Academic Press Dictionary of Science and Technology

Academic Press Dictionary of Science and Technology

Edited by
Christopher Morris

Academic Press
Harcourt Brace Jovanovich, Publishers
San Diego New York Boston London Sydney Tokyo Toronto

This book is printed on acid-free paper. ♾

Academic Press, Inc.
1250 Sixth Avenue, San Diego, California 92101-4311

United Kingdom Edition published by
Academic Press Limited
24–28 Oval Road, London NW1 7DX

Library of Congress Cataloging-in-Publication Data

Academic Press dictionary of science and technology / edited by
Christopher Morris
p. cm.
ISBN 0-12-200400-0
1. Science--Dictionaries. 2. Technology--Dictionaries.
I. Morris, Christopher G. II. Academic Press. III. Title:
Dictionary of science and technology.
Ref Q123.A33 ~~1991~~ 1992
503--dc20 90-29032
CIP

PRINTED IN THE UNITED STATES OF AMERICA
92 93 94 95 96 97 DO 9 8 7 6 5 4 3 2 1

BH 1/4/95

Table of Contents

Contributors to the Dictionary

EDITOR: Christopher Morris

SENIOR EDITORS: Daniel Hammer; Gail Rice; Elinor Williams

ASSOCIATE EDITOR: Amy Rosen

ASSISTANT EDITORS: Charles Allen; Margaret Syverson; Paul Walker

EDITORIAL ASSISTANT: Joseph Sheehan

SENIOR DEFINERS: Virginia Andersen; Robert Burnham; Judy Cannon; Robert Carlton; Karen Cooksey; Angela DeLong; Carola Dunn; Cheryl Farr; Anthony Gaudin; Cheryl Jeffrey; Bonnie Kenison; Phil Lauder; Jonathon Leifer; Fran Lesser; John McDonald; Roger McElmell; Lynn Meineke; Susan Olney; Italo Servi; Gloria Tierney; Carol Trueheart; Susan Van Benthuysen; Carol Wekesser

ASSOCIATE DEFINERS: Brian Alexander; Ellen Baskin; Brett Bernath; Eugene Bjerke; Steven Bloom; Grant Callaghan; Paul Cheatham; Flora Churnin; Lillian Cox; Kristen Crabtree; Carole Duebbert; Michelle Dvorak; Matthew Fogel; Sharon From; Meredith Garcia; Stanley Howard; Carolyn Jaquess; Irene Kirby; Agnes Leung; Lorraine Menger; Barbara Moran; Elizabeth Morris; Giovanni Parmigiani; George Patterson; Larry Peluso; Mitch Persons; Ken Reinstein; Joseph Ring; Andrew Robinson; Richard Robinson; Jeff Savage; Judy Sperling; Judith Swift; George Wilson

ASSISTANT DEFINERS: Richard Baker; Andy Bartlett; Mitch Becker; Joseph Bonanno; Michael Cathcart; Betsy De Fries-Burg; Tom Farrell; Dina Fogel; Mark Galicia; Michael Kilpatrick; Tina Lucas; Christopher Marler; Kent McCammon; Virginia Moran; Donna Okino; Ted Orahood; Wendy Perkins; Kate Peterson; Jeff Pham; Kathleen Phillips; Donna Potter; Carol Spirkoff Prime; John Romeo; Steve Sherwood; Steve Strickland; Asif Subedar; John Thomas; Ken Whaley; Peter Worland

Editorial Services

SPONSORING EDITOR: Marvin Yelles

COORDINATING EDITOR: Dean Irey

EDITORIAL CONSULTANT: Dolores Wright

CONTRIBUTING EDITORS: Chuck Arthur; Nikki Fine; Bob Kaplan; Lorraine Lica; Phyllis Moses; David Pallai

PHOTO RESEARCHER: Maggie Porter

Manuscript Preparation

COPY EDITING SUPERVISOR: Carolyn Conboy

CHIEF COPY EDITORS: Sandra Lee; Anne Schermer

COPY EDITORS: Heidi Davis; Cindy MacDonald

PROOFREADERS: John Coats; Teri Johnson; Diana Miller; Nancy Ostlie

TYPISTS: Steve Anderson; Lisa Casias; Suchin Yi; Chuck Yunkherr

Production

DIRECTOR OF PRODUCTION: Charles A. Goehring

ASSOCIATE DIRECTOR: Stuart Anderson

PRODUCTION EDITORS: Joni Benjamin; William LaDue

PHOTOCOMPOSITION SUPERVISOR: Katherine Pagaard

PHOTOCOMPOSITION ASSISTANTS: Cheryl Coates; Cathy Moore

Art and Design

ART DIRECTOR: Linda Shapiro

DESIGNER: Cathy Reynolds

ART COORDINATOR: Alli Spooner

GRAPHIC ARTISTS: Perry Dockins; Marji Drost; James Kenny

ILLUSTRATORS: Jacqueline Brown; Trevor Copenhaver; Jean Foster; John Foster; Trevor Moo-Young; Lauren Shavell

Preface to the Dictionary

"As by the cultivation of various sciences, a language is amplified; it will be more furnished with words deflected from their original sense . . ." Preface to A Dictionary of the English Language, *Dr. Samuel Johnson, 1755*

This dictionary, the *Academic Press Dictionary of Science and Technology,* is a collaborative effort of the company's Book Division, Journal Division, and Reference Division. It is a completely new and original work, not a revision or derivation of an older dictionary, and the entire text has been expressly written for this publication.

The dictionary consists of 2464 pages and contains a total of 133,007 entries, making it the largest scientific dictionary ever compiled in the English language. Included among these 133,007 entries are 112,227 main entry words and 20,780 secondary entries. This book is the product of four years of research and the efforts of more than four hundred different people.

As the largest U.S.-based publisher of scientific books and journals, Academic Press has recognized the need for an up-to-date description of the scientific vocabulary of the English language. English is not only the native language of Great Britain, the United States, Canada, and other countries, but it is also the world's most widely used second language, and as such is the international language of science.

The vocabulary range of this dictionary may be compared to the range of the visible spectrum. Beyond the dictionary's spectrum at the wider end are terms of the common vocabulary of English that have no special scientific meaning. Beyond the spectrum at the narrower end are terms that have scientific validity but are too rare and specialized to merit inclusion in a general-interest dictionary. Thus the dictionary includes those terms that are likely to be encountered in a scientific context by an informed general reader.

The entry list for this dictionary was compiled by examining the vocabulary found in four types of sources: (1) specialized encyclopedias, dictionaries, and glossaries, such as the *Encyclopedia of Human Biology, Dictionary of Cell Biology,* and *Dictionary of Behavioral Science*; (2) textbooks and professional books, such as *A Textbook of Physical Chemistry, University Physics, General Parasitology,* and *Invertebrates*; (3) journals, such as *Virology, Developmental Biology,* and *Toxicology and Applied Pharmacology*; and (4) scientific periodicals, such as *Scientific American* and *Science.*

The process of developing this dictionary was as follows. First, the dictionary's Advisory Board, in conjunction with the editors of the Academic Press Book Division, identified the areas of science and technology that the dictionary should cover. There are 124 of these scientific fields in all, ranging from Acoustical Engineering to Zoology. Some of these fields had their origin at the very beginning of human history, when people first speculated about their environment and attempted to understand it and control it; e.g., Agriculture, Astronomy, Botany, Engineering, Mathematics, and Medicine. On the other hand, some are entirely new fields that have recently developed as separate entities; e.g., Artificial Intelligence, Biotechnology, Chaotic Dynamics, Computer Programming, Oncology, and Robotics.

Second, the Reference Division developed a vocabulary list for each of the 124 individual fields. These lists were then verified by the Advisory Board, and the advisors provided additional terms as appropriate. Third, various outside experts in the different fields were enlisted to write the definitions for these vocabulary lists. Each definer was assigned a word list for a particular field in which he or she had expertise. For example, Anthony Gaudin, author of *Human Anatomy and Physiology,* was assigned definitions in these fields. The definers also added terms to the manuscript as appropriate. Fourth, all definitions were reviewed and edited by the Reference staff. They were then further reviewed by editors from Academic Press Books and by certain members of the Advisory Board.

Fifth, the definitions were independently verified by our Review Panel, a group of peer reviewers. Each reviewer read through the entire manuscript for a given field. He or she either checked off each definition as being acceptable as written, or else provided an edited version to replace the existing copy. Even at this stage, the reviewers recommended additional new terms for inclusion; the word list of the dictionary has been constantly evolving during the project.

Sixth, the entire text was given a final reading for correctness and consistency at the page proof stage, by editors from the Academic Press Journal and Book Divisions. This final reading was independent of all earlier stages of the project and provided a fresh eye on the manuscript.

Dr. Johnson said, ". . . no dictionary of a living tongue ever can be perfect, since while it is hastening to publication, some words are budding, and some falling away . . ." We have tried to deal with this classic lexicographic problem through the use of a computerized in-house typesetting facility that allows us to make substantial editorial changes up to the point of publication. For example, the final listing in our "Chronology of Modern Science" (page 2429) describes an event that occurred one month before press time.

We do not presume to describe our dictionary as a complete description of the vocabulary of contemporary science. What we can say is that this book of 2464 pages and 133,000 words has the largest and most current scientific vocabulary available anywhere, that the content has been thoroughly verified for accuracy, and that it provides a simple and readable format. We hope this book will be useful to its readers, and that it will in some small way contribute to the cause of science.

—*Christopher Morris, Editor*

Fields Covered by the Dictionary

Acoustical Engineering
Acoustics
Agriculture
Agronomy
Analytical Chemistry
Anatomy
Anthropology
Archaeology
Architecture
Artificial Intelligence
Astronomy
Astrophysics
Atomic Physics
Aviation
Bacteriology
Behavior
Biochemistry
Biology
Biotechnology
Botany
Building Engineering
Cardiology
Cartography
Cell Biology
Chaotic Dynamics
Chemical Engineering
Chemistry
Civil Engineering
Computer Programming
Computer Science
Computer Technology
Control Systems
Crystallography
Design Engineering
Developmental Biology
Ecology
Electrical Engineering
Electricity
Electromagnetism
Electronics
Endocrinology
Engineering
Entomology
Enzymology
Evolution
Fluid Mechanics
Food Technology
Forestry
Genetics
Geochemistry
Geodesy
Geography
Geology
Geophysics
Graphic Arts
Hematology
Histology
Horology
Hydrology
Immunology
Industrial Engineering
Inorganic Chemistry
Invertebrate Zoology
Linguistics
Materials
Materials Science
Mathematics
Mechanical Devices
Mechanical Engineering
Mechanics
Medicine
Metallurgy
Meteorology
Metrology
Microbiology
Military Science
Mineralogy
Mining Engineering
Molecular Biology
Mycology
Naval Architecture
Navigation
Neurology
Nuclear Physics
Nucleonics
Nutrition
Oceanography
Oncology
Optics
Ordnance
Organic Chemistry
Paleontology
Particle Physics
Pathology
Petroleum Engineering
Petrology
Pharmacology
Photogrammetry
Physical Chemistry
Physics
Physiology
Plant Pathology
Psychology
Quantum Mechanics
Radiology
Robotics
Science Terminology
Solid-State Physics
Space Technology
Spectroscopy
Statistics
Surgery
Systematics
Telecommunications
Textiles
Thermodynamics
Toxicology
Transportation Engineering
Vertebrate Zoology
Veterinary Medicine
Virology
Volcanology
Zoology

Scientific Fields in the Dictionary

A Foreword, by Harriet Zuckerman

Once united under the umbrella of natural philosophy, the sciences have become increasingly distinct and specialized. New scientific fields and specialties continue to appear. Some result from the discovery of new phenomena (radio-astronomy or virology, for example); others grow out of the development of new concepts (such as quantum physics or string theory), while still others have resulted from the invention of new technologies (computer science and X-ray crystallography are two among several examples).

New specialties sometimes grow up within the boundaries of older fields (nuclear physics, for example, or oncology), while others are the products of the merging of different fields or perspectives; molecular biology and bacterial genetics are two such combinations. Still others represent new perspectives on old phenomena; ecology, chaotic dynamics, and neuroscience are prominent examples here.

As specialization proceeds, groups of scientists come together who define themselves and are defined by others as working on the same sharply focused set of problems. Some of these aggregations persist, grow, and become formally organized, while others are transformed as scientists' understanding of what is important shifts and changes. Occasionally, established specialties cease to attract many new recruits, while others are entirely superseded and leave behind little more than grist for research by historians and sociologists of science.

Such changes in the social structure of the sciences are largely responses to alterations in their cognitive content, that is, to the creation and use of new ideas and new technologies and the assembling of new observations. Changes in cognitive content are also accompanied by the creation of new words and formulaic symbols as markers for scientific and technological novelty. Indeed, the rapid development of science and technology in the last hundred years can, in this sense, be gauged by changes in the size and content of their vocabularies.

Like other new dictionaries, a general dictionary of science is the outcome of three decisions: Which words should be included as entries in the dictionary, how should these words be defined, and which categories of words should be covered? In this case, what are the specific fields of science that a comprehensive dictionary must recognize and describe?

The *Academic Press Dictionary of Science and Technology* identifies 124 different fields or divisions of scientific knowledge. The divisions in this dictionary range from general classifications or disciplines such as medicine and geology, each of which includes thousands of entries, to highly specialized areas such as artificial intelligence and toxicology. This list must, of course, be time-bound and somewhat arbitrary, but the diversity represented here in the dictionary mirrors the diversity of science and technology today.

Further, the *Dictionary* categorizes each word it includes as belonging to one or more of these fields. Some individual terms, in fact, have definitions in a dozen fields or more—suggesting that the sciences are knit together by common ideas, techniques, and observations while also remaining highly specialized and distinct.

Early dictionaries could be and were products of single individuals, such as Samuel Johnson and Noah Webster, and expressed both their vast knowledge and their particular predilections. Individual authorship of dictionaries is now exceedingly rare; the language has become so diverse and complex as to make it unfeasible unless the subject of the dictionary is itself exceedingly narrow. But in those simpler times, mastery of the language was possible, and the classification of words into separate categories was apparently unnecessary. For example, the 1859 edition of *Webster's Unabridged Dictionary* provides both a definition of the word *mechanics* that is still serviceable today and a very thorough 17-line discussion of the word *fish*. But it does not identify these or other scientific words as belonging to any specific scientific field.

With the development of the monumental *Oxford English Dictionary* (1887–1928), it became an accepted principle that a large dictionary must be a collective effort. No longer did a single editor prepare the entire dictionary word list from A to Z, as Johnson and Webster had done. Instead, various editors worked on different sections of the dictionary, and their efforts were then combined as a single work. This change in lexicography was but an instance of the great shift in manufacturing from the preindustrial practice of a single worker producing a complete product to the more modern technique of a group of workers producing separate parts. It presaged parallel changes in science from research being done mainly by individuals to today's very large-scale collaborative efforts.

Scientific Fields in the Dictionary

This practice of collective, but essentially separate, efforts by a number of editors in the making of a dictionary inevitably led to the establishment of separate categories for words. From the beginning, this division tended to be arbitrary, with one editor being assigned the words beginning with the letter A, while others dealt with those beginning with B, C, D, E, and so on. Many general-vocabulary dictionaries are still produced in this fashion.

In a comprehensive dictionary such as the *Academic Press Dictionary of Science and Technology,* however, the content is too specialized for this approach. No single definer can be expected to be completely familiar with all the words in a consecutive alphabetical span; even a section of just one-half page in this dictionary can include words from 20 different scientific fields. Therefore, this dictionary was developed not as a single manuscript, but as 124 separate manuscripts, one for each different field covered in the book.

While this dictionary is not unique in identifying terms according to their specific areas of science, the total of 124 different fields represented here is the most comprehensive classification appearing in any scientific dictionary. The dictionary is also unique in that each of these scientific fields is represented not only by various entries appearing throughout the book, but also by a special article describing the field itself.

These articles or "window essays" are placed within a gray-tinted box to set them off from the definitions, which gives them a window-like appearance against the background of black text. They are intended to enlarge the dictionary's coverage of the field and to provide a context for the words specific to it.

For example, the term *chemistry* appears in alphabetical order on page 411, along with a definition of the word, as would be expected in any conventional dictionary. But it is also accompanied by the distinctly unconventional window essay "Chemistry," written by the Nobel Prize winner Glenn T. Seaborg of the University of California, Berkeley. Similarly, on page 559, there is not only a definition for *crystallography* but also an essay, "Crystallography," by the double Nobel Laureate Linus Pauling.

These essays vary according to the approach of the individual author and the nature of the field. Typically, they provide an expanded definition of the term itself, recount the history of the field, explain its current scope, and describe both the issues that it has historically addressed and the issues that it will continue to address.

Readers will encounter these "window essays" as they search for the appropriate term for the field (or, of course, another term on the same page). But the editors hope that readers will also browse through the book in search of these essays. (A complete list of the essays begins on the opposite page of this section, page xi.) For example, under the letter 'E' alone one will find, among other essays, "Electromagnetism," by Arno Penzias, "Endocrinology," by Rosalyn Yalow, "Entomology," by Edward O. Wilson, and "Evolution," by Stephen Jay Gould. Taken together, these window essays characterize the current state of the sciences, their problematics, and their prospects. This dictionary is thus much more than an inventory of words and their definitions.

Harriet Zuckerman
The Andrew W. Mellon Foundation and Columbia University

Authors of Essays on Scientific Fields

The following list includes the names and affiliations of the scientists who contributed the special "window" essays on the various scientific fields covered in this dictionary. Each of these essays accompanies the conventional dictionary definition of the term in question and is placed next to it in alphabetical order. The essays are set off from the rest of the text by a light gray background. In the list below, the page number in parentheses after the article title indicates the page of the book on which the essay will be found.

Acoustical Engineering **(page 26)**
ERIC E. UNGAR
Chief Consulting Engineer
BBN Systems and Technologies
(A Division of Bolt Beranek and Newman, Inc.)
Cambridge, Massachusetts

Acoustics **(page 28)**
ALLAN D. PIERCE
Leonhard Chair in Acoustics and Mechanical Engineering
Pennsylvania State University

Agriculture **(page 59)**
CHARLES J. ARNTZEN
Deputy Chancellor and Dean
The Agriculture Program
Texas A&M University

Agronomy **(page 60)**
DONALD L. SPARKS
Professor of Soil and Environmental Chemistry
Chairman of the Department of Plant and Soil Sciences
University of Delaware

Analytical Chemistry **(page 106)**
GEORGE H. MORRISON
Professor of Chemistry
Cornell University

Anatomy **(page 108)**
MICHAEL D. GERSHON
Professor and Chairman of Anatomy and Cell Biology
Columbia University

Anthropology **(page 127)**
DONALD JOHANSON
President, Institute of Human Origins
Berkeley, California

Archaeology **(page 145)**
MURIEL PORTER WEAVER
Shelter Island, New York
Author of *The Aztecs, The Maya, and Their Predecessors*

Architecture **(page 148)**
THOMAS HINE
Architecture Critic, The Philadelphia Inquirer
Author of *Populuxe* and *Facing Tomorrow*

Artificial Intelligence **(page 160)**
GORDON S. NOVAK, JR.
Associate Professor of Computer Sciences
University of Texas, Austin

Astronomy **(page 171)**
RICHARD J. TERRILE
Planetary Astronomer
Jet Propulsion Laboratory
California Institute of Technology

Astrophysics **(page 172)**
SUBRAHMANYAN CHANDRASEKHAR
Morton D. Hull Distinguished Service Professor, Emeritus
University of Chicago
Nobel Laureate in Physics

Atomic Physics **(page 177)**
BENJAMIN BEDERSON
Professor of Physics
New York University

Aviation **(page 194)**
KERMIT E. VAN EVERY
Design Consultant, Van Every & Associates
San Diego, California

Bacteriology **(page 205)**
THOMAS D. BROCK
E. B. Fred Professor of Natural Sciences, Emeritus
University of Wisconsin

Behavior **(page 240)**
GEORGE W. BARLOW
Professor of Integrative Biology
University of California, Berkeley

Biochemistry **(page 259)**
ARTHUR KORNBERG
Professor of Biochemistry
Stanford University School of Medicine
Nobel Laureate in Medicine

Biology **(page 261)**
R. C. LEWONTIN
Alexander Agassiz Professor of Zoology
Museum of Comparative Zoology
Harvard University

Essay Authors

Biomedical Engineering **(page 262)**
JOHN G. WEBSTER
Professor of Electrical and Computer Engineering
University of Wisconsin

Botany **(page 295)**
ROBERT W. PEARCY
Professor of Botany
University of California, Davis

Building Engineering **(page 323)**
ROBERT A. HELLER
Professor of Engineering Science and Mechanics
Virginia Polytechnic Institute and State University

Cardiology **(page 363)**
ELLIOT RAPAPORT
Professor of Medicine
University of California, San Francisco
Chief, Cardiology Division
San Francisco General Hospital

Cartography **(page 369)**
JOHN GARVER
Senior Assistant Editor and
Chief Cartographer
National Geographic Society
Washington, D.C.

Cell Biology **(page 384)**
ELIZABETH D. HAY
Professor and Chair of Anatomy and
Cellular Biology
Harvard Medical School

Chaotic Dynamics **(page 402)**
JOHN W. McDONALD
Weapons Technology Staff, retired
Los Alamos National Laboratory

Chemical Engineering **(page 409)**
W. W. GRAESSLEY
Professor of Chemical Engineering
Princeton University

Chemistry **(page 411)**
GLENN T. SEABORG
University Professor of Chemistry
University of California
Nobel Laureate in Chemistry

Civil Engineering **(page 438)**
T. H. LIN
Professor Emeritus of Civil Engineering
University of California, Los Angeles

Computer Technology **(page 489)**
BENNET P. LIENTZ
Professor of Computer Technology
University of California, Los Angeles

Computer Programming and Computer Science **(page 490)**
DONALD E. KNUTH
Professor of The Art of Computer Programming
Stanford University

Control Systems **(page 515)**
NADER M. BOUSTANY
Section Head, Systems Analysis & Information Management
Department
General Motors Corporation

NEIL A. SCHILKE
Technical Director, Systems Engineering Center
General Motors Corporation

Crystallography **(page 559)**
LINUS PAULING
Research Professor
Linus Pauling Institute of Science and Medicine
Nobel Laureate in Chemistry
Nobel Laureate in Peace

Design Engineering **(page 617)**
LAWRENCE D. MALONEY
Chief Editor, *Design News*

Developmental Biology **(page 622)**
JOHN GERHART
Professor of Molecular and Cell Biology
University of California, Berkeley

Ecology **(page 707)**
PETER M. VITOUSEK
Professor of Biological Sciences
Stanford University

Electrical Engineering **(page 719)**
ROBERT H. KINGSTON
Senior Lecturer
Electrical Engineering and Computer Science
Massachusetts Institute of Technology

Electricity **(page 721)**
S. H. DURRANI
Program Manager, Advanced Systems
NASA Office of Space Communication
Washington, D.C.

Electromagnetism **(page 726)**
ARNO PENZIAS
Vice President, Research
AT&T Bell Laboratories
Murray Hill, New Jersey
Nobel Laureate in Physics

Electronics **(page 730)**
RICHARD H. BUBE
Professor of Materials Science and Electrical Engineering
Stanford University

Essay Authors

Endocrinology (page 745)
ROSALYN S. YALOW
Solomon A. Berson Distinguished Professor-At-Large
Mt. Sinai School of Medicine
City University of New York
Nobel Laureate in Medicine or Physiology

Engineering (page 749)
EDWARD E. DAVID, JR.
Engineering Consultant
President, EED, Inc.
Bedminster, New Jersey

Entomology (page 753)
EDWARD O. WILSON
Baird Professor of Science
Harvard University

Enzymology (page 756)
PAUL D. BOYER
Professor of Biochemistry
University of California, Los Angeles

Evolution (page 785)
STEPHEN JAY GOULD
Paleontologist and Author
Professor of Geology
Harvard University

Fluid Mechanics (page 855)
JOHN L. LUMLEY
Willis H. Carrier Professor of Engineering
Cornell University

Food Science and Technology (page 863)
STEVE L. TAYLOR
Professor and Head of Food Science and Technology
University of Nebraska

Forestry (page 867)
DAVID M. SMITH
Morris K. Jessup Professor Emeritus of Silviculture
Yale University

Genetics (page 918)
ARNO G. MOTULSKY
Professor of Medicine and Genetics
University of Washington

Geochemistry (page 920)
DONALD D. RUNNELLS
Professor and Chair of Geological Sciences
University of Colorado

Geodesy (page 921)
WILLIAM E. STRANGE
Chief Geodesist
National Geodetic Survey
Rockville, Maryland

Geography (page 922)
GILBERT M. GROSVENOR
President and Chairman
National Geographic Society
Washington, D.C.

Geology (page 923)
RAYMOND SIEVER
Professor of Geology
Harvard University

Geophysics (page 925)
WILLIAM A. NIERENBERG
Director Emeritus
Scripps Institution of Oceanography
San Diego, California

Graphic Arts (page 952)
BARRY MOSER
Artist and Designer
The Pennyroyal Press
Winner of the American Book Award
North Hatfield, Massachusetts

Gynecology (page 975)
DAVID M. GERSHENSON
Professor and Deputy Chairman of Gynecology
M. D. Anderson Cancer Center
University of Texas, Austin

Hematology (page 1008)
DAVID G. NATHAN, M.D.
Robert A. Stranahan Professor of Pediatrics
Harvard Medical School
Physician-in-Chief
Children's Hospital
Boston, Massachusetts

Histology (page 1031)
RONALD A. BERGMAN
Professor of Anatomy
University of Iowa

Horology (page 1046)
JAMES JESPERSEN
Physicist
National Institute of Standards and Technology
Boulder, Colorado

Hydrology (page 1062)
CHIN-FU TSANG
Senior Scientist
Earth Sciences Division
Lawrence Berkeley Laboratory
Berkeley, California

Immunology (page 1088)
HELEN VAN VUNAKIS
Professor of Biochemistry
Brandeis University

Essay Authors

Industrial Engineering (page 1102)
HEWITT H. YOUNG
Emeritus Professor of Engineering
Arizona State University

Inorganic Chemistry (page 1113)
A. GEOFF SYKES
Professor of Inorganic Chemistry
University of Newcastle upon Tyne
Newcastle upon Tyne, England

Invertebrate Zoology (page 1138)
RALPH BUCHSBAUM
Emeritus Professor of Biology
University of Pittsburgh

Linguistics (page 1247)
MORRIS HALLE
Professor of Linguistics
Massachusetts Institute of Technology
Co-author of *The Sound Pattern of English*

Materials Science and Engineering (page 1324)
A. H. HEUER
Kyocera Professor of Ceramics
Case Western Reserve University

Mathematics (page 1325)
PAUL R. HALMOS
Professor of Mathematics
Santa Clara University

Mechanical Devices (page 1335)
CHARLES E. S. UENG
Professor of Civil Engineering
Georgia Institute of Technology

Mechanical Engineering (page 1336)
DANIEL C. DRUCKER
Professor of Mechanical Engineering
University of Florida

Mechanics (page 1337)
THEODORE YAOTSU WU
Professor of Engineering Science
California Institute of Technology

Medicine (page 1339)
MARY ELLEN AVERY, M.D.
Thomas Morgan Rotch Professor
Department of Pediatrics
Harvard Medical School
Harvard University

Metallurgy (page 1358)
ROBERT MADDIN
University Professor of Metallurgy, Emeritus
University of Pennsylvania

Meteorology (page 1362)
JAMES R. HOLTON
Professor of Atmospheric Sciences
University of Washington

Metrology (page 1369)
JAROMIR J. ULBRECHT
Director, Technical Programs
National Institute of Standards and Technology
Gaithersburg, Maryland

Microbiology (page 1372)
JOAN W. BENNETT
Professor of Cell and Molecular Biology
Tulane University
Past President, American Society for Microbiology

Military Science (page 1382)
LARRY H. ADDINGTON
Professor of History
The Citadel

Mineralogy (page 1386)
HATTEN S. YODER, JR.
Director Emeritus, Geophysical Laboratory
Carnegie Institution of Washington

Mining Engineering (page 1388)
JAAK DAEMEN
Professor of Mining Engineering
Mackay School of Mines
University of Nevada at Reno

Molecular Biology (page 1402)
GUNTHER S. STENT
Professor of Molecular Biology
University of California, Berkeley

Mycology (page 1432)
RICHARD P. KORF
Professor of Mycology
Cornell University

Naval Architecture (page 1445)
DONALD V. WALTER
Chief Naval Architect
National Steel and Shipbuilding Company
San Diego, California

Navigation (page 1446)
THOMAS D. DAVIE
(Deceased)
Rear Admiral, United States Navy
President, The Navigation Foundation

Neurology (page 1460)
JOHN W. OLNEY
Professor of Psychiatry and Neuropathology
Washington University School of Medicine
St. Louis, Missouri

Essay Authors

Nuclear Physics (page 1487)
HERMAN FESHBACH
Institute Professor Emeritus
Massachusetts Institute of Technology

Nucleonics (page 1489)
HAROLD M. AGNEW
Former Director
Los Alamos National Laboratory

Nutrition (page 1492)
WILLIAM J. DARBY
Professor Emeritus of Biochemistry (Nutrition)
Vanderbilt University

Oceanography (page 1499)
ROGER REVELLE
(deceased)
Founder, Scripps Institution of Oceanography
Former Chancellor
University of California, San Diego

Oncology (page 1511)
VINCENT T. DEVITA, JR.
Benno C. Schmidt Chair in Clinical Oncology
Attending Physician and Member
Program of Molecular Pharmacology and Therapeutics
Memorial Sloan-Kettering Cancer Center
New York, New York

Optics (page 1524)
BRIAN J. THOMPSON
Professor of Optics and Provost
University of Rochester

Ordnance (page 1529)
GEORGE LINDSEY
Senior Research Fellow
Canadian Institute of Strategic Studies
Ottawa, Ontario

Organic Chemistry (page 1530)
HARRY WASSERMAN
Eugene Higgins Professor Emeritus of Chemistry
Yale University

Paleontology (page 1559)
PETER ANDREWS
Department of Paleontology
Natural History Museum, London

Particle Physics (page 1580)
EDWARD FARHI
Associate Professor of Physics
Massachusetts Institute of Technology

Pathology (page 1587)
KIM SOLEZ
Professor and Chairman of Pathology
University of Alberta

Petroleum Engineering (page 1617)
J. E. WARREN
Director
Frontier Resources International, Inc.
Pittsburgh, Pennsylvania

Petrology (page 1618)
NORBERT BERKOWITZ
Professor Emeritus of Fuel Science
University of Alberta

Pharmacology (page 1622)
GEORGE H. HITCHINGS
The Wellcome Research Laboratories
Burroughs Wellcome Co.
Research Triangle Park, North Carolina

Photogrammetry (page 1636)
JACK B. EVETT
Professor and Associate Dean
College of Engineering
University of North Carolina at Charlotte

Physical Chemistry (page 1642)
ALLEN J. BARD
Hackerman/Welch Regents Chair
University of Texas, Austin

Physics (page 1643)
MARVIN L. MARSHAK
Professor of Physics
University of Minnesota

Physiology (page 1644)
JOSEPH F. HOFFMAN
Eugene Higgins Professor of Cellular and Molecular Physiology
Yale University School of Medicine

Plant Pathology (page 1665)
GEORGE N. AGRIOS
Professor and Chairman
Department of Plant Pathology
University of Florida

Plasmids (page 1666)
JOSHUA LEDERBERG
Professor and President-Emeritus
The Rockefeller University
Nobel Laureate in Physiology or Medicine

Psychology (page 1751)
RICHARD C. ATKINSON
Professor of Psychology and Chancellor
University of California, San Diego

Quantum Mechanics (page 1770)
JOHN D. McGERVEY
Professor of Physics
Case Western Reserve University

Essay Authors

Radiology (page 1787)
HAROLD G. JACOBSON
Emeritus Professor and Chairman
Department of Radiology
Albert Einstein College of Medicine at Montefiore Medical Center
Bronx, New York

Robotics (page 1872)
V. DANIEL HUNT
President, Technology Research Corporation
Springfield, Virginia

Science (page 1926)
CHARLES C. GILLISPIE
Professor of the History of Science Emeritus
Princeton University

Solid-State Physics (page 2031)
HENRY EHRENREICH
Clowes Professor of Science
Harvard University

Space Technology (page 2043)
STEWART W. JOHNSON
Principal Engineer
Advanced Basing Systems
BDM International, Inc.
Albuquerque, New Mexico

Spectroscopy (page 2050)
R. H. WOODWARD WAESCHE
Principal Scientist, Virginia Propulsion Division
Atlantic Research Corporation
Gainesville, Virginia

Statistics (page 2086)
JOSEPH B. KADANE
Leonard J. Savage Professor
Statistics and Social Sciences
Carnegie Mellon University

Surgery (page 2143)
MICHAEL E. DEBAKEY, M.D.
Chancellor and Chairman
Department of Surgery
Baylor College of Medicine

Systematics (page 2159)
ROBERT M. MAY
Royal Society Research Professor
Department of Zoology
Oxford University

Telecommunications (page 2177)
ALAN G. CHYNOWETH
Vice President, Applied Research
Bellcore (Bell Communications Research)
Morristown, New Jersey

Textiles (page 2197)
JACK LENOR LARSEN
Chairman and Design Director
Jack Lenor Larsen Incorporated
New York, New York

Thermodynamics (page 2204)
J. M. HONIG
Professor of Chemistry
Purdue University

Toxicology (page 2239)
I. GLENN SIPES
Professor and Head of Pharmacology and Toxicology
Health Sciences Center
University of Arizona

Transportation Engineering (page 2253)
MARK R. NORMAN
Deputy Executive Director
Institute of Transportation Engineers
Washington, D.C.

Vaccinology (page 2310)
JONAS SALK, M.D.
Founding Director and Distinguished Professor in International Health Sciences
The Salk Institute for Biological Studies
San Diego, California

Vertebrate Zoology (page 2326)
DAVID M. GREEN
Assistant Professor and Curator of Herpetology
Redpath Museum
McGill University
Montreal, Quebec

Veterinary Medicine (page 2329)
CHARLES CORNELIUS
Emeritus Professor of Veterinary Medicine
University of California, Davis

Virology (page 2335)
JANET E. MERTZ
Professor of Oncology
McArdle Laboratory for Cancer Research
University of Wisconsin Medical School

Volcanology (page 2342)
FRED M. BULLARD
Emeritus Professor of Geological Sciences
University of Texas, Austin

Zoology (page 2397)
DAVID J. RANDALL
Professor of Zoology
University of British Columbia
Vancouver, British Columbia

Board of Editorial Advisors

This list includes the names and affiliations of the eminent scientists who provided editorial guidance during the planning and making of this dictionary. The specific roles of individual Board members varied, depending on the demands of their particular field of expertise, but the general advisory role of the Board had four aspects.

First, the Board of Advisors identified and delimited the 124 specific scientific fields to be covered in the dictionary. The dictionary editors then prepared preliminary entry lists for each of these fields. Second, the Board reviewed these entry lists, either indicating approval for the inclusion of a given term or recommending its deletion. The Board also added many terms to the existing lists, particularly new words or new uses of existing terms. The dictionary editors then prepared a group of sample definitions for each field.

Third, the Board reviewed these sample definitions to provide each field with its own defining model, that is, a specimen entry indicating what information a proper definition for this field should include, the sequence in which the information should be presented, and the manner in which it should be worded. The dictionary definers then wrote definitions for each field following the model. Fourth, the Board reviewed these actual definitions, in many cases reviewing the entire manuscript for a given field, making comments, alterations, and additions as necessary. Individual Board members also dealt with other editorial issues that were specific to their own fields, such as the standard form of a definition for a given field.

Editorial Advisors

Editorial Advisors

Editoral Advisors

Panel of Manuscript Reviewers

All definitions in this dictionary were carefully reviewed for accuracy by our own editorial staff. After this initial review of the manuscript was completed, a second independent review was carried out by a panel of outside experts. The members of the Review Panel were chosen for their expertise in specific scientific fields.

These reviewers, who are listed below, read over the entire final manuscript for the field or fields indicated. For each individual entry, the reviewers either indicated their approval of the existing definition as written, or else they supplied new wording to improve or expand the definition. They also added important new terms to the manuscript that had not previously been included.

NEAL B. ABRAHAM
Professor of Physics
Bryn Mawr College
Chaotic Dynamics

LAURA A. ANDERSSON
Professor of Biochemistry
Kansas State University
Biochemistry; Spectroscopy

PETER ANDREWS
Department of Paleontology
Natural History Museum
London
Archaeology; Evolution; Paleontology

RODNEY ANDREWS
Professor of Chemistry
Stevens Institute of Technology
Materials Science

CHRISTIAN B. ANFINSEN
Professor of Biology
Johns Hopkins University
Biology

CHARLES J. BAER
Professor of Mechanical Engineering
University of Kansas
Engineering

ALLEN J. BARD
Hackerman/Welch Regents Chair
University of Texas, Austin
Analytical Chemistry; Chemistry; Physical Chemistry

FRANK BARNES
Professor of Electrical and Computing Engineering
University of Colorado
Electrical Engineering

NORBERT BERKOWITZ
Emeritus Professor of Fuel Science
University of Alberta
Edmonton, Alberta
Geochemistry; Petrology

WILLIAM BEYER
Professor of Mathematical Sciences
Associate Dean
College of Arts and Sciences
University of Akron
Statistics

FLOYD E. BLOOM
Professor of Neuropharmacology
The Scripps Research Institute
San Diego, California
Neurology

JOHN G. BOLLINGER
Dean
College of Engineering
University of Wisconsin
Design Engineering; Materials Science

STEFAN BOSHKOV
Henry Krumb Emeritus Professor
Henry Krumb School of Mines
Columbia University
Mining Engineering

RICHARD O. BUCKIUS
Professor of Mechanical and Industrial Engineering
University of Illinois
Thermodynamics

MICHAEL K. BUCKLAND
Professor
School of Library and Information Studies
University of California, Berkeley
Science Terminology

ALBERT W. BURGSTAHLER
Professor of Chemistry
University of Kansas
Organic Chemistry

ROBERT BURNHAM
Editor
Astronomy Magazine
Astronomy; Astrophysics

Manuscript Reviewers

RICHARD D. CAMPBELL
Professor of Developmental and Cell Biology
University of California, Irvine
Developmental Biology

JOEL E. COHEN
Professor, Laboratory of Populations
Rockefeller University
Vertebrate Zoology; Zoology

STEPHEN COOPER
Professor of Microbiology and Immunology
University of Michigan Medical School
Bacteriology; Microbiology

JOHN C. COURTNEY
Professor of Nuclear Engineering
Louisiana State University
Nucleonics

ARNOLD J. DAHM
Professor of Physics
Case Western Reserve University
Solid-State Physics

JOHN DASCHBACH
Department of Chemistry
University of Oregon
Physical Chemistry

RENATA DMOWSKA
Professor of Applied Science
Harvard University
Geophysics

CLAYTON DODGE
Professor of Mathematics
University of Maine
Mathematics

QUAN DONG
Research Associate in Biology
Vanderbilt University
Ecology

ANGELINE DOUVAS
Professor of Medicine
University of Southern California Medical Center
Medicine

W. RICHARD DUKELOW
Professor of Physiology and Animal Husbandry
Director
Endocrine Research Unit
Michigan State University
Developmental Biology

W. W. DULEY
Professor of Physics
University of Waterloo
Waterloo, Ontario
Electromagnetism

S. H. DURRANI
Program Manager, Advanced Systems
NASA Office of Space Communications
Electrical Engineering; Electricity

WILLIAM EARNSHAW
Professor of Cell Biology and Anatomy
Johns Hopkins University School of Medicine
Cell Biology

GEORGE ELLIS
Industrial Drives, Blacksburg, Virginia
Electricity

ROBERT E. FISCHER
President, Optics One, Inc.
Optics

CARITA FOCKLER
Professor of Genetics
University of California, Berkeley
Genetics

KATHLEEN A. FRENCH
Professor of Biology
University of California, San Diego
Anatomy

ANTHONY GAUDIN
Professor of Biology
California State University, Northridge
Physiology

DONALD A. GLASER
Professor of Physics, Molecular Biology, and Neurobiology
University of California, Berkeley
Biotechnology

JENNY P. GLUSKER
Senior Member
Institute for Cancer Research
Philadelphia, Pennsylvania
Crystallography

ROBERT E. GREENE
Professor of Mathematics
University of California, Los Angeles
Mathematics

JOSHUA E. GREENSPON
J. G. Engineering Research Associates
Acoustical Engineering; Acoustics

THOMAS P. GRUNDY
Professor of Geology
Texas A&M University
Geology

E. R. HART
Professor of Entomology and Forestry
Iowa State University
Entomology

Manuscript Reviewers

DAVID HARWOOD
Professor of Geology
Texas A&M University
Geology

DOUGLAS HENDERSON
Professor of Chemistry
University of Utah
Physical Chemistry

RICHARD H. HIMES
Professor of Biochemistry
University of Kansas
Enzymology

ARTHUR HOBBS
Professor of Mathematics
Texas A&M University
Mathematics

J. M. HONIG
Professor of Chemistry
Purdue University
Solid-State Physics; Thermodynamics

M. JAMSHIDI
AT&T Professor and Director
CAD Laboratory for Systems and Robotics
University of New Mexico
Robotics

RICHARD T. JOHNSON
Neurologist-in-Chief
Johns Hopkins University School of Medicine
Virology

JOSEPH B. KADANE
Leonard J. Savage Professor of Statistics and Social Sciences
Carnegie Mellon University
Statistics

JACK KEENE
Professor of Microbiology
Duke University
Virology

KENT KEETON
Professor of Pharmacology
University of Texas, Austin
Pharmacology

DONALD KELLY
Professor of Physics
Miami University
Physics

ROBERT C. KING
Professor of Biological Sciences
Northwestern University
Genetics

JACOB KLEINBERG
Professor Emeritus of Chemistry
University of Kansas
Chemistry; Inorganic Chemistry

KENNETH L. KOWALSKI
Professor of Physics
Case Western Reserve University
Nuclear Physics

T. T. KOZLOWSKI
Professor of Biological Sciences
University of California, Santa Barbara
Botany

ERHARD KREMPL
Professor of Mechanics
Rensselaer Polytechnic Institute
Mechanical Engineering

WILLIAM B. KRISTAN, JR.
Professor of Biology
University of California, San Diego
Anatomy

STEPHEN LAMBERT
Professor of Chemical Engineering
University of California, Berkeley
Chemical Engineering

MONTAGUE LANE
Professor of Pharmacology and Medicine
Baylor College of Medicine
Oncology

GLENN LANGER
Professor of Medicine and Physiology
Director
Cardiovascular Research Laboratory
University of California Medical Center, Los Angeles
Cardiology

TOM LEESON
Professor Emeritus of Anatomy
University of Alberta
Edmonton, Alberta
Histology

STANLEY P. L. LEONG
Associate Professor of Surgery
University of California, San Francisco
Immunology

GEORGE R. LINDSEY
Senior Research Fellow
Canadian Institute of Strategic Studies
Military Science; Naval Architecture; Ordnance

WARREN LOCKETTE
Professor of Endocrinology
Wayne State University School of Medicine
Endocrinology

Manuscript Reviewers

PHILIP W. MAJERUS
Professor of Medicine and Biochemistry
School of Medicine
Washington University
St. Louis, Missouri
Oncology

PHILIP MATTER, III
Consulting Geologist
Tucson, Arizona
Mineralogy

MICHAEL McGINNIS
Associate Director
Clinical Microbiology Laboratory
University of North Carolina
Mycology

HENRY M. McHENRY
Professor of Anthropology
University of California, Davis
Anthropology; Paleontology

CAROLYN MEIGS
Iowa Electric Light and Power Company
Nucleonics

STANLEY MIDDLEMAN
Professor of Engineering Sciences
University of California, San Diego
Fluid Mechanics

TERRY M. MIKITEN
Professor and Associate Dean
Graduate School of Biomedical Sciences
University of Texas Health Science Center, San Antonio
Physiology

WILLIAM A. NIERENBERG
Director Emeritus
Scripps Institution of Oceanography
San Diego, California
Geography; Geophysics; Oceanography

MARY L. NIERODZIK
Department of Hematology
New York University Medical Center
Hematology

GUSTAV NOSSAL
Director
The Walter and Eliza Hall Institute of Medical Research
Royal Melbourne Hospital
Victoria, Australia
Immunology

GORDON S. NOVAK, JR.
Associate Professor of Computer Science
University of Texas, Austin
Artificial Intelligence; Computer Science; Computer Programming; Computer Technology

ROBERT J. O'HARA
Postdoctoral Fellow
Division of Birds
National Museum of Natural History
Systematics

ANDREW P. PLOCHOCKI
Professor of Polymer Engineering
Stevens Institute of Technology
Materials Science

GARY POLIS
Professor of Environmental Sciences
Vanderbilt University
Ecology

DON POTTS
Professor of Forestry
University of Montana
Hydrology

A. RAVI P. RAU
Professor of Physics
Louisiana State University
Atomic Physics

HERBERT E. RAUCH
Staff Scientist and Senior Member
Lockheed Research Laboratory
Palo Alto, California
Aviation; Space Technology

WILLIAM S. RICHARDSON
Department of Surgery
Shriners' Hospital for Crippled Children
Boston Burns Unit
Boston, Massachusetts
Surgery

ARTHUR W. ROSE
Professor of Geosciences
Pennsylvania State University
Geochemistry

MIKE SAYRE
Professor of Biology
Stanford University School of Medicine
Molecular Biology

ANDREW V. SCHALLY
Professor of Medicine
Veterans Administration Hospital and
Tulane University School of Medicine
Endocrinology

NEVIN S. SCRIMSHAW
Director
Food Nutrition and Development Programme
The United Nations University of Tokyo
Harvard Center for Population Studies
Nutrition

Manuscript Reviewers

SATIMARU SENO
Professor Emeritus
Shigei Medical Research Institute
Okayama, Japan
Veterinary Medicine

ITALO S. SERVI
Consultant
Charles River Associates Incorporated
Boston, Massachusetts
Metallurgy

JACOB SHEKEL
Professor of Electrical and Computer Engineering
Northeastern University
Electronics

STEVEN N. SHORE
Astrophysicist
Goddard Space Flight Center
Astronomy; Astrophysics; Fluid Mechanics; Quantum Mechanics

CELIA D. SLADEK
Professor of Neurology and Neurobiology
University of Rochester Medical School
Neurology

ROBERT M. STERN
Professor of Psychology
Pennsylvania State University
Psychology

FRANK TAYLOR
AT&T Bell Laboratories, Retired
Telecommunications

GARETH THOMAS
Professor of Materials Science and Mineral Engineering
University of California, Berkeley
Materials; Materials Science

JENNIFER THORSCH
Professor of Biological Sciences
University of California, Santa Barbara
Botany

CHIA-HSIUNG TZE
Professor of Physics
Virginia Polytechnic Institute and State University
Particle Physics

CHARLES E. S. UENG
Professor of Civil Engineering
Georgia Institute of Technology
Building Engineering; Civil Engineering; Transportation Engineering

ARTHUR C. UPTON
Professor and Director
Institute of Environmental Medicine
New York University Medical Center
Pathology; Radiology

KERMIT E. VAN EVERY
Design Consultant
Van Every & Associates
San Diego, California
Aviation; Space Technology

JAMES L. WAY
Professor of Pharmacology and Toxicology
Texas A&M University
Toxicology

BARBARA WEBSTER
Professor of Agronomy and Range Science
University of California, Davis
Botany

GRADY L. WEBSTER
Professor of Botany
University of California, Davis
Botany

ROLF WEIL
Professor Emeritus of Materials Science and Engineering
Stevens Institute of Technology
Materials; Materials Science

THOMAS H. WELLER
Professor Emeritus of Tropical Public Health
Harvard University
Invertebrate Zoology

CRAIG WHITE
Professor of Geosciences
Boise State University
Petrology

THEODORE YAOTSU WU
Professor of Engineering Science
California Institute of Technology
Mechanics

HATTEN S. YODER, JR.
Director Emeritus
Geophysical Laboratory
Carnegie Institute of Washington
Cartography; Geodesy; Mineralogy; Photogrammetry

MICHAEL E. ZELLER
Professor of Physics
Yale University
Horology

Guide to the Dictionary

This guide to the *Academic Press Dictionary of Science and Technology* contains the following information:

A. Different Types of Main Entries
B. Alphabetical Placement of Entries
C. The Format of an Entry
D. Secondary Entries
E. Names of Scientific Fields
F. Definitions
G. Cross-References
H. Etymologies
I. Special Essays
J. Abbreviations Used in the Dictionary
K. Pronunciations

A. Different Types of Main Entries

Single-Word Entries

The typical entry in this dictionary is an individual word that begins with a lowercase letter, such as *life, output, electrolyte, cell, calcium,* or *ego*. However, although the single-word entry is the most obvious type of dictionary entry, it is only one of eight different types of main entries found in this book. The others are discussed below.

Compound Entries

The dictionary entry list includes not only single-word entries, such as *feedback, frequency, membrane, modulation, negative,* and *plasma,* but also compound entries that consist of two or more words, such as *frequency modulation, negative feedback,* and *plasma membrane*.

Compound entries appear whenever the meaning of a pair of words or a group of words cannot be readily understood by simply combining the meanings of the individual words. For example, *oxygen debt* and *cucumber mosaic* appear as compound entries in the dictionary, because *oxygen debt* does not mean "the fact of owing someone oxygen," and *cucumber mosaic* is not "a mosaic composed of cucumbers."

Self-Evident Compounds

On the other hand, phrases such as *aircraft propeller, boiler cleaning, engine oil,* and *mine superintendent* do not appear as main entries in this dictionary. They are self-evident compounds whose meanings are obvious from their individual components. An "aircraft propeller" is simply the propeller of an airplane, "boiler cleaning" is the process of cleaning a boiler, and so on. Including a series of obvious phrases such as *mine examiner, mine foreman, mine inspector, mine manager, mine surveyor,* and *mine superintendent* provides a dictionary with additional entries, but it does not really provide the reader with any additional information.

Abbreviations

This dictionary includes as main entries various abbreviations of words and phrases, such as *lat, ctn,* or *ALS*. A given abbreviation can vary widely in form; it may or may not have a period at the end, and it may or may not be capitalized. For example, *DC,* the abbreviation for "direct current," may also appear as *D.C., dc,* or *d.c*.

The form of a given abbreviation that we present in this dictionary is the one that is most commonly used or most widely accepted. If more than one form is given in the entry, the different forms are arranged in order of preference:

kWh or **kwh** kilowatt-hour. Also, **kW-hr.**

Regardless of the order of preference, any form of an abbreviation that is presented here is correct, and so are many other variations that we do not include here. For example, virtually any abbreviation that is written without a period could also appear with one.

Abbreviations are not listed in a separate section in the Appendix; they appear as part of the main A-to-Z list. The alternative practice of putting abbreviations in a separate list, while the acronyms remain in the main list, places a burden on the reader. He or she must know in advance that *AIDS* is an acronym and therefore is in the main word list, but that *AIC* and *AID* are abbreviations and therefore in the separate list.

Acronyms

This dictionary also includes acronyms as main entries. An acronym is a form composed of the first letter or letters of several words. *Radar* is the classic example of an acronym; it is formed from the phrase "radio detection and ranging." Other familiar scientific acronyms are *laser* (light amplification by stimulated emission of radiation) and *cyborg* (cybernetic organism). Such terms are spelled and pronounced just as if they were conventional words. In fact, as time goes on and the term continues to appear in print in lowercase form, the connection with the original phrase is often lost.

Many other acronyms retain the form of initialisms. That is, they are written in the form of a group of capital letters representing the first initial of each word of the relevant phrase. Acronyms of this type include *DOS* (disk operating system) and *CAT scan* (computerized axial tomography).

Proper Names

This dictionary is a compilation of the general vocabulary of science and technology. Most single-word entries or compound entries are listed as beginning with lowercase letters, indicating that they have a relatively generalized meaning. However, certain entries, such as *Orion, Canis,* or *Marfan's syndrome,* begin with capital letters, indicating that they have a more specialized meaning.

Geographical Entries

This dictionary is not intended to serve as a complete source of place names, but it does include a significant number of geographical entries, such as *Urals, Angel Falls, Northern Hemisphere,* and *Galapagos Islands.* The geographical terms that appear here are those that describe aspects of physical geography. Thus, major surface features of the earth, such as the names of rivers, seas, mountain ranges, and deserts, are included as entries. The dictionary does not include as entries political designations, such as the names of nations, states, provinces, or cities.

Biographical Entries

This dictionary includes as entries the names of several thousand people who have made significant contributions to the world of science, ranging chronologically from figures of the classical world such as Archimedes and Pappus of Alexandria, to contemporary figures such as Albert Sabin, Rosalyn Yalow, and Stephen Hawking. Biographical entries are listed alphabetically according to the person's last name.

Priestley, Joseph 1733–1804, English chemist; discovered oxygen (independently of Karl Scheele) . . . .

Skinner, B(urrus) F(rederic) 1904–1990, American psychologist and writer . . . .

Biographical entries appear in the main A-to-Z dictionary rather than in a separate section in the Appendix. We place the biographical entries in the main word list for two reasons. First, it is more convenient for the reader to have all terms beginning with the letter **n** in one place in the book, whether the term being looked up is *newt* or *Newton.* Second, this system means that the entry for a biographical figure appears in the same place in the book as any entries that are derived from his or her name, for example, in this case *newton, Newtonian, Newtonian relativity, Newton's first law,* and so on.

Prefixes, Suffixes, and Combining Forms

This dictionary also includes various entries that are not whole words, but rather word elements, such as *macro–, iso–, tele–, hypo–, proto–, –osis,* and *–graph.* These elements are used to form larger words, such as *macrocyst* from *macro–* and the root word *cyst, telegraph* from *tele–* and *–graph,* and so on.

A dictionary cannot anticipate all the possible vocabulary forms that can be created by adding prefixes or suffixes to root words. The value of word-element entries is that they enable the reader to grasp the meaning of many more terms than are entered in the dictionary. For example, the word *intralaryngeal* is a very specialized medical term and therefore is not included as an entry here. However, the forms *intra–* and *laryngeal* are entries. One can look these up and find that *intra–* means "within" and *laryngeal* means "of or relating to the larynx," and then understand that the term *intralaryngeal* means "within the larynx."

The use of this combining technique is especially useful in understanding unfamiliar terms in fields such as Medicine, Chemistry, Botany, Zoology, Ecology, and Pharmacology. In these fields, new words are usually coined according to established principles, using existing word elements. Thus, if a new life form were discovered and described as *phagolithic,* one could look up the prefix *phag–, phago–* and the entry *lithic* in this dictionary and from those terms derive the understanding that the term *phagolithic* means "stone-eating."

B. Alphabetical Placement of Entries

Alphabetization by Spelling

All the entries in this dictionary appear in a single alphabetical list. That is, there are no separate sections at the back of the book for abbreviations, biographical names, or any other category of entries. In the alphabetical list, each entry is alphabetized letter by letter, according to its spelling. In other words, factors such as spacing, hyphenation, and punctuation do not affect the placement of entries.

tidepool
tide prediction
tide-producing force

In the "whole-word" system of alphabetizing, which is used in telephone directories, in computerized sorting programs, and in some encyclopedias and dictionaries, *tide prediction* would come first in the list above, because it would be alphabetized only up to the word space after *tide.* We do not use this system, because many terms can be written either with or without word spaces or hyphens. For example, the first term cited here could correctly appear in context as either *tidepool* or *tide pool.* Under the whole-word system, this variation would affect the place in the book where the word would appear. With our letter-by-letter system, only the spelling of the word, not its form, determines its placement.

Different Entries with the Same Spelling

Occasionally the dictionary includes two or more different entries that have the same spelling. In such an instance, the entries are placed according to the following criteria:

1. A complete word precedes a word part (prefix or suffix) or an abbreviation or acronym.

gram *Metrology.* a basic unit of mass . . . .
-gram a combining form denoting something that is drawn . . . .

cent *Acoustics.* a unit of pitch interval . . . .
cent. centigrade; century.

2. A derived term follows an original word that is its source.

Hertz, Heinrich 1857–1894, German physicist . . . .
hertz *Metrology.* a standard unit of measurement for frequency, ...

leg *Anatomy.* **1.** the lower limb of the body . . . .
leg *Engineering.* any object or part that resembles a human or animal leg . . . .

3. In the case of two unrelated words, the more important or more common term comes first.

list a series of items written together or otherwise grouped in some manner; specific uses include: . . . .
list *Engineering.* to tilt toward the horizontal; lean to one side.

4. If all other factors are considered equal, then a capitalized form arbitrarily precedes a lowercase form.

DBP diastolic blood pressure.
dBp *Engineering.* decibels above one picowatt.
HD heavy-duty.
h.d. or **hd** at bedtime. (From Latin *hora decubitus.*)

Factors That Do Not Affect Alphabetization

As stated above, the alphabetization of entries is done on a letter-by-letter basis. This alphabetization system is not affected by other elements that are part of the entry word but that cannot be placed in a sequence, as letters can. Such elements are ignored in alphabetization. For example, chemical prefixes such as *cis–*, *sym–*, *tert–*, *ortho–* (*o–*), *para–* (*p–*), and *meta–* (*m–*) are not considered. Words beginning with such elements are placed according to the spelling of the term itself.

o-chlorobenzaldehyde
chlorobenzene
chlorobenzilate
o-chlorobenzoic acid

Similarly, numbers or symbols that begin an entry are not considered in alphabetization.

chlorobutane
1-chlorobutane
chlorobutanol

tocolysis
tocopherol
α-tocopherol

The same principles apply if the nonalphabetizing element occurs within the entry rather than at the beginning.

glucose oxidase
glucose 6-phosphatase
glucose phosphate
glucose 1-phosphate

The only exception to the above patterns is when the prefix, number, or symbol is the only distinguishing factor between two otherwise identical entries. Then this element is taken into account to place those two entries in sequence according to alphabetical or numerical order.

chlorobenzilate
o-chlorobenzoic acid
p-chlorobenzoic acid

butene
1-butene
2-butene

case I pointing
case II pointing
case III pointing

Guide Words

At the top of each dictionary page there are two items in larger boldface type. These are guide words that are useful in locating a word. The guide word at the top left corner of the page indicates the first main entry on the page. The guide word at the top right corner indicates the last main entry on the page. Thus the guide words show the alphabetical span of the page.

For example, suppose that a reader is looking for the word *minicell* and turns to page 1385. The guide words on this page are **miltonia** and **mineralogist.** This indicates that the desired word is not on this page, but on a later page, since *minicell* follows *mineralogist* in alphabetical order. In fact, the next page has the guide words **mineralogy** and **minigene,** indicating that the desired word *minicell* will be found on this page.

C. The Format of an Entry

The normal text typeface of this dictionary is Times Roman. This is the same typeface that is used in this Guide itself, i.e., the typeface of this sentence. Certain other typographic treatments are used to identify various elements of the dictionary.

The Main Entry Word

The main entry word or headword appears at the left margin of the column. It is printed in heavy black type and in a contrasting typeface, Helvetica, **which looks like this:**

JATO . . . .
jaundice . . . .

The Pronunciation

If a pronunciation is given for the entry, this follows next. It is enclosed in brackets and printed in Dictimes, which is a special phonetic variation of the normal Times Roman text type.

JATO [jā´tō] . . . .
jaundice [jôn´dis] . . . .

The Field Name

The description of the entry's meaning begins with the field name. This indicates the specific area of science or technology in which this term is used. It appears in Times Roman Italic, *which looks like this:*

JATO [jā´tō] *Aviation.* . . .
jaundice [jôn´dis] *Pathology.* . . .

The Definition

After the field label comes the definition itself. (If the entry has no specific field label, then the definition follows immediately after the headword.) The definition is printed in the normal Times Roman text type.

jaundice [jôn´dis] *Pathology.* a syndrome characterized by a yellowish appearance . . . .

Secondary Entries

If the entry contains a secondary entry, this item appears in Times Roman Bold, **which looks like this:**

JATO [jā´tō] *Aviation.* a takeoff utilizing one or more jet-producing units . . . . One rocket in such a system is called a **JATO engine.** . . .

Cross-References

If the entry has a cross-reference in place of or after a definition, this appears in small capitals, WHICH LOOK LIKE THIS:

jaundice [jôn´dis] *Pathology.* a syndrome characterized by . . . . *Invertebrate Zoology.* see GRASSERIE.

The Etymology

If the word is provided with an etymology, this comes at the very end of the entry. This information is enclosed in parentheses. It is printed in the normal Times Roman text type.

jaundice [jôn´dis] *Pathology.* a syndrome characterized by a yellowish appearance . . . . (From the French word for "yellow.")

D. Secondary Entries

Often a main entry contains one or more secondary entries within it. Like main entries, these items are printed in boldface type, but they are set in Times Roman to contrast with the Helevetica typeface of the main entry word.

pacemaker *Physiology.* **1.** any object, substance, or device that. . . . **2.** also, **cardiac pacemaker.** . . .

Variant Spellings and Variant Forms

A variant spelling is a term that is identical in meaning to the main entry word, but that has a slightly different spelling. Variant spellings are introduced by the term "Also."

aleurone [al´yə rōn´] *Botany.* the reserve store of protein . . . . Also, **aleuron.**

A variant form has the same meaning as the main entry word, but a somewhat different form. Variant forms are introduced by the term "Also."

taxonomic . . . . Also, **taxonomical.**
sensory memory . . . . Also, **sensory register.**

Plural Forms of the Main Entry

Another type of secondary entry is the plural form of a singular noun. This dictionary normally does not include plural forms, because they can be readily understood from the singular form. No secondary entry is provided here for plurals that follow the regular English pattern of adding the letters **s** or **es** to the end of the singular form. The plural form of such a word is obvious from the singular spelling. However, we do provide secondary entries for irregular plural forms that are not obvious from the form of the singular word.

alewife [āl´wīf´] *plural,* **alewives.** . . .

Irregular plurals often occur in fields that derive their terminology from Latin, such as Anatomy, Medicine, Bacteriology, and Botany. In Latin the plural of the neuter ending **–um** is **–a,** the plural of the feminine ending **–a** is **–ae,** and the plural of the masculine ending **–us** is **–i.**

staphylococcus *plural,* **staphylococci.** . . .

Extensions of the Main Entry

A third type of secondary entry is a term that is an extension of the main entry. It applies the same meaning as the main entry word to a different form or a different context. Extended terms are usually introduced by the term "Thus."

water-cooled *Mechanical Engineering.* describing an engine, machine gun, or other device that is cooled by water circulating in pipes or a water jacket. Thus, **water-cooled engine,** . . . .

The definition given here for *water-cooled* may also be applied to understand the extended term *water-cooled engine.*

An extended term may also be introduced by "Similarly." This indicates that the term is formed in a manner comparable to the main entry term and has an analogous meaning.

unfavorable current *Navigation.* a current that slows the speed of a vessel. Similarly, **unfavorable winds.**

Compounds Formed from the Main Entry

The fourth and final type of secondary entry is a compound term that is formed by using the main entry word as its root term. These terms are subtypes or specific versions of the more general terms, and can be best understood as part of a definition of the root term rather than as separate entries.

dose *Medicine.* the measured quantity of medicine . . . *Nucleonics.* the total or accumulated quantity of ionizing radiation. Also, DOSAGE. The **absorbed dose** in rads represents the amount of energy absorbed . . . .

E. Names of Scientific Fields

The entries in this dictionary are identified according to the particular field of science or technology in which they are used. A complete list of the fields covered in the dictionary appears on p. viii of this Introduction. The field name is printed in italic type and is always written in full, not abbreviated.

Terms with Additional Field Names

Many terms in the dictionary have different meanings in different fields. In such cases, the fields are not arbitrarily listed in alphabetical order but are placed in a logical sequence based on their relative importance. The field in which the term originated or in which it is most often used appears first.

intelligence *Psychology.* a general term encompassing various mental abilities, including the ability to remember and use what one has learned, in order to solve problems, ... (see p. 1120 for full entry)

Related fields are grouped together within an entry, so that they can be read together for a more general understanding of the term. Thus, the entry for *intelligence* begins with its basic meaning in the field of Psychology. This is immediately followed by a closely related meaning in the field of Behavior. The definition of *intelligence* then continues with meanings in the fields of Computer Technology, Telecommunications, and Military Science.

A term receives a field label only if it has a unique meaning within that field. For example, the term *intelligence* is used in Medicine, but no such field label appears, because this same use is fully covered by the definition in Psychology.

Terms without Field Names

Occasionally a term appears with no field label at all. This means that the word is used over a wide range of disciplines and cannot be limited to one specific field. Most terms of this type are word elements such as prefixes, or very general descriptive terms whose field is self-evident.

psychological **1.** of or relating to the science of psychology. **2.** relating to or involving mental processes.

We regard the field labels used in the dictionary as self-evident, since we have used the currently established name for each discipline. However, for readers who are uncertain as to the meaning of a field name or the scope that it covers, please see pages xi–xvi of this Introduction. Appearing there is a list that indicates the page of the dictionary on which each field is defined and then discussed in an essay.

F. Definitions

Even in this scientific dictionary, defining words is an art rather than a science, but there are certain systematic patterns followed in the definitions in this book.

Definitions of Nouns

Most terms that appear in the dictionary are nouns. The definition for a noun generally begins with an article, *a, an,* or *the.* The definition then typically identifies the item as to what it is and describes its physical qualities.

> **Tyrannosaurus** *Paleontology.* a genus of carnivorous saurischian dinosaurs in the family Dinodontidae; they attained a length of about 50 feet . . . .

This material is then followed by statements giving pertinent facts about location, historical context, and other such "encyclopedic" issues.

> . . . the most specialized of the carnivorous dinosaurs, and the largest land predator; found in North America and Asia; extant in the Upper Cretaceous.

Some noun terms describe entities, processes, or concepts, rather than concrete objects. These definitions vary in form according to the term being defined.

Definitions of Adjectives

Definitions of adjectives begin with a standardized participial phrase such as "describing" or "of or relating to" to indicate the adjective status of the word.

> **baric** *Meteorology.* of or relating to weight, particularly that of the atmosphere.
>
> **degenerative** *Medicine.* describing a disease . . . .

Definitions of Verbs

Definitions of verbs begin with the infinitive term "to" as an indication that the word is a verb.

> **calibrate** *Science.* **1.** to determine or mark a correct value using a meter or other measuring instrument. . . .

Standardized Definitions

In some scientific fields all or most of the pertinent terms are of the same type. Examples of such fields include Organic or Inorganic Chemistry (names of chemicals) and Mineralogy (names of minerals). In such cases the definitions tend to follow a standardized form.

> **tungsten carbide** *Inorganic Chemistry.* WS, black hexagonal crystals; insoluble in cold water; melts at approximately 2870°C and boils at 6000°C; the strongest structural material known; used in dyes, cutting tools, and electrical resistors.

The above definition follows a standard pattern for chemicals: formula; physical description; solubility; melting and boiling levels; behavior; uses.

Multiple Definitions

Many entries in this dictionary have more than one meaning for the same term in the same field. In such a case, the individual definitions are separately numbered.

The definitions are arranged so that the central meaning comes first—the one that is most common, most general, most basic, and so on. More specific meanings that derive from this sense then follow.

> **landslide** *Geology.* **1.** any perceptible downward mass movement of soil or rock under the influence of gravity. **2.** the mass itself.

Definitions in Different Fields

Many words are used in more than one area of science. In such a case, the different fields are placed in sequence using the same principle that is used for different definitions within the same field. The fields are arranged so that the core meaning comes first. See Section E, p. xxix, for more on this.

Generalized Meanings

Often a word has a broad meaning that can apply in more precise ways in specific areas of science. In this dictionary, such words are defined first by a general meaning.

> **joint** a connection between elements, or the place where this occurs; specific uses include:

Following the general description and the standard phrase "specific uses include," specific meanings are presented.

> *Anatomy.* the place of union or junction between two or more bones . . . . *Botany.* a part on a stem from which a leaf or branch grows . . . . *Engineering.* the surface where two or more structural or mechanical shapes are joined; . . . . *Robotics.* the point in a robotic arm . . . . *Graphic Arts.* either of two hinges . . . . *Electricity.* a temporary or permanent juncture of two wires . . . . *Geology.* a surface fracture, break, or parting plane . . . .

Note that the phrase is "specific uses include." It is not possible for a dictionary to anticipate all the possible uses of a general word. The technique of a general description followed by "specific uses include" allows the reader to understand new applications of the word by applying the general definition.

Usage Information

Many definitions in the dictionary include information about how a term is used. This usage information indicates that the use of the term is limited, as by being a nontechnical term.

> **armpit** *Anatomy.* the popular name for the axilla, the hollow area under the arm at the shoulder.

The usage information may indicate that the term is no longer in common use or no longer accepted as standard. Or, it may indicate that a certain term is used mainly in British English rather than in American English.

> **consumption** *Medicine.* . . . **2.** a former term for tuberculosis of the lungs.
>
> **metre** *Metrology.* the British spelling of meter. See METER.

In some cases, a separate usage note explains the distinction between terms that are similar in meaning.

> **gravity** *Mechanics.* . . . In popular usage the term *gravity* is often used to describe the force of attraction itself, but technically *gravitation* is the force and *gravity* is the observed effect of this force.

G. Cross-References

Some definitions in this dictionary do not provide a definition, but consist only of a cross-reference. A cross-reference is an entry that directs the reader to another entry that is a more common or more accepted form having the same meaning.

spotted cavy see PACA.

For such an item, the reference in small capital letters indicates the preferred entry where the full definition will be found, as in the case of *spotted cavy* above.

paca *Vertebrate Zoology.* a generally solitary nocturnal rodent, *Cuniculus paca,* . . . . Also, SPOTTED CAVY.

Please note in the example above that the definition of *paca* ends with the notation "Also, SPOTTED CAVY." If a reader first looked up *spotted cavy* and was directed here, this notation confirms that he or she has reached the correct form at *paca.*

The Use of Cross-References

In this dictionary we use the cross-reference technique sparingly; we list a term as a cross-reference only if it can be used interchangeably with the preferred term. That is, it must be exactly equivalent to the preferred term not only in its meaning, but also in the way it is used, so that if a definition were supplied for the term it would be nothing more than a repetition of the definition of the preferred term.

H. Etymologies

Etymologies, or word histories, are given selectively in this dictionary. Since the focus of this book is not the history of English, we do not etymologize every entry. But we do provide etymologies for certain words. The etymology is placed at the end of the entry, after the definition.

Mars *Astronomy.* the fourth planet . . . . (Named for *Mars,* the Roman god of war, because of its reddish appearance; the color red was associated with war.)

Most words in English gradually moved into the language over time, but there are some words that were explicitly coined at a specific point in history; for example, the names of chemical elements. Etymologies are given to explain such coinages.

lawrencium *Chemistry.* . . . (From the *Lawrence* Radiation Laboratory, Berkeley, California, founded by Ernest O. Lawrence.)

Etymologies are given for words that derive from people's names, from place names, or from other proper names.

sardine *Vertebrate Zoology.* a small marine fish . . . . (From its association with the island of *Sardinia.*)

Etymologies explain the source of acronyms and other such contrived words:

GIFT *Genetics.* an in vitro procedure . . . . (An acronym for gamete intra-Fallopian transfer.)

Also, there are etymologies for words having an anecdotal history that sheds light on the contemporary meaning.

orangutan *Vertebrate Zoology.* . . . (From pidgin Malay meaning "forest man.")

henbane *Botany.* . . . (Because its poison is especially destructive to domestic fowl.)

The etymologies in this dictionary deal only with the history of the word in English. Their purpose is to provide additional information about the meaning of the current English word.

I. Special Essays

This dictionary contains 124 special essays on the scientific fields covered in the book. The essays are placed alphabetically throughout the book. Thus the essay on Metrology appears on page 1369, in the 'M' section of the dictionary. Each essay appears on the page next to the appropriate main entry for the scientific field name it discusses. The Metrology essay is placed on the page just below the definition of the word *metrology* itself. A list of these essays and their authors appears on pages xi–xvi of this Introduction.

J. Abbreviations Used in the Dictionary

It is a policy of this dictionary to keep the use of abbreviations to a minimum. Therefore, we do not abbreviate the names of the scientific fields, and we do not use abbreviations within the context of a definition, such as *esp.* for "especially" or *usu.* for "usually."

The only use of a contextual abbreviation within a definition is the standard technique of abbreviating a genus name when it is immediately repeated with a species name. Thus in the example below, the term *"P. vulgaris"* indicates that the common primrose has the species name *Primula vulgaris.*

primrose *Botany.* any of various plants belonging to the genus *Primula* of the family Primulaceae; the **common primrose,** *P. vulgaris,* is a popular garden flower.

If there is an abbreviation used in a dictionary definition that is unfamiliar to you, please note the list below:

AC	alternating current	keV	kiloelectron volt
AD	anno Domini	kg	kilogram
atm	atmosphere	kHz	kilohertz
BC	before Christ	km	kilometer
c.	circa (about)	kw	kilowatt
°C	degrees Celsius	lat.	latitude
cm	centimeter	lb	pound
DC	direct current	long.	longitude
E	East	m	meter
e.g.	exempli gratia (for example)	min	minute
		ml	milliliter
eV	electron volt	mm	millimeter
°F	degrees Fahrenheit	μm	micrometer
ft	foot; feet	mw	milliwatt
g	gram	N	North
Hz	hertz	S	South
i.e.	id est (that is)	sec	second
in.	inch	sq	square
K	degrees kelvin	W	West

K. Pronunciations

This dictionary is intended mainly as a description of the written form of scientific and technical English, and thus it does not systematically include pronunciations, which apply to the spoken form of the language. Nevertheless, pronunciations are given here for certain terms in which the spoken form of the word is somehow in question.

macaque [mə kak´] *Vertebrate Zoology.* a member . . . .

Words That Are Pronounced

Pronunciations are given for certain words whose spoken form does not correlate with the spelling of the word, such as foreign names or foreign words that have entered English.

Descartes, Rene [dā kärt´]
gneiss [nīs]

Pronunciations are also given for terms that are often spoken but whose spoken form is unusually complex.

ichthyology [ik´thē äl´ə jē]
psyche [sīk´ē]

Occasionally, a word has more than one acceptable pronunciation and both versions are given. The pronunciation that appears first is more generally accepted in American English.

Himalayas [him ə lā´əz; hi mäl´yəz]

Pronunciation System

The pronunciation system of this dictionary is similar to the ones used in most general American dictionaries, such as the *Random House Dictionary, Webster's New World Dictionary, American Heritage Dictionary,* and *World Book Dictionary.* This system represents the sounds of English through the use of conventional letters of the alphabet and familiar symbols.

Syllable Division

Syllable division is indicated by either a space or an accent mark after the syllable. Note the word *ichthyology* above. If there is a space after a syllable, this means that this syllable gets no special emphasis in speaking. The syllables thē and ə are examples of this. If there is a bold accent mark **´** after a syllable, this indicates that it is spoken with more force than the other syllables of the word. The syllable äl´ is an example of this. If there is a lighter accent mark ´, this indicates that the syllable is spoken with some stress, but not as much as the main accented syllable. The syllable ik´ is an example of this.

Consonant Pronunciation

Consonants are pronounced just as they are spelled in their regular use, according to the following system:

b as in **b**oy	**l** as in **l**ive	**t** as in **t**ime
ch as in **ch**ip	**m** as in **m**ake	**th** as in **th**in
d as in **d**o	**n** as in **n**ot	<u>**th**</u> as in **th**ese
f as in **f**ix	**ng** as in ri**ng**	**v** as in **v**ery
g as in **g**o	**p** as in **p**ot	**w** as in **w**et
h as in **h**at	**r** as in **r**un	**y** as in **y**et
j as in **j**ob or **g**em	**s** as in **s**un or **c**ity	**z** as in **z**oo
k as in **k**ey or **c**ar	**sh** as in **sh**ow	**zh** as in mea**s**ure

Vowel Pronunciation

Vowels are pronounced according to the following system:

ə is the "schwa" sound that occurs in American English in many unstressed syllables. It may represent any of the vowels **a, e, i, o,** or **u.** It is the sound often described as "uh" that is heard for the spelling of **a** in **a**lone, **e** in it**e**m, **i** in penc**i**l, **o** in kingd**o**m, and **u** in circ**u**s.

a is the "short a" or "flat a" sound of the spelling **a** in words such as **a**pple and c**a**t.

ā is the "long a" sound of the spelling **a** in words such as **a**ge and d**a**y.

ä is the "broad a" sound of the spelling **a** in words such as **a**h, f**a**ther, and c**a**r. In American English, [ä] is also the sound of the spelling **o** in words such as **o**perate or c**o**llege. In British English, this **o** spelling is usually rendered as [o], the "short o" sound.

â is the "r-colored a" sound of the vowel that occurs before [r] in words such as **air, care, bear,** and **where.** In some American dialects, the [r] is suppressed or dropped.

e is the "short e" sound of the spelling **e** in words such as **e**nd and g**e**t.

ē is the "long e" sound of the spelling **e** in words such as m**e** and b**e**, and also of **ee** in fr**ee** and **ea** in **ea**sy.

i is the "short i" sound of the spelling **i** in words such as **i**t, sl**i**p, and p**i**ck. It is also the sound of an unstressed syllable in words such as d**e**ciduous [di sid´joo əs]. Many speakers render such sounds with a schwa [ə] sound instead of an [i], and this variation is equally acceptable.

ī is the "long i" sound of the spelling **i** in words such as **i**ce, t**i**me, and l**i**ght. It is also the sound of the spelling **y** in sk**y** and b**y**.

ō is the "long o" sound of the spelling **o** in words such as **o**pen, **o**wn, g**o**, and c**oa**t.

ô is the "aw" sound of the spelling **o** in words such as **o**rder and c**o**rn and of the spelling **aw** in **aw**ful and l**aw**. It is also heard in words such as **a**ll and c**au**ght.

oi is the "oy" sound that is heard in words such as **oi**l, b**oy**, and c**oi**n.

oo is the "ooh" or "long double o" sound of the spelling **oo** in words such as **oo**ze and t**oo**l and of **o** in d**o**. It is also heard in words such as **new.** It combines with the consonant [y] to form the sound [yoo], the "long u" sound heard in words such as **you, use,** and **few.**

ou is the "ow" sound that is heard in words such as **ou**t and **how.**

u is the "short u" sound of the spelling **u** in words such as **u**p and c**u**t, and also in "r-colored" words such as t**u**rn and b**i**rd. It is often considered to be a stressed variant of the schwa [ə] sound, the difference being that [u] occurs in one-syllable words such as **sun,** or in stressed syllables, as in **under** [un´dər].

u̇ is the sound that is heard for the spelling **u** in p**u**t and **oo** in g**oo**d. It is an intermediate sound between [u] and [oo]. Compare the progression in length of the sounds [u] in **luck,** [u̇] in **look,** and [oo] in **Luke.**

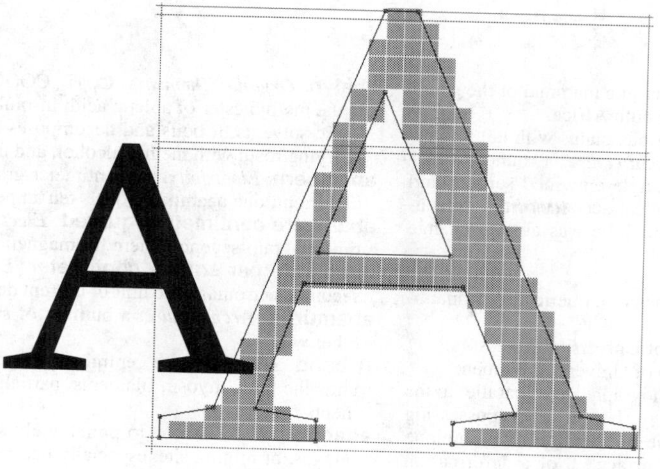

A *Science.* the first in a sequence or group. *Hematology.* a blood group in the ABO system, based on the presence of antigen A; persons with this blood type generally can donate blood safely to a person of type A or AB and receive blood from type A or O.

A or **A.** an abbreviation for: absolute; acre; adenine; alanine; ampere; angstrom; anode; area; argon.

A- *Aviation.* the general U.S. military designation for aircraft, as in A-20 or A-300; used mainly when a more specific designation, such as *B-* for "bomber," is lacking.

Å angstrom.

a or **a.** an abbreviation for: acre; are; area; atto-.

a- **1.** a prefix meaning "not" or "without," as in *asexual, amorphous.* **2.** a variant of the prefix *ad-*.

-a a suffix meaning "oxide of," as in *alumina, magnesia.*

α the Greek letter alpha (a).

aΩ abohm.

A1 time *Astronomy.* an atomic time scale in which one second equals 9,192,631,770 cycles of the cesium atom at zero field.

A-4 see SKYHAWK.

A-6 see INTRUDER.

A-7 see CORSAIR.

A-10 *Aviation.* a low-flying all-weather U.S. attack airplane designed as a tank fighter, equipped with bombs, missiles, and a cannon; popularly known as the Thunderbolt or Warthog.

A-10 II *Aviation.* a version of the A-10 attack plane equipped with a GAU-8A Avenger cannon and AGM-65 Maverick missiles; used to seek out and destroy enemy surface-to-surface missiles.

A-20 *Aviation.* a twin-engine attack bomber developed by Douglas Aircraft for use in World War II; an all-weather fighter version of this plane was designated the P-70.

A-300 *Aviation.* a large, wide-body European jetliner similar to the Boeing 767.

A-310 *Aviation.* a later version of the A-300 having a shortened fuselage and improved wing design.

A23187 *Biochemistry.* a calcium ionophore, a compound that facilitates movement of Ca^{2+} across membranes.

AA *Electricity.* a battery size for 1.5-volt dry cells, 1.4 cm in diameter and 5 cm in length.

AA *Aviation.* the airline code for American Airlines.

AA- *Ordnance.* the official designation for a Soviet missile series ranging from AA-1 to AA-13; see the specific code names such as ALKALI and ATOLL.

AA or **A.A.** **1.** Alcoholics Anonymous. **2.** anti-aircraft. **3.** author's alteration. **4.** achievement age.

AA-10 see ALAMO.

aA abampere.

aa or **a'a** [ä´ä´] *Volcanology.* a type of lava having a rough, jagged surface covered with clinkers and scoria. (From a Hawaiian word.)

aa. *Medicine.* an abbreviation for: **1.** arteries. (From Latin *arteriae.*) **2.** of each. (From Latin *ana.*)

AAA *Electricity.* a battery size for 1.5-volt dry cells, 1 cm in diameter and 4.3 cm in length.

AAA antiaircraft artillery.

AAA or **A.A.A.** American Association of Anatomists.

AAA pathway see AMINOADIPIC ACID PATHWAY.

aa channel *Volcanology.* a narrow, winding channel in which lava flows down from a central vent to feed an aa flow.

aA/cm² abampere per square centimeter.

AACP or **A.A.C.P.** American Academy of Child Psychiatry.

AAD or **A.A.D.** American Academy of Dermatology.

AADS or **A.A.D.S.** American Association of Dental Schools.

AAE or **A.A.E.** American Association of Engineers.

AAFP or **A.A.F.P.** American Academy of Family Physicians.

AAI or **A.A.I.** American Association of Immunologists.

aa lava see AA.

Aalenian [ô lēn´ē ən] *Geology.* in Great Britain, the lowermost Middle or uppermost Lower Jurassic period.

A* algorithm [ā´stär´] *Artificial Intelligence.* an algorithm for heuristic search, using an admissible heuristic function, that is guaranteed to find a minimum-cost solution if one exists.

Aalto, Alvar 1898?–1976, Finnish architect and designer; known for his modernist use of natural materials.

AAM air-to-air missile; antiaircraft missile.

AAM or **A.A.M.** American Academy of Microbiology.

AAMC or **A.A.M.C.** American Association of Medical Colleges.

AAN or **A.A.N.** American Academy of Neurology.

A & M or **A and M** Agricultural and Mechanical.

AAO or **A.A.O.** American Academy of Ophthalmology; American Association of Orthodontists.

AAOS or **A.A.O.S.** American Academy of Orthopedic Surgeons.

AAP or **A.A.P.** American Academy of Pediatrics; American Association of Pathologists.

AAPA or **A.A.P.A.** American Academy of Physician's Assistants.

AAPG or **A.A.P.G.** American Association of Petroleum Geologists.

AAPMR or **A.A.P.M.R.** American Academy of Physical Medicine and Rehabilitation.

Aardvark *Aviation.* a popular name for the F-111A fighter aircraft.

aardvark *Vertebrate Zoology.* a nocturnal, burrowing African mammal of the genus *Orycteropus* in the order Tubulidentata; it eats chiefly ants and termites and ranges from Ethiopia to South Africa. (From Afrikaans for "earth pig.")

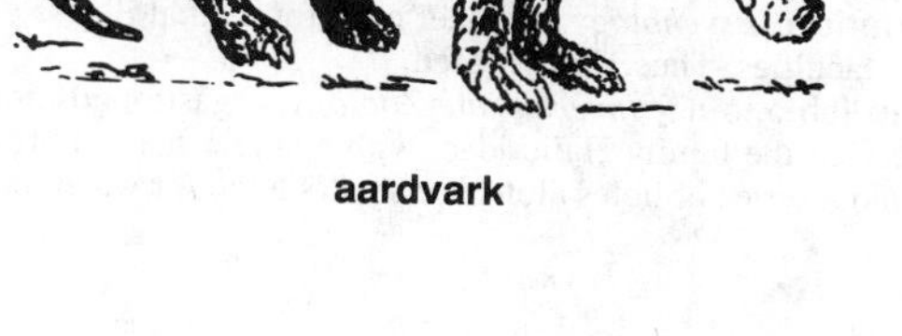

aardvark

aardwolf *Vertebrate Zoology.* a striped hyena-like mammal of the genus *Proteles* in the family Hyaenidae, native to South Africa.

Aaron's rod *Botany.* a popular term for various plants with tall flowering stems, such as a type of goldenrod. *Architecture.* a rounded ornamental molding that is decorated with a single entwined serpent and sometimes leaves and vines. (From the Biblical account of the miraculous rod used by the high priest *Aaron,* which he was able to change into a serpent.)

AAS Associate in Applied Science.

AAS or **A.A.S.** American Astronomical Society; American Astronautical Society.

AAUP or **A.A.U.P.** American Association of University Professors.

AAUW or **A.A.U.W.** American Association of University Women.

***a* axis** *Crystallography.* one of three crystallographic axes parallel to the edges of the unit cell of a crystal. *Petrology.* **1.** of a fabric possessing monoclinic symmetry, the axis lying at the intersection of the unique symmetry plane with a prominent fabric surface. **2.** of a deformation plan possessing monoclinic symmetry, the axis lying in the unique plane of symmetry and parallel to the movement plane; this represents the direction of tectonic transport.

AB *Hematology.* a blood type in the ABO system, based on the presence of both antigen A and antigen B; persons with this blood type generally can donate blood safely to a person of type AB only, but can receive blood from types AB, A, B, or O.

AB or **ab** an abbreviation for: **1.** Bachelor of Arts. (From Latin *artium baccalaureus.*) **2.** airborne.

ab- **1.** a prefix meaning "departing from" or "away from," as in *abnormal, abduction.* **2.** a prefix used to identify electromagnetic units of the centimeter-gram-second system.

ABA see ABSCISIC ACID.

abaca or **abacá** *Botany.* the Filipino name for the manila hemp plant, *Musa textilis.*

abactinal *Invertebrate Zoology.* in radially symmetrical animals, of or relating to the side opposite the mouth; in echinoderms this area lacks tube feet.

abacus *Mathematics.* a counting frame used to aid in arithmetic computation by manually sliding markers on rods or wires or in grooves. Two forms are especially common, both with markers divided into two sections: the biquinary Chinese **hsuan-pan,** in which each wire has two markers in the upper section and five in the lower, and the Japanese **soroban,** in which the upper section has one marker per wire and the lower has four per wire. *Architecture.* a slab that forms the uppermost section or division of the capital of a column.

abacus

abaft *Naval Architecture.* in a position or location toward the stern of a vessel.

abalienation *Psychology.* an older term for mental illness or the loss of mental faculties. Thus, **abalienated.**

abalone [ab´ə lo´nē] *Invertebrate Zoology.* a gastropod mollusk, genus *Haliotis,* of the family Haliotidae, with a single shell, a broad muscular foot, and a series of holes along the shell's edge for water excretion.

abalyn *Organic Chemistry.* $C_{19}H_{29}COOCH_3$, a rosin in liquid form that is the methyl ester of abietic acid; insoluble in water and miscible in organic solvents, it boils and decomposes at 360–365°C; prepared by esterifying rosin with methyl alcohol, and used as a plasticizer.

abampere *Electricity.* the centimeter-gram-second electromagnetic unit of current; one abampere equals ten amperes.

abampere centimeter squared *Electromagnetism.* a unit in the centimeter-gram-second system for magnetic moment.

abampere per square centimeter *Electricity.* the centimeter-gram-second electromagnetic unit of current density.

abamurus *Architecture.* a buttress or second wall built to support another wall.

A band *Histology.* the central region in a muscle sarcomere that contains the thick myosin filaments, partially overlapped by thin actin filaments.

abandon *Engineering.* to prematurely stop drilling or excavation at an oil, gas, or mining site, especially because such activity has become unprofitable or unsafe.

abandoned channel see OXBOW, def. 1.

abandoned cliff *Geology.* a cliff that is no longer subject to wave attack because of a relative drop of sea level.

abandoned field *Ecology.* a term used to describe an area that is in a transitional stage between active agricultural land and the native forest or other dominant vegetation of the region.

abandoned working *Mining Engineering.* any deserted excavation in which no further mining is intended. Also, **abandoned mine.**

abandonment *Petroleum Engineering.* **1.** the discontinued operation of an oil well for any of various reasons, such as the lack of oil, greatly reduced well pressure, a landslide, a damaged casing, or a flooded well hole. **2.** see ABANDONMENT CONTOUR. *Mining Engineering.* the voluntary surrender of legal rights or title to a mining claim, evidenced by failure to perform work, by conveyance, by absence, and by lapse of time.

abandonment contour *Petroleum Engineering.* a graph that shows the actual cumulative yield from an oil well in comparison to the estimated yield; an aid used in deciding upon the most financially opportune time to abandon an oil well.

abapertural *Invertebrate Zoology.* in mollusks, of or relating to a position away from the shell aperture.

abapical *Biology.* of or relating to the side opposite the apex; situated at the lower pole.

abarognosis *Medicine.* a loss of the sensation of weight.

abasement motive *Psychology.* a need to admit error or to accept blame or punishment. Also, **abasement need, abasement drive.**

abasia *Neurology.* an inability to walk, especially when due to muscular incoordination rather than paralysis.

abate *Engineering.* to carve or hammer down the surface of a material, especially so as to produce a relief figure.

abatement *Engineering.* **1.** the waste produced when a piece of material, such as metal or timber, is carved or shaped. **2.** a reduction in the quantity of some substance. **3.** a reduction of the effects of pollution, especially in relation to mine drainage. **4.** a lowering of a surface by erosion or evaporation.

abat-jour [ä´bä zhoor´] *Building Engineering.* a device that deflects daylight into a room from above; a skylight.

A battery *Electronics.* in a battery-operated vacuum-tube circuit, the battery that delivers power to heat the filaments of the tubes, as opposed to the one that furnishes the current (B battery).

abat-vent [ä´bä van´; vent´] *Building Engineering.* a louver designed to cut the wind while admitting light, air, and sound.

abaxial *Biology.* on the side opposite or facing away from the axis of an organism. *Botany.* **1.** the surface of a leaf facing away from the axis; the underside of an open leaf. **2.** of or relating to this side or surface.

abb *Materials.* coarse, low-quality wool from the edges and inferior parts of a fleece. *Textiles.* a warp yarn that is made of abb wool.

Abbé, Ernst [ä bā´] 1840–1905, German physicist and optician; developed refracting microscopes.

Abbé condenser *Optics.* a double lens that is located below the stage of a microscope, used to direct light onto the object being examined and through the focal plane of the objective.

Abbé constant see ABBÉ NUMBER.

Abbé number *Optics.* a number representing the ratio of the refractivity of an optical medium to the dispersion of that medium; it indicates the deviating effect of an optical medium on light of varying wavelengths. Also, ABBÉ CONSTANT, NU VALUE, V-NUMBER, V-VALUE, V-VALUE.

Abbé prism *Optics.* an optical system consisting of two prisms with perpendicular end faces that can be inserted between the objective and eyepiece of a telescope and used to invert an image.

Abbé refractometer *Optics.* an optical instrument that is used to determine the refractive index of a liquid, consisting of two glass prisms between which a liquid film is clamped and the angle of total internal reflection is measured.

Abbé sine condition *Optics.* a condition that must be met in order for an optical surface or any spherically corrected lens to be free from coma.

Abbé's theory *Optics.* a theory stating that a lens must be large enough to transmit the complete diffraction pattern of an object in order to form a true image of that object.

Abbevillian or **Abbevillean** *Archaeology.* a term for the period in European culture during which crude hand axes first appeared. (From *Abbeville,* in northern France, the site of early discoveries of these tools.) Also, CHELLEAN.

abbreviated dialing *Telecommunications.* an arrangement by which a telephone user can dial an abbreviated code of one or more digits that have been prearranged to represent longer telephone numbers; a feature of telephone station sets and automatic telephone switching systems and PBXs.

abb wool see ABB.

ABC or **A,B,C** *Ordnance.* atomic, biological, and chemical (weapons or missile warheads).

ABC or **abc** airborne control; automatic bass compensation; automatic brightness control.

ABC immunoperoxidase method *Immunology.* an immunocytochemical method using an antibody coupled to the enzyme peroxidase to stain tissue constituents; used to identify tissue antigens to aid diagnosis in surgical pathology.

abcoulomb *Electricity.* the centimeter-gram-second electromagnetic unit of electrical charge; one abcoulomb is the quantity of electricity that flows past any point in a circuit in one second when the current is one abampere; equal to ten coulombs.

abcoulomb centimeter *Electricity.* the centimeter-gram-second electromagnetic unit of electric dipole moment.

abcoulomb per cubic centimeter *Electricity.* the centimeter-gram-second electromagnetic unit of volume charge density.

abcoulomb per square centimeter *Electricity.* the centimeter-gram-second electromagnetic unit of surface charge density, electric polarization, and displacement.

ABC soils *Geology.* soils with distinctly developed profiles, including A, B, and C horizons.

ABC system *Geophysics.* a method of determining irregular weathering thickness, used in seismic surveying.

abd. abdomen.

Abderhalden reaction *Pathology.* a serum test formerly used to ascertain the presence of protective enzymes or "ferments" in the blood. (Named for Emil *Abderhalden,* 1877–1950, Swiss physiologist.)

abdom. abdomen.

abdomen *Anatomy.* the part of the body between the pelvis and the thoracic cavity, containing the stomach, intestines, and other digestive organs. *Invertebrate Zoology.* the elongated posterior region that is clearly separated in arthropods; it consists of up to ten segments in insects, seven segments or less in crustaceans, and it is usually unsegmented in arachnids.

abdomin- *Medicine.* a combining form meaning "abdomen" or "abdominal."

abdominal *Medicine.* relating to the abdomen or to diseases and conditions of the abdomen.

abdominal aorta *Anatomy.* the continuation of the thoracic part of the aorta, which gives rise to the inferior phrenic, lumbar, median sacral, superior and inferior mesenteric, suprarenal, renal, and testicular or ovarian arteries as well as the celiac trunk.

abdominal apoplexy *Medicine.* a sudden, severe hemorrhage within the abdomen caused by the rupture of a blood vessel, not by trauma to the abdomen.

abdominal cavity *Anatomy.* the body cavity within the abdomen, between the diaphragm and the pelvis. Also, ABDOMINAL REGION.

abdominal epilepsy *Medicine.* an epileptic seizure that presents abdominal pain, nausea, and headache rather than neurosensory symptoms as its primary symptoms.

abdominal hysterectomy *Surgery.* the removal of the uterus through an incision in the abdomen rather than through the vagina.

abdominal migraine *Medicine.* a vascular headache that is accompanied by recurrent abdominal pain and the associated symptoms of nausea, vomiting, and diarrhea.

abdominal muscle deficiency syndrome *Medicine.* a disorder marked by the absence of the muscles of the lower abdominal wall and by genital and urinary abnormalities.

abdominal pregnancy *Medicine.* implantation and development of the ovum in the abdominal cavity, most often following early rupture of a tubal pregnancy, or as the result of escape of a fertilized ovum through a fallopian tube.

abdominal quadrant *Anatomy.* any of four corresponding parts of the abdomen determined by drawing a vertical and a horizontal line bisecting the abdomen; so divided for descriptive and diagnostic purposes.

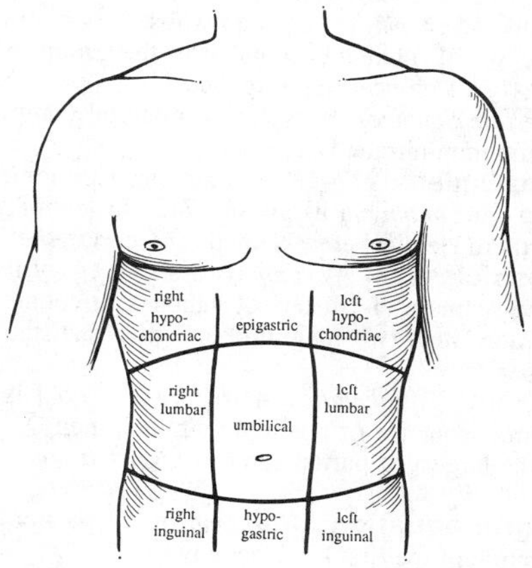

abdominal quadrants

abdominal region see ABDOMINAL CAVITY.

abdominal regions *Anatomy.* a theoretical formulation for descriptive and diagnostic purposes, dividing the abdomen into nine areas: the right hypochondriac, the epigastric, the left hypochondriac, the right lateral, the umbilical, the left lateral, the right inguinal, the pubic, and the left inguinal.

abdomino- *Medicine.* a combining form meaning "abdomen."

abdominocentesis see PARACENTESIS.

abdominocystic *Anatomy.* of or relating to both the abdomen and the urinary bladder.

abdominopelvic *Anatomy.* of or relating to the abdomen and the pelvis.

abdominoplasty [ab däm´ə nō plas´tē] *Surgery.* surgical repair of the abdomen, usually to remove excess skin after extreme weight loss; popularly known as a "tummy tuck."

abducens *Anatomy.* **1.** of or relating to abduction or to a structure that moves a body part away from the median plane of the body, such as the abducent nerve. **2.** see ABDUCENT NERVE. (A Latin term meaning "drawing away.")

abducent nerve *Anatomy.* the sixth cranial nerve, extending from the brainstem to the lateral rectus muscle of the eye. Also, ABDUCENS, NERVUS ABDUCENS.

abducent paralysis *Medicine.* a paralysis of the eye's lateral rectus muscle caused by a lesion on the sixth cranial nerve and marked by the absence of eye movement.

abduction *Physiology.* the movement of a body part away from the midline axis of the body, as in lifting an arm; the opposite of adduction. *Artificial Intelligence.* inference in which the result of a causal relationship is taken to imply the cause: given $P \rightarrow Q$, if Q is known, infer P. While not logically valid, this kind of reasoning is involved in areas such as medical diagnosis, in which a disease is inferred from the symptoms it causes.

abductor *Anatomy.* any muscle that acts to drag a part of the body away from the median line, such as the muscle in the foot that abducts the big toe.

ABE *Aviation.* the airport code for Allentown, Pennsylvania.

abeam *Navigation.* **1.** a bearing approximately 90 degrees to port or starboard of the craft's heading. **2.** of or relating to such a bearing; at right angles to the craft's fore-and-aft line.

Abegg's rule *Chemistry.* the statement that the sum of the positive and negative valencies for an element is equal to eight. (From Richard *Abegg,* 1869–1910, German chemist.)

ABEL *Artificial Intelligence.* an expert system for medical diagnosis of acid/base electrolyte disorders, developed at MIT.

Abel, Sir Frederick Augustus 1827–1902, English chemist; developed the Abel test.

Abel, John Jacob 1857–1938, American pharmacologist; first to isolate a hormone (epinephrine); also isolated adrenaline.

Abel, Niels Henrik 1802–1829, Danish mathematician; research in theory of elliptical functions, transcendental functions, and integrals.

Abelian *Mathematics.* relating to Niels Henrik Abel.

Abelian field *Mathematics* . **1.** an extension field whose Galois group is Abelian. **2.** a field of Abelian functions; i.e., a field of functions on an elliptic curve. Also, **Abelian domain.**

Abelian group *Mathematics.* a group whose binary operation is commutative; i.e., for all elements a and b in the group, $ab = ba$. (From Niels Henrik *Abel,* a pioneer in group theory.)

abelite *Materials Science.* an explosive material composed mainly of TNT and ammonium nitrate, NH_4NO_3.

Abell, George Ogden 1927–1983, American astronomer.

Abell catalog *Astronomy.* a listing of 2712 clusters of galaxies visible from the Northern Hemisphere. (Named for George *Abell.*)

Abell richness classes *Astronomy.* a six-category measure used in estimating the richness of clusters of galaxies; it counts the number of galaxies no more than two magnitudes fainter than the third brightest cluster member.

Abel's inequality *Mathematics.* if $\{b_1, b_2, \ldots, b_n\}$ is a finite monotone decreasing sequence of positive numbers, then $|\sum_{i=1}^{\infty} a_i b_i| \leq Mb_1$, where M is the largest of partial sums $|a_1|$, $|a_1 + a_2|$, $|a_1 + a_2 + a_3|$, . . . , $|a_1 + a_2 + \cdots + a_n|$.

Abel's integral equation *Mathematics.* a particular case of the Volterra equation of the first kind, namely,

$$f(x) = \int_0^x \phi(y)(x-y)^{-a}\,dy,$$

where $0 < a < 1$ and $x \geq a$, $f(x)$ is known, and $\phi(y)$ is to be determined.

Abel's problem *Mathematics.* the problem of describing the path a particle will follow if its movement is influenced only by gravity and its altitude–time function is given. In Abel's integral equation, suppose $f'(x)$ is continuous, and let $a = 1/2$ and $f(y) = (2g)^{-1/2}s(x)$, where $s(x)$ is the length of the path. Then

$$s(x) = \sqrt{(2g)}/\pi \; d/dx \int_0^x f(y)\,(x-y)^{-1/2}\,dy$$

is a solution to the equation and represents a solution to Abel's problem.

Abel summation *Mathematics.* a way of assigning a value to a series that may otherwise diverge; i.e., a special case of generalized sum for any series. Let $\sum_{n=1}^{\infty} a_n$ be a series of real numbers, not necessarily convergent, and let $f(x) = \sum_{n=1}^{\infty} a_n x^n$, where $0 \leq x < 1$. The series $\sum_{n=1}^{\infty} a_n$ is said to be **Abel summable** (**A-summable**) to S if $\lim_{x \to 1-} f(x) = S$; that is the limit as x approaches 1 from the left of $f(x)$ exists.

Abel test or **Abel process** *Physical Chemistry.* a process used to determine the flash point of petroleum and other highly volatile materials. Thus, **Abel tester.** (From Sir Frederick *Abel.*)

Abel theorem *Mathematics.* **1.** the theorem that if the complex power series $\sum_{n=1}^{\infty} a_n z^n$ has the radius of convergence equal to 1 and $\sum_{n=1}^{\infty} a_n$ converges, then

$$\sum_{n=1}^{\infty} a_n = \lim_{z \to 1-} \sum_{n=1}^{\infty} a_n z^n$$

Equivalently, if a power series $\sum_{n=1}^{\infty} a_n z^n$ converges to $f(z)$ for $|z| < 1$ and to A for $z = 1$, then

$$\lim_{z \to |1|} f(z) = A$$

2. the theorem that if a power series $\sum_{n=1}^{\infty} a_n z^n$ converges for $z = a$, then it converges absolutely for $|z| < |a|$. **3.** the theorem that if each of three series with nth terms a_n, b_n, and $c_n = a_0 b_n + a_1 b_{n-1} + \cdots + a_n b_0$, respectively, converges, then

$$\sum_{n=1}^{\infty} c_n = \sum_{n=1}^{\infty} a_n \sum_{n=1}^{\infty} b_n$$

abend *Computer Programming.* the abnormal termination of a task prior to completion because of an unrecoverable error condition, such as an attempt to divide by zero, encountered during task execution. (An acronym for abnormal end of task.)

abenteric *Medicine.* of or pertaining to organs or parts away from the intestine or to a process or disease occurring outside of the intestine.

ABE process *Microbiology.* an industrial fermentation process that utilizes the bacterium *Clostridium acetobutylicum* to produce acetone, butanol, and ethanol.

abequose *Biochemistry.* $C_6H_{12}O_4$, an unusual sugar present in the outer membrane of certain Gram-negative bacteria. Also, 3,6-DIDEOXY-D-XYLOHEXOSE.

Aberdeen Angus *Agriculture.* a breed of beef cattle having short, black hair and lacking horns. Also, ANGUS. (From the counties of *Aberdeen* and *Angus,* in Scotland, where this breed originated.)

aberrant *Biology.* being atypical or abnormal in structure or development; deviating from the usual.

aberration *Optics.* the failure of an optical system to produce a perfect image due to properties of the lens material, geometric defects in the reflecting or refracting surfaces, or other errors in the system; types of aberration include astigmatism, chromatic aberration, distortion, coma, curvature of field, and spherical aberration. Also, OPTICAL ABERRATION.

aberration of light *Optics.* an apparent displacement in the position of a celestial object due to the time it takes for light from the object to reach an earthbound observer and to the orbital motion of the earth during that time.

abetalipoproteinemia see FAMILIAL LIPOPROTEIN DEFICIENCY.

abfarad *Electricity.* the centimeter-gram-second electromagnetic unit of capacitance; one abfarad is the capacitance across which a charge of one abcoulomb produces a potential of one abvolt; equal to 10^9 farads.

abhenry *Electricity.* the centimeter-gram-second electromagnetic unit of inductance; one abhenry is the inductance across which a current that changes at the rate of one abampere per second induces a potential of one abvolt; equal to 10^{-9} henrys.

abherent *Materials Science.* a substance that inhibits or prevents a material from adhering.

Abies *Botany.* a genus of gymnosperm trees, including fir trees.

abietic acid *Organic Chemistry.* $C_{19}H_{29}COOH$, white, monoclinic plates that are insoluble in water and soluble in alcohol and dilute sodium hydroxide, melting at 173–174°C and boiling at 250°C (at low pressure); used in varnish manufacture, in esters for plasticizers, and in soaps.

ability *Psychology.* the actual power to perform a mental or physical act, whether innate or acquired by education or practice.

ab initio [ab´i nish´ē ō] *Science.* a phrase taken from Latin, meaning "from the beginning."

abiocoen [ab´ē ō sen´] *Ecology.* the sum of all the nonliving components of an environment or habitat. Also, **abiocen.**

abiogenesis [ab´ē ō jen´ə sis] *Biology.* the former theory that plant and animal life can spontaneously arise from nonliving organic matter in a relatively short time period. Also, AUTOGENESIS.

abiogenic *Biology.* not involved in or produced by living organisms. Also, **abiological.**

abioseston *Oceanography.* a collective term for dead suspended organic matter in the ocean.

abiosis *Medicine.* **1.** the absence of life. **2.** see ABIOTROPHY.

abiotic *Biology.* of or relating to nonliving things; independent of life or living organisms. *Ecology.* of or relating to the physical climatic and nonliving chemical aspects of an environment, as opposed to the biological aspects. Thus, **abiotic interaction.**

abiotic environment *Ecology.* the physical climatic and the nonliving chemical aspects of an environment.

abiotic factor *Ecology.* an environmental factor not associated with the activities of living organisms.

abiotrophy *Medicine.* premature loss of cell or tissue vitality, especially if caused by the late onset of a degenerative hereditary disease.

abjection *Mycology.* the process by which spores are cast off or discharged from the fungal structure that holds the spores.

ABL *Oncology.* a proto-oncogene that is translocated in chronic myelogenous leukemia.

ablastin *Immunology.* an antibody that inhibits the reproduction of parasites, particularly *Trypanosoma lewisi,* a rat parasite.

ablate *Aviation.* to carry out or undergo the process of ablation.

ablating material or **ablating agent** see ABLATIVE MATERIAL.

ablation *Aviation.* the loss or removal of material by an erosive process such as melting or vaporization, especially the intentional removal of material from a nose cone or spacecraft in order to provide thermal protection during reentry. *Medicine.* **1.** the removal of some part or growth by surgical excision or amputation. **2.** the removal of a noxious substance. *Geology.* the separation and removal of rocks and the formation of residual deposits, especially by erosion or weathering. *Hydrology.* **1.** the loss or wearing away of ice or snow from a glacier or a snowfield as a result of melting and evaporation. **2.** the quantity of ice or snow lost in this manner.

ablation area *Hydrology.* that part of the surface of a glacier or snowfield where annual ablation exceeds annual accumulation. Also, ZONE OF ABLATION.
ablation cone *Hydrology.* a cone of ice or snow covered by debris and formed by differential ablation.
ablation factor *Hydrology.* the rate at which a surface of ice or snow melts or evaporates.
ablation form *Hydrology.* a feature formed on the surface of ice or snow as a result of ablation.
ablation moraine *Geology.* a pile or layer of loosely consolidated rock debris that either overlies ice in the ablation area of a glacier or rests on ground moraine derived from the glacier.
ablatio placentae *Medicine.* the premature early separation of a normally situated placenta from its uterine attachment. Also, ABRUPTIO PLACENTAE.
ablative agent see ABLATIVE MATERIAL.
ablative cooling *Aviation.* the cooling of an object by means of ablation.
ablative material *Aviation.* a material, especially a coating material, that provides thermal protection to an object by burning away or disintegrating.
ablative shielding *Aviation.* the process of protecting an aircraft or spacecraft by the use of ablative materials.
ablatograph *Engineering.* an instrument used to record the amount of ablation (loss by melting or evaporation) on a glacial or ice surface.
A block *Civil Engineering.* a masonry unit with one end closed and the other open; a web separates the two ends so that when the unit is placed against a wall two cells are formed.
ablutomania *Psychology.* an irrational impulse to wash, or obsessive preoccupation with washing.
ABM antiballistic missile.
ABM-1 *Ordnance.* a Soviet antiballistic missile, code-named Galosh, designed for operation against ICBMs; it is powered by a multistage rocket motor and has a range of 185 miles; an improved version may have stop and restart capability, allowing it to distinguish warheads from decoys.
abmho *Electricity.* formerly, an electromagnetic unit of conductance; replaced by the absiemen.
abmortal *Neurology.* located or directed away from a dead or damaged part; applied especially to electric currents that are generated in injured tissue.
ABN or **abn** airborne.
Abney, William de Wiveleslie 1843–1920, English chemist; developed infrared-sensitive emulsion and used it to photograph the solar spectrum
Abney effect *Optics.* a shift in hue of a spectral color that arises from the addition of white light. Also, **Abney law.** (From William *Abney.)*
Abney level see HAND LEVEL.
abnormal *Psychology.* deviating from the normal; not conforming to the accepted societal rule or pattern.
abnormal anticlinorium *Geology.* an anticlinorium in which the axial planes of subsidiary folds converge upward. Also, **abnormal fold.**
abnormal behavior *Psychology.* a general term for behavior that is considered disruptive or inappropriate in a given situation and that is also statistically unusual.
abnormal end of task see ABEND.
abnormal glow discharge *Electronics.* a glow discharge that is in excess of that needed to surround the entire cathode completely with visible radiation.
abnormal grain growth *Materials Science.* the formation and growth of large grains, whose size is well outside the material's normal grain size distribution.
abnormality *Psychology.* an example or instance of abnormal behavior.
abnormal magnetic reading *Geophysics.* any aberrant magnetic compass reading caused by magnetic fields in a localized area that divert the needle from the magnetic pole.
abnormal magnetic variation *Geodesy.* an anomalous deflection of the compass needle from the magnetic meridian.
abnormal place *Mining Engineering.* a term for a particular area of a coal mine, usually associated with stalls or pillar methods of working, where geological or other conditions make it economically unfeasible for a miner to earn a wage equal to or above the minimum wage.
abnormal propagation *Telecommunications.* a phenomenon in which unstable atmospheric conditions prevent transmitted radio waves from traveling their normal route, thereby disrupting communications.
abnormal psychology *Psychology.* a branch of psychology that is concerned with the study of behavior and personality disorders, including their origins and treatment.
abnormal reading see ABNORMAL TIME.
abnormal reflections *Electromagnetism.* reflections of frequencies greater than the critical frequency of the ionized layer of the ionosphere, having sharply defined angles of reflections and high intensities.
abnormal synclinorium *Geology.* a synclinorium in which the axial planes of subsidiary folds converge downward.
abnormal tetragonality *Materials Science.* the distortion of the tetragonal lattice of ferritic martensite due to a partially disordered arrangement of carbon atoms in the lattice.
abnormal time *Industrial Engineering.* in time-motion studies, a time value that is far off the average, and therefore regarded as nontypical.
ABO blood groups *Hematology.* the major carbohydrate antigens that occur on red blood cells and are responsible for transfusion reactions.
ABO blood group system *Hematology.* one of the major human blood group systems; it includes two blood factors (A and B) and four blood types (A, B, AB, and O).
abohm *Electricity.* the centigrade-gram-second electromagnetic unit of resistance; one abohm is the resistance across which a steady current of one abampere will produce a potential difference of one abvolt; equal to 10^{-9} ohm.
abohm centimeter *Electricity.* the centimeter-gram-second electromagnetic unit of resistivity.
aboiement [äb´wə mant´] *Psychology.* the involuntary production of abnormal or unusual sounds, such as animal-like noises.
abomasitis *Veterinary Medicine.* inflammation of the abomasum.
abomasum [ab´ō mas´əm] *Vertebrate Zoology.* the fourth or true stomach of a ruminant (cud-chewing) animal. Also, **abomasus.**
A-bomb atomic bomb.
Abominable Snowman see YETI.
aboral *Invertebrate Zoology.* opposite to or away from the mouth.
aboriculture *Agriculture.* the cultivation of trees and shrubs.
aboriculturist *Agriculture.* an expert in the cultivation of trees and shrubs.
aboriginal *Ecology.* relating to or being one of the original inhabitants of a place.
aborigine [ab´ə rij´ə nē] *Anthropology.* one of the original or first-known human inhabitants of a place. *Ecology.* one of the original or native plants or animals of a region.

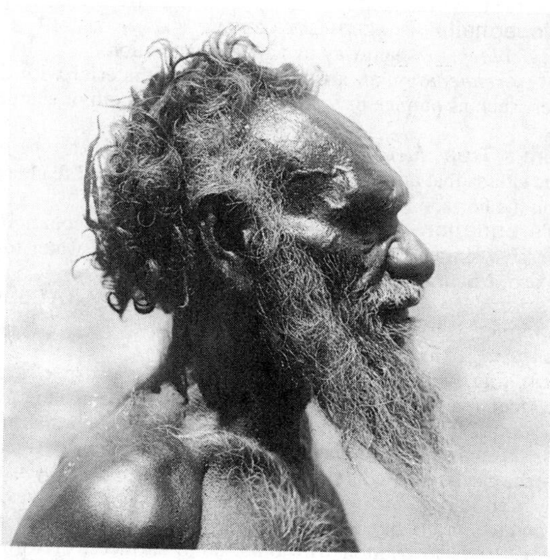

aborigine

abort *Medicine.* **1.** to give birth to a nonviable embryo or fetus; have a miscarriage. **2.** to remove or expel the embryo or fetus before it is viable outside the uterus. **3.** to terminate or stop a disease or other activity in its early stages. *Biology.* to stop or fall short of full-term growth and development. *Aviation.* **1.** to prematurely terminate an action, such as a launching or flight, especially because of an equipment failure or other technical problems. **2.** an incident in which such a termination takes place. *Ordnance.* to cut off the firing of a missile or other weapon after the order to fire has been given, but before the weapon actually fires. *Computer Programming.* **1.** to stop the execution of a program prior to its normal termination, because it is undesirable or impossible for the operation to continue. **2.** a procedure used by a computer operator or invoked automatically to execute such a premature termination.

abort branch *Robotics.* a branching instruction in the program of a robot to insure that the tool's center point does not drift away from its target.

abortifacient *Medicine.* **1.** causing or inducing abortion. **2.** an agent that will induce an abortion.

abortion *Medicine.* **1.** the premature expulsion, natural or artificially induced, of an embryo or of a nonviable fetus. **2.** the premature stopping of a natural or pathological process.

abortive *Biology.* incompletely or imperfectly developed; primitive.

abortive infection *Virology.* an infection that is terminated before it has completed its usual course or one that does not produce infectious progeny.

abortive transduction *Microbiology.* a process in which a fragment of bacterial DNA is introduced via a bacteriophage into another bacterial cell, where it remains and is expressed but does not integrate into the host DNA or replicate.

abortus *Medicine.* a fetus, weighing less than 500 grams (17 oz) or being less than 20 weeks gestational age, that is aborted.

abortus Bang reaction *Microbiology.* a test for the detection of brucellosis infection in cattle by assaying the milk for the presence of antibodies to the microorganism *Brucella abortus.*

abort zone *Aviation.* the period of time during which an abort may be successfully undertaken.

aboudikro see SAPELE.

aboulia see ABULIA.

about-sledge *Mechanical Devices.* a large hammer used by grasping it with both hands at one end of its handle.

ABP arterial blood pressure.

***ab* plane** *Geology.* the surface along which differential movement takes place; *a* represents the direction of maximum displacement, *b* the direction in the plane of movement at right angles to *a*.

AB power pack *Electricity.* **1.** an assembly that provides the required A and B direct-current voltages from an alternating-current power source. **2.** an assembly in one unit of the A and B batteries in a battery-operated vacuum-tube circuit.

ABQ *Aviation.* the airport code for Albuquerque, New Mexico.

ABR abortus Bang reaction.

abrachiocephalia see ACEPHALOBRACHIA.

abrade *Geology.* to wear away by abrasion or friction.

abrader *Archaeology.* an artifact made from material with abrasive qualities, such as pumice or sandstone; used to smooth or sharpen other objects.

Abraham's Tree *Meteorology.* the popular name for a type of cirrus radiatus clouds that appear to radiate in long plumes and feathers from a point on the horizon.

Abram's equation *Materials Science.* an empirical equation for the compressive strength of concrete as a function of its water-to-cement ratio. Also, **Abram's law.**

Abrams tank *Ordnance.* another name for the U.S. M-1A1 battle tank. (From former Army general Creighton *Abrams.*)

abranchiate *Zoology.* without gills.

abrasion the process of wearing or scraping away; specific uses include: *Medicine.* **1.** the wearing away of a substance or structure (such as the skin or the teeth) through some unusual or abnormal mechnical process. **2.** an area of skin that has been rubbed or scraped. *Geology.* the mechanical grinding, scraping, or wearing away of rock by friction resulting from impact with particles transported by wind, waves, moving water and ice, or gravity. *Engineering.* **1.** the wearing away or reduction of surfaces due to friction with other solid surfaces, or with liquids, gases, or a foreign substance. **2.** an area or surface where this process takes place. *Materials Science.* the removal of metal from parts that are difficult to machine.

abrasion drilling *Petroleum Engineering.* an oil-drilling method that uses a rock-cutting technique, rather than a conventional drill string and bit; an abrasive substance, such as blasting sand, is forced through nozzles at high velocity to erode away the rock.

abrasion mark *Graphic Arts.* **1.** a fine superficial mark on a print or film that does not penetrate to the base. **2.** a defect in a print or film caused by rubbing against another surface.

abrasion platform *Geology.* an extensive, submerged, nearly flat but sloping plain produced by continued wave erosion.

abrasion resistance *Materials Science.* the ability of a material to withstand wear from its surface.

abrasion-resistance index *Materials Science.* an expression of the abrasion resistance of materials relative to a standard material.

abrasion shoreline see RETROGRADING SHORELINE.

abrasion test *Materials Science.* any of various methods for determining the hardness of a material, as by measuring the pressure in grams and depth of a scratch from some abrasive stress.

abrasive *Geology.* **1.** a small, hard rock or mineral fragment involved in the natural abrasion of rock material or land surfaces. **2.** of or relating to such materials, or to the process of abrasion. *Materials Science.* **1.** any hard material used in the abrasion of other surfaces, such as sandpaper, emery, or the like. **2.** relating to or serving as an abrasive. Thus, **abrasive cloth, abrasive paper, abrasive belt,** and so on.

abrasive blasting *Mechanical Engineering.* a process that uses abrasive particles blown through a nozzle under air pressure to prepare a steel surface for painting.

abrasive cone *Mechanical Engineering.* a grinding wheel design whose operating surface is conical in shape.

abrasive cutoff *Materials Science.* the process of grinding a deep groove into a material with an abrasive wheel, to cause separation of a portion from the rest.

abrasive drilling *Mining Engineering.* a method of drilling characterized by the simultaneous abrasive action of the drilling medium as it rotates and is pressed against the rock.

abrasive grain *Materials Science.* hard refractory particles used as the abrasive medium in grinding operations. Similarly, **abrasive sand.**

abrasive ground *Geology.* see ABRASIVE.

abrasive jet cleaning *Engineering.* the cleaning of the surface of a solid by means of a jet stream (liquid or gas) containing abrasives.

abrasive machining *Materials Science.* the high-speed removal of metal through the abrasive action of a grinding wheel.

abrasiveness *Materials Science.* the fact or quality of being abrasive; being able to wear away another material.

abrasive wear *Materials Science.* the removal of surface material by moving contact with a harder material, which creates grooves in the softer material.

abreaction *Psychology.* a therapeutic technique for releasing repressed emotional tension by reexperiencing the original traumatic incident through words, feelings, and actions.

abreast mill *Mechanical Engineering.* a rotary disk attrition mill, with rotary steel disks that mount facing abrasive grinding plates; parts are set up in a row and are milled at the same time. Thus, **abreast milling.**

abreuvoir [ab′rə vwä′] *Civil Engineering.* in masonry, a space between stones to be filled with mortar.

abridge *Mathematics.* in an operation in which accuracy is required only to a certain extent, to drop digits that would not affect this accuracy. Thus, **abridged multiplication.**

Abrikosov-Suhl resonance *Solid-State Physics.* a scattering resonance of electronic states near the Fermi level that forms at temperatures below the Kondo temperature and accounts for the qualitative behavior of resistivity and magnetic susceptibility as functions of temperature in materials displaying the Kondo effect.

abrin *Toxicology.* a powerful poison found in the seeds of the rosary pea or jequirity, *Abrus precatorius,* formerly used topically to treat certain chronic eye disorders.

abrism *Toxicology.* poisoning due to ingestion of abrin; symptoms may include mucous membrane irritation, intestinal upsets, and internal bleeding.

Abrocomidae *Paleontology.* the rat chinchillas, a family of rodents in the infraorder Caviomorpha and the superfamily Octodontoidea; they arose in the Miocene and are still represented by two species of the central Andes.

abrosia *Medicine.* the total lack of food consumption; fasting.

abrupt *Botany.* of a plant or part, having or showing a sudden termination, as though cut off; truncate.

abruptio placentae see ABLATIO PLACENTAE.

abrupt junction *Electronics.* a transistor junction whose transition region between the P-type and N-type substrates is discontinuous.

ABS see ACRYLONITRILE BUTADIENE STYRENE RESIN.

ABS or **abs** *Computer Programming.* a shorter term for the ABSOLUTE VALUE FUNCTION.

abs or **abs.** absolute.

absarokite *Petrology.* a potassium-rich basalt with phenocrysts of olivine and augite, in a ground mass of labradorite rimmed with alkali feldspar. Ground mass may contain biotite, apatite, opaque oxides, and leucite.

abscess *Medicine.* a localized collection of pus caused by suppuration buried in soft tissues, organs teeth, or bones.

abscise *Surgery.* to cut off or remove.

abscisic acid *Biochemistry.* $C_{15}H_{20}O_4$, a plant hormone that plays a key role in growth and development by its antagonistic action to other plant hormones; for example, by inhibiting growth, bud formation, and seed germination under conditions of stress.

abscisin see ABSCISIC ACID.

abscissa [ab sis´ə] *Mathematics.* the first or horizontal coordinate of a two-dimensional Cartesian coordinate system on the Euclidean plane, usually denoted by *x*.

abscission *Botany.* the separation of leaves or other plant structures from an axis by the formation of an abscission layer.

abscission layer *Botany.* a layer of thin-walled cells that develop across the petiole at the base of a leaf, flower, or fruit when it is ready to fall from the tree.

abscopal *Radiology.* of or relating to the effect on nonirradiated tissues induced by the irradiation of other tissues within an organism.

absence seizure or **absence attack** see PETIT MAL.

absence status *Medicine.* a condition in which absence (petit mal) seizures continue for a period of minutes to hours without the presence of any normal neurological activity.

absenteeism *Behavior.* the behavior pattern among some animals of living apart from their newborn offspring and providing only minimal care.

absentmindedness *Psychology.*a habitual tendency to be overly occupied with one's own thoughts and inattentive to surrounding conditions or events.

absiemen *Electricity.* the centimeter-gram-second electromagnetic unit of conductance; one absiemen is the conductance at which a potential of one abvolt forces a current of one abampere; equal to 10^9 siemens.

absinthe [ab´sinth´] *Food Technology.* a bitter-tasting green liqueur that is flavored with wormwood or anise; it is no longer legally produced in most countries because of the extreme harmful effects of its prolonged use.

absolute having an ideal value as opposed to a conditional or relative one; specific uses include: *Chemistry.* free from other substances; not a mixture; pure; e.g., absolute alcohol. *Physics.* of or relating to measurement by means of fundamental, unvarying units. *Thermodynamics.* of or relating to the absolute temperature scale. *Geology.* referring to a date that is expressed in years or other units rather than in comparison with a similar entity. *Meteorology.* referring to the lowest or highest recorded value of a meteorological characteristic over a given period at a single station or over an area; most frequently refers to temperature.

absolute abundance *Ecology.* the precise number of individuals of a particular group in a given area, population, or community.

absolute accommodation *Physiology.* the ability of each eye to adjust separately for far or near vision.

absolute accuracy *Geodesy.* the result of evaluating all errors encountered while defining the position of a single feature or point in a geodetic datum or system. *Robotics.* **1.** the amount of tolerance allowed for each coordinate as a robot reaches a given point in space. **2.** the difference between the point where a robot was programmed to move and where it actually went.

absolute address *Computer Programming.* the number that uniquely identifies a location in computer memory. The absolute address is usually expressed in machine language. Also, MACHINE ADDRESS.

absolute addressing *Computer Programming.* the practice of using the actual number representing the memory location of the data to be processed as the address portion of the coded instruction.

absolute age *Science.* the age of an artifact, feature, or event as established by an absolute-dating process, such as radiometric dating; usually expressed in years before present.

absolute alcohol *Organic Chemistry.* a term for anhydrous ethanol that is at least 99% pure.

absolute altimeter *Navigation.* an instrument that measures and displays absolute altitudes by means of radio, sonic, or laser technology; used to find the true distance from an aircraft to the terrain.

absolute altitude *Navigation.* the distance from a height above the earth to the actual land or water surface below, as opposed to the height above sea level.

absolute angle of attack *Aviation.* the acute angle between the chord of an airfoil at any given moment during flight and the chord at zero lift. Also, AERODYNAMIC ANGLE OF ATTACK.

absolute blocking *Transportation Engineering.* a system by which a length of railroad track is divided into individual sections three or four miles in length; a train may not enter a section or block until that block is cleared of other trains.

absolute boiling point *Physical Chemistry.* the temperature at which a given substance boils, expressed in relation to absolute zero on the Kelvin temperature scale; e.g., at standard atmospheric pressure pure water has an absolute boiling point of 373.16 K, as compared to 100°C or 212°F.

absolute braking distance *Transportation Engineering.* the regulated interval between the rear of one vehicle and the front of the following vehicle.

absolute ceiling *Aviation.* the maximum altitude above sea level at which a particular aircraft with a given load could safely maintain horizontal flight under standard operating and atmospheric conditions.

absolute code *Computer Programming.* a source program code using machine instruction codes and absolute addresses that can be directly executed by the computer without prior translation by a compiler or interpreter. Also, DIRECT CODE.

absolute configuration *Biochemistry.* the spatial arrangement of four distinct substituent groups around an asymmetric carbon atom.

absolute convergence *Mathematics.* **1.** if the infinite series $\sum_{n=1}^{\infty}|a_n|$ converges (a_n real or complex), then the infinite series $\sum_{n=1}^{\infty}a_n$ is said to **converge absolutely** or to have the property of absolute convergence. **2.** if $\sum_{n=1}^{\infty}|a_n|$ converges, then the infinite product $\Pi(1 + a_n)$ is said to converge absolutely. **3.** the improper integral $\int_a^b f(x)dx$ is said to converge absolutely if $\int_a^b |f(x)|dx$ converges. Absolute convergence always implies convergence.

absolute dating *Science.* any of various processes, such as radiocarbon dating, by which an artifact, feature, or event can be assigned a fairly precise age that is not derived from the date of some other item or event. Thus, **absolute date.**

absolute datum *Geodesy.* a geodetic datum in which the reference ellipsoid is earth-centered, with its minor axis corresponding to the earth's axis of rotation.

absolute density see ABSOLUTE GRAVITY.

absolute deviation *Ordnance.* the distance between the center of the intended target and the point where a missile or projectile actually lands or bursts. *Statistics.* the magnitude of the deviation of a random variable, *a*, from a central value, *x*; i.e., the positive or absolute value of the difference between *x* and *a*.

absolute drought *Meteorology.* a term used in Britain to indicate a minimum period of 15 consecutive days with no measurable daily precipitation.

absolute efficiency *Acoustical Engineering.* the ratio of the useful acoustic power output from a transducer to the input.

absolute electrometer *Electricity.* an electrostatic instrument that determines electric potential by weighing the attraction of a charged disk against gravity. Also, ATTRACTED DISK ELECTROMETER.

absolute entropy *Thermodynamics.* the entropy of a system existing at a temperature of absolute zero; all pure substances can be assigned an entropy of zero at this point.

absolute error *Ordnance.* **1.** the distance between the center of impact or center of burst of a group of shots and the point where the farthest shot in the group impacts or bursts. **2.** the degree of error of a gunsight in relation to the accuracy of a master sight. *Mathematics.* $|x_a - x|$, where x_a is an approximation for the value of *x*.

absolute expansion *Thermodynamics.* the true expansion of a liquid with a change in temperature, allowing for the expansion of the container holding the liquid in calculating the measurement.

absolute frequency *Statistics.* the number of times that a given value appears in a sample.

absolute gain of an antenna *Electromagnetism.* the gain of an isotropic antenna isolated in space, for some specified direction.

absolute gravity *Chemistry.* the value given to denote the density or specific gravity of substances at standard conditions; e.g., at standard atmospheric pressure and a temperature of 0°C for gases.

absolute gravity station *Geodesy.* a marked point, usually in a laboratory, where the value of absolute gravity has been determined.

absolute humidity *Physics.* the ratio of the mass of water vapor present in a mixture to the volume it occupies. Also, VAPOR CONCENTRATION, VAPOR DENSITY.

absolute index of refraction see INDEX OF REFRACTION.
absolute instability *Meteorology.* the state of a column of air in the atmosphere whose lapse rate of temperature is greater than the autoconvective lapse rate, so that its air density increases with elevation. Also, AUTOCONVECTIVE INSTABILITY; MECHANICAL INSTABILITY.
absolute instruction *Computer Programming.* a computer instruction in its final binary form that specifies exactly the desired operation and can cause it to be executed. Also, BASIC INSTRUCTION.
absolute instrument *Engineering.* any instrument that measures a quantity in absolute units, such as pressure or temperature without the necessity of previous calibration.
absolute isohypse *Meteorology.* a line drawn on weather charts that indicates both constant pressure and height above mean sea level.
absolute luminosity *Optics.* the luminosity of an object as determined by comparing its brightness to the brightness of another standard object.
absolutely continuous *Mathematics.* **1.** a function f is absolutely continuous in the real interval $[a, b]$ if, for every $\varepsilon > 0$, there exists $\delta > 0$ such that $|\Sigma_{k=1}^{n}\{f(x_k) - f(x_{k-1})\}| < \varepsilon$, for all finite systems of nonintersecting intervals $I_k = (x_{k-1}, x_k)$ contained in $[a, b]$, and with

$$\Sigma_{k=1}^{n}(x_k - x_{k-1}) < \delta.$$

2. let μ and ν be two measures defined on a measure space X. ν is then said to be absolutely continuous with respect to μ if whenever $\mu(A) = 0$ for a set A in X, then $\nu(A) = 0$. Denoted $\nu << \mu$.
absolute magnetometer *Engineering.* a device able to measure the direction and force (vector) of a magnetic field, without the use of other magnetic instruments.
absolute magnitude *Astronomy.* the brightness that a star would have at a standard distance from earth of 10 parsecs (32.6 light-years). *Mathematics.* see ABSOLUTE VALUE.
absolute manometer *Engineering.* an instrument for measuring the elastic pressure of a gas or liquid, which is determined by using the constants of the apparatus.
absolute momentum *Meteorology.* the sum of a particle's momentum relative to the earth and its momentum due to the rotation of the earth. Also, **absolute linear momentum.**
absolute motion *Navigation.* motion relative to any fixed reference point. *Robotics.* a type of motion where the tool of a robot consistently returns to the same spot in the working envelope.
absolute number *Mathematics.* a number that has a single value expressed by a numeral, as opposed to a variable number represented by a letter or symbol. Similarly, **absolute term.**
absolute orientation *Navigation.* the orientation of a craft with respect to any fixed reference point. *Photogrammetry.* the act of scaling and leveling with a photogrammetric instrument to ground control of a relatively oriented stereoscopic model or group of models.
absolute parallax see ABSOLUTE STEREOSCOPIC PARALLAX.
absolute permeability *Electromagnetism.* the ratio of the magnetic flux density to the intensity of the magnetic field, measured in the number of webers per square meter in the meter-kilogram-second system of units. *Petroleum Engineering.* a measurement of the ability of a fluid, such as oil, gas, or water, to flow through a rock formation when the formation is at 100% saturation.
absolute pitch *Acoustics.* **1.** the exact pitch of a tone described in terms of vibrations per second. **2.** see PERFECT PITCH.
absolute porosity *Petroleum Engineering.* the ratio of the volume of the pore spaces or voids in a rock, available for the retention of fluid, to the total bulk volume of the rock.
absolute pressure *Physics.* the total pressure of a gas system measured with respect to zero pressure.
absolute pressure gauge *Engineering.* an instrument used to measure liquid pressure relative to a vacuum.
absolute programming *Computer Programming.* a method of source coding that uses no programming language but rather uses specific memory addresses and machine-coded (bit-pattern) operators. Also, ONE-LEVEL CODE; SPECIFIC CODE.
absolute reaction rate *Physical Chemistry.* the rate of a chemical reaction as calculated by the rate at which reactant molecules collide. Multiplied by an exponential; this gives reasonable agreement with experiment for a number of simple reactions.
absolute roof *Mining Engineering.* the entire mass of strata overlying a coal seam.
absolute scale see ABSOLUTE TEMPERATURE SCALE.
absolute space-time *Physics.* a concept in Newtonian mechanics stating that there exists a fixed frame of reference in which all measurements of space and time can be made. Also, ABSOLUTE TIME.
absolute specific gravity *Mechanics.* the ratio of an object's density to that of water, under conditions of equal temperature.
absolute stability *Meteorology.* the state of a column of air in the atmosphere whose lapse rate of temperature is less than the saturation-adiabatic lapse rate.
absolute standard *Physics.* a term for an object or quantity that is arbitrarily designated as an accepted standard and assigned a unit of one.
absolute stereoscopic parallax *Photogrammetry.* of a point on a pair of aerial photographs that are of equal principal distance, the algebraic difference of the distances of the two images from their photographic nadirs, as measured in a horizontal plane and parallel to the air base.
absolute stop *Transportation Engineering.* a signal indication requiring a vehicle to stop and remain completely stopped until the signal changes.
absolute system of units *Physics.* a system of fundamentally independent units of length, mass, time, and charge from which the meter-kilogram-second, centimeter-gram-second, Gaussian, and other systems are derived.
absolute temperature *Thermodynamics.* a temperature measurement that is made relative to the absolute zero temperature (which is 0 K on the Kelvin scale, or 0°R on the Rakine scale, –273.16°C on the Celsius scale, or –459.60°F on the Fahrenheit scale).
absolute temperature scale *Thermodynamics.* a temperature scale whose zero point corresponds to the absolute zero temperature; for example, the Kelvin scale or the Rankine scale.
absolute threshold *Physiology.* the minimum intensity and duration at which a stimulus will be perceived, usually established as a stimulus that will produce a response at least 50% of the time.
absolute time *Science.* see ABSOLUTE DATING. *Physics.* see ABSOLUTE SPACE-TIME.
absolute units *Physics.* the independent fundamental units necessary to compose the absolute system of units; specifically, units of length, mass, time, and charge.
absolute vacuum *Physics.* a completely evacuated system that has exactly zero pressure.
absolute value *Mathematics.* a norm (written as | |) on a vector space that is calculated as an ordinary Euclidean length without regard to its direction. Also, MAGNITUDE. For a real number x, $|x| = (x^2)^{1/2}$. For a complex number

$$z = x + iy, \quad |z| = (x^2 + y^2)^{1/2}$$

For a vector v with orthogonal coordinates

$$(v_1^2, v_2, \ldots, v_n), |v| = (v_1^2 + v_2^2 + \cdots + v_n^2)^{1/2}.$$

Computer Programming. a mathematical function provided by a programming language or software package that returns the positive or absolute value of the argument.
absolute value computer *Computer Technology.* a computer to which the magnitude of the quantities used is known but the algebraic sign is not relevant to the operation.
absolute viscosity *Fluid Mechanics.* the rate of deformation of a Newtonian fluid is directly proportional to the applied shear stress; the resulting constant of proportionality is the absolute or dynamic viscosity denoted by the Greek letter μ; shear stress is equal to the product of the absolute viscosity and the deformation.
absolute volume *Engineering.* the total volume of the particles of a granular material, excluding the spaces between those particles.
absolute vorticity *Fluid Mechanics.* a fluid's vorticity relative to a fixed coordinate system, such as the atmosphere's vorticity comparative to axes that do not rotate with the earth.
absolute wavemeter *Electromagnetism.* a device that measures a radiofrequency voltage by reading the length of a resonant line.
absolute weight *Engineering.* the mass (or weight) of a sample in a vacuum. Thus, **absolute weighing.**
absolute zero *Thermodynamics.* the zero point on a temperature scale of ideal gases, denoted by 0 K on the Kelvin scale or 0°R on the Rakine scale (–273.16°C on the Celsius scale, or –459.60°F on the Fahrenheit scale).
absorb *Chemistry.* **1.** to take up or receive matter. **2.** to lose intensity of light or radiation due to excitation of atoms or molecules within a substance. *Physics.* to take up energy or matter with no reflection or emission. *Electromagnetism.* to obtain energy from radiation.
absorbable suture *Surgery.* a surgical suture made of material that can be absorbed by the tissues of the body, hydrolyzed by proteolytic enzymes derived from inflammatory cells or water.

absorbance *Physical Chemistry*. the logarithm of the ratio of light intensity incident on a solution under analysis to the intensity transmitted by it; usually directly proportional to the concentration of the absorbing substance in a pure solution. Also, ABSORBANCY, EXTINCTION.

absorbancy see ABSORBANCE.

absorbed dose *Nucleonics*. the amount of energy from ionizing radiation that is absorbed by a unit mass of any material; it is expressed in rads or grays. One rad is defined as 100 ergs per gram; one gray is defined as one joule per kilogram. *Medicine*. the amount of medication that is absorbed into the bloodstream or into tissue within a given period of time.

absorbed-dose rate *Nucleonics*. the rate of energy deposition measured in rads or grays per unit time.

absorbency *Chemistry*. **1.** the taking up of a substance in bulk by other matter. **2.** the relative capacity of a substance to allow penetration by another substance. *Physics*. **1.** the taking in of radiation by the medium through which the radiation is passing. **2.** a value given to denote the amount of light or radiation absorbed by a particular substance.

absorbent *Materials Science*. **1.** any material or substance that absorbs. **2.** relating to or capable of absorption. Thus, **absorbent paper.**

absorbent cotton *Materials*. cotton from which the natural wax has been removed, for use in surgical dressings and for other medicinal or cosmetic purposes.

absorber any medium that is used to absorb something else, such as a sponge for liquid, rubber for vibration, or a liquid for vapor; specific uses include: *Engineering*. a device consisting primarily of a black surface or a system of fluids and pipes, used to absorb solar or other energy as heat. *Electronics*. any substance or device that collects and dissipates radiated energy. It may provide shielding from such energy or prevent its reflection. It may also allow particular components of radiated energy to be extracted from a material. *Nucleonics*. a material with a high neutron absorption cross section, such as cadmium or boron, used to adjust the neutron reaction rate in a reactor core.

absorber capacity *Chemical Engineering*. the maximum allowable vapor or liquid velocity in a packed or plate-type absorption tower at a given set of operating conditions.

absorber plate *Engineering*. a flat surface, usually black and sometimes incorporating the use of mirrors or transparent covers, used to collect solar energy.

absorbing boom *Civil Engineering*. a device that is floated on the surface of water in order to stop the spread of an oil spill and aid in its cleanup.

absorbing rod or **absorber rod** see CONTROL ROD.

absorbing set *Mathematics*. a subset A of a vector space X is an absorbing set if every vector in X is contained in αA for some positive scalar α.

absorbing state *Mathematics*. a special case of a recurrent state in a discrete Markov process in which the transition probability equals 1; so called because a process cannot exit the state once it enters.

absorbing well *Civil Engineering*. a well that is constructed for drainage purposes.

absorptance see ABSORPTIVITY.

absorptiometer *Biotechnology*. an instrument that measures the optical absorbance of a sample that is usually a colored liquid; used as a regulating device for waste treatment processes.

absorptiometric analysis *Analytical Chemistry*. a chemical analysis of a gas or liquid that is performed by measuring the peak electromagnetic absorption wavelengths characteristic of a specific element or compound.

absorption the process or fact of absorbing; specific uses include: *Physiology*. the passage of liquid and its solubles, such as digested food, through a cell wall into a cell by means of diffusion or osmosis. *Chemistry*. the process by which a liquid or gas is drawn into the permeable pores of a solid material. *Physics*. the action of energy or matter penetrating or being assimilated into a body of matter with no reflection or emission. *Electromagnetism*. the interaction between electromagnetic radiation and a material, in which energy is absorbed by the material by means of photoinduced transitions between energy levels. *Nucleonics*. a nuclear reaction process in which a neutron, proton, or other elementary particle is absorbed and retained by an atomic nucleus, raising the energy state of the absorbing nucleus with resultant fission or radiative capture. *Immunology*. a process in which an antibody or antigen is used to remove a corresponding antigen or antibody from a mixture. *Hydrology*. the passage of water into the lithosphere. *Telecommunications*. see ABSORPTION LOSS.

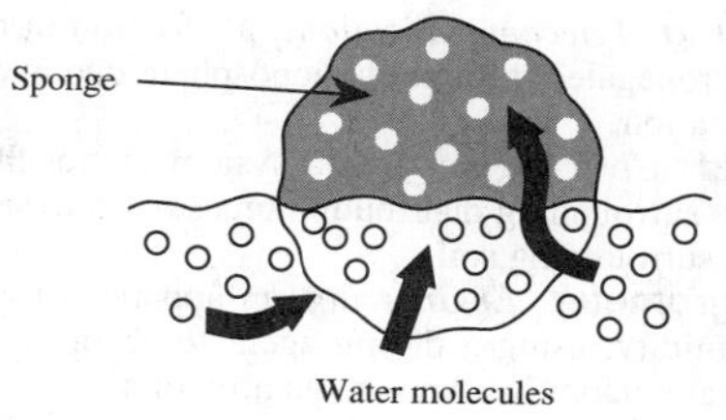

absorption: molecules actually penetrate

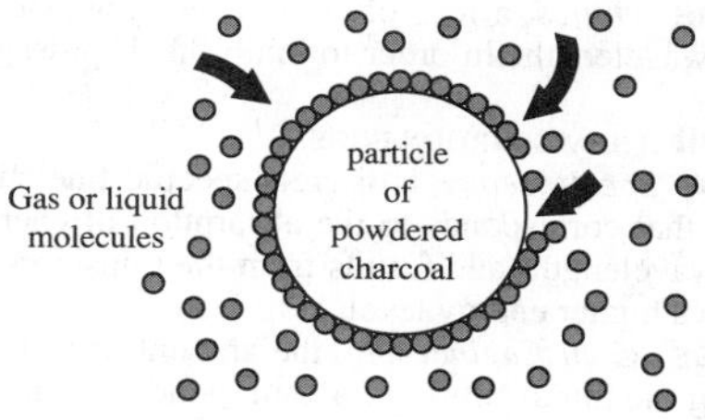

adsorption: molecules remain on the surface

absorption band *Physics*. a narrow range of frequencies, usually of electromagnetic radiation, that is absorbed by a given material while frequencies adjacent to the band are either reflected or transmitted through the material.

absorption bed *Civil Engineering*. a large pit used to absorb effluent from a septic tank; usually filled with coarse aggregate arranged about a distribution system.

absorption circuit *Electronics*. a resonant circuit that selects a particular frequency of electromagnetic energy to be absorbed. A wave trap is one example of such a circuit.

absorption coefficient *Optics*. the fraction of incident energy that is absorbed per unit pathlength. Also, ABSORPTION RATIO.

absorption column see ABSORPTION TOWER.

absorption constant see ABSORPTIVITY.

absorption control *Nucleonics*. in a nuclear reactor, the combined action of manual and automatic mechanisms for the insertion and withdrawal of neutron absorbers to control core reactivity. *Acoustical Engineering*. see ABSORPTION MODULATION.

absorption correction *Materials Science*. a correction for the photons or electrons that are lost in material, used especially in quantitative X-ray diffraction and electron microscopy.

absorption cross section *Electromagnetism*. an effective target area for aborption of radiation.

absorption current *Electricity*. a reversible component of measured current that is proportional to the voltage application time.

absorption curve *Physics*. **1.** a graphic representation of the amount of radiant energy absorbed by a material as a function of the wavelength. **2.** a graphical representation of the thickness of the absorbing material as a function of the intensity of radiant energy transmitted through the material.

absorption cycle *Mechanical Engineering*. a process within a refrigeration system during which the primary fluid (the refrigerant) and the secondary fluid (the absorbent) mix after the refrigerant leaves the evaporator.

absorption dynamometer *Engineering*. an instrument that uses the principles of electromagnetism to measure electric current, voltage, and power in a circuit, and that absorbs and dissipates the mechanical energy being measured.

absorption edge *Materials Science*. a wavelength characteristic of a filter, beyond which practically no absorption occurs in the filter. *Spectroscopy*. an abrupt change or discontinuity in the X-ray absorptivity of a given substance at a particular wavelength that corresponds to an energy jump accompanying ionization. Also, **absorption discontinuity.**

absorption-emission pyrometer *Engineering*. a thermometer used to determine the temperature of a gas by measuring a calibrated source of radiation before and after it is transmitted through and partially absorbed by the gas.

absorption factor *Nucleonics*. the probability that radiation from internally deposited radionuclides will undergo an interaction within the organ of reference.

absorption fading *Telecommunications.* a variation in the absorption of radio waves propagated through the ionosphere due to changes in the densities of ionization.

absorption field *Civil Engineering.* a system of trenches filled with coarse aggregate surrounding distribution pipes; used to seep septic tank effluent into the surrounding soil.

absorption hygrometer *Engineering.* an apparatus used to measure atmospheric humidity, using a drying agent to absorb and then weigh the amount of water vapor in a known quantity of air.

absorption index *Optics.* the quantity expressed as $a\lambda/4\pi n$, in which a is the absorption coefficient, λ is the wavelength, and n is the refractive index.

absorption lens *Optics.* a lens often used in eyeglasses, designed to absorb certain wavelengths in order to inhibit their passage through the lens.

absorption limit see ABSORPTION EDGE.

absorption line *Spectroscopy.* a discrete spectral line characteristic of atomic species that corresponds to the absorption of energy at a single frequency (or wavelength) and results from the transition of an electron from a lower to a higher energy level.

absorption loss *Civil Engineering.* the amount of water absorbed by the earth during the initial filling of a dam, pond, or canal. *Telecommunications.* **1.** in the transmission of electromagnetic waves over a wire or radio path, the loss of energy due to its conversion into other forms, such as heat. Also, ABSORPTION. **2.** the loss of energy experienced by a wave caused by intrinsic (material) absorption and by impurities consisting primarily of metal and OH^- ions in the transmission medium; also loss due to atomic defects in the transmission medium.

absorption meter *Engineering.* an instrument used to measure the amount of light transmitted through a transparent liquid or solid by means of a photocell or other light detector.

absorption modulation *Acoustical Engineering.* a type of amplitude modulation that couples a variable-impedance device, such as a microphone, to the output stage of a radio transmitter. Energy is thus absorbed from the transmitter according to the information captured by the microphone. Also, LOSS MODULATION.

absorption nebula *Astronomy.* a dark cloud of dust and gas that absorbs light from and impedes the view of background stars. Also, DARK NEBULA.

absorption oil *Materials.* a light oil that is used to absorb certain components from a gas and vapor mixture, as in the absorption of gasoline from wet natural gas. Thus, **absorption gasoline.**

absorption peak *Spectroscopy.* the wavelength at which the degree of absorption by a particular molecular substance is at a maximum.

absorption plant *Chemical Engineering.* a plant for conducting a gas-absorption process.

absorption process *Chemistry.* see ABSORPTION. *Chemical Engineering.* a process to separate gasoline from wet gas vapor by contacting a light oil with the gas to absorb the gasoline and then distilling the oil-gasoline mixture.

absorption rate *Building Engineering.* the amount of water absorbed by a brick or other porous building material, expressed in grams or ounces per minute. Also, SUCTION RATE.

absorption ratio see ABSORPTION COEFFICIENT.

absorption refrigeration *Mechanical Engineering.* a process that uses two fluids, a primary fluid (the refrigerant) such as ammonia, and a secondary fluid (the absorbent), water. When ammonia liquid is vaporized under pressure into an expansion tube, the vapors absorb heat, producing a refrigeration effect. Water absorbs the ammonia vapors to liquefy them to be distilled by a rectifying column. Thus, **absorption refrigerator**.

absorption spectrophotometer *Spectroscopy.* a device used to measure the relative intensity or brightness of the spectral lines or bands of an absorption spectrum.

absorption spectroscopy *Spectroscopy.* the study of radiant energy that is characteristically absorbed by a particular substance.

absorption spectrum *Spectroscopy.* a diagram, graph, or other display indicating the degree to which a substance absorbs radiant energy with respect to wavelength.

absorption system *Mechanical Engineering.* see ABSORPTION REFRIGERATION.

absorption tower *Engineering.* a tube in which a rising gas contacts a falling liquid so that part of the gas may be absorbed by the falling liquid. Also, ABSORPTION COLUMN.

absorption trench see ABSORPTION FIELD.

absorption wavemeter *Electronics.* a meter that displays the frequency of a circuit or device being examined, by employing a resonant circuit that absorbs maximum energy at the device frequency.

absorptive power see ABSORPTIVITY.

absorptivity *Thermodynamics.* the ratio of energy absorbed by a body to energy incident upon the same body. Absorptivity is usually denoted by the Greek letter alpha (α) and lies between zero and unity for a real body. In many practical applications, to simplify the analysis, absorptivity (α) is assumed to equal emissivity, the Greek letter epsilon (ε), although they are generally different for each body. *Analytical Chemistry.* a constant used in Beer's law that is a relative measure of the absorption intensity of a compound and is defined as the absorbance per unit concentration per unit length of light path.

absorptivity-emissivity ratio *Astrophysics.* for an object in space, the ratio between its absorption of solar radation and its infrared emission.

ABS resin see ACRYLONITRILE BUTADIENE STYRENE RESIN.

abstinence symptoms or **abstinence syndrome** *Medicine.* see WITHDRAWAL SYMPTOMS, WITHDRAWAL SYNDROME.

abstr or **abstr.** abstract.

abstract *Cartography.* a list of values of a certain quantity made for a field survey, derived directly from measurements of that quantity and recorded in the field book. *Computer Technology.* see ABSTRACTION.

abstract algebra *Mathematics.* the study of mathematical systems, each of which involves a set of elements, one or more operations on the set, and axioms for the interaction of elements and operations. It includes the theories of groups, rings, modules, fields, numbers, and categories, as well as linear algebra, representations, and Galois theory.

abstract automata theory *Computer Programming.* the mathematical discipline used in the design and study of abstract, idealized machines; usually abstractions of information-processing machines such as computers.

abstract data type *Computer Programming.* an abstraction that describes a class of data types; for example, a **linked list** describes data structures consisting of records, each of which contains a pointer to the succeeding record.

abstraction *Computer Technology.* a simplified description of a system that contains only significant information. *Hydrology.* **1.** the process of merging one or more streams into another stream having greater erosional activity. **2.** the portion of precipitation that does not become direct runoff.

abstract theory *Science.* any theory that describes a system in conceptual rather than concrete terms.

abstriction *Mycology.* the process by which dividing tissues called septa form and sever the spore from the spore-bearing structure.

abT abtesla.

abterminal *Biology.* of or relating to movement from an end toward the center; used especially in reference to the movement of electric current in a muscle.

abtesla see GAUSS.

Abt track *Civil Engineering.* a multiplate track that has teeth on different track plates staggered with respect to one another; sometimes used for railways in mountainous regions.

Abukuma-type facies *Petrology.* a type of dynathermal regional metamorphism characterized by low pressure. (Based on the metamorphic sequence of the *Abukuma* plateau, Japan.)

abulia *Psychology.* a disorder marked by the partial or total inability to make decisions.

abulic *Psychology.* **1.** of or relating to abulia. **2.** a person who is affected by abulia.

abundance *Nucleonics.* the amount of each constituent of the isotopic composition of an element.

abundance ratio *Nucleonics.* the proportion of various isotopes of an element in a given sample, usually expressed as a percentage. Also, RELATIVE ABUNDANCE.

abundant number *Mathematics.* a positive integer that is less than the sum of its factors, including 1 but not itself; for example, the factors of 12 (1, 2, 3, 4, and 6) are greater than 12.

abura *Materials.* the soft, light wood of a tropical African tree, *Mitragyne macrophylla,* which is used for construction where strength is not important.

aburton *Naval Architecture.* **1.** originally, a term used to describe casks stowed end-to-end across the width of a vessel's hold, as opposed to a fore-and-aft direction. **2.** of or relating to an object that is stowed in this manner.

abuse *Medicine.* the excessive use of a substance, such as alcohol or a drug, to an extent that is harmful; especially as distinguished from a more severe form of overuse that is characterized as addiction. Thus, **abuser**. *Psychology.* see CHILD ABUSE.

abut *Building Engineering.* to support with an abutment.

abutment *Architecture.* the part of a structure that directly receives thrust or pressure, as from an arch, vault, or strut. *Building Engineering.* an intersection between a roof surface and a wall rising above it. *Mining Engineering.* **1.** any point or surface constructed to absorb thrust. **2.** the structural portion of a furnace that is able to withstand the thrust of an arch.

abuttal *Cartography.* the boundary of a piece of land described in terms of the other pieces of land, or roads, rivers, etc., adjoining and bounding that land.

abutting joint *Building Engineering.* a joint connecting two pieces of wood so that the grains of the pieces are at an angle (usually 90°) to each other.

abutting tenons *Building Engineering.* two tenons that insert and connect within a common groove or slot made at opposite sides of a timber beam.

abvolt *Electricity.* the centimeter-gram-second electromagnetic unit of electromotive force; one abvolt is the difference of potential between any two points when one erg of work is required to move one abcoulomb of electricity between them; equal to 10^{-8} volts.

abvolt per centimeter *Electricity.* the centimeter-gram-second electromagnetic unit of electric field strength.

abwatt *Electricity.* the centimeter-gram-second unit of power.

abyss [ə bis´] *Oceanography.* **1.** an indefinite term referring to a deep part of the ocean, generally regarded as the area more than 1000 meters deep. **2.** see ABYSSAL ZONE.

abyssal *Geology.* **1.** of or relating to the region deep within the earth. **2.** see PLUTONIC. *Oceanography.* of or relating to the deepest regions of the ocean and the organisms inhabiting that environment. Also, **abysmal.**

abyssal-benthic *Oceanography.* of or relating to the ocean floor in the abyssal zone.

abyssal fan or **abyssal cave** see SUBMARINE FAN.

abyssal floor *Oceanography.* the ocean floor of the abyssal zone.

abyssal gap *Geology.* a passage in a ridge or rise that connects two abyssal plains at different levels.

abyssal hill *Geology.* a relatively low-relief hill rising up from an abyssal plain. Similarly, **abyssal knoll.**

abyssal plain *Geology.* a large, flat, nearly level deposit of sediments on the deep-ocean floor.

abyssal rock *Geology.* an igneous rock or intrusion occurring at considerable depths.

abyssal theory *Geology.* a theory that explains the formation of ore deposits as a result of the separation of minerals from the liquid stage during the cooling of the earth, and their subsequent transportation to and deposition in crust fractures.

abyssal zone *Oceanography.* the area of the great ocean depths, roughly between 2000 and 6000 meters deep (except as it includes abyssal hills or knolls, which may rise as high as 1000 meters); this area occupies about three-fourths of the total area of the Earth's oceans.

abyssolith see BATHOLITH.

abyssopelagic *Oceanography.* of or relating to organisms or phenomena in midwater, at great depths.

AC *Aviation.* the airline code for Air Canada.

AC or **A.C.** **1.** alternating current. **2.** air conditioning. Also, **ac** or **a.c.**

A.C. or **a.c.** **1.** before Christ. (From Latin *ante christum.)* **2.** before meals. (From Latin *ante cibum.)*

A/C or **a/c** **1.** alternating current. **2.** air conditioning.

Ac **1.** actinium. **2.** altocumulus.

Ac *Metallurgy.* a designation of the temperature at which a transformation takes place in the heating of steel, as indicated by the subscript number; e.g., Ac_1.

aC abcoulomb.

ac- a prefix meaning "toward," as in *accretion.*

-ac a suffix meaning "of" or "relating to," as in *cardiac.*

Ac_1 *Metallurgy.* in the thermal treatment of steel, the temperature at which the face-centered cubic phase (austenite) begins to form upon heating.

Ac_2 *Metallurgy.* in the thermal treatment of steel, a formerly designated critical transformation temperature, now recognized as the temperature above which a gradual magnetic change from ferromagnetic to paramagnetic behavior begins to occur upon heating.

Ac_3 *Metallurgy.* in the thermal treatment of steel, the temperature at which the transformation from the low-temperature body-centered cubic phase (alpha) to the face-centered cubic phase (gamma) is completed upon heating.

AC-130 *Aviation.* a U.S. Air Force cargo airplane; a modified version is used as a gunship.

ACA *Aviation.* the airport code for Acapulco, Mexico.

ACA or **A.C.A.** American College of Apothecaries.

acacia [ə kās´shə] *Botany.* any tree or shrub of the genus *Acacia,* of the mimosa family. All species are leguminous; most are spiny with clusters of small yellow flowers; many are economically important. **Acacia senegal** yields gum arabic (acacia gum). **Acacia catechu** and **Acacia suma** yield catechu, a powerful astringent. *Materials.* see GUM ARABIC.

acacia gum see GUM ARABIC.

acad. academy.

Acadian see ALBERTAN.

Acadian orogeny *Geology.* a Middle Paleozoic crustal deformation and accompanying igneous intrusion, climaxing in the late Devonian and found especially in the northern Appalachians.

acalcerosis *Medicine.* see ACALCICOSIS.

acalcicosis *Medicine.* a condition caused by calcium deficiency in the diet or by the loss of the mineral in the urine or other excreta.

acalculia [ə kal´kōō´ lē ə] *Medicine.* a form of aphasia that involves the inability to perform even the simplest mathematical operations.

acalyculate *Botany.* of a flower, without a calyx or outer covering.

Acalyptratae *Invertebrate Zoology.* a large group of shore flies, fruit flies, and leaf miners (among others), in the suborder Cyclorrhapha, with three-segment antennae and calypters (small lobes at the base of each wing) that are vestigial or absent.

Acalyptreatae see ACALYPTRATAE.

acanth- a combining form meaning "thorny" or "spiny."

acantha *Biology.* a thorn-like structure or spiny protrusion on a plant or animal.

Acanthaceae *Botany.* a family of dicotyledonous plants of which the acanthus is one type, typically perennial herbs and shrubs growing in tropical or subtropical areas with decorative, often spiny leaves and showy tubular flowers.

acanthaceous *Botany.* **1.** relating to or designating a spiny, prickly plant. **2.** relating to the Acanthaceae.

Acanthamoeba *Invertebrate Zoology.* an opportunistic free-living ameba capable of producing an amebic encephalitis, especially in chronically ill or weakened individuals, or those undergoing immunosuppressive therapy.

Acantharia *Invertebrate Zoology.* a subclass of essentially pelagic, marine protozoans, of the phylum Sarcodina, with skeletal rods constructed of strontium sulfate (celestite).

Acanthaster *Invertebrate Zoology.* a genus of Indo-Pacific sea stars in the order Spinulosida, including the crown-of-thorns sea star, *Acanthaster planci,* which eats the reef-building coral polyps.

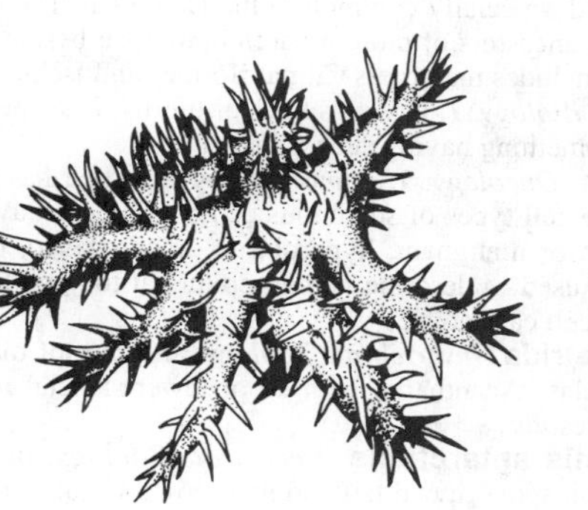

Acanthaster

acanthella *Invertebrate Zoology.* a transitional larva of the phylum Acanthocephala, which are parasitic, wormlike psuedocoelomates; it is intermediate between the acanthor (mature embryo previous to hatching) and the fully developed, infective cystacanth.

acanthesthesia *Neurology.* a perverted sensory perception by which a soft touch is sensed as the prick of a sharp object.

acanthite *Mineralogy.* Ag_2S, an opaque, metallic, blackish to dull gray monoclinic silver sulfide mineral that is dimorphous with argentite and is historically referred to as argentite, having a specific gravity of 7.2 to 7.3 and a hardness of 2 to 2.5 on the Mohs scale; an important ore of silver, found in low-temperature sulfide ore deposits associated with native silver and galena.

acantho- a combining form meaning "thorny" or "spiny."

Acanthobdellida *Invertebrate Zoology.* an order of leeches in the class Hirudinea, parasites of freshwater fish, with unusual characteristics that link true leeches with oligochaetes.

Acanthocephala *Invertebrate Zoology.* a phylum of spiny-headed, endoparastic worms with a reversible, elongated proboscis armed with hooks; juveniles develop within crustaceans and insects, and adults are parasitic in the digestive tracts of vertebrates.

acanthocephalan *Invertebrate Zoology.* any example of the Acanthocephala worms.

Acanthocheilonema perstans *Invertebrate Zoology.* a tropical filarial worm that is parasitic in humans.

acanthocheilonemiasis *Medicine.* a usually asymptomatic infection by the nematode *Acanthocheilonema perstans,* in which the adult worm can be found in the peritoneal, pleural, or pericardial cavity and occasionally in the mesentery.

acanthocladous *Botany.* **1.** any plant having thorny branches. **2.** of or relating to such a plant.

acanthocyte *Pathology.* a deformed erythrocyte distinguished by a multiple of irregularly spaced protoplasmic projections, giving the cell a thorny appearance.

acanthocytosis *Medicine.* a very rare genetic condition marked by red blood cells of distorted shapes; present in patients with abetalipoproteinemia and progressive neurologic deficits. Also, BASSEN-KORNZWEIG SYNDROME.

Acanthodes *Paleontology.* a genus of primitive fishes in the extinct family Acanthodidae, characterized by being toothless and having an elongated body and deeply embedded fin spines; extant from the Carboniferous to lower Permian.

acanthodian *Paleontology.* **1.** of or relating to the widespread class Acanthodii, primitive spiny fishes extant from the Silurian to the Permian, and the earliest known vertebrates with a movable, well-developed lower jaw. **2.** any member of the class Acanthodii.

Acanthodidae *Paleontology.* a family of spiny fishes in the extinct order Acanthodiformes; generally characterized by a single dorsal fin; extant from the Devonian to Permian.

Acanthodiformes *Paleontology.* an order of spiny fishes in the extinct class Acanthodii; it contains only one family, the Acanthodidae, but many genera are known only from isolated teeth, scales, or spines, and are therefore taxonomically imprecise; extant from the Devonian to Permian.

Acanthodii *Paleontology.* an extinct class of early fishes, the earliest known vertebrates with a movable and well-developed jaw; characterized by large eyes, a bony skeleton, and spines along the leading edge of the fins, becoming embedded in later forms; extant from the Silurian to Permian and especially common in the Devonian. The class may be related to the ancestors of more modern bony fish; besides the Acanthodiformes, it includes the orders Climatiiformes and Ischnacanthiformes.

acanthoid *Biology.* **1.** having spines or thorns. **2.** shaped like a spine or thorn. **3.** something having such a shape.

acanthoma *Oncology.* **1.** a mass of any type that has for its composition prickle cell types of squamous cells; while such a mass may be either benign or malignant, hyperplastic or neoplastic, the term is most commonly used to describe benign epithelial tumors. **2.** a well-defined squamous cell carcinoma.

Acanthometrida *Invertebrate Zoology.* an order of marine protozoans in the subclass Acantharia, with 20 or fewer skeletal rods of strontium sulfate (celestite).

Acanthophis antarcticus *Vertebrate Zoology.* the death adder, a snake with a spine-tipped tail and neurotoxic venom, found in Australia and New Guinea.

Acanthophractida *Invertebrate Zoology.* an order of marine protozoans in the subclass Acantharia; the skeleton has a latticework shell and skeletal rods.

acanthopodous *Botany.* relating to or designating a spiny, prickly plant part.

acanthopore *Paleontology.* a rod-like feature in some bryozoans, particularly the Trepostomata; acanthopores often accompany mesopores and autozooecia and are so named because they end in a spiny projection above the surface.

Acanthopteri *Vertebrate Zoology.* another name for the Perciformes, a very large order of bony fishes found worldwide in both fresh and salt waters.

Acanthopterygii *Vertebrate Zoology.* a superorder of saltwater spiny-finned fishes; often equated with the order Perciformes, although it contains other smaller orders; 60% of all living fishes belong to this superorder.

acanthor *Invertebrate Zoology.* the microscopic spiny larva of an acanthocephalan just previous to hatching, contained within the egg shell.

acanthosis *Medicine.* any abnormality of the prickle cell layer of the epidermis. Also, HYPERCANTHOSIS.

acanthosoma *Invertebrate Zoology.* the larval stage of the pelagic shrimps in the decapod family of Sergestidae.

Acanthosomatidae *Invertebrate Zoology.* a small subfamily of predaceous stink bugs in the order Hemiptera that produce a disagreeable odor.

acanthostegous *Invertebrate Zoology.* overlaid with two series of spines, as the ovicell or ooecium of certain bryozoans.

Acanthothiris *Paleontology.* an extinct genus of articulate brachiopods in the order Rhynchonellida.

acanthozooid *Invertebrate Zoology.* **1.** in bryozoan colonies, a specialized individual that secretes tubules which project as spines above the colony's outer surface. **2.** in cestode worms, the proscolex or head portion of a bladderworm.

Acanthuridae *Vertebrate Zoology.* the surgeonfishes, a family of fishes with movable knife-like spines, belonging to the suborder Acanthuroidei.

Acanthuroidei *Vertebrate Zoology.* a suborder of perch-like, largely herbivorous fish in the order Perciformes.

acanthus *Botany.* **1.** any of several prickly herbs native to the Mediterranean region. **2. Acanthus.** the genus that includes these plants. *Architecture.* the sculptured ornamentation characteristic of Corinthian capitals, consisting of a stylized representation of acanthus leaves.

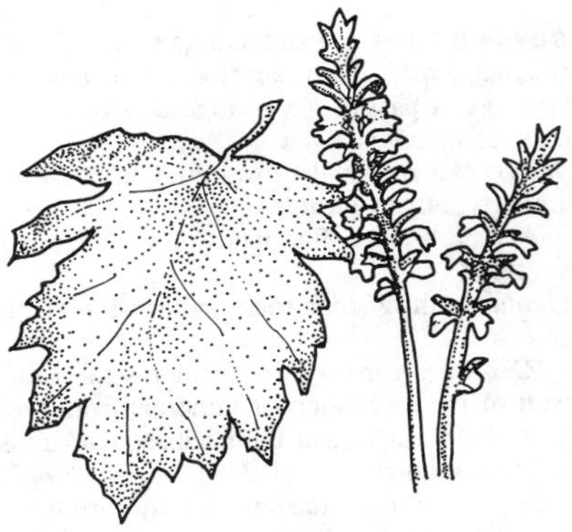

acanthus

acanthus family the common name for the Acanthaceae.

acapau see ACAPU.

acapnia *Medicine.* a low concentration of carbon dioxide in the blood.

acapsular *Medicine.* being without a capsule.

acapu *Materials.* a dark brown wood of trees of the genus *Andira,* which is used for flooring and heavy construction. Also, ACAPAU.

A-car *Transportation Engineering.* a self-propelled rail car with a control cab at one end.

acar- a combining form meaning "mite."

acardia [ə kär′dē ə] *Medicine.* the absence at birth of the heart.

acardiacus acephalus *Medicine.* a malformed fetus in which the head and heart are absent.

acardiotrophia *Medicine.* atrophy of the heart.

Acari see ACARINA.

acari- a combining form meaning "mite."

acariasis *Medicine.* an older term referring to dermatitis caused by mites.

acaricide *Toxicology.* **1.** destructive to mites. **2.** a pesticide or other agent that kills mites.

Acaridiae *Invertebrate Zoology.* a large and widely distributed family of mites (pale and weakly sclerotized) that feed on organic matter; serious pests of stored food products and skin parasites of mammals.

Acarina *Invertebrate Zoology.* the order of mites and ticks, in the class Arachnida; easily distinguished by a false head (gnathasoma) which has mouth parts, and by lack of distinct separation between the cephalothorax and the abdomen. Also, ACARI.

acaro- a combining form meaning "mite."

acaroid resin *Organic Chemistry.* a gum from a tree of genus *Xanthorrhoea*; consisting of a mixture of coumaric, cinnamic, and benzoic acids and their esters; used in varnishes and inks.

acarological *Invertebrate Zoology.* relating to mites and ticks or to the study of mites and ticks.

acarology *Invertebrate Zoology.* a branch of zoology that involves the study of mites and ticks. Thus, **acarologist.**

acarophilous *Biology.* of a plant, living in a symbiotic relationship with mites. Thus, **acarophile, acarophily.**

acarophobia *Psychology.* an irrational fear of mites, or of other tiny animals or objects.

acarpellous *Botany.* without carpels.

acarpous *Botany.* not producing fruit.

acarus *Biology. plural,* **acari.** another name for a mite.

ACAS airborne collision avoidance system.

acatalasia *Medicine.* a rare congenital disease, occurring primarily in the Japanese, in which the enzyme catalase is absent from the blood. Also, **acatalasemia.**

acatalepsia *Medicine.* an older term meaning: **1.** a mental deficiency characterized by an inability to reason or comprehend. **2.** uncertainty of diagnosis or prognosis. Also, **acatalepsy.**

acataleptic *Medicine.* characterized by acatalepsia.

acatamathesia *Medicine.* **1.** a loss of the ability to comprehend speech. **2.** a pathogenic degeneration of any of the perceptive faculties.

acataphasia *Neurology.* a speech disorder caused by a brain lesion and characterized by the inability to express one's thoughts in a logical manner, often involving the use of incorrect or nonexistent words. Thus, **acataphasic.**

acaulous *Botany.* not having a stem or a visible stem. Also, **acauline, acaulescent.**

acaustobiolith *Petrology.* a noncombustible rock that is organic or formed by organic accumulation of minerals; the category includes diatomite, radiolarite, phosphorite, and some limestones.

acaustophytolith *Petrology.* an acaustobiolith resulting from plant activity, such as a pelagic ooze containing diatoms.

ACC or **A.C.C.** American College of Cardiology.

accelerant see ACCELERATOR.

accelerate *Science.* to undergo or cause to undergo the process of acceleration; increase in velocity.

accelerated aging *Engineering.* a process in which a product or substance is placed under abnormal conditions such as intensified heat, pressure, or radiation, in order to determine the effects of long-term use or storage. *Electrical Engineering.* the testing of an electrical cable by using it with twice its normal working voltage in order to test its stability. Thus, **accelerated aging test.**

accelerated erosion *Geology.* **1.** erosion that occurs at a greater rate than is normal for a particular region, usually as a result of human activities that disturb or destroy the natural vegetation. **2.** soil erosion that occurs at a rate faster than the rate at which new soil horizons can form from parent material.

accelerated fatigue *Engineering.* a process in which equipment or a circuit is intentionally subjected to an extreme stress level, in order to simulate the effects of average use over an extended period of time. Thus, **accelerated fatigue test.**

accelerated hypertension see MALIGNANT HYPERTENSION.

accelerated life *Engineering.* the intentional overuse of a device, circuit, or system in order to evaluate the number of years the system would remain operable under usual or average circumstances. Thus, **accelerated life test.**

accelerated motion see ACCELERATION.

accelerated sulfur vulcanization *Materials Science.* the process of accelerating the polysulfide crosslinking of rubber through the use of organic catalytic intermediaries.

accelerated test *Engineering.* a test in which equipment, a material, or a circuit is intentionally subjected to an extreme stress level for a short time, in order to simulate the effects of its average use over an extended period of time.

accelerated testing *Engineering.* any process of equipment testing in which short-term experimental results are extrapolated to predict long-term in-service performance.

accelerated weathering *Engineering.* a test for wearability in which coatings or surfaces are exposed to intensified conditions in order to rapidly simulate and assess the effects of the weathering that would ordinarily occur over a greater period of time.

accelerating agent *Materials Science.* any material that increases the rate of a reaction.

accelerating electrode *Electronics.* in klystron, cathode-ray, and other electronic tubes, the electrode that increases electron velocity due to its high potential with respect to the cathode.

accelerating potential *Electronics.* the high positive potential in an electron tube that is applied to the electron accelerating electrode.

accelerating relay *Electricity.* a relay whose function is to assist in starting a motor or in increasing motor speed.

acceleration *Mechanics.* the vector representing the rate of change in velocity vector over time. It is expressed in meters (or feet) per second per second, and it involves an increase or decrease in speed and a change in direction. *Science.* in general, any increase in the speed or rate at which some process occur; in technical use *acceleration* and *speed* are not synonymous.

acceleration analysis *Mechanical Engineering.* a study to determine the type of acceleration—rectilinear, curvilinear, gravimetric, or angular—and the rate of change of the object in motion.

acceleration error *Navigation.* **1.** an error in a magnetic compass caused by the inclination of the earth's magnetic field. The after end of the compass card is tilted during acceleration or deceleration, causing an error, especially on east or west headings. **2.** a false reading in a sextant equipped with an artificial horizon, due to motion of the craft.

acceleration error constant *Control Systems.* the ratio of the acceleration of a controlled variable of a servomechanism to the actuating error, provided the actuating error is constant.

acceleration feedback *Aviation.* the transmission of data regarding the acceleration and gravitational forces acting upon an aircraft or missile from an accelerometer to the cockpit or mission control.

acceleration globulin *Biochemistry.* a plasma globulin that aids in blood clotting. Also, FACTOR V, ACCELERIN.

acceleration mechanisms *Astrophysics.* a term for the means by which cosmic rays and particles from solar flares are accelerated to high levels of energy.

acceleration of free fall see ACCELERATION OF GRAVITY.

acceleration of gravity *Mechanics.* the acceleration of an object due to the gravitational pull of a planet; on the earth, it has a standard MKS value of 9.80665 meters per second per second, but the actual magnitude varies with latitude and elevation. Also, ACCELERATION OF FREE FALL, APPARENT GRAVITY.

acceleration potential *Fluid Mechanics.* the partial derivative of the acceleration potential function of a flow field, with respect to some directed distance, is numerically equal to the acceleration vector of a fluid particle in that direction in the flow field; in an incompressible fluid, the product of the acceleration potential function and the fluid density is equal to the negative change of pressure of the fluid.

acceleration stress *Medicine.* the effect of acceleration on the physiological condition of the human body; used especially in reference to the effects of rapid acceleration or deceleration during aviation and space flights.

acceleration switch *Electricity.* a device that automatically opens or closes a circuit when the acceleration of a body to which the device is attached exceeds a predetermined rate in a given direction.

acceleration time *Computer Technology.* the time it takes for a magnetic tape drive or other such storage device to reach operating speed from a standing start.

acceleration tolerance *Physiology.* the limit of linear and centripetal gravitational force (expressed in G units) under which bodily systems will continue to function normally; used in reference to the effects of rapid acceleration or deceleration during aviation and space flights. *Engineering.* the maximum amount of force due to acceleration that an object or device can withstand.

acceleration voltage *Electronics.* in electron tubes, the potential difference existing between the electron accelerating electrode and the cathode.

accelerator something that causes acceleration or speeding up; specific uses include: *Mechanical Engineering.* a pedal or lever designed to control the speed of an engine by actuating the carburetor throttle valve or fuel-injection control. *Nucleonics.* an apparatus designed to accelerate charged particles to sufficient velocities (energy levels) to cause nuclear rearrangement upon bombarding a target so that the resulting reactions can be studied. Also, ATOM SMASHER, PARTICLE ACCELERATOR. *Materials Science.* any substance that is added to increase the rate at which some material process takes place, such as the hardening of cement or plaster or the development of a photograph. *Graphic Arts.* any alkali, such as sodium carbonate, that is added to a photographic developing solution to speed the developing process.

particle accelerator

accelerator jet *Mechanical Engineering.* a mechanical device used to inject additional fuel into a carbureted system to increase speed and power on demand by an operator.

accelerator linkage *Mechanical Engineering.* a device that transmits a command from an actuator to the accelerator pump or to a throttling valve.

accelerator pedal *Mechanical Engineering.* a device that controls the speed of an engine or electric motor by means of foot pressure exerted by an operator.

accelerator pump *Mechanical Engineering.* a piston that moves past an overrunning fuel port in a cylinder to pressurize a small quantity of fuel for injection into a carbureted system.

acceleratory reflex *Physiology.* an involuntary postural response that is elicited by the adequate stimulation of the labyrinthine sensory receptors in the inner ear by linear or angular acceleration.

accelerin see ACCELERATION GLOBULIN.

accelerogram *Engineering.* a graphic representation of the data recorded by an accelerograph.

accelerograph *Engineering.* an accelerometer that contains a pendulum for measuring the force of acceleration at a given point on earth, as during an earthquake or an underground explosion.

accelerometer *Engineering.* an instrument used to measure and record acceleration in a given direction; employed in aircraft, missiles, and spacecraft.

accelofilter *Chemistry.* a filtration device that increases the rate of filtration by forcing the liquid through the filter.

accent *Linguistics.* **1.** the relative prominence of a syllable within a word, based on its loudness, pitch, or length, or on some combination of these factors. In the word "access," the accent is on the first syllable: [ak´ses]. **2.** also, **accent mark.** a symbol indicating such prominence.

accent lighting *Civil Engineering.* light directed to highlight an object or to attract attention to an area.

accentor *Vertebrate Zoology.* any of several small Old World birds of the family Prunellidae, having a sparrowlike appearance, although with a more slender bill, especially the hedge sparrow. Also, **accenter.**

accentuation *Electronics.* the process of highlighting a particular band of frequencies by amplifying those frequencies.

accentuator *Electronics.* an electrical network or device that is used for preemphasis; i.e., to amplify a particular band of frequencies. Such devices include filters, equalizers, and tone controllers. Also, **accentuator circuit.**

accept *Computer Technology.* to receive an input with no detectable errors.

acceptability *Engineering.* the status of equipment, products, devices, or procedures that meet or exceed minimum standards.

acceptable daily intake *Food Technology.* a level of safety for chemicals used as food additives, usually equivalent to 1/100th of the maximum amount that causes no negative effect in humans. Also, ADI.

acceptable quality level *Industrial Engineering.* an established standard for the minimum percentage of output which must meet quality criteria. Also, **acceptable reliability level.**

acceptable risk *Geophysics.* the minimum level of expected seismic damage deemed tolerable when determining design requirements for structures.

acceptance criteria *Industrial Engineering.* the criteria established for determining if the offered output of a production process is acceptable; this may include physical and material specifications, performance standards, or other characteristics.

acceptance number *Industrial Engineering.* the maximum number of defective items per lot allowed for acceptance of the lot.

acceptance region *Statistics.* the range of values of the sample test statistic that will cause the null hypothesis to be accepted.

acceptance sampling *Industrial Engineering.* **1.** the sampling by the purchaser of a proportion of the output of some productive process to determine that it meets required standards; the results of the sampling test are then applied statistically to the whole production order to determine overall performance in meeting standards. **2.** the science or technique of sampling and acceptance decisions.

acceptance test *Industrial Engineering.* any test performed on a product or item to determine if it meets specifications or standards.

accepted depth *Oceanography.* the closest possible determination of the depth of a Nansen bottle or other such water-sampling device at the moment it is reversed.

accepting state *Computer Science.* a state of a finite automaton in which the input string is accepted as being a member of the language recognized by the automaton.

acceptor *Physical Chemistry.* a molecule, or part of a molecule's structure, that accepts an electron pair from a donor. Also, ELECTRON ACCEPTOR. *Chemistry.* **1.** any species that combines with or accepts other material. **2.** a chemical in which the reaction rate with another chemical increases because the other substance is undergoing another reaction. *Solid-State Physics.* a small quantity of impurities introduced into a crystalline semiconductor, and having a lower valency than the semiconductor from which they attract electrons so as to produce holes, creating *p*-type conduction. *Chemical Engineering.* a calcinated carbonate that absorbs the carbon dioxide evolved duringa process of coal gasification.

acceptor atom *Solid-State Physics.* an atom that has a hole for receiving an electron in a semiconductor material.

acceptor circuit *Electronics.* a circuit that passes a particular frequency with unity gain and attenuates all other frequencies.

acceptor control *Biochemistry.* the regulation of the rate of oxygen consumption by the concentration of ADP as phosphate acceptor; an element of cellular respiration in mitochondria.

acceptor material or **acceptor impurity** *Solid-State Physics.* see ACCEPTOR.

acceptor stem *Genetics.* the extended portion of a transfer RNA molecule that attaches to a specific amino acid.

access *Civil Engineering.* **1.** the freedom, ability, or legal right to pass from one point to another, as on a public right of way or on the sea. **2.** a means of entering or exiting a highway or expressway. *Computer Programming.* **1.** to locate a unit of code or data in memory and use it in a process. **2.** to identify a mass storage location and a memory location and initiate the transfer of data from mass storage into memory. **3.** to receive or derive two addresses and then transfer bits between them; e.g., between main memory and the ALU or a control unit. **4.** to write records to a file or read records from a file. **5.** any operation in which one or more of these processes takes place.

access arm *Computer Technology.* the element of a movable-head disk drive that supports one or more read/write heads and that is moved by a servo to position the read/write heads over the magnetic disk cylinders.

access code *Telecommunications.* **1.** the preliminary digits that a user must dial or key at the beginning of a call to be connected to a particular outgoing trunk, channel, service, or line. **2.** a numerical delineation of facility or internetwork switching. *Computer Technology.* a preliminary character sequence that a user must provide in order to gain entrance to a system; used to restrict the availability of data to certain authorized users.

access control *Transportation Engineering.* any system that limits the location or rate of access; e.g., an expressway with entrance available only at certain on-ramps. *Computer Technology.* **1.** special circuits in the central processing unit that inform memory or an interfacing input/output device as to whether the data transfer about to take place is a read or a write operation. **2.** a means of limiting the availability of proprietary or classified data or programs to only those individuals or programs permitted access to it. Thus, **access-control list, access-control register.**

access-control words *Computer Programming.* in microprogramming, a bit pattern stored in control memory that is used to transmit memory access information to or between operational programs.

access distance *Transportation Engineering.* the distance between a point of origin and a given station or stop.

access door *Building Engineering.* a temporary door fabricated during construction of a building to allow access to concealed areas, such as crawl spaces.

accessibility *Forestry.* **1.** the degree to which wood can be penetrated by a specific chemical under various conditions; used as a measure before cutting methods. **2.** the measured ease with which timber can be felled and cut into pieces during harvesting.

access line *Telecommunications.* **1.** a line connecting terminal equipment to a switching system. **2.** a circuit linking a PBX user with a switching center. **3.** any line giving access to larger system or network.

access management *Computer Technology.* the array of physical, hardware, and software techniques used to limit access to a computing system to authorized personnel, programs, or network nodes.

access matrix *Computer Programming.* a tabular representation of authorization granted to users of a given system's information and resources. Similarly, **access list.**

access mechanism *Computer Technology.* a device that moves one or more read/write heads into position so that certain data may be read to or written from a fixed data medium.

access method *Computer Technology.* any of various routines that connect programs with the information these programs transmit to and from memory for improved efficiency or flexibility; e.g., ISAM.

access mode *Computer Programming.* In COBOL, a clause that defines the manner in which the records of a file are to be accessed, such as sequential or random.

accessorius [ak sə sôr´ē əs] *Anatomy.* **1.** a muscle originating from one of the six lower ribs and inserting into one of the six upper ribs. **2.** a spinal accessory nerve. **3.** serving a supplementary function; accessory.

accessory something that serves a lesser or subsidiary function; specific uses include: *Anatomy.* **1.** relating to or designating a lesser feature that resembles some more significant organ in structure and function, such as the accessory pancreatic duct. **2.** see ACCESSORY NERVE. *Mechanical Engineering.* a device that is mounted in the peripheral area of a central machine and driven by belts, gears, vacuum, or electric motor to assist in the operation of that machine. *Nutrition.* ingested substances that are required for the maintenance of normal health but that are not themselves sources of energy, such as water, inorganic salts, and vitamins. *Geology.* see RESURGENT, def. 2.

accessory bud *Botany.* a bud located above or either beside or near a normal side of the main axillary bud.

accessory cell *Botany.* a morphologically distinct epidermal cell that is associated with guard cells. Also, SUBSIDIARY CELL.

accessory cells *Immunology.* the nonlymphocytic cells that act to regulate an immune response or lymphocyte processes and growth; frequently used to define antigen-capturing cells, e.g. macrophages.

accessory chromosome see SUPERNUMERARY CHROMOSOME.

accessory cloud *Meteorology.* a cloud whose formation and continuation depend upon the existence of a major cloud genus, either as an appendage or as an immediately adjacent cloudy mass.

accessory ejecta *Volcanology.* pyroclastic material formed from fragments of earlier lavas or preexisting solidified volcanic rocks originating from the same volcano. Also, RESURGENT EJECTA.

accessory element see TRACE ELEMENT.

accessory fruit see PSEUDOCARP.

accessory gland *Anatomy.* any gland of secondary importance in a system; a minor mass of glandular tissue that is situated near or at some distance from a similar structure. *Invertebrate Zoology.* one of the glands of varied structure and function associated with the male reproduction organs in insects.

accessory mineral *Mineralogy.* any mineral that is usually found in minor amounts in rocks and whose presence does not affect the proper classification of the rock.

accessory movement see SYNKINESIA.

accessory nerve *Anatomy.* the eleventh cranial nerve, serving structures in both the head and thorax.

accessory pigments *Biochemistry.* a class of pigments that are involved in photosynthesis by transferring absorbed energy to chlorophyll; found in chloroplasts and blue-green algae.

accessory plate *Optics.* a thin plate of quartz, gypsum, or mica used with a petrological microscope to determine the optical characteristics of translucent minerals.

access path *Computer Technology.* a sequence of steps, as from a dictionary to subdirectory or from one network node to another, that will reach a desired file, network node, or user.

access road *Civil Engineering.* **1.** a secondary road that allows movement on or off a highway or expressway. **2.** any thoroughfare that allows travel to an isolated area.

access time *Computer Technology.* **1.** the time interval between a request for data to be retrieved from memory and the time the data is available for processing. **2.** the time interval between generation of a read signal from the external interface of a storage device and the first bit or bit pattern to be read back out from the storage medium. Also, READ TIME. **3.** the time interval between generation of a mass storage address by a control unit or input/output controller and the initiation of data transfer between that location and the main memory. *Transportation Engineering.* the time elapsed in traveling from an origin to a station or stop and waiting there for a transport vehicle.

access tunnel *Civil Engineering.* a tunnel furnished for an access road.

access type *Computer Technology.* in a limited-access system, the allowable operations that a user may perform on a given file or program, such as read only, read/write, or execute. *Computer Programming.* in Ada programming language, the pointer fields embedded within a record that provide access to other records allowing user-defined dynamic data structures.

accident *Medicine.* **1.** an event that is unforeseen, especially one that is injurious. **2.** an unexpected complication that is not in the normal course of a disease or treatment. *Hydrology.* any occurrence that interrupts, interferes with, or ends the normal development of a river system.

accidental *Ecology.* **1.** relating to or being an individual organism found in a locale that is not a normal habitat for its species. **2.** describing a plant species that constitutes one-fourth to one-half of the vegetation of a given area. Thus, **accidental species.** *Anatomy.* see ACCIDENTAL PATTERN.

accidental degeneracy *Quantum Mechanics.* a special feature of the hydrogen atom characterized by degeneracy of the l quantum number related to angular momentum.

accidental ejecta *Volcanology.* pyroclastic rock material formed from fragments of preexisting rocks that are nonvolcanic or come from volcanic rocks unrelated to the erupting volcano. Also, NONCOGNATE EJECTA.

accidental error *Ordnance.* an unpredictable and random error in the operation of a weapon or instrument, caused by human or mechanical failure, as opposed to a consistently recurring error caused by an inherent fault in the system.

accidental inclusion see XENOLITH.

accidentalism *Medicine.* the theory that disease is only an accidental change from normal health and can be avoided or cured by correcting adverse external conditions; thus the focus is on treating disease symptoms rather than on examining causes.

accidental migration see DISPERSAL.

accidental pattern *Anatomy.* in fingerprint classification, one of the four basic types of patterns along with arch, loop, and whorl; an accidental pattern does not have a single defining form as the other types do, but combines features of these types.

accidental reinforcement *Behavior.* an event that follows a response and that seems to be a reward or punishment for it, but is actually a coincidence.

accidental rock *Geology.* igneous, metamorphic, or sedimentary rock that is derived from volcanic rock but not from the magma involved in an eruption.

accident block *Volcanology.* a rock fragment or chip that broke off from the subvolcanic basement and was ejected from a volcano.

Accident Cause Code *Industrial Engineering.* a standardized code number representing a given accident cause; used in the analysis of accident records to identify frequent causes of accidents.

accident frequency rate *Industrial Engineering.* the number of accidents at a facility or in a production process, normally stated as the number of accidents causing lost work time in a measuring period, multiplied by one million, and divided by the total man-hours worked during that period.

accident-prone *Psychology.* likely or inclined to injure oneself in accidents.

accident severity rate *Industrial Engineering.* a measure of accident severity effects, normally defined as the number of worker days lost due to accident injuries in a time period, multiplied by one million, and divided by the total man-hours worked during that period.

Accipitridae *Vertebrate Zoology.* a family of birds in the order Falconiformes; its nine diverse subfamilies include Old World vultures, kites, hawks, and eagles.

acciptor *Vertebrate Zoology.* any of various bird-eating, short-winged hawks of the family Accipitridae, including the North American goshawk and Cooper's hawk.

acclimatation see ACCLIMATION.

acclimate *Ecology.* to undergo the process of acclimation. Also, ACCLIMATIZE.

acclimated microorganism *Ecology.* any microorganism capable of acclimating to an environmental change, as in temperature.

acclimation *Ecology.* a reversible physiological or morphological change evinced by one individual in response to some alteration in its environment, such as a change in temperature or climate. Also, ACCLIMATIZATION, ACCLIMITATION.

acclimatization see ACCLIMATION.

acclimatize *Ecology.* to undergo the process of acclimation.

acclivity *Geology.* a slope that ascends from a reference point; the opposite of a declivity.

aC cm abcoulomb centimeter.

aC/cm^2 abcoulomb per square centimeter.

aC/cm^3 abcoulomb per cubic centimeter.

accolade *Architecture.* a decorative molding in the form of two ogee curves meeting above a window or doorway.

accommodate to carry out or undergo a process of accommodation; adjust to some new condition or circumstance.

accommodation the act or fact of adjusting to some different condition; specific uses include: *Physiology.* the contraction or relaxation of the lens of the eye to adjust for far or near vision. *Behavior.* any sensory adaptation that occurs in response to a change in stimulus. *Ecology.* the capacity of an organism to adjust to changes in its environment. *Psychology.* **1.** the modification of an existing perception and understanding as a result of new experiences. **2.** also, **social accommodation.** the adjustments that are made by people or groups in order to facilitate social harmony. *Robotics.* active or passive changes in a robot's motion in response to the robot's environment. *Cartography.* the range within which a stereoplotting instrument can adjust.

accommodation coefficient *Physics.* a measure of the ability of an incident gas to remove heat from a given surface, taken as a ratio of the actual heat that is lost from that surface to the heat that would be exchanged by the gas molecules if they were in thermal equilibrium with that surface.

accommodation ladder *Naval Architecture.* **1.** a ladder or set of steps hung over or attached to the side of a ship to permit boarding it from a small boat. **2.** a similar ladder that is used to move from one deck to another.

accommodation reflex *Physiology.* an involuntary change in the focal mechanism of the eye that accompanies the voluntary convergence of the two eyes onto an object in order to bring the image into clear focus.

accomplishment quotient see ACHIEVEMENT QUOTIENT.

accordant *Geology.* of two or more topographic features, matching or congruent, such as having the same elevation or a similar orientation. Thus, **accordant folds, accordant streams.**

accordant summit level *Geology.* a hypothetical surface where a horizontal plane intersects a hill or mountain tops over a broad region.

accordion cable *Electricity.* a flat, multiconductor cable prefolded in an accordion pattern and used to make connections to movable equipment.

accordion door *Building Engineering.* a series of movable solid panels that are suspended from a track and that fold and unfold like an accordion.

accordion fold *Graphic Arts.* an arrangement of parallel folds made in alternating directions on printed sheets; used in bookbinding.

accordion partition *Building Engineering.* a movable partition that is suspended from a track and that closes by folding flat like an accordion.

accordion roller conveyor *Mechanical Engineering.* a roller conveyor constructed in a spiral manner; the lightness of the spiral is determined by the size of material being transported and the transport time.

accouchement *Medicine.* a term for childbirth or the delivery of a child.

accoucher *Medicine.* **1.** a person, especially a man, who assists during childbirth. **2.** an obstetrician.

accounting machine *Computer Technology.* a noncomputerized, keyboard-activated machine that is used to perform calculations and produce accounting reports.

accouplement *Architecture.* the proximate placement of a pair of structural elements, such as two columns.

accrescent *Botany.* **1.** a plant part that continues to grow after flowering; usually the sepals **2.** of or relating to such a part.

accretion a gradual building up or enlargement; specific uses include: *Geology.* **1.** the gradual buildup of land on a shore over time as a result of natural forces, such as the deposition of sediment by wave action, tides, or currents. **2.** the process whereby the size of an inorganic mass increases as fresh particles are added to its surface. Also, AGGRADATION. *Meteorology.* the growth of a precipitation particle in which a frozen particle, such as a snowflake, collides with a supercooled liquid droplet that freezes upon contact. *Astronomy.* the process by which a star or other body increases in mass by the gravitational attraction of matter. *Civil Engineering.* **1.** the building up of sedimentation from water action as a result of a dam, jetty, or other construction. **2.** the sedimentation that results.

accretionary relating to or formed by accretion.

accretionary lava ball *Volcanology.* a ball of lava on the surface of an aa lava flow, formed by the molding of viscous lava around a solidified core.

accretionary limestone *Petrology.* a type of limestone formed by the gradual accumulation of organic remains.

accretionary ridge see ACCRETION RIDGE.

accretion disk *Astronomy.* a spinning, disk-shaped nebula surrounding and spiraling toward a black hole or other compact body, formed from particles attracted by the body's gravity.

accretion hypothesis see ACCRETION THEORY.

accretion line *Histology.* a microscopic line on a tooth indicating where a layer of enamel or dentin has been deposited.

accretion ridge *Geology.* a beach ridge found inland from the modern beach, indicating that the coast has been built out toward the sea.

accretion tectonics *Geology.* the collision and welding of continents resulting in continental growth as microcontinental material is added to the continental margins.

accretion theory *Astronomy.* the theory that the earth and the other planets were formed by accretion as a result of instabilities in the solar nebula.

accretion topography *Geology.* land-surface features that are built by the addition of sediment.

accretion vein *Geology.* a type of vein that is formed as channels are alternately filled with mineral deposits and then reopened by fractures.

accretion zone *Geology.* an area where accretion is occurring; i.e., a beach or continent.

acculturation *Anthropology.* **1.** a process by which cultural elements are exchanged between two societies in extensive contact, especially one in which a more dominant, complex culture influences a smaller or less complex group. **2.** any process by which individuals adapt to or become assimilated into another culture.

accumbent *Botany.* of a plant part, lying or folded against some other part.

accumulated discrepancy *Engineering.* a term for the total sum of the discrepancies occurring over the course of a survey.

accumulated divergence *Cartography.* the sum of the separate discrepancies for the sections of a line of levels, from the beginning of the line to any section end at which it is desired to compute the total divergence.

accumulated dose *Physiology.* a total dose resulting from repeated exposures to drugs or radiation. Also, CUMULATED DOSE.

accumulated total punching *Computer Technology.* in punch-card technology, an error detection technique that was used with card input to ensure that the complete card deck was read.

accumulating reproducer *Computer Technology.* in punch-card technology, a device that was used to reproduce a set of sorted cards and provide certain additional mathematical information on them.

accumulation *Hydrology.* **1.** the addition of ice or snow to the surface of a glacier or snowfield by condensation, snowfall, or any other process. **2.** the quantity of ice or snow that is added to a glacier or snowfield by such a process. **3.** see ALIMENTATION. *Mining Engineering.* **1.** in coal mining, the firedamp that tends to collect in higher parts of mine workings and at the edge of old workings and wastes. **2.** the collection of oil or gas in some form of trap.

accumulation area *Hydrology.* that part of the surface of a glacier or snowfield where annual accumulation exceeds annual ablation.

accumulation clock *Horology.* a system that measures long periods of time (thousands of years) by determining the ratio of the amount of a radioactive element in an object to the amount of its daughter element that has accumulated in the object as a result of radioactive decay; based on known, steady rates of decay.

accumulation factor *Mathematics.* $(1 + r)$ in the formula for compound interest, $A = P(1 + r)^n$, where A is the amount that is accumulated at the end of n periods from an an original principal P at the rate r per period.

accumulation model *Archaeology.* the theory that cultural change occurs gradually as societies accumulate behavioral traits.

accumulation point *Mathematics.* a point x_0 is an accumulation point for a subset A of a topological space X if, whenever U is an open set containing x_0, then U contains at least one point of A other than x_0. If A is a sequence of points $x_1, x_2, x_3, \ldots$, then for any integer n, each neighborhood of an accumulation point contains at least one member x_k of the sequence with $k > n$. Also, LIMIT POINT, CLUSTER POINT.

accumulation zone *Hydrology.* **1.** see ACCUMULATION AREA. **2.** the area where the main mass of snow that contributes to an avalanche was deposited originally.

accumulative see CUMULATIVE.

accumulative timing *Industrial Engineering.* a time-motion study result derived by repeated direct timings of successive work cycles.

accumulator a device that accumulates or stores something; specific uses include: *Mechanical Engineering.* any device used in hydraulic systems to store fluid under pressure in a container. *Petroleum Engineering.* a tank, chamber, or vessel for holding liquid or air under pressure for use in a hydraulic or air-actuated system. *Electricity.* a British term for STORAGE BATTERY. *Aviation.* in some gas-turbine engines, a device that stores fuel and releases it under pressure when needed, as when starting the engine. Also, FUEL ACCUMULATOR. *Mathematics.* a general term for any digital device that stores a number and, on receipt of another number, adds it to the stored number and stores the sum. *Computer Technology.* a special-purpose register and associated equipment within the arithmetic/logic unit of the processor, used for temporary storage of operands and intermediate results of arithmetic and logic operations.

accumulator battery see STORAGE BATTERY.

accumulator jump instruction *Computer Programming.* an instruction that directs a computer to leave the normal program sequence and branch to another part of the program when the accumulator has reached a certain status or value. Also, **accumulator transfer instruction.**

accumulator register *Computer Technology.* see ACCUMULATOR.

accumulator shift instruction *Computer Programming.* a computer instruction that causes the word stored in a register to be displaced a specified number of bit positions to the left or right.

accuracy *Metrology.* the extent to which a measurement approaches the true value of the measured quantity.

accuracy checking *Cartography.* **1.** the process of obtaining confirmation, based on sampling, of a map's relative conformity to specified standards of accuracy. **2.** see ACCURACY TESTING.

accuracy control character *Computer Programming.* a character used to identify data that is in error or that is to be disregarded.

accuracy control system *Computer Programming.* a means to detect and control errors in data input or in the results of processing.

accuracy life *Ordnance.* the number of rounds that a weapon can fire before it is affected by wear to the point where its accuracy level is unacceptable.

accuracy of fire *Ordnance.* the degree of precision of fire, calculated by the distance from the actual point of impact to the intended target.

accuracy testing *Cartography.* the process of obtaining confirmation, based on sampling, of a map's absolute conformity to specified standards of accuracy.

accurate contour *Cartography.* a contour line that is accurate to within one-half of the basic vertical interval. Also, NORMAL CONTOUR.

accustomization *Biology.* the process of adjusting to severe, unusual, or new conditions or a new environment; used especially in reference to the effects of space travel.

AC/DC or **ac/dc** alternating current/direct current.

AC/DC motor see UNIVERSAL MOTOR.

AC/DC receiver *Electronics.* a receiver that operates on an alternating-current or direct-current supply.

-acea *Systematics.* a suffix used in forming the names of classes and orders, as in *Crustacea,* the class of crustaceans.

-aceae *Systematics.* a suffix used in forming the names of families, as in *Liliaceae,* the lily family.

acebutolol *Pharmacology.* $C_{18}H_{28}N_2O_4$, a beta blocker used in the control of hypertension and angina.

acecarbromal *Pharmacology.* $C_9H_{15}BrN_2O_2$, a drug formerly used as a sedative; addictive and now replaced by newer drugs. Also, ACETYLCARBROMAL.

acellular *Biology.* not made up of or divided into cells.

acellular slime mold *Mycology.* the common term for the class of fungi called Myxomycetes.

AcEm actinium emanation.

acenaphthene *Organic Chemistry.* $C_{10}H_6(CH_2)_2$, colorless tricyclic needles derived from coal tar, insoluble in water and slightly soluble in alcohol; melts at 96.2°C; used as a dye intermediate and in pharmaceuticals, insecticides, and plastics. Also, **acenapthene.**

acenaphthenequinone *Organic Chemistry.* $C_{10}H_6(CO)_2$, yellow tricyclic needles that are insoluble in water and slightly soluble in alcohol; melts at 261°C; used in scarlet and red dyes. Also, **acenaphthequinone.**

-acene *Organic Chemistry.* a suffix indicating any condensed polycyclic compound that has a linear arrangement of three or more fused benzene rings.

acenesthesia *Psychology.* the loss of the physical awareness of one's body and bodily sensations.

acenocoumarin *Organic Chemistry.* $C_{19}H_{15}NO_6$, a white, crystalline powder, tasteless and odorless; melts at 197°C; slightly soluble in water and organic solvents; used medicinally as an anticoagulant. Also, **acenocoumarol.**

acentric *Biology.* not organized around a central point. *Genetics.* relating to or designating a chromosome or chromatid that does not have a centromere.

acentrous *Vertebrate Zoology.* having the notochord persistent throughout life and having no vertebral centra, as in certain primitive fishes.

-aceous a combining form meaning "having the nature or appearance of; being or having," as in *herbaceous, sebaceous.*

acephalia *Medicine.* the absence at birth of the head. Also, **acephalism.**

acephalic see ACEPHALOUS.

Acephalina *Invertebrate Zoology.* a suborder of parasitic protozoans of the order Eugregarinida; characterized by nonseptate trophozoites that do not undergo schizogony (asexual reproduction by multiple fission); e.g., earthworm parasites of the genus Monocystis.

acephalobrachia *Medicine.* the absence at birth of the head and arms. Also, ABRACHIOCEPHALIA.

acephalocardia *Medicine.* see ACARDIACUS ACEPHALUS.

acephalochiria *Medicine.* the absence at birth of the head and hands.

acephalocyst *Invertebrate Zoology.* in cestode flatworms, an abnormal hydatid or echinococcus cyst that fails to develop a head and brood capsules; parasitic in human organs.

acephalogastria *Medicine.* the absence at birth of the head, chest, and upper part of the abdomen.

acephalopodia *Medicine.* the absence at birth of the head and feet.

acephalorhachia *Medicine.* the absence at birth of both the head and the spinal column.

acephalostomia *Medicine.* the absence at birth of the greater part of the head, yet with a mouth-like opening present.

acephalothoracia *Medicine.* the absence at birth of the head and chest.

acephalous *Invertebrate Zoology.* lacking a head, or having a much reduced head-like structure. *Botany.* relating to or designating an ovary in which the style springs from the base rather than the apex.

acephaly see ACEPHALIA.

acephate *Organic Chemistry.* $C_4H_{10}NO_3PS$, white, moderately toxic crystals that are soluble in water and slightly soluble in alcohol; melts at 65–75°C; it is widely used as an insecticide, especially against foliage pests.

acephatemet *Organic Chemistry.* $CH_3OCH_3SPONH_2$, a white crystalline solid that is slightly soluble in water and melts at 39–41°C; used as an insecticide.

Acer *Botany.* a large genus of trees and shrubs that includes the maples and box elder.

Aceraceae *Botany.* a family of trees and shrubs in the order Sapindales and containing the maple genus *Acer*; most species accumulate quebrachitol and are often saponiferous and tanniferous.

aceramic *Anthropology.* nonceramic; relating to or designating a culture with no tradition of pottery making.

Aceraria *Invertebrate Zoology.* a genus of nematode parasites, including *Aceraria spiralis,* which affects the esophagus of birds.

acerate see ACEROSE.

acerbophobia or **acerophobia** *Psychology.* an irrational fear of sour tastes.

Acerentomidae *Invertebrate Zoology.* a family of minute, primitive insects belonging to the order Protura that lack eyes, wings, cerci, and antennae.

acerola *Botany.* **1.** an annual tree, *Malphigia glabra,* native to the West Indies and widely grown there and in adjacent areas for its cherrylike fruit. **2.** the fruit itself, a rich source of vitamin C used in various food preparations and as a diet supplement. Also, BARBADOS CHERRY, PUERTO RICAN CHERRY, WEST INDIAN CHERRY.

acerose *Botany.* having or growing in the shape of a needle; thin and sharp-pointed. Also, ACERATE, ACEROUS.

acerous *Zoology.* without horns or antennae. *Botany.* see ACEROSE.

acervate *Biology.* growing in a dense cluster or heap; used especially of certain fungi.

Acervularia *Paleontology.* an extinct cerioid massive coral of the Silurian, in the order Rugosa; unlike most rugose corals, it grew by axial increase.

acervulus *Mycology.* a saucer-shaped or flat mass of hyphae that carry certain spore-bearing structures called conidiophores, found in the fungal order Melanconiales.

Acetabularia *Botany.* the best-known genus in the green algae order Dasycladales, characterized by a unique naked stipe bearing one or more whorls of gametangia at its apex; includes mermaid's wine glass, *Acetabularia crenulata*.

acetabulum *Anatomy.* a cuplike socket in the pelvic bone into which the head of the femur fits. *Entomology.* the cavity or aperature on an insect's thorax into which a leg inserts and articulates.

acetal *Organic Chemistry.* **1.** $CH_3CH(OC_2H_5)_2$, a colorless, volatile liquid with a pleasant odor and nutty aftertaste; soluble in water and alcohol and decomposed by dilute acids; boils at 103–104°C; used in perfumes, cosmetics, and solvents. Also, DIETHYLACETAL, 1,1-DIETHOXYETHANE. **2.** see ACETALS.

acetaldehydase *Enzymology.* an enzyme that causes the white crystalline substance acetaldehyde to change to acetic acid.

acetaldehyde *Organic Chemistry.* CH_3CHO, a colorless liquid with a pungent, fruity odor, soluble in water and alcohol; melts at –121°C and boils at 20.8°C; an oxidation product of ethanol that is used to manufacture acetic acid. Also, ACETIC ALDEHYDE, ETHANAL.

acetaldehyde cyanohydrin see LACTONITRILE.

acetaldehyde dehydrogenase *Enzymology.* an enzyme that catalyzes the oxidation of acetaldehyde to acetic acid.

acetaldehyde reductase see ALCOHOL DEHYDROGENASE.

acetal resin *Organic Chemistry.* any of various synthetic, linear molecules obtained by the polymerization of formaldehyde. These resins are rigid and tough and have a high melting point; they are widely used as metal substitutes.

acetals *Organic Chemistry.* a class of compounds of the type $RCH(OR')_2$, where R and R' are organic radicals, and R may be hydrogen; prepared by adding alcohols to aldehydes.

acetamide *Organic Chemistry.* CH_3CONH_2, colorless, deliquescent, monoclinic crystals, slightly soluble in water and very soluble in alcohol; melts at 82.3°C and boils at 221.2°C; used in organic synthesis and as a general solvent.

acetamidine hydrochloride *Organic Chemistry.* $C_2H_6N_2{\cdot}HCl$, a crystalline solid, slightly deliquescent; soluble in water and alcohol; melts at 177–178°C; used in the synthesis of pyrimidines, imidazoles, and triazines.

acetaminophen *Organic Chemistry.* $CH_3CONHC_6H_4OH$, monoclinic, odorless crystals that melt at 168°C; slightly soluble in water and soluble in alcohol. Acetaminophen is used in azo dyes, in photographic chemicals, and as an analgesic; it is widely marketed as an over-the-counter pain reliever under brand names such as Tylenol and Panadol. Also, *p*-ACETYLAMINOPHENOL.

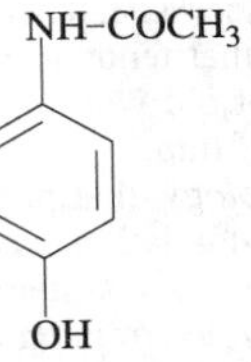

acetaminophen

acetanilide *Organic Chemistry.* $C_6H_5NHCOCH_3$, white, odorless crystalline leaflets or powder, soluble in hot water, alcohol, and ether; it melts at 114.3°C and boils at 304°C. It is used as a stabilizer for hydrogen peroxide, in dyestuffs, as an accelerator in rubber, and for various other purposes; it was formerly used as an analgesic.

acetarsone *Pharmacology.* $C_8H_{10}AsNO_5$, an arsenical compound used as a drug to destroy the parasite *Entamoeba histolytica* in the intestine and genitourinary tract; also used as a spirochetocide and antihistomonad in poultry. Also, **acetarsol; 3-acetamido-4-hydroxybenzenearsonic acid.**

acetate *Organic Chemistry.* either of two derivatives of acetic acid, CH_3COOH. **1.** a salt of acetic acid in which the terminal hydrogen atom is replaced by a metal, as in cupric acetate, $Cu(CH_3COO)_2$. **2.** an ester of acetic acid in which the replacement is by a radical, as in ethyl acetate, $CH_3COOC_2H_5$. *Textiles.* a synthetic fiber that is produced from partly hydrolyzed cellulose acetate. Also, **acetate rayon.** *Materials Science.* any of various other materials made from acetate or cellulose acetate. *Graphic Arts.* see ACETATE FILM.

acetate C-9 see NONYL ACETATE.

acetate dye *Chemistry.* **1.** one of a group of water-insoluble azo dyes capable of dyeing acetate fibers. **2.** one of a group of water-insoluble amino azo dyes made soluble with formaldehyde and bisulfite.

acetate film *Materials Science.* a type of photographic material that has a base of cellulose triacetate; it is widely used for motion picture film because of its low susceptibility to fire. *Graphic Arts.* a clear plastic film that is used as a protective overlay for photographs and other artwork.

acetate of lime *Organic Chemistry.* a commercial form of calcium acetate, $Ca(CH_3COO)_2$, made from pyroligneous acid and an aqueous suspension of calcium hydroxide.

acetate process *Chemical Engineering.* a process in which cellulose is combined with acetic acid or acetic anhydride and sulfuric acid catalyst to make cellulose acetate resin or fiber.

acetazolamide *Pharmacology.* $C_4H_6N_4O_3S_2$, an anhydrase inhibitor with diuretic properties; it occurs as a white or whitish crystalline powder, and is used to reduce the pressure in the eye in glaucoma, to change the level of excretion of certain substances in kidney disease, and formerly to treat edema associated with heart disease.

acetic *Chemistry.* **1.** relating to or producing acetic acid or vinegar. **2.** resembling vinegar; sour.

acetic acid *Organic Chemistry.* CH_3COOH, a clear, colorless, rhombic liquid, hygroscopic and crystallizing into deliquescent needles; miscible in water, alcohol, and acids; melts at 16.6°C and boils at 117.9°C. It is used in the production of plastics, dyes, food additives, and photographic chemicals. Also, ETHANOIC ACID.

acetic acid bacteria see ACETOBACTER.

acetic aldehyde see ACETALDEHYDE.

acetic anhydride *Organic Chemistry.* $(CH_3CO)_2O$, a colorless, combustible liquid with a strong pungent odor that reacts with water to form acetic acid and with ethanol to form ethyl acetate; melts at –73.1°C and boils at 139.55°C; widely used as an acetylating agent.

acetic ester or **acetic ether** see ETHYL ACETATE.

acetic fermentation *Microbiology.* the metabolic breakdown of an organic substrate to acetic acid by certain microorganisms, generating energy in the form of ATP.

acetic thiokinase *Enzymology.* an enzyme that catalyzes the formation of acetyl coenzyme A, which has an important function in metabolism.

acetidin see ETHYL ACETATE.

acetin *Organic Chemistry.* $C_3H_5(OH)_2O_2CCH_3$, a colorless, thick, hygroscopic liquid made by heating glycerol and acetic acid, and then distilling; soluble in water and alcohol and slightly soluble in ether; boils at 158°C; used as a dye solvent, in leather tanning, as a food additive, and as a gelatinizing agent in explosives. Also, MONOACETIN.

Acetivibrio *Bacteriology.* a genus of Gram-negative bacteria of the family Bacteroidaceae, composed of straight to curved rod-shaped cells, some of which exhibit flagellar motility.

acetoacetanilide *Organic Chemistry.* $CH_3COCH_2CONHC_6H_5$, a combustible white crystalline solid, slightly soluble in water and soluble in alcohol and ether; it is used in dyes and in organic synthesis. Also, ACETYLACETANILIDE.

acetoacetate *Organic Chemistry.* a salt of acetoacetic acid, containing the radical $CH_3COCH_2COO–$.

acetoacetic acid *Organic Chemistry.* CH_3COCH_2COOH, a colorless, oily liquid that is soluble in water, alcohol, and ether; decomposes at 100°C. It is derived from β-hydroxybutyric acid and is used in organic synthesis. Also, ACTYLACETIC ACID.

acetoacetic ester see ETHYL ACETOACETATE.

acetoacetyl coenzyme A *Enzymology.* a compound that plays an important role in metabolism, serving as fuel in certain metabolic processes in both the brain and nervous tissue, and affecting the oxidation of fatty acids. Also, **acetoacetyl-CoA.**

acetoamidoacetic acid see ACETURIC ACID.

acetobacter *Bacteriology.* **1.** any of a genus of ovoid or rod-shaped, nonmotile or flagellated, Gram-negative aerobic bacteria that occur on the surface of fruits, vegetables, and flowers, and that are used industrially in the manufacture of vinegar. **2. Acetobacter.** this genus itself.

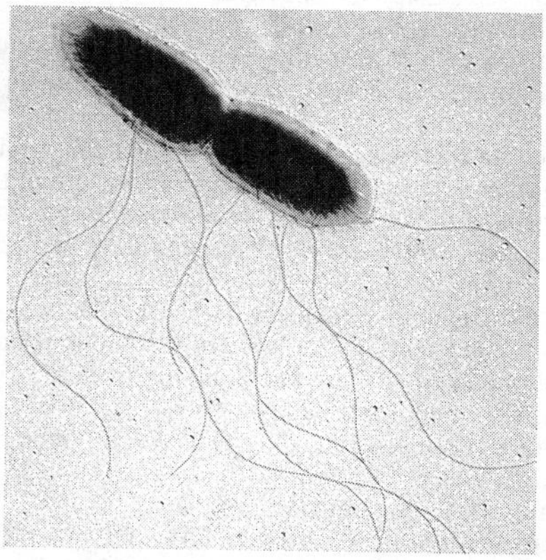

acetobacter

Acetobacteraceae *Bacteriology.* a family of Gram-negative, aerobic bacteria that typically oxidize ethanol to acetic acid and are found in acidic, ethanol-rich environments.

Acetobacter aceti *Bacteriology.* the type species for the bacterial genus *Acetobacter,* found typically on fruits and vegetables and able to transform wine or cider into vinegar.

Acetobacterium *Bacteriology.* a genus of Gram-negative, anaerobic bacteria that exist as rod-shaped cells, possess polar flagellation, and occur in freshwater and marine sediments.

Acetobacter suboxydans *Bacteriology.* a former name for the bacterial species *Gluconobacter oxydans.*

acetogen *Microbiology.* a bacterium that is capable of producing acetate as the major product of the oxidation of a number of substrates, including certain sugars.

acetoin *Organic Chemistry.* $CH_3CH(OH)COCH_3$, a slightly yellow liquid or crystals that are soluble in alcohol and water and slightly soluble in ether; it melts at 15°C and boils at 143°C. It is derived from the reduction of diacetyl, and is used as an aroma carrier in flavors and essences.

acetol *Organic Chemistry.* **1.** CH_3COCH_2OH, a colorless, combustible liquid that is soluble in water, alcohol, and ether; it freezes at –17°C and boils at 146°C. It is used as a solvent for nitrocellulose. Also, 1-HYDROXY-2-PROPANONE. **2.** a proprietary name for cellulose acetate. **3. Acetol.** a registered Canadian trade name for acetylsalicylic acid (aspirin).

acetolactic acid *Biochemistry.* $C_5H_8O_4$, a compound formed in plants and bacteria during the biosynthesis of the amino acid valine. Also, **α-acetolactic acid.**

acetolysis *Organic Chemistry.* the decomposition of an organic molecule through the use of acetic anhydride or acetic acid.

acetomeroctol *Pharmacology.* $C_{16}H_{24}HgO_3$, a crystalline substance formerly used as a topical antiseptic.

Acetomonas *Bacteriology.* the former name of *Gluconobacter*, a genus of Gram-negative aerobic rod-shaped bacteria of the family Acetobacteraceae.

acetone *Organic Chemistry.* CH_3COCH_3, a colorless, volatile liquid with a sweetish odor, miscible with water, alcohol, and most oils, that melts at –95.4°C and boils at 56.2°C. The simplest saturated ketone, it is extremely flammable and is frequently used as a solvent. It is found in the tissue, urine, and blood in very small amounts in healthy people and in larger amounts in diabetics and the starving. Also, 2-PROPANONE.

acetone–benzol process *Petroleum Engineering.* the use of acetone and benzol as solvents in a dewaxing process in petroleum refining.

acetone body see KETONE BODY.

acetone–butanol fermentation *Microbiology* the industrial production of acetone and *n*-butanol resulting from the fermentation of an organic substrate, such as molasses or corn, by certain species of the bacterial genus *Clostridium,* especially *C. acetobatylicum;* the first large-scale fermentation process to be used..

acetone chloroform see CHLOROBUTANOL.

acetone cyanohydrin *Organic Chemistry.* $(CH_3)_2C(OH)CN$, a colorless, highly toxic liquid that is soluble in water and organic solvents; melts at –20°C and boils at 82°C. Derived from condensing acetone with hydrocyanic acid, it is used in insecticides and as an intermediate for organic synthesis.

acetone fermentation see ACETONE-BUTANOL FERMENTATION.

acetonemia *Medicine.* the presence of acetone or acetone bodies in very large amounts in the blood. Also, KETONEMIA.

acetone number *Chemistry.* the numeric expression of a ratio that approximates the degree of polymerization of substances.

acetone pyrolysis *Organic Chemistry.* the decomposition by heat of acetone into ketene.

acetone sodium bisulfite *Organic Chemistry.* $(CH_3)_2CONaHSO_3$, a combustible crystalline material with a slight sulfur dioxide smell and a slightly fatty feel; soluble in water, slightly soluble in alcohol, and decomposed by acids; used in photography, textile dyeing, and printing.

acetone sugar *Organic Chemistry.* any reducing sugar that contains acetone; e.g., 1,2-monoacetone-D-glucofuranose. Also, **acetone glucose.**

acetonitrile *Organic Chemistry.* CH_3CN, a colorless liquid that is soluble in water and alcohol; melts at –41°C and boils at 82°C; used as a polar solvent, for the separation of fatty acids from vegetable oils, and in manufacturing synthetic pharmaceuticals. Also, METHYL CYANIDE.

acetonuria [as ə tə noor′ē ə] *Medicine.* large amounts of acetone in the urine indicating the incomplete oxidation of fats and most commonly occurring in diabetics and those suffering from fever, starvation, or cancer. Also, KETONURIA.

acetonylacetone *Organic Chemistry.* $CH_3COCH_2CH_2COCH_3$, a colorless liquid that is soluble in water; freezes at –5.5°C and boils at 192.2°C; used as a solvent for lacquers, as an intermediate for pharmaceuticals, and in photographic chemicals. Also, 2,5-HEXANEDIONE.

acetophenetidin see PHENACETIN.

acetophenone *Organic Chemistry.* $C_6H_5COCH_3$, a colorless liquid with a sweet, pungent odor and taste, slightly soluble in water, soluble in organic solvents and sulfuric acid; it melts at 20.5°C and boils at 202.6°C; used in perfumery, as an intermediate for pharmaceuticals, and in organic synthesis. Also, ACETYLBENZENE.

acetostearin *Organic Chemistry.* a general term for acetylated glyceryl monostearates. These solids are flexible but not greasy; used as plasticizers and as protective coatings for food.

acetous *Chemistry.* **1.** referring to acetic acid. **2.** referring to that which is sour to the taste.

acetovanillon see APOCYNIN.

acetoxime *Organic Chemistry.* $(CH_3)_2C{=}NOH$, colorless crystals with a chloral odor, soluble in water, alcohol, and ether; melts at 61°C and boils at 136°C; used as an intermediate in organic synthesis and as a solvent for celluose ethers.

acetphenetidin see PHENACETIN.

aceturic acid *Organic Chemistry.* $CH_3CONHCH_2COOH$, long, needlelike crystals that are soluble in water and alcohol and melt at 206°C. When combined with organic bases, it creates stable salts. Also, ACETYLGLYCINE.

acetyl *Organic Chemistry.* CH_3CO–, the radical of acetic acid, containing a methyl group and a carbonyl group.

acetylacetanilide see ACETOACETANILIDE.

acetylacetic acid see ACETOACETIC ACID.

acetylacetone *Organic Chemistry.* $CH_3COCH_2COCH_3$, a mobile, colorless to yellowish liquid, soluble in water, alcohol, and ether; freezes at –23°C and boils at 139°C. Light causes it to turn brown and to form resins. Acetylacetone is moderately toxic and is used as a solvent for cellulose acetate, a lubricant additive, a chelating agent for metals, and in pesticides.

***p*-acetylaminophenol** see ACETAMINOPHEN.

acetylase see ACETYLTRANSFERASE.

acetylate *Organic Chemistry.* to introduce one or more acetyl groups onto an organic molecule.

acetylated cotton *Materials.* a treated cotton fiber produced by the process of acetylation.

acetylating agent *Organic Chemistry.* a reagent such as acetic anhydride or acetyl chloride that is used to bond an acetyl group onto an organic molecule.

acetylation *Organic Chemistry.* the introduction of an acetyl group, CH_3CO–, onto the molecule of an organic compound having either –OH or $–NH_2$ groups.

acetylator *Organic Chemistry.* **1.** an organism that is capable of metabolic acetylation. **2.** an apparatus that is used in acetylation reactions.

acetylbenzene see ACETOPHENONE.

acetyl benzoyl peroxide *Organic Chemistry.* $C_6H_5CO{\cdot}O_2{\cdot}COCH_3$, white crystals that melt at 36.6°C and boil at 130°C; decomposed by water, alkaloids, and organic matter. It is toxic and a strong irritant, and is an explosion hazard if shocked or quickly heated; it is used as a germicide and disinfectant.

acetyl bromide *Organic Chemistry.* CH_3COBr, a colorless, fuming liquid that turns yellow in air; it is soluble in ether, chloroform, and benzene; freezes at –98°C and boils at 76°C; it decomposes violently in water and in alcohol; it is used in dye manufacture and for organic synthesis.

acetylcarbromal see ACECARBROMAL.

acetyl chloride *Organic Chemistry.* CH_3COCl, a colorless, fuming liquid that is highly refractive; soluble in ether, acetone, and acetic acid; it freezes at –112°C and boils at 51°C, and decomposes violently in water and alcohol; used in organic preparations as an acetylating agent, in dyestuffs, and in pharmaceuticals.

acetylcholine *Endocrinology.* $C_7H_{17}O_3N$, a derivative of choline that is a major chemical neurotransmitter of the autonomic nervous system; its release transmits impulses at the preganglionic synapse of the sympathetic and the parasympathetic systems and also at the postganglionic synapse of the parasympathetic system and sympathetic cholinergic fibers.

acetylcholinesterase *Enzymology.* an enzyme of the hydrolase class, that is found in skeletal muscle, red blood cells, and the gray matter of nerve tissue; it catalyzes a reaction that destroys acetylcholine released at neurohumoral junctions. Also, TRUE CHOLINESTERASE, CHOLINE ESTERASE I.

acetyl-CoA synthetase *Biochemistry.* an enzyme of bacteria and plants that catalyzes the reaction during which acetate enters metabolic pathways and acetyl coenzyme A is formed.

acetyl coenzyme A *Biochemistry.* $C_{21}H_{38}O_{16}N_7P_3S$, a coenzyme that plays a major role in intermediary metabolism, in which foodstuffs are transformed into simpler substances and synthesized by the cells. Also, **acetyl-CoA.**

acetylcysteine *Pharmacology.* $C_5H_9NO_3S$, an inhaled drug used in lung disorders to thin mucus and make it easier to expel.

acetylene [ə set´ə lēn] *Organic Chemistry.* HC≡CH, a colorless gas with a faint ethereal odor, soluble in water, highly flammable, and explosive when compressed; sublimes at –81°C. Acetylene is the simplest compound with a triple bond. It is typically produced by the reaction of calcium carbide with water, and it is used as a welding fuel (in combination with oxygen), for illumination, and in the synthesis of industrial compounds. *Engineering.* short for OXYACETYLENE; involving or using a mixture of oxygen and acetylene as a fuel. Thus, **acetylene welding, acetylene cutting.**

acetylene black *Organic Chemistry.* carbon black formed by the incomplete combustion of acetylene; used in dry cell batteries for its high electrical conductivity, and as a strengthener in rubber.

acetylene series *Organic Chemistry.* the series of unsaturated aliphatic hydrocarbons that have the general formula C_nN_{2n-2}. Each hydrocarbon contains at least one triple bond.

acetylene tetrabromide *Organic Chemistry.* $CHBr_2CHBr_2$, a yellowish liquid, insoluble in water and soluble in alcohol and ether; boils (decomposes) at 239–242°C; used for specific gravity separation of minerals, and as a solvent for fats, waxes, and oils. Also, *sym*-TETRABROMOETHANE.

acetylene torch see OXYACETYLENE TORCH.

acetylenic *Organic Chemistry.* relating to or like acetylene, especially in having a triple bond.

***N*-acetyl-D-galactosamine** *Biochemistry.* a derivative of the amino sugar galactosamine that has an acetyl group on the amino N; a common constituent of glycoproteins, heteropolysaccharides, and glycolipids.

***N*-acetyl-D-glucosamine** *Biochemistry.* $C_8H_{15}NO_6$, a derivative of the amino sugar glucosamine that has an acetyl group on the amino N; a common constituent of glycoproteins, heteropolysaccharides, and glycolipids.

acetylglycine see ACETURIC ACID.

acetylide *Organic Chemistry.* a carbide formed by bubbling acetylene through metal salts such as cuprous acetylide. Such carbides are violently explosive if heated or shocked.

acetylisoeugenol *Organic Chemistry.* C_6H_3 $(CH{=}CHCH_3)$ (OCH_3) $(OCOCH_3)$, white crystals with a spicy, clove-like odor that congeal at 77°C; used in perfumery and flavoring.

acetyl ketene see DIKETENE.

***N*-acetyllysine** *Biochemistry.* $C_8H_{16}N_2O_3$, a derivative of lysine that has an acetyl group on the ε-nitrogen and functions in translation in mammalian systems.

acetylmethadol see METHADYL ACETATE.

acetylmethylcarbinol see ACETOIN.

***N*-acetylmuramic acid** *Biochemistry.* the monosaccharide in bacterial peptidoglycan; muramic acid that has an acetyl substituent on its amino group.

acetyl number *Analytical Chemistry.* a number that indicates the extent to which a substance may be acetylated; a measure of the number of free hydroxyl groups in fats and oils, determined by the number of milligrams of potassium hydroxide used to neutralize one gram of acetylated product.

acetyl peroxide *Organic Chemistry.* $(CH_3CO)_2O_2$, colorless crystals that melt at 30°C and boil at 63°C; slightly soluble in cold water and soluble in alcohol and ether; flammable, highly irritating, and highly explosive; used as an initiator and catalyst for resins.

acetyl phosphate *Biochemistry.* $C_2H_5O_5P$, a compound formed during the metabolic activities of bacteria.

acetylpropionic acid see LEVULINIC ACID.

acetyl propionyl *Organic Chemistry.* $CH_3COCOCH_2CH_3$, a yellow liquid that melts at –52°C and boils at 106–110°C; slightly soluble in water; used in artificial flavorings of the butterscotch and chocolate type.

acetyl reduction assay *Biotechnology.* a sensitive technique that uses a flame ionization detector and a gas chromatography apparatus to measure the ability of an organism to fix nitrogen.

4-acetylresorcinol *Organic Chemistry.* $C_6H_3(OH)_2COCH_3$, light tan needle-like crystals, slightly soluble in water and alcohol; melts at 146–148°C; used as a dye intermediate, fungicide, and plant growth promoter. Also, 2,4-DIHYDROXYACETOPHENONE.

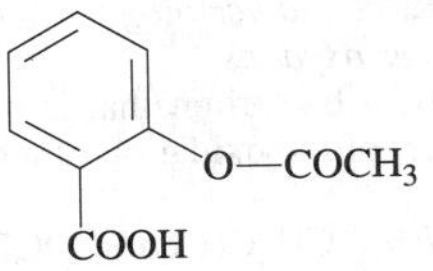

acetylsalicylic acid (aspirin)

acetylsalicylic acid [ə sēt´əl sal´i sil´ik as´id] *Organic Chemistry.* $CH_3COOC_6H_4COOH$, the chemical name for aspirin, an odorless white crystalline compound with a slightly bitter taste, derived from salicylic acid; soluble in water and alcohol, it melts at 132–136°C. It has widespread medicinal use as a pain reliever and to reduce inflammation and fever; it also inhibits blood clotting and has been shown to be effective in treating conditions associated with this.

***N*-acetylserine** *Biochemistry.* an acetylated serine that is thought to function for mammals as an initiator of translation, just as *N*-formylmethionine does for bacteria.

acetylserotonin *Endocrinology.* the acetylated form of 5-hydroxytryptamine, produced by the pineal gland along with serotonin and melatonin; of these products, only melatonin may be released by the pineal.

acetyl transacylase *Biochemistry.* an enzyme that catalyzes the transfer of the acetyl group from one substance to another; e.g., the transfer of acetyl group from acetyl CoA to orthophosphate to form acetyl phosphate.

acetyltransferase *Biochemistry.* an enzyme that mediates the attachment of acetyl groups to a variety of substances in intermediary metabolism. Also, ACETYLASE.

acetylurea *Organic Chemistry.* $CH_3CONHCONH_2$, colorless crystals that are slightly soluble in water and melt at 218°C.

acetyl valeryl *Organic Chemistry.* $CH_3COCOC_4H_9$, a combustible yellow liquid that is used in flavoring.

ACF diagram *Petrology.* a triangular diagram that shows the chemical character of a metamorphic rock, in which three principal points are plotted, including $A = Al_2O_2 + Fe_2O_3 - Na_2O + K_2O$, $C = CaO$, and $F = FeO + MgO + MnO$.

acfm actual cubic feet per minute.

AcG acceleration globulin.

a-c girdle *Geology.* in a fabric diagram, a concentration of points that forms a belt parallel to the plane containing the *a* and *c* fabric axes.

ACH *Medicine.* adrenal cortical hormone.

Achaenodontidae *Paleontology.* a family of dichobunoid piglike artiodactyl mammals in the extinct suborder Palaeodonta; from the Eocene.

achaetous *Invertebrate Zoology.* having no setae.

achalasia *Medicine.* **1.** the inability to relax the muscles of the gastrointestinal tract. **2.** see ACHALASIA OF THE CARDIA.

achalasia of the cardia *Medicine.* the inability to relax the esophagogastric sphincter muscle while swallowing.

Achariaceae *Botany.* a family of South African climbing or stemless monoecious herbs and shrubs of the order Violales; characterized by the absence of a latex system and alternate, palmately lobed leaves.

Achatocarpaceae *Botany.* a family of dioecious, sometimes spiny shrubs or small trees of the order Caryophyllales having berry fruits and alternate, simple, entire leaves without stipules; found in warm American climates.

ache *Medicine.* a continuous fixed pain, especially a dull pain, as opposed to a sharp, sporadic pain.

acheilia *Medicine.* the absence at birth of one or both of the lips.

acheilous *Medicine.* lacking one or both lips at birth; characterized by acheilia.

acheiropodia *Medicine.* the absence at birth of the hands and the feet.

achene [ā´kēn´] *Botany.* a dry, indehiscent, one-seeded fruit formed from a single carpel, and with the seed coat not fused with the ovary wall. Also, **achenium.**

Achernar *Astronomy.* Alpha (α) Eridani, a first-magnitude blue-white star in the southern constellation Eridanus, the River.

Acheson process *Materials Science.* a commercial method for producing silicon carbide, SiC, by reacting silica and coke at high electrically induced temperatures.

Acheulian or **Acheulean** [ə chōō´lē ən] *Archaeology.* **1.** a European culture of the Lower Paleolithic period, noted for the manufacture of the hand axe and the cleaver. **2.** the period when this culture existed. **3.** of or relating to this culture or period. (From Saint-*Acheul,* in northern France, the site of numerous significant findings from this period.)

Acheulian industry *Archaeology.* a progressive stage in the making of stone tools, characterized by the use of bifacial technique (chipping flakes off both sides of a stone rather than just one side).

achiasmatic *Cell Biology.* undergoing meiosis without the formation of chiasmata. Also, **achiasmate.**

achievement *Psychology.* an acquired ability, such as the ability to spell a list of vocabulary words correctly.

achievement age *Psychology.* the level of scholastic performance or accomplishment that is considered to be normal for a particular age, measured by academic achievement tests that are scored in terms of the chronological age of the average student. Also, EDUCATIONAL AGE.

achievement motive *Psychology.* a personal drive to succeed, to overcome obstacles, or to do difficult things quickly and well. Also, **achievement need, achievement drive.**

achievement quotient *Psychology.* the ratio of actual scholastic performance level to the expected level of performance for a particular age, calculated by dividing the achievement age by the mental age.

achievement test *Psychology.* a standardized test that is designed to measure an individual's level of knowledge in a particular area.

achilary *Botany.* having no lip or labellum, or having one that is undeveloped, as certain orchid flowers.

Achilles *Astronomy.* asteroid number 588, discovered in 1906 and measuring 116 kilometers in diameter. Of type D, it was the first asteroid in the Trojan group to be found.

Achilles group see TROJAN GROUP.

Achilles reflex *Physiology.* a brief, involuntary contraction of the calf muscles, produced by a sharp blow on the Achilles tendon. Also, **Achilles jerk.**

Achilles tendon *Anatomy.* a tendon in the leg that connects the calf muscle (triceps surae) with the calcaneus bone in the heel. (The association of the Greek hero *Achilles* with the heel comes from the legend that this was the only part of his body where he could be harmed.)

Achilles tendon reflex see ACHILLES REFLEX.

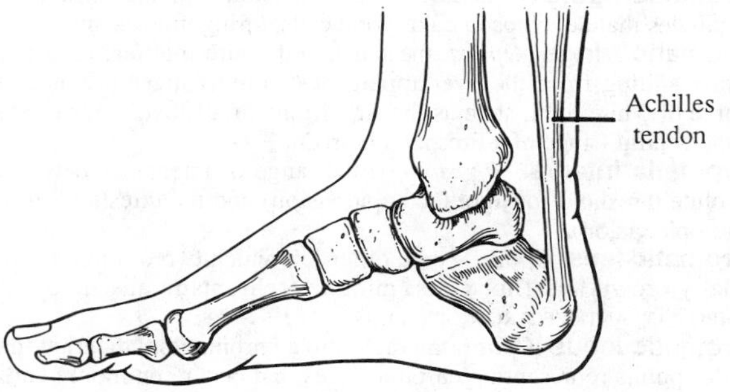

Achilles tendon

achillodynia *Medicine.* pain due to the inflammation of the Achilles tendon and the bursa. Also, **achillobursitis.**

achiote see ANNATTO.

achiral *Chemistry.* not chiral; able to be superimposed on its mirror image in a given configuration. Thus, **achiral molecule, achiral atom.**

achlamydeous *Botany.* **1.** a plant having no floral envelope. **2.** of or relating to such a plant.

achlorhydria *Medicine.* the absence of hydrochloric acid in the gastric juices.

achloropsia *Medicine.* the type of color blindness in which different shades of the color green cannot be distinguished.

achluophobia *Psychology.* an irrational fear of darkness.

Achlya *Mycology.* a genus of fungal mold belonging to the class Oomycetes, characterized by relatively wide hyphae and occurring in aquatic environments.

Achlyogetonaceae *Mycology.* a family of fungi belonging to the order Chytridiales, composed primarily of species that are parasites of freshwater algae.

Achnanthaceae *Botany.* a family of marine and freshwater diatoms constituting the suborder Monoraphidineae in the order Pennales, distinguished by a fully developed raphe on one of two valves.

Acholeplasma *Bacteriology.* a genus of small, spheroid bacteria of the family Acholeplasmataceae; facultatively anaerobic and distinguished by sterol-independent growth and the lack of a cell wall.

Acholeplasmataceae *Bacteriology.* a family of Gram-negative bacteria of the order Mycoplasmatales, containing the sole genus *Acholeplasma,* distinguished by an extremely small cell size, and requiring no sterols for growth.

acholia *Medicine.* the absence of bile or the failure to secrete bile. Thus, **acholic.**

acholuria *Medicine.* **1.** the absence of bile pigment in the urine. **2.** a type of jaundice that is characterized by this absence. Thus, **acholuric.**

achondrite *Astronomy.* a type of stony meteorite that lacks the small spherical bodies (chondrules) of the chondrites. Thus, **achondritic.**

achondroplasia *Genetics.* a genetically determined form of dwarfism due to a retarded growth of the long bones; in humans it is controlled by a dominant autosomal gene. Also, **achondroplasty.**

achondroplastic *Genetics.* relating to or affected by achondroplasia.

achordate *Zoology.* **1.** lacking a notochord. **2.** an organism lacking a notochord.

achroglobin *Biochemistry.* a colorless respiratory pigment found in certain mollusks and chordates.

achroite *Mineralogy.* a colorless variety of elbaite.

achromat see ACHROMATIC LENS.

Achromatiaceae *Bacteriology.* a family of schizomycete bacteria of the order Beggiatoales, containing the genus *Achromatirum* and occurring as large, motile cells in aquatic habitats.

achromatic *Science.* without color, or a color other than black, gray, or white. *Biology.* describing a cell structure that stains with difficulty. *Optics.* referring to or describing a lens or optical system that is capable of transmitting light without separating it into its constituent colors; used when chromatic aberration is corrected for at least two wavelengths.

achromatic color *Optics.* a color that has brightness, but does not have a hue, such as gray, black, or white.

achromatic condenser *Optics.* a condenser that eliminates chromatic and spherical anomalies, usually consisting of two achromatic lenses and two other optical elements; used in microscopes where high magnification is required.

achromatic figure *Cell Biology.* an older term for the spindle of microtubules that develops in eukaryotic cells during mitosis and meiosis.

achromatic fringe *Optics.* the black and white interference fringe of light resulting from the overlapping of colored fringes produced by a source of white light, such as the first fringe of a Lloyd's mirror system or the central fringe of a Fresnel's biprism.

achromatic interval *Physiology.* the range of intensities between the absolute threshold for light (scotopic vision) and the threshold for color (photopic vision).

achromatic lens *Optics.* a type of lens in which two or more elements, usually crown and flint glass, minimize chromatic aberration. Also, ACHROMAT, ANTISPECTROSCOPIC LENS.

achromatic locus *Optics.* the region in a chromaticity diagram that includes points representing chromaticities that occur commonly and may therefore be acceptable reference standards of illumination. Also, **achromatic region.**

achromatic prism *Optics.* an optical system consisting of two or more prisms with differing refractive indices, causing a minimum dispersion and maximum deviation of light that passes through; objects viewed through the device will not appear colored.

achromatic threshold *Physiology.* the minimum amount of light energy needed to stimulate the visual scotopic (or rod) system, which is sensitive to light but not to color.

achromatin *Cell Biology.* the part of a cell nucleus that does not stain with alkaline dyes.

achromatism the quality or fact of being achromatic; a lack of color. Also, **achromaticity.**

Achromatium *Bacteriology.* a genus of aerobic, gliding bacteria of the family Achromatiaceae, found in fresh water or marine mud, that are thought to contain sulfur and calcium carbonate inclusions.

achromatophilia *Biology.* a property of being resistant to common stains or dyes, such as those used to stain cells for microscopic study.

achromatous or **achromous** see ACHROMIC.

achromaturia *Medicine.* the lack of coloration in the urine, or the excretion of colorless urine.

achromia *Medicine.* the absence of normal color or pigmentation, as of the skin; pallor.

achromic *Biology.* without color; lacking the usual pigmentation.

achromin see ACHROMATIN.

Achromobacter *Bacteriology.* a genus of nonmotile rod-shaped bacteria within the family Achromobactereaceae, which formerly included some species now reclassified into the genera *Acinetobacter* and *Alcaligenes.*

Achromobacteraceae *Bacteriology.* a family of bacteria of the order Eubacteriales, occurring as rod-shaped cells in soil, fresh water, or salt water, and including some parasitic or pathogenic species.

achromogenic *Microbiology.* of or relating to an organism that does not produce any pigments or color.

A chromosome *Molecular Biology.* a normal chromosome of any eukaryotic cell, as distinguished from supernumerary chromosomes or B chromosomes.

Achroonema *Bacteriology.* a genus of gliding aquatic bacteria provisionally classified in the family Pelonemataceae, typically consisting of cylindrical cells in colorless unbranched filaments.

achylia *Medicine.* the absence of hydrochloric acid and pepsinogens in the gastric juices. Also, **achylia gastrica.**

ACI acoustic comfort index.

acicle see ACICULA.

acicula *plural,* **aciculae.** *Science.* any needle-shaped part, as of a plant or crystal. (From the Latin word for needle.)

acicular shaped like a needle, or having needle-shaped parts.

acicular ice *Hydrology.* freshwater ice containing air bubbles and having many long, needlelike crystals and hollow tubes in a layered arrangement. Also, FIBROUS ICE, SATIN ICE.

aciculate see ACICULAR.

aciculilignosa *Botany.* evergreen and deciduous needle-leaved vegetation.

Aciculoconidium *Mycology.* a genus of fungi belonging to the class Hyphomycetes, which produces asexual spores directly from its cells.

aciculum *plural,* **acicula** or **aciculums.** *Zoology.* a needle-like spine or bristle, as is present in certain types of worms or flagellates. *Science.* see ACICULA.

acid *Chemistry.* **1.** any of a fundamental category of compounds whose water solutions are identified by certain common characteristics, such as a sour or biting taste, the ability to turn blue litmus paper red, and the ability to react with bases and certain metals to form salts. Other definitions identify substances as acids by their activities rather than by their properties, as follows: **a. (Arrhenius acid)** any substance that increases the concentration of hydrogen (H^+) ions when added to a water solution. The greater the increase, the stronger the acid. (S. A. Arrhenius, 1887) **b. (Brønsted** or **Brønsted-Lowry acid)** any substance that serves as a proton donor to another substance which accepts a proton. (J. N. Brønsted and T. M. Lowry, 1923) **c. (Lewis acid)** any substance that accepts a pair of electrons to form a covalent bond. (G. N. Lewis, about 1915). **2.** containing acid-bearing pollutants: *acid* rain or snow. **3.** sour-tasting; acidic.

acid acceptor *Organic Chemistry.* a compound that stabilizes plastic and resin polymers by combining with trace acids that are formed when the polymer decomposes.

acid alcohol *Organic Chemistry.* a compound containing both a carboxyl (–COOH) group and an alcohol (–COH) group.

Acidaminococcus *Bacteriology.* a bacterial genus of the family Veillonellaceae, occurring as Gram-negative, anaerobic cocci that utilize amino acids as the main sources of carbon and energy; found in the intestinal tract of humans and pigs.

acidaminuria *Medicine.* an excess of amino acids in the urine. Also, AMINOACIDURIA.

acid anhydride *Chemistry.* an acid from which one or more water molecules have been removed.

acid azide *Organic Chemistry.* **1.** the unstable acyl or aroyl derivative of hydrazoic acids, formed when treated with nitrous acid, HNO_2. **2.** a compound where the azide group ($-N_3$) replaces the hydroxyl group of a carboxylic acid.

acid-base balance *Physiology.* an equilibrium of acids and alkalis that is maintained by a buffering system in the blood so that cells can function normally.

acid-base equilibrium *Chemistry.* a stable condition in a system of acids and bases; a reaction between an acid and base in which a state of equilibrium between reactants is reached.

acid-base indicator *Analytical Chemistry.* any substance that reveals the pH or degree of acidity or basicity of a solution through a color change.

acid-base pair *Chemistry.* an acid and a base that differ by one proton, in which the acid is the donor of the proton and the base is the acceptor of the proton.

acid-base titration *Analytical Chemistry.* a titration in which a specific volume of an acid of known concentration (the titrant) is added to a base of unknown concentration, or vice versa, for the purpose of determining the concentration of the unknown.

acid bronze *Metallurgy.* the formerly used name of various copper-base alloys containing about 5 to 10% tin, 10 to 20% percent lead, and 1% nickel.

acid calcium phosphate see CALCIUM PHOSPHATE.

acid cell *Histology.* a parietal cell of the stomach that secretes hydrochloric acid. *Physical Chemistry.* an electrochemical device in which an acid solution is used to generate electricity.

acid chloride *Organic Chemistry.* any of various compounds derived from acids, where chlorine replaces the hydroxyl group, such as benzoyl chloride.

acid clay *Geology.* a type of clay that produces hydrogen ions when dissolved in water.

acid conductor *Chemical Engineering.* the equipment used to heat and evaporate water from hydrolyzed acid or to distill water under a partial vacuum.

acid dilution *Petroleum Engineering.* the diluting of concentrated hydrochloric acid with water before oil-well acidizing.

acid disproportionation *Chemistry.* a state during a chemical reaction in which a single acid compound serves both as an oxidizing and as a reducing agent.

acid drift *Materials Science.* the tendency of ores and other materials to increase in acidity if left standing, through the acquisition of oxygen from the atmosphere.

acid dye *Organic Chemistry.* any of various dyes that contain the sodium salts of sulfonic or carboxylic acid, which gives them their specific color; frequently used with wool and silk.

acid egg or **acid blowcase** see BLOWCASE.

acid electrolyte *Physical Chemistry.* any acidic compound that will dissolve and dissociate into an acidic solution capable of transmitting an electric current.

acidemia *Medicine.* an increase in the acidity of the blood. Thus, **acidemic.**

acid-fast bacteria *Microbiology.* a category of bacterial cells, such as the mycobacteria, that resist decolorizing by mineral acids after treatment with an acid-fast stain.

acid-fast stain *Microbiology.* a dye, such as auramine-rhodamine stain, that is used to detect certain acid-fast bacteria, which cannot be decolorized or destained after application of the stain.

acid-fracture *Petroleum Engineering.* to open or enlarge a fracture by using a blend of oil and acid, or of water and acid under high pressure, to produce a hard limestone formation.

acid gases *Chemical Engineering.* the hydrogen sulfide and carbon dioxide emitted in refinery gases that form corrosive acids when combined with moisture.

acid halide *Organic Chemistry.* a compound that has the general formula RCOX, in which R is an alkyl or an aryl radical, and X is a halogen.

acid heat test *Analytical Chemistry.* a test that indicates the degree of unsaturation in hydrocarbons; performed by adding sulfuric acid to an organic compound and noting the increase in temperature.

acid hydrolysis see ACIDOLYSIS.

acidic *Chemistry.* **1.** relating to, being, or forming an acid or acids. **2.** having a pH value of less than 7.0. *Geology.* containing a high proportion of silica. *Nutrition.* having a sour or biting taste.

acidic group *Organic Chemistry.* the carboxyl radical (–COOH), which is found in organic acids.

acidic lava *Volcanology.* lava containing at least 66% silica; distinguished from basic and intermediate lava.

acidic oxide *Inorganic Chemistry.* an oxide of a nonmetal such as sulfur dioxide, SO_2, sulfur trioxide, SO_3, or phosphorus pentoxide, P_2O_5, that reacts with water to form acids.

acidic rock *Petrology.* igneous rock that is high in silica.

acidic titrant *Analytical Chemistry.* an acid solution of known concentration used as a standard solution for titration to determine the basicity of a solution of unknown concentration.

acidify *Chemistry.* to make acidic; convert into an acid. Thus, **acidification.**

acidimeter *Analytical Chemistry.* a device used to measure the amount of acid present in a solution.

acidimetry *Analytical Chemistry.* an analytical method for determining the amount of acid in a given sample by titration against a standard solution of a base.

aciding *Engineering.* **1.** a method of using acid to clean materials from surfaces, often so that a coating or paint may be added. **2.** the process of forming decorative designs on metal or cast stone by submersion in acid or treatment with acid on unprotected portions.

acidism *Toxicology.* poisoning due to ingestion of an acid. Also, **acidismus.**

acidity *Chemistry.* **1.** the fact of being acid, or the degree to which a given solution or substance is acid. **2.** the amount of acid present, as in an aqueous solution, which is dependent on the concentration of hydrogen ions.

acidity coefficient *Geochemistry.* the ratio of the number of atoms of oxygen in the alkaline compounds of a rock or mineral to the number of atoms of oxygen in silica. Also, ACIDITY QUOTIENT, COEFFICIENT OF ACIDITY, OXYGEN RATIO.

acidity quotient see ACIDITY COEFFICIENT.

acid jetting *Petroleum Engineering.* the jetting of an acid spray from a device lowered through oil-well tubing onto bottom-hole rock to clear away scale and mud that interfere with the flow of oil.

acid-loving *Agronomy.* of a plant, requiring relatively acidic soil for maximum productive growth; e.g., azaleas.

acid mine drainage *Mining Engineering.* drainage from bituminous coal mines characterized by a high concentration of acidic sulfates, usually ferrous sulfate.

acid mine water *Mining Engineering.* **1.** any mine water that contains free sulfuric acid, which results primarily from the weathering of iron pyrites. **2.** any mine drainage that has become acidic due to the breakdown of sulfide minerals under the chemical influence of oxygen and water.

acid mucopolysaccharide *Biochemistry.* chondroitin sulfuric acid, a glycosaminoglycan commonly found in cartilage.

acid number *Chemistry.* a measure of the free acid content of a substance, such as an oil, resin, or wax, as determined by the number of milligrams of potassium hydroxide, KOH, required to neutralize one gram of the substance. Also, ACID VALUE.

acidolysis *Organic Chemistry.* acid hydrolysis; the decomposition of a molecule with the addition of the elements of an acid to the molecule; similar to hydrolysis or alcoholysis.

acid open-hearth process *Metallurgy.* the processing of molten metals or alloys in a shallow hearth reverberatory furnace, in which bottom and lining are made of acid refractories such as silica. The slag covering the metal, if present, is also acid.

acidophil *Biology.* **1.** a cell that can be readily stained with an acid dye. **2.** an organism that lives or thrives in an acidic environment. *Histology.* **1.** see EOSINOPHIL. **2.** an acid-staining cell in the anterior lobe of the adenohypophysis.

acidophilia *Histology.* a structure or cell that stains readily with acid dyes.

acidophilic *Biology.* having an affinity for acid dyes.

acidophilus milk *Food Technology.* low-fat cow's milk that is lightly fermented with *Lactobacillus acidophilus* culture for nutritional or medicinal purposes, usually without a significant change in taste.

acidosis *Medicine.* a metabolic disturbance or significant change in the normal acid-base balance (pH 7.4), so that there is an excessive acidity of the blood or body tissues; classified as either **respiratory acidosis** or **metabolic acidosis,** depending on which type of malfunction causes the condition. *Veterinary Medicine.* a specific pathological condition of this type; may be brought on by grain overload, starvation, ischemia, or uremia, and aggravated by dehydration.

Acidothermus *Bacteriology.* a proposed genus of aerobic, thermophilic bacteria, existing as nonmotile, rod-shaped cells that occur in acidic hot springs.

acidotrophic *Biology.* requiring acid nutrients.

acid phosphatase *Enzymology.* an enzyme found in blood plasma that catalyzes the release of phosphate from more complex substances.

acid phosphate see SUPERPHOSPHATE.

acid polishing *Engineering.* the refining of glass or other surfaces through the use of acids.

acid potassium (hydrogen) sulfate see POTASSIUM BISULFATE.

acid precipitation *Meteorology.* **1.** see ACID RAIN. **2.** any precipitation having a pH of less than 5.6.

acid process *Metallurgy.* the processing of a molten metal or alloy in a furnace in which the bottom and lining are made of acid refractories such as silica. The slag covering the metal, if present, is also acid.

acid-producing *Mining Engineering.* describing a rock strata containing enough pyrite to result in acid formation when subjected to air and water weathering.

acid proteases *Enzymology.* enzymes that break down proteins in acid solution.

acid radical *Chemistry.* the remainder of a molecule of an acid after the removal of acidic hydrogen.

acid rain *Meteorology.* rain or other precipitation that has an excessive concentration of sulfuric or nitric acids, as a result of chemical pollution of the atmosphere from such sources as automobile exhausts and the industrial burning of oil and coal; generally defined as any precipitation with a pH of less than 5.6. Similarly, **acid snow, acid dust.**

acid reaction *Chemistry.* a process in which one species donates a proton and another accepts a proton.

acid-recovery plant *Chemical Engineering.* a plant, in certain refineries, for separating sludge acid into tar, acid oil, and weak sulfuric acid, with the capability of reconcentration.

acid refractory *Materials.* a refractory material with a high proportion of silica, used to resist high temperatures and attack by acid slag.

acid salt *Chemistry.* a salt of an acid and a base where only part of the hydrogen of the acid is replaced by a basic radical.

acid slag *Metallurgy.* the nonmetallic covering used in the processing of a molten metal or alloy, whenever such slag has an excess of acid constituents such as silica or alumina.

acid sludge *Chemical Engineering.* the gummy acid or oily residue that separates after petroleum oil has been treated with sulfuric acid to remove impurities.

acid soil *Geology.* any soil having a pH less than 7.

acid soot *Engineering.* particles of carbon that have absorbed acid as a result of combustion; often the cause of metal corrosion in incinerators.

acid spar *Mineralogy.* a fluorspar product that is more than 98% CaF_2 and no more than 1% SiO_2; used in the production of hydrofluoric acid.

acid steel *Metallurgy.* a steel processed in a furnace in which the bottom and lining are made of acid refractories, such as silica, and covered with a predominantly acid slag.

acid tartrate see BITARTRATE.

acid test *Science.* **1.** formerly, the use of nitric acid as a test to determine the presence of gold in a substance. **2.** in popular use, a conclusive test to determine the validity or quality of something.

acid treatment *Petroleum Engineering.* **1.** the use of acid to enlarge the pore spaces where reservoir fluids flow. **2.** a refining process whereby untreated petroleum products such as kerosene, gasoline, diesel fuel, and lubricating stocks are treated with sulfuric acid to enhance odor, color, and other characteristics.

acidulant *Food Technology.* any of a variety of chemicals added to a food product to increase its tartness or acidity. Also, **acidulent.**

acidulate *Chemistry.* to make acidic or sour.

acidulous *Chemistry.* acidic or sour. Also, **acidulated, acidulous.**

acidulous water *Hydrology.* mineral water that contains either dissolved carbonic acid or dissolved sulfur compounds.

aciduric *Microbiology.* of bacteria or other organisms, able to tolerate an acidic environment.

acid value see ACID NUMBER.

acid-water pollution *Engineering.* pollution caused by acidic water resulting from industrial processes, as in mining or the manufacturing of products such as chemical batteries and fibers.

acidylation see ACYLATION.

acierate *Metallurgy.* to convert iron into steel; carry out the process of acieration.

acieration *Metallurgy.* the conversion of iron into steel by refining the carbon content and adjusting the content of other chemical elements. Also, **acierage.**

aciform *Biology.* having a needlelike shape; acicular.

acinaciform *Biology.* having a shape like a scimitar, as certain leaves.

acinar *Anatomy.* **1.** having a grapelike form; clustered like grapes. **2.** relating to or affecting an acinus or acini. Also, **acinarious.**

acinar cell *Anatomy.* any of the cells lining an acinus, especially the zymogen-secreting cells of the pancreatic acini.

acinar gland see ACINOUS GLAND.

acinesia *Neurology.* see AKINESIS.

Acinetobacter *Bacteriology.* a widespread genus of aerobic, Gram-negative bacteria of the family Neisseriaceae, occurring as cocci or short rods and widely distributed in soil and water; some are human pathogens.

aciniform *Biology.* shaped like a grape cluster.

acinitis *Medicine.* an inflammation of the acini of a gland.

acinous *Biology.* **1.** consisting of or resembling a clustering of small globes, as on a blackberry. **2.** relating to or forming acini. Also, **acinose.**

acinous gland *Anatomy.* a gland made up of one or more acini.

acinus *plural,* **acini.** *Botany.* **1.** any of the small berries or drupelets making up a compound fruit in plants such as the blackberry. **2.** a berry that grows in clusters; e.g., a grape or current. *Anatomy.* **1.** a cluster of secretory cells that surround and empty into cavities. **2.** any small, saclike dilatation, especially one found in a gland. (A Latin word meaning "grape," so called because of its shape.)

Acipenser *Vertebrate Zoology.* a genus of the sturgeon family Acipenseridae.

Acipenseridae *Vertebrate Zoology.* the sturgeons, a family of bony fishes belonging to the order Acipenseriformes.

Acipenseriformes *Vertebrate Zoology.* an order of fishes that includes sturgeons and paddlefishes and belongs the the subclass Actinopterygii.

ACK *Computer Programming.* see ACKNOWLEDGE CHARACTER.

ackee *Botany.* a tropical fruit tree, *Blighia sapida,* native to West Africa, widely grown in tropical regions for its white, fleshy seed coverings, which are eaten as a vegetable. Also, AKEE.

Ackeret method *Fluid Mechanics.* a relationship, named for Jacob Ackeret, showing that for thin air foils at small angles of attack, the local pressure coefficient depends only on the local surface angle (denoted η) and the freestream Mach number (denoted M_f), as long as higher-order terms of the deflectional angle can be neglected.

Ackeret theory *Fluid Mechanics.* a theory, named for Jacob Ackeret, in which it is assumed that for thin air foils at small angles of attack, the oblique shock and the Mach angles are roughly similar. Under this theory, the wave angle is approximately equivalent to the Mach angle and the loss in total head is so small that the flow is considered to be at constant entropy.

Ackerman steering (gear) *Mechanical Engineering.* a type of automotive steering that uses an Ackerman compound axle which allows the inner axle to move through a greater angle than does the outer axle. Also, **Ackerman linkage.**

acknowledge *Telecommunications.* a statement transmitted by a sending location to a receiving location requesting confirmation that a previous message has been received and understood.

acknowledge character *Computer Technology.* a particular character or code transmitted by a receiving device to a sending device to represent an acknowledgment of data received without a detected error.

acknowledgment *Telecommunications.* a statement transmitted by a receiving location to a sending location confirming that a previous message has been received and understood.

aclasia *Medicine.* a pathological growth or continuity of a structure.

aclastic *Optics.* relating to or describing something that does not refract light. *Medicine.* relating to or affected by aclasia.

acleistocardia *Cardiology.* the condition of the foramen ovale cordis being open, especially an abnormal opening.

aclinal *Geology.* of a stratum or other feature, having little or no inclination or dip. Also, **aclinic.**

aclinic *Geophysics.* describing a state in which a freely suspended magnetic needle remains in a horizontal position; having no magnetic inclination.

aclinic line *Cartography.* the magnetic equator; a line on a map or chart drawn through all the points on the earth's surface where the magnetic inclination is zero.

ACM Association for Computing Machinery.

Acmaeidae *Invertebrate Zoology.* a family of limpets, in the class Gastropoda, that lack an opening at the apex (top) of the shell.

acme [ak´mē] *Science.* the highest point or level. *Medicine.* the crisis or critical stage of a disease. *Ecology.* the time of greatest diversification or abundance of a taxon.

acme harrow *Agriculture.* a type of harrow having long twisted blades that project toward the rear, used to stir the soil and to cut clods. Also, BLADE HARROW.

acme screw thread *Design Engineering.* a thread having a flat crest and a profile angle of 29°; used on lead screws on lathes and on power screws on presses, automobile jacks, and other devices. Also, **acme thread.**

acme zone *Paleontology.* a region or site having an abundance of a certain fossil form.

acmite see AEGIRINE.

acne *Medicine.* **1.** an inflammatory disorder involving the sebaceous glands, often occurring in adolescents; characterized by the presence of skin lesions in the form of pimples, blackheads, and sometimes cysts. Also, COMMON ACNE, ACNE VULGARIS. **2.** any of various other specific types of this condition, as indicated by a specific descriptive term; e.g., **acne cosmetica, acne artificialis, halogen acne, iodide acne.**

acne conglobata *Medicine.* a severe, chronic form of acne characterized by the presence of numerous comedones (blackheads) and large abcesses and cysts, and involving significant scarring. Also, CONGLOBATE ACNE.

acneform *Medicine.* relating to or having the appearance of acne. Also, **acneiform.**

acnegen *Medicine.* a substance that causes acne.

acnegenic *Medicine.* causing or capable of causing acne.

acne indurata *Medicine.* a severe form of acne that is a progression of acne papulosa, characterized by deep and extensive lesions that produce significant scarring. Also, INDURATE ACNE.
acne mechanica see MECHANICAL ACNE.
acnemia *Medicine.* atrophy of the calf muscles.
acne neonatarum see NEONATAL ACNE.
acne papulosa *Medicine.* a common form of acne in which the lesions are typically numerous small papules. Also, PAPULAR ACNE.
acne rosacea see ROSACEA.
acne venenata see CONTACT ACNE.
acne vulgaris *Medicine.* the common form of acne.
Acnidosporidia *Invertebrate Zoology.* an order of parasitic protozoans. Also, HAPLOSPOREA, HAPLOSPORIDEA.
acnode see ISOLATED POINT.
acoasm see ACOUSMA.
Acoela *Invertebrate Zoology.* an order of small (less than 2 mm), primitive marine flatworms, in the class Turbellaria, lacking a digestive tract.
Acoelea *Invertebrate Zoology.* an order of gastropod mollusks, commonly called sea slugs, subclass Opistobranchia.
Acoelomata *Invertebrate Zoology.* animals that lack a true coelom or body cavity, including sponges, coelentrates, and lower worms.
acoelomate [ā´sē´ lə māt´] *Invertebrate Zoology.* **1.** an organism lacking a true coelom or body cavity, such as a rotifer or a nematode. **2.** of or relating to such organisms.
acoelomous [ā´sə lō´məs] *Invertebrate Zoology.* relating to or describing an organism lacking a true coelom or body cavity. Also, **acoelomatous** [ā´sə lō´mə tis].
acoelous *Zoology.* **1.** without a true alimentary canal. **2.** without a true coelom or body cavity.
ACOG or **A.C.O.G.** American College of Obstetricians and Gynecologists.
acology *Medicine.* the science or study of remedies for disease.
Aconchulinida *Invertebrate Zoology.* an order of protozoans in the subclass Filosia, comprising a small group of naked amebas having filopodia.
aconitase *Enzymology.* an enzyme that catalyzes the transformation of citric acid into *cis*-aconitic acid and isocitric acid.
aconite *Botany.* **1.** a group of colorful plants of the crowfoot family, genus *Aconitum,* many of which are poisonous. **2.** any plant of this genus; also known by various popular names such as monkshood or wolfsbane. **3.** the European monkshood plant, *A. napellus;* its roots are the source of a powerful poison. *Pharmacology.* $C_{34}H_{47}NO_{11}$, the poison prepared from the dried root of *A. napellus;* formerly used as a sedative and to treat neuralgia, toothache, thematicism, and fever.
aconitic acid *Organic Chemistry.* $C_3H_3(COOH)_3$, a white to yellowish crystalline solid that is soluble in water and alcohol; melts (decomposes) at 195°C; found in sugarcane and sugar beets, and used in plasticizers, antioxidants, synthetic flavoring, and in organic synthesis.
aconitine *Pharmacology.* $C_{20}H_{25}NO_2$, the chief active principal of aconite; a highly poisonous substance when ingested or absorbed through the skin.
acontium *plural,* **acontia.** *Invertebrate Zoology.* in certain sea anemones, one of the free threads within the body cavity equipped with stinging cells.
aconuresis *Medicine.* involuntary urination.
acorea *Medicine.* the absence at birth of the pupil of the eye.
acoria *Psychology.* the inability to feel satiated, regardless of how much is eaten and even though the appetite may not be large. Also, AKORIA.
acorn *Botany.* the seed or fruit of oaks; an oval nut that grows in a hard, woody cup.

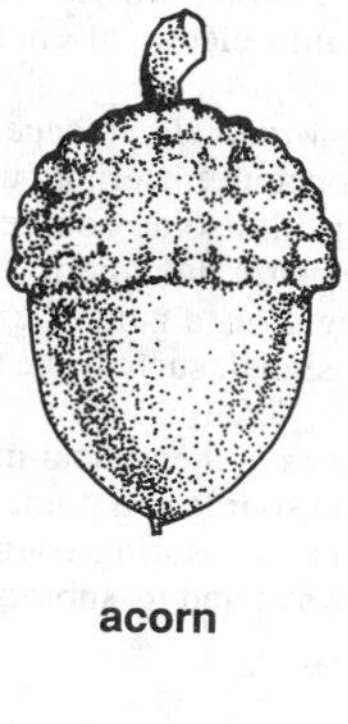

acorn

acorn barnacle *Invertebrate Zoology.* a small, conical sessile barnacle in the family Balanidae; common on intertidal rocks and ship bottoms.
acorn tube *Electronics.* an electron tube designed to function at UHF. The electrodes of this tube extend out of the tube's glass case radially. (So named because of its acorn-like shape.)
acorn worm *Invertebrate Zoology.* any of various worm-like marine animals in the class Enteropneusta of the Hemichordata phylum that burrow in sand or mud or live in algae or beneath stones.
acotyledon *Botany.* any plant in which cotyledons (seed leaves) are not present; e.g., ferns and lichens.
acotyledonous *Botany.* lacking cotyledons.
acou- or **acouo-** a combining form meaning "hearing" or "listening."
acoubuoy *Ordnance.* a listening device dropped by parachute onto land and water, used to detect sounds of enemy movements and transmit them to orbiting aircraft or land stations.
acouchi *Vertebrate Zoology.* a type of rodent that belongs to the family Dasyproctidae, which is classified as a dwarf variety of the agouti. Also, **acouchy.**
acousma *Medicine.* a simple auditory hallucination, such as buzzing or ringing. Also, **acoasma, acouasm.**
acousmatagnosis *Psychology.* an abnormal inability to understand spoken words and to recognize meaningful sounds.
acousmatamnesia *Psychology.* an abnormal inability to remember spoken words and to recognize meaningful sounds.
acoustic *Acoustics.* of or relating to sound, or the scientific study of sound. Many compound terms beginning with *acoustic* may also begin with *sonic* or *sound;* e.g., acoustic navigation = sonic navigation; acoustic energy = sound energy. *Biology.* relating to the sense or organs of hearing. *Materials Science.* of or relating to a material or substance that is designed to control the passage or quality of sound. Thus, **acoustic insulation, acoustic plaster, acoustic tile,** and so on. *Ordnance.* activated or directed by sound: an *acoustic* torpedo. *Electronics.* describing a guitar or other musical instrument whose sound is not electronically amplified or modified.
acoustic absorption *Acoustics.* the process by which sound energy is reduced as sound waves strike or pass through a surface. Also, **acoustic absorptivity.**
acoustic absorption coefficient *Acoustics.* the ratio of the amount of sound energy that is absorbed by a surface or medium to the amount that strikes it. Also, **acoustic absorption factor.**
acoustical of or relating to sound; acoustic. Many compound terms beginning with *acoustic* may also begin with *acoustical;* e.g., acoustical tile = acoustic tile.
acoustical ceiling *Building Engineering.* a ceiling covered with or built of material designed to control or absorb sound. Similarly, **acoustical door.**
acoustical cloud *Building Engineering.* a panel or other such device suspended from an auditorium ceiling, usually above the orchestra, to reflect sound.
acoustical Doppler effect see DOPPLER EFFECT.
acoustical engineering *Acoustics.* the practical application of the science of acoustics, including such elements as noise and vibration control, sound reproduction and broadcasting, and the use of sound instruments to make measurements and to examine or process various materials. *Building Engineering.* see ARCHITECTURAL ACOUSTICS.
See page 26 for ACOUSTICAL ENGINEERING essay.
acoustical holography see ACOUSTIC HOLOGRAPHY.
acoustical mass see ACOUSTIC MASS.
acoustical material *Materials Science.* any building material designed to reduce sound or absorb unwanted sound. Thus, **acoustical tile, acoustical plaster, acoustical (insulation) board.**
acoustical model see ACOUSTIC MODEL.
acoustical quality *Building Engineering.* see ACOUSTICS.
acoustical scintillation *Acoustics.* see SCINTILLATION.
acoustical stiffness see ACOUSTIC STIFFNESS.
acoustical surveillance *Ordnance.* any of various techniques involving the use of sound-detection devices to obtain information for intelligence purposes.
acoustical treatment see ACOUSTIC TREATMENT.
acoustic amplifier *Acoustical Engineering.* **1.** any device that increases the strength of an acoustic wave. **2.** see ACOUSTIC WAVE AMPLIFIER.
acoustic approximation *Fluid Mechanics.* an estimation that proceeds from a gas's nonlinear hydrodynamic equations to the linear wave equation for sound wave propagation.

Acoustical Engineering

Acoustical engineering deals with practical applications of the control of sound and with the control of sound and vibration. It encompasses a broad range of technical fields and impinges not only on physics and the classical engineering disciplines, but also on the arts, the life sciences, and the earth sciences. It is concerned not just with audible sound, but also with sound and vibration phenomena that range from barely measurable magnitudes to levels capable of inducing severe damage.

The traditional areas of acoustical engineering include the following, among others: architectural and musical acoustics, noise and vibration control, underwater acoustics, ultrasonics, communication engineering, shock and vibration engineering, and instrumentation engineering. Most of these areas have benefited greatly in recent years from advances in digital signal processing and micro-electronics.

Architectural acoustics deals with architectural spaces for musical and theatrical presentations, recording and broadcasting, as well as with acoustical environments in residences, offices, and other building areas. Musical acoustics relates to instruments and equipment for the production, recording, and reproduction of music. Noise and vibration control—keeping the noise and vibration of machines, of land, sea, and air vehicles, of industrial installations, and of highways and airports within acceptable limits—generally is motivated by safety, environmental, habitability, or machinery reliability concerns.

Underwater acoustics includes ship and submarine silencing and sonar engineering, as well as the use of sonar in oceanography and geology. Ultrasonics pertains to ultra-high-frequency sound and vibration; ultrasonic inspection and imaging systems, which in essence probe solid media much in the way sonar systems probe the ocean, have found wide use in industry, as well as in medicine.

Acoustical communication engineers are concerned primarily with the transmission of spoken information; they typically deal with such topics as speech synthesis, computer-aided recognition of spoken words and identification of the speakers, and speech compression for efficient transmission.

Acoustical engineering also deals with vibration and shock in structures and machinery, and thus it overlaps to some extent with mechanical and structural engineering. Acoustical engineers generally tend to focus on higher-frequency effects, whereas the more classical engineering concerns typically are associated with lower frequencies.

Instrumentation engineering supports all of the aforementioned areas. It relates to the design and application of sensors for sound and vibration, of sound and vibration generators, of recording systems, and of data analysis equipment.

Eric E. Ungar
Chief Consulting Engineer
BBN Systems and Technologies

acoustic array *Acoustical Engineering.* a group of acoustic transducers that are arranged to create a specific unified effect.

acoustic branch *Solid-State Physics.* the branch of the dispersion relation of waves in a crystal lattice that corresponds to equivalent acoustic waves propagating as crystal vibrations through an elastic medium.

acoustic center *Acoustical Engineering.* for a sound transmitter or receiver, the point from which sound appears to emanate or diverge.

acoustic clarifier *Acoustical Engineering.* a system composed of sound-absorbing cones loosely mounted on the baffle of a loudspeaker, used to absorb the energy of sudden bursts of loud sounds.

acoustic comfort index *Acoustics.* an arbitrary scale for sound level tolerance, in which a value of +100 is assigned to ideal conditions and a value of –100 is assigned to intolerable conditions.

acoustic compliance *Acoustics.* a medium's reactive capacitance as a result of its volume displacement, with little displacement of its center of gravity.

acoustic construction *Building Engineering.* a type of building construction designed to control or absorb sound by means of air spaces or acoustic material within the structure.

acoustic coupler *Acoustical Engineering.* a device that acts as an interface in the conversion of acoustic signals to electric signals, or vice versa.

acoustic delay *Acoustical Engineering.* an electronically or physically induced increase in the time between the transmission of a sound and reception of the sound, so as to produce some desired effect.

acoustic delay line *Electronics.* a transmission line or device that delays the transmission of acoustic signals. The delay is created by the sound pulses being reflected through some liquid or solid material; e.g., a quartz bar. Also, SONIC DELAY LINE, ACOUSTIC STORAGE.

acoustic detection *Acoustical Engineering.* **1.** the formation of the sound profile of an object, such as an underwater mountain range, by means of measuring the sound waves reflected off the object. **2.** a similar formation of a sound profile by measuring the sound radiated from the object. Thus, **acoustic detector.**

acoustic dispersion *Acoustics.* the separation of a sound wave into its frequency components as it passes through a particular medium. Also, refers to the fact that the velocity of the wave changes at it passes through the medium.

acoustic domain *Acoustics.* a term for a region of acoustic vibrational activity in a crystal lattice.

acoustic emission *Acoustics.* the transmission of sound pressure waves from a body due to a rapid energy release in response to internal or external mechanical stress placed upon the medium. *Materials Science.* the measurement of such sound waves as a means of detecting material flaws or failures.

acoustic energy see SOUND ENERGY.

acoustic fatigue *Mechanics.* the loss of strength in a material due to acoustic stresses.

acoustic feedback *Acoustical Engineering.* **1.** the return signal in signal processing that is used to adjust the output. **2.** an undesirable return signal that can produce excessive reinforcement and loud sounds, such as those produced when loudspeaker signals are picked up by a microphone.

acoustic filter *Acoustics.* a device used to cut off sound in a particular frequency or range of frequencies while admitting sound in another or other frequencies.

acoustic generator *Acoustical Engineering.* a device that creates sound energy, such as a noise generator that produces broadband noise or an oscillator that produces discrete frequencies.

acoustic grating *Acoustics.* a series of equally distant and equal-sized obstacles to sound waves, which cause wave diffraction, depending on the grating and wavelength.

acoustic heat engine *Engineering.* an engine that converts heat into sound energy and then into electrical energy, without the use of mechanical parts.

acoustic holography *Materials Science.* a nondestructive technique for evaluating flaws in opaque materials, utilizing phase and amplitude relationships between an ultrasonic source probe and its defect-generated reflection. Thus, **acoustic hologram.**

acoustic homing *Navigation.* a guidance system, as for a torpedo, that homes in on an acoustic signal, such as the sound of a ship's propellers.

acoustic horn see HORN, def. 1.

acoustic image *Acoustics.* a theoretical image, analogous to an optical image, of an actual sound source at a point on the opposite side of a barrier or boundary; used for measuring sound characteristics such as the combined strength of a wave and its subsequent reflections.

acoustic imaging *Acoustics.* the use of acoustic energy to form a representation of a physical object, such as ultrasound tomography on internal human organs or on microscopic images created with sound.
acoustic impedance *Acoustics.* the total reaction of a medium to sound transmission through it, represented as the complex ratio of the sound pressure to the effective flux (particle velocity times surface area) through the medium.
acoustic inertance or **acoustic inertia** see ACOUSTIC MASS.
acoustic intensity *Acoustics.* the average acoustic power transported across a unit area, usually expressed in watts per square meter.
acoustic interferometer *Acoustics.* a device that measures the velocity and attenuation of sound in a fluid medium by transmitting an ultrasonic beam through the medium to a reflector.
acoustic jamming *Acoustical Engineering.* the intentional saturation of an acoustic band with acoustic noise so that other sounds in the same band are masked.
acoustic labyrinth *Acoustical Engineering.* a speaker enclosure with special partitions and passages that dissipate cavity resonance and reinforce bass response.
acoustic lens *Acoustics.* a system of disks or other devices to spread or converge sound waves in a manner analogous to the way an optical lens refracts light.
acoustic levitation *Acoustics.* the use of intense sound waves to support an object in a fluid medium.
acoustic line *Acoustical Engineering.* a path along which acoustic signals are carried in such a way that resonant cavities, baffles, and labyrinths are interconnected for the desired transmission of bass frequencies; it is analogous to an electromagnetic transmission line.
acoustic logging *Petroleum Engineering.* a technique used to measure porosity in drill holes; a comparison of the depth to the travel time of a sonic impulse through a specific section of the borehole gives an indication of the rock composition and fluids found in the formation.
acoustic Mach meter *Aviation.* an instrument that indicates an object's speed by measuring the velocity of a sound pulse; used to calculate the Mach number.
acoustic mass *Acoustics.* the impedance of an acoustic volume, which for a small volume such as a tube can be derived by the formula $M_A = \rho l/S$, in which ρ is the density of air, l is the length of the volume, and S is the cross-sectional area.
acoustic mass reactance see ACOUSTIC REACTANCE.
acoustic measurement *Acoustics.* the logarithmic (base 10) measurement of acoustic energy, normally using decibel units by an equation such as the power equation: $dB = 10 \log(P/P_0)$ with P_0 representing the reference power, and P representing the nonreference power.
acoustic memory *Computer Technology.* an early type of computer memory using an acoustic delay line composed of a material such as mercury or quartz.
acoustic microscope *Optics.* an instrument that uses sonic waves at microwave frequencies to provide visual details of the microscopic and elastic properties of an object.
acoustic mine *Ordnance.* a naval mine that is activated by vibrations in the water caused by the noise of a ship's propeller, engine, and so on.
acoustic mode *Solid-State Physics.* a wave mode that is produced by vibrations in a continuous crystalline medium in which the long wavelengths behave like an acoustic wave and the shorter wavelengths approach the Debye frequency so that the phase velocity is observed to decrease.
acoustic model *Acoustical Engineering.* a scale model of a room, building, or other object used to measure qualities of sound distribution and noise control.
acoustic navigation see SONIC NAVIGATION.
acoustic nerve see AUDITORY NERVE.
acoustic noise *Acoustics.* **1.** any sound that is audible to human hearing. **2.** an unwanted sound that is subjected to some form of noise control.
acoustic ocean-current meter *Acoustical Engineering.* a device used to monitor currents by measuring the difference in travel time between each acoustic pulse transmitted in a direction opposite to the flow of a current and the return pulse.
acoustic ohm *Acoustics.* a unit of measurement based on Ohm's laws, used to measure acoustic impedance; equal to the impedance of a medium in which a sound pressure of one dyne per square centimeter produces a velocity of one centimeter per second.
acousticophobia *Psychology.* an irrational fear of sounds.
acoustic particle detection *Particle Physics.* a technique in particle detection in which a specimen is irradiated with an acoustic signal, and certain reflections of the signal indicate the presence of a particle.
acoustic perfume *Acoustical Engineering.* see WHITE NOISE.
acoustic phonetics *Linguistics.* the use of acoustic equipment and techniques to study human speech.
acoustic phonon *Solid-State Physics.* a quantum of acoustic mode excitation.
acoustic position reference *Petroleum Engineering.* a system used in offshore drilling that transmits ultrasonic signals to monitor the drilling ship's position in relation to the ocean floor.
acoustic power see SOUND POWER.
acoustic radar *Meteorology.* the use of pulsed acoustic waves to acquire various meteorologic information, such as humidity, wind speed, and turbulence.
acoustic radiation *Acoustics.* sound wave energy that travels from one point to another by exerting pressure on the medium, creating a pattern of compressions and rarefactions that can be described as a three-dimensional pattern.
acoustic radiation pressure *Acoustics.* the total pressure of a sound wave on any surface of interface.
acoustic radiator *Acoustical Engineering.* an element or object that radiates sound, such as the cone on a loudspeaker, the diaphragm on headphones, or an object under water.
acoustic radiometer *Engineering.* an instrument that detects and gauges the intensity of sound waves by measuring the pressure caused by the reflection or absorption of the waves.
acoustic ratio *Acoustical Engineering.* **1.** the ratio of the intensity of an acoustic signal that is radiated directly from a sound source to that of the same signal after it has reflected from a surface. **2.** the ratio of the intensity of an acoustic signal to that of a reference intensity; commonly used in the logarithm argument of a decibel rating.
acoustic reactance *Acoustics.* the imaginary portion of acoustic impedance (when impedance is expressed as a complex number), due to the inertia and elasticity of the medium and consisting of acoustic mass and acoustic compliance.
acoustic reactance unit see ACOUSTIC OHM.
acoustic receiver *Electronics.* any device that accepts an acoustic signal for the purpose of amplification or modulation.
acoustic reciprocity theorem *Acoustics.* a theorem used for calibration of a linear sound system, stating that if a force of specific magnitude is applied to any branch of the system and a response is measured on the branch, the ratio of the force to the magnitude of response will be unchanged if the points of application and measurement are reversed.
acoustic reflectivity *Acoustics.* **1.** the relative reflectivity of a specific material, that is, the tendency to deflect sound energy rather than absorb it, in a specific medium such as air, water, earth, and so on. **2.** the tendency of a surface to cause a sound wave to reflect at an angle equal to the angle of incidence, represented by the coefficient equation

$$R = (m \cos \theta_1 - n \cos \theta_2) / (m \cos \theta_1 + n \cos \theta_2)$$

with m representing the ratio of densities and n representing the ratio of velocities. Also, **acoustic reflection coefficient.**
acoustic reflex *Physiology.* the contraction of the stapedius muscle in the ear in response to intense sounds.
acoustic reflex enclosure *Acoustical Engineering.* a loudspeaker cabinet designed with certain materials and structure so as to direct low-frequency signals forward rather than attenuating these signals within the cabinet.
acoustic refraction *Acoustics.* the change in the direction of sound as it travels through a medium due to differences in temperature, pressure, and other characteristics of the medium, or a change in the medium.
acoustic regeneration see ACOUSTIC FEEDBACK.
acoustic resistance *Acoustics.* the real portion of acoustic impedance due to characteristics of the medium; the real part of the ratio of sound pressure at a specific point in the medium to particle velocity in a free plane.
acoustic resistance unit see ACOUSTIC OHM.
acoustic resonance *Acoustics.* a condition in which the amplitude of periodic sinusoidal motion in an acoustical system is at its maximum, due to the system being acted upon by sound waves at the same frequency as the natural oscillation frequency of the system.
acoustic resonator *Acoustics.* a device with an acoustic mass and compliance that mutually cancel each other's effects, thus rendering zero acoustic reactance at a specific natural frequency or frequencies for which it was designed.

Acoustics

Acoustics is the science of sound. The stem *acoust-* is related to Greek words pertaining to hearing, and the suffix *-ics* is an anglicized version of Greek and Latin suffixes implying "in the manner of." The term *acoustics* in the sense of a scientific discipline was first coined as the French term acoustique in an article published in 1701 by Joseph Sauveur, who stated (English translation) "I have formed the opinion that there is a higher science than music, and I call it acoustics; it has for its object sounds in general, whereas music has for its object sounds pleasing to the ear. To treat this science as other sciences, such as optics, it is necessary to explain the nature of sound, the organ of hearing, and all the properties of sound."

Up through the first third of the twentieth century, acoustics books and papers almost never had the word "acoustics" (or its foreign counterparts) in their titles. An important and influential exception was the book *Die Akustik* published by E. F. F. Chladni in 1802. Rayleigh, in his *Theory of Sound* which was first published in 1877, uses the term very sparingly with no explicit definition, but in the sense of implying the science of sound. Perhaps he and his contemporaries preferred the less-pretentious word "sound" to "acoustics."

The widespread modern use of the term probably stems from the desire to unambiguously refer to a science including more than that of sounds audible to the human ear. Advances in instrumentation made possible the systematic study of sounds (infrasound and ultrasound) with frequencies lower and higher than humans can hear, and it was recognized that the same physical principles are applicable. Thus, to many people, the term "sound" came to mean any mechanical disturbance with wave-like characteristics, and it is with this broadened usage that the definition of "acoustics" as the science of sound is now interpreted.

Allan D. Pierce
Leonhard Chair in Acoustics
Pennsylvania State University

acoustics *Physics.* the branch of physics that is concerned with the study of sound, including its production, propagation, and effects. *Building Engineering.* the overall acoustic characteristics of a room that determine the nature and quality of the sounds heard within it.

acoustic scattering *Acoustics.* a distribution of sound waves in many directions due to multiple reflections and bending (diffraction), such as can occur with music performed in a poorly designed auditorium.

acoustic seal *Acoustical Engineering.* a connection point or joint along an acoustic transmission line in which acoustic coupling is maximized; i.e., there are low losses of energy.

acoustic sensor *Robotics.* a sensor that uses sound waves to measure various phenomena.

acoustic shadow *Acoustics.* a space into which sound does not enter due to the refraction of sound waves; for example, a space immediately behind a large object that obstructs the sound wave.

acoustic shielding *Acoustics.* the use of barriers to prevent the transmission of sound into a space or past the barrier, such as barriers constructed to control unwanted noise.

acoustic signal processing *Acoustics.* the extraction of useful information from acoustic signals, using electronic techniques such as a fast Fourier transform.

acoustic signature *Engineering.* the characteristic pattern or profile of an object as detected by identification equipment that uses sound waves.

acoustic spectrograph *Oceanography.* a type of spectrograph that employs sound waves to study the transmission and reflection characteristics of ocean thermal layers and marine organisms.

acoustic spectrometer *Acoustical Engineering.* an electronic device that measures the intensities of acoustic frequencies for a selected frequency band.

acoustic spectrum *Acoustics.* **1.** the frequency range of acoustic signals. **2.** a graph showing such a range.

acoustic stiffness *Acoustics.* a quantity given by the product of the angular frequency $2\pi f$ and the acoustic stiffness reactance.

acoustic stiffness reactance *Acoustics.* the portion of the acoustic reactance of a medium associated with the potential energy, analogous to the capacitive reactance of an electrical system. *Materials Science.* the measurement of this resistance as a factor in the selection of materials for machinery.

acoustic storage see ACOUSTIC DELAY LINE.

acoustic strain gauge *Engineering.* a device that aids in assessing the amount of strain on portions of structures; a filament or wire is attached to the portion of the structure under strain, the filament is then triggered to vibrate, and the frequency of the vibration is measured by the device. This yields information about thermal currents and marine life.

acoustic streaming *Fluid Mechanics.* the production of unidirectional flow currents in a fluid because of the presence of sound waves.

acoustic suspension *Acoustical Engineering.* a method of sound reproduction in loudspeakers in which the speaker is loosely suspended in a sealed cabinet of sound-absorbent material; it allows for the reproduction of high-quality sound in relatively compact enclosures.

acoustic theodolite *Oceanography.* a device that produces a continuous vertical profile of an ocean current through the use of sound waves.

acoustic tile *Building Engineering.* a thin tile used as a covering on ceilings and walls to provide sound-absorbing properties.

acoustic tomography *Acoustics.* an imaging technique in which information is collected from beams of acoustic radiation that have passed through an object.

acoustic torpedo *Ordnance.* a torpedo that is directed toward its target by the noise of the target vessel or by a sonar signal.

acoustic transducer *Acoustical Engineering.* a device, such as an underwater hydrophone, that converts acoustic energy into electromagnetic energy, or a device, such as a loudspeaker, that converts electromagnetic energy into acoustic energy.

acoustic transformer *Acoustical Engineering.* a device in a sound system that changes an electroacoustical signal so that the reproduced sound is altered in some way from the original, as in its amplitude, frequency, or duration.

acoustic transmission *Acoustics.* the movement of acoustic energy by wave motion through a medium.

acoustic transponder *Navigation.* a transponder that responds with an acoustic code when interrogated by an acoustic signal.

acoustic treatment *Building Engineering.* the part of a building plan that provides for its acoustical environment.

acoustic velocity *Acoustics.* the speed at which sound travels.

acoustic wave *Acoustics.* a three-dimensional wave of compression and decompression of a medium, due to a disturbance of the medium by a source of acoustic energy.

acoustic wave amplifier *Electronics.* an amplifier that increases the strength of an acoustic wave by transferring the energy of semiconductor charge carriers to the wave, as the wave travels through a piezoelectric medium.

acoustic wave filter *Acoustical Engineering.* a filter designed to separate sound waves of different frequencies.

acoustic well logging *Oceanography.* a technique used in underground exploration that utilizes high-frequency pulsed sound waves.

acoustoelectric amplifier see ACOUSTIC AMPLIFIER.

acoustoelectric effect *Electronics.* the generation of a DC voltage in a crystal or in a metallic material, due to acoustic waves traveling along the surface of the material.

acoustoelectronics *Acoustical Engineering.* the use of acoustic energy to create electromagnetic waves, usually with crystals or metals that react when bombarded with acoustic waves, and the processing of such waves prior to reproduction of original sound.

acoustooptical cell *Electricity.* an electric-to-optical transducer in which an acoustic or ultrasonic electric input signal modulates or acts on a beam of light.

acoustooptical filter *Optics.* an optical filter tuned across the visible spectrum by acoustic waves ranging from 40 to 68 megahertz.

acoustooptic interaction *Optics.* the variation that occurs in the propagation of an optical wave when it passes through a medium with a low-frequency acoustic wave.

acoustooptic modulator *Optics.* a device that uses acoustic waves to vary the amplitude or phase of a light beam.
acoustooptics *Optics.* the study of interactions between sound waves and light waves in solid materials; important in laser and holographic technologies.
ACP acyl carrier protein.
ACP or **A.C.P.** American College of Physicians.
a-c plane *Crystallography.* the surface of a crystal whose edges are parallel to the *a* and *c* axes.
AC plate resistance see DYNAMIC PLATE IMPEDANCE.
acquire to come to have or get posession of; specific uses include: *Biology.* to develop some trait or quality after birth, as opposed to inheriting it. *Electronics.* to gather information from a transducer or a computer. *Ordnance.* to locate and track a radar target or satellite in order to obtain projectile firing data or other such data.
acquired *Biology.* of or relating to a trait developed by an organism after birth in response to the environment and not transmittable by inheritance; not genetic. *Medicine.* denoting some disease, predisposition, or habit that is not congenital, but is developed after birth.
acquired atelectasis see OBSTRUCTIVE ATELECTASIS.
acquired character *Genetics.* a modification in form or activity during an organism's lifetime that occurs as a result of environmental factors or use or disuse and that is not inheritable by future generations. Also, **acquired characteristic.**
acquired drive *Behavior.* a drive that is learned rather than innate.
acquired immune deficiency syndrome *Medicine.* the full name for AIDS, a fatal disease marked by the inability of the immune system to fight infection; see AIDS.
acquired immunity *Immunology.* the resistance of an organism to a disease or infection that is developed after birth, generally through the injection of an antigen or antibody, or through infection itself.
acquired immunodeficiency syndrome see AIDS.
acquired immunological tolerance *Immunology.* a condition in which an organism lacks the ability to form an immune response; produced by exposure to an antigen, frequently in embryonic life. Also, IMMUNOLOGICAL PARALYSIS.
acquired reflex see CONDITIONED REFLEX.
acquired trait see ACQUIRED CHARACTER.
acquisition the act or fact of acquiring something; specific uses include: *Behavior.* **1.** something that is gained by an individual, such as an idea, an item of information, or a new way of responding. **2.** an increase in the strength of a response after a behavior has been rewarded. *Engineering.* the detection of a desired radio signal or broadcast emission, usually by adjusting an antenna or telescope. *Ordnance.* the detection of a target by the human eye or a sensor system such as radar. *Space Technology.* the process of locating the path of a satellite or a space probe in order to gather tracking or telemetry data. *Linguistics.* the process of achieving competence or fluency in a language, especially in one's native language as a child.
acquisition and tracking radar *Engineering.* a radar system able to detect and then lock onto a target or onto an object emitting radiation or a radio signal.
acquisition tone *Computer Technology.* an audible signal that verifies user entry into a minicomputer, microcomputer, or other such device.
acquisitiveness *Psychology.* see HOARDING.
ACR or **A.C.R.** American College of Radiology.
acr- a combining form meaning "height" or "extremity."
Acrania *Zoology.* a subphylum of the Chordata, chordate animals without a true brain or skull. Also, **Acraniata.**
acrania *Medicine.* the absence at birth of the cranium, either partial or total. Thus, **acranial.**
Acrasia *plural,* **Acrasiales**. *Mycology.* a class of cellular slime molds noted for both plant and animal characteristics; usually classified in the division Myxomycota. (From a Greek word meaning "bad mixture.")
acrasia *Psychology.* the absence of self-control.
Acrasida plural, **Acrasidae**. *Mycology.* the order including the Acrasiales, cellular slime molds.
Acrasieae see ACRASIA.
acrasin *Biochemistry.* a substance produced by certain microorganisms that stimulates them to form fruiting bodies.
Acrasiomycetes *Mycology.* a class of fungi belonging to the division Myxomycota, believed to be related to protistans; it is composed of cellular slime molds that ingest nutrients through phagocytosis, and is similar to protozoans in its biochemical composition and its mode of sporulation.
acraspedote *Invertebrate Zoology.* describing or relating to tapeworm segments that are not overlapping
acratia *Medicine.* **1.** weakness, or the loss of strength. **2.** a lack of control, especially incontinence.
acraturesis *Medicine.* difficulty or weakness of urination.
acre *Metrology.* **1.** a unit of measure for land, equal to 43,560 square feet or 4046.86 square meters. **2.** any of several historic variations of this measure, such as the Scottish acre or Irish acre. (Originally based on the amount of land that a team of oxen could plow in one day.)
acreage *Agriculture.* the extent of an area of farmland as measured in acres.
Acree's reaction *Analytical Chemistry.* the addition of sulfuric acid and a formaldehyde solution containing ferric chloride to an unknown solution for the purpose of testing for proteins; the appearance of a violet-colored ring indicates a positive test.
acre-foot *Metrology.* a unit of capacity equal to 43,560 cubic feet, or the volume of water that is required to cover one acre to a depth of one foot.
acre-foot per day *Metrology.* a unit of volume rate of water flow.
acre-ft. acre-foot.
acre-ft./d acre-foot per day.
acre-in. acre-inch.
acre-inch *Metrology.* a unit of capacity equal to one-twelfth of an acre foot, or 3630 cubic feet.
Acremonium *Mycology.* a genus of fungi belonging to the class Plectomycetes, characterized by asexual, cone-shaped spores.
AC resistance see HIGH-FREQUENCY RESISTANCE.
acre-yield *Geology.* the average amount of oil, gas, or water that is recovered from one acre of a reserve.
acrid *Biology.* relating to or describing a substance that is sharp, biting, or irritating to the senses.
Acrid *Ordnance.* a Soviet air-to-air missile believed to be powered by a solid-propellant rocket motor and equipped with infrared or semiactive radar homing. It is the largest air-to-air missile in the world, with a launch weight of 1765 pounds; it probably carries a large conventional warhead at a speed of Mach 4 and a range of 15 to 50 miles. Officially designated **AA-6.**
acridine *Organic Chemistry.* $C_{13}H_9N$, small, colorless needles that are slightly soluble in hot water and soluble in alcohol and ether; melts at 111°C and boils at about 345°C; used in dyes and as an analytical reagent.

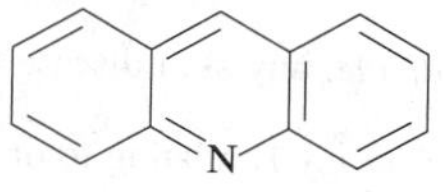

acridine

acridine orange *Organic Chemistry.* $C_{17}H_{19}N_3 \cdot HCl$, a dye that is used to identify cancer tumor cells. It complexes with DNA to produce a green color, and with RNA to produce an orange color, as observed by fluorescence microscopy.
acridophagous [ak′rə də fāj′is] *Biology.* feeding on grasshoppers or other related insects.
acriflavine *Organic Chemistry.* $C_{14}H_{14}N_3Cl$, brownish or orange odorless granules that are soluble in water and slightly soluble in alcohol; used in dyestuffs and as a powerful antiseptic and bacteriostat.
Acrilan *Textiles.* the trade name for a quick-drying, wrinkle-resistant synthetic textile fiber consisting of acrylonitrile copolymerized with vinyl acetate.
acrisia *Medicine.* an older term describing an event where both the diagnosis and the prognosis are uncertain.
acritarch *Paleontology.* a unicellular oceanic microfossil of unknown biological affinity; commonly found in Precambrian and Paleozoic strata. (From a Greek term meaning "uncertain beginning.")
acritical *Medicine.* without a crisis; describing a disease that does not have a crisis stage.
acro- a combining form meaning "height" or "extremity," as in *acrophobia* (the fear of heights) or *acroarthritis* (arthritis affecting the extremities).
acroagnosis *Medicine.* the loss or absence of sensation in a limb or extremity. Also, **acroanesthesia**.

acrobatholith *Geology.* a mineral deposit found in or near the rim of an exposed batholithic dome.
acrobatholithic *Geology.* of or relating to the stage in the erosion of a batholith during which such an area is exposed. *Mineralogy.* of a mineral deposit, found in or near the rim of an exposed batholithic dome.
acroblast *Cell Biology.* a granule within a spermatic that gives rise to the acrosome.
Acrobolbaceae *Botany.* a family of medium to robust liverworts of the order Jungermanniales, characterized by reddish to purplish pigmentation, lateral or ventral intercalary branches, scattered rhizoids, and absent or very reduced underleaves and bracteoles.
acrocarp *Botany.* a plant having the reproductive organs at the end of the main axis, as certain mosses do. Thus, **acrocarpous.**
acrocentric *Cell Biology.* describing a chromosome or chromatid whose centromere is located near one end, producing chromosome arms of unequal length. Thus, **acrocentric chromosome.**
acrocephalia see OXYCEPHALY.
acrocephalic or **acrocephalous** see OXYCEPHALIC.
acrocephalopolysyndactyly *Medicine.* the condition of acrocephalosyndactly also accompanied by polydactyly (the presence of extra fingers or toes). Also, **acrocephalopolysyndactylism.**
acrocephalosyndactylism see ACROCEPHALOSYNDACTYLY.
acrocephalosyndactyly *Medicine.* a congenital condition characterized by a peaked head and webbed fingers and toes; specific variations of the condition include **Alpert's syndrome, Apert-Crouzon syndrome, Chotzen syndrome,** and **Pfeiffer syndrome.**
acrocephaly see OXYCEPHALY.
Acroceridae *Invertebrate Zoology.* a family of two-winged dipteran flies with a very small head and a humpbacked thorax; the larvae are parasites of spiders.
Acrochaetiaceae *Botany.* a family of red algae belonging to the order Nemaliales, characterized by essentially filamentous simple thalli; also known as **Chantransiaceae** in older systems.
acrocinesis see ACROKINESIA.
acrocontracture *Medicine.* a condition in which the joints or fibrous tissue of an extremity, especially the hands and feet, are abnormally contracted.
acrocyanosis *Medicine.* a condition in which the extremities are blue, cold, and wet with perspiration.
Acrodelphidae *Paleontology.* a family of cetaceans in the suborder Odontoceti; represented by several genera in the Miocene and Pliocene.
acrodendrophilous [ak′rə den dräf′ə lis] *Ecology.* describing a species that lives or thrives in treetop habitats.
acrodermatitis *Medicine.* an inflammation of the skin of the hands or feet.
acrodermatosis *Medicine.* any skin disease involving the skin of the hands or feet.
acrodont *Vertebrate Zoology* **1.** relating to or describing a condition of tooth attachment in some lower vertebrates, such as lizards, in which the teeth sit on the edge of the jaw bone, rather than being held in sockets. **2.** an animal having such tooth attachment. Thus, **acrodontism.**
acrodrome *Botany.* of a leaf, having the main veins unite at the top.
acrodynia *Medicine.* a disease of infants and children, caused by ingestion of or contact with mercury and characterized by fatigue, irritability, rashes on the skin of the hands and the feet, generalized swelling of the extremities, gastrointestinal disturbances, itching of both the hands and the feet, and abnormal coloration of both the tip of the nose and the cheeks.
acrogen *Botany.* a plant growing and producing its reproductive structures only at the top or apex, such as mosses and ferns. Thus, **acrogenic.**
acrognosis *Neurology.* sensory perception of the limbs and knowledge of their different portions in relation to one another.
acrohypothermia *Medicine.* abnormal coldness of the hands or feet.
acrokinesia *Medicine.* excessive movement; abnormal freedom of movement of the limbs. Also, ACROCINESIS.
acroleic acid see ACRYLIC ACID.
acrolein *Organic Chemistry.* CH_2=CHCHO, a colorless or yellowish liquid with a pungent odor that is soluble in water, alcohol, and ether; it freezes at –86.9.°C and boils at 52.5–53.5°C. Unstable and readily polymerized by light, it is used in the production of plastics, perfumes, and colloidal metals, in organic synthesis, as a poison gas, and as an aquatic herbicide. Also, ACRYLALDEHYDE.
acrolein cyanohydrin *Organic Chemistry.* CH_2=CHCH(OH)CN, a liquid that is miscible in water and boils at 165°C. Ethylene and acrylonitrile will copolymerize with it; used as a modifier in synthetic resins.
acrolein dimer *Organic Chemistry.* $C_6H_8O_2$, a flammable liquid that boils at 151.3°C and is soluble in water; used as an intermediate in resins, pharmaceuticals, and dyes.
acrolein test *Analytical Chemistry.* a test for glycerin or fats, in which potassium bisulfate is added to the sample and heated; acrolein is released if the test is positive.
acromastitis *Medicine.* an inflammation of the nipple.
acromegalia see ACROMEGALY.
acromegalic *Medicine.* relating to or affected by acromegaly.
acromegaly *Medicine.* a chronic disease marked by the progressive enlargement of the hands, feet, face, head, and thorax due to abnormally high secretion of a growth hormone that is secreted by the anterior lobe of the pituitary gland.
acromelalgia see ERYTHROMELALGIA.
acromere *Histology.* the distal region of a rod or cone in the retina.
acrometer *Engineering.* an instrument used to determine the density of fluids.
acromial *Anatomy.* of or relating to the acromion.
acromicria *Medicine.* the converse of acromegaly; abnormal smallness of the head and extremities.
acromion *Anatomy.* the outer end of the scapula to which the collar bone is connected.
acromorph see SALT DOME.
acron *Invertebrate Zoology.* the head or nonsegmented preoral part of a segmented (metameric) animal, such as an annelid worm.
acronematic *Biology.* of or relating to flagella without hairs.
acronical *Astronomy.* relating to or occurring at sunset.
acronical rising *Astronomy.* the rising of a celestial object at sunset or just after sunset. Similarly, **acronical setting.**
acronym *Linguistics.* a word formed from the initial letter or letters of a group of words; e.g., *AIDS, radar.*
acroparalysis *Medicine.* the general paralysis of the muscles of one or more extremities.
acroparesthesia [ac′rō pâr′əs thēz′yə] *Medicine.* **1.** general abnormal sensations such as numbness or tingling in one or more extremities, due to nerve compression or polyneuritis. **2.** a disease marked by attacks of tingling, numbness, or stiffness of the extremities, sometimes with pain, skin discoloration, or slight cyanosis..
acropathy *Medicine.* a general term for any disease affecting the extremities. Thus, **acropathic.**
acropetal *Botany.* describing plant development from the base to the apex.
acrophobia *Psychology.* an irrational fear of heights or high places. Also, HYPSOPHOBIA.
acropleurogenous *Microbiology.* found at the tip and along the sides of a structure.
acropolis *Architecture.* **1.** the fortified upper part of an ancient Greek city, usually containing temples and other public buildings. **2. Acropolis.** a noted example of this in Athens. **3.** any elevated urban area containing public buildings.

Acropolis

Acrosaleniidae *Paleontology.* an extinct family of regular euechinoids in the superorder Stirodonta and order Salenioida; characterized by keeled teeth, gill slits, and solid spines; some genera evolved massive defensive spines; lower Jurassic to lower Cretaceous.

acrosin *Biochemistry.* an enzyme located in sperm and thought to play a role in egg penetration.

Acrosiphonia *Botany.* either of two genera of the family Acrosiphoniaceae, distinguished by multinucleate cells.

Acrosiphoniaceae *Botany.* a family of green algae of the order Acrosiphoniales, characterized by profusely branched filaments that are often matted by short hooked branches or rhizoids; restricted to cold marine waters, where they often grow on rocks or other algae.

Acrosiphoniales *Botany.* an order of cold marine and brackish-water green algae of the class Chlorophyceae; characterized by a simple or branched uniseriate thallus with filaments attached by rhizoids, multinucleate cells, and a perforated cylindrical chloroplast with many pyrenoids.

acrosome *Cell Biology.* the anterior caplike portion of the sperm, secreted by the Golgi apparatus of the spermatid.

acrospire *Botany.* the first sprout appearing after the germination of a grain.

acrospore *Mycology.* a fungal spore formed at the tip of a hypha.

acrosyndesis *Genetics.* a pairing activity of homologous chromosomes at meiosis that involves the terminal portions of the chromosomes.

acroteric *Anatomy.* of or relating to a tip or outermost part.

acroterion [ak´rə ter´ē än] *Architecture.* **1.** a pedestal at the peak or corners of a pediment, used to support a statue or other ornament. **2.** an entire ornamental element that includes such a pedestal. Also, **acroterium** [ak´rə ter´ē əm].

Acrothoracica *Invertebrate Zoology.* an order of naked burrowing barnacles in the subclass Cirripedia, that bore into and inhabit corals and the shells of mollusks and other barnacles.

acrotism [ə krät´iz əm] *Cardiology.* an older term for the condition of being without a pulse, or having a very weak pulse. Thus, **acrotic.**

Acrotretacea *Paleontology.* an extinct superfamily of small brachiopods in the order Acrotretida; characterized by cone-shaped pedicle valves and usually calcareo-corneous shells; lower Cambrian to Devonian.

Acrotretida *Invertebrate Zoology.* an order of branchiopods (small shrimp-like crustaceans); known from lower Cambrian fossils to the present living species.

Acrotretidina *Invertebrate Zoology.* a suborder of inarticulate branchiopods in the order Acrotretida, including only those species with shells composed of calcium phosphate.

Acrotrichaceae *Botany.* a monogeneric family of brown algae of the order Chordariales, characterized by trichothallic growth in which a single filament grows via an intercalary zone into a long, distal, colorless hair.

acrotropic [ak´rə trō´pik] *Biology.* of or relating to continued growth in the same direction as the original growth. Thus, **acrotropism.**

Acrotylaceae *Botany.* a family of red algae belonging to the order Gigartinales, having many-branched, flattened thalli; most species are restricted to the Southern Hemisphere.

acrozone see RANGE ZONE.

Acrux *Astronomy.* Alpha (α) Crucis, a first-magnitude double star in the southern constellation Crux, the Southern Cross.

acrylaldehyde see ACROLEIN.

acrylamide *Organic Chemistry.* CH_2=CHCONH$_2$, colorless, odorless crystals that melt at 84.5°C; soluble in water, alcohol, and acetone; stable at room temperature, but can polymerize readily while melting; used in dye synthesis, ore processing, sewage treatment, and in permanent press fabrics.

acrylamide copolymer *Organic Chemistry.* a combination of acrylamide and other resins, such as an acrylic resin, which forms a thermosetting resin.

acrylate *Organic Chemistry.* **1.** a salt or ester of an acrylic acid. **2.** see ACRYLATE RESIN.

acrylate resin *Organic Chemistry.* a polymer of acrylic acid or its esters with a $-CH_2-CH(COOR)-$ structure; used in surface coatings and finishes.

acrylic *Chemistry.* **1.** relating to or containing acrylic acid. **2.** describing a product derived from acrylic acid: *acrylic* sealants, *acrylic* adhesives. **3.** any of various synthetic products made with acrylic acid, such as an acrylic resin, fiber, plastic, or paint.

acrylic acid *Organic Chemistry.* CH_2=CHCOOH, a colorless, acrid liquid that is miscible with water, alcohol, and ether; melts at 13°C and boils at 141°C; it polymerizes readily and is used as a monomer for various acrylic polymers. Also, ACROLEIC ACID, 2-PROPENOIC ACID.

acrylic ester *Organic Chemistry.* an ester of acrylic acid.

acrylic fiber *Materials Science.* any of various quick-drying synthetic fibers produced by polymerizing acrylonitrile; specific types include Orlon, Lucite, Plexiglas, and Acrilan.

acrylic paint *Materials.* any of various quick-drying water-based paints made from acrylic resins, widely used both for industrial purposes and as an artistic medium.

acrylic plastic see ACRYLIC RESIN.

acrylic resin *Materials Science.* any of a large group of synthetic thermoplastic polymers created from various monomers (acrylic acid, methacrylic acid, esters of these acids, or acrylonitrile). These colorless liquids polymerize readily when exposed to light; they are widely used for fabrics, glass substitutes, paints, coatings, waxes, adhesives, etc.

acrylic rubber *Materials.* a synthetic rubber that contains acrylonitrile. Also, **acrylonitrile rubber.**

acrylonitrile *Organic Chemistry.* CH_2=CHCN, a colorless liquid that is soluble in all common organic organic solvents and partially miscible with water; freezes at –83°C and boils at 77.5–77.9°C, and is toxic, carcinogenic, and a dangerous fire risk; used as a monomer for acrylic fibers and in acrylic rubber, and for other industrial purposes.

acrylonitrile butadiene styrene resin *Materials Science.* a graft blend of acrylonitrile-styrene polymer and butadiene-acrylonitrile rubber. The resulting polymer offers both the hardness and strength of the vinyl resin, and the toughness and impact resistance of the nitrile rubber. Also, ABS RESIN.

acrylonitrile copolymer *Materials Science.* a polymer of acrylonitrile and compounds such as butadiene or acrylic acid; the resulting rubber is oil resistant.

acrylonitrile polymer *Materials Science.* a vinyl compound with a molecular structure oriented for strong fibers.

acrylophenone *Organic Chemistry.* CH_2=CHCOC$_6$H$_5$, a phenyl vinyl ketone and an isomer of cinnamaldehyde.

ACS or **A.C.S.** American Cancer Society; American Chemical Society; American College of Surgeons.

ACT American College Test.

Actaeonidae *Invertebrate Zoology.* the family of bubble shells, in the subclass Opisthobranchia, with the shell generally present but reduced or absent in some.

Actaletidae *Invertebrate Zoology.* a family of springtails, insects in the order Collembola, with simple tracheal systems.

act-frequency approach *Psychology.* a method of establishing the presence of a certain personality trait by the frequent occurrence of certain actions identified with this trait.

ACTH *Endocrinology.* adrenocorticotropic hormone, a peptide hormone secreted by the anterior pituitary gland; it stimulates the production of glucocorticoids by the adrenal cortex and is a trophic factor for the cortex as well. *Medicine.* a sterile preparation of this hormone, obtained from certain animals; used in treating arthritis, allergies, skin diseases, and other disorders. Also, ADRENOCORTICOTROPIN, CORTICOTROPIN.

actic *Ecology.* relating to or living in a rocky seashore environment.

Actidione *Microbiology.* the trade name for a preparation of the antibiotic cyclohexamide, produced by the bacterial species *Streptomyces griseus* and active against certain yeasts and fungi.

actin *Biochemistry.* a protein that plays an important role in the contraction and relaxation of striated muscle.

actinal *Invertebrate Zoology.* in radially symmetrical animals, the area from which tentacles or arms radiate and where the mouth is located; i.e., the oral side. *Zoology.* having rays or tentacles.

actin- or **actini-** a combining form meaning: **1.** ray or radial. **2.** of or relating to actinic radiation.

actin genes *Genetics.* a group of genes that code for various forms of the protein actin, some forms of which are responsible for muscle contraction.

acting-out *Psychology.* a term for actions or behavior patterns, especially negative or impulsive actions, that are a response to an earlier situation, but that occur in the context of a present, symbolically similar situation.

Actiniaria *Invertebrate Zoology.* the order of sea anemones, coelenterate subclass Zoantharia; solitary polyps with no skeleton.

actinic *Physics.* relating to the capability of electromagnetic radiation to initiate photochemical reactions, as in the fading of pigments.

actinic achromaticism *Optics.* a photographic lens system design that causes light sources of given wavelengths to meet at the same point and produce images of equal size. Also, FG ACHROMATISM.
actinic focus *Optics.* in an optical system, the point of convergence of the most chemically effective rays in the electromagnetic spectrum, usually involving rays that are in the ultraviolet range. Also, CHEMICAL FOCUS.
actinic glass *Optics.* glass that transmits most of the visible components of the light spectrum and absorbs most of the infrared and ultraviolet components.
actinic light *Photogrammetry.* the light or portion of the spectrum of short wavelengths that produces photochemical changes in light-sensitive emulsions.
actinide *Chemistry.* any element of the actinide series. (So called because all its members have properties similar to those of *actinium.*)
actinide series *Chemistry.* the series of heavy radioactive metallic elements extending from actinium (atomic number 89) or thorium (atomic number 90) through lawrencium (atomic number 103) on the periodic table. Also, **actinides.**
Actinidiaceae *Botany.* a family of woody, tanniferous trees, shrubs, and vines of the order Theales that are found in Asian tropics and mountains and characterized by raphide sacs in the parenchymatous tissues, scalariform vessels, and simple, alternate leaves.
Actiniscaceae *Botany.* a monogeneric family of nonphotosynthetic flagellates of the order Gymnodiniales, characterized by a lack of armor and by a flattened apical-antapical axis.
actinism *Chemistry.* the action or property of radiation that produces chemical effects, as in photography.
actinium *Chemistry.* a radioactive chemical element having the symbol Ac, the atomic number 89, an atomic weight (in its most stable isotope) of 227, a melting point of 1050°C and a boiling point of about 3200°C, and a half-life of 21.7 years; a rare silvery-white metal found in compound form in uranium ores or obtained from radium by neutron bombardment. (From a Greek term meaning "rays of light" or "radiance.")
actinium emanation see ACTINON.
actinium series *Nucleonics.* a series of 15 radioactive elements in the sixth period of the atomic system that form a sequence of parent-daughter relationships through α and β disintegrations, beginning with the longest-lived precursor, uranium-235, and ending with the stable isotope of lead-207. Also, **actinium decay series.**
actino- a combining form meaning: **1.** ray or radial. **2.** of or relating to actinic radiation.
actinobacillosis [ak tin´ō bas i lō´ sis] *Veterinary Medicine.* a bacterial disease of domestic animals, especially cattle or sheep, caused by *Actinobacillus lignieresii* and manifested by inflammation or lesions involving the skin, soft tissues of the head and neck, and internal organs, and by lymph node infection.
Actinobacillus *Bacteriology.* a genus of Gram-negative, mostly rod-shaped bacteria of the family Pasteurellaceae that are found both as pathogens and as commensal organisms in humans and domestic animals; several pathogenic species cause internal lesions.
Actinobifida *Microbiology.* a former genus of actinomycete bacteria containing mycelium-forming soil microbes, all species of which have now been reclassified into other genera.
actinobiology *Biochemistry.* the study of the effect of various forms of radiation on living organisms.
Actinobolina *Invertebrate Zoology.* a genus of freshwater ciliate protozoans, in the subclass Holotrichia, that are ovate or spherical with tentacles among the cilia.
Actinoceratoidea *Paleontology.* an extinct subclass of generally nektic Paleozoic cephalopods; Ordovician to Carboniferous; some species attained lengths of up to 10 meters.
actinochemistry *Chemistry.* a branch of chemistry that deals with the chemical effects of light and other forms of radiation.
actinochitin *Biochemistry.* a chitin (*N*-acetyl-D-glucosamine polymer) that occurs in mites as exoskeleton.
Actinochitinosi *Invertebrate Zoology.* a group name applied to two closely related suborders of mites, the Trombidiformes and the Sarcoptiformes.
actinodermatitis *Medicine.* a skin disease or inflammation resulting from excessive exposure to sunlight or other forms of radiation.
actinodielectric *Electricity.* exhibiting a temporary rise in electrical conductivity during electromagnetic radiation exposure.
actinodrome *Botany.* of a leaf, having veins extending out from a common center. Also, **actinodromous.**
actinoelectricity *Electricity.* the electromotive force produced in a substance by the action of radiant energy upon crystals during electromagnetic radiation exposure.
actinogestin see CONGESTIN.
actinogram *Engineering.* the recorded results of measurements made by an actinometer, which measures radiant energy.
actinograph *Engineering.* an actinometer that is equipped to record its measurements.
actinoid *Biology.* having a radiating or starlike shape.
actinoid elements see ACTINIDE SERIES.
actinolite *Mineralogy.* $Ca_2(Mg,Fe)_5Si_8O_{22}(OH)_2$, a bright to grayish-green, ferruginous monoclinic amphibole that forms a series with tremolite and ferro-actinolite, having a specific gravity of 3 to 3.3 and a hardness of 5.6 on the Mohs scale; found in crystalline, fibrous, and columnar varieties.
actinology *Physics.* a branch of physics dealing with the study of photochemical reactions.
Actinomadura *Bacteriology.* a genus of soil bacteria of the order Actinomycetales that form branching, substrate mycelia; some species are pathogenic in humans.
actinomere *Invertebrate Zoology.* one of the radial segments of a radially symmetrical animal.
actinometer *Engineering.* an instrument used to measure the intensity of radiation, especially that coming from a source causing photochemical reactions, such as the sun.
actinometry *Engineering.* the measurement of the intensity of radiation from a body, especially an astronomical body.
actinomorphic *Biology.* being radially symmetrical and capable of division into two identical halves by a longitudinal plane. *Botany.* describing flowers with their parts arranged in a radially symmetrical pattern. Also, **actinomorphous.**
Actinomyces *Bacteriology.* **1.** a genus of Gram-positive, anaerobic bacteria of the family Actinomycetaceae that are nonmotile and faculatively anaerobic, occurring as normal flora of the mouth and throat or as pathogens in humans and cattle. **2. actinomyces** *plural,* **actinomycetes** an organism of this genus.
Actinomycetaceae *Bacteriology.* a family of Gram-positive bacteria of the order Actinomycetales, characterized by nonmotility, non-acid-fastness, and a filamentous growth habit.
Actinomycetales *Bacteriology.* an order of Gram-positive, typically aerobic mycelial bacteria, in which the families are distinguished by the chemical nature of the cell wall and the lipid content of the cells.
actinomycete *plural,* **actinomycetes** *Bacteriology.* a member of the bacterial order Actinomycetales, characteristically Gram-positive, mycelium-forming species; some are found in soil or water, while others are plant or animal pathogens.
actinomycin *Pharmacology.* any of various antibiotics produced by *Streptomyces* species that are active against many Gram-positive and Gram-negative bacteria, inhibiting DNA-directed RNA synthesis by binding specifically to DNA.
actinomycin D *Molecular Biology.* an antibiotic that inhibits the synthesis of RNA by intercalating between bases on the DNA template, especially in regions rich in cytosine and guanine.
actinomycosis *Medicine.* a disease, common to cattle and swine but less common in humans, that affects the lymph nodes in the jaw in most cases but that can also affect the brain, lung, and gastrointestinal tract; caused by *Actinomyces israelii* in humans and *A. bovis* in animals.
actinomyosin *Biochemistry.* a term for a variety of proteins, such as hemoglobin, that have been elaborated with different strains of the bacteria *Streptomyces.*
Actinomyxida *Invertebrate Zoology.* an order of protozoan parasites, in the class Myxosporidea, with spores developed from several nuclei and enclosed in two or three valves. Also, **Actinomyxidia.**
actinon *Nuclear Physics.* the former name of a gaseous radioactive isotope of radon, now called radon-219, a member of the actinium series.
actinophage *Virology.* any virus that causes lysis of actinomycetes; a bacteriophage whose host is an actinomycete.
Actinophryida *Invertebrate Zoology.* an order of primarily freshwater protozoans in the class Actinopodea, widely distributed in stagnant water; they lack an organized test, a centroplast, and a capsule.
Actinophrys *Invertebrate Zoology.* a genus of sun animalcules, spherical freshwater protozoans in the order Heliozoa.
Actinoplanaceae *Bacteriology.* a family of bacteria of the order Actinomycetales, occurring as Gram-positive, spore-forming organisms in aquatic environments.

Actinoplanes *Bacteriology.* a genus of aerobic, sporogenous, Gram-positive bacteria of the family Actinoplanaceae, commonly found as branching mycelia in soil and aquatic habitats.

Actinopoda *Invertebrate Zoology.* a subclass of ameboid protozoans in the superclass Sarcodina, with axopodia.

Actinopolyspora *Bacteriology.* a genus of halophilic, mycelial bacteria of the order Actinomycetales, the sole species of which has been isolated from salt-rich bacteriological nutrient media.

Actinopteri *Vertebrate Zoology.* see ACTINOPTERYGII.

Actinopterygii *Vertebrate Zoology.* a large subclass of ray-finned fishes of the class Osteichthyes, distinguished by paired spiny fins supported by dermal rays; it includes most of the living bony fish. *Paleontology.* a fossil specimen of this type, first appearing during the Devonian.

actinorrhiza *Microbiology.* a symbiotic association between certain plant roots and certain bacteria, such as *Frankia* strains, in which nitrogen fixation occurs within specialized root nodules.

Actinosphaerium *Invertebrate Zoology.* a genus of sun animalcules, large freshwater protozoans in the order Heliozoa.

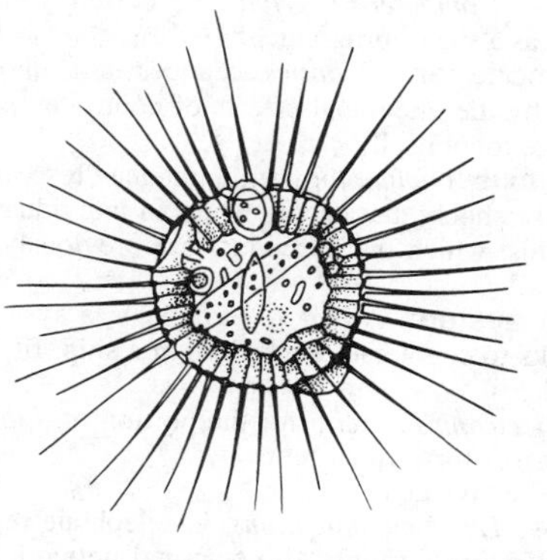

Actinosphaerium

actinostele *Botany.* a protostele that has vascular tissue arranged in radiating star-shaped arms, which in transection are interspersed with parenchyma.

actinostome *Biology.* **1.** the mouth of a radially symmetrical animal, such as the starfish. **2.** the area surrounding the mouth of an echinoderm.

Actinostroma *Paleontology.* a family of Paleozoic reef-building stromatoporoids similar to the Actinostromariidae.

Actinostromariidae *Paleontology.* a family of Mesozoic reef-building stromatoporoids in the order Actinostromatida; its laminar network is similar to Actinostroma, but the horizontal rods are heavier than the vertical pillars they radiate from and the structure is generally less regular; Jurassic to Cretaceous.

Actinosynnema *Bacteriology.* a genus of mycelial bacteria of the order Actinomycetales that occurs on vegetative matter in aquatic environments.

actinotherapy *Medicine.* the treatment of a disease with rays, light, photochemically actived compounds, X-rays, radium, and other radiochemicals.

actinotrocha *Invertebrate Zoology.* the free-swimming pelagic larvae of Phoronis, small marine tubicolous worms with characteristic ciliated lobes.

actinouranium *Nuclear Physics.* the former name of the radioactive uranium isotope of mass 235 that is now called uranium-235.

actinozoan see ANTHOZOAN.

actinula *Invertebrate Zoology.* an advanced larva in the life cycle of certain hydrozoans that possesses a mouth and tentacles, and may attach and develop into a new hydroid colony (as in Tubularia) or metamorphose into a medusa (as in Aglaura).

action *Military Science.* a specific encounter with the enemy, or combat in general. *Ordnance.* **1.** the operating mechanism in a gun, by which it is fired, loaded, and unloaded. **2.** a command ordering a unit or weapons crew to prepare to fire. *Artificial Intelligence.* a term referring to the clause in a production rule that indicates one or more conclusions that can be drawn if the premises are satisfied. *Mechanics.* **1.** according to Newton's third law of motion, an external force (the action) that is applied to a body and that is counteracted by an equal force in the opposite direction (the reaction). **2.** see ACTION VARIABLE.

actionable fire *Forestry.* any forest fire that requires suppression, particularly a fire started or allowed to spread in violation of the law.

action anthropology *Anthropology.* **1.** the application of strategies based on cultural information by an anthropologist acting as technical adviser and facilitator between two cultural groups. **2.** specifically, such techniques used to provide a minority culture with the opportunity to operate within a dominant culture while still preserving its own cultural integrity.

action at a distance theory *Physics.* the theory that two bodies separated in space will interact in some way, through gravitational forces, electromagnetic forces, and so on.

action chain *Behavior.* a set sequence of behaviors in which the completion of one response is the cue for the next response, as in certain courtship behaviors or in ritualized fighting. Also, CHAIN REFLEX, REACTION CHAIN.

action clause *Artificial Intelligence.* the part of a production rule that indicates one or more conclusions that can be inferred if the stated premises are satisfied.

action current *Physiology.* the electric current that flows along the axon of a nerve during the course of a nerve impulse or action potential.

action diagram *Computer Programming.* a software development tool that graphically illustrates the hierarchical structure and process flow of the program being developed.

action entry *Computer Programming.* the lower right-hand portion of a decision table that indicates which of the possible actions are associated with each set of conditions. *Computer Technology.* a list of all the types of actions that can be initiated by a microcomputer to operate an electromechanical machine; for example, motor speed and direction and solenoid actuation.

action integral see ACTION VARIABLE.

action portion *Computer Programming.* the lower portion of a decision table containing two sections, one that lists transformations to be done to data or materials (the action stub) and one that provides entries associating these actions with combinations of conditions (the action entry).

action potential *Physiology.* a rapid change in the polarity of the membrane of a neuron, gland, or muscle fiber that facilitates the interaction and transmission of impulses. Also, SPIKE POTENTIAL. *Behavior.* see TENDENCY.

action-reaction law *Physics.* Newton's third law of mechanics, stating that for every force imposed on a body, there is an equal and oppositely directed reaction.

action spectrum *Physiology.* a graphic plot of the correlation of an effect of electromagnetic radiation against the wavelength of the radiation.

action stub *Computer Programming.* the lower left-hand portion of a decision table consisting of a list of all possible actions to be taken or transformations to be made on data or materials.

action variable *Mechanics.* the time integral associated with the evolution of a physical system in the phase space; for a system of particles it equals the sum of the integrals of the system's generalized momenta over their canonically conjugate generalized coordinates.

actium *Ecology.* a rocky seashore environment or community.

activate to make active; start some process or motion; specific uses include: *Chemistry.* to cause or accelerate a reaction. *Materials Science.* to treat charcoal, carbon, or the like to improve their capacity for adsorbing impurities. *Physics.* to induce activity in a system that is static, as in neutron activation of radioactivity. *Biotechnology.* to purify sewage by treating it with air and bacteria. *Electricity.* **1.** to start the operation of an electrical device, usually by applying an enable signal or power to it. **2.** to use liquid as an additive in order to make a cell or battery operational. **3.** to apply material to the surface of a cathode to create or increase cathode emissivity. *Ordnance.* to cause a missile or explosive to be in an active state, ready for firing or explosion. *Military Science.* **1.** to assign a ship, troop unit, or the like to active service after it has previously been in inactive or reserve status. **2.** to bring to active status something that had not previously been in operation.

activated alumina *Inorganic Chemistry.* Al_2O_3, a highly porous, granular form of aluminum oxide with an absorptive capacity for moisture and odors found in some gases and liquids; used in the petroleum industry and in water purification.

activated carbon *Materials.* a highly porous form of charcoal treated so that it can readily adsorb large quantities of gases, vapors or undesirable impurities; widely used in filtering systems for absorbing gases and solids. Also, ACTIVATED CHARCOAL.

activated cathode *Electronics.* in an electron tube, a cathode that has been treated with a material such as thorium oxide, ThO_2. This treatment increases the emissivity of the cathode and thus the efficiency of the tube.

activated charcoal see ACTIVATED CARBON.

activated complex *Physical Chemistry.* a term for molecules in an unstable state intermediate to that of the reactants and products in a chemical reaction.

activated cross-slip *Materials Science.* the thermally sensitive process of a screw dislocation, transferring the direction of motion from one slip plane to another.

activated diffusion bonding *Materials Science.* a process for improving the strengths of joints in which a vacuum furnace braze is combined with subsequent annealing activated diffusion; used mostly for nickel-base superalloys.

activated graft *Surgery.* a graft characterized by the growth of nerves and blood supply for nourishment, following a period of denervation and tenuous vascularity.

activated mine *Ordnance.* a mine having a secondary fuse that will cause detonation of the mine if it is moved or disturbed.

activated plow *Mining Engineering.* any power-operated cutting blade used to shear coal seams that are unusually hard.

activated sintering *Metallurgy.* the sintering of a metal or alloy powder that has been subjected to a surface modification treatment in order to enhance its reactivity, thereby accelerating the agglomeration process.

activated sludge *Biotechnology.* the semiliquid, microbe-rich sediment that is added to secondary-stage sewage material in the activated-sludge process.

activated-sludge effluent *Biotechnology.* in the activated-sludge process, the liquid material that is transported to an aeration tank for secondary treatment.

activated-sludge process *Biotechnology.* a process used in sewage treatment that increases the level of biological activity by increasing the contact between the waste water and the actively growing microorganisms.

activated state *Physical Chemistry.* the higher energy state through which an atom passes in transforming from a metastable to a more stable state.

activated support *Biotechnology.* a chemically treated matrix material whose surface contains a large proportion of reactive groups.

activation the process of inducing or creating a state of activity; specific uses include: *Chemistry.* the process of treating a substance with heat or another form of energy in order to increase the internal energy of the original substance; this speeds any subsequent chemical or physical reaction. *Nucleonics.* the formation of radioisotopes by neutron irradiation, usually by thermal neutrons. *Electricity.* the process of using liquid as an additive in order to make a cell or battery operational. *Electronics.* in an electron tube, the process of applying a material to the surface of the cathode to create or increase the cathode emissivity. *Molecular Biology.* a process that starts with blast transformation and continues into cell division and differentiation. *Genetics.* the induction or stimulation of the transcription of a particular gene. *Physiology.* the process by which the central nervous system is stimulated to activity through regulation by the reticular activating system. *Neurology.* in electroencephalography, a deliberate electrical stimulation of the brain to produce activity. *Toxicology.* any metabolic reaction that alters the toxicity of a substance. *Metallurgy.* **1.** the change of a metal or alloy from a chemically or physically passive state to an active state. **2.** the increase in surface free energy or lattice binding energy of a solid.

activation analysis *Nucleonics.* the qualitative or quantitative determination of elements or specific isotopes in an unknown sample by comparing the characteristic energy spectra and half-lives of radioactive isotopes produced by neutron irradiation; useful in identifying trace amounts or impurities as small as several parts per billion. Also, NEUTRON ACTIVATION ANALYSIS.

activation energy *Physical Chemistry.* the least amount of energy that must be supplied to a system, over that of the ground state, in order to initiate a particular chemical reaction.

activation enthalpy *Atomic Physics.* the internal energy contribution to the free energy barrier that an atom in a metastable position must surmount in order to participate in a thermally activated process.

activation entropy *Atomic Physics.* the entropic contribution to the free energy barrier that an atom in a metastable position must surmount in order to participate in a thermally activated process.

activation record *Computer Programming.* the fresh storage that is allocated when a subroutine is called and that contains all the local storage for that procedure, including program block control information and variables. Also, STACK FRAME.

activation volume *Materials Science.* when referring to creep, a mechanism constant that relates changes in creep activation energies to the value of an applied stress.

activator *Chemistry.* a substance that causes another substance to become reactive, such as a catalyst. *Medicine.* any of various devices used in orthodontics to activate the force of the facial muscles in order to achieve a desired orthopedic correction, such as an enlargement of the upper jaw. *Genetics.* a molecule that activates or induces gene transcription. *Molecular Biology.* any molecule that activates or induces a biochemical reaction.

activator-dissociation system *Genetics.* a system of movable genetic elements in corn **(activator elements)** that cause chromosomal insertions and breaks.

activator RNA *Genetics.* specific ribonucleic acid molecules used to form a complex that links a receptor gene to a producer gene.

active *Electronics.* being a source of electrical energy. *Space Technology.* capable of transmitting a signal. *Volcanology.* relating to or being an active volcano. *Computer Programming.* currently in use or available for current use, as a file or program.

active accommodation *Robotics.* changes in a robot's programmed motion caused by the integrated effects of controllers, sensors, and the movement of the robot itself.

active anaphylaxis *Immunology.* an immediate response, including allergic reaction or shock, that occurs when an individual is reexposed to an antigen against which the individual has previously developed antibodies.

active antiroll system *Naval Architecture.* a system using fins or pumps and tanks to counteract the rolling of a ship; fitted on passenger-carrying vessels.

active area *Electronics.* a term for the portion of a metallic rectifying junction that passes forward current.

active arm see ACTIVE LEG.

active balance *Telecommunications.* in telephone repeater operation, the total of all return currents at a terminal network that is balanced against local impedance.

active biomass *Biotechnology.* the total weight or mass of living matter in a population of cells.

active center *Astronomy.* a nucleus or core of a galaxy that emits substantial amounts of energy.

active chaff *Ordnance.* see CHAFF.

active charcoal see ACTIVATED CHARCOAL.

active communications satellite see ACTIVE SATELLITE.

active component *Electricity.* an in-phase quantity (without reactance) in an AC circuit. *Electronics.* a component that can produce more power in the output signal than is present in the input signal. Also, ACTIVE DEVICE, ACTIVE ELEMENT.

active control(s) technology *Aviation.* the technology related to the automatic movement of the control surfaces of an aircraft in response to sensors.

active-cord mechanism *Robotics.* a manipulator that uses actuators along its body to move and make winding motions.

active countermeasures see ACTIVE ELECTRONIC COUNTERMEASURES.

active current *Electricity.* the current component that is in phase with the voltage in an alternating current; it equals the average power divided by the effective voltage.

active detection system *Engineering.* a system that emits energy, such as sound waves or radio waves, in order to detect or track targets or to profile objects or landscapes; sonar and radar are examples.

active device see ACTIVE COMPONENT.

active duty *Military Science.* the status of full-time service, especially in a combat environment. Also, **active service.**

active earth pressure *Civil Engineering.* the lateral pressure on a retaining structure from the soil when the structure yields and the soil is allowed to expand horizontally.

active ECM see ACTIVE ELECTRONIC COUNTERMEASURES.

active electric network *Electricity.* a network with at least one active component.

active electronic countermeasures *Electronics.* various actions, such as jamming, that are employed in disrupting or disabling electronic communications or surveillance by an enemy source; active countermeasures radiate energy and are thus detectable by the enemy.

active element *Nuclear Physics.* a substance that has one or more radioactive isotopes. *Electronics.* see ACTIVE COMPONENT.

active entry *Mining Engineering.* an entry in which coal is being mined from a point or connected section.

active file *Computer Programming.* a computer file that is in current use or to which entries or references are regularly made.

active filter *Electronics.* a filter with active components, such as amplifiers, that aid in signal selection and rejection. This filter usually employs negative feedback and tuning elements in the feedback path.

active fleet *Transportation Engineering.* in a transit system, the number of vehicles or cars that are in use or available for use on a regular basis.

active front *Meteorology.* a front or portion of a front that produces appreciable cloudiness and precipitation.

active galaxy *Astronomy.* a galaxy radiating unusually large amounts of energy from its nucleus.

active glacier *Hydrology.* **1.** a glacier having an accumulation area and exhibiting some degree of ice movement. **2.** a relatively rapid-moving glacier in a maritime environment at a low latitude where accumulation and ablation are both great.

active homing *Navigation.* a system in which a missile can lock in on a target, using an active detection system (such as radar), and steer directly toward the target.

active illumination *Engineering.* a lighting system whose intensity, orientation, or pattern is continuously controlled and altered by signals.

active image *Artificial Intelligence.* in a frame system or object-oriented system, a connection between a data value and a displayed image, such that the image reflects the value of the data as the program runs, and in some cases such that changes made interactively to the displayed image cause corresponding changes to the data.

active immunity *Immunology.* a resistance to an infection or disease of an organism, due to the organism's production of antibodies or a cellular immune attack, following an infection or inoculation.

active infrared detection *Engineering.* an active detection system that sends out probing infrared beams and receives reflected rays.

active jamming see JAMMING.

active layer *Geology.* the surface layer of soil above the permafrost that is subject to seasonal freezing and thawing. Also, FROST ZONE.

active leaf *Building Engineering.* in a two-leaf door, the leaf that carries the locking mechanism and that can be opened without opening the other leaf.

active leg *Electronics.* an electrical component of a transducer that varies its electrical properties in reaction to the transducer's input. Also, ACTIVE ARM.

active location system *Navigation.* a system that uses an active detection system to establish its own position; e.g., a terrain-matching radar system in a cruise missile.

active logic *Electronics.* a logic system that employs active components for the purpose of pulse width modulation, signal amplification, and signal phase shift.

active master file *Computer Programming.* a master file that contains elements that are accessed or updated frequently.

active master items *Computer Programming.* the most active elements in a master file based on amount of data usage.

active material *Electronics.* **1.** the material within a storage battery that is capable of reacting chemically to produce a flow of electric current. **2.** the fluorescent coating, such as calcium tungstate, on the screen of a cathode-ray tube. *Nucleonics.* the fissile and fissionable components required for a chain reaction in a nuclear reactor or a nuclear weapon.

active memory see ASSOCIATIVE MEMORY.

active metal brazing *Materials Science.* a metal-ceramic joining process in which surface wetting alloying elements in the brazing material promote a good bond between the braze and ceramic.

active mirror *Optics.* a mirror whose position, surface shape, and thus reflective qualities can be continually modified for maximum effectiveness as conditions in the surroundings change.

active oxidation method *Food Technology.* a measure of an oil's resistance to oxidation, defined as the number of hours needed for the oil to give a perioxide value of 100 under standard conditions.

active-passive transition *Materials Science.* the transition at some potential, at which the current density associated with anodic stainless steel or other passivate metal electrodes drops sharply and corrosion is minimized.

active permafrost *Geology.* permafrost that, once thawed by any means, reverts to its permanently frozen state under present climatic conditions.

active power *Electricity.* power, measured in watts, that can do work, including generating heat. For sinusoidal voltage and current, active power is $IV \cos \theta$, where θ is the phase difference between the current and voltage. In general, active power is $\int_0^t V(t)i(t)\,dt/t$. Also, REAL POWER.

active program *Computer Programming.* a program that is in current use or that is regularly used.

active prominence *Astronomy.* a rapidly changing arch of glowing gas erupting from the sun's photosphere.

active prominence region *Astronomy.* any region of the sun showing particularly spectacular prominences in the form of long flamelike filaments of glowing gas, sometimes extending beyond the sun's limb.

active region *Electronics.* in semiconductor material, the area that amplifies, rectifies, or in some other way dynamically changes a voltage or current. *Astronomy.* any region of the sun containing sunspots or other phenomena, such as prominences or flares, that are characteristic of intense localized magnetic fields.

active region filament *Astronomy.* a prominence from an active solar region that is seen in projection on the sun's disk.

active satellite *Space Technology.* a relay satellite that receives signals and then regenerates them before retransmission.

active site *Enzymology.* the specific portion of an enzyme molecule that binds with or interacts with a substrate, forming an enzyme-substrate complex. Also, BINDING SITE, CATALYTIC SITE.

active sleep see REM SLEEP.

active sludge *Biotechnology.* a sludge with abundant biodegrading bacteria, used to speed up the breakdown of raw sewage.

active sonar *Acoustics.* a sound navigation and ranging system that uses the time difference between a transmitted sound and the reception of its return echoes to determine ranges and bearings.

active substrate *Solid-State Physics.* a semiconductor or ferromagnetic substance in which radioactive elements are formed.

active sun *Astronomy.* a term describing the sun at those times when sunspots, flares, prominences, and similar phenomena are in abundance.

active system *Engineering.* any radar or radio system equipped to transmit as well as receive signals.

active technique *Psychology.* in psychotherapy, the practice by the therapist of moving away from an attitude of neutrality during the analysis to offer specific prohibitions or restrictions.

active tracking system *Navigation.* a system that uses an active detection system (e.g., radar or sonar) to track targets.

active transducer *Electronics.* a transducer that requires power from a source other than the input signal in order to operate; such transducers often contain transistors to supply the necessary power.

active transport *Physiology.* the direct participation of a cell to provide energy and carrier molecules for movement of specific chemical substances against a concentration gradient and usually across the cell membrane.

active value *Artificial Intelligence.* in a frame system or object-oriented system, a means of associating procedures with a data value so that the procedures will be called when the data is accessed or written.

active volcano *Volcanology.* a volcano that is either erupting or is capable of erupting and expected to do so.

active voltage *Electricity.* the component of voltage that is in phase with the current in an alternating-current circuit.

active window *Computer Technology.* a portion of the screen currently in use, visible to the user, and available for data entry or file editing.

active workings *Mining Engineering.* any part of a mine that is ventilated and inspected routinely.

activities of daily living *Medicine.* see ADL.

activity the process of being or becoming active; specific uses include: *Archaeology.* the customary use of a given artifact in a certain way, such as to prepare food. *Nuclear Physics.* the rate at which atoms disintegrate in a radioactive material per unit time; the common unit for measuring activity is the curie, defined to be 3.70×10^{10} disintegrations per second. *Physical Chemistry.* a thermodynamic quantity that represents the effective concentration of a solute in an ideal solution; if actual molar concentrations are replaced by activities, various exact equations can then be applied. *Computer Programming.* **1.** the act of using, altering, or referencing a given record. **2.** a measure of the extent to which such processing takes place for a given record. **3.** the lowest level function on a function chart. *Industrial Engineering.* in a PERT network, a single task that consumes time and must be completed in order to finish the project.

activity area *Archaeology.* a place where a certain archaeological activity is regularly carried out, such as food preparation.

activity chart *Industrial Engineering.* a chart used to measure progress in an industrial project by measuring each required operation against a time scale.
activity coefficient *Physical Chemistry.* a fractional number that when multiplied by the actual concentration of a substance in solution yields the chemical activity; it is a measure of the deviation from the the ideal state.
activity level *Computer Programming.* see ACTIVITY RATIO.
activity loading *Computer Programming.* the storing of records in a file so that the most frequently processed records can be most readily accessed. Also, ACTIVITY SEQUENCE METHOD.
activity ratio *Geology.* for a given sediment, the ratio of the plasticity index to the percentage of clay-sized minerals. *Computer Programming.* a measure of the degree of use of a file, in terms of the percentage of the total records in a file that are updated or inspected in a given period of time or during a given run. Also, ACTIVITY LEVEL.
activity sampling see WORK SAMPLING.
activity sequence method see ACTIVITY LOADING.
activity series *Chemistry.* an arrangement of the metals of the periodic table in the specific order of their tendency to react with water and acids. They are ordered so that each metal displaces from solution those below it in the series, and is displaced by those above it. Thus, K Na Mg Al Zn Fe Sn Pb H Cu Hg Ag Pt Au. Also, DISPLACEMENT SERIES, ELECTROMOTIVE SERIES.
actol see SILVER LACTATE.
actomyosin *Biochemistry.* a complex of the proteins actin and myosin that is found in muscle cells (myofibrils) and is responsible for muscle contraction.
actophilous *Ecology.* living or thriving in a rocky seashore environment. Thus, **actophile, actophily.**
actor-observer bias *Psychology.* the tendency of an individual to regard situations in which he or she is involved as caused by external factors, and to regard situations he or she observes as caused by the actions of those involved. Also, **actor-observer viewpoint, actor-observer indifference**.
actor-vs-observer (bias) see ACTOR-OBSERVER BIAS.
actual address *Computer Programming.* an absolute address that may be the result of an address computation.
actual age see ABSOLUTE AGE.
actual argument see ACTUAL PARAMETER.
actual code see ABSOLUTE CODE.
actual cost *Industrial Engineering.* the true cost of producing an item, including not only the manufacturing cost but also the allocated charges for overhead and other operating expenses.
actual cubic feet per minute *Chemical Engineering.* a measure of the volume of gas flow at operating pressure and temperature, as opposed to the volume of gas flow under standard conditions of pressure and temperature.
actual decimal point *Computer Programming.* a physical decimal point represented by a coded character that requires a storage location in a computer, as opposed to an implied decimal point at a fixed bit position in a number representation.
actual elevation *Meteorology.* the vertical distance of the ground at a meteorological station above mean sea level.
actual exhaust velocity *Aviation.* the velocity of an exhaust stream (including gases and other particles) as it moves past a precise point.
actual height *Electromagnetism.* the maximum altitude at which radio waves of a specified frequency are refracted.
actual hours see ACTUAL TIME.
actual instruction see EFFECTIVE INSTRUCTION.
actualization see SELF-ACTUALIZATION.
actual key *Computer Programming.* a specification in COBOL used to control the position of logical records in a file by identifying the actual or relative track and the specific record on the track.
actual mechanical advantage *Mechanical Engineering.* the ratio of output force to input force; i.e., the amount of force that must be supplied to a machine in order for it to do a given amount of work. For example, if a lever requires 10 pounds of force to raise a 40-pound load, it has an actual mechanical advantage of 4.
actual motion *Navigation.* the motion of a craft relative to the surface of the earth.
actual parameter *Computer Programming.* a value or address that is passed to a function, procedure, macroinstruction, or subroutine by a calling program at the time of call; it supplies the value for the formal parameter. Also, ACTUAL ARGUMENT.
actual pressure *Meteorology.* the atmospheric pressure reading obtained from a barometer, after correcting for current temperature, gravity, and instrumental errors.
actual relative movement *Geology.* see SLIP.
actual time *Industrial Engineering.* **1.** the time actually required by a worker to perform a given task or task element. **2.** the total hours actually worked by a worker in a particular time period. Also, OBSERVED TIME.
actual time of arrival *Aviation.* the precise time an aircraft touches down on the runway in landing. Similarly, **actual time of departure.**
actual time of interception *Navigation.* the precise time that a missile or interceptor impacts or reaches the immediate vicinity of a target craft.
actuate *Mechanical Engineering.* to put a device or mechanism into motion.
actuating system *Control Systems.* any system that supplies and transmits the electric, hydraulic, pneumatic, or other energy to drive a mechanism or system.
actuation *Robotics.* specific motions made by a robot using actuators.
actuator *Mechanical Engineering.* any device that is moved a predetermined distance to operate or control another mechanical device. *Acoustical Engineering.* a device that provides an external source of energy for an acoustic transducer, such as the voice for a sound-powered phone system. *Robotics.* a device that controls a robot's movement or articulation. *Computer Science.* see ACCESS ARM.
AcU actinouranium (uranium-235).
acu- a combining form meaning "with a needle" or "like a needle."
acuate *Biology.* having sharp points; needle-shaped.
acuity *Biology.* the keenness of a sense perception, as sight or smell.
Aculeata *Invertebrate Zoology.* the division of stinging bees, ants, and true wasps in the hymenopteran suborder Apocrita; the ovipositor is modified into a stinger.
aculeate *Biology.* **1.** bearing sharp points or spines. **2.** thin and sharp-pointed. Also, **aculeated.**
aculeus *Botany.* a prickle or thorn growing on the bark, as on the rose or raspberry. *Invertebrate Zoology.* **1.** an insect's ovipositor when modified into a sting. **2.** a spine on the wings of certain butterflies and moths. *Zoology.* any sharp, pointed process.
Aculognathidae *Invertebrate Zoology.* the family of ant-sucking beetles in the coleopteran superfamily Cucujoidea.
acuminate *Botany.* describing a leaf that has a long, projecting, sharp point.

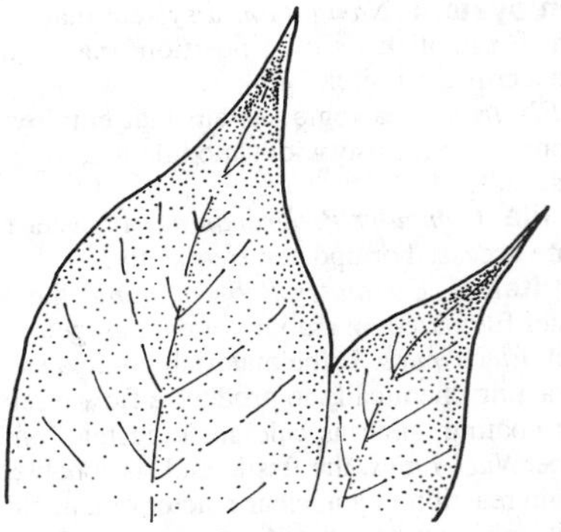

acuminate leaf

acupoint *Surgery.* a specific site along a body meridian where a needle is inserted in acupuncture.
acupressure *Medicine.* a therapeutic technique for relieving pain or regulating a body function by applying digital pressure to designated points on the body. *Surgery.* the compression of a bleeding vessel by the insertion of needles into adjacent tissue.
acupunctural *Medicine.* relating to or involving acupuncture.
acupuncture *Medicine.* a method of treating a disease or disorder with the use of long fine needles as a diagnostic and therapeutic tool, originating in China in ancient times and now also practiced in the West. It is accomplished by passing the needles though the skin at specific points **(acupuncture points)** along a series of lines called meridians. The needles are twirled, heated, or electrically energized, producing analgesia or altering sensory functions.

acupuncturist *Medicine.* a specialist who practices the procedure of acupuncture.

acus *Medicine.* a needle or needlelike part.

acusection *Surgery.* the process of cutting by means of an electrosurgical needle. Thus, **acusector.**

acutance *Optics.* the objective measure of the sharpness of a photographic image.

acute *Medicine.* **1.** referring to that which is sharp or severe. **2.** of a disease, having a relatively rapid onset with marked intensity and a short course. *Mathematics.* **1.** of an angle, less than 90°. **2.** of a triangle, containing only angles of less than 90°.

acute abdomen *Medicine.* a term for any abdominal condition that involves the sudden onset of severe pain in the abdominal cavity; requires immediate evaluation and may indicate the need for surgery.

acute alcoholism *Medicine.* the short-term intoxication caused by alcohol; drunkenness. The syndrome is characterized by muscle incoordination (including speech muscles) and impaired mental functions, and is often marked by such symptoms as blurred vision, memory loss, reduced motor control, decreased blood pressure, dehydration, nausea, headache, and stupor.

acute angle *Mathematics.* an angle smaller than a right angle; an angle of less than 90°.

acute angle block *Geology.* a fault block in which the strike on the down-dip side makes an acute angle with a diagonal fault.

acute anterior poliomyelitis see POLIOMYELITIS.

acute appendicitis *Medicine.* a sudden severe inflammation of the appendix, often requiring immediate surgical intervention, and characterized by severe abdominal pain, an elevated white blood cell count, fever, and nausea.

acute arch *Architecture.* a narrow arch with a sharply pointed apex. Also, LANCET ARCH.

acute arthritis *Medicine.* the sudden onset of inflammation of the joint, causing severe pain, swelling, and redness. Structural changes in the joint itself may result from persistence of this condition.

acute ascending myelitis *Medicine.* the sudden onset of inflammation of the spinal cord that may follow injury and that moves progressively up the spinal cord as the disease progresses.

acute bacterial endocarditis *Medicine.* the rapid and abrupt inflammation of the lining of the cardiac muscle, usually caused by pneumococci or staphylococci, that is generally limited to the external lining of a valve and sometimes of the chambers themselves.

acute benign lymphoblastosis see INFECTIOUS MONONUCLEOSIS.

acute berylliosis *Medicine.* the rapid onset of symptoms of beryllium poisoning due to the inhalation of this alkaline metal; the poisoning is characterized by acute pneumonitis, and, by chronic growths in the lungs, and also in the skin and other tissues.

acute bisectrix *Mineralogy.* in biaxial minerals, a line bisecting the acute angle of the optic axes.

acute bovine pulmonary emphysema and edema *Veterinary Medicine.* see FOG FEVER.

acute dermatitis *Medicine.* the rapid and severe onset of an inflammation of the skin producing redness, itching, and various skin lesions.

acute exposure *Radiology.* a short interval of usually heavy exposure to radiation.

acute glomerulonephritis *Medicine.* a rapid and severe onset of the inflammation of the glomeruli of the kidneys that frequently follows either infection or autoimmune disorders of an inflammatory nature, including systemic lupus erythematosus. These may be associated with inflammation of the respiratory tract. The inflammation is marked by edema, hematuria, hypertension, and in the more severe cases with dyspnea, delirium, convulsions, and coma.

acute granulocytic leukemia *Medicine.* a disease classified by the particular type of myelocyte involved and characterized by such symptoms as anemia, infection, bleeding, and fatigue; if untreated there is rapid progression of the disease, generally resulting in death within four to six months.

acute hemorrhagic conjunctivitis *Medicine.* specific acute endemic conjunctivitis that causes swelling of the eyelids, tearing, and conjunctival hemorrhages.

acute inflammation *Medicine.* a basic pathologic process having a rapid onset and consisting of recruitment of cells of the cellular immune system, antibodies, and lymphokines. Inflammation occurs in response to an injury or allergic reaction involving blood vessels and surrounding tissues. The cardinal signs are redness, warmth, swelling, pain, and impaired function.

acute leukemia *Medicine.* a rapidly progressing leukemia which if untreated will cause death within four to six months.

acute lymphocytic leukemia see LYMPHOBLASTOSIS.

acute monocytic leukemia *Medicine.* a variant of the acute leukemia in which the base abnormality is the progressive proliferation of the reticuloendothelial element invading the blood and the bone marrow; it is characterized by the symptoms present with other leukemias, with the addition of gum bleeding and inflammation.

acute necrotizing hemorrhagic encephalomyelitis *Medicine.* a degenerative disease of the brain and spinal cord, frequently fatal, characterized by edema, many minute hemorrhages in the affected area, and areas of dead tissue in blood vessel walls.

acute pericarditis *Cardiology.* the sudden onset of inflammation of the pericardium.

acute radiation syndrome *Radiology.* a syndrome resulting from a whole-body dose of ionizing radiation in excess of 1 gray, and characterized by many symptoms such as diarrhea, vomiting, fever, fatigue, erythema, bleeding of the mucous membranes, gastrointestinal hemorrhage, loss of hair, hypotension, tachycardia, dehydration, and reduction in lymphocyte, granulocyte, and platelet counts. Depending on the size of the dose, death may result within hours or weeks of exposure. Also, **acute radiation exposure.**

acute respiratory disease *Medicine.* the rapid and severe onset of any disease that affects the respiratory system, including pneumonia and adult respiratory distress syndrome.

acute rheumatic fever see RHEUMATIC FEVER.

acute rhinitis *Medicine.* the rapid and severe onset of inflammation of the mucous membranes of the nose marked by increased secretions and causing congestion and pain.

acute schizophrenia *Psychology.* a former term for a disorder characterized by the sudden onset of schizophrenic symptoms such as aberrant behavior and disorientation; since schizophrenia is now classified as chronic, such disorders are now usually described by other terms.

acute toxicity *Toxicology.* a toxic reaction that occurs in a short period of time.

acute transfection *Genetics.* the introduction of DNA into cells for a short time period.

acute triangle *Mathematics.* a triangle having angles that are all less than 90°.

acutifoliate *Botany.* **1.** a plant having sharp-pointed leaves. **2.** relating to or describing such a plant.

acutilobate *Botany.* **1.** a leaf that has acute lobes. **2.** relating to or describing such a plant.

ACV air-cushion vehicle.

acyclic *Physics.* having no cycles; continually varying without a regular pattern. *Organic Chemistry.* not cyclic; describing compounds, such as alkanes, that have their molecules arranged in open-chain structures instead of rings. *Botany.* having flowers arranged in spirals, not whorls. *Mathematics.* **1.** a chain complex whose homology groups are all zero. **2.** a transformation of a set to itself for which no nonzero power of the transformation leaves any element fixed. **3.** see ACYCLIC GRAPH.

acyclic

acyclic compound *Organic Chemistry.* see ACYCLIC.

acyclic feeding *Computer Technology.* a technique used by character readers that senses the trailing edge of the page just read and triggers the feeding of the next page, thus permitting character reading and recognition of various-sized documents.

acyclic graph *Mathematics.* a graph that has no circuits. An **acyclic directed graph** is one that has no directed circuits.
acyclic motion see IRROTATIONAL FLOW.
acyclovir *Virology.* a potent antiviral agent that is specific for alphaherpesviruses (herpes simplex and varicella zoster virus).
acyl *Organic Chemistry.* a general name for the radical of an organic acid, in which the OH of the carboxy group is replaced. The general formula is RCO–, where R is aromatic, aliphatic, or alicyclic.
acyl acid see ACID AZIDE.
acylated tRNA see CHARGED TRNA.
acylation *Organic Chemistry.* a substitution reaction where an acyl group is introduced into a molecule, usually to replace an active hydrogen. Also, ACIDYLATION.
acylcarbene *Organic Chemistry.* a carbene radical that has at least one acyl group bonded to its divalent carbon.
acyl carrier protein *Biochemistry.* a protein component of fatty acid synthetase, active in the biosynthesis of fatty acids.
acyl-CoA acyl-coenzyme A.
acyl-coenzyme A see FATTY ACYL-COENZYME A.
acyl derivative *Organic Chemistry.* an organic compound that contains an acyl group.
acyl halide *Organic Chemistry.* an organic compound that contains the halocarbonyl group, such as acyl chloride.
acylnitrene *Organic Chemistry.* a nitrene that has an acyl group covalently bonded to the nitrogen.
acyloin *Organic Chemistry.* an α-hydroxyketone condensation product of aldehydes, such as benzoin.
acyloin condensation *Organic Chemistry.* the process of heating an ester in the presence of metallic sodium and without oxygen. The intermediate formed then hydrolyzes into aliphatic α-hydroxyketones called acyloins.
acyltransferase *Enzymology.* any enzyme that catalyzes the transfer of an acyl group from one substance to another.
Acytoselium *Mycology.* the single genus in the family Acytosteliidae; a dictyostelid cellular slime mold characterized by aggregation of ameboid cells by irregular movement into centers or by radial stream.
AD or **A.D.** after Christ. (From Latin *Anno Domini,* year of the Lord.) *Medicine.* right ear. (From Latin *auris dextro.*) *Ordnance.* active duty. *Mathematics.* average deviation.
A/D *Electronics.* analog-to-digital (converter).
Ad *Mathematics.* a homomorphism from a Lie group G to the group of automorphisms of its Lie algebra $\boldsymbol{g}$. The element g of G is mapped to the automorphism of g that maps $X \in \boldsymbol{g}$ to $\mathrm{Ad}(g)X = gXg^{-1} \in \boldsymbol{g}$.
ad *Mathematics.* the derivative of Ad; that is, a homomorphism from a Lie algebra $\boldsymbol{g}$ to the Lie algebra of derivations of $\boldsymbol{g}$. The element X of $\boldsymbol{g}$ is mapped into the derivation of $\boldsymbol{g}$ that maps $Y \in \boldsymbol{g}$ to $\mathrm{Ad}(X)Y = [X, Y] \in \boldsymbol{g}$ $[X, Y]$ denotes the Lie bracket in $\boldsymbol{g}$.
ad *Medicine.* an instruction in prescriptions, indicating that a substance is to be added up to a certain amount. (From the Latin word for "to.")
ad- a prefix meaning "toward," as in *adhesion.*
ADA adenosine deaminase.
ADA or **A.D.A.** **1.** American Dental Association. **2.** American Diabetes Association.
ADA or **ada** air defense artillery.
Ada *Computer Programming.* a high-level programming language sponsored by the United States Department of Defense that supports modern structured design techniques such as structured programming, information hiding, abstract data types, and concurrent processing. (Named after Augusta *Ada* Byron, 1815–1852, daughter of Lord Byron; she assisted Charles Babbage in developing the forerunner of the modern computer.)
adalert *Geophysics.* an advance alert of abnormal solar activity given by a regional or national warning center.
Adam, Robert, 1728–1792, and his brother **James,** 1730–1794, Scottish architects; developed the popular **Adam style** of building and furniture design.
Adam-and-Eve see PUTTYROOT.
adamantane *Organic Chemistry.* $C_{10}H_{16}$, white crystals that sublime at 270°C; composed of four fused cyclohexane rings with the same arrangement of carbon atoms as the diamond lattice; used in synthetic lubricants and pharmaceuticals.
adamantine *Mineralogy.* of a mineral, resembling a diamond, especially in luster or extreme hardness. *Medicine.* relating to the enamel of the teeth.
adamantine spar *Mineralogy.* a type of corundum having a smoky brown tinge and silky luster.
adamantinoma [ad ə man´tə nōm´ə] see AMELOBLASTOMA.
adamantoblast see AMELOBLAST.
adambulacral [ad´am byə lāk´rəl] *Invertebrate Zoology.* in echinoderms, adjacent to the ambulacral areas (along which run the tube feet).
adamellite see QUARTZ MONZONITE.
adamite *Mineralogy.* $Zn_2(AsO_4)(OH)$, a zinc arsenate hydroxide often occurring as clear, white, or yellowish-green orthorhombic crystals, having a specific gravity of 4.34 to 4.45 and a hardness of 3.5 on the Mohs scale; a secondary mineral in the oxidation zone of ore deposits.
Adams, John Couch 1819–1892, English astronomer; discovered the planet Neptune, independently of Urbain Leverrier.
Adams, Roger 1889–1971, American chemist; analyzed molecular structures of gossypol, cannabinol, and other organic compounds.
Adam's apple *Anatomy.* the popular name for the protrusion at the front of the throat formed by the thyroid cartilage; more noticeable in men than in women. (From the Biblical legend that this feature originated when a piece of the forbidden fruit lodged in *Adam's* throat.)
Adams-Bashforth process *Mathematics.* a numerical integration method for recursively integrating a differential equation of the form $y'(x) = f[x, y(x)]$ by replacing f with a succession of linear interpolation polynomials.
Adam's catalyst *Chemical Engineering.* a platinum oxide produced by a fusion of hexachloroplatinic(IV) acid and sodium nitrate.
adamsite [ad´əm zīt´] *Organic Chemistry.* $C_6H_4(AsCl)(NH)C_6H_4$, canary-yellow crystals that are insoluble in water; sublimes readily and melts at 195°C; used in treating wood, in tanning leather, and in chemical warfare as a poisonous gas. Also, DIPHENYLAMINE CHLOROARSINE, PHENARSAZINE CHLORIDE. *Mineralogy.* a greenish-black variety of muscovite found in Vermont.
Adam's needle *Botany.* a yucca plant, *Yucca filamentosa,* having sword-shaped leaves with threadlike fibers along the edges and large white bell-shaped flowers; widely grown as an ornamental.
Adams-Stokes disease *Cardiology.* a syndrome marked by recurring episodes of slow or absent pulse, dizziness, fainting, and convulsions, usually due to heart block. Also, **Adams-Stokes syndrome.** (From Robert *Adams,* 1791–1857, and Robert *Stokes,* 1804–1878, Irish physicians.)
ada mud *Engineering.* a material added to drilling mud to aid in obtaining cores and samples.
Adanson, Michel 1727–1806, French naturalist; made a noted early classification of plants.
Adansonia *Botany.* a genus of tropical trees in the family Bombacacea, including the boabab, *A. digitata.* (From Michel *Adanson.*)
Adansonian taxonomy *Systematics.* an earlier taxonomic system that contributed to the modern practice of classification on the basis of overall similarities and differences rather than on a few subjective qualities. Also, **Adansonian classification.** (From Michel *Adanson.*)
adapertural *Invertebrate Zoology.* relating to or near the aperture, specifically the aperture of a conch.
Adapiformes *Paleontology.* an infraorder of small, lemur-like primates in the suborder Prosimii, extant from the late Paleocene to the Miocene.
Ada Program Support Environment *Computer Programming.* an extension of the minimal set of the Ada programming language design and development tools that provides more complete support for special applications.
adaptability *Ecology.* the capacity of an organism to adjust physiologically or behaviorially to changes in its environment.
adaptation the fact of changing in response to some condition; specific uses include: *Genetics.* a particular developmental, behavioral, anatomical, or physiological change in a population of organisms, based on genetic changes and occurring as a result of natural selection. *Evolution.* the general capacity of a species to undergo evolutionary change in response to its natural environment, so as to enhance its ability to survive. *Behavior.* any change in behavior patterns as an adjustment to environmental conditions. *Physiology.* the adjustment of the pupil of the eye to changes in light intensity. *Neurology.* a diminishing of sensitivity, such as a decrease in the frequency of neuron firing, in response to a repeated stimulus of constant intensity. *Psychology.* an individual's ability to adjust to new experiences and accept new information; the process of cognitive growth.
adaptation brightness or **adaptation level** see ADAPTATION LUMINANCE.
adaptation disease see ADAPTIVE DISEASE.
adaptation energy *Physiology.* Hans Selye's term for the individual's general ability to resist the physiological effects of stress.

adaptation luminance *Optics.* the brightness of an object in the proximity of a viewer approximating the visual range. Also, FIELD LUMINANCE.

adaptation syndrome see GENERAL ADAPTATION SYNDROME.

adapter something that allows for or causes a change; specific uses include: *Mechanical Devices.* any fitting or appliance that joins objects of different sizes or designs, such as hose connections or the like. *Optics.* any accessory or attachment that allows a camera to be used differently from the way in which it was designed, especially a device that allows for the interchange of lenses or accessories between different types of cameras. *Computer Technology.* a device that changes incoming signals or information from one characteristic to another, such as serial to parallel bit stream or different data transfer rates.

adapter skirt *Space Technology.* an attachment to or extension of a space vehicle section, providing a means of fitting it to some other object or section.

adapter transformer *Electricity.* a step-down transformer that provides low voltage usually to small electric equipment; often encased in plastic and designed to plug directly into a wall outlet.

adaption see ADAPTATION.

adaptive *Biology.* capable of or showing adaptation. *Evolution.* of or relating to genetic or environmental conditions that contribute to an organism's evolutionary success. *Computer Programming.* of a system, capable of some form of alteration or adjustment in response to existing conditions.

adaptive branch *Robotics.* a branching instruction in a robot's computer program; the decision made by the instruction usually depends on external conditions.

adaptive colitis see IRRITABLE BOWEL SYNDROME.

adaptive communications *Telecommunications.* a communications system that is designed to adjust automatically to changing inputs. Also, SELF-ADJUSTING COMMUNICATIONS.

adaptive control *Control Systems.* a method of control that uses data from feedback control signals to adjust system performance to meet specific goals. *Robotics.* a part of a robot's control program that allows it to adapt to external conditions by dynamically updating its performance data. Thus, **adaptive-control function** or **method**

adaptive convergence *Evolution.* the tendency of different organisms living under the same environmental conditions to assume certain similar characteristics; e.g., the fishlike shape of aquatic mammals such as the seal or whale.

adaptive disease *Medicine.* any disease that is thought to be caused or exacerbated by conditions of chronic stress, such as peptic ulcers or hypertension.

adaptive divergence *Evolution.* the tendency of related organisms living under different environmental conditions to develop certain divergent characteristics; e.g., polar bears and black bears.

adaptive enzyme *Enzymology.* an enzyme that is normally not present in a cell in significant quantities, but is produced in response to the presence of a certain substrate or inducer substance. Also, INDUCIBLE ENZYME.

adaptive equalization *Telecommunications.* any device or processing method that automatically compensates for the distortion experienced by a digital signal or digital modular signal in its passage through a transmission system.

adaptive immunity *Immunology.* a specific acquired resistance to infectious agents, produced by the introduction of an antigen through immunization.

adaptive integration *Mathematics.* an iterative integration algorithm that adjusts for irregularities in the function to be integrated, by adjusting the points at which the function is to be evaluated.

adaptive landscape *Evolution.* a graphic representation, in the form of a topographical map, of the fitness of specific genotypes or species in particular ecological niches.

adaptive norm *Genetics.* the array of genotypes possessed by a given population of a species which are compatible with its environment.

adaptive optics *Optics.* referring to or describing the design and use of optical systems whose performance is monitored and modified to meet the changing needs of a task, to compensate for aberrations, or to counteract thermal, mechanical, and acoustical disturbances.

adaptive peak *Evolution.* a peak or crest on an adaptive landscape, representing the fitness of a particular genotype or species in a particular ecological niche.

adaptive radiation *Evolution.* the diversification of a generalized ancestral form of an organism into several more specialized forms, each adapted to survive in slightly different ecological conditions in the former range, and each potentially a forerunner of a new form.

adaptive robot *Robotics.* **1.** a robot that can change its responses to fit the conditions of its environment. **2.** a robot that uses adaptive control.

adaptive system *Computer Programming.* a system that has the ability to learn, change its state, or otherwise react to a stimulus or a change in its environment.

adaptive system theory *Computer Science.* the branch of automata theory concerning systems that are capable of learning or adapting to outside stimuli.

adaptive valley *Evolution.* a valley or low point on an adaptive landscape, representing the low fitness of a genotype or species in a particular ecological niche.

adaptive value *Evolution.* a measurement of the ability of a given genotype to survive and reproduce, relative to other genotypes in the population. Also, SELECTIVE VALUE.

adaptive zone *Evolution.* **1.** a concept or graphic representation of an ecological continuum or pathway along which species or other taxonomic groups evolve over periods of time. **2.** the fundamental ecological niche or setting that will allow a species to survive and reproduce.

adaptor see ADAPTER.

adaptor RNA *Genetics.* a former term for transfer RNA, originally proposed by Francis Crick in 1958, when he hypothesized the existence of this molecule before its discovery.

adaptors *Molecular Biology.* a short sequence of DNA that serves to splice together two longer DNA molecules, one of which has cohesive ends.

ADAR see ADVANCED-DESIGN ARRAY RADAR.

adaxial [ad aks´sē əl] *Biology.* on the side facing toward the axis of an organism. *Botany.* **1.** the surface of a leaf facing toward the axis; the upper side of an open leaf. **2.** of or relating to this side or surface.

ADC analog-to-digital converter.

ADCC *Immunology.* antibody-dependent cell-mediated cytotoxicity; the capability of nonsensitized cells to rupture other cells that have been coated by a specific antibody.

ADCCP *Robotics.* Advanced Data Communication Control Procedure; a standard established by ANSI to deal with bit-oriented data communications; often used in robotic systems.

Adcock antenna *Electromagnetism.* an antenna that produces a radiation pattern in the form of a figure-eight, constructed of a pair of vertical antennas with a separation distance of less than one-half wavelength and opposite in phase. Thus, **Adcock direction finder.**

ADCON see ADDRESS CONSTANT.

adconductor cathode *Electronics.* in gas lasers, a cathode that has been treated with alkali metal atoms in order to emit electrons during glow discharge.

adcumulus *Petrology.* the late-stage growth phase of a cumulus crystal that maintains equilibration with the magma reservoir.

add *Mathematics.* to combine two or more numbers into one equivalent quantity, the sum of these numbers; carry out the process of addition.

ADD *Psychology.* see ATTENTION-DEFICIT DISORDER.

addax *Vertebrate Zoology.* a large, light-colored antelope of the subfamily Hippotraginae, having curved, widely spread horns; found in North Africa, Arabia, and Syria.

addax

addend *Mathematics.* **1.** of two numbers to be added together, the newly introduced number that is being added, as opposed to the previously existing one to which it is added (the augend). **2.** any of a group of numbers or quantities being added together.

addendum *Mechanical Engineering.* on a gear, the radial distance between the pitch line and the addendum circle.

addendum circle *Mechanical Engineering.* an imaginary circle positioned on a gear so as to pass through the top portion of the teeth; the outer circumference of a gear wheel.

adder *Vertebrate Zoology.* any of a wide variety of poisonous snakes belonging to the family Viperidae, such as the puff adder or the European viper. *Electronics.* **1.** any circuit in which two or more input signals are combined to give one output signal that is the sum of those signals. **2.** in a color TV receiver, a circuit that boosts the signal of the receiver's primary matrix. *Computer Technology.* a circuit in an arithmetic unit that can form the algebraic or arithmetic sum of two binary numbers held in two registers, the result of which is stored temporarily in the accumulator register. Also, **adder circuit.**

adder circuit *Electronics.* see ADDER, def. 1.

adder's tongue *Botany.* **1.** any of various ferns, genus *Ophioglossium,* having a single stalk with an upright fruiting spike thought to resemble a snake's tongue, especially *O. vulgarium,* which is found in the eastern half of North America. **2.** see DOGTOOTH VIOLET.

adder-subtracter *Computer Technology.* a combinational circuit that performs algebraic and arithmetic addition or subtraction and stores the result in the accumulator.

addict *Medicine.* a person who suffers from addiction; someone who is physically and emotionally dependent on the intake of a certain substance, such as alcohol or drugs.

addicted *Medicine.* suffering from addiction; dependent on a substance such as alcohol or drugs.

addiction *Medicine.* a physiological and psychological dependence on the intake of a certain substance, especially a harmful substance such as alcohol or narcotic drugs. *Psychology.* any habitual or obsessive practice of a harmful behavior: an *addiction* to gambling.

addictive *Medicine.* relating to or causing addiction: an *addictive* drug.

addictive disorder *Medicine.* a disorder that begins with the chronic use of a substance and results in tolerance and then physical dependence.

add-in *Computer Technology.* any of a variety of electronic components that can be added to an installed printed circuit board to increase a computer's capabilities or expand internal memory.

adding circuit *Electronics.* see ADDER, def. 1.

adding machine *Mechanical Devices.* a machine capable of adding numbers and performing other arithmetic functions; once widely used in accounting and other business applications but now largely replaced by computers and electronic calculators.

adding tape *Engineering.* a tape used in surveying, which is divided into 100 feet (or meters) and which also has one of these divisions or an additional foot (or meter) subdivided into smaller units.

Addis count *Pathology.* a numerical computation to determine the number of red blood cells, white blood cells, epithelial cells, casts, and protein content in a 12-hour urine sample; employed in the diagnosis and treatment of kidney disease. Also, **Addis test.** (Named after Thomas *Addis,* 1881–1949, American physician.)

Addison, Thomas 1793–1860, British physician; diagnosed and treated Addison's disease.

Addison's disease *Medicine.* a disease resulting from the reduction in the secretion of the adrenocortical hormones and characterized by increased pigmentation of mucous membranes and skin resulting in the appearance of black freckles on the head and neck, general weakness, loss of weight, hypotension, nausea, vomiting, confusion, and low blood sugar.

addition *Mathematics.* the process of combining two or more numbers to obtain one equivalent quantity, which is the sum; the binary operation denoted by the plus sign, +. The term usually refers to the operation of an Abelian group or the group operation in a ring, module, or vector space over which other operations distribute. *Chemistry.* a process or reaction in which two or more substances are directly combined to form a more complex compound.

addition agent *Physical Chemistry.* a substance added to a plating solution to improve the characteristics of the deposit formed.

addition item *Computer Programming.* an item that is to be added in a specific place in a previously established file.

addition mutation see INSERTION MUTATION.

addition of complex quantities *Mathematics.* given two complex numbers $z_1 = a_1 + b_1 i$ and $z_2 = a_2 + b_2 i$, addition is then defined as $z_1 + z_2 = (a_1 + a_2) + (b_1 + b_2)i$.

addition of vectors *Mathematics.* the combining of vectors by adding corresponding components; in particular, given two vectors $v_1 = (a_1, a_2, \ldots, a_n)$ and $v_2 = (b_1, b_2, \ldots, b_n)$, vector addition is then defined by $v_1 + v_2 = (a_1 + b_1, a_2 + b_2, \ldots, a_n + b_n)$.

addition polymer *Organic Chemistry.* a polymer formed by the chain addition of monomers, such as polyethylene and polystyrene. Also, **addition resin.**

addition polymerization *Materials Science.* the condensation product-free formation of a long-chain polymer from a monomer, involving an initiator to catalyze the reaction.

addition reaction *Organic Chemistry.* **1.** a chemical reaction in which an unsaturated molecule adds another atom or molecule so as to partly or completely saturate the original molecule. **2.** see ADDITION.

addition record *Computer Programming.* a record that causes creation of a new record in an established master file during updating.

addition rule *Statistics.* a method of combining the probability of occurrence of two or more events; i.e., event A_1 and event A_2.

addition solid solution *Crystallography.* a crystal with a random penetration of impurities into its interstices.

additive *Materials Science.* any substance that is added to another substance, usually in a small quantity, in order to produce a desired effect in the primary substance. *Food Technology.* a substance added in relatively small quantities to a food product, to facilitate its preparation or to improve its nutritional value, flavor, appearance, or keeping quality. *Petroleum Engineering.* any of various chemical compounds added to gasoline to improve engine performance; e.g., antiknock or antirust compounds. *Mathematics.* of or relating to addition. *Statistics.* see ADDITIVE MODEL.

additive color see ADDITIVE PRIMARY COLOR.

additive color viewer *Photogrammetry.* a projector for viewing positive transparencies obtained from multiband photography, in which one image is superimposed on another and each image is lit up with a different colored light.

additive function *Mathematics.* a function that satisfies the property $f(x + y) = f(x) + f(y)$ whenever f is defined on x, y, and $x + y$.

additive gene *Molecular Biology.* any of a series of gene sequences that affect the same phenotypic character in a synergistic fashion. Also, **additive factor.**

additive genetic variation *Genetics.* genetic variation in a population that results from the additive effects of multiple alleles that lack dominance and the added effects of epistasis from nonallelic genes.

additive inverse *Mathematics.* a number that when added to another given number yields a sum of zero; e.g., 6 is the additive inverse of –6.

additive model *Statistics.* a model in which the effect of a set of independent variables on a dependent variable is described as the sum of individual effects.

additive primary colors *Optics.* the three colors, usually blue-violet, green, and orange-red, that are combined in color photography and color television systems. Also, **additive primaries.**

additive process *Optics.* the process of combining various proportions of red, green, and blue light in a mosaic screen to closely approximate all color sensations and white; a technique used for most color television displays. Also, **additive color process, additive synthesis.**

additive recombination *Molecular Biology.* the insertion or splicing in of a new DNA sequence into an existing genome without any reciprocal loss of DNA; often achieved by insertion elements.

additive set function *Mathematics.* a set function f having the property $f(A \cup B) = f(A) + f(B)$ for any two disjoint sets A and B on which f is defined.

additive variance *Genetics.* genetic variance produced by the cumulative average effects of replacing an allele at a single locus or several alleles at different loci controlling a polygenic trait.

add list *Artificial Intelligence.* a term for a list of facts to be added to the world model after application of a STRIPS operator.

add-on *Computer Technology.* any of a variety of circuits, systems, or hardware devices that can be incorporated with a basic computer system to improve its capabilities; e.g., an extended memory unit, a modem to communication links, or a color graphic terminal. *Engineering.* any device added to a system or unit to improve its capabilities.

add-on memory *Computer Technology.* any of a variety of memory expansions to original storage capacity that are designed to enhance processing capabiity.

add operation *Computer Programming.* an arithmetic process in which two or more numbers are added together and the sum is stored in place of one of the original numbers.

add-overflow flip-flop *Computer Programming.* a one-bit register in a computer arithmetic unit that is set when the addition of two numbers results in a number too large to be properly represented in a computer word.

address *Computer Programming.* **1.** a bit pattern, representing a number, name, or label, that uniquely identifies a register, a location in memory, or an external data source or destination. **2.** the part of a fixed-form instruction that specifies the location of the operand. **3.** the coded portion of a message specifying the intended destination. **4.** to identify a storage location and obtain data from or store data in that location.

addressable *Computer Programming.* able to be directly accessed by the specification of a unique identifying location in memory or a specific input/output device.

addressable cursor *Computer Programming.* a facility of a visual display unit that permits varied positioning of the on-screen cursor by sending *X–Y* coordinates from an external source such as a keyboard or mouse.

address bus *Computer Technology.* the circuit that transmits the address of the operand referenced in an instruction from the CPU to main memory during the execution of an instruction, or to an external storage device during a read/write operation.

address computation *Computer Programming.* the process of calculating the effective address portion of an instruction.

address constant *Computer Programming.* a number used as the base or starting point from which subsequent addresses are to be modified.

address conversion *Computer Programming.* the process of translating a symbolic or relative address into an actual address, either manually or by means of a conversion program such as an assembler or compiler.

address counter *Computer Technology.* a binary incrementing counter in a direct-memory-access controller that is used to control data block transfer by storing the destination.

address field *Computer Programming.* the specific portion of a machine instruction that contains the operand, the address of the operand, or the information needed to derive the address of the operand.

address format *Computer Programming.* the arrangement of the fields and entries in the address portion of an instruction.

address-free program *Computer Programming.* a set of instructions written for an addressless machine that commonly uses stacks instead of memory locations for storing operands.

address generation *Computer Programming.* the process of obtaining an effective memory address by adding the contents of the base register to the relative address specified in the computer instruction.

addressing *Computer Programming.* a method of specifying the locations of operands and instructions in main memory or of specifying a particular input/output device.

addressing mode *Computer Programming.* a rule for interpreting or modifying the address field of an instruction to identify the operand; types include immediate, direct, indirect, and register.

addressing system *Computer Programming.* the hardware and software techniques by which specific locations in computer memory may be identified for storage and retrieval of data or instructions.

address interleaving *Computer Programming.* a method of assigning consecutive addresses to physically different memory modules that staggers memory access and reduces the effective memory cycle time.

addressless instruction (format) *Computer Programming.* an instruction format that has no address part, either because no address is required or because the address is implied by the instruction. Also, ZERO-ADDRESS INSTRUCTION.

address modification *Computer Programming.* the process of performing arithmetic operations on the address portion of an instruction during the execution of a program; used in a general loop control structure.

address part *Computer Programming.* the binary digits in a fixed-form instruction word that represent the location of the operand, the destination of the results of an operation, or the address of the next instruction, depending on the type of instruction.

address register *Computer Technology.* **1.** a special-purpose register in the central processing unit that is loaded with the address of the next memory location to be accessed. **2.** a register or one of a set of registers in a CPU that are specialized for use in address calculations. Also, MEMORY ADDRESS REGISTER.

address space *Computer Technology.* **1.** the set of memory addresses that a program may reference. **2.** the amount of memory allocated to a program or user. **3.** the amount of memory addressable by the address size of a machine instruction.

address translation *Computer Programming.* an operation performed by a compiler or assembler that changes the form of an address, often from symbolic to relative or virtual.

add-subtract time *Computer Programming.* the time required for the arithmetic/logic unit to perform addition or subtraction with two numbers after they have been obtained from memory.

add time *Computer Programming.* the time required for the arithmetic/logic unit to add two numbers after they have been obtained from memory.

adducent *Physiology.* relating to or performing adduction; moving toward a certain part or point. Also, ADDUCTIVE.

adduct *Chemistry.* a product of a process of chemical addition; a compound produced by the combination of two or more substances. *Physiology.* to draw inward; move toward the axis of the body or the axial line of a limb.

adduction *Physiology.* the movement of a body part toward the midline axis of the body, as in lowering an arm; the opposite of abduction.

adductive see ADDUCENT.

adductor *Anatomy.* a muscle that draws a part of the body toward the main axis of the body or of a limb.

Adelanthaceae *Botany.* a family of medium to robust liverworts of the order Jungermanniales; characterized by brown pigments, ventral intercalary branches, and erect leafy branches arising on rigid stems from a system of rhizoidous, prostrate stolons.

Adelard of Bath c. 1090–1150, English mathematician; translated classic Greek and Arabic works; wrote treatises on the abacus and astrolabe.

Adeleina [a də lē′ə nə] *Invertebrate Zoology.* a suborder of protozoan parasites in the order Eucoccida; sexual and asexual stages are in different hosts; they are parasitic on arthropods and have two sporozoites in each sporocyst.

Adélie penguin [ad′ə lē] *Vertebrate Zoology.* a small species of penguin, *Pygoscelis adeliae,* living in large colonies in Antarctica. (From the *Adélie* region of Antarctica; named for the wife of its discoverer.)

Adélie penguin

adelite *Mineralogy.* $CaMg(AsO_4)(OH)$, an orthorhombic transparent, colorless to gray, yellow, or green mineral occurring in manganese ore deposits; massive in habit, having a specific gravity of 3.74 and a hardness of 5 on the Mohs scale.

adelphous [ə del′fis] *Botany.* relating or describing a plant in which the stamens, instead of growing singly, combine by the filaments into one or more bundles.

aden- a combining form meaning "gland."

adenase see ADENINE DEAMINASE.

adendritic [ā′den drit′ik] *Zoology.* lacking dendrites.

adenectomy [ă′də nek′tə mē] *Surgery.* the surgical removal of a gland.

Aden, Gulf of *Geography.* the southwestern arm of the Arabian Sea, between southern Arabia and Somalia.

adenine *Biochemistry.* $C_5H_5N_5$, a purine base (6-aminopurine) that occurs in animal and plant tissues as a constituent of DNA and RNA; occurs chemically in the form of pearly white crystals.

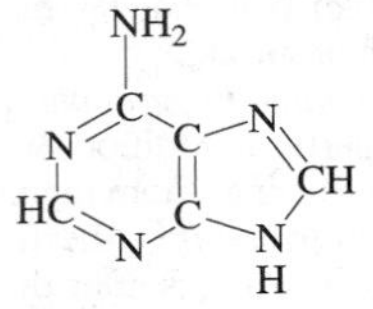

adenine

adenine deaminase *Enzymology.* an enzyme found in the liver, kidney, and spleen that catalyzes the breakdown of adenine into hypoxanthine and ammonia.
adenine deoxyriboside *Genetics.* a nucleoside produced when adenine bonds with deoxyribose.
adenitis *Medicine.* an inflammation of a gland or lymph node.
adeno- a combining form meaning "gland."
adenoacanthoma *Oncology.* an adrenocarcinoma containing some cells that have undergone squamous metaplasia. Also, **adenocancroid.**
adeno-associated virus see DEPENDOVIRUS.
adenocarcinoma [ad´ən ō´kar sə nō´mə] *Oncology.* a malignant adenoma in the epithelium of glandular tissue.
adenocele [ad´ən ō´sēl´] *Oncology.* a cystic tumor in a glandular structure.
adenocellulitis [ad´ən ō´sel yə lī´tis] *Medicine.* the inflammation of a gland and the surrounding tissue.
adenochondroma [ad´ən ō´kän drō´mə] *Oncology.* a tumor composed of both glandular and cartilaginous tissue, as in a mixed tumor of the salivary glands. Also, PULMONARY HAMARTOMA.
adenocystic carcinoma [ad´ən ō´sis´tik] *Oncology.* a malignant tumor consisting of cords or clusters of small epithelial cells arranged in a sievelike or nestlike pattern; typically found in the mammary glands, salivary glands, and mucous glands of the respiratory tract. Also, ADENOID CYSTIC CARCINOMA, ADENOMYOEPITHELIOMA.
adenocystoma [ad´ən ō´sis tō´mə] *Oncology.* a benign epithelial tumor containing visible lined cysts; typically found in the ovary, pancreas, and salivary glands. Also, **adenocyst.**
adenodynia [ad´ən ō´dī´nē ə] *Medicine.* glandular pain.
adenoepithelioma [ad´ən ō ep´ə thē lē ō´mə] *Oncology.* a tumor composed of glandular and epithelial elements.
adenofibroma [ad´ən ō fī brō´mə] *Oncology.* a tumor composed of fibrous connective tissue and glandular structures.
adenohypophysis [ad´ən ō´hī pä´fə səs] *Endocrinology.* the anterior lobe of the pituitary gland, consisting of a mixed population of cells that secrete a cell-specific complement of hormones, such as prolactin, growth hormone, and luteinizing hormone, in response to signals from the hypothalamus.
adenoid [ad´ə nōid; ad´noid] *Anatomy.* **1.** see ADENOIDS. **2.** of or relating to the lymphatic glands or to lymphoid tissue. **3.** resembling a gland; glandular.
adenoidal *Medicine.* **1.** having the adenoids swollen or enlarged. **2.** of the voice, having the nasal, obstructed tone associated with swelling of the adenoids. **3.** of or relating to the lymphatic glands or lymphoid tissue; adenoid.
adenoid cystic carcinoma see ADENOCYSTIC CARCINOMA.
adenoidectomy [ad´ə nōi dek´tə mē] *Surgery.* the surgical removal of the adenoids.
adenoiditis [ad´ə nōi di´tis] *Medicine.* an inflammation affecting the adenoids.
adenoids *Anatomy.* usually, **the adenoids.** the masses of lymphoid tissue at the back of the throat, behind the nose; during childhood they may become swollen and obstruct breathing. *Medicine.* a popular term for swelling of the adenoids.
adenolipoma [ad´ən ō lī pō´mə] *Oncology.* a tumor composed of fatty and glandular tissue elements.
adenolipomatosis *Oncology.* a condition that is characterized by the growth of multiple adenolipoma tumors.
adenolymphoma *Oncology.* a benign parotid tumor composed of glandular and cystic structures lined by eosinophilic epithelium. Also, WARTHIN'S TUMOR.
adenoma [ad ə nō´mə] *Oncology.* any of various benign epithelial tumors in which the cells form glandlike structures or develop from glandular epithelium.
adenoma sebaceum *Medicine.* an abnormal condition affecting the sebaceous glands, characterized by multiple yellowish wartlike growths on the face.
adenoma sudoriparum see SPIRADENOMA.
adenomatoid [ad´ə nō´mə toid] *Oncology.* relating to or resembling an adenoma. Thus, **adenomatoid tumor.**
adenomatosis [ad´ən ō´mə tō´səs] *Oncology.* a disorder marked by the formation of many glandular growths, especially when involving more than one gland.
adenomatous [ad´ə nō´mə təs] *Medicine.* **1.** relating to or resembling a gland. **2.** relating to an adenoma.
adenomatous goiter *Oncology.* an enlargement of the thyroid gland caused by the development of one or more nonmalignant encapsulated growths (adenomas) in the gland.
adenomegaly [ad´ən ō meg´ə lē] *Medicine.* an abnormal enlargement of a gland or glands.
adenomere [ad´ən ō´mir] *Developmental Biology.* the blind terminal part of a developing gland that eventually becomes the functional part of the organ.
adenomyoepithelioma [ad´ən ō mī´ō ep´ə thē lē ō´mə] see ADENOCYSTIC CARCINOMA.
adenomyoma [ad´ən ō mī ō´mə] *Oncology.* a benign tumor-like nodule formed by the growth of the endometrium into the muscle of the uterus.
adenomyometritis see ADENOMYOSIS.
adenomyosarcoma [ad´ən ō mī ō´sär kō´me] *Oncology.* a malignant tumor that is composed of both striated muscular elements and glandular elements.
adenomyosis [ad´ən ō mī ō´səs] *Oncology.* a condition characterized by benign growths that originate in the endometrium and invade the muscle layers of the uterus. Also, ENDOMETRIOSIS INTERNA or UTERINA, ADENOMYOMETRITIS.
adenopathy [ad´ən ä´pə thē] *Pathology.* a swelling or enlargement of the glands, particularly the lymph nodes. Thus, **adenopathic.**
Adenophorea *Invertebrate Zoology.* one of two classes of roundworms in the phylum Nematoda that lack phasmids (glands that function in chemoreception).
adenophyllous [ad´ən ä´fə ləs] *Botany.* relating to or being a leaf that bears glands.
adenosarcoma [ad´ən o sär kō´mə] *Oncology.* either of two malignant tumors: **1.** a tumor that is composed of sarcomatous and glandular elements. **2.** a tumor that originates in glandular epithelium and in mesodermal tissue.
adenosine [ə den´ə sēn] *Biochemistry.* $C_{10}H_{13}N_5O_4$, a nucleoside that is composed of adenine and ribose, which is found in muscle tissue and in the hereditary building block RNA (ribonucleic acid). Also, RIBOFURANOSYLADENINE.
adenosine arabinofuranoside see VIDARABINE.
adenosine 3',5'-cyclic phosphate see cAMP.
adenosine deaminase *Enzymology.* an enzyme of the hydrolase class that catalyzes the conversion of adenosine into inosine and other substances.
adenosine deaminase deficiency *Immunology.* a disorder of the immune system caused by the inability to produce adequate levels of the enzyme adenosine deaminase; genetically transmitted in individuals with certain immunodeficiency diseases.
adenosine diphosphatase *Enzymology.* an enzyme that catalyzes the release of adenosine diphosphate.
adenosine diphosphate *Biochemistry.* $C_{10}H_{15}N_5O_{10}P_2$, a substance involved in energy metabolism; formed by the breakdown of adenosine triphosphate.
adenosine diphosphate ribosylation *Biochemistry.* a reaction in which ADP-ribose is covalently attached to another compound. Also, ADP RIBOSYLATION.
adenosine monophosphate see ADENYLIC ACID.
adenosine phosphate *Biochemistry.* any of three types of compounds consisting of the nucleotide adenosine attached to either one, two, or three phosphoric acid molecules: adenosine monphosphate, adenosine diphosphate, or adenosine triphosphate. (See the specific entries for these terms.)
adenosine triphosphatase *Enzymology.* an enzyme that catalyzes the release of phosphate from adenosine triphosphate (ATP), which is the primary energy source in metabolism.

adenosine triphosphate *Biochemistry.* ATP, a nucleotide present in all living cells and acting as an energy source for many metabolic processes and required for ribonucleic acid synthesis.

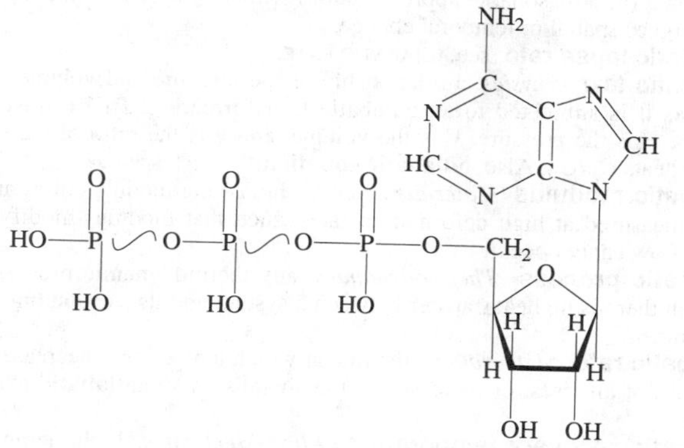

adenosine triphosphate (ATP)

adenosis [ad´ən ō´səs] *Medicine.* **1.** any disease of a gland, especially of the lymphatic glands. **2.** the abnormal development or enlargement of glandular tissue.

adeno-SV40 hybrid virus *Virology.* intramolecular recombinants that incorporate various amounts of SV40 DNA into an adenovirous DNA that had undergone deletions.

***S*-adenosylhomocysteine** *Biochemistry.* a compound produced when the methyl group of *S*-adenosylmethionine attaches to an acceptor such as phosphatidylethanolamine.

***S*-adenosylmethionine** *Biochemistry.* $C_{15}H_{22}N_6O_5S$, a reaction product of ATP and methionine in which the S atom of methionine is bound to the ribose of adenosine; a high-energy compound that serves as a methyl donor.

Adenoviridae *Virology.* a family of icosahedral, nonenveloped, linear dsDNA-containing viruses, usually found in the respiratory tract, that infect mammals or birds.

adenovirus [ad´ən ō vī´rəs] *Virology.* any virus of the Adenoviridae family; associated with various respiratory and eye diseases of humans and animals. Thus, **adenoviral.**

adenyl [ad´i nəl] *Chemistry.* see ADENYLYL.

adenylate [ə den´i lāt´] *Biochemistry.* **1.** an ionized form of adenosine monophosphate (AMP). **2.** to carry on the process of adenylation.

adenylate cyclase *Enzymology.* an enzyme occurring in plasma cell membranes that transforms adenosine triphosphate into adenosine 3',5'-cyclic phosphate (cAMP) and inorganic pyrophosphate; the resultant cAMP serves as a metabolic regulator.

adenylate energy charge *Biochemistry.* the measure for the amount of energy within the adenylate system (ATP-ADP-AMP) of a cell at a specified time.

adenylate kinase *Biochemistry.* an enzyme that interconverts two molecules of adenosine diphospate (ADP) into one molecule each of adenosine triphosphate (ATP) and adenosine monophosphate (AMP).

adenylation *Biochemistry.* the process by which the adenyl group of adenosine triphosphate (ATP) is transferred to an acceptor molecule.

adenyl cyclase see ADENYLATE CYCLASE.

adenylic acid *Biochemistry.* $C_{10}H_{14}N_5O_7P$, the ribonucleotide of the hereditary building block DNA, functions in energy metabolism to indicate the need to synthesize adenosine triphospate (ATP). Also, ADENOSINE MONOPHOSPHATE (AMP).

adenylyl [ad´ ə nə lil] *Biochemistry.* the radical resulting from the removal of an OH group from the phosphate group of adenosine monophosphate (AMP).

adenylylation *Biochemistry.* the process by which the adenylyl group of ATP is transferred to an acceptor molecule.

adenylyl transferase *Enzymology.* an enzyme that catalyzes the transfer of an adenylyl group.

Adephaga [ə def´ə gə] *Invertebrate Zoology.* a large suborder of predaceous beetles with filiform antennae in the order Coleoptera; the first abdominal segment is divided by coxae.

adequate contact *Medicine.* a term for the minimum amount of contact needed between one who is healthy and one who is a carrier of a disease, for the disease to be transmitted from one to the other.

adequate stimulus *Physiology.* any type of stimulus that is sufficient to produce a response from a particular receptor.

ader wax [äd´ər] see OZOCERITE.

ADF or **adf** automatic direction finder; aircraft direction finder.

ADF bearing indicator *Navigation.* **1.** the pointer on an automatic direction finder that indicates the relative bearing to the radio source. **2.** a separate instrument that does this.

adfreezing *Hydrology.* in permafrost studies, the process by which two objects adhere to each other as a result of the binding action of ice.

ADF reversal *Navigation.* the reversal in the bearing shown by an ADF bearing indicator as an aircraft passes over the ADF source station.

ADH antidiuretic hormone. See ARGININE VASOPRESSIN.

Adhara [ə där´ə] *Astronomy.* Epsilon (ε) Canis Majoris, a second-magnitude blue-white double star in the constellation Canis Major, the Greater Dog.

adhere *Science.* to stick together; become fastened together. *Physics.* of two or more dissimilar substances in contact, to be held together by molecular force acting at the surface.

adherend *Materials Science.* a material united with another substance by adhesion.

adherent *Science.* relating to or characterized by adhesion; sticking or clinging together. *Botany.* see ADNATE.

adherent cells *Immunology.* the cells from lymph tissues or inflamed areas (frequently macrophages) that cling to glass or plastic surfaces during laboratory testing.

adhesin *Microbiology.* a cell-surface appendage or extracellular macromolecular substance that facilitates adhesion of a cell to a surface or to other cells.

adhesion the sticking together of surfaces or parts; specific uses include: *Mechanics.* the static attractive force at the contacting surface between two bodies of different substances in contact with each other. *Physics.* a force that acts to hold two molecules of dissimilar substances together. *Materials Science.* the sticking together of structural parts by means of cement or glue. *Engineering.* the sticking together of two surfaces in contact as a result of bonds formed with stress or heat. *Biology.* an abnormal joining together of parts or organs that are normally separate. *Medicine.* **1.** the process by which two surfaces join to each other. **2.** a band of scar tissue binding two anatomic surfaces that are normally separate, such as those which commonly form in the abdomen following abdominal surgery.

adhesional theory of friction *Materials Science.* a description of the frictional coefficient between two materials in which welding of asperities on mating surfaces occurs; it is primarily dependent on the plastic parameters of the softer material, and also on the cleanliness of the mating surfaces.

adhesional work *Thermodynamics.* the amount of work that is required to separate two substances that are in contact with each other, particularly liquids at an interface, as calculated over an area of one square centimeter.

adhesion site *Microbiology.* any of a number of junctions or zones of apparent localized fusion formed between the cytoplasmic and outer membranes of Gram-negative bacteria; postulated to function as a site of transport to the outer membrane for both proteins and lipopolysaccharides.

adhesiotomy *Surgery.* the surgical cutting or division of adhesions, as in the intestines following earlier abdominal surgery.

adhesive *Science.* having the property of binding or sticking; adhering to another substance. *Materials Science.* **1.** a liquid, paste, powder, or dry film that bonds other materials together through physical or chemical means. **2.** of a material, coated with glue, paste, or another such sticky substance. *Physics.* of or relating to adhesion.

adhesive binding see PERFECT BINDING.

adhesive bonding *Materials Science.* the sticking together of solid surfaces by means of cement or glue. Thus, **adhesive bond.**

adhesive cell *Invertebrate Zoology.* a glandular thread-bearing cell that produces viscous adhesive secretions; found in ctenophore jellyfish, the pedal disk of Hydra, and the epidermis of Turbellaria worms.

adhesive gripper *Robotics.* an end effector that uses suction to grasp objects.

adhesive strength *Engineering.* the strength of an adhesive bond between two surfaces as measured by the tensile force needed to separate those surfaces.

adhesive tape *Materials.* any type of tape coated with an adhesive substance so that it will stick or hold to a surface, such as that used to hold bandages in place on the skin. Similarly, **adhesive bandage.**

adhesive water see PELLICULAR WATER.

adhesive wear *Materials Science.* the process of wear that occurs when two solid surfaces slide over one another under pressure and welding of asperities occurs.

ad hoc [ad häk´; ad hōk´] *Science.* for this specific situation; for the subject or task at hand. (A Latin phrase meaning "for this.")

ad hoc inquiry *Computer Programming.* a single request for information, such as a report, as opposed to periodic or routine requests.

ADI acceptable daily intake; average daily intake.

adiabat [ad´ē ə bat´] *Thermodynamics.* a line or curve on an adiabatic chart.

adiabatic [ad´ē ə bat´ik] *Thermodynamics.* **1.** describing a process in which there is no transfer of heat into or out of the system in question. **2.** without loss or gain of heat.

adiabatic atmosphere *Meteorology.* a model atmosphere characterized by a dry-adiabatic lapse rate throughout its vertical extent. Also, DRY-ADIABATIC ATMOSPHERE, CONVECTIVE ATMOSPHERE, HOMOGENEOUS ATMOSPHERE.

adiabatic calorimeter *Physical Chemistry.* a calorimeter that is heavily insulated to minimize heat loss or gain, so as to provide more accurate measurement of the thermal effect of the reaction.

adiabatic change *Thermodynamics.* any change in the thermodynamic state of a system that does not involve heat transfer between the system and its surrounding environment.

adiabatic chart *Thermodynamics.* a graphic representation of the change in volume and pressure of a substance undergoing an adiabatic change. Also, **adiabatic diagram.**

adiabatic compressibility *Thermodynamics.* an expression of the change in volume of a system resulting from a change in pressure without any heat transfer.

adiabatic compression *Thermodynamics.* a reduction in the volume of a system that does not involve heat transfer either into or out of the system.

adiabatic cooling *Thermodynamics.* a process in which the temperature of a system, or a portion of a system, is lowered without any heat transfer between the system and its surroundings.

adiabatic curing *Engineering.* a process of curing concrete in which there is no loss or gain of heat.

adiabatic curve *Thermodynamics.* the curve representing the relationship of pressure P to volume V on an adiabatic chart.

adiabatic demagnetization *Physics.* a process for cooling a paramagnetic salt to near absolute zero, by placing a salt sample in a strong magnetic field, drawing off the heat generated with a liquid helium bath, and then demagnetizing the sample.

adiabatic efficiency *Mechanical Engineering.* the ratio of the work done by a heat-driven machine to the total amount of heat made available to the machine.

adiabatic ellipse *Fluid Mechanics.* a quadrant of an ellipse that shows domains of compressible flow on the positive velocity, and positive sonic speed plane. This ellipse is arrived at by considering a stream tube in which the flow does not exchange heat with the fluid in neighboring stream tubes. The steady-flow energy equation for the flow in such a tube is then transformed into the equation of an ellipse in velocity and sonic speed.

adiabatic engine *Mechanical Engineering.* an idealized system in which an engine operates without any heat being added or taken away during the operating cycle.

adiabatic envelope *Thermodynamics.* a surface surrounding a thermodynamic system across which there is no heat transfer; disturbances to the envelope can only be made by long-range forces or motion of part of the envelope.

adiabatic equilibrium *Meteorology.* a vertical distribution of temperature and pressure in an atmosphere so that an air parcel displaced adiabatically will have the same temperature and pressure as its surroundings. Also, CONVECTIVE EQUILIBRIUM.

adiabatic expansion *Thermodynamics.* an increase in the volume of a system that does not involve any transfer of heat either into or out of the system.

adiabatic extrusion *Engineering.* the formation of rods, tubes, or sections of objects by forcing plastic through a shaped orifice and without the use of heat.

adiabatic flame temperature *Thermodynamics.* the temperature of the products in a combustion process that takes place with no heat transfer and no energy exchange; this is the maximum possible temperature for these products.

adiabatic flow *Fluid Mechanics.* a flow in which it can be assumed that the rate of energy transfer both into and out of the control volume is effectively zero.

adiabatic invariant *Physics.* a physical quantity that remains unchanged (to a reasonable approximation) while the system containing it undergoes spatial or temporal change.

adiabatic lapse rate see ADIABATIC RATE.

adiabatic law *Physics.* a relationship of the pressure and volume of a gas as it is subjected to an adiabatic transformation: $PV^g = \text{constant}$, where P is the pressure, V is the volume, and g is the ratio of the specific heats C_P/C_V. Also, **adiabatic equation.**

adiabatic modulus *Materials Science.* the Young modulus of a material measured at high deformation rates, such that modulus-modifying heat flow cannot occur.

adiabatic process *Thermodynamics.* any thermodynamic process in which there is no heat transfer between a system and its surrounding environment.

adiabatic rate *Meteorology.* the rate at which temperature decreases as a mass of air rises, or increases as the air falls. Also, **adiabatic gradient.**

adiabatic recovery temperature *Fluid Mechanics.* **1.** the temperature attained through the movement of fluid when it ceases to flow through an adiabatic procedure. Also, RECOVERY TEMPERATURE. **2.** in an adiabatic Carnot cycle, the beginning and ending temperatures.

adiabatic saturation pressure see CONDENSATION PRESSURE.

adiabatic shear *Materials Science.* a fracture process in metals whereby the heat associated with localized deformation encourages further deformation, leading to failure of the material. Similarly, **adiabatic softening.**

adiabatic system *Thermodynamics.* any system in which change can occur without any heat transfer with the environment, or with no appreciable gain or loss.

adiabatic theory *Astronomy.* a theory that superclusters of galaxies originally formed from flat, pancake-like primordial gas structures, which broke down into clusters due to internal shock waves and then condensed into individual galaxies.

adiabatic wall *Thermodynamics.* a boundary of a system that prevents any exchange of heat between the system and its surrounding environment.

adiabatic wall temperature *Fluid Mechanics.* **1.** a temperature that is used in place of the fluid freestream temperature in heat flux calculations for heat transfer between a wall and the stream flow at velocities approaching or exceeding the velocity of sound; this is generally used for applications such as high-speed aircraft, reentry vehicles, and missiles. **2.** the temperature that is assumed for a wall in a moving fluid stream in which no transfer of heat takes place between the stream and the wall.

adiactinic [a´dē ak tin´ik] *Optics.* describing a substance that does not transmit actinic light rays.

adiadochokinesia [ə dē´ə dō´kō kə nē´zhə] *Medicine.* the inability to perform rapid alternating movements with any accuracy, often symptomatic of cerebellar disease. Also, **adiadochokinesis.**

adiadokokinesia or **adiadokokinesis** see ADIADOCHOKINESIA.

adiagnostic *Petrology.* referring to a rock's texture, especially that of igneous rock, wherein individual components cannot be identified, even with the aid of a microscope.

Adiantaceae *Botany.* a large family of terrestrial ferns of the order Filicales, characterized by hairy or scaly creeping rhizomes and sori on the lower leaf surface along the end of vein lines; many of the ferns of dry regions belong to this family.

adiantum *Botany.* any of various ferns of the genus *Adiantum,* popularly called the maidenhair ferns.

adiaspiromycosis [ad´ē ə spēr´ ə mī kō´sis] *Pathology.* a noninfectious pulmonary disease that primarily affects animals and rarely affects humans; it is caused by inhaling spores from fungi of the genus Emmonsia.

adiaspore [ad´ē ə spôr´] *Bacteriology.* a spore produced by the soil fungus *Emmonsia parva* or *E. crescens,* the source of adiaspiromycosis.

adiathermancy [ad´ē ə thur´mən sē] *Physics.* the condition of being unaffected by heat waves. Also, **adiathermance.**

adiathermanous *Physics.* relating to the inability to transmit radiant heat. Also, **adiathermic.**

adiathermic [ad´ē ə thur´mik] *Physics.* not affected by heat; impervious to heat waves. Also, **adiathermanous.**

Adie-Holmes syndrome see ADIE'S SYNDROME.

Adie's syndrome [ā´dēz] *Medicine.* a condition marked by a tonic pupil **(Adie's pupil)** that responds slowly or not at all to light stimulus, and that changes slowly from near to distant vision; the condition sometimes includes the absence of reflexes in some of the tendons. (From W.J. *Adie,* 1886–1935, British neurologist.)

Adimeridae *Invertebrate Zoology.* a family of cylindrical bark beetles in the order Coleoptera. Also, COLYDIIDAE.

ad infinitum [ad´in fə nī´təm] *Mathematics.* to infinity; continuing without end. (A Latin phrase meaning "to the end.")

adinole [ad´i nōl] *Geology.* a clayey sediment that has undergone albitization as a result of contact metamorphism at the margin of a mafic intrusion.

adip- a combining form meaning "fat" or "fatty tissue."

adipate [ad´i pāt] *Organic Chemistry.* the salt that results from the reaction of adipic acid and a basic compound; an ester of adipic acid.

adipectomy see LIPECTOMY.

adiphenine [ə dif´ə nēn] *Pharmacology.* an anticholinergic smooth muscle relaxant formerly used to treat spasms of the gastrointestinal and genitourinary tracts.

adipic [ə dip´ik] *Biology.* of or relating to fatty tissue; adipose.

adipic acid *Organic Chemistry.* $COOH(CH_2)_4COOH$, a white crystalline solid that is soluble in alcohol and acetone and slightly soluble in water; melts at 152°C and boils at 337°C; used in nylon manufacture and polyurethane foams, and for various other purposes. Also, ADIPINIC ACID, HEXANEDOIC ACID.

adipinic acid see ADIPIC ACID.

adipo- a combining form meaning "fat" or "fatty tissue."

adipocellulose *Biochemistry.* a kind of cellulose that consists of fat and connective tissue, as is typically found in the cell walls of cork tissue.

adipocere [ad´ə pə sir] *Medicine.* a fatty, waxy substance found in the tissues of dead animals and humans and believed to be produced by the conversion of the proteins of these tissues into fat; it is found especially in moist places where the combination of favorable temperatures, moisture levels, and the lack of air promote the development of the substance.

adipocerite or **adipocire** see HATCHETTITE.

adipocyte [ad´i pə sīt´] *Cell Biology.* a large cell that is specialized for fat storage.

adipogenesis [ad´i pə jen´ə sis] *Physiology.* the formation of fat or fatty tissue. Thus, **adipogenic.**

adiponecrosis [ad´ə pō nə krō´sis] *Medicine.* necrosis (localized cell death) of fatty tissue in the body.

adiponecrosis (subcutaneous) neonatorum *Medicine.* a condition that occurs in otherwise healthy newborn infants, typically following a difficult labor, and that is marked by patchy subcutaneous fatty indurations and discolorations of the skin.

adiponitrile [ad´ə pō nī´trəl] *Organic Chemistry.* $(CH_2)_4(CN)_2$, a colorless to white, odorless, toxic liquid that is slightly soluble in water and soluble in alcohol; boils at 295°C; used in organic synthesis and as an intermediate in nylon manufacture.

adipose [ad´ə pōs] *Biology.* consisting of, relating to, or derived from fat; fatty. *Histology.* the animal fat stored in the cells of fatty tissue.

adipose fin *Vertebrate Zoology.* a small fatty fin posterior to the dorsal fin on some fishes, such as salmon and catfish.

adipose tissue *Histology.* a type of connective tissue that contains significant deposits of fat.

adiposis *Medicine.* an excessive accumulation of fat in the body.

adiposis dolorosa *Medicine.* a condition that is usually found in menopausal women, marked by deposits of symmetrical nodules and pendulous masses of fat in various areas of the body and accompanied by slight to moderate pain. Also, DERCUM'S DISEASE.

adiposogenital dystrophy [ə dip´sō jen´i təl] *Medicine.* a condition marked by excessive fat in the body, impaired development of the genital organs, and altered secondary sex characteristics. Also, FROHLICH'S SYNDROME.

adipsia [ə dip´sē ə] *Medicine.* the abnormal absence of thirst.

***a* direction** see *A* AXIS.

Adirondacks *Geography.* a mountain range in northeastern New York; highest peak: Mt. Marcy (5344 ft).

A display *Electronics.* a type of radar display that has its target signal deflection on a base perpendicular to that of the time base. The distance to the target is indicated by the horizontal position of the deflection, and the signal intensity is indicated by the vertical amplitude of the deflection. Also, A INDICATOR, A SCAN, A SCOPE.

adit *Mining Engineering.* a level, or nearly level, access passage from the surface used to excavate or drain the main tunnel.

aditus [ad´i təs] *Anatomy.* the entrance or channel leading to an organ or part.

ADIZ air defense identification zone.

adjacency *Computer Programming.* a situation in which modules execute in sequence, one right after the other. *Computer Technology.* a situation in character recognition in which two consecutive print characters are separated by less than the normal or desirable distance.

adjacency matrix *Mathematics.* for a (directed) graph with n distinct vertices $\{v_1, v_2, \ldots, v_n\}$, the $n \times n$ matrix (m_{ij}) where m_{ij} equals the number of edges from v_i to v_j.

adjacency structure *Mathematics.* a characterization of a graph by listing all vertices adjacent to each particular vertex. One format is the adjacency matrix of the graph.

adjacent *Mathematics.* of or relating to two vertices having a common edge, or two distinct edges having at least one common vertex. *Computer Technology.* of characters, showing adjacency.

adjacent angle *Mathematics.* either of two angles having a common vertex and lying on opposite sides of a common side.

adjacent channel *Telecommunications.* the channel with a frequency immediately above or below the desired signal.

adjacent-channel interference *Telecommunications.* extraneous power from a signal in an adjacent channel.

adjacent-channel selectivity *Electronics.* the process by which a receiver rejects signals that are one channel lower or higher than the frequency of the desired signal.

adjacent reentry *Materials Science.* the regular folding of chains in polymeric crystallization, in which they uniformly fold in upon themselves.

adjective dye *Chemistry.* a term for any dye for textile fibers that requires a mordant.

adjoint *Mathematics.* **1.** any operation $*$ defined on a collection of linear transformations that obeys the rule $(AB)^* = B^*A^*$ for all A, B on which $*$ is defined. **2.** see ADJOINT OF A MATRIX. **3.** see ADJOINT OPERATOR.

adjoint of a matrix *Mathematics.* the transpose of the matrix formed from a given matrix M by replacing each entry in M with its cofactor.

adjoint operator *Mathematics.* an operator A^* that is associated with a given linear operator A on a Hilbert space with inner product (,) such that $(A(x), y) = (x, A^*(y))$ for all elements x, y in the Hilbert space. A^* is said to be (Hermitian) adjoint to A. Also, (HERMITIAN) CONJUGATE OPERATOR, ASSOCIATE OPERATOR.

adjoint variable *Physics.* the canonically conjugate position and momentum, interpreted as generalized momenta in a dynamic classical system.

adjoint vector space see DUAL VECTOR SPACE.

adjoint wavefunction *Quantum Mechanics.* in Dirac's electron theory, the wavefunction that results from the application of the Dirac matrix b to the complex conjugate of the original function.

adjugate of a matrix see ADJOINT OF A MATRIX.

adjustable *Mechanical Devices.* describing a tool or device that is designed to be adjusted to various settings or configurations; used in many compound terms, such as **adjustable wrench, adjustable pliers, adjustable choke,** and so on.

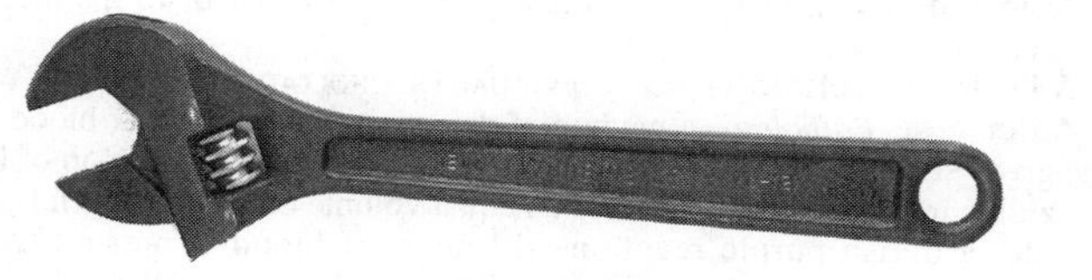

adjustable wrench

adjustable die *Mechanical Devices.* a die that is spring-tempered to allow flexibility when used in various sizes and shapes of specified components.

adjustable parallels *Engineering.* a device consisting of two wedge-shaped iron bars situated so that the thick end of each bar is positioned next to the thin end of the counterpart bar; the distance between the top face of one bar and the bottom face of the other bar may be adjusted, but the faces remain parallel.

adjustable propeller *Naval Architecture.* a screw propeller fitted with blades whose pitch angle can be adjusted to provide maximum efficiency at all speeds. Also, **adjustable-pitch propeller.**

adjustable resistor *Electricity.* a resistor with a mechanically changeable resistance. See POTENTIOMETER, REOSTAT.

adjustable square *Mechanical Devices.* a square with a sliding blade, used in L-shape or T-shape measuring or ruling operations. Also, DOUBLE SQUARE.

adjusted decibels *Electronics.* an expression of the decibel noise level in relation to some given reference level; e.g., –85 dBm referenced to one milliwatt. The measurement is usually done with a device that reads zero at the reference decibel level.

adjusted stream *Hydrology.* a stream that flows parallel to the strike of the least resistant rocks through which it cuts.

adjuster neuron *Anatomy.* any nerve supplying muscles that fit parts together.

adjusting point *Ordnance.* a physical feature or other point at or near the center of the target, used as the specific spot toward which fire at the target should be directed.

adjusting ring *Ordnance.* a device that is used to adjust the setting for an explosive fuse.

adjustive *Psychology.* relating to or contributing to adjustment.

adjustment the fact or process of adjusting; specific uses include: *Psychology.* the process by which behavior is altered in order to bring about a harmonious relationship with the environment. *Metrology.* the process of setting or compensating a measuring instrument so that the value indicated is as close as possible to the actual value. *Engineering.* in surveying, the practice of adjusting each observation in a series so that it becomes consistent with the others.

adjustment mechanism *Psychology.* a habitual or fairly permanent form of behavior that is adjustive.

adjustment of fire see FIRE FOR ADJUSTMENT.

adjustment reaction *Psychology.* a transient personality disorder that occurs in reaction to a stressful situation, such as divorce or business reverses. Also, **adjustment disorder.**

adjutage [aj′ə täzh] *Engineering.* a short tube placed at the opening of a container of liquid to aid or adjust the outflow of its contents.

adjutant [aj′ə tənt] *Military Science.* an administrative officer serving on the staff of a superior officer. *Vertebrate Zoology.* see ADJUTANT STORK.

adjutant stork *Vertebrate Zoology.* the largest stork, genus *Leptopilus,* including the East Asian species *L. dubius* and *L. javanicus* and the African species *L. crumeniferus.* Also, **adjutant bird.** (From its stiff, march-like style of walking.)

adjuvant [aj′ə vənt] *Medicine.* assisting or aiding in some treatment, remedy, or response. *Pharmacology.* a substance that, when added to a medication or other formulation, aids its action or otherwise affects it in some predictable way. *Immunology.* a substance capable of enhancing an immune response to an antigen.

ADK adenylate kinase.

adk gene *Genetics.* the genetic locus that encodes adenylate kinase. Also, DNAW GENE.

ADL *Medicine.* activities of daily living, a description of the routine activities and skills involved in carrying out a normal daily existence; used in evaluating the degree of care or intervention required by an outpatient, nursing home resident, or the like.

Adler, Alfred 1870–1937, Austrian psychiatrist; established the school of individual psychology.

Adlerian *Psychology.* of or relating to Alfred Adler or to his theories and techniques.

Adlerian psychology see INDIVIDUAL PSYCHOLOGY.

Adler test *Pathology.* a method of detecting the presence of blood in a given substance by adding to a 1-ml sample a saturated solution of benzidine in glacial acetic acid and an equal volume of 3% hydrogen peroxide; a bluish purple reaction will occur if blood is present. Also, BENZIDINE TEST. (From Oscar *Adler,* 1879–1932, and his brother Rudolph, 1882–1952, German physicians.)

ad lib or **ad lib.** *Medicine.* without restriction; freely; used as an instruction in prescriptions. (From Latin *ad libitum,* "at pleasure.")

adlittoral *Oceanography.* of or relating to shallow waters near the shore.

Adm. or **ADM** admiral.

administrative *Military Science.* relating to an item or operation that is not directly connected with combat or training. Thus, **administrative vehicle, administrative use, administrative movement, administrative unit,** and so on.

administrative map *Cartography.* **1.** a map containing administrative information relating to a tactical situation, such as evacuation and personnel sites, medical facilities, service areas, supply routes, traffic circulation, and boundaries. **2.** any map showing political and geographical boundaries of a country or countries.

administrative march *Military Science.* a movement of troops and vehicles made when no enemy attack or interference is expected.

Adminomonadaceae *Botany.* a monotypic marine planktonic family of unicellular algae of the order Desmocapsales, characterized by an apically inserted flagella.

admiral *Military Science.* **1.** a naval officer in command of a fleet of ships. **2.** the highest rank for an officer in the U.S. Navy and other navies. **3.** an admiral's flag, or the ship flying such a flag. *Invertebrate Zoology.* any of various large, brightly colored butterflies with reduced forelegs, of the genera *Vanessa* or *Limenitis,* including the well-known North American species *V. atalanta,* the **red admiral,** and *L. arthemis arthemis,* the **white admiral.**

admiralty *Military Science.* **1.** the rank or position of an admiral. **2.** a group or department in charge of naval affairs for a country.

admiralty brass *Metallurgy.* a copper-base alloy containing approximately 28% zinc and 1% tin. The arsenical variety also contains 0.02 to 0.60% arsenic; the antimonial and phosphorized varieties contain 0.02 to 0.10% antimony and phosphorus, respectively.

admiralty coal *Mining Engineering.* a high-grade, smokeless coal formerly used in steam-powered naval vessels.

admiralty constant *Naval Architecture.* the horsepower required to maintain a given speed by a vessel of a given displacement and midship section.

admissibility *Statistics.* in classical decision theory, the characteristic of a statistical procedure when no other procedure entails a risk that is equal to or smaller than that of the procedure in question.

admissible *Mathematics.* exhibiting the particular properties that are necessary for a theorem or procedure to be valid. For example, given a group G with operators, an admissible subgroup is a subgroup of G that admits the same operators. *Artificial Intelligence.* describing a heuristic function that does not overestimate the cost of reaching a goal from a given state.

admittance *Electricity.* the measure of how readily an alternating current flows through a circuit; the reciprocal of impedance, measured in siemens (formerly mhos).

admittance matrix *Electricity.* a matrix Y with elements that are the mutual admittances between the various meshes of an electrical network, satisfying the matrix equation $I = YV$.

admixture *Science.* anything added to form a mixture, or the mixture itself. *Materials Science.* an addition to wet concrete that affects properties such as curing time and flow characteristics. *Geology.* any of the subordinate grades of sediment, such as a fine or coarse admixture.

A DNA see A-FORM DNA.

adnate [ad′nāt′] *Biology.* united with or adhering to an organ of a different kind; referring to unlike parts of plants or animals that have grown together. Thus, **adnation.**

adnerval *Neurology.* **1.** directed toward a nerve; said especially of an electric current passing through muscle tissue toward a nerve's entrance point. **2.** located near a nerve. Also, **adneural.**

adnexa [ad neks′ə] *Anatomy.* subordinate or accessory anatomic parts attached to another or others, such as the Fallopian tubes in relation to the ovaries. Thus, **adnexal.**

adnexed *Mycology.* referring to the gill-like part of the mushroom joined to its stem or stalk.

adobe [ə dō′bē] *Geology.* **1.** a mixture of clay material and silt that can be sun-dried into bricks, commonly found in the southwestern United States and Mexico. **2.** clay soil that is derived from such deposits. *Building Engineering* a brick made from this material, or a house made from such bricks. Thus, **adobe brick, adobe construction.** (Derived from an Arabic word meaning "the brick.")

adobe flat *Geology.* a plain of sandy clay or adobe soil, formed by sheetflood deposition.

adolescence *Psychology.* the period of growth from puberty to maturity, lasting from about 12 to about 18 years of age. *Geology.* the stage in the erosion cycle occurring between youth and maturity.

adolescent *Psychology.* **1.** a person in the stage of adolescence. **2.** of or relating to adolescence. *Forestry.* a tree that is not yet mature or ready to be cut and sold for profit; a tree of the size that immediately precedes the marketable size. *Hydrology.* describing a river or stream having a well-cut, smoothly graded channel that may reach base level at its mouth. *Geology.* of or relating to the stage in the erosion cycle between youth and maturity. Thus, **adolescent river, adolescent stream.**

adolescent coast *Geology.* a shoreline marked by low, relatively continuous sea cliffs.

Adonis *Astronomy.* asteroid 2101; discovered in 1936, it measures 0.3 kilometers in diameter, has an eccentric orbit that comes close to the earth, and belongs to the Apollo group of asteroids.

adonitol *Biochemistry.* $C_5H_{12}O_5$, a pentahydric alcohol found in the plant *Adonis vernalis;* when oxidized, it yields the sugar ribose. Also, **adonite.**

adont hinge [ā´dänt´] *Invertebrate Zoology.* a type of ostracod hinge articulation that either lacks teeth and has overlapping valves, or has a ridge and a groove.

adoption *Behavior.* a term for a situation in which adult animals take over the care of young who are not their own offspring.

adoption study *Psychology.* a study that evaluates the inheritability of certain character traits or disorders by comparing the traits of adopted children with the traits of their biological and adoptive parents. Similarly, **adoption research.**

adoptive immunity *Immunology.* a resistance to disease or infection that is produced by the transfer of immunologically active cells from an immune individual to a nonimmune individual. Such cells from a nonimmune individual may be activated in vitro and transferred back to the same individual.

adoptive transfer *Immunology.* an immunity that is transferred by transferring immunocompetent cells from a primed donor to a nonimmune recipient.

adoral [a´dôr´əl] *Zoology.* **1.** near to or relating to the mouth. **2.** the side of an organism on which the mouth is located.

adoral zone of membranelles *Cell Biology.* along one side of the oral region of many protozoa, a series of membranes that are involved in food uptake.

Adoxaceae *Botany.* a monospecific family of delicate, circumboreal, perennial herbs of the order Dipsacales; characterized by trifoliolate leaves, five flowers in a compact cyme, and a dry drupe with stones.

ADP adenosine diphosphate.

ADP or **adp** automatic data processing.

ADPase adenosinediphosphatase.

ADP ribosylation see ADENOSINE DIPHOSPHATE RIBOSYLATION.

adradial canal *Invertebrate Zoology.* one of several gastrovascular canals extending radially and joining the circular canal at the rim of certain jellyfish.

Adrastea [ə dras´tē ə] *Astronomy.* the 15th moon of Jupiter; it was discovered in 1979 and measures 12.5 × 10 × 7.5 kilometers. Also, **Jupiter XV.**

adrenal [ə drē´nəl; ə dren´əl] *Anatomy.* **1.** of or relating to the kidneys. **2.** situated on or near the kidneys. **3.** see ADRENAL GLAND.

adrenal cortex *Anatomy.* the outer portion of the adrenal gland, consisting of columnar masses perpendicular to the surface of the gland; it secretes glucocortoids (e.g., hydrocortisone), mineralocorticoids (e.g., aldosterone), and various other hormones. Thus, **adrenal cortex hormone.**

adrenal-cortical see ADRENOCORTICAL.

adrenal-cortical insufficiency *Medicine.* a condition in which the activity of the adrenal gland is abnormally reduced.

adrenalectomy [ə drēn´ə lek´tə mē] *Surgery.* the surgical removal of an adrenal gland.

adrenal gland *Anatomy.* either one of the two flattened glands situated above each kidney, consisting of a cortex (outer wall) that secretes important steroid hormones and a medulla (inner part) that secretes adrenaline (epinephrine) and noradrenaline (norepinephrine). Also, **adrenal body.**

Adrenalin *Biochemistry.* a trademark for a commercially produced form of adrenaline (epinephrine).

adrenaline or **adrenalin** [ə dren´ə lin] *Biochemistry.* another name for EPINEPHRINE, a hormone secreted by the adrenal medulla that increases the heartbeat, blood pressure, and other bodily functions.

adrenal insufficiency *Medicine.* a condition in which the activity of the adrenal gland is abnormally reduced.

adrenalitis *Medicine.* an inflammation of the adrenal glands. Also, **adrenitis.**

adrenal medulla *Anatomy.* the inner portion of the adrenal gland largely made up of connective tissue and a plexus of veins; it secretes epinephrine and norepinephrine..

adrenalopathy see ADRENOPATHY.

adrenal virilism *Medicine.* the development of secondary male characteristics in a female due to the excessive secretion of androgens from the adrenal gland.

adrenergic [ad´rə nur´jik] *Physiology.* **1.** producing or activated by adrenaline (epinephrine). **2.** relating to nerve fibers that release certain adrenaline substances at their endings, specifically the sympathetic nerves of the autonomic nervous system.

adrenergic amine see CATECHOLAMINE.

adrenergic blocker *Biochemistry.* a compound that blocks the response to certain nerve impulses and thus the transmission of certain adrenaline or adrenalinelike substances; the two types are alpha blockers and beta blockers. Also, **adrenergic blocking agent.**

adrenine or **adrenin** see ADRENALINE.

adrenitis see ADRENALITIS.

adreno- a combining form meaning "relating to the adrenal glands."

adrenochrome *Biochemistry.* $C_9H_9NO_3$, a compound derived from the adrenal gland hormone epinephrine (catecholamine); affects capillary permeability and blood flow.

adrenocortical *Biochemistry.* relating to or coming from the cortex of the adrenal gland. Thus, **adrenocortical hormone.**

adrenocorticosteroid see CORTICOSTEROID.

adrenocorticotrophic see ADRENOCORTICOPTROPIC.

adrenocorticotropic [ə drēn´ō kôr´ti kō träp´ik] *Medicine.* capable of stimulating the adrenal cortex.

adrenocorticotropic hormone *Endocrinology.* ACTH, a peptide hormone that is secreted by the corticotrophs of the anterior pituitary gland, after cleavage from the prohormone product of the pro-opimelanocortin gene. It stimulates the production of glucocorticoids by the adrenal cortex and is a trophic factor for the cortex as well.

adrenocorticotropin see ADRENOCORTICOTROPIC HORMONE.

adrenodoxin *Biochemistry.* a protein produced by the adrenal glands that takes part in the transfer of electrons within animal cells.

adrenogenital [ə drēn´ō jen´ə təl] *Anatomy.* relating to the adrenal glands and the genitalia.

adrenogenitalism *Medicine.* a congenital condition caused by the excessive secretion of hormones from the adrenal cortex, marked by the development of secondary male characteristics in females and the accelerated growth and enlargement of the penis in males. Also, **adrenogenital syndrome.**

adrenokinetic *Physiology.* stimulating the action of the adrenal gland

adrenolytic [ə drēn´ə lit´ik] *Physiology.* **1.** inhibiting the action of the andrenergic nerves, or inhibiting the response to adrenaline (epinephrine). **2.** a drug or other substance that has this effect.

adrenomegaly [ə drēn´ə meg´ə lē] *Medicine.* the abnormal enlargement of one or both of the adrenal glands.

adrenopathy *Medicine.* any disease of the adrenal glands.

adrenotoxin [ə drēn´ə täks´in] *Toxicology.* any substance that is toxic to the adrenal glands.

adrenotropic [ə drēn´ə träp´ik] *Physiology.* of or relating to the action on or stimulation of the adrenal cortex. Also, **aderenotrophic.**

adrenotropic hormone see ACTH.

adret [ə drā´] *Agriculture.* a term for the more sunny side of a mountain or slope. (From a French word meaning "suitable" or "correct," with the idea that this is the better side for growing.)

Adrian, Baron Edgar Douglas 1889–1977, English physiologist; with Sherrington, Nobel Prize for research on neurons.

Adriatic Sea *Geography.* an arm of the Mediterranean separating Italy from Yugoslavia and Albania.

adromia [ə drō´ mē ə] *Neurology.* the absence of impulse conduction in a nerve of a muscle.

ADS antidiuretic substance.

ad sat. to saturation. (From Latin *ad saturandum.*)

adsorb *Chemistry.* to take up and hold another substance on the surface; carry on the process of adsorption. Thus, **adsorbable, adsorbability.**

adsorbate *Chemistry.* something that is adsorbed; material gathered on a surface by adsorption.

adsorbent *Chemistry.* **1.** a material that can hold or condense molecules of another substance on its surface by adsorption. **2.** relating to or capable of adsorption.

adsorption *Chemistry.* the taking up of the molecules from a gas or liquid on the surface of another substance; distinguished from *absorption,* a process where one substance actually penetrates into the inner structure of the other. (See ADSORPTION for illustration.) *Virology.* the adherence of a virus or other substance to particulate matter, such as cells, in solution or suspension.

adsorption chromatography *Analytical Chemistry.* a separation technique in which the components a liquid or a gas (mobile phase) are adsorbed to different extents on the surface of an adsorbent (stationary phase).

adsorption indicator *Analytical Chemistry.* an indicator, such as fluorescein, that shows a color change at the endpoint in a titration.

adsorption isobar *Analytical Chemistry.* a line that shows adsorption variations within a given parameter, such as temperature, while pressure remains constant.

adsorption isotherm *Physical Chemistry.* the relationship between solute concentration in the solid phase and the solute partial pressure or concentration in the fluid phase, while temperature conditions remain constant.

adsorption system *Mechanical Engineering.* a process that is used to purify a substance, by using the physical or chemical bonding that takes place on the surface of a solid or liquid to selectively absorb impurities from it .

ADT *Transportation Engineering.* average daily traffic, the average number of vehicles passing a given point or using a given highway per day.

adtidal [ad tīd′əl] *Ecology.* relating to an organism living immediately below low tide level.

adularescence *Mineralogy.* a milky-white or bluish sheen shown by certain gemstones, usually adularia and moonstones, when they are turned under light.

adularia *Mineralogy.* $KAlSi_3O_8$, a weakly triclinic variety of orthoclase occurring in transparent to white pseudo-orthorhombic crystals and having a specific gravity of 2.56 to 2.57.

adularization *Geology.* the process whereby the mineral adularia is introduced into a sediment or rock, sometimes replacing another mineral.

adult *Biology.* **1.** an organism that has reached full growth and development. **2.** of or relating to this phase of life.

adult phase *Forestry.* the period in the life of a tree during which flowering occurs.

adult rickets see OSTEOMALACIA.

adumbration [ad′əm brā′shən] *Optics.* the fact of being darkened or overshadowed. *Radiology.* the emergence of a shadow that conceals a structure being studied by radiography.

ad us. exter. *Medicine.* for external use. (From Latin *ad usum externum.*)

adv. or **Adv.** against. (From Latin *adversum.*)

advance to move forward or along some route or path; specific uses include: *Military Science.* **1.** a movement forward toward the enemy. **2.** to make such a movement. **3.** of or relating to a location that is closer to the enemy than the main force is. Thus, **advance force, advance element, advance detachment, advance (command) post, advance position.** *Engineering.* to cause some event to occur at an earlier time. *Navigation.* the distance a vessel moves in the original direction after the rudder is put over and before it begins to turn. *Geology.* **1.** the continued movement of a shoreline toward the sea. **2.** the net movement of such a shoreline over a given period of time. *Hydrology.* **1.** the forward, downslope movement of a glacier, which occurs when accumulation exceeds ablation. **2.** a time interval that is delineated by the general expansion of a glacier. *Surgery.* to surgically detach tissue, such as a muscle or tendon, and reattach it at an advanced point, especially as in strabismus surgery.

advance by bounds *Military Science.* an advance by troops consisting of a series of bounds, or separate movements, from one point to another.

advanced *Evolution.* of or relating to a form in a line of evolutionary development that displays a higher degree of complexity and specialization than the earlier forms. *Military Science.* of or relating to a location that is closer to the enemy than the main force is. Thus, **advanced base, advanced depot.**

advanced aluminide *Metallurgy.* an intermetallic compound consisting of aluminum and another metal such as titanium or nickel; primarily used in high-temperature applications.

advanced battery *Electricity.* a battery storage system used to harness solar or wind energy or to save excess electricity during low-demand periods for later use.

Advanced Cruise Missile *Ordnance.* an air-launched cruise missile, officially designated **AGM-129.**

advanced-design array radar *Ordnance.* a radar system that uses two antennas and a dataprocessing center to locate and identify enemy targets.

advanced gallery *Mining Engineering.* in tunnel excavation, a small heading driven before the main tunnel.

advanced gas-cooled reactor *Nucleonics.* a type of graphite-moderated power reactor that uses helium as coolant and graphite as a moderator, operating at substantial temperatures.

advanced line of position *Navigation.* a line of position that has been moved to account for the distance the craft has progressed since the line was established.

advanced potential *Electromagnetism.* an electromagnetic potential comparable to a retarded potential but existing on the future light cone of space-time; this potential has no physical interpretation.

advance growth *Forestry.* a growth of young trees that becomes established naturally, before any regeneration or clear-cutting techniques are used.

advance guard *Military Science.* a smaller body of troops who precede the main force to reconnoiter, clear the way, protect against surprise attacks, and so on. The **advance guard point** makes the farthest advance of this element, followed by the **advance guard support** and then the **advance guard reserve.**

advancement *Surgery.* the severance from its attachment and suture to a point farther forward of a tendon, muscle, or skin flap.

advance of the perihelion *Astronomy.* a slow movement of a planet's perihelion point forward in the same direction as its orbital motion; this movement is caused principally by gravitational perturbations from the other planets.

advance on *Military Science.* to advance toward a point when there is some doubt that this objective can be reached.

advance overburden *Mining Engineering.* an overburden, in open-cut mining, that is in excess of the average overburden-to-ore ratio and must be removed.

advance party *Military Science.* a part of an advance guard that is sent out ahead of the rest of the element.

advance signal *Transportation Engineering.* in a block system, a signal up to which a train is authorized to proceed within a block that is not completely clear.

advance slope grouting *Building Engineering.* in the placement of grouting, horizontal movement of the grout that is forced through set aggregate.

advance slope method *Building Engineering.* a way of placing concrete so that its fresh face slopes forward as it is laid.

advance stripping *Mining Engineering.* a process used to permit a minable grade of ore to be mined by removing barren or subore-grade earthy or rock materials.

advance to *Military Science.* to advance toward a point when it is reasonably certain that this objective can be reached.

advance wave *Mining Engineering.* the air pressure wave that occurs prior to the flame in a coal-dust explosion, resulting in dust suspension and violent eddies.

advancing *Mining Engineering.* the process of mining outward from the shaft toward the boundary.

advancing fire see ASSAULT FIRE.

advancing longwall *Mining Engineering.* the process of mining coal outward from the shaft pillar and maintaining roadways through the worked-out portion of the mine.

advect to undergo the process of advection.

advection *Meteorology.* **1.** the transfer of heat, cold, or other atmospheric properties by the horizontal motion of a mass of air. **2.** the rate of change of the value of an advected property, such as fog, at a given point. *Oceanography.* the generally horizontal mixing of bodies of water or air, on a large scale.

advectional inversion *Meteorology.* a departure from the usual decrease of temperature with increasing altitude, caused by advection.

advection fog *Meteorology.* a type of fog formed when warm, moist air moves horizontally over a cold surface and the air is consequently cooled to below its dew point; this is found especially along a coastline where the temperature of land and the temperature of water markedly differ.

advective *Meteorology.* relating to or caused by advection.

advective hypothesis *Meteorology.* an assumption that horizontal or atmospheric pressure changes are directly responsible for local temperature changes.

advective model *Meteorology.* a theoretical representation of an atmospheric property (typically of vertical temperature distribution), transported by the mass motion of the atmosphere.

advective thunderstorm *Meteorology.* a thunderstorm produced as a result of static instability during advection of relatively colder air at high altitudes, or relatively warmer air at lower altitudes, or by a combination of these conditions.

adventitia [ad´ven tish´ə] *Anatomy.* a membrane that covers an organ but is not part of that organ. Thus, **adventitial.**

adventitious *Biology.* not in the usual order or place; not natural or hereditary; e.g., roots that form on stems, a growth of hair where it usually does not grow, or the growth of a plant in a foreign habitat.

adventitious bud *Botany.* a leaf bud that occurs in an unusual position, such as on leaves or roots.

adventitious deafness *Medicine.* a loss of hearing that is either accidental or acquired, rather than congenital.

adventitious embryo *Medicine.* an embryo occurring outside the uterus.

adventitious reinforcement see ACCIDENTAL REINFORCEMENT.

adventitious root *Botany.* a root that occurs in an unusual position, such as on stems or leaves.

adventitious vein *Invertebrate Zoology.* a vessel appearing irregularly between the intercalary and accessory veins on certain insect wings.

adventitious virus *Virology.* a foreign virus that has invaded or established itself in another virus preparation or vaccine.

adventive *Biology.* **1.** an organism artificially or accidentally introduced into an environment where it is not native. **2.** not in the usual order or place; not natural or hereditary. *Geology.* see PARASITIC.

adventive cone *Volcanology.* a volcanic cone on the flank of, and subsidiary to, a larger cone.

adventive crater *Volcanology.* a crater on the flank of a large volcanic cone.

adverse *Botany.* of a leaf, turning inward toward the stem.

advertisement see DISPLAY BEHAVIOR.

advertising dress *Behavior.* a variation in appearance found in one sex of a species that serves to produce a response from the opposite sex, such as brighter coloration or a more conspicuous form.

advisory *Meteorology.* a report or bulletin providing information to the public on weather conditions, especially hazardous storm conditions. *Transportation Engineering.* a report providing flight or traffic safety information.

advisory area *Navigation.* the area within ten miles of an airport which has a flight service station, but not an operable control tower.

advolution *Biology.* growth or development that tends toward increasing similarity.

adynamia [a´də nā´mī ə] *Neurology.* a lack or loss of normal vital powers, such as extreme muscular weakness due to disease. Thus, **adynamic.**

adz *Mechanical Devices.* a manual cutting tool with a thin arched blade at right angles to the handle, used (especially in former times) for shaping timber and other wood applications. Also, **adze.**

adz block *Mechanical Engineering.* the section of a wood-planing machine that holds the cutters.

Ae *Metallurgy.* the temperature at which a solid-state transformation in iron or steel occurs at equilibrium; designated by various subscript numbers; e.g., **Ae_1**, the temperature at which the face-centered cubic phase (γ) forms under equilibrium conditions.

AE or **A.E.** **1.** Associate in Engineering. **2.** Agricultural Engineer.

AEC or **A.E.C.** Atomic Energy Commission.

Aechminidae *Paleontology.* a family of ostracods in the extinct order Palaeocopida and suborder Beyrichicopina; characterized by sulcated, lobate shells with a single horn-like dorsal spine; extant in the Ordovician to Carboniferous.

aecidioid [ēsh´ə doid´] *Mycology.* referring to a cluster-cup type of spore formed by a fruiting body called an aecium, which occurs in rust fungi.

aeciospore [ēsh´ē ə spôr] *Mycology.* a cylindrical or cup-shaped aecium (fruiting body) formed by rust fungi of the order Uredenales. Also, **aecidiospore.**

aecium [ēsh´ē əm] plural, **aecia.** *Mycology.* a cup-shaped fruiting body produced by rust fungi. Also, **aecidium.**

aedeagus *Invertebrate Zoology.* the chitinous copulatory organ of male insects; formed from a fingerlike evagination of the ventral body wall that encloses the terminal section of the ejaculatory duct.

Aedes [ā ē´dēz] *Invertebrate Zoology.* **1.** a large genus of mosquitoes, in the family Culicidae, including various species that carry diseases of humans and animals, such as yellow fever and dengue. **2. aedes.** any mosquito of this genus, such as *A. aegypti* or *A. africanus,* transmitters of yellow fever.

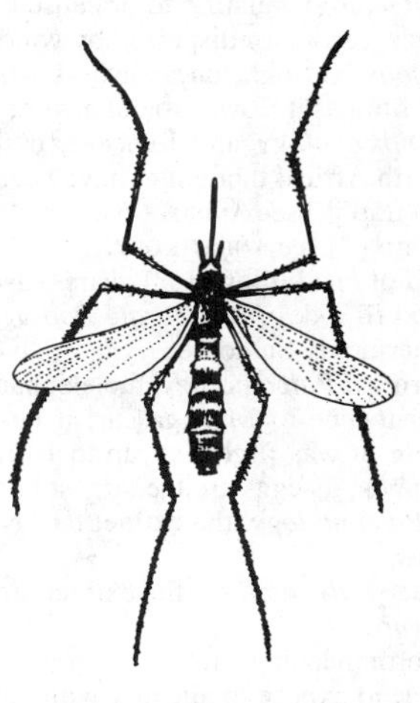

Aedes

Aedes cells *Virology.* cell lines that are established from mosquitoes of the genus *Aedes,* used to support the replication of a number of arboviruses.

Aeduellidae [ē´dü el´ə dī] *Paleontology.* a monogeneric family of primitive chondrostean fishes in the extinct suborder Palaeoniscoidea; possibly ancestral to higher neopterygian fish; lower Permian.

Aegean Sea *Geography.* an arm of the Mediterranean between mainland Greece, Crete, and Turkey.

Aegeriidae [ē´jə rē´ə dē] *Invertebrate Zoology.* a family of clear winged moths, in the suborder Heteroneura, that lack wing scales and resemble wasps and bees.

Aegialitidae [ē´jyə lid´ə dē] *Invertebrate Zoology.* a family of narrow-waisted bark beetles in the order Coleoptera. Also, SALPINGIDAE.

Aegidae [ē´jə dē] *Invertebrate Zoology.* a family of isopod crustaceans in the suborder Flabellifera with flat broad bodies; many are fish parasites.

aegirine *Mineralogy.* $NaFe^{+3}Si_2O_6$, a dark green to brown monoclinic mineral of the pyroxene group occurring as long, prismatic, often polysynthetic crystals having characteristic acute to blunt terminations, with a specific gravity of 3.50 to 3.55 and a hardness of 6 on the Mohs scale. Also, ACMITE, AEGIRITE.

aegirite see AEGIRINE.

Aegis [ē´jəs] *Ordnance.* **1.** a U.S. naval antiaircraft system with phased-array radar, computerized fire control, missile radar, and surface-to-air missiles. **2.** a cruiser or other ship equipped with such a system. (From the name of a shield in Greek mythology that protected Zeus and Athena.)

aegithognathous [ē´gə thäg nā´thəs] *Vertebrate Zoology.* a term used in classification to refer to birds with a palate in which maxillopalatine bones do not unite with each other or with the vomer, and in which the vomer is truncate in appearance. More than half of all living birds belong to this group.

Aegothelidae [ē´gə thel´ə dē] *Vertebrate Zoology.* a small family of birds belonging to the order Caprimulgiformes and including the owlet-nightjars or owlet-frog mouths of Australia and New Guinea.

Aegypiinae [ē´jə pī´ə nē] *Vertebrate Zoology.* the Old World vultures, a subfamily of large, strong scavengers in the family Accipitridae.

Aegyptianella [ə jip´shə nel´ə] *Bacteriology.* a genus of coccoid, Gram-negative bacteria of the family Anaplasmataceae, the sole species of which is a tick-borne, obligate parasite, infecting the erythrocytes of domestic and wild birds.

Aegyptopithecus [ə jip´tō pith´ə kəs] *Anthropology.* an Oligocene fossil primate found in the Fayum Beds of North Africa; thought to have been an ancestor of living African apes and of humans; probably arboreal and herbivorous, with dimorphic dentition.

Aelosomatidae *Invertebrate Zoology.* a family of microscopic fresh-water annelid worms, in the class Oligochaeta, with a ventrally ciliated prostomium and chaetae in four bundles on each segment.

aenigmatite *Mineralogy.* $Na_2Fe_5^{+2}TiSi_6O_{20}$, a black, triclinic mineral of the aenigmatite group, occurring as long prismatic crystals, and having a specific gravity of 3.74 to 3.85 and a hardness of 5.5 to 6 on the Mohs scale. Also, ENIGMATITE.

aeolian [ē´ōl ē ən] *Geology.* relating to or caused by the action of wind. *Agriculture.* of seeds, carried or dispersed by wind action. Also, EOLIAN.

aeolian tones *Acoustics.* high tones such as whistles or howls due to eddies formed in a fluid as it flows around a wire, cable, or pole.

Aeolopithecus *Anthropology.* an Oligocene fossil primate found in the Fayum Beds of North Africa; thought to have been ancestral to gibbons.

aeolotropic [ē´ə lō träp´ik] see ANISOTROPIC.

aeolotropy [ē´ə lä´trə pē] see ANISOTROPY.

aeon [ē´än] a period of one billion (10^9) years. Also, EON.

Aepophilidae [ē´pō fil´ə dē] *Invertebrate Zoology.* a family of bugs in the hemipteran superfamily Saldoidea.

Aepyornis [ē´pē ôrn´əs] *Paleontology.* the elephant bird, a genus known only from the Pleistocene of Madagascar; it became extinct after humans migrated there. It was flightless, up to 10 feet high, and weighed as much as 100 pounds; its eggs are the largest known bird eggs.

Aepyornithidae *Paleontology.* the extinct family of birds that is represented by *Aepyornis.*

Aepyornithiformes *Paleontology.* the extinct order of birds that is represented by *Aepyornis.*

aer- a combining form meaning "air."

aerate *Chemistry.* **1.** to expose to air; mix with air. **2.** to mix with a gas, such as carbon dioxide. *Physiology.* to expose to oxygen; oxygenate. *Agronomy.* to allow or cause air to enter the soil for the purpose of improving plant growth.

aerated flow *Physical Chemistry.* a flow of liquid that contains fine bubbles of air or gas.

aerating agent *Food Technology.* any gas used to carbonate beverages or to dispense liquids from charged containers.

A/E ratio see ABSORPTIVITY-EMISSIVITY RATIO.

aeration the process of exposing something to air or another gas; specific uses include: *Chemistry.* **1.** exposure to the chemical action of air. **2.** the passing of air through a substance, especially through a liquid. *Physiology.* the introduction of oxygen, especially the exchange of carbon dioxide for oxygen by the blood in the lungs. *Agronomy.* the process of allowing or causing air to enter the soil. *Food Technology.* a process used to make a liquid effervescent by charging or combining it with a gas such as carbon dioxide. *Engineering.* in air conditioning, the process of cooling or mixing by circulating air or through ventilation. *Mining Engineering.* a method of forming air bubbles in a flotation cell by introducing air into the pulp.

aeration cell *Physical Chemistry.* a device that generates electromotive force across electrodes that are made of the same material but located in different concentrations of dissolved air. Also, OXYGEN CELL.

aeration number *Biotechnology.* the mathematical ratio between gasflow rates and impeller speed and diameter in fermenters; used to analyze aeration-agitation data.

aeration tank *Engineering.* **1.** a tank holding fluid that is circulated or sprayed and thus exposed to the atmosphere. **2.** a tank designed to force air or gas through its fluid contents.

aerator *Mechanical Devices.* any of various devices used to saturate a substance with air or another gas. *Agronomy.* a device used to aerate soil.

aerenchyma [a´renk´ə mə] *Botany.* a type of tissue found in certain plants, especially aquatic plants, characterized by thin-walled cells and particularly large intercellular spaces; involved in internal gas exchange.

aerial relating to or in the air; specific uses include: *Electronics.* a radio or television antenna; i.e., a device of wires or metal rods that receives or transmits electromagnetic waves. In technical use *antenna* is the preferred term. *Botany.* **1.** a plant that lives above the surface of the ground or water. **2.** relating to or describing such a plant. *Military Science.* of or relating to air warfare or to operations involving aircraft. Thus, **aerial observation, aerial bombardment.** *Ordnance.* of a weapon or missile, launched, dropped, or fired from an aircraft. Thus, **aerial bomb, aerial mine, aerial torpedo, aerial dart,** and so on.

aerial archaeology *Archaeology.* the study of archaeological sites by observation or photography from the air.

aerial burst see AIRBURST.

aerial camera *Optics.* a camera devised for use in aircraft or space vehicles, designed to compensate for motion and to take high-resolution photographs. Thus, **aerial film.**

aerial cannon see AIRCRAFT CANNON.

aerial dusting see CROP DUSTING.

aerial exposure index *Photogrammetry.* the reciprocal of twice the exposure of the film, measured in meter-candle-seconds, at the point on the characteristic curve where the slope equals 0.6 gamma.

aerial ladder *Mechanical Devices.* a long extension ladder used especially in firefighting.

aerial logging *Forestry.* a system of transporting timber using helicopters or balloons to lift the logs.

aerial mapping *Cartography.* the making of plane and contoured maps and charts on the basis of ground surface photographs taken from an aircraft or spacecraft. Also, AEROCARTOGRAPHY.

aerial mosaic *Photogrammetry.* a group of aerial photographs whose edges have been cut and matched in such a way as to create a continuous view or representation of an area.

aerial mycelium *Mycology.* a mass of hyphae that extends above the surface of a medium or substrate.

aerial perspective *Optics.* an optical illusion produced by the atmospheric diffusion of light, in which distant objects appear less distinctive and are lighter in tone than those closer to the observer; appears commonly in landscape photography. *Graphic Arts.* a technique used to achieve this effect of depth and distance in a painting or other two-dimensional medium.

aerial photogrammetry *Photogrammetry.* the use of aerial photographs to aid in surveying and in making accurate measurements. Also, **aerial surveying.**

aerial photography *Aviation.* photographs of the earth's surface as taken from an aircraft or spacecraft, satellite, or rocket. Also, AEROPHOTOGRAPHY.

aerial reconnaissance *Military Science.* the use of aerial photography, as well as other data gathered from airborne sources, for the purpose of surveying a military objective or otherwise providing military intelligence. Also, **aerial photoreconnaissance, aerial photographic reconnaissance.**

aerial root *Botany.* a root growing partly or entirely above the ground; e.g., the roots of orchids. Similarly, **aerial stem.**

aerial sound ranging *Aviation.* the process of one aircraft determining another aircraft's location from the air by measuring the sound waves it emits.

aerial spud *Mechanical Engineering.* a cable that is used to move and anchor a dredge.

aerial survey *Engineering.* aerial photographs or other data, such as electronic information, gathered from an airborne source to aid in making topographical maps, charts, plans, and surveys.

aerial tramway or **aerial cableway** see TRAMWAY.

aerial triangulation see PHOTOTRIANGULATION.

aeriferous [âr if´ə ris] carrying or conveying air; e.g., the bronchial tubes.

aeriform [âr´ə fôrm´] relating to air; having the form of air.

aero- a combining form meaning "air."

aeroacoustics *Acoustics.* the study of sound transmission through the air, especially in terms of the effects of environmental noise from machinery, vehicles, aircraft, and so on.

aeroallergen [âr´ō al´ər jin] *Medicine.* any airborne substance that will cause the symptoms of an allergy, such as pollens, dust, smoke, fungi, or perfumes.

Aerobacter [âr´ō bak´tər] *Bacteriology.* a former genus of Gram-negative, bacillary bacteria belonging to the family Enterobacteriaceae, the species of which have now been reclassified into the genera *Enterobacter* and *Klebsiella.*

aeroballistics *Mechanics.* the study of the motion and behavior of projectiles and high-speed vehicles in the earth's atmosphere.

aerobe [âr´ōb] *Biology.* any organism that requires atmospheric oxygen to live, especially aerobic bacteria or other microorganisms.

aerobic [âr ō´bik] *Biology.* **1.** requiring atmospheric oxygen to live. **2.** relating to or occurring in the presence of oxygen. *Bacteriology.* relating to or caused by aerobic bacteria. *Medicine.* requiring supplemental oxygen for respiration. *Physiology.* relating to or being a form of aerobic exercise. Thus, **aerobic walking, aerobic dancing,** and so on.

aerobic-anaerobic *Biology.* relating to or involving both aerobic and anaerobic microganisms. *Biotechnology.* relating to the process of treating sewage material by means of both aerobic and anaerobic microorganisms.

aerobic-anaerobic interface *Biotechnology.* the point at which both aerobic and anaerobic microorganisms participate in a body of sewage sludge or compost; at this point the decomposition of the material goes no further.

aerobic-anaerobic lagoon *Biotechnology.* a pond containing partially treated or raw waste water where both aerobic and anaerobic stabilization occurs.

aerobic bacteria *Microbiology.* any bacteria that require oxygen for growth and are dependent on a respiratory metabolism to generate energy, with molecular oxygen usually serving as the terminal electron acceptor.
aerobic digestion *Microbiology.* the digestion of waste matter by aerobic microorganisms.
aerobic exercise *Physiology.* any of various forms of physical exercise intended to produce increased heart and lung activity and thus promote the body's use of oxygen; e.g., cycling, swimming, jogging, rowing, and fast dancing or walking. Also, AEROBICS.
aerobic lagoon *Biotechnology.* a pond in which sewage that is raw or only partially treated is decomposed by aerobic bacteria.
aerobic metabolism *Biochemistry.* the use of oxygen to generate energy from nutrients, as by humans and other mammals; carbon dioxide and water are by-products of this process. Also, RESPIRATORY METABOLISM.
aerobic respiration *Biochemistry.* a process of cellular respiration in which organic food, usually carbohydrates, is oxidized to form carbon dioxide and water, using atmospheric oxygen and resulting in a high output of energy. Also, **aerobic metabolism.**
aerobics see AEROBIC EXERCISE.
aerobic waste treatment *Biotechnology.* a commercial method of waste water and sewage treatment that alters the biochemical oxygen demand by the growth of aerobic microorganisms.
aerobiology *Biology.* the study of biological substances in the atmosphere, including airborne pollens, spores, and bacteria, and their effect on other organisms. Thus, **aerobiologist.**
aerobioscope [âr´ō bī´ə skōp] *Microbiology.* a device used to determine the bacterial composition of air.
aerobiosis [âr´ō bī ō´səs] *Biology.* life existing in or sustained by oxygen or air. Thus, **aerobiotic.**
aerocamera see AERIAL CAMERA.
aerocartography see AERIAL MAPPING.
aerochlorination *Civil Engineering.* the removal of grease from waste water by the use of compressed air and chlorine gas.
Aerococcus [âr´ō käk´əs] *Bacteriology.* a genus of Gram-positive bacteria of the family Streptococcaceae, occurring as microaerophilic, nonmotile cocci found in pairs or tetrads and displaying a characteristic alpha hemolysis pattern when cultured on a blood agar medium.
AERO code *Meteorology.* an international aviation operations code in which observable meteorological elements are encoded and transmitted in words five numerical digits in length.
aerodiscone antenna [e rō dis´kōn´] *Electromagnetism.* a small discone antenna used on aircraft for communication.
aerodontalgia [âr ō dän´tal´jē ə] *Medicine.* pain in the teeth caused by a change in atmospheric pressure, as during an airplane flight.
aerodrome [âr´ə drōm´] *Aviation.* another term for an airport, especially in British or older U.S. use.
aeroduct [âr´ə dukt´] *Aviation.* a ramjet engine designed to operate in the outer atmosphere by scooping up available ions and electrons and expelling particles derived from these as a jetstream.
aerodynamic [âr´ō dī nam´ik] *Fluid Mechanics.* of or relating to aerodynamics; having to do with the forces that operate on a body from air or another gas in motion.
aerodynamically *Fluid Mechanics.* in an aerodynamic manner; according to the laws or principles of aerodynamics.
aerodynamically clean *Fluid Mechanics.* relating to or describing a surface that produces a minimal amount of aerodynamic drag.
aerodynamically rough *Fluid Mechanics.* relating to or describing a surface whose irregularities are significant, so that it produces a turbulent flow of air even at the surface level.
aerodynamically smooth *Fluid Mechanics.* relating to or describing a surface whose irregularities are minimal, so that it produces a laminar (regular) flow of air at the surface level.
aerodynamic angle of attack see ABSOLUTE ANGLE OF ATTACK.
aerodynamic balance *Fluid Mechanics.* a balance that measures aerodynamic forces, commonly used in wind tunnels to test equipment that will be exposed to flowing air. Also, WIND-TUNNEL BALANCE. *Aviation.* see AERODYNAMIC CONTROL, def. 2.
aerodynamic center *Aviation.* **1.** in an airfoil or other aerodynamic configuration, the point about which a change of incidence will not cause a change of moment. **2.** the point about which the pitching moment remains constant despite changes in the angle of attack. Also, AXIS OF CONSTANT MOMENTS.
aerodynamic characteristics *Aviation.* a general term for the properties and performance of an airfoil or other body with respect to the forces acting upon it as it moves through air and other gaseous fluids.
aerodynamic chord *Aviation.* the line passing through the trailing edge of an airfoil and parallel to an airflow of less than Mach 1 as measured at zero lift; the resulting angle is used as a reference to measure the angle of attack of an airfoil profile.
aerodynamic coefficient *Fluid Mechanics.* a nondimensional expression of aerodynamic pressure, force, or moment, such as lift coefficient or drag coefficient, that indicates the features of a distinct shape at a specified incidence to the airflow.
aerodynamic configuration *Aviation.* a form or design for an aircraft or other such body, allowing it to utilize aerodynamic forces to maintain flight in an atmosphere.
aerodynamic control *Aviation.* **1.** any movable surface of an aircraft, whose movement can alter local aerodynamic forces. **2.** a control surface balanced by placing a part of the surface ahead of the hingeline in order to reduce hinge movements.
aerodynamic drag *Fluid Mechanics.* the force opposing motion encountered by a body moving relative to a fluid; this force is a function of the fluid density, the square of the fluid velocity, the planform area of the body, and the drag coefficient.

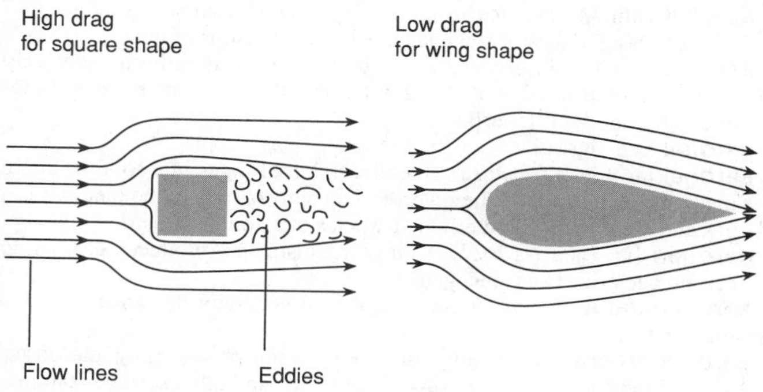

aerodynamic drag

aerodynamic force *Fluid Mechanics.* a force exerted on a body by the fluid surrounding it, either drag or lift, encountered due to a relative velocity between the body and the fluid. Also, **aerodynamic load.**
aerodynamic form drag see FORM DRAG.
aerodynamic friction drag see FRICTION DRAG.
aerodynamic heating *Fluid Mechanics.* the process by which a body is heated by air or other gases passing over its surface, due to friction and compression; this process is dominant mainly at high velocities.
aerodynamic lift *Fluid Mechanics.* an upward force encountered by a body moving relative to a fluid; this force is a function of the fluid density, the square of the fluid velocity, the planform area of the body, and the drag coefficient. Also, LIFT.
aerodynamic missile *Aviation.* **1.** a missile designed to utilize aerodynamic lift during flight. **2.** a missile capable of flight only within an atmosphere.
aerodynamic moment see MOMENT.
aerodynamic noise *Acoustics.* noise that is caused by a moving fluid, such as air or another gas, as it flows over a body. Also, **aerodynamic sound.**
aerodynamic phenomena *Fluid Mechanics.* the results produced by the flow of air over a body, such as mechanical, acoustic, electrical, and thermal effects.
aerodynamic resistance see AERODYNAMIC DRAG.
aerodynamics *Fluid Mechanics.* **1.** the study of gases in motion and the forces that affect this motion. **2.** specifically, the study of the effect of air in motion on an object; this can involve either objects moving through air, such as aircraft or automobiles, or stationary objects, such as bridges or tall buildings.
aerodynamic size *Physics.* the size of a sphere obtained by measuring the terminal velocity of a (nonspherical) particle as it passes through a fluid of known viscosity.
aerodynamic stability *Aviation.* the ability of an aircraft or other body to remain stable when acted upon by the forces of air, and thus maintain its course in flight.

aerodynamic trail *Fluid Mechanics.* a trail of condensed mixture in the air flowing over the exterior surfaces of aircraft at high speeds, created by adiabatic cooling to saturation or by minor supersaturation.

aerodynamic trajectory *Mechanics.* the path of a body significantly affected by aerodynamic forces, especially a rocket, missile, or the like.

aerodynamic turbulence *Fluid Mechanics.* a flow of a fluid in which the velocity at a given point shows fluctuations that are both random and irregular, such as is experienced by an airborne body with high aerodynamic drag.

aerodynamic vehicle *Aviation.* **1.** an airplane or other vehicle that utilizes aerodynamic forces to maintain its flight through the atmosphere. **2.** a vehicle capable of flight only within an atmosphere.

aerodynamic wave drag *Fluid Mechanics.* drag due to the formation of shock waves once any area of supersonic flow is encountered.

aerodyne [âr´ə dīn´] *Aviation.* any heavier-than-air craft; any vehicle that derives its lift from aerodynamic forces.

aeroelastic [âr´ō ē las´tik] *Fluid Mechanics.* **1.** relating to or resulting from aerodynamic forces. **2.** deformable due to aerodynamic forces.

aeroelasticity [âr´ō ē las tis´ ə tē] *Fluid Mechanics.* **1.** the elastic reaction of a body, such as an aircraft, to aerodynamic forces. **2.** the study of such aerodynamic movement or deformation. Also, **aeroelastics.**

aeroembolism *Medicine.* embolism due to air; it may occur as a result of surgery or during rapid, unchecked ascents to higher altitudes.

aerofall mill *Mechanical Engineering.* a large-diameter grinding mill with crushing bodies of steel balls, pebbles, or lumps of ore.

aerofilter *Civil Engineering.* a trickling filter containing relatively coarse material to filter at a high rate, which may be maintained by recirculation of the filter effluents.

aerofoil see AIRFOIL.

aerogel [âr´ə jel´] *Chemistry.* the dispersion of a gas in a solid or liquid medium, such as foam; the reverse of an aerosol, where the medium is a gas and the dispersed material is particles of a solid or liquid.

aerogen [âr´ə jen´ik] *Biology.* any organism that produces gas, as do certain bacteria. Thus, **aerogenic.**

aerogenerator *Electricity.* a wind-driven generator designed for commercial use.

aerogeography *Geography.* the use of aerial photography and observation in the study of landforms, topography, and other surface features.

aerogeology *Geology.* the study of the earth's geologic features through the use of aerial observation and aerial photography.

aerogram *Telecommunications.* an older term for a message that is transported by aircraft or communicated by radio waves.

aerograph *Meteorology.* a meteorological recording instrument designed to operate while airborne. Also, AEROMETEOROGRAPH.

aerography [âr äg´rə fē] *Meteorology.* **1.** the study of the atmosphere and its various phenomena. **2.** the practice of weather observations, map plotting, and keeping records of weather elements.

aerolite [ar´ō´līt] *Astronomy.* a common stony type of meteorite. Thus, **aerolitic.**

aerological day *Meteorology.* a specified day on which upper-air observations are made.

aerology *Meteorology.* **1.** the branch of meteorology concerned with the study of the atmosphere, especially the upper atmosphere. **2.** an older term for meteorology in general. Thus, **aerologic(al), aerologist.**

aeromagnetic *Geophysics.* relating to the earth's magnetic field as studied from the air. Thus, **aeromagnetic map, aeromagnetic surveying.**

aeromagnetic surveying *Geophysics.* the plotting of the earth's magnetic field through the use of airborne electronic magnetometers that are suspended from aircraft.

aeromarine *Navigation.* relating to navigation of aircraft above the ocean. Thus, **aeromarine radio beacon.**

aeromechanics *Fluid Mechanics.* the science of gases such as air in motion or in equilibrium, including the two distinct branches of aerodynamics and aerostatics. Thus, **aeroechanical.**

aeromedicine see AEROSPACE MEDICINE.

aerometeorograph see AEROGRAPH.

aerometer *Engineering.* an instrument used to measure the weight and density of air and other gases.

aeromonad [âr´ō mō´nad] *Bacteriology.* any Gram-negative bacterial species or strain belonging to the genus *Aeromonas.*

Aeromonas [âr´ə mōn´əs] *Bacteriology.* a genus of Gram-negative, rod-shaped, facultatively anaerobic bacteria possessing single polar flagella and frequently pathogenic for cold-blooded marine and freshwater animals; found in salt and fresh water, sewage, and soil.

aeromotor *Aviation.* an older term for an aircraft engine.

aeronaut *Aviation.* an older term for a pilot, especially the pilot of a lighter-than-air craft, such as a balloon or dirigible.

aeronautical [âr´ə nô´tə kəl] *Aviation.* relating to flight or to the design, construction, and operation of aircraft.

aeronautical beacon *Navigation.* an air navigation aid showing white or colored flashes of light to mark some significant feature of the terrain. Similarly, **aeronautical light.**

aeronautical chart *Cartography.* a specialized map of a given area that provides information necessary for air navigation, piloting, or the planning of air operations. Similarly, **aeronautical pilotage chart.**

aeronautical climatology *Meteorology.* the application of the data and techniques of climatology to aviation meteorological problems.

aeronautical engineering *Aviation.* the application of engineering principles and techniques to the design, construction, and operation of aircraft.

aeronautical flutter *Fluid Mechanics.* a self-activated aeroelastic vibration in which the airstream is the external origin of energy and the vibration relies upon the aerodynamic forces as well as the dissipative, inertial, and elastic forces of the system.

aeronautical meteorology *Meteorology.* the study of the effect of weather phenomena on aviation.

aeronautical mile see AIR MILE.

aeronautical mobile service *Telecommunications.* a radio service between aircraft stations or between aircraft stations and land stations.

aeronautical planning chart *Navigation.* any of various charts used to fulfill the requirements of preflight planning.

aeronautical radio beacon see AERONAUTICAL BEACON.

aeronautics *Fluid Mechanics.* the science of flight, including the design, construction, and operation of aircraft.

aeronavigation *Navigation.* the navigation of aircraft in flight.

aeronomy *Geophysics.* the study of the features and phenomena of the upper atmosphere. Thus, **aeronomical.**

aeronte see AEROLITE.

aerootitis see BAROTITIS.

aeropathy [âr´ō path´ē] *Medicine.* any disease or disorder caused by changes in atmospheric pressure; e.g, the bends.

aeropause *Geophysics.* the region of the upper atmosphere in which the air is too rarefied to support commercial aviation; thought of as a transition zone between the atmosphere and space.

aerophagia [âr´ō fāj´ē ə] *Medicine.* the excessive swallowing of air, especially as a manifestation of psychological anxiety. Also, **aerophagy.**

aerophare see RADIO BEACON.

aerophobia *Psychology.* **1.** an irrational fear of drafts or of fresh air, especially when associated with a fear of airborne diseases. **2.** a popular term for fear of flying in an aircraft.

aerophotography see AERIAL PHOTOGRAPHY.

aerophysics *Aviation.* the physics of flight, especially as applied to the design, construction, and operation of aircraft.

aerophyte see EPIPHYTE.

aeroplane *Aviation.* another term for an airplane, especially in British use or former U.S. use.

aeropulse see PULSEJET.

aeroresonator see RESOJET.

aeroservoelasticity *Aviation.* the performance of a servo-controlled elastic body when acted upon by aerodynamic forces.

aerosiderite see SIDERITE.

aerosinusitis see BAROSINUSITIS.

aerosol *Chemistry.* the suspension of very fine particles of a solid or droplets of a liquid in a gaseous medium. Fog, smoke, and volcanic dust are naturally occurring examples of aersols. *Materials Science.* **1.** a liquid substance stored under pressure in a container along with a propellant, usually a liquefied gas, and released in the form of a fine spray or foam. **2.** the container itself; an aerosol can.

aerosol can *Mechanical Devices.* a metal receptacle containing a pressurized liquid substance that is released in the form of a spray or foam when a valve is pressed; used as a dispenser for a wide variety of commercial products, such as insecticides, deodorants, paints, and cleaners. Also, **aerosol bomb, aerosol container.**

aerosol generator *Mechanical Devices.* any device, such as an aersol can, that can disperse a substance into the air in the form of dust or droplets. Also, **aersol propellant.**

aerospace *Meteorology.* of or relating to both the earth's atmosphere and space. *Aviation.* **1.** the two regions of the atmosphere and space considered as one area of flight activity. **2.** see AEROSPACE INDUSTRY.

aerospace engineering *Engineering.* the development and study of various technologies relating to aircraft, spacecraft, and missiles that are designed for flight in the earth's atmosphere and in outer space.

aerospace environment *Geophysics.* the conditions of the earth's atmosphere in space as encountered by airborne vehicles and missiles.

aerospace industry the various companies and enterprises involved in the design, development, manufacture, and sale of aircraft, spacecraft, missiles, and related products for aerospace flight.

aerospace medicine *Medicine.* the medical specialty that concerns itself with the physiological, pathological, and psychological effects of atmospheric and space flight on the body.

aerospace vehicle *Space Technology.* a vehicle capable of flight both within and outside an atmosphere.

Aerosporin *Microbiology.* the trademark for the polymyxin antibiotics, which are derived from the soil bacterium *Bacillus polymyxa* and possess antimicrobial activity against Gram-negative bacteria.

aerostat *Aviation.* a lighter-than-air craft, such as a balloon or dirigible.

aerostatic *Fluid Mechanics.* relating to air and other gases in equilibrium.

aerostatic balance *Engineering.* an instrument that weighs air.

aerostatics *Fluid Mechanics.* **1.** the study of air and other gases in equilibrium, and of bodies suspended or moving within such gases. **2.** the study of lighter-than-air aircraft.

aerotaxis *Biology.* the movement of an organism in response to the presence of oxygen, either toward or away from it; used especially in reference to aerobic and anaerobic bacteria. Thus, **aerotaxic.**

aerothermochemistry *Fluid Mechanics.* the study of gases as they are affected by chemical changes, heat, and motion.

aerothermodynamic border *Geophysics.* that region of the atmosphere above 100 miles (160 km), which is so rarefied as to create virtually no friction with the surface of a object moving through it.

aerothermodynamics *Fluid Mechanics.* **1.** a branch of thermodynamics that studies the effects of heating and the dynamics of gases. **2.** the analysis of aerodynamic phenomena at high gas speeds, incorporating the essential thermodynamic properties of the gas into the examination.

aerothermoelasticity *Fluid Mechanics.* the analysis of the integrated effects of aerodynamic loading and heating on elastic structures and their response.

aerotitis or **aerotitis media** see BAROTITIS.

aerotolerant *Biology.* able to live in air; used especially to refer to certain anaerobic bacteria that can survive in an aerobic environment.

aerotropism *Biology.* the growth or movement of an organism toward a source of oxygen. *Botany.* the deviation of roots from the natural direction of growth, by the action of oxygen or other gases. Thus, **aerotropic.**

aerozine *Materials.* an equal mixture of hydrazine and dimethylhydrazine, used as a rocket fuel.

aeschynite-(Ce) *Mineralogy.* (Ce,Ca,Fe,Th)(Ti,Nb)$_2$(O,OH)$_6$, a relatively rare, black or brown, orthorhombic mineral occurring in prismatic crystals, and having a specific gravity of 4.96 to 5.19 and a hardness of 5 to 6 on the Mohs scale.

aeschynomenous [es´kə näm´ə nəs] *Botany.* of or relating to a sensitive plant, such as those in the pea family.

Aesculus [es´kyü´ləs] *Botany.* the genus name for the the buckeyes, a common group of trees that includes the horse chestnut.

Aeshnidae [esh´nə dē] *Invertebrate Zoology.* a family of large dragonflies in the suborder Anisoptera, with partially fused eyes.

Ae star *Astronomy.* a type A star that shows pronounced lines of hydrogen emission in its spectrum.

aesthesia [es thē´zhə] see ESTHESIA.

aesthesiometer [es thē´zē äm´i tər] see ESTHESIOMETER.

aesthetes [es´thētz] *Invertebrate Zoology.* microscopic light sensory organs serving as simple eyes, located on the dorsal surface of the calcareous shell plates of certain Amphineura.

aesthetic [es thet´ik] see ESTHETIC.

aestidurilignosa *Ecology.* a deciduous broadleaf forest and bush community in which the trees are leafless in winter.

aestilignosa *Ecology.* a forest community of mixed evergreen and deciduous hardwoods.

aestivate [es´tə vāt] see ESTIVATE.

aestivation [es´tə vā´zhən] see ESTIVATION.

Aetosauria *Paleontology.* an extinct suborder of small, lizardlike thecodonts; moderately armored with dorsal and ventral scutes, descended from the Pseudosuchia; probably herbivorous; known only from the upper Triassic.

Aextoxicaceae *Botany.* a monospecific family of tanniferous trees native to Chile and belonging to the order Celastrales; having a dry drupe with a single stone and seed.

AF *Aviation.* the airline code for Air France.

AF or **af** absorption factor; audio frequency.

aF see ABFARAD.

af- a prefix meaning "toward," as in *afferent*.

A-factor *Biotechnology.* a bioregulator produced by *Streptomyces griseus* and other *Streptomyces* species; used to induce morphological differentiation and the production of streptomycin in mutant strains that lack this bioregulator.

afara [ə fä´rə] *Botany.* a West African tree, *Terminalia superba,* known for its straight-grained wood, which resembles teak.

AFB or **A.F.B.** air force base.

AFC *Immunology.* antibody-forming cells, especially to terminally differentiated B cells devoted to the production of secreted antibodies.

AFC or **afc** automatic frequency control.

afebrile [ā fēb´rəl] *Medicine.* without a fever; describing a condition that is not accompanied by fever.

affect *Psychology.* **1.** a specific emotion or feeling that is associated with some idea or mental representation. **2.** a general term for feelings, emotions, or mood.

affective *Psychology.* of or relating to affect; having to do with feelings, emotions, or mood.

affective disorder *Psychology.* a disorder characterized by wide fluctuation of moods, as from extreme depression to elation. Also, **affective reaction, affective psychosis**.

affectivity *Psychology.* the tendency to be susceptible to emotional stimuli or to react emotionally to experiences.

afferent [af´ər ənt] *Physiology.* moving or carrying inward or toward a central part. Thus, **afferent arteriole.** *Neurology.* **1.** a nerve that transmits sensory impulses toward the central nervous system **2.** relating or describing such a nerve. Thus, **afference.**

afferent nerve *Neurology.* see AFFERENT.

affiliation *Psychology.* **1.** the fact or experience of associating socially with other people. **2.** see AFFILIATION MOTIVE.

affiliation motive *Psychology.* an individual's need for social association with others, through friendship, cooperative relationships, groups and organizations, and so on. Also, **affiliation need** or **drive.**

affiliative *Psychology.* of or relating to affiliation.

affination *Food Technology.* the process of treating raw sugar crystals with a heavy sugar syrup to remove an adhering film of molasses.

affine [ə fīn´; af´īn´] of or relating to an affine transformation.

affine connection *Mathematics.* a connection on a manifold, whose form is unchanged under affine changes of parameter along curves, e.g., when the original parameter t of a curve $\gamma(t)$ is replaced by $t = as + b$.

affine coordinates *Mathematics.* if $(b_1, b_2, \ldots, b_n)$ is a basis of the n-dimensional vector space V and a given vector v is (uniquely) expressible as $(c_1, c_2, \ldots, c_n) = c_1b_1 + c_2b_2 + \cdots + c_nb_n$, then $(c_1, c_2, \ldots, c_n)$ are the affine coordinates of v with respect to the basis $(b_1, b_2, \ldots, b_n)$. Also, CONTRAVARIANT COORDINATES.

affine deformation *Cartography.* distortion in a map in which the scale along one axis or plane of reference differs from the scale along the other axis or plane. *Geology.* deformation in which preexisting linear structures remain linear after deformation. *Materials Science.* see AFFINE TRANSFORMATION.

affine geometry *Mathematics.* the study of those properties of space left unchanged by transformations that preserve only the relation of parallelism.

affine plane *Mathematics.* a set of lines and at least three noncollinear points satisfying the following two conditions: (i) for distinct points P and Q, there exists exactly one line l such that both P and Q lie on l; (ii) for a line l and a point P not on l, there exists exactly one line m such that P lies on m and m is parallel to l.

affine space *Mathematics.* suppose that W is an arbitrary extension field of a base field K. The affine space $An(W)$ is the set of n-tuples $(x_1, \ldots, x_n)$, where $x_i \in W$. $(x_1, \ldots, x_n) = x$ is called a point of $An(W)$. $An(W)$ may be viewed as an n-dimensional vector space with an affine connection defined on it, or as an n-dimensional vector space deprived of those vector space properties which depend on the location of the origin or the distance metric on the space.

affine strain *Geophysics.* a strain in the earth in which pressure is equal everywhere along the strain.

affine transformation *Cartography.* a transformation in which straight lines remain straight and parallel lines remain parallel, whereas angles may undergo changes and differential scale changes may be introduced. *Mathematics.* a transformation on a linear space to itself which can be expressed as the sum of a linear transformation and a fixed vector. The affine transformations form a group. In the plane, the group is six-dimensional, consisting of translations, rotations, stretchings and shrinkings, reflections, simple elongations and compressions, and simple shear transformations, as well as compositions of these.

affinial kin *Anthropology.* relatives related by marriage; in-laws.

affinity a connection or relationship; specific uses include: *Biology.* a structural resemblance between species or higher groups, indicative of common ancestry. *Chemistry.* the force by which particles and substances are attracted to others and held together in compounds. *Immunology.* the measurement of the strength with which an antibody combining site binds with a single unit or determinant of an antigen. *Computer Programming.* see AFFINITY FACTOR.

affinity chromatography *Analytical Chemistry.* a separation technique that utilizes the ability of biological molecules (mobile phase) to bind to certain ligands that are coupled to a matrix or support (stationary phase).

affinity constant *Chemistry.* a measure of the relative tendency of one substance to combine with another, expressed as F_d/F_b, where F_a is the tendency to decompose and F_b is the tendency to combine.

affinity factor *Computer Programming.* a measure used to categorize entities in a data base by examining their usage with respect to each other; i.e., if two entities are never used for the same activity, their affinity factor is zero, but if they are always used together for every activity, their affinity factor is 1.

affinity labeling *Biochemistry.* a method of labeling specific functional parts of a macromolecule, such as an enzyme, by inserting into the molecule a tagged synthetic substance similar to that to which the enzyme normally bonds; the labeled molecule is then more easily studied.

affinity partitioning *Biotechnology.* a procedure for liquid-liquid separation in which one of the liquids is chemically treated to produce a ligand that has a particular affinity for the molecule that is being separated.

affix [ā′fiks; af′iks] *Linguistics.* a syllable or syllables joined to a word or base to form a new word; any prefix, suffix, or combining form.

afforestation *Forestry.* the process of converting bare or cultivated land into forest, by planting seeds or transplanting seedlings or mature trees.

affricate [af′ri kət] *Linguistics.* a sound that is produced by beginning with a stop and ending with a fricative; in English these sounds are the [ch] in "church" and the [j] in "judge." (represented in IPA as [tʃ] and [dʒ]). Also, **affricative.**

afibrinogenemia [a′fī brin′ə jə nē′mē ə] *Medicine.* a rare blood disease marked by the absence of fibrinogen (coagulation factor I) in the blood, making it difficult for the blood to coagulate.

aflatoxicosis [a flāt′ō tāk sə kō′səs] *Medicine.* a toxic condition that is caused by the ingestion of nut or nut products that have been contaminated with *Aspergillus flavus* or other *Aspergillus* strains that will produce aflatoxin.

aflatoxin [af′lə täks′in] *Biochemistry.* $C_{17}H_{12}O_6$, a toxic substance produced by some strains of the fungi *Aspergillus flavus* and *A. parasiticus;* found on such plants as peanuts, corn, cottonseed, and soybeans, and thought to cause human hepatic carcinoma.

afocal [ā′fō′kəl] *Optics.* describing a lens or optical system with a convergent power of zero, whose object and image points stretch to infinity. Thus, **afocal lens, afocal system.**

A format *Computer Programming.* a format specification in FORTRAN that specifies the input or output of data as composed of alphanumeric characters.

A-form DNA *Genetics.* a right-handed double helical form of DNA that is similar to the common B form, but differs slightly in helical pitch and base pair orientation.

A-frame *Building Engineering.* a building having its structural frame in the shape of a triangle, usually with a steep double-pitched roof reaching to or near the ground.

Africa *Geography.* the second largest continent (area: 11,714,000 square miles), in the Eastern Hemisphere south of Europe.

African glanders see EPIZOOTIC LYMPHANGITIS.

African greenheart see DAHOMA.

African green monkey kidney cells *Virology.* cells taken from the green monkey, *Cercopithecus aethiops sabaeus,* and used to support the growth of certain viruses, such as poliovirus, that infect vertebrates.

African horse sickness *Veterinary Medicine.* an insect-borne, highly fatal infectious viral disease of horses and other equines, characterized by fever, swelling of the head and neck, thick yellow nasal discharge, and internal hemorrhaging.

African hunting dog *Vertebrate Zoology.* a long-legged dog, *Lycaon pictus,* a sub-Saharan species with a powerful short muzzle, large rounded ears, a splotchy coat, and four toes; highly social and cooperative, it hunts in great packs and can down almost any large game.

African sleeping sickness *Medicine.* a disease caused by the protozoa *Trypanosoma gambiense* or *T. rhodesiense,* marked by fever, chills, headache, vomiting, pain in the extremities, enlargement of the lymph glands, anemia, depression, fatigue, and eventually death if left untreated.

African swine fever see HOG CHOLERA.

African teak see IROKO.

African tick fever see RELAPSING FEVER.

African trypanosomiasis see SLEEPING SICKNESS.

African violet *Botany.* a group of tropical African plants of the genus *Saintpaulia,* with violet or pinkish flowers.

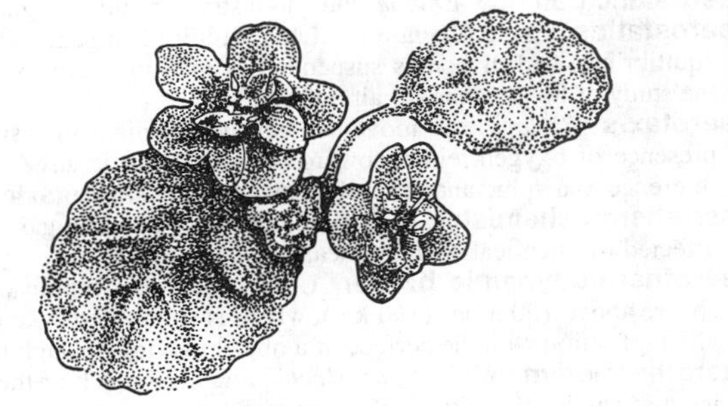

African violet

African walnut see ALONA.

Afrikaans [äf′rə känz′] *Linguistics.* a variation of the Dutch language spoken in South Africa, originating among the early Dutch settlers of the region.

Afro-Asiatic family *Linguistics.* a language family of Northern Africa and sections of the Middle East, including Semitic, Hamitic, and Chad.

Afropithecus [af′rō pith′i kəs] *Paleontology.* a genus of hominoid primates known from certain Miocene sites in Kenya.

Afrotarsius *Paleontology.* a genus of prosimians in the infraorder Tarsiiformes and the proposed family Tarsiidae, known from a single fossil jawbone found in an Oligocene deposit in North Africa.

AFSTN *Artificial Intelligence.* augmented finite state transition network.

aft *Naval Architecture.* at or toward the stern of a ship. *Aviation.* at or toward the rear of an aircraft.

AFT or **aft** automatic fine tuning.

afterbirth *Developmental Biology.* the placenta and membranes that are delivered from the uterus after birth of a mammal. Also, SECUNDINES.

afterblow *Metallurgy.* in the Bessemer steel-making process, the time interval from the endpoint (defined by the change in color of the visible flame) to the turning down of the air blow; the afterblow begins when the burning of the carbon contained in the steel is practically completed.

afterbody *Naval Architecture.* the part of a ship's hull that lies behind the middle of the vessel. Also, **after-end.** *Aviation.* the rear part of the body of an aircraft or other flight vehicle, after the main section. *Space Technology.* a companion object that follows a satellite or space vehicle. *Ordnance.* a part of a ballistic missile that reenters the atmosphere unprotected, behind the nose cone or some protected part of the missile.

afterboil *Mechanical Engineering.* an overheating of an automobile engine after the engine has been turned off.

afterbrain see METENCEPHALON (def. 2).

afterbreak *Mining Engineering.* a movement in which the material from the sides slides inward and follows a main break, positioning itself at right angles to the plane of the seam.

afterburner *Aviation.* an auxiliary combustion chamber placed behind a jet engine turbine to gain extra thrust by injecting additional fuel into the turbine's hot exhaust gases.

afterburning *Aviation.* a process in which fuel is injected into hot exhaust gases of a jet engine and reignited to provide additional thrust. *Space Technology.* the irregular combustion of leftover fuel in a rocket's firing chamber after fuel cutoff. Also, **afterflaming.**

afterburnt *Space Technology.* relating to or describing the period of time after a rocket propulsion system has exhausted its propellant.

afterburst *Mining Engineering.* **1.** a vibration that can follow a rock blast as a result of the ground adjusting to the new stress distribution. **2.** a delayed or sudden collapse of rock following a rock burst in an underground mine.

aftercare *Surgery.* the care and treatment of a convalescing patient, especially one who has undergone surgery. Also, AFTERTREATMENT.

after-chromed dye *Materials.* a dye that is improved in color or fastness by treatment with sodium dichromate, copper sulfate, or similar materials, after the fabric is dyed.

aftercondenser *Mechanical Engineering.* a small heat exchanger used in steam plants to condense motive steam, which allows noncondensable gases to be expelled to the atmosphere.

aftercooler *Mechanical Engineering.* a device or chamber for removing heat from compressed air. Thus, **aftercooling.**

after-cultivation *Agronomy.* harrowing or other cultivation carried out in a field after the crop has emerged.

afterdamp *Mining Engineering.* the residue mixture of gases in a mine following a mine fire or an explosion of firedamp.

aftereffect *Medicine.* a response to pharmacologic agents, drugs, traumatic events, or medical intervention that occurs after an interval of time.

afterfilter *Mechanical Engineering.* a filter used to separate fluid streams that contain only minute quantities of solids.

afterflaming *Space Technology.* see AFTERBURNING.

aftergases *Mining Engineering.* gases that are generated by mine fires or explosions.

afterglow *Meteorology.* a broad, high arch of radiant glow above the highest clouds at deepening twilight, caused by the scattering effect of very fine dust particles that are suspended in the upper atmosphere. Also, **afterlight.** *Atomic Physics.* luminescence that remains for more than 0.1 nsec after the energy that produces it disappears. *Physics.* any light that remains after or trails behind the object that is its source. Also, PHOSPHORESCENCE.

aftergrowth *Agronomy.* a second crop from the same land in the same growing season. Also, **aftercrop.**

afterheat *Nucleonics.* heat that is liberated by the residual radioactive decay of fission and activation products and by actinides after reactor shutdown. Also, DECAY HEAT.

afterimage *Physiology.* a sensory experience that continues after the original stimulus which produced it has ceased. *Computer Programming.* a file or record after updating.

afterloading *Medicine.* a radiotherapeutic technique in which an applicator or needle is placed within the patient during surgery and later loaded with a radioactive source under controlled conditions.

aftermath *Agriculture.* a second crop of grass that grows up after the first crop has been harvested or plowed under. Also, ROWEN.

afterpains *Medicine.* cramps in the uterus that occur in the first few days following childbirth.

afterpeak *Naval Architecture.* the aftermost compartment of a ship's hold.

after-perpendicular *Naval Architecture.* a line projected vertically from the point where the after side of the rudder post (or the midline of the rudder, if there is no post) intersects the load water line. The after-perpendicular, along with the forward-perpendicular, is one of the standard datum lines used in measuring the length of a vessel.

afterpotential *Physiology.* a small positive or negative electrical charge that follows one or more action potentials along a nerve.

afterripening *Botany.* the metabolic processes that must be undergone by certain seeds after harvest before germination can take place; although the embryo appears fully mature, the seed remains dormant until these processes occur.

afterrunning see DIESELING.

aftershaft *Vertebrate. Zoology.* in certain birds, an accessory feather growing out from the base of the shaft of a feather.

aftershock *Geology.* an earthquake that follows the main, or principal, shock of a large earthquake, and which originates at or near the epicenter of the original earthquake.

aftertack *Materials Science.* a residual stickiness remaining on paint or other coverings after the normal drying period.

after top dead center *Mechanical Engineering.* a reference to a timing mark used on engines when the number one piston is moving down after reaching the top of its stroke.

aftertreatment see AFTERCARE.

afterwind *Ordnance.* a strong inflowing wind that produces an updraft near an air or surface burst of a nuclear weapon, sometimes drawing dirt and debris toward the burst point to be carried up into the radioactive cloud.

Aftonian *Geology.* the first interglacial stage of the Pleistocene epoch in North America.

aftosa see FOOT-AND-MOUTH DISEASE.

afwillite *Mineralogy.* $Ca_3Si_2O_4(OH)_6$, a monoclinic mineral occurring as small colorless or white crystals having a specific gravity of 2.63 and a hardness of 3 on the Mohs scale.

Ag the chemical symbol for silver. (From Latin *argentum.*)

ag antigen.

ag- a prefix meaning "toward," as in *aggregate.*

AGA or **A.G.A.** American Gastroenterological Association.

agalactia [ā gə lak´ti kə]*Medicine.* the absence or failure of milk secretion following childbirth. Also, **agalactosis.**

agalactous *Medicine.* describing the absence or imperfect secretion of milk following childbirth.

agalite *Mineralogy.* a fine, fibrous variety of talc that is pseudomorphous after enstatite.

agalmatolite [ag´ əl mat´ ə līt] *Mineralogy.* any of several soft, waxy minerals or stones having a gray, green, yellow, or brown color and used by the Chinese for carving images. Also, FIGURE STONE, PAGODITE, LARDITE.

Agamemnon [ag ə mem´nän] *Astronomy.* asteroid 911, discovered in 1919 and measuring 154 kilometers in diameter; it belongs to type D. (Named for a noted king in Greek mythology.)

agameon [ā gam´ ē ən] *Biology.* an organism that does not produce gametes, and therefore reproduces only by asexual means.

agamete [ā ga´ mēt] *Biology.* an asexual reproductive cell.

agamic [ə gam´ ik] *Biology.* **1.** referring to organisms that do not reproduce sexually. **2.** of ova, not requiring fertilization by sperm. Also, **agamous, agametic.**

Agamidae [ə gam´ə dē] *Vertebrate Zoology.* a family of Old World lizards in the suborder Sauria, containing more than 200 species and found in Asia, Europe, Africa, and Australia.

agammaglobulinemia [ā gam´ə gläb´yə lə nē´mē ə] *Medicine.* a rare disease, either acquired or congenital, characterized by the near absence of gamma globulin from the blood plasma, resulting in an inability to produce immune antibodies; when congenital, the disease is sex-linked and inherited very much like hemophilia as a sex-linked recessive characteristic occurring only in male children.

agamogenesis [ā gam´ō jen´ə sis] *Biology.* the process of asexual reproduction, especially schizogony, or multiple fission. Thus, **agamogenetic.**

agamogony see AGAMOGENESIS.

agamospecies see AGAMEON.

agamospermy [ā´gam ə spurm´ē] *Botany.* a form of asexual reproduction in which seeds are produced, but neither fertilization nor meiosis occurs.

agamotropic *Botany.* of a flower, not closing again once it has opened.

aganglionic [ā´gang lē än´ik] *Neurology.* of or relating to the absence of ganglion cells.

Agaontidae *Invertebrate Zoology.* a family of wasps belonging to the hymenopteran suborder Apocrita, which pollinate certain varieties of figs. Also, **Agaonidae.**

agar [ä´gär; ag´är] *Materials.* a gelatinous product extracted from certain red algae, e.g., *Gelidium cartilagineum* or *G. robustum,* indigestible by almost all bacteria and therefore used as the basic ingredient in solid culture mediums. It also has a variety of industrial uses, as for thickening foods, making emulsions, as a laxative, as a substitute for gelatin, and so on. *Bacteriology.* a culture medium containing agar.

agar-agar see AGAR.

agar-gel reaction *Immunology.* an antigen-antibody reaction indicated by the formation of precipitin lines when the reactants are allowed to diffuse toward each other through an agar-gel medium.

agaric [ə gâr´ik] *Mycology.* **1.** a fungus that belongs to the family Agaricaceae. **2.** relating to mushrooms or fungi of this type. Also, **agaricaceous.**

Agaricaceae *Mycology.* a family of fungi belonging to the order Agaricales, which is composed of gill mushrooms, many of which are edible and commercially grown.

agaric acid *Organic Chemistry.* $C_{19}H_{36}(OH)(COOH)_3$, a colorless crystalline solid that is soluble in water and insoluble in benzene and melts at 141°C; used as an irritant. Also, **agaricic acid.**

Agaricales [ə gar´ə kā´lēz] *Mycology.* an order of fungi in the class Hymenomycetes, which contains mushrooms and toadstools; characterized by umbrella-like, gilled structures that are seldom pathogenic.

agaric mineral see ROCK MILK.

Agaricogastrales *Mycology.* an order of fungi belonging to the class Gasteromycetes, characterized by a hymenium that persists after maturity and a stipe similar to that of the fungi Agaricales.

agaricoid *Mycology.* referring to the fruiting body of a mushroom or toadstool of the order Agaricales, whose fruit-bearing surface, the hymenium, forms a layer on the gills and produces spores called ballistospores.

Agaricus *Mycology.* a genus of fungi belonging to the order Agaricales, which is characterized by gilled structures; most are edible.

agarose *Biochemistry.* a sugar found in the seaweed agar or marine algae that causes agar to gel.

agarose gel electrophoresis *Chemistry.* a method for fractionating DNA fragments under the influence of an electric field.

Agassiz [ag´ə sē] **1. Louis** 1807–1873, Swiss-American naturalist and geologist; studied marine fossils, glaciers, and zoological taxonomy. **2. Alexander** 1835–1910, American marine zoologist and mining engineer, son of Louis Agassiz.

Agassiz, Lake *Geology.* a late Pleistocene lake that occupied the site of the present Red River plain of Manitoba and other surrounding areas.

Agassiz trawl *Oceanography.* a sample-collecting device developed by Alexander Agassiz on the *Blake* expedition in 1877. A marked improvement on the collecting devices then available, it consisted of a cylindrical net with an iron hoop at each end, all suspended from an iron frame.

agate *Mineralogy.* SiO_2, a variety of chalcedony found in various low-intensity colors that are arranged in alternating bands, irregular clouds, or mosslike patterns, having a specific gravity of 2.6 and a hardness of 7 on the Mohs scale. *Materials.* something made of this quartz or of agate glass; specifically, a type of glass marble used in children's games. *Graphic Arts.* **1.** a type size measuring 5.5 points (or 14 lines to the inch), formerly the standard size for classified advertisements. **2.** a general term for small type.

agate glass *Materials.* a type of glass that has patterns of various color resembling the mineral agate. Similarly, **agateware.**

agate line *Graphic Arts.* a measure of space in advertising, one column wide and the depth of a line of agate type (0.07 inches).

agatized wood see SILICIFIED WOOD.

Agavaceae *Botany.* in some classifications, a family of desert and tropical plants related to the amaryllis and lily and including the genus Agave. Also, **agave family.**

Agave [ə gä´vā; ä´gə vā´] *Botany.* **1.** a genus of plants of the family Agavaceae, native to warm American regions and having spiny leaves and tall flower stalks. **2. agave.** any particular plant of this genus, especially *A. americana,* the century plant, noted for its fast growth and for blooming only once in its lifetime; cultivated for its sap, which is used in making tequila and pulque, and for the cordage in its leaves.

agave

agavose *Organic Chemistry.* $C_{12}H_{22}O_{11}$, a sugar from the agave tree; used medicinally as a purgative and diuretic.

agba *Materials.* the wood of an African tree, *Gossweilerodendron balsamiferum,* used in furniture making and for interior decoration.

AGC automatic gain control.

AGCT *Psychology.* Army General Classification Test, a group intelligence test widely used in the military in and after World War II.

AGE aerospace ground equipment.

age the period of time during which an organism or phenomenon exists or once existed; specific uses include: *Geology.* **1.** a subdivision of an epoch, corresponding to a particular stage of rock formation. **2.** an informal interval of geologic time corresponding to the formation of the rocks of any stratigraphic unit. **3.** in the earth's history, a time interval of unspecified length during which a particular life form was dominant or important, such as the Age of Mammals. **4.** an interval of time marked by specific physical conditions or geologic events, such as the Ice Age. *Biology.* a time of unspecified length during which a particular life form was dominant or important; e.g., the Age of Mammals. *Forestry.* **1.** the mean age of all trees in a forest, crop, or stand. **2.** of a tree, the time that has elapsed since the germination of the seed from which this tree developed.

age-area hypothesis *Anthropology.* the theory that the age or time period of a cultural trait can be determined by the distance it has traveled from its point of origin; therefore, the most widely distributed traits must be the oldest.

age class *Forestry.* **1.** one of the intervals, such as ten years, into which the age range of tree crops is divided for classification. **2.** the trees that fall into such an interval. *Anthropology.* see AGE GRADE.

age coating *Electricity.* the black deposit formed on the inner surface of an electric lamp bulb by material evaporated from the filament.

aged *Geology.* of a land-surface feature, nearly reduced to base level.

age determination *Geology.* the establishment of the geologic age of a specimen using any accepted method, such as dendrochronology or radiometric dating.

age distribution *Ecology.* the number or percentage of individuals in various age groups within a given population. Also, **age composition, age structure.**

age grade *Anthropology.* a culturally distinguished category that includes members of the same approximate age for whom there is a separate term, such as teenagers or the elderly. Also, **age group.**

age hardening *Materials Science.* a heat-treating process designed to produce a uniform dispersion of a fine, hard, coherent precipitate in a softer, more ductile matrix. Also, PRECIPITATION HARDENING.

ageing see AGING.

Agena [ə jē´nə] see BETA CENTAURI.

Agena rocket *Space Technology.* a restartable upper-stage launch vehicle widely used in the U.S. space program, especially in conjunction with the Atlas and Titan boosters.

agenda *Computer Programming.* **1.** a set of control program statements that specify the job run sequence and procedures which the computer is to follow. **2.** a sequence of processing operations to be performed in obtaining a computer solution. *Artificial Intelligence.* the order and priority in which to hypothesize and test uncertainties; the course of a session.

agenesis [á jen´ə sis] *Biology.* the failure to develop, or the incomplete development, of tissue, an organ, or any other part of a body.

agent *Science.* any substance, force, or organism capable of causing some change or reaction. *Artificial Intelligence.* a program that acts independently, as on a different computer, to attempt to accomplish a goal for a user or for another program.

Agent Orange *Toxicology.* the code name for a defoliant used by U.S. forces in the Vietnam War, composed of the herbicides 2,4,5-T and 2,4-D. After early use it was found that 2,4,5-T contains the highly toxic contaminant dioxin, which causes cancer and genetic damage in certain animals. The toxicity of Agent Orange to humans has not been precisely established, though it has been associated with severe health problems in military personnel exposed to it and birth defects in their offspring. **Agent Blue, Agent White,** and **Agent Purple** are other herbicides having similar use in the Vietnam War. (From the colored stripes used to identify containers of these herbicides.)

Age of Amphibians *Geology.* a term for the time roughly corresponding to the Carboniferous and Permian periods, when amphibians were the prominent life form on earth.

age of diurnal inequality *Geophysics.* the time elapsed between the maximum semimonthly declination of the moon (north or south) and the time at which that declination has its greatest effect on the range of the tides or the speed of tidal currents. Also, **age of diurnal tide.**

Age of Fish(es) *Geology.* a term for the time roughly corresponding to the Silurian and Devonian periods, when fish were the prominent life form on earth.

Age of Mammals *Geology.* a term for the time roughly corresponding to the Cenozoic period, when mammals became the prominent life form on earth.

Age of Man *Geology.* a term for the time roughly corresponding to the Quaternary period, when humans became the prominent life form on earth.

age of parallax inequality *Geophysics.* the time elapsed between the perigee of the moon's orbit and the time it takes that orbit to have its maximum effect on the range of the tides or speed of the tidal currents.

age of phase inequality *Geophysics.* the time elapsed between the new or full moon and the time it takes these phases to have their maximum effect on the range of tides or the speed of tidal currents.

Age of Reason *Science.* the period in European history extending roughly from the early 1600s to the late 1700s, in which philosophers emphasized the use of reason and the scientific method, and great advances were made in such fields as chemistry, anatomy, physics, and astronomy. Also, **Age of Rationalism.**

Age of Reptiles *Geology.* a term for the time roughly corresponding to the Mesozoic era, when reptiles were the prominent life form on earth.

age of the earth *Astronomy.* a conception estimated at 4.6 billion years; the earth is believed to have formed at the same time as the sun and the rest of the solar system. The oldest rocks thus far discovered in the earth's crust are estimated to be 3.96 billion years old.

age of the moon *Astronomy.* a conception estimated at 4.6 billion years, the same as the earth.

age of tide see AGE OF PHASE INEQUALITY.

ageostrophic wind [ə jē´ə sträf´ik] see GEOSTROPHIC DEPARTURE.

age ratio *Geology.* the ratio of daughter to parent isotope in a sample that is dated radiometrically.

age set *Anthropology.* a group of individuals of the same sex and approximate age who are initiated together and advance from one age grade to another.

age structure *Anthropology.* the cultural practice of grouping people according to life stages such as infants, children, teenagers, adults, and elderly.

ageusia *Neurology.* the absence of the sense of taste. Also, **ageustia.**

ageusic *Neurology.* **1.** of or relating to ageusia. **2.** a person who has an impaired sense of taste.

agger *Civil Engineering.* earthwork serving as a road, usually raised with sloping embankments. *Oceanography.* see DOUBLE TIDE.

agglomerate *Biology.* crowded together in a cluster or mass. *Geology.* a mass of coarse, angular fragments of volcanic rock formed during explosive volcanic eruptions. *Materials Science.* a collection of fine particles, resulting from attractive interparticle surface forces; especially important in powder processing of ceramics, Thus, **agglomerated.**

agglomerated settlement *Geography.* a settlement pattern characterized by dwellings packed closely together, in contrast to dispersed or scattered settlement.

agglomeration *Biology.* a cluster or mass, especially a dense cluster of flowers. *Metallurgy.* the intentional or unintentional gathering of particles, such as metal powder, into clusters or into a mass of larger size than the original particles. *Meteorology.* a process in which particles grow by colliding and assimilating with cloud particles or other precipitation particles. *Food Technology.* any chemical or physical treatment that improves the hydration of a powdered or granular food product. Also, INSTANTIZING.

agglomeration test *Mining Engineering.* a test to determine the binding qualities of coal. Also, BUTTON TEST.

agglutinant *Surgery.* **1.** promoting union by adhesion. **2.** a glue-like substance that holds parts together or causes parts to adhere during healing. *Materials Science.* any sticky substance used to hold materials together.

agglutinate cone see SPATTER CONE.

agglutination *Immunology.* the clumping or joining together of cells, due to the reaction of antigens to certain antibodies. *Medicine.* the process of union in wound healing.

agglutination test *Virology.* a virus recognition test based on the observation that some viruses, such as the influenza virus, attach to more than one cell and thus cause clumping (agglutination) of these cells.

agglutinin *Immunology.* a substance (usually an antibody) that causes antigens (such as bacteria) to form visible clumps.

agglutinogen *Immunology.* a particulate antigen that stimulates the production of the antibody agglutinin.

agglutinoid *Immunology.* an antibody that does not agglutinate, but is capable of joining with its agglutinogen.

aggradation *Geology.* **1.** the building up of any land surface, especially by the action of rivers and streams. **2.** see ACCRETION. *Hydrology.* **1.** the growth or spread of permafrost under present climatic conditions. **2.** the deposition of sedimentary material by a stream that serves to bring about or maintain uniformity of grade. Thus, **aggradational.**

aggradation recrystallization *Geology.* recrystallization that results in an increase in the size of a crystal.

aggrade *Geology.* to raise the level of a valley, stream bed, and so on by the deposit of silt, sediment, and the like.

aggraded valley floor *Geology.* the surface of a plain formed by a stream aggrading its valley, where the alluvial infilling is thicker than the depth of the stream channel.

aggregate *Geology.* a mass consisting of rock or mineral particles, or a mixture of both. *Materials.* the various small particles such as sand or gravel that form the basic constituents of concrete, along with water and cement. *Botany.* **1.** a flower that is formed by a cluster of carpels arranged in a dense mass. **2.** of or relating to such a flower.

aggregate bin *Engineering.* a container used to hold and dispense dry construction material such as broken stone, slag, gravel, and sand, and usually designed to funnel the material through a gate-like opening on the underside.

aggregate fruit *Botany.* a cluster of fruits developed from the many ovaries of a single flower, such as the raspberry or blackberry.

aggregate interlock *Engineering.* a situation in which the aggregate on one side of a concrete joint projects into the aggregate on the opposite side, facilitating load transfer.

aggregate recoil *Nuclear Physics.* the release of atoms from the surface of radioactive material, occurring primarily during α-particle decay when kinetic energy is transferred from the α-particle.

aggregate structure *Crystallography.* a mass composed of separate crystals or grains that are eclipsed when viewed under a polarizing microscope at different intervals while the stage is rotated.

aggregation *Science.* a grouping or clustering of individual items into a mass or group. *Behavior.* a gathering of members of a species that is not the result of social attraction, as when individual animals seek out the same watering place. *Mathematics.* the grouping together of several terms in an expression and treating them as the same term. *Astronomy.* see ACCRETION.

aggressin *Microbiology.* a substance that is produced by a pathogenic microorganism, and that facilitates the invasive spread of the pathogen and inhibits the defensive phagocytic process of the host.

aggression *Behavior.* an intentional attack or assault intended to cause harm or fear. Also, **aggressive behavior.**

aggression training *Psychology.* a behavior therapy technique in which aggressive responses are encouraged in order to improve the subject's sense of himself or herself as a forceful, effective individual.

aggressive *Behavior.* relating to or showing aggression. *Volcanology.* describing a magma or magmatic intrusion that forces its way into place. Also, INVASIVE.

aggressive carbon dioxide *Chemical Engineering.* the amount of carbon dioxide soluble in water beyond that needed to cause a specific concentration of calcium ions to precipitate as calcium carbonate; used as a measure of the scaling properties and corrosivity of water.

aggressive mimicry *Zoology.* a method of capturing prey by appearing to be something else, as seen in certain predatory fish or insects.

aggressiveness *Behavior.* the tendency to attack or to carry out actions of aggression..

aggressive water *Hydrology.* a term for any water that forces its way into place.

aggretope *Immunology.* the part of an antigen that interacts with an MHC molecule.

AGI or **A.G.I.** American Geological Institute.

aging the process of growing older or changing over time; specific uses include: *Food Technology.* **1.** the storing of food while it ripens, matures, or mellows, as with wine or cheeses. **2.** the artificial hastening or enhancement of this process, as when flour is treated with the oxidant ammonium persulfate in order to yield stronger and more resilient dough. *Materials Science.* the treatment of a material or product to stabilize or strengthen it by causing a coherent precipitate to form, increasing the permanency in a magnet, or the steaming of fabrics. *Nucleonics.* the continuous slowing down of neutrons, expressed as the spatial distribution of the slowing-down density of a nonabsorbing medium.

aging-lung emphysema *Medicine.* a pulmonary disease associated with aging, in which alveolar dilation occurs due to loss of tissue elasticity.

agitate *Mechanical Engineering.* to move or shake a material vigorously, especially so as to mix it with another material.

agitator *Mechanical Engineering.* a mechanical device used to maintain fluidity and plasticity, and to prevent segregation of liquids and solids in liquids, such as concrete and mortar.

agitator

agitator body *Mechanical Engineering.* a truck-mounted drum for transporting concrete or other semifluid materials, containing blades that are rotated to agitate the contents. Thus, **agitating truck.**

agit. a. us. *Medicine.* shake before using. (Latin *agita ante usum.)*

agium [á´jē əm] *Ecology.* a beach habitat or beach community. Also, AIGIALIUM.

AGK or **A.G.K.** *Astronomy.* Astronomische Gesellschaft Katalog (or Katalog der Astronomische Gesellschaft), a standard catalog of star positions.

AGL *Navigation.* above ground level; a term used in measuring altitude. *Medicine.* see ACUTE GRANULOCYTIC LEUKEMIA.

Aglaspida [ə glas´pə də] *Paleontology.* an extinct order of aquatic chelicerates in the subclass Xiphosura, resembling the modern horseshoe crab, *Limulus;* known only from the Cambrian.

aglomerular *Histology.* **1.** relating to a tissue or structure lacking glomeruli. **2.** such a tissue or structure.

Aglossa [á gläs´ə] *Vertebrate Zoology.* a suborder of tongueless, aquatic frogs in the family Pipidae.

aglycon *Biochemistry.* the portion of a glycoside compound that contains no carbohydrate molecules.

aglyphous [ā´glə fəs] *Vertebrate Zoology.* having solid teeth, used especially in reference to a group of snakes in the family Colubridae that includes garter, water, black, bull, and gopher snakes.

agmatine *Biochemistry.* $C_5H_{14}N_4$, a compound formed from the amino acid arginine.

agmatite *Petrology.* **1.** a migmatite with an inclusion layer such as a xenolith. **2.** fragmental plutonic rock with granitic cement.

agnate *Anthropology.* **1.** related only on the male side. **2.** a relative descended from a common male ancestor.

Agnatha [ag´nə thə] *Vertebrate Zoology.* the most primitive of the eight vertebrate classes; a class of jawless fish including ostracoderms and cyclostomes. Thus, **agnathan.**

agnathia [ag nath´ē ə] *Medicine.* a developmental anomaly characterized by the absence of the lower jaw.

agnathous [ag nath´əs] *Medicine.* relating to or affected with agnathia; lacking a lower jaw. *Zoology.* **1.** lacking jaws. **2.** belonging to the class Agnatha. Also, **agnathic.**

Agnesi, Maria Gaetana 1718–1799, Italian mathematician; developed the Witch of Agnesi.

agnosia *Medicine.* a loss of comprehension at the level of the central nervous system of any of the senses; the sensory sphere is intact, but the patient is unable to assimilate the meaning of the sense. Also, **agnosis.**

agnostid *Paleontology.* a trilobite of the extinct order Agnostida, generally small and eyeless; lower Cambrian to upper Ordovician; important in continental correlation because several species are geographically widespread but not long-lived.

Agnostus *Paleontology.* a genus of small, early Paleozoic trilobites known from widespread deposits, especially in the Cambrian.

agonal [ag´ə nəl] *Medicine.* relating to death, or to bodily processes occurring just before death. Thus, **agonal respiration.**

agonal clot *Hematology.* a blood clot formed in the heart during the death agony. Also, **agonal thrombus.**

agonic line [ā gän´ik] *Geophysics.* an imaginary line through all points on the earth's surface at which the magnetic declination is zero; the locus at which magnetic north coincides with true north.

Agonidae [ə gän´ə dē] *Vertebrate Zoology.* the poachers, a family of small marine fishes in the suborder Cottoidei; found in cold waters of the Northern Hemisphere.

agonist [ag´ə nist] *Biotechnology.* an agent capable of stimulating a biological response by occupying cell receptors. *Physiology.* any muscle whose contraction produces a desired reaction.

agonistic behavior *Behavior.* a class of behavioral patterns, including many types of attack, threat, defense, and escape, between members of the same species in response to social conflict.

agonistic buffering *Behavior.* a behavioral situation in which the presence of a child or infant serves to inhibit aggression between two potentially hostile adults.

Agonomycetacea [ə gän´ó mí sə tá´sé ə] *Mycology.* the single family in the order Agonomycetales, class Hyphomycetes.

Agonomycetales [ə gän´ó mí sə tal´éz] *Mycology.* a worldwide order comprising all mycelial imperfect fungi that do not produce spores; contains about 200 species, including several pathogens in the genus *Rhizoctonia* that are known to cause damping-off disease. Also, MYCELIA STERILIA.

agony *Medicine.* see AGONAL.

agora *Architecture.* a public meeting place, especially the main marketplace of an ancient Greek city.

agoraphobia [ə gôr´ə fó´bé ə] *Psychology.* an irrational fear of being in open or public places. (From *agora.)*

agoraphobic *Psychology.* **1.** also, **agoraphobe.** a person affected with the condition of agoraphobia. **2.** of or relating to this condition.

agouti *Vertebrate Zoology.* a stout-bodied nocturnal rodent of the genus *Dasyprocta*, family Dasyproctidae, ranging from Mexico to Peru and in the West Indies. *Zoology.* an irregular color pattern in the fur of various animals, in which an overall grayish or "grizzled" color is produced by alternating bands of light and dark on each individual strand of hair. *Genetics.* the gene that is responsible for this pattern of coloration.

agouti

agpaite [ag´pə īt] *Petrology.* a group of feldspathoid-bearing igneous rocks such as naujaite, lujavrite, and kakortokite.

agrad *Ecology.* a term for a cultivated plant, as opposed to one growing wild.

agraffe [ə graf´] *Surgery.* a clamplike instrument that holds the edges of a wound together. Also, **agrafe.**

agrammatism [ə gram´ə tiz´əm] *Neurology.* a cerebral disorder characterized by the inability to construct intelligible sentences using proper grammar. Also, **agrammatica**.

agranular [ā gran´yə lər] *Biology.* lacking granules; nongranular.

agranular leukocyte *Histology.* a white blood cell that lacks cytoplasmic granules and has a large nucleus.

agranular reticulum *Cell Biology.* intracellular membranes lacking attached ribosomes; smooth endoplasmic reticulum.

agranulocyte [ā gran´yə līt] *Immunology.* a white blood cell that lacks distinct cytoplasmic granules. Also, **agranular leukocyte.**

agranulocytosis [ā gran´yə lō sī tō´sis] *Medicine.* an acute disease, characterized by prostration, high temperature, and ulcerations of the mucous membranes, in which the white blood cell count drops to very low levels, causing neutropenia to become pronounced, usually in response to drugs, including analgesics, anticonvulsants, and anti-inflammatory agents. Also, SCHULTZ SYNDROME.

agraphia *Medicine.* an abnormal inability to express one's thoughts in writing, due to a lesion of the cerebral cortex. Thus, **agraphic.**

agrarian *Agriculture.* relating to agriculture or to cultivated plants.

A/G ratio see ALBUMIN-GLOBULIN RATIO.

agravic [ā grav´ik] *Geophysics.* having no gravitation.

agrellite *Mineralogy.* $NaCa_2Si_4O_{10}F$, a white to greenish white triclinic mineral found in alkalic igneous gneisses, having a hardness of 5.5 on the Mohs scale and a specific gravity of 2.9.

agrestal *Agriculture.* growing uncultivated in agricultural land, as weeds or other wild plants in a field. Also, **agrestial.**

agri- of or having to do with agriculture, as in *agribusiness.*

agricere *Geology.* a waxy or resinous coating of organic material found on soil particles.

agrichemical *Agriculture.* a chemical or chemical product used in farming, such as a pesticide or herbicide.

Agricola, Georgius (George Bauer) 1494?–1555, German physician and geologist; known as the father of mineralogy.

agricultural *Agriculture.* of, relating to, or used in farming.

agricultural chemical *Materials Science.* any chemical compound used as an aid to agriculture, including fertilizers, soil conditioners, insecticides, herbicides, and so on.

agricultural chemistry *Agriculture.* the use of chemical substances for agricultural purposes, such as enrichment of the soil, destruction of insect pests, and control of plant and animal diseases.

agricultural climatology *Agriculture.* the study of climatic conditions and their relationship to the production of specific strains or varieties of crops, as in the development of drought-resistant grasses for areas of low rainfall. Also, **agroclimatology.**

agricultural drain(pipe) see FRENCH DRAIN.

agricultural engineering *Agriculture.* the use of engineering principles for agriculture, as in the improvement of irrigation systems.

agricultural geology *Geology.* the use of geology in agricultural applications, for example in relation to the nature, formation, and distribution of soils.

agricultural lime *Materials.* lime (calcium oxide, CaO) used as a soil conditioner.

agricultural mechanics *Agriculture.* the use of principles of mechanics for agricultural purposes, as in the development of equipment such as automated feed mixers.

agricultural meteorology *Agriculture.* the use of meteorological information for agricultural purposes, as in the protection of crops from a predicted frost. Also, **agrometeorology.**

agricultural science *Agriculture.* the scientific study of all aspects of farm production, including soil management, crop production, animal husbandry, and the processing and marketing of farm products.

agriculture the science, art, or practice of raising crops and livestock for food or for other products useful to human activity.

agriculturist *Agriculture.* a person who studies or is involved with agriculture. Also, **agriculturalist.**

Agriochoeridae [ag´rē ō kir´ə dē] *Paleontology.* a family of piglike ungulates in the suborder Tylopoda and superfamily Merycoidodontoidea; restricted to North America; they arose in the Eocene, differentiated and spread in the Oligocene, and then dwindled, persisting in smaller numbers into the early Miocene.

agrioecology see AGROECOLOGY.

Agrionidae [ag´rē än´ə dē] *Invertebrate Zoology.* the family of large damselflies, in the suborder Zygoptera, usually having red and black markings on their wings.

agrium *Ecology.* a wildlife community on cultivated land or on other land subject to significant human activity.

agro- of or relating to the soil, as in *agronomy, agrobiology.* (From a Greek word meaning "field.")

Agrobacterium *Bacteriology.* a genus of Gram-negative, aerobic, rod-shaped bacteria of the family Rhizobiaceae, typically plant pathogens that form galls or tumors on roots or stems.

agrobiology *Agronomy.* the science of plant nutrition and plant growth. Thus, **agrobiological.**

agrochemical *Agriculture.* **1.** a chemical used to improve the quality of farm products. **2.** of or relating to the use of chemicals in agriculture.

Agriculture

The word "agriculture" is from the Latin words *ager,* meaning field, and *cultura,* meaning cultivation. The term has evolved to include raising livestock (animal husbandry) in addition to cultivating the soil to produce food and fiber from the land.

Broadly defined, agriculture encompasses the art and science of plant and animal production; the provision of machinery and materials for agricultural production; and the processing, manufacturing, and marketing of food and fiber products. As an applied science, agriculture integrates and interacts with other fields of inquiry including the biological sciences, chemistry, nutrition, human and veterinary medicine, engineering, social sciences, business, and economics.

Improvements in agricultural production were slow and incremental until the Industrial Revolution began the mechanization of agriculture. Throughout the first half of the 20th century, machines gradually replaced men and animals to plow, plant, and harvest more efficiently. During World War II and the decades that followed, new chemical fertilizers, pesticides, and herbicides helped farmers dramatically increase crop yields, producing a "Green Revolution" in the United States and parts of the developing world.

Agriculture has now entered an era of integrated, technology-intensive systems approaches. Computerized management systems and robotics are enabling greater precision in the use of water, soil, and agrichemicals. Integrated pest management systems combine biological pest control, genetically engineered resistant crops, and agrichemicals for an overall reduction in chemical use without compromising agricultural productivity.

Advances in molecular biology applied through various biotechnologies will make it possible to genetically engineer food crops to resist insects, weeds, diseases, and drought. Genetic improvement of livestock will produce animals that grow faster, are resistant to disease, and have desirable characteristics, such as leanness. These innovations will create a more efficient and environmentally benign agricultural system, new and improved products to meet consumer demands, and opportunities to increase worldwide food supplies.

Charles J. Arntzen
Deputy Chancellor and Dean
Agriculture Program, Texas A&M University

agrocinopine [ag´rō sin´ə pīn] *Biochemistry.* any of a class of opines present in crown gall, a type of plant tumor.

agroecology *Ecology.* the study of the relationship between the environment and agricultural crops. Thus, **agroecological.**

agroecosystem *Ecology.* the ecological relationships of agriculture in general or of a particular agricultural locale. Similarly, **agroenvironment.** Also, AGROSYSTEM.

agroforestry *Agronomy.* the practice of growing an integrated combination of forest and agricultural crops on the same land.

agroinfection *Genetics.* a technique for infecting plant cells with DNA from a plant virus, using the T-DNA portion of a Ti plasmid.

agrology *Agriculture.* the science of agricultural production.

agromania *Psychology.* an irrational desire to live alone in an open place, away from others.

Agromyces [ag´rō mī´sēz] *Bacteriology.* a genus of soil bacteria of the order Actinomycetales that grow as branched mycelia, multiply by fragmentation, and have an oxidative metabolism.

Agronomy

Agronomy is the study of crop and soil sciences. It is a broad field that includes the following areas: crop breeding, genetics, and cytology; plant molecular biology and plant molecular genetics; crop physiology and metabolism; crop ecology, production, and management; seed physiology, production, and technology; turf grass; crop quality and utilization, agroclimatology and agronomic modeling; soil physics; soil chemistry; soil microbiology; soil fertility and plant nutrition; soil genesis, morphology, and classification; soil mineralogy; and soil management.

It was not until 1840 that scientific investigations began in agronomy. Laboratory, field, and greenhouse studies were conducted on plant nutrition and the essentiality of plant nutrients. Advances were made over the next century in a number of areas, which included the discovery that soils could retain cations and anions; fundamental studies on soil water; understanding photosynthesis; studies on soil microorganisms and biological nitrogen fixation; the discovery that clays are crystalline; ways to reclaim saline and sodic soils; reduction of soil erosion through soil management and novel tillage systems; the creation of a modern soil classification system; and introduction of hybrid corn and improvement of crops through plant breeding and genetics.

Agronomy has changed significantly in the past decade. Agronomists, more than ever, are involved in interdisciplinary research, and there is increasing emphasis on improving plants through biotechnology and on environmental quality. They are making major accomplishments in the use of bioengineering to develop plants that are resistant to adverse environments, herbicides, diseases, and insects; genetic improvement of agronomic crops through gene characterization, germplasm acquisition and maintenance, plant breeding, and molecular markers; improving methods for manipulation of rhizosphere microorganisms to enhance plant growth; sustaining agriculture; preserving environmental quality by developing approaches for decontaminating soils, predicting and understanding inorganic and organic pollutant transport in soils and managing and disposing solid wastes; and using kinetics and in-situ surface spectroscopic techniques to elucidate reaction mechanisms at soil and soil component surfaces.

Donald L. Sparks
Chairman of Plant and Soil Sciences
University of Delaware

agronomist *Agriculture.* a person who studies or is an expert in crop production and soil science.

agronomy *Agriculture.* the scientific study of crop production and soil management.

agrostemmic acid [ag rə stem′ik] *Toxicology.* a poison found in the seeds of the corn cockle, *Agrostemma githagin.*

agrostology *Botany.* the branch of botany that deals with grasses. Thus, **agrostologic(al), agrostologist.**

agrosystem see AGROECOSYSTEM.

agrotechnology *Agriculture.* the use of technological devices and principles in agriculture to improve crop production. Thus, **agrotechnician, agrotechnologist.**

aground *Navigation.* on or into the bottom, dry land, a reef, and so on.

agrypnocoma *Neurology.* a state of coma characterized by lethargy with prolonged wakefulness and muttering delirium.

agrypnotic [ag rip nät′ik] *Medicine.* abnormally wakeful; affected by insomnia. Thus, **agrypnia.**

AGS *Aviation.* the airport code for Augusta, Georgia.

AGS or **A.G.S.** American Geriatrics Society.

AGU or **A.G.U.** American Geophysical Union.

aguaje [ä gwä′hä] *Toxicology.* a Spanish term used to refer to the poisoning that often results from rotting fish on beaches during an occurrence of the unusual current shift called *El Nino.*

ague [ág′yoo] *Medicine.* malarial fever or another disease marked by recurrent shaking and chills.

aguilarite *Mineralogy.* Ag_4SeS, an opaque, metallic black, orthorhombic mineral occurring as dodecahedral crystals, having a specific gravity of 7.586 and a hardness of 2.5 on the Mohs scale; found with argentite, silver, and calcite.

Agulhas Current [ə gool′häs] *Oceanography.* a current flowing southwestward along the southeast coast of Africa; well defined and narrow, it is one of the swiftest currents in the world.

agyiophobia [a′jē ə fōb′ē ə] *Psychology.* an irrational fear of streets or of crossing streets.

agyria *Neurology.* a congenital malformation in which the surface convolutions of the cerebral cortex are not normally developed and the brain is abnormally small, resulting in mental retardation. Also, LISSENCEPHALY.

aH abhenry.

AH or **ah** ampere-hour.

AHA or **A.H.A.** American Heart Association; American Hospital Association

ahaptoglobinemia [ä hap′tə glō bi nē′mē ə] the absence of haptoglobin in the blood.

Aharonov-Bohm effect see BOHM-AHARANOV EFFECT.

AHC acute hemorrhagic conjunctivitis.

ahermatypic [ā hər mə tip′ik] *Invertebrate Zoology.* referring to corals that are not reef-building.

AHF antihemophilic factor.

AHG antihemophilic globulin; aggregated human hemoglobin.

ahlfeldite *Mineralogy.* $NiSeO_3 \cdot 2H_2O$, a transparent monoclinic brownish to reddish mineral occurring as crystalline crusts, having a specific gravity of 3.37 and a hardness of 2 to 2.5 on the Mohs scale; forms a series with cobaltomenite.

aholehole [ä hó′lá hó′lá] *Vertebrate Zoology.* a silvery tropical fish of the family Kuhliidae, found in Indo-Pacific marine and fresh waters. Also, **aholeahole.**

-aholic *Psychology.* a combining form used to form compounds on the model of *alcoholic,* with the idea of compulsive or addictive behavior resembling alcoholism, as in *workaholic.*

A horizon *Geology.* the uppermost layer of soil, consisting of decayed organic matter and incorporating several transitional subsurface subdivisions that have lost minerals through leaching. Also, ZONE OF LEACHING.

AI *Aviation.* the airline code for Air India.

AI or **A.I.** **1.** artificial intelligence. **2.** artificial insemination. **3.** airborne intercept(ion).

AIA or **A.I.A.** American Institute of Architects.

AIAA or **A.I.A.A.** American Institute of Aeronautics and Astronautics.

AIC or **A.I.C.** American Institute of Chemists.

AIChE or **A.I.Ch.E.** American Institute of Chemical Engineers.

aichmophobia [āk′mō fōb′ē ə] *Psychology.* an irrational fear of pointed objects.

AID or **A.I.D.** artificial insemination donor.

aided tracking *Ordnance.* an automatic weapons system radar used to minimize error; the equipment is machine-controlled in conjunction with the operator. Thus, **aided-tracking mechanism.**

AIDS *Medicine.* acquired immunodeficiency syndrome, an epidemic retroviral disease due to infection with human immunodeficiency virus (HIV-1), transmissible via blood or semen, and characterized by an ineffective immune response; the disease follows a protracted and debilitating course and has a poor prognosis. Those at risk include homosexual or bisexual males, intravenous drug abusers, hemophiliacs and other blood transfusion recipients, all sexual contacts of males in at-risk groups, and newborn infants of mothers with AIDS. (An acronym for the full name acquired immunodeficiency syndrome.)

AIDS-related complex *Pathology.* a condition representing a less severe form of infection with the HIV virus than AIDS itself, generally characterized by enlarged lymph nodes and by fever, weight loss, diarrhea, and minor opportunistic infections; identified as a precursor to AIDS in some instances. Also, **AIDS-related condition** or **syndrome.**

AIDS virus *Virology.* the retrovirus identified as a cause of AIDS; HIV.

aid to navigation *Navigation.* a term used to describe any man-made device used to mark a location, generally to show a safe course or warn of danger.

AIEE or **A.I.E.E.** American Institute of Electrical Engineers.

aigialium see AGIUM.

aigialophilous *Ecology.* thriving in beach environments. Thus, **aigialophile, aigialophyte.**

aiguille [ā gwēl´] *Geology.* a sharply pointed rock peak or pinnacle, as found in intensely glaciated mountain regions. (A French word meaning "needle.")

AIHA or **A.I.H.A.** American Industrial Hygiene Association.

AIIE or **A.I.I.E.** American Institute of Industrial Engineers.

Aiken, Howard 1900–1973, American mathematician; designed and built the Mark I, an early digital computer.

aikinite *Mineralogy.* $PbCuBiS_3$, an opaque, lead-gray orthorhombic mineral with a metallic luster, massive in habit or as prismatic crystals with vertically striated faces, having a specific gravity of 7.1 to 7.2 and a hardness of 2 to 2.5 on the Mohs scale; found in veins with gold, galena, and white quartz.

ailanthus [ā lan´thəs; ī lan´thəs] *Botany.* any of various trees of the genus *Ailanthus,* native to Asia and Australia, especially *A. altissima,* known as the tree of heaven and widely cultivated in North America as an urban shade tree.

aileron [á´lə rän] *Aviation.* **1.** one of a pair of hinged control surfaces located near the trailing edge of each wing of an aircraft. Ailerons work as differential pairs, creating opposing lifting forces on opposite sides of the aircraft to produce a rolling motion about the craft's longitudinal axis. **2.** of or relating to ailerons, especially to their function of controlling roll. Used to form compounds such as **aileron angle, aileron balance,** and **aileron control.** *Architecture.* a half gable, as at the end of a penthouse roof. (From a French word meaning "little wing.")

aileron roll see SLOW ROLL.

aileron tab *Aviation.* a trim tab on an aileron.

ailsyte [al´sīt] *Petrology.* an alkalic microgranite, largely composed of riebeckite. Also, PAISANITE.

ailurophobe [i lür´ə fōb´] *Psychology.* a person who has an irrational fear or hatred of cats.

ailurophobia [i lür´ə fōb´ē ə] *Psychology.* an irrational fear or hatred of cats. Also, GALEOPHOBIA, GATOPHOBIA.

AIME or **A.I.M.E.** American Institute of Mechanical Engineers.

aiming circle *Engineering.* an instrument that measures angles in azimuth and at elevation; used for artillery firings and topographic mapping.

aiming point *Ordnance.* **1.** the exact point to which a weapon is sighted or an observer's instrument is directed in order to hit the target. **2.** the point used to determine the release of bombs in aerial bombardment.

aiming screw *Mechanical Engineering.* a screw on an automotive vehicle that secures a headlight and allows it to be aimed vertically and horizontally.

aiming stake *Ordnance.* a specially colored or marked stake, used as a reference point for aiming a weapon. Also, **aiming post.**

aimless drainage *Hydrology.* the discharge of water from a surface that does not have a defined drainage system.

A/in.² amperes per square inch.

A indicator see A DISPLAY.

ainhum [īn´yüm] *Medicine.* the gradual and spontaneous amputation of a toe due to a constricting fibrous ring that develops in the digitoplantar fold, seen chiefly in black adult males in Africa.

aiophyllous [ī ō fil´əs] *Botany.* describing a perenially green plant; evergreen.

AIP or **A.I.P.** American Institute of Physics.

aiphyllophilous *Ecology.* thriving in evergreen woodland environments. Thus, **aiphyllophile, aiphyllophyte.**

air *Chemistry.* the invisible, odorless, and tasteless mixture of gases forming the earth's atmosphere. At sea-level pressures, dry air consists of (percentage by volume) nitrogen 78.00%, oxygen 20.95%, argon 0.93%, carbon dioxide 0.033% (current percentage; thought to be increasing), neon 0.0018%, helium 0.0005%, methane 0.0002%, krypton 0.0001%, and smaller amounts of nitrous oxide, hydrogen, xenon, and ozone. *Engineering.* powered by or delivering air. See compound terms beginning with AIR and also with PNEUMATIC. *Aviation.* of or relating to aircraft or to flight in the air. See compound terms beginning with AIR and also with AIRBORNE, AIRCRAFT, AERIAL, and AERONAUTICAL. *Ordnance.* of an artillery shell, bursting in the air above the intended target.

AIR *Mechanical Engineering.* air-injection reactor, a system installed in an automotive engine that mixes fresh air with exhaust gases in the exhaust manifold, causing reaction with any escaped and unburned or partially burned fuel from the cylinders.

air-acetylene welding *Metallurgy.* a welding process in which welding is effected by the combustion of acetylene with air; filler metal may or may not be used.

Airacomet *Aviation.* a popular name for the F-59 jet fighter.

Airacomet

air-actuated *Engineering.* triggered or powered by compressed air.

air adit *Mining Engineering.* a passage driven expressly to ventilate a mine.

air adjustment *Ordnance.* the correction of fire of ground weapons based on aerial observation.

air alert *Military Science.* **1.** an alert against possible enemy air attack. **2.** a method of air defense in which fighter aircraft remain airborne, ready for immediate action. Also, **air-alert method.**

air-arc furnace *Engineering.* a furnace used to power wind tunnels by superheating air so that it expands greatly and emerges from containment at supersonic speeds.

air-aspirator valve *Mechanical Engineering.* a one-way air intake valve on some automotive engines, installed on the exhaust manifold, that provides extra oxygen to the exhaust to convert carbon monoxide to carbon dioxide.

air-assist forming *Engineering.* a method of using air pressure to preform a sheet of thermosetting plastic before the sheet is molded.

air-atomizing oil burner *Engineering.* an oil burner in which compressed air causes the stream of fuel to form fine droplets of oil.

airbag or **air bag** *Mechanical Engineering.* a passive-restraint safety device for automobile passengers, consisting of a large bag that instantaneously inflates to provide cushioning during the impact of a collision.

air barrage *Mining Engineering.* an airtight wall that partitions a mine's ventilation gallery by enabling air to enter through one part and return through the other.

air base *Aviation.* a center of operations for military aircraft. *Photogrammetry.* **1.** the line joining two air stations from which overlapping photographs have been taken, or the length of such a line. **2.** the distance in a stereoscopic model between adjacent perspective centers as reconstructed in the stereoscopic plotting instrument.

air battery *Electricity.* a connected group of air cells.

air belt *Mechanical Engineering.* a pressure-equalizing chamber in a cupola or furnace.

air bind *Engineering.* an air blockage that prevents the passage of liquid through a pipe.

air bladder *Biology.* **1.** a sac or cavity containing air, found in various animals and plants. **2.** an organ of this type found in most bony fish; more often called a SWIM BLADDER.

air blast *Engineering.* any strong jet of air produced mechanically. *Mining Engineering.* a disturbance in underground workings accompanied by a strong rush of air.

air-blast circuit breaker (switch) *Electricity.* an electric switch that, as it opens, uses a high-pressure gas blast to break the arc.

air-blast freezing *Food Technology.* a freezing system in which large quantities of very cold air (to –40°F) are circulated over a product at high speed.

airblasting or **air-blasting** *Engineering.* any technique of forcing a strong jet of air against a surface or through an aperture.

air bleeder *Mechanical Engineering.* **1.** a small air valve in an automotive engine that allows the air-to-fuel mixture to be varied. **2.** a valve or device for removing air from a hydraulic system. Also, **air bleed.**

airboat *Naval Architecture.* a shallow-draft boat propelled by a rear-facing aircraft propeller and steered by a rudder acting upon the airflow created by that propeller; used in swamps or other areas of very shallow water. *Aviation.* see SEAPLANE.

air-bond die *Metallurgy.* an angle-forging die in which the material touches the die only at three points; the material is formed without striking the bottom of the die.

airborne *Biology.* carried by or in the air, as seeds, pollen, bacteria, and so on. *Aviation.* of an aircraft, above the ground or surface and entirely supported by air. *Ordnance.* designating gear or equipment that is carried on an aircraft but is not an inherent part of the aircraft. *Military Science.* of or relating to an action in which personnel, equipment, or supplies are transported to the objective site by air. Used to form many compound terms, such as **airborne attack, airborne command (post), airborne force, airborne operation, airborne troops, airborne unit**, and so on.

airborne aircraft hour see AIRBORNE HOUR.

airborne assault *Military Science.* an airborne operation to attack the enemy or secure enemy territory, especially one in which troops land from helicopters rather than parachuting to the ground.

airborne collision avoidance (system) *Navigation.* a system that warns aircraft of collision hazards and advises evasive action, and that is actually installed aboard aircraft, as opposed to a ground-based warning system. Also, **airborne collision warning (system).**

airborne control system *Geodesy.* a method of surveying in which distances and horizontal and vertical angles are measured from two or more known points to an aiming point on a helicopter hovering above an unknown location on the ground.

airborne detector *Engineering.* a sensitive receiver carried in an aircraft and used to observe and measure infrared energy, in order to identify an air or surface object.

airborne early warning *Ordnance.* a term for any system that employs airborne radar equipment to detect the approach of enemy aircraft or missiles. Also, **airborne early warning and control.**

airborne expendable bathythermograph *Oceanography.* a small, buoyant, bomb-shaped device that is ejected from an aircraft into the water; it records temperature at different depths, transmitting the readings directly to the aircraft's recorder.

airborne hour *Aviation.* a standard measure of aircraft operational time based on elapsed time actually in the air per aircraft.

airborne intercept radar *Engineering.* a form of airborne radar used to detect and intercept enemy aircraft or missiles.

airborne magnetometer *Engineering.* an instrument designed to provide airborne measurements of the intensity and direction of a magnetic field.

airborne oceanography *Oceanography.* the study of ocean phenomena by the use of aircraft and helicopters.

airborne pollutants see AIRBORNE WASTE.

airborne profile *Cartography.* a terrestrial profile produced by an airborne profile recorder in an aircraft following a predetermined flight path.

airborne profile recorder *Cartography.* a synchronized system of altimeters and cameras used to record and measure constantly the altitude of an aircraft, thereby producing a profile of the terrain below. Also, TERRAIN PROFILE RECORDER.

airborne radar *Engineering.* any radar system carried by an aircraft to assess the distance, direction, and speed of other airborne objects to be targeted or avoided.

airborne radiation thermometer *Oceanography.* an infrared-sensing instrument that is carried on an aircraft to measure ocean surface temperature.

airborne sea and swell recorder *Oceanography.* an FM continuous wave radar system that measures and records wave height from an aircraft.

airborne speed *Aviation.* the average speed of an aircraft from takeoff to landing, equal to the distance flown divided by the time airborne.

airborne warning and control system see AWACS.

airborne waste *Ecology.* a collective term for various contaminating substances, such as dust, smoke, fumes, or vapors, set aloft into the atmosphere as a result of combustion and chemical processes. Also, **airborne pollutants.**

air-bound *Engineering.* describing a condition in which air pressure or an air pocket prevents a machine from operating properly, or prevents the liquid in a pipe from flowing smoothly.

air brake *Mechanical Engineering.* **1.** a mechanism operated by compressed air acting on a piston that is used to stop or slow a moving element. **2.** an absorption-type dynamometer that dissipates power through the rotation of a fan or like device.

airbrasive *Mechanical Devices.* relating to a device that cuts or cleans by the action of abrasive material propelled by an air blast. *Medicine.* a dental instrument for removing plaque deposits or preparing a cavity by the high-speed action of a stream of abrasive particles propelled by compressed air.

air breaking see AIRBLASTING.

air-break switch see AIR SWITCH.

air-breakup *Space Technology.* the breakup of some part of a reentry vehicle during or after its reentry into the atmosphere.

air-breathing *Mechanical Engineering.* of an engine or aerodynamic vehicle, requiring air for combustion purposes. Thus, **air-breathing missile, air-breather.**

air brick *Materials Science.* an open-sided ceramic or metal brick designed to allow air inside a building.

airbrush *Graphic Arts.* **1.** a small pencil-shaped gun that uses compressed air to fine-spray paint for adjusting tonal quality on drawings and photographs. **2.** to alter a piece of artwork with an airbrush, as by painting out some element of the artwork.

air bump *Aviation.* a local air disturbance, or the rise and fall of an aircraft caused by such a disturbance.

air burst *Ordnance.* the exploding of a shell in the air rather than on the ground or surface. *Petroleum Engineering.* a geophysical technique in which an air gun towed by a seismographic vessel releases bursts of compressed air to produce sound waves; air bursts are not detrimental to marine life as are explosive charges.

airbus *Aviation.* **1.** a term for a short-range or medium-range passenger aircraft carrying large numbers of passengers, especially on a regular shuttle route. **2. Airbus.** a series of commercial aircraft of this type, produced by the British-French consortium Airbus Industrie.

air bypass valve *Mechanical Engineering.* a valve through which air is allowed to flow as an alternative to passage through the normal piping. Also, DIVERTER VALVE.

air cap *Mechanical Engineering.* a device that forms and directs a pattern of air in the atomization of various spray materials.

air capacitor *Electricity.* a capacitor using air as the dielectric material between its plates. Also, AIR CONDENSER.

air carbon-arc cutting *Metallurgy.* a process in which a metal or alloy is cut by applying the intense heat of an electric carbon arc and removing the molten metal with a blast of air.

air cargo *Transportation Engineering.* any goods moved by air, including baggage, mail, or freight.

air carrier *Transportation Engineering.* any operator providing air transportation of passengers or goods over an established airway as a principal business.

air cavalry *Military Science.* a military unit transported to the combat area by helicopters or other aircraft; usually a body of troops landed from aircraft rather than paratroops.

air cell *Electronics.* a cell where a positive electrode is depolarized chemically by oxygen reduction.

air chamber *Mechanical Engineering.* any chamber or compartment filled with air, especially one installed on a piston pump or reciprocating pump to equalize the pressure and flow of the pumped fluid.

air change *Engineering.* a quantity of air equal to the volume of the enclosure to be ventilated. Ventilation is measured in air changes per hour.

air check *Acoustical Engineering.* a recording of a broadcast signal from a radio station in order to log the effective range of the signal.

air classification *Materials Science.* the forced-air separation of dry, typically ceramic powders by utilizing the relationship between particle size and settling velocity in air. *Food Technology.* a system of grading flour particles, using a controlled air system to separate them by size and weight.

air classifier *Mechanical Engineering.* a device that uses currents of air to sort particles by size, or to separate coarse material from finer dust. Also, **air elutriator.**

air cleaner *Engineering.* any of various devices that remove impurities such as dust, cinders, or fumes from the air to keep them from entering an instrument or space; e.g., the air filter of an automobile engine.
air composition *Meteorology.* the constituting substances of air, with the amounts expressed as a percentage of the total volume or mass.
air compression *Physics.* a decrease in volume of air in a closed container, due to an externally applied force.
air compressor *Mechanical Engineering.* a machine that draws in air at one pressure, compresses it by increasing its density, then delivers it at a higher pressure. Thus, **air-compressor valve.**
air condenser *Mechanical Engineering.* **1.** an air-steam heat exchanger that uses air to condense the steam. Also, **air-cooled condenser. 2.** a device that removes vapors from a compressed-air line. *Electricity.* see AIR CAPACITOR.
air-condition *Mechanical Engineering.* **1.** to supply a building, room, or other enclosure with air conditioning. 2. to treat air by means of air conditioning. Thus, **air-conditioned.**
air conditioner *Mechanical Engineering.* **1.** any device that modifies or controls one or more aspects of air, such as its temperature, relative humidity, purity, or motion. **2.** such a device used in a building, room, vehicle, or other enclosed area to maintain the air therein at a comfortably cool and dry level.
air conditioning *Mechanical Engineering.* the maintenance of air in enclosed spaces to control factors such as its temperature, relative humidity, and flow, with consideration given also for removing impurities, such as dust and contaminant gases. Also, CLIMATE CONTROL.
air content *Materials Science.* the amount of air space in concrete, cement, or similar materials, not including the pore space within the constituent particles.
air control *Aviation.* the direction of aircraft, especially military aircraft, by a surface communication center.
air-control center *Telecommunications.* an area, for example in a submarine, that is used exclusively for the control of aircraft.
air conveyor see PNEUMATIC CONVEYOR.
air-cooled *Mechanical Engineering.* cooled by a stream of air rather than by water or another liquid coolant. Thus, **air-cooled engine.**
air-core *Electronics.* describing a core that is not magnetic, as one of plastic or fiber. Thus, **air-core inductance.**
air-core coil *Electronics.* a coil that is wound on a nonmagnetic core.
air-core transformer *Electronics.* a transformer that has two or more air-core coils.
air corridor *Aviation.* a defined route through a country's airspace, which it is permissible for foreign aircraft to use.
air course *Mining Engineering.* see AIRWAY.
air cover *Military Science.* **1.** the protection given to surface forces against attack by enemy aircraft. **2.** aircraft providing such protection.
aircraft *Aviation.* **1.** any weight-bearing vehicle designed for navigation in and through the air, supported by the action of air upon its surfaces or by the vehicle's own buoyancy. **2.** of or relating to such a vehicle. Used to form many compound terms, such as **aircraft antenna, aircraft cannon, aircraft engine, aircraft fuel, aircraft gun, aircraft rocket.**
aircraft carrier *Naval Architecture.* a warship that has a large flat upper deck for the purpose of launching and recovering aircraft, and that is otherwise designed and equipped to serve as a base for combat aircraft.

aircraft carrier

aircraft ceiling *Meteorology.* the height between the ground and the lowest cloud base, cloud layer, or obscuring phenomenon aloft that is reported as broken, overcast, or obscured but not classified as scattered, thin, or partial. *Aviation.* the maximum altitude at which a given aircraft can be operated safely.
aircraft detection *Engineering.* a general term for any of the various methods used to detect the presence of an aircraft, as through the use of radar technology, acoustics, or optics.
aircraft electrification *Meteorology.* the accumulation of a net electric charge on the surface of an aircraft.
aircraft icing *Meteorology.* the accumulation of ice on exposed surfaces of an aircraft, formed as the aircraft is flown through supercooled water drops in clouds or rain.
aircraft impactor *Engineering.* a device carried by an aircraft to gather samples of airborne particles.
aircraft instrument *Aviation.* a term for an instrument in an aircraft other than the engine instruments; i.e., a device whose purpose is to analyze the craft's actions or to detect and measure data during flight.
aircraft mile *Transportation Engineering.* a measure of aircraft usage based on the airport-to-airport distance.
aircraft mile flown *Transportation Engineering.* a measure of aircraft usage based on the actual number of miles from takeoff to landing.
aircraft propeller see PROPELLER.
aircraft propulsion *Aviation.* a term for any action other than gliding that provides thrust to move an aircraft through the air.
aircraft thermometry *Meteorology.* the science of air temperature measurement taken from an aircraft.
aircraft weather reconnaissance *Meteorology.* detailed weather observations made from an aircraft during flight.
air crossing *Mining Engineering.* an airtight mine passage, in which a return airway passes over or under an intake airway. Also, **air bridge.**
air-cure *Chemical Engineering.* to harden or vulcanize at room temperature without the help of heat. *Materials Science.* see AIR-SEASON. Thus, **air-cured, air-curing.**
air current *Fluid Mechanics.* any moving stream of air. *Mining Engineering.* the flow of air that ventilates the workings of a mine.
air curtain *Mechanical Engineering.* a continuous broad stream of high-velocity, temperature-controlled air circulated across a doorway or other opening, to reduce airflow in or out of the space, exclude insects, and so on.
air cushion *Mechanical Engineering.* **1.** a shock-absorbing cushion of air. **2.** a mechanical device that arrests motion without shock, by trapping or compressing air.
air-cushion vehicle *Mechanical Engineering.* a transportation craft that rides on a cushion of air over water and relatively level land, with the air pressure under the vehicle maintained by rotors or fans. Also, GROUND-EFFECT MACHINE.
air-cut *Engineering.* of a liquid system, impeded by the inadvertent inclusion of air into the system.
air cycle *Mechanical Engineering.* a refrigeration cycle using air as the working fluid, characterized by alternate expansion and compression of the air itself, rather than by a conventional refrigerant. Also, ROVAC CYCLE.
air cylinder *Mechanical Engineering.* any cylinder that uses air to drive a piston, or that contains and compresses air.
air defense *Military Science.* any effort to protect against or repel attacks by enemy aircraft or guided missiles.
air deficiency *Chemistry.* an insufficient amount of air in an air-fuel mixture, which causes incomplete combustion or improper ignition.
air-depolarized battery *Electricity.* a primary battery that is kept depolarized by atmospheric oxygen instead of chemical compounds.
air diffuser *Building Engineering.* a device, often a louver, designed to introduce air from an air-conditioning duct in a ceiling and mix it with air already in the room. Also, AIR OUTLET.
air discharge *Geophysics.* **1.** a type of lightning discharge that branches toward the ground but dissipates in the subcloud layer. **2.** an infrequently occurring electrical discharge that develops above an electrical storm.
air diving *Engineering.* a term for scuba diving in which the breathing mix has the same proportion of oxygen to nitrogen as exists in the earth's atmosphere.
air door *Mining Engineering.* **1.** a door in a mine roadway constructed to inhibit the passage of air. **2.** a door that redirects the air of a ventilating system in a predetermined direction by closing part of the circulating system.

air drain *Civil Engineering.* a cavity designed to prevent damp air from reaching the interior of a building.

air drainage *Meteorology.* a gravity-induced downslope flow of relatively cold air.

air-dried *Materials Science.* subjected to drying; having a moisture content that is in equilibrium with the atmosphere. Thus, **air-dried lumber.**

air drift *Mining Engineering.* **1.** a roadway, usually inclined, that is driven in stone to promote ventilation. **2.** a drift that joins a ventilation shaft with a fan.

air drill *Mechanical Devices.* a drill driven by compressed air; a pneumatic drill.

air drilling *Mechanical Engineering.* a form of drilling in which compressed air or gas is the circulation medium. *Petroleum Engineering.* such a technique used instead of mud drilling because of its greater speed.

airdrome *Aviation.* another term for an airport or airfield, especially in British or older U.S. use.

air drop *Ordnance.* the delivery of troops, equipment, or supplies to a location on the ground from aircraft in flight. Also, **air delivery.**

air-dry *Materials Science.* **1.** of an object or material, having a moisture content that is in equilibrium with the moisture of the surrounding atmosphere. **2.** to dry an object or material by exposure to air. Thus, **air drying. 3.** see AIR-SEASON.

air duct *Engineering.* a pipe or passageway constructed to convey air, as for cooling or heating a room or building, to supply air to a pneumatic device, and so on.

air-earth conduction current *Geophysics.* the part of the air-earth current that is generated by the electrical conduction of the atmosphere.

air-earth current *Geophysics.* the natural transfer of electrical charge between the positively charged atmosphere and the negatively charged earth.

air ejector *Mechanical Engineering.* an air-removing device, as found in condensers, that uses a jet of fluid to entrain and thus remove the noncondensable gases.

air eliminator *Mechanical Engineering.* a device in a steam distribution system that opens when air or other noncondensables reach it, and closes if water or steam is present in the vent body, for the purposes of removing the noncondensables from the system.

air embolism see AEROEMBOLISM.

air endway *Mining Engineering.* a narrow roadway that is driven in a coal seam chiefly for ventilation, and that runs parallel and near to a winning headway.

air engine *Mechanical Engineering.* a heat engine actuated by compressed air.

air-entraining agent *Materials.* any of the resins or other agents used in air entrainment.

air entrainment *Materials Science.* the use of resins or other agents in the blending of cement or concrete in order to trap small bubbles of air in the mix, so as to improve cohesion, workability, and resistance to extreme weather conditions. Thus, **air-entrained cement.**

air environment *Telecommunications.* the total airborne paraphernalia of any communications system.

air equivalence *Nucleonics.* a property of material used in ionization chambers to measure photon radiation, formulated so that the energy spectrum of electrons that are produced in the material by photon-absorptive processes will be identical to the energy spectrum in an air chamber.

air exposure *Radiology.* the quantity of radiation measured in air at a given point, expressed in roentgens and omitting the radiation from affected material.

air express *Transportation Engineering.* the rapid movement of goods by air.

air-fall deposition *Volcanology.* the accumulation of layered airborne volcanic debris following an explosive eruption.

air feed *Metallurgy.* in the processes used to coat materials by thermal spraying, the conveyance of a coating material into the heat zone by a stream of air.

airfield *Aviation.* an area of ground used for the takeoff and landing of aircraft.

air filter *Engineering.* a device attached to an air intake mechanism to remove solid impurities from an airstream; may be used with ventilating mechanisms or to prevent pollutants from entering an instrument or engine.

air float *Mining Engineering.* a process of separating finer particles from coarser particles by the action of air.

airflow *Fluid Mechanics.* **1.** a rate of movement for air, computed by volume or mass for a time unit. **2.** an amount of air that flows through a wind tunnel, or a relative airflow that streams over sections of a moving craft. *Meteorology.* a natural movement of air in the atmosphere.

airflow pipe *Engineering.* a pipe through which air is passed from one point to another. Similarly, **airflow duct, airflow orifice.**

airflow stack effect *Fluid Mechanics.* the change of pressure with height, in air moving in a vertical duct, because of a variation in temperature between the air outside the duct and the moving air.

airfoil *Aviation.* a body, part, or surface designed to provide a useful reaction on itself, such as lift or thrust, when in motion through the air. On various air vehicles, airfoils may provide suspension (e.g., wings), stability (fins), or propulsion (propeller blades). (Originally so called because they are wide and thin like a leaf or *foil.*)

airfoil profile *Aviation.* the form or contour of an airfoil, especially in cross section.

airfoil-vane fan *Mining Engineering.* a fan used in mines, characterized by an airfoil-shaped blade, in which the vanes are curved backward to move air in the general direction of the axis about which it rotates.

air force *Military Science.* **1.** the branch of a country's armed forces that conducts military operations by air. **2. Air Force.** this branch of the U.S. armed forces.

airframe *Aviation.* the structure of an aircraft, as opposed to its engine and accessories, including the fuselage, wings, empennage, landing gear (minus tires), and engine mounts.

air freight *Transportation Engineering.* **1.** any air cargo other than baggage or mail. **2.** an air carrier service for transporting goods.

air-fuel ratio *Chemistry.* a method of expressing the composition of a mixture of fuel and air for combustion by the measurement of either weight or volume.

air gap the space existing between two elements or parts; specific uses include: *Electrical Engineering.* a small section of air or another nonmagnetic material between two magnetically related components; purposely designed so that a certain amount of voltage is required to overcome the gap. *Engineering.* the area between the lowest opening in a faucet or other such device that empties into a plumbing fixture, and the point at which this fixture will overflow. *Petroleum Engineering.* in an offshore drilling operation, the distance between the surface level of the ocean and the bottom of the base of the drilling platforms. *Geology.* see WIND GAP.

air gas see PRODUCER GAS.

air gauge *Engineering.* **1.** an instrument that measures air pressure. **2.** a device used to compare a machined surface to a standard surface by assessing the rate of passage of air between them. Also, **air gage.**

airglow *Astronomy.* a faint luminosity observed in the sky, caused by charged particles from the sun striking atoms and molecules in the earth's atmosphere. *Geophysics.* the fairly steady emission of light found in the middle and lower latitudes, caused by the absorption of energy by certain molecules of the upper atmosphere during the day that is then released at night in the form of radiant light.

air grating *Building Engineering.* in a ventilation system, a grille on a building's exterior through which air enters or leaves the building.

air-ground *Telecommunications.* of or relating to communication between aircraft and ground stations. *Ordnance.* see AIR-TO-GROUND.

air-ground communication *Telecommunications.* **1.** two-way communication between aircraft stations and land stations. **2.** in military usage, any communications method, system, or equipment used to transmit messages between air-to-ground and ground-to-air stations.

air-ground detection *Forestry.* any fire detection system that oversees significant areas by ground detectors and aerial patrol.

air hammer see AIRLIFT HAMMER.

air handling *Mechanical Engineering.* the circulation or mixing of air by an air conditioning unit.

air-hardening steel *Metallurgy.* a steel that hardens fully when cooled in a gaseous medium from a temperature above its transformation range, even in sections that are thicker than two inches. Also, SELF-HARDENING STEEL.

airhead *Military Science.* an area in hostile territory that, if taken and held, can be used as a base for the air landing of troops and equipment.

air heater or **air-heating system** see AIR PREHEATER.

air heave *Geology.* a plastic deformation of sediments that results from the growth and enlargement of air pockets.

air hoist *Mechanical Engineering.* a lifting tackle, tugger, or air winch, operated by compressed air lifting a piston in a cylinder. Also, PNEUMATIC HOIST.

airhole *Mining Engineering.* a small excavation or hole that optimizes the efficiency of ventilation by communication with other workings or with the surface.

air horn *Mechanical Engineering.* the air inlet of a carburetor in an automotive engine that is controlled by the choke plate and the throttle plate, usually streamlined or contoured to decrease inlet losses and turbulence.

air horsepower *Mechanical Engineering.* the theoretical minimum power required for air delivery by a fan, blower, compressor, or vacuum pump with no losses. Also, AIR POWER.

air hunger *Medicine.* a condition marked by rapid and labored breathing, seen in severe diabetic acidosis and coma.

air-injection (system) *Mechanical Engineering.* a system used primarily in diesel engines in which a fuel pump controls the amount of fuel flowing into the injection valve; the valve then opens mechanically, and compressed air drives the fuel charge and some air into the combustion chamber.

air inlet *Mechanical Engineering.* a device or opening through which air is exhausted from or discharged into a conditioned space, as in grilles, registers, diffusers, and slots in an air-conditioning system.

air-insulation *Electricity.* the use of ambient air for the insulation of part of an electric circuit. Thus, **air-insulated.**

air intake *Aviation.* an air duct opening designed to utilize an aircraft's motion to direct air to an engine or ventilator. *Mining Engineering.* a mechanism designed to direct clean air at the lowest possible temperature into a compressor.

air-jet process *Textiles.* a technique for increasing the bulk of filament yarns by exposing the yarn to a turbulent stream of air. Also, **air-jet spinning.**

air knife *Engineering.* a mechanism that uses a thin, powerful jet of air to remove surplus coating from newly coated paper. *Food Technology.* a series or assembly of forced-air blowers that are positioned along a moving food production line in order to remove excess batter or other materials.

air-knife coating *Engineering.* the thin film remaining on coated paper after it has been processed with an air knife.

air lance *Engineering.* **1.** a length of pipe used to add compressed air into a system to dislodge settled sand and recommence the unimpeded flow of water. **2. air-lance.** to cleanse a boiler wall by applying a stream of pressurized air to it. Thus, **air lancing.**

air-landed *Military Science.* of personnel or supplies, transported by air and then unloaded after the aircraft has landed, as opposed to being dropped from the air while it is still airborne.

air-launch *Ordnance.* to launch a missile from the air rather than from the surface. Thus, **air-launched missile.**

air layering *Botany.* a method of reproducing a plant by cutting a slit in the stem and inserting sphagnum moss; the cut surfaces are then dusted with an auxinlike growth substance and covered; after the roots are formed, the stem is excised and planted.

air leakage *Mechanical Engineering.* **1.** the uncontrolled, inward flow of air through cracks in a building element and around the doors of a building or structure, caused by the pressure effects of wind or the indoor-outdoor air density difference. **2.** the air that escapes from a joint coupling in ductwork. Also, **air infiltration.**

airless injection *Mechanical Engineering.* an internal-combustion system in which a high-pressure fuel pump injects fuel directly into the cylinder, without the aid of compressed air.

airless spraying *Engineering.* the spray application of paint at high pressure through an opening that forms the substance into a mist of minute droplets. Also, HYDRAULIC SPRAYING.

airlift *Aviation.* **1.** an operation to transport personnel or supplies by air rather than by surface means, when surface transportation is impractical or impossible. **2.** to carry out such an operation. *Mechanical Engineering.* **1.** a system or device for lifting dry, fine granular material or powder through pipes, using compressed air as the fluidizing medium. **2.** see AIRLIFT PUMP.

airlift fermenter *Biotechnology.* a bioreactor, used for aerobic fermentation, in which a plant or animal cell suspension is kept mixed and aerated by the introduction of air at the base of the central draft tube.

airlift hammer *Mechanical Engineering.* a double-acting, gravity-drop power hammer that uses an air cylinder to raise the ram; used in drop forging for roughing out heavy forgings in foundry work. Also, AIR HAMMER.

airlift pump *Mechanical Engineering.* a device for extracting water or another liquid from a well or borehole, consisting of two concentric tubes; the inner tube is charged with compressed air and submerged in the liquid, thus forcing a mixture of air and liquid up through the outer pipe.

airlight *Optics.* a term for the background light from the sun and sky that is scattered into the eyes of an observer by suspended atmospheric particles lying in the observer's cone of vision.

airlight formula *Optics.* an equation that relates the apparent luminance of a remote black object, the apparent luminance of the background sky, and the absorptivity of the air layer near the ground, fundamental to the visual-range theory.

airline *Aviation.* **1.** an established, usually commercial system of aerial transportation, including the equipment and facilities that are used in its operation. **2.** a company that owns or operates such a system. Also, **air line.**

air line *Engineering.* a hose, duct, or pipe that delivers air, compressed or otherwise, to a pneumatic tool or other such device. *Materials Science.* a defect in glass in the form of a long, thin air bubble.

air-line lubricator see LINE OILER.

air-line main *Mining Engineering.* a pipe column used to supply air to a quarry face from a compressor.

airliner *Aviation.* a large passenger aircraft operated by an airline.

air lines *Spectroscopy.* additional spectral lines that result from the emission of energy by molecules in air during electric discharge, and not from the molecules in the sample.

air load *Aviation.* the total weight carried by an aircraft excluding its own weight; i.e., the combined weight of on-board crew, passengers, cargo, and fuel. *Fluid Mechanics.* the aerodynamic force exerted on an aircraft surface during flight.

air lock *Civil Engineering.* an airtight chamber used to permit passage between two areas of differing air pressure; e.g., workers in an underwater tunnel enter the air lock through a surface door and the pressure is gradually raised until it equals the pressure of the compressed air in the work area, which they then enter through a second door. *Engineering.* a pocket of air or gas in a pipe that impedes or stops the flow of liquid. *Mining Engineering.* a casing at the top of an upcast mine shaft, used to inhibit the leakage of surface air into the fan.

air-lock strip *Building Engineering.* weatherstripping attached to the edges of each wing of a revolving door.

air log *Aviation.* a mechanical device that measures and records the air distance flown by an aircraft or missle.

airman *Military Science.* **1.** an enlisted man in an air force. **2.** a rank in the U.S. Air Force, below a sergeant. **3.** an enlisted man in the U.S. Navy whose duties are concerned with aircraft. *Aviation.* any person directly involved in air operations, including flight crew, ground service personnel, and air traffic control personnel.

air mass *Meteorology.* an extensive body of air whose temperature, humidity, and thermal structure have only slight variations.

air-mass analysis *Meteorology.* the theory and practice of surface chart analysis that consists of determining the extent, physical and stability properties, movements, and modifications of air masses; locating and analyzing the structure and movement of the fronts separating the masses; analyzing wave perturbations on the fronts; and using these factors to describe the weather.

air-mass climatology *Meteorology.* the representation of a regional climate by the frequency and characteristics of the air masses above the region.

air-mass precipitation *Meteorology.* any precipitation that is attributed to moisture and temperature distribution within an air mass that is not being influenced by a front or by any orographic lifting at the location.

air-mass shower *Meteorology.* a shower produced by local convection within an unstable air mass.

air-mass source region *Meteorology.* a large area of the earth's surface over which bodies of air remain long enough to acquire characteristic temperature and moisture properties imparted by that surface.

air meter *Engineering.* an instrument used to measure the rate of flow of air and other gases.

air mile *Navigation.* a distance unit in air navigation, equal in length to a nautical mile, but used for distance measurement over land as well as water.

air mileage *Navigation.* distance flown as measured in air miles. Thus, **air mileage indicator.**

air mileage unit *Navigation.* an instrument that calculates air distance flown by an airplane and feeds this information to the air mileage indicator and other instruments.

air-mixing plenum *Mechanical Engineering.* a compartment in which a recirculating air supply is mixed with incoming outdoor air before being discharged to an air-conditioning system.

air monitoring *Civil Engineering.* the practice of ongoing or continuous sampling of the air, especially in terms of its level of pollution or radioactivity.

air motor *Mechanical Engineering.* any device in which the rotation of a rotor or the motion of a piston is actuated by compressed air.

air movement *Military Science.* a term for the transporting of units, personnel, equipment, or supplies by air.

air mover *Mining Engineering.* a portable compressed-air device serving as a blower or exhauster; used for emergency ventilation in workings where auxiliary fans cannot be installed.

air navigation *Navigation.* the various techniques and skills used in determining, directing, or monitoring the position of an aircraft in flight.

air navigation facility *Transportation Engineering.* any ground facility serving air operations, including landing fields, lights, weather information stations, radar stations, and so on.

airometer *Engineering.* **1.** see AEROMETER. **2.** see AIR METER.

air outlet *Mechanical Engineering.* a device at the end of a duct that supplies air to a conditioned space in a ventilation system. *Building Engineering.* see AIR DIFFUSER.

air parcel *Meteorology.* an imaginary body of air to which some or all of the dynamic and thermodynamic properties of atmospheric air are assigned.

air permeability *Materials Science.* the degree to which a given material or substance will permit the passage of air. Thus, **air permeability test.**

air pickets *Ordnance.* aircraft dispersed around an area to detect and report approaching enemy aircraft.

airplane *Aviation.* **1.** an engine-driven, heavier-than-air aircraft supported by the dynamic reaction of airflow over fixed wings. **2.** more generally, any heavier-than-air vehicle designed to fly in an atmosphere, including also gliders and helicopters.

airplane flare *Engineering.* a flare that is released from an airplane as a signal or source of light, especially one that is attached to a small parachute.

air plant *Botany.* a plant that grows on another plant rather than having its own roots, but that is not a parasite; an epiphyte.

air plot *Navigation.* a visual or graphic display of the position and movement of an aircraft or other airborne object.

air pocket *Engineering.* any air-filled space in an area otherwise occupied by a liquid; e.g., trapped air within a sunken ship or air bubbles in a water pipe. *Meteorology.* a downdraft or nearly vertical air current that can cause an aircraft to suddenly lose altitude. *Aviation.* a popular term for a nearly vertical air current that can cause an aircraft to suddenly lose, or sometimes gain, altitude.

air pollution *Ecology.* a general term for the contamination of the atmosphere by wastes from such sources as industrial burning and automobile exhausts. Thus, **air pollutant.**

airport *Transportation Engineering.* a landing field and related areas providing for the arrival and departure of aircraft, including facilities for passenger service, aircraft repair and maintenance, air traffic control, and so on.

airport advisory service *Navigation.* a service provided at smaller airfields that do not have a control tower; it provides pertinent landing and takeoff information, but does not constitute official air traffic control clearance.

airport alternate *Transportation Engineering.* a secondary airport to which arriving aircraft may be diverted if the primary destination airport is not available for landing.

airport code *Transportation Engineering.* a standard system in which each of the world's major airports is given a unique three-letter designation; for example, LAX is Los Angeles International, and SNN is Shannon, Ireland.

airport engineering *Transportation Engineering.* all of the processes that are involved in the planning, design, construction, and operational maintenance of airports.

airport radar service area *Transportation Engineering.* the area over which an airport radar facility provides normal aircraft monitoring and advisory service.

airport surface detection *Navigation.* a radar system in an air traffic control tower that shows the position on the ground of aircraft and service vehicles.

airport traffic see AIR TRAFFIC.

airport traffic area *Navigation.* the airspace within a certain horizontal radius, usually five statute miles, of the center of an airport with an operable control tower, extending upward to 3000 feet above the elevation of the airport. This area is normally restricted to the landing or taking off of aircraft from that airport.

airport traffic control tower see CONTROL TOWER.

air position *Navigation.* the three-dimensional location of an aircraft, including both its horizontal position and its altitude. Thus, **air-position indicator.**

air power *Military Science.* the total strength of the military air forces of a nation or fighting group. *Mechanical Engineering.* see AIR HORSEPOWER.

air preheater *Mechanical Engineering.* a device in steam boilers that transfers heat from flue gases to the combustion air before entering the combustion chamber, thus increasing flame temperatures, returning utilizable heat to the furnace, and increasing furnace efficiency.

air pressure *Physics.* the pressure exerted by air due to the motions of air molecules.

air-pressure drop *Fluid Mechanics.* the loss of pressure due to friction encountered along an air passage.

air propeller see PROPELLER.

air properties *Physics.* the characteristics of air as a gas, such as density, average molecular weight, specific heats, and boiling points.

air pump *Mechanical Engineering.* a centrifugal or reciprocating pump used to add or remove air from an enclosed space, or to remove condensate from the condenser of a steam plant.

air puncher *Engineering.* an air-driven chisel or pick.

air purge *Mechanical Engineering.* **1.** elimination of particulates from air within an enclosed system by the displacement of air. **2.** removal of air from a piping system or hot water heating system.

air quantity see AIR CURRENT.

air raid *Military Science.* any attack on an objective by aircraft, such as bombing or strafing. Also, **air strike.**

air-raid shelter *Civil Engineering.* a heavily reinforced structure, often underground, designed to withstand an air attack and provide safety.

air register *Engineering.* a device that controls the dispensing of air from a duct into an area being ventilated, heated, or cooled. Often called a grille.

air regulator *Mechanical Engineering.* any device for regulating the flow or pressure in a system, as in the burner of a furnace. *Mining Engineering.* an adjustable door in permanent air stoppings that regulates ventilating current.

air reheater *Mechanical Engineering.* **1.** a device that heats the return air for reuse in an air-conditioning system. **2.** any device in a heating system that heats or reheats circulating air.

air resistance *Fluid Mechanics.* the drag experienced by a body passing through air.

air pollution

air rifle *Ordnance.* a low-powered rifle that fires small metal pellets or balls by means of the force of compressed air.

air ring *Engineering.* in thermoplastics, a circular manifold that dispenses an even stream of cool air through a tubular form that passes through it.

air route *Navigation.* a path regularly used by aircraft in transit between two points.

air-route surveillance radar *Navigation.* radar used by an air route traffic control center to track aircraft operating between terminal areas.

air-route traffic control *Navigation.* a traffic control system that serves aircraft en route, particularly those on IFR flight plans, as opposed to aircraft at or in the vicinity of an airport. Thus, **air-route traffic control center.**

air sac *Vertebrate Zoology.* in birds, any of various air-filled pouches located between certain organs of the body, connected to the lungs and aiding in respiration and in maintaining body temperature. *Entomology.* a thin-walled structure in the tracheal system of some species of fast-flying insects, increasing the efficiency of respiration and aiding in flight. *Anatomy.* any air-filled body cavity, especially an alveolus of the lungs.

air sacculitis *Veterinary Medicine.* an inflammation of the air sacs of poultry or other birds. Also, **air-sac disease.**

air sampling *Engineering.* the collection of samples of air, either from the atmosphere or from a contained area such as a factory or mine, in order to evaluate it for such conditions as pollen content, amount of pollutants, level of radioactivity, and so on. *Civil Engineering.* see AIR MONITORING.

air scoop *Mechanical Devices.* an air intake cowl that projects from the surface of a vehicle such as an automobile or aircraft; it is oriented so that airflow is directed into the vehicle, for purposes such as ventilation or combustion.

air screw *Mechanical Engineering.* an older term for an aircraft propeller, especially in British use.

air-season *Materials Science.* to prepare a product, such as lumber, for final use by exposing it to the action of atmospheric air at normal temperatures. Thus, **air seasoning.**

air seeding *Agriculture.* the process of sowing seeds by scattering them onto the soil surface, as opposed to placing them in holes in the ground. Thus, **air seeder.**

air-sensitive *Chemistry.* describing a substance that undergoes decomposition when exposed to air.

air separator *Mechanical Engineering.* a device that uses a current of air to separate particles by density or by size.

airshaft *Building Engineering.* an open space within a building, designed to admit air to windows or vents.

airship *Aviation.* an older term for a dirigible or other such lighter-than-air craft.

air shooting *Engineering.* **1.** the process of intentionally leaving or creating pockets of air in a blast hole in order to lessen the effect of rock shattering. **2.** an explosion triggered in the air over a rock formation for the purpose of analyzing the ensuing seismic wave that is produced. Thus, **air shot.** *Geophysics.* a technique for sending seismic pulses through a given area of the earth by detonating explosives in the air above the surface. *Ordnance.* the charging of shot with pockets of air in order to minimize the effect of shattering.

air shower *Astrophysics.* a short-lived burst of subatomic particles produced when a cosmic ray or gamma ray strikes an atomic nucleus in the earth's atmosphere.

air sickness *Medicine.* motion sickness during air flight, characterized in severe cases by nausea, vomiting, headaches, and vertigo. Thus, **airsick.**

air-slake *Chemistry.* to slake a substance such as lime by exposure to moist air.

air sounding *Meteorology.* a measurement of atmospheric phenomena or conditions at varying altitudes by means of scientific measuring devices carried aloft by a balloon or rocket.

airspace *Navigation.* an area of air above the land area of a country, considered to be an integral part of that country's sovereign territory. *Aviation.* the envelope occupied by an aircraft, formation, or aerial maneuver. *Meteorology.* of or relating to both the earth's atmosphere and space. *Building Engineering.* the space that serves as insulation between the inner and outer walls of a structure. *Botany.* any gas-filled space in the tissue of plants, especially a large space.

air-spaced *Electromagnetism.* describing coaxial cable whose dielectric material is air.

airspeed *Aviation.* the velocity of an aircraft or other airborne body relative to the velocity of the surrounding air; it differs from ground speed to the extent that the air is in motion. Thus, **airspeed indicator.**

air spring *Mechanical Engineering.* a type of shock absorber used in automobiles that consists of a plunger operating in a cylinder containing oil and compressed air.

air stack *Aviation.* a technique of air traffic control, formerly in widespread use, to provide a sequence for the landing of multiple aircraft by placing them in a holding pattern one above the other, separated by a fixed altitude.

air-standard cycle *Thermodynamics.* a thermodynamic cycle involving a working fluid that is considered to be an ideal gas with properties of dry air, such as a volume of 0.7756 cubic meters per kilogram at standard atmospheric pressure and a constant specific heat ratio of 1.4. It is an ideal cycle used to assess the performance of actual types of internal-combustion engines.

air-standard Diesel cycle see DIESEL CYCLE.

air-standard engine *Mechanical Engineering.* a heat engine that operates in an air-standard cycle, so that its performance or efficiency may be compared to other engines. Thus, **air-standard efficiency.**

air-standard refrigeration cycle *Thermodynamics.* an air-standard cycle used to assess the performance of actual refrigeration systems or substances.

airstart *Aviation.* the starting of an aircraft's engines while in flight, as after the flameout of a jet engine.

air station *Aviation.* a U.S. naval station where aircraft are based. *Photogrammetry.* the point occupied by the perspective center of an aerial camera at the instant a photograph is taken.

airstrip *Aviation.* a narrow surface, paved or unpaved, designed or used for the takeoff and landing of aircraft.

air superiority *Military Science.* a combat situation in which one air force has a tactical and numerical advantage to the extent that the enemy air force cannot prevent it from carrying out its operations.

air-supply mask see AIR-TUBE BREATHING APPARATUS.

air support *Military Science.* an air operation designed to assist a surface force in accomplishing its objective, as by bombing an enemy installation being attacked by ground troops.

air supremacy *Military Science.* a condition of air superiority in which the enemy air force is not capable of providing any effective interference whatsoever.

air surveillance *Transportation Engineering.* the systematic monitoring of airspace and air traffic by electronic, visual, or other means. Thus, **air surveillance radar**.

air-suspension encapsulation *Chemical Engineering.* a method for microencapsulation of different types of solid particles. The particles go through cycles in which they are suspended by a vertical current of air while they are sprayed with a coating material; they are then floated by the airstream to another zone for a drying treatment.

air-suspension system *Mechanical Engineering.* the intermediate parts of an automotive vehicle between the frame and wheels that support the vehicle's frame and body with a cushion of air to absorb road shock, as with an air spring.

air sweetening *Chemical Engineering.* a process to oxidize lead mercaptides to disulfides, using oxygen or air.

air switch *Electricity.* a switching device in which the electric circuit is broken in air. Also, AIR-BREAK SWITCH.

air syringe see CHIP BLOWER.

air system *Mechanical Engineering.* a refrigeration system using air as the working fluid, in which air is compressed and the heat of compression is dissipated, while the air is being chilled by expansion and the performance of work.

air tanker *Forestry.* an aircraft used for fire control, fitted with tanks and equipment for releasing fire retardants.

air temperature *Meteorology.* **1.** a representation of the average kinetic energy of molecular motion in a small region of air as measured by a calibrated thermometer in thermal equilibrium with the air. **2.** the temperature that the air outside an aircraft is assumed to have as indicated by a cockpit instrument.

air terminal *Transportation Engineering.* a building providing services for airline passengers as well as space for administrative functions.

air thermometer *Engineering.* a thermometer that measures temperature by means of a bulb whose air pressure or volume changes with the temperature.

airtight *Engineering.* constructed or sealed in such a way as to prevent the passage of air.

air-to-air *Ordnance.* describing a missile launched from an aircraft toward a target in the air. Similarly, **air-to-ground, air-to-space, air-to surface, air-to-underwater,** and so on.
air-to-air resistance *Civil Engineering.* the resistance that a wall provides to the passage of heat.
air-track drill *Mining Engineering.* a heavy drilling machine, equipped with caterpillar tracks and operated by independent air motors, used for quarry or open-cast blasting.
air traffic *Navigation.* the movement of one or more aircraft in a given area, especially in the vicinity of an airport, including aircraft moving on the ground at an airport.
air traffic clearance *Navigation.* authorization given by air traffic control for aircraft to proceed within controlled airspace under specified conditions.
air traffic control *Navigation.* a governmental service to ensure the safe and orderly flow of air traffic within and between airports; its personnel maintain radio contact with pilots and authorize or direct them to land, take off, or taxi in the appropriate manner according to conditions of air traffic, weather, and so on.
air traffic control center *Navigation.* a facility where air traffic control functions are carried out.
air traffic controller *Navigation.* a person who is trained and authorized to provide air traffic control service.
air transportation *Transportation Engineering.* the conveyance of passengers and cargo by means of aircraft. *Mining Engineering.* a method using pneumatic pipelines to move and stow filling material in a mine. Also, **air transport.**
air trap *Civil Engineering.* a trap that utilizes a water seal to prevent the rise of foul air from sinks, basins, or sewers. *Engineering.* bubbles of air that impede the flow of a liquid through a pipe.
air-tube breathing apparatus *Mining Engineering.* a device consisting of a helmet or mask that supplies the wearer with oxygen by means of a flexible tube connected to a mouthpiece; used for smoke protection in mining rescues. Also, AIR-SUPPLY MASK.
air turbulence *Meteorology.* a highly irregular atmospheric motion caused by rapid changes in windspeed and direction and associated updraft and downdraft air currents.
air valve *Mechanical Engineering.* **1.** any valve that serves to control air passage. **2.** specifically, a valve that is located at the highest point in a liquid-carrying pipeline, and that automatically lets air in or out when the internal pressure of the accumulated air has reached atmospheric pressure.
air-variable capacitor *Electricity.* a capacitor composed of one rotating set and one fixed set of metal plates, where the capacitance is adjusted by rotating the movable plates to alter the overlap with the fixed plates.
air-velocity measurement *Fluid Mechanics.* the measurement of air velocity by any of various meters, such as orifice flowmeters, flow nozzle flowmeters, or Venturi flowmeters; the choice of which is generally dependent upon cost, required accuracy, and head loss.
air vessel *Engineering.* a small chamber of air attached to the pipeline on the delivery side of a reciprocating pump, serving to cushion the vibration of the pump. It may also be at other locations in a piping system to minimize water hammer.
air void *Materials Science.* an airspace in a material such as cement or concrete.
air volcano *Volcanology.* a type of mud volcano from which large volumes of gas, in addition to mud and stones, are given off. *Geology.* a miniature crater produced by explosions of gas and mud and resembling a true volcano.
air wall *Nucleonic .* in an ionization chamber, a wall designed to produce an ionization .itensity inside the chamber that is equivalent to that of air.
air washer *Mechanical Engineering.* **1.** a device that sprays water into an air stream to heat, cool, humidify or dehumidify the air. **2.** a unit that removes particulates and impurities from the air by spraying the air stream with liquid.
air-water jet *Engineering.* a high-pressure jet stream of air and water mixed together; used to clean surfaces.
air-water storage tank *Engineering.* a water storage tank having compressed air above its primary contents.
air-water vapor mixture *Physics.* a mixture of dry air and water vapor, as in the atmosphere.
airwave *Meteorology.* a strong portion of the westerly current with a wavelike oscillation of the wind flow aloft.
airwaves *Electronics.* usually, **the airwaves.** an informal term for: **1.** the radio waves used in radio and television broadcasting. **2.** radio and television broadcasting itself. **3.** the frequency spectrum used for broadcasting.
airway a route or passage for air; specific uses include: *Antomy.* the route by which air passes into or out of the lungs. *Medicine.* a tube that is inserted in the windpipe to maintain the passage of air, as during general anesthesia or artificial respiration. *Navigation.* **1.** an air route established by national or international authority, for which traffic control and advisory services are available. **2.** specifically, in the United States, an established air route ten miles wide and extending to an altitude of 27,000 feet. *Building Engineering.* a ventilation passage between thermal insulation and the underside of a roof. *Mining Engineering.* an underground passage through which air is carried.
airway beacon *Navigation.* a lighted aid to air navigation that is used to mark an airway in remote mountain areas; the light flashes in Morse Code for identification.
airways forecast see AVIATION WEATHER FORECAST.
airways observation see AVIATION WEATHER OBSERVATION.
air wedge *Optics.* an arrangement of a wedge-shaped film of air between two flat reflective surfaces that produces light and dark bands of localized interference.
air well see AIRSHAFT.
airworthiness *Aviation.* the ability of an aircraft to operate safely in all the flight conditions to which an aircraft of its type will be subjected; the FAA and other national authorities issue airworthiness regulations and specifications for various classes of aircraft.
airworthy *Aviation.* of an aircraft, able to operate safely in all the flight conditions to which it will be subjected.
Airy, Sir George Biddell 1801–1892, English astronomer; discovered an inequality in the motions of Venus and Earth; calculated the earth's mean density.
Airy differential equation *Mathematics.* the differential equation $w'' - zw = 0$, where w is a function of the variable z. This arises in the study of light diffraction near a caustic surface.
Airy disc *Optics.* the image of a point source of light produced by an aberration-free lens, occurring as a bright, circular central area that is surrounded by concentric dark and light rings. Also, DIFFRACTION DISC.
Airy function *Mathematics.* a Bessel function of order $1/3$ which occurs as part of a pair of solutions to the Airy differential equation.
Airy isostasy [i säs´tə sē] *Geophysics.* a theory that explains the hydrostatic equilibrium of the Earth's surface by stating that mountains float on a base of fluid lava higher than do other features of the Earth's crust, and that the mass and roots of mountains increase along with their height.
Airy phase *Acoustics.* an acoustic wave produced by an explosion occurring in a bed of shallow water having a flat bottom.
Airy points *Engineering.* the optimal location for the support points of a bar being placed horizontally so as to minimize bending.
Airy stress function *Mechanics.* a biharmonic function whose second partial derivatives with respect to spatial coordinates give the components of stresses in a homogeneous linear elastic material.
AISI American Iron and Steel Institute.
AISI steel *Metallurgy.* any steel made to conform to the standards of the American Iron and Steel Institute.
aisle *Architecture.* **1.** a passage between sections of seats, as in a church or auditorium. Also, **aisleway. 2.** the sections flanking a church nave, usually separated from it by columns.
AIST or **A.I.S.T**. Agency for Industrial Science and Technology.
Aistopoda [ā´ə stäp´ə də] *Paleontology.* an order of small, snakelike amphibians in the extinct subclass Lepospondyli; characterized by the absence of limbs or girdles; formerly placed with the Stegocephalia; extant in the Permian to Carboniferous.
aithalophilous *Ecology.* thriving in evergreen thicket environments. Thus, **aithalophile, aithalophyte.**
Aitken, John 1839–1919, Scottish physicist; identified Aitken nuclei.
Aitken, Robert Grant 1864–1951, American astronomer; discovered approximately 3000 binary star systems
Aitken dust counter *Engineering.* an instrument that is used to determine the amount of dust present in the atmosphere. Also, **Aitken nucleus counter.**
Aitken nuclei *Meteorology.* a collection of microscopic particles in the atmosphere that serve as condensation nuclei for droplet growth under conditions of rapid adiabatic expansion produced by an Aitken dust counter.

Aitken's formula *Astronomy.* a mathematical formula for calculating the separation limit for true binary stars.

Aitoff equal-area map projection *Cartography.* a Lambert equal-area azimuthal projection of a hemisphere converted into a projection of the entire sphere; a projection bounded by an ellipse in which the line representing the equator is twice as long as the line representing the central meridian.

AIY *Aviation.* the airport code for Atlantic City, New Jersey.

Aizoaceae [ā īz ə wā´sē ē] *Botany.* a family of apetalous, mostly succulent plants, with opposite leaves, comprising 22 genera and 500 species, found primarily in South Africa.

Ajellomyces [ā´jel ə mī´sēz] *Mycology.* a genus of fungi belonging to the order Gymnoascales, occurring primarily in tropical climates and causing Gilchrist's disease in humans.

ajmaline [aj´mə lēn´] *Organic Chemistry.* $C_{20}H_{26}N_2O_2$, a yellowish, crystalline alkaloid found in the plant *Rauwolfia serpentina*; slightly soluble in water, and soluble in alcohol and ether; melts at 158–160°C; used medicinally.

AK-47 *Ordnance.* a widely used Soviet-made 7.62-mm assault rifle; it is lightweight, short-barreled, gas-operated and fires rifle-type cartridges at a rate of 600 rounds per minute and a muzzle velocity of 2330 feet per second; full name, **AK-47 Kalashnikov.**

Akabane virus disease *Veterinary Medicine.* a disease caused by the Akabane virus, carried and transmitted by mosquitoes, and resulting in congenital abnormalities of the central nervous system in ruminants.

akaganeite [a´kə gan´ē īt] *Mineralogy.* β-$Fe^{+3}O(OH,Cl)$, a tetragonal mineral that has a specific gravity of 3.555 and is polymorphous with feroxyhyte, goethite, and lepidocrocite; found in mines, brines, lunar rocks, and meteorites.

Akaniaceae [äk´ə nē ās´ē ē] *Botany.* a monospecific family of small trees in the order Sapindales that produce alkaloids and proanthocyanin, but not ellagic acid; native to eastern Australia and characterized by large, alternate, and pinnately compound leaves and offset leaflets.

akaryote *Cell Biology.* any cell that lacks a nucleus. Also, **akaryocyte.**

akathisia [ak´ə thiz´ē ə] *Medicine.* a pronounced inability to sit still or relax, either a physiological reaction to certain drugs or a psychological reaction to extreme anxiety.

akee see ACKEE.

akenobeite [a´kə nōb´ē it] *Petrology.* a light-colored form of aplite consisting of orthoclase and oligoclase with small amounts of quartz in the interstices.

akerite [ô´kə rīt´] *Petrology.* a syenite containing a variety of soda microcline, oligoclase, and augite.

Akerlund diaphragm *Radiology.* a spiral, adjustable lead device used in roentgenography to direct a radiation beam to a circle of variable size.

akermanite [äk´ər mə nīt´] *Mineralogy.* $Ca_2MgSi_2O_7$, a colorless or grayish tetragonal mineral of the melilite group that is isomorphous with gehlenite, has a specific gravity of 2.9 and a hardness of 5 to 6 on the Mohs scale, and is found in furnace slags and in calcium-rich eruptive and metamorphic rocks.

AKF diagram *Petrology.* a triangular diagram that shows the chemical character of a metamorphic rock in which the three principal points being plotted are $A = Al_2O_3 + Fe_2O_3 + (CaO + Na_2O)$, $K = K_2O$, and $F = FeO + MgO + MnO$.

akimbo [ə kim´bō] *Anatomy.* relating to the distance from one elbow to the other with the hands on the hips and the elbows out.

akinesia [ak´ə nēz´yə] *Medicine.* a partial or total loss of muscle movement due to peripheral or central nervous system abnormalities. Also, ACINESIA. *Entomology.* a temporary immobility of an insect caused by damage to or loss of sensory organs such as the antennae. Also, **akinesis.**

akinete [ak´ə nēt´; ə kī´nēt] *Botany.* a thick-walled spore formed during unfavorable conditions by certain blue-green algae.

akinetic [ak´ə net´ik] *Biology.* relating to or affected by akinesia.

akinetic autism see COMA VIGIL.

Akins' classifier *Mining Engineering.* a mechanism consisting of an interrupted-flight screw conveyor operating in an inclined trough; used to separate fine-size solids from coarser solids in a wet pulp.

AKO *Artificial Intelligence.* **1.** a kind of. **2.** see IS-A LINK.

akoria see ACORIA.

akrochordite *Mineralogy.* $Mn_4^{+2}Mg(AsO_4)_2(OH)_4{\cdot}4H_2O$, a reddish-brown monoclinic mineral occurring in aggregates of 0minute, nearly parallel crystals, having a specific gravity of 3.26 and a hardness of 3.5 on the Mohs scale.

aktology *Ecology.* the study of the conditions, sediments, or life of inshore or shallow water areas. Thus, **aktological.**

AL *Robotics.* Arm Language, a robotic language developed at Stanford University for research and development of prototype robotic systems.

Al the chemical symbol for aluminum.

al- a prefix meaning "toward," as in *alliance.*

ala [ā´lə] *plural,* **alae.** [ā´lē] *Biology.* a wing or winglike part, structure, or process. *Anatomy.* one of the flaring cartilaginous expansions of the nose. *Botany.* any of the side petals of certain flowers. *Invertebrate Zoology.* a finlike ridge on the cuticle of nematodes.

Ala alanine.

alabandite *Mineralogy.* $Mn^{+2}S$, an iron-black, opaque, cubic mineral with a submetallic luster having a specific gravity of 3.95 to 4.04 and occurring in granular or massive forms, mainly in metallic sulfide vein deposits.

alabaster *Mineralogy.* **1.** $CaSO_4{\cdot}2H_2O$, a fine-grained, compact and massive variety of gypsum, usually colorless to white, that is used as an ornamental building material. **2.** see ORIENTAL ALABASTER.

alae cordis *Entomology.* winglike bands of fibrous tissue that connect the heart to the pericardium in insects and other arthropods.

alalia [ə lā´lē ə] *Neurology.* an inability to speak, due to organic or functional impairment of the articulatory muscles. Thus, **alalic.**

alamalt *Food Technology.* a powder derived from sweet potatoes, used in making candy.

alamosite [al´ə mō´sīt] *Mineralogy.* $PbSiO_3$, a colorless to white, monoclinic lead silicate mineral occurring as fibrous crystals having a specific gravity of 6.5 and a hardness of 4.5 on the Mohs scale. (Named for the mining town of *Alamos,* Mexico, where it occurs as a vein mineral of the oxidized zone.)

Alangiaceae *Botany.* a monogeneric family of woody and sometimes thorny Old World plants in the order Cornales; characterized by articulated lactifers, spirally arranged leaves, and usually bisexual flowers growing on axillary cymes.

alanine *Biochemistry.* $C_3H_7NO_2$, an amino acid that plays a central role in the metabolism of amino acids; derivable from proteins and thus not essential in the diet.

alanine aminotransferase [ə mēn´ō tranz´tər ās] *Enzymology.* an enzyme that catalyzes the conversion of alanine and 2-oxoglutarate to glutamate and pyruvate; the reaction transfers nitrogen for excretion or for incorporation into other compounds.

alanyl *Organic Chemistry.* CH_3CHNH_2CO–, the acyl radical of alanine.

alar *Biology.* **1.** of or relating to a wing. **2.** wing-shaped.

Alariaceae [ə lär´i ā´sē ē] *Botany.* a family of kelp belonging to the order Laminariales and characterized by a blade that produces lateral outgrowths or pinnae; widely distributed in cold to temperate waters of the Northern and Southern Hemispheres.

alarm behavior see WARNING BEHAVIOR.

alarm gauge *Engineering.* a device that gives off a warning signal if the pressure of steam or level of water in a boiler reaches a crucial level.

alarm reaction *Physiology.* the first stage of the large-scale or complex response of an organism to stressful stimuli to which it is not adapted.

alarm song *Entomology.* a stress signal made by many families of beetles.

Alaska, Gulf of *Geography.* a broad inlet of the North Pacific off southern Alaska.

Alaska Current *Oceanography.* a current that flows generally northwestward along the northwest coast of North America; it contains water from the North Pacific Current and is therefore warmer than the water through which it flows.

Alaska Integrated Communications Exchange *Electronics.* see ALICE.

alaskite *Petrology.* a light-colored granite rock with only a few percent or less of mafic minerals.

alate [ā´lāt] *Biology.* having wings, or parts resembling wings. *Entomology.* undergoing a winged phase, as in ants that periodically swarm in mating flights to establish new colonies and lose their wings once the colony is established.

Alaudidae [ə laü´də dē] *Vertebrate Zoology.* the larks, a cosmopolitan family of passerine birds in the suborder Oscines, characterized as good fliers and runners and feeding on seeds, insects, and small invertebrates.

A layer *Geology.* in the classification of the earth's interior, the outermost seismic region, extending downward from the surface to the Mohorovicic discontinuity.

ALB *Aviation.* the airport code for Albany, New York.

alba *Anatomy.* referring to a white tissue or structure.

albacore [al´bə kôr´] *Vertebrate Zoology.* a long-finned species of tuna, *Thunnus alalunga,* found in tropical seas and important as a food and game fish.

Albada finder *Optics.* a viewfinder in which bright or white markings are reflected from the rear surface of the objective lens and form an outline framing the field of view of the camera lens.

albafite *Mineralogy.* a fusible variety of bitumen, containing as much as 15% oxygen, whose color is greenish to brownish but turns white when exposed to air; varies greatly in hardness and porosity.

albanite *Petrology.* a melanocratic leucitite found near Rome, Italy.

albarium *Materials.* a stucco made from powdered marble and lime, widely used in ancient buildings.

albatross *Vertebrate Zoology.* any of several large-winged soaring sea birds of the family Diomedieidae in the order Procellariformes, the largest oceanic birds; the **wandering albatross,** *Diomedea exulans,* is found in southern oceans and is the subject of many nautical legends.

albatross

al-Battani c. 850–929, Arab astronomer; discovered that the earth moves in a varying ellipse.

albedo [al´bē dō] *Optics.* the fraction of incident light that is reflected in all directions from an uneven surface, especially from a star.

albedometer [al´bi däm´ə tər] *Engineering.* a device that determines the albedo (reflecting power) of a surface.

albedo particles *Geophysics.* any particles freed from the earth's atmosphere by the effect of radiant energy from space. Also, **albedo neutrons.**

Alberger process *Chemical Engineering.* a process for manufacturing salt by heating brine under high pressure, then passing it to a graveler to remove calcium sulfate.

Albers (map) projection *Cartography.* a conical equal-area projection on which the meridians are straight lines that meet in a common point beyond the limits of the map, and the parallels are concentric circles having their center at the point of intersection of the meridians.

Albert, Lake *Geography.* a lake through which the Upper Nile River flows, between Uganda and Zaire.

Alberta low *Meteorology.* a body of air with lower pressure relative to that of the surrounding air that is centered on the eastern slope of the Canadian Rockies in the province of Alberta, Canada.

Albertan *Geology.* North American provincial series of the Middle Cambrian period, occurring after the Waucoban and before the Croixian. Also, ACADIAN.

Alberti, Leone Battista 1404–1472, Italian architect; formulated the elementary laws of perspective.

albertite *Mineralogy.* a black variety of bitumen having a specific gravity of 1.097 and a hardness of 1 to 2 on the Mohs scale, with a brilliant luster and a conchoidal fracture.

Albert's stain *Biotechnology.* a dye or stain used to detect metachromatic granules; a mixture of toluidine blue and malachite green is added to ethanol and acetic acid.

Albertus Magnus 1206–1280, German scholastic scientist; the preeminent medieval man of science; teacher of Thomas Aquinas.

Albian *Geology.* a European geologic stage of the uppermost Lower Cretaceous period, occurring after the Aptian and before the Cenomanian. Also, SELBORNIAN.

albic horizon *Geology.* a soil horizon from which clay and free iron oxides have been depleted or in which the iron oxides have been chemically rearranged after deposition.

albinism [al´bə niz´əm] *Biology.* the fact or condition of being an albino; an abnormal lack of coloration. Thus, **albinistic, albinotic.** *Medicine.* the congenital partial or total absence of pigment from the hair, skin, and eyes.

albino [al bí´nó] *Medicine.* a person having a congenital absence of all normal coloration in the skin, hair, and eyes, due to the inherited inability to produce melanin pigmentation. *Zoology.* an animal having an abnormal absence of pigmentation in the eyes, hair, and skin. *Biology.* a plant having a total or partial lack of natural pigments or chlorophyll.

albite *Mineralogy.* $NaAlSi_3O_8$, a colorless to white triclinic mineral of the feldspar group (plagioclase series), massive in habit or as commonly twinned crystals found in most rock groups, with a specific gravity of 2.61 to 2.63 and a hardness of 6 to 6.5 on the Mohs scale. Thus, **albitic.**

albite-epidote amphibolite facies *Petrology.* metamorphic rocks formed under intermediate temperature and pressure conditions by regional metamorphism or in the outer contact metamorphic zone.

albite law *Crystallography.* a twin law that specifies the orientation of alternating layers in a twin feldspar crystal in which the twinning plane is brachypinacoid, as in albite.

albitite *Petrology.* a coarse-grained porphyritic dike rock composed of a granular aggregate of albite and small amounts of muscovite, garnet, apatite, quartz, and opaque oxides.

albitization *Geology.* a process whereby the mineral albite is introduced into a sediment or rock, or replaces another mineral.

albitophyre *Petrology.* a porphyritic aphanitic rock containing albite phenocrysts in a ground mass composed mostly of albite.

albizia [al bits´ē ə] *Botany.* any of several trees or shrubs of the genus *Albizia,* native to warm or tropical climates, including several species grown as ornamental trees. Also, **albizzia.**

Alboll *Geology.* a suborder of the soil order Mollisol having an albic horizon and other distinct horizons above a seasonally perched water table.

albomycin [al´bə mīs´ən] see SIDEROMYCIN.

alb- or **alba-** a combining form meaning "white." (From Latin *albus.*)

alboranite *Petrology.* an olivine-free hypersthene basalt, which contains calcic plagioclase and hypersthene phenocrysts in a ground mass of calcic-labradorite, clinopyroxene, and magnetite.

Albuginaceae [al´byoo jin a´sē ē] *Mycology.* a family of fungi belonging to the order Peronsporales, composed of the white rusts, which are parasitic to plants.

albuginea [al´byoo jin´ē ə] *Histology.* a tunic or layer of dense white fibrous connective tissue that covers an organ or structure.

Albugo [al´byoo ´gō] *Mycology.* a genus of fungi belonging to the order Peronosporales; a plant parasite that produces the white rusts.

Albulidae [al´byool´i dē] *Vertebrate Zoology.* the bonefishes, a family of herringlike fish of the order Elopiformes; known for its eellike larval stage and found in all warm seas, but especially in shallow coastal waters, bays, and estuaries.

albumen [al´byoo´min] *Cell Biology.* the white of an egg, composed primarily of the water-soluble protein albumin. *Biochemistry.* see ALBUMIN.

albumen plate see SURFACE PLATE.

albumin [al´byoo´min] *Biochemistry.* a class of proteins in plants and animals, dissolving in water and coagulating when heated; found in egg white, milk, blood, and many other animal and vegetable tissues and secretions. In general, albumins from animal sources are of higher quality than those from vegetable sources, because animal proteins contain greater quantities of essential amino acids. *Hematology.* see SERUM ALBUMIN.

albumin A *Hematology.* the normal type of human serum albumin, as opposed to electrophoretic variants.

albuminate *Biochemistry.* a compound formed by the action of a base or an acid on albumin

albumin gene *Genetics.* a gene that codes for albumin, the major serum protein found in adult mammals.

albumin-globulin ratio *Medicine.* the ratio of albumin to globulin in blood serum, plasma, or urine, especially as measured for various types of kidney disease.

albuminoid *Biology.* relating to or resembling the animal and plant proteins called albumins. Also, **albuminous.** *Biochemistry.* **1.** resembling albumin. **2.** a fibrous protein, insoluble in water, that occurs in the hard parts of the body, which it supports and protects by virtue of its fibrous structure; examples are collagen and keratin. Also, SCLEROPROTEIN.

albumin suspension test *Pathology.* an analysis of the meconium (first feces of the newborn) for the presence of undigested albumin. The meconium is mixed with an equal portion of water, then shaken and centrifuged, and 10% trichloroacetic acid is then added. A white precipitate demonstrates the presence of undigested albumin, an indicator of meconium ileus.

albuminuria [al´byü mə nur´ ē ə] *Medicine.* the presence of albumin in the urine.

albumose [al´byə mōs´] *Biochemistry.* a protein derived by the action of certain enzymes, such as pepsin, on albumin.

alburnum *Botany.* the lighter, softer wood of a tree, between the bark and the harder center; sapwood.

ALC automatic level control.

alc. or **alc** alcohol.

Alcaligenes [al´kə lij´ə nēz´] *Bacteriology.* a widespread genus of Gram-negative, aerobic, rod-shaped, alkaline-producing bacteria of the family Achromobacteraceae, occurring in the intestinal tracts of vertebrates and as part of the normal skin flora, occasionally causing opportunistic infections.

Alcator [al´kād´ər] *Physics.* a device for confining a dense plasma within a toroidal chamber having a high toroidal magnetic field.

Alcedinidae [al´sə din´ə dē] *Vertebrate Zoology.* the kingfishers, a family of colorful, large-headed birds in the order Coraciiformes; found worldwide.

alchemist [al´kə mist] *Chemistry.* a person who practiced or studied alchemy.

alchemy [al´kə mē] *Chemistry.* a pseudoscience that existed from about 500 BC to about 1600 AD, and widely practiced in Europe in the Middle Ages. It was primarily concerned with attempts to transform base metals into gold and discover a universal cure for disease. Alchemy involved the experimental analysis of various materials and thus can be regarded as the forerunner of the modern science of chemistry.

Alcidae [al´sə dē] *Vertebrate Zoology.* a family of marine fish-eating birds in the order Charadriiformes, including the auks, puffins, razorbills, and guillemots. They are found chiefly on northern coasts.

Alciopidae [al´sē äp´ə dē] *Invertebrate Zoology.* a pelagic family of errantian worms in the class Polychaeta.

Alclad *Chemistry.* a trade name for a compound that is highly resistant to seawater corrosion, made from the aluminum alloy, duralumin, with a pure aluminum coating.

ALCM *Ordnance.* **1.** an acronym for air-launched cruise missile. **2.** specifically, a U.S. cruise missile designed to be carried and launched by the B-52 and B-1 bombers; it is powered by a turbofan engine, equipped with inertial guidance, and delivers a thermonuclear warhead at a speed of approximately 500 mph and a range of up to 1550 miles. Code name, **AGM-86B.**

alcogel *Chemistry.* a colloidal gel made with alcosol.

alcohol [al´kə hôl; al´kə häl] *Organic Chemistry.* **1.** any of a general class of organic compounds formed by the attachment of one or more hydroxyl (–OH) groups to carbon atoms in place of hydrogen atoms; e.g., methanol, ethanol, propanol, butanol. ROH is the general formula, with R indicating an aliphatic (hydrocarbon) radical and a hydroxyl group (–OH). **2.** specifically, C_2H_5OH, the transparent, colorless, volatile liquid that is the intoxicant in beverages such as wine, beer, or whiskey, known technically as ethyl alcohol or ethanol. See ETHANOL.

alcohol abuse *Medicine.* a term for a severe degree of alcoholism in which the individual's ability to function socially and occupationally is seriously affected.

alcoholate *Organic Chemistry.* a substance formed when an alkali metal reacts with an alcohol. Also, ALKOXIDE.

alcohol C-9 see NONYL ALCOHOL.

alcohol dehydrogenase *Biochemistry.* a zinc (Zn^{2+}) enzyme of the oxidoreductase class that catalyzes the breakdown of ethanol to acetaldehyde; this reaction is the first step in the metabolism of alcohols by the liver. Also, ACETALDEHYDE REDUCTASE.

alcohol dependence *Medicine.* a term for a less severe degree of alcoholism than alcohol abuse, in which the individual's ability to function is not as significantly affected.

alcohol fuel *Materials Science.* any automotive fuel consisting of or containing alcohol, especially gasohol.

alcoholic *Chemistry.* relating to, containing, or having the properties of alcohol. *Medicine.* **1.** a person suffering from alcoholism. **2.** relating to alcoholism or its effects.

alcoholic beverage *Food Technology.* any drink that contains ethyl alcohol, including fermented beverages such as wine or beer, and distilled beverages such as whiskey, vodka, gin, or rum.

alcoholic cirrhosis *Medicine.* the proliferation of fibrotic tissue in the liver (secondary to alcohol use) resulting in loss of viable hepatocytes and increased pressure in the flow of blood through the liver.

alcoholic fermentation *Microbiology.* a typical metabolic process of certain yeasts and other microorganisms, in which sugars are broken down to form carbon dioxide and ethyl alcohol under anaerobic or microaerobic conditions; used in making wine and beer and in baking.

alcoholism *Medicine.* a chronic condition characterized by the habitual consumption of alcoholic beverages to such an extent that it impairs physical or mental health and interferes with the activities of daily living; it may be progressive in nature and potentially fatal when producing pathological changes in organs.

alcoholometer *Engineering.* a hydrometer or other device that determines the amount of alcohol in an aqueous solution. Also, **alcoholimeter, alcoholmeter.**

alcohol oxidase *Enzymology.* an enzyme that catalyzes the oxidation of an alcohol.

alcoholysis *Organic Chemistry.* the reaction, similar to hydrolysis, that occurs between alcohol and another organic compound.

Alcor *Astronomy.* 80 Ursae Majoris, a fourth-magnitude star that forms a visual companion to Mizar in the handle of the Big Dipper.

alcosol *Chemistry.* a solution made with a mixture of a colloid and an alcohol.

alcove *Architecture.* **1.** a small, recessed space opening into a larger room. **2.** an arched opening in a wall; a niche. *Geology.* a large, deep niche formed by a spring or stream in a steep face of rock.

alcove hologram *Optics.* a hologram whose surface appears to be concave, which permits the viewing of images over a 180° angle.

alcove lands *Geology.* terrain characterized by terraced slopes consisting of hard, resistant beds of rock interstratified by softer rock beds.

Alcyonacea [al´sī ə nās´ē ə] *Invertebrate Zoology.* an order of soft corals in the class Anthozoa, with a skeleton of separate calcareous spicules.

Alcyonaria [al´sī ə ner´ē ə] *Invertebrate Zoology.* a subclass of soft corals, horny corals, sea fans, and sea pens in the class Anthozoa; they are polyps with eight branched tentacles and eight septa. Also, OCTOCORALLIA.

Aldebaran [al deb´ə rän] *Astronomy.* Alpha (α) Tauri, the first-magnitude reddish star in the Hyades cluster.

aldehyde [al´də hīd´] *Organic Chemistry.* a class of organic compounds containing the –CHO radical; e.g., formaldehyde, acetaldehyde. Thus, **aldehydic.**

aldehyde ammonia *Organic Chemistry.* $CH_3CH(OH)NH_2$, a white, crystalline solid that is very soluble in water and alcohol; melts (and partially decomposes) at 97°C; used in organic synthesis and as a vulcanizer accelerator for thread rubber.

aldehyde dehydrogenase *Enzymology.* an enzyme that catalyzes the conversion of aldehyde to its corresponding carboxylic acid.

aldehyde-lyase *Enzymology.* a class of enzymes that catalyzes the cleavage of carbon-carbon bonds producing an aldehyde. Also, ALDOLASE.

aldehyde polymer *Organic Chemistry.* any of the polymer plastics that contain an aldehyde, such as formaldehyde, acetaldehyde, or acrolein (acrylic aldehyde).

alder *Botany.* any of various trees and shrubs of the genus *Alnus,* usually growing in moist land; alder bark is used in dyeing and tanning, and the wood is used for bridges and piles because it resists underwater rot. The **red** or **Oregon alder,** *A. rubra,* is a commercially important hardwood in the Pacific Northwest.

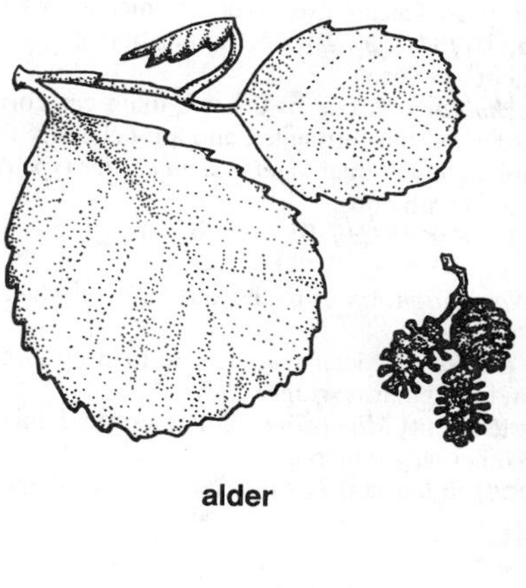

alder

Alder, Kurt 1902–1958, German chemist; with Diels, Nobel Prize for work in diene synthesis and the discovery of Diels-Alder reaction.

aldicarb *Organic Chemistry.* $C_7H_{14}N_2O_2S$, colorless crystals that are insoluble in water and slightly soluble in benzene, xylene, and acetone; melts at 100°C; used as a nematocide and insecticide.

aldohexose [al´də hek´sōs´] *Organic Chemistry.* any one of the six-carbon sugars (hexoses), such as glucose or galactose, that contain an aldehyde group.

aldol *Organic Chemistry.* $CH_3CH(OH)CH_2CHO$, a water-white or pale yellow thick liquid that is miscible with water, alcohol, ether, and organic solvents; boils at 83°C (20 mm); used in rubber accelerators and age resistors, perfumes, fungicides, and printing, and for various other purposes.

aldolase *Enzymology.* **1.** see ALDEHYDE-LYASE. **2.** the enzyme fructose-bisphosphate aldolase.

aldol condensation *Organic Chemistry.* the condensation of two aldehyde molecules, during which a hydroxyl group is formed, in addition to a carbonyl group. A catalyst is typically present, and water is eliminated.

aldol reaction *Organic Chemistry.* a reversible acid- or base-catalyzed addition of the α carbon atom of one carbonyl compound (or moiety of a molecule) to the carbonyl carbon of another molecule (or moiety of the original molecule) to form a β-hydroxy carbonyl compound; if dehydration occurs to form a conjugated unsaturated carbonyl compound, the reaction is then usually called an aldol condensation.

aldose *Organic Chemistry.* a group of monosaccharide sugars that contain an aldehyde group (–CHO).

aldose reductase *Enzymology.* an enzyme that dehydrogenates a class of monosaccharides known as aldoses.

aldosterone [al däs´tə rōn´] *Biochemistry.* $C_{21}H_{28}O_5$, a steroid hormone from the adrenal cortex that primarily functions to regulate sodium, potassium, and chloride metabolism. Also, **aldocortin.**

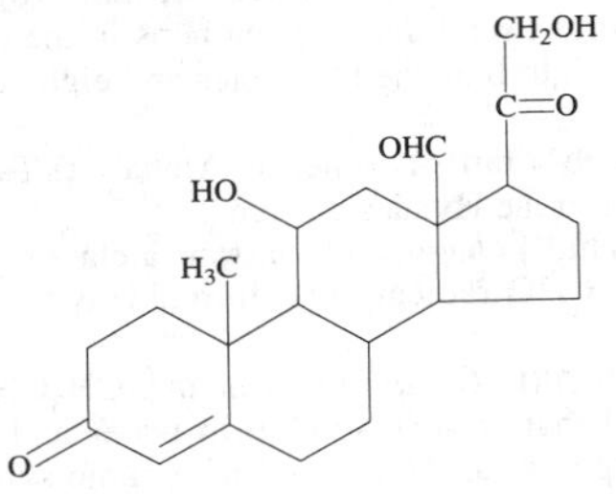

aldosterone

aldosteronism *Medicine.* a condition in which the blood contains unusually high levels of the hormone aldosterone, causing the retention of sodium and the loss of potassium though the excretion of urine and alkalosis; symptoms include paralysis, hypertension, certain types of cardiac irregularities, and polyuria.

Aldrich syndrome *Medicine.* an immunodeficiency syndrome marked by eczema, otitis media, anemia, and thrombocytopenic purpura and caused by a poor antibody response to antigens. Also, WISCOTT-ALDRICH SYNDROME. (From Robert A. *Aldrich,* American physician.)

aldrin *Organic Chemistry.* $C_{12}H_8Cl_6$, the common name of a brown to white crystalline solid, a chlorinated bismethylene bridged naphthalene; melts at 104–105.5°C; soluble in most organic solvents and insoluble in water; formerly used as an insecticide. (Named after Kurt *Alder.*)

Aldrovando, Ulyssi 1522–1605, Italian botanist; wrote a pioneering encyclopedia of science.

ale *Food Technology.* **1.** one of the two main categories of beer, along with lager. Ales are top-fermented and are generally darker, fuller-bodied, and more alcoholic than lagers. **2.** an earlier type of malt liquor that was not flavored with hops.

alecithal [ā les´ə thəl] *Cell Biology.* of an egg, possessing little or no yolk.

alee [ə lē´] *Navigation.* on or to the side of a ship away from the wind; leeward.

alee basin *Geology.* a deep-sea basin formed by turbidity currents deflected around a submarine ridge.

aleishtite [a lē ish´tīt] *Mineralogy.* a blue or green mixture consisting of dicktite and other clay minerals.

Alençon (lace) [ä´len sän] *Textiles.* a type of needlepoint lace that has a fine mesh background and a floral design outlined in heavy thread. (From *Alençon,* France, a city noted for the manufacture of this lace.)

aleph null *Mathematics.* the cardinal number of the set of positive integers, or of any set which can be assigned a one-to-one correspondence with the positive integers. Written $\aleph_0$. Also, **aleph-zero.**

Alepisauridae *Vertebrate Zoology.* the lancet fish family of the order Myctophiformes, having two species, one each in the Atlantic and the Pacific; they are characterized by a sail-like dorsal fin, a long body, long, fanged teeth, and a voracious appetite.

Alepocephaloidei *Vertebrate Zoology.* the slickheads, a suborder of deep-sea salmonlike fish in the order Salmoniformes.

Aleppo pine *Botany.* a pine tree, *Pinus halepensis,* native to Mediterranean areas and widely used since ancient times as a source of wood for shipbuilding; also used as a source of turpentine.

alert *Military Science.* **1.** a condition in which personnel are ready to launch or resist an attack, or to take other such offensive or defensive action. **2.** the time when such a condition is in effect, or a signal to begin this condition.

alerting signal *Telecommunications.* a ringing or other equivalent signal that is sent to customers to indicate they should answer their telephones.

-ales *Systematics.* a suffix used in forming the names of orders, as in Liliales.

aletophyte [ə let´ō fīt´] *Ecology.* a weed growing in an area where the natural vegetation has been cleared away, as along a roadside or railroad track. Thus, **aletophilous, aletophile.** Also, RUDERAL.

alette *Architecture.* **1.** a wing of a building. **2.** the wing on either side of an engaged column, often serving as the abutment of an arch. **3.** a door-jamb.

aleukemia [ā´lü kē´mē ə] *Medicine.* **1.** leukemia in which the total white blood cell count is normal or below normal. Also, **aleukemic leukemia. 2.** see ALEUKIA.

aleukemic [ā´lü kēm´ik] *Medicine.* characterized by aleukemia; lacking white blood cells.

aleukia [ə loo´kē ə] *Hematology.* the absence of white blood cells from the blood.

Aleuria [ə lûr´ē ə] *Mycology.* a genus of fungi belonging to the family Aleuriaceae of the order Pezizales, characterized by large, brightly colored spore cases and occurring on soil, dung, or wood.

aleurone [al´yə rōn´] *Botany.* the reserve store of protein (**aleurone grains**) found in the seeds of certain cereal plants. Also, **aleuron.**

aleurone layer *Botany.* the outer layer of the endosperm where aleurone is located.

aleuroplast *Botany.* a plastid that stores protein, often in granular form.

Aleutian Current *Oceanography.* a current that flows eastward between the North Pacific Current and the Aleutian Islands; a branch of the Kuroshio Extension.

Aleutian disease *Veterinary Medicine.* a fatal disease of mink believed to be caused by a virus; symptoms include weight loss, persistent diarrhea, bleeding from the mouth, anemia, degenerative changes in the liver and kidneys, and poor reproduction.

Aleutian Islands *Geography.* a chain of volcanic islands extending about 1200 miles from southwestern Alaska into the central North Pacific. Also, **Aleutians.**

Aleutian low *Meteorology.* a body of air with lower pressure relative to the surrounding air that is located near the Aleutian Islands on mean charts of sea-level pressure; it is one of the main centers of activity in the atmospheric circulation of the Northern Hemisphere.

alewife [āl´wīf´] plural, **alewives.** *Vertebrate Zoology.* a widely distributed North American oceanic fish, *Alosa pseudoharengus*; also found in fresh water; used extensively for food. It is part of the Clupeidae (herring) family.

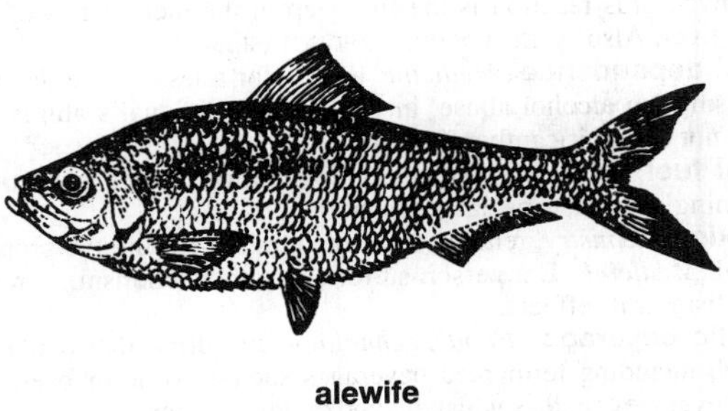

alewife

Alexander Archipelago *Geography.* a large group of islands off the coast of southeastern Alaska.

Alexanderson antenna *Electromagnetism.* an antenna designed for VHF (approximately 10 to 30 kHz), consisting of a horizontal wire that is grounded at regular intervals by vertical base-loaded wires. (From Ernst F. W. *Alexanderson,* 1878–1975, American engineer.)

Alexandrian *Geology.* a North American provincial series of the Lower Silurian period, occurring after the Cincinnatian of the Ordovician period and before the Niagaran.

alexandrite *Mineralogy.* an emerald-green variety of chrysoberyl that is valued as a gemstone, having a specific gravity of 3.75 and a hardness of 8.5 on the Mohs scale.

alexia [ə lek´sē ə] *Neurology.* the loss of the ability to understand written language as the result of a cerebral lesion. Also, WORD BLINDNESS, APHEMESTHESIA.

Aleyrodidae [al´ə räd´ə dē] *Invertebrate Zoology.* a family of minute whiteflies in the order Homoptera that are covered with white powdery wax and feed on plant juices; a serious plant pest.

alfa laval contractor *Biotechnology.* a centrifuge in which a vertically mounted bowl rotates around a central shaft; used for liquid-liquid extraction.

alfalfa *Botany.* a deep-rooted perennial plant of the pea family, *Medicago sativa,* having small divided leaves, purple cloverlike flowers, and spiral pods; probably originating in the Middle East and widely cultivated since ancient times as a livestock feed and cover crop. Also, LUCERNE.

alfalfa weevil *Invertebrate Zoology.* a reddish weevil, *Hypera postica,* originating in the Old World and now found in North America as a major pest of alfalfa crops.

alfa process *Food Technology.* a method of manufacturing butter by running high-fat cream through a three-cylinder cooler.

alfenol *Metallurgy.* a soft alloy with moderately high permeability at low field strength and with high electrical resistivity, consisting of iron and either 12% or 16% aluminum.

alfilaria or **alfileria** *Botany.* a European plant of the geranium family, *Erodium cicularium,* now widely grown in the western U.S. as a forage crop. Also, FILAREE.

alfin catalyst *Organic Chemistry.* a catalyst obtained from the reaction of alkali alcoholates and olefin halide; the resulting slurry is used to convert olefins into polyolefins, called **alfin rubbers.**

Alfisol *Geology.* an order of soils that develops in humid and subhumid climates and is characterized by a gray to brown surface horizon over a clay-rich subsurface layer.

Alford loop *Electromagnetism.* an antenna consisting of a multielement array driven so that the peripheral elements are approximately in phase; the antenna produces a circular radiation pattern in the polarization plane.

Alfrey-Price equation *Materials Science.* a rate equation for copolymerization in terms of monomer and radical reactivities and their electrostatic interaction.

Alfvén, Hannes Olof Gösta born 1908, Swedish physicist; Nobel Prize for research in magnetohydrodynamics.

Alfvénic of or relating to Hannes Alfvén or his research and theories.

Alfvén number *Physics.* a ratio that equals the speed of an Alfvén wave propagating in a fluid element divided by the speed of the fluid element.

Alfvén speed *Physics.* the propagation velocity of an Alfvén wave, given by $v = B_0(\rho\mu)^{-1/2}$, where B_0 is the magnetic field intensity, ρ is the mass density of the fluid, and μ is the magnetic permeability. Also, **Alfvén velocity.**

Alfvén theorem *Physics.* the statement that a perfectly conducting fluid contained within a tube of magnetic flux will be retained in that tube over time and motion.

Alfvén wave *Physics.* a transverse hydromagnetic wave that propagates along magnetic field lines and is generated by the low-frequency oscillation of ions in a fluid medium, typically a plasma.

alg- a combining form meaning "pain."

alg. algebra.

alga [al´gə] *Biology.* the singular form of algae; an individual organism of this type.

algae [al´jē] *Biology.* any of a large group of mostly aquatic organisms that contain chlorophyll and other pigments and can carry on photosynthesis, but lack true roots, stems, or leaves; they range from microscopic single cells to very large multicellular structures; included are nearly all seaweeds. (From Latin *alga,* "seaweed.")

algae bloom see ALGAL BLOOM.

algaecide see ALGICIDE.

algal [al´gəl] *Biology.* relating to or caused by algae. *Geology.* formed from or composed of algae.

algal ball see ALGAL BISCUIT.

algal biscuit *Geology.* a disk-shaped or spherical carbonate mass produced in fresh water as a result of precipitation by various types of blue-green algae. Also, ALGAL BALL, WATER BISCUIT.

algal bloom *Biology.* an unusual concentration of algae in or on a body of water, especially as a result of pollution from the runoff of fertilizers, industrial wastes, and so on. Also, ALGAE BLOOM.

algal coal see BOGHEAD COAL.

algal layer *Botany.* a layer of green or blue-green algal cells within the thallus of a lichen. Also, **algal zone.**

algal limestone *Petrology.* a type of limestone formed either from the remains of calcium-secreting algae or from formed algae that are bound together with the floccule remains of other lime-secreting organisms.

algal pit *Geology.* on sea ice or in the ablation area of a glacier, a small depression that contains algae.

algal reef *Geology.* an organic reef formed largely of algal remains in which algae are or were the main calcium carbonate-secreting organisms.

algal ridge *Geology.* at the outer edge of a coral reef, an elevated margin or ridge that is composed of calcium carbonate secreted by actively growing calcareous algae.

algal rim *Geology.* a low rim built by actively growing calcareous algae on the lagoonal side of a leeward reef or on the windward side of a reef patch in a lagoon.

algal structure *Geology.* a usually calcareous deposit formed by colonial secretion and precipitation by algae and having a definite form.

algebra *Mathematics.* **1.** the branch of mathematics that deals with the use of letters and other symbols, as well as numbers, to represent quantities and to express generalizations about them. For example, the relation $(x + y)^2 = x^2 + 2xy + y^2$ is the same for any two numbers x and y. **2.** the study of extending the rules of arithmetic to objects other than numbers, such as functions, linear transformations, symbols, and the like. **3.** a standard mathematical structure, such as a ring, enriched by one or more additional operations and axioms relating the new operations to the existing operations. In particular, if K is a commutative ring with identity, then a ring A is a K-algebra if A is a unitary (left) K-module and $k(ab) = (ka)b = a(kb)$ holds for all k in K and all a, b in A. Formerly called a hypercomplex system. If K and A are the same, the terms *algebra* and *ring* are often used interchangeably; thus a ring can be viewed as an algebra over itself. (From an Arabic term meaning "reduction.")

algebraic *Mathematics.* **1.** of or relating to algebra. **2.** in particular, of or relating to polynomials. For example, let F be an extension field of K. An element u of F is said to be algebraic over K if u is a root of some nonzero polynomial f in $K[x]$, the set of all polynomials with coefficients in K. If f is monic and has integer coefficients, then u is an **algebraic integer.** If every element of F is algebraic over K, then the field F is said to be algebraic over K, or F is an **algebraic extension** of K.

algebraic addition *Mathematics.* the combination of algebraic terms by addition or subtraction, such that subtraction of an element is defined as addition of the additive inverse. The result is called an **algebraic sum.**

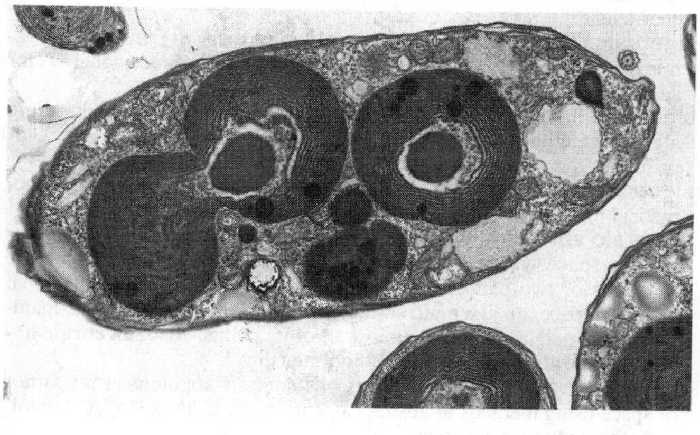

algae

algebraically closed *Mathematics.* a field (such as the complex numbers) in which all polynomial equations with coefficients in the field have roots in the field. A field F that has no algebraic closure except itself is said to be algebraically closed. This condition is equivalent to each of the following: (a) Every nonconstant polynomial f in polynomial ring $F[x]$ has a root in F. (b) Every nonconstant polynomial f in $F[x]$ splits over F; that is, f can be written as the product of linear factors in $F[x]$. (c) Every irreducible polynomial in $F[x]$ has degree 1. (d) There exists a subfield K of F such that F is algebraic over K and every polynomial in $K[x]$ splits in $F[x]$.

algebraic closure of a field *Mathematics.* **1.** an algebraic extension F of a field K that is algebraically closed; obtained by adjoining to K all the elements which are algebraic over K, then all elements algebraic over those, and so on, until there are no new elements. The resulting field F is called the **algebraic closure** of K and is said to be algebraically closed, since no new algebraic elements can be adjoined to F. **2.** equivalently, if F is a splitting field over K of the set of all irreducible polynomials in $K[x]$ (the ring of polynomials over K) then F is the algebraic closure of the field K and F is said to be algebraically closed if F exists.

algebraic curve *Mathematics.* a set of points in n-dimensional Euclidean space satisfying a polynomial in the n coordinate functions. If the polynomial is of degree m, then the curve is said to be of degree m.

algebraic equation *Mathematics.* an equation formed by setting an algebraic expression equal to zero.

algebraic expression *Mathematics.* an expression formed by performing a finite number of algebraic operations (addition, subtraction, multiplication, division, and raising to a rational power) on formal symbols. Also, RATIONAL EXPRESSION.

algebraic extension of a field *Mathematics.* F is an algebraic extension field of K if K is a subfield of F and if every element of K is algebraic over F.

algebraic function *Mathematics.* a function whose value is given by an algebraic expression of the argument of the function; i.e., the function values satisfy an algebraic equation having coefficients that are polynomials in the argument.

algebraic geometry *Mathematics.* the study of geometric properties of solutions of algebraic (polynomial) equations.

algebraic language *Mathematics.* the conventional rules of syntax for writing mathematical expressions and equations. *Computer Programming.* an algorithmic language many of whose statements are constructed to resemble algebraic expressions, such as ALGOL or FORTRAN.

algebraic manifold *Mathematics.* **1.** a subset M of an affine space that is defined as the set of common zeros of a finite number of polynomials $f_1, \ldots, f_r$; that is, the set of all solutions of the equations $f_1(\xi) = 0, \ldots, f_r(\xi) = 0$. **2.** a variety, especially one having singular points.

algebraic manipulation *Computer Programming.* the processing of symbolic mathematical expressions, resulting in new expressions, without regard for the numerical value of the symbols that represent numbers.

algebraic-manipulation language *Computer Programming.* a special-purpose language developed to create, combine, process, and simplify mathematical equations; used in manipulating complex physical equation sets and to generate programs for numerical simulation.

algebraic number *Mathematics.* a root of a polynomial with rational coefficients.

algebraic number field *Mathematics.* any extension field K of the rational numbers Q that can be represented as a finite-dimensional vector space over Q.

algebraic number theory *Mathematics.* the study of numbers using algebraic methods, including such topics as Fermat's last theorem, Diophantine equations, p-adic numbers, adeles, and class field theory.

algebraic simplification *Artificial Intelligence.* reduction of a mathematical formula to a simpler form by the application of algebraic laws.

algebraic surface *Mathematics.* the set of points in n-dimensional Euclidean space whose parametric representation consists of algebraic functions of two parameters. In a hypersurface, an $(n–1)$-dimensional complex space, an algebraic surface is usually required to be a 2-manifold in the neighborhood of almost all of its points. Also, **algebraic hypersurface, algebraic variety of codimension 1.**

algebraic topology *Mathematics.* the study of topological properties of space using methods of abstract algebra, such as homology, cohomology, and homotopy theories.

algefacient [al′jə fā′shənt] *Surgery.* cooling; refrigerant.

Algenib [al′jə nib] *Astronomy.* Gamma (γ) Pegasi, a bluish-white star that lies in the southeast corner of the Great Square in the constellation Pegasus, the Flying Horse.

algesi- or **algesio-** a combining form meaning "pain."

algesia [al jēz′yə] *Neurology.* sensitivity to pain, especially extreme sensitivity to pain.

algesic [al jēz′ik] *Neurology.* relating to or producing pain; painful. Also, **algetic.**

algesimeter [al′jə sim′ə tər] *Medicine.* an instrument for measuring the perceptual threshold for pain. Thus, **algesimetry** [al′jə sim′ə trē].

algesiogenic [al jēz′ē ō jen′ik] *Neurology.* producing pain.

algesiroreceptor see NOCICEPTOR.

algesthesia [al′jis thēz′ē ə] *Neurology.* **1.** the perception of pain; any painful sensation. **2.** the ability to sense pain. Also, **algesthesis**.

algetic *Neurology.* painful.

algi- a combining form meaning "pain."

algicide [al′jə sīd′] *Toxicology.* a substance or preparation that is used for killing algae, as in removing mustard algae growth from a swimming pool.

algin [al′jən] *Organic Chemistry.* a polysaccharide of D-mannuronic acid (alginic acid) found exclusively in brown algae; a cream-colored powder that is soluble in water and insoluble in alcohol, chloroform, and ether; it is used as a stabilizer in food products and in yarns and fibers.

alginate *Surgery.* a salt of alginic acid, extracted from marine kelp and used as a foam, clot, or gauze for absorbable surgical dressings, or as a gel for dental impressions. *Organic Chemistry.* any of various derivative salts of alginic acid, such as sodium alginate.

alginic acid *Organic Chemistry.* $(C_6H_8O_6)_n$, a colloidal polymannuronic acid derived from marine brown algae; soluble in alkaline solutions; used as a thickener and emulsifier, it can absorb up to 300 times its weight in water.

algio- a combining form meaning "pain."

algite [al′jīt] *Petrology.* the fundamental petrological algal unit found in algal or boghead coal. Also, **alginite.**

algo- a combining form meaning "pain."

algodonite [al′gə də nīt′] *Mineralogy.* Cu_6As, an opaque, metallic, white to gray mineral, usually massive in habit or as orthorhombic crystals, having a specific gravity of 8.38 and a hardness of 4 on the Mohs scale.

algogenesia [al′gō jə nēz′yə] *Neurology.* the origin or production of pain. Also, **algogenesis.**

algogenic [al′gə jen′ik] *Neurology.* **1.** relating to or causing pain. **2.** relating to or producing cold.

Algol [al′gäl; al′gōl] *Astronomy.* Beta (β) Persei, an eclipsing binary star, also known as the **winking star** or the **demon star,** in the constellation Perseus, the Hero. (From Arabic *al-ghūl* "ghoul" or "demon.")

ALGOL or **Algol** *Computer Programming.* a general-purpose computer language suitable for communicating algorithms to people and for executing them on a variety of computers. ALGOL is no longer widely used, but it has influenced many other computer languages, which are sometimes called **ALGOL-like languages**. (An acronym for <u>Algo</u>rithmic <u>L</u>anguage)

algolagnia [al′gə lag′nē ə] *Psychology.* sexual gratification derived from inflicting or enduring pain; **active algolagnia** is another term for sadism and **passive algolagnia** is another term for masochism.

algology [al gäl′ə jē] *Botany.* the study of seaweeds or algae; phycology. *Medicine.* the study of pain. Thus, **algologist, algological.**

Algol paradox *Astronomy.* the paradox that the binary star Algol seems to contradict accepted notions of stellar evolution, in that its more massive component star, Algol A, has evolved more slowly than the smaller companion, Algol B.

Algol symbiotic *Astronomy.* two stars in a binary pair whose evolution follows a path similar to Algol's.

Algol variable star *Astronomy.* an eclipsing binary star named for its prototype Beta Persei.

Algoman orogeny *Geology.* **1.** a Precambrian deformation of the earth's crust that affected the rocks of northern Minnesota and Ontario. **2.** see KENORAN OROGENY.

algometer *Medicine.* an instrument that measures sensitivity to pain. Thus, **algometry.**

Algonkian [al gong′kē ən] see PROTEROZOIC.

algophagous [al′gə fāj′is] *Biology.* feeding on algae.

algophobia *Psychology.* an irrational fear of experiencing pain or of witnessing pain in others.

algorithm [al´gə rith̲ əm] *Mathematics.* a set of well-defined directions to perform mathematical operations that lead to the solution or approximate solution of a given problem. If a potentially infinite number of steps is required, the process can still qualify as an algorithm if a stopping rule based on solution accuracy can be given. *Computer Programming.* an abstract procedure to carry out an operation by following a series of precise, unambiguous steps; often the term algorithm connotes a procedure that is guaranteed to produce a correct result in an efficient manner. *Science.* any formalized, step-by-step procedure for solving a problem, as in medical diagnosis. (From the Arab mathematical pioneer al-Khwarizmi.)

algorithmic error *Computer Programming.* an error that occurs when the algorithm used to solve a problem does not give the correct result, either because it is applicable only under certain conditions or because mathematical calculations have been imprecise.

algorithmic language *Computer Programming.* **1.** see ALGOL. **2.** any computer language that provides facilities for the solution of mathematical problems in a finite number of steps.

algor mortis *Pathology.* the cooling of the body after death.

Alhazen (ibn al-Haitham) 965–1038, Arab mathematician and astronomer; pioneered studies of optics and the atmosphere.

alias *Computer Programming.* **1.** an alternate name for a certain file, device, data element, or the like. **2.** an alternate entry point in main memory where program execution may begin. *Mathematics.* see ALIASING.

aliasing *Mathematics.* the condition that two or more functions are indistinguishable because they have the same values at a finite set of points. Such functions are said to be aliases of each other. The aliasing problem often occurs in an undersampled discrete Fourier transform. *Computer Programming.* the creation of an alternate name for a given piece of data.

ALICE or **Alice** *Electronics.* a system of radio stations that mainly use scatter-propagation technology to couple with early-warning radar stations. (An acronym for Alaska Integrated Communications Exchange.)

alicyclic [al´ə sī´klik] *Organic Chemistry.* **1.** describing a major class of compounds having both aliphatic and cyclic properties. The cyclic molecules do not have an aromatic benzene ring; for example, cyclohexane or cyclopentane. Most are derived from petroleum or coal tar. **2.** a compound of this type. Thus, **alicyclic hydrocarbon.**

alidade [al´ə dād] *Engineering.* a rule having a sight at each end, used in plane table surveying and topographic surveying. Also, SIGHT RULE. *Cartography.* the part of a surveying instrument that consists of a sighting device. *Navigation.* a gyrocompass repeater with a telescope mounted over it for measuring directions.

alienation *Psychology* **1.** an older term for mental illness. **2.** a feeling of being withdrawn or estranged from one's group or society.

alienist *Psychology* **1.** an older term for a psychiatrist. **2.** a psychiatrist who testifies in a court of law regarding a person's mental competency.

aliesterase *Biochemistry.* an enzyme that catalyzes the breakdown (by hydrolysis) of carboxylic acid esters. Also, CARBOXYLESTERASE.

aligning drift *Mechanical Devices.* a rod or bar for aligning parts during manufacturing or assembly.

aligning punch *Mechanical Devices.* a drift used to align holes in sheet metal for riveting or bolting operations.

alignment the fact of being in line or bringing into line; specific uses include: *Cartography.* the correct location, direction, and character of a line or feature on a map in relation to other lines or features. *Architecture.* the position of a building or one of its elements on a site. *Engineering.* **1.** a line of adjustment through two or more points. **2.** the fixing of points on the ground for the laying out of a railroad, wall, canal, and so on. *Electronics.* **1.** the process of varying the parameters of a tuned circuit to achieve the desired frequency response. **2.** the process of adjusting the synchronization of different components of an electrical system. *Mining Engineering.* the planned direction of a tunnel or other construction. *Nuclear Physics.* see NUCLEAR ALIGNMENT.

alignment correction *Engineering.* a correction in the length of measured line, used to compensate for not being able to apply the measuring instrument directly on the plane containing the line.

alignment pin *Mechanical Devices.* a metal pin positioned at the mouth of a tube configured with a single projecting rib, used to align it within a bore.

alimentary [al´ə men´trē; al´ə men´tə rē] *Biology.* of or relating to food, diet, nutrition, or digestion. Thus, **alimentary system.**

alimentary canal *Anatomy.* the gastrointestinal tract; the parts of the body through which food passes and is digested, and waste is eliminated.

alimentary castration *Entomology.* the infertility of certain castes of social insects due to nutritional deficiencies.

alimentary paste *Food Technology.* unleavened dough made from semolina or wheat flour, usually extruded and dried to make noodles, pasta, and other similar products.

alimentation *Biology.* **1.** the act of providing nutrients by feeding. **2.** the state of being nourished. *Hydrology.* the furnishing of a glacier with material that turns to ice.

Alioth [al´ē´äth] *Astronomy.* Epsilon (ε) Ursae Majoris, a spectroscopic binary star in the handle of the Big Dipper.

aliphatic [al ə fat´ik] *Organic Chemistry.* **1.** describing a major class of compounds in which carbon and hydrogen molecules are arranged in a straight or branched chain. They may be alkenes, alkanes, or alkynes. **2.** a compound of this type.

aliphatic acid *Organic Chemistry.* any organic acid that is derived from aliphatic hydrocarbons.

aliphatic acid ester *Organic Chemistry.* an organic ester that is derived from aliphatic acids.

aliphatic hydrocarbon see ALIPHATIC.

aliphatic polycyclic hydrocarbon *Organic Chemistry.* any hydrocarbon in which a minimum of two aliphatic structures are rings (cyclic).

aliphatic polyene compound *Organic Chemistry.* any unsaturated aliphatic or alicyclic substance with four or more carbons, and a minimum of two double bonds, such as hexadiene.

aliphatic series *Organic Chemistry.* a specific series of open- or branched-chain hydrocarbons. The two major classes are saturated and unsaturated.

aliquot [al´ə kwät´] *Mathematics.* a portion of a whole that is divided into equal parts without any remainder, each portion representing a known quantitative relationship to the whole and to each of the other portions; e.g., the number 2 is an aliquot of the number 6. Also, **aliquot part.** *Materials Science.* a sample of a material evaluated to determine the properties of the whole. *Medicine.* the division of a quantity of a specimen into equal parts, each portion representing a known quantitative relationship to the whole and to each of the other

Alismataceae [ə liz´mə tās´ē ē] *Botany.* a family consisting of monocotyledonous aquatic or marsh plants having elliptical leaves and white flowers.

alisphenoid [al´ə sfē´noid] *Anatomy.* either of two bones at the base of the skull that form wings of the sphenoid; remaining separate in some mammals and fused in others. Thus, **alisphenoidal.**

alive *Biology.* having life; living. *Electricity.* electrically connected to a voltage source. *Mining Engineering.* **1.** a term used to describe the productive component of a lode. **2.** a term used to describe coal when it bursts, cracks, and breaks off while under pressure, creating a rustling sound.

alivincular [al´ə ving´kyə lər] *Invertebrate Zoology.* in certain bivalves, a short ligament that has its longer axis transverse to the hinge line.

alizarin [ə liz´ə rin] *Organic Chemistry.* $C_6H_4(CO)_2C_6H_2(OH)_2$, combustible, orange-red, triclinic crystals or a brownish-yellow powder, slightly soluble in water and soluble in alcohol and acid; melts at 289–290°C and boils (sublimes) at 430°C. It occurs naturally in madder root and is synthesized from anthraquinone; used in dye and biological stain manufacture. Also, ALIZARINE, 1,2-DIHYDROXYANTHRAQUINONE.

alizarin dye *Organic Chemistry.* any of a series of dyes derived from alizarin, such as **alizarin red, alizarin crimson, alizarin yellow, alizarin blue,** and so on.

alizarine see ALIZARIN.

alk. alkaline.

alkalemia [al´kə lēm´ē ə] *Medicine.* a high alkalinity of the blood due to a decrease in the hydrogen ion concentration or to an increase in hydroxal ions. It is often the result of prolonged vomiting.

alkalescens-dispar group *Bacteriology.* a group of nonmotile strains of *Escherichia coli* distinguished by the anaerobic fermentation of glucose and the frequent absence or delay of lactose fermentation.

alkalescent *Chemistry.* having the property of being alkaline; having a pH greater than 7. Thus, **alkalescence.**

alkali *Chemistry.* a hydroxide of one of the alkali metals, producing solutions with pH greater than 7, soluble in water, neutralizing acids to form salts, and turning red litmus paper blue; e.g., sodium hydroxide or potassium hydroxide. In aqueous solutions, the term is equivalent to *base. Geology.* a bitter-tasting salt consisting of sodium or potassium carbonate, found in soils in arid or semiarid regions; generally unproductive for agriculture.

Alkali *Ordnance.* an early Soviet air-to-air missile powered by a solid-propellant rocket motor and equipped with radar homing, delivering a 60-pound high-explosive warhead at a speed of Mach 1 and a range of 5 miles; officially designated **AA-1.**

alkali-aggregate reaction *Chemistry.* in a cement, the chemical reaction between the aggregate and the alkali that causes a weakening.

alkali alcoholate *Organic Chemistry.* the product of a reaction between an alkali metal and an alcohol, with the metal replacing the hydroxyl hydrogen.

alkali blue *Organic Chemistry.* any of a series of sodium salts of triphenylrosanilinesulfonic acid, widely used as an indicator and a pigment. *Graphic Arts.* a blue printing ink using this pigment.

alkalic [al kal´ik] *Petrology.* **1.** or or relating to igneous rocks that have relatively high alkali content (K_2O and Na_2O), giving rise to feldspathoids or other minerals, such as acmite. **2.** of or relating to igneous rocks with a low alkali-lime index of 51 or less.

alkali-calcic series *Petrology.* the series of igneous rocks having an alkali-lime index between 51 and 55.

alkali cellulose *Materials.* the product formed by steeping wood pulp with sodium hydroxide, NaOH; the first step in the manufacture of cellulose derivatives such as rayon.

alkali chlorosis *Plant Pathology.* a condition in which plant foliage yellows due to an excess of soluble salts in the soil.

alkali denaturation test *Pathology.* a sensitive spectrophotometric method for ascertaining the concentration of fetal (F) hemoglobin molecule by alteration of its globin moiety when exposed to alkali.

alkali disease *Veterinary Medicine.* a chronic, toxic condition of livestock, characterized by incoordination, emaciation, and lameness, resulting from the consumption of forages and grains containing 5 to 40 parts per million of selenium. Also, (CHRONIC) SELENIUM POISONING.

alkali-earth metal see ALKALINE-EARTH METAL.

alkali emission *Geophysics.* an emission of light from energized lithium, potassium, and sodium molecules in the upper atmosphere that can be seen at twilight.

alkali feldspar *Mineralogy.* any feldspar that contains alkali metals and very little calcium, such as orthoclase, microcline, anorthoclase, albitic plagioclase, sanidine, and adularia.

alkali flat *Geology.* **1.** in an arid or semiarid region, a level area or plain where alkali salts have become concentrated as a result of evaporation and poor drainage. **2.** see SALT FLAT.

alkali grass *Botany.* a tall plant, *Zigadenus elegans,* having grasslike leaves and clusters of greenish-white flowers; widespread in western North America.

alkali hydroxide *Chemistry.* see ALKALI.

alkali lake *Hydrology.* a salt lake containing large quantities of dissolved sodium carbonate and potassium carbonate in addition to sodium chloride. Also, **alkaline lake.**

alkali-lime index *Petrology.* in a sequence of igneous rocks, the weight percentage of silica when the weight percentages of CaO (calcium oxide) and K_2O (potassium oxide) + Na_2O (sodium oxide) are equal.

alkali-loving *Botany.* of a plant, requiring relatively alkaline soil for optimal growth; not thriving in acidic soil.

alkali metal *Chemistry.* any of the elements of Group IA, which form highly alkaline solutions in water and burn vigorously in air. They all have a valence of one, and are softer and less dense than other metals; included are the elements lithium, sodium, potassium, rubidium, cesium, and francium.

alkalimeter *Analytical Chemistry.* an instrument that measures the amount of base in a solution; used to measure the amount of carbon dioxide liberated from a weighed sample.

alkalimetry *Analytical Chemistry.* an analytical method for determining the basicity of a solution, as by titration with a standard solution of acid.

alkaline [al´kə lin; al´kə līn´] *Chemistry.* **1.** relating to, containing, or having the properties of an alkali. **2.** solutions have a pH higher than 7.0.

alkaline cell *Electricity.* a primary cell in which the electrolyte consists of an alkaline solution, usually potassium hydroxide.

alkaline cleaner *Metallurgy.* a bath used to cleanse the surface of metals or alloys. Such a bath contains alkali hydroxides mixed with salts such as borates, silicates, or phosphates.

alkaline-earth metal *Chemistry.* any of the elements of Group IIA, which are divalent and strongly basic (though less so than the alkali metals). These include beryllium, magnesium, calcium, strontium, barium, and radium; some classifications do not include beryllium and may also omit magnesium.

alkaline flooding *Petroleum Engineering.* a form of oil recovery in which alkaline chemicals are combined with polymer flooding or are injected during a water flooding; surfactants form when the chemicals react with acids in the crude oil.

alkaline phosphatase *Enzymology.* an enzyme that occurs in a variety of normal and malignant tissues; its level in serum is useful in diagnosing certain diseases, such as bone diseases and hepatitis. Also, PHOSPHOMONOESTERASE.

alkaline protease *Enzymology.* an enzyme that catalyzes the breakdown of proteins in alkaline solution.

alkaline soil *Geology.* a soil having a pH greater than 7.

alkaline storage battery *Electricity.* a battery in which the electrolyte is composed of an alkaline solution.

alkaline tide *Physiology.* a temporary increase in the alkalinity of the urine during digestion.

alkaline wash *Chemical Engineering.* a process by which impurities are removed from kerosene by using a caustic soda or other alkaline solution.

alkalinity *Chemistry.* the fact of being alkaline; the extent to which a solution has excess hydroxide ions.

alkalipenia [al´kə lī pē´nē ə] *Medicine.* a condition of having low alkali reserves in the body.

alkali-resisting paint *Materials.* any of various paints intended to withstand exposure to akaline materials or conditions.

alkali soil *Geology.* a soil having a high degree of alkalinity (pH of 8.5 or higher) or containing greater than 15% exchangeable sodium.

alkalize *Chemistry.* to make or become alkaline. Also, **alkalanize.**

alkaloid *Chemistry.* **1.** any of a large group of organic nitrogenous bases found in certain plants as a defense against insects and herbivores and having pharmacological properties; usually crystalline solids, but also existing in gum or liquid form. Examples include caffeine, nicotine, cocaine, atropine, codeine, quinine, and morphine. **2.** a synthetic substance having similar properties, such as procaine. **3.** relating to or resembling an alkaloid compound. Thus, **alkaloidal.**

alkalometry *Analytical Chemistry.* an analytical method for determining the amount of alkaloids present in a solution.

alkalophile *Microbiology.* a microorganism that grows optimally in an alkaline environment.

alkalosis *Physiology.* an excessively high presence of alkali or decreased concentration of hydrogen ions in the blood, sometimes leading to a rise in pH of the blood, and often linked to persistent vomiting or gastric drainage.

alkamine *Organic Chemistry.* a compound having both amino and alcohol groups. Also, AMINO ALCOHOL.

alkane *Organic Chemistry.* any of various aliphatic hydrocarbons that have the general formula C_nH_{2n+2}. The first (lightest molecular weight) four are gases; higher members are liquids, and those above $C_{16}H_{34}$ are waxy solids. Thus, **alkane series.**

alkanet *Botany.* **1.** a European plant, *Alkanna tinctoria,* belonging to the borage family. **2.** the root of this plant. *Materials Science.* a red dye that is derived from the root of this plant.

alkannin *Organic Chemistry.* $C_{16}H_{16}O_5$, a red powder that is insoluble in water and soluble in alcohol, benzene, ether, and oils; it melts at 149°C; used as a coloring agent in fats, oils, wax, cosmetics, and wine. Also, **alkanin.**

alkanolamine *Organic Chemistry.* any of a group of amino alcohols in which the nitrogen is directly attached to the carbon of an alkyl alcohol. They are viscous, water-soluble liquids used in rubber accelerators and as absorbents for acidic gases. Also, ALKYLOLAMIDE.

alkapton (body) *Biochemistry.* homogenistic acid or another such substance with an affinity for alkali, found in the urine and associated with alkaptonuria.

alkaptonuria *Medicine.* a hereditary disease caused by an autosomal recessive gene that renders individuals incapable of producing the liver enzyme homogenistic acid oxidase. Homogenistic acid is not broken down in the body, but is excreted in the urine. Thus, **alkaptonuric.**

Alkar process *Chemical Engineering.* a process in which alkylaromatics are produced by the catalytic alkylation of aromatic hydrocarbons with olefins.

alkarsine see ALKYLOLAMIDE, COCODYL OXIDE.

alkene *Organic Chemistry.* any of various unsaturated aliphatic hydrocarbons having the general formula C_nH_{2n}. Thus, **alkene series.**

al-Khwarizmi c. 780–850, Arab mathematician; his *Arithmetic* and *Algebra* introduced Indian mathematics to the West. (His name is also transliterated as *al-Kwarizmi, al Khowarizmi, al-Khuwarizmi,* etc.)

Al-killed steel *Metallurgy.* steel that is dipped in molten aluminum to produce a coating resistant to oxidation and scaling at high temperatures. Also, ALUMINIZED STEEL, KILLED STEEL.

alkoxide see ALCOHOLATE.

alkoxy *Organic Chemistry.* an alkyl radical attached by an oxygen to a molecule; for example, the ethoxy radical.

alkyd resin *Organic Chemistry.* a thermosetting polymer made from ethylene glycol or glycerol and a polybasic alcohol such as phthalic anhydride. It is used as a coating and in various paints. Thus, **alkyd paint.**

alkyl *Organic Chemistry.* **1.** a monovalent hydrocarbon group formed by removing one hydrogen from an alkane. It is usually designated by the letter R, and has the the general formula of $-C_nH_{2n+1}$. **2.** relating to or describing such a group.

alkylamine *Organic Chemistry.* an alkyl group attached to an amine nitrogen, as in ethylamine, $C_2H_5NH_2$.

alkylaryl sulfonate *Organic Chemistry.* a sulfonate of both aliphatic and aromatic structure; for example, alkylbenzene sulfonate.

alkylate *Organic Chemistry.* **1.** to add an alkyl group to a compound. **2.** a substance produced in this way. *Petroleum Engineering.* a high-octane alkylate substance added to aviation fuel. Thus, **alkylated gasoline.**

alkylate bottom *Chemical Engineering.* the fraction remaining which boils at a higher temperature than the aviation gasoline which is collected after the distillation of alkylate.

alkylating agent *Organic Chemistry.* any of various substances that contain an alkyl radical and that therefore can replace a hydrogen atom in an organic compound; this type of reaction acts on DNA and interferes with cell replication. *Pharmacology.* a substance of this type used as a drug to destroy cells, especially cancer cells. Thus, **alkylating drug.**

alkylation *Organic Chemistry.* the introduction, by substitution or addition, of an alkyl radical into an organic compound. *Petroleum Engineering.* a refinery process in which a high-octane blending component for gasolines is created by chemically combining an isoparaffin and an olefin.

alkylbenzene sulfonate *Organic Chemistry.* a branched-chain sulfonate, such as dodecylbenzene or tridecylbenzene sulfonate, used as a detergent.

alkyl compound see ALKYL.

alkylene *Organic Chemistry.* an organic radical formed from unsaturated aliphatic hydrocarbons.

alkyl group see ALKYL.

alkyl halide *Organic Chemistry.* a compound formed from an alkyl group and a halogen, such as methyl bromide or ethyl bromide.

alkylolamide see ALKANOLAMINE.

alkyne *Organic Chemistry.* an unsaturated hydrocarbon compound containing a carbon-to-carbon triple bond, the simplest being ethyne or acetylene $HC{\equiv}CH$.

allachesthesia [al´ək əs thēzh´ə] see ALLESTHESIA.

allactite *Mineralogy.* $Mn_7(AsO_4)_2(OH)_8$, a reddish-brown, translucent, monoclinic mineral having one distinct cleavage, a specific gravity of 3.83, and a hardness of 4.5 on the Mohs scale; found with calcite, franklinite, willemite, and fluorite.

allalinite [ə lal´ə nīt´] *Petrology.* an altered olivine plutonic gabbro with the original texture but a chemically altered crystalline structure.

allanite *Mineralogy.* $(Ce,Ca,Y)_2(Al,Fe^{+2},Fe^{+3})_3(SiO_4)_3(OH)$, a brown to black cerium-bearing monoclinic mineral of the epidote group having a specific gravity of 4.0 to 4.2 and a hardness of 5.5 to 6 on the Mohs scale; often found as an accessory mineral in igneous or metamorphic rock. Also, ORTHITE.

allant- or **allanto-** a combining form meaning: **1.** of, relating to, or resembling a sausage, as in *allantiasis.* **2.** of or relating to the allantois, as in *allantotoxin.*

allantiasis [a lən tī´ə sis] *Toxicology.* sausage poisoning, especially poisoning due to ingestion of sausages contaminated with *Clostridium botulinum.*

allantoic *Developmental Biology.* of or relating to the allantois. Thus, allantoic fluid, allantoic vessel.

allantoic acid *Biochemistry.* $C_4H_8N_4O_4$, an acid that is produced by the breakdown of allantoin; it plays a role in nucleic acid metabolism.

allantoid [ə lan´toid] *Microbiology.* any slightly curved, sausage-shaped structure having rounded ends. Also, **allantoidal.** *Developmental Biology.* see ALLANTOIC.

allantoin [ə lan´tō in] *Biochemistry.* $C_4H_6N_4O_3$, a crystallizable substance that is found in many plants and in allantoic and amniotic fluid and fetal urine; it is also found as an excretory product in urine in many mammals other than primates. *Medicine.* a preparation of this substance, used (especially formerly) to promote healthy tissue growth in wounds and ulcers.

allantoinase [ə lan´tə wə nās´] *Enzymology.* an enzyme that catalyzes the conversion of allantoin into allantoic acid.

allantois [ə lan´tō is; ə lan´tois] *Developmental Biology.* an extraembryonic, saclike outgrowth of the ventral hindgut in mammals, reptiles, and birds; in humans its blood vessels develop into the vessels of the umbilical cord.

allantoxanic acid [a´lan´täk´san´ik] *Biochemistry.* $C_4H_3N_3O_4$, an acid derived from allantoin or uric acid and formed by oxidation.

allanturic acid [al´ən tur´ik] *Biochemistry.* $C_3H_4N_2O_3$, an acid derived from allantoin or uric acid and formed by oxidation.

Allard's law *Optics.* a formula that expresses the relationship between the intensity of a light source, conditions in the atmosphere, and the amount of light received from that source at a specific distance.

all-around traverse *Ordnance.* the capacity of a weapon to turn in a complete horizontal circle on its traversing mechanism.

all-burnt (time) *Space Technology.* the point in time at which a spacecraft or missile consumes all of its fuel.

allcharite see GOETHITE.

Alleculidae [Iol´ə kyu´lə dē] *Invertebrate Zoology.* a family of comb-clawed beetles in the order Coleoptera.

Allee's principle *Ecology.* the principle that a specific habitat will support an optimal population level for any given species if neither too few nor too many individuals of that species are present in the habitat. Also, **Allee's law.**

allège *Architecture.* a thinned portion of a wall, especially the area beneath a window.

alleghanyite [al ə gān´ē īt´] *Mineralogy.* $Mn_5^{+2}(SiO_4)_2(OH)_2$, a brittle, pinkish, monoclinic silicate mineral of the humite group that is found in zinc ores, having a specific gravity of 4.0 and a hardness of 5.5 on the Mohs scale.

Alleghenian of or relating to geological or biological features of the Allegheny Mountain region of eastern North America.

Alleghenian orogeny *Geology.* a late Paleozoic to early Triassic crustal deformation involving the Pennsylvanian and Lower Permian rocks of the Allegheny Plateau, central and southern Appalachians, and the Ridge and Valley provinces.

Alleghenian zone see EASTERN MIXED FOREST.

Allegheny subregion *Ecology.* a distinct zoogeographical region that includes the eastern half of the United States and adjacent areas of Canada.

allele [ə lēl´; al´ēl´] *Genetics.* any of two or more alternative forms of a gene occupying the same chromosomal locus; such as that which determines flower petal color in peas. Also, ALLELOMORPH.

allelic [ə lel´ik; ə lē´lik] *Genetics.* **1.** of or relating to an allele or alleles. **2.** produced by alternative genes. Thus, **allelism.**

allelic exclusion *Genetics.* the expression of only one of the two allelic forms of a gene present in a diploid cell.

allelic frequency see GENE FREQUENCY.

allelic gene *Genetics.* any of a series of two or more alternative forms of a gene that occur at the same site on homologous chromosomes.

allelic mutant *Genetics.* an organism or cell that differs from a parent due to mutations of its alleles.

allelo- a combining form meaning "relating to another," as in *allelochemistry.*

allelochemics *Biology.* chemical reactions between species that involve the release of active chemical substances such as scents, pheromones, or toxins. Thus, **allelochemical, allelochemic.**

allelochemistry *Chemistry.* the study of organic products that stimulate or inhibit other organisms.

allelomimicry *Behavior.* a characteristic behavior pattern in which members of the same species imitate the actions of another or others in the group. Thus, **allelomimetic.**

allelomorph see ALLELE.

allelomorphic see ALLELIC.

allelopathy *Plant Pathology.* the release of a toxic chemical by a plant that inhibits the growth of nearby plants of the same or other species, thus reducing competition. *Biology.* any effect or influence of one organism that is harmful to others of the same species. Thus, **allelopathic.**

allelotaxis *Developmental Biology.* the development of an organ from several different embryonic structures.

allelotoxin *Toxicology.* a toxic chemical released by a plant that inhibits the growth of other plants of the same or other species.

allelotropism *Biology.* a state of mutual attraction between two cells or organisms. Thus, **allelotropic.**

allelotype *Genetics.* the occurrence of an allele in a population; an example of a certain allele.

allemontite *Mineralogy.* a mixture of stibarsen with either arsenic or antimony.

Allen, Edgar V. 1892–1943, American physiologist; discovered estrogen; with Doisy, studied the role of hormones in reproduction.

Allen-Doisy test *Biochemistry.* a test used for the measurement of estrogenic substances. Similarly, **Allen-Doisy unit.**

allene *Organic Chemistry.* $CH_2{=}C{=}CH_2$, a colorless, easily liquefied gas that freezes at –136°C and boils at 34.5°C; used as an organic intermediate. Also, PROPADIENE.

Allen screw *Mechanical Devices.* a screw with a hollow hex-shaped socket design in its head to fit an Allen wrench.

Allen's rule *Zoology.* the principle that the extremities of warm-blooded animals (ears, limbs, tail) tend to be shorter in cold climates in order to conserve body heat. (From J. A. *Allen,* 1838–1921, American zoologist.) Also, **Allen's law.**

Allen's test *Medicine.* a practical test for the obstruction of the radial or ulnar arteries, in which the examiner compresses one of these arteries to determine if blood flow from the other one is adequate. (From Edgar *Allen.*)

Allen wrench *Mechanical Devices.* a hex-shaped head wrench that fits snugly into the matching socket in an Allen screw or other fitting; its bent shape allows for turning the screw to tighten or loosen it.

Allen wrench

allergen *Immunology.* an antigen that is capable of inducing an allergic reaction.

allergenic *Immunology.* **1.** causing an allergy. **2.** acting as an allergen.

allergic *Medicine.* **1.** having an allergy; sensitive to some allergenic substance. **2.** relating to or caused by an allergy; used as a descriptive name for various conditions, such as **allergic arteritis.**

allergic asthma *Medicine.* bronchial asthma caused by an allergen, typically by the inhalation of airborne allergens such as pollen.

allergic conjunctivitis *Medicine.* conjunctivitis caused by an allergen, typically by exposure to pollen, smoke, or air pollution.

allergic (contact) dermatitis *Medicine.* a skin rash caused by the contact of hypersensitive tissue with some allergen, involving inflammation, itching, and often surface lesions; its occurrence may be generalized or local.

allergic reaction see ALLERGY.

allergic rhinitis *Medicine.* inflammation of the nasal passages caused by an allergen, typically household dust or animal hair or dandruff; hay fever is a common form.

allergic vasculitis syndrome *Medicine.* the inflammation of a blood vessel due to a reaction to an allergen, typically a drug such as penicillin.

allergology [al′ər gäl′ə jē] *Medicine.* the study of allergy, including the causes, diagnosis, and treatment of various allergies. Thus, **allergologist, allergological.**

allergy *Immunology.* an exaggerated physical response to some antigen, typically a common environmental substance, that produces little or no response in the general population, resulting when histamine or histamine-like substances are released from injured cells. It involves various respiratory and dermatological symptoms, such as sneezing or itching. (From a Greek term meaning "other action.")

Allerød oscillation *Meteorology.* an increase in atmospheric temperature that occurred globally around the earth during the later Pleistocene epoch, approximately between 9850 and 8850 BC.

allesthesia [al′əs thēzh′yə] *Medicine.* a perception that tactile sensation is remote from the actual point of sensation. Also, ALLACESTHESIA.

allethrin [al′ə thrən] *Organic Chemistry.* $C_{19}H_{26}O_3$, a viscous liquid that is soluble in alcohol, carbon tetrachloride, and kerosene and insoluble in water. It is a synthetic pyrethroid used as an insecticide.

alliaceous [al′ē ā′shəs] *Science.* similar to garlic or onions in appearance, taste, or smell.

alliance *Systematics.* a taxonomic level pertaining to closely associated botanical families, usually occupying a rank below a class and above an order, and ending in *-ion* .

alliance theory *Anthropology.* the premise that marriage involving the exchange of women between groups in a society results in greater social solidarity in the form of cooperation and an exchange of obligations and resources.

allicin *Organic Chemistry.* $C_6H_{10}OS_4$, a yellow, oily liquid that is slightly soluble in water and soluble in alcohol, decomposing on heating; a substance that has antibacterial properties, extracted from garlic.

allidochlor [ə lid′ə klôr′] *Organic Chemistry.* $C_8H_{12}ClNO$, an amber liquid, slightly soluble in water and soluble in alcohol, that is used as a preemergence herbicide for vegetable crops.

alligator *Vertebrate Zoology.* a large reptile of the genus *Alligator,* having a broad head, a blunt snout, and teeth that do not protrude out over the jaw. The two species are the **American alligator,** *A. mississippiensis,* found in the southeastern United States, and the **Chinese alligator,** *A. sinensis,* found along China's Yangtze River.

alligator clip *Electricity.* a long, narrow spring clip with meshing jaws, primarily used with test leads to make temporary connections quickly.

alligator gar *Vertebrate Zoology.* a large North American freshwater fish, *Lepisoteus spatula,* having a prominent snout and a long tubular body.

Alligatoridae [al′ə gə tôr′i dē] *Vertebrate Zoology.* a reptile family in the order Crocodilia comprising seven species, including the American and Chinese alligators and the caimans.

alligatoring *Materials.* the condition of paint, varnish, or similar coating material that has developed cracks and lines from incorrect application or the effects of the elements. *Metallurgy.* **1.** during the rolling of metal or alloy slabs, the longitudinal splitting in a plane parallel to the rolling surface, and the curling of the resulting two portions of the slab. **2.** in a rolled metal or alloy, surface flaws resembling an alligator skin.

alligator lizard *Vertebrate Zoology.* any of several species of lizards of the genera *Algaria* and *Gerrhonotus,* ranging from the western United States to Panama and having bony-plated scales resembling those of an alligator.

alligator pear see AVOCADO.

alligator shears see LEVER SHEARS.

alligator snapping turtle *Vertebrate Zoology.* a North American turtle, *Macroclemys temmincki,* having a carapace with three prominent ridges and a wormlike mouth projection to attract prey; the largest freshwater turtle. Also, **alligator snapper.**

alligator wrench *Mechanical Devices.* a wrench having toothed jaws that form a V-point.

all-inertial guidance *Navigation.* a guidance system that relies entirely on inertial information to update its computed position.

Allisoniaceae *Botany.* a family of thallose liverworts of the order Metzgeriales, characterized by a thallus lacking central strands, having a poorly defined midrib, and thinning to wings, and by unicellular spores.

Allium *Botany.* **1.** a genus of bulbous plants of the lily family, having about three hundred species. **2. allium.** any plant of this genus, such as garlic, onions, or leeks.

allivalite [al′ə və līt′] *Petrology.* a holocrystalline rock composed mostly of equal proportions of anorthite and olivine, with lesser amounts of augite, apatite, and opaque iron oxides.

allo- a combining form meaning: **1.** another, different, or reversed, as in *allogrooming.* **2.** *Chemistry.* the more stable of two geometrical isomers. **3.** *Behavior.* other; another, as in *allomother.*

alloantibody *Immunology.* an antibody produced by one individual that has the ability to react with antigens of other members of the same species. Also, ISOANTIBODY.

alloantigen *Immunology.* an antigen that has the ability to stimulate the creation of a distinct antibody in other individuals within the same species. Also, ISOANTIGEN.

allobar *Meteorology.* a change in atmospheric pressure. *Physics.* a mixture of isotopes of an element proportionately differing from that naturally occurring. *Nuclear Physics.* an isotope with a different atomic weight than the naturally occurring form of the same element.

allobaric *Meteorology.* of or relating to change in atmospheric pressure. Thus, **allobaric wind.**

allocate *Computer Programming.* to assign, commit, or reserve specific parts of a computer system, such as areas of storage or storage systems, for a specific purpose. *Industrial Engineering.* to assign the use of resources, such as materials, capital, or time, among competing requirements. Thus, **allocation.**

allocentric *Psychology.* tending to focus on the thoughts and feelings of others; not egocentric. *Anthropology.* accepting or respecting the values and customs of other cultures; not ethnocentric.

allochem *Geology.* discrete calcareous particles or aggregates, including intraclasts, oolites, fossils, and pellets, that serve as framework grains in mechanically deposited limestone.

allochemical *Chemistry.* relating to or being a change in chemical composition.

allochemical metamorphism *Petrology.* metamorphism involving the addition or removal of material so that the bulk chemical composition of the rock is changed.

allochetite [a′lə ket′īt′] *Petrology.* a porphyritic igneous rock with phenocrysts of labradorite, orthoclase, titan-augite, nepheline, magnetite, and apatite in a dense ground mass of minerals such as augite, biotite, magnetite, hornblende, nepheline, and orthoclase.

allochiria [a′lō kī′rē ə] *Neurology.* a misperception of tactile stimuli such that a stimulus on one part of the body is felt at the corresponding point on the opposite side. Also, **allocheiria.**

allochoric *Ecology.* describing a species that occurs in two or more communities within a given geographical region. Also, **allochorous.**

allochromatic [al′ə krə mat′ik] *Science.* showing a variety of colors or changing in color, as a gem or crystal. *Physics.* of or relating to allochromy.

allochromatic crystal *Crystallography.* a crystal that exhibits various colors as a result of microimpurities within it. Therefore the crystal cannot be identified by its color.

allochromy [al′ə krōm′ē] *Physics.* the radiation emitted from a substance at a particular wavelength, resulting from the absorption of incident radiation of a different wavelength.

allochronic [al′ə krän′ik] *Science.* existing or occurring at different times; not contemporary.

allochronic speciation *Evolution.* the process of developing different species as a result of members of the population acquiring different breeding seasons or patterns, or differing in the use of time during the day or night, rather than through geographic separation.

allochronic species *Evolution.* **1.** any new species that results from allochronic speciation. **2.** species that constitute an evolutionary sequence or lineage, but that do not occur simultaneously.

allochthon [ə läk′thôn] *Geology.* a mass of rock that has been transported a great distance from its original place of deposition, usually by tectonic forces such as overthrusting or gravity sliding.

allochthonous [ə läk′thə nis] *Geology.* referring to materials whose present site is away from their place of origin. *Ecology.* living or growing away from the place of origin; not native.

allochthonous coal *Geology.* a type of coal formed from accumulated plant material that was transported from its place of growth and deposited elsewhere.

allochthonous stream *Hydrology.* a stream that is flowing in a channel it did not itself form.

allocryptic [al ə krip′tik] *Biology.* of or relating to an organism that conceals itself under living or nonliving material.

allocycly [al ə sik′lē] *Genetics.* a difference in the characteristics of coiling behavior of different chromosomes or portions of chromosomes.

allodiploid [al′ə dip′loid] *Genetics.* **1.** an organism that has two distinct chromosome sets derived from two different species, as in hybridization. **2.** of or relating to such an organism. Thus, **allodiploidic, allodiploidy.**

allodynia [al′ə dī′nē ə] *Neurology.* a sensation of pain as a result of a neutral stimulus on normal skin.

Alloeocoela [ə lē′ə sēl′ə] *Invertebrate Zoology.* an order of aquatic flatworms in the class Turbellaria, having a simple pharynx and a saclike intestine.

allogamy see CROSS-FERTILIZATION.

allogene *Geology.* a mineral or rock fragment that was derived from preexisting rock and transported to its present site of deposition. Also, ALLOTHIGENE.

allogeneic [al′ə jə nē′ik] *Genetics.* **1.** referring to genetically different cell lines in one individual, capable of producing different antigens. **2.** referring to individuals of the same species that are genetically different. Also, ALLOGENIC.

allogeneic effect *Immunology.* a nonspecific increase or depression in an immune response activated by T lymphocytes.

allogeneic graft see ALLOGRAFT.

allogenic *Genetics.* see ALLOGENEIC. *Ecology.* relating to the replacement of one community by another as a result of changes in the environment. Thus, **allogenic succession.** *Geology.* describing mineral and rock constituents that were formed at some place other than the site at which they are now found. Also, ALLOTHIGENIC, ALLOTHOGENIC, ALLOTHIGENOUS.

allograft *Surgery.* a graft from a donor of the same species as the recipient, but of a different genotype. Also, ALLOGENEIC GRAFT, HOMOGRAFT, HOMOPLASTIC GRAFT.

Allogromiidae [a′lə grə mī′ə dē′] *Invertebrate Zoology.* a family of protozoans in the order Foraminiferida.

Allogromiina [a′lə grə mī′ə nə] *Invertebrate Zoology.* a suborder of chiefly marine protozoans in the order Foraminiferida.

allogrooming *Behavior.* a behavior pattern in which an animal grooms the fur or skin of another animal of the same species. Similarly, **allopreening.**

allogyric birefringence [a′lō jī′rik] *Optics.* a phenomenon wherein passing plane polarized light through an optical substance produces two equal beams of circularly polarized light; one right-handed and the other left-handed, each having different velocities.

allokinesis [al′ō kə nē′sis] *Neurology.* any passive, involuntary, or reflex movement. Thus, **allokinetic.**

allolactose *Biochemistry.* a form of the disaccharide lactose that is produced by bacteria and is directly catalyzed by β-galactosidase formation.

allomarking *Behavior.* a behavior pattern in which an animal applies some scent-marking substance to another of its own species.

allomerism [ə läm′ə riz′əm] *Crystallography.* a variation in the chemical composition of a crystal that does not involve a change in its crystalline form. Thus, **allomeric.**

allomerize [ə läm′ə rīz′] *Crystallography.* to undergo allomerism.

allometric [al′ə met′ə rik] *Biology.* of or relating to allometry or to allometric growth.

allometric growth *Biology.* the phenomenon of different growth rates for different organs or parts of an organism.

allometry *Biology.* **1.** the relative growth rate of part of an organism in relation to another part or to the whole. **2.** the study or measure of such growth.

Allomonas [al′ə mōn′əs] *Bacteriology.* a proposed genus of Gram-negative, rod-shaped bacteria of uncertain classification, occurring in fresh water, sewage, and feces.

allomone [al′ə mōn′] *Biology.* a hormone or other substance that is produced by one species and has an effect on another, especially so as to benefit the emitting species. *Behavior.* a behavior pattern that influences the behavior of another species.

allomorph [al′ə môrf′] *Linguistics.* a variant form of any morpheme, such as the several endings that express the plural in English (*-s*, *-es*, *-a*, and so on); they occur in different contexts but their meaning is the same. *Chemistry.* see ALLOTROPE.

allomorphism [al′ə môr′fiz əm] *Crystallography.* a variation in crystalline form that does not involve a change in chemical composition. Thus, **allomorphic.**

allomorphite [al′ə môr′fīt] *Mineralogy.* **1.** a barite-containing mineral that is pseudomorphous after anhydrite. **2.** see BARITE.

allomorphosis [al′ə môr fō′sis] *Evolution.* an evolutionary change characterized by the rapid development or growth of one anatomical feature or organ in relation to the entire organism.

allomother *Behavior.* a female animal showing maternal behavior toward young that are not her own. Thus, **allomaternal.**

Allomyces [a lō mī′sēz′] *Mycology.* a genus of fungi belonging to the order Blastocladiales, characterized by spores possessing a single flagellum, and occurring in water, mud, and moist soil.

alloparent *Behavior.* an animal showing parental behavior toward young that are not its own. Thus, **alloparental.**

allopaternal *Behavior.* relating to or being a male alloparent.

allopath [al′ə path′] *Medicine.* a physician who uses allopathy as a method of treating disease; i.e., one who intervenes actively through medication, surgery, and so on. Also, **allopathic physician.**

allopathic [al′ə path′ik] *Medicine.* relating to or using allopathy as a method of treating disease.

allopathy [ə läp´ə thē] *Medicine.* a method of treating a disease by introducing a condition that is intended to cause a pathologic reaction which will be antagonistic to the condition being treated.
allopatric [al´ə pat´rik] *Ecology.* relating to or being two taxonomic entities or populations whose ranges are geographically separate and thus cannot interbreed.
allopatric speciation *Evolution.* an evolutionary change leading to the rise of new species resulting from the separation of a population into mutually exclusive geographic regions, thereby creating distinct gene pools. Also, GEOGRAPHIC SPECIATION.
allopatry [al´ə pat´rē] *Ecology.* the occurrence of allopatric (geographically separated) organisms or species
allopelagic [al´ō pə lā´jik] *Oceanography.* of a marine organism, living at various ocean depths.
allophanamide see BIURET.
allophane [al´ə fān´] *Geology.* an amorphous clay mineral having a variable composition of aluminum silicate, hydrated water, and traces of other minerals.
allophasis [al´ə fā´sis] *Medicine.* incoherent speech; delirium.
allophene [al´ə fēn´] *Genetics.* a phenotype due to other cells of the host rather than to mutation in the actual cells showing the characteristic; these will show a normal phenotype when transplanted to a normal host.
allophenic *Biology.* of or relating to single individuals originating from more than one conceptus; having an orderly coexistence of cells with different phenotypes ascribable to known allelic genotypic differences.
allopheny [al´ə fēn´ ē] *Genetics.* the interaction between the genes in host cells and transplanted cells in a host organism.
allophilous [ə läf´i ləs] *Biology.* living in or growing on sand.
allophone [al´ə fōn´] *Linguistics.* any of two or more alternate forms of the same phoneme that are not discerned as distinct sounds and thus do not affect meaning, e.g., the *t* sound in *steam* and the more aspirated *t* sound in *team* are allophones of the phoneme [t]. Thus, **allophonic.**
allophore [a´lō fôr´] *Histology.* a red pigment cell found in the skin of fishes, amphibians, and reptiles.
alloplast *Surgery.* **1.** a graft of an inert metal or plastic material. **2.** an inert foreign body used for implantation into tissues. Thus, **alloplastic.**
alloplasty *Surgery.* the replacement of a natural body part or tissue with synthetic material.
allopolyploid [al´ə päl´ə ploid´] *Genetics.* **1.** an organism, strain, or cell that is produced from the combination of two or more chromosome sets from different species. **2.** also, **allopolyploidic.** of or relating to such a form. Also, **alloploid.**
allopolyploid speciation *Evolution.* an evolutionary change leading to the rise of new species of polyploid plants resulting from hybrids between diploid ancestors.
allopolyploidy [al´ ə päl´ə ploi´dē] *Genetics.* the fact of having two or more distinct chromosome sets originating from different species.
allopsychic [al´ə sī´kik] *Psychology.* of or relating to the mind in its relation to the external world.
allopurinol *Pharmacology.* $C_5H_4N_4O$, a drug used to decrease the excessive amounts of uric acid in the blood caused by gout, certain blood disorders, and chemotherapy.

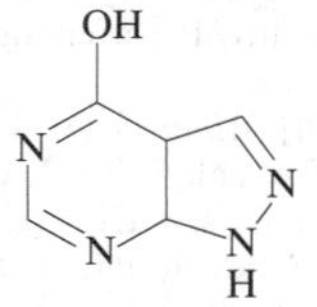

allopurinol

allorhythmia [al´ə rith´mē ə] *Cardiology.* an irregular heartbeat that recurs repeatedly.
all-or-none law *Cardiology.* the known principle that the heart muscle, when stimulated to a certain threshold, will either contract fully or not at all. *Physiology.* the principle that the strength of a nerve impulse is not dependent on the stimulus strength, since once the nerve threshold is reached, the fiber will respond completely at full capacity or not at all. *Behavior.* the principle that a behavioral stimulus either will be strong enough to produce a complete response or will produce no response at all. Also, **all-or-none principle, all-or-none responsiveness, all-or-nothing law.**
allosaur *Paleontology.* a large carnivorous dinosaur of the late Jurassic period, known from fossil remains in North America. Also, **allosaurus.**
allosome *Genetics.* **1.** a chromosome that deviates in size, form, and behavior from other chromosomes. **2.** a sex chromosome. Thus, **allosomic.**
allosteric *Biochemistry.* of or relating to allosterism.
allosteric effector *Biochemistry.* a molecule that changes the function of the molecule with which it reacts. Also, **allosteric modulator.**
allosteric enzyme *Enzymology.* a regulatory enzyme whose allosteric effectors cause its activity to increase or decrease. *Biochemistry.* any of a group of enzymes whose catalytic activity is varied by the noncovalent binding of a given metabolite at a site other than the catalytic one.
allosteric modulator see ALLOSTERIC EFFECTOR.
allosteric site *Biochemistry.* a region on an enzyme molecule to which an allosteric effector binds, either activating or inhibiting the enzyme.
allosteric transition *Biochemistry.* a structural change in certain parts of a protein molecule that occurs as a result of its interaction with an allosteric effector.
allosterism *Biochemistry.* a process in which the binding of a compound to a subunit of an enzyme or other protein molecule at an allosteric site changes the conformation of the protein. Also, **allostery.**
allostery [a´lō stir´ē] *Genetics.* a change from one conformation of a protein possessing two or more receptor sites to another conformation, when one of the receptor sites is occupied by a compound.
allosyndesis [al´ō sin´də sis] *Genetics.* a phenomenon in mitosis of allopolyploid cells, whereby homologous chromosomes from one parent pair only with one another, and do not pair with those from the other parent.
allotetraploid [al´ō tet´rə ploid] *Genetics.* an allopolyploid produced when a hybrid of two species doubles its chromosome number. Also, AMPHIDIPLOID.
Allotheria [a´lō thir´ē ə] *Paleontology.* an extinct infraclass of mammals that constitutes the single order Multituberculata; Holarctic distribution, extant in the upper Triassic to lower Oligocene.
allotherm *Biology.* an organism whose body temperature is determined by its surrounding conditions, rather than being internally consistent. Thus, **allothermic.**
allothigene see ALLOGENE.
allothigenic or **allothigenous** see ALLOGENIC.
allothimorph [ə läth´ə môrf´] *Geology.* a constituent of metamorphic rock whose original crystal outline was not altered during metamorphism.
allothogene [ə läth´ə jēn] see ALLOGENE.
allothogenic see ALLOGENIC.
allotransplant see ALLOGRAFT.
allotri- a combining form meaning "strange" or "foreign." Also, **allotrio-.**
allotrioblast [a´lə trē´ə blast] see XENOBLAST.
Allotriognathi *Vertebrate Zoology.* an order of the subclass Teleostei; a group of rare, bony deep-sea fishes with small teeth and large eyes, including the ribbon-fish, oar-fish, and moonfish. Also, LAMPRIDIFORMES.
allotriomorphic [ə lä´trē ə môr´fik] *Mineralogy.* of minerals in igneous or metamorphic rock, not bounded by their own rational crystal faces but whose forms are controlled by mineral grains adjacent to them. Also, ANHERAL, XENOMORPHIC.
allotriomorphism see ALLOTROPY.
allotrope *Chemistry.* a substance that exists and is metastable in two or more physical forms over a given temperature range.
allotrophic *Ecology.* **1.** of an organism, obtaining nourishment from other organisms. **2.** of an ecosystem, receiving nutrients from outside the system. **3.** see ALLOTROPIC. Thus, **allotroph, allotrophy.**
allotrophic lake *Ecology.* a lake that receives organic material by drainage from the surrounding land.
allotropic relating to or characterized by allotropy. Also, ALLOTROPHIC.
allotropy *Chemistry.* the property exhibited by an element that exists as two or more forms, but retains the same state; e.g., red and yellow phosphorus. *Materials Science.* the existence of different stable crystal structures of a material at different temperatures. *Ecology.* the existence of the same population or species in different habitats. Also, **allotropism.**
allotype *Immunology.* any of several allelic variants of a protein that are characterized by antigenic difference, especially allelic variants of immunoglobulin heavy and light chains. *Systematics.* a specimen in the type series that is of the opposite sex of the holotype and originally designated by the author. Thus, **allotypic.**
allowable bearing value see ALLOWABLE SOIL PRESSURE.

allowable burned area *Forestry.* the maximum acreage burned over a period of years that can be considered an acceptable loss for that region and forest type.
allowable soil loss *Agronomy.* the amount of soil that can be eroded from an area without long-term damage to soil productivity.
allowable soil pressure *Civil Engineering.* the maximum allowable loading on a soil with respect to the amount and type of settlement expected. Also, ALLOWABLE BEARING VALUE.
allowable stress *Mechanical Engineering.* the maximum force per unit area that can be applied to a structure or solid according to some saftey standard. Similarly, **allowable load.**
allowance *Design Engineering.* the intentional difference in dimensions between two mating parts, allowing clearance for a sliding fit or for a film of oil. *Industrial Engineering.* an amount of nonproductive time that is added to normal time to account for unexpected delays, worker fatigue, attendance to personal needs, and so on.
allowed energy bands *Solid-State Physics.* the range of energy levels that are permitted to electrons in a molecule or crystal.
allowed transition *Quantum Mechanics.* a transition from one quantum state to another, permitted by the selection rules of quantum mechanics.
alloxan *Biochemistry.* $C_4H_2N_2O_4$, a compound that is produced from uric acid and destroys cells of the pancreas, resulting in diabetes. Also, MESOXYALYUREA.
alloy *Metallurgy.* any of a variety of materials having metallic properties and composed of two or more intimately mixed chemical elements, of which at least one is a metal; e.g., brass is an alloy of copper and zinc. Alloys are produced to obtain some desirable quality such as greater hardness, strength, lightness, or durability.
Alloy 750 *Metallurgy.* the former designation of a casting aluminum alloy containing about 6.2% tin, 1% copper, and 1% nickel; the current designation is **UNS A08500.**
alloy cast iron *Metallurgy.* gray iron composed of 3% of the alloying elements silicon, nickel, chromium, copper, and aluminum, either alone or in combination.
alloyed nuclear fuel *Nucleonics.* a reactor fuel produced by adding such materials as silicon, zirconium, or chromium to metallic uranium or plutonium to offset the effects of irradiation damage and phase-change dynamics, and also to improve fabrication and operating characteristics.
alloying *Metallurgy.* any of various processes by which an alloy is manufactured, such as melting, sintering, electrochemical co-depositing, or diffusing.
alloy junction *Electronics.* a junction formed by alloying at least one impurity metal to a semiconductor. Depending on the type of impurity metal, the junction will be a P type or N type. Also, FUSED JUNCTION.
alloy junction diode *Electronics.* a diode whose P and N regions are fabricated by alloying different impurity metals to the semiconductor material. Also, **alloy diode.**
alloy junction transistor *Electronics.* a transistor in which opposite ends of the substrate are doped with an impurity metal to create a collector and an emitter at the wafer ends and a base in the middle.
alloy plating *Metallurgy.* the simultaneous deposition of two or more metallic elements onto a substrate.
alloy steel *Metallurgy.* steel in which other other elements as well as carbon have been added as alloys; e.g., chromium, nickel, or vanadium.
allozygote [al´ō zī´gōt] *Genetics.* an individual that is homozygous at a particular locus, but possesses homologous alleles that have originated from different ancestral sources.
allozyme [al´ə zīm´] *Enzymology.* any one of a group of functionally identical or similar enzymes that are produced by alleles of the same gene.
all-pass network *Electronics.* a network that provides equal attenuation (or gain) to signals of all frequencies, varying only the phase shift or delay depending on the frequency. Also, **all-pass filter.**
Allport, Gordon W. 1897–1967, American psychologist; studied the development of personality.
all-purpose computer *Computer Technology.* any computer that provides general-purpose processing capabilities as well as special-purpose business or scientific functions.
all-purpose flour *Food Technology.* an intermediate type of flour containing less gluten than bread flour but more than cake flour, thus being nearly as strong as the former and as tender and friable as the latter.
all-sky camera *Optics.* a camera that is directed downward toward a convex mirror in order to photograph the entire sky.
allspice *Food Technology.* an aromatic spice made from the berry of a tropical American tree, *Eugenia pimenta;* so called because its odor and flavor suggest a combination of cinnamon, cloves, and nutmeg. *Botany.* the tree that yields this spice, more often called PIMENTO.

allspice

all-terrain cycle see ATC.
all-terrain vehicle see ATV.
all-translational system *Robotics.* a simple robotic system in which the axes do not change while the robot's body is moving.
allulose *Organic Chemistry.* $CH_2OHCO(CHOH)_3CH_2OH$, a nonfermentable constituent of cane sugar molasses that is soluble in water.
alluvial *Geology.* **1.** relating to or consisting of any material that has been carried or deposited by running water. Thus, **alluvial valley, alluvial deposit. 2.** describing a placer or its associated mineral formed by the action of running water. (From a Latin term meaning "to wash.")
alluvial cone *Geology.* an alluvial fan having very steep slopes. Also, DEBRIS CONE, DEJECTION CONE, WASH.
alluvial dam *Geology.* a sedimentary deposit built up by an overloaded stream and obstructing the stream channel.
alluvial fan *Geology.* an outspread, relatively flat or gently sloping, fan- or cone-shaped mass of loose rock material that is deposited by a stream as its gradient decreases, as when it descends into a valley or upon a plain or where it junctions with the main stream. Also, FAN, DETRITAL FAN, DRY DELTA, TALUS FAN.
alluvial flat *Geology.* a small alluvial plain deposited alongside a river during floods. Also, RIVER FLAT.
alluvial mining *Mining Engineering.* the process of dredging, hydraulicking, or drift mining to exploit alluvial deposits.
alluvial ore *Geology.* valuable mineral particles that have been transported by a river or stream.
alluvial plain *Geology.* **1.** a flat or gently sloping tract of land alongside a periodically overflowing river that is produced by the deposition of alluvium. Also, WASH PLAIN, WASTE PLAIN. **2.** see BAJADA.
alluvial slope *Geology.* an alluvial surface that slopes down and away from a mountainside and merges with a plain or broad valley floor.
alluvial soil *Geology.* a soil developed on flood plains and deltas, having only the characteristics of the alluvium of which it is composed.
alluvial terrace *Geology.* a terraced embankment of loose, unconsolidated alluvial material that is built up adjacent to the sides of a river valley. Also, BUILT TERRACE, DRIFT TERRACE, FILL TERRACE.
alluviation *Geology.* the deposition or formation of alluvial sediments or features along the course of a river or stream.
alluvium *Geology.* clay, mud, sand, silt, gravel, and other unconsolidated detrital matter that is carried along and deposited by flowing water. Also, **alluvion, alluvial deposit.**
all-weather *Aviation.* **1.** of an aircraft, having special instrumentation enabling it to operate in the dark or in daylight weather conditions that limit visibility. **2.** taking place under any weather or visibility conditions, as in **all-weather flight. 3.** of instruments, equipment, or facilities, designed for use in all-weather flight. Thus, **all-weather airport, all-weather aircraft, all-weather fighter,** and so on.
ally *Biology.* any organism having a relationship with another, especially an evolutionary relationship.
allyl *Organic Chemistry.* an unsaturated radical ($-CH_2{=}CHCH_2$) that is derived from allyl alcohol by the removal of hydrogen.
allyl acetone *Organic Chemistry.* $CH_2{=}CHCH_2CH_2COCH_3$, a colorless liquid that is soluble in water and organic solvents; it boils at 127–129°C; used in perfumes, fungicides, and insecticides, and as an intermediate in pharmaceutical synthesis.

allyl alcohol *Organic Chemistry.* CH_2=$CHCH_2OH$, a colorless liquid with a mustardlike odor that is soluble in water, alcohol, and ether; freezes at –129°C, melts at –50°C, and boils at 97°C; used in herbicides, in esters for resins and plasticizers, in the manufacture of glycerol, and as a military poison.

allylamine [al´əl ə mēn´] *Organic Chemistry.* CH_2=$CHCH_2NH_2$, a colorless to light yellow liquid with an ammoniacal odor that is soluble in water, alcohol, and ether; boils at 58°C; attacks rubber and cork; used as a pharmaceutical intermediate and in organic synthesis.

allyl bromide *Organic Chemistry.* CH_2=$CHCH_2Br$, a colorless to light yellow liquid with an irritating odor that is soluble in alcohol, ether, and chloroform and insoluble in water; boils at 71.3°C; used in organic synthesis, resin preparation, and perfume intermediates.

allyl cation *Organic Chemistry.* a carbonium cation with the structure CH_2=$CHCH_2^+$; the bonding site is at the saturated carbon atom.

allyl chloride *Organic Chemistry.* CH_2=$CHCH_2Cl$, a colorless liquid with a pungent odor that is soluble in alcohol, chloroform, and ether, and slightly soluble in water; boils at 46°C; used in allyl compound synthesis.

allyl cyanide *Organic Chemistry.* CH_2=$CHCH_2CN$, a liquid with an onionlike odor that is soluble in alcohol and slightly soluble in water; it boils at 119°C; used in polymerization as a crosslinking agent.

allylene *Organic Chemistry.* $CH_3C{\equiv}CH$, a colorless gas that is soluble in ether and boils at –23.1°C; toxic and a dangerous fire risk; used as a specialty fuel. Also, PROPYNE, METHYLACETYLENE.

allyl group *Organic Chemistry.* the group of univalent compounds with the general formula CH_2=$CHCH_2^+$.

allylic [ə lil´ik] *Organic Chemistry.* having to do with allyl; involving the presence of allyl.

allylic hydrogen *Organic Chemistry.* in an allylic (three-carbon) system, the hydrogen attached to the carbon atom adjacent to a double bond.

allylic rearrangement *Organic Chemistry.* the migration of a double bond in an allylic (three-carbon) system from the 1,2 carbon position to the 2,3 carbon. The substituent migrates simultaneously from carbon 3 to carbon 1. Thus: C1=C2C3X ↔ XC1C2=C3.

allyl isothiocyanate [ī´sō thī ō sī´ə nāt´] *Organic Chemistry.* H_2C=$CHCH_2NCS$, a colorless to pale-yellow oily liquid with a pungent, irritating odor and biting taste; slightly soluble in water and soluble in alcohol and ether; boils at 152°C; used as a fumigant, in ointments, and as a poison gas. Also, MUSTARD OIL.

allyl mercaptan *Organic Chemistry.* CH_2=$CHCH_2SH$, a water-white liquid that darkens on standing and has a strong garlic odor; soluble in ether and alcohol and insoluble in water; boils at 67–68°C; used as a pharmaceutical intermediate and a rubber accelerator. Also, **allyl thiol.**

allyl resin *Materials Science.* a class of thermosetting polyester resins derived from esters of allyl alcohol and dibasic acids; used as laminating adhesives, varnishes, and heat-resistant finishes. Also, **allyl plastic.**

allyl sulfide *Organic Chemistry.* $(CH_2$=$CHCH_2)_2S$, a colorless liquid with a garlic odor that is insoluble in water and soluble in alcohol, ether, and chloroform; boils at 139°C; used in synthetic oil of garlic.

allylthiourea [al´il thī´ō yoo rē´ə; thī´ə yə rē´ə] *Organic Chemistry.* CH_2=$CHCH_2NHCSNH_2$, a white, crystalline solid with a slight garlic odor and bitter taste that is soluble in water, alcohol, and ether, and melts at 78°C; used as a corrosion inhibitor.

allyltrichlorosilane *Organic Chemistry.* CH_2=$CHCH_2SiCl_3$, a colorless, pungent, toxic liquid that boils at 117.5°C. It polymerizes easily, and is used in glass fiber finishes and as an intermediate for silicones.

allylurea [al´il yoo rē´ə] *Organic Chemistry.* $C_4H_8N_2O$, solid crystals that are soluble in water and alcohol and insoluble in ether, and melt at 85°C; used to manufacture corrosion inhibitors such as allylthiourea.

almandine *Mineralogy.* $Fe_3^{+2}Al_2(SiO_4)_3$, a cubic silicate mineral of the garnet group that has a specific gravity of 4.1 to 4.3 and a hardness of 7 to 7.5 on the Mohs scale; used as a gemstone and as an abrasive. Also, **almandite.**

Almeida's disease see SOUTH AMERICAN BLASTOMYCOSIS.

Al-Mg alloy *Metallurgy.* **1.** any of several cast aluminum-base alloys that contain from 4% to 10% magnesium, commonly designated as the 5xxx series. **2.** any of several wrought aluminum-base alloys that contain from 0.8% to 5.1% magnesium, commonly designated as the 5xxx series.

almond [ä´mənd; am´ənd] *Botany.* a tree of the rose family, *Prunus amygdalus,* widely grown in temperate regions for its edible nut, which is the kernel or stone of the fruit. The leaves and flowers of the almond tree resemble those of the peach.

almond oil *Materials.* a volatile essential oil distilled from the ground kernels of bitter almonds; used in cosmetic creams, perfumes, liqueurs, and food flavors. Also, OIL OF BITTER ALMOND.

almost everywhere *Mathematics.* if a given proposition concerning the points of a measure space is true for every point except for a set of points of measure zero, then the proposition is said to be true almost everywhere or for almost all points. Also, **almost all points.**

almost periodic (function) *Mathematics.* a continuous function $f(z)$ is said to be almost periodic if, for every $\varepsilon > 0$, there exists a number M such that for any real or complex number z, any interval of length M contains a nonzero number t with $|f(z+t) - f(z)| < \varepsilon$. This notion differs from that of a periodic function in that for a periodic function, a number T exists so that $f(z+T) = f(z)$ for all z independent of any choice of ε and M.

almost uniform convergence *Mathematics.* a sequence $\{f_n\}$ of almost everywhere finite-valued measurable functions defined on a measure space X is said to converge almost uniformly to the measurable function f if, for every $\varepsilon > 0$, there exists a measurable subset F of X of measure $< \varepsilon$ and such that $\{f_n\}$ converges uniformly on $X - F$.

Almquist unit *Biology.* a unit used for the standardization of vitamin K.

alnico magnet *Electromagnetism.* a ferromagnetic material commonly used for permanent magnets due to its high degree of magnetic intensity; the alloy composition is 54% iron, 18% nickel, 10% aluminum, 12% cobalt, and 6% copper.

Alnilam [al´ni ləm] *Astronomy.* Epsilon (ε) Orionis, a blue-white star that lies in center of the Belt in the constellation Orion.

alnoite [al´nə wit´] *Petrology.* a mafic lamprophyre characterized by melilite; it is feldspar-free but normally contains biotite, perovskite, olivine, and carbonate in the matrix.

aloe [al´ō] *Botany.* any of various plants of the genus *Aloe,* belonging to the lily family, grown in South Africa and other warm areas, and having various medicinal properties. *Pharmacology.* also, **aloes.** the juice of the leaves of an aloe plant, such as *Aloe barbadensis,* the **Barbados aloe,** used in compound benzoin tincture, a skin protectant, and formerly used as a purgative.

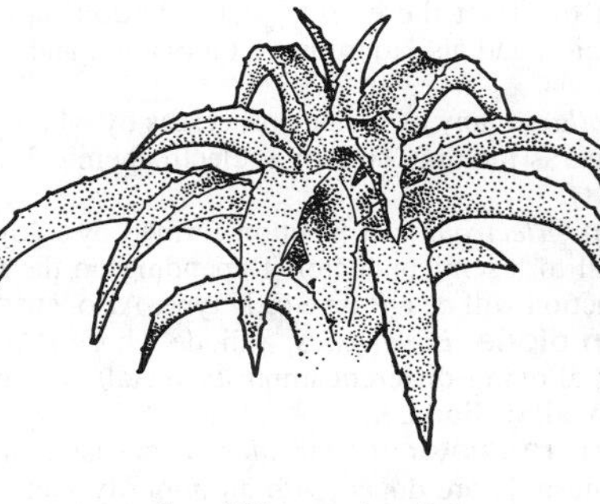

aloe (Aloe vulgaris)

Aloeaceae [a´lō ā´sē ē] *Botany.* a family of coarse, sparsely branched monocotyledonous plants of the order Liliales; characterized by narrow, sessile, strongly succulent, and often very firm leaves that usually produce anthraquinones and chelidonic acid.

aloe lace *Textiles.* a delicate lace fabric that is made from aloe plant fibers.

aloe wood *Botany.* the aromatic wood of an East Indian tree, *Aquilaria agallocha;* not a true aloe. Also, **aloes wood.**

aloe vera *Botany.* a species of aloe whose leaves provide a juice that is widely used as an ingredient in skin lotions and ointments.

alogia [ə lō´jē ə] *Neurology.* an inability to speak due to a central lesion.

aloin [al´ō in] *Pharmacology.* $C_{22}H_{21}O_9$, the bitter, purgative extract of certain aloes.

aloisiite [a´lə wis´ē īt´] *Mineralogy.* a brown to violet, amorphous silicate containing ferrous oxide, magnesium, sodium, and calcium.

alona *Materials.* the golden brown wood of a tree of West Africa, *Lovoa trichiliodes,* which is used in the making of high-quality furniture and veneers. Also, AFRICAN WALNUT.

Alongiaceae [ə lôn´gē ā´sē ē] *Botany.* a monogenetic family of woody and sometimes thorny Old World plants in the order Cornales; characterized by articulated lactifers, spirally arranged leaves, and usually bisexual flowers growing on axillary cymes.

alopecia [al′ə pē′sē ə] *Medicine.* the unnatural absence of hair; baldness. Used in the names of many specific conditions; e.g., **alopecia areata,** patchy loss of hair; **alopecia universalis,** complete loss of hair; **alopecia congenitalis,** congenital lack of scalp hair.

Alopiidae *Vertebrate Zoology.* the thresher sharks, a family in the order Isuriformes found worldwide in tropical and temperate seas.

alp *Geology.* any high mountain, especially a single peak of the Alps.

ALP alkaline phosphatase.

alpaca *Vertebrate Zoology.* a South American herbivorous mammal, *Lama pacos,* in the family Camelidae, resembling and related to the llama. It is raised commercially for its long, lustrous wool. *Materials.* the wool of this animal, or a fabric or garment made from it.

alpaca

alpenglow *Meteorology.* **1.** the reappearance of sunset colors on a mountain summit that occurs after the original colors have faded. **2.** a reddish glow that precedes the regular coloration of sunrise.

alpestrine *Botany.* living or found at high altitudes; alpine.

alpha *Science.* **1.** the first letter in the Greek alphabet; it is written as α. **2.** the first in a series or hierarchy. *Chemistry.* a prefix that indicates the position of a substituting carbon atom or group in an organic compound; on a carbon chain, the atom next to the functional group; in heterocyclic compounds, the carbon atom next to the heteroatom. *Behavior.* describing the dominant or highest-ranking individual in a group of animals. *Electronics.* the current gain from emitter to collector of a common-base configured transistor.

Alpha *Astronomy.* a term used to designate the brightest star in a constellation; e.g, Alpha Centauri.

alpha-adrenergic blocker see ALPHA BLOCKER.

alpha-amanitin *Mycology.* a type of amanatin poison found in mushrooms belonging to the genus *Amanita.*

alpha amylase *Enzymology.* an endoamylase that catalyzes the random hydrolysis of starch molecules at (1–4) glycosidic bonds.

alpha and beta cutoffs *Artificial Intelligence.* in alpha/beta search, threshold values that allow unpromising avenues of play to be identified and eliminated from the search.

Alpha Aquilae see ALTAIR.

Alpha Arietis see HAMAL.

Alpha Aurigae see CAPELLA.

Alpha axis *Robotics.* a turning axis that represents one degree of freedom in a robot.

alphabet *Linguistics.* any ordered set of characters that are used for writing a language. *Computer Programming.* the set of symbols used by a grammar or automation.

alpha/beta search *Artificial Intelligence.* a kind of game tree search that uses alpha and beta cutoffs to avoid unnecessary search of unpromising avenues of play.

alphabetical *Linguistics.* **1.** relating to or using an alphabet. **2.** in the order of the letters of the alphabet; A to Z. Also, **alphabetic.**

alphabetic character *Computer Programming.* a letter of the alphabet.

alphabetic coding *Computer Programming.* **1.** a system that uses alphabetic characters as a form of abbreviation to represent information. **2.** a binary code representation of the letters of the alphabet, for example by assigning A = 1, B = 2, etc.

alphabetic writing *Linguistics.* any form of writing that uses a letter or symbol to represent a sound, as opposed to systems that use symbols to represent whole words or larger concepts.

alpha blocker *Biochemistry.* any of various drugs or other substances that block chemical activity at the alpha receptors; e.g., the antihypertensive drug phenoxybenzamine. Also, **alpha-blocking agent.**

Alpha Bootis see ARCTURUS.

alpha brass *Metallurgy.* a copper-zinc alloy, the crystal structure of which is face-centered cubic. The maximum zinc content of alpha brass is approximately 37%.

Alpha Canis Majoris see SIRIUS.

Alpha Canis Minoris see PROCYON.

Alpha Carinae see CANOPUS.

alpha cell *Histology.* **1.** a cell type in the islets of Langerhans that produces glucagon. **2.** an acidophilic pigment cell found in the anterior lobe of the pituitary gland.

alpha cellulose *Organic Chemistry.* a highly refined cellulose from which all soluble materials (such as sugars and pectin) have been removed by a strong sodium hydroxide solution. It is the major component of wood and paper pulp.

Alpha Centauri [sen tôr′ē] *Astronomy.* a star lying 4.3 light-years away that is the sun's nearest stellar neighbor; it is a triple star also known as **Rigil Kentaurus,** found in the southern constellation of the Centaur.

Alpha Crucis see ACRUX.

alpha cutoff (frequency) *Electronics.* in a common-base configured transistor, the high frequency at which alpha drops 3 decibels from its low-frequency value.

Alpha Cygni [sig′nē] see DENEB.

alpha decay *Nuclear Physics.* a type of spontaneous disintegration that takes place within a radioactive nuclide in which the parent nucleus emits an alpha particle (identical with a helium nucleus) at high velocity in one or more discrete energy groups. Also, **alpha disintegration.**

alpha emission *Nuclear Physics.* the release of alpha particles at high velocity from an atom's nucleus as it undergoes radioactive transformation. Thus, **alpha emitter.**

Alpha Eridani see ACHERNAR.

alpha female *Behavior.* the dominant or highest-ranking female in a social group of animals.

alpha-fetoprotein [al′fə fēt′ō prō′tēn] *Biochemistry.* the fetal equivalent of albumin; a plasma protein with $alpha_1$ electrophoretic mobility, produced by the fetal liver, yolk sac, and gastrointestinal tract; used to monitor the response of hepatomas and neoplasms to treatment and in antenatal diagnosis of neural tube defects.

Alpha Geminorum see CASTOR.

alpha globulin *Biochemistry.* a kind of plasma globulin having alpha electrophoretic mobility.

alpha helix *Molecular Biology.* an element of secondary structure found in proteins; it consists of a right-handed helical shape in which the backbone is derived from peptide bonds between the carboxyl group of one amino acid and the alpha carbon amino group of the next. Thus, **alpha-helical.**

alpha hemolysis *Microbiology.* the appearance of a zone of greenish discoloration around a bacterial colony cultured on blood agar, resulting from incomplete decomposition of blood cell hemoglobin in the culture medium.

alpha-hemolytic streptococci see VIRIDANS STREPTOCOCCI.

Alphaherpesvirinae [al′fə her′pēz vēr′i nē] *Virology.* a subfamily of the family Herpesviridae, causing cell lysis and having a short replication cycle marked by rapid spreading in cell cultures.

alpha iron *Metallurgy.* crystalline iron having a body-centered cubic structure stable up to 910°C.

alpha irradiation *Nucleonics.* the bombardment of elements to produce artificial nuclear rearrangements, as distinguished from spontaneous nuclear transmutations.

Alpha Leonis see REGULUS.

Alpha Lyrae [lī′rē] see VEGA.

alpha male *Behavior.* the dominant or highest-ranking male in a social group of animals.

alphameric [al fə mer′ik] see ALPHANUMERIC.

alphanumeric [al′fə noo mer′ik] *Computer Programming.* **1.** a character that is either alphabetic or numeric; that is, a letter or a numeral. **2.** using both letters and numerals, and also punctuation marks and other symbols, as representations of data. Used in many compound terms, such as **alphameric character, alphanumeric data, alphanumeric display (device).**

alphanumeric code *Computer Programming.* a correspondence code for representing letters of the alphabet, numerals, and special symbols as integer values, such as the EBDIC and ASCII codes.

alphanumeric instruction *Computer Programming.* a computer instruction carried out on either alphabetic, numeric, or mixed data.

Alpha Orionis see BETELGEUSE.

alpha particle *Atomic Physics.* a positively charged atom, identical with the helium nucleus, consisting of two neutrons and two protons, so double charged; released at a very high speed in the disintegration of radium and other radioactive elements.

alpha-particle decay see ALPHA DECAY.

alpha-particle detector *Nucleonics.* typically, an ionization chamber having a counting-type meter and a thin window that can be penetrated by incident alpha particles; the instrument's pulse counter is adjusted to respond only to alpha particles and not to count beta particles or gamma rays that may enter the chamber.

alpha-particle emission see ALPHA DECAY.

alpha-particle scattering *Atomic Physics.* the movement of individual particles away from their initial paths when a beam of alpha particles passes through thin metallic foils.

Alpha Piscis Austrini see FOMALHAUT.

alpha position *Organic Chemistry.* a term used to designate a substituent positioned on the first carbon attached to a functional group in an organic compound.

alpha ray *Nucleonics.* a doubly ionized helium atom that is electrically charged and therefore is deflected in electric and magnetic fields and produces intense ionization in matter; the most easily absorbed of the three types of radiation emitted by radioactive substances. Also, **alpha radiation.**

alpha-ray vacuum gauge *Engineering.* an ionization gauge that emits alpha particles instead of electrons; primarily used to measure pressures in the 10^{-3} to 10 Torr range.

alpha receptor *Biochemistry.* a cell site that responds to adrenaline or adrenaline-like substances such as epinephrine, causing various physiological changes including pupil dilation, increased vascular resistance, and heightened muscle activity.

alpha rhythm *Physiology.* the waveforms of the electroencephalogram that occur at a frequency of 8–12 hertz in the normal relaxed adult.

alphascope *Computer Technology.* a computer device that displays alphanumeric characters on a video screen.

Alpha Scorpii see ANTARES.

alpha state *Psychology.* a state of relaxation and removal from sensory awareness, associated with alpha brain waves and characterized by a lack of anxiety and tension.

alpha system *Telecommunications.* any system that uses alphabetic characters to delineate a signaling code.

Alpha Tauri [tôr´ē] see ALDEBARAN.

alpha taxonomy *Systematics.* the first stage in the development of systematic understanding of a taxon, during which species are described and designated, based primarily on their morphological (phenotypic) characters.

alpha test *Science.* a term for the first phase of testing of a new scientific product or computer program by the developers of the product. Thus, **alpha test site.**

alphatron see ALPHA-RAY VACUUM GAUGE.

Alpha Ursae Majoris see DUBHE.

Alpha Virginis see SPICA.

Alphavirus *Virology.* a genus of viruses of the family Togaviridae; most species are transmitted through mosquitoes, and many are important agents of disease in humans and animals.

alpha waves *Physiology.* the pattern of slow brain waves of alpha rhythm.

Alpheidae [al´fē´ə dē´] *Invertebrate Zoology.* a family of snapping shrimps in the crustacean order of Decapoda; distinguishable from other snapping shrimps (Cragonidae) by the unequal size of the claws on the first pair of legs.

Alphonsus [al fän´səs] *Astronomy.* a lunar crater with a diameter of 70 miles.

Alpides [al´pə dēz] *Geology.* the great east-west system of young, folded mountains that includes the European Alps, the Himalayas, and related mountains of Asia. Also, MEDITERRANEAN BELT.

Alpine or **alpine** *Geology.* **1.** relating to or characteristic of the European Alps. **2.** relating to any mountain or mountain system that resembles the European Alps in topography, morphology, or structure. *Ecology.* **1.** relating to a mountain area above the timberline, but below the area of permanent snow. **2.** describing an organism, population, or community that lives or thrives in such an area. *Agriculture.* a breed of milk goat that is native to the Swiss Alps.

Alpine fir *Botany.* a fir tree, *Abies lasiocarpa,* native to western North America, whose wood is widely used for lumber.

alpine glacier *Hydrology.* a moving mass of ice that accumulates in a previously formed mountain stream valley; a snow catchment area. This alpine glacial erosion creates a rugged, sharp topography with U-shaped valleys; characteristics include horns, moraines, cirques, and aretes. Also, MOUNTAIN GLACIER, VALLEY GLACIER.

Alpine orogeny *Geology.* a Tertiary deformation of the earth's crust that affected the Alpides of southern Europe and Asia.

alpine tundra *Ecology.* an extensive region of lightly vegetated but treeless land above the timberline in a mountainous area.

alpine-type facies *Petrology.* dynamothermal metamorphism that occurs in high-pressure, low temperature (150–400°C) conditions; characterized by the pumpellyite and glaucophane schist facies.

alpinotype tectonics [al pē´nō tīp´] *Geology.* a deformational movement of alpine mountain belts characterized by deep-seated plastic folding, plutonism, and lateral thrusting of orthogeosynclines.

Alport's syndrome see HEREDITARY NEPHRITIS.

Alps *Geography.* a major mountain system of south central Europe, extending from southeastern France through northern Italy, Switzerland, southern Germany, and Austria into northern Yugoslavia; highest peak, Mont Blanc (15,771 ft).

ALS antilymphocyte serum; amyotrophic lateral sclerosis.

alsad *Ecology.* a plant that typically grows in a grove of plants of the same kind.

alsbachite [ôlz´bä kīt´] *Petrology.* a porphyritic plutonic rock composed of sodic plagioclase, quartz, and subordinate orthoclase, often with garnet, biotite, and muscovite.

Alseuosmiaceae *Botany.* a family of dicotyledonous shrubs of the order Rosales, characterized by regular, often highly scented flowers and bearing a bilocular berry; native to New Zealand.

Alsever solution *Medicine.* a solution consisting of citrate, dextrose, and sodium chloride, used to preserve reagent red blood cells.

Al-Si alloy *Metallurgy.* **1.** one of several cast aluminum-base alloys containing 5 to 12% silicon, commonly designated as the 4xxx series. **2.** one of several wrought aluminum-base alloys containing 5 to 12% silicon, commonly designated as the 4xxx series.

Al-Si bronze *Metallurgy.* one of several copper-base alloys containing aluminum up to 15% and silicon up to 3%.

alsium *Ecology.* an area growing with groves of trees or bushes.

ALT alanine aminotransferase.

alt *Graphic Arts.* see ALTERATION.

alt or **alt.** alternate.

Altai [al´tī] *Geography.* a high mountain range of central Asia, lying mostly along the border of northwestern China and southwestern Mongolia; highest peak, Belukha (White Mountain) (15,157 ft).

Altaic *Linguistics.* a language family that includes Turkic, Mongol, and Manchu.

Altair [al tīr´] *Astronomy.* Alpha (α) Aquilae, the first-magnitude white star in the constellation of the Eagle.

altaite [al tā´īt] *Mineralogy.* PbTe, an opaque, metallic tin-white cubic mineral, usually massively formed and only rarely occurring in cubic or octahedral crystals, having a specific gravity of 8.19 and a hardness of 3 on the Mohs scale; found in vein deposits associated with native gold, silver, antimony, sulfides, and tellurides.

altar *Architecture.* a raised structure, such as a table or slab, used in religious rites.

Altar see ARA.

altazimuth (telescope) [al taz´ə məth] *Optics.* a telescope that can be moved on the horizontal axis and vertical axis; it allows for the simultaneous observation of both horizontal and vertical directions or angles. Thus, **altazimuth mounting.**

alteration *Graphic Arts.* a change made in copy after it has been typeset, requested by the author or the publisher. *Petrology.* a weathering or hydrothermally induced change in the composition of a rock.

alteration enzyme *Enzymology.* a phage T4 enzyme that is injected with phage DNA into a host bacterial cell, where it inhibits the synthesis of host RNA by modifying the host RNA polymerase.

alteration switch *Computer Technology.* a manual switch on the computer console, or a programmed simulation of the switch, that can be set to control or issue instructions, or to input data to the computer at given points in the program run.

altered state (of consciousness) *Psychology.* any mental condition that varies from an individual's normal state of awareness, such as may be produced by a dream, hypnosis, meditation, hallucination, sensory deprivation, or intoxication by drugs or alcohol.

alter ego *Psychology.* an intimate friend who is considered by the individual to be a second or other self. Thus, **alter-egoism.**

alternant *Mathematics.* the determinant of an $n \times n$ matrix whose (i,j)th entry has the form $f_i(x_j)$, where $f_1(x), f_2(x), \ldots, f_n(x)$ are functions and $x_1, x_2, \ldots, x_n$ are particular quantities.

alternant hydrocarbon *Organic Chemistry.* a linear or cyclic conjugated polyene containing an even number of carbon atoms. If every other (alternate) carbon atom is identified by an asterisk, then no pair of adjacent carbon atoms bears the same designation. This situation is impossible to avoid in **nonalternant hydrocarbon** systems composed of chains or rings containing odd numbers of carbon atoms.

Alternaria [al´tər nâr´ē ə] *Mycology.* a genus of fungi belonging to the class Hyphomycetes, composed of many species that are pathogenic to plants or that are capable of causing dermatitis in humans.

alternate *Botany.* describing leaves, buds, or branches that are arranged at different heights along an axis; not opposite. *Electricity.* of a current, to reverse direction periodically.

alternate angles *Mathematics.* nonadjacent angles that lie on opposite sides of a transversal cutting two lines, each having one of the lines as a side. They are **alternate exterior angles** if neither lies between the two lines cut by the transversal, and **alternate interior angles** if both lie between the two lines.

alternate energy see ALTERNATIVE ENERGY.

alternate generations see ALTERNATION OF GENERATIONS.

alternate hypothesis *Statistics.* the statement about the distribution of a random variable that is accepted if the null hypothesis is rejected; the complement of the null hypothesis.

alternate immersion test *Metallurgy.* a test in which a specimen is alternatively immersed into, and withdrawn from, a corrosive environment at specified intervals of time.

alternate key *Computer Technology.* a special-purpose key similar to the shift key that is pressed simultaneously with another key to give the key a new meaning. Also, ALT KEY.

alternate routing *Telecommunications.* **1.** any technique used to direct telephone traffic to a substitute route within a network when the first choice route is unavailable for immediate use. **2.** any method of routing in which less desirable or less direct routes are selected at a switching point when congestion is encountered on the primary route.

alternate track *Computer Technology.* in disk storage, a track that is automatically substituted for a defective primary track.

alternating copolymer *Materials Science.* a polymeric chain consisting of two diverse monomer units that alternate regularly along the chain.

alternating current *Electricity.* an electric current that reverses direction of flow at regular intervals. Worldwide, virtually all electric power is distributed as sinusoidal alternating current. In North America, the frequency of alternation is 60 hertz.

alternating-current coupling *Electricity.* a coupling that allows the passage of AC signals but blocks DC signals; usually accomplished with a series capacitor.

alternating-current/direct-current *Electronics.* describing an electronic device capable of functioning from either an AC or DC power supply. Also, AC/DC.

alternating-current dump *Electronics.* the removal of all AC power from a system or component.

alternating-current erase *Electronics.* in magnetic recording, the removal of the magnetically represented data or sound by using alternating current to energize the erase head. Thus, **alternating-current erasing head.**

alternating-current generator *Electricity.* **1.** a machine that converts mechanical energy into alternating-current electrical energy. **2.** any device that produces an alternating current. Similarly, **alternating-current motor.**

alternating-current magnetic biasing *Electronics.* in magnetic recording, the application of high-frequency alternating current to the recording head to linearize operation.

alternating-current plate resistance see DYNAMIC PLATE IMPEDANCE.

alternating-current power supply *Electricity.* a power supply that yields alternating-current output voltage. Similarly, **alternating-current network.**

alternating-current resistance see HIGH-FREQUENCY RESISTANCE.

alternating-current transmission *Electronics.* **1.** the transmission of electrical energy between two points through the use of alternating current. **2.** a form of television transmission in which a particular signal magnitude always produces the same screen brightness for a very brief period.

alternating double filter *Biotechnology.* a highly effective wastewater treatment apparatus that consists of two cycling biological filters; after the biomass accumulated in the first filter consumes most of the biochemical oxygen demand, the order of filters is reversed.

alternating flashing light *Navigation.* a lighted navigational aid that shows flashes of two or more colors. The period of light is less than the period of darkness.

alternating function *Mathematics.* a scalar-valued function f of n variables, $n > 2$, for which $f(x_1, x_2, \ldots, x_n)$ changes sign if two of the variables are interchanged.

alternating gradient *Electromagnetism.* a magnetic field set up by a chain of magnets whose dipoles alternate in sequence.

alternating-gradient focusing *Electromagnetism.* the use of an alternating gradient to keep a beam of charged particles from dispersing.

alternating-gradient synchrotron *Nucleonics.* a particle accelerator in which protons are accelerated first by a Cockcroft-Walton accelerator and then by a linear accelerator before being injected into the main ring where the field of each of a series of magnets has a strong radial gradient, with alternating gradient directions in adjacent magnets.

alternating group *Mathematics.* the subgroup (of order $n!/2$) of the permutation group on n objects which contains only the even permutations.

alternating group-flashing light *Navigation.* a navigational light showing periodic groups of flashes with color variations; the period of light is less than the period of darkness.

alternating occulting light *Navigation.* a lighted navigational aid that flashes two or more colors and eclipses at regular intervals. The period of light is equal to or greater than that of darkness.

alternating ring *Mathematics.* a ring R in which the following restricted associative law holds: $a(ab) = (aa)b$ and $b(aa) = (ba)a$ for all elements a,b in R. Alternating rings can be viewed as nonassociative algebras; the algebra of Cayley numbers is an example.

alternating series *Mathematics.* a series of real numbers whose (nonzero) terms are alternately positive and negative.

alternating stress *Mechanics.* stress in a material produced by a force acting alternately in opposite directions.

alternating voltage *Electricity.* a voltage that changes polarity at regular intervals.

alternation *Electricity.* the process of periodically varying a current or voltage from zero to a maximum, back to zero, to a negative maximum, and then back to zero. *Physics.* the positive or negative variation of a waveform from zero to maximum and back to zero, equaling one-half a cycle. *Medicine.* an interrupted condition or occurrence, alternating with one that is different or opposite.

alternation of generations *Biology.* **1.** the regular alternative occurrence in an organism's life history of different forms of reproduction. **2.** specifically, the alternation of sexual and asexual reproduction in certain plants and lower animals.

alternation of multiplicities law *Chemistry.* the arrangement of the periodic table so that the ascending groups have alternating multiplicities, i.e., even and odd numbers of orbital electrons.

alternative energy *Engineering.* energy from a source other than the conventional fossil-fuel sources of oil, natural gas, and coal; e.g., wind, running water, the sun. Thus, **alternative fuel.**

alternative hypothesis *Statistics.* the second of two opposing hypotheses in a hypothesis test; often represents the proposition that is tentatively considered to be false.

alternative medicine *Medicine.* any of various healing techniques that are not part of the formal medical tradition and that typically do not involve the use of surgery or manufactured drugs, such as acupunture, chiropracty, or homeopathy.

alteRNAtive splicing see RNA SPLICING.

alternator *Electricity.* a machine or device that generates alternating current; an AC generator.

alterne *Ecology.* any of two or more ecological communities alternating in succession with each other in the same area.

Alteromonas *Bacteriology.* a genus of Gram-negative, aerobic bacteria that occur as rod-shaped cells, each with a single polar flagellum; they require sodium for growth and are found in marine habitats.

alt. h. or **alt. hor.** *Medicine.* an abbreviation meaning "every other hour." (Short for Latin *alternis horis.)*

altherbosa *Ecology.* a tall herb community growing up in an area where forest has been cleared away.

altigraph *Engineering.* a pressure altimeter designed to record changes in altitude on a graph.

altimeter [al tim´i tər; al´tə mē´tər] *Engineering.* an instrument for measuring altitude, typically an aneroid barometer that measures atmospheric pressure (which decreases with height) in relation to sea level or some other reference point; used in aircraft.

altimeter setting *Engineering.* the standard value of atmospheric pressure to which altimeters in aircraft are set in order to provide an accurate measurement of altitude.

altimetry [al tim´i trē] *Metrology.* the science or practice of measuring altitude, as with an altimeter.

altiplanation *Geology.* the erosional processes, such as solifluction, that result in the development of flat, terracelike surfaces in hillsides or summit locations. Thus, **altiplanation terrace, altiplanation surface.**

altiplane [al´ti plān´] *Geology.* a flattened summit, broad terrace, or other generally horizontal surface, produced by altiplanation and characterized by accumulations of loose rock material on a smooth slope.

altiplano [al´ti plän´ō] *Geology.* a high plateau or basin in a mountain area. *Geography.* **Altiplano.** a region of this type in the Andes Mountains of Bolivia, Peru, and neighboring countries.

altithermal *Geology.* **1. Altithermal.** a dry postglacial period extending from about 7500 to 4000 years ago, during which time temperatures were believed to be distinctly higher than present temperatures. **2.** of or relating to this period, especially to its climate, deposits, and geologic events. **3.** of or relating to any time period or climate characterized by high or rising temperatures.

altithermal soil *Geology.* a soil that belongs to or indicates a climate marked by high or rising temperatures.

altitude *Engineering* the vertical distance of something above a reference point, such as sea level or the earth's surface. *Astronomy.* the angular distance of a star, planet, or other celestial body above the observer's horizon. *Aviation.* a given, usually assigned height at which an aircraft travels. *Mathematics.* **1.** a line segment extending from a base and perpendicular to that base to a vertex or line (plane) parallel to the base at the top of a geometric figure (not necessary planar). **2.** the length of that line segment, indicating the height of the figure at that location. Altitude is not uniquely defined; it depends on choice of base and vertex or parallel line (plane).

altitude acclimatization *Physiology.* certain adjustments occurring at high altitudes (14,000–15,000 feet) to compensate for anoxia, such as increases in respiration rate and in the red blood cell count.

altitude azimuth *Engineering.* azimuth as determined by altitude, declination, and latitude.

altitude chamber *Engineering.* an airtight chamber having adjustable air pressure and temperature, thus enabling it to simulate the air pressure and temperature of any atmospheric altitude.

altitude circle *Astronomy.* a circle on the celestial sphere having equal altitude over the earth's surface and lying in a plane that is parallel to the horizon. *Electronics.* see ALTITUDE HOLE.

altitude curve *Astronomy.* the arc of a vertical circle that stretches between the horizon and a star, planet, or satellite, as measured upward from the horizon.

altitude datum *Engineering.* the selected horizontal level, usually mean sea level, from which altitude is measured.

altitude delay *Electronics.* in aircraft radar plan-position-indicating (PPI), a synchronous delay between the time at which the radar pulse is transmitted and the time at which the screen trace begins, thus keeping the altitude hole from appearing on the screen.

altitude difference *Engineering.* the discrepancy between computed or calculated altitude and observed altitude. Also, **altitude intercept.**

altitude hole *Electronics.* in aircraft radar plan-position-indicating (PPI), a small blank circle in the center of the display that represents the time it will take for a radar pulse to travel from an aircraft to the ground and back. Also, ALTITUDE CIRCLE.

altitude intercept see ALTITUDE DIFFERENCE.

altitude reservation *Navigation.* the temporary setting aside of a particular block of airspace within given altitude bounds, for a special purpose such as the movement of large numbers of aircraft.

altitude restriction *Navigation.* an altitude, or series of altitudes in a specific order, that an aircraft may be required to maintain until reaching a specified point or time.

altitude sickness *Medicine.* a condition affecting some persons at high altitudes, as during mountain climbing or flight in an unpressurized aircraft, caused by the diminished oxygen levels at these altitudes and involving such symptoms as breathlessness, dizziness, headache, and nausea. Also, HIGH-ALTITUDE SICKNESS.

altitude signal *Electronics.* a radar signal that is reflected back to an airborne radar device from a ground or water surface that is directly beneath the aircraft, and thus indicates the aircraft's altitude above that surface.

altitude valve *Mechanical Devices.* a valve that adjusts the air-fuel mixture in an engine carburetor to allow for changes in altitude, since at higher altitudes the air has less density.

altitude wind tunnel *Aviation.* a wind tunnel in which conditions at different altitudes can be simulated by varying the air temperature, pressure, and humidity.

altitudinal [al´ti too´də nəl] relating to or affected by altitude.

ALT key see ALTERNATE KEY.

altocumulus cloud *Meteorology.* a principal cloud type consisting primarily of small liquid water droplets and appearing as a partly fibrous or diffuse cloud, usually white or gray, occurring as a layer or patch with a waved, rounded, or rolled aspect at heights from 6,000–20,000 feet.

altocumulus

altostratus cloud *Meteorology.* a principal cloud type consisting of rain, snow, and ice pellets and appearing as a striated, fibrous, or uniform cloud in a gray or bluish sheet or layer usually covering most of the visible sky, with parts thin enough to reveal the sun's position at heights from 6,000–20,000 feet.

altrices [al´trə sēz´] *Vertebrate Zoology.* organisms, especially birds, that are altricial.

altricial [al´trish´əl] *Zoology.* of or relating to a species in which the young are at a relatively undeveloped stage of development when born. Also, **altricious.**

altrigenderism *Psychology.* an early stage of attraction to members of the opposite sex that does not involve any overt sexual activity or interests.

altruism *Behavior.* a type of behavior in which an organism benefits another member of its species, without concern for its own welfare and often to its own detriment. Also, **altruistic behavior.**

altruistic *Behavior.* relating to or showing altruism.

ALU arithmetic logic unit.

alula [al´yə lə] plural, **alulae.** *Vertebrate Zoology.* **1.** the portion of a bird's wing that corresponds to the human thumb. **2.** a group of three or four small feathers at this joint. *Invertebrate Zoology.* a small lobe separated from the wing base in certain insects. Thus, **alular.**

alum *Mineralogy.* **1.** $KAl(SO_4)_2 \cdot 12H_2O$, potassium aluminum sulfate, a colorless, isometric evaporite with an astringent taste, having a specific gravity of 1.76 and a hardness of 2 on the Mohs scale; used in medicine as an astringent and styptic and for certain industrial purposes such as tanning and dyeing. Also, **common alum. 2.** any hydrous alkali aluminum sulfate. *Inorganic Chemistry.* a general term for any of various double sulfates analogous to and including common alum, formed by the union of a trivalent metal (such as aluminum, chromium or iron) and a univalent metal (such as potassium or sodium).

alum. aluminum.

alum coal *Geology.* an argillaceous, iron-containing, brown coal that produces alum as a by-product of weathering.

alumel [al´yə mel] *Metallurgy.* a nickel-base alloy containing manganese, aluminum, and silicon; used as thermocouple wire.

alumetize see ALUMINIZE.

alumina [ə loo´mi nə] *Inorganic Chemistry.* the oxide of aluminum, which occurs in nature as corundum and which is used in its synthetic form for the production of aluminum metal. See ALUMINUM OXIDE. *Materials.* any of various industrial materials made from or containing alumina. Thus, **alumina brick, alumina porcelain.**

alumina cement *Materials.* a quick-setting cement that has a high alumina content, making it more resistant to heat and chemicals than Portland cement. Also, ALUMINATE CEMENT.

aluminate *Inorganic Chemistry.* a negative ion derived from aluminum hydroxide, strongly basic in solution; it is used to clarify and soften water.

aluminate cement see ALUMINA CEMENT.

alumina trihydrate see ALUMINUM HYDROXIDE.

aluminide *Metallurgy.* any of various intermetallic compounds containing aluminum, such as nickel or titanium aluminide; commercially used for high-temperature service.

aluminite *Mineralogy.* $Al_2(SO_4)(OH)_4{\cdot}7H_2O$, a monoclinic mineral found in white, reniform masses, having a specific gravity of 1.66 to 1.82 and a hardness of 1 to 2 on the Mohs scale.

aluminium [al´yə min´ē əm] *Chemistry.* the British term for aluminum.

aluminize *Metallurgy.* to coat or treat a material with aluminum or an aluminum alloy, as by dipping in a molten bath, thermal spraying, or diffusing. *Materials Science.* to add aluminum to a material; treat with aluminum. Thus, **aluminization.**

aluminized steel *Metallurgy.* a steel that is coated with aluminum or an aluminum alloy.

aluminon *Organic Chemistry.* $C_{22}H_{23}N_3O_9$, a brownish-red powder that is soluble in water and decomposes at 220–225°C; used as an organic reagent for the colorimetric detection of aluminum.

aluminosilicate *Inorganic Chemistry.* **1.** a compound of aluminum silicate with a metal oxide or other radical; used in refining petroleum, to soften water, and in detergents. **2.** see ALUMINUM SILICATE.

aluminosis *Medicine.* a chronic inflammation of the lungs due to the presence of aluminum; found especially in industrial workers who are exposed to aluminum-bearing dust in the air.

aluminothermic *Metallurgy.* relating to or produced by the process of aluminothermy.

aluminothermy *Metallurgy.* a process by which a metallic oxide is reduced by reaction of the oxide with finely divided aluminum particles at high temperature. Also, **aluminothermics, aluminothermic process.**

aluminotype *Graphic Arts.* a relief-type printing plate on which the raised surface is molded of aluminum.

aluminous cement see ALUMINA CEMENT.

aluminum *Chemistry.* a soft, lightweight, silver-white metallic chemical element, the third most common element, having the symbol Al, the atomic number 13, an atomic weight of 26.9815, a melting point of 650°C, and a boiling point of 2450°C. It is higly ductile, malleable, conductive, and resistant to corrosion and wear, and is widely used in alloys for beverage cans, household utensils, aircraft and automobile parts, electrical equipment, and many other products. (An altered form of a word coined by the English chemist Sir Humphry Davy, meaning "the earth of alum.")

aluminum acetate *Organic Chemistry.* $Al(CH_3COO)_3$, a white powder that is soluble in water and decomposes on heating; used in aqueous solution as an antiseptic, antiperspirant, and astringent.

aluminum acetylsalicylate *Pharmacology.* $C_{18}H_{15}AlO_9$, a form of aspirin, used to treat fever and pain.

aluminum alloy *Metallurgy.* an alloy based on at least 75% aluminum, and containing one or more metallic elements such as copper, manganese, magnesium, silicon, zinc, and lithium. The alloying elements enhance strength and mechanical properties, and often other properties such as corrosion resistance.

aluminum ammonium sulfate see AMMONIUM ALUMINUM SULFATE.

aluminum arrester *Electricity.* a lightning arrester made of a series of electrolytic cells formed from aluminum trays containing electrolyte. Also, **aluminum-cell arrester.**

aluminum borohydride *Inorganic Chemistry.* $Al(BH_4)_3$, a volatile liquid that boils at 44.5°C; it ignites spontaneously in air and reacts violently with water; it is used as a reducing agent and a jet fuel additive.

aluminum brass *Metallurgy.* a copper-base alloy containing about 20.5% zinc and 2% aluminum. The arsenical variety also contains 0.2 to 0.6% arsenic.

aluminum bronze *Metallurgy.* a copper-base alloy that contains 2.3% to 15.0% aluminum as an essential alloying element. Other elements such as iron, manganese, nickel, tin, and zinc may also be included.

aluminum cable steel-reinforced *Electricity.* referring to a type of power transmission line that consists of aluminum wires surrounding a steel core.

aluminum chloride *Inorganic Chemistry.* $AlCl_3$, a deliquescent compound in the form of white or yellowish crystals, soluble in water and melting at 190°C; an important catalyst in a variety of reactions involved in petroleum refining and manufacturing.

aluminum conductor *Electricity.* a conductor made solely of aluminum, often used in high-voltage transmission lines because of its light weight.

aluminum equivalent *Radiology.* an equivalent thickness of pure aluminum that would shield radiation to the same degree as a material being evaluated.

aluminum fluoride *Inorganic Chemistry.* AlF_3 or $AlF_3{\cdot}3.5H_2O$, a white, crystalline powder that exists in both anhydrous and hydrate forms, slightly soluble in water; added to aluminum production cells to lower the melting point and increase the electrical conductivity.

aluminum fluorosilicate *Inorganic Chemistry.* $Al_2(SiF_6)_3$, a white powder, slightly soluble in cold water and readily soluble in hot water; used for artificial gems, enamels, and glass. Also, ALUMINUM SILICOFLUORIDE.

aluminum foil *Materials.* a thin aluminum sheet, widely used as a food wrapping, cooking sheet, and insulation backing.

aluminum halide *Inorganic Chemistry.* a compound made up of aluminum plus a halogen element, such as aluminum chloride.

aluminum hydroxide *Inorganic Chemistry.* $Al(OH)_3$, a white powder, insoluble in water but soluble in mineral acids and caustic soda (sodium hydroxide); used in the manufacture of ceramic glasses and in paper coatings. *Medicine.* a form of this substance used as an antacid, in the form of either a powder or a gel. Also, ALUMINA TRIHYDRATE.

aluminum monopalmitate *Organic Chemistry.* $Al(OH)_2(C_{16}H_{31}O_2)$, a white powder that is insoluble in water and alcohol and melts at 200°C; used in waterproofing leather, in the production of high gloss for paper, and as a food additive.

aluminum monostearate [ə loom´i nəm män´ō ster´āt´] *Organic Chemistry.* $Al(OH)_2[OOC(CH_2)_{16}CH_3]$, a white to yellowish fine powder that is insoluble in water, alcohol, and ether, and forms a gel with aromatic and aliphatic hydrocarbons; melts at 155°C; it is used in paints, inks, waterproofing, and as a stabilizer in plastics.

aluminum-nickel-cobalt alloy see ALNICO.

aluminum nitrate *Inorganic Chemistry.* $Al(NO_3)_3{\cdot}9H_2O$, white deliquescent crystals, soluble in cold water and decomposing in hot water, soluble in alcohol and acetone; melts at 73°C; used as a mordant for textiles, in leather tanning, and as a catalyst in petroleum refining.

aluminum nitride *Inorganic Chemistry.* AlN, a yellowish crystalline solid, melting at 2150°C, decomposed by water, unless heated to high temperatures; used as a semiconductor in electronics.

aluminum oleate *Organic Chemistry.* $Al(C_{18}H_{33}O_2)_3$, a yellowish-white soaplike compound of aluminum and oleic acid that is soluble in alcohol, ether, and oil and insoluble in water; used in waterproofing and as a thickener for lubricating oils.

aluminum orthophosphate see ALUMINUM PHOSPHATE.

aluminum oxide *Inorganic Chemistry.* Al_2O_3, a compound also called alumina, commercially used for aluminum production and for the manufacture of paper, abrasives, ceramics, electrical insulators, and other products.

aluminum paint *Materials Science.* any of various paints to which aluminum is added to provide greater reflectivity, wear resistance, and heat retention.

aluminum phosphate *Inorganic Chemistry.* $AlPO_4$, white crystals, melting at 1500°C, insoluble in water and alcohol and slightly soluble in acids; highly corrosive to tissue in solution; used in ceramics, pharmaceuticals, paints, and paper. Also, ALUMINUM ORTHOPHOSPHATE, ALUMINUM TRIPHOSPHATE.

aluminum plate *Graphic Arts.* in offset lithography, a thin sheet of aluminum used for either deep-etch or surface-type plates.

aluminum potassium sulfate see POTASSIUM ALUMINUM SULFATE.
aluminum powder *Materials.* aluminum in the form of finely ground flakes or granules, used in paints and in making small machine parts. Similarly, **aluminum paste.**
aluminum silicate *Inorganic Chemistry.* any of the numerous types of clays that contain varying proportions of Al_2O_3 and SiO_2; synthetically produced for crystals or fibers of high strength that are used to reinforce plastics.
aluminum silicofluoride see ALUMINUM FLUOROSILICATE.
aluminum soap *Organic Chemistry.* any of the various salts of higher carboxylic acids and aluminum; insoluble in water and alcohol, but soluble in oils; used in oils, paints, lubricating greases, and waterproofing compounds.
aluminum solder *Metallurgy.* **1.** an alloy of gold, silver, copper, and sometimes zinc, formerly used for soldering aluminum bronze. **2.** the solder having the highest melting point (425°C), consisting of about 45% aluminum and 55% germanium.
aluminum stearate *Organic Chemistry.* $Al(C_{18}H_{35}O_2)_3$, a white powder that is soluble in oils and insoluble in water and alcohol; melts at 115°C; used as a paint drier, waterproofer, in cosmetics, and as a defoaming agent in beet sugar and yeast processing.
aluminumware *Materials Science.* cooking utensils and other such articles made from aluminum, offering good heat conduction with relatively light weight and low maintenance.
alum shale *Petrology.* a shale that is derived from the processes of oxidation and hydration of pyrite to sulfuric acid, which acts on potash and alumina constituents to yield alum. Also, **alum schist, alum slate.**
alunite *Mineralogy.* $KAl_3(SO_4)_2(OH)_6$, a white or reddish-gray trigonal mineral usually massive in form and rarely occurring as rhombohedral crystals. It is formed by sulfotaric processes on volcanic rock, has a specific gravity of 2.6 to 2.9 and a hardness of 3.5 to 4 on the Mohs scale, and is used in producing alum. Also, **alumstone, alum rock, alumite.**
alunogen *Mineralogy.* $Al_2(SO_4)_3 \cdot 17H_2O$, a white triclinic mineral, soluble in water, occurring as delicate fibrous masses having a specific gravity of 1.77 and a hardness of 1.5 to 2 on the Mohs scale; found in areas with volcanic activity.
alure [al′yoor] *Architecture.* a passage or gallery, especially behind a castle parapet, around a church roof, or along a cloister.
alurgite [ə lur′jīt] *Mineralogy.* **1.** a copper-red, manganiferous variety of muscovite mica found in Piedmont, Italy. **2.** a purple to red variety of manganese mica found in Quebec, Canada.
ALU sequence *Genetics.* a family of repetitive, short sequences of DNA that make up about 5% of the human genome. They appear to be related in sequence to a class of RNA associated with ribonucleoproteins that are involved in the secretion of proteins through membranes. Also, **ALU family.**
alvar *Ecology.* a plant community dominated by mosses and herbs, occurring on shallow limestone soils.
Alvarez, Luis Walter 1911–1988, American physicist; Nobel Prize for studies of subatomic particles.
alveator [al′vē āt′ər] *Invertebrate Zoology.* a type of pedicellaria (surface pincers) in echinoderms such as seastars and sea urchins.
alveolar [al vē′ə lər] *Biology.* **1.** of or relating to an alveolus. **2.** covered with pits; honeycombed. *Linguistics.* a sound that is made by placing the tongue against the hard ridge behind the upper teeth, such as the sounds of [t] and [d].
alveolar arch *Anatomy.* the dental arch formed by the alveolar process of the upper or lower jaw.
alveolar canal *Anatomy.* any of a number of canals in the maxilla for passage of the alveolar vessels and nerves that enter the root canals of the maxillary teeth.
alveolar-capillary block *Medicine.* the reduced ability of gases to pass though the pulmonary alveolar-capillary membrane.
alveolar cell carcinoma see BRONCHIOLAR CARCINOMA.
alveolar duct *Anatomy.* any of the small air passages in the lung connecting the respiratory bronchioles and the alveolar sacs.
alveolar gland see ACINOUS GLAND.
alveolar index see GNATHIC INDEX.
alveolar macrophage *Anatomy.* a type of cell within the alveoli of the lungs that ingests foreign particles inhaled from the air. Also, DUST CELL.
alveolar oxygen pressure *Physiology.* a measure of the pressure of oxygen in the alveoli of the lungs.
alveolar process *Anatomy.* the portion of bone in the upper or lower jaw that surrounds and supports the teeth.
alveolar prognathism *Paleontology.* a forward projection of the teeth-bearing portions of an animal's jaws.
alveolar ridge *Anatomy.* the bony ridge on the upper or lower jaw that contains the alveoli (sockets) of the teeth.
alveolar sac *Anatomy.* a group of alveoli having a common opening.
alveolate [al vē′ə lāt′] *Biology.* relating to or like an alveolus. Also, **alveolated.**
alveolated cell see EPITHELIOID CELL.
alveoli [al vē′ə lī′] the plural of ALVEOLUS.
Alveolites [al vē′ə lī′tēz] *Paleontology.* a genus of colonial corals in the extinct order Tabulata, suborder Favositina, and family Alveolitidae; extant in the middle to late Paleozoic.
alveolitis [al vē′ə līt′is] *Medicine.* the inflammation of the alveoli of the lungs, usually from the inhalation of some antigenic substance such as agricultural spores or fungi.
alveolitoid [al vē′ə lə toid] *Invertebrate Zoology.* a type of tabulate coral having a vaulted upper wall and a lower wall parallel to the surface attachment.
alveolus [al vē′ə lis] *plural,* **alveoli.** *Biology.* a small cavity, pit, or socket. *Anatomy.* **1.** any of the numerous tiny air cells of the lungs in which the exchange of oxygen and carbon dioxide takes place. Also, **pulmonary alveolus. 2.** any of the bony cavities or sockets in which the roots of the teeth are held in place in the upper and lower jaws. Also, **dental alveolus. 3.** any small cavity or socket. (From a Latin word meaning "hollow.")
Alydidae [ə lid′ə dē′] *Invertebrate Zoology.* a family of broad-headed bugs in the order Hemiptera that give off a strong odor if disturbed.
alymphocytosis [ā′lim fō sī′tō səs] *Hematology.* the absence or near absence of lymphocytes from the blood.
alyphite [al′ə fīt] *Mineralogy.* a bitumen that yields a high percentage of open-chain hydrocarbons, such as methane, when distilled.
Alysiella [ə lis′ē el′ə] *Bacteriology.* a genus of Gram-negative, gliding bacteria of the family Simonsiellaceae, occurring as square-ended, multicellular filaments in the oral cavities of warm-blooded vertebrates.
alysoid [al′ə soid′] see CATENARY.
alyssum [ə lis′əm] *Botany.* **1.** any of various plants of the genus *Alyssum,* native to the Mediterranean region and cultivated as ornamental ground covers. **2.** see SWEET ALYSSUM.
alytesin [ā līt′ə sin] *Endocrinology.* a biologically active peptide that is found in frog skin and is a member of the bombesin family of peptides.
Alzheimer's disease [ôlts′hī mərz; alts′hī mərz] *Medicine.* a brain disorder attributed to the accumulation of certain proteins in the brain (especially amyloid), characterized by the atrophy of the frontal and occipital lobes of the brain; a progressive, irreversible disease that typically develops after age 40 in both men and women and involves such symptoms as loss of memory, disorientation, emotional outbursts, speech and gait disturbances, and a general deterioration of mental ability. (From Alois *Alzheimer,* 1864–1915, German neuorologist, who first described it.)
AM *Aviation.* the airline code for Aeromexico. *Electronics.* amplitude modulation, the deliberate variation of the amplitude of a radio wave or carrier in order to transmit a signal. *Telecommunications.* **1.** a system of communication or broadcasting that uses amplitude modulation. **2.** using such a system: an *AM* radio.
Am americium; ammonium.
Am² ampere meter squared.
A/m ampere per meter.
A/m² ampere per square meter.
AMA actual mechanical advantage; automatic message accounting.
AMA *Aviation.* the airport code for Amarillo, Texas.
AMA or **A.M.A.** American Medical Association.
amacratic lens see AMASTHENIC LENS.
amacrine (cell) [am′ə krin] *Cell Biology.* a type of nerve cell found in the retina of the vertebrate eye that facilitates image processing.
amagat [ä′mə gä′] see AMAGAT DENSITY UNIT, AMAGAT VOLUME UNIT.
Amagat density unit *Physics.* the standard unit of the gas molecular density in the Amagat unit system, obtained when the pressure is one atmosphere and the temperature is 0°C; usually expressed in moles per cubic meter.
Amagat diagram *Physics.* a graphical representation of isothermal curves of the pressure-volume product as a function of pressure.
Amagat-Leduc rule *Physics.* the statement that the total volume of a gas mixture is equal to the sum of the volumes that would be occupied by the individual gases at the same pressure and temperature as the mixture. Also, **Amagat's law, Amagat's rule.**

Amagat system *Physics.* a unit system in which the unit of pressure is the atmosphere and the unit of volume is the gram-molecular volume; used in the study of the behavior of gases under pressure.

Amagat volume unit *Physics.* a unit of volume in the Amagat system, equal to 22.413 liters, the volume that one mole of an ideal gas occupies at a pressure of 1 atmosphere and a temperature of 0°C.

amagmatic [ā´mag mat´ik] *Volcanology.* having no magmatic activity.

amalgam *Mineralogy.* **1.** a silver-white, naturally occurring isometric alloy of silver and mercury having a brilliant metallic luster. **2.** any alloy of mercury with other metals. *Materials Science.* see DENTAL AMALGAM.

amalgamate *Metallurgy.* to mix one or more other metals with mercury, thus forming an alloy.

amalgamation *Metallurgy.* **1.** the treating of precious-metal ores with mercury for the purpose of extracting their values. **2.** the mixing of mercury with another metal or metals to form an alloy.

amalgamation pan *Metallurgy.* an iron container in which precious-metal ores can be ground with water and further treated with mercury and chemicals until amalgamation is completed; used especially, formerly, to separate gold from its ore.

amalgamation table *Metallurgy.* a flat metal surface on which mercury is spread so that it will amalgate with gold particles as gold-bearing ore is washed over it. Also, **amalgamation plate.**

amalgamator *Metallurgy.* a machine used to bring powdered ore in close contact with mercury to amalgamate metals contained in the ore.

amalgam treatment see AMALGAMATION.

Amalthea [am´əl thē´ə] *Astronomy.* a moon of Jupiter, possessing a somewhat distorted shape due to the gravitational pull of the planet; discovered in 1892. (From the name of a mythological character who nursed the infant Jupiter.) Also, JUPITER V.

Amaltheus [ə mal´thē əs] *Paleontology.* a genus of lower Jurassic ammonites in the class Cephalopoda; important as a zone fossil, occurring only in a stratum between the strata of its relatives Prodactylioceras and Dactylioceras.

Amanita [äm´ə nē´tə; äm´ə nī´tə] *Mycology.* **1.** a genus of fungal mushrooms having white spores and belonging to the order Agariales; it occurs in deciduous and coniferous forests and includes such poisonous types as the death cup, destroying angel, and deadly fly agaric. **2.** **amanita.** any fungus of this genus.

Amanitaceae *Mycology.* a family of fungi of the order Agaricales, composed of gill mushrooms, many species of which are poisonous.

amanitotoxin *Toxicology.* any of various powerful poisons found in certain mushrooms of the genus *Amanita.* Also, **amanitin, amantine.**

amantadine see AMANTADINE HYDROCHLORIDE

amantadine hydrochloride *Pharmacology.* $C_{10}H_{17}N \cdot HCl$, an antiviral compound used in the treatment of Type A influenza, and in the treatment of parkinsonism to promote dopamine release.

amara see BITTERS.

amaranth [am´ə ranth´] *Botany.* any of various plants of the genus *Amaranthus,* having small, brightly colored flowers with a lasting pigment. *Materials Science.* a dark-red dye derived from certain amaranth plants or prepared synthetically.

Amaranthaceae *Botany.* a family of herbs, trees, and vines, found widely in warm regions and including the amaranth, pigweed, and tumbleweed.

amaranthaceous [am´ə rən thā´shəs] *Botany.* of or relating to the Amaranthaceae family of plants. Also, **amaranthine.**

amaranth family see AMARANTHACEAE.

Amaranthus [am´ə ran´thəs] *Botany.* a genus of dicotyledenous plants of the family Amaranthacae. Some species are cultivated, such as love-lies-bleeding, *A. caudatus*; others are common weeds, such as the green amaranth, *A. retroflexus.*

amarantite *Mineralogy.* $Fe^{+3}(SO_4)(OH) \cdot 3H_2O$, a triclinic, amaranth-red mineral occurring in bladed or columnar masses, having a specific gravity of 2.2 and a hardness of 2.5 on the Mohs scale; decomposed by water and found with magnesium copiapite.

amaretto [am´ə ret´ō] *Food Technology.* a liqueur with a bitter-almond flavor.

amarillite *Mineralogy.* $NaFe^{+3}(SO_4)_2 \cdot 6H_2O$, a brilliant pale-green to yellowish monoclinic water-soluble mineral occurring as thick, tabular crystals, having a specific gravity of 2.19 and a hardness of 2.5 to 3.0 on the Mohs scale; found as veinlets in massive coquimbite.

Amaryllidaceae *Botany.* a family of plants related to and identified with the lily family, including several hundred species typically having flat, narrow leaves and large, lilylike flowers, such as the amaryllis, snowdrop, narcissus, and daffodil.

amaryllidaceous *Botany.* of or relating to the Amaryllidaceae family of plants.

amaryllis [am´ə ril´əs] *Botany.* **1.** any of several plants of the genus *Amaryllis* of the family Amaryllidaceae, especially the **true amaryllis,** *A. belladonna,* also called the belladonna lily. **2.** any of several other plants of related genera, such as *Hippeastrum.*

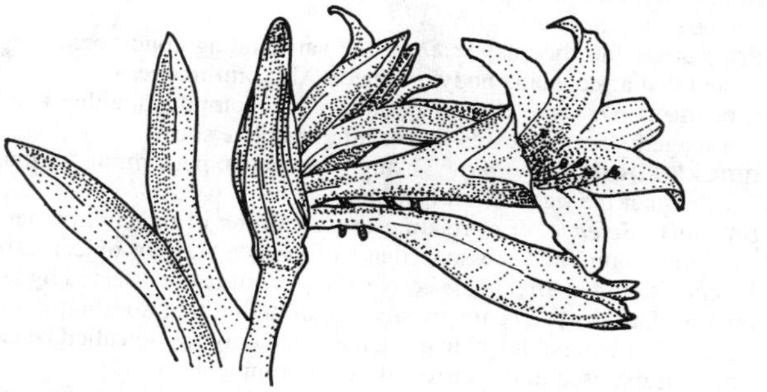

amaryllis

amaryllis family see AMARYLLIDACEAE.

A-mast *Engineering.* an A-shaped truss used to hold a machine designed for hoisting.

amasthenic lens *Optics.* a lens that refracts light rays into a single focus. Also, **amacratic lens.**

amastia [ə mas´tē ə] *Medicine.* the congenital absence of one or both breasts. Also, **amazia.**

amastigote *Pathology.* the morphologic phase in the life cycle of flagellate protozoan parasites such as Leishman-Donovan bodies.

amateur bands *Telecommunications.* bands of frequencies assigned to licensed amateur radio operators.

amateur radio *Electronics.* the practice of transmitting and receiving radio signals, building and operating radio equipment, and so on, as a hobby rather than as a business or a professional service. Also, HAM RADIO.

amathophobia [ə math´ə fō´bē ə] *Psychology.* an irrational fear of or aversion to dust.

amatol [am´i tôl] *Ordnance.* a powerful explosive mixture of ammonium nitrate and TNT.

amatoxin see AMANITIN.

amaurosis [am´ô ro´sis] *Neurology.* a condition of blindness, particularly when occurring with no apparent lesion of the eye, as from disease of the optic nerve, spine, or brain. (From Latin for "darkening.")

amaurosis fugax *Medicine.* a transient monocular blindness, lasting ten minutes or less, that may result from a transient ischemia caused by carotid artery insufficiency or from centrifugal force; it is often a warning sign of a stroke.

amaurotic [am´ô rät´ik] *Neurology.* of or relating to amaurosis, or to blindness in general.

amaurotic familial idiocy see TAY-SACHS DISEASE.

Amazon *Geography.* the longest river in South America, flowing nearly 4000 miles from the Peruvian Andes across northern Brazil to the Atlantic Ocean.

amazonite *Mineralogy.* a green to bluish variety of microcline, a triclinic silicate. Also, **amazon stone.**

amb- a prefix meaning "both," as in *amboceptor.*

amber *Mineralogy.* an oxygenated hydrocarbon derived from the fossil resin of certain coniferous trees, noted since ancient times as an accumulator of static electricity. It has a transparent yellow to reddish color, a specific gravity of 1.09, and a hardness of 2 to 2.5 on the Mohs scale; used in making jewelry. Extinct insects and other objects are often found trapped in amber.

amber codon *Molecular Biology.* the nucleotide triplet UAG, one of the three nonsense codons in messenger RNA that cause the termination of protein synthesis.

ambergris [am´bər gris] *Materials.* a grayish waxy substance originating in the intestines of sperm whales and found floating on the water or washed ashore; once widely used in perfumes.

amberjack *Vertebrate Zoology.* any of several amber-colored warfish of the genus *Seriola,* found in warm Atlantic waters; especially *S. dumerili,* a popular game fish. Also, **amberfish.**

amber mutation *Genetics.* a type of mutation to the UAG codon, a nonsense codon that prematurely stops the synthesis of a polypeptide chain.

amber suppressor *Genetics.* an allele coding for tRNA whose anticodon is altered in such a way that the tRNA inserts an amino acid at the site of UAG stop codons resulting from an amber mutation, thereby preventing premature termination of the polypeptide synthesis.

ambi- a prefix meaning "both," "on both sides," or "around," as in *ambidextrous, ambient.*

ambiance [am´bē əns] *Science.* the surrounding conditions or environment that affect some body or object. Also, **ambience.**

ambident *Organic Chemistry.* an organic compound, either a substrate or reagent, that possesses two or more reactive sites.

ambidextrous *Physiology.* using both hands; performing various tasks with either the right or left hand.

ambient *Science.* of or relating to a condition of the environment, such as temperature and pressure, that surrounds a body or object, especially a condition that affects the body or object but is not affected by it.

ambient diving *Oceanography.* ocean diving that does not involve the use of a submersible vehicle, such as scuba diving; so called because the diver is exposed to the surrounding environment.

ambient light *Optics.* the light generated from outside sources and present in the environment around an optical or vision system, such as the light from a nearby lamp that surrounds the picture on a television screen. Also, **ambient illumination.**

ambient noise *Acoustics.* background noise consisting of all noise sources, such as environmental noise, machinery, voices, and so on, other than a sound of interest, such as music being performed or a voice being recorded.

ambient noise level *Acoustics.* the uncontrollable background noise level present at a site, such as street traffic.

ambient pressure *Physics.* the pressure of a surrounding medium, such as air, that comes in contact with some specified body or object.

ambient stress field *Geophysics.* the distribution and numerical value of a stress field in the rock environment prior to its study and disturbance by humans.

ambient temperature *Physics.* the temperature of a surrounding medium, such as gas or liquid, that comes in contact with an apparatus and acts as a temperature reservoir.

ambiguity *Artificial Intelligence.* **1.** in understanding natural language, images, or other inputs, the superficial possibility of interpreting an input in more than one way. **2.** specifically, in computerized text analysis, a word that has more than one valid meaning; e.g., *left* as in "Turn left" (not right) and "Tom left" (departed). *Navigation.* a condition in which the elements of a fix utilizing circular or hyperbolic lines of position are consistent with two different positions. *Electronics.* a condition in which a synchro or servo system seeks a false null position in addition to the proper null position.

ambiguity error *Computer Technology.* an error, usually transient, that occurs during the reading of individual digits constituting a number, as when imprecise synchronization causes changes to occur in different digit positions of a given number at slightly different times.

ambiguity function *Electrical Engineering.* a function used to assess the suitability of radar waveforms resulting in the absence of ambiguities, reductions of clutter, and greater accuracy and resolution.

ambiguous codon *Genetics.* a type of codon that codes for more than one amino acid. Thus, **ambiguous genetic code.**

ambiguous grammar *Computer Science.* a grammar that allows some sentence or string to be generated or parsed in more than one way.

ambiguous name *Computer Programming.* a partially specified name that can be used to locate all files or other items to which it applies.

ambilineal descent *Anthropology.* a descent system in which individuals are voluntarily affiliated with numerous possible kin groups, and are related through either mother or father, and may also be related to any descent group of which their parents are members through the grandparents.

ambipolar *Electronics.* relating to or involving both positive and negative charges.

ambipolar coupling *Particle Physics.* the interaction between oppositely charged particles diffusing in a plasma, resulting from respective differences in diffusion rates.

ambipolar diffusion *Particle Physics.* the diffusion of charged particles in a plasma, resulting from the combined influences of a density gradient and the internal electric field set up by the separation of the electrons and the positive ions.

ambisense expression strategy *Virology.* the coding of viral proteins in both sense and antisense (complementary) mRNAs. Thus, **ambisense RNA.**

ambisexual *Biology.* relating to or showing the characteristics of both sexes. *Psychology.* sexually attracted to people of both sexes; bisexual.

ambisexuality the fact or condition of being ambisexual.

ambit *Cartography.* a boundary line thought of as enclosing a given area. Also, **ambitus.**

ambivalence or **ambivalency** *Psychology.* the simultaneous existence in an individual of opposing emotions, attitudes, or ideas toward a given person or situation. *Behavior.* see AMBIVALENT BEHAVIOR.

ambivalent relating to or showing ambivalence.

ambivalent behavior *Behavior.* a behavior pattern in which two differing or conflicting actions occur together or in rapid sequence, such as alternating movements of fleeing and attack.

ambly- a combining form meaning "blunt" or "dull," as in *amblyopia.*

amblygonite *Mineralogy.* $(Li,Na)Al(PO_4)(F,OH)$, a white, brittle, triclinic mineral with a pearly luster occurring as large, coarse crystals, having a specific gravity of 2.01 to 3.09 and a hardness of 5.5 on the Mohs scale; often mined as an ore of lithium, found in granite pegmatites.

amblyopia [am´blē ō´pē ə] *Medicine.* **1.** any of various conditions of impaired vision without any apparent defect or disease of the eye; types include **toxic, nutritional,** and **traumatic amblyopia. 2.** see SUPPRESSION AMBLYOPIA.

amblyopic *Medicine.* relating to or affected by amblyopia.

amblyopod [am´blē ə päd´] *Paleontology.* an extinct order or suborder of mammals, known from North American fossils of the Eocene, having a small brain and an elephant-like body.

Amblyopsidae *Vertebrate Zoology.* a family of small fishes in the order Percopsiformes. Most are blind, lack skin pigment, and live in swamps, caves, and subterranean streams in North America.

Amblyopsiformes [am´blē ōp´si fôr´mās] *Vertebrate Zoology.* another name for the order Percopsiformes.

amblyoscope [am´blē ə skōp´] *Optics.* an instrument that provides two separate images for individual viewing by each eye, used to train or stimulate vision in an amblyopic eye.

Amblypygi *Invertebrate Zoology.* an order of tailless whip scorpions, in the class Arachnida, with the tarsus of the first walking legs modified into an extremely long tactile organ.

Amblystegiaceae *Botany.* a family of glossy, matted, mostly aquatic mosses of the order Hypnobryales, mainly distributed in northern temperate and polar regions; characterized by branched prostrate or erect stems with lateral sporophytes.

amboceptor [am´bə sep´tər] *Immunology.* an antibody having two combining elements, one for a cellular antigen and one for complement.

amboceptor unit *Immunology.* the smallest amount of antibody that produces complete red blood cell destruction in the presence of an excess of complement.

ambonite *Petrology.* a collective name for hornblende-biotite andesites and dacites that contain cordierite. (Found on *Ambon* Island in Indonesia.)

Amborellaceae [am´bôr ə lās´ē e] *Botany.* a monospecific family of dicotyledonous evergreen arborescent shrubs in the order Laurales; noted for accumulating aluminum and for being the only family in the order that lacks vessels and ethereal oil cells.

Ambrosia [am brōz´yə] *Botany.* **1.** a genus of annual weeds that includes the common ragweed, *A. artemisiifolia,* and the giant ragweed, *A. trifida.* **2. ambrosia.** a plant of this genus.

ambrosia beetle *Invertebrate Zoology.* any of various beetles of the family Scotylidae that bore holes in wood, in which they cultivate a fungus that they use for food.

ambrosiaceous [am brōz´ē ā´sē əs] *Botany.* of or relating to the *Ambrosia* genus of plants.

ambrosia fungi *Mycology.* certain fungi that grow in tunnels formed by ambrosia beetles and serve as food for these insects and their larvae.

Ambrosiozyma *Mycology.* a genus of fungi belonging to the ambrosia family Saccharomycetaceae, developing budding yeast cells and a true mycelium, and consisting of intertwined hyphae divided into cells by septa.

ambrotype *Graphic Arts.* a positive image made on glass by the collodion process and viewed against a dark background; a common early type of photograph.

ambulacral *Invertebrate Zoology.* of or relating to the ambulacra. Thus, **ambulacral groove(s).**

ambulacrum [am´byə lā´krəm] *plural,* **ambulacra.** *Invertebrate Zoology.* in echinoderms, any of the five radial grooves along which the tube feet are located.
ambulant see AMBULATORY.
ambulatorial see AMBULATORY.
ambulatory *Zoology.* able to move, especially to walk, from one place to another; not stationary. *Medicine.* **1.** of a patient, able to walk; not confined to a bed, stretcher, wheelchair, and so on. **2.** of or relating to health services for patients who are not confined to a bed. *Architecture.* **1.** an aisle around the apse of a church. **2.** a walkway within a cloister.
ambulatory bud *Biology.* an asexually produced organism that breaks from its parent and is capable of independent movement, as in some yeasts and other fungi.
ambulatory care *Medicine.* a term for health services for persons who have realtively minor disorders and thus are able to visit a facility as outpatients and depart after their treatment. Thus, **ambulatory care center.**
ambulatory schizophrenia *Psychology.* a schizophrenic disorder that is not considered severe enough to require the individual to be institutionalized.
Ambystoma [am bis´tə mə] *Vertebrate Zoology.* a genus of common and widely distributed North American salamanders, having about 20 species.
Ambystomatidae *Vertebrate Zoology.* a family of North American salamanders and newts in the suborder Ambystomoidea that are usually terrestrial except during breeding season.
Ambystomoidea *Vertebrate Zoology.* a suborder of the amphibian order Caudata, which includes the family Ambystomatidae, or North American salamanders and newts.
AMC automatic modulation control.
ameba [ə mē´bə] *Invertebrate Zoology.* the common name for a number of species of one-celled, usually microscopic organisms of the order Amoebida of the class Sarcodina; found extensively in fresh and salt water, moist upper soil, and as parasites in humans and animals. An ameba has no shell or skeleton and constantly changes shape as it forms pseudopods (false feet), by which it moves, engulfs food particles, and forms food vacuoles; reproduction is usually by binary fission. (From a Greek word meaning "to change," in reference to its changeable form.) Also, AMOEBA.

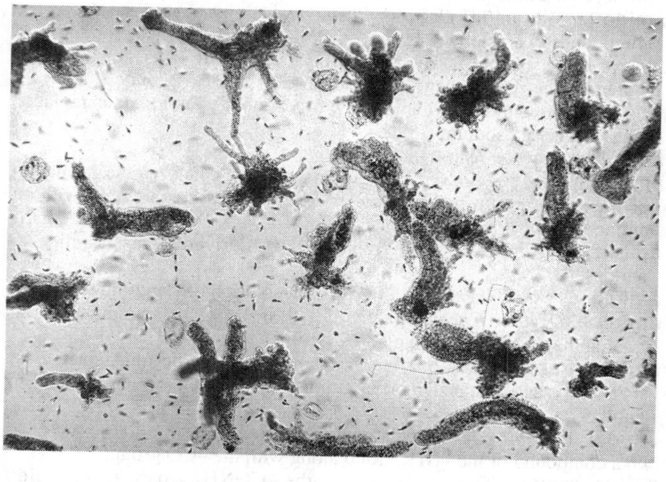

ameba

Amebelodon [äm´ə bel´ə dän] *Paleontology.* a genus of North American mastodons in the extinct family Gomphotheriidae; about 10 feet high and characterized by "shovel tusks" in the lower jaw; extant in the upper Miocene.
amebiasis [am ə bī´ə sis] *Pathology.* **1.** an infection of the intestine, liver, or other sites with *Entamoeba histolytica,* a pathogenic ameba acquired by ingesting contaminated water or food. **2.** in general, any infection caused by an amebic parasite. Also, AMOEBIASIS.
amebic *Invertebrate Zoology.* **1.** being an ameba, or related to the amebas. **2.** of or like an ameba, especially in reference to its shape or movement. *Medicine.* describing a disease or condition caused by amebas. Thus, **amebic colitis, amebic abscess.**
amebic dysentery *Pathology.* an inflammation of the intestine caused by infection with the amebic parasite *Entamoeba histolytica,* characterized by ulceration, especially of the colon, and by vomiting, nausea, severe diarrhea, and abdominal pain. Also, AMOEBIC DYSENTERY.
amebicide *Toxicology.* an agent that is destructive to ameba, especially a prepared substance used to kill ameba.
amebocyte [ə mēb´ə sīt´] *Invertebrate Zoology.* **1.** a wandering ameboid cell in the tissues and fluids of many animals that functions in excretion and other processes. **2.** any cell capable of moving like an ameba; i.e., by means of pseudopods. Also, AMOEBOCYTE.
ameboflagellate *Invertebrate Zoology.* describing amebas with flagella or the ability to assume a flagellate phase under certain conditions.
ameboid *Invertebrate Zoology.* relating to or resembling an ameba. Also, AMOEBOID.
ameboid movement *Cell Biology.* ameba-like movement; the locomotion of a cell that is accomplished by cytoplasmic streaming into pseudopods.
amebula *Invertebrate Zoology.* a small ameboid cell produced by reproduction in some protozoans.
ameiosis [ā´mī ō´sis] *Genetics.* a failure of both divisions of meiosis to occur, resulting in a form of nuclear division that does not involve a reduction in chromosome number.
ameiotic *Genetics.* relating to or characterized by ameiosis.
Ameiuridae [ām´i yûr´ə dē] *Vertebrate Zoology.* a family of North American freshwater catfishes in the suborder Siluroidei.
amelanosis [am´ə lə nō´sis] *Medicine.* the absence of pigmentation in tissue due to a lack of melanin. Thus, **amelanotic** [am´ə lə nät´ik].
amelia [ə mēl´yə] *Medicine.* the congenital absence of a limb.
amelification *Developmental Biology.* the process in which the ameloblasts (enamel cells) develop into dental enamel.
amelioration *Agronomy.* the process of improving soil for agricultural use, as by plowing, draining, fertilizing, and so on.
ameloblast *Developmental Biology.* a cell in the inner layer of the enamel organ of a developing tooth, related to the formation of enamel. Also, ADAMANTOBLAST, ENAMEL CELL.
ameloblastic *Anatomy.* relating to or located in the region of the ameloblasts. *Oncology.* of or relating to any of various tumors situated in this region. Thus, **ameloblastic fibroma, ameloblastic odontoma, ameloblastic sarcoma.**
ameloblastoma [am´ə lō blas tō´mə] *Oncology.* a malignant tumor of the jaw originating in or associated with the ameloblasts (enamel-forming cells). Also, ADAMANTINOMA.
amelogenesis *Developmental Biology.* the formation of the dental enamel. Thus, **amelogenic.**
amelogenesis imperfecta *Developmental Biology.* a congenital condition marked by the improper development of the dental enamel.
amelus *Medicine.* a person affected with amelia (the congenital absence of a limb).
amemolite *Geology.* a stalactite characterized by one or more changes in its axis of growth.
amendment *Agronomy.* a substance added to the soil to improve plant growth, such as lime.
amendment file see CHANGE FILE.
amenorrhea [ə men´i rē´ə] *Medicine.* the absence or abnormal stoppage of menstruation. **Primary amenorrhea** is the failure of menstruation to begin at puberty; **secondary amenorrhea** is the cessation of menstruation after the menstrual cycle has become established.
amensalism *Ecology.* a relationship between two species in which one is unaffected and the other is negatively affected, typically by a toxic chemical secreted by the first species.
ament [am´ənt; ə ment´] *Botany.* a tassellike spike of small, closely clustered flowers lacking petals and sepals, as on a willow, birch, or poplar. *Psychology.* a person affected by amentia.
amentia [ə men´shə] *Psychology.* an older term for mental retardation or dementia.
Amera *Invertebrate Zoology.* a major division of invertebrates comprising more or less wormlike, unsegmented forms (Platyhelminthes, Nemathelminthes, and certain Bryozoa).
American basement *Building Engineering.* a term for a basement wholly or partially above ground level, often containing the building's main entrance.
American bond *Building Engineering.* a widely used pattern of bricklaying in which four to six layers of bricks with the long sides exposed (stretchers) are placed between a single layer with the shorter ends exposed (headers). Also, COMMON BOND, SCOTCH BOND.

American boring system *Mining Engineering.* a system of percussive boring in which a string of boring tools is attached to a rope suspended from a derrick, thus allowing the tools to be raised clear of the hole in order to facilitate cleaning the hole with a sludger; widely used in the 19th-century American oil industry.

American caisson see BOX CAISSON.

American eagle see BALD EAGLE.

American filter see DISK FILTER.

American foulbrood *Entomology.* a bacterial disease of honeybee larvae that causes them to become glutinous.

American hemorrhagic fever viruses *Virology.* a group of viruses of the genus *Arenavirus* that infect wild rodents and sometimes mites.

American jade see CALIFORNITE.

American sign language *Linguistics.* a standardized system of communication through the use of hand gestures, used in North America to communicate by and with the deaf. Also, ASL, AMESLAN.

American spotted fever SEE ROCKY MOUNTAIN SPOTTED FEVER.

American Standard Code for Information Interchange see ASCII.

American standard screw thread *Design Engineering.* a U.S. standard for specified sizes of screw thread. Similarly, **American standard pipe thread.**

American wire gauge *Metallurgy.* a standard system used for designating the thickness of sheets or the diameter of wires. Also, **American standard wire gauge.**

americium [am´ə rish´ē əm] *Chemistry.* a crystalline silver-white transuranic element of the actinide series, produced by the neutron bombardment of plutonium, having the symbol Am, atomic number 95, an atomic weight of 243 in its most stable isotope, and a half-life of 475 years. It is used as a long-term alpha-particle emitter. (Named for *America* by one of its discoverers, the chemist Glenn T. Seaborg, to correlate with the rare-earth metal europium's being named for Europe.)

Amerosporae *Mycology.* according to the Saccardo classification system, a subgroup of imperfect fungi possessing unicellular spores.

amesite [ām´zīt´] *Mineralogy.* $Mg_2Al(SiAl)O_5(OH)_4$, an apple-green triclinic mineral of the kaolinite-serpentine group that occurs as pseudohexagonal plates or in prismatic crystals with a pearly luster on cleavage, having a specific gravity of 2.77 and a hardness of 2.5 to 3 on the Mohs scale; found with corundophilite, diaspore, and magnetite.

Ameslan see AMERICAN SIGN LANGUAGE.

Ames test *Pathology.* a test to determine if a substance is a carcinogen, performed by introducing a strain of *Salmonella typhimurium* that does not have the capacity to produce histidine into an agent lacking histidine but containing the test material; if mutation of the *Salmonella* bacteria occurs, there is proof that the material being tested causes DNA damage. (Devised by Bruce *Ames,* born 1928, American biochemist.)

ametabolous [ā´mə tab´ə lis] *Zoology.* not undergoing or showing evidence of metamorphosis. *Entomology.* relating to or being a growth stage of certain primitive insects, such as Thysanura, where there is an increase in size without distinct external changes. Also, **ametabolic.**

amethopterin or **amethopterine** see METHOTREXATE.

amethyst [am´ə thist] *Mineralogy.* a pale to deep purple, transparent to translucent variety of quartz whose color is derived from the presence of iron; used as a gemstone.

amethyst

ametria [am´ə trē´ə] *Medicine.* the congenital absence of the uterus.

ametropia *Medicine.* a defect in vision involving faulty refraction of images on the retina; for example, astigmatism or myopia. Thus, **ametropic.**

AM field signature *Electronics.* an alternating magnetic field's characteristic pattern as displayed on detection and classification equipment.

amherstite *Petrology.* a syenodiorite containing andesine and antiperthite with blue quartz, hypersthene, rutile, and traces of other minerals.

AMI air mileage indicator.

amianthus *Materials.* a variety of asbestos with long, flexible white fibers.

amiben see CHLORAMBEN.

amicable number *Mathematics.* either of a pair of positive integers, related to each other in such a way that each number is equal to the sum of the proper divisors of the other number (excluding the number itself). Also, FRIENDLY NUMBER.

Amici, Giovanni Battista [ä mē´chē; ə mē´chē] 1784–1863, Italian astronomer and biologist; improved mirrors for telescopes.

Amici prism *Optics.* a right angle prism where the hypotenuse is a root of two orthogonal reflective surfaces thereby giving a 90° beam angular change with two reflecting surfaces. A compound prism that is integral to many direct-vision spectroscopes, composed of three or more alternating layers of crown and flint glass and used to provide dispersion of light without deviation. Also, DIRECT-VISION PRISM, ROOF PRISM.

amicrobic [ā´mī krō´bik] *Pathology.* not caused by a microbe.

amicron [ā´mī´krän] *Physical Chemistry.* a term used to describe submicroscopic particles that are no larger than 10^{-7} centimeters.

amicroscopic [ā´mī krə´skäp´ik] *Optics.* too small to be seen through a microscope.

amictic *Invertebrate Zoology.* not capable of being fertilized, such as the diploid eggs of female rotifers that develop without fertilization.

amictic lake *Hydrology.* a perpetually frozen lake that does not undergo mixing of the water column.

amicyanin *Microbiology.* a copper-containing protein found in certain bacteria which gain energy from the oxidation of carbon monoxide.

amidase [am´i dās´] *Enzymology.* an enzyme that catalyzes the breakdown of an amide compound to a carboxylic acid and ammonia.

amidate *Organic Chemistry.* to add an amide group to a compound or produce an amide group within a molecule. Thus, **amidation.**

amide [am´id; am´id] *Organic Chemistry.* **1.** any of a class of organic compounds characterized by the $-CONH_2$ radical, formed from an acid by the replacement of a hydroxyl group, $-OH$, with $-NH_2$, or from ammonia by the replacement of hydrogen with an acyl group. **2.** any of a class of compounds in which the hydrogen atom of ammonia is replaced by a metal; e.g., sodium amide.

amide group see AMIDO GROUP.

amide synthase *Enzymology.* an enzyme that catalyzes the formation of the amide group by replacing the OH of a carboxylic acid with an amino group.

amidic *Organic Chemistry.* relating to or derived from an amide or amides.

amidin *Organic Chemistry.* a transparent solution of starch in water.

amidine *Organic Chemistry.* any of a class of organic compounds containing the $-CNHNH_2$ group.

amidinotransferase *Enzymology.* an enzyme of the transferase class that catalyzes the transfer of an amidino group from one compound to another.

amido- *Organic Chemistry.* a combining form referring to the presence in a compound of the $-NH_2$ group along with the $-CO$ group.

amido group *Organic Chemistry.* the $-CONH_2$ group. Also, **amido radical.**

amidohydrolase *Enzymology.* an enzyme of the hydrolase class that catalyzes the hydrolysis of an amido group. Also, DEAMIDASE.

amidol *Organic Chemistry.* $C_6H_3(NH_2)_2OH \cdot 2HCl$, a white crystalline solid that is soluble in water and slightly soluble in alcohol; melts at 205°C; used in organic synthesis and as a photographic developer. Also, 2,4-DIAMINOPHENOL HYDROCHLORIDE.

amidships *Naval Architecture.* at or near the middle of a vessel; midway between the bow and stern.

Amiidae [ə mī´ə dē] *Vertebrate Zoology.* a small family of fish in the order Amiiformes, comprising many extinct species and represented today by a single genus, *Amia,* the freshwater bowfin.

Amiiformes [ə mī´ə fôr´mēz] *Vertebrate Zoology.* an order of fish in the subclass Actinopterygii, containing the single family Amiidae.

amimia *Neurology.* a loss of the ability to express oneself with signs or gestures.

amination *Organic Chemistry.* **1.** the synthesis of amines. **2.** the process of adding the amino group, $-NH_2$, to an organic compound.

amine *Organic Chemistry.* a general term for a class of organic compounds that have been synthesized from ammonia, NH_3, by substituting organic groups for one or more of the ammonia hydrogens.

-amine *Chemistry.* a suffix indicating the presence of the amino group, $-NH_2$.

amine oxidase *Enzymology.* an enzyme that causes the ammonia derivatives called amines to oxidize to aldehyde; consisting of two types, copper-containing protein or **diamine oxidase** and flavin-containing protein or **monoamine oxidase.**

aminic *Organic Chemistry.* relating to or being an amine.

aminic acid see FORMIC ACID.

amino- or **amin-** *Chemistry.* a combining form referring to the presence in a compound of the amino group, $-NH_2$.

aminoacetic acid see GLYCINE.

amino acid *Biochemistry.* any organic compound containing an amino ($-NH_2$) and a carboxyl ($-COOH$) group; there are 20 α-amino acids from which proteins are synthesized during ribosomal translation of mRNA. An **essential amino acid** is any one of the amino acids that are essential for metabolism, health, and growth, but that are not synthesized by the body and thus must be obtained from food. For humans, these include isoleucine, leucine, lysine, methionine, phenyalanine, threonine, tryptophan, valine, and (during growth periods) arginine and histidine. A **nonessential amino acid** is an amino acid that is synthesized by the body and thus not specifically required in the diet. These include alanine, asparagine, aspartic acid, cysteine, glutamine, glutamic acid, glycine, proline, serine, and tyrosine.

amino acid activation *Biochemistry.* a coupled reaction catalyzed by a specific aminoacyl synthetase that attaches an amino acid to AMP and then to a transfer RNA (tRNA) in preparation for translation.

amino acid analyzer *Biochemistry.* an instrument used to determine the amino acid composition of a protein.

amino acid dating *Archaeology.* a method of dating bone or other organic material, in which specific changes in its amino acid structure (racemization) are measured after the death of the organism; aspartic acid is the compound most often used. *Geology.* a similar technique used in dating geological samples. Also, AMINOSTRATIGRAPHY.

amino acid oxidase see MONOAMINE OXIDASE.

amino acid racemization dating see AMINO ACID DATING.

amino acid requirement(s) *Nutrition.* the number of amino acids needed by an organism to provide the necessary proteins for a healthy diet, which may vary from zero in the case of an organism that synthesizes them all, to a complete requirement in the case of an organism in which all the biosynthetic pathways are blocked. Eight to ten amino acids are required by humans and certain other mammals.

amino acid sequence *Molecular Biology.* the precisely defined linear order of amino acids in a protein or polypeptide; the major determiner of protein structure.

aminoaciduria see ACIDAMINURIA.

α-aminoacylpeptide hydrolase *Enzymology.* an enzyme that hydrolyzes a peptide molecule by attacking the bond of a carboxylic acid.

aminoacyl transferase *Enzymology.* an enzyme that catalyzes the transfer of an aminoacyl group.

aminoacyl tRNA *Molecular Biology.* a transfer RNA molecule that is associated with a specific amino acid; the two are covalently bonded until the bond is broken during protein synthesis.

aminoacyl tRNA synthetase *Enzymology.* an enzyme that catalyzes the attachment of an amino acid to adenosine monophosphate (AMP), forming an aminoacyl adenylate, which is then attached to a transfer RNA molecule forming an aminoacyl-tRNA molecule.

aminoadipic acid pathway *Mycology.* a chemical pathway for the biosynthesis of the amino acid lysine, especially as found in such microorganisms as Blastocladiales, Chytridiales, Mucorales, and Ascomycetes. Also, AAA PATHWAY.

amino alcohol see ALKAMINE.

aminoanthraquinone *Organic Chemistry.* $C_6H_4(CO)_2C_6H_3NH_2$, either of two compounds: **a. 1-aminoanthraquinone.** red iridescent needles that melt at 252°C with sublimation; soluble in alcohol, benzene, and acetone, and insoluble in water; used in the manufacture of dyes and pharmaceuticals. **b. 2-aminoanthraquinone.** orange-brown or red needles that melt at 302°C with sublimation; soluble in alcohol, benzene and acetone, and insoluble in water; used in dye synthesis.

4-aminoantipyrine see AMPYRONE.

***p*-aminoazobenzene** see ANILINE YELLOW.

aminobenzoic acid *Organic Chemistry.* any of three isomers derived from benzoic acid, especially *p*-aminobenzoic acid (PABA).

***m*-aminobenzoic acid** *Organic Chemistry.* $C_6H_4NH_2CO_2H$, yellowish or reddish crystals that melt at 174°C; slightly soluble in water, alcohol, or ether; used in dye manufacture. Also, BENZAMINIC ACID.

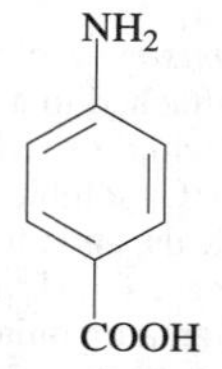

***p*-aminobenzoic acid**

***p*-aminobenzoic acid** *Biochemistry.* PABA, a compound of the vitamin B complex that is necessary for folic acid synthesis, but not essential for the human diet.

***o*-aminobenzoic acid** see ANTHRANILIC ACID.

aminobutane see BUTYLAMINE.

2-amino-1-butanol *Organic Chemistry.* $CH_3CH_2CHNH_2CH_2OH$, a colorless liquid that boils at 178°C; soluble in water and alcohol, and corrosive to copper, brass, and aluminum; used as an emulsifying agent for oils and waxes, as an absorbent for acidic gases, and as a vulcanizing accelerator.

γ-aminobutyric acid *Biochemistry.* $C_4H_9NO_2$, GABA, a fatty acid derivative that acts as an inhibitory neurotransmitter in the central nervous system.

ε-aminocaproic acid *Organic Chemistry.* $H_2N(CH_2)_5CO_2H$, a crystalline solid that melts at 204°C; soluble in water and insoluble in alcohol; used as an agent to counteract blood clot destruction, and as a spacer for affinity chromatography. ALSO, 6-AMINOHEXANOIC ACID.

aminocarb *Organic Chemistry.* $C_{11}H_{16}N_2O_2$, a brownish crystalline solid that melts at 93°C; slightly soluble in water and soluble in polar organic solvents; used as an insecticide.

3-amino-2,5-dichlorobenzoic acid see CHLORAMBEN.

aminodithioformic acid see DITHIOCARBAMIC ACID.

aminoethane see ETHYLAMINE.

2-aminoethanethiol see ETHANOLAMINE.

2-aminoethanesulfonic acid see TAURINE.

aminoglycoside [ə mēn′ō glī′kə sīd′] *Organic Chemistry.* a chemical compound containing one or more amino sugars in glycoside linkage. *Pharmacology.* any of a group of bacterial antibiotics having such a structure, widely used for the treatment of infections caused by Gram-negative bacilli; included are such common types as streptomycin, neomycin, gentamicin, and kanamycin.

amino group *Organic Chemistry.* the univalent group of one nitrogen atom and two hydrogen atoms, $-NH_2$; the essential component of amino acids. Also, AMINO RADICAL.

6-aminohexanoic acid see AMINOCAPROIC ACID.

***p*-aminohippuric acid** see PAHA.

α-aminohydrocinnamic acid see PHENYLALANINE.

aminohydrolase see DEAMINASE.

aminohydroxybenzoic acid see PAS.

2-amino-2-methyl-1,3-propanediol *Organic Chemistry.* $CH_2OHC(CH_3)NH_2CH_2OH$, colorless crystals that melt at 110°C and boil at 151°C, and that are soluble in water and alcohol; corrosive to copper, brass, and aluminum; used in organic synthesis, leather dressings, cleaning compounds, and as an emulsifier for cosmetics. Also, AMPD.

3-amino-2-naphthoic acid *Organic Chemistry.* $C_{10}H_6NH_2COOH$, yellow crystals that melt at 214°C; soluble in ether and alcohol; used as an indicator for copper, nickel, and cobalt.

1-amino-2-naphthol-4-sulfonic acid *Organic Chemistry.* $C_{10}H_5NH_2(OH)SO_3H$, white or gray needles that are soluble in hot sodium bisulfite; used to manufacture azo dyes.

2-amino-5-naphthol-7-sulfonic acid *Organic Chemistry.* $C_{10}H_5NH_2OHSO_3H$, white or gray needles that are soluble in hot water; used as a dye intermediate.

amino nitrogen *Chemistry.* nitrogen that is combined with hydrogen in an amino group.

2-amino-5-nitrothiazole *Pharmacology.* $C_3H_3N_3O_2S$, a drug used to treat the amebic disease histomoniasis (blackhead) in poultry and trichomoniasis in pigeons.

6-aminopenicillanic acid *Pharmacology.* $C_8H_{12}N_2O_3S$, a white crystalline substance that decomposes at 208°C; used in the manufacture of synthetic penicillins.

aminopeptidase *Enzymologyy.* an enzyme that catalyzes the cleavage by hydrolysis of amino acids from a polypeptide or protein, starting at the amino terminal.

aminophenol *Organic Chemistry.* a class of compounds possessing both $-NH_2$ and $-OH$ groups attached to a benzene ring; for example, ***p*-aminophenol,** $C_6H_4NH_2OH$, white or reddish-yellow crystals that turn violet in light and melt at 184°C; soluble in water and alcohol; used in dye manufacture, pharmaceuticals, and photography.

aminophylline *Pharmacology.* $C_{16}H_{24}N_{10}O_4$, a yellowish or white powder with a bitter taste and an ammoniacal odor; a drug used to dilate the air passages in asthma, bronchitis, and emphysema; formerly used as a heart and respiratory stimulant and as a diuretic.

aminoplast or **aminoplastic resin** see AMINO RESIN.

aminopolypeptidase see AMINOPEPTIDASE.

aminopterin *Pharmacology.* $C_{19}H_{20}N_8O_5$, a drug formerly used to check the proliferation of malignant cells, now largely replaced by methotrexate.

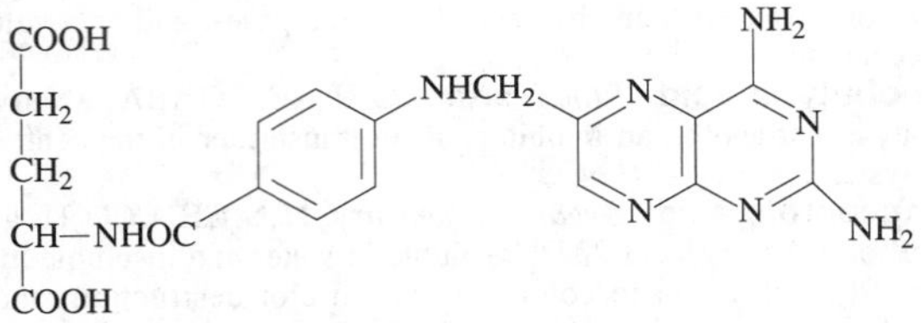

aminopterin

amino purine *Biotechnology.* a base analog that replaces adenine in the DNA of an organism; the result is incorrect base pairing in future cell generations and subsequent mutation.

2-aminopurine *Biochemistry.* a purine compound that can be used in DNA as a substitute for the purine adenine, but frequently causes mutations during DNA replication.

2-aminopyridine *Organic Chemistry.* $C_5H_4NNH_2$, white or colorless crystals, soluble in water, alcohol, ether, and benzene; melts at 58°C and boils at 210°C; used as an intermediate for antihistamines and other pharmaceuticals. Also, **α-aminopyridine.**

3-aminopyridine *Organic Chemistry.* $C_5H_4NNH_2$, a crystalline solid that melts at 64°C and boils at 274°C; soluble in water, alcohol, ether, and benzene; used in making dyes and pharmaceuticals. Also, **β-aminopyridine.**

4-aminopyridine *Organic Chemistry.* $C_5H_4NNH_2$, white crystals that melt at 158–159°C and are soluble in water; used as a chemical intermediate.

aminopyrine *Pharmacology.* $C_{13}H_{17}N_{3O}$, a painkiller and fever reducer that is now rarely used because of its dangerous side effects.

amino radical see AMINO GROUP.

amino resin *Organic Chemistry.* a general term for any of a group of resins resulting from the condensation polymerization of an aldehyde, such as formaldehyde, with a compound containing an amino group, such as urea.

***p*-aminosalicylic acid** see PAS.

Aminosol [ə mēn′i säl] *Medicine.* a trademark for a preparation of amino acids with glucose and mineral salts, used in intravenous injections.

aminostratigraphy see AMINO ACID DATING.

aminosuccinic acid see ASPARTIC ACID.

amino sugar *Biochemistry.* a simple sugar (monosaccharide) that contains an amino group in the place of hydroxyl groups; for example, glucosamine.

amino terminal *Biochemistry.* the end of a polypeptide or protein chain that contains a free amino group. Also, N-TERMINAL.

2-aminothiazole *Organic Chemistry.* $C_3H_4N_2S$, light yellow crystals that melt at 90°C; soluble in hot water and dilute mineral acids and slightly soluble in cold water, alcohol, and ether; used as an intermediate in the synthesis of sulfathiazole, a thyroid inhibitor.

aminothiourea see THIOSEMICARBAZIDE.

aminotoluene see TOLUIDINE.

aminotransferase see TRANSAMINASE.

aminotriazole *Organic Chemistry.* $C_2H_4N_4$, a crystalline solid that melts at 156–159°C, soluble in water, alcohol, chloroform, and methanol; used as a wide-spectrum herbicide against broadleaf and grassy weeds; not used on food crops because of suspected carcinogenic properties. Also, **amitrole.**

aminourea see SEMICARBAZIDE.

aminuria [am′i nûr′ē ə] *Medicine.* an excess of amines in the urine.

-aminyl *Chemistry.* a suffix indicating the formation of a free radical by the loss of a hydrogen atom from a base whose name ends in *-amine.*

amiodarone [ə′mē′ə də rōn′] *Pharmacology.* $C_{25}H_{25}I_2NO_3$, a drug used in the treatment of arrhythmia (abnormal heartbeat).

amitate [am′i tāt′] *Anthropology.* a prescribed social relationship between an aunt and her niece, in which the aunt is responsible for the training and discipline of the niece.

amitosis [am′i tō′sis; ā′mī tō′sis] *Cell Biology.* simple cellular fission or cell division in which the normal nuclear events of chromosome formation and separation do not occur. Thus, **amitotic.**

amitriptyline hydrochloride *Pharmacology.* $C_{20}H_{23}N{\cdot}HCl$, a white crystalline powder that is soluble in water; used as an antidepressant drug.

AML *Robotics.* a programming language developed by IBM for its 7535 robotic manufacturing system.

AML *Medicine.* acute myelocytic leukemia.

AMM or **amm** antimissile missile.

ammeter *Engineering.* an instrument used to measure and indicate the rate of flow of an electric current, usually in amperes.

ammine [am′ēn; ə mēn′] *Inorganic Chemistry.* a type of coordination compound in which a ligand is ammonia; it usually contains a metal ion that is connected directly to the nitrogen by coordinate linkage.

ammocoete [am′ə sēt] *Invertebrate Zoology.* the larval stage of the lamprey eel. Also, **ammocete.**

ammocolous *Ecology.* living or growing in sand. Also, **ammocole.**

Ammodiscacea *Invertebrate Zoology.* a superfamily of foraminiferal protozoans, in the suborder Textulariina, that possess simple labyrinthic test walls.

ammodyte *Botany.* a plant living or growing in sand.

Ammodytoidei *Vertebrate Zoology.* the sand eels or sand lances, an eel-like suborder of the order Perciformes. Its members swim in shoals near the shore and bury themselves in the sand.

ammonal *Materials.* a high explosive composed mainly of TNT, ammonium nitrate, and aluminum.

ammonation *Inorganic Chemistry.* a reaction between ammonia and a reactant resulting in the elimination of a small molecule such as water and the covalent union of $-NH_2$ or $=NH$ with the reactant residue. Example: $CO(OH)_2 + NH_3 \rightarrow CONH_2OH + H_2O$

ammonia *Inorganic Chemistry.* **1.** NH_3, a colorless gas that has a strong, highly irritating odor and an alkaline reaction in water, and is lighter than air and extremely soluble in water; formed in nature as an end product of animal metabolism by the decomposition of uric acid. It is commercially produced by the direct catalytic combination of nitrogen and hydrogen gases under high pressure and temperature, and has a wide variety of industrial uses; for example, as a fertilizer, household cleaning agent, and refrigerant. **2.** a solution of this gas in water; ammonium hydroxide. (From the Egyptian god *Amon* or *Ammon,* whose temple was near a noted early source of ammonium chloride.)

ammonia absorption refrigerator *Mechanical Engineering.* a refrigeration system that uses ammonia as the working fluid, and creates low temperatures by evaporating ammonia in a heat exchanger.

ammonia alum see AMMONIUM ALUMINUM SULFATE.

ammonia-beam maser *Physics.* a maser whose active medium is a beam of gaseous ammonia.

ammoniac *Materials Science.* a gum resin obtained from the stems of certain plants, used in concrete, adhesives, and formerly in medicine as an expectorant. *Botany.* any of several plants that are the source of this resin, especially *Dorema ammoniacum,* which is found in Iran and northern India. *Inorganic Chemistry.* see AMMONIACAL.

ammoniacal [am′ə nī′i kəl] *Inorganic Chemistry.* **1.** relating to or containing ammonia. **2.** like ammonia, especially in odor.

ammonia clock *Horology.* a type of atomic clock in which a maser separates ammonia molecules into two energy states; as the molecules fall into a lower energy state, they oscillate at a constant frequency that can be used for extremely precise timekeeping. Also, AMMONIA MASER CLOCK.

ammonia compressor *Mechanical Engineering.* a compressor unit that mechanically increases the pressure of gaseous ammonia in an ammonia refrigeration system.

ammonia condenser *Mechanical Engineering.* a tubular system used in refrigeration systems, in which compressed ammonia is cooled, either by spray or forced air, until it becomes a liquid.

ammonia dynamite *Materials Science* a common form of dynamite containing sensitized ammonium nitrate, NH_4NO_3, in place of or in combination with nitroglycerin; safer and cheaper than straight dynamite. Similarly, **ammonia gelatin dynamite.**

ammonia liquor *Chemical Engineering.* an aqueous solution that is obtained in the destructive distillation of soft coal; it contains ammonia, ammonium compounds, cyanogen, and hydrogen sulfide. Also, **ammoniacal liquid.**

ammonia lyase *Enzymology.* an enzyme that catalyzes the breakdown of a substrate, producing ammonia as one of the products.

ammonia maser clock see AMMONIA CLOCK.

ammonia meter *Engineering.* a type of hydrometer that measures the density of aqueous ammonia solutions.

ammonia nitrogen see AMINO NITROGEN.

ammonia print see DIAZO PAPER.

ammonia-soda process see SOLVAY PROCESS.

ammonia solution see AMMONIA WATER.

ammonia synthesis *Chemical Engineering.* the formation of ammonia by a combination of nitrogen and hydrogen gases at high temperature and pressure with the aid of a catalyst.

ammoniate *Chemistry.* **1.** to treat or combine with ammonia. **2.** a compound formed by a combination with ammonia. Thus, **ammoniated, ammoniation.**

ammoniated mercury *Inorganic Chemistry.* $HgNH_2Cl$, a white, odorless powder that darkens on exposure to light; insoluble in water and alcohol; used in pharmaceuticals. Also known by various other names, such as **ammoniated mercury chloride.**

ammoniated superphosphate *Inorganic Chemistry.* a fertilizer consisting of 5 parts of ammonia to 100 parts of superphosphate.

ammonia valve *Engineering.* a valve constructed of materials that are resistant to corrosion by ammonia.

ammonia water see AMMONIUM HYDROXIDE.

ammonification *Chemistry.* **1.** the natural formation of ammonia in soil by the action of bacteria on proteins in decaying organic matter. **2.** any similar process in which ammonia or an ammonia compound is produced.

ammonify *Chemistry.* **1.** to combine or be combined with ammonia. **2.** to produce or become ammonia. Thus, **ammonifier.**

ammonio- *Chemistry.* a combining form used to indicate the presence of ammonia or ammonium.

ammonioborite *Mineralogy.* $(NH_4)_2B_{10}O_{16}{\cdot}5H_2O$, a white monoclinic mineral that is soluble in water, having a specific gravity of 1.765; found especially in fumarolic deposits in certain lagoons of Tuscany, Italy.

ammoniojarosite *Mineralogy.* $(NH_4)Fe_3^{+3}(SO_4)_2(OH)_6$, a pale yellow trigonal mineral of the alunite group having a specific gravity of 3.0 and occurring in certain mercury deposits as microscopic, transparent, tabular grains in small earthy lumps or nodules.

ammonite *Paleontology.* **1.** a type of ammonoid fossil that is characterized by a thick, strongly ornamented shell with complex lines of junction, which are known as **ammonitic sutures**. **2.** the shell itself. **3.** see AMMONOID.

ammonitic *Paleontology.* relating to or characteristic of ammonite fossils.

ammonitid *Paleontology.* of or relating to a late type of ammonoid in the order Ammonitida, widespread in deeper waters in the Jurassic and Cretaceous; of uncertain classification.

ammonium *Chemistry.* the hypothetical radical NH_4^+; it does not appear in a free state, but it forms salts or compounds that are analogous to those of the alkali metals.

ammonium acetate *Organic Chemistry.* CH_3COONH_4, a combustible white crystalline solid that is soluble in water and alcohol; melts at 114°C; used as a chemical reagent, in textile dyeing, and for preserving meats.

ammonium alginate *Organic Chemistry.* $(C_6H_7O_6{\cdot}NH_4)_n$, a hydrophilic colloid; used as a stabilizer and thickening agent in cheese, canned fruits, ice cream, and other food products.

ammonium aluminum sulfate *Inorganic Chemistry.* $NH_4Al(SO_4)_2{\cdot}12H_2O$, colorless, odorless water-soluble crystals, melting at 94°C; used in manufacturing medicines and baking powder, and in dyeing, tanning, and papermaking. Also, AMMONIUM ALUM, ALUMINUM AMMONIUM SULFATE.

ammonium benzoate *Organic Chemistry.* $C_6H_5COONH_4$, a white crystalline or powdery solid that is soluble in water and alcohol; decomposes at 198°C; used as a preservative in rubber latex and in adhesives.

ammonium bicarbonate *Inorganic Chemistry.* NH_4HCO_3, a white crystalline salt, soluble in water and insoluble in alcohol; used in baking powder and fire extinguishers. Also, AMMONIUM HYDROGEN CARBONATE.

ammonium bichromate see AMMONIUM DICHROMATE.

ammonium bifluoride *Inorganic Chemistry.* NH_4HF_2, a corrosive salt that is soluble in water and alcohol; it is used as a preservative and sterilizer and in electroplating processes. Also, AMMONIUM HYDROGEN FLUORIDE.

ammonium binoxalate *Inorganic Chemistry.* $(NH_4)HC_2O_4{\cdot}H_2O$, colorless crystals, soluble in water and decomposed by heating; used as an analytical reagent and for ink removal.

ammonium bisulfate *Inorganic Chemistry.* NH_4HSO_4, colorless crystals, soluble in water; used as a reagent. Also, AMMONIUM HYDROGEN SULFATE.

ammonium bitartrate *Organic Chemistry.* $(NH_4)HC_4H_4O_6$, a colorless crystalline solid, soluble in water and insoluble in alcohol; used in making baking powder and as an indicator for calcium. Also, AMMONIUM HYDROGEN TARTRATE.

ammonium borate *Inorganic Chemistry.* $NH_4HB_4O_7{\cdot}3H_2O$, colorless crystals, soluble in water; used as a fire retardant on fabrics and as an herbicide.

ammonium bromide *Inorganic Chemistry.* NH_4Br, an ammonium halide, existing as both colorless crystals and a yellowish powder; soluble in water and alcohol; used in photography, pharmaceuticals, and textile finishing.

ammonium carbamate *Inorganic Chemistry.* $NH_4CO_2NH_2$, a white, rhombic, crystalline, highly volatile powder that decomposes in air to evolve ammonia; it forms an important unstable intermediate in the manufacture of urea and is used as a fertilizer.

ammonium carbonate *Inorganic Chemistry.* **1.** $(NH_4)_2CO_3$, the ammonium salt of carbonic acid. **2.** $NH_4HCO_3{\cdot}NH_2CONH_4$, the double salt of ammonium bicarbonate and ammonium carbamate; used in smelling salts, baking powder, and various other products.

ammonium chloride *Inorganic Chemistry.* NH_4Cl, a white crystalline salt derived as a by-product of the ammonia soda process; soluble in water and slightly soluble in alcohol; used in dry cells, galvanizing, and in medicine as an expectorant. Also, SAL AMMONIAC.

ammonium chromate *Inorganic Chemistry.* $(NH_4)_2CrO_4$, yellow toxic crystals, soluble in water and insoluble in alcohol; used in photography and as an analytical reagent.

ammonium chromium sulfate see CHROMIUM AMMONIUM SULFATE.

ammonium citrate *Organic Chemistry.* $(NH_4)_2HC_6H_5O_7$, a white granular solid that is soluble in water and slightly soluble in alcohol; used in pharmaceuticals and in analysis to determine phosphates in fertilizers.

ammonium dichromate *Inorganic Chemistry.* $(NH_4)_2Cr_2O_7$, orange needles that are soluble in water and alcohol and decompose with slight heating; a strong oxidizing agent whose dusts and aqueous solution are toxic; it is used in photography, lithography and as a mordant for dyeing.

ammonium fluoride *Inorganic Chemistry.* NH_4F, a white crystalline salt, soluble in cold water and decomposed by heat; used in analytical chemistry, glass etching, and wood preservation.

ammonium fluorosilicate *Inorganic Chemistry.* $(NH_4)_2SiF_6$, a toxic white crystalline powder that is soluble in water and alcohol; a strong irritant to the eyes and skin; used for mothproofing, glass etching, and electroplating. Also, AMMONIUM SILICOFLUORIDE.

ammonium formate *Organic Chemistry.* $HCOONH_4$, a colorless, deliquescent crystalline powder that melts at 115°C and is soluble in water and alcohol; used in metal separations.

ammonium gluconate *Organic Chemistry.* $NH_4C_6H_{11}O_7$, a white crystalline solid that is soluble in water and insoluble in alcohol; used as an emulsifier for cheese and salad dressing and as a catalyst in textile printing.

ammonium halide *Inorganic Chemistry.* a compound in which the ammonium ion is bonded to a simple ion formed from one of the halogen elements.
ammonium hydrate see AMMONIUM HYDROXIDE.
ammonium hydrogen carbonate see AMMONIUM BICARBONATE.
ammonium hydrogen fluoride see AMMONIUM BIFLUORIDE.
ammonium hydrogen oxalate *Inorganic Chemistry.* $(NH_4)HC_2O_4 \cdot H_2O$, colorless crystals that are soluble in water and decompose on heating; used as an analytical reagent and for ink removal.
ammonium hydrogen sulfate see AMMONIUM BIFSULFATE.
ammonium hydrogen tartrate see AMMONIUM BITARTRATE.
ammonium hydroxide *Inorganic Chemistry.* NH_4OH, a colorless liquid with a strong odor, formed by dissolving ammonia in water in concentrations up to 30%; widely used as a household cleanser and for a variety of manufacturing processes.
ammonium iodide *Inorganic Chemistry.* NH_4I, white hygroscopic crystals or powder, made from ammonia and hydrogen iodide; soluble in water and alcohol; used in photography and in pharmaceuticals.
ammonium iron sulfate see FERRIC AMMONIUM SULFATE.
ammonium lactate *Organic Chemistry.* $NH_4C_3H_5O_3$, a colorless to yellow viscous liquid that is soluble in water and alcohol; used in electroplating and in tanning and processing leather.
ammonium lineolate *Organic Chemistry.* $C_{17}H_{31}COONH_4$, a soft yellowish paste with an ammoniacal odor; soluble in water, ethanol, and methanol; used as an emulsifying agent and for waterproofing.
ammonium metatungstate *Inorganic Chemistry.* $(NH_4)_6H_2W_{12}O_{40}$, a white powdery solid that is soluble in water; used in electroplating.
ammonium metavanadate see AMMONIUM VANADATE.
ammonium molybdate *Inorganic Chemistry.* $(NH_4)_2MoO_4$, a white crystalline powder that is soluble in water and insoluble in alcohol; it is used as an analytical reagent, in pigments, and as a source of molybdate ions.
ammonium nickel sulfate see NICKEL AMMONIUM SULFATE.
ammonium nitrate *Inorganic Chemistry.* NH_4NO_3, a colorless crystalline powder that is soluble in water and alcohol, melting at 170°C and decomposing at 210°C; widely produced for various industrial uses, such as fertilizers, explosives, and freezing mixtures, and used in the manufacture of nitrous oxide.
ammonium oleate *Organic Chemistry.* $C_{17}H_{33}COONH_4$, a yellow to brownish mass, an ammonium soap; soluble in water and in hot alcohol; it melts at 70–72°C; used in cosmetics and as an emulsifying agent.
ammonium oxalate monohydrate *Organic Chemistry.* $(NH_4)_2C_2O_4 \cdot H_2O$, a colorless crystalline salt that is soluble in water and decomposed by heat; used in analytical chemistry, rust and scale removal, and safety explosives.
ammonium palmitate *Organic Chemistry.* $C_{15}H_{31}COONH_4$, yellowish granules, an ammonium soap that is soluble in water and in hot alcohol; it melts at 70–72°C; used in waterproofing and as a thickening agent.
ammonium perchlorate *Inorganic Chemistry.* NH_4ClO_4, white crystals that are soluble in water and decomposed by heat; a strong oxidizing agent that ignites violently with combustibles; used in explosives, pyrotechnics, and as a smokeless rocket and jet propellant.
ammonium permanganate *Inorganic Chemistry.* NH_4MnO_4, a purple crystal or powder that is soluble in water and decomposed by heat; used as an oxidizing agent and in explosives.
ammonium peroxydisulfate see AMMONIUM PERSULFATE.
ammonium persulfate *Inorganic Chemistry.* $(NH_4)_2S_2O_8$, white crystals that are soluble in water and decomposed by heat; used as an oxidizing and bleaching agent, and in electroplating, etching, and food preservation. Also, AMMONIUM PEROXYDISULFATE.
ammonium phosphate *Inorganic Chemistry.* $(NH_4)_2HPO_4$, a colorless to white salt of ammonia and phosphoric acid that is soluble in water; used as a fire retardant, in fertilizers, and for a wide variety of other industrial purposes. Also, DIAMMONIUM PHOSPHATE.
ammonium picrate *Organic Chemistry.* $NH_4OC_6H_2(NO_2)_3$, yellow crystals, slightly soluble in water and alcohol and decomposed by heat; used as a military explosive.
ammonium salt *Inorganic Chemistry.* any of various salts formed by the neutralization of ammonium hydroxide with acids, usually white and water-soluble and decomposed by heat.
ammonium silicofluoride see AMMONIUM FLUOROSILICATE.
ammonium soap *Organic Chemistry.* a cosmetic soap ingredient that is prepared from a reaction between ammonium hydroxide and a fatty acid.
ammonium stearate *Organic Chemistry.* $C_{17}H_{35}COONH_4$, a cream-colored, waxy solid that melts at 73°C and is soluble in alcohol and boiling water; used in making cosmetics and as a waterproofing agent, primarily for papers, textiles, and cement.
ammonium sulfamate *Inorganic Chemistry.* $NH_4OSO_2NH_2$, white hygroscopic crystals that are soluble in water and melt at 130°C; used for flameproofing textiles and paper, and as an herbicide.
ammonium sulfate *Inorganic Chemistry.* $(NH_4)_2SO_4$, grayish to white crystals that are soluble in water and insoluble in alcohol and melt at 140°C; a commercially important chemical produced mostly from the destructive distillation of coal; used as a fertilizer, a food additive, and in water treatment, tanning, and fermentation processes.
ammonium sulfate precipitation *Biotechnology.* a technique used to isolate and purify proteins by adding ammonium salts to a cell extract and then precipitating out proteins.
ammonium sulfide *Inorganic Chemistry.* $(NH_4)_2S$, yellow crystals that are soluble in water and alcohol; used in photographic developers and to provide synthetic flavors.
ammonium sulfite *Inorganic Chemistry.* $(NH_4)_2SO_3 \cdot H_2O$, colorless to white crystals that are soluble in water; used as a reducing agent and in photography.
ammonium tartrate *Organic Chemistry.* $C_4H_4O_6(NH_4)_2$, a colorless crystalline solid that is soluble in water and alcohol and decomposes on heating; used in processing textiles and in medicine.
ammonium thiocyanate *Organic Chemistry.* NH_4SCN, a colorless crystalline solid that is soluble in water, alcohol, acetone, and ammonia; melts at 149.6°C; used in analytical chemistry and for many industrial purposes, such as fabric dyeing, electroplating, and steel pickling, and in photographic chemicals and freezing solutions.
ammonium vanadate *Inorganic Chemistry.* NH_4VO_3, a white crystalline nonflammable powder that is slightly soluble in cold water and decomposes at 210°C; used in indelible inks and as a paint drier and textile mordant.
ammonoid *Paleontology.* **1.** one of the stratigraphically important cephalopod mollusks of the subclass Ammonoidea, noted for their distinctive shells. **2.** of or relating to these mollusks.
Ammonoidea *Paleontology.* an extinct subclass of mollusks in the class Cephalopoda, characterized by a large, chambered shell coiled in a spiral plane; the ammonoids are important as zone fossils in the Paleozoic, widespread and diverse from the Devonian to the Cretaceous.
ammonolysis [am´ə näl´ə sis] *Chemistry.* the solvolysis of a substance in a liquid ammonia solution. Thus, **ammonolytic.**
ammonolyze *Chemistry.* to subject to or undergo ammonolysis.
ammonotelic [am´ə nō tēl´ ik] *Zoology.* having ammonia as the chief nitrogenous waste product, as do certain freshwater fish.
Ammon's horn see HORN OF AMMON.
ammophilous [ə mäf´ə ləs] *Ecology.* thriving in sandy habitats. Thus, **ammophile, ammophily.**
ammophyte *Botany.* a plant living or growing in sand.
ammunition *Ordnance.* any object or material designed to be fired, exploded, or otherwise released by a weapon, especially bullets, shot, shells, and other such projectiles.
ammunition belt *Ordnance.* **1.** a band with loops or pockets for holding cartridges that are fed from it into a machine gun or other such automatic weapon. Also, **feed belt. 2.** see CARTRIDGE BELT.
ammunition box *Ordnance.* a box in which an ammunition belt can be folded for feeding cartridges into a machine gun.
ammunition carrier *Ordnance.* **1.** a vehicle that accompanies artillery guns and carries ammunition for them. **2.** a person in an artillery squad who carries ammunition and helps in loading.
ammunition day of supply *Ordnance.* the estimated amount of ammunition required per day to carry on operations in a specific combat area, expressed in terms of rounds fired per weapon per day.
ammunition dump *Ordnance.* an area for storing or holding supplies of ammunition. Also, **ammunition depot.**
amnesia [am nē´zhə] *Neurology.* the loss or absence of memory, caused by disease, physical injury, or emotional trauma.
amnesiac [am nē´zhē ak´] *Neurology.* **1.** a person affected by or displaying symptoms of amnesia. **2.** relating to, characterized by, or causing amnesia. Also, **amnesic.**
amnestic *Neurology.* characterized by or causing amnesia.
amnestic aphasia or **amnesic aphasia** see ANOMIC APHASIA.
amnestic apraxia *Neurology.* the inability to carry out some request or command even though the physical ability to do so is present, because of a failure of memory.

amnicolous [am´nə kā´ ləs] *Ecology.* living or thriving on sandy river banks. Also, **amnicole.**

amniocentesis [am´nē ō sen tē´sis] *Medicine.* a procedure for obtaining a sample of amniotic fluid from the uterus of a pregnant woman, usually involving the insertion of a hypodermic needle into the amnion and the withdrawal of amniotic fluid; used to detect fetal disorders and abnormalities or otherwise assess the condition or development of the fetus.

amniochorial *Developmental Biology.* pertaining to a combination of amnion and chorion. Also, **amniochorianic.**

amniocyte *Cell Biology.* a cell found floating free within the thin membranous sac, or amnion, surrounding an embryo.

amniogenesis *Developmental Biology.* the development of the amnion.

amniography *Radiology.* an X-ray examination of the amnion after the injection of an opaque solution, to permit viewing of the amniotic cavity and the fetus.

amnion [am´nē än] *Vertebrate Zoology.* **1.** in reptiles, birds, and mammals, the thin but tough membrane that forms the inner sac containing a developing embryo and its surrounding fluid (the amniotic fluid). **2.** the entire sac that is formed by this membrane. *Invertebrate Zoology.* a similar membrane found in insects and other invertebrates.

amnionic [am´nē än´ik] *Developmental Biology.* relating to or forming an amnion.

amniorrhea [am´nē ə rē´ə] *Medicine.* the escape of the amniotic fluid.

amniorrhexis *Medicine.* the rupture of the amnion.

amnioscope *Medicine.* an endoscope that permits the visual inspection of the fetus and of the color and amount of the amniotic fluid. Thus, **amnioscopy.**

Amniota *Vertebrate Zoology.* a collective term for higher land-living vertebrates whose embryos have an amnion, chorion, and allantois; i.e., reptiles, birds, and mammals.

amniote *Vertebrate Zoology.* any animal belonging to the Amniota; that is, a land-living vertebrate having an amnion during the embryonic stage.

amniotic [am´nē ät´ik] *Developmental Biology.* of or relating to the amnion.

amniotic cavity *Developmental Biology.* the area between the embryo and the amniotic membrane, containing the amniotic fluid.

amniotic fluid *Developmental Biology.* the watery fluid within the amnion, in which the embryo or fetus is suspended; it serves to cushion the fetus and also acts a medium of chemical exchange.

amniotic membrane *Vertebrate Zoology.* see AMNION, def. 1.

amniotic sac *Vertebrate Zoology.* see AMNION, def. 2.

amniotomy *Medicine.* the deliberate rupture of the fetal membranes to induce labor.

amobarbital *Pharmacology.* $C_{11}H_{18}N_2O_3$, a white crystalline powder, slightly soluble in water; a barbiturate used orally as a sedative and hypnotic; it may be addictive if abused.

A mode *Acoustics.* a method of ultrasonic medical tomography in which ultrasonic waves are transmitted on a single axial focus for specific applications, such as measurement of corneal thickness or brain development during pregnancy.

amoeba [ə mē´bə] *plural,* **amoebas** or **amoebae** [ə mē´bē] see AMEBA and related words.

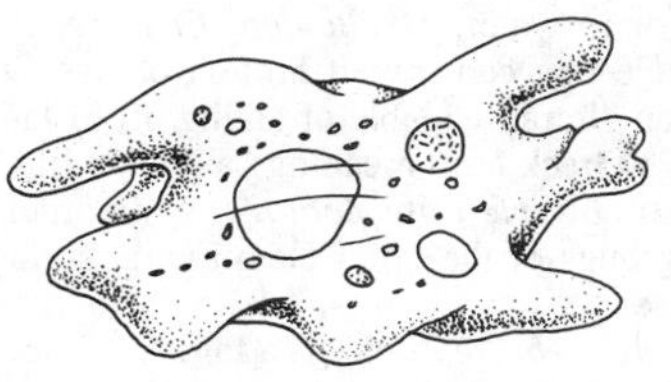

Amoeba

Amoeba *Invertebrate Zoology.* the genus of naked protozoans of the rhizopod order Amoebida. Asymmetrical with constantly changing body shape, they move by extruding pseudopodia and usually reproduce by cell fission.

amoebic see AMEBIC.

Amoebida *Invertebrate Zoology.* an order of naked protozoans belonging to the class Rhizopodea, containing all the amebas that are parasitic in animals.

Amoebidiaceae *Mycology.* a single family of fungi belonging to the order Amoebidiales, occurring in ponds and pools and on freshwater arthropods, cladocerans, copepods, amphipods, or mosquito larvae; it obtains its nutrients from water, rather than from its host.

Amoebidiales *Mycology.* an order of fungi belonging to the class Trichomycetes, classified as protozoans, that occur in ponds and pools and on freshwater arthropods and insect larvae.

Amoebobacter *Bacteriology.* a genus of purple sulfur bacteria of the family Chromatiaceae, consisting of spheroid, nonmotile photosynthetic cells that contain gas vacuoles and fix carbon dioxide in the presence of hydrogen sulfide.

amoebocyte see AMEBOCYTE.

amoeboid *Geology.* of a fold or other structure, having a shallow dip and no definite shape.

amoeboid glacier *Hydrology.* a glacier that is connected with its snowfield for only a part of the year.

Amoebophryaceae *Botany.* a monogeneric family of marine dinoflagellates of the order Syndiniales that are intracellular parasites on other marine dinoflagellates and other marine organisms, sometimes even attacking members of their own family.

Amor [ə môr´] *Astronomy.* asteroid 1221, discovered in 1932. It measures about half a kilometer in diameter and is located between Mars and Venus. Amor is the prototype of the class of Amor asteroids. (From *Amor* or Cupid, the Roman god of love and son of Mars and Venus.)

Amor asteroid *Astronomy.* any of a class of asteroids whose orbit has a perihelion between 1.0 and 1.3 astronomical units, crossing the orbit of Mars but not that of Earth. Also, **Amor object.**

amorph [ā´môrf; ä´môrf] *Genetics.* a gene that has no measurable effect; an inactive gene. Thus, **amorphic.**

amorphic allele *Genetics.* an allele that is not genetically active.

amorphism *Biology.* the fact or condition of being amorphous.

amorphization *Biology.* the process of making or becoming amorphous.

Amorphosporangium *Bacteriology.* a genus of rod-shaped sporogenous bacteria of the order Actinomycetales, found in leaf litter and other soil habitats; now usually classifed as the genus *Actinoplanes.*

Amorphothecaceae *Mycology.* a family of fungi belonging to the order Eurotiales, best known for its asexual state, *Hormconis resinae.*

amorphous not having a definite form or shape; specific uses include: *Biology.* not having a clearly defined structure or organization, such as amebas or certain bacteria. *Geology.* occurring in a continuous mass, without division into specific parts. *Crystallography.* not having crystals; not crystalline. *Materials Science.* describing a highly viscous liquid that does not possess the crystalline order normally characteristic of the solid state; e.g., glass.

amorphous frost *Hydrology.* a frost that lacks crystalline structure.

amorphous laser see GLASS LASER.

amorphous memory array *Computer Technology.* a physical unit of computer memory composed of a solid noncrystalline material.

amorphous peat *Geology.* fine-grained peat in which the original plant structures have been destroyed as a result of the decay of cellulose material.

amorphous ribbon *Metallurgy.* a narrow strip of a material that has no crystalline structure.

amorphous semiconductor *Solid-State Physics.* a semiconductor material that lacks crystalline (regular) structure, and tends to have lower electrical conductivities than crystalline semiconductors.

amorphous sky *Meteorology.* a state of the sky characterized by an abundance of clouds lacking any distinctive form, usually accompanied by precipitation falling from a higher cloud layer.

amorphous snow *Hydrology.* snow having an irregular crystalline structure.

amorphous solid *Materials Science.* a noncrystalline solid material whose microscopic arrangement exhibits no periodicity or long-range order, and whose shear viscosity is large enough that macroscopic shapes are maintained for relatively long periods.

amortizement *Architecture.* the sloping top of a buttress, pillar, or other such projecting feature.

amosite *Mineralogy.* a commercial name for iron-rich asbestiform varieties of the amphiboles grunerite, cummingtonite, anthophyllite, and ferro-gedrite. (From an acronym for Asbestos Mine of South Africa + the suffix *-ite.*)

amount of substance *Science.* a term for the amount of a given component that is present in a system, solution, medical preparation, or the like.

amoxicillin [ə mäks´i sil´in] *Pharmacology.* $C_{16}H_{19}N_3O_5S$, a type of oral penicillin, a derivative of ampicillin, used against a variety of infections caused by Gram-negative and Gram-positive bacteria.

AMP adenosine monophosphate.

amp or **amp.** **1.** ampere, amperes. **2.** amperage. **3.** amplifier.

ampacity *Electricity.* current-carrying capacity expressed in amperes.

ampangabeite see SAMARSKITE.

AMPD see 2-AMINO-2-METHYL-1,3-PROPANEDIOL.

Ampeliscidae *Invertebrate Zoology.* a family of tube-building amphipods in the crustacean suborder Gammaridea, with laterally compressed bodies.

ampelite *Petrology.* a former term for a black carbonaceous or bituminous shale.

amperage *Electricity.* the strength of an electric current expressed in amperes.

Ampère, André Marie 1775–1836, French mathematician and physicist; discovered several laws of electromagnetism.

ampere [am´pēr] *Metrology.* the basic unit of electric current in the meter-kilogram-second system; equivalent to one coulomb per second. One ampere is the current that, if held constant in two parallel conductors of infinite length at a distance one meter apart in a vacuum, will produce a force of 2×10^{-7} newtons per meter of length. (From André Marie *Ampère*.)

Ampère currents *Electromagnetism.* theoretical molecular ring currents that are thought to be responsible for magnetic phenomena as well as the apparent nonexistence of magnetic monopoles.

ampere-hour *Electricity.* the quantity of electricity that passes through a circuit in one hour when the rate of flow is one ampere, equal to 3600 coulombs.

ampere-hour capacity *Electricity.* the number of ampere-hours that can be delivered by a battery under specified conditions.

ampere-hour efficiency *Electricity.* a measure of the efficiency of a battery by comparing the ampere-hours delivered to the ampere-hours input during charge.

ampere-hour meter *Engineering.* a device that records the ampere-hours (current multiplied by time) for a given circuit at a given point.

Ampère law see AMPERE'S LAW.

ampere meter squared *Electromagnetism.* a unit of the SI or meter-kilogram-second system for electromagnetic moment.

ampere-minute *Electricity.* the quantity of electricity that passes through a circuit in one minute when the rate of flow is one ampere, equal to 60 coulombs.

ampere per meter *Metrology.* a unit of the SI or meter-kilogram-second system for magnetic field strength or magnetization.

ampere per square inch *Metrology.* a nonstandard unit of current density.

ampere per square meter *Metrology.* the standard international unit of current density.

Ampère rule see AMPÈRE'S RULE.

Ampère's law *Electromagnetism.* a formula that gives the magnetic effects at a certain point due to current elements, usually expressed in terms of a line integral surrounding a conductor.

ampere square meter per joule second *Metrology.* the SI or meter-kilogram-second unit for gyromagnetic ratio.

Ampère's rule *Electromagnetism.* a so-called "right-hand rule" stating that the direction of the magnetic field lines surrounding a conductor is counterclockwise when viewing the conductor with the current flow coming toward the observer and clockwise if the current is moving away from the observer.

Ampère's theorem *Electromagnetism.* a theorem stating that the magnetic field generated by a current in a conductor is equivalent to that which is generated by a magnetic shell whose boundary coincides with the conductor's surface and whose strength is equal to that of the current.

ampere-turn *Electromagnetism.* **1.** a unit of the SI or meter-kilogram-second system for the magnetomotive force of a circuit, equivalent to the force produced by a current of one ampere flowing through one complete loop or turn of a conducting coil. **2.** the loop or turn itself.

amperometric *Physical Chemistry.* relating to the measurement of the strength of an electric current by chemical analysis.

amperometric titration *Physical Chemistry.* the use of current to measure the change in the concentration of an analyte; generally the potential of the indicator eletrode is held fixed with respect to a reference electrode.

amperometry *Physical Chemistry.* any type of chemical analysis that involves measuring electrical currents; generally refers to a fixed applied potential.

amph- a prefix meaning "both" or "on both sides," as in *amphoteric.*

Ampharetidae *Invertebrate Zoology.* a family of tube-building worms belonging to the class Polychaeta, with retractable buccal (mouth area) tentacles.

Ampharetinae *Invertebrate Zoology.* a subfamily of polychaete worms of the family Ampharetidae.

amphetamine [am fet´ə mēn´; am fet´ə min] *Pharmacology.* **1.** $C_9H_{13}N$, a colorless, volatile liquid having a characteristic strong odor and burning taste; soluble in alcohol and slightly soluble in water; a nervous system stimulant that depresses blood pressure, appetite, and breathing. Abuse of this substance leads to severe side effects including dependence and possibly induced psychosis. **2.** any of various related compounds having similar effects, such as methamphetamine or dextroamphetamine.

amphi- a prefix meaning "both" or "on both sides," as in *amphimorphic.*

amphiarthrosis *Anatomy.* a cartilaginous joint in which the bony surfaces are joined by fibrocartilage.

amphiaster *Cell Biology.* the figure of two asters joined by a spindle that forms during cell division. *Invertebrate Zoology.* a type of sponge spicule that is stellate at both ends.

Amphibia *Vertebrate Zoology.* one of the eight classes of vertebrate animals, characterized by moist, smooth skin, cold-bloodedness, gills at some stage of development, and usually an initial stage of life as an aquatic larva and metamorphosis into a lunged adult that lives wholly or partly on land. Frogs, toads, salamanders, and newts are members of this class.

amphibian

amphibian [am fib´ ē ən] *Vertebrate Zoology.* **1.** a member of the class Amphibia. **2.** of or relating to this class. **3.** any other animal capable of living both on land and in water, such as a seal. *Botany.* a plant that grows on land or in water. *Mechanical Engineering.* a tank, truck, or other such vehicle that can travel through or over water as well as on land. *Aviation.* an aircraft capable of taking off or landing either on land or water. (From a Greek term meaning "a double life.")

Amphibicorisae *Invertebrate Zoology.* a suborder of semiaquatic or shore-inhabiting bugs of the order Hemiiptera; surface water bugs with exposed antennae.

Amphibioidei *Invertebrate Zoology.* a family of tapeworms in the order Cyclophyllidea.

amphibiotic *Zoology.* describing an animal that as a larva has gills and lives on water, and as an adult has lungs and lives on land.

amphibious *Biology.* able to live or grow both on land and in water. *Mechanical Engineering.* of or relating to crafts, vehicles, or equipment that may be used on land and on or in water. *Military Science.* **1.** of or relating to amphibious warfare. Thus, **amphibious force(s), amphibious operation.** **2.** specially equipped or designed for amphibious warfare. Thus, **amphibious tank.**

amphibious assault *Military Science.* an amphibious operation to land troops on an enemy beach.

amphibious assault ship

amphibious assault ship *Naval Architecture.* a ship designed to carry helicopters and helicopter-borne troops and equipment for sea-to-shore assault operations.

amphibious mine *Ordnance.* a mine designed especially for use against amphibious operations, as by damaging landing craft or temporary bridge structures.

amphibious tractor see AMTRAC.

amphibious warfare *Military Science.* the use of combined naval, air, and ground forces to achieve an objective on land, as by landing troops and equipment from naval vessels onto an enemy shore.

amphiblastic *Developmental Biology.* describing the complete but unequal cleavage of a telolecithal egg. Thus, **amphiblastic cleavage.**

amphiblastula *Developmental Biology.* a blastula in which there is complete but unequal cleavage, so that the cells of one hemisphere differ from those of the other. *Invertebrate Zoology.* the free-swimming larva of many sponges; a hollow ball composed of anterior flagellate cells and posterior nonflagellate cells.

amphibole [am´fə bōl´] *Minerology.* **1.** a large group of ferromagnesian silicate minerals that are important in rock formation, usually monoclinic but sometimes orthorhombic, and characterized by prismatic columnar or fibrous crystals with a structure of crosslinked double chains of tetrahedra. Also, **amphibole group. 2.** any mineral that belongs to this group, such as horneblende.

amphibolic *Geology.* of or relating to the amphibole group of minerals. *Zoology.* capable of being turned backward or forward, as the outer toe of certain birds. *Medicine.* of uncertain prognosis; doubtful; used to describe a stage of a disease **(amphibolic stage).**

amphibolic pathway *Biochemistry.* the central metabolic pathway, involved in both anabolic and catabolic processes.

Amphibolidae *Invertebrate Zoology.* a family of gastropod mollusks in the subclass Pulmonata; Amphibola is the only pulmonate with an operculum.

amphibolite *Petrology.* a crystalloblastic, coarse-grained metamorphic rock with fine prismatic cleavage qualities typically composed of an amphibole, such as hornblende, and a plagioclase, such as andesine, often with small amounts of quartz.

amphibolite facies *Petrology.* rocks produced by medium- to high-grade regional metamorphism.

amphibolization *Petrology.* the transformation of pyroxene and other minerals into amphibole in an igneous rock.

amphicarpic *Botany.* **1.** a plant that bears two kinds of fruit, differing either in shape or in the time of maturing. **2.** of or relating to such a plant. Also, **amphicarpous.**

Amphichelydia *Paleontology.* a former grouping at the suborder level of cryptodires and other primitive turtles; no longer in use. Also, **Amphichelida.**

amphichroic *Chemistry.* exhibiting either of two colors under varying conditions; e.g., litmus in reaction to an acid or an alkali.

amphichromatic *Chemistry.* see AMPHICHROIC. *Botany.* of an individual plant, producing flowers of different colors on the same stem. Thus, **amphichromatism.**

amphichrome *Botany.* an individual plant that produces flowers of different colors on the same stem.

amphiclinous *Genetics.* of a hybrid, exhibiting more of the characteristics of one parent than the other.

Amphicoela *Vertebrate Zoology.* a small amphibian order in the subclass Anura; it contains a single family, Ascaphidae, which are primitive toads with amphicoelous vertebrae.

amphicoelous [am´fə se´lis] *Vertebrate Zoology.* of or relating to a vertebra that is concave at both ends, as is the case in most fishes and certain reptiles.

amphicribral *Botany.* having a concentric vascular bundle in which phloem surrounds the central strand of xylem. Thus, **amphicribral bundle.**

amphicryptophyte *Botany.* a marsh plant with amphibious vegetative parts.

amphicyonid *Paleontology.* one of the Amphicyonidae, large dogs that appeared in the Oligocene and achieved a Holarctic distribution in the Miocene. The North American Daphoenodon was a typical amphicyonid, about five feet long.

Amphicyonidae *Paleontology.* a widespread and abundant family of carnivorous predators in the superfamily Arctoidea. Formerly placed in the Canidae, the amphicyonids are among the few post-Eocene carnivores that cannot be classified in living families; there is a possible relationship to the Ursidae.

amphicyte *Neurology.* a neuroglial cell that encapsulates the dorsal root ganglion cells. Also, CAPSULE CELL.

amphicytula *Developmental Biology.* a fertilized ovum that has the property of cleaving unequally by holoblastic means.

amphid *Invertebrate Zoology.* one of a pair of circular pits or tubes situated laterally at the anterior end of aquatic nematode worms, thought to be sense organs.

amphidetic *Invertebrate Zoology.* of a bivalve ligament, extending both before and behind the beak.

amphidiploid see ALLOTETRAPLOID.

Amphidiscophora *Invertebrate Zoology.* an order of deepwater glass sponges of the class Hyalospongia, with an anchoring root tuft and spicules having a stellate disk at both ends.

Amphidiscosa *Invertebrate Zoology.* an order or subclass of glass sponges in Amphidiscophora; in some classifications, equivalent to Amphidiscophora.

amphidromic point *Oceanography.* a point at which the variation of the tides is zero and from which the cotidal lines radiate, progressing through all phases of the tidal cycle. Similarly, **amphidromic region.**

amphigamy see AMPHIGENESIS.

amphigastrium *Botany.* a small, rudimentary leaflike growth on the underside of the stem of certain liverworts.

amphigastrula *Developmental Biology.* a gastrula composed of cells of unequal size in its upper and lower hemispheres.

amphigean *Ecology.* relating to or occurring in both the Old World and the New World; found around the earth. Also, **amphigeic.**

amphigene see LEUCITE.

amphigenesis *Biology.* normal sexual reproduction; the union of male and female gametes.

amphigenetic *Biology.* of or relating to sexual reproduction.

amphigenous *Microbiology.* of certain microorganisms, occurring on both surfaces or sides, or all over, a given structure.

amphigonadism see HERMAPHRODITISM.

amphigony see AMPHIGENESIS.

amphikaryon *Nucleonics.* a nucleus having the diploid number of chromosomes.

Amphilestidae *Paleontology.* a family of primitive Mesozoic mammals in the extinct order Triconodonta; represented by about five genera between the middle Jurassic and lower Cretaceous; its exact affinities are uncertain.

Amphilinidea *Invertebrate Zoology.* a family of cestodarian tapeworms, with flat leaflike bodies and an anterior extensible proboscis, inhabiting the coelom of sturgeon and other fishes.

Amphimerycidae *Paleontology.* an extinct family of artiodactyl mammals in the suborder Tylopoda, possibly related to the xiphodontids or protoceratids; known only from the Eocene and Oligocene of Europe.

Amphimerycoidea *Paleontology.* a former grouping of early Cenozoic tylopod ruminants at the superfamily level.

amphimixis *Biology.* the union of gametes of two organisms in sexual reproduction. *Psychology.* the union of oral, anal, and genital impulses in the development of sexuality.

Amphimonadaceae *Botany.* a family of colorless, mostly freshwater flagellate algae of the order Isochrysidales, characterized by flagella of equal length, the absence of hairs or other appendages on the flagella, and often thick organic scales on the body.

Amphimonadidae *Invertebrate Zoology.* a family of zooflagellate protozoans in the order Kinetoplastida, with a kinetoplast containing DNA.

amphimorphic *Geology.* **1.** any rock or mineral that was formed by more than one geologic process or factor, such as a deposit formed by a mineral-bearing thermal spring. **2.** relating to such a rock or mineral.

Amphineura *Invertebrate Zoology.* a class of marine chitons and related forms in the phylum Mollusca. The foot, if present, is broad and flat, the mantle is undivided, and the shell, if present, has eight plates.

Amphinomidae *Invertebrate Zoology.* a family of stinging or fire worms of the subclass Errantia, with brittle, venomous setae.

Amphinomorpha *Invertebrate Zoology.* a group name for three families of polychaete worms of the subclass Errantia: Amphenomidae, Euphrosinidae, and Spintheridae.

amphioxus *Zoology.* a popular name for the Branchiostoma, a family of small lance-shaped, translucent, boneless, fishlike animals.

amphipathic *Chemistry.* describing a molecule that has two groups with characteristically different properties, such as a detergent that has both a water-soluble (hydrophilic) polar end and a water-insoluble (hydrophobic) nonpolar end. Thus, **amphipath, amphipathic molecule.**

amphiphile *Chemistry.* a substance having amphiphilic properties.

amphiphilic *Chemistry.* describing a substance containing both polar, water-soluble groups and nonpolar, water-insoluble groups.

amphiphloic *Botany.* of or relating to a plant in which the phloem appears on both sides of the xylem.

amphiphyte *Botany.* a plant able to live either in water or in damp soil above the water level.

Amphipithecus *Paleontology.* an extinct genus of early primates in the suborder Anthropoidea, known from teeth and jaw fragments of the late Eocene.

amphiplatyan *Anatomy.* of a vertebra, having flat articular surfaces at each end.

amphiploid see ALLOPOLYPLOID.

amphipneustic *Vertebrate Zoology.* having both gills and lungs simultaneously throughout life, as do some amphibians.

amphipod *Invertebrate Zoology.* **1.** one of the Amphipoda. **2.** of or relating to the Amphipoda. **3.** having both walking and swimming appendages. Thus, **amphipodous.**

Amphipoda *Invertebrate Zoology.* a large order of beach hoppers, sand fleas, parasitic whale lice, and others in the crustacean division Peracarida, that have laterally compressed bodies with no carapace, biramous first antennae, gills on their thoracic limbs, and stalkless eyes.

amphiprostyle *Architecture.* **1.** having columns at the front and back but not on the sides. **2.** a building in this style. Also, **amphiprostylar.**

amphiprotic see AMPHOTERIC.

amphisarca *Botany.* a berrylike succulent fruit surrounded by a woody or crustaceous layer or rind forming a gourd.

amphisbaena *Vertebrate Zoology.* a worm lizard of the genus *Amphibaena* of the family Amphisbaenidae, having a head and tail that are not easily distinguishable and able to move either forward or backward. Thus, **amphisbaenic.** (From the name of a serpent in classical mythology that had a head at each end of its body.)

Amphisbaenidae *Vertebrate Zoology.* the worm lizards, a family of tropical, wormlike, burrowing, and usually legless lizards in the suborder Sauria.

Amphisoleniaceae *Botany.* a family of marine algae of the order Dinophysiales, characterized by laterally compressed cells and occurring in tropical oceanic plankton worldwide.

Amphisopidae *Invertebrate Zoology.* a family of isopod crustaceans in the suborder Phreatoicoidea.

amphispore *Mycology.* a type of rapidly germinating spore (urediospore), characterized by its thick, sometimes pigmented walls and by its ability to become a resting spore.

Amphissitidae *Paleontology.* a family of palaeocopid ostracods in the extinct suborder Beyrichicopina; known only from the Carboniferous of North America.

amphistome *Invertebrate Zoology.* **1.** one of the suborder Amphistomata, parasitic flukes in the order Digenea; somewhat conical in form with an anterior oral sucker and ventral posterior sucker. **2.** having a sucker at each end of the body. Thus, **amphistomous.**

amphistyle *Architecture.* **1.** having columns in porticoes either at both ends or at both sides and across the length of both sides or ends. **2.** a classical temple or other building having this style. Also, **amphistylar.**

amphistylic *Vertebrate Zoology.* having the upper jaw attached to the braincase and the hyomandibular cartilage, as in a small number of sharks.

amphitheater

amphitheater *Architecture.* **1.** an oval, circular, or semicircular auditorium with tiers of seats rising from a central open area. **2.** an outdoor theater in this style, especially in ancient Greece. *Geology.* a landform suggesting an amphitheater, having a relatively flat floor surrounding by steeply sloping sides. Also, **amphitheatre.**

amphithecium *Botany.* the outer layer of cells in the spore case of a moss.

Amphitheriidae *Paleontology.* a monogeneric family of primitive eupantotheres; one of the earliest mammals, known only from the middle Jurassic *Amphitherium*; in the line of evolution leading to the placentals and marsupials.

amphitopic *Ecology.* of a population or species, showing a tolerance to a wide variety of habitat and environmental conditions

amphitrichous *Microbiology.* having a single flagellum at each end of the cell, as do certain bacteria. Also, **amphitrichate.**

Amphitrite [am′ fə trī′ tē] *Astronomy.* asteroid 29, discovered in 1854; it has a diameter of 219 kilometers and belongs to type S.

amphitrite *Oceanography.* a type of inflatable ship used in diving operations as a tender. *Invertebrate Zoology.* a marine annelid worm characterized by branching gills and by many tentacles anterior to the mouth.

Amphitritinae *Invertebrate Zoology.* a subfamily of tube-building or sedentary burrowing worms in the family Terebellidae.

amphitropous *Botany.* having an ovule curved upon itself so that the ends are brought near to each other with the hilum in the middle.

Amphiumidae *Vertebrate Zoology.* a small family of eel-like salamanders in the suborder Salaman Salamandroidea, having tiny, weak limbs and both gills and lungs in the adult stage.

Amphiura *Invertebrate Zoology.* a genus of burrowing asterozoan echinoderms in the class Ophiuroidea, extant since the Miocene and still widespread.

amphivasal *Botany.* a concentric vascular bundle in which the xylem surrounds the phloem.

Amphizoidae *Invertebrate Zoology.* a small family of trout stream beetles in the order Coleoptera; aquatic insects with the unusual characteristic of not being able to swim.

amphogenic *Biology.* **1.** producing both male and female offspring. **2.** producing equal numbers of male and female offspring. Thus, **amphogenesis.**

ampholyte *Chemistry.* an electrolyte that acts as an acid or a base, depending on the acidity or basicity of the medium in which it is present.

ampholytic *Chemistry.* relating to or being an ampholyte.

ampholytic detergent *Chemistry.* a detergent that is both a cation (in acidic solutions) and an anion (in basic solutions), combining detergent and disinfectant qualities.

amphora [am fôr′ə] *plural* **amphorae** or **amphoras.** *Archaeology.* a jar having an oval body, two handles, and a narrow neck and mouth; widely found in ancient Greek and Roman sites.

Amphora *Botany.* a large, diverse genus of diatoms in the family Cymbellaceae, found primarily in marine areas ranging from deep waters to intertidal sands.

Amphoracrinus *Paleontology.* a genus of Carboniferous crinoids in the family Amphoracrinidae; distinguished by its flat, truncate base.

amphoric *Design Engineering.* having the shape of an amphora, or a similar tapering, narrow-necked shape. *Acoustics.* having a deep, hollow sound, as is produced by blowing across the mouth of a narrow-necked bottle or vase.

amphoric breathing *Medicine.* a condition in which the breath has an abnormal hollow, blowing sound, as perceived by means of a stethoscope; this sometimes indicates a cavity in the lung. Also, **amphoric breath sound.**

Amphoriscidae *Invertebrate Zoology.* a family of calcareous sponges in the order Sycettida.

amphoteric *Chemistry.* describing a substance that has both basic and acidic properties, such as water.

amphotericin [am′fə ter′ə sin] *Pharmacology.* either of two antifungal antibiotics, amphotericin A or amphotericin B.

amphotericin A *Pharmacology.* the less active form of the two complex antibiotics produced by *Streptomyces nodosus;* not used clinically.

amphotericin B *Pharmacology.* $C_{47}H_{73}NO_{17}$, a yellow to orange semicrystalline powder, insoluble in water; a polyene antibiotic produced by the bacterium *Streptomyces nodosus* and effective against a wide range of fungi; used systemically or topically in the treatment of fungal infections.

amphoterism *Chemistry.* the property of acting as either a base or an acid.

amphoterite *Geology.* a stony meteorite containing mainly pyroxene and olivine along with small amounts of oligoclase and nickel-rich iron.

amphoterous see AMPHOTERIC.

amphotony *Neurology.* a condition of increased tone in both the sympathetic and parasympathetic nervous systems.

amphotropic virus *Virology.* a retrovirus that can replicate both in its host of origin and in foreign host cells.

amp-hr ampere-hour.

ampicillin *Pharmacology.* $C_{16}H_{19}N_3O_4S$, a white, bitter crystalline powder, soluble in water; a semisynthetic penicillin used to combat a wide variety of Gram-negative and Gram-positive bacteria, especially those causing urinary and bronchial infections.

amplectant *Botany.* twisting or twining for support.

amplexicaul *Botany.* embracing or surrounding the stem, as the base of some leaves.

amplexus *Vertebrate Zoology.* the copulatory embrace of amphibians, especially frogs and toads. *Botany.* the overlap of the edges of a leaf on the edges of the leaf above it.

amplidyne *Electricity.* a rotating amplifier often used as a power amplification stage in electromechanical systems; a small increase in power input yields a large boost in power output.

amplification the fact of amplifying or being amplified; specific uses include: *Electricity.* an increase in the strength of a current, signal, or other electrical effect. *Physics.* an increase in the magnitude of some physical quantity or force. *Genetics.* **1.** the production of many DNA copies from one master region of DNA. **2.** the replication in bulk of a gene library. **3.** a process in which the amount of plasmid DNA is increased in proportion to the amount of bacterial DNA by means of treatment with certain antibiotics. *Psychology.* **1.** a technique of psychotherapy in which the client is instructed to exaggerate a particular feeling or behavior in order to become more conscious of it. **2.** a method of dream analysis that focuses on one element of a dream and the responses elicited by it.

amplification factor *Electronics.* in electron tubes, the incremental ratio of the plate potential to the grid potential with all other voltages and the plate current held constant; usually represented by the symbol μ. Also, *μ* FACTOR.

amplification noise *Electronics.* an unwanted signal introduced into an amplifier input signal by the amplifier; the unwanted signal is omnipresent within the amplifier.

amplified back bias *Electronics.* the feeding back of a falling voltage produced across a quick response circuit into the previous stage of an amplifier.

amplifier *Electronics.* a device that magnifies the strength of a relatively weak input signal without altering the characteristics (such as waveform) of that signal. Amplifiers are used in radio and television receivers and in stereophonic sound systems, often as a separate component of such systems. *Engineering.* any of various other devices used to increase the magnitude of some physical or mechanical effect.

amplify *Electronics.* to process an input signal so that the output signal has characteristics identical to those of the input, except that the signal intensity is increased.

amplifying delay line *Electronics.* in pulse compression, a delay line that amplifies signals of very high frequency.

amplitron *Electronics.* in microwave signal amplification, a cross-field amplifier with a reentrant electron beam and a circular spinning nonreentrant network.

amplitude the size, range, or extent of something; specific uses include: *Physics.* **1.** in any periodically varying function, such as a wave or vibration, the maximum absolute variation of the function from a given reference point through one cycle. **2.** specifically, the angle between the zero position and the peak position in the arc of a pendulum. *Electricity.* the amount that current or voltage varies from zero or from an average value. *Navigation.* the angle, measured north or south, between the prime vertical and some other point (or vertical circle) on the celestial sphere. *Astronomy.* the difference in magnitudes between the brightness of a variable star at maximum and minimum light. *Ecology.* the range of tolerance of a specific organism or species for a variety of environmental conditions. *Geology.* see AXIAL PLANE SEPARATION. *Mathematics.* **1.** the angle between a vector representing a complex number in an Argand diagram and the positive real axis. **2.** one-half of the distance between the greatest and least values of a periodic function. **3.** the angle component of the polar representation $re^{i\theta}$ of a complex number.

amplitude contrast *Optics.* in a transmission electron microscope, an imaging mode in which contrast is produced by imaging either the diffracted or transmitted electron beam.

amplitude discriminator see PULSE-HEIGHT DISCRIMINATOR.

amplitude fading *Telecommunications.* a type of signal variation in which the amplitudes of the frequency components of a radio wave are attenuated uniformly.

amplitude (frequency) distortion see FREQUENCY DISTORTION.

amplitude gate *Electronics.* a network or transducer that passes only that part of a wave signal which is between two close amplitude boundaries.

amplitude level *Physics.* the natural logarithm of the ratio of two amplitudes.

amplitude limiter see LIMITER.

amplitude-modulated indicator *Engineering.* a general category of radar indicators, in which an echo from a target is evident when the sweep of the electron beam is deflected from the baseline either vertically or horizontally.

amplitude modulation *Electronics.* the deliberate processing of a carrier signal so that its amplitude varies in accordance with the level of the modulating signal; used in ordinary radio broadcasting (AM radio) and in transmitting the video portion of a television signal. Also, AM.

amplitude-modulation noise *Telecommunications.* any noise signal produced by unwanted amplitude variations of a radio signal.

amplitude-modulation radio see AM RADIO.

amplitude modulator *Physics.* a device that is capable of periodically changing the amplitude of a signal of higher frequency.

amplitude noise *Telecommunications.* fluctuations in the amplitude of a signal that is returned by a target after a radar reflection.

amplitude of accommodation *Physiology.* the difference between the refractory power of the eye when it is adjusted for far-point vision and for near-point vision.

amplitude of convergence *Physiology.* the difference in the power required to turn the eyes from their far point of convergence to their near point.

amplitude peak *Physics.* the point of maximum absolute variation, either positive or negative, of any periodically varying function.

amplitude resonance *Physics.* the resonance that produces the maximum amplitude in a resonant system.

amplitude response *Electronics.* the gain or attenuation of a circuit as a function of the signal frequency.

amplitude selector see PULSE-HEIGHT SELECTOR.

amplitude separator *Electronics.* **1.** a circuit that isolates the portion of a waveform that is either above, below, or between given values. **2.** a circuit that separates control signals from video signals in a television receiver.

amplitude shift key(ing) *Telecommunications.* data signals that produce a number of different amplitude levels of a sine-wave carrier.

amplitude splitting *Optics.* a process used in studying and measuring radiation wherein a surface reflects part of a beam while transmitting another part of the beam and the parts are recombined to form interference.

amplitude-suppression ratio *Electronics.* in an FM receiver, the ratio between the amplitude variations in the input and output signals.

amplitude-vs-frequency distortion *Electronics.* the distortion of a waveform due to the variation of gain (or attenuation) with frequency.

ampule *Medicine.* a small glass or plastic container sealed to maintain its contents in a sterile condition; typically used to hold a specific dose of a drug or vaccine in solution form for injection. Also, **ampul, ampoule.**

ampulia *Botany.* a small membranous float attached to the leaves of some aquatic plants.

Ampulicidae *Invertebrate Zoology.* a family of wasps in the order Hymenoptera; black with a fairly elongated neck; predatory on roaches.

ampulla [am pul′ə; am poo′lə] *plural,* **ampullae.** *Anatomy.* **1.** a rounded, flasklike dilation of a duct, canal, or other tubular structure. **2.** the enlarged portion at the end of the semicircular canal of the ear. Also, AMPULLA OSSEA. **3.** see AMPULLA OF VATER. *Invertebrate Zoology.* **1.** an expandable sac at the base of the tube foot in some echinoderms, such as sea stars. **2.** either of a pair of small sacs forming part of the aboral sense organs in Ctenophora. **3.** a pit of the skeleton in Hydrocorallina. *Botany.* a flask-shaped organ or bladder on some aquatic plants.

ampullaceous see AMPULLARY.

ampulla of Lorenzini *Vertebrate Zoology.* a system of minute gelatinous canals in the head area of an elasmobranch fish, thought to function in sensing temperature. (From Stefano *Lorenzini,* a 17th-century Italian physician who described them.)

ampulla of Vater *Anatomy.* a dilation formed at the point of convergence of the pancreatic duct and the common bile duct as they enter the duodenum through the duodenal papilla. Also, **ampulla hepatopancreatica.** (From Abraham *Vater,* 1684–1751, German anatomist.)

ampulla ossea *Anatomy.* see AMPULLA, def. 2.

Ampullariella *Bacteriology.* a genus of bacteria of the family Actinoplanaceae that are found in soil and freshwater habitats; characterized both by cell wall chemotype and by the absence of true aerial mycelia.

ampullary *Biology.* **1.** of or relating to an ampulla. **2.** having the shape of an ampulla; flask-shaped. Also, **ampullar, ampullate.**

ampullary organ *Physiology.* a swelling or bulging at the end of each semicircular canal in the ear, where special hair cells are located that can detect changes in bodily acceleration and deceleration.

amputate *Surgery.* to carry out the procedure of amputation; cut off all or part of a limb or digit.

amputation *Medicine.* **1.** the surgical removal of all or part of a limb, digit, or other appendage of the body, as in the case of a severe infection, injury, or tumor. **2.** a similar spontaneouus removal as the result of an accident.

amputee *Medicine.* a person who has had all or part of a limb removed through surgical amputation.

ampyrone *Pharmacology.* $C_{11}H_{13}N_3O$, a drug used to treat fever and pain. Also, 4-AMINOANTIPYRINE.

AMRAAM *Ordnance.* an advanced U.S. medium-range air-to-air missile powered by an internal rocket motor and equipped with long-range infrared sensors, midcourse inertial guidance, and an active seeker for terminal homing; it carries a conventional warhead of less than 50 pounds at a speed of Mach 4 and a range of 30 miles. (An acronym for advanced medium-range air-to-air missile.)

AM radio *Telecommunications.* **1.** a system of radio communication that transmits information by means of amplitude modulation (deliberate variation of the amplitude of a carrier signal in accordance with the amplitude of the modulating signal). **2.** a radio using this system, or a receiver including such a system.

AMS *Aviation.* the airport code for Amsterdam (Schiphol), The Netherlands.

AMS automatic music search.

AMS or **A.M.S.** American Mathematical Society; American Meteorological Society.

ams amount of substance.

AM signature *Telecommunications.* a graphic delineation of the identifying characteristics of an AM radio signal.

Am star *Astrophysics.* a type A star that shows abnormally strong lines of metals in its spectrum; many are spectroscopic binaries.

amtrac *Ordnance.* a vehicle used for the landing of ships and cargo in an amphibious assault and for limited movement over water or land. (An acronym for amphibious tractor.)

AMU air mileage unit.

amu or **AMU** ATOMIC MASS UNIT.

Amur *Geography.* a river that flows about 1800 miles across northeastern Asia to the Sea of Okhotsk, mostly along the Sino-Soviet border. Also, HEILONG.

amusia *Neurology.* an inability to recognize or produce musical tones, usually as a result of a cerebral lesion. **Sensory amusia** is the inability to comprehend such sounds, and **motor amusia** is the inability to reproduce them.

AMV avian myeloblastosis virus.

amychophobia *Psychology.* an irrational fear of being scratched, as by the claws of a cat.

amyelia *Medicine.* the congenital absence of the spinal cord. Thus, **amyelic.**

amyelinic *Neurology.* without myelin; having no myelin sheath as a result of injury or developmental failure.

amygdala *Anatomy.* **1.** a general term for an almond-shaped structure or body. **2.** specifically, an almond-shaped body in the lateral ventricle of the brain. *Botany.* a seed of the almond tree, *Prunus amygdalus.*

amygdalaceous *Botany.* having to do with the almond or related fruits such as the peach or cherry.

amygdalate **1.** having to do with or related to almonds. **2.** having the shape of an almond. Also, **amygdaline.**

amygdale see AMYGDULE.

amygdalic *Biology.* having to do with or resembling an almond. *Chemistry.* relating to or dervived from amygdalin or amygdalic acid (mandelic acid).

amygdalic acid see MANDELIC ACID.

amygdaliform *Biology.* almond-shaped.

amygdalin *Biochemistry.* $C_6H_5CH(CN)OC_{12}H_{21}O_{10}$, a colorless, crystalline glucoside, soluble in water and alcohol and melting at 210°C; found especially in the seeds of the bitter almond tree, and also in parts of other plants of the genus *Prunus;* noted as the source of Laetrile, a compound alleged to have antineoplastic properties.

amygdaloid *Biology.* almond-shaped. *Geology.* **1.** any igneous or volcanic rock containing many amygdules. **2.** of or relating to such a rock. Also, **amygdaloidal.**

amygdaloidal lava *Geology.* see AMYGDALOID, def. 1.

amygdaloid body *Anatomy.* see AMYGDALA, def. 2.

amygdule *Geology.* **1.** a small cavity or vesicle formed by the expansion of gases in igneous rock during solidification and later filled with secondary mineral, such as chalcedony, opal, calcite, zeolite, and quartz. **2.** a pebble composed of agate.

amyl *Organic Chemistry.* any of the various forms of the pentyl group, $-C_5H_{11}$, occurring in eight isomeric arrangements. Also, PENTYL.

amyl- a combining form meaning "starch," as in *amylase.*

amyl acetate *Organic Chemistry.* $CH_3COOC_5H_{11}$, a colorless liquid that is soluble in water, alcohol, and ether and boils at 142°C; flammable and a dangerous fire risk; used as a solvent and flavoring agent and for industrial purposes. Also, ISOAMYL ACETATE.

amyl alcohol *Organic Chemistry.* **1.** a colorless, oily liquid, $C_5H_{12}O$, the chief component of fusel oil. Also, AMYLIC ALCOHOL. **2.** any of various isomers of amyl alcohol; eight isomers are possible and six are used commercially. **3.** see 1-PENTANOL.

***t*-amyl alcohol** see AMYLENE HYDRATE.

***n*-amylamine** *Organic Chemistry.* $C_5H_{11}NH_2$, a colorless liquid that is soluble in water, alcohol, and ether; freezes at –55°C and boils at 104°C; it has a variety of industrial uses, including the manufacture of pharmaceuticals, insecticides, dyestuffs, synthetic detergents, corrosion inhibitors, and gasoline. Also, PENTYLAMINE.

amylase *Enzymology.* any of a group of enzymes that catalyze the breakdown of starch into sugar in animals; different forms occur in saliva and in pancreatic juice, in plant parts, and in certain bacteria and molds.

β-amylase *Enzymology.* an enzyme that catalyzes the sequential removal of glucose residues from a starch molecule, starting at the nonreducing end of the starch molecule.

amyl benzoate see ISOAMYL BENZOATE.

***n*-amyl chloride** *Organic Chemistry.* $CH_3(CH_2)_3CH_2Cl$, a colorless liquid, insoluble in water and miscible with alcohol and ether, that freezes at –99°C and boils at 107.8°C; used as a solvent and chemical intermediate.

amylene *Organic Chemistry.* C_5H_{10}, any of five different isomeric forms of a highly flammable liquid hydrocarbon; nearly insoluble in water and miscible with alcohol and ether.

amylene hydrate *Organic Chemistry.* $(CH_3)_2C(OH)CH_2CH_3$, a clear, colorless liquid with a camphorlike odor and a burning taste; slightly soluble in water and miscible with alcohol and ether; freezes at –11.9°C and boils at 101.8°C; used as a pharmaceutical solvent and occasionally as a sedative. Also, *t*-AMYL ALCOHOL.

amyl ether *Organic Chemistry.* $C_{10}H_{22}O$, a colorless liquid, insoluble in water and soluble in alcohol and ether; it freezes at –70°C and boils at 186°C; used as a general organic solvent.

amyl group see AMYL.

amyl hydrate see AMYL ALCOHOL.

amyl hydride see PENTANE.

amylic *Organic Chemistry.* of or relating to amyl. Thus, **amylic fermentation.**

amylic alcohol see AMYL ALCOHOL, def. 1.

amylin see AMYLOPECTIN.

amylism *Toxicology.* poisoning due to the consumption of amylene hydrate.

amyl mercaptan *Organic Chemistry.* $C_5H_{11}SH$, a clear white to yellowish liquid, insoluble in water and soluble in alcohol, that has a boiling range of 104–130°C and a strong, offensive odor; used as an odorant for natural gas to detect gas-line leaks. Also, PENTANETHIOL.

amyl nitrate *Organic Chemistry.* $C_5H_{11}ONO_2$, a colorless liquid that boils at 145°C; used as an additive to raise the cetane number of diesel fuels.

amyl nitrite *Organic Chemistry.* $C_5H_{11}ONO$, a clear, yellowish, highly flammable liquid with a distinctive fruity odor; soluble in alcohol and nearly insoluble in water; boils at 96–99°C and is decomposed by air, light, and water; used in medicine as a vasodilator, and sometimes used abusively as a stimulative inhalant. Also, ISOAMYL NITRITE.

amylo- a combining form meaning "starch," as in *amyloplast.*

amylobarbitone or **amylobarbital** see AMOBARBITAL.

amylocellulose see AMYLOSE.

amylogen see AMYLOSE.

amyloglucosidase *Enzymology.* an enzyme that catalyzes the hydrolysis of starch, specifically breaking glucoside bonds between chains of glucose residues.

amyloid *Pathology.* **1.** starchlike or characterized by starchlike staining properties. **2.** the pathologic extracellular proteinaceous substance deposited in amyloidosis; a waxy, amorphous, hyalinelike material that exhibits a green birefringence under polarized light when stained with Congo red. Amyloid deposits are composed mainly of straight, nonbranching fibrils aggregated in bundles or interlocked in a meshwork; each fibril is composed of identical polypeptide chains arranged in stacked antiparallel beta-pleated sheets. In some chronic diseases, amyloids are present as extracellular deposits in the walls of blood vessels.

amyloidal *Pathology.* **1.** relating to or affected by amyloidosis. **2.** being or resembling an amyloid deposit.

amyloid body *Pathology.* a small mass of colloid substance frequently found in the tubules of the prostate gland.

amyloidosis [am´ə loi dō´sis] *Pathology.* any of a variety of diseases characterized by an accumulation of amyloid material in organs or tissues of the body, in which the accumulation impairs vital functions. In **primary amyloidosis,** there is no obvious predisposing condition; in **secondary amyloidosis,** the condition is associated with another chronic disease, such as tuberculosis.

amylolysis *Biochemistry.* the enzymatic breakdown (hydrolysis) of starch into soluble substances, such as sugar. Also, **amylohydrolysis.**

amylolytic *Biochemistry.* relating to or capable of amylolysis.

amylolytic enzyme *Biochemistry.* an enzyme that removes the starch molecules from more complex carbohydrates; used in textile manufacturing.

amylopectin *Biochemistry.* an insoluble complex carbohydrate that is a constituent of starch. Also, AMYLIN, α-AMYLOSE, STARCH CELLULOSE.

amylophagia [am´ə lə fāj´ē ə] *Medicine.* an abnormal craving to eat starch, especially laundry starch.

amyloplast *Botany.* the colorless plastid that forms starch granules. Thus, **amyloplastic.**

amylopsin *Biochemistry.* the α-amylase enzyme in pancreatic juice that converts starch into the sugar maltose.

amylose *Biochemistry.* a complex carbohydrate, such as cellulose or starch (polysaccharide), that is the soluble constituent of starch. Also, AMIDIN, AMYLOCELLULOSE, AMYLOGEN.

α-amylose see AMYLOPECTIN.

amylovorin *Biochemistry.* a polysaccharide produced by *Erinia amylovora,* a bacteria that induces fireblight in trees.

amyl propionate *Organic Chemistry.* $CH_3CH_2COOC_5H_{11}$, a colorless liquid that has a distillation range of 135–175°C, and is miscible with most organic solvents; a fire hazard; used in making food flavorings, perfumes, and lacquers.

amyl radical see AMYL.

amyl salicylate *Organic Chemistry.* $C_6H_4OHCOOC_5H_{11}$, a clear liquid that is insoluble in water and soluble in alcohol and ether; boils at 280°C; used in making soaps and perfumes. Also, ISOAMYL SALICYLATE.

amylum *Science.* the Latin word meaning "starch."

amylum star *Botany.* a star-shaped, starch-filled propagative body formed in certain stoneworts.

amyl xanthate *Organic Chemistry.* a salt produced by substituting a metal atom for the hydrogen atom attached to the sulfur in amylxanthic acid; used as a collector during the flotation of certain minerals.

Amynodontidae *Paleontology.* an extinct family of hornless rhinoceros-like mammals in the superfamily Rhinocerotoidea, characterized by canines in the form of curved tusks; analogous to hippopotamuses; extant in the Eocene.

amyostasia *Neurology.* a muscle tremor, especially when characteristic of locomotor ataxia. Thus, **amyostatic.**

amyotaxia or **amyotaxy** see ATAXIA.

amyotonia *Medicine.* an abnormal condition in which the muscles lack tone. Thus, **amyotonic.**

amyotonia congenita see OPPENHEIM'S DISEASE.

amyotrophic [am´ē ō trō´fik] *Medicine.* relating to or characterized by amyotrophy.

amyotrophic lateral sclerosis *Medicine.* a progressive, degenerative disease of the motor neurons of the brain stem and spinal cord, characterized by a general weakening and wasting of the voluntary muscles, and eventually complete paralysis. It is of unknown origin and presently incurable, and is usually fatal within two to five years. It is popularly known as Lou Gehrig's disease, after the baseball player who was a noted early victim.

amyotrophy *Medicine.* the progressive weakening and wasting of the muscles.

An the chemical symbol for actinon.

an- a prefix meaning: **1.** not or without, as in *anaerobic.* **2.** up or back, as in *anode.*

ana *Pharmacology.* of ingredients in a prescription, in equal quantities; so much of each. (From a Greek word meaning "of each.")

ana- a prefix meaning "up" or "back," as in *anamorphosis, anaplastic.*

ANA or **A.N.A.** American Nurses' Association.

Anab *Ordnance.* a Soviet air-to-air missile believed to be powered by a solid-propellant rocket motor and equipped with infrared or semiactive radar homing; it probably carries a large conventional warhead at a speed of Mach 2.5 and a range of 10 to 15 miles; officially designated **AA-4.**

Anabaena *Bacteriology.* **1.** a genus of freshwater, nitrogen-fixing blue-green algae that occurs in, and often contaminates, ponds, reservoirs, and other enclosed bodies of water. **2. anabaena.** any particular alga of this genus.

Anabantidae *Vertebrate Zoology.* a family of freshwater fishes in the order Perciformes that rely on atmospheric air, usually acquired at the surface of water; they are sometimes called labyrinth fish because of their complex accessory breathing organ.

Anabantoidei *Vertebrate Zoology.* a suborder of freshwater fishes in the order Perciformes, including the climbing perch or labyrinth fish.

Anabas *Vertebrate Zoology.* **1.** a genus of small freshwater fishes that resemble the perch and use their spiny fins for traveling on land; found in Africa and southeastern Asia. **2. anabas.** any fish of this genus.

anabasine *Organic Chemistry.* $C_{10}H_{14}N_2$, a colorless, toxic, liquid alkaloid produced by the plants *Anabasis aphylla* and *Nicotiana glauca;* it freezes at 9°C and boils at 105°C; soluble in alcohol and ether; used as an insecticide.

anabatic *Meteorology.* of air, moving upward.

anabatic wind *Meteorology.* a local wind blowing up a hill or mountain due to local surface heating; caused by the difference in density between the warm ground air and the cooler air in the free atmosphere.

Anabena *Botany.* a freshwater blue-green alga that sometimes imparts an unpleasant fishy odor and taste to water supplies. Also, **Anabaena.**

anabiosis *Biology.* **1.** a cyclic or seasonal state of greatly reduced metabolism, especially one in an aquatic invertebrate during a drought period; usually reversed by an increase in moisture. **2.** the restoration of vital processes after apparent death.

anableps see FOUR-EYED FISH.

anabohitsite *Petrology.* a variety of olivine-bearing pyroxenite containing about 65% hornblende and hypersthene and about 30% magnetite and ilmenite.

anabolic [an´ə bäl´ik] *Biochemistry.* of or relating to substances that increase the rate of metabolism in cells or organisms.

anabolic steroids *Endocrinology.* synthetic androgens that enhance the retention of nitrogen; used in certain clinical states, such as post-surgery or other trauma, to decrease muscle wasting; it is also used non-clinically by some athletes to build up muscle mass and enhance performance.

anabolism *Biochemistry.* the process of metabolism by which smaller molecules are joined to make larger molecules; the required energy is supplied by adenosine triphosphate (ATP).

anabolite *Biochemistry.* a product of anabolism.

anaboly *Evolution.* the addition of a new phase to the end of the ancestral pattern of embryonic development, resulting in evolutionary divergence from the ancestral form.

anabranch *Hydrology.* **1.** a stream that diverges from the main stream or river and later rejoins the main flow downstream. Also, VALLEY BRAID. **2.** a branch of a stream that gets lost in sandy soil.

Anacanthini *Vertebrate Zoology.* an order of bony fishes comprising the cod, haddock, pollack, and whiting. Also, GADIFORMES.

Anacardiaceae *Botany.* a large order of polypetalous trees or shrubs that are widely distributed throughout tropical America, Africa, and India; includes the cashew, pistachio, mango, and poisonous suma.

anacardiaceous *Botany.* belonging to the Anacardiaceae, or cashew family.

anacardium gum see CASHEW GUM.

anacidity *Medicine.* a lack of normal acidity, especially in the stomach.

anaclasis *Neurology.* a reflex action.

anaclinal *Geology.* of a stream or valley, having a downward slope opposite to that of the general dip of the underlying strata.

anaclisis *Psychology.* physical and emotional dependence on another person for security and gratification, especially the normal dependence of an infant on its mother.

anaclitic [an´ə klit´ik] *Psychology.* relating to or characterized by anaclisis; dependent.

anaclitic depression *Psychology.* in infants who have been separated from their mother or primary caretaker, a condition in which physical and intellectual development becomes seriously impaired.

anaclitic object choice *Psychology.* the choice of a loved one based on the person upon whom the individual was emotionally dependent as an infant.

anaconda *Vertebrate Zoology.* **1.** a nonvenomous constrictor snake, *Eunectes murinus,* native to tropical South America and belonging to the Boidae (boa) family. The largest living snake, it is partly aquatic and partly arboreal. **2.** any large constricting snake.

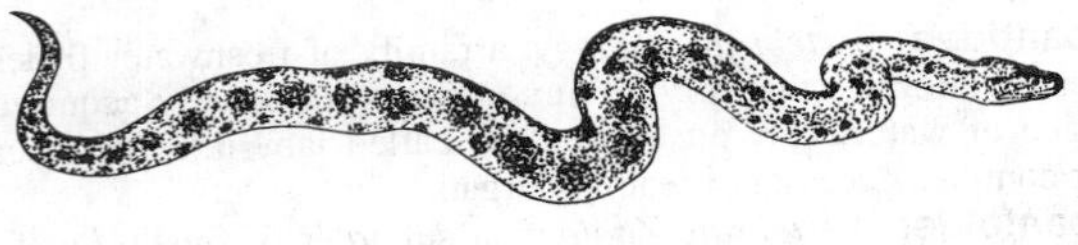

anaconda

anacoustic zone *Geophysics.* the part of outer space starting at 100 miles (160 km), at which the distance between gas molecules is so great that sound waves can no longer be carried. Also, ZONE OF SILENCE.

Anactinochitinosi *Invertebrate Zoology.* a group name for three related suborders of ticks and mites: Ixodides, Mesostigmata, and Onychopalpida.

anacusis *Medicine.* a condition of total deafness.

Anacystis *Bacteriology.* a genus of cyanobacteria.

anadromous *Vertebrate Zoology.* referring to fish, such as salmon, that live most of their lives in the ocean but migrate into freshwater streams to spawn.

anaerobe *Biology.* **1.** an organism, especially a bacterium, that does not require atmospheric oxygen to live. **2.** an organism that cannot survive in the presence of oxygen. Also, **anaerobium.**

anaerobic [an´ə rō´bik] *Biology.* **1.** occurring with little or no oxygen. **2.** of or relating to anaerobes. Also, **anaerobiotic, anerobic.**

anaerobic bacteria *Microbiology.* any microorganisms that grow only in the absence of molecular oxygen and that generate energy by fermentative reactions which do not involve molecular oxygen.

anaerobic digester *Biotechnology.* a bioreactor designed to facilitate anaerobic bacterial fermentation; commonly used in the treatment of sewage sludge.

anaerobic digestion *Biochemistry.* the process by which complex plant and animal compounds are broken down into simpler compounds in the absence of oxygen, producing a variety of gaseous and soluble products (including methane, commonly used in sewage treatment).

anaerobic filter system *Biotechnology.* a filtering procedure used in anaerobic bacterial fermentation when liquid effluents have a high level of dissolved organic matter and a low level of particulate solids.

anaerobic glycolysis *Biochemistry.* a process in which sugar is broken down into smaller molecules in the absence of oxygen; the conversion of glucose to lactic acid.

anaerobic metabolism *Biochemistry.* the metabolic process used by cells to produce energy from nutrients without using oxygen; termed *fermentation* when referring to microorganisms and *anaerobic glycolysis* when referring to humans and other animals.

anaerobic petri dish *Microbiology.* a glass dish used for growing cultures of anaerobic bacteria, usually in a thioglycollate agar medium with limited air space.

anaerobic respiration *Biochemistry.* a metabolic process in which cells produce chemical energy by breaking down organic compounds without using oxygen; this process is termed fermentation when referring to microorganisms and glycolysis when referring to humans. Also, **anaerobic metabolism.**

anaerobic sediment *Geology.* a highly organic sediment formed in hydrogen sulfide-rich water where limited circulation results in the absence or near absence of oxygen at the surface of the sediment.

anaerobiosis *Biology.* the condition of sustaining life without the presence of oxygen or air.

Anaerobiospirillum *Bacteriology.* a genus of Gram-negative, fermentative bacteria of the family Bacteroidaceae, consisting of helical, rod-shaped cells originally isolated from the colons and throats of dogs.

anaerogenic *Biology.* **1.** producing little or no gas. **2.** suppressing the production of gas by gas-producing bacteria.

anaerophyte *Botany.* a plant that grows without air, especially an anaerobic bacterium.

Anaeroplasma *Bacteriology.* a genus of obligately anaerobic bacteria of the class Mollicutes found in the rumen of cattle and sheep, distinguished by the lack of a cell wall and requiring sterols for growth.

Anaerovibrio *Bacteriology.* a genus of Gram-negative, lipolytic bacteria of the family Bacteroidaceae occurring in the rumen of cattle and sheep as curved, rod-shaped cells possessing polar flagellation.

anafront *Meteorology.* a warm or cold front at which the warm air is ascending relative to the frontal zone.

Anagalidae *Paleontology.* a family of early eutherian mammals generally classified in the order Anagalida; thought to be related to the ancestry of rodents, lagomorphs, and macroscelids, although some see them as related to the Dinocerata; extant from the Paleocene to the Oligocene.

anagenesis *Evolution.* **1.** an evolutionary change from simple toward complex forms of life, resulting in higher levels of organization or specialization. **2.** an evolutionary change in which modified forms replace one another in continuous succession without branching into new taxa. Also, PHYLETIC EVOLUTION.

anaglyph *Graphic Arts.* a stereoscopic picture made up of two images, one red and the other blue-green, that are viewed with corresponding color filters to produce a three-dimensional effect.

anagotoxic *Toxicology.* acting against a toxin.

ana-holomorph *Mycology.* a fungus whose sexual stage of reproduction is not yet known.

Anahuac *Geology.* the North American Gulf Coast stage of the Miocene epoch, after the Chickasawhay and before the Napoleonville.

anakinesis *Biochemistry.* a process in living organisms that produces energy-rich molecules such as adenosine triphosphate.

anal *Anatomy.* of, relating to, or near the anus. *Psychology.* **1.** of or relating to the anal stage of psychosexual development. **2.** having or displaying anal character traits.

analbite *Mineralogy.* an unstable triclinic polymorph of albite that becomes monoclinic at about 700°C; obtained by heating albite.

anal canal *Anatomy.* the terminal portion of the large intestine.

anal character *Psychology.* a personality type that derives from unresolved conflicts during the anal stage of psychosexual development, considered to be manifested by such traits as stinginess, inflexibility, and excessive neatness and orderliness. Also, **anal personality.**

analcime *Mineralogy.* $NaAlSi_2O_6{\cdot}H_2O$, a colorless to white cubic mineral with a vitreous luster, having a specific gravity of 2.22 to 2.29 and a hardness of 5 to 5.5 on the Mohs scale; found in basalts and other igneous rocks. Also, **analcite.**

analcimite *Petrology*. an extrusive or hypabyssal basalt that consists primarily of pyroxene and analcime. Also, **analcitite.**
analcimization *Geology*. in igneous rock, the replacement of feldspars or feldspathoids by the mineral analcime.
anal column *Anatomy*. a collection of longitudinal folds in the membrane of the anal canal.
analemma *Astronomy*. a figure-eight-shaped diagram, often placed on a terrestrial globe, that plots the sun's declination during the course of the year, as well as the equation of time.
analeptic *Pharmacology*. **1.** capable of restoring vigor; stimulating. **2.** any substance or medication that has an analeptic effect, such as caffeine or amphetamine.
anal-expulsive *Psychology*. of or relating to a phase of the anal stage in which sensual pleasure is obtained by expelling feces.
anal fin *Vertebrate Zoology*. an unpaired fin located in the middle of the posterior ventral part of a fish body, usually just behind the anus.
analgesia [an´əl jēz´yə] *Physiology*. **1.** the absence of a normal sensitivity to pain. **2.** the first stage below full consciousness through which an individual progresses after a general anesthetic has been administered, where he or she is still semiconscious but feels no pain.
analgesic [an´əl jēz´ik] *Pharmacology*. **1.** capable of relieving pain. **2.** any medication or other substance that relieves pain, such as aspirin.
anal gland *Vertebrate Zoology*. a gland that is located near the anus or rectum, usually serving an excretory or secretory function. *Invertebrate Zoology*. a gland in mollusks of the genus *Murex* that secretes a purple substance used in dyeing.
anallagmatic curve *Mathematics*. a curve that maps to itself under inversion in some circle. That is, every point on the curve can be considered to lie on a line containing a diameter of the circle (of radius r), and to each point is associated another point of the curve lying on the same line such that the product of the distances of the points from the center of the circle is the same as r^2.
analog a thing that is similar or comparable, but not identical, to another; specific uses include: *Evolution*. an organism that is similar to other organisms in function or behavior as a result of convergent evolution rather than common ancestry. *Chemistry*. a substance possessing a chemical structure and chemical properties similar to those of another substance. *Metrology*. representing information in a way that bears an exact relationship to the original information, by means of a continuous physical variable such as length, weight, voltage, or pressure. *Food Technology*. a substitute food that is manufactured from vegetable matter to look and taste like a meat or dairy product. Also, ANALOGUE.
analog adder *Electronics*. an analog signal whose output is the sum of two or more analog inputs.
analog clock *Horology*. a clock of the traditional type, in which time is represented by the position of hands that rotate on a dial, rather than by a numerical display.
analog communications *Telecommunications*. any system that uses a nominally continuous electrical signal.
analog comparator *Electronics*. a comparator that produces a high (binary 1) digital output signal when the sum of two analog voltages is positive or a low (binary 0) signal when the sum is negative.
analog computer *Computer Technology*. a computer in which information is stored and processed as physical values, such as voltages, that can vary smoothly between certain limits rather than having discrete, digital values.
analog data *Computer Technology*. the representation of information in a way that bears an exact relationship to the original information, so that it varies continuously rather than discretely as with digital data.
analogical control *Robotics*. control by signals from analog devices.
analog indicator *Electronics*. an indicator whose output is presented by pointer deflection or other continuously variable visual means. Similarly, **analog readout.**
analog multiplexer *Electronics*. a multiplexer that accepts only analog input signals.
analog multiplier *Electronics*. an analog circuit whose output is the product of two or more analog inputs.
analog network *Electronics*. a network whose voltage and current relationships are analogous to the relationships of some physical system.
analogous *Science*. relating to or being an analog; similar or comparable. *Biology*. similar in function and appearance, but not in structure or origin.
analogous pole *Solid-State Physics*. the positively charged pole of a crystal that arises when the crystal is heated.
analog radio system see AR SYSTEM.
analog recording *Electronics*. a method of recording material in which the recording signal varies in a manner analogous to the original signal.
analog signal *Electronics*. a signal whose parameters (such as amplitude, frequency, or phase) can change continuously over a given range, as distinguished from a digital signal where only some discrete values (usually 2) are considered significant.
analog simulation *Computer Programming*. the representation of physical systems that are describable by mathematical expressions, usually differential equations, containing variables which change continuously over time. *Electronics*. any electronic process that implements the transient behavior over time of a physical system.
analog states *Nuclear Physics*. the highly excited states of almost contiguous nuclear isobars that have identical nucleon wavefunctions except for the change of one or more neutrons into the same number of protons, resulting in a structure analogous to the original decaying state in the neighboring isobar. Also, ISOBARIC ANALOG STATES.
analog switch *Electronics*. a switching device that acts to pass the true analog signal of a transducer's output.
analog-to-digital converter *Electronics*. a device that changes a continuously variable quantity such as motion or electrical voltage into digital or discrete values. Thus, **analog-to-digital conversion.**
analog-to-frequency converter *Electronics*. a device that transforms an analog signal which is not in frequency form into a proportional change in frequency.
analogue see ANALOG.
analog voltage *Electronics*. a voltage representing an analog signal.
analog watch *Horology*. a watch of the traditional type, in which time is represented by the position of hands that rotate on a dial, rather than by a numerical display.
analogy a comparison of two things, based on their similarity in one or more respects; specific uses include: *Evolution*. a similarity in function or behavior among organisms or their anatomical structures resulting from convergent evolution rather than common ancestry. *Anthropology*. the practice or concept of using ethnographic information from recent cultures to make informed hypotheses about archaeological cultures, such as comparing the way of life of contemporary Kalahari Bushmen to that of Paleolithic African cultures.
anal plate *Vertebrate Zoology*. **1.** any of the plates at the rear of the ventral part of a turtle's shell. **2.** in snakes, a large scale just outside the anal opening.
anal-retentive *Psychology*. of or relating to a phase of the anal stage in which sensual pleasure is obtained by retaining feces.
anal sphincter *Anatomy*. a circular muscle surrounding and controlling the anal opening.
anal stage *Psychology*. in psychoanalytic theory, the second stage of psychosexual development, in the second year of life, during which the anus becomes the focus of sexual gratification through the sensations associated with carrying out or withholding bowel movements. Also, **anal period, anal phase.**
anal triangle *Anatomy*. the posterior half of the perineum.
analysand *Psychology*. a person who is undergoing psychoanalysis.
analysis *plural*, **analyses.** the separation of a thing into its constituent parts in order to study its nature; specific uses include: *Analytical Chemistry*. the detection and identification of the chemical composition of a substance, using classical laboratory techniques, microchemical interactions, and analytical instrumentation. *Psychology*. see PSYCHOANALYSIS. *Meteorology*. a detailed study of the state of the atmosphere based on actual observations. *Physics*. the separation of light into its prismatic components. *Mathematics*. the areas of mathematics that make use of the concepts of limits, convergence, and continuity. *Computer Programming*. see PROGRAM ANALYSIS.
analysis of covariance *Statistics*. a study of the effect of a set of both quantitative and qualitative variables on a quantitative response, with emphasis on the effect of the qualitative variables.
analysis of variance *Statistics*. a study of the effect of a set of qualitative variables on a quantitative response variable, based on a decomposition of the variance of the latter.
analyst *Psychology*. see PSYCHOANALYST.
analyte *Analytical Chemistry*. the substance being identified and measured in an analysis.
analytical of or relating to analysis. Also, **analytic.**
analytical balance *Engineering*. a precision balance that is designed to measure the mass of quantities to an accuracy ranging from 0.1 to 0.01 milligrams.

Analytical Chemistry

The philosophy of analytical chemistry is the application of physical phenomena to characterize the chemical composition and properties of matter in relation to a specific scientific question. Fifty years ago the question was largely a matter of amounts and determining the elemental composition of a sample. For example: How much iron is in this steel sample? How clean is my drinking water? Today, however, the questions are often much more complex: How old is this object? How does human tissue react to materials used in an artificial heart?

The scope of analytical chemistry now spans industry, agriculture, the environment, and medicine, as well as many research areas in chemistry, biology, biochemistry, geology, and other physical sciences. Contributing significantly to this growth has been the proliferation of technology: analytical instrumentation, computers, novel electronics and optics, and, more recently, robotics. Analytical methods now go beyond determining elemental composition in a sample to yield information about molecular composition, crystal structure, oxidation state, coordination state, surface phenomena, etc., in order to properly characterize the chemical forms of the species in a given system. Many techniques now used in analysis yield two- and three-dimensional information with high spatial resolution, resulting in a more realistic and useful view of that system.

Chemical analysis can now more accurately be described as being applied to a problem rather than to a sample, as one is more interested in the problem the sample represents rather than the sample itself. Analytical chemistry is the detective of the scientific world, one of its chief problem solvers.

George H. Morrison
Professor of Chemistry
Cornell University

analytical chemistry *Chemistry.* the branch of chemistry that deals with the identification of substances and the determination of the precise amount or composition of substances within a chemical system.

analytical distillation *Analytical Chemistry.* the vaporization and recondensation of a liquid mixture for the purpose of determining its components by comparing their respective boiling points.

analytical electron microscopy *Optics.* the use of characteristic X-rays produced by electron-specimen interaction in a scanning electron microscopy to obtain quantitative chemical composition data by analyzing their wavelengths. Also, ENERGY-DISPERSIVE X-RAY ANALYSIS.

analytical engine *Computer Technology.* an early device, conceived by Charles Babbage in 1833, that is considered to be the forerunner of the modern computer.

analytical extraction *Analytical Chemistry.* a separation technique in which soluble parts of a mixture are removed by dissolution; often used to extract pure metals from their ores.

analytical function generator *Electronics.* an analog computer device whose outputs are dependent on one or more inputs as defined by a function appearing in a physical law. Also, NATURAL (LAW) FUNCTION GENERATOR.

analytical mechanics *Mechanics.* the mathematical analysis of motions of particles or rigid bodies, considered as due to mutual interactions between the constituents and applied forces that can be expressed in terms of the Lagrangian and Hamiltonian functions.

analytical orientation *Photogrammetry.* the determination of position, tilt, direction, flight height, and angular and linear elements in rectifying aerial photographs.

analytical photogrammetry *Photogrammetry.* the mathematical determination of object size and shape from measurements made directly on the images, as opposed to mechanical determination from a stereoscopic model.

analytical ultracentrifuge *Engineering.* an ultracentrifuge that uses an optical system to determine sedimentation velocity or equilibrium.

analytic continuation *Mathematics.* a process that uses the properties of an analytic function to extend the function's definition beyond the original domain. If f is a function analytic in a domain D, and there exists a function F analytic in some domain containing D as a proper subset, such that $F(z) = f(z)$ for all z in D, then F is an analytic continuation of f. The process of extending f to F is also called analytic continuation.

analytic curve *Mathematics.* a curve in n-dimensional space whose parametric representation consists of n analytic functions of a single real variable.

analytic function *Mathematics.* a function that can be expanded in a convergent power series. Although this condition implies that the function has derivatives of all orders, analyticity is a stronger condition than infinite differentiability, since there exist functions which are infinitely differentiable, yet are not expandable in power series.

analytic geometry *Mathematics.* the representation of geometric figures and curves using coordinates, equations, and algebraic methods of reasoning.

analytic group *Geology.* a rock-stratigraphic unit having subdivisions that are classified as formations.

analytic number theory *Mathematics.* the study of numbers using analytic methods; includes such topics as distribution of primes, the Riemann zeta function, and formal power series.

analytic psychology *Psychology.* the system of psychology originated by Carl Jung, which distinguishes three layers of the psyche: the conscious mind, the personal unconscious, and the collective unconscious.

analyzer *Computer Programming.* any of various devices used to analyze and solve problems, such as a logic analyzer or differential analyzer. *Engineering.* an instrument system, usually composed of a number of basic instruments, that is used for making electronic measurements. *Mechanical Engineering.* **1.** in an absorption refrigeration system, a device that increases the concentration of refrigerant in the vapor entering the condenser and separates the water vapor from the refrigerant vapor. **2.** any device that assists in analyzing or interpreting the physical or chemical characteristics of a system. *Optics.* a device that transmits only plane polarized light; used in the eyepiece of a polariscope or similar instrument.

anamigmatization *Geology.* the remelting of preexisting rock under conditions of high temperature and pressure to yield magma. Also, **anamigmatism.**

anamnesis [an´am nē´sis] *Psychology.* the recollection of past things; memory.

anamnestic response *Immunology.* a response to antigenic stimulation that is characterized by large antibody production; it occurs during a subsequent or second introduction of the antigen. Also, BOOSTER RESPONSE.

Anamnia *Vertebrate Zoology.* a group of vertebrates that have no amnion during early development, including Agnatha, Fish, and Amphibia. Also, **Anamniota.**

anamniotic *Vertebrate Zoology.* describing vertebrates whose embryo has no amnion, such as fish or amphibians.

anamorph *Mycology.* the imperfect or asexual stage of a fungus.

Anamorpha *Invertebrate Zoology.* a division or subclass of Chilopoda (centipedes). The young insects are born with seven pairs of legs, and they develop more legs and body segments as they grow.

anamorphic *Science.* changing to a more complex form. *Optics.* describing a lens or optical system that produces unequal magnification of an image along the vertical and horizontal axes, resulting in a distorted image.

anamorphic lens *Optics.* a lens having one or more cylindrical surfaces; used in photography to produce images that are compressed in one dimension, such as horizontal, which can later be restored to true form. Also, **anamorphote lens.**

anamorphic system *Optics.* an optical system that produces a different magnification in the horizontal or vertical direction due to the incorporation of lenses, mirrors, or prisms that yield different magnifications and effects.

anamorphic zone *Geology.* the zone deep within the earth's crust in which anamorphism takes place.

anamorphism *Geology.* intense metamorphism deep within the earth, characterized by rock flow and the changing of simple, less dense mineral compounds into denser, more complex compounds through the processes of decarbonization, dehydration, and deoxidation.

anamorphoscope *Optics.* an instrument, usually consisting of a cylindrical lens or a convex viewing mirror, that reveals the true proportions of an image distorted by anamorphosis.

anamorphosis *Evolution.* a gradual and steady process of evolutionary change. *Optics.* the formation of a distorted image by using an optical system designed for this purpose. *Graphic Arts.* an image that is distorted unless viewed through a special device.

anamorphote lens see ANAMORPHIC LENS.

anamorphous solid see NONCRYSTALLINE SOLID.

Anancinae *Paleontology.* a subfamily of elephant-like proboscideans in the extinct family Gomphotheriidae; about 9 feet high, *Anancus* was characterized by its extremely long upper-jaw tusks, which grew as long as 12 feet; Miocene to lower Pleistocene.

anandrous *Botany.* of a flower, lacking stamens; female.

anankastic personality *Psychology.* an obsessive-compulsive personality syndrome. Also, **anancastic personality**.

Ananke *Astronomy.* a retrograde-orbiting moon of Jupiter, discovered in 1951; it has a diameter of about 30 kilometers. Also, JUPITER XII.

ananthous *Botany.* of a plant, lacking flowers.

anapaite *Mineralogy.* $Ca_2Fe^{+2}(PO_4)_2{\cdot}4H_2O$, a transparent, greenish white triclinic mineral occurring in crystalline, often aggregate forms, having a specific gravity of 2.8 and a hardness of 3.5 on the Mohs scale.

anapeirean *Petrology.* relating to igneous rocks of the Pacific series.

anaphase [an´ə fāz´] *Cell Biology.* a stage in cell division marked by the movement of separate chromatids (during the first meiotic division) or members of a homologous pair of chromosomes (during meiosis) away from one another on the metaphase plate and toward the opposite poles of the cell.

anaphase

anaphia *Neurology.* a loss, lack, or impairment of the sense of touch. Also, ANHAPHIA.

anaphor *Artificial Intelligence. plural,* **anaphora.** a word that refers to another word in a sentence, such as a pronoun.

anaphoresis [an´ə fə rē´sis] *Physical Chemistry.* the movement of charged particles, suspended in an electrolytic medium, toward the anode, following their exposure to an electric field. *Physiology.* diminished activity of the sweat glands.

anaphoria *Medicine.* an upward deviation of the visual axes of the eyes. Also, **anatropia.**

anaphylactic shock [an´ə fī lak´tik] *Immunology.* a life-threatening hypersensitivity reaction to a previously encountered antigen; symptoms usually include respiratory distress, vascular collapse, and shock.

anaphylatoxin *Immunology.* a toxic substance produced in animal tissue during an antigen-antibody reaction that causes anaphylaxis.

anaphylaxis see ANAPHYLACTIC SHOCK.

anaplasia *Evolution.* a progressive phase of development in an organism, prior to maturity, characterized by increased vigor and diversification. *Oncology.* a loss of differentiation of cells and of their orientation to one another and to their axial framework, a characteristic of tumor tissue. Also, **anaplastia.**

Anaplasma *Bacteriology.* a genus of coccoid, Gram-negative bacteria of the family Anaplasmataceae; they are obligate parasites that infect the erythrocytes of cattle, sheep, and deer.

Anaplasmataceae *Bacteriology.* a family of bacteria of the order Rickettsiales that infect vertebrate erythrocytes or plasma and are transmitted by ticks, fleas, and lice.

anaplasmosis *Pathology.* an infectious disease of ruminants varying from peracute to chronic, frequently caused by blood-feeding insects such as ticks, and characterized by anemia, icterus, and fever.

anaplastic *Oncology.* of or relating to anaplasia or reversed development. Thus, **anaplastic cells.** *Surgery.* of or relating to anaplasty.

anaplasty *Surgery.* the repair or replacement, especially by plastic surgery, of lost, defective, or damaged parts.

anaplerosis *Surgery.* the repair of a tissue defect by filling in missing or damaged parts. *Biochemistry.* an enzymatic reaction in intermediary metabolism that restores the concentration of a vital intermediate whose cellular function has been depleted.

anaplerotic *Biochemistry.* describing a reaction that replenishes the supply of intermediates in a metabolic pathway, as the citric acid cycle.

anaplerotic sequences *Biochemistry.* a series of reactions that replenish intermediates depleted by biosynthesis.

anapophysis *Anatomy.* an accessory process, especially of a vertebra.

Anapsida *Vertebrate Zoology.* one of the five subclasses of reptiles, characterized by a skull with a complete covering of dermal bones with no temporal openings, as in tortoises and turtles.

anaptic *Neurology.* suffering from an impaired sense of touch.

anarcestid *Paleontology.* of or relating to the earliest ammonoids, in the order Anarcestida; the anarcestids are known only from Devonian strata and were probably ancestral to all later ammonoids.

Anarhichadidae *Vertebrate Zoology.* the wolf fishes or wolf eels, a family of large marine fishes in the suborder Belnnioidei, characterized by a large head on a long body (up to nine feet), a long dorsal fin, and massive canines and molars.

Anasca *Paleontology.* a suborder of bryozoans in the order Cheilostomata; characterized by the absence of an ascus (a tubular hydrostatic compensation sac used in moving the tentacles). In anascans the tentacles are moved by internal muscle action of the frontal membrane; they first appeared in the Jurassic, and many genera are still extant.

anaseism *Geophysics* a movement, during an earthquake, of the earth away from the center of the quake.

anaspid *Paleontology.* **1.** a member of the extinct order Anaspida; small freshwater fish characterized by a single nostril, narrow rows of scales, and usually an armor-plated head. **2.** relating to this order.

Anaspida *Paleontology.* an order of small, armored ostracoderms in the class Agnatha and the extinct subclass Cephalaspidomorpha; first appeared in the late Silurian and became extinct in the Devonian; some have bony scales but none has more than a cartilaginous skeleton.

Anaspidacea *Invertebrate Zoology.* an order of primitive, shrimplike animals in the class Malacostracea; found in fresh water, they deposit eggs among water plants or under stones.

Anaspidea *Invertebrate Zoology.* an order of sluglike sea hares with a thin shell largely obscured by the mantle, and large parapodia, gill, jaws and storage crop; an ink gland produces copious purple secretions. Also, APLYSIACEA.

Anaspididae *Invertebrate Zoology.* a family of shrimplike crustaceans in the order Anaspidacea.

anastatic water *Hydrology.* underground water within the capillary fringe. Also, FRINGE WATER.

anastigmatic lens *Optics.* a combination of lenses that minimizes astigmatism and reduces chromatic and spherical aberration in an optical system. Also, **anastigmat (lens).**

anastomose *Physiology.* to open one part into another, either directly or by connecting channels, as blood and lymph vessels and hollow viscera. *Surgery.* to create, by surgical, traumatic, or pathological formation, a communication between two formerly separate structures, such as arteries, veins, or the gastrointestinal tract.

anastomosing stream see BRAIDED STREAM.

anastomosis *Science.* a union between multiple branches of a single system. *Botany.* the interconnection of cells or strands of cells with one another, as the veins in a leaf. *Medicine.* **1.** a connection between blood vessels that circumvents an obstruction in the usual channel. **2.** a connection or opening created by surgical, traumatic, or pathological means between two normally distinct organs or spaces. Also, INOSCULATION.

anastomotic ulcer see MARGINAL ULCER.

anastral *Cell Biology.* of a mitotic figure or process, lacking an aster.

anatabine *Chemistry.* $C_{10}H_{12}N_2$, a liquid alkaloid that is obtained from tobacco.

anatase *Mineralogy.* TiO_2, a transparent to nearly opaque tetragonal mineral found in colors ranging from black to brown to dark blue, having a specific gravity of 3.82 to 3.95 and a hardness of 5.5 to 6 on the Mohs scale; found mainly in gneiss, schists, and detrital deposits; trimorphous with rutile and brookite. Also, OCTAHEDRITE.

anatexis *Geology.* a high-temperature metamorphic process whereby existing deep-seated rock remelts and may be regenerated as magma.

anathermal *Geology.* **1. Anathermal.** a postglacial period extending from about 10,000 to 7500 years ago, during which temperatures generally rose following the last major continental glacier advance. **2.** of or relating to the climate, deposits, and events of this period. **3.** of or relating to any period during which temperatures are rising.

Anatidae *Vertebrate Zoology.* a large family of waterfowl in the order Anseriformes, including many species of ducks, geese, and swans.

anatine *Vertebrate Zoology.* **1.** of, relating to, or belonging to the family Anatidae. **2.** relating to or resembling a duck.

anatomical *Anatomy.* of or relating to anatomy.

anatomical dead space *Physiology.* the portions of the respiratory tract within the trachea, bronchi, and air passages containing air that does not reach the alveoli during respiration; the amount of dead space is increased by certain lung disorders, such as emphysema.

anatomical pathology *Medicine.* a branch of pathology in which morphologic changes in tissues and organs are studied to determine the cause of disease or death. Thus, **anatomical pathologist.**

anatomical position *Anatomy.* the conventional positioning of the body used in descriptive anatomy, in which the body is erect and facing the viewer, with the arms at the sides and the palms toward the front.

anatomical snuffbox *Anatomy.* a small depression on the back of the hand at the base of the thumb, formed by the tendons that reach toward the thumb and index finger when the wrist is flexed, the thumb is abducted, and the fingers are extended.

anatomicosurgical *Surgery.* relating to both anatomy and surgery.

anatomist *Anatomy.* a scientist who specializes in anatomy.

anatomize *Anatomy.* to dissect an animal or plant in order to examine the structure and relationship of its parts.

anatomy *Biology.* **1.** the science of the structure of the human body and the relationship of its parts. **2.** a formal dissection of a body. **3.** the structure of an organism.

anatropous *Botany.* of ovule orientation in the ovary, having the funiculus lengthened and the ovule inverted 180 degrees so that the micropyle is folded over and lies near the base of the funiculus.

anavenin *Toxicology.* a venom that has been inactivated, usually by the addition of formaldehyde, but that still retains its antigenic properties. Also, **anavenene, anavenom.**

Anaxagoras c. 500–428 BC, Greek philosopher and astronomer; studied eclipses, meteors, and the sun; inspired Aristotle.

anaxial *Biology.* lacking a distinct axis, resulting in an asymmetrical shape.

Anaximander 611–547 BC, Greek philosopher and astronomer; introduced the sundial to Greece; posited origins of all life from an infinite, indefinite mass.

Anaximenes late 6th century BC, Greek philosopher, student of Anaximander; posited air as the source of all matter.

Anbauhobel [an´bou´hō bəl] *Mining Engineering.* a fast-moving plough used on longwall faces, traveling at a rate of 75 feet per minute with a cutting depth ranging from 1.5 to 3 inches.

ANC *Aviation.* the airport code for Anchorage, Alaska.

Ancalomicrobium *Bacteriology.* a genus of Gram-negative, nonmotile bacteria, that typically bears prosthecae, is facultatively anaerobic, and is found in aquatic habitats, reproducing by bud formation.

ancestor *Evolution.* any taxonomically distinct progenitor of a more recent or existing species or other taxonomic group. *Computer Science.* a node in a tree that lies on a path between the given node and the root; a parent of a node or an ancestor of its parent.

ancestral *Evolution.* of or relating to a physiological feature or hereditary character derived from or originating with a phylogenic precursor. *Systematics.* of or relating to a character state found in an ancestor; the opposite of *derived.*

ancestroecium *Invertebrate Zoology.* a tube that encloses an ancestrula.

ancestrula *Invertebrate Zoology.* the first zooid of a bryozoan colony (Ectoprocta) from which other individuals bud asexually to form a new colony.

anchialine *Ecology.* describing coastal seawater habitats that do not have a surface connection to the sea.

anchieutectic [an´kē´yə tek´tik] *Geology.* of a rock or magma, having a mineral composition of such proportions that the system is almost at its minimum melting temperature.

Anatomy

Anatomy: The science of bodily structure; structure as discovered by dissection. The English word "anatomy" has been borrowed from Greek terms referring to cutting and dissection. In scientific use, however, anatomy refers to the knowledge derived from those activities and not to the procedures themselves. The particular type of knowledge is the definition of the structure of organized beings, animal or plant.

The use of the term (originally *Anothomia*) to convey this meaning dates to the 14th century. Structural information has traditionally been obtained by cutting, which includes both dissection and microtomy. In practice, modern investigation in the field of anatomy has involved much more than the mere description of structure and encompasses an understanding of function. In contrast to other disciplines, particularly that of physiology, however, anatomy seeks to use structure as a tool with which to comprehend function. It differs from physiology in the emphasis that is placed in anatomy on the parameter of distance, or spatial relation, as opposed to the parameter of time.

Recently, practitioners in the field of anatomy have suffered from the common tendency, even among biological scientists who should know better, to use the term anatomy to mean gross human anatomy. To many people the very word itself, "anatomy," conveys an image of a foul-smelling room, filled with the cold, still bodies of no longer living human beings, being worked upon by students who are passing through a ritual of admission to the medical profession. Anatomy, to these people, implies a huge body of knowledge, little of which remains to be discovered, but which needs instead to be gleaned from overly large tomes, and memorized.

In response to this unfortunate word association, departments of Anatomy in many universities have changed their names. Popular new titles include Anatomy and Cell Biology (or the reverse), Cell Biology, Anatomy and Neurobiology (or the reverse), and Structural Biology. These new titles are meant to distinguish the science that is carried out in these departments from gross human anatomy and to convey to prospective students, fellows, faculty members, and donors that the business of the departments is research in the modern sense. In that modern sense, departments of Anatomy (in the various permutations of words that have been incorporated into their titles) are currently undergoing a renaissance of investigative activity. The development of many new methods for the study of structure, particularly those that now utilize computers to obtain images, has breathed new life into the old discipline. Anatomy is very much a living science.

Michael D. Gershon
Professor and Chairman of Anatomy
Columbia University

anchimeric assistance *Organic Chemistry.* an interaction that occurs when a neighboring group participates in the rate increase of a reaction, such as reactions of carbocation intermediates.

achimonomineralic *Petrology.* referring to an igneous rock that consists almost entirely of a single mineral.

anchor a means by which something is held firmly in place; specific uses include: *Naval Architecture.* a heavy weight, usually of iron or steel and often having hooks or prongs, that is suspended from a chain or cable and lowered to the seabed to hold a ship in place. *Civil Engineering.* **1.** a tie rod connecting a retaining wall to an anchor block. **2.** the entire assembly of tie rod and anchor block. *Building Engineering.* a piling or block that serves as the base of a guy wire or similar support for holding building structures in place. *Mechanical Engineering.* any of various plowing devices that are similar in shape to a ship's anchor. *Metallurgy.* in the art of casting in sand molds, the metallic device that holds the mold core. *Invertebrate Zoology.* an anchor-shaped spicule found in sea cucumbers and certain sponges.

anchorage an object to which something is fastened; specific uses include: *Navigation.* an area used by a vessel to anchor. *Architecture.* a device used to attach one object to another, especially the lower members of a building to the foundation. *Civil Engineering.* a device that holds tendons in a pre- or post-tensioned concrete member. *Building Engineering.* see ANCHOR BLOCK. *Medicine.* **1.** in tissue cell culture, the attachment of proliferating cells to a solid surface. **2.** in dentistry, a tooth or an implanted tooth substitute to which a fixed or removable partial denture, crown, or restorative material is secured. **3.** in orthodontia, the resistance offered by an anatomical part that is used to effect movement of a tooth. *Surgery.* the surgical fixation of loose or prolapsed abdominal tissue.

anchorage deformation *Civil Engineering.* in prestressing concrete members, the deformation of an anchor or slippage of tendons when the prestressing force is transferred from the prestressing device to the member. Also, **anchorage slip.**

anchorage dependence *Biotechnology.* the necessity for some cell cultures to be in contact with a solid substrate, such as glass or plastic, in order to grow and multiply.

anchor and collar *Mechanical Devices.* a gate hinge arrangement consisting of a tube that is anchored in a masonry block at one end and fitted with a collar at the other.

anchor ball *Naval Architecture.* **1.** a flag with a black ball, displayed between the bow and foremast of a vessel that is anchored in or near a navigation channel. **2.** a grapnel attached to a cable and fired into the rigging of a stranded ship for use in rescue operations.

anchor bend *Naval Architecture.* a knot used to secure an anchor line to the anchor.

anchor block *Building Engineering.* a block set into a brick or masonry wall as a surface for fastening objects to the wall. Also, DEADMAN.

anchor bolster see HAWSE BOLSTER.

anchor bolt *Civil Engineering.* a bolt connecting a structure to its foundation to resist overturning from the lateral forces of seismic and wind loads.

anchor charge *Engineering.* the preloading of several charges in a blast hole at once, with the upper charges held down by an anchor.

anchor corners *Graphic Arts.* a layout technique in which illustrations or display type are placed at each corner of a page, especially in a newspaper.

anchored dune *Geology.* a sand dune whose movement or form has been stabilized by the growth of vegetation or the cementation of sand.

anchor escapement *Horology.* an escapement mechanism in a pendulum clock in which the pendulum continues to swing after the anchor-shaped pallet engages the escape wheel, causing the wheel to recoil slightly; a sturdier but less accurate escapement than the deadbeat escapement. Also, RECOIL ESCAPEMENT.

anchor gear *Naval Architecture.* **1.** an anchor and its accompanying chain. **2.** the equipment used to operate a vessel's anchors, including windlasses, hawsers, and related fittings.

anchor ice *Hydrology.* underwater ice that is formed on submerged objects or that is attached to the bottom of a body of water that itself is not frozen.

anchorite *Petrology.* a variety of diorite having nodules of mafic minerals and veins of felsic mineral.

anchor log *Civil Engineering.* a log, beam, or concrete block that is buried some distance from a retaining wall and connected to it by tie rods, in order to provide support for the wall by its weight and by the resistance of the passive earth pressure against it.

anchor pile *Civil Engineering.* a pile that is used to resist tension or lateral forces.

anchor plant *Botany.* a term for a plant that is grown to secure the soil against erosion, as on a slope or hillside.

anchor plate *Civil Engineering.* a heavy plate that is buried in concrete and secures a supporting cable, as in a suspension bridge.

anchor span *Civil Engineering.* in a cantilever or suspension bridge, a section that spans the distance between the anchorage and the closest pier or tower.

anchor station *Oceanography.* a site used by a research vessel to anchor while making scientific observations.

anchor tower *Civil Engineering.* a tower to which the leg of an anchor crane is anchored.

anchor well *Naval Architecture.* a recess in the side or deck of a ship, usually toward the forward end of the vessel, into which an anchor fits when it is raised.

anchor windlass *Naval Architecture.* the motor, machinery, and rotating drum used in raising and lowering an anchor.

anchovy *Vertebrate Zoology.* any fish of the family Engraulidae, especially the species *Engraulis encrasicholus*; a small, herringlike European fish that is widely used as a food fish or as a flavoring for other foods.

anchovy pear *Botany.* **1.** the fruit of a West Indian *Grias cauliflora* tree, which tastes like and resembles a mango. **2.** the tree on which this fruit grows.

Anchusa *Botany.* a genus of plants of the family Boraginaceae, having hairy stems and one-sided clusters of trumpet-shaped flowers in colors ranging from blue or purple to white.

anchylo- see ANKYLO-.

ancillary statistic *Statistics.* a statistic whose probability distribution does not depend on the unknown parameter to be estimated.

ancipital *Botany.* having two edges, as the flattened stems of some grasses.

Ancistrocladaceae *Botany.* a monogeneric family of dicotyledonous tropical Asian and African shrubs in the order Violales that climb by twining or hooked branch tips and sometimes produce alkaloids.

ancium *Ecology.* a canyon forest community.

ancon *Architecture.* a scrolled bracket or console that supports a cornice above a door or window.

anconeal *Anatomy.* pertaining to or located near the elbow.

anconeus *Anatomy.* an extensor muscle of the forearm, located at the back of the elbow.

ancophyte *Ecology.* a plant having a canyon forest habitat. Thus, **ancophilous, ancophile.**

ancora *Invertebrate Zoology.* the initial, anchor-shaped growth stage of graptolithinids.

ANCOVA *Statistics.* an acronym for analysis of covariance.

ancylite-(Ce) *Mineralogy.* $SrCe(CO_3)_2(OH) \cdot H_2O$, a light yellow to orange to gray orthorhombic mineral having a specific gravity of 3.95 and a hardness of 4 to 4.5 on the Mohs scale; found in hydrothermal veins with mafic and felsic alkalic igneous rocks.

ancylo- see ANKYLO-.

Ancylopoda *Paleontology.* a suborder of perissodactyl ungulates that includes the Eomoropidae as well as the widespread chalicotheres; extant in the Eocene to Pleistocene.

Ancylostoma *Invertebrate Zoology.* the type genus of the hookworm family Ancylostomidae, with teeth in the buccal cavity resembling hooks. They are parasites in the intestines of certain mammals, especially the species *Ancylostoma duodenale,* which causes microcytic hypochromic anemia in humans.

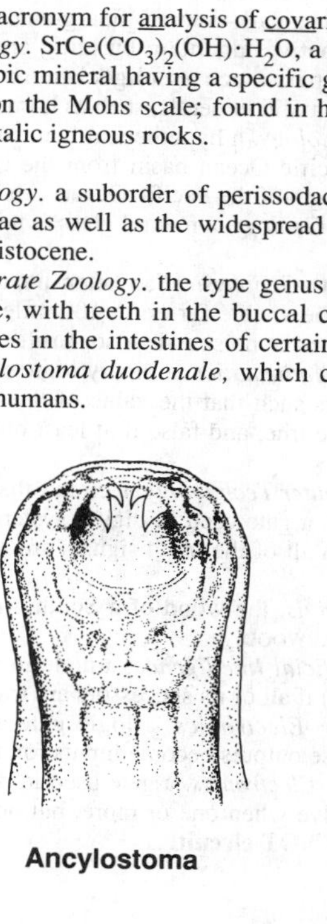

Ancylostoma

ancylostomiasis *Medicine.* an infection caused by hookworms.

Ancylostomidae *Invertebrate Zoology.* a family of hookworms in the order Strongyloidea. Also, **Ancylostomatidae.**

AND *Mathematics.* a logical operator used to connect elements in an AND function; if A is true and B is true, then (and only then) A AND B is true. *Computer Programming.* an instruction that computes the bitwise logical AND function on two computer words.

And see ANDROMEDA.

andalusite *Mineralogy.* Al_2SiO_5, a pinkish, reddish, or violet orthorhombic, orthosilicate mineral occurring in nearly square prisms, having a specific gravity of 3.13 to 3.16 and a hardness of 6.5 to 7.5 on the Mohs scale; found in slates and argillaceous schists as a contact mineral; trimorphous with kyanite and sillimanite.

Andaman and Nicobar Islands *Geography.* an island chain in the Andaman Sea.

Andaman Sea *Geography.* the eastern part of the Bay of Bengal, west of the Malay Peninsula.

Andean-type continental margin *Geology.* a continental margin where oceanic lithosphere descends beneath an adjacent continent, resulting in andesitic volcanism.

Andept *Geology.* a suborder of the soil order Inceptisol that develops chiefly from glassy material derived from volcanic explosions and has a low bulk density.

Andersen sampler *Biotechnology.* a device that collects airborne microflora by filtering air through a series of different-sized mesh screens and then having the collected material deposited on agar plates.

Anderson, Carl David 1905–1991, American physicist; studied cosmic rays; Nobel Prize for discovery of the positron.

Anderson, Philip Warren born 1923, American physicist; with Mott and Van Vleck, Nobel Prize for fundamental research on electronic structure of magnetic and disordered systems.

Anderson bridge *Electronics.* a six-branch bridge that measures self-inductance in terms of resistance and capacitance.

Anderson-Dayem bridge *Physics.* a junction within a superconducting film, formed by a narrow stricture (a few micrometers in width) that allows only electron pairs to cross.

andersonite *Mineralogy.* $Na_2Ca(UO_2)(CO_3)_3 \cdot 6H_2O$, a bright yellowish-green, secondary trigonal mineral with rhombohedral crystals, having a specific gravity of 2.8 and a hardness of 2.5 on the Mohs scale.

Andes *Geography.* the world's longest mountain range, extending approximately 4500 miles from Panama along western South America to Cape Horn; highest peak, Mount Aconcagua (22,831 ft).

andesine *Mineralogy.* a colorless or white triclinic plagioclase feldspar having a specific gravity of 2.68 and a hardness of 6 to 6.5 on the Mohs scale; commonly found as a rock-forming mineral in igneous rocks with intermediate silica content.

andesite *Geology.* a dark-colored, fine-grained, extrusive volcanic rock associated with active mountain-building areas and characterized by plagioclase feldspars, such as oligoclase or andesine, together with small amounts of mafic minerals, such as horneblende or olivine.

andesite line *Geology.* a hypothetical boundary separating the oceanic basalts of the Pacific Ocean basin from the continental andesite rocks that mark the margin of the Pacific Ocean. Also, MARSHALL LINE.

andesitic glass *Geology.* a natural glass that is the chemical equivalent of andesite.

Andes lightning *Geophysics.* an often intense electrical coronal discharge that may be visible for up to 100 miles and, because of the distance, appears to make no sound. Also, **Andes glow.**

AND function *Mathematics.* a binary logic operator that operates on a pair of statements such that the value of the operator is true if both of the statements are true, and false if at least one statement is false. Also, CONJUNCTION.

AND gate *Computer Technology.* a circuit that computes an AND function. *Electronics.* a gate used in digital systems, whose output signal is active only when all of the input signals are active. Also, **AND circuit, AND element.**

andiroba *Materials.* the wood of a South American tree, *Gymnanthes lucida.* Also, CRABWOOD.

AND node *Artificial Intelligence.* a node in an AND/OR graph that is true (or satisfied) if all of its successors are true.

AND/NOR gate *Electronics.* a logic gate array that consists of two AND gates whose outputs become inputs for a single NOR gate.

AND NOT gate *Electronics.* a gate used in digital systems, whose output signal is active when one or more, but not all, input signals are active. Also, **AND NOT circuit.**

AND-OR gate *Electronics.* a gate used in digital systems, whose output signal is active for many possible AND and OR combinations of the input signals. Also, **AND-OR circuit.**

AND/OR graph *Artificial Intelligence.* a graph or tree structure describing the decomposition of a goal in terms of alternative subgoals (OR nodes) or combinations of subgoals that must all be satisfied (AND nodes).

andorite *Mineralogy.* $PbAgSb_3S_6$, an opaque, metallic, dark gray to black orthorhombic mineral occurring as tabular crystals, having a specific gravity of 5.38 and a hardness of 3 to 3.5 on the Mohs scale; closely related to uchucchacuaite and ramdohrite.

ando soil *Geology.* a group of volcanic soils that contain a large percentage of the mineral allophane, which aids in storing water and inhibits the decay of organic matter.

Andr see ANDROMEDA.

andr- a combining form meaning "male," as in *android.*

Andrada e Silva, José Bonifácio d' 1763?–1838, Brazilian scientist and statesman; discovered new metals and alloys.

Andrade, Edward Neville da Costa 1887–1971, English physicist; discovered a formula that relates the viscosity of liquids to temperature.

Andrade's creep law *Mechanics.* a law stating that strain in a creeping solid is proportional to the cube root of time before its rate becomes constant.

Andrade's indicator *Chemistry.* $C_{20}H_{17}Na_2N_3O_9S_3$, a pH indicator having a molecular weight of 586.

andradite *Mineralogy.* $Ca_3Fe_2^{+3}(SiO_4)_3$, a variously colored cubic garnet containing calcium and iron, having a specific gravity of 3.7 to 4.1 and a hardness of 6.5 to 7 on the Mohs scale.

Andral's decubitus *Surgery.* the position of an individual lying on the sound or unaffected side, as assumed by a patient in early stages of pleurisy.

Andreaeales *Botany.* an order of bryophytes of the class Andreaeopsida, characterized by an unswollen tapering foot, very large thick-walled spores, the presence of seta, and a preference for limestone.

Andreaeobryaceae *Botany.* a monotypic moss family of the order Andreaeales, occurring in northwestern North America; contains only one genus, **Andreaea,** the granite mosses.

Andreaeopsida *Botany.* a small, ancient class of bryophytes, distinguished by capsule dehiscence along longitudinal slits and by the small, brittle, epilithic cushions or tufts of the gametophyte; distributed worldwide, particularly on rocks in arctic, antarctic, and cold temperate regions.

Andrenidae *Invertebrate Zoology.* a family of mining or burrowing bees in the order Hymenoptera; short-tongued solitary bees making complex underground nests.

andreoblastoma see ARRHENOBLASTOMA.

Andrews, Roy Chapman 1884–1960, American explorer and naturalist; made important discoveries in Asian fossil fields.

Andrews, Thomas 1813–1885, Irish chemist; discovered critical temperature in liquefication of gasses.

andrewsite *Mineralogy.* $(Cu^{+2},Fe^{+2})Fe_3^{+3}(PO_4)_3(OH)_2$, a silky, aquamarine-colored orthorhombic mineral having a specific gravity of 3.475 and a hardness of 4 on the Mohs scale.

andrite *Geology.* a meteorite composed primarily of augite together with small amounts of olivine and troilite.

andro- a combining form meaning "male," as in *androgenesis.*

androconium *Entomology.* any of certain modified scales on the wings of male butterflies and moths, which produces a scent that attracts the female.

androcyte *Cell Biology.* in many lower plants, a cell that develops into a spermatozoid upon differentiation.

androecium *Botany.* the male part of a flower, consisting of stamens and sometimes also staminodes.

androgen *Endocrinology.* a class of male sex hormones related to the steroid androstane and produced in the adrenal cortex and the testes; includes testosterone, androsterone, and androstenolone; responsible for the development of secondary male characteristics, such as a deep voice and facial hair. Also, **androgenic hormone.**

androgenesis *Developmental Biology.* the development of an embryo that contains only paternal chromosomes.

androgenetic merogony *Developmental Biology.* the development of a fragment of an egg that lacks female nuclei but that has been fertilized by a sperm cell.

androgenic *Biology.* describing genetic or hormonal conditions producing male characteristics.

androgenic gland *Invertebrate Zoology.* endocrine tissues around the sperm ducts in male crustaceans that control the development of testes and male sexual characteristics.

androgen unit *Biology.* a unit used for the standardization of male sex hormones.

androgynous [an´dräj´ə nəs] *Biology.* exhibiting characteristics of both sexes; especially, having both male and female sex organs. *Psychology.* showing or possessing both male and female traits.

androgyny [an´dräj´ə nē] *Biology.* a state of sexual ambiguity characterized by the presence of both male and female sex organs. *Psychology.* the presence in an individual of both male and female traits.

android *Biology.* relating to or resembling a human. *Robotics.* a robot having the form or features of a human being.

androma see ARRHENOBLASTOMA.

Andromeda [an dräm´ ə də] *Astronomy.* the Chained Woman, a constellation of the northern sky that is seen best in the autumn and whose brightest star is Alpheratz. (Named for a character in Greek mythology, an Ethiopian princess who married Perseus after he freed her from being chained to a rock.)

Andromeda Galaxy *Astronomy.* a spiral galaxy also known as M31 that lies at a distance of about 2.2 million light-years; it is visible to the naked eye and is thought to be similar to the Milky Way. Also, **Andromeda Nebula.**

Andromeda strain *Microbiology.* a hypothetical new type of microorganism that would cause widespread destruction of life on earth because of a lack of resistance to it. (From the 1969 novel *The Andromeda Strain,* in which a previously unknown and highly lethal form of bacteria is inadvertently brought to earth by a returning spacecraft.)

Andromedids *Astronomy.* a meteor shower peaking in late November whose members appear to come from the constellation Andromeda. Also, BIELIDS.

andromedotoxin *Toxicology.* a poison found in the leaves and bark of various ericaceous plants of the genera Andromeda, Kalmia, and Rhododendron; toxic to sheep and other grazing livestock.

andromerogony *Developmental Biology.* **1.** the formation of an egg fragment after the cutting, shaking, or centrifugation of a fertilized or unfertilized egg. **2.** the formation of a portion of an egg that contains only the male pronucleus because the nucleus of the egg has been removed before the male and the female pronuclei have fused.

andromonoecious *Botany.* of a plant, bearing both staminate and monoclinous flowers.

andropathy *Medicine.* any disease or disorder that affects only males rather than both sexes.

androphage *Microbiology.* a male-specific bacteriophage; a bacterial virus that will only infect bacteria with an F plasmid, i.e., male or donor cells.

androphilous *Ecology.* living or thriving in proximity to humans. Thus, **androphile, androphily.**

androphobia [an´drə fō´bē ə] *Psychology.* an irrational fear of men or of the male sex.

androphore [an´drə fôr´] *Botany.* a staminal column that is formed by a union of filaments, as in a monadelphous plant. *Invertebrate Zoology.* in coelenterates, a generative bud or modified medusa in which only male elements are developed.

Androsace *Botany.* **1.** a genus of herbs of the family Primulaceae that are found in the Northern Hemisphere and have tufted leaves, small pink or white flowers, and capsular fruits. **2. androsace.** a plant of this genus, especially the rock jasmine.

androsin *Biochemistry.* $C_{15}H_{20}O_8$, a glucoside sugar compound found in the herbal plant *Apocynum androsaemifolium.* When broken down through hydrolysis, it produces glucose and avetonvanillone, which is used as a drug to increase heart muscle tone.

androsperm *Biology.* a sperm carrying a Y chromosome.

androspore *Botany.* a zoospore that is typical of algae of the family Oedogoniaceae; develops into a small male plant that produces true spermatozoids.

androstane *Biochemistry.* $C_{19}H_{32}$, the source or parent hydrocarbon of the androgen hormones. Also, ETIOALLOCHOLANE.

androstenedione *Biochemistry.* one of three types of androgens secreted by the ovaries and adrenals that are involved in the synthesis of testosterone.

androsterone *Biochemistry.* $C_{19}H_{30}O_2$, a male sex hormone synthesized from progesterone and found in the urine and plasma of men and women; seven times weaker than testosterone.

anechoic [an´ e kō´ ik] *Acoustics.* echo-free; having the property of absorbing sound, as tiles in a sound room. *Radiology.* of or referring to a chamber or material that allows ultrasound waves to pass without reflecting back to their source.

anechoic chamber *Acoustical Engineering.* a room whose walls are constructed of sound-absorbing material in order to minimize internal sound reflection; used in acoustical testing. Also, **anechoic room, anechoic test chamber.**

anecumene *Geography.* see NEGATIVE AREA.

anelasticity [an´ē las tis´i tē] *Mechanics.* the failure of a material's strain to vary proportionately with applied stress due to small time effects or plastic deformation.

anelectric *Physics.* relating to the incapability of being electrically charged by friction.

Anelytropsidae *Vertebrate Zoology.* rare, legless, skinklike lizards of the family Amphisbaenidae, represented by a single species found in Mexico.

anemia *Pathology.* a condition in which the level of hemoglobin in the blood is below the normal range and there is a decrease in the production of red blood cells, often causing pallor and fatigue.

anemic *Pathology.* **1.** of or relating to anemia. **2.** affected by or suffering from anemia.

anemic infarct see WHITE INFARCT.

anemo- a combining form meaning "wind," as in *anemometer.*

anemobiagraph *Engineering.* a pressure-tube anemometer that has springs to keep the wind scale of the float manometer linear.

anemochore *Botany.* a plant whose seeds or spores are dispersed by the wind.

anemoclast *Geology.* a rock fragment that has been broken off and rounded by wind action.

anemoclinometer *Engineering.* an instrument used to measure the direction of wind in relation to the horizontal plane.

anemogram *Engineering.* a graphic representation of the record produced by an anemograph.

anemograph *Engineering.* an instrument designed to produce a permanent record of the speed (and sometimes the direction) of the wind, as measured by an anemometer.

anemology *Meteorology.* an older term for the study of winds.

anemometer *Engineering.* an instrument that measures wind speed, typically consisting of three or four hemispherical cups attached to the ends of arms that radiate from a vertical spindle, so as to be rotated by an air current.

anemometry *Meteorology.* the study of measuring and recording the direction and force of the wind, including its vertical component.

Anemone *Botany.* **1.** a genus of plants of the buttercup family Ranunculaceae, native to the northern United States and Canada, having a tall, slender stem and cup-shaped flowers in colors ranging from white to pink or purple. **2. anemone.** any particular plant of this genus, especially *A. quinquefolia,* the wood anemone, a delicate plant with deeply cut leaves, popularly known as WINDFLOWER.

anemonin *Toxicology.* a toxin found in plants of the genus *Anemone.*

anemonism *Toxicology.* poisoning by anemonin; symptoms may include blistering and acute gostrointestinal upset.

anemophily *Botany.* the pollination or dispersal of spores or pollen by means of wind. Thus, **anemophilous.**

anemophobia *Psychology.* an irrational fear of wind or drafts.

anemoscope *Engineering.* an instrument used to detect and measure the direction of slow-moving air currents.

anemosis *Botany.* a condition in which the annual layers of certain trees are separated from one another, frequently attributed to the effects of strong winds on the trunk, although some believe the condition results from exposure to frost or lightning. Also, WIND SHAKE.

anemotaxis *Biology.* the movement or orientation of an organism in response to wind.

anemotropism *Biology.* the movement of an organism away from or toward air currents or wind.

anemovane *Engineering.* an instrument that serves as both an anemometer and a wind vane.

anencephaly *Medicine.* a congenital malformation of the skull, in which the brain fails to develop and is either partially or completely absent. Thus, **anencephalic, anencephalous.**

anenterous *Zoology.* lacking intestines.

Anepitheliocystidia *Invertebrate Zoology.* a proposed superorder of digenetic Trematodes.

anergized culture see HABITUATED CULTURE.

anergy *Immunology.* the inability of an organism to produce an immune response to an antigen or antibody. *Medicine.* an abnormal lack of energy.

anerobic see ANAEROBIC.

anerobic sediment *Geology.* a highly organic sediment that is formed in hydrogen sulfide-rich water where limited circulation results in the absence or near absence of oxygen at the surface of the sediment. Also, **anaerobic sediment.**

aneroid *Engineering.* describing an instrument that operates without liquid. Thus, **aneroid calorimeter, aneroid flowmeter, aneroid liquid-level meter,** and so on.

aneroid altimeter *Engineering.* an altimeter whose readings depend on an aneroid barometer or on the movement of a pressure-sensitive element.

aneroid barograph *Engineering.* an aneroid barometer that continuously and automatically records changes in pressure on a rotating drum. Also, ANEROIDOGRAPH.

aneroid barometer *Engineering.* a small barometer having a sealed, bellowslike box that contracts or expands as air pressure rises or drops; used to measure the local atmospheric pressure and to estimate altitude.

aneroid barometer

aneroid capsule *Engineering.* in an aneroid barometer, a thin-walled, airtight, accordion-like compartment designed to expand and contract in response to air pressure. Also, **aneroid bellows.**

aneroid diaphragm *Engineering.* a thin metal plate that covers the end of an aneroid capsule and moves about an axis with the expansion or compression of that capsule.

aneroidograph see ANEROID BAROGRAPH.

anesthekinesia *Neurology.* motor paralysis accompanied by the loss of motor power or sensibility. Also, **anesthecinesia**.

anesthesia *Physiology.* a partial or complete loss of sensation or feeling, especially of pain.

anesthesiologist *Medicine.* a physician who specializes in the administration of anesthetics during surgery, labor and delivery, or other procedures.

anesthesiology *Medicine.* the branch of medicine that is concerned with anesthesia and anesthetics.

anesthesol see ETHYL *p*-AMINOBENZOATE HYDROCHLORIDE.

anesthetic *Pharmacology.* **1.** capable of producing anesthesia. **2.** any of various drugs that have an anesthetic effect, such as procaine or ether.

anesthetist *Medicine.* a nurse or technician who is trained to administer anesthetics.

anestrus *Vertebrate Zoology.* for a cyclically breeding female mammal, an interval of sexual inactivity between two estrus (sexually active) periods.

anethole [an´ə thōl´] *Organic Chemistry.* $CH_3OC_6H_4CH{=}CHCH_3$, white, sweet-tasting crystals that melt at 22.5°C and are soluble in alcohol, ether, acetone, and benzene; contained in the seeds of the anise herb, *Pimpinella anisum,* and used in making perfumes and food flavoring, and as a sensitizer in color-bleaching processes in color photography. Also, 4-PROPENYLANISOLE.

aneuploidy *Genetics.* a genetically unbalanced condition in which a cell or an organism has a number of chromosomes that is not an exact multiple of the haploid number for that species.

Aneuraceae *Botany.* a cosmopolitan family of liverworts of the order Metzgeriales, characterized by thalloid organization, the lack of leaf-like lateral lobes or a differentiated midrib, and sparing rhizoids usually restricted to stolons and basal thallus areas.

aneurin see THIAMINE.

aneurogenic *Neurology.* of, relating to, or characterized by the lack of formation of nerve fibers.

aneurysm [an´yə riz´əm] *Medicine.* a sac formed by the dilation of the wall of a vein or artery; sometimes congenital, but usually caused by disease and, occasionally, by trauma. Also, **aneurism.**

aneurysmectomy *Surgery.* the excision of an aneurysm by removal of the sac.

aneurysmoplasty *Surgery.* the plastic reconstruction of an aneurysmal artery by opening the sac of the aneurysm and suturing its walls to restore the normal dimension to the lumen of the artery. Also, ENDOANEURYSMORRHAPHY, ENDOANEURYSMOPLASTY.

aneurysmotomy *Surgery.* the incision into the sac of an aneurysm.

aneusomatic *Cell Biology.* of an organism or a group of cells, possessing variable numbers of each chromosome.

aneuspory *Botany.* the production of an unusual number of spores from each spore mother cell.

ANF antinuclear factor; atrial natriuretic factor.

Anfinsen, Christian Boehmer born 1916, American biochemist; shared Nobel Prize for structural analysis of RNA and other proteins.

angaralite *Mineralogy.* $Mg_2(Al,Fe)_{10}Si_6O_{29}$, a black mineral of the chlorite group.

Angara Shield *Geology.* a small shield area of exposed Precambrian rock located in Siberia.

angel *Engineering.* an informal term for a radar image coming from an unseen, undefined source; it may be caused by birds, swarms of insects, or atmospheric conditions. Also, **angel echo.**

angel dust *Pharmacology.* an informal name for phencyclidine (PCP).

angelellite *Mineralogy.* $Fe_4^{+3}(AsO_4)_2O_3$, a blackish-brown triclinic mineral occurring as globular and crystalline incrustations on andesite, with a specific gravity of 4.9 and a hardness of 5.5 on the Mohs scale.

Angel Falls *Geography.* the world's highest waterfall (3212 ft), located in southeastern Venezuela.

angelfish *Vertebrate Zoology.* any of various species of the order Perciformes that are thin and deep-bodied with elongated fins, primarily of the family Cichlidae and native to fresh waters of South America, including some tropical marine species of the family Pomacanthidae.

Angelica *Botany.* **1.** a genus of umbelliferous plants of the parsley family. **2. angelica.** any herb belonging to the genus *Angelica,* especially *A. archangelica.* Also, ARCHANGEL.

angelica oil *Materials.* a highly aromatic medicinal oil distilled from the roots and fruit of the herb *Angelica archangelica* and used in the production of certain perfumes, medicines, and food products.

angelique *Materials.* the wood of a South American tree, *Dicorynia paraensis,* which is resistant to fungi and insects; used in marine construction.

angi- a prefix meaning "vessel" (usually referring to a blood vessel), as in *angiectomy.*

angialgia *Medicine.* pain in a blood vessel.

angiectasis [an´jə ek´tas is] *Medicine.* the gross dilation and elongation of a blood vessel.

angiectomy [an´jə ek´tə mē] *Surgery.* the excision or resection of a blood vessel segment.

angiitis *Medicine.* inflammation of a blood vessel or lymph vessel. Also, VASCULITIS.

angina [an jī´nə] *Medicine.* **1.** a spasmodic, choking, or constricting pain, especially **angina pectoris,** sharp thoracic pain accompanied by a feeling of suffocation, usually due to a lack of oxygen of the myocardium and typically brought on by exertion, stress, or excitement. **2.** any inflammation of the throat.

angio- a prefix meaning "vessel" (usually a blood vessel), as in *angioplasty.*

angioaccess *Surgery.* the site of entry into a blood vessel, as in a blood vessel used for recurrent hemodialysis.

angioblast *Developmental Biology.* **1.** a cell involved in blood vessel formation. **2.** the primordial mesenchymal tissue from which embryonic blood cells and vascular endothelium differentiate.

angioblastoma *Oncology.* a vascular tumor of the brain or spinal cord.

angiocardiography *Radiology.* an X-ray examination of the heart and great vessels following the injection of a radiopaque contrast medium into a blood vessel or one of the cardiac chambers.

angiocardiopathy *Cardiology.* any disease of the heart and blood vessels.

angiocarditis *Cardiology.* inflammation of the heart and great blood vessels.

angiocarpic development *Mycology.* spore development of certain fungi, such as mushrooms, in which the spores develop in a chamber or cavity not exposed to the environment until the spores are ready to be released.

angiocarpous *Botany.* describing a fruit that is enclosed in a distinct covering, such as a shell or husk.

angiochondroma *Oncology.* a chondroma surrounded by an excessive number of blood vessels.

angioedema *Medicine.* an acute vascular reaction involving subcutaneous or subdermal swelling in the form of giant wheals; it may be hereditary or may be caused by allergy, infection, or emotional stress. Also, **angioneurotic edema.**

angiofibroma *Oncology.* a benign tumor composed of blood vessels and fibrous tissue. Also, TELANGIECTATIC FIBROMA.

angiogenesis *Developmental Biology.* the formation of new blood vessels.

angiogenin *Enzymology.* a polypeptide that stimulates the formation of endothelial cells.

angiogranuloma *Oncology.* a rapidly growing, ulcerated lesion with a substantial amount of vascular granulation tissue, representing an inflammatory response.

angiography *Radiology.* an X-ray examination of the blood vessels following the injection of a radiopaque contrast medium; used as a diagnostic tool in conditions such as stroke or heart attack. The X-ray that is produced is an **angiogram.**

angioleiomyoma *Oncology.* a benign tumor that consists of convoluted, thick-walled blood vessels and well-differentiated muscle elements; it usually occurs as a painful, nodular tumor on a lower extremity in middle-aged women. Also, **angiomyoma.**

angiolipoma *Oncology.* a benign tumor composed of blood vessels and fatty tissue.

angiology *Medicine.* the branch of medicine concerned with the study of the blood vessels and the lymph vessels.

angioma *Oncology.* a tumor of blood vessels or lymph vessels.

angiomatosis *Oncology.* a condition characterized by the formation of multiple angiomas; sometimes accompanied by an overgrowth of fat and tissue.

angiomyolipoma *Oncology.* a benign tumor consisting of a mixture of blood vessels, fatty tissue, and muscle elements, usually occurring in the kidney.

angiomyxoma *Oncology.* a chorioangioma of the placenta in which capillary-like blood vessels are evident; often extending into the myxomatous tissue of the umbilical cord.

angioneoplasm *Oncology.* a benign, vascular tumor composed of blood vessels.

angioneurectomy *Surgery.* the excision of the blood vessels and nerves of a part.

angioneuropathy *Neurology.* **1.** any neuropathy that is associated with vascular disease and principally affects blood vessels. **2.** a disorder of the vasomotor system, such as vasomotor paralysis.

angioparesis see VASOMOTOR PARALYSIS.

angioplasty *Surgery.* **1.** a surgical reconstruction of blood vessels. **2.** an angiographic procedure for eliminating areas of narrowing in blood vessels. Also, PERCUTANEOUS TRANSLUMINIAL ANGIOPLASTY.

angiopressure *Surgery.* the application of pressure on a blood vessel to control bleeding.

angiorrhaphy *Surgery.* the suture of a vessel or vessels.

angiosarcoma *Oncology.* a malignant tumor that is made up of endothelial and fibroblastic tissue.

angiospasm see VASOSPASM.

angiosperm *Botany.* a plant or tree belonging to the class Angiospermae, such as the apple or the oak; a flowering plant; a group of plants whose seeds are borne within a mature ovary (fruit).

Angiospermae *Botany.* a large class of flowering plants that bear their seeds in a closed seed vessel.

angiostenosis *Medicine.* a narrowing of the lumen of a vessel.

angiostomy *Surgery.* **1.** an operation in which an opening is made in a blood vessel. **2.** the opening so made.

angiostrophy *Surgery.* the twisting of a vessel to arrest hemorrhage. Also, **angiostrophe.**

angiotensin *Endocrinology.* any of a group of blood pressure-regulating peptides (angiotensins I, II, and III) formed by the sequential actions of three enzymes (renin, angiotensin converting enzyme, and a nonspecific peptidase) on angiotensinogen. **Angiotensin II** is a powerful vasopressor and a stimulator of aldosterone secretion by the adrenal cortex.

angiotensin converting enzyme *Enzymology.* an enzyme that catalyzes the conversion of angiotensin I to angiotensin II by removing a dipeptide from the C-terminal end. Also, DIPEPTIDYL I CARBOXYPEPTIDASE.

angiotensinogen *Endocrinology.* an α_2-globulin in plasma; the substrate from which angiotensin I is released by the action of renin. Also, RENIN SUBSTRATE.

angiotomy [an´jē ät´ə mē] *Surgery.* the cutting or severing of a blood vessel or lymph vessel.

angiotribe *Surgery.* a very powerful forceps that uses a screw to apply pressure, especially to crush tissue containing an artery, in order to control bleeding. Also, VASOTRIBE.

angiotripsy *Surgery.* the use of an angiotribe to stop bleeding by crushing the tissue containing an artery. Also, VASOTRIPSY.

angitis see ANGIITIS.

angle *Mathematics.* **1.** a measure of rotation about a fixed axis; commonly measured in degrees, radians, or revolutions. **2.** the region between two rays with a common vertex swept out by such a rotation. **3.** in higher dimensions, the angle between two hyperplanes is defined as the angle between the normals to the hyperplanes. The term *angle* may refer to a geometric figure, numeric quantity, or signed algebraic quantity.

angle azimuth indicator see RISER ANGLE INDICATOR.

angle bar *Civil Engineering.* see ANGLE IRON.

angle beam *Engineering.* ultrasonic waves that are directed at a specific angle for the purpose of inspecting a metallic surface.

angle blasting *Engineering.* sandblasting that is performed at an acute angle.

angle block *Engineering.* a small wood block used for joining wood pieces, especially at right angles, in order to make joints more rigid. Also, GLUE BLOCK.

angle board *Design Engineering.* a board having an angled face that allows it to be used as a guide for planing or cutting other boards at the same angle.

angle bond *Building Engineering.* a metal tie used in masonry work for bonding wall corners.

angle brace *Engineering.* **1.** a bar or support fixed across the inside angle in a framework to strengthen it. Also, ANGLE TIE. **2.** a tool used to drive a wood bit in a corner where lack of space makes it awkward to use an ordinary brace.

angle bracket *Architecture.* **1.** a bracket in an angle or corner of a molded cornice. **2.** a bracket set at an angle other than perpendicular to the wall from which it projects.

angle brick *Building Engineering.* a brick that has a corner with an oblique angle.

angle buttress *Architecture.* either of a pair of buttresses set at right angles to each other, forming the corner of a building.

angle capital *Architecture.* the capital of a corner column, especially an Ionic capital modified to project equally on both sides of the corner.

angle cleat *Building Engineering.* a bracket made of a short piece of angle bar, used to connect structural members at right angles. Also, **angle clip.**

angle closer *Engineering.* a brick cut at an angle and used to close the bond at the corner of a wall.

angle-closure glaucoma *Medicine.* a type of glaucoma that occurs when the angle between the iris and the cornea becomes closed due to contact between the iris and the inner surface of the trabecular meshwork, restricting the normal outflow of aqueous humor. Also, CLOSED-ANGLE GLAUCOMA, NARROW-ANGLE GLAUCOMA.

angle collar *Mechanical Devices.* a fitting made of cast iron, used to connect pipe ends that are not in alignment.

angle cut *Mining Engineering.* a drilling pattern in which the drill holes converge to blast out a core, creating an open free face for following shots, which are timed to follow rapidly.

angle cutter see ANGULAR CUTTER.

angle diversity *Telecommunications.* in the tropospheric scattering propagation of electromagnetic waves, the difference created by two or more feeders that produce multiple beams from the same reflector at slightly different launch angles.

angledozer *Mechanical Engineering.* a tractor having a broad steel blade that is set at an angle for pushing earth, snow, or debris to one side. Also, ANGLING DOZER.

angled stair *Architecture.* a stair whose successive flights are set at an angle other than 180° to one another.

angle fillet *Engineering.* a triangular wooden strip placed over an internal joint connecting two surfaces that meet at an angle of 180° or less.

angle fishplates *Civil Engineering.* railroad plates that join the rails and prevent the joint from sagging under heavy cars and locomotives.

angle float *Mechanical Devices.* a tool used for finishing plaster or concrete to adjacent 90° corner angles within a room.

angle gauge *Mechanical Devices.* an instrument used to lay out or check angles in carpentry, masonry, or bricklaying operations.

angle iron *Civil Engineering.* a rolled steel member with an L-shaped cross section. *Building Engineering.* an angled piece of steel or iron, especially right-angled, used to connect or reinforce two structural members. Also, ANGLE BAR, ANGLE SECTION.

angle jamming *Electronics.* radar jamming in which the return pulse is mixed with another pulse of erroneous bearing or phase.

angle joint *Engineering.* a joint used to connect two pieces of lumber at an angle.

angle lacing *Civil Engineering.* a system of laced angle iron that connects structural members.

angle-lighting luminaire *Optics.* an electric light that distributes light asymmetrically.

angle modillion *Architecture.* a modillion at the corner of a cornice.

angle modulation *Electronics.* a modulation technique that varies the angle of the carrier wave, as in phase modulation and frequency modulation.

angle noise *Electromagnetism.* interference in radar reception that results from variations in the angle at which an echo is received.

angle of action *Mechanical Engineering.* the angle of revolution of one of two meshed gears when at least one tooth is in contact.

angle of advance see ANGULAR ADVANCE.

angle of approach *Civil Engineering.* the maximum angle of incline onto which a vehicle can travel from the horizontal without hindrance. *Ordnance.* the angle between the line along which a target is moving and the line along which a gun is pointed to hit the target. *Mechanical Engineering.* the angle of revolution of one of a pair of wheels in gear, from the initial contact between two teeth until the pitch points of the same teeth fall together.

angle of arrival *Electromagnetism.* an angle between the propagation direction of a given form of electromagnetic radiation and the direction associated with a receiving antenna.

angle of attack *Aviation.* the angle between the wing chord or some reference axis of an airfoil and the direction of undisturbed airflow with respect to the airfoil; used in calculating and adjusting the lift of an aircraft. *Mining Engineering.* the angle created by the direction of air approach and the chord of the aerofoil section of a mine fan.

angle of bite see ANGLE OF NIP.

angle of cant see ANGLE OF PITCH.

angle of clearance *Ordnance.* the angle between the line along which a gun could be pointed directly at a target and the line along which it must be pointed to clear any obstruction between it and the target.

angle of climb *Aviation.* the angle between the path of a climbing aircraft and a horizontal plane.

angle of contact *Fluid Mechanics.* the angle between a vertical container surface and the tangent to the line of contact between two fluids in the container, as in the case of a graduated cylinder half-filled with water, where the angle of contact is measured within the liquid between the vertical cylinder wall and a tangent to the meniscus. Also, CONTACT ANGLE.

angle of convergence *Optics.* the vertical angle between the lines of sight of the two eyes of an optical instrument with two eyepieces. *Navigation.* see ANGULAR PARALLAX. *Transportation Engineering.* the angle at which traffic enters the traffic stream of a road; the narrower the angle of convergence, the higher the speed at which arriving traffic can enter the stream. *Ordnance.* the angle at which any gun parallel to the base gun in a battery must be turned in order for that gun to point at the target.

angle of current *Hydrology.* in gauging streams, the difference between 90° and the angle made by the current with a measuring section.

angle of cut *Navigation.* the smaller of the angles formed when two position lines cross; the greater this angle, the more reliable the fix thus created.

angle of deflection *Electronics.* the angle formed by an electron beam in a cathode-ray tube as it deflects from a straight path to a new position. *Ordnance.* the horizontal clockwise angle between the axis of the bore of a gun and the line of sighting when the gun is laid for direction.

angle of departure *Aviation.* the angle between the present path of an aircraft and some other designated path. *Ordnance.* the vertical angle between the line of sight and the axis of the bore of a gun at the instant the projectile leaves the muzzle.

angle of depression *Cartography.* the angle in a vertical plane between the horizontal and a descending line; the complement of tilt. Also, MINUS ANGLE, DESCENDING VERTICAL ANGLE, PLUNGE ANGLE. *Ordnance.* the vertical angle between the horizontal and the axis of the bore of a gun when the gun is pointed below the horizontal.

angle of descent *Aviation.* the angle between the path of a descending aircraft and a horizontal plane.

angle of dip *Geology.* see DIP, def. 1.

angle of divergence *Electronics.* the angle formed by the spread of an electron beam in a cathode-ray tube as it moves from the cathode to the screen.

angle of elevation *Cartography.* the angle in a vertical plane between the horizontal and an ascending line. Also, PLUS ANGLE, ASCENDING VERTICAL ANGLE. *Ordnance.* the vertical angle between the horizontal and the axis of the bore of a gun when the gun is pointed above the horizontal; i.e., the angle through which the axis of the bore must be raised so that the bullet or projectile will carry to a distant target.

angle of entry *Ordnance.* the degree of angle from the perpendicular at which a bomb enters a target.

angle of fall *Ordnance.* for a projectile, the vertical angle between the horizontal and the tangent to the trajectory when it returns to its altitude of firing.

angle of friction see ANGLE OF STATIC FRICTION.

angle of glide *Aviation.* the angle between the glide path of an aircraft and a horizontal plane.

angle of impact *Ordnance.* the acute angle with the horizontal at which a bomb or projectile strikes the ground or a target.

angle of incidence *Physics.* the angle between a ray that strikes a surface and the perpendicular to that surface at the point of incidence. *Acoustics.* the angle at which a wave strikes a reflective surface, which is normally the boundary of a different medium, and which results in a reflection from the surface, a transmission into the different medium, or a combination of the two. *Aviation.* the angle between a chord of a fixed or adjustable airfoil and the longitudinal axis of the aircraft. *Ordnance.* the angle with the vertical at which a bomb or projectile strikes the ground or a target; the complement of the angle of impact.

angle of inclination *Cartography.* an angle of elevation or an angle of depression.

angle of jump *Ordnance.* the angle between the line along which a gun is aimed and the line along which it is actually pointed at the instant when the projectile leaves the muzzle.

angle of nip *Mechanical Engineering.* the maximum included angle between two approaching faces of a crusher at which an object can still be gripped. Also, ANGLE OF BITE.

angle of obliquity *Mechanical Engineering.* the deviation in the direction of the force between two gear teeth that are in contact from the direction of their common tangent.

angle of orientation *Mechanics.* the angle between the plane containing the axis of a projectile and the tangent to the trajectory and the vertical plane containing the tangent to the trajectory.

angle of pitch *Aviation.* the angle between the longitudinal axis of an aircraft, as seen from the side, and some reference plane, usually a horizontal plane. Also, ANGLE OF CANT.

angle of pressure *Mechanical Engineering.* the angle between a gear tooth's profile and the radial line at its pitch point.

angle of radiation *Electromagnetism.* the angle between the surface of the earth and the direction of electromagnetic radiation propagating into the sky from a transmitting antenna.

angle of recess *Mechanical Engineering.* the angle of rotation of one of a pair of wheels in gear, from the coincidence of the pitch points of the two teeth until the last contact point of the same teeth.

angle of reflection *Physics.* the angle formed by the normal line to a surface and the direction of propagation of waves that are reflected from the surface.

angle of refraction *Physics.* the angle between the direction of propagation of a refracted ray and the perpendicular to the interface between the two media.

angle of repose *Engineering.* the steepest angle of a surface at which a mass of loose or fragmented material will remain standing in a pile on a surface, rather than sliding or crumbling away; the angle will vary according to the composition of the material. Also, **angle of rest.** *Mechanics.* see ANGLE OF STATIC FRICTION.

angle of roll *Aviation.* the angle between the lateral axis of an aircraft and some reference plane, usually a horizontal plane; this angle is positive when the port wing tip is higher than the starboard.

angle of safety *Ordnance.* the angle between the line along which a gun can be pointed directly at a target and the line along which it must be pointed in order to be certain of clearing friendly troops between the gun and the target.

angle of shift *Ordnance.* the horizontal angle that represents the distance a gun must be moved to shift fire from one target to another.

angle of site *Ordnance.* the vertical angle through which the axis of the bore must be moved to correct for the difference in altitude between gun and target.

angle of slide *Mechanics.* the angle, measured in degrees of deviation from the horizontal, at which loose or fragmented materials on a surface will begin to slide. Also, **angle of slip.**

angle of stall see CRITICAL ANGLE OF ATTACK.

angle of static friction *Mechanics.* for an object resting on an inclined plane surface, the maximum angle of inclination from the horizontal that can be reached before the object begins to slide down the plane under the action of gravity. Also, ANGLE OF FRICTION, ANGLE OF REPOSE.

angle of thread *Design Engineering.* the angle between the opposite sides of a screw thread, measured in an axial plane.

angle of torsion *Mechanics.* the angle through which a shaft or a similar body rotates when a torque is applied. Also, **angle of twist.**

angle of traverse *Ordnance.* **1.** a horizontal angle representing the extent to which a gun can be turned on its mount. **2.** the horizontal angle between the line from a gun to the left limit of its fire and the line to the right limit.

angle of view *Optics.* **1.** the breadth of the angle included by a camera lens and by the subject field recorded on the film; for photography it is the negative's diagonal, while in film or video it is the frame's width. **2.** in microscopy, the angle between two rays that come from opposite points on the edge of the visual field diaphragm of the eyepiece and pass through the exit pupil; the angle determines how much of the retina is covered by the image.

angle of visibility *Aviation.* the angle of vision from an aircraft that is not obstructed by any of the airplane surfaces, as seen from the position of the pilot or another crew member.

angle of yaw *Aviation.* the angle, as seen from above, between the longitudinal axis of an aircraft and a reference direction, usually the direction of travel. This angle is positive when the aircraft turns starboard.

angle plate *Mechanical Devices.* a cast iron connecting plate with one edge folded up at a right angle to the other, used for joining workpieces on a lathe face plate or machining tool table, or for marking off a workpiece on a surface plate.

angle-ply laminate *Materials Science.* an oriented fiber-polymer composite in which laminates of uniaxial composite are consolidated at specific angles to one another.

angle point *Cartography.* **1.** a point at which a stake is driven to indicate a change in the direction of a survey line. **2.** the stake itself.

angle post *Building Engineering.* **1.** a post set on a stair landing to support the handrail. **2.** in half-timber construction, the corner post.

angle press *Mechanical Engineering.* a hydraulic press having both vertical and horizontal rams that is used for molding and extruding plastics; specially designed for the production of complex moldings with deep undercuts, protrusions, or the like.

angler *Vertebrate Zoology.* any of several species of sluggish, deep-sea fishes of the family Lophiidae, many having rays on the tops of their heads ending in a luminous bulb. Also, **anglerfish.**

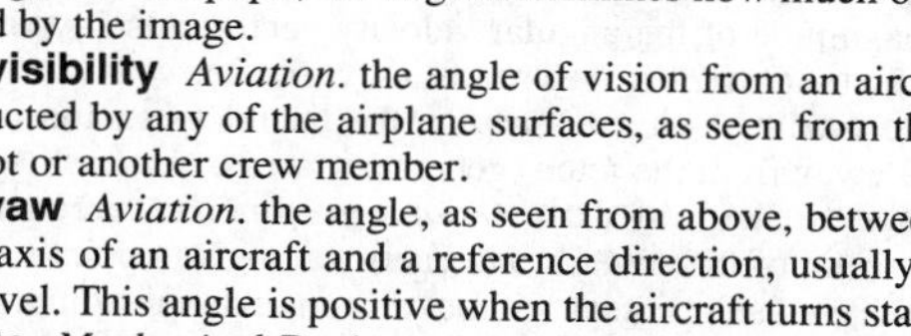

angler

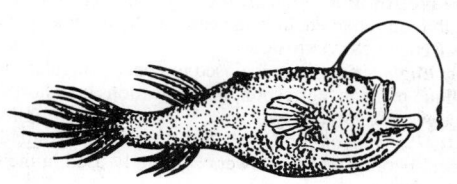

angle rafter see HIP RAFTER.

angle-resolved photoelectron spectroscopy *Spectroscopy.* the analysis of the kinetic energy of photoelectrons emitted from a solid surface, including the angles at which they are emitted, when the surface is bombarded by ultraviolet radiation.

angle rib *Architecture.* **1.** one of the diagonal ribs that divide the rectangles of a Gothic vaulting and form the main element of its structure. **2.** in decorative work, molding that ornaments an angle.

angle ridge see HIP RAFTER.

angle section see ANGLE IRON.

angle set *Mining Engineering.* **1.** a particular set in a series of sets, placed at angles to each other. **2.** a set manufactured with timber that utilizes an angle brace.

angle shot *Graphic Arts.* **1.** a still photograph taken by a camera that is tilted at an angle from the horizontal. **2.** a motion picture shot taken from a different angle than the previous shot.

anglesite *Mineralogy.* $PbSO_4$, a white or colorless orthorhombic mineral occurring as tabular or prismatic crystals, having a specific gravity of 6.38 and a hardness of 2.5 to 3 on the Mohs scale; found as a secondary mineral in lead deposits formed by the oxidation of galena.

angle staff *Building Engineering.* a vertical wooden or metal strip set flush at the exterior angle of two plastered surfaces to protect the plaster and serve as a guide for floating it.

angle-stem thermometer *Engineering.* a thermometer used to measure the temperature of the contents of a tank; fashioned at an angle to fit the shape of the tank.

angle stile *Building Engineering.* a plain, thin wooden molding that conceals the joint of a wall and a projecting wood member such as a cabinet.

angle structure *Civil Engineering.* a method of building a tower in which braces are placed at angles with respect to the vertical support rods to provide strength.

angle strut *Civil Engineering.* an angle iron designed to carry a compressive load.

angle tie see ANGLE BRACE.

angle to right *Cartography.* the horizontal angle measured clockwise from the preceding line to the following one. Also, CLOCKWISE ANGLE.

angle-to-right traverse *Cartography.* a surveying technique in which all angles are measured in a clockwise direction after the transit has been oriented by a backsight to the preceding station.

angle tracking noise *Electronics.* the variations of the angle of a servo tracking system around the correct angle.

angle valve *Mechanical Devices.* a spherically shaped, screw-down top valve with its outlet positioned 90° to its inlet; it is used as a fluid-flow regulator in a pipe.

angle variable *Mechanics.* the dynamical variable appearing in Hamiltonian mechanics, denoted w, that is conjugate to the action variable J; defined for periodic motion only.

angling dozer see ANGLEDOZER.

Angora *Agriculture.* **1.** a breed of domestic goat that has an outer coat of soft, shaggy wool called mohair. **2.** a breed of rabbit that produces fine silky white wool. (From the city of *Ankara*, in Turkey.)

angrite *Geology.* an achondritic stony meteorite composed mainly of augite together with small quantities of olivine and troilite.

angrogenic hormone see ANDROGEN.

Ångström, Anders Jonas [ang´strəm] 1814–1874, Swedish physicist and astronomer; made noted studies of light; mapped the solar spectrum; found hydrogen in the sun's atmosphere.

angstrom or **Angstrom** *Metrology.* a unit of length equal to one hundred-millionth of a centimeter (10^{-10} meter), used to measure extremely small distances, such as wavelengths of light or atomic diameters. (Named for Anders Jonas *Ångstrom*.) Also, **angström, Ångström.**

Ångström coefficient *Physics.* the coefficient A appearing in Ångström's formula.

Ångström compensation pyrheliometer *Engineering.* an instrument that measures levels of radiation by monitoring the amount of electrical current passed through a shaded manganic strip until its temperature matches that of another strip exposed to the radiation.

Ångström's formula *Physics.* a formula that describes the scattering of radiation from dust particles in the atmosphere: $S = A\lambda^{-B}$ where A is the Ångström coefficient, λ is the wavelength, and B is related to the size of the particle.

angstrom unit see ANGSTROM.

anguclast *Geology.* a large angular fragment composed of various-sized materials in a sediment or sedimentary rock.

Anguidae *Vertebrate Zoology.* a family of lizards in the suborder Sauria, composed of such diverse species as the alligator lizard, the glass snake, and the slow-worm.
Anguilliformes *Vertebrate Zoology.* a large order of fishes in the class Actinopterygii, including the true eel and other long bony fishes without scales and with reduced fins.
Anguilloidei *Vertebrate Zoology.* a suborder in the order Anguilliformes, including the congers, freshwater eels, and morays.
angular *Science.* of or relating to an angle or angles; having or forming angles or sharp corners.
angular acceleration *Mechanics.* the time rate of change in the angular velocity of a rotating body, expressed in radians per second squared.
angular accelerometer *Engineering.* an instrument that measures the rate of change of angular velocity between two objects.
angular advance *Mechanical Engineering.* the angle that the center of an eccentric sheave or valve actuator makes with a line set at 90° in advance of the crank pin, dependent on the lead and lap of the valve in a steam engine. Also, ANGLE OF ADVANCE, ANGULAR LEAD.
angular altitude *Photogrammetry.* a measurement in degrees of an object above the horizon, taken from a given point of observation and determined by the angle between the horizontal and the observer's line of sight.
angular aperture *Optics.* the breadth of the angle between the most widely separated rays that can pass through a lens to form an image.
angular bitstock *Mechanical Engineering.* a tool that is used to hold a bit in rotary motion, fitted with handles positioned so that it may be used in corners and cramped areas. Also, **angular bitstalk, angular brace.**
angular calibration constants *Photogrammetry.* in a multiple-camera assembly or muliple-lens camera, the values of angular orientation of the lens axes of the several lens-camera units to a common reference line.
angular clearance *Design Engineering.* a space below the straight of a die that allows the passage of slugs or blanks.
angular-contact bearing *Mechanical Engineering.* a rolling-contact antifriction bearing for heavy radial and thrust loads, designed with a high shoulder on one side of the outer race that takes the thrust.
angular correlations *Nuclear Physics.* an experimental technique for measuring how the intensity or the cross section of a nuclear reaction depends on its directional orientation with respect to incident and emitted radiation.
angular coverage *Photogrammetry.* the angle determined by a camera lens with the borders of the field of a vertical photo.
angular cutter *Mechanical Engineering.* a tool-steel milling cutter with the cutting face at an angle greater or less than 90° to the axis of the cutter; it is used to make flutes on taps, reamers, and the like. Also, ANGLE CUTTER.
angular diameter *Astronomy.* the apparent diameter of a celestial object, measured in degrees or radians.
angular displacement *Physics.* a change in angular position, described by a rotation about some axis.
angular dissymetry *Materials Science.* the asymmetric angular dependence of the intensity of light scattered by a solution; often used for polymeric molecular weight calculations.
angular distance *Astronomy.* the apparent distance between two celestial objects measured in degrees or radians. *Navigation.* the angle between two different directions, or the arc of the great circle joining two points on a sphere.
angular distortion *Cartography.* the distortion in a map projection resulting from nonconformality. Also, **angular alteration, angular change, angular deformation**.
angular distribution *Nucleonics.* the distribution of scattered particles or of the products of nuclear reactions relative to the incident beam.
angular error of closure see ERROR OF CLOSURE.
angular frequency *Physics.* the frequency of rotation or vibration, expressed in radians per second.
angular gear *Mechanical Engineering.* an arrangement of bevel gearing that transmits motion between two nonparallel rotating shafts, thus permitting changing the direction of motion. Also, **angle gear**.
angular height *Ordnance.* the vertical angle between the line along which a gun is sighted and the horizontal.
angular impulse *Mechanics.* the integral over time of all torques applied.
angularity *Science.* the fact or condition of having an angle or angles.
angular lead see ANGULAR ADVANCE.
angular leaf spot *Plant Pathology.* a plant disease that is characterized by foliage spots having sharply limited or angular outlines; in cucumbers, it is caused by the bacterium *Pseudomonas lachrymans,* and in cotton by the bacterium *Xanthomonas malvacearum.*
angular length *Mechanics.* a wavelength expressed in radians or an equivalent angular measure; equal to 2π radians or 360° times the length in wavelength units.
angular magnification *Optics.* a ratio used for optical systems that compares the angle subtended at the eye by an image of an object to the angle subtended at the eye by the object.
angular milling *Mechanical Engineering.* the milling of flat surfaces that are at an angle to the cutter axis of the milling machine.
angular momentum *Mechanics.* **1.** for a single particle moving about an axis, the moment of its linear momentum; i.e., the vector product of the particle's position and its linear momentum at the moment it passes a given point. Also, MOMENT OF MOMENTUM. **2.** for a system of particles, the vector sum of the individual angular momentum vectors of all the particles in the system.
angular momentum operator *Quantum Mechanics.* the operator that affects an infinitesimal rotation of coordinate axes.
angular parallax *Navigation.* the angular difference between two bearings to a given point, or observations of a given celestial body, as observed from different positions. If either the distance of the sighted object or the length of the baseline between observations is known, the other value can be determined trigonometrically.
angular perspective *Graphic Arts.* a form of plane linear perspective in which some of the principal lines of the picture are either parallel or perpendicular to the picture plane and some are oblique.
angular pitch *Design Engineering.* an angle derived from the distance along the pitch circle between successive teeth of a gear.
angular rate see ANGULAR SPEED.
angular resolution *Electromagnetism.* a measure of the ability of a radar to detect two targets exclusively by angle measurements.
angular resolver see RESOLVER.
angular separation see ANGULAR DISTANCE.
angular shear *Mechanical Engineering.* a metal-cutting process in which the shear is accomplished by two blade edges inclined to each other, effectively reducing the necessary shear force.
angular speed *Mechanics.* the scalar change in direction per unit time, i.e., the magnitude of the angular velocity vector. Also, ANGULAR RATE, ROTATION RATE. *Ordnance.* see ANGULAR VELOCITY.
angular spreading *Oceanography.* the lateral extension of waves as they move away from the fetch (generating area).
angular-spreading factor *Oceanography.* in forecasting ocean waves, the ratio of the actual wave energy at a given point to the energy that would have been present without any angular spreading.
angular travel *Ordnance.* the movement of a target, expressed by the angle from its original position to its new position, as judged from the observing point.
angular travel method *Ordnance.* a method of directing fire by calculating the rate of angular travel of the target.
angular unconformity *Geology.* an unconformity marking the boundary between two groups of rocks of different ages in which the bedding planes are not parallel or in which the older, underlying strata dip at a usually steeper angle than the younger, overlying strata. Also, DISCORDANCE, STRUCTURAL UNCONFORMITY, NONCONFORMITY.
angular velocity *Mechanics.* a vector whose magnitude equals the time rate of change of angular displacement, and which is directed along the axis of rotation of the body, i.e., perpendicularly to the radial and linear velocity vectors; it is usually expressed in radians per second or revolutions per minute. *Transportation Engineering.* the rate at which a turning vehicle changes direction, typically expressed in degrees per second. For a given angular velocity, turning radius is directly proportional to speed. *Ordnance.* the speed of a moving target, expressed by the rate of change of the line-of-sight as seen from the observing point.
angular wavenumber *Meteorology.* the number of waves at a given wavelength required to encircle the earth at the latitude of a disturbance. Also, HEMISPHERIC WAVENUMBER.
angulator *Engineering.* a device used to convert angular measurements on an oblique plane to corresponding projections on a horizontal plane.
Angus see ABERDEEN ANGUS.
anhalonin *Pharmacology.* $C_{12}H_{15}NO_3$, a poisonous alkaloid derived from mescal buttons (cactus flowers); formerly used in the treatment of asthma and angina.
anhaphia see ANAPHIA.

anharmonicity [an´här mə nis´ ə tē] *Physics.* **1.** relating to the mechanical vibration of a body in which the restoring force is nonlinear with the displacement. **2.** relating to the nonlinearity of the dipole moment with respect to the internuclear distance resulting in a nonlinear variation of intensities of radiation, particularly in infrared bands.

anharmonic oscillator *Physics.* a body that, when displaced from an equilibrium position, experiences a restoring force not linearly dependent on the displacement, so that motion is not simple harmonic.

anhedonia [an´hə dō´nē ə] *Psychology.* a lack of pleasure or satisfaction in experiences that would normally produce such a feeling.

anhedral see ALLOTRIOMORPHIC.

anhedron *Petrology.* rock that has the organized internal structure of a crystal without such a form externally.

anhidrosis *Physiology.* diminished activity of the sweat glands. Also, ANAPHORESIS.

Anhimidae *Vertebrate Zoology.* the screamers, a small family of goose-like birds with spurred wings, belonging to the order Anseriformes.

Anhingadae *Vertebrate Zoology.* the darters and anhingas, a family of birds in the order Pelecaniformes, characterized by their long slender necks and long pointed bills.

anhydrase *Enzymology.* an enzyme that triggers the removal of water from the substance upon which it is acting.

anhydremia *Medicine.* a deficiency of water in the blood.

anhydride *Chemistry.* a substance that is derived from an acid when water is removed, or that becomes an acid in the presence of water; e.g., carbon dioxide, CO_2, is the anhydride of carbonic acid, H_2CO_3.

anhydrite *Mineralogy.* $CaSO_4$, a colorless or white orthorhombic mineral occurring in granular and compact masses, having a specific gravity of 2.98 and a hardness of 3.5 on the Mohs scale; an important rock-forming mineral found with gypsum, limestone, or dolomite.

anhydrite evaporite *Petrology.* a compact, granular sedimentary rock harder than gypsum, composed essentially of calcium sulfate, resembling marble and deposited by evaporation.

anhydrobiosis *Biochemistry.* a state caused by dehydration, in which an organism's metabolism is reduced to an imperceptible level. *Physiology.* the state of living without water.

anhydrous alcohol see ABSOLUTE ALCOHOL.

anhydrous ammonia *Inorganic Chemistry.* liquid ammonia, a colorless substance liquefied under pressure, boiling at –33.5°C and freezing at –77.5°C; used as a refrigerant.

anhydrous ferric chloride see FERRIC CHLORIDE.

anhydrous phosphoric acid see PHOSPHORIC ANHYDRIDE.

anhydrous plumbic acid see LEAD DIOXIDE.

anhydrous sodium carbonate see SODA ASH.

anhydrous wolframic acid see TUNGSTIC OXIDE.

ani *Vertebrate Zoology.* any of several insectivorous cuckoos of the genus *Crotophaga,* having ranges from the southern United States to South America; characterized by shiny, black plumage and thin, blade-like bills, and known for their gregarious, communal nesting.

Anichkov's myocyte *Immunology.* a cardiac muscle cell occurring in a group of cells called Aschoff's bodies; characterized by striated nuclei.

anicteric *Medicine.* no-icteric; not associated with jaundice.

anidiomorphic see XENOMORPHIC.

anilazine *Organic Chemistry.* $C_9H_5Cl_3N_4$, a cream-colored solid compound that is insoluble in water; it melts at 150°C; used as a fungicide on vegetable crops and lawns.

anileridine *Pharmacology.* $C_{22}H_{28}N_2O_2$, a narcotic analgesic drug used in general anesthesia, as a postoperative sedative, and as an obstetric painkiller; abuse may lead to addiction.

anilide [an´ə lid; an´ə līd] *Organic Chemistry.* any of a class of compounds that are formed from aniline, $C_6H_5NH_2$, by the substitution of an organic acid group for the hydrogen of NH_2, such as benzanilide, $C_6H_5NHCOC_6H_5$, or acetanilide, $C_6H_5NHCOCH_3$.

Aniliidae *Vertebrate Zoology.* a small family of burrowing, nonvenomous snakes of tropical South America and Southeast Asia, in the order Squamata.

aniline [an´ə lin; an´ə līn] *Organic Chemistry.* $C_6H_5NH_2$, a colorless, oily, highly toxic liquid that turns brown on exposure to air; boils at 184°C; slightly soluble in water and soluble in alcohol, ether, and benzene. It is an important organic base for dyes and drugs and is used for various industrial purposes, as in manufacturing rubber and photographic chemicals, and in petroleum refining. Also, AMINOBENZENE.

aniline black *Organic Chemistry.* the black dye that is produced on certain textile fabrics when aniline or an aniline salt is oxidized on the fabric in the presence of certain metal salts.

aniline dye *Organic Chemistry.* any of various synthetic dyes produced from aniline or aniline salts.

aniline-formaldehyde resin *Chemistry.* a synthetic resin created by polymerizing aniline and formaldehyde; it is thermoplastic.

aniline hydrochloride *Organic Chemistry.* $C_6H_5NH_2 \cdot HCl$, a white to greenish crystalline solid that melts at 198°C and boils at 245°C; soluble in water, alcohol, and chloroform; used in making dyes and in printing. Also, **aniline chloride, aniline salt.**

aniline ink *Materials.* a fast-drying printing ink, originally a coal-tar dye solution, now usually employing pigments rather than dyes.

aniline point *Chemical Engineering.* the lowest temperature at which equal volumes of aniline and a test liquid are miscible; used as a test for components of hydrocarbon fuel mixtures.

aniline printing or **aniline process** see FLEXOGRAPHY.

aniline yellow *Organic Chemistry.* $C_6H_5N{=}NC_6H_4NH_2$, yellow to tan crystals that are slightly soluble in water and soluble in alcohol and ether; melts at 126–128°C and boils above 360°C; used in various yellow dyes. Also, *p*-AMINOAZOBENZENE.

anilinism *Toxicology.* poisoning due to exposure to aniline or any of its derivatives; characterized by cyanosis of the mucous membranes and lips and a rise in the blood levels of methemoglobin.

anima *Anthropology.* a soul or spirit thought to imbue some object in nature. *Psychology.* **1.** Carl Jung's term for the individual's inner or true self, which is closely connected to the unconscious. **2.** the feminine aspect of the male personality.

animal *Zoology.* any member of the kingdom that is generally characterized by the power of voluntary motion, specialized sense organs that provide rapid motor response to stimuli, limited capacity for regenerative growth, the lack of rigid cell walls, and the inability to manufacture nutrients from inorganic substances. These qualities and others distinguish the animal kingdom from the plant kingdom.

animal balance *Agriculture.* a balance scale designed to measure the weight of living, moving animals, providing relatively accurate readings in spite of any movement.

animal black *Chemistry.* a fine carbon pigment made by the calcination of animal bones, used as a decolorizer and purifier.

animal cell culture *Biotechnology.* the commercial isolation and growth of specific cell types for the production of antibodies and interferons.

animal charcoal *Chemistry.* the carbon residue left after the destructive distillation or carbonization of animal matter, such as blood, flesh, or bone.

animalcule *Microbiology.* a term formerly used to refer to any microscopic or minute animal organism; introduced by Leeuwenhoek.

animal fiber *Textiles.* a natural fiber of animal origin, such as wool or silk.

animal graft *Surgery.* a zooplastic graft.

animal husbandry *Agriculture.* a field of agriculture that involves the raising of domestic animals, including their breeding, care, and feeding.

Animalia [an´ə māl´yə] *Systematics.* one of the five basic categories, or kingdoms, in the taxonomic hierarchy, and including all extant and extinct eukaryotic, multicellular organisms that feed by ingesting other organic matter and are capable of spontaneous movement.

animal kingdom see ANIMALIA

animal locomotion *Zoology.* the ability of an animal to move under its own power from one place to another.

animal oil see BONE OIL.

animal pole *Cell Biology.* the region of an egg in which cytoplasm is concentrated; distinguished from the vegetal pole, which contains a high concentration of yolk.

animal power *Mechanical Engineering.* the rate of work done by a horse, bullock, or other work animal.

animal psychology *Behavior.* the study of animal behavior, especially of individualized or pathological patterns as opposed to normal species behavior.

animal virus *Virology.* any virus that can infect and replicate in the cells of an animal; one of three major types of viruses, along with plant viruses and phages. Their host range is all-inclusive, and modes of transmission include contact, aerosols, vectors, food, water, or placenta.

animate *Science.* **1.** characterized by life; alive. **2.** of or relating to animal life.

animatism see ANIMISM.

Animikean *Geology.* in the Canadian Shield, a provincial series formed during Proterozoic time. Also, ANIMIKIE, PENOKEAN.

Animikie see ANIMIKEAN.

animikite *Mineralogy.* white or gray granular masses of silver ore consisting of a mixture of sulfides, arsenides, and antimonides, and often containing nickel and lead.
animism *Anthropology.* a belief in spiritual beings that are embodied in nature, such as animals, plants, and natural objects, and that are thought to have a soul, or anima. *Psychology.* the belief that inanimate objects or forces of nature, such as rivers, rocks, or trees, have human qualities such as thought and emotion. Also, ANIMATISM.
animistic relating to or characterized by animism.
animus *Psychology.* the masculine aspect of the female personality.
anion [an´ē än; ə nī´än] *Chemistry.* a negatively charged ion, which is attracted to an anode, the positive electrode in an electrolytic cell.
anion analysis *Analytical Chemistry.* a test for negatively charged particles (anions) in an aqueous solution.
anion exchange *Chemistry.* a process in which anions in solution exchange with anions in an insoluble matrix or resin.
anion-exchange resin *Analytical Chemistry.* a type of resin used in chromatography; characterized by the ability of negative ions in its immobilized (stationary) phase to be exchanged for anions in its solute (mobile phase).
anionic *Chemistry.* of or relating to an anion.
anionic detergent *Materials.* any of a class of detergents that have negatively charged surface ions.
anionic polymerization *Materials Science.* the addition polymerization of negatively charged species with a monomer that contains a double bond. *Organic Chemistry.* a type of polymerization catalyzed by Lewis bases.
anionotropy *Chemistry.* the breaking off of an ion (such as OH– or X–) from a molecule to leave a positive ion in a state of dynamic equilibrium.
anis- a combining form meaning "unequal."
Anisakidae *Invertebrate Zoology.* a family of parasitic roundworms in the superfamily Ascaridoidea; infectious in humans as a result of eating uncooked fish.
anisaldehyde *Organic Chemistry.* $C_6H_4(OCH_3)CHO$, either one of two compounds: **1. *o*-anisaldehyde.** a white to light tan crystalline solid that melts at 37–39°C and boils at 238°C; insoluble in water and soluble in alcohol; used as a chemical intermediate in making perfumes, food flavorings, and antihistamine medicines. **2. *p*-anisaldehyde.** a colorless to pale yellow liquid that boils at 248°C; insoluble in water; used in perfumes, antihistamines, and food flavorings.
anisate *Organic Chemistry.* any salt of anisic acid.
anise *Botany.* a plant, *Pimpinella anisum,* of the parsley family, having small, umbelliferous flowers and aromatic seeds **(aniseed)**.

anise

aniseikonia *Medicine.* a condition in which the ocular image of an object as seen by one eye differs in size and shape from that seen by the other eye.
Anisian *Geology.* a European geologic stage of the lower Middle Triassic period, occurring after the Scythian and before the Ladinian.
anisic acid *Organic Chemistry.* $CH_3OC_6H_4COOH$, a white crystalline solid that melts at 184°C and boils at 275–280°C; slightly soluble in water and soluble in alcohol, ether, chloroform and benzene; used as an insecticide and in antiseptics. Also, *p*-METHOXYBENZOIC ACID.
anisic alcohol *Organic Chemistry.* $CH_3OC_6H_4CH_2OH$, a colorless liquid with a floral odor that is insoluble in water and soluble in alcohol and ether; it boils at 255–265°C; used in making perfumes and pharmaceuticals. Also, **anise alcohol, anisyl alcohol.**
anisidine *Organic Chemistry.* $CH_3OC_6H_4NH_2$, another name for *o-*, *m-*, or *p*-aminoanisole.
aniso- a combining form meaning "unequal," as in *anisotropic.*
anisocarpous *Botany.* of a flower, having fewer carpels than other floral parts, such as stamens.
anisochela *Invertebrate Zoology.* **1.** a type of sponge spicule with dissimilar ends. **2.** a chela (claw) in crustaceans with opposable parts of unequal size.
anisodesmic *Mineralogy.* of a compound or crystal, having ionic bonds of unequal strength.
anisogamete see HETEROGAMETE.
anisogamy see HETEROGAMY.
anisole [an´ə sōl´] *Organic Chemistry.* $C_6H_5OCH_3$, a toxic colorless liquid that freezes at –37.8°C and boils at 155°C, insoluble in water and soluble in alcohol, ether, acetone, and benzene; used as a solvent and in making perfumes. Also, METHYL PHENYL ETHER, METHOXYBENZENE.
anisomerous *Botany.* having an unequal number of parts in the floral whorls.
anisometric *Science.* not isometric; not being of equal size or measurement.
anisometric growth *Botany.* the unsymmetrical growth of a plant.
anisometric particle *Virology.* of a virus particle, not isometric, often rod-shaped rather than bullet-shaped.
Anisomyaria *Invertebrate Zoology.* an order of oysters, scallops, and mussels in the class Bivalvia, with the anterior adductor muscle highly developed and the posterior one only slightly developed.
Anisophylleaceae *Botany.* a family of dicotyledonous tropical trees and shrubs of the order Rosales, noted for accumulating aluminum; characterized by alternate simple leaves and by flowers borne on axillary spikes or racemes or on panicles on leafless shoots.
anisophyllous *Botany.* of a plant, having leaves of unequal sizes or shapes.
Anisoptera *Invertebrate Zoology.* a suborder of dragonflies in the order Odonata, having hind wings that are much wider than the front wings and are held horizontally at rest.
anisostemonous *Botany.* of a plant, having stamens that are unequal to the petals or sepals.
Anisotomidae *Invertebrate Zoology.* a family of fungus or carrion beetles in the order Coleoptera. Also, LEIODIDAE.
anisotropic *Physics.* relating to or having unequal physical properties along different directions. *Botany.* of a plant, having unequal dimensions along different axes. Also, **anisotropal.**
anisotropic inhibitor *Biochemistry.* a substance that reduces or prevents a metabolic or physiological process in only one direction.
anisotropic membrane *Chemical Engineering.* a filtration membrane consisting of a thin skin at the separating surface supported by a spongy sublayer.
anisotropy [an´ī sō´trə pē] *Materials Science.* the fact of being dependent on direction, especially in a crystalline lattice, of any mechanical, electrooptic, or magnetic property, such as elasticity, conductivity, or permeability. *Biology.* the condition of having unequal responses to external stimuli.
anisotropy constant *Electromagnetism.* a temperature-dependent parameter of a ferromagnetic material, associating magnitization in various directions to the anisotropy energy.
anisotropy energy *Electromagnetism.* the energy associated with a ferromagnetic crystal when the magnetization domain is rotated away from the direction of easy magnetization by an external field.
anisotropy factor see DISSYMMETRY FACTOR.
anisotropy of flow *Virology.* the orientation of different flows of particles in different directions.
Anitschkow's myocyte see ANICHKOV'S MYCOCYTE.
ankaramite *Petrology.* a basalt with abundant phenocrysts of pyroxene and olivine and typically alkaline affinity.
ankaratrite see OLIVINE NEPHELINITE.
anker *Metrology.* a unit of capacity equal to 10 U.S. gallons or 37.85 liters, used to measure liquids.
ankerite *Mineralogy.* $Ca(Fe^{+2},Mg,Mn)(CO_3)_2$, a white, gray, brown or pink trigonal, iron-rich mineral, with rhombohedral crystals, having a specific gravity of 2.97 to 3.02 and a hardness of 3.5 to 4 on the Mohs scale. Also, FERROAN DOLOMITE.
Ankistrodesmaceae *Microbiology.* a family of freshwater green algae of the order Chlorophyceae; also called Selenastraceae and used by some authorities to further segregate selected genera of the family Oocystaceae that have elongated or acicular cells.

Ankistrodesmus *Microbiology.* a common genus of freshwater green algae of the family Oocystaceae, having elongated cells united in colonies by entanglement or lateral cohesion; sometimes classifed as the separate family Ankistrodesmaceae.

ankle *Anatomy.* the part of the leg just above the foot, in the region of the ankle joint.

ankle bone see TALUS.

ankle jerk see ACHILLES JERK, ACHILLES REFLEX.

ankle joint *Anatomy.* the joint that connects the foot and the leg, formed by the articulation of the tibia and fibula with the talus.

ankylo- a combining form meaning "bent" or "joint." Also, ANCYLO-, ANCHYLO-.

ankylodactylia or **ankylodactyly** *Medicine.* adhesion of the fingers or toes to one another.

Ankylosauria *Paleontology.* a suborder of herbivorous ornithischian dinosaurs, extant in the Cretaceous. They were squat and quadrupedal, with stubby, hoofed feet, and were strongly armored, with thick bony plates completely covering the dorsal and lateral surfaces of body and tail. Some species grew to lengths of about 18 feet.

Ankylosaurus *Paleontology.* a genus of squat, quadrupedal, armored ornithischian dinosaurs in the suborder Ankylosauria and family Ankylosauridae, known only from upper Cretaceous deposits of North America and Asia.

ankylosed *Medicine.* stiffened or fused, as a joint.

ankylosing spondylitis *Medicine.* a form of rheumatoid arthritis of the spine that affects young adult males almost exclusively and that is marked by stiffening of the spinal joints and ligaments, so that movement becomes increasingly painful and difficult.

ankylosis *Physics.* the restriction of one or more degrees of freedom of motion by the use of mechanical restraints or friction. *Medicine.* an immobility and fixation of a joint, often in an abnormal position, due to injury, disease, or surgery; it often occurs in rheumatoid arthritis.

ankyrin *Biochemistry.* a globular protein that links spectrin and an integral membrane protein in the erythrocyte plasma membrane.

anlage [an´lə gə] *Developmental Biology.* a group of cells that can be initially identified as a future body part. Also, PRIMORDIUM.

annabergite *Mineralogy.* $Ni_3(AsO_4)_2 \cdot 8H_2O$, a white to light green, monoclinic mineral with vertically striated, prismatic crystals, having a specific gravity of 3.07 and a hardness of 1.5 to 2.5 on the Mohs scale; found as a secondary mineral in the oxidation zone of nickel-bearing ore deposits.

Annapurna *Geography.* a noted mountain in the Himalayas, in northern Nepal (26,504 ft).

annatto *Botany.* a small tropical American tree, *Bixa orellana,* that produces a reddish-yellow dye called **annotto.** Also, ANATTO, ACHIOTE.

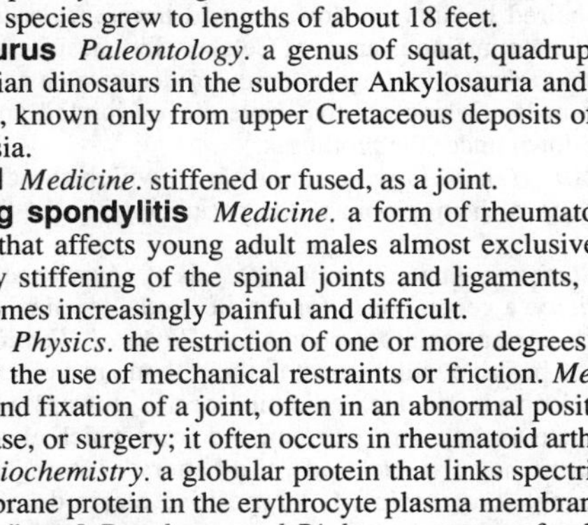

annatto

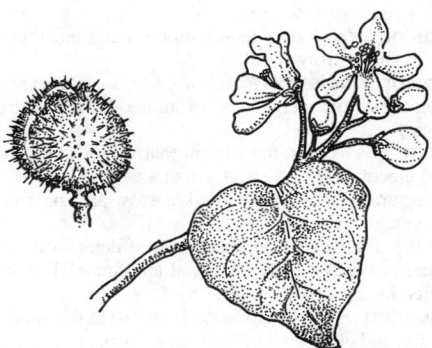

anneal *Engineering.* to carry out the process of annealing.

annealing *Engineering.* the sustained heating of a material, such as metal or glass, at a known high temperature, followed by the gradual cooling of the material; a process carried out in order to reduce hardness or brittleness, to eliminate various stresses and weaknesses, or to produce other qualities.

annealing oven *Engineering.* a furnace used to anneal materials such as metal and glass. Also, **annealing furnace.**

annealing point *Materials Science.* the temperature at which the viscosity of a piece of glass is low enough to allow for the relief of residual stresses. Also, **annealing temperature.**

annealing texture *Materials Science.* a preferred orientation of the grains of an annealed metal that might cause its properties to be anisotropic.

annealing twin *Materials Science.* a crystallographic twin formed during an annealing process subsequent to plastic deformation; found in certain metals and minerals.

Annedidae *Vertebrate Zoology.* the legless lizards, a burrowing, snakelike family of lizards in the suborder Sauria; composed of two species that are found in California and Baja California.

annelid *Invertebrate Zoology.* **1.** belonging to the phylum Annelida. **2.** one of the members of this phylum, such as earthworms or leeches.

Annelida *Invertebrate Zoology.* a phylum of ringed worms (metameric metazoans) that have a true coelom, a closed vascular system with hemoglobin-bearing blood, and a cuticle with chitinous setae from pits in the skin.

annex point *Cartography.* a point used to assist in the relative orientation of vertical and oblique photographs, selected in the overlap area between the vertical and its corresponding oblique about midway between the pass points.

annidation *Ecology.* the ability of a mutated genotype to survive, though competitively inferior, by occupying a niche its parent type is unable to inhabit.

Anniellidae *Vertebrate Zoology.* a family of limbless, snakelike lizards of California, belonging to the order Squamata.

annihilation *Particle Physics.* a nuclear event in which a particle and its corresponding antiparticle collide, converting into annihilation radiation.

annihilation operator *Quantum Mechanics.* **1.** an operator that, when applied to a wavefunction describing N particles, yields a wavefunction describing $N-1$ particles. **2.** an operator that, when applied to a known eigenfunction, yields the eigenfunction with the next lower eigenvalue; it yields zero if applied to the ground state.

annihilation radiation *Nuclear Physics.* the energy produced from the joining of a positron and an electron, in which the positive and negative charges neutralize each other and become electromagnetic radiation.

annihilator *Mathematics.* **1.** the annihilator of a set S is the class of all functions or operators on S which are zero at every point of S. **2.** in particular, the annihilator of an element r in a module over a ring R is the ideal A of i such that $ar=0$ for all a in A.

anniversary clock see FOUR HUNDRED-DAY CLOCK.

Annona *Botany.* a genus of trees and shrubs of the family Annonaceae, several species of which yield edible fruits as well as bark and leaves used in folk medicine. **Annona muricata** is the soursop, a popular tropical American fruit.

Annonaceae *Botany.* a family of tropical and subtropical woody plants having simple alternate leaves and sometimes edible fruits such as the cherimoya and sweetsop. Also, **annona family.**

annotation *Computer Programming.* any nonexecutable note or comment in a program that provides explanatory or descriptive information.

annual *Science.* happening once a year or relating to a period of one year. *Botany.* **1.** a plant that grows, flowers, produces seed, and then dies within one year or one growing season. Also, ANNUAL PLANT. **2.** relating to or describing such a plant.

annual aberration *Astronomy.* a minute change in a star's apparent position that stems from the orbital velocity of earth as it travels about the sun.

annual equation *Astronomy.* a periodic inequality in the moon's motion caused by the variable attraction of the sun due to the earth's orbit being an ellipse.

annual flood *Hydrology.* the highest flood peak of a stream during a given water year.

annual inequality *Oceanography.* the variation in water level or tidal current speed during the course of a year; it is more or less periodic and is caused mainly by meteorological conditions. *Astronomy.* see ANNUAL EQUATION.

annual labor see ASSESSMENT WORK.

annual layer *Geology.* **1.** a sedimentary layer that was deposited or was presumed to have been deposited during the course of one year. **2.** in a salt body, a dark band of disseminated anhydrite crystals.

annual magnetic change see MAGNETIC ANNUAL CHANGE.

annual magnetic variation see MAGNETIC ANNUAL VARIATION.

annual parallax *Astronomy.* the apparent displacement of a celestial body caused by the changing perspective of the observer as the earth follows its yearly path around the sun.

annual plant *Botany.* see ANNUAL.

annual ring *Botany.* the yearly increment of wood growth as seen in cross section of the stem of a plant, manifest as a ring with two layers, one of spring growth with large-diameter cells and one of summer growth with small-diameter cells; or, in secondary xylem (wood), the growth ring formed during one season. More than one ring may be formed during a single year as a result of environmental conditions. Also, GROWTH RING.

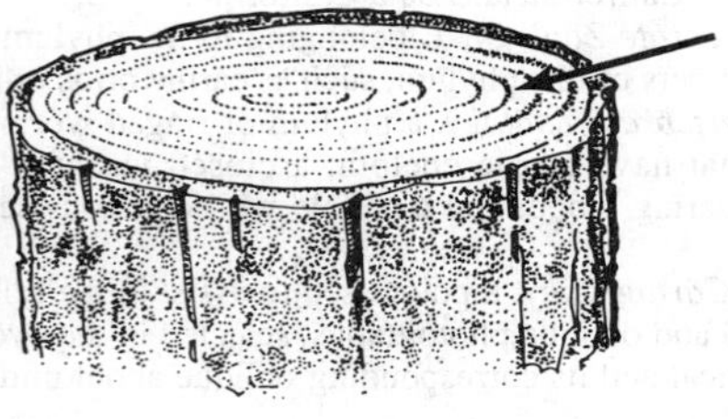

annual ring

annual storage *Hydrology.* the capacity of a reservoir that can contain the annual runoff of a watershed, but cannot carry over any water for longer than the year.

annual variation *Geophysics.* the observed changes in any geophysical field at a specific location over a period of one year.

annuation *Ecology.* a variation in a wildlife community, as in abundance or distribution, resulting from annual changes in its environment.

annular *Anatomy.* **1.** of or relating to an annulus. **2.** ring-shaped. *Mechanical Engineering.* **1.** the ringlike space between a cylindrical element and its surroundings, such as the space between the casing and the wall of a hole. **2.** having the shape or form of a ring, such as the *annular* fins of heat exchanger tubes.

annular atom *Organic Chemistry.* an atom that is part of the ring in a cyclic compound.

annular auger *Mining Engineering.* a ring-shaped tool that is designed to leave the core untouched by cutting an annular channel.

annular blowout preventer *Petroleum Engineering.* in a drilling operation, a type of blowout preventer which uses a sealing component made of a rubber ring that is compressed into the annular space between the pipe and the wellbore by hydraulically operated pistons. Also, SLEEVE BLOWOUT PREVENTER.

annular conductor *Electricity.* a number of wires stranded in three ring-shaped layers of alternating twists around a fiber or saturated hemp core.

annular drainage pattern *Hydrology.* a pattern of natural stream courses in which the tributaries follow a ringlike or concentric path along a belt of weak rocks.

annular eclipse *Astronomy.* a type of solar eclipse in which the moon's disk appears too small to fully cover the sun, thus leaving a ring (annulus) of sunlight visible around the moon's limb.

annular effect *Fluid Mechanics.* a phenomenon noted in fluid flow in a tube when the fluid movement quickly alternates, as in the propagation of sound waves, wherein the mean speed increases starting from the center of the tube toward the walls and then drops to zero, within a thin laminar boundary layer, at the wall itself.

annular gear *Mechanical Devices.* an annular ring with gear teeth attached to a base, commonly used in automobile starter hardware.

annular hernia see UMBILICAL HERNIA.

annular section *Engineering.* the open space between two concentric pipes or containers.

annular solid *Mathematics.* a solid generated by rotating a closed plane curve about an axis lying in the same plane as the curve but not intersecting the curve; a "doughnut" figure.

annular transistor *Electronics.* a transistor whose base, emitter, and collector are fabricated in concentric circular regions.

annular vault *Architecture.* a ring-shaped barrel vault.

annular velocity *Petroleum Engineering.* the rate of speed of the fluid carrying cuttings to the surface in the annular space of a drilling well; a typical velocity is 100 feet per minute.

annulated shaft or **annulated column** *Architecture.* a column made up of a cluster of shafts banded, at intervals, by rings.

annulate lamellae *Cell Biology.* stacks of pore-containing membranes that are found within the germ cells of many animals and may be a storage for nuclear pores.

annulene *Organic Chemistry.* any of a class of monocyclic conjugated hydrocarbons that have the general formula $(\text{–CH=CH–})_n$.

annulus *Anatomy. plural,* **annuli.** any ringlike or bandlike body part. Also, ANULUS. *Botany.* a band of plant tissue, such as the slender membrane that surrounds the stem of certain mushrooms after the spreading of the cap; in ferns, a row of specialized cells in a sporangium. *Mathematics.* **1.** the area in the plane between two concentric circles. **2.** the surface of the sphere between two parallel planes that intersect the sphere.

annulus conjecture *Mathematics.* if f and g are locally flat embeddings of S^{n-1}, the $n-1$ sphere, into R^n (n-dimensional real space), and if $f(S^{n-1})$ lies within the bounded component of $R^n - g(S^{n-1})$, then the closed region in R^n that is bounded by $f(S^{n-1})$ and $g(S^{n-1})$ is homeomorphic to the direct product of S^{n-1} and $[0, 1]$. This result has been proven true in all cases except $n = 4$.

annunciator *Engineering.* a remote communication device used to signal whether a current is flowing or has flowed.

anoa *Vertebrate Zoology.* the smallest buffalo (about three feet tall), a black sparse-haired bovid with down-curved horns, *Bubalus depressicornis;* found in mountainous areas of Celebes, Indonesia.

Anobiidae *Invertebrate Zoology.* a family of deathwatch and drugstore beetles in the order Coleoptera that have eleven-jointed antennae and the head bent down under the prothorax.

anochromasia *Hematology.* a condition in which the erythrocytes show a piling up of hemoglobin at the periphery, so that the center is pale.

ANOCOVA *Statistics.* an acronym for <u>an</u>alysis <u>o</u>f <u>cova</u>riance.

anode *Electricity.* a general term for the electrode, terminal, or element through which current enters a conductor. *Physical Chemistry.* **1.** the electrode that is positive with respect to the cathode in a electrochemical cell. It is the electrode at which oxidation occurs, toward which anions generally migrate as they carry current, and from which electrons leave the system. **2.** the negative electrode of a battery or storage cell that is delivering current. *Electronics.* **1.** the positive electrode of a electron tube. **2.** the positive plate of a capacitor.

anode balancing coil *Electricity.* a set of mutually coupled windings used to maintain approximately equal currents in parallel anodes operating from a common transformer terminal.

anode circuit *Electronics.* in an electron tube, the elements between the anode and the cathode, including the cathode source. Also, PLATE CIRCUIT.

anode circuit detector see ANODE DETECTOR.

anode copper *Metallurgy.* a slab of impure copper (at least 99.0% copper) produced by refining blister copper (containing at least 98.5% copper) and casting the refined product in a flat mold. Such slab is used as the anode in the electrolytic refining process that yields commercially pure copper (at least 99.95% pure).

anode corrosion *Metallurgy.* the dissolution of a metal or alloy acting as the anode in an electrolytic cell.

anode corrosion efficiency *Physical Chemistry.* an expression of the ratio between the actual weight loss of an anode from corrosion to the theoretical loss.

anode current *Electronics.* the current that flows into the anode to the cathode in an electron tube. Also, PLATE CURRENT.

anode dark space *Electronics.* a dark narrow area next to the surface of the anode in a gas tube.

anode detector *Electronics.* an electron tube detector in which current is proportional to variations in the signal amplitude. The anode current is thus rectified by the anode circuit.

anode dissipation *Electronics.* heat dissipated at the anode of an electron tube by ion and electron bombardment. Also, PLATE DISSIPATION.

anode drop see ANODE FALL.

anode effect *Physical Chemistry.* a condition that produces an abrupt increase in voltage and a decrease in current flow.

anode efficiency *Electronics.* the ratio of the (AC) load power to the (DC) anode input power in an electron tube. Also, PLATE EFFICIENCY.

anode fall *Electronics.* the drop in potential between the anode and the gas column near the anode in a gas tube. The value of the fall may be less than, equal to, or greater than zero. Also, ANODE DROP.

anode film *Chemistry.* a coating formed on the anode of an electrochemical cell.

anode furnace *Metallurgy.* in the processing of copper, the furnace from which anode copper is cast.

anode glow *Electronics.* a bright narrow area on the anode side of the positive column in a gas tube.

anode impedance *Electronics.* the total impedance seen between the anode and the cathode of an electron tube, exclusive of the electron stream. Also, PLATE IMPEDANCE, ANODE LOAD IMPEDANCE.

anode input power *Electronics.* the (DC) power consumed by the anode of a vacuum tube, determined by the product of the anode voltage and the anode current. Also, PLATE INPUT POWER.

anode load impedance see ANODE IMPEDANCE.

anode metal *Metallurgy.* a slab of impure metal such as copper or zinc, used as the anode in the electrolytic refining process.

anode modulation *Electronics.* modulation in an electron tube with a carrier signal present, due to the application of a modulating signal at the anode. Also, PLATE MODULATION.

anode mud *Metallurgy.* the insoluble residue that derives from the anodic dissolution of an impure metal such as copper during electrolytic refining. Also, **anode slime.**

anode neutralization *Electronics.* neutralization in an electron tube amplifier where an inverting network is placed in the anode circuit. Also, PLATE NEUTRALIZATION.

anode pulse modulation *Electronics.* modulation in an electron tube with a carrier signal present, where externally produced pulses are applied to the anode. Also, PLATE PULSE MODULATION.

anode rays *Electronics.* positive ions released at the anode, generally due to impurities in the metal of the anode. *Radiology.* beams of positively charged particles emitted from the anode of a partially evacuated tube when voltage is applied. Also, POSITIVE RAYS.

anode resistance *Electronics.* the ratio of an incremental change in the anode voltage to an incremental change in the anode current, with all other voltages held constant. Also, PLATE RESISTANCE.

anode saturation *Electronics.* in an electron tube, the valence of the anode current which does not further increase with an increase in anode voltage. Also, PLATE SATURATION, VOLTAGE SATURATION.

anode sheath *Electronics.* in a gas tube operated at high current levels, a layer of electrons surrounding the anode.

anodic[ə näd´ik] *Physics.* relating to an anode.

anodic control *Physical Chemistry.* the polarization at the anode of an electrochemical cell, in which a decrease of the anode potential dominates current density reduction.

anodic migration *Physical Chemistry.* the movement of a charged electric particle toward the anode in an electric field.

anodic polarization *Physical Chemistry.* the changes in an anode's energy levels caused by electron flow.

anodic protection *Materials Science.* corrosion protection of a metal or alloy achieved by impressing upon the metal an anodic current of sufficient magnitude to cause the formation of an adherent, protective oxide. Anodic protection is effective in certain metals such as stainless steel, but not in most others such as copper, zinc, or magnesium.

anodic reaction *Metallurgy.* in an electrochemical cell, the reaction occurring at the electrode that is connected with an external positive source of electricity.

anodize *Metallurgy.* to form a coating on a metal or alloy by oxidizing the surface in an electrolytic cell, in which the metal is connected with a positive source of electricity; most commonly applied to aluminum-base materials. Thus, **anodized aluminum.**

anodized dielectric film *Electricity.* a protective oxide film deposited on a conducting surface used for producing thin-film capacitors, trimming resistor values, and passivation in the manufacture of integrated circuits.

anole [ə nō´lē] *Vertebrate Zoology.* a tropical arboreal lizard of the genus *Anoles* in the family Iguanidae, having flattened, adhesive digits and changing color according to the intensity of sunlight.

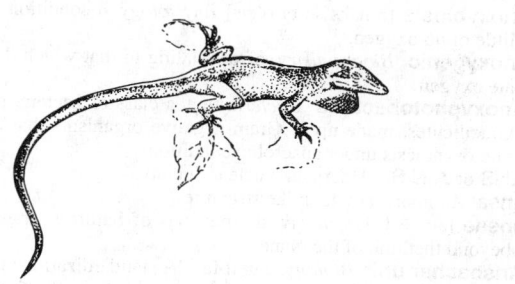

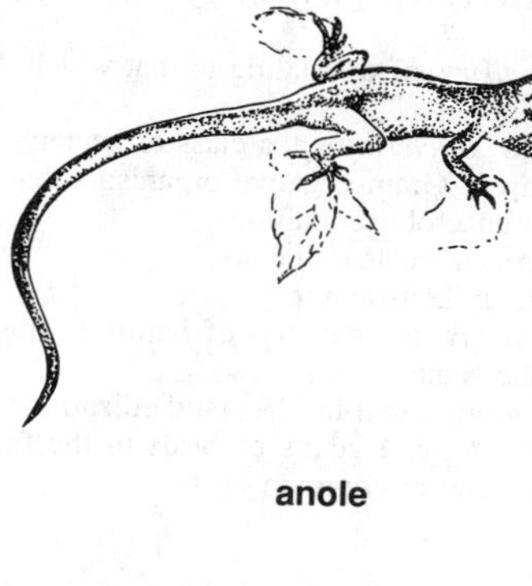

anole

anolyte *Chemistry.* the portion of an electrolytic solution that surrounds the anode.

anomalistic month [ə näm´ə lis´tik] *Astronomy.* the time it takes for the moon to make two successive passages through perigee: 27.55455 days. Also, **anomalistic period.**

anomalistic tide cycle *Oceanography.* the average period of about 27.55 days during which the moon makes one complete revolution around the earth.

anomalistic year *Astronomy.* the time it takes for the earth to make two successive passages through perihelion: 365.259635 days.

Anomalodesmata *Invertebrate Zoology.* a subclass of marine clams, burrowers with valves of unequal size.

anomalon *Nuclear Physics.* a fragment produced during a nuclear reaction, in which the projectile nucleus collides with a neighboring target nucleus at rest.

anomaloscope *Optics.* an instrument that detects and classifies defective color vision, wherein a subject mixes lights of two different colors (usually red and blue) in an attempt to match the color (usually yellow) of a third light.

anomalous [ə näm´ə ləs] *Science.* **1.** not usual or regular; abnormal. **2.** difficult to explain or classify. *Volcanology.* formed or changed by being incorporated into foreign material. Thus, **anomalous magma.**

anomalous dispersion *Optics.* an anomaly in the normal variation of refractive index versus wavelength, commonly observed near an absorption band. *Crystallography.* see ANOMALOUS SCATTERING.

anomalous expansion *Thermodynamics.* **1.** a characteristic property of water, in which its volume apparently increases at temperatures between 4°C and 0°C. **2.** any such increase in the volume of a substance as its temperature decreases.

anomalous magnetic moment *Particle Physics.* an observed difference between the experimentally measured magnetic moment of a particle and that which is predicted by Dirac's theory.

anomalous scattering *Crystallography.* a phase change that occurs upon the scattering of X rays by a crystal containing one or more atoms that strongly absorb the X rays. As a result, in crystals that lack symmetry elements of the second kind (rotation-inversion axes and glide planes), Bragg reflections from opposite faces of the crystal may have different intensities. These differences in intensity may be used to determine the absolute configuration of chiral crystals. Also, ANOMALOUS DISPERSION.

anomalous trichromatism *Physiology.* the deviation of visual color matching by a trichromat, who requires the three primary colors to match the whole spectrum but whose choices of each primary differ from the normal.

anomalous viscosity see NON-NEWTONIAN VISCOSITY.

anomalous water see POLYWATER.

anomalous Zeeman effect *Spectroscopy.* the splitting of spectral lines into more than three components for atoms that are not at ground state, when the source of radiation is brought into a magnetic field.

Anomaluridae *Vertebrate Zoology.* the scaly-tails, a family of African squirrel-like rodents in the order Rodentia; characterized by a wide gliding membrane on either side of the body and hard, pointed scales under the tail that give support in climbing.

anomaly [ə näm´ə lē] a deviation from the average or normal state; specific uses include: *Biology.* an abnormal deviation from the characteristic structure, function, or state of a group; an irregularity. *Medicine.* a congenital malformation, such as the absence of a limb. *Geodesy.* a deviation of an observed value (gravitational, magnetic, isostatic, etc.) from a theoretical value, due to a corresponding irregularity in the earth's structure at the area of observation. *Astronomy.* the angular distance between a planet and the apsidal line of its orbit, measured in the direction of its motion. *Oceanography.* the difference between actual conditions at a given location and those that would have existed if the water's temperature and salinity values had matched those of an arbitrary standard.

anomaly finder *Engineering.* a computer used on ships to record water depth, time, course, speed, and various geophysical data.

anomaly of dynamic height *Oceanography.* the excess of the actual geopotential difference, as measured between two given isobaric surfaces, over the geopotential difference in a homogeneous water column whose salinity is 35 parts per thousand and whose temperature is 0°C. Also, **anomaly of geopotential difference.**

anomaly of specific volume *Oceanography.* the excess of the actual specific volume of a sample of sea water over the specific volume of sea water whose salinity is 35 parts per thousand and whose temperature is 0°C, with pressure remaining constant. Also, STERIC ANOMALY.

anomer *Organic Chemistry.* one of a pair of isomers (designated α and β) of cyclic carbohydrates produced when a new point of symmetry arises after a rearrangement of atoms at the aldehyde or ketone position.
anomia [ə nōm´ē] *Neurology.* an inability to name objects or recognize their names.
anomic aphasia *Neurology.* a type of aphasia in which the individual cannot recall principal nouns and verbs due to brain damage.
anomite *Mineralogy.* a rarely occurring variety of biotite (differing with it only in optical orientation) found only in Orange County, New York, and Lake Baikal, Eastern Siberia.
Anomocoela *Vertebrate Zoology.* a suborder of toadlike amphibians of the order Anura, having procoelus sacral vertebrae and including spadefoot toads and horned frogs.
anomocoelous *Anatomy.* describing a vertebral column in which the vertebral centra are slightly biconcave or flat terminally, and the intervertebral cartilage is connected to and not subdivided between successive presacral vertebrae.
Anomphalus *Paleontology.* a genus of small aspidobranch gastropods characterized by an almost smooth rotelliform shell; extant in the Devonian and Carboniferous.
Anomura *Vertebrate Zoology.* a suborder of marine crustaceans in the order Decapoda, characterized by a tail fan and a forward-bending or soft asymmetrical thorax; it includes hermit crabs and sand crabs.
anonymous dimensionless group 1–4 *Chemical Engineering.* four dimensionless groups used to solve problems in laminar boundary-layer flow, gas absorption in wetted-wall columns, and transfer processes.
Anopheles [ə näf´ə lēz´] *Invertebrate Zoology.* **1.** a genus of mosquitoes that includes the species capable of transmitting malaria parasites to humans; characterized by the posture of holding the body at an angle with the head downward while resting or feeding. **2. anopheles.** any mosquito of this genus. (From a Greek word meaning "harmful.")

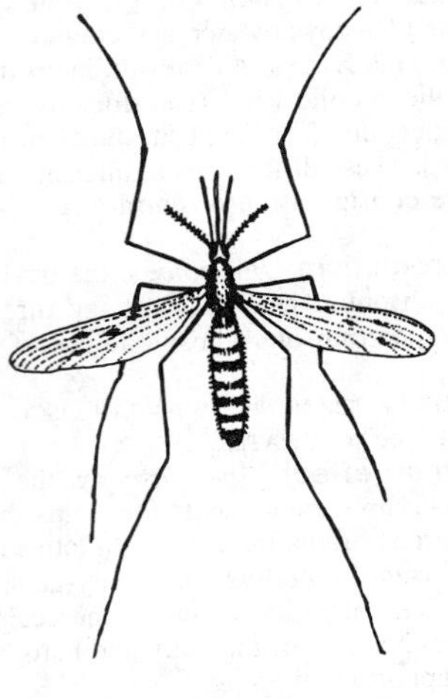

Anopheles

anopheline *Invertebrate Zoology.* relating to or involving mosquitoes of the *Anopheles* genus.
anophthalmia *Medicine.* a developmental abnormality characterized by the complete absence of the eyes or the presence of vestigial eyes.
Anopla *Invertebrate Zoology.* a subclass of worms in the phylum Nemertinea, with the mouth located below or posterior to the brain and a proboscis without stylets.
Anoplocephalidae *Invertebrate Zoology.* a family of tapeworms in the order Cyclophyllidea, with unarmed scolices that live as adults in the intestines of various herbivores and pass their larval stages in certain free-living mites.
Anoplotheriidae *Paleontology.* a family of primitive ruminants in the extinct superfamily Anoplotheroidea; exact affinities uncertain; extant in the upper Eocene to middle Oligocene.
Anoplotheroidea *Paleontology.* an extinct superfamily of selenodont artiodactyls in the suborder Tylopoda; extant in the Eocene to Miocene.
Anoplura *Entomology.* the sucking lice, an order of wingless blood-sucking insects that are parasites on mammals.
anorchidism *Medicine.* the congenital absence of testes. Also, **anorchism.**
anorectal *Anatomy.* of or relating to both the anus and rectum, or to their junction region.
anorexia [an´ə reks´sē ə] *Medicine.* a loss or absence of appetite. *Psychology.* see ANOREXIA NERVOSA. (From a Greek word meaning "without appetite" or "without desire.")
anorexia nervosa *Psychology.* a pathological loss of appetite and self-limited food intake, often associated with an overconcern about or misperception of one's body image, leading to severe emotional, physiological, and psychological disturbances, and sometimes to death through malnutrition or related effects.
anorexic [an´ə reks´ik] *Medicine.* pertaining to loss of appetite for food or to a substance that causes loss of appetite. Also, **anorectic.** *Psychology.* **1.** a person who has the condition of anorexia nervosa. **2.** of or relating to this condition.
anorgasmy *Medicine.* a failure to experience orgasm during coitus.
anorogenic *Geology.* of a feature, structure, event, or period, lacking or unrelated to any tectonic disturbance.
anorogenic time *Geology.* a geologic time period during which no significant deformation of the crust occurred.
anorthic crystal see TRICLINIC CRYSTAL.
anorthite *Mineralogy.* $CaAl_2Si_2O_8$, a white to grayish triclinic member of the feldspar group (plagioclase series), transparent to translucent, having a specific gravity of 2.74 to 2.76 and a hardness of 6 to 6.5 on the Mohs scale; found as a rock-forming mineral in basic plutonic and volcanic rocks.
anorthite-basalt *Petrology.* basalt in which the plagioclase phenocrysts are anorthite.
anorthoclase *Mineralogy.* $(Na,K)AlSi_3O_8$, a colorless or white triclinic alkali feldspar, commonly massive in habit, with a specific gravity of 2.56 to 2.62 and a hardness of 6 to 6.5 on the Mohs scale; found in volcanic rocks.
anorthopia *Medicine.* **1.** distorted vision in which straight lines appear as curves or angles, and symmetry is incorrectly perceived. **2.** any deviation of the eye that the patient cannot overcome; the visual axes assume a position relative to each other that is different from what is required physiologically.
anorthosite *Petrology.* a plutonic igneous rock composed almost entirely of plagioclase feldspar, normally labradorite, with minor amounts of pyroxene and olivine.
anorthositization *Geology.* the process whereby the mineral anorthosite is introduced into a sediment or rock, or replaces another mineral.
anoscope *Medicine.* a speculum for the examination of the lower rectum and anus.
anosmia *Neurology.* the absence of the sense of smell. Thus, **anosmic.** Also, ANOSPHRASIA, OLFACTORY ANESTHESIA.
anosphrasia see ANOSMIA.
Anostraca *Invertebrate Zoology.* the order of fairy shrimps, brine shrimps, and others in the subclass Branchiopoda; having an elongated carapace, stalked eyes, and long uniramous antennae, and swimming upside down.
anotia *Medicine.* the absence at birth of the external ears.
anotron *Electronics.* an electron tube rectifier that operates on the principle of cold-cathode discharge. It has a copper anode and a large cathode of sodium or some other material.
anotto see ANNATTO.
ANOVA *Statistics.* an acronym for analysis of variance.
anoxemia *Medicine.* a reduction of oxygen content of the blood below physiologic levels.
anoxia *Medicine.* a total lack of oxygen; sometimes used to indicate a reduced supply of oxygen to the tissues.
anoxybiosis [ə näks´sē bī ō´sis] *Physiology.* a condition of living with little or no oxygen.
anoxygenic *Biochemistry.* of or relating to that which does not generate oxygen.
Anoxyphotobacteria *Bacteriology.* a class of bacteria of the division Gracilicutes, made up of Gram-negative organisms that are capable of photosynthesis under anaerobic conditions.
ANS or **A.N.S.** American Nuclear Society.
ansa *Anatomy.* any looplike structure.
ansae [an sē´] *Astronomy.* the portions of Saturn's rings that protrude beyond the limb of the planet.
Ansbacher unit *Biology.* a unit for the standardization of vitamin K.
Anser *Vertebrate Zoology.* a genus of birds in the family Anatidae, comprising several species of common geese.

Anseranatini *Vertebrate Zoology.* a subfamily in the family Anatidae with one species, the magpie goose of Australia.
Anserifomes *Vertebrate Zoology.* an order of aquatic birds characterized by flattened, hard-edged bills and webbed feet, and including ducks, geese, and swans.
ANSI or **A.N.S.I.** American National Standards Institute.
Ansolpidaceae *Mycology.* a family of fungi belonging to the order Hypochytriales, occurring widely as a parasite of algae.
anstau *Oceanography.* a process that results in a piling up or sinking of water masses, as in convergence.
answerback *Telecommunications.* the ability of a communications device to identify itself in response to a specific signal received from a calling device.
answer extraction *Artificial Intelligence.* a technique for extracting a desired answer from a resolution proof by introducing an artificial answer predicate that serves to carry the variable bindings of the solution.
answering machine *Telecommunications.* a tape-recording device attached to a telephone to answer automatically; it responds to callers with a previously recorded message and allows them to record their own messages for later playback.
answer predicate *Artificial Intelligence.* an artificial predicate that serves to carry the variable bindings in a resolution proof to allow answer extraction. While the prover will return True if it proves a question (e.g., "Does there exist an *x* such that ..."), the desired answer may be the value of *x* that makes the statement true.
ant *Invertebrate Zoology.* the common name for a narrow-waisted, generally wingless insect of the family Formicidae (order Hymenoptera). Ants live in highly elaborate social communities of hundreds or thousands of individuals, and are prolific worldwide except in very cold regions.
Ant see ANTLIA.
ant- a prefix meaning "against," as in *antacid.*
anta *Architecture.* a pilaster formed by thickening a wall at its terminus.
Antabuse [an´tə byooz´] *Pharmacology.* the trade name for a commercial preparation of disulfiram, a drug used to treat alcoholism. A person will experience vomiting, dizziness, and other ill effects if alcohol is consumed after taking the drug.
antacid *Medicine.* any of various mildly alkaline substances used to treat excess stomach acidity, such as sodium bicarbonate, $NaHCO_3$. *Chemistry.* a substance that neutralizes acids or counteracts acidity.
antagonism *Biology.* an interaction between two organisms existing in close association, in which one may inhibit the growth or action of the other through competition for food or the release of toxic substances. *Medicine.* the opposing action between drugs and disease or bodily functions. *Physiology.* an opposing action between one muscle and another.
antagonist *Biology.* an agent or substance that counteracts the action of another, as one of two opposing muscles or a drug that acts to nullify a biological response.
antapex *Astronomy.* see SOLAR ANTAPEX.
antarafacial *Organic Chemistry.* a term used to describe the process of simultaneously making or breaking two bonds on opposite faces of a molecule during a concerted cycloaddition reaction.
Antarctic [ant´ärk´tik] *Geography.* **1. the Antarctic.** the continent of Antarctica and its surrounding oceans. **2.** also, **antarctic.** found in, coming from, or associated with this region. (Most compound entries using this term may be spelled either Antarctic or antarctic.)
Antarctica [ant´ärk´ti kə] *Geography.* an ice-covered continent lying mostly within the Antarctic Circle, with the South Pole near its center; area: about 5,100,000 square miles.
antarctic air *Meteorology.* an air mass that develops its characteristics over the antarctic region.
antarctic anticyclone *Meteorology.* a glacial anticyclone believed to overlie Antarctica. Also, ANTARCTIC HIGH.
Antarctic bottom water *Oceanography.* a cold, saline water mass that sinks below the 12,000-foot level in the South Atlantic and flows north from the Antarctic land mass, sometimes reaching latitudes of 45°N; characterized by temperatures around –0.4°C and salinity of around 34.66 parts per thousand.
Antarctic Circle *Geodesy.* in the Southern Hemisphere, the parallel of latitude that is equal to the complement of the winter solstice. Due to steady variations in the obliquity of the winter solstice, the position of the Antarctic Circle is constantly changing. *Cartography.* a parallel of latitude representing the geodesic antarctic circle, customarily drawn on maps at 66°33´S.
Antarctic circumpolar current *Oceanography.* a current that flows eastward completely around Antarctica, caused by the west-wind surface drift.
antarctic cod *Vertebrate Zoology.* a large, codlike fish of the family Nototheniidae, bottom dwellers along the Subarctic coast; they are able to survive freezing temperatures due to specialized blood proteins called "notothenoid antifreeze" and through skeletal reduction. Also, **antarctic blennie.**
Antarctic Convergence *Oceanography.* the confluence of subantarctic and subtropical surface waters at about 50°S; i.e., the point where colder polar waters and warmer waters meet and conditions of air temperature, weather, and marine life change significantly.
Antarctic Divergence *Oceanography.* the point near the continent of Antarctica where two prevailing surface currents meet, the clockwise west-wind drift and the counterclockwise east-wind drift.
Antarctic dragonfish see DRAGONFISH.
antarctic front *Meteorology.* the semipermanent, semicontinuous front that lies between the antarctic air of Antarctica and the polar air of the southern oceans.
antarctic high see ANTARCTIC ANTICYCLONE.
Antarctic Intermediate Water *Oceanography.* a cold (about 2.2°C) water mass of below-average salinity (33.8 parts per thousand), formed at the surface near the Antarctic Convergence, generally between 45° and 55°S.
Antarctic Ocean *Geography.* the waters surrounding Antarctica, composed of the southern Atlantic, Pacific, and Indian oceans.
Antarctic Peninsula *Geography.* a narrow, mountainous ridge of land extending about 1200 miles northward from West Antarctica toward South America; formerly called Palmer Land or the Palmer Peninsula.
Antarctic Zone see SOUTHERN FRIGID ZONE.
Antarctodolops *Paleontology.* a genus of primitive marsupials in the family Polydolopidae. The only marsupial found in an Antarctic deposit, Antarctodolops is known only from a recently discovered jawbone of *A. dailyi.* The site may have been part of South America in the Eocene, to which it is dated.
Antares [an tar´ ēz] *Astronomy.* Alpha (α) Scorpii, a first-magnitude reddish supergiant star in the constellation Scorpius, the Scorpion.
antazoline *Pharmacology.* $C_{17}H_{19}N_3$, an oral antihistamine used to relieve and treat allergies. Also, IMIDAMINE.
ant bear *Vertebrate Zoology.* another name for the giant anteater. See ANTEATER.
antbird *Vertebrate Zoology.* a common name for the family Formicariidae, medium-sized birds with weak flight that follow ant armies and catch flushed insects; found in the neotropics from southern Mexico to South America.
ante- a prefix meaning: **1.** before, as in *antecedent.* **2.** in front of, as in *antebrachium.*

anteater

anteater *Vertebrate Zoology.* the common name for any of various mammals in the family Myrmecophagidae, distinguished by their elongated snouts and diet of ants and termites; the best known is the **giant (great) anteater,** which is found in Central and South America and may grow to six feet long, including the tail.
antebrachium [ant´tə brak´ē əm] *Anatomy.* the part of the arm between the elbow and the wrist; the forearm.
antecedent [an´tə sēd´int] *Artificial Intelligence.* the left-hand side of an implication or rule; e.g., in the rule $P \rightarrow Q$, P is the antecedent.

antecedent conflict *Psychology.* the concept or theory that disturbing events in one's early life bring about an intensified reaction to conflict in one's adult life.

antecedent method *Artificial Intelligence.* see IF-ADDED METHOD.

antecedent platform *Geology.* a theoretically preexisting submarine platform, situated 50 meters or more below sea level, from which barrier reefs and atolls are described as building upward toward the water's surface.

antecedent precipitation index *Meteorology.* an index of soil moisture derived from a daily summation of precipitation amounts over a specified period of time.

antecedent reasoning *Artificial Intelligence.* see FORWARD CHAINING.

antecedent stream *Hydrology.* a stream that existed before the present topography and that has maintained its original course after and despite local geologic changes, such as uplift.

antecedent valley *Geology.* a stream valley that has maintained its original course despite subsequent uplift or other deformation.

ante cibum [an´tē sē´bəm] a Latin term meaning "before meals."

anteconsequent stream [an´tē kän´sə kwənt] *Hydrology.* a stream whose course, in its earlier stages, was determined by the general form and slope of the recently formed land surface on which the stream originated, and that maintained this course despite later geologic changes.

antecosta *Invertebrate Zoology.* in arthropods, the internal anterior ridge on the skeleton where longitudinal muscles attach.

antediluvian [an´tē di loov´ē ən] *Geology.* a term formerly used to describe the period before the great flood associated with the Biblical account of Noah. Also, **antediluvial.**

antefebrile [an´tē fē´brəl] *Medicine.* occurring or existing prior to the onset of fever; before a fever.

antefix *Architecture.* an ornament used to conceal the ends of the joint tiles on a roof. Also, ANTIFIX.

anteflexion *Medicine.* **1.** an abnormal forward curving of an organ or part. **2.** the normal forward curvature of the uterus.

antelocation *Medicine.* the forward displacement of an organ.

antelope *Vertebrate Zoology.* any of various species of hollow-horned, hoofed, ruminant mammals of the subfamily Antilopinae in the order Artiodactyla. Most,well-known species, such as the gnu, eland, gazelle, and impala, live in Africa; there are also some species in Asia. The North American antelope (pronghorn) is not a true antelope.

ante meridiem or **ante meridian** *Science.* before noon; relating to the portion of the day between midnight and the following noon. (Latin for "before midday.")

ante mortem [an´tē môr´təm] a Latin term meaning "before death."

antenatal [an´tē nā´təl] *Medicine.* prenatal; formed or occurring before birth.

antenna [an´ten´ə] *plural,* **antennas** or **antennae.** [an´ten´ē] *Electromagnetism.* a device that radiates or receives electromagnetic radiation in the radiofrequency or microwave range. *Invertebrate Zoology.* either one of a pair of long slender processes on the heads of certain insects, arthropods, and crustaceans, which act as feelers and sensory receptors.

satellite tracking antenna

antenna amplifier *Electromagnetism.* any device or component that is used to improve upon the signal-to-noise level in a receiving antenna.

antenna aperture *Electromagnetism.* an antenna that encloses some area and has a radiation field distributed over the area, while outside the area the field is negligibly small.

antenna array *Electronics.* see ARRAY.

antenna bearing *Navigation.* a bearing determined by the position of a rotating antenna, such as a radar or radio direction finder.

antenna chlorophyll *Biochemistry.* a kind of chlorophyll molecule that collects and passes energy from light to the photochemical reaction center; the predominant photosynthetic pigment.

antenna circuit *Electronics.* a circuit used to release a wave into free space in a transmitting system, or to capture a free-space wave in a receiver system.

antenna coil *Electromagnetism.* a coil connected in series with an antenna so that the antenna current passes through it.

antenna coincidence *Electromagnetism.* a situation in which two directional antennas have patterns that are pointed toward each other.

antenna counterpoise see COUNTERPOISE.

antenna coupler *Electromagnetism.* a device that is used to transfer energy from a transmitter to a transmission line or from a transmission line to a receiver.

antenna crosstalk *Electromagnetism.* a measurement of unwanted transfer of power between two antennas; equal to the ratio or the log of the ratio of the unwanted power received by one antenna to the power transmitted by another antenna.

antenna detector *Electromagnetism.* a device that detects radar signals and thus warns personnel that they are being observed by radar.

antenna directivity diagram *Electromagnetism.* a graphic representation of a quantity proportional to the gain of a particular antenna for various directions in a plane or cone.

antenna effect *Electromagnetism.* **1.** in a loop antenna, any negative effect resulting from the capacitance of the loop to ground. **2.** in an electronic navigation system, the presence of output signals having no directional information, caused by the directional array acting as a simple nondirectional antenna.

antenna effective area see APERTURE.

antenna efficiency *Electromagnetism.* the ratio of the amount of power radiated by an antenna to the amount of power that is fed to the antenna's terminals.

antenna field *Electromagnetism.* **1.** a group of antennas arranged in a given configuration so as to produce a particular radiation pattern. **2.** the region defined by these antennas. **3.** the effective free-space energy distribution produced by an antenna or group of antennas.

antennafier *Telecommunications.* a low-profile antenna/amplifier combination that is used mostly for portable communications systems.

antenna gain *Electromagnetism.* a measure of the effectiveness of a directional antenna as compared to a nondirectional antenna; for a transmitting antenna, usually expressed in decibels as the ratio of standard input power to directional antenna input power that will produce the same field strength in the desired direction; for a receiving antenna, usually expressed as the ratio of the signal power values produced at the receiver input terminals.

antennal gland *Invertebrate Zoology.* the principal excretory organs located at the base of the antennae in certain crustaceans. Also, GREEN GLAND.

antenna loading *Electronics.* the placement of a lumped inductor or capacitor in an antenna circuit, in order to increase or decrease the effective wavelength of the antenna.

antenna matching *Electromagnetism.* a matching of the impedance of an antenna to the characteristic impedance of its transmission line.

antennamitter *Telecommunications.* an antenna/oscillator combination.

antenna pair *Electromagnetism.* two antennas on a baseline of a fixed length, used in direction finding and exhibiting directional patterns.

antenna polarization *Electromagnetism.* the polarization of an electromagnetic wave that is radiated by a transmitting antenna.

antenna power *Electromagnetism.* of a transmitting antenna, the product of the square of the antenna current and the antenna resistance at the point where the current is measured.

antenna power gain *Electromagnetism.* of a transmitting antenna, the square of the antenna gain, expressed in decibels.

antenna resistance *Electromagnetism.* of a transmitting antenna, the power supplied to the antenna circuit divided by the square of the effective current measured at the power feeding point of the antenna; expressed in ohms.

Antennaridae or **Antennariida** *Vertebrate Zoology.* the frogfishes, a family of marine anglerfish in the order Lophiiformes, characterized by a short round body and a large mouth with numerous teeth; found in shallow water or clinging to floating seaweed in all tropical and subtropical seas.

antenna scanner *Electromagnetism.* a microwave feed horn that is designed to direct a signal over an antenna array in such a way so as to create a desired field pattern.

Antennata *Invertebrate Zoology.* a subphylum of the class Arthropoda; they have antennae and the first postoral appendages are mandibles. Also, MANDIBULATA.

antenna temperature *Astrophysics.* the apparent temperature of a celestial object as measured by the power of its radio signal.

antenna tilt error *Engineering.* a measure of the difference between the angle of the tilt of a radar antenna and the electrical center of the radar beam.

antennaverter *Telecommunications.* an antenna/converter combination that attaches to the intermediate-frequency amplifier of a receiver.

antennule *Invertebrate Zoology.* in crustaceans, one of a second pair of antennae, generally smaller than the first pair.

antenodal *Entomology.* in dragonflies, before or in front of the nodus (a cross vein near the middle of the costal border of the wing).

antepartal *Medicine.* occurring before parturition or childbirth, with reference to the mother. Also, **antepartum.**

anteport *Architecture.* an outer gate or door.

anter *Invertebrate Zoology.* part of a bryozoan operculum that serves to close off a portion of the operculum.

anteriad *Zoology.* toward the front or head of the body; anterior.

anterior *Zoology.* **1.** situated in front of an organ or in the front part of a body or structure. **2.** a position near the front or head.

anterior commissure *Anatomy.* an organized tract of fibers in the anterior part of the brain that connects corresponding parts of the right and left cerebral hemispheres.

anterior pituitary hormones *Endocrinclogy.* the group of hormones produced and released by the adenohypophysis, including lutropin, follitropin, somatotropin, prolactin, corticotropin, thyrotropin, and melanotropin.

antero- a prefix meaning "in front of" or "forward," as in *anteromedial, anterolateral.*

anterograde amnesia *Medicine.* the impairment of memory for events occurring after the onset of amnesia; an inability to form new memories.

anteroinferior *Anatomy.* situated in front and below.

anterointernal *Anatomy.* situated in front and to the inside.

anterolateral *Anatomy.* situated in front and away from the midline.

anteromedial *Anatomy.* situated in front and to the midline.

anteromedian *Anatomy.* situated in front and in the median plane.

anteroom *Architecture.* a small room serving as an entrance or waiting area for a larger room.

anteroposterior *Anatomy.* from front to back.

anteroseptal *Cardiology.* in front of the septum of the heart, specifically the atrioventricular septum.

anterosuperior *Anatomy.* situated in front and above.

antetheca *Paleontology.* the leading wall, representing the most recent stage of growth, of a fusulinid; when a new leading wall forms, the newly enclosed antetheca becomes a septum.

anteversion *Medicine.* a condition in which an organ is tipped forward but not bent at an angle, as is the case in anteflexion.

anth- a combining form meaning "flower," as in *anthesis.*

anthelate *Botany.* an elongated, flower-bearing branch, as in certain rushes.

Antheliaceae *Botany.* a monogeneric family of alpine-arctic liverworts belonging to the order Jungermanniales; characterized by isophyllous, prostrate plants that are often whitish and sometimes brown, rhizoids that are restricted to underleaf bases, and massive seta.

anthelic arc [ant hē´lik] *Astronomy.* a rare daytime luminous phenomenon that appears at the same elevation as the sun, but 180° away from it in azimuth.

anthelion [ant hēl´ yən] *Astronomy.* a bright white spot that sometimes appears in the sky directly opposite the Sun.

anthelminthic *Pharmacology.* any medication used to expel or destroy intestinal worms. Also, **anthelmintic.**

anther *Botany.* in flowering plants, the tip of the stamen, which contains the pollen sacs and in which the pollen grains are formed; usually elevated by means of a filament.

Antheraea eucalypti cells *Virology.* a line of cells that are isolated from insects and can support the growth of a number of arboviruses.

antheraxanthin *Biochemistry.* a yellow pigment found in plants and animals such as maize, algae, and bacteria.

anther culture *Biotechnology.* a method of producing haploid plants by inducing embryo formation in pollen, accomplished in solid or liquid culture. *Botany.* an aseptic culture of intact anthers or single pollen grains; used to induce embryo formation and produce haploid plants, most often in the Solanaceae and Gramineae genera.

antheridiophore *Botany.* in certain liverworts, an upright structure consisting of a stalk and cap that bears the antheridia.

antheridium *Botany.* the organ in which the male sex cells are developed in flowerless and seedless plants.

antheriferous *Botany.* of a plant, having anthers.

antherozoid *Botany.* in lower plants and some gymnosperms, a motile male gamete that moves by means of flagella.

anther smut *Mycology.* a kind of smut fungus that is a parasite to certain plants, in whose anthers it develops its spores.

anthesis *Botany.* the period, act, or state of full bloom in a flower; the maturing of the stamens.

Anthicidae *Invertebrate Zoology.* a family of antlike flower beetles in the superfamily Tenebrionoidea.

anthill *Entomology.* a mound of dirt carried by ants from their underground nest and heaped around its entrance.

antho- a combining form meaning "flower," as in *anthoblast.*

anthoblast *Invertebrate Zoology.* a developmental stage of some corals, produced by budding.

anthocarpous *Botany.* describing a plant that bears multiple fruit.

anthocaulus *Invertebrate Zoology.* the stalklike basal portion of some solitary corals, from which the oral portion is pinched off to form a new zooid.

Anthocerotae *Botany.* the class of hornworts.

Anthocerotales *Botany.* the hornworts, a worldwide order constituting the class Anthocerotae and composed of the single family Anthocerotaceae; characterized by prostrate plants often forming rosettes, a leafless thallus with lobate margins or a thick midrib and broad membranous wings, and a unique sporophyte with an erect or horizontal dehiscent capsule.

anthocodium *Invertebrate Zoology.* the free oral end of colonial sea anemones that have their basal portions united in a common mass.

Anthocoridae *Invertebrate Zoology.* the family of flower bugs in the order Hemiptera; small active bugs that are predacious on other insects.

anthocyanidin *Biochemistry.* a commonly occurring plant pigment derived from anthocyanin.

anthocyanin *Biochemistry.* a soluble glucoside plant pigment found in fruits, leaves, and blossoms of higher plants and responsible for their intense red, purple, blue, and black colors.

Anthocyathea *Paleontology.* a class of usually solitary archaeocyathids, characterized by a cylindrical or discoidal conical cup and by numerous dissepiments and tabulae; they were extant in the Lower to Upper Cambrian.

anthocyathus *Invertebrate Zoology.* in some solitary corals, the new free zooid formed by the budding of the anthocaulus (stalklike basal portion).

anthodite *Geology.* clusters of long needlelike or hairlike crystals of gypsum or aragonite that radiate from the roof or wall of a cave.

anthoinite *Mineralogy.* a white triclinic mineral with the approximate formula $WAl(O,OH)_3$, occurring as tabular crystals and having a specific gravity of 4.8 to 5.06 and a hardness of 1 on the Mohs scale; found in placer concentrates and quartz veins.

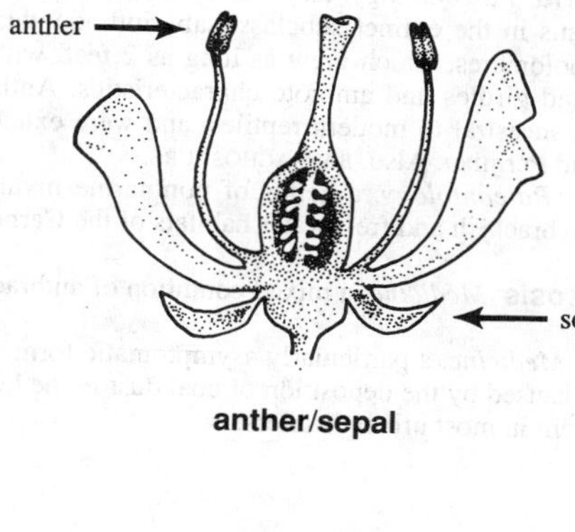

anther/sepal

Anthomedusae *Invertebrate Zoology.* an order of coelenterates in the class Hydrozoa. The medusoids (jellyfish stage) lack otocysts but may have ocelli while gonads are on the manubrium; the hydroid colony lacks a complete skeletal covering (athecate).

Anthomyzidae *Invertebrate Zoology.* a family of small flies in the order Diptera, with at least one pair of bristles on the front of the head bent upward.

anthophagous [an´thə fā´jəs] *Biology.* feeding mainly or exclusively on flowers. Thus, **anthophage, anthophagy.**

anthophilous [an´thə fil´əs] *Biology.* attracted to or feeding on flowers. Thus, **anthophile, anthophily.**

anthophobia *Psychology.* an irrational dislike or fear of flowers.

anthophore [an´thə fôr´] *Botany.* in some flowers, a lengthened internode below the receptacle that bears the pistils and corolla.

anthophyllite [an´thə fi´līt] *Mineralogy.* $(Mg,Fe^{+2})_7Si_8O_{22}(OH)_2$, a gray to green to brown, translucent, orthorhombic mineral with a vitreous luster, a specific gravity of 2.85 to 3.57, and a hardness of 5.5 to 6 on the Mohs scale; usually massive or rarely in prismatic crystals, found only in metamorphic rocks such as schists and gneisses.

Anthosomidae *Invertebrate Zoology.* a family of fish lice (copepods) in the crustacean suborder Caligoida; exoparasites on fish.

anthostele *Invertebrate Zoology.* the thick-walled, nonretractable, aboral section of certain coelenterates.

Anthozoa *Invertebrate Zoology.* a class of corals, sea anemones, and related forms in the phylum Cnidaria; the jellyfish stage is absent, and polyps have a septate gastrovascular cavity with stinging cells.

anthozooid *Invertebrate Zoology.* an individual zooid of a compound anthozoan.

anthracene [an´thrə sēn´] *Organic Chemistry.* $C_6H_4(CH)_2C_6H_4$, a colorless crystalline solid that melts at 217°C and boils at 340°C, and is insoluble in water and soluble in alcohol and ether; produced during the distillation of coal tar. It is used in making dyes.

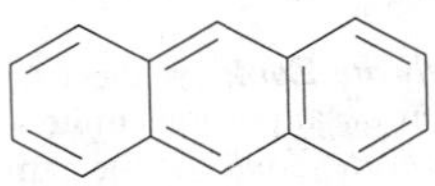

anthracene

anthracene oil *Materials.* an oil that is a coal-tar fraction, boiling at a range of 270–360°C; a source of anthracene and also used as a wood preservative and pesticide.

anthracene violet see GALLEIN.

anthraciferous coal *Mineralogy.* a type of hard coal containing anthracene.

anthracite [an´thrə sīt´] *Mineralogy.* a coal of the highest metamorphic rank, black with a bright, often submetallic luster, having a fixed carbon content of 92–98%, a specific gravity of 1.32 to 1.7, and a hardness of 2 to 2.5 on the Mohs scale. Also, HARD COAL, STONE COAL.

anthracite fines *Mining Engineering.* the produce from an anthracite coal preparation plant, usually less than one-eighth inch in diameter. Also, DUFF, FINES, GRAINS.

anthracitization *Geochemistry.* the transformation of bituminous coal into anthracite coal as a result of metamorphism.

anthracnose *Plant Pathology.* a plant disease caused by the fungi Melanconiales and characterized by black, usually sunken stem lesions.

Anthracosauria *Paleontology.* an order of small, salamandrine or eel-like amphibians in the extinct subclass Labyrinthodontia; best known from the embolomeres, which grew as long as 2 feet, with well-developed limbs and girdles and amniote characteristics. Anthracosaurians were directly ancestral to modern reptiles, and were extant in the Carboniferous and Permian. Also, BATRACHOSAURS.

Anthracosia *Paleontology.* a genus of nonmarine bivalves that was widespread in brackish and freshwater habitats of the Carboniferous and Permian.

anthracosilicosis *Medicine.* a mixed condition of anthracosis and silicosis.

anthracosis *Medicine.* a particularly asymptomatic form of pneumoconiosis that is caused by the deposition of coal dust in the lungs; this condition is present in most urban dwellers.

Anthracotheriidae *Paleontology.* an extinct family of large, piglike artiodactyls in the superfamily Hippopotamoidea; probably amphibious and possibly ancestral to the hippopotamuses; Holarctic distribution; extant in the Eocene to Pleistocene.

anthracoxene *Geology.* a brownish resin that occurs in brown coal and dissolves in ether to form an insoluble portion, anthrocoxenite, and a soluble portion, schlanite.

anthralin *Pharmacology.* $C_{14}H_{10}O_3$, a drug applied to the skin to treat chronic eczema, ringworm, athlete's foot, psoriasis, and other skin diseases.

anthranilate *Biochemistry.* a salt of anthranilic acid, a precursor of the amino acid tryptophan.

anthranilic acid *Organic Chemistry.* $C_6H_4NH_2COOH$, a white to yellow crystalline solid that melts at 144–146°C, and is soluble in hot water, alcohol, ether, and chloroform; used in making pharmaceuticals, dyes, and perfumes. Also, *o*-AMINOBENZOIC ACID.

anthranone see ANTHRONE.

anthrapurpurin *Organic Chemistry.* $C_6H_3OH(CO)_2C_6H_2(OH)_2$, a tricyclic orange-yellow crystalline solid; melts at 369°C and boils at 462°C; slightly soluble in hot water and soluble in alcohol and alkalies; used in dyeing textiles and in organic synthesis.

anthraquinone *Organic Chemistry.* $C_6H_4(CO)_2C_6H_4$, yellow needles that are insoluble in water and soluble in alcohol and ether; melts at 286°C and boils at 379–381°C; used as a cathartic and in the manufacture of certain dyes.

anthraquinone pigments *Biochemistry.* violet, red, orange, or yellow pigments occurring in certain plants and animals; used as a source for certain dyes.

anthrax [an´thraks´] *Veterinary Medicine.* a highly contagious bacterial disease of warm-blooded animals, especially cattle and sheep, caused by *Bacillus anthracis*; transmissible to humans by contaminated raw meat and other animal products, and giving rise to high fever, convulsions, spleen enlargement, lesions in the lungs and usually rapid death. Also, WOOLSORTER'S DISEASE.

anthraxolite *Geology.* a hard, black asphaltic material having a high fixed carbon content and often occurring in association with oil shales.

anthraxylon *Geology.* the vitreous constituents of coal that are derived from the woody tissues of plants.

Anthribidae *Invertebrate Zoology.* a small family of fungus weevils in the order Coleoptera that live and feed on woody fungi or decaying wood.

anthrone *Organic Chemistry.* $C_{14}H_{10}O$, a crystalline solid that melts at 155°C; insoluble in water and soluble in alcohol, acetone, and benzene; used in organic synthesis as a reagent for carbohydrates.

anthrop- a prefix meaning "human," as in *anthropic.*

anthropic *Anthropology.* relating to human beings or to the period of human existence on earth. Also, **anthropical.**

anthropic principle *Astronomy.* a principle of reasoning in cosmology which states that the present existence of life on earth implies certain limits on the ways that the early universe could have evolved.

anthropo- a prefix meaning "human," as in *anthropology, anthropoid.*

anthropocentric *Psychology.* **1.** of or relating to the belief that humans are the center of the universe. **2.** regarding or interpreting natural events or conditions in terms of human values. Thus, **anthropocentrism.**

anthropodesoxycholic acid see CHENODEOXYCHOLIC ACID.

anthropogenic *Ecology.* caused or influenced by the activities of humans. Also, **anthropogenous.**

anthropography *Anthropology.* a branch of anthropology that studies the ethnic and racial distribution of people over geographical areas.

anthropoid *Anthropology.* resembling a human; used especially to refer to the humanlike apes, including chimpanzees, gorillas, and many ancient apes. *Vertebrate Zoology.* of or relating to the great apes in the primate family Pongidae.

anthropoidal plate *Anthropology.* the primate ilium that grew into an enlarged plate of bone as primates evolved.

anthropoid ape see ANTHROPOID.

Anthropoidea *Vertebrate Zoology.* a mammalian suborder in the order Primates that includes humans, as well as monkeys and apes.

anthropological of or relating to anthropology.

anthropological linguistics *Anthropology.* a major branch of anthropology in which language is studied in a sociocultural context, with an emphasis on how language reflects the thought processes of particular cultures.

anthropologist *Anthropology.* a person who specializes in or studies anthropology.

Anthropology

Anthropology is the study of the social and biological aspects of humankind, both past and present. We humans are intrigued by who we are, where we came from, and why we behave the way we do. Anthropologists view the diversity in physical characteristics and cultural practices of modern peoples as variations on a common human theme. Traditionally anthropologists have deemphasized the study of European societies and concentrated their efforts on non-Western societies. The objective study of anthropology has led to the idea of cultural relativity, meaning that all societies must be evaluated within their own cultural framework.

Anthropologists attempt to understand human nature and humans' place in nature. A highly diverse discipline, anthropology is concerned with the sociocultural as well as the biological side of humanness. Sociocultural anthropologists study social, religious, political, and economic customs, as well as the language (linguistics) of living societies. Archaeologists attempt to understand past cultures by discovering and interpreting the material culture left behind by our forebears.

Physical anthropologists focus on understanding biological variation seen in extant human populations and in documenting the fossil record of human origins. Although human populations differ in physical characteristics such as skin color, physical anthropologists stress the essential biological unity of human species. Four million years ago our ancestors, *Australopithecus,* stood upright and launched the human career. Recent findings suggest that anatomically modern humans, *Homo sapiens,* also arose first in Africa nearly 200,000 years ago.

Donald C. Johanson
President, Institute of Human Origins
Berkeley, California

anthropology the science that deals with the study of human culture and evolution, consisting of such subdisciplines as physical anthropology, archaeology, anthropological linguistics, cultural anthropology, and social anthropology. Archaeology is sometimes regarded as a separate science rather than as a branch of anthropology.

anthropometer *Anthropology.* a complex measuring stick used to calculate the dimensions of the human body.

anthropometric *Anthropology.* relating to or based on anthropometry.

anthropometry [an′thrə päm′ə trē] *Anthropology.* the use of various techniques for taking measurements of the human body, employed in the study of human growth and variation.

anthropomorph *Archaeology.* a design element that portrays a human or humanlike figure, such as those found on ancient pottery.

anthropomorphic [an′thrə pə môr′fik] *Anthropology.* describing nonhuman objects in human form; e.g., rock art that depicts a god as having a human shape. *Behavior.* relating to or characterized by anthropomorphism; regarding animals as possessing human qualities.

anthropomorphism [an′thrə pə môr′fiz əm] *Anthropology.* the portrayal or conception of a human form or human characteristics in a deity, animal, or inanimate object. *Behavior.* the attribution of human abilities or emotions to animals, suggesting that their actions are conscious and intentional.

anthroponosis *Pathology.* an infectious disease confined to humans but originally derived from animals.

anthropophilic *Pathology.* human-seeking or selecting; referring to a parasite that prefers humans over animals as a source of blood meal..

anthropophobia *Psychology.* an irrational fear of people or of human society.

anthroposcopy *Anthropology.* the determination of characteristics or personality from the body shape and facial features; a technique used in earlier studies but now discredited.

Anthuridea *Invertebrate Zoology.* a small suborder of isopods in the crustacean class Malacostraca, with slender, elongated, dorsoventrally compressed bodies.

anti- a prefix meaning: **1.** "against," as in *anticoagulant.* Most "anti" words that are spelled with a hyphen can also be correctly spelled without the hyphen. **2.** used against a specified form of attack, as in *antitank, antisubmarine,* and so on.

antiacid bronze *Metallurgy.* the former designation of acid bronze, a corrosion-resistant, cast copper-base alloy.

antiadrenergic [an′tē ad rə nur′jik] *Neurology.* opposing or blocking adrenergic (sympathetic) neural activity. *Pharmacology.* any drug or agent that blocks adrenergic neural activity. Also, ANTISYMPATHETIC, SYMPATHOLYTIC.

antiagglutinin *Immunology.* a substance that neutralizes its corresponding agglutinin.

antiaggressin *Immunology.* a substance that neutralizes aggressin, which is produced by certain microorganisms to enhance virulence.

antiairborne *Ordnance.* of or relating to weapons or equipment used to resist airborne attacks.

antiaircraft *Military Science.* of or relating to an effort to resist attack by enemy aircraft and missiles. Thus, **antiaircraft artillery, antiaircraft defense, antiaircraft operations,** and so on. *Ordnance.* used to fire on or destroy enemy aircraft and missiles. Thus, **antiaircraft gun, antiaircraft missile, antiaircraft barrage,** and so on.

antianaphylaxis [an′tē an ə fə lak′sis] *Immunology.* a condition in which a sensitized animal is resistant to anaphylaxis.

antiandrogen *Endocrinology.* a substance, such as flutamide, that inhibits the biological actions of an androgenic hormone by binding to and thus activating the androgen receptor of target tissue.

antiantidote *Toxicology.* any substance that counteracts the action of an antidote. Thus, **antiantidotal, antiantidotic.**

Antiarchi *Paleontology.* an order of mainly freshwater placoderms, extant in the lower Devonian to lower Carboniferous, with the head and anterior body armored, the tail and posterior body fishlike. They retain the heavy armor of earlier placoderms, rather than becoming lighter as the contemporaneous arthrodires did; their antecedents are unclear.

antiatom [an′tē at′əm] *Atomic Physics.* an atom composed of antiparticles, such as antiprotons, antineutrons, and positrons. It thus has the same mass and spin as an ordinary atom, but the component parts are antiparticles with reversed charges.

antibacterial agent *Microbiology.* any agent that is either bacteriocidal or bacteriostatic, killing or inhibiting the growth of bacterial cells.

antiballistic *Ordnance.* used to intercept and destroy ballistic missiles.

antibaryon [an′té bâr′é än] *Atomic Physics.* the antiparticle of a baryon, including both antinucleons and antihyperons, with baryon number −1, strong interactions, and charge opposite that of the baryons.

antibiogram *Microbiology.* the sensitivity pattern of a given microorganism to a range of antibiotics.

antibiosis [an′tē bī ō′sis] *Biology.* a mutual relationship between two species in which one adversely affects the other, as by the production of a substance that is toxic to the other species.

antibiotic [an′tē bī ät′ik] *Biology.* **1.** having the ability to destroy life, or to interfere significantly with life processes. **2.** any substance or agent capable of having this effect. *Microbiology.* a chemical substance that is important in the treatment of infectious diseases, produced either by a microorganism or semisynthetically, having the capacity in dilute solutions to either kill or inhibit the growth of certain other harmful microorganisms. Antibiotics are widely used antibiotics include penicillin, streptomycin, and tetracycline.

antibiotic assay *Microbiology.* a test to determine the sensitivity of a microorganism to a range of antibiotics by examining the microbe's ability to grow in the presence of a standard dilution of each antibiotic.

antibody *Immunology.* a protein, produced as a result of the introduction of an antigen, that has the ability to combine with the antigen that caused its production. Also, IMMUNE BODY.

antibody combining site see ANTIGEN BINDING SITE.

antibody-dependent cell-mediated cytotoxicity *Immunology.* the capability of nonsensitized cells to rupture (kill) other cells that have been coated by a specific antibody.

antibody-forming cells *Immunology.* the terminally differentiated B cells (plasma cells) that are devoted to the production of secreted antibodies.

antibonding orbital *Physical Chemistry.* a relationship between an electron and a nucleus in which the electron's energy increases as it moves away from the nucleus, causing them to separate rather than bond. Also, **antibonding molecular orbital.**

anticapacitance switch *Electronics.* a switch designed to minimize capacitance between its open terminals.

anticarcinogen *Pharmacology.* any substance that counteracts the effects of a carcinogen, or cancer-causing agent.

anticatalyst *Chemistry.* an agent that inhibits the effects of a catalyst.

anticathode *Electronics.* in an X-ray tube, the electrode from which X rays are emitted and to which electrons are directed.

anticenter *Astronomy.* the point (which lies in Auriga) lying opposite the galactic center in the sky. *Geology.* thet point on the surface of the earth that is diametrically opposite the epicenter of an earthquake. Also, ANTIEPICENTER.

antichlor *Chemical Engineering.* any product that neutralizes and removes hypochlorite or free chlorine after the bleaching process in the manufacture of textiles or paper.

anticholinergic [an´tē kōl´ə nur´jik] *Neurology.* opposing or blocking cholinergic (parasympathetic) neural activity. *Pharmacology.* any drug or agent that blocks cholinergic neural activity. Also, PARASYMPATHOLYTIC.

anticholinesterase [an´tē kōl´ə nes´tə rās] *Organic Chemistry.* any compound, such as nerve gas, that inhibits the activity of the enzyme cholinesterase and thus blocks normal transmission of nerve impulses.

anticipatory control *Biotechnology.* the predictive regulation of a process or system based on anticipated events, a feed-forward rather than a feedback system.

anticipatory staging *Computer Programming.* the movement of blocks of data or instructions from slower auxiliary storage to a faster medium, such as main memory, so that the data is ready for access in the expectation that it will soon be needed by the program.

anticlastic *Mathematics.* if the two principal curvatures at each point of a surface have opposite signs, the surface is said to be anticlastic. Equivalently, at each point, one normal section is concave and the other is convex, so every point is a saddle point.

anticlastic bending *Physics.* an effect in which the cross section of a beam is distorted in bending, due to longitudinal contractions.

anticlinal [an´tē klī´nəl] *Botany.* at right angles to the surface; for example, the anti-clinal wall of a cell is arranged perpendicular to the surface of the plant body. *Geology.* of or relating to an anticline.

anticlinal fold

anticlinal axis *Geology.* **1.** in an anticline or folded structure, the common crest or median line from which the strata seem to dip away in opposite directions. **2.** a line representing the intersection of the axial plane with a given stratum in an anticline.

anticlinal bend *Geology.* a convex upward bend of rock strata in which one limb dips gently toward, and the other limb dips steeply away from, the crest.

anticlinal mountain *Geology.* a ridge or series of ridges formed by an anticlinal fold.

anticlinal nose *Geology.* see NOSE, def. 1.

anticlinal theory *Geology.* the theory that trapped underground petroleum and natural gas tend to accumulate in or near anticlinal structures.

anticlinal trap *Geology.* a formation at the top of an anticline where petroleum or natural gas accumulates.

anticlinal valley *Geology.* a valley that was carved from or follows the site of an anticline.

anticline [an´tē klīn] *Geology.* an upward fold of stratified rock in which the sides slope down and away from the crest; the oldest rocks are in the center, and the youngest rocks are on the outside.

anticlinorium *Geology.* a composite fold structure of great extent consisting of a series of minor folds and having the form of an anticline.

anticlutter gain control *Electronics.* gain control in a radar receiver that automatically raises the receiver gain slowly to maximum after a transmission pulse is received. This process minimizes amplification of clutter due to short-range echoes.

anticoagulant [an´tē kō´ag yə lənt] *Hematology.* **1.** acting against coagulation; tending to cause the blood not to coagulate. **2.** any substance or agent that tends to prevent blood clotting. *Pharmacology.* a drug that acts to prevent or suppress coagulation of the blood.

anticoagulant therapy *Cardiology.* a treatment program for reducing or preventing intravascular or intracardiac clotting through the use of anticoagulant drugs.

anticoding strand see ANTISENSE STRAND.

anticodon *Genetics.* the sequence of three nucleotides in a transfer RNA molecule that is complementary to the codon of a messenger RNA molecule.

anticoincidence *Nuclear Physics.* the occurrence of an event at one site without a simultaneous event at another.

anticoincidence circuit *Electronics.* in control systems, a circuit that produces an output pulse when one of its two inputs receives a pulse but the other does not, occurring within a given time interval.

anticollision radar *Engineering.* a radar system used specifically to warn about potential collisions between ships or between aircraft.

anticommutator *Mathematics.* the anticommutator of two linear operators A and B is the operator $AB + BA$.

anticommute *Mathematics.* two operators A and B anticommute if their anticommutator is zero, or, equivalently, if $AB = -AB$.

anticomplementary *Immunology.* referring to a substance capable of diminishing or eliminating the action of a complex system (known as a complement), which normally combines with antibodies to form a host defense mechanism.

anticonvulsant *Medicine.* acting or serving to prevent convulsions. *Pharmacology.* any drug that prevents or relieves the violent involuntary phasic or tonic contraction of muscles occurring during an epileptic seizure.

anticorona *Optics.* the rainbow-colored rings, complementary to the rings of a corona, that sometimes surround the shadow of an observer cast by the sun on a fog bank, mist, or cloud; often seen from mountaintops and airplanes. Also, BROCKEN BOW, GLORY.

anticorrosive *Materials Science.* describing a material that serves to inhibit or prevent corrosion.

anticreeper *Civil Engineering.* a device used to prevent the movement of a railroad rail in the lengthwise direction.

anticrepuscular arch see ANTITWILIGHT ARCH.

anticrepuscular rays [an´ tē kri pus´ kyə lər] *Astronomy.* crepuscular rays that lie more than 90° away from the sun.

anticryptic *Biology.* relating to the use of misleading body coloration for camouflage or concealment, especially as a means of defense against predators.

anticurl coating *Graphic Arts.* a thin film that is applied to the back of photographic material in order to prevent rolling or distortion.

anticusp *Invertebrate Zoology.* an anterior, downward projection in conodonts.

anticyclogenesis *Meteorology.* the process of strengthening or developing anticyclonic circulation in the atmosphere.

anticyclolysis *Meteorology.* a weakening or abatement of anticyclonic circulation in the atmosphere.

anticyclone *Meteorology.* the rotation of air about a concentrated center of higher atmospheric pressure in a direction opposite to that of the earth's rotation; it is clockwise in the Northern Hemisphere, counterclockwise in the Southern Hemisphere, and undefined at the equator.

anticyclonic *Meteorology*. of, relating to, or being an anticyclone.
anticyclonic winds *Meteorology*. the winds associated with the concentrated center of highest pressure in an anticyclone.
antidepressant *Pharmacology*. any drug that prevents or relieves depression or stimulates the mood of a depressive patient.
antiderivative see INDEFINITE INTEGRAL.
antidesiccant *Materials Science*. a material that serves to inhibit or prevent moisture loss, as from plants.
antidetonant see ANTIKNOCK.
antideuteron *Atomic Physics*. the antiparticle of the deuteron composed of an antiproton and an antineutron, thus having negative charge.
antidiabetic *Pharmacology*. any drug that prevents or relieves diabetes.
antidiarrheal *Pharmacology*. any agent that corrects or alleviates diarrhea.
antidisturbance fuse *Ordnance*. a sensitive fuse designed to detonate the explosive if it is moved or disturbed. Also, **antihandling fuse.**
antidiuretic [an′tē dī′yə ret′ik] *Pharmacology*. any drug that reduces or suppresses the formation of urine.
antidiuretic hormone see ARGININE VASOPRESSIN.
antidotal *Toxicology*. serving as an antidote. Also, **antidotic.**
antidote *Toxicology*. any substance that counteracts or prevents the action of a poison.
antidrag *Aviation*. **1.** of a force, acting against the force of drag. **2.** of an aircraft structural member, designed to resist the effects of drag.
antidune *Geology*. a temporary ripple or sand wave that is formed on a stream bed and is analogous to a sand dune, but migrates upcurrent.
antienzyme *Enzymology*. a polypeptide or protein that inhibits the action of an enzyme or destroys it.
antiepicenter see ANTICENTER.
antiestrogen *Endocrinology*. a substance, such as tamoxifen, that blocks the action of estrogens by binding to and thus inactivating the estrogen receptor of target tissue.
antifading antenna *Electronics*. an antenna that radiates at small angles of elevation to decrease fading caused by high-altitude radiation.
antiferroelectric crystal *Physics*. a crystalline substance composed of two interpenetrating sublattices having equal and opposite electric polarization when in a state of lower symmetry, and composed of unpolarized, indistinguishable sublattices when in a state of higher symmetry.
antiferromagnetic *Electromagnetism*. describing a substance composed of atoms or molecules whose spins are arranged antiparallel to their neighbors in the crystal lattice.
antiferromagnetic domain *Solid-State Physics*. a region within a solid substance in which the magnetic moments of the atoms or molecules are aligned antiparallel to each other.
antiferromagnetic resonance *Electromagnetism*. a magnetic resonant condition observed in antiferromagnetic materials when the material is placed in a rotating magnetic field.
antiferromagnetic susceptibility *Electromagnetism*. a measure of the magnetic response of an antiferromagnetic material when subjected to an external magnetic field.
antiferromagnetism *Solid-State Physics*. a phenomenon in which there is a lack of magnetic moment in certain metals, alloys, and transition element salts due to the antiparallel or spiral arrangement of the atomic magnetic moments.
antifertility agent *Pharmacology*. another term for a contraceptive drug.
antifertilizin *Biochemistry*. a substance produced by animal sperm that serves to attract the egg prior to fertilization.
antifibrinolysin *Biochemistry*. a substance that inhibits certain enzymes in plasma from catalyzing the breakdown of proteins.
antifix see ANTEFIX.
antifoaming agent see DEFOAMING AGENT.
antifogging *Materials Science*. describing a material that serves to inhibit or prevent fogging, as on a windshield or lens.
antiform *Geology*. an anticlinal structure in which the stratigraphic sequence is unknown.
antifouling *Materials*. **1.** describing a material applied to ship bottoms, underwater pilings, and the like, to protect them from barnacles and other marine organisms that accumulate on such surfaces. **2.** describing a material that prevents the accumulation of carbon on an engine part.
antifreeze *Chemistry*. **1.** a substance that lowers the freezing point of water, especially one that is used commercially for this purpose in automotive cooling systems, such as ethylene glycol. **2.** any substance that is added to a liquid to lower its freezing point; e.g., Drygas or another such mixture that is added to gasoline.
antifreeze proteins *Cell Biology*. a class of serum proteins produced by several species of marine fish found in the Arctic and Antarctic; they lower the freezing point of the blood.
antifriction *Mechanics*. **1.** serving to reduce friction. **2.** a substance or material that lessens friction. Thus, **antifriction bearing.**
antifriction alloy *Metallurgy*. an alloy that has a low coefficient of friction.
antifriction material *Engineering*. a general term for various metals, alloys, plastics, polyurethane rubber, and other complex compounds that produce little friction with themselves and other surfaces; used to line shafts, axles, bearings, and other such moving parts.
antifungal *Materials Science*. describing a material that serves to inhibit or prevent the growth of fungi.
antigen [an′ti jən] *Immunology*. **1.** a substance that causes the formation of an antibody or elicits a cellular response. **2.** see IMMUNOGEN.
antigen-antibody reaction *Immunology*. any reaction in which an antibody interacts with, combines with, or neutralizes an antigen.
antigen binding capacity *Immunology*. the measurement of the ability of an antibody to bind with a specific antigen.
antigen binding site *Immunology*. the area of an antibody that binds specifically with its related antigenic determinant. Also, ANTIBODY COMBINING SITE.
antigen deletion *Microbiology*. the loss of one or more antigenic determinants in a microorganism due to mutation or loss of a plasmid.
antigenemia *Hematology*. the presence of an antigen (such as hepatitis B antigen) in the blood. Also, **antigenaemia.**
antigenic [an′tə jen′ik] *Immunology*. relating to or being an antigen.
antigenic competition *Immunology*. a condition in which an immune response to an antigen becomes diminished due to the previous introduction of another, unrelated antigen.
antigenic determinant *Immunology*. the part of an antigen surface that is responsible for combining with the antibody or T-cell receptor combining site.
antigenic drift *Virology*. a slight change in the antigenic specificity of a virus over an extended period of time.
antigenicity *mmunology*. the ability of a substance to act as an antigen.
antigenic shift *Virology*. a major, usually sudden change in the antigenic specificity of a virus, brought about either by the introduction of a new virus or by a rearrangement of genes.
antigenic variation *Immunology*. a change in a cell or organism, usually genetic, in which new or altered antigens are produced or old antigens are eliminated.
antigen-presenting cell *Immunology*. an accessory or lymphocytic cell that bears an antigen or antigenic fragment and presents it to lymphocytes to provoke the production of an immune response.
antigen processing *Immunology*. the conversion of an antigen into a form that can be recognized by T lymphocytes in association with MHC molecules.
anti-glare screen *Computer Technology*. any of various nonreflective shields that are placed over or on a cathode-ray tube screen in order to reduce the reflection of room lighting and thus lessen eye strain.
antiglobulin *Immunology*. an antibody that is produced against the antigenic determinants of a serum globulin.
antiglobulin test see COOMBS' TEST.
Antigone complex [an tig′ ə nē] *Psychology*. a desire to sacrifice one's own life for the sake of a loved one. (From the Greek myth of *Antigone*, who buried the body of her brother despite an order not to do so and thus brought about her own death.)
antigorite [an tig′ə rīt] *Mineralogy*. $(Mg, Fe^{+2})_3Si_2O_5(OH)_4$, a monoclinic variety of serpentine that is greenish with a waxy luster, occurring as microscopic lamellar plates and as tiny, lathlike crystals and having a specific gravity of 2.61 and a hardness of 2.5 to 3.5 on the Mohs scale.
antigravity *Physics*. a theoretical phenomenon in which gravitational force causes two bodies to repel each other, rather than to attract. Also, GRAVITATIONAL REPULSION.
anti-g suit see G SUIT.
antihalation backing *Graphic Arts*. a light-absorbing material applied to the back of film to prevent the reflection of light rays from the base of the film into the emulsion.
antihemolytic *Hematology*. acting to prevent hemolysis.
antihemophilic factor *Hematology*. a relatively storage-labile coagulation factor participating only in the intrinsic pathway of blood coagulation. Deficiency of this factor, when transmitted as a sex-linked recessive trait, causes classical hemophilia. Also, ANTIHEMOPHILIC GLOBULIN, (COAGULATION) FACTOR VIII.

antihemophilic globulin see ANTIHEMOPHILIC FACTOR.
antihemophilic human plasma *Hematology.* normal human plasma that has been processed promptly to preserve the antihemophilic properties of the original blood; used for temporary correction of bleeding tendency in hemophilia.
antihemorrhagic vitamin *Biochemistry.* another term for vitamin K, so called because of its tendency to promote blood clotting.
antihistamine *Pharmacology.* any drug that counteracts the effects of histamine, a substance released in the body by allergic reaction and responsible for the symptoms of allergies.
antihunt circuit *Electronics.* a circuit that reduces hunting, or unwanted oscillation above and below a desired level. The circuit is often employed in a feedback arrangement in a control system. Similarly, **antihunt device.**
antihypercholesterolemic *Cardiology.* **1.** an agent that acts to decrease or prevent high blood cholesterol levels. **2.** the action of decreasing or limiting the blood cholesterol level.
antihyperon *Particle Physics.* the antiparticle of a hyperon, with a baryon number of –1, opposite strangeness (s), and charge Q.
antihypertensive agent *Pharmacology.* any drug that reduces high blood pressure.
anti-icer *Aviation.* an apparatus designed to prevent the formation of ice on an aircraft, as opposed to a de-icer, which removes ice.
anti-icing *Aviation.* of a material or procedure, preventing or retarding ice formation.
anti-idiotype *Immunology.* an antibody that reacts with the antigenic determinants on the V regions of other antibodies. Its ability to bind to antigen receptors of T and B cells makes it as efficient as antigens in stimulating these cells.
anti-idiotype antibody *Immunology.* an antibody that binds exclusively to an antigenic determinant characteristic of one individual animal.
anti-inflammatory *Pharmacology.* **1.** describing an agent that counteracts or suppresses inflammation without acting directly against the cause. **2.** any drug that so acts.
antijamming *Electronics.* **1.** any method used to opppose enemy efforts at jamming communication or radar. **2.** relating to such methods.
antiknock *Materials Science.* describing a substance that acts as an antiknock compound in petroleum. Thus, **antiknock additive, antiknock fluid, antiknock hydrocarbons,** and so on.
antiknock blending value *Engineering.* a numerical expression of the decrease in knocking achieved by an anitknock additive compound.
antiknock compound *Materials Science.* any of various organic compounds that are added to gasoline to reduce its tendency to detonate and cause a loud "knocking" sound. Tetraethyl lead has been the most widely used antiknock compound, but because of its contribution to air pollution it has been replaced by a nonmetallic compound in lead-free gasolines. Also, **antiknock agent.**
antiknock gasoline *Materials Science.* gasoline to which an antiknock compound has been added.
antiknock rating *Engineering.* a measure of the relative performance of a volatile liquid fuel in keeping an engine from knocking.
antileukocytic *Hematology.* acting to destroy leukocytes (white blood cells).
Antilles [an til´ēz] *Geography.* a chain of islands that separates the Caribbean Sea from the Atlantic Ocean. Cuba, Hispaniola, Jamaica, and Puerto Rico make up the **Greater Antilles;** the smaller islands extending southeast from Puerto Rico are the **Lesser Antilles.**
Antilles Current *Oceanography.* the northern branch of the North Equatorial Current; flows along the northern side of the Greater Antilles and joins the Florida Current to form the Gulf Stream.
Antillian subregion *Ecology.* a distinct zoogeographical region that includes the islands of the Caribbean Sea.
Antilocapridae *Vertebrate Zoology.* a family of ruminant mammals in the order Artiodactyla, containing a single living species, the pronghorn.
antilog see ANTILOGARITHM OF A NUMBER.
antilogarithm of a number *Mathematics.* the number whose logarithm is the given number. The base must be specified for the antilogarithm to be uniquely defined.
antilogous pole *Solid-State Physics.* the negatively charged pole of a crystal that arises when the crystal is heated.
Antilopinae *Vertebrate Zoology.* the antelopes, a diverse and wide-ranging subfamily of hollow-horned ruminant mammals belonging to the order Artiodactyla, including the oryx, eland, kudu, hartebeest, bongo, springbok, impala, gazelle, and many other species.
antilymphocyte serum *Immunology.* a powerful nonspecific immunosuppressive agent that causes destruction of circulating lymphocytes.
antilysin *Immunology.* a substance that counteracts the action of a lysin.
antimagnetic *Engineering.* able to avoid or diminish the influence of magnetic fields, usually by being constructed with magnetic shielding or nonmagnetic materials.
antimalarial *Pharmacology.* any drug used to prevent or treat malaria or to inhibit or destroy the malarial parasite.
antimatter *Physics.* any matter that is entirely composed of antiparticles, such as positrons, antiprotons, and antineutrons.
antimechanized *Ordnance.* used as a weapon or tactic against mechanized vehicles, such as tanks.
antimere *Invertebrate Zoology.* one of the radial segments of a radially symmetrical animal. Also, ACTINOMERE.
antimessage *Molecular Biology.* a single-stranded RNA molecule that is made artificially from a complementary DNA strand, and is thus capable of binding in a complementary way with normal messenger RNA.
antimetabolite *Pharmacology.* any substance that is similar in structure to an essential metabolite (a substance that is involved in normal metabolism) and that therefore interferes with the utilization of the metabolite in the body.
antimicrobial agent *Microbiology.* any substance that either destroys or inhibits the growth of a microorganism at concentrations tolerated by the infected host.
antimissile missile *Ordnance.* a defensive missile whose purpose is to intercept and destroy other missiles in flight.

antimissile missile

antimitotic drug *Pharmacology.* any substance that inhibits or prevents cell division and multiplication, often in malignant cancer cells.
antimolecule *Atomic Physics.* a molecule composed of antiparticles, such as antiprotons, antineutrons, and positrons, having the same mass and spin as their equivalents in an ordinary molecule, but the reverse of other properties, such as charge.
antimonate *Chemistry.* **1.** indicating antimony as the central atom in a anion. **2.** indicating the radical $[Sb(OH)_6]^-$ in antimony salts.
antimonic *Chemistry.* **1.** of or relating to antimony. **2.** describing various compounds of antimony, especially those in which it has a valence of 5.
antimonic acid see ANTIMONY PENTOXIDE.
antimonide *Inorganic Chemistry.* a compound made of antimony and a more positive element.
antimonite see STIBNITE.
antimonous *Chemistry.* **1.** of or relating to antimony. **2.** describing various compounds of antimony, especially those in which it has a valence of 3.
antimony [an´tə mō´nē] *Chemistry.* Sb, a metallic element of Group VA of the periodic table, a metallic, silver-white, opaque, trigonal solid having the atomic number 51, an atomic weight of 121.75, a specific gravity of 6.7, and a hardness of 3 to 3.5 on the Mohs scale; it melts at 630.5°C and boils at 1750°C. Antimony has low thermal conductivity and is used in lead alloys, especially for electric cable covers and storage batteries, and in semiconductors. (From a Medieval Latin word; originally from the Arabic name for this metal.)

antimony-124 *Nuclear Physics.* a radioactive isotope of antimony with a half-life of 60 days, commonly used as a tracer in solid-state and pipeline flow studies.
antimony black *Inorganic Chemistry.* antimony trisulfide occurring in the form of black crystals.
antimony bromide see ANTIMONY TRIBROMIDE.
antimony chloride see ANTIMONY TRICHLORIDE.
antimony fluoride see ANTIMONY TRIFLUORIDE.
antimony glance see STIBNITE.
antimony hydride *Inorganic Chemistry.* SbH_3, a colorless, poisonous gas that boils at –17°C and melts at –88°C; slightly soluble in water. Also, STIBINE.
antimony iodide see ANTIMONY TRIIODIDE.
antimonyl *Chemistry.* the monovalent radical SbO^-, which often occurs in formulas of antimony compounds.
antimonyl potassium tartrate see TARTAR EMETIC.
antimony needles *Inorganic Chemistry.* antimony trisulfide occurring in the form of needles.
antimony orange *Inorganic Chemistry.* antimony trisulfide occurring in the form of orange-red crystals.
antimony oxide see ANTIMONY TRIOXIDE.
antimony oxychloride *Inorganic Chemistry.* SbOCl, a white powder that decomposes at 170°C; insoluble in water, alcohol, and ether, and soluble in hydrochloric acid; used in flameproofing textiles.
antimony parchloride see ANTIMONY PENTACHLORIDE.
antimony pentachloride *Inorganic Chemistry.* $SbCl_5$, a reddish-yellow oily, hygroscopic liquid, decomposing in excess water; used in analytical testing and in dyeing.
antimony pentafluoride *Inorganic Chemistry.* SbF_5, a corrosive hygroscopic liquid that reacts violently with water; used in the fluorinating of organic compounds.
antimony pentasulfide *Inorganic Chemistry.* Sb_2S_5, an odorless orange-yellow powder, insoluble in water and soluble in alkali and concentrated hydrochloride acid; used as a red pigment. Also, ANTIMONY PERSULFIDE.
antimony pentoxide *Inorganic Chemistry.* Sb_2O_5, a white or yellowish powder, insoluble in water; used in the synthesis of antimony compounds.
antimony persulfide see ANTIMONY PENTASULFIDE.
antimony potassium tartrate see TARTAR EMETIC.
antimony red see ANTIMONY PENTASULFIDE.
antimony sodiate see SODIUM ANTIMONATE.
antimony sulfate see ANTIMONY TRISULFATE.
antimony sulfide see ANTIMONY TRISULFIDE.
antimony tribromide *Inorganic Chemistry.* $SbBr_3$, a yellow, deliquescent crystalline mass that is decomposed by water; used as a mordant and in manufacturing antimony salts.
antimony trichloride *Inorganic Chemistry.* $SbCl_3$, a hygroscopic, colorless crystalline mass, soluble in water; used as a mordant, in fireproofing textiles, and as a chlorinating agent.
antimony trifluoride *Inorganic Chemistry.* SbF_3, white to gray hygroscopic crystals, soluble in water; used as a fluorinating agent and in dyeing and pottery making.
antimony triiodide *Inorganic Chemistry.* SbI_3, red crystals, volatile at high temperatures and decomposing in water; insoluble in alcohol and chloroform.
antimony trioxide *Inorganic Chemistry.* Sb_2O_3, a white, odorless, crystalline powder, melting at 655°C; insoluble in water; used as a powerful reducing agent. Also, **antimony (III) oxide.**
antimony trisulfate *Inorganic Chemistry.* $Sb_2(SO_4)_3$, a white deliquescent powder, soluble in acids and decomposing in water; used in matches and explosives. Also, ANTIMONY SULFATE.
antimony trisulfide *Inorganic Chemistry.* Sb_2S_3, black or orange-red crystals, insoluble in water; used as a pigment and in matches.
antimony yellow see LEAD ANTIMONATE.
antimorph *Genetics.* a mutant gene that prevents or inhibits the expression of an ancestral or wild phenotype.
antimutagen *Genetics.* a compound that interferes with the action of another mutagenic agent or reduces the spontaneous mutation rate.
antiMüllerian factor or **hormone** *Endocrinology.* a glycoprotein that is produced by the embryonic testis and inhibits the development of the Müllerian ducts (precursors of the uterus and fallopian tubes). This factor thus inhibits the development of female phenotype, while the actions of testosterone are required to induce the male phenotype. Also, MÜLLERIAN REGRESSION FACTOR, MÜLLERIAN DUCT INHIBITORY FACTOR.
antimycin A *Biochemistry.* $C_{28}H_{40}N_2O_9$, an antibiotic that blocks respiration in the mitochondrial electron transport chain; used as a fungicide, insecticide, and miticide.
antimycotic *Medicine.* antifungal; destructive to fungi.
antineoplastic drug *Pharmacology.* any agent that inhibits the maturation and proliferation of benign or malignant tumor cells.
antineuralgic *Neurology.* counteracting neuralgia; said especially of drugs or therapies.
antineuritic *Neurology.* relieving or preventing nerve inflammation.
antineurotoxin *Toxicology.* any substance that counteracts the action of a neurotoxin.
antineutrino *Particle Physics.* the antiparticle of a neutrino; the particle has zero mass, spin 1/2, and positive helicity.
antineutron *Particle Physics.* an uncharged particle of mass equal to that of the neutron but with a magnetic moment in the opposite direction, relative to its spin.
antinode *Acoustics.* a point along a sinusoidal standing wave that is halfway between two nodes in the wave; such a point indicates a position of maximum intensity. *Astronomy.* either of two points in an orbit that lie 90° of orbital longitude away from the nodes.
antinoise *Acoustics.* a sound that is intentionally generated in order to mask the sound of another noise, by having its wavelength neutralize the wavelength of the unwanted noise. *Acoustical Engineering.* see WHITE NOISE.
antinoise microphone *Acoustics.* a microphone, such as a "lip button," that is designed to amplify only the voice or desired sound and to eliminate a high level of surrounding noise.
antinuclear *Nuclear Physics.* of or relating to an antinucleus.
antinuclear antibodies *Pathology.* antibodies targeting antigens arising in a cell nucleus; serologic measurements indicate antibody levels with the aid of immunofluorescent staining.
antinucleon *Particle Physics.* an antineutron or antiproton; i.e., a particle with the same mass as its nucleon counterpart but having opposite charge or opposite magnetic moment.
antinucleus *Nuclear Physics.* a type of nucleus that contains antineutrons and antiprotons in the same way that an ordinary nucleus contains neutrons and protons.
Antioch process *Metallurgy.* in the production of plaster molds for casting, a process that includes a dehydration step and a rehydration step, yielding a mold that improves the quality of the resulting casting.
antiodontalgic *Medicine.* relieving toothache.
antioncogene see TUMOR-SUPPRESSOR GENE.
antioncotic *Oncology.* a substance that can reduce swelling and inhibit the growth of tumors.
antioxidant *Materials Science.* any of various organic compounds that are added to materials such as paints, plastics, gasoline, rubber, and food products in order to reduce the effect of oxidation and the accompanying degradation of properties.
antiozonant *Chemistry.* a substance that is added to rubber to inhibit or prevent the severe oxidizing action of ozone on both natural and synthetic elastomers.
antiparallel *Molecular Biology.* describing an alignment of two polynucleotide strands that are linked together by base pairing so that the parallel linked strands have opposite polarity, as in native DNA. *Physics.* relating to two vectors that lie along a common line, but point in opposite directions.
antiparalytic *Neurology.* counteracting or preventing paralysis; said especially of drugs or therapies.
antiparasitic agent *Pharmacology.* a substance that destroys parasites.
antiparticle *Particle Physics.* an elementary particle that is identical to another elementary particle in mass and spin but opposite in electric and magnetic properties and that, when brought together with its counterpart, produces mutual annihilation.
Antipatharia *Invertebrate Zoology.* an order of black or thorny corals in the class Anthozoa, with a multibranched axial skeleton; found in deep tropical waters.
antipercolator *Mechanical Engineering.* in the carburetor of an automotive engine, a valve that vents vapor when the throttle is closed in order to prevent the boiling of fuel into the intake manifold and avoid flooding during restart.
antipersonnel *Ordnance.* of a bomb, missile, or other such weapon, intended to kill or injure people rather than to cause damage to structures, vehicles, and installations.
antipersonnel mine *Ordnance.* a land mine whose primary purpose is to harm enemy personnel rather than to destroy or disable vehicles.

antiperthite *Geology.* a variety of alkali feldspar composed of parallel or subparallel intergrowths and formed by the separation of sodium feldspar (albite) and potassium feldspar (orthoclase) during the slow cooling of molten mixtures.

antipetalous *Botany.* of a leaf, placed opposite to or in front of a petal.

antiphase boundary *Materials Science.* the high-energy boundary between two regions of an ordered metallic solid. Also, **antiphase domain.**

antiplasticization *Materials Science.* the process of offsetting the decrease in modulus or tensile strength, or the increase in molecular mobility, in a polymer by adding a plasticizer.

antipodal *Botany.* the cells, usually three, located in the end opposite the micropyle of the megagametophyte of flowering plants; often vestigial.

antipodes [an tip′ə dēz′] *Cartography.* two points on the earth's surface that are 180° from each other.

antiport *Biochemistry.* linked transport in opposite directions across a biologic membrane.

antiprincipal planes see NEGATIVE PRINCIPAL PLANES.

antiprincipal point see NEGATIVE PRINCIPAL POINT.

antiproton *Particle Physics.* an antiparticle of the proton having the equivalent mass of a proton, a unit negative charge, and a spin of 1/2; annihilating with a proton, it yields mesons.

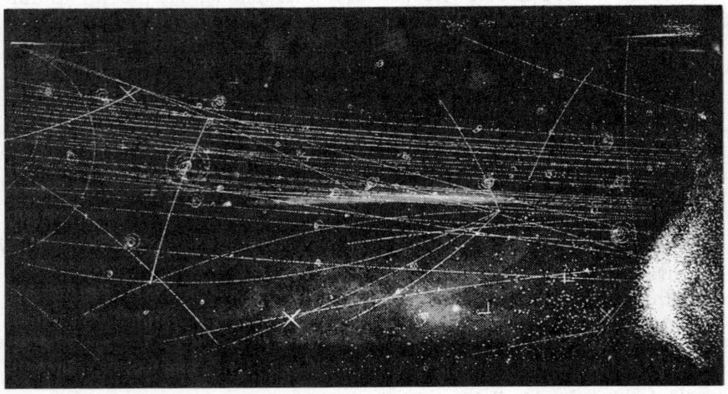

antiproton

antiprotonic atom *Atomic Physics.* an atom in which an antiproton orbits an ordinary nucleus.

antipruritic *Pharmacology.* any drug used to relieve or prevent itching; usually applied topically.

antipsychomotor *Neurology.* suppressing or preventing the motor effects of increased mental activity; said especially of drugs or therapies.

antipyretic *Pharmacology.* any drug that relieves or reduces fever.

antipyrine *Pharmacology.* $C_{11}H_{12}N_2O$, a drug that relieves pain and fever; seldom used because of its side effects.

antiquarian [an′tə kwâr′ē ən] *Archaeology.* **1.** a person who studies or collects ancient relics or artworks. Also, **antiquary. 2.** of or relating to such ancient objects.

antiquark [an′tē kwärk′] *Particle Physics.* the hypothetical antiparticle of a quark with the opposite electric charge, baryon number, and strangeness from its corresponding quark.

antiquing [an′tēk′ing] *Engineering.* **1.** the process of bringing a burnished finish to leather through the application of stain, wax, or oil. **2.** the process of exposing portions of the underlayer or underlayers of wet paint by combing, graining, or marbling techniques.

antiquities *Archaeology.* **1.** ancient relics, artworks, monuments, and the like. **2.** an older name for the study of ancient life and culture.

antirad *Chemical Engineering.* a substance that is added to rubber during processing to reduce the deteriorating affects of radiation.

antiradar coating *Engineering.* a coating that minimizes the reflection of electromagnetic waves; used to avoid detection by radar.

antiradiation missile *Ordnance.* a missile that homes passively on a radiation source.

antirattle spring *Mechanical Engineering.* a spring used in automotive vehicles to prevent undesired rattling and to hold together integral parts in the clutches and the disk brakes.

antireceptor *Virology.* a virion surface protein or an antibody that binds to a cell surface receptor.

anti-redeposition agent *Chemical Engineering.* any of various types of chemical agents that prevent or decrease the repeated deposition of low-solubility substances on metal surfaces in process streams.

antireflection coating *Engineering.* surface treatment with dielectric material to reduce the reflection of electromagnetic radiation while increasing light transmission.

antiresonance *Electricity.* that frequency at which the inductive reactance is equal to the capacitative reactance in a parallel resonant circuit. Also, PARALLEL RESONANCE. *Engineering.* the frequency at which the impedance of a tuned circuit approaches infinity and is a maximum.

antiresonance circuit see PARALLEL-RESONANT CIRCUIT.

anti-Rh agglutinin *Immunology.* an acquired antibody that acts against any Rh antigen.

anti-Rh immunoglobin *Immunology.* a serum protein, having antibody activity against Rh-positive fetal red blood cells, that prevents anti-Rh antibody formation in Rh-negative mothers when administered soon after delivery. Also, RH_O(D) IMMUNE GLOBULIN.

anti-Rh serum *Immunology.* a blood serum that contains high levels of anti-Rh antibodies.

antiroll bar see STABILIZER BAR.

antiroll fin *Navigation.* the external part of a gyro stabilizer on a large ship. It is in the form of a horizontal plate which acts on the moving water to minimize rolling. These fins can usually be retracted when not needed.

antirolling gyroscope *Navigation.* a gyroscope that acts as the sensing and controlling unit for a ship's stabilizer.

antiroll tank *Navigation.* a fluid-holding tank that is designed to counteract the rolling motion of a ship. Antiroll tanks are fitted in pairs on each side of the ship, and fluid (typically water or oil) is pumped between them; the shifting weight of the fluid acts against the vessel's rolling motion.

antirostrum *Invertebrate Zoology.* the terminal segment of the appendages of certain mites.

antisatellite missile *Ordnance.* a missile whose target is an artificial satellite in space.

antisense RNA *Molecular Biology.* a single-stranded RNA molecule that, by binding with a complementary sequence in another nucleic acid molecule, inhibits the function and/or completion of synthesis of the latter molecule. *Genetics.* an RNA molecule having a coded nucleotide sequence that is complementary to a specified mRNA.

antisense strand *Genetics.* the DNA strand that serves as the template for mRNA synthesis. Also, ANTICODING STRAND.

antisepsis *Medicine.* **1.** the prevention of infection by antiseptic means. **2.** any procedure that reduces the amount of infection-causing agents on skin or mucous membranes.

antiseptic *Medicine.* free or cleansed of infectious microorganisms. *Pharmacology.* **1.** any substance that inhibits or kills infectious microorganisms on environmental surfaces or a body surface (e.g., the skin). **2.** acting against infection.

antiseptic paint *Immunology.* secreted antibodies that wash over the mucous surfaces of the body.

antiserum *Immunology.* a type of serum containing antibodies that are specific to an antigen or antigens. Also, IMMUNE SERUM.

antisetoff powder *Graphic Arts.* a fine powder applied to freshly printed paper to prevent the transfer of wet ink to adjacent sheets.

antisideric *Pharmacology.* any substance that counteracts the physiological action of iron in the body.

anti-sidetone circuit *Electricity.* components or circuits, included in some telephone sets, with a balancing network that reduces sidetone interference.

antiskid plate *Engineering.* a metal sheet coarsened on both sides and positioned between objects in order to keep them from sliding.

antislip metal *Metallurgy.* a metal or alloy containing particles, usually abrasive, that increase its coefficient of friction.

antismudge ring *Building Engineering.* a metal frame that protects the surface on which a mechanical device is set; for example, around an air diffuser.

antisocial *Psychology.* hostile to society or to the established values and practices of society.

antisocial behavior *Psychology.* any behavior that violates rules or conventions of property and personal rights.

antisocial personality *Psychology.* a personality disorder characterized by an inability to feel certain emotions or to form personal relationships, by untruthfulness and insincerity, and by a lack of remorse or guilt.

antisolar point *Astronomy.* a point on the celestial sphere lying directly opposite the sun on the line from the sun through an observer.
antisound see ANTINOISE.
antispasmodic *Pharmacology.* any drug that relieves spasm or convulsion.
antispecificity factor *Molecular Biology.* a protein that is synthesized during the infection of bacteria *Escherichia coli* by T4 phage; prevents recognition of transcription initiation sites by RNA polymerase.
antispectroscopic lens see ACHROMATIC LENS.
antistat *Materials.* any antistatic agent or material.
antistatic *Materials Science.* of an agent or material serving to attract moisture from the air to a surface, thus improving the surface conductivity and reducing the likelihood of a spark or discharge
antistatic mat *Computer Technology.* a floor mat used in computer centers or manufacturing areas to prevent the buildup of static electricity, thereby reducing the risk of a discharge that would damage electronically stored data or electronic components.
anti-Stokes lines *Spectroscopy.* spectral lines produced by scattered light having shorter wavelengths (higher frequencies) than those of the original incident light; usually used in reference to Raman spectroscopy.
anti-Stokes transition *Materials Science.* a transition in which interactions between incident light and a material cause the scattered light to shift to shorter wavelengths.
antistreptolysin *Immunology.* a specific antibody that counteracts the activity of streptolysin, a toxic substance produced by certain streptococci.
antistress *Mineralogy.* of a mineral, not stable or unable to form in an environment of high shear stress.
antistructure disorder *Materials Science.* a defect in an ordered ionic crystal in which an anion and cation exchange sites.
antisubmarine *Ordnance.* used to attack or defend against submarines. Thus, **antisubmarine missile, antisubmarine torpedo.**
antisubmarine net *Ordnance.* a steel net stretched across the entrance to a harbor or waterway to prevent the passage of enemy submarines.
antisway bar see STABILIZER BAR.
antisymmetric matrix *Mathematics.* a matrix M that equals the negative of its transpose; that is, $M = -M^T$.
antisymmetric tensor *Mathematics.* if the sign of a tensor is changed when two covariant (or contravariant) arguments are interchanged, the tensor is said to be antisymmetric with respect to those arguments. If the tensor is antisymmetric with respect to every pair of covariant and contravariant arguments, it is said to be an antisymmetric tensor.
antisymmetric wavefunction *Physics.* the wavefunction of a multiparticle system that undergoes a sign change when any two identical particles in the system are interchanged.
antisympathetic see ANTIADRENERGIC.
antitactical (ballistic) missile *Ordnance.* a missile system that is designed to locate and intercept incoming enemy missiles; e.g., the U.S. Patriot system.
antitail *Astronomy.* an apparent sunward-extending tail of a comet; it is an optical illusion caused by the viewing geometry.
antitank *Ordnance.* **1.** used as a weapon against tanks and other armored vehicles. Thus, **antitank grenade, antitank gun, antitank mine, antitank (guided) missile. 2.** used to obstruct the movement of tanks and armored vehicles. Thus, **antitank ditch, antitank obstacle.**
antitank weapon *Ordnance.* any of various weapons used against tanks and other armored vehicles.
antitermination factor *Biochemistry.* a protein that blocks the termination of transcription. *Genetics.* a protein that permits an RNA polymerase molecule to ignore instructions to stop transcription of a genetic message.
antitheft device *Mechanical Engineering.* a system installed to prevent or hinder theft, especially on a motor vehicle; may include the sounding of an alarm or mechanical locks on the ignition or steering wheel.
antithetic variables *Statistics.* in Monte Carlo simulations, random variables that are highly negatively correlated.
antithrombin *Biochemistry.* a plasma protein that reduces the coagulation activity of thrombin and other clotting factors.
antitorpedo *Ordnance.* used to detect or intercept enemy torpedoes: an *antitorpedo* missile.
antitoxic *Toxicology.* **1.** of or relating to an antitoxin. **2.** an antitoxin.
antitoxin *Toxicology.* any substance that counteracts the action of a toxin or poison.
antitrades *Meteorology.* a deep layer of westerly winds found at upper levels above the surface trade winds of the tropics.
antitriptic wind *Meteorology.* any small-scale wind for which the pressure force exactly balances the viscous force and in which the vertical transfers of momentum predominate.
antitubercular *Pharmacology.* describing any agent that is therapeutically effective against tuberculosis. Also, **antituberculous, antituberculotic.**
antituberculin *Pharmacology.* the antibody produced by a skin test for tuberculosis.
antitumor antibiotic *Microbiology.* a substance produced by a bacterium that is cytotoxic to certain tumor cells, such as glidobactin A, which is produced by *Polyangium brachysporum.*
antitumorigenic *Oncology.* a substance that can inhibit the formation of tumors.
antitussive *Pharmacology.* any agent that relieves or suppresses coughing.
antitwilight arch *Meteorology.* a narrow band of about 3 degrees angular width, with a pink or purple cast, that rises with the antisolar point at sunset and sets with the antisolar point at sunrise. Also, ANTICREPUSCULAR ARCH.
antivenene *Toxicology.* any antitoxin used specifically to counteract animal venom, such as snake or spider venom. Also, **antivenin, antivenom.**
antiviral *Pharmacology.* any drug that destroys viruses or prevents their growth or replication.
antivitamin *Biochemistry.* a substance that is similar in structure to a given vitamin and inhibits the vitamin from engaging in its usual metabolic activities.
antiwithdrawal device *Ordnance.* a device used in an explosive to set off the charge if any attempt is made to withdraw the fuse.
antixerophthalmic vitamin see VITAMIN A.
ant. jentac. before breakfast. (From Latin *ante jentaculum.)*
antler *Vertebrate Zoology.* either of a pair of solid, bony, single or branched outgrowths on the heads of deer and similar animals; usually grown by the males only and shed annually.
antlerite *Mineralogy.* $Cu_3SO_4(OH)_4$, a translucent light green, orthorhombic mineral with a vitreous luster, having vertically striated, prismatic crystals, a specific gravity of 3.9, and a hardness of 3.5 on the Mohs scale; found as a secondary mineral in the oxidation zone of copper deposits in arid regions.
Antler orogeny *Geology.* a late Devonian and early Mississippian crustal deformation of Paleozoic rock that occurred in the Great Basin of Nevada.
Antlia [ant′ lē ə] *Astronomy.* the Air Pump, a small constellation of the southern sky with no star brighter than fourth magnitude.
ant lion *Invertebrate Zoology.* the common name for insects of genus *Myrmeleon* or related genera in the family Myrmeleontidae; the larvae dig small conical pits in sandy soil, lie buried in the bottom, and use their large jaws to catch ants and other small insects that fall in the pit.

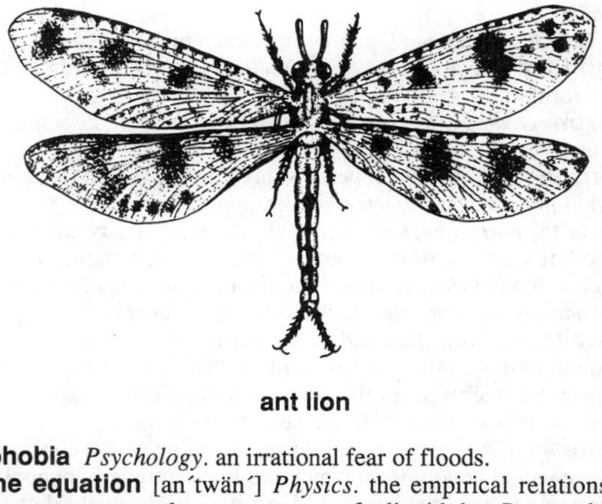

ant lion

antlophobia *Psychology.* an irrational fear of floods.
Antoine equation [an′twän′] *Physics.* the empirical relationship between temperature and vapor pressure of a liquid: $\log P = a - b/(c + T)$, where P is the vapor pressure, T is the temperature, and a, b, and c are constants that must be found experimentally.
Antonadi scale [än′ tō näd′ē] *Astronomy.* an evaluation measure for astronomical visibility, with a rating of I being given to excellent viewing conditions and V to very poor ones.

Antonoff's rule *Physics*. a rule stating that the surface tension at the interface of two liquids in equilibrium is equal to the difference of the surface tensions of the two liquids when exposed to the atmosphere.

ant pipet *Vertebrate Zoology*. an antbird of the genus Conopophaga of the South American forests. The ten species were formerly placed in their own family, Conopophagidae, but are now assigned to the Formicariidae. Also, GNATEATERS.

ant. prand. before dinner. (From Latin *ante prandium.)*

antrectomy *Surgery*. the excision of an antrum, especially of the antrum of the stomach.

antrorse *Biology*. turned or bent forward, or directed upward.

antrum *Anatomy*. a nearly closed cavity or chamber, especially in a bone.

antrum of Willis see GASTRIC ANTRUM.

ANTU see α-NAPHTHYLTHIOUREA.

anucleate *Cell Biology*. lacking a nucleus.

anucleolate *Cell Biology*. lacking a nucleolus.

anulus see ANNULUS.

Anura *Vertebrate Zoology*. the frogs and toads; a large order of about 2000 species in the class Amphibia.

anuran *Vertebrate Zoology*. **1.** a frog or toad. **2.** of or relating to frogs and toads.

anuresis *Medicine*. **1.** the retention of urine in the bladder. **2.** see anuria.

anuria *Medicine*. the complete suppression of urinary secretion by the kidneys.

anus *Anatomy*. the posterior outlet of the alimentary canal, through which waste is expelled.

anvil *Mechanical Devices*. a massive metal block used as a solid base or working surface by blacksmiths in forging operations. *Anatomy*. the middle of a group of three small bones in the middle ear; the incus. *Meteorology*. a supplementary cloud feature in which the upper portion of cumulonimbus spreads and takes on the form of an anvil.

anvil cloud *Meteorology*. a popular name for a cumulonimbus capillatus cloud, particularly one that embodies the supplementary anvil feature. Also, THUNDERHEAD.

anxiety *Psychology*. a strong and unpleasant feeling of nervousness or distress in response to a feared situation, often accompanied by physiological effects such as nausea, trembling, breathlessness, sweating, and rapid heartbeat.

anxiety attack *Psychology*. an experience of anxiety, especially one in which only the physical symptoms are present without conscious fear or apprehension.

anxiety disorder *Psychology*. any disorder in which anxiety is the central feature or in which anxiety appears when the individual tries to resist a phobia. Thus, **anxiety disorder of childhood** or **adolescence.**

anxiety hierarchy *Psychology*. a ranking of related experiences or situations in terms of the degree of anxiety they produce in an individual; used as part of a progressive therapy technique.

anxiety neurosis *Psychology*. a functional disorder characterized by persistent or ongoing anxiety and associated with apprehension, fear, and panic.

AOA or **A.O.A.** American Optometric Association.

AO* algorithm [ā´ō´stär´] *Artificial Intelligence*. an algorithm similar to an A* algorithm, for heuristic search of an AND/OR graph.

Ao horizon *Geology*. the layer of the A horizon that is composed only of humus.

Aoo horizon *Geology*. the uppermost layer of the A horizon, consisting of vegetation matter that has not yet decomposed.

aorta [ā ôr´tə] *plural*, **aortae.** *Anatomy*. the main artery carrying blood from the left ventricle of the heart to systemic circulation. *Zoology*. the large dorsal blood vessel leading from the heart in many invertebrates.

aortic *Anatomy*. of or relating to the aorta. Also, **aortal.**

aortic arch *Anatomy*. the bend of the aorta between its ascending and descending portions. Also, ARCH OF THE AORTA. *Vertebrate Zoology*. one of the paired branches from the aorta in aquatic and amphibious vertebrates that conducts blood from the heart to the gills.

aortic atresia *Cardiology*. the absence or closure of the aortic root orifice; a congenital condition in which the left ventricle is hypoplastic or nonfunctioning, with oxygenated blood passing from the left into the right atrium through a septal defect, and the mixed venous and arterial blood passing from the pulmonary artery to the aorta by way of a patent ductus.

aortic body see AORTIC PARAGANGLION.

aortic dissection see DISSECTING ANEURYSM.

aortic hiatus *Anatomy*. an opening in the diaphragm through which the aorta passes.

aortic notch *Cardiology*. a small decrease in the arterial pulse or pressure contour marked by a notch in an arterial pulse tracing; it immediately follows the closure of the semilunar valves and is sometimes used as a marker for the end of systole or the ejection period. Also, DICROTIC NOTCH.

aorticopulmonary window *Cardiology*. a radiolucent area below the aortic arch in the left anterior oblique view, in front of the spinal column, and behind the aorta and pulmonary artery.

aortic paraganglion *Anatomy*. a cluster of autonomic nerve cells adjacent to the aorta.

aortic valve *Anatomy*. a semilunar valve that regulates the flow of blood from the left ventricle into the aorta.

aortitis *Medicine*. inflammation of the aorta.

aortocoronary *Cardiology*. relating to or communicating with the aorta and coronary arteries.

aortocoronary bypass *Cardiology*. a surgical procedure to correct the effects of an obstructive lesion in a coronary artery, by the grafting of a section of saphenous vein between the aorta and a coronary artery distal to the affected area in the latter.

aortogram *Radiology*. a recording of the results of aortography.

aortography *Radiology*. radiography of the aorta after the intravascular injection of an opaque medium.

aortopathy *Medicine*. any disease of the aorta.

aortosclerosis *Cardiology*. hardening of the aorta associated with overgrowth of connective tissue.

aosmic *Neurology*. of, relating to, or suffering from anosmia.

aoudad *Vertebrate Zoology*. a wild sheep, *Ammotragus lervia,* native to northern Africa; characterized by large curved horns and a long fringe of hair on the throat, chest, and forelegs. Also, BARBARY SHEEP.

AP anterior pituitary gland, angina pectoris, arterial pressure.

ap- a prefix meaning "away from," as in *aphelion*.

apa *Botany*. a valuable timber tree, *Eperua falcata,* of tropical America that has pinnate leaves, clusters of red flowers, and reddish-brown wood used for palings and shingles.

APA or **A.P.A.** American Pharmaceutical Association, American Psychiatric Association, American Psychological Association.

Apache *Ordnance*. **1.** a U.S. Army attack helicopter designed for use in bad weather or darkness. **2.** a German/French air-to-surface modular weapon able to deliver several kinds of sub-munitions.

apandrous *Botany*. having functionless male sex organs.

apapane *Vertebrate Zoology*. a bright crimson-colored honeycreeper, *Himatione sanguinea,* that is native to Hawaii and characterized by black wings, tail, and bill.

aparalytic *Neurology*. without paralysis.

APAS *Robotics*. Adaptable Programmable Assembly System; a robotic system to automatically assemble and manufacture parts.

apastia *Neurology*. an abstention from eating that is not due to physiologic factors but is a neurologic symptom.

apastron [a´ pas´ trən] *Astronomy*. the point at which a binary star is farthest from its companion.

Apatemyidae *Paleontology*. an early family of rodentlike mammals that is now placed in its own order, the **Apatotheria;** it had been placed in many different eutherian orders and suborders, most recently in the suborder Soricomorpha; Paleocene to Oligocene.

apatetic *Zoology*. of or relating to the protective resemblance to some part of the environment or to the markings of another species for the purpose of disguise; serving to conceal by deceptive camouflage. Thus, **apatetic coloration.**

Apathornithidae *Paleontology*. a family of early flying birds usually placed in the order Ichthyornithiformes, characterized by a well-developed keel and other features that allowed it to fly; Cretaceous.

apatite *Mineralogy*. $Ca_5(PO_4CO_3)_3$ (F, Cl, OH), a group name including variously colored hexagonal and monoclinic phosphate minerals with the above general formula, having a vitreous luster and occurring in crystalline and massive forms; a major constituent of sedimentary phosphate rocks, of the bones and teeth in vertebrates, and present as an accessory mineral in igneous rock.

Apatosaurus *Paleontology*. a gigantic saurischian in the infraorder Sauropoda, reaching a length of 22 meters and a weight of around 30 tons; the preferred term for the group of dinosaurs commonly called Brontosaurus.

apatropine see APOATROPINE.

APC antigen-presenting cell, automatic phase control.

ape *Vertebrate Zoology.* any tailless primate having long arms and a broad chest in the family Pongidae, which includes the chimpanzee, gorilla, and orangutan, or in the family Hylobatidae, which includes the gibbon and the siamang.

apeirophobia *Psychology.* an irrational fear of infinity or of boundless space.

Apennines *Geography.* a mountain range running the length of Italy from the Gulf of Genoa to Sicily; highest peak: Monte Corno (9560 ft).

aperiodic [ā´pēr ē äd´ik] *Physics.* relating to or being of irregular occurrence; not periodic. Thus, **aperiodic wave.**

aperiodic antenna *Electromagnetism.* an antenna that is designed to maintain useful efficiency over a wide range of frequencies by means of suppressing reflections within the antenna system. Also, NONRESONANT ANTENNA.

aperiodic compass *Navigation.* a compass that returns directly, without oscillation, to its proper reading after deflection.

aperiodic damping *Physics.* overdamping that, when applied to an oscillatory system, brings the system to rest at equilibrium without allowing it to oscillate about the point of equilibrium.

aperiodicity *Chaotic Dynamics.* a state characterized by an irregular, nonrepeating evolution in time; temporal behavior that cannot be described as a superposition of a finite number of periodic signals.

aperiodic waves *Electricity.* **1.** waves that cannot be characterized by a period or frequency. **2.** the transient current wave in a series circuit having resistance R, inductance L, and capacitance C, when the equation $R^2C \geq 4L$ is satisfied.

apers *Ordnance.* an abbreviation for antipersonnel.

apertometer *Optics.* a device that measures the numerical aperture of a microscope objective.

aperturate *Botany.* of a pollen grain, having apertures or areas where the exine is either thinner or absent.

aperture [ap´ər chər] an opening or hole in an object or structure; specific uses include: *Optics.* **1.** the opening through which light enters an optical system. **2.** the diameter of a lens in an optical instrument that controls the amount of light traversing the system. The actual diameter of the opening in a camera for a given shot is often called the **working aperture. 3.** the diameter of a lens in a telescope. *Electromagnetism.* the effective area over which an antenna extracts power from an incident plane wave; equal to the product of the gain and the square of the wavelength divided by 4π. Also, ANTENNA EFFECTIVE AREA.

aperture aberration *Optics.* an error in optical imaging that results when two objects located at different distances from the axis of the system cannot be focused simultaneously due to the size of the aperture.

aperture antenna *Electromagnetism.* an antenna whose beam width is determined by the dimensions of a horn, lens, or reflector.

aperture conductivity *Acoustics.* a quantity given by the ratio of the density of a medium with an acoustic aperture to the acoustic mass measure at the aperture.

aperture diaphragm see APERTURE STOP.

aperture-grill picture tube *Electronics.* in television reception, a tube that has a perforated disk with many holes placed between the color beams and the screen. The disk ensures that the three beams strike the correct color dot on the screen.

aperture illumination *Electromagnetism.* the field profile over an aperture that indicates the amplitude and relative phase.

aperture plate *Electronics.* a magnetic memory array used in digital computers, consisting of a ferrite material perforated with uniform parallel rows of holes alternating with metal connecting strips.

aperture ratio *Optics.* the ratio of the diameter of a lens to its focal length, expressed in increments called *f*-numbers.

aperture sight *Ordnance.* a gun sight without a lens, having a hole or opening through which the target is viewed.

aperture slot *Optics.* the small rectangular opening through which light from a moving document enters a rotary camera. Also, **aperture slit.**

aperture stop *Optics.* a mechanical opening that physically limits the diameter of the light bundle that will pass into or through the lens of an optical system. Also, APERTURE DIAPHRAGM.

apetalous *Botany.* lacking petals.

apex *plural,* **apices** [ā´pə sēz´]. the highest point or level; specific uses include: *Biology.* the highest point, pointed end, or tip of an organ or structure. *Physiology.* the point of greatest response to any type of stimulation. *Geology.* **1.** the highest or uppermost point of a landform; summit. **2.** the highest point of a vein in relation to the surface. **3.** see CULMINATION. *Mathematics.* the highest point of a geometric figure or solid relative to some line or plane.

Apex *Ordnance.* a Soviet medium-range air-to-air missile powered by a solid-propellant rocket motor and probably equipped with infrared or semiactive radar homing.

apex stone *Architecture.* the usually ornamental top stone of a gable end. Also, SADDLE STONE.

Apgar score *Medicine.* an assessment of a newborn infant's physical condition, usually determined sixty seconds after birth, as indicated by a numerical expression composed of a value of zero to two assigned to each of five categories: heart rate, respiratory effort, muscle tone, reflex stimulation, and skin color; a score of ten indicates optimal condition. (From Virginia *Apgar,* 1909–1974, American anesthesiologist.)

aphagia [ə fāj´ē ə] *Medicine.* **1.** the fact of abstaining from eating; not eating. **2.** a loss of the ability to swallow.

aphakia [ə fāk´ē ə] *Medicine.* an absence of the lens of the eye, usually occurring as a result of cataract extraction, although the condition may be congenital or caused by trauma.

aphaniphyric *Petrology.* of or relating to a texture of porphyritic rocks with microaphanitic ground mass. Also, FELSOPHYRIC.

aphanite *Petrology.* a compact, fine-grained rock that is so uniform in texture that no distinct mineral crystals are visible to the naked eye.

aphanitic *Petrology.* of or relating to an aphanite.

Aphanizomenon *Bacteriology.* a genus of filamentous bacteria, growing as cylindrical cells that aggregate into filaments and form floating colonies in fresh or brackish water.

Aphanochaetaceae *Botany.* a family of marine and freshwater green algae of the order Chlorophyceae, in which ultrastructural details are known only for the genus *Aphanochaete*.

Aphanochaete *Botany.* a genus of cosmopolitan freshwater green algae of the family Aphanochaetaceae that is a common epiphyte of freshwater algae and aquatic angiosperms.

Aphanomyces *Mycology.* a genus of fungi of the order Saprolegniales that is a parasite to certain algae and causes root rot in higher plants.

aphasia [ə fāz´ē ə; ə fāz´yə] *Neurology.* a loss or impairment of the power of speech or writing, or of the ability to understand written or spoken language, due to a brain injury or disease.

aphasic *Neurology.* **1.** of, relating to, or affected with aphasia. **2.** a person who is affected with aphasia. Also, **aphasiac.**

aphasiology *Neurology.* the study of speech and language disorders.

Aphasmida see ADENOPHOREA.

Aphasmidea *Invertebrate Zoology.* a subclass of roundworms lacking minute sensory structures called phasmids, including most free-living marine and freshwater nemotodes; also includes some terrestrial dwellers and parasites.

Aphelenchoidea *Invertebrate Zoology.* a superfamily of soil-inhabiting nemotodes; includes families Aphelenchidae and Paraphelenchidae.

aphelion [ə fēl´ yən] *Astronomy.* the point in the orbit of a planet or comet that lies farthest from the sun.

apheliotropic *Botany.* turning away from the sun.

Aphelocheiridae *Invertebrate Zoology.* a family of bugs in the order Hemiptera.

aphemesthesia see ALEXIA.

aphemia *Neurology.* aphasia that is due to a central brain lesion.

aphid *Invertebrate Zoology.* the common name for a harmful plant parasite in the family Aphididae, including any of the insects that suck plant sap and exude sugary secretions that are favored by ants; bisexual or parthenogenic reproduction, viviparous or egg-bearing. The newly born females of some species contain embryos; other species are important as vectors of plant virus diseases. Also, PLANT LOUSE.

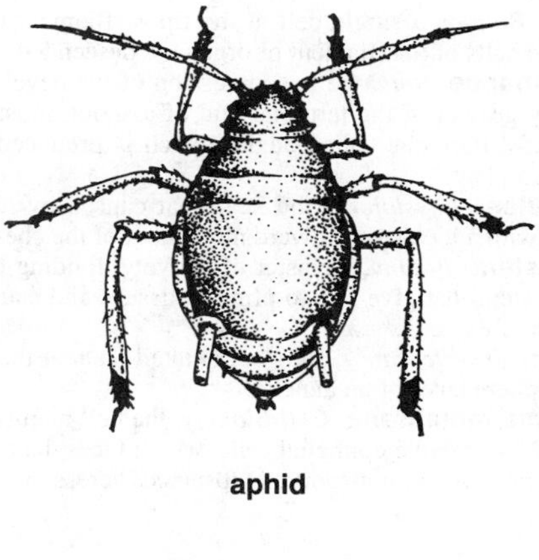

aphid

Aphid *Ordnance.* a Soviet close-range air-to-air missile powered by a solid-propellant rocket motor and equipped with infrared or possibly semiactive radar homing; one of the smallest guided air-to-air missiles in the world, with a launch weight of around 120 pounds.

Aphididae *Invertebrate Zoology.* a family, in the order Homoptera, that consists of over 3000 species of true plant lice or aphids.

Aphidoidea *Invertebrate Zoology.* a superfamily of aphids or greenflies in the order Homoptera; most species are polymorphic, parthenogenic or bisexual, viviparous or egg-laying; phytophagous parasites.

Aphis *Invertebrate Zoology.* an aphid genus.

aphonia *Medicine.* loss of the voice; this may be due to overuse, disease, or psychological causes such as hysteria.

aphotic zone *Oceanography.* the deeper parts of the ocean, where there is insufficient light for plants to carry on photosynthesis.

Aphredoderidae *Vertebrate Zoology.* a family of fishes in the order Percopsiformes, represented by only one small species, the pirate perch.

aphrodisiac [af′rō dēz′ē ak] *Physiology.* **1.** a drug or agent that arouses or increases sexual responses. **2.** causing or increasing sexual desire.

Aphroditidae *Invertebrate Zoology.* a family of bristle worms and sea mice, large polychaetous annelids covered with long lustrous setae; wide-ranging scavenger marine worms favoring sandy bottoms.

aphtha *Medicine.* a small white or grayish ulcer, usually occurring in the mouth or gastrointestinal tract due to a virus or fungal infection.

aphthitalite *Mineralogy.* $(K,Na)_3Na(SO_4)_2$, a white, soluble, trigonal mineral with tabular, commonly twinned crystals, having a specific gravity of 2.7 and a hardness of 3 on the Mohs scale.

aphthous fever see FOOT-AND-MOUTH DISEASE.

Aphthovirus *Virology.* a genus of viruses of the family Picornaviridae that cause foot-and-mouth disease in animals and are marked by unpredictable antigenic activity that renders vaccination control difficult.

aphylactic map projection *Cartography.* a map projection that has none of the three special properties of equivalence, conformality, or equidistance. Also, ARBITRARY MAP PROJECTION.

Aphylidae *Invertebrate Zoology.* a family of apparently phytophagous hemipteran insects inhabiting eucalyptus bark.

Aphyllophorales *Mycology.* an order of fungi in the class Hymenomycetes; it occurs on land and is characterized by tough and sometimes crustlike fruiting bodies, having spores that are exposed to the environment throughout their development; contains over 1000 species.

aphyllous *Botany.* without leaves.

aphyric *Petrology.* of or relating to an igneous rock characterized by a fine-grained texture and an absence of any phenocrysts.

aphytic zone *Ecology.* a lake floor that has no plant growth because it is too deep to be reached by light. Also, **aphytal zone.**

Apiaceae *Botany.* the umbels, a family of aromatic dicotyledonous herbs, shrubs, and trees in the order Apiales. Some species such as the hemlock are poisonous, others are cultivated for food or spice, including dill, celery, carrot, parsnip, caraway, and parsley. Also, UMBELLIFERAE.

Apiales *Botany.* an order in the subclass Rosidae, consisting of dicotyledonous woody or herbaceous plants with well-developed secretory cells or cavities. Also, UMBELLALES.

apian *Agriculture.* of or relating to bees or to the raising of bees. (From *apis,* Latin for *bee.*). Also, **apiarian.**

apiarist *Agriculture.* a person who keeps bees. Also, **apiculturist.**

apiary *Agriculture.* a place in which bee colonies and hives are kept, especially for the purpose of honey production and collection.

apical of, relating to, or located at an apex.

apical angle *Mechanics.* the projected angle formed by the pointed tip of a projectile.

apical cell *Botany.* a single cell at the tip of filament or organ from which all the cells of the filament or organ are descended.

apical dominance *Botany.* the suppression of the development of lateral buds by growth of the terminal bud of a shoot, most likely due to the action of a hormone called auxin, which is produced in the apical meristem of a plant.

apical impulse *Physiology.* the heartbeat caused by left ventricular contraction with a localized outward movement of the chest wall.

apical meristem *Botany.* a cluster of actively dividing cells at the tip of a root or stem that give rise to primary tissues and cause an increase in the length of the axis.

apical organ *Invertebrate Zoology.* a ciliated plate at the anterior apex of a trochosphere larva of an annelid.

apical plasma membrane *Cell Biology.* the cell membrane found on the surface of vertebrate epithelial cells, which faces the external lumen, and may be involved in transport of substances across the cell.

apical plate *Invertebrate Zoology.* a cluster of specialized cells, usually with cilia, at the apex or anterior end of certain pelagic larvae, that performs sensory and nervous functions.

apiculate *Botany.* tipped with a short, abrupt point.

apiculture *Agriculture.* the raising of honeybees, especially on a large scale for commercial purposes.

Apidae *Invertebrate Zoology.* a family of insects containing bumblebees and honeybees.

API gravity *Petroleum Engineering.* the gravity, or weight per unit of volume, of petroleum products or other liquids as determined by the recommended procedure of the American Petroleum Institute.

Apioceridae *Invertebrate Zoology.* a family of noisy, long, bushy flies in the suborder Orthorrhapha, that frequent flowers and consume nectar; includes over 100 species, most of which are Australian.

Apiocrinites *Paleontology.* a genus of long-stemmed crinoids with a rigid, pear-shaped calyx; common in shallow-water Jurassic deposits; Apiocrinites survives today only in an abyssal form.

apioid *Physics.* the pear-shaped form of a rapidly revolving mass of liquid due to the influence of gravity.

apiology *Invertebrate Zoology.* the scientific study of honeybees.

apiphobia *Psychology.* an irrational fear of bees or of being stung by a bee. Also, MELISSOPHOBIA.

Apis *Invertebrate Zoology.* a genus in the family Apidae containing the common honeybee.

API scale *Chemical Engineering.* a scale of relative density used by the American Petroleum Institute to measure the specific gravity of liquids such as crude oil. On this scale, the larger the number, the lighter the oil, as with a light crude at 40 degrees API and a heavy crude at 20 degrees API.

apisination *Toxicology.* poisoning from a bee sting; in mild cases symptoms may include skin irritation; in severe cases, symptoms may include anaphylactic shock.

Apis mellifera *Entomology.* the honeybee, a member of the family Apidae that has been domesticated and is kept for the commercial production of honey.

Apistobranchidae *Invertebrate Zoology.* a family of polychaete, burrowing, deposit-feeding worms favoring sandy or muddy bottoms, from shallow to deep water; cosmopolitan.

apitoxin *Toxicology.* the toxic protein constituent of bee venom. Also, **apisin.**

apivorous *Zoology.* feeding on bees, such as certain birds native to Africa.

apjohnite *Mineralogy.* $Mn^{+2}Al_2(SO_4)_4{\cdot}22H_2O$, a white monoclinic mineral having a specific gravity of 1.8 and a hardness of 1.5 on the Mohs scale; occurring in fibrous masses, crusts, and efflorescences.

APL *Computer Programming.* a high-level interactive programming language with an extensive set of operations and data structures, well suited to handling complex operations on arrays. (An acronym for A Programming Language.)

aplacental *Zoology.* having no placenta, such as the marsupials.

Aplacophora *Invertebrate Zoology.* a small group of primitive, sluggish, wormlike carnivorous mollusks common to all water depths, favoring muddy substrates; various forms feed on coelenterates, protozoans, or other minute organisms. Also, **Aplacophorea.**

aplanatic lens *Optics.* a lens designed using aplanatic points so that it is corrected for spherical aberration and coma.

aplanatic points *Optics.* two points on the axis of an optical system located in such a way that all rays radiating from one point converge to, or appear to diverge from, the other point.

aplanetic *Biology.* referring to organisms in which there is no motile stage.

aplanogamete *Biology.* a nonmotile gamete.

aplanospore *Mycology.* a nonmotile, asexual spore produced either within the sporangium in certain fungi or within a cell in certain algae, in which it develops a new cell wall distinct from that of the parent cell.

aplasia *Immunology.* a condition in which cell or organ development is partially or totally lacking.

aplastic *Biology.* referring to an organism having defective development, such as incomplete organ development.

aplastic anemia see MYELOPHTHISIS.

aplite *Petrology.* a light-colored, fine-textured igneous rock composed mainly of quartz and feldspar.

Aplodontidae *Vertebrate Zoology.* a monotypic family of primitive rodents of the suborder Sciuromorpha, found in cool wet regions of the Pacific Northwest. Also, SEWELLEL; MOUNTAIN BEAVER.

Aplysia *Invertebrate Zoology.* the sea hares; a genus of large sluglike mollusks, often colorful, with finlike swimming lobes on the foot, four tentacles, and a rudimentary internal test; some forms emit dark fluid when disturbed. Also, TETHYS.

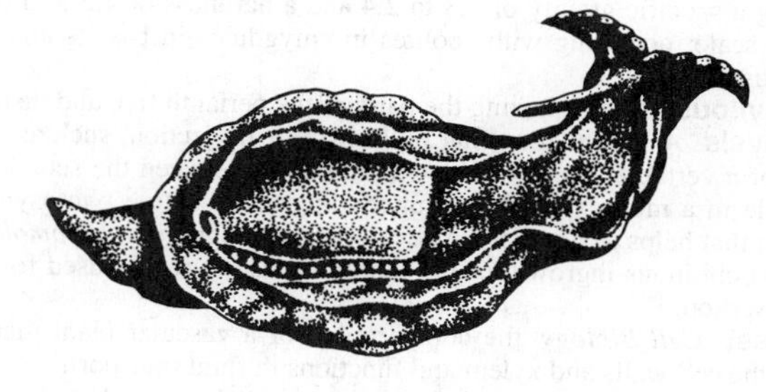

Aplysia

aplysiatoxin *Biochemistry.* a toxic substance found in blue-green algae.
apnea *Physiology.* a temporary cessation of breathing, occurring in some newborns and in some adults during sleep. *Zoology.* a temporary suspension of breathing in hibernating animals.
Apneumonomorphae *Invertebrate Zoology.* in some classifications, a suborder of Arachnida consisting of arachnomorph spiders that lack book lungs.
apneusis *Physiology.* sustained inspiration that is uninterrupted by expiration.
apneustic center *Zoology.* the part of the brain that controls the inflation of the lungs in the higher vertebrates.
apneustic tracheal system *Entomology.* the closed spiracle respiratory system found in many aquatic insects; oxygen passes into the system from the surrounding medium, adaptations include tracheal gills and spiracular gills.
apo- a prefix meaning: **1.** away; apart. **2.** separated or derived from another chemical compound, as in *apomorphine.*
apoapsis [ap´ ō ap´ sis] *Astronomy.* the point in an orbit where the moving body lies farthest from the celestial body around which it orbits.
apoatropine *Organic Chemistry.* $C_{17}H_{21}NO_2$, a poisonous crystalline alkaloid, derived from atropine; melts at 62°C; insoluble in water and soluble in alcohol, ether, acetone, and benzene. It is used as an antispasmodic agent. Also, ATROPAMINE, APATROPINE.
apob *Meteorology.* an aircraft sounding taken aloft with an aerometeorograph. (An acronym for airplane observation.)
apocarpous *Botany.* having separate or distinct carpels, as in buttercups.
apocenter *Astronomy.* the point on an elliptic orbit at the greatest distance from the principal focus or center of attraction. Also, APOFOCUS.
apochromatic lens *Optics.* a compound lens that is virtually free from chromatic error and that has been corrected for spherical aberration for two wavelengths and for central coma for one wavelength. Also, **apochromat.**
apochromatic system *Optics.* an optic system that has been corrected chromatically so that three colors come to a common focus.
apocodeine *Pharmacology.* $C_{18}H_{19}NO_2$, a monomethyl ether of apomorphine, used to induce vomiting.
apocope *Surgery.* another term for amputation.
apocrine gland *Physiology.* any of certain sweat glands that occur in hairy areas of the body, and whose secretion contains part of the secreting cells.
Apocrita see CLISTOGASTRA.
Apocynaceae *Botany.* the dogbane family of herbs, shrubs, trees, and some vines; characterized by simple opposite leaves, flowers without a corona, and milky and often poisonous juice; fruit is usually a berry, drupe, or dry pod.
apocynin see ASAPOCYNIN.
Apocynum *Botany.* a genus of poisonous North American apocynaceous plants noted for their digitalislike cardioactive properties; includes Canadian hemp, black Indian hemp, and dogbane.
Apoda *Vertebrate Zoology.* the Caecilians, an order of wormlike, burrowing animals in the class Amphibia, having reduced eyes and an anterior sensory tentacle.
Apodacea *Invertebrate Zoology.* a subclass of burrowing sea cucumbers with simple tentacles, digitate or pinnate, greatly reduced or absent tube feet, and missing anterior body retractor muscles.
apodeme *Invertebrate Zoology.* an internal support plate for muscles and certain internal organs of arthropods.
Apodes see ANGUILLIFORMES.
Apodi *Vertebrate Zoology.* a suborder of birds in the order Apodiformes, composed of the swifts.
Apodida *Invertebrate Zoology.* an order in subclass Apodacea containing three families of burrowing, wormlike sea cucumbers.
Apodidae *Vertebrate Zoology.* a widely distributed family of birds in the order Apodiformes, composed of the swifts, having flat skulls and toes pointing forward, but appearing and behaving swallowlike.
Apodiformes *Vertebrate Zoology.* an order of fast-flying insectivorous birds, including the swifts and hummingbirds.
Apodiniaceae *Botany.* a monogeneric family of unicellular algae of the order Blastodiniales, being nonphotosynthetic ectoparasites native to the Mediterranean.
apodization *Optics.* the process of blacking out the center portion of an aperture, or the systematic variation of the transmission of a lens element, in order to reduce diffraction, as in a reflecting telescope with a secondary mirror.
apodous *Zoology.* having no legs or feet. Also, **apodal.**
apodous larvae *Invertebrate Zoology.* a type of insect larvae in which the trunk appendages are completely suppressed; occurs in Diptera, Hymenoptera, and Lepidoptera.
apoenzyme *Enzymology.* the protein portion of an enzyme, which is catalytically inactive. Also, **apoferment.**
apoferritin *Biochemistry.* a protein in the mucous cells of the small intestines that functions to facilitate the absorption of iron; ferritin without its iron component.
apofocus see APOCENTER.
apogamia see APOMIXIS.
apogamy *Biology.* **1.** the asexual development of an embryo. **2.** the development of a sporophyte directly from the cells of a gametophyte without fertilization, resulting in a constant chromosome number.
apogean [ap´ ə jē´ən] *Astronomy.* of or connected with the apogee, or occurring when the moon is at the apogee of its orbit.
apogean range *Oceanography.* the average semidiurnal tidal range at the time of apogean tides.
apogean tides *Oceanography.* the tides of decreased range that occur when the moon is farthest from the earth, at apogee. Also, **apogean tidal currents.**
apogee [ap´ ə jē´] *Astronomy.* the point in an orbit around the earth at which the moon, or any other satellite, lies farthest from the earth's center of attraction. *Space Technology.* the point in the trajectory of a missile that is farthest from the earth. *Medicine.* the state of the greatest severity of a disease.

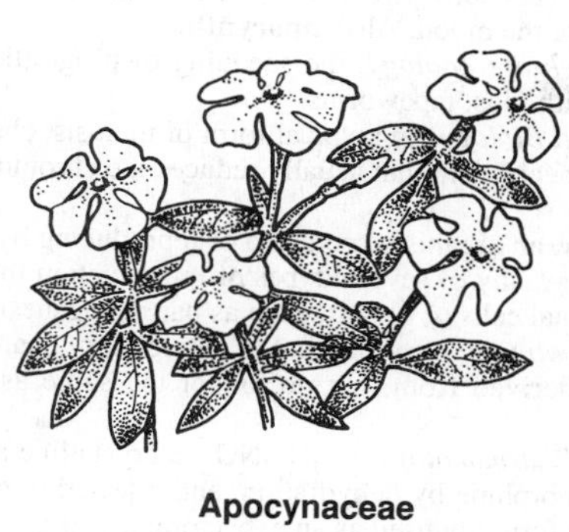

Apocynaceae

apogeny *Botany.* a loss of the reproductive function; the inability to reproduce.
apogeotropic *Botany.* turning away from the earth.
apogeotropism *Botany.* negative geotropism; the tendency of leaves or other parts of a plant to turn upward or away from the earth.
Apogonidae *Vertebrate Zoology.* the cardinal fishes, a family of tropical marine fishes in the order Perciformes, some of which are bioluminescent.
Apoidea *Invertebrate Zoology.* the insect group or superfamily consisting of bees.
apoinducer *Biochemistry.* a regulatory protein that activates a gene by binding to control regions in the DNA and stimulating transcription.
apolegamy *Biology.* the forces of selection, especially sexual selection in breeding.

apolipoprotein *Biochemistry.* a protein that plays an critical role in the transportation of liposides, minus the lipid component.

Apollo *Astronomy.* asteroid 1862, discovered in 1932 and measuring about 1.5 kilometers in diameter. Of type S, it is the prototype of a class of asteroids whose perihelia lie at less than 1.0 astronomical unit. *Space Technology.* see APOLLO PROGRAM.

Apollonian personality *Anthropology.* a categorization by the anthropologist Ruth Benedict, describing an ideal society with an even temperament and calm social structure exemplified by the Zuñi.

Apollonius of Perga c. 262–205 BC, Greek mathematician; formulated basic laws of conic sections; established the terms *ellipse, parabola,* and *hyperbola.*

Apollo object *Astronomy.* any asteroid whose orbit comes closer to the Sun than 1.0 astronomical unit at perihelion.

Apollo Program *Space Technology.* the U.S. space project (1961–1975) that was designed to place astronauts on the moon's surface for exploration and return them to earth. In July 1969, **Apollo 11** took the first astronauts to the moon.

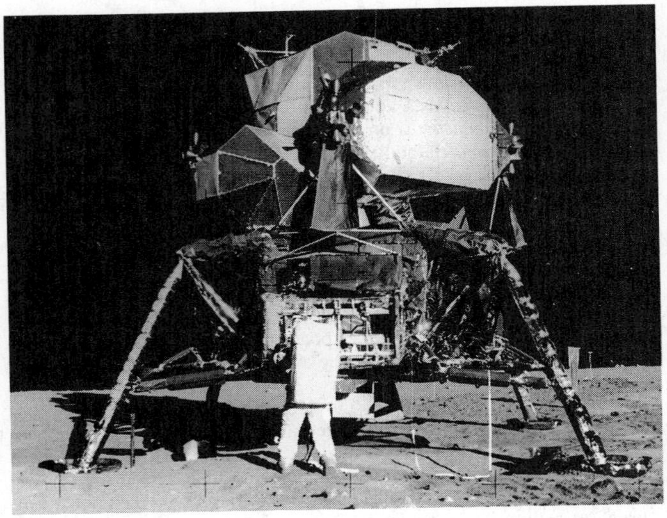

Apollo 11

apolune *Space Technology.* the point in a lunar orbit that is farthest from the center of the moon. Also, **apocynthion.**

apolysis *Invertebrate Zoology.* the shedding of proglottids (segments) during life, as with most tapeworms.

apomeiosis *Cell Biology.* an atypical form of meiosis, characterized by a failure of the mechanism that usually reduces the chromosome number by one-half.

apomict *Biology.* an organism produced or reproducing by apomixis.

apomixis *Biology.* any of several types of reproduction that occur without union of sexual cells or organs, such as parthenogenesis.

apomorphic *Evolution.* of or relating to a hereditary character in an organism that is derived from, but no longer the same as, an ancestral character.

apomorphine *Pharmacology.* $C_{17}H_{17}NO_2$, a crystalline alkaloid that is obtained from morphine by dehydration, and injected to effect instantaneous vomiting; formerly used as an expectorant.

apomyoglobin *Biochemistry.* the oxygen-binding protein of muscle, minus its heme group (which is the site of O_2 binding).

aponeurology the scientific study of aponeuroses and fasciae.

aponeurosis *Anatomy.* **1.** a fibrous sheet or expanded tendon connecting two or more muscles or organs. **2.** the whitish, fibrous membrane that serves as the attachment of certain nerves.

Aponogetonaceae *Botany.* a monogeneric family of perennial, glabrous herbs that secrete oil, tannin, or latex and float or grow submerged in fresh water; characterized by leaves that are all basal and an inflorescence that is raised above the water on a leafless scape.

apopathetic behavior *Behavior.* behavior that is influenced by the presence of others but is not directed toward them.

apophorometer *Engineering.* an instrument that identifies minerals through a process which condenses the minerals from a vapor.

apophyge [ə päf′ə jē] *Architecture.* a small section at the top or bottom of a column where the shaft curves out to meet the capital or base. Also, ESCAPE.

apophyllite *Mineralogy.* $KCa_4Si_8O_{20}(F,OH)\cdot 8H_2O$, a group name for white to grayish tetragonal minerals with a pearly to vitreous luster, having a specific gravity of 2.3 to 2.4 and a hardness of 4.5 to 5 on the Mohs scale; occurring with zeolites in amygdules in basalts and other igneous rocks.

apophyllous *Botany.* having the parts of the perianth free and distinct.

apophysis *Anatomy.* a normal outgrowth or projection, such as a process of a vertebra. *Botany.* the swollen region between the seta and the capsule in a moss sporophyte, functioning as an active photosynthetic region that helps to nourish the developing sporogonium. *Entomology.* a ventral chitinous ingrowth of the thorax in insects that is used for muscle insertion.

apoplast *Cell Biology.* the compartment of a vascular plant that contains the cell walls and xylem and functions in fluid transport.

apoplectic *Neurology.* of, relating to, or affected by apoplexy.

apoplexy [ap′ə pleks′ē] *Neurology.* a sudden neurologic or cerebral impairment due to a cerebrovascular accident, such as a hemorrhage into the brain or rupture of an artery. Also, **apoplexia, apoplectic stroke.**

apoprotein *Biochemistry.* the protein on a polypeptide chain that is detached from the ligands and prosthetic groups.

apopyle *Invertebrate Zoology.* an opening in the radial canal or flagellated chamber of a sponge through which water passes.

aporepressor *Biochemistry.* a polypeptide that when activated permits a gene to bind to the operator locus and regulate the activity of nearby genes.

Aporidea *Invertebrate Zoology.* in some classifications, an order of Cestoda consisting of a single genus, *Nematoparataenia,* of unsegmented tapeworms that are parasitic primarily in swans.

aporogamy *Botany.* fertilization that results when a pollen tube penetrates an ovule at some point other than through the micropyle.

aposematic *Biology.* relating to or characterized by aposematism. Thus, **aposematic coloration.**

aposematic resemblance *Entomology.* the similar appearance of a harmless species to a poisonous or dangerous one; serves to provide protection from predation.

aposematism *Biology.* the protective coloration or structures possessed by certain species to warn potential predators of poisonous or distasteful defense mechanisms.

apospory *Mycology.* the development of a gametophyte from a sporophyte cell without meiosis and the formation of spores, with the resulting gametophyte having the same chromosome number as the parent sporophyte.

apostatic selection *Ecology.* the tendency of a predator to select the most abundant form of prey available in a given area, which tends to equalize the population of the various forms in the area. Also, FREQUENCY-DEPENDENT SELECTION.

apostilb *Optics.* a unit of luminance expressed as $1/\pi$ candles per square meter; equivalent to one ten-thousandth of a lambert. Also, BLANDEL.

Apostomatida *Invertebrate Zoology.* an order of ciliated protozoans of the subclass Hypostomatica that are mainly parasitic or symbiotic on or in marine invertebrates.

apothecaries' measure *Metrology.* a system of units, including drams, ounces, and pounds, used in pharmacy for the preparation of liquid medicines. Also, **apothecaries' liquid measure.**

apothecaries' weight *Metrology.* a system of weights used in pharmacy for the preparation of medicines, based on a 12-ounce pound. Also, **apothecaries' system.**

apothecioid *Mycology.* referring to a kind of fungal spore-bearing body whose surface is exposed to the environment.

apothecium *Mycology.* an open or dish-shaped part of lichens and certain fungi that holds the spores exposed to the environment throughout their development.

apothem *Mathematics.* the perpendicular distance from a center.

apotreptic behavior *Behavior* behavior toward another animal of the same species that tends to cause that animal to withdraw.

apotypic state *Evolution.* the character state diverging from the basic or typical form in a transformation series.

apozymase *Enzymology.* the protein portion of the enzyme zymase.

Appalachian relating to the Appalachians or to Appalachian orogeny.

Appalachian orogeny *Geology.* a late Paleozoic deformation of the earth's crust that affected the southeastern portion of the North American continent.

Appalachians *Geography.* a mountain system of eastern North America, extending from southeastern Canada to Alabama.

apparatus of Perroncito [per′ən sēt′ō] *Neurology.* a random formation and proliferation of fibrils in the form of spirals and networks with newly formed axons, occurring at the stump of a divided nerve during regeneration. Also, PERRONCITO'S PHENOMENON, PERRONCITO'S SPIRALS. (Named for Aldo *Perroncito,* 1882–1929, Italian histologist.)

apparent *Navigation.* of a phenomenon, actually observed, rather than tabulated or calculated.

apparent additional mass *Fluid Mechanics.* an imaginary amount of fluid added to the body mass to achieve the force necessary to advance the body through the fluid.

apparent altitude *Navigation.* sextant altitude corrected for instrument errors and dip (and Coriolis, if applicable); the altitude above the horizon at which a given body appears.

apparent candlepower *Optics.* the amount of candlepower emitted by a point source of light having the same luminous intensity as an extended source of light at a specified distance.

apparent cohesion *Geology.* the resistance to separation exhibited by particles of soil, silt, or sand resulting from surface tension of the water surrounding each particle.

apparent density *Metallurgy.* **1.** in powder metallurgy technology, the weight of a powder per unit volume, measured according to a standard procedure. **2.** the weight of any porous object per unit volume. Apparent density is invariably lower than the theoretical density of the material.

apparent depth *Optics.* the seeming depth or thickness of a transparent optical medium when viewed by an observer; differs from the actual depth due to refraction.

apparent diameter *Astronomy.* the angular size of a celestial object as viewed from the earth.

apparent diffusion coefficient *Materials Science.* an overall coefficient of diffusion obtained by experiment, in which more than one diffusion mechanism is contributing to the total flux. Also, EFFECTIVE DIFFUSION COEFFICIENT.

apparent dip *Geology.* the angle made with the horizontal by an exposed rock body or structural surface, measured in any vertical section that is not perpendicular to the strike.

apparent distance see ANGULAR DISTANCE.

apparent expansion *Thermodynamics.* the expansion of a liquid with a change in temperature.

apparent force *Mechanics.* a fictitious force that appears to exist from physical experience or observation made in a noninertial frame of reference; e.g., the fact that a force seems to pull passengers forward if a car stops suddenly, or pull them outward as the car rounds a curve. Also, INERTIAL FORCE, OBSERVED FORCE.

apparent fracture surface energy see FRACTURE SURFACE ENERGY.

apparent gravity see ACCELERATION OF GRAVITY.

apparent magnitude *Astronomy.* the magnitude of a celestial object as seen from the earth.

apparent motion see RELATIVE MOTION.

apparent noon *Astronomy.* noon as shown on a sundial.

apparent oxygen utilization *Oceanography.* an expression of the amount of change in the oxygen content of a sample of sea water since the time that the sample had been at the surface; the difference between the observed oxygen concentration in a sample of subsurface water and the theoretical saturation concentration.

apparent photosynthesis *Botany.* the rate of photosynthesis minus the rate of respiration.

apparent place *Astronomy.* the apparent position of a celestial body on the celestial sphere, determined by right ascension and declination.

apparent porosity *Materials Science.* the volume of open pores divided by the volume of the material.

apparent power *Electricity.* the power, measured in voltage-amps, composed of active (real) power and reactive power. It is equal to the product of RMS current and RMS voltage: apparent power = $[(\text{active power})^2 + (\text{reactive power})^2]^{1/2}$.

apparent procession see APPARENT WANDER.

apparent relative movement see SEPARATION.

apparent shoreline *Photogrammetry.* the outer limit of marine vegetation, as shown in aerial photographs where the actual shoreline cannot be seen.

apparent slope *Graphic Arts.* a true slope that appears vertically distorted or exaggerated in air photographs viewed with a stereoscope.

apparent solar day *Astronomy.* the time between two successive passages of the sun over a given meridian.

apparent solar time *Astronomy.* the time of day shown by a sundial or expressed by the position of the real sun (rather than a fictitious mean sun) relative to some reference point. Also, **apparent time.**

apparent sun *Astronomy.* in timekeeping and navigation, the actual sun in the sky.

apparent velocity see RELATIVE VELOCITY.

apparent vertical *Geophysics.* the direction of all accelerations, including resultant gravitational acceleration.

apparent visual angle *Optics.* the angle subtended at a source, determined by the size of the source and its distance from the eye of the observer.

apparent volume *Physics.* the difference between the volume of a solution composed of dissolved matter in a pure solvent and the volume of the pure solvent.

apparent wander *Geophysics.* the apparent directional change in the axis of rotation of a spinning body, such as a gyroscope, due to the rotation of the earth. Also, APPARENT PRECESSION.

apparent water table see PERCHED WATER TABLE.

apparent weight *Mechanics.* the measured weight of an object immersed in a fluid; its true weight minus the weight of the displaced fluid.

apparent wind see RELATIVE WIND.

apparition *Astronomy.* the period when a comet or other celestial body is in view from earth.

appearance potential *Physics.* the minimum potential required to accelerate an electron beam in order to ionize a source and thus produce ions of a particular species.

appeasement behavior *Behavior.* a pattern of submissive behavior that serves to inhibit aggression or attack by another animal of the same species.

appeasement display *Behavior.* an action or posture that serves to inhibit aggression or attack by another animal of the same species.

append *Computer Programming.* to add a record, text, or element to the end of a file, string, or linked list.

appendage *Biology.* an external or subordinate structure protruding from the main body, such as limbs, tentacles, fins, or wings, used for locomotion, sensory reception, feeding, and other purposes. Also, APPENDIX. *Naval Architecture.* any fitting that projects underwater from a vessel's hull, such as a rudder, propeller, or shaft.

appendectomy [ap′en dek′tə mē] *Surgery.* the surgical removal of the vermiform appendix.

appendicitis [ə pen′də sī′tis′] *Medicine.* **1.** any inflammation of the vermiform appendix. **2.** see ACUTE APPENDICITIS.

appendicitis obliterans *Medicine.* appendicitis with hardening and shrinking of the submucous tissue and plastic peritonitis, leading to obliteration of the lumen of the appendix. Also, OBLITERATIVE APPENDICITIS, PROTECTIVE APPENDICITIS.

appendicular skeleton *Anatomy.* the skeleton of the girdles and limbs of the body.

appendiculate *Biology.* forming or having limbs or appendages.

appendix *Anatomy.* a small, fingerlike projection from the cecum of the large intestine. *Biology.* see APPENDAGE.

appersonation see APPERSONIFICATION.

appersonification *Psychology.* a delusion in which an individual identifies with and then takes on the characteristics of another person, typically a famous or important figure.

appertization *Food Technology.* **1.** a method of sterilizing food in hermetically sealed containers. **2.** other similar heat-sterilization processes. (From the French chemist Francois *Appert,* 1749–1840, who invented the process.)

appestat *Neurology.* the nerve center of the brain that controls the amount of food intake and the sensation of appetite; probably located in the hypothalamus.

appetite *Biology.* a natural longing or desire, particularly the recurring desire for food.

appetitive behavior *Behavior.* behavior by an animal that is directed toward achieving a specific goal and that varies according to the situation.

appinite *Petrology.* a plutonic rock rich in prismatic hornblende phenocrysts with a high feldspar content.

apple *Botany.* **1.** any of various small deciduous, insect-pollinated and self-sterile trees or occasionally shrubs of the genus *Malus* in the Rosaceae family, having white to pink flowers borne on short lateral branches and an edible fruit that is a pome. **2.** the fruit of an apple tree, particularly *Malus sylvestris,* the common cultivated variety.

apple coal *Geology.* any soft coal that is easily mined, so called because it breaks into fragments the size of apples.
apple essence see ISOAMYL VALERATE.
Applegate diagram *Electronics.* a two-dimensional plot of the positions of drift electrons versus time, for an electron tube with velocity modulation.
apple gum *Botany.* a gum tree native to Australia that resembles the apple tree; valued for its hard brown timber.
apple honey see APPLE SYRUP.
apple maggot *Invertebrate Zoology.* the larva of certain insects from the same family as the fruit fly, feeding on the flesh of apples and carrying the bacterial rot organisms. Also, RAILROAD WORM.
apple of Peru *Botany.* an annual plant of the nightshade family, *Nicandra physalodes,* having large, pale-blue flowers.
apple oil see ISOAMYL VALERATE.
Appleton, Sir Edward V. 1892–1965, British physicist; Nobel Prize for discovery of the Appleton layer (F_2 layer).
Appleton layer see F_2 LAYER.
appliance *Engineering.* **1.** in general, any tool or machine that is used to carry out a specific task or produce a desired result. **2.** specifically, an electrical device that is used for some household purpose, such as a washing machine, dishwasher, toaster, or food processor. *Transportation Engineering.* any aircraft fitting or item of aircraft gear apart from the airframe, including instruments, survival gear, etc., but not including engines or propellers.
appliance panel *Engineering.* a metal protective housing containing devices to prevent excessive current in circuits that feed portable electric appliances.
application *Computer Technology.* any specific purpose for which a computer system or program is employed; payroll records, sales summaries, and inventory control are typical business applications.
application development *Computer Programming.* the process of creating or adapting software products for the specific needs of their users.
application-development language *Computer Programming.* **1.** a high-level language that simplifies programming by generating segments of code in a conventional language. **2.** an added capability of a database management or spreadsheet package that supports high-level programming of those functions.
application-development system *Computer Programming.* a collection of software products that enables efficient development of application programs, e.g., debuggers, syntax checkers, and code formatters.
application generator *Computer Programming.* a program that constructs another program by selecting instructions from a specifically designed set of instructions and arranging them in accordance with prescribed rules and specifications.
application layer *Computer Technology.* the highest layer in an open-system network architecture, developed by the using organization and providing the functions needed for that user to access and utilize the network.
application of fire *Ordnance.* in gunnery, the placing of fire on a specific target or in a specific area.
application program see APPLICATIONS PROGRAM.
applications package *Computer Programming.* a commercially available, predeveloped set of integrated programs for standard scientific or business applications, such as linear programming or spreadsheets.
applications program *Computer Programming.* a computer program written for or by a specific end user to accomplish a specific purpose. Thus, **applications programmer.**
applications satellite *Space Technology.* any earth satellite designed for a practical use, such as a communications or weather satellite.
applications software *Computer Programming.* software created for a specific purpose, such as word processing, as opposed to software used in the running of a system.
applications technology satellite *Space Technology.* a high-orbit satellite that carries some new technology into space in order to facilitate experiments into the practical uses of such technology.
application study *Computer Programming.* a detailed analytic process of determining the feasibility of using a computer for a specific application and establishing the requirements and specifications for the application system.
application system *Computer Programming.* the set of programs and computer equipment that together perform a specific function.
applicative language see FUNCTIONAL LANGUAGE.
applicative programming see FUNCTIONAL PROGRAMMING.
applied *Science.* dealing with practical issues rather than research and theory; based on actual experience or used to solve actual problems.
applied anthropology *Anthropology.* a branch of anthropology in which aspects of cultural knowledge are utilized in conjunction with modern techniques, and in which the anthropologist is an advocate for the culture that he or she studies by directing culture change.
applied archaeology *Archaeology.* the use of archaeological methods and techniques to obtain information about contemporary society rather than about past cultures.
applied climatology *Meteorology.* the scientific analysis of climatic data for application in manufacturing, agriculture, or technology.
applied ecology *Ecology.* the area of ecology concerned with management of the ecological character of a certain area, as by attempts to preserve an endangered species or to introduce natural predators of an overpopulated species.
applied epistemology *Artificial Intelligence.* the study of the origin, nature, and limits of human knowledge as it relates to the design and development of machines intended to simulate human perception and cognition.
applied ethology *Psychology.* the study of behavior in animal species whose behavior patterns are of direct practical concern to humans, such as pets, livestock, and zoo animals.
applied linguistics *Linguistics.* the study of applications of linguistic theory to psychology, lexicography, or the teaching of language, as well as applications to the fields of data processing and telecommunications.
applied meteorology *Meteorology.* the application of current weather data, analyses, and forecasts to specific practical problems.
applied psychology *Psychology.* the attempt to deal with practical issues and problems by applying the principles and techniques of psychology, as in education, industry, or the law.
applied research *Science.* research that is conducted for a practical purpose, as distinguished from basic research.
applied science *Science.* the use of scientific findings and scientific practices to deal with practical problems.
applied shock *Electronics.* the excitation that produces shock motion within a system.
appliqué *Graphic Arts.* a piece of material that is cut out and attached to the surface of another as a decoration. *Optics.* a combination of lenses that offers the same focal length for three or more wavelengths.
appliqué armor *Ordnance.* an additional layer of armor that can be installed on a tank on top of its structural armor.
appliqué circuit *Electricity.* a circuit for adapting existing equipment for special usage.
apposition *Cell Biology.* the deposition of additional layers of cell wall material upon a preexisting plant cell wall. *Botany.* the deposition of successive layers of cellulose upon those already present on the inner surface of a plant cell wall, a process that strengthens the overall cell stucture. *Surgery.* the process of bringing or placing together, as the edges of a wound.
apposition beach *Geology.* any of a series of successively formed beaches lying parallel to an older beach on its seaward side.
apposition eye *Invertebrate Zoology.* the compound eye of diurnal insects in which collected points of light are combined to form visual images.
apposition fabric *Petrology.* a primary orientation of the elements of most sedimentary rocks developed or formed at the time of deposition of the material. Also, PRIMARY FABRIC.
apposition suture *Surgery.* a superficial suture used for the exact approximation of the cutaneous edges of a wound. Also, COAPTATION SUTURE.
Apprentice *Robotics.* an all-purpose robot developed by Unimation, Inc., that has five axes of freedom and a spherical coordinate system.
appressed *Biology.* pressed close to or flattened against something, but not united to it.
appressorium *Mycology.* a flattened hyphal structure, formed by many parasitic fungi, that permits penetration of the host epidermis.
approach *Navigation.* the process of an aircraft descending for a landing. Thus, **approach course.** *Oceanography.* the part of the sea that is adjacent to a shoreline.
approach-approach conflict *Psychology.* a conflict situation in which the individual must choose between equally attractive but incompatible goals.
approach-avoidance conflict *Psychology.* a conflict in which the individual feels both attracted to and repelled by the same goal.
approach behavior see EPITREPTIC BEHAVIOR.

approach-control radar *Electronics.* any radar set used in a ground-controlled approach system, especially an airport surveillance radar that shows on a screen the positions of aircraft in the traffic control area.

approach fix *Aviation.* a specified point, often marked on government charts, from which the final approach to an airport is begun.

approach gate *Aviation.* an imaginary point, at least five miles from an airport, from which air traffic control provides headings for landing aircraft.

approach gradient *Psychology.* a gradual yet steady increase in the drive to approach a positive goal as one gets closer to the goal, e.g., when excitement about a forthcoming vacation trip increases as the day of departure gets closer.

approach lights *Navigation.* a set of lights along the extended center line of a runway indicating the desired approach to it, usually laid in a precise pattern with crossbars at set distances.

approach march *Military Science.* the advance of troops when direct ground contact with the enemy is about to be made or the attack position is about to be reached.

approach path *Navigation.* **1.** the portion of a flight path in the immediate vicinity of a landing area that terminates at the touchdown point. **2.** the path actually followed by an aircraft during its landing approach.

approach speed *Navigation.* the indicated air speed at which an aircraft approaches for landing.

approach vector *Robotics.* a vector that describes the orientation of a robot's gripper in relation to the direction from which it approaches a workpiece.

appropriate technology *Science.* the use of methods and equipment that are suited to available conditions and local resources, especially in underdeveloped areas.

approval motive *Psychology.* an individual's need for his or her behavior to be praised or approved by parents, peers, and others. Also, **approval need.**

approximate *Science.* [ə präks´ə mət *for 1, 3;* ə präks´ə māt´ *for 2, 4*] **1.** describing a value or result that is not exactly correct, but is close enough for some predetermined purpose. **2.** to obtain a value or result approaching the actual or desired one. **3.** close in position or location; near. **4.** to bring near or into apposition.

approximate absolute temperature *Physics.* a temperature scale that approximates the Kelvin temperature scale by rounding off the ice point of water at 1 atmosphere of pressure to 273 K and the boiling point to 373 K.

approximation *Science.* **1.** a result that is not exactly correct but is close enough for some purpose. **2.** the process of obtaining such a result. **3.** the process of bringing together, as the edges of a wound.

approximation suture *Surgery.* a deep suture that draws the deep tissues of a wound together.

appulse *Astronomy.* **1.** an apparent, very close approach of two celestial bodies, especially the close approach of a planet to a star without the occurrence of an eclipse. **2.** the approach or actual incident of conjunction between two celestial bodies.

apractic *Neurology.* of, relating to, or characterized by apraxia. Also, **apraxic.**

apraxia [ə prak´sē ə] *Neurology.* an inability or loss of ability to perform purposeful, coordinated motor activities in the absence of paralysis or other sensory or motor dysfunction, particularly the inability to use an object properly. (From a Greek word meaning "not acting.")

apricot *Botany.* **1.** any of various small trees of the genus *Prunus* in the Rosaceae family with ovate, closely serrate leaves; pinkish or white flowers; and smooth-skinned fruit. **2.** the edible fruit of one of these trees, *Prunus armeniaca.*

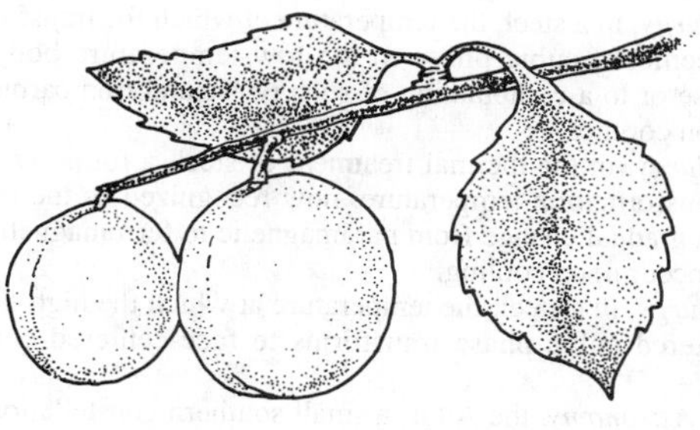

apricot

aproctous [ā´präk´təs] *Zoology.* having no anal pore or opening.

apron a device or area thought to resemble the familiar garment used to protect the front of the body; specific uses include: *Geology.* a continuous, broad, blanketlike accumulation of alluvial, glacial, or other unconsolidated material commonly formed at the base of a mountain or in front of a glacier. *Hydrology.* **1.** see RAM. **2.** see ICE APRON. *Aviation.* a usually paved area adjacent to a runway or other airfield facility where aircraft can park, load or unload, refuel, or undergo servicing. *Civil Engineering.* any device made to protect a surface of the earth from the damaging power of moving water, such as a platform that receives water falling over a dam. *Building Engineering.* **1.** an extension of an interior trim, especially a flat piece of finished wood set directly beneath a windowsill. **2.** a vertical extension at the back of a sink or lavatory. **3.** a concrete slab extending on grade from a building. **4.** see APRON FLASHING. *Mining Engineering.* **1.** a canvas-covered frame on a miner's rocker. **2.** an amalgamated copper plate placed below a stamp battery over which the pulp passes. Also, COPPER PLATE. **3.** a hinged extension of a loading chute. *Ordnance.* a metal plate or cover for an artillery piece to protect its operator.

apron conveyor *Mechanical Engineering.* a conveyor used for transporting packages or bulk materials, made up of a series of metal or wood plates attached to a continuous belt.

apron flashing *Building Engineering.* flashing along the joint between the lower side of a vertical surface (especially a chimney) and a sloping roof.

apron lining *Building Engineering.* in a staircase, the wooden casing that covers an apron piece.

apron pan *Mining Engineering.* one of a series of pans which, when attached to a chain or pivotally attached one to another, forms the conveying medium for a traveling apron.

apron piece *Building Engineering.* a beam that supports a staircase landing. Also, PITCHING PIECE.

apron rail *Building Engineering.* on a door, a horizontal center rail having a decorative molding.

apron wall see SPANDREL.

aprosody *Neurology.* a lack of normal fluctuations in the pitch, stress, and tempo of speech.

A-protein *Virology.* in the cell wall or in extracts of *Staphylococcus aureus,* a protein that binds the Fc portion of immunoglobulins and is useful in gathering antigen-antibody complexes; it occurs as a coat protein in cells at high pH and low ionic strength.

aprotic *Chemistry.* describing a substance that neither accepts nor donates protons. Thus, **aprotic solvent.**

aprotinin *Biochemistry.* a substance present in tissues and blood that inactivates kallikrein, trypsin, chymotrypsin, plasmin, and papain.

APS or **A.P.S.** American Physical Society, American Physiological Society.

apse *Architecture.* a semicircular or polygonal end of a building, especially the east end of a church designed to house the altar. *Astronomy.* an apsis.

apsidal [ap´ sə dəl] *Astronomy.* describing the apsides of an orbit. *Architecture.* of or relating to the apse of a church.

apsidal motion *Astronomy.* the rotation of an eccentric orbit's major axis in the plane of the orbit, caused primarily by gravitational perturbations.

apsides [ap´ sə dēz´] the plural form of APSIS.

Apsidospondyli *Vertebrate Zoology.* an amphibian subclass in which members have a vertebral centra formed from two blocks of bone or cartilage called the intercentrum and pleurocentrum. This group includes present-day frogs.

apsis *plural,* **apsides.** *Astronomy.* the point in an orbit that is farthest from or closest to the body about which it revolves. Also, APSE.

AP site *Genetics.* a site on a DNA molecule that lacks either purines (apurinic) or pyrymidines (apyrimidinic).

Ap star *Astronomy.* a type A star that shows unusually strong lines of ionized metals and rare-earth elements in its spectrum.

APT automatic picture transmission, Automatic Programming Tool.

apterium *Vertebrate Zoology.* the bare or down-covered space between rows of contour feathers on a bird's skin.

apterous *Biology.* not winged; lacking wings or winglike expansions.

Apterygidae *Vertebrate Zoology.* the kiwis, a family composed of three species of primitive, nocturnal, flightless birds from New Zealand, in the order Apterygiformes.

Apterygiformes *Vertebrate Zoology.* a ratite order of flightless birds with vestigial wings, containing one family, Apterygidae or the kiwis.

Apterygota *Invertebrate Zoology.* a subclass or group of primitively wingless Insecta whose members undergo little or no metamorphosis; includes bristletails, springtails, and silverfish. Also, **Apterygote.**

Aptian *Geology.* a European geologic stage of the Lower Cretaceous period, after the Barremian and before the Albian. Also, VECTIAN.

aptitude *Psychology.* the potential ability to acquire a new skill or perform a previously unlearned task.

aptitude test *Psychology.* a test that is intended to predict an individual's capacity to become proficient in a given activity.

aptyalism *Medicine.* an absence or deficiency of saliva. Also, **aptyalia.**

APU auxiliary power unit.

APUD *Oncology.* an acronym for amine percursor uptake and decarboxylation.

APUD cells *Oncology.* a system of apparently unrelated endocrine cells sharing a number of cytochemical and ultrastructural characteristics and found scattered throughout the body; this system synthesizes structurally related peptides that function as hormones or neurotransmitters, and produces various substances, such as epinephrine, norepinephrine, dopamine, serotonin, and neurotensin.

apudoma *Oncology.* any tumor composed of APUD cells that secrete one or more hormones or amines.

Apus [ā´ pəs] *Vertebrate Zoology.* a genus of Old World swifts in the bird family Apodidae. *Astronomy.* the Bird of Paradise, a small and faint constellation of the southern sky.

apyrase *Enzymology.* an enzyme that catalyzes the breakdown of adenosine triphosphate, releasing phosphate.

apyrexia *Medicine.* the absence or intermittent cessation of fever.

AQ *Psychology.* achievement quotient.

aq. or **Aq.** water. (From Latin *aqua.*)

aq. bull. boiling water. (From Latin *aqua bulliens.*)

aq. dest. distilled water. (From Latin *aqua destillata.*)

aq. ferv. warm water. (From Latin *aqua fervens.*)

AQL see ACCEPTABLE QUALITY LEVEL.

Aql see AQUILA.

aq. pur. pure water. (From Latin *aqua pura.*)

Aqr see AQUARIUS.

aq. tep. tepid water. (From Latin *aqua tepida.*)

aqua [äk´wə] *Science.* the Latin word for water. *Chemistry.* relating to or being an aqueous solution.

aqua ammonia see AMMONIUM HYDROXIDE.

aquaculture see AQUICULTURE.

Aquadag *Electronics.* a trade name for a conductive graphite material coated to the interior and exterior of some cathode-ray tubes. On the interior it serves to collect any secondary electrons emitted by the screen, while on the exterior it serves as the final capacitor of the high voltage filtering circuit. *Materials.* a trade name for a colloidal suspension of graphite in water, used as a lubricant and a conductive coating.

aqua fortis *Inorganic Chemistry.* another name for nitric acid, HNO_3. See NITRIC ACID.

aquagene tuff see HYALOCLASTITE.

Aqualf *Geology.* a suborder of the soil order Alfisol that develops in water-saturated areas and is characterized by a gray or mottled color.

aqualung *Engineering.* a self-contained underwater breathing apparatus (scuba) in which the air supply is automatically regulated by a two-stage valve or by the demand regulator.

aquamarine *Mineralogy.* a gem variety of beryl.

aquametry *Analytical Chemistry.* analytical procedures such as oven drying, distillation, or Karl Fischer titration, used to measure the quantity of water present in materials.

aquapuncture *Surgery.* the injection of water beneath the skin.

aqua regia *Inorganic Chemistry.* a highly corrosive, suffocating liquid made by mixing one part nitric acid to three or four parts hydrochloric acid; used to dissolve metals, including gold and platinum.

Aquarius *Astronomy.* the Water Bearer, a zodiacal constellation located between Pisces and Capricornus.

Aquaspirillum *Bacteriology.* a genus of Gram-negative bacteria that grow in freshwater habitats as helical cells, exhibiting a distinct corkscrewlike motility by means of bipolar flagellation.

aquatic *Biology.* living or growing in or on water.

aquatint *Graphic Arts.* an etching process on a copper plate that produces halftones resembling watercolors.

aquation *Chemistry.* the process of forming a complex of water molecules with ions or other molecules.

aquatone *Graphic Arts.* an offset printing process using a gelatin-coated zinc plate.

aqueduct *Civil Engineering.* a conduit used for carrying water over long distances.

Aquent *Geology.* a suborder of the soil order Entisol that develops in areas periodically saturated with water and is characterized by a bluish- or greenish-gray color.

aqueous [äk´wē əs] *Science.* of or relating to water. *Chemistry.* of a solution, containing water. *Geology.* of rocks, formed of matter deposited by or in water.

aqueous humor *Physiology.* a clear watery fluid between the cornea and the iris of the eye that helps the cornea keep its shape.

aqueous lava *Volcanology.* mud lava formed as a result of the mixture of volcanic ash with water or condensing volcanic vapor.

aqueous rock *Petrology.* a sedimentary rock that is deposited in or by water. Also, HYDROGENIC ROCK.

aqueous solution *Chemistry.* a solution with water as the solvent.

aqueous vapor see WATER VAPOR.

Aquept *Geology.* a suborder of the soil order Inceptisol that develops in areas where restricted natural drainage results in water saturation; characterized by a dark surface horizon over a mottled or gray subsoil.

aquiclude *Geology.* a body of relatively impermeable rock that serves to restrict the flow of groundwater as it cannot transmit enough water to directly supply a well or spring.

aquiculture *Biology.* **1.** the cultivation of aquatic plants and animals for human food consumption or other human use. **2.** specifically, freshwater cultivation, as opposed to marine cultivation (mariculture). Also, AQUACULTURE.

aquifer [ak´wə fər] *Hydrology.* a permeable body of rock or other geologic structure that contains and conducts economically significant quantities of groundwater to supply wells and springs.

Aquifoliaceae *Botany.* the holly family of trees and shrubs, characterized by alternate, usually evergreen leaves; often having spiny-toothed margins, small, greenish-white flowers in axillary clusters, and red or purple berrylike drupes.

aquifuge [äk´wə fyooj´] *Geology.* a body of rock that is incapable of absorbing or transmitting water, thus rendering it impermeable.

aquiherbosa *Ecology.* the herbaceous plant communities of ponds and swamps.

Aquila [ak´ wə lə] *Astronomy.* the Eagle, a constellation lying south of Cygnus and Lyra that contains the bright star Altair.

aquiprata *Ecology.* the plant communities influenced by the presence of groundwater in damp meadows.

Aquitanian *Geology.* a European geologic stage of the lowermost Miocene epoch, occurring after the Chattian of the Oligocene and before the Burdigalian.

aquitard *Geology.* a leaky confining bed that transmits water at a very slow rate to or from an adjacent aquifer.

Aquod *Geology.* a suborder of the soil order Spodosol that develops in water-saturated areas as a result of fluctuating water tables or humid climates; characterized by a black surface horizon, a white albic horizon, or a cemented spodic horizon.

aquo ion *Chemistry.* a term for an ion that contains one or more water molecules.

Aquoll *Geology.* a suborder of the soil order Mollisol that develops in areas periodically saturated with water; characterized by a black surface horizon over a mottled or gray subsoil.

Aquox *Geology.* a suborder of the soil order Oxisol that develops in areas periodically saturated with water and is characterized by a mottled subsoil.

Aquult *Geology.* a suborder of the soil order Ultisol that develops in areas periodically saturated with water and is characterized by a gray or mottled color.

Ar see ARGON.

Ar_1 *Metallurgy.* in a steel, the temperature at which the transformation of the face-centered cubic phase to the low-temperature body-centered cubic phase or to a combination of such phase and iron carbide is completed upon cooling

Ar_2 *Metallurgy.* in the thermal treatment of steel, a formerly designated critical transformation temperature, now recognized as the temperature at which a gradual change from paramagnetic to ferromagnetic behavior begins to occur upon cooling.

Ar_4 *Metallurgy.* in a steel, the temperature at which the high-temperature body-centered cubic phase transforms to face-centered cubic during cooling.

Ara [ā´ rə] *Astronomy.* the Altar, a small southern constellation with few bright stars.

Arabellidae *Invertebrate Zoology.* a family of long, thin, burrowing polychaete worms, generally carnivorous; includes some endoparasitic forms that exist in other invertebrates until maturity.

Arabia *Geography.* a large desert peninsula in southwestern Asia, between the Red Sea and the Persian Gulf. Also, **Arabian Peninsula.**

Arabian Sea *Geography.* the northwestern part of the Indian Ocean, between Arabia and India.

Arabic numerals *Mathematics.* the numerals 0, 1, 2, 3, 4, 5, 6, 7, 8, and 9. They were introduced to Europe from Arabia but are thought to have originated in India. Also, **Hindu-Arabic numerals.**

arabinose *Biochemistry.* $C_5H_{10}O_5$, a simple five-carbon sugar obtained from some vegetable gums.

arabinosis *Toxicology.* a toxic reaction to arabinose, a pectin sugar found in many plants.

arabinosyl nucleoside *Enzymology.* a molecule formed by bonding a purine or pyrimidine base to the pentose sugar arabinose, using an *N*-glucoside bond.

arabinoxylan *Biochemistry.* a polysaccharide constituent of hemicellulose of angiosperm cell walls.

arabite see ARABITOL.

arabitol *Organic Chemistry.* $CH_2OH(CHOH)_3CH_2OH$, a stereoisomeric crystalline solid that is derived from arabinose; melts at 106°C and is soluble in water. Also, ARABITE, D-LYXITOL.

arable [âr´ə bəl] *Agriculture.* of or relating to farm land that is plowed or that is suitable for plowing.

arable crop *Agriculture.* a crop that requires the land to be plowed and seeds to be sown each year.

2-araboketose see RIBULOSE.

aracanga *Materials.* the durable, hard, heavy wood of timber trees of the species *Aspidosperma* of South America; used in the construction of such objects as furniture, flooring, boat frames, and railroad crossties.

Araceae *Botany.* the arum family of monocotyledenous herbs, characterized by an inflorescence of numerous tiny flowers on a spadix sheathed by a large spathe.

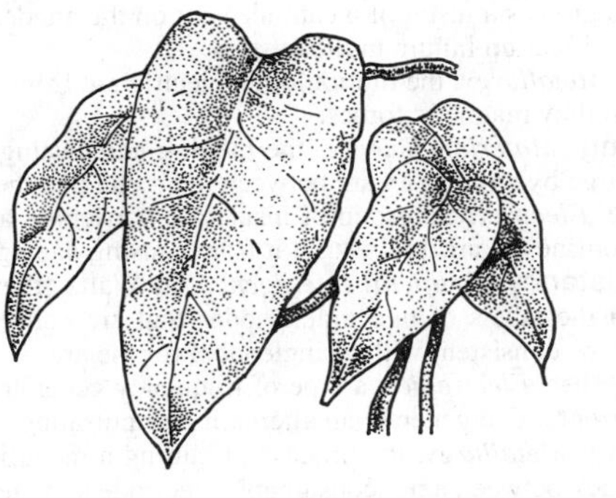

Araceae (philodendron)

arachic acid or **arachidic acid** see EICOSANOIC ACID.

arachidonic acid *Biochemistry.* $C_{20}H_{32}O_2$, an unsaturated fatty acid necessary in the human diet for its role in the production of regulatory substances such as leukotrienes, thromboxanes, and prostaglandins; found in animal fats such as liver and egg yolk. Also, **arachidonate.**

arachis oil see PEANUT OIL.

Arachnia *Bacteriology.* a genus of Gram-positive, faculatively anaerobic bacteria of the family Actinomycetaceae, occurring as irregularly shaped rods having a fermentative metabolism.

Arachniaceae *Mycology.* a family of fungi belonging to the order Lycoperdales which is characterized by its distinct hymenium.

arachnid [ə rak´nid] *Invertebrate Zoology.* **1.** an organism belonging to the class Arachnida. **2.** of or relating to this class.

Arachnida *Invertebrate Zoology.* a class of arthropods allied to insects and crustaceans, but with eight legs, no wings or antennae, two body regions, and a breathing mechanism of tracheal tubes or pulmonary sacs; the class includes spiders, mites, ticks, daddy longlegs, and scorpions.

arachnidism *Toxicology.* poisoning resulting from an arachnid bite, especially a spider bite. Also, ARANEISM.

arachno- or **arach-** a combining form meaning: **1.** of or relating to a spider or other arachnid. **2.** resembling a spider's web.

arachnodactyly see MARFAN'S SYNDROME.

arachnoid *Invertebrate Zoology.* **1.** pertaining to or resembling an arachnid. **2.** resembling a spider's web. *Anatomy.* **1.** weblike. **2.** the arachnoid membrane. *Botany.* formed of or covered with long, delicate, entangled hairs or fibers, giving a cobwebby appearance.

arachnoidal granulations *Anatomy.* small elevations, thought to be enlargements of arachnoid villi, that project into the superior sagittal sinus and create slight depressions on the inner surfaces of the brain; the structures through which cerebrospinal fluid is reabsorbed into the blood. Also, PACCHIONIAN BODIES.

arachnoidal villi *Anatomy.* numerous microscopic projections of the arachnoid membrane into some venous sinuses. Also, **arachnoid villi.**

Arachnoidea *Invertebrate Zoology.* a subclass of Arachnida.

arachnoiditis *Neurology.* an inflammation of the arachnoid membrane. Also, **arachnitis.**

arachnoid membrane *Anatomy.* a delicate membrane that lies between the dura mater and pia mater and envelops the brain and spinal cord; the middle of the three meninges.

arachnology *Invertebrate Zoology.* the study of spiders and other arachnids.

arachnophobia *Psychology.* an irrational fear of spiders.

Aradidae *Invertebrate Zoology.* fungus bugs or flat bugs, a family of small, flat, narrow-headed bugs with suckers, generally found under bark where most feed on fungus.

Aradoidea *Invertebrate Zoology.* a superfamily of hemipteran, mostly fungus feeding insects, true bugs with conspicuous antennae.

Araeoscelidia *Paleontology.* an artificial grouping of euryapsid tetrapods of the Permian and Triassic; thought to be ancestral to the nothosaurs and plesiosaurs; sometimes called Proterosauria; neither classification is now accepted, but still *incertae sedis.*

araeostyle *Architecture.* an intercolumniation of four or more diameters. Also, **areostyle.**

Arago, Dominique Francois 1786–1853, French physicist; discovered magnetism of rotation and basic laws of light polarization.

Arago distance [a´ rə gō] *Astronomy.* the angular distance between the antisolar point and the nearest Arago point.

aragonite *Mineralogy.* $CaCO_3$, a white to pale yellow orthorhombic mineral with a vitreous luster, having a specific gravity of 2.95 and a hardness of 3.5 to 4 on the Mohs scale; found in limestone caverns and near hot springs and geysers; trimorphous with calcite and vaterite.

Aragon spar see ARAGONITE.

Arago point *Optics.* a bright spot that appears in the center of the shadow of a circular disk illuminated by a point source, due to diffraction. Also, **Arago spot.** *Astronomy.* one of three points on the sky where the diffuse sky radiation is unpolarized.

Arago's disk *Electromagnetism.* an electrically conductive disk that, when rotated in a horizontal plane about a vertical axis, can induce a bar magnet suspended above the disk to rotate in a likewise direction due to the mutual interaction between the magnetic field and the eddy currents generated by the disk.

Arales *Botany.* an order of monocots that includes the families Araceae and Lemnaceae; characterized by radical or cauline alternate leaves, flowers borne on a spadix, and fruit that is usually a berry.

Araliaceae *Botany.* the ginseng family of herbs, shrubs, and trees; characterized by alternate lobed to bipinnate toothed leaves, dense clusters of small whitish or greenish flowers in umbels, and berrylike fruit.

aralkyl *Organic Chemistry.* an organic group in which an alkyl hydrogen atom is replaced by an aryl group.

Aral Sea *Geography.* a large, slightly saline lake in the southwestern USSR, east of the Caspian Sea.

aramayoite *Mineralogy.* $Ag(Sb,Bi)S_2$, an iron-black, triclinic, pseudotetragonal mineral having a specific gravity of 5.6 and a hardness of 2.5 on the Mohs scale; found with tetrahedrite, stannite, and pyrite.

aramid *Materials.* any of a class of strong, flame-retardant synthetic materials composed of long-chain polyamides, which are frequently used in the manufacturing of such products as cables, ropes, and protective clothing. Thus, **aramid fibers.**

Aramidae *Vertebrate Zoology.* a family of cranelike birds in the order Gruiformes, containing only one species, the limpkin or courlan.

Araneae *Invertebrate Zoology.* commonly called spiders, the largest of all arachnid orders, containing over 40,000 known species distributed worldwide.

Araneida see ARANEAE.

araneism *Toxicology.* see ARACHNIDISM.

araneology *Invertebrate Zoology.* the study of spiders.

araneose see ARACHNOID.

Araphidineae *Botany.* a suborder of pennate diatoms of the order Pennales, characterized by a lack of a raphe.

arapahite [ə rap´ə hīt] *Petrology.* a dark-colored, porous, fine-grained basalt composed of magnetite, bytownite, and pale green augite.

Araqil see AQUILA.

Ararat, Mount *Geography.* a mountain in eastern Turkey (16,946 ft); traditionally regarded as the landfall of Noah's Ark.

Aratus of Soli c. 315–245 BC, Greek didactic poet; stimulated Greek and Roman interest in astronomy.

Araucariaceae *Botany.* a family of tall evergreen trees of the class Pinatae; characterized by symmetrically whorled branching, spirally arranged leaves, and large and woody pistillate cones; an important timber and hard resin source.

Arbacioida *Invertebrate Zoology.* an order of sea urchins, in the class Echinoidea, that is primarily herbivorous; wide-ranging distribution.

Arber, Werner born 1929, Swiss molecular biologist; shared Nobel Prize for discovery and study of restriction enzymes.

Arber's law *Evolution.* a law stating that any structure that disappears from a phylogenetic lineage during the course of evolution is never regained by descendants of that line.

arbiter *Computer Technology.* a computer device that sets the priority sequence for two or more devices competing for access to a single computer resource, such as memory or the central processor.

arbitrary access *Computer Technology.* equal access to all memory locations, without regard for the location of the previous reference.

arbitrary level *Archaeology.* an excavation that is dug by prescribed levels, such as ten-centimeter levels, rather than by following the natural layers of rock or earth.

arbitrary map projection see APHYLACTIC MAP PROJECTION.

arbitrary precision arithmetic *Computer Science.* arithmetic, usually in terms of integer or rational values, that employs as many words of computer memory as necessary to represent exactly the value of a result.

arbitration bar *Metallurgy.* in casting, a bar cast from a sample of molten metal to determine whether its composition, hardness, and mechanical properties are within specifications.

arbor *Mechanical Engineering.* **1.** an axle or spindle on which cutting tools, such as boring tools, milling cutters, and reamers, are mounted. **2.** the driving shaft carrying the workpiece in a lathe that enables machining to be made on the work surface. *Metallurgy.* in casting, a metal component embedded in sand mold cores to support the sand or to support the load applied. *Horology.* in a timepiece, a shaft or spindle that functions as an axle for a wheel, pinion, or gear attached to it.

arboreal [är bôr´ē əl] *Botany.* of, relating to, characteristic of, or resembling a tree or trees. *Zoology.* pertaining to animals that live primarily in trees. Also, **arboricolous, arbicole.**

arboreous *Botany.* wooded, especially heavily wooded.

arborescence *Biology.* the state of being branched, or treelike, in structure, appearance, growth, or other properties.

arborescent *Botany.* like a tree in growth, structure, or appearance.

arboretum [är´bə rē´təm] *Botany.* a living collection of various trees and shrubs that are grown and usually displayed publicly for scientific investigation, education, and aesthetic enjoyment.

arbor hole *Design Engineering.* the hole in the center of a grinding or cutting wheel that allows the wheel to be mounted on an arbor.

arboricity *Mathematics.* the arboricity of a graph G is the smallest number of subsets that partition the edge set $E(G)$ such that each subset induces a forest. A theorem of Nash-Williams is that, if G is an undirected graph in which at least one edge is not a loop, then the arboricity of G is max $[|E(H)|/|v(H)|]-1$, where the maximum is taken over all nontrivial vertex-induced subgraphs of G and $[x]$ is the least–integer function.

arboriculture *Botany.* the cultivation of shade and ornamented trees and shrubs. *Forestry.* the growing and tending of trees and shrubs, especially for purposes of ornament or instruction, rather than for profit or for preservation of parkland.

arborization *Biology.* a branching or treelike configuration, as in the bronchial tubes or at nerve dendrites.

arbor press *Mechanical Engineering.* a machine that forces arbors or mandrels into or out of drilled or bored parts before turning or grinding, by means of a screw press or hydraulic power. Also, MANDREL PRESS.

arbor vitae or **arbor vitae** [är´bər vī´tē] *Anatomy.* a treelike arrangement of white and gray nerve tissue that is seen in a median section of the cerebellum. *Botany.* any of several ornamental shrubs and timber-producing trees of the genus *Thuja,* family Cupressaceae; characterized by bright green leaves, overlapping scales, and small cones. (A Latin phrase meaning "tree of life.")

arbovirus *Virology.* any of a nontaxonomic group of viruses that are transmitted by arthropods and can replicate in either the host vertebrate or the arthropod vector. Also, ARTHROPOD-BORNE VIRUS.

Arbuckle orogeny *Geology.* a late Pennsylvanian crustal deformation that produced the Witchita and Arbuckle mountains of Oklahoma.

arbuscules *Mycology.* tuftlike haustorial growths occurring in certain fungi called mycorrhiza.

arbutin *Organic Chemistry.* $C_{12}H_{16}O_7$, a colorless to white crystalline solid that melts at 199°C, and is soluble in water and alcohol; derived from the cranberry, blueberry, most types of pear, and other plants, and also prepared synthetically; used as a urinary antiseptic. Also, HYDROQUINONE-D-GLUCOPYRANOSIDE.

Arbutus *Botany.* a genus of shrubs and small trees, such as *A. unedo,* the strawberry tree of Ireland and southern Europe, or *A. menziesii,* the madrone tree of western North America.

arc *Mathematics.* **1.** a continuous portion of a circle or other curve. **2.** either of the two parts of a circle cut off by two points on the circle. The longer arc is the **major arc** and the shorter is the **minor arc. 3.** in graph theory, a directed edge of a graph. *Navigation.* the graduated scale of an angle-measuring instrument, such as a sextant. *Electricity.* a luminous, sustained discharge of electricity across an insulating medium. *Geology.* a line of islands, mountains, or other landforms that lie along a great curve. *Robotics.* the trajectory of a robot.

ARC or **arc** see AIDS-RELATED COMPLEX.

arcade *Invertebrate Zoology.* a type of cell found in the pharynx of nematodes, joined to similar cells by an arching structure. *Architecture.* **1.** a row of arches supported by columns. **2.** a covered walkway lined with such arches on one or both sides.

arcanite *Mineralogy.* K_2SO_4, an orthorhombic mineral whose chemical, optical, and crystallographic properties are closely related to those of mascagnite.

Arcanobacterium *Bacteriology.* a genus of bacteria of the order Actinomycetales that are capable of growth under both anaerobic and aerobic conditions as irregular rod-shaped cells.

arcature *Architecture.* **1.** a blind arcade. **2.** a small arcade.

arcback *Electronics.* in a gas tube, reverse current flow in the anode-cathode region as a result of a cathode spot on the anode. This condition causes rectification failure in the tube.

arc blow *Metallurgy.* the motion of an electric arc from its normal location, caused by magnetic forces.

arc casting *Materials Science.* the melting and casting of a powdered metal charge by striking an arc between the metal and the mold wall.

arc chute *Electricity.* a group of insulating barriers in a circuit breaker used to confine the arc and restrict it from causing damage.

arc consistency *Artificial Intelligence.* in Waltz filtering, a requirement that the labels of two graph nodes that are connected by a constraint arc be consistent with a single label for the arc.

arc converter *Electronics.* a type of frequency generator that utilizes a direct-current arc to generate an alternating or pulsating current.

arc cutting *Metallurgy.* the process of cutting a metallic material with an arc struck between a nonconsumable electrode and the workpiece.

arc discharge *Electricity.* an arc between anode and cathode of a gas-filled tube.

arc-discharge method *Materials Science.* a way of growing ceramic single crystals, in which an arc is struck between a single crystal anode and a source cathode of the same material, resulting in anode growth.

arc doubleau [ärk´ də blō´] *Architecture.* a massive arch used to support a vault across a wide space.

Arcella *Invertebrate Zoology.* a genus of Arcellidae (or Testacida), protozoans resembling amoebas with chitinous umbrellalike shells; found on decaying plants in freshwater.

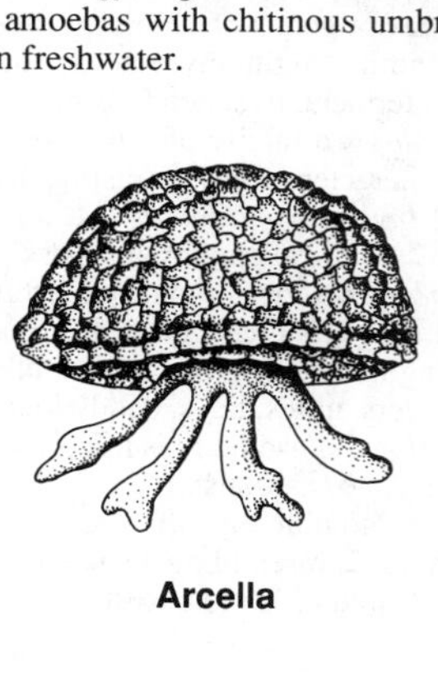

Arcella

Arcellinida *Invertebrate Zoology.* an order of protozoa that usually has a rigid test with a single opening; primitive freshwater organisms.
arc cosecant, arc cosine, arc cotangent see ARC FUNCTION.
arc excitation *Atomic Physics.* a process in which current from an electric arc is used to boost electrons into a higher energy state.
arc function *Mathematics.* **1.** an inverse function of a given function in the family of trigonometric functions. The arc function of a number is an angle whose (trigonometric) function is the given number, where the function is cosine, sine, secant, cosecant, tangent, or cotangent. Also, INVERSE TRIGONOMETRIC FUNCTION. **2.** in order to have an arc function be single-valued, it is customary to restrict the function to a particular standard range. The resulting value is called the *principal value* and is distinguished by capitalizing the "a" in "arc."

arc function	abbreviation	principal value interval
Arc sine	Arc sin x or $\text{Sin}^{-1}x$	$[-\pi/2, +\pi/2]$
Arc cosecant	Arc csc x or $\text{Csc}^{-1}x$	$[-\pi/2, 0)$ u $(0, +\pi/2]$
Arc cosine	Arc cos x or $\text{Cos}^{-1}x$	$[0, \pi]$
Arc secant	Arc sec x or $\text{Sec}^{-1}x$	$[0, \pi/2)$ u $(\pi/2, \pi]$
Arc tangent	Arc tan x or $\text{Tan}^{-1}x$	$(-\pi/2, +\pi/2)$
Arc cotangent	Arc cot x or $\text{Cot}^{-1}x$	$(0, \pi)$

arc furnace *Metallurgy.* a melting furnace in which the charge is heated by an electric arc struck either between an electrode and the charge, or between two electrodes near the charge.
arch *Civil Engineering.* a curved structure designed to exert horizontal forces on its supports when subjected to vertical loads.
arch- a combining form meaning "first, principal" or "most basic."
Archaebacteria *Bacteriology.* a categorization of bacteria of the division Mendosicutes, typically found in unusual environments and distinguished from the rest of the procaryotes by several criteria, including the number of ribosomal proteins and the lack of muramic acid in cell walls. Also, **Archaeobacteria.**
archaeo- **1.** ancient; original; primitive. **2.** of or relating to archaeology. (From a Greek word meaning "ancient" or "beginning.")
archaeoastronomy *Archaeology.* the study of evidence of prehistoric knowledge of astronomical events, such as calendars, observatory sites, and astronomical images in rock art.
archaeocete *Paleontology.* of or relating to the ancestral whales, the extinct suborder Archaeoceti, which were fish-eating carnivores about the size of modern porpoises.
Archaeoceti *Paleontology.* an extinct suborder of fish-eating whale-like mammals in the order Cetacea; among the first ancestral whales to return to the ocean; characterized by peglike teeth, an elongated snout, nostrils on top of the head, and nonprotruding vestigial hind legs; Eocene to Oligocene.
Archaeocidaridae *Paleontology.* a family of Paleozoic echinoids in the long-ranged order Cidaroida; characterized by a flexible test and from two to ten columns of plates; large primary as well as secondary tubercles; Devonian to Permian.
Archaeocidaris *Paleontology.* a genus of endocyclic echinoids in the extinct order Cidaroida and family Archaeocidaridae; Carboniferous.
Archaeocopida *Paleontology.* an order of thin-shelled ostracods, including about 20 genera; characterized by a primarily chitinous shell, only weakly calcified or phosphatized; probably ancestral to all later ostracods, but otherwise of uncertain affinity; Cambrian to Ordovician.
Archaeocyatha *Paleontology.* an extinct phylum of primitive, sponge-like, reef-building marine organisms, they were generally conical and had perforated calcareous walls; their skeletons' height was generally between 20 and 150 mm, with some giant individuals even taller; they are among the earliest known reef animals, invading stromatolite reefs in the lower Cambrian; their ancestors are unknown but are presumed to be protozoans; after their fossils disappear in the middle Cambrian, there is no evidence of any reef animal life for about 60 million years.
archaeogastropod *Paleontology.* relating to the Archaeogastropoda, a large order of prosobranch gastropods characterized by bipectinate ctenidia and a nonsiphonate shell; they arose in the Cambrian and became widespread and diverse in the middle Paleozoic, and are represented today by about half of their Paleozoic families and genera.
Archaeogastropoda *Invertebrate Zoology.* an order of primitive gastropods including topshells, abalones, and limpets, usually found on rocky shores; algae feeders.
archaeological of or relating to archaeology.
archaeological chemistry *Archaeology.* the use of chemical processes to obtain archaeological data; includes laboratory analysis of materials or objects and the removal of surface deposits on artifacts.
archaeological context *Archaeology.* the status of an artifact or feature that no longer is in active use by humans and exists as part of the archaeological record.
archaeological culture *Archaeology.* the artifacts that are typical of a specific region at a particular time.
archaeological recovery see RECOVERY.
archaeological sequence *Archaeology.* the placement of a group of similar objects in a chronological sequence, according to stylistic changes that have occurred over time.
archaeologist a person who specializes in or studies archaeology.
archaeology the scientific study of materials and objects that remain as evidence of the life and culture of prehistoric or ancient peoples, such as artifacts, structures, and settlements.

Archaeology

Archaeology is a science that has as its goal the reconstruction of culture history before there was a written record. In the United States, archaeology is taught in universities in the Anthropology Department along with its related fields of Physical Anthropology, Linguistics, and Ethnology. Today a qualified archaeologist normally has a doctorate most often based on experience in the field, for there is no substitute for the shovel, paintbrush, trowel, and tweezers.

A major excavation today is an interdisciplinary undertaking using anthropologists as well as specialists in a variety of fields such as agronomy, zoology, botany, and geology. A career path in archaeology usually leads to teaching at a university that provides the opportunity for fieldwork in the summer or on sabbaticals. Research institutions and museums seek archaeologists as curators for their collections and also sponsor field expeditions.

The establishment of chronology is a prime concern for understanding spatial and temporal interrelationships. To this end, stratigraphy and seriation are long-standing reliable techniques of investigation. Other methods used with success are tree-ring counts, obsidian hydration, and fluorine analysis. The greatest boon for archaeology since 1949 has been radiocarbon dating, employing a variety of organic materials. In the last few decades aerial photography and satellite images have been used to locate archaeological sites and detect ancient raised field agriculture. Certain geophysical technologies have successfully identified structures beneath the surface, and in Egypt cosmic radiation detectors have been helpful in finding chambers in the Giza pyramids.

Computers too have a role, especially for handling quantitative and statistical analyses. The potential is there for the formulation of databases to store and share material that will be more easily used than the traditional site reports.

Archaeology is not without unique problems. Faking and looting continue unabated as huge monuments are broken and carted away and murals are wrenched from walls to find their way to the ever-present marketplace, thus leaving an irreplaceable void in the history of a site. But the future is bright with conferences swelling with participants and the increasing pride of the public in its heritage. With new discoveries in the field of technology for exploration and dating, archaeology is a growing and expanding science.

Muriel Porter-Weaver
Shelter Island Heights, New York
Author of *The Aztecs, The Maya, and Their Predecessors*

archaeomagnetic dating *Archaeology.* a method of dating ancient objects containing magnetic materials; used mainly with stationary ceramic objects such as earthen fireplaces, kilns, and brick walls.

archaeomagnetism *Geophysics.* the principle that the earth's magnetic field has changed over time; this permits the dating of certain ancient materials by comparing their magnetic orientation to the current local magnetic orientation.

archaeometric *Archaeology.* relating to or denoting the use of analytical scientific techniques from fields such as chemistry, geology, and physics for the analysis of archaeological data.

Archaeopteridales *Paleontology.* an order of pteridophytes in the class Progymnospermopsida; the archaeopterids were primitive plants that displayed many characteristics of gymnosperms; they arose in the middle Devonian and persisted until the early Carboniferous.

Archaeopteris *Paleontology.* a genus of Devonian plants in the order Archaeopteridales that is one of the best known genera of its time, having been widely studied since 1871; it was significant in the development of the higher plants, especially because of its heterospory and its gymnospermic secondary wood.

Archaeopterygiformes *Paleontology.* the order of birds that is represented by *Archaeopteryx.*

Archaeopteryx *Paleontology.* a genus of primitive birds known only from specimens found in 1861 at the Solnhofen limestone quarry in Germany, along with the bird-like dinosaur *Compsognathus*; the size of a large pigeon, *Archaeopteryx* was probably a weak flyer; its feathers are its only strictly avian characteristic; upper Jurassic.

Archaeornithes *Paleontology.* the subclass of birds that is represented by *Archaeopteryx.*

Archaeosphaerodiniopsidaceae *Botany.* a rare monotypic family of marine planktonic dinoflagellates of the order Peridiniales, thought to be based on the encysted stage of another dinoflagellate.

Archaeothyris *Paleontology.* a genus of lizard-like synapsids in the extinct order Pelycosauria and family Ophiacodontidae; one of the earliest known pelycosaurs, *Archaeothyris* was less than two feet long and resembled a small, short-legged iguana; upper Carboniferous.

archaic *Archaeology.* in New World chronology, a period during which there was a shift from hunting and gathering to agriculture; characterized by the use of food-processing tools and the beginnings of village organization. Also, **Archaic.** *Psychology.* of or relating to elements in the psyche, usually unconscious, that endure from humans' prehistoric experience and that may reappear in dreams and other phenomena.

archangel see ANGELICA.

Archangiaceae *Bacteriology.* a family of gliding soil bacteria of the order Myxococcales that grow vegetatively as rods, produce miocysts, and develop fruiting bodies.

Archangium *Bacteriology.* a genus of Gram-negative, gliding bacteria of the family Archangiaceae that occur in soil and decaying vegetation and produce fruiting bodies.

Archanodon *Paleontology.* a genus of Devonian freshwater bivalves.

Archanthropinae *Paleontology.* an anthropological subdivision of hominoids in the direct line of descent to *Homo sapiens;* this includes *Pithecanthropus* and related forms.

arch bar *Building Engineering.* **1.** a curved bar in a window sash. **2.** a curved chimney bar. **3.** a bar supporting brickwork over an opening.

arch beam *Civil Engineering.* a beam curved in the vertical plane, used to support heavy loads over long spans.

arch bridge *Civil Engineering.* a bridge that uses long span arches for support.

arch bridge

arch center *Civil Engineering.* a temporary structure used to support masonry or concrete arches during construction.

arch corner bead *Building Engineering.* a corner bead fabricated on location in bridge construction; used to reinforce the curved area of arch openings.

arch dam *Civil Engineering.* a dam that utilizes an arched shape in the horizontal plane to transfer the forces of the retained water to the sides of a canyon.

arche- a combining form meaning "first, original," as in *archetype.*

Archean *Geology.* of or relating to the Archeozoic or its rocks. Also, **Archaean.**

arc heating *Metallurgy.* the process of heating a material by using the energy of an electric arc.

arched construction *Building Engineering.* construction using arches and vaults rather than beams and lintels for support.

archegoniophore *Botany.* in bryophytes and ferns, the specialized structure that bears archegonia.

archegonium *Botany.* **1.** a female gamete-producing structure of the Bryophyta; a flask-shaped structure consisting of an elongated neck and a swollen venter that produces a single egg and in which fertilization and development of the sporophyte occurs. **2.** a female gamete-producing structure of the Pteridophyta, borne on the prothallus, in which fertilization and development of the embryonic sporophyte occurs.

archencephalon *Developmental Biology.* the anterior portion of the embryonic brain from which the forebrain develops.

archenteron *Developmental Biology.* the primitive gastric cavity of the gastrula. Also, GASTROCOELE, PRIMITIVE GUT, PRIMORDIAL GUT.

archeo- a combining form meaning "ancient," as in *archeology.*

archeocortex see ARCHIPALLIUM.

archeocyte *Invertebrate Zoology.* an undifferentiated itinerant cell, as in the sponge embryo, that can develop into more specialized cell types; useful in reproduction, regeneration, or repair. Also, **archaeocyte.**

archeological see ARCHAEOLOGICAL and related words.

archeology see ARCHAEOLOGY and related words.

Archeozoic *Geology.* early Precambrian geologic time. Also, **Archaeozoic.**

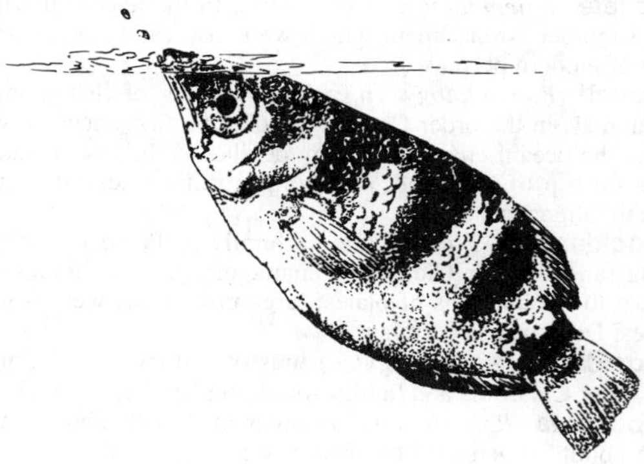

archer

archer *Vertebrate Zoology.* a small freshwater fish that shoots a stream of water out of its mouth to kill insects for food; any member of the family Toxotidae in the order Perciformes. Also, **archerfish.**

Archer *Astronomy.* see SAGITTARIUS. *Ordnance.* a Soviet air-to-air missile, officially designated **AA-11.**

Archeria *Paleontology.* a genus of primitive fish-eating labyrinthodont amphibians in the extinct suborder Embolomeri and family Archeriidae; Lower Permian.

archespore *Botany.* the cell or group of cells that give rise to a spore mother cell, from which spores are ultimately derived.

archetype [är´kə tīp´] *Evolution.* a hypothetical ancestral form of a given organism, developed by systematically eliminating complexity and specialization from that organism. *Psychology.* Carl Jung's term for the structural part of the collective unconscious, directing one's perception of the world and the development of one's personality.

arch filament *Astronomy.* a solar prominence that connects regions of opposite magnetic polarity.

arch girder *Civil Engineering.* a normal steel girder bent in a curved shape.

arch-gravity dam *Civil Engineering.* an arch dam whose large mass provides stability due to gravity.

archi- a combining form meaning "first, principal" or "most basic," as in *architect, archinephridium.*

Archiacanthocephala *Invertebrate Zoology.* an order of spiny-headed worms in the phylum Acanthocephala; parasites in terrestrial birds and mammals.

Archiannelida *Invertebrate Zoology.* a small group of marine worms living primarily in crevices in a seabed; one species lives in underground water caverns and some forms are parasitic.

archibenthic zone *Oceanography.* the upper part of the benthic region, including the continental shelf and extending from the sublittoral zone to the abyssal zone.

archiblast *Biology.* the parts of an ovum that actively form an embryo, as distinguished from the yolk.

archicarp *Mycology.* the part of certain fungi that develops into its fruiting body.

archicarpous *Mycology.* relating to or being an archicarp.

Archichlamydeae *Botany.* a group of dicots in which the petals of the flowers are separate or absent.

archicoel *Zoology.* the primary space between the ectoderm and the alimentary canal in the development of certain animals.

Archidiales *Botany.* an almost cosmopolitan order of mosses of the class Bryophyta, comprising the single family Archidiaceae and characterized by small, delicate plants with a unique sporophyte developmental pattern; these plants grow as perennials or emphemerals on sandy soil.

archidictyon *Entomology.* an irregular network of wing venation that is evident in primitive fossil insects and present-day mayflies and dragonflies.

archigastrula *Developmental Biology.* the gastrula in its most basic developmental stage.

Archigregarinida *Invertebrate Zoology.* a small order of primitive, parasitic gregarines without septa, found in the intestines of marine invertebrates and lower chordates.

Archimedean [är′kə mē′dē ən] *Science.* of or relating to Archimedes or his methods, discoveries, or principles. *Mathematics.* an ordered field F is said to be Archimedean over a subfield K (with the induced ordering inherited from F) if no element a of F exists such that $|a| \geq x$ for all x in K. (Equivalently, K satisfies Archimedes' axiom as a subfield of F.) An intermediate field F_1, with $F \supset F_1 \supset K$, is said to be maximal Archimedian over K in F if it is Archimedean over K and no other intermediate field containing F_1 is Archimedean over K. K is maximal Archimedean in F if it is maximal Archimedean over itself in F. K may also be called an **Archimedean ordered field.**

Archimedean principle *Physics.* see ARCHIMEDES' PRINCIPLE.

Archimedean solid *Mathematics.* one of thirteen possible solid figures whose faces are all regular polygons (not necessarily of the same type) and whose polyhedral angles are all congruent. Sometimes called semiregular solids to distinguish them from the Platonic solids, whose faces are all congruent.

Archimedean spiral *Mathematics.* the curve spiraling into the origin whose polar equation is $r = a\theta$. Geometrically, it is the plane curve that is the locus of a point moving at a constant velocity v, starting at the origin, along the radius vector while the radius vector moves about the origin with uniform angular velocity ω, and $a = v/\omega > 0$.

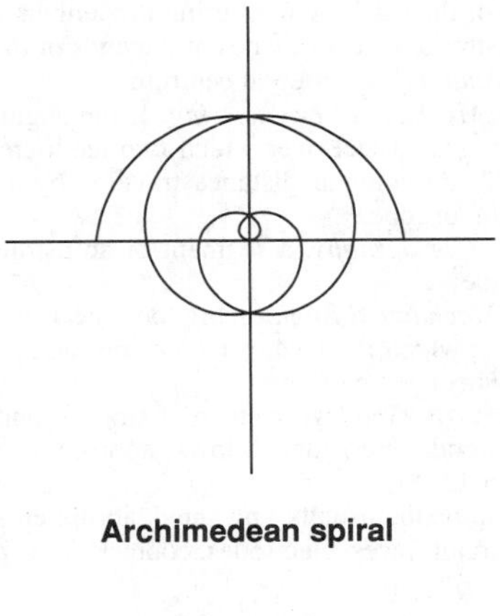

Archimedean spiral

Archimedes [är′kə mē′dēz] c. 287–212 BC, Greek mathematician; discovered basic laws of hydrostatics and of lever and pulley; invented the catapult and Archimedean screw; his mathematical discoveries included the doctrine of limits.

Archimedes

Archimedes' axiom *Mathematics.* if x is any real number and $>$ is a linear ordering, there exists an integer n such that $n > x$.

Archimedes' method *Materials Science.* a technique for measuring the apparent density of a particle by measuring its mass and volume while it is suspended in a liquid of known density; this is based on Archimedes' principle.

Archimedes' number *Fluid Mechanics.* a dimensionless group that expresses the ratio of viscous force to gravitational force.

Archimedes' principle *Physics.* the explanation by Archimedes of the principle of buoyancy. He noted that an object placed in water apparently loses an amount of weight equal to the weight of the water displaced by the object; this led to the statement that a body immersed in a fluid is buoyed up by a force equal to the weight of the fluid it disperses.

Archimedes' problem *Mathematics.* the problem of dividing a hemisphere into two sections of equal volume, using a plane parallel to the base of the hemisphere. This problem is not solvable using Euclidean methods, i.e., using a compass and straightedge.

Archimedes' screw *Mechanical Devices.* a device conceived by Archimedes to raise water against the force of gravity, by means of an inclined cylinder containing a well-fitted screw that can be rotated to lift the water upward.

archinephridium *Invertebrate Zoology.* the excretory organ of certain annelid larvae.

archinephros *Vertebrate Zoology.* an excretory organ of primitive vertebrates, consisting of long, segmented tubules. Also, HOLONEPHROS.

arching *Geology.* the folding of rocks or sediments into anticlines. *Mining Engineering.* **1.** a curved support for the roof of an opening in a mine. **2.** the development of peripheral cracks around an excavation due to the difference in stress between the skin rock and the rock in the stress ring. *Civil Engineering.* the bridging or transferring of shear stress from a soil of low shear strength to an adjacent area of soil with higher shear resistance.

archipallium *Anatomy.* a portion of the cerebral cortex having little or no lamination and constituting one-twelfth of the cortical area corresponding to the olfactory lobe. Also, ARCHEOCORTEX.

archipelagic apron *Geology.* a smooth, gently sloping, fanlike bank or cone that surrounds a volcanic island or a seamount.

archipelago [är′kə pel′ə gō] *Geography.* **1.** a cluster or chain of islands. **2.** an area of an ocean or sea containing many islands.

archistome see BLASTOPORE.

architect *Architecture.* a licensed professional trained in the art of designing buildings and overseeing their construction.
architectonic *Geology.* of or relating to forces that determine the structure of a landform. *Architecture.* of or conforming to the technical principles of architecture.
architect's scale *Graphic Arts.* a ruler marked with a scale that converts the dimensions of a reduced-size drawing to the dimensions of the actual object drawn.
architectural acoustics *Acoustical Engineering.* **1.** the process or technology of designing a room or building to provide for optimal sound conditions within it. **2.** specifically, the design of architectural spaces for recording and broadcasting and for musical and theatrical productions.
architectural bronze *Metallurgy.* a wrought copper-base alloy containing about 2–3% lead and 37–46% zinc.
architectural concrete *Materials.* a term for concrete that is used for finishing or decoration on a building surface, as opposed to a structural use.
architectural engineering *Civil Engineering.* engineering involved with the materials, components, and design of the structural systems of buildings, as opposed to heavy construction such as bridges or dams.
architectural millwork *Civil Engineering.* millwork fabricated to meet the specifications of a particular job, in contrast to stock millwork.
architectural photogrammetry *Photogrammetry.* the use of photogrammetry for the documentation of specific information regarding the sizes and shapes of historic buildings and monuments.
architectural volume *Civil Engineering.* the volume of a building calculated by multiplying the floor area by the building height.
architecture **1.** the art and science of designing and constructing buildings. **2.** the buildings of a particular place or time period.

Architecture

Although the word *architect* comes from ancient Greek roots meaning "master builder," it was adopted to describe not an artisan but rather someone responsible for the design of a building or place. The words *architect* and *architecture* did not appear in English until the mid-16th century, and while building and the making of monuments is an ancient human activity, these words reflect a renaissance idea that reason can be applied on a very large scale.

Many of the earliest uses of the word refer to God as the original architect, and they reflect a confidence that His creation can be understood, like any other well-drawn set of plans. It is worth noting that the name for the field derives from the word for the practitioner; architecture is what architects do. There have been master builders throughout history—and we retrospectively label what they built architecture—but the word *architect* emerged only as individual creativity began to be generally celebrated. One possible definition of architecture is that it is building with ego.

Long-standing uncertainty about how to classify architecture is reflected in the Oxford English Dictionary's first definition: "The art or science of building or constructing edifices of any kind for human use." In 1848, the English critic John Ruskin wrote that architecture unites "the technological and imaginative elements as essentially as humanity does the soul and body," though he deplored what he saw as a materialist bias. In fact, since Ruskin's time, technological innovation in construction has been primarily the province of engineers, while architects have been more concerned with how people understand, experience, and use buildings.

Thomas Hine
Architecture Critic, *The Philadelphia Inquirer*
Author of *Populuxe* and *Facing Tomorrow*

Architeuthis *Invertebrate Zoology.* a genus of giant squid, deepwater cephalopod mollusks up to 55 feet in length including body and tentacles; the largest invertebrate.
architrave [är′kə trāv′] *Architecture.* **1.** the lowest section of an entablature, resting directly on the capitals of two or more columns. **2.** a set of decorative moldings around a doorway, window, or other opening.
architrave cornice *Architecture.* an entablature with an architrave and a cornice but no frieze.
archival storage *Computer Programming.* storage media, such as offline magnetic tapes or disks, used for long-term storage of data seldom accessed, such as transaction history or backup copies of working programs.
archive or **archives** *Computer Programming.* a storage location separate from the main location, used for long-term storage.
archiving *Computer Programming.* the process of storing data or programs in long-term storage such as microfiche, magnetic tape or disk, or optical disk.
arch of the aorta see AORTIC ARCH.
Archonta *Paleontology.* a superorder proposed to include primates with bats, tree shrews, and dermopterans; the evidence for such a relationship has not yet been considered compelling enough to make such a major reclassification.
Archosauria *Vertebrate Zoology.* the "ruling lizards"; the superorder of advanced diapsids that includes the modern crocodiles as well as the thecodonts, pterosaurs, and dinosaurs; the archosaurs first appeared in the earliest Triassic.
archosaurs *Paleontology.* a general term for the diapsids of the superorder Archosauria.
Archostemata *Invertebrate Zoology.* a small suborder of beetles with about 30 species worldwide; the larvae are specialized wood feeders; fossils of some species date to the Permian age.
arch pattern *Anatomy.* a distinctive pattern of ridges at the fingertip; used to classify fingerprints.
arch press *Mechanical Engineering.* a machine used for forming metals or punching holes in metals by means of a punch, moved by a ram with an arch-shaped frame to allow work on wide pieces.
arch rib *Civil Engineering.* a protruding molded member subdividing the underside of an arch.
arch ring *Civil Engineering.* the part of an arch that carries the load.
arch truss *Civil Engineering.* a truss utilizing the shape of an arch or series of arches.
arc hyperbolic function *Mathematics.* an inverse function of one of the hyperbolic functions; usually called inverse hyperbolic function.
arcing contacts *Electricity.* special contacts between which an arc occurs when they are separated.
arcing ring *Electricity.* a metal ring attached to an insulator for protection from a power arc.
arcing time *Electricity.* **1.** the interval between the parting of arcing contacts and the extension of the arc. **2.** the interval between the severance of the fuse link to the final interruption of the circuit under a specified condition.
arc jet engine *Aviation.* a rocket engine whose propellant gas is heated as it passes through an electric arc.
arc lamp *Electricity.* an electric lamp in which light is generated by an arc produced when current flows through ionized gas between two electrodes.
Ar cm *Metallurgy.* in a hypereutectoid steel, the temperature during cooling at which the iron carbide begins to separate.
arc measurement *Geodesy.* a surveying method used to determine the size and figure of the earth by measuring the lengths of arcs of triangulation and the astronomic coordinates of the ends of the arc.
arcocentrum *Anatomy.* a vertebral centrum.
arc of contact *Mechanical Engineering.* **1.** the angular distance on the pitch circle of a gear wheel over which two teeth are in contact. Also, **arc of action. 2.** the angular distance traveled by a pulley in contact with a belt, chain, or rope.
arc of parallel *Cartography.* a segment of an astronomic or geodetic parallel of latitude.
arc of recess *Mechanical Engineering.* on a gear wheel, the arc on the pitch circle along which the face of the driving wheel is in contact with the side of the wheel being driven.
arc of triangulation *Geodesy.* a chain of single, connected figures (triangles, quadrilaterals, etc.) that follows, approximately, an arc on the reference ellipsoid.
arc-over *Electricity.* the usually unwanted, abrupt creation of an arc between printed circuit traces, electrodes, contacts, or capacitive plates.

arc resistance *Electricity.* the ability of a material, usually a dielectric, to resist formation of arcing; expressed as the ratio of the voltage that yields an arc discharge to the arc current.

arc secant see ARC FUNCTION.

arc spectrum *Spectroscopy.* the spectrum produced by a neutral atom as a result of vaporization within an electric arc generated between two electrodes.

arc-suppressor coil *Electricity.* a device, usually a diode, resistor-capacitor network, or coil, that extinguishes an arc discharge.

arc tangent see ARC FUNCTION.

arc-through *Electronics.* a condition in a gas tube, in which current flows in the forward direction during what should be the negative half cycle of the AC voltage.

Arctic [ärk´tik] *Geography.* **1. the Arctic.** the region of continuous cold surrounding the North Pole, usually regarded as including all areas within the Arctic Circle or all areas north of the tree line; thus it includes the northernmost parts of Alaska, Canada, Scandinavia, and the Soviet Union, as well as Greenland. **2.** also, **arctic.** found in, coming from, or associated with this region. Most compound entries using this term may be spelled either as Arctic or arctic.

arctic air *Meteorology.* an air mass, cold aloft and extending to great heights, whose characteristics develop mostly in winter over arctic surfaces of ice and snow.

arctic-alpine *Ecology.* describing areas above the timberline in mountainous northern regions.

arctic anticyclone see ARCTIC HIGH.

Arctic Archipelago *Geography.* a large group of islands in northeastern Canada between the Canadian mainland and Greenland.

Arctic Circle *Geodesy.* in the Northern Hemisphere, the parallel of latitude that is equal to the complement of the summer solstice; due to steady variations in the obliquity of the summer solstice, the position of the Arctic Circle is constantly changing. *Cartography.* a parallel of latitude representing the geodesic arctic circle, customarily drawn on maps at 66°33´N.

arctic climate see POLAR CLIMATE.

Arctic Convergence *Oceanography.* the ill-defined polar convergence zone of the Northern Hemisphere.

arctic desert see POLAR DESERT.

Arctic fox *Vertebrate Zoology. Alopex lagopus,* a carnivorous mammal of the dog family known for its thick, white winter coat as well as a gray phase; arctic adaptations include rounded ears, a short muzzle, and fur-covered footpads. Also, BLUE FOX, WHITE FOX, POLAR FOX.

Arctic fox

arctic front *Meteorology.* the semipermanent, semicontinuous front between deeper, colder arctic air and shallower, less cold polar air of northern latitudes.

arctic haze *Meteorology.* a condition in which visibility is reduced in horizontal and slant directions but remains unimpeded vertically; encountered by aircraft over arctic regions.

arctic high *Meteorology.* a weak high appearing on mean charts of sea-level pressure over the Arctic Basin during late spring, summer, and early autumn. Also, ARCTIC ANTICYCLONE, POLAR ANTICYCLONE, POLAR HIGH.

Arctic Intermediate Water *Oceanography.* the relatively small intermediate water masses of the North Pacific and North Atlantic, respectively located north of a line between the Aleutian Islands and Japan and north of a line between Iceland and Labrador.

arcticization [ärk´tik ə zā´shən] *Engineering.* the construction or treatment of machinery so it can be used in areas of very low temperatures.

arctic mist *Meteorology.* a light ice fog that appears as a mist of ice crystals.

Arctic Ocean *Geography.* the ocean that surrounds the North Pole, lying north of Europe, Asia, and North America.

arctic pack see POLAR ICE.

arctic sea smoke *Meteorology.* a steam fog rising from areas of open water within sea ice.

arctic smoke see STEAM FOG.

Arctic suite *Petrology.* a group of igneous rocks characterized by alkali and alkali-calcic rocks intermediate in composition between Atlantic and Pacific suites.

Arctic Zone see NORTHERN FRIGID ZONE.

Arctiidae *Invertebrate Zoology.* a family of moths, including tiger moths and ermine moths; some caterpillars appear woolly; the larvae feed mainly on lichens although a few species are injurious to plants.

arc time *Metallurgy.* the time an electric arc is maintained uninterrupted during the arc-welding process.

arc-to-chord correction see CONVERSION ANGLE.

Arctocyonidae *Paleontology.* a large family of primitive condylarths in the infraorder Caviomorpha; unspecialized, some genera being bear-sized and omnivorous, and others smaller and herbivorous; upper Cretaceous to Eocene.

Arctogea *Ecology.* the zoogeographical area including the Palaearctic, Nearctic, Ethiopian, and Oriental regions.

Arctolepis *Paleontology.* a genus of primitive arthrodire placoderms in the family Phlyctaeniidae; only about 20 cm long, *Arctolepis* was characterized by very strong pectoral fins, which were about 15 cm from tip to tip; extant in the Devonian.

arc triangulation *Engineering.* in surveying, a type of triangulation whereby two distant points are connected by tracing an arc of a large circle over the earth's surface.

Arcturidae *Invertebrate Zoology.* a blind or sighted species of Valvifera (class Crustacea) existing in seashore algae and on deep-sea organisms.

Arcturus [ärk´ tür´ əs] *Astronomy.* Alpha (α)-Boötes, a first-magnitude reddish star in the constellation Boötes, the Herdsman.

arcuale *Developmental Biology.* any of the eight paired, segmental cartilages from which a vertebra develops.

arcuate *Anatomy.* arched or curved like a bow.

arcuated *Architecture.* having arches.

arcuate delta *Geology.* a curved delta whose convex outer margin faces the body of water. Also, FAN-SHAPED DELTA.

arcus *Meteorology.* a dense, roll-shaped cloud with tattered edges, appearing as an accessory to the lower front part of a cumulonimbus cloud.

arcus palmaris profundus see DEEP PALMAR ARCH.

arcwall coal cutter *Mining Engineering.* an electric or compressed-air cutter designed for undercutting or overcutting a coal seam in narrow work.

arc-welder's disease see SIDEROSIS.

arc welding see ELECTRIC-ARC WELDING.

arcwise connected *Mathematics.* a topological space X is arcwise connected if, given any two points a and b in X, there exists a continuous path between them; i.e., there is a continuous map q: $[0,1] \to X$ such that $q(0) = a$ and $q(1) = b$. If X is arcwise connected, then it is also connected; the converse is not necessarily true. X can be arcwise connected without being locally arcwise connected, and vice versa.

arcwise connected set see CONNECTED SET.

Arcyria *Mycology.* a genus of fungi belonging to the class Myxomycetes which comprises slime molds.

ard *Archaeology.* an ancient plow form with a simple blade and share that scratches the surface of the soil rather than producing a furrow.

ARD see ACUTE RESPIRATORY DISEASE.

ardealite *Mineralogy.* $Ca_2(SO_4)(HPO_4){\cdot}4H_2O$, a white to pale yellow monoclinic mineral that occurs as fine-grained powdery masses, having a specific gravity of 2.3; found with gypsum in phosphate deposits.

Ardeidae *Vertebrate Zoology.* the herons, bitterns, and egrets, a widely distributed family of wading birds in the order Ciconiiformes.

Ardennian orogeny *Geology.* a brief, late Silurian rock deformation that occurred in western Europe.

ardennite *Mineralogy.* $Mn_4^{+2}(Al,Mg)_6(SiO_4)_2(Si_3O_{10})[(As,V)O_4](OH)_6$, a yellow to yellowish-brown orthorhombic mineral occurring as prismatic crystals, having a specific gravity of 3.62 to 3.7 and a hardness of 6 to 7 on the Mohs scale.

ARDS see ACUTE RESPIRATORY DISEASE SYNDROME.

are [âr; är] *plural,* **area** *Metrology.* a unit of measure for land, equal to 100 square meters or 119.6 square yards.

area *Mathematics.* a measure of the size of a surface or region. *Computer Programming.* an abstract representation of a region of secondary storage, the exact nature of which is determined by the system. *Military Science.* see THEATER.

area bombing *Ordnance.* a method of bombing in which bombs are dropped on a general area, with no attempt to hit a particular target or to bomb in a specific pattern.

area control *Navigation.* an air traffic control procedure in which aircraft are assigned to fly through a specified area, within which they are free to choose their specific flight paths.

area defect *Materials Science.* a two-dimensional material defect, such as the boundary between neighboring grains, a twin and the matrix, and other abrupt changes in the crystal structure.

area defense *Ordnance.* a defense against air attack that is intended to protect a general area rather than a specific point.

area detector *Crystallography.* an electronic device for measuring at one time the intensities of a large number of X-ray beams diffracted by a crystal. Such a device may, for example, involve a multiwire proportional counter coupled to an electronic device for recording the data in computer-readable form.

area excavation *Archaeology.* an excavation technique in which the full horizontal extent of a site is cleared while a gradual vertical probe is made.

area fire *Ordnance.* firing directed at a certain area rather than at an individual target.

area forecast *Meteorology.* a weather forecast for a specified geographical area. Also, REGIONAL FORECAST.

areal eruption *Volcanology.* a type of volcanic eruption caused by the collapse of the roof of a batholith.

areal geology *Geology.* the study of the structural features, surface forms, and stratigraphic units over a relatively large region.

area light *Civil Engineering.* a light source used to illuminate a significant area with large dimensions in two directions, such as a bay window.

arealization *Cell Biology.* the process by which cells are committed to particular pathways of differentiation based on their different locations in a developing organism.

areal pattern *Petroleum Engineering.* the distribution pattern of water- or gas-injection wells and oil-production wells in a specific oil reservoir.

areal velocity *Astrophysics.* the rate at which the radius vector of an object sweeps across a given area.

area meter *Engineering.* an instrument that uses a piston or float to measure the rate of fluid flow through an open channel in a designated area.

area navigation *Navigation.* a navigational method that allows aircraft to follow any desired course within specified limits.

area of error *Navigation.* the area within which a craft is located as determined by a fix or by an estimated position with a given degree of probable error.

area opaca *Developmental Biology.* the outer, opaque, portion of the blastodisk of a bird embryo.

area pellucida *Developmental Biology.* the central, clear portion of the blastodisk of a bird embryo.

area placentalis *Developmental Biology.* that part of the trophoblast that touches the uterine mucosa in the embryos of early placental vertebrates.

area redistribution *Physics.* a method of measuring an irregular pulse by constructing a rectangular pulse with the same peak amplitude and the same area on a graph of amplitude versus time.

area rule *Aviation.* an airfoil design that minimizes zero-lift drag at the aircraft's intended speed by distributing its total cross-sectional area along the direction of flight; also used in the design of particular aerodynamic configurations such as wing-body configuration.

area search *Computer Programming.* a computerized search to locate a set of files or records that belong to a specified category or area of interest.

area supplementary control *Electrical Engineering.* the control that is applied to area generator speed in order to maintain scheduled system frequency and established net interchanges when changes occur in system frequency or tie-line loading.

area target *Ordnance.* a target for gunfire or bombing that occupies a considerable area, such as a large factory complex or an airport, as opposed to a single building or other specific point.

area tie line *Electrical Engineering.* a transmission line connecting one control area to another.

area triangulation *Engineering.* triangulation that extends in various directions from a control point, covering the surrounding region.

area vitellina *Developmental Biology.* the yolk area outside the area vasculosa in meroblastic embryos.

area wall *Civil Engineering.* a retaining wall around an open area, particularly one below grade such as an areaway or entrance to a basement.

areaway *Civil Engineering.* an open area below grade used to provide light, air, or access to an adjacent building.

area weighted average resolution *Photogrammetry.* a single average value for the resolution over the picture format for any specified focal plane.

area-wide service *Transportation Engineering.* a system that provides comparable levels of service throughout an area.

Arecaceae *Botany.* the palm family of tropical or subtropical evergreen trees and shrubs; characterized by thick, unbranched stems, fruit that is a fleshy or fibrous drupe borne in leaf axils, and a spiral crown of large, long-stalked, fan-shaped, or pinnate leaves. Also, PALMAE.

arecaidine methyl ester see ARECOLINE.

Arecales *Botany.* an order of monocots that includes the single family Arecaceae. Also, PALMALES.

Arecidae *Botany.* a subclass of monocots in the class Liliopsidae, characterized by numerous small flowers, often in a spadix, surrounded by a prominent spathe and broad, petiole leaves without typical parallel venation.

arecoline *Organic Chemistry.* $C_8H_{13}NO_2$, a colorless liquid that boils at 209°C and is miscible with water, alcohol, and ether and soluble in chloroform; an oily alkaloid extract from betel nuts; formerly used in making pharmaceuticals.

areflexia [âr′ə fleks′ē ə] *Neurology.* an absence of the reflexes.

arelem *Robotics.* the arithmetic element of an APT program that uses machine motion instruction and geometric inputs to calculate cutter locations.

arena *Behavior.* a defined area used by an animal group for courtship displays and mating.

arenaceous *Geology.* of a sediment or sedimentary rock, composed primarily of sand or sand-sized particles of rock or minerals, or having a sandy texture or appearance. Also, ARENARIOUS, PSAMMITIC, SABULOUS.

arenarious see ARENACEOUS.

Arenaviridae *Virology.* a family of single-stranded RNA viruses containing a single genus, *Arenavirus.*

Arenavirus *Virology.* a genus of viruses that can infect through both vertical and horizontal transmission and can replicate in a wide range of mammals, particularly rodents; a persistent infection results from slow or insufficient immune response.

arendalite *Mineralogy.* a dark green variety of the monoclinic mineral epidote, occurring in crystalline and massive forms and found in the Arendal valley of Norway.

arene *Organic Chemistry.* another term for an aromatic hydrocarbon.

Arenicolidae *Invertebrate Zoology.* a family of lugworms, polychaete annelids that ingest mud.

arenicolite *Geology.* a U-shaped hole or groove in sedimentary rock believed to be a burrow or trail made by the marine worm *Arenicola* or the trail of a crustacean or mollusk which later became filled with sand.

arenicolous *Zoology.* living on or in the sand.

Arenigian or **Arenig** *Geology.* European stage of the Lower Ordovician period, occurring after the Tremadocian and before the Llanvirnian. Also, SKIDDAVIAN.

arenite or **arenyte** *Petrology.* a consolidated sand-texture sedimentary rock irrespective of composition. Also, PSAMMITE.

Arent *Geology.* a suborder of the soil order Entisol whose soil horizons have been disrupted or have disappeared as a result of mechanical mixing of the soil by earth-moving operations such as deep plowing.

areocentric *Astronomy.* a coordinate system centered at Mars.

areodesy [ar′ ē ō des′ ē] *Astronomy.* the geodesy of Mars.

areogeography *Ecology.* the study of the geographical distribution of plant and animal life.

areography [ar′ ē äg′ rə fē] *Astronomy.* the geography of Mars.

areola [[ə rē´ə lə; âr´ē ō´lə] *plural,* **areolae** or **areolas.** *Anatomy.* **1.** a ring of pigmented epithelium surrounding the nipple of a mammary gland. Also, **areola mammae. 2.** any circular or bulbous structure.

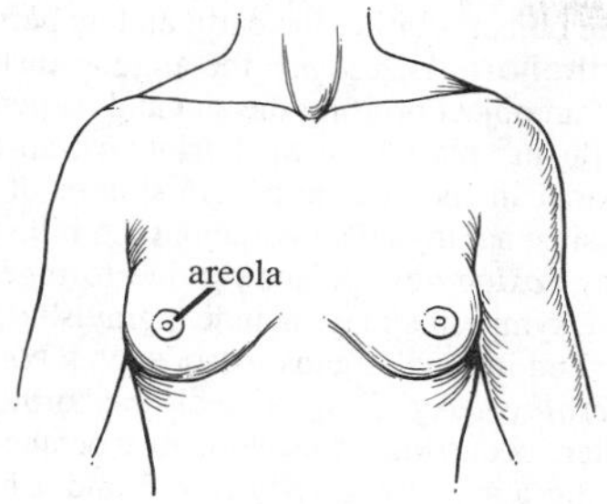

areola

areolar tissue [ə rē´ə lər; âr´ē ō´lər] *Histology.* the connective tissue that serves as a loose packing for practically all organs and tissues.
areolate [ə rē´ə lāt´] *Biology.* **1.** pitted or divided into small patches. **2.** of or relating to an areola.
areolitis [ə rē´ə lī´tis] *Medicine.* an inflammation of the pigmented area surrounding the nipple of the breast.
areostyle see ARAEOSTYLE.
Ares [êr´ēz] *Astronomy.* another name for the planet Mars. (*Ares* was the ancient Greek god of war; he was identified with the Roman god Mars.)
arete or **arête** [ə rāt´] *Geology.* a jagged, narrow rocky ridge or spur created by glacial erosion between cirques, usually located above the snowline in rugged mountains. Also, ARRIS, CRIB, SERRATE RIDGE.

arete

aretic behavior *Behavior.* behavior that is destructive, hostile, and aggressive.
ARFOR *Meteorology.* an international code word indicating an area forecast.
ARFOT *Meteorology.* an international code word indicating an area forecast in English system units.
arfvedsonite *Mineralogy.* $Na_3(Fe^{+2},Mg)_4Fe^{+3}Si_8O_{22}(OH)_2$, a black, monoclinic amphibole occurring as long prismatic or tabular crystals, having a specific gravity of 3.37 to 3.50 and a hardness of 5 to 6 on the Mohs scale; found in plutonic alkaline igneous rocks.
Argand, Jean Robert 1768–1822, French mathematician; developed Argand diagram.
Argand diagram [är´gand; är´gänd] *Mathematics.* the standard rectangular representation of the complex plane, in which the horizontal axis represents the real part of a complex number and the vertical axis represents the complex part.
Argand lamp *Engineering.* a gas lamp having a tubular wick enclosed by two concentric metal tubes, designed so that air flows both outside and inside the flame.
Argasidae *Invertebrate Zoology.* a family of soft ticks including a number that are parasitic on and carry diseases to humans and other mammals; all lack a scutum and display little sexual dimorphism.
Argelander, Friedrich [är´gə land´ər] 1799–1875, German astronomer; compiled catalog of northern stars; pioneered study of variable stars.
Argelander method *Astronomy.* a technique for determining the brightness of a star relative to two others that straddle it in brightness. Also, STEP METHOD.
argentaffin cell *Histology.* a cell whose granules are readily stained by silver salts. Such cells are widespread and are associated with the formation of polypeptides and proteins with hormonal acitivity; e.g., secretin, gastrin.
argentaffinoma *Oncology.* a tumor that originates from argentaffin cells in the terminal ileum or appendix. Also, CARCINOID.
argentic [är jen´ tik] *Chemistry.* **1.** of or relating to silver, which is also called argentum. **2.** describing various compounds of silver, especially those in which the element has a valence of 2.
argentic oxide see SILVER PEROXIDE.
Argentinoidei *Vertebrate Zoology.* a family of marine, often deep-sea, fish in the order Salmoniformes, including the deep-sea smelts.
argentite *Mineralogy.* Ag_2S, a blackish to gray cubic mineral that is dimorphous with acanthite; now discredited as a mineral not found in nature, stable only above 179°C.
argentocyanides *Inorganic Chemistry.* complexes formed when silver cyanide reacts with solutions of soluble metal cyanides.
argentojarosite *Mineralogy.* $AgFe_3^{+3}(SO_4)_2(OH)_6$, a brilliant yellow to brown trigonal mineral of the alunite group occurring as tiny, fine-grained crystals, having a specific gravity of 3.66, and found as a secondary mineral with anglesite, barite, and quartz.
argentometer *Engineering.* an instrument that measures the amount of grains per ounce of silver nitrate in a silver bath.
argentometry *Analytical Chemistry.* a volumetric analysis using the precipitation of insoluble silver salts.
argentophil *Biology.* a cell, tissue, or other structure that has an affinity for silver, especially one that can be stained by silver solutions.
argentous [är jen´ təs] *Chemistry.* **1.** of or relating to silver, which is also called argentum. **2.** describing various compounds of silver, especially those in which the element has a valence of 1.
argentous chlorate see SILVER CHLORATE.
argentous oxide see SILVER OXIDE.
argentum *Chemistry.* the Latin word for silver; this is the derivation of the symbol Ag to represent silver.
Argid *Geology.* a suborder of the soil order Aridisol that develops in arid climates; characterized by an argillic or a natric horizon.
Argidae *Invertebrate Zoology.* a family of mostly tropical small sawflies with three-jointed antennae; there are 400–500 known species worldwide.
argillaceous [är´jə lā´ shəs] *Geology.* **1.** of sediment or sedimentary rock, largely composed of or containing clay minerals or clay-size particles. Also, ARGILLOUS, CLAYEY, PELOLITHIC. **2.** containing or composed of the rock argillite.
argillation *Geology.* a process by which the weathering of aluminum silicates forms clay minerals.
argillic *Geology.* of, relating to, containing, composed of, or characteristic of clay; argillaceous.
argillic alteration *Geology.* a process of rock alteration whereby certain minerals in a rock are changed into clay minerals.
argillic horizon *Geology.* a soil horizon in which silicate clays have percolated down from an overlying layer and accumulated.
argilliferous *Geology.* of a structure, producing or containing clay.

argillite [är´jə līt´] *Petrology.* massive, fine-grained rocks formed from compact siltstone, shale, or claystone but intermediate in degree of induration and structure between them and slate; argillite is more indurated than mudstone but lacks the fissility of shale.

argillite

argillophilous [är´jə läf´ə ləs] *Ecology.* living or thriving in clay or mud. Thus, **argillophile, argillophily.**

argillous see ARGILLACEOUS.

arginase *Enzymology.* an amino acid that is essential for human nutrition; it catalyzes the formation of urea from the amino acid arginine.

arginine [är´ jə nēn´] *Biochemistry.* $C_6H_{14}N_4O_2$, an amino acid found in plants and animals that is essential for the human diet; also produced by the breakdown of proteins.

arginine vasopressin *Endocrinology.* a peptide hormone of the posterior pituitary (neurohypohysis) that regulates the resorption of water in the distal convoluted tubule of the kidney; so named because it has the amino acid arginine in the eighth position (vs Lys-vasopressin). Also, ANTIDIURETIC HORMONE.

arginine vasotocin *Endocrinology.* a peptide hormone produced by the pineal gland and considered to be involved in the regulation of the reproductive system during development.

argol *Food Technology.* a crude tartar deposit formed in wine casks during aging.

argon *Chemistry.* a nonmetallic chemical element, one of the noble gases, having the symbol Ar, the atomic number 18, and an atomic weight of 39.948; freezes at –189.2°C and boils at –185.7°C. It is a colorless, odorless, inert gas that forms 0.93% of the atmosphere and that is not known to form any chemical compounds. It is used to fill light bulbs, as a gas shield in welding, and in lasers. (From a Greek term meaning "not working; idle;" because of its being completely inert.)

Argonauta *Invertebrate Zoology.* the paper nautilus, a genus of cephalopods in the order Octopoda, similar to octopuses, but the females live in a thin white spiral shell, coming and going at will.

Argo Navis *Astronomy.* a large constellation south of Canis Major, now split into Carina, Puppis, Vela, and Pyxis.

argon ionization detector *Nucleonics.* an ionization chamber for measuring nuclear radiation in which argon is used as the detecting medium because the mean energy expended per ion pair produced in it is almost identical for α particles, protons, and other light particles.

argon laser *Optics.* a laser that uses singly ionized argon, requires high-power input, and emits several watts of laser energy in the blue-green region of the visible spectrum. Also, **argon-ion laser.**

argon oxygen decarburization *Metallurgy.* a refining process in steelmaking, in which carbon impurities are removed while minimizing the loss of desirable alloying elements.

Argovian *Geology.* in Great Britain, a geologic substage of the Upper Jurassic period, occurring after the Oxfordian stage and before the Rauracian substage.

Arguloida *Invertebrate Zoology.* fish lice in the subclass Branchiura, found on both freshwater and marine fish throughout much of the world.

argument *Mathematics.* **1.** the object upon which a function acts, that is, the value at which the expression describing the function is to be evaluated. **2.** the angle between a vector representing a complex number in an Argand diagram and the positive real axis. Also, AMPLITUDE. *Computer Programming.* see PARAMETER, def. 1.

argument of latitude *Astronomy.* the angular distance between an orbit's ascending node and the object measured in the orbital plane.

argument of perigee *Astronomy.* the angular distance between the ascending node of an object orbiting the earth and its perigee.

argument of perihelion *Astronomy.* the angular distance between the ascending node of an object orbiting the sun and its perihelion.

Argyre Planitia [är´jīr plə nish´ē ə] *Astronomy.* an 800-kilometer-diameter impact basin in the southern hemisphere of Mars that is the youngest of its basins, having an age of about 3.5 billion years.

argyria [är jī´rē ə] *Toxicology.* poisoning due to ingestion of silver or one of its salts; the symptoms may include a grayish discoloration of the skin, conjunctiva, and internal organs. Also, SILVER POISONING.

argyrodite *Mineralogy.* Ag_8GeS_6, a steel-gray orthorhombic mineral with metallic luster, occurring as pseudocubic octahedral or dodecahedral crystals, having a specific gravity of 6.2 and a hardness of 2.5 on the Mohs scale; found in vein deposits with acanthite, other silver minerals, and sulfides.

arheol see SANTALOL.

Arhynchobdellae *Invertebrate Zoology.* freshwater, amphibious, or terrestrial leeches, including well-known blood-sucking leeches; worldwide distribution, no marine species are known.

Arhynchodina *Invertebrate Zoology.* a suborder of protozoans, in the order Thigmotrichida, with a mouth and rows of cilia.

arhythmicity [ā´rith mis´ə tē] *Biology.* a condition in which an organism deviates from an expected behavioral or psychological rhythm.

ariboflavinosis [ā´rī bō flav´ə nō´sis] *Medicine.* a deficiency of riboflavin in the diet, producing a syndrome characterized by dryness, cracking, or inflammation of the lips, mouth, and tongue, scaliness of the skin in the areas of the cheeks, eyelids, nose, and earlobes, vascularization of the corneas, and anemia.

arid *Meteorology.* very dry; lacking moisture. *Ecology.* lacking vegetation.

arid climate *Meteorology.* any extremely dry climate, more specifically, a climate with a value of –60 to –40 on the moisture index scale.

arid erosion *Geology.* in arid regions, the wearing away of rock primarily by the action of wind.

aridification *Meteorology.* a natural or manmade climatic change causing a formerly humid region to become increasingly dry.

Aridisol *Geology.* an order of soils that develops in arid climates, characterized by being low in organic matter and high in calcium and magnesium.

aridity *Meteorology.* the fact of being arid; a measure of the degree to which a climate lacks life-promoting moisture.

aridity coefficient *Meteorology.* a function of precipitation and temperature that represents the aridity of a location, given by the latitude factor × temperature range × precipitation ratio.

aridity index *Meteorology.* a measure of aridity, given as an index of the degree of water deficiency below water need at a given station.

ariegite *Petrology.* a group of pyroxenites composed principally of clinopyroxene, orthopyroxene, and spinel, devoid of feldspar and olivine, but may contain some garnet or amphibole.

Ariel *Space Technology.* any of a series of British satellites launched (beginning in 1962, with U.S. Scout boosters) as scientific probes.

Ariel (Uranus 1) *Astronomy.* the first moon of Uranus; discovered in 1851, it has a diameter of 1174 kilometers.

Aries [âr´ēz] *Astronomy.* the Ram, a small zodiacal constellation located between Pisces and Taurus that is seen best in the autumn.

arietiform *Vertebrate Zoology.* resembling or shaped like a ram's horns, as the black markings on the face of a kangaroo rat.

Ariidae *Vertebrate Zoology.* a family of tropical marine catfishes in the order Siluriformes.

Arikareean *Geology.* the North American Lower Miocene geologic stage.

aril *Botany.* an additional integument that is formed on some seeds after fertilization; an outgrowth that develops from the funiculus or the chalaza to form a fleshy bag which may completely or partially surround the seed; may be spongy, fleshy, or hairy and is often brightly colored.

arilode *Botany.* a false aril; an aril that arises from the micropyle rather than the funiculus or chalaza.

A ring *Astronomy.* the outermost of the three rings of Saturn visible from Earth in a modest telescope; with its outer edge at a distance from Saturn's center of 136,200 kilometers, it is approximately 15,200 kilometers wide.

Arionidae *Invertebrate Zoology.* a family of prominent, primarily vegetarian European slugs possessing a shell vestige; includes the red slug *(Arion rufus)* and black slug *(A. ater),* both of which vary in color.

arista *Invertebrate Zoology.* the bristly structure near or at the tip of the antenna of many two-winged flies.

Aristarchus (of Samos) c. 310–230 BC, a scientist of ancient Greece who is known as the first to propose that the earth moves around the sun rather than vice versa.

Aristarchus [ar ə stär´ kəs] *Astronomy.* a bright lunar crater 45 kilometers in diameter.

aristogenesis see ORTHOGENESIS.

Aristolochiaceae *Botany.* the birthwort family of herbs, vines, and shrubs; characterized by alternate, heart-shaped leaves and showy, strong-smelling flowers; includes snakeroots and wild ginger.

Aristolochiales *Botany.* an order of dicots that includes the families Aristolochiaceae, Rafflesiaceae, and Hydnoraceae; distinguished by a petaloid perianth and inferior ovary.

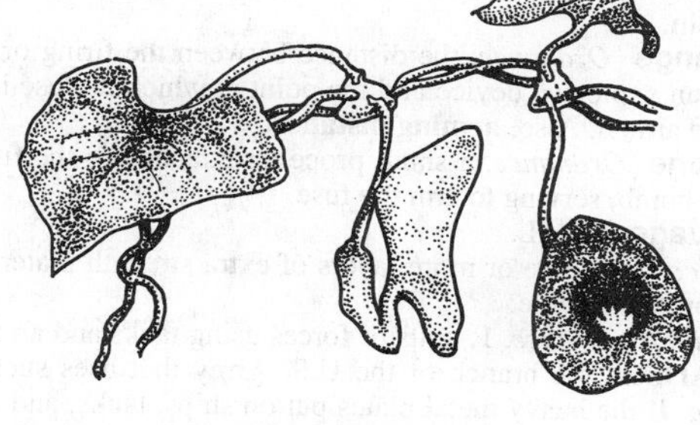

Aristolochiales

aristolochic acid *Organic Chemistry.* $C_{17}H_{11}NO_7$, a brownish crystalline solid that decomposes at 281°C, slightly soluble in water and soluble in alcohol, ether, and acetone; used as an aromatic bitter. Also, **aristolochine.**

aristopedia *Invertebrate Zoology.* the replacement of an arista by a relatively perfect leg; a fairly common anomoly in flies of the genus *Drosophila* (includes fruit flies). Also, **aristapedia.**

Aristotle [âr´ə stät´əl] 384–322 BC, Greek philosopher who extended many areas of knowledge, including biology, physics, and astronomy.

Aristotle's anomaly *Neurology.* a phenomenon that causes a person to sense two objects when the first and second fingers are crossed and a single object, such as a pencil, is placed between them.

Aristotle's lantern *Invertebrate Zoology.* a chewing mechanism in Echinoidea consisting of a framework of muscles and ossicles that supports the teeth and encloses the lower part of the oesophagus; named for its resemblance to ancient oil lamps.

arithmetic [ə rith´mə tik *for 1;* âr´ith met´ik *for 2*] *Mathematics.* **1.** the processes of addition, subtraction, multiplication, and division; usually of integers, but also of real or complex numbers. **2.** having to do with or involving the use of arithmetic.

arithmetical *Mathematics.* relating to or involving arithmetic.

arithmetical instruction *Computer Programming.* a machine instruction whose operation code is one of the arithmetic operations that, with logic and shift instructions, provide complete computational processing capabilities.

arithmetical operation *Computer Programming.* any of the basic mathematical processes such as addition, subtraction, multiplication, division, and negation.

arithmetical unit *Computer Technology.* a part of arithmetic logic unit that performs arithmetical operations.

arithmetic check *Computer Programming.* a process for verifying the results of an arithmetic operation by performing a different operation that should yield the same result and comparing the outcome with the original result.

arithmetic circuitry *Computer Technology.* the electronic circuits that implement the operations in the arithmetical unit of the central processor.

arithmetic logic unit *Computer Technology.* a digital circuit or collection of circuits within the central processing unit that performs a set of arithmetic micro-operations and a set of logic microoperations.

arithmetic mean *Mathematics.* the average of a set of terms, computed by dividing their sum by the number of terms. Given the arithmetic mean A of n positive real numbers and the geometric mean G of the same n positive real numbers, $A > G$.

arithmetic progression *Mathematics.* a sequence in which the difference between any term and its predecessor is some constant. In particular, if a is the first term, then the nth term is $a + (n - 1)d$. Also, **arithmetic sequence.**

arithmetic register *Computer Technology.* special purpose registers in the arithmetic-logic unit that function as the source or destination of operands and in which arithmetic and logic operations can be performed.

arithmetic scan *Computer Programming.* a procedure that examines an expression to determine the intended order of arithmetic operations.

arithmetic series *Mathematics.* a series whose terms are an arithmetic progression.

arithmetic shift *Computer Programming.* the left or right movement within a fixed-length word that effectively multiplies or divides the number by two, once for each bit position the number is shifted.

arithmetic symmetry *Electronics.* a situation in which a filter's frequency response is symmetrical about the center frequency when frequency is plotted linearly. see LOGARITHMIC SYMMETRY.

arithmetization *Mathematics.* the study of a mathematical system using the operations of arithmetic.

arity *Artificial Intelligence.* the number of arguments of a function.

Arizona ruby *Mineralogy.* a ruby-colored pyrope garnet found in the southwestern United States.

Arizonite *Petrology.* a dike rock composed of about 80% quartz, 18% orthoclase, and traces of mica and apatite. *Mineralogy.* $Fe_2^{+3}Ti_3O_9$, a brown to black, hexagonal mineral having a specific gravity of 4.13 to 4.81 and a hardness of 3.5 on the Mohs scale, occurring as irregular thin plates. Also, PSEUDORUTILE.

arizonosis *Veterinary Medicine.* an egg-transmitted infection affecting chiefly turkey poults and often characterized by nervous symptoms, blindness, peritonitis, and infection of the intestinal tract.

Arkansas stone *Petrology.* a form of novaculite that is quarried in the Ozark Mountains of Arkansas and used as a whetstone for sharpening tools and instruments.

arkite *Petrology.* a feldspathoid-rich rock consisting of pseudoleucite, nepheline, melanite, pyroxene, and traces of orthoclase, apatite, and sphene.

arkose *Petrology.* a feldspar-rich, coarse-grained sandstone derived from the erosion of granite and composed of angular to subangular, poorly to moderately sorted grains; contains quartz and other detrital minerals.

arkosic *Petrology.* of or relating to rocks that are wholly or partly characterized by arkose.

arkosic bentonite *Petrology.* bentonite derived from volcanic ash, which contains between 25 and 75% unaltered detrital crystalline grains. Also, SANDY BENTONITE.

arkosic limestone *Petrology.* an impure clastic limestone containing relatively high proportions of feldspar.

arkosic sandstone *Petrology.* **1.** a sandstone in which large amounts of feldspar are present, such as one containing unassorted products of granular disintegration of fine- or medium-grained granite. **2.** a general term for arkose, subarkose, and various arkosic rocks.

arkosic wacke see FELDSPATHIC GRAYWACKE.

arkosite *Petrology.* a feldspar-rich quartzite. Also, **arkose quartzite.**

Arkwright, Sir Richard 1732–1792, English inventor; developed the water frame, revolutionizing the machine spinning of cotton.

arkyochrome *Neurology.* any nerve cell in which the chromatic substance arranges itself in the form of a network.

arkyostichochrome *Neurology.* any nerve cell in which the chromatic substance appears partly as a network and partly as parallel threads, i.e., as both an arkyochrome and a stichochrome.

arm *Anatomy.* **1.** the part of the upper extremity that extends between the shoulder and the elbow. **2.** loosely, the part of the upper extremity between the shoulder and the hand, including both the arm and the forearm. **3.** any armlike branch of an organ or structure.

arm a structure, device, area, or other entity thought to resemble the human arm in some way; specific uses include: *Zoology.* a forelimb of an animal. *Ordnance.* **1.** any weapon for use in combat or warfare. **2.** to cause ammunition or explosives to be ready to detonate, as by removing a safety device, setting a fuse or other firing device, and so on. *Robotics.* an interconnecting set of links and joints that move with one or more degrees of freedom and support a wrist socket and an end effector. *Acoustical Engineering.* a freely suspended portion of a record player that holds the cartridge stylus over the record. *Geology.* **1.** a narrow inlet or cove that extends inland from a body of water. **2.** a spur or ridge that extends from a mountain. *Mathematics.* one of the sides of an angle. *Physics.* the perpendicular distance between the axis and a line along which a force acts to cause a rotation about the axis. *Naval Architecture.* **1.** the sections of an anchor that project at an angle from either side of the shank; each arm ends in a fluke. **2.** the end of a yard or boom. *Electricity.* a portion of a network consisting of one or more two-terminal elements in series. *Military Science.* a combat branch of a military force, especially of the U.S. Army, such as the infantry, artillery, or armored cavalry.

armadillo *Vertebrate Zoology.* any member of the mammalian family Dasypodidae in the order Edentata; characterized by a bony, platelike armor over the head, back, and sides, divided by transverse furrows that permit movement.

armadillo

Armalite *Ordnance.* an American assault rifle.

armament *Military Science.* the fact of being or becoming armed with weapons. *Ordnance.* **1.** the weapons that a vehicle, aircraft, or combat unit is armed with: a fighter plane whose *armament* includes cannons and missiles. **2.** also, **armaments.** any form of war equipment, such as weapons, ammunition, and combat vehicles.

armament error *Ordnance.* the deviation of any shot from the center of impact of a series of shots, after all possible human errors and adjustments of sighting have been accounted for.

armangite *Mineralogy.* $Mn^{+2}_{26}As^{+3}_{18}O_{50}(OH)_4(CO_3)$, a black trigonal mineral occurring as short prismatic crystals, having a specific gravity of 4.43 and a hardness of 4 on the Mohs scale, and found with hematite in calcite-barite veinlets.

armature *Architecture.* the structural iron bars used to frame tracery or to reinforce building features such as slender columns or hanging canopies. *Electromagnetism.* the movable part in an electromagnetic mechanical device, such as the winding in which the electromotive force is induced in an electric motor.

armature chatter *Electromagnetism.* vibrations of an armature caused by irregularities in the coil current.

armature contact see MOVABLE CONTACT.

armature reactance *Electromagnetism.* the inductive reactance associated with the magnetic flux produced by the current in the armature conductors.

armature reaction *Electromagnetism.* the interaction between the magnetic field produced by the current in an armature and the main magnetic field in an electric motor.

armature resistance *Electricity.* the ohmic resistance in the armature windings of an electric generator or motor.

arm conveyor *Mechanical Engineering.* a vertical conveyor consisting of armlike projections or shelves connected to a belt or chain, best suited for moving barrels, drums, and the like. Also, RIGID ARM CONVEYOR.

armed *Ordnance.* of or relating to an explosive device that has all safety devices removed and is ready to be detonated: an *armed* mine.

armed forces *Military Science.* **1.** all the military forces of a nation as a group. Also, **armed service** or **armed services. 2. Armed Forces.** the U.S. military forces as a group; the Army, Navy, Air Force, and Marine Corps.

armed merchantman *Naval Architecture.* a ship designed primarily for trade, but fitted with armament. Under international law, a neutral merchant ship may carry a defensive armament without compromising its status as a merchantman.

armed reconnaissance *Military Science.* a reconnaissance mission that also searches for and attacks targets of opportunity.

arm elevator *Mechanical Engineering.* a chain elevator for carrying fixed-shape objects, such as barrels or drums, and equipped with projecting curved arms that cradle the objects during transport. Also, BARREL ELEVATOR.

armenite *Mineralogy.* $BaCa_2Al_6Si_9O_{30}{\cdot}2H_2O$, a colorless to green, translucent hexagonal mineral occurring as prismatic crystals, having a specific gravity 2.76 and a hardness of 7.5 on the Mohs scale; associated with axinite, pyrrhotite, and quartz.

ARMET *Meteorology.* an international code word indicating an area forecast in metric system units.

Armillaria *Mycology.* a genus of fungi belonging to the order Agaricales living in or on wood and causing root rot.

Armillaria root rot *Plant Pathology.* a disease of forest and orchard trees in which the fungus *Armillaria melea* first invades the roots and then the lower trunk. Also, BARK-SPLITTING DISEASE.

Armilliferidae *Invertebrate Zoology.* a family of bloodsucking parasitic arthropods in the suborder Porocephaloidea.

arming *Ordnance.* describing a device used to arm a bomb, mine, or other such explosive device, such as an **arming fuse, arming wire, arming pin.**

arming range *Ordnance.* the distance between the firing or launching point for an explosive device and the point at which its fuse is expected to become armed. Also, **arming distance.**

arming vane *Ordnance.* a small propeller attached to the fuse mechanism of a bomb, serving to arm the fuse.

arm language see AL.

armor *Electricity.* one or more layers of extra strength material to reinforce or protect a cable.

armor *Military Science.* **1.** units or forces using tanks and armored vehicles. **2. Armor.** the branch of the U.S. Army that uses such vehicles. *Ordnance.* **1.** the heavy metal plates put on ships, tanks, and other vehicles to resist shells and projectiles. **2.** any similar protective covering put on vehicles or worn by people. *Electricity.* one or more layers of extra strength material to reinforce or protect a cable.

armored *Ordnance* **1.** having protective armor: an *armored* tank. **2.** using or involving vehicles with protective armor: *armored* combat, an *armored* unit.

armored artillery *Military Science.* an artillery unit that is part of an armored division or that uses weapons mounted on self-propelled armored vehicles.

armored cable *Electricity.* cable with a wrapping of metal for mechanical protection.

armored car *Ordnance.* an armored vehicle that moves on wheels, usually having a mounted gun, used to provide security or for reconnaissance.

armored cavalry *Military Science.* a combat unit that uses tanks and other armored vehicles to carry out swift, concentrated attacks.

armored infantry *Military Science.* an infantry unit attached to or supporting an armored unit.

armored mud ball *Geology.* a large, roundish mass of clay or silt that collects a coating of coarse sand and fine gravel as it rolls downstream. Also, PUDDING BALL.

armored personnel carrier *Ordnance.* an armored vehicle that is used to transport troops to or from a combat area

armored vehicle *Ordnance.* any vehicle that has armor protection and is used by a military unit for combat, security, or transportation, such as a tank or armored car.

armorer *Ordnance.* a person whose work is servicing and repairing small-arms weapons and equipment.

armorhead see BOARFISH.

Armorican orogeny see VARISCAN OROGENY.

armor-piercing *Ordnance.* designed to penetrate the protective armor of a ship, tank, or other such vehicle, such as an **armor-piercing bomb, armor-piercing bullet, armor-piercing incendiary, armor-piercing tracer.**

armor-piercing discarding sabot *Ordnance.* a long, thin armor-piercing projectile fired with high velocity by the aid of a sabot fitting the gun and discarded on leaving the muzzle.

armory *Military Science.* a building where a reserve military unit trains or conducts operations. *Ordnance.* a place where arms and military equipment are manufactured or stored.

armpit *Anatomy.* the popular name for the axilla, the hollow area under the arm at the shoulder.

arms *Ordnance.* military weapons, such as rifles, artillery, bombs, and the like. Used in combinations such as **arms chest, arms rack, arms locker.**

arm solution *Robotics.* the calculations by which a robot controller translates joint positions into desired tool positions.

Armstrong, Edwin Howard 1890–1954, American electrical engineer; invented frequency modulation (FM).

Armstrong frequency-modulation system *Electronics.* a phase-shift modulation system developed by E.H. Armstrong, in which a low-frequency carrier is modulated at a low level, and the signal is passed through several amplifying stages to reach the desired high level and high carrier frequency.

Armstrong oscillator *Electronics.* an oscillator circuit designed by E. H. Armstrong utilizing inductive feedback. It has a tuned grid coil and an untuned plate coil whose coupling determines the feedback ratio.

Armstrong's acid *Organic Chemistry.* $C_{10}H_6(SO_3H)_2$, a white crystalline solid that melts at 240°C and is soluble in water and alcohol; used to make dyes. Also, NAPHTHALENE-1,5-DISULFONIC ACID.

armure *Textiles.* a fabric with an irregular surface characterized by small fancy designs on a ribbed or woven background.

army *Military Science.* **1.** any large military force that fights on land. **2.** **Army.** the branch of the U.S. Armed Forces trained and equipped to fight on land. **3.** also, **Army.** a major unit of the U.S. Army, made up of two or more corps.

army artillery *Military Science.* artillery assigned or attached to an army and under its command. Similarly, **army aircraft, army aviation.**

army corps see CORPS.

army depot see DEPOT.

army group *Military Science.* a major army unit consisting of several field armies.

army tactical missile system *Ordnance.* a tactical surface-to-surface ballistic missile for battlefield delivery of precision-guided sub-munitions.

army worm *Invertebrate Zoology.* the larvae of certain species of Noctuidae moths, so named because they attack crops and other vegetation in large numbers.

Arnaldus of Villanova c. 1235–1313, Spanish-French physician; revived the Hippocratic method of medical observation.

Arnaudon's green see PLESSY'S GREEN.

Arndt-Eistert synthesis *Organic Chemistry.* a reaction that converts an acyl halide to a carboxylic acid with one additional carbon by reacting an acid chloride with excess diazomethane, followed by rearrangement of the resulting diazoketone to a ketene in the presence of water to form the homologated acid.

Arneth's index *Histology.* a system for classifying human blood granulocytes according to the number of lobes present in the nucleus. Also, **Arneth's count, Arneth's formula.**

arnimite *Mineralogy.* $Cu_5^{+2}(SO_4)_2(OH)_6 \cdot 3H_2O$, a hydrous copper sulfate hydroxide mineral that may be equivalent to antlerite.

2-arninopropane see ISOPROPYLAMINE.

Arnold sterilizer *Microbiology.* an apparatus for sterilizing objects, or freeing them of living microorganisms, by the use of live steam at atmospheric pressure.

Arodoidea *Invertebrate Zoology.* a superfamily of hemipteran land bugs with conspicuous antennae.

arolium *Invertebrate Zoology.* a padlike projection between the tarsal claws of numerous insects.

aromatase *Enzymology.* a microsomal enzyme complex that catalyzes the conversion of testosterone to estradiol.

aromatic *Science.* having a strong but pleasant odor; spicy, fragrant, and so on. *Organic Chemistry.* **1.** describing a major class of unsaturated cyclic hydrocarbons characterized by the presence of one or more rings; they are so called because they have a strong odor. The class is typified by benzene, which has a six-carbon ring containing three double bonds. Other aromatic series include the naphthalenes and the anthracenes. **2.** a compound of this type.

aromatic alcohol *Organic Chemistry.* an aromatic compound having a hydroxyl group located in a side chain on a benzene ring.

aromatic aldehyde *Organic Chemistry.* an aromatic compound having the –CHO group; for example, benzaldehyde.

aromatic amine *Organic Chemistry.* an aromatic organic compound that contains at least one amino group.

aromatic amino acid *Biochemistry.* a kind of amino acid characterized by the presence of a so-called aromatic group or benzene ring.

aromatic hydrocarbon *Organic Chemistry.* see AROMATIC.

aromatic ketone *Organic Chemistry.* an aromatic compound containing the –CO group; for example, acetophenone.

aromatic nucleus *Organic Chemistry.* the six-carbon, core ring structure that characterizes benzene, naphthalene, anthracene, and related compounds.

aromatic series *Organic Chemistry.* see AROMATIC.

aromatic spirit of ammonia *Pharmacology.* smelling salts or sal volatile, a substance used to stimulate breathing after fainting. Also, SPIRIT OF HARTSHORN.

aromatic sulfuric acid *Pharmacology.* sulfuric acid diluted with alcohol and flavored with ginger and cinnamon; formerly used as an astringent to treat diarrhea, night sweats, and spitting blood.

aromatization *Chemical Engineering.* the conversion of nonaromatic hydrocarbons to aromatic hydrocarbons.

aromorphosis *Evolution.* an evolutionary change that is characterized by the development of complexity in the organism, usually involving larger size and anatomical integration, without any increase in specialization.

arostat process *Chemical Engineering.* a process in which aromatic molecules are saturated by catalytic hydrogenation to yield high-quality jet fuels, high-purity cyclohexane from benzene, and solvents which are nearly free of aromatics.

arousal *Behavior.* the general level of an animal's excitability or responsiveness to stimulation, as shown by the type of behavior exhibited or by physiological indicators.

aroyl *Organic Chemistry.* the –ArCO group, where Ar is an aromatic substituent, such as phenyl or naphthyl.

aroylation *Organic Chemistry.* a reaction in which the aroyl group is substituted for another atom or radical in a molecule.

ARPANET *Computer Technology.* a wide area packet-switched network established by the U.S. Defense Department that links many universities, military laboratories, and other computer centers together. (An acronym for Advanced Research Projects Agency Network.)

arquerite *Mineralogy.* a silver-rich (86.6%) form of amalgam that is malleable and very soft and has a specific gravity of 10.8.

arrastre [ə rä′strə] *Mining Engineering.* a circular rock-lined pit in which broken ore is pulverized by stones attached to horizontal poles fastened to a central pillar and dragged around the pit. Also, **arrastra.**

array *Science.* a set of identical (or very similar) elements arranged in a regular pattern. *Computer Programming.* a collection of data elements, such as numbers or character strings, arranged so that each item in the array can be located; the entire array is often given a single name with subscripts or indices referring to individual elements. *Statistics.* see ORDERED ARRAY.

array antenna *Electronics.* an antenna that has a group of individual radiating elements positioned for some desired directional property.

array processor *Computer Technology.* a computer having many processing elements that are capable of carrying out basic arithmetic operations in parallel in a single step under control of a master program.

array radar see PHASED-ARRAY RADAR.

array sonar see PHASED-ARRAY SONAR.

arrested evolution *Evolution.* an evolutionary change that occurs extremely slowly in one taxonomic group in comparison to the observed rates of change in similar taxonomic groups.

arrester *Electricity.* a device used to protect an installation from lightning. *Engineering.* a protective screen above a chimney or incinerator that keeps the burning material within the stack.

arrester gear *Aviation.* an apparatus designed to halt an aircraft's forward motion, as when landing on an aircraft carrier; usually includes arresting wires on the landing surface and an arresting hook on the aircraft.

arresting hook *Aviation.* a device fitted to an aircraft to engage arresting wires (as on an aircraft carrier) and absorb the aircraft's forward momentum during routine landings or emergencies.

arresting wires *Aviation.* a series of cables stretched across the flight deck of an aircraft carrier, used to decelerate landing aircraft. An arresting hook at the rear of the aircraft engages one of the arresting wires, pulling it out against friction releases and bringing the aircraft to a quick stop.

Arrhenius, Svante August [är rān′ē əs] 1859–1927, Swedish physicist and chemist; developed the theory of electrolytic dissociation.

Arrhenius acid see ACID, def. 1a.

Arrhenius base see BASE, def. 1a.

Arrhenius equation *Physical Chemistry.* a rate equation that is used for many chemical transformations and processes, in which the rate is exponentially related to temperature; one version is $k = Ae^{-Ea/RT}$, in which k is the rate constant of the chemical reaction, A is a constant called the preexponential factor or frequency factor, Ea is the activation energy, R is the gas constant, and T is the absolute temperature. (From S. A. *Arrhenius.*)

Arrhenius-Guzman equation *Physics.* a relationship between the viscosity of a liquid and the temperature of the liquid, usually expressed in kelvin degrees: $h = Ae^{B/RT}$, in which A and B are constants and R is the universal gas constant.

Arrhenius theory *Chemistry.* the theory that an acid can be defined as any substance that increases the concentration of hydrogen ions when added to a water solution.

Arrhenius viscosity formulas *Physics.* three equations that relate the viscosity of a liquid to its temperature, the viscosity of a solution to its concentration and to the viscosity of the solvent, and the viscosity of a colloidal solution to the viscosity of the liquid.

arrhenoblastoma *Oncology.* a neoplasm of the ovary that arises from the ovarian stroma, mimicking derivatives of the sex chord mesenchyme in the testis; may causes defeminization and virilization. Also, ANDREOBLASTOMA, ANDROMA, ARRHENOMA, SERTOLI-LEYDIG CELL TUMOR.

arrhenoma *Oncology.* see ARRHENOBLASTOMA.

arrhenotoky *Biology.* a form of nonsexual reproduction in which a parthenogenetic female, such as an unfertilized queen bee, produces only male offspring.

arrhythmia [ə ri̲t̲h̲′mē ə] *Cardiology.* any variation from the normal rhythm of the heartbeat, as in atrial fibrillation or paroxysmal tachycardia. Thus, **arrhythmic.**

arris *Architecture.* an edge at the intersection of two planes or curves, especially in moldings or between two flutes of a Doric column. Also, **aris.** *Geology.* see ARETE.

arris fillet *Building Engineering.* a triangular piece of wood used to raise the roof tiles or slates on the upper side of a chimney to shed rainwater.

arris gutter *Building Engineering.* a V-shaped gutter.

arris hip tile *Building Engineering.* an L-shaped tile made to fit over the hip of a roof. Also, ANGLE HIP TILE.

arris rail *Building Engineering.* a wooden rail of triangular section.

arrissing tool *Engineering.* a tool used for rounding the edges on freshly poured concrete.

arris tile *Building Engineering.* an angular tile purposely fabricated to cover intersections of hips and ridges in slated and tile roofs. Also, BANNET TILE.

arrisways *Building Engineering.* **1.** of timber, cut or sawed diagonally. **2.** of bricks, tiles, slates, or boards, laid diagonally or at an angle, or with an angle or edge presented. Also, **arriswise.**

arrojadite *Mineralogy.* $KNa_4CaMn_4^{+2}Fe_{10}^{+2}Al(PO_4)_{12}(OH,F)_2$, a translucent, dark green monoclinic mineral occurring as cleavable masses and crystals, with a specific gravity of 3.56 and a hardness of 5 on the Mohs scale; found with granitic pegmatite; forms a series with dickinsonite.

Arrow *Ordnance.* an Israeli weapons system under development as a high-altitude surface-to-air missile defense against incoming surface-to-surface missiles. *Astronomy.* see SAGITTA.

arrowhead *Archaeology.* the pointed end of an arrow; often found at sites of prehistoric peoples. *Botany.* an aquatic or marsh plant of the genus *Sagittaria,* having arrowhead-shaped leaves and clusters of white, cuplike flowers.

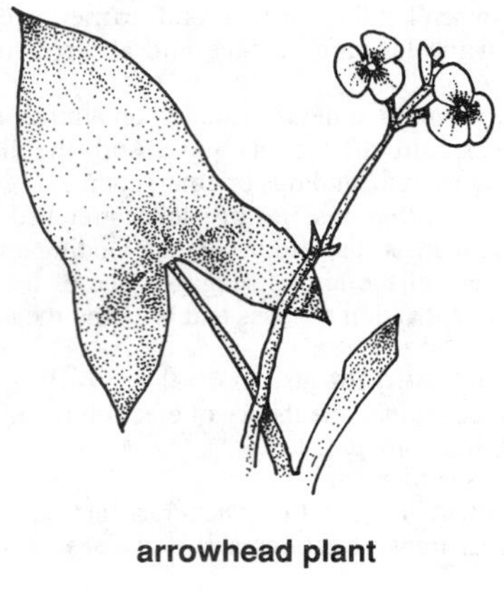

arrowhead plant

arrowhead wing *Aviation.* a wing configuration in which the wings are swept sharply aft, thus suggesting an arrow's head in planform.

arrowroot *Botany.* a tropical American plant, *Maranta arundinacea,* the rhizomes of which produce and store starch. *Food Technology.* nutritive starch from the roots of *Maranta arundinaceae* or any of several plants yielding a similar starch, used in cooking to thicken or bind sauces and stocks.

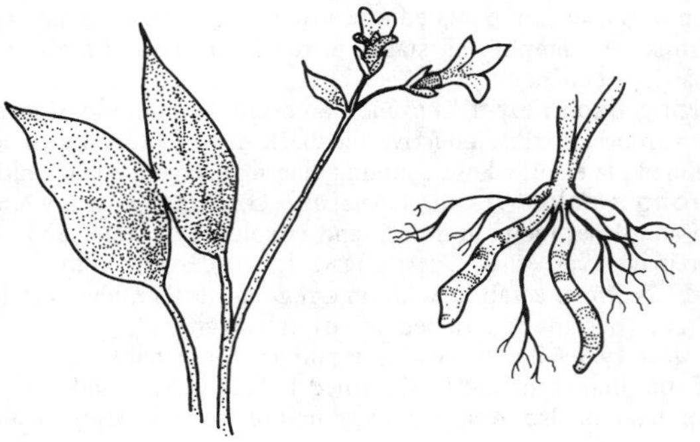

arrowroot

arrowroot starch *Food Technology.* a nutritive starch made from the underground stems of arrowroot.

arrowworm *Invertebrate Zoology.* a small, slender, mostly transparent worm, in the phylum Chaetognatha, with prehensile hooks or teeth on each side of the mouth; among the most abundant animals in the marine plankton, includes some forty species.

arroyo [ə roi′ō] *Geology.* a deep, usually dry channel or gully carved by an ephemeral or intermittent stream; common in the semiarid regions of the southwestern United States and Latin America. (From a Spanish word meaning "shaft" or "pit.")

ARS autonomously replicating sequence.

arsenal *Ordnance.* a place in which arms and military equipment are manufactured or stored.

arsenate *Inorganic Chemistry.* **1.** $(AsO_4)^{3-}$, a negative ion derived from arsenic acid. **2.** a salt or ester of arsenic acid.

arsenic *Chemistry.* a chemical element having the symbol As, the atomic number 33, an atomic weight of 74.9216, and a boiling point (sublimes) of 613°C. It is a silver-gray crystalline solid forming compounds that are highly toxic and carcinogenic. It is used as an alloy, especially in lead shot, and in electrical circuits. *Mineralogy.* this element in its native form, a tin-white, opaque, trigonal mineral usually occurring in granular massive form, or as small, rhombohedral crystals with a nearly metallic luster and highly perfect cleavage, having a specific gravity of 5.7 and a hardness of 3.5 on the Mohs scale; found in hydrothermal veins, in limestones, and in the anhydrite cap rock of salt domes. (Thought to go back to a Persian word meaning "gold; golden;" also influenced by a Greek term meaning "manly" or "strong;" with reference to its powerful properties.)

arsenic acid *Inorganic Chemistry.* **1.** $H_3AsO_4 \cdot 1/2H_2O$, white translucent crystals, soluble in water and alcohol; used in insecticides, glassmaking, and defoliants. Also, ORTHOARSENIC ACID. **2.** see MAGNESIUM ARSENATE.

arsenical *Chemistry.* **1.** of or relating to arsenic. **2.** a compound containing arsenic, especially a drug or insecticide.

arsenicalism *Toxicology.* poisoning caused by the consumption of arsenic or one of its derivatives; symptoms may include severe gastrointestinal disturbances and failure of the liver, kidneys, and blood-forming tissues. Also, **arsenism.**

arsenic bromide see ARSENIC TRIBROMIDE.

arsenic chloride see ARSENIC TRICHLORIDE.

arsenic disulfide *Inorganic Chemistry.* As_2S_2, red, orange, or black crystals that melt at 307°C; soluble in acids and alkalis, insoluble in water; used as a pigment, in the leather industry, and in fireworks.

arsenic fluoride see ARSENIC TRIFLUORIDE.

arsenic hydride see ARSINE.

arsenic oxide see ARSENIC PENTOXIDE.

arsenic pentasulfide *Inorganic Chemistry.* As_2S_5, yellow or orange crystals, soluble in nitric acid and alkalis, insoluble in water; used as a pigment.

arsenic pentoxide *Inorganic Chemistry.* As_2O_5, a white, deliquescent, amorphous solid; soluble in water and alcohol and decomposed by heat; used in insecticides and herbicides and for dyeing and printing.

arsenic sulfide see ARSENIC DISULFIDE.

arsenic tribromide *Inorganic Chemistry.* $AsBr_3$, yellowish-white, hygroscopic crystals, decomposed by water; used in analytical chemistry and medicine.

arsenic trichloride *Inorganic Chemistry.* $AsCl_3$, an oily, colorless to pale yellow liquid, highly toxic; soluble in concentrated hydrochloric acid and most organic solvents, decomposed by water; used in ceramics and as an intermediate for pharmaceuticals and insecticides.

arsenic trifluoride *Inorganic Chemistry.* AsF_3, a yellow liquid that fumes in air and is decomposed by water; highly toxic.

arsenic trioxide *Inorganic Chemistry.* As_2O_3, a white, odorless, poisonous powder, slightly soluble in water, soluble in acids and alkalis; used in ceramics, glass, and pigments, in insecticides and herbicides, and for various other purposes. Also, ARSENOUS ACID.

arsenic trisulfide *Inorganic Chemistry.* As_2S_3, yellow or red crystals or powder, insoluble in water and melting at 300°C; used as a pigment and reducing agent.

arsenide *Chemistry.* a compound of arsenic and a metal.

arsenin *Organic Chemistry.* a heterocyclic organic compound composed of a six-atom ring system with no nitrogen atoms, in which the carbon atoms are unsaturated and the unique heteroatom is arsenic.

arseniopleite *Mineralogy.* $NaCaMn^{+2}(Mn^{+2},Mg)_2(AsO_4)_3$, a monoclinic mineral of the alluaudite group.

arseniosiderite *Mineralogy.* $CaFe_3^{+3}(AsO_4)_3O_2{\cdot}3H_2O$, an opaque, yellowish-brown monoclinic mineral occurring in fibrous aggregates and granular pseudomorphs after scorodite, having a specific gravity of 3.6 and a hardness of 4.5 (1.5 fibrous) on the Mohs scale.

arsenious see ARSENOUS.

arsenite *Inorganic Chemistry.* a salt or ester of arsenous acid (arsenic trioxide).

arseno- *Organic Chemistry.* relating to or describing an arseno compound.

arsenobenzene *Organic Chemistry.* $(C_6H_5As{=}AsC_6H_5)_3$, a white crystalline solid that melts at 212°C, soluble in hot water, alcohol, and benzene; used in making medicines to treat bacterial diseases.

arsenobismite *Mineralogy.* Near $Bi_2(AsO_4)(OH)_3$, a yellowish-brown mineral occurring in microcrystalline aggregates, having a specific gravity of 5.7; found with oxidized bismuth ore.

arsenoclasite *Mineralogy.* $Mn_5^{+2}(AsO_4)_2(OH)_4$, a red orthorhombic mineral found in Swedish dolomite.

arseno compound *Organic Chemistry.* a compound having the general formula $(RAs)_n$, where R is a functional group and the As=As bond is present.

arsenolamprite *Mineralogy.* As, a dark-gray orthorhombic dimorph of arsenic occurring in fibrous foliated masses, having a specific gravity of 5.3 to 5.5 and a hardness of 2.0 on the Mohs scale; found with native silver.

arsenolite *Mineralogy.* As_2O_3, a rare transparent, colorless or white cubic mineral of secondary origin occurring in octahedrons, having a specific gravity of 3.87 and a hardness of 1.5 on the Mohs scale; dimorphous with claudetite.

arsenopyrite *Mineralogy.* FeAsS, an opaque, silver-white to steel-gray monoclinic (pseudo-orthorhombic) mineral with a metallic luster and distinct cleavage, having a specific gravity of 5.9 to 6.2 and a hardness of 5.5 to 6 on the Mohs scale; the most abundant arsenic mineral, commonly found in metasomatic and metamorphic rocks and in medium- to high-temperature veins.

arsenous *Chemistry.* **1.** of or relating to arsenic. **2.** describing various compounds of arsenic, especially those in which the element has a valence of 3. For compounds of this type, see the entries ARSENIC TRIOXIDE, ARSENIC TRICHLORIDE, and so on.

arsenous acid or **arsenious acid** see ARSENIC TRIOXIDE.

arsine *Inorganic Chemistry.* AsH_3, a colorless, highly toxic gas released during electrochemical reactions involving arsenic compounds; soluble in water and slightly soluble in alcohol; used in organic synthesis and as a military poison. Also, ARSENIC HYDRIDE.

arsoite *Petrology.* an olivine-bearing diopside trachyte with a groundmass of sanidine, oligoclase, diopside, magnetite, and sodalite.

arsonic acid *Inorganic Chemistry.* an acid derived from arsenic acid, having various industrial uses.

arsonium *Inorganic Chemistry.* AsH_4, a radical that forms compounds such as AsH_4OH (**arsonium hydroxide**).

Arsonval, (Jacques) Arsene d' [də är´ sōn väl´] 1851–1940, French biophysicist; invented d'Arsonval galvanometer; a pioneer in physical therapy. See also D'ARSONVAL.

arsphenamine *Pharmacology.* $C_{12}H_{12}N_2O_2As_2{\cdot}2HCl{\cdot}2H_2O$, a yellow hygroscopic powder, the first drug discovered to treat syphilis, yaws, and other spirillum infections; now replaced by antibiotics. *Organic Chemistry.* see SALVARSAN.

Artacaminae *Invertebrate Zoology.* a subfamily of annelid polychaete worms with a mass of matted tentacles at the anterior end.

Artemia salina *Invertebrate Zoology.* the brine shrimp, a genus of crustaceans in the order Anostraca, found in salty lakes and saltpans.

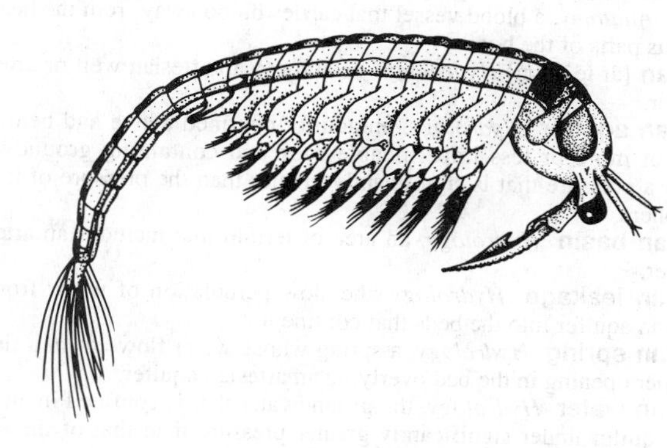

artemia salina

artenkreis *Evolution.* a group of closely related species that replace one another geographically.

arteri- a combining form meaning "artery."

arterial *Anatomy.* of or relating to an artery or arteries.

arterial aneurysm *Cardiology.* a localized saclike or fusiform bulging of an artery.

arterial blood *Hematology.* oxygenated blood, found in the pulmonary veins, the left chambers of the heart, and the systemic arteries; it is bright red in color.

arterial canal or **arterial duct** see DUCTUS ARTERIOSUS.

arterial cone see CONUS ARTERIOSUS.

arterial transport line *Transportation Engineering.* a high-speed, high-volume, long-distance transport route; a main line.

arteriectasis *Medicine.* the dilation and, usually, lengthening of an artery.

arterio- a combining form meaning "artery," as in *arteriogram.*

arteriogram *Medicine.* the filmed record produced by arteriography.

arteriography *Medicine.* the use of X-rays to produce a film of an artery or arteries after a radiopaque material has been introduced into the bloodstream.

arteriole [är tēr´ē ōl´] *Anatomy.* a small artery (8–50 μm in diameter in humans) leading into capillaries.

arteriolitis *Medicine.* an inflammation of the arterioles.

arteriolosclerosis [är tēr´ē ō´lō sklə rō´sis] *Medicine.* the sclerosis and thickening of the walls of the smaller arteries (arterioles).

arteriorrhexis *Medicine.* the rupture of an artery.

arteriosclerosis [är tēr´ē ō´sklə rō´sis] *Medicine.* a disease of the arteries characterized by thickening, loss of elasticity, and calcification of arterial walls, resulting in a decreased blood supply particularly to the cerebrum and lower extremities; it often develops with aging, and in hypertension and diabetes.

arteriosclerosis obliterans *Medicine.* arteriosclerosis in which the lumen of the artery has been completely obliterated by the proliferation of the intima of small vessels.

arteriotomy *Surgery.* the incision of an artery.

arteriovenous anastomosis *Anatomy.* a cross connection between a vein and an artery.

arteriovenous angiorrhaphy *Surgery.* the joining by suture of an artery and a vein in order to divert arterial blood into the vein. Also, ARTERIOVENOUS SHUNT

arteriovenous fistula *Medicine.* an abnormal communication between an artery and a vein, caused by either injury (**traumatic arteriovenous fistula**) or congenital abnormality (**congenital arteriovenous fistula**) and usually resulting in arteriovenous aneurysm.

arteriovenous shunt *Surgery.* see ARTERIOVENOUS ANGIORRHAPHY.

arterite *Petrology.* **1.** a migmatite produced by the injection of residual magmas from extrinsic sources. **2.** a veined gneiss derived from magmatic processes. **3.** a veined gneiss whose veins are formed from intrusions of molten granite secretions.
arteritic migmatite *Petrology.* a composite rock probably formed by the introduction of granitic magma parallel to the foliation in a schist or other layered rock.
arteritis *Medicine.* the inflammation of an artery.
Arterivirus *Virology.* a genus in the family Togaviridae that currently includes only equine arteritis virus, which infects horses by unknown vectors and causes marked necrosis of small arteries, often leading to spontaneous abortion in pregnant mares.
artery *Anatomy.* a blood vessel that carries blood away from the heart to various parts of the body.
artesian [âr tē´zhən] *Hydrology.* relating to an artesian well or artesian aquifer.
artesian aquifer *Hydrology.* an aquifer confined above and below by beds of more or less impermeable rock and containing groundwater under a pressure that is significantly greater than the pressure of the atmosphere.
artesian basin *Hydrology.* an area of terrain that includes an artesian aquifer.
artesian leakage *Hydrology.* the slow percolation of water from an artesian aquifer into the beds that confine it.
artesian spring *Hydrology.* a spring whose water flows from a fissure or other opening in the bed overlying an artesian aquifer.
artesian water *Hydrology.* the groundwater that is confined in an artesian aquifer under significantly greater pressure than that of the atmosphere, so that the water is capable of rising if given an opportunity to do so.
artesian well *Hydrology.* a well that taps into water that is confined in an artesian aquifer.
Arthoniaceae *Botany.* a family of lichens belonging to the order Hysteriales.
Arthoniales *Mycology.* an order of fungi belonging to the subdivision Ascomycotina, including certain fungi and lichens, that has been reclassified within the order Myriangiales.
Arthopyrenia *Mycology.* a genus of lichenized fungi in the family Arthopyreniaceae, most often found in cool-temperate regions.
Arthopyreniaceae *Mycology.* a family of fungi belonging to the order Pleosporales composed of species that live off decaying organic matter, as lichens on woody plant material or as parasites on algae.
arthral *Medicine.* of or relating to a joint. (From Greek *arthron,* "joint.")
arthralgia *Medicine.* pain in a joint.
arthrectomy [är threk´tə mē] *Surgery.* the surgical removal of a joint.
arthresthesia *Neurology.* the ability to perceive movement of the joints.
arthritic [är thrit´ik] *Medicine.* **1.** of, relating to, or afflicted with arthritis or gout. **2.** a patient afflicted with arthritis.
arthritide *Medicine.* any of various skin eruptions that originate with arthritis or gout.
arthritis [är thrī´tis] *Medicine. plural,* **arthritides.** rheumatism in which the inflammatory lesions are confined to the joints; any of numerous inflammatory conditions of the joints, most notably osteoarthritis and rheumatoid arthritis.

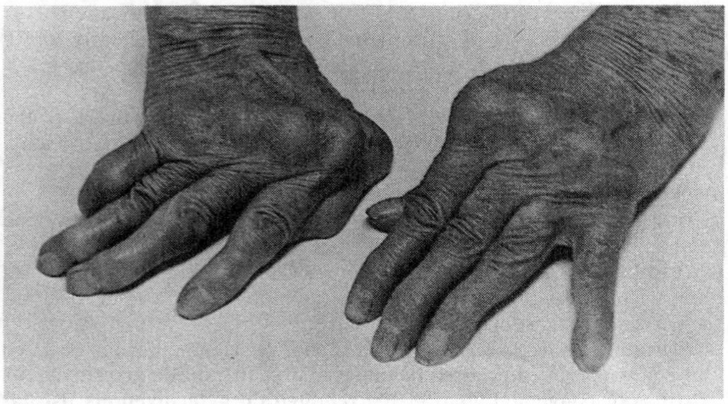

arthritis

arthritis deformans see RHEUMATOID ARTHRITIS.
arthro- a prefix meaning "joint," as in *arthropod, arthritis.* Also, **arthr-**
Arthroacus *Mycology.* a genus of fungi belonging to the family Saccharomycetaceae that develops yeastlike cells and a true hyphae.
Arthrobacter *Bacteriology.* a genus of Gram-variable, obligately aerobic bacteria of irregular cell form, occurring in soil.
Arthrobotrys *Mycology.* a genus of fungi belonging to the imperfect fungi (subdivision Deuteromycotina); it is most commonly encountered in soils and obtains nutrients from feeding off certain worms called nematodes.
arthrobranch *Invertebrate Zoology.* a malacostracan gill attached to the articular membrane between the body and basal joint of the leg of a crustacean.
arthrocentesis [arth´rō sen tē´sis] *Surgery.* the surgical puncture and removal of fluids from a joint.
arthrochondritis *Medicine.* the inflammation of the cartilage of a joint.
Arthrocladiaceae *Botany.* a monotypic family of multibranched brown algae of the order Desmarestiales, characterized by axes covered with fuzzy whorls of tufted filaments; occurs on most European coasts and on North America's eastern coast.
arthroclasia *Medicine.* the surgical breaking down of an ankylosis to enable free movement of a joint.
Arthroderma *Mycology.* a genus of fungi of the family Gymnoascaceae that comprises skin parasites called dermatophytes.
arthrodesis see ARTIFICIAL ANKYLOSIS.
arthrodia *Anatomy.* a joint in which articulating surfaces are flat, permitting free gliding movement in every direction, as between the patella and femur.
Arthrodira *Paleontology.* an order of primitive bottom-dwelling placoderms; characterized by a primitive, joint-necked structure, strong armor (especially the head shield), and a tuberculate surface; ranged in length from a few centimeters to 5 meters; the order includes several large suborders and numerous genera of uncertain affinities; Lower to Upper Devonian.
Arthrodonteae *Botany.* a family of mosses characterized by thin membranous peristome teeth surrounding the rim of the sporangium opening.
arthrodynia *Medicine.* any pain in a joint.
arthrogram *Medicine.* an X-ray record of a joint into which an opaque contrast medium has been introduced.
arthrography *Medicine.* an X-ray of a joint after injection of opaque contrast mateial.
arthrogryposis *Medicine.* **1.** the persistent flexing or contracting of a joint. **2.** tetanoid spasm.
arthrolysis [ar thräl´ə sis] *Surgery.* the surgical loosening of adhesions in an ankylosed joint.
Arthromitaceae *Bacteriology.* in former classifications, a family of parasitic bacteria found in the intestinal tracts of insects and crustaceans.
arthropathy *Medicine.* any joint disease.
arthroplasty *Medicine.* **1.** plastic surgery of a joint or joints. **2.** the formation of movable joints.
Arthropleura *Paleontology.* a genus of Carboniferous millipedes in the class Myriapoda and the extinct subclass Arthropleurida; *Arthropleura* attained lengths of up to six feet.
arthropod [ärth´rə päd] *Invertebrate Zoology.* **1.** any member of the phylum Arthropoda. **2.** of or relating to this phylum.
Arthropoda [ärth´rə pō´də] *Invertebrate Zoology.* the largest animal phylum, including insects, arachnids, and crustaceans, typified by segmented bodies and paired, jointed antennae, wings, or legs.
arthropod-borne virus see ARBOVIRUS.
arthropodin *Biochemistry.* a kind of protein that is soluble in water and forms part of the inner layer of the external protective covering of insects.
arthrosclerosis *Medicine.* the stiffening or hardening of a joint.
arthroscope [ärth´rə skōp´] *Medicine.* a tubelike instrument used to examine the interior of a joint and carry out diagnostic and therapeutic procedures within the joint, including minor surgical procedures.
arthroscopic [ärth´rə skäp´ik] *Medicine.* relating to or involving the use of an arthroscope. Thus, **arthroscopic surgery.**
arthroscopy [är thräs´kə pē] *Medicine.* the use of an arthroscope to examine the interior of a joint or to perform minor surgery.
arthrosis *Anatomy.* a degenerative condition of a joint.
arthrospore *Botany.* an isolated, thick-walled vegetative cell in a resting state, formed by segmentation of the filament in certain blue-green algae and fungi.

Arthrotardigrada *Invertebrate Zoology.* minute water bears in the phylum Tardigrada, usually with a median cirrus; marine and freshwater species.
arthrotomy [är thrät´ə mē] *Surgery.* an incision into a joint.
Arthus reaction *Immunology.* an immediate hypersensitive reaction occurring in blood vessel walls after the introduction of an antigen to an organism that already has a precipitating antibody to that antigen.
artichoke *Botany.* a tall thistlelike plant, *Cynara scolymus,* having enlarged fleshy involucral bracts; the bracts and the receptacle of the immature flower head are eaten as a vegetable.

artichoke

article *Invertebrate Zoology.* in arthropods, a segment of any articulated (jointed) structure or appendage. *Anatomy.* any one of the sections of a jointed series.
articulamentum *Invertebrate Zoology.* the inner, pearly layer of a chiton's shell plate, which is hinged with the plates on either side.
articular [är tik´yə lər] *Anatomy.* of or relating to a joint.
articular cartilage *Anatomy.* the cartilage of a joint or articulation.
articular disk *Anatomy.* a disk of cartilage between two bones in a joint.

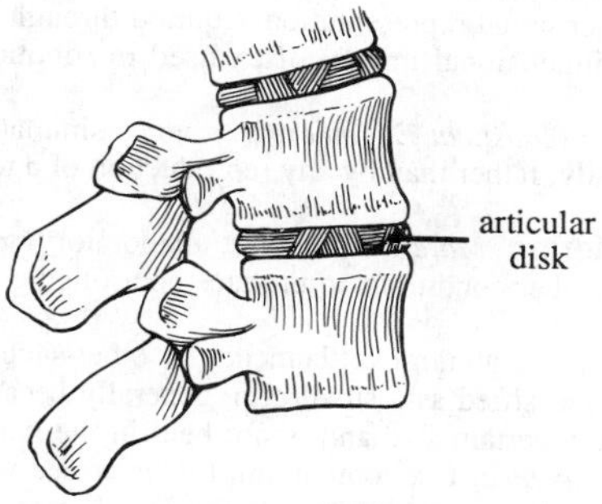

articular disk

articular membrane *Invertebrate Zoology.* a flexible layer of tissue acting as a joint between segments of the exoskeletons of arthropods.
Articulata *Invertebrate Zoology.* one of two classes of Brachiopoda (lamp-shells) having hinged valves and shells containing calcium carbonate, typically found in deep water.
articulate *Anatomy.* **1.** to form a joint. **2.** divided into or united by joints.
articulated *Science.* **1.** connected by movable joints. **2.** consisting of separate segments jointed so as to be movable with respect to each other.
articulated arm *Robotics.* a manipulator arm with several joints.
articulated drop chute *Engineering.* a structure that serves as a channel for a flow of concrete, made up of an arrangement of metal cylinders that are tapered so that the bottom end of each one fits into the top end of the cylinder below.
articulated leader *Mechanical Engineering.* a wheel-mounted unit with a pivoting loading element for moving soil, sand, and the like.
articulated train *Transportation Engineering.* a railroad train having cars that are permanently connected, usually with flexible hingelike connectors.
articulatio *Anatomy.* a term used in anatomical nomenclature to designate a place of junction of two or more bones of the skeleton, as indicated by the following term; e.g., **articulatio humeri** is the joint of the humerus (long bone of the arm) with the scapula (shoulder blade).
articulation the process of joining two or more parts, or the point where this occurs; specific uses include: *Anatomy.* a place of union or junction between two or more bones; a joint. *Botany.* a joint or point of union between two separate parts of a plant, as between the segments of a stem or fruit. *Robotics.* the manner in which the various parts of a robot are joined together to give it certain degrees of movement or freedom. *Telecommunications.* in telephony, a measure of the success with which sounds can be transmitted over a given assembly of apparatus, usually expressed as the fraction or percentage of the sounds correctly recognized by the listener.
articulation equivalent *Telecommunications.* a measure of the articulation achieved in a telephone connection or system, obtained by comparing the articulation of speech reproduced over it with a working reference system. Also, **articulation reference equivalent.**
artifact *Archaeology.* **1.** any object made by humans or affected in some way by human behavior. **2.** such an object used to identify or characterize a certain people, culture, or stage of development. *Radiology.* any feature or structure that appears as an image on a radiograph but is not naturally present in living tissue; such an image may result from the measuring instrument itself. *Histology.* any structure visible in a microscopic preparation caused by the activities of the preparator or the reagents used, and not characteristic of the living tissue. Also, **artefact.**
artifactitious *Radiology.* relating to or resulting from an artifact.
artifact theory *Paleontology.* a theory that provides an explanation for the absence in the fossil record of a long succession of pre-Cambrian ancestors for modern life forms by claiming that the record is incomplete and has not preserved these forms.
artifact type *Archaeology.* a category of artifacts whose attributes are similar; spoons or tables are artifact types.
artificial aging *Metallurgy.* aging caused by a deliberate thermal treatment performed above ambient temperature.
artificial ankylosis *Surgery.* the surgical stiffening of a joint by fusion of joint surfaces. Also, ARTHRODESIS, SYNDESIS.
artificial antenna see DUMMY ANTENNA.
artificial asteroid *Space Technology.* any object made and placed in solar orbit by humans.
artificial atmosphere *Chemical Engineering.* a gas or mixture of gases used in place of air in industrial or laboratory operations; classified as an active, or process, atmosphere, or as an inactive, or protective, atmosphere.
artificial carborundum see MOISSANITE.
artificial cardiac valve *Cardiology.* a synthetic valve surgically implanted to replace a dysfunctional heart valve.
artificial classification *Systematics.* a classification that is designed for convenience of use, or for identification purposes, and that is not based on any underlying concept of natural order.
artificial drive *Petroleum Engineering.* a technique for producing oil from a reservoir when natural solution-gas, drives-gas-cap, and water are not present or have been exhausted; waterflood, recycling or repressuring, and in situ combustion are types of artificial drives.
artificial echo *Electromagnetism.* a radar signal that is reflected from an artificial target such as a corner reflector or echo box.
artificial feel *Space Technology.* a control feel mechanically simulated in an aircraft's control system in which the craft controls are not directly connected to control surfaces and thus do not receive true feedback.
artificial fiber *Textiles.* a fiber that is made from synthetic materials such as nylon or rayon.
artificial gold see STANNIC SULFIDE.
artificial gravity *Space Technology.* the simulation of gravity within a spacecraft by means of rotation or acceleration.
artificial ground *Electricity.* an operating ground for antennas that is not directly connected to earth.
artificial heart *Medicine.* a pumping mechanism that duplicates the output, rate, and blood pressure of the natural heart; it may replace the function of the entire heart or any portion of it, and may be inside, outside, or alongside the body.
artificial horizon *Navigation.* any device for indicating the horizontal; used for taking celestial observations when a natural horizon is not available.
artificial hypothermia *Medicine.* the reduction of body temperature to counteract a prolonged fever or as a means of decreasing the need for oxygen during various surgical procedures, especially on the heart.
artificial insemination *Medicine.* the deposit of seminal fluid into the vagina or cervix by artificial means; the procedure is timed to coincide with the expected time of ovulation so that fertilization can occur.

Artificial Intelligence (AI)

The term AI, coined by John McCarthy in 1956, was preceded by an excellent essay on the subject by Alan Turing in 1950; mechanical thought was considered by Ada Lovelace, assistant to Charles Babbage, in 1842.

I define AI as the study of the computation required for intelligent behavior and the attempt to duplicate such computation using computers. Intelligent behavior connects perception of the environment to action appropriate for the goals of the actor. Intelligence, biologically costly in energy, pays for itself by enhancing survival. It isn't necessary to understand perfectly, but only to understand well enough to act appropriately in real time. Different points on the scale of intelligence versus cost, from insects to humans, are viable. Intelligence is computation in the service of life, just as metabolism is chemistry in the service of life. Airplanes fly with different hardware than birds; AI uses different hardware than humans, but will duplicate the essence of intelligence: appropriate connection between perception and action.

Current goals of AI include visual and speech perception, robotics, understanding human languages, representation of knowledge, reasoning, and learning. Expert systems have begun to capture human expertise and reproduce it in narrow domains. AI is a hard problem that will take much time and many advances; real progress has been made. Cognitive Science attempts to fuse AI with understanding of human cognition from psychology and neuroscience; this has occurred for early vision.

AI is controversial: it may redefine our place in the universe, as did astronomy and evolution, by removing our claim to uniqueness by virtue of superior intelligence. Critics have made false arguments against AI: since a Turing Machine has limits, it cannot be intelligent (as if humans had no limits!); or embodiment is required for intelligence. Ignore them: we have an existence proof, and we are it.

Gordon S. Novak, Jr.
Associate Professor, Computer Sciences
University of Texas at Austin

artificial intelligence *Computer Science.* a field of study concerned with the development and use of computer systems that have some resemblance to human intelligence, including such operations as natural-language recognition and use, problem solving, selection from alternatives, pattern recognition, generalization based on experience, and analysis of novel situations.

artificial ionization *Electronics.* the introduction of an artificial reflecting layer into the atmosphere for long-distance communication.

artificial kidney *Medicine.* a device to remove elements usually excreted in the urine from the blood while it is circulated outside the body. An intracorporeal artificial kidney using intestinal or pulmonary tissue as a filtration membrane is under development. Also, HEMODIALYZER.

artificial language *Artificial Intelligence.* a language that has been explicitly developed at a specific time, rather than evolving naturally over time through use by a community of speakers; for example, Esperanto.

artificial lift *Petroleum Engineering.* a procedure for producing oil from wells by injecting gas into the sand or rock formation, or by pumping with a rod, tubing, or bottom-hole centrifugal pump.

artificial load *Electricity.* **1.** a dissipative but largely nonradiating device which essentially has the impedance characteristics of an antenna or a transmission line for testing. **2.** a dissipative device used to terminate a generator or amplifier for testing.

artificial malachite see CUPRIC CARBONATE.

artificial monument *Engineering.* a manmade, relatively permanent structure that marks the site of a certain survey site.

artificial nourishment *Civil Engineering.* a process of replenishing a beach by depositing dredged material from another site, or by some other artificial means.

artificial oil of bitter almond *Organic Chemistry.* a popular name for benzaldehyde, C_6H_5CHO, an almond flavoring agent. See BENZALDEHYDE.

artificial radiation belt *Geophysics.* a belt of high-energy electrons trapped in the earth's magnetic field, as a result of nuclear explosions.

artificial radioactivity see INDUCED RADIOACTIVITY.

artificial radio aurora see RADIO AURORA.

artificial reality see VIRTUAL REALITY.

artificial recharge *Civil Engineering.* the use of injection wells or other techniques to recharge an aquifer depleted by abnormally large withdrawls.

artificial respiration *Medicine.* respiration that is maintained by manual or mechanical means when normal breathing has stopped.

artificial satellite *Space Technology.* an object that is made and placed in orbit by humans, usually around the earth but sometimes around another body.

artificial sea water *Oceanography.* a prepared solution that approximates the chemical composition of sea water.

artificial selection *Genetics.* a process of selective breeding by humans of another species, with the intention of producing economically or aesthetically desirable characteristics in the offspring.

artificial sky *Architecture.* a device used to study or display an architectural model under daylight conditions.

artificial sweetener *Food Technology.* any substitute for sugar, such as saccharin.

artificial upwelling *Oceanography.* the concept of mechanically inducing (probably by production of heat) the upwelling of water rich in nutrients toward the surface.

artificial vision *Artificial Intelligence.* computer technology that interprets a two-dimensional representation acquired through a sensor into a two- or three-dimensional image; often used in robotics. Also, COMPUTER VISION.

artificial voice *Acoustical Engineering.* a voice simulation that is created electronically, rather than by any reproduction of a recorded human voice.

artificial weathering *Materials Science.* a laboratory process that simulates actual weather conditions to produce controlled changes in materials.

artillery *Ordnance.* a gun, rocket launcher, or other such weapon that is too large to be classified as a small arm, generally because it fires ammunition above a certain size and is not held in the hand or hands for firing. *Military Science.* **1.** a combat unit that is armed with weapons of this kind. **2. Artillery.** the branch of the U.S. Army that uses such weapons.

artilleryman *Military Science.* a member of an artillery unit.

artillery sled *Ordnance.* a flat steel conveyance used to transport artillery weapons over snow, ice, and other such terrain where traction for wheels is difficult.

artillery survey *Ordnance.* the process of determining the relative locations of artillery pieces and their targets, and of providing accurate data for firing.

artillery train *Ordnance.* a group of field artillery pieces being transported together.

Artinian *Mathematics.* a ring or module is left (resp. right) Artinian if it satisfies the descending chain condition on left (resp. right) ideals. A ring or module is Artinian if it is both left and right Artinian.

artinite *Mineralogy.* $Mg_2(CO_3)(OH)_2 \cdot 3H_2O$, a white monoclinic mineral with a vitreous to silky luster, occurring in minute prismatic crystals and fibrous aggregates, having a specific gravity of 2.0 and a hardness of 2.5 on the Mohs scale.

Artinskian *Geology.* a European geologic stage of the Lower Permian period, occurring after the Sakmarian and before the Kungurian.

Artiodactyla *Vertebrate Zoology.* a large, diverse order of herbivorous, even-toed mammals (two- or four-toed hooves), including the antelope, pig, hippopotamus, giraffe, bison, and sheep.

ARTS *Meteorology.* a navigational air-traffic control system used in the vicinity of airports, utilizing both airport surveillance radar and an air-traffic radar beacon system. (An acronym for automated radar terminal system.)

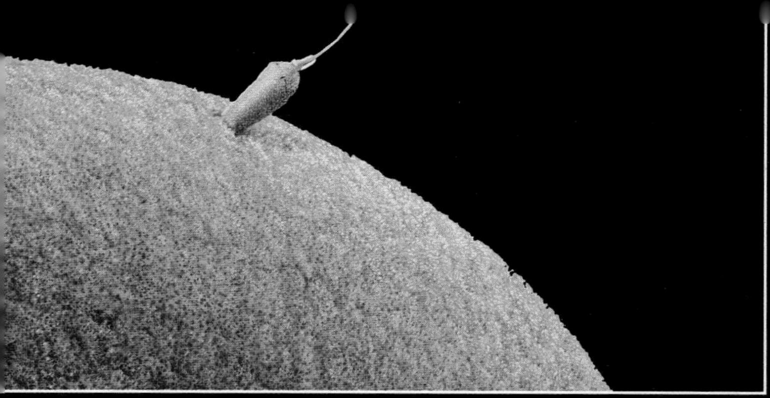

-luman spermatozoon fertilizing a human ovum.
-rancis Leroy/SPL/Photo Researchers.

Trail of hominid footprints, fossilized in volcanic ash, dating from 3.0 to 3.7 million years ago.
John Reader/SPL/Photo Researchers.

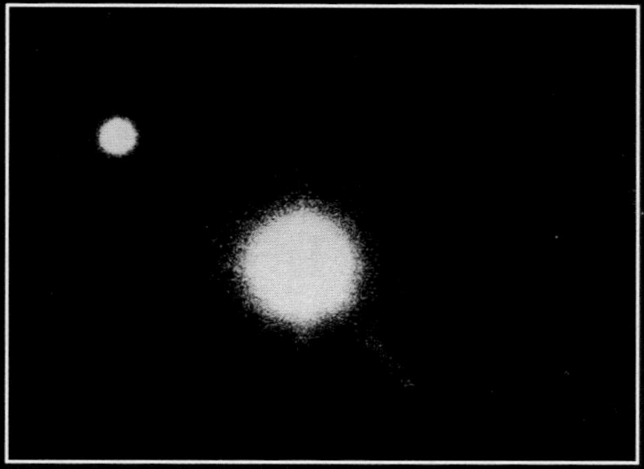

Quasar with associated jet. Courtesy of NASA.

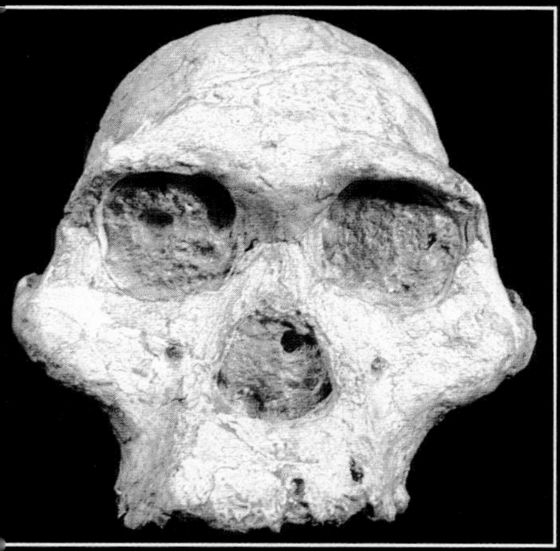

\dult skull, toothless and perhaps female, of *\ustralopithecus africanus* approximately 2.6 nillion years old. Eric Delson/City University of lew York.

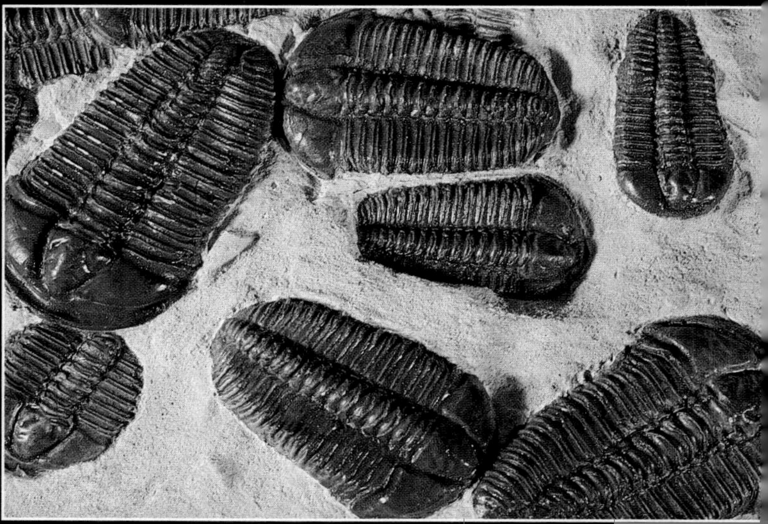

raphy

Standard Model of
FUNDAMENTAL PARTICLES AND INTERACTIONS

The "Standard Model" is a term used to describe the quantum theory that includes the theory of strong interactions (quantum chromodynamics or QCD) and the unified theory of weak and electromagnetic interactions (electroweak). Gravity is included on this chart because it is one of the fundamental interactions even though not part of the "Standard Model."

FERMIONS

matter constituents
spin = 1/2, 3/2, 5/2,...

Leptons spin = 1/2		
Flavor	Mass GeV/c^2	Electric charge
ν_e electron neutrino	$< 2 \times 10^{-8}$	0
e electron	5.1×10^{-4}	−1
ν_μ muon neutrino	$< 3 \times 10^{-4}$	0
μ muon	0.106	−1
ν_τ tau neutrino	$< 4 \times 10^{-2}$	0
τ tau	1.784	−1

Quarks spin = 1/2		
Flavor	Approx. Mass GeV/c^2	Electric charge
u up	4×10^{-3}	2/3
d down	7×10^{-3}	−1/3
c charm	1.5	2/3
s strange	0.15	−1/3
t top (not yet observed)	> 89	2/3
b bottom	4.7	−1/3

Spin is the *intrinsic* angular momentum of particles. Spin is given in units of $\hbar$, which is the quantum unit of angular momentum, where $\hbar = h/2\pi = 6.58\times10^{-25}$ GeV s $= 1.05\times10^{-34}$ J s.

Electric charges are given in units of the proton's charge. In SI units the electric charge of the proton is 1.60×10^{-19} coulombs.

The **energy** unit of particle physics is the electron volt (eV), the energy gained by one electron in crossing a potential difference of one volt. **Masses** are given in GeV/c^2 (remember $E = mc^2$), where 1 GeV = 10^9 eV = 1.60×10^{-10} joule. The mass of the proton is 0.938 GeV/c^2 = 1.67×10^{-27} kg.

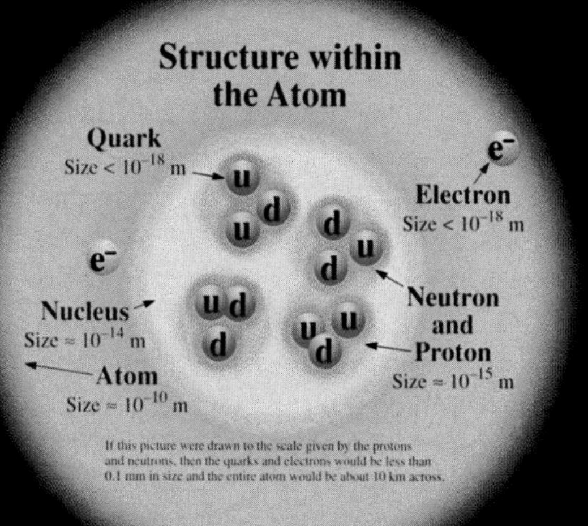

BOSONS

force carriers
spin = 0, 1, 2,...

Unified Electroweak spin = 1	Mass GeV/c^2	Electric charge
γ photon	0	0
W^-	80.6	−1
W^+	80.6	+1
Z^0	91.16	0

Strong or color spin = 1	Mass GeV/c^2	Electric charge
g gluon	0	0

Color Charge

Each quark carries one of three types of "strong charge," also called "color charge." These charges have nothing to do with the colors of visible light. There are eight possible types of color charge for gluons. Just as electrically charged particles interact by exchanging photons, in strong interactions color-charged particles exchange gluons. Leptons, photons, and *W* and *Z* bosons have no color charge and hence no strong interactions. One cannot isolate quarks and gluons; they are ***confined*** into color-neutral hadrons. This confinement (binding) results from multiple exchanges of gluons among the color-charged objects.

Confinement

As color-charged particles (quarks and gluons) are separated, the color force between them approaches a constant value and the energy in the color-force field increases. This energy eventually is converted into additional quark-antiquark pairs (see the figures below). The objects that finally emerge are color-neutral combinations called **hadrons** (mesons and baryons)

Residual Strong Interactions

The strong binding of the color-neutral protons and neutrons to form nuclei is due to residual strong interactions between their color-charged constituents. It is similar to the residual electrical interaction which binds electrically neutral atoms to form molecules. It can be viewed as the exchange of mesons between the hadrons.

PROPERTIES OF THE INTERACTIONS

Sample Fermionic Hadrons: Baryons qqq and Antibaryons $\bar{q}\bar{q}\bar{q}$					
Symbol	Name	Quark content	Electric charge	Mass GeV/c^2	Spin
p	proton	uud	1	0.938	1/2
$\bar{p}$	anti-proton	$\bar{u}\bar{u}\bar{d}$	−1	0.938	1/2
n	neutron	udd	0	0.940	1/2
Λ	lambda	uds	0	1.116	1/2
Ω^-	omega	sss	−1	1.672	3/2

Property \ Interaction	Gravitational	Weak (Electroweak)	Electromagnetic (Electroweak)	Strong: Fundamental	Strong: Residual
Acts on:	Mass – Energy	Flavor	Electric Charge	Color charge	See Residual Strong Interaction Note
Particles experiencing:	All	Quarks, Leptons	Electrically charged	Quarks, Gluons	Hadrons
Particles mediating:	Graviton (not yet observed)	W^+ W^- Z^0	γ	Gluons	Mesons
Strength for two u quarks at 10^{-18} m (relative to electromagnetic)	10^{-41}	0.8	1	25	Not applicable to quarks
Strength for two u quarks at 3×10^{-17} m (relative to electromagnetic)	10^{-41}	10^{-4}	1	60	Not applicable to quarks
for two protons in nucleus	10^{-36}	10^{-7}	1	Not applicable to hadrons	20

Sample Bosonic Hadrons: Mesons $q\bar{q}$					
Symbol	Name	Quark content	Electric charge	Mass GeV/c^2	Spin
π^+	pion	$u\bar{d}$	+1	0.140	0
K^-	kaon	$s\bar{u}$	−1	0.494	0
ρ^+	rho	$u\bar{d}$	+1	0.770	1
D^+	D^+	$c\bar{d}$	+1	1.869	0
η_c	eta-c	$c\bar{c}$	0	2.980	0

Matter and Antimatter

For every particle type there is a corresponding antiparticle type, denoted by a bar over the particle symbol. Particle and antiparticle have identical mass and spin but opposite charges. Some electrically neutral bosons (e.g., Z^0, γ, and $\eta_c = c\bar{c}$, but not $K^0 = d\bar{s}$) are their own antiparticles.

Figures

These diagrams are an artist's conception of physical processes. They are *not* exact and have *no* meaningful scale. Green shaded areas represent the cloud of gluons or the gluon field, red lines the quark paths, and black lines the paths of leptons.

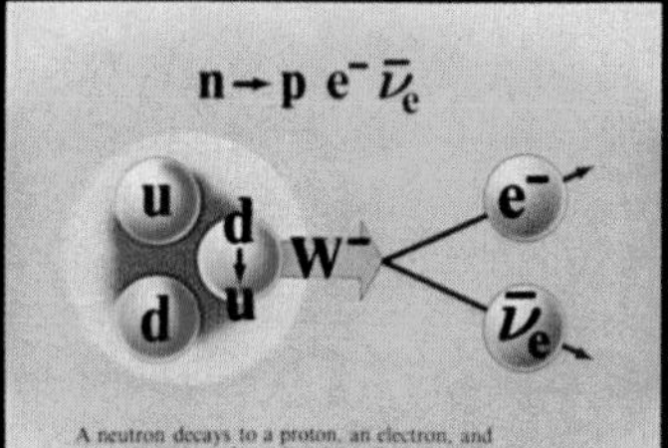

A neutron decays to a proton, an electron, and an antineutrino via a virtual (mediating) W boson. This is neutron β decay.

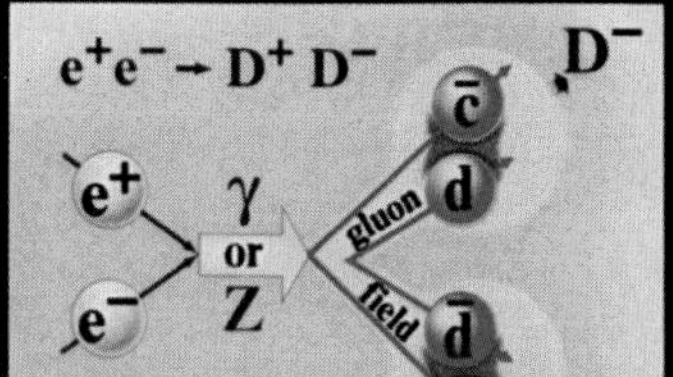

An electron and positron (antielectron) colliding at high energy can annihilate to produce D^+ and D^- mesons via a virtual Z boson or a virtual photon.

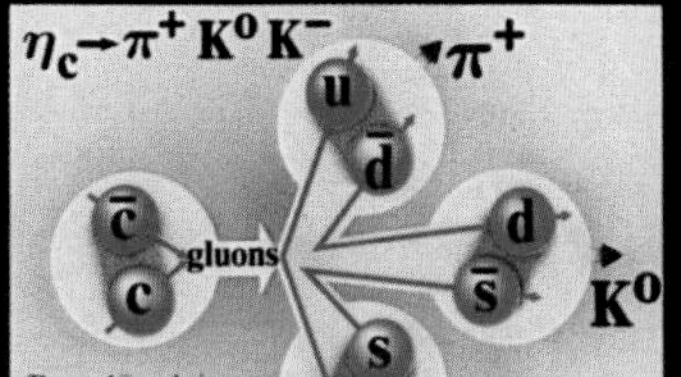

The c and $\bar{c}$ quarks in an η_c meson annihilate into virtual gluons. Quark pair production in the gluon cloud sometimes gives π^+, K^0, and K^- as the final products.

Contemporary Physics Education Project

This chart was created by the Contemporary Physics Education Project. For information on materials prepared to supplement this chart (book, computer software, film), write: CPEP, MS 50-308, Lawrence Berkeley Laboratory, Berkeley, CA 94720. Production of this chart was supported by funding from:

U.S. DEPARTMENT OF ENERGY
LAWRENCE BERKELEY LABORATORY
STANFORD LINEAR ACCELERATOR CENTER
SUPERCONDUCTING SUPER COLLIDER LABORATORY
CERN-MICROCOSM
IBM
ROCKWELL INTERNATIONAL
MARTIN MARIETTA ASTRONAUTICS GROUP
AMERICAN ASSOCIATION OF PHYSICS TEACHERS
BURLE Electron Tubes

Arundel method *Cartography*. a method, based on radial triangulation and graphic and analytical techniques, for point-by-point topographic mapping from aerial photographs.

aryl *Organic Chemistry*. **1.** an organic group produced by removing one hydrogen atom from an aromatic hydrocarbon. **2.** of or relating to such a group.

aryl acid *Organic Chemistry*. an organic acid that contains an aryl group.

arylalkyl see ARALKYL.

arylamine *Organic Chemistry*. an aromatic compound that contains at least one amine group.

aryl compound *Organic Chemistry*. a compound whose molecules have the six-carbon ring structure that is characteristic of aromatic compounds, that of either benzene or another aromatic series; e.g., phenyl, C_6H_5-.

aryldiazo compound *Organic Chemistry*. a diazo group attached to the ring structure of benzene, a benzene derivative, or some other aromatic derivative.

arylene *Organic Chemistry*. a bivalent radical formed by the removal of hydrogen from two carbon sites on an aromatic compound.

aryl group see ARYL.

aryl halide *Organic Chemistry*. an aromatic compound in which a halide atom has replaced one of the ring hydrogens.

arylide *Organic Chemistry*. an aromatic compound consisting of a metal and an aryl group.

aryloxy compound *Organic Chemistry*. a group of aryl compounds having an aryl radical and oxygen, such as (2,4-dichlorophenoxy)acetic acid (2,4-D), widely used as weed killers.

arylsulphatase *Enzymology*. an enzyme that catalyzes the hydrolysis of a phenyl sulfate to a phenol and sulfate.

arylsulphatase test *Microbiology*. an assay for the identification of species of *Mycobacterium* that, if grown in a liquid medium containing an indicator dye, will cause a color change in this medium upon addition of an alkali.

aryne *Organic Chemistry*. a class of very reactive aromatic compounds in which two adjacent atoms of a ring lack substituents, with two orbitals each missing an electron.

arytenoid cartilage *Anatomy*. either of two cartilages of the larynx to which the vocal cords are attached.

AS *Aviation*. the airline code for Alaska Airlines.

AS-1 *Ordnance*. an early Soviet air-to-surface antishipping missile powered by a turbojet engine, delivering a conventional warhead at high subsonic speed and a maximum range of around 65 miles. A surface-to-surface version used for coastal defense is code-named **Samlet.**

AS.12 *Ordnance*. a French air-to-surface missile powered by a two-stage solid-propellant rocket engine and equipped with a wire-transmitted optical-tracking guidance system, delivering a conventional 63-pound armor-piercing charge at a speed of 210 miles per hour and a maximum range of 5 miles; alternately, it may carry an antitank-shaped charge or prefragmented antipersonnel charge.

AS-15 *Ordnance*. a Soviet air-launched cruise missile capable of carrying a single nuclear warhead at a range of up to 1900 miles.

AS.20 *Ordnance*. a French air-to-surface missile powered by a dual-thrust solid-propellant rocket engine and equipped with a radio-command guidance system, delivering a 66-pound conventional warhead at a speed of Mach 1.7 and a range of 1.8 to 4 miles.

AS.30 *Ordnance*. a French air-to-surface tactical missile powered by a two-stage solid-propellant rocket engine and equipped with a radio-command or TCA automatic guidance system, delivering a 529-pound conventional warhead at a speed of Mach 1.5 and a maximum range of 7 miles. The **AS.30L** is equipped with an automatic tracking laser illumination guidance system.

ASA or **A.S.A.** American Surgical Association; American Society of Anesthesiologists; Acoustical Society of America.

Asaphus *Paleontology*. a genus of large, smooth trilobites in the suborder Asaphina, reaching lengths of almost 2 feet; *Asaphus* first appeared in the Middle Cambrian but is especially characteristic of the Lower Ordovician.

asapocynin *Pharmacology*. $C_9H_{10}C_3$, a substance that is derived from Canadian hemp and other plants of the genus *Apocynum*, used as a heart stimulant similar to digitalis. Also, ACETOVANILLON, APOCYNIN.

asarone *Organic Chemistry*. $C_{12}H_{16}O_3$, a crystalline solid that melts at 67°C, insoluble in water and soluble in alcohol; produced by plants of the genus *Asarum*; used as a constituent of essential oils.

ASB or **A.S.B.** American Society of Bacteriologists.

asbestos *Mineralogy*. a general term for any of various silicate minerals that can be separated into long, thin heat-resistant, flexible fibers used for a number of commercial purposes. Asbestos can be divided into two general structural groups, amphibole and serpentine, each of which contains an array of fibrous species. *Materials Science*. relating to or containing such materials. Thus, **asbestos board, asbestos felt, asbestos insulation, asbestos pipe, asbestos plaster, asbestos roofing, asbestos shingles,** and so on.

asbestos cement *Building Engineering*. a mixture of asbestos fibers, Portland cement, and water; used to make the wallboard known as **asbestos sheeting** or **asbestos-cement board.**

asbestos-cement cladding *Building Engineering*. the formation of a wall or wall facing using asbestos boards and component wall pieces that are directly supported by framing elements.

asbestosis *Medicine*. a lung disease (pneumoconiosis) caused by inhaling asbestos fibers; marked by intersititial fibrosis of the lung, and associated with pleural mesothelioma and bronchogenic carcinoma.

asbolite *Mineralogy*. a soft, black mineral aggregate containing hydrated oxides of manganese; classed as a type of wad, and rich (as much as 32%) in cobalt. Also, BLACK COBALT.

as-brazed *Metallurgy*. the condition of a metal or alloy that has been brazed, but not subjected to later thermal or mechanical treatments.

A scale *Acoustics*. **1.** a diatonic scale that starts on the standard musical reference note A, having a frequency of 440 hertz. **2.** a scale of noise measurement, represented as L_A, to which most sound measuring devices are calibrated, so that the instrument has a response approximating the 40-phon loudness contour for an 8000-hertz tone.

A-scan *Optics*. an ultrasonic technique used in nondestructive evaluation, in which a stationary transducer is used in conjunction with an oscilloscope to detect internal cracks or other flaws in a body. *Radiology*. a visual display of ultrasonographic echoes on a cathode ray tube in which one axis represents the elapsed time of echo return and the other corresponds to echo strength. *Electronics*. see A DISPLAY.

Ascaphidae *Vertebrate Zoology*. a small family of primitive frogs in the order Anura, found in New Zealand and the northwest United States.

ascariasis *Medicine*. an infection by roundworms, especially *Ascaris lumbricoides,* found in the small intestine and causing colicky pain and diarrhea, usually in children.

ascarid *Invertebrate Zoology*. any nematode worm of the family Ascaridae, which includes common intestinal parasites of terrestrial mollusks, arthropods, and vertebrates.

Ascaridata see ASCARIDINA.

Ascarididae *Invertebrate Zoology*. a family of nematodes in the superfamily Ascaridoidea, with a well-developed lip region separated from the cervical region; parasites of the intestinal tract in land mammals, reptiles, and amphibians.

Ascaridina *Invertebrate Zoology*. a suborder of Rhabditida comprising sizable nematode worms, with no stylet; parasites of various arthropods, mollusks, and vertebrates (including humans).

Ascaridoidea *Invertebrate Zoology*. a superfamily in the suborder Ascaridina; medium to large nematode worms, intestinal parasites of vertebrates, with a highly muscular esophagus. Also, ASCAROIDEA.

ascaridole *Organic Chemistry*. $C_{10}H_{16}O_2$, an unstable terpene peroxide liquid that will explode when heated to 115°C, soluble in alcohol, acetone, and benzene; used to initiate polymerization reactions and formerly used in medicine.

Ascaris

Ascaris *Invertebrate Zoology*. a genus of nematode worms having three-lipped mouths and looking like earthworms; includes the common roundworm.

Ascaroidea see ASCARIDOIDEA.

ascaroid nematode *Invertebrate Zoology*. roundworms of the superfamily Ascaridoidea, parasitic in vertebrates, including humans; *Ascaris lumbricoides* is a typical species.

ascarylose *Biochemistry.* a sugar present in the lipopolysaccharide of certain strains of a Yersinia pseudotuberculosis.
ascend *Aviation.* to move upward in a flight vehicle, often directly upward in a near-vertical trajectory, as opposed to the more gradual trajectory of a climb.
ascender *Graphic Arts.* a part of a lowercase letter that extends above the main element of the letter; the letters b, d, f, h, k, l have ascenders.
ascending aorta *Anatomy.* the proximal part of the aorta, arising from the left ventricle and giving rise to the right and left coronary arteries before continuing into the arch of the aorta.
ascending branch *Mechanics.* the portion of an object's trajectory prior to reaching the summit, along which its altitude increases constantly with time.
ascending central series *Mathematics.* let $C_1(G)$ denote the center of a group G, i.e., the (normal) subgroup of G consisting of all elements of G which commute with every element of G. Let $C_2(G)$ denote the inverse image of $C(G/C_1(G))$ under the canonical projection $G \to G/C_1(G)$. It can then be shown that $C_2(G)$ is a normal subgroup of G which contains $C_1(G)$. Then continue this process inductively: $C_i(G)$ is the inverse image of $C(G/C_{i-1}(G))$ under the canonical projection $G \to G/C_{i-1}(G)$. The resulting sequence of normal subgroups of G, denoted $(e) < C_1(G) < C_2(G) < \cdots$, is the ascending central series of G (e is the identity element of G and $<$ means "is a normal subgroup of"). If $C_n(G) = G$ for some n, then G is said to be **nilpotent.**
ascending chain condition *Mathematics.* a module A is said to satisfy the ascending chain condition (ACC) on submodules (or to be **Noetherian**) if for every chain $\cdots A_3 \supset A_2 \supset A_1$ of submodules of A there exists an integer n such than $A_i = A_n$ for all $i \geq n$.
ascending chromatography *Analytical Chemistry.* a separation technique in which the mobile phase of a mixture rises through the stationary phase due to capillary action.
ascending colon *Anatomy.* the part of the colon lying between the cecum and the hepatic flexure.
ascending node *Astronomy.* the point in orbit at which a body enters the northern celestial hemisphere. *Space Technology.* the reference point on the earth's surface that an orbiting satellite passes as it moves north across the equator.
ascending series *Mathematics.* a series whose terms form an increasing sequence.
ascending sort *Computer Programming.* a sort process that results in a sequence of records in which the record with the lowest-valued key field appears first, and each subsequent record has a higher key value.
Aschelminthes *Invertebrate Zoology.* in some classifications, a phylum of pseudocoelomate animals (roundworms, rotifers, and allies) including the classes Rotifera, Gastrotricha, Kinorhyncha, Nematoda, Nematomorpha, and sometimes Priapuloidea and Acanthocephala, all of which have at times been regarded as separate phyla.
Aschersonia *Mycology.* a genus of fungi belonging to the order Sphaeropsidales, and occurring in warm habitats; it is parasitic on scale insects and white flies, and is used commercially as an insecticide.
Aschheim-Zondek test *Pathology.* a test for pregnancy, involving the injection of the urine of women thought to be in the first trimester of pregnancy into immature female mice; the indication of pregnancy is the swelling and hemorrhages of the ovaries of the mice and premature maturation of the ovarian follicles. (Named after Selmar *Aschheim,* 1878–1965, and Bernhardt *Zondek,* 1891–1966, German gynecologists.)
aschistic *Geology.* of rocks of a minor igneous intrusion, having a composition equivalent to the parent magma and showing no significant differentiation into light and dark portions.
Aschoff bodies *Medicine.* characteristic collections of cells and leukocytes that occur in the interstitial tissues of the heart in rheumatic myocarditis. Also, **Aschoff nodules.**
Ascidiacea *Invertebrate Zoology.* the sea squirts, a class or order of simple or compound tunicates, usually sedentary at maturity, having a saclike covering through which water channels.
ascidiform *Botany.* pitcher-shaped.
ascidium *Botany.* any pitcher-shaped plant organ or appendage.
ascigerous *Mycology.* referring to fungi having spore sacs called **asci.**
ASCII [as´kē´] *Computer Programming.* a widely used standard 8-bit code for representing characters as binary numbers; used for information interchange among data processing systems, data communication systems, and associated equipment; consists of seven information bits and one parity bit, which can be combined to represent 128 letters, numerals, punctuation, and special control characters. (An acronym for American Standard Code for Information Interchange.)
ASCII file *Computer Programming.* a file consisting of ASCII characters with no nonprinting characters or commands, so that it can be read by various text processors. Also, TEXT FILE.
ascites [ə sit´ēz] *Medicine.* an effusion and accumulation of serous fluid in the abdominal cavity.
ascitic tumor *Oncology.* a tumor found in the peritoneal cavity which results in the production of hydroperitoneum or ascites.
Asclepiadaceae *Botany.* the milkweed family of herbs, shrubs, and vines, characterized by an umbrellalike cluster of small flowers and long pods that split open to release numerous flattened airborne seeds, each having a plume of long silky hairs. Most species have milky juice; many are medicinal or poisonous.
asco- a combining form meaning "sac," as in *ascospore.*
ascocarp *Mycology.* in fungi belonging to the subdivision Ascomycotina, the reproductive portion of the microorganism that holds the spore sacs or asci. Also, **ascoma.**
Ascochyta *Mycology.* a genus of fungi of the order Sphaeropsidales; one species, *Ascochyta pisi,* causes leafspot disease in the pea plant.
ascogenous *Mycology.* referring to fungi that form asci.
ascogonium *Mycology.* the female sexual organ or cell in fungi belonging to the subdivision Ascomycotina. Also, **ascogone.**
Ascolichenes *Botany.* a group of lichens in which the fungus is an ascomycete.
ascomycetes *Mycology.* a former term for a class of fungi, still commonly used to describe members of the subdivision Ascomycotina characterized by producing asci. Also, SAC FUNGI.
Ascomycotina *Mycology.* a subdivision of fungi characterized by its ascospores, which are produced sexually in asci; it includes such fungi as edible truffles and morels, most yeasts, and some mildews, and is composed of five classes: Hemiascomycetes, Loculoascomycetes, Plectomycetes, Pyrenomycetes, and Discomycetes. Also, SAC FUNGI.
ascon *Invertebrate Zoology.* a sponge or sponge larva having inhalant canals leading directly to the paragaster, and connecting with the exterior directly through a vent.

ascon

A scope *Electronics.* a radar scope that produces an A display.
ascorbic acid *Biochemistry.* $C_6H_8O_6$, vitamin C, a water-soluble vitamin that is found in numerous fruits and vegetables and is necessary for human growth; its absence leads to scurvy and poor wound repair.
Ascoseirales *Botany.* a monotypic order of brown algae of the class Phaeophyceae, composed of the family Ascoseiraceae and occurring in Antarctica and subantarctic regions; characterized by a simple blade that continuously splits into equal segments until multiple thick, narrow blades are attached to a branched stipe.
Ascosphaeraceae *Mycology.* a family of fungi belonging to the order Ascosphaerales that occurs in association with bees and beehives; some species are pathogenic to bees.
Ascosphaerales *Mycology.* an order of fungi belonging to the class Plectomycetes that is characterized by asexual reproduction and occurs in association with pollen and bees.

ascospore *Mycology.* a fungal spore of the subdivision Ascomycotina that is produced sexually within an ascus.
ascostroma *Mycology.* the ascocarp or reproductive structure of certain fungi that bears the spore sacs within cavities called locules.
Ascothoracica *Invertebrate Zoology.* an order of parasitic cirripedes, marine crustaceans with larvae living on sea anemones or sea urchins and maturing into sessile adults.
Ascovirus *Virology.* a genus of large, enveloped dsDNA viruses that have been isolated from lepidopteran larvae and replicated rapidly in the nuclei of various tissues.
ASCP or **A.S.C.P.** American Society of Clinical Pathologists.
ascus *plural,* **asci.** *Mycology.* in fungi of the subdivision Ascomycotina, the spore sac within which ascospores are sexually produced.
asdic *Electronics.* a British term for underwater acoustic detection equipment. (An acronym for Anti-Submarine Detection Investigation Committee.)
aseismic area *Geophysics.* a geographic area in which earthquakes do not usually occur.
Asellariaceae *Mycology.* the only family of fungi belonging to the order Asellariales; it occurs in freshwater, marine, and land environments as a parasite of isopods and insects.
Asellariales *Mycology.* an order of fungi belonging to the class Trichomycetes, characterized by branched, septate thalli.
Aselloidea *Invertebrate Zoology.* a group of free-living isopod crustaceans belonging to the suborder Asellota, found in fresh water.
Asellota *Invertebrate Zoology.* a suborder of Isopoda with over 20 families found in marine and freshwater habitats, often considered to contain the most primitive isopods.
asemasia *Neurology.* a loss or lack of ability to communicate by speech or by signals.
asemia *Neurology.* an inability to use or understand either speech or signs, due to a central lesion.
asepsis *Surgery.* the absence of, or prevention of contact with, living pathogenic organisms; a condition of sterility.
aseptic *Medicine.* **1.** free from infection. **2.** free from septic material; sterile.
asepticism *Surgery.* the theory and method of aseptic surgery.
aseptic meningitis *Medicine.* a mild form of meningitis, usually viral meningitis, characterized by malaise, fever, headache, nausea, stiffness in the neck and back, and a short, uncomplicated course.
aseptic surgery *Surgery.* surgery that is conducted in sterilized conditions in order to prevent infection by pathogenic organisms.
asexual [ā´seks´yoo əl] *Biology.* **1.** not involving sexual processes, as some types of reproduction. **2.** lacking functional sex organs. *Psychology.* lacking sexual interest or identity. Thus, **asexuality.**
asexual reproduction *Biology.* a form of reproduction in which new individuals are formed without the involvement of gametes and usually by a single parent. In plants, it includes vegetative propagation and the production of spores, and in animals it includes budding and fission.
ash *Botany.* any of several species of deciduous timber trees of the genus *Fraxinus,* characterized by opposite, pinnately compound leaves and a samara fruit. *Materials.* the wood of various species of ash trees. *Chemistry.* the residual inorganic matter left after an organic substance has been completely burned. *Engineering.* the residue from the burning of diesel fuel, used as a quantitative indicator for fuel cleanliness. *Volcanology.* fine or very fine pyroclastic particles, less than 4 millimeters in diameter, that are blown out from a volcanic explosion.

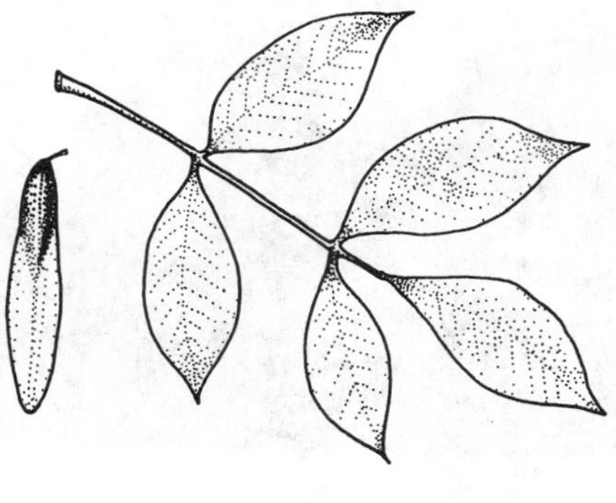

ash

Ash *Ordnance.* a Soviet air-to-air missile believed to be powered by a solid-propellant rocket motor and equipped with infrared or semiactive radar homing; it probably carries a large conventional warhead and cruises at a speed of Mach 3 and a range of 15 to 35 miles; officially designated **AA-5.**
ASH or **A.S.H.** American Society of Hematology.
ASHA or **A.S.H.A.** American Speech and Hearing Association.
Ashby *Geology.* a North American geologic stage of the Middle Ordovician period, occurring after the Marmor and before the Porterfield.
ash collector see DUST CHAMBER.
ash cone *Volcanology.* a conical hill composed mainly of unconsolidated volcanic ash that forms around and above a volcanic vent.
ash dump *Engineering.* a covered opening in a fireplace floor that permits accumulated ashes to drop down into the ash pit for later removal.
ashen light *Astronomy.* a faint glow, possibly an optical illusion, on the unilluminated side of Venus.
ash fall *Volcanology.* a shower of volcanic ash that falls from an eruption cloud, or a deposit of such material lying on the ground. Also, ASH SHOWER.
ash field *Volcanology.* an area covered by a thick, extensive accumulation of unconsolidated volcanic ash.
ash flow *Volcanology.* an extremely hot mixture of ash and volcanic gases ejected by an explosion of magma from a volcanic crater or fissure. Also, PYROCLASTIC FLOW, INCANDESCENT TUFF FLOW, GLOWING AVALANCHE.
ash furnace *Engineering.* a furnace used to frit materials for glassmaking.
ash fusibility *Geology.* the ability of coal ash to gradually soften and melt with an increase in temperature as a result of chemical reactions and the melting of its components.
Ashgillian *Geology.* a European geologic stage of the Middle Ordovician period, occurring after the Caradocian and before the Llandoverian of the Lower Silurian.
ashing *Analytical Chemistry.* the heating of a substance to leave only noncombustible ash, which is analyzed for elemental composition.
ashlar *Civil Engineering.* **1.** stones as they come from a quarry. **2.** stones that are cut and dressed.
ash oven see ASH FURNACE.
ash pan *Building Engineering.* a pan under a fireplace grate, used for collecting and removing ashes.
ash pit *Engineering.* an area beneath a fireplace in which ashes are accumulated for removal.
ash rock *Volcanology.* rock composed of sandy material that is produced from volcanic explosions.
ash shower see ASH FALL.
ashstone *Petrology.* a rock composed of lithified fine-grained volcanic ash.
ashy grit *Geology.* **1.** a deposit of fine pyroclastic material that is sand-sized or smaller. **2.** a mixture consisting of volcanic ash and sand.
Asia *Geography.* the world's largest continent (area: 17,012,000 square miles) in the Eastern Hemisphere north of the equator; part of the Eurasian land mass, separated from Europe by the Ural Mountains.
Asia Minor *Geography.* a peninsula in southwestern Asia, bounded by the Mediterranean, Aegean, and Black seas; the Asian part of Turkey. Also, **Anatolia.**
Asian flu *Medicine.* **1.** a pandemic outbreak of influenza A that occurred in 1957 and was thought to have originated in China. **2.** any similar acute viral respiratory infection originating with an influenza A virus, especially an A-2 virus.
Asiatic ash see MANCHURIAN ASH.
ASIC *Computer Technology.* an array of special-purpose processor components. (An acronym for Application Specific Integrated Circuit.)
asiderite see STONY METEORITE.
Asilidae *Invertebrate Zoology.* robber flies, a family of sizable, slender two-winged flies with strong legs and wings and a hardened beak (proboscis) used for sucking the body fluids of insects captured in flight.
ASIM or **A.S.I.M.** American Society of Internal Medicine.
A site *Molecular Biology.* one of two tRNA binding sites on the ribosome; holds the incoming tRNA molecule charged with the next amino acid waiting to be attached to the growing polypeptide chain.
A size *Engineering.* one of a series of specific sizes to which items are cut in manufacturing paper or board.
ASM or **A.S.M.** American Society for Microbiology; American Society for Metals.
ASM or **asm** air-to-surface missile.

Asobolaceae *Mycology.* a family of fungi belonging to the order Pezizales, occurring primarily on dung and sometimes on wood, soil, charcoal, or decaying plant material.

asocial [ā´sō´shəl] *Psychology.* indifferent to or unaware of social codes and values.

Asocorticeae *Mycology.* a family of fungi belonging to the order Helotiales, composed of parasites forming a crustlike growth on the bark, branches, or leaves of trees.

Asoideceae *Mycology.* a family of fungi belonging to the order Endomycetales, occurring in soil, grasses, insects, and plants.

Asopinae *Invertebrate Zoology.* a family of hemipteran insects, including some that are predators on caterpillars.

A-space resin see RESOLE.

asparaginase *Enzymology.* an enzyme that triggers the breakdown of the amino acid asparagine, releasing ammonia; used in treating childhood leukemia.

asparagine [ə spâr´ə jēn´] *Biochemistry.* $C_4H_8N_2O_3$, a nonessential amino acid that is found in certain vegetables, such as asparagus and beets; active in the metabolism of cells and brain and nerve tissue. Also, **asparagin.**

asparagus *Botany.* any plant of the genus *Asparagus,* liliaceaous herbs having reduced, scalelike or bristlelike leaves and small leaflike branches called cladophylls that photosynthesize, especially **Asparagus officinalus,** which is cultivated as a garden vegetable.

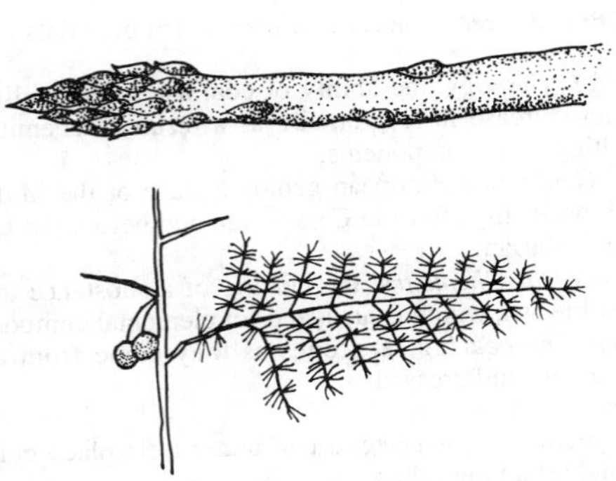

asparagus

asparagus stone *Mineralogy.* a yellowish-green variety of apatite, occurring in crystalline form and granular or cleavable masses. Also, **asparagolite.**

aspartame [ə spär´tam´; as´pər tām´] *Organic Chemistry.* $C_{14}H_{18}N_2O_5$, a dipeptide ester that is soluble in water; melts at 246–247°C; it is 150–200 times sweeter than natural sugar, and is widely used as a low-calorie artificial sweetener in soft drinks and packaged foods.

aspartase *Biochemistry.* a bacterial enzyme that catalyzes the breakdown of L-aspartate, producing ammonia and fumarate. Also, **aspartate ammonia-lyase.**

aspartate *Biochemistry.* a salt or ion of aspartic acid.

aspartate aminotransferase *Enzymology.* an enzyme that catalyzes the reaction of aspartate with 2-oxoglutarate to form glutamate and oxaloacetate; the elevation of this enzyme in serum is used as a diagnostic test. Also, GLUTAMIC OXALOACETIC TRANSAMINASE.

aspartate kinase see ASPARTOKINASE.

aspartate transcarbamylase *Biochemistry.* the enzyme that initiates movement of a carbamoyl group from carbamoyl phosphate onto the amino group aspartate.

aspartic acid *Biochemistry.* $C_4H_7NO_4$, a nonessential amino acid found in plants and animals that has an important role in the metabolism of amino acids.

aspartic acid dating see AMINO ACID DATING.

aspartic proteinase *Enzymology.* an enzyme that catalyzes the hydrolysis of peptide bonds in a protein and has an aspartic acid in its active site.

aspartokinase *Biochemistry.* an enzyme that causes aspartic acid to react with adenosine triphosphate and yield aspartyl phosphate. Also, ASPARTATE KINASE.

aspartyl *Biochemistry.* the acyl radical derived from aspartic acid.

aspartyl phosphate *Biochemistry.* $H_2O_3POCOCH_2NH_2COOH$, a substance that serves as a precursor in the biosynthesis of the compound pyrimidine, the bases of which are sometimes found in nucleic acid.

aspartyl proteinase *Enzymology.* an enzyme catalyzing the hydrolysis of peptide bonds in a protein and having aspartic acid in its active site.

aspect *Architecture.* the direction that a building faces in relation to the points of the compass. *Astronomy.* the apparent position of any heavenly body with respect to another. *Ecology.* **1.** the extent to which a site is exposed to environmental factors. **2.** a seasonal change in the physical appearance of a species. **3.** the direction the slope of a hill or mountain faces; e.g., north-facing aspect.

aspect angle *Engineering.* an angle formed between the lengthwise axis of a moving airborne object and the axis of a radar beam.

aspection *Ecology.* the periodic physiological changes observed in certain species that occur with periodic changes in the environment, such as seasonal changes. Also, **aspectation, aspect diversity.**

aspect ratio the ratio of one dimension to another, such as width to height; specific uses include: *Aviation.* the ratio of the span of an airfoil to its mean chord, or of the square of its span to its total area. An airfoil with a high aspect ratio has a relatively long span and short chord; one with a low aspect ratio has a relatively short span and long chord. *Materials Science.* the ratio of the length of a whisker or fiber of a material to its diameter. *Mechanical Engineering.* the ratio of the height of an automobile tire to its width. Also, TIRE ASPECT RATIO, TIRE PROFILE. *Computer Technology.* in computer graphics, the ratio of the vertical to horizontal dimensions of a displayed image or frame.

aspen *Botany.* any of various trembling-leaved poplar trees of the genus *Populus,* characterized by soft wood, long-stalked broad leaves, flowers in drooping catkins, and airborne seeds attached to siky hairs.

Aspergillaceae *Mycology.* a term formerly used for the fungus family Eurotiaceae belonging to the subdivision Ascomycotina; most are now assigned to the molds of the genera *Aspergillus* and *Penicillium* or the fungus *Thielavia basicola.*

Aspergillales *Mycology.* a term formerly used for the fungal order Eurotiales belonging to the subdivision Ascomycotina; some genera are pathogenic to humans, for example, *Aspergillus.*

aspergillic acid *Biochemistry.* $C_{12}H_{20}O_2N_2$, an antifungal antibiotic produced by *Aspergillus flavus* that blocks the growth of tuberculosis bacteria.

aspergillin [as´pər jil´ən] *Biochemistry.* **1.** a black pigment occurring in the spores of the common bread mold, *Aspergillus niger.* **2.** an antibacterial substance obtained from the molds *A. flavus* and *A. fumigatus* that works effectively against a wide variety of bacteria.

aspergillosis [as´pər jə lō´sis] *Medicine.* a diseased condition caused by an imperfect fungus (mold) of the genus *Aspergillus,* marked by inflammatory granulomatous lesions in the skin, ears, nasal sinuses, lungs, and sometimes bones and meninges. Also, **aspergillomycosis.**

Aspergillus *Mycology.* a genus of imperfect fungi, of the class Hyphomycetes, including species that cause food spoilage and diseases.

aspermia *Medicine.* a failure of production or emission of sperm. Also, **aspermatism.**

asperomagnetic state *Solid-State Physics.* an antisymmetric speromagnetic (scattered) state of matter.

asphalt *Materials.* **1.** a dark, tarry, bituminous material found naturally or distilled from petroleum. **2.** a mixture of this material with sand, gravel, or similar additives, used in paving. **3.** to apply any of these materials. **4.** relating to or containing such materials. Thus, **asphalt block, asphalt cement, asphalt paint, asphalt paper, asphalt paving, asphalt roofing, asphalt shingle, asphalt tile,** and so on.

asphalt paver

asphalt cutter *Mechanical Engineering.* a power-operated machine with a rotating abrasive disk or blade that is designed to cut asphalt or other bituminous surfacing material.

asphaltene [as fôl′tēn; as′fəl tēn′] *Materials Science.* any component of the bitumen in petroleum that is soluble in carbon disulfide but not in paraffin naphthas.

asphalt heater *Engineering.* a device used in the paving process to increase the temperature of bitumen.

asphaltic *Materials Science.* relating to, containing, or derived from asphalt. Thus, **asphaltic concrete, asphaltic base oil.**

asphaltic sand *Geology.* a naturally occurring mixture of sand and asphalt in varying proportions.

asphaltite *Mineralogy.* a naturally occurring black to dark brown solid hydrocarbon that is highly soluble in carbon disulfide and less fusible than native asphalt.

asphalt leveling course *Civil Engineering.* a course of asphalt paving that is laid to smooth and eliminate the irregularities of an existing pavement.

asphalt mastic see MASTIC ASPHALT.

asphalt overlay *Civil Engineering.* courses (or layers) of asphalt paving placed over existing pavement.

asphalt rock *Geology.* any porous sedimentary rock that is infused naturally with asphalt. Also, ROCK ASPHALT.

asphalt soil stabilization *Civil Engineering.* the treatment of soil with an asphalt penetration mixing, used to improve the load-bearing capacity of the soil.

aspherical [ā′sfēr′ə kəl] *Optics.* describing a lens or mirror whose surface is not precisely spherical,in order to minimize spherical aberration.

aspherical lens *Optics.* a lens having one or more surfaces that deviate slightly from a true spherical shape.

asphyxia [as fiks′ē e] *Medicine.* an extreme condition that is caused by a lack of oxygen and excess of carbon dioxide in the blood due to interference in respiration or to an insufficiency of oxygen in respired air, resulting in an impending or actual cessation of apparent life; suffocation.

asphyxiant [as fiks′ē ent] *Toxicology.* any poison that exerts its effects by depriving tissues of oxygen.

asphyxiate [as fiks′ē āt′] *Medicine.* to enter or put into a state of asphyxia; suffocate.

asphyxiation [as fiks′ē ā′shən] *Medicine.* a state of asphyxia; suffocation.

aspiculate *Invertebrate Zoology.* without spicules.

Aspide *Ordnance.* an Italian air-to-air missile that is powered by a solid-propellant rocket motor and equipped with semiactive radar homing; it carries a 75-pound conventional fragmentation warhead at a speed of Mach 4 and a range of 30 to 60 miles.

Aspidiphoridae *Invertebrate Zoology.* a family of coleopteran insects that feed on spores of slime molds.

Aspidisca *Invertebrate Zoology.* a genus of small aquatic ciliate protozoans in the order Hypotrichida, with large cirri, a flat ventral surface, and a horseshoe-shaped macronucleus.

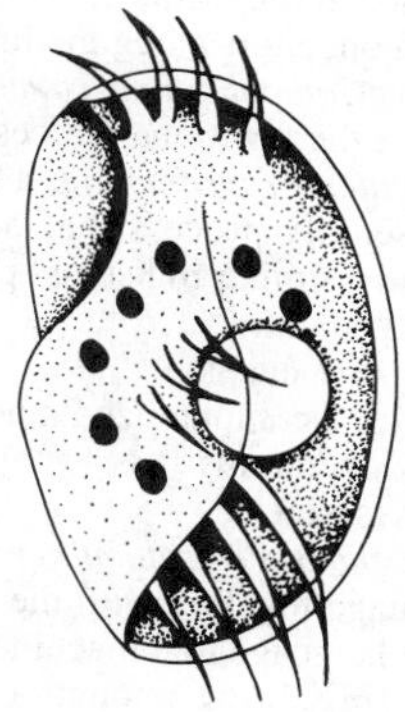

Aspidisca

Aspidobothria see ASPIDOGASTREA.

Aspidobranchia *Invertebrate Zoology.* marine gastropods with slightly concentrated nervous systems and generally showing clear traces of ancient bilateral symmetry, with two kidneys and two auricles; includes limpets, periwinkles, and conches.

Aspidochirotacea *Invertebrate Zoology.* a subclass of sea cucumbers having 10 to 30 shield-shaped tentacles and lacking anterior retractor muscles; the body is bilaterally symmetrical, with tube feet (podia); a toxin in the body wall of some types is used as a poison.

Aspidochirotida *Invertebrate Zoology.* an order of echinoderms in the subclass Aspidochirotacea; includes some of the largest species of holothurians (sea cucumbers).

Aspidocotylea see ASPIDOGASTREA.

Aspidodiadematidae *Invertebrate Zoology.* a family of sea urchins, in the order Diadematoida, favoring fine mud substrates; cosmopolitan species characterized by long, slender, downward-curving spines (outrigger type) and a fragile test.

Aspidogastrea *Invertebrate Zoology.* a small order or subclass of Trematoda, comprising some 40 freshwater and marine species of flukes characterized by a large complex ventral sucker or row of suckers; internal or external parasites of aquatic animals, sometimes with a variety of hosts (snails, clams, and, later, fish or turtles). Also, ASPIDOCOTYLEA, ASPIDOBOTHRIA.

Aspidogastridae *Invertebrate Zoology.* a family of trematode worms, in the subclass Aspidogastrea, both adult and developmental forms of which are found inside gastropod and lamellibranch mollusks; adults are also found in fish and turtles, with the common United States host being the freshwater mussel.

Aspidorhynchidae *Paleontology.* a family of slender, predaceous fish in the extinct order Aspidorhynchiformes of the Jurassic and Cretaceous periods; they resembled the modern gars in shape and size but were probably not related to them.

Aspidorhynchiformes *Paleontology.* an extinct monofamilial order of neopterygian fish of the Jurassic and Cretaceous periods.

aspidospermine *Pharmacology.* $C_{22}H_{30}N_2O_2$, an alkaloid derived from the bark of the tree *Aspidosperma quebracho-blanco*, formerly used to stimulate breathing in asthma and dyspnea.

Aspinothoracida *Paleontology.* a term once used to refer to the extinct placoderm suborder Brachythoraci.

aspirate *Surgery.* **1.** to draw in or take away by suction. **2.** the material removed by suction.

aspirated sound *Linguistics.* an explosive push of air in the formation of certain sounds: [p] [b] [k].

aspirating see DEDUSTING.

aspirating burner *Engineering.* a burner in which fuel and air are sucked into a combustion chamber, where the fuel is burned in suspension.

aspirating needle *Surgery.* a long, hollow needle used to withdraw liquid or air from a cavity.

aspiration *Science.* the process of removing substances by suction; the product of such a process. *Engineering.* **1.** the removal of air from a chamber or enclosure by means of suction. **2.** describing a device that is ventilated by a suction fan. Thus, **aspiration meteorograph, aspiration psychrometer, aspiration thermograph,** and so on. *Microbiology.* the use of a pipette to withdraw a fluid sample by means of suction. *Physiology.* the act of inhaling. *Surgery.* the withdrawal of liquid or gases from a cavity by suction.

aspiration biopsy *Surgery.* the removal of a biopsy specimen by aspirating it through a needle or trocar; the instrument first pierces the skin or the external surface of the organ and then the underlying tissue to be examined. Also, NEEDLE BIOPSY.

aspirator *Engineering.* **1.** a tubular device that draws up gases, granular materials, or fluids by suction. **2.** a device used to protect the mouth and lungs from dust by means of wire gauze, cloth, or fibrous material that covers the mouth and nose.

aspirin *Pharmacology.* the common name for acetylsalicylic acid, a widely used pain reliever. See ACETYLSALICYLIC ACID.

aspite *Geology.* of a volcano, having a crater whose base is wider than its height.

Aspleniaceae *Botany.* a large, cosmopolitan family of ferns of the order Filicales, including spleenworts, wood ferns, holly ferns, and lady ferns; characterized by a creeping or erect scaly rhizome, a terrestrial or epiphytic habit, and sori occurring on lower leaf surfaces.

asporogenic mutant *Microbiology.* a microorganism that has become genetically incapable of producing spores.

asporogenous *Botany.* not producing spores. Also, **asporogenic.**

Aspredinidae *Vertebrate Zoology.* a family of marine catfish in the order Siluriformes, found near South America.

asRNA see ANTISENSE RNA.

AS/RS automated storage/retrieval system.

ass *Vertebrate Zoology.* a member of any of several horselike species of the family Equidae in the order Perissodactyla; usually smaller than the horse, with longer ears.

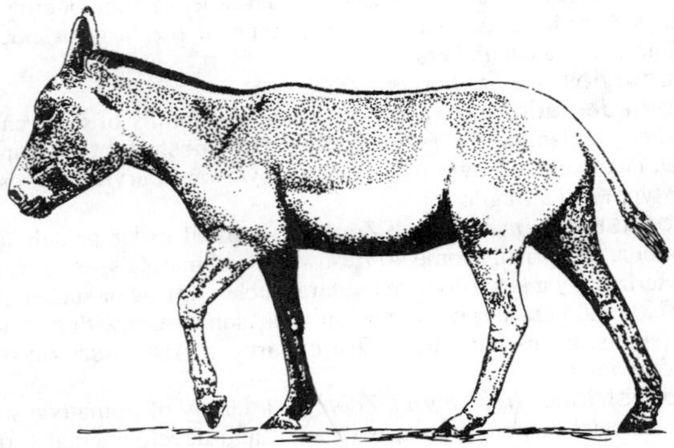

ass

assault *Military Science.* **1.** a swift and sudden attack on a fortification, gun position, or other such specific objective. **2.** see AIRBORNE ASSAULT. **3.** see AMPHIBIOUS ASSAULT. **4.** the final stage of a ground attack, where hand-to-hand combat may take place. **5.** to make a swift, sudden attack.

assault aircraft *Aviation.* aircraft, often helicopters, used to move troops and cargo into combat areas and keep them resupplied.

assault boat *Ordnance.* a boat used to carry assault troops. Similarly, **assault craft, assault glider, assault shipping, assault (transport) aircraft.** *Naval Architecture.* a light boat used in amphibious or river-crossing assault operations.

assault course *Military Science.* an area used to prepare troops for ground assaults, as by training in hand-to-hand combat or in the use of a rifle or bayonet.

assault echelon *Military Science.* **1.** see ASSAULT WAVE. **2.** see ASSAULT FORCE.

assault fire *Ordnance.* **1.** fire delivered by attacking troops at close range to the enemy; usually fired from the hip or a standing position at a rapid, sustained rate. **2.** short-range artillery fire concentrated on a specific target.

assault force or **forces** *Military Science.* in an amphibious or airborne assault, the unit or units assigned to take control of the initial beachhead or airhead.

assault gun *Ordnance.* any artillery weapon used for direct fire from close range at a specific target.

assault model *Ordnance.* a vehicle used to provide direct fire in an assault operation. Also, **assault-landing model, amphibious assault-landing model.**

assault phase *Military Science.* the period between the time assault units enter the combat area and the time that their initial objective is judged to have been secured.

assault position *Military Science.* the final position from which an assault unit moves to close with the enemy and reach its objective.

assault wave *Military Science.* the unit or vessels that lead the attacking force in an assault. Also, ASSAULT ECHELON.

assault wire *Ordnance.* a light field telephone wire wound on reels small enough to be carried by one person under assault conditions.

assay *Analytical Chemistry.* a quantitative or qualitative analysis of the composition of a material. *Mining Engineering.* **1.** to test ores or minerals by chemical or blowpipe examination. **2.** a means of ascertaining the commercial value of a mineral substance.

assay balance *Engineering.* a sensitive balance used in the assaying of precious metals.

assay bar *Metallurgy.* a standard specimen of pure gold or silver, used as a reference for establishing the purity of a commercial noble metal.

assay plan *Mining Engineering.* a map marking the positions of the assay, stope, width, etc., of samples removed.

assay pound *Mining Engineering.* a varying weight used by assayers as proportionately representing a pound.

assay ton *Mining Engineering.* a unit of weight of ore equal to the number of troy ounces in a short ton.

assay value *Mining Engineering.* the amount of gold or silver, in ounces per ton of ore, of any given sample as indicated by assay.

assay walls *Mining Engineering.* the planes to which an ore body can be profitably mined, with the metal content of the country rock determined from assays as the limiting factor.

Assel elongation *Metallurgy.* the hot working process in which metal tubing is finish rolled to a reduced diameter and wall thickness.

assemblage a general collection of objects or items; specific uses include: *Ordnance.* the collecting of various supplies and items of equipment for a single purpose. *Archaeology.* a group of objects found in association with each other and thus considered to be the product of one people from one period of time; the objects may be all of the same type or of different types. *Paleontology.* a group of fossils that is unified in some way in respect to assemblages in neighboring strata.

assemblage zone *Paleontology.* the stratigraphic units or periods of time characterized by a given fossil assemblage, usually named for a taxon characteristic of the assemblage. Also, CENOZONE.

assemble *Industrial Engineering.* in micromotion studies, a term for the elemental motion of bringing together two connecting parts.

assembled stone *Materials Science.* describing an artificial gem that is composed of two or more components, at least one of which is a precious stone.

assembler *Computer Programming.* a computer program that translates assembly language code into machine language instructions. Also, SYMBOLIC ASSEMBLY SYSTEM, ASSEMBLER PROGRAM.

assembler directive *Computer Programming.* a pseudo-operation in an assembly language that provides information to control the assembly process or provides a way of defining data words in a program; an assembler directive does not translate to a machine instruction.

assembler language see ASSEMBLY LANGUAGE.

assembler program see ASSEMBLER.

assembly the act or result of assembling; specific uses include: *Mechanical Engineering.* a fabricated product containing the component parts of a machine, mechanism, or the like. *Military Science.* the act or fact of assembling, especially the grouping of units or aircraft at a specific point. Thus, **assembly area, assembly point, assembly position.** *Computer Programming.* the process of translating a source program written in assembly language into an object program consisting of machine instructions.

assembly costs *Industrial Engineering.* that portion of the cost of manufacturing a product associated with the process of assembling its parts, as opposed to the cost of manufacturing or obtaining the parts.

assembly economics *Industrial Engineering.* the financial analysis and evaluations associated with the cost of assembling parts.

assembly language *Computer Programming.* a low-level programming language that makes use of mnemonic instruction codes, labels, and names to refer directly to their binary equivalents. Typically, one line of assembly language corresponds to one machine instruction.

assembly line *Industrial Engineering.* a production process in which the production stages are arranged in sequential order, with unfinished items moving "along the line" through each production stage.

assembly-line balancing *Industrial Engineering.* the process of arranging either machine and operator assignments on an assembly line, or the product mix passing through the line, in order to minimize the idle time of machines and operators along the line.

assembly-line diagram *Computer Programming.* a type of Warnier-Orr diagram used to define the functional processes of a program.

assembly listing *Computer Programming.* an optional printed listing produced during the assembly process that documents details of the source and object programs; similar to listings produced by a compiler.

assembly mark see PIECE MARK.

assembly program see ASSEMBLER.

assembly robot *Robotics.* a computerized manufacturing tool that performs specified assembly operations.

assembly routine see ASSEMBLER.

assembly system *Computer Programming.* a system consisting of a symbolic assembly language together with the assembler that translates a program written in that language into machine language.

assembly time *Engineering.* **1.** the amount of time it takes to assemble a mechanism or product. **2.** the time interval between the spreading of an adhesive and the point when its strength becomes effective.

assembly unit *Computer Technology.* a portion of a program that is incorporated into a larger program using an assembler.

assert *Artificial Intelligence.* to add a fact to the database of true or currently believed facts.

assertion *Artificial Intelligence.* an element in a database that is assumed to be true.

assertive behavior *Psychology.* a type of interpersonal behavior that involves the straightforward expression of thoughts and feelings as well as concern for the rights of others, in a form that is regarded as socially appropriate.

assertiveness training *Psychology.* a behavior therapy technique in which assertive responses are reinforced in order to reduce passive behavior in interpersonal relationships. Also, **assertion training.**

assessment *Psychology.* **1.** the systematic gathering of information for the purpose of identifying and describing individual differences in human behavior. **2.** the area of psychology that is concerned with measuring personality, achievement, and ability differences.

assessment drilling *Mining Engineering.* the drilling performed to fulfill the prescribed amount of work that must be done yearly on an unpatented mining claim in order to retain title.

assessment work *Mining Engineering.* the work accomplished in a year on an unpatented mining claim on the public domain necessary under United States law for the maintenance of the possessory title. Also, ANNUAL LABOR.

assignable cause *Industrial Engineering.* an identifiable cause of a variation in the quality of an output.

assignment operator *Computer Programming.* an operator that causes the variable on the left to assume the value of the expression on the right; often an = sign.

assignment problem *Computer Programming.* a quantitative analysis method related to linear programming that optimizes the assignment of personnel to jobs in such a way as to minimize the total cost or time required; every job is assigned to a worker and every worker is assigned exactly one job.

assignment statement *Computer Programming.* a computer program instruction that computes the value of an arithmetic or logical expression and assigns that value to a specified variable name.

assimilation *Psychology.* the process by which an individual incorporates new stimuli into his or her existing world view; the ability to apply existing concepts to new experience. *Geology.* the process by which magma is incorporated with solid rock, other magma, or foreign gas. Also, MAGMATIC DIGESTION. *Linguistics.* the phonological process in which sounds are changed to resemble surrounding sounds, making words easier to pronounce, as in "gramma" for "grandma." *Anthropology.* the gradual merging of two cultures to form one.

assimilative nitrate reduction *Microbiology.* a process by which microorganisms reduce nitrate to ammonia.

assimilative sulfate reduction *Microbiology.* a process by which microorganisms reduce sulfate to sulfide.

assist *Aviation.* **1.** to provide supplemental thrust to an aircraft, as in a jet assist, rocket assist, or assisted takeoff. **2.** the process of assisting.

assist(ed) panel *Computer Technology.* a display explaining instructions the computer needs in order to proceed, along with the alternatives available and the expected format for response.

assisted takeoff *Aviation.* an aircraft launching boosted by an extra propulsive source such as rockets.

assize *Civil Engineering.* **1.** a cylindrical block of stone forming a section of a column. **2.** a layer of stone in a building.

associate *Psychology.* a specific item that is linked to another in an individual's mind.

associate curve see BERTRAND CURVE.

associated automatic movement see SYNKINESIA.

associated corpuscular emission *Geophysics.* all the secondarily charged particles that are associated with the passage of a gamma-ray or X-ray beam through the air.

associated flow rule *Materials Science.* a plastic flow relationship in which the function describing the yield surface is itself a plastic potential.

associated gas *Petroleum Engineering.* gas that occurs with oil; gaseous hydrocarbons present in a free gas phase under original temperature and pressure conditions in an oil reservoir in the early stages of production. Also, GAS-CAP GAS, SOLUTION GAS.

associated prime ideal *Mathematics.* A prime ideal P of a Noetherian ring R is said to be associated with a module M over R if P is the annihilator of some nonzero element of M.

associated production *Particle Physics.* a phenomenon whereby, in any nuclear reaction that produces strange particles, those particles are invariably produced in pairs (a strange particle and one of opposite strangeness), never one particle alone.

associate matrix see HERMITIAN CONJUGATE OF A MATRIX.

associate operator see ADJOINT OPERATOR.

association *Psychology.* a bond or connection between two images, thoughts, ideas, or other psychological phenomena, whereby the occurrence of one tends to bring to mind the other. *Behavior.* the forming of a learned connection between a stimulus and a response, or between one stimulus and another. *Neurology.* a coordination of functions of similar body parts that involves consciousness as well as a high degree of modifiability. *Genetics.* the joint occurrence of two genetically determined characteristics in individuals of a given population at a higher frequency than would be expected based on their independent frequencies. *Archaeology.* the relationship of various objects that share the same general location and stratigraphic level and that are thought to have been deposited at the same time. *Astronomy.* a group of stars that form within a fairly short time from a cloud of dusty gas and that are loosely bound by gravity.

association area *Physiology.* any of various regions of the cerebral cortex that are identified with visual discrimination, learning, perceptual-motor activity, and complex processes of thought and language.

association center *Invertebrate Zoology.* an invertebrate nervous center in which stimuli from sensory receptors are coordinated and distributed.

association constant *Immunology.* a mathematical measurement of the affinity of bonding between two molecules at equilibrium (such as an antigen and an antibody).

association fiber *Anatomy.* any of certain nerve fibers that interconnect portions of the cerebral cortex within a brain hemisphere. Also, **association nerve fiber, association neurofiber.**

associationist theory of memory *Artificial Intelligence.* the theory that human memory is organized in terms of concepts that are related by associational links.

association key *Computer Programming.* the specific data element value or content used to compare with data stored in associative memory, so that matching data may be retrieved or accessed.

association list *Artificial Intelligence.* in Lisp, a list composed of pairs of a symbolic name and associated value, used to represent variable bindings, substitutions, and data sets.

association neuron *Anatomy.* a neuron that lies within the central nervous system between afferent and efferent neurons.

association test *Psychology.* a technique that is designed to assess an individual's response to specific stimuli such as colors or isolated words.

association theory *Psychology.* a theory that concepts are learned by a simple, reinforced connection between a stimulus and a desired response.

associative algebra *Mathematics.* an associative ring that is also an algebra. If the requirement of associativity is dropped, then the result is a **nonassociative algebra.** Examples are Lie algebras and alternating rings.

associative dimensioning system *Robotics.* in manufacturing process control, a computerized tool that automatically adjusts to variations in workpiece dimensions.

associative facilitation *Psychology.* an ability to establish new associations with ease due to previous associations.

associative inhibition *Psychology.* an impaired ability to establish new associations due to previous associations.

associative language *Computer Programming.* any of a class of nonprocedural languages whose basic operations are entry, removal, and retrieval of entity associations in memory.

associative law *Mathematics.* a binary operation $*$ is said to satisfy the associative law if $(a * b) * c = a * (b * c)$ for all a, b, and c on which $*$ operates.

associative learning *Psychology.* the process of learning through the formation of associations between ideas and events that were experienced together and thus are mentally linked.

associative lookup *Computer Programming.* the process of retrieving data from an associative or content-addressable memory.

associative memory *Computer Technology.* a content-addressable computer memory in which stored data is identified by properties of its own value, rather than by an address as in conventional memories.

associative processor *Computer Technology.* a type of parallel processor that performs identical operations simultaneously on different portions of the data and that has an associative or content-addressable memory.

associative ring *Mathematics.* a ring in which the ring multiplication obeys the associative law.

associative storage see ASSOCIATIVE MEMORY.

assortative mating *Genetics.* a type of sexual reproduction in which the tendency for males of a certain kind to breed with females of a certain kind is not random. In **positive assortative mating,** the two parents of a pair tend to be more alike than would be expected by chance; in **negative assortative mating,** they tend to be less alike.

assortment see INDEPENDENT ASSORTMENT.

assumed decimal point *Computer Programming.* the position in a numeric word at which the decimal point is implied; no physical storage space is required as with an actual decimal point.

assumed force see CORRECTION FORCE.

assumed ground elevation *Cartography.* the elevation assumed to prevail in the vicinity of a critical point, such as a mountain peak, in the local area covered by a photograph or group of photographs.

assumed plane coordinates *Engineering.* a system of plane coordinates that is set up specifically for the needs of a surveyor.

assumed position *Navigation.* a point at which a craft is assumed to be located, generally used as a starting point for determining the actual location of the craft.

assumption-based truth maintenance system see TRUTH MAINTENANCE.

assured mineral see DEVELOPED RESERVES.

assyntite *Petrology.* an antiquated term for plutonic rock consisting largely of orthoclase and pyroxene, with small amounts of sodalite and nepheline, biotite, sphene, apatite, and opaque oxides.

AST aspartate aminotransferase.

astable circuit [ā´stā´bəl] *Electronics.* a circuit that bounces between two unstable states at a rate determined by the circuit's time constants; an example of this is a blocking oscillator.

astable multivibrator *Electronics.* a multivibrator circuit that requires no external control input; it operates between two continuously alternated states at a rate dependent on the circuit's time constants. Also, FREE-RUNNING MULTIVIBRATOR.

Astacidae *Invertebrate Zoology.* a family of crayfish, dominant freshwater decapods in the order Decapoda, found in the temperate Northern Hemisphere.

astacin *Biochemistry.* a red pigment found in the shells of some crustaceans, such as boiled lobsters.

Astacinae *Invertebrate Zoology.* a subfamily of crayfish common to Europe and the far western United States.

A-stage resin *Organic Chemistry.* an early stage in a thermosetting resin reaction; at this stage the resin is fully soluble in common solvents and is fusible at less than 150°C.

A star *Astrophysics.* a white star whose spectrum is characterized by strong absorption lines of hydrogen; it has a temperature in the range of 7500 to 10,000 kelvins.

Astartian see SEQUANIAN.

astasia [ə stāz´yə] *Neurology.* an inability to stand due to motor incoordination, despite the presence of strength and sensation in the legs.

astatic [ā´stat´ik] *Physics.* being without orientation or directional characteristics; having no tendency to change position. *Engineering.* describing an instrument in which a negative restoring force has been applied to enhance any deflecting force, thus making the instrument highly sensitive. Thus, **astatic galvanometer, astatic gravimeter, astatic magnetometer, astatic wattmeter,** and so on.

astatic coils *Electromagnetism.* two similar coils that are connected in series and suspended from an axis so that an external magnetic field exerts no net torque on the system.

astatic governor see ISOCHRONOUS GOVERNOR.

astatic gravimeter *Engineering.* an extremely sensitive instrument that is used to measure minute changes in specific gravity.

astatic pair *Electromagnetism.* two magnets of equal field strength, arranged in such a manner that their polarity directions are opposite and perpendicular to the line that bisects the pair, resulting in no net force or torque on the system.

astatic pendulum *Physics.* a pendulum that is not restricted to swinging in a plane.

astatine [as´tə tēn´] *Chemistry.* a nonmetallic chemical element having the symbol At, the atomic number 85, and an atomic weight of 211 in its most stable isotope. It is the heaviest element of the halogens, and all of its known isotopes are radioactive. Astatine occurs in infinitesimal amounts in nature and is produced artificially, and is thought to have properties similar to those of iodine. (From a Greek word meaning "unstable," because of the fact that it decays so rapidly; it has a half-life of 7 to 8 hours.)

A station *Navigation.* the main radio transmitting station in a system of two or more synchronized stations used for position finding. *Cartography.* a subsidiary station established between principle stations of a traverse, for convenience in measuring the distance between the principle stations.

astatized see ASTATIC.

astaxanthin *Biochemistry.* a violet pigment found in certain bird feathers and crustacean shells.

Asteidae *Invertebrate Zoology.* a small family of two-winged flies, in the order Diptera.

astel *Mining Engineering.* an overhead board or arching in a mine gallery.

astelic *Botany.* lacking a central stele; having discontinuous vascular bundles.

asteltoxin *Mycology.* a kind of poison found in the fungal species *Aspergillus stellatus.*

aster *Botany.* any composite plant of the genus *Aster,* family Asteraceae, having a pappus of bristles and rays of pink, white, or blue surrounding a yellow central disk. *Cell Biology.* the array of microtubules that radiates out from each of the two centrosomes of a cell during the process of mitosis.

aster

Asteraceae *Botany.* the daisy family of herbaceous plants of the order Asterales; characterized by numerous small flowers arranged on a common receptacle in a disk-shaped head surrounded by a ray of petals extending from the rim of the disk. Also, COMPOSITAE.

Asterales *Botany.* an advanced order of herbaceous to woody dicots of the subclass Asteridae, including the families Asteraceae and Cichoriaceae; characterized by inflorescences arranged in heads called capitula surrounded by an involucre of bracts.

Asteridae *Botany.* a large subclass of dicots in the class Magnoliopsida, characterized by sympetalous flowers.

Asteriidae *Invertebrate Zoology.* a large and important family of echinoderms in the order Forcipulatida; generally slow-moving starfish; includes common North American and European species, with *Asterias* the best known genus.

Asterinaceae *Mycology.* a family of fungi, belonging to the order Asterinales, that forms blotches on leaves and stems of plants growing in tropical to warm regions.

Asterinales *Mycology.* an order of fungi, belonging to the class Loculoascomycetes, that occurs primarily in tropical regions and is composed of members living off dead organic matter or as parasites of higher plants.

asterism *Astronomy.* any small, noticeable group of stars such as the Big Dipper or the Summer Triangle.

asterixis *Neurology.* a recurrent muscular tremor of the upper limbs, characterized by an intermittency of sustained muscle contractions; the result is a motion similar to a bird flapping its wings. Also, FLAPPING TREMOR.

astern *Navigation.* at or to the rear of a ship, aircraft, or other vessel.

asternal *Anatomy.* not attached to the sternum, such as the floating ribs.

Asterniidae *Invertebrate Zoology.* a widely distributed family of starfish, in the order Spinulosida; usually pentagonal, and quite flat with short arms.

Asterococcaceae see PALMELLOPSIDACEAE.

asterognosis *Medicine.* a loss of power or ability to recognize objects by touch. Also, TACTILE AMNESIA.

asteroid *Astronomy.* a rocky body, less than 1000 kilometers across, in orbit around the sun. Also, MINOR PLANET.

asteroid belt *Astronomy*. the region of the solar system between the orbits of Mars and Jupiter, where most of the asteroids orbit.
Asteroidea *Invertebrate Zoology*. a class of Echinodermata comprising starfishes.
Asterolampraceae *Microbiology*. a family of advanced marine diatoms of the order Centrales, having valves bearing large, modified areolae that open to the interior by a slitted foramen.
Asteroschematidae *Invertebrate Zoology*. a family of brittle stars, in the order Phrynophiurida, with small disks and stout, unbranched arms.
Asterozoa *Invertebrate Zoology*. a subphylum of echinoderms comprising the starfishes (Asteroidea) and brittle stars (Ophiuroidea); includes one extinct and three living classes.
aster wilt *Plant Pathology*. the most common disease of asters, caused by the fungus *Fusarium oxysporum f. callistephi* and characterized by brown streaks on the stems.
aster yellows *Plant Pathology*. a viral disease of asters and other plants, transmitted by leafhoppers and characterized by yellowing and dwarfing.
asthen- or **astheno-** a combining form meaning "weak" or "thin."
asthenia [as thēn´ē ə] *Medicine*. any loss or lack of strength or energy; a condition of weakness.
asthenic [əs then´ik] *Medicine*. relating to or characterized by asthenia; weak. *Psychology*. a body-type classification for people with a thin build.
asthenolith *Geology*. a body of magma that is formed by local melting within the solid portion of the earth's crust in response to heat generated by natural activity.
asthenopia [as´thə nōp´ē ə] *Medicine*. weakness or fatigue of the eyes. Thus, **asthenopic.**
asthenosphere *Geology*. the zone or layer of the earth's upper mantle that lies below the lithosphere, that is capable of plastic deformation, and in which magmas may be generated and the velocity of seismic waves reduced.
asthma [az´mə] *Medicine*. a generally chronic condition that is characterized by recurring attacks of wheezing, coughing, and labored breathing; it may be caused by allergies, physical exertion, chemical irritation, or emotional stress. (From a Greek word meaning "panting.")
asthma crystals see CHARCOT-LEYDEN CRYSTALS.
asthmatic [az mat´ik] *Medicine*. **1.** relating to, caused by, or affected with asthma. **2.** a person affected with asthma.
Astian *Geology*. a European geologic stage of the Upper Pliocene epoch, occurring after the Plaisancian and before the Calabrian (Villifranchian) of the Lower Pleistocene.
Asticcacaulis *Bacteriology*. a genus of Gram-negative, aerobic soil bacteria, similar to those of the Caulobacter group, occurring as rod-shaped cells with a single prostheca at one pole.
asticity Neurology. a state of increased muscular tone marked by exaggerated reflexes and increased resistance to the stretching of muscles.
astigmatic [as tig mat´ik] relating to or affected by astigmatism.
astigmatic lens *Optics*. a lens used in eyeglasses to correct for astigmatism in the eyes, such as a planocylindrical, spherocylindrical, or pherotoric lens. Also, **astigmat.**
astigmatism [ə stig´mə tiz´əm] *Physiology*. a condition of the eye in which there is an unequal curvature of the surface of the cornea, so that light rays are not brought into a precise focus at one point on the retina, but are spread over a diffuse area. This condition causes the eye to perceive blurred or indistinct images. *Optics*. an aberration in a lens or optical system that causes rays that are in one plane to focus at a different location from that of rays that are in an orthogonal plane. *Electronics*. in an electron beam, a beam divergence where electrons of different axial planes fail to focus at the same point.
astigmatizer *Optics*. a lens that stretches a single point of light into a line or band, commonly attached to range finders.
astigmometer *Optics*. an instrument that measures the degree of astigmatism present in an optical system.
ASTM or **A.S.T.M.** American Society for Testing and Materials.
ASTM grain size *Materials Science*. a standardized number that quantifies a material's grain size, based on the number of grains in a flat surface of defined area.
astogeny *Invertebrate Zoology*. a change in size or form shown by all the zooids in colonizing animals as the colony matures.
astomatal *Botany*. without stomata.
Astomatida *Invertebrate Zoology*. an order of protozoans in the subclass Holotricha, with no mouth; parasites of invertebrates.
astomatous *Invertebrate Zoology*. having no mouth.
astomous *Botany*. without a stomium or other line of dehiscence; bursting irregularly.
Aston, Francis William 1877–1945, English physicist; awarded the Nobel Prize for the invention of the mass spectrograph.
Aston dark space *Electronics*. in a gas tube, a dark region in the immediate vicinity of the cathode, where an electron's velocity is insufficient to excite the gas.
Aston process *Metallurgy*. a process for making controlled quality wrought iron. Its most critical step consists of pouring a continuous stream of refined metal into a large volume of molten slag.
Aston whole number rule *Physics*. a rule stating that the atomic weight of an element is very close to an integral value and that any small departure from the integral value is due to the presence of isotopes of the element.
ASTOR torpedo see ANTISUBMARINE TORPEDO.
astraean *Invertebrate Zoology*. a star coral.
astraeid *Invertebrate Zoology*. reef-building corals with massive forms; includes some perforate corals and brain corals.
astragal *Architecture*. a small, semicircular molding, often decorated with a string of beads. *Building Engineering*. a vertical molding along the edge of one of a pair of doors or windows, covering the joint between them and serving as a weather seal.
astragaloid bone see TALUS.
astragalus see TALUS.
astrakanite see BLOEDITE.
astral *Astronomy*. of or relating to the stars.
astral dome see ASTRODOME.
astral lamp *Engineering*. a type of Argand lamp having a flattened annular oil reservoir which causes the lamp to cast a shadow across a table when placed centrally.
astraphobia *Psychology*. an irrational fear of lightning and thunderstorms.
Astrapotheria *Paleontology*. an extinct order of South American ungulates, including the Trigonostylopidae and the Astrapotheriidae; Paleocene to Miocene.
Astrapotheriidae *Paleontology*. a family of large South American ungulates in the extinct order Astrapotheria; characterized by well-developed forequarters, slender hindquarters, a protruding lower jaw, and large and ever-growing canines; Eocene to Miocene.
Astreaceae *Mycology*. a family of fungi belonging to the order Sclerodermatales; members live in sandy areas of open woods, in symbiotic relationship to forest trees.
Astrephomenaceae *Botany*. a monogeneric family of biflagellates of the order Volvocales; characterized by a coenobium of 16–128 subspherical cells arranged at the periphery of a sphere, all the same except for a few at the posterior that are smaller with stiff, rudderlike flagella.
Astrephomene *Botany*. the single genus in the Astrephomenaceae (or Volvocaceae) family of green algae, composed of 16 to 128 biflagellate cells arranged at the periphery of a sphere.
astringent *Medicine*. **1.** causing contraction; constrictive. **2.** an agent causing contraction, especially locally after topical application.
astro- *Astronomy*. a prefix meaning "star" or "heavenly body"; used in relation to celestial objects or space flight.
astroballistics *Mechanics*. the study of the motion of meteroids and other solid bodies entering the earth's atmosphere.
astrobiology *Biology*. the study of life outside the earth's atmosphere, such as on planets or celestial bodies other than earth.
astroblast *Neurology*. a cell that develops into an astrocyte.
astroblastoma *Oncology*. a rare, malignant tumor composed of cells with abundant cytoplasm and two or three nuclei.
astrobleme *Geology*. an ancient circular crater on the earth's surface that was produced by the impact of a cosmic body.
astrochanite see BLOEDITE.
astrochemistry *Astronomy*. the study of the chemical composition and evolution of the universe.
astrocompass *Navigation*. a device for determining direction by celestial observations.
astrocyte [as´trə sīt´] *plural*, **astrocytes** or **astroglia.** *Neurology*. a star-shaped cell of the nervous system, that provides nutrients, support, and insulation for neurons of the central nervous system; one of the major categories of neurological cells.
astrocytoma [as´trō sī tō´mə] *Medicine*. a tumor made up of astrocytes.
astrocytosis [as´trō sī tō´sis´] *Medicine*. a proliferation of astrocytes owing to the destruction of nearby neurons during an episode of hypoxia or hypoglycemia.

astrodome *Aviation.* a transparent, usually plastic dome mounted on top of an aircraft or spacecraft, used chiefly for celestial navigation. Also, NAVIGATION DOME.

astrodynamics *Space Technology.* the systematic planning and directing of spacecraft trajectories through the application of celestial mechanics and allied fields.

astrogation see ASTRONAVIGATION.

astrogeodetic *Geodesy.* relating to direct measurements of the earth.

astrogeodetic coordinate system *Cartography.* a coordinate system having its origin at a point with known geodetic coordinates and its axes oriented like those of an astronomic coordinate system.

astrogeodetic datum orientation *Geodesy.* the position of a reference ellipsoid in relation to the geoid in a specified area of a geodetic network.

astrogeodetic deflection *Geodesy.* a deflection of the vertical obtained by comparing the astronomical and geodetic coordinates at a given point on the surface of the earth (or on the geoid).

astrogeodetic leveling *Geodesy.* a method in which astrogeodetic deflections of the vertical are used to determine the separation of the geoid and spheroid in studying the figure of the earth.

astrogeodetic undulation *Geodesy.* the separation or difference between an astrogeodetic geoid, as defined for a particular datum, and a specified ellipsoid surface.

astrogeology *Astronomy.* the study of the geology of extraterrestrial solid objects.

astroglia *Histology.* the astrocytes of the central nervous system, which function as connective tissue.

astrograph *Astronomy.* a photographic telescope, usually with a field of view at least 1° across.

astrographic position see ASTROMETRIC POSITION.

astrogravimetric leveling *Geodesy.* a method in which a gravimetric map is used for the interpolation of the astrogeodetic deflections of the vertical to determine the separation of the geoid and the ellipsoid in studying the figure of the earth.

astrogravimetric points *Geodesy.* astronomical positions (points) corrected for the deflection of the vertical by gravimetric methods.

astroid *Mathematics.* the graph in the plane of the equation $x^{2/3} + y^{2/3} = a^{2/3}$. It has the shape of a four-pointed star centered at the origin with cusps at $(\pm a, 0)$ and $(0, \pm a)$.

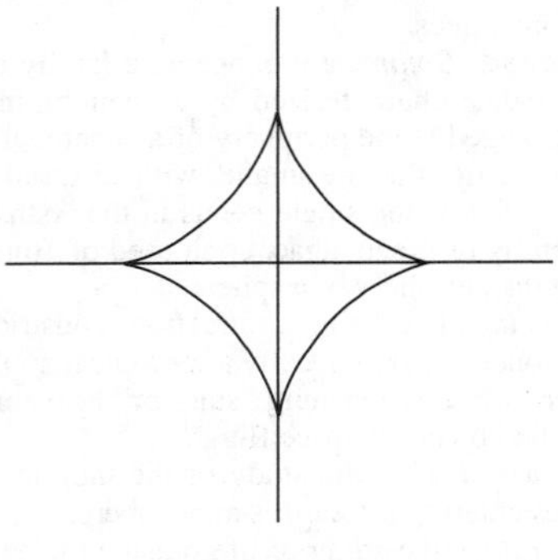

astroid

astrolabe *Engineering.* an instrument for measuring the altitude of the sun and stars, widely used by astronomers until replaced by the sextant.

astrometric binary star *Astronomy.* a double-star system in which the presence of an unseen component star is shown by irregularities in the proper motion of its primary star.

astrometric position *Astronomy.* the apparent position of an object, for which the annual abberation has been corrected.

astrometry *Astronomy.* the precise measurement of the motions and positions of celestial bodies.

astron *Nuclear Physics.* an early form of an experimental nuclear fusion reactor using field-reversing particle rings for the magnetic confinement of a fusion plasma.

astronaut *Space Technology.* a crew member or passenger in a space vehicle.

astronautics *Aviation.* the science and practice of space flight.

astronavigation *Navigation.* the plotting of a flight vehicle's position from within the vehicle by means of sighting on celestial objects.

astronomical *Astronomy.* of or relating to the science of astronomy. Also, **astronomic.**

astronomical atlas *Astronomy.* an atlas of the sky; it plots the positions of stars, galaxies, nebulae, and so on, but not planets.

astronomical azimuth *Astronomy.* the angle between the astronomical meridian plane of the observer, and the plane containing the observed point and the true normal of the observer, measured in the plane of the horizon.

astronomical azimuth mark *Cartography.* a signal or target having an astronomical azimuth from the survey station that can be determined by direct observation of a celestial body.

astronomical bearing see TRUE BEARING.

astronomical camera *Optics.* a camera designed to record astronomical phenomena, such as stars, nebulae, galaxies, or planets, and the spectra of such phenomena.

astronomical camera

astronomical catalogue *Astronomy.* a list of the celestial positions of stars; some catalogues also list galaxies, nebulae, and so on.

astronomical clock *Horology.* **1.** a precise pendulum clock with separate dials for seconds, minutes, and hours; it was originally used by astronomers to calculate astronomical time. **2.** a clock indicating the movement of the sun, planets, and other astronomical data.

astronomical constants *Astrophysics.* precisely measured fundamental quantities, such as solar parallax, the constant of aberration, and the obliquity of the ecliptic.

astronomical coordinates *Geodesy.* quantities defining a point on the surface of the earth, or of the geoid, in which the local direction of gravity is used as a reference. Also, GEOGRAPHIC COORDINATES.

astronomical coordinate system *Astronomy.* coordinates (most often right ascension and declination) used for determining positions of celestial objects.

astronomical date *Astronomy.* a date designated by year, month, day, and decimal fraction of a day.

astronomical day *Astronomy.* the mean solar day that begins at mean noon.

astronomical distance *Astronomy.* the distance of a celestial object from earth, given in any standard unit of astronomical measurement.

astronomical eclipse see ECLIPSE.

astronomical ephemeris see EPHEMERIS.

astronomical equator *Geodesy.* the line on the earth's surface on which the astronomical latitude at every point is 0°.

astronomical instrument *Engineering.* any item used for the study of the position, composition, and movement of celestial bodies.

astronomical latitude *Geodesy.* the angle between a plumb line and the celestial equator; a north or south position on the earth reckoned directly from observations of celestial bodies and uncorrected for deflection of the vertical.

astronomical leveling see ASTROGEODETIC LEVELING.

astronomical longitude *Geodesy.* the angle between the plane of the celestial meridian and the plane of an initial, arbitrarily chosen meridian.

astronomical meridian *Geodesy.* a line on the earth's surface that has the same astronomical longitude at every point. Also, TERRESTRIAL MERIDIAN.

astronomical meridian plane *Geodesy.* a plane that contains the vertical of the observer and is parallel to the instantaneous rotation axis of the earth.

astronomical nutation see NUTATION.

astronomical observatory *Astronomy.* **1.** a building constructed to house astronomical instruments. **2.** an organization that conducts astronomical research through the use of such a building.

astronomical parallel *Geodesy.* a line on the earth's surface having the same astronomical latitude at every point; because deflection of the vertical is not the same at all points on the earth, an astronomical parallel is an irregular line, not lying in a single plane.

astronomical photography *Optics.* the use of photographs to record extraterrestrial objects in order to study their surface features, positions, motions, radiation, and spectra.

astronomical position *Geodesy.* **1.** a point on the earth having coordinates determined by observation of celestial bodies. Also, ASTRONOMICAL STATION. **2.** a point on the earth defined in terms of astronomical longitude and latitude.

astronomical refraction *Geophysics.* the bending of light or a ray of celestial radiation as it passes into the atmosphere from space.

astronomical scintillation *Astrophysics.* the twinkling of starlight caused by variations of refractivity in high-altitude layers of the earth's atmosphere.

astronomical spectrograph *Spectroscopy.* a spectroscope in which the spectra of stars and other celestial objects are recorded on photographic film.

astronomical spectroscopy *Spectroscopy.* the analysis of radiant energy emitted by celestial objects in order to gather data about physical and chemical properties of the celestial objects, such as density, temperature, and chemical composition.

astronomical station see ASTRONOMICAL POSITION.

astronomical surveying *Geodesy.* a method of determining longitude and latitude by observation of celestial bodies, with separations calculated by computing distances corresponding to measured angular displacements along the reference spheroid.

astronomical telescope *Optics.* **1.** a telescope used for observing extraterrestrial objects. **2.** a telescope that collects, detects, or records electromagnetic radiation emitted from extraterrestrial sources.

astronomical theodolite see ALTAZIMUTH.

astronomical tidal constituent *Oceanography.* one of the harmonic elements in mathematical expressions and formulas for the tide-producing force, tides, and tidal currents. Also, HARMONIC CONSTITUENT, TIDAL CONSTITUENT, CONSTITUENT NUMBER, PARTIAL TIDE.

astronomical tide *Oceanography.* a tide caused by the attractive forces of the sun and moon, as opposed to a meteorological tide caused chiefly by wind and atmospheric pressure.

astronomical time *Astronomy.* the solar time in an astronomical day. *Horology.* **1.** any standard of time based on astronomical observations. **2.** a system of time measurement formerly used by astronomers, based on solar time in a mean solar day beginning at noon, and superseded by civil time.

astronomical traverse *Engineering.* a line running across an area defined in surveying, located by observations of celestial bodies and subsequent computations.

astronomical triangle *Astronomy.* the spherical triangle on the sky formed by the three great circles that connect a celestial object, the zenith point, and the nearest celestial pole.

astronomical twilight *Astronomy.* the time between sunset or sunrise and the moment when the sun's center lies 18° below the horizon.

astronomical unit *Astronomy.* the average distance from earth to the sun; about 150 million kilometers.

astronomical year see TROPICAL YEAR.

astronomic coordinate system *Cartography.* **1.** a coordinate system consisting of the plane of the celestial equator and another plane passing through an arbitrarily selected celestial meridian. **2.** a coordinate system consisting of a plane perpendicular to the earth's axis of rotation and another parallel to that axis and passing through the vertical at a specified point.

astronomic north *Cartography.* the perpendicular direction to the left of an observer facing the direction of the earth's rotation. Also, TRUE NORTH.

astronomic tide see ASTRONOMICAL TIDE.

Astronomy

Astronomy is the study of the nature of our universe. This field applies our current knowledge of physical, chemical, and biological principles to our observations of the total physical world or cosmos. It is a way of putting into context our place, origin, and uniqueness in the universe.

Astronomy had its beginnings in man's early curiosity about the puzzling lights in the night sky. Why the night sky was different than the day sky was a profound mystery, and many mythologies were developed to incorporate the sky into a model for the nature of the world. The suggestion that the sun is a star and that all the stars are suns was an insightful intellectual connection which expanded man's view of his universe. The real physical world was found to be organized on a much grander scale than the earth and was rich in puzzling observable phenomena. Phases of the moon, eclipses, and motion of the planets against the background stars were all problems to be explained and incorporated into this astronomical view.

Astronomers found that the same laws of physics which explained local phenomena could be applied to the entire universe. Today most fields of science have direct applications to astronomy and have helped give us our current understanding. Observations indicate that the universe was created about 10 to 15 billion years ago in an explosion referred to as the "big bang."

Life on earth appears to be a natural consequence of the events that took place after the big bang. Heavy elements were forged in the nuclear furnaces of stars, released in supernova explosions, and complex organic molecules grew out of them in interstellar space. We now find that the raw materials which were required for the origin of life on earth are common on other planets in our own solar system and in interstellar space.

Modern astronomy incorporates many fields of study, ranging from particle physics experiments, which are helping to understand the first moments of time after the big bang, to biology, which gives us insight into the origin of life on earth and the possibility of its origin on other worlds.

Astronomy has shown that the earth is but one infinitesimal component in the cosmos and that we have evolved from the universe to the point where we now try to comprehend it as astronomers.

Richard J. Terrile
Planetary Astronomer, Jet Propulsion Laboratory
California Institute of Technology

astronomy *Science.* the scientific study of celestial bodies, including their size, features, composition, motion, relative position, and so on. (From a Greek term meaning "star arrangement.")

Astropectinidae *Invertebrate Zoology.* a family, in the order Paxillosida, comprising many starfish species with large marginal plates and long, well-developed spines; they are generally quick-moving.

astrophobia *Psychology.* an irrational fear of heavenly bodies and outer space.

astrophyllite *Mineralogy.* $(K,Na)_3(Fe^{+2},Mn)_7Ti_2Si_8O_{24}(O,OH)_7$, a bronze-yellow to gold-yellow triclinic mineral with perfect cleavage, bladed crystals up to 15 cm long, and a submetallic, pearly luster, having a specific gravity of 3.3 to 3.4 and a hardness of 3 on the Mohs scale; found with quartz, feldspar, riebeckite, and zircon.

astrophysical *Astronomy.* relating to or used in astrophysics.

astrophysical camera see ASTRONOMICAL CAMERA.

Astrophysics

Astrophysics, as the name implies, is devoted to the understanding of astronomical phenomena in terms of the laws of physics. Defined in this way, the line of demarcation between "astronomy" and "astrophysics" largely fades. Indeed, there has been no real distinction from the very beginning; the first and greatest book on the physical sciences, Newton's *Principia,* laid the foundations both of physics and of astronomy. While Newton's *Principia* has commonly been regarded as devoted to the understanding of the motions in the solar system (as encapsulated in Kepler's laws), Newton himself based his formulation of the universal law of gravitation, principally, on his precise measurements of the periods of oscillations of "pendulous bodies" which established the equality of the inertial and the gravitational mass or "quantity of matter" and "weight" in Newton's terminology.

It is customary, however, to date the birth of astrophysics with Kirchhoff's interpretation in 1859 of the Fraunhoffer lines in the solar spectrum as revealing the presence of the familiar metals, such as sodium and potassium, as glowing vapors in the sun's atmosphere; and this only a few decades after the idealistic philosopher Bishop Berkeley had claimed that the sun, the moon and the stars are but "so many sensations in our minds" and that it would be "meaningless" to enquire into the composition of the stars.

After Kirchhoff's discovery, to speak of the composition of the stars was no longer in the realm of idle dreams; it became a problem of intense practical interest. And in the development during the twenties of the quantum theory of atomic and molecular spectra, leading astrophysicists interested in the unravelling of the spectra of the sun and the stars (for example, M.N. Saha and H.N. Russell) played leading roles. Three discoveries in this context stand out: that of the identification of helium in the chromosphere of the sun before its terrestrial identification; the identification of the negative ion of hydrogen as the source of the opacity in the solar photosphere before its isolation in its free-state in the laboratory; and the identification of the Coronal lines of the sun as arising from very high stages of ionization of iron.

More recently, the composition of the distant galaxies like the Quasars derived from spectral analysis—that is, the unravelling of the so-called "forest" of hydrogen Lyman α lines—has led to an understanding of the evolution of galaxies during the past ten or more billion years. And the parallel development following the discovery of the universal 3° K black-body radiation led to the application of the latest advances in particle physics to the very beginnings of the astronomical universe in the "big bang."

Alongside the foregoing developments, the understanding of the source of energy of the stars as due to nuclear reactions resulting in the burning of hydrogen into helium followed on the earlier foundations on the structure of stars laid by Emden, Schwarzschild, and Eddington. Besides, the application of the developments in quantum statistics, based on Pauli's exclusion principle, led to our understanding of the structure of the highly dense white dwarf-stars. These developments in turn led to an understanding of the supernova phenomenon and of the structure of neutron stars.

In a different area, continuing the development of gravitational astronomy along the lines of Newton, one has incorporated into the theory of the formation of star clusters and galaxies the ideas of Einstein and of Smoluchowski on Brownian motion which provided the molecular basis for the ideas of Maxwell and Boltzmann.

A new branch of astrophysics was founded early in this century when it was discovered that in the spectra of distant stars one found "fixed lines," i.e., lines which did not partake in the motions of the stars. These fixed lines were soon interpreted as due to the absorption of starlight by matter in the intervening space between the stars. Besides common atoms (such as sodium, calcium, and iron) and molecules (such as carbon dioxide, carbon monoxide, and water vapor), large molecules with atomic weights exceeding one hundred have been identified. The study of these interstellar lines has played central roles for the delineation of the spiral arms in the milky way and in the location of the seats of formation of stars.

And finally, reference should be made to recent advances in "relativistic astrophysics" in which one has applied the principles of general relativity to deduce the last stages in the evolution of stars, the structure of pulsars, and the formation of black holes.

Subrahmanyan Chandrasekhar
Morton D. Hull Distinguished Service Professor Emeritus
University of Chicago; Nobel Laureate in Physics

astrophysics *Astronomy.* the physics of stars and nebulae.

astrosphere *Cell Biology.* the center of an aster.

Asturian orogeny *Geology.* a brief Pennsylvanian crustal deformation involving Phanerozoic rock.

astylar *Architecture.* without columns or pilasters.

asulcal *Biology.* not having normal or characteristic grooves and furrows, such as on the surface of the cerebrum.

A-summable see ABEL SUMMATION.

A supply *Electronics.* a source of current used to heat filaments of an electron tube. Also, A POWER SUPPLY.

as-welded *Metallurgy.* the condition of a metal or alloy that has been welded, but not subjected to subsequent thermal or mechanical treatments.

asyllabia *Neurology.* a form of aphasia characterized by an inability to form or understand syllables, even while recognizing individual letters.

asymbolia *Neurology.* a form of aphasia characterized by an inability to understand and interpret symbols, including words, gestures, signs, and figures.

asymmetrical [ā´si met´trə kəl] *Science.* not symmetrical; characterized by asymmetry. Also, **asymmetric.**

asymmetrical bedding *Geology.* a sequence in the layering of sedimentary rock in which the rock's distinctive features succeed each other in a circuitous arrangement, as in the sequence 1-2-3-1-2-3-1-2-3.

asymmetrical cell *Electronics.* a photoelectric cell whose impedance for current flow in one direction is greater than that in the other direction.

asymmetrical conductivity *Electricity.* a change in the conductivity of a conductor over its cross section that is not symmetric with respect to the central axis of the conductor.

asymmetrical fold *Geology.* a rock fold whose axial plane is not vertical, causing its limbs to dip at different angles. Also, **asymmetric fold.**

asymmetrical laccolith *Geology.* a lens-shaped circular mass of intrusive igneous rock in which the beds dip at distinct angles in different sectors.

asymmetrical ripple mark *Geology.* the normal form of a ripple mark with short, steep downstream slopes and long, gentle upstream slopes.

asymmetrical vein *Geology.* a deposit of crustified geologic materials within a rock fracture or joint having unlike layers on each side.

asymmetric carbon atom *Organic Chemistry.* a carbon atom that is bonded to four different atoms or groups.

asymmetric unit *Crystallography.* the smallest part of a crystal structure from which the complete structure can be obtained from the space group symmetry operations.

asymmetry [ā´sim´ə trē] *Science.* a condition in which the parts of a shape or configuration that are on opposite sides of a central line are not symmetrical. Also, DISSYMMETRY, NONSYMMETRY. *Physical Chemistry.* a structure or design of a molecule, atom, or ion that cannot be duplicated by reversing or reorienting the pattern. Also, MOLECULAR ASYMMETRY.

asymptomatic [ā´sim tō mat´ik] *Medicine.* showing or causing no symptoms.

asymptote [ə sim´tōt´] *Mathematics.* **1.** a line (or a curve) approached arbitrarily and closely by a curve as the distance from the origin increases; the line approaches closer and closer to the curve but does not touch it. **2.** the line which is the limit of the tangents to a curve as the point of tangency approaches infinity.

asymptotic cone of acceptance [ā´sim tät´ik] *Geophysics.* a solid angle that particles must enter in order to be picked up and counted by a neutron monitor on earth.

asymptotic curve *Mathematics.* a curve on a surface with the property that the osculating place at each point of the curve is the same as the tangent plane to the surface at that point.

asymptotic direction of arrival *Geophysics.* the direction of positively charged particles at a set point on the earth's surface, after they have passed through the geomagnetic field.

asymptotic expansion *Mathematics.* a (possibly divergent) series of the form $a_0 + (a_1/x) + (a_2/x^2) + (a_3/x^3) + \cdots$ is an asymptotic expansion of the function $f(x)$ if $\lim_{x\to\infty} x^n [f(x) - S_n(x)] = 0$ for any fixed n, and where $S_n(x)$ is the sum of the first n terms of the series. Also, ASYMPTOTIC SERIES.

asymptotic formula *Mathematics.* if $f(x)/g(x)$ approaches 1 as x approaches infinity or some constant value x_0, then $f(x) = g(x)$ is said to be an asymptotic formula (and $f(x)$ and $g(x)$ are **asymptotically equal** at infinity or at x_0, as the case may be), even if $f(x)$ does not actually equal $g(x)$ for all values of x.

asymptotic freedom *Particle Physics.* a concept that the strong force between quarks becomes weaker as the quarks move closer together.

asymptotic series see ASYMPTOTIC EXPANSION.

asymptotic stability *Mathematics.* the solution $x(u_0, t)$ of a system of ordinary differential equations with initial conditions u_0 is said to exhibit asymptotic stability or to be asymptotically stable if:
(i) $\lim_{u\to u_0} |x(u,t) - x(u_0,t)| = 0$ for each $t > 0$; and (ii) there exists a neighborhood U of u_0 such that $\lim_{t\to\infty} |x(u_1,t) - x(u_0,t)| = 0$ for any $u_1 \in U$.

asynapsis *Cell Biology.* the failure of homologous chromosomes to pair during meiosis.

asynchronism *Neurology.* a disturbance of coordination marked by a lack of synchrony of movement.

asynchronous [ā´sin krän´əs] *Science.* not at the same time. *Computer Technology.* **1.** operating at a rate determined by the system rather than at a regular rate of chronological time; without a fixed time pattern. **2.** describing the relationship of two or more systems that run at their own rates and interact at unpredictable times.

asynchronous communication *Computer Technology.* a communication that takes place when one party transmits data, as when a user types a letter on a keyboard.

asynchronous computer *Computer Technology.* a computer in which there is no clock or external timing mechanism, but rather the logic is arranged in stages; when the output of one stage is stabilized, new input to that stage is accepted. Thus, **asynchronous data.**

asynchronous control *Control Systems.* a method of control based on real time rather than predetermined portions of a fixed machine cycle. Thus, **asynchronous device.**

asynchronous input/output *Computer Technology.* the transfer of data between a computer and peripheral input/output devices that have independent internal timing.

asynchronous logic *Electronics.* in digital systems, logic whose input and output relationships are independent of a clock.

asynchronous network control *Transportation Engineering.* a system in which headways are controlled but not prescheduled.

asynchronous operation *Computer Technology.* a mode of character-by-character data transfer that requires the transmission of control signals (start and stop bits) to separate transmitted characters. Also, **asynchronous working.** *Electronics.* an operation initiated and controlled by the input signals only, with its speed and timing determined by the circuit and the signals without the intervention of a clock signal.

asynchronous timing *Industrial Engineering.* a queue simulation procedure in which the simulation is updated for each arrival or departure, resulting in variable updatings of the queue timing clock.

asynchronous transfer mode *Telecommunications.* a method for the switching of channels or groups of channels in packetized format and contained in broadband (150–600 megabits per second) channels for BISDN communications.

asynchronous transmission *Telecommunications.* data transmission in which time of occurrence of a specified significant instant in each byte, character, word, block, or other unit of data is arbitrary and occurs without necessarily being dependent on preceding signals on the channel.

asyndetic *Linguistics.* relating to **asyndeton,** a form of expression in which conjunctions are omitted, as in "I came, I saw, I conquered." *Computer Technology.* omitting connectives between operands.

asynergia *Neurology.* the incoordination of the various muscular contractions involved in an action, resulting in jerky, nonsynchronous movement.

asystole [ā sis´tə lē] *Cardiology.* cardiac arrest or the absence of a heartbeat. Also, **asystolia.**

asystolic [ā´sis täl´ ik] *Cardiology.* characterized by asystole.

At see ASTATINE.

at see TECHNICAL ATMOSPHERE.

AT-1 *Ordnance.* an early Soviet antitank missile powered by a solid-propellant motor and equipped with wire guidance; it was replaced by the AT-2 in the late 1960s.

AT-2 *Ordnance.* a Soviet antitank missile powered by a solid-propellant motor and equipped with heat-seeking homing; it delivers a hollow-charge warhead capable of penetrating two feet of armor plate and has a speed of 335 miles per hour and a range of 7200 ft. Code name, **Swatter.**

AT-3 *Ordnance.* a small Soviet antitank missile that can be transported and launched by a two-man infantry team; it delivers a 6.6-lb armor-piercing hollow charge at high subsonic speed and a range of 9800 ft. Code name, **Sagger.**

AT-5 *Ordnance.* a Soviet antitank missile.

AT-6 *Ordnance.* an advanced Soviet antitank missile that is probably powered by a dual-thrust solid motor and equipped with laser guidance, delivering a large high-explosive antitank warhead at 620 miles per hour and a range of 150 to 16,500 ft. Code name, **Spiral.**

ATA see AMINOTRIAZOLE.

Atabrine *Pharmacology.* a trademark for a preparation of quinacrine hydrochloride, an anthelmintic agent that is effective against flatworms (Cestodes) and has replaced the older, more toxic agents in the therapy of tapeworm; repeated treatment is necessary for *Hymenolepsis nana.*

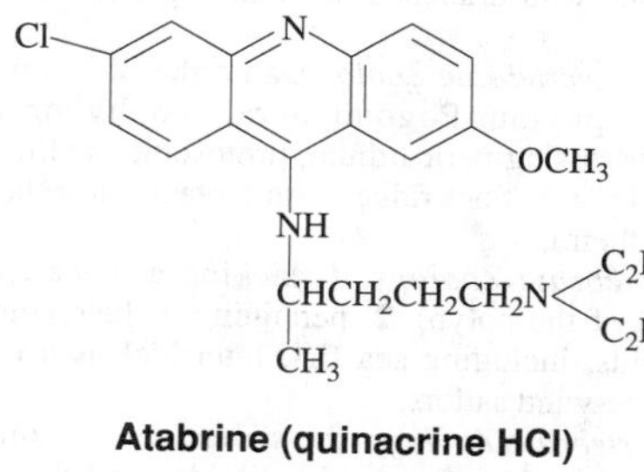

Atabrine (quinacrine HCl)

atacamite *Mineralogy.* $Cu_2^{+2}Cl(OH)_3$, a brittle, bright to dark green orthorhombic mineral with a highly perfect cleavage and an adamantine to vitreous luster, occurring as slender prismatic crystals having a specific gravity of 3.76 and a hardness of 3 to 3.5 on the Mohs scale; found as a secondary mineral in oxidation zones of copper deposits.

atactic *Neurology.* **1.** of muscle movements, irregular; lacking in coordination. **2.** of, relating to, or characterized by ataxia. Also, **ataxic.** *Organic Chemistry.* **1.** a polymer molecule in which the side groups attached to opposite sides of a carbon chain are distributed at random in an unsymmetrical pattern on the two sides. **2.** relating to or describing such a molecule.

atactic polymer *Materials Science.* a polymer in which side groups are randomly arranged on either side of the main chain.

atactic structure *Materials Science.* the structure of a thermoplastic in which the pendant methyl group is randomly arranged on either side of the main carbon chain.

atactostele *Botany.* a type of stele in which the vascular bundles are scattered randomly throughout ground tissue, typical of monocots.

atavic see ATAVISTIC.

atavism *Evolution.* the reappearance of ancestral forms or characters in a contemporary organism as a result of a reactivation of genes inherited from that ancestral form.

atavistic *Evolution.* relating to or showing atavism.

atavistic mutation *Genetics.* a mutation that returns a tissue or organism to a more primitive state.

ataxaphasia *Neurology.* an inability to form phrases and sentences despite the ability to enunciate individual words. Also, ATAXIAPHASIA.

ataxia [ə taks´ē ə] *Neurology.* a failure of muscular coordination or irregularity of muscular action. Also, **ataxy.**

ataxiaphasia see ATAXAPHASIA.

ataxia telangelectasia *Pathology.* a familial autosomal recessive disorder marked by a loss of coordination or failure in muscular movement and by a progressive cerebellar immunodeficiency associated with both B and T lymphocytes. Also, LOUIS-BAR SYNDROME.

ataxic *Geology.* of a mineral deposit, lacking stratification.

ataxiophobia *Psychology.* an irrational fear of disorder.

ataxite *Geology.* **1.** a variety of fine-grained iron meteorite containing more than 10% nickel and lacking the structure of hexahedrite or octahedrite. **2.** a taxitic rock that resembles a breccia.

ATBM antitactical ballistic missile.

ATC *Mechanical Devices.* all-terrain cycle; a three-wheeled motorcycle designed for off-road use. Also, ATV.

ATC air-traffic control; automatic train control.

atectonic *Geology.* of an event, occurring in the absence of widespread crustal deformations.

atel- or **atelo-** a prefix meaning "imperfect" or "incomplete."

Ateleopoidei *Vertebrate Zoology.* an order of deep-sea fishes in the class Cetomimiformes distinguished by a long body, lack of dorsal fin, and continuous anal and caudal fins.

atelestite *Mineralogy.* $Bi_8(AsO_4)_3O_5(OH)_5$, a transparent to translucent yellow monoclinic mineral occurring in minute, tabular crystals, having a specific gravity of 6.82 and a hardness of 4.5 to 5 on the Mohs scale.

Atelopidae or **Atelopodidae** *Vertebrate Zoology.* a family of small, brightly colored Central American tree frogs in the suborder Procoela.

Atelostomata *Invertebrate Zoology.* a superorder of echinoderms, in the subclass Euechinoidea; members are irregular echinoids lacking any chewing mechanism in adulthood.

Aten *Astronomy.* asteroid 2062, discovered in 1976; it has a diameter of less than 1 km, belongs to type S, and orbits mostly inside the earth's orbit.

ATH *Aviation.* the airport code for Athens, Greece.

Athalamida *Invertebrate Zoology.* an order of amoebas, in the subclass Granuloreticulosia, with branched, threadlike pseudoposia and no shell or skeleton.

Athecanephria *Invertebrate Zoology.* an order of beard worms, in the class Frenulata, phylum Pogonophora, subphylum Perviata; anthecanephrids possessing pericardium, protosome, and mesosome separated externally by a distinct ridge, with tentacles attached individually and not fused at their bases.

athecate *Invertebrate Zoology.* **1.** lacking a theca (protective cup around the head of the polyp). **2.** pertaining or belonging to the gymnoblastic hydroids, including sea firs, intertidal hydroids, freshwater hydras, and by-the-wind sailors.

Atherinidae *Vertebrate Zoology.* the silversides, a family of omnivorous marine fishes in the suborder Mugiloides and the order Atheriniformes; members are distinguished by the lack of a lateral line, and by having a small mouth and feeble teeth.

Atheriniformes *Vertebrate Zoology.* a large order of marine fishes in the infraclass Teleostei, including such diverse types as needlefishes, silversides, and killifishes.

athermalize *Engineering.* to render a system independent of temperature or thermal effects. Thus, **athermalization.**

athermal solution *Physical Chemistry.* a solution that resembles an ideal solution in that its heat of mixing is zero, but that exhibits some internal energy change and thus is nonideal; this may occur when components are similar in chemical nature but very different in molecular size.

athermal transformation *Physics.* any change, either physical or chemical, in which there is no change in the temperature of the system. *Metallurgy.* in an alloy, a solid-state transformation, such as the martensitic, that rapidly changes the alloy structure without the help of thermal activation and without diffusion.

athermancy [ā´thər mən sē] *Electromagnetism.* a property whereby a substance is incapable of transmitting infrared radiation.

athero- a combining form meaning "fatty," and often denoting degeneration. (From a Greek word meaning "gruel.")

atheroma *Medicine.* a mass of plaque of degenerated, thickened arterial intima that occurs in atherosclerosis.

atheromatous plaque *Cardiology.* a clearly defined yellow area or swelling on the inner surface of an artery due to a fatty deposit.

atherosclerosis [āth´ə rō´sklə rō´sis] *Medicine.* a common form of arteriosclerosis in which deposits of yellowish plaques containing cholesterol, lipoid substances, and lipophages are formed within large and medium-sized arteries.

athetosis *Neurology.* a disturbance marked by the involuntary repetition of slow, writhing, movements, especially of the hands and feet.

Athiorhodaceae see RHODOSPIRILLACEAE.

athlete's foot *Medicine.* a dermatophytosis of the feet, particularly the interdigital spaces and soles; often caused by *Trichophyton rubrum* or *Epidermophyton floccosum,* and characterized by intensely pruritic lesions varying from mild, chronic, and scaling to acute, pustular, and bullous. Also, TINEA PIDIS.

athletic heart *Cardiology.* hypertrophy (or enlargement) of the heart that is sometimes seen in athletes and is not caused by valve disease.

athodyd [āth´ə did] *Aviation.* a jet engine resembling a tube with both ends open. Air is admitted at one end, compressed by the forward movement of the aircraft, heated by fuel combustion, and expelled at the other end, producing thrust. (An acronym for aerothermodynamic duct.)

athrocyte *Histology.* a phagocytic cell of the venal tubule that ingests macromolecules by a process similar to phagocytosis.

athwartship *Naval Architecture.* at a right angle to the fore-and-aft line of a vessel.

Athyrididina *Paleontology.* a suborder of articulate brachiopods in the extinct order Spiriferida; biconvex and bistrophic, characterized generally by smooth shells and strong beaks; upper Ordovician to Jurassic.

Atkinsiella *Mycology.* a genus of fungi, belonging to the order Saprolegniales, that includes some species that are parasites of the eggs of certain marine crustacea.

ATL *Aviation.* the airport code for Atlanta, Georgia.

Atlantacea *Invertebrate Zoology.* a superfamily of free-swimming gastropod mollusks, appearing near sea coasts during fall or winter storms; characteristics include back-swimming and complete transparency.

Atlantic *Geography.* found in or associated with the Atlantic Ocean.

Atlantic guitarfish see GUITARFISH, def. 1.

Atlantic Ocean *Geography.* the ocean extending from the Arctic to Antarctica between the Americas and Europe and Africa.

Atlantic series *Petrology.* a great group of igneous rocks, largely of alkalic or alkali-calcic affinity, based on tectonic settings found in nonorogenic areas, and often associated with block sinking and great crystal instability; such rocks will erupt along faults and fissures or through explosion vents. Also, **Atlantic suite.**

Atlantic time *Astronomy.* the zone of civil time that is the fourth west of Greenwich. Also, **Atlantic standard time.**

Atlantic-type continental margin *Geology.* an aseismic edge of a continent, similar to that of the Atlantic, where oceanic and continental lithospheres are coupled.

atlantite *Petrology.* an olivine-bearing nepheline tephrite in which dark minerals predominate over the light.

atlas *Cartography.* a book or collection of maps. *Anatomy.* the first cervical vertebra, which supports the skull. *Mathematics.* a collection of open sets on a differentiable n-manifold, with coordinates defined on each set. The coordinates are required to be compatible in that the composition of the coordinates on one open set with the inverse coordinates on an overlapping open set is a differentiable function from R^n to R^n.

Atlas *Astronomy.* the fifteenth moon of Saturn, discovered in 1980 and measuring about 40 × 30 km. *Geography.* a mountain range running across northwestern Africa, mostly in Morocco and Algeria; highest peak: Jebel Toubkal (13,671 ft). *Space Technology.* a missile originally developed as an ICBM, later used to launch a wide variety of space vehicles, including the Mercury, Agena, and Centaur.

Atlas-Centaur *Space Technology.* a combination of an upper-stage Centaur space vehicle with a modified Atlas missile used as a first-stage rocket booster.

atlas grid *Cartography.* a reference system consisting of two perpendicular axes, one divided into units labeled with letters, the other with numbers, used to designate points or areas on a map or photograph. Also, ALPHANUMERIC GRID.

atlatl [at´lät əl] *Archaeology.* a spear-throwing device used by the Aztecs and other peoples of the Americas.

ATM *Electronics.* automated teller machine, an electronic banking device that allows a customer to automatically carry out transactions such as deposits and cash withdrawals by placing an identification card in the machine and entering a personal code. *Ordnance.* antitactical missile. *Telecommunications.* asynchronous transfer mode.

atm ATMOSPHERE.

atmidometer see ATMOMETER.

atmo- a combining form meaning "air" or "vapor," as in *atmosphere, atmometry.*

atmoclast *Geology.* a rock fragment that is broken off in places by chemical or mechanical atmospheric weathering.

atmoclastic *Petrology.* referring to a clastic rock composed of rock fragments broken by atmospheric weathering and recemented in the same pattern as the previous arrangement.

atmogenic *Geology.* of a rock, mineral, or deposit, originating in the atmosphere and formed through the action of condensation, wind, or deposition of volcanic vapors.

atmolith *Geology.* any atmogenic rock.

atmolysis *Fluid Mechanics.* a procedure for separating the components of a blend of two gases that relies upon the differing rates of gas diffusion through a porous membrane or partition. *Toxicology.* a destruction of body tissues caused by contact with fumes of toxic volatile fluids.

atmometer *Engineering.* an instrument that measures the rate at which water evaporates into the air.

atmometry *Metrology.* the science or practice of measuring evaporation. Thus, **atmometric.**

atmophile element *Meteorology.* any of the atmospheric elements that exist singly, in an uncombined state, or as a volatile, concentrated compound in the gaseous primordial atmosphere.

atmosphere *Meteorology.* the envelope of gases surrounding the earth and held to it by the force of gravity. It consists of four distinct layers, whose bondaries are not precise: the troposphere (extending from sea level to about 5–10 miles above the earth), the stratosphere (up to about 30 miles), the mesosphere (up to about 60 miles), the thermosphere (up to about 300 miles or more). The upper region of the troposphere is often regarded as a separate region, the exosphere. *Astronomy.* **1.** the gas bound gravitationally to a planet. **2.** the outer layers of a star. *Mechanics.* a unit of pressure that is taken to be the standard pressure of the earth's atmosphere at sea level; equal to the pressure of a column of mercury 760 mm high and expressed as 101.325 kilopascals (1.01325×10^5 newtons per square meter), or about 14.7 pounds per square inch.

atmospheric *Meteorology.* relating to or taking place in the earth's atmosphere.

atmospheric absorption *Geophysics.* the absorption of radiation by the gases and moisture in the atmosphere.

atmospheric acoustics *Acoustics.* the propagation of sound through the atmosphere, which affects sound in predictable ways according to atmospheric conditions such as temperature and precipitation.

atmospheric attenuation *Geophysics.* the depletion of electromagnetic energy in the atmosphere due to absorption or diffusion.

atmospheric boil see TERRESTRIAL SCINTILLATION.

atmospheric boundary layer see SURFACE BOUNDARY LAYER.

atmospheric braking *Space Technology.* the slowing of the speed of an object as it encounters the drag of a planetary atmosphere. This may be initiated or enhanced deliberately, as in landing a space vehicle.

atmospheric chemistry *Meteorology.* the study of the atmospheric constituents in the troposphere and the stratosphere.

atmospheric composition *Meteorology.* the chemical constituents in the earth's atmosphere, including nitrogen, oxygen, carbon, argon, water vapor, neon, carbon dioxide, helium, krypton, methane, hydrogen, and nitrous oxide.

atmospheric condensation *Meteorology.* the transformation of water vapor in the air to dew, fog, or cloud.

atmospheric convection current *Geophysics.* the vertical movement of air currents due to temperature variations.

atmospheric cooler *Mechanical Engineering.* a cooler for fluids that uses air circulation obtained by natural convection to cool hot, fluid-filled tubes. Also, NATURAL-DRAFT COOLER.

atmospheric density *Meteorology.* a ratio of a portion of atmospheric mass to the volume it occupies.

atmospheric diffusion *Meteorology.* an exchange of fluid parcels between atmospheric regions in random movements too small to be treated by equations of motion.

atmospheric distillation *Chemical Engineering.* a distillation procedure performed at atmospheric pressure, as distinguished from vacuum or pressure distillation.

atmospheric disturbance *Meteorology.* any disruption of the atmospheric steady state.

atmospheric drag *Fluid Mechanics.* a critical perturbation of the orbits of closely adjacent low-orbit artificial satellites due to atmospheric resistance; the effects extending over ages of time are semidiameter, period, and decreasing eccentricity.

atmospheric duct *Geophysics.* a layer of the troposphere in which refractive properties are such as to trap a large proportion of certain high frequency radiations.

atmospheric electric field *Geophysics.* a measure, in volts per meter, of the electrical energy in a given portion of the earth's atmosphere at a given time.

atmospheric electricity *Geophysics.* a collective term for all electrical phenomena occurring in the atmosphere, including lightning, St. Elmo's fire, atmospheric ionization, and air-earth currents.

atmospheric entry *Space Technology.* the penetration of a planetary atmosphere by an object approaching from space, especially of the earth's atmosphere by a reentering spacecraft.

atmospheric evaporation *Hydrology.* the passage of water in a gaseous state from the earth's surface, including oceans, lakes, rivers, ice, snow, and soil, into the atmosphere.

atmospheric general circulation *Meteorology.* a statistical representation of the mean global flow pattern of the atmosphere.

atmospheric impurity *Meteorology.* any foreign material that mixes with and contaminates the air in the atmosphere.

atmospheric interference *Geophysics.* the radio-frequency electromagnetic radiation, originating for the most part in lightning discharges, that interferes with radio communications. Also, ATMOSPHERICS.

atmospheric inversion see INVERSION.

atmospheric ionization *Geophysics.* the charging of neutral particles in the atmosphere through violent contact with charged particles.

atmospheric lapse rate see ENVIRONMENTAL LAPSE RATE.

atmospheric layer see ATMOSPHERIC SHELL.

atmospheric noise *Electronics.* noise heard in a radio receiver because of interference from the atmosphere.

atmospheric optics see METEOROLOGICAL OPTICS.

atmospheric physics *Geophysics.* a branch of science dealing with the physical phenomena of the atmosphere.

atmospheric pressure *Physics.* the pressure of the earth's atmosphere. See ATMOPSPHERE..

atmospheric pressure cure *Petroleum Engineering.* the preparation of petroleum specimens for testing purposes by aging them at regular atmospheric pressure for a specific time period at a particular humidity and temperature.

atmospheric radiation *Geophysics.* the radiation emitted by the atmosphere either upward into space or downward toward the earth, consisting mainly of long-wavelength terrestrial radiation plus the small amount of short-wavelength solar radiation absorbed in the atmosphere.

atmospheric radio wave *Electromagnetism.* a radio wave that propagates through the atmosphere by reflections and refractions occurring in the atmosphere.

atmospheric refraction *Geophysics.* the refraction of light passing through the earth's atmosphere, including both astronomical refraction and terrestrial refraction. *Astronomy.* an apparent upward displacement of celestial objects relative to the horizon as light from them is bent toward the vertical by the decreasing density with altitude of the earth's atmosphere; it is greatest for objects on the horizon and negligible at elevations higher than about 45°.

atmospheric region see ATMOSPHERIC SHELL.

atmospherics see ATMOSPHERIC INTERFERENCE.

atmospheric scattering *Geophysics.* a diffusion or alteration in the direction of the propogation, frequency, or polarization of electromagnetic radiation through contact with atoms in the atmosphere.

atmospheric shell *Meteorology.* any one of a number of layers of the atmosphere, most commonly distinguished by temperature distribution. Also, ATMOSPHERIC LAYER, ATMOSPHERIC REGION.

atmospheric shimmer see TERRESTRIAL SCINTILLATION.

atmospheric sounding *Meteorology.* the measuring of atmospheric conditions above the effective range of surface weather observations.

atmospheric structure *Meteorology.* the constituting elements that characterize the atmosphere, including wind direction, velocity, altitude, air density, and velocity of sound.

atmospheric suspensoids *Meteorology*. any particles, such as dust, that are finely divided and suspended in the atmosphere.

atmospheric tide *Geophysics*. the rhythmic, periodic oscillation of the earth's atmosphere due to the gravitational effects of the earth, sun, and moon and to the absorption of radiation by the atmosphere. *Oceanography*. a tidal movement of the atmosphere resembling an ocean tide but caused principally by diurnal temperature changes.

atmospheric turbulence *Meteorology*. apparently random fluctuations of the atmosphere often causing deformations of its fluid flow.

atmospheric window *Astrophysics*. a range of wavelengths in which the atmosphere is partly or largely transparent.

ATMS *Artificial Intelligence*. short for assumption-based truth maintenance system.

ATN augmented transition network.

ATO *Transportation Engineering*. automatic train operation.

atocia [ə tō´sē ə] *Medicine*. a condition of sterility in a female.

Atokan *Geology*. a North American provincial series of the lower Middle Pennsylvanian, occurring after the Morrowan and before the Desmoinesian.

atoke *Invertebrate Zoology*. the anterior part of a marine polychaete worm from which the sexual part develops during breeding season.

atoll *Geology*. a low, ring-shaped coral island surrounding or almost surrounding a lagoon.

Atoll *Ordnance*. a Soviet air-to-air missile powered by a solid-propellant rocket motor and equipped with infrared or semiactive radar homing; it delivers a 6-kg blast-fragmentation warhead at a speed of Mach 2.5 and a range of 4 miles; officially designated **AA-2**.

atoll texture *Geology*. a ring of one or more minerals that surrounds another mineral within a deposit. Also, CORE TEXTURE.

atom *Chemistry*. **1.** the smallest unit of a chemical element that can still retain the properties of that element. Atoms combine to form molecules, and they themselves contain several kinds of smaller particles. An atom has a dense central core (the nucleus) consisting of positively charged particles (protons) and uncharged particles (neutrons). Negatively charged particles (electrons) are scattered in a relatively large space around this nucleus and move about it in orbital patterns at extremely high speeds. An atom contains the same number of protons as electrons and thus is electrically neutral and stable under most conditions. **2.** in Dalton's theory and other earlier use, the smallest possible particle of matter, which cannot be further divided and which makes up all larger forms of matter. *Computer Programming*. a class of primitive data elements or elementary objects. *Artificial Intelligence*. **1.** in propositional or predicate calculus, a predicate symbol with the appropriate number of arguments. For example, MORTAL (Socrates) is an atom, where MORTAL is the predicate and Socrates is a constant term that is its argument. Also, ATOMIC FORMULA. **2.** in Lisp, any data that is not a cons cell, such as a symbol or a number. *Mathematics*. **1.** a nonzero element A of a measure ring with the property that if $A \supset E$, then either $E = A$ or E is the zero element of the measure ring. That is, there is no nonzero element of the measure ring which is properly contained in A. **2.** a nonzero element a of a lattice with the property that no nonzero element of the lattice precedes a.

atomic *Chemistry*. **1.** of or relating to atoms. **2.** separated into individual atoms. *Ordnance*. of or relating to atomic or nuclear weapons: *atomic* ammunition, *atomic* missiles, an *atomic* blast.

atomic absorption coefficient *Physics*. a coefficient derived by dividing the absorption coefficient of a particular medium by the atomic density.

atomic absorption spectrophotometry *Analytical Chemistry*. a method to determine the elemental composition of a substance by vaporizing the sample and measuring at specific wavelengths the amount of radiation absorbed, which is proportional to the concentration of the element in the sample.

atomic absorption spectroscopy *Spectroscopy*. the study of absorption spectra by means of passing electromagnetic radiation through an atomic medium that is selectively absorbing; this produces pure electronic transitions free from vibrational and rotational transitions.

atomic battery see NUCLEAR BATTERY.

atomic beam *Physics*. a narrow beam of atoms or molecules.

atomic beam resonance *Physics*. a phenomenon in which transitions between energy states in the nuclei of atoms in an atomic beam are induced when the beam is passed through a region in which an oscillating magnetic field is imposed at right angles to a uniform magnetic field, and the oscillation frequency is a resonant frequency characteristic of the nuclear energy transition.

atomic bomb

atomic bomb or **atom bomb** *Ordnance*. a powerful bomb that derives its explosive force from the sudden release of huge amounts of atomic energy.

atomic charge *Atomic Physics*. the electric charge of an ion, calculated by the number of electrons the atom gains or loses during ionization times the charge of an individual electron.

atomic clock *Horology*. any extremely precise electronic clock that uses the oscillations of individual atoms or molecules, especially the cesium-133 atom, to regulate its movement.

atomic clock

atomic cloud see RADIOACTIVE CLOUD.

atomic coordinates *Crystallography*. a set of numbers that specifies the position of an atom in a crystal structure with respect to the axial directions of the unit cell of the crystal. Coordinates are generally expressed as the dimensionless quantities x, y, z (fractions of unit-cell edges).

atomic device see ATOMIC WEAPON.

atomic diamagnetism *Atomic Physics*. the susceptibility of electrons in an atom to the influence of a magnetic field.

atomic emission analysis see FLAME EMISSION SPECTROPHOTOMETRY.

atomic emission spectroscopy *Spectroscopy*. the study of emission spectra by means of producing and measuring the light emitted by a specimen when the sample is vaporized by an electric arc.

atomic energy *Atomic Physics*. the energy that exists in atoms; nuclear energy.

atomic energy level *Atomic Physics*. the amount of energy in an atom at various levels of excitation, ranging from the ground level to its most energized state.

atomic fallout see FALLOUT.

atomic form factor see ATOMIC SCATTERING FACTOR.

atomic formula *Artificial Intelligence*. see ATOM, def. 1.
atomic fusion *Physics*. see FUSION.
atomic gas laser *Optics*. a laser that employs an electric field to raise the energy level of atoms undergoing radioactive decay.
atomic ground state *Atomic Physics*. the lowest energy state; the energy state in which the electron normally resides. Also, ATOMIC UNEXCITED STATE.
atomic heat *Physical Chemistry*. the amount of heat required to raise the temperature of a gram-atomic weight of an element by one degree. Also, **atomic heat capacity.**
atomic hydrogen *Chemistry*. hydrogen in a gaseous form whose molecules are separated into atomic form.
atomic hydrogen maser *Physics*. a maser in which the stimulation medium is a beam of hydrogen atoms.
atomic hydrogen welding *Metallurgy*. an arc-welding process of limited industrial significance, based on an electric arc operating in a hydrogen atmosphere.
atomicity *Chemistry*. **1.** the fact of being atomic or existing in the form of atoms. **2.** the number of atoms in one molecule of a substance. **3.** the number of electrons that an atom will lose, add, or share when it reacts with other atoms; valence.
atomic magnet *Atomic Physics*. an atom that exhibits magnetic properties in the ground state or in an excited state.
atomic magnetic moment *Atomic Physics*. a measure of an atom's magnetic properties, expressed in magnetons.
atomic mass *Physics*. the mass of an atom or molecule as expressed in atomic mass units, using a scale in which the most abundant isotope of carbon has a mass of 12.
atomic mass unit *Physics*. a unit of mass convenient for describing the masses of atoms and molecules; the standard atomic mass unit is 1/12 of the mass of a carbon atom with mass number 12.
atomic module *Computer Programming*. in structured program design, a bottom-level module, one that has no subordinates.
atomic moisture meter *Engineering*. a device that measures the amount of moisture in coal instantaneously and continuously by bombarding it with neutrons and measuring the neutrons that rebound after they strike hydrogen atoms of water.
atomic number *Nuclear Physics*. the number, denoted by the letter *Z*, of protons in the nucleus of an atom; this number uniquely characterizes a nuclear species and determines its place in the periodic table; the atomic number is written as a subscript before the elemental symbol, thus $_{92}U$.
atomic number factor *Optics*. a correction factor in analytical electron microscopy originating in average atomic number differences between the sample and a standard.
atomic orbital *Atomic Physics*. the set of quantum numbers for an electron in an atom. *Physical Chemistry*. the region of high probablity that is occupied by an individual electron as it travels with a wavelike motion in the three-dimensional space around a nucleus; an indicator of the electron's energy level.
atomic packing factor *Atomic Physics*. the ratio of the volume that is occupied by the atoms in a unit cell to the volume of the cell.
atomic paramagnetism *Electromagnetism*. a form of paramagnetism that results from a permanent magnetic moment of an atom.
atomic parameters *Crystallography*. a set of numbers that specifies the position of an atom in the unit cell of a crystal structure and the extent of vibration of that atom. Usually three parameters define position; one parameter can be used to define isotropic vibration, or six to define simple (harmonic) anisotropic vibration.
atomic particle *Atomic Physics*. one of the elementary particles, such as an electron, proton, or neutron, that makes up an atom.
atomic percent *Chemistry*. the number of atoms of a given single element that are present in 100 specimen atoms of a certain substance.
atomic photoelectric effect see PHOTOIONIZATION.
atomic physics a branch of physics primarily devoted to the study of the structures and energies of atoms and molecules.

Atomic Physics

Atomic physics is concerned with the study of atoms and simple molecules and their interactions with each other, including their several constituents, and with radiation fields. The subject is a very broad one, since it impacts on virtually all physical phenomena occurring outside the nucleus (and sometimes even inside). For example, condensed matter physics builds on the properties of individual atoms and molecules, physical phenomena in the gas phase are determined in the first instance by the behavior of the constituent atoms and molecules, chemical reactions are examples of what could be characterized as atomic and molecular collision phenomena, atmospheric science is a complex multidisciplinary subject in which fundamental processes that govern the atmosphere are determined by atom-molecule interactions in the presence, generally, of solar radiation.

Atomic and molecular processes range from the most fundamental to the very applied. On the fundamental level, precision spectroscopic studies of isolated atoms are used to perform tests of fundamental concepts in quantum mechanics, quantum electrodynamics, and symmetry concepts. Atomic physics studies of "exotic" species such as muonium, positronium, pionium, and even antihydrogen are also fruitful testing grounds of these fundamental concepts as well as of our understanding of the structure of subnuclear particles.

Despite the fact that the basic atomic interactions, which are overwhelmingly electromagnetic in nature, are well understood, virtually no atomic problem is exactly solvable, possibly excepting atomic hydrogen (although even this problem isn't exactly solvable if one includes quantum electrodynamic and particle theory effects). As a consequence, atomic physics is a superb testing ground for many (i.e., more than two!) -body theory. Atomic physics has therefore become a prime user of advanced computational techniques.

In parallel with the enormous theoretical and computational advances which have occurred in atomic physics in recent years has been spectacular developments in our ability to manipulate atoms in the laboratory. They can literally be made to jump through hoops, through laser control of atoms. Trapping of atoms in laser fields and of ions in combined electric and magnetic fields enables one to perform spectroscopic measurements to a precision far beyond what was possible even a decade or two ago. Precision spectroscopy using lasers has pushed the field to as much as twelve or more figure accuracy.

Sophisticated beam techniques have made it possible to study details of atomic interactions such as the effect of atomic orientation on chemical reactions. For example, molecules can be induced to possess differing reaction rates by selectively orienting collision partners. Dependence of interactions upon spin and orbital angular momentum orientations of collision partners are now routinely studied.

Coincidence experiments (*e*, 2*e*) or (*e*, *hv*) can yield detailed information concerning electron correlations and atomic electron momentum distributions. Indeed, atomic and molecular physics, among the oldest of all subfields of physics, has undergone a complete renaissance, almost as though it was starting over again.

Benjamin Bederson
Professor of Physics
New York University

atomic pile see NUCLEAR REACTOR.
atomic polarization *Physical Chemistry.* the distortion of the electronic charge clouds between unlike atoms in molecules.
atomic power plant see NUCLEAR POWER PLANT.
atomic radius *Physical Chemistry.* half the distance from center to center between two like atoms that are not bonded together.
atomic reactor see NUCLEAR REACTOR.
atomic rocket *Aviation.* **1.** a proposed means of spacecraft propulsion in which nuclear bombs would be detonated behind the space vehicle as a propellant. **2.** a space vehicle propelled by such methods.
atomic scattering factor *Physics.* a factor that describes the directional distribution of the scattering of X rays of a given wavelength by an atom. Also, ATOMIC FORM FACTOR.
atomic second *Horology.* the amount of time taken by 9,192,631,770 cycles of radiation resulting from an energy-level transition of the electrons in a cesium-133 atom at zero magnetic field; the basic unit of atomic time.
atomic spectroscopy *Spectroscopy.* the study of radiant energy either absorbed or emitted by atoms.
atomic spectrum *Spectroscopy.* the spectrum produced as a result of energy level transitions within an atom as radiant energy is either absorbed or emitted.
atomic standard *Physics.* a property associated with an atom or molecule that is strictly reproducible, such as the wavelengths of spectral lines observed in a hydrogen discharge tube.
atomic stopping power see STOPPING POWER.
atomic structure *Atomic Physics.* the organization of an atom, consisting of a small but massive and positively charged nucleus, containing protons and neutrons, surrounded by negatively charged extranuclear electrons.
atomic surface burst *Ordnance.* the explosion of an atomic weapon at the surface of the ground. Other similar compounds include **atomic air burst, atomic underground burst, atomic underwater burst.**
atomic susceptibility *Electromagnetism.* the magnetization of a substance per atom per unit of applied magnetic field strength; a dimensionless quantity.
atomic theory *Chemistry.* the theory that all matter is composed of minute, distinct particles called atoms, and that these particles cannot be further subdivided; first developed by the Greek philosopher Democritus in about 400 BC and formally proposed by the English chemist John Dalton in 1808; later modified as it became established that atoms consist of even smaller particles.
atomic time *Horology.* a standard of time based on the number of oscillations of the cesium-133 atom as its electrons move from high to low energy levels.
atomic unexcited state see ATOMIC GROUND STATE.
atomic vibration *Atomic Physics.* a periodic change in the position of atoms within a molecule that gives rise to a number of physical phenomena, including heat conduction.
atomic volume *Physical Chemistry.* the value obtained by dividing the atomic weight of an element by its specific gravity in the solid condition.
atomic weapon see ATOMIC BOMB, NUCLEAR WEAPON.
atomic weight *Chemistry.* the averaged mass of one atom of an element, based on a scale where the isotope carbon-12 weighs exactly 12.00 atomic mass units.
atomism *Chemistry.* another name for the atomic theory; i.e., the theory that all matter is composed of atoms.
atomist *Chemistry.* a supporter of the theory of atomism.
atomistic groups *Anthropology.* the organization of hunters and gatherers into small and scattered bands or families during certain seasons to efficiently exploit resources in a limited resource area; practiced by groups such as the Great Basin Shoshoneans and the Eskimos.
atomization *Mechanical Engineering.* **1.** the process of reducing a liquid or meltable solid to fine particles or spray by forced passage through a nozzle or jet. **2.** the reduction of liquid fuel to a fine spray or mist that readily ignites in an automotive engine.
atomizer *Mechanical Devices.* any of various mechanical devices that reduces a bulk liquid into a fine spray or mist, as by steam, air pressure, or the like.
atomizer burner *Mechanical Engineering.* a liquid-fuel burner that sprays the unignited fuel into a readily igniting fine mist before it enters the combustor.
atomizer mill *Mechanical Engineering.* a solids grinder that reduces materials to a fine powder through fracturing and abrasion.
atomizing *Metallurgy.* the process by which metallic powders are produced by impinging water, steam, or an inert gas at high velocity on a stream of molten metal.
atomizing humidifier *Mechanical Engineering.* a device that adds a fine spray of water to a stream of air.
atom probe *Engineering.* a device used to detect a single atom or molecule in a substance, in which the atom is isolated in a probe hole after being separated from the specimen by pulsed field evaporation, and then examined in a mass spectrometer.
atom set see HERBRAND BASE.
atom smasher see ACCELERATOR.
atonia *Neurology.* a weakness or lack of normal tone in a tissue or organ. Also, **atony.**
atopite *Mineralogy.* a yellow to brown variety of romeite containing fluorine.
atopy *Immunology.* an immunological hypersensitive condition against common environmental antigens that is believed to be hereditary.
atoxic *Toxicology.* **1.** not toxic. **2.** not caused by or associated with a toxin.
atoxigenic *Toxicology.* not producing toxins.
ATP adenosine triphosphate.
ATPase adenosinetriphosphatase.
ATP synthetase *Enzymology.* an enzyme that catalyzes the formation of adenosine triphosphate.
A trace *Electronics.* the first (higher) trace on a radar scope display.
Atractidae *Invertebrate Zoology.* a family of nematodes in the superfamily Oxyuroidea; parasites of fish, amphibians, and land vertebrates, including primates.
atractyloside *Biochemistry.* a toxic substance that causes convulsions and hinders nerve conduction by blocking the ATP-ADP carrier system; formed by a species of thistle found in some Mediterranean countries.
atraumatic *Medicine.* not traumatic; not causing damage or injury.
atrazine *Organic Chemistry.* $C_8H_{14}ClN_5$, a white crystalline solid that melts at 173°C; widely used as an herbicide. Also, 2-CHLORO-4-ETHYLAMINO-6-ISOPROPYLAMINE-*S*-TRIAZINE.
atremia *Neurology.* **1.** the absence of tremor. **2.** a hysterical paralysis of the legs resulting in the inability to walk.
atresia *Medicine.* a congenital absence or closure of a normal body orifice or tubular organ.
atreto- a combining form denoting an abnormal closure or the absence of a normal opening, as in *atretocephalus.*
atrial [ā´trē əl] *Anatomy.* of or relating to the atrium.
atrial fibrillation *Cardiology.* a rapid, highly irregular heart rhythm.
atrial flutter *Medicine.* a condition of cardiac arrhythmia in which the atrial contractions are rapid (200 to 320 per minute), but regular. The ventricles are unable to respond to each atrial impulse, so that a partial block is usually present.
atrial natriuretic factor *Endocrinology.* one of a family of peptide hormones, cleaved from a single precursor peptide and produced in the cardiac atria, the physiological effects of which include increased urine output, increased sodium excretion, and a receptor-mediated vasodilation, the net result of which is lowered blood pressure. Also, **atrial natriuretic peptide.**
atrial septum *Anatomy.* the muscular septum separating the atria of the heart.
Atrichornithidae *Vertebrate Zoology.* a family containing one species of bird, the wrenlike Australian scrub-bird, in the suborder Menurae, distinguished by its acute powers of mimicry.
atrichous *Cell Biology.* the lack of hair or hairlike structures.
atriomegaly *Medicine.* the abnormal enlargement of one of the atria.
atriopeptin see ATRIAL NATRIURETIC FACTOR.
atriopore *Zoology.* **1.** an opening from the atrial cavity to the exterior in Cephalochordates. **2.** a spiracle in a tadpole.
atrioseptoplasty [ā´trē ō sep´tō plas´tē] *Surgery.* a surgical repair of the septum between the atria of the heart as in atrial septal defect repair.
atrioventricular [ā´trē ō ven´trik´yə lər] *Anatomy.* of or relating to an atrium and ventricle of the heart.
atrioventricular block *Cardiology.* a condition in which transmission of impulse from atrium to ventricle through the atrioventricular node is impaired. Also, **atrioventricular heart block.**
atrioventricular canal *Developmental Biology.* the canal in the embryonic heart leading from the common sinuatrial chamber to the ventricle.
atrioventricular groove see CORONARY SULCUS.
atrioventricular heart block see ATRIOVENTRICULAR BLOCK.

atrioventricular node *Anatomy.* tissues located in the wall of the right atrium, specialized to generate the pacemaker activity that produces heartbeat.

atriplicism *Toxicology.* poisoning caused by ingestion of the leaves of a kind of spinach, *Atriplex littoralis*; symptoms may include painful swelling and gangrene of the fingers and toes.

atrium *Architecture.* **1.** an open courtyard within a house. **2.** a multistoried court or hall within a hotel or other building, usually having a skylight. *Medicine.* a chamber that gives entrance to another structure or organ; usually used alone to refer to an atrium of the heart.

atropamine see APOATROPINE.

atropine *Pharmacology.* $C_{17}H_{23}NO_3$, an alkaloid used for such purposes as to relieve Parkinsonism, increase the heart rate, dilate the pupils, counteract toxic agents, and reduce secretions such as sweat.

atropine flush *Toxicology.* a reaction to atropine; symptoms may include a reddening of the skin of the face and neck.

atropine sulfate *Pharmacology.* $C_{34}H_{48}N_2O_{10}S$, a drug derived from atropine, used to cause paralysis of the ciliary muscle in the eye and to dilate the pupils.

atropinism *Toxicology.* poisoning caused by the ingestion of atropine or belladonna, or of parts or preparations of any of the plants from which these drugs are derived; symptoms may include dryness of the mouth and throat, fever, and hallucinations. Also, **atropism.**

atropisomer *Organic Chemistry.* one of two conformational isomers that can be readily separated, isolated, and studied.

atropous see ORTHOTROPOUS.

Atrypa *Paleontology.* a Middle Paleozoic genus of articulate brachiopods in the extinct suborder Atrypidina and family Atrypidae.

Atrypidae *Paleontology.* a family of articulate brachiopods in the extinct order Spiriferida and suborder Atrypidina; Middle Ordovician to Upper Devonian.

Atrypidina *Paleontology.* a suborder of articulate brachiopods in the extinct order Spiriferida (in some classification systems given its own order, Atrypida); among the first spire-bearing brachiopods, typically biconvex; Ordovician to Devonian.

ATS automatic train supervision.

attached dune *Geology.* a dune of any size or shape that accumulates around a rock or other geological structure in the path of windblown sand.

attached groundwater *Hydrology.* groundwater that adheres to the walls of the spaces between subsurface particles of soil and rock.

attached processing *Computer Technology.* a multiprocessor architecture designed to improve system efficiency and throughput; two or more computers share the same functions or are assigned support roles. An attached processor may perform specialized functions such as matrix processing at high speed.

attached processor *Computer Technology.* one of two or more processors connected together to improve processing efficiency.

attached shock wave *Fluid Mechanics.* a conical or oblique shock wave that seems to be touching the nose of a body in a highly supersonic flow field or the leading edge of an airfoil.

attached thermometer *Engineering.* a term for a thermometer connected to an instrument to measure temperature while in operation.

attached X chromosome *Genetics.* a genetic mutation in *Drosophila* that is expressed as a failure of the two X chromosomes to separate during oogenesis, resulting in the production of eggs that contain either two or no X chromosomes.

attachment *Behavior.* an intangible connection between an animal or human and a particular place, object, or companion.

attachment cord see PATCH CORD.

attachment plaque *Cell Biology.* a specialized site on the nuclear envelope to which the chromosomal ends are attached during meiosis.

attachment plug *Electricity.* a device, with an attached flexible cord containing conductors, used to form an electrical connection between the conductors in the cord and conductors permanently connected to a receptacle into which the device is inserted.

attack see ANGLE OF ATTACK.

attack heading *Navigation.* the heading of an aircraft after allowance is made for its angle of attack; it differs from the physical heading only if the aircraft is banking, and then differs by the sine of the angle of bank times the angle of attack.

attack plane *Aviation.* the reference plane used for determining an aircraft's angle of attack.

attack time *Forestry.* the time elapsed from the end of the report of a fire to the first organized attack on it.

attapulgite *Mineralogy.* $(Mg, Al)_2Si_4O_{10}(OH)\cdot 4H_2O$, a group of tough, lightweight, fibrous minerals characterized by a distinctive rodlike shape.

attapulgite

attar of roses see ROSE OIL.

attemperation of steam *Mechanical Engineering.* the control of any excessive superheat in a steam boiler, either by admixture between the superheated steam and cooling steam, or by forcing cooling steam across superheated steam tubes, thereby regulating the final steam temperature.

attemporate *Engineering.* to control the temperature of a material or system. Thus, **attemporation.**

attended time *Industrial Engineering.* a term for the time during which a machine has an operator present and is operating, or is idle but capable of operation.

attention *Behavior.* the selective direction of an animal's interest toward a particular stimulus.

attention-deficit disorder *Medicine.* a condition observed in children in which the attention span is markedly decreased, and activity and excitability are increased; the existence of an actual brain dysfunction is still unproven. Also, MINIMAL BRAIN DYSFUNCTION SYNDROME.

attention-deficit hyperactivity disorder see ATTENTION-DEFICIT DISORDER.

attention hypothesis see SELECTIVE ATTENTION.

attention-seeking *Psychology.* the process of carrying out certain actions, especially socially unacceptable actions, in order to gain attention and recognition from others. Also, **attention-getting.**

attention span *Psychology.* the length of time that a person can concentrate on one event or thing.

attenuate *Botany.* of a plant part, gradually tapered.

attenuated *Control Systems.* of a signal, having decreased in passing through a control system or control element.

attenuated strain *Virology.* a virus strain that through replication has diminished potency but still mimics a virulent parent virus, thus affording an immune response.

attenuated total reflectance *Spectroscopy.* an infrared spectroscopic technique for measuring the absorption spectra of opaque materials, such as paints and varnishes, by determining the energy reflected at the place where two media of different refractive indexes are in optical contact with each other. Also, FRUSTRATED INTERNAL REFLECTANCE, INTERNAL REFLECTANCE.

attenuated vaccine *Immunology.* a live vaccine that contains organisms whose virulence for a host has been diminished or abolished; it is administered to produce active immunity.

attenuated virus *Virology.* any virus that has undergone attenuation by serial passage or other means.

attenuation *Physics.* a reduction in amplitude, density, or energy as the result of such effects as friction, absorption, or scattering. *Electricity.* an exponential reduction of signal amplitude due to distance. *Control Systems.* a decrease in a signal as it passes through a control system or control element; usually expressed in decibels or as a ratio. *Microbiology.* a procedure for reducing the virulence of a human pathogen, usually by adaptation to another host species or to a different culture medium.

attenuation coefficient *Electromagnetism.* the fraction of a beam of electromagnetic energy that is lost by scatter or absorption in passing through a given thickness of a material; used in relation to light.

attenuation constant *Physics.* a quantity that represents the exponential rate of spatial decay of the amplitude of a wave at a given frequency, usually expressed in decibels per unit length or nepers per unit length. Also, **attenuation factor.**

attenuation distortion *Telecommunications.* a significant distortion in a cable, attenuator, coupling, or other device when passing electrical signals; usually expressed in decibels.

attenuation equalizer *Electronics.* a device that equalizes transmission loss for a line or circuit for all frequencies of interest.

attenuation length *Physics.* **1.** the inverse of the attenuation constant. **2.** the distance over which a wave travels while its amplitude is attenuated by a factor of $1/e$, or approximately 37%.

attenuation network *Electronics.* a passive network that provides equal attenuation throughout a frequency band while maintaining nearly constant phase.

attenuation ratio *Physics.* the magnitude of the ratio of the complex electric field strength at the destination point to that of the beginning point for an electromagnetic wave propagating from one point to another.

attenuator *Electronics.* a resistive or capacitative circuit designed to lower a signal amplitude to some desired value without distorting the signal waveform.

attenuator region *Molecular Biology.* a region within the gene where RNA polymerase molecules will stop elongation of a transcript unless a specific molecular signal is received.

Atterberg scale *Geology.* a geometric and decimal grade scale used in soil classification and in the classification of particles in sediments. Also, **Atterberg grade scale.**

attic *Architecture.* in classical architecture, a low story built above a cornice. *Building Engineering.* the space immediately below a roof, located wholly or partly within the roof frame; a garret.

Attican orogeny *Geology.* a brief Late Miocene crustal deformation involving Phanerozoic rock.

attic tank *Building Engineering.* a water tank placed above the highest fixture in a house, using gravity to provide water pressure.

attic ventilator *Building Engineering.* **1.** a louver set in a gable to allow hot air to escape from an attic. **2.** a vent pipe running from an attic to the roof. **3.** a mechanical fan placed in an attic.

attitude *Behavior.* a learned tendency to react consistently in a given manner to certain individuals, objects, or concepts. *Geology.* the relation of a structural surface or rock bed to the horizontal plane, as determined by measuring both its strike and dip. *Graphic Arts.* the position of a camera or photograph relative to a given external reference system. *Aviation.* the position or orientation of a flight vehicle as determined by the relationship between its axes and some reference line or plane.

attitude control *Aviation.* **1.** the adjustment of the attitude of a flight vehicle. **2.** a device that automatically adjusts attitude.

attitude (control) jet *Aviation.* **1.** a jet stream of gas used to adjust the attitude of an aircraft or spacecraft. **2.** a fixed or movable nozzle used to direct such a jet stream.

attitude gyro *Aviation.* a gyro-operated instrument that indicates the attitude of a flight vehicle, especially one utilizing a reference system of 360° of rotation about each axis of the vehicle.

attitude indicator see ATTITUDE GYRO.

atto- a combining form meaning one quintillionth, or 10^{-18}.

attracted-disk electrometer see ABSOLUTE ELECTROMETER.

attracting object see COUNIVERSAL OBJECT.

attraction *Behavior.* **1.** any characteristic of an object or individual that produces approach responses in others. **2.** the inclination to approach an object or individual.

attraction cone see FERTILIZATION CONE.

attraction force see DISPERSION FORCE.

attraction gripper *Robotics.* a robotic end effector that uses adhesion, magnetism, or suction to grasp a workpiece.

attractor *Chaotic Dynamics.* a set of points in the phase space of a dissipative dynamical system that are visited in the asymptotic (infinitely long time) evolution of a trajectory. *Mathematics.* A closed set A on a manifold M with a flow is an attractor (or **attracting set**) for the flow if there exists a neighborhood U of A such that for each neighborhood V of A , the image of U under the flow is eventually contained in V. A is also required to be preserved under the flow. A maximal neighborhood U is called the **basin of attraction.**

attribute *Archaeology.* a distinct characteristic used to classify artifacts into groups; used to describe objects in terms of their physical traits, such as size, shape, weight, and color. *Computer Programming.* **1.** any characteristic of a data variable, such as length or format. **2.** a characteristic or data field of a record that may take on a variety of values, such as the address or telephone number in a customer file record. *Statistics.* a qualitative random variable.

attribute sampling *Industrial Engineering.* a procedure of quality control sampling by observation of the output for desired or undesired attributes. Similarly, **attribute testing.**

attribution *Psychology.* the process by which people assign causes or explanations to their behavior and that of others.

attribution error *Psychology.* the tendency to overestimate the influence of psychological factors in behavior, while underestimating environmental and circumstantial factors that have equal or greater effect.

attribution theory *Psychology.* the study of the way in which people explain their own behavior and that of others, in terms of whether psychological or external causes are cited as the determining factor.

attribution therapy or **attribution training** see REATTRIBUTION THERAPY.

attributive or **attributional** *Psychology.* of or relating to attribution.

attrital coal *Geology.* a coal having a ratio of anthraxylon to attritus ranging from 1:1 to 1:3.

attrition *Geology.* the wearing away by friction and the subsequent reduction in size of rock fragments as they grind, rub, and scrape against one another while being moved about by water, wind, or gravity. *Military Science.* the fact of suffering losses in personnel, equipment, and supplies over a given period of time, as through combat casualties.

attrition milling *Materials Science.* a size-reduction process for extractive ceramic processing, in which a ceramic powder is combined with a grinding medium and agitated by stirring arms.

attrition minesweeping *Ordnance.* minesweeping designed to minimize the number of mines in an area when complete removal of all mines or closing of the area is not feasible.

attrition rate *Ordnance.* a figure expressing the rate at which attrition takes place in forces or materials.

attritus [ə trī´təs] *Geology.* any dull gray to black, translucent to opaque constituent of coal that is formed primarily from finely divided fragments of relatively resistant plant residue.

att site *Genetics.* the specific place in the genomes of a bacterium and an infecting bacteriophage at which the phage can integrate and excise by recombination.

atü *Physics.* a unit of pressure equal to 1 atmosphere, referring to the amount of pressure above standard atmospheric pressure.

ATV *Mechanical Devices.* **1.** all-terrain vehicle; a jeep or other such vehicle designed for off-road use. **2.** see ATC.

Atwater coefficient *Nutrition.* numerical factors used to calculate and reflect the number of kilocalories of energy physiologically available from a gram of food.

Atwood, George 1746–1807, English mathematician; studied acceleration of motion; invented the Atwood machine.

Atwood machine *Mechanical Engineering.* a device consisting of a pulley over which is passed a stretch-free cord connecting two weights; can be used to determine the acceleration of gravity.

at. wt. atomic weight.

Atyidae *Invertebrate Zoology.* a family of thin-shelled gastropod crustaceans in the section Caridea, including many forms of shrimp.

A-type inclusion body *Virology.* an eosinophilic intranuclear inclusion that is found in cells infected with a virus such as herpes viruses or morbilliviruses.

A-type virus particle *Virology.* any of a group of intracellular, noninfectious, spherical RNA virus particles having dual shells.

atypical [ə tip´ə kəl] *Medicine.* not typical; not having the usual signs or symptoms.

atypical interstitial pneumonia *Veterinary Medicine.* various lung diseases of cattle, having in common atypical clinical signs, usually with acute or chronic respiratory distress, little evident toxemia, and a progressive course that does not respond to treatment.

atypical mycobacteria *Bacteriology.* any species of the bacterial genus *Mycobacterium* that can cause mild infection in humans, but does not cause the more severe diseases associated with *Mycobacterium,* such as tuberculosis.

atypical verrucous endocarditis see LIBMAN-SACKS ENDOCARDITIS.

AU astronomical unit.

Au the chemical symbol for gold. (From Latin *aurum.)*

aucuparious *Ecology.* attracting birds or having abundant bird life.
audio- a combining form meaning "hearing" or "sound," as in *audiogram.*
audio component *Telecommunications.* the part of any wave or signal whose frequencies are within the audible range.
audio device *Computer Technology.* any of a wide variety of computer components that accept or produce sound.
audio frequency *Acoustics.* a term for any frequency within the audio frequency range; i.e., from about 20 to 20,000 hertz.
audio-frequency choke *Electromagnetism.* a coil used in electrical circuits to impede audio frequency currents.
audio-frequency meter *Acoustical Engineering.* any of various instruments used to measure the frequencies of sound waves that are audible to the human ear.
audio-frequency oscillator *Electronics.* an oscillator that produces AF signals (20 to 20,000 hertz).
audio-frequency peak limiter *Electricity.* a circuit that restricts signal peaks to a predetermined value.
audio frequency range *Acoustics.* the portion of the acoustic energy spectrum, from about 20 to 20,000 vibrations per second, that can be detected by a human with normal hearing.
audio-frequency shift modulation *Telecommunications.* in radio, a system of facsimile transmission in which the required frequency shift is achieved through an 800-hertz shift of the audio signal.
audio-frequency transformer *Electricity.* an iron-core transformer designed for use with audio-frequency circuits.
audiogenic seizure *Neurology.* a form of reflex epilepsy resulting from exposure to sound, usually a sudden loud noise.
audiogram *Acoustics.* a graphic representation of hearing ability, especially a representation that indicates hearing loss at various frequency levels.
audio image *Acoustics.* a specific sound that a sound-reproduction system attempts to record, or to reproduce from a recording.
audio-impedance measurement *Acoustics.* the direct assessment of acoustic impedance over a given range of audible frequencies.
audiologist *Acoustics.* a person who is trained or skilled in audiology. *Medicine.* a person skilled in the science of hearing, including the rehabilitation of patients whose hearing cannot be improved medically or surgically.
audiology *Acoustics.* the scientific study or measurement of hearing ability. *Medicine.* the study of hearing loss or impairment, and of techniques or methods for dealing with such a condition.
audio masking see MASKING.
audiometer *Acoustical Engineering.* **1.** an instrument that generates sounds of known frequency and intensity; used to measure an individual's hearing ability, especially the level at which a sound becomes audible at a given frequency. **2.** an instrument used to measure the intensity of sounds.
audiometry *Acoustics.* the scientific measurement of hearing ability, through the use of audiometers and similar devices. Thus, **audiometric.**
audio-modulated radiosonde *Engineering.* a miniature radio transmitter that is carried aloft, having a carrier wave that is regulated by audio-frequency signals; the frequency of the signals is operated according to the sensing devices of the instrument.
audio oscillator *Acoustics.* an electronic device that produces a signal of a specified frequency or range of frequencies in the audio spectrum, for the calibration of a sound-measurement device such as a real-time analyzer. *Electronics.* see AUDIO-FREQUENCY OSCILLATOR.
audio output *Computer Technology.* computer output in the form of simulated or recorded spoken words.
audio patch bay *Acoustical Engineering.* an electronic device that interfaces two or more audio channels for multichannel recordings and playback.
audio peak limiter see AUDIO-FREQUENCY PEAK LIMITER.
audio range see AUDIO FREQUENCY RANGE.
audio response unit *Telecommunications.* a device designed to provide a voice response to digital enquiries that are sent, for example, to a computer via a keyboard and modem.
audio signal *Acoustics.* any signal within the audio frequency range; i.e., from about 20 to 20,000 hertz.
audio spectrometer see ACOUSTIC SPECTROMETER.
audio system see SOUND-REPRODUCING SYSTEM.
audio taper *Acoustical Engineering.* the logarithmic change in volume and tone with the rotation of a control potentiometer.
audio transformer see AUDIO-FREQUENCY TRANSFORMER.
audiovisual *Telecommunications.* of or relating to the use of electrical, chemical, mechanical, and optical media in the reproduction of audible signals and visual images.
audiphone *Acoustical Engineering.* a device used by individuals with certain types of hearing impairment; it consists of a diaphragm that picks up vibrations on a tooth, rather than the eardrum, and transmits the signal to the inner ear.
audit *Computer Programming.* **1.** a process used to detect accidental input or processing errors as well as fraud, often using test data and special-purpose software. **2.** a set of procedures established to ensure the quality and integrity of a data base. **3.** to carry out such a process or procedure.
audition *Physiology.* the sensation and perception of sounds produced by stimulation of nerve receptors in the ear; hearing.
audition coloree see CHROMATIC AUDITION.
auditory *Physiology.* of or relating to the sense of hearing.
auditory adaptation *Neurology.* the ability to adapt to changes in sound, as evidenced by changes in the auditory threshold resulting from exposure to sound.
auditory association area *Physiology.* the auditory sensory center located bilaterally in the temporal lobes of the brain.
auditory capsule see OTIC CAPSULE.
auditory impedance *Physiology.* a mechanical factor that determines the amount of sound energy that is absorbed or reflected at the eardrum; related to loss of transmission of sounds to the middle ear and the cochlea.
auditory nerve *Anatomy.* the eighth cranial nerve, or vestibulocochlear nerve, which innervates the ear and carries impulses relating to both sound stimuli and balance to the brain. Also, ACOUSTIC NERVE.
auditory perspective *Acoustics.* **1.** a three-dimensional sound field in which sound is produced. **2.** the creation by a sound reproduction system of the illusion of this three-dimensional reality.
auditory placode *Developmental Biology.* an epidermal thickening, next to the hindbrain in the early embryo, from which the internal ear develops. Also, OTIC PLACODE.
auditory saucer see AUDITORY PLACODE.
auditory synesthesia *Neurology.* an auditory sensation that occurs when another sense is stimulated. Also, PHONISM, SUBJECTIVE SOUND.
auditory threshold see THRESHOLD OF AUDIBILITY.
auditory tube see EUSTACHIAN TUBE.
audit total *Computer Programming.* a known quantity or sum that is used to verify intermediate or final results of data processing, usually in an accounting or other financial application.
audit trail *Computer Programming.* the procedure of keeping a record of transactions entered or steps in processing data to be able to reconstruct how the results were obtained.
Audubon, John James [ô′də bən] 1785–1851, Haitian-born American artist and naturalist; known for highly accurate paintings of birds.
Auerbach's plexus see MYENTERIC PLEXUS.
Aufbau principle [ôf′bou′] *Chemistry.* a description of the building up of the elements, in which the structure of each successive element is obtained by adding one proton to the atomic nucleus and one electron to an atomic orbital simultaneously. The electrons are placed in their orbitals in order of increasing energy.
aufwuch [ôf′wük′] *Ecology.* **1.** an aquatic organism that clings to submerged rocks, plants, and other objects. **2. aufwuchs.** a community of such organisms forming a layer on underwater objects.
auganite *Petrology.* basalt essentially composed of augite and plagioclase, but no olivine.
augelite *Mineralogy.* $Al_2(PO_4)(OH)_3$, a colorless to white monoclinic mineral occurring in tabular to prismatic crystals, having a specific gravity of 2.7 and a hardness of 4.5 to 5 on the Mohs scale; found in veins with lead sulfosalts and in andalusite deposits.
augen *Petrology.* large lenticular grains or aggregates of minerals, commonly alkali feldspars, that are found in schists and gneisses and have an eye-shaped appearance in cross section.
augend *Mathematics.* a quantity to which an addend is added; that is, the quantity being augmented by an addend.
augen gneiss [ô′gən nīs′] *Petrology.* a gneiss with eye or almond-shaped porphyroblasts, typically quartz or feldspar.
augen schist [ô′gən shist′] *Petrology.* **1.** a rock that is associated with mylonite and composed of granulated, recrystallized minerals in aggregates surrounded by, and alternating with, schistose streaks and lenticles of completely recrystallized minerals. **2.** a schist having small augen or eyes of feldspar or other minerals.

augen structure *Petrology.* a structure characteristic of certain gneisses, schists, and granites, with elliptical or lens-shaped constituents often outlined by mica flakes, thus resembling eyes.

auger [ôg´ər] *Mechanical Devices.* a manually operated boring device for wood or earth, consisting of a bit and transverse handle.

Auger, Pierre V. [ō´zhā´] born 1899, French physicist; discovered the Auger effect.

auger bit *Mechanical Devices.* a spiral bit shaped much like an auger with a square tang at its upper end used in wood and earth boring. Thus, **auger drilling, auger boring.**

Auger coefficient *Atomic Physics.* the relationship between the number of Auger electrons and X-ray photons emitted during any change in energy levels.

auger conveyor see SCREW CONVEYOR.

Auger effect *Atomic Physics.* a process in which an atom ionizes without emitting radiation; this occurs when an electron in the outer regions fills a vacancy in an inner orbit, and at the same time another outer electron is ejected by the atom. Also, AUTOIONIZATION, AUGER TRANSITION.

Auger electron *Atomic Physics.* an electron that is emitted from an atom as a result of the Auger effect.

Auger electron spectroscopy *Spectroscopy.* the analysis of the energy of secondary (Auger) electrons ejected from a solid surface upon irradiation with accelerated electrons or X-ray photons.

auger mining *Mining Engineering.* an inexpensive method of mining used by strip miners when the overburden gets too thick to be removed economically.

auger packer *Mechanical Engineering.* a feed mechanism in which a rotating screw or auger enclosed in a pipe is used to transport granulated solids into containers for shipping.

Auger recombination *Atomic Physics.* a process in which an electron and a hole recombine without generating radiation, by passing the excess energy and momentum onto another electron or hole.

Auger shower *Astronomy.* a shower of cosmic rays.

Auger transition see AUGER EFFECT.

auget [ô´jet´] *Engineering.* a priming tube that is used in blasting. Also, **augette.**

augite [ô´jīt´] *Mineralogy.* $(Ca,Na)(Mg,Fe,Al,Ti)(Si,Al)_2O_6$, a dark-green to black monoclinic mineral of the pyroxene group, occurring as prismatic crystals, having a specific gravity of 3.2 to 3.5 and a hardness of 5.5 to 6 on the Mohs scale; found as a component of basalts, dolerites, and gabbros.

augitite [ô´jə tīt´] *Petrology.* an alkalic volcanic rock with phenocrysts of augite and iron ore, and often some biotite or hornblende, in a base of brown glass.

augitophyre [ô´jīt´ə fī ər] *Petrology.* a form of volcanic lava with phenocrysts of augite in a groundmass of leucite or potash feldspar.

augmentation *Navigation.* the apparent increase in the semidiameter of a celestial body as its altitude increases.

augmentation correction *Navigation.* a correction that is applied to a celestial observation to account for the increase in semidiameter of a body with the increase in altitude. It is generally only applied to observations of the moon.

augmentation distance *Nuclear Physics.* the extrapolation distance from the physical boundary of a nuclear reactor to an imaginary point where the neutron flux is zero.

augmentation mammaplasty *Surgery.* breast enlargement, accomplished by inserting an autologous or artificial implant.

augmentation system *Aviation.* any vehicle system that boosts propulsive thrust by an auxiliary device such as afterburning or an augmenter tube, improves flight stability, or enhances sensor tracked target signatures.

augmented matrix *Mathematics.* in a set of simultaneous linear equations in variables, a matrix whose entries are the coefficients and constant terms of the equations. In particular, the ith row of the matrix consists of the coefficients and then the constant term of the ith equation.

augmented operation code *Computer Programming.* a particular machine instruction operation code to which further definition or limitation is specified in another part of the instruction.

augmented transition network *Artificial Intelligence.* a formalism for describing parsers, especially for natural language. Similar to a finite automaton, but augmented in that arbitrary tests may be attached to transition arcs, subgrammars may be called recursively, and structure-building actions may be executed as an arc is traversed. Also, **augmented finite state transition network.**

augmenter *Aviation.* any device, such as an **augmenter tube,** through which the exhaust gases of an aircraft's engines can be directed to provide extra thrust.

augmenting factor *Oceanography.* a correction factor used in harmonic analyses of tides and tidal currents.

Aujeszky's disease *Veterinary Medicine.* an encephalomyelitis affecting a variety of domestic animals, caused by *Herpesvirus suis,* often transmitted through wounds, and characterized by trembling, incoordination, and paralysis; swine that remain relatively resistant to the virus serve as vectors to cattle.

auk [ôk] *Vertebrate Zoology.* any of several species of flightless, black and white diving seabirds of the northern coasts, belonging to the family Alcidae, and including the extinct **Great Auk.**

auk

auklet [ôk´lit] *Vertebrate Zoology.* any of six species of black and white seabirds of the family Alcidae whose range includes the Bering Sea and North Pacific waters, and who winter as far south as Mexico. Also, SEA SPARROW.

Aulacoceratida *Paleontology.* the oldest coleoids; an order of cephalopods in the subclass Coleoidea that arose in the Carboniferous and became extinct in the Upper Jurassic.

aulacogen *Geology.* a narrow, elongated basin bounded by normal faults and located above hot mantle spots in cratons.

Aulacomniaceae *Botany.* a family of large, yellow-green mosses of the order Bryales that form mats on soil, humus, peat, tree trunks, and logs; characterized by erect stems with terminal sporophytes, loosely erect to flexuose leaves, a strong costa at or near the apex, and a bipolar distribution.

Aulodonta *Invertebrate Zoology.* an order of sea urchins having a rigid test, gills and grooved teeth; a typical species is Diadema.

Aulolepidae *Paleontology.* a monogeneric family of early teleost fishes in the extinct order Ctenothrissiformes; possibly ancestral to both major groups of spiny teleosts; Upper Cretaceous.

Aulophyllum *Paleontology.* a Carboniferous genus of solitary rugose corals in the extinct superfamily Zaphrenticae and family Aulophyllidae; Aulophyllum is distinguished from its close relative Dibunophyllum by a haphazardly arranged axial structure.

aulophyte [ôt´ə fīt´] *Botany.* a nonparasitic plant that lives within a hollow cavity of another plant.

Aulopodidae *Vertebrate Zoology.* a family of bottom-living marine fish of the order Myctophiformes, found in warm coastal waters of the North Atlantic, Pacific, and near southern Australia. Also, **Aulopidae.**

Auloporidae *Paleontology.* a family of tabulate corals in the extinct order Auloporida, occurring in colonies that were flat and encrusting or shrubby-looking, sometimes with funnel-shaped platforms; Ordovician to Permian.

Aulorhynchidae *Vertebrate Zoology.* the tubesnouts, a family of marine fishes of the suborder Gasterosteioidei, having a protractile upper jaw and well-developed premaxillaries.

Aulostomidae *Vertebrate Zoology.* the trumpetfishes, a monogeneric family of tropical and subtropical reef fishes in the suborder Syngnathoidei, characterized by a small mouth at the end of a long, tubular snout.

Aur Auriga.

aur. the ear or ears. (From Latin *auris, aures.)*

aura *Neurology.* a subjective sensation preceding and marking the onset of a paroxysmal attack, such as an epileptic seizure or a migraine.

aural [ô´rəl] *Biology.* of or relating to the ear or to the sense of hearing.
aural masking see MASKING.
aural null *Navigation.* the point at which the sound of a signal is at a minimum or completely absent; normally it is the point at which a radio direction-finding bearing is taken.
aural radio range *Electronics.* a radio range whose tracking is determined by aural signal interpretation.
aural signal *Acoustics.* **1.** any acoustic signal that can be identified by listening to its acoustic characteristics such as pitch, beats, and so on. **2.** the audio portion of a television signal. *Electronics.* **1.** any signal that is audible. **2.** the sound portion of a TV signal.
aural transmitter *Telecommunications.* any radio equipment that is used for the transmission of sound signals from a television broadcasting station.
auramine *Organic Chemistry.* $C_{17}H_{21}N_3 \cdot HCl$, a yellow to orange crystalline solid that melts at 267°C, soluble in water, alcohol, and chloroform; used as a fabric dye and an antiseptic. Also, **auramine hydrochloride, auramine O**.
aurantia *Organic Chemistry.* $C_{12}H_8N_8O_{12}$, an orange aniline dye that is used as a biological stain and in making certain photographic filters.
aurantlin see NARINGIN.
Aurelia *Invertebrate Zoology.* a genus of large, common jellyfish.
aureofacin *Microbiology.* a potent antifungal antibiotic that is a mixture of two related compounds, vacidin A and gedamycin.
aureole [ôr´ē ōl´] *Meteorology.* a poorly defined, bluish-white disk with a reddish-brown outer edge around the sun or the moon, produced by a cloud composed of droplets distributed over a wide size range. *Geology.* an area of metamorphosed country rock surrounding an igneous intrusion. Also, CONTACT AUREOLE, CONTACT ZONE, METAMORPHIC ZONE.

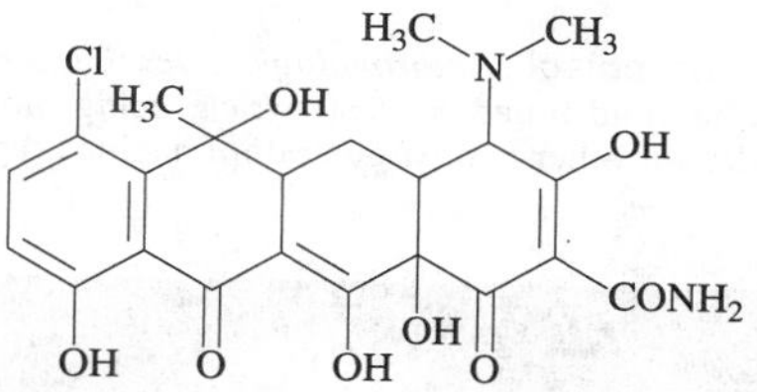

Aureomycin

Aureomycin [ôr´ē ō mī´sin] *Microbiology.* a trade name for the broad-spectrum antibiotic chlortetracycline hydrochloride.
aureothricin *Microbiology.* a broad-spectrum antibiotic produced by certain species of Streptomyces. Also, THIOLUTIN.
aureusidin *Biochemistry.* a yellow pigment found in certain plants, such as the yellow snapdragon. Also, 4,6,3',4'-TETRAHYDROXYAURONE.
Auri see AURIGA.
auri- a combining form meaning: **1.** ear, as in *auricular.* (From Latin *auris.)* **2.** gold, as in *auriferous.* (From Latin *aurum.)*
auriasis see CHRYSIASIS.
auric chloride see GOLD CHLORIDE.
aurichalcite *Mineralogy.* $(Zn,Cu^{+2})_5(CO_3)_2(OH)_6$, a transparent, orthorhombic green to blue mineral with a pearly luster, having a specific gravity of 3.96 and a hardness of 1 to 2 on the Mohs scale.
auric hydroxide see GOLD HYDROXIDE.
auricle *Anatomy.* **1.** the portion of the external ear not contained within the head. **2.** a muscular flap in the walls of a portion of the atria of the heart. Also, PINNA.
auric oxide see GOLD OXIDE.
Auriculaceae *Botany.* a monogeneric family of marine diatoms of the order Pennales, characterized by semicircular to semielliptical valves that slope up from the straight side to the convex side.
auricular *Anatomy.* of or relating to the ear or auricle.
auricular flutter see ATRIAL FLUTTER.
Auricularia *Mycology.* a genus of fungi belonging to the order Eutremellales, having species ranging from parasites (e.g., *A. auricula-judae)* to edible fungi (e.g., *A. polytricha,* cultivated on oak trees in China).
Auriculariaceae *Mycology.* a family of fungi belonging to the class Phragmobasidoiomycetes, characterized by its exposed fruiting bodies and obtaining its nutrients primarily from dead organic matter; some species are edible.
auricularia (larvae) *Invertebrate Zoology.* the larvae of sea stars and certain other echinoderms.
Auriculariales *Mycology.* a former term for an order of fungi, some of which live from inorganic matter, others of which are parasites of fungi such as mosses.
auricularis *Anatomy.* any of three small muscles beneath the skin around the ear, **auricularis anterior, a. posterior,** or **a. superior.**
auriculin see ATRIAL NATRIURETIC FACTOR.
auriculoventricular node *Cardiology.* a former term for atrioventricular node.
auriferous *Geology.* of a substance, containing or yielding gold.
Auriga [ô rī´ gə] *Astronomy.* the Charioteer, a northern constellation that lies between Perseus and Gemini, and is visible on late summer and autumn evenings.
Aurignacian [ôr´ig nā´sē ən] *Archaeology.* **1.** a European culture of the Upper Paleolithic period, noted for the use of characteristic stone and bone tools. **2.** the period when this culture existed, from about 35,000 to about 20,000 years ago. **3.** also, **Aurignac.** of or relating to this culture or period. (From *Aurignac,* a village in southern France, the site of numerous findings of this type.)
aurin *Organic Chemistry.* $C_{19}H_{14}O_3$, a red crystalline solid that decomposes at 308°C; insoluble in water and soluble in alcohol and ether; used in making dyes. Also, ROSOLIC ACID.
aurintricarboxylic acid *Biochemistry.* a dye that prevents protein synthesis in prokaryote and eukaryote cells by preventing mRNA from binding to ribosomes; it also chelates metal ions, and in higher concentrations, hinders chain elongation.
auris see EAR.
Auriscalpiaceae *Mycology.* a family of fungi of the order Agaricales, occurring primarily on wood.
aurora [ə rôr´ə] *Geophysics.* a form of sporadic radiant emission occurring in the upper atmosphere over the middle and high latitudes and seen most often in the Arctic and Antarctic regions; thought to be caused by charged particles from the sun that collide with and excite atoms in the upper atmosphere, which then emit light as they return to their ground state.
Aurora *Astronomy.* asteroid 94, a type C asteroid that was discovered in 1867 and has a diameter of about 420 kilometers.
aurora australis *Geophysics.* the aurora of the Southern Hemisphere.
aurora borealis [ə rôr´ə bôr ē al´is] *Geophysics.* the aurora of the Northern Hemisphere.

aurora borealis

auroral [ə rôr´əl] *Astronomy.* of or pertaining to the aurora.
auroral absorption event *Geophysics.* a large increase of electric and radio wave density in the D-level of the atmosphere during an aurora or a magnetic storm.
auroral caps *Geophysics.* the polar regions within the auroral zones.
auroral electrojet *Geophysics.* a strong current of electricity that flows through the auroral zone during a polar substorm.
auroral form *Geophysics.* any of various shapes of auroral emissions: **arcs** (bands of arching light extending acrosss the sky), **rays** (beams of light, appearing singularly or in bundles), **draperies** (sheets of light spreading across the sky), **crowns** (rays that seem to emanate from a common point), and **diffuse aurora** (diffuse cloudlike surfaces).

auroral frequency *Geophysics.* the frequency of auroral activity, determined by figuring the percentage of nights an aurora can be seen from a given location.
auroral isochasm *Geophysics.* an imaginary line that connects areas of equal auroral intensity observed over a number of years.
auroral line *Spectroscopy.* a prominent green or red line observed in the spectra of auroras that corresponds to certain forbidden transitions of the neutral oxygen atom.
auroral oval *Geophysics.* an oval-shaped region, centered on the Northern or Southern magnetic pole, within which auroral emissions occur.
auroral poles *Geophysics.* the sites at which the auroral isochasms coincide with the magnetic-axis poles of the geomagnetic field.
auroral region *Geophysics.* the geographic area in either hemisphere from which auroral activity can normally be observed, usually within 30 degrees magnetic latitude from each magnetic pole.
auroral storm *Geophysics.* a series of auroral substorms coming in rapid succession during a ge magnetic storm.
auroral substorm *Geophysics.* a characteristic pattern of an auroral episode, consisting of the intensification of auroral emission that occurs about midnight in which auroral arcs move rapidly toward the pole, causing a bulge in the auroral oval.
auroral zone *Geophysics.* the region of maximum auroral activity in either hemisphere that lies 10–15 degrees geomagnetic latitude from the geomagnetic poles.
auroral zone blackout *Geophysics.* a term for the disruption of communications within the auroral zone due to increased ionization in the atmosphere.
aurora polaris *Geophysics.* the auroral emissions of the Northern or Southern Hemisphere that occur at higher than usual altitudes.
aurothioglucose *Pharmacology.* $C_6H_{11}AuO_5S$, a gold salt used in the treatment of early active rheumatoid arthritis that does not respond to other therapies.
aurovertin *Mycology.* any of certain poisons found in the fungus *Calcarisporium arbuscula.*
AUS *Aviation.* the airport code for Austin, Texas.
auscultation *Medicine.* the act of listening for sounds within the body, especially to ascertain the condition of the lungs, heart, pleura, abdomen, or other organs, or to detect fetal heart sounds.
ausforming *Metallurgy.* the hot deformation of the face-centered cubic phase of steel within specified temperature and time ranges, in order to avoid the formation of nonmartensitic transformation products.
ausrolling *Metallurgy.* ausforming by a rolling process.
austausch coefficient see EXCHANGE COEFFICIENT.
austempered nodular iron *Metallurgy.* iron containing a small amout of magnesium or cerium that is austenitized and quenched to obtain spherical nodular graphite. Also, NODULAR IRON.
austempering *Metallurgy.* in a steel, a thermal treatment consisting of cooling from above the transformation temperature at a sufficiently high rate so that structural changes are prevented, and of holding at a temperature that is below the temperature at which pearlite forms, but above the temperature at which martensite forms.
austenite *Metallurgy.* an interstitial solid solution of carbon in gamma iron, having a face-centered cubic structure and a much higher solid solubility for carbon than alpha ferrite (2.08% at 1148°C to 0.8% at 723°C).
austenitic *Metallurgy.* of or relating to a steel that has a face-centered cubic crystal structure. Thus, **austenitic steel.**
austenitic cast iron *Metallurgy.* a cast iron that has a face-centered cubic structure; it is alloyed with nickel or another suitable element.
austenitic manganese steel see HADFIELD MANGANESE STEEL.
austenitic stainless steel *Metallurgy.* any of several nickel-bearing stainless steels that are austenitic and therefore nonmagnetic. Also, **austenite stainless steel.**
austenitize *Metallurgy.* to heat treat a steel so that the resulting structure is either partially or fully face-centered cubic.
auster see OSTRIA.
Austinian *Geology.* a North American Gulf Coast stage of the Upper Cretaceous period, after the Eaglefordian and before the Tayloran.
austinite *Mineralogy.* $CaZn(AsO_4)(OH)$, a colorless to yellow orthorhombic mineral occurring as tiny, prismatic or acicular crystals, having a specific gravity of 4.13 and a hardness of 4 to 4.5 on the Mohs scale; found with adamite on limonite.
austral *Geography.* of or related to the south.
austral axis pole *Geophysics.* the location, in the Southern Hemisphere, at which the magnetic pole intersects the earth's surface.
Australia *Geography.* an island continent southeast of Asia with an area of about 2,940,000 square miles, entirely occupied by the nation of Australia.
Australia antigen *Immunology.* a polymorphic infectious agent that causes hepatitis in some individuals. Also, HEPATITIS B ANTIGEN.
Australian region *Ecology.* a distinct zoogeographical region that includes the subregions of Australia, New Zealand, Austro-Malaya, and Polynesia.
Australian subregion *Ecology.* a distinct zoogeographical subregion that includes the continent of Australia.
australite *Mineralogy.* a jet-black variety of tektite that is often round or button-shaped; found in southern Australia.
Australopithecinae *Paleontology.* a subfamily of hominids that includes the several species of *Australopithecus.*
australopithecine *Anthropology.* the earliest-known hominid in human evolution, small-brained but walking erect; fossils are from East and South Africa and date to 3.6 million years.
Australopithecus *Paleontology.* a genus of primates in the family Hominidae of the Pliocene and Early Pleistocene, 1.5–3 million years B.P.; they resemble apes in some ways but also have human characteristics, with cranial capacity around 600 cc; some authorities subdivide them into "robust" *(Paranthropus)* and "gracile" *(Homo africanus)* types.
Australopithecus afarensis *Anthropology.* a form of australopithecine hominid that is the oldest known to date, offering evidence of early group organization. *Paleontology.* a possible new species of hominoid primates in the family Hominidae, known from remains of several individuals found in the Laetolil Beds near Laetoli, Tanzania; the remains have been dated as about 3.5 million years old; the most complete skeleton is commonly known as "Lucy."
Australopithecus africanus *Anthropology.* a small, slender form of australopithecine hominid, appearing primarily in the Pliocene in South and East Africa.
Australopithecus boisei *Anthropology.* a very large and robust australopithecine hominid found in East Africa, being more massive and larger-toothed than other robust australopithecines, appearing in the Pleistocene.

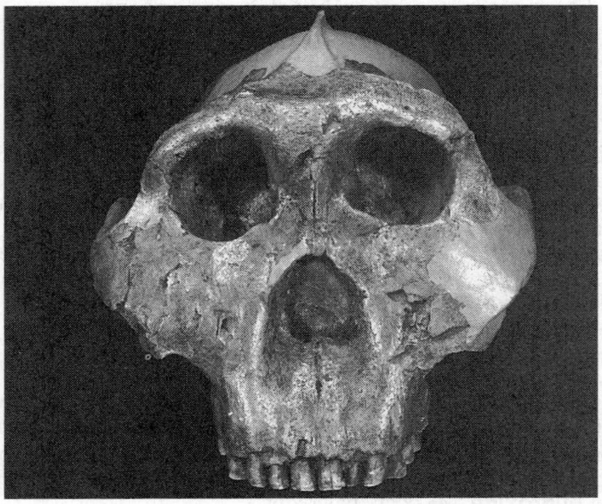

Australopithecus boisei

Australopithecus prometheus *Anthropology.* a name given to a group of australopithecine hominids found at Makapansgaat in South Africa; formerly thought to show evidence of fire use.
Australopithecus robustus *Anthropology.* a large and robust form of australopithecine hominid; primarily from the Pleistocene.
Austrian orogeny *Geology.* a brief Early Cretaceous crustal deformation involving Phanerozoic rock.
Austro-Asiatic family *Linguistics.* a major language family that includes Munda of eastern India and Vietnamese.
Austroastacidae *Invertebrate Zoology.* a family of Australian crayfish-type crustaceans.
Austrobaileyaceae *Botany.* a monospecific family of evergreen woody vines of the order Magnoliales, characterized by scattered spherical cells containing volatile oils and by malodorous flowers; native to northeastern Australia.

Austrodecidae *Invertebrate Zoology.* a family of marine arthropods in the subphylum Pycnogonida.

Austro-Malayan subregion *Ecology.* a distinct zoogeographical region that includes a small portion of northern Australia and various oceanic islands to the north of this.

autacoid *Endocrinology.* describing any of various physiologically active, endogenous substances, such as histamine, serotonin, and angiotensin, that are not yet placed in existing classifications.

autallotriomorphic *Petrology.* of or relating to an aplitic texture in which all the constituents crystallized simultaneously, thus preventing formation of euhedral crystals.

autapomorphic *Evolution.* of or relating to an apomorphic character unique in a species or other taxonomic group in a single phylum.

autarchic genes *Genetics.* genes in a mosaic organism that are not inhibited by cellular products from genetically different neighboring tissues.

autecology *Ecology.* the ecology of an individual organism or species. Thus, **autecological.** Also, AUTOECOLOGY.

authalic latitude *Cartography.* a latitude based on a sphere having the same area as the spheroid, and such that the areas between successive parallels of latitude are exactly equal to the corresponding areas on the spheroid. Also, EQUAL-AREA LATITUDE.

authalic map projection see EQUAL-AREA MAP PROJECTION.

authentication *Telecommunications.* **1.** in a communications system, the process of checking the validity of a transmission, message, or originator in order to protect against accepting a fraudulent transmission. **2.** any procedure used to identify stations, users, and equipment and to verify eligibility to receive special categories of information.

authenticator *Telecommunications.* **1.** a sequence of characters arranged in a predetermined manner that is inserted in a message for the purposes of authentication. **2.** the means used to identify or verify the eligibility of a station, originator, or person to access special categories of information.

authigene *Mineralogy.* any mineral found at the place it was formed.

authigenic *Geology.* of mineral and rock constituents, formed in place at the same time as, or after, the formation of the rock of which they are a part. Also, **authigenous, authigenetic.**

authoring language *Computer Programming.* a programming language and accompanying dialect expressly designed for programming instructional materials for computer-assisted instruction.

authoritarian *Psychology.* **1.** of or relating to the authoritarian personality type. **2.** a person of this type.

authoritarianism *Psychology.* a tendency to prefer situations in which one leader has unquestioned authority over a group and the group members willingly follow the leader's direction.

authoritarian personality *Psychology.* a personality type characterized by strong group allegiance and negative attitudes toward people outside the group, rigid adherence to convention, and submission to those in authority.

authorized carrier frequency *Telecommunications.* **1.** a frequency that is allocated and assigned (along with maximum permitted deviations) by a regulatory authority to a specific user for a specific purpose. **2.** the portion of the radio spectrum assigned to a specific user.

authorized library *Computer Programming.* a list of programs to which a particular user is permitted access.

authorized program *Computer Programming.* a single computer program to which a user is permitted access.

author's alteration *Graphic Arts.* a copy change made by the author after a manuscript has been typeset; distinguished from printer's errors in billing for corrections. Also, **author's addition.**

author's proof *Graphic Arts.* the master proof; the set of galleys given to the author, who makes corrections and alterations and returns the proof to the compositor.

autism *Psychology.* a severe disorder originating in early childhood, characterized by extreme self-absorption, language disturbances, highly structured, unvarying behavior, an inability to form personal relationships, and a pattern of repetitive body movements.

autistic *Psychology.* relating to or affected by autism.

autistic disorder see AUTISM.

auto- a combining form meaning: **1.** self or same, as in *autochthonous.* **2.** automatic, as in *autofocus.*

autoabstract *Computer Programming.* a process of automatically selecting key words and phrases that represent the basic content of a document; often using statistical or counting techniques. Also, AUTOMATIC ABSTRACTING.

autoacceleration *Organic Chemistry.* a self-actuated increase in the rate of bulk polymerization and molecular weight that occurs in certain vinyl monomers.

autoadaptivity *Control Systems.* the ability of a machine to adapt to its environment by accepting commands, analyzing the input of sensors, and using the data to carry out preplanned operations.

autoagglutination *Immunology.* the spontaneous clumping together of particular antigens that occurs when the antigens are suspended in saline. Also, AUTOHEMAGGLUTINATION.

autoagglutinin *Immunology.* an antibody that is a normal constituent of an individual's blood serum and that causes the clumping together of the individual's red blood cells.

autoalarm see AUTOMATIC ALARM RECEIVER.

autoanalysis *Psychology.* psychoanalysis of oneself; examination of one's own psychic components.

autoantibody *Immunology.* an antibody produced by an organism that works against its own antigens.

autoantigen *Immunology.* a type of antigen present in the tissues of an organism that causes the development of corresponding antibodies. Also, SELF-ANTIGEN.

autoasphyxiation *Physiology.* a cessation or failure of the respiratory process due to internal metabolic activities.

autoaudible *Cardiology.* describing the condition in which the sounds of an individual's heart are audible to himself or herself.

autobarotropy *Meteorology.* the state of a fluid with equal coefficients of barotropy and piezotropy.

autobasidium *Mycology.* a basidium found in the fungal subdivision Basidiomycotina, characterized by its not dividing after septation.

autobrecciation *Geology.* a process of rock formation in which portions of the first solidified crust of a lava flow are mixed into the portion that is still fluid.

autocall *Telecommunications.* the automatic dialing of a telephone number by a computer or automatic calling unit.

autocarp *Botany.* fruit resulting from self-fertilization.

autocarpic *Botany.* producing fruit by self-fertilization.

autocarpy *Botany.* the process of producing fruit by self-fertilization.

autocatalysis *Chemistry.* a catalytic reaction produced by the products of the reaction.

autochrome plate *Graphic Arts.* a photographic plate formerly used for color photography, consisting of a sheet of glass coated with dyed starch grains, with the image formed directly on the glass, which was then viewed as a transparency.

autochthon *Geology.* a large rock body that remains rooted to its basement at its site of origin. *Paleontology.* an organism that is indigenous to the locality in which it was found. Also, **autochthone.**

autochthonous *Science.* occurring in the place where originally formed; indigenous. *Ecology.* produced within a given habitat or system. *Pathology.* of a disease or growth, found in an individual at the place where it originated. Thus, **autochthony.**

autochthonous coal *Geology.* coal or peat that is derived from decayed plant material and found at the site of its origin.

autochthonous sediment *Geology.* a deposit of mineral or organic material formed in place by decomposition.

autochthonous stream *Hydrology.* a stream that is flowing in the channel in which it originated.

autoclastic *Geology.* of rock, fragmented in place as a result of mechanical processes or orogenic forces.

autoclastic schist *Geology.* a well-foliated metamorphic rock formed in place by the pressure of one rock mass upon another or by the impact of massive rock movement.

autoclave *Engineering.* an airtight steel vessel that heats substances under high pressure; used for chemical manufacturing and industrial processing. *Medicine.* **1.** a machine that sterilizes instruments with pressurized steam. **2.** to sterilize in an autoclave.

autoclave curing *Engineering.* a process in which certain materials are steam cured in an autoclave at temperatures generally between 170° and 215°C.

autoclave molding *Engineering.* a process in which reinforced plastics are cured through the use of an autoclave that sets the resin by means of high steam pressure.

autoclaving *Engineering.* the process of heating or sterilizing using an autoclave.

Autoclip *Surgery.* the trademark for a stainless steel surgical clip used to close a wound; the clip is inserted with a mechanical applicator that automatically feeds a series of clips.

Autocoder *Computer Programming.* an early IBM programming language that permitted symbolic and mnemonic coding for the IBM 1401 computer; analogous to assembly language.

autocollimation *Optics.* a technique for collimating an optical instrument by projecting an illuminated object to infinity, then reflecting its image from a flat mirror surface, and adjusting the instrument until both the object and image are in focus at the same plane.

autocollimator *Optics.* **1.** a device used to detect and measure small angular displacements in a reflective surface, with a high degree of precision. **2.** a device that detects and measures small deviations in a light beam and makes the diverging light parallel. **3.** a telescope equipped with an autocollimator.

autocolony *Invertebrate Zoology.* an offspring colony formed within a cell of a colony and duplicating in miniature its parent colony.

autoconsequent stream *Hydrology.* a stream whose course is guided by the slopes of alluvium it has itself deposited.

autoconsequent waterfall *Hydrology.* one of a series of waterfalls that develop at particular sites along a stream course where part of the solution load is precipitated as a result of warming, evaporation, or other factors.

autoconvection *Meteorology.* a phenomenon in which the initiation of convection occurs spontaneously in an atmospheric layer in which the lapse rate is equal to or greater than the autoconvective lapse rate.

autoconvection gradient see AUTOCONVECTIVE LAPSE RATE.

autoconvective instability see ABSOLUTE INSTABILITY.

autoconvective lapse rate *Meteorology.* the environmental lapse rate of temperature in an atmosphere in which the density is constant with height, equal to g/R, where g is the acceleration of gravity and R the gas constant.

autocopulation *Invertebrate Zoology.* the process of self-copulation, which occurs occasionally in certain hermaphroditic worms.

autocorrelation *Statistics.* a linear relation between successive measurements of a random variable over time. Also, SERIAL CORRELATION. *Electronics.* extracting information from a signal by integrating the product of the signal waveform and an identical waveform with a variable delay.

autocorrelation function *Mathematics.* the autocorrelation function of a given function $f(t)$ is: $\lim_{T\to\infty} {}^1/_{2T} \int_{-T}^{T} f(t)f(t-\tau)\,dt$, where τ is a time-delay parameter. The autocorrelation function reveals whether the value of $f(t)$ is, on the average, influenced by the value of f at $t-\tau$ by exhibiting a peak at τ.

autocorrelator *Electronics.* an electronic network designed to distinguish an intelligible signal from noise by means of autocorrelation.

autocrine *Endocrinology.* a hormonal pathway characterized by the production of a biologically active substance by a cell; the substance then binds to receptors on that same cell to initiate a cellular response.

autodecrement addressing *Computer Programming.* a register-indirect addressing technique for accessing a table in memory, in which the highest address is loaded into the register and then automatically decremented after each data access until the entire table has been accessed.

autodeme *Ecology.* a plant population that is made up mainly of self-fertilizing individuals.

autodermic *Surgery.* of or relating to the patient's own skin; especially, describing autodermic skin grafts.

autodermic (skin) graft *Surgery.* a skin graft taken from the patient's own body.

autodictionary *Computer Programming.* in a language translation system, the table look-up function that indicates a word-for-word substitution from one programming language to another; or substitutes codes for longer words or phrases. Also, AUTOMATIC DICTIONARY.

autodrainage *Surgery.* the surgical or spontaneous drainage of an abscess or cavity by diversion of the fluid into a new channel within the patient's own body.

autodyne circuit *Electronics.* a circuit that acts both as an oscillator and a heterodyne detector, and whose output signal frequency is the difference between that of the received signal and that of the oscillator.

autodyne reception *Telecommunications.* radio reception that is implemented by an oscillator/detector device.

autoecious *Mycology.* of rust fungi, requiring only one host to complete all stages of the life cycle. Also, AUTOICOUS.

autoecology see AUTECOLOGY.

autoexec *Computer Programming.* in microcomputers using MS-DOS, the file that contains the instructions to be executed when the computer is turned on; it contains such commands as setting the system date/time and customizing the screen prompt message.

autofocus rectifier *Optics.* a photoenlarger that corrects for distortions in negatives caused by tilt during aerial photography.

autofrettage *Engineering.* a process in which a favorable distribution of initial or residual stress in a tube is induced, as in the manufacturing of gun barrels.

autogamous [ô täg´ə məs] *Biology.* relating to or characterizd by autogamy.

autogamy [ô täg´ə mē´] *Biology.* any of various types of self-fertilization, especially the fertilization of a flower by its own pollen, or the union of two closely related cells or nuclei of protozoans or fungi. *Invertebrate Zoology.* see PAEDOGAMY.

autogenesis see ABIOGENESIS.

autogenetic *Geology.* of a landform, developed solely under the action of local conditions, such as rains or streams, without interference by crustal movements. Also, **autogenous, autogenic.**

autogenic drainage *Hydrology.* a drainage system that has developed only under the conditions of the surface over which the constituent streams flow.

autogenous [ô täj´ə nəs] *Science.* generated without external influence. *Invertebrate Zoology.* able to produce eggs without requiring a meal of blood.

autogenous electrification *Physics.* the accumulation of charge by an object, generated by friction between the object and its surroundings.

autogenous grinding *Mechanical Engineering.* the secondary grinding of a material achieved by tumbling the material alone in a revolving drum, without the use of balls or bars.

autogenous healing *Engineering.* the closing up of cracks in concrete that occurs when concrete sections are kept damp and in contact.

autogenous mill *Mechanical Engineering.* a ball mill grinder that uses the coarse incoming material as the grinding medium. Also, **autogenous tumbling mill.**

autogenous regulation *Genetics.* a type of gene regulation in which a protein is able to control the synthetic activity of the gene that synthesizes that protein.

autogenous vaccine *Immunology.* the immunizing material made from cultures of an infected person that is subsequently injected back into the same person.

autogenous welding *Metallurgy.* the process of joining metallic components by fusion welding without using a filler metal.

autogeosyncline *Geology.* a geosyncline within a stable area that has developed without the adjoining highlands and is filled primarily with carbonate sediments.

autogiro [ôt´ō jī´rō] *Aviation.* a rotorcraft in which, except for the initial start, the rotors are driven solely by the action of air upon them while moving. Modern autogiros are equipped with propellers or other propulsive means independent of the rotor system. Also, **autogyro.**

autograft *Immunology.* a graft transplanted from one part of the body to another in the same individual. Also, AUTOTRANSPLANT.

autograph *Photogrammetry.* a stereoplotting instrument that uses negatives or glass positives to produce a trace of an object; the image is formed by the mechanical, chemical, or radiation effects of the object itself and made visible by development.

autohemagglutination see AUTOAGGLUTINATION.

autohemolysis *Hematology.* a breakdown of the red blood cell membrane by factors in the organism's own serum that cause the release of hemoglobin.

autohemorrhage *Entomology.* in certain insects, the automatic expulsion of blood that is nauseating or poisonous to would-be predators.

autohemotherapy *Medicine.* treatment by reinjection of an individual's own blood.

autohypnosis *Psychology.* self-induced hypnosis.

autoicous see AUTOECIOUS.

autoignition *Mechanical Engineering.* in an internal combustion engine, the spontaneous ignition of fuel when introduced into the combustion chamber, either due to the heat of compression or to glowing carbon in the chamber. *Chemistry.* see SPONTANEOUS COMBUSTION.

autoimmune [ôt´ō i myoon´] *Immunology.* of, relating to, or characteristic of autoimmunity or an autoimmune disease.

autoimmune disease *Immunology.* a disease that is caused when an individual produces an immune reaction against its own tissues.

autoimmunity *Immunology.* a condition in which an individual produces an immune reaction against its own tissues.

autoimmunization *Immunology.* a condition in which antibodies are naturally produced against the constituents of an individual's own cells or tissues.

autoincrement addressing *Computer Programming.* a register-indirect addressing technique for the purpose of accessing a table in memory, in which the lowest address is loaded into the register and then automatically incremented after each data access until the entire table has been accessed.
autoindexing *Computer Programming.* the automatic incrementing or decrementing of an index value during program execution.
autoinducible enzyme *Enzymology.* a drug-metabolizing enzyme whose production is stimulated by the chronic administration of that drug.
autoinfection *Medicine.* infection by an agent that is already present in the body, often by the transfer of the agent from one part of the body to another.
autoinoculation *Medicine.* inoculation with microorganisms from one's own body.
autointoxicant see AUTOTOXIN.
autointoxication *Medicine.* intoxication by some poison generated within the body.
autointrusion *Geology.* a process in which residual fluids from a solidifying magma are injected into rifts formed in the crystallized fraction. Also, **autoinjection.**
autoionization see AUGER EFFECT.
autokinesis *Physiology.* the power of voluntary motion. Thus, **autokinetic.**
autolith *Petrology.* in a granitoid rock, an accumulation of iron-magnesium minerals in oval or elongated structure.
autolithography *Graphic Arts.* a lithography process in which the artist draws a design directly on the printing surface, with no transfer of images.
autoload *Computer Technology.* a key on a computer console that, when pressed, boots the operating system into memory and initializes the system.
autologous *Immunology.* produced from the subject's own tissues or derived from the subject's body, such as skin taken from one part of the body and grafted to another part.
autoluminescence *Atomic Physics.* a glow that arises from energy within a material, such as that generated by radioactive materials.
autolysin *Enzymology.* an enzyme that hydrolyzes and destroys the components of a cell.
autolysis [ô täl´ə sis] *Chemistry.* **1.** the return of a substance to solution as a result of natural processes, such as decay or decomposition. **2.** the process by which a substance dissolves or digests itself. *Pathology.* **1.** the automatic dissolution of cells or tissue by the enzymes contained within them, occurring upon death or under particular pathological conditions. **2.** the destruction of cells due to the presence of a lysin.
autolytic [ôt´ə lit´ik] relating to or causing autolysis.
autolytic enzyme *Enzymology.* an enzyme found in the cell wall that causes the breakdown of the cell following traumatic events such as injury or death.
Autolytus *Invertebrate Zoology.* a genus of marine annelids in the family Syllidae that reproduce asexually by developing new living segments near the posterior end.
automanual system *Transportation Engineering.* a railroad signal system that automatically moves to the danger position as a train passes but must be reset manually.
automarking *Behavior.* a behavior pattern in which an animal applies some scent-marking substance to itself as a signal to others of its own species.
automata [ô täm´ə tə] the plural of AUTOMATON.
automata theory *Control Systems.* **1.** a theory concerning the operating principles, applications, and behavioral characteristics of automatic devices. **2.** a theory concerning models of objects and processes such as computers, digital circuits, nervous systems, cellular growth, and reproduction.
automate *Engineering.* **1.** to convert a facility or operation to automation, especially one that formerly required a significant amount of human labor. **2.** to operate or control any system by means of automation.
automated *Engineering.* done by automation; involving the principles or techniques of automation. Thus, **automated inventory, automated system,** and so on.
automated assembly *Design Engineering.* the use of a computer system that directs operations to be performed by machines in a manufacturing plant; the system monitors the quality and production levels at each assembly operation.
automated diagramming *Computer Programming.* a software engineering tool for interactive diagramming on a computer screen that speeds up software development, aids in enforcing software standards, and automates software documentation.
automated diffractometer *Crystallography.* an instrument for automatically measuring and recording the intensities of X-ray beams diffracted by a crystal. The mutual orientations of the crystal and of the detector with respect to the source of radiation are computed from some initial data on a few selected reflection. The computer calculates these orientation angles and drives the gears that move the crystal orienter and detector to the desired angular settings.
automated flight service station *Aviation.* an unmanned, automated station from which pilots can obtain flight briefings.
automated guided vehicle system *Control Systems.* a computer-controlled transport system used to move workpieces to predetermined locations within a flexible manufacturing system.
automated guideway transit *Transportation Engineering.* a system in which fully automated, unmanned vehicles travel on fixed tracks.
automated machining *Design Engineering.* the use of computer-controlled machines to create part geometry by providing tool path motions or other machine functions.
automated machining cells *Design Engineering.* a cluster of machine tools organized to provide an optimum machining environment for parts production.
automated manufacturing system *Design Engineering.* a computerized system of manufacturing machines to produce products with reduced human intervention.
automated office *Computer Technology.* the application of computer and network technology to improve office productivity in such areas as text processing, electronic mail, information storage and retrieval, and task scheduling and management. Similarly, **automated management.**
automated radar plotting aid *Navigation.* an automated system that converts raw radar data input into radar plotting information.
automated radar terminal system *Navigation.* a sophisticated identification system used in air traffic control in which aircraft identification and other flight information is displayed along with a given aircraft's radar image.
automated reasoning *Artificial Intelligence.* the derivation of conclusions from observed or given data, often by a proof procedure operating on the data and a set of axioms.
automated storage/retrieval system *Design Engineering.* a combination of equipment and controls used to store and retrieve materials with precison, accuracy, and speed under a defined degree of automation with systems varying from relatively simple, manually controlled machines to massive computer-controlled storage/retrieval systems fully integrated into a manufacturing and distribution process.

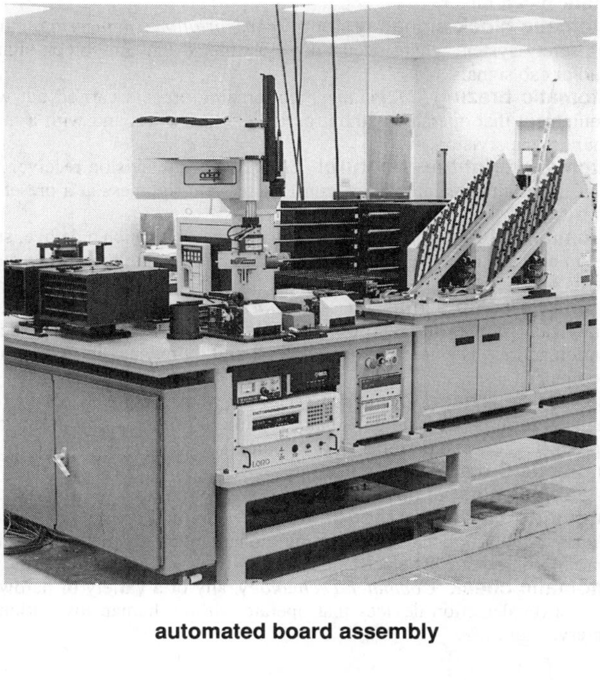

automated board assembly

automated tape library *Computer Technology.* an early method of massive bulk storage in which the computer itself had the capability of selecting the desired tape reel from a rack, mounting the reel, accessing data stored thereon, and returning the reel to the rack when the run was complete.

automated teller machine see ATM.

automated welding *Design Engineering.* the process of providing automated positioning of a welding apparatus during the process of welding parts.

automatic *Engineering.* describing any instrument or device that has the capacity to operate on its own, without human control, through a self-regulating mechanism. Used to form many compound terms involving the idea of self-initiated or self-controlled action, including the specific entries listed below and numerous others. *Ordnance.* **1.** of a weapon, firing continuously as long as pressure on the trigger is maintained: an *automatic* rifle. **2.** of or relating to such weapons: *automatic* fire. **3.** a pistol that is fired in this manner.

automatic abstracting see AUTOABSTRACT.

automatic alarm receiver *Electronics.* a component of an international network of alarm receivers actuated by a radio-frequency signal to indicate an international emergency. Also, AUTOALARM.

Automatically Programmed Tools *Computer Programming.* an early programming language used to produce control information for numerically controlled machine tools.

automatic altitude reporting *Transportation Engineering.* a transponder system that continually reports aircraft altitude when interrogated by a control radar in Mode C. Ground controllers are given continual altitude information in 100-foot increments and are not dependent upon pilot reports.

automatic back bias *Electronics.* a technique used in radar reception to prevent a receiver from being overloaded by a jamming signal or radar echo. It is accomplished by gain control feedback loops.

automatic background control see AUTOMATIC BRIGHTNESS CONTROL.

automatic balance *Engineering.* a type of balance that can weigh objects without operator assistance.

automatic bar and chucking machines *Design Engineering.* an automatic screw machine with the feature of rapid screw production through the use of a variety of tool slides which are automatically sequenced, each performing a portion of necessary machining on a rotating bar, then cutting off finished pieces in rapid succession.

automatic bass compensation *Electronics.* a circuit used in sound receivers and amplifiers to boost the bass signal, to compensate for the human ear's relative inability to perceive lower-frequency sounds.

automatic batcher *Mechanical Engineering.* a batcher, generally used for concrete, that blends all the ingredients of a mixture at predetermined weights and is actuated by a single starter switch. Also, **automatic batch mixer.**

automatic block signal system *Transportation Engineering.* a traffic safety system where occupancy of a block activates block signals and/or cab signals.

automatic brazing *Metallurgy.* any brazing process carried out with equipment that operates without periodic adjustments and with a minimum of supervision.

automatic brightness control *Electronics.* a television receiver circuit that maintains an average constant screen brightness at a preset intensity.

automatic calibration *Engineering.* a process by which a device, such as an electronic balance, automatically recalibrates the measuring range of its scale.

automatic calling unit *Telecommunications.* a device that automatically calls another terminal over a public data network when the destination address is entered into the calling unit by manual entry or from a computer.

automatic carriage *Computer Technology.* any device in which continuous form paper is automatically fed into a printer, as by means of a sprocket wheel and perforated paper borders.

automatic C bias see SELF-BIAS.

automatic celestial navigation see CELESTIAL-INERTIAL GUIDANCE.

automatic character recognition *Computer Technology.* a technique, usually optical, of converting alphanumeric character input data into machine-readable electronic form.

automatic check *Computer Technology.* any of a variety of hardware data error detection devices that operate without human invocation or intervention.

automatic choke *Mechanical Engineering.* on the carburetor of an automotive engine, an automatic device that opens and closes the choke to satisfy air requirements of the engine, preventing excessive richness of the air-fuel mixture by opening the choke when the normal operating temperature for the engine is reached.

automatic coding *Computer Programming.* a process of using a computer to convert a program written in a high-level or symbolic language or, in the case of software development tools, a program design into machine executable instructions.

automatic color control *Electronics.* a color television receiver circuit that maintains a constant color intensity by automatically adjusting the gain of the chrominance bandpass amplifier. Also, **automatic chroma control, automatic chrominance control.**

automatic computer *Computer Technology.* a computer that can function largely without human intervention, such as in military automatic weapons systems.

automatic connection *Electronics.* a connection made between users of communicating devices by automatic electronic switching.

automatic contrast control *Electronics.* a television receiver circuit that maintains a constant average screen contrast by adjusting the gain of the radio frequency amplifier and the video intermediate frequency amplifier.

automatic controller *Control Systems.* an instrument that continuously measures the value of a variable quantity or condition and automatically takes action to correct any deviation from a preprogrammed value. Thus, **automatic control (system).**

automatic counter see REVOLUTION COUNTER.

automatic cutout *Electricity.* a protective device that automatically removes or shorts part of a circuit at a particular moment; for example, a circuit breaker.

automatic data processing *Computer Technology.* the use of computer systems to process information with a minimum of human assistance or intervention. *Computer Programming.* a processing of (typically business) data by computer. Used in various compounds such as **automatic data-processing system, automatic data-processing department.**

automatic data-processing equipment *Computer Technology.* the computers and related electronic devices used in business data processing. Similarly, **automatic data-processing auxiliary equipment.**

automatic degausser *Electronics.* a magnetic circuit that automatically demagnetizes a color television picture tube.

automatic dialer *Electronics.* a device that automatically dials a certain telephone number in its memory when it is activated by a preset button or other such device.

automatic dictionary see AUTODICTIONARY.

automatic direction finder *Navigation.* a radio receiver that senses and indicates the direction toward a nondirectional radio beacon.

automatic door bottom *Building Engineering.* a horizontal moveable device attached to the bottom of a door, which moves downward when the door is shut, forming a seal along the threshold. Also, THRESHOLD CLOSER.

automatic drill *Mechanical Devices.* a bit shank containing a coarse-pit screw that engages and holds the bit when the handle is pushed down.

automatic driller or **automatic drilling control unit** see DRILLING CONTROL.

automatic error correction *Telecommunications.* **1.** a characteristic of a transmission system that enables a certain proportion of errors in the received signal to be detected and automatically corrected. **2.** a system in which errors are detected and corrected automatically through the use of error detection equipment and codes without initiating a request for repetition.

automatic exchange *Electronics.* any type of communication exchange in which operator action is not required.

automatic exposure *Optics.* a control system that automatically regulates and adjusts the exposure setting in a camera; it is powered by an electric current that is produced by the action of light falling on a photocell.

automatic fine-tuning control *Electronics.* a color television receiver circuit that automatically maintains a constant tuner frequency by correcting drift and incorrect tuning.

automatic firearm see AUTOMATIC WEAPON.

automatic flight control *Aviation.* an automated system that is used for controlling an aircraft, intended to combine safety with on-time performance.

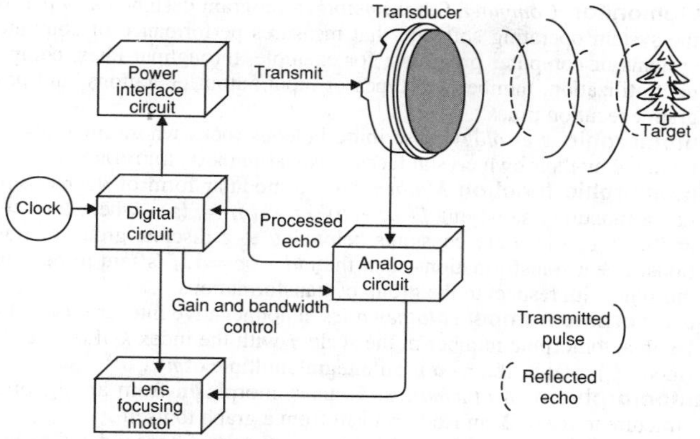

Polaroid automatic focusing system

automatic focus *Optics.* a control system in a camera or enlarger that automatically keeps an object in focus; the system is equipped with a sensor that is capable of estimating distance or image contrast, and a drive mechanism that adjusts the lens.

automatic frequency control *Electronics.* any device that automatically maintains the frequency response of a circuit; used in superheterodyne receivers, radar receivers, and television receivers.

automatic gain control *Electronics.* a device that automatically adjusts a receiver's gain to maintain a constant output amplitude, regardless of the amplitude of the input signal.

automatic grid bias see SELF-BIAS.

automatic gun see AUTOMATIC WEAPON.

automatic horizon see ARTIFICIAL HORIZON.

automatic ignition *Engineering.* **1.** the ignition of a charge in an internal combustion engine by the heat of compression, rather than by an ignition device. **2.** the process of igniting a fuel mixture by turning on a gas burner valve.

automatic indexing *Computer Programming.* the process of searching a document for specified key words and recording their locations for entry in the document index; a feature of many commercial word-processing programs. *Robotics.* a method of finding the alignment and position of a workpiece so that a robot can perform some operation on it.

automatic intercept *Telecommunications.* a subsystem designed to process intercepted calls automatically and connect them to a vacant number announcement or to an announcement machine that gives correct numbers or transfers calls to an intercept operator for handling.

automatic interrupt *Computer Programming.* a program-controlled interrupt system that causes transfer of program control to a specified location where the interrupt handling program is stored.

automatic level compensation *Telecommunications.* a circuit or subsystem that makes an automatic adjustment for deviations in the output levels of a circuit or system.

automatic level control *Electronics.* in magnetic recording, a device that automatically adjusts the recording level, in order to stop intermittent loud sounds from creating distortion and to pick up quiet sounds. *Mechanical Engineering.* **1.** in an automotive vehicle, a suspension system that compensates for variations in load at the front, rear, or both ends of the car by means of electrically controlled air-chamber shock absorbers that position the car at a predetermined level, regardless of load. **2.** a self-activated system that automatically controls a fluid level in a vessel, container, or the like.

automatic message accounting *Telecommunications.* a system operating as part of, or in conjunction with, an automatic telephone switching system to create records from which billing information can be derived for individual calls, such as long-distance calls. Also, AUTOMATIC TOLL TICKETING.

automatic message-switching center *Telecommunications.* a method of automatically handling messages through a switching center, either from local users or from other switching centers, whereby a distant electrical connection is established between the calling and called stations.

automatic mode *Transportation Engineering.* in a rapid-transit system, an operating condition in which vehicle movement and headways are controlled electronically.

automatic modulation control *Electronics.* a transmitter circuit that prevents overmodulation by lowering the gain of overly strong signals.

automatic modulation limiting *Telecommunications.* in citizen-band radio transmitters, a circuit that curtails overmodulation.

automatic music search *Acoustical Engineering.* a control mechanism in a tape player that allows it to move forward or in reverse from any given point on a tape to the nearest gap in sound, thus enabling the listener to search for the beginning of a selection.

automatic noise limiter *Electronics.* a circuit that defines the maximum amplitude of any received signal, including noise; loud noise signals thus are clipped.

automatic pagination *Computer Programming.* the process of placing successive page numbers on a document automatically; a feature of many commercial word-processing programs.

automatic part inspection *Design Engineering.* the use of sensors, data bases, and automatic handling devices to determine the acceptability of a part or product.

automatic peak limiter see LIMITER.

automatic phase control *Electronics.* a color television receiver circuit that synchronizes the phase of the 3.58 megahertz color carrier with a color burst signal.

automatic picture control *Electronics.* in a color television receiver, a switching device that disconnects one or more manual controls and connects the corresponding presets.

automatic picture-transmission system *Electronics.* a television satellite system that transmits pictures during the day and at night (infrared) from weather satellites. It utilizes the slow-scan (200 s per image) technique.

automatic pilot *Navigation.* a mechanism which allows an aircraft to maintain a set course and level flight without human control.

automatic pistol *Ordnance.* see AUTOMATIC, def. 3.

automatic positioning machine *Design Engineering.* a machine that is capable of providing a position in space from a defining data base.

automatic press *Mechanical Engineering.* a press equipped with a mechanical feeding device that synchronizes the feed of the workpiece with the press action.

automatic process *Psychology.* any learned process that occurs without the intervention of deliberate or conscious thought.

automatic programming *Computer Programming.* **1.** the use of a computer to generate computer programs from a human understandable form or language. **2.** an application of list processing techniques to generate computer programs from axiomatic specifications, information specifications, or from examples.

automatic ranging see AUTORANGING.

automatic regulator see AUTOMATIC CONTROLLER.

automatic relay *Telecommunications.* a system having the means of switching for automatic equipment to record and retransmit messages.

automatic repeat request *Telecommunications.* a signal automatically transmitted by the receiving unit when it detects an error in data, which causes the sending unit to retransmit.

automatic routine *Computer Programming.* a program module that automatically executes if certain conditions or values occur during processing; for example, error processing or end-of-file instructions.

automatic search see AUTOMATIC MUSIC SEARCH.

automatic sensitivity control *Electronics.* a receiver circuit that maintains the sensitivity of a receiver at a prescribed level.

automatic short-circuiter *Electricity.* a device used to automatically short circuit the commutator bars in single phase commutator motors.

automatic shutdown *Computer Programming.* the orderly emergency termination of computer operations; the procedure includes recording the contents of all registers and status words in order to be able to recover and restart with a minimum of data lost.

automatic spider *Mining Engineering.* a drill-rod clamping device actuated by foot or hydraulically.

automatic stability *Aviation.* **1.** the tendency of a body to resist displacement and maintain its attitude or to return to its original conditions after being displaced. **2.** the use of automatic stabilization equipment to accomplish this.

automatic stabilization equipment *Aviation.* any device, such as automatic pilot, that automatically operates a craft's control systems in order to maintain stability.

automatic stoker *Mechanical Engineering.* a device that supplies solid fuel continuously by gravity; it may also carry the fuel on an endless chain through a boiler furnace and deposit the ash. Also, MECHANICAL STOKER.

automatic stop *Computer Technology.* automatic cessation of computer operations upon detection of certain specified types of processing errors.
automatic switchboard *Telecommunications.* a system that has the capability of automatically processing traffic without operator intervention; it can recognize dial and pushbutton signaling, automatically route calls, handle precedences, dispose of incorrect calls, and switch incoming calls from line-to-line, trunk-to-trunk, and trunk-to-line.
automatic threshold closer see AUTOMATIC DOOR BOTTOM.
automatic tint control *Electronics.* a color television receiver circuit that corrects phase errors in the chroma signal before demodulation in order to maintain correct skin tones.
automatic titrator *Analytical Chemistry.* **1.** a titration with a measured flow of the reactant. **2.** electrically generated reactant with potentiometric, amperometric, or colorimetric end-point determination.
automatic toll ticketing see AUTOMATIC MESSAGE ACCOUNTING.
automatic tool identification *Design Engineering.* the process of using a bar code or other coded identifier on individual tools in order to provide for automated selection and retrieval of tooling.
automatic train control *Transportation Engineering.* a rail system in which all operations, protection, and supervision are controlled electronically.
automatic tuning system *Control Systems.* an electrical or mechanical system for automatically tuning a radio transmitter or receiver to a predetermined frequency.
automatic typewriter *Engineering.* an electric typewriter that has the capability to record information as it produces conventional typed copy, so that it later can automatically retype the stored information.
automatic voltage regulator see VOLTAGE REGULATOR.
automatic volume compressor see VOLUME COMPRESSOR.
automatic volume control *Electronics.* in a radio receiver, a device that maintains a constant output volume by adjusting the gain control automatically. The volume is thereby independent of the input signal.
automatic volume expander see VOLUME EXPANDER.
automatic weapon *Ordnance.* a weapon that repeatedly fires, extracts, ejects, and loads new rounds without power from an outside source, so long as pressure on the trigger or firing mechanism is maintained. Also, AUTOMATIC GUN, AUTOMATIC FIREARM.
automatic weather station *Meteorology.* an unmanned weather station equipped with weather element recording and telemetric apparatus for transmitting data at predetermined times.
automatic wet-pipe sprinkler system *Engineering.* a sprinkler system whose pipe network is filled with water at a pressure that is sufficient to provide an immediate discharge of water when the system is activated.
automatic writing *Psychology.* the writing of meaningful material unconsciously, as during hypnosis.
automation *Engineering.* **1.** the use of mechanical, electronic, or computerized systems to replace or augment human labor. **2.** an industrial process in which machines operate automatically with minimal human involvement.
automatism *Biology.* **1.** spontaneous, automatic activity of body tissues. **2.** a behavior or act that is not under control of the will, such as a reflexive response or an epileptic seizure.
automaton [ô täm´ə tän´] *plural,* **automata** or **automatons.** *Computer Technology.* an abstract mathematical machine that performs actions in response to its preprogramming, memory contents, and inputs; used as formal models of real or theoretical computers. Kinds of automata include finite automata, pushdown-stack automata, linear-bounded automata, and turning machines.
automechanism *Control Systems.* any machine or device that operates under automatic control or under control of a servomechanism.
automecoic *Cartography.* of or related to true scale dimensions.
autometamorphism *Petrology.* the metamorphic process in an igneous rock caused by the action of its own volatile water-rich liquid trapped within its impervious chilled border.
autometasomatism see AUTOMETAMORPHISM.
automimicry *Biology.* mimicry within a species, as when some members of the species are unpalatable, and palatable members of the same species imitate them.
automobile *Mechanical Engineering.* **1.** a mechanical, four-wheeled, trackless passenger vehicle, generally designed for use on paved roads and propelled by electricity, diesel fuel, or gasoline. **2.** of or relating to such a vehicle. Also, CAR.
automobile industry see AUTOMOTIVE INDUSTRY.
automonitor *Computer Programming.* a program designed as a part of the system operating software that measures performance of computer systems or computer programs; for example, throughput rates, component utilization, number and types of input/output operations, and program execution times.
automorphic *Petrology.* describing igneous rocks whose minerals are bounded by their own crystal faces. Also, EUHEDRAL, IDIOMORPHIC.
automorphic function *Mathematics.* a modular form of degree zero; i.e., a function f satisfying $f((az+b)/(cz+d)) = f(z)$, where Im $z > 0$ and a, b, c, and d are constants belonging to a discrete group of fractional linear transformations such that $ad - bc = 1$. f is said to be automorphic with respect to the group of transformations.
automorphic number *Mathematics.* a nonnegative integer a is said to be an automorphic number in the scale n with the index k if $a^2 - a \equiv 0$ (mod n^k); that is, if $a^2 - a$ is an integral multiple of n^k.
automorphism *Mathematics.* **1.** an isomorphism from an algebraic structure to itself. **2.** an isomorphism from a graph to itself.
automorphosis *Petrology.* the metamorphosis of consolidated, solidified, igneous rock by intrinsic liquid heating.
automotive *Mechanical Engineering.* **1.** of or relating to a self-propelled vehicle designed for land transportation, such as an automobile, truck, motorcycle, or bus. Thus, **automotive engine, automotive fuel, automotive transmission,** and so on. **2.** of or relating to any vehicle that is propelled by a self-contained engine or motor, such as a motor boat, helicopter, or airplane.
automotive engineering *Engineering.* the development and study of various technologies relating to automobiles and other motor vehicles. The design and management of the production of these vehicles.
automotive industry *Mechanical Engineering.* the various companies and enterprises involved in the design, development, manufacture, and sale of motor vehicles and related products.
automotive vehicle *Mechanical Engineering.* **1.** an automobile or other motor vehicle. **2.** any other self-propelled vehicle used for the transportation of people or materials, including aircraft and motor boats.
autonomic *Physiology.* self-controlling; functionally independent.
autonomic movement *Botany.* any movement in response to internal rather than external stimulation; includes growth and chromosome movement.
autonomic nervous system *Physiology.* the part of the nervous system that controls involuntary actions, including the smooth muscles, cardiac muscle, and glands.
autonomic reflex system *Physiology.* the autonomic fibers that innervate involuntary action of the smooth muscles, glands, and conductive tissues of the heart.
autonomously replicating sequence *Molecular Biology.* a DNA molecule that is found in the cytoplasm or in the nucleus but is not attached to a chromosome and replicates independently of the chromosomal DNA. *Genetics.* any of several DNA sequences required for chromosome replication in yeast cells.
autonomous system *Mathematics.* a system of linear first-order differential equations of the form $dx_i/dt = A_i(x_1, \ldots, x_n)$, where the functions A_i are independent of t.
autonomy stage *Psychology.* the second of Erik Erikson's eight stages of human development, at about the age of two to three, when children develop a sense of independence and begin to control their impulses and actions.
autonomy vs doubt *Psychology.* the conflict that can arise during the autonomy stage of development, when the child may doubt his ability to become independent if his parents are overly critical or overprotective. Also, **autonomy vs shame and doubt.**
autooxidation *Chemistry.* **1.** a self-catalyzed oxidation reaction that occurs spontaneously in the atmosphere, initiated by light, heat, and so on. **2.** an oxidation reaction initiated only by an inductor.
autoparasitization *Entomology.* an unusual form of intraspecific interaction in which one sex lives as a parasite on the other sex, as with certain scale insects.
Autopass *Robotics.* a robot language developed by IBM for their automatic programming system for computerized mechanical assembly.
autopatch *Electronics.* a remote-controlled device that incorporates a radiocommunications system into a land-line telephone network.
autopath *Neurology.* a person who suffers from allergic disorders due to a sensitive autonomic nervous system. Also, **autophil.**
autopatrol *Mechanical Engineering.* a self-propelled machine with a blade for shaping excavated surfaces to the desired shape or slope. Also, MOTOR GRADER.

autophagic vacuole *Cell Biology.* a large digestive vacuole that contains digestive enzymes, and in which cellular organelles are digested.

autophagocytosis *Cell Biology.* a process occurring within a cell in which the cell's cytoplasm is digested by enzymes contained within vacuoles in the cell.

autophagy [ôt´ō faj´ē] *Cell Biology.* intracellular digestion of a cell's own cytoplasm by enzymes produced within the cell. Also, **autophagia.**

autophobia [ôt´ō fō´bē ə] *Psychology.* an irrational fear of oneself or of being alone.

autophyte [ôt´ə fīt´] *Botany.* any plant capable of manufacturing its own food from inorganic substances; an autotrophic plant.

autopilot see AUTOMATIC PILOT.

autoplaque *Microbiology.* a clear, circular area seen on certain bacterial lawns, similar to that caused by a lytic bacteriophage but occurring in the absence of bacteriophage; thought to result from autolysis of bacterial cells.

autoplast *Cell Biology.* a spheroplast that is formed from a cell by the action of lytic enzymes of that cell. *Surgery.* see AUTOGRAFT.

autoplastic graft see AUTOGRAFT.

autoplasty *Surgery.* the use of tissue from one part of the body to repair defects in another part.

autoplotter *Computer Technology.* see PLOTTER.

autopneumatolysis *Geology.* a process of igneous rock alteration whereby new minerals are formed through the action of volatiles that originate in the magma or rock itself.

autopoisonous see AUTOTOXIC.

autopolarity *Electronics.* on a digital meter display, the ability to display the correct polarity for the quantity measured, by means of a plus sign for positive polarity and a minus sign for negative polarity.

autopolymerizing resin *Materials Science.* the room temperature polymerization of a monomer due to a catalytic addition to the monomer liquid.

autopolyploid *Genetics.* a polyploid that contains multiple copies of a basic set of chromosomes.

autopositive *Graphic Arts.* a film or paper that is processed in a single development stage, producing a positive image when exposed to a positive or a negative image from a negative.

autoprothrombine see THROMBOKINASE.

autoprotolysis *Chemistry.* the transfer of a proton from one molecule to another molecule of the same substance.

autopsy [ô´täp sē] *Pathology.* **1.** an analytical examination of the internal organs of the body after death to ascertain the cause of death or the nature of pathological changes present; a postmortem examination. **2.** in the empirical school of ancient Greek studies, the intentional replication of symptoms for analytical purpose.

autoradiogram *Pathology.* a recording of the radiation given off by a tissue sample after introduction of radioactive agents into a body area.

autoradiograph *Engineering.* an image produced by autoradiography. Also, **autoradiogram.**

autoradiography *Engineering.* a technique in which the image of a radioactive specimen is obtained using a photographic or X-ray sensitive emulsion. Thus, **autoradiographic.** Also, RADIOAUTOGRAPHY.

autorail *Mechanical Engineering.* a self-propelled vehicle equipped with flanged wheels and pneumatic tires in order to adapt it to both railways and roads.

autoranging *Engineering.* in electric meters, automatic range switching that continues to increasingly higher ranges until it reaches a range for which the full-scale value is not exceeded.

autoreducing tachymeter *Engineering.* a surveying instrument that simultaneously determines distances and elevations of distant objects.

autoregressive series *Mathematics.* a difference equation of the form

$$f(t) = a_1 f(t-1) + a_2 f(t-2) + \cdots + a_n f(t-n) + k.$$

auto-repeat *Computer Technology.* a keyboard feature that repeats a character as long as the key is held down.

auto-restart *Computer Technology.* the capability of some computers to reinitialize the operating system automatically following the recovery from a power or equipment failure.

autorotation *Mechanics.* the turning of a rotor due solely to the action of passing air, as in windmills and other wind-energy devices. *Aviation.* the movement of an aircraft propeller or rotor in this manner to permit flight in the absence of engine power.

autoscreen film *Photogrammetry.* a photographic film embodying a halftone screen that automatically produces a halftone negative from continuous-tone copy.

autoserum *Immunology.* a type of serum that is injected into the individual from whom it was previously derived.

autosexing *Biology.* the characteristic or process of exhibiting different characteristics at birth or hatching, according to sex; especially used in reference to fowl.

autoshaping *Behavior.* a conditioning pattern in which an animal that is repeatedly exposed to a stimulus paired with a certain reward begins to behave toward that stimulus as if it were the reward.

autoskeleton *Invertebrate Zoology.* an internal skeleton; specifically, the endoskeleton of a sponge, made up of spicules or spongin fibers secreted by the cells.

autosled *Mechanical Engineering.* a propeller-driven vehicle equipped with four retractable runners and wheels, for travelling on packed snow, ice, and bare roads.

autosome *Genetics.* any chromosome other than the X or Y sex chromosome.

autospore *Botany.* a daughter cell that has the same shape as the parent cell; formed by the division of a single cell, especially in single-celled algae such as diatoms.

autostability *Control Systems.* the ability of a device to maintain its position under or through the control of a servomechanism.

autostarter *Electricity.* **1.** an automatic starting and switchover generating system composed of a standby generator linked to the station load through an automatic power transfer control unit. **2.** a motor starter with an autotransformer that furnishes a reduced voltage for starting.

autostoper *Mining Engineering.* a stoper or light compressed-air rock drill mounted on an air-leg support that both supports the drill and exerts pressure on the drill bit.

autostylic *Vertebrate Zoology.* relating to the attachment of the upper jaw to the neurocranium, as in amphibians, reptiles, and some fishes.

autosuggestion *Psychology.* a suggestion that originates within the person, rather than with another person. Also, SELF-SUGGESTION.

autosyndesis *Cell Biology.* the pairing of homologous chromosomes from the same parent that occurs in polyploid and aneuploid organisms. *Molecular Biology.* see DIFFERENTIAL SPLICING.

autotest program *Computer Programming.* a set of instructions that reside in a computer's operating system and are executed in order to test various system components, such as main memory and the hard disk drive, usually when the system is booted.

autotetraploid *Genetics.* an autopolyploid somatic cell that contains four sets of identical chromosomes.

autotomy *Zoology.* the ability of certain lower animals, such as lizards and starfish, to cast off injured body parts and, usually, to regenerate new ones.

autotoxic *Toxicology.* of a substance, poisonous to the body that produces it. Also, AUTOPOISONOUS.

autotoxin *Toxicology.* any poisonous substance that is produced within the body, such as urea. Also, AUTOINTOXICANT.

autotransductor *Electromagnetism.* a reactor in which the main current and the control current are driven through the same windings.

autotransformer *Electricity.* **1.** a power transformer with one continuous winding, part of which serves as the primary winding and all of which serves as the secondary winding, or vice versa. **2.** a device designed to transform voltage, current, or impedance and having parts of one winding common to both the primary and secondary circuits.

autotransfusion *Surgery.* the reinfusion of blood or blood products into the body of the patient from which the blood was originally removed.

autotransplant see AUTOGRAFT.

autotroph *Biology.* any organism, especially photosynthetic green plants and chemosynthetic bacteria, that obtains its nutrients by synthesis from the environment, rather than by consuming other organisms.

autotrophic *Biology.* of an organism, able to manufacture food from inorganic sources and to live independent of outside sources of complex organic molecules such as vitamins and amino acids. *Ecology.* of an ecosystem, receiving little or no nutrient materials from outside the system. Thus, **autotrophy.**

autotrophic lake *Ecology.* a lake in which all or most of the organic material is derived from within the lake, rather than by drainage from the surrounding land.

autotropic *Biology.* tending to grow in a straight line, regardless of external factors.

autoxidation see AUTOOXIDATION.

autozooid *Invertebrate Zoology.* any member of a zooid colony that is capable of feeding itself.

autumn *Astronomy.* in the Northern Hemisphere, the season that begins with the September equinox and ends with the December solstice; in the Southern Hemisphere, the season between the March equinox and the June solstice.
autumnal [ô tum´nəl] *Astronomy.* of or relating to autumn.
autumnal equinox *Astronomy.* the moment when the sun crosses the celestial equator from north to south; this occurs approximately on September 23rd.
autumn ice *Oceanography.* young sea ice in the first weeks of formation; crystalline and comparatively salty.
Autunian *Geology.* a European geologic stage of the Lower Permian period, occurring after the Stephanian of the Carboniferous and before the Saxonian.
autunite *Mineralogy.* $Ca(UO_2)_2(PO_4)_2 \cdot 10–12H_2O$, a yellow to green tetragonal mineral formed in the oxidation zones of uranium deposits, having a highly perfect cleavage, a specific gravity of 3.05 to 3.19, and a hardness of 2 to 2.5 on the Mohs scale.
Auversian *Geology.* a European geologic stage of the middle Eocene epoch, occurring after the Lutetian and before the Bartonian. Also, LEDIAN.
aux- a combining form meaning "growth" or "increase," as in *auximone.*
auxanogram *Microbiology.* a plate culture used in auxanography to define the effects on growth of certain environmental conditions.
auxanography *Microbiology.* a technique in which a given microorganism is exposed to a variety of substances, such as different carbon sources, and subsequent growth on each source is examined in order to determine the most suitable growth medium.
auxanometer *Engineering.* a device that is used to measure the rate of plant growth.
aux channel see AUXILIARY CHANNEL.
auxesis [ôk sē´sis] *Physiology.* an increase in the size or volume of an organism from the growth of individual cells without cell division.
auxiliary *Military Science.* **1.** assisting or supporting a main unit in carrying out its duty or assignment. **2.** a person or group that does this.
auxiliary aiming point *Ordnance.* a point or object used to sight a gun toward a target that cannot be seen, so that when the sight is aimed at that point, fire will be directed to the target.
auxiliary cell *Botany.* in some algae, a single-celled female sex organ having a prolonged, tubular distal end that receives the male gamete; a carpogonium.
auxiliary channel *Telecommunications.* in data transmission, a secondary channel whose direction of transmission is independent of the primary channel and is controlled by an appropriate set of secondary control interchange circuits.
auxiliary circle *Astronomy.* a circle circumscribing an orbital ellipse.
auxiliary contacts *Electricity.* supplementary contacts that are actuated by and function with the main contacts in switches or relays, usually for control functions such as interlocks.
auxiliary dead latch *Mechanical Devices.* a dead latch that locks the main latch bolt of a door automatically when it is closed. Also, **auxiliary latch bolt.**
auxiliary electrode *Physical Chemistry.* in a electrochemical cell, such as a battery, the electrode that transfers current to the test electrode. Also, SECONDARY ELECTRODE.
auxiliary equipment *Computer Technology.* a term for any peripheral device not under direct control of a computer. Also, OFF-LINE EQUIPMENT.
auxiliary fault *Geology.* a minor fault that extends from or abuts a major fault. Also, BRANCH FAULT.
auxiliary fluid ignition *Space Technology.* a means of igniting a liquid propellent rocket engine in which combustion is initiated by injecting a liquid that ignites spontaneously when mixed with either the fuel or the oxidizer in the rocket's combustion chamber.
auxiliary instruction buffer *Computer Technology.* an area in memory containing a block of consecutive instructions that permits the instruction fetch and execute cycles to overlap. Also, INSTRUCTION LOOK-AHEAD; INSTRUCTION PIPELINE.
auxiliary landing gear *Aviation.* a part of a landing gear that does not bear any significant weight during landing but is useful for stabilizing the craft.
auxiliary memory *Computer Technology.* secondary storage units, usually slower and less expensive than main or primary memory, intended for storing large amounts of data for relatively long periods of time. Also, AUXILIARY STORAGE, EXTERNAL MEMORY, SECONDARY STORAGE.
auxiliary mineral *Mineralogy.* any unimportant or infrequently occurring light-colored mineral that is found in igneous rock.
auxiliary operation *Computer Technology.* an operation performed off-line by a peripheral device while not under direct control of the main processor.
auxiliary plane *Geology.* a plane that is perpendicular to the slip of a fault plane.
auxiliary plant *Mechanical Engineering.* accessory equipment, such as condenser pumps, mechanical stokers, feedwater pumps, and fans, that are used with the main boiler, turbine, or engine at a power-generating station.
auxiliary processor *Computer Technology.* a special-purpose processor that increases processing speed by performing concurrently with the main processor; for example, input/output servers and array processors.
auxiliary relay *Electricity.* **1.** a relay that supports another relay or device in the performance of a function. **2.** a relay that is actuated by another relay.
auxiliary routine *Computer Programming.* a routine that is designed to assist in computer operation or in debugging other routines.
auxiliary ship see AUXILIARY VESSEL.
auxiliary station *Cartography.* any surveying station connected to a network of main stations and dependent on that network for its position.
auxiliary storage *Computer Technology.* **1.** data storage devices and media under the control of but external to the computer, such as magnetic tape or disk and optical disks. Also, EXTERNAL STORAGE. **2.** see AUXILIARY MEMORY.
auxiliary switch *Electricity.* **1.** a switch in series or parallel with another switch. **2.** a switch that is actuated by another switch.
auxiliary target *Ordnance.* a secondary target used in adjusting fire before directing fire at the actual target, in order to surprise the enemy when fire at the actual target begins.
auxiliary vessel *Ordnance.* a vessel that takes part in a military operation but is not a combat vessel, such as a transport or hospital ship.
auximone *Biochemistry.* a substance thought to be similar to a vitamin and to influence growth in plants.
auxin *Biochemistry.* a hormone that promotes growth in plant cells and tissues by enlarging or lengthening the cells rather than increasing their number.
auxo- a combining form meaning "growth" or "increase," as in *auxotrophic.*
auxoautotrophic *Biology.* requiring no growth factors introduced from or produced outside the organism for metabolic synthesis.
auxocardia [ôks´ə kär´dē ə] *Cardiology.* **1.** enlargement of the heart. **2.** diastole.
auxochrome *Chemistry.* a group that, in the presence of a chromophore, enhances coloration and the ability of the color to act as a dye. An auxochrome group contains lone pairs of electrons.
auxocyte [ôk´sə sīt´] *Biology.* any gamete-producing cell, such as an oocyte, spermatocyte, or sporocyte, during its growth period.
auxograph *Engineering.* an instrument used for the automatic recording of the changes in volume of a body.
auxoheterotrophic *Biology.* requiring growth factors introduced from or produced outside the organism for metabolic synthesis.
auxometer *Engineering.* an instrument that locates the position of optical axes; used particularly to determine the degree to which a lens system is magnified.
auxospore *Invertebrate Zoology.* the reproductive cell of a diatom, formed by the union of two small cells resulting from repeated cell division.
auxotonic *Botany.* induced by growth rather than by external stimulus.
auxotroph *Biology.* any organism (e.g., a bacterium) that, as a result of mutation, can no longer synthesize a substance that is necessary for its own nutrition (usually an amino acid), and thus requires an external supply of that substance.
AV *Aviation.* the airline code for Avianca.
aV abvolt.
av see ARITHMETIC MEAN.
availability *Nutrition.* the degree to which an ingested nutrient is present in a form that can be absorbed and utilized. *Physics.* the difference between the enthalpy per unit mass of a substance and the product of the entropy per unit mass and the lowest temperature available to the substance for heat discard; used to determine the ratio of work actually performed to work that theoretically should have been performed.
availability bias *Psychology.* the tendency to reach conclusions or make judgments based on the information available in one's memory.

availability ratio *Industrial Engineering.* the proportion of production time that a given process or piece of equipment is actually ready for service, as opposed to time spent undergoing repair or awaiting parts; the proportion of "up time" to total time.

available *Chemistry.* of an electron, able to be utilized in a chemical reaction. *Computer Science.* of an expression, having been computed in the computation path preceding the current location.

available energy *Mechanical Engineering.* the maximum amount of energy that can be converted to mechanical work. *Nutrition.* those ingested nutrients that can be absorbed from the intestine into the blood and converted into energy. In some foodstuffs, nutrients shown to be present by chemical testing may be unavailable, or only partly available to the animal.

available fuel *Forestry.* the part of the total potential fuel component in a specific forest area that will actually burn under various prescribed conditions.

available heat *Mechanical Engineering.* the maximum amount of heat energy that can be obtained in the combustion of a given fuel under ideal conditions.

available line *Electronics.* the portion of a scanning line in a facsimile network that can accommodate picture signals. It can be expressed as a percentage of the scanning line length.

available power *Electronics.* the maximum power that can be delivered from the terminals of a circuit or a source to a load. This power will be delivered if the load impedance is conjugate to the circuit or source impedance at these terminals.

available power gain *Electronics.* the ratio of the available power from the output port of a transducer to the available power from the source connected to the input port of the transducer.

available process time *Industrial Engineering.* the portion of a time period during which a system or agent can perform productively.

available relief *Geology.* a measurement used to determine the amount of headwater erosion and depth to which a valley can be cut in a given area, equal to the vertical distance between the altitude of the original upland surface and the level of the valley floor.

available signal-to-noise ratio *Electronics.* the ratio of the signal power available at a given point in a circuit to the existing random noise power.

available space list *Computer Programming.* a dynamic memory allocation technique that maintains a list of all currently available memory cells; usually used when demands for space and releases are interspersed and random.

available time see UPTIME.

available water *Hydrology.* soil or vadose water that is available for use by plants. Also, **available moisture.**

avalanche *Hydrology.* a large mass of snow or ice, often mixed with rock or soil, sliding or falling rapidly down a slope under the force of gravity. *Electronics.* a term for a condition in which a highly reverse-biased semiconductor junction generates large currents due to high energy electrons that ionize atoms and create new charge carriers with a snowball effect.

avalanche breakdown *Electronics.* **1.** in a semiconductor junction, a nondestructive breakdown caused by multiplicative ion liberation that creates large reverse current flow. **2.** a breakdown in which new charge carriers are created by existing charge carriers in a strong electric field that gains sufficient energy to free valence electrons.

avalanche chute *Hydrology.* the central track or path along which an avalanche has moved.

avalanche conduction *Physiology.* a widespread conduction of nerve impulses when several different axons interact with relatively little input.

avalanche cone *Hydrology.* the mass of material deposited where an avalanche has fallen, consisting of snow, ice, rock, soil, and all other material carried along by the force of the fall.

avalanche diode *Electronics.* a silicon diode that is capable of nondestructive breakdown, in which the junction breakdown voltage is nearly constant for different current levels. Also, BREAKDOWN DIODE.

avalanche effect *Electronics.* see AVALANCHE.

avalanche impedance *Electronics.* in an avalanche diode, the lowered junction impedance during breakdown.

avalanche-induced migration *Electronics.* an interconnection technique for a field-programmable logic array in which certain base-emitter junctions are shorted by the application of an appropriate voltage.

avalanche noise *Electronics.* noise created in an avalanche diode at the beginning of avalanche breakdown.

avalanche oscillator *Electronics.* an oscillator used to convert direct current to microwave frequencies by using an avalanche diode as negative resistance.

avalanche photodiode *Electronics.* a photodiode that uses avalanche breakdown to achieve internal multiplication of photocurrent.

avalanche track see AVALANCHE CHUTE.

avalanche transistor *Electronics.* a transistor that uses avalanche breakdown at high reverse voltage to produce chain generation of charge-carriers.

avalanche voltage *Electronics.* the voltage across a P-N junction diode at which breakdown occurs.

avalanche wind *Meteorology.* a rush of air that precedes a dry-snow avalanche or landslide.

avant-corps *Architecture.* a part of a building that projects out from the main mass, such as a pavilion from the facade.

AVC automatic volume control.

aV/cm abvolt per centimeter.

aven see POTHOLE, def. 2.

Avena *Botany.* a genus of mostly annual grasses of the family Poaceae, known as oats; characterized by flat leaves, loose panicles, and hanging, stoutly awned spikelets.

avena celeoptile test *Botany.* a test of the effect of a growth hormone (e.g., auxin) on plant growth, as judged by its affect on a growing oat of the genus *Avena.*

avenin *Nutrition.* the glutelin protein obtained from oats *(Avena sativa).* Also, PLANT CASEIN, LEGUMIN.

aventurine *Mineralogy.* a greenish to yellowish to brownish-red type of allochromatic quartz that contains glistening scales of hematite or mica.

average *Mathematics.* the sum of a group of items divided by the total number of items.

average acceleration *Mechanics.* the ratio of change in velocity to elapsed time for an accelerating body as it moves from one given point to another point.

average branching factor *Artificial Intelligence.* the average number of possible alternatives from any given state in a search tree.

average calculating operation *Computer Programming.* a technique for estimating a computer's calculation speed by taking the average time for nine additions and one multiplication.

average current see MEAN CURRENT.

average cycle time *Industrial Engineering.* an average that is obtained by dividing the sum of observed work times by the number of observations.

average daily traffic *Transportation Engineering.* the mean number of vehicles per day passing a given point or using a given stretch of road. Also, **average daily traffic volume.**

average delay *Transportation Engineering.* the average time by which vehicles are delayed by a choke point or congestion. For each vehicle this is given by the time in the congested zone minus the time it would take to pass through that zone if it were not congested. The mean of these delay times is the average delay.

average deviation *Mathematics.* **1.** for a discrete random variable with values $\{x_i\}$, mean value μ, and probability function p, the average (or mean) deviation is $\sum |x_i - \mu| p(x_i)$. If the sum converges, sometimes the median (rather than the mean) is used for μ. **2.** more generally, given a random variable X with mean μ and probability density function f, the average or mean deviation of X is the expected value of $|X - \mu|$, that is, $\int_{-\infty}^{\infty} |X - \mu| f(x)\,dx$, if the integral converges.

average discount factor see DISCOUNT FACTOR.

average driver *Transportation Engineering.* a hypothetical driver of average visual acuity, reaction times, and driving skill.

average-edge line *Computer Technology.* in optical character recognition, the imaginary line that traces and regularizes the form of printed or handwritten characters for easier recognition.

average effectiveness level see EFFECTIVENESS LEVEL.

average gradient *Graphic Arts.* a measure of contrast in a photographic image, expressed as the slope of a straight line joining two density points on the sensitometric curve.

average heading *Navigation.* the time-averaged heading maintained by a yawing craft.

average igneous rock *Petrology.* a hypothetical rock whose chemical composition corresponds to the average chemical composition of the outermost 10-mile (16-kilometer) shell of the earth.

average information content *Telecommunications.* the mean entropy per character for all possible messages from a stationary message source.

average limit of ice *Oceanography.* the average distance, in normal winters, that solid ice extends out from the land.

average molecular weight *Materials Science.* the averaged value of the molecular weight of a single polymeric chain in a bulk polymer, reflecting a distribution of chain lengths present in all polymers.

average noise factor *Electronics.* the ratio of the total noise at the output terminal of a circuit to the thermal noise at the input terminal, at a standard noise temperature of 290 K. The ratio is taken over the entire frequency spectrum and is usually expressed in decibels. Also, **average noise figure.**

average outgoing quality limit *Industrial Engineering.* in quality control, the maximum percent of defectives in outgoing material.

average path length *Computer Programming.* in a binary search algorithm, the simple or weighted average of the number of comparisons needed to search the different keys in the list.

average power output *Electronics.* the radio-frequency power, averaged over a modulation cycle, at the output terminals of a transmitter.

average sample number *Industrial Engineering.* in quality control testing, the average number of items sampled per lot.

average speed *Transportation Engineering.* the distance covered in a trip divided by the time elapsed. Average speed may be computed in various ways for different purposes; for example, the time spent at stopovers may be included or excluded.

average strength *Military Science.* the official average strength of a unit as computed from daily morning reports.

average terrestrial pole *Geodesy.* the average position of the instantaneous axis of rotation of the earth, averaged over a specified time period.

average total inspection *Industrial Engineering.* in quality-control testing, the average number of items inspected per lot.

average velocity *Mechanics.* the ratio of displacement to elapsed time for an accelerating body as it moves from one given point to another point. Similarly, **average speed.**

average waiting time *Transportation Engineering.* the arithmetic mean of the waiting times experienced by individual vehicles passing through a queue.

average wind *Navigation.* the mean distribution of wind direction. For geographical locations, it is often represented on charts by a compass-rose diagram in which the frequency with which the wind blows from a given direction is represented by the length of a vector drawn in that direction from the center.

averaging *Robotics.* a method of using feedback from a robot sensor to reduce interference.

averaging device *Engineering.* an instrument that is used to calculate the arithmetic mean of several readings.

Averröes 1126–1198, Arab-Spanish philosopher; revived Aristotelian cosmology.

aversion *Psychology.* the fact of avoiding some painful or unpleasant experience or situation.

aversion therapy *Psychology.* a form of therapy that eliminates undesirable behavior by associating it with something that causes the person to react with aversion, as by pairing the taste and smell of alcohol or cigarettes with nausea.

aversion zone *Mycology.* an area between two fungal or bacterial colonies in which no growth takes place.

aversive *Psychology.* of or relating to aversion; avoiding an experience or situation.

aversive behavior *Behavior.* see AVOIDANCE.

aversive conditioning *Behavior.* a type of counterconditioning in which an undesirable response is eliminated by pairing it with a negative stimulus, such as pain or nausea.

aversive stimulus *Behavior.* a stimulus that causes the individual to withdraw or retreat. Also, PUNISHER.

Aves [āv´ēz´] *Vertebrate Zoology.* the class of vertebrates that includes all birds, characterized by being warm-blooded and egg-laying and having feathers and wings.

avgas aviation gasoline.

Aviadenovirus *Virology.* a genus of viruses in the family Adenoviridae that have been isolated from birds; various species cause diseases such as hemorrhagic enteritis and inclusion hepatitis in birds, and some can induce tumors in rodents.

avian [āv´ē ən] *Vertebrate Zoology.* **1.** of or relating to birds. **2.** occurring in or affecting birds.

avian enteroviruses *Virology.* a group of viruses of the genus Enterovirus, family Picornaviridae, that includes avian encephalomyelitis.

avian (infectious) encephalomyelitis *Veterinary Medicine.* a disease of young domestic fowl that is caused by a picornavirus and marked by ataxia, trembling of the head and neck, leg weakness followed by paralysis, and finally, death. Also, EPIDEMIC TREMOR.

avianize *Virology.* to attenuate the virulence of a virus strain by passing it through chicks or their embryos.

avianized vaccine *Immunology.* a vaccine that contains microorganisms whose virulence has been lessened by adaptation to and passage through chicken embryos.

avian leukosis *Veterinary Medicine.* the general name for a group of widespread viral diseases of domestic fowl that result in leukemia and tumors of internal organs.

avian monocytosis see BLUE COMB.

avian pneumoencephalitis or **avian pseudoplague** see NEWCASTLE DISEASE.

avian tuberculosis *Veterinary Medicine.* an infectious disease of farm birds and captive birds caused by *Mycobacterium avium*; symptoms include gradual weight loss and liver enlargement.

aviation *Aviation.* **1.** the art, science, technology, and process of operating heavier-than-air aircraft. **2.** a corporation or other organization engaged in the development and manufacture of aircraft. **3.** a term for military aircraft.

Aviation

Aviation is the development, manufacture, and operation of heavier-than-air aircraft. Aviation includes airplanes, gliders, and helicopters. Helicopters can fly vertically and hover, but they have low forward speeds compared with that of airplanes. Some compromises have been developed, combining unique vertical and hover capabilities of helicopters with the high speed of airplanes.

People have always dreamed of flying like birds. Gliding flights preceded the Wright brothers' powered flight on December 17, 1903, when their airplane took off and achieved a speed of 31 miles per hour. They had studied the results of previous glider flights by Octave Chanute and Otto Lilienthal, pioneers in gliders. Shortly after the Wrights proved that powered flight was possible, World War I erupted, and airplanes were used as observation platforms, and later as aggressive vehicles using machine guns firing through the propellers without hitting the blades.

When World War II ended, a period of "barn-storming" or stunt demonstration flying, air races, and long distance flights took place, Charles Lindbergh's solo flight from New York to Paris in 1927 being the most famous.

World War II brought military aircraft into their own. These craft were powered by piston engines driving propellers, had retractable landing gears, flaps, and closed cockpits, and were made primarily of steel or aluminum alloy structures.

The development of the jet engine near the end of World War II changed the future of aviation. Most of our military strategy now revolves around this highly maneuverable, high-speed aircraft.

Commercially, jet engines have made air travel between countries economically feasible and physically comfortable.

Supersonic flight was first made possible by jet engines with afterburners and configuration changes, e.g., thin swept wings. The future may find flight possible for more people for longer distances in shorter times.

K. E. Van Every
Design Consultant
Van Every & Associates, San Diego, CA

aviation electronics see AVIONICS.
aviation medicine see AEROSPACE MEDICINE.
aviation mix *Materials Science.* an antiknock mixture used in aviation fuel.
aviation weather forecast *Meteorology.* a forecast of weather elements of particular interest to aviation, such as ceiling, upper winds, icing, turbulence, precipitation, and storms. Also, AIRWAYS FORECAST.
aviation weather observation *Meteorology.* an evaluation given according to a set procedure of those weather elements most important for aircraft operations. Also, AIRWAYS OPERATION.
Avicenna 980–1037, Persian philosopher and physician; his *Canon of Medicine* was the most influential medical text of the Middle Ages.
avicularium *Invertebrate Zoology.* a small prehensile process on many bryozoans that resembles a bird's head; the stalk contains a flexible, muscular mandible for seizing or swallowing prey.
aviculture *Agriculture.* the raising of birds, especially wild birds, in captivity.
avidin *Biochemistry.* a protein isolated from raw egg white that acts as a vitamin antagonist to biotin, binding tightly to it and thereby rendering it unavailable to the body (producing the syndrome known as biotin deficiency).
avidin-biotin technique *Immunology.* a system designed to detect antigens using the strong binding affinity of avidin and biotin being coupled to specific antibodies.
avidity *Immunology.* the strength of the bond between an antibody and an antigen.
aviolite *Petrology.* a variety of hornfels that is composed of mica and cordierite.
avionic *Engineering.* of or relating to avionics.
avionics *Engineering.* a general term for the development and production of electrical and electronic equipment for use in aircraft, spacecraft, and missiles.
aviophobia *Psychology.* an irrational fear of flying.
Avipoxvirus *Virology.* a genus of viruses in the family Poxviridae that infect birds and are usually transmitted by arthropod vectors.
avirulent *Virology.* of a pathogen, lacking in virulence.
avitaminosis see HYPOVITAMINOSIS.
avivement *Surgery.* the surgical debridement of the edges of a wound, especially to aid in healing.
AVMA or **A.V.M.A.** American Veterinary Medical Association.
A-V node *Cardiology.* a small mass of interwoven conducting tissue under the right atrial endocardium. Also, ATRIOVENTRICULAR NODE.
avocado *Botany.* a tropical tree, *Persea americana,* of the Lauraceae family, having large oval to elliptical leaves, small yellowish flowers, and large pear-shaped edible fruits.

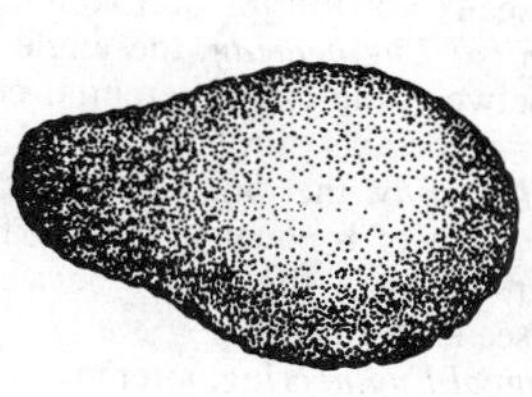

avocado

avocado oil *Materials.* an edible oil high in unsaturated fatty acids, used in cosmetics, hair and skin conditioners, and salad oils.
avocet [av´ə set´] *Vertebrate Zoology.* large shorebirds of the family Recurvirostridae, having long legs, contrasting plumage, and a distinctive long and slender recurved bill that is swept side to side in the mud to uncover crustaceans, seeds, and insects; found worldwide.
avodire *Materials.* a type of decorative, light mahogany wood from a tropical West African tree, *Turraeanthus african.*
avogadrite *Mineralogy.* $(K,Cs)BF_4$, a colorless or white orthorhombic sublimate mineral found in the fumaroles of Mt. Vesuvius, occurring as eight-sided tabular crystals and having a specific gravity of 3.5.
Avogadro, Amadeo [ä´vō gä´drō] 1776–1856, Italian physicist; formulated Avogadro's law; distinguished atoms from molecules.
Avogadro's law *Physics.* a hypothesis stating that equal volumes of different gases under the same conditions of temperature and pressure contain the same number of molecules. Also, **Avogadro's hypothesis.**
Avogadro's number *Physics.* 6.02×10^{23}, the number of atoms or molecules in one mole.
avoidable delay *Industrial Engineering.* a delay in a production process that could be avoided by proper management or work practices, as opposed to delay due to external causes or uncontrollable circumstances.
avoidance *Behavior.* the process of withdrawing or retreating from an object or situation as a consequence of an associated negative experience, such as pain, fright, or nausea. Also, **avoidance behavior.** *Psychology.* a refusal to encounter situations, activities, or objects that would produce anxiety or conflict.
avoidance-avoidance conflict *Psychology.* a conflict between two negative goals or alternatives, in which the avoidance of one requires choosing the other.
avoidance conditioning *Behavior.* conditioned training by which an individual learns to withdraw from or avoid a negative stimulus. Also, **avoidance training, avoidance learning.**
avoidance gradient *Psychology.* a gradual yet steady increase in the drive to avoid a negative goal as one gets closer to the goal; e.g., fear of going to the dentist increases as the actual day of the appointment approaches.
avoidant disorder of childhood *Psychology.* a childhood personality disorder characterized by an extreme aversion to contact with strangers.
avoidant personality disorder *Psychology.* a personality disorder characterized by extreme sensitivity to rejection and criticism, social withdrawal, and low self-esteem.
avoirdupois [av´ər doo pwä´] **1.** of or relating to the avoirdupois system of weights. **2.** an avoirdupois weight: 1 kg is about 2 lb *avoirdupois.*
avoirdupois ounce see OUNCE, def. 1.
avoirdupois pound see POUND, def. 1.
avoirdupois weight or **avoirdupois system** *Metrology.* a system of weights traditionally used in Great Britain, the U.S., and other English-speaking countries, based on a pound of 16 ounces. (From a Middle English phrase meaning "goods sold by weight.")
Avonian see DINANTIAN.
AVP *Aviation.* the airport code for Wilkes-Barre/Scranton, Pennsylvania.
Avrami equation *Materials Science.* an equation describing the kinetics of crystallization, relating the fraction of transformed material to variables of time, temperature, and mechanism.
avulsion *Hydrology.* the cutting off or separation of a portion of land by a flood or by a sudden change in the course of a stream.
avunculate *Anthropology.* a prescribed social relationship between uncle and nephew in which the uncle is responsible for the training and discipline of the nephew.
avunculocal residence *Anthropology.* a marital residence pattern in which the family resides with the husband's maternal uncle or kinsmen.
AWACS [ā´waks´] *Aviation.* **1.** a U.S. Air Force program that uses airborne radar to detect enemy aircraft and provide data for interceptor forces. **2.** an aircraft equipped for use in this system. (An acronym for airborne warning and control system.)
awareness *Psychology.* a conscious knowledge of some sensation, experience, or event.
awaruite *Mineralogy.* Ni_2Fe to Ni_3Fe, an opaque, silver to grayish-white metallic cubic mineral having a hardness of 5 on the Mohs scale; found in gold washings and in serpentinized ultramafic rocks.
awash *Oceanography.* just level with the surface of the water and alternately covered and exposed by waves.
AWG American wire gage.
awl *Mechanical Devices.* a tool with a long sharpened point, used to mark or puncture hard material such as wood or leather.

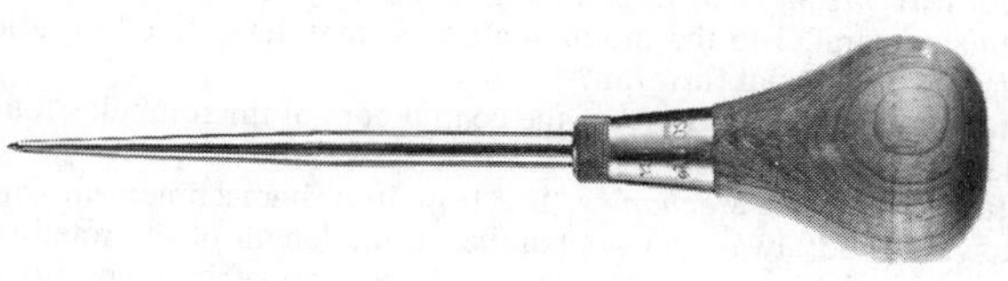

awl

awn *Botany.* a bristlelike structure projecting from the lemma of certain grasses.
awning *Building Engineering.* an often adjustable rooflike covering that is made of canvas, placed over a door, window, or deck to provide protection from the sun, wind, or rain.

awning deck *Naval Architecture.* a light full-size deck running above a ship's main deck, especially on passenger ships; it is used for protection from bad weather and as a walking area for passengers. Also, HURRICANE DECK.

awning window *Building Engineering.* a window containing a vertical series of top-hinged sections that are opened by a crank or other control device, swinging outward from their bottom edges.

axe or **ax** *Mechanical Devices.* a heavy woodworking tool with a wedge-shaped metal hood used in cutting, cleaving, hewing, or chopping materials such as wood or stone.

axes

axed brick *Materials Science.* a term for brick having a rough appearance from being shaped by an ax.

Axelrod, Julius born 1912, American pharmacologist; Nobel Prize for physiology of humoral transmitters in sympathetic nerves.

axenic culture *Biology.* the culture of organisms of a single species in a sterile environment, in the absence of any other living cells or organisms. Also, PURE CULTURE.

axes **1.** the plural of AXE. **2.** the plural of AXIS.

axes of an aircraft *Aviation.* any one of a set of reference axes on an aircraft, usually mutually perpendicular and intersecting at the craft's center of gravity.

axes of inertia *Physics.* a set of axes for a rigid body in which the moment of inertia is maximum about one axis, the moment of inertia is minimum about another axis, and the third axis is perpendicular to both.

axhammer *Mechanical Devices.* a tool having an ax blade on one side of its head and a hammer face on the other.

axial *Science.* of, relating to, or along an axis.

axial compression *Geology.* in experimental work with rock cylinders, a procedure whereby pressure is applied to a rock parallel to its cylinder axis.

axial correlation *Materials Science.* short-range order along the length of a polymeric chain.

axial culmination *Geology.* a fold distortion in which the fold axis curves upward.

axial dipole field *Geophysics.* an ideal dipole field with its axis coincident with the earth's rotational axis, located at the earth's center in a theoretical magnetic field.

axial fan *Mechanical Engineering.* a fan in which the direction of flow remains parallel to the motor shaft axis; may have fived or adjustable blades. Also, **axial flow fan.**

axial filament *Cell Biology.* the central core of microtubules found in a flagellum or cilium.

axial flow *Fluid Mechanics.* **1.** a flow in turbomachinery in which the flowing fluid always moves parallel to the length of the rotating shaft, as in an axial flow compressor or turbine. **2.** in general, any flow parallel to the axis.

axial-flow compressor *Mechanical Engineering.* a fluid turbocompressor that accelerates the flow of the compressed fluid generally in a direction parallel to the rotation axis. Also, **axial compressor.**

axial-flow jet engine *Aviation.* a jet engine containing a compressor through which the flow of air is predominantly along the longitudinal axis of the engine.

axial-flow pump *Mechanical Engineering.* a pump with a propeller-type impeller to induce axial flow; operates with minimum head and maximum capacity. Also, PROPELLER PUMP.

axial force diagram *Civil Engineering.* in statics, a graphical representation of the axial forces acting on a structural member.

axial gradient *Developmental Biology.* a graded difference in some activity along an embryonic axis, which influences the subsequent development.

axial hydraulic thrust *Mechanical Engineering.* the sum of unbalanced axial impeller forces in single and multistage pumps.

axial jet *Fluid Mechanics.* **1.** a jet engine that uses an axial compressor to compress some or all of its products of combustion (usually air). **2.** a flowing turbulent stream that mixes in three dimensions with quiescent fluid.

axial lead *Electricity.* a lead coming from the end or from along the axis of a resistor, capacitor, or other component.

axial length *Crystallography.* the lengths of the vectors **a**, **b**, **c** defining the edges of a unit cell.

axial load *Mechanics.* **1.** a concentrated force that is normal to a sectional plane and that is applied at the centroid of the plane. **2.** a distributed force whose resultant acts at the centroid of a plane to which it is perpendicular.

axial mixing *Biotechnology.* a technique in which the addition of various compounds to a bioreactor is accelerated or slowed down relative to the compounds' average mixing time in fluid flow; used to improve bioreactor efficiency.

axial moment of inertia see MOMENT OF INERTIA.

axial musculature *Anatomy.* the muscles of the head and trunk.

axial observation *Ordnance.* the observation of gunfire from a point that is nearly on the line between the gun and the target.

axial organ *Invertebrate Zoology.* an organ that is thought to help circulate the blood in starfish and other echinoderms.

axial period *Astronomy.* the period in which an object rotates once on its axis.

axial plane *Geology.* the plane or surface determined by the hinge lines of layers in a fold. Also, AXIAL SURFACE. *Crystallography.* a plane that includes two of the axes of a crystal.

axial plane cleavage *Geology.* a cleavage that is more or less parallel to the axial plane of a fold.

axial plane foliation *Geology.* a foliation that is formed parallel to the axial plane of a fold and at right angles to the direction of chief deformational pressure.

axial plane schistosity *Geology.* a distinct foliation type in metamorphic rock that is formed parallel to the axial plane of a fold.

axial plane separation *Geology.* the distance between the axial planes of adjacent folds. Also, AMPLITUDE.

axial quadrupole see LONGITUDINAL QUADRUPOLE.

axial rake *Mechanical Engineering.* the angle of a milling cutter or reamer measured between the circumferential cutting edge and a line parallel to the center of the axis of rotation.

axial ratio *Crystallography.* the ratio of the axial lengths, customarily expressed with the value of **b** equal to unity. These ratios may be deduced from measurements of the angles between faces on a crystal. *Electromagnetism.* see ELLIPTICITY.

axial relief *Mechanical Engineering.* a removed portion behind the end cutting edge of a milling cutter that prevents dragging in the metal being cut.

axial runout *Mechanical Engineering.* the eccentricity of a milling cutter in the direction of the rotational axis.

axial segregation *Materials Science.* solidification-induced concentration of rejected solute in the center of a cast metal slab or billet.

axial skeleton *Anatomy.* the bones of the skull and vertebral column.

axial stream *Hydrology.* **1.** in a mountain valley, the main stream that flows through the deepest part and parallel to the longest dimension of the valley. 2. a stream whose course follows the axis of an anticline or a syncline.

axial surface *Geology.* see AXIAL PLANE.

axial symmetry *Mathematics.* a symmetry with respect to a line. In particular, an aspect of a geometric configuration is said to have axial symmetry if it remains unchanged when rotated about a given line. (The line is called the *axis of symmetry.*)

axial tilt *Astronomy.* the inclination of an object's rotation axis relative to its orbit.

axial trace *Geology.* the intersection of the axial plane of a fold with another surface, usually the earth's, indicating the trend of the fold.

axial trough *Geology.* a fold distortion in which the fold axis curves downward.
axial vector *Mathematics.* a vector that reverses direction under linear transformations that are reflections. Axial vectors exist only in 3-dimensional space, where both differential 2-forms and 1-forms can be mapped by duality to vectors. Axial vectors correspond to differential 2-forms and usually occur as vector cross products. *Physics.* see PSEUDOVECTOR.
axiation *Developmental Biology.* the formation of an embryonic axis, or the determination of polarity of an egg.
Axiidae *Invertebrate Zoology.* a family of decapod crustaceans, including hermit crabs, of the suborder Reptantia.
axil *Biology.* the angle created by a structure and the axis from which it arises; usually used in relation to the angle between the upper side of a leaf or stem and the stem or branch that supports it. *Botany.* **1.** the angle formed by the adaxial surface of a leaf or branch and the shoot or trunk that bears it. **2.** the place at which a leaf or branch attaches to a stem.
axilla *Anatomy.* the concave area between the arm and the wall of the thorax. Also, ARMPIT.
axillary *Anatomy.* of or relating to the axilla or to some structure within the axilla. *Botany.* of, relating to, or growing from an axil, such as an **axillary bud**.
axillary sweat glands *Anatomy.* glands of excretion located in the axilla.
Axinellina *Invertebrate Zoology.* a genus of sponges.
axinite *Mineralogy.* a group name for several triclinic borosilicates with the general formula $A_3Al_2BSi_4O_{15}(OH)$, where A=Ca,Fe^{+2},Mg,Mn^{+2}, occurring as transparent brown, blue, yellow, or green wedge-shaped to platy crystals with a vitreous luster and a hardness of 6 to 6.5 on the Mohs scale; a minor gemstone found in contact zones of granitic and diabasic intrusions.
axinitization *Geology.* the process whereby an axinite replaces another mineral in a rock.
axiolite *Mineralogy.* a type of spherulitic aggregate in which the crystals radiate perpendicularly from a central axis.
axiom *Mathematics.* any of the assumptions or postulates upon which a particular mathematical system or theory is based. All other propositions of the theory can be derived from the axioms. *Artificial Intelligence.* in theorem proving, an initially given fact, rule, or theorem.
axiomatic **1.** relating to or having the nature of an axiom. **2.** like an axiom; self-evident or obvious.
axiomatic S-matrix theory *Particle Physics.* a theory that is based on several fundamental axioms including unitarity, Lorentz invariance, analyticity for energies and momenta near the corresponding physical values, and singularities; the theory is used to construct the S-matrix.
axiom of choice *Mathematics.* the Cartesian product of a nonempty family of nonempty sets is nonempty. Equivalently, if $\{A_i\}$ is a family of sets with i in some index set $I \neq \varnothing$, and $A_i \neq \varnothing$ for each i in I, then there exists at least one choice function for the family $\{A_i\}$. Intuitively, a "choice function" can be described as a simultaneous choice of an element from each of many sets. The axiom of choice, also known as **Zermelo's axiom,** is equivalent to the well-ordering principle, Zorn's lemma, the Hausdorff maximality principle, and Tukey's lemma.
axiom of continuity *Mathematics.* to every point on the real line, there corresponds a real number.
axiom of infinity *Mathematics.* there exists a set that contains $\varnothing$ (the empty set) and the successor of each of its elements; used in the set theoretic construction of the natural numbers.
axiom of specification *Mathematics.* to every set A and to every condition $S(x)$ there corresponds a set B whose elements are exactly those elements x of A for which $S(x)$ holds. Often referred to by its German name, *Aussonderungsaxiom,* this axiom is one of the major principles of set theory.
axion *Particle Physics.* a hypothetical neutral pseudoscalar boson having a mass of approximately 1 MeV; postulated to preserve parity and time-reversal invariance of strong interactions.
axis *Mechanics.* an imaginary straight line through a body, regarded as a coordinate line or as a line about which the body rotates. *Anatomy.* **1.** an imaginary central line extending the length of the body, such as the line that coincides with the vertebral column in a vertebrate. **2.** the second cervical vertebra. *Mathematics.* **1.** any of the reference lines in a coordinate system. **2.** a line that determines axial symmetry for a geometric configuation; a line of symmetry. **3.** a line or hyperplane left invariant by a 1-dimensional group of rotations. *Ordnance.* the imaginary central line through the bore of a gun. Also, **axis of the bore, bore axis.** *Botany.* a central part or support on which parts are arranged, such as a main stem or root. *Geology.* **1.** any line that indicates the general trend of a landform or other large geologic structure. **2.** the main or dominant area of a mountain range. **3.** in a fold, the line representing the intersection of the axial plane with each bed.
axis interchange *Robotics.* the use of memory area that is ordinarily reserved for one axis of a robotic system to store information concerning another axis.
axis inversion *Robotics.* the process of using the mirror image of an axis for left-handed manufacturing operations.
axis of abscissas *Mathematics.* the horizontal or x-axis of a two-dimensional Cartesian coordinate system.
axis of constant moments see AERODYNAMIC CENTER.
axis of freedom *Mechanical Devices.* an axis around which a gimballed gyro is free to rotate for a specified number of degrees.
axis of homology *Cartography.* the intersection of the plane of a photograph with the horizontal plane of the map or the plane of reference of the ground; corresponding lines in the photograph and map planes intersect on the axis of homology. Also, MAP PARALLEL, PERSPECTIVE AXIS, AXIS OF PERSPECTIVE.
axis of movement *Military Science.* the line along which combat troops move either toward the front line or toward the rear. Similarly, **axis of advance, axis of evacuation, axis of supply, axis of (signal) communication.**
axis of ordinates *Mathematics.* the vertical or y-axis of a two-dimensional Cartesian coordinate system.
axis of perspective see AXIS OF HOMOLOGY.
axis of rotation *Mechanics.* the axis of a rotating rigid body, which may change its direction during the body's rotation.
axis of symmetry *Crystallography.* an imaginary line about which a body or two-dimensional figure can be rotated to produce an object indistinguishable from the first. If such a situation occurs for a rotation of $360°/n$, the object is said to have an ***n*-fold axis of symmetry.**
axis of the pelvis *Anatomy.* **1.** a curved line corresponding to the curve of the sacrum and coccyx between the coccyx and the front of the pelvis. **2.** the path that a fetus travels in passing through the pelvis.
axis of thrust see THRUST AXIS.
axis of tilt *Graphic Arts.* a line through the perspective center perpendicular to the principal plane.
axis of trunnions *Ordnance.* see TRUNNIONS.
axisymmetric flow see AXIAL FLOW.
axle *Mechanical Engineering.* a supporting member, beam, or shaft designed to carry a wheel that may be attached to it, driven by it, or freely mounted on it.
axle box *Engineering.* the box-shaped bearing arrangements of a railway through which an axle passes in the hub of a wheel.
axle ratio *Mechanical Engineering.* the ratio of the rotational speed of the drive shaft to the rotational speed of the drive wheels in an automotive vehicle.
axoaxonic *Physiology.* a type of synaptic junction, formed between the axons of two neurons.
axocoel [aks′ə sēl′] *Invertebrate Zoology.* the part of the coelom in starfish and other echinoderms that forms a canal open to the exterior.
axodendritic *Physiology.* referring to a type of synaptic junction formed between the axon of one neuron and a dendrite of another.
axolemma *Neurology.* the delicate plasma membrane that covers the axon of a neuron. Also, **axilemma**.
axolotl [aks′ə lät′əl] *Vertebrate Zoology.* **1.** a dark, speckled salamander of Mexico and the western United States, known for its permanent external gills and other larval-like features. **2.** in the genus *Amystoma*, the full-grown larval stage in which external gills are still present, resembling the adult Mexican axolotl.

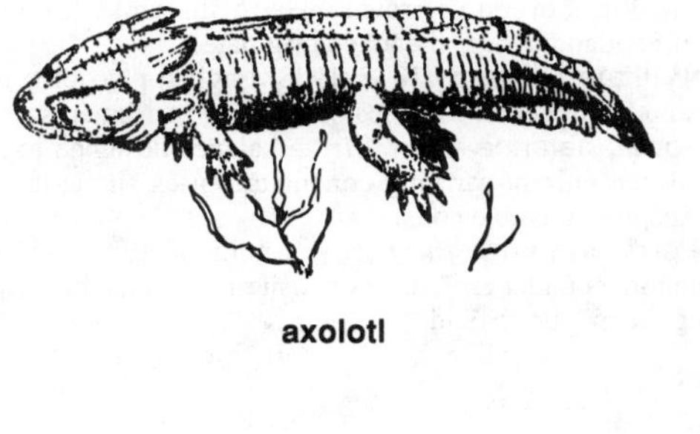

axolotl

axolotl unit *Biology.* a unit used for the standardization of thyroid extracts.

axolysis *Neurology.* the degeneration and dissolution of an axon of a nerve cell due to direct injury or death of parent neurons.

axometer *Engineering.* a device used to mark the position of optical axes, particularly one used to adjust eyeglass lenses with respect to the axes of a person's eyes.

axon *Anatomy.* an extension of a neuron that carries propagated impulses.

axoneme *Cell Biology.* the characteristic bundle of 9 + 2 microtubules found in the central core of a eukaryotic cilium or flagellum.

Axonolaimoidea *Invertebrate Zoology.* a superfamily of nematode worms containing one family, Axonolaimidae; amphids that are diverse in form and exist in marine and brackish waters.

axonometric projection *Graphic Arts.* the representation of a three-dimensional object on a single plane, placing the object at an angle to the plane of projection.

axoplasm *Cell Biology.* the cytoplasm of an axon. Also, HYALOPLASM.

axopodium *Invertebrate Zoology.* a semipermanent, long, narrow pseudopodium consisting of an axial rod enclosed in an ectoplasmic envelope, typical in Radiolaria and Heliozoa.

axosomatic *Neurology.* **1.** of or relating to the synapse of an axon on a nerve cell. **2.** describing the synaptic junction formed between the axon of one neuron and the cell body of another.

axospongium *Neurology.* the fine fibrillar network that constitutes the substance of an axon.

axostyle *Invertebrate Zoology.* a slender axial rod acting as a stiffening skeletal support in many parasitic flagellates.

ayan *Materials.* the bright yellow wood of a tree of West Africa, *Distemonanthus benthamianus,* which is used in interior decoration, flooring, and paneling. Also, SATINWOOD.

aye-aye *Vertebrate Zoology.* a nocturnal, arboreal lemur found only in Madagascar; it constitutes the only species in the primate family Daubentoniidae.

aye-aye

Aylesbury [ālz′ber′ē] *Agriculture.* a large breed of duck that is characterized by white feathers and skin and orange feet; an excellent meat producer that is similar to the Pekin breed. (Named after *Aylesbury,* England.)

Ayre method *Naval Architecture.* the technique of estimating the effective horsepower of a vessel by analysis of scale-model tests.

Ayrshire [âr′sher′] *Agriculture.* a sturdy breed of dairy cattle characterized by markings of red or brown mixed with white. (From the county of *Ayr,* in Scotland, where the breed originated.)

Ayrton, William Edward 1847–1908, English physicist; invented the ammeter and many other electrical measuring instruments.

Ayrton-Jones balance *Electricity.* a balance designed to measure the force between current-carrying conductors; uses single-layer solenoids as the fixed and movable coils.

Ayrton-Perry winding *Electricity.* a noninductive winding in which two inductors conduct current in opposite directions, the opposing flow canceling the magnetic field.

Ayrton shunt *Electricity.* a high-resistance universal shunt for increasing the range of a galvanometer without changing the damping.

Aytoniaceae *Botany.* a cosmopolitan family of typical liverworts of the order Marchantiales; characterized by dichotomous or ventral branching, air chambers in several layers in the thallus, and sporophytes on raised stalked archegoniophores.

AZ *Aviation.* the airline code for Alitalia.

azacrown ether *Organic Chemistry.* a crown ether containing both nitrogen donor atoms and oxygen donor atoms.

9-azafluorene see CARBAZOLE.

azaserine *Pharmacology.* $C_5H_7N_3O_4$, an antibiotic drug derived from *Streptomyces* that is used to treat infections, both bacterial and fungal, and to inhibit proliferation of abnormal cells; interferes with the production of urine.

6-azauridine *Pharmacology.* $C_8H_{11}N_3O_6$, a drug that inhibits the proliferation of malignant cells, especially white blood cells; used in the treatment of acute leukemia.

azelaic acid *Organic Chemistry.* $HOOC(CH_2)_7COOH$, a colorless crystalline solid that melts at 106.5°C and boils at 287°C; soluble in hot water, alcohol, and ether; used in organic synthesis and in making polyamides, lacquers, and alkyd salts. Also, NONANEDIOIC ACID.

azelate *Organic Chemistry.* a salt or ester of azelaic acid.

azel display *Electronics.* a type of plan-position-indicating (PPI) radar monitor that displays azimuth and elevation information.

azalea *Botany.* **1.** any of various flowering shrubs or trees, genus *Rhododendron,* of the heath family, having deciduous leaves and funnel-shaped flowers. **2.** the flower of this plant.

azel mounting see ALTAZIMUTH MOUNTING.

azeotrope [az′ē ō trōp′] *Chemistry.* **1.** a compound that forms part of an azeotropic mixture. **2.** see AZEOTROPIC MIXTURE.

azeotropic [az′ē ō trō′pik] *Chemistry.* relating to or being an azeotropic mixture.

azeotropic copolymerization *Materials Science.* copolymerization when the ratio of monomer units in the polymer and the ratio of monomer units in the precursor feed are equivalent and constant.

azeotropic drying *Chemistry.* a method of removing water from a liquid at temperatures lower than 100°C; a second liquid that forms an azeotropic mixture with water is added to the sample liquid.

azeotropic mixture *Chemistry.* a liquid mixture whose boiling point is constant, so that the vapor produced in distillation or partial evaporation has the same composition as the liquid phase. The boiling point of an azeotropic mixture will be at a minimum or maximum level compared to those of other mixtures of the same substances. Also, AZEOTROPE.

2-azetidinecarboxylic acid *Biochemistry.* $C_4H_7NO_2$, a crystal used to produce certain polypeptides of unusually high molecular weight.

azide *Inorganic Chemistry.* any of a large class of compounds that contain the $-N_3$ group and are synthesized from hydrazoic acid, HN_3. All the heavy metal azides are explosive, as are most light metal azides and organic azides.

azidothymidine see AZT.

azimidobenzene see 1,2,3-BENZOTRIAZOLE.

azimuth [az′ə məth] *Astronomy.* the angular distance of an object along the horizon measured from north toward east. *Cartography.* a horizontal angle measured from a vertical reference plane, usually a meridian. *Navigation.* the true bearing of a celestial body from the observer, generally measured through 360°.

azimuthal chart *Cartography.* a chart on an azimuthal projection. Also, ZENITHAL CHART.

azimuth alignment *Acoustical Engineering.* in electromagnetic tape recording, a 90° alignment of the gap between the recording or reproducing heads and the tape as the tape moves past the heads.

azimuthal map projection *Cartography.* a map projection on which the azimuths or directions of all lines radiating from a central point or pole are the same as the azimuths or directions of the corresponding lines on the sphere. Also, ZENITHAL MAP PROJECTION.

azimuthal quantum number *Physics.* a quantity denoted by l assigned to the angular momentum of an electron in a given orbit; the value of l controls the general spatial distribution of electron probability.

azimuth angle *Navigation.* the horizontal angle between the observer and a celestial body. It is measured either from the elevated pole through 180° or from the nearest north or south point through 90°. It is labeled with the reference point, the number of degrees, and the direction from the reference point, as: N 38° W.

azimuth blanking *Electronics.* a momentary interruption of a radar screen signal as the antenna sweeps an azimuth sector.

azimuth circle *Navigation.* a device oriented 360° concentrically in the horizontal plane on a compass and scribed with degree marks for reading bearings or directions.

azimuth compass *Navigation.* a compass designed to determine directly the amount of magnetic variation for a given location by comparing the magnetic bearing of a celestial body with the calculated bearing.

azimuth deviation *Ordnance.* the difference in azimuth between the line from the gun to the target and the line from the gun to the point at which a shell strikes or bursts.

azimuth dial *Engineering.* a horizontal sundial that casts a shadow marking the sun's azimuth.

azimuth equation *Cartography.* a condition equation that expresses the relationship between the fixed azimuths of two lines that are connected by triangulation or traverse.

azimuth gating *Electronics.* in a plan-position-indicating (PPI) radar display, the brightening of a chosen azimuth sector.

azimuth instrument *Engineering.* any device used for the measurement of azimuths, especially one that envelops the central pivot in the glass cover of a magnetic compass. Also, AZIMUTH BAR.

azimuth line *Engineering.* a line that extends from a central point in a photograph, which corresponds with a similar line in an adjacent photograph in the same flight line.

azimuth mark *Geodesy.* a geodetic monument carrying a mark that has an azimuth known by definition or by measurement from a given point.

azimuth marker *Engineering.* a line on a radar display that passes through a target so that the bearing may be identified.

azimuth of the principal plane *Photogrammetry.* a horizontal angle that is calculated clockwise from the meridian.

azimuth resolution *Electromagnetism.* the minimum angular separation between two objects at equidistant range that is required by a radar system to differentiate the objects.

azimuth scale *Engineering.* a device that is a part of certain instruments or gun carriages that measure angles to identify azimuths.

azimuth tables *Astronomy.* tables used in navigation that list objects' azimuths for a given latitude, hour angle, and declination.

azimuth transfer *Engineering.* the act of joining the nadir points of two vertical photographs of overlapping flights with a straight line.

azimuth traverse *Engineering.* a line drawn across an area in surveying that is established by azimuth and verified by back azimuth.

azine *Organic Chemistry.* a six-membered, benzenelike ring compound, containing at least one nitrogen atom.

azine dye *Organic Chemistry.* any of a diverse group of dyes, such as the nigrosines and safranines, that are derived from phenazine, oxazine, and thiazine.

azino- *Organic Chemistry.* a combining form indicating the presence of the =N–N= group.

aziridine see ETHYLENEIMINE.

azo- *Organic Chemistry.* a combining form referring to the presence of the –N=N– group.

azobenzene *Organic Chemistry.* $C_6H_5N{=}NC_6H_5$, an orange-red crystalline solid occurring as stereoisomers; the *cis* form melts at 71°C, and the *trans* form melts at 68.5°C; both forms are insoluble in water and soluble in alcohol, ether, benzene, and acetic acid; used in making rubber accelerators and dyes.

azobisisobutyronitrile *Organic Chemistry.* $C_8H_{12}N_4$, a crystalline solid that melts at 105°C, insoluble in water and soluble in alcohol; used as an initiator of free radical reactions and as a blowing agent for plastics and elastomers.

azo dye *Organic Chemistry.* any of a diverse group of commercially important dyes manufactured from amino compounds, containing the –N=N– chromophore. They may be acidic, basic, direct, or mordant dyes, according to the chemical composition. They are subdivided into monoazo, diazo, triazo, and tetraazo dyes by the number of –N=N– groups present.

azoic *Ecology.* of any environment, devoid of life.

Azoic *Geology.* the part of Early Precambrian time for which the corresponding rocks show no trace of life.

azoic dye *Organic Chemistry.* a water-insoluble azo dye that is produced by combining the components of the dye on the fabric fiber. Also, INGRAIN COLOR, ICE COLOR.

azoic printing *Graphic Arts.* the method whereby azoic compositions, that is, mixtures of naphthols and diazotized products temporarily inhibited from color development, are printed on cloth; when the printed material is passed through steam containing formic acid vapor, the coupling reaction occurs and development takes place.

azole *Organic Chemistry.* a member of a class of five-membered heterocyclic organic compounds that contain nitrogen atoms and two double bonds.

Azolla *Botany.* a genus of minute water ferns of the family Saviniaceae, characterized by a sporophyte with pinnately branched stems and small distichous bilobed leaves.

Azollaceae *Botany.* a family of minute floating ferns of the order Salviniales that are found in tropical and warm-temperate regions; characterized by rhizomes that creep along the water surface, dangling roots, and sporocarps on the leaf bases underwater. A blue-green alga lives in small pouches in the leaves.

Azomonas *Bacteriology.* a genus of Gram-negative, aerobic bacteria of the family Azotobacteraceae, occurring in soil or water and distinguished within the family by an ability to form cysts.

azomycin *Microbiology.* an antimicrobial agent that is produced synthetically or by *Streptomyces,* and is active against anaerobes.

azonal peat see LOCAL PEAT.

azonal soil *Geology.* any soil that is characterized by the lack of well-developed horizons and a resemblance to the parent material. Also, IMMATURE SOIL.

Azores *Geography.* an island group in the Atlantic, about 900 miles west of mainland Portugal.

Azores high *Meteorology.* a semipermanent subtropical high situated over the North Atlantic Ocean; it is one of the principal centers of action in northern latitudes.

Azospirillum *Bacteriology.* a genus of Gram-negative bacteria, occurring as singly flagellated rods in soil and plant roots, with a respiratory metabolism and the ability to fix nitrogen.

azotemia *Medicine.* an excess in the blood of urea or other nitrogen compounds. Thus, **azotemic.**

Azotobacter *Bacteriology.* a genus of Gram-negative, aerobic bacteria of the family Azotobaceraceae, found chiefly in soil and water.

Azotobacteraceae *Microbiology.* a family of Gram-negative, nitrogen-fixing bacteria, the genera of which are distinguished on the basis of DNA base composition.

azotometer see NITROMETER.

azoturia *Medicine.* an excess in the urine of urea or other nitrogen compounds. Thus, **azoturic.**

azoxybenzene *Organic Chemistry.* $C_6H_5NO{=}NC_6H_5$, a yellow crystalline solid occurring as stereoisomers; the *cis* form melts at 87°C, and the *trans* form melts at 36°C and is insoluble in water and soluble in alcohol, ether, and ligroin; both are used in organic synthesis.

AZT *Virology.* azidothymidine, an analog of thymidine that blocks DNA replication and is thus useful in the treatment of AIDS.

azulene *Organic Chemistry.* $C_{10}H_8$, a blue or grayish-black crystalline ring-structure isomer of naphthalene that melts at 99–100°C and decomposes at 270°C; soluble in alcohol, ether, and acetone. It is derived from certain plants of the genus *Artemisia* and is used in making cosmetics.

azulite *Mineralogy.* a pale-blue, translucent variety of smithsonite occurring most often in large masses.

azure blue see COBALT BLUE.

azurin *Biochemistry.* **1.** a turquoise-blue protein that contains copper, is present in some bacteria such as *Alcaligenes denitrificans* and *Paracoccus denitrificans,* and functions in electron transfer. **2.** a solution sprayed on crops to kill or prevent the growth of fungi.

azurite *Mineralogy.* $Cu_3^{+2}(CO_3)_2(OH)_2$, an azure-blue, transparent to translucent monoclinic mineral having a vitreous luster, a specific gravity of 3.77, and a hardness of 3.5 to 4 on the Mohs scale; found as a secondary mineral in oxidation zones of copper deposits with malachite, and used as a gemstone.

azurmalachite *Mineralogy.* an azurite-malachite mixture characterized by concentric banding and used ornamentally or as a gemstone.

Azusa *Engineering.* a tracking system that measures the space position and velocity of a moving object.

Azygia *Invertebrate Zoology.* a genus of parasitic flatworms in the group Distomata, found in fish intestines.

azygospore or **azygosperm** *Mycology.* a spore that is produced from a reproductive cell called a gamete without being fertilized. Also, PARTHENOSPORE.

azygote *Biology.* any organism produced by haploid parthenogenesis, and therefore having only half the number of chromosomes characteristic of organisms produced by sexual reproduction.

azygous vein *Anatomy.* a vein in the thorax that receives blood from the intercostal veins in the rib cage and delivers it to the superior vena cava.

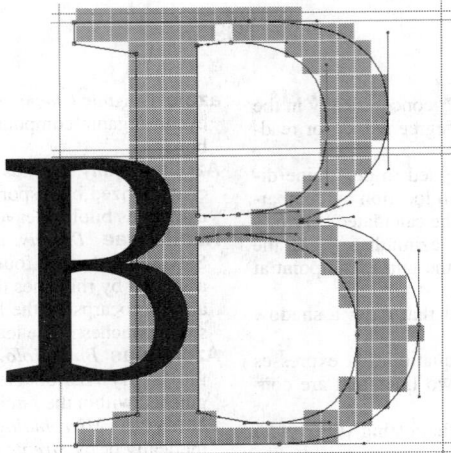

B *Science.* the second in a sequence or group. *Hematology.* the blood group in the ABO system in which the red cells carry the B antigen; individuals of this group have anti-A natural antibodies.

B the chemical symbol for boron.

B or **B.** magnetic field.

b barn.

b or **b.** an abbreviation for: bacillus; back; before; billion; born.

B- *Aviation.* the U.S. military designation to identify a bomber aircraft, as in B-17 or B-52.

B-1 *Aviation.* a long-range supersonic bomber developed by Rockwell for the U.S. Air Force and first flown in 1974.

B-2 *Aviation.* a twin-engine night bomber used in the 1930s and made famous in the Battle of Britain; popularly known as the Condor.

B-17 *Aviation.* a four-engine midwing bomber flown extensively in Europe and the Mediterranean during World War II; popularly known as the Flying Fortress.

B-24 *Aviation.* a four-engine midwing bomber used with a variety of modifications in World War II; popularly known as the Liberator.

B-25 *Aviation.* a twin-engine bomber used extensively in the Pacific theater and the Mediterranean area during World War II; later used for training and liaison; popularly known as the Mitchell (after Colonel Billy Mitchell).

B-26 *Aviation.* a twin-engine bomber used in World War II; popularly known as the Marauder.

B-29 *Aviation.* a four-engine bomber used extensively in the Pacific during World War II, particularly in the bombing of Japan; popularly known as the Superfortress.

B-47 *Aviation.* a bomber having six jet engines, sweptback wings, and a double-wheel bicycle landing gear.

B-52 *Aviation.* a heavy bomber having eight jet engines, sweptback wings, and a double-wheel bicycle landing gear; widely used by the U.S. Air Force.

BA *Aviation.* the airline code for British Airways.

BA or **B.A.** Bachelor of Arts; British Academy.

Ba the chemical symbol for barium.

BAA or **B.A.A.** Bachelor of Applied Arts; British Astronomical Association.

Baade, Walter [bä´ də] 1893–1960, German-born American astronomer; increased the distance scale of the universe.

Baade's star [bä´ dəz] *Astronomy.* the massive star, now a pulsar, whose explosion as a supernova created the Crab Nebula in Taurus about 950 years ago.

Baade's window *Astronomy.* a region of sky in the Milky Way close to the galactic center that is unusually free of obscuration from interstellar gas and dust.

BAAE or **B.A.A.E.** Bachelor of Aeronautical and Astronautical Engineering.

babassu *Botany.* a palm of northeastern Brazil, *Orbignya barbosiana,* with hard-shelled nuts that yield babassu oil.

babassu oil *Materials.* an oil extracted from the nuts of the babassu; used primarily in the manufacture of soaps and cosmetics, and as a cooking oil.

Babbage, Charles 1791–1871, English mathematician; he designed early calculating machines that were forerunners of the modern computer.

Babbitt metal *Metallurgy.* any of several antifriction bearing alloys composed of tin or lead, and containing lesser amounts of copper or antimony. (From the American inventor Isaac *Babbitt.*)

babble *Telecommunications.* a confused mixture of intermingling sounds in a circuit, resulting from crosstalk from a number of mutually interfering channels.

babbler *Vertebrate Zoology.* any of a family of passerine birds with a loud, chattering cry; especially the songbird forming the subfamily Timaliinae of the family Muscicapidae, having short rounded wings, and known for its rapid vocalizations.

Babcock, Horace Welcome born 1912, American astronomer; with his father, **Harold Babcock,** developed a method of measuring the magnetic fields of stars.

Babcock, Stephen 1843–1931, American chemist; developed the Babcock test.

Babcock coefficient of friction *Fluid Mechanics.* an estimation of the coefficient of friction for the flow of steam in a circular pipe having a diameter of *d* inches; mathematically, $0.0027[1 + (3.6/d)]$.

Babcock magnetograph *Astrophysics.* a photoelectric detector used to record small magnetic fields on the sun's photosphere. (Named for its inventor, Harold *Babcock.*)

Babcock test *Food Technology.* a method for measuring the butterfat content of milk by adding acid to a test sample and using centrifugal force to cause the fat to rise to the top. (From the American chemist Stephen *Babcock.*)

Babes-Ernst granules *Microbiology.* intracellular metachromatic bodies found in certain bacteria that, upon treatment with a basic dye, develop a color different from the color of the dye that is used to stain them.

Charles Babbage

Babesia [bə bē´zhə] *Invertebrate Zoology.* a genus of protozoa of the order Piroplasmida that are parasitic on the blood of mammals and are transmitted by ticks.

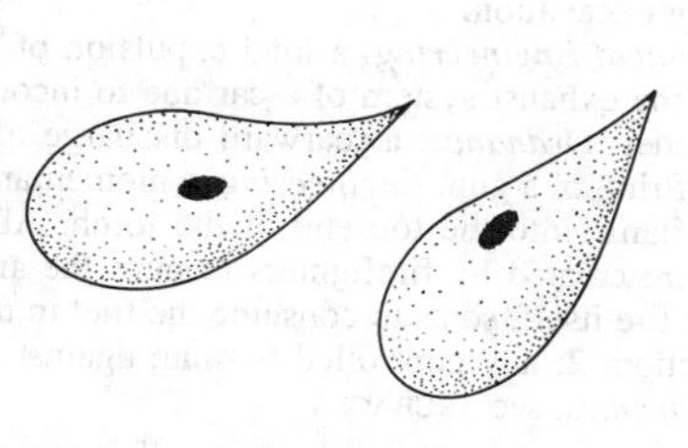

Babesia

babesiasis [bab´i sī´ ə sis] *Veterinary Medicine.* a tickborne parasitic disease of mammals caused by protozoa of the genus *Babesia,* and characterized by high fever, anemia, and reddened urine or hemoglobinuria. Also, BABESIOSIS, PIROPLASMOSIS.

Babesiidae *Invertebrate Zoology.* a family of parasitic, one-celled animals that reproduce within mammal red blood cells; transmitted by tick bite, the ticks serving as intermediate hosts. Also, **Babesidae.**

Babinet, Jacques [bab´i nāt´] 1794–1872, French physicist; invented a polariscope, hygrometer, and goniometer.

Babinet compensator *Optics.* a device made of two quartz prisms fashioned into a rhombus that is used for the analysis of polarized light and to determine the optical retardation and the degree of birefringence of crystal plates.

Babinet point *Optics.* a neutral point that is situated 15° to 20° above the sun.

Babinet's principle *Optics.* a principle stating that two complementary diffraction screens, in which the opaque portions of one correspond to the transparent parts of the other, and vice versa, will produce identical diffraction patterns.

babingtonite *Mineralogy.* $Ca_2(Fe^{+2},Mn)Fe^{+3}Si_5O_{14}(OH)$, a greenish or brownish black triclinic mineral having a specific gravity of 3.36 and a hardness of 5.5 to 6.0 on the Mohs scale, occurring in short prismatic crystals with a perfect cleavage; found in cavities and fracture coatings in trap rock, gneiss, and granite.

babirusa [bab´ə roo´ sə] *Vertebrate Zoology.* an East Indian wild pig, *Babirousa babirussa,* of the family Suidae that is characterized by its hairless, rough, gray hide. The male has four long canine teeth that grow upward through the snout and curve toward its forehead. Also, **babirussa.**

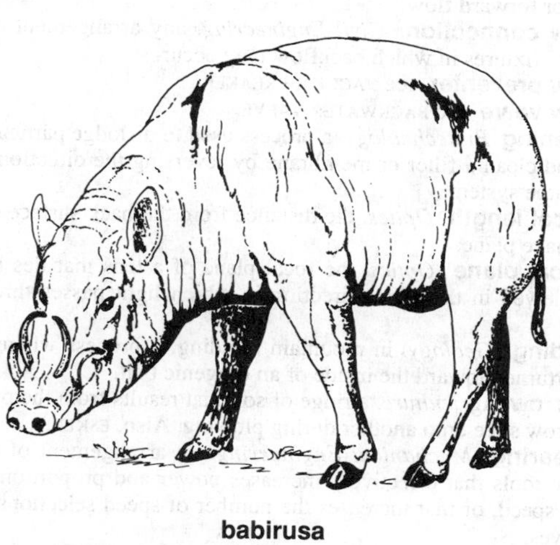

babirusa

baboon *Vertebrate Zoology.* any of several African and Asian terrestrial primates of the genus *Papio* in the family Cercopithecidae, characterized by a doglike muzzle, highly colored callosites on the buttocks, and a short, nonprehensile tail.

Babo's law [bä´ bōz] *Physical Chemistry.* the principle that when a substance is dissolved in a solvent, the vapor pressure of the solvent is lowered by an amount proportional to the concentration of the solution, regardless of temperature.

babs blind approach beacon system.

babul *Botany.* **1.** any of the leguminous trees of the genus *Acacia* that yield gum and tannin; especially the African *A. nilotica.* **2.** the gum or bark of the tree.

baby hamster kidney cells *Virology.* a line of fibroblastic cells that are taken from the kidneys of day-old hamsters and used to grow many viruses in culture. Also, **BHK.**

baby-pig disease *Veterinary Medicine.* an acute hypoglycemia in piglets up to two weeks old, probably caused by the failure of a sow's milk supply or the inability to suckle; symptoms include weakness, loss of appetite, and diarrhea. Also, HYPOGLYCEMIA OF PIGLETS.

baby's-breath *Botany.* a tall plant, *Gypsophila paniculatta,* of the pink family, with stalks bearing clusters of small, fragrant flowers, usually white or pink. Also, **babies'-breath.**

baby spot *Electricity.* a small, usually hooded spotlight used to concentrate light on an area or object a short distance from the spotlight.

baby teeth see DECIDUOUS TEETH.

bac. bachelor. (From Latin *baccalaureus.)*

bacca [bak´ə] *Botany.* a berry.

baccate *Botany.* resembling a berry.

bacci- *Botany.* a combining form meaning "berry."

bacciferous *Botany.* producing berries.

baccivorous *Botany.* feeding on berries.

Bacillaceae *Bacteriology.* a family of mostly Gram-positive bacteria made up of endospore-forming rods and cocci, within the order Eubacteriales, comprising the genera *Bacillus* and *Clostridium.*

Bacillariophyceae *Botany.* a class of mostly unicellular algae known as diatoms; characterized by a silica cell wall of two halves, one over the other, by chlorophyll, xanthopyll, and carotenoid pigments, and by oil and leucosin as storage products.

Bacillariophyta *Botany.* a division of unicellular algae coexistent with the Bacillariophyceae.

bacillary [bas´ ə lär´ē] *Microbiology.* relating to bacilli or rod-shaped microorganisms.

bacillemia *Pathology.* the presence of bacilli in the blood.

bacilliform [bə sil´ ə fôrm´] *Bacteriology.* having the shape or appearance of a bacillus; rod-shaped.

bacillophobia see BACTERIOPHOBIA.

bacilluria *Pathology.* the presence of bacilli in the urine.

Bacillus [bə sil´əs] *Bacteriology.* a genus of bacteria of the family Bacillaceae, including large aerobic or facultatively anaerobic, spore-forming, rod-shaped cells, mostly Gram-positive and motile.

bacillus *Microbiology.* a rod-shaped bacterial cell.

Bacillus Calmette-Guérin vaccine *Immunology.* an attenuated strain of *Mycobacterium bovis* used for immunization against tuberculosis.

bacitracin *Microbiology.* a polypeptide antibiotic produced by *Bacillus subtilis,* active against Gram-positive bacteria by inhibiting cell wall synthesis.

back *Anatomy.* the dorsal part of the body, from the neck to the pelvis. Also, DORSUM.

back acter *Mining Engineering.* the front-end equipment, consisting of a jib with an arm and bucket, that is fitted to an excavator and designed primarily for vertically sided trenching.

backarc basin *Oceanography.* an area between a continent and an offshore subduction zone, especially common in the Western Pacific Ocean.

back arch *Architecture.* an arch that supports the unexposed inner face of a wall, whose exterior facing is supported by a lintel.

back azimuth *Cartography.* an azimuth taken from a previously observed azimuth, back to the point of the original observation; because of the convergence of the meridians the two measurements are rarely complementary, or they differ by exactly 180°.

backband *Building Engineering.* the outside member of a door or window casing formed by a piece of millwork surrounding the trim at the top and sides. Also, **backbend.**

back bay *Oceanography.* a bay separated from the sea by barrier islands, with narrow passes between them through which inland runoff flows into the sea.

backbeach see BACKSHORE.

back bearing *Navigation.* the reciprocal of a bearing, differing from it by 180 degrees, and equivalent to the bearing of the observer as measured from the point observed.

backbending *Nuclear Physics.* a discontinuity in the plot of the nuclear moment of inertia against the rotational frequency squared in some rare-earth elements.

back-bent occlusion see BENT-BACK OCCLUSION.

back bias *Electronics.* **1.** a negative or positive feedback signal. **2.** the illumination of the rear surface of the mosaic in a television tube, which results in greater sensitivity.

backblast *Ordnance.* a rearward discharge of gases from the breech of a recoilless weapon when the weapon is fired.

backblast area *Ordnance.* a cone-shaped area at the rear of a recoilless weapon or rocket launcher where there is danger of injury from the weapon's backblast.

backbone *Anatomy.* the collection of bones along the center of the back; the vertebrae. *Geology.* **1.** a ridge forming the principal axis of a mountain. **2.** the primary mountain ridge, range, or system of an area. *Molecular Biology.* the supporting spine forming the axis of a polymer molecule from which side chains project. *Naval Architecture.* the central fore-and-aft assembly of a ship's keel that reinforces the bottom.

backbone structure *Materials Science.* a term describing the main chain configuration in a polymer.

back boxing *Building Engineering.* see BACK LINING.

backbreak see OVERBREAK.

back bulb *Botany.* a protocorm in certain orchids that may remain on the plant for as long as two years and can be used to propagate the plant.

backburn *Forestry.* see BACKFIRE.

backcast stripping *Mining Engineering.* a stripping method that uses one dragline to strip and cast the overburden, while a second recasts a portion of the overburden.

back check *Mechanical Devices.* a mechanism within a hydraulic door closer that limits the speed at which the door can be opened.

back-coated mirror *Optics.* a glass substrate with a reflective coating on the rear surface. Also, **back-surface mirror.**

backcoating *Graphic Arts.* a sensitized layer applied to the back of a film base material to absorb light.

back contact *Electricity.* a stationary contact on a relay that is normally closed but opens when the relay is energized. Also, BREAK CONTACT.

back course *Navigation.* the reciprocal of a course, differing from it by 180 degrees.

backcross *Genetics.* a breeding technique in which a hybrid of two parental strains is bred with either one of the parental strains.

back cut *Forestry.* a notch that is cut on the backside of a tree, after and higher than the level of the undercut.

backdeep see EPIEUGEOSYNCLINE.

backdigger see BACKHOE.

back diode *Electronics.* a semiconductor diode in which maximum current flows in the reverse direction (when the N-type region of the junction is positive).

backdirt *Archaeology.* the soil removed from an archaeological excavation.

backdoor cold front *Meteorology.* a frontal weather system in which a cold air mass moves in a south and southwesterly direction along the Atlantic seaboard of the United States.

back echo *Electromagnetism.* a signal on a radar screen resulting from the echo of a signal transmitted by a minor lobe of a search radar beam.

backed cloth *Textiles.* any cloth woven with extra warp or weft threads to add weight or reversibility. Also, DOUBLE CLOTH.

backedging *Engineering.* a process of cutting a tile or brick by first chipping away the glazed surface and then the portion beneath this.

backed-up value *Artificial Intelligence.* in a game tree search, the value of a position as determined by starting with approximate evaluations of positions at terminal nodes of the tree (given by a static evaluation function), then calculating the values of higher nodes by assuming that each player will select the most advantageous of the available moves at each point.

back electromotive force see COUNTERELECTROMOTIVE FORCE.

back emission *Electronics.* the emission of electrons in the reverse direction from anode to cathode. Also, REVERSE EMISSION.

back-emission electron radiography *Electronics.* a technique used in microradiography to visualize the presence of materials with different atomic numbers on the specimen surface.

back end see THRUST YOKE.

back-end processor *Computer Technology.* a processor used for specialized functions such as database management, as distinguished from a general purpose I/O front-end processor. Also, **back-end system**.

backfill *Building Engineering.* **1.** to refill an excavation. **2.** the material used to refill an excavation. **3.** a trench around a building, bridge, abutment, or similar structure.

back fillet [fil′ ət] *Building Engineering.* the return of the margin of a door jamb or window jamb that projects beyond a wall.

backfilling *Building Engineering.* the process of moving backfill into an open trench or excavation.

backfire *Mechanical Engineering.* a loud expulsion of gases that have accumulated in the exhaust system of a car due to incomplete combustion in the cylinder. *Ordnance.* a rearward discharge of gases or fragments from the firing of a gun. *Engineering.* a momentary recession of a welder's torch flame into the top end of the torch. Also, FLASHBACK. *Forestry.* **1.** a fire created by firefighters to burn the area between the fire line and the fire itself, so as to consume the fuel in the fire's path or change its direction. **2.** any controlled burning against the wind. Also, BACKBURN. *Electronics.* see ARCBACK.

backfire

backflap hinge *Mechanical Devices.* a flat plate or strap, consisting of two square leaves, screwed to the side of a door or shutter that is too thin to permit the use of a butt hinge. Also, FLAP HINGE.

backflash *Chemistry.* the burning of a gas in a place where combustion is not supposed to occur, such as at the air inlet of a Bunsen burner.

backflooding *Hydrology.* the reversal of water flow at the water table as a result of changes in precipitation.

backflow *Fluid Mechanics.* the direction of a flow that differs from the normal or forward flow.

backflow connection *Civil Engineering.* any arrangement of pipes, drains, or fixtures in which backflow may occur.

backflow preventer see VACUUM BREAKER.

backflow valve see BACKWATER VALVE.

backflushing *Biotechnology.* a process used to dislodge particulate material and clean a filter or membrane by reversing the direction of flow through the system.

back focal length *Optics.* the distance from the back surface of a lens to its image plane.

back focal plane *Optics.* the focal plane of a lens that lies behind it when viewed in the same direction in which light passes through the lens.

backfolding *Geology.* in mountain building, a process whereby folds are overturned toward the inside of an orogenic belt.

back furrow *Agriculture.* a ridge of soil that results from the overlap of one furrow slice onto another during plowing. Also, ESKER.

back gearing *Mechanical Engineering.* an arrangement of gears on machine tools that effectively increases power and proportionately decreases speed, or that increases the number of speed selections on cone belt drives.

background the conditions that form the setting in which something of greater significance is experienced or studied; specific uses include: *Physics.* the sum of all extraneous effects with which a phenomenon under investigation might be confused. *Acoustics.* the sum of all noise from unidentifiable sources. *Telecommunications.* the sum of the effects from which a signal must distinguish itself in order to be identified. *Computer Technology.* the part of a display screen on which characters and images are not displayed. *Geochemistry.* a high concentration of an element or a chemical property of any material that occurs naturally in an area.

background count *Physics.* the level of radioactivity due to sources other than the source being measured.

background density *Radiology.* the opaqueness of a processed radiograph not attributable to the radiation passed through the object.

background discrimination *Engineering.* the capability of a measuring instrument to separate background noise from signal.

background job *Computer Technology.* a program that has lower priority, such that it can be executed when system resources are not required by higher priority jobs. Also, **background program.**

background luminance *Optics.* the brightness of the background behind an object being viewed.

background noise *Acoustics.* the sum of all noises that interfere with the production, reception, detection, or recording of a specific sound, whether the sound of interest is present or not. *Telecommunications.* any undesired signals that enter a frequency band and cannot be separated from desired signals. *Computer Technology.* an energy, such as electrical noise generated by thermal motion of atoms, that is added to a desired signal and may cause errors in the received signal.

background processing *Computer Technology.* **1.** the execution of lower priority programs when higher priority programs or real-time entries are not using system resources. **2.** in a multiaccess sytem, processing that does not use on-line or interactive facilities.

background radiation *Nucleonics.* ionizing radiation that arises naturally from cosmic rays and from radioactive isotopes in the soil and air.

background reflectance *Computer Technology.* in optical character recognition, the reflection produced in the areas surrounding inked characters.

background returns *Engineering.* a term for irrelevant signals that appear on a radar screen, caused by reflection from objects other than the target.

background signal *Engineering.* the output of residual gas from a leak detector, to which the detector device responds.

background temperature *Astrophysics.* the antenna temperature of the diffuse galactic background radiation

back gutter *Building Engineering.* a gutter installed on the upsloping side of a chimney to divert water around the chimney.

backhand welding *Metallurgy.* a welding process in which the flame or electrode hand faces the direction of travel, and consequently postheats the existing weld.

back-haul *Telecommunications.* **1.** the routing of a call that appears to take an illogical path through a network. **2.** a link between a central office and a main microwave or multiplex station.

back hearth *Building Engineering.* that part of the hearth (the floor) that lies under the grate. Also, INNER HEARTH.

back heating *Electronics.* the excess heating of a cathode caused by bombardment by high-energy electrons returning to the cathode.

backhoe or **back hoe** *Mechanical Engineering.* a powered excavating machine that cuts trenches by a boom-mounted bucket drawn through the ground toward the machine. Also, DRAGSHOVEL, HOE SHOVEL.

backing something forming an aid or support; specific uses include: *Materials.* the base of sand paper or other coated abrasive products. *Electronics.* a flexible material, such as cellulose acetate or polyester, found on magnetic tape and used as the carrier for the oxide coating. *Mining Engineering.* the timbers across the top of a level, supported in notches cut in the rock. *Graphic Arts.* the process of rolling and shaping the spine of a bound book. *Meteorology.* a change in wind direction to counterclockwise in the Northern Hemisphere and to clockwise in the Southern Hemisphere.

backing board *Building Engineering.* a flat gypsum board used in a suspended acoustical ceiling to which soundproofing tiles with an adhesive backing are attached.

backing cloth *Graphic Arts.* a fiber material used on the back side of photographic paper to add strength.

backing-off *Engineering.* a process in which metal is removed from the edge of a worn cutting tool to reduce friction.

backing pump *Engineering.* a mechanical vacuum or roughing pump consisting of two pumps in tandem; used to establish a low pressure into which a vapor diffusion pump can operate. Also, FORE PUMP.

backings *Building Engineering.* the battens affixed to rough walls to which wood or other linings are attached.

backing storage see AUXILIARY STORAGE.

backing up *Graphic Arts.* the process of printing on the reverse side of a printed sheet. *Building Engineering.* the use of inferior bricking for the inner face of a wall.

back jamb see BACKLINING.

backjoint *Civil Engineering.* a rabbet in masonry to receive a slip, such as a wood nailer over a fireplace.

backlands or **back lands** *Geology.* the lowlands along the sides of a river in a floodplain, lying behind the natural levee.

backlash *Electronics.* **1.** a reverse current flowing in a grid circuit, caused by the loss of electrons by the grid. **2.** the persistence of oscillation in a resonant circuit, when the stimulus is reduced below the value that is necessary to commence oscillation. *Engineering.* **1.** unwanted play or looseness between two parts of a mechanism due to imperfect connection. **2.** the lost motion that occurs when one moving part has to travel a certain distance before engaging and activating another part. **3.** a discrepancy.in the dial readout of a designated quantity, according to whether the dial is moved clockwise or counterclockwise to obtain the reading.

backlight *Graphic Arts.* a light that illuminates the subject of a photograph from behind, separating the subject from the background.

backlimb *Geology.* in an asymmetrical anticline, the limb that is less steep.

backlining or **back lining** *Building Engineering.* **1.** a thin strip next to the wall and opposite the pulley on a window casing. Also, BACK JAMB. **2.** the portion of a frame that forms the back of a recess for boxing shutters. *Graphic Arts.* a strip of very durable paper **(backlining paper)** that is glued to the spine of a hardcover book in order to bind the signatures and strengthen the spine.

back lintel *Building Engineering.* a lintel used to support the backing portion of a masonry wall.

back lobe *Electromagnetism.* the portion of an electromagnetic radiation pattern of a directional antenna transmitted in a direction opposite to the intended direction.

back marsh *Ecology.* marshland formed in poorly drained areas of an alluvial plain.

back matter *Graphic Arts.* the material following the main text of a book, including the index, appendix, end notes, and glossary. Also, END MATTER.

back-mixing *Chemical Engineering.* the mixing of already reacted chemicals with the unreacted feed to a chemical reactor.

back mutation *Genetics.* a mutation that reverses a previous mutation, usually causing a mutant gene to regain its wild-type function.

back nailing *Building Engineering.* a technique used to fortify a roof against slippage by nailing plies to the substrate.

back nut *Mechanical Devices.* **1.** a nut with one dish-shaped side that fits a grommet in watertight fittings. **2.** a locking nut on the shank of a pipe or valve.

backout *Space Technology.* a cancellation or reversal of steps already taken, usually in reverse order, after a countdown has been halted. *Computer Technology.* to remove a previously made change in a computer program. *Metallurgy.* to nullify the effect of positive electrical potentials that occur in the anodic area of a cathodic protection system.

backpatching *Computer Science.* the process of filling in the address of a label, which has just become defined, in preceding parts of the program that made forward references to it.

backplane *Electronics.* a wiring board used in microcomputers and minicomputers to provide the required connections between logic, memory, and input-output modules; it is usually constructed as a printed circuit. *Computer Technology.* see MOTHERBOARD.

backplastering *Building Engineering.* **1.** the use of lath and plaster partitions between the inner and outer surfaces of a stud wall to add to its insulating properties. **2.** a coat of plaster applied to exterior brickwork in order to keep moisture from seeping from within the building.

backplate *Building Engineering.* a plate made out of metal or wood that serves as a backing for a structural member. *Electronics.* the electrode in a television camera tube that receives the stored charge image via capacitance coupling.

back porch *Electronics.* the portion of a composite television picture signal located between the trailing edge of a horizontal-sync pulse and the trailing edge of the corresponding blanking pulse.

back-porch effect *Electronics.* the extension of the collector current in a transistor for a short time after the input signal has decreased to zero.

back pressure *Engineering.* in a drainage pipe, any air pressure exceeding atmospheric pressure. *Mechanical Engineering.* **1.** the resistance encountered by the drill stem in a rock drill that causes the bit to be fed faster than it can cut. **2.** reverse pressure opposing forward flow.

back prop *Building Engineering.* a raking strut that is inserted under every second or third frame in order to transfer the weight from the timbering of a deep trench to the ground.

back propagation *Artificial Intelligence.* the adjustment of the weights of connections in a neural network, using reward and punishment based on training data for which the desired outcome is known, so that the output of the network will eventually approximate the desired output.

backquote *Artificial Intelligence.* in Lisp, a character macro, specified by the backward-quote or accent-grave symbol, that allows a symbolic structure of specified form to be created, with values substituted in certain positions.

back rake *Mechanical Devices.* the backsloping angle of the blade of a cutting tool, such as a plane or lathe, from its top surface to a plane parallel to its base.

back range *Navigation.* a range measured between the stern of a seacraft to an object to guide the craft as it is moving away.

back reef *Geology.* the area between a reef and the adjacent shore.

back-reflection photography *Crystallography.* an X-ray diffraction technique in which photographic film is placed between the X-ray source and a crystal specimen; used in the analysis of large specimens.

back resistance *Electronics.* the contact resistance opposing the inverse current of a contact rectifier.

back-run process *Chemical Engineering.* a method for manufacturing water gas in which the material is passed first through a superheater, then up through a carburetor, down through a generator, and finally purified in a scrubber.

back rush *Oceanography.* the return of water seaward after the waves rush onto the beach.

backsaw *Mechanical Devices.* a short crosscut saw with a stiff metal rib backing to the blade; used in precision work.

backsawing *Forestry.* the act of sawing wood at right angles to the medullary rays. Also, CROWN-CUT, SLASH-SAWING.

backscatter *Physics.* radioactive waves or particles deflected by scattering processes at angles greater than 90° to the original direction of the beam of radiation.

backscattering *Physics.* the scattering of radiation, particles, or light waves at an angle greater than 90°. *Electromagnetism.* the undesired radiation of energy to the rear by a directional antenna. *Telecommunications.* any radio waves or light waves produced by propagating in a direction approximately reverse to that of the incident wave.

backset *Building Engineering.* the horizontal distance between the face on the centerline of a lock or latch to the center of the keyhole, lock cylinder, or lock mechanism through which a bolt passes.

backset bed *Geology.* a crossbed that inclines against the direction of the depositing wind or water current.

backshore *Geology.* the usually dry inner or upper section of a shore, lying between the coastline and the high spring tide mark. Also, BACKBEACH.

backshore terrace see BERM.

backsight *Engineering.* **1.** a survey point or line that has been previously determined. **2.** a survey reading taken on a previously determined point from a new position.

backsighting *Ordnance.* a method of orienting and synchronizing two pieces of equipment by sighting them directly at each other.

backsight method *Engineering.* **1.** a plane-table traversing process in which the table's line of sight is positioned so that it corresponds with the ground line. **2.** the placing of two pieces of equipment adjacent to each other so that they correspond in azimuth and elevation.

back siphonage *Civil Engineering.* the flowing of used, contaminated, or polluted water back into the potable water system, caused by reduced pressure in the pipe or vessel.

backslant *Graphic Arts.* any slanted type that is tilted to the left, rather than to the right.

backslope or **back slope** *Geology.* **1.** the dip slope when the underlying rock diverges from the angle of the land surface. **2.** the less steep side of a cuesta or of a fault block.

back solution *Robotics.* the use of calculations to convert tool-coordinated positions into specified robotic joint positions.

backspace *Computer Technology.* **1.** to move a printing or display element backward one unit at a time. **2.** to move one record backward in a file. *Graphic Arts.* to shift a typewriter carriage or typing element one space backward by depressing a particular key for this purpose.

backstay *Engineering.* a support that prevents a standing object from moving or falling forward.

back stoping see OVERHEAD STOPING.

backstress *Materials Science.* a stress opposing the applied stress used in rolling to reduce the tensile stresses; on a slip plane, this can be caused by dislocation pileup.

back stroke *Cardiology.* the recoiling action of the ventricle that follows the pumping of blood into the aorta.

back-surface mirror see BACK-COATED MIRROR.

backsweep see SWEEPBACK.

back-to-back connection *Electronics.* the parallel connection of tubes, anode to cathode, or transistors in parallel in opposite directions, to control AC current without rectifying. Also, **back-to-back circuit.**

back-to-front ratio *Electromagnetism.* a ratio used to compare the signal strength of an antenna, metal rectifier, or other device, in one direction with that in the opposite direction.

backtracking *Artificial Intelligence.* a procedure that enables an inference engine to retreat from a dead-end solution, forget conflicting information, and examine another path. Also, TRUTH MAINTENANCE.

back up *Computer Technology.* to duplicate hardware, software, or data for retrieval in the event of system failure or user error.

backup *Computer Technology.* any hardware, software, or data that can be retrieved in the event of system failure or user error. *Petroleum Engineering.* the holding of one section of a pipe while another section is screwed into or out of it. *Building Engineering.* the part of a masonry wall behind the exterior facing. *Civil Engineering.* an overflow or an accumulation, usually in a drain, due to water flow stoppage.

backup file *Computer Programming.* a complete or partial copy of a file created for later retrieval in the event of the loss or damage to files.

backup strip *Building Engineering.* a wood slat affixed at the corner of a wall or partition to which the ends of a lath can be nailed.

Backus-Naur form *Computer Programming.* a metalanguage used to specify the syntax of a programming language. Also, **Backus normal form.**

backward chaining see TOP-DOWN REASONING.

backward conditioning *Behavior.* in classical conditioning, the presentation of an unconditioned stimulus before the conditioned stimulus, so that a conditioned response comes after, rather than before, the unconditioned response.

backward diode *Electronics.* a semiconductor diode having no forward tunnel current; used as a low-voltage rectifier.

backward folding see BACK FOLDING.

backward reasoning see TOP-DOWN REASONING.

backward wave *Electronics.* a wave in a traveling-wave tube with group velocity in the opposite direction to the electron beam.

backward-wave oscillator *Electronics.* an oscillator containing a special vacuum tube in which electrons are bunched by an RF magnetic field as they flow from cathode to anode. The bunching action produces a backward wave that grows larger as it moves toward the end of the tube with the electron gun.

backward-wave tube *Electronics.* a type of traveling wave tube in which energy on a slow-wave circuit flows in the opposite direction to the electrons; used as stable, low-noise amplifiers or as oscillators.

backwash *Oceanography.* **1.** water thrown back by an obstruction, such as a ship, breakwater, or cliff. **2.** back rush of waves.

backwater *Hydrology.* **1.** a body of water that is backed up or turned back from its course by an obstruction, an opposing current, or a tide movement. **2.** any tranquil body of water lying parallel to a river or joined to a main stream and fed by backflow. **3.** a creek or series of lagoons parallel to the coast and separated from the sea, but fed through barred outlets.

backwater valve *Building Engineering.* a check valve in a drainage pipe for which a reversal of flow causes the valve to close, thus cutting off flow.

Bacon, Francis 1561–1626, English philosopher, statesman, and essayist; early advocate of empiricism and scientific methods.

Bacon, Roger c.1214–1294, English philosopher and man of science; a pioneer in experimental science and the study of optics.

BACON *Oncology.* an acronym for a combination of the drugs bleomycin, Adriamycin, CCNU (lomustine), Oncovin, and nitrogen mustard; used in cancer chemotherapy treatment.

BACOP *Oncology.* an acroynm for a combination of the drugs bleomycin, Adriamycin, cyclophosphamide, Oncovin, and prednisone, used in a malignant lymphoma chemotherapy treatment.

bact. bacterial; bacterium; bacteriology. Also, **Bact.**

BACTEC *Microbiology.* a commercial assay for the detection of microorganisms in a given sample of body fluid, such as blood, from which microorganisms are normally absent.

bacter- a combining form meaning "bacteria," as in *bacteroid.*

bacteremia *Pathology.* the presence of bacteria in the blood.

bacteri- a combining form meaning "bacteria," as in *bacteritic.*

bacteria *singular,* **bacterium** *Bacteriology.* extremely small, unicellular microorganisms that multiply by cell division and whose cell is typically contained within a cell wall, occurring in spherical, rodlike, spiral, or curving shapes and found in virtually all environments; some types are important agents in the cycles of nitrogen, carbon, and other matter, while others cause diseases in humans and animals.

bacterial [bak´tēr´ ē əl] *Bacteriology.* relating to or caused by bacteria.

bacterial blight *Plant Pathology.* any bacteria-caused plant disease that results in withering of the plant or death of plant parts without rotting, such as common bacterial blight, fuscous blight, and halo blight.

bacterial brown spot *Plant Pathology.* a plant disease caused by the bacteria *Pseudomonas syringae,* characterized by water-soaked, rust-colored spots or cankers on the plant foliage or woody stems. Also, **bacterial canker.**

bacterial conjugation *Microbiology.* the process of sexual reproduction in bacteria in which genetic material is transferred from a donor bacterial cell to a recipient cell during direct cell-to-cell contact.

bacterial endocarditis *Pathology.* a bacterial infection of the inner lining of the heart; characterized by fever, heart murmur, and enlarged spleen.

bacterial endospore *Microbiology.* a dormant body formed within certain Gram-positive bacterial cells as a consequence of nutrient limitation; highly resistant to adverse environmental conditions and capable of developing into a vegetative cell.

bacterial lawn *Microbiology.* a confluent growth of one bacterial strain that covers the entire surface of a petri plate.

bacterial leaf spot *Plant Pathology.* a plant disease caused by bacteria and characterized by well-defined, discolored spots on the plant foliage; includes angular leaf spot and leaf blotch.

bacterial luminescence *Microbiology.* the emission of light by certain bacteria in the presence of oxygen, catalyzed by the enzyme luciferase, and often seen in marine species.

bacterial metabolism *Microbiology.* the sum of all processes in a bacterial cell that transform energy into a biologically useful form, including catabolism, or degradative reactions, and anabolism, or biosynthetic reactions.

bacterial motility *Microbiology.* the movement or locomotion of a bacterial cell, including flagellar-mediated swimming, a gliding movement along a solid substrate, or movement mediated by cilia.

bacterial photosynthesis *Microbiology.* photosynthesis by the cells of green and purple bacteria; distinct from the plant process in having only one type of photosynthetic reaction center.

bacterial pigmentation *Microbiology.* a coloration apparent in bacterial colonies due to the deposition of organic coloring matter, or pigment, within the cells.

bacterial pustule *Plant Pathology.* a plant blight characterized by blisters on the plant foliage and caused by the bacterium *Xanthomonas phaseoli.*

bacterial slant *Microbiology.* a bacterial culture inoculated on a nutrient medium that has been allowed to solidify in a culture tube in a diagonal, sloping orientation.

bacterial soft rot *Plant Pathology.* a bacterial disease of plants that causes the breakdown and decay of tissues.

bacterial speck *Plant Pathology.* a bacterial plant disease characterized by the appearance of small lesions on plant stems and leaves.

bacterial spot *Plant Pathology.* any of various bacterial plant diseases characterized by small, scabby lesions on leaves, stems, or fruits.

bacterial transposition *Microbiology.* a specific sequence of DNA capable of translocating from one position on a bacterial genome to another, which carries genes encoding its own transposition as well as genes encoding other functions, such as antibiotic resistance.

bacterial vaccine *Immunology.* a preparation containing antigens produced by killed or living bacteria, used to induce immunity to a specific bacterial disease or to enhance immunity in an infected individual.

bacterial virus see BACTERIOPHAGE.

bacterial wilt disease *Plant Pathology.* a plant disease common to cucumber and muskmelon, characterized by wilting and shriveling of leaves and stems; caused by the bacteria *Erwinia tracheiphila.*

bactericide [bak tēr´ə sīd] *Biology.* an agent that kills bacteria.

bactericidin *Immunology.* a substance containing both antibody and certain nonantibody components in the serum, that kills bacteria.

bacteriform *Microbiology.* resembling a bacterium in form.

bacterin *Immunology.* a vaccine consisting of suspended bacteria cells that have been killed or weakened by heat or chemical means.

bacterio- a combining form meaning "bacteria," as in *bacterioscopy.*

bacteriochlorophyll *Biochemistry.* any of a group of pigments occurring in bacteria and functioning in anaerobic photosynthesis.

bacteriocin *Microbiology.* a protein substance released by certain bacteria that kills but does not lyse closely related strains of bacteria.

bacteriocinogen *Genetics.* a bacterial plasmid that controls the synthesis of bacteriocin.

bacteriocyte *Entomology.* a modified fat cell in certain insects, containing groups of bacterium-shaped rods thought to be symbiotic bacteria.

bacteriogenic *Microbiology.* caused or produced by bacteria.

bacterioid *Microbiology.* resembling bacteria. Also, **bacterioidal.**

bacteriological warfare *Military Science.* military operations using harmful bacteria as a weapon to infect enemy personnel.

bacteriologist *Microbiology.* a person who specializes in the study of bacteria; an expert in bacteriology.

bacteriology *Microbiology.* the scientific study of bacteria.

Bacteriology

Bacteriology is the study of bacteria, the smallest living creatures. In the living world, two basic cell types have been recognized, which are called procaryote and eucaryote. Bacteria are the only procaryotes. There are several groups of eucaryotes, including the fungi, algae, and protozoa. In addition, all higher life forms (plants and animals) are eucaryotes. The term *bacteriology* is often used as a synonym for *microbiology,* which is the study of all microorganisms, including not only the bacteria, but also the fungi, algae, and protozoa.

Bacteria are thought to represent the first organisms to have arisen on earth, from which all other organisms are descended. Studies using the techniques of molecular biology have shown that bacteria can be divided into two major lines of descent, called eubacteria and archaebacteria, which differ profoundly in many of their fundamental biochemical properties. The archaebacteria in particular are distinguished by their ability to grow in extreme environments, such as in the presence of high salt concentrations or at high temperatures.

Bacteria are extremely important in the overall economy of the earth and have profound effects on human activities. Many of the most important diseases of humans are caused by bacteria, including tuberculosis, scarlet fever, syphilis, gonorrhea, diphtheria, and plague. Throughout human history, some of these diseases have profoundly affected the manner in which society has developed. However, many bacteria are beneficial, carrying out important natural processes, such as nitrogen fixation and methane formation. Bacteria often find wide practical use in agriculture, industry, and medicine.

Many of the medically useful antibiotics are produced by bacteria, and the biotechnology industry makes use of bacteria as their principal agents in the production of new pharmaceuticals by genetic engineering. Bacteria are the most important living organisms involved in the biodegradation of toxic wastes and the sewage treatment process is linked to their activities. Petroleum formation and recovery depend importantly on the activities of bacteria, which are also principally responsible for ecosytem recovery from oil spills.

Thomas D. Brock
E. B. Fred Professor of Natural Sciences, Emeritus
University of Wisconsin at Madison

bacteriolysin *Microbiology.* any substance capable of joining with and subsequently killing bacteria.

bacteriolysis [bak´tēr ē äl´ə sis] *Microbiology.* a disruption of the structural integrity of a bacterial cell, causing release of the cell contents.

Bacterionema *Bacteriology.* a genus of bacteria of the family Actinomycetaceae occurring as facultatively anaerobic, Gram-positive, pleomorphic organisms, comprising nonseptate and septate filaments and bacilli.

bacteriophaeophytin *Biochemistry.* a derivative of bacteriochlorophyll that is found in certain photosynthetic bacteria, including several purple and green bacteria.

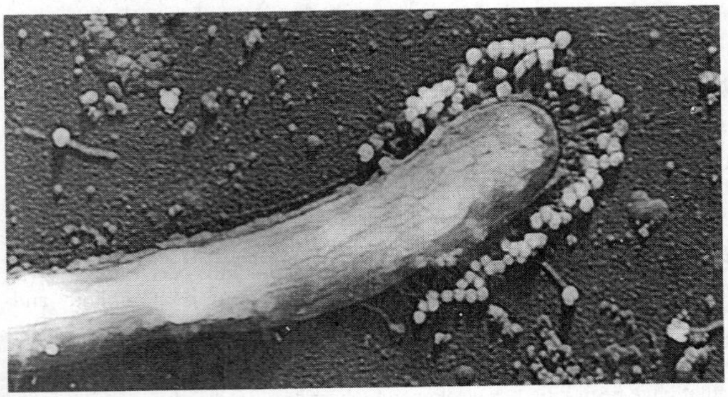

bacteriophage

bacteriophage [bak´tēr´ē ə faj´] *Virology.* any virus that infects and replicates within a bacterium, producing transmissible lysis or becoming lysogenic; usually species-specific. Also, BACTERIAL VIRUS.

bacteriophage lambda *Molecular Biology.* a virus that infects and replicates within the host cells of a bacteria; it does not always elicit the lytic response but may enter into a stable relationship with the host bacteria. Also, TEMPERATE PHAGE.

bacteriophagous [bak´tēr ē ə fāj´əs] *Microbiology.* of or relating to to any virus that consumes bacteria.

bacteriophobia [bak´tēr ē ə fō´bē ə] *Psychology.* an irrational fear of becoming infected by bacteria or other microorganisms. Also, BACILLOPHOBIA.

bacteriophytoma *Pathology.* a reactive, tumorlike lesion caused by bacteria.

bacteriorhodopsin *Microbiology.* a pigmented protein involved in proton transport across the cell membranes of certain photosynthetic bacteria, such as *Halobacterium,* allowing illuminated cells to synthesize ATP energy.

bacterioruberin *Microbiology.* a carotenoid pigment that functions in gathering light energy in certain photosynthetic Halobacteriaceae.

bacteriosis *Plant Pathology.* any type of plant disease that is caused by bacteria.

bacteriostasis *Microbiology.* the inhibition, but not extermination, of bacterial growth and reproduction.

bacteriostat *Biotechnology.* a chemical agent that stops or inhibits the multiplication of bacteria. Also, **bacteriostatic agent.**

bacteriostatic *Microbiology.* referring to any agent that reversibly inhibits the growth and proliferation of bacteria.

bacteriotoxin [bak´tēr ē ō täks´ən] *Microbiology.* any toxin that is produced by or toxic to bacteria.

bacteriotropin [bak´tēr ē ō trōp´ən] *Immunology.* a substance (usually an antibody) that combines with bacteria, making the bacteria more susceptible to phagocytosis.

bacterium [bak´tēr´´ē əm] *Microbiology.* the singular of bacteria. (From a Greek word meaning "little rod.") See BACTERIA.

Bacterium *Microbiology.* in former systems of classification, a genus of bacteria made up of non-spore-forming, rod-shaped bacteria.

bacteriuria *Pathology.* the presence of bacteria in the urine. Also, **bacteruria.**

bacterization *Botany.* any process in which bacterial action alters the composition of a plant.

bacteroid *Microbiology.* **1.** a cell that resembles a bacterium. **2.** a modified bacterial cell, such as the bacterial endosymbionts found in nodules of leguminous plants.

Bacteroidaceae *Bacteriology.* a family of Gram-negative, mainly anaerobic bacteria comprising the genera *Bacteroides, Fusobacterium, Sphaerophorus,* and *Streptobacillus.*

Bacteroides *Bacteriology.* a genus of Gram-negative bacteria of the family Bacteroidaceae, which grow as obligately anaerobic, rod-shaped cells or filaments, possess a fermentative metabolism, and are found in the human oral cavity and intestinal tract.

bactofugation [bak´tō fū gā´shən] *Food Technology.* a milk pasteurization method that uses a centrifuge to remove most spore-forming bacteria; it also extends the life of the milk and can be applied to remove yeast and mold cells.

bactoprenol *Microbiology.* a bacterial membrane-bound, isoprenoid compound that serves to facilitate the transfer of a number of molecules across the inner membrane, including lipopolysaccharide and peptidoglycan.

Bactrian camel *Vertebrate Zoology.* the two-humped camel, *Camelus bactrianus,* used extensively for transportation and food in desert regions of central Asia.

Bactrites *Paleontology.* a genus of early ammonoids in the family Bactridae of the Silurian and Devonian, having a shell that curves only slightly.

baculite *Mineralogy.* a minute crystal that looks like a dark rod.

Baculoviridae *Virology.* a family of double-stranded DNA-containing insect viruses comprising the genus *Baculovirus.*

Baculovirus *Virology.* **1.** the single genus of the family Baculoviridae; isolated from arthropods, and often used for the species-specific control of insects. **2. baculovirus.** a member of this family, especially a granulosis virus.

baculum *Vertebrate Zoology.* a heteroptic bone forming the skeleton of the penis in all insectivores, bats, rodents, carnivores, and pinnipeds, and in primates except humans. Also, OS PRIAPI.

bad break *Graphic Arts.* a term for a line break in typesetting that fails to follow normal syllable division or that falls too near the beginning or the end of a word, such as the break at the end of this line: this is an e-xample of a bad break.

baddeleyite *Mineralogy.* ZrO_2, a colorless, yellow, brown, or black monoclinic zirconium oxide having a specific gravity of 5.74 to 5.82 and a hardness of 6.5 on the Mohs scale; found in Brazil, Ceylon, Sweden, and the United States.

badge meter see FILM BADGE.

badger *Vertebrate Zoology.* a small burrowing animal of the family Mustelidae in the order Carnivora, characterized by black lines running from the eyes over the ears. *Materials.* the stiff fur of this animal, used especially for brushes. *Mechanical Devices.* a tool used to clean mortar out of drain joints.

badger

badge reader *Computer Technology.* an optical-recognition device designed to scan data that is magnetically coded on a plastic badge; used especially for building security.

badger plane *Mechanical Devices.* a plane with its mouth set at an oblique angle, used for planing corners.

badger softener *Building Engineering.* a brush filled with badger hair; used in water color graining.

Badger's rule *Physical Chemistry.* a rule that establishes the observable relationship for the maximum distance a molecular bond can be stretched.

Badlands

Badlands *Geography.* a region in southwestern South Dakota and northwestern Nebraska known for its barren, eroded landscape.
bad top *Mining Engineering.* a term for a roof in a coal mine that is thought to be precarious, sometimes resulting from a blast.
BaE barium enema.
BAE or B.A.E. Bachelor of Aeronautical Engineering; Bachelor of Agricultural Engineering; Bachelor of Architectural Engineering.
baeckeol *Organic Chemistry.* $C_{13}H_{18}O_4$, a pale-yellow, crystalline phenolic ketone found in the oils of certain plants of the myrtle family.
BAeE or **B.Ae.E.** Bachelor of Aeronautical Engineering.
Baekeland, Leo 1863–1944, Belgian-born American chemist; invented Bakelite.
bael *Botany.* a spiny citrus tree, *Aegle marmelos,* that grows in India; or the hard-shelled fruit from this tree.
baeocyte *Invertebrate Zoology.* an immotile or motile blue-green bacterial endspore.
Baer, Karl Ernst von 1792–1876, Estonian-born German biologist; discovered the human ovum; founded the study of embryology.
Baeyer, Adolf von 1835–1917, German chemist; analyzed and synthesized indigo.
Baeyer strain theory *Organic Chemistry.* a theory stating that in cyclic compounds, the deflection from the symmetrical tetrahedral angles (109° 28') found between the four valences of sp^3-hybridized carbon produces strain, causing the unstable molecule to undergo ring-opening reactions.
Baeyer-Villiger reaction *Organic Chemistry.* the process of oxidizing ketones (open chain, cyclic, and aromatic) to esters or lactones by the use of peracids.
Baffin Bay *Geography.* a long (700 miles), icy finger of the North Atlantic between Greenland and Baffin Island.
Baffin Island *Geography.* a large island (area: 195,927 square miles) northeast of Hudson Bay in the Northwest Territories of Canada.
baffle a barrier or obstruction; specific uses include: *Acoustical Engineering.* **1.** a small acoustic screen attached to a microphone in order to reduce air flow that might cause local variations in the acoustic field around it. **2.** a partition in a loudspeaker cabinet that is used to reduce the interaction between the sound produced by the speaker and the walls of the cabinet. *Acoustics.* any partition or barrier that blocks an incoming sound. *Building Engineering.* an artificial obstruction for checking or deflecting light flow. *Electricity.* an instrument used for shielding a circuit breaker from gas or oil.
baffle plate *Electromagnetism.* a conductive plate inserted into a waveguide to reduce the cross-sectional area of the guide for filtering purposes or wave conversion.
baffling wind *Meteorology.* a randomly shifting wind over the sea that impedes movement by sailing vessels.
bag *Computer Science.* a collection of items, analogous to a set, but allowing multiple occurrences of an item. Also, MULTISET.
bagasse [bə gas´] *Food Technology.* **1.** the crushed fibrous plant residue that is a by-product of sugarcane refining. **2.** similar fibrous by-products from winemaking or other food processing.
BAg or **B.Ag.** Bachelor of Agriculture.
bag filter *Engineering.* a cloth bag filter used for the recovery of impurities that are suspended in gas.
bagged charges *Ordnance.* any propelling charges for heavy guns that are enclosed in a fabric container and used in separate-loading or semifixed ammunition.
bagging *Textiles.* a material woven of hemp or jute and used for bags.
bagherra [bä gēr´ə] *Textiles.* a crush-resistant velvet fabric with an uncut loop pile.
baghouse *Engineering.* a chamber or room in which bag filters trap solids that have temporarily combined with gases.
bag molding *Materials Science.* a process in which plastics are molded into curved shapes by applying fluid pressure through a flexible cover that forces the material to be molded against a rigid die.
bag plug *Engineering.* a type of drain plug made up of a cylindrical canvas bag that is inserted in the drain pipe and inflated.
Bagridae *Vertebrate Zoology.* the bagrid catfish family of the order Siluriformes, found in fresh waters of Asia and Africa and characterized by nasal barbs.
bag trap *Engineering.* a trap shaped in the form of the letter S, having vertical inlet and outlet pipes that are aligned.
baguette [bə get´] *Mineralogy.* a small, narrow, rectangular-shaped gemstone. *Building Engineering.* see BEADED MOLDING.
baguio [bäg´yō] *Meteorology.* a severe tropical cyclone in the Philippines. Also, **bagio, bagyo.**
bahada see BAJADA.
bahamite *Petrology.* a fine-grained, consolidated limestone of high purity, without abundant fossils; formed of sediment similar to a type currently accumulating in the Bahama Islands.
bahiaite *Petrology.* a holocrystalline igneous rock formed mainly of hypersthene with hornblende and sometimes minor amounts of olivine, ferroan spinel, and other minerals.
bahut [bä´hoot´] *Architecture.* **1.** a low wall that rises above the cornice of a building and carries the roof. **2.** the rounded upper course of masonry on a parapet wall.
bai *Meteorology.* a yellow mist of Eastern China and Japan that falls during the spring and fall, caused by moisture collected in clouds of sand blown eastward from the interior of China. Also, SAND MIST.
Baikal, Lake [bī käl´] *Geography.* a large (12,160 square miles), very deep freshwater lake in southeastern USSR.
Baikalian orogeny *Geology.* a late Precambrian crustal deformation that affected the area near Lake Baikal, USSR.
bail *Engineering.* **1.** to remove unwanted liquid from a place, especially water from a ship or boat. **2.** a steel loop that snap-fits around the parts of a device to hold them together.
bailer *Mining Engineering.* a long pipe with a foot valve that is used to extract water, sand, or other such unwanted material from a well or drilling hole, thus allowing the desired material in the hole to flow or be removed.
Bailey, Liberty 1858–1954, American botanist and horticulturist; pioneered agricultural education; wrote noted horticultural encyclopedia.
Bailey bridge *Civil Engineering.* a temporary bridge that can be assembled quickly from portable prefabricated sections of welded steel in a lattice pattern; first widely used by the Allies in WWII. (Named after its designer, British engineer Sir Donald *Bailey.*)
Bailey meter *Engineering.* an instrument that determines the rate of flow in ducts, spouts, or pipes by means of a counter that measures the weight of granular material as it moves through the structure.
bailing *Engineering.* the process of clearing a well or hole of extraneous material with the use of a bailer.
bailing bucker see BAILER.
bailout *Aviation.* an emergency procedure in which a pilot and crew exit an aircraft in flight and descend via parachute.
bailout bottle *Aviation.* a small container of oxygen attached to a parachute harness or ejection seat for individual use during a bailout or other emergency.
Baily's beads *Astronomy.* bright points of light, resembling a string of beads, seen around the edge of the moon immediately before and after a total eclipse; caused by sunlight streaming through valleys on the moon's limb. Also, **Bailey's beads.**
Bainbridge reflex *Physiology.* an increase in heart rate caused by pressure in or distension of the veins of the right atrium.
Bain correspondence *Materials Science.* the relationship between two specific lattices when a particular structural unit of one becomes the unit cell of the second; used specifically in reference to martensite transformations.

Bain distortion *Materials Science.* the distortion of one lattice type into another by contraction or expansion along specific crystallographic axes. Also, BAIN STRAIN.

bainite [bā´nīt] *Metallurgy.* a steel intermediate between pearlite and martensite created when eutectoid steels in the austenitic condition are isothermally transformed by hot-quenching to temperatures in the 550°C to 250°C range.

bainite nose *Materials Science.* the section on a TTT curve for steel representing a maximum in the bainite transformation rate.

Bain strain see BAIN DISTORTION.

Bairdiacea *Invertebrate Zoology.* a superfamily of bivalve aquatic crustaceans in the subclass Podocopa.

Baird, John Logie 1888–1946, Scottish electrical engineer; pioneered many developments in television, including color television, large-screen TV, and stereoscopic TV.

Baire σ-algebra *Mathematics.* the smallest σ-algebra of subsets of a topological space X that contains all sets of the form $\{x \in X: f(x) = 0\}$, where f is any continuous function on X. Equivalently, the Baire σ-algebra of a topological space X is the smallest σ-algebra containing all closed, compact subsets of X. Also, **Baire sigma algebra.**

Baire class *Mathematics.* **1.** all continuous functions are said to be of the first Baire class (or sometimes Baire class zero). **2.** a function is said to be of Baire class α if it is not of any class $\beta < \alpha$ and if the function is a pointwise limit of functions of Baire classes β, for any $\beta < \alpha$ (α and β are ordinal numbers, not necessarily finite).

Baire function *Mathematics.* the family of all Baire functions on a topological space X is the smallest class of functions defined on X that contains all the continuous functions on X and also contains the limit of every pointwise (but not necessarily uniformly) convergent sequence of functions; that is, Baire functions are functions of the first Baire class on X. In order for an extended real-valued function f defined on a *Borel* set to be a Baire function, it is necessary and sufficient that, for every finite real number a, $f^{-1}([a, \infty])$ is a Borel set. (If $f^{-1}([a, \infty])$ is a Borel set, then f is a Borel function; that is, it is the inclusion of a which distinguishes a Borel function from a Baire function.)

Baire measure *Mathematics.* any finite-valued measure defined on the Baire sets of a topological space.

Baire's category theorem *Mathematics.* let X be a complete metric space, and suppose that a subset A of X is of the first category (i.e., A is a countable union of nowhere dense sets) in X. Then $X \cap A^c$ is dense in X; that is, $X \cap A^c$ is of the second category in X, where A^c is the complement of A.

Baire set *Mathematics.* for a given topological space X, a member of the Baire σ-algebra of X.

Baire's theorem *Mathematics.* let M be a complete metric space and $\{A_i\}$ a countable sequence of close sets in M whose union is M. Then at least one of the sets A_i contains an interior point.

Bairstow number *Fluid Mechanics.* a term formerly used to signify a Mach number.

baiting *Zoology.* the process of providing food plants, water, shelter, or other necessities in order to attract animals to an area, so as to control their numbers or modify their normal routine.

bait shyness *Behavior.* the avoidance by an animal of a food that previously caused it to become ill.

baize *Textiles.* a soft woolen or cotton fabric that resembles felt and is usually green; used to cover billiard tables and bulletin boards.

bajada [bä hä´də] *Geology.* in desert and semiarid regions, a wide, uninterrupted, gently sloping alluvial surface extending inland from the base of a mountain range into a basin, and formed by the lateral merging of adjacent alluvial fans. Also, BAHADA, COMPOUND ALLUVIAL FAN, ALLUVIAL PLAIN.

bajada breccia [bä hä´də brek´ē ə] *Petrology.* an imperfectly stratified accumulation of coarse, angular rock fragments that is mixed with mud; it is usually found in arid climates, and is formed from a mudflow with considerable water.

Bajocian [bə jō´shən] *Geology.* a European geologic stage of the Middle Jurassic period, occurring after the Toarcian and before the Bathonian.

Bakanae disease *Plant Pathology.* a disease of rice occurring especially in Japan, caused by the fungus *Gibberella fulikurae,* which results in rotting of the plant stem.

baked core *Metallurgy.* a core in sand casting that has been heated to harden and set its shape.

baked finish see BAKING FINISH.

baked permeability *Metallurgy.* the gas permeability of a mold in casting that has been heated above the boiling point of water.

baked strength *Metallurgy.* the strength of a mold in casting that has been baked above the boiling point of water.

Bakelite [bāk´līt´; bāk´ə līt´] *Materials Science.* a trademark for various synthetic resins created by the condensation of formaldehyde with phenols; for example, vinyl, polypropylene, phenolic, perylene, polysulfone, ABS, and other resins and compounds.

bakeout *Engineering.* a procedure of heating a vacuum system to drive out gases from the surfaces of the system.

baker bell dolphin *Civil Engineering.* a dolphin with a heavy bell-shaped cap; used to absorb the impact of a ship.

bakerite *Mineralogy.* $Ca_4B_4(BO_4)(SiO_4)_3(OH)_3{\cdot}H_2O$, a colorless, transparent monoclinic mineral having a specific gravity of 2.9 and a hardness of 4.5 on the Mohs scale, and occurring in compact masses.

Baker-Nunn camera *Optics.* a widefield camera consisting of three aspheric correcting elements and a spherical primary mirror; designed to track earth-orbiting satellites.

Baker-Schmidt telescope *Optics.* a type of Schmidt telescope designed to produce an image that is free of astigmatism and distortion.

Baker's rule *Ecology.* the principle that self-pollinating species will have a greater tendency to colonize remote habitats because the presence of only a single individual is required.

bakers' yeast *Food Technology.* the yeast *Saccharomyces cerevisiae,* commonly used as a leavening agent because of its rapid growth and low alcohol-producing properties.

baking *Food Technology.* the process of heating any food in an oven. *Metallurgy.* the process of heating a metallic component at a moderate temperature in order to remove gases. *Engineering.* a process in which heat applied to fresh paint films accelerates the evaporation time of its thinners and enhances the reaction of its binder components, resulting in a hard polymeric film. Also, STOVING.

baking finish *Engineering.* **1.** a finish that is considerably changed when heated, in contrast to a finish that undergoes normal or forced drying. **2.** a finish obtained by baking paint or varnish at temperatures above 66°C, which provides greater durability.

baking powder *Food Technology.* a powdered leavening agent consisting of a carbonate and acid agents that release carbon dioxide from the carbonate when moistened.

baking soda *Inorganic Chemistry.* a name for sodium bicarbonate, $NaHCO_3$, or for potassium bicarbonate, $KHCCO_3$, especially when used in baking powder. See SODIUM BICARBONATE, POTASSIUM BICARBONATE.

bal. balance; balancing.

Balaenicipitidae *Vertebrate Zoology.* the shoebills, a tropical African family of large, wading, storklike birds in the order Ciconiiformes, characterized by short necks, large heads, and large shoe-shaped bills.

Balaenidae *Vertebrate Zoology.* a family of large-headed, toothless whales in the suborder Mysticeti, comprising the right and bowhead whales; characterized by the absence of a dorsal fin and by grooves on the throat and chest.

Balaenopteridae *Vertebrate Zoology.* a family of the baleen whale order Mysticeta, including the blue whale, fin whale, sei whale, and minke whale. Also, RORQUAL, FINBACK WHALE.

balance a condition or means of achieving equality; specific uses include: *Engineering.* an instrument for determining mass typically consisting of a bar attached to a central support so that it has free horizontal movement, with a pan or tray suspended from each end of the bar. If a substance of unknown mass is placed on one tray, its mass can be determined by the amount of known mass that must be placed on the other tray in order to bring the two into exact horizontal alignment. *Mining Engineering.* the counterpoise or weight attached by cable to the drum of a winding engine to balance the weight of the cage and hoisting cable. *Electricity.* **1.** to adjust a bridge circuit for zero volts. **2.** to equalize voltage, current, or impedance between output terminals of a network. *Computer Programming.* in data structures, a measure of the relative size of the left and right subtrees of a node in a tree structure notation. *Chemistry.* an exact equality of the number of atoms of various elements entering into a chemical reaction, with the number of atoms of those elements present in the final products of the reaction. *Aviation.* the maintenance of equilibrium and steady flight by a flight vehicle. *Acoustics.* a condition for which each parallel-processing segment of an acoustic amplifier system produces an equivalent average output from the same acoustic or electrical input signal. *Hydrology.* **1.** the change in mass of a glacier over some specified time interval. **2.** see NET BALANCE. *Horology.* the vibrating member of a watch, chronometer, or clock. *Electronics.* see BALANCE CONTROL.

Balance see LIBRA.

balance arm *Building Engineering.* a support for the side of a projected window, constructed so as to allow the center of gravity of the sash to be constant when the window is open or closed. *Horology.* the part of the balance that connects the rim to the staff.

balance beam *Civil Engineering.* a long, heavy beam that allows a canal-lock gate to be swung on its pintle, and also helps to balance the outer end of the gate. Also, **balance bar.**

balance coil *Electricity.* an iron-core solenoid with adjustable taps that provides a neutral terminal to convert a two-wire circuit to a three-wire circuit.

balance control *Electronics.* a control for the loudspeakers of a stereo sound system, which can vary the volume of one speaker relative to the other while the overall volume remains relatively constant; i.e., the sound will be perceived as coming primarily from the left speaker if the control is adjusted in this direction.

balanced amplifier *Electronics.* an amplifier circuit with two identical signal branches connected to operate in phase opposition, with each input and output connection balanced to ground.

balanced armature unit *Acoustical Engineering.* an electromagnetically balanced armature on a record cutter or player that houses a cutting stylus or a needle cartridge, producing fluctuations representative of recorded sound as the stylus or needle moves to the right or left over a rotating disk.

balanced bridge *Electricity.* a bridge circuit adjusted for zero output voltage.

balanced chromosome *Genetics.* a chromosome containing several inversions that prevent recombination with its homologue.

balanced circuit *Electricity.* **1.** a circuit in which the two conductors are at equal and opposite potentials with reference to ground at all times. **2.** a balanced bridge.

balanced cloth *Textiles.* a fabric that has the same number and size of yarns in both warp and weft.

balanced construction *Building Engineering.* a sandwich panel with an odd number of plies laminated together so that the construction is identical on both sides of a plane through the center of the panel.

balanced converter see BALUN.

balanced currents *Electricity.* any currents flowing in the two conductors of a balanced line that are equal in amplitude and opposite in polarity.

balanced design *Statistics.* an experiment design in which the experimental units are distributed equally among the different levels of the factors.

balanced detector *Electronics.* a symmetrical demodulator used in FM receivers, in which the audio output is the rectified difference between the voltages produced across two resonant circuits, one tuned slightly above the carrier frequency and the other slightly below.

balanced differences *Statistics.* an estimate of error in a systematic sample based on differences between values falling within a specified area.

balanced digit system *Mathematics.* a number system of positional representation in which the digits in each position range from $-n$ to n, where n is a nonzero integer such that $(n + 1)$ is greater than half the number base.

balance delay see BALANCING DELAY.

balanced error *Computer Technology.* a state in which all values in an error range have an equal probability, and the maximum and minimum values in the ranges are equal in value but opposite in sign; the positive errors offset the negative errors, resulting in a net error of zero.

balanced growth *Microbiology.* the growth of cells in culture with all components increasing at the same rate so that cell composition is invariant during growth.

balanced input *Electronics.* a two-terminal input circuit whose electrical midpoint is grounded.

balanced lethal system *Genetics.* a population containing recessive nonlinked alleles that are lethal when homozygous and nonlethal when heterozygous.

balanced line *Electricity.* a transmission line composed of two conductors with equal impedance, in which the currents are equal in amplitude and opposite in polarity. *Industrial Engineering.* an assembly line that has been arranged to provide for a steady flow of output with minimal idle time for operators and machines on the line.

balanced load *Electricity.* a load of equal impedance at both terminals with respect to ground.

balanced merge *Computer Programming.* in a disk sort program, a process that uses an equal number of input and output lists.

balanced method *Engineering.* a system of instrumentation that uses a central point of zero and makes other readings in relation to zero.

balanced modulator *Electronics.* a circuit that accepts carrier and modulating signals as input, and outputs a suppressed-carrier, amplitude-modulated signal.

balanced motion pattern *Industrial Engineering.* a series of similar motions that are evenly distributed between the right and left hands in order to achieve coordination and minimize idle time for either hand.

balanced network *Electricity.* a part of a balanced system; a network with one or more terminal pairs (usually input and/or output) in which both terminals of each pair have the same impedance to ground.

balanced oscillator *Electronics.* an oscillator in which the impedance centers of the tank circuits are at ground potential, and the voltages between the opposite ends and their centers are equal in magnitude and opposite in phase.

balanced output *Electronics.* an output circuit whose electrical midpoint is grounded.

balanced polymorphism *Genetics.* genetic polymorphism that is maintained in a population because the heterozygotes for the alleles involved have a greater adaptive value than either homozygote.

balanced range of error *Statistics.* a range in which the maximum and minimum possible errors are equal in number and opposite in sign.

balanced reading see BALANCED METHOD.

balanced reciprocity *Anthropology.* the practice within a group of exchanging equivalent goods or services in an immediate or limited time period.

balanced ring modulator *Electronics.* a modulator that suppresses the carrier signal using tubes or diodes and provides double-sideband output.

balanced rock see PERCHED BLOCK, def. 1.

balanced rudder *Naval Architecture.* a rudder with part of its surface area in front of the rudder post, so that the rudder axis is near the center of side water pressure when the rudder is turned.

balanced salt solution *Chemistry.* a solution of salts in which the ratio of salts is balanced proportionally so that the toxic effects of each salt are eliminated.

balanced sash see SLIDING SASH.

balanced set *Electronics.* two or more components with identical, or nearly identical, gain and load characteristics, connected in parallel or push-pull configuration. *Mathematics.* a subset B of a vector space is balanced if αB is contained in B for every scalar α such that $|\alpha| \leq 1$.

balanced step *Building Engineering.* a winder with a tread width at the narrow end equal to the tread width of the adjacent stair flight. Also, DANCING STEP, DANCING WINDER.

balanced stock *Genetics.* a genetic strain that can be maintained in a heterozygous condition without artificial selection by a balanced combination of recessive genes that are lethal to hemizygous males or confer sterility to homozygous females. *Ordnance.* a condition in which the supplies needed and the quantity available are equal to each other. Also, **balanced supply.**

balanced surface *Aviation.* a control surface whose hinge or pivot moment is completely or nearly self-balanced, for example, by tabs or by a mass located in front of the hinge axis.

balanced translocation *Genetics.* a type of chromosome aberration in which a precise exchange of chromosomal segments occurs between two nonhomologous chromosomes. Also, RECIPROCAL TRANSLOCATION.

balanced transmission line *Electricity.* a transmission line in which each conductor has the same impedances to ground.

balanced valve *Engineering.* a valve in which pressure forces from the fluid being controlled oppose one another, ensuring that resistance to opening and closing the valve is negligible. Also, EQUILIBRIUM VALVE.

balanced voltages *Electricity.* any voltages having equal magnitude and opposite polarity with respect to ground.

balanced weave *Textiles.* a weave in which the length of the yarn is the same in both the weft and warp directions and on both sides of the fabric.

balanced wire circuit *Electricity.* a circuit or conductor system with two identical sides that are electrically identical and symmetrical with respect to ground and other conductors.

balance equation *Meteorology.* a diagnostic equation expressing a balance between the pressure field and the horizontal field of motion.

balance escapement *Horology.* an escapement mechanism in a timepiece that relies on the regular oscillations of a spring-loaded balance wheel to stop and release the escape wheel, thereby releasing the motion of the going train in steady timekeeping impulses.

balance lug *Navigation.* a lugsail that has a section forward of the mast and a long foot, usually with a boom.

balance method see NULL METHOD.

balance of nature *Ecology.* a state of balance tending to produce populations of relatively constant size in the natural environment, resulting from the constant interaction and interdependence of various organisms. Also, ECOLOGICAL BALANCE.

balancer *Invertebrate Zoology.* one of a pair of filaments, found in place of hind wings in diptera, that help the insect to balance in flight. Also, HALTERE. *Vertebrate Zoology.* one of a pair of lateral appendages on the heads of some larval salamanders. *Electricity.* the part of a direction finder that improves the sharpness of the direction indicator.

balancer set *Electricity.* **1.** a pair of linked direct-current generators that are used to equalize the voltage on each side of a three-wire system. **2.** a device used for equalizing the loads on the outer lines of a three-wire power distribution system.

balance sheet *Nutrition.* a measurement indicating that the amount of energy coming into the body via ingested nutrients equals the amount of energy the body expends in its activities.

balance spring see HAIRSPRING.

balance staff *Horology.* the staff that carries the balance wheel in a timepiece.

balance theory *Psychology.* the theory that people tend to prefer a balanced relationship with others and with their environment, and that they will behave in a manner that is intended to achieve this balance.

balance tool *Mechanical Engineering.* a tool used primarily for taking rough cuts on the surface of a workpiece in a lathe; it is held in a tool holder at an unvarying angle that is determined before grinding the cutting edge.

balance-to-unbalance transformer *Electricity.* a device used to match a pair of unbalanced transmission lines to a balanced pair of transmission lines.

balance wheel see FLYWHEEL.

balancing *Design Engineering.* in rotating machines, the balancing of centrifugal forces to counteract any vibration in any plane. Also, DYNAMIC BALANCING.

balancing a survey *Engineering.* a surveying process in which adjustments are made along a traverse, so that there are no errors of closure and an adjusted position is established for each traverse station. Also, TRAVERSE ADJUSTMENT.

balancing band *Naval Architecture.* a band around the shank of an anchor at its center of gravity with a ring attached on one side; when hung from the ring, the anchor will hang on its side.

balancing capacitor *Electronics.* a variable capacitor used to increase the accuracy of a radio direction finder. Also, CONDENSING CAPACITOR.

balancing delay *Industrial Engineering.* the process of keeping one hand idle in hand work processes while the other hand catches up. Also, BALANCE DELAY.

balancing selection *Evolution.* a process of natural selection in which heterozygotes are selected far more than either homozygote.

balancing unit *Electricity.* a device that converts balanced to unbalanced transmission lines, or the reverse, by placing discontinuities at the junction between the lines.

balancing valve *Engineering.* a valve that controls the flow of liquid in a pipe. Also, **balancing plug cock.**

Balanidae *Invertebrate Zoology.* a highly evolved sessile family of acorn barnacles in the suborder Balanomorpha.

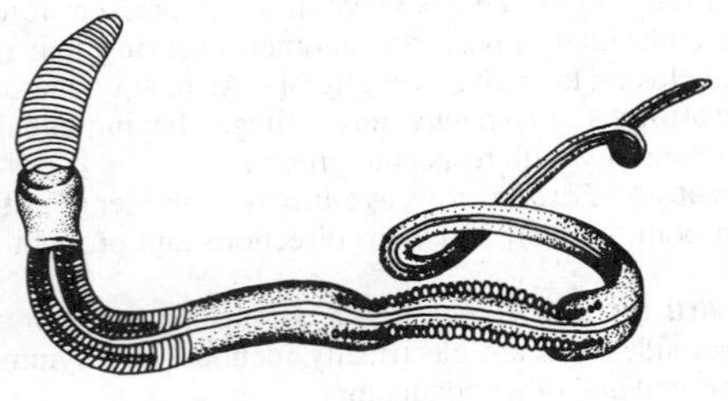

Balanoglossus

Balanoglossus *Invertebrate Zoology.* a genus of marine acorn worms, in class or order Enteropneusta, with an acorn-shaped proboscis. They are one of several in this class that, because of certain vertebrate-like characteristics, afford a link between invertebrate and vertebrate life forms.

Balanomorpha *Invertebrate Zoology.* a suborder of order Thoracica, subclass Cirripedia (barnacles), which includes the rock barnacle.

Balanopaceae *Botany.* a small Australasian family of dioecious evergreen shrubs or trees with scalelike proximal leaves and normal distal leaves on each branch, and bearing fruit that is a drupe in a persistent involucre, resembling an acorn.

Balanopales *Botany.* an order of trees coexistent with the family Balanopaceae. Also, **Balanopsidales.**

Balanophoraceae *Botany.* a family of dicots that are chlorophyll-less root parasites with scalelike leaves and a funguslike appearance.

Balanophyllia *Paleontology.* a genus of solitary cup corals in the order Scleractinia that arose in the Tertiary and still persists in British coastal waters.

Balantidium *Invertebrate Zoology.* a genus of large parasitic ciliate protozoans. The species *Balantidium coli* infests the intestines of amphibians and mammals, and may cause ulcerative dysentery in humans.

Balantiopsaceae *Botany.* a family of dioecious liverworts of the order Jungermanniales occurring in south temperate regions; characterized by a well-developed marsupium, reduced bracteoles, and rhizoids rising from the stem near the underleaves.

Balanus *Invertebrate Zoology.* a genus of common nonparasitic barnacles, in which a calcareous shell encases cirri (limbs) that emerge and then retract with collected food particles.

balanus see GLANS PENIS.

balata [bə lä′tə] *Materials.* a nonelastic, rubberlike gum obtained from the latex of the tropical American tree, *Manikara bidentata*; used in making gaskets, golf ball covering, and chewing gum and as a substitute for gutta-percha.

balau [bə lou′] *Materials.* the wood of the trees of the *Shorea* species of Southeast Asia; used as a structural timber, for flooring, barrels, and shipbuilding. Also THITYA, TENG, PHCHEK.

Balbiani chromosome *Genetics.* a banded polytene chromosome found in the larvae of certain diptera, which exhibits Balbiani rings.

Balbiani rings *Genetics.* giant chromosome puffs present on certain polytene chromosomes in the midge *Chironomus* that code for proteins secreted in the larval salivary glands.

balbriggan *Textiles.* a knitted lightweight cotton fabric often used for making hosiery or underwear.

balconet *Building Engineering.* a low ornamental railing on a door or window that projects slightly beyond the threshold or sill.

balcony *Building Engineering.* a platform that projects from a wall, either inside or out, usually supported by pillars and enclosed by a railing.

bald *Zoology.* having white on the head. *Geography.* a treeless hilltop or mountaintop. *Mining Engineering.* a term meaning without framing, in reference to a flat-ended mine timber.

baldachin [bal′də kin] *Textiles.* a fabric of silk brocade interwoven with gold or silver threads. *Architecture.* a permanent canopy suspended above a throne or altar, usually supported by pillars and surrounded by a railing. Also, **baldacchino.**

bald eagle *Vertebrate Zoology.* a fish-eating eagle of the family Accipitridae, species *Haliaetus leucocephalus*, characterized by a white head and darkly feathered adult body; it is considered an endangered species.

bald eagle

baldheaded anticline *Geology.* an anticline whose crest has been eroded, exposing the stratigraphically older rocks of its core.

Baldwin spot see BITTER PIT.

bale *Agriculture.* a large, compressed and bound bundle of produce, such as cotton or hay. *Industrial Engineering.* a large package of tightly pressed material secured with rope or wire, usually wrapped in paper.

Balearic Islands [bal´ē âr´ik] *Geography.* an archipelago in the western Mediterranean off the eastern coast of Spain; its principal islands are **Majorca (Mallorca), Minorca, Ibiza,** and **Formentera.**

baleen [bə lēn´; bă´lēn´] *Vertebrate Zoology.* the horny plates hanging from the roof of the mouth in toothless whales; used as a strainer to eject water but retain solids while feeding. Also, WHALEBONE.

baleen whale *Vertebrate Zoology.* any toothless whales of the suborder Mysticeta, with whalebone plates on the sides of the jaw for straining shrimp and plankton from the water. Also, WHALEBONE WHALES.

baler *Agriculture.* a farm machine that picks up hay from a swath or windrow and forms it into bales. *Mechanical Engineering.* a machine that compresses large quantities of loose bulky material and secures it for convenient transport.

baler

Balfour, Francis 1851–1882, British biologist; wrote the first comparative embryology.

Balfour's law *Developmental Biology.* a postulate stating that the speed at which any portion of an ovum divides is proportional to the amount of protoplasm in that area.

Bali [bä´lē; bal´ē] *Geography.* an island in southern Indonesia, east of Java (area: 2905 square miles).

baling *Civil Engineering.* a technique used to compact loose refuse into dense heavy blocks.

balisage [bal´ə säzh] *Military Science.* a method of marking a land route with dim lights so that it can be traveled at faster speeds during a blackout.

Bali wind *Meteorology.* a strong easterly wind prevalent at the eastern end of Java.

balk or **baulk** *Archaeology.* an area or band of unexcavated earth left between excavated units or pits to reveal the stratigraphy of an excavation. *Building Engineering.* the use of squared timber in building construction. *Mining Engineering.* a sudden thinning out of a bed of coal for a certain distance.

balking *Industrial Engineering.* the refusal of a customer or input item to enter a queue.

ball a round or roundish body or mass; specific uses include: *Mechanical Engineering.* one of the crushing bodies, usually made of ceramic or steel, used in a ball mill for fine grinding or ore crushing. *Ordnance.* a general-purpose bullet, as opposed to special forms of ammunition such as explosive or armor-piercing bullets. *Geology.* **1.** a spherical mass of sedimentary material. **2.** see LONGSHORE BAR. **3.** see BALLSTONE, def. 1. *Oceanography.* a low ridge of sand parallel to a beach, below high tide and separated by a trough from the beach. *Mathematics.* in a metric space X with distance function $d(x, y)$, the *open ball* centered at x with radius r is defined by $\{y \in X: d(x, y) < r\}$. The *closed ball* is defined similarly but with $\leq$ replacing $<$.

ball-and-race-type pulverizer *Mechanical Engineering.* a grinding machine consisting of balls rotated by a motor that crushes incoming material, such as coal, against the races, reducing it to a fine consistency. Also, BALL-BEARING PULVERIZER, BALL-RACE MILL.

ball-and-socket joint *Anatomy.* a joint formed when the head of one bone is inserted into a hollow in another bone, as in the hip joint of the femur with the pelvis. Also, ENARTHROSIS. *Geology.* a transverse fracture in a basalt column that is curved either upward or downward. Also, CUP-AND-BALL JOINT. *Mechanical Engineering.* a joint in which a member with a spherical end is placed within a socket recessed to fit it, thus permitting relative movement within a given cone, or a cutout in the socket.

ballas *Mineralogy.* a dense, spherical aggregate of minute diamond crystals used for industrial purposes.

ballast *Naval Architecture.* weight carried low in a hull to lower a ship's center of gravity and increase stability. *Electricity.* **1.** a component used to limit the current flow through, or stabilize the operation of, a circuit, stage, or device. **2.** the choke used in series with fluorescent lamps. *Civil Engineering.* **1.** a layer of coarse stone or gravel used as a base for concrete. **2.** the rock upon which the ties of a railroad are set.

ballast lamp *Electricity.* a resistance lamp that maintains a nearly constant current by increasing resistance when current increases; used as a ballast resistor, current limiter, or alarm, or to stabilize a discharge lamp.

ballast line *Naval Architecture.* the waterline at which a ship floats when ballasted.

ballast movement *Navigation.* a voyage or part of a voyage by a tanker with no commercially valuable cargo in its tanks. Also, **ballast leg, ballast passage.**

ballast reactor *Electricity.* a coil wound on an iron core and linked in series with a fluorescent lamp to adjust for the negative-resistance characteristics of the lamp by yielding an increased voltage drop as the current through the lamp increases.

ballast resistor *Electricity.* a resistor that maintains a nearly constant current by compensating for fluctuations in alternating-current power-line voltage by increasing resistance when current increases. Also, BARRETTER.

ballast tank *Naval Architecture.* **1.** a space between the inner and outer hulls of a submarine that is filled with water when submerged and with air when surfaced. **2.** any tank used to hold water ballast.

ballast tube *Electricity.* a ballast resistor in a vacuum tube. The vacuum improves voltage-regulating action by reducing heat radiation from the resistance element.

ball bearing *Mechanical Engineering.* a friction-reducing bearing that consists of hardened steel balls that roll in spherical grooves in the surfaces of two concentric rings, so that the load is carried by the balls.

ball-bearing hinge *Mechanical Engineering.* a friction-reducing hinge fitted with ball bearings between the hinge knuckles.

ball-bearing pulverizer see BALL-AND-RACE-TYPE PULVERIZER.

ball-breaker see WRECKING BALL.

ball burnishing *Metallurgy.* **1.** see BALL SIZING. **2.** the process of removing burrs and polishing small metallic parts by tumbling.

ball bushing *Mechanical Engineering.* a ball bearing that allows axial motion of the shaft.

ball catch *Mechanical Devices.* a door latch fitted with a spring-loaded ball that engages a small hole in the striking plate to keep the door closed until it is forced open.

ball check valve *Engineering.* a valve that allows flow in one direction only, composed of a ball that is held against a seat and controls the flow.

ball clay see PIPE CLAY, def. 1.

ball coal *Geology.* a type of coal that occurs in spherical masses, probably formed by a jointing process.

ballcock *Mechanical Engineering.* a system for regulating the supply of water in a toilet tank or similar vessel, consisting of a floating ball attached to a valve; the rise and fall of the ball, which corresponds to the rise and fall of the water level, opens and shuts the valve.

Ballerup-Bethesda group *Bacteriology.* the strains of the Gram-negative bacterium *Citrobacter freundi* that ferment lactose slowly and are typically found in feces.

ball float *Mechanical Engineering.* a floating, spherical device that is used to operate a ball valve by moving or raising the cup-shaped seat of the valve.

ball flower [flou´ər] *Architecture.* an ornament resembling a ball enclosed within the three or four petals of a flower, usually set at regular intervals along a hollow molding.

ball grinder see BALL MILL.

ballhead *Mechanical Engineering.* the part of the governor containing the rotating balls or flyweights, whose force is at least partially counteracted by a speeder spring.

ball ice *Oceanography.* an unusual sea ice phenomenon consisting of soft, spongy spheroids of ice about 2.5 to 5 cm in diameter.

Balling hydrometer *Engineering.* an instrument that measures the density of sugar in a solution, usually by determining its specific gravity.

balling up *Metallurgy.* the undesirable formation of metallic beads in brazing, caused by inadequate wetting of the base metal.

ballism *Neurology.* a condition of the nervous system characterized by crude, violent, flinging motions of one or more limbs; caused by contractions of the proximal limb muscles as a result of destruction of the subthalamic nucleus of Luysii or its fiber connections. Also, **ballismus.**

ballista [bə lēs´tə] *Ordnance.* an artillery piece that was used by the ancient Roman army to hurl large boulders at a fortification.

ballistic [bə lis´tik] *Mechanics.* of or relating to the science of ballistics.

ballistic area *Ordnance.* the space between the center of impact of a group of shots beyond the target and the center of impact of another group short of the target.

ballistic body *Engineering.* **1.** any object whose movement and physical attributes are affected by surrounding forces, conditions, or substances, such as by gravity, temperature, or pressure. **2.** specifically, a missile, bullet, shell, rocket, or other such object.

ballistic camera *Optics.* a camera that employs multiple exposures to record the trajectory of a projectile.

ballistic cap see FALSE OGIVE.

ballistic coefficient *Mechanics.* a numerical measurement that expresses a projectile's efficiency to overcome air resistance, such as the drag coefficient or deceleration coefficient.

ballistic conditions *Ordnance.* the various factors that affect the motion of a projectile, including its weight, size, and shape, its muzzle velocity, and the existing wind and air conditions.

ballistic correction *Ordnance.* an adjustment in firing data based on conditions affecting the flight of a projectile, such as wind or air density, rather than on observation of fire.

ballistic curve *Mechanics.* the curve followed by a bullet, bomb, missile, or other projectile, determined by the ballistic conditions, the propulsive force, and the action of gravity.

ballistic deflection *Mechanics.* the deflection of a missile as a result of its aerodynamic properties.

ballistic density *Mechanics.* a representative effective air density that would produce the same range for a projectile as the actual density distribution, expressed as a percentage of the density according to the standard artillery atmosphere.

ballistic director *Ordnance.* an instrument that computes firing data for the future position of a moving target, allowing for such factors as wind, air density and temperature, and muzzle velocity. Also, PREDICTOR.

ballistic efficiency *Ordnance.* a measure of the effect of air resistance on the flight of a projectile, depending mainly on its weight, size, and shape.

ballistic entry *Mechanics.* the entry into the atmosphere by an unpropelled body.

ballistic galvanometer *Electricity.* a galvanometer used to measure the quantity of electricity in a current pulse of short duration.

ballistic instrument *Engineering.* any instrument used to measure the effect of moving energy on an object, or to measure the energy itself.

ballistic lead [lēd] *Ordnance.* the allowance made for wind and other ballistic factors when sighting a moving target.

ballistic limit *Ordnance.* the velocity limit of a projectile at which a plate of a certain thickness will just resist complete penetration by the projectile.

ballistic magnetometer *Engineering.* an instrument designed to measure the magnetic intensity in a coil that is subjected to transient voltage when either the magnetized sample or the coil itself is moved relative to the other.

ballistic measurement *Mechanics.* the measurement of the velocity and drag of a projectile by means of a ballistic pendulum, ballistic galvanometer, camera, or other such device.

ballistic missile *Ordnance.* a long-range missile that is self-propelled and guided in the early part of its flight and free-falling in a ballistic trajectory in the remaining part of the flight.

ballistic pendulum *Engineering.* a device consisting of a heavy block suspended by rods that measures the velocity of a projectile, such as a bullet, by calculating the angle of the pendulum's swing.

ballistic range *Ordnance.* a firing range for testing weapons and projectiles. Also, **ballistics range.**

ballistics [bə lis´tiks] *Mechanics.* the study of the motion and behavior of projectiles, missiles, rockets, bullets, and similar objects, and of the accompanying phenomena.

ballistic separator *Civil Engineering.* a device that separates noncompostable material, such as rocks, glass, or metal, from solid waste, by using a rotor to fling the material into the air so that the lighter compostable material travels a shorter distance than the heavier noncompostable material.

ballistics of penetration *Ordnance.* the branch of terminal ballistics concerning the motion of a projectile as it forces its way into targets of solid or semisolid substances, such as concrete, steel, earth, and so on.

ballistic table *Mechanics.* the compilation of ballistic data from which trajectory characteristics are obtained.

ballistic temperature *Ordnance.* an assumed constant temperature that would have the same effect on the trajectory of a projectile as the varying temperatures actually encountered in its flight.

ballistic trajectory *Mechanics.* the path of an unpowered object that moves only under the influence of gravity and atmospheric friction, with its surface providing no significant lift to alter its course.

ballistic tube *Ordnance.* a gun tube used for ballistic tests.

ballistic uniformity *Ordnance.* the ability of a propellant to bring about uniform muzzle velocity and produce similar interior ballistic results when fired under identical conditions from round to round.

ballistic vehicle *Engineering.* a vehicle that does not leave the ground surface and that follows a path affected only by gravitational forces and by its resistance to the medium through which it moves.

ballistic wave *Mechanics.* the audible acoustic disturbance produced by the compression of the air ahead of a missile in its flight.

ballistic wind *Ordnance.* an assumed constant wind that would have the same overall effect on the trajectory of a projectile as the varying winds it actually encounters in its flight.

ballistite *Materials Science.* a powerful smokeless explosive consisting of nitrocellulose and nitroglycerin; used as a solid fuel for rockets.

ballistophobia *Psychology.* an irrational fear of missiles or projectiles.

ballistospore *Mycology.* a fungal spore that is forcibly discharged from a fungus when it reaches maturity. Also, BALLISTOCONIDIUM, SHOT SPORE.

ball joint see BALL-AND-SOCKET JOINT.

ball lightning *Geophysics.* a rare form of white or reddish-colored lightning that occurs in the shape of a sphere approximately 30 cm in diameter, moves relatively slowly, and usually disappears without a detonation. Also, GLOBE LIGHTNING.

ball mill *Mechanical Engineering.* a low-speed pulverizer consisting of a horizontal, rotating cylinder that contains a charge of steel balls or pebbles that tumble with the material to be pulverized.

ballonet *Aviation.* an auxiliary air bag or compartment within the main envelope of an airship or balloon; used to maintain trim and altitude control. Also, **ballonnet.**

balloon *Aviation.* a vehicle or device incorporating a nonporous bag filled with heated air or lighter-than-air gas so that it rises and floats in the atmosphere. *Surgery.* an inflatable sac that is used to keep tubes or catheters in the body, to inflate or distend a cavity or overlying structures, or to provide support to body structures.

balloon angioplasty see PERCUTANEOUS TRANSLUMINAL ANGIOPLASTY.

balloon barrage *Ordnance.* a series of moored balloons, usually strung together with steel cables, for impeding a low-level attack by enemy aircraft; used over cities during the European campaign of World War II.

balloon catheter *Surgery.* a catheter with an inflatable sac at its tip; the inflated balloon may ease passage of the tube through a blood vessel or may help occlude the vessel or may keep the catheter in place. Also, **balloon-tip catheter.**

balloon ceiling *Meteorology.* a ceiling height classification determined by timing the ascent and disappearance of a pilot balloon.

balloon chuck *Mechanical Devices.* a lathe chuck with a hollow hemisphere for holding small parts, such as the balance staffs of watches, with only their ends exposed.

balloon drag *Meteorology.* a small balloon, attached to a larger balloon, that is loaded with ballast and inflated to explode at a predetermined altitude; used to slow the ascent of a radiosonde during its initial ascent so that more detailed measurements may be obtained.

balloon framing *Building Engineering.* framing for a wooden building that is composed of machine-sawed scanterlings fastened with nails; each stud is one piece from roof to foundation with joists nailed to the studs and supported by sills or ribbons placed into the studs.

ballooning *Aviation.* the art of operating a balloon. *Surgery.* the distention of a body cavity with air or gas for therapeutic purposes, especially in balloon angioplasty.

balloon-type rocket *Space Technology.* a rocket vehicle that obtains its structural integrity from the pressure of the propellants and other gases within it; the best-known example is the Atlas.

balloon vine *Botany.* a tropical climbing plant, *Cardiospermum halicacabum,* bearing large bladderlike pods and black seeds with a white heart-shaped spot.

balloting *Ordnance.* the tossing of a projectile within the limits of a bore diameter, when moved through the bore by propellant gases.

ballotini [bal´ə tēn´ē] *Cell Biology.* a mass of small glass beads used for cell disruption; available in various sizes.

ballottement [bə lät´mənt] *Medicine.* a technique of physical examination to estimate the size and mobility of a deeply situated floating object. When used as a diagnostic maneuver in pregnancy, the examining finger is used to sharply tap the uterine wall, sending the embryo tossing upward and falling back against the wall.

ball-peen hammer *Mechanical Devices.* a type of hammer whose head has one conventional flat side and one ball-shaped side, used in riveting and metal-forming operations.

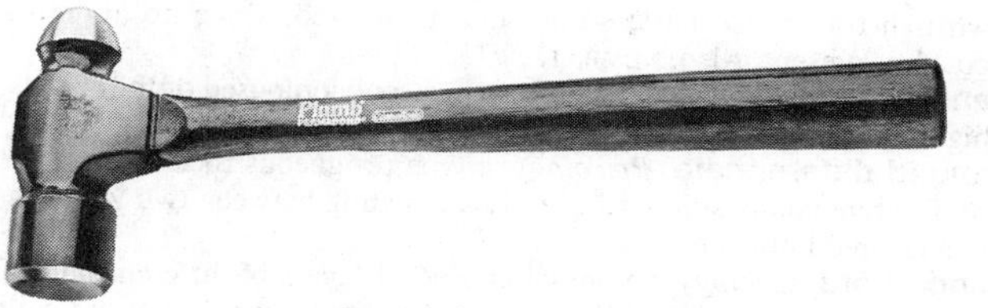

ball-peen hammer

ball-pendulum test *Engineering.* a test to determine the effect of explosives by measuring the degree to which a pendulum swings when a specific weighted charge is exploded.

ball planting *Forestry.* the setting out for growing of young trees after their roots have been enclosed in rough balls of earth.

ball race *Mechanical Devices.* a steel ring-shaped groove, track, or channel in which ball bearings move.

ball-race mill see BALL-AND-RACE-TYPE PULVERIZER.

ball screw *Mechanical Engineering.* a device that converts rotation to linear motion, consisting of a threaded nut linked to a threaded shaft by friction-reducing ball bearings constrained to move in the threads.

ball sizing *Metallurgy.* the process of finishing and sizing a hole by forcing a suitably sized steel ball through it.

ballstone *Geology.* **1.** a large spherical nodule in a stratified structure, especially an ironstone nodule in coal. Also, BALL. **2.** a rounded concretion consisting of unstratified limestone surrounded by calcareous shale or bedded limestone. Also, CROGBALL, WOOLPACK.

ball test *Civil Engineering.* **1.** a test to determine the consistency of concrete by measuring the penetration of a weight dropped into it. **2.** the process of rolling a specifically sized ball through a drain to determine whether the drain is circular and unobstructed.

ball-up *Engineering.* the accumulation of a viscous consolidated material by a section of drilling equipment while it is operating. *Navigation.* the pulling away of an anchor from a soft bottom with a large clump of mud attached to it.

ballute [bal´oot; bə loot´] *Space Technology.* a balloonlike parachute used as a braking system for a sounding rocket or other spacecraft.

ball valve *Mechanical Engineering.* **1.** a nonreturn valve consisting of a ball held against a cup-shaped seat having a circular opening of smaller diameter than the ball. When fluid flows from the opening toward the sphere, it pushes the ball away, thus allowing unimpeded flow; flow from the other direction pushes the sphere against the opening, thus sealing the valve. **2.** a valve that controls fluid flow by the rotation of a ball on a spindle; the ball has a hole in its center which permits or stops fluid flow depending upon its position.

balm *Botany.* a tall, fragrant herb of the mint family, Labiatae, genus *Melissa,* used for seasoning and flavoring foods. *Geology.* a concave cliff or overhanging rock that forms a shelter or a cave.

Balmer, Johann Jakob 1825–1898, Swiss mathematician; discovered that the line frequencies of the hydrogen spectrum form a convergent series.

Balmer discontinuity *Astrophysics.* a relatively abrupt decrease in a continuum spectrum at about 3650 Ångstroms caused by hydrogen absorption lines in the Balmer series crowding to their series limit. Also, **Balmer jump.**

Balmer formula *Spectroscopy.* a mathematical equation for determining the wavelengths of the lines that appear in the spectrum of hydrogen, $1/\lambda = R$.

Balmer jump *Spectroscopy.* the sudden decrease in the intensity of the spectral lines of hydrogen at the Balmer limit. Also, **Balmer discontinuity.**

Balmer limit *Spectroscopy.* the limiting wavelength near which the distance between consecutive lines in the Balmer series decreases and approaches a continuum.

Balmer series *Astrophysics.* a series of hydrogen spectral lines that lies in the visible part of the spectrum and is associated with the atom's second energy level; most prominent as absorption lines in the spectra of A-type stars. Also, **Balmer lines.**

balsa [bôl´sə] *Botany.* a large, fast-growing tree, *Ochroma pyramidale* (*lagopus*), in the Bombacaceae family, characterized by simple, angled or lobed leaves and yellowish to brownish terminal flowers. *Materials.* the wood of this tree, the lightest in commercial use; used for insulation, floats, paneling, and models.

balsam bôl´səm] *Materials Science.* a semifluid, resinous, and fragrant substance from certain trees, especially trees of the genus *Commiphora,* that contain benzoic or cinnamic acid.

Balsaminaceae *Botany.* a family of herbs in the order Geraniales that are almost always glabrous, sometimes tanniferous but without ellagic acid, and characterized by raphide sacs in the stems and leaves.

balsam of Peru see PERU BALSAM.

Baltic Sea *Geography.* an inland arm of the North Atlantic, located in northern Europe between Sweden and the USSR.

Baltimore, David born 1938, American microbiologist; shared Nobel Prize for research on tumor viruses.

Baltimore classification *Virology.* a system of classifying viruses on the basis of gene arrangement and nucleic acid type. (Named for David Baltimore, who proposed the system.)

Baltimore oriole *Vertebrate Zoology.* a songbird of North and South America, *Icterus galbula galbula,* a subspecies of the northern oriole. (From the black and orange plumage of the male, the colors of Lord *Baltimore's* coat of arms.)

balun or **BALUN** [bal´ən]*Electronics.* a quarter-wavelength, cylindrical sleeve placed over the end of a coaxial cable feed to an antenna, which matches an unbalanced coaxial transmission line or system to a balanced, two-wire line or system. It isolates the outer conductor of the cable from ground. (An acronym for balanced to unbalanced.) Also, BALANCED CONVERTER, LINE-BALANCE CONVERTER.

balustrade *Architecture.* a row of short columns (**balusters**) topped by a handrail.

Bamberga *Astronomy.* the asteroid 324, which has a diameter of almost 500 kilometers and belongs to type CP; it has one of the darkest surfaces of any asteroid, reflecting only about 3% of the sunlight striking it.

Bamberger's formula *Organic Chemistry.* a structural formula for naphthalene that shows the valencies of the benzene rings pointing toward the centers.

bamboo *Botany.* any of several woody, treelike grasses in the family Gramineae that have jointed hollow stems with solid nodes; often used in furniture making.

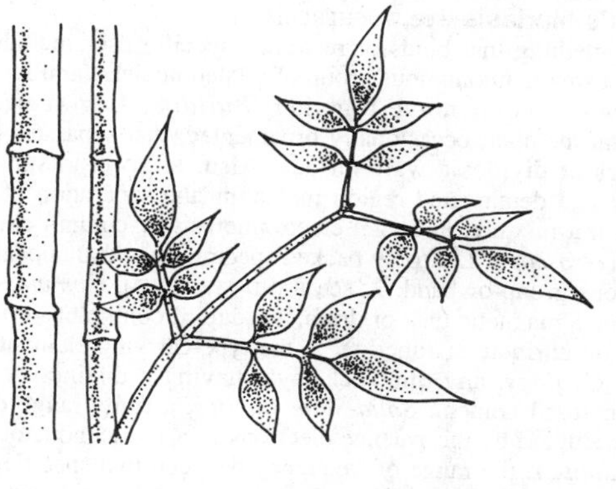

bamboo

Bambusoideae *Botany.* a subfamily of the family Gramineae that includes bamboo species; characterized by usually broad and pseudopetiolate leaves.

Banach, Stefan [ban´äk; ban´ik] 1892–1945, Polish mathematician; founder of modern functional analysis; formulated theory of topological vector spaces.

Banach algebra *Mathematics.* an algebra that is also a Banach space and for which $\|u \cdot v\| \leq \|u\| \cdot \|v\|$ for all vectors u, v in the algebra and norm $\| \, \|$.

Banach's closed graph theorem see CLOSED GRAPH THEOREM.

Banach space *Mathematics.* a normed linear (or vector) space that is also complete.

Banach-Steinhaus theorem *Mathematics.* let X be a Banach space, Y a normed vector space, and $\{T_i\}$ a sequence of bounded operators from X to Y. Then a number $M > 0$ exists such that $\|T_i\| \leq M$ for all i. That is, if the sequence $\{T_i\}$ is pointwise bounded, then it is also uniformly bounded.

banak *Botany.* a Central American timber tree of the genus *Virola.*

banakite *Petrology.* an alkalic basalt made up of plagioclase, sanidine, and biotite, with small quantities of analcime, augite, olivine, and possibly traces of quartz or leucite.

banana *Botany.* **1.** a herbaceous treelike tropical monocot of the genus *Musa* that has a false stem formed by leaf bases that grow concentrically and through which the stalk develops. **2.** the edible fruit of this plant.

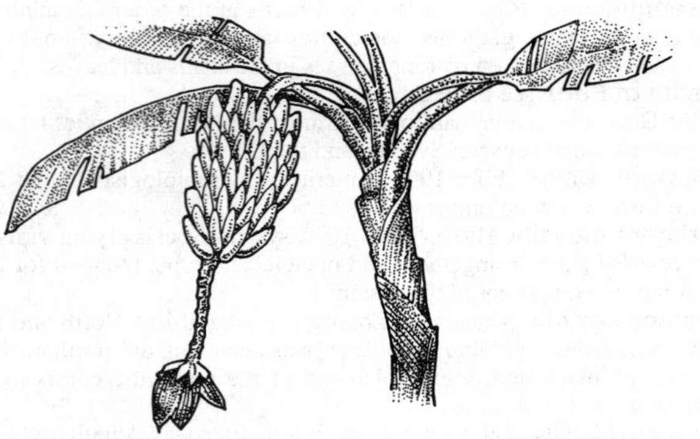

banana plant

banana freckle *Plant Pathology.* a disease of banana plants caused by the fungus *Macrophoma musae,* producing black spots on the fruit and leaves.

banana oil *Organic Chemistry.* **1.** a solution of cellulose nitrate, usually in amyl acetate or a similar solvent; so named because of its banana-like odor. **2.** a popular name for isoamyl acetate, $CH_3COOC_5H_{11}$.

banana plug *Electronics.* a single conductor plug with a spring-metal tip, resembling a banana, used on test leads or as terminals for plug-in components. Thus, **banana jack.**

banatite *Petrology.* a genetic derivative of shoshonite containing orthoclase, plagioclase, and quartz proportionately similar to quartz monzonite but at a lower silica content.

banco *Hydrology.* a meander, oxbow lake, or other part of a stream channel or flood plain that has been cut off and left dry by a change in the course of a stream or a river.

Bancroft's filariasis see WUCHERERIASIS.

band something that binds or restrains; specific uses include: *Anthropology.* a small, autonomous group of related nuclear families, often nomads, who live or move together. *Building Engineering.* a flat horizontal member, occasionally ornamented, that separates a series of moldings or divides a wall surface. Also, BAND COURSE. *Analytical Chemistry.* a demarcated region indicating the separation of a component from a mixture, as in gel electrophoresis or column chromatography. *Spectroscopy.* **1.** closely packed spectral lines that appear to form a continuous group or band. **2.** see BAND SPECTRUM. *Computer Technology.* **1.** on a magnetic disk or drum, a group of contiguous tracks having a common attribute or function. **2.** the type-carrying element of a band printer. *Geology.* any thin rock layer having a distinctive lithology, color, or fossil content. *Solid-State Physics.* a finite range of energies that is occupied by the valence electrons in a semiconductor. *Telecommunications.* **1.** the range of frequency between two specified points or limits. **2.** a specified area of broadcast transmission frequency.

bandage *Medicine.* a strip of protective gauze or other material, used to wrap or bind a part of the body, as for a cut or or other injury.

band brake *Mechanical Engineering.* a brake consisting of a flexible band anchored at one end and wound partially around a drum; frictional force is applied by increasing tension in the band to tighten it around the drum.

band chain *Engineering.* a strong steel or Invar tape that is at least 100 feet long and graduated in feet; used for accurate surveying.

band clamp *Mechanical Devices.* a two-piece flexible metal clamp or band fitted with bolts at each end; used to secure riser pipes or shape chair backs.

band clutch *Mechanical Engineering.* a friction-type clutch, consisting of a fabric-lined steel band that is contracted at the periphery of the clutch rim by means of engaging gear.

band conveyor see BELT CONVEYOR.

band course *Building Engineering.* see BAND.

banded *Petrology.* describing the appearance of rocks with thin and nearly parallel bands of varying textures, colors, and minerals.

banded agate *Mineralogy.* an agate whose colors alternate within parallel bands or stripes that may vary in thickness, and that are formed by silica deposits in irregular rock cavities.

banded anteater *Vertebrate Zoology.* an Australian marsupial of the family Myrmecobiidae that feeds on termites and is marked with whitish transverse bars; somewhat misnamed, since no abdominal pouch is present. Also, NUMBAT.

banded coal *Geology.* a heterogeneous coal composed of thin bands of highly lustrous coalified wood within layers of striated coal.

banded differentiate *Petrology.* a type of igneous rock having layers of different composition, frequently alternating between two varieties as in a layered intrusion.

banded ore *Geology.* an ore composed of layers of different minerals, or of one mineral that is different from layer to layer in color or texture.

banded peat *Geology.* peat composed of alternating bands of decayed vegetable matter.

banded structure *Meteorology.* the appearance of precipitation echoes on a radar plan position indicator (ppi) scope. *Metallurgy.* the structure of a metal or alloy, consisting of parallel bands, typically oriented along the rolling direction. *Petrology.* an outcrop of igneous or metamorphic rocks having nearly parallel layers, strips, flat lenses, or streaks that differ in mineral composition and texture.

banded vein *Geology.* a vein composed of alternating bands of differing minerals lying parallel to the walls. Also, RIBBON VEIN.

band-elimination filter see BAND-STOP FILTER.

band gap *Solid-State Physics.* the minimum energy separation between the highest occupied state and the lowest empty state, that determines the temperature dependence of the electrical conductivity of a pure semiconductor.

band girdle *Forestry.* a band of bark that is removed all around a living trunk to weaken or kill the tree because of disease or to implement forest management goals. Also, RING BARKING.

band head *Spectroscopy.* the wavelength at which the spectral lines of a particular band are more concentrated, making one edge of the band appear sharper than the other.

bandicoot [ban´di koot´] *Vertebrate Zoology.* any of several species of lively nocturnal marsupials belonging to the family Peramelidae, found in Australia, Tasmania, and New Guinea.

bandicoot

banding *Science.* a pattern of differentiated material that is striped in color, texture, or structure. *Forestry.* a method of protecting a tree by encircling the stem with a band of any material that shields it against the passage of insects. *Electronics.* any visible horizontal bands in a video picture that are caused by a defect in videotape recording heads. *Molecular Biology.* the process of staining chromosomes to identify patterns of bands. *Petrology.* the alternate layers of different materials arranged in a thin bed within sedimentary rocks. *Surgery.* the process of encircling and binding with a thin strip of material. *Genetics.* see CHROMOSOME BANDING.

bandit *Military Science.* an enemy aircraft, especially an attacking fighter plane.

bandit problems *Statistics.* problems in sequential allocation of resources to alternative experiments.

band lightning see RIBBON LIGHTNING.

band of fire *Ordnance.* a very dense pattern of fire from automatic guns, used to protect a final line or position.

band of position *Navigation.* an area extending to either side of a line of position within which a craft is believed to be located.

bandolier [ban′də lēr′] *Ordnance.* an ammunition belt worn over one shoulder and across the chest, having loops or pockets for cartridges. Also, **bandoleer.**

bandpass *Electronics.* a range, usually expressed in hertz, that indicates the difference between the limiting frequencies at which a desired fraction of maximum output is obtained.

bandpass amplifier *Electronics.* an amplifier that maintains an essentially uniform response when passing a selected band of frequencies.

bandpass filter *Electronics.* a filter that freely passes signals having frequencies within specified nominal limits and rejects signals with frequencies outside those limits.

bandpass response *Electronics.* an amplitude response that shows an essentially uniform value in a band between two frequencies and high attenuation outside the band. Also, FLAT-TOP RESPONSE.

band printer *Computer Technology.* a type of impact line printer in which the font is carried as embossed characters on a thin steel or polyeurethane band.

band III protein *Biochemistry.* a protein spanning the red blood cell membrane that forms an anion channel through which HCO_3^- is exchanged for Cl in the lungs.

band-rejection filter see BAND-STOP FILTER.

band saw *Mechanical Engineering.* a power-driven endless toothed steel band running on revolving pulleys; used for cutting wood. Also, **band mill, bandsawing machine.**

band scheme *Solid-State Physics.* the identification of allowed energy bands, bandwidths, and band spacings for electron states in a solid.

band selector *Electronics.* a switch that chooses any one of the bands in which a device, such as a receiver, is designed to operate, usually containing two or more sections to allow simultaneous changing in all tuning circuits. Also, BANDSWITCH.

band shape *Spectroscopy.* a distribution of spectral emission or absorption line intensities as a function of frequency; the distribution is typically Gaussian in nature and is due to the motion of atoms or molecules in the specimen.

band spectrum *Spectroscopy.* the spectrum characteristic of molecular gases and chemical compounds consisting of a series of closely spaced emission or absorption lines.

band spreading *Telecommunications.* **1.** in electromagnetic transmission, the artificial dispersion of a signal over a bandwidth that is substantially wider than that of the original signal. **2.** in heavily used radio bands, a widening of tuning scales in order to facilitate tuning.

band-spread tuning *Electronics.* a tuning control on some receivers, in which stations are spread in a single band of frequencies over an entire tuning dial; it controls small variable capacitors connected in parallel with each main tuning capacitor.

band-stop *Electronics.* the ability to suppress signals of a desired frequency or band of frequencies while passing all others.

band-stop filter *Electronics.* a filter that passes signals with all frequencies except those between two specified cutoff frequencies. Also, **band-rejection filter.**

bandswitch see BAND SELECTOR.

bandswitching *Electronics.* a system of inductance or capacitance switching that allows a number of frequency bands to be covered by a single tuning dial.

band theory of solids *Solid-State Physics.* a theory stating that there exist certain restricted energy ranges for the electrons in a solid.

band wheel *Mechanical Engineering.* a large flat pulley over which the main drive belt runs, transmitting power from the engine to the main crankshaft on a churn or cable-system drill.

bandwidth *Telecommunications.* the range or difference between the limiting frequencies of a continuous frequency band. *Mathematics.* let $L(G)$ be the set of all labelings of the vertices of graph G with distinct integers. The bandwidth of G is

$$\min_{l \in L(G)} \; \max_{uv \in E(G)} \; |l(u) - l(v)|$$

bandylite *Mineralogy.* $CuB(OH)_4Cl$, a translucent tetragonal mineral, dark-blue with greenish highlights, with tabular or equant crystals, having a specific gravity of 2.81 and a hardness of 2.5 on the Mohs scale; found as a secondary mineral with atacamite and eriochalcite.

Bang, Bernhard 1848–1932, Danish veterinarian; discovered bacterial agents of Bang's disease and brucellosis.

bang-bang circuit *Electronics.* in an analog computer, an operational amplifier with double feedback limiters driving a high-speed relay.

bang-bang control *Robotics.* a type of control that uses mechanical stops to change the direction of movement.

bang-bang robot *Robotics.* a simple robot that can move only in two directions, both restricted by a mechanical stop.

Bangiaceae *Botany.* a family of red algae belonging to the order Bangiales, having a duo-phased life history of filamentous and parenchymatous stages and including the important genus *Porphyra.*

Bangiales *Botany.* an order belonging to the red algae division Rhodophyta, characterized by genera with more than one morphological phase and filamentous thalli that often have a basal disk.

Bangiophyceae *Botany.* a subclass of mostly marine red algae, characterized by having a single star-shaped chloroplast in each cell.

Bang's disease see CONTAGIOUS ABORTION.

bang-zero-bang control *Robotics.* a type of control using only three values: maximum, minimum, or zero. Also, **bang-bang-off control.**

bank *Geology.* **1.** any elevated slope of earth that borders a body of water, especially the rising ground that confines a stream or a river to its channel. **2.** any steep slope of sand, gravel, or other unconsolidated material. **3.** see EMBANKMENT. **4.** see ORGANIC BANK. *Oceanography.* a level plateau or shelf in the ocean, usually near the continental margin, that is shallow but navigable. *Mining Engineering.* the surface around the mouth of a shaft. *Civil Engineering.* a ridge or mass of earth constructed to carry a railroad or roadway above the natural grade. *Aviation.* to incline an aircraft laterally, particularly when making a turn. *Industrial Engineering.* a quantity of materials on hand and awaiting further processing. *Electricity.* a collection of similar devices used in conjunction with each other as a single device.

bank-and-turn indicator *Aviation.* an instrument combining a bank indicator and a turn indicator in a single housing; used during execution of a turn to avoid slipping or skidding by indicating the rate at which the aircraft is turning and the position of the wings. Also, **bank indicator.**

bank-and-wiper switch *Electricity.* a switch in which electromagnetic ratchets or other mechanisms move wipers first to a desired group of terminals and then over these terminals to the desired bank contacts.

bank cushion *Navigation.* a stretch of water adjacent to the banks of a channel that is sufficiently deep for navigation, but that should be avoided by vessels because of the danger of bank suction.

banked winding *Electronics.* a method of winding radio-frequency coils in which single turns are wound successively in two or more layers in a flat outward spiral, with the entire winding proceeding from one end of the coil to the other without being returned.

banker *Engineering.* a stone or wood bench that is used by masons for dressing stones or bricks.

banket *Geology.* a compact conglomerate consisting of vein-quartz pebbles embedded in a quartzite matrix.

bankfull stage *Hydrology.* the elevation of the water surface of a stream that fills its channel to the maximum capacity without overflowing its banks.

bank gravel *Geology.* any naturally occurring deposit of gravel that is intermixed with fine sand or clay. Also, BANK-RUN GRAVEL.

bank height *Mining Engineering.* the height of a bank measured vertically between its highest point or crest and its toe at the digging level or bench. Also, BENCH HEIGHT; DIGGING HEIGHT.

bank-inset reef *Geology.* a coral reef that rises from a submarine flat at a distance back from its outer margin of rimless shoals.

bank material *Civil Engineering.* any soil or rock in its natural position before excavation.

bank reef *Geology.* **1.** a large reef that is characterized by an irregular shape and the absence of reef-forming organisms. **2.** see BANK-INSET REEF.

bank-run gravel see BANK GRAVEL.

Banks, Sir Joseph 1743–1820, English naturalist; scientific coordinator for Captain James Cook's expeditions.

banksia [bangk´sē ə] *Botany.* any Australian tree or shrub of the genus *Banksia,* with alternate leaves and cylindrical flowers. (Named after Sir Joseph *Banks.*)

bank slope *Mining Engineering.* the angle, measured in degrees of deviation from the horizontal, at which the earthy or rock material will stand in an excavated, terracelike cut in an open-pit mine or quarry. Also, BENCH SLOPE.

banksman *Mining Engineering.* the person in charge of the shaft and cage or skip at the surface of a colliery.

banks oil see COD-LIVER OIL.

bank storage *Hydrology.* water that has been absorbed and retained in the permeable material which forms the banks and bed of a stream. Also, LATERAL STORAGE.

bank suction *Navigation.* an effect encountered especially in narrow channels in which a vessel moving parallel to the bank of the channel is drawn toward the bank, resulting in the risk of grounding.

bank winding see BANKED WINDING.

Banneker, Benjamin 1731–1806, American astronomer, mathematician, and surveyor; helped to survey and lay out the original city of Washington, D.C.

banner *Botany.* the uppermost petal standing at the back of a papilionaceous or other irregular flower. Also, STANDARD, VEXILLUM.

banner cloud *Meteorology.* a cloud trail or plume that extends downwind from an isolated mountain peak and is often observed even on clear days. Also, CLOUD BANNER.

bantam tube *Electronics.* a compact tube that has a standard octal base, but a glass envelope considerably smaller than the standard glass octal tube.

Banting, Sir Frederick Grant 1891–1941, Canadian physician; shared Nobel Prize (with Macleod) for discovery of insulin.

banyan *Botany.* a very large tree, the *Ficus bengalensis* of East India, whose branches send out shoots that root in the soil to form secondary trunks, enabling the tree to cover large areas.

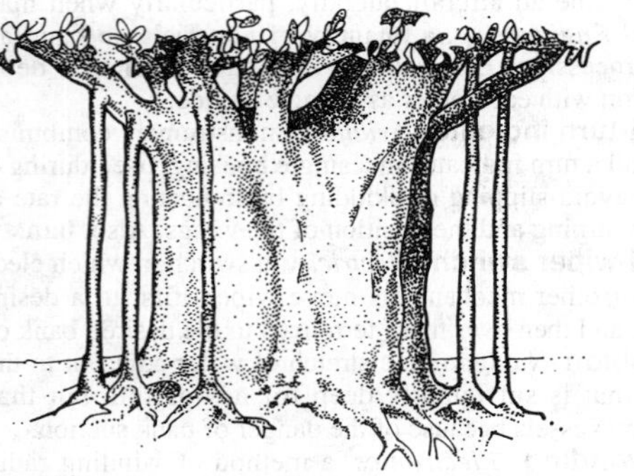

banyan

baobab [bā´ō bäb] *Botany.* any large tree of tropical Africa, genus *Adansonia,* having a thick trunk and bearing a gourdlike fruit.

BAR Browning automatic rifle.

bar a comparatively long, evenly shaped piece of some solid substance; specific uses include: *Oceanography.* an offshore ridge, bank, or mound of sand, gravel, mud, or mollusk shells, submerged at least at high tide; located generally at the mouth of a river or estuary but sometimes parallel to a beach. *Geology.* **1.** any similar deposit built up along the banks of a river or stream. **2.** a mass of hard, barren rock that cuts across a stream bed. **3.** a vein of hard rock that runs through a lode. *Metallurgy.* a metallic mill product, long in proportion to its thickness, and, for round or square bars, having a width-to-thickness ratio of about one. *Mining Engineering.* a drilling or tamping rod. *Metrology.* a unit of pressure equal to one million dynes per square centimeter, or 100,000 newtons per square meter.

bar. barometer; barometric; barrel.

baraboo *Geology.* an isolated remnant of upstanding resistant rock that had been buried and was later uncovered as a result of the partial erosion of overlying strata.

baragnosis *Pathology.* the absence of the ability to recognize weight.

Baragwanathia *Paleontology.* the oldest known lycopod club moss, known from lower Devonian deposits in Australia; one of the earliest vascular plants, with some specimens dating as far back as the upper Silurian.

bararite *Mineralogy.* $(NH_4)_2SiF_6$, a white, low-temperature, hexagonal sublimate mineral having a specific gravity of 2.15 and a hardness of 2.5 on the Mohs scale; occurring most frequently in tabular crystals, but also found in arborescent and mammillary forms; dimorphous with cryptohalite.

barat *Meteorology.* a strong, squall-type west or northwest wind, occurring from December to February over Menado on the north coast of Celebes.

barathea [bar´ə thē´ə] *Textiles.* a tightly woven fabric of silk, cotton, rayon, or wool with a pebbly texture.

Bárány, Robert [bä´rän´ē] 1876–1936, Austrian physician; Nobel Prize for research on the physiology and pathology of the ear.

Bárány chair *Engineering.* a chair that tests the effects of circular motion on humans, especially airplane pilots.

barb *Botany.* a hooked or a sharp bristle. *Vertebrate Zoology.* one of the parallel hairlike projections on the shaft of a feather. *Meteorology.* a short, straight line drawn obliquely toward lower pressure from the end of a wind-direction staff to represent windspeed on a synoptic chart. Also, FEATHER.

Barbados cherry see ACEROLA.

Barbados earth *Geology.* a fine-grained siliceous accumulation of remains of radiolarian fossils occurring in Barbados, West Indies.

barban *Organic Chemistry.* $C_{11}H_9Cl_2NO_2$, a white, crystalline compound that is insoluble in water and soluble in benzene; it melts at 75–76°C; used as an herbicide and plant growth regulator.

barbarism [bär´bə riz´əm] *Anthropology.* in the 19th-century evolutionary view of culture, the intermediate stage of social development, following savagery and preceding civilization.

Barbary ape [bär´bə rē] *Vertebrate Zoology.* a tailless monkey, *Macaca sylvanus,* inhabiting mountain ranges in northwestern Africa. A small colony of unknown origin lives on the Rock of Gibralter and is protected by the British government.

barb bolt *Mechanical Devices.* a bolt with jagged or barbed edges to inhibit withdrawal from the object into which it is driven. Also, RAG BOLT.

bar beach see BARRIER BEACH.

barbed tributary *Hydrology.* a stream that forms a sharp bend pointing upstream as it joins the main flow in an upstream direction.

barbed wire *Materials Science.* a wire with sharp points or barbs twisted on it every few inches; widely used for fencing. Also, **barbwire.**

barbel *Vertebrate Zoology.* **1.** a fleshy whiskerlike sensory growth on the mouth or nostril of some fishes, such as catfishes. **2.** a European freshwater fish of the genus *Barbus* in the family Cyprinidae.

barbellate *Biology.* having short, stiff, hooked hairs or bristles.

bar bending *Civil Engineering.* the bending of reinforcement steel into the shapes required for a reinforced concrete structure.

barber *Meteorology.* **1.** a severe storm at sea during which ocean spray and precipitation freeze on contact with the decks and riggings of ships. **2.** see FROST SMOKE.

barber chair *Forestry.* a high, thick splint, resembling a chair back, left standing on a stump above the undercut because of faulty felling or the heavy lean of a tree.

barberite *Metallurgy.* the name once used for a corrosion-resistant copper-base alloy containing approximately 5% nickel and 2% tin.

barberry *Botany.* a shrub of the genus *Berberis,* having yellow flowers in an elongated cluster, or the fruit of this shrub.

barbertonite *Mineralogy.* $Mg_6Cr_2(CO_3)(OH)_{16}{\cdot}4H_2O$, a pink to purple hexagonal mineral having a specific gravity of 2.1 and a hardness of 1.5 to 2 on the Mohs scale; found in serpentine rock with stichtite and chromite; dimorphous with stichtite.

barbet *Vertebrate Zoology.* a tropical bird of the family Capitonidae, having a stout bill with bristles at its base.

barbette *Naval Architecture.* a nonrotating cylindrical armored bulwark protecting a rotating gun mount or the base of a turret. *Ordnance.* a platform or mound in a fort from which a mounted gun can fire.

Barbeyaceae *Botany.* a monospecific family of northeastern African and Arabian small dioecious trees of the order Urticales; having small, wind-pollinated, apetalous flowers and opposite, simple leaves.

barbicel *Vertebrate Zoology.* a tiny hook on the end of each barbule on a contour feather, but not on a down feather, that holds rows of barbules together. Also, HAMULUS.

barbital *Organic Chemistry.* $C_8H_{12}N_2O_3$, a white crystalline compound that is soluble in hot water, alcohol, ether, and acetone; melts at 187–192°C. A habit-forming hypnotic, it is usually administered orally as a sedative. Also, BARBITONE.

barbital sodium *Pharmacology.* $C_8H_{11}N_2NaO_3$, the soluble sodium salt of barbital; a long-acting barbiturate used as a sedative and to induce sleep. Also, SOLUBLE BARBITAL.

barbitone see BARBITAL.

barbiturate [bär bich´ə rit; bär bich´ə rāt´] *Pharmacology.* any of a class of sedative-hypnotic agents derived from barbituric acid or thiobarbituric acid and classified into long-, intermediate-, short-, and ultrashort-acting classes: ultrashort-acting barbituates, such as thiopental, are used as intravenous anesthetics; long-acting phenobarbital is an important anticonvulsant used in the treatment of epilepsy. Other barbituates have been used as sedatives and hypnotics, but they have generally been replaced by drugs with fewer side effects.

barbituric *Pharmacology.* relating to or or derived from barbituric acid.

barbituric acid *Organic Chemistry.* $C_4H_4N_2O_3$, the parent compound of barbiturates, consisting of large colorless to white crystals that melt at 248°C and are slightly soluble in water and alcohol; derived by condensing malonic acid and urea. Also, PYRIMIDINETRIONE, MALONYLUREA.

barbule [bär´byool] *Vertebrate Zoology.* one of a series of parallel filaments fringing the barbs of a feather and held in place by a small hook at the distal end.

bar buoy *Navigation.* a buoy that marks the location of a bar at the mouth of a river or the approach to a harbor.

barchan *Geology.* a gently sloping, moving sand dune having a crescent shape with horns pointing downwind and lying transverse to the direction of the prevailing winds. Also, BARKHAN, CRESCENTIC DUNE, HORSESHOE DUNE.

bar chart see BAR GRAPH.

bar clamp *Mechanical Devices.* a clamping device consisting of a screw clamp and an adjustable stop mounted on a wood or metal bar; used for clamping two parts together when applying glue or epoxy.

Barclayaceae *Botany.* a monogeneric family of dicotyledonous aquatic herbs of the order Nymphaeales, native from tropical southeastern Asia to New Guinea; characterized by parenchymatous tissue with conspicuous air passages, floating bladelike leaves rising from a rhizome, and solitary extra-axillary flowers.

bar code *Computer Technology.* a code consisting of magnetic ink lines of varying widths that can be read optically with a scanning device; commonly used on product labels at the point of sale in retail stores, using the Universal Product Code System.

bar-code scanner *Computer Technology.* an optical character reader that can read data represented by bar codes and translate the codes into digital signals.

Bardeen, John born 1908, American physicist; twice shared Nobel Prize, for invention of transistor and for theory of superconductivity.

Bardeen-Cooper-Schrieffer theory *Solid-State Physics.* a theory stating that superconductivity is achievable because the conduction electrons in the conductive medium tend to move in spin-paired states, and thus can move at low temperatures through a solid without resistance.

Bardeen-Herring source *Materials Science.* a dislocation source activated by the interaction of climbing and immobilized dislocations.

bar diggings *Mining Engineering.* any gold-washing claims on the shallows of a stream that are worked when the water is low, using cofferdams.

bar drawing *Metallurgy.* the process of producing a bar by drawing it through a die.

bar drill see DRILL COLUMN.

bare board *Electronics.* a printed circuit board that has conductors but not electronic components.

bare charge *Ordnance.* an explosive charge without a casing; used in testing blast characteristics.

bare charm *Particle Physics.* the quality of a particle in which the charm property is clearly evident (as in a 0 meson), as distinguished from hidden charm, in which the states are a combination of charm and anticharm quarks.

bare disk *Electronics.* a floppy disk drive that has no electronic circuit controls.

bare electrode *Metallurgy.* the filler metal in consumable arc welding, consisting of an uncoated wire or rod.

barefaced tenon *Engineering.* a projecting member in a piece of material that is inserted into a mortise to make a joint, characterized by having a shoulder on one side only.

bareface fabric *Textiles.* a smooth fabric with no nap.

bare fallow *Agronomy.* land not sown for a season, but kept in cultivation to keep the vegetation down.

barefoot completion see OPEN-HOLE COMPLETION.

barefoot well *Petroleum Engineering.* a well that has no casing, screen, or perforated pipe and shows no need of one; limestone or sandstone wells that give no indication of disintegration or caving in are often completed without a casing or liner, so that reservoir fluids flow unrestricted into the open wellbore.

barege [bə rezh´] *Textiles.* a sheer fabric in a leno weave of silk warp and a cotton or worsted filling; used to make veils and dresses. Also, **barége.**

bare rock *Navigation.* a seabed with no mud or sand to hold an anchor; provides poor holding ground, and is avoided by anchoring vessels.

bare-root *Agronomy.* of or relating to bare-root planting.

bare-root planting *Agronomy.* a transplanting technique in which the plant is placed in the ground with very little surrounding soil on the roots. Also, NAKED-ROOT PLANTING, OPEN-ROOT PLANTING.

baresthesia *Neurology.* the sensibility of weight or pressure. Also, BARYESTHESIA.

bare value *Quantum Mechanics.* the value of some property of a particle, such as mass, in the absence of any external interactions.

bar-finger sand *Geology.* a deposit of coarse sediment that forms an elongated body underlying a distributary channel in a bird-foot delta, and grows seaward in the form of long fingers. Also, **bar finger.**

Barfoed's test *Analytical Chemistry.* a test for monosaccharides using an acid solution; a red precipitate indicates a positive test.

barge *Naval Architecture.* **1.** a flat-bottomed, towed, unpowered cargo vessel. **2.** the largest boat of a battleship.

barge board *Building Engineering.* a normally ornamental board that is positioned at the gable end of a roof to hide the ends of horizontal timbers and protect them from weathering. Also, VERGEBOARD.

barge couple *Building Engineering.* **1.** either of a pair of rafters that carry the part of a gable roof that projects beyond the gable wall. **2.** a rafter under the barge course that serves as a base for the barge boards and carries the plaster for the soffits. Also, BARGE RAFTER.

barge course *Building Engineering.* **1.** the tiling on a gable roof beyond the exterior surface of a gable wall. **2.** the tiles, bricks, or slates that are put upon and project over the raking edges of a gable roof. **3.** a coping course of bricks placed on edge and arranged transversely on a wall.

bar generator *Electronics.* a generator of pulses that are equally separated in time, synchronized by the synchronizing pulses of a television system; used to produce a stationary bar pattern as a test pattern for a television picture.

barge rafter see BARGE COUPLE.

barge spike see BOAT SPIKE.

barge stone *Building Engineering.* one of a number of stones that form the sloping edge of a gable.

bar graph *Statistics.* a graphic display in which rectangles are used to portray data, usually with the bases of the rectangles representing class intervals and their heights representing the corresponding class frequencies. Also, BAR CHART, HISTOGRAM.

bar hole *Engineering.* a small hole that runs underground along the path of a gas pipe in a bar test survey.

baric *Meteorology.* of or relating to weight, particularly that of the atmosphere.

baric topography see HEIGHT PATTERN.

baric wind law see BUYS-BALLOT'S LAW.

barines *Meteorology.* westerly winds in eastern Venezuela.

baring *Mining Engineering.* the surface soil and useless strata overlying a seam of coal, clay, or ironstone that has to be removed in preparation for working the mineral. *Geology.* see OVERBURDEN, def. 1.

Bari-Sol process *Chemical Engineering.* a method of removing waxes from liquid hydrocarbons by extracting the wax with a solvent of mixed ethylene dichloride and benzene, and then separating it from the hydrocarbons in a centrifuge.

barite *Mineralogy.* $BaSO_4$, a white, yellow, blue, red, brown, or gray orthorhombic mineral with perfect cleavage and a vitreous luster, having a specific gravity of 4.5 and a hardness of 3 to 3.5 on the Mohs scale; found in medium- to low-temperature hydrothermal vein deposits; it forms a series with celestine. Also, BARYTE.

barium [bâr′ē əm] *Chemistry.* a chemical element of the alkaline-earth group, having the symbol Ba, the atomic number 56, an atomic weight of 137.34, a melting point of about 725°C and a boiling point of about 1640°C. It is an extremely reactive, silver-white, somewhat malleable metal. (From a Greek word meaning "heavy.")

barium-140 *Nuclear Physics.* a radioactive isotope of barium with a half-life of 12.8 days that emits beta particles during decay.

barium acetate *Inorganic Chemistry.* $Ba(C_2H_3O_2)_2{\cdot}H_2O$, white crystals, soluble in water and insoluble in alcohol; a barium salt made by adding acetic acid to a solution of barium sulfide; used as a paint and varnish dryer.

barium azide *Inorganic Chemistry.* $Ba(N_3)_2$, a crystalline solid that is soluble in water and slightly soluble in alcohol; used in high explosives.

barium bromate *Inorganic Chemistry.* $Ba(BrO_3)_2{\cdot}H_2O$, white poisonous crystals or powder, slightly soluble in water and insoluble in alcohol; decompose at 260°C; used for preparing other bromates and as an analytical reagent and corrosion inhibitor.

barium bromide *Inorganic Chemistry.* $BaBr_2{\cdot}2H_2O$, colorless poisonous crystals, soluble in water and alcohol; used in photography and formerly used in medicine.

barium carbonate *Inorganic Chemistry.* $BaCO_3$, a white powder, insoluble in water and soluble in acids; occurring in nature as witherite; used to remove sulfates from brines before they are introduced into electrolytic cells,and also used in rodent poisons, ceramics, optical glass, and television picture tubes.

barium chlorate *Inorganic Chemistry.* $Ba(ClO_3)_2{\cdot}H_2O$, colorless prisms or white powder, soluble in water and melting at 414°C; a strong oxidizer that is used in explosives and fireworks.

barium chloride *Inorganic Chemistry.* $BaCl_2{\cdot}H_2O$, colorless, flat crystals, soluble in water and insoluble in alcohol; a toxic salt that is used as a rodent poison and as a reagent; formerly used in medicine.

barium chromate *Inorganic Chemistry.* $BaCrO_4$, a heavy yellow crystalline powder that is soluble in acids and insoluble in water; used as a pigment, in safety matches, as a corrosion inhibitor and metal primer, and for various other purposes.

barium citrate dihydrate *Organic Chemistry.* $Ba_3(C_6H_5O_7)_2{\cdot}2H_2O$, a grayish-white toxic crystalline powder that is soluble in water and hydrochloric and nitric acid; used for stabilizing latex paints.

barium cyanide *Organic Chemistry.* $Ba(CN)_2$, a very posionous white crystalline powder that is soluble in water and alcohol and decomposes in air; used in metallurgy and electroplating.

barium dioxide see BARIUM PEROXIDE.

barium enema *Radiology.* a contrast solution of barium sulfate that is injected into the rectum to facilitate X-ray examination of the intestines.

barium fluoride *Inorganic Chemistry.* BaF_2, a white powder that is slightly soluble in water; derived from barium sulfide and hydrofluoric acid; used as a dry-film lubricant and in spectroscopy.

barium fluorosilicate *Inorganic Chemistry.* $BaSiF_6H$, a white crystalline powder that is insoluble in water; used in ceramics and insecticides. Also, BARIUM SILICOFLUORIDE.

barium fuel cell *Electricity.* a fuel cell that uses barium with either oxygen or chlorine to convert chemical energy into electrical energy.

barium hydroxide *Inorganic Chemistry.* $Ba(OH)_2$, a compound that exists both in an anhydrous form and in several hydrated forms; used in a wide variety of chemical processes, such as sugar refining, fat saponification, and water softening or purification.

barium iodide *Inorganic Chemistry.* $BaI_2{\cdot}H_2O$, colorless crystals, decomposing and reddening on exposure to air; soluble in water and slightly soluble in alcohol; used in the manufacture of other iodides.

barium manganate *Inorganic Chemistry.* $BaMnO_4$, an emerald green powder that is insoluble in water; used as a paint pigment.

barium mercury iodide see MERCURIC BARIUM IODIDE.

barium molybdate *Inorganic Chemistry.* $BaMoO_4$, a white powder that is slightly soluble in water and acids; used in electronic and optical equipment and as a paint pigment.

barium monoxide see BARIUM OXIDE.

barium nitrate *Inorganic Chemistry.* $Ba(NO_3)_2$, a crystalline salt, soluble in water and insoluble in alcohol; often used in fireworks because it gives off a green light on ignition; also used in explosives and ceramics.

barium nitrite *Inorganic Chemistry.* $Ba(NO_2)_2{\cdot}H_2O$, a white to yellowish crystalline powder, soluble in water and alcohol; used in explosives.

barium oxide *Inorganic Chemistry.* BaO, a white to yellow powder that melts at 1923°C, absorbs carbon dioxide readily from the air, and reacts with water to form the hydroxide; used as a dehydrating agent and as a detergent for lubricating oils.

barium perchlorate *Inorganic Chemistry.* $Ba(ClO_4)_2{\cdot}4H_2O$, colorless crystals, soluble in water and methanol; used in explosives and as a rocket fuel.

barium permanganate *Inorganic Chemistry.* $Ba(MnO_4)_2$, a brownish to violet crystalline compound that is soluble in water; used as a strong disinfectant and to depolarize dry cells.

barium peroxide *Inorganic Chemistry.* BaO_2, a grayish-white powder that is insoluble in water, and melts at 450°C; a strong oxidizing agent, used in bleaching and in the manufacture of hydrogen peroxide.

barium protoxide see BARIUM OXIDE.

barium silicate *Inorganic Chemistry.* $BaSiO_3$, a colorless powder, melting at 1604°C; insoluble in cold water and decomposes in hot water; used in ceramics.

barium silicide *Inorganic Chemistry.* $BaSi_2$, a gray solid that evolves hydrogen gas upon exposure to water; used to deoxidize steel.

barium silicofluoride see BARIUM FLUOROSILICATE.

barium star *Astronomy.* a red giant or supergiant star whose spectrum shows abnormally strong lines of barium.

barium stearate *Organic Chemistry.* $Ba(C_{18}H_{35}O_2)_2$, a white crystalline solid that melts at 160°C and is insoluble in water and alcohol; used as a lubricant, in plastic and rubber compounds, and as a waterproofing agent.

barium sulfate *Inorganic Chemistry.* $BaSO_4$, a white powder that is odorless and tasteless, having a specific gravity of 4 to 4.5; most of its industrial applications are due to its unusually high density, including its widespread use in oil well drilling muds.

barium sulfide *Inorganic Chemistry.* BaS, a compound that exists both as yellowish-green crystals and as a gray powder; decomposes in water to the hydrosulfide; used in luminous paints and as a flame retardant.

barium sulfite *Inorganic Chemistry.* $BaSO_3$, a white powder that is decomposed by heat; insoluble in water and soluble in dilute hydrochloride acid; used in paper manufacturing.

barium superoxide see BARIUM PEROXIDE.

barium swallow *Radiology.* a contrast solution of barium sulfate ingested by a patient for an X-ray examination of the upper gastrointestinal area.

barium tetrasulfide *Inorganic Chemistry.* $BaS_4{\cdot}H_2O$, red or yellow rhombic crystals, soluble in water.

barium thiocyanate *Inorganic Chemistry.* $Ba(SCN){\cdot}2H_2O$, white deliquescent crystals, soluble in water and alcohol; used in dyeing and in photography.

barium thiosulfate *Inorganic Chemistry.* $BaS_2O_3{\cdot}H_2O$, a white powder that is decomposed by heat; slightly soluble in water and insoluble in alcohol; used in matches, explosives, luminous paints, and varnishes.

barium titanate *Inorganic Chemistry.* $BaTiO_3$, a light gray powder that is insoluble in water and soluble in concentrated sulfuric and hydrofluoric acids; single crystals are often doped with iron and used as a ferroelectric ceramic.

barium tungstate *Inorganic Chemistry.* $BaWO_4$, a white powder that is insoluble in water; used in X-ray photography and as a pigment.

barium yellow see BARIUM CHROMATE.

bar joist *Building Engineering.* a welded steel joist with an open web consisting of a single bent bar running in a zig-zag pattern between roof and floor supports.

bark *Botany.* a protective corky tissue of dead cells on the outer part of the stems of woody plants. *Forestry.* to remove the bark from a tree or a piece of round timber. *Naval Architecture.* a three-masted ship with all masts square-rigged but the aftermost, which is fore-and-aft-rigged.

bark bar *Forestry.* the strip of bark left between two adjacent faces. Also, INTERSPACE, INTERFACE.

bark cloth *Textiles.* a cloth made by soaking and pounding certain kinds of tree bark to a desired thinness.

bark crepe *Textiles.* a crepe fabric textured to resemble tree bark.

bar keel *Naval Architecture.* a keel constructed of long flat iron bars overlapped and joined lengthwise; it connects the transverse members and bulkhead.

barker *Engineering.* **1.** a machine that removes bark from logs. **2.** a person who removes bark and dirt from wood using a steam barker or drum barker.

Barker method *Crystallography.* a procedure based on the recognized constancy of interfacial angles that allows identification of crystals by morphological means alone.

barkevikite *Mineralogy.* a term formerly used for a monoclinic sodic amphibole that is close to arfvedsonite but more calcic. Also, FERROHORNBLENDE.

bark graft *Botany.* a plant graft made by slitting the bark of the stock and slipping the scion beneath it.

barkhan see BARCHAN.

Barkhausen, Heinrich [bärk´hou´zən] 1881–1956, German physicist and electrical engineer; discovered Barkhausen effect in ferrous metals.

Barkhausen effect *Electromagnetism.* a succession of small, abrupt steps in the magnetization of a ferromagnetic material as observed when the material is subjected to a steadily increasing or decreasing external magnetic field.

Barkhausen interference *Telecommunications.* interference that is caused by unwanted oscillations in the horizontal output tube of a TV monitor. (So called because such interference may be produced by a Barkhausen-Kurz oscillator.)

Barkhausen-Kurz oscillator *Electronics.* a triode oscillator having a positively biased grid and a negatively biased anode; oscillation depends on the transit time of electrons between cathode and anode. Also, **Barkhausen oscillator.**

Barkhausen oscillation *Electronics.* an unwanted oscillation in the horizontal-output tube of a television receiver, resulting in narrow, dark, ragged vertical lines near the left side of the picture or raster.

Barkla, Charles 1877–1944, English physicist; showed that the number of electrons in an atom is about half its atomic weight.

bark louse *Inveterbrate Zoology.* any of the insects of the order Psocoptera that live on tree bark and other plants.

bark pocket *Botany.* a patch of bark that is partially or completely enclosed in the annual rings of a tree.

bark slip *Forestry.* the separation of bark from the wood of a tree at a functioning layer of cambium.

bark-splitting disease see ARMILLARIA ROOT ROT.

bark spud *Mechanical Devices.* a tool used to remove bark from logs.

barley *Botany.* a grass of the genus *Hordeum* and its edible grains, characterized by long slender awns and spikelets at the node of a flattened rachis; its principal use is as a cereal and in the manufacture of malt.

barley coal *Mining Engineering.* any stream-sized anthracite sized on a round punched plate with approximately quarter-inch holes. Also, BUCKWHEAT NO. 3.

barley scald *Plant Pathology.* a disease of the barley plant caused by the fungus *Rhynchosporium secalis* that produces yellowish to blue-green blotches and blighted foliage.

barley smut *Plant Pathology.* a destructive fungus disease of barley characterized by the appearance of dusty masses of spores on the plant organs; having either a loose form (caused by the fungus *Ustilago nuda)* or a covered form (caused by the fungus *Ustilago hordei).*

barley stripe *Plant Pathology.* a barley disease caused by the diffusible toxin of the fungus *Helminthosporium gramineum* and characterized by the appearance of light-colored stripes on the leaves.

barley stripe mosaic virus group see HORDEIVIRUS GROUP.

barley yellow mosaic virus group *Virology.* a nontaxonomic group of plant viruses with flexuous rod-shaped particles; transmitted in nature by fungi.

bar linkage *Mechanical Engineering.* a system of rigid members or bars joined at pivots with pins; used to transmit power or change direction. Also, LINK MECHANISM.

Barlow lens *Optics.* a lens system used in telescopes, in which one or more strongly negatively powered lens elements are used to increase the effective focal length and thereby increase magnification.

Barlow's equation *Mechanical Engineering.* an equation used to compute the strength of a cylinder under internal pressure, $t = DP/4S$, where t is the minimum thickness of the cylinder, D is its diameter, P is the internal pressure in excess of the external, and S is the allowable tensile strength.

Barlow's rule *Physical Chemistry.* a rule that the space occupied by atoms in a given molecule is proportional to their lowest valence value.

bar magnet *Electromagnetism.* a straight bar-shaped magnet that has been strongly magnetized to serve as a permanent magnet.

barmine *Ordnance.* an antitank mine constructed of plastic, having a long and narrow shape that is well adapted to mechanical laying.

bar mining *Mining Engineering.* the alluvial mining of sandbanks, river bars, or submerged deposits, usually between low and high waters.

barn *Agriculture.* a farm building used to store hay, grain, and farm implements, or to house farm animals. *Nuclear Physics.* a unit of measure for nuclear reaction cross sections, proportional to the probabilities of nuclear reaction, and equal to 10^{-24} cm^2 (or 10^{-28} m^2); the term is derived from a wartime code word describing a particular cross section as "as big as a barn."

barnacle *Invertebrate Zoology.* the common name for any of various marine crustaceans of the order Cirripedia; the larvae are free-swimming, but adults form a hard outer shell and affix to an underwater surface. They may be stalked or sessile, parasitic or nonparasitic.

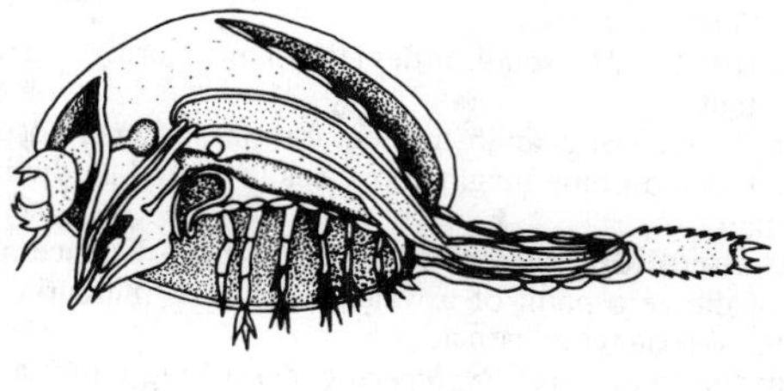

barnacle

Barnard, Christiaan [bär´närd´] born 1922, South African surgeon; performed first human heart transplant.

Barnard, Edward E. [bär´nərd] 1857–1923, American astronomer; discovered fifth satellite of Jupiter and dark nebulae of the Milky Way.

Barnard's loop [bär´nərdz] *Astronomy.* a gigantic shell of hydrogen gas visible in long-exposure photographs, roughly centered on the Belt and Sword region of Orion.

Barnard's star [bär´nərdz] *Astronomy.* a 10th-magnitude star in Ophiuchus that has the greatest known apparent motion on the celestial sphere, 10.3 seconds of arc per year; at 5.9 light-years' distance, it is also the second-nearest star to the sun.

Barnett effect *Electromagnetism.* an observed slight magnetization developed in an initially unmagnetized iron rod when the rod is rotated at high speeds about its longitudinal axis.

Barnett method *Electromagnetism.* the determination of the gyromagnetic moment of a ferromagnetic material by means of the Barnett effect.

Barnette's conjecture *Mathematics.* the conjecture that, if G is a 3-connected cubic bipartite planar graph, then G is Hamiltonian.

barney *Mining Engineering.* a small car or truck attached to a rope or cable that pushes cars up a slope or inclined plane. Also, BULLFROG, DONKEY, GROUNDHOG, RAM.

barn owl *Vertebrate Zoology.* a white and golden brown owl, *Tyto alba,* with a mostly white heart-shaped face; found in hollow trees in forests near open country as well as in barns and silos in farmlands.

Barnum effect *Psychology.* a tendency for people to accept very general or vague characterizations as applying specifically and accurately to themselves, as in personality profiles, astrological predictions, and so on. (Referring to the comments of P.T. *Barnum,* 1810–1891, American showman, about public gullibility.)

baro- a combining form meaning "pressure," as in *barotaxis,* or sometimes "weight," as in *baromacrometer.*

baroceptor *Physiology.* a pressure-sensitive receptor organ of the nervous system, found, for example, in the walls of blood vessels.

baroclinic *Physics.* of, relating to, or having the property of baroclinity.

baroclinic disturbance. *Meteorology.* the cyclone migration associated with a strong baroclinity of the atmosphere, as evidenced on synoptic weather charts by temperature gradients in the constant-pressure surfaces, vertical wind shear, tilt of pressure troughs with height, and concentration of solenoids in the frontal surface near the ground. Also, BAROCLINIC WAVE.

baroclinic field *Meteorology.* a distribution of atmospheric elements of pressure and mass in such a manner that the specific volume or density of the air is not solely a function of pressure.

baroclinic instability *Meteorology.* an atmospheric hydrodynamic instability arising from the existence of a meridional temperature gradient and resulting in a thermal wind in an atmosphere in quasi-geostrophic equilibrium with static stability.

baroclinic wave see BAROCLINIC DISTURBANCE.

baroclinity *Physics.* a state of fluid stratification in which isobaric surfaces and isosteric surfaces are not parallel, but intersect. Also, **baroclinicity, barocliny.**

baroduric see BAROTOLERANT.

barodynamics *Mechanics.* the study of the mechanics of heavy structures that are liable to collapse under their own weight.

barognosis *Neurology.* the conscious perception of weight; the faculty for recognizing weight.

barogram *Engineering.* a graphic representation of changes in atmospheric pressure, as measured by a barograph.

barograph *Engineering.* **1.** an instrument that continuously and automatically records changes in pressure on a rotating drum. Also, BAROMETROGRAPH. **2.** see ANEROID BAROGRAPH.

Barolo [bə rō´lō] *Food Technology.* a heavy, acidic Italian red wine made with Nebbiolo grapes.

baromacrometer *Medicine.* a device for measuring the weight and length of infants.

barometer *Engineering.* an instrument for measuring atmospheric pressure; used in determining height above sea level and predicting changes in the weather.

barometer elevation *Meteorology.* the vertical distance above mean sea level of the zero point of a weather station's mercurial barometer. Also, ELEVATION OF IVORY POINT.

barometric [bâr´ə met´rik] *Engineering.* describing information derived from the use of a barometer. Thus, **barometric pressure.** *Physics.* relating to pressure, particularly atmospheric pressure.

barometric altimeter see PRESSURE ALTIMETER.

barometric condenser *Mechanical Engineering.* a condenser used in steam jet refrigeration systems, in which steam inside a long vertical pipe is condensed at reduced pressure by direct contact with a countercurrent rain of cooling water.

barometric draft regulator *Mechanical Engineering.* a mechanically balanced damper that rotates with pressure changes within the breeching or chimney connector of a boiler, so as to bleed air into the breeching to maintain a steady overfire draft in the combustion chamber. Also, **barometric damper.**

barometric elevation *Meteorology.* the actual elevation, above mean sea level, of a weather service station as determined by the station's mercurial barometer.

barometric error *Horology.* an error in a timepiece caused by the fluctuations in density of the atmosphere through which the balance or pendulum moves.

barometric gradient see PRESSURE GRADIENT.

barometric hypsometry *Engineering.* the use of mercurial or aneroid barometers to determine elevations.

barometric leveling *Cartography.* a process of surveying, in which barometric readings assist in determining the approximate differences in the elevation of an area; generally used for surveying large areas.

barometric surface *Physics.* an isobaric surface

barometric switch see BAROSWITCH.

barometric tendency see PRESSURE TENDENCY.

barometric tide *Geophysics.* the daily rising and falling of atmospheric pressure caused by the gravitational effects of the sun and the moon.

barometric wave *Meteorology.* any wave in the atmospheric pressure field; usually short-term variations not associated with cyclonic scale motions or with atmospheric tides.

barometrograph see ANEROID BAROGRAPH.

barometry [bə räm´ə trē´] *Engineering.* the scientific study of the measurement of atmospheric pressure.

baromil *Mechanics* a unit of measure in the CGS system used in a graduated mercury barometer.

baropacer *Cardiology.* an electronic unit implanted in the necks of dogs for continuous stimulation of the carotid artery.

barophilic *Microbiology.* relating to or describing a microorganism that grows optimally or obligately at high atmospheric pressure; said especially of certain deep-sea bacteria. Thus, **barophile.**

barophobia *Psychology.* an irrational fear of the force of gravity.

Baroque [bə rōk´] *Architecture.* the predominant style in 17th-century European architecture and design, characterized by its elaborate use of curved surfaces, sculpturing, decoration, and color.

baroque dolomite *Mineralogy.* a white, opaque dolomite with large, curved, saddle-shaped crystals that is found in sulfide ore deposits and veins.

baroscope *Engineering.* an instrument that shows changes in the pressure of the atmosphere.

barostat *Engineering.* a device that regulates and maintains pressure at a constant value within a chamber.

baroswitch *Engineering.* a switch that is operated by changes in atmospheric pressure.

barotaxis *Biology.* the stimulation of living matter by change of the pressure relations under which it exists.

barothermogram *Engineering.* a graphic representation of pressure and temperature readings, as made by a barothermograph.

barothermograph *Engineering.* an automatic instrument used to record temperature and pressure.

barothermohygrogram *Engineering.* a graphic representation of readings produced by a barothermohygrograph.

barothermohygrograph *Engineering.* an automatic instrument that simultaneously records the temperature, pressure, and humidity of the atmosphere.

barotolerant *Microbiology.* relating to or describing a microorganism that grows optimally at standard atmospheric pressure but is able to grow at higher pressures as well. Also, BARODURIC.

barotropic [bâr´ō trōp´rik] *Meteorology.* relating to or characterized by an atmospheric condition in which surfaces of equal pressure coincide with surfaces of equal density. *Physics.* having a density that is a function solely of pressure.

barotropic atmosphere see BAROTROPIC MODEL.

barotropic disturbance *Meteorology.* a cyclonic scale atmospheric wave in which areas of low pressure (troughs) and areas of high pressure (ridges) are approximately vertical. Also, **barotropic wave.**

barotropic field *Meteorology.* a distribution of atmospheric pressure and mass so that the specific volume is solely a function of pressure.

barotropic model *Meteorology.* a model atmosphere with zero horizontal temperature gradient at all levels so that the isopleths of density and pressure coincide, and the thickness chart has no pattern. Also, BAROTROPIC ATMOSPHERE.

barotropy *Physics.* a state of a fluid in which the density is a single-valued function of the pressure, in which isosteric surfaces coincide everywhere with isobaric surfaces.

bar pattern *Electronics.* a color pattern of repeating bars or lines produced by a bar generator for adjusting color television receivers.

bar plain *Geology.* a smooth plain formed by a braided stream, characterized by a web of interconnected, elongated, irregularly sized bars.

bar printer *Computer Technology.* an impact printer in which the type slugs are carried on a type bar that moves vertically to select the desired character; type bars are positioned side-by-side in every print position.

barracuda [bâr´ə kood´ ə] *Vertebrate Zoology.* any of numerous species of highly aggressive carnivorous marine fish of the family *Sphyraenida* in the order Perciformes; characterized by very strong, sharp teeth.

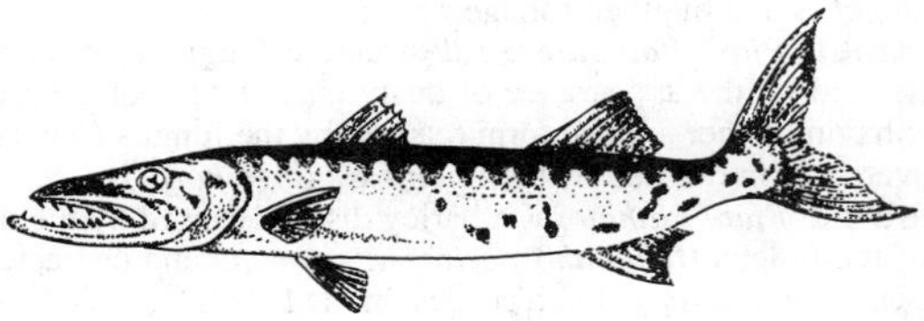

barracuda

barracudina *Vertebrate Zoology.* a deep-sea fish of the family Paralepididae in the order Myctophiformes, having a long, slender body, large eyes, a long slender snout, and a large mouth with fanglike teeth.

barrage *Ordnance.* a concentration of artillery fire to protect friendly troops or halt the movement of enemy forces. *Civil Engineering.* a low dam, with gates running its full length, that increases the depth of a river or water course, or diverts it for irrigation or navigation. *Mycology.* an aversion response of sexually incompatible fungus cultures growing in proximity; characterized by a persistent growth gap between them.

barrage balloon *Ordnance.* a balloon or blimp moored to the ground by cables; used as part of an antiaircraft defense, by presenting a threat to low-flying aircraft.

barrage jamming *Telecommunications.* intentional signal jamming accomplished by transmitting a band of frequencies that is large in relation to the bandwidth of a single emitter, either by presenting multiple jammers on adjacent frequencies or by using a single wideband transmitter.

barrage reception *Telecommunications.* reception in which interference of radio signals from any particular direction is minimized by selecting from among a number of directional antennas the one that gives the maximum signal/interference ratio.

barrage-type spillway *Civil Engineering.* a route for excess water with a sluice gate across the width of the entrance.

barranco [bə rän´kō] *Geology.* **1.** a deep, steep-sided hole or break formed by heavy rains in the southwestern United States. Also, **barranca. 2.** a deep, steep-sided drainage valley formed by erosion on the side of a volcanic cone.

Barr body *Genetics.* a condensed, inactivated X chromosome found in the nuclei of somatic cells of females, which condenses into heterochromatin and is visible during mitotic interphase; the total number of Barr bodies is always one less than the total number of X chromosomes present in the cell or organism.

barré [bä rā´] *Textiles.* a bar, stripe, or defect in fabric that runs crosswise, or parallel to the weft.

barred basin see RESTRICTED BASIN.

barred gate *Civil Engineering.* a gate with horizontal timber rails.

barred spiral galaxy *Astronomy.* a type of spiral galaxy whose nucleus is crossed by a bright elongated bar from which the spiral arms extend.

barrel *Mechanical Devices.* a circular container that is widest at its middle and is held together by hoops and staves. *Ordnance.* the tubular part of a gun through which the projectile is fired. *Optics.* the housing that contains the optical devices used in camera lenses and iris diaphragms. *Naval Architecture.* the cylindrical central portion of a windlass or capstan, around which a line or chain is wound. Also, BARREL OF CAPSTAN, DRUM. *Horology.* a cylinder in a spring-driven timepiece that holds the mainspring and is connected by gear teeth or a fusee to the gear trains. *Metrology.* any of various units of measure based on the capacity of a standard barrel, typically in the U.S. 31.5 gallons for liquids and 105 quarts for dry materials. *Mathematics.* a closed, convex, balanced, absorbing subset of a topological vector space.

barrel arch *Architecture.* an arch with a cross section in which the length is greater than the span, resembling a barrel.

barrel assembly *Ordnance.* the barrel of a gun and the parts that attach it to the rest of the gun.

barrel bolt *Mechanical Devices.* a sliding cylindrical door bolt between one side of the door edge and the socket on the door frame.

barrel cactus *Botany.* any of several large, cylindrical, ribbed, spiny cacti of the genera *Echinocactus* and *Ferocactus.*

barrel ceiling *Architecture.* a ceiling that is semicylindrically shaped.

barrel chest see EMPHYSEMATOUS CHEST.

barrel compressor *Mechanical Engineering.* a centrifugal compressor with a barrel-shaped housing that partially depends upon centrifugal forces for pressure increase.

barrel distortion *Optics.* a distortion produced by a lens or lens system in which lateral magnification is greater at the central area of an image than at its periphery, causing the image of a square grid to appear barrel-shaped. Also, POSITIVE DISTORTION.

barrel drain *Building Engineering.* a cylindrical drain.

barreled space *Mathematics.* a topological vector space in which every barrel in the space contains a neighborhood of zero.

barrel elevator see ARM ELEVATOR.

barrel fitting *Mechanical Devices.* a short threaded fitting used to connect pipe ends.

barreling *Metallurgy.* the formation of convex surfaces in a cylindrical or conical body, caused by compression.

barrel life *Ordnance.* the number of rounds that may be fired through a gun barrel before it loses its effectiveness.

barrel of capstan *Naval Architecture.* see BARREL.

barrel plating *Metallurgy.* a process of plating articles in a rotating container, usually a perforated cylinder, that operates at least partially submerged in a solution.

barrel printer *Computer Technology.* an impact line printer that uses a complete font of characters embossed on a drum in each printing position; a print hammer at each position strikes paper and inked ribbon against a character as it rotates by the drum.

barrel roll *Aviation.* a maneuver in which a plane performs a complete roll by revolving once around its longitudinal axis.

barrels per day *Chemical Engineering.* a unit of measurement that indicates the rate of petroleum production at a refinery.

barrels per stream day *Chemical Engineering.* a unit of measurement that indicates the rate of oil or oil-product flow at a fluid-processing unit in continuous operation.

barrel vault or **barrel roof** *Architecture.* a semicylindrical vault supported by parallel walls or arcades. Also, CRADLE VAULT; TUNNEL VAULT; WAGON VAULT.

Barremian *Geology.* a European geologic stage of the Lower Cretaceous period, occurring after the Hauterivian and before the Aptian.

barren *Biology.* **1.** unable to support life. **2.** unable to produce seeds or offspring.

barren-ground caribou *Vertebrate Zoology.* the race of caribou found in the Barren Ground region, somteimes considered a separate species.

Barren Grounds *Geography.* a sparsely inhabited tundra region in northern Canada, especially west of Hudson Bay. Also, **Barren Lands.**

barrens *Geography.* an area of poor, often sandy soil supporting scant vegetation.

barretter [bar´et ər] *Electronics.* an iron-wire resistor with a positive temperature coefficient of resistivity; sometimes used in microwave power measuring units, but more suited as a microwave signal detector. *Electricity.* see BALLAST RESISTOR.

barricade *Military Science.* an obstacle placed across a road or path to impede or prevent the movement of enemy personnel and vehicles. *Engineering.* a structure made for the purpose of lessening or confining the effects of an explosion.

barricade shield see BARRIER SHIELD.

barrier anything that prevents passage or hinders; specific uses include: *Navigation.* an obstruction that prevents passage of a craft. *Electrical Engineering.* the solid insulating material in transformers that provides the main insulation, apart from the oil. *Ecology.* any physical, biological, or climatic factor that hinders or prevents the spread of a population. *Physics.* **1.** a region where the potential energy is greater than the total energy of a particle, rendering the region inaccessible to the particle. **2.** the depletion layer of a P-N junction in a diode.

barrier bar see LONGSHORE BAR.

barrier basin *Geology.* a basin produced through the action of a natural obstacle such as a dam, landslide, or moraine.

barrier beach *Geology.* an elongated offshore ridge composed of sand or other unconsolidated material that lies roughly parallel to the shore but separated from it by a lagoon or other water area, and emerges slightly above the high-tide level. Also, BAR BEACH.

barrier capacitance see DEPLETION-LAYER CAPACITANCE.

barrier chain *Geology.* a succession of barrier beaches, barrier spits, and barrier islands that extends along a considerable length of coastline.

barrier curb *Civil Engineering.* a curb high enough to hold back vehicles.

barrier flat *Geology.* a relatively level region, often occupied by pools of water, lying between the seaward edge of a barrier and the lagoon on the landward side.

barrier-grid storage tube see RADECHON.

barrier ice *Hydrology.* **1.** the ice that makes up the Antarctic ice shelf. **2.** see SHELF ICE.

barrier injection transit-time diode *Electronics.* a microwave diode in which the carriers that cross the drift region are generated by minority carrier injection from a forward-biased junction.

barrier island *Geology.* **1.** a broadened barrier beach or a dynamic, long, narrow, wave-built island lying parallel to the shore. **2.** a separate section of barrier beach lying between two inlets.

barrier lagoon *Oceanography.* a shallow body of water between the mainland and a barrier reef.

barrier lake *Hydrology.* **1.** a lake or other body of water that is retained or enclosed by a natural dam or barrier, such as a landslide or ice dam. **2.** a freshwater lagoon that is separated from its lake by a sandbank or shore dune.

barrier layer see DEPLETION LAYER.

barrier-layer capacitance see DEPLETION-LAYER CAPACITANCE.

barrier-layer cell or **barrier-layer photocell** see PHOTOVOLTAIC CELL.

barrier light *Ordnance.* a searchlight used to throw a fixed beam on a water area, to detect enemy vessels, or to cordon off a certain sector.

barrier material *Materials.* a material used to withstand oil, water, or other liquids and gases. *Ordnance.* a nonexplosive material placed in an explosive charge to shape the detonation wave.

barrier penetration see TUNNELING.

barrier pillar see COAL BARRIER.

barrier reef *Geology.* an active coral reef growing roughly parallel to the shore of a continent or volcanic island but separated from it by a deep, wide lagoon.

barrier separation *Chemical Engineering.* a method used to separate a two-component gaseous mixture by selective diffusion of one component through a separative barrier which may be microporous metal or a nonporous polymeric matrix.

barrier shield *Engineering.* a wall that can withstand ionizing radiation; used to protect an operator from a radioactive source.

barrier spit *Geology.* a sand barrier connected at one end to the mainland.

barrier strip *Electronics.* a terminal strip in which each terminal is separated from another by a raised insulating partition or barrier.

barrier tactics *Military Science.* any tactics that involve the use of fortified lines supported by gunfire.
barrier theory of cyclones *Meteorology.* a cyclone-development theory stating that a slow-moving cold air mass in the path of a warm air mass moving rapidly in an eastward direction will trigger the formation of low pressure on the leeward side of the cold air. Also, DROP THEORY.
barrier voltage *Electronics.* the minimum voltage needed for conduction through a P-N junction.
barrier zone *Botany.* a zone that prevents the distribution of plants.
Barrovian metamorphic zone *Geology.* a zone of regional metamorphism based on the first appearance of successive index minerals within the metamorphosed rock.
barrow *Agriculture.* **1.** a male pig that has been castrated. **2.** see WHEELBARROW. *Archaeology.* a mound of earth or rocks, used to cover a burial place; often found in prehistoric sites in Britain.
Barrow, Isaac 1630–1677, English mathematician; mentor of Isaac Newton.
barrow run *Civil Engineering.* a temporary ramp for wheeled transport of materials on a construction site.
bar scale see GRAPHIC SCALE.
bar screen *Mechanical Devices.* a large screen or mesh formed of metal bars, placed in a drain or sewer to catch large solid objects. Also, **bar strainer.**
bar sight *Ordnance.* a rear sight for a firearm, consisting of a movable bar with a notch for sighting through.
bar steel *Metallurgy.* a bar-shaped steel produced by solid-state diffusion of carbon into bar-shaped iron.
Barstovian *Geology.* a North American geologic stage of the Lower Miocene, occurring before the Clarendonian.
bar tack *Textiles.* a series of close stitching across a piece of cloth to reinforce it at a point of concentrated strain.
barter *Anthropology.* a form of trade or commerce between groups in which certain goods are exchanged for other goods regarded as equivalent in value. Also, **bartering.**
bar test *Engineering.* a process in which bar holes are driven in the ground periodically along the path of a gas pipe so that air in the holes can be tested with a gas detector; used to test for gas leakage in the pipe.
bar theory *Geology.* a theory that explains the origin of large thick deposits of salt, gypsum, and other evaporites in a lagoon separated from the ocean by a bar.
Bartholin, Caspar Thomasen 1655–1738, Danish anatomist; discovered Bartholin's glands and a duct of the sublingual gland.
Bartholin, Erasmus 1625–1698, Danish mathematician; discovered the double refraction of light.
Bartholin, Thomas 1616–1698, Danish anatomist; best known for observations on the lymphatic system; father of *Erasmus Bartholin.*
Bartholin's duct *Zoology.* an excretory duct of the sublingual gland.
Bartholin's gland *Zoology.* either of a pair of small, oval, mucus-secreting glands in some female mammals, lying on each side of the upper end of the vagina.
Bartlett force *Nuclear Physics.* a force that arises from interaction between two nucleons in which the spin coordinates are exchanged.
Bartlett's test *Statistics.* a test for equality or homogeneity of variances in a set of normal populations.
Barton, Derek born 1918, English chemist; shared Nobel Prize for applying Hassel's concept of conformation to organic molecules
Bartonella *Bacteriology.* a genus of Gram-negative, parasitic bacteria of the family Bartonellaceae, occurring within or on human erythrocytes and endothelial cells; transmitted by insects in South America.
Bartonellaceae *Bacteriology.* a family of Gram-negative, aerobic parasitic bacteria of the order Rickettsiales, including the genera *Bartonella* and *Grahamella.*
bartonellosis see CARRION'S DISEASE.
Bartonian *Geology.* a European geologic stage of the Upper Eocene epoch, after the Auversian and before the Ludian. Also, MARINESIAN.
Bartram, John 1699–1777, American botanist; planted the first American botanical garden.
Bartramiaceae *Botany.* a cosmopolitan family of tufted and matted dull-colored mosses of the order Bryales, occurring on moist soil and in calcareous seeps and crevices of calcareous rocks; characterized by erect stems with abundant tomentum and leaves that often appear to arise from an expanded basal portion.
Bart reaction *Organic Chemistry.* the formation of an aromatic arylarsonic acid by treating the aromatic aryldiazonium compound with an alkali arsenite in the presence of cupric salts or powdered silver or copper.
Bartter's syndrome *Medicine.* a rare hereditary disorder in which hypertrophy and hyperplasia affect the juxtaglomerular cells, which produces hypokalemic alkalosis and hyperaldosteronism. Usually, there is an absence of hypertension in the presence of markedly increased plasma renin concentrations and insensitivity to the pressor effects of angiotensin. It affects children and is sometimes accompanied by mental retardation and short stature.
bary- a combining form meaning "heavy," as in *baryon.*
barycenter [bâr´ə sen tər] *Astronomy.* the center of mass in any system of celestial objects moving under mutual gravity; used primarily with reference to the earth-moon system. *Mathematics.* the center of gravity of a finite number of masses. The barycenter of a simplex spanned by the points $\{P_0, P_1, \ldots, P_n\}$ is the point $(P_0 + P_1 + \cdots + P_n)/(n+1)$.
barycentral dynamical time *Astronomy.* a system of dates and times used for planetary positions relative to the solar system's barycenter.
barycentric coordinates *Mathematics.* the barycentric coordinates of a point P are the positive or negative weights $\{m_i\}$ assigned to the points $P_i (i = 0, 1, \ldots, n)$ of the barycentric coordinate system so that P is their centroid. The barycentric coordinates $\{m_i\}$ are not unique; if they are all multiplied by the same number, the point determined by them remains the same. Sometimes it is required that $\sum_{i=1}^{n} m_i = 1$, in which case the representation is unique; in this case the coordinates are normal.
barycentric coordinate system *Mathematics.* a coordinate system based on points that are the vertices of a simplex, i.e., redundant coordinates on affine n-space based on a set of $n + 1$ points $\{P_0, P_1, \ldots, P_n\}$, not all lying in the same hyperplane.
barycentric elements *Astronomy.* a set of orbital elements for a pair of gravitating celestial objects (such as a star-star, star-planet, or planet-satellite system) that uses the system's barycenter as the reference point.
barycentric energy *Mechanics.* the energy of a system in its center-of-mass coordinate frame.
Barychilina *Paleontology.* a genus of Paleozoic ostracods in the extinct suborder Platycopina; of uncertain order, Platycopida or Podocopida; Silurian to Carboniferous.
barye see MICROBAR.
baryesthesia see BARESTHESIA.
Barylambdidae *Paleontology.* a family of Paleocene amblypod mammals in the extinct order Pantodonta; characterized by a compact build, heavy limbs and tail, and a small skull.
baryon [bâr´ē än] *Particle Physics.* any of the class of the heaviest subatomic particles that includes protons, neutrons, and other particles whose eventual decay products include protons. Baryons form a subclass of the hadrons and are subdivided into nucleons and hyperons.
baryon number *Particle Physics.* a quantum number that is assigned to elementary particles; a baryon is given a baryon number 1, an antibaryon –1, and nonbaryons 0. The baryon number remains the same in every particle reaction.
baryon octet *Particle Physics.* a configuration of six hyperons (one lambda, three sigma, and two xi hyperons) and two nucleons to form a symmetrical pattern; all particles having positive parity and spin 1/2.
baryon resonance *Particle Physics.* any of the highly unstable subatomic particles considered to be composites of several relatively stable particles or very short-lived high-energy or mass states of these stable particles. Baryon resonances decay into one of the more stable baryons.
baryon spectroscopy *Particle Physics.* the study of energy levels in baryon particles by means of spectroscopic techniques.
baryon-to-photon ratio *Astrophysics.* the ratio of the number of heavy particles (such as protons) to the number of photons in the universe; the ratio, 10*E*–9, is thought to have remained unchanged since the beginning of the universe.
barysphere **1.** see CENTROSPHERE. **2.** see PYROSPHERE.
baryta see BARIUM OXIDE.
baryta feldspar see HYALOPHANE.
baryte see BARITE.
Barytheriidae *Paleontology.* an extinct upper Eocene family of proboscideans represented solely by *Barytherium;* formerly thought ancestral to *Deinotherium* but now considered only distantly related.
Barytherioidea *Paleontology.* an extinct upper Eocene suborder of proboscideans represented solely by *Barytherium.*
barytine see BARITE.
barytocalcite *Mineralogy.* $BaCa(CO_3)_2$, a white, gray, greenish, or yellowish monoclinic mineral with a vitreous to resinous luster; prismatic, striated crystals having a specific gravity of 3.66 to 3.71 and a hardness of 4 on the Mohs scale; found with calcite, barite, and fluorite in veins within limestone.

barytropic gas *Physics.* a gas whose pressure is solely dependent on its density.

BAS or **B.A.S.** Bachelor of Applied Science; Bachelor of Arts and Sciences.

bas- a combining form denoting a relationship to a base or foundation, or to a chemical base.

basal *Science.* **1.** of, related to, or forming the base of a structure. **2.** essential; vital; primary. *Physiology.* of the lowest value in a measurement, usually of primary importance, such as the lowest rate needed to maintain a function.

basal area *Forestry.* the area of the cross section of a tree stem near its base, generally taken at standard breast height and including bark.

basal arkose *Geology.* a consolidated feldspathic residue forming a thin, discontinuous deposit at the base of a sedimentary sequence overlying granitic rock.

basal body *Cell Biology.* a cylindrical structure that forms the base of a flagellum or cilium; it contains sets of microtubules and is structurally similar to a centriole. Also, BLEPHAROPLAST.

basal canker *Plant Pathology.* any of several fungus-caused diseases that appear as a crack in tree bark where the fungus has killed the wood.

basal cell carcinoma *Oncology.* an epithelial tumor that rarely metastasizes but has the capacity for local invasion and destruction. Also, BASALOMA, HAIR MATRIX CARCINOMA.

basal cleavage *Crystallography.* a break along a naturally occurring cleavage plane that is parallel to the basal plane of a crystal.

basal clinker *Petroleum Engineering.* a hard mass of lava or other material situated at the bottom of a drill hole in such a way that it actually or potentially obstructs a drilling operation.

basal conglomerate *Geology.* a conglomerate of homogenous substances that rests on an erosional surface and forms the initial stratigraphic unit or bottom member of a sedimentary series.

basal disk *Invertebrate Zoology.* the relatively flat base area in anthozoans and other coelenterate polyps where they attach to substrate.

basal ganglia *Anatomy.* masses of cell bodies located deep in each cerebral hemisphere; includes the caudate nucleus, puntamen and globus pallidus (together the lentiform nucleus), and the claustrum. These nuclei participate in motor control. Also, CEREBRAL NUCLEI.

basal groundwater *Hydrology.* a major body of groundwater that floats on and is in equilibrium with a body of salt water.

basalia *Vertebrate Zoology.* a collective term for the cartilaginous rods that support the pectoral and pelvic fins in elasmobranch fishes.

basalis *Histology.* the deepest layer of the endometrial lining of the uterus.

basal lamina *Developmental Biology.* the thin layer of extracellular matrix produced by, and underlying, an epithelium.

basal mental age *Psychology.* in a standardized intelligence test, the mental age at which all items are successfully passed.

basal metabolic rate *Physiology.* the rate of basal metabolism as measured by a calorimeter, with the subject at absolute rest 14–18 hours after eating, and expressed in calories per hour per square meter of body surface.

basal metabolism *Physiology.* the minimal energy expended for vegetative functions such as respiration, circulation, peristalsis, muscle tone, body temperature, and glandular activity.

basaloma see BASAL CELL CARCINOMA.

basal orientation *Crystallography.* the orientation parallel to the basal plane of a crystal.

basal pinacoid *Mineralogy.* a crystal form that consists of two parallel facies that cut the vertical *c* axis and lie parallel to the lateral axis planes *a* and *b*.

basal plane *Crystallography.* the plane in a crystal that is perpendicular to its *c* axis.

basal rot *Plant Pathology.* any disease that produces a breakdown or decay of the roots or bulbs of a plant.

basalt *Petrology.* a dark, fine-grained igneous rock originating from a lava flow or minor intrusion, composed mainly of plagioclase clinopyroxene and sometimes olivine, and often displaying a columnar structure; the most common volcanic rock in the earth's crust.

basalt glass see TACHYLITE.

basaltic *Petrology.* of, relating to, containing, situated in, or resembling basalt.

basaltic dome see SHIELD VOLCANO.

basaltic hornblende or **basaltine** *Petrology.* a black or brown variety of hornblende that is rich in ferric iron and low in hydroxyls, found chiefly in basalts and other volcanic rocks. Also, OXYHORNBLENDE.

basaltic lava *Petrology.* lava of basaltic composition; normally very fluid compared to lavas richer in silica.

basaltic rock *Petrology.* a general term for any dark-colored fine-grained igneous rock including basalt, diabase, dolerite, and sometimes andesite.

basalt obsidian see TACHYLITE.

basal tunnel *Engineering.* a water-supply tunnel along the basal water table of an area.

basaluminite *Mineralogy.* $Al_4(SO_4)(OH)_{10}{\cdot}5H_2O$, a white mineral occurring in compact masses and formed in the crevices of ironstone associated with allophane, gypsum, and aragonite.

basal water see BASAL GROUNDWATER.

basal water table *Hydrology.* the water table of a body of unconfined basal groundwater.

basanite *Petrology.* a group of basaltic rocks composed of olivine, clinopyroxene, calcic plagioclase, and a feldspathoid, such as nepheline or leucite; any rock in this group.

basculating fault see WRENCH FAULT.

bascule *Engineering.* a device that operates on the principle of a balance or seesaw, so that a shift in the weight of one part of the structure is counterbalanced by an opposing weight.

bascule bridge *Engineering.* a drawbridge whose sections can be raised or lowered by pivoting on a horizontal shaft; as the roadway section is elevated, the counterweight section descends into a pit.

base *Chemistry.* any of a fundamental category of compounds that are identified by certain common characteristics, such as a bitter taste, a slippery feeling in water solution, the ability to turn litmus paper blue, and the ability to react with acids to form salts. Other definitions identify substances as bases by their activities rather than by their properties, as follows: **a.** also, **Arrhenius base.** any substance that increases the concentration of hydroxide (OH^-) ions when added to a water solution. The greater the increase, the stronger the base. **b.** also, **Brønsted** or **Brønsted-Lowry base.** any substance that accepts a proton from another substance which serves as a proton donor. **c.** also, **Lewis base.** any substance that donates a pair of electrons to form a covalent bond. *Biology.* the part of a plant or organ that is nearest its point of attachment. *Engineering.* the lower part of a structure, especially one upon which an instrument rests or to which it is attached. *Building Engineering.* the lowermost part of a wall or other building member. *Architecture.* the lower part of a column or pier, thicker than the shaft and resting on a plinth or pedestal. *Robotics.* the end of a robotic arm opposite to the end that grasps or manipulates objects. *Graphic Arts.* the flexible transparent material over which an emulsion is layered to produce photographic film. *Mathematics.* **1.** the number of digits available in each position of a number system; in the binary system, for example, the base number is 2, so there are only two digits available for each position (0, 1). Also, RADIX. **2.** in geometry, the line or surface most near the horizontal, upon which a figure rests. *Military Science.* **1.** a usually protected or fortified location from which a military force operates or military supplies are obtained. Thus, **base command, base services. 2.** see BASE OF OPERATIONS. **3.** see BASE UNIT. *Ordnance.* **1.** see BASELINE. **2.** the part of a projectile below the rotating band; the rear part of a projectile. Thus, **base fuse.** *Chemical Engineering.* the main substance in a solution of crude oil that remains after distillation. *Electronics.* in a bipolar transistor, the area between the emitter region and the collector region, into which minority carriers are injected from the emitter region.

base address *Computer Programming.* **1.** a numeric value that is used as a reference for the calculation of memory addresses in the execution of a computer program. **2.** a reference address added to the relative address of an operand or instruction to compute its absolute address.

base analog *Molecular Biology.* **1.** a chemical whose molecular structure mimics that of a DNA base. **2.** a purine or pyrimidine that can be substituted for a normal nitrogenous base despite slight structural differences. Also, **base analogue**.

base anchor *Building Engineering.* a metal tie used to attach a doorframe to the floor beneath it.

base apparatus *Engineering.* in surveying, any measuring instrument used to determine the length of a baseline.

baseball *Physics.* a machine for confining plasma during controlled fusion reactions, so called because its current carriers resemble the seams of a baseball.

baseband *Telecommunications.* **1.** a frequency band containing signals that are used to modulate a carrier prior to its transmission by line or radio. **2.** a frequency band associated with or comprising an original signal from a modulated source.

base bias *Electronics.* the direct voltage applied to the base of a bipolar transistor to ensure that signal excursions operate over the desired region of the transistor characteristic.
base bleed *Ordnance.* a method of reducing the drag of a large projectile by burning a small quantity of propellant, which bleeds gas into the space behind the base.
baseboard *Building Engineering.* a sometimes ornamented board forming the base of an interior wall and designed to cover the joint between the wall and floor.
baseboard heater *Building Engineering.* a heating element or outlet located near the foot of an interior wall.
base box *Metallurgy.* in measuring tin plate or terneplate, a unit equal to 31,360 square inches (the total area of 112 sheets, each measuring 14 by 20 inches).
base bullion *Metallurgy.* **1.** smelted lead that contains a recoverable impurity such as gold, zinc, or especially silver. **2.** any base metal containing a recoverable amount of precious metal as an impurity.
base camp *Military Science.* a main camp from which a unit operates, as opposed to a smaller camp set up temporarily in the field.
base-centered lattice *Crystallography.* a crystal lattice in which the lattice points occur at the vertices and the centers of opposite faces.
base circle *Mechanical Devices.* a design circle from which involuted gear teeth are constructed.
base color *Graphic Arts.* in multicolor printing, the color printed on the first run through the press and serving as the background upon which other colors are printed.
base conditions *Petroleum Engineering.* the standard conditions (i.e., 60°F and 14.65 psia pressure) used in estimating the percentage of gas found in oil from a well.
base construction line *Cartography.* the bottom line of a map projection, perpendicular to the central meridian, along which other meridians are established.
base correction *Engineering.* in surveying, an adjustment designed to make field measurements conform to base station values.
base course *Building Engineering.* the lowest course of masonry in a building. *Civil Engineering.* in road construction, the surfacing layers other than the wearing course; in flexible pavements, a layer of chosen and compacted soil covered with a thin layer of asphalt.
base density *Graphic Arts.* the optical transmission density of a film base, excluding the emulsion layer.
base detonation see BASE INITIATION.
base development *Military Science.* the setting up of facilities to support a military operation, such as a supply or repair station or a training site.
base direction *Cartography.* the direction of the vertical plane containing the air base, which may be expressed as a bearing or an azimuth.
Basedow's disease see GRAVES' DISEASE.
base drag *Fluid Mechanics.* the drag due to a base pressure lower than the ambient pressure, and part of the pressure drag.
base elbow *Mechanical Devices.* a cast-iron pipe elbow that includes at its base a flange or plate designed to support the fitting or receive a supporting member.
base element *Electronics.* in a transistor, a contact that carries current to the base region. Also, **base electrode**.
base exchange *Geochemistry.* see ION EXCHANGE.
base flashing *Building Engineering.* flashing that is designed to cover and protect the joint between a roof and a wall.
base flow *Hydrology.* the regular, steady, or fair-weather flow of water into stream channels.
base fracture *Mining Engineering.* a major break or crack made by blasting in a quarry.
base frequency see FUNDAMENTAL TONE.
base-height ratio *Cartography.* the ratio of the length of an air base to the average altitude from which a stereoscopic pair of photographs was taken. Also, K-FACTOR.
base initiation *Ordnance.* the detonating of a charge from the base or rear of the charge. Also, BASE DETONATION.
base insulator *Electricity.* a heavy-duty insulator used to support an antenna mast.
base level *Geology.* the theoretical limit approached by the erosion of the earth's surface, or the lowest level to which a stream can erode the land, i.e., sea level.
base-leveled peneplain see PENEPLAIN.
base-level plain *Geology.* a flat surface of land produced by the erosion of a region to or near its base level.
baseline or **base line** *Science.* a value that represents the background level of a measurable quantity and is used for comparison with values that represent responses to environmental stimuli or an experimental intervention; a set of data used as a basis for subsequent comparison or control. *Ordnance.* a line used as a reference point for plotting firing data. *Cartography.* a line extending east and west along an astronomical parallel passing through an initial point, along which standard township, section, and quarter-section corners are established. *Navigation.* a line connecting two radio transmitters. *Graphic Arts.* in typesetting, the bottom alignment of capital letters in a font, used as a guide for aligning other letters and characters. *Electronics.* **1.** a horizontal or vertical line created by the movement of a scanning dot across a cathode-ray tube. **2.** specifically, the line that appears on a radar display in the absence of an echo, from which the pulse waves seem to emanate. *Engineering.* a survey line measured very precisely to form a starting point and reference for subsequent measurements and triangulation.
baseline break *Electronics.* a break in the baseline of a radar display due to a strong pulse signal.
baseline technique *Analytical Chemistry.* a technique for determining the absorbance of a sample, in which the distance from a line drawn tangent to the spectrum background to the height of the absorption peak is measured.
Basellaceae *Botany.* a family of tropical and subtropical dicotyledonous herbs of the order Caryophyllales; characterized by twining or scrambling stems growing from rhizomes and simple, somewhat succulent leaves. A common example is the Madeira vine.
basellaceous *Botany.* relating to or resembling the family Basellaceae.
base load *Electricity.* the constant load that a power station or system must provide in order to cover minimum or normal needs, as distinguished from the peak load generated during intermittent periods of heavy usage.
base-loaded antenna *Electromagnetism.* a vertical antenna whose electrical height is increased by adding inductance in series at the antenna's base.
base magnification *Optics.* the ratio of the distance between the centers of the objectives to the distance between the centers of the eyepieces in a pair of binoculars.
base map *Cartography.* a reference map against which other maps and data can be compared or from which more detailed maps can be prepared by the addition of information.
basement *Building Engineering.* a story of a building that is wholly or partly below street or ground level. *Geology.* the portion of the earth's crust lying between rocks of interest and the Mohorovicic discontinuity. Also, BASEMENT ROCK, BASEMENT COMPLEX, FUNDAMENTAL COMPLEX.
basement membrane *Histology.* a thin, delicate layer of tissue beneath the epithelium of mucous membranes and secreting glands; produced by the extracellular condensation of mucopolysaccharides and proteins. Also, BASILEMMA.
basement rock *Geology.* see BASEMENT.
base metal *Chemistry.* any of the metals found at the lower end of the electrochemical series, such as zinc, copper, lead, or tin; the opposite of a noble metal. Base metals oxidize and corrode readily and are thus relatively low in value.
base modulation *Electronics.* amplitude modulation from the application of the modulating voltage to the base of an amplifier transistor.
base molding *Building Engineering.* a decorative baseboard, or a strip of molding used to decorate the top of a baseboard.
base net *Engineering.* in surveying, a set of triangles and quadrilaterals that start with a measured base line and connect with a line of the main scheme of a triangulation net.
basenji [bə sen′jē] *Vertebrate Zoology.* a breed of small, terrierlike dogs having a chestnut coat with a curly tail and noted for their inability to bark; originally bred in central Africa as hunting dogs.
base of a logarithm *Mathematics.* the number of which the logarithm is the exponent; that is, if $\log_b a = c$, then $b^c = a$, and b is the base of the logarithm.
base of a matroid *Mathematics.* any of the sets in the set B of bases of a matroid; a maximal independent set in a matroid.
base of a number system *Mathematics.* the number whose powers determine place value in a positional representation of a number system.
base of a topological space *Mathematics.* a family B of open subsets of a topological space X such that every open subset of X can be expressed as a (not necessarily finite) union of sets in B.
base of drift *Geology.* a seismic-velocity discontinuity between glacial drift materials and the formation beneath them.

base of fire *Ordnance.* a force that gives support to the advance or attack of other units with its fire.
base of operations *Military Science.* any area used by a military force as a central point for its activities; the place from which a unit begins an advance or to which it makes a retreat.
base of projectile *Ordnance.* see BASE, def. 2.
base ore see COMPOUND ORE.
base pair *Genetics.* a pair of purine-pyrimidine bases, each in a separate nucleotide, in which each base is hydrogen-bonded to the other. In DNA, the two such pairs are guanine with cytosine and adenine with thymine. In RNA, adenine is paired with uracil.
base-pairing *Genetics.* the bonding of bases in DNA and RNA.
base-pairing rules *Genetics.* the rules stating that in DNA, adenine necessarily pairs with thymine and cytosine pairs with guanine, and that in RNA, adenine pairs with uracil.
base-pair ratio *Genetics.* in a given sample of DNA, the ratio between the number of adenine-thymine base pairs and the number of guanine-cytosine base pairs.
base-pair substitution *Genetics.* a breakage of a DNA molecule that allows the replacement of one base pair by another.
base period *Transportation Engineering.* any time of day other than a peak period.
base piece *Ordnance.* a single gun in a battery or group that is used to calculate initial firing data, and as a reference for firing data for the other guns. Also, **base gun.** Similarly, **base mortar.**
base plate or **baseplate** *Medicine.* in dentistry, a preformed piece of plastic or wax designed to represent the base of a denture; used for arranging denture elements and for trial placement in the mouth. *Mechanical Devices.* a metal plate or flange that serves as a support for a machine, device, or part. *Building Engineering.* a metal plate that forms the foundation or base of a wall stud or other partition. Also, SOLE PLATE. *Virology.* a structural formation of bacteriophage particles in which tail pins and fibers connect to the terminal end of a phage tail.
base-plus-fog density *Optics.* the optical transmission density of an unexposed area of processed film on paper; incorporates both the density of the film base and the density of the developed emulsion layer.
base point *Ordnance.* a central point in a firing area that has a known location, used as a reference for calculating firing data.
base post see BASE STAKE.
base pressure *Fluid Mechanics.* the pressure forced upon the foundation or aft of a body in fluid flow. *Mechanics.* standard atmospheric pressure or another such pressure used as a reference for other values. Also, GAUGE PRESSURE.
base quantity *Metrology.* a fundamental physical quantity in a system of measurement that is defined, independent of other physical quantities, by means of a physical standard and by comparing the quantity to be measured with the standard. Also, FUNDAMENTAL QUANTITY, BASIC QUANTITY.
base rate or **baserate** *Psychology.* the normal occurrence of any response, statistic, or other phenomenon in a given population. *Industrial Engineering.* **1.** an established rate of pay per piece or per unit of time. Also, BASIC RATE. **2.** the standard rate at which a given task is expected to be performed.
base rate area *Telecommunications.* in telephone exchanges, the region in which all services are provided without charge for mileage.
base region *Electronics.* the interelectrode region of a transistor into which minority carriers are injected.
base register *Computer Technology.* a register containing a base address that can be added to an address during the execution of an instruction; the actual address is determined by adding the contents of the base register to the address specified by the instruction. A base register contains the base address of a large block of memory. Also, INDEX REGISTER, RELOCATION REGISTER.
base runoff *Hydrology.* runoff that represents the natural flow of a stream, including that delayed by slow movement through lakes or swamps. Also, FAIR-WEATHER RUNOFF, SUBSTANTIAL RUNOFF.
base screed *Building Engineering.* **1.** a piece of plaster, wood, or metal used as a level guide in pouring a concrete slab. **2.** a board or metal strip that is dragged across freshly poured concrete in order to set the proper level.
base sequence *Genetics.* the order of the base pairs in a given DNA molecule.
base sheet *Building Engineering.* in building up layered roofing, a sheet of coated felt that is stretched over the framing before other materials are applied.
base shoe *Building Engineering.* a narrow strip of molding, often a quarter round, that is applied along the joint between a baseboard and the floor. At corner joints, a wedge-shaped piece of base shoe (**base-shoe corner**) may be used.
base spray *Ordnance.* fragments of a bursting shell that are thrown to the rear rather than to the front or the side.
base-spreading resistance *Electronics.* the resistance of the base region caused by the resistance of the bulk material of the base region in a transistor.
base stake *Ordnance.* a marker used to indicate the line from a gun to the base point. Also, BASE POST.
base station *Engineering.* the location that serves as the starting point of a survey. *Telecommunications.* **1.** a land station in a land mobile service carrying on a communication service with land mobile stations and other base stations. **2.** a radio station in mobile service providing communication with mobile stations and other base stations.
base stock *Graphic Arts.* untreated paper that will later be laminated, coated, or otherwise finished.
base substitution *Genetics.* in a DNA molecule, the replacement of one base by another base.
base surge *Volcanology.* a high-velocity, ring-shaped cloud of gas or solid debris that accompanies a volcanic eruption. *Physics.* radioactive mist that spreads rapidly into the atmosphere from a point in the sea directly above an underwater nuclear explosion.
base tee *Mechanical Devices.* a pipe tee fitting that includes a base plate or flange for support.
base tilt *Cartography.* the inclination of an air base with respect to the horizontal.
base time *Industrial Engineering.* a standard amount of time that is needed to perform a particular task, with no allowance included for incidental factors affecting a worker's performance.
base unit *Metrology.* a unit based on a physical standard such as kilogram, meter, second, or coulomb, all of which are independent units; the fundamental unit in a system of measure, from which other units are derived. Also, FUNDAMENTAL UNIT, BASIC UNIT. *Military Science.* the unit or group around which an operation is planned and performed. Also, BASE ELEMENT.
bashing *Mining Engineering.* the process of closing down and sealing a section of a mine after operations have terminated, sometimes involving the explosive destruction of workings or roadways.
Bashkirian *Geology.* a European geologic stage of the Middle Carboniferous period, after the Namurian and before the Moscovian.
basi- a combining form denoting a relationship to a base or foundation, or to a chemical base.
basic *Chemistry.* **1.** relating to, being, or forming a base or bases. **2.** having a pH value greater than 7.0. **3.** describing a form of a compound that is more alkaline than the parent compound from which it is derived, such as basic lead carbonate. *Military Science.* **1.** see BASIC TRAINING. **2.** of lowest rank, such as **airman basic.** *Petrology.* of or relating to igneous rock having a low silica content, usually between 45 and 52%.
BASIC *Computer Programming.* a high-level programming language originally designed for conversational mode in an on-line program. (An acronym for Beginner's All-purpose Symbolic Instruction Code.)
basic access see BASIC DIRECT ACCESS.
basic airspeed *Aviation.* an airspeed value calculated by correcting indicated airspeed for instrument error.
basic anxiety *Psychology.* anxiety or insecurity that originates when a child senses indifference or coldness from its parents, making the child feel isolated and insecure.
basic block *Computer Science.* a sequence of program statements such that if any one of them is executed, all of them are; a sequence of statements with a label (if any) only at the beginning and a branch (if any) only at the end. Also, BLOCK.
basic chromosome number see BASIC NUMBER.
basic converter see BASIC-LINED CONVERTER.
basic cover *Military Science.* an aerial photograph that is used as the standard to compare to later photographs to identify changes in an area.
basic direct access method *Computer Programming.* an access method used to retrieve or directly update specific blocks of data set on a direct-access storage device.
basic dye *Materials Science.* any of many usually synthetic dyes that are salts of colored organic bases, consisting mostly of amino or imino compounds; used mainly in coloring inks, typewriter ribbons, and chemical testing materials.
basic element see ELEMENTAL MOTION.

basic food group *Nutrition.* any of four to seven categories of food sources commonly used in planning nutritionally adequate diets; the most common groupings are (1) dairy, (2) protein, including meat or substitutes, (3) fruits and vegetables, and (4) bread and cereals.

basic front *Geology.* in rocks undergoing granitization, an advancing zone rich in calcium, magnesium, iron, and other elements in excess of those needed to form granite. Also, MAFIC FRONT, MAGNESIUM FRONT.

basic granulite *Petrology.* a metamorphic rock produced by the total recrystallization of a basalt or gabbro.

basic group *Chemistry.* a group of atoms that form part of a compound and react together as a base, such as the hydroxide group OH–.

basic hornfels *Petrology.* hornfels that is the recrystallized product of a basic igneous rock.

basichromatic *Biology.* of or relating to a substance that can be stained with a basic dye, such as the particles in the nucleus of a cell.

basic indexed sequential access method *Computer Programming.* an access method in which a key within the record or an index of such keys is used to locate the data on a direct-access device.

basic input/outout system *Computer Programming.* in an operating system, the element that transfers information from a program to a peripheral device such as a monitor or printer.

basic instruction see ABSOLUTE INSTRUCTION.

basic lava *Volcanology.* lava containing less than 52% silica, as distinguished from acidic and intermediate lava.

basic lift *Aviation.* the lift of a given airfoil section when the overall lift is zero; calculated according to the shape of the airfoil.

basic-lined *Metallurgy.* having a floor and walls lined with basic refractories, as in a **basic-lined converter.**

basic linkage *Computer Technology.* any linkage that is used repeatedly in a routine, program, or system and follows the same set of rules each time.

basic load *Ordnance.* the total amount of ammunition carried by a unit, including the ammunition carried by individual soldiers and that which is loaded in weapons or transported on vehicles.

basic motion *Industrial Engineering.* see ELEMENT.

basic motion times see SYNTHETIC BASIC MOTION TIMES.

basic motion-time study *Industrial Engineering.* a study of the times required for basic motions and groups of motions that cannot be timed with a stopwatch, obtained by filming a variety of operations with a motion-picture or video camera paired with a timing device.

basic number *Genetics.* the haploid number of chromosomes in a species, organism, or cell.

basiconic receptor *Entomology.* an insect sense organ consisting of a small projecting cone connected to an internal nerve fiber.

basic open-hearth process *Metallurgy.* a steelmaking process, now becoming obsolete, performed in a furnace with a shallow hearth and a low roof, in which the bottom, lining, and slag are basic.

basic oxide *Inorganic Chemistry.* a metallic oxide that is a base, or that forms a hydroxide when combined with water.

basic process *Metallurgy.* any steelmaking process using a basic-lined converter, including Bessemer, open-hearth, and electric processes.

basic processing unit *Telecommunications.* in a communications system, the primary controller and data processor. Also, CENTRAL PROCESSING UNIT.

basic pulse repetition rate *Navigation.* the minimum repetition rate in a group of pulses.

basic Q see NONLOADED Q.

basic quantity see BASE QUANTITY.

basicranial flexure see PONTINE FLEXURE.

basic rank *Behavior.* an animal's position in a rank order that is not dependent on or influenced by the status of other group members. Also, **basic ranking.**

basic rate *Industrial Engineering.* see BASE RATE.

basic refractory *Metallurgy.* any basic, heat-resistant material used to line a converter in metal processing, such as magnesia, lime, or magnesite.

basic research *Science.* any research that is conducted for the purpose of investigation into the nature of something in order to obtain knowledge about the subject, as distinguished from *applied research,* which has more practical goals; however, knowledge gained through basic research may be put to practical uses through applied research.

basic rock see BASIC.

basic salt *Inorganic Chemistry.* a compound that is both a base and a salt because it contains hydroxide or oxide as well as the usual positive and negative radicals of normal salts.

basic schist *Petrology.* a schistose rock derived from the metamorphism of a basic igneous rock.

basic sediment *Petroleum Engineering.* foreign matter and impurities that are found in oil produced from a well. Also, BUSHWASH, BOTTOM SETTLINGS.

basic sediment and water *Petroleum Engineering.* the impurities such as soil, water, and foreign matter that settle and collect in the bottom of petroleum storage tanks; the residue of the first stage of an oil separation process. Also, BOTTOMS, BOTTOM SETTLINGS AND WATER.

basic sequential access *Computer Technology.* an access method that stores and retrieves data blocks in continuous sequences, using either a sequential or a direct-access device; it always stores or retrieves the next record with no regard to key values.

basic slag *Metallurgy.* a by-product of smelting containing phosphorus and other basic materials; commonly used as fertilizer or as basic lining for furnaces.

basic steel *Metallurgy.* steel produced by any basic process.

basic tactical unit *Military Science.* in the U.S. Army, the fundamental unit capable of carrying out an independent tactical mission, such as a rifle company in the Infantry or a battery in the Artillery.

basic titanium sulfate see TITANIUM SULFATE.

basic titrant *Analytical Chemistry.* a basic solution of known concentration used as a standard solution for titration to determine the acidity of another solution.

basic training *Military Science.* an initial training period in which new recruits are given instruction in general principles of military conduct and fundamental combat skills, prior to specialized training in specific skills. Also, BASIC.

basic unit see BASE UNIT.

Basidiobolaceae *Mycology.* a family of fungi belonging to the order Entomophthorales, found primarily in dung, plant debris, or soil; some are pathogenic to animals and humans.

basidiocarp *Mycology.* the spore-bearing part of fungi belonging to the subdivision Basidiomycotina.

basidiole *Mycology.* **1.** an immature basidium or cell that bears the spores of fungi belonging to the subdivision Basidiomycotina. **2.** a basidium that has been aborted. **3.** a cell that resembles a basidium but is sterile.

Basidiolichenes *Botany.* a group of lichens in which the fungus is a basidiomycete.

Basidiomycotina *Mycology.* a subdivision of fungi belonging to the division Eumycota, characterized by the formation of basidiospores that are sexually produced on basidia and comprising such fungi as rust and smut fungi, wood-rotting fungi, and edible mushrooms. Also, **Basidiomycetes.**

Basidiophora *Mycology.* a genus of fungi of the order Peronosporales, characterized by sporangiospores that have a swollen base.

basidiophore *Mycology.* a kind of spore that bears a basidium.

basidiospore *Mycology.* an asexual spore formed by a basidium.

basidium *Mycology.* in fungi belonging to the subdivision Basidiomycotina, a terminal fungal structure that produces spores called basidiospores.

basification *Geology.* the natural enrichment of rock with calcium, magnesium, iron, manganese, and other elements through the assimilation of country rock contaminated by granitic magma.

basifixed *Botany.* of a leaf or stem, attached by the base; not capable of independent movement.

basil [baz´əl; bāz´əl; bās´əl] *Botany.* any of several aromatic herbs of the genus *Ocimum* of the Labiateae family having purplish-green ovate leaves; commonly used in cooking and flavoring. Also, SWEET BASIL.

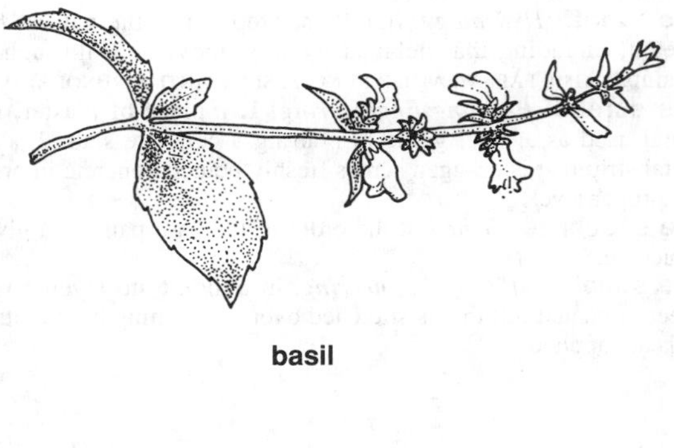

basil

basilar [bas´ə lär] *Biology.* of, relating to, located near, or growing from a base, especially the base of the skull. Also, **basilaris, basilary.**

basilar membrane *Anatomy.* a membrane in the cochlea of the inner ear, stretching from the spiral laminae to the outer wall of the cochlea and supporting the organ of Corti.

basilar papilla *Vertebrate Zoology.* in tetrapod vertebrates, an embryonic sensory area that develops into the area where hairlike cells respond to sound waves and other environmental vibrations.

basilar plate *Developmental Biology.* in vertebrates, an embryonic plate derived from the parachordals and anterior notochord that forms several bones of the skull.

basilar process *Anatomy.* the basal portion of the occipital bone, which extends from the foramen magnum and articulates with the sphenoid.

basilemma see BASEMENT MEMBRANE.

basilica [bə sil´ə kə] *Architecture.* **1.** in ancient Rome, a large oblong building used as a court of justice, usually lit by a clerestory and having double colonnades and a semicircular apse. **2.** a similar building used as a Christian church.

basilic vein *Anatomy.* a large vein on the inner side of the biceps muscle.

basilisk [bas´ə lisk´] *Vertebrate Zoology.* any of the water lizards of the genus *Basiliscus*, family Iguanidae, native to riverine areas of tropical America and characterized by a broad crest on the back of the head said to resemble the mythological basilisk monster; noted for running upright and skimming the water when provoked.

basilisk

Basilosaurus *Paleontology.* a genus of early whales in the family Basilosauridae; known only in the Eocene, they reached lengths of up to 16 meters.

basimesostasis *Geology.* a process whereby plagioclase crystals in a diabase become partly or wholly enclosed by augite.

basin *Geology.* **1.** an area in which rock beds dip inward, toward a central spot. **2.** a sediment-filled area, or an area in which sediments accumulate. **3.** a depressed or low-lying area of land surrounded by higher land. *Geography.* **1.** an area drained by one river and its tributaries, such as the Mississippi basin. Also, DRAINAGE BASIN. **2.** an area surrounding an ocean or sea, such as the Mediterranean basin. *Oceanography.* a large, deep area of the ocean, generally 4–5 km deep and lying between the continental margin and oceanic ridges. *Agronomy.* a shallow, circular depression dug around the trunk of a tree to hold surface water in place for absorption by the roots. *Mechanical Devices.* a wide, shallow, circular container used to hold liquids. *Metallurgy.* see POURING BASIN. *Anatomy.* the pelvis.

basin accounting see HYDROLOGIC ACCOUNTING.

basin-and-range *Geology.* a regional topographic structure that is characterized by parallel fault-block mountain ranges separated by broad valleys filled with alluvium.

basin area *Geology.* the total area of a drainage basin that contributes overland flow to a given stream or river.

basin budget see HYDROLOGIC BUDGET.

basining *Geology.* a process whereby the settling of the earth's surface results in the formation of a basin caused by erosion or by absorption and transport of salt and gypsum deposits.

basin irrigation *Agronomy.* an irrigation technique in which a dike or levee holds water in place, allowing it to soak into the soil. Also, **basin check method, basin method.**

basin length *Geology.* the length of a basin, measured in a straight line from the farthest point on its drainage divide to the mouth of its stream.

basin of attraction *Physics.* a region of phase space from which trajectories lead to the same attractor.

basin order *Geology.* a system of classification in which a drainage basin is assigned the same number as the order of the stream which drains into it.

basin peat see LOCAL PEAT.

basin range *Geology.* a long, narrow mountain range shaped primarily by the tilting and faulting of a solid block of strata and isolated by alluvium-filled valleys.

basin relief *Geology.* the difference in elevation between the mouth of a stream and the highest point on the perimeter of or within its drainage basin.

basin valley *Geology.* a wide, shallow valley having gently inclined sides.

basio- a combining form denoting a relationship to a base or foundation, to a chemical base, or to the basion.

basion *Anatomy.* a cranial landmark located at the midpoint of the front border of the foramen magnum in the midsagittal plane.

basion-bregma height *Anthropology.* the span between the basion and the bregma, a craniometric measure now used mainly for fossil identification.

basionym [bas´ē ə nim] *Systematics.* a revised or modified version of an original, valid taxon name that retains the original name's root. Also, **basinym, basionymn.**

basiophitic *Geology.* of or relating to the texture of an ophitic rock that is composed of augite.

basipetal *Biology.* **1.** descending or otherwise moving toward the base. **2.** growing or developing toward the base, as with some spores.

basiphil see BASOPHIL.

basipodite *Invertebrate Zoology.* **1.** the proximal joint of an arthropod limb. **2.** the second joint of certain crustacean limbs.

basipterygium *Vertebrate Zoology.* a bone or cartilage supporting either of the paired fins in most fishes.

basis *Science.* the physical, factual, or theoretical foundation upon which something rests. *Mathematics.* a set of linearly independent vectors in a vector space V such that each vector in V can be represented as a finite linear combination of vectors from the basis set. Vectors in the basis are called base vectors. The number of base vectors is the dimension of the vector space and may be infinite.

basis metal see BASE METAL.

basisternum *Invertebrate Zoology.* the anterior of the two sternal skeletal plates in insects.

basistyle *Invertebrate Zoology.* either of a pair of relatively flexible processes on the hypopygium of some two-winged male flies.

basis weight *Graphic Arts.* the weight of a standard-size ream of a given paper; used in printing to compute manufacturing costs.

basitarsus *Invertebrate Zoology.* in arthropods, the often enlarged or otherwise conspicuously differentiated basal segment of the tarsus.

basket *Mechanical Devices.* a usually small perforated container made of woven wood strips, twigs, or other lightweight flexible material. *Mining Engineering.* a cylindrical tool, with catching teeth at its base, designed to be inserted into a bore to pick up rock samples or to retrieve objects such as dropped bolts. Also, **basket barrel.** *Ordnance.* **1.** a structure within a tank turret that carries the personnel who operate the turret. **2.** a general area into which a homing missile must be delivered before its guidance system can acquire its more precise target. *Aviation.* a passenger or cargo compartment suspended beneath a balloon.

basket cell *Histology.* a cell of the cerebellar cortex whose axon gives off brushes of fibrils, forming the basketlike nest in which each Purkinje cell rests.

basket coil *Electronics.* a method of winding wire in a loose crisscross coil, thereby limiting distributed capacitance. Also, BASKET WINDING.

basketfish *Invertebrate Zoology.* any of numerous echinoderms of the class Opiuroidea, related to the starfishes and having slender, complexly branched interlacing arms that serve to entrap the fish on which they feed. Also, BASKET STAR, SEA SPIDER.

basket flower *Botany.* a composite annual plant, *Centaurea americana,* native to Mexico and the southwestern United States; distinguished by its pink or purple rayed flower heads, each surrounded by a basketlike involucre.

basket-handle arch *Architecture.* a low-crowned (flattened) arch drawn from three or more centers. Also, SEMIELLIPTICAL ARCH, MULTICENTERED ARCH.

basketlike structure *Materials Science*. a term for a microstructure having a woven appearance, associated with crystallographically influenced grain growth.

basket-of-gold *Botany*. a widely cultivated perennial herb of the mustard family, *Alyssium saxatile,* having gray leaves and clusters of small yellow flowers.

basket planting *Forestry*. the setting out for growing of young trees in loosely woven baskets, in which they have been raised.

basket star see BASKETFISH.

basketweave *Textiles*. of or relating to a fabric composed of two or more yarns worked together as warp and weft in a simple weave resembling a checkerboard woven basket pattern. *Building Engineering*. an arrangement of bricks in a checkerboard pattern.

basket winding see BASKET COIL.

basking shark *Vertebrate Zoology*. a large shark, *Cetorhinus maximus,* found in cold to temperate seas, and often floating on the surface

baso- a combining form meaning "relating to a base or foundation."

Basommatophora *Invertebrate Zoology*. an order of aquatic snails of the class Gastropoda, subclass Pulmonata, having paired tentacles with eyes at the base; various species breathe air or water.

basophil *Histology*. **1.** a white blood cell having a two-lobed nucleus and cytoplasmic granules that readily stain with basic dyes and containing vasoactive amines such as histamine and serotonin. **2.** any structure, cell, or other element that readily stains with basic dyes. Also, BASIPHIL.

basophile *Biology*. **1.** a basophil. **2.** basophilic.

basophilia *Biology*. the response of relatively undeveloped red corpuscles to basic dyes. *Pathology*. a condition occurring in the bloodstream in which there is a predominance of basophilic leukocytes or an increase in the equivalence of parenchymatous cells in an organ, often indicating leukemia, advanced anemia, or malaria.

basophilic *Immunology*. referring to tissues that have an affinity for basic dyes. Also, **basophilous.**

basophilism see CUSHING'S SYNDROME.

basophilous *Ecology*. living or thriving in alkaline habitats. Thus, **basophily.**

basophobia *Psychology*. an irrational fear of standing or walking.

Basov, Nikolai born 1922, Russian physicist; shared Nobel Prize for theory and development of masers.

Basquin relation *Materials Science*. a power law expression for metal fatigue, relating the applied elastic strain range and the number of cycles to failure.

bas-relief [bä´rē lēf´] *Architecture*. carving or scupture in which the figures project only slightly from the background and are not undercut. Also, LOW RELIEF, BASSO-RELIEVO.

bass [bās] *Acoustics*. **1.** the lower portion of the audio-frequency spectrum, below about 250 hertz. **2.** notes of musical works written in the **bass clef,** which starts at middle C and progresses downward. **3.** low in pitch: a *bass* singing voice. **4.** lowest in pitch of its type: a *bass* guitar.

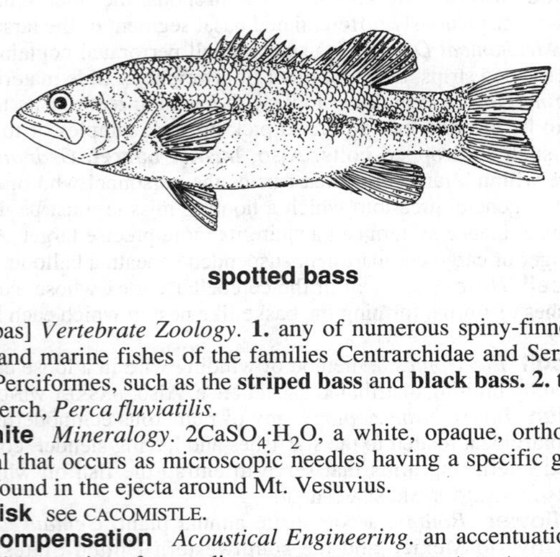

spotted bass

bass [bas] *Vertebrate Zoology*. **1.** any of numerous spiny-finned freshwater and marine fishes of the families Centrarchidae and Serranidaea, order Perciformes, such as the **striped bass** and **black bass. 2.** the European perch, *Perca fluviatilis*.

bassanite *Mineralogy*. $2CaSO_4 \cdot H_2O$, a white, opaque, orthorhombic mineral that occurs as microscopic needles having a specific gravity of 2.70; found in the ejecta around Mt. Vesuvius.

bassarisk see CACOMISTLE.

bass compensation *Acoustical Engineering*. an accentuation of the lower frequencies of an audio system or component, as to compensate for speakers with weak bass response or a listening environment that deadens the bass. Also, **bass extension, bass boost.**

bass control *Acoustical Engineering*. a manual tone control that adjusts the bass frequencies in an audio amplifier.

Bassen-Kornzweig syndrome see ACANTHOCYTOSIS.

basset *Vertebrate Zoology*. a breed of dog having long ears, short legs, and a long body with a black, tan, and white coat; originally bred in France as a hunter. Also, **basset hound.** *Geology*. an edge of a geologic structure that is exposed at the earth's surface; an outcrop edge. Also, **basset edge.**

bassetite *Mineralogy*. $Fe^{+2}(UO_2)_2(PO_4)_2 \cdot 8H_2O$, a transparent, yellow monoclinic mineral found in the Basset mines of Cornwall, England, occurring as tabular crystals, having a specific gravity of 3.63 and a hardness of 2.5 on the Mohs scale.

basso-relievo see BAS-RELIEF.

bass reflex *Acoustical Engineering*. in a loudspeaker cabinet, a baffle designed to enhance sound reproduction of low frequencies by directing bass frequencies from the rear of the cabinet to the front so as to reinforce the signals emerging directly from the speaker. Also, **bass reflex baffle**.

bass reflex enclosure *Acoustical Engineering*. a speaker cabinet equipped with a bass reflex baffle.

bass response *Acoustical Engineering*. the extent to which a loudspeaker, amplifier, or other such audio device handles low frequencies.

basswood *Botany*. any tree of the genus *Tilia,* especially the American linden, *Tilia americana;* characterized by asymmetrical buds with a large outer scale; fragrant, creamy white small flowers; and small round fruits borne in drooping clusters. Also, LINDEN, WHITEWOOD.

bassy *Acoustical Engineering*. of sound reproduction, having excessive resonance at low frequencies; overemphasizing the bass.

bast *Botany*. see PHLOEM. *Materials*. see BAST FIBER.

bastard *Mineralogy*. an inferior or impure rock or mineral, usually of little or no commercial value. *Mining Engineering*. **1.** a vein that is located near a main deposit but contains a lower grade of ore. **2.** a metal or ore that gives misleading assays.

bastard coal *Mining Engineering*. see BATT.

bastard pointing *Building Engineering*. in masonry, an imitation tuck pointing in which a slight projection, which casts a shadow, is given to the stopping to each joint. Also, **bastard-tuck pointing.**

bastard sawing see BACKSAWING.

bastard thread *Design Engineering*. a screw thread not conforming to any standard thread.

bast fiber *Materials*. any of several strong fibers (such as flax, hemp, ramie, or jute) that are obtained from the inner bark of certain trees and used in manufacturing textiles and paper. Bast fibers include phloem fibers, pericyclic fibers, and cortical fibers; they are composed of long, slender thick-walled cells with tapering ends.

bastion *Geology*. a mass of bedrock that juts out from the mouth of a hanging glacial trough into a main glacial valley.

bastite *Mineralogy*. a green to green-black to brown variety of serpentine formed by the alteration of orthorhombic pyroxene found as foliated masses in igneous rocks.

bastnäsite-(Ce),(La),(Y) *Mineralogy*. $(Ce,La,Y)CO_3F$, three hexagonal rare-earth minerals with the above general formula, often found in pegmatites and associated with syenitic intrusives; valued chiefly as a source of cerium and other rare earths. (From *Bastnäs,* Sweden, where these minerals are mined.) Also, **bastnaesite.**

bast ray see PHLOEM RAY.

bat *Vertebrate Zoology*. any of more than 1000 species of Chiroptera, a worldwide order of nocturnal mammals having forelimbs modified with webbing to form wings; the only true flying mammals. *Building Engineering*. a piece of brick that is large enough to use in wall construction and is cut transversely so as to leave one end whole. *Materials Science*. a slab or lump of clay, plaster, or similar material. *Textiles*. see BATTING.

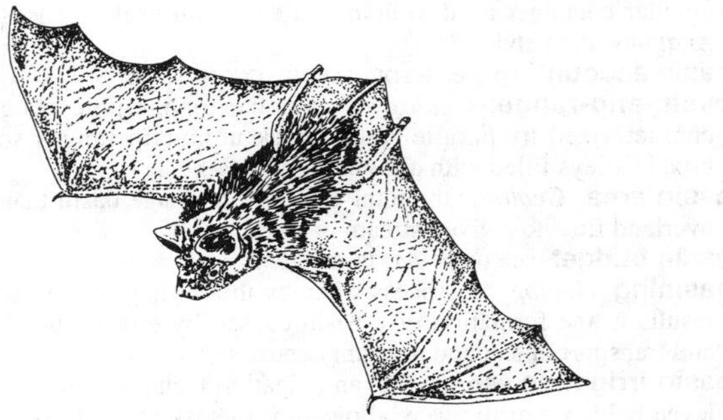

bat

Bataceae *Botany.* a monogeneric family of dicotyledonous shrubs of the order Batales that bear mustard oil and are characterized by slender, succulent leaves and a drupe fruit adapted to flotation in salt water.

batai *Materials.* the pale pinkish brown wood of a tree of Southeast Asia, *Albizia falcataria,* which is easily worked and is used to make such objects as plywood cores, fiberboard, and pulp.

Batales *Botany.* a small order of dicots with anatropous ovules and reduced flowers; contains a single family with only the genus *Batis.*

bat bolt *Mechanical Devices.* a bolt that is jagged or rough at its butt or tang.

batch *Engineering.* **1.** the amount of material that is required for or produced in a single operation. **2.** the amount of material that is needed for a chemical or physical process to make a generally uniform end product. *Computer Programming.* a collection of computer transactions to be processed as a unit, such as a group of instructions or programs with associated input data.

batch-and-forward system *Computer Technology.* a data-processing system in which data is collected at the source or central location for a specified period of time, then transmitted in bulk to the computer for processing in one batch.

batch box *Engineering.* a box of established volume used for measuring materials, such as cement or sand, for subsequent testing or mixing. Also, GAUGE BOX.

batch compiler *Computer Programming.* a compiler program that translates the complete program at one time, sometimes making two or more passes through it.

batch culture *Microbiology.* a closed culture system in which the nutrients are gradually consumed, allowing the microorganisms to undergo balanced growth for only a few generations or to the point of harvest; the opposite of continuous culture.

batch distillation *Chemical Engineering.* a distillation in which the entire distillation charge is placed in the still at the beginning of the operation, as opposed to continuous distillation, in which the still is fed continuously.

batched water *Engineering.* water that is added to concrete or mortar just before or at the start of the mixing process.

batcher *Mechanical Engineering.* a machine that measures and combines the ingredients of a mixture, such as concrete, into batches.

batch furnace *Metallurgy.* a heat-treatment furnace in which all materials are loaded and unloaded at the same time through a single door.

batch growth *Microbiology.* see CONTINUOUS CULTURE.

batching *Engineering.* the process of measuring the volume of the components of a batch of concrete or mortar prior to mixing. *Petroleum Engineering.* **1.** in a pipeline operation, the pumping of different grades of oil or gasoline that are next to one another and of different densities to prevent mixing of deliveries. **2.** in a refining operation, the mixing of two grades of petroleum in an effort to improve the distillation of one. *Computer Programming.* see BATCH PROCESSING.

batching plant *Engineering.* a manufacturing plant or station where concrete is mixed prior to its delivery to a construction site.

batching sphere *Petroleum Engineering.* a large rubber ball that is placed in a pipeline and inflated to separate consecutive batches of petroleum products.

batch interface *Petroleum Engineering.* the narrow zone where two different products or densities of crude oil converge in a pipeline as they are pumped in succession, one after the other.

batch job *Computer Programming.* **1.** any application program that is executed, often with other batch jobs, from beginning to end with no human interaction. **2.** any transaction that is accumulated or processed in batch processing.

batch manufacturing see BATCH PRODUCTION.

batch mixer *Mechanical Devices.* any mixing device that is used to prepare, mix, or contact batches of material, such as concrete, during a designated time cycle.

batch plant see BATCHING PLANT.

batch process *Industrial Engineering.* any manufacturing process in which a product is produced in separate batches, as opposed to mass production.

batch processing *Computer Programming.* **1.** a method of data processing that accumulates transactions and processes them periodically as a unit rather than as they occur. **2.** an older kind of operating system designed to execute a succession of batch jobs. Also, **batching**.

batch production *Industrial Engineering.* any manufacturing process in which all the materials to be used are accumulated and routed through each stage of production. Also, BATCH MANUFACTURING.

batch-push system *Industrial Engineering.* a system of inventory control in which each work station produces at a constant rate that is independent of the input requirements of the next station.

batch rectification *Chemical Engineering.* a batch distillation process in which the boiled-off vapor recondenses into liquid and a portion is refluxed back to the still to contact the rising vapors.

batch retort *Food Technology.* a type of autoclave designed to can food under superimposed steam pressure.

batch separator *Petroleum Engineering.* **1.** equipment for separating batches of oil-and-water mixtures and emulsions which may be produced from time to time in oil well operations. **2.** a machine for separating, with the aid of water or air, materials of different gravities.

batch system *Computer Technology.* a computer system that processes applications in a batch mode, as opposed to interactively.

batch terminal *Computer Technology.* a set of I/O devices connected via data communication links to the central computer and a control unit that allows the remote user the same access to the computer as local users; often includes a card reader and a printer.

batch terminal simulation *Computer Technology.* a method of testing connectivity to remote batch terminals by using either hardware or software to simulate the messages and responses of the terminal.

batch total *Computer Programming.* any of a variety of check sums used to verify the operation of a batch process or to assess the accuracy of the operation.

batea *Mining Engineering.* a wide, shallow pan made of wood or iron and used to sift gravel or sand for ore.

Bates, Henry Walter 1825–1892, English explorer and naturalist; discovered Batesian mimicry in Amazon insects.

Batesian mimicry *Behavior.* a form of mimicry in which the morphology and coloration of a certain species imitate those of another species that is less vulnerable to attack because of being dangerous, unpalatable, difficult to capture, and so on.

batfish *Vertebrate Zoology.* any fish of the primarily deep-water family Ogcocephalidae, characterized by a thin flat body, lumpy scaleless skin, a dangling "fishing pole" apparatus that lures prey, and movement along the ocean bottom by limblike pectoral and ventral fins; found in tropical and temperate oceans.

bath *Chemical Engineering.* **1.** an immersion of materials for a special purpose, such as a **chemical bath** designed to precipitate a salt. **2.** a chemical solution used in such a treatment.

bat-handle switch *Electricity.* a toggle switch with an elongated lever that resembles a baseball bat.

bathesthesia see BATHYESTHESIA.

Bathian see BATHONIAN.

batho- a combining form meaning "deep" or "depth," as in *bathometer.*

bathochrome *Physical Chemistry.* an atom or radical that when added to a compound causes its absorption band to shift toward the red end of the visible spectrum.

bathochromy *Physical Chemistry.* a process in which a compound shifts its absorption band toward the red end of the spectrum. Also, **bathochromatic shift.**

batholith or **batholite** *Geology.* a large mass of intrusive igneous rock having an exposed surface area of more than 40 square miles, with no apparent base or floor of older rock. Also, BATHYLITH, BATHYLITE, ABYSSOLITH, INTRUSIVE MOUNTAIN.

bathometer *Engineering.* an instrument that measures the depth of water.

Bathonian *Geology.* European geologic stage of the Middle Jurassic period, occurring after the Bajocian and before the Callovian. Also, BATHIAN.

bathophobia *Psychology.* an irrational fear of depths and deep places.

Bathornithidae *Paleontology.* an extinct family of birds in the order Gruiformes, suborder Cariamae, extant from Eocene to Miocene.

bathtub curve *Industrial Engineering.* a statistical curve representing the typical pattern of equipment failure of a given machine over time: the initially high failure rate drops sharply after the machine is broken in, levels out over extended use, and then rises again as the machine wears out.

bathvillite *Mineralogy.* a light brown, amorphous, brittle, woody resin found in Bathville, Scotland, and occurring as porous lumps in algal coal.

bathy- a combining form meaning "deep" or "depth," as in *bathymetry.*

bathyal *Oceanography.* of or relating to the **bathyal zone,** the biogeographic region of the ocean bottom between the sublittoral and abyssal regions, with depths of 660 to 13,000 feet (200–4000 meters).

bathyanesthesia *Neurology.* a loss of sensibility of internal parts of the body, such as muscles and joints.

bathybic *Oceanography.* of, relating to, or found in the deep sea.

bathycardia *Cardiology.* a condition in which the heart is fixed at a low position in the body because of the nature of the anatomy, not because of illness.

bathyclinograph *Oceanography.* an instrument designed for measuring vertical deep-sea currents.

bathyconductograph *Engineering.* an instrument that operates from a moving ship to measure the degree of electrical conductivity of ocean water at selected depths.

Bathyctenidae *Invertebrate Zoology.* a family of deep-sea coelenterates, of the order Cydippida; little is known of their behavior or development.

Bathydraconidae *Vertebrate Zoology.* the deep dragonfishes, a family of Antarctic marine fishes of the suborder Notothenioidei.

Bathyergidae *Vertebrate Zoology.* a family of burrowing mammals in the order Rodentia, suborder Hystricomorpha, including the African mole rats and sand rats.

bathyesthesia *Neurology.* consciousness or sensibility of the internal parts of the body, such as muscles and joints. Also, BATHESTHESIA.

bathygram *Oceanography.* a graphic record of bathymetric measurements, for example, from an echo sounder.

bathyhyperesthesia *Neurology.* extreme sensitivity of internal body structures, such as muscles, bones, and tendons.

Bathylaconoidei *Vertebrate Zoology.* a suborder of deep-sea salmon of the order Salmoniformes.

bathylimnetic *Biology.* living in the depths of lakes.

bathylith or **bathylite** see BATHOLITH.

bathymetry *Oceanography.* **1.** the science of measuring the depths of oceans and other large bodies of water. **2.** the information gathered from such measurements. Thus, **bathymeter, bathymetric.**

Bathynellacea *Invertebrate Zoology.* an order of primitive crustaceans, blind and generally colorless, in the superorder Syncarida; found among sand grains in groundwater streams and occasionally in hot water springs.

bathyorography *Geodesy.* the measurement of the heights of mountains and the depths of oceans.

bathypelagic *Oceanography.* of, relating to, or living in the ocean depths, especially in a bathyal region.

bathypelagic zone *Oceanography.* the open ocean environment at depths between about 1000 and 4000 meters.

bathyphotometer *Oceanography.* an instrument designed to measure and record the intensity of light at various ocean depths.

bathypitometer *Oceanography.* an instrument designed to measure and record the movements, depth, and temperature of deep-sea currents.

bathyscaphe [bath´ə skaf´] *Naval Architecture.* a diving vessel used for observation at great depths, consisting of a thick-walled steel sphere attached to a large hull, and having a fluid lighter than water for buoyancy and steel weights for ballast. Also, **bathyscaph.**

bathysphere *Naval Architecture.* a spherical manned vessel designed to be suspended by a cable and lowered to great depths.

Bathysquillidae *Invertebrate Zoology.* the mantis shrimps, a family of stomatopods living in deep, usually tropical waters.

bathythermogram *Oceanography.* a graphic representation of the information recorded by a bathythermograph.

bathythermograph *Oceanography.* an instrument that records water temperature in relation to ocean depth. Also, **bathythermosphere**.

batik *Textiles.* **1.** an originally Javanese method of hand-dyeing fabrics: wax is applied to areas not to be colored, the fabric is immersed in dye solution, and then the wax is dissolved in boiling water. **2.** a design or fabric produced by this process.

bating *Chemical Engineering.* the treatment of delimed animal skins with pancreatin or other tryptic enzymes in order to produce softer, smoother, and more flexible leather products.

batiste *Textiles.* a soft, sheer, plain-woven fabric made of fine fibers, usually cotton or linen.

Batoidea *Vertebrate Zoology.* the rays, a large suborder of cartilaginous freshwater or marine fishes in the order Squaliformes, characterized by a dorsoventrally flattened body and very large pectoral fins; includes the sawfishes, guitarfishes, skates, stingrays, and mantas.

batophobia *Psychology.* an irrational fear of being on or near high objects, such as tall buildings or trees.

batrachian *Vertebrate Zoology.* of or relating to Salientians (frogs and toads).

batrachoid *Vertebrate Zoology.* resembling a frog or toad.

Batrachoididae *Vertebrate Zoology.* the toadfishes and midshipmen, the only family in the marine order **Batrachoidiformes;** characterized by a large flat head and wide mouth, lack of scales, and carnivorous and voracious nature.

Batrachospermaceae *Botany.* a family of freshwater red algae belonging to the order Nemaliales, characterized by a thallus with a prostrate basal portion and an upright filamentous or pseudoparenchymatous portion, with gametangia formed only in the upper part of the latter.

batrachotoxin *Toxicology.* a highly toxic secretion from the skin glands of several species of tropical frogs of the genus *Phyllobates.*

bat ray *Vertebrate Zoology.* a marine fish, *Myliobatis californicus,* of the family Myliobatidae, that inhabits the Pacific coast; characterized by a large, liplike pad around the front of the head, a long, spiny tail spine, and a grinding tooth plate; known for its voracious appetite for shellfish. Also, BATFISH.

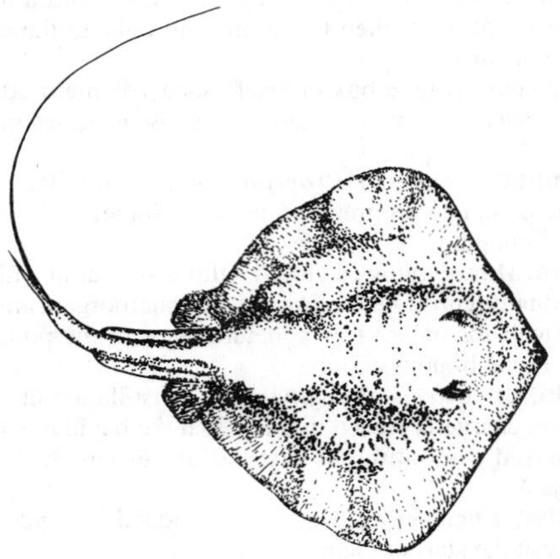

bat ray

batt *Mining Engineering.* a thin parting of coal, sometimes immediately overlying a coal bed, occurring in the lower part of shale strata. Also, BASTARD COAL. *Materials.* any hardened clay, but not fire clay. Also, BEND, BIND. *Textiles.* see BATTING.

batt. battalion; battery.

battalion *Military Science.* a ground force consisting of a headquarters and two or more companies, batteries, or other such units; usually commanded by a lieutenant colonel.

batted work *Engineering.* a stone surface covered with narrow parallel striations that have been scored by hand with a batting tool.

batten *Building Engineering.* a strip of square-sawn timber, usually 2 to 4 inches thick and 5 to 8 inches wide, used for a wide variety of building purposes such as supporting laths, reinforcing doors or other parallel-board constructions, or covering joints between floorboards. *Naval Architecture.* **1.** a thin iron bar used to secure the hatch cover on a ship. **2. batten down.** to secure a hatch so as to make watertight. **3.** a thin piece of wood or other material slipped into a pocket along the edge of a sail in order to make it stiffer. *Aviation.* **1.** on an airplane wing or tail, a stiffening member consisting of a strip attached (usually temporarily) to prevent movement of the control surface. **2.** in the nose cone of a nonrigid airship, any of a radially arranged group of metal, wood, or plastic strips used to provide rigidity to the nose or an attachment for mooring.

battenberg *Textiles.* a coarse linen lace made of braid or tape. Also, RENAISSANCE LACE.

batten door *Building Engineering.* a door formed with battens oriented side to side and reinforced with additional battens fastened horizontally.

battened wall *Building Engineering.* a plasterboard or lath-and-plaster wall that is reinforced with battens on its internal faces.

batten plate *Civil Engineering.* an iron or steel plate used to join and reinforce the parts of a composite strut or girder.

batter *Building Engineering.* a uniform slope running backward and upward, especially on the face of a wall or other building member; usually expressed as a ratio of one foot horizontal per *x* feet vertical.

batter board *Building Engineering.* one of a set of boards that are placed horizontally to support strings as a guide in laying out the foundation and excavation lines of a building.

battery *Electricity.* a multicelled direct-current voltage source that converts chemical, thermal, nuclear, or solar energy into electric energy; commonly applied to single cells. *Medicine.* a series of related procedures, especially a **battery of tests.** *Military Science.* an artillery unit corresponding to a company in the infantry. *Ordnance.* **1.** a group of artillery pieces, mortars, machine guns, or other such weapons set up in one place under one commander. **2.** the personnel who operate such a group of guns. **3.** a fortification or emplacement equipped with artillery. **4. in battery.** of a gun, in firing position and ready to fire. Similarly, **out of battery.** *Chemical Engineering.* two or more pieces of processing equipment operating as a single unit, such as distillation columns.

battery acid *Inorganic Chemistry.* a popular name for sulfuric acid, H_2SO_4. See SULFURIC ACID.

battery cable *Electricity.* an apparatus that is composed of connecting wire with two spring-jawed clamps **(battery clips)** at each end; used to transfer a charge from one bipolar battery to another. Also, BOOSTER CABLE, JUMPER CABLES.

battery charger *Electricity.* a device that applies power to charge a battery. Usually the voltage and current are controlled so the battery is charged rapidly, but without undue stress.

battery electrolyte see ELECTROLYTE.

battery eliminator *Electronics.* a device that provides direct-current energy from a power line to a radio receiver, electron tube, or other circuit that would otherwise require battery power.

battery limits *Chemical Engineering.* the portion of a chemical plant where the actual processes are carried out, as distinguished from offices, storage buildings, and other off-site structures.

battery separator *Electricity.* an insulating plate inserted between the positive and negative plates of a battery in order to prevent contact.

batting *Textiles.* a loosely composed sheet of cotton, wool, or synthetic fiber used for felt and other layered, nonwoven fabrics. Also, BATT, BAT.

batting tool *Engineering.* a type of chisel, generally 3 to 4.5 inches wide, that is used to make striated stone surfaces.

batt insulation *Building Engineering.* a blanket made of synthetic batting and used to insulate building walls and roofs; its standard size is 16 inches wide by 3 to 6 inches thick.

battle *Military Science.* **1.** see COMBAT. **2.** a component of a campaign, more substantial than a skirmish or engagement.

battle casualty *Military Science.* a person who is killed, wounded, injured, or missing in action, or rendered ill by combat conditions.

battle fatigue *Psychology.* an earlier term for post-traumatic stress disorder bought on by the experience of wartime combat.

battlefield recovery *Ordnance.* the removal of disabled or abandoned equipment and supplies from a combat area.

battle map *Cartography.* a type of military map showing sufficient ground detail to provide information for all types of operations.

battlement *Architecture.* a parapet having indentations for defense or decoration.

battle position *Military Science.* in ground combat, the position on which the main effort of the defense is concentrated.

battleship *Naval Architecture.* **1.** a ship designed primarily to engage in fleet actions with similar enemy ships. **2.** a heavily gunned and armored ship; the largest surface warship other than aircraft carriers.

battleship

battle short *Electricity.* a switch designed for the purpose of short-circuiting safety interlocks and also lighting a red warning light.

battle sight *Ordnance.* a sight setting that is determined beforehand for use at battle range if existing combat conditions do not permit a more exact setting.

battle tank *Ordnance.* see TANK.

batture *Geology.* an elevated section of a river bed formed by the gradual accumulation of sediment.

baud *Telecommunications.* in telegraphy, a unit of signaling speed, equal to the number of pulses and spaces that are transmitted per second. If the pulses have only one level, then the bits per second is equivalent to bauds; if the pulses are multilevel, then the rate in bauds contains additional information, and the number of bits per second is a greater number. *Computer Science.* in a computer system, the number of signal elements, usually bits, that can be transmitted per second. Also, **baud rate.** (Named for Émile *Baudot.*)

Baudot, J. M. Émile [bô´dō´] 1845–1903, French engineer; pioneer in telegraphy and inventor of the Baudot code.

Baudot code *Telecommunications.* a data transmission code in which five bits of equal length represent one character; now replaced by the start-stop asynchronous ITA Number 2.

Bauhaus [bou´hous´] *Architecture.* **1.** a German school of art, architecture, and design founded by Walter Gropius; its members applied traditional functional craftsmanship to modern industrial materials and techniques. **2.** relating to or in the style of this school.

baulk see BALK.

Baumé, Antoine [bō mā´; bō´mā] 1728–1804, French chemist; devised the Baumé hydrometer and calibration scale. Also, BÉ.

Baumé (hydrometer) scale *Physical Chemistry.* a numerical scale used with a hydrometer to calibrate the specific gravity of liquids; it is based on the extent to which the given liquid is heavier or lighter than water.

baumhauerite *Mineralogy.* $Pb_3As_4S_9$, an opaque, gray triclinic mineral occurring as short, prismatic crystals with a metallic luster, having a specific gravity of 5.3 and a hardness of 3 on the Mohs scale; found in crystalline dolomite.

Bauschinger effect *Metallurgy.* in a metal or alloy, a change in stress-strain behavior that is caused by microscopic stresses inside the material. A common repercussion of this effect is the decrease in yield strength in one direction when plastic deformation occurs in the opposite direction.

Bautz-Morgan classification *Astronomy.* a classification system that divides clusters of galaxies into three groups: type I has a single supergiant galaxy, type III has no dominant bright galaxies, and type II is intermediate between types I and III.

bauxite *Petrology.* the principal ore that is used in producing aluminum; a residual sedimentary rock of a dirty-white, yellow, brown, gray, or reddish color, formed by the leaching of groundwater and consisting of one or more aluminum hydroxides (chiefly gibbsite and boehmite), along with variable amounts of clay (usually kaolinite) and some iron and titanium oxides. (From *Les Baux,* in southern France, where it was discovered.)

bauxite treating *Chemical Engineering.* a petroleum-refinery treatment process in which a vaporized petroleum fraction is passed through beds of bauxite to convert sulfur compounds, such as mercaptans, into hydrogen sulfide.

bauxitization *Geology.* the process whereby bauxite is formed from primary aluminum silicates or secondary clay minerals under heavy tropical and subtropical weathering conditions.

bavenite *Mineralogy.* $Ca_4Be_2Al_2Si_9O_{26}(OH)_2$, a brittle, white orthorhombic mineral occurring as prismatic to fibrous crystals, having a specific gravity of 2.71 to 2.74 and a hardness of 5.5 on the Mohs scale; found in granite pegmatites with beryl.

Baveno twin law *Crystallography.* a twin law, described as a normal twin, that is found in both monoclinic and triclinic crystals in which the twin axis is perpendicular to the twin plane.

***b* axis** *Crystallography.* in a triclinic system, one of three axes constructed by bisecting each angle, the *b* axis being the axis that extends horizontally (usually oriented from right to left) from the point of intersection of the three axes and is the shorter of the two axes that are at right angles with the third. *Geology.* the fabric axis that indicates the direction of the fold axis, usually perpendicular to the direction of transport. *Petrology.* of a fabric or deformation plan possessing monoclinic symmetry, the axis that is perpendicular to the unique plane of symmetry. Also, *B* DIRECTION.

bay

bay *Botany.* **1.** any of several evergreen trees with shiny dark-green leaves of the family Lauraceae, especially the **bay laurel,** *Laurus nobilis.* **2.** any of various laurel-like trees or shrubs.

bay *Geology.* **1.** a recess in the shoreline of a sea or lake. **2.** the body of water formed by such a recess; an inlet of a sea or lake that is larger than a cove but smaller than a gulf. *Architecture.* **1.** a repeated structural unit in a building, especially between beams or columns. **2.** a protruding part of a building, containing a bay window. *Aviation.* a space formed by partitions between structural members in a flight vehicle, used to house something such as an engine, ordnance, or the cargo. *Engineering.* **1.** a small enclosure such as a machine housing or storage compartment. **2.** the space separating two principal beams or trusses. **3.** a compact, well-defined section where concrete is laid at one point during the construction of a large ground area, such as a pavement or a runway. *Geophysics.* a small magnetic disturbance of short duration, usually an hour, so called because of the V- or bay-shaped mark it makes on a magnetic record. *Electromagnetism.* a single segment of an antenna array.

Bayard-Alpert ionization gauge *Electronics.* an apparatus used to measure very low gas pressure by collecting ions on a fine wire inside a helical grid.

bay barrier *Geology.* a sandy shoal or spit that extends across the mouth of a bay, serving to separate the bay from the main body of water.

bayberry *Botany.* **1.** any of several aromatic trees or shrubs of the genus *Myrica.* Also, WAX MYRTLE. **2.** the waxy, resinous berry of such plants, the source of bayberry wax. **3.** see BAY RUM TREE.

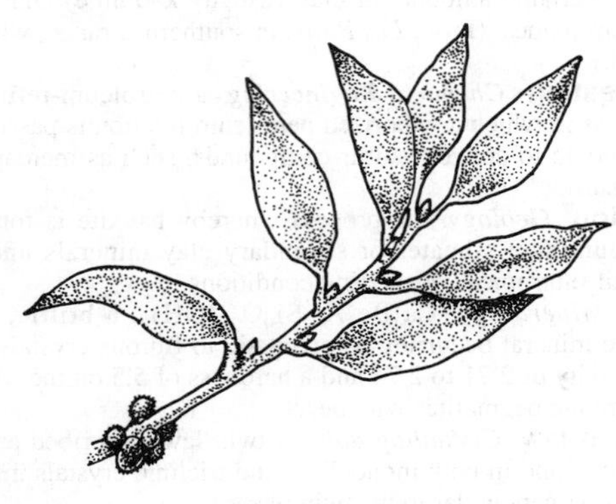

bayberry

bayberry wax *Materials.* a thick paste made by boiling bayberries; used to make aromatic candles and as a source of vegetable tallow, myrtle wax, and ocuba wax.

bay delta *Geology.* a delta that occurs where the mouth of a stream enters and fills or partially fills the head of a bay or drowned valley.

Bayer, Johann 1572–1625, German astronomer; simplified stellar nomenclature; charted 12 newly discovered constellations.

Bayer letter *Astronomy.* Greek letters assigned to the naked-eye stars of each constellation by Johann Bayer in his 1603 star atlas *Uranometria;* usually (though not always) α was given to the brightest star in the constellation, β to the next brightest, and so on.

Bayer name *Astronomy.* a star designation comprising a Greek letter plus the genitive form of a constellation's name; for example, Spica, the brightest star in Virgo, is called α Virginis.

Bayer process *Metallurgy.* the first step in the extraction of aluminum from bauxite, consisting of digesting the ore in a sodium hydroxide solution.

Bayer's constellations *Astronomy.* the twelve constellations in the southern celestial hemisphere described by the German astronomer Johann Bayer for his 1603 star atlas *Uranometria*: Apus the Bird of Paradise, Chamaeleon the Chameleon, Dorado the Goldfish, Grus the Crane, Hydrus the Lesser Water Snake, Indus the Indian, Musca the Fly, Pavo the Peacock, Phoenix the Firebird, Triangulum Australe the Southern Triangle, Tucana the Toucan, and Volans the Flying Fish.

Bayes, Thomas 1702–1761, English mathematician; established first method of probability inference; defended Newtonian calculus.

Bayesian [bāz′ē ən] *Mathematics.* relating to an approach to probability that views probabilities as degrees of belief that are modified by experience in accordance with Bayes' rule; this contrasts with the frequentist approach, which represents probabilities as relative frequencies of occurrence.

Bayesian statistics *Statistics.* the application of Bayesian decision theory and estimation to the analysis of sample data for the purpose of deriving chartacteristics of the underlying population; contrasts utilizing prior information with classical statistical methods of hypothesis testing and statistical inference.

Bayes' decision rule *Statistics.* $(a_1) = Sx_i\,P(x_i)$; the criterion of selecting from among several alternatives the course of action with the highest expected payoff, where the expected payoff for each action is calculated by multiplying each payoff by its respective probability and summing all such products.

Bayes' rule *Statistics.* a rule by which the predetermined probability of an event E is modified according to the actual outcomes of the event; that is, the prior probability (the probability determined *before* the event takes place) is modified to the posterior probability (the conditional probability value *given* the information that event E has occurred) in order to take into account the impact of new information.

Bayes' theorem *Statistics.* a theorem used to interconvert conditional probabilities. If $x_1, x_2, \ldots, x_n$ form a set of mutually exclusive and collectively exhaustive events, and if E is some other event, the conditional probability of x_i occurring, given that E occurs, can then be expressed as $P(x_i \mid E) = P(E \mid x_i)P(x_i) \mid P(E)$, which is the product of the conditional probability of E occurring, given that x_i occurs, and the marginal probability of x_i, divided by the marginal probability of E. $P(x_i)$ is considered the prior probability, which is modified to the posterior probability $P(x_i \mid E)$. *Genetics.* in genetic counseling, the use of this theorem along with pedigree data to predict the risk of inheriting an X-linked recessive allele or the likelihood of the late onset of an autosomal dominant trait.

bayhead or **bay head** *Geology.* a generally swampy section of a bay, located closest to the shore and farthest from the main body of water with which the bay connects.

bayhead bar *Geology.* an embankment of sand, silt, or gravel stretching across the head of a bay near the shore. Also, **bayhead barrier.**

bayhead beach *Geology.* a crescent-shaped beach that is formed near the head of a bay by materials eroded from nearby headlands and transported to the bayhead by longshore currents or storm waves. Also, POCKET BEACH, COVE BEACH.

bayhead delta *Geology.* a delta formed at the head of a bay or an estuary into which a river flows.

bay ice *Oceanography.* a mass of new, smooth sea ice that forms in a sheltered arctic bay.

bayldonite *Mineralogy.* $PbCu_3(AsO_4)_2(OH)_2 \cdot H_2O$, a massive to finely granular, green monoclinic mineral having a specific gravity of 5.50 to 5.65 and a hardness of 4.5 on the Mohs scale; found as a secondary mineral in oxidation zones of copper-bearing ore deposits.

bay leaf *Food Technology.* the dried leaf of a laurel tree, used as an herb.

bayleyite *Mineralogy.* $Mg_2(UO_2)(CO_3)_3 \cdot 18H_2O$, a transparent, yellow, water-soluble monoclinic mineral occurring in minute prismatic crystals having a specific gravity of 2.05.

Bayliss, William 1860–1924, English biologist; isolated secretin (with Starling); wrote *The Nature of Enzyme Action.*

baymouth bar *Geology.* a long, narrow embankment of sand or gravel that is deposited by waves or currents and connects or nearly connects the headlands of a bay. Also, **bay bar.**

bayonet *Ordnance.* a knifelike steel weapon, designed to be attached to the end of a rifle for use in hand-to-hand combat.

bayonet base *Electricity.* a base for a device, such as a lamp, consisting of two pins that project from either side of a smooth cylindrical surface and fit into corresponding slots in a bayonet socket in order to secure the base and socket.

bayonet coupling *Mechanical Devices.* a coupling with projections from its face that engage a bayonet socket.

bayonet fitting *Mechanical Devices.* a socket with a spring-loaded base into which a plug is pushed and twisted until the plug's two (or more) pins engage a corresponding L-shaped slot in the socket.

bayonet socket *Mechanical Devices.* an attachment socket with J-shaped grooves into which the bayonet coupling of a device, such as a lamp socket, is inserted and slightly turned; the coupling is then held firm in the short ends of the "J" grooves by a spring.

bayou [bī´yoo´] *Hydrology.* **1.** a marshy inlet, outlet, or arm of river or lake in flat country, usually sluggish or stagnant due to floods and lack of drainage. **2.** an estuarial creek or inlet. (A Louisiana French word thought to be derived from the Choctaw *bayuk.*)

bay-rum tree *Botany.* a West Indian shrub, *Pimenta racemosa,* having aromatic leaves used in making bay oil and bay rum. Also, WEST INDIAN BAYBERRY.

bayside beach *Geology.* a beach formed along the side of a bay, primarily by materials from adjacent headlands that are transported by longshore currents.

bay window *Architecture.* **1.** the window of a protruding bay. **2.** an entire bay, including the window.

bazooka *Ordnance.* a 2.36-inch rocket launcher used by U.S. forces as an antiarmor weapon, especially during World War II, firing a 3.5-pound projectile at a range of 400 yards; a similar 3.5-inch rocket launcher was called a **super bazooka.** *Electronics.* an informal term for a balun.

bazzite *Mineralogy.* $Be_3(Sc,Al)_2Si_6O_{18}$, a transparent, azure-blue, hexagonal mineral, with acicular crystals, having a specific gravity of 2.77 to 2.82 and a hardness of 6.5 on the Mohs scale, that is found in drusy cavities and as inclusions in quartz in pegmatites.

B battery *Electronics.* a battery that supplies the required power to the anode in battery-operated vacuum tube equipment.

B-B fraction *Chemical Engineering.* a blend of butenes and butanes from the distillation of a solution of light liquid hydrocarbons.

BB gun see AIR RIFLE.

bbl or **bbl.** bushel; barrel; barrels.

bbls or **bbls.** bushels; barrels.

BBT basal body temperature.

BC or **B.C.** before Christ.

B-CAVe *Oncology.* an acronym for a combination of the drugs bleomycin, CCNU (lomustine), and vinblastine, used in chemotherapy.

BCC metals *Materials Science.* a collective term for metals whose crystal structures are classified as body-centered cubic.

BCD or **bcd** binary-coded decimal.

BCE or **B.C.E.** before the Common Era; Bachelor of Chemical Engineering; Bachelor of Civil Engineering.

B cell *Immunology.* a lymphocyte produced in the bone marrow that is an active component of an immune response.

B-cell domain *Immunology.* the area of a lymphoid organ that contains the highest proportion of B cells.

B-cell growth and differentiation factors *Immunology.* any substances originating from T lymphocytes that enhance B-lymphocyte growth in culture or cause B lymphocytes to differentiate into antibody-secreting cells.

bcf billion cubic feet.

BCG Bacillus Calmette-Guérin.

BCh or **B.Ch.** Bachelor of Chemistry.

BChE or **B.Ch.E.** Bachelor of Chemical Engineering.

B chromosome see SUPERNUMERARY CHROMOSOME.

BCLS basic cardiac life support.

BCN *Aviation.* the airport code for Barcelona, Spain.

B complex see VITAMIN B COMPLEX.

BCS British Computer Society.

BCS or **B.C.S.** Bachelor of Chemical Science.

Bd baud.

bd. board; boundary; bundle.

BDA bomb damage assessment; British Dental Association. *Aviation.* the airport code for Bermuda.

BDAM basic direct-access method.

BDC bottom dead center.

Bdella *Invertebrate Zoology.* a genus of mites, including **Bdella cardinalis,** a species that is parasitic on other insects.

bdellium *Materials.* the fragrant gum resin of several species of *Commiphora* trees, used chiefly as an adulterant of myrrh because of its similar appearance and aroma.

bdelloid *Invertebrate Zoology.* leechlike.

Bdelloidea *Invertebrate Zoology.* an order of rotifers, 16-jointed forms that swim freely using the ciliated disk and also creeplike leeches; primarily freshwater, with a wide distribution including Antarctic lakes and hot-water springs

Bdellonemertea *Invertebrate Zoology.* a small order of ribbon worms in the class Enopla having an unarmed proboscis. Also, **Bdellonemirtea, Bdellonemertini, Bdellomorpha.**

Bdellovibrio *Bacteriology.* a genus of small, aerobic, motile Gram-negative bacteria occurring as intracellular parasites of other Gram-negative bacteria, reproducing between the cell wall and plasma membrane of the cell bacterium and ultimately killing the host cell; found worldwide in soil and in both fresh and marine waters.

bd. ft. or **bd ft** board foot; board feet.

***b* direction** see *B* AXIS.

B display *Electronics.* a type of radar display that has a rectangular scope which shows targets as bright spots, and indicates target bearing and distance by means of the horizontal and vertical coordinates. Also, B INDICATOR, B SCAN.

BDL *Aviation.* the airport code for Bradley International, Hartford.

bdl. bundle. Also, **bdle.**

B-DNA *Biochemistry.* the biologically important form of double-helical DNA in which the helix is right-handed and the helical axis is in the center of the molecule; modeled by Crick and Watson.

B-DOPA *Oncology.* an acronym for a combination of the drugs bleomycin, dacarbazine, Oncovin, prednisone, and Adriamycin, used in cancer chemotherapy.

bds. boards; bundles.

BDS or B.D.S. Bachelor of Dental Surgery.

BDSc or B.D.Sc. Bachelor of Dental Science.

BE or **B.E.** Bachelor of Engineering; Bachelor of Education.

Be the chemical symbol for beryllium.

Bé or **Bé.** Baumé.

beach *Geology.* **1.** a gently sloping zone of unconsolidated material along a shore, ranging from the low-water line to the line of permanent vegetation or to where there is a distinct change in material or in physiographic form. **2.** the shore of a body of water. **3.** the material in transit or deposited along the shoreline, usually consisting of sand, pebbles, rock, and shell fragments, and accompanied by mud. *Navigation.* to draw a vessel up onto a beach. *Military Science.* short for BEACHHEAD, as in *beach* defense, *beach* dump, *beach* obstacle, *beach* support area.

beach aster *Botany.* a perennial seaside herb, *Erigeron glaucus,* having solitary heads of violet-colored flowers. Also, SEASIDE DAISY.

beach capacity *Military Science.* an estimate of the amount of cargo that can be landed on a certain strip of shore in a given time.

beach cycle *Geology.* the periodic retreat and outbuilding of a beach caused by tides and waves.

beach drift *Geology.* **1.** material (such as gravel, sand, and shell fragments) that is transported by waves breaking at an angle with the shore. **2.** the movement of such material along the shore.

beach exit *Military Science.* a route for personnel and land vehicles to leave a beachhead area.

beach face see FORESHORE.

beach firmness *Geology.* the ability of a beach sand to resist pressure, depending mainly on its degree of packing, moisture content, and particle size.

beachfront *Geography.* land along a beach or facing a beach.

beach grass *Botany.* any of several tough, erect, strongly rooted grasses commonly found on exposed sandy beaches, especially *Ammophilia arenaria.*

beach group see SHORE PARTY.

beachhead *Military Science.* an area on an enemy shoreline that is the first objective of a landing force; a beach area that if held will allow the landing of troops and equipment and movement inshore.

beach markings *Materials Science.* in fatigued samples, fracture surface striations associated with cyclic crack growth.

beachmaster *Military Science.* the officer in command of a beach or shore party. *Vertebrate Zoology.* the dominant bull of a territory within a fur-seal breeding ground.

beach party see SHORE PARTY.

beach plain *Geology.* an area formed by successive embankments of wave-deposited beach material that is added uniformly to a shoreline.

beach platform see WAVE-CUT BENCH.

beach plum *Botany.* a shrub of the rose family, *Prunus maritima,* having sharp-toothed leaves, white flowers, and a purplish acidic fruit used in making jams and preserves; grows wild along seashores of the northeastern U.S.

beach pool *Hydrology.* **1.** a small, usually temporary body of water between two beaches or two beach ridges. **2.** a pool adjoining a lake, formed by wave action from the lake.

beach profile *Geology.* the intersection of a beach surface with a vertical plane perpendicular to the shoreline.

beach ridge *Geology.* a low, continuous mound of beach material formed by the action of waves and currents and lying more or less parallel to the shoreline.

beach scarp *Geology.* a steep slope formed by wave erosion on the seaward side of a berm.

beacon *Navigation.* a fixed, usually elevated device that serves as an aid to marine or aeronautical navigation, usually by emitting a signal warning of its location, such as a flashing light or radar pulse.

beaconage *Navigation.* **1.** a system of beacons. **2.** a tax or user fee for maintaining beacons.

beacon delay *Electronics.* in a responder beacon, the time required to convert an incoming message into an outgoing signal.

beacon presentation *Electronics.* the radarscope presentation produced by radio-frequency waves sent out by a radar beacon.

beacon stealing *Electronics.* in radar tracking, the interference between radar beacons, causing the loss of beacon tracking.

beacon tracking *Electronics.* the tracking of a moving object by means of signals emitted by a transmitter or transponder located on or within the object.

bead *Building Engineering.* a small convex molding formed on wood or other material. *Analytical Chemistry.* a globule of borax or other flux used in a bead test. *Geochemistry.* a drop of a fused material used as a solvent in color testing for various metals. *Electromagnetism.* a ceramic, glass, or plastic insulator used to position and support the inner and outer conductors of a coaxial cable. *Metallurgy.* **1.** in welding, the deposit of filler metal made in one pass. **2.** in the assaying of noble metals, the beads of noble metals obtained by cupellation.

bead and butt *Building Engineering.* framed work having a flush panel, with a bead run along the two edges in the direction of the grain. Also, **bead butt.**

bead and flush see BEADFLUSH.

bead and quirk *Building Engineering.* a bead that is separated by a narrow groove from the surface that it decorates. Also, QUIRK BEAD.

bead and reel *Building Engineering.* a convex molding having a pattern of elongated beads that alternate with one or more disks placed edge-on, with spherical beads, or with both.

bead-bed reactor *Biotechnology.* a reactor used for the rapid commercial propagation of mammalian cells; a growth medium is pumped through a glass column packed with glass beads on which the cells are growing.

beaded cable *Electromagnetism.* a coaxial cable that uses beads to position the inner and outer conductors so that the dielectric material between the conductors is (effectively) air as opposed to a continuous sleeve of material. Also, **beaded transmission line.**

beaded esker *Geology.* an esker having many bulges along its length, indicating differences in the rates at which fluvioglacial sediments were deposited.

beaded lake *Hydrology.* **1.** a narrow lake between sand dunes. **2.** see PATERNOSTER LAKE.

beaded lightning see CHAIN LIGHTNING.

beaded lizards *Vertebrate Zoology.* a large, venomous lizard, *Heloderma horridum,* with black, beadlike scales with yellow to pinkish spots, found in western Mexico. Also, MEXICAN BEADED LIZARD.

beaded molding see BEAD MOLDING.

beader *Mechanical Devices.* a tool used to carve beads for decorative molding.

beadflush *Building Engineering.* of a panel or paneling work, flush with its stiles and rails and finished with a flush bead on all its edges. Also, BEAD-AND-FLUSH.

beading *Building Engineering.* **1.** bead molding. **2.** all of the bead moldings constituting the ornamentation of a given surface or set of surfaces. *Metallurgy.* the process of raising a ridge or projection on a metal sheet.

bead-jointed *Building Engineering.* of a carpentry joint, ornamented with a bead along one of the butting edges.

Beadle, George Wells 1903–1989, American geneticist; with E. L. Tatum, Nobel Prize for proof that genes act by controlling chemical and enzymatic reactions in cells.

Beadle-Tatum experiment *Molecular Biology.* a series of experiments with mutant colonies of the red bread mold *Neurospora crassa,* performed in the early 1940s by George Beadle and Edward Tatum and leading to their hypothesis that one gene coded for a single enzyme.

bead mill *Biotechnology.* an apparatus that bursts cells by using a ball-type mill loaded with glass or ceramic beads; used for laboratory and large-scale processing.

bead molding *Building Engineering.* a molding edge or cornice decorated with a string of beads, often made of cast plaster.

bead plant *Botany.* any of several creeping plants of genus *Nertera* having tiny leathery leaves and transparent orange-colored berries; native to New Zealand and South America.

bead test *Analytical Chemistry.* the formation of a blister or bead to test for the presence of certain metals; oxides of these metals, when fused with borax on a platinum wire, produce a characteristic color. Also, BORAX BEAD TEST.

bead thermistor *Electricity.* a thermistor consisting of a small bead of resistive material attached to terminal wires.

beak *Vertebrate Zoology.* **1.** the bill or projecting jaws of a bird, having a horny epidermal covering and variously adapted for different modes of feed. **2.** a bill-like projection, as on a turtle or sawfish. *Invertebrate Zoology.* a horny protrusion or process resembling a bird's bill, as on an octopus or certain insects. *Botany.* a narrowed or prolonged tip on certain fruits or carpels. *Architecture.* a small pendant molding, usually serving as a drip. Also, **beak molding.**

beaked whales *Vertebrate Zoology.* any of the toothed whales of the family Hyperoodontidae (Ziphiidae) having beaklike jaws and found in all the oceans.

beaker *Science.* a deep, wide-mouthed glass container usually having a pointed lip for pouring. *Archaeology.* a simple clay drinking vessel without handles, widely used in prehistoric Europe. Also, BELL BEAKER.

Beaker folk *Archaeology.* a probably late Neolithic people living in various parts of Europe, characterized by the beakers found in their burial sites. Also, **Beaker people.**

beakhead *Naval Architecture.* a platform, with supporting knees and rails, that projects from the bow of a sailing ship; on early ships it was often covered with a grate and used as a latrine. Also, HEAD.

beaking joint *Building Engineering.* a resultant straight joint made by several members, such as flooring planks, having the same endpoint.

beakiron or **beak iron** *Mechanical Devices.* **1.** the tapered, projecting horns of an anvil. **2.** an anvil having such horns. Also, BICKIRON, BICKERN. (From Latin *bicornis,* "two horned.")

beam *Physics.* a concentrated, unidirectional stream of particles or radiation, as of light or sound. *Electromagnetism.* a signal transmitted in a certain direction, for example, by a beacon or radio tower. *Forestry.* a long, heavy piece of timber that is shaped for use as a supporting member. *Naval Architecture* the width of a ship's hull at its widest point. *Navigation.* **on the beam.** at right angles to a ship's keel or heading. *Textiles.* a large flanged spool designed to hold warp yarns before they are inserted into a loom. *Ordnance.* the side of a targeted vehicle, such as a ship, aircraft, or tank.

beam angle *Electromagnetism.* the angular measure of a directional electromagnetic beam, usually congruent with the area in which the intensity of the beam is half of its maximum value. Also, BEAM WIDTH.

beam antenna *Electrical Engineering.* an antenna, especially a VHF or short-wave antenna, that is designed to transmit in a certain direction.

beam attenuator *Spectroscopy.* an attachment used to adapt a spectrophotometer for small samples by reducing the energy of the reference beam.

beam bender see ION TRAP.

beam brick *Building Engineering.* an often triangular face brick used for bonding to a concrete lintel that is poured in place.

beam building *Mining Engineering.* a method for stabilizing overlying bedded rock in flat-lying deposits by bolting together the strata, which then act as a single beam capable of supporting the roof of the mine section.

beam compass *Graphic Arts.* an instrument used for drawing large arcs, consisting of a horizontal bar (or beam) having two adjustable sliding heads that hold marking points.

beam-condensing unit *Spectroscopy.* an attachment for a spectrophotometer that provides reduced radiation at the sample by concentrating and remagnifying the incident beam.

beam coupling *Electronics.* a method of producing an alternating current by modulating the density of a direct-current electron beam. The ratio of the alternating current to the direct beam current is called the **beam-coupling coefficient.**

beam current *Electronics.* **1.** the electric current in an electron beam. **2.** in a cathode-ray tube, the part of the gun current that projects on the screen.

beam damage *Optics.* the alteration or contamination of specimens due to interactions induced by the beam of an electron microscope.

beam-deflection tube *Electronics.* an electron beam tube in which the transverse movement of an electron beam controls current to an output electrode.

beam diameter *Optics.* the distance between two diametrically opposed points on a light beam of a specified irradiance; usually refers to a beam whose cross section is nearly circular.

beam divergence *Optics.* the angular spread occurring in a light beam at a specified fraction of peak irradiance.

beam drop *Electromagnetism.* a distortional fall in transmission altitude occurring in one section of an electromagnetic beam, particularly in detection radar.

beam efficiency *Electromagnetism.* the degree to which a beam antenna radiates in the desired direction, usually expressed as a ratio of the energy in a single beam to the total energy radiated by the antenna.

beam expander *Optics.* an optical system designed to increase the diameter of a beam of radiation, such as a laser beam.

beam fill *Building Engineering.* brick, masonry, or concrete work used to fill spaces between the joists or beams of a wall. Also, **beam filling.**

beam holding *Electronics.* the regenerating of charges stored on a cathode-ray tube screen by a diffused beam of electrons.

beam hole *Nucleonics.* a closable port in the protective shielding of a research reactor that allows the passage of a beam of neutrons for a variety of physical studies, including reactor diagnostics, direct measurement of nuclear reactions, ionization effects, and the like.

beam-indexing tube *Electronics.* a color picture tube that uses a single electron gun and whose screen is composed of vertical stripes of red, green, and blue phosphors arranged in sequence.

beam lead *Electronics.* a flat, thick-film lead placed on a semiconductor chip chemically or by evaporation as a connecting lead for a semiconductor device or integrated circuit.

beam-lobe switching *Electronics.* a method of locating the direction of a source of a radio signal by comparing the received signal while changing the lobe pattern of the receiving antenna.

beam magnet *Electromagnetism.* a magnet used to unite two or more electromagnetic beams.

beam network *Building Engineering.* a lattice of beams used in housing construction, especially to frame a roof.

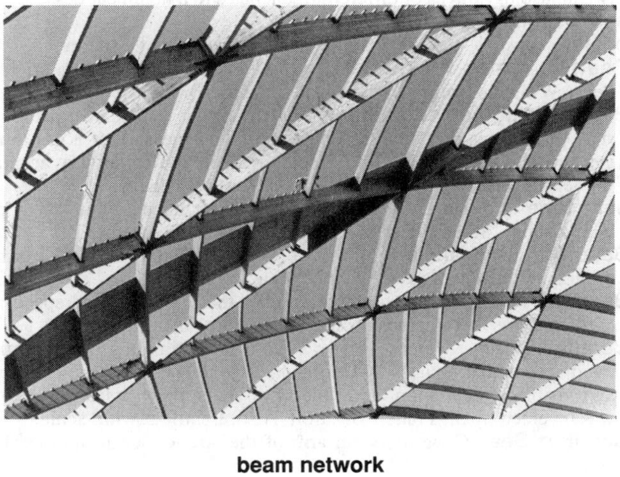

beam network

beam power tube see BEAM TETRODE.

beam recording *Electronics.* the recording of computer data directly onto microfilm by means of an electron beam.

beam resonator *Optics.* an instrument devised to isolate a beam of electromagnetic radiation, such as a laser beam, to a given area without requiring guidance along the entire length of the beam.

beam rider *Space Technology.* a missile designed to maintain or return to a course set by a radar beam; such a navigation system is called **beam-rider guidance**; the process of following a radar beam is called **beam riding.**

beam sea *Navigation.* wave action that strikes a vessel on the beam, that is, at right angles to its keel or heading.

beam search *Artificial Intelligence.* a search method that maintains a predetermined number of the best search paths found thus far at any given point. Thus, it considers more possibilities than depth-first search, but avoids the exponential number of possibilities of breadth-first search.

beam shaping *Electrical Engineering.* the act of controlling the radiation pattern emitted from an antenna by controlling the shape of the antenna, and the amplitude and phase of the driving signal.

beam splitter *Optics.* **1.** a mirror that allows some incident light to pass through it while reflecting the remainder. **2.** an optical device that uses a mirror or prisms to divide a light beam into two or more paths; especially, the prism system used in a color television camera to separate red, green, and blue light.

beam storage *Computer Technology.* an earlier type of storage device in which electron beams were used to enter or retrieve information in storage cells.

beam switching see BEAM-LOBE SWITCHING.

beam-switching tube *Electronics.* an electron tube that is designed as a multiposition electronic switch. It contains several (usually ten) identical arrays of electrodes around a central cathode, which act with a ring-shaped permanent magnet surrounding the glass envelope to provide crossed electric and magnetic fields.

beam tetrode *Electronics.* an electron-beam tube having an auxiliary pair of plates that focus the electron beams so as to augment power-handling capability and reduce secondary emission effects. Also, BEAM (POWER) TUBE, BEAM VALVE.

beam therapy *Radiology.* a treatment using a beam of light or other source of radiant energy.

beam tide *Navigation.* a tidal current setting at right angles to a vessel's keel or heading.

beam tube see BEAM TETRODE.

beam valve see BEAM TETRODE.

beam width see BEAM ANGLE.

beam wind *Navigation.* a wind that blows at approximately right angles to the heading of a vessel or aircraft.

bean *Botany.* **1.** any of the large, edible, kidney-shaped seeds of the Leguminoseae family, especially those of the genus *Phaseolus*, and the plants that produce them. **2.** any of various other beanlike seeds and the plants that produce them, such as coffee beans. *Engineering.* a small protruding device used to slow the flow of a liquid from a pipe or a well. *Mining Engineering.* a piece of coal that is small enough to pass through a screen or mesh.

bean anthracnose *Plant Pathology.* a disease of bean plants that is similar to bean blight, caused by the fungus *Colletotrichum lindemuthianum;* characterized by dark lines in the veins on the underside of the leaves and pinkish to brown lesions on the pods and seeds.

bean blight *Plant Pathology.* a disease of bean plants caused by the bacteria *Xanthomonas phaseoli* and characterized by irregular water-soaked spots and blotches on the plant parts.

bean caper *Botany.* any plant of the genus *Zygophyllum,* especially *Z. fabago*, a small tree of the eastern Mediterranean whose flower buds are used as a caper substitute.

bean curd see TOFU.

bean ore *Geology.* a loose, coarse-grained iron ore occurring in lens-shaped aggregations.

bean shot *Metallurgy.* a term for granulated copper having a beanlike appearance from being poured in hot water.

bean tree *Botany.* any of several trees having beanlike pods, such as the carob and catalpa.

bear *Navigation.* **1.** to tend to move in a certain direction. **2. bear down.** to approach something, especially another vessel, from the windward. **3. bear off.** to turn away; avoid. *Mining Engineering.* to drive in at the side or top of a working.

bear *Vertebrate Zoology.* any member of the widespread mammalian family Ursidae, the largest and least carnivorous animals in the order Carnivora; characterized by heavy hairy bodies, short strong legs, and plantigrade feet.

brown bear

bear cat see PANDA, def. 2.

beard *Biology.* a hairy, bristly, or fibrous growth, such as the tuft on a goat's chin, the stiff hairs around a bird's beak, the silk on a cob of corn, or the awns on the head of oats, barley, and other grains. *Textiles.* a flexible hook on the end of certain sewing needles. *Graphic Arts.* on a piece of type, the slope between the face and shoulder.

bearded needle *Textiles.* a needle having a flexible hook at one end, used in machine knitting to produce a tight and even texture. Also, SPRING NEEDLE.

bearded seal *Vertebrate Zoology.* a large grayish or yellowish Arctic seal, *Erignathus barbatus,* having square foreflippers and a thick growth of bristles on either side of the muzzle.

beardfish see BARBUDO.

bearding *Naval Architecture.* **1.** the slanting edge along which the longitudinal members of a hull meet the sternpost or the stempost. **2.** the forward edge of a rudder. **3.** the process of trimming wood from the hull of a ship in order to improve its performance.

beard-lichen *Botany.* a greenish-gray lichen, *Usnea barbata,* that grows on and hangs from trees, resembling a beard.

Bear Driver see BOÖTES.

bearer *Botany.* a tree or plant that produces fruit or flowers. *Building Engineering.* **1.** a joistlike crosspiece supporting the boards of a scaffold. **2.** see BEARING MEMBER. *Graphic Arts.* a support that protects a printing surface from excess pressure or prevents the application of ink to a nonimage area.

bearing a direction, orientation, or relative position; specific uses include: *Navigation.* the horizontal direction of one terrestrial point from another; normally measured clockwise through 360°. *Cartography.* an angle measured clockwise from north.

bearing a supporting member; specific uses include: *Mechanical Engineering.* any part of a machine or device that supports or carries another part that is in motion in or upon it, such as a journal bearing. *Building Engineering.* **1.** the part of a beam that actually rests on its supports. **2.** the compressive stress on a beam or its supports. **3.** the joint or area of contact between a bearing member and a wall.

bearing angle *Navigation.* a directional bearing expressed as an angle right or left from a baseline, normally a north-south line.

bearing area *Engineering.* the part of a structure or member that effectively carries a load.

bearing bar *Engineering.* **1.** a device used in a magnetic compass to determine bearing. **2.** a load-carrying bar that supports a grating or cover.

bearing cap *Mechanical Engineering.* a metal piece designed to fit over the top of a journal bearing to hold the shaft in place.

bearing capacity see BEARING STRENGTH.

bearing cursor *Engineering.* in radar, the radial line on a transparent disk that can be rotated to determine bearing.

bearing distance *Engineering.* the unsupported span of a beam between two bearings.

bearing loss *Mechanical Engineering.* in a machine, a loss of power due to friction between a bearing and the part that it supports.

bearing member *Engineering.* a supporting part of a structure, such as a column or joist.

bearing metal *Mechanical Engineering.* any low-friction alloy, such as bronze, used in a journal bearing at the points of contact with the shaft.

bearing pile *Engineering.* a column, usually made of prestressed concrete or reinforced timber, that is driven into bedrock or subsoil and used to transmit the load of a structure into the firm foundation.

bearing plate *Engineering.* a flat, heavy steel plate designed to receive and distribute weight, as from a wall, column, or one end of a truss.

bearing pressure see BEARING STRESS.

bearing resolution *Electronics.* the minimum angular difference in bearing at which a given radar system is able to differentiate between two adjacent targets having the same range.

bearing strain *Engineering.* the deformation of a bearing part under a load.

bearing stratum *Engineering.* a geologic bed or formation that is used to support a structure, serving as a natural foundation.

bearing strength *Mechanics.* the maximum load that a structure such as a column, wall, footing, or a joint can support, divided by its effective bearing area. Also, BEARING CAPACITY.

bearing stress *Mechanics.* the stress on a surface in contact with a load divided by the bearing area; used to predict the wear expected in a moving part. Also, BEARING PRESSURE.

beat *Physiology.* the rhythmic throb or pulsation of the heart or an artery. *Horology.* **1.** the audible ticking of a timepiece. **2.** the rhythm produced by the oscillations of an escapement mechanism, especially with regard to its regularity. A timepiece that ticks irregularly is **out of beat.** *Physics.* the periodic variation of amplitude resulting from the superposition of two sinusoidal signals of slightly different frequencies.

beat Cepheid *Astronomy.* a Cepheid variable star in which two or more nearly equal periods of variability pass into and out of phase with each other, producing a "beat."

beaten zone *Ordnance.* an area on the ground that is hit by fire from a gun battery.

beater an apparatus used to beat some material; specific uses include: *Agriculture.* a revolving blade mechanism on a threshing machine or combine harvester that loosens the kernels from ears of cereal. *Food Engineering.* a revolving device equipped with blades to crush and mix ingredients. *Graphic Arts.* in papermaking, a machine that initiates the preparation of the fibers by cutting, bruising, fibrillating, and hydrating the stock. *Mining Engineering.* a tool used to pack material around a charge of powder in a blasthole. *Textiles.* the part of a loom that drives the newly inserted threads against the edge of the woven fabric.

beater mill see HAMMER MILL.

beat frequency *Electronics.* the frequency of a signal at the output of a nonlinear circuit, equal to the sum or difference of the frequencies of two or more signals applied to the input.

beat-frequency oscillator *Electronics.* an oscillator that produces a signal frequency by combining two generating frequencies. Also, HETERODYNE.

beating-in *Electronics.* the process of manipulating the frequency of one of two interconnected oscillators until no beat frequency is heard in a connected receiver.

beat note *Electronics.* the beat frequency resulting from the feeding of two sinusoidal waves of different frequencies to a nonlinear device.

beat oscillator see LOCAL OSCILLATOR.

beats *Acoustics.* periodic variations in the loudness of a sound.

Beattie, James Alexander 1895–1981, American chemist; with Percy Williams Bridgman, formulated Beattie-Bridgman equation.

Beattie-Bridgman equation *Thermodynamics.* a standard equation of state that relates the volume, temperature, and pressure of a gas to the gas constant while taking into account molecular effects; very accurate for densities less than about 0.8 times the critical density.

beat tone *Acoustical Engineering.* a tone produced when two tones within about fifteen cycles of each other are played together.

Beaufort, Sir Francis [bō′fərt] 1774–1857, English admiral and hydrographer; developed the Beaufort scale.

Beaufort force *Meteorology.* a number that denotes wind speed according to the Beaufort scale. Also, BEAUFORT NUMBER.

Beaufort scale *Meteorology.* a system for estimating and reporting wind speeds that equates Beaufort force, wind speed, descriptive term, and visible effects upon land objects or sea surface; its original form was invented by Sir Francis Beaufort. Also, **Beaufort wind scale.**

Beaufort Sea *Geography.* an arm of the Arctic Ocean, northeast of Alaska.

beau gregory *Vertebrate Zoology.* a blue-and-yellow damselfish, *Pomacentrus leucostictus*, found in tropical waters off Florida and the West Indies. Also, **beau gregoire.**

Beaujolais [bō´jə lā´] *Food Technology.* **1.** a light, fruity French red wine made with Gamay grapes. **2.** a similar wine made in California or elsewhere. Also, GAMAY BEAUJOLAIS. (From the *Beaujolais* district, in the Burgundy region of central France.)

Beaumont, William [bō´mänt´t] 1785–1853, American physician; known for studies of gastric juices and the physiology of digestion.

beauty *Particle Physics.* the qualities of a bottom quark, which is informally called a **beauty quark.**

Beauveria *Mycology.* a genus of imperfect fungi of the family Moniliaceae, order Moniliales. *B. bassiana* causes muscardine in silkworms; *B. tenella* was formerly used for pest control.

beaver *Vertebrate Zoology.* a semiaquatic rodent of the family Castoridae found throughout North America and Europe and characterized by soft brown fur, webbed hind feet for swimming, a broad, flat, scaly tail, and strong, prominent front teeth.

beaver

beaverboard *Building Engineering.* a construction sheeting made of wood fiber and resembling a heavy cardboard; used especially in building partitions, ceilings, and temporary structures.

beaver cloth *Textiles.* a thick-napped cotton or woolen cloth resembling beaver fur.

beaverite *Mineralogy.* $Pb(Cu^{+2},Fe^{+3},Al)_3(SO_4)_2(OH)_6$, a bright-yellow trigonal mineral of the alunite group, occurring in earthy masses of microscopic plates and having a specific gravity of 4.36; found as a secondary mineral in the oxidation zones of lead-copper deposits.

beavertail *Electronics.* a broad, flat antenna pattern signal that is directed up and down in order to determine the altitude of an object.

bebeerine *Organic Chemistry.* $C_{36}H_{38}N_2O_6$, a crystalline alkaloid derived from the bark of the bebeeru tree and related species, that is soluble in benzene and melts at 215°C; used as a tonic in malaria.

bebeeru *Botany.* a South American tree, *Nectandra rodiei,* of the laurel family; its hard, durable wood is used in shipbuilding and heavy construction; its bark is the chief source of bebeerine, a tonic used in malaria. Also, GREENHEART.

Béchamp reduction *Organic Chemistry.* a method of using ferrous salts or iron in aqueous acid to reduce aromatic nitro groups to amino groups.

bêche [besh] *Mining Engineering.* a tool used in well boring to retrieve drill rods or other items from inside the bore.

bêche-de-mer see SEA CUCUMBER.

Becke line *Mineralogy.* a thin line of light seen under a microscope at the junction of two substances with different refractive indices. It is used to perform the **Becke test,** comparing the refractivity of two such minerals or of a mineral and an immersion liquid: when the microscope tube is raised, the Becke line appears to move toward the more refractive mineral.

Becker and Kornetzki effect *Physics.* the reduction of the internal friction of a ferromagnetic material that is placed in a saturating magnetic field.

beckerite *Mineralogy.* a variety of retinite that has a high concentration (20–23%) of oxygen.

Becklin-Neugebauer object *Astronomy.* a bright, relatively small object, probably a star in the process of formation, that has been detected in the Orion Nebula at infrared wavelengths.

Beckmann, Ernst Otto 1853–1923, German chemist; discovered Beckmann transformation; developed Beckmann thermometer.

Beckmann thermometer *Engineering.* a small-range, high-precision mercury thermometer.

Beckmann transformation *Organic Chemistry.* a type of reaction in which oximes are converted to amides by the action of acids, as when benzophenone oxime, $(C_6H_5)_2C{=}NOH$, is converted to benzanilide on treatment with a catalyst, usually phosphorus pentachloride; used especially to determine the configuration of ketoximes. Also, **Beckmann rearrangement.**

beclomethasone dipropionate *Pharmacology.* $C_{28}H_{37}ClO_7$, a glucocorticoid that is administered by aerosol inhalation to treat bronchial asthma symptoms.

Becquerel, Antoine César [bek´ə rel´; bə krel´] 1788–1878, French physicist; pioneer of electrochemistry; invented constant-current electric pile, electromagnetic balance, and thermoelectric needle.

Becquerel, Antoine Henri 1852–1908, French physicist; shared Nobel Prize for discovery of radioactivity in uranium.

Antoine Henri Becquerel

becquerel [bek´ə rel´] *Nucleonics.* a unit of radionuclide activity, equivalent to 1 disintegration per second; this term is now usually used in place of the term *curie.*

Becquerel cell *Electricity.* a photoelectric cell consisting of two identical, unequally illuminated electrodes that are placed in an electrolyte, causing a voltage to flow between them. Also, PHOTOCHEMICAL CELL.

becquerelite *Mineralogy.* $Ca(UO_2)_6O_4(OH)_6{\cdot}8H_2O$, a brownish-yellow orthorhombic mineral with a resinous luster, prismatic crystals, a specific gravity of 5.14 and a hardness of 2.5 on the Mohs scale; found as a secondary mineral with uraninite.

Becquerel rays *Nuclear Physics.* a term formerly used to identify all rays emitted by radioactive substances, now designated by the more specific terms alpha rays, beta rays, and gamma rays.

bed *Agriculture.* **1.** a small plot of soil used to grow seedlings, flowers, shrubs, or vegetables. **2.** see SEEDBED. *Forestry.* **1.** the flat and horizontal surface of the undercut. **2.** to level and then construct a cushion of brushwood on the ground along the line on which a tree is to be felled, in order to minimize the shattering of its timber. *Geology.* **1.** a subdivision of a stratified rock series consisting of relatively homogeneous material separated from the layers above and below by visually or physically well-defined boundary planes. **2.** any layer of rock or ore deposit found in a stratified sequence of rocks. **3.** the bottom or ground upon which a body of water, such as a stream, lake, or ocean rests, or the land marking the site of a former body of water. *Graphic Arts.* on an offset press, the flat surface to which the type or plates are attached and over which the rollers and impression cylinders pass. *Chemistry.* a solid, porous mass of material that acts as a reactant, catalyst, or adsorbent in a chemical reaction. *Mechanical Engineering.* the part of a machine having a planed, precisely machined surface that supports or aligns other machine parts, as in a lathe *bed.*

BEd or **B.Ed.** Bachelor of Education.

bedbug *Entomology.* the common name for *Cimex lectularius* and related species of flat, wingless, mostly nocturnal parasites of the family Cimicidae (order Heteroptera) that feed by sucking the blood of mammals, especially humans.

bedded *Geology.* of or relating to material that is formed, arranged, or deposited in layers or beds.

beddedness index see STRATIFICATION INDEX.

bedded plant see BEDDING PLANT.

bedded vein *Geology.* a vein or other mineral deposit that follows a bedding plane.

bed detector *Mining Engineering.* any instrument designed to locate and measure the scope of underground mineral deposits.

bedding

bedding *Geology.* the arrangement of sedimentary rock in beds, strata, or layers of varying thickness and character. *Agronomy.* the plowing of a field by alternating back furrows with dead furrows. *Petrology.* the division of certain crystalline rocks, such as granite, into well-defined planes lying parallel to the land surface. *Building Engineering.* any base layer upon which something rests, such as putty laid beneath a window pane (**bed glazing**) or concrete beneath a pipe. *Mining Engineering.* a firm but flexible material that serves as a filler or cushion, such as soft copper wire that is handset under or around a diamond in a drill bit. Also, CALKING.

bedding cleavage *Geology.* the tendency of a rock to split parallel to the bedding plane. Also, **bedding-plane cleavage.**

bedding fault *Geology.* a fault whose surface follows along or is parallel to the bedding plane of the component rocks.

bedding fissility *Geology.* the tendency of fine-grained sedimentary rock to split easily more or less parallel to the bedding.

bedding glide *Geology.* overthrusting in which a bed thrusts laterally along the roof or floor of a mine.

bedding joint *Geology.* a joint that follows along or is parallel to the bedding plane.

bedding plane *Geology.* a surface or division plane that visibly separates each successive layer of stratified rock from its neighboring layers. Also, STRATIFICATION PLANE.

bedding-plane slip *Geology.* the relative displacement of sedimentary rock strata along bedding planes that occurs during folding. Also, FLEXURAL SLIP.

bedding plant *Agriculture.* a usually ornamental plant that is especially suited for growing in open-air beds. Also, **bedded plant.**

bedding schistosity *Geology.* foliation in coarse-grained, metamorphic rock that is parallel to the original rock bedding.

bedding surface *Geology.* a surface within a mass of stratified rock that represents the original surface of deposition or the interface between two neighboring sedimentary beds.

bedding thrust *Geology.* a nearly horizontal thrust fault that follows along or is parallel to the bedding plane. Also, **bedding glide.**

bedding void *Geology.* an open space formed between successive lava flows.

bedeguar or **bedegar** *Plant Pathology.* a mosslike gall that is produced on roses (especially the sweetbrier) by a gall wasp.

Bedford cord *Textiles.* a heavy woven fabric that is ribbed lengthwise like corduroy.

Bedford limestone see SPERGENITE.

bediasite or **Bediasite** *Petrology.* a jet-black to brown tektite found in east-central Texas. (From the town of *Bedias,* Texas.)

bed joint *Building Engineering.* **1.** a horizontal joint in masonry or brickwork. **2.** a radiating joint in an arch.

bedload or **bed load** *Geology.* a heavy sediment that is transported on or very near a streambed by the natural flow of the stream.

bed molding *Architecture.* **1.** in a cornice, the molding immediately above the frieze and below the corona. **2.** any decorative molding below a projecting member.

Bednorz, J. Georg born 1950, German physicist; with K. Müller, received Nobel Prize for discovery of high-temperature superconductors.

Bedoulian *Geology.* in Switzerland, a geologic substage of the lower Aptian stage of the Lower Cretaceous period.

bedpan *Medicine.* a receptacle for the urinary and fecal discharges of a bedridden patient.

bedplate *Mechanical Engineering.* a metal plate that serves as the base or foundation of a machine or of certain machine parts.

bedrock or **bed rock** *Geology.* the solid, undisturbed rock underlying the soil or other unconsolidated surface material.

bedrock mortar *Archaeology.* a deep basin in granite or other large rock outcroppings, formed by the action of grinding or crushing foods with a stone. A flat or shallow surface of this same type is called a **bedrock grinding slick.**

bed separation *Mining Engineering.* thin cavities that are formed along bedding planes by the differential lowering of strata over mine workings.

Bedsonia *Bacteriology.* a former name for CHLAMYDIA.

bedsore see DECUBITUS ULCER.

bedstraw *Botany.* any of several rubiaceous herbs of the genus *Galium* having square stems, whorled leaves, and small white or brownish flowers; once used in cheesemaking to curdle milk. (From their former use as straw to stuff mattresses.)

bee *Invertebrate Zoology.* the common name for the members of a large group of membranous-winged, segmented, often stinging insects of the family *Apidae,* many of which are major plant pollinators and honey producers. They are widely distributed throughout the world except in extremely cold climates.

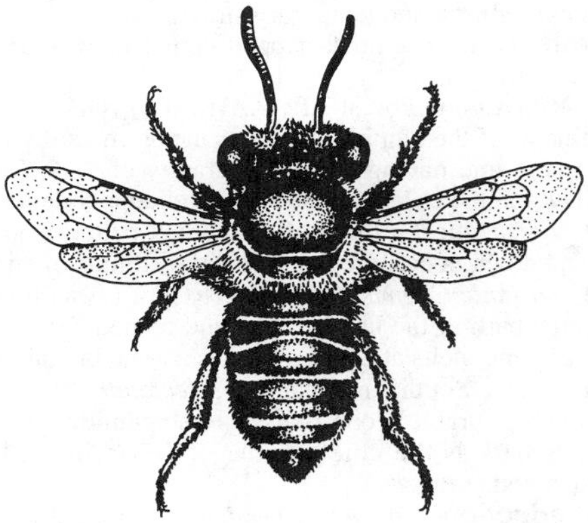

bee

BEE or **B.E.E.** Bachelor of Electrical Engineering.

bee balm *Botany.* **1.** any of several perennials of the mint family, especially *Monarda didyma,* having thin leaves and white, lavender, or deep red flowers that are attractive to bees and hummingbirds. Also, OSWEGO TEA. **2.** see LEMON BALM.

Beebe, William 1877–1962, American naturalist; a leading ornithologist, known as explorer of exotic regions.

bee beetle *Entomology.* a European beetle, *Trichodes apiarius,* whose larvae (bee wolf) sometimes infest beehives.

bee bird *Vertebrate Zoology.* any bird that feeds on bees, such as the European flycatcher.

bee bread *Entomology.* a bitter, brownish mixture of nectar and pollen that is often prepared, stored in the hive, and used by bees as food for their larvae and young.

beech *Botany.* **1.** any deciduous hardwood tree of the genus *Fagus* in the Fagaceae family, characterized by smooth gray bark; hard, straight-grained wood **(beechwood);** glossy leaves; long, slender, scaly winter buds; and triangular edible nuts **(beechnut). 2.** or **beech family.** the Fagaceae family, including beeches, chestnuts, and oaks.

beech bark disease *Plant Pathology.* an often fatal disease of beech trees caused by the **beech scale** *(Cryptococcus fogi)* and the fungus *Nectria coccinea faginata,* which act together in destroying bark and causing foliage to wilt.

beechdrops *Botany.* a low North American plant without green leaves, *Epifagus virginiana,* that is parasitic on the roots of the beech.

beech fern *Botany.* either of two ferns of the genus *Thelypteris (Dryopteris)* that grow throughout beech forests in moist northern temperate regions such as northeastern North America.

bee eater *Vertebrate Zoology.* any of several brightly colored tropical birds of the family Meropidae that feed on bees and other insects.

beef *Agriculture.* **1.** the meat that comes from adult cattle. **2.** *plural,* **beeves** or **beefs.** any cow, bull, or steer that is fattened to be slaughtered for food. *Mineralogy.* fibrous calcite that occurs in thin veins or layers in sedimentary rock.

beefalo *Agriculture.* a hybrid animal (typically 5/8th domestic cow and 3/8th buffalo) that is bred for disease resistance and low-fat meat.

beefly *Vertebrate Zoology.* any of many dipterous insects of the family Bombyliidae, resembling bees but having only one pair of wings.

Beefmaster *Agriculture.* a breed of beef cattle crossbred from Brahman, Hereford, and Shorthorn cattle. *Botany.* a widely grown variety of beefsteak tomato.

beef stearin see OLEOSTEARIN.

beefwood *Forestry.* the heavy, hard reddish wood of any of several species of casuarina tree; used in making furniture and cabinets.

beehive *Agriculture.* a domicile of bees that is either man-made or created by a bee colony within a tree crevice or other hollow cavity. *Entomology.* a bee colony's population (queens, drones, and workers).

Beehive *Astronomy.* a large open (or galactic) star cluster in Cancer, just visible to the naked eye. Also, PRAESEPE, MANGER, M44.

beehive oven *Engineering.* an arched oven used to carbonize coal into coke.

bee hummingbird *Vertebrate Zoology.* a variety of hummingbird found in Cuba and the Isle of Pines; the world's smallest bird and among the smallest of the warm-blooded vertebrates, weighing about 2 grams and measuring 2 inches in length including tail and bill.

bee hummingbird

beekeeping see APICULTURE.

beekite *Mineralogy.* a variety of chalcedony often incrusting or pseudomorphous after coral or shell fossils.

bee louse *Entomology.* a tiny wingless dipteran insect of the family Braulidae that infests the bodies of honeybees; its larvae burrow into the hive to feed on pollen and honey.

beeper see PAGER.

Beer, August 1825–1863, German physicist; formulated Beer's law.

beer *Food Technology.* **1.** an alcoholic beverage produced by the slow fermentation of malted cereal grains, usually barley and hops; the two main types of beer are top-fermented ales and bottom-fermented lagers. **2.** in common use, a lager as opposed to an ale.

beerbachite *Petrology.* fine-grained basic hornfels with poikiloblastic crystals of olivine, derived from xenolithic inclusions in gabbro and composed of labradorite, clinopyroxene (diallage), hornblende, hypersthene, and others.

beer scale *Food Technology.* a deposit of calcium oxalate and other salts that forms on the insides of an apparatus used in the brewing process. Also, **beer stone.**

Beer's law *Physical Chemistry.* a law stating that the amount of light absorbed by a solution varies exponentially with the concentration of the solution and the length of the light path in the solution.

beerwort see WORT.

beeswax see WAX, def. 2.

beet *Botany.* any of various biennial plants of the genus *Beta* in the goosefoot (Chenopodiaceae) family having a large, tapering taproot swollen with stored food, especially *Beta vulgaris,* with a red-rooted variety grown as a vegetable and a white variety (the sugar beet) often used a source of sugar.

beetle *Invertebrate Zoology.* a member of the large insect order Coleoptera (about 330,000 known species), characterized by forewings that are thickened to form a shell-like cover that protects the membranous hind wings. *Mechanical Devices.* a heavy, usually wooden mallet used for beating, mashing, or hammering; a maul. *Mining Engineering.* **1.** a remote-controlled, cable-hauled propulsion unit with the power to move a train of wagons at the mine surface. Also, CHARGER. **2.** a small compressed-air locomotive.

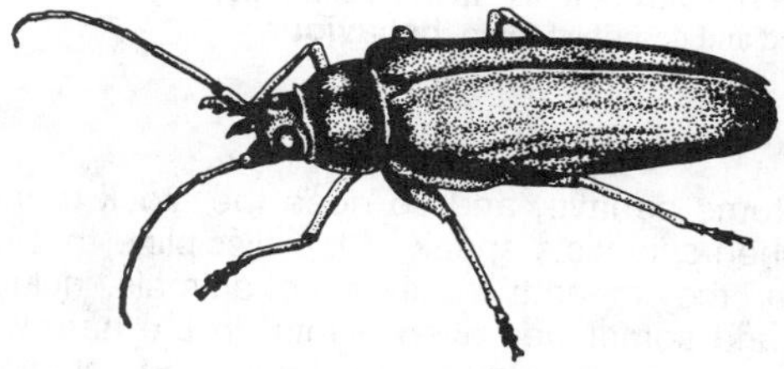

beetle

beet leafhopper *Invertebrate Zoology.* a leafhopper, *Circulifer tenellus,* of the western U.S. that feeds on the leaves of sugar beets and acts as a vector of curly-top disease.

beetle stone see SEPTARIUM.

beetleweed see GALAX.

beetling *Textiles.* the process of pounding cotton or linen cloth with a beetle or other tool in order to produce a smooth, glossy finish.

beetree *Botany.* any hollow tree in which bees hive, especially basswood (American linden).

beet sugar *Food Technology.* a sugar produced by the extraction and crystallization of juice from sugar beets.

bee wolf *Entomology.* a term for a bee larva.

Before Present or **before present** see BP.

before the wind *Navigation.* describing a sailing vessel that is sailing with the wind coming from astern.

Beggiatoa *Bacteriology.* a genus of sulfur-oxidizing gliding bacteria of the family Beggiatoaceae that consist of colorless cells in unattached filaments, occur in sulfide-rich aquatic environments, and deposit sulfur granules in the presence of hydrogen sulfide.

Beggiatoaceae *Bacteriology.* a family of Gram-negative gliding bacteria in the order Cytophagales that consist of colorless filaments containing cells in flexing chains, live freely in fresh or salt water or in soil, and may accumulate sulfur granules in the presence of hydrogen sulfide.

Beggiatoales *Bacteriology.* a former taxonomic order of gliding bacteria that are now included in the order Cytophagales.

begging *Behavior.* a term for various behavior patterns by which young animals induce their parents to provide food for them.

Beginner's All-purpose Symbolic Instruction Code see BASIC.

beginning-of-file label *Computer Programming.* a record at the beginning of a file that contains information about the file such as name, length, and block or record size. Also, HEADER RECORD.

beginning-of-information marker *Computer Technology.* a reflective strip on a magnetic tape that is sensed by the tape drive to indicate the point where recording can begin. Also, **beginning-of-tape marker.**

beginning-of-volume label *Computer Programming.* a record that is used for multivolume file processing at the beginning of a flexible disk, magnetic tape reel, or other data storage unit. The record contains information about the volume such as file name, volume number within the file, size, and first and last records.

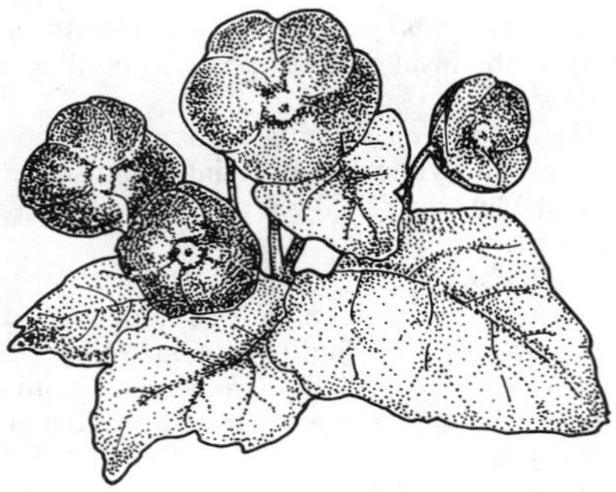

begonia

begonia *Botany.* any tropical plant of the family Begoniaceae, especially those of the genus *Begonia*, which are cultivated for their waxy flowers and large, richly colored succulent leaves.

Begoniaceae *Botany.* the begonia family of succulent herbs, characterized by thick rhizomes or tubers; radical or alternate, usually asymmetrical leaves; and a winged capsule fruit with numerous tiny seeds.

behavior *Psychology.* any action that an individual carries out in response to a stimulus or to its environment, especially an action that can be observed and described. Also, **behaviour.**

behaviorism *Psychology.* a psychological school that regards as its subject matter only overt actions that can be directly observed, measured, and manipulated, and that ignores unobservable mental experiences such as ideas and emotions. Thus, **behaviorist, behavioristic.**

behaviorist approach *Behavior.* an approach to psychology that focuses on observable behavior.

behavior mechanism see DEFENSE MECHANISM.

behavior modification *Behavior.* the use of techniques of conditioning to control or change learned behaviors.

behavior potential *Behavior.* the likelihood that a certain behavior will occur in a particular situation.

behavior ratio *Behavior.* a measurement consisting of the number of responses to one alternative divided by the total number of responses to all alternatives.

behavior rehearsal *Behavior.* a technique of assertiveness training in which the individual acts out appropriate and effective ways to handle troublesome situations.

behavior repertoire *Behavior.* all the behaviors that are possible for a species.

behavior therapy *Behavior.* a type of psychological treatment that uses conscious learning techniques to help individuals make constructive changes in their behavior. Thus, **behavior therapist.**

beheaded stream *Hydrology.* a stream whose upper part has been cut off and diverted into the channel of another stream that has greater erosional activity and that flows at a lower level.

Behavior

Atoms behave, and so does the stock market. But when scientists speak of the discipline that is called *behavior,* they mean the activities of animals, including humans, and sometimes even plants. In the narrow sense, behavior refers to the reactions of animals to change in the environment. But animals also behave spontaneously, in the absence of any obvious stimulus. Therefore, behavioral scientists attempt to understand how an animal takes in information from its environment, processes that information, and then acts. Processing information in the central nervous system may entail integrating information over time, including stimuli coming from the organism's own internal environment, such as hormones. Thus the connection between stimulus and response can be delayed and indirect.

The study of behavior dates back to Aristotle and even earlier. Charles Darwin considered the behavior of animals and humans in an evolutionary context, foreshadowing attempts of comparative psychologists and ethologists. Ethology arose in Europe shortly before World War II, centered around the Dutch biologist Niko Tinbergen and the Austrians Konrad Lorenz and Karl von Frisch; these three received the Nobel Prize in Medicine in 1974 for their contributions. They studied naturally occurring behavior and how it evolved. The views of Lorenz and Tinbergen on the genetic basis of behavior resulted in a widely aired controversy with North American comparative psychologists, led by Daniel Lehrman; he argued that ethologists were genetic determinists who failed to appreciate the delicate interplay of environment and genes in the development of behavior.

That issue was later amicably resolved; however, it burst into the open again briefly with the advent of E.O. Wilsons's book *Sociobiology,* published in 1975. Today animal behavior is studied by biologists, biological psychologists, and anthropologists. Some analyze the immediate causes of behavior whereas others address the adaptiveness and evolution of behavior.

George W. Barlow
Professor of Integrative Biology
University of California, Berkeley

behavioral *Behavior.* of or relating to animal or human behavior.

behavioral anthropology *Anthropology.* another name for evolutionary anthropology.

behavioral fatigue *Behavior.* a pattern in which a certain behavior, after having been repeated several times, can no longer be elicited again for some time, or can be elicited only by a much stronger stimulus. Also, RESPONSE FATIGUE.

behavioral genetics *Science.* the scientific study of the effects of heredity on behavioral patterns or traits.

behavioral inventory see ETHOGRAM.

behavioral medicine an area of research and practice that combines behavioral techniques with medical science in order to promote health and to manage disease, as in the use of biofeedback for treating medical disorders.

behavioral science a field of knowledge, such as anthropology, psychology, or sociology, that studies the actions of humans and other organisms in order to form conclusions about social behavior.

behavioral therapy or **psychotherapy** see BEHAVIOR THERAPY.

behavior disorder *Psychology.* any persistent behavioral pattern that is undesirable or destructive, especially one that is not part of a well-defined neurosis or psychosis.

behenic acid *Organic Chemistry.* $CH_3(CH_2)_{20}COOH$, a combustible solid that is soluble in alcohol and ether; it melts at 80°C and boils at 306°C (60 Torr); a saturated fatty acid that is obtained from plant seed oils, especially black mustard; used in the manufacture of cosmetics and waxes. Also, DOCOSANOIC ACID.

behenyl alcohol *Organic Chemistry.* $CH_3(CH_2)_{20}CH_2OH$. a colorless, waxy solid that melts at 65–72°C and boils at 180°C (0.2 Torr); a long-chain saturated fatty alcohol used in synthetic fibers, lubricants, and evaporation retardants on water surfaces. Also, 1-DOCOSANOL.

Behnken's unit *Radiology.* a unit of X radiation that yields an ionic charge equal to one electrostatic unit when applied to 1 cubic centimeter of air at 18°C and standard atmospheric pressure.

Behrens-Fisher problem *Statistics.* a problem that considers statistical data of random samples from two different normally distributed populations with unknown unequal variances. Also, FISHER-BEHRENS PROBLEM.

Behrens-Fisher test *Statistics.* a test of significance for the difference between the means of random samples in a Behrens-Fisher problem.

Behring, Emil Adolf von 1854–1917, German physician and bacteriologist; Nobel Prize for producing the first antitoxins (against tetanus and diphtheria).

beidellite *Mineralogy.* $(Na,Ca_{0.5})_{0.3}Al_2(Si,Al)_4O_{10}(OH)_2 \cdot nH_2O$, a white, reddish, or brownish-gray monoclinic clay mineral of the smectite group.

bei function *Mathematics.* the imaginary part of $J_n(ze^{\pm 3\pi i/4})$, where J_n is the *n*th Bessel function.

Beijerinckia *Bacteriology.* a genus of Gram-negative, nitrogen-fixing bacteria of uncertain affiliation, occurring in acidic soils, having a respiratory metabolism, and dividing by cell constriction.

Beilby layer *Metallurgy.* a flow layer on a metallic or mineral surface that has been heavily disturbed by mechanical deformation, as by polishing, such that the normal crystal structure is obliterated.

Beilstein, F. P. 1838–1906, German chemist; compiled *Handbuch der Organischen Chemie*, describing the properties and reactions of organic compounds.

bejel see NONVENEREAL SYPHILIS.

Békésy, Georg von [bā´kā shē] 1899–1972, Hungarian-born American physicist; Nobel Prize for discoveries on stimulation within the cochlea.

Békésy audiometry *Acoustics.* a technique of testing hearing in which the subject controls the procedure by pressing a signal button; the intensity of a tone diminishes as long as the button is depressed, and increases when it is released.

Bekhterev-Mendel reflex see MENDEL-BEKHTEREV REFLEX.

Bekhterev's sign *Neurology.* a loss of pain sensation in the area behind the knee, characteristic of patients suffering from tabes dorsalis (a form of neurosyphilis).

bel *Physics.* a dimensionless quantity used to describe a comparison of power levels, the number of bels being the logarithm of the ratio of the powers.

belat *Meteorology.* a strong land wind from the north or northwest that blows across the southeast coast of Arabia and is accompanied by sand blown from the interior desert areas.

belay *Navigation.* to secure the smaller ropes and lines aboard a ship by wrapping them around a cleat or belaying pin; used especially in reference to the running rigging.

belemnite *Paleontology.* **1.** an extinct cephalopod resembling an octopus or cuttlefish, traditionally placed in the subclass Coleoidea but proposed for inclusion in Belemnoidea. **2.** the dart-shaped fossil guard piece of such an animal.

Belemnoidea *Paleontology.* a proposed extinct subclass of mollusks in the class Cephalopoda, characterized by an internal shell consisting of an elongated phragmocone and a calcite rostrum, with soft body parts surrounding the shell much as in modern squids; they arose in the Carboniferous, proliferated in the Mesozoic, and then disappeared in the Eocene.

belfry *Architecture.* **1.** an attached or freestanding belltower. **2.** the part of a steeple in which bells are hung.

Belgian *Agriculture.* a large, sturdy breed of draft horse, usually roan or chestnut in color, that originated in Belgium.

Belgian endive *Agriculture.* a young chicory plant cultivated in darkness so that it forms a tight, narrow head of whitish leaves, which are used raw in salads or cooked as a vegetable.

Belgian hare *Agriculture.* a breed of domestic rabbit having reddish-brown fur.

belian *Materials.* the wood of a tree of Southeast Asia, *Eusideroxylon zwageri*, which is used for heavy construction projects such as shipbuilding and flooring.

belief function *Artificial Intelligence.* a function that determines how strongly an uncertain conclusion is believed.

belief system *Anthropology.* the portion of a culture that encompasses religion, spiritualism, magic, ritual, and origin and culture-hero mythologies.

B eliminator see BATTERY ELIMINATOR.

Belinurus *Paleontology.* an extinct Carboniferous genus of xiphosuran arthropods in the order Limulida (or, in some classifications, order Xiphosurida); similar to the modern "horseshoe crab" Limulus, but with a partly segmented abdomen.

belite see LARNITE.

bell *Acoustics.* a hollow, flared-mouth metallic vessel that vibrates at a fixed pitch when struck by a clapper or hammer. *Navigation.* the sounding of a bell as a signal, especially a system used to sound the time every half hour onboard a ship. *Botany.* the corolla of a flower. *Architecture.* the underside of a foliated capital between the abacus and the neck molding. *Metallurgy.* a cap used to seal the top of a blast furnace; using a double bell and hopper allows one to charge materials without fully opening the furnace.

Alexander Graham Bell

Bell, Alexander Graham 1847–1922, Scottish-born American anatomist and educator of the deaf; inventor of the telephone.

Bell, Sir Charles 1774–1842, Scottish anatomist; discovered distinct roots of sensory and motor nervous systems; described function of facial nerves; formulated Bell's law and described Bell's palsy.

belladonna *Botany.* a poisonous perennial herb, *Atropa belladonna,* of the family Solanaceae, characterized by dull green leaves, purple flowers, and cherrylike red to black fruits and containing several medicinal alkaloids, such as atropine and belladonine. Also, DEADLY NIGHTSHADE. *Toxicology.* any of several poisonous alkaloids produced by *Atropa belladonna.*

belladonnine *Organic Chemistry.* $C_{34}H_{42}N_2O_4$, an alkaloid derived from belladonna that is slightly soluble in water and soluble in alcohol; melts at 129°C; used in the preparation of anticholinergic compounds.

Bellatrix *Astronomy.* γ Orionis, the star on the northwest corner of the constellation Orion that marks the Hunter's left shoulder.

bell barrow see BARROW.

bell beaker see BEAKER.

bellbird or **bell bird** *Vertebrate Zoology.* any of various birds having a clear, bell-like call, especially the wood thrush *(Hyocichla mustelina)* of North America, the cotingas (family Cotingidae) of tropical America, the three-wattled bellbird *(Procnias tricarunculata)* of Central America, and the honey eater *(Anthornis melanura)* of New Zealand.

bell buoy *Navigation.* a floating aid to navigation that includes a bell which is rung by wave action, compressed gas, or electrically operated hammers. A bell buoy sounds a single tone, unlike a gong buoy, which can sound several tones.

bell cap *Chemical Engineering.* a triangular or hemispherical metal casting installed on distillation-column trays to force upflowing vapors to bubble through downcoming liquid.

bell character *Computer Technology.* an ASCII code character, control-G, that causes the computer or terminal to make a bell or beeping sound to attract the user's attention.

bell crank *Mechanical Devices.* a lever having two arms set at a right angle and joining at a common fulcrum.

bell curve *Statistics.* a symmetrical frequency curve resembling the outline of a bell. Also, BELL-SHAPED CURVE, NORMAL CURVE.

belled *Design Engineering.* resembling a bell in some way, usually being flared at one end, such as a **belled caisson.**

Bellerophon *Paleontology.* an extinct genus of aspidobranch symmetrical gastropods in the order Archaeogastropoda, widespread in the Carboniferous. (In Greek mythology, *Bellerophon* was a Corinthian hero who mounted Pegasus and slew the Chimera.)

Bellerophontacea *Paleontology.* an extinct superfamily of snail-like mollusks that was long classified with gastropods but is now widely thought of as monoplacophorans; the shell coils symmetrically in one plane, with a flaring pouter lip; Ordovician to Permian.

bellflower *Botany.* any plant of the family Campanulaceae (the bellflower family) that has bell-shaped flowers, especially any of the genus *Campanula* having showy, usually blue flowers.

bellflower

bell gable *Architecture.* in a church having no belfry, a pierced gable that is built or extended above the roof to house a bell.

bellingerite *Mineralogy.* $Cu_2^{+2}(IO_3)_6 \cdot 2H_2O$, a green triclinic mineral occurring in crusts with tiny prismatic crystals, having a specific gravity of 4.89 and a hardness of 4 on the Mohs scale; found as a secondary mineral with gypsum.

Bellini, Lorenzo 1643–1704, Italian anatomist; discovered Bellini's ducts and Bellini's ligaments.

bell jar *Engineering.* a bell-shaped glass jar used to enclose a vacuum, hold gases, and cover delicate objects.

bell-jar testing *Engineering.* a test for leakage from a sealed vessel, by filling it with gas and placing it in a vacuum chamber to determine if gas is drawn into the vacuum.

bell metal *Metallurgy.* an alloy containing about 80% copper, 20% tin, and often small amounts of lead and zinc; used for making bells.

bell metal ore see STANNITE.

bell-mouthed *Design Engineering.* designed with a bell-shaped opening at one or both ends. Thus, **bell-mouthed jar, bell-mouthed nozzle, bell-mouthed pipe.**

bellows *Mechanical Devices.* **1.** a flexible, accordion-sided device designed for pumping air, consisting of a chamber that expands to draw in air through a valve and contracts to expel it through a tube. **2.** any device resembling or incorporating an air bellows, especially one designed to allow for expansion and contraction. Thus, **bellows expansion joint, bellows gas meter, bellows gauge, bellows seal.** *Optics.* an accordion-like structure situated between the lens and the film of a camera that allows variation of the distance between those two components in order to focus and also to prevent exposure of the film by surrounding light.

bellows expansion joint *Design Engineering.* a joint formed in a run of piping that allows for expansion or contraction of the piping.

bell pepper *Botany.* the bell-shaped fruit of a sweet pepper.

bell-shaped curve see BELL CURVE.

Bell's inequalities *Quantum Mechanics.* a family of inequalities, implied by hidden-variables theory, that assert the existence of discrepancies between probabilities of simultaneous events in separated parts of a composite system.

Bell's law *Physiology.* a statement that the anterior roots of spinal nerves are motor in function, while the posterior ones are sensory.

bells of Ireland *Botany.* a hardy annual plant, *Molucella laevis,* having small white flowers, each surrounded by an enlarged, bell-shaped green calyx. Also, SHELLFLOWER.

Bell's palsy *Medicine.* a sudden paralysis and characteristic distortion of one side of the face, caused by a lesion of the facial nerve. (From Sir Charles *Bell.*)

Bell's theorem *Quantum Mechanics.* a theorem stating that there exist no hidden-variable theories of the second kind capable of making statistical predictions in complete agreement with quantum mechanics for some experiments.

bell tap *Petroleum Engineering.* a tool used to cut into, hold, and retrieve lost drilling equipment from a well.

bell transformer *Electricity.* a small step-down transformer designed to supply current to a low-voltage signaling device such as a doorbell, buzzer, or alarm.

bell-type manometer *Engineering.* a device used to measure differing pressures in which a bell floating in liquid is filled with one pressure and the other pressure is exerted down on the top of the bell. The measurement is taken from the level of the bell in the liquid.

bellwether *Agriculture.* a male sheep (wether) that leads a flock, usually wearing a bell.

Bell-Wigner inequality *Quantum Mechanics.* an extension of Bell's inequalities that implies conflicts in anticorrelations predicted by quantum mechanics.

bell wire *Electricity.* a solid copper wire that is used in low-voltage applications.

belly *Anatomy.* **1.** the lower front part of the body between the chest and the thigh. **2.** the abdominal cavity. **3.** the fleshy portion of any skeletal muscle, where its diameter is the greatest. *Graphic Arts.* on a piece of moveable type, the side bearing the nick, which points upward when placed in a setting stick.

beloeilite *Petrology.* a granular plutonic rock that contains sodalite, potassium feldspar, and a small amount of mafic minerals.

Belondiroidea *Invertebrate Zoology.* a superfamily of nematodes having a muscular spiral sheath enveloping the postcorpus of the esophagus; though often found near plant roots, its feeding habits are unknown.

belonephobia *Psychology.* an irrational fear of pins and needles, or other sharp pointed objects.

belongingness *Psychology.* the sense of being accepted by another individual or by a group. *Behavior.* the principle that associations between items are formed more readily if the properties of one item are similar to those of the other.

Beloniformes *Vertebrate Zoology.* a former name of the order Atheriniformes.

belonite *Mineralogy.* an elongated minute embryonic crystal having rounded or pointed ends.

Belontiidae *Vertebrate Zoology.* the gouramis, small freshwater labyrinth fishes of the suborder Anabantoidei found in tropical Africa, India, and Asia.

Belostomatidae *Invertebrate Zoology.* a family of large, water-dwelling insects that feed on tadpoles, young fish, mollusks, and other insects; includes several North American species.

below minimums *Meteorology.* weather conditions that are below operational weather limits, prohibiting the operation of aircraft.

belt *Mechanical Engineering.* an often endless flexible band made of leather, plastic, fabric, or the like and used to convey materials or to transmit rotary motion between shafts by running over pulleys with special grooves. *Ecology.* any definable strip or area of vegetation. *Geography.* an often elongated zone having distinctive characteristics. *Hydrology.* a long strip of pack ice whose width may range from 1 kilometer to 100 kilometers or more.

belt conveyor *Mechanical Engineering.* a travelling endless band designed for conveying materials, in which the motion of the band is driven by horizontal drums or pulleys. Also, BAND CONVEYOR.

belt course see STRING COURSE.

belt desmosome *Cell Biology.* a type of intercellular adhering junction that occurs as a band encircling each cell, binding adjacent cells together; commonly found in animal epithelial cells.

belt drive *Mechanical Engineering.* **1.** power transmission from one shaft to another by means of an endless belt which connects pulleys mounted on the shafts. **2.** a device used to transmit power in this mode.

belted *Design Engineering.* reinforced with a band of material, such as a **belted tire.** *Zoology.* marked with a band of color, such as a **belted sandfish.**

belted kingfisher *Vertebrate Zoology.* a North American kingfisher, *Ceryle alcyon,* having a grayish-blue body, a crested head, and a white breast with a blue band.

belted plain *Geology.* a plain whose surface displays alternating, parallel bands or belts of rock worn down to different levels.

belteroporic *Mineralogy.* of mineral crystals in rocks, oriented in the direction of easiest growth.

belt feeder *Industrial Engineering.* a short belt conveyor for transporting solid materials from storage to a processing area or machine.

belt-feeding or **belt-fed** *Ordnance.* of an automatic weapon, supplied with cartridges by an ammunition belt. Also, **belt-loading.**

belt grinding *Metallurgy.* grinding with an abrasive-coated belt.

Beltian orogeny *Geology.* a Precambrian crustal deformation that supposedly occurred in western North America, the existence of which is not generally accepted.

belt line *Transportation Engineering.* a railroad or other transportation system that wholly or nearly encircles a city or region.

belt of cementation see ZONE OF CEMENTATION.

belt of soil water *Geology.* the upper subdivision of the zone of aeration from which water is released into the air by the action of plants or by evaporation.

belt printer *Computer Technology.* an early impact printer whose fonts are carried on a flexible belt; similar to a chain printer.

belt sander *Mechanical Devices.* an electrical portable sanding tool consisting of a continuous abrasive belt that is driven in one direction.

belt shifter *Mechanical Devices.* an iron bar having a pair of forks or prongs, used to move a belt backward or forward between loose and fast pulleys or to replace a slipped belt.

belt slip *Mechanical Engineering.* the slipping or the difference in speed between a belt and a pulley or drum due to insufficient frictional grip between them, as between a belt conveyor and a driving drum.

Beltsville small white *Agriculture.* a breed of small turkeys with a white body and wings and orange feet and legs; it has a large proportion of white meat.

belt tightener *Mechanical Devices.* **1.** a device for pulling belt ends together for coupling up. **2.** a device that maintains uniform tension on driving belts, or causes them to fit more closely to the pulley circumference.

beltway *Transportation Engineering.* a highway built around the perimeter of an urban area, such as the interstate **Beltway** around Washington, D.C.

beluga *Vertebrate Zoology.* **1.** a white sturgeon, *Acipenser huso,* of the Black and Caspian seas; a valuable source of caviar. **2.** or **beluga whale.** see WHITE WHALE.

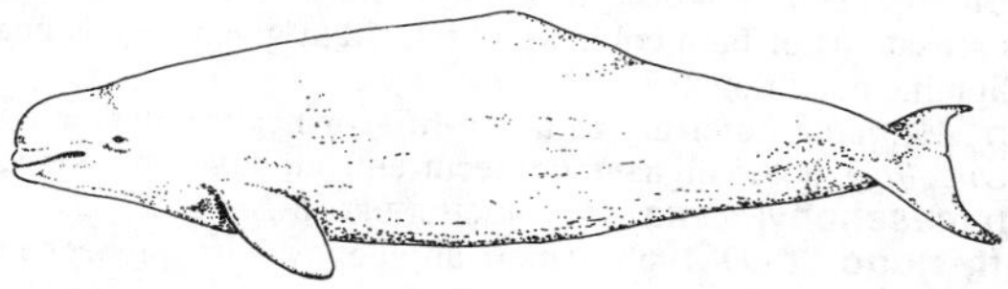

beluga whale

Beluzov-Zhabotinsky reaction *Chaotic Dynamics.* a chemical reaction in which a variety of stable, periodic, quasiperiodic, and chaotic oscillations in the amounts of the reaction products (and colors) in a well-stirred mixture with a fixed flow rate of the components are observed for different flow rates; without stirring, the reaction leads to the formation of concentric spiral patterns of different reaction products.

belvedere *Architecture.* a rooftop pavilion built to command a view. (An Italian word whose root meaning is "fine view.")

BEM or **B.E.M.** Bachelor of Engineering of Mines.

BEMA Business Equipment Manufacturers Association.

bema *Architecture.* **1.** in ancient Greece, a speaker's platform or stage. **2.** in a Christian church or basilica, the open (Western) or enclosed (Eastern) area between the apse and the nave.

bemegride *Pharmacology.* $C_8H_{13}NO_2$, a respiratory stimulant that has been used to treat barbiturate poisoning.

ben *Botany.* a tree of tropical Asia whose winged leaves are the source of ben oil.

Benacerraf, Baruj born 1920, Venezuelan-born American immunologist; Nobel Prize for pioneering work on the major histocompatibility complex and genetic control of immune response.

Benadryl [ben´ə dril´] *Pharmacology.* a trade name for preparations of the antihistamine diphenhydramine hydrochloride.

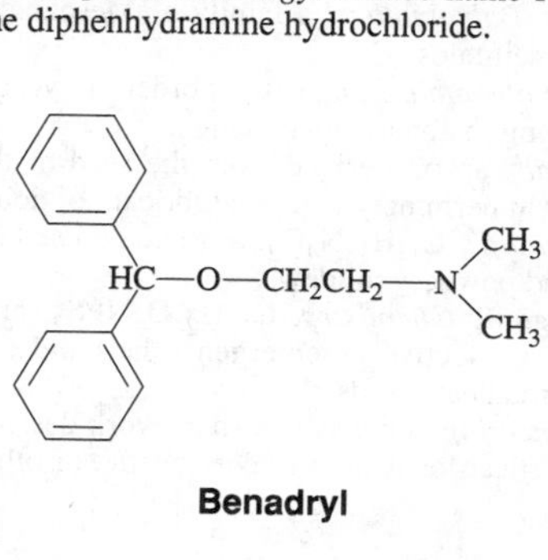

Benadryl

Bénard convection cells *Physics.* an array of regular hexagonal cells that appears in a liquid heated from below; the liquid tends to rise in the center and fall near the walls of the cell.

Bence Jones protein *Pathology.* a protein, characterized by low molecular weight and unique properties of thermosolubility, present in the urine of patients with multiple myeloma and sometimes in patients with other reticuloendothelial diseases. (Named after Henry *Bence Jones,* 1814–1873, English physician.)

Bence Jones proteinuria *Pathology.* the condition in which Bence Jones protein is present in the urine.

bench *Geology.* **1.** a long, narrow, relatively flat or gently inclined surface that is bounded by steeper slopes above and below. **2.** a small terrace or steplike ledge of earth or rock that breaks the continuity of a slope. **3.** a nearly horizontal area on the ocean side of an artificial dike that marks the level of maximum high water. *Mining Engineering.* a layer of coal that is mined separately from other layers in the same seam. *Civil Engineering.* see BERM.

bench and pipe vice *Mechanical Devices.* a bench vice that is fitted with pipe jaws and sometimes a small anvil, used to engage a pipe.

bench assembly *Engineering.* a term for the fitting and joining of parts.

bench blasting *Mining Engineering.* any of various processes used to reach and remove ore or waste by blasting away a succession of horizontal layers (benches).

bench check *Mechanical Engineering.* a test of a machine or device that is made in the workshop rather than under field conditions; the opposite of a field test.

bench clamp *Mechanical Devices.* a vise that is mounted on a work bench, used to force parts together or hold work in place.

bench dog *Engineering.* a wooden or metal peg that, when placed in a slot on a work bench, prevents a work piece from slipping.

bench face *Mining Engineering.* a mass, usually inclined, of any soil or rock material rising above the digging level in an open-pit mine or quarry.

bench gravel *Geology.* a gravel bed that is located on the side of a valley above the present stream bottom, representing part of the stream bed when it was at a higher level.

bench grinder *Mechanical Devices.* a motor grinder that is mounted on a work bench, having a double-ended shaft for mounting polishing wheels or buffers.

bench height see BANK HEIGHT.

bench hook *Mechanical Devices.* any of several devices used to keep work from slipping toward the rear of a work bench, such as a metal hook or a flat piece of wood having blocks fixed to the back edge of the top and the front edge of the bottom.

benching *Engineering.* the process of building or cutting in stepped levels or benches. *Building Engineering.* **1.** a berm that is located above a ditch. **2.** concrete that is poured on earth surfaces to prevent sliding. *Petroleum Engineering.* concrete that is used in conjunction with a pipeline for additional strength as it is laid on its sides.

bench irrigation *Agronomy.* an irrigation method for a sloping field in which low ditches run parallel to the slope. Also, BENCH-BORDER IRRIGATION; CONTOUR IRRIGATION.

bench lands *Agronomy.* terracesor shelf-like land forms representing former water levels of lakes, rivers, or seas.

bench lathe *Mechanical Devices.* a small lathe that is suitable for attachment to a workbench.

bench lava *Volcanology.* a lava that forms raised platforms and crags about the edge of a lava lake. Also, **bench magma.**

bench mark or **benchmark** *Science.* a measurement or standard against which similar units can be compared. *Engineering.* in surveying, a fixed point of reference whose elevation is known; often a permanent mark embedded in an enduring object indicating its elevation above or below some standard point, such as sea level. *Geology.* a permanent mark embedded in an enduring object indicating elevation above or below some standard datum, such as sea level, that is used as a reference point in topographic or tidal surveys.

benchmark program *Computer Technology.* a program selected from a specific class of problems or area of applications used to evaluate and measure the performance time and cost of computers relative to each other.

benchmark test *Computer Technology.* a program or group of programs executed on several different computers for the purpose of comparing their performance and cost.

bench mesa see MESA.

bench placer *Geology.* a bench gravel formed by mechanical concentration of mineral particles from weathered debris that is mined at or near the surface of the earth.
bench plane *Mechanical Devices.* any of various planes (such as a jack plane or block plane) that are designed for work on flat surfaces, especially on a work bench.
bench-scale testing *Engineering.* the practice of examining materials, methods, or chemical processes on a scale that can be performed on a work bench.
bench screw *Mechanical Devices.* a usually metal screw that is mounted on a work bench and fitted with various jaws to form a vise.
bench slope see BANK SLOPE.
bench stop *Mechanical Devices.* a type of bench hook that is set flush with a work bench, with an adjustable head designed to hold materials in place.
bench surgery *Surgery.* surgery performed on an organ that has been removed from the body, after which it is reimplanted.
bench table *Building Engineering.* **1.** a course of masonry that forms a seat, table, or bench at the foot of an interior wall around a column. **2.** a work bench.
bench terrace *Agronomy.* an embankment of earth having a flat top and a steep or vertical downhill side; usually built on the contour of sloping land in order to control erosion or improve land stability.
bench test see BENCH CHECK.
benchwork or **bench work** *Mechanical Engineering.* work that is performed at a bench or similar work station in a factory or laboratory, often using hand tools or small machines; distinguished both from machine work (using larger machines) and from field work (performed outdoors or on a job site). Also, CRADLE WORK.
bend *Geology.* **1.** a curve or turn in the course, bed, or channel of a stream. **2.** a curved part of a lake, inlet, or coastline. *Navigation.* to turn a vessel, especially in a relatively narrow waterway. *Design Engineering.* a curved section of a pipe or machined part. *Mining Engineering.* any hardened clayey substance that is encountered in mining operations. *Mechanical Engineering.* **1.** a curved or bent length of tubing or conduit that is used to connect the ends of two adjacent straight lengths set at angles to one another. **2.** a knot used to fasten two ropes together, or a piece of rope to something else; to fasten with such a knot. *Electronics.* a change in the direction of a rigid or flexible waveguide.
bend allowance *Design Engineering.* in any material, the length of the arc of the neutral axis between the axis points along a bend.
Benday process *Graphic Arts.* a method of shading line drawings for reproduction by etching line, dot, or texture patterns onto a metal printing plate. A plate so treated is known as a **Benday plate.** (Named for *Ben*(jamin) *Day,* 1838–1916, an American printer who developed this process.)
bend contours *Materials Science.* image contrast in a transmission electron microscope due to bending of the thin foil.
benderboard *Materials.* a lightweight flexible board often cut in strips, used for fencing, as a landscape or garden divider, and so on.
bender element *Electronics.* a union of two different thin strips of piezoelectric material connected in such a way that one strip increases in length and the other becomes shorter when voltage is applied.
bend glide *Materials Science.* the bending deformation of a crystal under an applied moment, due to the formation and movement of dislocations within its compressive and tensile regions.
bending *Engineering.* the act of forming a metal or softened wooden part into a curved shape. *Hydrology.* the first stage in ice deformation, in which lateral pressure produces an upward or downward motion in sea ice.
bending brake *Mechanical Devices.* a metal-forming press brake that makes linear bends with sharp angles in sheet metal.
bending capacity see BENDING STRENGTH.
bending iron *Mechanical Devices.* a tool used to either expand or straighten lead pipe. Also, **bending pin.**
bending moment *Mechanics.* the first moment of stress integrated over any cross section of a structural member, equal to the algebraic sum of the moments of all forces to either side of the section; the moment is about the sectional neutral axis along which the bending stress is equal to sectional mean bending stress; a positive bending moment bends the beam convex downward, while a negative one bends the beam convex upward.
bending-moment diagram *Mechanics.* a graph plotting the variations of bending moment (about the *y*-axis, the neutral axis) along a beam (the *x*-axis). Also, MOMENT DIAGRAM.
bending slab *Naval Architecture.* a thick cast-iron plate having holes through which iron pins called dogs are inserted; the dogs form a template for bending metal hull frames to their proper curvature.
bending strength *Mechanics.* the ability of a beam or other structural member to withstand a bending moment. Also, BENDING CAPACITY.
bending stress *Mechanics.* the internal longitudinal stress that is developed in a beam in response to curvature due to external loadings.
bend radius *Design Engineering.* the radius that corresponds to the curve of a bent part, as measured at the interior surface of the bend.
bendroflumethiazide *Pharmacology.* a thiazide diuretic used in treatment of edema and hypertension. Also, **bendrofluazide.**
bene- a combining form meaning "well."
Beneckea *Bacteriology.* a former genus of bacteria, whose species are now reclassified in the genus *Vibrio.*
Benedicks effect *Physics.* an electromotive force produced in a circuit containing only one metal, but with impurities or internal strains, when the circuit is subjected to an asymmetrical temperature distribution.
Benedict, Ruth 1887–1948, American anthropologist; explored cultural influences on personality formation.
Benedict's solution *Analytical Chemistry.* a solution of sodium citrate, copper sulfate, and sodium carbonate used to test for reducing sugars; the formation of a red color indicates a positive test.
beneficiate *Metallurgy.* to remove impurities from or otherwise refine an ore prior to smelting, a process known as **beneficiation.**
Bengal, Bay of *Geography.* an arm of the Indian Ocean, east of India and west of Burma.
bengaline *Textiles.* a woven fabric having distinct crosswise ribs.
BEng. or **B.Engr.** Bachelor of Engineering.
Benguela Current *Oceanography.* the northward-flowing current off the southwest coast of Africa, between 35°S and 15°S.
Benham top *Optics.* a disk made up of black and white arcs that produces sensations of faint color when rotated at given speeds under specific lighting conditions.
benign *Medicine.* referring to a non-life- or health-threatening condition. *Oncology.* not malignant, not recurrent, and favorable for recovery.
benign mesenchymoma see LIPOCHONDROMA.
Benioff, Hugo 1899–1968, American geophysicist; pioneer in instrumental seismology.
Benioff extensometer *Geophysics.* a linear-strain seismograph used to measure variations in distance between two linked reference points.
Benioff zone *Geophysics.* a plane of seismic activity associated with dipping of a moving lithospheric plate underneath another plate and into the earth's mantle along a subduction zone (e.g., along the circum-Pacific belt). (Named for Hugo *Benioff,* an American seismologist.)
benitoite *Mineralogy.* $BaTiSi_3O_9$, a rare dichroic hexagonal mineral occurring in clear-to-blue pyramidal or tabular crystals. (Named for San *Benito* County, California, where it was discovered.)
benjamin see BENZOIN RESIN.
benjamin bush *Botany.* any of several North American shrubs of the genus *Lindera* having an aromatic bark that is used as a stimulating tonic. Also, SPICEBUSH, BENZOIN.
benjaminite *Mineralogy.* $(Ag,Cu)_3(Bi,Pb)_7S_{12}$, a gray monoclinic mineral occurring in granular masses, having a metallic luster, a specific gravity of 6.34 to 6.70, and a hardness of 3.3 to 3.5 on the Mohs scale.
benjamin tree *Botany.* any of several southeast Asian trees of the genus *Styrax* that are the source of benzoin resin.
Benlate *Toxicology.* the trade name for benomyl.
benmoreite *Petrology.* an igneous rock that is intermediate in composition between mugearite and trachyte.
benne *Botany.* a name used in the southeastern U.S. and West Indies for the sesame plant or its seeds or oil. Also, **benny.**
Bennettitaceae *Paleontology.* a family of Mesozoic cycadeoid plants in the order Bennettitales.
Bennettitales *Paleontology.* an extinct order of cycadeoid plants of the Mesozoic, growing to about 2 meters high.
ben oil *Materials.* an oil derived from the seed of the ben, a tropical Asian tree; used in perfumery and as a lubricant of delicate machinery.
benomyl *Toxicology.* $C_{14}H_{18}N_4O_3$, a pesticide used against such fungi as apple scab and powdery mildews.
bensulide *Organic Chemistry.* $C_{14}H_{24}O_4NPS_3$, an amber liquid or crystal used as a selective preemergence herbicide, primarily against crabgrass and broadleaf weeds.
bent *Civil Engineering.* a transverse framework designed to carry either horizontal or vertical loads on a bridge, trestle, or other spanning structure.

bent-back occlusion *Meteorology.* an occluded front whose direction of motion has reversed because of the development of a new cyclone or, less often, the displacement of the old cyclone along the front. Also, BACK-BENT OCCLUSION.

bent grass *Botany.* any of several hardy perennial lawn or pasture grasses of the genus *Agrostis*, such as redtop.

benthal *Oceanography.* **1.** see BENTHIC. **2.** see BENTHOS.

Bentham, George 1800–1884, English botanist; wrote the important botanical work *Genera Plantarum* (with J. Hooker).

benthic *Oceanography.* **1.** of, relating to, or living in a benthos. **2.** of or relating to a benthon.

benthiocarb *Organic Chemistry.* $C_{12}H_{16}NOCl$, an amber liquid that boils at 126–129°C; used as an herbicide to control aquatic weeds in rice crops.

benthon *Oceanography.* the organisms living in a benthos. Thus, **benthonic.**

benthos *Oceanography.* **1.** the biogeographic region at the bottom of a sea or ocean (or, more broadly, at the soil-water interface of an ocean, sea, or lake). **2.** the organisms living in such a region; benthon.

bent housing *Petroleum Engineering.* a drill housing that is curved to facilitate directional drilling, for example in a positive-displacement downhole turbodrill.

bentonite *Mineralogy.* a soft, porous rock ranging from white to light green or light blue in color, composed chiefly of the clay mineral montmorillonite and having the ability to absorb water with an increase in volume; used chiefly as a suspending agent and to thicken oil-well drilling mods. (Named for Fort *Benton,* Montana.)

ben tree see BEN.

bentwood *Design Engineering.* **1.** wood that is brought to a curved shape by bending it with moist heat, then set by cooling and drying. **2.** furniture made with such wood.

Benz, Karl 1844–1929, German engineer; pioneer in automobile design and production.

benz- a combining form denoting the presence of benzene, benzoic acid, or one or more of the phenyl groups, as in *benzaldehyde.*

benz. benzene; benzoic.

benzadox *Organic Chemistry.* $C_6H_5CONHOCH_2COOH$, white crystals that melt at 140°C; used as an herbicide.

benzal *Organic Chemistry.* C_6H_5CH, a bivalent radical derived from benzaldehyde. Also, BENZYLIDENE.

benzalacetone see BENZYLIDENE ACETONE.

benzal chloride *Organic Chemistry.* $C_6H_5CHCl_2$, a colorless, highly refractive oily liquid that freezes at –16°C and boils at 207°C; insoluble in water and soluble in alcohol. It is made by chlorinating toluene and is used in dyes and as an intermediate in producing benzaldehyde.

benzaldehyde *Organic Chemistry.* C_6H_5CHO, a colorless or yellowish volatile oil that freezes at –56°C and boils at 178°C; slightly soluble in water and miscible with alcohol and ether; used in the production of dyes and perfumes, as a solvent, as an almond flavoring agent, and for various other purposes. Also, BENZOIC ALDEHYDE.

benzaldoxime *Organic Chemistry.* $C_6H_5CH{=}NOH$, either of two stereoisomeric crystalline oximes formed from benzaldehyde and hydroxylamine; soluble in alcohol and ether; the *syn* form melts at 34°C and the anti form at 130°C; used in the production of other organic compounds.

benzalkonium *Organic Chemistry.* $C_6H_5CH_2N(CH_3)_2R^+$, an organic radical in which the alkyl (R) groups range from C_8H_{17} to $C_{18}H_{37}$.

benzalkonium chloride *Organic Chemistry.* $C_6H_5CH_2N(CH_3)_2RCl$, a highly toxic white to yellow-white salt that is soluble in water, alcohol, and acetone; used as an antiseptic and germicide applied topically to the skin and mucous membranes. The mixture of alkyls (R) may range from C_8H_{17} to $C_{18}H_{37}$.

benzamide *Organic Chemistry.* $C_6H_5CONH_2$, a colorless crystalline compound that melts at 132–133°C and boils at 290°C; soluble in hot water, alcohol, and ether. It is derived from the action of benzoyl chloride on ammonia or ammonium carbonate and is used in organic synthesis. Also, BENZOIC ACID AMIDE.

benzaminic acid *Organic Chemistry.* $C_6H_4NH_2COOH$, yellowish or reddish crystals (colorless when pure) that melt at 174°C, slightly soluble in water, alcohol and ether; used in dye manufacture. Also, *m*-AMINOBENZOIC ACID.

benzanilide *Organic Chemistry.* $C_6H_5CONHC_6H_5$, a white, crystalline compound that melts at 160°C; insoluble in water and soluble in alcohol; an anilide of benzoic acid or benzoyl chloride that is used in the manufacture of pharmaceuticals and dyes.

benzanthracene *Organic Chemistry.* $C_{18}H_{12}$, a crystalline cyclic hydrocarbon that melts at 162°C and sublimes at 435°C; insoluble in water and soluble in alcohol, ether, acetone, and benzene. It is carcinogenic and is found in small amounts in coal tar. Also, NAPHTHANTHRACENE.

benzanthrone *Organic Chemistry.* $C_{17}H_{10}O$, pale yellow needles, insoluble in water and soluble in alcohol; melting at 170°C; used as a dye intermediate.

Benzedrine [ben´zə drēn´] *Pharmacology.* the trade name for a preparation of amphetamine.

benzene *Organic Chemistry.* **1.** C_6H_6, a colorless, volatile, flammable, toxic liquid that freezes at 5.5°C and boils at 80°C. It is soluble in alcohol, ether, and acetone and insoluble in water, and has an aromatic odor and toxic fumes. Benzene is obtained chiefly from petroleum and also from coal distillation; it is a widely produced chemical and is used as a solvent and in the manufacture of dyes, synthetic rubber, and commercial chemicals. Also, BENZOL. **2.** see BENZENE SERIES.

benzene

benzene 1-allyl-4-hydroxy see CHAVICOL.

benzenecarboxylic acid *Organic Chemistry.* an aromatic acid derived from benzene, especially benzoic acid, C_6H_5COOH.

benzenediazonium chloride *Organic Chemistry.* $C_6H_5N_2Cl$, an explosive, highly toxic crystalline ionic salt that is soluble in water; produced by the reaction of aniline hydrochloride and nitrous acid; used as a dye intermediate.

benzene hexachloride *Organic Chemistry.* $C_6H_6Cl_6$, a water-insoluble crystalline compound prepared by the chlorination of benzene in actinic light; its γ isomer is a powerful insecticide. Also, HEXACHLOROCYCLOHEXANE.

benzene nucleus *Organic Chemistry.* **1.** the group of six carbon atoms that, together with the hydrogen atoms, make up the benzene ring. **2.** the benzene ring itself.

benzenephosphorus dichloride *Organic Chemistry.* $C_6H_5PCl_2$, a highly reactive colorless liquid that freezes at –51°C and boils at 234.6°C; strongly irritating; used in organic synthesis for derivation of plasticizers, polymers, antioxidants, and oil additives.

benzene ring *Organic Chemistry.* the structure of benzene, consisting of a closed hexagonal chain of six carbon atoms, each attached to a hydrogen atom and attached to each other by alternating single and double bonds. In derivatives of benzene, the structure remains the same, but one or more of the hydrogen atoms are replaced by other atoms or radicals. Also, **benzene formula.**

benzene series *Organic Chemistry.* a family of liquid and solid aromatic hydrocarbons based on the general formula C_nH_{2n-6}, in which benzene is the simplest, C_6H_6, and toluene is the next highest, C_7H_8.

benzene sulfonate *Organic Chemistry.* any salt or ester of benzenesulfonic acid.

benzenesulfonic acid *Organic Chemistry.* $C_6H_5SO_3H$, a colorless, crystalline aromatic acid that is soluble in water and alcohol; it melts at 65–66°C for the anhydrous form and 45–46°C for the hydrate form; derived by sulfonating compounds of the benzene series and used as a catalyst in the manufacture of detergents and phenols.

1,2,4-benzenetricarboxylic acid see TRIMELLITIC ACID.

1,2,4-benzenetriol *Organic Chemistry.* $C_6H_3(OH)_3$, a crystalline solid that melts at 141°C; soluble in water, alcohol, and ether; used in the analysis of gases.

benzenoid *Organic Chemistry.* **1.** describing an organic compound that resembles the structure of carbon atoms of benzene. **2.** a compound whose structure is similar to that of benzene.

benzestrol *Pharmacology.* $C_{20}H_{26}O_2$, a synthetic nonsteroidal estrogen compound in crystalline powder form that is administered orally; used in oral contraceptives, to ease discomfort in breast and prostate cancers and menopause, to treat osteoporosis, and to inhibit milk production.

benzethonium chloride *Organic Chemistry.* $C_{27}H_{42}ClNO_2$, colorless, odorless plates with a very bitter taste, melting at 164°C and soluble in water, alcohol, and acetone; a synthetic ammonium compound having various applications as an antiseptic, disinfectant, or preservative.

benzhydrol *Organic Chemistry.* $(C_6H_5)_2CHOH$, a colorless, crystalline solid that melts at 69°C and is slightly soluble in water and soluble in alcohol and ether; a secondary alcohol produced by reducing benzophenone with zinc and aqueous alcohol alkali; used in the preparation of certain antihistamines and insecticides. Also, **benzohydrol.**

benzidine *Organic Chemistry.* $NH_2C_6H_4{\cdot}C_6H_4NH_2$, a white to pinkish crystalline compound that melts at 128°C and boils at 400°C; soluble in water and alcohol; usually derived from nitrobenzene and used as an intermediate in the production of azo dyes. It is a known carcinogen

benzidine test *Biotechnology.* a test using anaeromatic amines such as benzidine to detect the presence of sulfates, as in water analysis or for blood identification.

benzil *Organic Chemistry.* $C_6H_5COCOC_6H_5$, large, yellow, six-sided prisms that melt at 94–95°C and boil at 346–348°C; insoluble in water and soluble in alcohol and ether; used in organic synthesis and insecticides.

benzilic acid *Organic Chemistry.* $(C_6H_5)_2C(OH)COOH$, a combustible white to tan powder with a distinctive odor; melts at 151–152°C and is soluble in hot water and alcohol; used in organic synthesis.

benzimidazole *Organic Chemistry.* $C_7H_6N_2$, tabular crystals that melt at 170.5°C and decompose above 360°C; slightly soluble in water and soluble in alcohol. It is toxic to yeast and animals in the absence of its homologous compounds, adenine and guanine.

benzin or **benzine** *Organic Chemistry.* a clear, colorless, highly flammable, volatile liquid mixture of hydrocarbons (mostly of the methane series) that is obtained from the distillation of petroleum and used as a solvent for many organic compounds. Also, PETROLEUM BENZIN(E), PETROLEUM ETHER.

benzo- a combining form denoting the presence of benzene, benzoic acid, or one or more of the phenyl groups, as in *benzonitrile*.

benzoate *Organic Chemistry.* any salt or ester of benzoic acid.

benzoated *Organic Chemistry.* of a compound, containing or combined with benzoic acid.

benzoate of soda see SODIUM BENZOATE.

benzocaine *Pharmacology.* $C_9H_{11}NO_2$, a local anesthetic occurring as a white crystalline powder that is applied topically to the skin and mucous membranes.

benzodiazepine *Pharmacology.* any of a group of minor tranquilizers (such as diazepam, flurazepam, and chlordiazepoxide) having a common molecular structure and similar pharmacological actions, including antianxiety, muscle relaxing, and sedative and hypnotic effects.

benzoic acid *Organic Chemistry.* C_6H_5COOH, an aromatic compound of colorless to white plates or needles that melt at 122.1°C and boil at 249.1°C (60 Torr); slightly soluble in water and soluble in alcohol, ether, acetone, and benzene. Benzoic acid is found especially in benzoin and cranberries and is used as a fruit preservative, in pharmaceuticals and cosmetics, for ringworm treatment, and in organic synthesis. Also, PHENYLFORMIC ACID, BENZENECARBOXYLIC ACID.

benzoic acid amide see BENZAMIDE.

benzoic acid nitrile see BENZONITRILE.

benzoic aldehyde see BENZALDEHYDE.

benzoic anhydride *Organic Chemistry.* $(C_6H_5CO)_2O$, colorless rhombic prisms that melt at 42°C and boil at 360°C; insoluble in water and soluble in most organic solvents; used in dyes, for pharmaceuticals, and in organic synthesis.

benzoin *Botany.* see BENJAMIN BUSH. *Materials Science.* see BENZOIN RESIN. *Organic Chemistry.* $C_{14}H_{12}O_2$, a highly toxic white crystalline compound that is slightly soluble in water and soluble in alcohol and chloroform; it melts at 137°C and is combustible and capable of being optically active; used in organic synthesis. Also, BENZOYLPHENYLCARBINOL.

α-benzoin oxime *Organic Chemistry.* a crystalline, light-sensitive solid that melts at 151–152°C, slightly soluble in water and soluble in alcohol, ether, and acetone; used as a reagent to determine the presence of copper, molybdenum, and tungsten.

benzoin resin *Materials.* a fragrant balsamic resin obtained from the bark of various Asian trees (particularly certain Javanese trees of the genus *Styrax*) that is widely used in cosmetics, perfumery, and medicine; e.g., as a topical protectant, an antiseptic, an irritant expectorant, and an inhalant. Also, BENJAMIN, GUM BENJAMIN, GUM BENZOIN.

benzol see BENZENE.

benzol-acetone process *Chemical Engineering.* a solvent dewaxing method, using benzol-acetone as a solvent, in which the mixture of solvent and oil with wax is cooled until the wax solidifies and is then removed by filtration.

benzolism *Toxicology.* poisoning that is caused by exposure to benzene; the symptoms may include anemia, in chronic cases, and coma, in acute cases.

benzonatate *Pharmacology.* $C_{30}H_{53}NO_{11}$, a clear, pale yellow viscous liquid that is administered orally as an antitussive. Also, **benzononatine.**

benzonitrile *Organic Chemistry.* C_6H_5CN, a colorless, oily, highly toxic compound that melts at –13°C and boils at 190.7°C;it is soluble in hot water, alcohol, and ether. It has an almondlike odor and a sharp taste. It is used chiefly as a solvent for synthetic resins. Also, PHENYL CYANIDE.

benzophenone *Organic Chemistry.* $(C_6H_5)_2CO$, colorless to white prisms with a sweet, roselike odor, insoluble in water and soluble in alcohol and ether. The α **benzophenone** form melts at 48.1°C and the β **benzophenone** form melts at 26°C; both boil at 305.9°C. Benzophenone is used chiefly in perfumery and organic synthesis. Also, DIPHENYL KETONE.

benzopurpurine *Organic Chemistry.* any of a series of scarlet-colored azo dyes that are used as a contrast stain with hematoxylin and other blue stains.

benzopyrene *Organic Chemistry.* $C_{20}H_{12}$, a five-ring aromatic hydrocarbon occurring in the form of pale yellow crystals; insoluble in water and soluble in benzene; melts at 176–179°C and boils at 310–312°C. It is highly carcinogenic and is found in coal tar, cigarette smoke, and as a product of incomplete combustion. It is a procarcinogen that requires metabolic activation to have a mutagenic effect. Also, BENZO[*a*]PYRENE, 3,4-BENZPYRENE.

benzo[*a*]pyrene see BENZOPYRENE.

5,6-benzoquinoline *Organic Chemistry.* $C_{13}H_9N$, a crystalline solid that melts at 94°C and is soluble in alcohol, ether, acetone, and benzene; used as an analytical reagent for cadmium. Also, β-NAPHTHOQUINOLINE.

benzoquinone see QUINONE.

benzosulfimide see SACCHARIN.

benzothiazole *Organic Chemistry.* C_6H_4SCHN (bicyclic), a yellow liquid that boils at 227°C; slightly soluble in water and soluble in alcohol; combustible and highly toxic; used chiefly as a rubber softener or a dyestuff intermediate.

benzothiofuran see THIANAPHTHENE.

1,2,3-benzotriazole *Organic Chemistry.* $C_6H_5N_3$, a highly toxic, explosive, crystalline solid that melts at 100°C and is soluble in alcohol, benzene, and chloroform; used in organic synthesis and in producing ultraviolet absorbers. Also, AZIMIDOBENZENE.

benzotrifluoride *Organic Chemistry.* $C_7H_5F_3$, a water-white liquid with a noticeable odor; freezes at –29.1°C and boils at 102.1°C; miscible with alcohol, ether, acetone, and benzene. This flammable and highly toxic compound is used as an intermediate for dyes, in pharmaceuticals, as a vulcanizing agent and in insecticides.

benzoyl *Organic Chemistry.* the univalent radical C_6H_5CO– of benzoic acid and its derivatives. Also, BENZOYL GROUP, BENZOYL RADICAL.

benzoylate *Organic Chemistry.* to introduce a benzoyl radical into a compound. Thus, **benzoylated, benzoylation.**

benzoyl chloride *Organic Chemistry.* C_6H_5COCl, a colorless liquid with a pungent odor that boils at 197.2°C and melts at –1.0°C; soluble in ether; highly toxic and a strong irritant; used as an intermediate in organic synthesis, as in dyes.

***N*-benzoylglycin** see HIPPURIC ACID.

benzoylglycine see HIPPURIC ACID.

benzoyl group see BENZOYL.

benzoyl-pas calcium *Pharmacology.* $C_{28}H_{20}CaN_2O_8{\cdot}5H_2O$, an antibacterial occurring as a whitish powder; used as an oral tuberculostatic.

benzoyl peroxide *Organic Chemistry.* $(C_6H_5CO)_2O_2$, white crystals that are slightly soluble in water and soluble in organic solvents; melts at 103–106°C; highly toxic and explosive if heated; used as a bleaching agent, a catalyst for free radical reactions, in acne treatment, and for other purposes.

benzoylphenylcarbinol see BENZOIN.

benzoyl radical see BENZOYL.

3,4-benzpyrene see BENZOPYRENE.

benztropine mesylate *Pharmacology.* $C_{21}H_{25}NO \cdot CH_4O_3S$, a drug occurring as a white crystalline powder having anticholinergic, antihistiminic, and local anesthetic actions.

benzyl *Organic Chemistry.* $C_6H_5CH_2^-$, a univalent radical formed by removing one hydrogen atom from the side chain of toluene, as for example in benzyl alcohol, $C_6H_5CH_2OH$.

benzyl acetate *Organic Chemistry.* $C_6H_5CH_2OCOCH_3$, a colorless fragrant liquid ester that boils at 212°C; insoluble in water and soluble in alcohol and ether. It is found in jasmine oil and other essential oils, and is used in perfumery, as an additive in tobacco, soaps, and cosmetics, and for various other purposes. Also, PHENYLMETHYL ACETATE.

benzyl alcohol *Organic Chemistry.* $C_6H_5CH_2OH$, a colorless, slightly aromatic liquid; soluble in water and alcohol; freezes at –15.3°C and boils at 205.3°C; combustible and highly toxic. It is the simplest homologue of the aromatic alcohols, and is used as a solvent, in perfumery, medicine, and ester making, and for other purposes. Also, PHENYLCARBINOL, PHENYLMETHANOL.

benzylamine *Organic Chemistry.* $C_6H_5CH_2NH_2$, a toxic, combustible, colorless liquid base that is soluble in water, alcohol, and ether; boils at 185°C; used as a chemical intermediate in dye production.

benzyl benzoate *Organic Chemistry.* $C_6H_5COOCH_2C_6H_5$, a colorless oily liquid ester that occurs naturally in balsams; insoluble in water and soluble in alcohol and ether; freezes at 18.8°C and boils at 325°C; used in medicine and perfumery.

benzyl bromide *Organic Chemistry.* $C_6H_5CH_2Br$, a colorless liquid that freezes at –3°C and boils at 201°C; insoluble in water and soluble in acohol and ether. It is highly toxic and highly irritating to the skin and eyes, and is used to manufacture poison gas, as a foaming agent, and in organic synthesis.

benzyl butyrate *Organic Chemistry.* $C_{11}H_{14}O_2$, a combustible liquid having a fruity odor; boils at 240°C; used as a plasticizer and flavoring agent.

benzyl chloride *Organic Chemistry.* $C_6H_5CH_2Cl$, a colorless liquid with a pungent odor, insoluble in water and soluble in alcohol and ether; freezes at –39°C and boils at 179.3°C. It is highly toxic and irritating, and is used as an intermediate in the synthesis of benzyl derivatives. Also, CHLOROMETHYLBENZENE.

benzyl chlorocarbonate *Organic Chemistry.* $C_6H_5CH_2OCOCl$, an oily liquid with an acrid odor causing the eyes to tear; reacts with water to form hydrochloric acid and decomposes above 100°C; used to block the amino group in peptide synthesis. Also, **benzyl chloroformate.**

benzyl cinnamate *Organic Chemistry.* $C_{16}H_{14}O_2$, an aromatic white crystalline ester that melts at 39°C and boils at 244°C (18 Torr); insoluble in water and soluble in alcohol. It is found naturally in balsam and is used in perfumery. Also, CINNAMEIN.

benzyl cyanide *Organic Chemistry.* $C_6H_5CH_2CN$, a toxic, oily aromatic compound that freezes at –23°C and boils at 230°C, insoluble in water and soluble in alcohol; found in some essential oils, such as garden peppergrass, and used in organic synthesis.

benzyl dichloride see BENZAL CHLORIDE.

benzyl disulphide see DIBENZYL DISULFIDE.

benzyl ether *Organic Chemistry.* $(C_6H_5CH_2)_2O$, a colorless, oily liquid; insoluble in water and soluble in alcohol and ether; unstable at room temperature; boils at 297°C; used in perfumes and as a plasticizer for nitrocellulose.

benzyl ethyl ether *Organic Chemistry.* $C_6HCH_2OC_2H_5$, a colorless, oily liquid with a boiling point of 298°C; insoluble in water and miscible with alcohol and ether; used in organic synthesis and as a flavoring

benzyl fluoride *Organic Chemistry.* $C_6H_5CH_2F$, a toxic, irritating colorless liquid that freezes at –35°C and boils at 139.8°C; used in organic synthesis.

benzyl formate *Organic Chemistry.* $C_6H_5CH_2OOCH$, a colorless liquid with a fruity-spicy odor; boils at 203.4°C and is insoluble in water and miscible with alcohol and oils; used in perfumes and as a flavoring.

benzyl group see BENZYL.

benzylidene see BENZAL.

benzylidene acetone *Organic Chemistry.* $C_6H_5CH{=}CHCOCH_3$, a colorless, lustrous crystalline compound that melts at 42°C and boils at 260–262°C; insoluble in water and soluble in alcohol, ether, and benzene; used in perfumery and organic synthesis. Also, BENZALACETONE.

benzylidene chloride see BENZAL CHLORIDE.

benzyl isoeugenol *Organic Chemistry.* $CH_3CH{=}CHC_6H_3(OCH_3)OCH_2C_6H_5$, a white, crystalline solid with a floral odor, soluble in alcohol and ether; used in perfumery.

benzyl mercaptan *Organic Chemistry.* $C_6H_5CH_2SH$, a colorless liquid that boils at 195°C; insoluble in water and soluble in alcohol and carbon disulfide; it is toxic, combustible, and an irritant; used as an odorant and for flavoring. Also, THIOBENZYL ALCOHOL.

benzylmorphine hydrochloride *Pharmacology.* $C_{24}H_{26}ClNO_3$, a narcotic pain killer; abuse may lead to addiction. Also, PERONINE.

benzylpenicillin or **benzyl penicillin** see PENICILLIN G.

benzyl propionate *Organic Chemistry.* $C_2H_5COOCH_2C_6H_5$, a colorless liquid with a sweet odor, that boils at 220°C and is insoluble in water; combustible and highly toxic; used in perfumes and for flavoring.

benzyl radical see BENZYL.

benzyl salicylate *Organic Chemistry.* $C_6H_4(OH)COOCH_2C_6H_5$, a thick liquid with a faint sweet odor; slightly soluble in water and miscible with alcohol; it boils at 208°C (25 Torr); used as a fixer in perfumery and in sunburn preparations.

benzyl sulfide *Organic Chemistry.* $(C_6H_5CH_2)_2S$, colorless plates that melt at 49–50°C and decompose on heating; insoluble in water and soluble in alcohol and ether; used in organic synthesis.

benzyl thiocyanate *Organic Chemistry.* $C_6H_5CH_2CNS$, a colorless, crystalline solid that is insoluble in water and soluble in alcohol and ether; melts at 43°C and boils at 256°C; used as an insecticide.

benzyne *Organic Chemistry.* C_6H_4, an unstable intermediate compound formed by removing two adjacent hydrogens from a benzene ring.

bephenium *Pharmacology.* an anthelmintic that is effective against hookworm and other intestinal nematodes.

bequest *Artificial Intelligence.* any data that is developed or acquired by an object or rule in a knowledge system and is made available to other objects within the system.

BER *Aviation.* the airport code for Berlin, Germany.

Beranek scale [bə ran′ik] *Acoustics.* a scale of noise-level measurement that is based on a large band of analysis rather than several smaller bands; used in a given environment to determine whether conversation will be easy, difficult, or impossible.

beraunite *Mineralogy.* $Fe^{+2}Fe_5^{+3}(PO_4)_4(OH)_5 \cdot 4H_2O$, a reddish monoclinic mineral commonly found in iron ore deposits.

berbamine *Organic Chemistry.* $C_{37}H_{40}N_2O_6$, a crystalline alkaloid that melts at 197–200°C (anhydrous) and 156°C (hydrous); insoluble in water and soluble in alcohol and ether, found especially in barberry.

Berberidaceae *Botany.* the barberry family of shrubs and herbs, characterized by alternate leaves that vary from spiny to those reduced to spines and by a berry or capsule fruit; many species are cultivated as hedges.

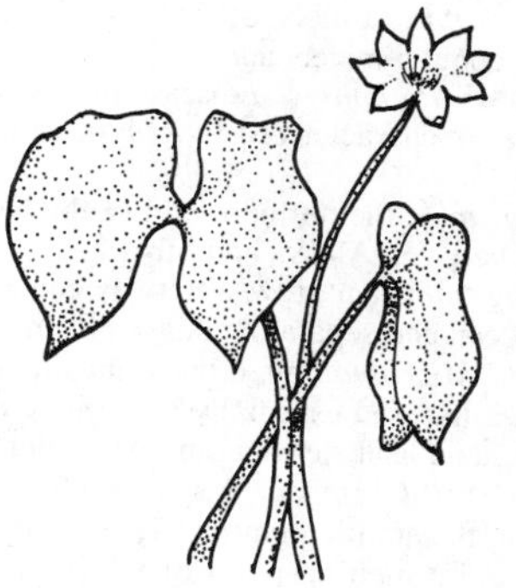

Berberidaceae (barberry)

berberidaceous *Botany.* of, relating to, belonging to, or resembling the barberry family.

berberine or **berberin** *Organic Chemistry.* $C_{20}H_{19}NO_5$, a toxic, white to yellow crystalline alkaloid that melts at 145°C (anhydrous); insoluble in water and soluble in alcohol and ether. It is present in the roots of barberry, golden seal, and other plants, and is used as an amebicide in the treatment of cholera.

berdache *Anthropology.* a cultural institution of some North American Indian societies in which a male assumes the role of a female.

Berenice's Hair or **Berenice's Locks** see COMA BERENICES.

berenil *Organic Chemistry.* $C_{22}H_{29}N_9O_6$, a yellow solid that decomposes at 217°C; soluble in 14 parts water, and slightly soluble in alcohol, ether, and chloroform; used in veterinary medicine.

Beretta *Ordnance.* any of several weapons manufactured by the Italian firm Beretta, including various 9-mm semiautomatic pistols and 9-mm submachine guns.

ber function *Mathematics.* the real part of $J_n\,(ze^{\pm 3\pi i/4})$, where J_n is the nth Bessel function.

Berg, Paul born 1926, American biochemist; Nobel Prize for method of mapping DNA.

bergall see CUNNER.

bergamot [bur´gə mät´] *Botany.* **1.** a small citrus tree, *Citrus aurantium bergamia,* having an acidic, pear-shaped fruit whose rind is the source of bergamot oil. **2.** also, **bergamot mint.** any of various fragrant plants of the mint family that yield an oil similar to bergamot, especially *Mentha citrata* and *Monarda fistulosa.*

bergamot oil *Materials.* an essential oil produced from the rind of the bergamot tree, used widely in perfumery. Also, **bergamot essence.**

Bergeron-Findeisen theory *Meteorology.* a theoretical explanation of the process by which precipitation particles form within a cloud composed of ice crystals and liquid water drops, stating that if the total water content is sufficiently high within an admixture of these particles, the ice crystals would gain mass by sublimation at the expense of the liquid, which would then lose mass by evaporation. Also, ICE CRYSTAL THEORY.

Bergius, Friedrich 1884–1949, German chemist; shared Nobel Prize with Carl Bosch for hydrogenation of coal.

Bergius process *Chemical Engineering.* a method of producing an oil similar to crude petroleum by the hydrogenation and liquefaction of coal or cellulosic materials at high hydrogen pressures in the presence of a catalyst.

Bergman, Tobern Olaf 1735–1784, Swedish chemist and naturalist; wrote a pioneering study of chemical affinities.

Bergmann's rule *Ecology.* the principle that within a species of warm-blooded animals, those living in colder climates tend to be larger than those in warmer climates.

bergmehl *Mineralogy.* a fine, white, mealy calcareous deposit on limestone cave walls that has been eaten in times of great scarcity. (A German word meaning "mountain meal.")

Bergoldiavirus see GRANULOSIS VIRUS.

bergschrund *Hydrology.* a deep gap or crevasse, or a series of closely spaced fissures, formed at or near the head of a mountain glacier or snowfield and separating the moving ice or snow from the relatively stationary ice and snow. Also, SHRUND.

Bergström, Sune Karl born 1916, Swedish biochemist; Nobel Prize for discovering the structure of prostoglandins and related substances.

berg till see FLOE TILL.

bergy bit *Oceanography.* a piece of floating ice that is smaller than an iceberg, generally about 5 meters across.

beriberi or **beri-beri** *Pathology.* a disease that is caused by vitamin B_1 deficiency and that is characterized by polyneuritis, cardiac pathology, and edema.

Bering Sea *Geography.* the northern arm of the North Pacific, north of the Aleutians and between Alaska and Siberia.

Bering Strait *Geography.* a narrow waterway connecting the Bering Sea with the Arctic Ocean and separating Alaska from Siberia.

Berkefeld candle *Microbiology.* a bacterial filter consisting of a tube of diatomaceous earth, used especially to sterilize heat-labile fluids that are unable to withstand heat sterilization. Also, **Berkefeld filter.**

berkelium [burk´lē əm] *Chemistry.* a synthetic radioactive element having the symbol Bk, the atomic number 97, an accepted atomic weight of 247 (uncertain), and a melting point of 990°C; a chemically reactive, crystalline metal. It is produced by the bombardment of americium isotope 241 with positively charged helium atoms (alpha particles); its most easily produced isotope has a half-life of 314 days. (First produced at the radiation laboratory of the University of California, *Berkeley.*)

berkeyite see LAZULITE.

Berkshire *Agriculture.* a breed of hog having a black body with white markings on its feet, snout, and tail. (From *Berkshire,* England.)

berlin *Textiles.* a soft woolen yarn used for knitting and embroidery; made from a fine, dyed wool called **Berlin wool.**

berlina *Materials.* the durable, pale brown wood of the trees of the *Berlinia* species of West Africa, used for decorative fittings and heavy structural work.

berlinite *Mineralogy.* $AlPO_4$, a colorless to gray to red trigonal mineral occurring in compact masses, with a specific gravity of 2.64 and a hardness of 6.5 on the Mohs scale; found in iron mines with augelite and other phosphates.

Berl saddle *Chemical Engineering.* a type of column packing used in distillation columns.

berm *Geology.* **1.** on the backshore of a beach, a temporary bench, shelf, ledge, or narrow terrace formed by the action of waves. **2.** a benchlike remnant of a surface resulting from the interruption of an erosion cycle. *Civil Engineering.* a stabilizing earthwork, especially the shoulder of a road. Also, BENCH. *Building Engineering.* a mound of earth placed against a building wall for stabilization or insulation.

bermanite *Mineralogy.* $Mn^{+2}Mn_2^{+3}(PO_4)_2(OH)_2 \cdot 4H_2O$, a reddish brown monoclinic (pseudo-orthorhombic) mineral that occurs in crystalline and massive form, having a specific gravity of 2.84 and a hardness of 3.5 on the Mohs scale. Bermanite is found in pegmatite with other phosphates.

Berman-Moorhead locator *Surgery.* an instrument that is used to locate metallic fragments embedded in body tissues.

berm crest *Geology.* the seaward or outer limit and generally the highest part of a beach berm.

Bermuda or **Bermudas** *Geography.* a group of islands (total area: 21 square miles) in the Atlantic Ocean, 580 miles east of North Carolina.

Bermuda grass *Botany.* a long-lived perennial grass, *Cynodon dactylon,* that is commonly grown for lawns and pastures, having extensive creeping rhizomes and stolons and flowers borne on slender spikes.

Bermuda high *Meteorology.* the semipermanent subtropical high of the western area of the North Atlantic ocean.

Bernal chart *Materials Science.* an overlay chart used in conjunction with a rotating crystal diffraction pattern that relates reciprocal lattice coordinates to diffraction spots on the film.

Claude Bernard

Bernard, Claude [ber´närd´] 1813–1878, French physiologist; discovered that liver converts sugar to glycogen and stores it.

Bernard's canal see DUCT OF SANTORINI.

Bernard-Soulier disease *Pathology.* an autosomal recessive disorder characterized by platelets having a wide range in size and morphology; its variable clinical signs include mucocutaneous and visceral hemorrhaging, purpura, and prolonged bleeding. Also, **Bernard-Soulier syndrome.** (Named for Jean *Bernard* (born 1907) and Jean-Pierre *Soulier* (born 1915), French hematologists.)

Berne virus *Virology.* an enveloped RNA virus that has been isolated from horses, believed to be of the family Toroviridae.

Bernini, Gianlorenzo 1598–1680, Italian Baroque architect and sculptor; designed St. Peter's Church and Square at the Vatican.

Bernouilli, Daniel [bur noo´lē] 1700–1782, Swiss physician and mathematician; formulated Bernouilli's Principle.

Bernouilli, Jacques (Jacob) 1654–1705, Swiss mathematician; worked in calculus, analytic geometry, and theory of probability.

Bernouilli, Jean (Johann) 1667–1748, Swiss mathematician; discovered exponential calculus.

Bernoulli distribution see BINOMIAL DISTRIBUTION.

Bernoulli effect *Fluid Mechanics.* the force exerted upon a fluid stream is diminished as the velocity is augmented.

Bernoulli equation *Fluid Mechanics.* an equation that expresses the law of conservation of energy for an incompressible irrotational flowing fluid; the terms in the Bernoulli equation represent energy in a flowing fluid due to pressure, velocity, and height, and it assumes steady, frictionless flow along a streamline. *Mathematics.* the differential equation $y'f = (x)y + g(x)y^n$, $n \neq 1$. Also, **Bernoulli differential equation.**

Bernoulli-Euler law *Mechanics.* a linear elastic law stating that the radius of the curvature of a beam under bending is proportional to the product of its Young's modulus and the moment of inertia of a cross section about the tranverse neutral axis, divided by the bending moment.

Bernoulli number *Mathematics.* the coefficients B_n in the expansion $xe^x/(e^x - 1) = \sum_{n=0}^{\infty} B_n x^n/n!$. The B_n are rational, the odd numbered B_n are zero and, in general

$$B_n = (2n)!\,(2^{1-2n})(\pi^{-2n}) \sum_{n=1}^{\infty} (1/i)^{2n}.$$

Bernoulli polynomial *Mathematics.* the polynomials $B_n(x)$ in the expansion $te^{xt}/(e^t - 1) = \sum_{n=0}^{\infty} B_n(x)t^n/n!$, where $|t| < 2\pi$. The Bernoulli numbers are the Bernoulli polynomials evaluated at $x = 0$.

Bernoulli process *Statistics.* a sequence of n identical and independent Bernouli trials of a random experiment.

Bernoulli's lemniscate *Mathematics.* the curve described by the equation $(x^2 + y^2)^2 = a^2(x^2 - y^2)$. It is shaped like a figure eight, centered at the origin and intersecting the axes at $\pm a$; the area of each loop is $a^2/2$.

Bernoulli theorem *Fluid Mechanics.* the law of variation of pressure along a stream line assuming steady motion and constant fluid density. *Statistics.* a theorem stating that if p is the probability of a given event and m is the number of occurrences of the event in n trials, then the probability that, for any $e>0$, $|(m/n) - p| < e$ has a limit of 1 as n approaches infinity.

Bernoulli trial *Statistics.* a trial or experiment with two possible outcomes, conventionally termed "success" and "failure," if the probability of success is p and the probability of failure is q, then $p + q = 1$, since one of the two outcomes must occur. Also, **Bernoulli experiment.** (Named for Swiss mathematician Jacques *Bernoulli.*)

Beroida *Invertebrate Zoology.* an order of thimble-shaped marine comb jellies, in phylum Ctenophora, that are compressed in the tentacular plane.

Berriasian *Geology.* a European geologic stage of the Lower Cretaceous period, after the Portlandian and before the Valangianian.

berry *Botany.* any indehiscent fruit, such as a gooseberry, grape, or tomato, with a succulent pericarp and stoneless, embedded seeds. *Invertebrate Zoology.* an egg of a lobster, crayfish, or other crustacean.

berth *Navigation.* **1.** the space assigned to a ship at a wharf or at anchor. Also, **berthage. 2.** to bring a ship into such a space. **3.** the safe distance maintained between a ship and any other object. *Naval Architecture.* the living quarters on a ship, particularly a sleeping space or a cabin assigned to a ship's officer.

Berthelot, Marcellin [ber´tə lō] 1827–1917, French chemist; pioneered thermochemistry and synthesis of organic compounds.

Berthelot equation *Physical Chemistry.* an equation that demonstrates the relationship of the physical properties of a gas, such as temperature, pressure, and volume, to its constant. Retains the same crude representation of the contribution of the repulsive forces as the van der Waals equation but uses a less well founded representation of the contribution of the attractive forces.

Berthelot principle *Physical Chemistry.* the principle that of all possible chemical processes that can proceed without the aid of external energy, the one that will occur is the one accompanied by the greatest evolution of heat; this applies at low temperatures only and does not account for endothermic reactions. Also, THOMSON-BERTHELOT PRINCIPLE.

Berthelot relation *Physical Chemistry.* a relationship that expresses the constants of molecular attraction between like and unlike species.

berthierite *Mineralogy.* $FeSb_2S_4$, an opaque, metallic, dark-gray orthorhombic mineral occurring in granular and fibrous masses, having a specific gravity of 4.64 and a hardness of 2 to 3 on the Mohs scale.

Berthollet, Claude Louis [ber´tə lā] 1748–1822, French chemist; author of the first systematic study of physical chemistry and chemical affinity.

berthollide *Chemistry.* any compound whose composition does not conform to the law of definite proportions. Also, NONSTOICHIOMETRIC COMPOUND.

Berthon, Edward Lyon [bur´thän´] 1813–1899, English clergyman and inventor; invented a collapsible boat and a number of instruments used in steamship navigation.

Berthon dynamometer *Engineering.* an instrument consisting of two metal straightedges set at an angle and joined together, used to measure the diameter of objects inserted between the straightedges.

Bertiella *Invertebrate Zoology.* a genus of tapeworms of the family Anoplocephalidae; some species are occasionally parasitic in man and higher apes.

Bertrand, Joseph [ber´tränd´] 1822–1900, French mathematician; proposed Bertrand's postulate.

Bertrand curve *Mathematics.* one of a pair of curves whose principal normals are the same. Also, ASSOCIATE CURVE, CONJUGATE CURVE.

bertrandite *Mineralogy.* $Be_4Si_2O_7(OH)_2$, a colorless or pale-yellow orthorhombic mineral occurring in pegmatites as tabular or prismatic crystals.

Bertrand's postulate *Mathematics.* the theorem, proved by Chebyshev in 1852, that there is always at least one prime number between n and $2n$ for $n > 3$.

Bertrand's rule *Microbiology.* a rule stating that in compounds containing *cis* secondary alcoholic groups with at least one carbon atom of D configuration, the atom or atoms are dehydrogenated by the *Acetobacter suboxydans,* thus producing a ketone.

Beryciformes *Vertebrate Zoology.* an order of spiny-finned, mostly deep-sea fishes in the superorder Acanthopterygii, having an air bladder with a pneumatic duct. Also, **Berycomorphi.**

beryl *Mineralogy.* $Be_3Al_2Si_6O_{18}$, a green, blue, yellow, or pink hexagonal mineral found in metamorphic rocks and granitic pegmatite; used as a gemstone and also the principal ore of beryllium.

beryl

beryllate *Inorganic Chemistry.* a salt formed by the reaction of a strong base with beryllium oxide, BeO.

beryllide *Inorganic Chemistry.* an intermetallic compound formed by chemically combining beryllium with metals such as zirconium and tantalum.

berylliosis *Pathology.* beryllium poisoning, usually an inflammation involving the lungs and sometimes the skin, subcutaneous tissues, lymph nodes, liver, or other structures.

beryllium [bə ril´ē əm] *Chemistry.* a chemical element having the symbol Be, the atomic number 4, an atomic weight of 9.01218, and a melting point of about 1280°C. It is a brittle, hard, gray-white metal with a high ratio of strength to weight; the lightest structural metal known. It is widely used in spacecraft construction and is also used in electronic devices and lasers and for windows in X-ray tubes. (From the mineral *beryl,* where it is often found.)

beryllium alloy *Metallurgy.* an alloy that either contains beryllium or is based on beryllium.

beryllium carbide *Inorganic Chemistry.* BeC, yellow hexagonal crystals that decompose above 2100°C; used in nuclear reactor cores.

beryllium chloride *Inorganic Chemistry.* $BeCl_2$, white or colorless deliquescent needles, very soluble in water; used as a catalyst for certain organic reactions.

beryllium copper *Metallurgy.* any of several cast or wrought copper-base alloys containing up to 2.7% beryllium, usually along with other elements such as cobalt or nickel. Also, **beryllium bronze.**

beryllium detector see BERYLOMETER.

beryllium fluoride *Inorganic Chemistry.* BeF_2, a colorless hygroscopic solid that is very soluble in water and melts at 800°C; used in metallurgy and in glass manufacturing; a known carcinogen.

beryllium nitrate *Inorganic Chemistry.* $Be(NO_3)_2 \cdot 3H_2O$, a white to yellowish deliquescent compound, highly toxic, that is soluble in water and alcohol; melts at 60°C; used to introduce beryllium oxide into materials and as a hardening agent in incandescent mantles.

beryllium nitride *Inorganic Chemistry.* Be_3N_2, colorless to white crystals that melt at 2200°C; used to make rocket fuels and to produce radioactive tracers.

beryllium oxide *Inorganic Chemistry.* BeO, a white powder that is insoluble in water and soluble in acids and alkalis, having a hardness of 9 on the Mohs scale and melting at 2530°C; a unique ceramic material characterized by high electrical resistivity and thermal conductivity, resistance to damage by nuclear radiation, and transparency to microwave radiation; used in electron tubes, resistor cores, transistor mountings, and for various other purposes.

beryllonite *Mineralogy.* $NaBePO_4$, a rare colorless or yellow, transparent to translucent monoclinic mineral occurring in short prismatic or tabular crystals, having a specific gravity of 2.8 and a hardness of 5.5 to 6 on the Mohs scale; found with albite, tourmaline, and herderite.

berylometer *Engineering.* an instrument that uses gamma-ray activation analysis to detect and analyze beryllium.

Berytidae *Invertebrate Zoology.* the stilt bugs, a cosmopolitan family of fragile, slender, hemipteran bugs having long, thin legs.

berzelianite *Mineralogy.* Cu_2Se, an opaque, metallic, lead-gray cubic mineral that occurs as thin veinlets and dendritic crusts, having a specific gravity of 6.65 to 6.71 and a hardness of 2 on the Mohs scale; usually found in highly weathered siliceous gangue with native gold and lead.

Berzelius, Jöns Jakob [bər zā´lē əs] 1779–1848, Swedish chemist; founder of electrochemical theory; established chemical nomenclature; discovered elements cerium, selenium, and thorium; first to isolate calcium, silicon, and tantalum.

Bessel, Friederich 1784–1846, German astronomer and mathematician; found first authentic star parallax; invented Bessel functions.

Bessel ellipsoid of 1841 *Geodesy.* a reference ellipsoid having an ellipticity of 1/299.15, a semimajor axis of 6,377,397.2 meters, and a semiminor axis of 6,356,078.9 meters.

Bessel equation *Mathematics.* the second-order linear ordinary differential equation $z^2w'' + zw' + (z^2 - v^2)w = 0$, where w is a function of z and v is a complex number.

Bessel-Fourier transform see HANKEL TRANSFORM.

Bessel function *Mathematics.* the best-known solution $J_v(z)$ of the Bessel equation for v. An independent solution $Y_v(z)$, called the Weber function or Bessel function of the second kind, is often adjoined to express the general solution of the Bessel equation as a linear combination of the two. Numerous other selections of pairs of independent solutions exist (Hankel functions, modified Bessel functions, MacDonald functions, and so on)

Besselian elements *Astronomy.* numerical values used in calculating solar eclipses that establish the position of the moon's shadow relative to the earth.

Besselian star numbers *Astronomy.* numerical values used to convert a star's celestial position from a tabulated mean position to its actual apparent position on a given date and time.

Besselian year *Astronomy.* the year that begins when the mean sun reaches 18h40m in right ascension.

Bessel inequality *Mathematics.* given a vector space V with an inner product (,) and an orthonormal set $\{e_1, e_2, \ldots, e_m\}$, Bessel's inequality then states that

$$(v, v) \geq \sum_{k=1}^{m} |(v, e_k)|^2 \quad \text{for all } v \text{ in } V.$$

Bessel transform see HANKEL TRANSFORM.

Bessemer, Henry 1813–1898, English steel manufacturer; invented the Bessemer process.

Bessemer converter *Metallurgy.* a large, refractory-lined, barrel-shaped furnace used to make steel by the Bessemer process.

Bessemer iron *Metallurgy.* pig iron having a sufficiently low phosphorus content so that it can be refined by the Bessemer process. Also, **Bessemer pig iron.**

Bessemer ore *Metallurgy.* an iron ore that is low in phosphorus and therefore suitable for the Bessemer process.

Bessemer process *Metallurgy.* a steel refining process in which air is blown through the molten charge to oxidize impurities such as carbon, silicon, and manganese. Any steel produced by this process is called **Bessemer steel.**

Best, Charles Herbert 1899–1978, Canadian physiologist; discovered histaminase and assisted in the discovery of insulin.

Be star *Astronomy.* a star of spectral type B that is irregularly variable in brightness and shows emission lines of hydrogen in its spectrum due to a shell of gas around the star.

best commercial practice *Engineering.* a standard applied in the manufacturing of an item that was not designed according to established codes.

best-first search *Artificial Intelligence.* a search in which a queue of nodes that have been generated is kept; nodes are evaluated by a heuristic function, and the best node found so far is examined and expanded at each step.

beta *Science.* Β or β, the second letter in the Greek alphabet, often used to denote the second item in a group or series. *Astronomy.* usually, **Beta.** the second-brightest star of a constellation. *Electronics.* the current transfer ratio (gain) of a transistor in the common-emitter circuit arrangement; it is expressed as the ratio of AC collector current to AC base current. *Crystallography.* the intermediate index of refraction in a biaxial crystal.

Beta *Botany.* a genus of glabrous succulent herbs of the goosefoot family, including the beet *(Beta vulgaris).*

beta alumina *Engineering Materials.* $Na_2 \cdot 11Al_2O_3$, an aluminum oxide used for refractories, having hexagonal crystals that are heat-stabilized with sodium.

beta blocker *Physiology.* any agent, such as propranolol, that blocks activity of the beta-adrenergic receptors in the sympathetic nervous system; used especially in the treatment of angina and hypertension.

beta brass *Metallurgy.* any of several alloys based on copper and zinc and having a body-centered cubic crystal structure.

beta carotene *Biochemistry.* $C_{40}H_{56}$, the precursor to vitamin A; commonly found in plants in association with chlorophylls; cleaved into two molecules of vitamin A in the liver of numerous animals.

beta cell *Histology.* **1.** a basophilic pigment cell found in the anterior lobe of the adenohypophysis. **2.** any cell that secretes insulin in the pancreatic islets of Langerhans.

Beta Centauri *Astronomy.* see HADAR.

Beta Cephei *Astronomy.* any hot variable star of spectral type 09 to B3 that varies in brightness by less than a third of a magnitude. Also, **Beta Canis Majoris.**

beta-chain *Biochemistry.* one of the two types of polypeptide present in adult hemoglobin; a hemoglobin chain consists of two alpha-chains and two beta-chains.

betacin *Microbiology.* a bacteriocin-like substance that is produced by certain lysogenic strains of *Bacillus subtilis* and that inhibits growth of nonlysogenic bacterial strains.

beta circuit *Electricity.* a circuit designed to transmit a portion of the amplifier output back to the input.

beta coefficient *Statistics.* in multiple regression analysis, the number of standard deviations that the dependent variable changes in response to a change of one standard deviation in an independent variable; the transformed net regression coefficients of the regression equation.

Beta Crucis *Astronomy.* a first-magnitude star in the Southern Cross.

beta cutoff frequency *Electronics.* the frequency at which the beta of a transistor is three decibels below the low-frequency value.

betacyanin *Biochemistry.* a pigment found in certain plants of the Caryophyllales plant order.

beta decay *Nuclear Physics.* a type of radioactivity in which the parent nucleus emits either a negatively charged electron (negatron) or a positively charged electron (positron), thus raising or lowering the atomic number by one while leaving the atomic mass unchanged. Also, **beta disintegration, beta-particle decay.**

Betadine *Pharmacology.* a trade name for preparations of povidone-iodine, used as an antiseptic.

beta distribution *Statistics.* a flexible probability distribution having finite upper and lower bounds, used in representing uncertainty on the parameter of a binomial data process.

beta diversity *Ecology.* the diversity of species among differing habitats or communities.

beta emitter *Nuclear Physics.* a nuclide that decays by emitting electrons.

beta factor *Physics.* a ratio comparing the kinetic pressure of plasma to its magnetic pressure.

beta function *Mathematics.* the function of the parameters x and y defined by

$$\beta(x, y) = \int_0^1 t^{x-1}(1-t)^{y-1}dt.$$

The integral converges when both x and y are positive and diverges otherwise. Also, EULER'S INTEGRAL OF THE FIRST KIND.

beta-gamma survey meter *Nucleonics.* a portable instrument that provides a quantitative measure of low-level beta and gamma radiation in decontamination and for general survey of personnel and equipment; the most common type is the Geiger-Müller counter.

Beta Geminorum see POLLUX.

beta globulin *Biochemistry.* the portion of the plasma globulin that includes transferrin.

beta hemolysis *Microbiology.* in the medium surrounding a bacterial colony cultured on blood agar, the appearance of a colorless zone characteristic of certain pathogenic bacteria, caused by decomposition of hemoglobin in the blood cells of the medium.

Betaherpesvirinae *Virology.* a subfamily of the family Herpesviridae that has a long replication cycle and narrow host range; causes infection of the kidneys and salivary glands in a wide range of vertebrates, including humans. Also, CYTOMEGALOVIRUS GROUP.

betaine *Organic Chemistry.* $C_5H_{11}NO_2$, colorless crystals that decompose at 293°C and are soluble in water; an oxidation product of choline found first in sugar beet molasses and later in many plants and animals, and also synthesized from glycine. It is used in the treatment of muscular weakness and degeneration.

beta interaction see WEAK INTERACTION.

betalain *Biochemistry.* a class of nitrogen-containing red or yellow pigment that occurs in exclusive plants from the family Caryophyllales, including beets and chards, where it serves the same functions as the anthocyanin plant pigments.

beta-lipoprotein see LOW-DENSITY LIPOPROTEIN.

Betamax *Telecommunications.* a trade name for a system of recording-playback format for video cassette recorders using half-inch tape and a 3–3.5 MHz frequency response. Also, **Beta.**

betamethasone *Pharmacology.* $C_{22}H_{29}FO_5$, a synthetic glucocorticoid that is the most active of the anti-inflammatory steroids, occurring as a white crystalline powder used orally or applied to the skin; also used to derive several medicinally important esters, such as **betamethasone acetate** and **betamethasone sodium phosphate.**

betanaphthol *Pharmacology.* $C_{10}H_8O$, one of two isomers of naphthol, occurring as a colorless or pale-buff crystalline compound and formerly used as a topical antiseptic, especially in fungal infections.

betanin *Biochemistry.* a nitrogen-containing red betalain pigment found in beets.

Beta Orionis see RIGEL.

beta particle *Nuclear Physics.* an electron or positron that is emitted from a nucleus during beta decay.

Beta Persei star *Astronomy.* an eclipsing binary. Also, ALGOL VARIABLE.

beta plane *Geophysics.* a spherical model of the earth with a rate of rotation that changes linearly with the north-south location.

beta-pleated sheet *Biochemistry.* a common secondary structure in proteins, in which polypeptide chains lying side by side are linked by hydrogen bonds between peptide C=O and N–H groups; occurs as two types, *parallel* and *antiparallel.*

beta quartz see HIGH QUARTZ.

beta ray *Nuclear Physics.* a stream of beta particles.

beta-ray spectrometer *Spectroscopy.* a device used to measure the energy of emitted beta particles. Also, **beta spectrometer.**

beta rhythm *Physiology.* the waveforms of the electroencephalogram that range in frequency from 18 to 30 Hz.

beta rule see REDUCTION RULE.

beta sheet *Molecular Biology.* a structure resulting from the regular folding of polypeptide chains; the chief alternative to the alpha helix.

beta taxonomy *Systematics.* the second stage in the development of systematic understanding of a taxon during which the interrelationships of its component taxa are determined.

beta test *Computer Programming.* the second stage of testing of a new software product, usually conducted outside the developer's facility and in a near-real working environment, called a **beta test site.**

betatron *Nucleonics.* a device for accelerating charged particles by continuously increasing the magnetic flux within the particle's orbit.

betatropic *Nuclear Physics.* of two atoms, differing in atomic number by one unit, as if one atom could produce the other by ejecting a beta particle.

beta waves see BETA RHYTHM.

beta weight see BETA COEFFICIENT.

betaxanthin *Biochemistry.* a yellow betalain plant pigment found in the family Caryophyllales.

betazole *Pharmacology.* $C_5H_9N_3$, a substance that is used in medicine to produce various compounds, notably **betazole hydrochloride,** $C_5H_9N_3{\cdot}2HCl$, a white crystalline powder used in diagnostic tests to stimulate gastric secretion of hydrochloric acid.

betel *Pharmacology.* an East Indian masticatory preparation consisting of a **betel nut** (the dried seed of the betel palm) wrapped in a **betel leaf** (the leaf of a betel pepper) with quicklime; it contains a narcotic called arecaine, which produces mild stimulation and a sense of well-being.

Betelgeuse [bēt′əl jooz′; bet′əl jooz′] *Astronomy.* α-Orionis, a first-magnitude orange-red star in the northeast corner of Orion, the Hunter. (From an Arabic phrase meaning "shoulder of the giant.")

betel nut *Botany.* the astringent, dried seed of the palm tree *Areca catechu,*that is chewed as a masticatory stimulant by natives of Ceylon and Malaya.

betel palm *Botany.* a tall tropical Asian palm tree, *Areca catechu,* that produces the betel nut.

betel pepper *Botany.* a climbing East Indian pepper plant, *Piper betel,* whose leaves are used in preparing betel. Also, **betel vine.**

BET equation see BRUNAUER-EMMETT-TELLER EQUATION.

Bethe, Hans Albrecht [bā′tē′] born 1906, German-born American physicist; Nobel Prize for theory of solar and astral energy production.

Bethe-ansatz technique *Physics.* a method for solving one-dimensional many-body problems; first applied to one-dimensional magnets, then generalized to many-body problems with point interactions.

Bethell's process *Forestry.* a method of preserving dried timber by impregnating it with creosote under the pressure of a partial vacuum.

Bethe range *Materials Science.* the average distance that incident electrons travel in a sample, calculated using the Bethe relationship for energy loss per unit distance.

Bethe-Salpeter equation *Particle Physics.* an equation that is analogous to the integral form of the Schrödinger equation for two bodies.

Bethe theory *Materials Science.* a description of ordering in metal alloys as a function of temperature, which predicts that short-range interactions contribute principally to the temperature-dependent degree of disorder.

Bethune, Norman 1890–1939, Canadian surgeon; invented surgical tools and techniques; set up Chinese hospitals and medical schools.

Bethylidae *Invertebrate Zoology.* a family of small wasps in the order Hymenoptera, whose females oviposit on other insects that they sting and paralyze.

Bethyloidea *Invertebrate Zoology.* a superfamily of wasps in the suborder Apocrita, containing over 5000 known species worldwide.

betonite *Petrology.* a montmorillonite-type clay formed by the alteration of volcanic ash that varies in composition and is usually highly colloidal and plastic; it occurs in thin deposits in the western U.S. and is used in making refractory linings, decolorizing of oils, and thickening drilling muds.

betony *Botany.* a labiate plant, *Stachys* (formerly *Betonica*) *officianalis,* having purple flowers and hairy, aromatic leaves; once used medicinally as an astringent and emetic.

betrunked river *Hydrology.* a river whose lower course has been removed by the submergence of a valley or by the recession of the coast.

betrunking *Hydrology.* the disappearance of the lower part of a stream course as a result of the submergence of a valley or the recession of a coast, thus causing the upper branches of the stream system to enter the seas as smaller, independent streams.

betta *Vertebrate Zoology.* a tropical freshwater labyrinth fish of the family Belontiidae, the single species, *Betta splendes,* having beautiful color and long, delicate, flaglike fins; the males fight aggresively for territory.

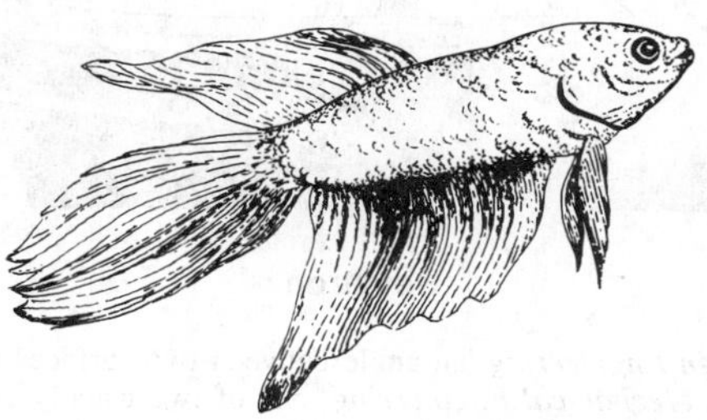

betta

Bettelheim, Bruno 1903–1990, Austrian-born American psychiatrist and author; noted for his work with emotionally disturbed children.

Betterton-Kroll process *Chemical Engineering.* an extractive process for obtaining bismuth and purifying desilverized lead that contains bismuth by adding metallic calcium or magnesium to the molten lead.

Betti number *Mathematics.* the kth Betti number of a manifold is the vector space dimension of the kth homology group of the simplicial complex for the manifold; on a differentiable manifold, also equal to the vector space dimension of the kth cohomology group of the complex of alternating differential forms on the manifold (by deRham's theorem).

Betti reciprocal theorem *Mechanics.* a generalization of Maxwell's reciprocal theorem for linear elastic response: for two systems of loads, the work that would be done by the forces in the first system acting through the displacements produced in the second system is equal to the work that would be done by the forces in the second system through the displacements of the first system. The terms *load* and *displacement* are used in their general sense; for example, the load can be a moment, and the displacement a rotation.

Bettsia *Mycology.* a genus of fungi in the order Ascophaerales that lives off of the pollen stored in the honeycombs of bee hives.

Betts process *Metallurgy.* an electrochemical process for refining lead and producing bismuth in which naturally occurring bismuth impurities in lead are electrolytically removed from the melt and recovered as anode slimes.

Betula *Botany.* the birch trees, a genus of deciduous trees or shrubs of the Betulaceae family, characterized by leaves with prominent veins, noticeably toothed margins, and conelike strobiles that disintegrate and scatter tiny seeds.

Betulaceae *Botany.* the birch family of deciduous trees, characterized by simple, alternate-toothed leaves, flowers in catkins, and a nut or nutlet fruit in a conelike strobile, cluster, or husk; includes birches, alders, and hazelnuts.

betulaceous *Botany.* relating to or resembling the birch family of trees.

betulinic acid *Organic Chemistry.* $C_{29}H_{46}(OH)COOH$, a crystalline triterpenoid acid formed from betulinol by oxidation; slightly soluble in water, alcohol, and acetone. Also, **betulic acid.**

betweenbrain see DIENCEPHALON.

Betz cell *Histology.* any of a group of large pyramidal ganglion cells found in the internal pyramidal layer of the cerebral cortex. (Named for V. A. *Betz,* 1834–1894, Russian anatomist.)

beudantite *Mineralogy.* $PbFe_3^{+3}(AsO_4)(SO_4)(OH)_6$, a green to brown or black mineral occurring as trigonal crystals, having a specific gravity of 4 to 4.48 and a hardness of 3.5 to 4.5 on the Mohs scale; found as a secondary mineral in the oxidized zone of ore deposits.

BeV *Physics.* one billion electron volts, 10^9 eV. Also, GEV.

Bevatron *Nucleonics.* a type of proton synchrotron at the University of California, Berkeley. (So named because of its ability to produce particles in the billion electron volt energy range.)

Bevatron

bevel *Design Engineering.* an angle between two surfaces that is not a right angle. *Mechanical Engineering.* one of two usually conical gears used to connect two shafts whose axes are set at right angles to each other. *Building Engineering.* to cut or join at an angle. *Graphic Arts.* a drafting instrument having two arms that are jointed so they can open to any angle; used for drawing angles and measuring bevels.

beveling *Geology.* the cutting across or planing by erosion of a geologic structure or landform. *Mechanical Engineering.* a machining process of cutting off a sharp corner of a machined part in order to make a surface not at right angles with the rest of the part. Also, CHAMFERING.

bevel joint *Building Engineering.* a miter joint formed by two wood pieces or other materials that meet at an angle other than 90°.

bevel square *Mechanical Devices.* an adjustable tool used to measure angles between surfaces, especially in carpentry.

Beverage antenna see WAVE ANTENNA.

beyerite *Mineralogy.* $(Ca,Pb)Bi_2(CO_3)_2O_2$, a white or yellow tetragonal mineral occurring in compact masses and flattened reactangular crystals, having a specific gravity of 6.56 and a hardness of 2 to 3 on the Mohs scale; found as a secondary mineral with bismutite.

Beyrichia *Paleontology.* a genus of dimorphic Paleozoic ostracods in the extinct order Palaeocopa and family Beyrichiidae.

Beyrichiacea *Paleontology.* a superfamily of equivalve ostracods in the extinct order Palaeocopida and suborder Beyrichicopina; characterized by a long, straight hinge and generally lobate shells; Ordovician to mid-Permian.

Beyrichicopina *Paleontology.* the suborder of ostracods whose type genus is *Beyrichia.*

Beyrichiidae *Paleontology.* the family of ostracods whose type genus is *Beyrichia.*

bezel *Design Engineering.* **1.** the sloped surface of a cutting tool. **2.** a grooved rim used to hold a window, lens, or other transparent covering, such as on a clock or headlight. *Horology.* the metal flange in the case of a timepiece into which a clock glass or watch crystal is set. *Mineralogy.* on a cut gem, the sloping surface just above the girdle or between the girdle and the table.

BF board foot; board feet; blastogenic factor.

BF or **B.F.** Bachelor of Forestry.

bf or **b.f.** boldface.

BFA or **B.F.A.** Bachelor of Fine Arts.

BFL *Aviation.* the airport code for Bakersfield, California.

BFO beat frequency oscillator.

B form DNA *Genetics.* the double-helical DNA molecule in its normal structural configuration; the most common form of DNA. It differs from the other right-handed helical form of DNA (called the A form) and from the left-handed helical form (the Z form) in terms of base stacking and helical pitch.

BFP biologic false-positive.

BFRA *Metallurgy.* a composite material consisting of aluminum reinforced with boron fibers.

BG brigadier general.

bGH bovine growth hormone.

BGI *Aviation.* the airport code for Barbados.

BGR *Aviation.* the airport code for Bangor, Maine.

BH Brinell hardness.

Bhabha scattering *Particle Physics.* the physical processes that occur in scattering in which positrons are deflected by electrons.

bhd. bulkhead.

BHK cells see BABY HAMSTER KIDNEY CELLS.

BHM *Aviation.* the airport code for Birmingham, Alabama.

BHN Brinell hardness number.

B horizon *Geology.* the layer of soil directly beneath the surface layer (or A horizon); characterized by an accumulation of clay, iron, aluminum, and organic matter and by a blocky or prismatic structure. Also, ZONE OF ACCUMULATION, ZONE OF ILLUVIATION.

bhp brake horsepower. Also, **b.h.p., BHP.**

BHX *Aviation.* the airport code for Birmingham, England.

Bi the chemical symbol for bismuth.

bi- a prefix meaning "two" or "twice," as in *bidirectional, biennial.*

BIA or **B.I.A.** Bachelor of Industrial Arts.

Bial's test *Pathology.* a method of detecting the presence of pentose in urine by the use of five ml of a reagent, consisting of 1.5 grams of orcinol, 500 grams of fuming hydrochloric acid, and 10% ferric chloride, which is boiled before several drops of urine are added. The immediate appearance of a green color indicates pentose. (Named after Manfred *Bial,* 1870–1908, German physician.)

biamperometry *Analytical Chemistry.* a measurement of the current through two polarizing electrodes to detect the end point of a titration.

Bianchi cosmology *Astrophysics.* a model cosmology in which the universe appears different to different observers, unlike the standard (Friedmann) mode, which looks the same to all observers within it.

Bianchi identity *Mathematics.* viewing the curvature tensor of a symmetric connection as a vector-valued 2-form, the Bianchi identity asserts that the covariant exterior derivative of the curvature tensor is zero. To this is sometimes added **Bianchi's first identity,** asserting that the exterior product of the curvature tensor with the basis-valued 1-form is zero.

bianchite *Mineralogy.* $(Zn,Fe^{+2})SO_4 \cdot 6H_2O$, a colorless or white, transparent monoclinic mineral with a specific gravity of 2.03 that is soluble in water and forms a crystalline crust.

Biarritzian *Geology.* a European geologic stage of the upper Middle Eocene epoch, considered to be the equivalent of the Bartonian stage.

biarticulate *Biology.* having two joints, as in the antennae of some insects.

bias *Acoustical Engineering.* the addition of sound to a recording, especially a high-frequency signal, to produce full frequency coverage and prevent distortion due to the lack of a signal. Also, **bias compensation.** *Electronics.* an electrical, mechanical, or magnetic force, voltage or current applied to a device to establish a desired level of operation. *Statistics.* **1.** the systematic distortionary effect of a measurement or sampling process on a statistical result that prevents the result from being representative of its population. **2.** the expected magnitude of the error incurred by a biased estimator, equal to the difference between the expected value of the estimator and the parameter of interest.

bias binding *Textiles.* a nonfraying, narrow strip of material used for binding curved seams and garment edges.

bias cell *Electronics.* a dry cell that provides the necessary C bias voltage in the grid circuit of a vacuum tube.

bias current *Electronics.* **1.** the steady, constant current that presets the operating threshold of a circuit. **2.** the current flowing into each of the two input terminals of a differential amplifier.

bias distortion *Electronics.* distortion caused by the operation of a tube or transistor with incorrect bias. *Telecommunications.* **1.** a distortion affecting a two-condition (binary) modulation in which all signal intervals correspond to only one of the two signals. **2.** also, **bias telegraph distortion.** in a telegraph or teletype system, an improper lengthening of either the mark or space elements that is caused by a lack of symmetry in transmitting or receiving equipment. **3.** a distortion due to improper biasing of amplifying device to cause the operating point to be moved from a linear region of amplification to one that is nonlinear.

biased relay see PERCENTAGE DIFFERENTIAL RELAY.

biased statistic *Statistics.* a statistical measure whose expected value is not equal to the population statistic: $E(q) \neq /q$. Also, **biased estimator.**

bias error *Statistics.* the magnitude of the error incurred by a biased estimator, equal to the difference between the mean of the estimator and the true parameter.

bias oscillator *Electronics.* an oscillator used in a magnetic recorder to generate an AC signal having a frequency in the range required to erase prerecorded material and bias the system to give a linear recording characteristic.

bias register *Computer Programming.* a register used in the relocation approach to memory mapping, in which the contents of the register are added to a logical memory address to produce a physical address.

bias resistor *Electronics.* a resistor used to produce the voltage drop necessary to provide a desired biasing voltage.

bias voltage *Electronics.* **1.** the steady voltage that presets the operating threshold of a circuit. **2.** the non-signal or mean potential of any electrode in a thermionic tube, as measured with reference to the cathode.

bias winding *Electricity.* a DC control winding in a magnetic amplifier or other magnetic device.

biaxial crystal *Crystallography.* a crystal with two directions along which the wave-normal velocity for monochromatic light is constant, so that as a result, the crystal has principal refractive indices; true for triclinic, monoclinic, and orthorhombic crystals.

biaxial indicatrix *Crystallography.* an ellipsoidal with the three principal indices of refraction of a bixial crystal for light waves (of a given wavelength) in their direction of vibration as the lengths of the three mutually perpendicular semiaxes.

biaxial stress *Mechanics.* a condition in which two of three mutually perpendicular principal stresses are in the same plane, and one is zero.

Biazzi process *Chemical Engineering.* a continuous-flow process used to combine glycerin with a nitrate to form nitroglycerin, glycol dinitrate, and diethylene glycol nitrate.

bib. *Medicine.* an abbreviation meaning "drink" or "to drink." (From Latin *bibe.*) Also, **Bib.**

bibcock. *Mechanical Devices.* any faucet, tap, or plug with a downturned opening. Also, **bibb cock.**

bibenzyl *Organic Chemistry.* $C_6H_5CH_2 \cdot CH_2C_6H_5$, a crystalline hydrocarbon consisting of two benzene rings attached to ethane; insoluble in water and soluble in alcohol and carbon disulfide; melts at 52°C and boils at 284°C. Also, DIBENZYL.

Bibionidae *Invertebrate Zoology.* a family of flies, in the order Diptera, comprising stout insects with short, thick-set antennae; some adult species act as orchard pollinators, while larvae tend to damage crops.

Bible paper *Graphic Arts.* a paper that is lightweight (14–30 pounds) and low-bulk but fairly opaque; used for printing Bibles, dictionaries, and other books with a large number of pages. Also, INDIA PAPER.

bibliophobia *Psychology.* an irrational fear or hatred of books.

bicameral *Biology.* having two branches, chambers, or divisions.

bicapsular *Anatomy.* consisting of two capsules, as does a body joint. *Botany.* referring to plants having a capsule containing two chambers or cells within which specialized structures develop, as within an anther.

bicarbonate *Inorganic Chemistry.* a salt made by the neutralization of one hydrogen atom in carbonic acid; a salt containing the $-HCO_3$ group.

bicarbonate of soda *Inorganic Chemistry.* another name for sodium bicarbonate, $NaHCO_3$, especially when used as an antacid. See SODIUM BICARBONATE.

bicarinate *Biology.* having two keel-shaped projections or appendages.

bicaudate *Zoology.* having two tails or tail-like appendages. Also, **bicaudal.**

bicellular *Biology.* having or made up of two cells.

bicentric *Biology.* **1.** of or relating to a taxon having two centers of evolution. **2.** relating to a plant or animal having two centers of distribution.

bicephalous *Biology.* having two heads.

biceps *Anatomy.* any muscle that has two heads or two points of origin, especially the biceps brachii.

biceps brachii *Anatomy.* a muscle of the upper arm that flexes and supinates the forearm.

biceps femoris *Anatomy.* a hamstring muscle of the thigh that rotates the leg laterally and flexes the leg at the knee.

Bichat, Marie 1771–1802, French anatomist, physiologist, and pathologist; founder of histology; wrote landmark *Descriptive Anatomy.*

bichir *Vertebrate Zoology.* any of several African primitive freshwater fishes of the genus *Polypterus* that have functional lungs as well as gills.

biciliate *Biology.* having two cilia.

bicipital *Anatomy.* **1.** two-headed. **2.** relating to a biceps muscle. *Botany.* having two heads or supports, as the stamens of certain flowers.

bicipital groove see INTERTUBERCULAR SULCUS.

bicollateral bundle *Botany.* a vascular bundle that has two strands of phloem, one on each side of the xylem, on the same radius.

bicompact set see COMPACT SET.

bicomponent fiber *Textiles.* a synthetic fiber made from two polymers, arranged to lie side-by-side or as a sheath surrounding a core.

biconcave lens see DOUBLE-CONCAVE LENS.

biconditional *Mathematics.* an if and only if statement; the biconditional sentence $P \leftrightarrow Q$ means Q is a necessary and sufficient condition for P; equivalently, if P then Q, and if Q then P. The symbol $\leftrightarrow$ is called the biconditional operator.

biconditional operator see EQUIVALENCE.

biconditional statement see IF AND ONLY IF.

biconnected graph *Mathematics.* a graph that is a block. In that same usage, a block of a graph is called a **biconnected component.**

biconstituent fiber *Textiles.* a fiber composed of dissimilar fibers that are combined and then extruded. Also, MATRIX FIBER.

bicontinuous function see HOMEOMORPHISM.

biconvex lens see DOUBLE-CONVEX LENS.

bicorn *Biology.* **1.** describing a plant or animal that has two horns or hornlike parts. **2.** shaped like a crescent. *Mathematics.* the graph of the equation $(x^2 + 2ay - x^2)^2 = y^2(a^2 - x^2)$, where a is a constant.

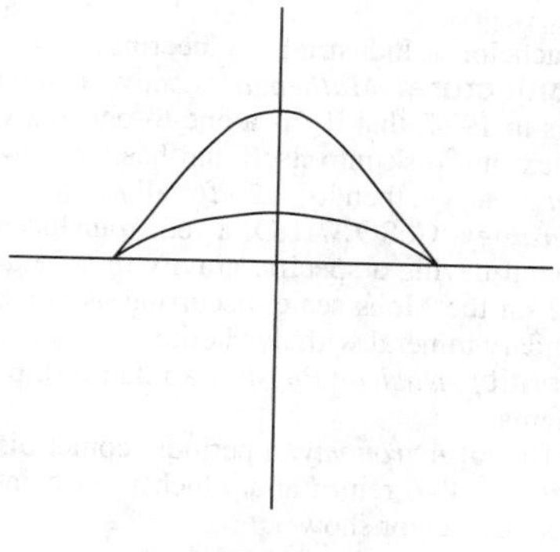

bicorn

bicornuate *Biology.* having two horns or having horn-shaped branches.
Bicosoecida *Invertebrate Zoology.* an order of free-living, colorless protozoans, in the class Zoomastigophorea, with two flagella.
bicostate *Botany.* of a leaf, having two main ribs running longitudinally.
bicron see STIGMA.
bicrystal *Materials Science.* two crystals joined or grown onto one another, often at specific orientations.
biculturalism *Anthropology.* the participation of a minority culture in two separate culture spheres, as in British colonial Africa where the native population participated in the British culture in cities and in the traditional culture in their villages.
bicuspid *Anatomy.* having two points, cusps, or flaps. *Botany.* having two points or cusps.
bicuspid teeth *Anatomy.* the teeth between the molars and the canines.
bicuspid valve *Anatomy.* the valve between the left atrium and left ventricle. Also, MITRAL VALVE.
bicycle *Mechanical Engineering.* a two-wheeled, human-powered vehicle, with one wheel in front of the other, usually driven by a rider turning foot pedals attached to the rear wheel by a chain.
bicyclic *Botany.* having or arranged in two whorls.
bicyclic compound *Organic Chemistry.* a compound that contains two (usually fused) ring structures, as in naphthalene.
bicyclo[2.2.1]heptane see NORBORNANE.
BID or **B.I.D.** Bachelor of Industrial Design.
bidalotite see ANTHOPHYLLITE.
Bidder's organ *Vertebrate Zoology.* a vestigial ovary found in front of a developed testis or ovary in some species of toads of both sexes.
Biddulphiaceae *Botany.* a family of marine diatoms of the order Centrales having bi-, tri-, or polypolar valves with pseudoocelli-bearing elevations at the valve poles.
Biddulphiineae *Botany.* a suborder of mostly marine diatoms of the order Centrales, characterized by special structures at the poles or circular valves and ocelli or labiate processes.
bidentate *Biology.* having two teeth or toothlike structures. *Chemistry.* describing a molecule whose atoms are joined to a metal atom or ion to form a coordination complex.
bidentate ligand *Inorganic Chemistry.* a chelating agent that has two groups capable of attachment to a metal ion.
bidenticulate *Biology.* having two teeth or toothlike parts or processes.
bidirectional *Engineering.* having the ability to function in two directions, usually two opposite directions. Thus, **bidirectional transistor, bidirectional transducer,** and so on.
bidirectional antenna *Electromagnetism.* an antenna whose radiation pattern has two principle lobes of different directions.
bidirectional data bus *Computer Technology.* a single conductor that is used to transmit digital data in either direction, as between the central processing unit and the memory.
bidirectional microphone *Acoustical Engineering.* a microphone, such as a ribbon microphone, that is responsive toward the front and back, but dead toward the sides, thus resulting in a figure-eight response pattern.
bidirectional printer *Computer Technology.* a line printer that eliminates the carriage return and speeds up the printing by alternately printing a line from left to right, then from right to left.
bidirectional replication *Molecular Biology.* a mechanism of DNA replication, commonly found in both prokaryotes and eukaryotes, in which two DNA replication forks move out in either direction from a single site of initiation.
Bidrin see DICROTOPHOS.
BIE or B.I.E. Bachelor of Industrial Engineering.
Bieberbach conjecture *Mathematics.* now a theorem, proved by Louis deBranges in 1984, that if f is a one-to-one analytic function that maps the complex unit disk into itself and has the power series expansion $z + a_2x^2 + a_3x^3 + \cdots$, then $|a_n| \leq n$ for all n.
bieberite *Mineralogy.* $CoSO_4{\cdot}7H_2O$, a red, translucent, water-soluble monoclinic mineral having a specific gravity of 1.95 and a hardness of approximately 2 on the Mohs scale, occurring as crusts and stalactites; found as a secondary mineral with pyrrhotite.
Biedenharn identity *Nuclear Physics.* a relationship among the six-*J* Wigner coefficients.
Biela's Comet [bē´lə] *Astronomy.* a periodic comet discovered in 1826 that broke up on its 1846 return and which is associated with the Andromedid (or Bielid) meteor shower.
Bielids [bē´ lidz] *Astronomy.* see ANDROMEDIDS.
Bienayme-Chebyshev inequality *Statistics.* a special case of Chebyshev's inequality such that the function of the random variable is the square of the deviation from the mean: for any $k>1$, at least $1 - 1/k^2$ of the distribution lies within k standard deviations from the mean.
biennial *Botany.* **1.** a plant having a two-year life cycle, with germination in one year or growing season, and flowering, fruit production, and death in the next year or growing season. Also, **biennial plant**. **2.** of or relating to such a plant.
Bierbaum hardness test *Engineering.* a test designed to measure the hardness of a sample by measuring the width of a scratch made by a diamond point.
biface (tool) *Mechanical Devices.* any cutting tool with symmetrical, V-shaped faces, such as an ax. *Archaeology.* a stone tool having both sides or surfaces flaked rather than just one.
bifacial *Botany.* having distinct dorsal and ventral surfaces that differ in structure from each other. *Design Engineering.* of or relating to a tool in which both sides are alike. *Archaeology.* of a stone tool, having flakes chipped from both sides or surfaces.
bifarious *Botany.* in two vertical rows. (From a Latin word meaning "twofold; double.")
bifenox *Organic Chemistry.* $C_{14}H_9Cl_2NO_5$, a tan, crystalline compound that melts at 84–86°C; insoluble in water; used as an herbicide.
bifid *Anatomy.* divided into two parts or branches, often at the apex. *Biology.* separated or cleft into two equal parts or lobes.
Bifidobacterium *Bacteriology.* a genus of Gram-positive, anaerobic bacteria of the family Actinomycetacea, occurring as irregular rod-shaped cells among the normal flora of the urogenital and gastrointestinal tracts.
bifilar *Science.* fitted or furnished with two filaments or threads.
bifilar electromagnetic oscillograph *Electromagnetism.* an apparatus consisting of a galvanometer with a bifilar mirror suspension that is deflected in proportion to a current passing through the galvanometer; the mirror reflects a beam of light onto a screen and thus can record low-frequency oscillations.
bifilar micrometer *Astronomy.* an instrument attached to the eyepiece end of a telescope in order to measure the angular separation and orientation of a visual double star.
bifilar resistor *Electricity.* a resistor wound with a wire that doubles back in order to reduce residual inductance.
bifilar suspension *Engineering.* the suspension of a body by two parallel vertical wires that give a considerable controlling torque.
bifilar transformer *Electricity.* a transformer in which unity coupling is approached by interwinding the primary and secondary coils; the turns of the coils are wound side by side in the same direction.
bifilar winding *Electronics.* a winding used for noninductive coils, in which the current passes through two side-by-side wires, in opposite directions, so that their outer magnetic field is largely balanced.
biflabellate *Entomology.* of or relating to insect antennae that have short joints with long flat processes positioned on opposite sides.
biflagellate [bī flaj´ə lāt´] *Biology.* having two flagella.
bifluoride *Chemistry.* an acid fluoride having the formula MHF_2.
bifocal *Optics.* **1.** having two foci. **2.** describing eyeglasses or contact lenses having two separate sections, one for near vision and one for far.
bifocal lens *Optics.* a lens having two parts, each with a different focal length; used in eyeglasses to correct for both near and far vision.
bifocals *Optics.* eyeglasses having bifocal lenses.
bifoliate *Botany.* having two leaves.
biforate *Biology.* having two pores or openings.
bifunctional monomer *Materials Science.* a monomer that utilizes two active bonds for the polymerization of long chains.
bifurcate [bī´fər kāt´] *Biology.* divided into two branches; forked.
bifurcated contact *Electricity.* a forked contact acting as two contacts in parallel for increased reliability.
bifurcate-merging system *Anthropology.* a kinship system in which there are no distinctions based on certain lineal and collateral relatives, as when a father and the father's brother are both called "father."
bifurcation [bī´fər kā´shən] *Science.* the process of dividing into two parts, or a product of this process. *Chaotic Dynamics.* a charge in the stability or type of solutions of a dissipative dynamical system as a parameter is varied. When a single property of the solutions is plotted versus the parameter, there is a branching of the solutions or a change of stability at the bifurcation point (the critical value of the parameter).
bifurcation diagram *Chaotic Dynamics.* a plot of some characteristics of the stable solutions of a system (such as the peak value of one quantity) versus a parameter.

bifurcation ratio *Hydrology.* a measure of the degree of branching within a drainage basin and equal to the ratio of the number of streams of any given order to the number of streams of the next higher order.

big bang theory *Astronomy.* a widely held theory for the origin of the present universe. According to this theory, about 10 to 20 billion years ago a highly concentrated mass of gaseous matter at a single point in space underwent a gigantic explosion (the **big bang**). The universe then began to expand away from this point of explosion, and the process of expansion continues today, with galaxies moving away from one another at great speed. Also, **Big Bang theory.**

Big Bertha or **Big Berta** *Ordnance.* a nickname for various large German cannons of World War I, especially the 232-mm guns that bombarded Paris at long range. (After *Bertha* Krupp of Germany's Krupp steel works, the manufacturer of such weapons.)

big brown bat see BROWN BAT.

big bud *Plant Pathology.* any of several varieties of plant diseases that result in an abnormal swelling of the bud; common to such plants as currants and tomatoes.

Big Dipper *Astronomy.* a prominent group of seven bright stars in the constellation Ursa Major, resembling a dipper in outline.

bigeminal pulse *Physiology.* a pulse having a sequence of two heartbeats in rapid succession, a pause, two more beats, a pause, and so on; related to the occurrence of ventricular premature beats.

bigeminy *Medicine.* the occurrence of pairs of premature atrial or ventricular heartbeats.

bigeye *Vertebrate Zoology.* one of two widely distributed fishes of the genus *Priacanthus,* having a reddish to silvery color, a short, flattened body, and large eyes.

big gastrin see PROGASTRIN.

bighead *Veterinary Medicine.* **1.** a bulging of the skull bones of an animal, due to osteomalacia. **2.** an acute infectious disease of sheep, characterized by intense edematous swelling of the head, face, and neck. Also, SWELLED HEAD.

bighorn *Vertebrate Zoology.* a wild North American mountain sheep of the family Bovidae, species *Ovis canadensis,* ranging from Canada to Mexico, and characterized by a compact, muscular body, large, ribbed and spiral horns, and sharp, elastic hooves for gripping rocky mountainsides. Also, ROCKY MOUNTAIN SHEEP.

bighorn sheep

bight *Geology.* **1.** a long, gradual bend or gentle curve in a shoreline that forms an open coast or bay. **2.** a bend or curve in a river or in a mountain chain. *Hydrology.* a crescent-shaped indentation formed in an ice edge by currents or winds.

big inch (pipe) *Petroleum Engineering.* a pipeline 24 inches in diameter that carries gas or oil for long distances; first built during World War II, and used to carry crude oil from Texas to Pennsylvania.

bigleaf maple *Botany.* a tree of North America, *Acer macrophyllum,* having large, deeply lobed leaves and fragrant yellow flowers that grow in drooping clusters. Also, OREGON MAPLE.

Bignoniaceae *Botany.* a family of dicotyledonous tropical woody plants, vines, and herbs in the order Scrophulariales, having opposite or whorled leaves and large, showy flowers; included are the bignonia, catalpa, princess tree, and trumpet creeper. (From Abbé *Bignon,* librarian of Louis XIV of France.)

bignum *Computer Science.* a result of arbitrary precision arithmetic. (An acronym for big number.)

biguanide *Organic Chemistry.* $C_2H_7N_5$, crystals; melts at 136°C and decomposes at 142°C; soluble in water and alcohol and insoluble in ether, benzene, and chloroform; used as sulfate to determine the presence of copper and nickel.

biguttulate *Mycology.* of or relating to a fungus with two oil globules in its spores that resemble nuclei.

big vein *Plant Pathology.* a soil-borne viral disease of lettuce that results in swollen, yellow leaf veins and stunted plant growth.

bigwoodite *Petrology.* a medium-grained alkali syenite consisting of microcline, microcline-microperthite, sodic plagioclase, and hornblende, aegirine-augite, or biotite.

Biharian *Geology.* a European geologic stage of the Middle Quaternary era, occurring after the Villafranchian and before the Olderburgian.

biharmonic function *Mathematics.* any solution to the differential equation $\Delta^2 u = 0$, where Δ is the Laplacian operator and u is a function of a finite number of independent variables.

bihor. *Medicine.* during two hours; every two hours. (Short for Latin *bihorium.)* Also, **Bihor.**

bijection *Mathematics.* a one-to-one correspondence between two sets; that is, a function that is both an injection and a surjection.

bijou bottle [bē´joo´] *Biotechnology.* a glass bottle often used in centrifuges for filtration; two bottles can be clamped together mouth to mouth with a filter between them.

bijugate *Botany.* having two pairs of leaflets.

Bijvoet differences *Crystallography.* the differences in the intensities of pairs of Bragg reflections, *I(hkl)* and *I(−h,−k,−l)*,obtained by X-ray diffraction of a crystal containing an anomalously scattering atom. It is experimentally possible to determine the absolute configuration of an optically active molecule by use of this effect.

bikaverin *Organic Chemistry.* $C_{20}H_{14}O_8$, a red crystalline solid that melts at 320–325°C with decomposition. It is a hemolytic agent and an antibiotic with antiprotozoal action.

bilabial *Linguistics.* the sound that is formed by bringing the lips close together, as in the production of *w,* or touching, as in *b* and *p.*

bilabiate *Botany.* having two lips, as the corolla of certain flowers.

bilateral [bī´lat´ər əl] *Biology.* relating to or having two symmetrical sides or parts. *Electronics.* operating in two different directions. *Ordnance.* relating to a system of plotting fire that employs two separate instruments or observers at a distance from each other. Also, **bilateral observation, bilateral spotting.**

bilateral amplifier *Electronics.* an amplifier that can amplify signals traveling in opposite directions; the signal coming into any one of its two ports comes out amplified at the other port.

bilateral antenna *Electromagnetism.* an antenna having a radiation pattern in which the two principal lobes are 180° apart and of equal intensity.

bilateral circuit *Electricity.* a circuit with equipment at opposite ends that is managed, operated, and maintained by different services.

bilateral cleavage *Developmental Biology.* a pattern of division of a zygote that is bilaterally symmetrical.

bilateral descent *Anthropology.* a kinship system in which descent is figured from both parents equally.

bilateral hermaphroditism *Zoology.* a type of hermaphroditism in which there is a testis and an ovary on each side of the body.

bilateral Laplace transform *Mathematics.* a nonconventional generalization of the Laplace transform, in which the path of integration includes both the positive and negative real axis, not just the positive.

bilateral manipulator *Robotics.* **1.** a two-armed manipulator. **2.** a master-slave manipulator with symmetric force reflection.

bilateral network *Electricity.* a network in which the current magnitude remains constant when the voltage polarity is reversed.

bilateral symmetry *Biology.* a condition in which parts are arranged in mirror halves on either side of a central line so that each half is the counterpart of the other, divisible in one plane only. *Botany.* a flower having two seeds that can be divided by a longitudinal plane through the floral axis resulting in a mirror image.

bilateral tolerance *Design Engineering.* the allowable variance above and below the base dimension of a machine part.

bilateral trama *Mycology.* certain tissues located and arranged bilaterally in the gills of mushrooms and other fungi of the class Phrgmobasidiomycetes.

Bilateria *Zoology.* all those phyla of the animal kingdom whose members have developed a taxonomy that is bilaterally symmetrical.

bilayer *Chemistry.* a structure consisting of two layers that are each one molecule thick, such as the phospholipids in cellular membranes.
bile *Physiology.* a secretion of the liver that is stored in the gallbladder and facilitates digestion by emulsifying fats.
bile acid *Biochemistry.* a steroid acid produced by the liver and a component of bile; plays a significant role in digesting fats such as cholic acid.
bile duct *Anatomy.* a tube that carries bile from the liver to the duodenum.
bile pigment *Biochemistry.* any of seven colored components of bile, such as orange (bilirubin) and green (biliverdin); derived from the breakdown of the heme of hemoglobin, myoglobin, and other heme proteins. Also, BILIARY PIGMENT.
bile salt *Biochemistry.* a salt of bile that is produced in the liver and aids in the digestion of fats by emulsification; consists of a bile acid coupled to glycine or taurine.
bilge *Naval Architecture.* **1.** the curved part of the shell joining the bottom and the sides of a hull. **2.** a space or well inside a double-bottomed hull into which seepage drains to be pumped out. Also, **bilge well.** **3.** see BILGE WATER.
bilge board *Naval Architecture.* a covering placed over the bilge to protect the cargo from bilge water and to keep the bilge free of objects that might block the bilge pump.
bilge keel *Naval Architecture.* either of two projecting flanges extending along the bilge of a ship to reduce rolling. Also, **bilge piece.**
bilge keelson *Naval Architecture.* an internal longitudinal member resembling a keelson, but running along a ship's bilge.
bilge pump *Naval Architecture.* a pump used to discharge collected water from the lower portion of a ship's hull.
bilge water *Naval Architecture.* the water, usually foul-smelling, that collects in the lower portion of a ship's hull; a rise in its level may indicate a leak.
bilharzia *Invertebrate Zoology.* a group of flatworms parasitic in the veins or the pelvic region and urinary organs of humans, usually in Africa and other tropic regions.
bilharziasis see SCHISTOSOMIASIS.
biliary of or relating to bile, to bile ducts, or to the gallbladder.
biliary pigment see BILE PIGMENT.
biliary system *Anatomy.* the system in the body that produces and transports bile.
bilicyanin *Biochemistry.* a blue pigment derived from the bile pigments bilirubin and biliverdin and found in gallstones. Also, CHOLECYANIN.
bilification *Physiology.* the formation and secretion of bile.
bilinear [bī´lin´ē ər] *Science.* having or relating to two lines.
bilinear concomitant *Mathematics.* **1.** given a homogeneous differential equation $L(y)$ and its adjoint $L^*(y)$, a bilinear concomitant for L is an expression $B(u,v)$ such that $vL(u) - vL^*(v) = d/dx\,[B(u,v)]$. It can be shown that $B(u,v)$ is linear and homogeneous in $u, u', \ldots, u^{(n-1)}$ and in $v, v', \ldots, v^{(n-1)}$, where exponents denote derivatives. **2.** in invariant theory, a concomitant that is bilinear.
bilinear expression *Mathematics.* an expression that is linear in each of two variables separately.
bilinear function *Mathematics.* a function from the Cartesian product of two vector spaces into a third vector space that is linear in each of its two arguments separately; a multilinear function of two variables. If the range is one-dimensional (i.e., a field), the term **bilinear form** is sometimes used.
bilinear map *Mathematics.* a map $f(x,y)$ that is linear in each of its two variables; i.e., a mapping $f(x,y)$ from the Cartesian product $E \times F$ of two R-modules into R, such that for each $x \in E$, the function f_x that takes y to $f(x, y)$ is linear in y, and for each $y \in F$, the function f_y which takes x to $f(x, y)$ is linear in x.
bilinear transformation see FRACTIONAL LINEAR TRANSFORMATION.
bilinite *Mineralogy.* $Fe^{+2}Fe_2^{+3}(SO_4)_4 \cdot 22H_2O$, a white to yellow, radiating fibrous monoclinic mineral having a specific gravity of 1.87 and a hardness of 2 on the Mohs scale.
bilious [bil´ē əs; bil´yəs] *Pathology.* of or relating to bile or an excess secretion of bile.
biliprotein *Biochemistry.* an organic compound involved in photosynthesis that is found in certain algae; contains the pigments phycocyanin, phycoerythirn, and allophycocyanin conjugated to a protein.
bilirubin *Biochemistry.* $C_{33}H_{36}N_4O_6$, a red-orange bile pigment formed from the bile pigment biliverdin and resulting from heme catabolism (mainly the breakdown of aged red blood cells); high concentrations lead to jaundice.
biliverdin or **biliverdinic acid** *Biochemistry.* $C_{33}H_{34}N_4O_6$, a green bile pigment that results from heme catabolism and is transformed to the bile pigment bilirubin; present in the bile of amphibians, birds, and humans. Also, DEHYDROBILIRUBIN.
bill *Vertebrate Zoology.* the horny part of the mouth of a bird; the beak. *Invertebrate Zoology.* the flattened area of the shell margin of an oyster's broad end. *Mechanical Devices.* a sharp, hooked instrument for pruning or cutting. Also, **billhook.** *Naval Architecture.* the point of a fluke of an anchor.
billabong *Hydrology.* an Australian term referring to a channel leading from a river to a dry stream bed that may be filled seasonally, or to a long, stagnant pool that is the result of a temporary stream overflow.
billboard array *Electromagnetism.* an antenna array of stacked dipoles spaced from 1/4 to 3/4 of a wavelength from a sheet metal reflector.
billet *Military Science.* **1.** a lodging or shelter for troops in a private or nonmilitary public building. **2.** an official order to provide such lodging. *Building Engineering.* a piece of timber in which three sides are sawn and one side is left round. *Metallurgy.* a generic term describing the starting stock of a forging or extrusion operation. *Forestry.* see BOLT.
Billet split lens *Optics.* a lens that has been divided along the optical axis into two halves, with the two halves being slightly displaced; this lens system produces two coherent images and is used in interferometry.
billfish *Vertebrate Zoology.* any of several fishes with a long, sharp bill or snout, such as the gar, needlefish, or saury.
billiard cloth *Textiles.* a wool felt cloth with a closely cropped dress-face finish that makes it smooth and moisture-resistant; used on billiard tables.
Billingsellacea *Paleontology.* a superfamily of articulate brachiopods in the extinct order Orthida and suborder Orthidina, occurring in the lower Cambrian to lower Ordovician; among the first articulate brachiopods, they are important phylogenetically as ancestors of the very diverse and important order Strophomenida.
billion *Mathematics.* **1.** in the system used in the United States and France, a number represented as 10^9, or 1,000,000,000. **2.** in the system used in Great Britain and Germany, 10^{12}.
bill of materials *Design Engineering.* a descriptive listing of all the materials necessary to manufacture an assembly, subassembly, or end-item unit.
billow cloud *Meteorology.* broad, nearly parallel, bands of cloud oriented perpendicular to the wind direction with cloud bases near an inversion surface.
bilobate *Biology.* divided into or having two lobes.
bilobular *Biology.* divided into or having two lobules.
bilocal residence *Anthropology.* a residence pattern in which a couple lives with the husband's or wife's relatives with equal occurrence.
bilocular *Biology.* having two chambers, cells, or cavities.
bilophodont *Zoology.* the grinding teeth, as in a tapir, with ridges joining the two anterior and two posterior cusps.
bimaceral *Geology.* a microscopically visible band of a coal mass consisting of two organic constituents.
bimaculate *Zoology.* marked with two spots.
bimag core see BISTABLE MAGNETIC CORE.
bimanous *Zoology.* having two hands, especially as distinct from feet. Also, **bimanal.**
bimaxillary *Zoology.* of or relating to the two halves of the maxilla, the laterally moving appendages behind the mandibles.
bimetal *Metallurgy.* a material made by bonding two sheets of different metals, each metal with a different coefficient of thermal expansion.
bimetallic *Metallurgy.* relating to or being a bimetal.
bimetallic balance *Horology.* a compensation balance in which a bimetallic strip is used to obtain temperature compensation.
bimetallic corrosion *Metallurgy.* the corrosion caused by the electrolytic dissolution of a less noble metal that is in contact with a more noble metal.
bimetallic strip *Engineering.* a strip consisting of two metals in which one side expands at a different rate than the other; used to control or measure temperature.
bimetal plate or **bimetallic plate** *Graphic Arts.* an extremely durable lithographic plate made of two electroplated metal layers, one ink receptive (such as copper) and the other water receptive (such as aluminum); used for long press runs.
Bimini Islands *Geography.* a group of small islands in the western Bahamas, covering nine square miles; supposed site of the Fountain of Youth. Also, **Biminis.**
bimodal *Statistics.* having two modes, as in a distribution.

bimolecular [bī´mə lek´yə lər] *Chemistry.* consisting of or relating to two molecules.

bimolecular reaction *Chemistry.* the most common form of chemical reaction, occurring when two molecules interact with each other.

bimorph cell *Electronics.* a unit in which two piezoelectric plates are cemented together so that application of voltage causes one to expand and the other to contract, bending the cell in proportion to the applied voltage; used in microphones and vibration detectors. Also, **bimorph, bimorphous cell.**

bimorphemic *Linguistics.* containing two morphemes, as in the words *waited* or *dogs.*

binapacryl *Organic Chemistry.* $C_{15}H_{18}N_2O_6$, a toxic crystalline compound; insoluble in water and soluble in alcohol and acetone; melts at 70°C; used to kill mites and ticks and as a fungicide.

binary [bī´nâr ē; bīn´ə rē] *Science.* composed of or containing two parts or elements. *Computer Programming.* of or relating to a two-state system based on only two possible alternative states such as "on" and "off" or "true" and "false"; often used as a short term for the binary number system that uses only two values, 0 and 1. Groups of binary digits or bits can represent all characters and numerals.

binary alloy *Metallurgy.* an alloy containing only two elements; for instance, cupronickel which contains only copper and nickel.

binary arithmetic operation *Computer Programming.* any arithmetic operation in which all operands and results are in binary notation.

binary cell *Computer Programming.* an electronic cell of one binary digit capacity.

binary chain *Computer Technology.* a series of separate binary circuits existing in one of two possible states and arranged so that each circuit can affect the conditions of adjacent circuits.

binary circuit see DIGITAL CIRCUIT.

binary code *Computer Programming.* a system in which the encoding of any data is done through the use of bits, such as 0 or 1.

binary coded character *Computer Programming.* any alphabetic, numeric, or special character that is represented by a prespecified arrangement of a fixed number of binary digits.

binary coded decimal system *Computer Programming.* a form of numeral representation using binary numerals to represent each decimal digit in a decimal numeral where each decimal digit is separately and individually coded rather than the numeral as a whole.

binary component *Electronics.* a component that can be in either one of two states at any given time. Also, **binary device.**

binary compound *Chemistry.* a term for a chemical compound that consists of only two elements; e.g., sodium chloride (table salt), NaCl; gallium arsenide, GaAs; or boron trifluoride, BF_3.

binary conversion *Computer Technology.* the process of converting a number or other character into or out of a binary representation.

binary decision *Computer Programming.* a choice between only two courses of action.

binary dump *Computer Programming.* a printout of the contents of a computer main memory and registers in binary form, often used during the debugging of a program.

binary encoder *Electronics.* an encoder that transforms angular, linear, or other forms of input into binary-coded numbers.

binary fission *Biology.* a form of asexual cellular reproduction in which the nucleus divides into two cells, essentially equal in size.

binary galaxy *Astronomy.* a pair of galaxies that are bound to each other gravitationally.

binary incremental representation *Computer Programming.* a type of variable representation in which the rate of change in the variable is limited to one of two values at each quantum step: plus one for a positive change and minus one for a negative change.

binary large object *Computer Programming.* in object-oriented information organization and management, a large object stored in binary form.

binary loader *Computer Programming.* a program that transfers an object or machine-readable binary program held on some external storage medium into main memory in a form suitable for execution.

binary logic *Electronics.* an assembly of digital logic components that process binary signals.

binary magnetic core *Solid-State Physics.* a ferromagnetic core that is designed to remain in one of two different stable magnetic states unless switched by an external agent.

binary number *Mathematics.* a number expressed in base 2.

binary number system *Mathematics.* a positional representation for numbers using 2 as the base and only the digits 0 and 1.

binary operation *Computer Programming.* an operation in which one operator involves two operands, such as addition ($X + Y$) and subtraction ($X - Y$); in contrast to a unary operation that involves only one operand, such as a square root. *Mathematics.* any rule for combining two elements of a set to obtain a third element of some set. Examples of binary operations include addition and multiplication, as well as set union and intersection.

binary phase diagram *Materials Science.* a diagram that indicates the phases and their compositions present in a mixture of two pure substances, i.e., elements or compounds, over any range of composition and temperature.

binary picture *Graphic Arts.* a black-and-white illustration with no shades of gray, such as a line drawing.

binary point *Computer Programming.* the point, implied or actual, that separates the integer portion of a binary number from the fractional portion; it is analagous to the decimal point in a decimal number.

binary pulsar *Astronomy.* a pulsar that exists as one member of a binary star system.

binary representation see DYADIC EXPANSION.

binary scaler *Electronics.* a counting circuit whose output appears as a binary coded number.

binary search *Computer Programming.* a method of searching for a particular record by successively testing for the desired record, examining the half of the remaining records in which the desired record would normally be found. Also, DICHOTOMIZING SEARCH.

binary separation *Chemical Engineering.* the separation by distillation or solvent extraction of a miscible liquid mixture of two chemical compounds.

binary signal *Electronics.* voltage or current that conveys information by varying between two values that correspond to the binary values 0 and 1.

binary star *Astronomy.* a system of two stars bound gravitationally and moving in elliptical orbits around a common center of mass. Also, **binary star system.**

binary system *Engineering.* any system involving only two main components, such as 0 and 1 or yes and no. *Mathematics.* a number system using the base 2.

binary-to-decimal conversion *Mathematics.* the process of changing a base 2 or binary representation of a number into a base 10 or decimal representation.

binary touch sensor *Robotics.* a touch sensor that provides binary feedback and sensory input to a robotic controller.

binary tree *Mathematics.* a directed tree such that each vertex is the initial point for at most two edges.

binary weapon *Ordnance.* a weapon in which two chemicals that are separately nontoxic are combined in a shell or bomb to produce a lethal mixture.

binary word *Computer Technology.* a fixed-length group of bits that occupies one storage location in a computer memory and is treated and transferred as a unit.

binate *Botany.* produced or growing in pairs or couples.

binaural [bī nôr´əl] *Medicine.* **1.** having two ears. **2.** relating to or affecting both ears. *Acoustics.* relating to or producing binaural sound.

binaural hearing *Physiology.* the perception of a sound arriving simultaneously at the two ears.

binaural intensity effect *Acoustics.* an effect in which the angle between the direction of sound and the median plane that bisects the line joining both ears is proportional to the logarithm of the ratio of the sound intensities arriving at each ear separately, provided that the sound at each ear is of the same frequency and phase.

binaural phase effect *Acoustics.* an apparent shift or displacement in the direction of a sound source that results from a change in the phases of two otherwise identical sources; the angular displacement is proportional to the amount of phase shift.

binaural sound *Acoustics.* sound recorded through two separate microphones and transmitted through two separate channels to produce a stereophonic effect.

binche lace *Textiles.* a bobbin lace with flat floral scroll designs on a coarse mesh background interspersed by scattered snowflake motifs.

bin cooler *Food Technology.* a covered, vented box or chamber in which dried food is cooled between drying and packing.

bind *Textiles.* to loop one stitch over another to form a strong edge on a knitted fabric. *Medicine.* to form a weak, reversible chemical bond, such as an antigen to an antibody or a hormone to a receptor. *Mining Engineering.* see BATT.

binder *Materials Science.* any substance that causes a mixture to adhere. *Metallurgy.* any additive that increases the green strength of the powder compact or facilitates sintering. *Building Engineering.* a girder that supports the ends of two sets of floor joists. *Geology.* a mineral cement that holds loosely aggregated sediments together by filling in the spaces between grains. *Agriculture.* an attachment to a reaper that binds cut grain.

binder course *Mining Engineering.* any coarse bituminous aggregate containing a small percentage of asphalt; used as an intermediate connecting link between the concrete foundation and the top wearing course of asphalt pavement. *Civil Engineering.* a progression of masonry unit sandwiched between two walls in order to bind them.

binderless briquetting *Engineering.* the briquetting of coal by the application of pressure without the addition of binders.

binder's board *Graphic Arts.* a very high quality paperboard used in case-made book covers, usually covered with woven cloth.

binder's die see STAMPING DIE.

bindheimite *Mineralogy.* $Pb_2Sb_2O_6(O,OH)$, a gray to brownish or yellowish cubic mineral, occurring as reniform masses, having a specific gravity of 4.6 to 7.32 and a hardness of 4 to 4.5 on the Mohs scale; found as a secondary mineral in oxidized zones of lead-antimony ore deposits. (Named after J.J. *Bindheim,* German chemist.)

B indicator see B DISPLAY.

bindin *Endocrinology.* a glycoprotein that is formed in sea urchin spermatozoa and believed to be involved in the sperm-egg recognition necessary for fertilization.

binding *Graphic Arts.* **1.** the process of folding, gathering, and fastening together the printed signatures or sections of a book and securing them in a cover. **2.** the product of this process. *Computer Programming.* the connection between a variable or expression and its current value.

binding energy *Physics.* **1.** the minimum amount of energy required to extract an individual particle from a system of particles. **2.** the minimum amount of energy required to disassemble a system of particles and separate them at infinite distances so that there is no interaction. Also, BOND ENERGY, TOTAL BINDING ENERGY. *Physical Chemistry.* see BOND ENERGY. *Nuclear Physics.* the difference, expressed in units of energy according to the mass-energy equivalence relation $E = mc^2$, between the isotopic mass of an atom and the sum of the masses of its constituent protons, neutrons, and orbital electrons.

binding paper *Graphic Arts.* paper that is stretched over boards to form covers for hardcover books.

binding screw *Electricity.* a screw for holding a conductor to a terminal. Also, **binding post.**

binding site see ACTIVE SITE.

binding strake *Naval Architecture.* a strong heavy strake of planking, especially one next to a sheer strake.

binding time *Computer Programming.* the moment at which a symbolic expression in a program is reduced to a form that can be directly interpreted.

bindweed *Botany.* a perennial weedy plant that has a slender, twining stem, belonging to the morning-glory family, Convolvulaceae, and comprising the genus *Convolvulus.*

bine *Botany.* the slender, twining stem of certain climbing plants, especially the hop. Also, **bind.**

Binet, Alfred [bi nā´] 1857–1911, French psychologist.

Binet age *Psychology.* a child's mental age as computed from his or her score on the Binet-Simon intelligence scale.

Binet's formula *Psychology.* a maxim stating that children under the age of nine who are mentally underdeveloped by two years are possibly mentally deficient, while those over the age of nine who are mentally underdeveloped by three years are definitely mentally deficient

Binet-Simon (intelligence) scale *Psychology.* a system of rating the intellectual development of children; from test scores, the child's mental age can be computed and compared to the chronological age. Also, **Binet-Simon (intelligence) test.** (From its developers, Alfred *Binet* and Theodore *Simon.*)

Bingham equation *Materials Science.* an empirical mechanical relationship for unfired ceramic bodies, in which shear strain rate is linear with stress above a critical level.

Bingham number *Fluid Mechanics.* a dimensionless group that indicates the magnitude of yield stress relative to viscous stress for Bingham plastic fluids.

Bingham plastic *Fluid Mechanics.* a plastic non-Newtonian material that deforms elastically but does not flow until a critical shear stress is reached; beyond that shear stress, the shear stress increases linearly with the rate of deformation.

bing ore *Mineralogy.* **1.** the purest ore of lead, containing the largest crystals of galena. **2.** an unassimilated heap or pile of anything, especially ore or slag.

binistor *Electronics.* a semiconductor device having two stable states used in switching and storage circuits, depending largely on an external voltage supply for its negative-resistance characteristic.

binnacle *Naval Architecture.* the housing of a ship's steering compass, usually with storage space for the logbook and other materials needed by the watch officer; originally called a bittacle.

Binnig, Gerd born 1947, German physicist; with Rohrer, received the Nobel Prize for development of scanning tunneling microscope.

binocular *Optics.* **1.** describing an optical instrument that requires the use of both eyes simultaneously, to give a stereoscopic effect to an image or to enhance observation. **2.** see BINOCULARS. *Biology.* **1.** having to do with or the use of both eyes simultaneously. **2.** of or relating to visual perception of depth based on the angular difference between retinal images.

binocular accommodation *Physiology.* the changes in the lenses of each eye in binocular vision that are needed to bring the object of fixation into clear focus.

binocular microscope *Optics.* a compound microscope that presents a separate image to each of the observer's eyes simultaneously, thereby yielding stereoscopic viewing.

binoculars *Optics.* an optical instrument that allows the magnified viewing of distant objects with good depth perception, consisting of two small, identical telescopes joined together side by side, so that the viewer can look through one with each eye.

binocular tube *Optics.* a viewing tube that is designed to accommodate two eyepieces, for binocular viewing.

binocular vision *Physiology.* the use of both eyes together without diplopia (double vision).

binodal curve *Materials Science.* the molecular weight dependent curve representing equilibrium between phases in a multicomponent polymer system.

binode *Electronics.* a thermionic double diode having one cathode and two anodes.

binomen *Systematics.* the two-word phrase that is used as the scientific name of a species; it is composed of the genus name, which is capitalized, and the species name, which is usually lowercase. Also, BINOMIAL NAME.

binomial *Mathematics.* a polynomial with exactly two terms. *Electromagnetism.* a bidirectional antenna array designed for suppressed minor lobes and oppositely directed principal lobes.

binomial array see PASCAL'S TRIANGLE.

binomial coefficient *Mathematics.* combinatorially, the number of ways of choosing k objects from n objects without regard to order. The numbers $\binom{r}{k}$, defined for all real numbers r and for all integers $k \geq 0$, by $\binom{r}{k} = 1$ for $k = 0$ and $\binom{r}{k} = r(r-1)(r-2)\cdots(r-k+1)/k!$ for $k > 0$. If r is a nonnegative integer and k is an integer such that $0 \leq k \leq r$, then $\binom{r}{k} = r!/(r-k)!k!$. The binomial coefficients satisfy the recursion relation $\binom{r}{k} = \binom{r-1}{k} + \binom{r-1}{k-1}$ for all real r and for all positive integers k, and the symmetry relation $\binom{r}{k} = \binom{r}{r-k}$ for all nonnegative integers r and $k \leq r$, as well as numerous other identities. These are called binomial coefficients because they are the coefficients in the expansion of $(+y)^r$. When n is a positive integer and $0 \leq k \leq n$, $\binom{r}{k}$ make up the entries in Pascal's triangle. Also denoted by $C_{r,k}$.

binomial differential *Mathematics.* a differential having the form $x^m(a + bx^n)^p dx$, where m, n, and p are rational and a and b are constants.

binomial distribution *Statistics.* the probability distribution of the number X of successes in n identical and independent Bernoulli trials: $P(X = x) = C_{n,x}p^x(1-p)^{n-x}$ for $x = 0, 1, 2, \ldots, n$, where p is the probability of success in each trial and $C_{n,x}$ is the binomial coefficient. Also, BERNOULLI DISTRIBUTION.

binomial equation *Mathematics.* an equation of the form $x^n - a = 0$.

binomial law *Mathematics.* if p is the probability that a given event will occur in a Bernoulli trial, the binomial law states that the probability of the event occurring k times in n independent trials is $\binom{n}{k}p^k(1-p)^{n-k}$.

binomial name see BINOMEN.

binomial nomenclature *Systematics.* the system of nomenclature, devised by Linnaeus, in which the name of an organism is made up of two words, a genus name and a species name.

binomial series *Mathematics.* the infinite series expansion of $(x + y)^r = \Sigma_{k=0}^{\infty}\binom{r}{k}x^{r-k}y^k$ for any real number r. This converges to $(x + y)^r$ if $|y/x| < 1$ or if r is a nonnegative integer. In the case that r is a nonnegative integer, the series has only finitely many nonzero coefficients.

binomial surd *Mathematics.* a binomial of the form $ac^{1/m} + bd^{1/n}$ or $a + bd^{1/n}$, where m and n are integers greater than 1 and the indicated roots are irrational numbers. **Conjugate binomial surds** are pairs of the form $a + bd^{1/n}$ and $a - bd^{1/n}$ or $ac^{1/m} + bd^{1/n}$ and $ac^{1/m} - bd^{1/n}$.

binomial theorem *Mathematics.* for an integer $n \geq 1$ and a and b nonzero real numbers, $(a + b)^n = \sum_{k=1}^{n} \binom{n}{k} a^k b^{n-k}$.

binomial trials *Statistics.* a sequence of trials in which the particular result for each trial may or may not happen. Also, BERNOULLI PROCESS.

binormal *Mathematics.* the binormal to a curve at a point P in Euclidean 3-space is the vector passing through P that is normal to the osculating plane of the curve at P. The direction of the vector is chosen so that, together with the positive tangent and principal normal to the curve at P, it forms a right-handed Cartesian system.

bin sort see RADIX SORT.

binucleate *Cell Biology.* having two nuclei.

binucleolate *Cell Biology.* containing two nucleoli.

bio- a combining form meaning "life" or "living organism," as in *biomechanics, bioluminescence.*

bioacoustics *Behavior.* the study of sound production and reception in organisms other than humans. *Biology.* the study of the effects of sounds on living things.

bioaeration *Biotechnology.* a system of purifying sewage by oxidation, in which crude sewage is passed through special centrifugal pumps.

bioaffinity sensor *Biotechnology.* a sensor in which molecular recognition generates a biochemical signal, so that immobolized hormone receptors or antibodies can be used to detect hormones or antigens.

bioaltruism see ALTRUISM.

bioamine see BIOGENIC AMINE.

bioanthropology see BIOLOGICAL ANTHROPOLOGY.

bioassay *Analytical Chemistry.* a determination of the concentration of a substance by its effect on the growth of a test organism under controlled conditions. *Toxicology.* the use of a living organism or cell culture to test for the presence of a substance. *Virology.* a test to quantify a virus by measuring its infectivity for a host.

bioastronautics *Biology.* the study of the effects of space and interplanetary travel on living organisms.

bioautography *Analytical Chemistry.* a bioassay of certain compounds, usually antibiotics or vitamins, by evaluating their ability to enhance the growth of some organism and to repress that of others.

bioavailability *Nutrition.* the extent to which a drug or other substance becomes available to the target tissue after administration.

Bio-beads S *Biotechnology.* a polystyrene support material used to fractionate molecular compounds in gel filtration chromatography with lipophilic solvents.

bioblast *Biology.* **1.** a formative cell. **2.** an elementary unit of protoplasmic structure.

bioburden *Microbiology.* the number of contaminating organisms found on a given amount of material prior to undergoing an industrial sterilization procedure.

biocatalyst *Biochemistry.* a catalyst involved in chemical processes of living organisms; e.g., an enzyme.

biocellate *Biology.* having two ocelli or eyelike parts.

biocen see BIOCOEN.

biocenology [bī´ō sə näl´ə jē] *Ecology.* the study of communities of organisms and of the relationship among the members of such communities. Also, CENOBIOLOGY, BIOCOENOLOGY.

biocenosis [bī´ō sə nō´sis] *Ecology.* a self-sufficient community of organisms naturally occupying and interacting within a specific biotope. Also, BIOCOENOSIS. *Paleontology.* a taphonomic association of a functional community of fossils.

biochemical *Chemistry.* of or relating to biochemistry.

biochemical deposit *Geology.* a sedimentary rock deposit, such as bacterial iron ore, resulting directly or indirectly from the chemical actions and vital activities of living organisms. Also, **biochemical rock.**

biochemical fuel cell *Electricity.* an electrochemical energy source in which electricity is generated chemically by the oxidation of biological substances.

biochemical oxygen demand *Microbiology.* the amount of oxygen required by aerobic organisms to carry out oxidative metabolism in water containing organic matter, such as sewage. Also, BIOLOGICAL OXYGEN DEMAND.

biochemical oxygen demand test *Biochemistry.* a test measuring the amount of oxygen consumed per liter of water containing organic matter, such as sewage, after a five-day incubation at 20°C. Also, BIOLOGICAL OXYGEN DEMAND TEST, BOD TEST.

Biochemistry

Biochemistry embraces the chemical reactions of living cells. It is based on the proposition that virtually all of life can be understood as chemistry. Standing between biology and chemistry, the field of biochemistry interdigitates with more than twenty disciplines named in this Dictionary and relates to all branches of chemistry and biology, ranging from genetics to physical chemistry, from medicine to agriculture, from nutrition to biotechnology.

From its early origins in organic chemistry and physiology in the mid-19th century, modern biochemistry emerged at the turn of the 20th century with the recognition that a complex cellular process could be observed in the juices extracted from the cell. The fermentation of sugar to alcohol by a yeast cell was reconstituted by a succession of twelve enzyme-catalyzed reactions which capture the fuel energy of sugar for cell growth; essentially the same series of reactions in a muscle cell supplies the energy for its contraction. The universality of biochemistry throughout Nature, displayed in all aspects of metabolism, is a triumph of evolution and one of the great revelations of this century.

Recent advances in genetic chemistry have profoundly affected all the biosciences and hastened their coalescence. All these disciplines now communicate in the language of chemistry. Applications of the new knowledge, using recombinant DNA technology, have revolutionized the diagnosis, treatment and prevention of diseases. Previously unavailable hormones, interferons and vaccines are in clinical use. Advances in agriculture and industry based on biochemistry are imminent. Mind as well as matter may soon be the subject of biochemistry.

Arthur Kornberg
Professor of Biochemistry
Stanford University Medical Center
Nobel Laureate in Medicine

biochemistry *Chemistry.* the science concerned with all aspects of the chemistry of living organisms.

biochemorphology *Biochemistry.* the study of the chemical structure of nutrients and pharmaceuticals and their effect on living organisms.

biochore *Ecology.* a group of similar biotopes.

biochrome *Biochemistry.* a natural pigment that is produced by living organisms.

biochron *Paleontology.* a relatively short-lived fossil flora or fauna.

biochronology *Geology.* the study of the age of the earth based on the relative dating of rocks and geologic events by the use of fossil evidence.

biocide *Ecology.* any substance that is toxic or lethal to living organisms, such as a pesticide, herbicide, or fungicide.

bioclastic rock *Petrology.* a rock composed of broken or calcareous remains of organisms, such as limestone composed of shell fragments. *Geology.* a rock composed mainly of fragments of preexisting rock that have broken off as a result of the actions of living organisms.

bioclimatology *Ecology.* the study of the relationship between climate and the activities and characteristics of plants and animals. Also, **bioclimatics.**

biocoen [bī´ə sen´] *Ecology.* the sum of all the living components of an environment or habitat. Also, BIOCEN.

biocoenology see BIOCENOLOGY.

biocoenosis see BIOCENOSIS.

biocompatibility *Biology.* the capability of being harmonious with biological life without causing toxic or injurious effects.

biocontrol see BIOLOGICAL CONTROL.

biocultural functionalism *Anthropology.* a theory of Bronislaw Malinowski proposing that culture and its various institutions are responses to the fulfillment of human biological and psychological needs.

biocybernetics [bī´ō sī´bər net´iks] *Biology.* the study of communication and control in living organisms, especially physiological feedback mechanisms and central nervous system control.

biocycle *Ecology.* the rhythmic repetition of certain phenomena observed in living organisms.

biocytin *Biochemistry.* $C_{16}H_{28}N_4O_4S$, a naturally occurring complex of biotin and lysine. Also, BIOTIN COMPLEX OF YEAST.

biocytinase *Enzymology.* an enzyme found in the blood and liver that changes biocytin into biotin and lysine.

biodegradable *Biology.* susceptible of decomposition by natural biological processes, as by the action of bacteria, plants, and animals.

biodegradation *Toxicology.* the destruction of one or more chemicals by biological processes.

biodiversity *Zoology.* the existence of a wide range of different species in a given area or during a specific period of time.

biodynamics *Biology.* the study of the nature and determinants of all organismic (including human) behavior.

bioecology *Ecology.* the study of the relationships of organisms to their natural environments.

bioelectric current *Physiology.* a current that flows between parts of a cell membrane having different electrical potential.

bioelectricity *Physiology.* the presence of electrical currents that arise and flow within muscular and neural tissues.

bioelectric model *Physiology.* a representation, in the form of an electrical current, of the electrical energy flow in living tissues.

bioelectrochemistry *Physiology.* the use of techniques, tools, and knowledge gained in the study of the electrochemistry and physiology of living organisms.

bioelectronics *Biology.* the study of the role of intermolecular transfer of electrons in biological regulation and defense.

bioenergetics *Biochemistry.* the study of the transformations of energy in living organisms.

bioengineering *Biotechnology.* the science that specializes in the manufacture of artificial replacements for various parts or organs of the body. *Chemical Engineering.* the application of engineering methods for achieving biosynthesis of animal and plant products, such as fermentation processes. *Engineering.* any of various other applications of engineering methods and technology to the fields of medicine or biology.

bioequivalence *Pharmacology.* the condition of having the same strength and similar bioavailability in the same dosage as another specimen of a given substance. Thus, **bioequivalent.**

biofacies [bī´ō fā´sēz] *Geology.* **1.** a subdivision of a stratigraphic unit that is distinguished from adjacent subdivisions solely on the basis of its fossils. **2.** a lateral variation in the biologic aspect of a stratigraphic unit, considered to be an expression of local biologic conditions.

biofeedback *Psychology.* information about certain physiological functions, such as pulse or blood pressure, that is supplied to an individual in order to help him or her learn how to affect or control those functions.

biofilm *Biotechnology.* a method of cell immobilization in which a microbe population grows in a thin layer of a living or nonliving surface.

bioflavonoid *Biochemistry.* a group of compounds widely occurring in plants, such as citrus fruits, and acting to maintain cell wall permeability of small blood vessels. Also, CITRUS FLAVONOID COMPOUND, VITAMIN P.

biofog *Meteorology.* a steam fog caused by contact between very cold air and the warm moist air that surrounds human or animal bodies.

biofuel *Biotechnology.* a solid, liquid, or gaseous fuel that is obtained from biological raw material; the conversion is accomplished through thermochemical or biological methods.

biogas *Biotechnology.* a mixture of methane and carbon dioxide along with traces of other gases, such as nitrogen, hydrogen, and water vapor, that is produced during anaerobic digestion.

biogenesis [bī´ō jen´ə sis] *Biology.* the theory that living matter always arises from preexisting living matter, and not through spontaneous generation. Thus, **biogenetic.**

biogenetic law *Biology.* a theory claiming that the development of the animal embryo and young traces the evolutionary development of the species. Also, RECAPITULATION THEORY.

biogenic [bī´ō jen´ik] *Biology.* **1.** resulting from the actions of living organisms. **2.** necessary for life processes.

biogenic amine *Neurology.* a class of chemical substances, such as amines, generated by cells that alter cerebral, vascular, and other body functions. Also, BIOAMINE.

biogenic amine hypothesis *Neurology.* a theory that biogenic amines play a significant role in the development of major affective disorders and schizophrenia.

biogenic chert *Petrology.* chert derived from the shells of pelagic silica-secreting organisms.

biogenic reef *Geology.* a moundlike layered structure built by and composed predominantly of organic remains such as shells and skeletons of sedentary organisms.

biogenic rock *Geology.* an organic rock produced by the physiological activities of plant or animal organisms. Also, **biogenetic rock, biogenous rock.**

biogenic sediment *Geology.* a sediment that directly results from the physiological activities of organisms.

biogenous [bī´äj´ə nəs] *Biology.* originating from life or producing life.

biogeochemical cycle *Geochemistry.* the circulation of chemical components through the biosphere from or to the lithosphere, atmosphere, and hydrosphere.

biogeochemical prospecting *Geochemistry.* the chemical analysis of plants in a particular area, in order to detect concentrations of certain elements that might reflect the presence of hidden mineral deposits.

biogeochemistry *Geochemistry.* a branch of geochemistry dealing with the interactions and relationships between plant and animal organisms and the global distribution of chemical elements.

biogeography *Ecology.* the study of the distribution of plant and animal life in the earth's environment, and of the biological and historical factors that produced this distribution. Also, AREOGEOGRAPHY.

bioglasses see SURFACE ACTIVE GLASSES.

bioglyph see TRACE FOSSIL.

biohazard *Ecology.* any risk or harm that results from exposure to infectious bacteria, viruses, or other harmful agents or their products, particularly those found in a clinical microbiology laboratory or used in genetic recombination studies.

bioherm *Geology.* a mound, dome, or reeflike mass of rock that is composed almost exclusively of the remains of sedentary marine organisms and is embedded in rock of different physical character.

biohermite *Petrology.* any limestone formed of debris from a bioherm, often found in reef cores.

biohydrology *Ecology.* the study of the interaction between plant and animal life and the water cycle.

bioimplant *Surgery.* a prosthesis made of biosynthetic material.

bioinstrumentation *Engineering.* the use of instruments, such as sensors, to record and transmit physiological data from living humans and animals.

biokinetics *Developmental Biology.* the study of the movements within developing organisms.

biol. biology; biologic; biological; biologist.

biolinguistics *Linguistics.* the study of language functions as related to or derived from biological characteristics of an organism.

biolite *Mineralogy.* any rock or group of minerals formed from organic material or by the action of organisms.

biolith *Petrology.* a rock that is formed from or by organic material, either by noncombustible, inorganic processes that form an acaustobiolith, or by combustible, organic processes that form a caustobiolith Also, **biolite.**

biolithite *Petrology.* any limestone of an organic nature such as a reef rock or stromatolite.

biological *Biology.* of or relating to living organisms or life processes. *Pharmacology.* any medicinal preparations made from living organisms and their products, such as serums, vaccines, antigens, and antitoxins.

biological agent *Ordnance.* a virus, microorganism, or toxic organic substance that is used in warfare to cause death or disease.

biological anthropology *Anthropology.* a branch of anthropology that deals with humans as biological organisms, including areas such as primatology, human genetics, human ecology, paleoanthropology, and fields of applied anthropology such as anthropometrics and forensic anthropology. Also, PHYSICAL ANTHROPOLOGY, BIOANTHROPOLOGY.

biological clock *Biology.* an internal physiological mechanism that controls certain biological rhythms in plants and animals, including humans, such as metabolic changes, sleep cycles, and photosynthesis.

biological community *Ecology.* all the organisms inhabiting a given area.

biological control *Ecology.* the deliberate use by humans of one species of organism to eliminate or control another, as in the control of undesirable plants or insects by the use of natural parasites, diseases, or predators, rather than by herbicides and pesticides. Also, BIOCONTROL.

Biology

Biology is the study of the properties and history of living organisms and of their interactions with the non-living world. Beginning in the 17th Century with Descartes' machine model of the organism, biologists have viewed living organisms as physico-chemical machines to be understood by breaking them down into component parts and elementary functions. Each level of organization of a living organism is explained by reference to the properties of lower levels. So, the properties of populations are seen as the aggregate of the properties of individuals. Individuals in turn are understood as assemblages of organs or cells with different physiological properties which, in turn, derive from the chemical properties of the molecules of which they are composed.

Using physics as a model, biologists have attempted to find universal processes and properties of all organisms that can be applied to all forms, despite the apparent diversity of life. So, Mendel's laws of inheritance and the molecular processes of gene duplication and function, processes of cell division, the movement of cells and the movement of tissues in development, and evolutionary change by natural selection are all seen as universal processes that apply in detail to all of life.

Biological systems differ from simpler physical systems, however, in three important respects. First, they are intermediate in size between atoms and celestial bodies and are, therefore, in a range of physical phenomena that does not allow simple rules of behavior such as Kepler's Laws of Motion for planets or the Wave Equation for particles. Second, living organisms are the nexus of very large numbers of different effective causal pathways as a consequence both of their size and their complex multi-level structure. Third, living organisms have evolved and all of life on earth is the result of a single historical sequence. Since nearly none of the kinds of organisms that might have arisen from different combinations of genes have in fact ever existed, actual organisms have been a historically contingent, partly accidental, assemblage of objects. So, if biology has any parallel with physics, it is with hydrodynamics, rather than with atomic physics.

There are no completely universal rules of biological organisms. While nearly all organisms share the same genetic code that specifies proteins from DNA, the code is not absolutely universal and some organisms like Trypanosoma have alternative code elements presumably as an accident of history. None of Mendel's "laws" is universal and many exceptions exist to Mendelian inheritance. Not all aspects of organisms are the result of natural selection for better function. For many aspects of organisms, the question of their function is inappropriate. They are products of random mutation and random chance fixation of these mutations in the species. Even maladaptive characteristics can be fixed in a species by chance.

Living organisms are affected by and affect the non-living world strongly. The evolution of organisms has caused an evolution of the earth's surface properties and the chemical composition of its atmosphere. Therefore, the study of the history of the earth and its atmosphere is inseparable from the study of biology.

R. C. Lewontin
Alexander Agassiz Professor of Zoology
Museum of Comparative Zoology, Harvard University

biological corrosion *Metallurgy.* the corrosion of a metallic component in an oxygen-free environment caused by anaerobic bacteria.

biological engineering *Biology.* the artificial selection of different strains of plant or animal species to improve an organism, especially an agriculturally useful plant or animal.

biological half-life *Physiology.* the time required by a living body to eliminate one-half of the quantity of an administered substance, such as a radioisotope, through normal metabolic channels.

biological magnification *Ecology.* an increasing concentration of toxic substances within each successive link of the food chain. Also, **biological amplification, biomagnification.**

biological oceanography *Oceanography.* the study of oceanic plant and animal life in relation to the marine environment.

biological oxidation *Biochemistry.* the aerobic biochemical processes in which food is metabolized, hydrogen atoms are transferred from one molecule to another, and ATP is synthesized via oxidative phosphorylation, thus releasing energy.

biological oxygen demand see BIOCHEMICAL OXYGEN DEMAND.

biological response modifiers *Immunology.* any proteins, such as interferons or interleukins, that modify an immune or other biological response, usually enhancing it.

biological shield *Nucleonics.* a structure of dense material, such as concrete or lead, around a nuclear reactor to protect against radiation.

biological specificity *Biology.* the orderly patterns of developmental and metabolic reactions that characterize an individual and its species.

biological therapy *Psychology.* any form of treatment for abnormal behavior that alters the individual's physiological processes, such as electric shock treatment or surgery.

biological value *Biochemistry.* a relative measurement of a protein's nutritional value, and thus of how it can function to maintain and aid in the growth of an individual's body tissues; usually measured in terms of egg protein, which has the highest value (BV = 0.9–1.00).

biological warfare *Military Science.* tactics and strategy in warfare that involve the use of biological agents.

biologist an expert or specialist in biology.

biology *Science.* the scientific study of living organisms and their interactions with the nonliving world.

bioluminescence *Biology.* the emission of visible light caused by catalytic chemical reactions that produce a form of electricity in certain organisms, such as glowworms, fireflies, jellyfish, and some fungi.

biolysis *Biology.* the chemical decomposition of organic matter by the action of living organisms.

biolytic *Biology.* destructive to life.

biomarker *Toxicology.* a physiological or pharmacological measure that is used to predict a toxic event in an animal.

biomass fuel machine

biomass *Ecology.* **1.** plant material or vegetation that can be converted to useful fuel and that is considered as a potential energy source. **2.** a quantitative estimate of the entire assemblage of living organisms, both animal and vegetable, of a given habitat, considered collectively and measured in terms of mass, volume, or energy in calories.

biomaterial *Surgery.* any nondrug substance that can be used as a system or part of a system to treat, enhance, or replace any tissue, organ, or function of the body.

biomathematics *Biology.* the application of mathematical methods to biology and medicine.

biome *Ecology.* a recognizable complex biotic community of a given region, produced by the interaction of climatic factors, living organisms, and substrate; especially a community that has developed to climax vegetation, such as tundra, coniferous forest, or grassland.

biomechanics *Biology.* the application of mechanical laws to living structures, particularly to the locomotor system of the human body.

biomedical engineering *Biotechnology.* the use of engineering methods, instrumentation, and technology to solve medical problems, including the manufacture of artificial limbs and organs, the design and construction of hospitals, the development of community health programs, and the study of ways to control the environment.

Biomedical Engineering

Biomedical engineering is the application of the methods of engineering analysis to solve problems in clinical medicine and the life sciences. For example, electrical engineers design cardiac pacemakers to stimulate the heart when the natural pacemaker fails. They design probes that beam ultrasound into the body to determine the position of the fetus or to determine the quantity of blood flow. Computer engineers design programs that monitor and analyze the heart rate, blood pressure, and temperature of patients in the intensive care ward.

Chemical engineers design dialysis systems that cleanse the blood of waste products when kidney function fails. To replace blocked blood vessels, they design replacement vessels of special plastics that inhibit clotting. Mechanical engineers design powered wheelchairs and other support surfaces for patients with spinal cord injury. Clinical engineers manage the increasingly complex equipment in the hospital so it is effective and safe. Since biomedical engineers learn physiology and medicine, they can communicate with physicians, evaluate patients' needs, and develop appropriate devices to alleviate the disease.

For the life sciences, engineers develop instruments to improve our understanding of the working of the body. Improved electrodes probe into the brain and improved computer techniques analyze the brain signals to unravel the intricacies of how the brain works. Engineers also develop mathematical models of the functioning of tiny nerve cells and large organs such as the heart, lungs, and kidneys. These mathematical models permit analysis and predictions of the outcome of therapy, for example, during the use of drugs to control blood pressure.

John G. Webster
Professor of Electrical & Computer Engineering
University of Wisconsin at Madison

biomedical photogrammetry see BIOSTEREOMETRICS.

biometeorology *Ecology.* the study of the effects on living organisms of both natural and artificial atmospheric phenomena, such as temperature, humidity, barometric pressure, rate of air flow, and air ionization.

biometer *Biology.* an instrument by which minute quantities of carbon dioxide can be measured; used in measuring the carbon dioxide given off from functioning tissue.

biometrical genetics *Genetics.* the mathematical analysis of the inheritance of variable phenotypic characteristics in animal and plant breeding.

biometrics *Statistics.* **1.** the statistical study of biological observations and events. **2.** the calculation of life expectancy. Also, **biometry.**

biomicrite *Petrology.* a limestone composed of skeletal remains in a matrix of carbonate mud.

biomicroscopy *Pathology.* **1.** the microscopic examination of living tissue in the body. **2.** the examination of the cornea or the lens using a combination of a slit lamp and a corneal microscope.

biomicrosparite *Petrology.* **1.** biomicrite in which the carbonate-mud matrix has recrystallized to microspar. **2.** microsparite containing fossil fragments or fossils.

biomicrudite *Petrology.* biomicrite containing fragments or whole fossils greater than 1 millimeter in diameter.

biomimetic catalyst *Organic Chemistry.* a model synthetic catalyst that mimics natural organic processes at ambient conditions.

biomimetics *Biochemistry.* a branch of biology that uses information from biological systems to develop synthetic systems.

biomolecule *Biochemistry.* a compound that occurs as a component of a living organism, such as a protein or nucleic acid.

bion *Biology.* an individual living organism. *Ecology.* see BIONT.

bion. bionic.

bionavigation *Vertebrate Zoology.* the instinctual ability of some animals to return to a given site without the use of landmarks, as birds to their roost or salmon to spawning grounds.

bionic *Engineering.* **1.** of or relating to bionics. **2.** in popular use, describing electronic or mechanical devices that replace human body parts, especially so as to greatly enhance strength or performance.

bionics *Engineering.* the study of the functions, characteristics, and phenomena of the living world, and the application of this knowledge to create and improve electronic devices and mechanical parts.

biont *Ecology.* a term for a single individual organism. Also, BION.

bionucleonics *Radiology.* the study of the application of radioactive and rare stable isotopes to biologic systems.

biopack *Engineering.* a container used to house a living organism during space flight and to monitor physiological functions. Also, **biopak.**

biopelite see BLACK SHALE.

biopelmicrite *Petrology.* a limestone similar to biopelsparite except that the microcrystalline matrix exceeds calcite cement.

biopelsparite *Petrology.* a limestone that is intermediate in content between biosparite and pelsparite, and is similar to biosparite except that the ratio of fossils and fossil fragments to pellets ranges between 3:1 and 1:3.

biophagous *Biology.* feeding on living organisms.

biophile *Biochemistry.* any element occurring in living organisms or organic matter, such as carbon, oxygen, or nitrogen.

biophysics *Biology.* the study of biological structures and processes in terms of the principles of physics.

bioplast *Biology.* **1.** a functional independent mass of living protoplasm. **2.** an ameboid cell.

biopolymer *Biochemistry.* a macromolecule occurring in a living organism such as a protein, polysaccharide, or nucleic acid.

biopotency *Biochemistry.* the capability of or extent to which a chemical substance, such as a hormone, acts within a biological system.

biopotential *Biology.* the quiescent form of energy in biological material before it is transformed into an active state that results in the manifestations of life.

bioprocess *Biology.* a method used to prepare a biological product, particularly one of genetic engineering, for commercial use.

biopsy *Pathology.* **1.** the removal and microscopic examination of tissue taken from the living body and performed to establish an exact diagnosis. **2.** a specimen removed by biopsy.

biopterin *Biochemistry.* 2-amino-4-hydroxy-6-(1',2'-dihydroxypropyl)-pteridine, a pteridine derivative for which the reduced form functions as a coenzyme in hydroxylation reactions of amino acids; also functions as a growth factor in the absence of high concentrations of folic acid.

bioptic *Surgery.* of or relating to a biopsy.

bioptome *Surgery.* a cutting instrument used to take biopsy specimens.

biopyribole *Mineralogy.* a general term for rocks containing biotite, pyroxene, or amphibole, or for those minerals themselves.

bioreactor *Biotechnology.* any of a variety of small-scale or large-scale closed containers used for both fermentation and enzyme reactions.

bioregion *Ecology.* an area that constitutes a natural ecological community.

biorheology *Biology.* the study of the deformation and flow of matter in living systems and in materials directly derived from them.

BIOS basic input/output system.

biosatellite *Space Technology.* an artificial satellite that can support plant or animal life forms.

bioscience *Biology.* the study of biology in which all the sciences are applied.

biosensor *Engineering.* any sensor that transmits data relating to a biological process, such as blood pressure. *Biotechnology.* a device or system that uses an immobilized biological material to detect and measure a chemical compound.

bioseries *Evolution.* a recognizable sequence of evolutionary changes of a single hereditary character.

biosocial theory *Psychology.* the theory that personality is formed by the interaction of biological factors with social and environmental influences.

biosonar *Physiology.* the use of ultrasonic instruments to analyze and map parts of the body. *Behavior.* the use by certain animals of echoes of sounds they produce to locate objects in the perceptual field.

biosparite *Petrology.* a limestone consisting of less than 25% intraclasts, less than 25% oolites, with a volume ratio of fossils and fossil fragments to pellets more than 3:1, and calcite cement exceeding microcrystalline calcite.

biosphere *Ecology.* **1.** that part of the earth's atmosphere, hydrosphere, and lithosphere in which living organisms are known to exist. **2.** the ecosystem composed of the earth and the living organisms inhabiting it.

biostabilizer *Civil Engineering.* a composting system that tumbles moistened soil waste in a drum until the waste biodegrades into a fine dark compost.

biostasis *Ecology.* the capacity of an organism to tolerate changes in its environment without undergoing adaptive change itself.

biostat *Biotechnology.* a vessel in which the biomass of a continuous culture is monitored by parameters other than turbidity.

biostatics *Biology.* the study of the structure of organisms in relation to their functions, as opposed to biodynamics.

biostatistics *Statistics.* the application of statistical methods and processes to the analysis of biological data.

biostereometrics *Photogrammetry.* the spatial analysis of biological form based on the principles of analytical geometry. Also, BIOMEDICAL PHOTOGRAMMETRY.

biostratigraphic unit *Geology.* a layer of rock that is classified solely on the basis of its fossil content.

biostratigraphy *Paleontology.* the study of the stratigraphic distribution of fossils.

biostratinomy *Paleontology.* the preservation history of fossils, including their transportation, any physical, chemical, or biological damage, the attachment of epifauna, and postburial disturbance by other organisms or geological processes.

biostromal limestone *Geology.* a blanketlike mass of limestone, composed mainly of loosely aggregated shells and shell fragments.

biostrome *Geology.* **1.** a broad, bedded, blanketlike mass of rock built by sedentary organisms and composed mainly of their remains. **2.** an organic layer, such as a bed of shells or corals.

biosynthesis *Biochemistry.* the building up of a chemical compound in the physiologic processes of a living organism.

biosystematics *Systematics.* a botanical term designating the systematic study of populations and species, especially by means of ecological or experimental methods.

Biot, Jean Baptiste [bī´ō] 1774–1862, French physicist and mathematician; known for work on polarization of light.

biota *Biology.* **1.** the plant and animal life characteristic of a specific region or biosphere or of a given time period. **2.** plant and animal life in general.

Biot and Savart's law *Physics.* a law stating that the magnetic field caused by a short current element is directly proportional to the current and inversely proportional to the distance of the observation point from the conductor.

biotechnical *Engineering.* **1.** of or relating to biotechnology. **2.** applying techniques or principles of biotechnology.

biotechnical robot *Robotics.* a robot that can work only in conjunction with a human operator.

biotechnology *Engineering.* the application of advanced biological techniques in the manufacture of industrial products, including the production of antibiotics, insulin, and interferon, or for environmental management, such as waste recycling.

biotelemetry *Engineering.* the study of the behavior and physiology of living things at a distance from the observer, using telemetry techniques such as radio waves.

Biot-Fourier equation *Thermodynamics.* a differential equation that relates the temperature gradient in a body to the thermal diffusivity, used to represent the nonsteady heat conduction through solids.

biotic *Biology.* of or relating to living things; caused or produced by living organisms. *Ecology.* of or relating to the biological aspects of an environment, as opposed to its geological or meteorological aspects. Thus, **biotic environment, abiotic substance.**

biotic community *Ecology.* an aggregation of various species coexisting in the same habitat. *Paleontology.* a group of species whose fossil remains are frequently found in the same site.

biotic district see BIOTIC PROVINCE.

biotic environment *Ecology.* the living organisms of a community habitat or environment.

biotic factor *Ecology.* an environmental factor associated with or resulting from the activities of living organisms.

biotic potential *Ecology.* the capacity of a population of living organisms to increase under optimal environmental conditions.

biotic province *Ecology.* a specific area in which a particular combination of organisms live.

biotin *Biochemistry.* $C_{10}H_{16}N_2O_3S$, a B vitamin that functions as a coenzyme and is ubiquitous in nature. Also, VITAMIN H, COENZYME R.

biotin carboxylase *Enzymology.* an enzyme that transforms biotin to the compound carboxybiotin.

biotin complex of yeast see BIOCYTIN.

biotinylation *Molecular Biology.* the labeling of a probe with conjugated biotin, whose high affinity for anti-biotin antibodies is exploited to mark the place at which the probe binds by indirect immunoassay. Also, **biotin labeling.**

biotite *Mineralogy.* $K(Mg,Fe^{+2})_3(Al,Fe^{+3})Si_3O_{10}(OH,F)_2$, a dark brown to green monoclinic, magnesium-iron mica with perfect basal cleavage having a specific gravity of 2.7 to 3.4 and a hardness of 2.5 to 3 on the Mohs scale; widely distributed in igneous and metamorphic rocks. (Named after Jean Baptiste *Biot*.) Also, BLACK MICA.

Biot number *Fluid Mechanics* a dimensionless constant, named after Jean Baptiste Biot, that is equivalent to the ratio of the heat transfer coefficient for convection at the surface of the solid to the specific conductance of the solid.

biotope *Ecology.* the smallest region of a habitat uniform in environmental conditions and in distribution of life forms; e.g., a tidal pool or a forest canopy. *Biology.* the location of a parasite in the body of its host.

biotoxication *Toxicology.* intoxication that is caused by exposure to a biotoxin.

biotoxicology *Toxicology.* the branch of toxicology concerned with biotoxins.

biotoxin *Toxicology.* any poison produced by and derived from the cells or secretions of a living organism, either plant or animal.

biotransformation *Biotechnology.* **1.** the sequence of chemical reactions that is undergone by a compound in a biological system, usually due to enzymatic metabolic reactions. **2.** the transformation of matter from one form to another by living organisms or enzymes. Also, **bioconversion.**

biotron *Biology.* a chamber with a completely controlled environment, designed to study organisms for use in experiments. *Botany.* see PHYTOTRON.

biotroph *Biology.* an organism that can live and multiply only on another living organism.

Biot's law *Optics.* a law stating that the rotation exhibited in an optically active medium is proportional to the length of the light path, to the concentration of a solution, and to the inverse square of the light's wavelength.

bioturbation *Geology.* the agitation and disruption of a sediment by the activities of organisms.

biotype *Genetics.* **1.** a population of genetically identical organisms. **2.** a specific anatomical or physiological configuration of an organism due to its adaptation to specific environmental conditions, such as a high-altitude plant growing at sea level.

bioultrasonics [bī´ō ul´trə sän´iks] *Acoustics.* the use of ultrasonic sound for biological applications, such as ultrasonic medical tomography, ultrasonic microscopy, and physical therapy.

biozone *Paleontology.* 1. in terms of time, the period of existence of a species. 2. in lithic terms, an abstract zone that includes all strata deposited during the period of time that a given taxon lived.

Biozzi mice *Immunology.* any selectively bred mice that produce high or low antibody responses to a variety of antigens.

bipack *Graphic Arts.* a type of film composed of two emulsions that are exposed simultaneously.

bipara *Medicine.* a female who has had two children at separate births.

biparental *Biology.* relating to or having traits or characteristics derived from both parents, male and female.

biparietal *Anthropology.* relating to the two parietal eminences or bones, especially as they are determinants in measurements of the breadth of the skull.

biparous *Vertebrate Zoology.* producing two ova or offspring at one time. *Botany.* having dichotomous branches or axes.

bipartite *Zoology.* having two parts. *Botany.* divided into two parts nearly to the base, as a leaf.

bipartite cubic *Mathematics.* the two separate branches of the graph of the cubic equation $y^2 = x(x - a)(x - b)$. The graph is symmetric about the x-axis and intersects the x-axis at 0, a, and b.

bipartite graph *Mathematics.* a graph G whose vertices can be divided into two disjoint sets V_1 and V_2 such that every edge of G joins a vertex of V_1 to a vertex of V_2. G is said to have **bipartition** V_1, V_2.

bipectinate *Invertebrate Zoology.* **1.** bearing comblike toothed projections on each side, as in the antennae of certain moths. **2.** branched like a feather along both sides of a main shaft, as in the organ of smell in certain snails.

biped *Vertebrate Zoology.* **1.** any animal having two feet, such as humans or birds. **2.** having two feet.

bipedal *Biology.* **1.** having two feet. **2.** walking on two feeet.

bipedal dinosaur *Paleontology.* one of many dinosaurs that walked primarily on the hind legs; the weight of the body was pivoted on the hip bone which became fused to the backbone giving more rigid support. Many had shortened forelimbs adapted for grasping.

bipedalism *Anthropology.* the condition of being two-footed or of using two feet for standing and walking; noted in human evolution by the upright vertebral column and head position, the anthropoidal plate with flared attachments for flexors and extensors, and the arched foot with its strengthened great toe.

bipenniform *Anatomy.* having muscle fibers arranged like feathers on both sides of a median tendon.

biphasic *Botany.* having both a gametophyte phase and a sporophyte phase in the life cycle.

biphenyl *Organic Chemistry.* $C_6H_5{\cdot}C_6H_5$, a pleasant-smelling, white or slightly yellow crystalline solid, insoluble in water and soluble in alcohol and ether; melts at 71°C and boils at 255.9°C. It is used as a heat transfer agent, as a fungistat in packaging citrus fruits, in plant disease control, and in organic synthesis. Also, DIPHENYL, PHENYLBENZENE.

biphyletic *Evolution.* of or relating to species in different phyla that are descended from a common ancestor.

Biphyllidae *Invertebrate Zoology.* the false skin beetles, a family of coleopterans generally found in forests living under tree bark.

bipinnaria *Invertebrate Zoology.* the bilaterally symmetrical larva of certain starfish; swimming by means of ciliated bands, the larvae may move about freely for several weeks before settling into metamorphosis.

bipinnate *Botany.* pinnate, with the divisions also pinnate, as in a leaf.

biplane *Aviation.* an airplane having two wings on each side of the fuselage, one usually slightly forward and above the other.

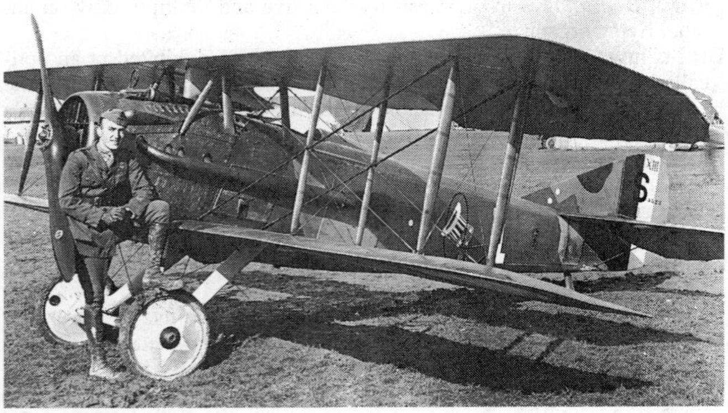

biplane

bipolar *Science.* having or relating to two poles, directions, or extremities. *Anatomy.* of a neuron, having two processes.

bipolar affective disorder see MANIC-DEPRESSIVE PSYCHOSIS.

bipolar amplifier *Electronics.* an amplifier that can supply a pair of output pulse signals corresponding to the positive or negative polarity of the input signal.

bipolar circuit *Electronics.* a logic circuit where 0's and 1's are treated in a bipolar manner, rather than by the presence or absence of a signal.

bipolar disorder see MANIC-DEPRESSIVE PSYCHOSIS.

bipolar electrode *Electrical Engineering.* an electrode in an electroplating bath that is not connected to either the anode or the cathode. Also, SECONDARY ELECTRODE.

bipolar flagellation *Microbiology.* the occurrence of flagella at both poles, or ends, of a bacterial cell.

bipolar format *Computer Technology.* a method of transmitting a binary data stream in which a binary 0 is sent as no pulse and a binary 1 is sent as a pulse that alternates in sign (plus or minus) for each successive binary 1 sent.

Bipolarina *Invertebrate Zoology.* a suborder of parasitic protozoa, phylum Mixozoa, class Myxosporea, order Bivalvulida, with spores usually elongated in a sutural plane or zone.

bipolar integrated circuit *Electronics.* an integrated circuit whose principal element is a bipolar junction transistor.

bipolar memory *Computer Technology.* a semiconductor memory that uses integrated circuit bipolar junction transistors as memory cells, in contrast with MOS memories that use metal-oxide semiconductor transistor technology. Bipolar uses more energy than MOS but is faster.

bipolar outflow *Astronomy.* gas jets from a young star that are ejected at right angles to the nebula of dusty gas surrounding the star.

bipolar signal *Telecommunications.* in digital transmission, a pseudoternary signal conveying binary digits, in which successive marks or spaces are of alternate polarity (positive and negative) but equal in amplitude, and in which spaces or marks are of zero amplitude.

bipolar transistor *Electronics.* a transistor that uses both positive and negative charge carriers and that can be manufactured as a separate device or incorporated into integrated circuits.

bipotential *Biology.* relating to or characterized by bipotentiality.

bipotentiality *Biology.* **1.** the capacity to function or develop in either of two possible ways; usually referring to sexual differentiation. **2.** an abnormal condition in which both male and female reproductive organs are present in one individual. Also, HERMAPHRODITISM.

biprism *Optics.* a very obtuse prism whose apex forms an angle of almost 180°; it produces double images and interference fringes used in measuring the wavelength of light. Thus, **biprism interference.**

bipropellant *Space Technology.* a rocket propellant consisting of two liquids, a fuel and an oxidizer, that are kept separate before combustion.

biquadratic *Mathematics.* **1.** a polynomial expression of degree 4. Also, QUARTIC. **2.** a polynomial in one variable in which each term has degree 2 or 4 only.

biquartz *Optics.* an optical device consisting of two adjoining sections of quartz, one section right-handed and the other left-handed, that is used with analyzers, such as a Nicol prism, to study polarized light.

biquinary see QUIBINARY.

biquinary abacus *Mathematics.* a Chinese-style abacus, in which the markers are separated into two- and five-part sections.

biquinary notation *Mathematics.* a mixed-based notation of numbers as represented on a biquinary abacus. Digits are grouped in pairs, the first of which indicates 0 or 1 unit of 5, and the second 0, 1, 2, 3, or 4 units of 1.

biradial *Biology.* having a combination of radial and bilateral symmetry of parts, as in some marine invertebrates. Also, DISYMMETRY.

biradical *Chemistry.* any radical having two free electrons.

biramous *Biology.* having two branches or being forked, such as a crab claw. Also, **biramose.**

Biraphidineae *Botany.* a suborder of diatoms of the order Pennales, characterized by a fully developed specialized raphe on both valves and no labiate processes.

birch *Botany.* a slender, hardy deciduous tree of the genus *Betula,* with a smooth, laminated outer bark that often peels into thin papery strips as in *B. papyrifera. Materials.* the strong, close-textured wood of the birch tree, grown in North America and northern Asia; used for furniture, paneling, and toys.

bird *Vertebrate Zoology.* any of the many vertebrates in the class Aves; all are warm-blooded, lay eggs, and have wings and feathers, and most can fly.

birdcage *Engineering.* the rotating fan for a hot-air furnace.

birdcage clock see LANTERN CLOCK.

bird cherry *Botany.* any of several species of wild cherry, especially *Prunus padus* of Europe and Asia, and *P. pennsylvanica* of the U.S., or the fruit of these trees. Also, PIN CHERRY.

bird-foot delta *Geology.* a river delta whose seaward-extending distributaries form a pattern resembling the outstretched claws of a bird. Also, **bird's-foot delta.**

bird-footed dinosaur *Paleontology.* a bipedal carnivorous dinosaur of the order Theropoda, whose three-clawed toes left birdlike tracks.

bird grass *Botany.* a forage and lawn grass, *Poa trivialis,* introduced into North America from Europe. Also, KNOTGRASS.

bird-hipped dinosaur *Paleontology.* a dinosaur belonging to the order Ornithischia, whose pelvis resembled that of a bird.

Birdiavirus *Virology.* a former name for nuclear polyhedrosis viruses.

bird louse *Invertebrate Zoology.* any of a number of wingless insects of the order Mallophaga (biting lice), mostly bird parasites that cause injury by feeding on feathers or skin.

bird of paradise *Vertebrate Zoology.* a brilliantly colored oscine bird of the family Paradisaeidae, ranging primarily from New Guinea to Australia; the males have beautiful courtship plumage, including long head plumes, expandable caped wings, and elaborate tails. *Botany.* a small shrub, *Caesalpinia gilliesii,* native to South America and grown in California and Florida, having bright, showy flowers that resemble a bird of paradise in flight.

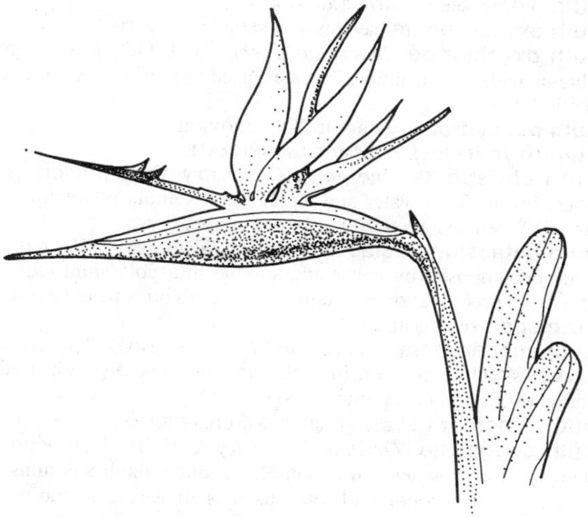

bird of paradise

bird of prey *Vertebrate Zoology.* any of a group of carnivorous birds that hunt their prey, including the orders Falconiformes (eagles, hawks, vultures, etc.) and Strigiformes (owls).

birds-eye *Botany.* any of various small plants with small, round, bright-colored flowers, especially the speedwell, *Veronica chamaedrys. Textiles.* a fabric that has a woven pattern of diamonds with center dots.

bird's-eye primrose *Botany.* a primrose, *Primula mistassinica,* having pink or lilac flowers with yellow centers and leaves clustered at the base of the stalk; found in meadows and on moist cliffs.

bird's-eye rot *Plant Pathology.* a disease of grape plants, caused by the fungus *Elsinoe ampelina,* characterized by fruit marked by small, dark sunken spots with light centers.

bird's-foot trefoil *Botany.* a perennial plant, *Lotus corniculatus,* of the pea family, with pods that spread like a crow's foot; a European import that now grows across the U.S.

bird's-foot violet *Botany.* a violet, *Viola pedata,* with divided leaves that resemble a bird's foot and single, pansylike flowers.

bird's mouth *Building Engineering.* a term for a notch cut at 90° on the underside of a rafter to allow it to rest over the arris of a cross timber.

bird's-nest fungus *Plant Pathology.* a fungus, *Cyathus stercoreus,* whose peridia resemble a bird's nest containing eggs; often found in gardens fertilized with manure.

birectangular *Mathematics.* any geometric figure, not necessarily planar, having two right angles.

bireflection *Optics.* a variation in the color or brightness of optically anisotropic reflecting objects when they are viewed with plane-polarized light as the direction of their vibration changes.

birefringence *Optics.* **1.** the separation of a ray of light, on passing through a crystal, into two unequally refracted, plane-polarized rays (of orthogonal polarizations). This effect occurs in crystals in which the velocity of light is not the same in all directions; that is, the refractive index is anisotropic. Uniaxial crystals have one direction in which double refraction does not occur; biaxial crystals have two. Also, DOUBLE REFRACTION. **2.** the difference between the greatest and smallest index of refraction in a birefringent crystal.

birefringent *Optics.* having or showing birefringence.

birefringent filter *Optics.* a filter composed of alternating layers of polarizing film and plates of birefringent crystal that transmits light in a series of widely spaced wavelength bands; often used for photographing solar flares. Also, LYOT FILTER, MONOCHROMATIC FILTER.

birefringent plate *Optics.* a thin, flat piece of material capable of producing birefringence.

birimose *Botany.* opening by two slits.

Biringuccio, Vanocchio 1460–1539, Italian engineer; wrote early studies on metallurgy, mining engineering, and industrial chemistry.

Birkeland, Kristian Olaf Berhard 1867–1927, Norwegian physicist; explained aurora borealis and discovered a process for fixing nitrogen.

Birkeland-Eyde process *Chemical Engineering.* a method of nitrogen fixation in which air passes through an alternating-current arc flattened by a magnetic field, resulting in the formation of about 1% nitric oxide.

Birkhoff, George David 1884–1944, American mathematician; formulated ergodic theorem; worked in dynamics and differential equations.

Birkhoff's theorem *Physics.* a theorem stating that if space-time contains centrally symmetrical matter that satisfies Einstein's general relativity equations, it is necessarily static and, when under a coordinate transformation, becomes the equivalent of the Schwarzchild solution.

Birmingham wire gauge *Design Engineering.* a standard for sizing wire, metal tubing, steel sheeting, and other metal materials. Also, **BWG.**

Birnaviridae *Virology.* a family of bisegmented dsRNA viruses that cause infectious pancreatic necrosis in fish and infectious bursal disease in chickens.

Birostrina *Paleontology.* a genus of subtriangular Mesozoic bivalves with a larger convex left valve; of some stratigraphic importance in the Cretaceous.

birotulate *Invertebrate Zoology.* a sponge spicule with two wheel-shaped ends.

birth canal *Anatomy.* the path through which a fetus passes during birth, from the uterus through the cervix, vagina, and vulva.

birth defect *Pathology.* any defect present at birth, such as a morphological defect or an inborn error of metabolism. Also, CONGENITAL DEFECT.

birth order *Psychology.* the position of a child in age in relation to his or her siblings; often regarded as having an effect on personality and behavior patterns.

birth rate *Biology.* the ratio of the number of live births in a year in a given area in relation to a given portion of the population in that area; often expressed as births per one hundred or one thousand people.

bis- *Chemistry.* a prefix meaning twice or again, used in the names of compounds to indicate that a chemical grouping or radical occurs twice in the molecule, e.g., **bisphenol.**

Biscay, Bay of *Geography.* a broad arm of the Atlantic off the coast of western France and northern Spain.

2,2-bis(*para*-chlorophenyl)-1,1-dichloroethane *Organic Chemistry.* $C_{14}H_{10}Cl_4$, a colorless, crystalline solid that melts at 109°C; used as a wide-spectrum insecticide on many food crops. Also, DDD, TDE.

bischofite *Mineralogy.* $MgCl_2 \cdot 6H_2O$, a colorless to white, monoclinic mineral, occurring in crystalline-granular or fibrous forms, having a hardness of 1.5 on the Mohs scale, found in salt deposits.

biscuit *Materials.* unglazed ceramic ware that has been fired. *Metallurgy.* **1.** the starting stock of a drop forging operation. **2.** a small cake of any primary metal. *Acoustical Engineering.* a lump of vinylite record compound that is placed between the two sides of the record mold during the stamping of a record. Also, PREFORM.

biscuit cutter *Mining Engineering.* a core barrel, six to eight inches in diameter, with a sharp bottom that is forced into the rocks by the jars.

bisdimethylthiocarbamyl sulfide see TETRAMETHYLTHIURAM MONOSULFIDE.

bise *Meteorology.* a cold, dry wind blowing in a northerly direction in or from the mountain regions of France and Switzerland; most frequent in spring and winter. Also, BIZE.

bisection *Surgery.* the process of dividing into two parts by cutting.

bisector *Mathematics.* **1.** a line or hyperplane that divides a given angle into two equal angles. **2.** a point, line, or plane that passes through the midpoint of a line segment.

bisectrix *Crystallography.* either of two lines bisecting the acute and obtuse angles formed by the optic axes in a biaxial crystal.

biserial *Biology.* arranged in two rows, series, or cycles. *Statistics.* relating to the correlation between two sets of measurements, in which one set is limited to one of two values.

biserrate *Biology.* **1.** having a serrated edge on which the straight edges are themselves serrated, such as on some leaves. **2.** being serrated on both sides, such as on antennae.

bisexual *Biology.* **1.** of or relating to two sexes. **2.** having gonads of both sexes. *Psychology.* **1.** sexually attracted to one's own sex as well as to the opposite sex. **2.** a person who is sexually attracted to both sexes.

bisexuality *Psychology.* the fact or condition of being bisexual.

bishop *Vertebrate Zoology.* a small, terrestrial weaverbird of the family Ploceidae, genus *Euplectes*; found in African grasslands and known for its long, well-woven globular nest. Also, **bishop bird.**

bishop

Bishop, J. Michael born 1936; American microbiologist; shared 1989 Nobel Prize for research on oncogenes.

Bishop's ring *Meteorology.* a faint, broad reddish-brown corona most often seen in dust clouds that arise from volcanic eruptions.

bisilicate *Metallurgy.* a slag having a silicate degree of 2.

bismanol *Metallurgy.* a highly ferromagnetic alloy consisting of bismuth and manganese, with a high degree of magnetic force.

bismite *Mineralogy.* Bi_2O_3, a yellowish to greyish-green, translucent monoclinic mineral often found in an earthy, impure state, having a specific gravity of 8.64 to 9.22 and a hardness of 4.5 on the Mohs scale. Also, **bismuth ocher.**

bismuth *Chemistry.* an element, symbol Bi, atomic number 83, atomic weight 208.98, having a melting point of 271°C; a brittle metal with a reddish tinge. Though not generally considered radioactive, it has four naturally radioactive isotopes. It is used in the manufacture of pharmaceuticals, alloys, and cosmetics. *Mineralogy.* the hexagonal rhombohedral mineral form of this element, having a specific gravity of 9.7 to 9.83 and a hardness of 2 to 2.5 on the Mohs scale. (From a Latinized version of a German word; original source unkown.)

bismuth alloy *Metallurgy.* any of several low-melting alloys containing 33% to 67% bismuth, various amounts of lead, tin, or cadmium, and, at times, indium or antimony; some of these alloys expand upon freezing.

bismuthate *Inorganic Chemistry.* a compound of bismuth containing the BiO_3^- ion, in which the element has a valence of +5.

bismuth blende see EULYTITE.

bismuth carbonate see BISMUTH SUBCARBONATE.

bismuth chloride *Inorganic Chemistry.* $BiCl_3$, white deliquescent crystals volatized by heat; melts at 230–232°C and decomposes in water to the oxychloride; used as a catalyst and to produce bismuth salts.

bismuth chromate *Inorganic Chemistry.* $Bi_2O_3 \cdot Cr_2O_3$, an orange-red powder that is insoluble in water; derived from bismuth nitrate and potassium chromate; used as a pigment.

bismuth citrate *Inorganic Chemistry.* $BiC_6H_5O_7$, a white powder that is slightly soluble in water and alcohol; decomposed by heat; used in pharmacy as an astringent and in the preparation of other bismuth remedies.

bismuth glance see BISMUTHINITE.

bismuth hydrate see BISMUTH HYDROXIDE.

bismuth hydroxide *Inorganic Chemistry.* $Bi(OH)_3$, a white amorphous powder that is soluble in acids and insoluble in water, precipitated by the action of sodium hydroxide on bismuth salt solutions; often used in plutonium separations.

bismuthinite *Mineralogy.* Bi_2S_3, an opaque, gray to white orthorhombic mineral with a metallic luster and perfect cleavage, occurring in massive form or as prismatic crystals, having a specific gravity of 7.8 and a hardness of 2 on the Mohs scale; found in low-to-high temperature hydrothermal vein deposits. Also, BISMUTH GLANCE.

bismuth iodide *Inorganic Chemistry.* BiI_3, grayish to black metallic crystals that are soluble in alcohol and insoluble in water; melting at 408°C; used to analyze materials.

bismuth nitrate *Inorganic Chemistry.* $Bi(NO_3)_3 \cdot 5H_2O$, clear lustrous hygroscopic crystals that decompose in water; used in luminous paints and enamels, and as an astringent and antiseptic.

bismuth oleate *Organic Chemistry.* $Bi(C_{17}H_{33}COO)_3$, a yellowish-brown, granular mass, insoluble in water and soluble in benzene and ether; used as a catalyst.

bismuthosis *Toxicology.* poisoning due to chronic exposure to bismuth; symptoms may include skin rashes and degeneration of the liver and kidneys. Also, **bismuthism.**

bismuth oxide see BISMUTH TRIOXIDE.

bismuth oxycarbonate see BISMUTH SUBCARBONATE.

bismuth oxychloride *Inorganic Chemistry.* BiOCl, a white powder, soluble in acids and insoluble in water; used in pigments, cosmetics, and dry cells.

bismuth oxyhydrate see BISMUTH HYDROXIDE.

bismuth oxynitrate see BISMUTH SUBNITRATE.

bismuth phosphate *Inorganic Chemistry.* $BiPO_4$, odorless white crystals; insoluble in water and alcohol and soluble in weak acids; decomposes when heated; used in plutonium recovery.

bismuth potassium iodide *Inorganic Chemistry.* BiI_7K_4, red crystals that are decomposed by water and soluble in a potassium iodide solution; used in precipitation of vitamins and antibiotics from solution.

bismuth spar see BISMUTITE.

bismuth subcarbonate *Inorganic Chemistry.* $(BiO)_2CO_3$, an odorless and tasteless white powder, insoluble in water and alcohol; used in X-ray diagnostics and in ceramic glass.

bismuth subchloride see BISMUTH OXYCHLORIDE.

bismuth subgallate *Organic Chemistry.* $C_7H_5BiO_6$, an odorless and tasteless yellow powder that is soluble in dilute alkali solutions but insoluble in water, alcohol, and ether; used as an astringent and protective in the treatment of conditions such as ulcerative colitis, dysentery, and diarrhea.

bismuth subnitrate *Inorganic Chemistry.* $4BiNO_3(OH)_2 \cdot BiO(OH)$, a heavy white powder used in perfumes, pharmaceuticals, and ceramic enamels.

bismuth subsalicylate *Inorganic Chemistry.* $Bi(C_7H_5O)_3Bi_2O_3$, an odorless and tasteless white powder that is sensitive to light; insoluble in water and alcohol; used as a surface coating for plastics and in copying paper.

bismuth sulfate *Inorganic Chemistry.* $Bi_2(SO_4)_3$, white needles or powder, insoluble in water and alcohol; decomposes at 405°C; used in the analysis of metallic sulfates.

bismuth sulfide *Inorganic Chemistry.* Bi_2S_3, a blackish-brown powder that is insoluble in water; it decomposes at 685°C; used in manufacturing bismuth compounds.

bismuth telluride *Inorganic Chemistry.* Bi_2Te_3, gray hexagonal platelets; melts at 573°C; used for semiconductors and for thermoelectric cooling.

bismuth trichloride see BISMUTH CHLORIDE.

bismuth trioxide *Inorganic Chemistry.* Bi_2O_3, a heavy yellow powder that is insoluble in water and soluble in acids; melts at 820°C; used in enamels and to color ceramics.

bismutite *Mineralogy.* $Bi_2(CO_3)O_2$, a massive to earthy, white to yellowish to gray tetragonal mineral, having a specific gravity of 8.5 and a hardness of 2.5 to 3.5 on the Mohs scale; found as a secondary mineral in oxidized zones of bismuth bearing veins and pegmatites. Also, BISMUTH SPAR.

bismutotantalite *Mineralogy.* $Bi(Ta,Nb)O_4$, a black, orthorhombic mineral occurring as large, imperfect crystals having a specific gravity of 8.26 to 8.84 and a hardness of 5 on the Mohs scale; found in pegmatites with black tourmaline and muscovite.

bisnaga *Botany.* any of various thorny cactuses belonging to the genera *Echinocactus, Ferocactus,* and *Astrophytum;* found in the southwestern U.S. Also, BIZNAGA.

bison *Vertebrate Zoology.* **1.** a large, wild, cattle-like mammal, *Bison bison,* of the family Bovidae, characterized by enormous forequarters, a hump just behind the neck, and thick shaggy fur. **2.** a related mammal, *Bison bonasus,* of Europe, larger and less shaggy than the American bison. Also, WISENT.

American bison

bisphenoid *Crystallography.* a crystallographic form bounded by eight scalene triangles arranged in pairs.

bisphenol A *Organic Chemistry.* $(CH_3)_2C(C_6H_5OH)_2$, a white or brownish crystalline solid that has a mild odor of phenol; insoluble in water and soluble in alcohol; melts at 153°C; used as a plastics intermediate for the manufacture of epoxy resins and as a fungicide.

bisporangiate *Botany.* having both microsporangia and megasporangia; having both functional male and female parts.

bispore *Botany.* an asexual spore produced in pairs in bisporangium of certain of the Rhodophyta.

bisque firing [bisk] *Materials Science.* a low-temperature firing of a ceramic prior to high-temperature firing and glazing.

bistability *Chaotic Dynamics.* the coexistence in phase space of two different attractors for the same values of the parameters; which attractor is observed depends on the initial conditions.

bistable *Science.* having or capable of having two stable states, such as on/off or 0/1. Thus, **bistable unit, bistable (optical) device,** and so on.

bistable circuit *Electronics.* a circuit having two stable states that can be decided by input signals; used in counters and scalers.

bistable device *Computer Technology.* any device with two stable states, usually "on" and "off"; such a device is commonly used in memory and registers to store a bit. Also, **bistable latch.**

bistable multivibrator *Electronics.* a circuit capable of assuming either one of two stable states and requiring two input pulses to complete a cycle. Also, FLIP FLOP.

bistatic radar *Engineering.* a radar system based on two locations at great distance from each other; signals are transmitted to a body in outer space from one location, and signals reflecting back to earth are received at the other.

bistatic reflectivity *Optics.* the property of a reflecting surface to reflect light or energy along one or more paths that are different from that of the incident ray.

bistoury *Surgery.* a long, narrow-bladed knife used to open abscesses, slit sinuses and fistulas, and the like.

bisulfate *Inorganic Chemistry.* a compound that has the $-HSO_4^-$ ion, derived from sulfuric acid; an acid sulfate.

bisulfide see DISULFIDE.

bisulfite *Inorganic Chemistry.* a compound that has the $-HSO_3^-$ ion, derived from sulfurous acid; an acid sulfite.

bit *Mechanical Devices.* **1.** a drilling or boring tool for use in a brace or drill press. **2.** the cutting iron of a plane. **3.** a removable boring head used on certain kinds of drills, such as a rock drill.

bit *Computer Programming.* the basic unit of information in a digital computing system, with a value of either 1 or 0. *Mathematics.* a place value in a binary number; it can equal 0 or 1. (An acronym for the term binary digit).

bitartrate *Organic Chemistry.* any salt that contains the univalent radical $-C_4H_5O_6$. Also, ACID TARTRATE.

bit blank *Mechanical Devices.* a drill bit in which diamonds or other cutting elements can be inset by hand or attached by mechanical means.

bitblt [bit′blit′] *Computer Science.* **1.** an instruction or subroutine for rapid transfer of blocks of memory from one area of memory to another, such that the transfer may begin and end at specified bit positions; useful especially for bitmap graphics. **2.** to make such a transfer. (An acronym for bit block transfer.)

bit brace *Mechanical Devices.* a tool to which a boring bit may be attached and by means of which the bit is rotated. Also, **bitstock.**

bit buffer unit *Telecommunications.* a unit that terminates incoming and outgoing bit-serial communications lines and temporarily stores information for retiming and/or reformation.

bit density *Computer Technology.* a measure of the number of bits recorded per unit of length or area.

bit drag *Mechanical Devices.* a bit with serrated teeth, used in rotary drilling. Also, DRAG BIT.

bite *Biology.* **1.** to grasp, tear, or seize with the teeth. **2.** to wound, sting, or pierce for bloodsucking with a proboscis, as does a mosquito or bee. **3.** a wound or puncture made by the teeth or other parts of the mouth. *Anatomy.* the alignment of the upper and lower teeth. *Graphic Arts.* **1.** the relative ability of a given paper to accept ink or other impressions, based on thickness, porosity, and other surface characteristics. **2.** one of a series of acid-etching steps in photoengraving.

bite biopsy *Surgery.* the removal of a fragment of tissue using an instrument.

bitegmic *Botany.* having two integuments.

biternate *Botany.* ternate, with the divisions also ternate.

bit gauge see BIT STOP.

bithionol *Organic Chemistry.* $C_{12}H_6Cl_4O_2S$, a white or grayish-white crystalline powder that melts at 187°C; insoluble in water and soluble in alcohol, ether, and acetone; used as an ingredient in deodorants, germicides, fungistats, and pharmaceuticals, and especially effective against Gram-positive cocci. It is a skin irritant and cannot be used in cosmetics.

bithorax complex *Genetics.* a complex of at least nine genes in *Drosophila* that contributes to normal thoracic segmentation.

biting angle *Ordnance.* the angle at which a shell or projectile penetrates armor.

biting-in *Graphic Arts.* the corrosive action of acid in etching a design onto a metal plate.

biting louse see BIRD LOUSE.

bit location *Computer Programming.* a storage location in a record capable of storing one binary digit.

bit manipulation *Computer Programming.* the altering of bits from one state to another, often to control the logical sequence of computer operations. Also, BIT FLIPPING.

bitmap *Computer Science.* **1.** a specification of a binary (black-and-white) image in which each pixel is represented by a single bit. **2.** a data structure or file containing a bitmap.

bit mapping *Computer Technology.* a graphics display technique in which one or more memory bits are assigned to each pixel (picture element) of the displayed image. Also, **bit map display.**

bit matrix *Mechanical Devices.* a metal or metal alloy that forms the material in which the diamonds in a bit crown are embedded. Also, DIAMOND MATRIX.

bit parallel *Telecommunications.* a method of sending, processing, or receiving digital code information in which as many paths are used as there are bits in the word being considered. Also, BYTE SERIAL.

bit pattern *Computer Programming.* the specific arrangement of n bits used in a system to represent 2^n possible choices: a 3-bit pattern represents 8 possible combinations; an 8-bit pattern represents 256 possible combinations, and so on.

bit position *Computer Programming.* the relative position of a single bit within a binary word.

bit rate *Telecommunications.* **1.** the speed at which bits are generated or transmitted. **2.** in a binary digital communication system, the number of bits transmitted per unit time, usually bits per second.

bit serial *Telecommunications.* the transmission, in sequence, of character-forming bits.

bit shank *Mechanical Devices.* the threaded part of a bit.

bit-sliced microprocessor *Computer Technology.* a processor unit in a microcomputer that is constructed of multiple large-scale integration chips, with each chip one, two, or four bits wide, and interconnected to form the desired word length; each bit-slice chip contains all the circuits required to perform arithmetic, logic, register, and some I/O functions on its segment of the word. Also, **bit slice architecture.**

bits per inch *Computer Technology.* a measure of binary information recorded on a magnetic tape storage medium.

bits per second *Telecommunications.* the number of bits occurring at or passing a point each second.

bit stop *Mechanical Devices.* an attachment that stops a bit at a desired depth. Also, BIT GAUGE.

bit stream *Computer Programming.* a binary signal without regard to groupings in character; often used in connection with synchronous transmission.

bit string *Computer Programming.* a sequence of binary digits in which each digit is considered an independent unit and its significance is determined by its position.

bit sub *Petroleum Engineering.* a short length of pipe created to be placed between the drill collar and the drill bit or between the shock sub and the drill collar.

bit synchronization *Telecommunications.* the part of a message header that synchronizes all bits and characters.

bitt *Naval Architecture.* **1.** an upright post on a ship's hull used to secure a rope or anchor cable; often placed in pairs. **2.** to secure a rope or cable around a bitt.

bitter almond *Botany.* a variety of the almond tree, *Prunus amygdalus amara,* which bears inedible fruit but which is the source of a volatile oil and of the glucoside amygdalin (Laetrile).

bitter cress *Botany.* any plant of the genus *Cardamine*, of the mustard family, bearing white, pink, or purple flowers; the small leaves are often used in salads.

bitter end *Naval Architecture.* **1.** the part of the anchor cable that lies behind the bitts to which it is secured, thus remaining on board while the ship rides to her anchor. **2.** the extreme end of any important rope or chain.

bitterling *Vertebrate Zoology.* a small, carplike freshwater fish, *Rhodeus sericeus,* found in central and eastern Europe; the female has a long, yellow or red ovipositor with which she deposits eggs into the mantle cavity of mussels.

bittern *Vertebrate Zoology.* any of several species of tawny, long-legged, nocturnal herons, such as *Botaurus lentiginosus* of North America, and *B. Stellaris* of Europe, noted for their bellowing cry. *Chemical Engineering.* the bitter liquor remaining after the removal of sodium chloride crystals from concentrated sea water or brine; used as a source of bromides, magnesium, and calcium salts.

bittern

bitternut *Botany.* a hickory, *Carya cordiformis,* that grows in the swamps and moist woods of the eastern and southern U.S. and bears a smooth, gray, thin-shelled nut with a bitter kernel.

Bitter pattern *Solid-State Physics.* a pattern of ferromagnetic particles that indicates the magnetic domains on the surface of a ferromagnetic crystal.

bitter pit *Plant Pathology.* a disease of unknown cause affecting the pear, apple, and quince; characterized by brown, sharply defined corky flecks in the flesh of the fruit and discolored surface depressions.

bitterroot *Botany.* a small plant, *Lewisia rediviva,* of the purslane family, with thick, edible roots and pink or white flowers; found in the northern Rocky Mountains. It is the state flower of Montana.

bitter rot *Plant Pathology.* a disease of grapes, apples, and other fruits caused by the fungus *Glomerella cingulata;* characterized by the appearance of cankers on the woody plant parts and spotting and decay of the fruit.

bitters *Food Technology.* a medicinal agent that has a bitter taste; used as a tonic or appetizer.

bittersweet *Botany.* **1.** a climbing plant, *Solanum dulcamara,* of the nightshade family, with purple, blue, or white flowers and small, poisonous berries. Also, WOODY NIGHTSHADE. **2.** a climbing shrub of North America, *Celastrus scandens,* with greenish flowers and orange seed cases that open and show red seeds.

bit test *Computer Programming.* a test performed by a program to determine the state of a particular bit, usually in order to control the sequence of operations.

bitumen *Geology.* any naturally occurring flammable substance consisting mainly of a mixture of hydrocarbons, such as petroleum or asphalt. *Materials.* **1.** originally, a type of asphalt occurring naturally in Asia Minor. **2.** any similar black, sticky mixture of hydrocarbons occurring naturally or pyrolytically in the atmosphere and completely soluble in carbon disulfide; obtained mainly from natural oxidized petroleum products or from a petroleum distillation process.

bitumenite see TORBANITE.

bitumenize *Building Engineering.* to change into, cover, or mix with bitumen.

bituminization see COALIFICATION.

bituminol see ICHTHAMMOL.

bituminous [bi too´mi nəs] *Mineralogy.* of, relating to, containing, or characteristic of bitumen. Thus, **bituminous cement, bituminous coating, bituminous concrete, bituminous paint,** and so on.

bituminous coal *Mineralogy.* a dark brown to black coal that has a heat value greater than 10,500 Btu/lb, contains 15–20% volatile matter, and yields pitch or tar on heating. Also, SOFT COAL.

bituminous concrete *Materials.* a pavement made up of aggregates, such as crushed stone, gravel, or slag, combined with a bituminous binder that is used instead of cement.

bituminous distributor see BLACKTOP PAVER.

bituminous lignite *Mineralogy.* a brittle, lustrous brownish-black coal having a heat value less than 8300 Btu/lb. Also, PITCH COAL.

bituminous limestone *Mineralogy.* a dark, dense limestone that contains a high amount of organic matter and emits a foul odor when broken or rubbed vigorously.

bituminous rock *Mineralogy.* **1.** any natural or rock asphalt. **2.** a rock in which the percentage of impregnation is comparatively low.

bituminous sand *Geology.* sand that is naturally impregnated with bituminous substances.

bituminous sandstone *Petrology.* any sandstone containing bituminous material.

bituminous shale *Petrology.* a shale containing hydrocarbons or bituminous material; some yield oil or gas on distillation, such as oil shale. Also, BLAE.

bituminous stabilization *Materials Science.* the process of mixing a bituminous material with soil to act as a binder or as a waterproofing agent; the type and amount of bituminous material depend on the required function, the soil type, and the climatic conditions. Also, SOIL STABILIZATION.

bituminous wood *Geology.* a variety of lignite having a fibrous structure that resembles wood.

bitunicate ascus *Mycology.* an ascus possessing two walls: an inelastic outer wall that breaks open at maturity and an elastic inner wall that stretches as the spores develop within it until maturity.

bitypic *Evolution.* of or relating to a taxon comprising only two immediately subordinate subtaxa, as in a genus comprising two species.

biuret *Organic Chemistry.* $NH_2CONHCONH_2$, colorless, odorless needles that decompose at 190°C; soluble in hot water and alcohol; used to identify urea, from which it is derived. Also, ALLOPHANAMIDE, CARBAMOYLUREA.

bivalence *Chemistry.* the fact or condition of having a valence of 2.

bivalent *Chemistry.* having a valence of 2.

bivalent chromosome *Cell Biology.* either of a pair of homologous chromosomes that are grouped together during the zygotene stage of meiosis.

bivalve *Invertebrate Zoology.* **1.** mollusks of the class Lamellibranchia (clams, mussels, and oysters), usually with two shell sections hinged at the base and controlled by adductor muscles. **2.** any organism with a two-part shell.

Bivalvia *Invertebrate Zoology.* **1.** an order of mollusks with a laterally compressed body and a hinged double shell, each half with a valve. Also, LAMELLIBRANCHIA. **2.** sometimes, the Brachiopoda and Lamellibranchia considered as a group.

bivane *Engineering.* a sensitive wind vane that measures both the horizontal and vertical components of wind direction.

bivariate analysis *Statistics.* the study of the joint occurrence of two random variables.

bivariate normal distribution *Statistics.* a vector random variable (X, Y) in which the conditional probability of Y given X is normally distributed.

bivariate regression model *Statistics.* a statistical model describing the relationship between a dependent variable Y and an independent variable X as $Y = a + bX + e$, where a and b are parameters and e is a random error term.

biventer *Anatomy.* any organ or muscle that has two bellies.

biventricular *Cardiology.* relating to or affecting both ventricles of the heart.

Bivine's method *Toxicology.* a method of administering chloral hydrate to treat convulsions caused by ingestion of strychnine.

bivinyl see BUTADIENE.

bivittate *Zoology.* **1.** having two stripes. **2.** having two oil receptacles.

bivium *Invertebrate Zoology.* the two posterior rays of a starfish on either side of the sieve plate (madreporite).

bivoltine *Entomology.* having two broods during a single breeding season or producing two generations a year, as certain silkworm moths.

bivouac [biv´ə wak´] *Military Science.* a temporary camp in the field, usually with tents or improvised shelter.

Bixaceae *Botany.* a family of dicotyledonous tropical trees and shrubs of the order Violales, characterized by red or orange juice carried in abundant secretory cells or canals, and large alternate leaves.

bixbyite *Mineralogy.* $(Mn^{+3},Fe^{+3})_2O_3$, an opaque, metallic black cubic mineral having a specific gravity of 4.95 and a hardness of 6 to 6.5 on the Mohs scale; occurs with topaz in cavities of rhyolite.

bixin *Organic Chemistry.* $C_{25}H_{30}O_4$, orange crystals that decompose at 217°C; soluble in alcohol and acetone; a carotenoid obtained from the seeds of *Bixa orellano;* used to color margarine.

bize see BISE.

biznaga see BISNAGA.

Bjerknes, Vilhelm 1862–1951 and **Jakob** 1897–1975, [byerk´nəs] Norwegian meteorologists; developed theory of the polar front.

BJS *Aviation.* the airport code for Beijing, China.

Bk the chemical symbol for berkelium.

BKK *Aviation.* the airport code for Bangkok, Thailand.

bl. black; block.

BL or **B-L** breadth-length.

Blaauw mechanism [blō] *Astrophysics.* a mathematical explanation of the disruption of a binary star system as one component throws off a shell of gas and the resulting loss of gravitational attraction changes the orbit of the companion star.

black *Optics.* describing an object of low reflectance that evenly absorbs most of the light of all visible wavelengths. *Chemistry.* any of several forms of finely divided carbon, either pure or mixed with waxes, oils, or fats, such as bone black or lampblack. *Telecommunications.* see BLACK SIGNAL.

Black, Davidson 1884–1934, Canadian anatomist and anthropologist; discovered Peking Man *(Sinanthropus pekinensis).*

Black, Joseph 1728–1799, Scottish physician and chemist; made first systematic study of a chemical reaction; isolated carbon dioxide; developed theory of latent heat.

black alder *Botany.* a holly, *Ilex verticillata,* of eastern and midwestern North America, bearing red berries that remain into early winter. Also, WINTERBERRY.

black alkali *Geology.* a deposit of sodium carbonate that lies on the surface in arid to semiarid areas and tends to blacken organic matter.

black-and-white *Graphic Arts.* describing a monochromatic picture done with black and white only.

black-and-white television *Telecommunications.* television in which the final reproduced picture is in shades of gray ranging between black and white. Also, MONOCHROME TELEVISION.

black annealing *Metallurgy.* box or pot annealing of ferrous alloy sheet, strip, or wire. Also, **box annealing.**

black area *Telecommunications.* a region that contains only an encrypted signal.

black ash *Mining Engineering.* an impure sodium carbonate containing sodium sulfide with some carbon; produced especially for recovery of its soda content.

blackband *Geology.* a dark, earthy, carbonaceous variety of clay ironstone that is frequently found together with coal. Also, **blackband ore, blackband ironstone.**

black bass *Vertebrate Zoology.* any of various freshwater American game fish of the genus *Micropterus,* such as the largemouth bass.

black bear *Vertebrate Zoology.* the most common bear of North America, *Ursus americanus,* of the family Ursidae; a forest-dweller with an omnivorous diet, it weighs up to 500 pounds, stands about five feet, and varies in color from black to light brown; noted for having the smallest mammalian young (about 8 ounces) in proportion to adult size.

black-bellied plover *Vertebrate Zoology.* a large, gray, black, and white plover, *Pluvialis squatarola,* that breeds in the Arctic regions of North America and Siberia; it has black underparts in nuptial plumage.

blackberry *Botany.* any of several long-stemmed upright or trailing shrubs of the genus *Rubus,* family Rosaceae, characterized by fruit that is an aggregate of many drupelets which cling closely to the axis.

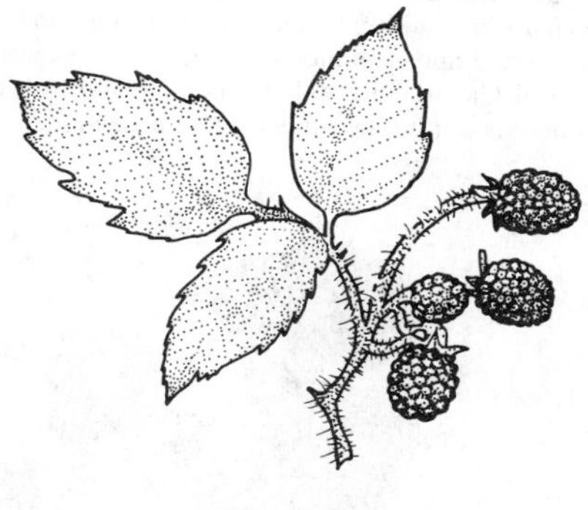

blackberry

blackberry lily *Botany.* a garden perennial, *Belamcanda chinensis,* of the iris family, having orange, lilylike flowers with red spots and clusters of black seeds that resemble blackberries.

black-billed cuckoo *Vertebrate Zoology.* a North American cuckoo with a black bill, *Coccyzus erythropthalmus,* that builds its own nest and raises its own young.

black bindweed *Botany.* a twining, weedy vine, *Polygonum convolvulus,* native to Europe but found in America as a troublesome weed.

blackbird *Vertebrate Zoology.* one of several species in the family Icteridae (American blackbirds) and the subfamily Turdinae (Old World blackbirds) in which the males are usually all black with yellow beaks and the females are mottled brown.

black blight *Plant Pathology.* any of several diseases of tropical plants caused by sooty molds on plant surfaces.

black blizzard see DUST STORM.

blackboard *Materials.* a sheet of slate or other hard material, widely used in classrooms, meeting rooms, and the like for writing with chalk; originally black but now usually light green. *Artificial Intelligence.* **1.** a generally accessible data set or control mechanism. **2.** a shared work station.

blackboard architecture *Artificial Intelligence.* a modular artificial intelligence architecture in which independent knowledge sources solve a problem cooperatively by posting hypotheses in a globally accessible data structure.

blackboard model *Artificial Intelligence.* a representation of the control structure of human planning in which processing is asynchronous and opportunistic.

black body or **blackbody** *Physics.* a theoretical object that is simultaneously a perfect radiator and absorber, and whose radiation is governed solely by its temperature. Most stars behave approximately like black bodies.

blackbody radiation *Thermodynamics.* the radiation that would be emitted from an ideal black body so that the energy distribution is dependent only on the temperature; the emission spectrum is based on Planck's equation and is expressed by the Stefan-Boltzmann law.

blackbody temperature *Thermodynamics.* the temperature at which a perfect black body would emit the same radiation per unit area as that emitted by a given body at a given temperature.

black box *Engineering.* **1.** a self-contained, removable electronic device, originally black, that records data on the flight and functions of an aircraft. **2.** an informal term for any compact electronic device that controls or monitors a system, and that is considered to have internal functions or processing sequences which are intricate (and mysterious to some persons). *Computer Technology.* a type of specialized hardware for converting one code into another.

blackbuck *Vertebrate Zoology.* a medium-sized, blackish-brown antelope, *Antilope cervicapra,* common in India.

black-bulb thermometer *Engineering.* a type of thermometer in which the sensing element, when covered with a shade, approximates a body capable of absorbing all incident radiation without reflecting any.

black buran see KARABURAN.

blackbutt *Materials.* the wood of several species of the eucalyptus tree (such as *Eucalyptus patens),* having a trunk base that resembles charred wood; cut for timber in Australia.

blackcap *Biology.* **1.** a black raspberry that grows wild in many parts of the U.S., and that is also cultivated. **2.** any of various birds with black at the top of the head, such as the Old World blackcap, *Sylvia atricapilla.*

black-capped chickadee *Vertebrate Zoology.* a mostly gray chickadee, *Parus atricapillus,* with a black cap and throat, and a white cheek patch; found in mixed and deciduous forests, suburbs, and parks of the northern U.S. and Canada. Its call is the familiar "chick-a-dee," in which the first note is a full tone higher than the second.

black-capped chickadee

black chaff *Plant Pathology.* a wheat disease that is caused by the bacteria *Xanthomonas translucens undulosa;* characterized by dark stripes running lengthwise along the chaff.

black cherry *Botany.* **1.** an American wild cherry, *Prunus serotina,* with drooping clusters of fragrant white blossoms and a sour, but edible fruit (drupe). **2.** the fruit and the wood of this tree.

black coal *Mineralogy.* any bituminous coal or anthracite, as distinct from brown coal. *Geology.* see NATURAL COKE.

black cobalt see ASBOLITE.

black compression *Electronics.* the reduction in the gain of a television picture signal at those levels that correspond to dark areas in the picture; its overall effect is to reduce the contrast in the low lights of the picture. Also, BLACK SATURATION.

black copper *Metallurgy.* impure copper originating from an oxide ore smelted in a blast furnace.

black coral *Invertebrate Zoology.* a coral, in the order Antipatharia, with polyps arranged around an axial skeleton composed of a black horny material and bearing thorns; mainly located at deep ocean depths.

black coring *Materials Science.* a condition usually resulting from the premature firing of the exterior of a ceramic piece that prevents the oxidation of carbonaceous material and sulfur compounds, so that the interior remains in a reduced state, causing a black color.

blackdamp *Mining Engineering.* a nonexplosive, heavier-than-air, mixture of carbon dioxide and other gases, especially nitrogen, which is unable to sustain life. Also, CHOKEDAMP.

Black Death see PLAGUE.

black diamond see CARBONADO.

black disease *Veterinary Medicine.* an infectious, usually fatal, necrotic hepatitis of sheep and sometimes of cattle or pigs caused by an organism *(Clostridium novyi)* that multiplies in parts of the liver previously damaged by the common liver fluke; it can be controlled by vaccination.

black drop *Astronomy.* an optical illusion that creates an apparent extension of the disks of Mercury or Venus when seen telescopically in transit across the face of the Sun.

black durain *Geology.* a brownish-black band of a coal mass, characterized by a dull luster and high hydrogen content and containing many microspores, volatiles, and some vitrain.

black dwarf *Astronomy.* the final stage of a white dwarf when it is devoid of light or heat; black dwarfs are difficult to detect because they emit no radiation.

black earth *Materials Science.* finely ground brown coal that is used as a pigment. Also, COLOGNE EARTH, CASSEL BROWN, VANDYKE BROWN.

black end *Plant Pathology.* **1.** a nonparasitic disease of pears, believed to be caused by an irregular water supply; characterized by a blackening of the skin and flesh near the calyx. **2.** a disease of bananas caused by several fungal species, especially *Gloesporium musarum;* characterized by the darkening of the fruit stem.

blacker-than-black *Electronics.* a voltage value that is greater than the value representing the black portions of the image; used to control impulses in a composite television signal. *Graphic Arts.* in four-color printing, the addition of an extra black ink as a fifth color.

Blackett, Baron Patrick 1897–1974, English physicist; Nobel Prize for research on cosmic rays and improvement of cloud chamber.

black-eyed pea see COWPEA.

blackfellow's bread *Mycology.* the edible part of the fungus *Polyporus mylittae,* which occurs in Australia.

blackfire *Plant Pathology.* a tobacco disease caused by the bacterium, *Pseudomonas angulata;* characterized by dark, angular spots on the leaves that may eventually fall out, leaving ragged holes.

black fish *Vertebrate Zoology.* **1.** any of several dark-colored fishes, such as the sea bass or the tautog. **2.** a small, edible fish, *Dallia pectoralis,* of Alaska and Siberia; characterized by its ability to survive below-freezing water temperatures.

black fly *Invertebrate Zoology.* a minute gnat of the dipterous family Simulidae, having a black body and aquatic larvae; the female can inflict a painful bite.

black frost *Meteorology.* an intense cold with no deposit of hoar frost, causing vegetation to turn black.

black granite see DIORITE.

black gunpowder see BLACK POWDER.

black haw *Botany.* **1.** a shrub or small tree of the eastern and southern United States, *Viburnum prunifolium,* of the honeysuckle family; characterized by large clusters of white flowers and bunches of bluish-black berries. Also, STAGBUSH. **2.** a similar shrub, the sheepberry.

blackhead *Medicine.* an accumulation of keratin and sebum within the dilated opening of a hair follicle, having a dark appearance because of the effect of oxygen on sebum; the basic lesion of acne vulgaris. *Veterinary Medicine.* an infectious disease affecting turkeys and occasionally other domestic and wild fowl caused by *Histomonas meleagridis;* characterized by lesions of the intestine and liver and a dark discoloration of the comb. Also, HISTOMONIASIS.

blackhead disease *Plant Pathology.* **1.** a disease afflicting bananas, caused by parasitic eel worms of the family Tylenchidae. **2.** a disease of banana rootstock caused by the fungus *Thielaviopsis paradoxa;* characterized by breakdown and decay of the plant.

blackheart *Botany.* a variety of cherry with a sweet taste, a heart shape, and a black skin. *Plant Pathology.* a disease of plants, especially of potatoes and various trees; characterized by a blackening of the internal tissues, usually as a result of extremes in temperature.

blackheart malleable see MALLEABLE CAST IRON.

black hole *Physics.* an object in space so dense that no light or radiation can escape from it. Black holes, which cannot be observed directly, are believed to be formed when a star collapses inward upon itself.

black ice *Hydrology.* **1.** a thin layer of clear ice formed on land, freshwater, or saltwater, which appears dark because of its transparency. **2.** dark glacier ice formed by the freezing of salt water. **3.** see GLAZE.

blacking *Metallurgy.* a carbon-base material, such as graphite, used to coat metallurgical equipment to prevent the molten metal from sticking to the surfaces of the equipment.

black iron oxide see FERROUS OXIDE.

black jack see SPHALERITE.

black kernel *Plant Pathology.* a disease of rice plants caused by the fungus *Curvularia lunata;* characterized by the appearance of abnormally darkened kernels.

black knot *Plant Pathology.* a disease of plants, especially of cherries, plums, apricots, and some nuts, caused by the fungus *Dibotryon morbosa;* characterized by black knotlike swellings on the twigs and limbs.

black lead see GRAPHITE.

blackleg *Veterinary Medicine.* an acute infectious disease of cattle and sheep, caused by the bacterium *Clostridium chauvaei;* characterized by a crepitant swelling of the heavy muscles. Also, **blackquarter**.

black level *Electronics.* the level of a television picture signal that corresponds to the maximum limit of black peaks.

black light *Optics.* invisible forms of light such as ultraviolet radiation that cause fluorescent materials to emit visible light.

black lignite *Geology.* a variety of brown coal having a heat value between 6300 and 8300 Btu/lb. Also, LIGNITE A.

black line *Plant Pathology.* a disorder of grafted walnut trees, characterized by a black line of dead tissue at the graft point, that can eventually cause the entire tree to die.

black liquor *Materials.* the residual liquid from wood chips treated by a sulfite process.

black measles *Medicine.* a rare, severe, often fatal form of measles; characterized by hemorrhage into the skin lesions and mucous membranes, convulsions, delirium, and marked respiratory distress. *Plant Pathology.* a disease of unknown cause affecting California grapevines; characterized by black spots on the grapes, brown, dry leaves, and a dying back of the canes from the tip.

black membrane *Cell Biology.* a planar lipid bilayer that is synthetically formed at the interface of two aqueous compartments, and typically used to study the permeability properties of artificial membranes.

black mica see BIOTITE.

black mold *Plant Pathology.* a fungus disease affecting a wide variety of plants; characterized by superficial dark brown or black blotches or coatings on leaves, stems, and fruit. *Mycology.* a species of fungi belonging to the genus *Aspergillus,* having black spore heads; for example, *A. niger*.

black mud *Geology.* a dark, sulfide-containing mud formed under anaerobic conditions.

black nightshade *Botany.* a common weed, *Solanum nigrum,* of the nightshade family, having white flowers, poisonous leaves, and edible berries that are poisonous when not ripe.

blacknose *Plant Pathology.* a disease afflicting dates in which the distal end of the fruit becomes dark, brittle, and withered.

black oak *Botany.* any of several North American oak trees, such as *Quercus velutina,* with a dark or blackish bark, common in the eastern United States.

black opal *Mineralogy.* any opal that displays red or green highlights against a dark body color.

blackout *Military Science.* an order to put out or conceal all lights in an area at night so that they will not be seen by enemy aircraft. *Electricity.* a sudden loss of alternating-current line power as a result of deliberate means or power failure. *Telecommunications.* **1.** a ban on live local broadcasting. **2.** a complete disruption of communications due to a faulty transmission medium; sometimes referred to as **radio blackout**. *Medicine.* a condition due to diminished circulation to the brain, characterized by a temporary loss of vision and momentary unconsciousness.

blackout lamp *Ordnance.* a light for use on vehicles during blackouts, designed to be difficult to observe from the air. Also, **blackout light.**

black oxide *Metallurgy.* the oxide intentionally produced on a metallic material by immersing it in a suitable solution for the purpose of creating a protective or aesthetic coating.

black patch or **blackpatch** *Plant Pathology.* a disease of red and white clover caused by an unidentified fungus; marked by the simultaneous appearance of brown or blackish lesions on a group of the plants.

black peak *Telecommunications.* the peak excursion of a black signal at any time in a transmission.

black pepper *Food Technology.* the dried berries of the *Piper nigrum* plant, used as a spice; or the vine itself.

black piedra *Medicine.* a fungus disease that affects the hair and that can be identified by black, gritty nodules on the hair shafts of the scalp; caused by *Piedraia hortae,* the condition usually occurs in tropical climates of the Americas, Africa, Southeast Asia, and Indonesia.

black pit *Plant Pathology.* a disease of citrus fruits, particularly lemons, caused by the bacteria *Erwina citrimaculans;* characterized by depressed dark brown spots on the fruit.

black pod *Plant Pathology.* a disease of the cacao plant caused by the fungus *Phytophthora palmivora;* characterized by the darkening and subsequent decay of the plant pod.

black point *Plant Pathology.* a disease of cereal grains caused by bacteria of the *Pseudomonas* and *Xanthomonas* genera, and by various fungi; characterized by a darkening of the grain embryo ends.

black powder *Materials.* an explosive powder consisting of saltpeter, sulfur, and charcoal; used primarily for fireworks and in old guns fired for sport. Also, BLACK GUNPOWDER.

black pox *Veterinary Medicine,* an ulcerative mastitis of cow teats exhibiting a black spot in the center of each ulcer.

black printer *Graphic Arts.* a black film component in color reproduction that is combined with the three primary colors in order to sharpen neutral tones and details.

black ring *Plant Pathology.* **1.** a virus disease of members of the cabbage family, characterized by black concentric rings or spots on plant leaves and often black flecks in the roots as well. **2.** a virus disease of various plants characterized by small dark rings or ring patterns on young leaves.

black root rot *Plant Pathology.* **1.** any of several plant diseases caused by various fungi and characterized by dark lesions on the root and reduced vigor and stunted growth aboveground. **2.** an apple disease caused by the fungus *Xylaria mali.* **3.** a disease of tobacco plants and other plants caused by the fungus *Thielaviopsis basicola*.

black rot *Plant Pathology.* any plant disease caused by bacteria, such as *Xanthomonas campestris,* or fungi, such as *Guignardia bidwellii;* characterized by the darkening and decay of the plant.

black rust *Plant Pathology.* any of various diseases of plants, especially of wheat and other cereals and grasses, caused by rusts and characterized by black discoloration.

black sand *Geology.* a sand that consists mainly of grains of heavy, dark-colored rocks or minerals such as magnetite and ilmenite, and that occasionally yields gold and platinum.

black saturation see BLACK COMPRESSION.

black scope *Electronics.* a cathode-ray tube operating on the threshold of luminescence without the application of video signals.

black scour *Veterinary Medicine.* a hemorrhagic enteritis of sheep, swine, or cattle, mainly affecting young animals; associated with heavy worm burden; it may also be caused by bacterial infection or poor diet.

Black Sea *Geography.* a large inland sea in southeastern Europe, north of Turkey and south of the USSR; connected with the Mediterranean by the Bosporus, Sea of Marmara, and Dardanelles.

black sea bass *Vertebrate Zoology.* a black-striped sea bass, *Centropristes striata,* found off the coast of eastern North America; an important food fish.

black shale *Petrology.* a thinly bedded shale that is rich in carbon, sulfide, and organic material; formed by anaerobic decay of organic matter.

black shank *Plant Pathology.* a tobacco disease caused by the fungus *Phytophthora parasitica nicotianae,* characterized by a darkening and rotting of plant leaves and stems, and a damping off of the seedlings.

black signal *Telecommunications.* a signal that corresponds to the darkest part of a transmitted image. Also, BLACK.

black silver see STEPHANITE.

blacksmith's drill *Mechanical Devices.* a drill fitted with a one-half inch diameter flat shank to fit a certain kind of holder.

black smoke *Engineering.* smoke containing a relatively large amount of solid carbon particles, resulting from inefficient burning.

blacksnake *Vertebrate Zoology.* **1.** the American black racer, *Coluber constrictor,* found in the eastern U.S. that grows as long as six feet. Also, **black racer. 2.** a large, semiaquatic, venomous snake of Australia, whose body is black on top and red underneath.

blacksnake

black snow *Hydrology.* a term for dark-colored snow that results from falling through a particulate-laden atmosphere.

black spot *Plant Pathology.* a plant disease caused by bacteria or fungi, such as *Diplocarpon rosae;* characterized by black spots or blotches on plant parts, defoliation, and rotting.

black stem *Plant Pathology.* any of various plant diseases caused by fungi such as *Ascochyta imperfecta* or *Mycosphaerella lethalis;* characterized by darkening of the plant stem and defoliation.

blackstrap molasses *Food Technology.* a syrupy liquid that remains after sugar has been completely refined; used in the production of alcohols and cattle feed.

black-surface enclosure *Thermodynamics.* an enclosure surrounding a thermodynamic system in which the interior surface of the enclosure has the properties of a black body.

black-surface field *Electronics.* a layer of p^+ material applied to the back surface of a solar cell to reduce hole-electron recombinations and increase the cell's efficiency.

black-tailed deer *Vertebrate Zoology.* a variety of the mule deer, *Odocoileus hemionus columbianus,* having a black tail; found on the western slope of the Rocky Mountains.

black thread *Plant Pathology.* a disease of the para rubber tree caused by the fungus *Phytophthora meadii*; it causes black stripes at the exposed bast where the tree is tapped that extend into the cambium or wood. Also, **black stripe, black canker.**

black tip *Plant Pathology.* any of various plant diseases characterized by darkened, dead areas at the tip of a fruit or seed.

blacktongue *Veterinary Medicine.* a now rare pellagra of dogs characterized by black discoloration of the tongue and lesions of the mouth and digestive tract; caused by a deficiency of niacin in the diet.

blacktop *Materials Science.* **1.** to cover a road or other surface with asphalt or a similar dark bituminous material. **2.** the material used in such a process, or a road covered in this manner.

blacktop paver *Mechanical Engineering.* a vehicle, such as a tank truck, that spreads a desired thickness of heated bituminous mixture over the prepared surface of a road, parking lot, or the like. Also, BITUMINOUS DISTRIBUTOR.

black transmission *Telecommunications.* a type of transmission in which the lowest transmitted frequency is equivalent to the darkest part of the object whose image is being transmitted.

blackwater fever *Pathology.* a severe form of malaria characterized by kidney failure and bloody urine, occurring primarily in tropical areas.

Blackwell, Elizabeth 1821–1910, British-American physician; first woman in the U.S. to receive a medical degree.

black whale see SPERM WHALE.

black widow *Invertebrate Zoology.* a venomous spider, *Latrodectus mactans,* found widely in the U.S. and in Canada; the female has a shiny black body with a red or yellow, hourglass-shaped mark on the underside of its abdomen.

Black Widow *Aviation.* a popular name for the F-61 night fighter.

black yeast *Mycology.* a fungus that produces black yeastlike cells; occurs primarily on dairy products.

bladder *Anatomy.* **1.** a musculomembranous sac in the anterior part of the pelvic cavity that serves as a reservoir for urine. **2.** any membranous sac that serves as a receptacle for a secretion.

bladder cell *Invertebrate Zoology.* a blood or an ameboid cell located in the outer layer of the tunics of certain tunicates.

bladder press *Mechanical Engineering.* a machine for building pneumatic tires that both shapes and cures (vulcanizes) the rubberized material at the same time.

bladder worm see CYSTICERCUS.

blade the flat, cutting part of an object; specific uses include: *Mechanical Devices.* the flat arm of a fan, turbine, or propeller. *Aviation.* a propeller arm or rotary wing, especially the part that cleaves the air and has an efficient airfoil shape. *Botany.* **1.** the flat, expanded portion of a leaf. **2.** the flat, leaflike thallus of certain brown algae, such as kelp. *Electrical Engineering.* the moving part of a knife-switch that makes contact with the fixed jaws and conveys the current. *Archaeology.* a long, thin flake of stone that is detached from a prepared core; generally characterized by a length at least twice the width. *Vertebrate Zoology.* a single plate of the bony growths hanging down from the upper jaws or palates of some whales.

blade angle *Aviation.* **1.** the angle between the zero-lift chord of a propeller blade and a plane perpendicular to its angle of rotation. **2.** the angle between the zero-lift chord of a rotor blade and a plane perpendicular to a reference angle through the rotor.

blade back *Aviation.* the side of a propeller that faces forward, corresponding to the top of a wing.

blade-coated *Materials Science.* describing paper or other material subjected to a blade-coating process.

blade coating *Materials Science.* a process in which a polymer is applied to a moving substrate, and the coating thickness is determined by a movable blade.

blade face *Aviation.* the side of a propeller opposite the blade back.

blade harrow see ACME HARROW.

blade loading *Aviation.* the load sustained by the rotor blades of a helicopter; determined by dividing the gross weight of the aircraft by the total area of its blades.

blade root *Aviation.* the part of a propeller blade nearest the hub; the point at which a blade shank joins the hub.

blade shank *Aviation.* the part of a propeller arm or rotary wing between the blade proper and the hub. The shank of a propeller blade usually has a round or oval cross section.

blade stall *Aviation.* a condition on a rotary-wing aircraft that occurs when the blade operates at an angle of attack greater than the angle of maximum lift.

blae see BITUMINOUS SHALE.

Blagden's law *Physical Chemistry.* a rule stating that the lowering of a solution's freezing point is proportional to the molar concentration of the dissolved substance.

Blair-Brown knife *Surgery.* a knife with a long, sharp blade that is used to cut skin grafts.

blairmorite *Petrology.* a porphyritic volcanic rock consisting mainly of analcite phenocrysts in a groundmass of sanidine, analcite, alkalic pyroxene with traces of sphene, melanite, and nepheline.

Blaise reaction *Organic Chemistry.* the combination of α-bromocarboxylic esters with zinc, in the presence of nitriles, giving ß-oxoesters. An organozinc intermediate reacts with the nitrile, then the complex is hydrolyzed with 30% potassium hydroxide.

Blake breaker see BLAKE JAW CRUSHER.

blakeite *Mineralogy.* $Fe_2(TeO_3)_3$, a ferric tellurite mineral, reddish to deep brown, found in Goldfield, Nevada, and occurring as microcrystalline crusts; crystal system unknown.

Blake jaw crusher *Mechanical Engineering.* a heavy-duty rock-breaking machine, having one fixed-jaw plate and an inclined swing jaw pivoted at the top, between which large lumps of material are crushed; this contrasts with the Dodge jaw crusher, in which the movable jaw is pivoted at the bottom. Also, ALLIGATOR, BLAKE BREAKER, JAW BREAKER.

Blake number *Fluid Mechanics.* a dimensionless group showing the ratio of inertial force to viscous force for flow of a fluid through beds of solids.

Blancan *Geology.* a North American geologic stage of the lowermost Pleistocene epoch, after the Hemphillian and before the Irvingtonian.

blanc fixe [blängk´ fēks´] *Inorganic Chemistry.* a synthetic form of barium sulfate, $BaSO_4$, a white precipitate formed by reacting barium chloride with sodium sulfate; used as a pigment extender. (From French; literally, "fixed white.")

blanching *Food Technology.* the process of scalding or parboiling raw foods in steam or boiling water prior to canning; used to soften texture, remove air, and destroy bacteria and enzymes.

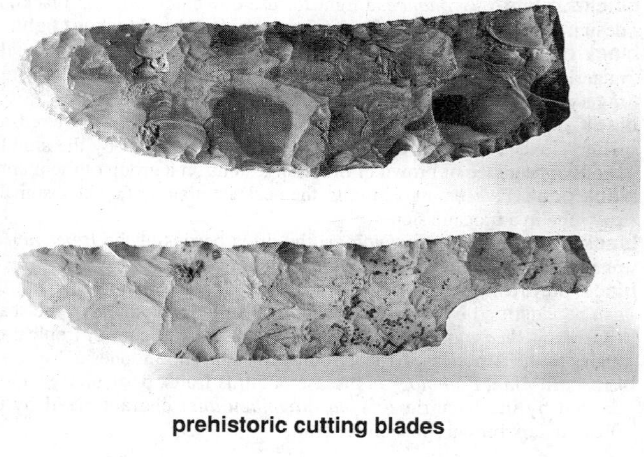

prehistoric cutting blades

Blanc rule [blängk] *Organic Chemistry.* a rule stating that glutaric and succinic acids will yield cyclic anhydrides on pyrolysis, while adipic and pimelic acids yield cyclic ketones; there are certain exceptions.

blandel see APOSTILB.

blank unfinished or empty; special uses include: *Metallurgy.* **1.** a cut piece of sheet metal before it is shaped by a forming or finishing operation. **2.** a pressed and sintered semifinished component in powder metallurgy. *Mechanical Devices.* a key that has not yet been cut to a specific shape. *Electronics.* **1.** describing an audiotape or videotape that has no recorded material on it. **2.** to cut off the electron beam of a cathode-ray tube. *Computer Programming.* describing a computer disk that has no data stored on it. *Archaeology.* any usable, unmodified flake of stone that has a size and form suitable for working into a tool. *Ordnance.* a cartridge that contains an explosive charge but no projectile, designed to produce a noise without inflicting damage; used in training, signaling, and so on. Also, **blank ammunition** or **blank cartridge.**

blank carburizing *Metallurgy.* the process of simulating the carburizing process without introducing carbon into a ferrous workpiece; usually accomplished by using an inert material instead of the carburizing agent, or by applying a suitable protective coating to the ferrous alloy.

blank character *Computer Programming.* any character or bit pattern that produces a physical space in stored data or on an output medium. Also, SPACE CHARACTER.

blank column detection *Control Systems.* the collator function of checking for a blank column in a data field and signaling an error condition if one is found.

blank disk *Computer Programming.* see BLANK.

blanket a device or material that covers or hides something in the manner of a blanket; specific uses include: *Telecommunications.* to obliterate or override a weak radio signal by transmitting a stronger one. *Nucleonics.* in a breeder reactor, an area surrounding the core containing fertile material (thorium-232 or uranium-238) that is converted into a fissile species (uranium-233 or plutonium-239) by neutron capture and subsequent radioactive decay. *Graphic Arts.* a rubber-coated fabric that covers the blanket cylinder of an offset press. *Mining Engineering.* **1.** a textile material used in ore-treatment plants for catching coarse free gold, and associated minerals, such as pyrite. **2.** any soil or broken rock left or placed over a blast to confine or direct throw of fragments.

blanket bog *Ecology.* an extensive layer of peat over flat and rolling land, formed by bog vegetation and deriving its nutrients primarily from rainfall; found in cold, wet parts of the world.

blanket cloth *Textiles.* a thick heavyweight fabric of wool or cotton that provides good thermal insulation.

blanket cylinder *Graphic Arts.* in an offset press, the rubber-covered cylinder that receives the inked design from the plate cylinder and then transfers (offsets) it onto the paper.

blanket deposit *Geology.* a flat, horizontal ore body or sedimentary deposit of relatively uniform thickness, whose length and width greatly exceed its thickness.

blanket gas *Chemical Engineering.* a gas, such as inert nitrogen, used to blanket the vapor space of a vessel or storage tank containing a liquid feed or product in order to prevent or reduce vapor loss, contamination, fire, or vapor explosion. Also, CUSHION GAS.

blanket grouting *Building Engineering.* a form of ground treatment used beneath concrete and earth dams to consolidate the foundation to a depth of 20–50 feet; holes are drilled using a low-pressure grouting technique on a 10–20 foot grid that controls and prevents the upheaval of the foundation block.

blanketing *Telecommunications.* the broadcasting of a signal, usually close to the antenna, which interferes with the reception of other signals being broadcast at different frequencies.

blanket insulation *Building Engineering.* an insulation used in construction that has been vaporproofed by the addition of a paper backing; usually taken from a rolled sheet.

blanket sand *Geology.* a blanket deposit of sand or sandstone covering an unusually large area.

blanket sluice *Mining Engineering.* a gently down-sloping sluice with a bottom lining of coarse blankets on which the fine, but heavy particles of gold, amalgam, etc., are caught.

blanket-to-blanket press *Graphic Arts.* an offset press having two units, each containing a blanket cylinder that serves as the impression cylinder for the other.

blank flange *Mechanical Devices.* **1.** a solid disk used to close off a pipe end. Also BLIND FLANGE. **2.** a flange that is finished except for a drilled hole or threads.

blank hole. *Mining Engineering.* a borehole in which no minerals or other valuable substances were encountered. Also, BARREN HOLE, DRY HOLE.

blanking *Engineering.* **1.** the process of shearing or punching shapes from metal or plastic sheets. **2.** the insertion of a circular barrier in a pipe joint in order to cut off the flow of liquid during repair. Also, BLINDING. *Electronics.* the process of disabling or blocking a circuit for a specified period of time.

blanking circuit *Electronics.* in television scanning, a circuit that prevents the transmission of brightness variations during horizontal and vertical retrace intervals.

blanking die *Mechanical Devices.* a press that uses dies of a specified size to cut out a strip of sheet metal.

blanking level *Electronics.* the level in a composite picture signal that separates the range of the composite signal containing picture information from the range containing synchronizing information.

blanking pulse *Electronics.* a positive or negative square wave used to switch off a part of a television or radar set electronically for a predetermined period of time.

blanking signal *Electronics.* a wave of recurrent pulses that are related in time to the scanning processes and are used to effect blanking; they occur at both the line and field frequencies, cutting off the electron beam during retrace at both transmitter and receiver.

blanking time *Electronics.* the period of time in which the electron beam of a cathode-ray tube is cut off.

blank nitriding *Metallurgy.* the process of simulating the nitriding operation without introducing nitrogen; usually accomplished by using an inert material instead of the nitriding agent, or by applying a suitable protective coating to the ferrous alloy.

blank stamping see BLIND STAMPING.

blank tape *Electronics.* see BLANK.

Blasiaceae *Botany.* a cosmopolitan family of branched liverworts of the order Metzgeriales, characterized by a distinct midrib that merges gradually with the lamina, colorless rhizoids on the midrib's ventral surface, and distinct ventral scales borne in two rows.

blast *Physics.* an explosion or other such phenomenon that causes a sudden shock to the surrounding air or ground. *Engineering.* **1.** the act of setting off an explosion. **2.** a sudden gust of air or jet of water released under pressure. **3.** air that is forced into a furnace by a blower to increase the rate of combustion. *Computer Programming.* to release areas of main or secondary memory storage when they are no longer required by the operational program, making them available for reallocation.

-blast see BLASTO-.

blastaea *Invertebrate Zoology.* a hypothetical metazoan ancestral form corresponding in organization to the blastula larva, yet existing as an abundant adult organism.

blast area *Ordnance.* **1.** a scorched area of earth around the muzzle of a gun, caused by the effect of repeated firing. Similarly, **blast mark. 2.** any ground area affected by the force of an explosive blast.

blast bomb see LIGHT-CASE BOMB.

blast burner *Engineering.* a burner in which a continuous flow of pressurized oxygen or air fuels the illuminating gas.

blast cell *Histology.* one of the undifferentiated cells in the human reticuloendothelial system that produces blood cells. *Immunology.* a cell, usually a primitive precursor cell, that is capable of division.

blast chamber *Aviation.* the combustion chamber of a gas turbine, jet, or rocket engine. Also, COMBUSTION CHAMBER.

blast ditching *Civil Engineering.* the process of using explosives as a part of trench excavation, as in laying pipelines.

blast effect *Physics.* **1.** the violent air movement and pressure changes that are associated with an explosion. **2.** the damage that is inflicted by an explosion.

blastema *Developmental Biology.* **1.** the embryonic cellular mass that develops into an organ or part. **2.** the cluster of cells that consolidates at a site of tissue regeneration and then forms the new structure.

blaster *Engineering.* a device used to detonate an explosive. Also, **blasting machine.**

blast furnace *Metallurgy.* a vertical shaft furnace in which solid fuel, such as coke, is burned with an air blast to smelt ore in a continuous operation

blast gate *Metallurgy.* a gate in the blow pipe of a cupola furnace, used to control air flow.

blasthole or **blast hole** *Engineering.* **1.** a hole used to hold an explosive charge. **2.** the hole through which water enters the bottom of a pump.

blasthole drill

blasthole drill *Engineering.* any type of drilling machine used to produce holes in which an explosive charge is placed. Thus, **blasthole drilling.**

-blastic a combining form meaning "having many buds, germs, cells, or cell layers," or "undergoing a given type of development."

blasticidin S *Organic Chemistry.* $C_{17}H_{26}N_8O_5$, crystals that are soluble in water and insoluble in alcohol; decomposes at 235–236°C; used as a fungicide for rice crops.

blasting *Mining Engineering.* the operation of breaking coal, ore, rock, or other hard material by making a hole in it, inserting an explosive charge, and detonating or firing it. Also, SHOT FIRING. *Metallurgy.* the process of cleaning a metallic surface by the impact of high-velocity particles of an abrasive material, such as sand or hard metal shots. *Geology.* the abrasion or erosion of a stationary surface caused by the force of fine particles of sand and dust driven by wind or water. *Acoustics.* **1.** the act of producing or especially reproducing sounds with excessive volume. **2.** sound distortion that is caused by the overloading of an amplifier or receiver.

blasting agent *Materials Science.* a cap-insensitive chemical composition that contains no explosive ingredient and that can be made to detonate when initiated with a high-strength explosive primer.

blasting cap *Engineering.* **1.** a copper shell closed at one end and containing a charge of detonating compound, ignited by electric current or the spark of a fuse; used for detonating high explosives. **2.** a small sensitive charge placed in a larger explosive charge to detonate the larger charge.

blasting fuse *Engineering.* a length of slow-burning material containing or saturated with a combustible substance; used to lead to and set off an explosive charge from a distance. It has a given burning rate in feet per minute.

blasting gelatin *Materials Science.* a high-explosive material consisting of nitroglycerin and nitrocotton, whose rubberlike, elastic form is unaffected by water; used chiefly in underwater work.

blasting powder *Materials Science.* a form of gunpowder using sodium nitrate, $NaNO_3$, instead of potassium nitrate, KNO_3; used chiefly for blasting rock and ore.

blast lamp see BLOWTORCH.

blasto- *Developmental Biology.* a prefix meaning "relating to a bud or budding," particularly to an early embryonic stage, as a primitive element, cell, or layer. Also, -BLAST. *Petrology.* a prefix used to indicate the presence of metamorphosed residual structures in a rock.

Blastobacter *Bacteriology.* a genus of Gram-negative, aerobic bacteria that reproduce by budding and that are made up of rod-shaped or wedge-shaped cells attached in a rosette formation; found in ponds and lakes.

Blastobasidae *Invertebrate Zoology.* a family of moths, in the order Lepidoptera.

blastocarpous *Botany.* germinating and developing while still in the pericarp.

blastochyle *Developmental Biology.* the fluid contained in the blastocoele.

Blastocladiaceae *Mycology.* a family of fungi belonging to the order Blastocladiales that occurs on land and in water, living off dead plants and animal debris; widely used in biological research.

Blastocladiales *Mycology.* an order of fungi belonging to the class Chytridiomycetes that occurs in soil or water and lives off plants, invertebrates, or dead organic matter.

blastocoel or **blastocoele** *Developmental Biology.* the fluid-filled cavity of the blastula that develops from the cleavage of a fertilized egg. Also, CLEAVAGE CAVITY, SEGMENTATION CAVITY.

blastocone *Developmental Biology.* an undeveloped blastomere.

Blastocrithidia *Invertebrate Zoology.* a genus of protozoa of the order Kinetoplastida, parasitic in arthropods and other invertebrates, that pass through epimastigote, amastigote, and presumably promastigote developmental stages in their life cycle.

blastocyst *Developmental Biology.* the modified blastula stage of mammalian embryos, made up of the inner cell mass and a thin trophoblast layer that encloses the blastocele. Also, BLASTODERMIC VESICLE.

Blastocystis *Invertebrate Zoology.* a wide-ranging genus of commensal yeasts common in human feces; may be confused with pathogenic protozoans during fecal exams.

blastocyte *Developmental Biology.* an undifferentiated blastomere of the morula or blastula stage of an embryo.

blastocytoma see BLASTOMA.

blastoderm *Developmental Biology.* the thin, disk-shaped cell mass of a young telolecithal embryo and its extraembryonic extensions over the surface of the yolk; when complete, all three of the primary germ layers (ectoderm, endoderm, and mesoderm) are present. Also, GERMINAL MEMBRANE.

blastodermic vesicle see BLASTOCYST.

Blastodiniaceae *Botany.* a monogeneric family of parasitic algae of the order Blastodiniales, occurring in the stomach of marine planktonic Copepoda.

Blastodiniales *Botany.* an order of ectoparasitic algae of marine invertebrates and fish, belonging to the class Dinophyceae.

blastodisc *Developmental Biology.* **1.** the disc of active cytoplasm at the animal pole of a telolecithal ovum. **2.** the blastoderm, especially in very early stages.

blastoff *Aviation.* **1.** the launching of a missile, rocket, or spacecraft. **2.** the time at which this occurs.

blastogenesis *Biology.* **1.** the development of an individual from a blastema, that is, by asexual reproduction. **2.** the transmission of inherited characters by the germ plasm. *Immunology.* a process in which small lymphocytes are transformed into primitive blastlike cells; caused by agents such as phytohemagglutinin, certain antigens, or leukocytes from an unrelated individual.

blastogranitic rock *Petrology.* a metamorphic granitic rock in which parts of the original granitic texture remain intact.

Blastoidea *Paleontology.* a class of Paleozoic echinoderms in the subphylum Pelmatozoa that were sessile, with a relatively simple cup at the end of a stem; occurring in the Silurian to Lower Permian.

blastokinesis *Invertebrate Zoology.* the rotational displacement of the developing embryo within the egg of certain insects.

blastoma *Oncology.* a neoplasm composed of embryonic cells derived from the blastema of organs and tissues. Also, BLASTOCYTOMA.

blastomere *Developmental Biology.* one of the cells into which the ovum separates after fertilization. Also, CLEAVAGE CELL, SEGMENTATION SPHERE.

blastomerotomy see BLASTOTOMY.

Blastomyces *Mycology.* a genus of thermal dimorphic imperfect fungi of the family Moniliaceae that grow as mycelial forms at room temperature and as yeastlike forms at body temperature; pathogenic for humans and animals.

Blastomycetes *Mycology.* formerly a class of fungi belonging to the subdivision Deuteromycotina, now included in the class Hyphomycetes; comprising yeast fungi that reproduce by budding or ballistospores.

blastomycosis *Pathology.* **1.** an infection caused by organisms of the genus *Blastomyces.* **2.** any infection caused by a yeastlike organism.

blastomylonite *Petrology.* metamorphosed rock that has recrystallized after granulation.

blastopelitic *Petrology.* of or relating to the structure of metamorphosed argillaceous rocks dominated by relict sedimentary textures that identify the kindred of the parent rock.
blastophere see BLASTULA.
blastophitic *Petrology.* of or relating to the relict texture of a metamorphic rock in which traces of an original ophitic texture remain.
blastophore *Cell Biology.* the cytoplasmic portion of a spermatid that is not converted into a spermatozoon.
blastopore *Developmental Biology.* the opening leading into the archenteron developed by the invagination of the blastula. Also, ARCHISTOME, BLASTOPORE LIP, PROTOSTOMA.
blastopore lip *Developmental Biology.* the tissue at the edge of the blastopore.
blastoporphyritic *Petrology.* relating to the relict texture of a metamorphic rock in which traces of an original porphyritic texture remain.
blastopsammite *Geology.* a metamorphosed sandstone in which relicts of the parent rock remain. Thus, **blastopsammitic.**
blastopsephite *Geology.* a metamorphosed conglomerate or breccia in which relicts of the original rock remain. Thus, **blastopsephitic.**
blastosphere see BLASTULA.
blastospore *Mycology.* a kind of fungal spore produced as a bud in yeasts.
blastostyle *Invertebrate Zoology.* a process or modified feeding polyp in certain thecate hydroids from which medusae bud and are produced asexually.
blastotomy *Developmental Biology.* the destruction of cleavage cells during embryo formation. Also, BLASTOMEROTOMY.
blastozooid *Invertebrate Zoology.* an individual developed as a result of asexual reproduction.
blast pressure *Physics.* the air pressure resulting from an explosion.
blast roasting *Mining Engineering.* a process of oasting by blasting to maintain internal combustion within the charge of finely divided ores; done in a Dwight-Lloyd machine, in which roasting is accompanied by sintering.
blast transformation *Immunology.* a process of activating small lymphocytes in a culture so that they increase in size and form cells resembling lymphoblasts.
blastula *Developmental Biology.* the stage of an embryo following cleavage and preceding gastrulation, usually consisting of a spherical blastoderm surrounding a blastocoele. Also, BLASTOSPHERE.
blastulation *Developmental Biology.* the development of the blastula or blastocyst.
blast wall *Engineering.* a barrier used to shield an area from the effects of an explosion.
blast wave *Physics.* a pressure wave produced by an explosion, radiating outward from the center of the blast.
Blattabacterium *Bacteriology.* a genus of Gram-negative, nonmotile, rod-shaped bacteria of uncertain affilliation that occur as intracellular endosymbionts of the cockroach.
Blattidae *Invertebrate Zoology.* a family of cockroaches, in the suborder Blattaria, including the cosmopolitan domestic cockroaches.

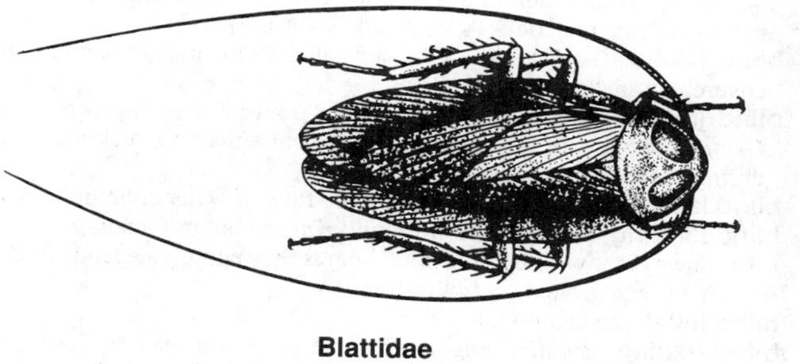

Blattidae

B layer *Geology.* in the classification of the earth's interior, the seismic region equivalent to the low-velocity zone of the upper mantle, extending from the Mohorovicic discontinuity to 410 km below the surface.
blaze *Cartography.* a highly visible mark made on a tree trunk at about chest height, used to determine the elevation of the surrounding terrain.
blaze-of-grating technique *Optics.* the use of a diffraction grating with ruled grooves to reflect and concentrate light.
blazing star *Botany.* **1.** a group of wild flowers of the genus *Liatris,* having thick clusters of small purple or rose-red blossoms along the stem on thin spikes that resemble a shooting star; found on prairies and meadows in the eastern and midwestern U.S. Also, **rough blazing star.** **2.** a plant, *Mentzelia laevicaulis,* having large, light-yellow flowers with conspicuous stamens. Also, **giant blazing star.**
bld. blood.
bleach *Chemistry.* **1.** to whiten or make lighter by exposure to light or by using a chemical agent. **2.** any chemical used in bleaching, such as hydrogen peroxide, chloride of lime, or sodium hypochlorite.
bleaching *Graphic Arts.* the process of removing the negative silver in developing a positive photographic image. *Building Engineering.* a preparatory treatment for wood to remove exposure stains or to balance color variations before staining or varnishing. *Optics.* a decrease in the optical absorption of an object that arises from a chemical process or from exposure to radiation.
bleaching agent *Chemistry.* a substance that causes whitening or decolorizing of fibers by chemical action.
bleach spot *Geology.* a greenish or yellowish area in a red rock resulting from the reduction of ferric oxide around an organic particle.
bleb *Mineralogy.* a small, usually rounded inclusion of one mineral in another.
Blechnaceae *Botany.* a family of tropical ferns of the order Filicales; characterized by a creeping or erect scaly rhizome, simple or pinnate fronds, free or netted veins, and sori located in a line along the pinna midveins.
bleed *Medicine.* to lose blood from the vascular system, either internally or externally. *Botany.* to lose or remove sap or other liquid from a cut or scratched surface on a plant. *Chemistry.* **1.** to release pressure gradually by withdrawing gas or liquid from a container through a valve. **2.** of a dye, to run or become mixed. *Graphic Arts.* **1.** to extend an illustration beyond the margins to the edge of a printed page after trimming. **2.** an illustration so extended. *Computer Technology.* in optical character recognition, the capillary flow of ink beyond the specified boundaries of a printed character.
bleeder *Medicine.* **1.** a person who bleeds freely or is subject to hemorrhagic diathesis. **2.** any blood vessel cut during a surgical procedure that requires clamping, cautery, or ligature. **3.** one who draws blood. Also, PHLEBOTOMIST. *Electronics.* a resistor connected across a power source to regulate voltage when the load is disconnected or to discharge capacitors when the power is switched off. Also, BLEEDER RESISTOR. *Engineering.* a valve to drain off unwanted fluid or air from a tank or system.
bleeder resistor *Electronics.* see BLEEDER.
bleeder turbine *Mechanical Engineering.* a multistage turbine in which high-pressure steam is expanded to the required pressure and then extracted or bled from the turbine casing for process or feedwater heating purposes.
bleeding *Medicine.* **1.** the escape of blood from an injured vessel. **2.** the letting of blood. *Engineering.* **1.** the purposeful and controlled draining off of steam, water, air, and other fluids through an outlet in a closed or hydraulic system. **2.** the separation and surfacing of liquid from a liquid-solid mixture, as with the water that separates and rises from cement paste after it is poured. *Materials Science.* the segregation of water to the surface of a wet concrete mix. *Textiles.* the running or diffusion of colored dyes in fabric when it is wet.
bleeding canker *Plant Pathology.* a disease of hardwoods caused by the fungus *Phytophthora cactorum;* characterized by sunken areas in the bark that ooze reddish-brown fluid and eventually cause wilting and branch dieback.
bleeding cycle *Mechanical Engineering.* a regenerative steam cycle in which the feedwater is heated by steam extracted from the turbine at one or more stages.
bleeding disease *Plant Pathology.* a disease of the coconut palm caused by the fungus *Ceratostomella paradoxa;* characterized by the oozing of reddish-brown fluid from cracks on the stem.
bleeding time *Physiology.* a measure of blood coagulation time, made by pricking the skin and recording the time required for the drop of blood to stop flowing.
bleed valve *Engineering.* the mechanical device through which undesired fluids in a container or line may be drained off.
bleep *Electronics.* **1.** a short, high-pitched beeping sound inserted on the recorded soundtrack of a videotape or film. **2.** to remove a section of a soundtrack and replace it with such a sound, especially so as to delete obscenities or other objectionable material.
blende *Mineralogy.* **1.** any metallic sulfide mineral having a bright or resinous luster, such as zinc blende or cadmium blende. **2.** see SPHALERITE.

blended unconformity *Geology.* an interruption in the continuity of a sequence of rocks having no distinct surface of separation or sharp contact.

blended whiskey *Food Technology.* a whiskey that is a mixture of two or more whiskeys, or a mixture of whiskey with neutral spirits, and that contains at least 20% of 100-proof straight whiskey by volume after blending.

blending *Metallurgy.* in powder metallurgy, the thorough mixing of powders having the same composition, but different particle sizes and shapes. *Textiles.* the thorough mixing of two or more kinds of fibers to produce a single yarn.

blending inheritance *Genetics.* a supposed pattern of inheritance, according to which traits that are contrasted in the parents will appear as an intermediate blend in the offspring; e.g., a short parent and a tall parent will produce offspring of medium height. Blending inheritance was a pre-Mendelian proposal, now discredited, to account for the observation that offspring typically show some characteristics that are similar to each parent.

blending stock *Chemical Engineering.* any material used to compound petroleum products.

blending value *Chemical Engineering.* the number calculated and utilized to indicate the ability of an additive (such as tetraethyllead, isooctane, or aromatic hydrocarbons) to enchance the octane rating of a gasoline

blend stop *Building Engineering.* a thin wood strip used to hold a sash in position; it is fastened on the exterior vertical edge of a pulley stile or door jamb.

Blenniidae *Vertebrate Zoology.* the combtooth blennies, a family of carnivorous marine fishes in the suborder Blennioidea.

Blennioidei *Vertebrate Zoology.* a suborder of bony marine fishes in the order Perciformes, including the blennies, wolffishes, gunnels, and eelpouts.

blenny *Vertebrate Zoology.* any of several fishes of the family Blenniidae and related families, especially of the genus *Blennius,* characterized by a long tapering body and small pelvic fins in front of the pectoral fins.

bleomycin *Oncology.* a glycopeptide antibiotic that is isolated from *Streptomyces verticellus* and used to treat various cancers, including Hodgkin's disease, squamous cell cancers of the head and neck, and cancers of the testicles.

blepharal *Anatomy.* of or relating to the eyelid.

blephar- or **blepharo-** a combining form meaning "eyelid," as in *blepharitis, blepharostat.*

blepharedema *Medicine.* a swelling of the eyelids.

Blephariceridae *Invertebrate Zoology.* a family of aquatic two-winged insects, in the suborder Orthorrhaphia; the larvae have a row of suckers for clinging to stones in rapidly flowing streams.

Blepharisma *Invertebrate Zoology.* a genus of marine and freshwater ciliate protozoans in the order Heterotricha, spindle-shaped, pyriform, or ovoid.

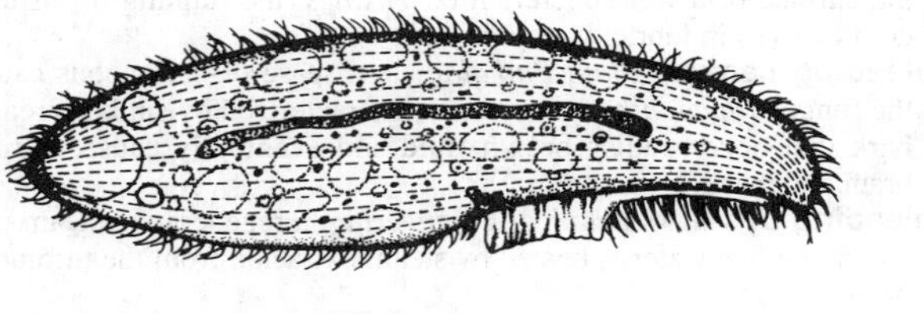

Blepharisma

blepharitis *Medicine.* an inflammation of the eyelid.

blepharoplast see BASAL BODY.

blepharoplasty *Surgery.* a plastic surgery of the eyelids.

blepharoptosis *Medicine.* the drooping of the eyelid. Also, PTOSIS.

blepharospasm *Pathology.* a tonic spasm of the orbicularis oculi muscle, producing more or less complete closure of the eyelids.

blepharosphincterectomy *Surgery.* removal of part of the skin and muscle of the eyelid to relieve pressure on the cornea.

blepharostat *Surgery.* an instrument that holds the eyelids apart during eye surgery.

blewit or **blewitt** *Mycology.* the pale-bluish, edible part of the field mushroom, *Tricholoma personatum,* and wood mushroom, *T. nudum.* Also, BLUETTE, BLUE LEG.

blight *Plant Pathology.* **1.** a plant disease characterized by general and rapid browning of leaves, flowers, and stems, resulting in their death. **2.** the bacterium, fungus, or virus that causes such disease.

blight canker *Plant Pathology.* a stage of fire blight, during which cankers appear on branches or other plant parts; if the canker enlarges and encircles the branch, the part of the branch above the infection dies.

blimp *Aviation.* a nonrigid airship. (Originally a British colloquial expression for "dirigible type b, limp," *blimp* became an official U.S. Navy term in 1939.)

blind *Medicine.* not having the sense of sight; unable to see. *Anatomy.* closed at one end, as a *blind* fistula. *Aviation.* done under conditions of poor visibility with the aid of instruments, radar, or radio aids; used to form compounds such as **blind flying** and **blind landing.** *Graphic Arts.* without ink or other coloring agent. *Engineering.* a solid disk inserted in the joint of a pipe to block fluid flow during repair. Also, BLANK. *Geology.* of a mineral deposit, terminating below the surface. *Hydrology.* describing a lake or pond that has no inflow or outflow.

blind arcade *Architecture.* an arcade whose arch supports are filled in with masonry.

blind arch *Architecture.* a closed arch that does not penetrate the structure; used for ornamentation.

blind coal *Geology.* any type of coal that burns without a flame.

blind-code dating *Food Technology.* a product labeling using codes so that the retailer knows when products should be removed from the shelf, but the consumer does not.

blind creek *Hydrology.* **1.** a creek that is dry except during rainfall. **2.** see CLOSED DRAINAGE.

blind drift *Mining Engineering.* **1.** a horizontal passage in a mine that is not yet connected with the other workings. Also, BLIND LEVEL. **2.** an inverted siphon for water in a mine.

blind drilling *Engineering.* a term for drilling in which the liquid used to aid in boring does not return to the surface.

blind embossing *Graphic Arts.* a raised design on a material, made without ink or leaf.

blindfish *Vertebrate Zoology.* any of several kinds of fishes, such as the cavefish of the family Amblyopsidae; some have rows of small projections with a keen sense of touch in place of eyes; they are found in the waters of caves in the eastern U.S.

blind flange *Building Engineering.* a flange that closes the end of a pipe, producing a blind or dead end.

blind folio *Graphic Arts.* a page that is counted as part of the total page count but does not have a page number (folio) printed on the page.

blind hole *Design Engineering.* a hole that does not extend completely through a workpiece. *Engineering.* **1.** a borehole where the liquid used to aid in drilling does not return to the surface. **2.** a rivet hole drilled inaccurately, so that it does not coincide with the corresponding hole.

blind image *Graphic Arts.* a lithographic image that is no longer ink-receptive.

blinding *Engineering.* **1.** the layering of a clean bed of concrete over soil or clay so that other materials can be applied or reinforcements attached. **2.** the spreading of sand or stone chips to fill in or cover the tar layer of a road being surfaced. **3.** see BLANKING, def. 2. *Mining Engineering.* an obstruction of a screen mesh during screening, created by a matting of fine materials. Also, BLOCKING; PLUGGING.

blind island *Geology.* in a lake, a small area of organic matter that is covered by shallow water.

blind joint *Engineering.* a joint hidden from view at any perspective. *Geology.* a plane of potential fracture where a massive rock may break during excavation.

blind lake *Hydrology.* a lake that has no flow of water either in or out.

blind landing *Transportation Engineering.* an aircraft landing in which the runway is invisible to the pilot, who is thus entirely dependent on instruments and ground control instructions.

blind level see BLIND DRIFT.

blind nailing *Building Engineering.* the process of nailing floorboards or other wood pieces or beams, so that no nail heads appear when construction is completed. Also, SECRET NAILING, TOSH NAILING.

blind nipple *Mechanical Engineering.* a short section of pipe or tube in a boiler that has one end sealed or closed.

blind passage *Microbiology.* an isolation procedure for the enrichment of a small amount of virus, in which the viral suspension is passed through a series of test cell populations to the extent that host cell symptoms can be detected.

blind roller *Oceanography.* a long, high swell that increases in height almost to breaking as it passes over shoals.

blinds *Building Engineering.* a panel or series of panels used to cover a window frame to block out exterior light, provide security, obscure vision, or serve as an ornament or decoration. Also, SHUTTERS.

blind search *Artificial Intelligence.* a search in which no heuristic information is available to distinguish one search path from another, so that exhaustive examination of possibilities must be done.

blind seed *Plant Pathology.* a disease of forage grasses (such as rye grass), caused by the fungus *Phialea temalenta,* characterized by shriveled, soft seeds.

blind shaft *Mining Engineering.* a sublevel shaft that is connected to the main shaft by a transfer station.

blindsnake *Vertebrate Zoology.* a burrowing, wormlike snake of the superfamily Typhlopoidea, having fused head shields and a short tail; found throughout subtropical and tropical areas of the world. Also, WORM SNAKE.

blind spot *Physiology.* the place on the retina of the eye where the optic nerve enters and there are no sensory receptors. *Optics.* any specific area that cannot be seen within a larger area that is visible. *Telecommunications.* a point within normal range of a transmitter at which signals are weak and reception is poor. Also, DEAD SPOT. *Engineering.* an area on a filter where no filtration can occur, as in the blockage of a portion of the screen of a sieve with particles. Also, DEAD AREA.

blind stamping *Graphic Arts.* the process of impressing a design or lettering on a book cover using a heated relief die, without ink or leaf. Also, **blind tooling.**

blindstory *Architecture.* a story or major horizontal division of a wall without exterior windows or other major openings, especially the triforium of a Gothic church.

blind trial *Statistics.* an experiment, typically designed to study the relative effectiveness of health-care treatments, in which the subjects are not aware of the type of treatment they are receiving; in a **double-blind trial,** neither the subjects nor the experimenters are aware of the type of treatment.

blind valley *Geology.* a steep-walled valley of Karst terrain that ends abruptly at the point downstream where the stream disappears underground.

blindworm *Vertebrate Zoology.* a legless lizard, *Anguis fragilis,* having tiny eyes and movable eyelids, and smooth, round scales covering its body; found in Europe, western Asia, and Algeria.

blind zone *Telecommunications.* a region in which the strength of a broadcast signal is abnormally weak due to geographical interferences in the line of sight of a broadcast station.

blink *Metrology.* a unit of time equal to the one hundred-thousandth part of a day, or 0.864 seconds. *Meteorology.* the brightening of a cloud base layer caused by the reflection of light from a snow-covered or ice-covered surface.

blink comparator *Optics.* an optical instrument commonly used in astronomy to detect subtle differences between two very similar photographs by alternating them rapidly.

blinker light *Telecommunications.* **1.** a blinking light that is used to attract the attention of studio personnel. **2.** a light aboard a ship that is used to send coded messages.

blinking *Telecommunications.* a periodic change in the intensity of a group of characters on a display terminal, usually used to emphasize an important message to the user. *Navigation.* a modification of a loran signal to indicate incorrect operation.

blink microscope *Optics.* a type of blink comparator that magnifies the images being viewed and compared.

Blinksiaceae *Botany.* a monospecific family of red algae of the order Gigartinales, known from a single locality and having thalli comprised of a horizontal layer of large, irregularly arranged cells.

blip *Electronics.* **1.** on a radar screen, a spot of light or a base-line irregularity representing the radar reflection from a target object. **2.** any small spot of light on a CRT screen. **3.** see BLEEP.

blister *Medicine.* a thin-walled swelling on the skin containing watery matter or serum, often caused by a burn or rubbing. *Metallurgy.* a subsurface delamination that causes a small raised area on the surface; can occur between a coating and the base material, or in a rolled product. *Graphic Arts.* a raised spot, on a photographic film or plate, at which the emulsion has lifted off the base. *Engineering.* **1.** a raised, rounded area on the surface of metal or plastic, caused by gas escaping from within while the material was molten. **2.** a bubble in a surface coating, such as paint or varnish. *Geology.* a downward protrusion of roof rock into a coal seam. *Volcanology.* a hollow, domelike swelling in the crust of a lava flow.

blister beetle *Entomology.* any of various beetles of the family Meloidae; some emit a secretion that can blister the skin.

blister blight *Plant Pathology.* a tea plant disease caused by the fungus *Exobasidium vexans,* characterized by translucent, sometimes pinkish, circular, concave lesions on the upper sides of the leaves.

blister canker *Plant Pathology.* a disease of apple trees caused by the fungus *Numularia discreta,* resulting in rough, black lesions on the large limbs and trunk of the tree. Also, APPLE POX.

blister copper *Metallurgy.* a matte of from 96% to 99% copper, having a black, blistered surface after smelting due to gases generated during smelting; usually further refined for commercial uses.

blister hypothesis *Geology.* a theory of orogeny stating that heat from radioactive decay causes the melting and expansion of the overlying crust, producing a blister, or regional uplift, in the near-surface rocks.

blistering *Engineering.* the formation of raised, rounded pockets of air, gas, or moisture, sealed within a material such a metal, plastic, paint, or varnish.

blister rust *Plant Pathology.* any of several plant diseases caused by the rust fungi of the genus *Cronartium;* characterized by large orange blisters and cankers that may spread until the tree dies.

blister spot *Plant Pathology.* a disease of apple trees caused by the bacteria *Pseudomonas papulans,* characterized by rough cankers on the tree limbs and dark blisters on the fruit.

blitz *Military Science.* **1.** short for BLITZKRIEG. **2.** an intensive aerial bombing attack: the London *blitz* of World War II.

blitz can see JERRY CAN.

blitzkrieg *Military Science.* a form of offensive warfare developed by the Germans in World War II, involving swift, large-scale attacks by mechanized troops and armored units supported by dive bombers. (From German for "lightning war.")

blizzard *Meteorology.* a severe weather condition characterized by very cold temperatures and strong winds bearing a heavy amount of mostly fine, dry snow picked up from the ground.

blk. black; block; bulk.

BL Lacertae object [lə sur´ tē] *Astronomy.* a celestial object with rapidly varying radio emission that may be a "missing link" between galaxies and quasars. Also, LACERTIDS.

bloat *Veterinary Medicine.* a digestive disorder in ruminants, especially cattle and sheep, where the rumen becomes overdistended due to an accumulation of gas, resulting from dietary factors or impeded eructation.

bloating *Materials Science.* the evolution of gas in a hot pressed ceramic during subsequent annealing, due to the presence of trapped impurities.

BLOB binary large object.

blob *Meteorology.* a small-scale temperature and moisture inhomogeneity produced by turbulence in the atmosphere as seen on an oscilloscope.

Bloch, Felix 1905–1983, Swiss-American physicist; shared Nobel Prize for method of studying magnetic properties of nuclei.

Bloch, Konrad Emil born 1912, German-born American biochemist; shared Nobel Prize for discovering how cells produce cholresterol.

Bloch equations *Solid-State Physics.* a set of equations that approximate the rate of change in magnetization in a solid due to spin relaxation and gyroscopic precession.

Bloch function *Solid-State Physics.* a wavefunction for an electron in a crystal having periodic structure; the function has the form $y=u(r)e^{ikr}$, where k is the wave vector, r is a position vector, and $u(r)$ is a function of the position that has the lattice periodicity.

Blochman bodies *Entomology.* intracellular, prokaryotic symbionts occurring in mycetomes in insects.

Bloch theorem *Solid-State Physics.* a theorem stating that every electronic wave in a periodic crystal lattice can be represented by a Bloch function.

Bloch wall *Solid-State Physics.* in a crystal lattice, a layer of thickness on the order of a few hundred lattice constants, separating two magnetic domains.

block a compact piece of solid material; specific uses include: *Building Engineering.* **1.** a solid mass of wood or stone with a number of flat faces. **2.** a hollow masonry building unit made of cement or terra cotta. **3.** a mold or piece on which a material is shaped or kept in shape. *Mechanical Devices.* **1.** the basic metal form of an engine, into which cylinder bores and other openings are machined, and to which other engine components are fitted or attached. **2.** a metal or wood housing that holds one or more pulleys fitted with a hood to grasp objects. *Mechanical Engineering.* see CYLINDER BLOCK.

block a group of things considered as a single unit; specific uses include: *Forestry.* the primary subdivision of a forest area, generally bounded by natural features and given a proper name or Roman numeral and then divided into smaller compartments. *Transportation Engineering.* a specified section of track or guideway. *Mining Engineering.* **1.** a mine division usually created by workings, but sometimes by survey lines or other arbitrary limits. **2.** a rectangular panel of ore allowing access for sampling, testing, and mining purposes. Also, **blocked-out ore.** *Computer Programming.* **1.** a group of characters, words, or records that is handled as a single unit. **2.** the area or tape position where such a group of data is stored. *Computer Technology.* see BASIC BLOCK. *Statistics.* a group of experimental units that share a characteristic believed to affect the response under study. *Mathematics.* **1.** a component of a block design; one of the k-element subsets of a (balanced incomplete) block design. There are b blocks in a (b, n, r, k, λ) - configuration. **2.** a connected graph with no cut vertices. **3.** in a graph G, a subgraph that is a block and is maximal with respect to that property is a **block of G.**

block an obstruction or stoppage; special uses include: *Military Science.* an obstacle or barrier to prevent the advance of enemy ground troops. Thus, **blocking force, blocking position.** *Behavior.* an inability to respond or perform that cannot be explained in terms of the immediate circumstances. *Psychology.* **1.** an inability to express or release emotions in an adequate manner. **2.** see BLOCKING. *Chemistry.* to make a chemical substance inactive or hinder its effect.

blockade *Pathology.* an agent that interferes with or prevents a specific action in an organ or tissue, such as the interruption of a nerve impulse or a heart muscle-contraction impulse.

block and tackle *Mechanical Engineering.* a mechanical system of pulley blocks and ropes or cables, used for hoisting or moving heavy loads. Also, **block and fall.**

block body *Computer Programming.* the list of executable statements that follows the block head in a block-structured program.

block brake *Mechanical Engineering.* a vehicle brake in which a block of material is forced against the rim of the revolving wheel, either manually, electromagnetically, or hydraulically.

block caving *Mining Engineering.* mining in which a large block of ore is undercut and forced to cave in; its fragments then gravitate to withdrawal points. Also, CAVING SYSTEM; CUMBERLAND METHOD.

block clay see MÉLANGE.

block coal *Mining Engineering.* **1.** a type of tough, usually semisplint bituminous coal having a tendency to break into sizable lumps or cubical blocks. **2.** any coal that will not pass through a certain-sized screen.

block coefficient *Naval Architecture.* a measure of how completely an underwater hull occupies a space; equal to displacement in tons divided by [(length × beam × depth)/35]. Also, COEFFICIENT OF FINENESS.

block copolymer *Materials Science.* a linear polymer consisting of two or more comonomer units arranged in extended sequences of each monomer type. Also, BLOCK POLYMER.

block cutting

block cutting *Forestry.* a cutting method that involves removing timber in several stages over an extended period, which can last up to 20 years. A new stand is established as the old one is removed; this method is used with trees requiring shade during their first years of growth, including ponderosa pine, red pine, and Douglas fir.

block design *Mathematics.* **1.** let $X = \{x_1, x_2,..., x_n\}$ be a set of n objects. A **complete block design** of X is b (b an integer greater than zero) replications of X, with the objects of X arranged according to given specifications. **2.** a **balanced incomplete block design** is a collection of bk-element subsets of X (blocks) such that: (i) $k < n$. (When $k = n$, the incomplete design degenerates to a complete block design.) (ii) Every object of X appears in exactly r of the b blocks. (iii) Every two objects appear simultaneously in exactly λ of the b blocks. Since the parameters b, n, r, k, and λ completely characterize the block design, it is called a (b, n, r, k, λ)-configuration. **3.** a **randomized complete block design** is one in which the objects in each replication of X are randomly ordered.

block diagram *Engineering.* a pictorial representation of a mechanical or electronic system, using rectangles and other simple geometric shapes to represent components or processes of the system, and lines to signify steps in the process or flow of materials, energy, or data.

blocked impedance *Electricity.* the input impedance of a device, often a transducer, when its output load is infinite.

blocked operation *Chemical Engineering.* the alternate use of a single refinery process unit in more than one operation.

blocked resistance *Acoustical Engineering.* the resistance due only to electrical loss by an audio-frequency transducer measured when its elements are unable to move.

block encryption *Telecommunications.* the process of using cryptography techniques to encode a block of data in order to provide data security during transmission.

blocker-type forging *Engineering.* a type of design forging characterized by soft contours with expanded radii and draft angles.

blockette *Computer Programming.* a relatively small block of information or a subset of a block transferred or stored as a unit, usually in the form of consecutive machine words.

block faulting *Geology.* **1.** a type of normal faulting in which a region of the earth's crust is divided into blocks of differing elevations and orientations. **2.** the process by which block mountains are formed.

block field see FELSENMEER.

block gap *Computer Technology.* the physical space that is left blank between data blocks recorded on magnetic tape or other storage media.

block glide *Geology.* a translational landslide in which the slide mass essentially moves outward and downward as a unit.

block-grid keying *Telecommunications.* a method of keying a continuous-wave transmitter by operating the amplifier stage as an electronic switch.

block head *Computer Programming.* a list of variable, procedure, and function declarations that appears at the beginning of a program block prior to the executable statements in block-structured programming.

block hole see BLASTHOLE, def. 1.

block hour *Transportation Engineering.* a measure of aircraft use based on the total time in motion from one stop to the next; so called from wheel chocks or blocks. Also, BLOCKS OFF-BLOCKS ON TIME.

blockhouse *Building Engineering.* a protective structure, usually constructed of reinforced concrete, with small openings; used for the protection of personnel and controls, direction of gunfire, observation, and operation.

block identifier *Computer Programming.* a means of identifying an area of memory so that the storage locations can be shared by a program and its subroutines.

block ignore character *Computer Technology.* a control character that indicates an error has been detected in data preparation or transmission and that some data should be ignored.

blocking obstructing or supporting by the use of a block or blocks; specific uses include: *Building Engineering.* **1.** the use of numerous small wood pieces to fill interstices or spacing joints for reinforcing members. **2.** the undesirable adhesion of layers of plastic or metallic materials during storage and use. *Electronics.* **1.** the process of stopping or impeding current flow. **2.** the process of overloading a receiver by an unwanted signal so that the automatic gain control reduces the response to the desired signal. *Psychology.* the sudden, unexplained inability of a speaker to continue with a certain train of thought, usually regarded as a reaction to an unpleasant association. *Histology.* the process of embedding tissue in a block of solid material, such as paraffin, for the purpose of thin sectioning. *Solid-State Physics.* the introduction of an impurity into a solid to block the movement of dislocations, which results in a subsequent hardening of the solid. *Meteorology.* a large-scale obstruction of the normal west-to-east movement of cyclones and anticyclones.

blocking a grouping together into a block or blocks; specific uses include: *Computer Programming.* the grouping of two or more logical records into a single physical record so that they are jointly read or written by one machine instruction. Also, BATCHING. *Statistics.* the process of dividing components of an experiment into sections, or blocks, of similar items in order to measure total variability of data through comparisons between blocks; often an effective means of isolating sources of heterogeneity in the total data.

blocking antibody *Immunology.* a type of antibody capable of interfering with the binding of another antibody or of sensitized cells.

blocking factor *Computer Programming.* the number of logical records per physical record or block on a magnetic tape or disk. *Immunology.* any substance (generally antibodies or serum proteins) that causes a weakening or suppression of the immune response.

blocking group *Organic Chemistry.* a group in peptide synthesis that is reacted with a free amino or carboxyl group on an amino acid to prevent its taking part in subsequent formation of peptide bonds.

blocking high *Meteorology.* a high-pressure system or anticyclone that remains nearly stationary and effectively blocks the movement of typical west-to-east migratory cyclones across its latitudes. Also, **blocking anticyclone.**

blocking oscillator *Electronics.* an electron-tube oscillator that operates intermittently as its grid bias increases during oscillation, then decreases until oscillation resumes.

blocking-oscillator driver *Electronics.* a circuit that develops a square pulse to drive the modulator tubes; it usually contains a line-controlled blocking oscillator that shapes the pulse into the square wave.

blocking test *Immunology.* a test in which the specific reactions between an antibody and antigen are stopped by adding an antibody or antigen of the same nature (but with dissimilar biological behavior).

block lava *Volcanology.* a lava whose surface consists of angular blocks with relatively smooth faces.

block length *Computer Programming.* the total number of characters, words, or records contained in one block. Also, **block size.**

block loading *Computer Programming.* a method for loading a program into main memory so that the control segments are loaded into adjacent locations.

block mark *Computer Programming.* a special character used when handling variable length data blocks to indicate the end of the block.

block mountain *Geology.* a mountain that is formed by block faulting.

block movement *Mining Engineering.* a general failure of a mine's hanging wall.

block multiplexer *Computer Technology.* a peripheral transfer facility in high-performance devices that permits simultaneous data transmission by interleaving large blocks of data.

block out *Graphic Arts.* to remove an unwanted part of a photograph (such as a background figure) in photoengraving, usually by opaquing the negative.

block parity *Telecommunications.* a form of parity checking applied to an entire block of transmitted data.

block plane *Mechanical Devices.* a short plane with no cap iron and the cutting bevel reversed; used for planing down end grain and cutting across the grain of the wood.

block planting *Forestry.* a patterned method of planting in which young trees are set out in specific groups.

block polymer *Organic Chemistry.* a polymer that has a large number of monomers, arranged in such a way that relatively long sequences, or blocks, of one monomer unit alternate with blocks of another, such as M1M1M1M1M2M2M2M2. *Materials Science.* see BLOCK COPOLYMER.

block printing *Graphic Arts.* a printing done from engraved or carved blocks of wood or metal; used extensively on textiles.

block protector see PROTECTOR BLOCK.

block sequence *Metallurgy.* a sequence of multiple-pass welding in which separate segments are partially or completely built up in cross section before intervening lengths are deposited.

block signal system *Control Systems.* an automated traffic control system for railroads that uses the tracks as electrical circuits to detect the presence of train engines or cars. *Transportation Engineering.* a system of controlling headways using fixed signals at the entrance of each block on a line.

blocks off-blocks on time see BLOCK HOUR.

block sort *Computer Programming.* a coarse sort on the most significant characters in the key for the purpose of breaking a large file into more manageable subfiles.

block splitting see CELLULAR SPLITTING.

block stream *Geology.* an accumulation of boulders or angular blocks at the edge of a cliff that forms the uppermost layer over solid or weathered bedrock. Also, ROCK STREAM.

block structure *Computer Programming.* a programming language concept first introduced in ALGOL 60 that allows related declarations and statements to be grouped together in blocks.

blocks world *Artificial Intelligence.* a microworld used in early artificial intelligence research, involving the perception, manipulation, or planning for manipulation of toy blocks on a table.

block system *Control Systems.* an automated railroad traffic control system that uses the tracks as electrical circuits to detect the presence of train engines or cars. *Mining Engineering.* a pillar mining system used to divide coal into uniform blocks that are then extracted on retreat.

block transfer *Computer Programming.* the process of moving one or more blocks of data in a computer.

blocky iceberg *Oceanography.* an iceberg with very steep sides and a nearly flat upper surface.

blödite *Mineralogy.* $Na_2Mg(SO_4)_2 \cdot 4H_20$, a colorless to greenish, yellowish, or red monoclinic mineral, having a specific gravity of 2.25 and a hardness of 2.5 to 3 on the Mohs scale, occurring in short prismatic crystals and granular or compact masses; found in oceanic salt deposits. Also, **bloedite.**

Bloembergen, Nicolaas born 1921, Dutch-born American physicist; with Schawlow, Nobel Prize for developing laser spectroscopy.

blomstrandine see PRIORITE.

Blondel-Rey law *Optics.* a law that is used to determine the apparent point brilliance of a flashing light source of less than 5 hertz.

blood *Hematology.* the fluid that circulates through the heart, arteries, veins, and capillaries in humans and other vertebrates, serving chiefly to carry oxygen and nutrients to the body cells and to remove carbon dioxide and other waste products from the cells. In humans it consists of a pale yellow liquid, plasma, in which the solid elements (red blood cells, white blood cells, and platelets) are suspended.

blood agar *Hematology.* an agar medium containing heart infusion, peptone, and sodium chloride, autoclaved and enriched by the addition of sterile defibrinated blood; it is used for primary plating and subculturing, especially to determine bacterial hemolysis. The blood used may be sheep (for group A *Streptococcus*), rabbit (for *Haemophilus parahemolyticus*), or horse. Also, **blood agar culture medium.**

blood albumin *Hematology.* see SERUM ALBUMIN.

blood-brain barrier *Physiology.* a barrier, made up of capillary walls and surrounding neuroglia, that limits the passages of substances between the blood and brain tissue.

blood cell *Hematology.* an erythrocyte (red blood cell) or leukocyte (white blood cell).

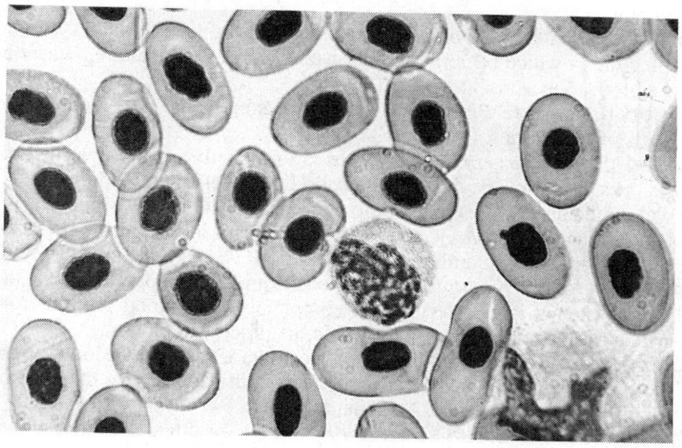

blood cells

blood clot *Hematology.* a coagulum formed of blood, either in or out of the body. Also, COAGULUM.

blood count *Hematology.* a numerical computation to determine the number of formed elements (red or white blood cells, or platelets), found in a cubic millimeter of blood, made by calculating the cells in an accurate volume of diluted blood.

blood culture *Hematology.* a laboratory preparation of a blood sample in order to analyze the blood or propagate various blood elements.

blood disease *Plant Pathology.* a disease of banana plants affecting the vascular tissues and characterized by blighted leaves and rust colored rot on the fruit; thought to be caused by the bacteria *Xanthomonas celebensis.*

blood dyscrasia *Pathology.* a condition in which one or more of the constituents of the blood are abnormal or are present in abnormal quantity, as in leukemia or hemophilia.

blood feud *Anthropology.* the cultural institution of retaliatory murder between two warring clans or lineages, sometimes lasting through several generations.

blood fluke *Invertebrate Zoology.* a flatworm belonging to the genus *Schistosoma;* parasitic on humans and domestic animals, attacking the liver, spleen, intestines, and urinary systems, and occasionally the brain.

blood group *Hematology.* **1.** a classification of blood defined by one or more cellular antigenic structural groupings under the control of allelic genes. Blood groups, especially for humans, are identified by agglutination supported by specific antiserums and by lectins extracted from certain plants. Two important types are the ABO and Rh blood groups. **2.** any characteristic, function, or trait of a cellular or fluid component of blood, considered as the expression of the actions and interactions of dominant genes, and useful in studies or analyses of human inheritance (e.g., legal determination of paternity). Also, BLOOD TYPE.

blood group chimerism *Genetics.* the occurrence of blood cells of two genetic types in one organism, as when dizygotic twins exchange hematopoietic stem cells *in utero* and continue to form blood cells of both types after birth.

blood group substance *Hematology.* any of the antigenic substances responsible for blood group specificities.

blood group system see BLOOD GROUP.

bloodhound *Vertebrate Zoology.* a breed of medium to large hound dogs known for an extremely keen sense of smell.

Bloodhound *Ordnance.* a British surface-to-air missile designed as the standard RAF home defense missile; the Bloodhound Mk 2 is powered by two ramjet engines and four solid-propellant rocket boosters, equipped with semiactive radar homing, and carries a conventional warhead at a maximum range of over 50 miles.

blood island *Developmental Biology.* a collection of mesenchyme cells in the angioblast of the young embryo that becomes vascular endothelium and blood corpuscles.

blood level *Pathology.* the concentration of a drug or other substance in a specified amount of plasma, serum, or whole blood; often expressed as a percent or in milligrams or micrograms per deciliter of blood.

blood plasma *Hematology.* the liquid portion of the blood in which the particulate components are suspended.

blood-plate hemolysis *Microbiology.* the decomposition of red blood cell hemoglobin in a blood agar medium that surrounds a bacterial colony.

blood poisoning *Pathology.* a popular term for septicemia, a systemic infection in which bacteria or other pathogenic substances are present in the blood. See SEPTICEMIA.

blood pressure *Physiology.* the force exerted by the blood on the walls of the arteries.

blood pump *Surgery.* a machine used to move blood through the tubing of circulation devices outside the body; designed to move blood without damaging the blood components, especially the erythrocytes.

blood rain *Meteorology.* a reddish-colored rain involving the presence of dust particles containing iron oxide.

blood serum *Hematology.* the clear liquid that separates from the blood when it is allowed to clot completely; i.e., blood plasma from which fibrinogen has been removed in the process of clotting.

bloodstone *Mineralogy.* a subtranslucent variety of chalcedony, bright to leek green with spots of jasper that have the appearance of drops of blood; a variety of quartz. Also, HELIOTROPE.

bloodstream or **blood stream** *Physiology.* the circulation throughout a living body of the blood flowing in the arteries, veins, and capillaries.

blood sugar *Biochemistry.* a popular term for the glucose present in the blood.

blood test *Pathology.* any test of the physical or chemical composition of a blood sample.

blood transfusion *Pathology.* the administration of whole blood or a component to replace blood lost through trauma, surgery, or disease.

blood type see BLOOD GROUP.

blood typing *Immunology.* the determination of a blood group through laboratory testing that monitors the agglutination reactions of the blood in relation to one or more blood groups. Also, **blood grouping.**

blood vessel *Anatomy.* any tube in the body through which blood flows; an artery, vein, or capillary.

bloodwood *Botany.* any of various trees with red wood or red sap, such as the Australian *Eucalyptus gummifera* or *E. ptychocarpa,* or the African *Pterocarpus angolensis.*

blooey line *Petroleum Engineering.* a pipe or flexible tube that conducts cuttings-laden air or gas from the collar of a borehole far enough from a drill rig to keep the air around the drill dust free.

bloom a blossoming or blooming; specific uses include: *Botany.* **1.** to produce blossoms or flowers. **2.** the state of flowering. **3.** a delicate, whitish, waxy or powdery coating on the fruit or leaves of certain plants. **4.** a sporadic seasonal increase in a population of phytoplankton.

bloom a surface coating; specific uses include: *Optics.* **1.** a coating applied to the surface of a lens, mirror, prism, or filter in order to minimize reflection. **2.** the color of oil in reflected light, differing from its color in transmitted light. *Metallurgy.* **1.** a semifinished product, similar to a billet, but usually larger. **2.** in electroplating, an exudation on the surface of the electrolyte bath. **3.** a corrosion product, looking like a flower, formed on certain metallic materials in humid environments. *Engineering.* a semifinished thick block of steel approximately 6 inches square; an intermediate stage in the rolling process. *Geology.* the oxidized, decomposed, or weathered outcrop of a mineral vein or coal bed. *Mineralogy.* a whitish powder that is found on rocks in arid regions, produced by loss of water of crystallization or by the evaporation of water brought to the rock surface by capillary action.

bloomer *Metallurgy.* a mill used to convert ingots into blooms.

Bloomfield, Leonard 1887–1948, American linguist; a pioneer in the systematic study of language.

blooming *Electronics.* **1.** an increase in the size of the scanning spot on a radar scope or other CRT screen, because of an increase in signal intensity or duration. **2.** a loss of focus in an area of a television picture in which brightness is excessive.

blooming mill *Metallurgy.* in steelmaking, a rolling mill used to produce blooms.

Bloom's syndrome *Medicine.* an inherent autosomal recessive trait resulting in dwarfism, accompanied by a hypersensitivity of the skin and a butterfly telangiectatic erythema primarily of the face but occasionally of the hands and forearms, with numerous defects of skin pigmentation, keratinization, dentition, and development.

blossom *Botany.* **1.** the flower of a plant, especially one that produces edible fruit. **2.** to produce blossoms or flowers. *Geology.* the oxidized, decomposed, or weathered outcrop of a mineral vein or coal bed.

cactus blossom

blossom-end rot *Plant Pathology.* **1.** any fruit rot that occurs at the point of the fruit at which the blossom ends. **2.** a disease of tomatoes, peppers, and other fruits and vegetables that is caused by adverse moisture conditions during growth, and which is characterized by discoloration and withering of the tip end of the fruit.

blotch *Plant Pathology.* any abnormal, irregular, and superficial spot on the leaves, shoots, or fruits of plants.

blotch print *Textiles.* cloth that is printed with both a ground color and a design.
blotter *Engineering.* a disk placed between a grinding wheel and its flanges to prevent stress.
blotter press *Chemical Engineering.* a plate-and-frame filtering assembly in which the filter medium is blotting paper.
blotting *Genetics.* a widely used procedure for transferring proteins, DNA, or RNA to a solid supporting matrix, either by electrophoresis or capillary action.
blow *Electricity.* to forcefully open a circuit by overloading it with excess current or voltage.
blowback *Chemical Engineering.* **1.** an intentional reverse flow of a liquid through a filter in order to remove caked solids. Also, BACKWASH. **2.** a continuous flow of gas or liquid that bleeds through air lines from instruments and onto the monitored process line, preventing the process fluid from backing up and contacting the instrument. *Ordnance.* **1.** the escape of gases to the rear during the firing of a gun. **2.** a method of operating an automatic gun in which the force of gases expanding to the rear provides the energy to carry out the firing cycle. Thus, **blowback gun.**
blowball *Botany.* the downy seed ball of a plant such as the dandelion.
blowby or **blow-by** *Mechanical Engineering.* **1.** in an internal-combustion engine, the leakage of the air-fuel mixture or combustion gases past a piston and into the crankcase during periods of maximum pressure in a particular cylinder; caused by an excessive gap or seized rings. **2.** a device fitted to the crankcase and designed to route such leaking substances back to the cylinders for combustion.
blowcase *Chemical Engineering.* a cylindrical or egg-shaped container used in moving acids or other liquids by air pressure. Also, ACID EGG, ACID BLOWCASE. *Petroleum Engineering.* **1.** a pumping system for transferring water mixtures and crude oil with no agitation to prevent emulsions. **2.** a discharge mechanism used to collect and remove liquid accumulations from a gas transmission line.
blow defect *Materials Science.* a defect in cast metal due to gas evolution on a wet mold wall.
blowdown *Chemical Engineering.* the use of pressure in order to remove solids or liquids from a process vessel, such as the removal of sludge or concentrated feedwater from the boiler of a steam-generating plant. *Mechanical Engineering.* **1.** the difference between the opening and closing pressures of a safety valve. **2.** see BLOWOFF. **3.** see COMPRESSION TEST.
blowdown line *Chemical Engineering.* a large conduit that receives and restrains liquids expelled by pressure from process vessels.
blowdown stack *Chemical Engineering.* a vertical chimney into which the materials of a chemical process unit are emptied in an emergency.
blowdown tunnel see BLOWDOWN WIND TUNNEL.
blowdown turbine *Aviation.* a turbine attached to and powered by the exhaust gases of a reciprocating engine.
blowdown wind tunnel *Aviation.* an open-circuit wind tunnel into which compressed gas is introduced and allowed to escape to the atmosphere or to another chamber by way of the test section. Also, BLOWDOWN TUNNEL.
blower *Mechanical Engineering.* a fan that is enclosed in such a way that the inlet gas is compressed to a higher pressure at discharge. *Aviation.* a rotary supercharger built into an aircraft engine and geared to its crankshaft; called a **high blower** when set at high rpm and a **low blower** at low rpm.
blow forming *Materials Science.* a superplastic forming technique for metals, in which the material to be shaped is forced by air pressure into a die, under controlled temperature and strain rate conditions.
blowhole *Vertebrate Zoology.* the nostril on top of the head of a whale or other cetacean, through which a compressed air and vapor stream from the lungs is released as the animal surfaces. *Geology.* a nearly vertical hole or natural chimney in the roof of a sea cave or a wave-formed tunnel, through which incoming waves and rising tides force air to rush upward or water to spout intermittently. *Volcanology.* a tiny crater on the surface of a thick lava flow. *Hydrology.* an opening that passes through a snowbridge into a crevasse or system of crevasses that are otherwise sealed by snowbridges. *Metallurgy.* in a casting or weld, a cavity caused by evolution of gas during solidification.
blow-in *Metallurgy.* the starting operation of a blast furnace. *Petroleum Engineering.* an oil well that begins to eject gas and oil.
blowing *Chemical Engineering.* the introduction of compressed air at the bottom of a container or tank to mix the liquid inside the container. *Engineering.* see BLOW MOLDING.
blowing agent *Materials Science.* an additive used in the production of polymer foams that breaks down into a gas at elevated processing temperatures. Pellets (granules) of polymer are made containing the blowing agent; then the pellets are heated, the polymer softens or melts, and the blowing agent decomposes to form a gas within the pellet, and the walls of the pellet expand.
blowing cave *Geology.* a cave in which air is alternately blown out and sucked in at the entrance. Also, BREATHING CAVE.
blowing pressure *Engineering.* the air or gas pressure used to pump up a mass of molten plastic into a mold during the formation of a hollow plastic object.
blowing snow *Meteorology.* snow lifted from the surface of the earth by the wind to a height of six feet or more above the surface and blown about in quantities that restrict visibility at and above that height.
blowing still *Chemical Engineering.* a still in which oxidized or blown asphalt is produced.
blowing-up furnace *Metallurgy.* a furnace in which ore is sintered and zinc and lead are volatized.
blowlamp see BLOWTORCH.
blow-lifting gripper *Robotics.* an end effector for workpieces that should not be touched, using compressed air to lift and manipulate objects.
blow molding *Engineering.* the formation of a hollow plastic object by forcing air or gas into a mass of molten plastic located within the cavity of a split mold. Also, BLOWING.
blown-fuse indicator *Electricity.* a neon warning light that is connected across a fuse and lights when the fuse is blown.
blown glass *Engineering.* articles of glass made in part by air forced into a mass of molten glass.
blowoff *Mechanical Engineering.* the discharge of steam or water from a cooling tower, boiler, evaporative condenser, or similar apparatus in order to decrease the pressure within the apparatus or the concentration of dissolved solids in the circulating water. Also, BLOWDOWN. *Space Technology.* the separation of a payload or package section from the remaining part of a rocket vehicle or reentry body; done by explosive or other enforced means, usually to facilitate retrieval of an instrument or flight record.
blowoff valve *Mechanical Engineering.* **1.** a valve that maintains the required pressure in a vessel by discharging or relieving excess pressure from the system. Also, RELIEF VALVE. **2.** in boiler piping, a valve that facilitates removal of solids from the circulating boiler water.
blowout *Electricity.* **1.** the usually forceful opening of a circuit caused by excess current or voltage. **2.** the extinguishing of an arc. *Engineering.* a rupture of a container that suddenly expels or exhausts the air, gas, or liquid contained within. *Geology.* **1.** any small depression, basin, valley, or cuplike hollow that is formed by wind erosion on a dune or other sand deposit, or where vegetation has been destroyed by fire or by overgrazing. **2.** any weathered surface exposure of altered, discolored rock thought to be indicative of a mineral deposit. *Hydrology.* see SAND BOIL. *Petroleum Engineering.* a sudden, unplanned, and dangerous eruption of gas or oil from a well being drilled; a wild well. *Aviation.* a flameout, especially in a jet engine due to excessive air velocity. *Electromagnetism.* the termination of an electric arc by means of a magnetic field applied perpendicularly to the arc and thus deflecting it.
blowout coil *Electromagnetism.* a coil that produces a magnetic field in order to deflect and extinguish an electric arc, particularly at a point in a circuit where the current may be interrupted by opening of the circuit.
blowout dune *Geology.* a shallow hollow in the land surface of an arid region resulting from wind blowing on the loose surface material.
blowout magnet *Electromagnetism.* a permanent magnet or an electromagnet used to recognize and extinguish the arc formed when a high-current switch is opened.
blowout preventer *Petroleum Engineering.* an arrangement of heavy-duty valves secured to the top of the casing to control well pressure; this prevents loss of pressure in the annular space between the drill pipe and casing or in the open bore hole during drilling operations.
blowpipe *Biology.* a small tube that concentrates a flow of air into a body cavity to reveal or clean it. *Engineering.* **1.** a long, straight tube used to force air into a mass of molten glass when making blown glassware. **2.** a small tube used to carry a stream of air to a flame; used in soldering and in laboratory flame tests. **3.** see BLOWTORCH.
Blowpipe *Ordnance.* a British shoulder-fired, close-range surface-to-air missile; it is powered by a solid-propellant rocket motor, equipped with radio command guidance, and carries a conventional warhead.
blow pressure see BLOWING PRESSURE.

blow rate *Engineering.* the speed at which air or gas pressure is forced into a molten mass of material during the process of blow molding.
blowtank *Chemical Engineering.* a papermaking pit or tank into which the pulp charge from a digester is blown when a cook is completed.
blowtorch *Engineering.* a gaseous-fueled, portable blast burner used for producing flame and intense heat.
blow tube see BLOWPIPE.
blow up *Graphic Arts.* **1.** to enlarge all or part of a photographic image. **2. blowup.** such an enlargement. *Forestry.* a sudden increase in the intensity of a fire that precludes direct control or thoroughly changes the existing control plans.
blowup ratio *Engineering.* **1.** in blow molding, the ratio of the diameter of the mold cavity to the width of the mass of molten plastic to be molded in the cavity. **2.** in blow tubing, the ratio between the diameter of the manufactured product and the diameter of its die.
BLQ *Aviation.* the airport code for Bogotá, Colombia.
BLS Bureau of Labor Statistics.
bls. bales; barrels.
blubber *Vertebrate Zoology.* the thick subdermal layer of insulating fat in whales and other large marine animals that may constitute up to one-quarter of the animal's total weight.
blue *Optics.* **1.** the color sensation that corresponds to radiation in the 455 to 492 nanometer wavelength region of the light spectrum and lies between violet and green on the color spectrum. **2.** describing a filter that absorbs less blue light than red and green light.
blue annealing *Metallurgy.* softening of a steel by heating a hot-rolled sheet in an atmospheric environment, followed by cooling in air. A blue oxide is formed on the surface, but does not affect properties.
blue asbestos *Mineralogy.* a common name for crocidolite.
blue baby *Medicine.* a term used to describe an infant who is born with cyanosis due to a congenital defect of the heart or lung.
blue band *Geology.* a thin bed of bluish clay found throughout the Illinois-Indiana coal basin. *Hydrology.* a layer of relatively bubble-free glacier ice or a band marking the appearance of such a layer on the surface of a glacier.
bluebell *Botany.* a general name for various plants having blue, bell-shaped flowers, such as the **bluebell of Scotland,** *Campunula rotundifolia,* or the **Virginia bluebell,** *Mertensia virginica.*
blueberry *Botany.* any of several species of the genus *Vaccinium* and their edible berries, which are usually black with a pale bluish bloom and numerous tiny seeds.

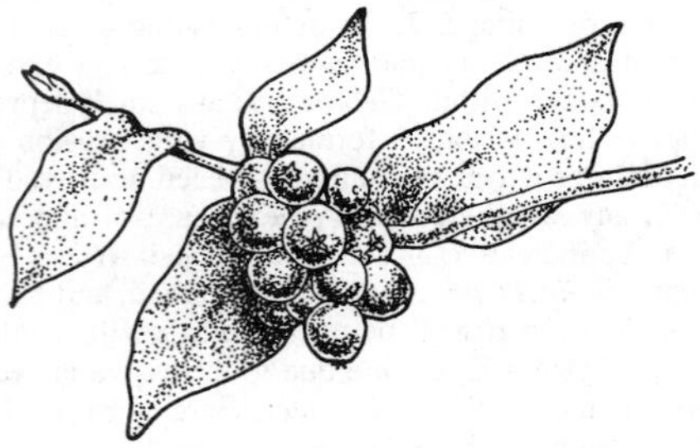

blueberry

bluebird *Vertebrate Zoology.* any of several North American songbirds in the family Turdidae, genus *Sialia,* having generally blue plumage and dwelling in open country or woodlands
blue brittleness *Metallurgy.* the embrittlement of some steels that have been heated between 200°C and 370°C. This phenomenon does not affect killed steels and is more prevalent in steels that have been worked at elevated temperatures.
blue cap *Mining Engineering.* a blue halo or tip of the flame of a safety lamp that indicates the presence of firedamp in the air.
blue cheese *Food Technology.* any semisoft cheese with blue mold, such as Roquefort, Gorgonzola, or Stilton; American blue cheese is similar to Roquefort but is usually made from cow's rather than sheep's milk.
blue comb *Veterinary Medicine.* an acute or subacute disease of unknown cause affecting domestic fowl and turkeys, characterized by a drop in egg production, loss of appetite, watery diarrhea, and sometimes by a bluish-purple discoloration of the comb. Also, AVIAN MONOCYTOSIS, PULLET DISEASE.
blue copper ore *Mineralogy.* a popular name for azurite and coveline.

blue crab

blue crab *Vertebrate Zoology.* an edible crab, *Callinectes sapidus,* with a dark green body and bluish legs; found in North America along the Atlantic and Gulf Coasts.
blue edge *Astronomy.* a location on the H-R diagram showing the maximum temperature at which an RR Lyrae star is stable.
bluette see BLEWIT.
bluefish *Vertebrate Zoology.* a widely distributed predatory fish that is the only species of the family Pomatomidae in the order Perciformes, distinguished by its bluish color.
blue flash *Meteorology.* a bluish color appearing in the upper portion of the sun, resulting from atmospheric refraction; visible occasionally when the sun rises above or sinks below the horizon. Also, GREEN FLASH.
blue fox see ARCTIC FOX.
blue gas see WATER GAS.
bluegill *Vertebrate Zoology.* a common sunfish of the family Centrarchidae, species *Lepomis macrochirus,* native to the freshwaters of the eastern U.S. and introduced to other regions of the U.S.; characterized by sharp dorsal and anal fins and brightly colored gill covers.
blue glow *Electronics.* a glow caused by the ionization of the molecules of mercury vapor in electron tubes that contain mercury vapor. *Metallurgy.* a luminescence emitted by certain ores.
blue granite see LARVIKITE.
bluegrass *Botany.* any of various grasses of the genus *Poa,* characterized by blades that have their margins folded together at the tip to resemble the bow of a boat, dense tufts of bluish-green blades, and creeping rhizomes; grown for lawns and pastures.
blue ground *Geology.* **1.** unoxidized slate-blue or blue-green kimberlite containing plutonic rock fragments and diamonds. **2.** a group of coal seams consisting mainly of hard clay or shale beds.
blue haze *Astronomy.* a Martian atmospheric haze that renders the atmosphere opaque to violet and blue light.
blue ice *Hydrology.* **1.** pure, clear, coarse-grained glacier ice having a bluish to greenish color. **2.** on the Antarctic ice sheet, an ablation area where bare glacier ice shows at the surface as a result of wind erosion.
blue iron earth see VIVIANITE.
blue iron ferrocyanide see PRUSSIAN BLUE.
bluejay *Vertebrate Zoology.* a crested jay, *Cyanocitta cristata,* common in eastern North America, characterized by a bright blue back and a gray breast.
blue lead *Metallurgy.* a term for metallic lead, to distinguish it from other lead compounds such as white lead, red lead, and orange lead. *Mineralogy.* **1.** the gold-bearing gravel deposits of the Sierra Nevada mountains. **2.** the bluish auriferous gravel deposit in the ancient river channels of California.
blue leg see BLEWIT.
blueline *Graphic Arts.* **1.** a prepress proof printed in blue on white paper treated with iron-salt compounds. Also, BLUEPRINT. **2.** a blue outline that does not reproduce when photographed, used as a guide for laying out a printed page.
blue magnetism *Geophysics.* a magnetism that is characteristic of the northern pole of a magnet or of the earth's north magnetic pole as shown by the south-seeking end of a freely suspended magnet.
blue malachite *Mineralogy.* a common name for azurite.
blue metal *Geology.* a hard, fine-grained bluish-gray shale or mudstone found at the base of many coal beds in England.

blue mold *Mycology.* a fungal species of the genus *Penicillium* that possesses blue spores.

blue mold rot *Plant Pathology.* **1.** any fungal plant disease caused by fungus of the genus *Penicillum* (also known as blue mold). **2.** a fungus disease of tobacco plants, characterized by yellow spots and a white to light-violet moldy growth on the undersides of the leaves. Also, DOWNY MILDEW.

blue mud *Geology.* a deep-sea sediment containing a small amount of terrigenous material, colored bluish-gray by the presence of finely divided iron sulfides and organic matter.

Blue Nile *Geography.* a tributary of the Nile, flowing 1000 miles from Lake Tana through northern Ethiopia and eastern Sudan and joining the White Nile at Khartoum to form the Nile's trunk stream.

blue ocher see VIVIANITE.

blueprint *Graphic Arts.* **1.** a photographic reproduction of a plan or drawing, consisting of white lines on a blue background. **2.** see BLUELINE.

blueprint paper *Graphic Arts.* a paper used for printing blueprints, characterized by its uniform surface and absorbency as well as its lack of chemicals that might react to blueprint developing solution.

blue protein *Biochemistry.* any of a class of proteins that contain copper and are present in bacteria, fungi, and in the plasma of mammals; some blue proteins called cupredoxins play a role in the movement of electrons.

blue pus *Medicine.* pus with a bluish tint, indicating the presence of an antibiotic pigment, pyocyanin, which is a product of *Pseudomonas aeruginosa.*

Blue Ridge Mountains *Geography.* the southeastern range of the Appalachians, extending from southern Pennsylvania to northern Georgia; highest peak: Grandfather Mountain (5964 ft).

blue rot *Plant Pathology.* a conifer disease caused by fungi of the genus Ceratostomella and characterized by a bluish discoloration of the wood. Also, BLUING.

blues *Graphic Arts.* a set of blueline prepress proofs.

blueschist facies *Petrology.* a facies of metamorphic rocks that is formed under high pressure and moderate to low temperatures associated with subduction zones, producing a broad mineral association including glaucophane, actinolite, jadeite, aegirine, lawsonite, and pumpellyite.

blueshift *Astrophysics.* the apparent shift of spectral lines toward the blue end of the spectrum that occurs when a source is moving toward the observer.

blue-sky scale see LINKE SCALE.

blue spar see LAZULITE.

blue stain *Plant Pathology.* an abnormal bluish cast that occurs on the sapwood of many trees and is caused by various fungi, including Fusarium, Ceratostomella, and Penicillium.

blue star *Astronomy.* a star emitting mostly blue light.

Blue Steel *Ordnance.* a British air-to-surface strategic stand-off missile powered by a twin-chamber liquid-propellant rocket engine and equipped with inertial guidance; it carries a thermonuclear warhead at supersonic speed with a range of approximately 200 miles.

bluestem grass *Botany.* any of serveral perennial grasses of the genus *Andropogon,* usually densely tufted, with bluish leaf sheaths and young shoots strongly flattened.

bluestone *Petrology.* **1.** a highly argillaceous sandstone with an even texture and bedding. **2.** a dark-bluish gray feldspathic sandstone that is easily split into thin slabs and used commercially as flagstone. *Mineralogy.* see CHALCANTHITE.

blue stragglers *Astronomy.* the hottest stars belonging to a cluster that lie just above the main sequence.

bluetongue *Veterinary Medicine.* a noncontagious, insect-transmitted viral disease of mostly sheep and sometimes cattle or wild game animals, manifested by fever, rhinitis, enteritis, lameness, swelling and blue coloration of the tongue, and eruptions on mouth membranes.

blue verdigris see CUPRIC ACID.

blue vitriol see CHALCANTHITE.

blue water gas *Mining Engineering.* a gas obtained in a cyclic process in which steam is passed over red-hot coke; called blue water gas because of its blue flame, a characteristic of the combustion of carbon dioxide.

blue whale *Vertebrate Zoology.* the largest animal ever to have lived on earth, weighing about 150 tons and measuring up to 100 feet long; it has remarkable speed and agility for its size and ranges in all oceans from polar seas to the equator; an endangered species.

bluffs

bluff *Geography.* **1.** a cliff or headland with a broad, steep face. **2.** a high, steep bank rising from a plain or shore.

bluff body *Space Technology.* **1.** a reentry vehicle or other body having a broad, flattened nose. **2.** the shape of such a body.

bluff-bowed ship *Naval Architecture.* a ship with a full, blunt bow, usually associated with high capacity and low speed.

bluffing *Behavior.* see DEIMATIC BEHAVIOR.

bluing *Metallurgy.* **1.** a process used to enhance corrosion resistance and improve the appearance of a steel sheet, strip, or component. This process consists of heating in air, steam, or other oxidizing environment. **2.** heat treating a spring to improve its properties. *Plant Pathology.* see BLUE ROT.

Blumberg, Baruch S. born 1925, American physician; Nobel Prize for discoveries in diagnosis and treatment of hepatitis.

blunderbuss *Ordnance.* a firearm of the 16th and 17th centuries, having a large bore and a wide, flaring muzzle, used to scatter shot at short range.

blunger *Engineering.* a vat having a rotating shaft equipped with rotors for mixing clay with water in making slip (clay having the consistency of cream) for ceramics. Also, **blunging machine.**

blunging *Engineering.* the agitated mixing of a material such as clay with liquid, in order to make a substance suitable for manufacturing ceramics.

blunt-end DNA *Molecular Biology.* a fragment of DNA that is cut evenly so that neither strand of the duplex is longer than the other.

blunt file *Mechanical Devices.* a file of rectangular form, with straight edges and sides.

B lymphocyte *Immunology.* a type of cell found in blood, tissues, and lymph that originates from the bone marrow and becomes transformed through antigenic stimulation, becoming either a memory cell or a plasma cell that forms antibodies against a given antigen.

Blytt-Sernander climatic classification *Geology.* a system of late-glacial and postglacial climate classification that has been inferred from bog stratigraphy and megascopic plant remains.

BM basal metabolism; bench mark.

BM or **B.M.** Bachelor of Medicine.

bm or **b.m.** board measure (board feet).

BMA or **B.M.A.** British Medical Association.

BMD ballistic missile defense.

BME or **B.M.E.** Bachelor of Mechanical Engineering; Bachelor of Mining Engineering.

B meson *Particle Physics.* a particle with mass 1237 MeV and baryon number $B = 0$.

BMet. or **B.Met.** Bachelor of Metallurgy.

BMetE. or **B.Met.E.** Bachelor of Metallurgical Engineering.

BMEWS [bē´myōōz´] *Ordnance.* a radar system with linked bases in Alaska, Greenland, and the United Kingdom, designed to detect intercontinental ballistic missilies before they reach the horizon. (An acronym for Ballistic Missile Early Warning System.)

B mode *Acoustics.* a method of ultrasonic medical tomography in which a display is formed by scanning on a single plane focus to produce a two-dimensional image; used for applications such as assessing fetal anatomy or identifying pelvic mass.

BMR or **B.M.R.** basal metabolic rate.

BMS or **B.M.S.** Bachelor of Medical Science; Bachelor of Marine Science.

BMT or **B.M.T.** Bachelor of Medical Technology.

BN or **B.N.** Bachelor of Nursing.

BNA *Aviation.* the airport code for Nashville, Tennessee.

BNC connector *Electricity.* a small bayonet coaxial-cable connector, often used in low-power radio-frequency and test applications.

BNF Backus-Naur Form.

BNOA ß-naphthoxyacetic acid.

BN object Becklin-Neugebauer object.

BOA or **B.O.A.** British Orthopaedic Association.

boa *Vertebrate Zoology.* **1.** broadly, any constricting snake of the family Boidae in the order Squamata, including the boa constrictor, anaconda, and python **2.** specifically, snakes of the subfamily Boinae, characterized by bearing live young, as opposed to the egg-laying pythons.

boa

boa constrictor *Vertebrate Zoology.* **1.** a tropical American snake, *Constrictor constrictor,* that may reach ten feet or more in length; it crushes and suffocates its prey by coiling around it. **2.** broadly, any large snake of the boa family

boar *Vertebrate Zoology.* **1.** a wild hog. Also, WILD BOAR. **2.** broadly, a mature male of certain other mammals, such as a boar raccoon. *Agriculture.* a mature, uncastrated male swine.

board *Forestry.* a long, flat, rectangular piece of cut wood that is relatively wide in comparison to its thickness. *Building Engineering.* a composition material fabricated in large sheets; for example, plasterboard or fiberboard. *Electronics.* a sheet of insulated material on which components are mounted and interconnected for installation in electronic equipment; used especially in reference to printed circuits. Also, CIRCUIT BOARD. *Graphic Arts.* either of two stiff foundation pieces placed inside the cover material of a hardcover book. *Computer Technology.* an electrical panel whose circuitry can be changed by adding or deleting external wiring. Also, PLUGBOARD, WIRE BOARD, PANEL. *Naval Architecture.* the side of a ship; used in compound terms such as star*board* and over*board*. *Navigation.* the distance traversed in one windward leg or tack.

board and batten *Building Engineering.* siding or other construction having wide boards or plywood sheets mounted vertically with the butt joints covered by battens.

board coal see BITUMINOUS WOOD.

board computer *Computer Technology.* a computer consisting of a single circuit board that holds all the electronic components.

board drop hammer *Mechanical Engineering.* a hammering device consisting of a metal battering head attached to a wooden frame that moves between two rollers; friction between the rollers raises the head to its high point, from which it drops and strikes the surface being worked. Also, **board hammer.**

board foot *Engineering.* a unit of measure equal to the volume of a board 1 foot square and 1 inch thick (144 cubic inches); used to measure the volume of lumber.

boarding *Engineering.* **1.** a collection of boards, or a structure made with boards. **2.** the process of covering something with boards.

board measure *Engineering.* a system of measuring lumber in which the basic unit is the board foot.

board sheathing *Building Engineering.* the reinforcement of studs with plywood, fiberboard, gypsum, or other wood boards in order to increase rigidity, lateral stability, and insulation properties, or to form a base for a finish.

boarfish *Vertebrate Zoology.* any of various marine fishes that have a projecting snout similar to that of a boar, including fish of the family Pentacerotidae of the Indo-West Pacific region, and also *Capros aper*, a brilliant red, deep-bodied, spiny-rayed fish of the Mediterranean region.

Boas, Franz [bō´az] 1858–1942, German-born American anthropologist; pioneer in physical and cultural anthropology.

Boas, Ismar Isidor 1858–1938, German physician; developed a number of tests relating to stomach contents and motility.

boast *Graphic Arts.* **1.** to roughly carve or shape material, especially stone or wood, in preparation for further embellishment with finer detail. **2.** material that has been prepared in this manner.

boaster *Graphic Arts.* a broad-faced chisel used in rough-cutting stone. Also, **boasting chisel.**

Boas' test *Pathology.* a method of detecting the presence of hydrochloric acid in stomach contents, using a solution of resorcinol, sugar, and dilute alcohol. (Named after Ismar I. *Boas.*)

boat *Naval Architecture.* **1.** a small craft capable of being carried aboard a larger vessel. **2.** any relatively small vessel capable only of limited independent operation. *Physical Chemistry.* see BOAT FORM. *Analytical Chemistry.* a dish made of a material such as ceramic or platinum, used to hold a substance to be analyzed by combustion.

boat boom *Naval Architecture.* a horizontal spar extended from the side of the ship in order to secure smaller boats and keep them from bumping into the larger vessel. Also, BOAT SPAR, RIDING SPAR.

boat configuration or **boat conformation** see BOAT FORM.

boat deck *Naval Architecture.* a partial deck above the main deck, used mainly for stowage of boats.

boat form *Organic Chemistry.* a configuration of atoms in a molecule with more than five atoms, in which the bond angles form an irregular boatlike pattern, as in a less stable modification of cyclohexane where the hydrogen atoms are staggered in relation to the carbon atoms.

boat spar see BOAT BOOM.

boat spike *Mechanical Devices.* a square, chisel-headed spike used for heavy timber construction. Also, BARGE SPIKE.

boatswain [bō´sən] *Navigation.* a naval warrant officer in charge of seamanship.

boattail *Design Engineering.* on a rocket vehicle or other elongated body, the aft portion that is tapered to reduce drag.

bob *Mechanical Devices.* a weight, generally on the end of a line or rod, that is used as a balance or counterbalance, such as the weight on the end of a pendulum.

Bobasatranidae *Paleontology.* a family of chondrostean fishes in the extinct order Palaeonisciformes and suborder Platysomoidei; they lived in the upper Permian and lower Triassic periods and were characterized by a deep body and fully heterocercal tail fin.

bobbed *Genetics.* describing a mutant allele located on the X and Y chromosomes in *Drosophila* that, when homozygous, leads to the formation of dwarf bristles and reduces the number of copies of genes for ribosomal RNA.

bobbin *Textiles.* a small hollow cylinder, flanged at one or both ends, around which yarn or thread is wound. *Electromagnetism.* **1.** also, **bobbin core.** an insulated spool-like foundation around which a coil of wire or ferromagnetic tape is wound. **2.** the coil wound in this manner.

bobbinet *Textiles.* a net of hexagonal mesh, manufactured by machine from cotton, silk, nylon, or other materials.

bobbing *Electronics.* the fluctuation of radar echo strength due to alternate interference and reinforcement of reflected waves.

bobbin lace *Textiles.* a lace that is made by hand from threads wound on small bobbins; the thread is twisted around pins placed on a pillow or pad in order to create a particular design. Also, PILLOW LACE.

bobcat *Vertebrate Zoology.* a solitary, nocturnal, carnivorous cat of the family Felidae, *Lynx rufus*; native to the mountains and deserts of west-central North America; characterized by a dark-spotted brown coat, white underbelly, and tufted ears.

bobcat

bobierrite *Mineralogy.* $Mg_3(PO_4)_2 \cdot 8H_2O$, a colorless to white monoclinic mineral occurring in minute crystals and in massive form, having a specific gravity of 2.2 and a hardness of 2 to 2.5 on the Mohs scale.

bobolink *Vertebrate Zoology.* a common American migratory songbird, *Dolichonyx oryzivorus,* of the family Icteridae, characterized by a rollicking melodic song, and by yellow, white, and black plumage. Also, REEDBIRD, RICEBIRD.

bobtail curtain antenna *Electromagnetism.* a bidirectional, vertically polarized, phase-array antenna containing two horizontal sections attaching with three vertical sections.

bobwhite *Vertebrate Zoology.* a game bird of the family Phasianidae, genus *Colinus,* ranging from Canada to Central America and living in brush or piney woods; characterized by brown and white speckled plumage and a black "necklace" marking on males.

bocca *plural,* **bocche.** *Volcanology.* any opening through which lava and gases escape from a volcano, especially on the side or at the foot. (An Italian word meaning "mouth.") Also, **boca.**

bock *Food Technology.* a dark, rich, full-bodied beer, usually drunk in the spring; originally brewed in Germany.

Bock's ganglion see CAROTID GANGLION.

BOD biochemical oxygen demand; biological oxygen demand.

Bodansky unit *Biology.* a standardized unit, equal to the quantity of phosphatase in 100 ml of serum required to liberate 1 mg of phosphorus as phosphate ion from sodium ß-glycerophosphate, in a time of 1 hour at a temperature of 37°C.

Bode, Johann Ehlert 1747–1826, German astronomer; published Bode's law in 1772 (which was formulated by the mathematician Johann Titius).

Bode diagram *Electronics.* a diagram in which the gain or phase angle of a device is plotted against the logarithm of the frequency to determine the stability of the device and to show what degree of negative feedback can safely be applied.

bodenite *Mineralogy.* a steel-gray metallic composite of nickel, iron, arsenic, bismuth, and cobalt. (Named for *Boden,* Sweden, where it is mined.)

Bodenstein number *Fluid Mechanics.* a dimensionless group used when the diffusive force is important in reactor analysis or design.

Bode's law *Astronomy.* an empirically derived statement, now believed to be a coincidence, that describes the spacing of the planets out through Uranus; it breaks down in the case of Neptune and Pluto. Also, TITIUS-BODE LAW.

Bodonidae *Invertebrate Zoology.* a family of aquatic protozoans in the order Kinetoplastida, with two flagella of unequal length; found in both free-living and parasitic forms.

body *Biology.* the complete physical substance and structure of an organism, living or dead. *Zoology.* the trunk of an animal, excluding the head, limbs, and tail. *Anatomy.* **1.** the largest, most important part of any organ. **2.** any mass or collection of material, such as the adrenal body. *Physics.* a mass, especially one that is complete and independent. *Geography.* a separate mass of water, such as an ocean, sea, lake, or stream, as distinguished from other water masses. *Astronomy.* a moon, planet, star, or other heavenly object. *Mechanical Engineering.* the part of a vehicle in which passengers ride or a load is carried, as opposed to the wheels, engine, and so on. *Naval Architecture.* the hull of a ship. *Aviation.* **1.** the main structure or central part of a flight vehicle; a fuselage or hull. **2.** broadly, any aerodynamically designed structure or form. *Ordnance.* **1.** the central part of a shell between the rotating band and bourrelet. **2.** broadly, any ballistically designed object. *Materials Science.* **1.** the quality, consistency, or density of a substance, especially a liquid. **2.** the unformed material from which a ceramic object is made. *Geology.* **1.** the property that makes coal combustible. **2.** a pocket of ore or mineral deposit. *Graphic Arts.* **1.** the main reading matter in a printed text, as opposed to the headlines, captions, and other display lines. **2.** the solid shank supporting the printing face on a piece of type.

body angle *Aviation.* the acute angle between a given line and the longitudinal axis of an airframe.

body axis *Aviation.* the longitudinal, horizontal, or vertical axis of a flight vehicle; these three mutually perpendicular axes form a reference system that moves with the vehicle.

body build index see BODY MASS INDEX.

body burden *Nucleonics.* the total amount of radioactive material that is present in the body of a human or an animal.

body capacitance *Electricity.* capacitance between the hand or body of an operator and a piece of electronic equipment, often causing detuning or interference signals or noise.

body cavity *Anatomy.* any visceral cavity, such as the thoracic, abdominal, or pelvic cavity.

body-centered cubic *Crystallography.* one of the basic lattice types, in which lattice points are arranged on the vertices and at the center of a cubic unit cell. There are two lattice points per unit cell, one corner point because eight corners are each shaped by eight unit cells and one point in the center. Also, **body-centered lattice.**

body centrode see CENTRODE.

body clock see BIOLOGICAL CLOCK.

body cone *Mechanics.* the cone that is traced out by the angular velocity vector of a rotating rigid body. Also, POLHODE CONE.

body dysmorphic disorder see DYSMORPHIC DISORDER.

body equilibrium *Physiology.* a maintained balance between the intake of materials into the body and the excretion of materials from the body.

body force *Mechanics.* any external force that acts on a volume element of a body and is proportional to the volume, such as gravity.

body frame *Mechanics.* a reference frame rotating with a body, which allows equations for rotational motion to be applied to this body.

body language *Behavior.* the expression of thoughts or feelings through posture or body movements. Also, KINESICS.

body mass index *Nutrition.* a measurement of the amount of fat and muscle in a given human body; determined by weight (kg) divided by height (m) squared, and used as an index of obesity. Also, BODY BUILD INDEX, QUETELET'S INDEX.

body of revolution *Mathematics.* the three-dimensional object formed by rotating a planar curve or planar region about a line (axis of revolution) in the same plane. If the curve is closed and the axis of revolution does not intersect the curve, the resulting special case is an annular solid.

body rhythm *Physiology.* the rhythmic patterns of biological functions that may range in time periods from seconds to months, such as the patterns of the electroencephalogram, or the circadian rhythms of sleep and waking.

body-righting reflex *Physiology.* the normal upright orientation of the head in space and the positioning of the body in a normal relationship to the head.

body rotation Robotics. the main axis of motion for pick-and-place robots.

body wave *Geophysics.* a seismic wave that travels within the earth, as opposed to one that travels along its surface.

boehmite see BÖHMITE.

boehm lamella *Geology.* any of a group of thin layers or bands of inclusions in quartz that are subparallel to the basal plane.

Boerhaave, Hermann 1668–1738, Dutch physician and educator; wrote seminal textbooks on chemistry and physiology.

Boethius, Anicius Manlius Severinus 480?–524?, Roman philosopher; translated and preserved classical mathematical works.

Boettger's test *Analytical Chemistry.* a test for saccharides; a positive result is indicated when bismuth subnitrate is reduced to a precipitate of metallic bismuth.

bog *Ecology.* a wet, spongy swamp area made up of peat and decaying vegetable materials.

bog asphodel *Botany.* any of several North American or European plants of the lily family, genus *Narthecium,* found in swampy environments; especially *N. americanum,* a grassy yellow-blossomed wildflower native to New Jersey.

bog butter *Geology.* "fossil" peat that has been buried for storage, forgotten, and found much later, preserved in Irish peat bogs.

bogen structure *Geology.* a structure of vitric tuffs composed mainly of shards of glass.

bogey *Military Science.* a hostile or unidentified aircraft or missile. Also, BOGIE, BOGY.

bog harrow *Agriculture.* a type of harrow having large notched disks for breaking up very hard soil or heavy plant growth.

boghead cannel *Geology.* a cannel coal that is rich in organic residues of algae.

boghead cannel shale *Geology.* a coaly shale that is rich in waxy or fatty algae.

boghead coal *Geology.* a variety of nonbanded, translucent coal derived from organic residues of algae, containing a high percentage of volatile matter, and yielding tar and oil upon distillation. Also, ALGAL COAL.

bogie or **bogey** *Mechanical Engineering.* a small truck, cart, wagon, or the like that is used for various purposes, as in mining or railroading.

Ordnance. the portion of an artillery weapon that supports the load of the weapon when it is being transported. *Aviation.* a landing-gear assembly having two sets of wheels set around a central strut. *Industrial Engineering.* any of the supporting wheels or rollers on the inside of an endless track. *Military Science.* see BOGEY.

bog lake *Hydrology.* a small, bog-surrounded body of water with a false bottom of organic material and vegetation on a floating mat of peat.

bog ore *Mineralogy.* any of several spongy, impure minerals found in sandy soil that has been submerged under a swamp or marsh, including **bog iron ore, bog manganese ore,** and **boglime.**

bog spavin see SPAVIN.

Bohemian gemstone *Mineralogy.* any of three gemstones found in Bohemia; they include a garnet (pyrope), a false ruby (rose quartz), and a false topaz (citrine).

Bohm-Aharanov effect *Quantum Mechanics.* the phase shift of an electron wave function resulting in an alteration in the interference pattern of a double-slit electron diffraction experiment in the presence of a potential magnetic field even if the magnetic field is shielded so that diffracted electrons do not pass through it.

böhmite *Mineralogy.* AlO(OH), a grayish-brown or red orthorhombic mineral occurring as microscopic tabular crystals with good basal cleavage, having a specific gravity of 3.03 to 3.07 and a hardness of 3 on the Mohs scale; forms a major component of some bauxites; dimorphous with diaspore. Also, BOEHMITE. (After the German scientist J. *Böhm.*)

Bohr, Aage N. born 1922, Danish physicist; with Mottelson, Nobel Prize in 1975 for discovering connection between collective motion and particle motion in atomic nuclei.

Bohr, Niels 1885–1962, Danish physicist; developed Bohr theory, the foundation of quantum mechanics; Nobel Prize in 1922.

Niels Bohr

Bohr effect *Biochemistry.* a relationship between the amount of carbon dioxide or pH in the blood and the amount of oxygen that hemoglobin can release: as the amount of carbon dioxide increases, the amount of oxygen bound by hemoglobin decreases.

Bohr magneton *Nuclear Physics.* a unit of magnetic moment defined as being equal to the moment of one electron spinning about its own center, or $\mu_B = eh/4\pi m_e c$, where e = charge, h = Planck's constant, m_e = rest mass, and c = the velocity of light.

Bohr-Sommerfeld quantization rule *Quantum Mechanics.* a rule stating that atomic energy levels are quantized such that the classical orbit circumference accommodates exactly an integral number of de Broglie wavelengths for the electron; equivalently, that the action integral be an integral multiple of $\hbar$ (Planck's constant divided by 2π).

Bohr-Sommerfeld theory *Nuclear Physics.* a modification of the Bohr theory that allows for the possibility of elliptical electron orbits. Similarly, **Bohr-Sommerfeld atom.**

Bohr theory *Nuclear Physics.* an early theory in quantum mechanics that postulated the structure of an atom as consisting of a positively charged nucleus around which revolve one or more electrons in discrete circular orbits of constant energy; an increase or decrease in an electron's energy must be accompanied by a transition to another orbit. Similarly, **Bohr atom.**

Bohr-van Leeuwen theorem *Quantum Mechanics.* a theorem claiming that magnetism is not explicable in classical physics, but is a quantum phenomenon.

Bohr-Wheeler theory of fission *Nuclear Physics.* an explanation for the stability of a nucleus against fission based on the concept that the nucleus is an evenly charged incompressible liquid droplet possessing surface tension.

BOI *Aviation.* the airport code for Boise.

Boidae *Vertebrate Zoology.* the boas, a family of chiefly topical, nonpoisonous constricting snakes in the order Squamata, noted for bearing live young.

boil *Physical Chemistry.* to undergo or cause to undergo the process of boiling; change from a liquid to gaseous state. *Civil Engineering.* an accidental flow of water and solid material into an excavation due to excessive water pressure outside the excavation. Also, BLOW. *Medicine.* see FURUNCLE.

boil disease *Veterinary Medicine.* a protozoal disease of fish caused by *Myxobolus pfeiffer*, characterized by tumors in the muscles and connective tissues and often resulting in death.

boiler *Mechanical Engineering.* a closed, pressurized vessel in which water or other liquid is heated, either to be utilized in its liquid state or to generate steam energy.

boiler compound *Chemistry.* any chemical used to treat the water in a boiler, as to prevent corrosion, mineral deposit, and so on.

boiler draft *Mechanical Engineering.* the difference between the atmospheric pressure outside a boiler unit and the pressure within the system.

boiler furnace *Mechanical Engineering.* the heating unit in a boiler system for the mixing of air and fuel and for the fuel's combustion; consists of a rectangular-shaped steel casing lined with refractory material to protect the casing and to prevent loss of heat from the furnace.

boiler horsepower *Mechanical Engineering.* an older method of rating a boiler's output, defined as the evaporation into dry saturated steam of 34.5 lb of water per hour at a temperature of 212°F.

boilerplate *Metallurgy.* a flat-rolled steel plate used to fabricate boilers; also used in shipbuilding. Also, **boiler steel.** *Geology.* a fairly smooth surface on a cliff having little or no foothold. *Hydrology.* frozen, crusty, hard-packed snow.

boilerplate *Computer Programming.* previously written and formatted material, such as business letters, announcements, forms, and the like, that accompanies a word-processing program; it can be adapted to the specific needs of the individual user. (From the use of the term *boilerplate* in journalism to refer to syndicated or previously written copy.)

boilerplate model *Space Technology.* a metal replica conforming to the outer configuration of a flight vehicle model, usually made of heavier material than the original and designed to withstand stress tests.

boiler setting *Mechanical Engineering.* the foundation structure of a boiler system.

boiler trim *Mechanical Engineering.* the external tubes and pipes that connect the control instruments and gauges to a boiler unit.

boiling *Physical Chemistry.* the process in which heat converts a liquid into a gas or vapor; for a given pressure, this occurs at a fixed temperature in a pure substance and over a range of temperatures in mixtures of substances.

boiling point *Physical Chemistry.* the temperature at which a liquid boils; the temperature at which the liquid's vapor pressure equals the atmospheric pressure of its environment. The boiling point of water at sea level is 212°F or 100°C. Also, BP or BP.

boiling-point elevation *Physical Chemistry.* the observation that the boiling point of a liquid in solution is higher than its standard boiling point in a pure state; the amount of elevation is directly dependent on the amount of solute particles present. Also, ELEVATION OF BOILING POINT.

boiling range *Physical Chemistry.* the range of temperature in a laboratory distillation from the point when a liquid begins to boil until the point when it is entirely evaporated.

boiling spring *Hydrology.* **1.** a spring whose waters are agitated by heat action. **2.** a rapidly flowing stream characterized by strong vertical eddies. **3.** a stream rising from the bottom of a residual clay basin and located at the head of an interior valley.

boiling-water reactor *Nucleonics.* a nuclear power reactor in which water passing as coolant through the core is turned to steam by direct use of fission heat from the uranium oxide fuel; steam for driving the turbogenerator is formed within the reactor vessel itself rather than in an external heat exchanger and, after being condensed, returns as feedwater to the reactor vessel.

boil-off *Thermodynamics.* the process by which a liquid is allowed to reach its boiling point and is then vaporized, as when liquid oxygen is exposed to room temperature. *Space Technology.* a loss of liquid fuel from a rocket during countdown due to this process. *Textiles.* a process of removing impurities from a fabric by boiling it in a solution.

boil smut *Plant Pathology.* a common corn disease caused by the fungus *Ustilago maydis,* in which grayish white galls containing black spores appear on all parts of the plant.

bojite *Petrology.* a type of gabbro with primary brown hornblende, diallage, and some biotite and plagioclase.

Bok globule *Astronomy.* a small, roughly spherical dark nebula about a third of a light-year in diameter, believed to be an area of star formation. (Named for the American astronomer Bart J. *Bok.*)

Bol. or **bol.** *Medicine.* an abbreviation for "pill." (From Latin *bolus.*)

Bolbitiaceae *Mycology.* a family of fungi of the order Agaricales, occurring mainly on wood and dung, and including some gill mushrooms.

bold *Graphic Arts.* short for BOLDFACE.

boldface *Graphic Arts.* type that is heavier and darker than its normal companion type: **this is boldface.**

boldface italic *Graphic Arts.* type that is both heavy (bold) and slanted (italic): ***this is boldface italic.***

Boldiaceae *Botany.* a monotypic family of red algae in the order Bangiales, characterized by a saccate thallus rising from a discoid base.

bole *Geology.* any of several varieties of fine, compact, red, yellow, or brown earthy clay composed mainly of hydrous aluminum silicates. Also, BOLUS. *Forestry.* the trunk of a tree, especially the highly marketable lower portion of a tree trunk.

boleite *Mineralogy.* $Pb_{26}Ag_{10}Cu^{+2}_{24}Cl_{62}(OH)_{48}\cdot 3H_2O$, an indigo-blue, translucent cubic mineral with perfect cleavage and pseudocubic crystals, having a specific gravity of 5.05 and a hardness of 3 to 3.5 on the Mohs scale; found as a secondary mineral with cumengite.

Boletaceae *Mycology.* a family of fungi of the order Agaricales, composed of fleshy, edible mushrooms, commonly known as the **boletes.**

Boletales *Mycology.* a term formerly used for an order of fungi belonging to the subclass Holobasidiomycetidae.

boletic acid see FUMARIC ACID.

Boletogastraceae *Mycology.* a family of fungi belonging to the order Cribbiaceae, which is related to the family Boletaceae and includes many widespread North American species. Also RHIZOPOGONACEAE.

Boletus *Mycology.* **1.** a genus of basidiomycetous Hymenomycetes. **2. boletus.** any mushroomlike fungus of this genus, characterized by an easily separable layer of tubes or spores on the underside of the pileus; includes edible and toxic varieties.

bolide [bō´ līd] *Astronomy.* a very bright meteor, especially one that explodes, sometimes audibly, and falls to earth in the form of meteorites.

boll *Botany.* a dry, rounded pod or capsule of a plant; used especially in reference to cotton or flax.

bollard *Naval Architecture.* a thick, short post on a dock or ship used to secure mooring lines. *Civil Engineering.* a post designed to prevent vehicular infringement.

boll rot *Plant Pathology.* a cotton disease caused by the fungi *Glomerella gossypii* and *Xanthomonas malvacearum* that produces the decay of cotton bolls.

boll weevil *Invertebrate Zoology.* a small, snouted beetle injurious to young cotton bolls, *Anthonomus grandis,* that is found in Mexico and the southern United States.

bolometer *Electrical Engineering.* an instrument that measures small amounts of microwave or infrared radiation by detecting changes of electrical resistance on a thin, heat-sensitive metal conductor; applications include standing wave detectors and infrared search and guidance systems. Also, THERMAL DETECTOR. *Cardiology.* an older instrument for measuring the heartbeat's force, as opposed to blood pressure.

bolometric correction *Astronomy.* a photometric adjustment to an object's measured magnitude to include the energy from its nonvisible radiation.

bolometric magnitude *Astronomy.* a magnitude for a heavenly body that includes its nonvisible radiation, either from all wavelengths or from those wavelengths that reach the earth.

bolometric neutrino detection *Nucleonics.* the use of a bolometer's sensitive thermometer, which registers small changes in electrical resistance induced by incident radiation to detect and measure transient temperature rises caused by neutrino interactions.

bolson *Geology.* **1.** in desert regions of the U.S., a broad, saucer-shaped valley or basin almost or completely surrounded by mountains and having no surface outlet. **2.** a temporary lake that forms in such an area.

bolster a cushion, frame, or support; specific uses include: *Building Engineering.* **1.** a horizontal timber used on a post to lessen the middle free span section of a beam. **2.** a timber that connects two ribs of a centering. *Civil Engineering.* a structural member that supports the end of a bridge truss. *Mechanical Engineering.* a metal plate to which dies are fastened in a pressing or punching machine. Also, **bolster plate.** *Naval Architecture.* **1.** a round casting on the side of a vessel, through which the anchor chain passes. **2.** a block of wood or canvas serving to lessen friction on the rigging.

bolt *Mechanical Devices.* **1.** a flat-headed, partially threaded cylindrical fastener commonly used with a nut. **2.** a sliding, cylindrical or bar-shaped fitting forming the tongue of a lock. *Ordnance.* a sliding bar or rod in some firearms that serves to move a cartridge into place, close the breech for firing, and then eject the spent cartridge. *Textiles.* a roll of fabric of a specified length. *Forestry.* **1.** a short round section of a log. Also, BILLET. **2.** a length of timber that is to be cut or sawed into smaller pieces. *Building Engineering.* a block of timber used to make stair treads or shingles. *Graphic Arts.* any of the three folded edges of a sheet before cutting. *Mining Engineering.* see BOLTHOLE.

bolt-action *Ordnance.* of a rifle, firing by means of a bolt.

bolted joint *Engineering.* a connection of two or more structural parts by means of threaded bolts.

bolt face *Ordnance.* the portion of a gun bolt that contacts the base of the cartridge.

bolthole *Mining Engineering.* a short, narrow opening that connects the main workings with the airhead or ventilating drift of a coal mine.

bolting *Engineering.* the joining of parts by using nuts, bolts, or studs; the usual means of fixing together steel parts. *Mining Engineering.* a method for separating different-sized particles using vibrating sieves. *Food Technology.* the process of sifting flour or meal through a sieve or fine-meshed cloth. Thus, **bolting cloth.** *Botany.* the premature production of flowers and seeds, especially of a biennial plant in its first year.

bolt mechanism *Ordnance.* the entire mechanical assembly that includes a gun bolt and its related parts.

boltwoodite *Mineralogy.* $HK(UO_2)SiO_4\cdot 1.5H_2O$, a pale yellow monoclinic mineral having a specific gravity of 3.6 and a hardness of 3.5 to 4 on the Mohs scale; found in silicate alteration zones surrounding zones of hydrated uranyl oxides.

Boltzmann, Ludwig [bōlts´män; bōlts´mən] 1844–1906, Austrian physicist; laid the foundation of statistical mechanics.

Boltzmann constant *Physics.* a constant equal to 1.38×10^{-23} joules per kelvin, used in statistical physics.

Boltzmann distribution *Physics.* a relation describing the distribution of energies for particles when in thermal equilibrium.

Boltzmann distribution law see DISTRIBUTION LAW.

Boltzmann entropy hypothesis *Physics.* a hypothesis stating that the entropy of a system of particles is a linear function of the logarithm of the statistical probability of the distribution.

Boltzmann equation *Physics.* a fundamental equation based on particle conservation, showing that the rate of losses, including leakage from the region and disappearance by reactions of all kinds, is equal to the rate of production from sources within the region and the rate of scattering into the region. The equation is usually given in the form

$$\partial f/\partial t + v \cdot \partial f/\partial x + \partial f/\partial v = (df/dt)_{\text{other}}$$

where the left-hand side is the total time derivative of particle density f due to motion in phase space, and the right-hand side is the total derivative of f due to physical processes such as particle collisions, annihilations, and so on.

Boltzmann factor *Physics.* the factor $\exp(-E/kT)$ that appears in the expression giving the probability for atoms to have an energy E at temperature T, where k is the Boltzmann constant.

Boltzmann H theorem *Physics.* a theorem stating that the entropy of a system always tends to increase.

Boltzmann statistics see MAXWELL-BOLTZMANN STATISTICS.

Boltzmann superposition principle *Materials Science.* a theory allowing for prediction of the stress state of a linear viscoelastic material, through the superposition of the effects of all applied strains.

Boltzmann transport equation *Physics.* an equation that describes the rate of change of the properties of a fluid subject to external fields and gradients of pressure and temperature. Also, MAXWELL-BOLTZMANN EQUATION.

Boltzmann-Vlasov equations *Physics.* the equations governing a high-temperature plasma in which the collisional mean free path is much larger than all other characteristic lengths of the system.

bolus *Pharmacology.* any oral medication formed into a large soft pill. *Medicine.* a concentrated mass of medication injected to aid in diagnosis. *Radiology.* a mass of scattering material, such as paraffin, bags of water, or a rice-flour mixture, placed between the radiation source and the skin to achieve a precalculated isodone pattern in the tissue irradiated. *Physiology.* a small lump of food that has been chewed and swallowed. *Geology.* see BOLE.

bolus alba see KAOLIN, def. 1.

Bolyai, János 1802–1860, Hungarian mathematician; developed hyperbolic geometry, the first successful form of non-Euclidean geometry (independently of Gauss and Lobachevsky).

Bolyai geometry see HYPERBOLIC GEOMETRY.

Bolzano, Bernard 1781–1848, Czech mathematician; defined noninfinitesimal continuous functions and properties of infinite sets.

Bolzano's theorem *Mathematics.* the theorem that if f is real-valued and continuous on the real interval $[a, b]$ and $f(a) > 0$ and $f(b) < 0$, then there exists some number x_0 in (a,b) for which $f(x_0) = 0$.

Bolzano-Weierstrass property *Mathematics.* a topological space X has the Bolzano-Weierstrass property if every infinite subset of X has a limit point in X. Also, FRÉCHET COMPACTNESS.

Bolzano-Weierstrass theorem *Mathematics.* the theorem that every bounded infinite set in R^n possesses at least one accumulation point, which may or may not be an element of the given set.

BOM *Aviation.* the airport code for Bombay, India.

Bomarc *Ordnance.* a U.S. long-range surface-to-air missile powered by two ramjet engines and a solid-propellant rocket booster and equipped with radio command guidance and active radar homing; it delivers a nuclear warhead at a speed of Mach 2.8 and a range of 440 miles; officially designated **CIM-10B.**

bomb *Ordnance.* **1.** a projectile filled with explosive material, designed to be dropped from an aircraft to cause damage to enemy targets. **2.** any similar explosive device used as a weapon to kill or destroy, employed in various ways such as being thrown by hand, fired from artillery, concealed in the ground or in a building or vehicle, and so on. *Engineering.* the thick-walled steel vessel of a bomb calorimeter. *Volcanology.* a large individual fragment of volcanic material that was viscous when ejected but assumed a rounded form as it cooked and solidified before striking the earth. *Computer Programming.* **1.** to process a computer program incompletely or incorrectly due to incorrect instructions or bad input data. **2.** a catastrophic failure of a program or computer system due to hardware or software failure. *Geochemistry.* a vessel used in geochemistry and petrology that can withstand high temperature and pressure. *Radiology.* a heavy metal-shielded apparatus containing a quantity of radium or other radioactive element for use in teleradiation therapy.

Bombacaceae [bäm´bə kā´sē ē] *Botany.* the bombax family of tropical deciduous trees, characterized by palmate leaves, large solitary or clustered flowers, and large dry or fleshy fruit with a woolly or cottony fiber surrounding the seeds; includes the balsa, baobab, and silk-cotton trees.

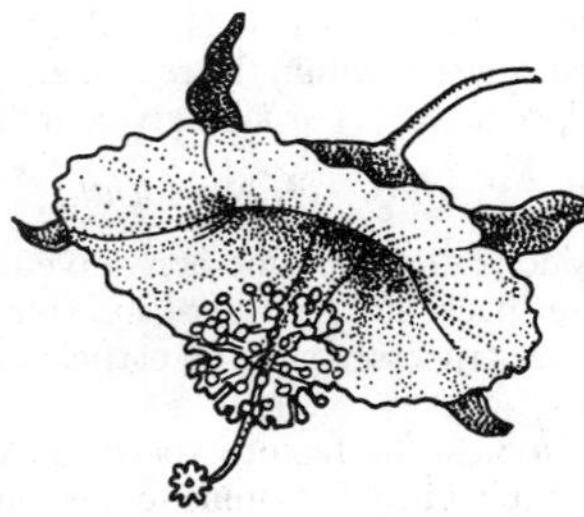

Bombacaceae (baobob)

bombard *Nucleonics.* to subject a target to the impact of an intense stream of high-energy particles, such as electrons, nucleons, alpha particles, or other atomic nuclei, for the experimental study of resultant reactions.

bombardier *Military Science.* **1.** a member of a bomber crew who operates the bombsight and releases the bombs. **2.** a former term for an artilleryman.

bombardier beetle *Invertebrate Zoology.* any of various ground beetles, especially of the genus *Brachinus,* characterized by the ejection of a volatile fluid from its abdomen when disturbed; the fluid vaporizes with a popping sound upon contact with the air.

bombardment *Military Science.* the process of bombarding a target. Thus, **bombardment aviation, bombardment aircraft, bombardment photography,** and so on.

bombax see BOMBACACEAE.

Bombay blood group *Genetics.* a rare genetic variation of the human ABO blood groups produced by a recessive mutation in a locus other than that directly involved in producing the A and B antigens; this mutation prevents production of a precursor substance necessary for the synthesis of the A and B antigens. Also, **Bombay phenotype.** (For *Bombay,* India, where it was first discovered.)

bombazine *Textiles.* a plain or twilled fabric with a silk warp and wool filling, often dyed black.

bomb bay *Aviation.* the compartment in the fuselage of a bomber where bombs are kept in readiness for release.

bomb calorimeter *Engineering.* a container into which an organic material of a known mass is sealed with oxygen and then electrically ignited under water, so that the subsequent rise in water temperature from the ignition can be used to calculate the calorific value of the sample; uses include the measurement of the value for fuels and foods.

bomb carpet see CARPET BOMBING.

bomb casing *Ordnance.* see CASING.

bomb disposal *Ordnance.* the process of handling an unexploded bomb so as to disarm it or detonate it harmlessly.

bomber *Aviation.* **1.** a military aircraft designed to carry and drop bombs; often distinguished by size (for example, *light bomber*) or function (for example, *attack bomber*). **2.** of or relating to such an aircraft. Thus, **bomber attack, bomber squadron.**

bombesin *Endocrinology.* a tetradecapeptide originally isolated from frog skin; on infusion into dogs, it stimulates gastric acid secretion, gallbladder contraction, pancreatic secretion, and relaxation of the choledochoduodenal junction. It is a pressor agent, and it is found in the brain tissue and gut of humans and classified as a neuropeptide.

Bombidae *Invertebrate Zoology.* the bumblebees, a family of hairy medium to large social bees, usually black and yellow.

bombing run *Ordnance.* the flight path of a bombing aircraft just before it releases its bombs. Also, **bomb run.**

bombing through overcast see BTO.

bomblet *Ordnance.* any of a number of relatively small, separately detonating bombs that are encased and delivered in a larger bomb.

bomb line *Military Science.* an imaginary line designated by ground forces, beyond which bombing attacks can be carried out against the enemy without endangering friendly troops. Also, **bomb safety line.**

bombload *Ordnance.* the bomb or bombs carried by an aircraft, or the total weight of such bombs.

bombproof *Ordnance.* describing a structure or building that is able to withstand the effects of most explosive bombardment, especially an underground facility specifically constructed for this purpose. *Computer Science.* see BULLETPROOF.

bomb rack *Ordnance.* a mechanical device in or on an aircraft that holds bombs for release over the target. Similarly, **bomb dispenser.**

bomb reconnaissance *Ordnance.* the act of reconnoitering an area to locate an unexploded bomb and determine its nature.

bomb-release line *Ordnance.* an imaginary line around a target area that represents the point at which an attacking aircraft should release its bombs to hit the target. Similarly, **bomb-release point.**

bomb sag *Geology.* a depression formed by a bomb that fell on layers of tuff or fine-grained ash deposit and was buried there.

bombshell *Ordnance.* see BOMB.

bomb shelter *Civil Engineering.* a structure or area designed to protect people during bombings; often underground and/or specially reinforced.

bombsight *Ordnance.* a device used in a bomber aircraft to guide the release of bombs so that they will strike the target.

bomb test *Engineering.* a test in which items being checked for leaks are placed in pressurized fluid; the fluid is forced into any holes or ruptures present in the item.

Bombycidae *Invertebrate Zoology.* a family of moths, in the order Lepidoptera, comprising a single species, *Bombyx mori,* whose larvae are silkworms.

Bombycillidae *Vertebrate Zoology.* a North American passerine songbird family of the suborder Oscines, characterized by variable soft plumage, a short hooked bill, and long wings; it includes the waxwings, silky flycatchers, and palm chats.

Bombyliidae *Invertebrate Zoology.* a family of bee flies, in the suborder Orthorrhapha, named for their beelike appearance and frequent presence upon flowers; certain larvae are parasitic on bee larvae.

Bombyx mori *Invertebrate Zoology.* a creamy-white domesticated silkworm moth whose larvae secrete a delicate cocoon thread that is unravelled by humans and spun into silk cloth.

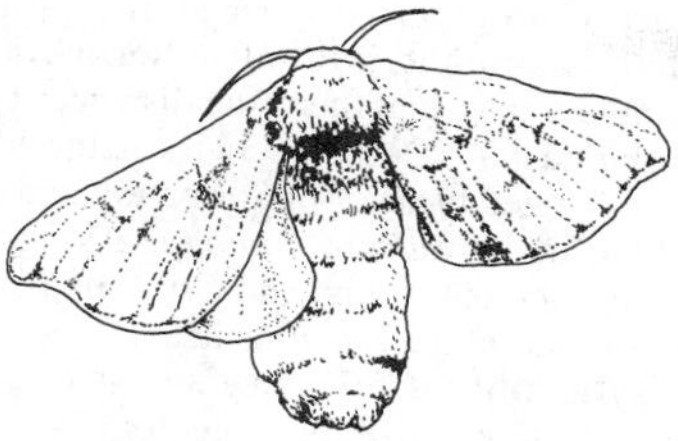

Bombyx mori (silkworm)

bonaci *Vertebrate Zoology.* any of several edible fishes of the genus *Mycteroperca,* commonly known as groupers, found in the waters off of Florida and the West Indies. (From *bonasi,* an American Spanish term for fish.)

bond something that connects or holds together; specific uses include: *Chemistry.* the attractive force that links atoms together in a molecule or an ionic structure, so that they act as a unit. Also, CHEMICAL BOND. *Engineering.* any substance used to hold together the abrasive grains of material in a grinding wheel, such as ceramics, rubber, or resin. *Medicine.* see BONDING. *Civil Engineering.* adhesion between concrete and steel reinforcement due to shrinkage of the concrete and the natural adhesion between the particles. *Metallurgy.* in welding, brazing, or soldering, the interface between the joined parts, or between a part and the filler. *Building Engineering.* any of various generally overlapping configurations of bricks, stones, or similar materials designed to increase the strength or enhance the appearance of building construction. *Electricity.* a conductor that provides a continuous path of electric current between metal parts or other equipotential points. Thus, **bonding wire.** *Behavior.* a close social attachment, usually occurring between mated pairs or between parents and their young. *Materials.* see BOND PAPER.

Bond, George Phillips 1825–1865, American astronomer; discovered Saturnian ring and satellite; pioneer in astronomical photography.

Bond albedo *Optics.* the ratio of the total light reflected from an illuminated sphere to the incident light upon that sphere.

Bond and Wang theory *Mechanical Engineering.* a theory used to calculate the amount of energy in horsepower hours needed to pulverize a short ton of solid material.

bond angle *Physical Chemistry.* the angle between the lines connecting the nucleus of one atom to the nuclei of two other atoms that are bonded to it. Also, VALENCE ANGLE.

Bondarzewiaceae *Mycology.* a family of fungi belonging to the order Agaricales, comprised of root parasites of living trees.

bond axis *Physical Chemistry.* a central line between the nucleus of one atom and the nucleus of another atom that is bonded to it, bisecting the area of high electron density between the two nuclei.

bond dissociation energy *Physical Chemistry.* the actual bond energy required to break a single bond of one type in one molecule. Also, SINGLE BOND ENERGY.

bond distance see BOND LENGTH.

bonded fabric *Textiles.* **1.** a fabric made from fibers that are held together by a chemical adhesive rather than being woven or spun. **2.** a fabric composed of two fabrics that have been joined together, usually a face fabric and a lining fabric.

bonded NR diode *Electronics.* an N+ junction semiconductor device that combines avalanche breakdown and current flow conductivity modulation to form negative resistance.

bonded-phase chromatography *Analytical Chemistry.* a form of high-pressure liquid chromatography that uses a stable stationary phase.

bonded strain gauge *Engineering.* a strain gauge in which a fine wire functions as the resistance element, generally in a zigzag shape; it is impregnated into an insulating material and bonded to the pressure-sensing element.

bond electron see BONDING ELECTRON.

bond energy *Physical Chemistry.* the force that maintains a chemical bond, described in terms of the average amount of energy per mole required to break bonds of the same type per molecule; in general, the larger the bond energy, the stronger the bond. Also, BINDING ENERGY, TOTAL BOND ENERGY. *Physics.* see BINDING ENERGY.

bonderize *Metallurgy.* to apply a phosphate anticorrosion coating to a steel surface.

bond hybridization *Chemistry.* the manner in which atomic orbitals mix or change to accommodate different types of chemical bonds.

bonding *Chemistry.* the combining of atoms to form molecules or ionic compounds. *Engineering.* the structural joining of two components by means of an adhesive, especially under high temperature and pressure. *Medicine.* the process of fixing orthodontic brackets or other attachments directly to the enamel surface of the teeth with orthodontic adhesives. Thus, **tooth bonding.** *Behavior.* the act of forming or maintaining a close social attachment between two or more individuals. Thus, **bonding behavior, bonding drive.**

bonding electron *Physical Chemistry.* an electron whose charge density resides predominately between two atoms, so that it helps hold the molecule together.

bonding layer *Building Engineering.* a thin layer of cement mortar that is spread on a moist, prepared, hardened concrete surface before applying fresh concrete.

bonding orbital *Physical Chemistry.* a relationship between an electron and a nucleus in which the electron's energy decreases as it draws closer to the nucleus, causing them to unite. Also, **bonding molecular orbital.**

bonding pad *Electronics.* a metallized region on the surface of a semiconductor device, to which connections can be made.

bonding strength *Mechanics.* the strength of adhesives, welds, solders, glues, and the like in response to shear or tension. *Materials Science.* a measure of the effectiveness of a commercial product in terms of this strength.

bonding type *Materials Science.* a description of the interatomic bonding characteristics of a material, based on the nature of the electron distribution about the atoms.

bond length *Physical Chemistry.* the average distance between the nuclei of two atoms bonded together in a stable molecule. Also, BOND DISTANCE. *Civil Engineering.* that length of a reinforcing bar which corresponds to its grip ability.

Bond number *Fluid Mechanics.* a dimensionless group that gives the ratio of gravitational force to surface-tension force.

bond paper *Graphic Arts.* any of a wide variety of fine papers, usually having a high rag content, characterized by their strength, rigidity, and absorbency. (Originally, such paper was used to print stock certificates and government *bonds.*) Also, BOND.

Bond's law *Mechanical Engineering.* a theory used to explain the relationship between the amount of energy needed to crush solid materials, and the size and shape of the end-product; intermediate between Kick's law and Rittinger's law. Also, **Bond's third theory.**

bondstone *Building Engineering.* **1.** a stone used for bonding masonry to a masonry backing by joining the coping of a gable above a wall. **2.** a stone that passes through a masonry wall so as to support it.

bond stress *Building Engineering.* a shear stress at the surface of a reinforcing bar that prevents relative movement between the bar and concrete.

bonduc *Botany.* a semitropical tree of the legume family, *Caesalpinia bonducella,* that produces hard, gray seeds used as rosary beads and in making jewelry.

Bondy-Chvatal theorem *Mathematics.* the closure $c(G)$ of an undirected graph G is Hamiltonian if and only if G is Hamiltonian. Dirac's theorem and Ore's theorem are corollaries of this theorem.

bone *Anatomy.* **1.** the dense connective tissue that makes up the majority of the skeleton of most vertebrates, consisting of a mineralized matrix surrounding living osteocytes. **2.** one of the individual parts of the skeleton of a vertebrate.

bone age *Physiology.* the stage of the skeletal development of an individual expressed in terms of the average chronologic age at which such development is normally attained.

bone bed *Geology.* any sedimentary layer or layers that are characterized by a concentration of fossil bones, bone fragments, and other such organic remains.

bone chert *Petrology.* a white or red weathered residual chert with a chalky and porous texture.

bone coal *Geology.* **1.** a hard, compact coal with a high ash content. **2.** a thin, argillaceous sedimentary layer found in coal seams.

bone dating *Archaeology.* a method of dating in which dates are found by utilizing stratigraphy and F-U-N (fluorine, uranium, nitrogen) techniques to analyze bone samples; human remains are compared with animal bone or fossils found in the same strata.

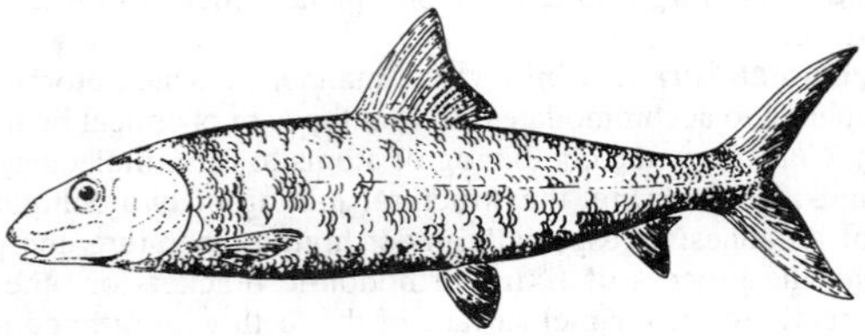

bonefish

bonefish *Vertebrate Zoology.* any of several tropical or semitropical marine fishes of the family Albulidae, especially *Alpula vulpes,* a game and food fish with a slender silvery body and a skeleton composed of many fine bones. Also, LADYFISH.

bone graft *Surgery.* a graft using a piece of bone to replace another bone that is lost due to trauma or disease.

Bonellia *Invertebrate Zoology.* a genus of marine worms, in the order Echiuroidea, of which the males are parasitic on the females.

Bonellidae *Invertebrate Zoology.* a family of wormlike marine animals, in the order Echiuroidea.

bone marrow *Histology.* soft, higly vascular connective tissue that occurs primarily in certain flat bones and serves as the main area of red and white blood cell production.

bone phosphate of lime see TRICALCIUM PHOSPHATE.

bone seekers *Neutronics.* radioisotopes of calcium and its chemical analogues strontium and barium that tend to replace normal calcium in calcifying tissue of bone, as well as radium and radioistopes of the rare-earth elements, such as cerium, that are deposited to a lesser extent in other parts of the bone; the resultant deposition of alpha- and beta-particle radiations in small amounts may cause serious bone and bone marrow damage.

bone spavin see SPAVIN.

bone turquoise *Mineralogy.* fossil bone or ivory that has been colored blue with phosphate of iron; it is used as a gemstone, but loses its color over time. Also, FOSSIL TURQUOISE, ODONTOLITE.

bone wax *Surgery.* a waxy substance that is used to pack small bone cavities and to control bleeding from them, e.g., in the bones of the skull.

bongkrekic acid *Organic Chemistry.* $C_{28}H_{38}O_7$, a white solid that melts between 50 and 60°C; it is a toxic antibody produced by *Pseudomonas cocovenenans.*

bongo *Vertebrate Zoology.* a large forest antelope, *Taurotragus eurycerus,* of the family Bovidae, native to tropical Africa and having spiraled, broad horns, a very short mane down the length of the back, and thin white stripes on a heavy, reddish-brown body.

bongo

boninite *Petrology.* a glassy andesite containing abundant phenocrysts of bronzite, less of olivine and augite, and no feldspar.

bonito *Vertebrate Zoology.* **1.** any of three marine food and game fishes of the mackeral family, including the Atlantic or common bonito, *Sarda sarda,* the Pacific or California bonito, *S. chiliensis,* and the Indo-Pacific bonito, *S. orientalis.* **2.** any of several related fishes, including the leaping bonito, *Cybiosarda elegans,* and the oceanic bonito, the skipjack tuna.

Bonnemaisoniaceae *Botany.* a family of red algae of the order Nemaliales, characterized by differing gametangial and tetrasporangial thalli and often producing phenolic compounds; some species are considered to be an edible delicacy in the South Pacific.

Bonne's projection *Cartography.* a conical map projection in which the central meridian is straight, with all other meridians shown as curved lines and parallels spaced at their true distances apart.

Bonnet, Charles [bō´nā´] 1720–1793, Swiss naturalist; rediscovered parthenogenesis; popularized ideas of preformation and catastrophism.

bonnet monkey [bän´it] *Vertebrate Zoology.* a long-tailed macaque, *Macaca radiata,* found in south India and Sri Lanka and characterized by a bonnetlike tuft of hair on the top of its head.

Bonnevie-Ullrich syndrome see TURNER'S SYNDROME.

Bononian *Geology.* in Great Britain, a geologic stage of the Upper Jurassic period, occurring after the Kimmeridgian and equivalent to the lower Portlandian stage.

bonsai [bän´sī´] *Botany.* **1.** the Japanese art of growing dwarf trees by pruning, pinching, and training to produce a desired shape or effect. **2.** a plant dwarfed by these methods.

bontebok [bän´tə bäk´] *Vertebrate Zoology.* a large South African antelope, *Damaliscus dorcus,* that has a purplish-red body and a white face and rump; it is threatened with extinction.

bony fish *Vertebrate Zoology.* a general term for fishes in the vertebrate class Osteichthyes, characterized by a bony skeleton.

bony labyrinth *Anatomy.* the network of passages within the temporal bone that forms the vestibule, cochlea, and semicircular canals of the inner ear.

Boo see BOÖTES.

booby *Vertebrate Zoology.* any of several tropical seabirds of the family Sulidae, genus *Sula,* characterized by sharp bills and webbed feet. (So called by sailors because they were very easy to capture.)

booby trap *Ordnance.* **1.** a concealed explosive device set to go off when moved or disturbed by an unsuspecting person. **2.** any similar device concealed to cause injury to enemy personnel. **3. booby-trap.** to conceal such a device within or beneath a seemingly harmless object.

Boodleaceae *Botany.* a family of green marine algae belonging to the order Siphonoclades.

bookbinding see BINDING.

book capacitor *Electricity.* a variable capacitor in which the metal plates are hinged along one edge like the pages of a book.

book gill *Invertebrate Zoology.* in king crabs, a respiratory gill with membranous folds arranged like pages in a book. Also, GILL BOOK.

bookkeeping *Computer Programming.* see HOUSEKEEPING.

book louse *Invertebrate Zoology.* any minute, soft-bodied wingless insect of the order Psocoptera, including around 1000 species; often found feeding among old books and papers.

book lung *Invertebrate Zoology.* the respiratory organ of spiders and scorpions, containing parallel membranous layers arranged like pages in a book. Also, LUNG BOOK.

book mold *Metallurgy.* in casting, a permanent mold that is split longitudinally and hinged.

book paper *Graphic Arts.* any of a wide variety of good, sturdy papers traditionally used in book publishing and commercial printing.

book structure *Geology.* a rock structure in which parallel sheets of ore alternate with valueless rock, usually quartz.

Boole, George 1815–1864, English mathematician; developed Boolean algebra.

Boolean algebra *Mathematics.* **1.** a nonempty class of sets that is closed under union and the taking of set complements. **2.** a Boolean ring in which the ring elements are sets, ring addition is defined to be symmetric difference of sets, and ring multiplication is defined to be set intersection. **3.** in logic theory, a Boolean algebra is formulated from a nonempty set S_0 of statements. The set is extended to the smallest set S of statements (or formulas) such that the negation of each member of S is a member of S and each of the conjunction, disjunction, conditional, and biconditional of any two members of S is a member of S. Two elements of S are indistinguishable if they yield equal truth tables. Since it can be shown that each formula is equivalent to one involving no connectives other than conjunction and negation, the statement calculus under conjunction, negation, and a convention of truth equivalence is a Boolean algebra. **4.** let B be any nonempty set, and suppose that to each $a \in B$ there corresponds a unique element $a^* \in B$ and to each pair of elements $a, b \in B$ there is a unique element $a \vee b \in B$ such that the following are true: (i) $a \vee b = b \vee a$; (ii) $(a \vee b) \vee c = a \vee (b \vee c)$; and (iii) $(a^* \vee b^*)^* \vee (a^* \vee b)^* = a$. Then B is a Boolean algebra.

Boolean function *Mathematics.* a function that takes only two values (often referred to as T and F), and whose domain is a collection of propositional statements. *Computer Programming.* an expression in Boolean algebra that uses logical operators such as AND, OR, NOT, NAND, NOR, and exclusive OR, to combine binary variables or words of binary values. Also, LOGIC FUNCTION.

Booleanization *Mathematics.* let X be a topological space. The Booleanization of X is a topological space X^b on X that has the compact open sets of X and their complements as a subbase for open sets.

Boolean operator *Computer Programming.* a logic operator whose operands and result are variables that can assume one of only two states, or are words of multiple binary values.

Boolean ring *Mathematics.* a ring R such that $a^2 = a$ for all elements a in R.

Boolean σ-ring *Mathematics.* a Boolean ring S (whose elements are usually thought of as sets with symmetric difference and intersection taking the roles of ring addition and multiplication, respectively) that is closed under countable unions; i.e., the union of every countable collection of elements of S is also in S. A Boolean σ-ring is also closed under countable intersections; for example, the σ-ring of subsets of a set X is a Boolean σ-ring. Also, **Boolean sigma ring.**

Boolean search *Computer Programming.* a search for information that meets multiple conditions combined by logical (Boolean) operators such as AND, OR, or NOT.

Boolean value *Computer Programming.* either of two possible values, true or false. A Boolean variable and a Boolean function may only assume one of the Boolean values. Also, LOGICAL VALUE.

boom *Naval Architecture.* **1.** a horizontal spar attached at one end to the lower part of a mast, used to control the sheet of a fore-and-aft sail. **2.** a horizontal spar attached to a mast at one end, used to hoist boats or cargo. **3.** a horizontal spar extending from the side of a ship and used to support a small hull, called an outrigger, attached to the main hull to provide stability. *Navigation.* a barrier designed to prevent the passage of a vessel. *Forestry.* a barrier designed to confine floating logs. *Aviation.* **1.** a structural member connecting the tail to the fuselage on some aircraft. **2.** a retractable refueling pipe on a tanker aircraft. *Mechanical Engineering.* a beam member of a crane used to project the upper end of the hoisting tackle. *Robotics.* a device that allows a robotic arm to be extended telescopically. *Acoustical Engineering.* see BOOM MICROPHONE.

boom crutch *Naval Architecture.* a Y-shaped fitting used to secure a boom in place when it is not being used. Also, **boom cradle.**

boomer *Engineering.* a device used to tighten the chains securing a load of equipment, such as pipe or timber.

booming *Mining Engineering.* in placer mining, the accumulation and sudden discharge of a volume of water. Also, GOUGING.

boom microphone *Acoustical Engineering.* an apparatus widely used in moviemaking and television broadcasting, consisting of a movable mechanical arm from which a microphone is suspended; it can be positioned so that it is close above the performer's head but still high enough to be out of camera range.

boomslang *Vertebrate Zoology.* a rear-fanged, venomous snake of the Colubridae family, *Dispholidus typus,* found in the sub-Saharan savanna; an arboreal hunter of chameleons, frogs, and birds.

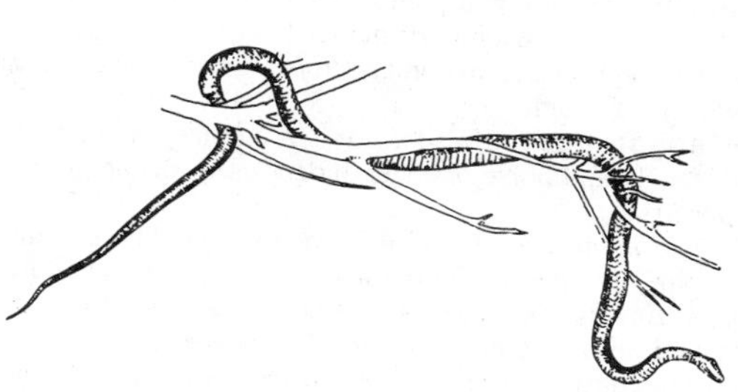

boomslang

boom stop *Mechanical Engineering.* a device that is used to stop a boom when it has reached a maximum angle.

boongary a small arboreal kangaroo, *Dendrolagus lumholtzi,* found in Queensland, Australia. Also, **Lumholtz's kangaroo.**

Boophilus *Invertebrate Zoology.* a genus of blood-sucking ixodid cattle ticks comprising many species that are vectors of bovine anaplasmosis and babesiosis.

Boopidae *Invertebrate Zoology.* a family of lice in the order Mallophaga that are exoparasites on Australian marsupials.

Boord synthesis *Chemical Engineering.* a technique for making α-olefins by reducing α-bromo ethers with zinc.

boorga see BURGA.

boost *Aviation.* **1.** to supply additional thrust or air pressure. **2.** the additional thrust or air pressure supplied, or any action or device that supplies this. **3.** see BOOSTER PRESSURE. *Space Technology.* to raise a space vehicle from the surface to a programmed altitude by rocket propulsion. *Mechanical Engineering.* a rise in the atmospheric pressure in the induction of a supercharged or turbocharged piston engine. *Electronics.* to augment the total voltage acting in a circuit by the series connection of an additional supply source. *Acoustical Engineering.* see BASS BOOST.

booster *Mechanical Engineering.* any of various devices used to increase the power or thrust of an engine. Thus, **booster charge, booster pump, booster system,** and so on. *Space Technology.* see BOOSTER ENGINE, BOOSTER ROCKET. *Electronics.* **1.** a radio-frequency amplifier used to amplify and rebroadcast a television or communication radio signal at higher power without altering the carrier frequency. **2.** any device that increases the amplitude of a signal or of an energy source. *Ordnance.* a small explosive charge used to augment the fuse in setting off the main explosive charge. *Immunology.* a vaccine given after a primary immunization to renew or increase an immune response.

booster brake *Mechanical Engineering.* an air chamber connected to the brake pedal that uses the vacuum produced by the engine to force hydraulic fluid or compressed air to the wheel brake, thus requiring less pressure on the brake pedal to stop the vehicle.

booster cable see BATTERY CABLE.

booster engine *Space Technology.* the engine of a booster rocket or similar device used to provide auxiliary thrust.

booster fan *Mechanical Engineering.* a fan used to increase the amount of air pressure or air flow.

booster pressure *Aviation.* manifold air pressure that is greater than the prevailing atmospheric pressure. Also, BOOST.

booster pump *Mechanical Engineering.* an auxiliary pump used to increase pressure in a liquid or compressed-air pipe.

booster response see ANAMNESTIC RESPONSE.

booster rocket *Space Technology.* **1.** a rocket used to launch a vehicle before another engine takes over. **2.** a rocket engine used along with a sustainer engine or other propulsive system to provide additional thrust.

booster station *Engineering.* one of a series of facilities located along a long-distance pipeline carrying oil, gas, or other liquid products, having pumps or compressors to restore lost pressure.

booster unit *Forestry.* a water supply and pump outfit used for fire control that consists of a tank, a hose reel, and a self-contained power pump mounted on a truck or trailer.

boot *Computer Technology.* to load (an operating system, program, or the like) into the memory of a computer. *Electricity.* a protective boot-shaped jacket pulled over a cable or connector. *Petroleum Engineering.* a cylindrical separator through which well fluids sometimes pass before flowing into the actual separator; the boot aids the separation process by working as a surge column to allow gas to separate out.

Boötes [bō ō´ tēz] *Astronomy.* a conspicuous northern constellation containing the bright star Arcturus; common names include the Hersdman and the Bear Driver.

boothite *Mineralogy.* $CuSO_4 \cdot 7H_2O$, a blue monoclinic water-soluble mineral with a specific gravity of 2.1 and a hardness of 2 to 2.5 on the Mohs scale; found in massive forms with copper and iron sulfates.

bootjack *Engineering.* a device that retrieves objects from inaccessible locations, such as a drilling well.

bootleg *Mining Engineering.* a hole created by a blast that has shattered the rock improperly, resulting in a bootleg shape. Also, GUN.

bootstrap *Engineering.* a general term for a self-sustaining system that is started by external forces and continued by means of its own actions. *Computer Technology.* **1.** the process by which a programmed loader, whose job it is to load other software into a computer, gets itself into the computer. **2.** a small segment of machine code that is either stored permanently in nonvolatile memory or activated by pushing a console button; it loads the operating system from disk or external memory storage into random access memory and calls it when the computer is turned on. *Space Technology.* **1.** a self-sustaining rocket ignition system in which the initiation of one operation sets a sequence in motion. Thus, **bootstrap process. 2.** a rocket that lifts off and ascends rapidly. *Statistics.* a procedure to compute the distribution of a statistic of interest based on random resampling from the observed data.

bootstrap button *Computer Technology.* a button on a computer console that, when pressed, initializes the computer and loads the operating system.

bootstrap circuit *Electronics.* a circuit in which part of the output is fed back across the input, providing a higher effective input impedance and unity gain.

bootstrap loader *Computer Technology.* a binary loader that is used for the initial loading of a program into an empty computer.

bootstrap memory *Computer Technology.* a device that is built into a computer, usually in read-only memory or hard-wired, in order to perform the initial loading of software into the machine.

bootstrapping *Computer Technology.* see BOOTSTRAP. *Electronics.* a method of lifting a generator circuit above ground by a voltage value derived from its own output signal.

bootstrap scheme *Particle Physics.* the aspect of the strong interaction theory that is concerned with the self-consistency of the theory, whereby no single particle can be distinguished as being responsible for the interactions between two hadrons. The fact that all hadrons can take part in strong interactions, either as initial and final particles or as the exchanged particle responsible for the interaction force, provides self-consistency constraints on the scattering amplitudes describing those reactions. Also, **bootstrap theory.**

Bopyridae *Invertebrate Zoology.* a family of isopod crustaceans; the females are parasites attached to shrimp and other decapods, while males remain free-swimming.

Bopyrina *Invertebrate Zoology.* a genus of isopod crustaceans in the suborder Epicaridea that are parasitic on marine shrimp.

Bopyroidea see EPICARIDEA.

bora *Meteorology.* **1.** a violent, cold, and dry wind that blows from the north or northeast. **2.** a moist wind with a source so cold that when the air reaches the lowlands or coast, the dynamic warming is insufficient to raise the air temperature to a normal level for that region.

boracic acid see BORIC ACID.

boracite *Mineralogy.* $Mg_3B_7O_{13}Cl$, a brittle, white, orthorhombic pseudocubic mineral with a vitreous luster inclining to adamantine, having a specific gravity of 2.95 and a hardness of 7 to 7.5 on the Mohs scale; found in the water-insoluble residue from rock salt.

bora fog *Meteorology.* a dense fog caused when a bora lifts a spray of droplets from the surface of the sea.

Boraginaceae *Botany.* a family of dicots, characterized by simple alternate hairy leaves and inflorescences that are a spirally coiled cyme, unfurling as the flowers bloom. Also, **borage family.**

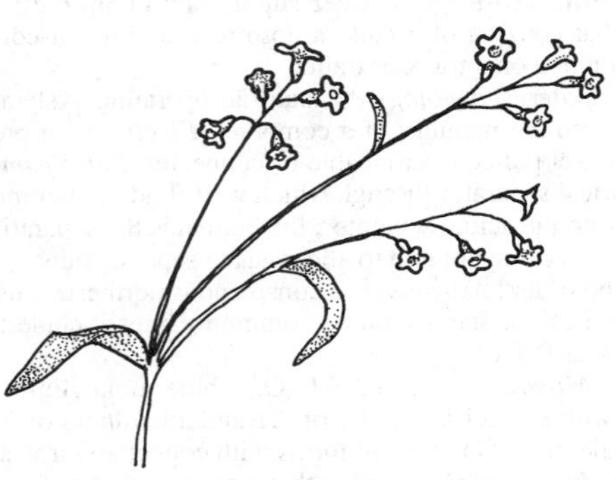

Boraginaceae (forget-me-not)

Boralf *Geology.* a suborder of the soil order Alfisol having a dull brown or yellowish-brown color, characterized by a mean annual soil temperature between 0°C and 8°C and by not being completely dry for either 90 consecutive days or for 60 consecutive days in the 90-day period following the summer solstice.

borane *Inorganic Chemistry.* any of various compounds of boron and hydrogen (boron hydrides), including the simplest, B_2H_6 (diborane), and higher boranes such as pentaborane, B_5H_9, hexaborane, B_6H_{10}, and decaborane, $B_{10}H_{14}$; used as a high-energy fuel.

borate *Chemistry.* **1.** any of the salts or esters of boric acid, H_3BO_3. **2.** to treat with boric acid.

borate mineral *Mineralogy.* a group of minerals having as a principal chemical constituent boron combined with oxygen in a basic structure of BO_3^{-3}.

borax *Mineralogy.* $Na_2B_4O_5(OH)_4 \cdot 8H_2O$, a brittle, white, translucent to opaque, monoclinic mineral with a vitreous to resinous luster, occurring in prismatic, densely grouped crystals, and having a specific gravity of 1.71 and a hardness of 2 to 2.5 on the Mohs scale; found in evaporated saline lake and playa deposits; uses include cleansing agents, enamels, pottery, glass, textiles, and tanning. Also, SODIUM BORATE HYDRATE.

borax bead test see BEAD TEST.

borazole *Inorganic Chemistry.* $B_3N_3H_6$, a colorless, highly toxic liquid, boiling at 55°C, that hydrolyzes with water to form boron hydrides. It is the inorganic analogue of benzene, with similar physical properties.

Borda, Jean Charles 1733–1799, French naval officer; contributions in hydrodynamics (the Borda mouthpiece) and nautical astronomy.

Borda mouthpiece *Fluid Mechanics.* a re-entrant orifice or mouthpiece that eliminates flow along tank walls so that the pressure there is nearly hydrostatic.

Bordeaux [bôr´dō´] *Food Technology.* any wine from the Bordeaux region of southwestern France, such as claret or Sauterne.

Bordeaux mixture *Toxicology.* an insecticide and fungicide mixture made by adding slaked lime to copper sulfate and water, used against garden pests.

border disease *Veterinary Medicine.* a congenital viral disease affecting newborn lambs, characterized by an excessive birth coat, low birth weight, tremors, and abnormal gait and head shape; death usually occurs within the first few weeks after birth. Also, HAIRY SHAKER DISEASE.

bordered *Botany.* of a leaf, having a margin differentiated by its structure or marking.

bordered pit *Botany.* a pit on the cell wall of tracheids and wood vessels in which the secondary cell wall arches over the pit cavity.

border facies *Geology.* the outer edge of an igneous intrusion, differing in texture and composition from the main body of the intrusion.

border irrigation *Agronomy.* an irrigation method for a sloping field in which water flows down the slope in low ditches. Also, **border method, border strip irrigation.**

borderland *Geology.* a crystalline landmass once presumed to have bordered certain orogenic belts near the edges of the North American continent.

borderline defective *Psychology.* a person in the category of borderline mental retardation.

borderline (personality) disorder *Psychology.* see BORDERLINE SYNDROME.

borderline mental retardation *Psychology.* a mild form of mental retardation, with an IQ of about 71 to 85, marked by a slight impairment of adaptive behavior. Also, **borderline intellectual functioning.**

borderline psychosis *Psychology.* a condition of an individual who is potentially psychotic and who displays some psychotic traits, but who has not yet broken with reality. Thus, **borderline schizophrenia.**

borderline syndrome *Psychology.* a personality disorder involving a variety of nonpsychotic but serious symptoms, such as long-standing repressed anger, impulsive and unpredictable behavior, lack of a sense of personal identity, and chronic feelings of loneliness and emptiness.

Bordet, Jules [bôr´dā´; bôr´de´] 1870–1961, Belgian bacteriologist; Nobel Prize for work in immunology; discovered antibody in blood serum, hemolysis, and agent of whooping cough (with Gengou).

Bordetella *Bacteriology.* a genus of Gram-negative, strictly aerobic bacteria possessing a respiratory metabolism, occurring as parasitic or pathogenic coccobacilli in mammals; *B. pertussis* is the causative organism of whooping cough.

Bordet-Gengou agar *Microbiology.* a culture medium, containing glycerol, starch, and horse or sheep blood, used for the growth of *Bordetella pertussis.*

bore *Design Engineering.* **1.** a circular or ellipsoidal hole or interior of a cylinder, pipe, or tube. **2.** the diameter of a hole or a hollow interior. *Mechanical Engineering.* to enlarge a hole to a specific diameter, using a cutting tool inside the hole. *Mining Engineering.* a tunnel in the process of being constructed. *Ordnance.* **1.** the inside part of a gun barrel. **2.** the diameter of this area. *Oceanography.* a turbulent, wall-like advancing wave of water caused by a very rapid rise of the tide, usually in a long, shallow, and narrowing estuary. Also, TIDAL BORE. *Volcanology.* an outlet of a geyser at the surface of the earth.

boreal *Geography.* of or relating to the north.

boreal forest *Ecology.* a vast area of forest vegetation that lies adjacent to the arctic tundra, and stretches southward from the arctic timberline down to the north temperate forest; in North America it covers large areas of Canada and parts of the U.S. Northeast and upper Midwest. Also, TAIGA.

boreal period *Ecology.* in northern Europe, a period extending from 7500 to 5500 BC, characterized by cold winters, warm summers, and the growth of boreal forests.
boreal zone *Ecology.* a climactic and biotic zone in North America and northern Eurasia characterized by short summers and long cold winters.
bore axis *Ordnance.* see AXIS.
borecole see KALE.
bore diameter *Ordnance.* see BORE, def. 2.
bore erosion *Ordnance.* the enlargement of a gun bore due to a wearing away of the surface.
bore evacuation *Ordnance.* see EVACUATION.
bore evacuator *Ordnance.* see EVACUATOR.
bore expansion *Ordnance.* the enlargement of a gun bore due to expansion from interior pressure.
borehole *Mining Engineering.* a hole made in the ground by a drill, auger, or other device to explore strata in search of minerals, for water supply, for blasting, or to relieve underground pressures caused by accumulation of gases or water. Also, WELLBORE. *Geology.* see BORING.
borehole mining *Petroleum Engineering.* the extraction of minerals in the form of gas or liquid from the earth, through the use of suction pumps and boreholes.
borehole survey *Mining Engineering.* **1.** a check to determine any deviation from the vertical plane or from the planned target in a borehole, by means of a compass, plumbing line, camera, or gyroscope lowered into the hole. **2.** the record of information thus obtained. Also, DRILL-HOLE SURVEY.
bore impression *Ordnance.* a plastic impression of the bore of a gun to determine the condition of the rifling.
Borel, Félix Édouard Émile 1871–1956, French mathematician; demonstrated importance and uses of Heine-Borel Theorem; pioneer of game theory.
Borel σ-algebra *Mathematics.* **1.** for a topological space X, the smallest σ-algebra of subsets of X that contains every open subset of X. **2.** for X a locally compact Hausdorff space, the σ-ring generated by the class of all compact subsets of X. Also, **Borel sigma algebra.**
Borelli, Giovanni Alfonso 1608–1679, Italian mathematician and physiologist; founder of the iatrophysical school, which uses mechanical principles to explain movements of the animal body.
Borel measurable function *Mathematics.* an extended real-valued function f defined on a topological space X with Borel σ-algebra $B(X)$ is said to be **Borel measurable** (or f is a **Borel function**) if $f^{-1}([a, \infty]) \in B(X)$ for all real numbers a; that is, the preimage of any interval $[a, \infty]$ (or, equivalently, of any open interval) is a Borel set. If, in addition to this, $f^{-1}([a, \infty])$ is a Borel set, then f is a Baire function.
Borel set *Mathematics.* for a given topological space X, a member of the Borel σ-algebra of X.
Borel subalgebra *Mathematics.* a maximal solvable subalgebra of a Lie algebra; corresponds to a Lie algebra of upper-triangular matrices.
bore premature *Ordnance.* the premature explosion of a projectile within the bore.

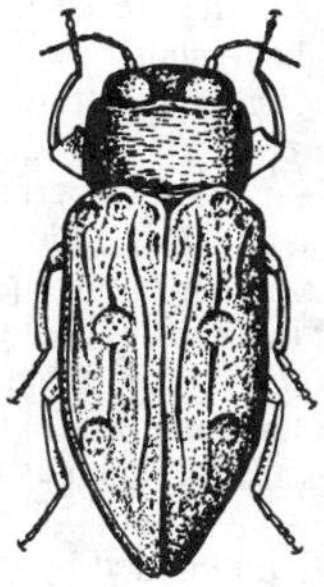

borer (metallic wood boring)

borer *Mechanical Engineering.* a tool or apparatus that is used to bore a hole. *Invertebrate Zoology.* any of various animals or their larvae that bore into wood, rock, or other surfaces, as in certain bivalves, sponges, and insects. *Vertebrate Zoology.* any of various jawless fishes of the subclass Cyclostomata that bore into other fishes to feed on their flesh, such as the hagfish.
bore riding pin *Ordnance.* see RIDING PIN.
borescope *Engineering.* a slender straight-tube periscope, usually provided with illumination, that is used to inspect the interior of machinery or artillery weapons for defects.
bore sight *Ordnance.* a device used to align the axis of the bore of a gun with an aiming point. Also, **boresight.**
boresight camera *Optics.* a camera used in conjunction with radar to photograph a stationary target or objects being tracked, in order to assess the need for realignment of the radar.
boresighting *Ordnance.* a process by which the axis of a gun bore and the line of sight are made parallel to each other or are made to converge on a certain aiming point, by aligning a sight at the bore of the gun with one at the muzzle. *Engineering.* the alignment of a microwave or radar antenna toward a fixed target at a specific location.
Borhyaenidae *Paleontology.* a family of extinct carnivorous marsupials in the superfamily Borhyaenoidea, with fossils dating to the Miocene Epoch; they were about the size of a wolf with sharp teeth and claws.
Borhyaenoidea *Paleontology.* a superfamily of extinct carnivorous marsupials, with fossil remains dating to the Miocene Epoch.
boric acid *Inorganic Chemistry.* H_3BO_3, colorless to white crystals, soluble in boiling water and alcohol; melts at about 169°C; a comparatively weak acid derived from borax, widely used as an eyewash or a mild antiseptic and for various industrial purposes.
boric acid ester *Organic Chemistry.* any compound that is readily hydrolyzed to yield boric acid and the respective alcohol; for example, trimethyl borate hydrolyzes to boric acid and methanol. They are colorless to yellow liquids that boil above 230°C; used as catalysts, solvents, dehydrating agents, and for various other purposes.
boric oxide *Inorganic Chemistry.* B_2O_3, a trioxide of boron, a colorless powder or crystals; soluble in alcohol and hot water and slightly soluble in cold water; melts at about 450°C. It is the principal ingredient in highly heat-resistant glass. Also, BORON OXIDE.
boride *Inorganic Chemistry.* an interstitial compound having a hardness of 8 to 10 on the Mohs scale and a melting point above 2000°C, formed by heating a mixture of boron and a metal; the boron atoms link together in zigzag chains. It is used in high-temperature applications such as rocket nozzles.
boring *Mechanical Engineering.* the process of making or enlarging a cylindrical hole using a lathe, boring machine, or boring mill. *Petroleum Engineering.* the process of creating deep holes for the exploration or exploitation of oil fields. *Geology.* a cylindrical sample of earth strata obtained from a borehole.
boring log see DRILL LOG.
borings see CUTTINGS.
boring sponge *Invertebrate Zoology.* any sponge of *Cliona* or related genera that penetrates shells, some Atlantic species are injurious to oysters.
BORIS *Artificial Intelligence.* a commercially available electronic chess machine for microcomputers.
borism *Toxicology.* poisoning due to ingestion of boron compounds; symptoms may include nausea, vomiting, diarrhea, and abdominal pain.
Born, Max 1882–1970, German physicist; Nobel Prize in 1954 for fundamental work in quantum mechanics.
Borna disease *Veterinary Medicine.* an infectious viral encephalitis occurring primarily in horses, but also affecting sheep and cattle; characterized by slight fever, depression, nervous symptoms, then convulsions and paralysis preceding death. (Named for *Borna,* a district in Saxony where an epidemic occurred.) Also, ENZOOTIC ENCEPHALITIS, EQUINE ENCEPHALITIS, CRAZY DISEASE.
bornane see CAMPHANE.
Born approximation *Quantum Mechanics.* an approximate perturbation method used in scattering calculations in which the interaction energy is assumed to be much smaller than the kinetic energy; the scattered wavefunction is assumed to be composed of plane waves.
Borna virus *Virology.* an unclassified RNA virus that causes disease in horses.
Borneo *Geography.* the world's third-largest island, located in the Malay Archipelago on the equator; area, 287,023 square miles.
Borneo camphor see BORNEOL.
borneol *Organic Chemistry.* $C_{10}H_{18}O$, a secondary alcohol occurring as white, translucent lumps with a sharp, camphorlike odor; melts at 208°C; insoluble in water and soluble in alcohol and ether. It is synthesized from camphor or obtained naturally from *Dryobalanops aromatica,* a tree growing in Borneo, Sumatra; used in perfumery. Also, BORNEO CAMPHOR, BORNYL ALCOHOL.
borneol acetate see BORNYL ACETATE.

Born equation *Physical Chemistry.* an equation that determines the free energy of solvation of an ion in terms of the Avogadro number, the ionic radius, the ionic valency, the ion's electronic charge, and the dielectric constant of the electrolytic.

Born-Haber cycle *Solid-State Physics.* a four-stage cycle of chemical and physical processes acting on a substance, commonly a crystalline metallic halide, to determine the lattice energy of sonic crystals.

Bornholm disease see EPIDEMIC PLEURODYNIA.

bornite *Mineralogy.* Cu_5FeS_4, an opaque, brittle, copper-colored cubic mineral that develops a brilliant iridescent purple tarnish, usually occurring in granular or compact masses and having a specific gravity of 5.07 and a hardness of 3 on the Mohs scale; it is a valuable ore mineral in many copper deposits.

Born-Madelung model *Solid-State Physics.* a model used in classical theory for cohesive energy, compressibility, and lattice spacing in ionic crystals.

Born-Mayer equation *Solid-State Physics.* an equation, based on the summation of coulomb interaction energies, repulsive potentials, and van der Waals potentials for nearest neighbor interaction, that specifies the binding energy in an ionic crystal.

Born-Oppenheimer approximation *Physical Chemistry.* a system for calculating electronic states, such as wavefunctions and energy levels, in which nuclei are assumed to have infinite masses and be in a fixed position in space, because their vibrations, compared to those of the electrons, are exceedingly slow. Also, **Born-Oppenheimer method.**

Born-von Kármán theory *Solid-State Physics.* a modification to the Debye theory of specific heat of solids in which the medium is considered to be a lattice of elastically coupled discrete particles rather than a continuous medium; the heat capacity depends on the frequency and frequency distribution of acoustical vibrations in the medium.

bornyl acetate *Organic Chemistry.* $C_{10}H_{17}OOCCH_3$, a colorless liquid that forms crystals at 10°C. It has a characteristic piney, camphorlike odor; used in perfumes and for flavorings. Also, BORNEOL ACETATE.

bornyl alcohol see BORNEOL.

bornyl chloride *Organic Chemistry.* $C_{10}H_{17}Cl$, a white crystalline compound derived from borneol or pinene; insoluble in water and soluble in alcohol, ether, and benzene; melts at 132°C and boils at 207–208°C.

bornyl formate *Organic Chemistry* $C_{10}H_{17}OOCH$, a colorless, combustible liquid with a pine odor; used in soaps and disinfectants.

bornyl isovalerate *Organic Chemistry.* $C_{10}H_{17}OOC_5H_9$, an aromatic, limpid fluid that is a constituent of valerian oil; boils at 255–260°C; insoluble in water and soluble in alcohol and ether. It is used in medicine and flavorings.

boroarsenate *Mineralogy.* any borate mineral that contains arsenic. Similarly, **borofluoride.**

borolanite *Petrology.* a hypabyssal rock having a granitoid texture and consisting mainly of orthoclase and melanite with subordinate nepheline, biotite, and pyroxene.

Boroll *Geology.* a suborder of the soil order Mollisol having fairly thick, nearly black A horizons, dark grayish-brown B horizons, and paler C horizons; characterized by a mean annual soil temperature of less than 8°C and by never being dry for 60 consecutive days during the 90-day period following the summer solstice.

boron *Chemistry.* a highly reactive nonmetallic chemical element, having the symbol B, the atomic number 5, an atomic weight of 10.811, a specific gravity of 2.34 (amorphous) or 2.46 (crystalline), a valence of 3, and a melting point of 2300°C. It occurs as a brown amorphous powder or as black monoclinic crystals with a hardness of 9.3 on the Mohs scale; it is found only in compounds such as borax. (A blend of the words borax and carbon.)

boron-10 *Nuclear Physics.* a stable isotope of boron that emits γ rays after neutron capture and, because of this property, is widely used in radiation detectors.

boron alloy *Metallurgy.* a uniformly dispersed mixture of boron with another metal or metals, such as iron or manganese.

boronatrocalcite see ULEXITTE.

boron bromide see BORON TRIBROMIDE.

boron carbide *Inorganic Chemistry.* B_4C, a very hard, black, water-insoluble, crystalline solid formed by heating boron oxide, B_2O_3, with coke in an electric furnace; melts at 2350°C and boils above 3500°C; used as an abrasive in cutting tools and for other purposes.

boron chloride see BORON TRICHLORIDE.

boron counter *Nucleonics.* a proportional counter containing boron to measure low-level neutron flux during early stages of reactor start-up. Also, **boron counting tube, boron chamber.**

boron fiber *Inorganic Chemistry.* a fiber made by a deposition of boron on a heated tungsten wire; resistant to chemicals, high temperatures, and electricity; used for yarns and woven products. Also, **boron filament.**

boron fluoride see BORON TRIFLUORIDE.

boron hydride see BORANE.

boron nitride *Inorganic Chemistry.* BN, a white powder that melts (sublimes) at about 3000°C. It is made up of extremely small particles, about one micrometer in diameter. Because of its high melting point and superior electrical resistance, it is often used in furnace insulation and aerospace shielding, and for various other industrial purposes.

boron oxide see BORIC OXIDE.

boron polymer *Organic Chemistry.* macromolecules formed by the polymerization of certain compounds containing boron-nitrogen, boron-phosphorus, or boron-arsenic bonds.

boron thermopile *Nucleonics.* a nonionizing sensor for measuring neutron flux during in-core experiments that contains a series of thermocouples with alternate junctions coated with boron-10; the junctions absorb neutrons and generate heat, which is converted to voltage by the thermocouples, indicating radiation intensity on an external meter.

boron tribromide *Inorganic Chemistry.* BBr_3, a colorless, fuming liquid that is decomposed by alcohol and water; boils at about 91°C; used in the manufacture of diborane.

boron trichloride *Inorganic Chemistry.* BCl_3, a colorless, fuming and corrosive liquid that is decomposed by alcohol and water; boils at 12.5°C; used in refining aluminum, magnesium, and copper, and in the manufacture of diborane.

boron triethoxide see ETHYL BORATE.

boron triethyl see TRIETHYLBORANE.

boron trifluoride *Inorganic Chemistry.* BF_3, a colorless gas that does not support combustion; soluble in cold water; liquid BF_3 boils at about –100°C; used in industry to catalyze chemical reactions.

boron trifluoride counter *Nucleonics.* a type of boron counter that is filled with boron trifluoride gas, BF_3; used to to determine slow-neutron densities.

boron trifluoride etherate *Organic Chemistry.* $(C_2H_5)_2OBF_3$, a toxic, fuming liquid immediately hydrolyzed by air and boiling at 125.7°C; used as a catalyst.

borosilicate glass *Materials.* a heat-resistant glass with at least 5% boron oxide, B_2O_3; used for chemical glassware, telescope mirrors, ovenware, and pump impellers because of its durability, low coefficient of thermal expansion, and high softening point (about 600°C).

Borrelia *Bacteriology.* a genus of Gram-negative, helical bacteria of the family Spirochaetaceae, occurring as parasites in humans and animals.

Borrelinavirus see NUCLEAR POLYHEDROSIS VIRUS.

borreliosis [bə rel´ē ō´sis] *Medicine.* an infectious disease transmitted by lice and ticks and characterized by fever alternating with apyrexia, chills, headache, neuromuscular pains, and occasionally vomiting. Also, RELAPSING FEVER.

Borrman effect *Physics.* an irregular transmission of X-rays when a single crystal of high perfection is placed in a reflecting position in a monochromatic X-ray beam.

Borrmann method *Materials Science.* a topographical method using transmitted X-rays to evaluate the presence and extent of imperfections in thick, nearly perfect single crystals.

borrow *Civil Engineering.* earth material used as fill in one location after being excavated in another.

borrowing *Mathematics.* a method of performing subtraction by modifying the digits of the minuend; 10 is added to a lower-order digit and 1 is subtracted from the next higher-order digit, allowing the corresponding digit of the subtrahend to be subtracted from a number that was previously too small. *Linguistics.* a word from one language that has been incorporated into another language through cultural contact; e.g., English has many borrowings from French as a result of the Norman Conquest of England.

borrow pit *Archaeology.* a pit created by the removal of dirt for fill by the prehistoric builder rather than by modern excavation. *Civil Engineering.* a pit created to provide fill material in another location.

BORSCHT *Telecommunications.* an acronym for the seven functions and components used in telecommunications: battery feed, overvoltage protection, ringing, supervision, coding, hybrid, and testing.

bort *Mineralogy.* **1.** an aggregate of granular to fine, imperfectly crystallized diamonds or of fragments produced in cutting diamonds. **2.** a flawed, off-color diamond of the lowest quality, suitable only for industrial use.

BOS *Aviation.* the airport code for Boston.

Bosanquet's law *Electromagnetism.* a statement analogous to Ohm's law: the magnetomotive force divided by the magnetic flux in a magnetic circuit will give the reluctance.

Bosch, Karl 1874–1940, German chemist; Nobel Prize in 1931; developed commercial processes for production of ammonia and hydrogenation of coal and oil.

Bosch fuel-injection pump *Mechanical Engineering.* a fuel pump in which the plunger and barrel are closely lapped to reduce leakage; used in internal-combustion engines.

Bose, J. Chandra 1858–1937, Indian physicist and plant physiologist; founded Bose Research Institute; studied stimulus response in plants.

Bose, Satyendranath 1894–1974, Indian physicist; developed Bose-Einstein theory of quantum statistics; bosons named in his honor.

Bose-Einstein condensation *Physics.* the condensation of an ideal Bose gas in momentum space when its temperature is lowered below a characteristic temperature.

Bose-Einstein distribution *Physics.* a function that specifies the number of particles in each of the allowed energy states for an assembly of independent bosons.

Bose-Einstein statistics *Particle Physics.* the statistical mechanics law obeyed by a system of particles whose wavefunction is unchanged when two identical particles are interchanged. It is assumed that all identical particles be regarded as absolutely indistinguishable, and that any number of identical particles can have the same set of quantum numbers in a given system. Particles obeying Bose-Einstein statistics have symmetric eigenfuctions and are known as bosons.

Bose gas *Physics.* any assembly of noninteracting or weakly interacting bosons.

bosh *Metallurgy.* **1.** in ironmaking, the part of a blast furnace from the tuyeres to the plane of maximum diameter. **2.** in copper smelting, the quartz build-up formed during operation. **3.** a tank for washing metallic components or for holding clean components.

boson *Particle Physics.* a particle that obeys Bose-Einstein statistics and has zero or integral spin. Unlike fermions, bosons are not conserved in number: They can be generated or destroyed singly, rather than in particle-antiparticle pairs.

Bosporus *Geography.* a strait connecting the Black Sea with the Sea of Marmara, near Istanbul.

bosque see TEMPERATE AND COLD SCRUB.

boss a raised, usually rounded area; specific uses include: *Biology.* a rounded protruberance or excrescence on the body or on an organ of an animal or plant. *Medicine.* a rounded eminence on the surface of a bone or tumor. *Design Engineering.* a protrusion on a machine part for mounting, support, or ornamentation, or for machining to a higher tolerance. *Naval Architecture.* **1.** the swollen portion of a ship's hull around the propeller shaft. **2.** the rounded hub of the propeller. *Geology.* a knoblike, circular, or elliptical mass of plutonic igneous rock on the surface of the earth. *Graphic Arts.* a piece of raised metal laid over the corner of a book for protection or decoration

bossage *Building Engineering.* stones that have been cut roughly and often laid into position for later finishing.

bostonite *Petrology.* a rock with coarse trachytic texture formed almost wholly of albite and microcline and with subordinate mafic minerals that are usually altered. (Found in *Boston,* Massachusetts.)

Boston ridge *Building Engineering.* the use of shingles in an alternating, overlapping configuration to the roof of a house from one side of the ridge to the other.

Bostrichidae *Invertebrate Zoology.* false powder-post beetles, a family of small cylindrical beetles having a hoodlike thorax; both larvae and adults bore into wood, living plants, stored products, and cables.

Bostrichoidea *Invertebrate Zoology.* a superfamily of beetles in the suborder Polyphaga.

BOT beginning-of-tape marker.

bot *Invertebrate Zoology.* the larva of a botfly. Also, BOTT.

bot. botany; botanist; botanical; bottle; bottom.

botallackite *Mineralogy.* $CU_2^{+2}Cl(OH)_3$, a powdery, blue-green to green monoclinic mineral having a specific gravity of 3.6 that is found with atacamite and paratacamite as a secondary mineral.

Botallo's duct see DUCTUS ARTERIOSUS.

botan. botanical.

botanical *Botany.* of or relating to plants or to the science of botany. *Pharmacology.* any drug that is derived from a plant, especially such a drug in its unrefined state.

botanical garden *Botany.* a diverse garden that is developed and maintained for exhibition, scientific, and educational purposes.

Botany

Botany is the subdiscipline of biology that is concerned with the study of plants. Although an intimate knowledge of plants and their uses has certainly been with man since the beginning, the study of plants from a scientific perspective had its origins in ancient Greece, especially with the scholar Theoprastes. Modern botany began in the seventeenth and eighteenth centuries when the development of the microscope made possible the study of the detailed structure of cells and tissue and the scientific method led to experiments on how plants function. The Italian Malphigi and the Englishman Grew described the anatomy of plants and the modern classification system now applied to all organisms was developed by the Swedish botanist, Linnaeus. Hales, Priestley and Ingenhouse conducted experiments on the ascent of sap and on the production of oxygen by plants.

Modern botany is concerned with the form and function of plants, their evolutionary relationships, and their interactions with other organisms and with their physical environment. It uses experimental methods to study the regulation of growth and development, photosynthesis, transport of water, nutrients and sugar, and other processes necessary for the survival of plants.

Plant systematics is a branch of botany that is concerned with the evolutionary relationships between plants that form the basis of the modern classification. Plant ecology is concerned with the relationship between plants and their physical and biological environment.

Botany involves ever more detailed studies of the biochemical and physical mechanisms underlying the growth of plants. Increasingly, research in botany uses techniques from molecular biology to understand how the genes controlling plant processes are regulated. Studies in botany are the basis of plant biotechnology that may lead to new ways to increase crop production, protect crops from pests or protect and enhance the environment.

Robert W. Pearcy
Professor and Chair of Botany
University of California, Davis

botany *Biology.* **1.** the branch of biology that is concerned with the study of plants and plant life. **2.** the properties exhibited by a specific plant, plant group, or plant community. (From a Greek word meaning "herb.")

Botany Bay *Geography.* a bay on the southeast coast of Australia, near Sydney.

botfly or **bot fly** *Invertebrate Zoology.* any of various insects belonging to the order Diptera, family Oestridae, whose larvae are parasitic in animals, especially horses and sheep.

Bothe, Walther [bō´tə] 1891–1957, German physicist; Nobel Prize for using coincidence method to prove the particle nature of cosmic rays.

bothridium *Invertebrate Zoology.* one of the four leaflike suckers asymmetrically placed on the head of a tapeworm of the order Tetraphyllidea. Also, PHYLLIDEA.

Bothriocidaris *Paleontology.* an unusual genus of echinoderms that is known only from Ordovician and Silurian deposits found in Estonia and Scotland; possibly an echinoid, but also has been classified in the past as a diploporite cystoid and as a holothuroid; characterized by a radial madreporite.

Bothriocidaroida *Paleontology.* an order of early Paleozoic echinoderms whose type genus is *Bothriocidaris.*

Bothriolepsis

Bothriolepsis *Paleontology.* a genus of Devonian ostracoderms of the family Asterolepidae.

bothrium *Invertebrate Zoology.* a groovelike sucker on certain tapeworms, such as that found on either side of the head of *Diphyllobothrium latum.*

Bothrops *Vertebrate Zoology.* a genus of South American serpents, including *B. atrox,* the fer-de-lance.

Botriocephaloidea *Invertebrate Zoology.* see PSEUDOPHYLLIDA.

Botrydiaceae *Botany.* a monogeneric family of terrestrial yellow-green algae of the order Vaucheriales, consisting of plants with macroscopic aerials, globular parts, and subterranean colorless branched rhizoids; usually found on drying mud in freshwater environments.

Botrydiopsidaceae *Botany.* a family of freshwater yellow-green algae of the order Mischococcales; characterized by the cells' ability to sustain growth without division, producing multinucleate cells with many chloroplasts.

Botryochloridaceae *Botany.* a family of freshwater yellow-green algae of the order Mischococcales; characterized by cells that remain attached after reproduction, forming colonies that may or may not be embedded in mucilage.

botryogen *Mineralogy.* $MgFe^{+3}(SO_4)_2(OH){\cdot}7H_2O$, a deep red to orange monoclinic mineral occurring in reniform or botryoidal masses, or as striated prismatic crystals, having a specific gravity of 2.14 and a hardness of 2 to 2.5 on the Mohs scale.

botryoid *Geology.* a formation of calcium carbonate that resembles a bunch of grapes and usually occurs in caves.

botryoidal *Science.* shaped like a cluster of grapes; used to describe minerals, plants, and animal parts. (From Greek *botrys,* a bunch of grapes.)

botryoidal tissue *Invertebrate Zoology.* loosely arranged mesenchyme or "packing" tissue in certain leeches; darkly pigmented tubular cells traversed by intracellular capillaries containing red bloodlike fluid.

botryomycosis *Veterinary Medicine.* a chronic bacterial infection in domestic animals, often occurring after castration, caused by *Staphylococcus aureus* and generally characterized by a weeping fibromatous tumor that forms at the site of the infected wound or injury.

Botrytis *Mycology.* a genus of fungi of the class Discomycetes that includes some species which are plant pathogens and may cause a gray mold on the hosts.

botrytis disease *Plant Pathology.* any plant disease caused by the fungus *Botrytis,* generally characterized by leaf blight, mold, and the breakdown and deterioration of plant parts; affects many plants including apples, potatoes, lettuce, onions, and roses. Also, **botrytis rot.**

bott *Metallurgy.* a plug of clay attached to a long rod and used to stop the flow of molten metal from a cupola or blast furnace. Thus, **bott plug, bott stick.** *Invertebrate Zoology.* see BOT.

Böttger, Johann Friedrich 1682–1719, German chemist; first European to produce porcelain.

bottle *Engineering.* **1.** a portable container used to hold liquids, typically made of glass or plastic and having a narrow neck that can be closed with a cap, cork, or stopper. **2.** a similar container used to store and transport gases.

bottle cast *Oceanography.* a set of water samples and readings of ocean temperature made by activating the mechanism to operate a series of bottles fixed to a line.

bottle cell *Cell Biology.* in an amphibian gastrula, a bottle-shaped cell that, due to its shape, may facilitate invagination and subsequent formation of the blastopore during development.

bottle centrifuge *Engineering.* a machine designed to rapidly spin a group of test tubes or bottles filled with various mixtures, in order to separate each mixture into substances of different densities.

bottled gas see LIQUEFIED PETROLEUM GAS.

bottle graft *Botany.* a graft in which the scion is protected from wilting by keeping its base in a bottle of water until it unites with the stock.

bottleneck *Transportation Engineering.* a point or section in a road that experiences traffic congestion, due to a lower traffic capacity than preceding and following sections of the road; it may be permanent, due to road design or traffic patterns, or it may be temporary, due to road work or an accident. *Petroleum Engineering.* an area of reduced diameter in a drill pipe, caused by inordinate longitudinal strain or weight or a combination of strain and abnormal swaying of the mechanism.

bottleneck assignment problem *Industrial Engineering.* in linear programming, the problem of maximizing the efficiency of the least efficient stage of a production process.

bottleneck period *Genetics.* a segment of time when a large population becomes reduced to only a few individuals, then expands again with an altered gene pool. Thus, **bottleneck effect.**

bottlenose **1.** see BOTTLE-NOSED DOLPHIN. **2.** see BOTTLE-NOSED WHALE.

bottle-nosed dolphin *Vertebrate Zoology.* any of various relatively large North Atlantic and Mediterranean dolphins of the genus *Tursiops,* especially *T. truncatus,* characterized by a prominent beak.

bottle-nosed whale *Vertebrate Zoology.* any of several whales of the family Hyperoodontidae, including the North Atlantic *Hyperoodon ampulatus* and the South Atlantic *H. planifrons,* characterized by a prominent beak and bulbous forehead.

bottle test *Chemistry.* an analytical procedure to determine the quantity of a chemical required to separate an emulsion into clear oil and water fractions, performed by adding that chemical to samples of a water-oil emulsion.

bottom *Geology.* **1.** the ground beneath a body of water. **2.** see BOTTOM LAND. *Mining Engineering.* **1.** the floor of an underground mining cavity. **2.** to strike bedrock or clay when sinking a mine shaft. *Petroleum Engineering.* see BOTTOM FRACTION. *Naval Architecture.* the section of a ship's hull between the bilges; broadly, the part of a ship's hull that lies below the water line. *Agriculture.* the part of a plow that breaks up the soil. Also, BASE.

bottom blow *Engineering.* a machine used in blow molding that forces air upward through a mass of molten plastic and into the mold.

bottom-blown converter *Metallurgy.* a pollution-causing metal refining process in which oxygen is blown into the bottom of the melt, providing heat and oxidizing impurities.

bottom break *Botany.* a branch or shoot growing from the base of a plant rather than from an axillary bud on a branch. *Mining Engineering.* a break or crack that separates a block of stone from a quarry floor.

bottom chord *Civil Engineering.* any member from the bottom series of members of a truss extending parallel to the roadway of a bridge.

bottom cut *Mining Engineering.* **1.** a machine cut made in the floor of a seam before shot firing. **2.** the lower of two converging lines of horizontally spaced blasting holes in drilling and blasting a tunnel.

bottom dead center *Mechanical Engineering.* the point at which a piston is at the end of its downstroke.

bottom-discharge bit see FACE-DISCHARGE BIT.

bottom drive *Petroleum Engineering.* any of several methods used to stimulate and drive oil from a difficult formation, including water, steam, or CO_2 drive, pressure maintenance by gas injection, or in situ combustion in the formation.

bottom-dump *Mechanical Engineering.* describing a vehicle that dumps its load through gates or doors in the underside of its body.

bottom-dump truck *Mechanical Engineering.* a trailer or semitrailer that can dump its contents through doors in the underside of its body. Also, DUMP WAGON.

bottomed *Mining Engineering.* describing a shaft or hole that has been driven to completion, or that has been abandoned.

bottom fauna see BENTHOS.

bottom-fermented *Food Technology.* of beer, brewed with yeast strains that sink to the bottom during fermentation; characteristic of lagers.

bottom flow *Hydrology.* a current flowing along the bottom of a body of water that is denser than any section of the surrounding water.

bottom fraction *Petroleum Engineering.* in petroleum distillation, the heaviest, least volatile fraction; the final cut. Also, BOTTOM, BOTTOMS.

bottom hole *Petroleum Engineering.* the section of an oil well with the greatest depth.

bottom-hole pressure *Petroleum Engineering.* **1.** the recorded gas-drive pressure, expressed in pounds per square inch, at the base of an oil well shaft; used to determine oil-reservoir performance and downhole equipment productivity. **2.** the well pressure determined at a point opposite the producing formation. *Geology.* see RESERVOIR PRESSURE.

bottom-hole pump *Petroleum Engineering.* a compact, high-volume electric or hydraulic pump, placed near the bottom of a well to lift well fluids.

bottom-hole separator *Petroleum Engineering.* a unit that separates natural gas and crude oil at the bottom of a well in order to increase the volumetric efficiency of the pumping system.

bottom ice see ANCHOR ICE.

bottoming *Geology.* the thinning out or ending of an ore body, either structurally or as a lessening of valuable minerals. *Civil Engineering.* the lowest layer of material in the construction of a road. *Electronics.* a limiting action in positive grid-voltage peaks in a beam-power tube or pentode, caused by the formation of a virtual cathode under certain operating conditions; the anode current is limited by the virtual cathode and flattens the corresponding anode voltage peak.

bottoming drill *Mechanical Devices.* a square-headed drill used to bore out the bottom of a drilled hole from a conical to a cylindrical form.

bottoming tap *Mechanical Devices.* the final tap applied to the finish of an internal thread of a blind hole without bevelling. Also, PLUG TAP. *Metallurgy.* a tap with a chamfer of one to one and one-half threads in length.

bottomland *Geology.* **1.** any low-lying, level, highly fertile land. **2.** in the Mississippi Valley region, a grassy lowland formed by deposition along the edge of a watercourse. **3.** an alluvial or a flood plain.

bottom mine *Ordnance.* a mine with negative buoyancy designed to rest on the seabed; used in shallow waters.

bottom moraine see GROUND MORAINE.

bottomonium *Particle Physics.* a combination of two bottom quarks to form an upsilon particle.

bottom pillar *Mining Engineering.* a large, unworked block of solid coal left around the shaft.

bottom quark *Particle Physics.* the fifth type of quark; having electric charge –1/3 times the elementary charge, bottom quantum number –1, and a mass three times that of a charmed quark. Also, B QUARK.

bottom rot *Plant Pathology.* **1.** a lettuce disease caused by the fungus *Pellicularia filamentosa* and characterized by decay on the lower leaves that gradually spreads upward. **2.** a disease of tree trunks caused by pore fungi.

bottoms *Engineering.* nickel sulfide in the Orford process of separating nickel and copper as sulfides. *Chemical Engineering.* the process stream leaving the bottom of a contacting column. *Petroleum Engineering.* see BASIC SEDIMENT AND WATER.

bottom sample *Petroleum Engineering.* a core removed from the bottom of a borehole for reservoir evaluation.

bottom sampler *Oceanography.* an instrument having a sounding lead with adhesive on its underside, used to obtain sample materials from the sea bed.

bottom sediment see BASIC SEDIMENT AND WATER.

bottomset bed *Geology.* any of a number of horizontal or gently sloping layers of fine-grained silts and clays that have been slowly deposited on the bottom of a sea or lake at the front of a growing delta.

bottom settlings see BASIC SEDIMENT AND WATER.

bottom terrace *Geology.* a landform that is produced by streams and characterized by a broad, gently sloping surface in the direction of flow and a steep cliff facing downstream.

bottom-up analysis *Computer Programming.* a type of syntactic analysis or parsing that starts with a string and attempts to develop a tree that converges on a root symbol.

bottom-up design *Computer Programming.* a design process that begins by identifying the components to be used, then constructs from them other components that are progressively closer to the system goal.

bottom water *Oceanography.* the deepest, densest water mass of a body of sea water. *Hydrology.* water that lies immediately below oil or gas within a stratum of rock.

botulin [bäch′ə lən] *Toxicology.* an extremely potent neurotoxin causing botulism disease, produced by the bacterium *Clostridium botulinum.* Also, **botulinus toxin, botulismotoxin.**

botulinum cook *Food Technology.* the heating of food for a specified time at 121.1°C to destroy the endospheres of *Clostridium botulinum.*

botulinus [bäch′ə lin′əs] *Microbiology.* a Gram-positive anaerobic bacterium, *Clostridium botulinum,* that is responsible for the symptoms of botulism.

botulism [bäch′ə liz əm] *Toxicology.* food poisoning by the neurotoxin botulin, characterized by vomiting, abdominal pain, difficulty of vision, central nervous symptoms, disturbances of secretion, dyspepsia, a barking cough, and ptosis. Also, **botulism disease.**

boturon *Organic Chemistry.* $C_{12}H_{13}N_2OCl$, a white solid that melts at 145–146°C; used as an herbicide.

bouclé [boo′klā′] *Textiles.* **1.** an uneven loop-textured yarn, usually consisting of a longer thread continuously looped by two shorter threads. **2.** a fabric of this yarn. **3.** a fabric with a rough, textured, looplike surface.

boudin [boo′dan′] *Geology.* any of a series of sausage-shaped segments occurring in a boudinage. (A French word for a type of sausage.)

boudinage [boo′də nazh′] *Geology.* a structure often found in deformed sedimentary and metamorphic rocks in which an original layer or bed has been stretched and broken into regular, elongated segments like sausages.

Boudouard reaction *Chemistry.* a term for the chemical reaction in which carbon dioxide is reduced by carbon.

bougainvillea [boo′gən vēl′ē ə] *Botany.* any of several plants of the genus *Bougainvillea,* small shrubs or vines having small flowers with bracts of various bright colors; native to South America, widely cultivated in warm regions as ornamentals.

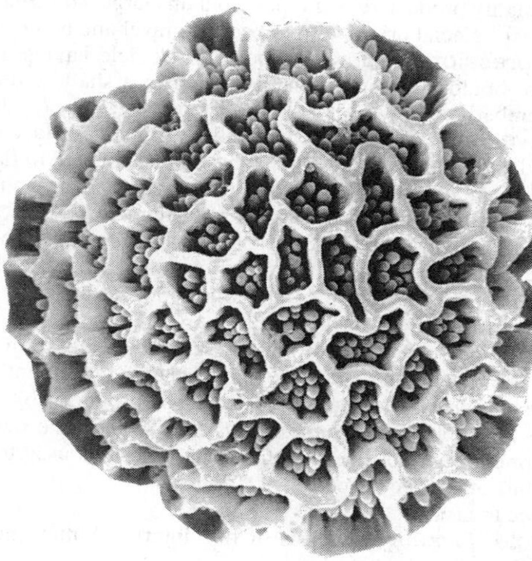

bougainvillea pollen grain

bough *Botany.* any large or main branch of a tree.

bougie *Surgery.* a cylindrical instrument used to calibrate or dilate constricted areas in tubular organs such as the urethra or the rectum.

bougie decimale *Optics.* a former unit of measure of luminous intensity, equivalent to 0.96 international standard candle.

bougienage *Surgery.* the process of passing a bougie through a tubular structure or organ in order to increase its caliber, as in the treatment of a constricted esophagus.

Bouguer, Pierre [boo′gâr′] 1698–1758, French mathematician; founder of photometry; made major contributions in hydrography and geodesy.

Bouguer anomaly *Geophysics.* a gravity anomaly calculated after corrections for latitude, elevation, and terrain have been made, without taking into account isostasy corrections. Also, **Bouguer gravity anomaly.**

Bouguer-Lambert-Beer law *Analytical Chemistry.* a law stating that the intensity of a beam of monochromatic radiation in an absorbing medium decreases exponentially with penetration.

Bouguer-Lambert law *Analytical Chemistry.* a law stating that the change in the intensity of light transmitted through an absorbing substance is related exponentially to the thickness of the absorbing medium and a constant dependent on sample and wavelength.

Bouguer plate *Geology.* an imaginary layer of infinite length and width that is at the height of the observation point above the geoid or datum surface.

Bouguer reduction *Geology.* a correction made in gravity work to account for the attraction effect of latitude, elevation, and type of terrain. Also, **Bouguer correction.**

Bouguer's halo *Meteorology.* an infrequently observed, faint, white circular arc or ring with a radius of about 39° centered on the antisolar point. Also, ULLOA'S RING.

boulangerite *Mineralogy.* $Pb_5Sb_4S_{11}$, an opaque, bluish-gray monoclinic mineral with a metallic luster, occurring as long prismatic crystals or in fibrous masses, having a specific gravity of 6.23 and a hardness of 2.5 to 3 on the Mohs scale; occurring in low- to medium-temperature hydrothermal vein deposits.

boulder *Geology.* a detached rock mass having a diameter greater than 256 mm that has been rounded or otherwise shaped by chemical weathering, mechanical weathering, or abrasion in the course of transport.

boulder barricade *Geology.* an accumulation of large boulders that is visible along a coast between low and half tide.

boulder barrier *Geology.* a shore ridge that has been created by pressure from floating ice moved by strong winds.

boulder bed *Geology.* **1.** a glacial deposit that contains many different sizes of particles. **2.** a conglomerate that contains boulders.

boulder belt *Geology.* a long, narrow accumulation or zone of glacial boulders lying across the direction of glacier movement.

boulder buster *Mechanical Devices.* a heavy, pointed steel device attached to the bottom of a drill rod; used to pound and break up boulders obstructing a borehole. Also, **boulder cracker.**

boulder clay *Geology.* **1.** a glacial deposit composed of striated boulders of different sizes that are embedded in hard, pulverized clay or rock flour. Also, DRIFT CLAY. **2.** see TILL. *Agronomy.* parent material for a number of highly productive soil types; contains large, compact deposits of gritty clay of glacial origin with imbedded gravel and boulders.

boulder depression *Geology.* a type of block field having a flat surface of pure boulder material, usually found in a shallow depression below the timberline.

boulder pavement *Geology.* **1.** a relatively smooth surface scattered with striated and polished boulders that have been abraded to flatness by glacier movement or by local and temporary erosion. **2.** an accumulation of boulders left after finer material has been removed by the action of wind, waves, or river currents. **3.** a somewhat inclined surface of randomly spaced, flattened blocks resulting from soil flow or other mass movement. **4.** a thin, smooth, or sheetlike concentration of polished boulders covering a desert surface.

boulder train *Geology.* a line or belt of glacial boulders from the same bedrock source extending in the direction of glacier movement.

boule *Crystallography.* a small rounded mass produced by fusion of alumina, as in sapphires and rubies. *Materials Science.* the raw product initially formed in any of the manufacturing processes used to produce single crystals of desired shapes.

boulimia see BULIMIA.

Bouma cycle *Petrology.* a series of five intervals, often incomplete, making up the sequence of a turbidite.

bounce *Electronics.* **1.** an abrupt abnormal variation in television picture brightness, independent of the actual brightness of the original subject. **2.** a similar variation in the vertical position of the picture.

bounce cast *Geology.* a mark or depression, originally made in a soft stream bed and preserved on the underside of a layer, consisting of a short ridge that fades out gradually in the downstream direction.

bounce table *Mechanical Engineering.* a device that tests the resilience of objects, especially metal components, by subjecting them to high impacts caused by dropping.

boundary *Military Science.* a line designating the limits of a combat zone or an area of fire. *Mathematics.* the set of all boundary points of a given subset of a topological space. Also, FRONTIER. *Chaotic Dynamics.* an area of infinite complexity between orderly and chaotic behavior, the simplest example of which is generated by plotting the real and imaginary roots of the polynomial equation $x^3 - 1 = 0$ that divide the complex plane into three identical regions which act as attractors with boundaries having the peculiar property that every point borders all three regions; magnified portions of the boundaries repeat the basic pattern on ever-smaller scales. *Electronics.* an area of meeting of P-type and N-type semiconductor materials where the donor and acceptor concentrations are equal. *Robotics.* a line that defines the edge of an image in machine vision.

boundary condition *Mathematics.* **1.** a constraint on the solutions of a system of differential equations, involving a specified set of values of the independent variable(s). **2.** a constraint on one or more values of a difference equation or recurrence relation, necessary to initiate the computation.

boundary current *Oceanography.* a deep ocean current that develops as a result of sudden changes in temperature and salinity.

boundary current system *Oceanography.* the series of currents that form the eastern and western segments of the oceanic gyres in the three tropical oceans, carrying water poleward in the stronger western boundary currents and toward the equator in the weaker eastern currents.

boundary fault *Mining Engineering.* a fault occurring where displacement has limited the size of a coalfield and truncated the coal-bearing strata. Also, MARGINAL FAULT.

boundary films *Mineralogy.* films of one constituent of an alloy that surround the crystals of another constituent.

boundary layer *Botany.* a layer of air along the surface of a leaf, whose width varies with wind speed, thus controlling the rate of transpiration. *Fluid Mechanics.* **1.** a layer of fluid in the immediate vicinity of a bounding surface. **2.** the region adjacent to the boundary dominated by shear stresses in a flow that is over a flat plate or through a pipe.

boundary-layer control *Aviation.* the design or control of slotted or perforated wings with suction to reduce undesirable aerodynamic effects caused by the boundary layer.

boundary-layer flow *Fluid Mechanics.* a flow in the region adjacent to the boundary where shear stresses dominate in the flow over a flat plate or through a tube.

boundary-layer separation *Fluid Mechanics.* the laminar boundary layer that eventually separates from a surface, leaving a turbulent boundary layer adjacent to the surface in the flow over a flat plate or through a pipe.

boundary-layer theory see FILM THEORY.

boundary line *Cartography.* a line of demarcation between two adjoining politically or legally distinct areas of land. *Geochemistry.* the line where any two phase areas meet in a binary system, or the line where any two liquidus surfaces intersect in a ternary system.

boundary lubrication *Engineering.* partial lubrication that may occur between solid surfaces because of adsorbed monofilm layers of lubricant on those surfaces.

boundary monument *Cartography.* an object placed on or near a boundary line to preserve and identify the ground location of the line.

boundary point *Mathematics.* in a topological space X, a boundary point of a subset A of X is a point x such that any open set in the topology of X that contains x also contains at least one point of A and at least one point of the complement of A. Equivalently, x is a boundary point of A if it belongs both to the closure of A and to the closure of the complement of A. The set of all boundary points of A is called the *boundary* or *frontier* of A.

boundary stratotype *Geology.* a type section that is used as a standard for definition and recognition of a time-stratigraphic boundary.

boundary value problem *Mathematics.* the problem of solving a given system of differential, integral, or difference equations together with boundary conditions; that is, the solution must not only satisfy the system of equations, the solution or its derivatives may also be required to take on specified values along portions of the boundary of the solution domain in time and space.

boundary vista *Cartography.* land along a boundary line cleared of all trees and shrubbery for the ready identification of that line.

boundary wave *Geophysics.* a seismic wave that moves along a free surface or the boundary of clearly defined layers.

bound barrel *Ordnance.* a rifle barrel that contacts the stock in such a way that expansion due to the heat of firing will cause the barrel to bend and impair accuracy.

bound charge *Electricity.* an electrostatic charge that is held by the attraction of an inducing charge of opposite polarity, as opposed to a free charge. Also, POLARIZATION CHARGE.

bounded depth-first search *Artificial Intelligence.* a depth-first search in which a bound is placed on the number of operator applications that can be considered; any sequence of operations that exceeds the depth bound is considered a failure.

bounded function *Mathematics.* a real function f for which there is a positive real number r, called a **bound** for f, such that $f(x) < r$ for all x in the domain of f. More generally, a function whose range is a metric space is bounded if an open ball exists that entirely contains the range of the function.

bounded growth *Mathematics.* a real-valued function f defined on the positive real numbers is said to have bounded growth if there exist real numbers M and a such that $|f(t)| \leq Ma^t$ for all $t > 0$.

bounded linear transformation *Mathematics.* a linear transformation between topological vector spaces is bounded if it takes bounded sets to bounded sets. Note that this differs from the usual notion of a bounded function, which has bounded range. No linear transformation except the zero transformation has bounded range.

bounded set *Mathematics.* **1.** a set of real numbers M is said to be **bounded above** if there exists a real number B such that $x \leq B$ for all x in M; B is called an **upper bound** of M. **2.** the set M is said to be **bounded below** if there exists a real number b such that $b \leq x$ for all x in M; b is called a **lower bound** of M. **3.** a subset of a metric space is bounded if it can be contained in some open ball of finite radius.

bounded variation *Mathematics.* a function has bounded variation over a given closed interval of its domain if its total variation on that interval is finite.

bound glue state see GLUEBALL.

bounding mine *Ordnance.* an antipersonnel mine that has a smaller charge which sends the case up into the air, at which point the main charge explodes at a height of three or four feet.

bound morpheme *Linguistics.* a morpheme that always occurs attached to another morpheme; e.g., the prefix *non-* or the plural form *-s.*

bound occurrence *Artificial Intelligence.* in a predicate calculus formula, an occurrence of a variable within the scope of a quantifier (\for all or \exists) of that variable.

bound particle *Physics.* a particle that does not possess enough energy to escape a system and is therefore confined to the system.

bounds register *Computer Technology.* in a multiprogramming environment, a special-purpose register that controls access to main memory by storing the memory locations of the upper and lower limits of the current user's memory block.

bound state *Quantum Mechanics.* a discretely determined stationary state in which the probability of exactly determining a quantum particle's wavefunction approaches zero at infinity in all directions and for all time.

boundstone *Geology.* a sedimentary rock having components that were bound together during deposition which have stayed essentially in the same position.

bound symbol *Telecommunications.* a contextual symbol that is neither preceded nor followed by space.

bound variable *Mathematics.* a variable that occurs within the scope of a quantifier or is explicitly cited by the quantifier.

bound vector *Mechanics.* a vector whose direction, line, and point of application are all prescribed.

bound water *Chemistry.* molecules of water held tightly by chemical groups in a larger molecule; proteins tend to hold water in this way. It may not freeze until as low as −40°C.

bourbon *Food Technology.* a type of whiskey distilled from a mash that is at least 51% corn combined with malt and rye. (From *Bourbon* County, Kentucky, where it was first produced.)

Bourdon (pressure) gauge [bur′dən] *Engineering.* an instrument used to measure the pressure of gases or liquids, consisting of a curved hollow metal tube closed at one end that tends to expand and straighten under pressure, moving a needle on an attached indicator. (From Eugene *Bourdon,* 1808–1884, French inventor.) Also, **Bourdon tube.**

Bourges process [bu̇rzh] *Graphic Arts.* in color lithography, a fake color process in which the four primaries are laid down as ink films on a transparent support, then adjusted by hand and shot through color filters.

Bourne shell *Computer Science.* a shell program for Unix systems; the standard shell for Unix System V.

Bourneville's disease see TUBEROUS SCLEROSIS.

bournonite *Mineralogy.* $PbCuSbS_3$, a relatively brittle, opaque, gray to black orthorhombic mineral with a metallic luster, occurring as short prismatic crystals, having a specific gravity of 5.8 and a hardness of 2.5 to 3 on the Mohs scale.

bourrelet *Ordnance.* the cylindrical portion of a projectile that contacts the bore while in the weapon.

Boussinesq, Joseph [boo′sə nesk′] 1842–1929, French physicist; contributions in hydrodynamics, including Boussinesq approximation.

Boussinesq approximation *Fluid Mechanics.* **1.** the approximate treatment of buoyancy in an incompressible fluid in which the density perturbation is a function of temperature only. **2.** a modification of the Stokes' approximation of solitary waves of permanent type progressing in a channel, wherein a fluid velocity is impressed on the wave.

Boussinesq coefficient *Hydrology.* a dimensionless constant that is computed based upon the momentum passing through a section in a channel. Also, MOMENTUM COEFFICIENT.

Boussinesq equation *Engineering.* an equation defining soil as an elastic solid and showing that the lines of equal vertical stress in soil under the point of a load or foundation are approximately circular.

Boussinesq number *Fluid Mechanics.* a dimensionless number applied to the examination of wave action in open passages.

Boussingault, Jean Baptiste 1802–1887, French chemist and mining engineer; important research on plant nutrition and respiration.

boussingaultite *Mineralogy.* $(NH_4)_2Mg(SO_4)_2 \cdot 6H_2O$, an easily fusible, water soluble, colorless to yellow-pink monoclinic mineral occurring as short prismatic crystals, having a specific gravity of 1.7 and a hardness of 2 on the Mohs scale; found as crusts and stalactites with other sulfate minerals.

bouton [bo′ tōn′] *Anatomy.* **1.** a pimple, knoblike swelling, or pustule. **2.** the terminal of an axon.

Bouvealt-Blanc method *Organic Chemistry.* the formation of alcohols by the reduction of esters with sodium dissolved in an alcohol.

Bouwers-Maksutov system see MAKSUTOV SYSTEM.

Boveri, Theodor 1862–1918, German embryologist; studied centrosomes; devised diagram of spermatogenesis and oogenesis.

Bovet, Daniel born 1907, Swiss-born Italian pharmacologist; Nobel Prize for development of antihistimines and muscle relaxants.

bovid *Vertebrate Zoology.* **1.** of or relating to animals of the family Bovidae. **2.** any animal of this family.

Bovidae *Vertebrate Zoology.* a large family of even-toed ungulates in the suborder Ruminantia, characterized by paired, unbranched hollow horns usually in both the male and female; includes cattle, sheep, goats, and antelopes.

bovine *Vertebrate Zoology.* **1.** of or relating to ruminant mammals of the subfamily Bovinae, including cattle, buffalo, and kudus. **2.** any animal belonging to this subfamily; often applied more specifically to animals belonging of the genus *Bos,* such as the ox, cow, or water buffalo. *Immunology.* referring to a substance or process that is related to or derived from a cow, ox, or other closely related species.

bovine atypical interstitial pneumonia see FOG FEVER.

bovine mastitis *Veterinary Medicine.* any of various inflammations of the mammary gland in cows.

bovine staggers *Veterinary Medicine.* a disease of cattle in South Africa caused by ingestion of the poisonous herb *Matricaria nigellaefolia* and manifested by incoordination, weight loss, and paralysis.

Bovoidea *Vertebrate Zoology.* the pronghorns and bovids, a superfamily of ruminants in the order Artiodactyla.

bow *Architecture.* a curved or polygonal section projecting from a flat exterior wall. *Naval Architecture.* the forwardmost part of a vessel. *Ordnance.* the front of a tank or other armored vehicle. Thus, **bow gun.** *Aviation.* the forward end of an airship or, rarely, of an airplane.

bow compass *Graphic Arts.* a small instrument for drawing circles, consisting of two legs that are joined by a bow-shaped spring. Similarly, **bow divider.**

Bowden cable *Mechanical Engineering.* a spring-steel wire in a flexible metal casing, used to pull loads over long distances.

Bowden-Tabor theory *Materials Science.* a description of the plastic deformation of asperities when two rough surfaces come in contact.

Bowditch, Henry Pickering 1840–1911, American physiologist; discovered all-or-none law and Bowditch's law.

Bowditch, Nathaniel 1773–1838, American mathematician and astronomer; wrote *The New American Practical Navigator.*

Bowditch curve *Mathematics.* a simple harmonic curve defined by $x = \sin(nT + d)$, $y = b \sin t$, bounded by the rectangle $|x| \leq a$, $|y| \leq b$.

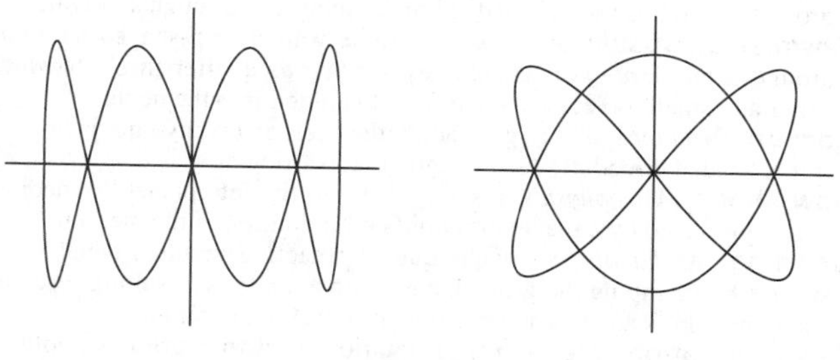

Bowditch curve

bowels *Anatomy.* an older term for the intestines. See INTESTINE.
Bowen, Norman Levi 1887–1956, Canadian-born American geologist; a pioneer in petrology; studied phase; proposed Bowen's reaction series.
bowenoid *Oncology.* a condition of Bowen's disease in which lesions are formed from structural breakdown of cells of the epidermis which have turned neoplastic.
Bowen's disease *Oncology.* an intraepidermal squamous-cell carcinoma, often occurring in multiple primary sites. Also, PRECANCEROUS DERMATITIS.
Bowen's reaction series *Mineralogy.* a series of minerals in which any early formed phase tends to react with the melt during differentiation and form a new mineral lower in the series. Also, REACTION SERIES.
bower anchor *Naval Architecture.* one of the two heavy anchors carried on either side of a vessel's bow and used in normal anchoring; it is permanently attached to a cable running through the hawsepipe.
Bower-Barff process *Metallurgy.* a process for applying an anticorrosion coating to a steel surface by heating the steel at 800°C, first in air, then in steam, and finally in producer gas.
bowerbird *Vertebrate Zoology.* any of various passerine birds of the family Ptilonorhynchidae, found in Australia and New Guinea; males build bowers, or arched chambers, that they often decorate with bright objects to attract females.

bowerbird

bowfin *Vertebrate Zoology.* a North American carnivorous freshwater fish, *Amia calva,* distinguished by its well-developed lungs.
Bowie formula *Geophysics.* a formula that calculates the local gravity anomaly on earth.
bowl classifier *Chemical Engineering.* a shallow, concave-bottomed bowl in which a liquid-solid suspension is admitted to the center. Larger particles drop to the bottom and are moved to a discharge point; small particles and liquid are collected as they flow over the edge of the bowl.
bowl crater *Astronomy.* any small crater whose interior walls lack terraces or other structure.
Bowles-Mackenzie theory *Materials Science.* a phenomonological description of martensitic transformations that relates operative strains to lattice deformation.
bowline *Naval Architecture.* **1.** a line running forward from the sheet of a square sail. **2.** a common type of nonslipping knot.
bowl mill *Mechanical Engineering.* a crushing device that consists of a rotating bowl and stationary rollers, with the grinding pressure being produced by means of springs; it grinds and pulverizes coal to the preferred size range and then passes the crushed fuel directly to a furnace. Also, **bowl-mill pulverizer.**
bowl scraper *Mechanical Engineering.* a steel bowl hung on a wheeled frame; as it is pulled forward by a tractor, its bottom edge digs into the ground; the soil is then ejected at the dumping site through a tailgate.
Bowman's capsule *Anatomy.* a double walled, cup-shaped structure around a glomerulus of the kidney serving as a filter through which water and small molecules move from the blood into the nephron.
bow-on *Ordnance.* of a target, facing the firer; an enemy tank is *bow-on* if it is headed directly toward the gun that is firing at it.
bow shock *Astrophysics.* a shock wave in the interplanetary medium that is produced by a planet or moon's motion through the medium.
bowsing *Archaeology.* a technique of detecting features beneath the surface by thumping the ground, e.g., with a mallet; a resulting resonant sound may indicate a buried chamber or pit. Also, **bosing.**
Bow's notation *Mechanics.* a notation used to represent coplanar forces graphically, where forces are represented by line segments.
bowsprit *Naval Architecture.* a fixed spar projecting from the bow of a vessel.
bowstring girder *Civil Engineering.* a structural truss member that has a curved upper portion and a flat horizontal bottom connecting the ends. Also, **bowstring beam, bowstring truss.** Similarly, **bowstring bridge.**
bow thruster *Naval Architecture.* a small propeller mounted near the bow with its axis transverse to the length of the ship; used for maneuvering.
bowtie antenna *Electromagnetism.* a dipole antenna with a bowtie shape having the property of a nonuniform distribution of inductance and capacitance, which gives rise to a wider bandwidth over that of a simple dipole antenna.
bow wave *Fluid Mechanics.* a shock wave produced at the fore or the front portion of a body. Also, **bow shock.**
box *Engineering.* **1.** a container or case, generally square or rectangular, made of cardboard, wood, or metal, and often having a cover. **2.** broadly, any protective housing, such as a fuse *box. Mechanical Devices.* **1.** the threaded nut for an auger drill screw. Also, BOXING. **2.** the part of a die that is attached to the ejector half and may also include the core mechanism. *Artificial Intelligence.* in logic or resolution theorem proving, the empty clause; from its mathematical symbol, \Box.
box *Botany.* **1.** an evergreen shrub or small bushy tree of the genus *Buxus,* especially *B. sempervirens,* characterized by small, leathery leaves and hard, durable wood; also applied to the wood itself. Also, BOXWOOD. **2.** of or relating to evergreen shrubs and trees of the family Buxacae in the order Euphorbiales.
Box and Cox transformation *Statistics.* a power transformation of the observed data, often used to achieve normality in linear models.
Box and Jenkins model *Statistics.* any of various models for the analysis of times series data; based on autoregression, moving averages, and accommodation of trend.
box annealing *Metallurgy.* annealing a metal or alloy in a container, so that oxidation is minimized. Also, CLOSE ANNEALING, POT ANNEALING.
box barrage *Ordnance.* an antiaircraft barrage in which shots are delivered in a boxlike pattern.
box caisson *Civil Engineering.* a steel or concrete box with an open top that may be floated to and sunk at a foundation site in a river or seaway. Also, AMERICAN CAISSON.
box camera *Optics.* the simplest type of camera, consisting of a box that contains a simple lens, shutter, and viewfinder, as well as some type of mechanism for winding a roll of film.
box canyon *Geology.* a canyon having vertical walls, usually closed upstream.
boxcar *Transportation Engineering.* a railroad car with a flat roof and vertical sides, used to transport freight. *Telecommunications.* a stream of pulses having a long duration in contrast to the spaces separating them.
boxcar circuit *Electronics.* a circuit used in radar to sample voltage waveforms and store the latest value sampled. (So called because of the flat, boxlike shape of the waveform.)
boxcar function *Mathematics.* a function (step function) whose value is zero except for a finite interval of the domain, where the function has a nonzero constant value.
boxcar spectrum *Spectroscopy.* a spectrum of wavelengths over an interval in which there are definite cutoff limits and all wavelengths between these limits have a constant nonzero intensity.
box chronometer *Horology.* see CHRONOMETER, def. 2.
box culvert *Civil Engineering.* a culvert of rectangular or square cross section used to carry a small flood flow under roads or embankments.
boxer's encephalopathy see TRAUMATIC ENCEPHALOPATHY.
box fold *Geology.* a fold having a broad, flat crest or trough and steep limbs.
box girder *Building Engineering.* a hollow girder with a square or rectangular cross section, generally made of light steel or cast iron. Also, **box beam.**
box-girder bridge *Civil Engineering.* a permanent bridge constructed from box section girders.
box heading *Building Engineering.* a heading that is close-timbered on both the roof and the building sides.
boxing *Design Engineering.* the material used to make boxes or casings. *Building Engineering.* the portion of a window frame that receives a folded shutter. *Metallurgy.* a fillet weld that is continued around a corner of the workpiece as a continuation of the main weld. *Mechanical Devices.* see BOX, def. 1. *Medicine.* in forming dental restorations and appliances, the creation of vertical walls, generally of wax, around an impression to preserve the desired shape and size of the base of the cast.

boxing-frame construction *Building Engineering.* a building method using a long, thin block of flats, offices, or the like with concrete slab floors mounted on load-bearing walls of brick or concrete across the width of the building.

boxing shutter *Building Engineering.* a window shutter that folds into a recess at the side of a window frame.

boxing the compass *Navigation.* naming off the points of an old-fashioned compass in clockwise order, starting from north.

boxlike bacteria *Bacteriology.* a nontaxonomic group of angular bacteria that contain pigments resembling bacteriorhodopsin, occurring in salt-rich environments.

box loom *Textiles.* a loom with more than one shuttle box, for weaving with several types or colors of filling yarns.

box nail *Mechanical Devices.* a flat-headed nail with a long shank and sharp point.

box nut *Mechanical Devices.* a nut with a closed end that protects the end of a bolt. Also, CAP NUT, DOME NUT.

box piles *Civil Engineering.* pile foundations constructed by combining two sections of sheetpiling, beams, channels, or plates.

box planting *Forestry.* the process of setting out young trees for continued growth in the various types of slatted wooden boxes in which they have been raised.

box plot *Statistics.* a rectangular graph representing the frequency distribution of a set of values, with the maximum and minimum values represented by the ends of the rectangle, and the median and quartile values represented by lines across the width of the rectangle parallel to the ends.

box saw *Mechanical Devices.* a thin-bladed saw held in place by rods and turnbuckles; used in making curved cuts.

box turtle *Vertebrate Zoology.* any of various chiefly terrestrial North American turtles of the genus *Terrapene,* characterized by a hinged shell that can be closed for protection. Also, **box tortoise.**

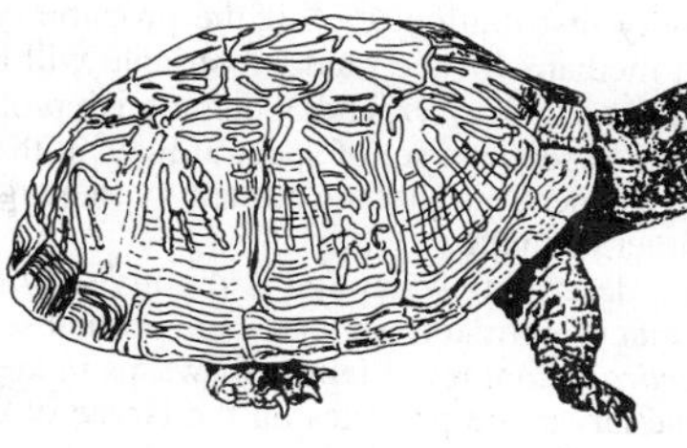

box turtle

boxwood *Botany.* see BOX.

boxwork *Geology.* a network of open, boxlike spaces consisting of intersecting blades or plates of minerals that had been deposited along fracture planes and left after the host rock dissolved.

box wrench *Mechanical Devices.* a wrench with a closed end that fits over and around the head of the nut or bolt.

box wrench

Boyden chamber *Immunology.* a piece of equipment made up of two compartments divided by a micropore filter, used to test for chemotaxis.

Boyden procedure *Immunology.* a process in which erythrocytes are treated with tannic acid so that their surfaces can readily adsorb soluble proteins.

Boyle, Robert 1627–1691, Irish-born British chemist and physicist; founder of modern chemistry; formulated Boyle's law.

Boyle's law *Physics.* a law stating that the product of the pressure of a gas and its volume is a constant while the temperature of the gas remains constant.

Boyle's temperature *Thermodynamics.* the temperature at which the virial coefficient B vanishes in the equation of state for a given gas:

$$Pv = RT\,[1 + (B/v) + (C/v^2) + \cdots].$$

Boy's camera *Optics.* a camera used to photograph lightning flashes by means of gyrating lenses that capture separate strokes.

BP or **B.P.** blood pressure; before the present; blueprint; Bachelor of Pharmacy.

bp base pair.

bp or **BP** *Physical Chemistry.* see BOILING POINT. Also, **b.p.**

BPA British Paediatric Association; British Parachute Association.

B particle see B MESON.

BPD or **bpd** barrels per day.

BPetE or **B.P.E.** Bachelor of Petroleum Engineering.

BPH benign prostatic hypertrophy.

BPH or **B.P.H.** Bachelor of Public Health.

B. Ph. British Pharmacopoeia.

BPharm or **B.Pharm.** Bachelor of Pharmacy.

BPI or **bpi** bits per inch; bytes per inch.

BPMC *o-sec*-butylphenyl-*N*-methyl carbamate.

BPO benzoyl peroxide; butyl peroxide.

B power supply see B SUPPLY.

bps bits per second.

BPT *Aviation.* the airport code for Beaumont-Port Arthur, Texas.

Bq becquerel.

b quark or **b-quark** see BOTTOM QUARK.

BR *Medicine.* an abbreviation for "bed rest."

Br the chemical symbol for bromine.

Br1 mycoplasmavirus group *Virology.* a proposed genus of phages that are isolated from Mycoplasma.

BRA ß-resorcylic acid.

bra and ket notation *Quantum Mechanics.* in the Dirac notation, the use of left-hand < and right-hand > members of the bracket < | > to denote the scalar product of two wavefunctions, where the scalar product of a bra vector <*B*| and a ket vector |*A*> is written as a complete bracket <*B* |*A*>.

Braarudosphaeraceae *Botany.* a family of flagellate marine algae of the order Coccosphaerales, having flattened coccoliths composed of five variously shaped elements radiating from a central point and angular coccospheres.

Brabender plastograph *Materials Science.* a device for mixing, as of polymers and additives, with the proviso for recording the torque required to turn a rotor in the melt, also used for estimating viscosity during mixing, typically of polymer melts.

brace *Building Engineering.* a stiffening member, often diagonal, designed to create or withstand tension and/or compression on a structure. Also, especially in describing braces collectively, **bracing.** *Medicine.* an orthopedic appliance designed to support, align, or hold parts of the body in correct position. *Naval Architecture.* a rope extending from a yard to the deck and used to swing the yard into position. *Mechanical Devices.* the U-shaped handle of a hand drill used to turn a drill bit. Thus, **brace and bit.**

braced frame *Civil Engineering.* a structural frame composed of solid girts mortised into solid posts with full-story studs between the posts. Thus, **braced framing.**

braced-rib arch *Civil Engineering.* a form of steel arch that has a system of diagonal bracing, typically used in bridge construction.

bracehead *Mining Engineering.* a cross handle attached to the top of a column of drill rods, used to turn the column and bit. Also, **brace key.**

brace root see PROP ROOT.

brachi- a combining form meaning "arm," as in *brachial.*

brachia *Invertebrate Zoology.* tentacles. *Anatomy.* the plural of BRACHIUM.

brachial *Zoology.* of or relating to the arm or an armlike process.

brachial artery *Anatomy.* an artery in the upper arm.

brachial cavity *Invertebrate Zoology.* a frontal space inside the valves of brachiopods where the brachia are retracted.

brachialgia *Neurology.* a sensation of pain in the arm.

brachial plexus *Anatomy.* an interwoven network of lower cervical and upper thoracic nerves serving the arm and hand.

Brachiata see POGONOPHORA.

brachiate *Botany.* having opposite, widely spreading branches resembling arms.

brachiation *Zoology.* a method of locomotion using the arms to swing from one hand hold to another; found in primates, e.g., the gibbon ape.

brachidium *Invertebrate Zoology.* in some brachiopods, a calcareous skeletal support of the lophophore.

brachio- a combining form meaning "arm," as in *brachiopod.*

brachiocephalic artery see INNOMINATE ARTERY.

brachiolaria *Invertebrate Zoology.* a larval stage of starfish found when a bipinnaria develops three small projections on its anterior end that it affixes to an object on the ocean floor during metamorphosis into the adult stage.

brachiopod [brak′ē ə päd′; brä′kē ə päd′] *Invertebrate Zoology.* any marine invertebrate of the phylum Brachiopoda. Also, LAMPSHELL.

Brachiopoda *Invertebrate Zoology.* a phylum of marine animals characterized by a bivalve shell and by a pair of cilia-covered arms used to collect food; includes some 250 living species and 30,000 fossil species.

brachioradialis *Anatomy.* a muscle of the lateral forearm that flexes the forearm.

brachiosaur [brak′ē ə sôr′; brä′kē ə sôr′] *Paleontology.* a Late Jurassic sauropod of the genus *Brachiosaurus,* characterized by a small head, massive body, and forelegs longer than its hind legs; the heaviest of the dinosaurs, it reached up to 70 feet in length and weighed up to 80 tons.

brachistochrone *Mechanics.* the curved path taken by a body, descending without friction under the force of gravity alone, from one point to another in minimal time.

brachistochrone problem *Mathematics.* the problem, posed by Johann Bernoulli in 1696 as a challenge to European mathematicians, of finding the equation of the path along which a particle falls from one point to another in the shortest time. The solution, a cycloid through the two points, involves minimizing the integral expressing this path and is the earliest problem in the calculus of variations.

brachium [brak′ē əm; brä′kē əm] *Anatomy.* the part of the upper limb from the shoulder to the elbow; the arm.

brachy- a combining form meaning "short," as in *brachycephalic.*

brachyanticline *Geology.* an anticline that is wider than it is long.

Brachyarchus *Bacteriology.* a genus of nonmotile, Gram-negative, rod-shaped bacteria, found in lakes.

brachyaxis *Crystallography.* the shorter of the horizontal axes of an orthorhombic crystal.

Brachybasidiales *Mycology.* an order of plant-parasitic fungi belonging to the class Hymenomycetes.

brachyblast *Botany.* a short branch that occurs on the same plant with normal branches.

brachycephalic [brak′ə sə fal′ik] *Anthropology.* having a relatively short, broad head, in whch the skull width is 81–85.4% of the length; e.g., American Indians, Malayans, or Burmese. Also, **brachycephalous.**

brachycerous *Entomology.* having short antennae, as do some Diptera.

brachydactylia *Medicine.* a condition in which the fingers and toes are abnormally short.

Brachydiniaceae *Botany.* a family of marine algae comprising the order Brachydiniales.

Brachydiniales *Botany.* a monofamilial order of marine, unicellular algae of the class Dinophyceae, characterized by a reduced cell body with four to five mobile, elongated arms.

Brachygastra *Entomology.* the honey-pot ant; some individuals in this genus serve as living food-stores for other ants by enormously inflating their stomachs with sugary honeydew secreted by aphids.

brachyodont *Zoology.* **1.** of or relating to mammals with a primitive type of grinding tooth characterized by a low crown. **2.** the tooth itself.

Brachypteraciidae *Vertebrate Zoology.* a family of colorful birds in the order Coraciiformes, native to Madagascar. Also, GROUND ROLLERS.

brachypterous *Entomology.* having short underdeveloped wings, usually preventing flight; a common phenomenon among many orders of insects that often serves to limit their natural range.

brachysclereid *Botany.* an isodiametric sclereid, typical in the pith, cortex, and bark of many stems and in certain fruits such as the pear. Also, STONE CELL.

brachysm *Botany.* a type of dwarfism caused by shortening of the internodes in certain plants.

brachysyncline *Geology.* a syncline that is wider than it is long.

Brachytheciaceae *Botany.* a family of shiny, loosely matted mosses of the order Hypnobryales that grow on soil, logs, humus, rocks, and tree trunks; characterized by branched, prostrate and creeping or erect stems and lateral sporophytes, and common to Northern Hemisphere boreal and temperate regions.

brachytherapy *Radiology.* treatment with ionizing radiation in which the radioactive source is close to the body area being treated; it may be implanted, in physical contact, or located a short distance away.

Brachythoraci *Paleontology.* the largest suborder of arthrodire placoderms, lighter and more advanced than the actinolepids and phlyctaenaspids; extant in the Lower to Upper Devonian.

Brachyura *Invertebrate Zoology.* a group of crustaceans in the order Decapoda, with a short abdomen hidden under the cephalothorax; approximately 4000 mostly marine species, including the true crabs.

brachyury *Genetics.* a mutant phenotype in the mouse, characterized by a short tail.

bracing *Engineering.* **1.** the process of adding strength and stability to a component or structure by using rods, ties, or other means of support. **2.** the support or supports so used.

brackebuschite *Mineralogy.* $Pb_2(Mn^{+2},Fe^{+2})(VO_4)_2 \cdot H_2O$, a dark brown to black, strongly pleochroic monoclinic mineral occurring as striated prismatic crystals, having a specific gravity of 6.05; found in oxidized zones of lead-zinc deposits in Cordoba, Argentina.

bracket *Building Engineering.* **1.** an L-shaped support for a shelf or shelflike load. **2.** a vertical board that supports the tread of a stair. *Architecture.* an often decorative structural member projecting from a wall and supporting an overhang. *Medicine.* in orthodontic work, a small metal attachment that fastens the arch wire to an orthodontic band or directly to a tooth. Also, ORTHODONTIC BRACKET, ORTHODONTIC ATTACHMENT. *Ordnance.* **1.** a distance between two shots or series of shots, one of which is over the target and one of which is short, or one of which is to the right and the other to the left. **2.** to deliver artillery fire that straddles a target in this manner.

bracket fungus *Mycology.* fungi belonging to the order Agaricales, growing on tree trunks in the shape of shelves or brackets. Also SHELF FUNGUS, CONK.

bracketing *Ordnance.* a method of adjusting artillery fire by establishing a bracket and then successively reducing its size until the target is hit. Also, **bracketing method.**

bracketing salvo *Ordnance.* a group of shots in which an equal number fall short of the target and go over the target.

bracket median *Statistics.* a value of the random variable that corresponds to the midpoint of one of several equal probability groups of the cumulative probability distribution; e.g., if the probability groups are each .2, the bracket medians for the random variable will be those values corresponding to the .1, .3, .5, .7, and .9 cumulative probabilities.

Brackett series *Spectroscopy.* a set of spectral lines in the far infrared region of the hydrogen spectrum that are related by making certain substitutions in the Balmer formula.

brackish *Hydrology.* describing water that is slightly salty; water whose salinity is between that of normal fresh water and normal sea water.

Braconidae *Invertebrate Zoology.* a family of wasps in the superfamily Ichneumonoidea; the larvae are parasites on the larvae of other insects, and one caterpillar may feed more than one hundred braconids; adults frequent flowers.

bract *Botany.* **1.** a small modified leaf with a relatively undeveloped blade, in the axil of which grows an inflorescence or flower. **2.** a leaf or leaflike structure associated with a sporangium.

bracteate *Botany.* having bracts. Also, **bracteose**.

bracteolate *Botany.* having bracteoles.

bracteole *Botany.* small secondary bracts at the base of certain flowers.

brad *Mechanical Devices.* a thin wire nail with a small head, either flush with the shank or with a projection to one side.

bradawl *Mechanical Devices.* a straight awl with a chisel edge used to make small holes for brads, nails, or screws.

bradding *Mechanical Engineering.* the loss of sharpness and subsequent deformation of a bit tooth when too much weight is placed on it.

Bradfordian *Geology.* **1.** in North America, a stage of the uppermost Devonian period, occurring after the Cassadagan and before the Mississippian. Also, CONEWANGOAN. **2.** in Britain, a substage of the Middle Jurassic period.

Bradley, James 1693–1762, English astronomer; discovered the aberration of light and nutation of earth's axis.

Bradley aberration *Astrophysics.* a small apparent displacement in the position of a star created by the effect of Earth's orbital motion on the starlight received from the object.

bradleyite *Mineralogy.* $Na_3Mg(PO_4)(CO_3)$, a light gray monoclinic mineral occurring as fine-grained masses and having a specific gravity of 2.7; found in oil shale.

bradshot or **bradsot** see BRAXY.

brady- a combining form meaning "slow," as in *bradycardia.*

bradyauxesis *Biology.* the slower growth or development of one part of a structure or organism relative to the whole structure or organism.

bradycardia *Medicine.* a slower than normal contraction of the heart muscle causing a pulse of less than 60 beats per minute.

bradyesthesia *Neurology.* an abnormal slowness or dullness of sensation perception.

bradykinesia *Medicine.* an abnormal slowness of movement; sluggishness of physical or mental responses.

bradykinesia [brā´dē kə nēz´yə] *Neurology.* an abnormal slowness of movement, often involving restricted range; a slowness of mental or physical responses.

bradykinin *Biochemistry.* $C_{50}H_{73}N_{15}O_{11}$, a kinin (a type of plasma hormone) that causes blood vessels to dilate, resulting in a drop in blood pressure, the contraction of muscles in the lungs, intestines, and uterus, and pain. Also, CALLIDEIC I, KALLIDIN I. *Endocrinology.* a peptide autacoid that is formed in tissue or plasma and is a potent vasodilator, hyperalgesic agent, and mediator of the inflammatory response.

bradylalia *Neurology.* abnormal slowness of speech due to a brain lesion. Also, BRADYPHASIA.

bradymenorrhea *Medicine.* a slow or prolonged menstrual flow.

bradyodont *Paleontology.* a general term referring to a variety of early cartilaginous fishes in the subclass Holocephali, including psammodonts and helodonts.

bradypepsia *Medicine.* an abnormal slowness of digestion.

bradyphasia see BRADYLALIA.

bradyphrenia *Neurology.* sluggish thinking that results from organic causes, such as epidemic encephalitis.

bradypnea *Medicine.* an abnormal slowness of breathing.

Bradypodidae *Vertebrate Zoology.* the sloths, a family of slow-moving arboreal mammals in the order Edentata, known for their habits of hanging upside-down from tree branches.

Bradyrhizobium *Bacteriology.* a genus of Gram-negative bacteria of the family Rhizobiaceae that cause the formation of plant root nodules in certain legumes; some strains carry out nitrogen fixation under microaerobic conditions.

bradytely *Evolution.* a significantly slow rate of evolutionary change for a given species within a group of related species.

bradytherapy *Medicine.* a form of radiotherapy in which the agent used is close to, on the surface of, or implanted in the body.

bradytocia *Medicine.* prolonged labor and slow delivery.

Bragg, Sir William Henry 1862–1942, and **Sir William Lawrence** 1890–1971, English physicists; developed the X-ray spectrometer and used it to analyze the structure of crystals; shared Nobel Prize in 1915.

Bragg angle *Physics.* an angle q satisfying the relation $\sin q = nl/2d$ for the scattering of X-rays in Bragg scattering, where n is a positive integer, l is the wavelength, and d is the characteristic atomic spacing.

Bragg cell see ACOUSTOOPTIC MODULATOR.

Bragg equation *Crystallography.* in the diffraction of X-rays by crystals, each diffracted beam is considered as a **Bragg reflection** from a lattice plane in the crystal. If the angle between the nth order of diffraction of X-rays, wavelength l, and the normal to a set of crystal lattice planes is $(90° - q)$, and the perpendicular spacing of the lattice planes is d, then $nl = 2d \sin(q)$. Thus when X-rays strike a crystal they will be diffracted, when, and only when, this equation is satisfied. With this equation, Bragg first identified the integers, h, k, and l of the Laue equation with the Miller indices of the lattice planes.

braggite *Mineralogy.* (Pt,Pd,Ni)S, a steel-gray tetragonal mineral with a dull external luster, occurring as minute grains, having a specific gravity of 9 to 10 and a hardness of 5 on the Mohs scale; found in norites of the Bushveld complex, Transvaal, South Africa.

Bragg-Pierce law *Physics.* a relationship for determining the atomic absorption coefficient of an element for X-rays when the atomic number of the element and the wavelength of the X-rays are known.

Bragg scattering *Solid-State Physics.* the scattering of X-rays or neutrons from the atomic layers of a periodic crystal lattice; constructive interference produces characteristic diffraction patterns of diffraction at definite angles. Also, **Bragg diffraction, Bragg reflection.**

Bragg's law *Solid-State Physics.* a law specifying the relation between the reflection angle of maximum intensity q, the wavelength of incident radiation l, and the interatomic spacing d: $\sin(q) = nl/2d$, where n is a positive integer.

Bragg spectrometer *Engineering.* a spectrometer used to analyze the structure of crystals and to measure the wavelengths of X-rays and gamma rays.

Bragg-Williams theory *Materials Science.* a description of ordering in metal alloys as a function of temperature, in which long-range interactions contribute mainly to the temperature-dependent degree of disorder.

Tycho Brahe

Brahe, Tycho [brä; brä´hē] 1546–1601, Danish astronomer; made precise observations of stars and planets; discovered first supernova.

Brahman *Agriculture.* a breed of beef cattle having a grayish body, a large hump over the neck and shoulders, drooping ears, and a prominent dewlap; it originated in India. Also, **Brahma.** (From a Sanskrit word meaning "prayer.")

Brahmaputra *Geography.* a river flowing from the Tibetan Himalayas through northeastern India and Bangladesh to the Bay of Bengal.

braid *Textiles.* **1.** to interweave three or more strands of yarn, rope, or similar material by laying one strand on top of another strand; to make an article, such as a rug, by such interweaving. **2.** a thin strip of fabric used as a decorative trimming, often on clothing. *Hydrology.* to branch and rejoin many times, forming a network of interlacing channels.

braided stream *Hydrology.* a stream that divides into a system of many interlacing, branching, and reuniting channels separated by islands. Also, ANASTOMOSING STREAM.

braided wire *Electricity.* a tube of fine wires used as a grounding strap or woven around a cable for shielding purposes.

brain *Vertebrate Zoology.* a mass of nerve tissue, consisting of an outer cortex of gray matter and an inner white matter, contained within the cranium and forming an extension of the spinal cord; it is the part of the nervous system that receives impulses and transmits messages, thus controlling mental and physical actions. Also, ENCEPHALON. *Invertebrate Zoology.* in many invertebrates, the pre-oral ganglia that roughly corresponds in position and function to the vertebrate brain.

braincase see CRANIUM.

brain coral *Invertebrate Zoology.* any of several corals of the genus *Meandrina* whose convolutions resemble the human brain.

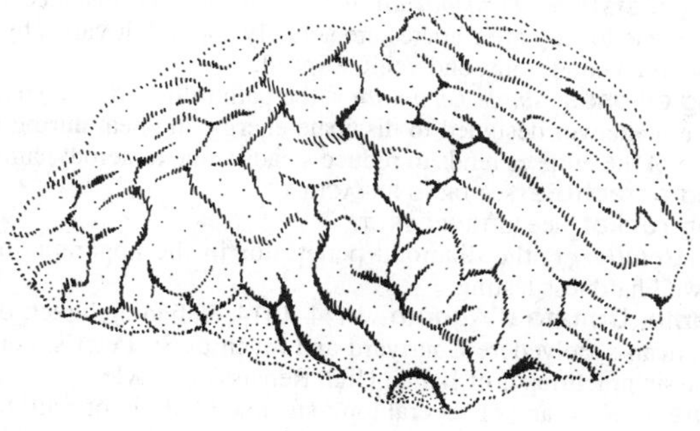

brain coral

brain death *Medicine.* irreversible brain damage characterized by the absence of both responsiveness to stimuli and spontaneous muscle activity (respiration, shivering, etc.), with an isoelectric electroencephalogram for thirty minutes, all in the presence of cardiac activity and in the absence of hypothermia or intoxication by central nervous system depressants. Also, IRREVERSIBLE COMA, CEREBRAL DEATH.

brain-heart infusion medium *Microbiology*. a microbial growth medium that contains infusions of both calf brain and beef heart and is used to culture a range of bacteria and fungi.

brain sand see CORPORA ARENACEA.

brainscan *Radiology*. a diagnostic imaging technique in which radioisotopes are injected intravenously, circulating to the brain where they gather in abnormal tissue; they are then traced and photographed by a scintiscanner.

brainstem *Anatomy*. the base of the brain caudal to the cerebrum and cerebellum, consisting of the midbrain, pons, and medulla oblongata.

brainstorming *Psychology*. a technique of group interaction in which all participants are encouraged to offer new ideas freely and spontaneously.

brainwashing *Psychology*. **1.** a technique of employing intensive stress and coercion to persuade an individual to reject his existing beliefs and attitudes, especially political views, and accept opposing doctrines. **2.** in popular use, any propagandistic or persuasive technique that influences people to accept new or different beliefs without evaluation.

brain waves *Physiology*. the changes in the electrical potential recorded by electrodes placed on the scalp of the human subject; usually designated as alpha, beta, and delta rhythms, depending on their wave frequencies.

brake *Mechanical Engineering*. **1.** a device designed to slow or stop the motion of a vehicle or machine by the use of friction in a controlled manner. **2.** a similar device that controls the tension on the reel of a rotary printing press. *Mechanical Devices*. a toothed device used to break up flax or hemp into fibers. Also, BREAKER. *Ecology*. an area that is overgrown with bushes, briers, cane, or similar vegetation.

brake band *Mechanical Engineering*. the flexible band of a band-type brake that is tightened around a brake drum to stop the motion of a wheel.

brake block *Mechanical Engineering*. the part of the brake where the brake shoe is held.

brake drum *Mechanical Engineering*. the revolving cylindrical component with which either the brake band or the brake shoe comes into frictional contact.

brake fluid *Materials*. an automotive fluid designed to transmit pressure from the brake pedal to the brake pistons.

brake horsepower *Mechanical Engineering*. the actual horsepower of an engine measured at the flywheel by the use of a dynamometer.

brake lining *Mechanical Engineering*. the asbestos-based facing material bonded to brake bands or brake shoes to reduce heat and increase friction.

brake mean-effective pressure *Mechanical Engineering*. the theoretical constant pressure that, if exerted on an engine during the power stroke, would produce the actual power produced during the power stroke.

brake shoe *Mechanical Engineering*. the curved metal structure containing the brake lining on the outside surface that comes into frictional contact with the inside of a brake drum.

brake thermal efficiency *Mechanical Engineering*. the ratio of a brake's energy output to its input.

braking distance *Transportation Engineering*. the distance required for a vehicle to stop after brakes are actually applied; it varies by type of vehicle, road conditions, and vehicle speed.

braking ellipses *Space Technology*. the path followed by a space vehicle in a maneuver designed to dissipate energy and heat during reentry; the size of the ellipses tends to reduce steadily due to aerodynamic drag.

braking parachute see DRAG PARACHUTE.

braking rocket see RETROROCKET.

brale *Metallurgy*. the diamond penetrator in the apparatus used for Rockwell hardness testing.

Bramante, Donato d'Agnolo 1444–1514, Italian architect; designed the Vatican's Belvedere Courtyard and much of St. Peter's; considered the classic practitioner of Italian High Renaissance style.

bramble *Botany*. any of several long-stemmed upright or trailing shrubs of the genus *Rubus*, family Rosaceae, characterized by a fruit that is an aggregate of many drupelets.

brammallite *Mineralogy* $(Na,H_3O)(Al,Mg,Fe)_2(Si,Al)_4O_{10}[(OH)_2,H_2O]$, a white monoclinic sodium-rich variety of the illite group of mica-clay minerals, having a specific gravity of 2.88, found as a fissure filling and surface coating on shale.

bran *Food Technology*. fibrous residue from the milling of grain husks with some endosperm.

branch *Botany*. **1.** an offshoot of a main stem or trunk. **2.** to spring off or out. *Hydrology*. **1.** any small stream that flows into or out of another, usually larger, stream. **2.** any small stream, tributary, or distributary. *Electronics*. a part of a network, consisting of one or more two-terminal elements in series. *Electricity*. see ARM. *Computer Programming*. **1.** to depart from normal sequential instruction processing. **2.** to transfer control to another part of the program, sometimes based on computer results. **3.** an instruction that can cause a transfer of control. Also, JUMP. *Mathematics*. in graph theory, an edge in a (spanning) tree.

branch-and-bound technique *Industrial Engineering*. in nonlinear programming, a procedure of "branching" the problem into elements, then "bounding" these for feasible alternative solutions.

branch circuit *Electricity*. the wiring system segment which extends beyond the final overcurrent device that guards the circuit.

branch-circuit distribution center *Telecommunications*. a communications distribution center at which branch circuits emanate.

branch cut *Mathematics*. a line or curve C joining branch points on a sheet of a Riemann surface of an analytic, multivalued function f. The same curve is drawn on each sheet of the function's Riemann surface, and the various sheets are connected to each other along C. A variable point z crossing C is considered to have passed from one sheet to another, and $f(z)$ has passed from one of its branches to another.

branch cutout *Electricity*. the fuse holder that guards a branch circuit in an interior wiring system.

branched acinous gland *Anatomy*. a gland consisting of several small secretory lobules, each drained by ducts that fuse into a single main duct.

branched polymer *Materials Science*. a polymer whose backbone structure repeatedly branches in many directions.

branched tubular gland *Anatomy*. a gland in which the terminal secretory portion is divided into two or more tubular branches.

brancher see GLUCAN BRANCHING ENZYME.

branch fault see AUXILIARY FAULT.

branch gain see BRANCH TRANSMITTANCE.

branchial *Zoology*. of or relating to gills.

branchial arch *Vertebrate Zoology*. any of the series of paired cartilaginous arches on either side of the pharynx and supporting a gill in a fish or amphibian. Also, GILL ARCH, VISCERAL ARCH, BRANCHIAL BASKET.

branchial basket see BRANCHIAL ARCH.

branchial cleft *Vertebrate Zoology*. **1.** any of a series of slits in the pharynx through which aquatic vertebrates draw water. **2.** any of the branchial ectodermal grooves of mammalian embryos.

branchial heart *Invertebrate Zoology*. in Cephalopoda, a muscular dilation of a vein that contracts to drive blood through a set of capillaries in the gills.

branchial plume *Invertebrate Zoology*. an accessory respiratory organ in some gastropods.

branchial pouch *Zoology*. a respiratory cavity found in primitive animals and in some sharks.

branchial sac *Invertebrate Zoology*. the pharynx of a tunicate, functioning as a gill.

branchial segment *Developmental Biology*. any or all of the paired portions of the pharynx of vertebrate embryos.

branchiate *Vertebrate Zoology*. relating to or having gills.

branchiating robot *Robotics*. **1.** a robot capable of moving over the surface of an object. **2.** a robot with legs or other equipment that allows a manipulator to move around within its working environment.

branching *Computer Programming*. a method of selecting the next instruction to be executed, either based on actual computer results, called **conditional branching,** or specified in the branch instruction itself, called **unconditional branching.** *Nuclear Physics*. the tendency for a radioactive element to decay by two or more competing modes of disintegration.

branching coefficient *Materials Science*. the probability that a terminating trifunctioning unit is connected to an end of a polymer chain that already possesses a similar unit at its other end.

branching diagram *Systematics*. a treelike diagram, such as a cladogram, phenogram, or phylogenetic tree, that depicts relationships among taxa. Also, DENDOGRAM.

branching enzyme see GLUCAN BRANCHING ENZYME.

branching factor see AVERAGE BRANCHING FACTOR.

branching fraction *Nuclear Physics*. the percentage of atoms that decay according to a particular mode in branching.

branching process *Statistics.* a stochastic model describing the growth of a population; based on the branching property, according to which each individual reproduces independently of others.

branching ratio *Nuclear Physics.* the relative intensities of competing disintegration modes determined by the number of atoms on average that follow one mode of decay versus the number of atoms that follow another.

branch instruction *Computer Programming.* an instruction in a program that indicates the conditions under which the transfer of control is to take place and the address of the next instruction to be executed. Also, **branching instruction.**

Branchiobdellidae *Invertebrate Zoology.* aberrant annelids in the family Oligochaeta, intermediate between leeches and earthworms; parasitic or commensal on freshwater crayfish.

branchiocranium *Vertebrate Zoology.* the part of a fish skull that includes the hyal and mandibular regions and the branchial arches.

branchioma *Oncology.* a tumor that originates from bronchial cysts or epithelium.

branchiomere see BRANCHIAL SEGMENT.

branchiomeric musculature *Vertebrate Zoology.* the muscles derived from gill segments in vertebrates.

Branchiomycetes *Mycology.* a taxonomically undetermined fungus that is considered to be closely related to the order Saprolegniales and is found only on fish gills, where it causes gill rot.

Branchiopoda *Invertebrate Zoology.* a subclass of crustaceans with flattened trunk limbs forming the basis of a complex filter feeding mechanism; mostly freshwater, but a few marine species; includes anostracans (brine shrimp), cladocerans (water fleas), conchostracans, and notostracans.

branchiostegal ray see GULAR, def. 2.

branchiostegite *Invertebrate Zoology.* in certain malacostracan crustacea, the expanded pleural part of the carapace forming the gill cover and chamber.

Branchiostoma *Zoology.* a genus of lancets commonly found along U.S. coasts, having many gill slits that are segmentally arranged myotomes and a tubular, dorsal nerve cord.

Branchiotremata *Invertebrate Zoology.* a branch of the subphylum Oligomera; free-living primitive worms, including acorn worms, with gill slits.

Branchiura *Invertebrate Zoology.* a subclass of fish lice, in the class Crustacea, found in both freshwater and marine environments; all species are parasitic on fish.

branch joint *Electricity.* a joint used for connecting a branch conductor or cable that continues beyond the branch.

branch line *Transportation Engineering.* a subsidiary route that connects with a main line. *Engineering.* a horizontal drainage pipe stemming off from another pipe.

branch migration *Molecular Biology.* the ability of a partially paired DNA strand to displace one resident strand of a duplex and so increase the length of its base pairing to the other strand of that duplex.

branch of a function *Mathematics.* a restriction of a multivalued analytic function to a particular region (which may include the point at infinity) of the complex plane, with a choice of value at each point so that the resulting single-valued function is continuous.

branch point *Electricity.* a point at which two or more conductors, components, or circuits join in an electrical connection. Also, JUNCTION POINT. *Transportation Engineering.* the place where a route branches into two or more lines. *Computer Programming.* a step in a computer program that contains a branch instruction. *Mathematics.* a point z_0 on a Riemann surface of a multivalued analytic function f is a branch point of order m of f if: (i) $w = f(z)$ has the inverse function $z = f^{-1}(w)$ in some neighborhood of $w_0 = f(z_0)$, and (ii) $[f^{-1}(w)]'$ has a zero of order m or a pole of order $m + 2$ at $f(z_0)$. It is at z_0 that $m + 1$ sheets of the Riemann surface are joined. Also, RAMIFICATION POINT.

branch sewer *Civil Engineering.* a sewer that carries discharge to a larger sewer main from a relativly smaller collection area.

branch transmittance *Control Systems.* the increase of current or voltage in a branch of an electrical network. Also, BRANCH GAIN.

brand *Agriculture.* **1.** an identifying mark that is burned onto the hide of an animal, using a hot iron. **2.** to burn such a mark on an animal.

branding iron *Agriculture.* an iron rod with a metal stamp at one end that is used to brand livestock.

Brandt *Ordnance.* one of several mortars manufactured by the French company Hotchkiss-Brandt, including various 120-mm mortars and a 81-mm light mortar in a short- and long-barreled version.

brandtite *Mineralogy.* $Ca_2(Mn^{+2},Mg)(AsO_4)_2 \cdot 2H_2O$, a fusible, colorless to white monoclinic mineral occurring as prismatic crystals, having a specific gravity of 3.67 and a hardness of 3.5 on the Mohs scale.

brandy *Food Technology.* an alcoholic beverage distilled from wine or fermented juice after the wine has aged in casks.

Brangus *Agriculture.* a breed of beef cattle cross-bred from the Brahman and Angus breeds.

Branhamella *Bacteriology.* a subgenus of Gram-negative, coccoid bacteria of the family Neisseriaceae, distinguished on the basis of cell shape; occurring as parasites in the mucous membranes of mammals. Also, MEISSERIA CATARRHALIS.

brannerite *Mineralogy.* $(U,Ca,Y,Ce)(Ti,Fe)_2O_6$, a highly radioactive, opaque, black monoclinic mineral occurring as prismatic crystals or in granular masses, having a specific gravity of 4.2 to 5.4 and a hardness of 4.5 to 5.5 on the Mohs scale; found as a primary mineral in pegmatites.

Brans-Dicke theory *Physics.* a gravitation theory that includes a scalar field in addition to the tensor field of Einstein's general relativity; the scalar field is determined by the distribution of mass-energy in the universe.

brash ice *Hydrology.* an accumulation of small floating fragments of ice, resulting from the breakup of larger masses. Also, DEBRIS ICE.

brashness *Forestry.* sawed wood that is low in shock resistance and shows failure under bending stress.

brass *Metallurgy.* any of several copper-zinc alloys, often containing other elements. *Geology.* yellowish iron pyrite that is found in coal.

brass chill see METAL-FUME FEVER.

Brassica *Botany.* a genus of the Brassicaceae that includes cabbages, turnips, brussels sprouts, rutabagas, black mustard, and other odoriferous vegetables.

Brassicaceae *Botany.* the mustard family of dicot herbs, characterized by a cruciform arrangement of sepals and petals, and by fruit that is a podlike capsule; many species are cultivated for food.

brassin *Biochemistry.* a plant hormone that plays a role in the elongation and division of plant leaf and stem cells.

Brathinidae *Invertebrate Zoology.* the rove beetles, in the superfamily Staphylinoidea; predacious, cosmopolitan coleopterans dwelling in moist leaf litter.

Brattain, Walter Houser 1902–1987, American physicist; shared Nobel Prize for invention of the transistor.

brattice *Mining Engineering.* **1.** a board or other partition used to divert air into a particular working place or section of a mine. **2.** an airtight partition in a mine shaft, designed to separate intake from return air. **3.** planking used to support a wall or roof. Also, **brattish.**

brattice cloth *Mining Engineering.* any fire-resistant canvas or duck from which a brattice is made.

Braulidae *Invertebrate Zoology.* a family of bee lice, including three European and one rare North American species in the order Diptera; the adults are external parasites of bees, and the larvae feed on pollen in honeycomb wax.

Braun, Karl Ferdinand 1850–1918, German physicist; with Marconi, Nobel Prize for development of wireless telegraphy.

braunite *Mineralogy.* $Mn^{+2}Mn_6^{+3}SiO_{12}$, a brittle, gray or brownish-black tetragonal mineral with a submetallic luster, occurring as striated pyramidal crystals, having a specific gravity of 4.72 to 4.83 and a hardness of 6 to 6.5 on the Mohs scale; found as a secondary mineral formed under weathering conditions in veins.

Braun lipoprotein *Biochemistry.* a lipoprotein present in enterobacteria such as *Escherichia coli.*

Braun MSK tissue disintegrator *Biotechnology.* a cooled container filled with glass or plastic beads that oscillates rapidly; used to rupture cells in a tissue culture.

Bravais, Auguste 1811–1863, French chemist and mathematician; made important advances in crystallography and group theory.

Bravais lattice *Crystallography.* any one of the fourteen possible arrays of points repeated periodically in three-dimensional space such that the arrangement of points about any one of the points is identical in every respect to that about any other point in the array.

bra vector *Quantum Mechanics.* a state vector in Hilbert space analogous to a row vector in algebra; the dual of a ket vector.

brave west winds *Meteorology.* a term for the strong and persistent westerly winds over the oceans in temperate latitudes.

bravoite *Mineralogy.* another name for nickeloan pyrite, $(Fe,Ni)S_2$.

Braxton-Hicks contraction *Medicine.* a light, usually painless uterine contraction that occurs intermittently during pregnancy.

braxy *Veterinary Medicine.* a highly fatal, acute disease of younger sheep caused by *Clostridium septicum* ingested by the animal in a "braxy pasture" or in frozen grass or feed; characterized by inappetence, diarrhea, and inflammation of the abomasum. Also, BRADSHOT, BRADSOT.

brayer *Graphic Arts.* a small roller used for hand inking of printing blocks or plates.

Brayton cycle *Thermodynamics.* a thermodynamic cycle that consists of four processes: a reversible adiabatic (no heat transfer) compression at constant entropy; a heat transfer at constant pressure up to the maximum temperature; an adiabatic expansion at constant entropy back to the original pressure; and a heat transfer at constant pressure back to the original volume and entropy. This is the ideal cycle for the actual performance of a simple gas turbine. Also, JOULE CYCLE.

brazed shank tool *Mechanical Engineering.* a cutting device in which the metal blade is soldered to a different, nonferrous alloy.

braze joint *Metallurgy.* a joint produced by brazing.

braze welding *Metallurgy.* the joining of metallic components with a filler that melts above 450°C but below the melting point of the base.

brazier's chill see METAL-FUME FEVER.

Brazil Current *Oceanography.* a warm saline current that flows southward along the coast of Brazil; it is a continuation of the South Equatorial Current.

brazilianite *Mineralogy.* $NaAl_3(PO_4)_2(OH)_4$, a yellowish-green monoclinic mineral occurring as short prismatic to equant crystals of gem quality, having a specific gravity of 2.98 and a hardness of 5.5 on the Mohs scale; found in Brazil as a hydrothermal mineral in pegmatites.

Brazilian rosewood *Materials.* the hard, coarse wood of the jacaranda tree of northern South America, used for high-quality furniture and cabinetry and for novelty items; it is dark brown to purple in color with a characteristic fragrance.

Brazilian subregion *Ecology.* a distinct zoogeographical region that includes the country of Brazil and other areas within the Amazon River basin.

Brazil nut *Botany.* the large, three-angled seed of the broadleaf evergreen tree *Bertholletia excelsa;* develops in a large pod that contains 18 to 24 closely packed seeds.

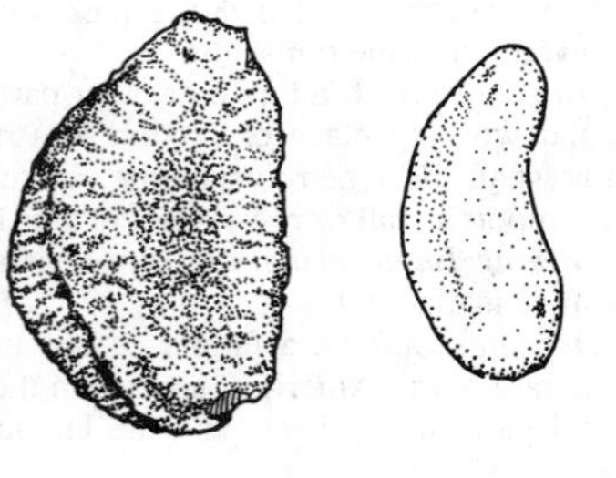

Brazil nut

brazing *Materials Science.* the joining of materials by a layer of an alloy that has a lower melting temperature than the materials to be joined and that wets upon melting; its melting point is higher than that of solders.

brazing alloy *Metallurgy.* an alloy suitable as a filler for brazing or braze welding. Similarly, **brazing metal, brazing sheet.**

breach *Geology.* any deep opening in a landform, especially one cut by erosion. *Military Science.* **1.** a gap or opening made by force in a fortification or troop position. **2.** to make such a gap or opening.

breached anticline *Geology.* an anticline whose crest has been greatly eroded, thus producing a valley more or less along the axis of the fold.

breached cone *Volcanology.* a volcanic cone or crater whose rim has been broken through and carried away by the outpouring of lava.

bread *Food Technology.* a food made of flour or meal that is mixed with water or milk, usually kneaded into dough with a leavening agent, and then baked.

breadboard *Electronics.* **1.** a developmental or prototype implementation of an electronic circuit; often temporary arrangements of electronic components laid out on a board in a rough manner in order to experiment with the circuit design. **2.** any basic framework, such as a perforated board, on which electronic components can be quickly mounted and wired. **3.** to construct an electronic circuit for experimental or demonstration purposes.

breadboarding *Electronics.* the construction of an electronic circuit on a breadboard to test or demonstrate its feasibility.

breadboard model *Electronics.* a working assemblage of the parts of an instrument or electrical system, wherein components are unhoused and rest flat on a board for the purposes of inspection, experimentation, and demonstration.

breadcrust *Geology.* a surface of salt concretions that resembles a crust of bread.

breadcrust bomb *Volcanology.* a volcanic bomb whose surface is characterized by a network of opened cracks.

bread flour *Food Technology.* a wheat flour with high gluten content, giving it high tensile strength, great expandability, and light density.

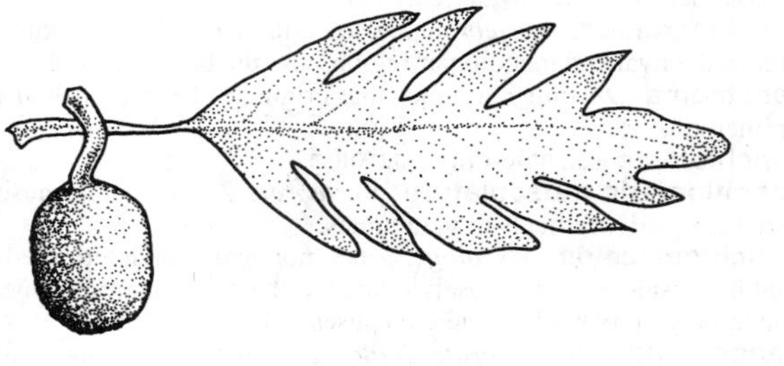

breadfruit

breadfruit *Botany.* **1.** a tall, large-leafed tree, *Artocarpus altilis,* of the Moraceae (mulberry) family; native to the Pacific islands and grown commercially in the Caribbean. **2.** the large, round edible fruit of this tree, composed of a perianth, pericarp, and receptacle joined together.

bread mold *Mycology.* any of several black fungi of the order Mucorales that grow on spoiled bread, especially *Rhizopus nigricans.* Also, BLACK MOLD.

breadnut *Botany.* the round, brownish fruit of the ramon, a West Indian tree of the mulberry family, which is roasted or boiled and widely eaten as a bread substitute.

breadroot *Botany.* the edible starchy root of a North American plant, *Psoralea esculenta,* of the pea family.

bread spinning see ROTOR SPINNING.

breadth-first search *Artificial Intelligence.* a knowledge system behavior that considers all information and possible options at a given level before proceeding to a deeper level.

break an interruption in the continuity of a structure or action; specific uses include: *Geology.* **1.** a marked variation of topography, such as an abrupt change in a slope or profile. **2.** a marked change or interruption in the lithology of a normal geologic sequence, such as a soft layer of rock between harder layers; this usually signifies an unconformity or hiatus. *Meteorology.* **1.** a sudden, marked weather change after a period of persistent dry, hot, wet, or cloudy weather. **2.** a hole or gap in a layer of clouds. *Electronics.* **1.** an open circuit. **2.** a circuit fault. **3.** in a circuit-opening device, the shortest distance between the contacts when they are in the open position. *Telecommunications.* **1.** a temporary transmission interruption. **2.** a signal sent by a receiver to a transmitter indicating that the receiver desires to transmit. **3.** in radio or television, a given time scheduled for commercial announcements. *Graphic Arts.* the end of a line, column, page, or other unit of printed copy. *Computer Programming.* **1.** to interrupt computer processing, usually by manual means. **2.** a separating point in a file at which a significant change in a key value, such as the initial alphabetic character, occurs. *Artificial Intelligence.* in Lisp, the halting of execution due to occurrence of an error.

breakage and reunion *Cell Biology.* the interchange of DNA between homologous chromosomes during meiosis, involving a physical break in each of two chromatids at corresponding points, a crossing over of the DNA strands, and a subsequent rejoining of the broken chromatids in a reciprocal manner.

breakaway *Design Engineering.* designed to shatter under stress as a safety feature, such as a *breakaway* traffic sign.

breakaway action *Robotics.* a safety feature that allows a wrist socket to cease movement when it senses that too much force is being applied to the end effector; a socket so equipped is called a **breakaway wrist.**

breakaway height *Aviation.* under given conditions, the altitude below which an aircraft landing approach is unsafe, requiring the aircraft to climb and make another landing attempt.

breakaway phenomenon see BREAKOFF PHENOMENON.

break-before-make *Electricity.* a switch or relay in which a closed contact must open before an open one closes, thus breaking one circuit before making another.

breakbone fever or **breakbone** see DENGUE.

break contact see BACK CONTACT.

breakdown *Chemistry.* a process of chemical decomposition or analysis. *Electricity.* **1.** a disruptive discharge through insulation, insulators, or other materials separating circuits, involving a sudden large increase in current due to complete failure under electrostatic stress. **2.** an undesired runaway increase in an electrode current in a gas tube. *Metallurgy.* **1.** in rolling or drawing, the preliminary operation that breaks down the structure of the starting stock. **2.** a preliminary operation in press-forging. *Materials Science.* see DIELECTRIC BREAKDOWN. *Psychology.* see NERVOUS BREAKDOWN.

breakdown diode *Electronics.* a Zener diode, avalanche diode, or any other semiconductor diode having a reverse-voltage breakdown mechanism based on either the Zener or the avalanche effect.

breakdown impedance *Electronics.* a small-signal impedance at a specified direct current in the breakdown region of a semiconductor diode.

breakdown law see FAILURE DISTRIBUTION.

breakdown maintenance *Industrial Engineering.* unscheduled or emergency maintenance that involves diagnosing a problem and repairing a machine in response to a malfunction.

breakdown region *Electronics.* that part of a semiconductor-diode characteristic curve lying beyond the initiation of breakdown.

breakdown torque *Electrical Engineering.* the maximum torque that an electric motor will develop with rated voltage applied at rated frequency without an abrupt drop in speed.

breakdown voltage *Electronics.* **1.** the voltage produced at a particular current in the electrical breakdown region of a semiconductor diode. Also, ZENER VOLTAGE. **2.** the voltage at which breakdown occurs in a dielectric or in a gas tube.

breaker *Oceanography.* **1.** a wave that has become too steep to remain stable and is therefore breaking into foam near the shore. **2.** a wave that is breaking against a rock or other solid object. *Electricity.* see CIRCUIT BREAKER. *Mechanical Devices.* see BRAKE. *Mining Engineering.* **1.** a building where coal is broken, sized, and cleaned for market. Also, COALBREAKER. **2.** a machine designed to carry out this process. *Textiles.* a machine designed to separate foreign matter from a fiber to be carded. *Agriculture.* a tomato or other vegetable that is picked and shipped to market before it is ripe.

breaker cam *Mechanical Engineering.* in an internal combustion engine, an engine-driven rotating device that opens the breaker points of the ignition system.

breaker depth *Oceanography.* the stillwater depth of the ocean at the point where a wave breaks. Also, BREAKING DEPTH.

breakerless ignition see ELECTRONIC IGNITION.

breaker plate *Engineering.* a perforated plate that is attached to the extruder head of a plastics die form, used to hold a screen that filters out foreign particles.

breaker points *Electricity.* a pair of low-voltage contacts designed to make or break the flow of current in the primary circuit of a gasoline engine's ignition system; replaced by electronic ignition in most modern engines. Also, POINTS.

breaker zone *Oceanography.* the shore area between the outermost breakers and the extent of wave uprush. Also, SURF ZONE.

break-even or **breakeven** *Industrial Engineering.* **1.** having income equal expenses. **2.** see BREAK-EVEN POINT. *Nucleonics.* **1.** the point or stage at which a fission or fusion reaction becomes self-sustaining. **2.** in a nuclear reactor, a condition in which as much fissile material is produced as is consumed, resulting in a conversion ratio equal to 1.

break-even point *Industrial Engineering.* the production level at which the value of the output just matches the cost of producing it; the minimum economical level of production.

break for color see COLOR BREAK.

break frequency *Control Systems.* the frequency at which there is an abrupt change in slope in the graph of the logarithm of the amplitude of the frequency response of a system versus the logarithm of the frequency. Also, CORNER FREQUENCY, KNEE FREQUENCY.

breakfront *Design Engineering.* **1.** having a front set on more than one plane, such as a central section set forward of those on either side or an upper section set back from the base. **2.** a cabinet, bookshelf, or other object having such a front.

breaking *Plant Pathology.* any discoloration of a flower, such as streaking or blotching, that is caused by a viral infection.

breaking capacity see BREAKING STRENGTH.

breaking circulation *Petroleum Engineering.* the resumption of drilling fluid, flowing down the drillpipe through the bit and up through the annulus to the surface.

breaking current *Electricity.* the maximum current that a circuit breaker, switch, or other device can interrupt without being damaged.

breaking depth see BREAKER DEPTH.

breaking-drop theory *Geophysics.* a theory, based on the separation of electric charge due to the breakup of water drops, that explains the bipolar charge distribution characteristic of thunderclouds.

breaking joint *Building Engineering.* any method of laying bricks in staggered courses such that the vertical joints are not continuous.

breaking load *Mechanics.* the minimum load or stress that, when applied steadily, will break or fracture a certain material or member.

breaking piece *Engineering.* an easily replaceable machine part that is designed to break when subject to a certain level of sudden force, thus protecting the entire machine from such stress; comparable to the fuses used in electrical circuits. Also, **breaking device, breaking-pin device.**

breaking strength *Mechanics.* the ability of a material to withstand tensile stress without breaking. Also, BREAKING CAPACITY.

breaking stress *Mechanics.* any stress, such as compression or tension, that will cause a material to rupture. Also, FAILURE STRESS, FRACTURE STRESS.

break-in operation *Telecommunications.* a procedure in which a receiving station may interrupt a transmission to request the transmitting station to perform a specific function, such as wait, change frequency, or repeat transmission.

break key *Computer Technology.* a key that interrupts the incoming message and holds the transmission line as long as it is depressed; used in some half-duplex data communications systems in which data can flow in only one direction at a time.

breakoff phenomenon or **breakoff feeling** *Aviation.* a sudden feeling, experienced by some high-altitude flyers and spaceflight crews, of being totally cut off from earth and humanity. Also, BREAKAWAY PHENOMENON.

breakout *Electricity.* a junction at which one or more conductors extend from a multiconductor cable to complete circuits along the main cable. *Metallurgy.* an unwanted escape of molten metal due to failure of its containing system. *Petroleum Engineering.* the process of withdrawing and disassembling equipment used in drilling a well, such as drill rods, bits, or collars.

breakout block *Petroleum Engineering.* a block of steel that fits the square of a drill bit and holds the bit while it is unscrewed from a drill collar.

breakoutput *Computer Programming.* an operation in the ALGOL language that causes a buffer to empty to an output device before the buffer is filled.

breakout tongs *Petroleum Engineering.* a heavy wrench, usually mechanically actuated, used to couple or uncouple drill rods, drill pipe, casing, or drive pipe. Also, MAKEUP TONGS.

breakover *Electronics.* in a circuit component, the transition from a forward-blocking to a forward-conducting state.

breakover voltage *Electronics.* the value of positive anode voltage at which a circuit component changes into the conductive state.

break period *Telecommunications.* **1.** the duration of a given break. **2.** in telephone usage, the span of time that the circuit contacts remain open.

breakpoint *Chemical Engineering.* **1.** an abrupt change in the slope of a mathematical relation, caused by some physical event; for example, the pressure drop-gas velocity relation for gas-liquid flow in a packed column will show a distinct change in slope at the loading point, and another abrupt change at the flooding point. **2.** see BREAKTHROUGH. *Computer Programming.* **1.** a place in a program at which the programmer has arranged for normal execution to be interrupted so that the values of the variables, the registers, and the counters can be analyzed or saved for later resumption of the program; used mainly in debugging new or revised programs. **2.** a step in the processing that requires an external action by the operator, such as mounting a tape, that causes a pause in execution and displays a message for the operator. *Robotics.* a point in the program of a robot at which changes can be made. *Industrial Engineering.* an identifiable point in a work cycle at which one element is completed and the next one begins. Also, ENDPOINT, READING POINT.

breakpoint switch *Computer Technology.* a bistable switch used during debugging that may be set externally and tested by branching statements within the program.

breakpoint symbol *Computer Programming.* an optional symbol added to an instruction identifying it as a breakpoint; useful in ensuring complete removal of breakpoints after debugging is complete.

break roll *Food Technology.* a grooved steel roller, usually one of a pair, used in wheat milling to cut open the kernels.

breaks in overcast *Meteorology.* a condition in which the cloud cover exceeds 0.9 but is less than 1.0.

break spinning see OPEN-ENDED SPINNING, ROTOR SPINNING.

break sprag *Mining Engineering.* a wooden or iron bar that is located between the spokes of a tram's wheels and is used to control its speed on an incline.

breakthrough *Chemical Engineering.* **1.** a break in a section of filter cake that allows unfiltered fluid to pass through. Also, BREAKPOINT. **2.** in certain ion-exchange systems, the initial appearance of unadsorbed ions, indicating that the resin bed is exhausted. *Metallurgy.* in the ion-exchange recovery of metal from solution, the point at which the first traces of metal appear in the last of a series of resin-filled stripping columns. *Mining Engineering.* **1.** a ventilation passage that is cut through a pillar to allow air to travel from one room to another. Also, CROSSCUT, ROOM CROSSCUT. **2.** an opening that is made, either accidentally or deliberately, between two underground workings. **3.** the point at which a drill bit leaves the rock and enters either a natural or a manmade opening. *Computer Technology.* in optical character recognition, a break in the intended character stroke.

break thrust *Geology.* a deep-angle reverse fault that cuts across a limb of a fold which formed during deformation of the fold.

breakup *Hydrology.* **1.** the melting, loosening, or destruction of winter snow or ice that occurs during springtime. **2.** the period when such a thaw occurs.

breakwater *Civil Engineering.* a barrier constructed out into the sea to break the force of the waves and provide a safe harbor behind it.

bream *Vertebrate Zoology.* **1.** any of various deep-bodied European freshwater fishes, especially of the genus *Abramis*. **2.** any of several freshwater sunfishes of the genus *Lepomis*. **3.** any of the several porgies, especially the **sea bream,** *Archosargus rhomboidalis*.

breast *Anatomy.* **1.** collections of milk-secreting glands in the anterior ventral part of a female body. **2.** a corresponding undeveloped gland in the male. *Building Engineering.* **1.** a section of wall between a window and the floor. **2.** a projection from a wall, such as the interior part of a chimney. *Metallurgy.* the front of an open-hearth furnace. *Mining Engineering.* any place currently being mined underground, such as a working face or heading; used especially in coal mining.

breastbeam *Naval Architecture.* the deck beam at the break of a partial deck such as a forecastle or poop. *Textiles.* on a loom, a horizontal bar over which the woven fabric passes to the cloth roll. *Building Engineering.* see BREASTSUMMER.

breast board *Civil Engineering.* one of a set of movable wooden sheets or boards that are used to support the face of an excavation.

breast drill *Mechanical Devices.* a hand drill turned by bevel gears, with added driving force supplied by the **breast plate,** a concave face at the top of the handle that can be pressed against the operator's chest.

Breasted, James Henry 1865–1935, American archeologist; excavated Megiddo (Armageddon); established ancient Egyptian historical periods; founded University of Chicago Oriental Institute.

breast height *Forestry.* a standard height that is measured on standing trees from ground level, usually taken at 4 feet 3 inches or 4 feet 6 inches; used for recording diameter, girth, or basal area.

breast line *Navigation.* a line used to secure a ship that is moored alongside a pier.

breast pump *Medicine.* a device used to withdraw milk from the breast.

breast stope *Mining Engineering.* an underground working face (breast) from which detached ore will not gravitate in flat lodes without assistance.

breastsummer *Building Engineering.* a horizontal beam that supports an exterior wall over a large opening such as a shop window. Also, BREASTBEAM.

breast wall *Mining Engineering.* a wall built to support a bank of earth, especially a working face in an underground mine.

breastwork *Ordnance.* a defensive wall erected at about chest height, designed to allow defenders to fire over the top from a standing position.

breather *Mechanical Engineering.* a small vent in the top of a covering or container, to allow circulation of air, escape of excess liquid or fumes, equalization of pressure, and the like. Similarly, **breather pipe.**

breather pipe *Mechanical Devices.* a length of pipe that directs air circulation in parts of an engine such as the oil tank or crankcase.

breathing *Physiology.* the process of alternately inhaling and exhaling air; respiration. *Genetics.* the formation of regions in a double-strained DNA molecule where the double helix unwinds into localized regions of single-stranded DNA. *Mechanical Engineering.* **1.** the process of ventilating a container, tank, or other closed system. **2.** in the molding of plastics, the opening and closing of a mold to permit the escape of accumulated gases. Also, DEGASSING. *Hydrology.* see RIVER BREATHING.

breathing apparatus *Mechanical Devices.* any device designed to allow an individual to breathe in conditions where respiration would not otherwise be possible, as in an area permeated with gases or smoke.

breathing cave see BLOWING CAVE.

breathing line *Building Engineering.* a level, usually five feet above the floor, used especially in the construction of public rooms as a reference for determining air-conditioning requirements, maximum occupancy, and the like.

breccia [bresh´ē ə; bresh´yə] *Petrology.* a coarse clastic rock composed of angular fragments of older rocks that are cemented together in a fine matrix; formed in a number of ways from a variety of materials. (An Italian word meaning "broken stones" or "rubble.")

breccia dike *Geology.* an intrusive tabular mass of breccia that cuts across a pre-existing rock bed or formation.

breccial *Petrology.* of or relating to breccia.

breccia marble *Petrology.* any marble that contains angular rock fragments.

breccia pipe see PIPE, def. 3.

brecciated *Petrology.* formed into breccia; having the form of breccia; conglomerated.

brecciation *Petrology.* **1.** a brecciated formation. **2.** the process of forming breccia; conglomeration.

breccia vein *Geology.* a fissure that contains many fragments of wall rock with mineral deposits in the interstices.

Breda's disease see YAWS.

Breda virus *Virology.* an ssRNA-containing virus that causes diarrhea in calves.

breech *Anatomy.* the lower posterior of the trunk; buttocks. *Ordnance.* the rear part of the bore of a gun, especially the opening and related mechanism for inserting the projectile.

breech birth *Medicine.* the emergence of the infant's lower extremities or buttocks first during the birth process.

breechblock or **breech-block** *Ordnance.* a movable metal piece for closing the breech opening of a gun and holding it closed against the force of the charge.

breechbolt *Ordnance.* see BOLT.

breech bore sight *Ordnance.* a sight located at the breech of a gun, used in conjunction with a muzzle bore sight to align the axis of the gun bore with the axis of the sight.

breech delivery see BREECH BIRTH.

breeches buoy *Naval Architecture.* an apparatus consisting of a life preserver or belt fitted with a canvas sling resembling a pair of short trousers (breeches) and pulled along lines strung between ships or between a ship and shore; used for rescues and for transferring personnel.

breeching *Mechanical Engineering.* a pipe or tubing system that conducts combustion by-products from a furnace or boiler up through a chimney.

breech interlock *Ordnance.* a device used in a breech-loading weapon to prevent a round from being loaded when the breechblock is not fully open.

breech-loading *Ordnance.* of or relating to weapons in which the ammunition is inserted at the rear of the bore: a *breech-loading* rifle.

breech presentation *Medicine.* an intrauterine position of the fetus in which the buttocks or feet lie closest to the cervix at the time of labor.

breech pressure *Ordnance.* the pressure of exploding gases against the interior of the breechblock when a weapon is fired.

breed *Biology.* **1.** a strain or variety of animals or plants that have a common origin and distinguishing characteristics. **2.** to reproduce sexually, especially under controlled conditions in order to select certain characteristics to be transmitted to plant or animal offspring. *Genetics.* a group of individuals, derived from a common ancestor and having similar genetic characteristics, that is artificially maintained. *Nucleonics.* to produce more fissile fuel than is consumed in a reaction.

breeder reactor *Nucleonics.* a nuclear reactor specifically designed to produce fissionable fuel from fertile material at a faster rate than it consumes fuel for energy production.

breeding *Biology.* the use of controlled reproduction to improve certain characteristics in plants and animals. *Nucleonics.* the production of fissile material from fertile material, usually in excess of the amount consumed; specifically, the production of plutonium-249 and uranium-233 from, respectively, uranium-238 and thorium-232.

breeding gain *Nucleonics.* in a breeder reactor, the net increase in the number of fissile atoms per fuel atom consumed.

breeding ground *Zoology.* a place where animals breed, especially one to which a certain group returns each breeding season.

breeding ratio *Nucleonics.* the ratio of the total number of fissile nuclei produced from fertile material to the number of fissile nuclei consumed in fission and nonfission reactions.

breeding stock *Agriculture.* farm animals or plant specimens of high quality that are kept to produce offspring.

breeze *Meteorology.* **1.** a light wind. **2.** a wind speed ranging from 4 to 31 mph on the Beaufort wind scale.

breezeway *Architecture.* a roofed, open-sided passageway connecting two buildings, such as a house and a garage.

Brefeldiellaceae *Mycology.* a family of fungi belonging to the order Asterinales and occurring on the surfaces of leaves, primarily in the tropics of the Southern Hemisphere.

bregma *Anatomy.* the point of union of the coronal and sagittal sutures of the skull.

Bréguet spring *Horology.* a type of balance spring whose outer coil is raised above the plane of the spiral and curved back toward the center.

brei *Biology.* tissue that is ground to a pulp and sometimes suspended in an isotonic medium for use in research or experimentation.

Breinl strain *Bacteriology.* the virulent strain of the microorganism *Rickettsia prowazekii* that is the causative agent of the acute, systemic disease typhus fever.

Breit, Gregory 1899–1981, Russian-born American physicist; worked in quantum theory and electrodynamics; in the first practical application of radar, determined the height of the ionosphere.

breithauptite *Mineralogy.* NiSb, a bright copper-red hexagonal mineral having a specific gravity of 7.59 to 8.23 and a hardness of 5.5 on the Mohs scale, usually occurring in massive form and found in calcite veins with native silver, niccolite, and cobaltite.

Breit-Wigner formula *Nuclear Physics.* a mathematical expression that defines the cross section for resonant absorption as a function of collision energy; derived from the Breit-Wigner theory.

Breit-Wigner theory *Nuclear Physics.* a theory that explains neutron cross sections in terms of neutron absorption to form a compound nucleus at or near resonance peaks with a high probability of neutron capture. Also, DISPERSION THEORY.

Bremia *Mycology.* a genus of fungi in the order Peronosporales that causes downy mildew on lettuce.

bremsstrahlung *Electromagnetism.* **1.** a stream of electromagnetic radiation produced by the rapid deceleration of a high-speed charged particle (usually an electron or beta particle) in the electrical field of another high-speed charged particle (usually a nucleus). **2.** the deceleration that produces such radiation. (German for "braking radiation.")

Bren gun or **bren gun** *Ordnance.* a .303-caliber, gas-operated, air-cooled light machine gun used by the British Army during World War II; although obsolete in Britain, it is still used in other countries.

brennschluss *Space Technology.* **1.** flameout in a rocket engine or rocket missile. **2.** the moment when this occurs. (A German word meaning "combustion termination.").

Brentidae *Invertebrate Zoology.* a family of straight-snouted beetles belonging to the order Coleoptera, widespread in warm regions, with wood-boring larvae; the adults may steal burrows and eject other wood-borers to lay eggs.

brepho- a combining form denoting a relationship to an embryo, fetus, or infant.

brephotrophic *Nutrition.* of or relating to the nourishment of infants.

Bretonian orogeny *Geology.* a late Devonian deformation in Cape Breton, Nova Scotia.

breton lace [brə tän′] *Textiles.* a type of lace having heavily embroidered designs, often in colored thread, on a net background. (From the *Breton* people of France, who originated this lace.) Also, **bretonne lace.**

Bretschneideraceae *Botany.* a dicotyledonous and monospecific family of deciduous trees of the order Sapindales, noted for producing mustard oil and native to the mountains of western China.

Brettanomyces *Mycology.* a widespread genus of fungal yeasts belonging to the class Hyphomycetes and found in such fermented beverages as grape must, wine, and beer.

brettice or **brettis** see BRATTICE.

Breuer, Josef [broi′ər] 1842–1925, Austrian neurologist; pioneered treatment of hysteria; with Freud, cofounder of psychoanalysis.

Breuer, Marcel 1902–1981, Hungarian-born architect and furniture designer; taught at Bauhaus; best known for his tubular steel chair.

Breuil, Abbé Henri 1877–1961, French priest; made major contributions to archeology of Upper Paleolithic cave art.

breunnerite *Mineralogy.* a white, yellowish, or brownish variety of magnesite, sometimes bituminous, that contains about 9% FeO.

brevet *Military Science.* a promotion giving an officer higher honorary rank but no increase in pay and no increase or a limited increase in authority. Also, **brevet rank.**

brevi- a combining form meaning "short," as in *breviductor.*

Brevibacteriaceae *Bacteriology.* in former systems of classification, a family of Gram-positive bacteria, occurring as non-spore-forming, irregularly shaped rods.

Brevibacterium *Bacteriology.* a genus of Gram-positive, aerobic, coryneform bacteria of uncertain taxonomical classification, found in soil, water, dairy products, and decomposing matter; some species are of industrial importance.

brevicaudate *Zoology.* having a short tail or tail-like appendage.

brevipennate *Vertebrate Zoology.* having short wings.

brevirostrate *Vertebrate Zoology.* having a short beak or bill.

brevitoxin *Biochemistry.* a toxin formed by the unicellular flagellate protozoan *Ptychodiscus brevis.*

brevity code *Telecommunications.* any noncryptic code designed to shorten the time and space required for message transmission.

brew *Food Technology.* **1.** to carry out the process of brewing an alcoholic beverage; a beverage so prepared. **2.** to prepare a liquid by steeping a solid food product such as tea in hot water; a liquid so prepared.

Brewer jar *Microbiology.* a gas-tight cylindrical chamber designed to establish an atmosphere containing no oxygen; used to cultivate anaerobic bacteria on petri dishes. Also, **Brewer anaerobic jar.**

brewers' yeast *Food Technology.* dried yeast cells recovered from the brewing of beer, used for medicinal purposes or as a diet supplement.

brewing *Food Technology.* the process of preparing alcoholic beverages such as beer by fermentation of a mash consisting of malted grains, especially barley and hops.

Brewster, Sir David 1781–1868, Scottish physicist; discovered polarization of biaxial crystals and double refraction by compression.

Brewster angle *Optics.* **1.** the angle of incidence at which a reflected ray of unpolarized light obtains maximum plane polarization. **2.** the polarizing angle in Brewster's law.

brewsterite *Mineralogy.* $(Sr,Ba,Ca)Al_2Si_6O_{16}\cdot 5H_2O$, an easily fusible, white to yellowish to grayish monoclinic mineral having a specific gravity of 2.45, a hardness of 5 on the Mohs scale, a perfect cleavage, and short, prismatic, often twinned crystals.

brewsterlinite *Mineralogy.* liquid CO_2 that occurs as inclusions in the cavities of minerals such as quartz and topaz.

Brewster point *Optics.* a neutral point situated approximately 15–20° below the sun.

Brewster's law *Optics.* a law stating that when light is reflected at the polarizing angle (Brewster angle), the reflected and refracted rays are perpendicular; used to determine the refractive index of a solid.

Brewster window *Optics.* a glass window positioned at a Brewster angle to incident light in order to minimize reflection losses that would result from the use of external mirrors; used in gas lasers.

Brianchon's theorem *Mathematics.* if a conic section is circumscribed by a hexagon, then the three diagonals of the hexagon are concurrent. The point of intersection is called the **Brianchon point.** This theorem is the dual of Pascal's theorem.

briar see BRIER.

brick *Materials.* a building or paving material composed of clay that has been hardened by heat, either in the sun or in a kiln; it is generally rectangular, and the traditional U.S. dimensions are 2-1/4 inches by 3-3/4 inches by 8 inches.

brickfielder *Meteorology.* a dusty, hot, and dry northerly wind that blows from the interior of Australia across to its southern coast.

bricking *Building Engineering.* work made to resemble brickwork, usually on plaster, stucco, or concrete surfaces.

bricklayer's hammer *Mechanical Devices.* a hammer having both a flat head and a sharp peen, used to shape and set bricks.

brickwork *Building Engineering.* any structural or decorative work done in brick.

bridal-veil fall *Hydrology.* a waterfall of great height and low water volume, such that the falling water is largely dissipated in a fine spray, resembling a white veil, before it reaches the base. Yosemite Valley in California has the most noted example of this type.

bride price *Anthropology.* an amount given in goods or money to the bride's kin before the marriage.

bride service *Anthropology.* work performed for the bride's kin by the groom for a period of time before or after marriage.

natural bridge

bridge *Civil Engineering.* a structure that connects two points and carries pedestrian or vehicle traffic over an obstacle such as a body of water, a declivity, or another road. *Geology.* a natural structure resembling this, especially an archlike rock formation that spans a ravine or valley. *Anatomy.* **1.** the upper part of the external nose formed by the junction of the nose bones. **2.** a structure connecting two points such as parts of an organ. *Medicine.* a dental appliance, usually a fixed partial denture supported in place by attachments to adjacent teeth. *Naval Architecture.* the raised command position of a ship; so called because on early steamers it had the form of a span between the paddlewheel boxes. *Chemistry.* an atom or group of atoms that connects other atoms or groups, usually within a ring. *Horology.* any of several partial plates in a timepiece that mount on the main plate and serve to cover the wheels of the trains and to bear their upper pivots. *Mining Engineering.* a natural rock slab that spans a passage from wall to wall and inclines less than 45° from the horizontal. *Electricity.* see BRIDGE CIRCUIT. *Mathematics.* in graph theory, a cutset with only one edge. Also, ISTHMUS.

bridge abutment *Civil Engineering.* one of two terminal supporting structures of a bridge.

bridge bearing *Civil Engineering.* a fixed or adjustable support that carries part of the load of a bridge to a pier.

bridgeboard *Building Engineering.* a notched board that carries the steps of a wooden stair.

bridge cable *Civil Engineering.* a cable that suspends the roadway or truss of a suspension bridge, made up of wires that are laid parallel or coiled in ropelike strands.

bridge circuit *Electricity.* a type of electrical network, circuit, or instrument that is used in electrical measurements, usually having four extremities arranged in a diamond, three with known impedances and the fourth attached to the source of unknown impedance. Also, BRIDGE.

bridge deck *Naval Architecture.* a deck that runs from side to side in the center of a vessel but does not extend to the bow or stern; often the deck from which the vessel is operated.

bridged-T network *Electricity.* a T network having an additional (fourth) branch bridged across the two series arms of the T between an input terminal and an output terminal.

bridge foundation *Civil Engineering.* the ensemble of piers and abutments that support the superstructure of a bridge.

bridge graft *Botany.* a means of healing or overcoming wounds (for example, when a tree is girdled by rodents or winter injury) in which scions are inserted with one end below and the other end above the gap in the bark; eventually damaged and new tissues unite to fill in the gap.

bridgehead *Military Science.* a position on the hostile side of a river or similar body of water that is established by advancing troops to allow the main force to cross safely.

bridge house *Naval Architecture.* an enclosed structure whose roof forms the bridge deck of a ship.

bridge hybrid see HYBRID JUNCTION.

bridge network see LATTICE NETWORK.

bridge of Varolius see PONS.

bridge oscillator *Electronics.* an oscillator in which positive feedback, limitation of amplitude, and frequency of oscillation are determined by a bridge circuit.

bridge pier *Civil Engineering.* a supporting structure that carries a bridge span; in multispan bridges, one or more piers are usually set between two abutments.

bridge rectifier *Electronics.* a full-wave rectifier having four rectifiers in the form of a bridge, with the AC supply connected across one diagonal and the DC output taken from the other.

Bridgerian *Geology.* a North American geologic stage of the Middle Miocene epoch, occurring after the Wasatchian and before the Uintan.

Bridges, Calvin Blackman 1889–1938, American geneticist; worked on chromosome mapping and chromosome theory of heredity.

bridge vibration *Mechanics.* any mechanical vibration of a bridge or other such suspended structure that is caused by an external stimulus, as opposed to vibration from failure of the structure.

bridgewall *Mechanical Engineering.* in a furnace or boiler, a transverse wall, usually of firebrick, over which combustion by-products flow.

bridgeware *Computer Technology.* any software or hardware that assists in making a transition to an upgraded computer system by converting programs developed for an earlier system.

bridging *Building Engineering.* **1.** a method of supporting and holding in place the joists in a roof or floor by setting short crosswise braces between them. **2.** the wooden braces used in such an arrangement. *Electricity.* **1.** a process of making a solder bridge. **2.** the process of connecting two or more circuits in parallel. *Cartography.* the use of stereoscopic plotting instruments and images in map production. *Photogrammetry.* in a photogrammetric survey, the extension and adjustment of unknown land between regions with established ground control. *Mining Engineering.* an obstruction that develops through the accumulation of small pieces of rock or other material, as in a drillhole, shaft, well, or material-crushing device. *Metallurgy.* **1.** the formation of a solid metal mass above the surface of a molten metal pool. **2.** in metal melting or ore smelting, the unintentional joining of the charge, causing downfeed impedance.

bridging amplifier *Electronics.* an amplifier whose input is bridged across a transmission line and whose output provides a signal to a branching line. Its input impedance has to be high enough so it will not affect the signal level in the main transmission circuit.

bridging connection *Electronics.* a parallel connection designed to withdraw some of the signal energy in a circuit without a noticeable effect on the normal operation of the circuit.

bridging gain *Electronics.* the ratio between the power delivered by a transducer to a specified load impedance and the power dissipated in the reference impedance across which the input of the transducer is spanned; usually expressed in decibels.

bridging ligand *Organic Chemistry.* a ligand that is simultaneously bonded to two or more metal atoms.

bridging loss *Electronics.* the reciprocal of bridging gain, the ratio between the power dissipated in the reference impedance and the power delivered by the transducer.

bridging point *Geography.* a place where a river is or can be bridged, especially the lowest of such points in the course of a river.

Bridgman, Percy Williams 1882–1961, American physicist; Nobel Prize for research in high-pressure physics.

Bridgman anvil *Physics.* a device, consisting of two large pistons with small contact heads that act in opposing directions, used for applying high static pressure.

Bridgman effect *Solid-State Physics.* the liberation or absorption of heat in an anisotropic crystal with passage of an electric current due to the nonuniform current distribution.

Bridgman relation *Solid-State Physics.* a relation between the Ettingshausen coefficient P, the Nernst-Ettingshausen coefficient Q, the temperature T, and the thermal conductivity s in a transverse magnetic field: $P = QTs$.

Bridgman technique *Solid-State Physics.* a method of growing crystals under a controlled temperature gradient, with nucleation of the crystal at the tip of a cone-shaped container.

bridled *Mechanical Engineering.* of a device or instrument, restrained in range or motion, as by a spring mechanism.

bridled pressure plate *Meteorology.* an inductance-type transducer in which the pressure on a plate exposed to the wind is balanced by the force of a spring; used to measure air velocity.

bridle joint *Building Engineering.* a heading joint in which the first member is cut away in the center with a tenon projecting on each side, and the second member is cut away at the sides to receive the tenons; the converse of a mortise-and-tenon joint.

briefing *Military Science.* a summary, usually oral, of the plans for a future mission or operation.

brier *Botany.* the white heath tree, *Erica arborea,* of southern Europe; its woody root **(brierroot** or **brierwood)** is used to make tobacco pipes. Also, BRIAR.

brig *Naval Architecture.* a two-masted ship that is square-rigged on both sides. *Military Science.* a place of imprisonment, especially on a warship. *Physics.* a dimensionless quantity used to describe a comparison of two quantities, the number of brigs being the logarithm of the ratio of the two quantities; it is similar to a bel except that a bel is used for power ratios only.

brigade *Military Science.* a force consisting of a headquarters and two or more regiments or similar units; usually commanded by a brigadier general.

brigadier general *Military Science.* the lowest rank of general officer, above a colonel and below a major general; the insignia is one star. Also, **brigadier.**

brigantine *Naval Architecture.* a two-masted ship that is square-rigged on the foresail and fore-and-aft rigged on the mainsail.

Briggs, Henry 1556–1630, English mathematician; revised Napier's logarithmic scale; compiled logarithmic tables.

Briggs stretcher carriage *Mining Engineering.* an easily maneuverable, cushioned stretcher used to transport injured miners from underground workings.

bright *Optics.* describing an area or object that appears to radiate or reflect light, especially in large amounts.

Bright, Richard 1789–1858, English physician; diagnosed Bright's disease.

bright annealing *Metallurgy.* the technique or process of heating and slow cooling (annealing) an alloy in a protective atmosphere in order to minimize oxidation and surface discoloration.

bright band *Meteorology.* the enhanced echo of snow melting to rain as displayed on a range-height indicator (RHI) scope.

bright coal *Geology.* a type of jet-black, pitchlike, banded coal that contains more than 5% anthraxylon and less than 20% opaque matter.

bright dipping *Metallurgy.* immersing a metallic material in a chemical solution that produces a clean and bright surface.

bright-field *Optics.* in microscopy, of or relating to the brightly lit background behind a target specimen or object that consequently appears dark.

bright-line spectrum *Spectroscopy.* an emission spectrum consisting of colored or bright lines or bands against a dark background.

bright nebula *Astronomy.* a cloud of interstellar gas that shines by fluorescence.

brightness *Optics.* **1.** the amount of light that appears to emanate from an object or celestial body. **2.** the intensity of a color, apart from its hue and saturation; pure white has maximum brightness, pure black minimum. **3.** see LUMINANCE, def. 1. *Astrophysics.* the measured luminosity of an astronomical body located in some particular region of the spectrum.

brightness control *Electronics.* the control in a television receiver or other CRT device that governs the average brightness of the reproduced image.

brightness scale *Photogrammetry.* the ratio of the luminance of highlights to that of the deepest shadows in the terrain of a particular field of view.

brightness temperature *Astrophysics.* the apparent temperature of a celestial object, assuming that it radiates like a black body.

bright points *Astronomy.* bright regions in the sun's corona that come and go, radiating X rays and extreme ultraviolet radiation.

bright rim structures *Astronomy.* the bright edges of dusty nebulae.

Bright's disease *Medicine.* a former general term describing kidney disease with increased secretion of protein in the urine.

bright segment *Geophysics.* a faint band of light appearing over the horizon just before sunrise or after sunset, caused by sunlight reflected from particulate matter in the upper atmosphere.

bright spots *Petroleum Engineering.* on a seismographic recording strip, white dots that may indicate the presence of hydrocarbons.

Bright Star Catalog *Astronomy.* a star catalog published by Yale University Observatory that includes stars with magnitudes of 6.5 or brighter.

bril *Optics.* a unit of luminance, with 1 bril representing a doubling of the level of luminance; a luminance of 100 brils is equivalent to 1 millilambert.

brill *Vertebrate Zoology.* a European flatfish, *Scophthalmus rhombus,* that is closely related to the turbot.

brilliance *Optics.* the luminance of a body, including saturation and brightness. *Electronics.* **1.** the degree of brightness and clarity of the image on a cathode-ray tube. **2.** the degree to which higher audio frequencies are produced or reproduced, as for example in a radio broadcast or the playback of a recording.

brilliant cut *Mineralogy.* a method of cutting gemstones so as to enhance sparkle while retaining weight; a **brilliant-cut stone** has 18–104 facets (typically 58) and takes the form of two pyramids set base to base, with the upper (the crown) truncated to give it a broad table.

Brillouin, Léon-Nicolas [brē´wän] 1889–1969, French-born American physicist; worked in quantum theory; discovered Brillouin scattering.

Brillouin, Marcel-Louis 1854–1948, French physicist; worked in kinetic theory, structure of crystals, and viscosity of liquids and gases.

Brillouin function *Solid-State Physics.* the quantum-mechanical analog to the Langevin equation for paramagnetism.

Brillouin scattering *Solid-State Physics.* the scattering of light by acoustic phonons.

Brillouin zone *Solid-State Physics.* a fundamental polyhedron in wave vector space (k-space) whose geometry plays an important role in band theory and the specification of diffraction condition; it is bounded by a Wigner-Seitz primitive cell in the reciprocal lattice.

Brill's disease *Medicine.* a recurrence of typhus many years after the initial infection. Also, **Brill-Zinsser disease.**

brimstone *Mineralogy.* a common term for sulfur, especially commercial roll sulfur.

brine *Hydrology.* **1.** a general term for sea water. **2.** sea water that is unusually high in salt content because of evaporation or freezing. **3.** any water with a high content of dissolved salts. *Mechanical Engineering.* an aqueous solution, often containing sodium chloride, that is cooled and circulated in a refrigeration system.

brine cell *Hydrology.* **1.** a small enclosure in sea ice, containing residual liquid that is higher in salt content than ordinary seawater. Also, **brine pocket.**

brine lake see SALT LAKE.

Brinell, Johann 1849–1925, Swedish metallurgist; invented the Brinell test for measuring the hardness of metals and alloys.

Brinell hardness *Materials Science.* the relative hardness of a material as measured by the Brinell test. Thus, **Brinell (hardness) number.**

Brinell test *Engineering.* a procedure that establishes the relative hardness of a material by pressing a 10-mm steel ball under a known load into the material being tested; the Brinell number is calculated by dividing the load in kilograms by the curved area in square millimeters of the resulting indentation.

brine pit *Hydrology.* an opening at the mouth of a salt spring, from which water is obtained to be evaporated for salt.

brine shrimp *Invertebrate Zoology.* any small branchipod crustacean of the genus *Artemia,* especially *A. salina;* commonly found in saline lakes and ponds.

brine spring see SALT SPRING.

B ring *Astronomy.* the middle and widest of the three rings of Saturn that are visible from Earth in a telescope; it is 25,800 kilometers wide, between 92,200 and 118,000 kilometers from Saturn's center.

Brinkman number *Fluid Mechanics.* a dimensionless number whose magnitude indicates the degree to which viscous friction causes a temperature rise in a specific flow.

brinolase *Enzymology.* a fibrinolytic enzyme produced by the mold *Aspergillus oryzae.*

briolette *Mineralogy.* any pear-shaped gem whose entire surface is cut with triangular facets.

briquette [bri ket′] *Engineering.* a molded lump of burnable fuel that is made by compressing coal dust or sawdust, often with a binder, at high temperature. Also, **briquet.**

brisa *Meteorology.* a northeasterly wind blowing off the sea. Also, BRIZA.

brisance *Ordnance.* the shattering effect or crushing force of a high explosive.

brisance index *Ordnance.* a scale used to measure the power of an explosive by comparing its ability to shatter graded sand with the standard power of trinitrotoluene (TNT).

brisant *Ordnance.* of materiel, having a shattering effect; explosive.

brise-soleil [brēz′sō lā′] *Architecture.* a louvered structure, such as a screen or trellis, that is constructed in front of windows in order to shield a building from the sun's glare while admitting fresh air and light. (A French term meaning "sun break.")

Brisingidae *Invertebrate Zoology.* a cosmopolitan family of deep-sea starfish in the subphylum Asterozoa, having a small central disk and 9 to 44 upward-arching arms.

brisote *Meteorology.* a strong brisa blowing over the island of Cuba.

bristle *Biology.* a short, stiff, coarse hair or hairlike structure on a plant or animal. *Building Engineering.* a similar structure on a paintbrush, usually made of hog's hair or of a synthetic material such as nylon.

bristlecone pine *Botany.* either of two evergreen trees of the family Pinus that grow at high altitudes in the western U.S., bearing cones with long, curving spines on the scales as well as short needles that grow in long, thick bundles. The **Great Basin bristlecone pine,** *Pinus longaeva,* lives longer than any other tree.

bristol *Graphic Arts.* any of a group of very sturdy high-fiber papers characterized by their strength, hardness, and foldability, such as **printing bristol** and **wedding bristol**.

Brit. British; Britain.

Britannia metal or **britannia metal** *Metallurgy.* a white alloy consisting of 80–90% tin with antimony, copper, and sometimes zinc, lead, and bismuth; similar to pewter, it is used to make tableware and as an antifriction material.

britholite-(Ce) *Mineralogy.* $(Ce,Ca)_5(SiO_4,PO_4)_3(OH,F)$, a brown hexagonal mineral, having a specific gravity of 3.86 and a hardness of 5 on the Mohs scale; found in nepheline syenites and contact metasomatic deposits.

British absolute system of units *Physics.* a system of measurement in which the fundamental units are the foot, the second, and the pound mass. Also, FPS SYSTEM.

British Anti-Lewisite or **British antilewisite** see DIMERCAPROL.

British engineering system of units see BRITISH GRAVITATIONAL SYSTEM OF UNITS.

British gravitational system of units *Physics.* a system of measurement in which the fundamental units are the foot, the second, and the slug mass. Also, BRITISH ENGINEERING SYSTEM OF UNITS, ENGINEER'S SYSTEM OF UNITS.

British hundredweight see LONG HUNDREDWEIGHT.

British imperial see IMPERIAL.

British Isles *Geography.* an island group comprising Great Britain, Ireland, and adjacent smaller islands.

British measure or **system** see ENGLISH SYSTEM.

British thermal unit *Thermodynamics.* **1.** an energy unit that is equivalent to 1054.5 joules or 252 calories; 10^5 BTU = 1 therm. **2.** originally, a measure of heat defined as the amount of heat energy required to raise the temperature of one pound of water by one degree Fahrenheit (usually from 60 to 61°F) at standard atmospheric pressure.

British ton see LONG TON.

Britten-Davidson model *Genetics.* a former hypothetical model of eukaryotic gene regulation in which four distinct types of DNA genes (sensor genes, integrator genes, receptor genes, and producer genes) work to sense a stimulus and eventually produce a messenger RNA that directs protein synthesis. (Proposed in 1969 by Roy *Britten* and Eric *Davidson.*)

brittle *Materials Science.* having hardness and rigidity but little tensile strength; tending to fracture easily from low stress.

brittlebush *Botany.* any of several desert plants of the genus *Encelia* having a brittle stem, alternate leaves, and yellow ray flowers; found in Mexico and the southwestern U.S.

brittle fracture of metals *Materials Science.* a fracture that usually proceeds with little or no prior plastic deformation.

brittle fracture of polymers *Materials Science.* the fracture of polymeric materials in which a part of the fracture energy goes into forming distorted localized regions called crazes before cracking occurs; fracture occurs at small strains.

brittle mica *Mineralogy.* any of a group of micaceous minerals that, unlike true micas, contain calcium instead of potassium as an essential ingredient and cleave to brittle flakes.

brittleness *Materials Science.* the quality of being brittle; the tendency of a material to fracture under low stress without much prior elastic deformation.

brittle silver ore see STEPHANITE.

brittle star *Invertebrate Zoology.* any marine echinoderm of the class Ophiuroidea, related to and resembling the starfish but having long, fragile, serpentine arms. Also, SERPENT STAR.

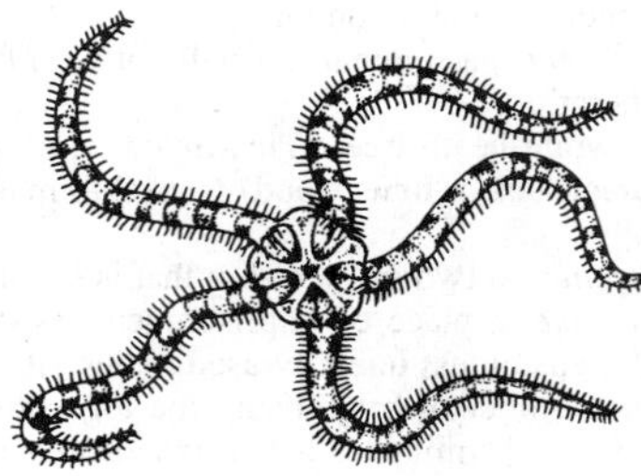

brittle star

brittle temperature *Materials Science.* the temperature below which a material, especially a metal, is brittle; that is, the temperature below which it will fracture without prior plastic deformation.

brittle temperature of polymers see GLASS TRANSITION TEMPERATURE.

Brix degree *Chemical Engineering.* a unit of the Brix scale; the percentage by weight of soluble solids in a syrup at 68°F.

Brix scale *Chemical Engineering.* a unit for sugar solutions that indicates the percentage of sugar by weight in the solution at a specific temperature; frequently used to express the sugar level of wines. (Named for the German inventor A. F. W. *Brix.*)

briza see BRISA.

brl. barrel.

BRM biological response modifier.

broach *Mechanical Devices.* a pointed hand tool used for gouging and enlarging holes and in masonry for the rough dressing of stone. *Architecture.* any of four short, sloping pyramidal members that form the transition from a square tower to a broach spire.

broaching *Engineering.* **1.** the process of enlarging or smoothing a borehole or drilled hole. **2.** the breaking down of the walls between two adjacent drill holes.

broach spire *Architecture.* an octagonal spire that rises from a square tower without any intervening structure, the transition being made at the corners of the tower by broaches that blend into the spire.

broad band *Telecommunications.* **1.** a high-speed transmission communications channel that uses a greater bandwidth than that of a voice-grade channel. **2.** also, **broadband.** referring to the use of coaxial cable for data transmission of analog signals.

broad-band amplifier *Electronics.* an amplifier having essentially flat response over a wide range of frequencies.

broad-band antenna *Electromagnetism.* an antenna that is designed to operate efficiently over a wide range of frequencies, commonly on the order of at least 10% of its center frequency.

broad-band channel *Telecommunications.* a transmission path that is capable of transmitting analog signals of greater frequency than a voice-grade channel.

broad-band klystron *Electronics.* a klystron having three or more resonant cavities that are loaded externally and stagger-tuned to broaden the bandwidth.

broad-band path *Telecommunications.* a path having a bandwidth of at least 20 kilohertz.

broad-breasted bronze *Agriculture.* a breed of very large turkeys that have dark brown feathers with white tips and a large amount of breast meat.

broadcast *Agriculture.* to sow seeds in all directions by scattering them over the ground. *Telecommunications.* a transmission method whereby any station within the range of the transmitting station can receive radio and television signals on a given frequency. Thus, **broadcast station, broadcast transmitter,** and so on.

broadcast band *Telecommunications.* a frequency band that broadcast stations may use to send and receive radio signals.

broadcast burning *Forestry.* a technique whereby a controlled fire is allowed to burn over a specific area within designated boundaries in order to reduce the fuel hazard.

broadcaster *Agriculture.* a device designed for scattering seeds, typically consisting of a hopper from which seed is supplied to a revolving fanlike mechanism that disperses it.

broadcasting *Agriculture.* the distribution of seeds uniformly over an area, as opposed to seeding in rows. Also, **broadcast seeding.** *Telecommunications.* **1.** the process of transmitting analog signals over radio or television. **2.** the art, business, or profession of radio or television.

broadcast mode see DAISY CHAIN.

broadcast network *Computer Science.* a network in which a message is sent to all nodes connected to the network; the intended recipient reads the message, while others ignore it.

broadcloth *Textiles.* a lightweight, lustrous fabric with a closely woven crosswise rib.

broad gauge *Transportation Engineering.* **1.** of a railroad track, set further apart than the standard gauge of 56.5 inches. **2.** or **broad-gauge.** designed to run on such a track. Also, **broad gage.**

broadleaf *Agriculture.* any of several tobacco strains having broad leaves, used especially for cigars.

broadleaf tree *Botany.* any tree having broad, flat leaves rather than needlelike leaves, including all deciduous trees and certain evergreens. Also **broadleafed tree, broadleaved tree.**

broadloom *Textiles.* of a carpet, rug, or fabric, woven on a wide loom, especially one wider than 54 inches.

broad on *Navigation.* describing the bearing of a vessel: **1. broad on the bow.** about 45° to starboard or port from the heading of the vessel. **2. broad on the beam.** about 90° to starboard or port from the heading. **3. broad on the quarter.** about 135° to starboard or port.

broadside *Naval Architecture.* the whole side of a ship above the waterline. *Military Science.* the simultaneous firing of all of a ship's heavy guns. *Electromagnetism.* the fact of being perpendicular to an axis or a plane. *Graphic Arts.* a sheet of paper that is printed on one or both sides, now often an advertising circular.

broadside array *Electromagnetism.* an antenna array that is designed so that the direction of maximum radiation is directed perpendicular to the plane or line of the array.

broad-spectrum antibiotic *Microbiology.* an antibiotic that is effective against a wide range of bacteria, both Gram-positive and Gram-negative.

broad tuning *Electronics.* poor selectivity in a radio receiver, so that two or more stations are audibly received at a single dial setting.

Broca, Pierre Paul 1824–1880, French surgeon and anatomist; pioneer in the study of the brain.

Broca's aphasia see EXPRESSIVE APHASIA.

brocade *Textiles.* a luxurious woven fabric, usually made of silk or polyester and having elaborate, slightly raised designs.

brocatel *Textiles.* a stiff fabric having patterns woven into it in high relief; used for tapestry and upholstery. *Petrology.* a type of varicolored ornamental marble, quarried mostly in Spain and Italy. Also, **brocatelle.**

broccoli *Botany.* a cruciferous plant, *Brassica oleracea,* whose green leafy stalks and clustered buds are eaten as vegetables.

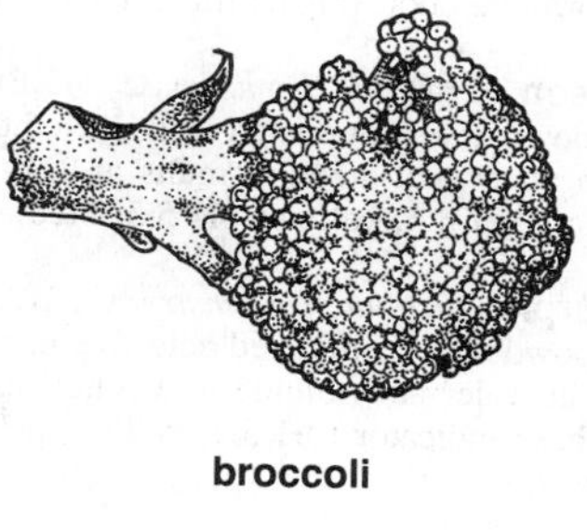

broccoli

brochantite *Mineralogy.* $Cu_4^{+2}(SO_4)(OH)_6$, a bright to blackish-green monoclinic mineral with a vitreous luster and a perfect cleavage, having a specific gravity of 3.97 and a hardness of 3.5 to 4 on the Mohs scale; found as a secondary mineral in oxidized zones of copper deposits. (Named for *Brochant* de Villiers, 1772–1840, French mineralogist.) Also, **brochanite, brochanthite, brochantit.**

Brochothrix *Bacteriology.* a genus of Gram-positive, facultatively anaerobic bacteria that grow as rod-shaped cells joined together into chains, typically found in meat and meat products.

Brocken *Geography.* a mountain in central Germany; the highest peak of the Harz Mountains (3745 ft).

Brocken bow see ANTICORONA.

Brocken specter *Meteorology.* **1.** an optical illusion that occurs at high altitudes when the image of an observer placed between the sun and a cloud is projected on the cloud as a greatly magnified shadow. **2.** an illusory appearance of a gigantic shadow, frequently observed on the Brocken Peak in the Harz Mountains of Saxony.

Brock's syndrome see MIDDLE LOBE SYNDROME.

brocot suspension *Horology.* a type of clock-pendulum suspension that can be adjusted from the face of the clock. (Developed by French horologist Achille *Brocot.)*

brodiaea or **Brodiaea** *Botany.* any bulbous plant of the genus *Brodiae,* in the Amaryllis family, having grasslike basal leaves and clusters of variously colored flowers. (Named for James *Brodie,* 1744–1824, Scottish botanist.)

Brodmann, Korbinian 1868–1918, German neurologist; distinguished 47 areas of the cerebral cortex by the arrangement of their six cellular layers, and assigned a number to each.

Brodmann's areas *Physiology.* the cytoarchitectural areas of the cerebral cortex, distinguished by differences in the arrangement of their six cellular layers. (Named for Korbinian *Brodmann.)*

broeboe *Meteorology.* a dry, strong, gusty easterly wind that blows over the island of Celebes.

Broenner acid see BRÖNNER ACID.

Broensted see BRÖNSTED.

broiler *Agriculture.* a chicken that is tender enough to be broiled, especially one that weighs less than 2½ pounds.

broke *Graphic Arts.* of paper, unsuited for commercial use; to be re-pulped.

broken *Meteorology.* of or relating to a sky coverage of between 0.6 and 0.9, expressed to the nearest tenth.

broken-back transit *Optics.* a transit telescope containing a prism with a 90° apex that is located where the optical and rotational axes intersect and that interrupts the path of light.

broken belt *Hydrology.* a band of broken ice that forms a transition between solid ice and open water.

broken-color work *Building Engineering.* any decorative effect, such as antiquing, that is achieved by manipulating color washes or glazes with a sponge, rag, or other device in order to vary and break up the color blocks.

broken ice *Oceanography.* sea ice that covers 50–75% of the surface in a given area.

broken line *Mathematics.* a sequence of line segments lying end to end so that the endpoints are not all collinear; useful in approximating (the length of) a curve.

broken pediment *Architecture.* a pediment that is interrupted at the apex.

broken-stick model *Ecology.* a model of the proportional representation of a species in a community or environment, obtained by arbitrarily dividing a line into segments, representing the resources of the environment.

broken stream *Hydrology.* a stream that disappears and reappears over and over again, as in an arid region.

broken twill *Textiles.* fabric in which the diagonal line formed by the weft threads (the twill) is broken or reverses direction at intervals.

broken water *Oceanography.* an area of water whose surface shows ripples or eddies, usually surrounded by calm water.

broken wind see HEAVES.

broken-wing display *Behavior.* a behavior pattern in which an adult bird feigns a broken wing or similar injury to appear easy to catch and thus draw a predator's attention away from its eggs or chicks.

brolga *Vertebrate Zoology.* a large crane, *Grus rubicunda,* found in Australia and New Zealand, having silver plumage and a bare red patch on the head. It is noted for its elaborate mating dance. Also, NATIVE COMPANION.

brom- a combining form denoting the presence of a bromine atom in a molecule, as in *bromphenol.*
bromacetone see BROMOACETONE.
bromacil *Organic Chemistry.* $C_9H_{13}BrN_2O_2$, a crystalline solid that is soluble in water and alcohol; melts at 158–159°C; used as an herbicide.
bromargyrite *Mineralogy.* AgBr, a bright to amber-yellow, sectile, cubic mineral that occurs as massive crusts and waxy coatings, having a specific gravity of 6.47 and a hardness of 2.5 on the Mohs scale. Also, BROMYRITE.
bromate *Chemistry.* any salt of bromic acid, $HBrO_3$.
bromato- a combining form meaning "food," as in *bromatotherapy.* Also, **broma-** .
bromatotoxin *Toxicology.* a poison that is produced in decaying food by fermenting bacteria.
bromatotoxism *Toxicology.* poisoning due to chronic consumption of decayed food that contains bromatotoxin. Also, **bromatotoxismus.**
bromcresol green see BROMOCRESOL GREEN.
bromcresol purple see BROMOCRESOL PURPLE.
bromegrass *Botany.* any of numerous grasses of the genus *Bromus* having flat blades and open clusters of spikelets. Also, **brome.**
bromelain or **bromelin** *Biochemistry.* an enzyme that breaks down proteins and clots milk; used in the food industry to tenderize meat. Also, PLANT PROTEASE CONCENTRATE.
Bromeliaceae *Botany.* the bromeliad family of xerophytic or epiphytic monocots, characterized by thick, stiff, narrow leaves and including the pineapple. (Named for Olaf *Bromel* (Olaus Bromelius), 1639–1705, Swedish botanist.)
bromeliad *Botany.* any plant of the family Bromeliaceae, especially an ornamental house plant having large lance-shaped leaves and showy flowers.

bromeliad

Bromeliales *Botany.* the order of monocots coexistent with the family Bromeliaceae.
bromellite *Mineralogy.* BeO, a white hexagonal mineral having a specific gravity of 3.017 and a hardness of 9 on the Mohs scale, occurring as minute crystals in calcite veinlets in skarns.
brome mosaic virus group see BROMOVIRUS GROUP.
bromhidrosiphobia *Psychology.* an irrational fear of offensive odors.
bromic acid *Inorganic Chemistry.* $HBrO_3$, a colorless to yellowish liquid, known in solution only; that is, it is very soluble in cold water, decomposes at 100°C, and is unstable except in very dilute solutions. It is used in pharmaceuticals and dyes and as an oxidizing agent.
bromide *Chemistry.* any binary compound derived from hydrobromic acid, HBr.
bromide paper *Graphic Arts.* a sensitive photographic paper that is coated with a silver bromide emulsion.
brominate *Chemistry.* to treat or combine with bromine.
brominating agent *Chemistry.* any compound that can be used to insert a bromine atom into a molecule, e.g., phosphorus tribromide, PBr_3.
bromination *Chemistry.* the process of inserting a bromine atom into a molecule.
bromine *Chemistry.* a nonmetallic halogen element, having the symbol Br, the atomic number 35, an atomic weight of 79.909, a melting point of –7.2°C, and a boiling point of 58.78°C. It is a dark, reddish-brown, very reactive liquid that is found in nature only in combination with other elements; it fumes at room temperature with an irritating odor. It is used in manufacturing antiknock gasoline, in photography, and in water purification. (From a Greek word meaning "foul odor; stench.")
bromine azide *Inorganic Chemistry.* BrN_3, crystals or red liquid that melts at about 45°C; a strong oxidizing agent, explosive when heated or shocked; used in detonators and other explosive devices.
bromine iodide see IODINE BROMIDE.
bromine monochloride *Inorganic Chemistry.* BrCl, a reddish-yellow liquid or gas that readily hydrolyzes; soluble in water or ether; it freezes at –66°C and decomposes at 10°C ; it is used in industrial disinfectants.
bromine number *Analytical Chemistry.* the number that indicates the degree of unsaturation of an oil, determined by the number of centigrams of bromine that are absorbed by one gram of oil. Also, BROMINE VALUE.
bromine pentafluoride *Inorganic Chemistry.* BrF_5, a colorless, fuming liquid that reacts with every known element except nitrogen, oxygen, and the noble gases; freezes at –61.3°C and boils at 40.5°C. It is corrosive to skin and tissue and explodes on contact with water; used in liquid rocket fuels and in chemical synthesis.
bromine test *Chemical Engineering.* a laboratory test to determine unsaturated hydrocarbons found in crude oil by mixing a sample with bromine and measuring the absorption rate.
bromine trifluoride *Chemistry.* BrF_3, a colorless to yellowish liquid, very reactive and corrosive, that decomposes violently on contact with water; melts at 8.8°C and boils at 135°C; used as a fluorinating agent and electrolytic solvent.
bromine value see BROMINE NUMBER.
bromine water *Chemistry.* a saturated solution of bromine in water, used as a laboratory reagent.
brominism *Medicine.* chronic bromide intoxication, a toxic condition caused by extended use of bromides. Also, **bromism.**
bromo- **1.** a combining form denoting the presence of a bromine atom in a molecule, as in *bromoform.* **2.** a combining form meaning "foul smelling."
***N*-bromoacetamide** *Organic Chemistry.* $CH_3CONHBr$, needlelike white crystals; melts at 102–105°C; soluble in warm water and cold ether; emits toxic bromine fumes on heating; used as a brominating agent and in the oxidation of primary and secondary alcohols.
***p*-bromoacetanilide** *Organic Chemistry.* C_8H_8BrNO, a crystalline solid; melts at 168°C; soluble in benzene, chloroform, and ethyl acetate, and insoluble in cold water; used as an analgesic and antipyretic.
bromoacetic acid *Organic Chemistry.* $BrCH_2COOH$, colorless crystals; melts at 50°C and boils at 208°C; soluble in water, alcohol, and ether. It is a powerful irritant and is used in organic synthesis.
bromoacetone *Organic Chemistry.* $BrCH_2COCH_3$, a colorless liquid that rapidly turns violet; freezes at –54°C and boils at 136°C; slightly soluble in water and soluble in alcohol, ether, acetone, and benzene. It is toxic and a powerful irritant, and is used as a tear gas and in organic synthesis.
bromo acid see EOSIN.
bromoalkane *Organic Chemistry.* an aliphatic hydrocarbon with bromine bonded to it. Also, BROMOHYDROCARBON.
***p*-bromoaniline** *Organic Chemistry.* $BrC_6H_4NH_2$, rhombic crystals; insoluble in water and soluble in alcohol and ether; melts at 66–66.5°C; used in the preparation of azo dyes.
bromobenzene *Organic Chemistry.* C_6H_5Br, a colorless oily liquid that freezes at –30.5°C and boils at 156°C; insoluble in water and miscible with most organic solvents. It is usually obtained by the bromination of benzene and is used as a solvent, in motor fuels and compounds, and to make other chemicals. Also, PHENYL BROMIDE.
***p*-bromobenzyl bromide** *Organic Chemistry.* $BrC_6H_4CH_2Br$, a crystalline solid with an aromatic odor; melts at 61°C; soluble in water, alcohol, and ether; used to identify aromatic carboxylic acids.
bromochloromethane *Organic Chemistry.* CH_2BrCl, a clear, colorless liquid that freezes at –86.5°C and boils at 68°C; insoluble in water and soluble in organic solvents. Its vapor is moderately toxic and has an irritating chloroformlike odor. It is nonflammable and is used in fire extinguishers.
bromocresol green *Organic Chemistry.* $C_{21}H_{14}Br_4O_5S$, a yellowish-gray crystalline powder; a brominated acid dye of the sulfon-phthalein series; soluble in water and alcohol; melts at 218–219°C; used as an acid-base indicator (pH 4.5, yellow; pH 5.5, blue). Also, BROMCRESOL GREEN.
bromocresol purple *Organic Chemistry.* $C_{21}H_{16}Br_2O_5S$, a pinkish-gray crystalline powder; a brominated acid dye of the sulfon-phthalein series; insoluble in water and soluble in alcohol; melts at 241–242°C; used as an acid-base indicator (pH 5.2, yellow; pH 6.8, purple). Also, BROMCRESOL PURPLE.

bromocriptine *Pharmacology.* $C_{32}H_{40}BrN_5O_5$, an ergot alkaloid that is used to suppress prolactin secretion, thereby inhibiting lactation and stimulating ovulation; also used in treatment of Parkinson's disease and acromegaly.
bromoderma *Pathology.* a skin eruption caused by a hypersensitive reaction to bromide ingestion.
bromodiethylacetylurea *Pharmacology.* $C_7H_{13}BrN_2O_2$, the chemical name for the drug carbromal, a sedative and weak hypnotic.
bromofenoxim *Organic Chemistry.* $C_{13}H_7Br_2N_3O_6$, a creamy, crystalline powder that melts at 196–197°C; used as an herbicide.
bromoform *Organic Chemistry.* $CHBr_3$, a heavy, colorless, addictive liquid with a chloroformlike odor; freezes at 9°C and boils at 151°C, and is slightly soluble in water and soluble in alcohol, ether, chloroform, and benzene. It is used for the separation of minerals, by floating relative densities of less than 2.9 and sinking densities greater than 2.9.
bromohydrocarbon see BROMOALKANE.
bromol *Organic Chemistry.* $Br_3C_6H_3OH$, a crystalline solid, insoluble in water and soluble in alcohol and ether; melts at 95°C and boils at 244°C. Also, TRIBROMOPHENOL.
α-bromonaphthalene *Organic Chemistry.* $C_{10}H_7Br$, a thick, colorless liquid that freezes at 6.2°C and boils at 281°C, and is slightly soluble in water and soluble in alcohol, ether, acetone, and benzene; used in refractometric fat determination and in organic synthesis.
1-bromooctane *Organic Chemistry.* $CH_3(CH_2)_6CH_2Br$, a colorless liquid; insoluble in water and soluble in alcohol and ether; boils at 198°C; used in organic synthesis. Also, *N*-OCTYL BROMIDE.
bromophos *Organic Chemistry.* $C_8H_8BrCl_2O_3PS$, yellowish crystals; soluble in water; melts at 53–54°C; used as an insecticide.
bromophosgene see CARBONYL BROMIDE.
bromopicrin *Organic Chemistry.* CBr_3NO_2, prismatic crystals that melt at 103°C and explode if rapidly heated; slightly soluble in water and soluble in alcohol, benzene, and ether. It is a strong irritant and an explosion risk, and is used for poison gas. Also, NITROBROMOFORM.
bromopnea *Medicine.* bad breath; halitosis.
***N*-bromosuccinimide** *Organic Chemistry.* $C_4H_4BrNO_2$, fine white crystals; soluble in water and acetone; decomposes at a range of 172–178°C; used for controlled low-energy bromination.
bromotrifluoromethane *Organic Chemistry.* $CBrF_3$, a colorless gas, toxic and an irritant, that freezes at –168°C and boils at –59°C; it is nonflammable and is used as a fire-extinguishing agent and chemical intermediate.
bromouracil *Biochemistry.* $C_4H_3BrN_2O_2$, the bromine derivative of the DNA base uracil; has the capacity to create a mutable polymer in its reaction with DNA, and is useful in genetic research as a marker because it causes no harm to the chromosome.
Bromovirus group *Virology.* a group of ssRNA-containing plant viruses with a narrow host range; transmitted mechanically and by beetles. Also, BROME MOSAIC VIRUS GROUP.
α-bromo-*meta*-xylene *Organic Chemistry.* $CH_3C_6H_4CH_2Br$, a liquid that is insoluble in water and soluble in alcohol and ether; boils at 212–215°C; used in organic synthesis and as a tear gas. Also, XYLYL BROMIDE.
bromoxynil *Organic Chemistry.* $C_7H_3Br_2NO$, a colorless solid that melts at 194–195°C; slightly soluble in water; used as an herbicide.
bromoxynil octanoate *Organic Chemistry.* $C_{15}H_{17}Br_2NO_2$, a pale brown liquid that is insoluble in water; melts at 45–46°C; used for weed control.
brompheniramine *Pharmacology.* $C_{16}H_{19}BrN_2$, an antihistaminic drug that is the bromine analog of chlorpheniramine; often prepared as a maleate salt, **brompheniramine maleate,** $C_{16}H_{19}BrN_2 \cdot C_4H_4O_4$, a white crystalline powder that is administered orally or by injection.
bromphenol blue *Organic Chemistry.* $C_{19}H_{10}Br_4O_5S$, prisms or crystals that are slightly soluble in alcohol and benzene and slightly soluble in water; used as an acid-base indicator (pH 3.0, yellow; pH 4.6, purple).
Brompton's mixture *Pharmacology.* an analgesic mixture administered primarily to terminally ill patients, containing alcohol, morphine or heroin, and sometimes cocaine. (Developed at *Brompton* Hospital, London.) Also, **Brompton's cocktail.**
bromthymol blue *Organic Chemistry.* $C_{27}H_{28}Br_2O_5S$, white to yellow crystals that are soluble in alcohol and alkaline solutions and slightly soluble in water; used as an acid-base indicator (pH 6.0, yellow; pH 7.6, blue).
bromuration *Histology.* the process of treating a tissue with bromine or a bromine-containing reagent.
Bromwich contour *Mathematics.* the contour of integration usually employed to ensure convergence of the integral for inverting the Laplace transform of a function $f(t)$ of a single variable. In the Argand diagram, it is the line $\mathrm{Re}(s) = c$, where c is larger than the exponent b in the growth condition $|f(t)| < Ae^{bt}$ used to ensure the analyticity of the Laplace transform itself.
bromyrite see BROMARGYRITE.
bronch- a combining form meaning "of or relating to the bronchi," as in *bronchitis.*
bronchi *Anatomy.* the plural of *bronchus.*
bronchial *Anatomy.* of or relating to a bronchus or bronchi.
bronchial adenoma *Medicine.* a benign, gland-like tumor of the bronchus.
bronchial asthma *Medicine.* a condition marked by recurring attacks of paroxysmal dyspnea, caused by constriction of the bronchi and bronchioles, increased mucus, spasms of the smooth muscles, and edema of the lung membranes.
bronchial tree *Anatomy.* the network of tubes of the bronchial system leading to the lung.
bronchiectasis *Medicine.* chronically enlarged bronchi with inflammation, most commonly occurring in the lower portion of the lung; marked by fetid breath and paroxysmal coughing.
bronchio- a combining form meaning "of or relating to the bronchi," as in *bronchiolitis;* often an alternate to *broncho-*, though the latter is usually preferred, as in *bronchospasm (bronchiospasm).*
bronchiolar carcinoma *Oncology.* a rare type of adenocarcinoma of the lung, where elongated tumor cells line the alveolar septa. Also, ALVEOLAR CELL CARCINOMA.
bronchiole *Anatomy.* any of the small subdivisions of a bronchus.
bronchiolectasis *Medicine.* dilation of the bronchioles.
bronchiolitis *Pathology.* an inflammation of the bronchioles, such as bronchopneumonia.
bronchiolitis obliterans *Medicine.* a buildup of scar tissue blocking the bronchioles; it sometimes follows pneumonia, but the cause of the scarring is often unknown. Also, **bronchiolitis fibrosa obliterans.**
bronchitic *Pathology.* relating to or affected with bronchitis.
bronchitis [brän kī´təs] *Medicine.* an inflammation of the bronchi, broadly classified as **acute bronchitis,** a bronchitic attack having a relatively short and severe course, or as **chronic bronchitis,** a long-continued form that tends to recur after stages of quiescence.
broncho- a combining form meaning "of or relating to the bronchus," as in *bronchospasm.*
bronchoalveolar *Anatomy.* of or relating to a bronchus and alveoli.
bronchocele *Pathology.* a localized dilatation of a bronchus.
bronchoconstriction *Medicine.* the process or fact of decreasing the caliber of a bronchus.
bronchodilator *Medicine.* **1.** an enlarging of the air spaces in the lungs. **2.** a substance that causes expansion of the air spaces of the lungs.
bronchogenic *Medicine.* originating in a bronchus.
bronchogram *Medicine.* the radiographic record produced by bronchography.
bronchography *Medicine.* the X-ray of a lung after filling a bronchus with an opaque liquid.
broncholithiasis *Medicine.* a condition marked by the presence of calculi **(broncholiths)** in the lumen of the tracheobronchial tree.
bronchopathy *Medicine.* any disease of the bronchi.
bronchoplasty *Surgery.* plastic surgery of a bronchus; surgical closure of a bronchial fistula.
bronchopneumonia *Pathology.* an inflammation of the bronchioles, especially one beginning in the terminal bronchioles.
bronchopulmonary *Anatomy.* of or relating to the lungs and air passages; both bronchial and pulmonary.
bronchorrhea *Pathology.* an excessive discharge of mucus from the air spaces of the lungs.
bronchoscope *Medicine.* a device used to look into the lung to diagnose lung conditions and to remove biopsy tissue or foreign bodies.
bronchoscopic *Medicine.* referring to either bronchoscopy or the bronchosope.
bronchosinusitis *Pathology.* a coexisting infection of the paranasal sinuses and the lower respiratory passages.
bronchospasm *Pathology.* an abnormal spasmodic contraction of the smooth muscle of the bronchi, as occurs in asthma.
bronchospirometry *Medicine.* the means by which the air capacity of a bronchus is measured.
bronchovesicular see BRONCHOALVEOLAR.

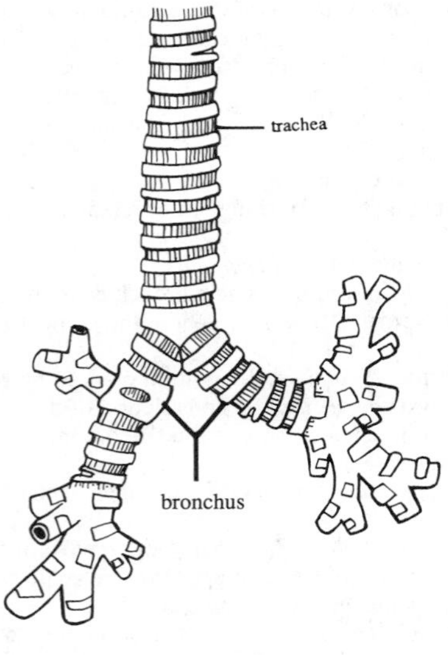

bronchus

bronchus *Anatomy. plural,* **bronchi.** either of the two main branches of the trachea.
Bronco *Aviation.* a popular name for the OV-10 reconnaissance plane.
Brongniart, Alexandre 1770–1847, French mineralogist and geologist; pioneer in study of strata; delineated four orders of reptiles.
Bronner's acid *Organic Chemistry.* $C_{10}H_6(NH_2)SO_3H$, a colorless mixture of crystalline monosulfonic acids; used as an intermediate for dyestuffs.
Brönsted, Johannes Nicolaus 1879–1947, Danish physical chemist; known for defining acids and bases in terms of protein loss or gain.
Brönsted acid *Chemistry.* see ACID.
Brönsted base *Chemistry.* see BASE.
Brönsted-Lowry theory *Chemistry.* the theory that any molecule or ion that can produce a proton is an acid, and any that can take up a proton is a base. Thus all reactions between acid and base are simply the transfer of a proton from one base to another. Also, **Brönsted theory**.
brontides *Geophysics.* the short-lived rumbling sounds from the earth that can be heard in active seismic regions.
brontophobia *Psychology.* an irrational fear of thunder.
brontosaurus *Paleontology.* any of several huge herbivorous dinosaurs of the genus *Apatosaurus* having a small head, long neck, massive body, and thick limbs; widespread in the Jurassic period. Also, **brontosaur.**
Brontotheriidae *Paleontology.* a family of herbivorous, large-hoofed, rhinoceroslike perissodactyl mammals in the superfamily Brontotherioidea, characterized by horny projections at the end of the snout; originated in North America in the Eocene, spread to Asia in the Oligocene.
Brontotherioidea *Paleontology.* a monofamilial superfamily of perissodactyl mammals in the suborder Hippomorpha, related to the horses of Equoidea, the other superfamily of the suborder.
bronze *Metallurgy.* any of various brown metals, alloys composed mostly of copper with up to 11% tin and sometimes small amounts of zinc or other metals.
Bronze Age *Archaeology.* a time period, about 3000 BC, that is designated in the Three-Age System as following the Stone Age and preceding the Iron Age; defined by a shift from the use of stone tools to the use of bronze.
bronzed grackle *Vertebrate Zoology.* a bird, *Quiscalus quiscula versicolor,* the western subspecies of the North American common grackle, having black plumage that on the male is tinted with bronze on the back and purple and blue-green around the neck.
bronzing *Metallurgy.* **1.** the process of forming a colored chemical coating on a bronze alloy. **2.** the process of applying bronze as plating on another material. *Materials Science.* any agent, such as a varnish, that is used to give a bronze finish to another material. *Graphic Arts.* the process of applying metallic powder to a printed surface or proof.
bronzite *Mineralogy.* $(Mg,Fe^{+2})_2Si_2O_6$, a green or brown, iron-containing variety of enstatite.
bronzitite *Petrology.* a pyroxenite that is composed almost entirely of bronzite with subordinate amounts of olivine, hornblende, or spinel. Also, **bronzitfels.**
brood *Zoology.* **1.** the offspring of animals, usually a number of young that are produced or hatched at one time. **2.** to sit on and cover eggs in order to warm and hatch them; incubate. *Agriculture.* of an animal, kept and used for breeding. *Botany.* of a plant, heavily infested with insects.
brood capsule *Invertebrate Zoology.* in a larval tapeworm, a section of infolded germinal epithelium that holds buds from which new adults will grow and constitutes an infective agent if eaten by a suitable host.
broodmare or **brood mare** *Agriculture.* a female horse that is used for breeding.
brood parasite *Biology.* an organism that uses a host species to brood the young of its own species. Also, NEST PARASITE.
brood parasitism *Biology.* the use of a host species by a parasite to brood the young of its own species.
brood pouch *Vertebrate Zoology.* a body pouch where eggs or embryos undergo certain stages of development, such as a sacklike appendage on a male seahorse or pipefish in which a female lays her eggs, or a pouch on the backs of some frogs and toads where eggs are carried.
brook *Hydrology.* a natural stream of flowing water, regarded as smaller than a creek, especially a stream that flows swiftly over a rocky bed.
brookite *Mineralogy.* TiO_2, a brown, yellowish, reddish, or iron-black orthorhombic mineral occurring in crystals of varied habit with adamantine to submetallic luster, having a specific gravity of 4.14 and a hardness of 5.5 to 6 on the Mohs scale; trimorphous with anatase and rutile.
brooklime *Botany.* **1.** any of various plants of the genus *Veronica* that grow in marshes or along streams, such as **American brooklime,** *Veronica americana*, a creeping plant having leafy stems and clusters of small blue flowers. **2.** see WATERCRESS.
brook trout *Vertebrate Zoology.* a speckled game fish, *Salvelinus fontinalis,* commonly found in freshwater streams of eastern North America. Also, SPECKLED TROUT.
brookweed *Botany.* either of two herbs of genus *Samolus* in the primrose family having small white flowers and growing in damp places. Also, WATER PIMPERNEL.
broom *Botany.* any of numerous shrubby plants of the genera *Cytisus* and *Genista* in the pea family, having long, slender branches, small leaves, and yellow flowers. *Mechanical Devices.* a tool used for sweeping, made originally from this plant but now of straw or synthetic material. *Building Engineering.* a head of a wooden pile that has been crushed and splayed by pounding.
Broom, Robert 1866–1951, Scottish anatomist and paleontologist; discovered and analyzed fossils of the genus *Australopithecus*.
broomcorn *Botany.* any of several varieties of sorghum resembling corn and having long, stiff panicles that are used in making brooms.
brooming *Civil Engineering.* a method of imparting a grooved texture to a concrete slab by brushing the freshly poured concrete surface with a broom.
broomrape *Botany.* **1.** any of various parasitic plants of the genus *Orobanche*, which live on the roots of broom. **2.** or **broomrape family.** the family Orobanchaceae of parasitic plants, including broomrape, beechdrops, and squawroot.
broomweed *Botany.* any of a wide variety of wild herbs or shrubs that can be used in making brooms.

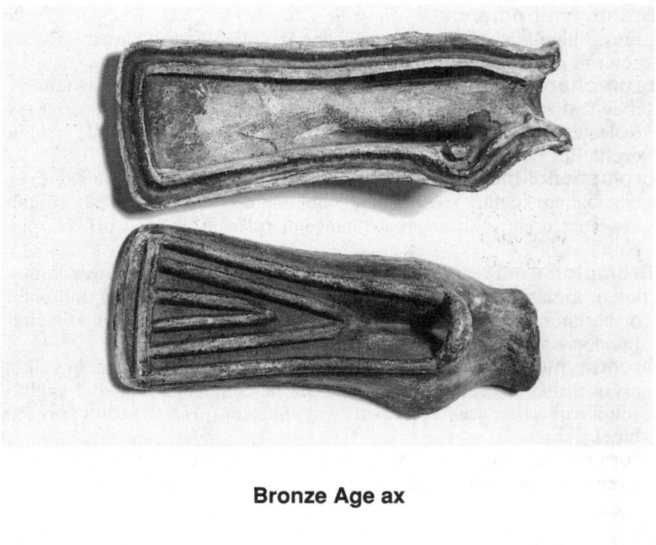

Bronze Age ax

broomy flow *Fluid Mechanics.* a fluid that eddies in a pipe after it has passed through a tight area or makes an alteration in its direction.

brosse *Invertebrate Zoology.* a brushlike organelle of cilia that is found on the anterodorsal surface of certain ciliate protozoa.

broth *Food Engineering.* a thin soup made by boiling meat or vegetables, often used as a stock or base in preparing foods. *Microbiology.* a liquid culture medium for the cultivation of microorganisms, such as a **carbohydrate broth** or **nitrate broth.**

brotula *Vertebrate Zoology.* any of several deep-sea or cave-dwelling fishes in the family **Brotulidae,** order Perciformes.

Brouncker's continued fraction *Mathematics.* the continued fraction $4/\pi = 1 + 1/2 + 9/2 + 25/2 + 49/2 + \cdots$.

Brouncker's series *Mathematics.* the infinite series $\ln 2 = 1/1\cdot 2 + 1/3\cdot 4 + 1/5\cdot 6 + 1/7\cdot 8 + \cdots$.

Brouwer fixed-point theorem *Mathematics.* any continuous map of a closed ball in R^n to itself leaves at least one point fixed.

Brown, Herbert C. born 1912, British-born American chemist; Nobel Prize for discovery and study of borohydrides and organoboranes.

Brown, Michael Stuart born 1941, American biochemist; with J. Goldstein, Nobel Prize for discovering low-density lipoprotein receptors.

Brown, Robert 1773–1858, Scottish physician and botanist; discovered Brownian movement; collected 4000 Australian plant species.

brown acid *Chemical Engineering.* oil-soluble petroleum sulfonate found in sludge after petroleum products are treated with sulfuric acid.

brown algae *Botany.* algae of the class Phaeophyceae; the largest and most structurally complex of the algae, characterized by large holdfasts, fucoxanthin pigment, and laminarin starch as a storage material; includes kelp and gulfweed.

Brown and Sharpe wire gauge see AMERICAN WIRE GAUGE.

brown bast *Plant Pathology.* a disease of the para rubber tree characterized by a stoppage of the flow of latex and discolored inner bark near the tapping cut.

brown bat *Vertebrate Zoology.* any of several small- to medium-sized brown-colored bats, including the **little brown bat** *(Myotis lucifugus)* and **big brown bat** *(Eptesicus fuscus)* of North America.

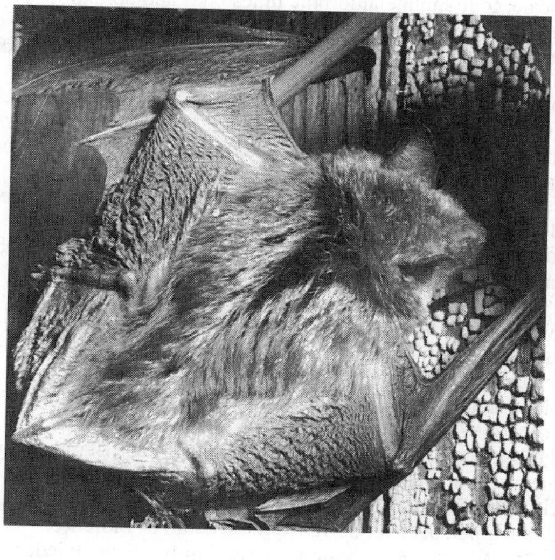

brown bat

brown bear *Vertebrate Zoology.* any bear of the species *Ursus arctos,* characterized by tan to dark brown fur, a hump high on the back, and an upturned muzzle; includes the European and Asian brown bears as well as the grizzly and Kodiak bears, formerly considered separate species.

brown bent *Botany.* a sturdy North American grass, *Agrostis canina,* that is commonly used for lawns and putting greens.

brown blight *Plant Pathology.* a virus disease of lettuce characterized by pale yellow or brown spots on inner leaves, stunted plants, and gradual browning that begins at the base.

brown blotch *Plant Pathology.* **1.** a mushroom disease caused by the bacterium *Pseudomonas tolaasi* and characterized by the appearance of irregular brown spots. **2.** a pear disease caused by a fungus and characterized by irregular brown spots on the surface of the fruit.

brown body *Invertebrate Zoology.* **1.** a brown mass formed in certain bryozoans by a degenerating polyp or polypide; thought to function as an excretory organ. **2.** any of numerous flat, ovoid dark masses found in the hind region of earthworms, resulting from the breakdown of amoebocytes and related debris.

brown bullhead *Vertebrate Zoology.* a brown freshwater catfish, *Ictalurus nebulosus,* of eastern North America.

brown canker *Plant Pathology.* a disease of roses caused by the fungus *Cryptosporella embrina* and characterized by stem cankers having a reddish-purple border and by leaf and stem lesions that change in color from purple to white and finally to buff.

brown clay ironstone *Geology.* limonite that occurs in the form of concretionary masses.

brown coal *Mining Engineering.* lignite, especially loosely consolidated lignite that is mined and used as a fuel, intermediate in fuel value between peat and the true coals.

brown creeper *Vertebrate Zoology.* a small brown-and-white bird, *Certhia americana,* of North America having a curved bill and stiff tail feathers.

brown dwarf *Astronomy.* a hypothetical object whose mass is greater than that of a large planet (like Jupiter) and smaller than the smallest known star, a red dwarf.

Browne, Sir Thomas 1605–1682, English author and physician; coined the term "electricity."

brown-eyed Susan *Botany.* a composite plant, *Rudbeckia triloba,* of the southeastern U.S., having trilobate leaves and a single flower with a dark brown center and brown-to-yellow rays; a coneflower similar to the black-eyed Susan.

brown fat *Histology.* in humans and many other mammals, brownish-yellow adipose tissue composed of spherical **brown fat cells,** which contain small fat droplets in the cytoplasm and are a major source of heat energy.

brown felt blight *Plant Pathology.* a disease in conifers caused by various Ascomycetes fungi, especially *Herpotrichia nigra* and *Neopeckia coulteri,* characterized by brown matted needles covered with a dense, dark, feltlike growth.

Brown Forest soil *Geology.* any of a group of calcium-rich intrazonal soils that develop in a warm climate under deciduous forests and are characterized by a mull horizon.

brownheart *Plant Pathology.* a brownish discoloration of the flesh of apples in storage caused by a buildup of carbon dioxide.

brown hematite see LIMONITE.

Brown-Hopps modification *Microbiology.* a procedure for staining bacterial cells in tissue sections, using a modification of the Gram stain technique.

Brownian movement *Physics.* the random movement of particles suspended in a fluid, caused by the interaction of these particles with the molecules of the fluid. *Mathematics.* a stochastic process that is the limit of random walks on the sample space of non-negative real numbers. Also, **Brownian motion.** (From Robert *Brown.*)

brown induration *Pathology.* **1.** a pigmented area in the lung seen in pneumonia. **2.** a thickening and pigmentation of the connective tissue of the lung from chronic congestion caused by heart valve disease or anthracosis.

browning *Plant Pathology.* any plant abnormality resulting in a brownish discoloration of a plant part. Also, STEM BREAK.

Browning, John M. 1855–1926, American inventor and designer of firearms, notably the Browning automatic rifle (BAR).

Browning *Ordnance.* any of various weapons produced by the Browning Arms Company, founded by American gunsmith John M. Browning; the most famous are the Browning automatic rifle (BAR) and the Browning .50-caliber heavy machine gun; others include automatic pistols, rifles, shotguns, .30-caliber machine guns, and aircraft cannons.

browning agents *Food Technology.* any food ingredient or constituent that gives a darker color to a food during either processing or cooking, such as milk, eggs, and many sugars.

Browning automatic rifle *Ordnance.* a gas-operated, portable automatic weapon capable of firing 200 to 300 rounds per minute; usually fired from a bipod.

brown iron ore see LIMONITE.

brown lead oxide see LEAD DIOXIDE.

brown leaf rust *Plant Pathology.* a disease of rye plants caused by the fungus *Puccinia dispersa.*

brown lignite *Geology.* a variety of brown coal with a heat value less than 6300 BTU/lb and a high moisture content. Also, LIGNITE B.

brownline *Graphic Arts.* a prepress proof printed in brown on white paper treated with silver nitrate. Also, VANDYKE.

brown lung see BYSSINOSIS.

brownout *Electricity.* **1.** a deliberate reduction of line voltage in order to lessen load demands. **2.** a reduction of electric lighting in a city during wartime as a precaution against air attack. **3.** any reduction or curtailment of electric power, as occurs during a storm, for example.

brown patch *Plant Pathology.* a turf disease characterized by roughly circular patches of brown grass bordered by grayish-black mycelia and caused by soil-borne fungi that multiply under conditions of high temperatures and humidity.

brown rat see NORWAY RAT.

brown recluse spider *Invertebrate Zoology.* a small, brown venomous spider, *Loxosceles reclusa,* having a violin-shaped marking on its back. Also, BROWN SPIDER, FIDDLEBACK SPIDER, VIOLIN SPIDER.

brown rice *Food Technology.* unpolished rice; that is, rice from which the bran and germ have not been removed.

brown-ring test *Analytical Chemistry.* a test for the nitrate ion using a solution of sample and dilute ferrous sulfate layered on top of concentrated sulfuric acid; a brown ring between the layers indicates a positive test.

brown root *Plant Pathology.* a disease of tropical plants caused by the fungus *Hymenochaete noxia,* in which the roots become incrusted with earth and stones bound together by brown masses of fungi filaments.

brown root rot *Plant Pathology.* **1.** a disease caused by the fungus *Thielava basicola* and characterized by the decay and darkening of the roots and lower stem, occurring most frequently in plants of the pea, cucumber, and potato families. **2.** a similar disease occurring in tobacco and caused by attacks of meadow nematodes.

brown rot *Plant Pathology.* **1.** any fungal or bacterial disease that results in browning and the breakdown of plant tissue, especially in plums and other fruit trees. **2.** in timber trees, decay caused by a fungal attack on cellulose.

browns *Graphic Arts.* a set of brownline prepress proofs.

brown seaweed *Botany.* any brown algae, especially the larger species.

brown smoke *Engineering.* a term for smoke containing a noticeable amount of solid carbon particles, but a lesser amount than black smoke.

brown snow *Meteorology.* snow that is mixed with dust particles.

brown soil *Geology.* any of a group of zonal soils having a brown surface grading into a light-colored subsurface over a layer of calcium carbonate.

brown spar *Geology.* any light-colored crystalline carbonate mineral that is colored brown by the presence of iron.

brown spider see BROWN RECLUSE SPIDER.

brown spot *Plant Pathology.* a fungus disease of corn, soybeans, and other plants that is characterized by irregular brown lesions on the leaves and fruit and by cankers on the stems; caused by any of several fungi, such as *Certophorum setosum.*

brown stem rot *Plant Pathology.* a disease of soybeans that is caused by the fungus *Cephalosporium gregatum* and characterized by a brownish discoloration and decay of the internal tissue of the stem and leaves.

brownstone *Petrology.* a ferruginous sandstone composed of quartz grains coated with iron oxide derived from the Triassic of the Connecticut River Valley. *Building Engineering.* a building, especially a multistory house, whose exterior walls are constructed of this sandstone.

brown stringy rot *Plant Pathology.* a fungal disease of conifers characterized by brown or rusty fibrous linear markings in the inner tissues.

brown sugar *Food Technology.* **1.** unrefined or partially refined sugar that retains all or some of its natural molasses content. **2.** refined white sugar to which molasses has been added.

Brown Swiss *Agriculture.* a breed of dairy cattle varying from light to dark brown in color; originally bred in Switzerland.

browse *Agriculture.* **1.** of livestock, to feed or nibble on tender portions of trees or shrubs, such as buds, sprouts, leaves, stems, and vines. **2.** the tender portions of trees or shrubs that are fed upon in this way. *Computer Programming.* **1.** to examine information stored in a data base. **2.** to examine parts of a program in order to understand it better.

browser *Vertebrate Zoology.* an animal that feeds on those parts of plants which are above ground, that is, shoots, twigs, and the leaves attached to them. *Computer Programming.* a program that is designed to facilitate browsing of data or programs by finding the desired data, displaying it in understandable form, and so on.

browsing *Agriculture.* the act of feeding or nibbling by livestock on tender portions of trees or shrubs, as opposed to grazing.

BRU *Aviation.* the airport code for Brussels, Belgium.

brubru *Meteorology.* a squall in the East Indies.

Bruce, Sir David 1855–1931, British surgeon; discovered bacillus of undulant fever; isolated germ and agent of sleeping sickness.

Brucella *Bacteriology.* a genus of Gram-negative, nonmotile, aerobic, rod-shaped or ovoid bacteria of uncertain affiliation, occurring as intracellular parasites or pathogens in humans and animals.

Brucellaceae *Bacteriology.* in former systems of classification, a family of Gram-negative, aerobic bacteria occurring as small, nonmotile coccoid or rod-shaped cells.

brucellergen *Biochemistry.* a protein found in the aerobic bacterium *Brucella*; used in a skin test to detect the presence of the organism.

brucellergen test *Immunology.* a skin test that detects the presence of the pathogenic bacteria *Brucella.*

brucellin *Immunology.* a substance derived from the bacteria *Brucella,* used either in skin tests to determine exposure to or infection caused by *Brucella,* or as a vaccine against brucellosis.

brucellosis [broo´sə lō´sis] *Medicine.* a chronic systemic disease characterized by fever, weakness, and general malaise; it is caused by infection with a *Brucella* organism, and is transmitted to humans by direct or indirect contact with infected animals or their milk.

Bruchidae *Invertebrate Zoology.* the seed weevils, a family (type genus *Bruchus*) of small beetles with larvae that infest the seeds of peas and other legumes.

Bruch's membrane *Anatomy.* a transparent membrane that is a part of the vascular tunic of the eye, lying between the pigmented epithelium of the retina and the choroid.

brucine *Organic Chemistry.* $C_{23}H_{26}N_2O_4$, a white crystalline solid that melts at 178°C (anhydrous) or 105°C (hydrate); slightly soluble in water and soluble in alcohol, chloroform, and benzene; a poisonous alkaloid found in the seeds of the plants *Strychnos ignatii* and *Strychnos nux-vomica,* used in denaturing alcohol and as a lubricant additive. It also forms **brucine hydrochloride, brucine nitrate dihydrate, brucine phosphate,** and **brucine sulfate heptahydrate.**

brucite *Mineralogy.* $Mg(OH)_2$, a whitish, sectile, transparent to translucent trigonal mineral occurring in tabular crystals and fibrous forms, having a specific gravity of 2.38 to 2.4 and a hardness of 2.5 on the Mohs scale; found in metamorphic limestones and dolomitic schists.

Bruckner cycle *Meteorology.* an episodic climatologic cycle of a region in which relatively cool-damp and warm-dry periods alternate over a period of approximately 35 years.

brugnatellite *Mineralogy.* $Mg_6Fe^{+3}(CO_3)(OH)_{13}\cdot 4H_2O$, a micaceous, pink to brownish-white hexagonal mineral having a specific gravity of 2.14 and a hardness of 2 on the Mohs scale; found as coatings and crusts in hydrothermally altered serpentinic rocks.

bruise *Medicine.* a discoloration of an area of skin or mucus membrane caused by blood leaking into the tissues under the skin.

bruissement *Cardiology.* a heart tremor that feels, to the touch, like a cat's purring. Also, PURRING TREMOR.

bruma *Meteorology.* an afternoon haze that appears on the coast of Chile when sea air is transported inland.

Brunauer-Emmett-Teller equation *Physical Chemistry.* an equation that determines the adsorption rate of a film more than one molecule thick; it computes the surface area of the monolayer or top layer. Also, BET EQUATION.

Brunel, Isambard 1806–1859, English engineer; famous for design and construction of railroads, bridges, and steamships.

Brunelleschi, Fillipo [broo´nə les´kē] 1377–1446, Italian architect; invented linear perspective; designed the Duomo of Florence.

Brunelliaceae *Botany.* a monogeneric family of dicotyledonous evergreen trees of the order Rosales that are native to tropical America; some species are covered by a dense down.

Brunfels, Otto 1489–1534, German herbalist; wrote and published *Living Portraits of Plants,* with fine illustrations by Hans Weiditz.

Bruniaceae *Botany.* a family of dicotyledonous South American shrubs and small trees of the order Rosales, characterized by small tough leaves and sessile flowers in spikes or heads.

Brunizem see PRAIRIE SOIL.

Brunner's glands *Anatomy.* small, flattened mucus-secreting glands in the lining of the duodenum.

Bruno, Giordano c. 1548–1600, Italian philosopher; postulated an infinite universe; anticipated relativity.

Brunoniaceae *Botany.* a monospecific family of dicotyledonous perennial herbs of the order Campanulales, native to Australia and characterized by flowers on a leafless stalk.

Brunschwig, Hieronymus c. 1452–1512, Dutch physician; introduced medical applications for distillates and improved methods of producing them.

Brunt-Douglas isallobaric wind see ISALLOBARIC WIND.

Brunton compass *Engineering.* a type of field compass affixed with sights and a reflecting device; used in surveying. Also, **Brunton, Brunton pocket transit.**

brüscha *Meteorology.* a wind that blows from the northeast in the Engadin, Switzerland, bringing fine weather.

brush *Electricity.* **1.** a conductor that maintains contact between stationary and moving parts of a generator or motor. **2.** the moving member of a selector, or a similar device, that makes contact with the terminals of a bank. *Computer Technology.* see PAINTBRUSH.

brush biopsy *Surgery.* a biopsy that is obtained by passing a bristled catheter into suspected areas of disease and removing cells caught on the bristles.

brush border *Cell Biology.* the large number of microvilli that line the surface of an intestinal epithelial cell, increasing the membrane surface area for nutrient absorption.

brush discharge *Electricity.* a luminous discharge of electricity into the air from a conductor when its potential exceeds a predetermined value but remains too low for spark formation.

brush encoder *Electronics.* an encoder that converts positional information to digitally encoded data, using brushes that make contact with conductive segments on a moving surface.

brush fire *Forestry.* a fire in an area that has scrub trees, bushes, and shrubs, rather than a fire of full-size trees.

brush harrow *Agriculture.* a simple harrow consisting of a log or pole with short branches attached, used to cover seeds. Also, **brush drag.**

brushing *Textiles.* the process of rubbing a fabric against a rough surface to produce a thick nap. Also, NAPPING, POLISHING.

brushite *Mineralogy.* $CaHPO_4{\cdot}2H_2O$, a colorless to yellow, monoclinic, easily fusible mineral occurring as needlelike, prismatic crystals, having a specific gravity of 2.32 and a hardness of 2.5 on the Mohs scale.

brush lag *Electricity.* the distance that brushes on a motor are displaced opposite to the rotation of the motor to adjust for armature reaction.

brush lead *Electricity.* the distance the brushes on a motor are displaced in the direction of the motor rotation to adjust for armature reaction.

brush machine *Food Technology.* a machine that smooths the grains in rice processing.

brush rocker *Electricity.* an adjustable yoke to which the brush holders in an electrical machine are attached.

brushwood *Forestry.* **1.** shrubs and twigs that have fallen or been cut and remain on the forest floor. **2.** shrubby vegetation and groups of trees that do not produce commercial timber.

brussels sprouts *Botany.* a plant, *Brassica olearacea gemmifera,* of the family Brassicaceae, having small, headlike buds that grow in the leaf axils along the stem and are eaten as a vegetable.

brute-force filter *Electricity.* a type of powerpack filter that depends on large values of capacitance and inductance to smooth out pulsations.

brute-force technique *Computer Technology.* the use of great processing power to do a job.

brute supply *Electricity.* a power supply that is fully unregulated, employing no circuitry to hold output voltage constant with the variations in the input line or load changes.

Bruxellian *Geology.* a European geologic stage of the lower Middle Eocene epoch, occurring after the Ypresian and before the Auversian.

Bryaceae *Botany.* a large family of mosses of the order Bryales, growing either in bare soil or humus or (when epiphytic or saxicolous) in loose to dense cushions; characterized by large, thin-walled hexagonal cells, stout costa, and inclined, smooth capsules.

Bryales *Botany.* an order of mosses of the subclass Bryidae in which the gametophyte is perennial and erect.

Bryant, Gridley 1789–1867, American engineer; a pioneer in railway construction; invented the portable derrick.

Bryidae *Botany.* a subclass of mosses in which the sporophyte consists of a seta, a capsule that opens by a peristome, and a filamentous protonema.

Bryobartramiaceae *Botany.* a monotypic family of tiny mosses of the order Pottiales that form small tufts on exposed soil; characterized by an extensive rhizoidal system and terminal sporophytes; found in South Africa and Australia.

bryology *Botany.* the branch of botany that deals with mosses and other bryophytes.

Bryophyta *Botany.* a division of the plant kingdom that includes small, simple nonvascular plants: the mosses, liverworts, and hornworts. The gametophyte is the visible, free-living generation on which the sporophyte is nutritionally dependent.

bryophyte [brī′ə fīt′] *Botany.* a moss or liverwort; any member of the division Bryophyta.

Bryopsida *Botany.* the moss family, the largest and most highly developed group of Bryophyta; characterized by a leafy gametophyte with multicellular rhizoids and spores that produce a filamentous protonema.

Bryopsidaceae *Botany.* a family of marine green algae in the order Bryopsidales (Siphonales), characterized by a rhizomelike prostrate base and an erect pinnately branching portion.

Bryopsidales *Botany.* an order of green algae of the class Chlorophyceae, characterized by a thallus made up of branched tubes or siphons that often constrict but are seldom septate.

Bryopsis *Botany.* the type genus of tropical marine green algae of the family Bryopsidaceae, characterized by nonseptate filaments that form a horizontal rhizomatous base and an upright pinnately branched portion.

Bryoxiphiaceae *Botany.* a monotypic family composed of a slender, rigid moss of the order Dicranales that grows on moist, acid cliff faces; characterized by erect to hanging stems with terminal sporophytes, closely spaced complanate leaves, and no peristome.

Bryozoa *Invertebrate Zoology.* the moss animals, a phylum of minute, mosslike marine and freshwater creatures, with distinct alimentary canals, that form permanent colonies and reproduce by budding. Also, POLYZOA.

bryozoan *Invertebrate Zoology.* any member of the phylum Bryozoa.

bryozoan colony *Invertebrate Zoology.* a community of aquatic animals that reproduce by budding and live united in branching, flat, or mosslike groups permanently attached to stones, seaweed, or shells.

bryozoan zooid *Invertebrate Zoology.* a complex single aquatic entity having an alimentary canal with a distinct mouth and anus; consists of a zooecium (body wall) and a polypide which is retractable into the zooecium; lives attached to other bryozoan zooids.

BS *Petroleum Engineering.* bottom settlings, the emulsions of oil, water, and mud that settle out of crude oil in storage tanks. *Medicine.* blood sugar; bowel sounds; breath sounds.

BS or **B.S.** Bachelor of Science; Bachelor of Surgery; British Standard.

BSA body surface area.

BSc or **B.Sc.** Bachelor of Science.

B-scan *Radiology.* a visual two-dimensional display of ultrasonographic echoes on a CRT screen in which the position of a dot on the screen corresponds to the time elapsed, and the brightness of the dot corresponds to the strength of the echo.

BSCh or **B.S.Ch.** Bachelor of Science in Chemistry.

B scope *Electronics.* a radar scope that produces a B display.

bsh. bushel; bushels.

BSI British Standards Institution.

B size *Engineering.* one of a series or range of specific sizes to which items are cut in manufacturing paper or board.

BSO or **bso** blue stellar object.

B-splines *Design Engineering.* the representation of curves when considering trajectory generation and control of a robotic arm in mechanical design applications.

BSS *Computer Science.* a pseudo-operation for some assemblers, used to specify the reservation of a block of storage. (An abbreviation of block starting with symbol.)

B-stage resin *Organic Chemistry.* an intermediate stage in a thermosetting resin reaction, when the resin has been thermally reactive beyond the A stage, so that it is partially soluble in common solvents and is not fully fusible even above 150°C. It has limited commercial use.

B star *Astronomy.* a star whose spectrum displays absorption lines of hydrogen and neutral helium and which has a surface temperature ranging from about 11,000 to 25,000 kelvins.

B supply *Electronics.* the supply of positive voltage to the anodes and other positive electrodes of electron tubes. Also, B POWER SUPPLY.

BS & W *Petroleum Engineering.* an abbreviation for basic sediment and water, the impurities of oil, water, and foreign matter that collect in the bottom of petroleum storage tanks. Also, BOTTOMS, BOTTOM SEDIMENT, SEDIMENT AND WATER, BASE SEDIMENT AND WATER.

BT body temperature; bleeding time.

B tectonite *Petrology.* **1.** a tectonite with a fabric dominated by linear elements without an S-plane that indicates an axial direction rather than a slip surface. **2.** a tectonite whose point diagram is girdled about the *b*-axis.

BTO *Military Science.* bombing through overcast; the process of bombing a ground target that is obscured by clouds, through the use of radar, infrared equipment, or other such devices.
BTR *Aviation.* the airport code for Baton Rouge, Louisiana.
B trace *Electronics.* the second (lower) trace on a radar scope display.
B-Tree *Computer Science.* a method of organizing large files of data for rapid access to desired records kept on secondary storage. Each node contains several sorted key values. If the desired key value is not found, it will fall between two key values in the node; a pointer stored between these values gives the child node of the tree in which (or below which) the desired record can be found.
btry. battery.
BTU or **B.T.U.** British thermal unit or units. Also, **Btu, B.t.u.**
BTV *Aviation.* the airport code for Burlington, Vermont.
B-type inclusion body *Virology.* cytoplasmic sites of virus replication found in vertebrate Poxvirus-infected cells. Also, GUARNIERI BODY.
B-type virus particle see TYPE B ONCOVIRUS GROUP.
bubble *Physics.* a small cavity or thin film, usually filled with a fluid, whose boundary is roughly spherical and is completely immersed in another fluid. *Solid-State Physics.* a cylindrical region of magnetization that is produced in a thin film of magnetic material by an externally applied field. Also, MAGNETIC BUBBLE. *Meteorology.* see BUBBLE HIGH.
bubble cap *Chemical Engineering.* a metal cap that covers a riser in the plate of a distillation tower; designed to permit vapors to rise through the cap to make contact with the liquid on the plate.
bubble-cap plate *Chemical Engineering.* a device designed to produce a bubbling action in a distillation column that provides intimate contact between the vapor bubbles flowing up the column and the downflowing liquid.
bubble cavitation *Fluid Mechanics.* the formation and collapse of vapor cavities in a flowing liquid, resulting whenever the local pressure is reduced to that of the liquid vapor pressure.
bubble chamber *Physics.* a device used for tracking the trajectories of ionizing particles by photographing the trails of bubbles in superheated liquids, such as liquid hydrogen.
bubble domains *Materials Science.* very small, reversely magnetized cylindrical domains perpendicular to the surface of a thin magnetic material.
bubble high *Meteorology.* a localized high-pressure area, with anticyclonic circulation, from 50 to 300 miles across; often induced by precipitation and vertical currents associated with thunderstorms.
bubble horizon *Navigation.* a device attached in front of the horizon glass of a sextant to provide an artificial horizon in the form of a bubble.
bubble impression *Geology.* a shallow depression having a smooth surface without a raised rim, usually formed on a sedimentary deposit by a bubble of gas after it has escaped into the air or the water above.
bubble memory *Computer Technology.* a nonvolatile storage technology using small integrated circuit chips in which microscopic magnetic areas called "bubbles" float in a film of magnetic material where they can be manipulated, by means of electromagnetic signals, to represent data stored in memory.

bubble memory

bubble-mold cooling *Engineering.* the cooling of plastic formed within a mold by means of a liquid stream that flows through a cavity surrounding the mold.
bubble point *Physical Chemistry.* the temperature at which gas bubbles first appear in a solution with more than one component.
bubble pulse *Geophysics.* an extraneous pulse created by a bubble formed when detonating a charge in water during seismic prospecting.
bubble raft *Solid-State Physics.* a phenomenon in which a layer of bubbles floating on a soap solution simulates a two-dimensional crystal.
bubble sextant *Navigation.* a type of sextant that uses a bubble to indicate the horizontal.
bubble sort *Computer Programming.* a method of sorting information that compares consecutive pairs of elements in a list, determines whether each pair is in order, and interchanges those that are not; involves repeated passes through the list until the entire list is in order. Also, EXCHANGE SORT, RIPPLE SORT.
bubble test *Engineering.* a test to determine the largest aperture in the grating of a filter, by forcing air through the grating while it is submerged in liquid.
bubble tower *Chemical Engineering.* a bubble-cap plate distillation column.
bubble train *Volcanology.* a string of vesicles in lava that reveals the path that rising gas has taken as it escaped.
bubble tray *Chemical Engineering.* a circular plate with bubble caps, installed in a vapor-liquid contacting tower. Also, **bubble-cap tray.**
bubble wall fragment *Geology.* a glassy volcanic shard showing part of a curved or flat vesicle surface.
Bubnoff unit *Geology.* a measure of the rate of geologic movements, equal to 1 μm/year.
bubo [boo´bō´; byoo´bō´] *Medicine.* a swollen and tender lymph node, especially in the axilla or groin, seen in plague, syphilis, gonorrhea, tuberculosis, and other infections. Also, **bubon.**
bubonic [boo´bän´ik; byoo´bän´ik] *Medicine.* **1.** relating to or characterized by buboes. **2.** relating to or affected by bubonic plague.
bubonic plague *Medicine.* the most common form of plague, characterized by an abrupt onset of fever, chills, weakness, and headache followed by pain, tenderness, and buboes associated with a marked hemorrhagic tendency, the development of disseminated coagulation, necrotic purpura, and extensive symmetrical gangrene; severe complications include pneumonia and septicemia. The notorious **Black Plague** that devastated the population of Europe and Asia in the 1300s was a form of bubonic plague.
bucc- a combining form meaning "cheek," as in *buccal.* Also, **bucco-.**
buccal *Anatomy.* of or relating to the mouth or cheek.
buccal cavity *Anatomy.* the cavity of the mouth.
buccal gland *Anatomy.* any of numerous small glands in the mucous membrane of the mouth that secrete saliva.
Buccinacea *Invertebrate Zoology.* a superfamily of marine gastropod mollusks, in the order Prosobranchia, including dove shells and typical whelks.
Buccinidae *Invertebrate Zoology.* a family of neogastropods containing some species that are marine scavengers; includes the whelk.
Bucconidae *Vertebrate Zoology.* the puffbirds, a family of arboreal forest birds in the order Piciformes, native to the tropical Americas.
Bucerotidae *Vertebrate Zoology.* the hornbills, a family of large tropical birds of Africa and Asia in the order Coraciiformes, characterized by a downwardly curved beak and a hornlike head ornamentation.
Buch, Baron Leopold von 1774–1853, German geologist; a founder of volcanology.
Buchan-type facies series *Geology.* a dynamothermal regional metamorphism that occurs in high-pressure regions.
Bucherer reaction *Organic Chemistry.* **1.** a method of preparing hydantoin from carbonyl compounds by reaction with potassium cyanide and ammonium carbonate. **2.** a method for the interconversion of α- or β-naphthylamine by heating with bisulfate salts. Also, **Bucherer-Bergs reaction.**
Buchholz protective device *Electricity.* a protective relay connected to an oil-filled tank that is equipped with a transformer; engaged either by gas produced by faults or by oil surges produced by explosive faults in the transformer.
buchite *Petrology.* a partially vitrified inclusion of sandstone in igneous rock.
Buchner, Eduard 1860–1917, German chemist; Nobel Prize for identifying zymase as the agent of alcoholic fermentation.
Buchner's zymase see ZYMASE.

buck *Vertebrate Zoology.* a male animal, especially a male deer, goat, antelope, rabbit, hare, or rat. *Building Engineering.* the frame for a finished door. *Electronics.* to reduce the total voltage in a circuit by the series connection of an additional voltage of opposite polarity. *Mining Engineering.* **1.** to break up or crush a material, such as ore. **2.** to push coal down a chute toward a mine car, in anthracite coal regions. **3.** a large quartz reef void of accessory minerals such as gold. Also, BUCK QUARTZ, BULL QUARTZ.

bucket *Engineering.* **1.** a deep, wide container with a flat bottom, used to hold and carry water and other liquids. **2.** a scooping device, usually designed to open and close to take on or let go of a load. **3.** a water outlet in a turbine. *Botany.* see CALYX. *Computer Programming.* a group of items that have been arranged together for some logical reason, such as a bucket of symbols that all begin with the same letter.

bucket-brigade device *Electronics.* a semiconductor device in which carriers transfer charges from point to point in sequence.

bucket conveyor *Mechanical Engineering.* a bulk-material carrier that consists of a series of buckets attached by pivots to two endless roller chains; the buckets pivot in such a manner that they remain upright except when tilted into a dumping position.

bucket dredge *Mechanical Engineering.* a dredge resting on two pontoons, between which passes a series of buckets on a conveyor chain or belt; the buckets excavate soil material from the bottom of a pond or lake and deposit the soil on the deck of the dredge.

bucket drill *Mining Engineering.* an auger stem drill with a steel cylinder instead of a drill bit to confine the cutting; especially effective in obtaining samples of clay deposits.

bucket thermometer *Oceanography.* a water-temperature thermometer whose bulb is surrounded by an insulated container. Thus, **bucket temperature.**

bucket-wheel excavator *Mechanical Engineering.* a power-operated shovel consisting of a series of buckets attached to a wheel, which allows for continuous digging.

buckeye *Botany.* any of various deciduous trees or shrubs of the genus *Aesculus;* characterized by opposite, palmately compound leaves, large erect panicles of tubular or bell-shaped flowers, and large leathery capsules that split open to release round, nutlike seeds.

bucking coil *Electromagnetism.* a coil that is positioned and connected so that the magnetic field generated by the coil opposes the field generated by another coil.

Buckingham's equations *Mechanical Engineering.* a series of equations for calculating the durability of gears and the dynamic loads to which they are subjected in relation to their composition, dimensions, hardness, and surface endurance.

Buckingham's pi theorem *Physics.* a theorem stating that if there is one and only one mathematical expression connecting n physical quantities x_j that can be expressed in terms of m fundamental quantities, such that $f(x_1, x_2, \ldots, x_n) = 0$, then f can be expressed in the form $F(p_1, p_2, \ldots, p_{n-m}) = 0$ where the p's are $n-m$ dimensionless products of the x_i.

bucking voltage *Electricity.* a voltage of opposite polarity compared to that of another voltage against which it acts.

buckle *Metallurgy.* **1.** undesirable waviness perpendicular to the rolling direction in a rolled metallic product, such as a steel bar or sheet. **2.** in casting, undesirable formation of a surface indentation caused by the expansion of the molding sand into the cavity where the metal is poured.

buckle defect *Materials Science.* a metal-casting defect in which a flaw exists on the mold wall surface prior to spalling off.

buckle fold *Geology.* a fold caused by a lateral compression of sedimentary rock beds. Also, TRUE FOLD.

buckle plate *Civil Engineering.* a steel floor plate that is slightly arched in order to increase its rigidity.

Buckley gauge *Engineering.* a device that measures extremely low levels of gas pressure by detecting the degree of ionization generated in the gas by a given amount of electric current.

buckling *Mechanics.* the bending or abrupt lateral deflection of a column, plate, or similar structure under longitudinal compression which causes instability of one mode of deformation bifurcating to another stable mode. *Nucleonics.* a measure of neutron density distribution in a reactor, relative to the thermal diffusion length in a core lattice, the magnitude of which is inversely proportional to the size that a reactor core must have to be critical; the just-critical size is attained when **geometric buckling** (determined by reactor shape) is equal to **material buckling** (dependent on reactor materials).

buckling load *Mechanics.* the minimum load or stress that, when applied steadily, will cause a certain structure to buckle.

buckling strength *Mechanics.* the ability of a material to withstand stress without buckling. Also, BREAKING CAPACITY.

buckminsterfullerene see BUCKY BALL.

buck quartz *Mining Engineering.* see BUCK.

buckstay *Mechanical Engineering.* a supporting rack or frame for a furnace wall.

buckthorn toxin *Toxicology.* any of a group of poisons found in the fruit of the desert shrub *Karwinskia humboldtiana* and the fruit and bark of plants of the genus *Rhamnus.*

buckwheat *Botany.* any low sprawling herb or shrub of the genus *Fagopyrum,* family Polygonaceae, characterized by alternate leathery evergreen leaves and small white flowers in racemes or corymbs.

buckwheat

buckwheat coal *Geology.* anthracite coal of a size that will pass through a 5/16- to 9/16-inch mesh screen.

buckwheat no. 3 see BARLEY COAL.

bucky ball *Chemistry.* a molecule made up of 60 atoms of carbon that are shaped like a geodesic dome; the third form of pure carbon (the other two are ordinary graphite and precious diamonds). This molecule is extremely durable and can be blasted apart only by powerful lasers; bucky balls are thought to be in the forefront of such applications as powerful superconductors and as carriers of radioactive atoms in cancer therapy. (Named after the American inventor Buckminster Fuller, 1885-1983, who designed geodesic domes.) Also, BUCKMINSTERFULLERENE.

Bucky-Potter diaphragm *Radiology.* a grid containing a series of thin lead strips that moves between the patient and the film to filter out scattered radiation during a radiographic exposure, thus increasing detail and contrast. Also, POTTER-BUCKY GRID.

bud *Botany.* a compact, undeveloped shoot or flower having overlapping, immature leaves folded about the growing tip; usually enclosed within protective bud scales. *Invertebrate Zoology.* **1.** a protruberance from the body of an organism that develops into a new individual, as in bryozoans. **2.** to reproduce asexually from a parent cell or zooid.

BUD *Aviation.* the airport code for Budapest, Hungary.

Budan's theorem *Mathematics.* the number of roots of an nth degree polynomial $f(x)$ that lie in an open interval (a, b) is $V(a) - V(b)$, where $V(x)$ is the number of sign changes in the sequence $f(x), f'(x), f''(x), \ldots, f^{(n)}(x)$, evaluated at a and b, respectively. Vanishing terms in the sequence do not increase $V(x)$.

budbreak *Botany.* the beginning of growth from a bud.

budded virus *Virology.* any virus that envelops by budding through the membrane of an infected cell. Also, **budding virus.**

budding *Botany.* **1.** the production of buds. **2.** reproduction by developing bud outgrowths, which may or may not be set free. **3.** asexual reproduction in which an outgrowth of a parent organism detaches and forms a new individual of the same species; most common among yeast plants. Also, GEMMATION. **4.** see BUD GRAFTING. *Virology.* a process of viral assembly in which the nucleocapsid is enveloped as it passes through the virus-modified membrane of the host cell.

budding bacteria *Microbiology.* any of a number of bacterial species that reproduce by developing an outgrowth from the bacterial cell that ultimately separates, resulting in two cells.

buddy system *Military Science.* a system of grouping people in pairs to provide protection and support; each member of a pair is required to remain near and watch out for the other member.

bud fission *Mycology.* in fungi, a certain way of reproducing by budding. Also, GEMMATION.

budgerigar [buj´ə ri gär˝] *Vertebrate Zoology.* a small, seed-eating Australian parrot of the family Psittacidae, species *Melopsittacus undulatus,* having many color mutations and the ability to mimic speech; a popular cage bird. Also, BUDGIE, SHELL PARAKEET.

budgerigar

budget year *Meteorology.* a one-year period that begins at the first sign of the accumulation season at the firn line of a glacier or ice cap and extends through the ablation season of the following summer.

budgie see BUDGERIGAR.

bud grafting *Botany.* a method of vegetative propagation in which a single bud with a small amount of adjacent tissue is inserted into a slit in the bark of the stock. Also, BUDDING.

budling *Botany.* a shoot that develops from the scion in bud grafting.

bud rot *Plant Pathology.* any plant abnormality resulting in the breakdown and decay of bud tissue.

bud scale *Botany.* a tough, scalelike, modified leaf that forms a protective covering for a bud until it opens; may be covered with hairs or water-repellent resin, gum, or wax.

bud sport *Botany.* a branch, flower, or shoot that differs markedly from the rest of the plant and passes on differences to vegetatively propagated offspring. Also, **bud variation**.

bud stick *Botany.* a shoot of a plant from which buds are cut for propagation of the plant.

BUE *Aviation.* the airport code for Buenos Aires, Argentina.

Buerger precession method see PRECESIUM METHOD.

Buerger's disease *Medicine.* an inflammation and destruction of the blood vessels, primarily found in young men; characterized by a swelling or festering of the lower extremities leading to impaired circulation and gangrene.

buetschlite *Mineralogy.* $3K_2CO_3{\cdot}2CaCO_3{\cdot}6H_2O$, a probably hexagonal mineral that is found as clinkers formed by the fusion of wood ash.

BUF *Aviation.* the airport code for Buffalo, New York.

Buffalo *Aviation.* **1.** a popular name for the F2A fighter. Also, **Brewster Buffalo. 2.** a popular name for a small civilian cargo aircraft made by DeHaviland of Canada.

buffalo *Vertebrate Zoology.* **1.** a general name for several species of wild oxen in the family Bovidae, including Old World buffaloes such as the water buffalo or Cape buffalo. **2.** in popular use, the American bison.

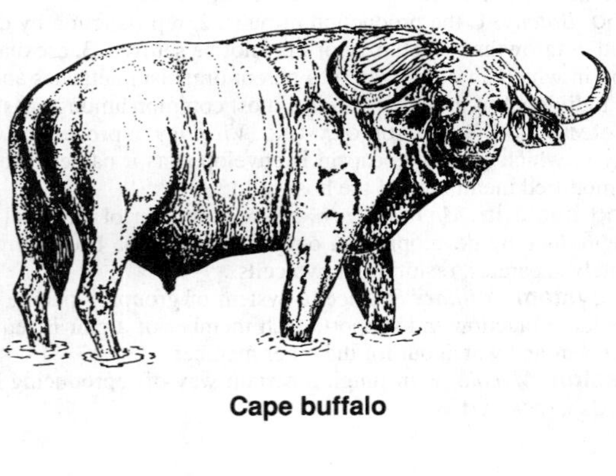

Cape buffalo

buffer *Computer Technology.* an intermediate storage area used to compensate for differences in rates of data flow when information is being transmitted from one computer device to another. *Electricity.* a circuit or other component used to prevent undesirable electrical interaction between circuits or components. *Chemistry.* a solution containing a weak acid and a conjugate base of this acid (or, less commonly, a weak base and its conjugate acid); it resists change in its pH level when an acid or a base is added to it, because the acid neutralizes an added base and vice versa. Also, BUFFER SOLUTION. *Agronomy.* organic matter or a carbonate and phosphate compound in the soil that preserves hydrogen-ion concentrations and resists change in pH value. *Ecology.* see BUFFER SPECIES.

buffer amplifier see ISOLATION AMPLIFIER.

buffer capacitor *Electronics.* a capacitor used to suppress voltage surges that might damage other parts in the circuit. It is connected across the secondary of a vibrator transformer or between the anode and cathode of a cold-cathode rectifier tube.

buffer capacity *Chemistry.* the amount of an acid or base that must be added to a solution to produce a given change in pH.

buffered computer *Computer Technology.* a computer system that makes use of buffers in conjunction with its input/output control system to accommodate the different rates at which data is transferred.

buffered device *Computer Technology.* a peripheral device, such as a printer or teleprinter, that is capable of serving as a buffer.

buffered I/O channel *Computer Technology.* a control channel for most peripheral devices that has associated temporary storage areas for input and output data; the channel controls the input/output operation, freeing the CPU for other processing.

buffered terminal *Computer Technology.* **1.** any of a variety of user terminals that can accommodate differing data transmission rates by temporarily storing incoming data. **2.** a character display, interactive terminal with a local memory holding the text displayed on the screen.

buffer element *Electricity.* an inverting driver circuit element of low impedance.

buffering action *Chemistry.* resistance to change in either acidity or alkalinity.

buffering agent *Food Technology.* any chemical added to a processed food to prevent marked changes in its pH. The best buffering agents combine a weak acid with an acid salt of a strong base.

buffer level *Design Engineering.* the level to which an intermediate storage unit has been filled.

buffer pooling *Computer Programming.* a buffering technique in which a number of buffers are available to the input/output control system; as a physical record is produced, a buffer is drawn from the pool and used to hold the record; after the record has been transmitted, the buffer is returned to the pool.

buffer solution *Chemistry.* see BUFFER.

buffer species *Ecology.* an animal that is introduced into an area to provide a food supply for predators and thus reduce the effect of their predation on a more desirable species.

buffer storage *Computer Programming.* any storage device or register used for temporary storage of data or for synchronization between external and internal storage forms.

buffer strip *Agronomy.* a strip of grass, clover, or another such low crop, grown between strips of a row crop such as corn to control erosion.

buffer time *Transportation Engineering.* the margin of safety between actual headway and minimum allowable headway.

buffer utilization *Design Engineering.* the percentage of capacity of a storage buffer.

buffer zone *Aviation.* an airspace safety zone between adjacent traffic lanes or holding patterns. *Forestry.* a strip of land along roads, trails, or waterfronts that is maintained to enhance aesthetic values. *Computer Programming.* an area in computer memory temporarily assigned to store input or output data.

buffet boundary *Aviation.* a point, expressed in terms of the lift coefficient and speed of a given aircraft, beyond which buffeting will occur.

buffeting *Aviation.* an irregulatr beating or oscillation of an aerodynamic body caused by turbulence, an excessive angle of attack, or an attempt to fly at a Mach number exceeding the specifications for a subsonic aircraft.

buffing *Engineering.* the smoothing, polishing, or removing of layers from a surface, usually by means of a rotating wheel of soft cloth impregnated with a liquid or an abrasive substance.

Buffon, Comte de [boo´fōn˝] 1707–1788, French naturalist; wrote first comprehensive natural history; pioneer of evolutionary theory.

Buffon's needle problem *Mathematics.* the problem of determining the likelihood that a needle dropped on a equidistant pattern of parallel lines will intersect one of the lines. If the needle is chosen to have length $c - \delta$, where c is the distance between the lines, the probability is given by $2\delta/(\pi c)$. This is a classic problem in what is now known as geometrical measure theory.

buffy coat *Hematology.* the thin yellowish layer of leukocytes overlying the packed red cells in centrifuged blood. Also, **buffy crust.**

Bufonidae *Vertebrate Zoology.* a family of amphibians in the order Anomocoela, including the true toads.

bufotenine *Organic Chemistry.* $C_{12}H_{16}N_2O$, a hallucinogenic, colorless crystalline solid that boils at –4°C; insoluble in water and soluble in alcohol. It is a toxic alkaloid found in the skin secretions of the common toad and also made synthetically; it is hallucinogenic.

bufotoxin *Toxicology.* a poison produced in the skin glands of the European toad, *Bufo vulgaris,* and other toads.

bug *Invertebrate Zoology.* a common name for any creeping or crawling insect in the order Hemiptera; true bugs have beaks for sucking and piercing. *Computer Programming.* an error in either the mechanics or the logic of a computer program. *Engineering.* any malfunction in a system that is as yet untraced or uncorrected.

bug dust *Mining Engineering.* the fine coal or other material left over from a boring or cutting of a drill, a mining machine, or a pick.

bug strip *Building Engineering.* the vertical weatherstripping at the edge of a screen door that seals the opening when the door is latched at the opposite side.

buhrstone *Petrology.* a porous, silicified, fossiliferous limestone with abundant cavities previously occupied by fossil shells; formerly used as a millstone. Also, MILLSTONE.

buhrstone mill *Mechanical Engineering.* a grain-processing machine in which the grinding surfaces are of buhrstone.

Build *Robotics.* a programming language to control robots.

building *Building Engineering.* a fixed structure built for human use and occupancy.

building code *Civil Engineering.* governmentally adopted specifications for the minimum requirements regarding the design and construction of buildings.

building dock *Civil Engineering.* a type of graving dock or basin in which ships are constructed, and which is filled after construction to float out the ship.

building engineering *Engineering.* the use of engineering principles and practices in the construction of various types of buildings.

building line *Civil Engineering.* a setback line that establishes by law the limits of a building envelope.

building noise *Acoustics.* a term for background noise that is inherent in a particular sound-recording location, such as the noise of a central air-conditioning unit.

building-out circuit *Electricity.* a short section of transmission line that shunts another line for matching and tuning. Also, **building-out section**.

building-out network *Electricity.* a network connected to a base network so that the combination will simulate the sending-end impedance of a line that has a termination other than that for which the basic network was designed.

buildup or **build-up** *Military Science.* the process of bringing a unit or force up to its required personnel strength and level of supply. *Metallurgy.* in electrodeposition, the excessive deposit occurring at loci of high current density.

buildup pressure *Petroleum Engineering.* the increase of bottom-hole pressure up to an equilibrium value in a shut-in gas or oil well.

buildup sequence *Metallurgy.* the order in welding in which the beads are deposited.

buildup test *Petroleum Engineering.* a test to measure the effective drainage radius of a wellbore and the presence of permeability barriers or other deterrents to production; for a certain period of time, the well is closed and a bottom-hole pressure bomb is used to record pressure.

built-in antenna *Electromagnetism.* an antenna that is enclosed inside the cabinet of a radio receiver unit or television.

built-in check *Computer Technology.* an automatic function performed by the hardware circuitry of a computer, including input/output checks, instruction checks, and parity checks.

built-in function *Computer Programming.* a procedure provided with a programming language that can be invoked by the programmer by naming the function and providing the argument(s); examples are square root and trigonometric functions. Also, **built-in procedure.**

built terrace see ALLUVIAL TERRACE.

built-up *Ordnance.* of or relating to a gun barrel formed by two or more cylinders shrunk one over the other rather than by a single cylinder.

built-up beam *Engineering.* a steel beam fashioned and built up in layers by welding, riveting, or gluing rather than being formed by rolling.

built-up fraction *Graphic Arts.* a fraction, such as 3/4, whose numerator and denominator are typeset on two separate blocks of type, with half the slash typeset on each block. Also, SPLIT FRACTION or PIECE FRACTION.

Building Engineering

Building Engineering, as the term implies, is involved with the construction of every type of structure from single-family dwellings to skyscrapers, bridges, tunnels, roads and even space stations. The building engineer is part architect and part civil and mechanical engineer.

This discipline requires a thorough understanding of construction techniques, materials, and structural, mechanical, and electrical systems, as well as familiarity with building codes and health and environmental regulations.

The building engineer's most demanding problem is scheduling both the delivery of components and the work assignments of various trades in order to produce a smooth and uninterrupted progress of the project. These aspects are closely related to the engineer's ability to control the costs of construction, because incorrect scheduling may be responsible for laborers standing idle or possibly for the destruction and eventual rework of prematurely completed assignments.

Though the building engineer is seldom required to design the various parts of the project, it is her/his responsibility to know how these parts fit together and to advise the designers of problem areas. It is a frequent occurrence that the beams and columns of the structural engineer interfere with the placement of the heating, air conditioning, and plumbing systems of the mechanical engineer.

The building engineer is also responsible for the health and safety of the construction crew. Accidents caused by poorly designed scaffolding, falling objects, or unsafely operated equipment are all too frequent in this industry. Within this context, the engineer must also be cognizant of the dangers to nearby structures produced by excavation and blasting as well as of the inconvenience to neighbors resulting from the noise, commotion, and traffic that are inseparable parts of construction projects.

The recent emphasis on protection of the environment rests on the shoulders of the engineer when choosing between materials previously approved but currently considered harmful, such as lead or asbestos, and newer, safer substitutes. Another environmental issue arises from the disposal of the waste material produced by the project.

It can be seen from the above discussion that building engineering encompasses a wide range of analytical and practical information. Unfortunately, few universities offer degree programs in building engineering. Those desiring to work in this field usually earn degrees in civil engineering and obtain much of the required background from "on the job" experience.

Robert A. Heller
Professor of Engineering Science and Mechanics
Virginia Polytechnic Institute and State University

bulb *Botany.* a modified shoot, consisting of a reduced stem bearing concentric layers of fleshy leaves protected by scale leaves; it functions for storage, reproduction, or carrying the plant through seasons unfavorable for growth. *Electricity.* the glass envelope that encloses the elements of an incandescent lamp, vacuum tube, gas tube, or photocell.

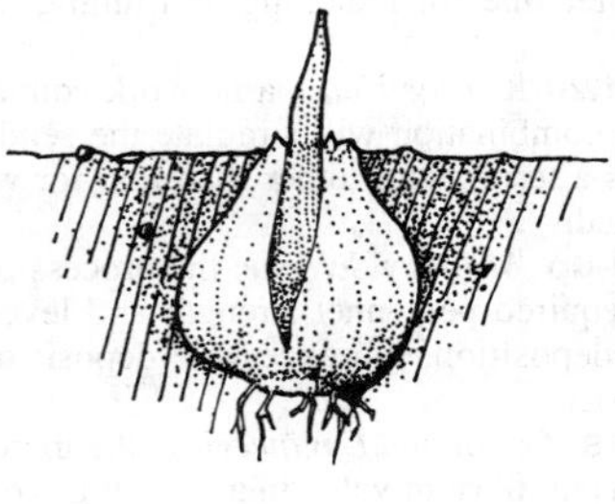

bulb

bulb angle *Design Engineering.* an angle iron with a bulbous thickening at one end.

bulbar paralysis *Medicine.* a degenerative condition marked by progressive paralysis and atrophy of muscles of the lips, tongue, mouth, pharynx, and larynx.

bulb glacier see EXPANDED-FOOT GLACIER.

bulbil *Botany.* a small, bulblike organ that may grow from a leaf axil, an inflorescence, or a stem base, typically swollen with reserve food materials.

Bulbochaete *Botany.* a genus of freshwater green algae of the family Oedogoniaceae, characterized by multibranched filaments in which each cell bears a long hair with a bulbous base.

bulb of force *Archaeology.* a swelling on a stone flake at the point where it has been struck to detach it from a core. Also, **bulb of percussion.**

bulb of the penis *Anatomy.* a rounded enlargement of the corpus spongium that encloses the base of the male urethra.

bulbospongiosus *Anatomy.* a voluntary muscle in the area between the anus and the male genitals that can contract to expel urine and semen from the part of the urethra in the penis.

bulbourethral gland *Anatomy.* either of two pea-shaped glands below the prostate gland that add fluid to the semen during ejaculation.

bulbous *Science.* having the form or nature of a bulb; resembling a bulb. *Botany.* having or growing from bulbs.

bulbous bow *Naval Architecture.* a projecting bulbous fairing of a ship's bow below the waterline; employed in large, fast ships.

bulbul *Vertebrate Zoology.* a small, gregarious bird of the family Pycnonotidae, having short wings and squared or forked tails, and sometimes destructive to crops; native to the Old World tropics.

bulbus *Medicine.* a rounded mass or enlargement.

bulbus oculi see EYEBALL.

bulgecin *Biochemistry.* a glycopeptide that produces a bulge in the cells of susceptible bacteria when joined with β-lactam antibiotic.

bulgur *Food Technology.* wheat that has been boiled, dehydrated, lightly milled, and cracked. Also, **bulghur.**

bulimia [bə lēm′ē ə] *Psychology.* a mental disorder occurring predominantly in adolescent females, characterized by episodes of binge eating that continue until terminated by abdominal pain, sleep, or self-induced vomiting; such binges usually alternate with periods of normal eating or fasting.

bulimic [bə lēm′ik] *Psychology.* relating to or affected by bulimia.

Buliminacea *Invertebrate Zoology.* a superfamily of deepwater amoeboid, protozoan formaniferans in the suborder Rotalina.

Bulitian *Geology.* a North American geologic stage of the Paleocene epoch, occurring after the Ynezian and before the Penutian.

bulk *Graphic Arts.* **1.** the thickness of a particular type of paper, expressed in sheets per inch. **2.** the thickness of a book, excluding the cover, expressed in inches.

bulk acoustic wave *Acoustics.* a total acoustic wave, consisting of the first propagation and all subsequent propagations of the wave, which can be measured for ultrasonic sound in a superfluid such as helium.

bulk carrier *Naval Architecture.* a vessel designed to carry undifferentiated bulk cargo, such as grain or ore.

bulk density *Materials Science.* the weight per unit volume of a material; used especially in reference to powdered or granulated substances.

bulk diode *Electronics.* a semiconductor microwave diode that utilizes the bulk effect.

bulked yarn *Textiles.* yarn that has been treated chemically to introduce air spaces between the fibers and increase its loft. Also, TEXTURED YARN.

bulk effect *Electronics.* an effect that takes place within the entire bulk of a semiconductor material, rather than in a localized region or junction. Thus, **bulk-effect device.**

bulk eraser *Electromagnetism.* a device used to delete an entire reel of recorded magnetic tape.

bulk factor *Engineering.* the ratio of the volume of the loose powdered material used in molding a plastic article to the volume of the finished article.

bulk flotation *Mining Engineering.* the process of intentionally raising a multimineralized froth in order to concentrate more than one valuable mineral in a single operation.

bulk fractionating *Materials Science.* the process of separating bulk polymers by exploiting solubility differences among its fractions differing in mole mass.

bulkhead *Naval Architecture.* **1.** an inboard vertical partition capable of providing structural strength or preventing flooding. **2.** any vertical wall or partition aboard ship. *Aviation.* a partition or frame perpendicular to the longitudinal axis of a flight vehicle and serving to divide, support, or give shape to the fuselage, hull, or float. *Building Engineering.* **1.** a partition built in a subterranean environment such as a conduit or tunnel to impede water flow. **2.** a retaining structure of timber, steel, or reinforced concrete used to shore up land areas adjacent to water bodies, especially harbors. **3.** the horizontal or incline outside a door above a stairway which leads to a cellar. *Mining Engineering.* a tight wood, rock, mud, or concrete partition for protection against gas, fire, and water hazards.

bulkhead deck *Naval Architecture.* the deck laid along the top of a ship's watertight bulkheads.

bulkhead line *Civil Engineering.* the farthest line offshore to which a structure may be placed and not interfere with navigation.

bulking book paper *Graphic Arts.* any paper designed to provide maximum bulk for a given number of sheets.

bulking power *Textiles.* the ability of a textile fiber to increase the volume of a finished fabric.

bulking value *Chemical Engineering.* the relative capacity of a pigment or other substance to increase the volume of paint.

bulk insulation *Engineering.* a type of insulation that impedes the flow of heat by the use of many air pockets or by its imperviousness to heat radiation.

bulk lifetime *Solid-State Physics.* the average length of time that elapses before a minority charge carrier recombines with the carriers of opposite charge in a semiconductor material.

bulk memory *Computer Technology.* storage devices, such as direct-access disks and sequential-access magnetic tapes, that are used to store large volumes of information in conjunction with high-speed memory.

bulk mining *Mining Engineering.* a mining process that indiscriminately collects both low- and high-grade ore.

bulk modulus *Mechanics.* the ratio of the change in pressure applied on a body to the corresponding fractional change in volume that this pressure produces; usually denoted by *B* or *K*. Also known by various other names, such as MODULUS OF VOLUME ELASTICITY, HYDROSTATIC MODULUS, COMPRESSION MODULUS.

bulk modulus of elasticity see BULK MODULUS.

bulk photoconductor *Electronics.* a photoconductor having high-power handling capabilities and other unique properties depending on the semiconductor and doping materials used, such as cadmium, selenite, germanium, and silicon.

bulk plant *Petroleum Engineering.* a wholesale facility for receiving and distributing petroleum products, equipped with tank car unloading facilities, warehouses, truck loading racks, and railroad sidings. Also, **bulk terminal.**

bulk polymerization *Materials Science.* polymer production in the absence of a solvent, in which the monomer, initiator, and polymer are all mutually soluble in a single phase.

bulk provenience *Archaeology.* the location (provenience) of a group of similar objects by type of material and by level or surface.

bulk resistor *Electronics.* an integrated-circuit resistor that uses the N-type layer of a semiconducting substrate as a noncritical high-value resistor.

bulk strain *Mechanics.* the fractional change in volume of a body under compression.

bulk strength *Mechanics.* the strength per unit volume of a solid.

bulk stress *Mechanics.* an external force that tends to cause a change in the volume of a body.
bulk temperature *Thermodynamics.* a cross-sectional average temperature of a flowing fluid over its flow area, which varies with different specific tube or duct geometries. Also, **bulk mean temperature.**
bulk transport *Mechanical Engineering.* any mechanical system for conveying bulk materials, such as cement, coal, or grain.
bulk volume *Materials Science.* the total volume of a material, including voids.
bull *Agriculture.* a mature male bovine animal. *Vertebrate Zoology.* the mature male of certain other large mammals, such as the moose, elephant, or whale.
bull. let it boil. (From Latin *bulliat.*) Also, **Bull.**
bullate *Botany.* having a blistered or puckered appearance.
bull bit *Mining Engineering.* a flat drill bit.
bulldozer *Mechanical Engineering.* a tractor-driven machine with a broad, blunt, horizontal, front-mounted blade that is used to push soil, rock, and other excavated material. *Metallurgy.* a machine used for bending, forming, and punching narrow plates and bars.
Bullera *Mycology.* a genus of fungi belonging to the class Hyphomycetes; found in certain plant material and characterized by respiratory metabolism, budding cells, and ballistospores.
bullet *Ordnance.* **1.** a cone-shaped or round piece of lead, steel, or other metal designed to be fired from a rifle, pistol, or other such small-caliber weapon. **2.** this piece together with its casing; a cartridge. *Engineering.* **1.** a weight, usually bullet-shaped, on a wire rope or line that snags retrievable devices, such as core barrels, from a borehole. **2.** a scraper with an adjustable spring blade that removes debris from the walls of a pipeline. **3.** a bullet-like cylinder containing nitroglycerin that is exploded in a borehole. **4.** see TORPEDO.
bullet drop *Ordnance.* the vertical drop of a bullet in flight.
bullet group *Ordnance.* see GROUP.
bulletin board *Computer Technology.* a service in a computer network that provides a means for members of the network to enter and store brief messages to be posted for other members to read.
bullet jacket *Ordnance.* see JACKET.
bullet nose *Mathematics.* the graph in the plane of $a^2/x^2 - b^2/y^2 = 1$.
bullet perforator *Petroleum Engineering.* perforating equipment in the shape of a tubular gun that fires bullets through the casing and cement at specific depths in a well; holes are formed through which formation fluids are able to enter the wellbore.
bullet planting *Forestry.* the setting out for growing of young trees grown in bullet-shaped plastic tubes, from which they are ejected by a spring-loaded gun into prepared holes in the ground.
bulletproof *Ordnance.* able to resist or withstand the impact of a bullet. Thus, **bulletproof glass, bulletproof armor,** etc. *Computer Programming.* of a program, resistant to users' mistakes. Also, BOMBPROOF.
bulletproof vest *Ordnance.* a bulletproof garment worn to protect the chest and upper body.
bullet splash *Ordnance.* the scattering of small metal fragments produced by the impact of a bullet against armor plate or another hard object.
bulletwood *Materials.* **1.** the very hard and durable wood of the balata tree of Guyana. **2.** the wood of the gutta-percha tree of Malaya, which is deep red in color with a fine, open grain.
bullfinch *Vertebrate Zoology.* a European songbird, genus *Pyrrhula* of the family Fringillidae, including several short, stocky, congenial species from Eurasia.
bullfrog *Vertebrate Zoology.* a deep-voiced, heavy-bodied frog, especially *Rana catesbeiana* of the family Ranidae, the largest North American frog.

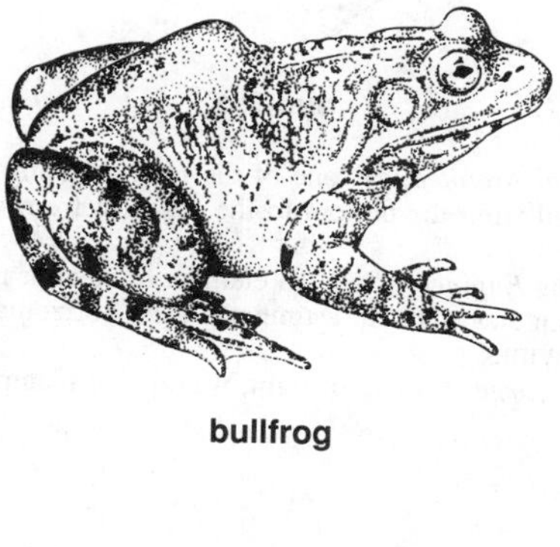

bullfrog

bullhorn *Mechanical Devices.* a portable, usually battery-operated loudspeaker.
bulliform *Science.* bubble-shaped.
bulliform cell *Botany.* any of the large, thin-walled cells in the epidermis of most monocot leaves, which change turgor to unroll developing leaves as well as to roll and unroll mature leaves in order to regulate water loss. Also, MOTOR CELL.
bullion *Metallurgy.* **1.** unwrought refined gold or silver. **2.** in the refining of precious metals, an intermediate product containing recoverable metal values.
bullnose *Building Engineering.* a rounded or obtuse exterior angle such as a corner between two walls, or at a window return or door frame.
bullous emphysema *Medicine.* a type of emphysema in which fluid-filled blisters, or **bullae,** are found in the lung tissue.
Bullpup *Ordnance.* a type of air-to-surface command-guided rocket-propelled missile capable of carrying both conventional and nuclear warheads; officially designated **AGM-12.**
bull rake see SWEEP RAKE.
bull's-eye *Ordnance.* **1.** the center of a practice target, usually a dark, solid circle. **2.** the center or central area of a military target. **3.** a shot, bomb, missile, or the like that hits the center of the target; a direct hit.
bull's-eye rot *Plant Pathology.* a disease of apples, caused by the fungus *Neofabraea malicorticis* or *Gloesporium perennans* and characterized by the appearance of spots that resemble eyes on the fruit.
bull's-eye squall *Meteorology.* a fair-weather squall off the coast of South Africa, named for the rather peculiar appearance of a small isolated cloud marking the top of the invisible vortex of the storm.
bull wheel *Mechanical Engineering.* the largest and usually strongest wheel or gear of a machine. Similarly, **bull gear.**
bulwark *Naval Architecture.* a solid portion of a ship's side rising above the main deck; it provides protection against heavy seas and helps prevent sailors and passengers from washing overboard.
Bulygen number *Thermodynamics.* a dimensionless number used in the analysis of evaporative heat transfer.
Bumastus *Paleontology.* a genus of large (up to 15 cm) trilobites in the suborder Illaenina and family Scutelluidae; distinguished by a smooth and strongly convex cephalon and pygidium.

bumblebee

bumblebee *Invertebrate Zoology.* any of several large, hairy, hole-dwelling social bees of the genus *Bombus* in the family Apidae.
bumblefoot *Veterinary Medicine.* **1.** a dietary or trauma-based disease of adult birds that is often caused by *Staphylococcus aureus* and manifested by abscess formation on the soft portions of the foot and consequent lameness. **2.** a foot abscess occurring in sheep.
bummock *Oceanography.* a downward projection from a sheet of ice; the underwater counterpart of a hummock.
bump *Aviation.* a local air disturbance, or the rise and fall of an aircraft caused by such a disturbance. Also, AIR BUMP. *Geology.* see CRUMP.
bump Cepheid *Astronomy.* a Cepheid variable star that shows a secondary bump in its light curve.
bump contact *Electronics.* a large-area contact for alloying directly to the substrate of a transistor for mounting or interconnecting uses.
bumper *Engineering.* **1.** a horizontal bar of metal or other material across the front and back of a car, truck, or other motor vehicle, serving to protect the body of the vehicle from collision damage. **2.** a device used in drilling as a stay or to dislodge cable tools. *Metallurgy.* in casting, a machine used to pack molding sand into a container by repeated blows.

bumping *Chemistry.* the uneven boiling of a liquid mixture caused by large bubbles of highly volatile components that escape rapidly and irregularly. *Mechanical Engineering.* a low-frequency knocking within a furnace or other device, indicating unstable combustion. *Metallurgy.* **1.** the process of forming a dish in a metallic material by repeated blows. **2.** the process of forming a head, as of a rivet. **3.** in sheet metal fabrication, separating the sheet at a seam. **4.** in casting, packing molding sand.

BUN blood urea nitrogen.

buna rubber *Materials.* a synthetic rubber that is a copolymer of acrylonitrile and butadiene and has excellent fuel and oil resistance.

bunched pair *Electricity.* a set of wire pairs tied together or marked for identification.

buncher gap see INPUT GAP.

buncher resonator *Electronics.* the resonator in a velocity-modulated tube, next to the cathode, where the faster electrons catch up with the slower ones to produce bunches of electrons.

bunching *Electronics.* the succession of electron groups during the flow of electrons from cathode to anode of a velocity-modulated tube, resulting from the differences of electron transit time produced by the velocity modulation. *Transportation Engineering.* a buildup of vehicles beyond the number needed or scheduled.

bunching voltage *Electronics.* in a velocity-modulated tube, the peak value of a radio-frequency voltage between the grids of the buncher resonator.

bunchy top *Plant Pathology.* a viral disease of plants, characterized by a crowding of twigs and leaves at the upper stem and a shortening of the stem internodes.

bund *Civil Engineering.* an embankment or embanked thoroughfare along a body of water.

bundle *Botany.* a group of cells that act in conduction and support in the stems and leaves of plants. *Computer Science.* see BUNDLED PROGRAM. *Optics.* a concentrated group of light rays. *Mathematics.* a triple (E, B, π) consisting of two topological spaces E and B and a continuous surjective mapping $\pi: E \to B$. E is called the total space and B is called the **base of the bundle.** $\pi^{-1}(x)$ is a topological space for all $x \in B$, and is called the fiber at x. The simplest bundle is the **Cartesian bundle** $(B_1 \times B_2, B_1, \pi_1)$, where π_1 is the first projection defined by $\pi_1(x_1, x_2) = x_1$ for all $x_1 \in B_1$ and $x_2 \in B_2$. Used to generalize topologcal products; e.g., a Möbius strip cannot be described globally as a topological product.

bundle branch *Anatomy.* either of the two branches of the bundle of specialized impulse-conducting tissue that extend from the atrioventricular bundle through the myocardium of the ventricle.

bundle burial *Archaeology.* a secondary burial practice in which bones are collected after the flesh has decayed and then reburied in a vessel or other grave.

bundled program *Computer Science.* an application program developed by a computer manufacturer and sold along with hardware at no extra cost.

bundle of His *Anatomy.* the collection of impulse-conducting fibers located in the atrioventricular septum of the heart between the atrioventricular node and the bundle branches. (Named for Wilhelm *His,* Jr., 1863–1934, German physician.)

bundle scar *Botany.* a mark within a leaf scar that indicates the end of a vascular bundle.

bundle sheath *Botany.* one or more layers of cells surrounding a vascular bundle.

bundling machine *Mechanical Engineering.* a machine that can be preset to count items, load them into cartons, and wrap the cartons for shipping.

bungarotoxin *Toxicology.* a neurotoxin component of the venom of certain poisonous snakes of the genus *Bungarus* (the kraits). Three forms have been identified, α, β, and γ; **α-bungarotoxin** binds with and immobilizes acetylcholine receptors, thereby blocking neuromuscular transmission.

bunion *Medicine.* an irregular protrusion at the first metatarsal caused by thickening of the joint, shifting the great toe to the side.

bunker *Ordnance.* a fortified structure, either wholly or partly underground, that is used to defend or conceal a gun position, protect personnel, and so on.

bunkering *Engineering.* the storage of fuel in a receptacle from which it can later be withdrawn in order to stoke a furnace or engine or feed a fuel tank.

bunker silo *Agriculture.* an above-ground horizontal silo consisting of two parallel walls with movable dividers, between which cattle pass for feeding.

Bunn chart *Materials Science.* a series of calculated curves used for the graphical indexing of crystal lattices from X-ray diffraction data.

bunodont *Anatomy.* having molars with small, rounded cusps.

bunolophodont *Vertebrate Zoology.* **1.** of teeth, having cone-shaped outer cusps and transversely ridged inner cusps. **2.** having such teeth, as do tapirs.

Bunonematoidea *Invertebrate Zoology.* a superfamily comprising some of the more unusual nematodes, bacterial or fungal feeders, common in compost and other decaying matter.

bunoselenodont *Vertebrate Zoology.* **1.** of teeth, having blunt, cone-shaped inner cusps and longitudinal, crescent-shaped outer cusps. **2.** having such teeth, as did the extinct titanotheres.

Bunsen, Robert Wilhelm 1811–1899, German chemist; with Kirchhoff, discovered spectrum analysis and the elements cesium and rubidium; invented Bunsen burner.

Bunsen burner *Engineering.* a gas burner having an adjustable air inlet that allows the heat of the flame to be modified; widely used in laboratories.

Bunsen ice calorimeter *Engineering.* a device used to determine the amount of heat released during the melting of a compound, by measuring the increase in volume of an ice-water solution surrounding the compound.

bunsenite *Mineralogy.* NiO, a green, transparent, cubic mineral occurring in octahedral crystals, having a specific gravity of 6.7 to 6.9 and a hardness of 5.5 on the Mohs scale; found in oxidized zones of nickel-uranium veins.

Bunsen-Kirchhoff law *Spectroscopy.* a law stating that every atomic species has a characteristic emission spectrum consisting of bright lines and a characteristic absorption spectrum consisting of dark lines.

Bunsen photometer or **Bunsen screen** see GREASE SPOT PHOTOMETER.

bunt *Plant Pathology.* a disease of wheat caused by either of two species of the fungus *Tilletia* in which the wheat grains are replaced by odorous smut spores.

buntal *Textiles.* a fine white fiber that is made from the talipot palm. Also, BURI.

Bunter *Geology.* a European geologic stage of the Lower Triassic period, occurring after the Permian period and before the Muschelkalk stage.

Bunte salts *Organic Chemistry.* $RSSO_2ONa$, water-soluble salts of various alkyl or aralkyl thiosulfuric acids; they are used against coccidia parasites.

bunting *Vertebrate Zoology.* a stout-billed bird of the family Fringillidae, distinguished by strong head markings, an angular gape, and a bony knob on the palate; found in both the Old World and the U.S.

bunting

buntline *Naval Architecture.* one of the ropes running across the front of a square sail from the bottom to the yardarm from which the sail is hung.

bunton *Mining Engineering.* **1.** an element of steel or timber in the lining of a rectangular shaft. **2.** a timber placed horizontally in a circular shaft. Also, DIVIDER.

Bunyan bag *Surgery.* a bag of light, waterproof material used to cover wet dressings.

Bunyaviridae *Virology.* a large family of enveloped, ssRNA-containing viruses that infect a wide host range, mainly of vertebrates and arthropods, with transmission by numerous vectors including aerosol infection.

Bunyavirus *Virology.* a genus in the family Bunyaviridae with a host range of various vertebrates, for which mosquitoes serve as the primary vector.

buoy *Navigation.* a floating, anchored object used to aid in navigation by marking navigable courses or the position of natural obstructions.

buoyage *Navigation.* a system of buoys.

buoyancy *Fluid Mechanics.* a force, due to fluid pressure, that acts on objects immersed in or floating on the surface of a liquid.

buoyancy parameter *Fluid Mechanics.* a dimensionless number consisting of the ratio of the Grashof number divided by the square of the Reynolds number.

buoyancy pontoons *Petroleum Engineering.* pontoons that hold up offshore pipelines while they are welded together on the water, then are removed to allow the pipeline to sink into the correct position.

buoyancy tank *Naval Architecture.* a sealed air tank aboard a boat, intended to provide buoyancy if the boat is swamped.

buoyant density *Physics.* the density of a solute as determined by observing a sedimentation gradient. *Virology.* the density at which a virus floats as a band when suspended in a viscous sugar or a heavy metal salt.

buoyant force *Fluid Mechanics.* a force, due to fluid pressure, that holds up an object immersed or floating on the surface of a liquid; it is equal in magnitude to the weight of the object that is above the level of the liquid.

buoy sensor *Acoustical Engineering.* a buoyant device that has one or more hydrophones for picking up sound in the ocean and then transmitting radio frequencies with information regarding these sounds.

Buprestidae *Invertebrate Zoology.* a mainly tropical family of various, usually green, blue, or copper-colored metallic beetles, many being destructive wood borers as larvae.

Buprestoidea *Invertebrate Zoology.* the jewel beetles, a superfamily of metallic, colorful wood-boring insects; there are 15,000 known species, many of which are serious fruit tree pests.

BUR *Aviation.* the airport code for Burbank, California.

bur. bureau.

Burali-Forti paradox *Mathematics.* a contradiction based on the faulty assumption that there exists a set that comprises all the ordinal numbers. If such a set exists, form the supremum of all ordinal numbers, which would then be an ordinal number greater than or equal to every ordinal number. But this is impossible, because for each ordinal number there exists a strictly greater one.

buran *Meteorology.* a strong northeast wind of Russia and Siberia, most frequent in winter.

Burbank, Luther 1849–1926, American horticulturalist; developed hybrid plant species.

burble *Fluid Mechanics.* a separation or breakdown of the laminar flow past a body and the eddying or turbulent flow resulting from this occurrence.

burden *Electricity.* the power drawn from the circuit that connects the secondary terminals of an instrument transformer, expressed in volt-amperes. *Geology.* any rock or earthy material overlying a deposit. *Metallurgy.* **1.** the material melted in a direct arc furnace. **2.** in an iron blast furnace, the ratio of ferrous materials to coke. Also, COVER.

Burdigalian *Geology.* a European geologic stage of the Lower Miocene epoch, occurring after the Aquitanian and before the Helvetian.

buret *Chemistry.* a glass tube marked with a scale, usually open at the top and with a stopcock at the bottom, used for measuring the volume of a liquid as it is dispensed in a titration. Also, **burette.**

burga *Meteorology.* a northeasterly storm in Alaska that brings sleet or snow. Also, BOORGA.

Burger's circuit *Materials Science.* a series of unit translations about a dislocation, in which an addition step is required to complete the circuit due to the presence of the dislocation.

Burger's vector *Materials Science.* the addition step to complete the Burger's circuit; its orientation with the respect to be dislocation line indicates the dislocation type.

Burgess Shale *Geology.* a Middle Cambrian bed of black shale in western Canada that is the site of major discoveries of fossil fauna. *Paleontology.* the well-preserved fossil fauna of soft-bodied Cambrian animals contained in this site, one of the most important fossil finds in paleontological history.

Bürgi, Jobst 1552–1632, Swiss mathematician; developed logarithmic tables and mathematical notation; made astronomical instruments.

burglar alarm *Engineering.* a security system that sets off an alarm when a door or window is moved or opened, or a certain object is tampered with; typically the alarm is activated if there is a break in an electric circuit or a light beam.

Burgundy *Food Technology.* **1.** any wine from the Burgundy region of central France. Most red Burgundy wines are made with Pinot Noir grapes; the best-known white wines are made from Chardonnay grapes. **2.** also, **burgundy.** a wine similar to a red Burgundy, produced in California or elsewhere.

Burhinidae *Vertebrate Zoology.* the stone-curlews or thick-knees, a family of wide-ranging nocturnal shorebirds in the order Charadriiformes, characterized by their seemingly swollen knees.

buri see BUNTAL.

burial goods see GRAVE GOODS.

burial ground *Archaeology.* an area that is set aside by a certain group or culture for the burial of the dead. *Nucleonics.* a land area that is set aside for the burial of sealed drums containing low-level radioactive waste, consisting mostly of contaminated air filters, cleaning rags and paper, tools, glassware, containers, and protective clothing, as well as immobilized wet solids and condensates from laundry effluent and waste concentration.

burial metamorphism *Geology.* a type of metamorphism in which the fabric of the original rock is largely preserved but the mineral composition is generally altered.

burial mound *Archaeology.* a large mound of earth placed over a grave at the time of burial.

burial orientation *Archaeology.* the direction or alignment of the body at the time of burial, especially the direction toward which the head is positioned; this may vary according to the culture involved.

buried *Geology.* not visible; beneath the surface.

buried hill *Geology.* an ancient hill of resistant rock that was later buried by sediments.

buried placer *Geology.* an ancient ore deposit that was buried beneath lava flows, soil, or other material.

buried river *Geology.* a river bed that was buried beneath alluvial deposits, lava, pyroclastic rocks, or till.

buried set-point method *Robotics.* a method of using soft servos to guide a robot's manipulator along the edge of a template.

buried soil see PALEOSOL.

buried suture *Surgery.* a suture that is placed so that it is covered by skin.

burin *Mechanical Devices.* a tool that is used for etching or engraving metal or wood. *Archaeology.* a specialized flake or blade with a chisel edge; used to carve or engrave softer materials such as bone, wood, or antler.

burkeite *Mineralogy.* $Na_6(CO_3)(SO_4)_2$, a white to buff to grayish orthorhombic mineral having a specific gravity of 2.57 and a hardness of 3.5 on the Mohs scale.

Burkitt's lymphoma *Oncology.* a cancerous growth arising in the lymphatic system and spreading to the jaw, the eye socket, the abdominal organs, and the ovaries.

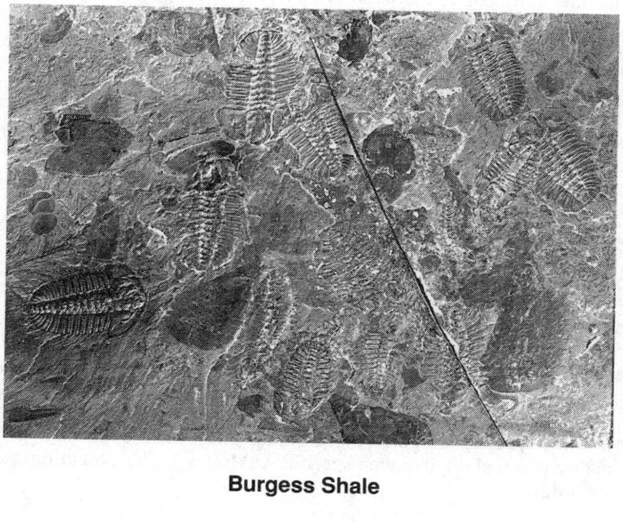

Burgess Shale

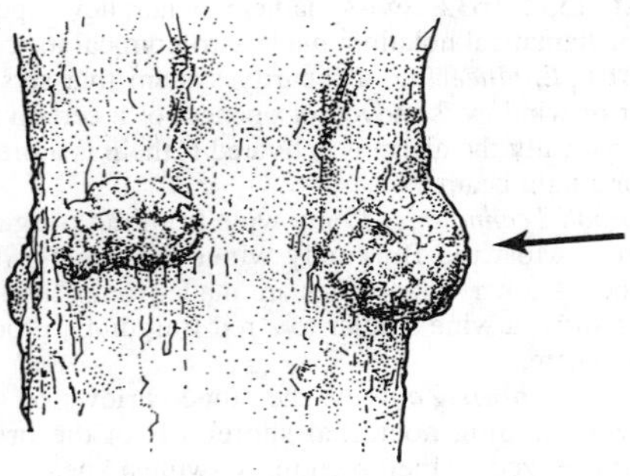

burl

burl *Botany.* on a tree trunk or branch, an abnormal hard, woody outgrowth sometimes due to stimulation by insects.

burlap *Textiles.* a coarse woven fabric made of jute, hemp, or flax. Also, HESSIAN.

burling *Textiles.* a process for removing knots or lumps in the finishing of wool fabrics.

Burmanniaceae *Botany.* a family of mostly saprophytic monocots having scalelike leaves, tuberous rhizomes, and numerous tiny seeds, and usually lacking chlorophyll.

burn *Chemistry.* to undergo fast or slow combustion; oxidize. *Physics.* to undergo fission or fusion. *Medicine.* an injury to tissues that is caused by contact with heat, chemicals, electricity, friction, or radiant and electromagnetic energy; classified as first, second, or third degree, depending upon the severity. *Computer Technology.* to write data into a programmable read-only memory (PROM); so called because some PROMs were written by blowing small fuses in the PROM.

burnable poison *Nucleonics.* a nuclide of large neutron absorption cross section, such as boron, that is incorporated into a nuclear reactor's fuel or fuel cladding to compensate for loss of reactivity as fuel is consumed and fission-product poisons accumulate.

burn cut see PARALLEL CUT.

burner *Mechanical Engineering.* the part of a fuel-burning device, such as a furnace or boiler, in which the fuel and air are mixed and combustion occurs. *Chemical Engineering.* a furnace in which sulfur or sulfide ore is burned to produce sulfur dioxide and other gases. *Aviation.* a combustion chamber or fuel-injection nozzle therein, especially in a jet engine.

Burnet, Sir Frank Macfarlane 1899–1985, Australian physician and virologist; Nobel Prize (with Medawar) for developing theory of acquired immunological tolerance.

burnettize *Engineering.* to douse fabric or timber thoroughly with a zinc chloride solution to prevent decay; an obsolescent process.

Burnett's hypercalcemia see MILK-ALKALI SYNDROME.

Burnham, Daniel Hudson 1846–1912, American architect, designed early skyscrapers including the Flatiron Building (New York City).

burn-in *Engineering.* the preliminary operation of a device or system prior to its actual on-line operation, to stabilize its characteristics and identify early failures. *Graphic Arts.* a process for giving extra exposure to certain areas of a photographic image to increase the density of those areas.

burning block *Forestry.* a designated area in a fire control plan that has sufficiently uniform conditions of trees and other fuel so as to be treated in one particular way under given burning prescriptions.

burning constant *Ordnance.* a figure expressing the relative burning rate of a given type of propellant.

burning glass *Optics.* a lens that converges sun rays onto a small area, producing intense heat at the focus.

burning line *Petroleum Engineering.* a pipeline used for carrying refinery fuel gas, as opposed to gas designated for later processing.

burning off *Forestry.* the process of setting fire under regulated conditions to areas of unwanted vegetation or fuel.

burning period *Forestry.* that part of each 24-hour period when fires spread most rapidly; usually the interval between 10 AM and sundown.

burning point *Engineering.* the lowest temperature at which an inflammatory oil will light when a burning match is held near its exposed surface; used in safety tests.

burning quality *Engineering.* a rating given for the burning performance of a fuel oil, based on specific ASME tests. Thus, **burning-quality index.**

burning velocity *Chemistry.* the normal velocity of the area of burning gas toward the unburnt gas in the combustion of a flammable mixture.

burn-in test *Computer Technology.* a method of testing a computer or other device by applying power to it and running it for a period of time; many failures occur early in the life of a chip, so a chip that survives the burn-in test is likely to be good.

burnish *Engineering.* to rub a surface with pressure, usually by means of a special tool and lubricant, to obtain a smooth, polished finish.

burnisher *Engineering.* any tool designed for burnishing. *Graphic Arts.* a tool designed for polishing gold leaf, used in printing book covers.

burnoff *Meteorology.* the dissipation of fog or low stratus cloud layers by daytime heating from the sun.

burn-off *Metallurgy.* **1.** in the electrodeposition on a nonmetallic base, the unintentional removal of a preexisting electroless deposit. **2.** the removal of a lubricant from a powder compact prior to sintering.

burnout *Engineering.* **1.** the abrupt breakdown of a device, often due to excessive current and subsequent overheating, usually resulting in serious damage. **2.** the moment in the flight of a rocket when the flow of propellant is exhausted or halted. *Aviation.* in a reaction engine, the termination of fuel combustion due to either fuel (or oxidizer) exhaustion or fuel shutoff; the moment when this occurs. *Nucleonics.* in water-cooled reactors, a rupture in fuel cladding, with release of fission products into the coolant, caused by localized heat buildup which results from film boiling that insulates a heat transfer surface. *Psychology.* a popular term for a condition of extreme fatigue and apathy, especially as a result of prolonged stress and overwork.

burn out *Graphic Arts.* **1.** to overexpose, often unintentionally, a photosensitive material. **2.** to intentionally overexpose an area on a press plate so that no tints come up, often done when a line of type is printed over a photographic image.

burnout velocity *Space Technology.* the speed of a rocket vehicle at the moment of burnout. Also, **burnt velocity.**

Burnside boring machine *Mechanical Engineering.* a drilling machine equipped with an apparatus to contain any water that may be tapped.

burnt deposit *Metallurgy.* in electrodeposition, an undesirable deposit caused by excessive current density.

burn-through see JAMMER FINDER.

burnt shale *Mining Engineering.* any carbonaceous shale that has experienced spontaneous combustion after remaining in a colliery tip for a long time, and then has been converted into a coppery slag material. Also, OXIDIZED SHALE, RED DOG.

burnup *Nucleonics.* a measure of the consumption of nuclear fuel in a reactor, usually expressed as the percentage of heavy metal atoms fissioned or the total heat released per unit weight of fuel.

burr *Botany.* the rough, prickly envelope of certain fruits. *Metallurgy.* **1.** the rough edge resulting from a mechanical operation such as cutting or grinding. **2.** a rotary tool with a surface similar to that of a hand file.

burrfish see PORCUPINEFISH.

burr mill *Food Technology.* a type of grinding mill that crushes cereal grain between two ribbed plates.

burro *Vertebrate Zoology.* a small donkey of the family Equidae, used as a pack animal.

burrow *Mining Engineering.* a refuse heap at a coal mine.

bursa *Anatomy.* a sac or pouch containing fluid, located between a tendon and a bone, or between other structures where friction may occur.

bursa of Fabricus *Vertebrate Zoology.* a saclike projection connected to the cloaca in birds, in which lymphocytes develop.

bursectomy *Surgery.* the excision of a bursa.

Burseraceae *Botany.* the bursera family of deciduous tropical and subtropical dicot trees; having alternate, pinnately compound leaves, flowers in clusters, and resin chambers in the bark; many are economically important.

bursicle *Botany.* the purselike or pouchlike structure of the receptacle in some orchids.

bursitis [bur sīt′əs] *Medicine.* inflammation of the bursa.

bursolith *Medicine.* a calculus or concretion in a bursa.

burst *Astrophysics.* an abrupt increase in nonthermal radio emission from a solar flare or from Jupiter. *Ordnance.* **1.** the explosion of a projectile in the air or as it strikes a target or surface. **2.** of a projectile, to explode in this way. **3.** a sustained series of shots fired by an automatic weapon. *Computer Technology.* **1.** to separate sheets of continuous feed paper produced as computer output. **2.** in data communications, an unbroken bit stream. *Electronics.* a sudden increase in the strength of a signal. *Telecommunications.* a short block of transmitted signals or data.

burst amplifier *Electronics.* in a color television receiver, an amplifier stage keyed into conduction and amplification by a horizontal pulse at the arrival of the color-burst signal

burst center see CENTER OF BURST.

burster *Botany.* an abnormal double flower in which the calyx splits or fragments. *Computer Technology.* an off-line mechanical device used to separate pages of single- or multiple-copy continuous-form printed output; may also strip the perforated edges and remove carbon paper. *Ordnance.* **1.** a charge used in chemical ammunition, such as tear gas bombs, to break open the casing and disperse the contents. **2.** see BURSTER COURSE. *Meteorology.* see SOUTHERLY BURSTER.

burster block *Ordnance.* a slab of concrete or other such material used to form a burster course.

burster course *Ordnance.* a layer of hard material used on the surface of a fortification to detonate projectiles before they can penetrate deeply enough to cause great damage. Also, BURSTER, BURSTING LAYER.

burst-forming unit *Cell Biology.* any undifferentiated bone marrow stem cell, such as an erythroblast, that undergoes several cell divisions to develop into the end-stage cell of its pathway.

bursting layer see BURSTER COURSE.

bursting strength *Mechanics.* the ability of a material to withstand pressure without rupture. *Materials Science.* the minimim hydraulic pressure required to burst or rupture a given vessel.

burst mode *Computer Technology.* the mode of operation of a channel that connects input and output devices with the central processor and memory, during which only one device at a time may transmit data over the channel.

burst pedestal *Telecommunications.* a television signal of roughly rectangular shape that is part of a color burst.

burst(ing) point see POINT OF BURST.

burst pressure *Mechanical Engineering.* the highest internal pressure that a vessel can support without bursting.

burst separator *Electronics.* the circuit that separates the color burst from the composite video signal in a color television receiver.

burst size *Virology.* the average number of virus particles per cell released during a lytic one-step growth infection of a host cell culture.

burst source *Astronomy.* a binary star that emits bursts of X-rays.

burst transmission *Telecommunications.* the transmission of data as a sequence of bursts, rather than a continuous stream, in order to economize transmission capacity or equipment.

burton *Mechanical Engineering.* a light hoisting tackle, usually with a single block and a double block.

Burton's sign *Toxicology.* a blue line that develops along the gums following chronic poisoning caused by an ingestion of lead or copper. Also, **Burton's line.** (Named for Henry *Burton,* 1799–1849, British physician.)

Buruli ulcer *Medicine.* a skin infection that is caused by *Mycobacterium ulcerans* and characterized by a small, hard, painless, movable subcutaneous nodule that grows and becomes fluctuant and ulcerates, leaving an undetermined edge.

bus *Transportation Engineering.* **1.** a large, elongated motor vehicle that is fitted with seats and used as a public conveyance. **2.** a passenger automobile that resembles such a vehicle. *Electricity.* **1.** a noninsulated conductor used to carry a large current or to make a common connection between several circuits. **2.** see BUSBAR. *Computer Technology.* a channel or path for transferring data or power from one of many sources to one or more of many destinations.

bus architecture *Computer Technology.* a computer system configuration that uses a shared or multiplexed communication path between the processor and the peripheral devices, permitting multiple devices to transmit data at the same time.

busbar *Electricity.* a heavy bar or strap that is used as a bus.

Buschke-Lowenstein tumor *Oncology.* a tumor that clinically resembles squamous cell carcinoma but microscopically represents a form of condyloma acuminatum; it usually occurs on the uncircumcised penis, but sometimes elsewhere in the anogenital area. Also, GIANT CONDYLOMA ACUMINATUM.

Busch lemniscate *Meteorology.* a point of concentration in the sky of all points at which the polarization plane of diffuse sky radiation is inclined 45° to the vertical. Also, NEUTRAL LINE.

bush *Botany.* a low-growing woody plant, especially one having many separate branches growing from or near the ground. *Ecology.* wild, uncultivated land or open forest.

Bush, Vannevar 1890–1974, American electrical engineer; invented differential analyzer.

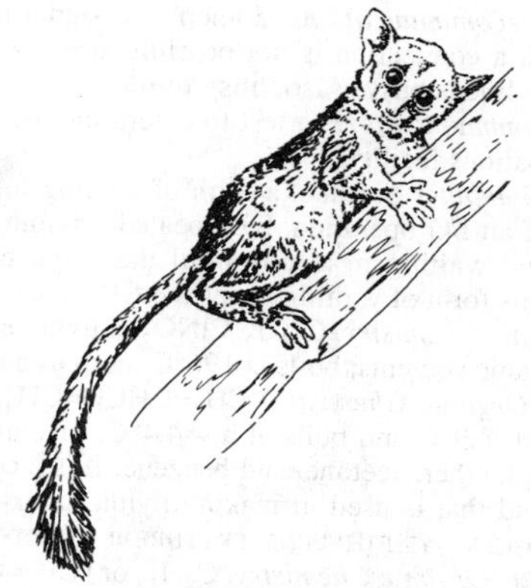

bushbaby

bushbaby *Vertebrate Zoology.* a small arboreal African primate of the family Lorisidae and the genera *Galago* and *Euoticus,* having a long tail, large eyes and ears, and long hind limbs that make it an agile leaper. Also, NIGHT-APE, GALAGO.

bush bear see POTTO.

bushbuck *Vertebrate Zoology.* a sub-Saharan African antelope of the family Bovidae, having a dark coat with white markings under the neck and strips or spots on the sides, straight and spiralled horns, and a compact body.

bush dog *Vertebrate Zoology.* a wild dog, *Speothos vanaticus,* of the family Canidae, found in forests of South America; it is short in stature with a long, stocky body and is largely nocturnal.

bush hammer *Mechanical Devices.* a hammer with one or two serrated faces used for dressing stone.

bushing *Mechanical Engineering.* **1.** a removable, soft-metal lining of a shaft used as a bearing or to limit the size of the opening. **2.** a removable hardened steel tube used as a guide for certain tools or parts, as for a drill rod or valve rod.

Bushnell, David c. 1742–1824, American inventor; invented "Bushnell's Turtle," a forerunner of the submarine.

bush pig *Vertebrate Zoology.* a wild African swine, *Potamochaerus porcus* of the family Suidae, characterized by long, reddish hair, tassled ears, and white cheek patches; known for its habit of ranging for food at night in large groups called sounders.

bushtit *Vertebrate Zoology.* an insectivorous songbird of the family Paridae that is native to western North America, having dull gray-brown plumage.

bus network *Telecommunications.* a network topology in which computers or terminals are connected to one main cable over which data or control signals are passed.

bus rapid transit *Transportation Engineering.* a high-speed bus system operating on an exclusive right-of-way.

bus reactor *Electricity.* a current-limiting reactor for connection between two different buses or two sections of the same bus for limiting and localizing disturbance due to a fault in either bus.

bustard *Vertebrate Zoology.* a large, long-legged Old World or Australian gamebird of the family Otididae in the order Gruiformes, related to cranes and plovers.

buster *Metallurgy.* in forging, a pair of shaped dies, used in preliminary operations. *Mining Engineering.* an expanding wedge used to break down coal or rock.

bustle pipe *Metallurgy.* an annular pipe designed to distribute heat in a blast furnace.

busulfan *Pharmacology.* $C_6H_{14}O_6S_2$, an akylating agent that is used primarily for the treatment of chronic granulocytic leukemia, polycythemia vera, and myeloid metaplasia; its major side effect is bone marrow depression.

busway *Electricity.* a standardized assembly of busbars rigidly supported by solid insulation and enclosed in a sheet-metal housing. *Transportation Engineering.* a roadway or lane reserved for bus rapid transit.

busy *Telecommunications.* of a line, in use and not able to acccept another incoming call. *Computer Science.* **1.** of a device, occupied or unable to perform another operation because the current operation is still in progress. **2.** of the central processing unit, requiring execution of CPU instructions. **3.** describing a variable whose value will be needed later during program execution.

busy signal *Telecommunications.* a telephone signal indicating to the calling party that a connection is not possible at a given time because the called party's line is busy. Also, **busy tone.**

busy test *Telecommunications.* a test to determine the availability of a given communications facility.

busy waiting *Computer Science.* a form of waiting for an event, such as completion of an I/O operation, by repeatedly issuing instructions to test its status and waiting in a loop until the event has occurred; so called because this form of waiting keeps the CPU busy.

butachlor *Organic Chemistry.* $C_{17}H_{26}ClNO_2$, a light yellow oil; soluble in water and organic solvents; boils at 196°C; used as an herbicide.

1,3-butadiene *Organic Chemistry.* CH_2=CHCH=CH_2, a colorless gas that freezes at –108.9°C and boils at at –4.4°C; insoluble in water and soluble in alcohol, ether, acetone, and benzene. It is a commercially important compound that is used in making nylon, latex paints, and synthetic rubber. Also, VINYL ETHYLENE, ERYTHRENE, BIVINYL.

butadiene dimer *Organic Chemistry.* C_8H_{12} or $(CH_2{=}CHCH{=}CH_2)_2$, a flammable liquid diolefin that is very reactive and polymerizes readily; the third ingredient in ethylene-propylene-terpolymer (EPT) synthetic rubbers.

butamben *Pharmacology.* $C_{11}H_{15}NO_2$, a topical anesthetic used to treat painful skin conditions and the discomfort of hemorrhoids.

Butamer process *Chemical Engineering.* a trademark for the process of isomerization of normal butane to isobutane, using hydrogen and a solid, noble metal catalyst of unrevealed components.

butane *Organic Chemistry.* C_4H_{10}, a colorless, highly flammable gas; soluble in water, alcohol, and chloroform; flash point –138.4°C and boils at –0.5°C. It is extremely stable and does not corrode metals or react with moisture. It occurs in natural gas and is produced by cracking petroleum; widely used as a fuel for lighters and other household products, as a component of LPG (liquefied petroleum gas), as a gasoline additive, and for other industrial purposes.

butane dehydrogenation *Chemical Engineering.* the removal of two or four hydrogen atoms from butane to yield butene or butadiene.

butanedioic acid see SUCCINIC ACID.

butanedioic anhydride see SUCCINIC ANHYDRIDE.

butanediol see BUTYLENE GLYCOL.

butanethiol see BUTYL MERCAPTAN.

butane vapor-phase isomerization *Chemical Engineering.* a method of isomerization of normal butane into isobutane, using an aluminum chloride catalyst and hydrogen chloride promoter.

butanoic acid see BUTYRIC ACID.

butanol *Organic Chemistry.* C_4H_9OH, a colorless toxic liquid that is known in four isomeric forms; soluble in water and miscible with alcohol and ether; used as a solvent, paint remover, and for other purposes. Also, BUTYL ALCOHOL, BUTYRIC ALCOHOL.

butanol fermentation see ACETONE-BUTANOL FERMENTATION.

butanone see METHYL ETHYL KETONE.

Butazolidin *Pharmacology.* a brand of phenylbutazone, $C_{19}H_{20}N_2O_2$, a nonsteroidal anti-inflammatory agent; administered to race horses and associated with heightened performance.

butcherbird *Vertebrate Zoology.* any bird that impales its prey on spines or twigs, such as the shrike (genus *Lanius*) or the Australian butcherbird of the family Cracticidae, a stocky, heavy-billed, strongly territorial bird known for beautiful song performed in duet.

butcherbird

butcher linen *Textiles.* a heavy, plain-woven linen or similarly textured synthetic fabric.

butcher marks *Archaeology.* marks made on animal bone by stone tools during butchering; used to associate humans with animal remains for a relative date; they are classified according to form and function as **cut marks, chop marks, and scrapes.**

bute [byoot] *Pharmacology.* a popular name for Butazolidin. See BUTAZOLIDIN.

Butenandt, Adolph born 1903, German chemist; Nobel Prize for discoveries and study of sex hormones.

butene see BUTYLENE.

1-butene *Organic Chemistry.* CH_2=$CHCH_2CH_3$, a colorless, highly flammable gas that is insoluble in water and soluble in most organic solvents; freezes at –185.3°C and boils at –6.3°C; a liquefied petroleum gas that is used in making butadiene and as a solvent and cross-linking agent. Also, α-BUTYLENE, ETHYLETHYLENE.

2-butene *Organic Chemistry.* CH_3CH=$CHCH_3$, a colorless, highly flammable gas that is insoluble in water and soluble in most organic solvents; a liquefied petroleum gas that is used in making butadiene and in the synthesis of four- and five-carbon organic molecules. It occurs in both *cis* and *trans* forms, the *cis* being known as **high-boiling 2-butene** (boiling point 3.7°C) and the *trans* as **low-boiling 2-butene** (boiling point 0.3°C).

***cis*-butenedioic acid** see MALEIC ACID.

Buteoninae *Vertebrate Zoology.* in some systems, a subfamily of the family Acciptridae, including the red-tailed hawk, common buzzard, and bald eagle.

Buthus *Invertebrate Zoology.* a genus of scorpions of the family Buthidae, order Scorpionida, containing a number of dangerous species that occur in warm or tropical regions, including *B. carolinianus,* of the southern United States.

butlerite *Mineralogy.* $Fe^{+3}SO_4(OH){\cdot}2H_2O$, an orange monoclinic mineral occurring in tabular to octahedral crystals, having a specific gravity of 2.55 and a hardness of 2.5 on the Mohs scale; dimorphous with parabutlerite.

Butler oscillator *Electricity.* an oscillator based on a two-stage amplifier with a quartz crystal placed from output to input.

Butler-Volmer equation *Physical Chemistry.* an equation for the rate of an electrochemical reaction, where the relation between current density and activation overpotential is given.

Butomaceae *Botany.* the flowering rush family of emergent aquatic perennial monocots; characterized by a horizontal rhizome, linear and erect parallel-veined leaves, and axillary inflorescences.

butopyronoxyl *Organic Chemistry.* $C_{12}H_{18}O_4$, a yellow to reddish-brown liquid that has an aromatic odor; it is insoluble in water and is miscible with alcohol and ether; it boils at 256–270°C. It is used mainly as an insect repellent. It is toxic when ingested, and may cause liver damage.

2-butoxyethanol *Organic Chemistry.* $HOCH_2CH_2OC_4H_9$, a liquid that boils at 171°C; it is soluble in water, alcohol, ether, acetone, and benzene; it is used in dry cleaning, and as a solvent for oils, greases, and resins.

Butschliellaceae *Botany.* a family of freshwater flagellates usually placed in the order Cryptomonadales, distinguished by the presence of four flagella of equal length and brownish-green to blue-green chloroplasts.

butt the end of something, especially the rear or bottom end; specific uses include: *Ordnance.* **1.** the rear end of the stock of a rifle or the handle of a pistol. **2.** on a firing range, a wall or mound of dirt or other material behind the target area to serve as a backstop for bullets. *Forestry.* **1.** the base of a tree trunk. **2.** the bottom or lower end of a log. **3.** to cut off the end of a log in order to eliminate defects and raise the log's commercial value. *Mechanical Devices.* the enlarged portion of a lever or connecting rod at the end to which a load is applied. *Mining Engineering.* **1.** any coal exposed at right angles to the face and generally having a rough surface. **2.** the point in a slate quarry at which the overlying rock comes in contact with an inclined stratum of slate rock. *Botany.* **1.** the end of a plant from which the roots spring. **2.** the end of a stalk or twig that is opposite the flowering end. **3.** either of two myrtaceous trees: *Eucalyptus pilularis,* **black butt,** and *E. longifolia,* **wooly butt.**

butt chisel *Mechanical Devices.* a woodworking chisel whose blade does not exceed four inches.

butt contact *Electricity.* a dome-shaped contact designed to press against a similarly shaped contact.

butte

butte *Geography.* a detached flat-topped hill, usually an eroded remnant of a mesa or a conical volcano.

butter *Food Technology.* the yellowish product of churned butterfat, usually obtained from cream. By law, butter must contain at least 80% butterfat by weight; it also contains water (approximately 15%) and sometimes salt (2.5% in salted butter) and vegetable coloring (such as carotene).

butterfat *Food Technology.* the natural fat of milk and butter, consisting of a mixture of glycerides.

butter finish *Metallurgy.* a semigloss finish, produced with a mildly abrasive wheel.

butterfly *Invertebrate Zoology.* the common name for any of various insects of the order Lepidoptera having four broad, usually brightly colored wings. Butterflies are found in virtually all habitats, especially in tropical rain forests.

butterfly

butterfly capacitor *Electricity.* a variable capacitor used as a tuner in nondigital VHF and UHF circuits; so named because the stator and rotor plates resemble butterfly wings.

butterfly effect *Chaotic Dynamics.* the phenomenon of sensitive dependence on initial conditions; so named for the influence exerted on a dynamical system by a minuscule difference (error or uncertainty) in input that evolves into an overwhelming difference in output. (From a 1979 Lorenz paper entitled "Does the Flap of a *Butterfly's* Wings in Brazil Set Off a Tornado in Texas?")

butterfly hinge *Mechanical Devices.* a hinge that runs through the center of a valve plate, allowing the plate to move like the wings of a butterfly.

butterfly nut see WING NUT.

butterfly valve *Engineering.* a circular disc having hinges that allow fluid to flow in only one direction; used inside a pipe or ventilating system to regulate the flow of its contents; often used with a controller. Also, **butterfly damper.**

buttering *Metallurgy.* in welding, the process of depositing a transitional material when joining two dissimilar metals or alloys.

buttermilk *Food Technology.* **1.** the whitish fluid residue of butter production; called **churned buttermilk. 2.** a fermented milk made by adding bacteria cultures to sweet milk; called **cultured buttermilk.**

butter oil *Food Technology.* a concentrated, high-fat product of separated cream that is used in making ice cream.

butter rock see HALOTRICHITE.

Butterworth filter *Electronics.* a filter with a maximally flat response in its passband. Mathematically, maximally flat means that at the center of the passband as many derivatives of the response shape as possible have a value of 0.

butt fusion *Engineering.* the process of bonding two like plastic or metal objects by melting the ends of each and applying pressure.

butt gauge *Engineering.* a gauge designed to mark the place on a door where mortise cuts for hinges will be made.

buttgenbachite *Mineralogy.* $Cu_{19}Cl_4(NO_3)_2(OH)_{32}\cdot 2H_2O$, an azure-blue hexagonal mineral occurring in minute acicular crystals, having a specific gravity of 3.42 and a hardness of 3 on the Mohs scale.

butt hinge *Mechanical Devices.* any type of door hinge other than a strap hinge.

butt joint *Engineering.* a joint formed by two pieces of wood or metal joined end to end without overlapping; it is usually used with one or more cover plates or other strengthening devices. *Electricity.* a connection achieved by welding, brazing, or soldering the ends of two conductors together. *Electromagnetism.* a waveguide connection in which the physical contact is made between the ends of the waveguide, thus preserving electromagnetic continuity.

buttock *Anatomy.* either of the two fleshy parts formed by the gluteal muscles at the back of the hips. *Naval Architecture.* **1.** also, **buttocks.** the rounded part of a vessel's stern above the waterline. Also, COUNTER. **2.** also, **buttock lines.** in ship drafting, the set of lines formed by the intersection of the underwater hull form and vertical planes running parallel to the keel. *Mining Engineering.* **1.** a corner at which two coal faces come together at approximately right angles. **2.** the rib of coal exposed at one or both ends of a longwall face. **3.** any coal that has been undercut and is ready to be broken.

buttock lines *Engineering.* the lines of intersection of the surface of a solid body, such as an aircraft fuselage or ship hull, with its vertical planes.

button *Electricity.* a small knob used to activate an electrical circuit. *Electronics.* **1.** the metal container that holds the carbon granules of a carbon microphone. **2.** a piece of metal joined to the base wafer of an alloy transistor. *Metallurgy.* **1.** in metal assaying, the globule obtained after fusion is completed. **2.** in the destructive testing of a welded specimen, the part that tears out. *Mycology.* a small, young mushroom.

button bit *Mechanical Devices.* a drilling bit composed of tungsten carbide inserts.

button die *Mechanical Devices.* a mating member, generally replaceable, that fits onto a piercing punch. Also, DIE BUSHING.

buttonhead *Mechanical Devices.* a circular bolt or screw head with a slot for a screwdriver and an oval top used with a flexible gasket fitted under it. Also, HALF-ROUND SCREW.

buttonhole *Surgery.* **1.** a short, straight cut made into an organ or through the wall of a cavity. **2.** the contraction of an opening down to a narrow slit, as in **buttonhole stenosis.**

buttonhook contact *Electricity.* a hook-shaped contact used on feed-through terminals of headers to aid soldering or unsoldering of leads.

button knife *Surgery.* a small knife used to cut cartilage.

button suture *Surgery.* a suture that is passed through a material and then tied, in order to prevent the threads from severing the flesh.

button test see AGGLOMERATION TEST.

buttress *Architecture.* **1.** an exterior masonry mass attached to and supporting a wall or vault. **2.** see FLYING BUTTRESS. *Civil Engineering.* an exterior pier, often sloped, used to provide support for the lateral forces, particularly for tall walls. *Botany.* on certain trees, the branch roots or other aboveground protuberant portions of the trunk or root that arch away from the trunk before they reach the soil, forming additional props; prop roots.

buttress dam *Civil Engineering.* a concrete dam utilizing a series of buttresses for stability.

buttress sand *Geology.* a sandstone body that intersects an underlying unconformity, often forming a trap for oil.

buttress thread *Design Engineering.* a screw thread used for strength and efficiency in transmitting power; the drive face of the thread is perpendicular to the axis and the back face is sloped for bracing it.

buttress unconformity *Geology.* a surface on which onlapping rocks rest against a regional scarp.

butt rot *Plant Pathology.* a fungus disease of woody plants caused by polyspores that attack and cause tissue decay at the base of the trunk.

butt weld *Metallurgy.* a welded joint between two metallic parts lying approximately in the same plane.

butvar see POLYVINYL BUTYRAL.

butyl [byoot′əl] *Organic Chemistry.* the group $-C_4H_9$, which includes four isomeric alkyl radicals: $CH_3CH_2CH_2CH_2-$, $(CH_3)_2CHCH_2-$, $CH_3CH_2CHCH_3-$, or $(CH_3)_3C-$.

***n*-butyl acetate** *Organic Chemistry.* $CH_3COOC_4H_9$, a colorless, flammable liquid with a fruity odor; freezes at –75°C and boils at 126.3°C; slightly soluble in water and soluble in alcohol and ether. It is toxic and a skin irritant..

***s*-butyl acetate** *Organic Chemistry.* $CH_3COOCH(CH_3)C_2H_5$, a colorless, flammable liquid that boils at 112.2°C; insoluble in water and miscible with alcohol and ether; used as a solvent.

***t*-butyl acetate** *Organic Chemistry.* $CH_3COOC(CH_3)_3$, a colorless, flammable liquid that boils at 96°C; insoluble in water and soluble in alcohol and ether; used as a solvent and as an anti-knock additive to nonleaded gasoline.

butyl acetoacetate *Organic Chemistry.* $CH_3COCH_2CO_2C_4H_9$, a colorless, combustible liquid that boils at 213.9°C; insoluble in water and soluble in alcohol and ether; used as an intermediate in the synthesis of metal derivatives, dyestuffs, and pharmaceuticals.

***n*-butyl acrylate** *Organic Chemistry.* $CH_2{=}CHCOOC_4H_9$, a colorless, flammable liquid; soluble in water; freezes at –64°C and boils at 145.7–148°C; used as an intermediate for organic synthesis and polymers.

butyl alcohol see BUTANOL.

***n*-butylamine** *Organic Chemistry.* $CH_3(CH_2)_3NH_2$, a colorless, flammable liquid that is miscible with water, alcohol, and ether; it freezes at –49.1°C and boils at 77.8°C; used as an intermediate in organic synthesis and to make insecticides, emulsifying agents, and pharmaceuticals. Also, 1-AMINOBUTANE.

***s*-butylamine** *Organic Chemistry.* $CH_3CH_2CH(NH_2)CH_3$, a colorless, flammable liquid that is miscible with water, alcohol, and ether; boils at 63°C; used as a fungicide. Also, 2-AMINOBUTANE.

***t*-butylamine** *Organic Chemistry.* $(CH_3)_3CNH_2$, a colorless, flammable liquid that is miscible with water, alcohol, and ether; freezes at –67.5°C and boils at 44.4°C; used as an insecticide and fungicide, and for dyestuffs and pharmaceuticals.

***n*-butyl-*p*-aminobenzoate** *Pharmacology.* $C_{11}H_{15}NO_2$, the chemical name for butamben, a topical anesthetic used to treat painful skin conditions and the discomfort of hemmorhoids.

butylate *Organic Chemistry.* **1.** to introduce the butyl group into a compound. **2.** $C_{11}H_{23}NOS$, a colorless liquid; soluble in water; boils at 138°C (at 21 torr); used as an herbicide.

butylated hydroxyanisole *Organic Chemistry.* $(CH_3)_3CC_6H_3OH(OCH_3)$, a white or yellowish waxy solid; insoluble in water and soluble in alcohol and ether; melts at 48–55°C and boils at 264–270°C; toxic if ingested; used as an antioxidant for fats and oils in food packaging.

butylated hydroxytoluene *Organic Chemistry.* $[(CH_3)_3C]_2C_6H_2OH(CH_3)$, a crystalline solid that melts at 72°C, insoluble in water and soluble in methanol and ethanol; used in petroleum products.

***n*-butylbenzene** *Organic Chemistry.* $C_6H_5C_4H_9$, a colorless, toxic liquid; insoluble in water and miscible in alcohol and ether; freezes at –87.9°C and boils at 183.2°C; used in organic synthesis.

***s*-butylbenzene** *Organic Chemistry.* $C_6H_5CH(CH_3)C_2H_5$, a colorless, toxic liquid; insoluble in water and miscible in alcohol and ether; freezes at –75.7°C and boils at 170.6°C; used as a solvent and plasticizer and for organic synthesis.

***t*-butylbenzene** *Organic Chemistry.* $C_6H_5C(CH_3)_3$, a colorless, toxic liquid, insoluble in water and soluble in alcohol; freezes at –57.8°C and boils at 169.1°C; used in organic synthesis and polymerization.

***n*-butyl chloride** *Organic Chemistry.* C_4H_9Cl, a colorless, flammable liquid that is insoluble in water and miscible with alcohol and ether; freezes at –123.1°C and boils at 78.44°C. It is toxic on prolonged inhalation, and is used as an alkylating agent in organic synthesis. Also, 1-CHLOROBUTANE.

***s*-butyl chloride** *Organic Chemistry.* $CH_3CHClCH_2CH_3$, a colorless, flammable liquid; slightly soluble in water and miscible with alcohol and ether; freezes at –131.3°C and boils at 68.2°C; used in organic synthesis. Also, 2-CHLOROBUTANE.

butyl diglycol carbonate *Organic Chemistry.* $(C_4H_9OCO_2CH_2)_2O$, a colorless liquid of low volatility that boils at 164–166°C; insoluble in water and soluble in organic solvents; compatible with many resins and plastics; used as a high-boiling-point solvent and as a plasticizer.

butylene [byoot′ə lēn′] *Organic Chemistry.* the alkene hydrocarbon group C_4H_8, including three known isomeric forms; all are flammable and easily liquefied gases. Also, BUTENE.

α-butylene see 1-BUTENE.

1,3-butylene glycol *Organic Chemistry.* $CH_2HOCH_2CH(OH)CH_3$, a syrupy, colorless liquid that is soluble in water and alcohol, and boils at 207.5°C; used as a solvent, as a food flavoring and food additive, and as a plasticizer. Also, 1,3-BUTANEDIOL.

1,4-butylene glycol *Organic Chemistry.* $HO(CH_2)_4OH$, a colorless oily liquid that boils at 230°C; miscible with water and soluble in alcohol; used as a solvent and humectant, and in making plastics and pharmaceuticals. Also, 1,4-BUTANEDIOL.

2,3-butylene glycol *Organic Chemistry.* $CH_3CH(OH)CH(OH)CH_3$, a colorless crystalline solid that boils at 178–182°C; miscible with water and alcohol, soluble in ether and acetone; a compound produced during the fermentation of sugar; used in resins and dyes and as a chemical intermediate. Also, 2,3-BUTANEDIOL.

butylene group see BUTYLENE.

1,2-butylene oxide *Organic Chemistry.* C_4H_8O, a colorless, flammable liquid; soluble in water and miscible in organic solvents; boils at 63°C; used in synthesizing polymers. Also, 1,2-EPOXYBUTANE.

butyl ether *Organic Chemistry.* $C_8H_{18}O$, a colorless liquid; insoluble in water and miscible with alcohol and ether; freezes at –95.3°C and boils at 142.2°C; toxic and flammable; used as a solvent for hydrocarbons and fatty materials, and as an extracting agent.

butylethylacetic acid see 2-ETHYLHEXANOIC ACID.

butyl formate *Organic Chemistry.* $HCOOC_4H_9$, a colorless liquid that freezes at –90°C and boils at 107°C; a dangerous fire risk; used as a solvent for cellulose compounds and in lacquers, perfumes, and synthetic flavorings.

butyl group see BUTYL.

***t*-butyl hydroperoxide** *Organic Chemistry.* $(CH_3)_3COOH$, a water-white, highly reactive liquid that freezes at –8°C and decomposes at 75°C; soluble in water and organic solvents. It is used in polymerization, oxidation, and deodorizing.

***n*-butyl lactate** *Organic Chemistry.* $CH_3CHOHCOOC_4H_9$, a water-white, stable liquid; slightly soluble in water; freezes at –43°C and boils at 188°C; used as a solvent and in lacquers and varnishes.

butyl mercaptan *Organic Chemistry.* C_4H_9SH, a colorless liquid with a strong skunklike odor; slightly soluble in water and soluble in alcohol and ether; boils at 97–101°C. It is toxic by inhalation and flammable, and is used as a gas-odorizing agent. Also, BUTANETHIOL.

butyl oleate *Organic Chemistry.* $CH_3(CH_2)_7CH{=}CH(CH_2)_7COOC_4H_9$, a light-colored oleaginous liquid that boils at 223–227°C (14–15 torr), insoluble in water and miscible with alcohol and ether; used as a plasticizer, particularly for PVC, and as a solvent and lubricant.

***p*-*t*-butylphenol** *Organic Chemistry.* $(CH_3)_3CC_6H_4OH$, needlelike white crystals with a distinctive odor; melts at 100°C and boils at 239°C; insoluble in water and soluble in alcohol and ether. It is used as a plasticizer and a chemical intermediate, and for other purposes.

***n*-butyl propionate** *Organic Chemistry.* $C_2H_5CO_2C_4H_9$, a water-white liquid with an applelike odor; freezes at –89°C and boils at 146°C; slightly soluble in water and soluble in alcohol and ether. It is used as a solvent and as an ingredient in perfumes and flavors.

butyl rubber *Materials.* a copolymer of isobutylene and isoprene that displays the lowest permeability among rubbers; it is resistant to abrasion, tearing, flexing, sunlight, and chemicals; used for thermoplastic rubber production, encapsulation, inner tubes, hose, and seals for food jars. Also, ISOBUTYLENE-ISOPRENE ISOMER.

butyl stearate *Organic Chemistry.* $C_{22}H_{44}O_2$, colorless crystals; freezes at 27°C and boils at 343°C; soluble in alcohol and ether and insoluble in water; used as a lubricant, in polishes, as a plasticizer, and for industrial purposes.

butynediol *Organic Chemistry.* $HOCH_2C{\equiv}CCH_2OH$, white rhombic crystals that melt at 58°C and boil at 238°C; soluble in water, alcohol, and acetone; used as a corrosion inhibitor, electroplating brightener, and defoliant.

butyraldehyde *Organic Chemistry.* $CH_3(CH_2)_2CHO$, a water-white liquid having a pungent aldehyde odor; freezes at –99°C and boils at 75.7°C; slightly soluble in water and soluble in alcohol and ether; used in plasticizers and rubber accelerators.

butyrate *Organic Chemistry.* a salt or ester of butyric acid that contains the radical $C_4H_7O_2$.

butyric acid *Organic Chemistry.* C_3H_7COOH, a colorless liquid with an obnoxious odor, occurring in spoiled butter; freezes at –4.5°C and boils at 165.5°C; miscible with water, alcohol, and ether. It is used in the synthesis of butyrate ester perfume and flavor ingredients and in disinfectants and pharmaceuticals.

butyric alcohol see BUTANOL.

butyric aldehyde see BUTYRALDEHYDE.

butyric anhydride *Organic Chemistry.* $C_8H_{14}O_3$, a colorless liquid that is soluble in either, boils at 199–201°C, and decomposes in water to form butyric acid.

butyric fermentation *Biochemistry.* a bacterial fermentation in which butyric acid is produced as a major product of organic matter decomposition.

butyrinase *Enzymology.* an enzyme that hydrolyzes the glyceryl ester butyrin, which is present in blood serum.

Butyrivibrio *Bacteriology.* a genus of Gram-negative, anaerobic bacteria of the family Bacteroidaceae occurring as curved rods that are motile by means of polar or subpolar flagella; found in the rumen of animals.

butyrolactone *Organic Chemistry.* $C_4H_6O_2$, a combustible oily liquid that freezes at –43.4°C and boils at 204°C; miscible with water, alcohol, and ether. It is used chiefly as a solvent for resins.

butyronitrile *Organic Chemistry.* $CH_3(CH_2)_2CN$, a colorless liquid that freezes at –112.6°C and boils at 116–117°C; slightly soluble in water and soluble in alcohol and ether; used in various industrial and pharmaceutical intermediates and products.

Buxaceae *Botany.* a family of cosmopolitan, usually evergreen trees, shrubs, and herbs of the order Euphorbiales; many species produce steroid alkaloids, and all are characterized by dicotyledonous embryos, the lack of a nectary disk, and shiny black seeds.

Buxbaumiales *Botany.* an order of very small mosses, characterized by a large asymmetrical capsule.

Buys-Ballot, Christoph 1817–1890, Dutch meteorologist; published Buys-Ballot's law.

Buys-Ballot's law *Meteorology.* a law that describes the relationship of the horizontal wind direction in the atmosphere to the pressure distribution. Also, BARIC WIND LAW.

buzz *Electronics.* a low-pitched sound with high-frequency components resulting from nonsinusoidal voltage interference. *Control Systems.* a force with a controlled amplitude and frequency, continuously applied to a servomotor-driven device to prevent it from sticking in the null position. Also, DITHER. *Aviation.* **1.** a humming noise produced by vibration of a control surface. **2.** to fly low over or dive at a place or object.

buzzard *Vertebrate Zoology.* **1.** a collective name for several large hawklike birds of prey of the subfamily Buteonidae, characterized by a curved beak and featherless lower legs. **2.** a name applied to New World vultures, a group of largely scavenger birds in the family Cathartidae.

buzz bomb *Ordnance.* a nickname for the German V-1 pulsejet-powered cruise missile of World War II; so called because of the sound it made over a target.

bv budded virus.

B virus *Virology.* a herpesvirus of the subfamily Alphaherpesvirinae that naturally infects certain monkeys and can be transmitted to humans through monkey bites or scratches, causing encephalomyelitis.

BW biological warfare; bacteriological warfare; black and white.

BWI *Aviation.* the airport code for Baltimore-Washington International.

BWO backward wave oscillator.

BWR boiling-water reactor.

Bw star *Astronomy.* a type B star with weak spectral lines of helium.

BX or **bx** biopsy.

BX cable *Electricity.* insulated electrical wires sheathed in flexible metal tubing; used for power transmission to electronic equipment.

Bx curve *Photogrammetry.* a graphic representation of vertical errors in a stereotriangulated strip, in which errors in the *x*-direction on horizontal control points are plotted as ordinates versus *y*-coordinates of these points.

by or **b.y.** billion years.

Byblidaceae *Botany.* a family of dicotyledonous perennial herbs and shrubs of the order Rosales, noted for insectivorous species; native to Australia and South Africa.

By curve *Photogrammetry.* a graphic representation of vertical errors in a stereotriangulated strip, in which errors in the *y*-direction on horizontal control points are plotted as ordinates versus *x*-coordinates of these points.

byerite *Geology.* a dark brown to black bituminous coal that melts and expands upon heating.

byon *Geology.* gem-bearing gravel, especially one containing a brownish-yellow clay in which rubies and sapphires occur.

bypass *Transportation Engineering.* a road that diverts traffic around a congested area or temporary hindrance. *Engineering.* any device that serves to divert the flow of a fluid around a fixture, obstruction, pipe, or valve, instead of through it. *Surgery.* a shunt or detour through an alternate vessel, created surgically to provide an auxiliary flow for blood or other fluids. *Electricity.* a low-impedance path carrying current around a component or circuit instead of through it. Thus, **bypass capacitator.**

bypass channel *Civil Engineering.* **1.** a channel designed to carry excess water or flood water from a stream. Also, FLOOD RELIEF CHANNEL. **2.** a channel built to divert water from a primary channel or canal.

bypass filter *Electronics.* a filter that provides a low-attenuation path around some other circuit or equipment.

bypass valve *Engineering.* in a fluid dynamic system, a valve opened to reduce pressure or to direct the fluid in a direction in which it does not normally flow.

Byrrhidae *Invertebrate Zoology.* a family of pill beetles in the superfamily Byrrhoidea having strongly arched bodies and flattened legs, and feeding on algae and mossy plants.

Byrrhoidea *Invertebrate Zoology.* the superfamily of pill beetles, in the order Coleoptera, usually black or metallic green; some 300 species are found in temperate regions; the soil-dwelling larvae feed on mosses, liverworts, lichens, and grass roots.

bysmalith *Geology.* a roughly vertical igneous intrusion along a steep fault, having a more or less cylindrical shape.

byssaceous *Botany.* consisting of or having fine threads or filaments.

byssinosis *Pathology.* a pulmonary disease caused by the breathing of cotton dust; found among textile workers. Also, BROWN LUNG.

byssocausis see MOXIBUSTION.

byssochlamys *Mycology.* a genus of fungi of the order Eurotiales that sometimes cause damage to canned foods.

byssoid *Biology.* made up of fine, silky threads; having a delicate, filamentous structure.

byssus *Invertebrate Zoology.* a tuft of strong, silky filaments secreted by a gland in the foot of certain bivalve mollusks; used for attachment to surfaces.

bystander apathy *Psychology.* a tendency of bystanders who witness an emergency not to offer aid to the afflicted person or persons; research indicates that the larger the number of onlookers, the less likely it is that anyone will offer aid.

bystander phenomena *Immunology.* the nonspecific effects of an immune response on tissues that are not directly involved with the response, but may be affected by it.

byte *Computer Programming.* a group of adjacent binary digits operated on as a unit; in many computers, a byte is an 8-bit group that represents one alphanumeric character or two decimal digits.

byte-addressable *Computer Programming.* of or relating to a characteristic of a computer in which each byte in main memory has a unique address and can be accessed individually by program instructions.

byte mode *Computer Technology.* a mode of operation of an I/O channel in which the bytes of data from several input and output devices are interleaved and then directed to the proper destination, as when several low-speed devices are connected to a multiplexer channel. Also, **byte multiplexing.**

byte multiplexer channel *Computer Technology.* an approach to connecting peripheral devices to a processor and main memory using a shared channel that transmits one byte of data at a time and interleaves bytes of data from several devices; used with slower I/O devices.

byte serial see BIT PARALLEL.

bytownite see PLAGIOCLASE.

Byturidae *Invertebrate Zoology.* the raspberry fruit worms, insects in the order Coleoptera; larvae are fruit pests.

Byzantine *Architecture.* the style of architecture that evolved at Constantinople, capital of the Eastern Roman Empire (330–1453); characterized by round arches and large domes with vast interior spaces.

bz. benzene.

Bz curve *Photogrammetry.* a graphic representation of the vertical errors in the aerotriangulation of a strip of photographs.

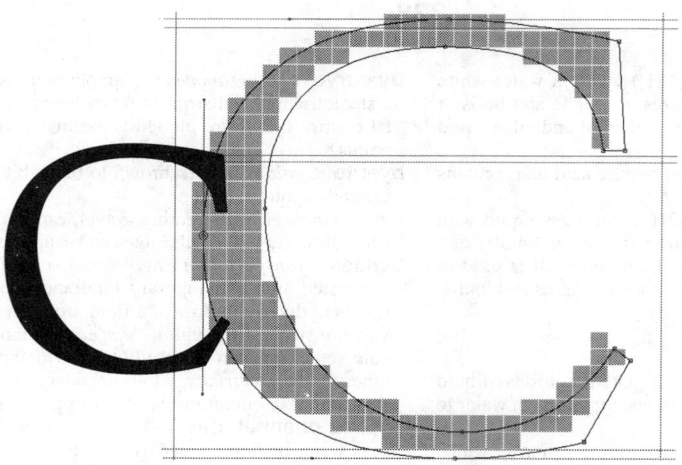

C *Science.* the third in a sequence or group. *Metrology.* the symbol for the Celsius (centigrade) temperature scale. *Biotechnology.* a symbol that describes the saturation concentration of dissolved oxygen; often used in fermentation investigations and procedures. *Computer Programming.* **1.** a general-purpose programming language used, among other applications, to develop the UNIX operating system. **2.** the character that represents the hexadecimal equivalent of a decimal 12.

C the chemical symbol for carbon.

C or **C.** an abbreviation for: capacity; capacitance; capacitor; coulomb.

C- *Aviation.* the U.S. military designation for a cargo aircraft, as in *C-47* or *C-5A*.

C++ *Computer Programming.* an object-oriented extension to the C programming language that adds to it the class concepts of the language Simula 67.

c3 spiroplasmavirus group see SPIROPLASMAVIRUS GROUP.

C-5A *Aviation.* a very large, long-range military transport; popularly known as the *Galaxy*.

C-9A *Aviation.* a military version of the DC-9, adapted as a troop transport; popularly known as the Nightingale.

^{14}C see CARBON-14.

C-17 *Aviation.* a long-range military cargo and transport aircraft that is specially designed to operate on very short runways.

C-17

C-47 *Aviation.* a military version of the DC-3, adapted to carry troops and heavy loads and used extensively in World War II; popularly known as the Skytrain (U.S.) or Dakota (British).

C-53 *Aviation.* a military version of the DC-3, adapted as a hospital and troop transport and used extensively in World War II; popularly known as the Skytrooper.

C-97 *Aviation.* a double-decked four-engine heavy military transport airplane; popularly known as the Stratofreighter.

C-141 *Aviation.* a heavy, long-range military transport airplane powered by four turbofan engines; popularly known as the Starlifter.

CA *Aviation.* the airline code for Air China.

CA or **C.A.** *Psychology.* chronological age.

Ca the chemical symbol for calcium.

CAB Civil Aeronautics Board.

cab *Transportation Engineering.* the part of a locomotive, rail car, bus, or truck that houses the operator and operating controls.

caballing *Oceanography.* the mixing of waters of equal density but different temperatures and salinities, so that the resulting mixture is denser than its two components and therefore sinks.

cabane *Aviation.* a series or arrangement of struts used to carry a load above a fuselage or wing, as to support the upper wing of a biplane.

Cabannes' factor *Analytical Chemistry.* a correcting factor for the depolarizing effect of horizontal components of scattered light during the determination of molecular weight by optical methods.

cabbage

cabbage *Botany.* any of several varieties of *Brassica olearacea capitata,* biennial plants of the mustard family whose green or purple leaves are compacted into an edible head.

cabbage yellows *Plant Pathology.* a disease of cabbage, caused by the fungus *Fusarium conglutinans* and characterized by yellowing and stunting of part or all of the infected plant.

Cabernet Sauvignon [cab´ər nā sô vin yōn´] *Botany.* a premium red wine grape, grown in the Bordeaux region of France and elsewhere. *Food Technology.* **1.** a dry red Bordeaux wine made mostly or entirely with this grape. **2.** a similar wine produced in California, Australia, and elsewhere. Also, **Cabernet.**

cabezon [kab´ə zōn´] *Vertebrate Zoology.* a large Pacific sculpin, *Scorpaenichthys marmoratus,* found along rocky bottoms in shallow waters.

Cabibbo theory *Particle Physics.* the theory that the up quark will change into down or strange quarks under the action of the weak force.

cabin configuration *Aviation.* the basic arrangement and fittings of an aircraft cabin. The three types of cabin configuration are passenger, air cargo, and convertible between the two.

cabinet file *Mechanical Devices.* a coarse woodworking file with one flat face and one convex face, used by joiners and cabinetmakers. Similarly, **cabinet scraper.**

cabinet saw *Mechanical Devices.* a short saw with one rip edge and one crosscut edge.

cable *Mechanical Devices.* a rope, fiber, or heavy steel gauge wire that is used for support, lifting, hauling, and other load-supporting functions. *Telecommunications.* **1.** see COAXIAL CABLE. **2.** see CABLE TELEVISION. *Electricity.* a group of electrical conductors bound together and insulated from each other within a protective sheath. *Architecture.* a convex molding set within a flute of a column or pilaster.

cable armor see ARMORED CABLE.
cable bend *Naval Architecture.* a knot, or a short length of line, used to secure an anchor line to the anchor.
cable buoy *Navigation.* a floating buoy attached to the anchor of a ship or boat, in order to mark its location.
cablecar or **cable car** *Transportation Engineering.* a vehicle that moves on a cable railway or on an aerial tramway.
cable code see MORSE CABLE CODE.
cable complement *Electricity.* a group of cable pairs having a common distinguishing feature.
cable conveyor *Mechanical Engineering.* a conveying machine consisting of a bulk-transport car that rides on a flexible cable.
cable delay *Computer Technology.* the time it takes for an electrical pulse to pass through a cable or wire, approximately 1.5 nanoseconds per foot, depending on cable characteristics and other factors.
cable drilling *Mining Engineering.* a percussive method of drilling in which a sharp, heavy tool suspended in the borehole from a wire cable continually pounds the rock.
cable fill *Electricity.* the ratio of the number of pairs in use to the total number of pairs in the cable.
cablegram *Telecommunications.* a telegraph message that is sent via a submarine cable.
cable lacquer *Materials.* a lacquer that is made from synthetic resins and has a high dielectric strength; used for flexible, durable coatings.
cable-laid rope *Design Engineering.* a rope consisting of several individual strands twisted together so that the twist of the strands is in the direction opposite that of the rope.
cable layer *Naval Architecture.* a vessel equipped for laying underwater cables.
cable length *Oceanography.* a nautical unit of horizontal distance, generally considered to be about 600 feet.
cable noise *Electricity.* electrical noise that is received by cable conductors.
cable railway *Mechanical Engineering.* an engine-driven continuous track used to carry coupled bulk-transport cars.
cable release *Engineering.* a device for releasing a camera shutter, in which the trigger is actuated by a plunger of stiff wire cable in a flexible tube.
cable run *Electricity.* the path used by a cable on cable racks from one termination to another.
cable skidding *Forestry.* the act of dragging or hauling logs from a stump to a landing, using chains or heavy cables. Thus, **cable skidder.**

cable skidder

cable stopper *Naval Architecture.* a short line with a gripping device attached to an anchor cable to take strain off the anchor windlass. There are four main types: **Blake Slip, Screw Slip, Senhouse Slip,** and **Devil's Claw.** Also, CHAIN STOPPER, SLIP.
cable system *Mining Engineering.* a drill system in which the bottom of a hole is penetrated by a heavy bit suspended from a flexible manila or steel cable; a reciprocating motion is imparted to the bit by an oscillating beam through the suspension cable. Thus, **cable-system drill.**
cable television *Telecommunications.* a television broadcasting system that uses giant antennae to receive signals, which are relayed to individual subscribers via coaxial cable or other broadband distribution media.
cable tier *Naval Architecture.* the storage space for a ship's cables, usually located on the lowest deck or hold.
cable tool *Mining Engineering.* any bottom-hole tool in which the drilling bit is connected by cable with the machine on the surface, allowing a percussive action to drill the boreholes. Thus, **cable-tool drilling.**
cable trough *Electricity.* an enclosed channel that provides a cable path.
cable twist *Design Engineering.* a construction or design in which each successive twist is opposite to the preceding twist.
cabochon [kab′ə shän′] *Mineralogy.* a cut, but unfaceted gemstone, usually dome-shaped and highly polished.
Cabombaceae *Botany.* a family of dicotyledonous aquatic herbs of the order Nymphaeales having creeping rhizomes, solitary aerial flowers, and elongated, floating, leafy stems.
caboose *Engineering.* a car, usually attached to the rear of a train, used for personnel.
Cabot, John 1450?–1498?, English navigator, born in Genoa, Italy; discovered North American mainland.
Cabot, Sebastian 1484?–1557, English navigator and mapmaker, born in Venice, Italy; explored and charted American coasts; son of John Cabot.
Cabot's ring *Cell Biology.* a ringlike structure found in immature red blood cells and erythrocytes that are involved in severe anemias.
Cabral, Pedro Alvarez [kə bräl′] 1467?–1528?, Portuguese navigator; discovered Brazil.
CABRA numbers *Metallurgy.* a former standard designation of commercial copper-base products, superseded by CDA numbers and by UNS, the Universal Numbering System of metals and alloys.
cabretta *Materials.* sheepskin leather that is made from sheep that grow hair and not wool; a very strong leather used for shoes and gloves.
cab signal *Transportation Engineering.* an indicator on a locomotive that notifies the engineer about conditions influencing the operations and movements of the train.
cab signal control system *Transportation Engineering.* in a locomotive or rail car, a control system that indicates track conditions and stops the train if the operator does not respond to adverse conditions.
cabuya see MAURITIUS HEMP.

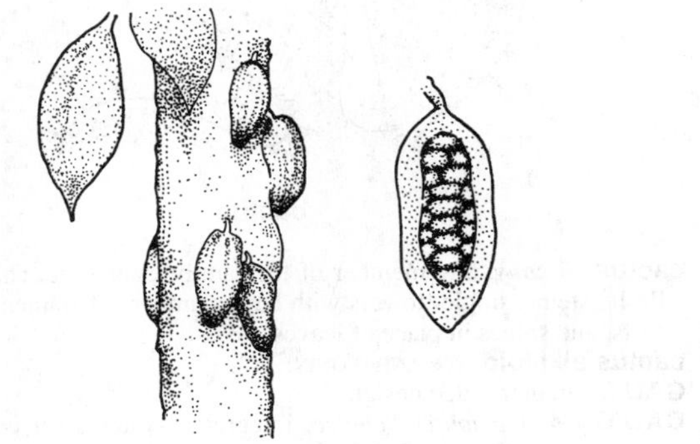

cacao

cacao *Botany.* **1.** a small tropical tree, *Theobroma cacao,* from whose seeds cocoa and chocolate are derived. **2.** the seeds or fruit of this tree.
cacao butter see COCOA BUTTER.
Cacao swollen shoot virus group *Virology.* a nontaxonomic group of plant viruses, consisting of bacilliform particles that have narrow host ranges and are transmitted by aphids, leafhoppers, and mealybugs.
cache *Archaeology.* a hidden deposit of a group of artifacts, often discovered in burials or caves; it usually consists of ceremonial objects or emergency food supplies. *Computer Technology.* a very high-speed memory mechanism in the memory hierarchy between slower-speed main memory and the CPU; used to improve effective memory transfer rates. Also, MEMORY CACHE.
cachectin *Biochemistry.* a protein that inhibits lipid utilization by enzymes.
cachexia *Medicine.* a profound and marked state of poor health and malnutrition, usually associated with a serious disease such as cancer.
cachinnation *Psychology.* unrestrained laughter without any appropriate cause; a symptom of a form of schizophrenia.

Cachonellaceae *Botany.* a family of marine parasitic algae of the order Blastodiniales, occurring on the ectoderm of siphonophores.
cacimbo [kə sim´bō] *Meteorology.* wet fogs and drizzles of West Africa, observed with onshore winds from the Benguela current.
caco- or **cac-** a combining form meaning "bad," as in *cacotrophy, cachexia.*
cacodyl *Organic Chemistry.* $(CH_3)_2AsAs(CH_3)_2$, crystalline plates that are slightly soluble in water and soluble in alcohol; melts at –6°C and boils at 165°C. Also, TETRAMETHYLBIARSINE.
cacomistle [kak´ə mis´əl] *Vertebrate Zoology.* a small nocturnal mammal of the southwest United States, belonging to the carnivorous family Procyonidae and resembling a raccoon. Also, CIVET CAT, RINGTAIL, BASSARISK.
caconym *Systematics.* a taxonomic name that is considered unacceptable for linguistic, rather than scientific, reasons.
cacotheline *Organic Chemistry.* $C_{21}H_{21}N_3O_7$, yellow crystals; slightly soluble in water; used to indicate metals in chelometric titrations.
cacoxenite *Mineralogy.* $(Fe^{+3},Al)_{25}(PO_4)_{17}O_6(OH)_{12}\cdot 75H_2O$, a yellow or brown hexagonal mineral occurring as minute crystals in tufted aggregates or fibrous crusts, having a specific gravity of 2.3 and a hardness of 3.4 on the Mohs scale; found as a secondary mineral with other phosphates and with limonite.
Cactaceae *Botany.* the cactus family of leafless, spiny succulent plants, containing over 2000 species found mainly in the arid regions of North and South America.
cactolith *Geology.* an irregular mass of igneous rock shaped somewhat like a cactus.

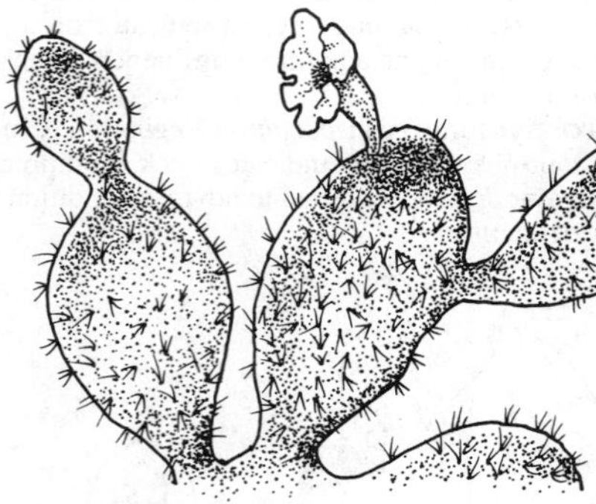

cactus

cactus *Botany.* any member of the family Cactaceae, characterized by fleshy stems, single flowers with large numbers of stamens and perianth parts, and spines in place of leaves.
cactus alkaloid see ANHALONIN.
CAD computer-aided design.
CAD/CAM *Computer Science.* integrated systems for computer-aided design and manufacturing.
cadalene *Organic Chemistry.* $C_{15}H_{18}$, an aromatic liquid; insoluble in water and soluble in oils; boils at 291–292°C.
cadang-cadang *Plant Pathology.* an infectious viral disease of the coconut palm, especially in the Philippines; characterized by yellowish bronzing of the older leaves.
cadastral *Cartography.* of or relating to surveys or maps used to determine the location and ownership of real estate, usually for the purposes of taxation.
cadastral map *Cartography.* a large-scale map indicating the boundaries and dimensions of subdivisions and individual tracts of land in order to show ownership.
cadastral survey *Cartography.* a land survey made to delimit and identify property lines.
cadaver *Medicine.* a dead body, especially a human corpse being used for medical purposes.
cadaverine *Biochemistry.* $C_5H_{14}N_2$, a toxic, foul-smelling liquid that is produced as a result of certain chemical reactions in decomposing flesh.
caddis fly *Invertebrate Zoology.* an insect of the order Trichoptera that generally inhabits stream sides; its aquatic larvae often spin silken retreats fortified by bits of twig, shell, and the like.
cade oil *Materials.* a brown, viscous oil that has a tar odor and is slightly soluble in water. It is derived from the wood of a Mediterranean juniper, *Juniperus oxycedrus,* and is used in treating skin diseases and in perfumes. Also, JUNIPER TAR OIL.
cadinene *Organic Chemistry.* $C_{15}H_{24}$, an alicyclic sesquiterpene that is insoluble in water and slightly soluble in alcohol; boils at 274°C; present in camphor, oil of savin, oil of cubeb, and cade oil. Also, **3,9-cadinadiene.**
cadmiosis *Toxicology.* poisoning due to ingestion or inhalation of cadmium dust or fumes; symptoms may include emphysema.
cadmium [kad´mē əm] *Chemistry.* a rare element having the symbol Cd, the atomic number 48, an atomic weight of 112.4, a melting point of 320.9°C, and a boiling point of 767°C. It is a white, ductile metal obtained from zinc ores, and is used as an anticorrosive and in making alloys. Its compounds are highly toxic. (From a Greek term meaning "earth of Cadmus.")
cadmium acetate *Inorganic Chemistry.* $Cd(OOCCH_3)_2\cdot 3H_2O$, a colorless crystalline compound that is soluble in water and alcohol; melts at 256°C; used in chemical testing for sulfides, selenides, and tellurides, and for producing iridescent effects on porcelain.
cadmium blende see GREENOCKITE.
cadmium bromate *Inorganic Chemistry.* $Cd(BrO_3)_2\cdot H_2O$, colorless to white crystals or powder, soluble in water; a strong oxidizer, highly toxic and irritating; used as an analytical reagent.
cadmium bromide *Inorganic Chemistry.* $CdBr_2$, a white to yellowish crystalline solid that is soluble in water and alcohol; melts at 567°C and boils at 863°C; used in photography and lithography.
cadmium carbonate *Inorganic Chemistry.* $CdCO_3$, a white trigonal solid, insoluble in water and soluble in dilute acids; decomposes below 500°C; used as a chemical reagent.
cadmium cell *Electricity.* a voltage reference cell; at 20°C it has a voltage of 1.0186 volts.
cadmium chlorate *Inorganic Chemistry.* $Cd(ClO_3)_2\cdot 2H_2O$, a colorless prismatic solid that is deliquescent; soluble in water, alcohol, and acetone and melts at 80°C. It is highly toxic.
cadmium chloride *Inorganic Chemistry.* $CdCl_2$, small, colorless to white hexagonal crystals, soluble in water and acetone; melts at 586°C and boils at 960°C; used in photography and analytical chemistry and for various industrial purposes.
cadmium cyanide *Inorganic Chemistry.* $Cd(CN)_2$, white crystals, decomposing above 200°C; a precipitate obtained when potassium or sodium cyanide is added to a concentrated solution of a cadmium salt; used in electroplating copper.
cadmium fluoride *Inorganic Chemistry.* CdF_2, a white crystalline compound, soluble in water and acids; melts at 1100°C and boils at 1758°C; used as a starting material for laser crystals.
cadmium hydrate see CADMIUM HYDROXIDE.
cadmium hydroxide *Inorganic Chemistry.* $Cd(OH)_2$, a white amorphous powder that is insoluble in water and soluble in dilute acids; decomposes at 300°C; used to make cadmium-nickel storage batteries.
cadmium iodate *Inorganic Chemistry.* $Cd(IO_3)_2$, a fine white powder that is slightly soluble in water and soluble in nitric acid; decomposed by heat; used as an oxidizing agent.
cadmium iodide *Inorganic Chemistry.* CdI_2, a white compound that slowly turns yellowish green when exposed to air; soluble in water, alcohol, and acids; melts at 387°C and boils at 796°C; used in photography and process engraving.
cadmium molybdate *Inorganic Chemistry.* $CdMoO_4$, yellow plates or crystals; slightly soluble in water and soluble in acids; used in optics and electronics.
cadmium nitrate *Organic Chemistry.* $Cd(NO_3)_2$, a salt found in the form of white, amorphous pieces or hygroscopic needles; soluble in water, alcohol, and ammonia; melts at 350°C. It is a fire and explosion risk, and is used in photography and in coloring glass and porcelain.

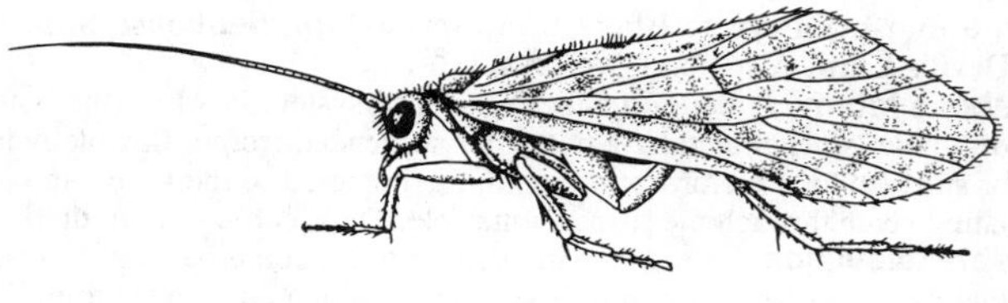

caddis fly

cadmium ocher see GREENOCKITE.

cadmium oxide *Inorganic Chemistry.* CdO, an amorphous brown powder that is insoluble in water and soluble in acids and alkalis; melts above 1500°C and decomposes at 900–1000°C. It is highly toxic and carcinogenic and is used in battery electrodes and cadmium plating baths.

cadmium potassium iodide *Inorganic Chemistry.* $CdI_2 \cdot 2KI \cdot 2H_2O$, a deliquescent white powder that yellows with age; soluble in water, alcohol, ether, and acids and decomposes on heating; toxic and carcinogenic; used in analytical chemistry and medicine. Also, POTASSIUM TETRAIODOCADMATE, POTASSIUM CADMIUM IODIDE.

cadmium red *Materials.* a pigment that is made of the sulfide and selinide of cadmium, having a vibrant red color and excellent film-forming properties; used in painting.

cadmium selenide *Inorganic Chemistry.* CdSe, a red or greenish-brown powder that is insoluble in water; melts above 1350°C; used as a red pigment and in photoelectric cells.

cadmium selenide cell *Electronics.* a photoconductive cell that uses a cadmium selenide semiconductor to measure electromagnetic radiation, characterized by a fast response time and high sensitivity to waves emitted by incandescent lamps and some infrared light sources.

cadmium-silver oxide cell *Electricity.* an alkaline-electrolyte cell that can be used without recharging in primary batteries.

cadmium sulfate *Inorganic Chemistry.* $CdSO_4$, colorless, odorless crystals, soluble in water and insoluble in alcohol; melts at 1000°C; used in fluorescent screens, pigments, and electroplating. It is a known carcinogen.

cadmium sulfide *Inorganic Chemistry.* CdS, a brown or yellowish-orange powder that is insoluble in cold water and forms a colloid in hot water; melts at 1750°C under a pressure of 100 atmospheres; used in ceramic glazes, scintillation counters, and solar cells.

cadmium sulfide cell *Electronics.* a photoconductive cell that uses a small wafer of cadmium sulfide, characterized by an extremely high dark to light resistance ratio, to measure electromagnetic radiation.

cadmium telluride *Inorganic Chemistry.* CdTe, brownish-black cubic crystals that are insoluble in water; oxidizes on prolonged exposure to air and melts at 1091°C. It is toxic on inhalation, and it is used in semiconductors.

cadmium telluride detector *Electronics.* a device that measures electromagnetic radiation continuously at ambient temperatures of up to 400°C; used in solar cells and in infrared nuclear radiation and gamma-ray detectors.

cadmium tungstate *Inorganic Chemistry.* $CdWO_4$, white or yellow crystals or powder, very slightly soluble in water and toxic by inhalation; used in fluorescent paint, X-ray screens, and scintillation counters.

cadmium yellow see CADMIUM SULFIDE.

CADUCEUS [kə doo´sē əs] *Artificial Intelligence.* an expert system used for differential diagnosis in internal medicine, developed at the University of Pittsburgh. (From *caduceus,* an ancient symbol that is used to represent the medical profession.)

caducicorn *Vertebrate Zoology.* having horns that are shed and regrown on a regular basis, as is the case with some deer.

caducous *Botany.* of or relating to those plant parts, usually floral, that readily and naturally fall off the plant.

cadwaladerite [kadwäl´ə də rīt´] *Mineralogy.* $Al(OH)_2Cl \cdot 4H_2O$, a lemon-yellow transparent mineral, having a specific gravity of 1.66 and an undetermined hardness; a transparent mineral occurring as amorphous grains and small masses in sulfate deposits with halite.

CAE *Aviation.* the airport code for Columbia, South Carolina.

CAE computer-aided engineering.

Caecilian *Vertebrate Zoology.* a name for various members of the order Apoda, a legless, wormlike amphibian of the tropics.

caecum see CECUM.

Caedibacter *Bacteriology.* a genus of Gram-negative, nonmotile bacteria of uncertain classification, occurring as obligate endosymbionts in the microorganism *Paramecium aurelia.*

Caenolestidae *Vertebrate Zoology.* the opossum rats, a small family of marsupial mammals occurring outside Australia, including many extinct forms. *Paleontology.* a family of marsupials in the extinct suborder Caenolestoidea; rat-sized and mostly insectivorous, they were the most abundant of all small marsupials in South America in the early Miocene; Eocene to Pliocene.

caenostylic *Vertebrate Zoology.* of or describing a condition in which the first two visceral arches are attached to the cranium and function in the intake of food; found in sharks, amphibians, and chimeras.

caeoma [sē ō´mə] *Mycology.* **1.** a former genus of fungi belonging to the order Uredenales, characterized by a fruiting body that lacks an outer wall (or peridium). **2.** a fruiting body or sporocarp lacking an outer wall.

caeomatoid [sē´mə toid´] *Mycology.* referring to a fruiting body or aecium that is irregularly shaped and lacks an outer wall, or peridium.

Caerfaian *Geology.* a European geologic stage of the Lower Cambrian period, occurring after the Precambrian and before the Solvan.

caerulein *Endocrinology.* a peptide hormone that is isolated from frog skin and has sequence homology with cholecystokinin. It is a constrictor of the gallbladder.

Caesalpinoideae *Botany.* a subfamily of tropical and subtropical trees and shrubs of the family Leguminosae, having alternate, pinnate leaves and flowers in panicles or racemes.

caesarean or **Caesarean** see CESAREAN.

caesium chloride density centrifugation [sēz´ē əm] *Molecular Biology.* a type of analytical ultracentrifugation in which DNA or RNA molecules sink through a cesium chloride solution of differential density until they accumulate at characteristic levels in the solution.

caesium chloride density gradient *Molecular Biology.* a tube containing a cesium chloride solution that is highly concentrated at the bottom and progressively less concentrated toward the top; used to separate heavier DNA and RNA molecules from lighter ones.

caespitose see CESPITOSE.

caffeic acid *Organic Chemistry.* $C_9H_8O_4$, yellow prisms or plates, soluble in water and alcohol; decomposes at 223°C.

caffeine *Organic Chemistry.* $C_8H_{10}N_4O_2$, white odorless, hexagonal needles or prisms with a bitter taste, slightly soluble in water and alcohol; it melts at 238°C and sublimes at 178°C; derived by extraction from coffee beans, tea leaves, or kola nuts, and used in beverages and medicine. Caffeine stimulates the central nervous system, has a diuretic effect on the kidneys, and also has many effects on the cardiovascular system.

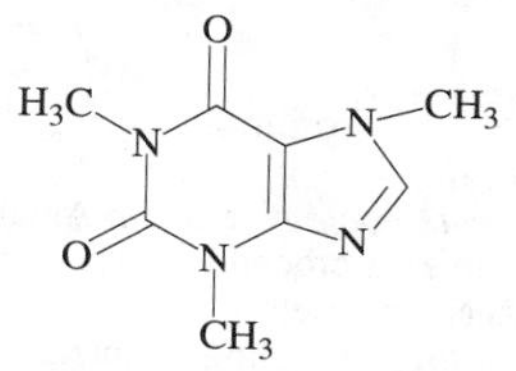

caffeine

caffeinism *Toxicology.* poisoning caused by the consumption of excessive amounts of caffeine; symptoms may include tachycardia, dyspepsia, irritability, and insomnia.

cage *Mechanical Engineering.* **1.** an enclosing device within a bearing that uniformly separates the individual balls or rollers. **2.** the enclosed platform or carriage of an elevator, used especially in mine shafts. *Navigation.* to set up a gyro or lock it in place. *Petroleum Engineering.* **1.** in rod-boring, a structure of elastic iron rods that is placed into the borehole to prevent the rods from vibrating. **2.** in a ball valve, the container that holds the ball, such as the pump often used in oil production. *Mining Engineering.* a usually multidecked structure in a vertical mine shaft used to transport personnel and equipment. *Crystallography.* a void occurring in a crystal structure that is capable of trapping one or more foreign atoms. *Mathematics.* for positive integers g and r, a graph with the least possible number of vertices that is r-regular and has girth g is called an **(r, g)-cage.**

cage antenna *Electromagnetism.* a broadband dipole antenna whose elements are supported such that at each pole the structure resembles a conical or cylindrical cage of wires.

cage compound *Organic Chemistry.* any large organic molecule having an internal space that is completely enclosed.

cage effect *Materials Science.* the recombination of free radicals, a process that limits their efficiency as initiators in addition polymerization.

cage guide *Mining Engineering.* a conductor used to guide cages in a mine shaft to keep them from swinging and colliding with one another. Also, FIXED GUIDE, ROPE GUIDE.

cage mill *Mechanical Engineering.* a grinding machine used to pulverize moist, sticky materials, such as clay, talc, and asbestos. Also, STEDMAN DISINTEGRATOR.

cage paralysis *Neurology.* a complex nutritional deficiency that produces softening and tenderness of the bones; sometimes seen in captive primates.

Cagniard de la Tour, Charles [kän´yärd´ dä lä toor´] 1777–1859, French chemist and engineer; liquefied several gases; determined the critical temperature of water.

Cahn-Hilliard equation *Materials Science.* an equation describing the rate of spinodal decomposition in terms of a free energy gradient.

cahnite *Mineralogy.* $Ca_2B(AsO_4)(OH)_4$, a colorless to white, transparent, brittle tetragonal mineral occurring as usually sphenoidal twinned crystals with a crosslike appearance, having a specific gravity of 3.156 and a hardness of 3 on the Mohs scale; found in cavities in axinite veinlets or implanted on garnet crystals in franklinite ore.

CAI *Aviation.* the airport code for Cairo, Egypt.

CAI computer-aided instruction; computer-assisted instruction.

Cailletet, Louis Paul [kī´ə tā´] 1832–1913, French chemist; the first to liquefy oxygen, hydrogen, and nitrogen.

Cailletet-Mathias law *Physical Chemistry.* the principle that the relationship between the density of a pure liquid at a given temperature and the density of its saturated vapor at the same temperature is a linear function of that temperature.

caiman [kā´mən] *Vertebrate Zoology.* any of several Central and South American reptiles of the genus *Caiman,* some species of which are endangered. Also, CAYMAN.

caiman

caiman lizard *Vertebrate Zoology.* a South American lizard, *Dracaena guianensis,* that resembles a crocodile, having powerful jaws used to crush its food of snails and mussels.

Cain complex *Psychology.* a feeling of intense rivalry and hatred toward a brother. (From the Biblical figure *Cain,* who murdered his brother Abel.)

cainophobia see NEOPHOBIA.

Cainotheriidae *Paleontology.* a family of highly specialized ruminant artiodactyls belonging to the suborder Tylopoda and extinct superfamily Anoplotheroidea; rabbit-sized and characterized by a strongly curved back, they were similar in several ways to modern hares and rabbits, though unrelated; extant in the Eocene and Oligocene.

Cainotherium *Paleontology.* a genus of rabbitlike selenodont artiodactyls belonging to the extinct family Cainotheriidae; extant in Europe in the Eocene to Oligocene.

cairn [kârn; kern] *Archaeology.* **1.** a burial mound constructed chiefly of stone. **2.** a stone pile set up as a landmark, monument, or the like. *Cartography.* an artificial mound of stones or masonry used to designate a point of surveying or of cadastral importance. Also, **carn.**

cairngorm *Mineralogy.* a smoky-yellow to dark brown or black type of quartz, often used as a gem. Also, SMOKY QUARTZ.

Cairns' mechanism *Genetics.* a mechanism for the replication of a double-stranded circular DNA molecule, as in bacterial DNA, in which replication begins at a fixed point and progresses around the circle in opposite directions, with both strands being replicated at the same time, ending at a defined terminus. (After John *Cairns,* its discoverer.)

caisson [kā´sän´] *Ordnance.* **1.** a two-wheeled vehicle used to carry artillery ammunition. **2.** an ammunition chest. **3.** formerly, a chest filled with explosives that detonated in the path of the enemy.

caisson disease *Medicine.* an illness caused by decompression after deep-sea diving.

cajeput oil *Materials.* a greenish-blue aromatic oil that is derived from the leaves and twigs of a tree in the myrtle family, *Melaleuca leucadendron,* native to Australia and New Guinea; used in medicine and perfumes.

cajeputol see EUCALYPTOL.

cajuput oil see CAJEPUT OIL.

cajú rains *Meteorology.* the light showers that occur during October in the northeastern region of Brazil.

CAK *Aviation.* the airport code for Akron/Canton, Ohio.

cake *Metallurgy.* **1.** a rectangular casting of a copper-base product, used as a starting stock for rolling into sheet. **2.** the coalesced portion of an unpressed metal powder. *Mining Engineering.* any drill sludge that has hardened, especially on the wall of a borehole. *Hydrology.* see ICE CAKE.

cake flour *Food Technology.* a wheat flour having a low gluten content, thus yielding tender and friable dough.

cake of gold *Metallurgy.* unmelted gold created by removing mercury from a gold-mercury amalgam.

caking *Engineering.* the fusing of a powdery substance into a solid mass by heat, pressure, or water.

caking coal *Geology.* a type of coal that softens and cakes upon heating and produces a hard, gray cellular coke.

Cal or **Cal.** large calorie.

cal. or **cal** small calorie; caliber.

Calabar swelling *Medicine.* an egg-sized swelling in the tissues caused by the body's allergic reaction to the organism *Loa loa,* a nematode found in Central and West Africa.

Calabrian [kə lāb´rē ən] *Geology.* **1.** a European geologic stage of the Lower Pleistocene epoch, following the Astian stage of the Pliocene epoch. **2.** in France and Italy, the marine equivalent of the terrestrial Villafranchian stage.

calaite see TURQUOISE.

calamine *Pharmacology.* a protectant and mild astringent consisting of zinc oxide, ZnO, with a small proportion of ferric oxide, Fe_2O_3, occurring as a fine pink powder or in solution as **calamine lotion;** used as a topical treatment for skin disorders. *Mineralogy.* a name used in the United States for hemimorphite, a commercial mining and metallurgical term for oxidized ores of zinc.

calamine brass *Metallurgy.* an alloy of copper and zinc carbonate, formerly used to imitate gold.

Calamitales *Paleontology.* an extinct order of arborescent Carboniferous sphenopsids; they were reedlike plants that grew to 30 meters in height and 30 centimeters in diameter.

calamus see QUILL.

calamus oil *Materials.* a yellow aromatic oil that is derived from the roots of the sweet flag plant, *Acorus calamus;* used in medicine and perfumes.

calandria *Chemical Engineering.* a closed vessel through which tubes carry the heating fluid in a vacuum-evaporating system.

Calanoida *Invertebrate Zoology.* a suborder of free-living plankton, in the crustacean Copepoda; vast numbers in the oceans form an important part of the marine food chain.

Calappidae *Invertebrate Zoology.* a family of box crabs, in the subsection Oxystomata of the section Brachyura.

calathiform *Biology.* shaped like a vase or cup; being almost hemispherical.

calaverite *Mineralogy.* $AuTe_2$, a brass-yellow to tin-white opaque, metallic, very brittle monoclinic mineral that often contains silver as well as gold, and is found in vein deposits that contain native gold, fluorite, quartz, sulfides, sulfosalts, and tellurides, having a specific gravity of 9.1 to 9.4 and a hardness of 2.5 to 3 on the Mohs scale; one of the principal gold ore minerals of Colorado and California.

Calberla's solution *Microbiology.* a medium composed of glycerol, ethanol, and distilled water that is used to mount specimens of airborne microbes for examination with a rotorod instrument.

calbindin *Endocrinology.* a member of a family of intracellular calcium binding proteins, believed to be induced by vitamin D, and found in a variety of cell types; these proteins are thought to serve as buffers that prevent large changes in intracellular calcium concentration.

calc- a combining form meaning "calcium," as in *calcaroid.*

calc-alkalic *Petrology.* **1.** describing igneous rock that has an alkali-lime index between 55 and 61. Thus, **calk-alkalic series. 2.** describing igneous rock that contains plagioclase feldspar.

calcaneal *Anatomy.* of or relating to the calcaneus, or heel bone.

calcaneitis *Medicine.* an inflammation of the heel bone.

calcaneocuboid ligament *Anatomy.* any of the ligaments attaching the calcaneus with the cuboid bone in the foot

calcaneodynia *Medicine.* a pain in the heel.

calcaneofibular *Anatomy.* of or relating to the heel bone and the fibula.

calcaneoplantar *Anatomy.* of or relating to the heel bone and the sole.

calcaneotibial *Anatomy.* of or relating to the heel bone and the tibia.

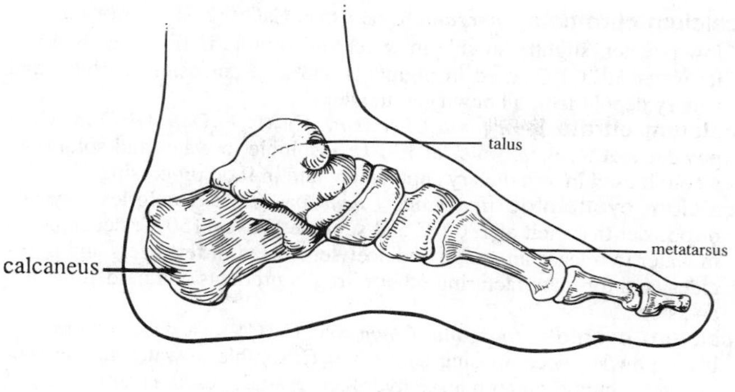

calcaneus

calcaneus *Anatomy.* the tarsal bone that forms the heel; the heel bone. Also, **calcaneum**.

calcar *Biology.* **1.** a spur or spurlike process projecting from the leg of a bird. **2.** a bony or cartilaginous process on the heel bone of bats, which helps to support the portion of the wing membrane lying between the legs. **3.** in insects, a tibial spine.

calcarate *Botany.* having a spur.

Calcarea *Invertebrate Zoology.* a class of the phylum Porifera, that includes sponges with calcium carbonate skeletons.

calcarenite *Petrology.* a type of limestone or dolomite composed of carbonate sand of a chemical or biochemical origin, consisting of oolites, coral or shell fragments, or debris formed by erosion of older limestones, with particle size ranging from .0625 to 2 millimeters.

calcareous [kal kâr′ə əs] *Science.* composed of, containing, or resembling calcium carbonate.

calcareous algae *Botany.* algae that thrive in environments permeated with lime.

calcareous crust see CALICHE.

calcareous ooze *Geology.* a deep-sea sediment ooze containing calcium carbonate from skeletal remains.

calcareous schist *Petrology.* a coarse-grained metamorphic rock derived from calcareous sediment containing impurities.

calcareous sinter see TRAVERTINE.

calcareous soil *Agronomy.* soil that contains enough calcium carbonate to foam when treated with hydrochloric acid.

calcareous tufa see TUFA.

calcarine sulcus *Anatomy.* a fissure on the dorsal median surface of each cerebral hemisphere, running horizontally and separating the occipital lobe from the lingual gyrus. Also, **calcarine fissure.**

Calcaronea *Invertebrate Zoology.* a subclass of sponges, in the class Calcarea, with free-swimming flagellate larvae.

Calceola *Paleontology.* a genus of solitary rugose corals in the suborder Cystiphyllina, that lay on the sea floor with the convex side down; extant only in the Devonian.

calci- a combining form meaning "calcium," as in *calcification.*

calcia [kal′sē ə] *Materials.* a chemical name for lime; a calcium oxide that occurs naturally in limestone, coral, shells, and chalk; used in mortar, cement, as an absorbent, and for liming acid soils.

calcic *Science.* derived from or containing calcium or lime. *Geochemistry.* describing minerals and rocks having a high calcium content. *Petrology.* describing igneous rock having an alkali-lime index greater than 61.

calcicole [kal′sə kōl′] *Botany.* of or relating to a plant that thrives in lime-rich, chalky soil. *Ecology.* a plant that thrives in soils rich in calcium salts. Also, CALCIPHILE.

calcicolous *Botany.* growing on chalky soil or on limestone.

calcicosis [kal′si kō′sis] *Medicine.* a disease of the lung caused by breathing marble dust.

calciferol see ERGOCALCIFEROL.

calciferous [kal sif′ə rəs] *Biology.* containing or producing calcium, calcite, or calcium carbonate.

calciferous gland *Invertebrate Zoology.* a gland that secretes calcium carbonate into the esophagus of certain oligochaete worms.

calcification [kal′sə fi kā′shən] *Geochemistry.* a process in which the colloids in surface soil become highly saturated with calcium, making them relatively fixed and more or less neutral in reaction. *Physiology.* the deposition of calcium salts in various body tissues such as the muscles or bones.

calcifuge [kal′sə fyooj′] *Botany.* of or relating to a plant that grows in lime-deficient soils. *Ecology.* a plant that does not tolerate soils rich in calcium salts. Also, CALIPHOBE.

calcilutite *Petrology.* **1.** consolidated mud of calcium carbonate formed of calcite grains or crystals with a diameter less than .0625 millimeter. **2.** a nonsiliceous dolomite or limestone formed of calcerous rock flour.

calcimeter [kal sim′ə tər] *Engineering.* a device used to determine the approximate amount of lime contained in a soil.

calcination *Chemical Engineering.* the heating of a solid to a high temperature, below its melting point, to create a condition of thermal decomposition or a phase transition other than melting or fusing.

calcine [kal′sīn′] *Engineering.* **1.** to reduce a substance to calx or powder by means of heat; also, of or relating to such a calx or powder. **2.** to roast metallic ore in order to eliminate sulfur or carbon dioxide; also, the ore so treated. **3.** a ceramic material or mixture fired, but not fused, for use as a constituent in a ceramic composition.

Calcinea *Invertebrate Zoology.* a subclass of sponges, in the class Calcarea, with free-swimming ciliate larvae.

calcined gypsum see PLASTER OF PARIS.

calcined soda see SODA ASH.

calcining furnace *Engineering.* a furnace in which ores or metallurgical products are calcined. Also, **calciner.**

calcinosis *Medicine.* a disease in which calcium salts accumulate in body tissues. Also, CALCIUM GOUT.

calcioferrite *Mineralogy.* $Ca_4Fe^{+2}(Fe^{+3}, Al)_4(PO_4)_6(OH)_4 \cdot 13H_2O$, a yellowish white to yellow-green, opaque, brittle monoclinic mineral of the montgomeryite group, having a specific gravity of 2.53 and a hardness of 2.5 on the Mohs scale;

Calciosoleniaceae *Botany.* a family of flagellate algae of the order Coccosphaerales that occur in marine phytoplankton and are characterized by diamond-shaped coccoliths and cylindrical cells.

calciovolborthite *Mineralogy.* $CaCu(VO_4)(OH)$, a greenish-yellow subtranslucent orthorhombic mineral; having a specific gravity of 3.75 and a hardness of 3.5 on the Mohs scale; found as a secondary mineral in oxidation zones of deposits containing vanadium minerals.

calcipenia *Medicine.* a deficiency of calcium.

calciphile see CALCICOLE.

calciphobe see CALCIFUGE.

calcirudite *Petrology.* dolomite or limestone formed for the greater part of worn or broken pieces of coral or shells or of limestone fragments over 2 mm in diameter; the interstices are filled with sand, calcite, or mud, and the whole is bound together with a calcareous cement.

calcisiltite *Petrology.* limestone consisting of silt-sized grains of calcite.

Calcisol *Geology.* a zonal soil characterized by a neutral or calcareous A horizon, no B horizon, and a calcareous C horizon.

calcite [kal′sīt′] *Mineralogy.* $CaCO_3$, a colorless to white trigonal mineral of the calcite group that occurs in highly varied crystal forms, having a specific gravity of 2.71 and a hardness of 3 on the Mohs scale; as stalactites and stalagmites in caverns, in geodes, and as a petrifying material replacing fossil animal and plant remains; the chief ingredient of all limestones and marble, and one of the most widely distributed minerals in the world; trimorphous with aragonite and vaterite, and forms a series with rhodochrosite. Also, CALCSPAR.

calcite

calcite compensation depth *Geology.* the depth in the ocean below which calcium carbonate dissolves at a faster rate than it is deposited. Also, COMPENSATION DEPTH, CARBONATE COMPENSATION DEPTH, CALCIUM CARBONATE COMPENSATION DEPTH.

calcite dolomite *Petrology.* a carbonate rock composed of 10–50% calcite and 50–90% dolomite.

calcitonin *Biochemistry.* a polypeptide hormone that is produced in the thyroid and parathyroid glands as a result of a condition called hypercalcemia, one of whose effects is to lower the level of calcium and phosphate in plasma. Also, THYROCALCITONIN.

calcitonin gene-related peptide *Biochemistry.* a 37-amino-acid-long neuropeptide that is structurally homologous to salmon calcitonin.

calcium *Chemistry.* a common alkaline earth element having the symbol Ca, the atomic number 20, an atomic weight of 40.08, a melting point of 845°C, and a boiling point of 1480°C; a fairly soft, silver-white metal found in milk, bone, chalk, and limestone and used as an alloying agent and deoxidizer; compounds are used in cooking, as a bleaching agent, and in making plaster. (From the Greek word for "lime" or "limestone.")

calcium-45 *Nuclear Physics.* a radioactive isotope of calcium with a half-life of 165 days; commonly used as a tracer in the study of calcium metabolism in humans.

calcium acetate *Organic Chemistry.* $Ca(CH_3COO)_2$, brown or gray hygroscopic crystals, soluble in water, slightly soluble in methanol, and insoluble in ethanol, acetone, and benzene; decomposes at 160°C; used in the manufacture of acetic acid and acetone, in dyeing, tanning, and curing, and as a food stabilizer and corrosion inhibitor.

calcium acrylate *Organic Chemistry.* $Ca(CH_2{=}CHCOO)_2$, a white powdery solid, deliquescent, soluble in water; used in making foundry molds, as a soil stabilizer and sealant for oil wells, as a binder for clay products, and as an ion exchanger.

calcium arsenate *Inorganic Chemistry.* $Ca_3(AsO_4)_2$, a white powder that is slightly soluble in water and soluble in dilute acids and that decomposes on heating; used as an insecticide and a germicide.

calcium arsenite *Inorganic Chemistry.* $CaAsO_3H$, white granules or powder, insoluble in water and soluble in acids; used as an insecticide and a germicide.

calcium ATPase *Enzymology.* an enzyme that catalyzes the removal of phosphate from an adenosine triphosphate molecule, and requires calcium.

calcium biphosphate see MONOBASIC CALCIUM PHOSPHATE.

calcium bisulfite *Inorganic Chemistry.* $Ca(HSO_3)_2$, a yellowish solid with a strong odor, corrosive to metal and irritating to skin and tissue; used as a bleach and preservative.

calcium bromate *Inorganic Chemistry.* $Ca(BrO_3)_2{\cdot}H_2O$, a white, crystalline powder; loses its water at 180°C; very soluble in water; used as a maturing agent.

calcium bromide *Inorganic Chemistry.* $CaBr_2$, colorless, highly deliquescent crystals or needles, soluble in water and alcohol; melts at 730°C and boils at 806–812°C; used in photography, freezing mixtures, food preservatives, fire retardant devices, and for various other industrial purposes; formerly used as a sedative and anticonvulsant.

calcium carbide *Organic Chemistry.* CaC_2, a colorless or grayish-black solid crystal that decomposes in water and melts at approximately 2300°C; used in the generation of acetylene gas for welding, and as a reagent.

calcium carbonate *Inorganic Chemistry.* $CaCO_3$, a white powder or colorless crystals, very slightly soluble in water. One of the most abundant inorganic substances in nature, it is the major component of sedimentary rocks, occurring in a relatively pure state as calcite, aragonite, marble, limestone, and chalk, as well as in animal shells and bones. Calcium carbonate has a wide variety of important chemical uses, as in antacids, tooth powders, white paints, cement, and lime, and in the smelting of iron ore.

calcium carbonate compensation depth see CALCITE COMPENSATION DEPTH.

calcium chlorate *Inorganic Chemistry.* $Ca(ClO_3)_2{\cdot}2H_2O$, white to yellowish crystals that decompose rapidly when heated; soluble in water and alcohol; valuable as an herbicide and also used in photography.

calcium chloride *Inorganic Chemistry.* $CaCl_2$, colorless to white deliquescent crystals or cubes, often occurring as a hydrate; used in pulp and paper treatment, in the deicing of highways, and as an antidust agent.

calcium chlorite *Inorganic Chemistry.* $Ca(ClO_2)_2$, white crystals or cubes; decomposes in water; a strong oxidizer.

calcium chromate *Inorganic Chemistry.* $CaCrO_4{\cdot}2H_2O$, a bright yellow powder, slightly soluble in water and soluble in dilute acids; loses its water at 200°C; used in pigments and as a corrosion inhibitor and battery depolarizer; a known carcinogen.

calcium citrate *Inorganic Chemistry.* $Ca_3(C_6H_5O_7)_2{\cdot}4H_2O$, a white powder that loses its water at 120°C; insoluble in water and soluble in alcohol; used in as a dietary supplement and in food processing.

calcium cyanamide *Inorganic Chemistry.* $CaCN_2$, colorless crystals or powder that melt at 1300°C and sublime above 1150°C; decomposes in water to release ammonia and acetylene; used in fertilizers and pesticides, and for manufacturing other nitrogen products and hardening iron and steel.

calcium cyanide *Inorganic Chemistry.* $Ca(CN)_2$, a white or grayish-black powder, decomposing above 350°C; soluble in water and decomposing in humid air to release hydrogen cyanide; used as an insect and rodent poison and in fumigation, and also for the leaching of gold and silver ore.

calcium cyclamate dihydrate *Organic Chemistry.* $Ca(C_6H_{11}NHSO_3)_2{\cdot}2H_2O$, white crystals, soluble in water and almost insoluble in alcohol; used as a food sweetener.

calcium dihydrogen phosphate see CALCIUM PHOSPHATE.

calcium dihydrogen sulfite see CALCIUM BISULFITE.

calcium dioxide see CALCIUM PEROXIDE.

calcium fluoride *Inorganic Chemistry.* CaF_2, a colorless to white powder that occurs in nature as fluorite or fluorspar; melts at 1423°C and boils at about 2500°C; used in etching glass, lasers, and electronic spectroscopy.

calcium fluoride structure *Materials Science.* a crystal structure in which the unit cell has calcium (Ca^{2+}) ions located at all four of the face-centered cubic, unit-cell sites and eight fluoride (Fl^-) ions occupying all of the tetrahedral interstitial sites.

calcium folinate see LEUCOVORIN CALCIUM.

calcium gluconate *Organic Chemistry.* $Ca(C_6H_{11}O_7)_2$, a white powder or granules, soluble in water and insoluble in alcohol; loses water at 120°C; used as a food additive and in vitamin tablets.

calcium gout see CALCINOSIS.

calcium hardness *Chemistry.* a term for hardness of water that is caused by the presence of calcium ions from dissolved carbonates and bicarbonates.

calcium hydrate see CALCIUM HYDROXIDE.

calcium hydride *Inorganic Chemistry.* CaH_2, white crystals that decompose when heated to 600°C; decomposes in water; used in the production of titanium and zirconium and as a cleaner for blocked oil wells.

calcium hydrogen phosphate see CALCIUM PHOSPHATE.

calcium hydrogen sulfite see CALCIUM BISULFITE.

calcium hydroxide *Inorganic Chemistry.* $Ca(OH)_2$, a soft white crystalline powder with an alkaline taste, derived from the action of water on calcium oxide. It absorbs carbon dioxide from the air, and it melts and loses its water at 580°C. It is used in mortar, plasters, cements, as a whitewash and disinfectant, and in soil conditioning, water softening, and many other industrial processes. Also known by various other names, such as SLAKED LIME, HYDRATED LIME, and CAUSTIC LIME.

calcium hypochlorite *Inorganic Chemistry.* $Ca(OCl)_2$, a white crystalline solid that decomposes at 100°C; soluble in cold water; a stable chlorine carrier. It is widely used as a bleaching agent, disinfectant, algicide, and bactericide.

calcium iodate *Inorganic Chemistry.* $Ca(IO_3)_2$, colorless to white crystals or powder; decomposes at 540°C; soluble in water and insoluble in alcohol; used as a food additive and as an antiseptic.

calcium iodide *Inorganic Chemistry.* $CaI_2{\cdot}6H_2O$, a deliquescent yellowish-white powder that is soluble in water; melts at 784°C and boils at about 1100°C; used in photography and medicine.

calcium iodobehenate *Organic Chemistry.* $Ca(OOCC_{21}H_{42}I)_2$, a white powder, soluble in cold water and insoluble in alcohol; used in medicine and pharmaceuticals.

calcium lactate pentahydrate *Organic Chemistry.* $Ca(C_3H_5O_3)_2{\cdot}5H_2O$, white needles that are soluble in water and insoluble in alcohol; used as a calcium supplement and as a blood coagulant.

calcium leucovorin see LEUCOVORIN CALCIUM.

calcium metabolism *Biochemistry.* chemical processes that act to maintain a constant level of calcium concentration in plasma and produce enough calcium to mineralize bones.

calcium metasilicate *Inorganic Chemistry.* $CaSiO_3$, a colorless, absorbent white powder that is insoluble in water; used in the manufacture of glass and Portland cement.

calcium molybdate *Inorganic Chemistry.* $CaMoO_4$, a colorless to white crystalline powder, insoluble in water, alcohol, and ethanol; used as an alloying agent.

calcium naphthenate *Organic Chemistry.* a light, sticky, water-insoluble calcium compound derived from a cycloparaffin hydrocarbon (usually a cyclopentane or cyclohexane); used as a hardening agent in plastic compounds, in waterproofing adhesives, and in wood fillers and varnishes.

calcium nitrate *Inorganic Chemistry.* $Ca(NO_3)_2 \cdot 4H_2O$, a white deliquescent mass, soluble in water and alcohol; melts at 42.7°C and decomposes at 132°C; a strong oxidizer used as a fertilizer and in explosives.

calcium nitride *Inorganic Chemistry.* Ca_3N_2, brown hexagonal crystals that melt at 1195°C and decompose in water; soluble in dilute acids and insoluble in alcohol.

calcium nitrite *Inorganic Chemistry.* $Ca(NO_2)_2 \cdot H_2O$, colorless to yellowish deliquescent crystals, soluble in water and slightly soluble in alcohol; loses its water at 100°C; used as a corrosion inhibitor.

calcium orthoarsenate $Ca_3(AsO_4)_2$, a white powder that is insoluble in water; used as an herbicide and insecticide.

calcium orthophosphate see MONOBASIC CALCIUM PHOSPHATE, DIBASIC CALCIUM PHOSPHATE, TRIBASIC CALCIUM PHOSPHATE.

calcium oxalate *Inorganic Chemistry.* CaC_2O_4, a colorless to white crystalline powder, a salt of oxalic acid; insoluble in water; a strong irritant, used in rare earth separations.

calcium oxide *Inorganic Chemistry.* CaO, an odorless, white or grayish-white solid that is very slightly soluble in water; melts at 2614°C and boils at 2850°C. It reacts with water to form calcium hydroxide, and is prepared commercially by heating limestone (calcium carbonate) until the carbon dioxide is expelled. It has a number of important uses; for example, in steel manufacturing and in the refining of aluminum, copper, and zinc; in fertilizers, insecticides, and fungicides; as a water softener and as a soil conditioner and stabilizer; in sewage treatment, and in sugar refining and glass manufacture. Also, LIME, QUICKLIME, BURNT LIME, CALX.

calcium pantothenate *Organic Chemistry.* $Ca(C_9H_{16}NO_5)_2$, white powder, insoluble in alcohol, slightly hygroscopic; it melts at 170–172°C and decomposes at 195–196°C; it is used in medicine, animal feeds, and dietary supplements.

calcium peroxide *Inorganic Chemistry.* CaO_2, a white or yellowish odorless powder that is slightly soluble in water and soluble in acids; decomposes at 275°C; it is used in dentifrices, antiseptics, and detergents.

calcium phosphate *Inorganic Chemistry.* **1.** see MONOBASIC CALCIUM PHOSPHATE, DIBASIC CALCIUM PHOSPHATE, TRIBASIC CALCIUM PHOSPHATE. **2.** any of various other phosphates of calcium, such as calcium metaphosphate, $Ca(PO_3)_2$, or calcium hypophosphate, $Ca_2P_2O_6 \cdot 2H_2O$.

calcium phosphate transfection *Biotechnology.* a method for the integration of viral DNA into a eukaryote cell's chromosomes in which calcium phosphate serves as the facilitator.

calcium phosphide *Inorganic Chemistry.* Ca_3P_2, red-brown crystals or gray lumps that melt at about 1600°C; used as a rodent poison and in explosives and fireworks.

calcium plumbate *Inorganic Chemistry.* Ca_2PbO_4, orange to brown crystals that are decomposed by heat; insoluble in cold water and decomposes in hot water; used in storage batteries and in the manufacture of glass and matches.

calcium pyrophosphate *Inorganic Chemistry.* $Ca_2P_2O_7$, a white powder, insoluble in water and soluble in dilute acids; melts at 1230°C; used as a polish and abrasive and in food supplements.

calcium resinate *Organic Chemistry.* a flammable, yellowish-white, amorphous powder that is insoluble in water and soluble in acid; prepared by boiling rosin with calcium hydroxide and filtering; it is used in waterproofing, tanning leathers, and in making paint driers and enamel. Also, LIMED ROSIN.

calcium reversal lines *Spectroscopy.* in the spectra of certain stars, narrow, bright calcium emission lines that appear in the center of broad calcium absorption bands.

calcium star *Astronomy.* a former name for a star of spectral type F.

calcium stearate *Organic Chemistry.* $Ca(C_{18}H_{35}O_2)_2$, a white powder, insoluble in water and soluble in alcohol; melts at 179°C; used as a water repellant, in paints, emulsions, cements, and cosmetics, and as a food additive.

calcium sulfate *Inorganic Chemistry.* $CaSO_4$, a grayish-white dense powder that occurs in nature in both an anhydrous form (anhydrite) and hydrated form (gypsum); it is also the by-product of many chemical reactions. It has many industrial uses; for example, as a source of sulfur and sulfuric acid, in cements, tiles, and plaster, as a soil conditioner, in paints, dyes, and polishes, and as a food additive.

calcium sulfide *Inorganic Chemistry.* CaS, a yellow to light-gray powder with an unpleasant odor and taste, decomposed by acids and partly decomposed by water; used as a base for luminescent materials and as a flotation agent.

calcium sulfite *Inorganic Chemistry.* $CaSO_3 \cdot 2H_2O$, a white powder that loses its water at 100°C; derived from the action of sulfurous acid on calcium carbonate; used in the manufacture of wood pulp and as a disinfectant in the sugar industry.

calcium tungstate *Inorganic Chemistry.* $CaWO_4$, white tetragonal crystals, insoluble in water and decomposed by hot acids; used in the manufacture of luminous paints and fluorescent lamps.

calciuria *Medicine.* the presence of calcium in the urine.

Calcivirus *Virology.* the sole genus of the family Calciviridae, including multiple strains of feline calcivirus, San Miguel sea lion virus, vesicular exanthema of swine virus, and possibly Norwalk virus.

calclacite [kalk´lə sīt´] *Mineralogy.* $CaCl_2 \cdot Ca(C_2H_3O_2)_2 \cdot 10H_2O$, a white monoclinic or triclinic mineral; having a specific gravity of 1.5 and an undetermined hardness on the Mohs scale; occurs as hairlike efflorescences, found as a formation on pottery, and on calcareous rocks and fossils kept in oak museum cases.

Calclamna *Paleontology.* an extinct genus of holothuroids in the order Dendrochirotida; relatives of the modern sea cucumbers; extant in the Jurassic to Eocene.

calclithite *Petrology.* a litharenite with 50% or more fragments of older eroded limestone.

calcrete [kal´krēt´] *Geology.* **1.** a conglomerate of surface sand and gravel cemented into a hard mass by calcium carbonate. **2.** in a semiarid climate, a hard crust containing calcium carbonate formed on the surface of a soil.

calc schist *Petrology.* a schistous marble derived from metamorphism of an impure limestone.

calc-silicate *Geology.* describing a metamorphic rock that is largely composed of calcite and calcium-bearing silicates. Also, LIME-SILICATE.

calc-silicate hornfels *Petrology.* a contact metamorphic rock with a fine grain of calcium silicate minerals such as lime-garnet, vesuvianite, tremolite, wollastonite, and epidote.

calc-silicate marble *Petrology.* marble with conspicuous calcium silicate or magnesium silicate minerals.

calc-sinter see TRAVERTINE.

calcspar see CALCITE.

calc-tufa see TUFA.

calculated address *Computer Programming.* an address computed by a program instruction, often based on intermediate results or prespecified criteria.

calculator *Computer Technology.* **1.** any device that is used to perform arithmetic operations, usually under direct operator intervention but sometimes programmable. Also, **calculating machine. 2.** specifically, a small hand-held electronic device performing arithmetic operations.

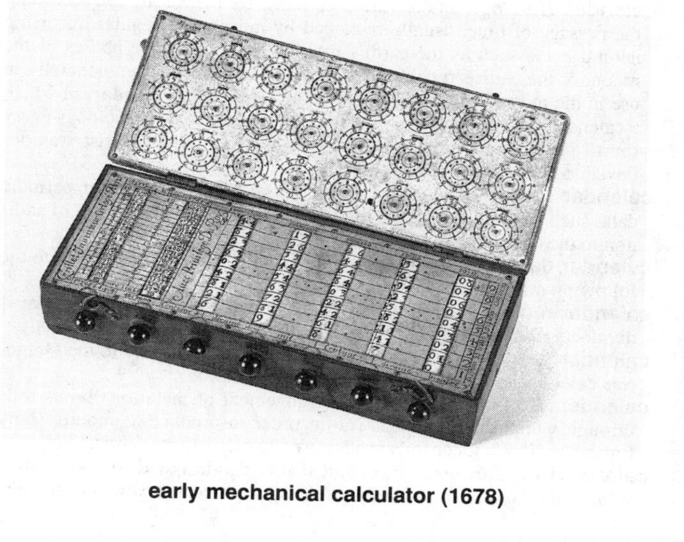

early mechanical calculator (1678)

calculus *Mathematics*. the branch of mathematics that deals with differentiation and integration of functions and related topics. Historically, the calculus arose from attempts to calculate precisely the slope of a curved line and the area of the region bounded by a curve. *Artificial Intelligence*. a formal mathematical treatment of logic, knowledge, and thought, using predicate and propositional calculus. *Pathology*. a mass of crystallized or precipitated salts or other material such as cholesterol that is formed within a body chamber such as the gallbladder, kidney, or urinary bladder. Also, STONE.

calculus of finite differences *Mathematics*. a method of interpolation using difference operators that evaluate functions at points of an arithmetic progression. Analogs of differentiation and integration are defined. Also, **calculus of enlargement.**

calculus of residues *Mathematics*. a method of applying the Cauchy residue theorem and related theorems of complex analysis to perform calculations such as the evaluation of contour integrals, computation of residues of meromorphic functions, and series expansions of meromorphic functions.

calculus of tensors *Mathematics*. those techniques in differential geometry that deal with differentiation of tensors in which the tensor components are represented by symbols with multiple subscripts and superscripts. The chief aim of tensor calculus is the investigation of relations that remain valid after coordinate transformations; significant applications are the theory of relativity, differential geometry of space, and many areas of theoretical physics. Also, RICCI CALCULUS.

calculus of variations *Mathematics*. the study of problems concerned with determining maxima or minima of a given functional or definite integral, relative to a class of functions or to the dependent variables of the integrand function. Historically, this area of mathematics arose from the brachistochrone problem posed by Johann Bernoulli in 1696.

calculus of vectors *Mathematics*. the area of calculus concerned with integration and differentiation of vector-valued functions; a subarea of tensor calculus since vectors are the simplest tensors.

caldera *Volcanology*. a large, more or less circular volcanic crater usually resulting from the collapse of underground lava reservoirs, and having a diameter that is many times greater than that of the vent.

Calderone, Mary born 1904, American physician; promoter of sex education and family planning.

caldesmon *Biochemistry*. a calcium-binding protein that is abundant in smooth muscle and that may be involved in the calcium-dependent control of contraction; it also binds calmodulin.

Caledonian orogeny *Geology*. a late Silurian to early Devonian deformation of the earth's crust, extending from Ireland and Scotland through Scandinavia.

Caledonides *Geology*. the mountain system formed by the Caledonian orogeny.

caledonite [kal′ə də nīt′] *Mineralogy*. $Pb_5Cu_2(CO_3)(SO_4)_3(OH)_6$, a brittle, green, transparent to translucent orthorhombic mineral occurring as small, elongated prismatic crystals, having a specific gravity of 5.76 and a hardness of 2.5 to 3 on the Mohs scale; it is found in small amounts in the oxidization zone of copper-lead deposits.

calefaction *Engineering*. the process of warming or the state of being warmed.

calendar *Horology*. any of various systems for measuring and recording the passage of time, usually reckoned by means of a regular, recurring natural cycle such as the earth's rotation on its axis, the phases of the moon, or the earth's revolution around the sun. The system generally in use in the modern industrial world is the **Gregorian calendar,** in which a calendar year consists of 365 days (with a leap year of 366 days every fourth year); years are numbered from the year Jesus Christ was believed to have been born.

calendar clock *Horology*. a clock that indicates some other periodic data, such as the date, day of the week, or phases of the moon, in addition to the time of day.

calendar day *Horology*. a period of 24 hours, from one midnight to the following midnight.

calendar month *Horology*. one of the periods into which a calendar is divided, ordinarily twelve but ten or thirteen in some earlier systems.

calendar year *Horology*. a period of one year according to the Gregorian calendar or some other system. See CALENDAR.

calender *Engineering*. a vertical arrangement of metal or fibrous rolls through which dried paper is run in order to produce a smooth, shiny finish. *Textiles*. a mechanical roller used in calendering fabric.

calendering *Textiles*. a mechanical process designed to give a stiff, shiny finish to fabrics. *Materials Science*. a process in which a softened molten or solid plastic is pulled through rips in a set of rolls rotating with a speed differential; the rolls squeeze the plastic into a thin oriented sheet or film of the polymer.

calendrical age determination *Archaeology*. a method of absolute dating used when an object has been inscribed with a date from an ancient calendrical system that can be correlated with a modern calendar.

calf *Agriculture*. a young cow or bull. *Vertebrate Zoology*. the young of certain other large mammals, such as the moose, elephant, or whale. *Oceanography*. see CALVED ICE.

calf diphtheria *Veterinary Medicine*. an infection due to a species of the bacterium *Sphaerophorus* that causes ulceration and necrosis of the mouth and pharynx of calves. Also, NECROTIC SOMATITIS, MALIGNANT SOMATITIS.

calf scours *Veterinary Medicine*. a contagious bacterial disease affecting calves within the first few weeks of life; caused by *Escherichia coli*, probably aggravated by stress and poor nutrition, and marked by whitish diarrhea, emaciation, and weakness. Also, SCOURS, WHITE SCOURS.

caliber *Ordnance*. **1.** the outer diameter of a projectile or the inner diameter of a gun barrel or launching tube. **2.** a measure of the length of a cartridge case, gun barrel, or launching tube.

calibrate *Science*. **1.** to determine or mark a correct value using a meter or other measuring instrument. **2.** to determine a measured value by adjusting for instrument error or environmental conditions.

calibrated airspeed *Aviation*. an airspeed value obtained from a barometric airspeed indicator corrected for instrument error and attitude but not (as with true airspeed) for altitude.

calibrated altitude *Navigation*. the measured angular altitude of a celestial body after correction is made for calibration error in the sextant.

calibrating tank *Engineering*. a tank that verifies the volumetric accuracy of liquids with the aid of positive displacement meters.

calibration *Science*. **1.** a measurement or comparison against a standard. **2.** the determination of any equipment deviation from a standard source so as to ascertain the proper correction factors.

calibration baseline *Cartography*. a line on which markers are placed at intervals that are accurately measured so that they can be used for calibrating distance-measuring instruments or equipment.

calibration constants *Photogrammetry*. the numerical results of calibration of an aerial camera, giving the calibrated focal length of the camera-lens unit, the relationship of a principal point of a photograph to the fiducial marks of a camera, and significant corrections for lens distortions.

calibration curve *Engineering*. a line that plots the value for each reading calibrated from a meter or control dial.

calibration fire *Ordnance*. test fire intended to determine the corrections necessary for the individual guns or launchers of a battery to hit the desired target or targets.

calibration markers *Engineering*. the marks presented digitally on a radar screen that show the values for navigational factors, such as bearing, distance, or time.

calibration plate *Photogrammetry*. a glass negative exposed with its emulsion side in the same position as that occupied by the service emulsion at the time of exposure, giving a record of the distance between the camera's fiducial marks. Also, MASTER GLASS NEGATIVE, FLASH PLATE.

calibration point *Ordnance*. the point at which calibration fire is directed.

calibration radio-beacon *Navigation*. a radio beacon used to calibrate radio direction-finding equipment.

calibration reference *Science*. a standard that indicates whether an analytical instrument or procedure is working within prescribed limits.

calibration template *Cartography*. a template made in accordance with the calibration constants of an aerial camera to show the relationship of the principal point of a camera to its fiducial marks, for rapid and accurate marking of the principal points on a series of photographs.

caliche or **caliché** [kə lē′chē] *Archaeology*. a salt encrustation or lime deposit that appears on the surface of materials such as stone, bone, or ceramic after they have been buried or exposed to moisture for an extended time. *Geology*. **1.** in Chile and Peru, a composite of gravel, rock, soil, or alluvium cemented with soluble sodium salts found in nitrate deposits. **2.** in Peru, a thin layer of clayey soil capping a gold vein; in Chile, a whitish clay in the selvage veins; in Mexico, feldspar, white clay, or a compact limestone; in Columbia, a mineral vein recently discovered, or a bank composed of clay, sand, and gravel in placer mining. **3.** in the arid and semiarid regions of the United States, opaque concretions of calcium carbonate and other minerals found in layers at or near the surface of stony soils. Also, CALCAREOUS CRUST.

Caliciviridae *Virology.* a family of RNA viruses that cause gastroenteritis in pigs, cats, calves, sea lions, and human infants.
calico *Textiles.* an inexpensive cotton fabric, heavier than muslin, usually printed with a figured pattern. (Originally made in India and named for *Calicut,* the city from which it was imported.)
calicole see CALCICOLE.
California, Gulf of *Geography.* an inlet of the eastern North Pacific between Baja California and mainland Mexico. Also, SEA OF CORTEZ.
California Achievement Test(s) *Psychology.* a series of standardized tests given to elementary and high school students, including four different levels with similar but increasingly difficult content.
California bearing ratio *Geology.* for a given soil, a measure of its relative resistance to penetration under controlled conditions of density and moisture content.
California Current *Oceanography.* an ill-defined current flowing southward along the west coast of the United States; the major branch of the Aleutian Current.
California encephalitis virus *Virology.* a virus of the family Bunyaviridae that has been isolated from mosquitoes in the southwestern U.S. and associated with encephalitis in humans.
California fog *Meteorology.* **1.** the fog formed on the California coast due to ocean winds that create a displacement effect on the relatively warm surface water, causing the colder, underlying water currents to rise to sea surface. **2.** a morning fog formed in the California coastal valleys due to moist ocean air blowing inland in the afternoon that is subjected to radiational cooling overnight.
California method *Hydrology.* a type of frequency analysis that uses as a parameter the average time between the occurrence of a hydrological quantity and that of an equal or greater quantity.
California Nebula *Astronomy.* an emission nebula of ionized hydrogen in Perseus, so called because its outline resembles that of the state. Also, NGC 1499.
Californian subregion *Ecology.* a distinct zoogeographical region that includes the state of California and adjacent areas north of it along the Pacific Coast.
California Personality Inventory *Psychology.* a set of personality scales used to provide descriptions for normal personalities.
California polymerization *Chemical Engineering.* a catalytic polymerization method to convert C3 and C4 olefins into motor fuel.
California sampler *Mining Engineering.* a type of drive sampler with a retractable, mechanical piston that can be set at any point within the barrel of the sampler.
California-type dredge *Mining Engineering.* a single-lift dredge with a stacker.
californite *Mineralogy.* a compact, green, translucent to opaque variety of vesuvianite that resembles jade; used as an ornamental stone. Also, **California jade.**
californium *Chemistry.* a synthetic radioactive element of the actinide series, having the symbol Cf, the atomic number 98, and an atomic weight of 252. It has several isotopes, and is found only in various compounds. (Named after *California*; it was discovered in 1950 by Glenn T. Seaborg and his associates at the University of California, Berkeley.)
Caligidae *Invertebrate Zoology.* a family of fish ectoparasites, in the crustacean suborder Caligoida, with sucking mouths.
Caligoida *Invertebrate Zoology.* a suborder of the crustacean order Copepoda; all members are fish ectoparasites.
calina *Meteorology.* a summer haze prevalent in Spain and Ecuador, caused by air filled with dust that is swept up from dry ground by strong winds.
caliper *Mechanical Devices.* a two-pronged measuring instrument with an attached scale used to measure linear dimensions between them, such as thickness, or inner and outer diameters of pipes, bores, or other circular hardware. *Graphic Arts.* the thickness of a single sheet of paper, expressed in thousandths of an inch.
caliper gauge *Mechanical Devices.* a micrometerlike device used to take measurements from a caliper. Also, CALLIPER.
caliper log *Mining Engineering.* a continuous record of the variations in diameter or in cross-sectional area of a borehole with depth.
caliper rule *Mechanical Devices.* a graduated scale with a fixed head and a sliding extension used to measure the thickness of boards or the diameters of a shaft or hole.
caliper splint *Medicine.* a splint for the leg consisting of two rods running from the thigh to a metal plate under the foot.
caliper square *Mechanical Devices.* a less accurate version of a vernier caliper.
calite *Metallurgy.* the former name of an oxidation-resistant iron-nickel alloy containing some aluminum and chromium, used as a casting.
calix see CALYX.
calixarene *Organic Chemistry.* basket-shaped molecules derived from the condensation of formaldehyde with phenolic compounds.
calk see CAULK.
calking see BEDDING, def. 2.
call *Computer Programming.* to invoke and transfer control to a subroutine or procedure with the provision that on completion of the subroutine, control is returned to the next sequential instruction in the calling program. *Cartography.* a reference to, or statement of, an object, course, distance, or other matter of description in a land survey, requiring or calling for a corresponding object on the land.
calla [kal′ə] *Botany.* **1.** any of several plants of the genus *Zantedeschia* having arrow-shaped leaves and a yellow spike enclosed in a large white spathe. Also, **calla lily. 2.** a related plant, *Calla palustris,* having heart-shaped leaves.
call announcer *Telecommunications.* an instrument that receives pulse signals and translates them into words that may be heard by a manual operator.
Callavia *Paleontology.* a genus of lower Cambrian trilobites in the family Olenellidae; Callavia has short spines and no axial spine.
call-back security *Computer Technology.* a security measure designed to prevent unauthorized access to computers via dial-up lines. A user calls the computer by telephone and enters the user's identification; the computer then hangs up and calls the user back at the user's prespecified number and only then allows log-in.
call by name *Computer Programming.* a method of transferring arguments from the calling program to the called subroutine by passing an expression representing the argument in the calling sequence; a subprogram created by the compiler that computes the value of the expression.
call by reference *Computer Programming.* a method of transferring arguments from the calling program to the called subroutine by providing the address of the memory location where the argument can be found. The subroutine uses or modifies this memory location directly.
call by value *Computer Programming.* a method of transferring arguments from the calling program to the called subroutine by providing the value of the argument itself in the calling sequence.
call by value/result *Computer Programming.* a transfer of arguments to a subroutine using call by value, with results transferred back in a similar manner.
call circuit *Electricity.* a communications circuit between switching points for transmitting switching instructions; used by traffic forces.
Callendar, Hugh 1863–1930, English physicist; invented the platinum resistance thermometer, compensated air thermometer, and continuous-flow calorimeter.
Callendar-Barnes' continuous-flow calorimeter *Engineering.* a calorimeter in which the heat to be measured is absorbed in water that flows through a tube, and the quantity of heat is then determined by the difference of temperature at each end of the tube and by the rate of the flow.
Callendar's equation *Thermodynamics.* **1.** an equation of state for water vapor whose temperature is well above the boiling point but well below the critical temperature. This equation does not apply at the critical point or in the liquid-vapor region. **2.** an equation that relates the temperature and the electrical resistivity of platinum.
Callendar's thermometer see PLATINUM RESISTANCE THERMOMETER.
call fire *Ordnance.* fire delivered at a specific target in response to a request from the supported unit.
call forwarding *Telecommunications.* a telephone service feature in which calls can be rerouted from one line to another.
Callichthyidae *Vertebrate Zoology.* the armored catfishes, a family of tropical freshwater fishes in the suborder Siluriformes, characterized by heavy, armorlike scales.
callideic I see BRADYKININ.
call in *Computer Programming.* to bring the called subroutine instructions into the processor for execution, temporarily transferring control to the subroutine.
call indicator *Electronics.* an instrument that receives signals from an automated switching system and displays the corresponding called number before an operator at a manual switchboard.
calling device *Electronics.* an instrument that produces the signals that control connections in an automated telephone switching system.
calling parameter *Computer Programming.* a value supplied to a subroutine by the calling program, usually as part of the calling sequence.

calling sequence *Computer Programming.* a set of program instructions and data used to call a subroutine; it consists of an instruction that transfers control, parameters to be passed to the subroutine, and one or more locations for the subroutine to transfer results back when its execution is complete.

call instruction *Computer Programming.* a program instruction that transfers control to the first statement of a called subroutine; upon completion of the subroutine, control returns to the instruction immediately following the call instruction in the calling program.

Callionymoidei *Vertebrate Zoology.* the dragonets, a suborder of colorful marine bottom fishes in the order Perciformes.

Callipallenidae *Invertebrate Zoology.* a family of marine arthropods in the subphylum Pycnogonida that have no palpi.

Calliphoridae *Invertebrate Zoology.* a family of blow flies in the subsection Calypteratae that lay their eggs on meat and other foods, and in open wounds; includes the bluebottle.

callipic cycle *Astronomy.* four 19-year metonic cycles, or 76 years.

Callipus c. 380 BC, Greek astronomer; studied the movement of celestial bodies.

Callisto [kə lis′tō] *Astronomy.* the fourth Galilean moon of Jupiter, discovered in 1610; 4840 kilometers in diameter; its composition is about half water and half silicate rock.

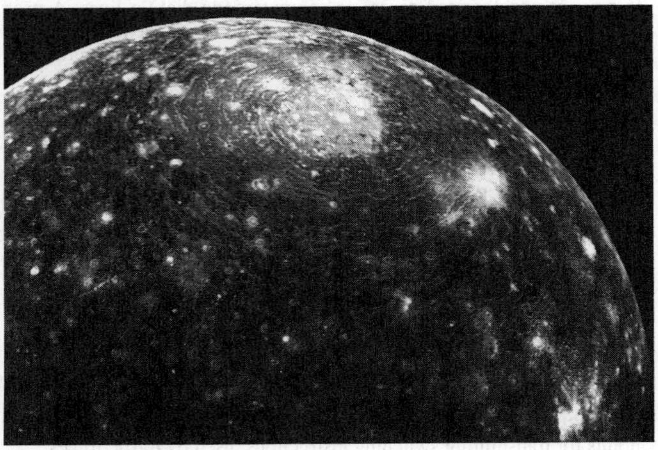

Callisto

Callithricidae *Vertebrate Zoology.* the marmosets and tamarins, a primate family of small Central and South American arboreal monkeys in the suborder Simiae, often characterized by tufts of hair on the sides of their faces.

Callitrichaceae *Botany.* the water starworts, a family of monogeneric cosmopolitan aquatic herbs of the order Callitrichales.

Callitrichales *Botany.* an order of mostly aquatic flowering plants that are typically low, slender, and multibranched with distally floating stems.

call letters *Telecommunications.* a sequence of alphanumeric characters that uniquely identify a transmitting station.

call mission *Military Science.* an air support mission requested on short notice, thus preventing detailed planning and briefing of pilots before takeoff; aircraft used for such a mission are kept on alert and armed with a prescribed load.

Callon's rule *Mining Engineering.* a rule stating that when a pillar is left in an inclined seam for support in a shaft or surface structure, a greater width should be left on the rise side of the shaft or structure than on the dip side

Callorhinchidae *Vertebrate Zoology.* a family of wide-ranging cartilaginous carnivorous fishes in the order Chimaeriformes, distinguished by a rounded snout with a hoe-shaped proboscis.

callose *Botany.* a complex branched carbohydrate, which is a common wall component in the sieve areas of sieve elements; it also may develop in reaction to injury in sieve elements and parenchyma cells; it also occurs in paller tubes.

Callovian *Geology.* a European geologic stage of the lowermost Upper Jurassic or uppermost Middle Jurassic period, occurring after the Bathonian and before the Oxfordian.

callow *Entomology.* an insect just emerged from the pupa, having a soft body and incomplete coloration; refers primarily to worker ants.

Callow flotation cell *Mining Engineering.* a pneumatic flotation cell into which air is blown at the bottom of the tank at low pressure through a porous septum, and mineralized froth overflows along the sides of the tank while the tailings progress to the discharge end.

Callow process *Mining Engineering.* a flotation process in which the agitation is generated by air forced into the pulp through the canvas-covered bottom of the cell.

Callow screen *Mining Engineering.* a mechanism that separates fine solids from coarser ones by a continuous belt of fine-screen wire.

call setup time *Telecommunications.* the total length of time needed to establish a connection between entities.

call sign *Telecommunications.* the radio identification code assigned to an aircraft. For a commercial airliner, it normally consists of the carrier name and flight number.

callus *Medicine.* a buildup of the outer layer of skin in a certain area that receives friction or pressure. *Botany.* **1.** a mass of plant cells that form around a wound on a plant. **2.** in some grasses, an extension of the base of the lemma where it attaches to the rachilla.

callus culture *Biotechnology.* the growth of undifferentiated or parenchymatous plant tissue on semisolid agar; the result is a mass of cells with no regular form.

call waiting *Telecommunications.* a telephone service feature that alerts a party already engaged in conversation that a new incoming call has arrived. The notified party can choose to ignore the alerting signal or respond with a signal requesting that the existing connection be placed on hold and the incoming call be connected. A second signal from the alerted party will cause the original connection to be reestablished and the second incoming call to be dropped.

calm *Meteorology.* a condition in which the wind has a speed of less than 1 mile per hour.

calmagite *Organic Chemistry.* $C_{17}H_{14}N_2O_5S$, a red crystalline solid, soluble in water; used as an acid-base indicator in the titration of calcium or magnesium with EDTA (pH 7.1–9.1, red; pH 9.1–11, blue).

calm belt *Meteorology.* a band of latitude in which the winds are light and variable; principally the horse latitudes and the doldrums.

calmodulin *Biochemistry.* a protein that binds with calcium and is present in all cells having a nucleus; controls many other enzymes.

calms of Cancer *Meteorology.* the light, variable winds and calms that occur in the centers of the subtropical high pressure belts over the oceans, at about 30°N and S, and, with the calms of Capricorn, make up the horse latitudes.

calms of Capricorn *Meteorology.* the light, variable winds and calms that occur in the centers of the subtropical high-pressure belts over the oceans, at about 30°N and S, and, with the calms of Cancer, make up the horse latitudes.

Calobryales *Botany.* an order of the class Hepaticae (liverworts) having stalked anthers, simple stems, and erect leafy branches.

Calocera *Mycology.* a genus of fungi belonging to the subclass Metabasidiomycetidae; its species, some of which are yellow, grow on various kinds of wood.

calomel [kal′ə mel′] *Mineralogy.* Hg_2Cl_2, a colorless or gray, transparent to translucent tetragonal mineral that occurs as tabular, pyramidal, or prismatic crystals, having a specific gravity of 7.6 and a hardness of 1.5 on the Mohs scale. It is found as a secondary mineral formed by the alteration of minerals containing mercury. Calomel is often used as a cathartic or fungicide.

calomel electrode *Physical Chemistry.* an electrode that consists of a pool of mercury connected to a layer or paste of mercurous chloride (calomel) and a solution of potassium chloride; used as a reference in measuring the electrical potential of other electrodes. Also known by various other names, such as CALOMEL REFERENCE ELECTRODE, MERCURY POOL ELECTRODE, SATURATED CALOMEL ELECTRODE, STANDARD CALOMEL ELECTRODE.

calomel reference electrode see CALOMEL ELECTRODE.

Calomniaceae *Botany.* a monogeneric family of light-green, slender, delicate mosses of the order Tetraphidae, often forming loose mats on the trunks and aerial roots of tree ferns in New Zealand, Tahiti, and Samoa.

calorescence [kal′ə res′əns] *Physics.* the production of visible light by means of heating with invisible infrared radiation.

caloric [kə lôr′ik] *Thermodynamics.* of or relating to heat transfer. *Nutrition.* of or relating to calories or to the measurement of the calorie content of specific foods.

calorie *Thermodynamics*. a unit of energy, originally defined as the heat transfer required to raise the temperature of one gram of pure water by one degree Celsius, from 14.5°C to 15.5°C, at standard atmospheric pressure; equivalent to 4.1840 joules. This is also called a **small calorie** to contrast it with the unit known as the **kilocalorie** or **large calorie,** which is 1000 small calories. *Physiology*. a unit of energy output for an organism, equal to one kilocalorie. *Nutrition*. the quantity of a given food that has an energy content of one kilocalorie.

calorific value *Engineering*. the number of heat units obtained by the combustion of unit mass of a fuel, expressed in heat units per unit mass. Also, HEATING VALUE.

calorifier *Engineering*. an apparatus for heating water in a tank by the immersion of a separate coil of heated pipes in the water.

calorimeter [kal´ə rim´ə tər] *Engineering*. an apparatus used to measure quantities of heat generated or emitted by a body during processes such as burning, change of state, and friction, often by observing the quantity of a solid liquefied or of a liquid vaporized under given conditions.

calorimetry [kal´ə rim´ə trē´] *Engineering*. the measurement of thermal constants, such as specific heat, latent heat, or calorific value.

Caloris Basin *Astronomy*. a ring-shaped topographic structure on Mercury, measuring 1300 kilometers in diameter and created by the impact of an asteroid.

calorizator *Food Technology*. in sugar processing, an apparatus used to heat the juice of sugarbeets.

calorize *Metallurgy*. to treat the surface of a ferrous material with aluminum powder at 800 to 1000°C to improve oxidation resistance.

Calosiphoniaceae *Botany*. a small family of rare red algae of the order Gigartinales, having cylindrical, branched, multiaxial gametangial thalli and superficially produced spermatangia.

Calostomataceae *Mycology*. a family of fungi belonging to the order Tulostomatales that is characterized by vibrant-colored fruiting bodies and occurs in the Americas, Australasia, and southern Asia.

Calothrix *Bacteriology*. a genus of cyanobacteria, or blue-green algae, that grow as filaments with heterocysts occurring at the bases of the trichomes; typically found on coastal rocks.

calotte *Biology*. a cap or a domelike structure. *Invertebrate Zoology*. a cap or structure like a cap, as in certain parasitic Mesozoa (Dicyemidae).

calpain *Biochemistry*. a proteinase that requires calcium for activity.

calsequestrin *Biochemistry*. a protein that stores calcium and releases it in muscle fiber upon contraction.

calthrop *Invertebrate Zoology*. a sponge spicule that has four axes in which the rays are equal or nearly equal in length.

Calutron *Nucleonics*. an early apparatus used for the elecromagnetic method of isotope separation. (So called because it used a magnet from one of the University of California cyclotrons.)

calvaria *Anatomy*. the upper dome of the skull. Also, **calvarium.**

calved ice *Oceanography*. a fragment of floating ice that has separated from a coast glacier or an iceberg.

Calvin, Melvin born 1911, American chemist; shared Nobel Prize for tracing chemical reactions of carbon in photosynthesis.

Calvin-Benson cycle *Biochemistry*. the cycle of reactions that occurs in green plants and photosynthetic bacteria, in which carbon dioxide is converted into glucose using ATP.

calving *Geology*. the falling or caving in of a mass of earth or rock. *Hydrology*. the breaking away of a mass of ice from a glacier, an ice shelf, or an iceberg. *Vertebrate Zoology*. the process of giving birth to a calf.

calx [kalks] *plural,* **calxes** or **calces.** *Mineralogy*. **1.** the oxide or ashy residue that is left after metals or minerals have been thoroughly roasted or burned. **2.** another term for lime (calcium oxide, CaO).

Calycanthaceae *Botany*. a family of aromatic shrubs and small trees in the order Laurales, native to North America and China and characterized by short leafy branches and solitary flowers that are pollinated by beetles.

calycanthemy *Plant Pathology*. an abnormal development that occurs in the outermost portion of a flower, in which petals or petallike structures are produced.

Calyceraceae *Botany*. a family of dicotyledonous herbs that are native to Central and South America and that constitute the order Calycerales; characterized by the storage of carbohydrates as inulin and commonly producing iridoid compounds.

Calycerales *Botany*. an order of flowering tropical plants of the subclass Asteridae, native to the western hemisphere.

calyciform *Botany*. resembling or having the form of a calyx.

calyculate *Botany*. having bracts that simulate the calyx.

Calymene *Paleontology*. a genus of phacopid trilobites in the suborder Calymenina; Ordovician to Silurian.

calymma *Invertebrate Zoology*. the gelatinous outer layer of certain radiolarians.

Calymmatobacterium *Bacteriology*. a genus of Gram-negative, nonmotile, rod-shaped bacteria, occurring as pathogens in humans.

Calymnidae *Invertebrate Zoology*. a family of echinoderms, in the order Holasteroida, with ovoid tests.

Calymperaceae *Botany*. a large, tropical family of erect tufted or matted mosses of the order Pottiales that grow on trees, rocks, and soil; characterized by terminal sporophytes, large clear cells in the basal sheath that form the cancellinae, and sometimes by spongy and multistratose leaves.

Calypogeiaceae *Botany*. a cosmopolitan family of medium-size, prostrate, creeping liverworts of the order Jungermanniales; characterized by linear and spirally twisted capsule valves, vestigial bracts, a general lack of secondary pigmentation, and reduced or obsolete leaf and underleaf lobing.

Calypso *Astronomy*. the fourteenth moon of Saturn, discovered in 1980 and measuring 15 km by 13 km by 8 km.

calypter *Entomology*. a secondary lobe on the wing of a house fly or other similar dipteran insect, often sufficient to hide the haltare. Also, ALULA, SQUAMA.

Calyptoblastea *Invertebrate Zoology*. a suborder of coelenterates, in the order Hydroida, whose life cycle alternates between free-swimming jellyfish and sessile colonies. Also, LEPTOMEDUSAE.

calyptra *Botany*. any hood or cap of cells protecting a plant part, such as cells protecting a root tip, the sporangia of mosses and liverworts, and the filament tip of a myxophyceae.

Calyptratae *Invertebrate Zoology*. a subsection of dipteran insects in the suborder Cyclorrhapha that have calypters associated with their wings.

calyptrate *Botany*. having a cap.

calyptrogen *Botany*. the meristematic cells from which the root tip forms and grows.

Calyptrosphaeraceae *Botany*. a family of usually flagellate marine algae of the order Coccosphaerales, characterized by coccoliths composed entirely of rhombohedric microcrystals.

Calyssozoa *Invertebrate Zoology*. the only class of the subphylum Entoprocta; bryozoans with an anal opening surrounded by tentacles.

calyx [kal´iks] *Botany*. the outermost, protective group of floral parts; a usually green outer whorl of sepals that protects the flower. *Invertebrate Zoology*. a cuplike structure in crinoids to which the arms are attached. *Anatomy*. an organ or body cavity shaped like a cup. Also, CALIX.

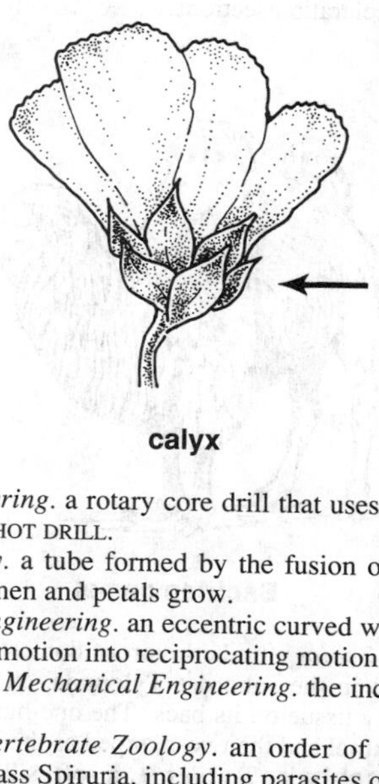

calyx

calyx drill *Engineering*. a rotary core drill that uses hardened steel shot to cut rock. Also, SHOT DRILL.

calyx tube *Botany*. a tube formed by the fusion of a flower's sepals, from which the stamen and petals grow.

cam *Mechanical Engineering*. an eccentric curved wheel on a shaft used to transform rotary motion into reciprocating motion.

cam acceleration *Mechanical Engineering*. the increased speed of the cam follower.

Camallanida *Invertebrate Zoology*. an order of phasmid nematode worms, in the subclass Spiruria, including parasites of mammals.

camarodont dentition *Invertebrate Zoology*. in echinoderms, keeled teeth meeting the epiphyses, so that the foramen magnum of the jaw is closed.

Camaroidea *Paleontology.* a minor order of encrusting graptolites known only from the Ordovician; a representative genus is Bithecocamara.
Camarosporium *Mycology.* a genus of fungi belonging to the class Coelomycetes.
Cambaridae *Invertebrate Zoology.* a family of crayfish, decapod crustaceans in the section Macrura, with well-developed tail fans.
Cambarinae *Invertebrate Zoology.* a subfamily of crayfish in the family Astacidae, including all eastern United States species.
camber *Design Engineering.* a slight convexity, arching, or curvature in an assembly or component. *Aviation.* **1.** the curvature of an airfoil section relative to its chord; usually expressed as the ratio of the height of the curved line (mean line) between the leading and trailing edges to the length of a straight line between the same two points. **2.** an inclination of landing wheels away from the vertical plane. *Naval Architecture.* an upward curvature of a deck in cross section, promoting water drainage. *Geology.* **1.** a structure resembling an arch or ridge where the bed sags downward toward topographically lower areas. **2.** a terminal shoulder of the continental shelf.
camber angle *Mechanical Engineering.* the angled tilt of the steerable wheels of a motor vehicle, which are closer together at the bottom than at the top; current car models have very little camber compared to older models.
camber arch *Architecture.* a flat arch having a slightly curved intrados and flat or slightly curved extrados.
cambered *Fluid Mechanics.* referring to an airfoil with a curved mean line.
camber keeled *Naval Architecture.* describing a vessel whose keel curves up fore and aft.
cambic horizon *Geology.* a soil horizon in which the alteration or removal of mineral matter causes a mottled or gray and red color.
cambium *Botany.* in most vascular plants, a layer of meristematic tissue that gives rise to the xylem, phloem, and (in woody plants) bark.
Cambrian *Geology.* **1.** the earliest geologic period, occurring during the Lower Paleozoic era from about 570 to 500 million years ago. **2.** referring to the rocks formed during that time.
Cambrian explosion *Paleontology.* a term for the fact that nearly all contemporary animal phyla first appear in the fossil record during the Cambrian period.
cambric *Textiles.* **1.** a high-quality, tightly woven white linen fabric. **2.** a usually cotton fabric resembling true cambric.
camcorder *Electronics.* equipment combining a video camera and a video tape recorder in one portable unit. (Short for <u>cam</u>era-re<u>corder</u>.)
cam cutter *Mechanical Engineering.* a grinding machine that uses a rotating master cam to cut and contour a duplicate.
cam dwell *Design Engineering.* the surface of a cam between the opening and closing acceleration sections.

Bactrian camel

camel *Vertebrate Zoology.* an herbivorous domesticated mammal of the family Camelidae in the suborder Tylopoda, characterized by one or two humps of fatty tissue on its back. The one-humped camel is known as the **Arabian camel** and the two-humped as the **Bactrian camel.**
camelhair or **camel hair** *Textiles.* **1.** a soft, silky fiber obtained from the underside of a camel, frequently blended with wool and used to make warm outerwear. **2.** a cloth made with this fiber, or resembling cloth made with this fiber.
Camelidae *Vertebrate Zoology.* a family of domesticated mammals in the suborder Tylopoda and the order Artiodactyla, including the Arabian or dromedary camel (one-humped), the Bactrian camel (two-humped), the llamas, the guanacos, and the vicunas.
Cameloidea *Vertebrate Zoology.* a superfamily of ruminant mammals in the order Artiodactyla, including the camels and llamas.
Camelopardalis *Astronomy.* the Giraffe, a large dim constellation identified in 1624, lying near the north celestial pole. Also, **Camelopardus.** (From the earlier use of the term *camelopard,* "camel leopard," as a name for a giraffe.)
Camembert or **camembert** [kam´əm bâr´] *Food Technology.* a cream-colored soft cheese ripened from the surface inward by the mold *Penicillium camemberti.* (From *Camembert,* a town in Normandy known for such cheeses.)
cam engine *Mechanical Engineering.* a reciprocating engine that uses a cam mechanism to create rotary motion.
camera *Optics.* any of a wide variety of devices through which light from an object is focused onto a light-sensitive material, such as film, in order to record the image as a photograph. *Electronics.* **1.** a device that converts images into electric signals. **2.** specifically, the device in a television broadcasting facility in which the image to be transmitted is displayed before it is converted into electrical impulses.
camera base *Graphic Arts.* the focal point of a lithographic camera.
camera cable *Electronics.* the cable that carries picture signals from the television camera to the control room.
camera calibration *Photogrammetry.* the process of determining the calibrated focal length, the location of the principal point in relation to the fiducial marks, the point of symmetry, the resolution of the lens, the degree of flatness of the focal plane, and the effective lens distortion in the focal plane of the camera.
camera chain *Telecommunications.* the ensemble of equipment needed to transform images of objects into television signals, typically including a television camera, camera control unit, sync generator, power supply, monitor, and connecting cables.
camera copy *Graphic Arts.* in making plates for offset lithography, an original image serving as a subject to be photographed, such as reproduction proofs, line art, or halftones.
camera copyboard see COPYBOARD.
camera lucida *Optics.* an accessory for a microscope or other optical instrument that employs a reflecting prism or mirrors to cast a duplicate image of an object onto a plane surface so that an outline of the object can be traced. Also, DRAWING APPARATUS.
camera obscura *Optics.* an enclosure, usually darkened, in which the light from an object passes through a lens and forms an image of the object at the back of the enclosure without producing a photographic reproduction.
camera-ready *Graphic Arts.* describing a reproduction proof or other item that is ready to be photographed for platemaking.
camera station *Cartography.* the point occupied by the perspective center of a camera at the instant a photograph is taken. In aerial photography, the AIR STATION. Also, EXPOSURE STATION.
Camerata *Paleontology.* a Paleozoic subclass of crinoids that includes two orders, the Diplobathrida and the Monobathrida; characterized by long and flexible stems, the incorporation in the theca of the proximal arm ossicles, and a tightly sutured calyx; extant in the middle Ordovician to upper Permian.
camera tube *Electronics.* a tube that scans an optical image to generate an electrical signal.
cam follower *Mechanical Engineering.* **1.** the output link or plunger of a cam mechanism. **2.** any part to which movement is imparted by a cam.
Camilla *Astronomy.* asteroid 107, discovered in 1868 and measuring 252 kilometers in diameter; it belongs to type C.
cam mechanism *Mechanical Engineering.* a device that consists of a rotating cam linked to a roller or pin.
Cammett table *Mining Engineering.* a side-jerking table for concentrating ore; similar to the Wilfley table.
camniume *Ecology.* an ecological succession caused by cultivation.
cam nose *Mechanical Engineering.* the highest point of a cam, used to keep valves open or closed in a reciprocating engine.
camouflage [kam´ə fläzh´] *Military Science.* the process of concealing or disguising troops, equipment, and installations by making them blend with the natural background.
camouflet *Ordnance.* **1.** a small explosive detonated underground to destroy an enemy's mines without disturbing the surface. **2.** the underground cavity created by such an explosive.

camp *Military Science.* **1.** a group of tents, huts, or other shelters set up temporarily for troops; more permanent than a bivouac. **2.** any military post, whether temporary or permanent. *Archaeology.* a habitation site at which a few individuals have stayed for a limited time to perform a specialized task, such as seasonal hunting or gathering. Also, **campsite.**

CAMP *Oncology.* an acronym for a cancer chemotherapy regimen that includes the drugs cyclophosphamide, Adriamycin (doxorubicin), methotrexate, and procarbazine.

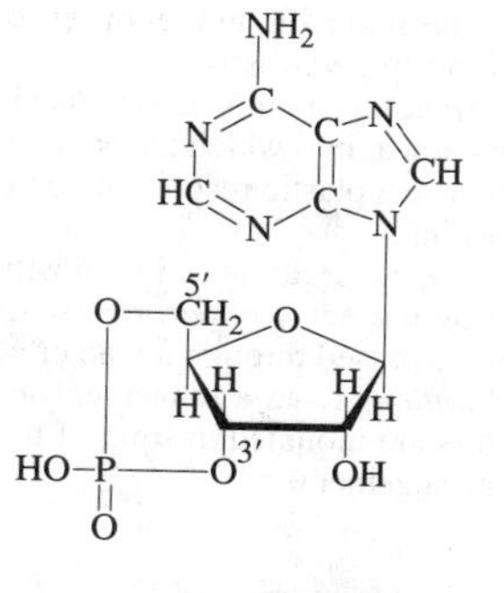

c-AMP (cyclic AMP)

c-AMP or **cAMP** *Biochemistry.* 3',5'-cyclic adenosine monophosphate, a nucleotide that is formed by adenylate cyclase when stimulated by specific hormones; functions as a second messenger within cells to initiate the activation of glycogen phosphorylase. Also, ADENOSINE 3',5'-CYCLIC PHOSPHATE, CYCLIC AMP.

campaign *Military Science.* a plan for a series of related military operations designed to accomplish a common objective, usually within a specific area and period of time.

Campanella *Invertebrate Zoology.* a genus of colonial freshwater ciliate protozoans, characterized by an inverted bell shape and long noncontracting stalks.

Campanian [kam pan′ē ən] *Geology.* a European geologic stage of the Upper Cretaceous period, occurring after the Santonian and before the Maestrichtian.

campaniform receptors *Entomology.* sense organs in the form of bell-like projections from the cuticle that detect strains in the integument and enable an insect to move in a coordinated fashion.

campanile [kam′pə nēl′] *Architecture.* a freestanding bell tower.

Campanulaceae *Botany.* a family of flowering plants of the order Campanulales, characterized by bisexual, bell-shaped flowers having petals that are usually united. Also, SPENOCLEACEAE.

Campanulales *Botany.* an order of dicots of the subclass Asteridae having alternate leaves without stipules, an inferior ovary, and bisexual flowers, usually in racemes but sometimes solitary or in spikes.

campanulate *Botany.* of a flower, bell-shaped.

cam pawl *Mechanical Engineering.* a device that allows a cam to rotate in only one direction.

Campbell-Baker-Hausdorff formula *Mathematics.* a formula for Z when $e^Z = e^X e^Y$, and X and Y may not commute. It applies, for example, when X and Y are matrices or elements of a Lie algebra. The first few terms are

$$Z = X + Y + (1/2)[X, Y] + (1/12)\{[[X, Y], Y] + [[Y, X], X]\} + \cdots$$

where $[X, Y]$ denotes the Lie bracket.

Campbell bridge *Electricity.* a bridge designed for mutual inductance comparison.

Campbell model *Genetics.* a model proposed to simulate the mechanism by which the λ bacteriophage can be inserted into the *Escherichia coli* chromosome without a reciprocal loss of bacterial DNA.

Campbell process *Metallurgy.* in steelmaking, an obsolete process based on a charge of ore and pig iron; a precursor of the open-hearth process.

Campbell's formula *Electromagnetism.* a formula expressing a general relationship between the propagation constant of a loaded transmission line and the propagation constant and characteristic impedance of an unloaded transmission line and the impedance of each loading coil.

Campbell's law *Geology.* a law stating that drainage divides tend to migrate toward an axis of uplift and away from an axis of subsidence.

Campbell-Stokes recorder *Engineering.* a recorder consisting of a glass sphere that focuses the sun's image onto a bent strip of card on which the hours of the day are marked; as the focused heat burns through the card, the duration of sunshine can be read from the burnt track.

Campephagidae *Vertebrate Zoology.* the songbird family of cuckoo-shrikes and minivets of the suborder Oscines, found throughout Africa, Asia, Australia, and the Pacific Islands.

camphane [kam′fān′] *Organic Chemistry.* $C_{10}H_{18}$, crystalline needles that are insoluble in water and very soluble in alcohol; melts at 51–52°C and boils at 158.5–159.5°C. Also, BORNANE.

camphene [kam′fēn′] *Organic Chemistry.* $C_{10}H_{16}$, a bicyclic terpene found in plant oils; insoluble in water and soluble in alcohol and ether; melts at 51–52°C and boils at 158.5–159.5°C.

camphor [kam′fər] *Organic Chemistry.* $C_{10}H_{16}O$, colorless or white plates that are insoluble in water and very soluble in alcohol; melts at 179.8°C and sublimes at 204°C; obtained mainly from the evergreen tree *Cinnamomum camphora*, and used in medicine, insecticides, tooth powders, and flavorings.

camphoric acid [kam′fôr′ik] *Pharmacology.* $C_{10}H_{16}O_4$, an oxidation derivative of camphor that formerly was used to stimulate breathing.

camphorism *Toxicology.* poisoning due to consumption of camphor; symptoms may include severe gastrointestinal disturbances, convulsions, coma, and respiratory failure, and may cause death.

camphor oil *Materials.* an essential oil that is obtained from the wood of the camphor tree, *Cinnamomum camphora*; used in the manufacture of camphor and safrole.

***d*-camphorsulfonic acid** *Organic Chemistry.* $C_{10}H_{16}O_4S$, prismatic crystals that are very soluble in water; melts at 194–195°C.

camphor tree *Botany.* an East Asian evergreen tree, *Cinnamomum camphora,* from whose leaves camphor is derived.

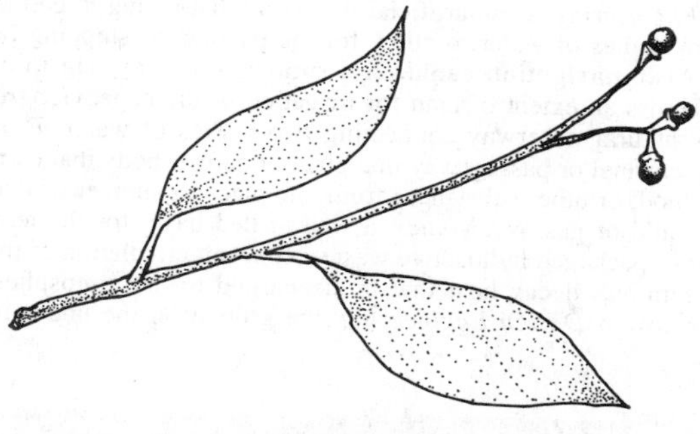

camphor tree

campo *Geography.* a region of grassy plains with scattered bushes and small trees, as is found in large areas of South America.

Campodeidae *Invertebrate Zoology.* a family of mostly wingless insects in the order Diplura, found in the northern temperate zones.

campodeiform *Invertebrate Zoology.* shaped like a bristletail, a wingless insect of genus *Campodea* with several segmented hairs at the tail end of the body.

campodeiform larva *Invertebrate Zoology.* any of various legged, active larval forms, including a common form of beetle larva that is usually long-legged, predatory, and lacking in abdominal appendages other than cerci.

camp-on *Telecommunications.* a telephone service feature that allows a new incoming call to a line or station already engaged in communication to be connected to the new caller when the present communication has ended and disconnected.

cam profile *Design Engineering.* the outline of a cam whose shape is determined by the form of its flanks and tip, and by means of which motion is communicated to the follower.

CAMP test *Microbiology.* a test for the identification of group B streptococci, in which a characteristic zone of hemolysis occurs between a streak of colonies of the group B streptococcus and colonies of *Staphylococcus aureus.* (An acronym for Christie-Atkins-Munch-Petersen, the discoverers of this phenomenon.)

camptonite *Petrology.* a lamprophyre composed of pyroxene, sodic hornblende, and olivine as dark constituents and labradorite as the light constituent, possibly interlaced with sodic orthoclase.

Campylobacter *Bacteriology.* a genus of Gram-negative, motile, microaerophilic bacteria, possessing a single flagellum at one or both cell poles, and found in the reproductive and intestinal tracts of animals and humans.

Campylobacteriosis *Pathology.* any of a number of diseases in animals that are caused by bacterial strains of *Campylobacter*, including genital diseases in cattle and sheep as well as human infections.

campylotropous *Botany.* of an ovule, curved horizontally so that the chalaza and the micropyle lie at nearly right angles to each other.

camshaft *Mechanical Engineering.* a rotating shaft to which a cam is fastened.

camwood *Materials.* the coarse, hard, and dense-grained wood of the *Baphia nitida* tree of West Africa; used for tool handles.

can *Mechanical Devices.* a metal container having a removable cover and less than ten gallons capacity. *Aviation.* any of the individual flame tubes in the cannular combustion chamber of a turbojet engine. *Nucleonics.* see CLADDING, def. 1.

Canaceidae *Invertebrate Zoology.* the seashore flies, a family of dipteran insects in the subsection Acalypteratae.

Canada goose *Vertebrate Zoology.* a common North American wild goose, *Branta canadensis.* Also, **Canadian goose.**

Canada lynx see LYNX, def. 2.

Canadian *Geology.* a North American provincial series of the Lower Ordovician period, occurring after the Croixian of the Cambrian and before the Champlainian.

Canadian Shield *Geology.* a plateau formed during the Precambrian period, extending over about two million square miles of Canada and the northeastern U.S., from Labrador southwest along the Hudson Bay and northwest to the Arctic Ocean; composed of granite, gneiss, marble, and other igneous and metamorphic rocks. Also, LAURENTIAN PLATEAU, LAURENTIAN SHIELD.

Canadian subregion *Ecology.* a distinct zoogeographical region that includes most of the country of Canada and also the state of Alaska.

canal *Engineering.* **1.** an artificial waterway that is dug to connect two adjacent bodies of water to allow for the passage of shipping between them. Also, **navigation canal. 2.** a similar waterway dug to conduct water across an extent of land for irrigation or drainage. *Geography.* a narrow natural waterway connecting two bodies of water. *Biology.* a tubular channel or passageway in a plant or animal body that carries air, water, food, or other substances from one place to another or that holds some liquid or gas. *Nucleonics.* a water-filled basin for the temporary storage of packaged radioactive wastes; constant circulation of the basin water removes decay heat that is discharged to the atmosphere by a cooling tower. *Design Engineering.* the groove on the underside of a corona.

irrigation canal

canal boat *Naval Architecture.* a long, narrow, shallow-draft boat designed for use along a canal.

canal cell *Botany.* any of several initial cells in the central cavity of the neck of an archegonium that disappear as the archegonium matures.

canalete *Materials.* the durable, flexible brown wood of the tropical American timber trees *Cordia gerascanthus* and *Cordia alliodora.*

canaliculate *Biology.* relating to or having very small longitudinal canals or grooves.

canaliculus [kan′əl ik′yə ləs] *plural,* **canaliculi.** *Histology.* a narrow tubular passage or small channel, such as the numerous microscopic canals connecting osteocytes in bone tissue, or **bile canaliculi,** which form a three-dimensional network between liver cells (hepatocytes).

canalis vertebralis see NEURAL CANAL.

canalization [kan′əl ′zā′shən] *Engineering.* a system of channels used for navigation, flood prevention, irrigation, or conduits for power generation. *Physiology.* the formation of new channels, such as capillaries in a blood clot or neural pathways in the central nervous system formed by repeated passage of nerve impulses. *Surgery.* the surgical formation of canals for drainage without the use of tubes. *Behavior.* the restriction of a variable behavior pattern to a limited set of responses.

canal of Cuvier see DUCTUS VENOSUS.

canal of Shlemm *Anatomy.* a circular venous channel, at the junction of the sclera and the cornea, into which the aqueous humor drains.

canaloplasty *Surgery.* the plastic reconstruction of a passage, as of the external auditory opening.

canal rays *Radiology.* rays composed of positive ions in a vacuum tube; so named following early experiments in which particles discharged from the anode passed through a hole or canal in the cathode.

canard [kə närd′] *Aviation.* **1.** an arrangement in which the horizontal stabilizer and elevators are mounted in front of the wing or wings. **2.** an aircraft having this arrangement.

canary

canary *Vertebrate Zoology.* **1.** a small songbird of the finch family Fringillidae, species *Serinus canarius*; having a melodic song and bright yellow plumage and often kept caged as a pet; originating in the Canary Islands, Azores, and Madeira Islands. **2.** any of various related birds of North America, Europe, and Africa..

Canary Current *Oceanography.* the cool current that flows southwestward along the northwest coast of Africa, from the Canary Islands to the Cape Verde Islands.

Canastotan *Geology.* in New York State, a geologic stage of the Upper Silurian period.

canavanine *Biochemistry.* $C_5H_{12}O_3N_4$, an amino acid that is present in only certain legumes and thus is useful in legume taxonomy.

can buoy *Navigation.* a floating aid to navigation in the shape of a cylinder.

cancel *Graphic Arts.* **1.** to cut out blank pages from a signature at the end of a book. **2.** to cut out a printed page, especially one containing an error. **3.** a sheet printed to correct an error and replace a canceled page. **4.** a sheet printed to replace a page in a looseleaf binder, as to update material in a legal reference book.

cancellate *Biology.* latticed; cancellous.

cancellation circuit *Electronics.* a circuit, used in radar, that processes moving-target signals by rejecting all fixed-target signals.

cancellation law *Mathematics.* **1.** the rule that if $x * z = y * z$, then $x = y$, for x, y, and z elements of an algebraic structure such as a group and $*$ the group operation. This rule may hold in a ring with no division operations, such as an integral domain. **2.** the process of dividing like (nonzero) factors out of the numerator and denominator of a fraction; for example, $ac/bc = a/b$ for $b, c \neq 0$.

cancellous *Biology.* having a latticelike or porous structure.

cancellous bone *Histology.* a spongy type of bone filled with marrow, found in the interior of flat bones and at the ends of long bones.

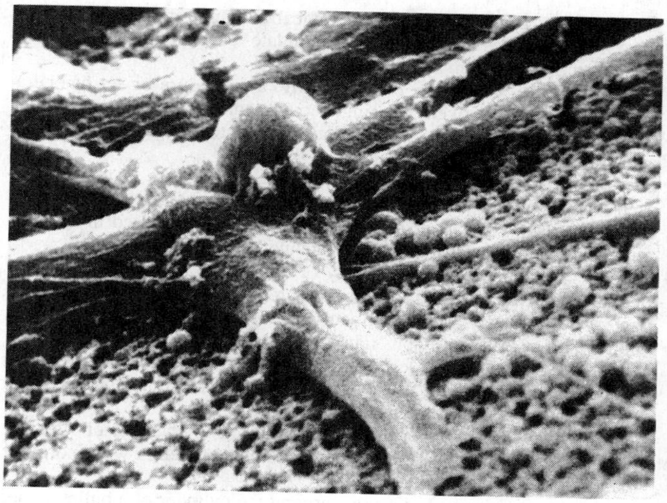

cancer cell

cancer *Oncology.* a malignant tumor whose cells have the properties of endless replication, loss of contact inhibition, invasiveness and the ability to metastasize and whose result, generally, if left untreated, is fatal.

Cancer *Astronomy.* the Crab, a faint northern constellation dating from antiquity that lies in the zodiac between Gemini and Leo.

cancer bodies see RUSSELL'S BODIES.

cancer embolus see MALIGNANT EMBOLUS.

canceremia [kan´sə rē´mē ə] *Oncology.* the presence of cancerous cells in the blood. Also, **cancerous cachexia.**

cancericidal *Oncology.* of any agent, destructive to cancer or to malignant cells.

cancerous [kan´sə rəs] *Oncology.* **1.** relating to or affected by cancer. **2.** being a form of cancer.

cancerous cachexia see CARCINEMIA.

cancerphobia see CARCINOPHOBIA.

cancriform *Oncology.* resembling a cancer. Also, CANCROID.

cancrinite *Mineralogy.* $Na_6Ca_2Al_6Si_6O_{24}(CO_3)_2$, a colorless, orange or pink to reddish, transparent to translucent hexagonal mineral, occurring as rare prismatic crystals or more commonly in massive form; having a specific gravity of 2.42 to 2.51 and a hardness of 5 to 6 on the Mohs scale; found as a primary mineral in some alkali rocks.

cancroid [kan´kroid] *Oncology.* **1.** see CANCRIFORM. **2.** a tumor similar in appearance and constituency to a cancer. **3.** a skin cancer of moderate malignancy.

candela *Optics.* the base unit of luminous intensity, equivalent to 1/60th of the typical intensity of one square centimeter of a blackbody radiator, at the temperature at which platinum solidifies (2042 K) at a pressure of 1 bar.

Candida *Mycology.* a genus of yeastlike imperfect fungi belonging to the family Cryptococcaceae which is found in soils, plants, and foods and is pathogenic to humans and other animals.

candidate virus *Virology.* **1.** a virus that might be the causal agent of an infection. **2.** a virus that has not been categorized but reveals an indication for a particular taxonomic group.

candidiasis *Medicine.* a disease produced by the invasion of a form of *Candida* fungus and attacking almost exclusively the mucosa of the mouth, respiratory tract, and vagina.

candle jar *Microbiology.* an oxygen-depleted culture incubation chamber produced by placing a lighted candle in the container prior to sealing it.

Candlemas Eve winds *Meteorology.* the heavy winds occurring in England in February or March. (Because of their association with the church festival of *Candlemas,* February 2nd.) Also, **Candlemas crack.**

candlepower *Optics.* a measure of the illuminating capacity of a light source, such as a lamp or candle, expressed in candelas.

candling *Agriculture.* the process of examining eggs in the shell, using a bright light to detect blood spots, bubbles, or shell cracks. *Virology.* a process by which the internal details of an embryonated egg are examined through transmitted light.

cane *Botany.* **1.** the long, hollow, jointed stem of certain plants including some palms and larger grasses and reeds. **2.** a plant having such a stem, such as sugar cane. **3.** the stem of a raspberry, blackberry, rose, or young grape vine.

cane blight *Plant Pathology.* a fungus disease of currants, raspberries, and other bush fruits that produces rows of wartlike fungus fruiting bodies on the canes.

Canellaceae *Botany.* a family of tropical aromatic trees and shrubs of the order Magnoliales; characterized by scattered spherical cells containing volatile oils, regular perfect flowers, and a compound pistil.

cane molasses *Food Technology.* a heavy syrup that remains after cane sugar has been crystallized.

canescent *Botany.* covered with fine, usually grayish hairs.

cane sugar see SUCROSE.

Canes Venatici [kān´ēz´ ven´ə tē´sē] *Astronomy.* the Hunting Dogs, a dim northern constellation introduced in 1536 that lies south of the Big Dipper's handle.

Canes Venatici I cloud *Astronomy.* a loose and sparse group of galaxies, mostly irregulars and spirals, lying about 15 million light-years away.

canfieldite *Mineralogy.* Ag_8SnS_6, an opaque, black or steel gray orthorhombic (pseudocubic) mineral, having a specific gravity of 6.276 and a hardness of 2.5 on the Mohs scale; found in vein deposits associated with acanthite, other silver minerals, and sulfides.

canga *Geology.* in Brazil, an erosion-resistant, unstratified, iron-rich rock occurring as a surficial deposit overlying older rocks.

can hoisting system *Mining Engineering.* a hoisting operation controlled at the top of the shaft; used in shallow lead and zinc mines.

Canidae *Vertebrate Zoology.* a family of carnivorous mammals that includes dogs, wolves, foxes, and coyotes and belongs in the superfamily Canoidea. *Paleontology.* a family of carnivores in the order Carnivora, represented today by dogs and wolves; Miocane to present.

Caniformia *Vertebrate Zoology.* a suborder of the mammalian order Carnivora, including the dogs, bears, mustelids, procyonids, and pinnipeds.

canine *Anatomy.* the sharp pointed tooth on either side of the upper or lower jaw, between the incisors and the bicuspids *Vertebrate Zoology.* **1.** of or relating to dogs or doglike animals, or to the family Canidae. **2.** a member of the family Canidae, especially a domestic dog.

canine diastema *Anthropology.* a gap present in the dentition of many primates other than hominids to accommodate a long canine tooth.

canine distemper *Veterinary Medicine.* a highly contagious, often fatal disease primarily striking young dogs, but also affecting other members of the families Canidae and Mustelidae; characterized by depression, nervous symptoms, fever, and disorder of the respiratory and intestinal tracts.

canine tooth *Anatomy.* a sharp pointed tooth on either side of the upper or lower jaw between the incisors and the bicuspids.

Caninia *Paleontology.* a genus of large, solitary rugose corals of the Carboniferous.

Canis *Vertebrate Zoology.* the type genus of the family Canidae, including all domestic dogs, wolves, and coyotes.

Canis Major *Astronomy.* the Greater Dog, a bright southern constellation lying in the Milky Way and containing Sirius, the sky's brightest star.

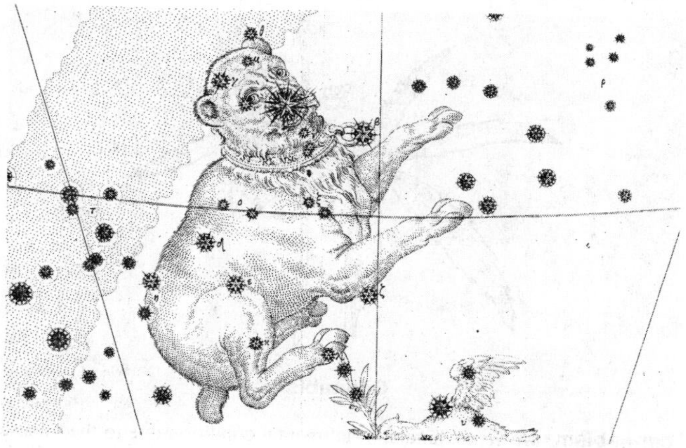

Canis Major

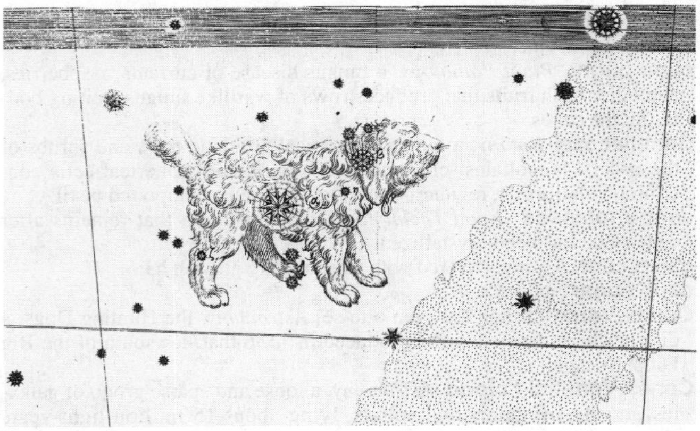

Canis Minor

Canis Minor *Astronomy.* the Lesser Dog, a faint constellation on the celestial equator whose brightest star is Procyon.

canister *Ordnance.* an antipersonnel projectile consisting of a light metal casing filled with steel balls, fragments, or slugs; the casing opens as it leaves the muzzle of the gun, and the projectiles scatter in a cone-shaped pattern similar to that of a shotgun shell.

canker *Plant Pathology.* a sunken or cracked dead area on a woody plant stem, caused by a bacteria, virus, or toxin; hardened tissue forms over the area, eventually causing the death of the plant stem.

canker sore *Medicine.* a lesion of the mouth or lips.

canker stain *Plant Pathology.* a disease of plane trees that is caused by the fungus *Endoconidiophora fimbriata plantani* and characterized by dark blue or red discolorations under dark cankers on the trunk and branches.

cankerworm *Invertebrate Zoology.* the common name for any of several lepidopteran insect larvae in the family Geometridae that are serious plant pests.

Cannabidaceae *Botany.* a family of erect or twining, usually dioecious herbs in the order Urticales, lacking latex and producing quebrachitol, pyridine alkaloids, and proanthocyanins; cultivated in different forms for hemp fiber and for psychotropic drugs such as marijuana. Also, **Cannabaceae.**

cannabidiol *Organic Chemistry.* $C_{21}H_{31}O_2$, a rodlike solid that is insoluble in water and soluble in alcohol; it melts at 67°C and boils at 187–190°C (2 torr).

cannabinol *Organic Chemistry.* $C_{21}H_{26}O_2$, a compound in the form of yellow plates, insoluble in water and soluble in alcohol and acids; melts at 76–77°C and sublimes at 185°C (0.05 torr); produced by the hemp plant *Cannabis sativa*, which causes hallucinogenic and intoxicating effects when ingested or inhaled.

Cannabis *Botany.* a genus of the family Cannabidaceae having green flowers and alternate lobed leaves; grown commercially for its fibers, which are used to produce rope, and its flowers and leaves, from which marijuana and hashish are produced.

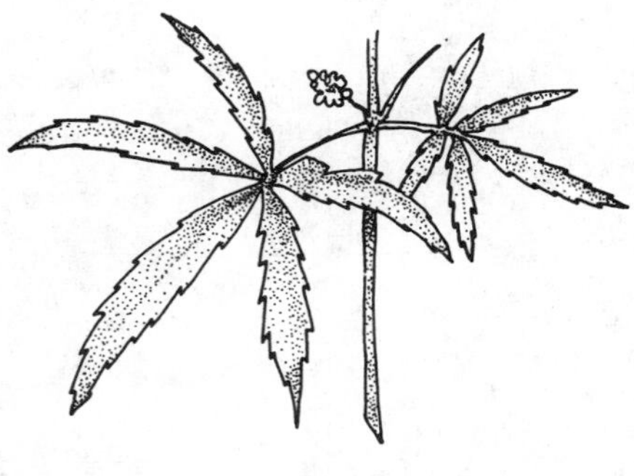

Cannabis

cannabism *Medicine.* an abnormal mental condition due to the misuse of cannabis, or the hemp plant, varieties of which are known as marijuana or hashish.

Cannaceae *Botany.* a family of tropical and subtropical plants of the order Zingiberales having alternate petiolate leaves and a single operative stamen and pollen sac.

canned cycle *Computer Programming.* a procedure or set of operations that can be invoked by a single command.

canned motor *Mechanical Engineering.* a motor encased with a pump that lubricates the motor's bearings.

canned program *Computer Programming.* a program that is prewritten for a specific application, stored, and available to users.

canned pump *Mechanical Engineering.* a pump enclosed in a watertight casing for use in underwater machinery and locations.

cannel coal *Geology.* **1.** a variety of nonbanded, fine-grained, highly volatile coal derived mainly from spores, characterized by a dull to greasy luster and conchoidal fracture. **2.** a former term for a coal that burns with a steady luminous flame. **3.** see JET COAL.

canneloid *Geology.* **1.** describing cannel coal or coal that resembles cannel coal. **2.** referring to a type of coal that is intermediate between bituminous coal and cannel coal.

cannel shale *Geology.* a black shale formed in stagnant or standing water by the accumulation of organic sediments together with inorganic matter.

cannelure *Ordnance.* **1.** a groove around the cylinder of a bullet, either containing a lubricant or into which the cartridge case is crimped. **2.** a groove around the band of a gun projectile that reduces the resistance against the rifling. **3.** a groove around the base of a cartridge that allows the extractor to take hold. **4.** a circular groove that locks the jacket of an armor-piercing bullet to the core.

cannibal *Behavior.* an organism that feeds on members of its own species. Thus, **cannibalism.**

cannibalize *Engineering.* to remove one or more components from a machine in order to use them for the repair or operation of another machine.

canning *Food Technology.* a process of food preservation in which the food is packed into glass jars or metal cans and then subjected to high temperatures that sterilize both the food and the container. *Metallurgy.* the process of enclosing a reactive metal, such as titanium, in a less reactive can, such as one made of iron, before hot-working. *Nucleonics.* see CLADDING, def. 2.

Cannizzaro, Stanislao 1826–1910, Italian chemist; distinguished atomic and molecular weights; discovered the Cannizaro reaction.

Cannizzaro reaction *Organic Chemistry.* a base-catalyzed dismutation of aromatic aldehydes or aliphatic aldehydes with no α-hydrogen into the corresponding acids and alcohols.

Annie Jump Cannon

Cannon, Annie Jump 1863–1941, American astronomer; compiled star catalogs and bibliographies; discovered new and variable stars.

Cannon, Walter 1871–1945, American psychologist and physiologist; noted for his studies of the nervous system and of emotional stimuli.

cannon *Ordnance.* **1.** originally, a type of artillery with a large caliber and a short barrel. **2.** in modern usage, a component of a gun, howitzer, or mortar, consisting of a tube, a breech mechanism, and a firing mechanism or base cap. **3.** a family of guns intermediate between small arms and artillery, usually of calibers between 20 mm and 40 mm, for use against aircraft or lightly armored vehicles.

cannonball *Ordnance.* **1.** originally, a round, solid projectile fired from a cannon. **2.** any projectile fired from a cannon.

cannon cradle *Ordnance.* a device that supports a cannon, allowing it to recoil and counterrecoil.

cannon pinion *Horology.* a tube terminating in a pinion, which fits over and rotates with the center arbor, and usually carries the minute hand of a timepiece.

cannon primer *Ordnance.* a primer that is used with separate-loading ammunition.

Cannon theory *Psychology.* the theory that emotions and the physical reactions associated with them are both caused by the brain's reaction to the same external stimulus, but that they are produced by different neural messages and are essentially independent of each other. Also, **Cannon-Bard theory.** (From Walter *Cannon* and Philip *Bard,* American psychologists.)

cannula *Medicine.* a pipette used in a trocar, an instrument used to remove fluid from body cavities. *Surgery.* a tube to be inserted into a cavity or duct.

cannular combustion chamber *Aviation.* the combustion chamber of a turbojet engine, consisting of a series of individual flame tubes (cans) enclosed in a common casing.

cannulated drill *Surgery.* a drill with a hole through the center of the long axis; used over a guidewire in surgery.

cannulation *Surgery.* the insertion of a cannula. Also, **cannulization.**

Canoidea *Vertebrate Zoology.* a mammalian superfamily in the suborder Caniformia and the order Carnivora, including only one family, Canidae, the dogs, wolves, foxes, and coyotes.

canonical [kə nŏn´i kəl] *Science.* of a statement, expression, procedure, or the like, in the simplest form that conforms to a general rule or accepted methodology.

canonical analysis *Statistics.* a methodology for the study of the relationship between two sets of random variables, based on summarizing each set through appropriately chosen canonical variables.

canonical basis *Mathematics.* the basis $\{e_1, e_2, \ldots, e_n\}$ of R^n, as a vector space over the real numbers R, where e_i is the column or row vector that has 1 in the ith position and 0 everywhere else.

canonical change *Astronomy.* a periodically repeating variation in one of the elements of an orbit.

canonical coordinates *Mathematics.* **1.** a coordinate system, defined in a neighborhood of the identity of a Lie group G. The coordinates $(t_1, \ldots, t_n)$ of a point $p \in G$ are given by

$$p = \exp(t_1 e_1)\exp(t_2 e_2)\cdots\exp(t_n e_n) \text{ or } p = \exp(t_1 e_1 + \cdots + t_n e_n)$$

for $\{e_\alpha\}$ a basis of the tangent space $T_e(G)$ at the origin e. Also, **normal coordinates.** It can be shown that, given an arbitrary system of coordinates in G, a canonical system of coordinates exists and is unique modulo any change of basis in $T_e(G)$. **2.** a coordinate system $(q^1, \ldots, q^n, p_1, \ldots, p_n)$ on a symplectic manifold (e.g., the cotangent bundle of a differentiable manifold) in which the **canonical 2-form** is expressed as $\sum_{i=1}^{n} dq^i \wedge dp_i$.

canonical coordinates and momenta *Quantum Mechanics.* the representation of the number of particles and corresponding momentum components by Cartesian coordinates and components of velocity, respectively.

canonical correlation *Statistics.* in canonical analysis, any of the correlations between canonical variables.

canonical distribution *Ecology.* a distribution plot in which the Y axis is the number of species and the X axis is an environmental variable calibrated logarithmically.

canonical ensemble *Physics.* an ensemble used to represent a system that can exchange energy but not particles with its environment.

canonical equations *Mathematics.* the equations $q_r' = \partial H/\partial p_r$ and $p_r' = -(\partial H/\partial q_r)$. Also, HAMILTON'S EQUATIONS.

canonical equations of motion see HAMILTON'S EQUATIONS OF MOTION.

canonical form *Science.* the simplest expression of an equation, statement, or rule. *Control Systems.* the simple, standardized matrix of rows and columns used to represent a given dynamic system. *Artificial Intelligence.* a convention for representing a symbolic expression that allows easier manipulation of such expressions. *Organic Chemistry.* each of the various possible electronic structures that a resonance-stabilized molecule may have.

canonically conjugate variables *Mechanics.* a generalized coordinate and its conjugate momentum; the action and angle variable.

canonical momentum see CONJUGATE MOMENTUM.

canonical random variables *Statistics.* in canonical analysis, two sets of random variables, S and T, that satisfy the following four conditions: (a) any two members of the same set have zero correlation; (b) each X_i of set S and Y_i of set T have zero correlation for all $i \neq j$; (c) each X_i is a linear combination of members of set S and each Y_i is a linear combination of members of set T; (d) each X_i and Y_i has mean 0 and variance 1.

canonical schema *Computer Programming.* in data base design, a map of the overall inherent structure of the data base that is independent of individual applications and computer systems.

canonical transformation *Mathematics.* **1.** any transformation of the coordinates on a symplectic manifold that leaves the canonical 2-forms in invariant form. Such a coordinate change also leaves Hamilton's equations unchanged. **2.** any function or transformation that has a standard form in some context.

Canopus *Astronomy.* the second-brightest star seen in the sky, a supergiant about 100 light-years away in Carina with spectral type F0 and an apparent brightness of –0.7 magnitude; used as guide star for many deep space probes.

canopy *Aviation.* **1.** the transparent overhead cover of an aircraft cockpit. **2.** the umbrella-like fabric body that supports a parachute. *Forestry.* the leafy cover or top layer formed by the uppermost branches of trees in a forest. Also, CROWN CANOPY, CROWN COVER.

cant *Ordnance.* the sideways tilt of a gun.

cantala *Materials Science.* a tough fiber taken from the agave; used to make twine.

cantaloupe *Botany.* a small melon, the fruit of the *Cucumis melo* plant. Also, **cantaloup.**

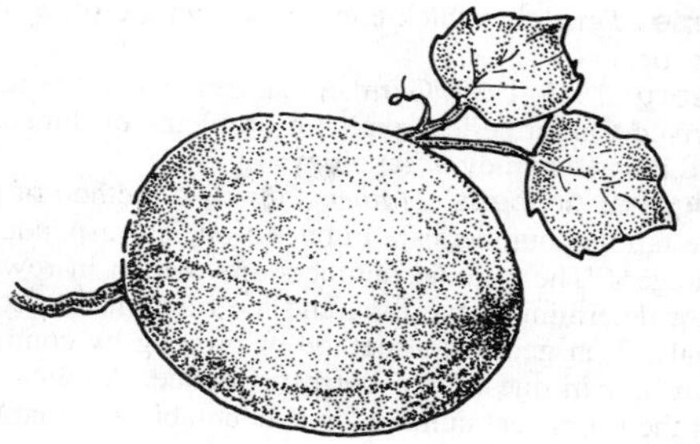

cantaloupe

cant body *Naval Architecture.* the sections of a hull that are supported by cant frames and curve toward the bow and stern.

Canterbury northwester *Meteorology.* a strong northwest foehn that descends the New Zealand Alps to the Canterbury Plains of South Island, New Zealand.

cant file *Mechanical Devices.* a file used to sharpen saw teeth with its surface serrations running at an angle to the length and edges.

Cantharellaceae *Mycology.* a family of fungi belonging to the order Agaricales which occurs primarily on humus and is well known for its edible species, the Chanterelle mushroom.

Cantharellales *Mycology.* a term formerly used for an order of fungi belonging to the subclass Holobasidiomycetidae.

canthariasis *Medicine.* an infestation of beetles within the body of a mammal after consumption of larvae or adult insects.

Cantharidae *Invertebrate Zoology.* soldier beetles, a family of carnivorous coleopteran insects in the superfamily Cantharoidea.

cantharides *Entomology.* a diuretic, blister treatment or purported aphrodisiac manufactured from the crushed, dried bodies of the blister beetle, Spanish fly, or others of the family Meloidae.

cantharidic acid *Organic Chemistry.* $C_8H_{12}O(COOH)_2$, an organic acid once used as a diuretic and stimulant.

cantharidin *Organic Chemistry.* $C_{10}H_{12}O_4$, rhombic plates or scales; insoluble in water and slightly soluble in alcohol and acid; melts at 218°C and sublimes at 110°C at 12 torr; used in veterinary medicine.

cantharidism *Toxicology.* poisoning caused by the ingestion or absorption of powdered cantharides (Spanish fly); symptoms may include severe gastrointestinal disturbances, nephritis, and death.

Cantharoidea *Invertebrate Zoology.* a superfamily of beetles in the suborder Polyphaga.

cant hook *Mechanical Devices.* a tool consisting of a stout wooden lever shaft with a pointed head and a pivoted iron hook near the pointed end; used for grasping, canting, and turning over legs.

canthus [kan´thəs] *plural,* **canthi.** [kan´thī´] *Anatomy.* the angle at either end of the fissure between the eyelids; often designated as **nasal** (inner) **canthus** or **temporal** (outer) **canthus.**

cantilever *Architecture.* any structural member that projects from a vertical support. *Building Engineering.* a beam or member built in or held down by weight fixed at one end while hanging freely at the other end.

cantilever aerotriangulation *Geodesy.* the process of determining the coordinates of ground points in a strip of aerial photographs when coordinates are known for ground points at only one end of the strip.

cantilever arch see CORBEL ARCH.

cantilever arm *Building Engineering.* on a cantilever bridge, the overhang from the support into the central span.

cantilever design *Robotics.* a type of robotic architecture consisting of a pedestal with a beam that projects out from the top of it.

cantilever extension *Cartography.* the extension and adjustment of a photogrammetric survey from a region with ground control into a region without ground control.

cantilever spring *Mechanical Engineering.* a spring supported at one end and holding a load at the other end.

cantilever vibration *Mechanics.* the transverse oscillation of a beam that is fixed (clamped) at one end, the other being free.

cantilever wall *Building Engineering.* a reinforced-concrete retaining wall stabilized by the weight of the retained material on its heel end.

canting *Mechanics.* the process of subjecting a cantilever beam to a sideways force, just smaller than the fracture load, in order to displace the free end of the beam.

Canton, John 1718–1772, English educator; investigated aurora borealis and terrestrial magnetism.

canton crepe *Textiles.* a thick crepe fabric with crosswise ribs, made of silk or rayon.

Cantor, Georg 1845–1918, German mathematician; the founder of set theory; derived Cantor series; explored problems of dimension and infinities; laid bases for study of topology.

Cantor diagonal process *Mathematics.* **1.** a method of proving that the positive rational numbers are in one-to-one correspondence with the positive integers. The rational numbers are written in rows, with each denominator determining one row, and the fractions are counted along the diagonals. **2.** in general, a method of proving by contradiction that two sets are not in one-to-one correspondence. Applications include proofs that the set of real numbers is uncountable and that the set of algebraic numbers is countable.

Cantor discontinuum See CANTOR TERNARY SET.

Cantor function *Mathematics.* the continuous monotone increasing function ψ that maps the Cantor ternary set onto the interval [0, 1], defined as follows: For every x in [0, 1], write

$$x = \sum_{i=1}^{\infty} \alpha_i / 3^i = \alpha_1 \alpha_2 \alpha_3 \cdots$$

where $\alpha_i = 0, 1,$ or 2. (Note that if $\alpha_i \neq 1$ for all i, then x is a member of the Cantor set C.) Let $n = n(x)$ be the first index i for which $\alpha_n = 1$. (If there is no such n, i.e., x is in C, then define $n(x) = \infty$.) Then define the function ψ by the equation

$$\psi(x) = \sum_{1 \leq i < n} \alpha^{i+1}_{i/2} + 1/2^n + \cdots.$$

The graph of $\psi(x)$ is known as the **devil's curve,** and the area bounded by it and the x-axis is known as the **devil's staircase.**

Cantor's axiom *Mathematics.* the assertion that the cardinal number of the set of points on a line extending indefinitely in both directions equals the cardinal number of the set of real numbers; i.e., there exists a one-to-one correspondence between the points on the line and the set of real numbers.

Cantor set *Chaotic Dynamics.* a fractal set of points created on a line segment by first removing the central third and then successively removing the central third of the remaining segments. The dimension of the Cantor set is log 2/log 3.

Cantor's paradox *Mathematics.* the contradiction that there is a largest cardinal number, based on the (incorrect) assumption that there exists a set containing all sets.

Cantor ternary set *Mathematics.* the set C of numbers in the closed interval [0, 1] whose ternary representations do not use the digit 1. Geometrically, this set can be constructed as follows: let X be the interval [0, 1]; X_1 = (1/3, 2/3) the open middle third of X; X_2 and X_3, the open middle thirds of $X - X_1$, i.e., (1/9, 2/9) and (7/9, 8/9) respectively; X_4, X_5, X_6, and X_7 the open middle thirds of the four closed intervals which make up $X - (X_1 \cup X_2 \cup X_3)$; and so on *ad infinitum.* Then the Cantor ternary set is

$$C = X - \bigcup_{n=1}^{\infty} X_n$$

It can be shown that C has Lebesgue measure = 0; C is perfect and nowhere dense; and C has the cardinal number of the continuum. Also, CANTOR DISCONTINUUM.

Cantor theorem *Mathematics.* the theorem that there is no one-to-one correspondence between the members of a set and the set of its subsets, and therefore that the cardinal number of a set is strictly less than the cardinal number of the set of its subsets (and that there is no largest cardinal number). Also, **Cantor subset theorem**.

cant strip *Building Engineering.* **1.** an inclined or bevelled wood strip used to change the pitch of a roof slope so as to round out the angle between the flat roof and an adjoining parapet to prevent bending. **2.** a strip positioned under the edge of the lowest row of tiles on the base of a tiered or gently sloping roof so as to equalize the slope.

canvas *Textiles.* a strong, heavy, plain-woven fabric made of linen, hemp, or cotton.

canyon *Geology.* a deep, steep-sided gorge or ravine usually cut by a river or stream.

canyon

canyon bench *Geology.* a relatively narrow, flat landform along a canyon wall resulting from differential erosion of horizontal strata.

canyon dune *Geology.* a dune that forms in a box canyon.

canyon fill *Geology.* loose, unconsolidated material consisting of permanently or temporarily deposited sediment that fills a canyon.

canyon wind *Meteorology.* **1.** a night-time flow of air down a canyon, caused by cooling at the canyon walls. **2.** any wind that is modified as it is forced to flow through a canyon or gorge. Also, GORGE WIND.

caoutchouc *Materials.* another term for raw rubber.

cap *Engineering.* **1.** a top or cover for an object or opening. **2.** to seal or cover a borehole. *Petroleum Engineering.* a cover for an oil well. *Mining Engineering.* **1.** a piece of timber placed on top of a prop or post in a mine. **2.** the horizontal section of timber used as a roadway suppport in a mine. *Building Engineering.* a finished cement top on a wall or chimney. *Genetics.* **1.** the addition of a methylated guanoside residue to the 5' end of much of the hnRNA and mRNA synthesized in eukaryotic cells. **2.** the concentration of a specific protein at a localized region on the surface of a cell. Also, CAPPING. *Mycology.* the pileus or spore-bearing portion of mushroom-shaped fruiting bodies found in fungi belonging to the class Hymenomycetes.

CAP Civil Air Patrol; catabolite activator protein.

5' cap see CAP OF MRNA.

capability list *Computer Programming.* **1.** a list of the permitted operations a user may perform on a given file. **2.** a row in an access matrix.

capable fault *Geology.* a term used by the Nuclear Regulatory Commission to describe a fault that is capable of movement in the relatively near future; the designation was developed for the purpose of selecting suitable sites for nuclear power plants.

Atoms & Molecules

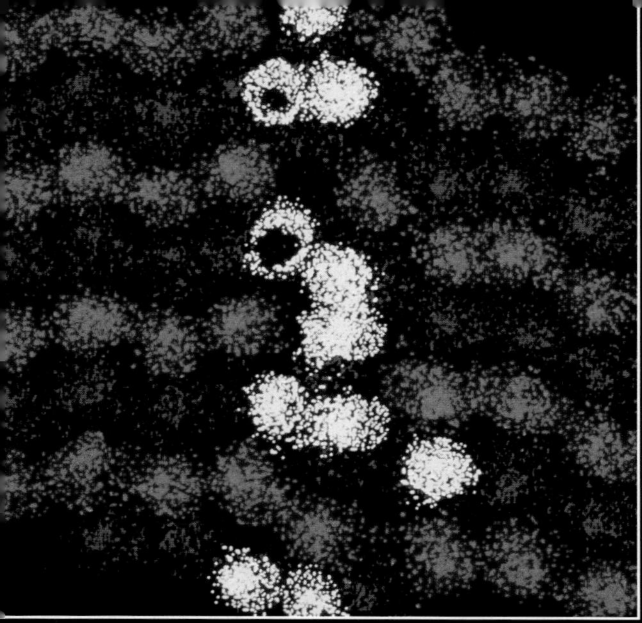

Scanning tunneling microscope image of atoms in a solar cell. Science Source/Photo Researchers.

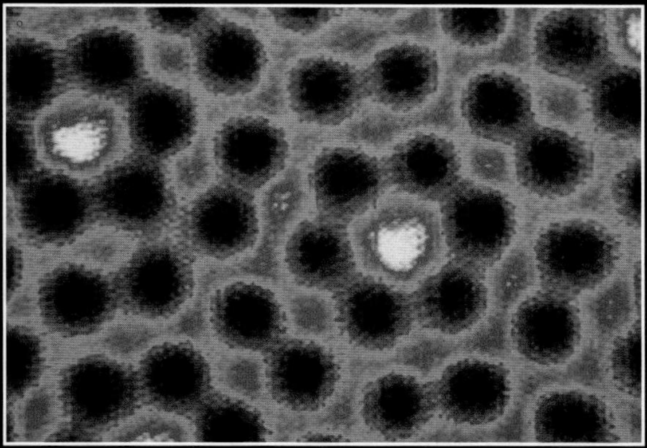

Scanning tunneling microscope image of the topmost atoms of a silicon surface. Courtesy of International Business Machines Corporation.

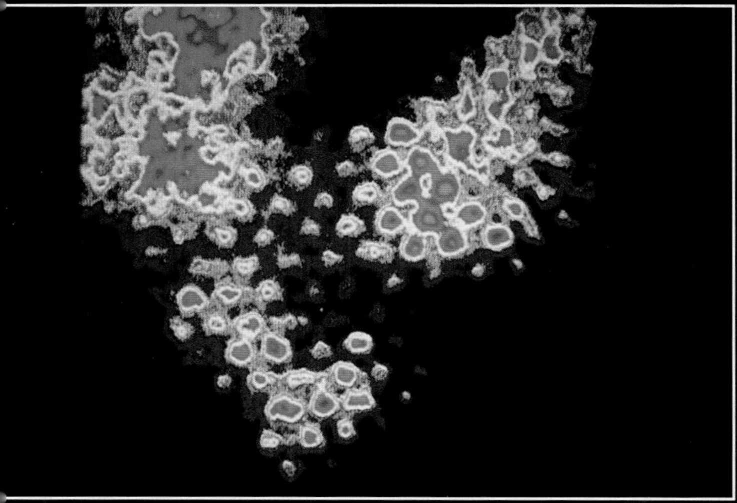

False-color scanning transmission electron micrograph of uranyl microcrystals from a uranyl acetate solution. Dr. M. Ohtsuki/SPL/Photo Researchers.

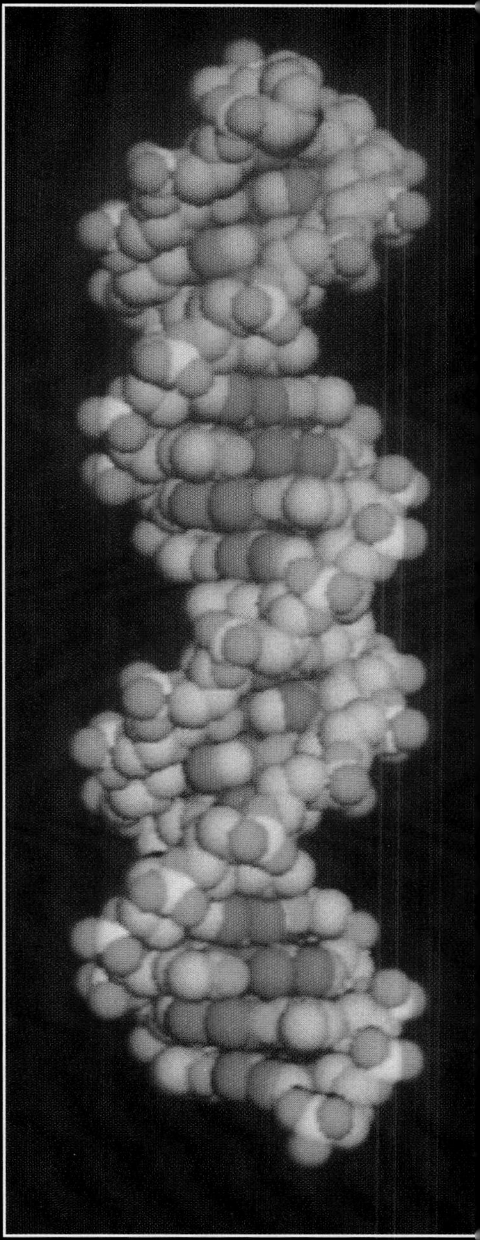

Molecular computer graphic of the DNA molecule, showing its double helix structure. Dr. A. Lesk/SPL/Photo Researchers.

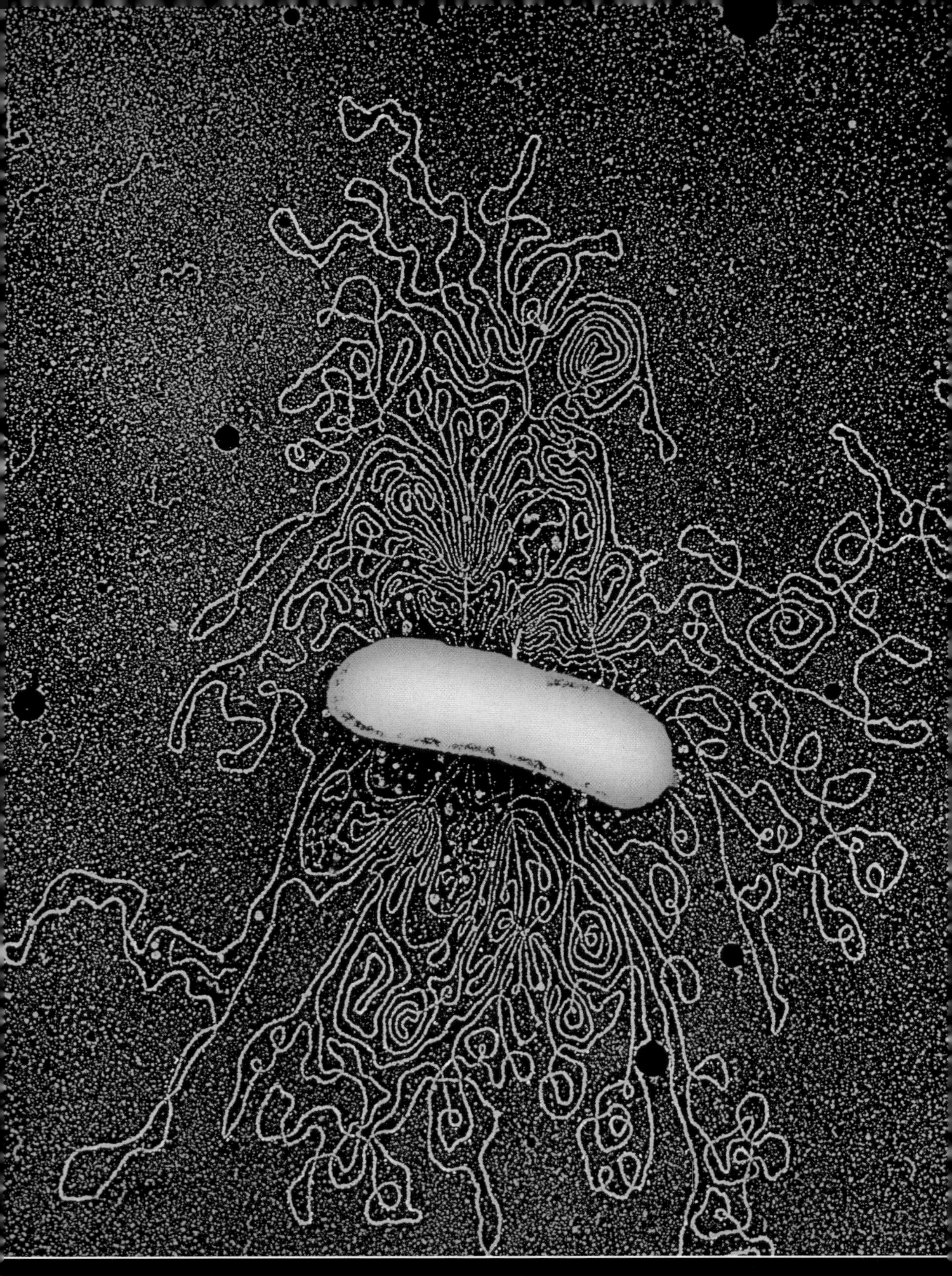

Transmission electron micrograph of DNA extruded from *Escherichia coli* cell (center). Dr. G. Murti/SPL/Photo Researchers.

capacitance *Electricity.* the ability of conductors separated by dielectric material to store energy in the form of electrically separated charges; a value described as the ratio of a quantity of electricity to a potential difference.

capacitance altimeter *Engineering.* an altimeter that measures the altitude of an aircraft by measuring the alterations in capacitance between two conductors on the aircraft when the ground is close enough to serve as a third conductor.

capacitance box *Electricity.* a device having capacitors and switches to allow capacitance adjustment at the terminals in uniform steps.

capacitance bridge *Electricity.* a bridge designed for measuring capacitance by comparison with a known capacitor.

capacitance hat *Electromagnetism.* a network of wire that is placed on top of an antenna for increasing the bandwidth or lowering its resonant frequency.

capacitance meter *Engineering.* an instrument used to measure the ratio between the charging current and the rate of change of voltage over time of a circuit.

capacitance probe *Petroleum Engineering.* a sensing instrument that measures the dielectric constants of water and oil constituents in an oil-water emulsion.

capacitance relay *Electronics.* a circuit with a relay that responds to a small change in capacitance, as when a hand or body nears a pickup wire.

capacitation *Physiology.* the processes that enable spermatozoa to fertilize ova in the presence of fluids of the female genital tract.

capacitator transducer *Acoustical Engineering.* a reciprocal transducer that creates sound due to electrical excitation between two capacitor plates, one stationary and the other flexible. Also, CONDENSER TRANSDUCER.

capacitive coupling *Electricity.* the connecting of two or more circuits by means of capacitance mutual to the circuits.

capacitive diaphragm *Electromagnetism.* a plate conductor that is inserted into a waveguide for the purpose of introducing capacitive reactance at the transmitted frequency.

capacitive-discharge ignition *Electronics.* a system in which energy stored in a capacitor is discharged across the gap of spark plug through the transformer and distributor each time a control rectifier is triggered; used in some types of automotive ignition systems.

capacitive divider *Electricity.* two or more capacitors in series across a source.

capacitive feedback *Electronics.* the process of returning part of the energy generated in the anode of a tube back to its cathode using a capacitance common to both circuits.

capacitive load *Electricity.* a load that can be represented as a capacitor or a combination of a capacitor and a resistor. Also, LEADING LOAD. *Electromagnetism.* a load whose capacitive reactance exceeds the inductive reactance and thus causes the current to lead the voltage.

capacitive loading *Electromagnetism.* the process of raising or lowering the resonant frequency of an antenna by installing a fixed capacitor in series.

capacitive post *Electromagnetism.* a conductive screw or post that is mounted on a waveguide perpendicular to the electric field in order to provide parallel capacitive susceptance with the waveguide for tuning or matching purposes.

capacitive-pressure transducer *Engineering.* a pressure-measuring device that monitors the strength of an electric signal emitted by the device when the capacitor experiences a change in pressure.

capacitive reactance *Electromagnetism.* the reactance associated with the capacitance of a capacitor circuit element, equivalent to the inverse of the product of the capacitance and the angular frequency; measured in ohms.

capacitive tuning *Electronics.* a system that adjusts the frequency of a signal by varying the capacitor in a circuit.

capacitive window *Electromagnetism.* a pair of conductive diaphragms inserted into a waveguide that are used to provide capacitive susceptance in parallel with the waveguide.

capacitor *Electricity.* a device composed of two conducting surfaces separated by a dielectric; it has the ability to store electrical energy and block the flow of direct current.

capacitor antenna *Electromagnetism.* an antenna consisting of two horizontal conductive elements and characterized by the capacitance associated with the elements.

capacitor box *Electronics.* a chamber that holds a capacitor in a heat-absorbing substance, such as water. Also, CONDENSER BOX.

capacitor bushing *Electricity.* a bushing in which cylindrical conducting layers are configured coaxially with the conductor within the insulating material. Also, CONDENSER BUSHING.

capacitor color code *Electricity.* color dots or bands placed on a capacitor to indicate the values of capacitance, tolerance, voltage rating, and other related information.

capacitor hydrophone *Acoustical Engineering.* a type of capacitator microphone that is used for detecting or responding to sounds under water.

capacitor-input filter *Electronics.* a power-supply filter in which a capacitor is connected across or parallel with the rectifier output.

capacitor loudspeaker *Acoustical Engineering.* any type of loudspeaker in which the mechanical energy is produced by the action of two electrically charged conductors; for example, a moving-coil microphone.

capacitor microphone *Acoustical Engineering.* a widely used type of microphone that has two electrically charged metal plates set slightly apart, one a flexible plate and the other rigid; sound waves make the flexible plate vibrate, which causes variations in the electric current from the plates. Such variations can be detected and amplified for sound reproduction. Also, CONDENSER MICROPHONE.

capacitor motor *Electricity.* a single-phase induction motor that has the main winding connected to an AC power source and the auxiliary winding connected in series with a capacitor.

capacitor-start motor *Electricity.* a capacitor motor in which the auxiliary winding is closed as a means of starting. The auxiliary winding is opened during normal operation.

capacity the maximum amount that a container can hold or that a system can receive; specific uses include: *Transportation Engineering.* the maximum number of vehicles per hour which a highway can carry; attempts to exceed capacity will normally produce congestion and delays rather than greater traffic volume. *Psychology.* see CAPABILITY. *Analytical Chemistry.* a measurement of the adsorption ability of ion-exchange materials in chromatography. *Computer Technology.* **1.** the maximum rate at which a computer system can process work. **2.** the total amount of data that a computer memory component can store. *Electricity.* **1.** the current-output capability of a cell or battery over a period of time, usually expressed in ampere-hours. **2.** capacitance. *Mathematics.* the nonnegative real number $\psi(a)$ assigned to each arc a of a network is the **arc capacity.** A network may be defined as a pair $N = (D, \psi)$, where D is a directed graph and ψ is a function from the arc set of D to the nonnegative real numbers.

capacity correction *Engineering.* a correction added to certain mercury barometers to allow for a difference in the cistern's level as atmospheric pressure varies.

capacity-demand relationship *Transportation Engineering.* the relationship between the capacity of a transport system and the number of trips demanded by users.

capacity-design procedure *Transportation Engineering.* a highway design procedure in which the capacities of each section, access, egress, etc., are mutually correlated so that, ideally, no potential capacity is wasted due to upstream bottlenecks.

capacity loading *Design Engineering.* the degree to which a system is operated at capacity.

capacity models *Design Engineering.* mathematical or computer models describing the loading of a manufacturing system.

capacity of the wind *Geology.* the total weight of particles of a given size, shape, and specific gravity that can be carried per unit volume of air by wind that is blowing at a given speed.

capacity planning *Design Engineering.* the setting of limits or levels for manufacturing operations at some point in the future, based on sales forecasts and on the requirements and availability of equipment, time, materials, and human resources. Similarly, **capacity expansion planning.**

capacity-restrained assignment *Transportation Engineering.* a method of adjusting minimum-time paths according to the volume, density, and speed on a route.

capacity test *Microbiology.* an assay to determine the ability of a disinfectant to remain active upon the addition of increasing numbers of bacteria.

cap-binding protein *Molecular Biology.* a specific protein that accumulates at one location on the surface of a cell to form a cap.

cap cloud *Meteorology.* **1.** a cloud that hovers above or over an isolated mountain peak, formed by the cooling and condensation of moist air forced up and over the peak. Also, CLOUD CAP. **2.** see PILEUS.

cap crimper *Engineering.* a tool, similar to pliers, that presses the open end of a blasting cap on the safety fuse before placing it in the primer.
cape *Geography.* a point of land jutting into the sea; a headland or promontory.
cape chisel *Mechanical Devices.* a cold chisel with a narrow blade, used to cut keyway grooves, form slots, and for other confined work.
Cape doctor *Meteorology.* a term for a strong, invigorating wind from the southeast on the southern coast of Africa.
cape foot or **Cape foot** *Metrology.* a unit of measure used in the Republic of South Africa, equal to 1.033 English feet.
Cape Horn Current *Oceanography.* the section of the west-wind surface drift that flows eastward near Cape Horn and then northeastward to become the Falkland Current.
Capella *Astronomy.* a double-star system 42 light-years away in Auriga, with spectral types G5 and G0 and having a combined apparent brightness of 0.1 magnitude.
Capell fan *Mining Engineering.* a centrifugal fan used in a mine shaft.
capillarity [kap´ə lâr´ə tē] *Fluid Mechanics.* the general behavior of fluids acting with surface tension on interfaces or boundaries.
capillarity correction *Engineering.* a correction for capillarity effects in order to determine the true inclination angle of a borehole, made by figuring the specific angular value from the apparent angle, as indicated by the plane of the etch line in an acid-survey bottle.
capillaroscope *Medicine.* an instrument for the microscopic examination of capillaries.
capillary *Anatomy.* **1.** any of the tiny blood vessels through which exchange of materials between the blood and surrounding tissues takes place; serve to join the arteries with the veins. **2.** any tube having a very small bore. *Geology.* describing a mineral that forms hairlike or threadlike crystals. *Petroleum Engineering.* a fine crack or fissure in a formation, through which oil and other fluids flow.
capillary action *Fluid Mechanics.* the attraction of the surface of a liquid to the surface of a solid, which either elevates or depresses the liquid depending upon molecular surface forces; for example, crude oil clings to the surface of each pore in a rock formation, making it difficult to recover oil.
capillary attraction *Fluid Mechanics.* the attractive force arising from surface tension effects at interfaces.
capillary bed *Anatomy.* an array of capillaries joining an artery to a vein.
capillary collector *Engineering.* an instrument used to collect moisture from the atmosphere; it is covered with a porous material that allows water, but not air, to pass into a capillary system.
capillary column *Analytical Chemistry.* a very fine-bore tube used to measure surface tension of various liquids by comparing the ratio of wetted surface of the tube to the volume of the liquid.
capillary condensation *Materials Science.* a process to evaluate pore size and distribution within a material in which surface adsorption of an intruding vapor is dependent upon pore diameter. *Physical Chemistry.* a term for the liquid that appears within the fine pores of an adsorbate.
capillary conductivity *Geology.* the ability of unsaturated rock or soil to transmit liquids.
capillary drying *Engineering.* the evaporation of moisture from the surface of a porous mass and the subsequent capillary movement of moisture from the interior of the mass to its surface; eventually the amount of moisture on the surface and interior of the mass stabilizes.
capillary ejecta see PELE'S HAIR.
capillary electrometer *Engineering.* an instrument in which small electric currents are detected by movement of a mercury meniscus in a capillary tube. Also, LIPPMANN ELECTROMETER.
capillary electrophoresis *Analytical Chemistry.* a separation technique in which a sample is introduced into a capillary tube and the components are separated by the application of a high voltage.
capillary equilibrium method *Petroleum Engineering.* a test procedure to predict gas and oil flow through an oil reservoir core; the flow is reduced to keep capillary flow in equilibrium between gas and oil within the reservoir.
capillary fringe *Hydrology.* the lower subdivision of the zone of aeration, located just above the water table and containing water under less than standard atmospheric pressure. Also, **capillary-moisture zone.**
capillary gas chromatography *Analytical Chemistry.* a separation technique in which the gaseous phase of a sample (mobile phase) passes through capillary tubes whose inner walls contain a thin film of an adsorbing medium (stationary phase) and the components are detected instrumentally as they are eluted.
capillary interstice *Hydrology.* a pore space that is small enough to hold water at a height above a free water surface by surface tension, but large enough to prevent molecular attraction from extending across the opening.
capillary migration *Hydrology.* the movement of water into small spaces in rock or soil as a result of surface tension. Also, **capillary flow.**
capillary pressure *Fluid Mechanics.* the pressure due to capillary action; i.e., the pressure of fluids under the influence of surface tension and adhesion. *Physiology.* the blood pressure in the capillaries, which are the smallest vessels of the circulatory system.
capillary rise *Fluid Mechanics.* the phenomenon by which surface tension causes the meniscus to rise in a liquid-filled manometer.
capillary tension see WATER TENSION.
capillary tube *Engineering.* a tube with a bore so fine that the rise or fall of a liquid in it due to capillary attraction is perceptible to the eye.
capillary viscometer *Materials Science.* a device for measuring shear viscosity of polymers (solutions or melts). The viscosity is calculated from the time required for a solution or melt to flow through a precisely dimensioned capillary tube under known pressure (stress) differential.
capillary water *Hydrology.* **1.** water that is held in or moving through small spaces between soil particles as a result of surface tension. **2.** water in the capillary fringe.
capillary wave *Fluid Mechanics.* a wave of short wavelength formed on the free surface of a liquid as a result of surface tension.
capillary yield *Geology.* the amount of capillary water that rises parallel to the water table at a specified distance below the land surface.
capillitium [kap´i lish´ē əm] *Mycology.* a network of filaments or hyphae threading through the spores of the fruiting body of certain slime molds belonging to the fungal class Myxomycetes and of certain fungi belonging to the class Gasteromycetes.
Capillovirus group *Virology.* a genus of plant viruses that resemble the closterovirus group, have a narrow host range, and are transmitted mechanically and by seed.
capital *Architecture.* the topmost member of a column or pilaster, crowning the shaft and carrying the entablature.
capitalist *Ecology.* an informal term for a plant that has significant reserves of stored food.
capital ship *Naval Architecture.* a warship of one of the largest and most powerful classes, such as an aircraft carrier or battleship.
capitate *Anatomy.* head-shaped. *Botany.* having a head-shaped appearance, usually referring to clusters of flowers.
capitellate *Botany.* having a small, bulbous end.
Capitellidae *Invertebrate Zoology.* a family of mud-swallowing annelid worms, sometimes called blood worms, in the group Sedentaria.
capitellum *Anatomy.* a rounded surface on the distal end of the humerus that articulates with the head of the radius.
Capitonidae *Vertebrate Zoology.* the barbets, a family of brilliantly colored pantropical birds in the order Piciformes, distinguished by the bristly nostril feathers around their bills.
capitonnage [kap´ə tə nazh´] *Surgery.* the surgical closure of a cyst cavity by applying sutures to approximate the opposing surfaces of the cavity.
capitular *Botany.* growing in or otherwise related to a capitulum.
capitulum [kə pich´yə ləm] *Biology.* a rounded, headlike protruberance on an anatomical structure. *Botany.* any inflorescence made up of many small, densely packed sessile or subsessile flowers, giving the whole a knobby appearance. *Invertebrate Zoology.* **1.** a protuberance at the tip of a process or part, as in the enlarged tip of the proboscis of a fly, the end of a capitate antenna, or the body portion of a hydroid polyp that bears the tentacles. **2.** the false head, or beak, of a tick, composed of palps and mouthparts.
cap molding *Architecture.* **1.** molding at the top of a dado. **2.** ornamental molding above the casing of a door or window.
Capnocytophaga *Bacteriology.* a genus of Gram-negative, gliding bacteria of the family Cytophagaceae, occurring in the human mouth and implicated in the pathogenesis of periodontal disease.
Capnodiaceae *Mycology.* a family of fungi belonging to the order Asterinales that forms sooty mold masses on the surfaces of plants and trees in warm to temperate regions.
cap of mRNA *Molecular Biology.* a modified guanine nucleotide that forms the 5' end of a eukaryotic mRNA molecule and protects the mRNA from degradation by nucleases. Also, 5' CAP.
capomo *Materials.* the wood of the trees of the *Brosimum* species of Central and South America; used for construction in flooring and veneers, and in the making of tool handles.

capon *Agriculture.* a male chicken that has been castrated to tenderize its flesh for market.

Caponidae *Invertebrate Zoology.* a family of arachnid arthropods, in the order Araneida, that have tracheae rather than book lungs.

CAPP *Design Engineering.* a computer application program that assists in the development of a process production plan for manufacturing. (An acronym for computer-aided process (or production) planning.)

Capparidaceae *Botany.* a family of dicotyledonous plants of the order Capparidales; characterized by alternate, simple, or palmately compound leaves, mostly bisexual flowers, and parietal placentation. Also, **Capparaceae.**

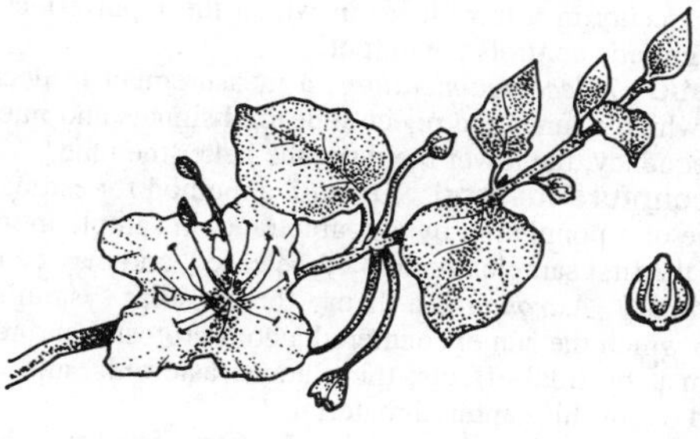

Capparidaceae (caper)

Capparidales *Botany.* an order of dicotyledonous shrubs, herbs, and trees of the subclass Dilleniidae, noted for their production of mustard-oil glucosides and alkaloidal amines. Also, **Capparales.**

capped column *Hydrology.* a hexagonal-shaped, columnar crystal of ice or snow having thin plates or stars at each end.

capped fuse *Engineering.* a length of safety fuse with the cap or detonator crimped on before it is taken to be used.

capped steel *Metallurgy.* steel cast in an ingot mold onto which a mechanical or chemical cap is placed after filling the mold with liquid metal; properties are intermediate between rimmed and semikilled steel.

cappelenite-(Y) *Mineralogy.* $Ba(Y,Ce)_6Si_3B_6O_{24}F_2$, a greenish-brown, brittle, trigonal mineral having a specific gravity of 4.407 and a hardness of 6 to 6.5 on the Mohs scale; occurs in nepheline syenite pegmatites with woehlerite and rosenbuschite.

cap piece *Mining Engineering.* a piece of wood that is fitted over a straight post or timber to provide more bearing surface for the support.

capping the process of placing a cap or top on something; specific uses include: *Petroleum Engineering.* **1.** the process of sealing or closing in a borehole, such as a spouting gas or oil well, to prevent the escape of fluids. **2.** the device used to cover or seal a borehole. *Mining Engineering.* **1.** the process of attaching a shackle or a swivel to the end of a winding rope. **2.** the overburden atop a valuable seam or bed of mineral. *Engineering.* to prepare a capped fuse. *Genetics.* **1.** the concentration of a specific protein at a localized region on the surface of a cell. **2.** the addition of a methylated guanoside residue to the 5' end of much of the hnRNA and mRNA synthesized in eukaryotic cells. *Geology.* **1.** see CAP ROCK, def. 1. **2.** see GOSSAN.

cap product *Mathematics.* **1.** a set intersection. **2.** a Boolean operation, denoted $\cap$ or Λ.

capraldehyde see DECANAL.

caprate *Organic Chemistry.* a salt of capric acid, containing the group $C_9H_{19}CO_2^-$.

Caprellidae *Invertebrate Zoology.* skeleton shrimps, a family of crustaceans, in the suborder Caprellidea, with slender cylindrical bodies.

Caprellidea *Invertebrate Zoology.* a suborder of crustaceans, in the order Amphipoda, living in marine or brackish water.

capric acid *Organic Chemistry.* $CH_3(CH_2)_8CO_2H$, combustible, white plates, slightly soluble in water and soluble in alcohol and acid; melts at 31.5°C and boils at 270°C; used to manufacture esters for perfumes and fruit flavors, and as an intermediate for food-grade additives. Also, DECANOIC ACID, DECYLIC ACID.

capric anhydride *Organic Chemistry.* $[CH_3(CH_2)_8CO]_2O$, a white crystalline solid that is insoluble in water and soluble in alcohol; melts at 24.7°C; used in organic synthesis. Also, DECANOIC ANHYDRIDE.

Capricornus *Astronomy.* the Sea-Goat, a large dim southern constellation lying on the zodiac between Sagittarius and Aquarius.

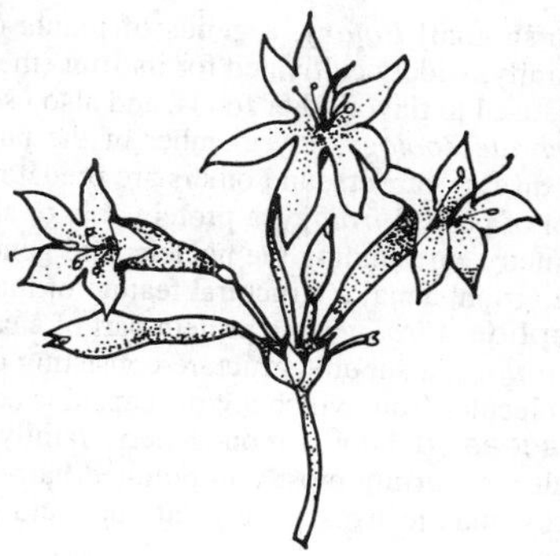

Caprifoliaceae (honeysuckle)

Caprifoliaceae *Botany.* the honeysuckle family of dicotyledonous plants of the order Dipasacales, having simple or pinnate leaves and bisexual epigynous flowers.

caprilydene see OCTYNE.

Caprimulgidae *Vertebrate Zoology.* the night jars, a family of nocturnal woodland birds having large mouths, belonging to the order Caprimulgiformes.

Caprimulgiformes *Vertebrate Zoology.* a diverse, widespread order of nocturnal birds with large mouths and weak feet, including nighthawks, whippoorwills, frogmouths, potoos, and others.

caprine arthritis-encephalitis *Veterinary Medicine.* a viral disease of goats resulting in chronic arthritis in the adult animal and encephalitis in the young; when fatal, death is preceded by paralysis and seizures.

caprinized vaccine *Immunology.* a vaccine that contains microorganisms whose virulence has been lessened by adaptation to and passage in goats.

Capripoxvirus *Virology.* a genus of viruses of the family Poxviridae that infect ungulates and are transmitted by arthropods.

caproamide *Organic Chemistry.* $CH_3(CH_2)_4CONH_2$, a crystalline solid that is slightly soluble in water and soluble in in alcohol; melts at 101°C and boils at 255°C; used as a chemical intermediate.

cap rock *Geology.* **1.** consolidated barren rock material overlying a mineral or ore deposit, which must be removed before mining. **2.** a layer of hard rock overlying the shale above a coal bed. **3.** in a salt dome, an impervious body of anhydrite and gypsum overlying the top of the salt body, or plug. **4.** an impervious layer of rock overlying a natural gas or oil deposit.

caproic acid see HEXANOIC ACID.

Caproidae *Vertebrate Zoology.* the boarfishes, a family of plankton-eating marine fish of the order Zeiformes, found in coastal waters around the world.

caprolactam *Organic Chemistry.* $C_6H_{11}NO$, a toxic, leaflike solid that is soluble in water and very soluble in alcohol; melts at 70°C and boils at 139°C (12 torr); used in manufacture of synthetic fibers. Also, 2-OXO-HEXAMETHYLENEIMINE.

ε-caprolactone *Organic Chemistry.* $C_6H_{10}O_2$, a liquid compound that is soluble in water and alcohol; boils at 98–99°C (2 torr). Also, 2-OXEPANONE.

caprylamide *Organic Chemistry.* $CH_3(CH_2)_6CONH_2$, leaflike plates; slightly soluble in water and very soluble in alcohol; melts at 105.9–110°C; used in organic synthesis.

capryl group *Organic Chemistry.* a terminal group found in many organic compounds. Also, OCTYL GROUP.

caprylic acid see OCTANOIC ACID.

caprylic anhydride *Organic Chemistry.* $[CH_3(CH_2)_6CO]_2O$, a white solid that decomposes in water and is soluble in alcohol; melts at –1°C and boils at 280–285°C; used in organic synthesis. Also, OCTANOIC ACID ANHYDRIDE.

capsaicin *Organic Chemistry.* $C_{18}H_{27}NO_3$, monoclinic plates or scales; insoluble in water and very soluble in alcohol; melts at 65°C and boils at 210–220°C (0.01 torr).

Capsaloidea *Invertebrate Zoology.* a superfamily of ectoparasitic trematodes, in the subclass Monogenea, with a sucker-shaped, hooked holdfast.

capsanthin *Biochemistry.* a red pigment found in paprika.

cap screw *Mechanical Devices.* a machined screw with threads along the length of its shank held by a threaded hole in the part to which it is joined, without the need for a nut.

capsicum [kap′sə kəm] *Botany.* a genus of plants of the Solanaceae (nightshade) family; widely cultivated for its fruit (the pepper), which is eaten as a food, used to flavor other foods, and also used medicinally.

capsid *Invertebrate Zoology.* any member of the bug family Miridae; some are plant disease carriers, and others are predators useful in the biological control of pests. *Virology.* a protein coat or shell of icosahedral or helical symmetry surrounding the nucleic acid genome or nucleoprotein core of the virion; a major structural feature of many viruses.

capsid polypeptide *Virology.* the protein part of a capsid.

capsomere *Virology.* a subunit structure consisting of a group of identical protein molecules from which a virus capsid is constructed.

Capsosiphonaceae *Botany.* a monogeneric family of green algae of the order Ulvales, occurring mostly in polluted harbors, in estuaries, or on muddy rocks; mature fronds are gelatinous and tubular and sometimes flattened.

capstan *Engineering.* a spoonlike drum mounted on a vertical axis, used for various forms of hoisting or pulling, and operated by steam, by electric power, or manually. Also, WINDLASS, WINCH. *Computer Technology.* the rotating spindle on a magnetic tape drive that spins the tape reel at a constant speed.

capstan bar *Mechanical Devices.* a long lever used to manually turn a capstan.

capstan nut *Mechanical Devices.* a nut with transverse holes through which a tightening bar can be inserted and the nut can be turned.

capstan screw *Mechanical Devices.* a screw with transverse holes in its head through which a tightening bar can be inserted and the screw can be turned. Also, **capstan-head screw.**

capstone *Architecture.* a coping stone.

capsular ligament *Anatomy.* any ligament that is attached to an articular or joint capsule.

capsular swelling reaction see QUELLUNG REACTION.

capsulate *Biology.* enclosed in or having a capsule.

capsule *Pharmacology.* a hard or soft soluble shell, often made of gelatin, containing a single dose of medication, usually to be administered orally. *Anatomy.* any sac or membrane that envelops an organ. *Botany.* **1.** any dry dehiscent fruit that is formed from more than one carpel. **2.** the spore-containing part of the sporophyte generation of mosses. *Engineering.* a small sealed compartment used to protect sensitive instrumentation. *Space Technology.* **1.** a small spacecraft or compartment capable of supporting life during a spaceflight, reentry, or high-altitude flight. **2.** a sealed encasement for instrumentation in space. *Microbiology.* a loose, extracellular layer of polysaccharide or protein, surrounding many procaryotic cells, and serving various functions including permeability resistance and ion exchange. *Invertebrate Zoology.* a sheath of tissue or an organically produced substance enclosing body organs or eggs. *Virology.* a proteinaceous layer of external material, produced during granulosis virus infections, that envelops the cell wall.

capsule cell see AMPHICYTE.

capsulectomy *Surgery.* the excision of a capsule, especially the capsule of a joint or of the lens.

capsuloma *Oncology.* a capsular or subcapsular tumor, especially in the kidney.

capsulorrhaphy *Surgery.* the suture of a capsule, especially a joint capsule.

capsulotomy *Surgery.* incision through a capsule, especially the capsule of the lens as in a cataract operation.

captacula *Invertebrate Zoology.* in some mollusks and annelids, sensory filaments or tentacles used for catching food.

captain *Military Science.* **1.** in the Navy, an officer ranking above a commander and below a rear admiral; equivalent to a colonel in the Army, Air Force, or Marine Corps. **2.** in the Army, Air Force, or Marine Corps, an officer ranking above a 1st lieutenant and below a major.

captan *Organic Chemistry.* $C_9H_8Cl_3NO_2S$, white to cream crystals, partially soluble in water; melts at 172°C; used as a fungicide in paints, plastics, fabrics, and fruit preservation.

caption *Graphic Arts.* the title or description of an illustration.

captive balloon *Aviation.* a towed or moored balloon, usually spherical and secured by steel cables; used chiefly for military observation.

captive fastener *Mechanical Devices.* a screw-type fastener that is restrained when unscrewed from the matching machine member.

captive mine *Mining Engineering.* a mine that produces coal for use by the same company that owns the mine.

captive test *Engineering.* a test performed by firing a rocket, usually one that is unmanned, while it is securely fastened to the test stand. Thus, **captive firing.**

Captor *Ordnance.* a U.S. self-propelled marine mine; it remains moored to the sea bottom until its acoustic sensor detects a suitable ship or submarine target, at which time it launches a small homing torpedo.

Captorhinomorpha *Paleontology.* a suborder of primitive anapsids in the extinct order Captorhinida; may be ancestral to all modern reptiles; includes the earliest of the amniotes, the family Protorothyrididae; Carboniferous to Permian.

capture *Nuclear Physics.* the process by which a neutron is absorbed (captured) by a target nucleus to form a compound nucleus, with the resulting excess energy being released as gamma radiation.

capture area *Acoustical Engineering.* a term for the frequency range over which a receiver can pick up signals.

captured rotation see SYNCHRONOUS ROTATION.

capture effect *Electronics.* **1.** the tendency of the stronger of two signals of the same frequency to diminish the weaker signal to zero. **2.** an effect that occurs in a transducer in which the input effect having the largest magnitude controls the output.

capture ratio *Telecommunications.* a measurement in decibels of the degree to which a tuner can reject undesired signals and interference on a given frequency; the lower the ratio, the better the tuner.

capture-recapture method *Statistics.* a method for estimating the unknown size of a population by a double random sample in which the elements of the first sample are marked, released, and may be resampled.

capture theory *Astronomy.* a theory for the solar system's origin, according to which the sun encountered a cool protostar and removed material from it by tidal effects; the planets, asteroids, and comets then condensed out of this captured material.

capturing *Engineering.* **1.** a process of using a torquer to hold the spin axis in a gyro to an exact position in relation to the spin reference axis. **2.** a process in which a missile is taken over by its guidance system.

capuchin

capuchin *Vertebrate Zoology.* a monkey of the genus *Cebus* that inhabits forests from Nicaragua to Paraguay; known to be very intelligent and easily trained (once popular as "organ-grinder" monkeys) and named for the crown of hair resembling cowls of capuchin monks.

capybara [kap′ə bär′ə] *Vertebrate Zoology.* a large South American rodent resembling a guinea pig and distinguished as an excellent swimmer, having partly webbed feet. Also, WATER PIG.

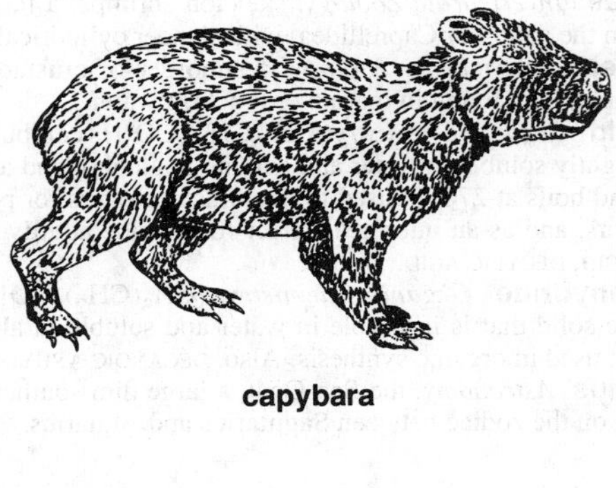

capybara

car *Mechanical Engineering.* **1.** a four-wheeled motor vehicle carrying passengers; an automobile. **2.** a vehicle running on rails, such as a railroad *car* or street*car*. **3.** the passenger compartment in an elevator, balloon, or other conveyance. *Mining Engineering.* see MINE CAR. *Artificial Intelligence.* in Lisp, the function that extracts the first half of a cons cell, usually containing a value that is an element of a list.

CAR civil air regulations.

Carabidae *Invertebrate Zoology.* ground beetles, a family of predatory coleopteran insects in the suborder Adephaga.

caraboid larva *Invertebrate Zoology.* the active and predacious larva of certain beetles.

caracara *Vertebrate Zoology.* a long-legged hawk of the falcon subfamily Polyborinae, characterized by red cheek and throat patches and long legs; often feeds on carrion; found from the southern U.S. to the tip of South America.

Caracarinae *Vertebrate Zoology.* the caracaras, a subfamily of South American carrion-eating hawks in the family Falconidae and the order Falconiformes.

caracolite *Mineralogy.* $Na_3Pb_2(SO_4)_3Cl$, a colorless or grayish, transparent monoclinic mineral, having a specific gravity of 5.1 and a hardness of 4.5 on the Mohs scale; occurs in prismatic crystals as crystalline incrustations with percylite, anglesite, and galena.

Caracul or **caracul** see KARAKUL.

Caradocian *Geology.* a European geologic stage of the Middle and Upper Ordovician period, occurring after the Llandeilian and before the Ashgillian.

caramel *Materials.* **1.** a flavoring and coloring agent derived by burning sugar until it changes color. **2.** a chewy candy made from milk, butter, and sugar; usually made into small blocks.

Carangidae *Vertebrate Zoology.* a family of marine fishes in the suborder Percoidea and the order Perciformes, including horse mackerels, jacks, scads, and pompanos.

carapace *Geology.* **1.** the upper limb of a fold having an almost horizontal axial plane. **2.** the frozen surface of a lava dome, especially its top part. *Invertebrate Zoology.* the tough, shieldlike plate covering the cephalothorax of decapods and other crustaceans.

Carapidae *Vertebrate Zoology.* the pearlfishes, a family of eel-like tropical fishes in the suborder Ophidioidei that sometimes live as parasites in other organisms. Also, CARAPODIDAE.

Carathéodory, Constantin [kär´ə tā´ə dôr´ē] 1873–1950, Greek mathematician; worked in calculus of variations and theory of real functions; formulated Carathéodory's principle.

Carathéodory outer measure see OUTER MEASURE.

Carathéodory's principle *Thermodynamics.* a principle associated with the second law of thermodynamics, which claims that in the neighborhood of an equilibrium state of a thermodynamic system, there exist states which are not accessible by either reversible or irreversible adiabatic processes.

caravel *Naval Architecture.* a small sailing vessel of Portugal and Spain, usually lateen-rigged on two or three masts; used in the Middle Ages and later. Also, **carvel.**

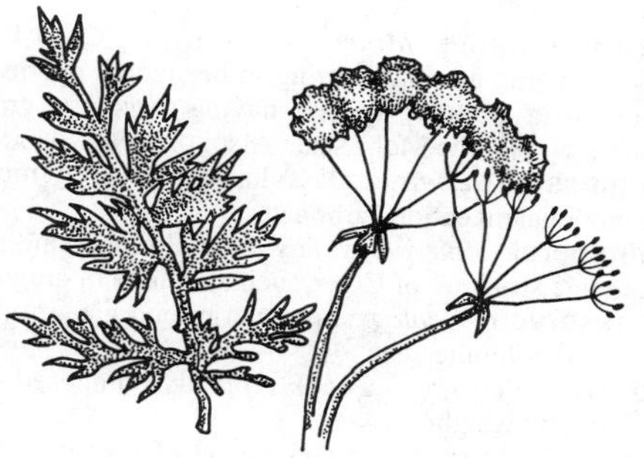

caraway

caraway *Botany.* a perennial herb, *Carum carui,* of the family Umbelliferae, whose seeds are cultivated as a spice.

carb- a combining form meaning "carbon," as in *carbide.*

carbachol *Pharmacology.* $C_6H_{15}ClN_2O_2$, a cholinergic drug used to contract the pupils in glaucoma treatment and ocular surgery. Also, CARBAMYLCHOLINE CHLORIDE.

carbalkoxylation *Organic Chemistry.* the addition of a –COOR group to a cyclic compound by treatment with esters and α-keto acids.

carbamate *Biochemistry.* an ester of the chemical carbamic acid.

carbamate esters *Organic Chemistry.* compounds with the formula ROCONHR´.

carbamic acid *Biochcmistry.* NH_2COOH, a simple amino acid that is active in metabolism.

carbamide see UREA.

carbamide peroxide see UREA PEROXIDE.

carbamino *Biochemistry.* a compound formed by the linkage of carbon dioxide with an aliphatic amine.

carbamoyl *Organic Chemistry.* the group NH_2CO–.

carbamoyl phosphate *Biochemistry.* $NH_2COPO_4H_2$, an organic compound termed an ester that is produced from phosphoric acid and carbamyl acid and, as a high-energy compound, is involved in certain types of biosynthesis. Also, **carbamyl phosphate.**

carbamoyltransferase *Enzymology.* an enzyme that catalyzes the transfer of the carbamoyl group H_2NCO–.

carbamylcholine chloride see CARBACHOL.

carbamylhydrazine hydrochloride see SEMICARBAZIDE HYDROCHLORIDE.

carbamylurea see BIURET.

carbanilide *Organic Chemistry.* $C_6H_5NHCONHC_6H_5$, rhombic, bipyramidal prisms, slightly soluble in water and alcohol; melts at 238°C and decomposes at 260°C; used in organic synthesis. Also, DIPHENYLUREA.

carbanion *Chemistry.* an anion formed by the removal of one or more protons from a carbon-containing compound. *Organic Chemistry.* **1.** a negatively charged ion of the type R_3C–, having a complete octet of electrons around the carbon atom with two unshared electrons. **2.** see METHYL ANION.

carbanion polymerization *Materials Science.* the successive polymerization of negatively charged species with a monomer that contains a double bond.

carbaryl *Organic Chemistry.* $C_{10}H_7OCONHCH_3$, a toxic solid; insoluble in water and soluble in acetone; melts at 145°C; used as an insecticide.

carbazole *Organic Chemistry.* $C_{12}H_9N$ white, leafy plates, insoluble in water and soluble in alcohol; melts at 247–248°C and boils at 355°C; used in the manufacture of dyes, reagents, explosives, and insecticides.

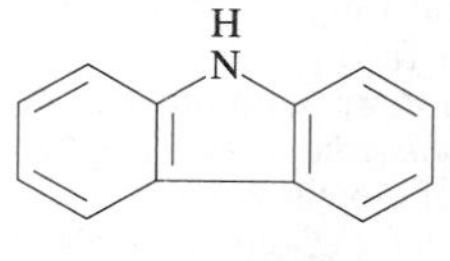

carbazole

carbene *Organic Chemistry.* an organic radical containing divalent carbon, making it very reactive.

carbenium ion *Organic Chemistry.* a cation that contains a carbon atom bearing only six electrons, and a positive charge.

carbenoid species *Organic Chemistry.* any reagent or material that is similar to carbene.

carbide *Inorganic Chemistry.* any of various compounds made up of carbon and another element, other than hydrogen; typically a metal such as iron, calcium, tungsten, silicon, or boron; usually produced by heating the reacting substances in an electric furnace. Many carbides are characterized by extreme hardness and are used as abrasives.

carbide lamp *Mining Engineering.* a lamp that burns acetylene that forms as a result of charged calcium carbide and water.

carbide miner *Mining Engineering.* an automated coal-mining machine that is controlled from outside the coal seam and operates as a continuous miner.

carbide nuclear fuel *Nucleonics.* a ceramic formulation of uranium in the form of uranium or plutonium monocarbide or dicarbide that has desirable thermal conductivity characteristics; used in advanced reactors that operate at very high temperatures.

carbide tool *Mechanical Devices.* a machining, forming, or cutting tool consisting of carbides embedded in a metallic matrix. A common type consists of tungsten carbide embedded in a cobalt-nickel matrix; it is used in cutting hard materials at high temperatures.

carbidization *Materials Science.* the surface reaction between carbon-containing molecules and metals, typically transition elements, to form surface carbides.
carbine [kär´bīn´; kär´bēn´] *Ordnance.* a general term for a short-barreled, lightweight military rifle.
carbinol see METHANOL.
Carbitol *Organic Chemistry.* $HOCH_2CH_2OCH_2CH_2OC_2H_5$, a trade name for a combustible, hygroscopic, colorless liquid used as a solvent and in textile manufacturing; miscible with water; boils at 195–201°C. Also, DIETHYLENE GLYCOL MONOETHYL ETHER.
carbo- a combining form meaning "carbon," as in *carbohydrate.*
carbocation *Organic Chemistry.* an unstable cation whose positive charge is associated with one or more carbon atoms.
carbocationic polymerization see CATIONIC POLYMERIZATION.
carbocyclic compound *Organic Chemistry.* any compound whose molecular structure includes rings on carbon atoms.
carbodihydrazide *Organic Chemistry.* colorless needles, very soluble in alcohol and water; melts at 154°C; used in photography.
carbodiimide *Organic Chemistry.* **1.** HN=C=NH, an unstable tantomer of cyanamide. **2.** any compound having the formula RN=C=NR.
carbofuran *Organic Chemistry.* $C_{12}H_{15}NO_3$ a white solid, soluble in water; melts at 150°C; used to control insects, mites and nematodes.
carbohumin see ULMIN.
carbohydrase *Enzymology.* an enzyme that catalyzes the breakdown of complex carbohydrates such as disaccharides and polysaccharides into simple sugars.
carbohydrate *Biochemistry.* an organic compound present in the cells of all living organisms and a major organic nutrient for human beings; consists of carbon, hydrogen, and oxygen, and makes up sugar, starch, and cellulose. *Organic Chemistry.* a compound of carbon, hydrogen, and oxygen, with the latter two usually in a 2:1 proportion. It is the most abundant class of organic compounds, including mono-, oligo-, and polysaccharides.
carbohydrate gum *Organic Chemistry.* a gelatinous gum produced by dispersing a thick solution of a polysaccharide in water at low concentrations, e.g., gum arabic.
carbohydrate metabolism *Biochemistry.* the chemical processes by which carbohydrates are broken down and used by the body.
carbolic acid see PHENOL, def. 1.
carbolism *Toxicology.* poisoning due to consumption of phenol; symptoms may include nausea, vomiting, convulsions, heart failure, and death. Also, PHENOL POISONING.
carbometer *Engineering.* an instrument that determines the amount of carbon in steel by measuring the magnetic properties of the sample within a given magnetic field.
carbomycin *Microbiology.* an antibiotic produced by the bacterial species *Streptomyces halstedii,* effective against Gram-positive bacteria through the inhibition of protein synthesis.
carbomycin B *Microbiology.* an antibiotic produced by the bacterial species *Streptomyces halstedii;* differs from carbomycin in that there is one less oxygen atom in its molecule.
carbon *Chemistry.* a very common nonmetallic element having the symbol C, the atomic number 6, an atomic weight of 12.01115, and melting point about 3600°C; found in the crystalline forms of diamond and graphite and in various amorphous forms including charcoal, coal, and coke; the defining element of organic compounds, and the active element of photosynthesis. The natural, dominant isotope of carbon, *carbon-12,* is the basis of the atomic weights of other elements (^{12}C = at. wt. 12). (From the Latin word for "coal" or "charcoal.")
carbon-12 *Nuclear Physics.* an isotope of carbon that has a stable nucleus; it occurs naturally and makes up most of the carbon found in nature.
carbon-14 *Nuclear Physics.* a naturally occurring, radioactive isotope of carbon with a half-life variously described as 5380 to 5730 years; commonly used in dating archaeological materials and in demonstrating the metabolic path of carbon in photosynthesis.
carbon-14 dating *Nucleonics.* a common method for estimating the age of a carbonaceous archaeological artifact by measuring the radioactivity of its carbon-14 content; this will determine how long ago the specimen was separated from equilibrium with the atmosphere-plant-animal cycle. Carbon-14 is continuously produced in the atmosphere by cosmic-ray bombardment and decays with a half-life typically described as 5568 years; dating is accomplished by comparing the carbon-14 activity per unit mass of the artifact with that in a contemporary sample.
carbonaceous *Science.* containing or composed of carbon.
carbonaceous chondrite *Astronomy.* a rare type of stony meteorite that is rich in carbon compounds and is thought to be relatively unaltered since the beginning of the solar system; its spectrum (and probably also its composition) closely resembles that of the C-type asteroids. Also, **carbonaceous meteorite.**
carbonaceous chondrite, type C1 *Geology.* a variety of carbonaceous chondrite having a relatively low density and strong magnetic properties; contains approximately 3.5% carbon and some sulfates.
carbonaceous chondrite, type C2 *Geology.* a variety of carbonaceous chondrite having a relatively high density and low water content, and generally composed chiefly of olivine.
carbonaceous chondrite, type C3 *Geology.* a variety of carbonaceous chondrite that is low in magnetic properties, and contains approximately 2.5% carbon along with some sulfur, which generally occurs as free sulfur.
carbonaceous rock *Petrology.* rock containing a substantial proportion of organic material.
carbonaceous sandstone *Petrology.* sandstone rich in carbon; commonly contains organic elements such as peat or lignite.
carbonaceous shale *Geology.* a dark gray or black shale with a high carbon content.
carbon acid *Organic Chemistry.* an organic molecule with an acidic proton attached.
carbonado *Mineralogy.* a dark, opaque aggregate of minute diamond particles that forms a rounded mass with superior toughness and is used as an industrial diamond. Also, BLACK DIAMOND.
carbon arc *Electricity.* an electrical discharge between two carbon electrodes; used in welding and high-intensity lamps.
carbon arc cutting *Metallurgy.* the process of cutting a metallic material with an electric arc struck between a carbon or graphite electrode and the workpiece.
carbon arc lamp *Electricity.* a lamp that employs an arc discharge between carbon electrodes. Also, CARBON LAMP.
carbon arc welding *Metallurgy.* welding with an electric arc struck between a carbon or graphite electrode and the workpiece.
carbonate *Chemistry.* **1.** any compound formed by the reaction of carbonic acid with either a metal (yielding a salt, calcium carbonate) or an organic compound (yielding an ester, such as diethyl carbonate). **2.** of, relating to, or containing such a compound. **3.** containing the carbonate ion, CO_3^{2-}.
carbonate compensation depth *Oceanography.* the depth below which the calcareous skeletons of marine animals dissolve as fast as they accumulate. Also, CALCITE COMPENSATION DEPTH, CALCIUM CARBONATE COMPENSATION DEPTH.
carbonate cycle *Geochemistry.* the circulation of carbonates and carbon dioxide through the biosphere, atmosphere, hydrosphere, and lithosphere via biogeochemical pathways, including the solution, deposition, and metabolism of carbonate compounds.
carbonated beverage *Food Technology.* any of various drinks that are impregnated with carbon dioxide gas, thereby becoming effervescent; carbonated beverages may be sweetened, flavored, acidified, or colored.
carbonated spring *Hydrology.* a spring containing carbon dioxide gas in solution.
carbonate-fluorapatite *Mineralogy.* $Ca_5(PO_4,CO_3)_3F$, a colorless mineral of the apatite group occurring in hexagonal prisms; also stalactitic, fibrous, or in granular masses; having a specific gravity of 3.12 and a hardness of 5 on the Mohs scale Also, DEHRNITE, FRANCOLITE.
carbonate glass *Materials.* a glass that is formed by mixing divalent cation and monovalent cation carbonates.
carbonate mineral *Mineralogy.* any mineral compound having a fundamental anionic structure of CO_3^{-2}, such as calcite or aragonite.
carbonate reservoir *Geology.* an underground gas or oil trap formed by limestones or dolomite.
carbonate rock *Petrology.* a rock typically composed of more than 50% carbonates by weight.
carbonation *Chemistry.* **1.** the conversion of a compound into a carbonate. **2.** the impregnation of a liquid with carbon dioxide (CO_2) under pressure.
carbonatite *Petrology.* **1.** igneous rocks that are composed primarily of carbonate minerals and that vary in composition from almost pure carbonate to carbonate-silicate; commonly associated with alkaline magmas. **2.** a sedimentary rock containing 80% calcium or magnesium.
carbon bit *Mechanical Devices.* a diamond bit with a brazen or set carbon-cutting medium.

carbon black *Chemistry.* an amorphous form of finely divided carbon that is produced by the partial combustion of hydrocarbons. Also, GAS BLACK. *Materials.* an amorphous powdered carbon produced by the incomplete combustion of gas; used as a color pigment and to increase the durability of tires.

carbon brush *Electricity.* a current-carrying brush made of amorphous carbon, carbon and graphite, or carbon and copper; used in motors, generators, variable resistors, and variable autotransformers.

carbon burning *Nuclear Physics.* the synthesis that takes place in stars by the fusion of two carbon-12 nuclei at temperatures that exceed billions of degrees Celsius.

carbon burning rate *Chemical Engineering.* the weight of carbon burned per unit time from the catalytic-cracking catalyst in a regenerator.

carbon button *Acoustical Engineering.* the enclosure that holds the grains of carbon in a carbon microphone.

carbon-carbon lyase *Enzymology.* an enzyme that catalyzes the cleavage of carbon-carbon bonds.

carbon cycle *Geochemistry.* the circulation and movement of carbon atoms through the biosphere, atmosphere, hydrosphere, and lithosphere as a result of various chemical reactions. *Nuclear Physics.* the process in which a carbon-12 atom engenders a succession of thermonuclear reactions which results in the release of massive amounts of energy and the creation of the other carbon-12 atom, believed to be the source of energy in the sun and other stars. Also, CARBON-NITROGEN CYCLE. *Astrophysics.* see CARBON-NITROGEN-OXYGEN CYCLE.

carbon dating see RADIOCARBON DATING.

carbon-detonation supernova model *Astrophysics.* a model for the explosive disruption of a 6- or 7-solar-mass star involving the nuclear ignition of its carbon-oxygen core.

carbon dioxide *Inorganic Chemistry.* CO_2, a colorless, odorless, noncombustible gas that is slightly more than 1.5 times as dense as air; becomes a solid (dry ice) below –78.5°C. It is present in the atmosphere as a result of the decay of organic material and the respiration of living organisms, and it represents about 0.033% of the air. Carbon dioxide is produced by the burning of wood, coal, coke, oil, natural gas, or other fuels containing carbon, by the action of an acid on a carbonate, or naturally from springs and wells. It has a wide range of uses, as in carbonated beverages, fire extinguishers, refrigeration systems, and aerosols.

carbon dioxide absorption tube *Analytical Chemistry.* an absorbent-packed tube used to capture the carbon dioxide formed during the Pregl combustion procedure.

carbon dioxide fire extinguisher *Chemical Engineering.* a type of chemical fire extinguisher with liquid carbon dioxide as the extinguishing agent, stored under pressure of 800 to 900 pounds per square inch at normal room temperature.

carbon dioxide gas laser *Physics.* a laser that operates with gaseous carbon dioxide as the active medium, capable of delivering several hundreds of watts of infrared power in a continuous wave mode.

carbon dioxide indicator *Mining Engineering.* a carbon dioxide detector, which is portable and operates on the absorption of the base by potassium hydroxide.

carbon dioxide laser *Optics.* a gas laser in which the transitions of energy states in carbon dioxide molecules produce powerful infrared wavelengths of 10.6 micrometers.

carbon dioxide process *Metallurgy.* a casting process based on a sand mold with a silicate binder that has reacted with carbon dioxide.

carbon dioxide sensitivity virus see SIGMA VIRUS.

carbon disulfide *Organic Chemistry.* CS_2, a toxic, clear, colorless, or faintly yellow, highly flammable liquid; soluble in water and alcohol; melts at –111.53°C and boils at 46.25°C; used in the manufacture of carbon tetrachloride, rayon, and soil disinfectants, and as a solvent.

carbon electrode *Metallurgy.* an electrode that is made up of a coated or uncoated carbon or graphite rod; used for arc brazing, cutting, or welding.

3'-carbon end *Molecular Biology.* the end of a DNA or RNA molecule in which the C-3 of the pentose is not involved in phosphodiester bond formation.

5'-carbon end *Molecular Biology.* the end of the DNA or RNA molecule in which the C-5 of the pentose is not involved in phosphodiester bond formation.

carbon equivalent *Materials Science.* an empirical relationship in which the effect of alloying elements on the hardenability of steels is related to the equivalent effect of a percentage of carbon. *Metallurgy.* **1.** in a cast iron, the sum of its carbon content and one third of its silicon and phosphorus contents. **2.** in welding, a rating of weldability related to carbon, chromium, copper, manganese, molybdenum, nickel, and vanadium content.

carbon fiber *Materials.* a tough, thin fiber of nearly pure carbon that is made by exposing various organic raw materials to high temperatures, combined with synthetic resins to make a strong, lightweight material that is used in the construction of aircraft and spacecraft.

carbon film *Analytical Chemistry.* a thin layer of carbon deposited by evaporation onto a specimen; the carbon protects and prepares the sample for electron microscopy.

carbon-film hygrometer element *Electricity.* a hygrometer element composed of a plastic strip coated with a film of carbon black dispersed in a hygroscopic binder.

carbon-film resistor *Electricity.* a resistor produced by depositing a carbon film on a ceramic form.

carbon fixation *Biochemistry.* the synthesis of organic compounds from carbon dioxide, such as in photosynthesis.

carbon-halide lyase *Enzymology.* an enzyme that catalyzes the cleavage of a bond between a carbon atom and a halogen.

carbon-hydrogen analyzer *Analytical Chemistry.* an instrument used to analyze the carbon and hydrogen content of organic compounds.

carbon hydrophone *Acoustical Engineering.* a type of carbon microphone used for detecting or responding to sounds under water.

carbonic acid *Inorganic Chemistry.* H_2CO_3, a weak acid formed by combining carbon dioxide and water; inorganic carbonates (salts of carbonic acid) are formed from it by reaction with metals or metal oxides, and organic carbonates (esters of carbonic acid), by reaction with organic compounds.

carbonic anhydrase *Enzymology.* an enzyme involved in the transport and release of carbon dioxide by acting on carbonic acid to produce CO_2 in physiological processes; it occurs primarily in the blood and in the mucous membrane of the stomach lining. Also, **carbonate dehydratase, carbonic acid anhydrase, carboanhydrase.**

Carbonicola *Paleontology.* a genus of vagrant Carboniferous bivalves in the order Unionida that lived in fresh or brackish water.

Carboniferous *Geology.* **1.** a geologic division of the Upper Paleozoic era from about 345 to 280 million years ago. **2.** referring to the rocks formed during that time.

carbonification see COALIFICATION.

carbonifics *Materials.* a polyhydric material, rich in carbon, that is often a component of intumescent coatings.

carbon isotope ratio *Geology.* see CARBON RATIO, def. 3.

carbonite see NATURAL COKE.

carbonitrided steel *Metallurgy.* a steel that has undergone carbonitriding.

carbonitriding *Metallurgy.* a case-hardening process in which both carbon and nitrogen diffuse into the surface layer of a ferrous material.

carbonitrile *Organic Chemistry.* a compound consisting of a nitrile group attached to a ring.

carbonium ion *Organic Chemistry.* a positively charged organic ion that lacks a pair of electrons and acts in subsequent chemical reactions as though the positive charge was localized on the carbon atom.

carbonization *Chemistry.* the conversion of a carbon-containing substance into carbon, as in the destructive distillation of coal. *Geochemistry.* **1.** in the formation of coal, the collection of carbon residue as a result of changes in and decomposition of organic matter. **2.** the buildup of carbon as a result of the slow decay of organic matter underwater. **3.** the conversion of a carbon-containing substance into carbon or carbon residue, either by natural processes or by heating to remove other components. *Materials Science.* a high-temperature (1500–2000°C) method of producing very fine high-strength graphite fibers. *Organic Chemistry.* the heating of coal in the absence of oxygen to produce coke and several gaseous and liquid petroleum by-products essential to industrial organic chemistry.

carbon knock *Mechanical Engineering.* the abnormal ignition in an internal combustion engine, usually caused by low octane-number fuel and resulting in carbon buildup and a loud metallic noise.

carbon lamp see CARBON-ARC LAMP.

carbon log *Petroleum Engineering.* in a well, the record of hydrocarbon pressure shown by determining the carbon content.

carbon microphone *Acoustical Engineering.* a type of microphone in which a flexible metal diaphragm presses against a container filled with grains of carbon; the diaphragm vibrates in response to sound waves, which causes variations in the electric circuit coupled to the carbon. Such variations can be detected and amplified for sound reproduction.

carbon monoxide *Inorganic Chemistry.* CO, a colorless, almost odorless, highly poisonous, and flammable gas; widely produced for industrial applications when a carbon-containing fuel is burned with an insufficient amount of oxygen; it is also evolved as an exhaust product from automobile engines. Carbon monoxide, if inhaled, can cause asphyxiation by combining with blood hemoglobin to prevent the flow of oxygen.

carbon monoxide laser *Optics.* a gas laser in which the transitions of energy states in carbon monoxide molecules produce wavelengths of up to 5.7 micrometers.

carbon monoxide poisoning *Toxicology.* poisoning caused by exposure to carbon monoxide, a gas that competes with oxygen in binding with hemoglobin, thereby decreasing the carrying capacity of oxygen by hemoglobin.

carbonmonoxyhemoglobin *Biochemistry.* hemoglobin that has carbon monoxide bound to the iron atom of its four heme prosthetic groups; blocks the binding of hemoglobin with oxygen. Also, CARBONYLHEMOGLOBIN, CARBOXYLHEMOGLOBIN.

carbon-nitrogen cycle *Nuclear Physics.* the process in which carbon, nitrogen, and oxygen nuclei act as catalysts in certain thermonuclear reactions, often referred to as the *carbon cycle.*

carbon-nitrogen lyase *Enzymology.* an enzyme that catalyzes the cleavage of a bond between carbon and nitrogen atoms.

carbon-nitrogen-oxygen cycle *Astronomy.* a series of high-temperature thermonuclear reactions in stars that use carbon as a catalyst to fuse four hydrogen nuclei into one helium nucleus and release energy as a by-product.

carbon-nitrogen-phosphorus ratio *Oceanography.* the relatively constant relationship between concentrations of (a) carbon, nitrogen, and phosphorus in marine algae and protozoa, and of (b) nitrogen and phosphorus in sea water; there is a conservative balance between the amounts of the elements absorbed by the organisms and the amounts returned to the sea upon their decomposition.

carbon number *Analytical Chemistry.* the number of carbon atoms in a material that is being analyzed.

carbon-oxygen lyase *Enzymology.* an enzyme that catalyzes the cleavage of a bond between carbon and oxygen atoms.

carbon oxychloride see CARBONYL CHLORIDE.

carbon paper *Materials.* **1.** a thin paper coated on one side with a dark material, such as carbon black, so that when it is placed between two sheets of paper, the writing or typing on the top sheet is transferred to the bottom sheet. **2.** a paper used in making photographic prints by the carbon process.

carbon pile *Electricity.* a variable electrical resistor that is composed of a number of carbon disks stacked between a fixed metal plate and a movable one. The resistance changes when the carbon is compressed.

carbon-pile pressure transducer *Engineering.* a transducer that measures pressure changes in a conductive carbon core by monitoring the differences in the strength of the electrical signals emitted.

carbon potential *Metallurgy.* in the process of steelmaking, the ability of a carbon-bearing environment to change or maintain the carbon content of a steel.

carbon ratio *Geology.* **1.** the percentage of fixed carbon in coal. **2.** the ratio of fixed carbon in coal to fixed carbon plus volatile hydrocarbons. Also, FIXED CARBON RATIO. **3.** the ratio of carbon-12 to either of the isotopes, carbon-13 or carbon-14, or the reciprocal of one of these ratios. Also, CARBON ISOTOPE RATIO.

carbon ratio theory *Geology.* the hypothesis that the gravity of the oil in any area is inversely proportional to the carbon ratio of the coal in the area.

carbon replication *Analytical Chemistry.* a carbon-film mold of a specimen surface thin enough to be studied using an electron microscope.

carbon residue *Chemical Engineering.* the carbon produced by heating lubricating oil under standard conditions in a closed system.

carbon residue test *Chemical Engineering.* a destructive-distillation method to determine the quantity of carbon residue in lubricating oils and fuels. Also, CONRADSON CARBON TEST.

carbon resistance thermometer *Engineering.* a highly sensitive resistance thermometer that is used to measure temperatures ranging from 0.05 to 20 K and can register changes of the order of 10^{-5} degrees.

carbon resistor *Electronics.* a resistor in which carbon particles mixed with a ceramic binder and baked into a cylindrical shape cause resistance to current flow to decrease as temperature increases. Also, COMPOSITION RESISTOR.

carbon restoration *Metallurgy.* a process by which the decarburized surface layer of a steel is restored to its original carbon level.

carbon sequence see CARBON-NITROGEN-OXYGEN CYCLE.

carbon star *Astronomy.* a star of spectral type R or N whose spectrum shows strong bands due to molecular carbon and carbon compounds. Also, C STAR.

carbon steel *Metallurgy.* any of numerous steels containing only incidental amounts of alloying elements, except for minimum quantities of copper, manganese, and silicon.

carbon suboxide *Inorganic Chemistry.* C_3O_2, a colorless gas with a strong pungent odor that causes irritation to the eyes and impairs breathing.

carbon-sulfur lyase *Enzymology.* an enzyme that catalyzes the cleavage of a bond between carbon and sulfur atoms.

carbon tetrachloride *Organic Chemistry.* CCl_4, a toxic, colorless liquid that is insoluble in water and soluble in alcohol; melts at –22.99°C and boils at 76.54°C; it is used in dry cleaning (especially formerly), in fumigation, and for the manufacture of chlorofluoromethane refrigerants. Also, PERCHLOROMETHANE.

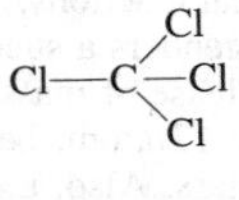

carbon tetrachloride

carbon tetrafluoride *Organic Chemistry.* CF_4, a toxic, colorless gas, slightly soluble in water; melts at –150°C and boils at –129°C; used as a refrigerant and gaseous insulator. Also, TETRAFLUOROMETHANE.

carbon tissue *Graphic Arts.* a thin sheet of paper coated with photosensitive gelatin, used in gravure platemaking.

carbon transducer *Engineering.* a type of transducer that has a movable electrode and carbon granules in contact with a fixed electrode, so that the motion of the movable electrode will vary the resistance of the granules.

carbon transfer recording *Telecommunications.* a method of facsimile recording in which carbon particles are placed on a record sheet based on the received signal.

carbonyl *Organic Chemistry.* the divalent group CO that occurs in a wide range of chemical compounds, such as aldehydes and ketones.

carbonylation *Chemistry.* the introduction of carbon monoxide into a molecule or species to create a carbonyl.

carbonyl bromide *Organic Chemistry.* CBr_2O, a heavy, colorless, toxic liquid; hydrolyzed by water, decomposed by light and heat; boils at 64–65°C; used as a poison gas. Also, BROMOPHOSGENE.

carbonyl chloride *Organic Chemistry.* $COCl_2$, a highly toxic, colorless to light yellow liquid or easily liquefied gas; slightly soluble in water; boils at 8.2°C and freezes at –128°C; used as a pesticide, herbicide, and war gas, and in dye manufacture. Also, PHOSGENE.

carbonyl compound *Organic Chemistry.* any compound containing a carbonyl group.

***N,N′*-carbonyldiimidazole** *Organic Chemistry.* $C_7H_6N_4O$, off-white powder or crystals; hydrolyzed by water; melts at 118–120°C; used as an enzyme crosslinking agent.

carbonyl fluoride *Organic Chemistry.* COF_2, a highly toxic, colorless, hygroscopic gas that decomposes in water and in alcohol; it melts at –114°C and boils at –83°C; it is used in organic synthesis. Also, FLUOROSPHOSGENE.

carbonylhemoglobin see CARBONMONOXYHEMOGLOBIN.

carbonyl process *Metallurgy.* the process by which a powdered metal, such as nickel, cobalt, or iron, is produced by thermally decomposing a metal carbonyl.

carbophenothion *Organic Chemistry.* $(C_2H_5O)_2P{=}SSCH_2S(C_6H_4)Cl$, an amber liquid; insoluble in water and miscible in common solvents; boils at 82°C; used as an insecticide and acaricide.

carborane *Organic Chemistry.* a class of compounds containing carbon, boron, and hydrogen; for example, $B_{10}C_2H_{12}$.

carboxamide *Organic Chemistry.* a compound containing the amide group –$CONH_2$.

carboxin *Organic Chemistry.* $C_{12}H_{13}NO_2S$, a white solid that melts at 91°C; used as a fungicide. Also, DCMO.

carboxybenzene see BENZOIC ACID.

carboxybiotin *Biochemistry.* a biotin derivative produced when it reacts with adenosine triphosphate and carbon dioxide.

Carboxydobacteria *Bacteriology.* a nontaxonomic group of bacteria distinguished by the ability to utilize carbon monoxide as the sole source of both carbon and energy, typically occurring in soil, sewage, and polluted water.

γ-carboxyglutamic acid *Biochemistry.* an amino acid that binds calcium; it forms when carboxylation of glutamic acid comes into contact with vitamin K.

carboxy group see CARBOXYL GROUP.

carboxylase *Enzymology.* an enzyme that causes a molecule of carbon dioxide to join with another compound, thus forming a carboxyl group.

carboxylate anion *Organic Chemistry.* a negatively charged ion having the structure $RCOO^-$.

carboxylation *Organic Chemistry.* the introduction of a carboxyl group into a compound.

carboxylesterase *Enzymology.* an enzyme that catalyzes the hydrolysis of carboxylic acid esters. Also, ALI-ESTERASE, CARBOLIC ESTER HYDROLASE.

carboxyl group *Organic Chemistry.* the –C(=O)OH group, characteristic of weak organic acids, and including fatty acids and amino acids.

carboxylhemoglobin see CARBONMONOXYHEMOGLOBIN.

carboxylic *Organic Chemistry.* containing one or more of the carboxyl groups (COOH), which are common to organic carboxylic acids.

carboxylic acid *Organic Chemistry.* **1.** any acidic substance possessing a carboxyl group. **2.** see PROLINE.

carboxylic ester hydrolase see CARBOXYLESTERASE.

carboxyl terminal *Biochemistry.* the end of a peptide chain that contains a free carboxyl group. Also, C-TERMINAL.

carboxyltransferase *Enzymology.* an enzyme that catalyzes the transfer of a carboxyl group from one molecule to another.

carboxymethyl *Organic Chemistry.* any carboxylic acid derivative that contains the group $-CH_2CO_2H$.

carboxymethyl cellulose *Organic Chemistry.* a semisynthetic, water-soluble, white polymer in which some of the hydrogen atoms of hydroxyl groups have been replaced by the carboxymethyl group.

carboxypeptidase *Enzymology.* an enzyme that frees amino acids from the carboxyl terminals of proteins; it is involved in protein digestion in the small intestine.

carboxysomes *Cell Biology.* a body of unknown function found in many autotrophic prokaryotes.

carbro process *Graphic Arts.* the making of color reproduction prints by the use of carbon tissues to transfer and superimpose pigments lifted from enlarged bromide prints of color-separation negatives. (From car-bon and bromide.)

carbuncle *Medicine.* a group of boils in the skin and in the tissue beneath, usually caused by *Staphylococcus aureus,* with the formation of drainage ducts and necrotic tissue.

carbuncular *Medicine.* relating to or resembling a carbuncle.

carburetion *Mechanical Engineering.* the process of mixing air and fuel in a carburetor. *Chemical Engineering.* the process of enriching a combustible gas by adding volatile carbon compounds.

carburetor [kär′bə rāt′ər] *Mechanical Engineering.* a device that is used to provide and regulate the mixture of air and fuel that is burned inside the cylinders of an internal-combustion engine.

carburetor icing *Mechanical Engineering.* the effect when the temperature of an air-fuel mixture in a carburetor drops below the freezing point of water because of volumetric expansion, causing ice to form in the carburetor.

carburization *Materials Science.* a high-temperature process in which carbon is diffused into the surface region of a steel piece, while it is austenite, producing a hardened martensitic surface upon quenching.

carburize *Metallurgy.* to case harden a ferrous material by diffusing carbon into its surface layer.

Carbyne *Organic Chemistry.* **1.** a trade name for the herbicide barban. **2.** see METHYLIDYNE.

carcenet *Meteorology.* a very cold gorge wind occurring in the eastern Pyrenees, particularly in the upper Aude valley.

Carcharhinidae *Vertebrate Zoology.* a family of sharks in the order Carcharhiniformes, including the black-tipped shark, the hammerhead shark, the tiger shark, and others.

Carcharhiniformes *Vertebrate Zoology.* the largest order belonging to the shark subclass Selachii, including the requiem sharks, the hammerhead sharks, the tiger sharks, and others.

Carchariidae *Vertebrate Zoology.* the sand sharks, a family of almost extinct primitive sharks that live near sandy shores and belong to the suborder Galeoidea.

Carcharodon *Paleontology.* the great white sharks; a living genus of sharks in the suborder Lamnoidea and family Lamnidae; Carcharodon first appeared in the Paleocene.

carcharodont *Vertebrate Zoology.* having very sharp, triangular teeth with serrated edges, such as those on a white shark.

Carchesium *Invertebrate Zoology.* a genus of colonial ciliate protozoans, in the order Peritricha, with contractible branched stalks, each individual having its own stalk muscle for independent contraction; marine and freshwater.

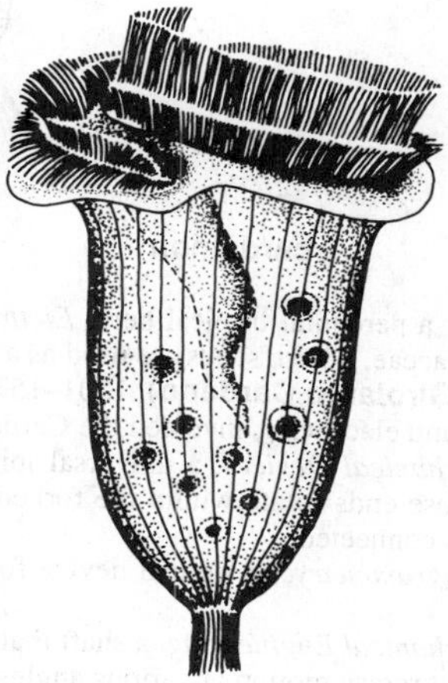

Carchesium

carcin- or carcino- a combining form meaning "cancer," as in *carcinemia, carcinogenic.*

carcinemia [kär′sə nēm′ē ə] *Oncology.* the state of emaciation, debility, and malnutrition that is associated with cancer and other disease conditions. Also, CANCEROUS CACHEXIA.

carcinogen [kär′sən ə gin; kär′sin′ə gin] *Oncology.* a chemical, physical, or biological substance that is capable of causing cancer.

carcinogenesis [kär′sə nō gen′ə sis] *Oncology.* the origin or production of a carcinoma.

carcinogenic [kär′sən ə gen′ik; kär′sə nō gen′ik] *Oncology.* causing cancer; producing a carcinoma.

carcinogenicity [kär′sə nō gə nis′ə tē] *Oncology.* the tendency or ability to produce a carcinoma.

carcinoid [kär′sə noid′] *Oncology.* a well-defined, yellowish benign tumor of the gastrointestinal tract.

carcinoid syndrome *Oncology.* a group of symptoms caused by the secretion of substances from a carcinoid tumor and including transient diarrhea, dusky appearance of the skin, attacks of asthma, heart murmurs, and an enlarged liver.

carcinolysis [kär′sə näl′ə sis] *Medicine.* the process of destroying cancerous cells.

carcinoma [kär′sə nō′mə] *Oncology.* a malignant epithelial cell tumor, which can occur anywhere in the body and spread through the bloodstream. (From the Greek word for "crab" or "cancer.")

carcinoma in situ *Oncology.* a malignant epithelial tumor which has not yet spread to underlying tissue; the term is commonly used for such tumors found in the uterus. Also, **cancer in situ.**

carcinomatosis [kär′sə nō mə tō′sis] *Medicine.* the extensive spread of cancer over a major portion of the body. Also, **carcinosis.**

carcinophilic [kär′sə nō fil′ik] *Oncology.* of a substance, particularly attracted to cancerous tissue.

carcinophobia [kär′sə nə fō′bē ə] *Psychology.* an irrational fear of cancer or of becoming afflicted with cancer.

carcinosarcoma *Oncology.* a mixed tumor containing carcinosarcoma and sarcomatous components. Also, SARCOCARCINOMA.

Carcinus viruses *Virology.* an unclassified group of viruses that have been isolated from crabs.

card *Electronics.* **1.** a thin phenolic panel on which components are mounted and wired to produce a circuit. **2.** the thin insulating strip around which a resistor element is wound. *Computer Technology.* a general term for any of various machine-processable data storage media.

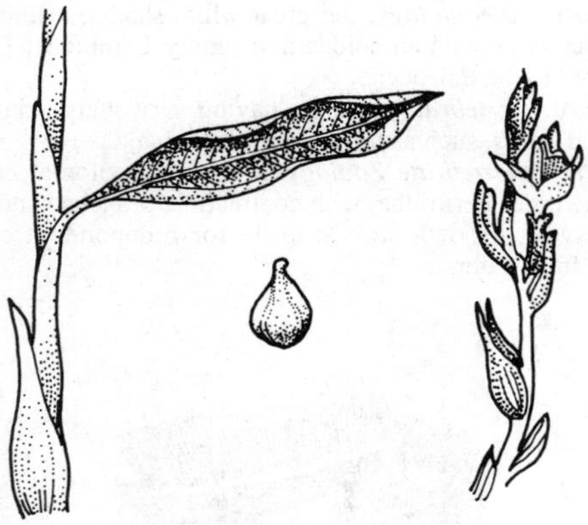

cardamon

cardamon *Botany.* a perennial tropical herb, *Elettaria cardmomum,* of the family Zingiberaceae, whose seeds are used as a spice.

Cardan, Jerome (Girolamo Cardano) 1501–1576, Italian physician; pioneer in algebra and electricity; invented the Cardan shaft.

Cardan joint *Mechanical Devices.* a universal joint that consists of a crosslike piece whose ends rotate within the forked end of each of two shafts to which it is connected.

cardan link *Photogrammetry.* an optical device for universal scanning around a point.

Cardan shaft *Mechanical Engineering.* a shaft that has a universal joint at its end to transmit rotary motion at various angles.

Cardan's suspension *Mechanical Devices.* a gimballed mounting that concentrates the weight of its load on a point; used in a mariner's compass.

cardboard *Materials.* a relatively thick, stiff material made of paper pulp; widely used for packing and other purposes.

card dialer *Telecommunications.* a device that uses a card with telephone numbers recorded on it to automatically dial a telephone number.

card-edge connector *Electricity.* a connector that mates with wiring leads from the edge of a printed circuit-board.

cardi- a combining form meaning "heart," as in *cardiac.*

cardia *Anatomy.* the part of the stomach surrounding the orifice at the junction of the esophagus and the upper part of the stomach.

cardiac [kär′dē ak] *Anatomy.* **1.** relating to or affecting the heart. **2.** relating to the cardia.

cardiac arrest *Medicine.* the cessation of heart function.

cardiac arrythmia *Cardiology.* irregularity of the heartbeat. Also, DYSRHYTHMIA.

cardiac block *Cardiology.* the impairment of impulse transmission from atria to ventricles. Also, ATRIOVENTRICULAR BLOCK.

cardiac catheterization. *Cardiology.* the passage of a tube, by way of a blood vessel, into the heart in order to introduce or remove fluids, determine intracardiac pressure, or detect cardiac anomalies.

cardiac cirrhosis *Medicine.* fibrosis of the liver, usually associated with congestive heart disease.

cardiac cycle *Physiology.* one complete sequence of events during heart action.

cardiac edema *Medicine.* the retention of fluid in the extremities as a secondary effect of heart disease.

cardiac electrophysiology *Physiology.* the branch of physiology that records and analyzes the electrical activity of the heart using electrocardiography.

cardiac failure *Medicine.* a condition in which the heart cannot pump enough blood to meet the circulatory requirements of the body. Also, HEART FAILURE.

cardiac gland *Anatomy.* any of the gastric glands near the orifice at the junction of the esophagus and the stomach.

cardiac impulse *Cardiology.* the palpable or recorded movement of the chest due to the heartbeat.

cardiac index *Cardiology.* the heart output per unit of time over body surface, usually expressed in terms of liters per minute per square meter.

cardiac input *Physiology.* the volume of venous blood entering the right atrium of the heart in a specified time period, usually one minute.

cardiac insufficiency *Cardiology.* the inability of the heart to perform its normal function. Also, HEART FAILURE.

cardiac loop *Developmental Biology.* the embryonic heart, created by the coiling of the cardiac tube.

cardiac massage *Medicine.* the stimulation of the heart by manually pressing on the chest (**closed cardiac massage**) or by actually massaging the heart itself after the chest has been surgically opened (**open cardiac massage).**

cardiac murmur *Medicine.* an abnormal sound produced by vibrations of the heart muscle and the flow of blood. Also, HEART MURMUR.

cardiac muscle *Histology.* the specialized, syncytial mass of involuntary striated muscle constituting the vertebrate heart.

cardiac output *Physiology.* the volume of blood that is pumped in one minute by the left ventricle.

cardiac plexus *Anatomy.* a network of autonomic nerves serving the heart.

cardiac reserve *Cardiology.* the heart's potential ability; the ability of the heart to perform beyond basal functioning.

cardiac septum *Anatomy.* the membranous partition that separates the left and the right sides of the heart.

cardiac skeleton *Cardiology.* the dense, mainly fibrous structure that supports the musculature of the heart. Also, SKELETON OF HEART.

cardiac souffle *Cardiology.* a soft, blowing heart or vascular sound.

cardiac sphincter *Anatomy.* an ordinary circular esophageal muscle located at the juncture of the esophagus and the stomach.

cardiac tamponade *Cardiology.* the acute compression of the heart caused by accumulation of blood or fluid in the pericardium, from rupture of the heart or penetrating wound. Also, **cardiac compression.**

cardiac thrombosis *Cardiology.* an aggregation of blood platelets, other blood factors, and other trapped materials into a clotlike formation in the heart, frequently creating an obstruction.

cardiac valve *Anatomy.* **1.** a valve created by the cardiac sphincter that prevents food from moving back from the stomach into the esophagus. **2.** any of the valves of the heart, between the atria and ventricles and between the ventricles and arteries.

cardialgia [kär′dē alj′ə] *Cardiology.* **1.** formerly, a disagreeable burning sensation in the lower chest, stomach, or heart area. Also, HEARTBURN. **2.** pain in the heart.

cardias dyspnea *Cardiology.* labored breathing due to heart disease.

cardiectomy [kär′dē ek′tə mē] *Surgery.* the removal of the cardiac portion of the stomach, which is near the esophagus.

cardinal *Vertebrate Zoology.* a finch, *Cardinalis cardinalis,* of the family Fringillidae, that is native to U.S. woodlands east of the Rockies; characterized by its crested head, clear whistles, and (in males) bright red plumage and black face.

cardinal

cardinal fish *Vertebrate Zoology.* any of various perchlike fish of the family Apogonidae, including species that are bright red with black markings.

cardinal heading *Navigation.* a heading directly toward one of the four cardinal points or directions.

cardinal number *Mathematics.* the number of elements in a given set. A cardinal number is finite if it is the cardinal number of a finite set; otherwise it is infinite. The cardinal number of a set X is formally defined as being the least member of the (well-ordered) set of the ordinal numbers equivalent to X. Also called the **power of** X.

cardinal point *Cartography.* any of the four principal astronomical directions on the surface of the earth: north, south, east, or west. Thus, **cardinal winds.** *Optics.* see GAUSS POINTS.

cardinal teeth *Invertebrate Zoology.* the ridges and grooves on the inner surfaces of the valves, near the anterior end of the hinge, of a bivalve mollusk.

cardinal vein *Zoology.* any of four major systemic venous channels in adult primitive vertebrates and in embryos of higher vertebrates.

carding *Materials Science.* the process of cleaning and preparing fibers for use in nonwoven articles.

cardio- a combining form meaning "heart," as in *cardiovascular.*

cardioangiology *Cardiology.* a medical specialization that involves the study of the heart and blood vessels.

Cardiobacterium *Bacteriology.* a genus of Gram-negative, nonmotile, rod-shaped or filamentous bacteria, that are anaerobic or facultatively aerobic and may cause endocarditis in humans.

cardioblast *Invertebrate Zoology.* an early embryonic cell in insects, from which the heart develops.

cardiocentesis *Surgery.* the surgical puncture or incision of the heart.

Cardioceras *Paleontology.* a genus of large ammonoids in the superfamily Stephanocerataceae, characterized by strong, curved ribs and an evolute and compressed shell; found in Jurassic strata worldwide.

cardiodiaphragmatic *Anatomy.* relating to the heart and diaphragm.

cardiodilatin *Cardiology.* one of two polypeptide hormones that are produced by the atrium or isolated from the atria of several animals, including humans; used as a vasodilator and for control of blood pressure.

cardiodynia *Cardiology.* an instance or condition of heart pain.

cardiogenic *Cardiology.* relating to or originating in the heart or caused by a dysfunctional heart.

cardiogenic plate *Cardiology.* an area of splanchnic mesoderm, first cephalad, then pharyngeal, from which the heart originates.

cardiogenic shock *Medicine.* the collapse of the body due to failure of the heart muscle, commonly the result of myocardial infarction.

cardiogram *Cardiology.* a graphic recording of the heart's action, made with a cardiograph; an electrocardiogram.

cardiograph *Cardiology.* an instrument producing a graphic recording of the heart's action. Also, ELECTROCARDIOGRAPH.

cardiography *Cardiology.* the process of recording the heart's functions for study. Thus, **cardiographic.**

cardioid *Science.* having the shape of a heart. *Mathematics.* a heart-shaped curve which is the locus of points generated by a fixed point on a circle of diameter a as the circle rolls around a second fixed circle of diameter a. This epicycloid of one loop and special case of the limaçon is the graph in the plane of the equation $r = a(1 - \cos \phi)$.

cardioid condenser *Optics.* a condenser used in dark-field microscopy that allows only the light diffracted or dispersed from the specimen to enter the microscope.

cardioid microphone *Acoustical Engineering.* a microphone that has a cardioid (heart-shaped) pickup pattern due to unequal pickup of the frequencies within its audio range.

cardiolipin *Biochemistry.* a phospholipid found in cell nuclei, in inner mitochondrial membranes, and in the plasma membranes of bacteria; used as an antigen in syphilis tests. Also, DIPHOSPHATIDYL GLYCEROL.

cardiologist *Cardiology.* a physician specializing in the cardiovascular system.

cardiology *Cardiology.* the discipline concerned with the functions and diseases of the heart.

Cardiology

Cardiology, defined literally, means the study of the heart. Although its origins lie in antiquity, particularly in the writings of Aristotle and Galen, modern cardiology began with the observations of William Harvey. It was Harvey who, in *De Motu Cordis,* published in 1628, first established the concept that blood flowed in a circular pathway propelled by the forces of the heartbeat from arteries to veins and back to the heart. Importantly, it was a conclusion achieved by observation, experimentation, and deliberation, setting an example for the conduct of future research. The foundation was laid for the discovery of a number of fundamental anatomic and physiologic principles over the next two centuries as well as the recognition of the effects of disease on the heart and circulation.

Scientific cardiology now began to flourish. The pulse was felt, the blood pressure taken, and the human thorax was percussed. However, the greatest impact on the teaching of clinical cardiology at this time was the introduction of the stethoscope. It led to enormous interest and study of heart sounds and murmurs. By the end of the nineteenth century, the genesis and description of most of the abnormal sounds and murmurs produced by diseases of the heart and circulation had been published.

The turn of the century also witnessed the introduction of two fundamental discoveries which were to revolutionize the diagnostic approach to the cardiac patient. These were the electrocardiogram and the X-ray examination of the heart and aorta. In a number of countries, a few physicians began to devote their practice largely to cardiology. Cardiac clubs and associations were formed; these became the precursors of the professional cardiac societies which exist in almost all countries today. They became a forum for discussion and the presentation of clinical research. They led to scientific journals devoted to the specialty. Hospitals began to establish special units for the study of heart cases. By the 1930s cardiology was clearly emerging as a subspecialty of internal medicine.

Since World War II, new knowledge and the number of physicians specializing in cardiology have grown exponentially. Years of additional training after certification as internal medicine specialists are required of those aspiring to take the examination needed to be certified as cardiac specialists. The growth of cardiology has paralleled the importance of heart disease itself, which in the industrialized world is the leading cause of death. This growth has also reflected the fact that effective therapy has emerged, particularly cardiac surgery and modern cardiovascular pharmacotherapy.

The past two decades have also seen the introduction of a number of new technologic advances in both the diagnostic and therapeutic arenas. These advances include ultrasonic examination of the heart including Doppler evaluation of intracardiac flow, ambulatory monitoring of the electrocardiogram, coronary angiography, stress electrocardiography, nuclear cardiology, balloon angioplasty, laser angioplasty, and percutaneously introduced cardiac stents, to name but some. It is clear that these additional ways to diagnose and treat cardiac problems have great potential for improving health care. But a word of caution is also appropriate. The cardiac specialist must not be transformed into a cardiac technician ordering a battery of tests which are costly and serve as a major contributor to the crisis of health care cost which we face today. He must remain a physician grounded in an understanding of the natural history of disease and one who continues to be a clinician devoted to holistic treatment of his patient.

Elliot Rapaport
Professor of Medicine, U. of California, San Francisco
Chief of Cardiology, San Francisco General Hospital

cardiomegaly see MEGALOCARDIA.
cardiometer *Cardiology.* an instrument formerly used to measure the force of the heart's action.
cardionatrin *Cardiology.* one of two polypeptide hormones that are produced in response to stretching by the atrium or are isolated from the atria of several animals, including humans. It increases excretion of sodium by the kidneys, controls blood pressure, and acts as a vasodilator. Also, ATRIAL NATRIURETIC FACTOR.
cardionephric *Anatomy.* of or relating to the heart and kidney.
cardiopaludism *Cardiology.* heart disease brought on by malaria, resulting in dilation of the right heart and arrhythmias.
cardiopathy *Cardiology.* any disease or disorder of the heart.
cardiophobia *Psychology.* an irrational fear of heart disease.
cardioplasty see ESOPHAGOGASTROPLASTY.
cardioplegia [kär´dē ō plāj´ē ə] *Cardiology.* **1.** paralysis of the heart. **2.** the use of any of a variety of techniques such as chemical injection, hypothermia, or electrical stimuli to temporarily arrest cardiac activity for the purpose of performing surgery on the heart.
cardioplegic *Cardiology.* having the effect of arresting cardiac activity.
cardiopneumatic [kär´dē ō´noo mat´ik] *Cardiology.* of or relating to the heart and respiration.
Cardiopteridaceae *Botany.* a monogeneric family of twining herbs of the order Celastrales that are native to southeast Asia and northeast Australia; noted for bearing a milky juice in the leaves and stems.
cardiopulmonary [kär´dē ō pul´mə nâr´ē] *Anatomy.* of or relating to the heart and lungs.
cardiopulmonary bypass *Cardiology.* the artificial maintenance of circulation by machine, as used in some cardiac surgeries.
cardiopulmonary resuscitation *Medicine.* CPR, the effort to revive a victim of cardiac arrest by systematic application of sternal pressure and by artificial respiration.
cardiorrhaphy [kär´dē ôr´ ə fē] *Surgery.* the operation of suturing the heart muscle.
cardiorrhexis [kär´dē ō rek´sis] *Cardiology.* a tear or break in the heart tissue.
cardioscope *Cardiology.* a device formerly used to look into the interior of the heart.
cardiospasm *Medicine.* **1.** an involuntary contraction of the heart muscle. **2.** a malfunction of the cardiac sphincter muscle of the stomach, preventing food from passing into the stomach from the esophagus.
cardiotachometer *Medicine.* a device that provides a continuous record of the heart rate.
cardiotocography *Medicine.* the process of recording the fetal heart rate and the force, frequency, and duration of uterine contractions during labor. Also, **cardiotokography.**
cardiotomy *Surgery.* **1.** a surgical incision of the heart. **2.** a surgical incision into the cardiac end of the stomach.
cardiotonic **1.** any drug or other agent that has a tonic effect on the heart, such as strengthening its action or muscle contraction. **2.** having a tonic effect on the heart.
cardiovalvulitis *Cardiology.* inflammation of the cardiac valves.
cardiovalvulotomy *Surgery.* an incision into or excision of part of a heart valve, to relieve stenosis.
cardiovascular [kär´dē ō vask´yə lər] *Anatomy.* of or relating to the heart and blood vessels.
cardiovascular system *Anatomy.* the heart and blood vessels, by which blood is pumped and circulated through the body.
Cardiovirus *Virology.* a genus of viruses of the family Picornaviridae that cause myocarditis and encephalitis in rodents.
carditis *Medicine.* an inflamed condition of the heart.
card key access *Engineering.* an unlocking device that deciphers magnetically coded information on the strip of a card.
card reader *Computer Technology.* a device to interpret the data represented by holes in punched cards by using photoelectric techniques or, formerly, sensing pins or reading brushes. Also, PUNCHED-CARD READER.
card sorter *Computer Technology.* a machine, little used at present, that physically reorders a deck of cards based on a specified field or column. Also, PUNCHED-CARD SORTER.
caregiver or **care-giver** *Medicine.* any person who provides health care services to a patient, such as a physician, nurse, psychologist, or social worker. *Psychology.* a person who cares for an infant or young child, especially someone other than the biological parent.
δ^3-carene *Organic Chemistry.* $C_{10}H_{16}$, a clear, colorless, combustible liquid, bicyclic monoterpene, insoluble in water and soluble in acid; boils at 167°C; used as a solvent and an intermediate.
caret *Graphic Arts.* a proofreading symbol (∧) used to indicate the location of an insert or correction. *Computer Technology.* a similar character used in computers, especially as a cursor blinking on a display screen to indicate where the next character typed by the operator will appear.
Carettochelyidae *Vertebrate Zoology.* the pitted-shell turtles, a family of aquatic turtles in the infraorder Cryptodira and the order Testudines, characterized by webbed feet and a soft skin rather than horny plates covering the bony carapace.
car float *Naval Architecture.* a lighter fitted with railroad track and used to ferry a railroad car.
car-following control *Transportation Engineering.* any system of headway control in which a car determines its speed and position relative to the preceding car and maintains a safe separation.
car-following theory *Transportation Engineering.* a model devised to show the relationship among various motor vehicles based on relative speed, absolute speed, and separation.
cargo boom *Mechanical Engineering.* a long arm that extends from the top of a derrick; used to support or guide objects being handled.
cargo carrier *Ordnance.* **1.** a vehicle designed to carry military cargo. **2.** in the U.S. armed forces, an air-transportable, amphibious, unarmored, full-tracked vehicle used to carry cargo and to accompany and resupply self-propelled artillery weapons; officially designated **M548.**
cargo cult *Anthropology.* a revitalization movement in Melanesian culture that was a reaction to white domination; the belief was that whites would be destroyed by a deluge and all their goods would arrive by ship; cult members ceased working because they believed that the event would only happen when their own supplies were exhausted.
cargo system *Anthropology.* a Meso-American institution in which positions of prestige were rotated throughout the community due to the great cost to an individual, resulting in a redistribution of wealth.
cargo winch *Mechanical Engineering.* an engine-driven machine that has one or more rotating drums on which is coiled a chain or cable for hoisting or hauling.
Cariamidae *Vertebrate Zoology.* the seriemas, a family of long-legged, terrestrial, predacious birds of South America in the order Gruiformes; noted for their snake-killing habits and often domesticated for this.
Caribbean [kâr´ə bē´ən; kə rib´ē ən] of or relating to the Caribbean Sea.
Caribbean Current *Oceanography.* the strong current flowing westward in the Caribbean Sea through the Yucatan Channel, where it turns north and then east toward the Straits of Florida.
Caribbean Sea *Geography.* an arm of the North Atlantic between the West Indies and Central and South America.
Cariboo orogeny *Geology.* an early Paleozoic crustal deformation believed to have affected the Cordillera of British Columbia, especially in the Selkirk and Omineca mountains.
caribou *Vertebrate Zoology.* any of four North American species of large deer of the genus *Rangifer*, family Cervidae, adapted to cold climates; characterized by large velvety antlers with flattened tines in both sexes and by an audible click made by the tendon in walking; they travel in migratory herds of tens of thousands. They are closely related to and identified with the Old World reindeer.

caribou

Caricaceae *Botany.* a family of mostly soft-stemmed shrubs and small tropical American trees of the order Violales, typically having an unbranched trunk topped by a cluster of leaves and producing mustard oil or an alkaloid; one species, **Carica papaya,** produces the edible papaya fruit.

Caridea *Invertebrate Zoology.* a section of decapod crustaceans, in the suborder Natantia, that includes many shrimps and prawns.

caries *Medicine.* the decomposition and decay of a tooth or a bone, causing discoloration, softening, and porosity.

carina *Vertebrate Zoology.* the large platelike projection on the center of the sternum of a bird or bat to which wing muscles are attached.

Carina *Astronomy.* the Keel, a bright southern constellation lying in the Milky Way that was formerly part of the now-obsolete constellation Argo Navis, the Ship.

Carina arm *Astronomy.* a spiral arm of the Milky Way Galaxy that lies inward of the spiral arm the sun belongs to and is seen in the direction of the constellation Carina.

Carina Nebula see ETA CARINAE NEBULA.

carinate *Biology.* shaped like a ridge or keel of a ship, or having a structure or projection thus shaped.

carination *Archaeology.* in an ancient jar or pot, a sharply angled shoulder dividing the neck from the body.

Carinomidae *Invertebrate Zoology.* a family of shore-dwelling, ribbonlike worms in the order Palaeonemertini.

Carinthian furnace *Metallurgy.* a zinc distillation furnace. *Mining Engineering.* a small reverberatory furnace, with an inclined hearth and fueled by wood, in which lead ore is treated by roasting and reaction.

cariostatic *Physiology.* relating to inhibition of the formation of dental caries, such as the cariostatic action of fluorides.

Carius method *Analytical Chemistry.* a procedure for analyzing organic compounds for halogens, phosphorus, and sulfur, that involves heating the sample with concentrated nitric acid in a sealed tube; the organic matter is oxidized and the elements can be determined.

Carlavirus group *Virology.* a group of plant viruses that infect a narrow host range, usually transmitted by aphids.

Carl Gustav or **Carl Gustaf** *Ordnance.* **1.** a type of Swedish 9-mm submachine gun. **2.** a Swedish 84-mm recoilless antitank gun.

carling *Naval Architecture.* one of the large pieces of squared timber running fore and aft between the beams of a wooden ship; used to support the deck or frame an opening in the deck. Also, CARLINES.

Carlsbad law *Crystallography.* a twin law, described as a parallel twin, found in both monoclinic and triclinic crystals in which the twin-axis lies in the plane perpendicular to the *b* axis.

car maneuvering *Transportation Engineering.* the movement of a car relative to the overall traffic stream; for example, by changing lanes.

Carme [kär´mā´] *Astronomy.* the eleventh moon of Jupiter; it was discovered in 1938 and is about 20 kilometers in diameter.

carminic acid *Organic Chemistry.* $C_{22}H_{20}O_{13}$, red, monoclinic prisms, soluble in water and alcohol; decomposes at 136°C; used in pigments and color photography. Also, COCHINILIN.

carminite *Mineralogy.* $PbFe_2^{+3}(AsO_4)_2(OH)_2$, a carmine-red orthorhombic translucent mineral occurring as lathlike minute crystals; having a specific gravity of 5.22 and a hardness of 3.5 on the Mohs scale; found in microcrystals with scorodite, anglesite, and cerussite.

Carmovirus group *Virology.* a genus of ssRNA-containing plant viruses having a wide host range; transmitted naturally through soil.

carnallite *Mineralogy.* $KMgCl_3 \cdot 6H_2O$, a milk-white to red, transparent to translucent, shiny orthorhombic mineral occurring in granular massive form or as pyramidal crystals; it is found as a component of thick, sedimentary saline deposit, and is used as a raw material for fertilizer manufacture.

carnassial *Anatomy.* a tooth that is specialized for tearing or shearing, such as the last premolars of the upper jaw and the first molars of the lower jaw.

carnation ringspot virus group see DIANTHOVIRUS GROUP.

carnaubic acid *Organic Chemistry.* $CH_3(CH_2)_{22}CO_2H$, a long-chain, fatty acid found in many plant oils and resins, that forms the basic component of carnauba wax.

carnegieite *Mineralogy.* a synthetic compound that is the high-temperature equivalent of nepheline.

Carnian *Geology.* a European geologic stage of the Upper Triassic period, occurring after the Ladinian and before the Norian. Also, KARNIAN.

carnitine *Biochemistry.* $C_7H_{15}NO_3$, an amino acid found in muscle tissue and the liver. Also, VITAMIN B_7.

Carnivora [kär niv´ə rə] *Vertebrate Zoology.* a large order of placental mammals that are almost all flesh eaters and quadrupeds; characterized by large canine teeth and sharp molars and premolars; includes the suborder Caniformia (dogs, bears, sea lions, walrus, pandas, weasels, and seals) and the suborder Feliformia (civets, hyenas, and cats).

carnivorous *Biology.* **1.** of animals, meat eating **2.** of plants, subsisting on nutrients obtained from animal protein.

carnosaur *Paleontology.* a member of the group of large, carnivorous saurischian dinosaurs in the suborder Theropoda, characterized by a large head and a short neck; *Tyrannosaurus rex* is the representative carnosaur.

Carnosauria *Paleontology.* a grouping of carnivorous saurischian dinosaurs in the suborder Theropoda, having short forelimbs, a short neck, and a large head; some authorities do not recognize this grouping.

carnosine *Biochemistry.* $C_9H_{14}N_4O_3$, a dipeptide that is composed of β-alanine and histidine, and is present in vertebrate muscle tissue.

Carnot, Sadi [kär´nō´] 1796–1832, French physicist; discovered Carnot thorem, a foundation of the theory of thermodynamics.

Carnot cycle *Thermodynamics.* a thermodynamic cycle that consists of four successive reversible processes: a constant-temperature expansion and heat transfer to the system from a high-temperature reservoir; an expansion with no heat transfer; a constant-temperature compression and heat transfer from the system to a low-temperature reservoir; and a compression with no heat transfer that restores the system to its original state. This is a hypothetical cycle of ideal efficiency that is used as a standard of comparison for actual heat engine cycles.

Carnot efficiency *Thermodynamics.* the highest efficiency possible for any engine, real or imaginary, operating between temperatures T_1 and T_2. It is a quantity given by the expression $(T_1 - T_2)/T_1$, where T_1 is the absolute temperature for the heat transfer from a high-temperature reservoir in the initial step of a Carnot cycle, and T_2 is the absolute temperature for the heat transfer to a low-temperature reservoir.

Carnot engine *Mechanical Engineering.* an ideal heat engine that functions according to the Carnot cycle, with no energy loss from friction or other factors.

carnotite *Mineralogy.* $K_2(UO_2)_2V_2O_8 \cdot 3H_2O$, a highly radioactive, yellow to greenish-yellow monoclinic mineral occurring as pearly, microscopic crystals or powdery aggregates in sandstones; having a specific gravity of 4.7 to 4.95 and an undetermined hardness on the Mohs scale; ore of uranium and vanadium.

Carnot number *Thermodynamics.* the expression of the Carnot efficiency of a heat engine.

Carnot's principle see CARNOT THEOREM.

Carnot's reagent *Chemistry.* an alcoholic solution of sodium bismuth thiosulfate that is made from sodium thiosulfate and bismuth subnitrate; used in the determination of potassium.

Carnot theorem *Thermodynamics.* **1.** the principle that no heat engine can be more efficient than an ideal, perfectly reversible Carnot engine, whose efficiency is dependent only on the temperatures of the thermal reservoirs. **2.** the principle that all Carnot engines operating between the same two reservoirs will have the same efficiencies, regardless of the working substance.

carob wood [kâr´əb] *Botany.* a large Brazilian tree, *Jacaranda copaia. Materials.* also, **carob wood.** the wood of this tree.

Caro's acid *Inorganic Chemistry.* H_2SO_5, white crystals that decompose in water; melts (decomposes) at 45°C; a strong oxidizer, used in dyeing and bleaching and in treating woolens to prevent felting and shrinking. Also, PEROXYSULFURIC ACID, PERSULFURIC ACID.

carotenase *Biochemistry.* an enzyme that catalyzes the breakdown of pigments called carotenoids.

carotene *Biochemistry.* $C_{40}H_{56}$, a carotenoid pigment widely distributed in nature, especially in green, leafy, and yellow vegetables and in yellow fruit; converted to vitamin A in animals. Also, CAROTIN, PRE-VITAMINE H.

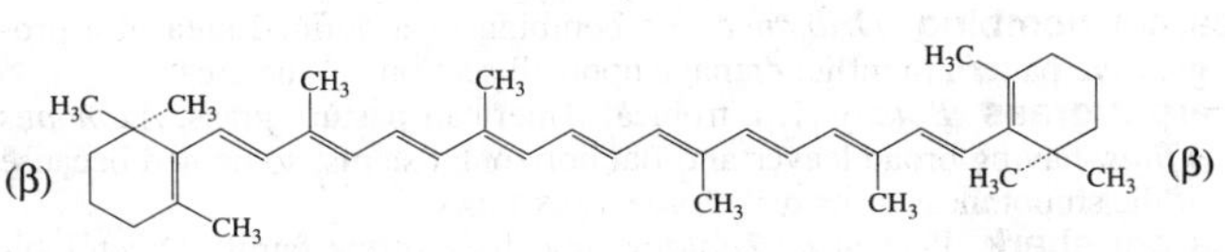

carotene (β-carotene)

β-carotene cleavage enzyme *Enzymology.* an enzyme occurring in the intestinal mucosa that breaks down β-carotene.

carotenemia *Hematology.* the excessive accumulation of carotene in the blood, sometimes causing a jaundiced appearance of the skin.

carotenoid *Biochemistry.* a class of red, orange, purple, and yellow pigments occurring widely in nature, synthesized in plants, and characterized by their solubility in fats; they are referred to as lipochromes when concentrated in animal fat.

carotenoid band shift *Biochemistry.* a change in the range of light a membrane absorbs during photosynthesis as a result of a slight change (measured in nanometers) in the absorption spectra of its carotenoids.

Carothers equation *Materials Science.* a relationship between the number-averaged degree of polymerization and the extent of polymerization reaction.

carotid artery *Anatomy.* either of the two great arteries of the neck that carry blood from the aorta to the head.

carotid body *Anatomy.* a body of tissue at the bifurcation of the carotid artery that is sensitive to oxygen and carbon dioxide levels and from which messages reach the medulla to regulate heartbeat.

carotid ganglion *Anatomy.* a small swelling on the undersurface of the carotid artery that communicates with the Gasserian ganglion and the sixth cranial nerve. Also, BOCK'S GANGLION, LAUMOMIER'S GANGLION.

carotid sinus *Anatomy.* a dilation of the common carotid artery at the point where it bifurcates to form the internal and external carotid arteries; contains receptors that help regulate blood pressure.

carotin see CAROTENE.

carotol *Biochemistry.* $C_{15}H_{25}OH$, an alcohol found in carrots.

carousel *Mechanical Engineering.* a circular conveyor on which objects are kept in continuous motion.

carp *Vertebrate Zoology.* any of various freshwater fishes of the family Cyprinidae in the order Cypriniformes, found all over the world and characterized by soft fins, a suckerlike mouth, and pharyngeal teeth.

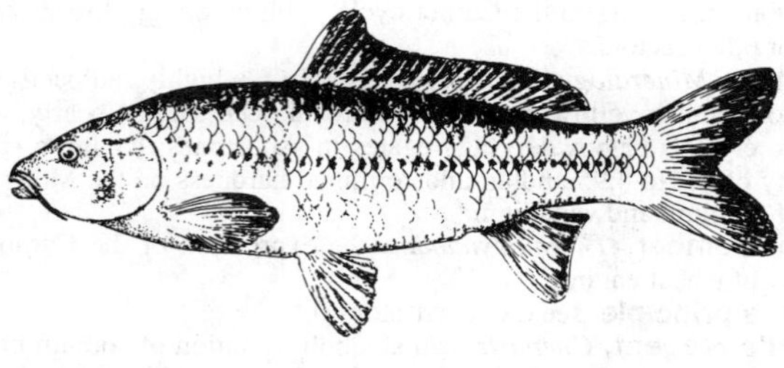

carp

carpal *Anatomy.* of or relating to the carpus or wrist.

carpal tunnel *Anatomy.* the osseofibrous passage for the median nerve and flexor tendons, formed by the flexor retinaculum and the carpal (wrist) bones.

carpal-tunnel syndrome *Medicine.* a complex resulting from compression of the median nerve of the carpal tunnel, characterized by pain with burning or tingling in the fingers, hand, and lower arm.

Carpathians *Geography.* a mountain range of central and eastern Europe, mostly in southern Poland, eastern Czechoslovakia, and Romania; highest peak: Gerlachovka (8711 ft).

carpectomy *Surgery.* the surgical removal of a wrist bone.

carpel *Botany.* the structure that bears the ovule in flowering plants.

carpenter *Building Engineering.* a person whose work is building and reparing items of wood.

carpenter's level *Mechanical Devices.* a bar-shaped spirit level made of aluminum or wood.

carpenter's square *Mechanical Devices.* a flat steel square used in framing construction, consisting of a 24-inch body and a 16-inch tongue marked by equal fractions in inch increments. Also, STEEL SQUARE.

carpentry *Building Engineering.* **1.** the profession or work of a carpenter. **2.** an item or items produced by a carpenter.

carpet bombing *Ordnance.* the bombing of a defined area in a progressive pattern to inflict damage upon all sections of the area.

carpet grass *Botany.* **1.** a tropical American pasture grass, *Axonopus affinis,* having broad leaves and flat horizontal stems; so named because of the smoothness of its turf. **2.** see SMUT GRASS.

carpet shark *Vertebrate Zoology.* any shark of the family Orectolobiformes, with mottled skin and fleshy lobes on the sides of the head; found in tropical Indo-Pacific waters.

carpholite *Mineralogy.* $Mn^{+2}Al_2Si_2O_6(OH)_4$, a yellow orthorhombic mineral occurring as fibrous radiated tufts; having a specific gravity of 2.9 to 3.02 and a hardness of 5 to 5.5 on the Mohs scale; found in mountainous regions of Germany, Yugoslavia, Czechoslovakia, and Belgium.

car pincher *Mining Engineering.* a mine worker who positions cars under loading chutes, at breaker or tipple, inserting a pinch bar under the car wheels and bearing down or pulling up on it to force the car forward. Also, **car shifter, car spotter.**

carpogonium *Botany.* the female sex organ of the red algae.

Carpoidea *Paleontology.* a term formerly used to denote a class of echinoderms distinguished by a laterally flattened calyx; now generally subdivided into three classes: Stylophora, Homoiostelea, and Homostelea; Cambrian to Devonian.

carpoids *Paleontology.* the echinoderms of the former class Carpoidea, now divided into three classes.

carpology *Botany.* the branch of botany concerned with the structure of fruits and seeds.

carpomycetes *Mycology.* fungi that form fruiting bodies, as in fungi belonging to the subdivisions Ascomycotina and Basidiomycotina.

carpopedal *Anatomy.* of or relating to the wrist and foot.

carpophagous *Zoology.* feeding mainly on fruit.

carpophalangeal *Anatomy.* of or relating to the wrist and finger bones.

carpophore *Botany.* the thin wiry stalk that supports each half of the dehiscent fruit of plants of the parsley family (Umbelliferae). (A Greek word meaning "fruit bearer," also used as an epithet of the goddess Demeter.)

carpophyte *Botany.* a thallophyte that, after fertilization, has formed a sporocarp.

carpoptosis see WRISTDROP.

carposporangium *Botany.* in some red algae, a sporangium that develops directly from a zygote and depends on its parent for nutrients.

carpospore *Botany.* a haploid or diploid spore produced by red algae.

carpus *Anatomy.* **1.** the eight bones of the wrist. **2.** the wrist itself.

carrageen *Botany.* a red alga, *Chondrus crispus,* found in the North Atlantic and gleaned for carrageenan, used chiefly as an emulsifier.

carrageenan *Organic Chemistry.* an aqueous, usually gel-forming, cell-wall polysaccharide mucilage that is found in the red algae *Chondrus crispus* and *Gigartina mammillosa*; used as an emulsifier in food products, cosmetics, and pharmaceuticals. Also, **carrageenin.**

Carrara marble *Petrology.* all marble quarried near Carrara, Italy; characterized by a white to bluish color or a white color with blue veins; used in tiles and statuary.

Carrel, Alexis 1873–1944, French surgeon and biologist; Nobel Prize for vascular surgery and organ and tissue transplants.

Carrel flask *Biotechnology.* a short, squat flask with an upwardly sloping, open-ended tube projecting from its neck on one side.

Carrel-Lindbergh pump see LINDBERGH PUMP.

Carrel's method *Surgery.* **1.** a method of end-to-end suture of blood vessels. **2.** a method of determining when to make secondary closure of wounds by taking a loop of material from the wound, spreading the material on a slide, staining it, and counting the number of bacteria. (After Alexis *Carrel.*)

Carrel's treatment *Surgery.* a process for treating wounds that involves exposing the wound, removing all foreign material and dentalized tissue, cleansing meticulously, and irrigating repeatedly with a dilute sodium hypochlorite solution, while protecting the adjacent skin with petrolatum gauze. Also, **Carrel-Dakin treatment.** (After Alexis *Carrel* and Henry *Dakin,* American chemist, 1880–1952.)

car retarder *Engineering.* a device for reducing or controlling the speed of mine cars on a track.

carriage *Engineering.* a mechanism that moves in a specified path in a machine and carries another part, as a recorder head. *Ordnance.* see GUN CARRIAGE. *Graphic Arts.* the mechanical or electronic part of a typesetting machine that places the type characters onto a galley.

carriage bolt *Mechanical Devices.* a square bolt with an oval head used with a nut in through bolt applications to steady the bolt while the nut is tightened.

carriage-control character *Computer Programming.* on a line to be printed, especially on an impact printer, an initial character that determines the vertical space to be skipped before the next line is printed.

carriage return *Computer Programming.* a command sent to a printer or terminal to move to the first position of the line.

carriage stop *Mechanical Engineering.* a device on a lathe used to mark off positions on the workpiece for turning and cutting.

carriage tape see CONTROL TAPE.

carrier a person or thing that carries; specific uses include: *Immunology.* **1.** a symptomless individual who is host to a pathogenic microorganism and who has the potential to pass the pathogen to others. **2.** a molecule (usually a protein) that binds with a partial antigen to make the antigen produce an immune response. *Veterinary Medicine.* an animal that bears and may transmit the causative agent of a disease, especially an animal that has recovered from or is immune to the illness. *Genetics.* a stable isotope of an element mixed with a small quantity of radioisotope of the element sufficient for chemical tracing to occur. *Radiology.* a material used to convey traces of another material through the body so that a physical process may be studied. *Histology.* a particle that transports substances through cell membranes in a reversible process. *Mechanical Engineering.* a wheeled chassis that is the base mounting for rough terrain cranes. *Naval Architecture.* see AIRCRAFT CARRIER. *Solid-State Physics.* a mobile entity that carries a charge, such as a conduction electron or a hole in a semiconductor crystal. Also, CHARGE CARRIER. *Telecommunications.* a continuous frequency or tone that is capable of being modulated by an information-bearing signal.

carrier aircraft *Aviation.* **1.** an aircraft designed to operate from an aircraft carrier; often equipped with folding wings, arresting hooks, or other special features. **2.** an aircraft designed to carry another aircraft, a guided missile, or other very large loads.

carrier air group *Military Science.* two or more aircraft squadrons placed under one command for administrative and tactical control of operations from an aircraft carrier.

carrier amplifier *Electronics.* an amplifier that operates through a capacitor with a capacitance that varies with the applied voltage. Also, DIELECTRIC AMPLIFIER.

carrier amplitude regulation *Telecommunications.* the amplitude variation of an amplitude-modulated carrier wave when symmetric modulation is used.

carrier beat *Telecommunications.* in facsimile transmission, an unwanted sequence of signals, each synchronous with a different stable oscillator, producing a pattern in received copy.

carrier-cable logging see SKYLINE LOGGING.

carrier channel *Telecommunications.* **1.** a path along which modulated carrier waves can be sent. **2.** the equipment that constitutes a complete carrier-current circuit.

carrier culture *Virology.* a cell culture that is persistently infected with a virus, but in which only a small proportion of the cell population is actually infected at any given time.

carrier current *Telecommunications.* a current in the form of an electromagnetic wave that is modulated by an information-bearing signal.

carrier density *Solid-State Physics.* the volume density of mobile charge carriers in a semiconductor material.

carrier detect *Computer Technology.* an interchange signal between a modem and its data terminal indicating that the modem has responded to a distant modem and is about to accept data.

carrier frequency *Telecommunications.* the frequency of a carrier wave that can be modulated by an information-bearing signal of a different frequency. Also, CENTER FREQUENCY.

carrier gas *Metallurgy.* in thermal spraying, the gas that carries the metallic powder to the gun.

carrier leak *Telecommunications.* the remainder of a carrier wave after carrier suppression.

carrier level *Telecommunications.* the power level of a carrier wave at a given time in a transmission system, in relation to a reference level; usually expressed in decibels.

carrier line *Electricity.* a transmission line for multiple-channel carrier communication.

carrier loading *Electromagnetism.* the addition of lump inductance to a coaxial cable transmission line used in the carrier transmission of frequencies up to about 35 kilohertz; minimizes cable attenuation and reduces mismatch of cable impedance and open wire impedance.

carrier mobility *Solid-State Physics.* a measure of the average speed of a carrier in a given semiconductor, given by the average drift velocity of the carrier per unit electric field.

carrier noise *Telecommunications.* any noise produced from unwanted variations of a carrier in the absence of any intended modulation.

carrier pipe *Engineering.* a pipe that conducts fluids or other materials.

carrier power output rating *Telecommunications.* the unmodulated power that is nominally supplied to an antenna transmission line by a radio transmitter, averaged during one frequency cycle during which time there is no modulation; unless noted, this is the normal rating of the transmitter.

carrier repeater *Electronics.* a circuit that boosts the power of a carrier signal level without increasing the noise level.

carrier rig *Petroleum Engineering.* a self-contained mobile unit created to service gas and oil wells with many pieces of equipment, such as engines, hoists, masts, and the like.

carrier rocket *Space Technology.* a rocket designed to carry a satellite or other large load.

carrier shift *Telecommunications.* **1.** the transmission of information by shifting the carrier frequency in one direction for a mark signal and in the other direction for a space signal. **2.** the variation between the mark and space frequencies in a transmission system using a frequency shift modulation technique.

carrier signaling *Telecommunications.* signaling methods that are used in the transmissions of more than one channel.

carrier striking force *Military Science.* a naval task force composed of aircraft carriers and supporting warships capable of conducting strike operations.

carrier suppression *Telecommunications.* a method of transmission in which the carrier frequency is not transmitted.

carrier swing *Telecommunications.* the complete deviation of a frequency or phase-modulated wave from the smallest to the largest frequency at any given time.

carrier system *Telecommunications.* a method of deriving a group of channels over a single path by modulating each channel on a different carrier frequency.

carrier telegraphy *Telecommunications.* a method of telegraphy in which the information-bearing signals modulate a carrier wave.

carrier telephony *Telecommunications.* a method of telephony in which the information-bearing signals modulate a carrier wave. Also, WIRED RADIO.

carrier terminal *Electronics.* an apparatus attached to one end of a carrier transmission system in which such transmission functions as modulation, filtering, and amplification are performed.

carrier test *Microbiology.* a test to determine the ability of a disinfectant to cleanse a given object that is contaminated with microorganisms.

carrier-to-noise ratio *Telecommunications.* the ratio of the amplitude value of a carrier to the received noise level.

carrier transfer filter *Electronics.* a group of filters that act as a carrier-frequency bridge between two transmission circuits.

carrier transmission *Telecommunications.* a common method of transmission in which a carrier is modulated by an information-bearing signal and transmitted.

Carrington, Richard Christopher 1826–1875, English astronomer; studied sunspots; discovered systematic drift of photosphere.

Carrington rotation number *Astronomy.* a system for counting rotations of the sun based on the average synodic rotation period of sunspots (27.2753 days) and beginning November 9, 1853.

Carrion's disease see OROYA FEVER.

Carr medium *Microbiology.* a bacteriological culture medium for distinguishing between strains of *Acetobacter* and *Gluconobacter,* using a color indicator to demonstrate that oxidation of the same substrate by the two genera yields different metabolic products.

carrot *Botany.* a plant, *Daucus carota,* of the parsley family, widely cultivated for its edible orange root.

Carrpaceae *Botany.* a monotypic family of specialized minute liverworts of the order Marchantiales found only in desert saltpans of Australia; characterized by delicate plants, smooth and colorless rhizoids that arise from the ventral thallus surface, and no air pores, although there are compound barrel-type pores on the archegoniophores.

carry *Mathematics.* the process in arithmetic addition that occurs when the sum of the digits in a given position is $mb + a$, where b is the number base, $a < b$, and $m > 0$; a is entered in the given position of the sum, and m is **carried** or added to the next-higher-order digit.

carry-complete signal *Computer Technology.* a signal produced by a parallel binary adder circuit, indicating the completion of the carry-and-add operation.

carry flag *Computer Technology.* **1.** a flip-flop circuit that holds the carry bit during a parallel binary add operation. **2.** a bit in the condition code register that indicates a carry from the most significant bit of the word following an add or subtract operation. Also, EXTEND FLIP-FLOP.

carrying capacity *Electricity.* the capability of a wire or other component to handle power or current; the maximum current or power that can be handled by a wire or component. *Ecology.* the maximum population that can be supported by the resources of a certain habitat at any given time. Also, ECOLOGICAL CAPACITY.

carrying contour *Cartography.* a single contour line on a map representing two or more contours, used to show vertical or near-vertical topographical features.

carry lookahead *Computer Technology.* a technique used to speed the execution of parallel add operations by predicting the final carry bit from the output of each partial adder.

carryover or **carry-over** *Chemical Engineering.* any undesired solid or liquid substance entrained by the overhead effluent from a reaction vessel, absorber, or fractionating column. *Hydrology.* the portion of stream flow during a given month or year that is derived from precipitation from earlier periods.

carry-save adder *Computer Technology.* a binary adder, used for fast summation of three or more operands, that outputs the sum and a second word consisting of all the carry bits from the add operation; the two output words are then added to obtain the final sum.

carry signal *Computer Technology.* a signal produced by a binary adder when the sum of bits in the same bit position of the operands exceeds 1, indicating that a 1 must be added to the next-higher significant bit position.

carry time *Computer Technology.* **1.** the time required to transfer a carry bit to the next-higher bit position and add it to the operands. **2.** the total time required to transfer all the carry bits to higher bit positions and add them to the operand bits.

carsickness *Medicine.* a condition induced by the movement of an automobile, similar to other types of motion sickness; characterized by vertigo, nausea, and sometimes vomiting.

Carson, Rachel 1907–1964, American marine biologist and science author; wrote influential early works on environmental concerns.

Rachel Carson

car stop *Engineering.* a device that arrests the movement of a mine car.

Cartan calculus *Mathematics.* a former name for the algebra of alternating differential forms.

Cartan connection *Mathematics.* an assignment of a matrix $\omega = (\omega^i_j)$ of 1-forms to every moving frame F on a manifold M such that $\omega' = A^{-1}dA + A^{-1}\omega A$ holds whenever moving frames F and F' are related by $F' = FA$.

Cartan integer *Mathematics.* for the ith and jth roots in a given root system, the quantity $2(\alpha_i \cdot \alpha_j)/(\alpha_j \cdot \alpha_j)$, where $(\cdot)$ denotes the usual (Euclidean) dot product. By theorem, these numbers are integers.

Cartan matrix *Mathematics.* for a given root system, the matrix whose (i, j)th entry is the Cartan integer relating the ith and jth roots.

Cartan's criterion *Mathematics.* in Lie algebra, the theorem that if $\mathrm{trace}(x\,y) = 0$ for all y in a Lie algebra L and for all x in the commutator subalgebra of L, then L is solvable.

Cartan subalgebra *Mathematics.* a nilpotent subalgebra of a Lie algebra L which equals its normalizer in L. If L is semisimple over a field of characteristic zero, its Cartan subalgebra is a maximal Abelian subalgebra of semisimple elements of L. The standard structure theory of semisimple Lie algebras is based on the properties of the Cartan subalgebra.

Carter, Howard 1873–1939, English archaeologist; discovered the tomb of Tutankhamen.

Carter chart *Electromagnetism.* an Argand diagram of the complex reflection coefficient of a waveguide junction, including lines of constant magnitude and phase of the impedance.

Carteria *Botany.* a genus of unicellular freshwater green algae belonging to the family Chlamydomonadaceae, characterized by its quadriflagellate structure.

Carterinacea *Invertebrate Zoology.* a superfamily of deep-sea foraminiferan protozoa in the suborder Rotaliina.

Cartesian [kär tē´zhən] of or relating to René Descartes or his mathematical methods, particularly regarding his logical analysis and mechanistic interpretation of nature.

Cartesian axis *Mathematics.* the x, y, or z axis in Euclidean space. More generally, any of a set of n $(n \geq 2)$ mutually perpendicular lines intersecting at a single point, called the *origin.*

Cartesian-coordinate robot *Robotics.* a robot with a nonrotary base that uses the Cartesian coordinate system of three perpendicular and intersecting straight lines.

Cartesian coordinates *Mathematics.* the x and y coordinates of a point in the plane, written (x, y). More generally, the coordinates of a point in R^n given by the directed distances of the point from the origin along the n Cartesian axes. The coordinates of a given point on the ith coordinate axis are zero in every position except the ith, and equal to the directed distance of the point from the origin in the ith position.

Cartesian coordinate system *Mathematics.* in n-space, the coordinate system determined by the n Cartesian axes. Also, RECTANGULAR COORDINATES.

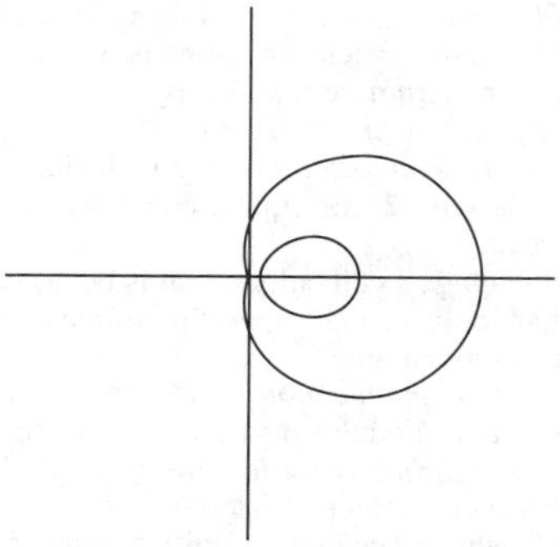

Cartesian oval

Cartesian oval *Mathematics.* the set of points P in the plane satisfying $\alpha(PF_1) + \beta(PF_2) = \gamma$, where α, β, and γ are scalar constants, F_1 and F_2 are fixed points, and (XY) is the length of the line segment from X to Y; used by Descartes in his study of optics and development of analysis of conics.

Cartesian product *Mathematics.* **1.** the Cartesian product $A \times B$ of two sets A and B is the unique set consisting of all ordered pairs (a, b) where a is an element of A and b is an element of B. The operation of forming the Cartesian product is neither commutative nor associative. **2.** more generally, the Cartesian product $\mathrm{X}_{\iota \in I} A_\iota$ of a family of sets $\{A_\iota\}_{\iota \in I}$ is the set of all (choice) functions x having domain I such that $x_\iota = x(\iota)$ is an element of A_ι for each ι in the index set I. For each such x, the value x_ι is the ιth coordinate of x. The axiom of choice ensures that the Cartesian product of the family of sets $\{A_\iota\}_{\iota \in I}$ exists if the A_ι and index set I are nonempty.

Cartesian surface *Mathematics.* a surface in three-dimensional space obtained by rotating the curve

$$b_0(x^2 + y^2)^{1/2} \pm b_1[(x - a)^2 + y^2]^{1/2} = c$$

about the x-axis.

Cartesian tensor *Mathematics.* a Cartesian tensor of the $(s + p)$th order in n-dimensional Euclidean space is a mixed tensor of n^{s+p} components that is contravariant of the sth order and covariant of the pth order when the coordinates undergo a positive orthogonal transformation. (The components may not transform correctly when a more general transformation is used.)

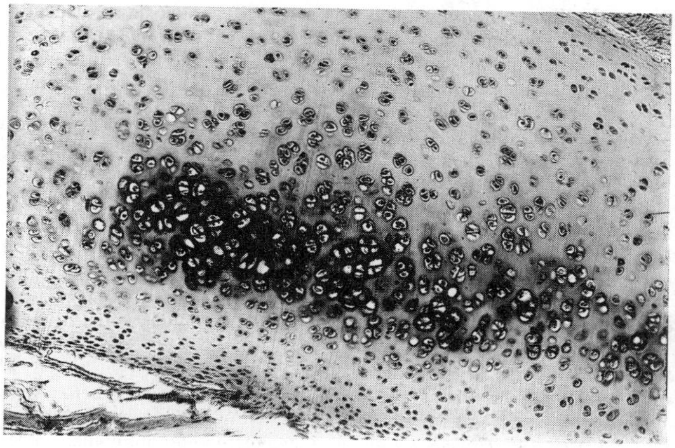

cartilage

cartilage *Histology.* a dense connective tissue consisting of cells embedded in a tough but flexible matrix.

cartilaginous [kär′tə läj′ə nəs] *Zoology.* **1.** relating to or composed of cartilage. **2.** having a skeleton composed of cartilage.

cartilaginous fish *Vertebrate Zoology.* the common name for all members of the vertebrate class Chondrichthyes, or those fishes with skeletons composed of cartilage and skin covered with dermal denticles, including sharks, dogfish, rays, skates, and so on.

cartogram *Cartography.* a type of single-topic map of no constant scale, in which features are deliberately distorted to show their relative value to one another in terms of that single topic; e.g., a map in which states are made large or small according to their agricultural production.

cartographic *Cartography.* of or relating to the tools and techniques of surveying land surfaces for the purposes of making maps, or the graphic tools and techniques used to produce maps.

cartographic annotation *Photogrammetry.* the delineation of additional new data such as airfields and cities, or the deletion of destroyed or dismantled features such as roads, bridges, and cultural landmarks, on a mosaic in order to portray current details.

cartographic license *Cartography.* the freedom to adjust, add, or omit features on a map, within stated allowable limits, to obtain the best cartographic expression.

cartographic satellite *Space Technology.* an artificial satellite designed to take photographs and collect data used in preparing maps of the earth's surface.

cartography *Geography.* the science of mapmaking.

cartometric scaling *Cartography.* the accurate measurement of geographic or grid coordinates on a map or chart by means of a scale.

cartouche [kär′toosh′] *Cartography.* a panel on a map, often decorated, enclosing the title or other legends, the scale, and other information. *Graphic Arts.* a device often having the appearance of scrollwork, used to enclose names or titles.

cartridge *Ordnance.* **1.** originally, a case containing all the components necessary to fire a gun once, including powder, shot, and so on. **2.** in modern usage, the ammunition necessary to fire a gun once, contained in a single unit and loaded in a single operation. *Engineering.* **1.** a container holding photographic film that allows the film to be inserted into, and removed from, a camera without exposing it to light. **2.** a container that permanently houses a continuous loop of motion picture film, videotape, or magnetic tape. *Electronics.* **1.** a unit attached to the arm of a record player that houses the pickup transducer. **2.** a closed-loop reel and housing developed primarily for eight-track recordings, which operate at 4-3/4 inches per second and are now obsolete due to their poorer quality recording and reproduction as well as durability, compared to cassette recordings. *Nucleonics* see CLADDING.

cartridge brass *Metallurgy.* an alloy containing 70% copper and 30% zinc, except for incidental impurities.

cartridge case *Ordnance.* an ammunition container holding the primer and propellant, and to which the projectile may be connected.

cartridge disk *Computer Technology.* a data storage medium consisting of a single rigid magnetic disk housed in a plastic casing.

cartridge filter *Engineering.* a clarifying filter having a vertical stack of metal disks spaced to trap particles contained in process liquids poured between the disks.

cartridge fuse *Electricity.* a fuse having a current-responsive element inside a fuse tube with a ferrule on each end for plug-in connections.

cartridge lamp *Electricity.* a pilot or dial lamp that has a cylindrical glass envelope with metal terminals at each end.

cartridge magazine *Ordnance.* a metal, spring-loaded container that holds cartridges and is loaded into the magazine well of a pistol, rifle, or machine gun; when empty, it may be removed from the gun, refilled with cartridges, and reloaded for further use.

cartridge starter *Mechanical Engineering.* a combustible device that detonates in an engine, moving a piston and starting the engine.

cartridge tape drive *Computer Technology.* an encased tape unit that contains magnetic tape and associated reels to permit automatic loading of the magnetic tape.

Cartwheel Galaxy *Astronomy.* the nickname given to a ring galaxy, once a normal spiral, that has been tidally distorted by an encounter with another galaxy into a ring-and-hub structure.

caruncle *Anatomy.* any small, fleshy process. *Botany.* in seeds, a fleshy outgrowth of the seed coat found at or near the hilum.

Carus, Karl Gustav 1789–1869, German physiologist; studied skull structure and connection between cell structure and psychology.

carvacrol *Organic Chemistry.* $C_{10}H_{14}O$, a colorless liquid; slightly soluble in water and soluble in alcohol; melts at 1°C and boils at 237°C; used in perfumes, fungicides, disinfectants, and flavoring.

Carvallo paradox *Optics.* the tenet that since light contains endless waves of energy at various frequencies, a spectrograph should present the spectrum of a light source both before and after it is lit.

Cartography

Cartography is the science and art of mapmaking; it also includes the study and interpretation of maps. Maps are graphic simplifications of reality, portraying relationships on the surface of the earth (or other celestial body) by means of points, lines, areas, symbols, colors, and typography. Some maps are charts that display data specifically for nautical and aeronautical navigation. A map can be either two- or three-dimensional.

One of the first known true maps, a clay tablet locating recognizable features in Mesopotamia (ca. 2300 BC), was unearthed at Yorghan Tepe near Kirkuk, Iraq. Ancient China benefited from the efforts of its mapmakers. The Greeks and Romans were competent cartographers, too; Ptolemy, a Greek astronomer and geographer, describes a map of the world (ca. AD 150) that shows remarkable knowledge of the size and shape of Eurasia and northern Africa. The Age of Discovery spanning the 15th and 16th centuries gave mapmakers wide new horizons. More complete and accurate views of the earth included Martin Waldseemuller's world map "Universalis Cosmographia" (1507), on which the name *America* appeared for the first time, and Gerardus Mercator's famous world map (1569), drawn on the mathematical projection named after him.

Cartography came of age during the 19th and 20th centuries with the development of lithography, photography, photogrammetry, and high speed color printing. The electronic era of computer-assisted cartography marks the most recent great leap forward. Already digital maps and atlases are available as personal-computer software and videodiscs, and before long the marriage of satellite imagery and advanced computer technology will enable the cartographer to make, on demand, instant, interactive maps for many uses.

John B. Garver
Senior Assistant Editor and Chief Cartographer
National Geographic Society

George Washington Carver

Carver, George Washington 1864–1943, American botanist; best known for research on various uses of the peanut.

carvone *Organic Chemistry.* $C_{10}H_{14}O$, a pale-yellowish or colorless liquid, slightly soluble in water and very soluble in alcohol; boils at 231°C; used in flavoring, liqueurs, perfumes, and soaps.

caryinite *Mineralogy.* $Na(Ca,Pb)(Ca,Mn)(Mn,Mg)_2(AsO_4)_3$, a brown, translucent, brittle, monoclinic mineral of the alluaudite group, having a specific gravity of 4.29 and a hardness of 4 on the Mohs scale; found as veinlets in limestone skarn with calcite, berzellite, and hausmannite.

caryo- a combining form meaning "seed" or "nucleus."

Caryocaraceae *Botany.* a family of dicotyledonous tropical American trees and shrubs of the order Theales, having evergreen leaves, large perfect flowers, and seeds that are edible in some species but toxic in others.

Caryophanaceae *Bacteriology.* a former classification for a family of bacteria in the order Caryophanales, consisting of filamentous or rod-shaped cells found in the oral cavity and intestinal tract of ruminants.

Caryophanales *Bacteriology.* a former classification for an order of bacteria in the class Schizomycetes, occurring in water, decomposing organic materials, and the intestinal tracts of arthropods and vertebrates.

Caryophanon *Bacteriology.* a genus of Gram-positive, obligately aerobic bacteria characterized as motile filaments able to use acetate as the only major source of carbon; found in cow dung, but its affiliation and true habitat are uncertain.

Caryophyllaceae *Botany.* a family of dicotyledonous herbs of the order Caryophyllales having opposite, sometimes connate-perfoliate leaves, flowers with long, tapered petals, and an ovary with free-central placentation.

Caryophyllaceae (carnation)

Caryophyllales *Botany.* an order of herbaceous, dicotyledonous plants, characterized by regular, bisexual flowers, axile to free central placenation, and a curved embryo. Also, CENTROSPERMALES.

caryophyllene *Organic Chemistry.* $C_{15}H_{24}$, an unstable sesquiterpenoid liquid; boils at 123°C (14 torr); forms the chief hydrocarbon component of clove oil.

Caryophyllidae *Botany.* a subclass of the Magnoloipsida that have a multilayered nucleus and ovules with two integuments. *Invertebrate Zoology.* solitary corals with goblet or horned-shaped thecae, found in colder ocean waters, including those off the coast of Great Britain; *Caryophyllia smithii* **(Devonshire Cup Coral)** is a typical example. Also, **Caryophyllidea.**

caryophyllin *Organic Chemistry.* $C_{30}H_{48}O_3$, needles or prisms, insoluble in water and slightly soluble in alcohol and acid; melts at 310°C. Also, OLEANOLIC ACID.

caryopilite *Mineralogy.* $(Mn^{+2},Mg)_3Si_2O_5(OH)_4$, a brownish-red, easily fusible monoclinic mineral that is closely related to friedelite, having a specific gravity of 2.8 to 2.9 and a hardness of 3 to 3.5 on the Mohs scale.

caryopsis *Botany.* in grasses, an achenelike fruit in which the pericarp is fused to the seed.

CAS *Aviation.* the airport code for Casablanca, Morocco.

casaba *Botany.* a winter melon, the fruit of the *Cucumis melo.* Also, CASSABA.

Casale process *Chemical Engineering.* a method for ammonia synthesis from hydrogen and nitrogen at high pressure, similar to the Huber-Bosch and Claude processes.

cascade *Hydrology.* a small, steep waterfall, especially one that is part of a series of such falls descending over steeply slanting rocks. *Geology.* **1.** a short, rocky downward slope in a stream bed over which water flows faster than through a rapids. **2.** a glacial valley whose floor resembles a broad staircase with irregular benches separated by steep risers. Also, GLACIAL STAIRWAY. *Electricity.* **1.** an arrangement of two or more often similar circuits or components in which the output of one is the input of another. Also, TANDEM. **2.** of or relating to such an arrangement. Thus, **cascade converter, cascade network.** *Computer Programming.* a mode of processing in which two or more components or actions are arranged in sequence, and the output of one is input to the next. *Engineering.* any group of similar devices connected or arranged in sequence so that each operates the one following and multiplies the effect of the one preceding.

cascade amplifier *Electronics.* a circuit arranged in stages, allowing the output produced by one to serve as input for the next, with power increasing at each stage.

cascade carry *Computer Technology.* in parallel binary addition, a process in which the carry bit from each bit position is carried to the next-higher bit position and combined with the operand bits until the most significant bit position is reached.

cascade compensation *Control Systems.* a method of compensation where the compensator is in series with the forward transfer function. Also, SERIES COMPENSATION, TANDEM COMPENSATION.

cascade connection *Electronics.* a tandem arrangement of two or more component devices in which the output of one is connected to the input of the next.

cascade control *Computer Technology.* an automatic control system in which the components are linked sequentially, with each unit controlling the operation of the subsequent unit.

cascade cooler *Chemical Engineering.* a fluid-cooling apparatus in which fluid flows through a series of horizontal tubes, one above the other, while cooling water runs from a trough over each tube.

cascaded feedback canceler *Electronics.* a device that eliminates interference frequently accompanying radar signals produced by moving targets. Also, VELOCITY-SHAPED CANCELER.

cascade fermentation *Biotechnology.* a system in which a fermenting liquor is fed through a series of up to ten fermenters; the yeast is retained and settled in intermediate tanks. Also, OVERFLOW CONTINUOUS FERMENTATION.

cascade fold *Geology.* one of a series of minor folds that are produced by the buckling and collapse of a larger fold under the influence of gravity.

cascade gamma emission *Nuclear Physics.* an event in which a nucleus emits two or more gamma rays in succession.

cascade image tube *Electronics.* a tube in which a number of sections are joined so that the image generated by one section is projected into the next section; used for extremely low-level light detection.

cascade impactor *Engineering.* a device for sampling aerosols or dusty air that automatically separates the particles or droplets by drawing the gas stream through a series of jet impactors of decreasing nozzle size.

cascade junction *Electronics.* a network of semiconductor devices joined in tandem.

cascade limiter *Electronics.* a circuit that continues to emit signals at predetermined amplitude no matter how widely the incoming signals may vary, thus improving the operation for both weak and strong signals. Also, DOUBLE LIMITER.

cascade liquefication *Physics.* a method of liquefying gas with a high critical temperature by applying pressure, then allowing the gas to evaporate; the evaporate then cools another gas below its critical temperature, and that gas is liquefied by pressure, and so on.

cascade mechanism *Endocrinology.* a series of intracellular events, initiated by a single signal (usually the binding of an agonist to its specific receptor on or in the cell) and then transduced and amplified by the subsequent activation of other intracellular signalling pathways, resulting in a specific terminal event such as secretion, protein synthesis, or cell division.

cascade merge *Computer Programming.* a complicated sort/merge routine devised to reduce the number of devices needed to hold input and output media.

cascade mixer-settler *Chemical Engineering.* a multiple-stage liquid-liquid contacting system, with each stage consisting of an agitated mixing vessel connected with an unstirred settling tank.

cascade noise *Electronics.* the sounds emanating from a communications receiver when an incoming signal is processed simultaneously through two separate amplifiers.

cascade pulverizer *Mechanical Engineering.* a crushing machine that pulverizes materials by the repeated rolling action of hard objects such as balls or pebbles within a tumbler.

cascade sequence *Metallurgy.* in multiple-pass welding, the sequence in which the weld is continuously produced by depositing overlapping layers of beads.

cascade shower *Particle Physics.* a cosmic-ray shower in which high-energy mesons, protons, and electrons create high-energy photons, that produce further electrons and positrons, thus increasing the number of particles until the energy is dissipated.

cascade system *Mechanical Engineering.* a series of pieces of equipment, such as evaporators, condensers or reactors, that are used to complete a process.

cascade transformer *Electricity.* a high-voltage source consisting of a limited number of step-up transformers.

cascade tray *Chemical Engineering.* a fractionating device with a series of parallel troughs placed so that one tray spills onto the tray below.

Cascadian orogeny *Geology.* a post-Tertiary deformation of the earth's crust found in western North America.

cascading *Electricity.* **1.** an application in which the devices nearest the source of power have interrupting ratings equal to, or in excess of, the available short-circuit current, while devices further from the source have successively lower interrupt ratings. **2.** the connection of two or more circuits in a cascade. *Mechanical Engineering.* a milling process that involves crushing balls that continually fall onto the feed material, such as a grain load, pulverizing the top layer.

cascading errors *Computer Science.* a situation, as in compiling a program, in which one error causes many subsequent errors. For example, failure to declare a variable may cause an error every time that variable is referenced.

cascading glacier *Hydrology.* a glacier having many crevasses as a result of passing over a steep, irregular bed.

cascode amplifier *Electronics.* a circuit that increases a signal's power by passing it through a grounded cathode then through a grounded-grid, resulting in high gain, high impedance, and low noise.

case *Engineering.* an item that holds a given object in an exact location because it conforms to the size of the object, but that may be detached from the object. *Metallurgy.* **1.** the surface layer of a ferrous material, having a composition intentionally altered from that of the base. **2.** the hardened exterior of steel that has been strengthened on the outside by carburizing, cyaniding, or nitriding, as well as by heating and quenching. *Petroleum Engineering.* to place a steel or iron pipe into a borehole to prevent caving of the walls or the ingress of water. *Graphic Arts.* **1.** the cover of a hardbound book. Also, **case binding. 2.** a compartmentalized wooden tray designed to hold metal type. *Mining Engineering.* a small break that allows water to enter into the mine workings. *Artificial Intelligence.* see DEEP CASE.

CASE computer-aided software (or systems) engineering.

caseation necrosis *Pathology.* a condition, often seen in patients with tuberculosis, in which tissue is transformed into a dry, amorphous mass resembling cheese, containing a mixture of protein and fat that is assimilated very gradually.

case-based reasoning *Artificial Intelligence.* reasoning based on similarity of a given set of data to a known case stored in a library, as in the use of legal precedents.

case bay *Building Engineering.* a section between two principal rafters and the joints between them, forming a division of a roof or a floor.

casebound see HARDBOUND.

cased charges *Ordnance.* a U.S. naval term for powder charges enclosed in a metal case and used in guns larger than 1 inch in diameter.

cased hole *Petroleum Engineering.* a wellbore that has been lined with casing of tough, thick-walled steel pipe.

case frame *Artificial Intelligence.* in natural language processing, a lexical entry that specifies a single sense meaning of a verb, together with information about related phrases such as case relation slots, selection restrictions on the phrases that can fill the slots, prepositions that are used for each case, and so on.

case hardening or **casehardening** *Materials Science.* the hardening of the outer layer of an object, such as a machine part, by carburizing; the end result is a part with a hard martensotic outside case for wear resistance and a tough inner core for fracture resistance. *Metallurgy.* any of several processes for hardening the surface of a ferrous material by diffusing into it an element such as carbon or nitrogen, and cooling at a controlled rate. *Geology.* a process by which a cementlike mineral coating forms on the surface of porous rock as a result of evaporation.

case history *Medicine.* a complete record of available data relating to an individual, consisting of observation, examination, test results, interviews, and professional findings; used in determining the individual's diagnosis and prognosis.

casein [kā´sēn´] *Organic Chemistry.* a white, amorphous, colloidal aggregate, soluble in acids, that is found in milk and is composed of several proteins together with phosphorus and calcium.

casein-formaldehyde *Organic Chemistry.* a formaldehyde-modified natural polymer.

casemate *Ordnance.* **1.** a chamber built into the walls of a fortress, used to house troops and guns and ammunition. **2.** a bombproof structure used to hold guns or ammunition. **3.** a type of artillery emplacement in which the gun fires through an opening in the wall of a fortress.

casement window *Building Engineering.* a window that opens on hinges attached to its sash, usually at the top.

Caseobacter *Bacteriology.* a genus of rod-shaped, aerobic bacteria of the order Actinomycetales that occur on certain cheeses.

caseous lymphadenitis *Medicine.* an inflammation of the lymph nodes in which the tissue breaks down into a dry, cheeselike mass.

case I pointing *Ordnance.* direct pointing, laying, or firing of a gun; both the direction and elevation of the gun are set by sighting the target through a sight or telescope. Also, **case I firing.**

case II pointing *Ordnance.* a combination of direct and indirect pointing, laying, or firing of a gun; the direction of the gun is set by sighting through a sight or telescope, while the elevation is set by using an elevation quadrant, range quadrant, or range disk. Also, **case II firing.**

case III pointing *Ordnance.* indirect pointing, laying, or firing of a gun; the direction of the gun is set with an azimuth circle or with the sight or telescope aiming at a point other than the target, and the elevation is set by using an elevation quadrant, range quadrant, or range disk. Also, **case III firing.**

case relation *Artificial Intelligence.* see DEEP CASE.

case shot *Ordnance.* **1.** originally, a hollow, cast-iron projectile filled with musket balls that was designed to explode a short distance in front of enemy troops. **2.** a hollow shell loaded with shrapnel. **3.** a British term for CANISTER.

case structure *Computer Programming.* a control structure provided by many programming languages that defines multiple conditions and corresponding program sections to be executed, depending on which condition is true; used in place of nested "if . . . then" constructs.

case study *Psychology.* the in-depth study of a single individual, family, social unit, or social situation. Thus, **case-study method.**

case-study approach *Anthropology.* a technique in which individuals are studied rather than groups as a whole; it evolved from the influence of psychology through culture and personality studies.

CASE tool *Computer Programming.* a CASE program to aid in constructing, maintaining, or documenting programs.

casework *Psychology.* direct involvement by a social worker with an individual case involving one person or family, with the goal of improving a negative psychological or social situation. Thus, **caseworker.**

cash crop *Agronomy.* a crop that is grown to be sold, rather than to be eaten on the farm or fed to farm animals.

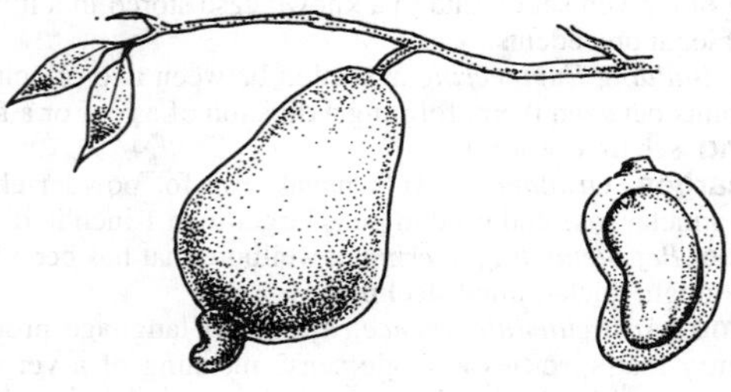

cashew

cashew *Botany.* a tropical evergreen tree, *Anacardium occidentale,* grown for its edible nut.

cashew gum *Materials.* a hard, light brown gum that is obtained from the bark of the cashew tree; used for inks, varnishes, insecticides, and bookbinding gum.

cash machine *Electronics.* another name for an ATM (automated teller machine). See ATM.

Cashmere [kazh´mēr´] *Agriculture.* a breed of small goat that is raised in the Himalayas, known for its fine, silky wool. Also, KASHMIR. *Textiles.* also, **cashmere.** a fine, soft fiber obtained from the undercoat of the Cashmere goat, often blended with wool and used in making high-quality sweaters and coats.

cash register *Mechanical Engineering.* a mechanical or electronic business machine designed to calculate, record, and display the purchase prices and total receipts from a series of commercial sales.

Casimir-du Pré theory *Solid-State Physics.* a theory stating that the lattice and spin systems in a crystal can be treated as distinct thermodynamic systems in thermal contact with each other.

Casimir element *Mathematics.* an element of the universal enveloping algebra of a Lie algebra *g* which commutes with every element of *g*. Always exists when *g* is semisimple.

casing *Engineering.* heavy steel pipe or tubing that is screwed or welded together, lowered and secured into a borehole by cementing; used to stop liquids, gas, or rocks from entering the hole and to prevent the loss of circulation liquid into crevassed or porous ground. *Mechanical Engineering.* a protective covering that encloses machinery, such as that of a boiler. *Building Engineering.* **1.** the formwork for concrete. **2.** a finishing member, covering, or housing surrounding a window or door.

casing centralizer *Petroleum Engineering.* see CENTRALIZER.

casing cutter *Petroleum Engineering.* a heavy, cylindrical, internal-cutting fishing tool bearing knives and run on a string of tubing or drill pipe into a well; as the pipe string is rotated, the knives are forced outward to cut the inner walls of the pipe to release a particular section.

casinghead *Petroleum Engineering.* the top of a casing where the control valves and flow pipes are attached; it protrudes above the surface of the well, thus permitting the pumping operation to occur along with the separation of gas and oil.

casinghead gas *Materials.* the natural gas obtained from an oil well.

casinghead gasoline *Petroleum Engineering.* liquid hydrocarbons that are removed from casinghead gas through refrigeration, absorption, or compression.

casing nail *Mechanical Devices.* a fine nail used in wooden trims, moldings, or the laying of floorboards.

cask see COFFIN.

CASNET *Artificial Intelligence.* a medical expert system applied to the treatment of glaucoma by using a causal network to associate treatments with various diagnostic observations. (An acronym for causal association network.)

casparian strip *Botany.* a bandlike region of primary wall that contains suberin and lignin, typical of endodermal cells in roots.

Caspian Sea *Geography.* the world's largest saltwater lake (about 144,000 sq mi in area), in the USSR and Iran on the traditional border between Europe and Asia.

cassaba see CASABA.

Cassadagan *Geology.* a North American geologic stage of the Upper Devonian period, occurring after the Chemungian and before the Bradfordian.

cassava *Botany.* perennial shrubs of the genus *Manihot,* which are cultivated for their tuberous edible roots. *Food Technology.* a nutritious starch that is derived from the root of this plant; used to make tapioca.

Cassegrain antenna [kas´grān´; kas´ə grān´] *Electromagnetism.* a directional microwave antenna whose feed radiator radiates forward toward a semispherical reflector located at the focus of a dish-shaped main reflector.

Cassegrain focus *Optics.* the pricipal focus in a Cassegrain telescope situated behind the primary mirror.

Cassegrain telescope *Optics.* a telescope in which light rays are reflected from two mirrors, a concave large primary mirror and a smaller convex secondary mirror, causing the image to fold back onto itself.

cassel brown see BLACK EARTH.

Casselian see CHATTIAN.

cassette *Acoustical Engineering.* a small cartridge, usually made of plastic, containing magnetic tape that passes from one reel to another for recording or playback on an audio or video tape recorder. Also, TAPE CASSETTE. *Radiology.* an aluminum or plastic housing for unexposed X-ray film, containing fluorescent intensifying screens between which the film is placed. *Graphic Arts.* a lightproof camera cartridge used to load film into a camera, especially during bright daylight.

cassette mechanism *Genetics.* a mechanism to explain the switching of mating type in yeast; genetic information for both mating types is assumed to be present on a yeast chromosome as silent *cassettes*; a copy of either type of cassette may be transposed to the mating-type locus, where it is *played,* or transcribed.

cassette memory *Computer Technology.* auxiliary magnetic tape storage media; often used to load programs into main memory and to capture local data.

Cassiar orogeny *Geology.* a late Paleozoic deformation of the earth's crust found in the Cordillera of British Columbia.

Cassidulinacea *Invertebrate Zoology.* a superfamily of deep-sea foraminiferan protozoa in the suborder Rotaliina.

Cassiduloida *Invertebrate Zoology.* an order of sea urchins in the subclass Euechinoidea, with five petal-shaped ambulacra around the mouth.

cassidyite *Mineralogy.* $Ca_2(Ni,Mg)(PO_4)_2 \cdot 2H_2O$, a pale to bright green fibrous triclinic mineral of the fairfieldite group, having a specific gravity of 3.1 and an undetermined hardness on the Mohs scale; found as thin crusts and spherules in cavities and cracks of weathered meteorites.

Cassini, Giovanni Domenico 1625–1712, Italian-French astronomer; found four satellites of Saturn and Cassini's Division in its ring.

Cassini projection *Cartography.* a map projection in which arc lengths are plotted as rectangular coordinates on a plane, and are scaled along the central meridian and along a great circle perpendicular to the meridian.

Cassini's Division *Astronomy.* the 2600-kilometer-wide gap in Saturn's rings that separates the A Ring from the B Ring; a tidal resonance with the moon Mimas keeps it relatively free of material.

Cassiopeia [kas´ ē ə pē´ ə] *Astronomy.* the Queen, a bright northern constellation in the Milky Way that lies on the opposite side of the pole from the Big Dipper.

Cassiopeia A *Astronomy.* a strong source of radio emission in Cassiopeia that is believed to be the remnant of a supernova of unknown age lying about 90,000 light-years distant.

cassiterite *Mineralogy.* SnO_2, a brown or black, transparent to opaque tetragonal mineral commonly occurring as twinned short prismatic crystals, having a specific gravity of 6.99 and a hardness of 6 to 7 on the Mohs scale. It is the principal ore of tin, and is found in medium- to high-temperature hydrothermal veins, as well as in metasomatic and alluvial deposits.

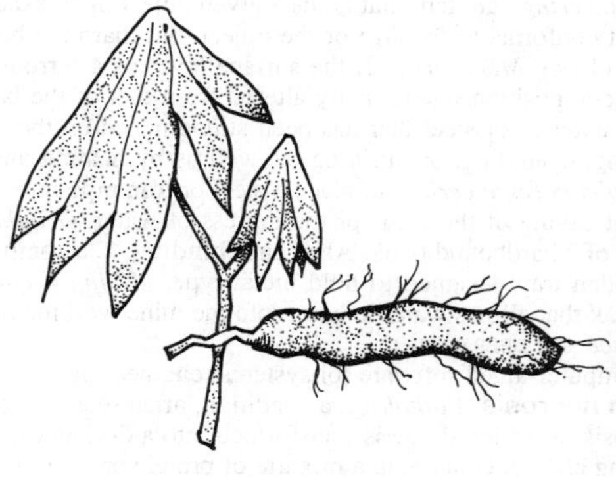

cassava

cassowary [kas´ə wär´ē] *Vertebrate Zoology.* a large flightless bird of Australia and New Guinea with dark, glossy plumage, an unfeathered neck and head, and a bony crest on the head; it belongs to the family Casauriidae in the order Casauriiformes.

cassowary

cast *Medicine.* **1.** a rigid bandage used to prevent movement of a body part after a fracture, dislocation, or other trauma. **2.** a mold of the jaw created to produce dentures and dental replacements. **3.** see STRABISMUS. *Physiology.* a molded substance that forms in the cavities of certain diseased tissues or organs, such as a **renal cast.** *Engineering.* **1.** an object formed by a mold. **2.** to form such an object. **3.** the mold used to form such an object. *Paleontology.* **1. external cast.** a complete replica of an entire fossil organism, usually formed by mineral replacement after the outer shell has dissolved and left a cavity in sediment where it had been buried. **2. internal cast.** a complete internal replica of a fossil organism, composed of sediment or minerals and generally formed by a shell or bones that enclose a void left by the disappearance of the soft parts of the living animal. *Geology.* a sedimentary structure representing the filling in of a mark or depression made on a soft bed and preserved as a solid form on the underside of the overlying stratum. *Optics.* a modification in the appearance of a color when another hue is added.

castability *Materials Science.* a qualitative assessment of the ease in which a molten material can be cast.

caste *Anthropology.* **1.** originally, one of the strict hereditary classes into which the Hindu society of India was divided. **2.** a distinct social group having a certain rank in a hierarchical society; often associated with a particular occupation; e.g., a warrior caste. *Entomology.* a specific group of mature social insects in a colony that carry out special functions; e.g., workers or soldiers.

castellanus *Meteorology.* a cloud type in which a part of the upper portion consists of vertically developed cumulus projections that are taller than they are wide, giving the cloud a turreted appearance.

castellated beam *Mechanical Devices.* a rolled metal beam with a welded web for added strength and depth, used in building construction.

castellated bit *Mechanical Devices.* **1.** a bit with sawtooth or prong-like projecting cutters. **2.** a coring bit with varying length prongs whose tips are coated with diamonds or other hard metal. Also, PADDED BIT.

castellated nut *Mechanical Devices.* a tall lock nut with approximately six projections forming a radial slit pattern wherein a cotter pin, key, or safety wire may be inserted to provide a lock for a correspondingly formed hole for a bolt or screw. Also, CASTLE NUT.

caster *Engineering.* **1.** the tilt of the kingpins in the front wheels of an automobile. **2.** a wheel that turns on a swivel at right angles to its axis, placed on the bottom of refrigerators, heavy furniture, carts, and so on, to provide support and maneuverability.

cast-film extrusion see CHILL-ROLL EXTRUSION.

Castigliano's theorem [kə stig´lē än´ō; kas´til yä´nō] *Mechanics.* a theorem related to the principle of virtual work, applying to a mechanical system of an elastic material obeying Hooke's law and subjected to loads. It states that the displacement of the system in the direction of each load equals the partial derivative of the system's complementary energy with respect to that load. Also, **Castigliano's principle.**

Castile soap *Materials.* **1.** a mild soap made from olive oil and sodium hydroxide. **2.** any of various similar soaps made from fats and oils.

casting *Metallurgy.* **1.** the process of pouring molten metal into a mold and letting it solidify for the purpose of producing a shaped component. **2.** a component produced by casting.

casting alloy *Metallurgy.* an alloy used exclusively for making components that are commercially used in their cast conditions.

casting copper *Metallurgy.* a fire-refined copper used to produce components that are utilized commercially in their cast conditions.

casting out nines *Mathematics.* a method of checking arithmetic computations, based on the fact that the residue of an integer (mod 9) equals the residue (mod 9) of the sum of the digits of the integer.

castings *Geology.* see FECAL PELLETS.

casting shrinkage *Metallurgy.* the shrinkage of a metal or alloy occurring during solidification. Liquid shrinkage occurs at temperatures down to the liquidus, solidification shrinkage from the liquidus to the solidus, and solid shrinkage below the solidus.

casting stress *Metallurgy.* residual stress present in a casting, caused by impeded contraction during cooling. Similarly, **casting strain.**

casting wheel *Metallurgy.* a turning table to which several flat molds are attached and filled in sequence; mainly used to cast anode metals.

cast iron *Metallurgy.* any of many iron-based cast alloy containing more than 2% carbon; i.e., more carbon than steels; known for hardness and heaviness; widely used to make engine blocks, machine parts, stoves, cookware, and various other products..

cast-iron *Mtallurgy.* made of cast iron.

cast-iron front *Architecture.* a prefabricated structural building facade of cast iron, often elaborately molded; widely used in late 19th-century American commercial architecture.

castle *Architecture.* **1.** a large structure characteristic of the European Middle Ages, consisting of a fortified building or buildings surrounded by high, thick walls and usually also a water-filled moat; used by feudal lords as a fortress and residence. **2.** any massive, imposing building that is modeled after or thought to resemble a medieval castle.

castle wheel see CLUTCH WHEEL.

cast loading see MELT LOADING.

Castner cell *Chemical Engineering.* a type of mercury cell used in the commercial manufacture of sodium and chlorine.

Castner process *Chemical Engineering.* an industrial method to create high-purity sodium cyanide by combining sodium, charcoal, and dry ammonia gas to form sodamide; the sodamide is converted to cyanamide which, in turn, is converted to cyanide.

Castniidae *Invertebrate Zoology.* a small family of diurnal butterfly-like moths in the superfamily Castnioidea.

Castnioidea *Invertebrate Zoology.* a superfamily of lepidopteran insects in the suborder Heteroneura.

castoff *Graphic Arts.* the process of determining or estimating the number of pages or lines that a typewritten manuscript will occupy when it is set in type.

Castor *Astronomy.* a multiple star system about 45 light-years away in Gemini, with a spectral type of A and a total apparent brightness of 1.6 magnitude.

castor arabia see SEN.

castoreum gland *Vertebrate Zoology.* a scent gland near the sex organs of a beaver.

Castoridae *Vertebrate Zoology.* the beavers, a family of semiaquatic mammals in the order Rodentia, characterized by their long front teeth, webbed feet, flattened tails, and den and dam building habits.

castorite *Mineralogy.* a transparent variety of petalite.

castor oil *Materials.* a yellow or brown, syrupy, nondrying liquid that melts at –10°C; obtained from the seeds of *Ricinus communis;* used in cosmetics and hydraulic fluids, and as a drying oil in paint; formerly in wide use in medicine as a cathartic.

castor-oil plant *Botany.* a tall tropical shrub, *Ricinus communis,* whose poisonous seed, the **castor bean,** is the source of castor oil.

castration *Medicine.* the removal of the male or female gonads, with the loss of ability to reproduce offspring.

castration anxiety *Psychology.* **1.** anxiety caused by fear of losing the genital organs. **2.** see CASTRATION COMPLEX.

castration complex *Psychology.* **1.** the fear of a male child that his father will castrate him because of the child's sexual feelings for the mother. **2.** the fantasy of a female child that her penis has been removed as a punishment.

cast steel *Metallurgy.* a steel that is commercially used in cast conditions, such as a centrifugally cast steel seamless pipe.

cast stone *Materials.* **1.** any of various precast, artificially manufactured building components such as blocks, sills, and lintels. **2.** precast cement with a facing of a fine material and cement that is meant to look like natural stone.

cast structure *Metallurgy.* the metallographic structure of a cast material.

cast-weld *Metallurgy.* to join metallic parts by pouring molten metal onto them.

casual find *Archaeology.* a nonscientific discovery of an archaeological object, as by a hunter or explorer.

casualty *Military Science.* **1.** a person lost to a military unit as a result of combat with the enemy, including those who are killed, wounded, captured, or listed as missing in action. **2.** any person who is lost to active service by other means, as through disease, noncombat injury or accident, unauthorized absence, and so on.

Casuariidae *Vertebrate Zoology.* the cassowaries, a family of large, flightless birds in the ratite order Casuariiformes, found in Australia and New Guinea, and characterized by black body plumage and an unfeathered but brilliantly colored head and neck.

Casuariiformes *Vertebrate Zoology.* the emus, cassowaries, and kiwis; an order of flightless birds of Australia and New Guinea having only the vestiges of wings and simple, hairlike feathers.

Casuarinaceae *Botany.* a family of trees and shrubs native to Australia and the Pacific Islands, having slender branchlets, reduced flowers in cones, and minute reduced leaves.

Casuarinales *Botany.* a single-family order of trees and shrubs characterized by reduced leaves and reduced unisexual flowers having neither calyx nor petals.

cat *Vertebrate Zoology.* any member of the family Felidae, including the common domestic cat as well as the tiger, lion, leopard, moumtain lion, jaguar, lynx, bobcat, ocelot, and others. *Naval Architecture.* **1.** to hoist an old-fashioned anchor to a vessel's cathead from where it can be stowed or lowered. **2.** the system of blocks and ropes used in hoisting the anchor to the cathead. **3.** a small, shallow-draft, broad-beamed boat, usually with a single large fore-and-aft sail. Also, **catboat.**

cat catalyst.

CAT computerized axial tomography; computed axial tomography; computer-aided testing; California Achievement Test; clear-air turbulence.

cat- or **cata-** a prefix meaning: **1.** down; lower; under. **2.** against. **3.** along with. **4.** very. Also, KAT- or KATA-.

catabiosis *Physiology.* the normal senescence of cells. (A Greek word meaning "a passing life.")

catabolism *Biochemistry.* the part of metabolism involving the breakdown of complex compounds into simpler ones, accompanied by the release of energy. Also, DEVOLUTION, DISSIMULATION, REGRESSION.

catabolite *Biochemistry.* any product of catabolism.

catabolite activator protein *Molecular Biology.* a dimeric protein composed of two identical subunits that, along with cAMP, regulates certain sites on bacterial chromosomes to heighten the frequency with which regions near that binding site are transcribed to messenger RNA. Also, **catabolite gene activator protein.**

catabolite repression *Molecular Biology.* a slowdown or reduction in the synthesis of enzymes involved in the catabolism of sugars, caused by glucose or one of its breakdown products (a catabolite).

cataclasis *Geology.* the process of rock deformation involving the fracture or crushing and rotation of mineral grains without recrystallization or the growth of new minerals.

cataclastic flow *Geology.* a rock deformation consisting of intergranular movement that causes displacement of particles relative to each other.

cataclastic metamorphism *Petrology.* local metamorphism in a region of faults and overthrusts caused by mechanical forces (cataclasis).

cataclastic rock *Petrology.* rock having angular fragments formed by cataclasis. Also, **cataclastite.**

cataclastic structure see MORTAR STRUCTURE.

cataclastic texture *Petrology.* in a dynamically metamorphosed rock, a texture characterized by granular, fragmentary, flattened, crushed crystals.

cataclysmic variable *Astronomy.* a class of variable star that is subject to sudden and unpredictable increases in brightness; it is believed to be a close binary star system in which gas from one star is transferred to the other, causing eruptions.

catacomb or **catacombs** *Architecture.* an underground burial chamber or cemetery that consists of rooms and passageways, such as those of the early Christians of Rome.

cata-condensed polycyclic *Organic Chemistry.* an aromatic compound that has no more than two rings sharing a single carbon atom.

catadicrotism *Cardiology.* an abnormal pulse characterized in a pulse tracing by two small additional waves or notches in the phase of the pulse following the main beat, and indicating two minor expansions of the artery following the main beat.

catadioptric *Optics.* describing an optical system that operates by both reflection and refraction; used to reduce aberrations in telescopes. Thus, **catadioptric telescope, catadioptric imaging system.**

catadromous *Vertebrate Zoology.* of fish, living in fresh water but migrating downriver to spawn at sea.

catagelophobia *Psychology.* an irrational fear of ridicule.

catagen *Physiology.* the brief portion of the hair growth cycle in which growth (anagen) stops and resting (telogen) begins.

catagenesis *Evolution.* an evolutionary change toward less complexity and specialization, resulting in an organism's loss of independence from, and control over, the environment. Also, KATAGENESIS.

Catalan forge *Metallurgy.* a blast furnace for remelting iron, formerly used exclusively in Catalonia and nearby areas.

Catalan numbers *Mathematics.* the numbers $c_n = \binom{2n}{n}/(n + 1)$, with generating function $C(x) = (2x)^{-1}[1 - (1 - 4x)^{1/2}]$. The c_n are among the most frequently occurring combinatorial numbers. They have numerous combinatorial interpretations, including: (a) the number of ways of dissecting a convex polygon of $n + 2$ sides into n triangles by drawing nonintersecting diagonals, (b) the number of ways of completely parenthesizing a product of $n + 1$ terms so that there are two factors inside each set of parentheses, (c) the number of bifurcated rooted planar trees with $n + 1$ endpoints, and (d) in an election with two candidates A and B, the number of ways n votes can come in so that A is never behind B.

catalase *Enzymology.* an enzyme found in the cells of almost all organisms except anaerobic and lactic acid bacteria; it causes the removal of the poisonous compound hydrogen peroxide from the cell.

catalepsy *Psychology.* a trancelike condition in which the person's limbs will remain in any position in which they are placed and there is an apparent loss of sensation and awareness; a condition associated with schizophrenia, epilepsy, hysteria, etc.

cataleptic *Psychology.* **1.** of or relating to catelepsy: *cataleptic* states or conditions. **2.** a person affected by catelepsy.

cataleptic attack *Psychology.* an episode of catalepsy.

catalog or **catalogue** a list of data; specific uses include: *Computer Programming.* a file, usually maintained in auxiliary storage, that contains an ordered list of names and other information about files stored in a computer system or on removable storage media. *Archaeology.* an inventory of archaeological data in which an artifact is labeled with a reference number and then is described in complete detail. *Astronomy.* see STAR CATALOG.

cataloged procedure *Computer Programming.* a set of job control statements associated and cataloged with an application program.

cataloging *Science.* the process of compiling a catalog or of entering data into a catalog. Also, **cataloguing.**

catalog-order device *Electronics.* a device that can be obtained from a manufacturer to upgrade or enhance circuits commonly used in computers.

catalogue equinox *Astronomy.* the date for which the information in a catalog of celestial objects is tabulated.

catalufa *Vertebrate Zoology.* a nocturnal, carnivorous fish of temperate and tropical Atlantic of the family Priacanthidae, having large eyes, rough scales, and a bright red color.

catalysis *Chemistry.* the acceleration of a chemical reaction by a catalyst.

catalyst *Chemistry.* any substance that notably affects the rate of a chemical reaction without itself being consumed or essentially altered. **Positive catalysts** (by far the majority of catalysts) accelerate reactions; **negative catalysts** retard them.

catalyst carrier *Chemistry.* a neutral material, such as alumina or activated carbon, that is used to support a catalyst by increasing its surface area. Also, **catalyst support.**

catalyst selectivity *Chemistry.* a process in which a catalyst affects one compound in a mixture more than (or rather than) others.

catalyst stripping *Chemical Engineering.* the removal of hydrocarbons retained on a catalyst by adding steam at the point where the used catalyst exits the reactor.

catalytic activity *Chemical Engineering.* the space velocity of a test catalyst, relative to that of a standard catalyst, required to achieve a given conversion in a particular reactor.

catalytic converter *Chemical Engineering.* an antipollution device in an automotive exhaust system that uses a catalyst to chemically convert pollutants in the exhaust gases, such as carbon monoxide and unburned hydrocarbons, into harmless compounds.

catalytic cracking *Petroleum Engineering.* a process of breaking down the heavy hydrocarbons of petroleum, using silica or alumina as a catalyst; used to convert heavy oils into lighter and more valuable products. Also, CAT CRACKING.

catalytic dehydrogenation *Organic Chemistry.* a dehydrogenation reaction accomplished by heating a hydroaromatic compound with a catalyst such as platinum, palladium, or nickel.

catalytic hydrogenation *Chemical Engineering.* hydrogenation using catalysts, such as nickel or palladium.

catalytic polymerization *Chemical Engineering.* the use of catalysts to convert low-molecular-weight molecules into polymer molecules of higher molecular weight.

catalytic reforming *Organic Chemistry.* an industrial process for converting alkanes and cycloalkanes into benzene and other aromatic compounds, thus providing the chief raw materials for large-scale synthesis of other classes of compounds.

catalytic site see ACTIVE SITE.

catalyze *Chemistry.* to change or bring about by catalysis.

catamaran *Naval Architecture.* **1.** a ship or boat with twin side-by-side hulls. **2.** a platform set between two pontoons or floats.

catamenia *Medicine.* the menses; menstruation.

catamount *Vertebrate Zoology.* a popular name for the mountain lion, *Felis concolor,* or for various other wild cats of North America.

cat-and-mouse engine *Mechanical Engineering.* an engine in which the pistons move in a circular direction, rather than vertically.

cataphoresis [kat′ə fə rē′sis] *Physical Chemistry.* the movement of charged particles, suspended in an electrolytic medium, toward the cathode, after their exposure to an electric field. Also, ELECTROPHORESIS.

cataplectic *Neurology.* of, relating to, or affected by cataplexy.

catapleiite *Mineralogy.* $Na_2ZrSi_3O_9{\cdot}2H_2O$, a yellow or brown, transparent to opaque hexagonal mineral commonly occurring as twinned thin hexagonal crystals, having a specific gravity of 2.7 to 2.8 and a hardness of 5 to 6 on the Mohs scale; found in alkali rocks and pegmatites with feldspars, nepheline, and other minerals.

cataplexy *Neurology.* an abrupt temporary loss of voluntary muscular function and tone, evoked by an emotional stimulus such as laughter, pleasure, anger, or excitement. Also, **cataplexis.**

Catapochrotidae *Invertebrate Zoology.* a family of coleopteran insects in the superfamily Cucujoidea.

catapult *Ordnance.* **1.** an ancient weapon used to hurl heavy stones, incendiaries, or other objects at enemy fortifications. **2.** a similar mechanical device to hurl grenades or bombs. *Aviation.* **1.** a device providing an auxiliary thrust to help propel a flight vehicle to a safe flying speed in a short distance, especially from an aircraft carrier. **2.** an emergency device for ejecting a crew member safely from the cockpit of an aircraft.

cataract *Medicine.* a partial or complete clouding of the lens of the eye, especially one that impairs vision. *Hydrology.* **1.** a large-volume waterfall whose vertical descent is concentrated in one sheer drop. **2.** a series of steep rapids within a large river. **3.** a large rush of water; a flood.

cataracting *Mechanical Engineering.* the movement of feed material, such as crushed grain, from the top of a load to the bottom after the impact of crushing balls falling from above.

catarrh [kə tär′] *Medicine.* an older term for an inflammation of the mucous membranes of the respiratory tract, causing an effusion of mucus.

catarrhal conjunctivitis *Medicine.* an inflammation of the lining of the eye marked by excessive production of mucus.

catarrhal jaundice *Medicine.* an obsolete term for infectious hepatitis.

catarrhine *Anthropology.* a human or humanoid member of the order Catarrhini. Also, **catarrhinian.**

Catarrhini *Paleontology.* the infraorder of anthropoids that includes humans, apes, and the Old World monkeys; the New World monkeys are now classified in the infraorder Platyrrhini.

catastrophe theory *Mathematics.* a branch of dynamical systems theory dealing with singularities; its best-known result is Thom's theorem.

catastrophic error *Computer Programming.* a programming error that causes the computer to abandon program compilation or execution.

catastrophic failure *Engineering.* **1.** an abrupt breakdown of a machine or system, without prior warning. **2.** a failure within one portion of a system that jeopardizes the operation of the entire system. *Materials Science.* a rapid crack growth in which more energy is available for crack advancement than is needed.

catastrophic waves *Oceanography.* sudden large and violent waves caused by nonperiodic events such as earthquakes or volcanic activity.

catastrophism *Paleontology.* **1.** the view that the earth's geological features, and observable differences in the form and distribution of fossils and archaeological artifacts, were caused by cataclysms that no longer occur, followed by new periods of creation. This idea was generally discredited by the end of the nineteenth century because of increasing knowledge of the fossil record, which led to the acceptance of the principle of uniformitarianism, the view that geological and biological changes of the past were caused by the same physical forces observable in the present. **2.** any of various contemporary theories that attempt to explain certain mass extinctions and geological phenomena as the result of such cataclysmic events as the impact of very large meteorites.

catastrophizing *Psychology.* the tendency of an individual to overreact to mild symptoms of anxiety and to regard them as a prelude to a severe panic attack.

catatonia [kat′ə tōn′ē ə] *Psychology.* a type of schizophrenia that is characterized by such symptoms as stupor, mutism, a complete resistance to instructions and attempts to be moved, rigidity and bizarre posturing, and sometimes excessive or apparently unprovoked excitement. Also, **catatonic schizophrenia.**

catatonic [kat′ə tän′ik] *Psychology.* **1.** of or relating to catatonia. **2.** a person affected by catatonia.

catatricrotism *Cardiology.* an abnormal pulse that is characterized in a pulse tracing by the appearance of three small additional waves or notches in the descending limb, indicating three minor expansions of the artery following the main beat.

catazone *Geology.* the deepest or most intense zone of metamorphism within the earth, marked by high temperatures (500–700°C), strong hydrostatic pressure, and little or no shearing stress. Also, KATAZONE.

catback *Naval Architecture.* a rope used to assist in hooking onto a traditional anchor in the process of catting it.

catbird *Vertebrate Zoology.* any of several species of songbirds known for their catlike mewing, usually dark gray with a black cap and white or red coverts beneath the tail; primarily of the family Mimidae.

catbird

cat block *Naval Architecture.* a heavy block having a hook for engaging the ring on an anchor; used for hoisting the anchor to the cathead.

cat box *Molecular Biology.* a sequence of DNA usually consisting of the sequence CAAT, believed to act as a recognition site for a class of DNA-binding proteins that stimulate transcription of adjacent genes by RNA polymerase II.

catch crop *Agronomy.* **1.** a fast-growing crop that is planted and harvested between two regular growing seasons. **2.** a crop planted after another has failed or when it is too late in the season for the usual crop.

catcher *Electronics.* an element in a velocity-modulated electron tube that abstracts the energy in a bunched electron stream as it passes through.

catching diode *Electronics.* a diode that short-circuits whenever its cathode begins to attract more electrons than its anode, preventing the voltage from rising above the voltage at the cathode.

catching-up *Graphic Arts.* the point in a press run when a printing plate is beginning to accept ink in nonimage areas.

catchment *Archaeology.* the process by which the inhabitants of a village or camp obtain resources from the surrounding area.

catchment area see DRAINAGE BASIN.

catchment glacier see SNOWDRIFT GLACIER.

catch pit *Mining Engineering.* in mineral processing, a pit at the lowest point of the floor in a mill, to which spillage gravitates or is hosed.

cat cracker *Petroleum Engineering.* a petroleum refinery vessel in which catalytic cracking takes place.

cat cracking see CATALYTIC CRACKING.

cat cry syndrome see CRI-DU-CHAT SYNDROME.

catechol [kat′ə kōl′] *Organic Chemistry.* $C_6H_4(OH)_2$, combustible, colorless prisms; soluble in alcohol; melts at 104–106°C and boils at 240–245°C; used as an antiseptic and in photography, electroplating, and organic synthesis.

catecholamine *Biochemistry.* an amine compound derived from the amino acid tyrosine that affects the activities of the sympathetic nervous system. Also, ADRENERGIC AMINE.

categorical data *Statistics.* any data arising from counts rather than measurements.

category *Mathematics.* **1.** a class C of objects together with (a) a class of disjoint sets, denoted hom(A, B), for each pair A and B in C, the elements of which are morphisms, and (b) a composition rule for the morphisms that is associative and has an identity. Examples of categories include (a) the class S of all sets; for A, B in S, hom(A, B) is the set of all functions from A to B. (b) the class G of all groups; hom(A, B) is the set of all group homomorphisms from A to B. (c) the one-element class consisting of any multiplicative group G; hom(G, G) is the set of elements of G and composition is the group operation. (d) the class of all partially ordered sets under a given order; hom(A, B) consists of all functions f from A to B such that if $x \leq y$ then $f(x) \leq f(y)$, for x and y in A. **2.** a subset A of a metric space M that is the union of a countable family of nowhere dense subsets is said to be of the **first (Baire) category**. Any set not of the first category is of the **second (Baire) category.**

category theory *Mathematics.* the study of categories and their morphisms. Concepts from category theory are often expressed in terms of commutative diagrams.

catelectrotonus *Physiology.* the local depolarization and increased irritability of a nerve that is proximate to a negatively charged electrode.

catena [kə tēn′ə] *Computer Programming.* a listed series or chain, usually consisting of data items. *Geology.* a group of soils in a given area that are derived from a common parent material, but show differences resulting from variations in local topology or drainage patterns.

catenane *Organic Chemistry.* a compound with interlocking rings, resembling the links of a chain, that are not chemically bonded but that cannot be separated without breaking at least one valence bond.

catenary *Mathematics.* the graph of the equation $y = a \cosh(x/a)$. Physically interpreted, a catenary is the plane curve under the force of gravity (the y-direction) in which a uniform flexible cable hangs when suspended from two points, where $a > 0$ is the y-intercept.

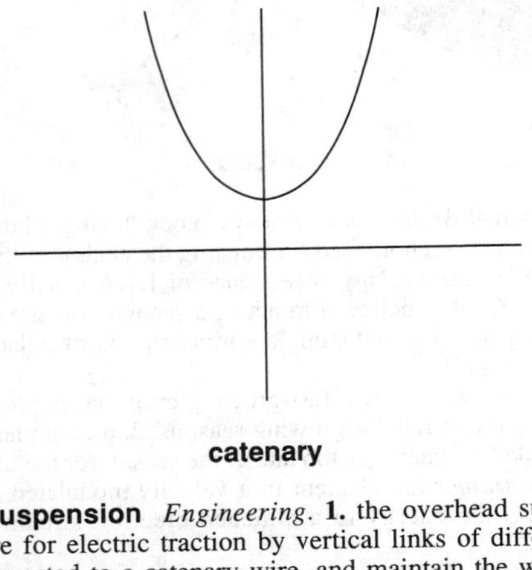

catenary

catenary suspension *Engineering.* **1.** the overhead suspension of contact wire for electric traction by vertical links of different lengths that are connected to a catenary wire, and maintain the wire at a constant height, especially as used in airships to fasten the cabin to the bag. **2.** any suspension that carries items such as pipelines across space.

catenation *Chemistry.* the self-linking of certain elements, especially carbon, to form chains or rings.

Catenipora *Paleontology.* a genus of tabulate corals in the family Halysitidae and subfamily Cateniporinae; associated with dasyclad algae in middle Ordovician to upper Silurian strata.

catenoid *Mathematics.* the surface of revolution generated by the rotation of a catenary about the y-axis.

catenulate *Biology.* **1.** chainlike or ringed in form, as the arrangement of some chromosomes. **2.** composed of individual units joined in a line.

Catenulida *Invertebrate Zoology.* an order of colorless, threadlike freshwater animals.

caterer problem *Mathematics.* a linear programming problem that seeks the optimal policy for a caterer who may choose to buy new napkins as needed or to send used napkins to either a fast or slow laundry service.

caterpillar

caterpillar *Invertebrate Zoology.* **1.** the prolegged wormlike larval stage of a butterfly or moth. **2.** the larvae of certain insects, such as scorpion flies and sawflies.

Caterpillar *Mechanical Engineering.* a trade name for a tractor that runs on two continuous belts, which wrap around the vehicle's toothed driving wheels.

caterpillar chain *Mechanical Devices.* a driving chain fitted with teeth to engage a conveyer chain.

catfish *Vertebrate Zoology.* the common name for several families of fish in the suborder Siluriformes of the order Cypriniformes, having sensory barbels at or near the mouth; mostly freshwater but with some ocean species.

catforming *Chemical Engineering.* a naphtha-reforming method using a catalyst composed of platinum, silica, and alumina.

cat gene *Genetics.* a gene that encodes chloramphenicol acetyltransferase.

catgut *Materials.* a strong cord made from the intestines of animals, originally cats but now especially sheep and other animals; used for surgical sutures, and cords for violins and tennis rackets.

cath cathode.

cath- a prefix meaning: **1.** down; lower; under. **2.** against. **3.** along with. **4.** very.

catharsis [kə thär′sis] *Psychology.* the expression of repressed thoughts, feelings, and experiences from the unconscious, accompanied by emotional relief and release.

cathartic [kə thär′tik] *Pharmacology.* **1.** any substance that causes evacuation of the bowels, whether by stimulating peristaltic action or by increasing bulk. **2.** causing evacuation of the bowels. *Psychology.* relating to or producing a catharsis.

Cathartidae *Vertebrate Zoology.* the New World vultures, including the king vultures, condors, and turkey buzzards; a family of carnivorous birds in the order Falconiformes, having hooked beaks and powerful claws and usually feeding on carrion.

cathead or **cat head** *Naval Architecture.* a beam or girder projecting from either side of a wooden sailing ship's bow, used in "catting" or securing the anchor when raised. *Petroleum Engineering.* an extension of the drawworks drum used for lifting items of great weight at the rig.

cathead line see CATLINE.

cathepsin *Biochemistry.* an enzyme that breaks down proteins into smaller substances, such as proteoses and peptones, in animal tissue.

catheter [kath′ə tər] *Medicine.* a hollow, usually flexible tube used to drain a body cavity, commonly the bladder, or to instill liquids for diagnostic testing or treatment.

catheterization [kath′ə tər ə zā′shən] *Medicine.* the placement of a catheter into a body sinus such as the bladder.

catheterostat *Surgery.* a stand for holding and sterilizing catheters.

cathetometer [kath´ə täm´i tər] *Optics.* a comparator having a telescope or microscope that slides up and down a vertical column carrying vernier markings; used to measure vertical distances of near objects and to determine the precise inclination of the telescope. Also, READING MICROSCOPE, READING TELESCOPE.

cathode *Electricity.* **1.** a negative electrode, such as the negative electrode of a battery or electrolytic capacitor. **2.** an electrode through which a primary stream of electrons enters the interelectrode space. *Physical Chemistry.* the electrode in an electrochemical cell at which reduction occurs; usually negative with respect to the anode. *Electronics.* the terminal of a forward-biased semiconductor diode that is negative with respect to the other terminal. Also, NEGATIVE ELECTRODE.

cathode bias *Electronics.* a method of restricting the flow of electrons into an electron tube so that the rate of electric current stays near ground level.

cathode copper *Metallurgy.* commercially pure copper that is deposited at the cathode by electrorefining anode copper, or by electrowinning copper from a solution.

cathode corrosion *Metallurgy.* **1.** in an electrochemical operation, the corrosion occurring at the cathode. **2.** corrosion occurring at the cathodic element of a bimetallic couple immersed in an electrolyte.

cathode-coupled amplifier *Electronics.* an amplifier having two vacuum tubes whose cathodes are connected to a common resistor, with the input applied to the grid of one tube and the output obtained at the anode of the other (or at both anodes).

cathode coupling *Electronics.* the process of adding input or output elements to a cathode circuit for coupling energy to another stage.

cathode dark space *Electronics.* the relatively nonluminous region of a cathode between the cathode glow and the negative glow in a glow-discharge cold-cathode tube. Also, DARK SPACE, CROOKES' DARK SPACE.

cathode disintegration *Electronics.* a process in which a cathode is stripped of its emissive coating.

cathode drop *Electronics.* the difference of potential due to the space charge near the cathode. Also, CATHODE FALL.

cathode efficiency *Chemical Engineering.* the current needed at the cathode compared to the theoretical current needed in an electrochemical process.

cathode emission *Electronics.* the process by which electrons flow from the cathode.

cathode fall see CATHODE DROP.

cathode follower *Electronics.* a circuit in which the input is applied to the grid, and the output is taken from the cathode, which may be at ground potential. Also, GROUNDED-PLATE AMPLIFIER.

cathode glow *Electronics.* the luminosity surrounding the cathode in a gas-discharge tube, such as a neon light, as it operates at low pressure.

cathode interface capacitance *Electronics.* a capacitance value that appears as part of the cathode interface impedance. Also, LAYER CAPACITANCE.

cathode interface impedance *Electronics.* a restriction of the flow of electric current between the cathode base and coating, which often results from a poor mechanical bond. Also, LAYER IMPEDANCE.

cathode modulation *Electronics.* a method of changing the frequency of a signal by applying a modulated voltage to the cathode circuit.

cathode ray *Electronics.* a beam of electrons produced when the filament in a tube is heated or bombarded by positive ions.

cathode-ray oscilloscope *Electronics.* an instrument that uses a cathode-ray tube to display signal waveforms.

cathode-ray output *Computer Technology.* any alphanumeric or graphic computer output that is displayed on a CRT screen.

cathode-ray tube *Electronics.* a tube whose electron beam can be focused to present alphanumeric or graphical data on an electroluminescent screen; widely used in television receivers and cameras, computer monitors, radar screens, and the like. Also, CRT.

cathode resistor *Electronics.* a resistance opposing the flow of electrons from the cathode in an electron tube to generate an adequate voltage level.

cathode spot *Electronics.* the region of a cathode from which electrons are emitted in an arc, and in which the current density is much higher than in simple thermionic emission.

cathodic [kə thäd´ik; kə thōd´ik] relating to or located at a cathode.

cathodic coating *Materials Science.* an electrodeposited polymeric coating in which the electrolyte is a cationic polymer.

cathodic control *Materials Science.* the polarization of the cathode of a galvanic cell, in which the cathode potential primarily determines the reaction rate.

cathodic inhibitor *Chemical Engineering.* a compound used to coat a metal surface with a thin film at the cathodes in order to protect the entire surface against corrosion in a conducting medium.

cathodic migration *Physical Chemistry.* the movement of a positively charged electric particle toward the negative pole (cathode) in an electric field.

cathodic polarization *Physical Chemistry.* a reduction in current from the cathode in an electrochemical cell, such as a battery, usually caused by excessive layers of hydrogen that increase resistance in the cell.

cathodic polymer see CATIONIC POLYMER.

cathodic protection *Metallurgy.* the protection of a metallic material from corrosion, effected either by coupling such material with a less noble metal or by impressing a current.

cathodoluminescence *Electronics.* a luminescence produced when metal in an electron tube emits radiation after the metal has been bombarded by high-velocity electrons.

cathodophosphorescence *Electronics.* an afterglow that follows the emission of radiation in an electron tube.

catholyte *Chemistry.* the part of an electrolyte that is immediately adjacent to the cathode.

cat ice see SHELL ICE.

cation [kat´ī ən] *Chemistry.* any positively charged ion, a species that moves to the cathode during electrolysis. Also, KATION.

cation analysis *Analytical Chemistry.* a test for positively charged particles (cations) in an aqueous solution.

cation exchange *Chemistry.* a reversible chemical reaction in which positive ions in a solution are exchanged for the equivalent ions in a solid. *Geochemistry.* see ION EXCHANGE.

cation-exchange resin *Analytical Chemistry.* a type of resin used in chromatography in which the positive ions on the immobilized (stationary) phase can be exchanged for cations in the solute (mobile phase).

cationic *Chemistry.* relating to or resembling a cation.

cationic hetero atom *Chemistry.* a positively charged atom of an element other than carbon within a chain made up largely of carbon atoms.

cationic polymer *Materials Science.* a positively charged polymer in solution, which coats on the cathode during electrodeposition. Also, CATHODIC POLYMER.

cationic polymerization *Materials Science.* the growth of a polymer chain by the addition of monomer units to a negatively charged growing chain. *Organic Chemistry.* a chain reaction polymerization initiated by a Lewis acid. Also, CARBOCATIONIC POLYMERIZATION.

cationic reagent *Chemistry.* any surface-active substance whose active principle is the positive ion.

cationic softener *Materials.* an ion-exchange agent used to soften water by substituting sodium for calcium and magnesium ions.

cationic starch *Materials.* a starch with a stable negative polarity on the molecules, giving greater adhesion to cellulose fibers in paper or textiles.

cationic surfactant *Organic Chemistry.* a large molecule whose positively charged polar end interacts with water, while its nonpolar end does not.

cationtrophy *Chemistry.* the removal of a positive ion from a molecule, leaving a negative ion in equilibrium.

cativo *Materials.* the nondurable, fine-textured, rust-colored wood of the *Prioria copaifera* tree of Central and South America; used for interior trim, veneering, plywood, and furniture.

catkin *Botany.* a large, drooping spike of unisexual flowers found in woody plants and adapted for aerial pollination.

catline *Petroleum Engineering.* in oil-well drilling, a rope attached to a cable used on the cathead to lift heavy objects. Also, CATHEAD LINE.

catlinite see PIPESTONE.

catoptric *Optics.* **1.** describing an optical system that operates only by reflection. **2.** referring to light reflected from a mirror, such as light from a filament focused onto a parallel beam with a reflector. Also, KATOPTRIC.

catoptric system *Optics.* an optical system in which images are produced by curved-surface mirrors.

catoptrite see KATOPTRITE.

Catoscopiaceae *Botany.* a monotypic family of small, circumboreal mosses of the order Bryales; characterized by tristichous leaves, abundant brown tomentum, long setae, and small, curved black capsules.

Catostomidae *Vertebrate Zoology.* the suckers, buffalo fish, and redhorse fish; a family of mostly bottom-dwelling freshwater fishes in the order Cypriniformes, found in North and Central America and characterized by a ventral mouth with fleshy lips.

CAT scan *Radiology.* the use of a computer to process data from a tomograph for a reconstruction display of the body in cross section on a screen. (Short for computerized axial tomography scan.)

cat-scratch disease *Medicine.* an infectious disease that spreads locally around the site of a cat scratch or bite; it is usually mild and may cause a slight fever and inflamed lymph nodes. Also, **cat-scratch fever.**

catshaft *Petroleum Engineering.* the shaft or axle on the drawworks where the catheads are mounted; one cathead is a drum acting as a capstan for light hoisting, and the other is a drum with a manual or air-actuated, quick-release friction clutch.

cat's paw *Meteorology.* a light breeze that affects a small area, such as one causing slight ripples on the surface of still water.

CATT or **CATTT** controlled avalanche transit-time triode.

Cattanach's translocation *Genetics.* a translocation in the mouse discovered by B. M. Cattanach, in which a segment of chromosome 7 is inserted into an X chromosome, resulting in partial inactivation of the autosomal genes during inactivation of a copy of the X chromosome.

Cattell, James M. 1860–1944, American psychologist and editor of scientific journals.

Cattell, Raymond born 1905, American psychologist; noted for studies of intelligence and of personality traits.

Cattell infant intelligence scale *Psychology.* an intelligence test, intended for children of 2 to 30 months of age, based on the revised Stanford-Binet test.

cattle *Agriculture.* a term for bovine animals as a group. (From a Middle English word meaning "personal property" or "valuable property." In early times, a person's wealth was often measured by the amount of livestock owned.)

cattle plague see RINDERPEST.

cattle tick fever see BABESIASIS.

CATV cable television; community antenna television.

catwalk *Engineering.* a narrow pathway, usually of wood or metal, that gives access to parts of large machines, as from the midship portion of a ship to the bow or stern, over a large printing press, along the inside keel of an airship, or in large, open building areas.

catwhisker *Electronics.* a small, sharply pointed and flexible wire that is used to establish contact with a sensitive point on the surface of a semiconductor.

Caucasian or **caucasian** see CAUCASOID.

Caucasoid or **caucasoid** *Anthropology.* **1.** a term used to describe a wide variety of groups of people residing or originating in Europe, the Middle East, and North Africa, typically having relatively light skin coloring. **2.** of or relating to such people.

Caucasus *Geography.* a mountain range in the southwestern USSR, between the Black and Caspian seas on the traditional border between Europe and Asia; highest peak: Mount Elbrus (18,481 ft).

Cauchy, Baron Augustin 1789–1857, French mathematician; made advances in wave propagation, convergency, elasticity, and calculus.

Cauchy-Bunyakovskii-Schwartz inequality see CAUCHY-SCHWARZ INEQUALITY.

Cauchy condensation test *Mathematics.* a monotone decreasing series $\sum_{n=1}^{\infty} a_n$ of positive terms converges if and only if $\sum_{n=1}^{\infty} p^n a\binom{n}{p}$ converges for each positive integer p.

Cauchy dispersion formula *Optics.* the relationship between the refractive index n of a medium and the wavelength λ of light.

Cauchy distribution *Statistics.* a symmetric probability distribution, generally given as $f(x) = 1/[\pi(1 + x^2)]$, where x is a real variable. It can be seen as a special case of Student's distribution with one degree of freedom.

Cauchy principal value *Mathematics.* **1.** let f be bounded on an interval (a, b) except possibly at some point c in the interval. Then the Cauchy principal value of the (improper) integral of f from a to b is $\lim_{\delta \to 0^+} [\int_a^{c-\delta} f(x)dx + \int_{c+\delta}^{a} f(x)dx]$ if this limit does exist. Even when the Cauchy principal value of $\int_a^b f(x)dx$ exists, the improper integral itself may not be finite; also denoted by $(\text{CPV})\int_a^b f(x)dx$. **2.** if f is Riemann integrable over every finite real interval, the Cauchy principal value of the infinite integral of f over R is defined as $\lim_{r \to \infty} \int_{-r}^{+r} f(x)dx$ provided this limit exists. Also denoted by $(\text{CPV})\int_{-\infty}^{\infty} f(x)dx$.

Cauchy problem *Mathematics.* **1.** the problem of finding a solution $y(x)$ of the ordinary differential equation $F[x, y(x), y'(x), \ldots, y^{(r-1)}(x)]$ of order r, given the initial conditions $y(x_0) = y_0$, $y'(x_0) = y_0^{(1)}, \ldots, y^{(r-1)}(x_0) = y_0^{(r-1)}$. **2.** more generally, the problem of finding a solution to a system of partial differential equations of order m, given a surface S, the initial conditions on the values of the solution, and initial conditions on the derivatives of order less than m on S.

Cauchy product *Mathematics.* the formula for the product of two infinite series: $(\sum_{n=0}^{\infty} a_n)(\sum_{n=0}^{\infty} b_n) = \sum_{n=0}^{\infty}(\sum_{k=0}^{n} a_k b_{n-k})$. This formula will hold when one of the two series $\sum_{n=0}^{\infty} a_n$ or $\sum_{n=0}^{\infty} b_n$ is absolutely convergent, although the other need only be convergent. In this case, the product series is not necessarily absolutely convergent.

Cauchy product theorem *Mathematics.* if each of the series $\sum_{n=0}^{\infty} a_n$ and $\sum_{n=0}^{\infty} b_n$ is absolutely convergent with the sums A and B, respectively, then their product converges absolutely to $A \cdot B$.

Cauchy radical test See CAUCHY ROOT TEST.

Cauchy ratio test *Mathematics.* the theorem that the series $\sum_{n=0}^{\infty} a_n$ converges absolutely or diverges as $\lim_{n \to \infty} |a_n/a_{n-1}| < 1$ or > 1. If this limit equals one, then the test fails. Also, RATIO TEST.

Cauchy relations *Solid-State Physics.* a set of six relations that are satisfied by solids whose atoms interact with forces acting along the line connecting them, and in which each atom is considered as a source of symmetry.

Cauchy residue theorem *Mathematics.* if f is complex-valued and analytic in a simply connected domain D of the complex plane containing a curve C except at a finite number of poles $a, b, c, \ldots$ within C, then the line integral of f around C is the sum of the residues of f at those poles; that is, $(1/2)\pi i \int_C f(z)\,dz = a_{-1} + b_{-1} + c_{-1} + \cdots$ (finite sum).

Cauchy-Riemann equations *Mathematics.* the pair of partial differential equations $\partial u/\partial x = \partial v/\partial y$ and $\partial u/\partial y = -\partial v/\partial x$. The Cauchy-Riemann equations are satisfied in a domain D if $f(z) = f(x+iy) = u(x,y) + iv(x, y)$ is analytic in D. If the partial derivatives $\partial u/\partial x$, $\partial v/\partial y$, $\partial u/\partial y$, and $\partial v/\partial x$ are continuous in D, then the Cauchy-Riemann equations are sufficient conditions that $f(z)$ be analytic in D.

Cauchy root test *Mathematics.* the theorem that the series $\sum_{n=0}^{\infty} a_n$ converges absolutely or diverges as $\lim_{n \to \infty}(|a_n|)^{1/n} < 1$ or > 1. If the limit equals 1, then the test will fail. Also, CAUCHY RADICAL TEST, CAUCHY'S TEST FOR CONVERGENCE, ROOT TEST.

Cauchy-Schwarz inequality *Mathematics.* the inequality $(ac + bd)^2 \leq (a^2 + b^2)(c^2 + d^2)$ for real numbers a, b, c, d. More generally, in a normed vector space with inner product (•), and norm $\| \|$, the inequality $|(x, y)| \leq \|x\| \cdot \|y\|$, for all vectors x and y. Equality holds if, and only if, x and y are linearly dependent. Also, CAUCHY-BUNYAKOVSKII-SCHWARTZ INEQUALITY.

Cauchy sequence *Mathematics.* a sequence $\{a_n\}$ with the property that for every $\varepsilon > 0$, there is an integer N such that $|a_n - a_m| < \varepsilon$ for all $m, n > N$.

Cauchy's method *Mathematics.* a method of solving the inhomogeneous linear differential equation of order r on an open interval: $f(x) = y^{(r)}(x) + f_{r-1}(x)y^{r-1}(x) + \cdots + f_1(x)y'(x) + f_0(x)y(x)$, where the coefficients $f_i(x)$ are assumed to be continuous on the open interval in question. If $y = c_1y_1 + \cdots + c_ry_r$ is the the general solution to the corresponding homogeneous equation, one determines the coefficients c_i so that $y(a) = 0$, $y'(a) = 0, \ldots, y^{r-2}(a) = 0$, and $y^{r-1}(a) = f(a)$, where a is an arbitrary parameter. If $y(x,a)$ is a solution thus obtained, then a particular solution to the inhomogeneous equation is $y(x) = \int y(x, a)\,da$, where the integral is evaluated from x_0 to x and for which $y(x_0) = y'(x_0) = \cdots = y^{r-1}(x_0) = 0$.

Cauchy's test for convergence see CAUCHY ROOT TEST.

Cauchy's theorem *Mathematics.* if G is a finite group whose order is divisible by a prime p, then G contains an element of order p.

cauda *Anatomy.* any structure resembling a tail.

cauda equina *Anatomy.* the base of the spinal cord and the associated roots of the spinal nerves that emanate below the first lumbar nerve.

caudal *Zoology.* relating to or having a tail.

caudal artery *Vertebrate Zoology.* a large dorsal artery in the tail of a vertebrate.

caudal vertebrae *Vertebrate Zoology.* the vertebrae that support the tail in a tailed vertebrate.

Caudata *Vertebrate Zoology.* the salamanders, an order of amphibians with a long body, a head, trunk, and tail, and four limbs, and usually having aquatic eggs and larva with gills.

caudate *Vertebrate Zoology.* relating to or having a tail or cauda.

caudate lobe *Anatomy.* one of four lobes of the liver, located just in front of the vena cava and connecting the right and left lobes of the liver.

caudate nucleus *Anatomy.* the most frontal of the basal ganglia; damage to caudate neurons is seen in Huntington's chorea and other motor disorders.

caudex *Botany.* **1.** the persistent, usually underground base stem and root system of a plant. **2.** the trunk of a fern or palm.

caudicle *Botany.* on many orchids, the threadlike stalk that attaches the pollen mass to the sticky disk.

caul- or **-caul** *Botany.* a combining form meaning "stem."
cauldron *Volcanology.* **1.** a caldera formed by magmatic erosion along a ring fault at the base of an intrusion. **2.** broadly, any volcanic subsidence structure.
cauldron subsidence *Volcanology.* **1.** a structure formed by the sinking of a cylindrical mass of crust into a magma chamber along a ring fault, sometimes accompanied by the rising of magma to form a ring dike. **2.** the process by which such a structure is formed.
caulerpa *Botany.* a genus of green algae constituting the family Caulerpaceae, occurring on tropical sea bottoms and having a thallus branching into rhizoids below and foliage-like expansions above.
Caulerpaceae *Botany.* a monogeneric family of green algae in the order Siphonales, having a single-celled thallus and reproducing both sexually and asexually.
caulescent *Botany.* having a stem above ground.
caulid *Botany.* the stemlike supporting member of a moss or liverwort.
cauliflorous *Botany.* bearing flowers on the trunk or on the main stems.
cauliflory *Botany.* the production of flowers on thickened tissue such as the trunk or branch of a plant, rather than at the apex of a meristem; seen especially in plants growing below a tropical canopy, such as *Theobroma cacao* (cocoa).

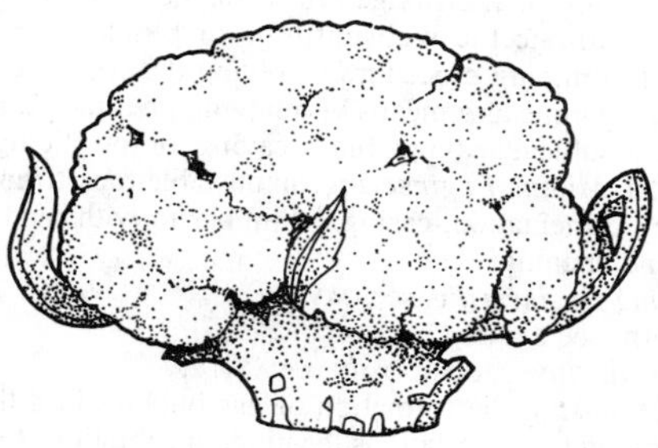

cauliflower

cauliflower *Botany.* a biennial plant, *Brassica oleracea,* of the mustard family that is cultivated for its edible parts, which are modified flowers.
cauliflower disease *Plant Pathology.* **1.** a disease of strawberries caused by the attack of the eelworm and characterized by the formation of abnormally shaped leaves that form clusters resembling cauliflowers. **2.** a plant disease, particularly common to strawberries, that is caused by the bacterium *Corynebacterium fascians.*
Caulimovirus group *Virology.* a group of dsDNA-containing plant viruses, having a narrow host range, associated primarily with cytoplasmic proteinaceous inclusion bodies, and transmitted mechanically and by aphids.
cauline *Botany.* relating to or growing out from a stem, especially the top of the stem.
caulk or **calk** *Building Engineering.* **1.** to apply a puttylike material in order to seal the seams, cracks, or similar openings of a window frame, plumbing fixture, etc. **2.** see CAULKING COMPOUND. *Naval Architecture.* to apply oakum between the planks of a boat in order to make the vessel watertight. *Engineering.* to close the spaces between overlapping riveted plates or other joints by hammering the exposed edge of one plate into close contact with the other or by packing it with a filler material.
caulking *Building Engineering.* **1.** the process of lining a seam or joint, as of tile on a tub or shower, with a caulking compound, or with lead, oakum, or dry-pack paste, in order to make it moistureproof, waterproof, or airproof. **2.** see CAULKING COMPOUND.
caulking compound *Building Engineering.* a puttylike material applied in certain caulking processes. Also, CAULKING, CAULK.
caulking gun *Mechanical Devices.* a caulking device consisting of a hollow cylinder filled with caulking material, which is dispensed by pressing a trigger or lever at the base.
caulking tool *Mechanical Devices.* a steel-step tool like a cold chisel but with a blunt edge; used with a hammer in deforming metal operations, such as closing the plates of a boiler. Similarly, **caulking iron.**
Caulobacter *Bacteriology.* a genus of Gram-negative, aerobic bacteria distinguished by the possession of a stalk or holdfast body; found in soil and fresh water containing organic matter.
Caulobacteraceae *Bacteriology.* a former classification for a family of rod-shaped bacteria of the order Pseudomonadales, occurring in fresh or salt water either as single cells or in short chains.
caulocarpic *Botany.* of a stemmed plant, flowering and bearing fruit annually.
Caulococcus *Bacteriology.* a genus of manganese-depositing bacteria of uncertain classification that are typically found in mud and lake bottoms.
caulome *Botany.* a plant's stem structure.
causalgia *Neurology.* neuralgia accompanied by an intense burning sensation, redness, and swelling of the affected area, due to injury of a peripheral nerve. Also, THERMALGIA.
causality *Science.* the relationship between a cause and its effect, or between correlated events, particularly natural phenomena. *Physics.* **1.** a hypothesis stating that a set of precisely determined conditions will always repeat the same effects. **2.** a principle stating that an event cannot precede its cause. *Mechanics.* a principle stating that if the values of the dynamical variables of a system are specified at a given time, and if the actions of all external forces on the system at all later times are known, then the values of the system variables at later times are exactly determined. Also, LAPLACE DETERMINISM. *Quantum Mechanics.* the principle that the occupation probablilities of future quantum states and the expected values of observables can be determined exactly from knowledge of the present state of the system.
causal system *Control Systems.* a system in which the response to an input signal is not dependent upon future input values. Also, NONANTICIPATORY SYSTEM, PHYSICAL SYSTEM.
caustic *Chemistry.* **1.** caustic soda (sodium hydroxide), NaOH. **2.** any hydroxide of a light metal, or any other compound that is chemically similar to sodium hydroxide. **3.** of an alkaline substance, being corrosive or irritating to living tissue. *Physics.* a curve or surface tangent to adjacent orthogonals of waves that are reflected or refracted from a surface. *Optics.* a surface fashioned from the outer envelope of light rays near an image. *Mathematics.* the image of a Lagrangian submanifold of the cotangent bundle of a manifold M in which the projection onto M has become singular.

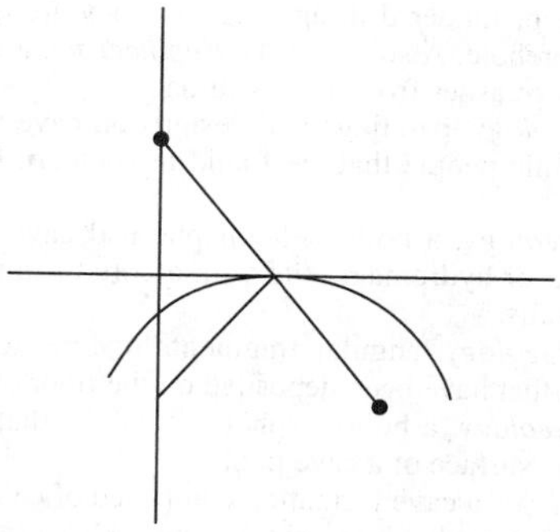

caustic

caustic alcohol see SODIUM ETHYLATE.
caustic dip *Metallurgy.* an alkaline bath used to neutralize acidity present on the surface of a metallic product, for etching or for removing contaminants such as grease or paint.
caustic embrittlement *Metallurgy.* the embrittlement of a stressed steel, caused by a caustic solution at moderate temperatures.
caustic injection *Petroleum Engineering.* the injection of aqueous solutions of caustic and polymer down a well to help increase oil recovery, as when oil production has declined due to reduced production from the formation.
causticity *Chemistry.* the property of being caustic (corrosive).
causticization *Chemical Engineering.* the process of producing an alkaline material such as sodium hydroxide.
caustic lime see CALCIUM OXIDE.
caustic metamorphism *Volcanology.* a geomorphological change caused by the passage of a lava flow or dike.
caustic potash *Inorganic Chemistry.* a popular name for potassium hydroxide, KOH, a widely used commercial caustic. See POTASSIUM HYDROXIDE.
caustic soda *Inorganic Chemistry.* a popular name for sodium hydroxide, NaOH, a widely used commercial caustic. See SODIUM HYDROXIDE.
caustic treater *Chemical Engineering.* a vessel in which acidic gases or liquids are contacted and neutralized by an alkaline solution.

caustic wash *Chemistry.* **1.** a water solution of caustic soda. **2.** the use of such a solution to remove impurities.
caustobiolith *Geology.* a combustible organic rock, such as coal peat, of plant origin.
caustolith *Geology.* a combustible rock derived from organic and/or inorganic materials.
cauterization *Surgery.* the electrical or chemical means of destroying tissue in a lesion, or of sealing blood vessels to prevent bleeding.
cauterize *Surgery.* to carry out a process of cautery.
cautery *Surgery.* **1.** the application of heat, electric current, or chemicals in order to scar, burn, or cut skin or tissues. **2.** an agent or instrument used for cautery.
cautery knife *Surgery.* a device, usually electric, designed to incise tissue with minimal bleeding, and used to cauterize vessels.
caution area *Navigation.* an area that is navigable, but in which special care should be taken due to the presence of nearby hazards.
cautious control *Control Systems.* a method of control that uses hedging and lower gain when dealing with uncertain conditions.
cav cavity.
cavaburd *Meteorology.* a thick fall of snow in the Shetland Islands. Also, KAVABURD.
cavaliers *Meteorology.* a local name in the vicinity of Montpelier, France, for the days near the end of March or the beginning of April when the mistral is usually at its peak.
cavalry *Military Science.* **1.** originally, a military unit operating on horseback. **2.** a mobile military unit utilizing armored vehicles.
caval valve *Anatomy.* the valve of the inferior vena cava, extending from the junction of the inferior vena cava with the right atrium toward the anterior part of the fossa ovalis of the heart.
cavascope *Surgery.* a device used to light and examine a cavity. Also, CELOSCOPE.
cave *Geology.* **1.** a naturally or artificially formed hollow area, chamber, or series of chambers beneath the surface or in the side of a mountain or hill, often having an opening to the surface. **2.** to collapse, as an excavation or mine. *Mining Engineering.* any fragmented rock materials that obstruct the hole or hinder drilling progress; they are derived from the sidewalls of a borehole. Also, CAVINGS. *Engineering.* a cavity that holds the accumulation of ashes from a glass furnace.
cave art *Archaeology.* paintings and designs on cave walls, especially those by Paleolithic people that are found in southern France and elsewhere in Europe.
cave blister *Geology.* a hollow, hemispherical cave formation composed of gypsum or hydromagnesite with a core of mud, and found attached to cave walls.
cave breccia *Geology.* angular fragments of limestone produced by cave breakdown that have been deposited on the floor of a cave.
cave bubble *Geology.* a hollow sphere of calcite that forms around a gas bubble on the surface of a cave pool.
cave coral *Geology.* a cave formation composed of calcite and having a knobby appearance, so that it resembles coral. Also, CORALLOID.
cave cotton *Geology.* thin, flexible filaments of gypsum or epsomite that are found projecting from a cave wall.
cave dwelling *Archaeology.* a prehistoric living place inside a cave or rock shelter, often inhabited in colder periods by Paleolithic hunters and gatherers. Thus, **cave dweller.**
cave earth *Geology.* an unconsolidated deposit of detrital materials that accumulates in a cave.
cave flower *Geology.* a curved, elongated deposit of gypsum or epsomite found attached to a cave wall.
cave-in *Mining Engineering.* a collapse that results from the ineffectiveness of borehole sidewalls or mine workings.
Cavellinidae *Paleontology.* a genus of small ostracods in the order Platycopida, characterized by a ridged shell surface and sexually dimorphic; Carboniferous to Permian.
caveman *Archaeology.* a popular term for a prehistoric cave dweller.
Cavendish, Henry 1731–1810, English chemist and physicist; determined properties of hydrogen and carbon dioxide, density of earth, composition of water, and weights of gases.
caveola *plural,* **caveolae.** *Cell Biology.* on the surface of a cell, a small invagination that facilitates the uptake of molecules or fluid.
cave painting see CAVE ART.
cave pearl *Geology.* in limestone caves, an unattached, rounded deposit formed by uniform precipitation of calcite about a nucleus.
caver *Meteorology.* a gentle breeze occurring in the Hebrides west of Scotland. Also, KAVER.
cavern *Geology.* **1.** a vast subterranean hollow or underground chamber in a rock. **2.** same as a cave, but implying large size and indefinite extent.
cavernoscope *Surgery.* a device used to view any cavity, as one that is inserted between the ribs and into the lung cavity.
cavernoscopy *Surgery.* the examination of a cavity with the aid of a cavernoscope.
cavernostomy *Surgery.* the surgical opening of any cavity in order to permit drainage, especially in pulmonary abscess.
cavernous *Geology.* **1.** describing a cavern or other geologic formation that contains caverns or caves. **2.** describing a rock texture that is honeycombed or filled with cavities, cells, or large interstices from erosion.
cavernous sinus *Anatomy.* a venous sinus located on each side of the sella turcica at the base of the skull, through which pass the carotid artery and branches of the third, fourth, and sixth cranial nerves.
caves see HOT CELLS.
cavetto *Architecture.* a simple concave molding having a curve of at least a quarter circle. Also, GORGE.
Caviidae *Vertebrate Zoology.* the cavies, guinea pigs, and capybaras; a herbivorous, diurnal or nocturnal rodent family distinguished by a vestigial tail and a reduced number of toes.
caving *Petroleum Engineering.* **1.** the collapsing of the walls of a wellbore. Also, SLOUGHING. **2.** dislodged rock fragments that fall into a wellbore and contaminate the well cuttings or block the hole. *Mining Engineering.* **1.** a mining procedure involving the controlled collapse of the roof in a deep mine as a means of relieving pressure. **2.** any undercut stoping, such as sublevel caving, block caving, or top slicing.
caving ground *Mining Engineering.* an unstable rock formation requiring support by cementation, casing, or timber to withstand the walls of an underground opening.
cavings *Mining Engineering.* see CAVE.
caving system see BLOCK CAVING.
cavitas abdominalis see ABDOMINAL CAVITY.
cavitation *Chemistry.* the formation of gas bubbles in a liquid, caused for example by pressure variations, heating, or vibration. *Fluid Mechanics.* **1.** the formation and instantaneous collapse of innumerable tiny cavities within a liquid that is subjected to rapid and intense changes in pressure. **2.** the formation and casting off of small vapor-pressure bubbles assembled on the surface of an object moving through degassed liquid. *Materials Science.* localized corrosion due to the stress produced by vapor bubbles bursting on a surface. *Pathology.* the formation of abnormal cavities, as in the lungs when pulmonary tuberculosis is present. Thus, **cavitation damage.**
cavitation erosion *Metallurgy.* the surface erosion of a material caused by the formation and scallops of cavities when the material is immersed in a liquid environment; for example, the erosion of an operating bronze ship propeller.
cavitation number *Fluid Mechanics.* the ratio of pressure forces to inertia forces in the study of cavitation phenomena where the cavitation number = $(\Delta P)/(0.5\rho v^2)$, where ΔP is the difference between the pressure in the liquid stream and the liquid vapor pressure, ρ is the fluid's density, and v^2 is the square of the fluid's velocity; essentially, this is the Euler number evaluated at cavitation conditions.
caviton *Physics.* a region of a plasma with reduced mass density and enhanced wave energy density.
cavity *Anatomy.* a hollow place or space within the body or an organ; it may be actual or potential, and normal or abnormal. *Medicine.* a lesion or area of deterioration in a tooth, caused by dental caries. *Electromagnetism.* see CAVITY RESONATOR.
cavity charge see SHAPED CHARGE.
cavity coupling *Electromagnetism.* the process of removing electromagnetic energy from a cavity resonator by using probes, apertures, or other means.
cavity filter *Electromagnetism.* a microwave filter consisting of a resonant cavity and other coupling devices inserted in a waveguide or coaxial cable.
cavity frequency meter *Engineering.* a device used for coaxial or waveguide systems that measures microwave frequencies by means of a cavity resonator.
cavity impedance *Electronics.* an impedance appearing across the gap between the cathode and the anode, which blocks the flow of current in a microwave tube.
cavity magnetron *Electronics.* a magnetron used as a microwave-transmitting oscillator in which the motion of electrons in relation to the magnetic field generates the microwave radiation.

cavity oscillator *Electronics.* a circuit that generates alternating current at a frequency determined by the magnetic energy stored in a cavity resonator.

cavity radiator *Thermodynamics.* an enclosure that has a black interior surface with a small opening through which radiation may enter or escape; this radiation approximates blackbody radiation.

cavity resonance *Electromagnetism.* a phenomenon whereby a conductive cavity supports a resonant condition of electromagnetic radiation, particularly at microwave frequencies.

cavity resonator *Electromagnetism.* a metal-lined enclosure whose interior geometry is designed to induce and support resonance at certain frequencies, commonly in the microwave range. Also, CAVITY.

cavity slide *Biotechnology.* a microscope slide with a small depression or cavity in the center about one centimeter across; used to hold drops of water or other liquids.

cavity tuning *Electromagnetism.* the process of changing the resonant frequencies of a cavity resonator by inserting a metal plunger to a certain depth, thus altering the geometry of the chamber's interior.

cavity wall *Building Engineering.* a wall built with an enclosed inner space to provide thermal insulation. Also, HOLLOW WALL.

caviuna *Materials.* the heavy, hard wood from trees of the genus *Machaerium* of Central and South America; used for fine furniture, decorative veneers, and cabinets.

CAVU *Aviation.* an operational term for "ceiling and visibility unlimited"; a condition in which the ceiling is in excess of 10,000 feet and visibility is in excess of 10 miles.

cavy *Vertebrate Zoology.* any of the tailless rodents in the family Caviidae, originating in South America and including the Patagonian cavy, the guinea pig, and the capybara.

cavy

cay *Geology.* **1.** a small, flat, coastal island or emergent reef composed of sand or coral. **2.** a flat mound of coral fragments and sand built up on a reef at or slightly above high-tide level.

Cayley, Arthur 1821–1895, English mathematician; wrote extensively on quantics, matrices, groups, and abstract geometry.

Cayley, Sir George 1773–1857, English aeronautical pioneer; built an early glider; wrote on biplanes, helicopters, and parachutes.

Cayley algebra *Mathematics.* an eight-dimensional, nonassociative algebra C derived from an underlying field K by forming a sum of two copies of K^3 with two copies of K as follows: Let (e_1, e_2, e_3) be the usual orthonormal basis of K^3, endowed with its usual inner (or dot) product $v \cdot w$. K^3 also has a vector (cross) product $v \times w = -(w \times v)$, satisfying $e_i \times e_i = 0$ $(i = 1, 2, 3)$, $e_1 \times e_2 = e_3$, $e_2 \times e_3 = e_1$, and $e_3 \times e_1 = e_2$. As a convenience, elements C can be written as 2×2 matrices $\begin{pmatrix} a & v \\ w & b \end{pmatrix}$, where $a, b \in K$ and $v, w \in K^3$. Addition and scalar multiplication are done in the conventional manner, but the product of two elements of C is a bit more complicated:

$$\begin{pmatrix} a & v \\ w & b \end{pmatrix}\begin{pmatrix} d & v' \\ w' & b' \end{pmatrix} = \begin{pmatrix} c & y \\ z & d \end{pmatrix}$$

where $c = aa' - v \cdot w'$; $d = bb' - w \cdot v'$; $y = av' + b'v + w \times w'$; and $z = a'w + b'v + v \times v'$. A basis $(c_1, \ldots, c_8)$ for C is given by:

$$c_1 = \begin{pmatrix} 1 & 0 \\ 0 & 0 \end{pmatrix},\ c_2 = \begin{pmatrix} 0 & 0 \\ 0 & 1 \end{pmatrix},\ c_{2+i} = \begin{pmatrix} 0 & e_i \\ 0 & 0 \end{pmatrix},\ c_{5+i} = \begin{pmatrix} 0 & 0 \\ e_i & 0 \end{pmatrix}$$

for $i = 1, 2, 3$. The identity element is $\begin{pmatrix} 1 & 0 \\ 0 & 1 \end{pmatrix} = c_1 + c_2$. If K is the field of real numbers, C may be viewed as being composed of pairs of quaternions. Also, OCTONIAN ALGEBRA.

Cayley-Hamilton theorem *Mathematics.* a linear transformation satisfies its own characteristic polynomial; i.e., let $\phi: E \to E$ be an endomorphism (linear transformation) of an n-dimensional vector space over a field K with characteristic polynomial $p(x) \in K[x]$. Then $p(\phi) = 0 \in E$.

Cayley-Klein parameters *Mathematics.* a set of four complex numbers that specify a rotation in R^3 by virtue of the fact that the group SU(2, C) is a covering group of SO(3, R). An element of SO(3, R) acting on a rigid body determines the body's (rotated) position relative to some reference frame, so the Cayley-Klein parameters may be thought of as coordinates which describe the possible orientations of a rigid body. Alternative coordinates on SO(3, R) are the Euler angles.

Cayley numbers *Mathematics.* an older term for the elements of a Cayley algebra.

Cayley's formula *Mathematics.* a specialization of the formula for the Cayley transform for M, an $(n \times n)$ orthogonal matrix. If $\det(M + I) \neq 0$, $M = (I - T)(I + T)^{-1}$, if T is an antisymmetric matrix and I is the $n \times n$ identity matrix.

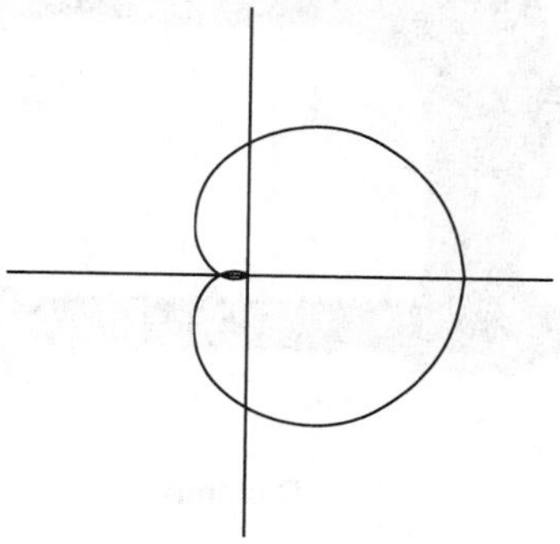

Cayley's sextic

Cayley's sextic *Mathematics.* the plane curve with polar equation $r = 4a \cos^3(\theta/3)$, where a is a constant.

Cayley's theorem *Mathematics.* if G is a group, then there is a monomorphism mapping G to $P(G)$, the permutation group of G. In particular, every finite group of order n is isomorphic to some subgroup of S_n, the symmetric group of order n.

Cayley transform *Mathematics.* let V be a finite-dimensional complex inner product space with identity I_V, and S be a symmetric linear operator on V. It can be shown that $(I_V + iS)^{-1}$ exists. The Cayley transform of S is the operator $U = (I_V - iS)(I_V + iS)^{-1}$. S is self-adjoint if and only if its Cayley transform is unitary. The transform may be viewed as providing a convenient (nonsurjective) parameterization of the group of unitary operators on V.

cay sandstone *Geology.* a coral sand that has been cemented by calcium carbonate and formed near the base of a coral-reef cay.

Caytoniales *Paleontology.* an order of pteridospermous gymnosperms of the Mesozoic; the Caytoniales were once thought to be a link between earlier plants and the angiosperms (flowering plants), but a direct relationship is now considered doubtful.

Cayugan *Geology.* a North American provincial series of the Upper Silurian period, occurring after the Niagaran and before the Ulsterian.

Cazenovian *Geology.* a North American geologic stage of the Middle Devonian, occurring after the Onesquethawan and before the Tioughniogan.

CB see CITIZENS' BAND.

C.B. Bachelor of Surgery. (An abbreviation of the Latin *Chirurgiae Baccalaurens.*)

Cb the chemical symbol for columbium, now called niobium.

Cb the meteorological symbol for cumulonimbus.

C battery *Electronics.* an energy source that supplies the voltage for biasing the grid of an electron tube. Also, GRID BATTERY.

CBI computer-based instruction.

C bias see SELF-BIAS.

CBM see COGNITIVE BEHAVIOR MODIFICATION.

CBW chemical-biological warfare.

Cc cirrocumulus.

cc cubic centimeter.

ccc cathodal closure contraction.

CCD calcite compensation depth, charge-coupled device.

CCD camera *Robotics.* charge-coupled-device camera, a device that reads the relative brightness of each picture element as the light coming into it is altered by a self-scanning semiconductor imaging device.
CCFAS compact colony-forming active substance.
CCIT 2 code *Telecommunications.* a telegraph code in which each character is delineated by five binary digits. Also, INTERNATIONAL TELEGRAPH ALPHABET.
CCITT Comité Consultatif International Telegraphique et Telephonique, (International Telegraph and Telephone Consultative Committee), a U.N. communications standards-making organization.
CCK cholecystokinin.
C clamp *Mechanical Devices.* a C-shaped clamp having a screw that threads through the bottom arm of the C up to the top.

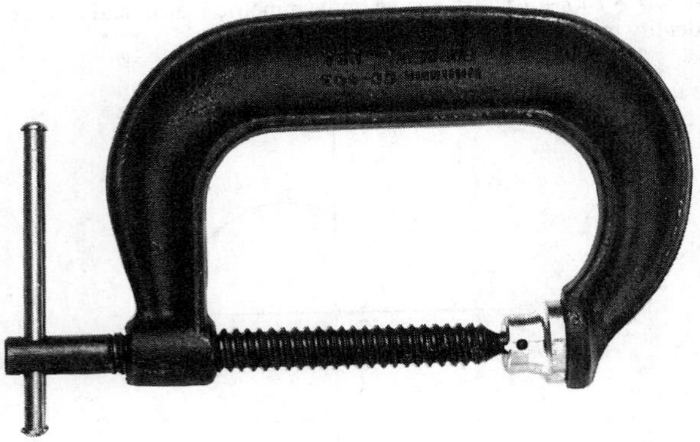

C clamp

CCS *Aviation.* the airport code for Caracas, Venezuela.
CCTV closed-circuit television.
CCU central control unit. *Medicine.* cardiac care unit; coronary care unit; critical care unit.
CCU *Aviation.* the airport code for Calcutta, India.
ccw counterclockwise. Also, **cckw**.
CD *Acoustical Engineering.* short for compact disk. See COMPACT DISK.
CD civil defense.
Cd the chemical symbol for cadmium.
cd candela.
CD-3 *Acoustical Engineering.* a miniature compact disk measuring 3 inches in diameter (as opposed to the standard 4.75 inches) and having a capacity of up to 20 minutes of recorded sound, rather than the longer capacity of standard CDs.
C damage *Ordnance.* damage that prevents an aircraft from completing its mission.
C dating or **C-14 dating** see RADIOCARBON DATING.
C-day *Military Science.* the unnamed day on which an operation deploying troops, cargo, or weapons commences or will commence.
CDG *Aviation.* the airport code for Charles de Gaulle Airport, Paris.
cD galaxy *Astronomy.* the class of supergiant elliptical galaxies, usually found in rich clusters of galaxies and thought to be the largest objects in the universe visible at optical wavelengths.
C display *Electronics.* the representation of targets on a radar screen that appear as blips whose bearings are indicated by the horizontal coordinate and whose angles of elevation are indicated by the vertical coordinate. Also, C INDICATOR, C SCAN.
CDM code division multiplex.
CDMA code-division multiple access.
CD markers *Immunology.* the cell surface molecules of leukocytes and platelets that are identified by this cluster designation (CD) with monoclonal antibodies, and that can be used to differentiate cell populations.
cDNA *Molecular Biology.* complementary DNA synthesized by RNA-directed DNA polymerase using RNA as a template; may be used as a probe for the presence of a gene code.
cdr [koud´ər] *Artificial Intelligence.* **1.** in Lisp, the function that extracts the second half of a cons cell, usually pointing to the remainder of a list following the first element. **2.** to move down a list of elements by taking successive cdr's.
cdr coding *Artificial Intelligence.* a method of reducing the amount of storage required for a cons cell, taking advantage of the fact that the next cell is usually located close to its predecessor in memory.
CD-ROM *Computer Technology.* a storage technology using a laser beam to write information on an optical disk, which may then be read by a lower-power laser beam; used for archival storage of a very large volume of information. (An acronym for compact disk read-only memory.)
CDT Central Daylight Time.
CDTA *Organic Chemistry.* $C_6H_{10}[N(CH_2CO_2H)_2]_2 \cdot H_2O$, a white, crystalline solid; very slightly soluble in water; melts at 200–220°C; used as a chelating agent.
CE chemical engineer; civil engineer.
Ce the chemical symbol for cerium.
Cebidae *Vertebrate Zoology.* a family of South American arboreal monkeys in the order Primates, including the spider monkeys, howlers, and capuchins; having opposable thumbs and large toes, long limbs and usually prehensile tails, small heads, and bare, rounded ears.
cebid herpesvirus *Virology.* a herpesvirus species that infects monkeys.
Cebochoeridae *Paleontology.* a family of primitive, piglike artiodactyls in the extinct superfamily Entelodontoidea; related to the Choeropotamidae; extant in the Eocene and Oligocene.
cebollite *Mineralogy.* $Ca_2(Mg,Fe^{+2},Al)Si_2(O,OH)_7$, a colorless, transparent, orthorhombic fibrous mineral, having a specific gravity of 2.96 and a hardness of 5 on the Mohs scale; found as an alteration product of melilite.
Cech, Thomas born 1947, American chemist; shared Nobel Prize for discovery that RNA can act as an enzyme and can replicate itself.
cecidium *Plant Pathology.* a swelled growth on a plant, caused by insects laying eggs in the plant or by a fungus invasion after an infection.
cecilite *Petrology.* a basaltic rock having few phenocrysts, composed of about 50% leucite with smaller amounts of augite, melilite, nepheline, olivine, anorthite, magnetite, and apatite.
cecoileostomy *Surgery.* **1.** the creation of a new opening between the cecum and the ileum. **2.** the opening itself. Also, ILEOCECOSTOMY.
Cecropiaceae *Botany.* a tropical American family of dicotyledonous trees, shrubs, and woody vines of the order Urticales, characterized by usually deeply lobed leaves with pinnate or palmate veins, small unisexual flowers, and a pistil composed of a single carpel, style, and stigma.
Cecropidae *Invertebrate Zoology.* a family of crustaceans in the suborder Caligoida; ectoparasites of fish.
cecum [sē´kəm] *Anatomy.* a pouch at the beginning of the large intestine. *Vertebrate Zoology.* a digestive cavity or outpouching open at one end. Also, CAECUM, BLIND GUT.
cedar *Botany.* **1.** any of several coniferous evergreen trees of the genus *Cedrus* having fragrant wood and widely spreading branches; one species, *Cedrus libani,* is known as the **Cedar of Lebanon**. **2.** any of various junipers having reddish-brown bark, such as the **red cedar,** *Juniperus virginiana.* **3.** any of several trees of the mahogany family, such as the **Spanish cedar,** *Cedrela odorata. Materials.* any of several woods of the genus *Cedrus* that are characterized by a yellow color, fragrant odor, and durability; used for construction and furniture.

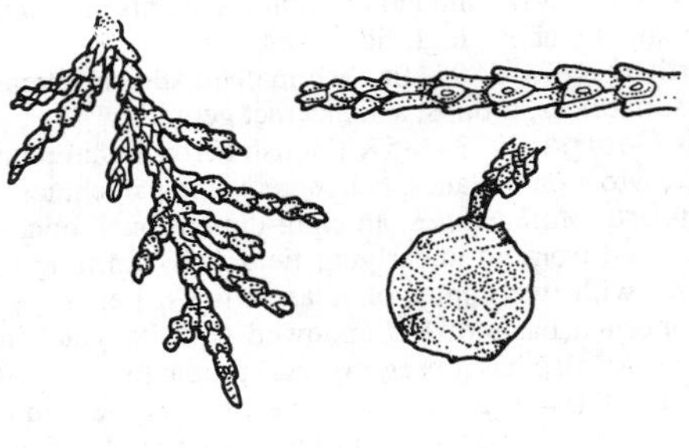

cedar

Cedecea *Bacteriology.* a genus of Gram-negative, facultatively anaerobic bacteria of the family Enterobacteriaceae that are found in the human respiratory tract and that may be opportunistic pathogens.
ceiba *Materials.* the soft, lightweight wood of the *Ceiba pantandra* tree of Central and South America; used for plywood, packages, and paper.
ceiling *Meteorology.* the measurable distance between the earth's surface and the base of the lowest layer of clouds reported as broken, overcast, or obscured. *Aviation.* **1.** the clarity of the air for looking upward, as in "ceiling and visibility unlimited." **2.** the absolute ceiling of a flight vehicle. **3.** the maximum height at which a vehicle or its crew can fly under given conditions.

ceiling balloon *Aviation.* a small free balloon used to determine the height of the meteorological ceiling by rising into the cloud cover at a steady, known rate of ascent.

ceiling classification *Meteorology.* a letter designating the ceiling height in aviation weather observation.

ceiling function see LEAST INTEGER FUNCTION.

ceiling light *Engineering.* a cloud height indicator that vertically projects a narrow beam from a searchlight onto a cloud base. Also, **ceiling projector.**

ceiling temperature *Materials Science.* the temperature for each polymerization system above which the formation of a high polymer is impossible at an equilibrium concentration.

ceilometer *Engineering.* a device that measures and records the altitude of a cloud formation.

celadonite *Mineralogy.* $K(Mg,Fe^{+2})(Fe^{+3},Al)Si_4O_{10}(OH)_2$, a green monoclinic mineral of the mica group, having a specific gravity of 3 and a hardness of 2 on the Mohs scale, occurring in earthy form or as small micaceous scales in formations of altered volcanic rocks of basaltic composition.

Celastraceae *Botany.* a family of evergreen and deciduous trees, shrubs, and woody vines of the order Celastrales having opposite or alternate leaves, small bisexual or unisexual flowers, and erect or basal ovules.

Celastrales *Botany.* an order of dicotyledonous plants of the subclass Rosidae having regular flowers and simple leaves.

Celebes *Geography.* a large island in central Indonesia, east of Borneo on the equator. Also, SULAWESI.

celery *Botany.* a biennial plant, *Apium graveolens dulce,* of the Umbelliferae family that is grown commercially for its edible leaf stalks.

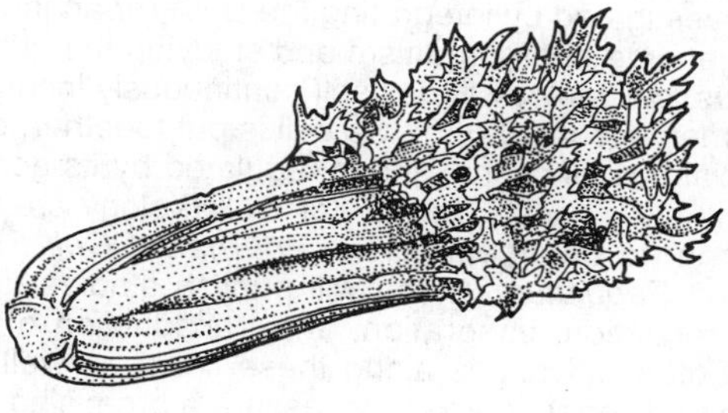

celery

celestial *Astronomy.* **1.** of the sky or visible heavens. **2.** of or relating to navigation of the sky.

celestial axis *Astronomy.* the projection of the earth's rotation axis north and south onto the celestial sphere.

celestial coordinates *Astronomy.* any set of coordinates, such as ecliptic coordinates, equatorial coordinates, or galactic coordinates, that locate an object's position on the celestial sphere.

celestial declination *Navigation.* the angular distance of a body north or south of the celestial equator; it is measured through 90° and labeled North or South.

celestial equator *Astronomy.* an imaginary great circle on the sky, lying directly above the earth's equator and separating the northern celestial hemisphere from the southern.

celestial fix *Navigation.* an exact position determined by observation of celestial bodies.

celestial geodesy *Geodesy.* the branch of geodesy that utilizes observations of close celestial bodies, including earth satellites, to determine the size and shape of the earth.

celestial globe *Astronomy.* a globe on which the positions and brightnesses of stars are plotted, often in reverse from their actual positions on the sky, as if the celestial sphere were being viewed from the outside.

celestial guidance *Navigation.* an automatic guidance system, as for missiles, that uses celestial fixes to determine position.

celestial horizon *Astronomy.* a great circle lying 90° from both the zenith and the nadir.

celestial-inertial guidance *Navigation.* an automatic guidance system, as for missiles, that uses both celestial fixes and inertial data to determine position.

celestial latitude *Astronomy.* the distance in degrees north or south measured from the ecliptic to a celestial object.

celestial line of position *Navigation.* a line, somewhere along which the craft is located, determined by observation of celestial bodies.

celestial longitude *Astronomy.* the distance in degrees to an object measured eastward along the ecliptic from the vernal equinox.

celestial mechanics *Astronomy.* the mathematical science that studies and predicts the motions and gravitational interactions of celestial objects by means of Newtonian physics.

celestial meridian *Astronomy.* the great circle passing through the north and south points on the horizon and through the zenith and nadir.

celestial navigation *Navigation.* the plotting of a moving flight vehicle's position from within the vehicle by sighting on celestial objects.

celestial observation *Navigation.* the measurement of heavenly bodies for the purpose of determining position.

celestial pole *Astronomy.* the point on the sky, north or south, at which the earth's rotation axis would touch the celestial sphere if sufficiently projected.

celestial sphere *Astronomy.* an imaginary sphere completely enveloping the earth, on which the stars and other objects appear to lie, all at the same infinite distance, and on which celestial coordinates are projected.

celestine *Mineralogy.* $SrSO_4$, a white, gray, blue, green, red, or brown, transparent to translucent orthorhombic mineral occurring as tabular or lathlike crystals, having a specific gravity of 3.97 and a hardness of 3 to 3.5 on the Mohs scale; found in sedimentary rocks, especially limestones, and in hydrothermal vein deposits. Also, **celestite.**

celiac [sēl´ē ak] *Anatomy.* relating to or located in the abdominal cavity.

celiac artery *Anatomy.* the first major branch of the abdominal aorta below the diaphragm; supplies blood to the stomach, duodenum, liver, spleen, and pancreas.

celiac disease *Pathology.* an enlargement of the intestines resulting in failure of digestion, seen in children and infants.

celiac syndrome *Medicine.* a metabolic syndrome characterized by an inability to digest glutenous foods such as wheat; symptoms include failure to absorb nutrients, malnutrition, and an enlarged abdomen.

celiocentesis see PARACENTESIS.

celiohysterectomy *Surgery.* the removal of the uterus through an abdominal incision. Also, CESAREAN HYSTERECTOMY.

celioma *Oncology.* an abdominal tumor, especially mesothelioma of the peritoneum.

celiomyomotomy *Surgery.* an incision into a benign muscle tumor through the abdominal wall.

celiorrhaphy *Surgery.* the suture of a wound in the abdominal wall.

celioscopy see LAPAROSCOPY.

celiotomy see LAPAROTOMY.

cell *Biology.* **1.** the fundamental microscopic unit of which all living things except viruses are composed, consisting of a nucleus and cytoplasm and bounded by a membrane; the minimal structural unit of life that is capable of functioning independently. *Engineering.* any small, limited space or compartment. *Physical Chemistry.* any vessel, such as a tube or a jar, that produces an electrical charge by passing electric current from one metal terminal (cathode) through a chemical solution to another metal terminal (anode). A battery is a group of connected cells. *Electricity.* **1.** any single unit designed to receive, store, and deliver electrical energy. **2.** a unit designed to convert radiant energy into electrical energy; for example, a unit that converts solar energy. *Nucleonics.* a unit or compartment in a reactor lattice that contains an individual fuel rod, its cladding, and other materials like coolant and moderator. *Computer Technology.* an address, a location in memory, or a register that is capable of holding one character or one word of information. *Telecommunications.* the geographical area served by a cellular phone system. *Mathematics.* **1.** an **open *n*-cell** is any homeomorphic image of the open unit ball $B_n = \{(x_1, \ldots, x_n) : \sum_{i=1}^{n} x_i^2 < 1\}$. A **closed *n*-cell** (also called an ***n*-disc** or **solid *n*-sphere**) is any homeomorphic image of the closed unit ball $C_n = \{(x_1, \ldots, x_n) : \sum_{i=1}^{n} x_i^2 \leq 1\}$. **2.** if S is any set and $p(S)$ is a partition of S, then the elements of $p(S)$ are the cells of $p(S)$.

cella *Architecture.* the sanctuary of a classical temple.

cell affinity chromatography *Biotechnology.* a method used to obtain functionally defined homogeneous cell populations from mixed cultures by pouring the sample in a column packed with absorbents and then washing out impurities.

cellar *Petroleum Engineering.* an area for collection of drainage water and other fluids for later disposal. *Computer Programming.* see PUSH-DOWN STACK.

cell-associated virus *Virology.* **1.** any virus that is found in cells in culture, rather than in superatant media. **2.** any virus in which infection is transmitted by infected cells.

Cell Biology

Today, cell biology is a distinct scientific discipline, centered on the study of the structure and function of the individual units called cells that comprise the multicellular organism. Cells vary greatly in size; the average diameter is about 40 μm (0.04 mm). Each contains a nucleus rich in genetic information surrounded by cytoplasm and a cell membrane. During development of the embryo into an adult organism, cells specialize in many different ways. Cell biologists and developmental biologists pride themselves in bringing to bear a variety of research techniques on questions of how the cell is put together, how it works, and how it specializes.

The structure and function of cells are studied by morphological (anatomical) techniques including light and electron microscopy, biochemical techniques, and molecular biological techniques used in conjunction. In some cases, investigators have turned to unicellular organisms like yeast to focus better on generalized cell functions that are difficult to study in multicellular plants and animals. The technique of culturing isolated living cells and organs outside the multicellular body has also made possible experimental manipulations of individual cells that would be impossible in whole organisms.

The existence of the cell as the basic structural unit of the living organism was widely accepted by the end of the 1930s, but the term *cell biology* was not used to describe a research discipline until the 1960s. Light microscopists, who called themselves cytologists and histologists, made sections of organisms embedded in wax to describe cells and their assembly into tissues that, in turn, form the organs of digestion, reproduction, etc. The resolution of the light microscope is so poor (0.2 μm), however, that an understanding did not emerge of where the molecules extracted from cells by biochemists might localize. In 1945, Keith R. Porter and his colleagues examined thin parts of whole cells in the newly developed electron microscope and discovered a cytoplasmic organelle they named the endoplasmic reticulum. A few years later, methods were devised to section cells thin enough for observation in the electron microscope, where image resolution is about 1 nm. In 1954, the macromolecular structure of motile cell processes called cilia was deciphered by Don W. Fawcett and Keith R. Porter, and by 1960, George E. Palade and others had combined cell fractionation, biochemistry, and electron microscopy to learn how a number of cytoplasmic organelles actually work. In 1960, a group led by Porter and Palade met in New York to found the American Society for Cell Biology, whose first President in 1961 was Fawcett. Shortly thereafter, the name of the newly created *Journal of Biophysical and Biochemical Cytology* was changed to the *Journal of Cell Biology,* and cell biology emerged as a distinct scientific discipline dedicated to the concept that the simultaneous application of multiple techniques is the best approach to solve the complex workings of the various cells that make up the body.

A great deal of recent excitement in the field of cell biology centers on the development of molecular techniques that can be used in combination with anatomy and chemistry to investigate cell structure, development, and function. It is possible using a retroviral enzyme called reverse transcriptase to create a complementary DNA (cDNA) from any isolated mRNA. Subsequent sequencing of the DNA has led to the discovery of many entirely new and previously unidentifiable proteins, the function of which can be investigated by integrating the DNA (or an inhibitor) into the genome of an organism and studying the effect on cells. Thus, cell biology finds itself continuously incorporating new methods to learn how a cell is put together, develops specialized functions, and is regulated by its genome. In 1989, the American Society for Cell Biology created a new journal to complement the *Journal of Cell Biology* called *Cell Regulation,* which will publish studies dealing with the reception, integration, and transduction of information. Future advances along these lines, as well as in the more traditional approaches, assure a promising future for this exciting and relatively new science that seeks to unravel the secrets of the cell.

Elizabeth D. Hay
Professor and Chair of Anatomy and Cellular Biology
Harvard Medical School

cell biology *Biology.* the scientific study of the structure, function, behavior, growth, and reproduction of cells and cell components.

cell-bound antibodies *Immunology.* any antibodies that are fixed to the surfaces of cells.

cell-chain theory *Neurology.* a former theory holding that nerve fibers consisted of chains of special cells and were not processes of a single cell.

cell complex *Mathematics.* a member of an inductively constructed sequence of topological spaces, each formed by sewing a cell to the boundary of its predecessor.

cell constancy *Biology.* a condition in which an adult organism has the same number of cells in its body or in a certain body part as other adults of the same species.

cell constant *Physical Chemistry.* a normalization factor used to obtain the specific conductance of a solution from the measured conductance in a particular cell.

cell control *Control Systems.* in the control hierarchy for controlling a cell, a module controlled by the central control module or by a factory control level.

cell cycle *Cell Biology.* a repeating sequence of events in eukaryotic cells consisting of two periods: first, a **cell-growth period** comprising the first gap or growth phase (G1), the DNA synthesis phase (S), and the second gap or growth phase (G2); and second, a **cell-division period** comprising mitosis (M).

cell determination *Cell Biology.* the process by which a cell that has not yet initiated differentiation becomes committed to a particular pathway of differentiation.

cell differentiation *Cell Biology.* the process by which a cell becomes specialized for a specific structure or function by selective gene expression of some genes and selective repression of others.

cell disruption *Cell Biology.* the lysis or breakage of cells.

cell division *Cell Biology.* the process by which two cells are produced from a single cell, including both nuclear division and cytoplasmic division.

cell envelope *Cell Biology.* the bacterial cell membrane and all surrounding external layers.

cell-free extract *Cell Biology.* a fluid preparation that is fractionated from lysed whole cells and contains the soluble portion of cell contents and organelles.

cell-free system *Biochemistry.* **1.** a system in which all the cells are incomplete, because material has been extracted from them. **2.** of or relating to material with incomplete cells.

cell-free translation system see CELL-FREE EXTRACT.

cell fusion see FUSION OF CELLS.

cell growth factor *Cell Biology.* any substance, typically obtained from the surrounding medium, that is required by a particular cell or cell type for proliferation.

cell inclusion see INCLUSION.

cell line *Biotechnology.* a population of animal cells that develops on repeated secondary cultivations for an indefinite time; this arises from a cell strain wherein some cells become altered in animal virology. *Molecular Biology.* a homogeneous group of cells derived through a tissue culture of a single sample of cells from a tissue or organ.

cell lineage *Developmental Biology.* the mitotic ancestry of blastomeres from their first cleavage division to their ultimate development into tissue cells.

cell-mediated cytotoxicity *Immunology.* the disintegration or dissolution of body cells by cytotoxic T lymphocytes and other natural killer cells, or by activated macrophages.

cell-mediated immunity *Immunology.* a specific immunity that is dependent upon the presence of reactive T lymphocytes.

cell membrane see PLASMA MEMBRANE.

cell memory *Cell Biology.* the phenomenon whereby a cell that is committed to a particular differentiation pathway will continue this specialized differentiation through several generations, even in the absence of influences that initially induced this differentiation.

cell movement *Cell Biology.* the locomotion of a cell by means of cilia, flagella, or protoplasmic flow.

cellobiase *Enzymology.* an enzyme that catalyzes the conversion of cellobiose into glucose.

cellobiose *Organic Chemistry.* $C_{12}H_{22}O_{11}$, colorless crystals, soluble in water and insoluble in alcohol; decomposes at 225°C; a disaccharide produced by the partial hydrolysis of cellulose, composed of two D-glucose residues; used in bacteriology.

cellodextrin *Organic Chemistry.* the water-soluble portion of cellulose, derived from prolonged exposure of cellulose to concentrated mineral acids; precipitated by alcohol. Also, CELLULOSE DEXTRIN.

cellophane *Materials.* a thin, transparent material manufactured from cellulose; used to wrap foods and other products in order to preserve freshness and protect against contamination.

cellosolve *Organic Chemistry.* **1.** $C_2H_5OCH_2CH_2OH$, a colorless liquid, miscible with water and alcohol; melts at –70°C and boils at 135.4°C; used in industrial solvents. Also, 2-ETHOXYETHANOL. **2.** a trade name for ethers of ethylene glycol.

cell pathology *Pathology.* a branch of scientific medicine that interprets the origin of disease in terms of cellular mutations, working from the premise that every cell recessively descends from some preexisting cell.

cell permeability *Cell Biology.* a measure of the ease with which a molecule can penetrate the plasma membrane and gain entry into a cell.

cell plate *Cell Biology.* the precursor of a new plant cell wall, forming during cell division in most higher plants and partitioning a single cell into two daughter cells.

cell potential *Materials Science.* the electromotive force developed in a galvanic cell, an indication of the tendency of a metal to corrode. *Physical Chemistry.* the potential energy difference between the two half-cells of the same electrochemical cell.

cell reaction *Physical Chemistry.* the overall chemical reaction that occurs in an electrochemical cell; the sum of the two half-reactions of oxidation and reduction in the cell.

cells and systems *Design Engineering.* a manufacturing unit consisting of at least two work stations and material-transport or conveyor systems and storage areas, all of which are accessible to robots.

cell sap *Cell Biology.* the fluid content of a plant cell vacuole, composed of an aqueous colloidal mixture of organic solutes.

cell separator *Cell Biology.* an apparatus that can differentially separate cells based on a particular distinguishing feature, such as the DNA content or presence or absence of a bound antibody coupled to a fluorescent marker.

cells of Paneth *Histology.* secretory cells in the crypts of Lieberkuhn of the small intestine, containing coarse cytoplasmic granules.

cell sorting out *Cell Biology.* a phenomenon whereby a mixed population of cell types separates, forming groups of discrete cell types; believed to result from differential attractions and adhesions between cells of the same type.

cell-surface differentiation *Cell Biology.* any change in the morphology or composition of the molecules expressed on a cell surface.

cell-surface ionization *Physiology.* the distribution of ions inside and outside a permeable cell membrane; ionic movement across a membrane upon stimulation is the main generator of an action potential.

cell synchronization *Cell Biology.* any of a number of techniques by which a population of cells is induced to assume the same stage of the cell cycle.

cell theory *Biology.* a theory stating that the cell is the fundamental structural and functional unit of life, and that the properties of an organism are the sum of the properties of its component cells.

cell-type tube *Electronics.* a radio-frequency switching tube that is filled with gas and that operates in an external resonant circuit.

cellular *Biology.* **1.** of or relating to cells; consisting of cells. **2.** composed of numerous chambers or compartments, like a honeycomb. *Petrology.* pertaining to igneous rock having numerous cavities that are larger than pore size, but smaller than caverns, and which may or may not be connected.

cellular affinity *Biology.* the tendency of the cells of certain sponges, slime molds, and vertebrates to selectively adhere to one another.

cellular automaton *Computer Technology.* a theoretical representation of a parallel computer consisting of simple computer processors that are arrayed as a pattern of cells and are connected to their immediate neighbors.

cellular chain *Computer Programming.* in data base structure, a linked list or chain that is not permitted to cross cell boundaries.

cellular control *Design Engineering.* a control system providing the direction to two or more machines in a manufacturing cell.

cellular convection *Meteorology.* an organized fluid motion characterized by areas of convection cells in which air moves upward away from the heat source in the center of the cell and downward at its edges.

cellular design *Design Engineering.* the design of a machining system in a cell configuration.

cellular immunity *Immunology.* **1.** the increased ability of phagocytes to destroy or digest parasites. **2.** see CELL-MEDIATED IMMUNITY. **3.** immunological processes that have a cellular basis or can be demonstrated by observing the behavior of cells.

cellular infiltration *Medicine.* the infusion of cells remote from their normal source into other tissues.

cellular manufacturing see GROUP TECHNOLOGY.

cellular phone *Telecommunications.* **1.** a mobile communications facility having radio transmission and telephone switching capabilities that provide communication between mobile users. This technique reuses mobile frequencies in nonadjacent geographical areas or cells to greatly expand total system capacity. **2.** an individual terminal operated by a mobile user in such a system, such as a telephone set and receiver in an automobile. Also, **cellular telephone, cellular mobile radio.**

cellular respiration *Biology.* the process within living cells in which carbohydrates are acted on by enzymes to produce energy, atmospheric oxygen combines with hydrogen to form water, and carbon dioxide is released as a waste product.

cellular slime molds *Biology.* a class of funguslike protozoans that form a slimy, mobile mass on decaying organic matter, while retaining their individual cellular characteristics rather than becoming an ameba-like mass (as does true slime mold).

cellular soil see POLYGONAL GROUND.

cellular splitting *Computer Programming.* a way of adding information to a full cell or block by splitting the cell into two half-full cells and adding the information to one of the new cells. Also, BLOCK SPLITTING.

cellular striation *Engineering.* a layer of cells within a cellular-plastic object that differs significantly from the cell formation of the remainder of the object.

cellulase *Enzymology.* an enzyme found in certain fungi and lower animals, such as insects and bacteria, that catalyzes the breakdown of cellulose into cellobiose.

cellulin granules *Mycology.* cells made up of chitin and gluten that occur in the vegetative hyphae of fungi belonging to the order Leptomitales. Also, **cellulin, cellulin grains.**

cellulitis *Medicine.* a condition marked by the severe inflammation of the connective tissue underlying the skin.

cellulolytic *Biology.* having the capacity to hydrolyze cellulose; used in reference to some protozoans and certain bacteria.

Cellulomonas *Bacteriology.* a genus of Gram-variable soil bacteria of the family Corynebacteriaceae, order Eubacteriales, having an aerobic or facultatively anaerobic metabolism; capable of hydrolyzing cellulose.

celluloneuritis *Neurology.* an inflammation of neurons, resulting from the spread of inflammation in surrounding soft tissue.

cellulose *Biochemistry.* $(C_6H_{10}O_5)_n$, a polysaccharide that is the major complex carbohydrate in plants, especially their cell walls.

α-cellulose *Organic Chemistry.* a highly refined cellulose from which all soluble materials, such as sugars and pectin, have been removed by a strong sodium hydroxide solution. It is the major component of wood and paper pulp.

cellulose acetate *Organic Chemistry.* a tough, flammable, easily fabricated thermoplastic resin occurring in white flakes or powder; melts at about 260°C; used in acetate fiber, lacquers, photographic film, and magnetic tapes.
cellulose acetate butyrate *Organic Chemistry.* a combustible, thermoplastic resin with a high resistance to oil and grease; produced by the reaction between a mixture of acetic acid, butyric acid, and their anhydrides on purified cellulose. It is an extremely durable and resistant compound that is used in making lenses, lacquer, plastic film, and outdoor signs.
cellulose acetate rayon *Textiles.* a rayon fabric made by spinning the acetic ester of cellulose.
cellulose dextrin see CELLODEXTRIN.
cellulose diacetate *Organic Chemistry.* a compound produced by treating cellulose with acetic acid and acetic anhydride; used in making safety film and rayon.
cellulose ester *Organic Chemistry.* cellulose in which the free hydroxyl groups have been replaced wholly or in part by acidic groups.
cellulose ether *Organic Chemistry.* etherified cellulose; a combustible, white, granular thermoplastic solid; insoluble in water; used in the production of textiles, films, and plastics.
cellulose fiber *Organic Chemistry.* elongate polymers of cellulose esters or ethers; found as threadlike structures in plants.
cellulose nitrate *Organic Chemistry.* a pulpy, cottonlike, amorphous solid, derived by treating cellulose with a mixture of nitric and sulfuric acids; used in high explosives, photographic film, and lacquers. Also, NITROCELLULOSE, PYROXYLIN.
cellulose propionate *Organic Chemistry.* an unusually stable ester of cellulose and propionic acid. Also, **cellulose acetate propionate.**
cellulose triacetate *Organic Chemistry.* $(C_{12}H_{16}O_8)_n$, nonflammable yellowish flakes that are insoluble in water and alcohol; derived by treating cellulose with acetic acid and acetic anhydride; used in protective coatings, textile fibers, and as a base for magnetic tape.
cellulose xanthate *Organic Chemistry.* a compound derived by treating cellulose with carbon disulfide and aqueous sodium hydroxide; used in making rayon and cellophane.
cellulosic *Organic Chemistry.* any compound derived from cellulose, such as cellulose acetate.
cellulosic resin *Organic Chemistry.* any resin produced from cellulose compounds.
cellulosome *Microbiology.* an organelle found at the cell surface of certain cellulolytic bacteria, which functions in the binding of cellulose and contains cellulolytic enzymes.
cell wall *Cell Biology.* a structure that surrounds a plant, fungal, or bacterial cell and maintains cell shape and rigidity.
Celor lens system *Optics.* an anastigmatic lens system in which two air-spaced achromatic doublet lenses are situated on either side of the stop.
celsian *Mineralogy.* $BaAl_2Si_2O_8$, a colorless, white, or yellow transparent monoclinic mineral having a specific gravity of 3.1 to 3.4 and a hardness of 6 to 6.5 on the Mohs scale; commonly occurring as twinned short prismatic crystals in the contact zones of manganese deposits.
Celsius, Anders 1701–1744, Swedish astronomer; developed the Celsius temperature scale.
Celsius or **celsius** *Thermodynamics.* relating to or expressed by the Celsius temperature scale. Thus, **Celsius thermometer.**
Celsius degree *Thermodynamics.* a unit of temperature increment used in the Celsius (centigrade) and Kelvin temperature scales.
Celsius temperature scale *Thermodynamics.* **1.** a temperature scale, also called the centigrade scale, defined so that zero degrees corresponds to the freezing point of water and 100 degrees corresponds to the boiling point, at standard atmospheric pressure. Absolute zero on this scale occurs at –273.16°C. **2.** a redefinition of this scale based on a single fixed point, the triple point of water (the state in which the solid, liquid, and vapor phases of water exist in equilibrium), which has a value of 0.01°C. This provides essentially the same freezing and boiling points as the other scale. Also, **Celsius scale.**
Celsus, Aulus Cornelius c. 30 AD, Roman natural philosopher; follower of Hippocrates; wrote the major medical text *De re medica.*
Celyphidae *Invertebrate Zoology.* a family of two-winged insects in the subsection Acalypteratae.
cement *Materials.* **1.** a powdered substance composed primarily of burned clay and limestone that is mixed with water, sand, and gravel to form concrete. **2.** the concrete itself. **3.** a soft adhesive substance that hardens when dry. **4.** to apply any of these substances. *Geology.* any chemically precipitated material or ore mineral that binds together loose particles of sediment into coherent rock. *Histology.* see CEMENTUM. *Invertebrate Zoology.* an adhesive secretion produced by certain invertebrates, that hardens on exposure to water or air and is used to bind objects.
cementation *Engineering.* the process of filling or packing a hole or cavity with cement. *Geology.* the process by which sediments are consolidated into hard, compact rock through the precipitation of binding material around the individual particles or grains. *Metallurgy.* the process of coating a metal, usually steel or iron, by heating it with another substance, usually a powdered metal such as zinc or aluminum, so that the coating substance diffuses into the surface of the metal. *Petroleum Engineering.* a diagnetic process in which a coarse, clastic substance becomes consolidated into hard rock; lithification is accomplished by precipitation of minerals in the interstices between the particles or grains of sediment.
cementation factor *Petroleum Engineering.* in oil-reservoir analysis, a mathematical expression for the degree that grains, such as sand, have been bound together by precipitated minerals.
cementation sinking *Mining Engineering.* a technique involving the injection of liquid cement or chemicals into the ground to facilitate the sinking of a shaft through strata containing water.
cement bond *Petroleum Engineering.* the adhesion to the cement of a well casing in a drill hole.
cement carrier *Naval Architecture.* a ship designed to carry dry cement, which can be on- or off-loaded through pneumatic hoses.
cement channeling *Petroleum Engineering.* a failure during the process of cementing the casing to the formation in which the slurry does not rise uniformly and leaves open areas, creating a weak bond.
cement copper *Metallurgy.* industrially significant impure copper, produced from a copper solution by a replacement reaction caused by ferrous scrap. This solution is usually produced by leaching mine overburden dumps.
cement gland *Invertebrate Zoology.* a structure in many invertebrates that produces an adhesive cementlike secretion.
cement gun *Mechanical Engineering.* a mortaring device used for mixing and applying cement.
cementicious bonding *Materials Science.* the addition of an inorganic ceramic adhesive to an aggregate of ceramic particles; used to produce reaction, hydraulic, and precipitation cements.
cementing basket *Petroleum Engineering.* a cone made of collapsible metal, created to fit next to a wellbore wall in order to prevent the passage of cement. Also, METAL-PETAL BASKET.
cementite *Materials Science.* a hard and brittle intermetallic compound that has negligible solubility limits and has the formula Fe_3C. *Mineralogy.* see COHENITE.
cementitious coatings *Materials Science.* coatings composed of inorganic bonding systems based on alkali-silicate, Portland and calcium aluminate cements; used to protect surfaces in very corrosive environments, such as those in the chemical and power industries.
cement kiln *Engineering.* a rotating, refractory-line, horizontal, steel shell for burning Portland cement.
cement mill *Mechanical Engineering.* a pulverizing machine used for grinding rock into a powder for cement.
cement rock *Petrology.* an argillaceous limestone containing variable proportions of lime, silica, and alumina as well as, usually, some magnesia; suitable for manufacture of natural hydraulic cement.
cementum *Histology.* the thin layer of calcified tissue that covers the dentin of the root of a tooth and anchors the tooth in its bony socket. Also, CEMENT.
CEMF counterelectromotive force.
ceno- a combining form meaning "new" or "recent," as in *Cenozoic.*
cenobiology see BIOCENOLOGY.
Cenomanian *Geology.* a European stage of the lower Upper Cretaceous period, occurring after the Albian and before the Turonian.
cenophobia see KENOPHOBIA.
Cenophytic *Geology.* a paleobotanic division of geologic time, beginning with the development of angiosperms in the middle or late Cretaceous period.
cenosite see KAINOSITE-(Y).
cenote see POTHOLE, def. 2.
Cenozoic *Geology.* **1.** the most recent geologic era, extending from the beginning of the Tertiary period (about 65 million years ago) to the present. **2.** referring to the rocks formed during that time.
cenozone see ASSEMBLAGE ZONE.

censored data *Statistics.* data in which the values of some elements in the sample are unobserved, but are known to lie in some interval; for example, studies on survival times that are concluded before the death of all experimental units.

cent *Acoustics.* a unit of pitch interval equivalent to 1/100th of tempered semitone, 1/200th of a tempered tone, and 1/1200th of an octave, which facilitates the adding and subtracting of musical intervals. *Nucleonics.* a reactivity unit equivalent to 1/100th of a dollar.

cent. centigrade; century.

cental see HUNDREDWEIGHT.

Centaurus [sen tôr´ əs] *Astronomy.* the Centaur, a bright constellation in the southern Milky Way that contains Alpha Centauri, the nearest star system to the Sun.

Centaurus A *Astronomy.* a strong radio source associated with the peculiar elliptical galaxy NGC 5128.

Centaurus cluster *Astronomy.* an irregular cluster of several hundred galaxies about 260 million light-years away that is a source of X-rays.

Centaurus X-3 *Astronomy.* an eclipsing binary star system that emits X-rays as gas from one star falls onto the other and is heated to immensely high temperatures by compresssion.

center *Mathematics.* **1.** the **center of a group** G (or ring) is the set $C = \{x \in G : xg = gx \ \forall \ g \in G\}$; that is, the elements which commute with every element of the group (or ring) . The center of a group is a normal subgroup. **2.** the point about which an ellipsoid or hyperboloid is symmetric. In the particular case of the n-sphere, it is the point which is equidistant from all the points of the n-sphere. **3.** for a regular polygon, the center of the inscribed circle. Also, CENTER OF SYMMETRY. **4.** the barycenter of an area or volume.

center arbor *Horology.* the arbor at the center of the works of a timepiece, which carries the center wheel and the cannon pinion, and which revolves once each hour. Also, CENTER STAFF.

centerboard *Naval Architecture.* a retractable fin projecting down from the centerline of a sailing boat, used to increase stability and weatherliness.

center burst *Materials Science.* a defect inside a metal bar when it is rolled at temperatures below the necessary standard; this occurs because of the rapid deformation of the material next to the rolls relative to the less deformed metal at the center.

center control *Robotics.* the control of a robot's tool center.

center-coupled loop *Electronics.* a coil in the energy-reflecting chamber of a microwave unit that transfers AC current electromagnetically.

center drill *Engineering.* a twist drill having a conical section with an angle of 60° and a countersink for drilling center holes in workpieces to prepare them for mounting in a lathe.

centered lattice *Crystallography.* the lattice of a crystal that has lattice points at the corners and at the center of the unit cell or at the centers of faces.

center-feed tape *Computer Technology.* a class of punched paper tape in which the center line of the sprocket or feed holes is aligned with the larger holes representing data in order to facilitate synchronization.

centerfire or **center-fire** *Ordnance.* **1.** relating to or describing a cartridge in which the primer is located in the center of the base. **2.** describing a firearm that uses such cartridges.

center frequency see CARRIER FREQUENCY.

center-gated mold *Engineering.* an injection mold having a filling orifice connected to the center of the mold cavity and the filling nozzle.

center gauge *Mechanical Devices.* a measuring device to check angles when setting a tool or tool points by means of its V-shaped notches.

centering *Building Engineering.* a curved temporary framework used to support a masonry arch or dome during the layup construction phase of a building.

centering control *Electronics.* one of two controls that position the image on a cathode-ray tube screen, either horizontally or vertically.

centering machine *Mechanical Engineering.* a device used for positioning work on a lathe before drilling.

center jump *Meteorology.* the formation of a new low-pressure center within a well-developed center that diminishes as the new center deepens, so that the cyclone appears to jump from the old to the new point of low pressure.

centerless grinder *Mechanical Engineering.* a metal-grinding machine that supports the workpiece between two abrasive wheels.

center line *Cartography.* **1.** the line connecting opposite corresponding quarter corners, or their theoretical positions, in a land survey. **2.** a line extending from the true center of overlapping aerial photographs through each of the transposed center points.

center loading *Electromagnetism.* the process of altering the resonant frequency of a transmitting antenna by inserting an inductance or capacitive halfway between the antenna's feedpoint and the end.

center of action *Meteorology.* **1.** any of the semipermanent highs and lows that appear on mean charts of sea-level pressure. **2.** a region in which the variation of a meteorological element is related to weather of the following season in other regions.

center of area *Mathematics.* given a figure in the plane, the center of mass of a uniform or homogeneous plate having the same boundaries as the given figure. Also, CENTER OF FIGURE, CENTER OF GIVEN FIGURE, CENTROID OF GIVEN FIGURE .

center of attraction *Mechanics.* a point toward which a force on a particle or a body is always directed; the force's magnitude depends only on the distance between its point of application and this center point.

center of buoyancy *Mechanics.* for a body fully or partly submerged in a fluid, the centroid of the volume of the displaced fluid; it is also the point at which the resultant buoyant force acts on the body.

center of burst *Ordnance.* see MEAN POINT OF IMPACT.

center of compliance *Robotics.* a point, determined by the characteristics, position, and orientation of compliance elements, where the compliance of the system is located, or around which rotation occurs.

center of curvature *Mathematics.* **1.** given a point p on a planar curve C, the center of the circle of curvature of C at p. **2.** in general, given a point p on a curve or surface in n-dimensional space, the center of an oscillating n-sphere at p.

center of figure see CENTER OF AREA.

center of force *Mechanics.* the effective point from which or toward which a central force acts.

center of given figure **1.** see CENTER OF AREA. **2.** see CENTER OF VOLUME.

center of gravity *Mechanics.* **1.** a fixed point in a body, such that the total force exerted by gravity on the body can be assumed to act on this point. **2.** in popular use, a theoretical central point in a body about which its weight is uniformly distributed.

center of inertia see CENTER OF MASS.

center of inversion *Crystallography.* a point through which an inversion operation is performed, converting an object into its enantiomorph. Also, CENTER OF SYMMETRY.

center of lift *Aviation.* the mean of all centers of pressure of an airfoil.

center of mass *Mechanics.* the point in a body which moves as though the body's entire mass were concentrated there, and as though all forces were applied there. Also, CENTER OF INERTIA.

center-of-mass system *Mechanics.* a reference coordinate frame that moves with the same velocity as the center of mass of a system, so that the center of mass can be thought of as being at rest in the system. Also, **center-of-mass coordinate system**.

center-of-momentum system *Mechanics.* a reference frame that is used in kinematics, having the same essential property as a center-of-mass system; i.e, the total linear momentum of the system is zero. Also, **center-of-momentum coordinate system**.

center of oscillation *Mechanics.* the point on a pendulum such that if the pendulum's entire mass were concentrated there, it would execute the same motion.

center of percussion *Mechanics.* the point in a body about which it begins to rotate upon experiencing an impulse force applied at a particular point elsewhere on the body.

center of pressure *Aviation.* the point at which the chord line of an airfoil section (extended, if necessary) intersects the line of action of the combined or resultant aerodynamic forces on the airfoil, and about which the air pressures balance.

center-of-pressure coefficient *Aviation.* the ratio of the distance of a center of pressure from the leading edge of an airfoil to the chord length of the airfoil.

center-of-pressure travel *Aviation.* a movement of the center of pressure of an airfoil along the chord depending on the angle of attack of the airfoil.

center of similitude *Mathematics.* the (fixed) point from which two properly placed similar figures are central projections of each other; i.e., the single point through which lines joining corresponding points on the figures pass, and such that the lines are divided in a constant ratio by this point. If similar figures can be placed in this way, they are said to be radially related or homothetic figures. The ratio of the line segments is the homothetic ratio and is well-defined (constant and independent of the configuration). Also, HOMOTHETIC CENTER.

center of suspension *Mechanics.* the center of the circle whose arc is described by a pendulum bob.
center of symmetry see CENTER OF INVERSION.
center of thrust see THRUST AXIS.
center of twist *Mechanics.* the point at which a transverse force must act in order not to produce any twisting in the structure. Also, SHEAR CENTER.
center of volume *Mathematics.* given a three-dimensional figure, the center of mass of a uniform or homogeneous solid having the same boundaries as the given figure.
center staff see CENTER ARBOR.
center tap *Electricity.* a terminal at the electrical midpoint of a winding, resistor, filament, or capacitor pair.
center wheel *Horology.* in a timepiece, a large geared wheel on the center arbor that transmits power from the barrel or fusee through the rest of the going train to the escapement mechanism.
centesis *Medicine.* the insertion of a needle into a body sinus to draw out fluid or gas to treat or diagnose an illness.
centi- *Metrology.* a metric combining form meaning: **1.** one-hundredth (.01), as in *centimeter;* symbolized as c. **2.** one hundred, as in *centipede.*
centiare [sen´ tə âr´] *Metrology.* a unit of measure equal to one square meter.
centibar *Metrology.* a unit of pressure equal to one-hundredth of a bar.
centigrade or **Centigrade** *Thermodynamics.* relating to or expressed by the centigrade (Celsius) temperature scale. Thus, **centigrade thermometer.** *Metrology.* divided into 100 degrees or parts.
centigrade degree see CELSIUS DEGREE.
centigrade (temperature) scale see CELSIUS TEMPERATURE SCALE.
centigram *Metrology.* a unit of mass equal to 0.01 gram. Also, **centigramme.**
centigray *Radiology.* a unit of absorbed ionizing radiation equivalent to one-hundredth of a gray.
centiliter *Metrology.* a unit of capacity equal to 0.01 liter. Also, **centilitre.**
centimeter *Metrology.* a basic unit of length in the metric system, equivalent to 0.01 meter or 0.39 inch. Also, **centimetre.** *Electricity.* see ABHENRY.
centimeter candle see PHOT.
centimeter-gram-second *Metrology.* of or relating to a system of measurement in which the centimeter is the basic unit of length, the gram is the basic unit of mass, and the second is the basic unit of time.
centimeter-gram-second system *Physics.* a system of measurement in which the fundamental units are the centimeter, gram, and second. Also, CGS SYSTEM.
21-centimeter line *Spectroscopy.* in the spectrum of neutral atomic hydrogen, a hyperfine spectral line having a wavelength of approximately 21 centimeters.
centimeter of mercury *Metrology.* a unit of pressure equal to the amount of pressure that will support a column of mercury one centimeter high.
centimetric waves *Telecommunications.* an electromagnetic wave in the superhigh-frequency range with a wavelength between 1 and 10 centimeters, yielding a frequency of 3 to 30 GHz and a digital data rate of up to 100 Mbps.
centiMorgan *Genetics.* a unit of distance on the chromosome map that is equivalent to 0.01 of a Morgan unit.
Centipeda *Bacteriology.* a genus of Gram-negative, anaerobic bacteria that occur as rod-shaped cells; the type species occurs in human periodontal lesions.
centipede *Invertebrate Zoology.* the common name for a many-footed carnivorous arthropod in the class Chilopoda.

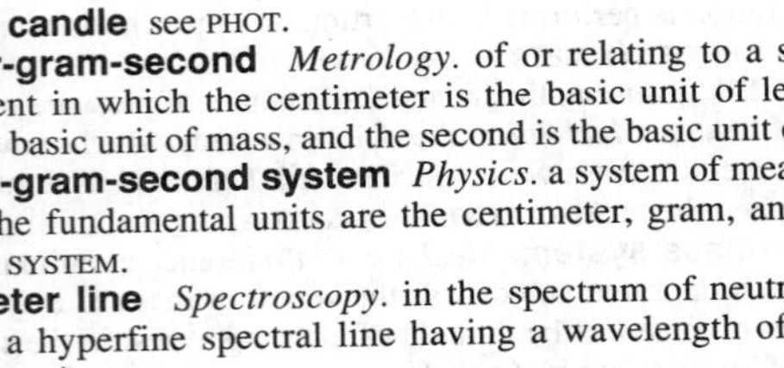

centipede

centipede grass *Botany.* a species of lawn grass, *Eremachloa ophiuroides,* native to China but grown widely in the Americas; named for the appearance of its many creeping runners.
centipoise *Fluid Mechanics.* one-hundredth of a poise, the basic unit of viscosity in the absolute metric system; 1 poise = 0.1 kg/m sec = 1 g/cm sec.
centistoke *Fluid Mechanics.* one-hundredth of a stoke, the unit of kinematic viscosity in the absolute metric system of units; 1 stoke = 0.0001 m^2/sec = 1 cm^2/sec.
centner *Metrology.* **1.** a unit of weight equal to 50 kilograms, or about 1 long hundredweight. **2.** see METRIC CENTNER.
centr- a combining form meaning "center."
centrad *Mathematics.* 0.01 radian, or approximately 0.573 degree.
central apparatus *Cell Biology.* the centrosome of a cell, composed of one or two centrioles surrounded by a region of differentiated cytoplasm.
central-battery system *Telecommunications.* a telegraph or telephone system that uses a single battery as its energy source. Batteries or other forms of remote energy sources are not required for the telephones served by the system.
central breaker *Mining Engineering.* a breaker that receives and prepares coal from several mines in a single district.
central canal *Anatomy.* a tube that runs through the gray commissure throughout the length of the spinal cord.
central collision see HEAD-ON COLLISION.
central condensation *Astronomy.* any kind of central grouping or massing in a celestial object, either apparent or real.
central control *Ordnance.* see CENTRALIZED CONTROL. *Space Technology.* the hub of a coordinated operation governing the entire process of a space flight, from preflight preparations at the launch site to the end of the flight, at the landing site.
central control unit *Control Systems.* a computer that provides the primary control for an entire control system.
central deadlock detector *Computer Technology.* in a star network distributed data system, a device installed at the central node that can communicate with the local deadlock detectors to resolve deadlocks by instructing some transactions to abort, shift some resources, reprioritize, or carry out other actions.
central deafness *Medicine.* diminished hearing due to malfunction of the cochlea or auditory nerve.
central difference operator *Mathematics.* the difference operator $\partial f(x) = [f(x+h) - f(x-h)]/2$, where h is fixed and is usually taken to be small. If $f(x)$ is known at a series of equally spaced values, repeated application of ∂ at those values gives a series of quantities called **central differences.** This technique is useful for numerical calculations and integration. $\mu f(x) = \partial f(x)$ is called the **central mean operator.**
central dogma *Biochemistry.* the concept describing the interrelations between DNA, RNA, and protein; in essence, DNA serves as a template for the replication of RNA, which in turn is translated into protein.
central eclipse *Astronomy.* an eclipse in which the obscuring body passes directly in front of the one to be hidden; usually involving a total concealing of the second body.
Centrales *Botany.* an order of diatoms, mostly circular in form, having protoplasts that contain a number of chromatophores, and valves with radial markings.
central field approximation *Physics.* an approximation in which the electric field due to the positive charge in the nucleus and the surrounding electrons of an atom is assumed to be spherically symmetric throughout the atom.
central force *Mechanics.* a force that is always directed toward or from a fixed central point (the center of attraction), and whose magnitude depends only on the distance between that point and the point of application.
central gear *Mechanical Engineering.* see SUN GEAR.
centralized configuration see STAR NETWORK.
centralized control *Control Systems.* a method of control in which all control decisions are made from a central location. *Ordnance.* a control mode used in antiaircraft batteries whereby target assignments are made by a higher echelon rather than by individual gunners.
centralized data base *Computer Programming.* a data base located at a single central site that contains all the information required by the organization.
centralized data processing *Computer Programming.* the performance of all data processing activities at a single physical location; data may be obtained locally or from remote sites.

centralizer *Mathematics.* the set of elements in a group G that commute with a given set S of elements of G. If a subgroup H of a group G acts on G by conjugation, the subgroup $C_H(S) = \{h \in H : hxh^{-1} = x \text{ for all } x \in S\} = \{h \in H : hx = xh \text{ for all } x \in S\}$ is called the centralizer of S in H. If $H = G$, then $C_G(S)$ is simply called the centralizer of S. S is often taken to be a single element of G. *Petroleum Engineering.* a cylindrical, cagelike device secured around the casing during drilling to keep the pipe centered in the borehole, allowing a more uniform cement case to be formed around the pipe. Also, CASING CENTRALIZER.

central-limit theorem *Statistics.* a theorem stating that the average of a large number of independent and identically distributed random variables, appropriately standardized, is approximately normally distributed. Also, LAW OF ERRORS.

central meridian *Astronomy.* an imaginary line drawn on a planet, the sun, or the moon that passes through the object's rotation poles and the center of its visible disk. *Cartography.* the line of longitude at the center of a map projection, usually the basis for constructing the projection.

central-meridian transit *Astronomy.* the instant when a feature on the surface of an object crosses its central meridian.

central nervous system *Anatomy.* the brain and spinal cord.

central office *Telecommunications.* a telephone-switching office at which a common carrier terminates and interconnects user lines with other user lines or trunk circuits to connecting offices or other switching systems.

central paralysis *Medicine.* any immobility caused by central nervous system damage.

central peak *Astronomy.* a mountain or group of mountains lying inside a crater; usually produced by elastic rebound of the target rocks immediately after the crater-forming inpact, but in some cases the central peak may be due to volcanism.

central-peak phenomenon *Materials Science.* a factor in the study of the structure and dynamics of crystals encountered by measuring the scattering intensity of monoenergetic beams of X-rays, neutrons, or light; the scattering peaks centered at zero energy transfer that might occur as a result of certain scattering mechanisms, such as critical opalescence, defect and impurity scattering at the critical point, or domain wall scattering.

central placentation *Botany.* a structure with ovules positioned in the center of the ovary.

central-place theory *Geography.* a theoretical model of the spatial relationships of settlements in a region; formulated in the early 1930s by the German geographer Walter Christaller. It is based on the idea that a settlement is centrally located in the region it serves, and that smaller settlements will arrange themselves around a larger central place.

central pressure *Meteorology.* the atmospheric pressure at the center of a high or low at any given moment.

central processing unit *Computer Technology.* the part of a computer that controls the interpretation and execution of machine language instructions; consists of a control unit, an arithmetic and logic unit, and a main memory. Also, **central processor.**

central-processing-unit time *Computer Programming.* the time expended in executing a set of instructions in the arithmetic/logic unit of a central processor; the amount of time spent in computation.

central-processor queue *Computer Science.* a queue of processes that are ready to run and waiting for execution by the central processor. Also, READY QUEUE.

central quadric *Mathematics.* ellipsoids and hyperboloids; that is, quadrics that have centers.

central rays *Radiology.* the straight lines passing through the center of a radiation source and through the center of a beam-limiting diaphragm.

central sulcus *Anatomy.* a fissure that separates the frontal and parietal lobes of the brain.

central terminal *Computer Technology.* in a distributed processing environment, a device that controls communication between the user terminals and the central processing unit.

central trait *Psychology.* a strong personality characteristic that is highly influential in a person's behavior but not as fundamental as a cardinal trait. Also, **central disposition.**

central transform *Computer Programming.* in structured program design, the logical process that converts the input to the system into the output; one of the three basic parts of a data flow diagram.

central valley see RIFT VALLEY, def. 2.

central water *Oceanography.* a subdivision of the surface group of water masses, found in the temperate climatic zones and characterized by high salinity and fairly warm temperatures.

Centrarchidae *Vertebrate Zoology.* the sunfishes, a family of North American freshwater fishes in the order Perciformes, characterized by spiny rays and nest-building habits.

centration *Psychology.* the tendency to focus one's attention exclusively on one feature of a situation to the exclusion of others; regarded as a trait of young children from the ages of 2 to 6.

centri- a combining form meaning "center," as in *centrifugal.*

centric *Anatomy.* of, relating to, in, or near a center; central.

centric fission *Genetics.* the formation of two chromosomes by the splitting of one chromosome at (or near) the centromere.

centric fusion *Genetics.* the fusion of two acrocentric or telocentric chromosomes into one metacentric chromosome. *Molecular Biology.* see WHOLE-ARM FUSION.

centrifugal [sen trif′ə gəl] *Mechanics.* acting or moving in a direction away from the axis of rotation. *Neurology.* moving away from the cerebral cortex. (Coined by Sir Isaac Newton; in Latin it literally means "fleeing or avoiding the center.")

centrifugal atomization *Materials Science.* the process of atomizing liquids by means of a device that feeds the liquids into the center of a spinning disk, which forces the liquids outward in sheets, where they become atomized. Thus, **centrifugal atomizer.**

centrifugal barrier *Mechanics.* in a rotating coordinate system, a steep, unbounded rise in potential around the axis of rotation. This results from the centrifugal force on the particle, and it prevents the particle from reaching the axis of rotation. Also, CENTRIFUGAL POTENTIAL ENERGY.

centrifugal brake *Mechanical Engineering.* a braking mechanism on a drum that is automatically applied if the drum's speed should exceed a set limit.

centrifugal casting *Metallurgy.* **1.** the process of casting molten metals in a rapidly rotating mold, in which centrifugal forces help to attain a sound product; extensively used for the production of seamless pipes. **2.** a product made by such a process.

centrifugal clarification *Mechanical Engineering.* the use of centrifugal force to rapidly separate fine solid matter or suspended liquid mists from a second liquid.

centrifugal classification *Mechanical Engineering.* the use of centrifugal force to separate only large particles within a liquid without affecting smaller particles; may involve reducing either the speed of the device or the duration of the operation.

centrifugal classifier *Mechanical Engineering.* a machine that uses centrifugal force to separate particles into fractions according to size.

centrifugal clutch *Mechanical Engineering.* a clutch mechanism activated by centrifugal force produced from a rapidly rotating shaft.

centrifugal collector *Mechanical Engineering.* a device that uses centrifugal force to separate fine solid particles or liquid droplets from an airstream.

centrifugal compressor *Mechanical Engineering.* a machine that uses centrifugal force to compress and discharge large volumes of gas; depending upon the pressure rise required, up to six or seven stages may be installed in a single compressor casing. Also, TURBOBLOWER.

centrifugal discharge elevator *Mechanical Engineering.* a fast-moving bucket elevator in which bulk materials are carried to the top and discharged by means of centrifugal force.

centrifugal distortion *Physics.* a slight radial elongation of a rotating body.

centrifugal drainage pattern see RADIAL DRAINAGE PATTERN.

centrifugal extractor *Chemical Engineering.* an apparatus consisting of a series of concentric cylinders contained in a cylindrical drum that rotate around a shaft; used to contact and separate components of different density in a liqiud-liquid extraction system.

centrifugal fan *Mechanical Engineering.* a fan used for mild compression of large volumes of gas, constructed in one of three general designs: straight-blade or steel-plate fan, forward-curved-blade fan, and backward-curved-blade fan.

centrifugal filter *Engineering.* a rotating container that separates heavy materials from light materials, eliminating the heavy materials through a nozzle.

centrifugal filtration *Mechanical Engineering.* the process of removing liquid from a slurry by feeding the mixture into a sieve, screen, or perforated basket in which solid matter is captured by the netting and the liquid is propelled outward by centrifugal force.

centrifugal flow *Fluid Mechanics.* the manner in which flowing fluid moves radially away from the rotating shaft in turbomachinery such as a centrifugal flow compressor.

centrifugal force *Mechanics.* **1.** a postulated inertial force in a rotating coordinate system, directed outward from the axis of rotation and, for a given magnitude of velocity, becoming weaker with increased distance (inversely proportional to the distance from the axis of rotation). **2.** in popular use, the force that seems to pull an object outward as it moves in a circle, thus acting as an equal opposing effect to the actual centripetal force.

centrifugal governor *Mechanical Engineering.* an engine speed-control device in which rotating balls or weights sense the rate of speed and adjust the flow of fuel accordingly.

centrifugal inflorescence *Botany.* a cluster of flowers opening in descending order from the tip.

centrifugal molecular still *Chemical Engineering.* an apparatus used for molecular distillation, in which material is fed into the center of a hot, rapidly rotating cone sited in a chamber at a high vacuum; centrifugal force spreads the material over the hot surface, sending the evaporable material to the condenser as vapor.

centrifugal moment *Mechanics.* the product of the magnitude of the centrifugal force acting on a body and the distance between the body and the axis of rotation.

centrifugal potential energy see CENTRIFUGAL BARRIER.

centrifugal pump *Mechanical Engineering.* a machine that pumps water and other liquids by using centrifugal force to spin the liquid radially outward, after which it is caught in the surrounding casing.

centrifugal sedimentation *Chemical Engineering.* a process to selectively fractionate suspensions according to density difference using a centrifuge.

centrifugal separation *Mechanical Engineering.* the process of using centrifugal force to separate two liquids, or solid particles from a liquid, or a liquid from a gas. *Nucleonics.* a method of isotopic enrichment in which centrifuges separate molecular species with different masses by rotating them at high speed so that the heavier molecules move toward the periphery of the centrifuge for collection, leaving the lighter species behind.

centrifugal settler *Chemical Engineering.* a rotating container that separates liquids from solid particles by centrifugal force.

centrifugal stretching *Physics.* the elongation of molecular bonds in a molecule due to centrifugal distortion, resulting in modification to the moment of inertia and energy levels.

centrifugal switch *Mechanical Engineering.* a switch that opens or shuts by centrifugal force; used within certain internal-combustion engines to maintain speed.

centrifugal tachometer *Mechanical Engineering.* a device that measures the rotational speed of a driveshaft by determining the amount of centrifugal force that is acting on a mass which is attached to the shaft and rotating with it.

centrifuge *Mechanical Engineering.* **1.** a rotating device that uses centrifugal force to separate substances of different densities, to remove moisture, or to simulate gravitational effects. **2.** to separate substances by means of such a device. Thus, **centrifugation.**

centrifuge microscope *Optics.* a microscope that allows a specimen to be magnified and observed while it is being centrifuged, having a stationary ocular and an objective that rotates; used to observe living cells.

centrifuge refining *Chemical Engineering.* the use of a centrifuge for purifying a process stream.

centrifuge tube *Analytical Chemistry.* a small, calibrated, cylindrical glass container that is used to hold samples for separation by centrifugation.

centrilobular emphysema *Medicine.* emphysema affecting the medial secretory cells of the lungs.

centriole *Cell Biology.* small, cylindrical, cellular organelle, composed of microtubules and thought to be active in formation of the spindle during mitosis.

centripetal [sen trip´ə təl] *Mechanics.* acting or moving in a direction toward the axis or center of rotation. *Neurology.* moving toward the cerebral cortex. (Coined by Sir Isaac Newton; in Latin it literally means "seeking the center.")

centripetal acceleration *Mechanics.* the acceleration of a particle moving in a circular path; it is always directed toward the center of the circle.

centripetal force *Mechanics.* the force that is required to keep an object moving around a circular path; it is directed toward the center of the circle. In the absence of this effect, the object would move in a straight line tangential to the circle.

centripetal selection see STABILIZING SELECTION.

Centriscidae *Vertebrate Zoology.* the shrimpfishes, a family of marine fishes of the suborder Syngnathoidei, characterized by a compressed body with a sharp ventral edge and encased in a thin, bony armor; often living among the spines of a sea urchin.

Centritractaceae *Botany.* a family of yellow-green algae of the order Mischococcales having elongate, solitary, uninucleate, free-living cells.

centro- a combining form meaning "center," as in centrosome.

centrobaric *Mechanics.* relating to the center of gravity of an object or system, or to the process of locating this center.

centroclinal *Geology.* describing a geologic stratum or structure that dips toward a common center.

centrode *Mechanics.* the path traced by the instantaneous center of a rigid body undergoing plane motion. If the path is within the body, it is called a **body centrode;** if it is in space, it is termed a **space centrode.**

Centrohelida *Invertebrate Zoology.* an order of protozoans in the subclass Heliozoia, with pseudopodia and siliceous spines but no central capsule.

centroid *Mechanics.* the point that may be considered as the center of mass for a one- or two-dimensional figure; i.e., the point in a system whose motion can be taken to represent the motion of the whole system. *Transportation Engineering.* in traffic and transit planning, a point selected as representative of activity within a zone.

centroid of a measure *Mathematics.* the centroid of a measure μ on a compact convex subset C of a topological space X is a point $x_0 \in X$ such that $f(x_0) = \int_C f(x)\mu(dx)$ for every continuous linear functional f on X; corresponds to center of mass if μ is ordinary Lebesgue measure in R^n.

centroid of asymptotes *Control Systems.* the point in a root-locus diagram where asymptotes intersect.

centroid of given figure **1.** see CENTER OF AREA. **2.** see CENTER OF VOLUME.

centroids of areas and lines *Mathematics.* points that are, respectively, the center of a given area or midpoint of a given line segment.

centrolecithal ovum *Cell Biology.* any egg having a central yolk.

Centrolenidae *Vertebrate Zoology.* the leaf frogs, a family of Central and South American arboreal frogs in the order Anura, noted for their transparent green skin and for depositing their eggs on moist vegetation above streams rather than in the water.

Centrolepidaceae *Botany.* a family of small, tufted, monocotyledonous grassy or mosslike annual herbs of the order Restionales, characterized by short, erect rhizomes and growing mostly in nutrient-poor soils or on tropical mountains.

centromere *Cell Biology.* a specialized region of the chromosome at which the two sister chromatids are held together and which is the site of attachment of the chromosome to the spindle fibers during mitosis.

centromeric DNA *Genetics.* short sequences of apparently nonfunctional DNA that are found in highly repetitive groups in the centromere region of the chromosome. Also, SIMPLE SEQUENCE DNA.

Centronellidina *Paleontology.* an extinct suborder of archaic articulate brachiopods in the order Terebratulida; characterized by a rhomboidal shape, by an often strongly incurved ventral beak, and generally by smooth or faintly striate shells; lower Devonian to Permian.

Centropomidae *Vertebrate Zoology.* the snooks or robalos, a huge, generalized family in the suborder Percoidei, inhabiting tropical American oceans.

centrosome *Cell Biology.* an amorphous region of differentiated cytoplasm that contains the centrioles and functions as a microtubule-organizing center for a cell. Also, MICROCENTRUM.

Centrospermales see CARYOPHYLLALES.

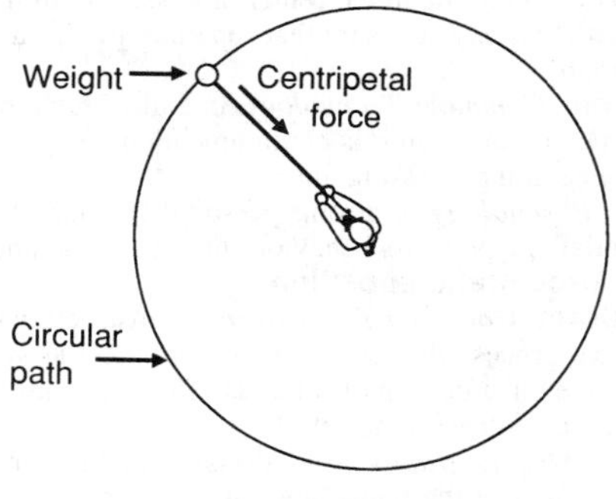

centripetal force

centrosphere *Geology.* the central core of the earth, below a depth of about 2900 kilometers. Also, BARYSPHERE. *Cell Biology.* the area of differentiated cytoplasm surrounding the centrioles within the centrosome.

centrosymmetric crystal structure *Crystallography.* a crystal structure that crystallizes in a space group with a center of symmetry. When the origin is chosen at the center of symmetry, the phase angle for each Bragg reflection, when the crystal diffracts X-rays, is either 0° or 180°.

centrosymmetry *Materials Science.* a property by which a system or body retains the same space inversion up to a specified point.

centrotype *Systematics.* in phenetics, the operational taxonomic unit (OTU) that falls closest to the center of a cluster of OTUs.

centrum *Botany.* **1.** the central air space in hollow stems. **2.** the asci and other associated cells found in the perithecium of some pyrenomycetes.

centrum of a vertebra *plural,* **centra.** *Anatomy.* the body of a vertebra, as distinct from the vertebral arches.

CEP circular error probable.

cephaeline *Organic Chemistry.* $C_{28}H_{38}N_2O_4$, one of the three major alkaloids produced by the plant ipecac, *Cephaelis ipecacuanha;* needles that are soluble in acid and alcohol and insoluble in water; melts at 115–116°C; used as an emetic and amebicide.

cephal- or **cephalo-** a prefix meaning "head" or "anterior," as in *cephalic, cephalopathy.*

cephalad [sef´ə läd] *Biology.* relating to the head; toward the head.

cephalalgia *Neurology.* pain in the head; headache. Also, **cephalgia, cephalodynia.**

Cephalaspidae *Paleontology.* a family of small, primitive ostracoderms in the extinct order Osteostraci; about one foot long and characterized by a well-developed pineal opening between the eyes; Silurian and Devonian.

Cephalaspidomorphi *Vertebrate Zoology.* the lampreys and hagfishes, a class in the superclass Agnatha, characterized by an unusually large number of gills.

Cephalaspis *Paleontology.* a genus of primitive armored fish in the subclass Monorhina and extinct order Osteostraci (Cephalaspida); the cephalaspids were characterized by a flattened head shield and pairs of pectoral fins and spines.

cephaledema *Neurology.* edema in the intracellular tissue spaces of the head.

cephalic [sə fal´ik] *Biology.* relating to the head or the head region.

cephalic birth *Medicine.* a birth in which the fetal head appears first.

cephalic flexure see CRANIAL FLEXURE.

cephalic index *Anthropology.* the ratio of maximum head length to head breadth; expressed as maximum head breadth times 100, divided by maximum head length.

cephalic presentation *Medicine.* the appearance of the head of the fetus first during birth.

cephalic vein *Anatomy.* a large, superficial vein in the arm running from the hand to the shoulder.

cephalin *Biochemistry.* an acid, phosphatidylethanolamine or phosphatidylserine, that is present in many living tissues, especially those of the central nervous system.

Cephalina *Invertebrate Zoology.* a suborder of protozoans in the order Eugregarinida; parasites of invertebrates.

cephalitis *Neurology.* a former term for encephalitis.

cephalization *Zoology.* the tendency in animals to concentrate neural, sensory, and feeding functions in the anterior or cephalic region of the body.

Cephalobaenida *Invertebrate Zoology.* an order of primitive parasitic arthropods in the class Pentastomida; the larvae have six legs.

Cephalocarida *Invertebrate Zoology.* a subclass of Crustacea, minute shrimplike animals with a horseshoe-shaped head, no eyes, and two pairs of short antennae.

Cephalochordata *Zoology.* a subphylum of the phylum Chordata, containing the lancelets and a few other related marine fishes; they are very primitive and are characterized by a simple cartilaginous skeleton, a notochord in adults, no definite head, no true brain or heart, and no paired fins. Also, **Cephalochorda.**

cephalogyric *Neurology.* of or relating to turning motions of the head.

Cephaloidae *Invertebrate Zoology.* the false longhorn beetles, a small family of coleopteran insects in the superfamily Tenebrionoidea.

cephalomere *Invertebrate Zoology.* one of the somites composing the head of an arthropod.

cephalometer *Radiology.* **1.** an instrument for measuring the dimensions of the head. **2.** a device for positioning the head correctly for radiographic examination and measurement.

cephalomotor *Neurology.* of, relating to, or generating movements of the head.

cephalon *Paleontology.* the "head" of a trilobite.

cephalont *Invertebrate Zoology.* a sporozoan immediately before spore formation.

cephalopathy *Neurology.* any disease of the brain or head.

cephalopelvic [sef´ə lō pel´vik] *Medicine.* relating to the relationship of the fetal head to the maternal pelvis.

cephalopelvic disproportion *Medicine.* a condition in which the head of the fetus is too large for the maternal pelvis.

Cephalopoda *Invertebrate Zoology.* the group of marine animals, including the squid, octopus, and nautilus, that constitute the most advanced class of the Mollusca.

cephalosporin *Microbiology.* any of a class of broad-spectrum antibiotics derived from the fungus *Cephalosporium*; structurally related to penicillin.

cephalostat *Radiology.* an instrument used to position the head precisely; used in dental radiology.

Cephalotaceae *Botany.* a family of small, tanniferous, insectivorous bog herbs in the order Rosales that are native to Australia; characterized by short rhizomes, pitcherlike leaves spiraled in basal rosettes, and a scape bearing clusters of small, apetalous flowers.

Cephalothecaceae *Mycology.* a family of fungi belonging to the order Eurotiales, which live on decaying wood, compost, soil, or dung.

cephalothin *Microbiology.* a semisynthetic derivative of the natural antibiotic **cephalosporin C,** effective against Gram-positive bacteria and certain Gram-negative species.

cephalothorax [sef´ə lō thôr´aks] *Invertebrate Zoology.* the part of the body of an arachnid or higher crustacean that is made up of the head and thorax.

Cephalothrididae *Invertebrate Zoology.* a family of ribbonlike worms in the order Palaeonemertini.

cephalotrichous flagellation *Cell Biology.* the localization of flagella at one end of a cell.

cephalotropic *Neurology.* having an affinity for brain tissue; said especially of a centrally active drug.

Cephaloziaceae *Botany.* a family of cosmopolitan liverworts of the order Jungermanniales, creeping or prostrate plants having bilobed leaves, biseriate antheridial stalks, and brownish, purplish, or reddish pigmentation.

Cephaloziellaceae *Botany.* a cosmopolitan family of minute to small liverworts of the order Jungermanniales, characterized by creeping to erect plants with reddish or brownish pigments, irregular branches, and highly reduced seta.

Cepheid instability strip [sef´ē id; sēf´ē id] *Astronomy.* the region of the Hertzsprung-Russell color-luminosity diagram where most pulsating yellow variable stars (including Cepheid variables) are found.

Cepheid variable *Astronomy.* a supergiant star of spectral type F or G that is characterized by continuous changes in brightness which repeat over periods of from 5 to 30 days, and which have an amplitude of between 0.5 and 1 magnitude; named for the prototype star Delta Cephei.

Cepheus [sē´ fē əs] *Astronomy.* the King, a conspicuous constellation lying close to the north celestial pole.

Cephidae *Invertebrate Zoology.* the stem sawflies, the only family in the hymenopteran insect superfamily Cephoidea; a plant pest whose larvae tunnel into the stems of grasses and other plants.

Cephoidea *Invertebrate Zoology.* a superfamily of hymenopteran insects in the suborder Symphyta.

CER conditioned emotional response.

cer- a combining form meaning "wax," as in *cerate.*

Ceractinomorpha *Invertebrate Zoology.* a subclass of horny sponges in the class Demospongiae.

ceramagnet *Electromagnetism.* a ferromagnetic material having the chemical structure $BaO{\cdot}6Fe_2O_3$.

Cerambycidae *Invertebrate Zoology.* the longhorn beetles, a family of coleopteran insects within the superfamily Chrysomeloidea.

cerambycoid larva *Invertebrate Zoology.* the larvae of Cerambycidae beetles, plant pests that bore into the wood and roots of trees.

Ceramiaceae *Botany.* a family of predominantly marine red algae of the order Ceramiales, having very regularly branched, filamentous or pseudoparenchymatous uniaxial thalli, and single or clustered spermatangia occurring superficially.

Ceramiales *Botany.* an order of red algae characterized by very regularly branched thalli of uniaxial construction, sporangial and gametangial thalli of similar form, and superficially clustered spermatangia.

ceramic *Materials.* **1.** of or relating to products, such as pottery, porcelain, or tile, that are made from nonmetallic mineral substances. **2.** an object made of such a material.

ceramic amplifier *Electronics.* a circuit that boosts a signal by using the piezoelectric properties of ceramics, such as barium titanate, and the piezoresistive properties of semiconductors, such as silicon.

ceramic analysis *Archaeology.* the study of artifacts made from fired clay to obtain archaeological data.

ceramic-based microcircuit *Electronics.* a microminiature circuit fired on a waferlike piece of ceramic.

ceramic bond *Materials Science.* a glassy phase that results from the solidification of a liquid phase at the grain boundaries, serving as a binder; the liquid phase results from a lowering of the melting range due to impurities or intentional aided substances.

ceramic capacitor *Electricity.* a capacitor whose dielectric is a ceramic material; the electrodes are generally silver coatings.

ceramic cartridge *Acoustical Engineering.* the element in a ceramic microphone or ceramic pickup containing the piezoelectric material.

ceramic coating *Metallurgy.* **1.** the process of applying a layer of ceramic material, such as alumina, onto a metallic product in order to protect against extremely high temperatures. **2.** the material thus applied.

ceramic crazing *Materials Science.* the formation of a network of fine surface cracks on ceramic ware; usually occurs during cooling or later when ware is in service.

ceramic filter *Electronics.* a filter consisting of electrically coupled, two-terminal piezoelectric ceramic resonators in ladder and lattice configurations, designed to impede the flow or attenuate the amplitude of specific frequency bands.

ceramic glaze *Engineering.* the glossy, vitreous finish on porcelain and some pottery, achieved by coating the clay object with metallic oxide or other substance, and firing in a kiln.

ceramic magnet *Electromagnetism.* a permanent magnet that is manufactured from mixtures of ferromagnetic and ceramic powders.

ceramic microphone *Acoustical Engineering.* a microphone that employs the piezoelectric properties of certain ceramic materials, such as quartz or Rochelle salts; the pressure of sound waves against the ceramic material, or an object contacting the material, produces an electric current.

ceramic mold casting *Metallurgy.* a precision casting process that uses permanent patterns and a slurry of zircon and high alumina mullite; molds are not monolithic as in investment casting.

ceramic packaging *Materials Science.* the manufacture and use of ceramic materials to protect products against air, water, relative humidity, microorganisms, insects, light, compression, and impact.

ceramic pickup *Acoustical Engineering.* a phonograph pickup that uses a ceramic element to convert sound energy to electromagnetic waves.

ceramic radiant *Engineering.* a component in a gas heating device that emits heat when illuminated with a flame.

ceramic rod flame spraying *Metallurgy.* a process for thermally spraying a ceramic material, such as alumina or zirconia, onto a metallic surface, using an oxygen-assisted flame.

ceramics *Materials Science.* the art and science of making and using solid articles composed primarily of inorganic, nonmetallic materials, including purification of the raw material, study and production of the chemical compounds, formation into components, and study of structure, composition, and properties of these materials, which are subjected to temperatures of 540°C or greater during manufacturing.

ceramic tile *Materials.* a flat piece of clay with a surface glaze; used for decorative purposes, especially in kitchens, bathrooms, and outdoor patios.

ceramic-to-metal seal *Materials Science.* a mechanically strong hermetic bond between carefully matched metal and ceramic components; these seals have the important properties of vacuum tightness, high mechanical strength, refractoriness, thermal shock resistance, good dielectric quality, and radiation stability; widely used in the electronic, nuclear, and mechanical fields.

ceramic tube *Electronics.* a tube in which electrons are encased in ceramic that can withstand temperatures in excess of 500°C; used on guided missiles to withstand reentry heat.

ceramide *Biochemistry.* an N-acylated sphingoid commonly produced from a fatty acid with 18 to 26 atoms and found along with phosphocholine in sphingomyelin.

ceramoplastic *Materials.* a high-temperature insulating material that is made from the bonding of synthetic mica with glass.

Ceramoporidae *Paleontology.* a family of stenolaemate bryozoans in the extinct order Cystoporata; Ordovician and Silurian.

Cerapachyinae *Invertebrate Zoology.* a subfamily of predacious ants, such as the army ant, within the family Formicidae.

Ceraphronidae *Invertebrate Zoology.* a family of tiny hymenopteran insects in the superfamily Proctotrupoidea; parasites on the eggs and larvae of other insects.

cerargyrite see CHLORARGYRITE.

cerat- or **cerato-** a combining form meaning "horn," as in *ceratine, Ceratopsia.*

cerata *Invertebrate Zoology.* the bright-colored, branching, external gills on the mantle of certain nudibranchs.

cerate *Pharmacy.* any medical preparation containing wax to make it flexible for external application. *Organic Chemistry.* a metal salt or soap of a fatty acid.

Ceratiaceae *Botany.* a monogeneric family of marine and freshwater dinoflagellates of the order Peridiniales, members of which are photosynthetic and biflagellate and have a prominent theca.

ceratine *Invertebrate Zoology.* the hornlike material produced by some anthozoans.

Ceratiomyxaceae *Mycology.* the sole family of fungi belonging to the order Ceratiomyxales; includes the species *C. fruticulosa,* which causes wood decay.

Ceratiomyxales *Mycology.* an order of fungi belonging to the subclass Ceratiomyxomycetidae.

Ceratiomyxomycetidae *Mycology.* a subclass of fungi belonging to the class Myxomycetes, which is characterized by spores that are borne externally on the surface of the column-shaped sporophores.

ceratitic *Paleontology.* of or relating to the suture pattern typical in ceratitic ammonoids, which is more complex than the goniatitic suture but less complex than the ammonitic suture.

ceratitid *Paleontology.* of or relating to the ammonoids of the order Ceratitida, which were characterized by marginally frilled partitions (ceratitic suture lines); widespread in the Triassic.

Ceratitida *Paleontology.* a Mesozoic order of ammonoids, generally characterized by a ceratitic suture; extant in the Triassic.

Ceratium *Invertebrate Zoology.* a genus of elaborately armored marine and freshwater flagellates in the order Dinoflagellata, some species having importance in the plankton of northern seas.

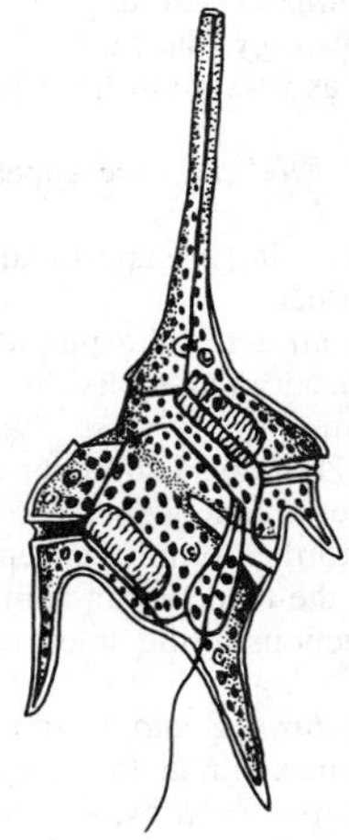

Ceratium

Ceratobasidiaceae *Mycology.* a family of fungi belonging to the order Metatremellales, occurring on decaying organic matter in tropical to temperate regions.

Ceratocoryaceae *Botany.* a monogeneric family of tropical and subtropical marine planktonic flagellates of the order Peridiniales, members of which are thecate and photosynthetic, with one species being bioluminescent.

Ceratocystis *Mycology.* a genus of fungi belonging to the order Eurotiales, causing serious plant diseases.

Ceratolithaceae *Botany.* an ancient monogeneric family of aflagellate marine algae of the order Coccosphaerales, characterized by ellipsoidal cells enclosed in an asymmetrical horseshoe-shaped coccolith.

Ceratomorpha *Vertebrate Zoology.* the tapirs and rhinoceroses, a suborder of the mammalian order Perissodactyla; odd-toed ungulates with pachydermatous skin.

Ceratomycetaceae *Mycology.* a family of fungi belonging to the suborder Laboulbeniineae, which is composed of some mole cricket parasites and some aquatic species.

Ceratophyllaceae *Botany.* a family of aquatic, dicotyledonous plants having no roots, floating leafy branches, and single unisexual flowers at separate nodes.

Ceratopsia *Paleontology.* a small suborder of ornithischian dinosaurs that includes the Psittacosauridae and Ceratopsidae; all were large and stocky, and most genera had the characteristic ceratopsian horns and frill over the back of the neck; extant in the Cretaceous.

Ceratosaurus *Paleontology.* a genus of large, carnivorous saurischian dinosaurs in the suborder Theropoda, widespread in the middle to upper Jurassic; 6 meters long, distinguished by thick supraorbital ridges and a short, hornlike structure on the snout.

cercaria *Invertebrate Zoology.* the final larval generation of the liver fluke, living in the freshwater snail; when mature it passes out and encysts, developing to the adult stage only if subsequently swallowed by a suitable host.

cerci *Entomology.* a pair of extensions on the hind end of the abdomen that often serve as sense organs; found on many insects, including crickets and cockroaches.

Cercidiphyllaceae *Botany.* a monogeneric family of dioecious, tanniferous trees of the order Hamamelidales that are native to Japan and China, having simple deciduous leaves, flowers with separate carpels, and producing both proanthocyanins and ellagic acid.

Cercopidae *Invertebrate Zoology.* the frog hoppers, a family of homopteran insects in the series Auchenorrhyncha, characterized by a froglike appearance and hopping ability; nymphs of many species produce and live in "cuckoo spit."

Cercopithecidae *Vertebrate Zoology.* the Old World monkeys, including baboons, mandrills, and macaques; a family of mostly arboreal monkeys in the infraorder Catarrhini and suborder Haplorhini, usually having a nonprehensile tail and four limbs almost equal in size.

cercopod *Invertebrate Zoology.* either of two filamentous projections found on the posterior end of notostracan crustaceans.

Cercospora *Mycology.* a genus of fungi belonging to the class Hyphomycetes, that causes numerous plant diseases, such as leaf spot.

cercospora leaf spot *Plant Pathology.* any of several plant diseases caused by the fungi of the genus Cercospora and characterized by discolored areas on the leaves.

cercus *Invertebrate Zoology.* either of two appendages found on the last abdominal segment of many insects and certain other anthropods.

cere *Vertebrate Zoology.* a soft swelling of tissue at the base of the upper beak of birds of prey and parrots, bearing the bird's nostrils.

cereal *Botany.* any grass whose grains are grown for food, such as rice, wheat, corn, oats, or barley. *Food Technology.* a breakfast food made from processed cereal grains.

cerebellar *Anatomy.* of or relating to the cerebellum.

cerebellar ataxia *Medicine.* the loss of motor coordination due to a cerebellar tumor.

cerebellifugal *Neurology.* tending or moving away from the cerebellum. Also, **cerebellofugal**.

cerebellipetal *Neurology.* tending or moving toward the cerebellum.

cerebellitis *Neurology.* inflammation of the cerebellum.

cerebellum [ser ə bel´əm] *Anatomy.* the section of the brain behind and below the cerebrum, consisting of two lateral lobes and a middle lobe that function as a coordinating center for muscle movement.

cerebr- a combining form meaning "brain," as in *cerebral.*

cerebral [sə rē´brəl; ser´ə brəl] *Anatomy.* of or relating to the cerebral hemispheres of the brain.

cerebral aneurysm *Medicine.* a thin sac filled with blood, formed by the dilatation of the walls of an artery in the brain.

cerebral angiography *Medicine.* an X-ray of the cerebral blood vessels.

cerebral coma see BRAIN DEATH.

cerebral cortex *Anatomy.* the gray outer portion of the cerebrum.

cerebral embolism *Neurology.* the blockage of a cerebral artery by a solid or gaseous embolism.

cerebral hemisphere *Anatomy.* either of the two lobes of the cerebrum.

cerebral hemorrhage *Medicine.* the rupture of a blood vessel, usually an artery, within the brain.

cerebral localization *Anatomy.* the process of locating the functional centers of the cerebrum. *Neurology.* **1.** a concept that designates a certain portion of the brain as the area that controls a given physiologic action or faculty. **2.** the designation of a certain region in the brain as the site of a lesion.

cerebral meningitis *Medicine.* an inflammation of the membranes of the brain.

cerebral palsy *Medicine.* any of several persistent, nonprogressive brain dysfunctions found in young children and marked by faulty motor ability.

cerebral peduncle *Anatomy.* either of the two large masses of substance that descend from each cerebral hemisphere, converging where they enter the pons; they form the ventral part of the mesencephalon.

cerebral seizure *Neurology.* a minor epileptic attack in which the seizures are predominantly one-sided or local; caused by a neuronal discharge in a discrete location in one hemisphere of the brain.

cerebrate *Psychology.* to use the mind; think. Thus, **cerebration.**

cerebrifugal *Neurology.* tending or moving away from the cerebrum or brain.

cerebripetal *Neurology.* tending or moving toward the cerebrum or brain.

cerebritis *Neurology.* inflammation of the cerebrum.

cerebro- a combining form meaning "brain," as in *cerebrovascular.*

cerebrocranial *Anatomy.* of or relating to the brain and skull.

cerebrogalactoside see CEREBROSIDE.

cerebrology *Neurology.* the study of the cerebrum or brain.

cerebromacular *Neurology.* of, relating to, or affecting the brain and the macula of the retina.

cerebropathy *Neurology.* any disease or dysfunction of the brain. Also, CEREBROSIS.

cerebroside *Biochemistry.* a kind of sugar compound occurring in nerve and brain tissue. Also, GALACTOLIPID, CEREBROGALACTOSIDE.

cerebrosis *Neurology.* see CEREBROPATHY.

cerebrospinal *Anatomy.* of or relating to the brain and spinal cord.

cerebrospinal axis *Anatomy.* the brain and spinal cord.

cerebrospinal fluid *Physiology.* a fluid within the brain and the subarachnoid space of the spinal cord that cushions the brain inside the skull and protects the spinal cord from mechanical shocks.

cerebrospinal meningitis *Medicine.* a disease characterized by the inflammation of the brain and spinal cord membranes; it is transmitted by direct human contact with infected persons and is marked by acute headaches, fever, rash, convulsions, and sometimes coma and death.

cerebrotonic *Psychology.* a personality type characterized by rigidity in posture and movement, inhibition in social situations, and extreme awareness of one's external environment and internal state.

cerebrovascular *Anatomy.* of or relating to the blood vessels of the cerebrum.

cerebrovascular accident see STROKE.

cerebrum *Anatomy.* the upper part of the brain, consisting of left and right hemispheres; receives conscious sensation and controls voluntary motor activities.

cerelose see GLUCOSE.

ceremedullary tube see NEURAL TUBE.

ceremonial center *Archaeology.* a complex of buildings that served as the focus of religious and civic activities in certain early New World societies.

ceremonial object *Archaeology.* an artifact that is associated with a ritual or ceremony or that functions only in a symbolic sense, as opposed to a household tool or other such practical device.

Cerenkov counter *Nucleonics.* a device used to identify beam particles and particles produced in collisions of high-energy particles, in which Cerenkov radiation is detected usually by photomultiplier tubes and the optical collection systems often are arranged to accommodate only one angle of emission.

Cerenkov radiation *Electromagnetism.* radiation of blue light that is emitted when an energetic charged particle passes through a transparent nonconductive material at a velocity greater than the velocity of light within the material.

Cerenkov rebatron radiator *Electronics.* a device with a small aperture that generates radio-frequency signals by passing an electron beam through a piece of nonconductive material.

Ceres [sēr´ ēz] *Astronomy.* asteroid 1, the largest and earliest-known asteroid, discovered in 1801 and having a diameter of 1025 kilometers; it belongs to type C. (Named for the Roman goddess of grain.)

ceria see CERIC OXIDE.

Ceriantharia *Invertebrate Zoology.* an order of solitary polyps in the subclass Zoantharia, with an elongated, anemonelike body.

Ceriantipatharia *Invertebrate Zoology.* a subclass of corals including the anthozoan orders Antipatharia and Ceriantharia.

ceric oxide *Inorganic Chemistry.* CeO_2, a heavy, brown or yellowish to white powder; soluble in sulfuric acid and insoluble in water and dilute acids; melts at about 2600°C; used in nuclear fuels, in ceramics, and as a polish for glass.

ceric sulfate *Inorganic Chemistry.* $Ce(SO_4)_2{\cdot}4H_2O$, white or reddish-yellow crystals that are very soluble (with decomposition) in water; used in waterproofing, in mildewproofing, and in dyeing and printing textiles.

cerite *Mineralogy.* $(Ce,Ca)_{10}(SiO_4)_6(OH,F)_5$, a brown, red, or gray trigonal mineral occurring in massive form or as pseudo-octahedral crystals, having a specific gravity of 4.75 to 4.78 and a hardness of 5 to 5.5 on the Mohs scale; found with allanite, fluorite, monazite, uraninite, and quartz.

Cerithiacea *Invertebrate Zoology.* a superfamily of gastropod mollusks in the order Prosobranchia.

cerium [sēr′ē əm] *Chemistry.* a rare-earth lanthanide element having the symbol Ce, the atomic number 58, an atomic weight of 140.12, a melting point of 795°C, and a boiling point of 3257°C; a gray, ductile metal that is highly reactive, oxidizing in moist air at room temperature. (Named for the asteroid *Ceres,* whose discovery was a notable event in science at the time this element was identified.)

cerium-140 *Nuclear Physics.* an isotope of cerium that has a stable nucleus and that makes up over 88% of the cerium found in nature.

cerium-142 *Nuclear Physics.* a stable isotope of cerium that makes up over 11% of the cerium found in nature.

cerium-144 *Nuclear Physics.* a radioactive isotope of cerium that is a fission product with a half-life of 285 days and that emits beta particles when it decays.

cerium chloride *Inorganic Chemistry.* $CeCl_3$, colorless to white deliquescent crystals; soluble in water, alcohol, and acids; melts at 848°C and boils at 1727°C; used in incandescent gas mantles, in the preparation of cerium metal, and in spectrography.

cerium dioxide see CERIC OXIDE.

cerium fluoride *Inorganic Chemistry.* CeF_3, white hexagonal crystals; insoluble in water and acids; melts at 1460°C and boils at 2300°C; used in the preparation of cerium metal and in arc carbons to increase their brilliance.

cerium oxide see CERIC OXIDE.

cerium stearate *Organic Chemistry.* $Ce(C_{18}H_{35}O_2)_2$, a white, waxy solid that melts at 100°C; used in making waterproofing materials.

cerium sulfate see CERIC SULFATE.

Cermak-Spirek furnace *Engineering.* a rectangular, reverberatory kiln separated into two portions by a barrier.

cermet *Materials.* a composite material produced by powdered metallurgy techniques, and made of ceramic grains dispersed in a metal matrix; used as a tool material for hard-to-machine materials, high-speed cutting, mechanical seals, bearings, and pump rotors. (An acronym for <u>cer</u>amic-<u>met</u>al.)

cermet resistor *Electricity.* a resistor consisting of finely powdered metals and insulating materials fired onto a ceramic substrate.

cernuous *Botany.* of a plant part, nodding or drooping.

cero- a combining form meaning "wax," as in *ceroma.*

cerography *Graphic Arts.* a form of printing using an engraved, wax-coated metal plate from which a duplicate printing plate is made.

cerolite see KEROLITE.

ceroma *Oncology.* a tissue tumor that has degenerated into an amyloid waxy state.

Cerophytidae *Invertebrate Zoology.* a family of beetles in the superfamily Elateroidea.

cerotic acid *Organic Chemistry.* $CH_3(CH_2)_{24}CO_2H$, combustible, white, crystalline needles; insoluble in water and very soluble in alcohol; melts at 88–89°C; a fatty acid obtained from beeswax, carnauba wax, or Chinese wax. Also, HEXACOSANOIC ACID.

cerotin see CERYL ALCOHOL.

cerous chloride see CERIUM CHLORIDE.

cerous fluoride see CERIUM FLUORIDE.

cers *Meteorology.* another name for the mistral wind in southern France and northeastern Spain.

certainty equivalence control *Control Systems.* a method of optimizing a control system by using known parameters to solve the control problem for unknown parameters.

certainty factor *Artificial Intelligence.* a percentage expression provided by an expert system that evaluates the likelihood that the conclusion reached is correct; consists of a mathematical composite of the certainties of all "if . . . then" rules applied in reaching the conclusion.

Certhiidae *Vertebrate Zoology.* the tree creepers, a family of small insect-eating birds of the passerine suborder Oscines, characterized by streaked dark plumage, a pointed curved bill, and a long tail with generally stiff feathers; found in forests of the Northern Hemisphere, Africa, and the Philippines.

ceruloplasmin *Biochemistry.* a plasma enzyme that aids in copper transport in the blood and is important in copper metabolism. Also, FERROXIDASE.

cerumen *Physiology.* a waxlike secretion of the ceruminous glands in the exterior canal of the ear that limits the entry of foreign matter into the ear. Also, EAR WAX.

ceruminous gland *Anatomy.* a wax-secreting gland of the outer ear.

cerussite *Mineralogy.* $PbCO_3$, a colorless, smoky, transparent to subtranslucent, pearly orthorhombic mineral commonly occurring as twinned tabular crystals, having a specific gravity of 6.55 and a hardness of 3 to 3.5 on the Mohs scale; found as a secondary mineral in the oxidation zone of ore deposits.

cervantite *Mineralogy.* $Sb^{+3}Sb^{+5}O_4$, a yellow orthorhombic mineral occurring as fine-grained masses, having a specific gravity of 6.5 and a hardness of 4 to 5 on the Mohs scale; found with other antimony oxides in Yugoslavia, Hungary, and Bolivia.

cervic- or **cervico-** a combining form meaning: **1.** neck, as in *cervical, cervicofacial.* **2.** cervix uteri, as in *cervicitis, cervicolabial.*

cervical *Anatomy.* of or relating to the neck, or to a necklike region within an organ.

cervical canal *Anatomy.* the canal through the cervix of the uterus.

cervical flexure *Developmental Biology.* the ventrally concave bend in the neural tube, located at the point at which the brainstem and spinal cord meet.

cervical ganglion *Anatomy.* any of three nerve centers located in the neck.

cervical plexus *Anatomy.* a network of nerves formed by the anterior divisions of the four upper cervical nerves.

cervical sinus *Developmental Biology.* a temporary depression caudal to the embryonic hyoid arch in humans; the depression contains the succeeding branchial arches, is overgrown by the hyoid arch, and closes off as the cervical vesicle.

cervical vertebra *Anatomy.* any of the vertebrae of the neck.

cervicitis *Medicine.* an irritation of the cervix of the uterus.

cervico- see CERVIC-.

cervicocolpitis *Medicine.* an inflammation of the cervix of the uterus and the vagina.

cervicodorsal *Anatomy.* of or relating to the neck and back.

cervicofacial *Anatomy.* of or relating to the neck and face.

cervicoplasty *Surgery.* plastic surgery of the neck.

Cervidae *Vertebrate Zoology.* the deer, including true deer, elk, moose, and caribou; a large family of herbivorous mammals in the suborder Ruminantia, inhabiting virtually all habitats worldwide except in Australia, and distinguished by males with solid calcareous antlers.

cervix *Anatomy.* a necklike portion of an organ, especially the constricted inferior portion of the uterus.

cervix uteri *Anatomy.* the necklike part of the uterus.

Cervoidae *Vertebrate Zoology.* a superfamily in the infraorder Pecora and the suborder Ruminantia, including deer, giraffes, okapis, and related species.

ceryl alcohol *Organic Chemistry.* $CH_3(CH_2)_{25}OH$, combustible, colorless, rhombic plates; insoluble in water and soluble in alcohol; melts at 80°C and boils at 305°C (70 torr, with decomposition). Also, 1-HEXACOSANOL, CEROTIN.

Cesalpino, Andrea 1519–1603, Italian botanist; formulated early system of plant classification according to roots and fruit organs.

cesarean [sə zâr′ē ən] *Medicine.* **1.** relating to or involving a cesarean section. **2.** a shorter term for a cesarean section. Also, CAESAREAN. (From a Latin word meaning "to cut"; associated with *Caesar* because of the belief that Julius Caesar was delivered in this way.)

cesarean hysterectomy *Medicine.* a hysterectomy immediately following a cesarean section.

cesarean section *Medicine.* the surgical delivery of the fetus and placenta by making an incision into the lower abdomen; performed when abnormal maternal or fetal conditions exist and are judged likely to make a normal vaginal delivery dangerous.

cesarolite *Mineralogy.* $PbH_2Mn_3^{+4}O_8$, a steel-gray, opaque mineral of unknown crystal structure, having a specific gravity of 5.29 and a hardness of 4.5 on the Mohs scale, and occurring as botryoidal crusts and friable masses in cavities of galena at Sidi-amor-ben-Salem, Tunisia.

Cesàro summation *Mathematics.* **1.** a way of assigning a value to a series which may otherwise diverge; i.e., a special case of generalized sum for any series. Let $\sum_{n=1}^{\infty} a_n$ be a series of real numbers, not necessarily convergent,

$$s_n = \sum_{k=1}^{n} a_k$$

its *n*th partial sum, and

$$\sigma_n = (1/n) \sum_{k=1}^{n} s_k$$

the sequence of arithmetic means of the partial sums. The series $\sum_{n=1}^{\infty} a_n$ is said to be **Cesàro summable** if $\lim_{n\to\infty} \sigma_n$ exists. An important application is Fejér's theorem in Fourier analysis. **2.** a generalization of the concept of limit for any sequence. Let $\{x_n\}$ be any sequence of points in R^p and $\{\sigma_n\} = \{1/n \sum_{k=1}^{n} x_k\}$ the sequence of arithmetic means of $\{x_n\}$. If $\lim_{n\to\infty} \sigma_n = x \in R^p$, then the sequence $\{x_n\}$ is said to be Cesàro summable to *x*, and *x* is the **Cesàro limit** of $\{x_n\}$. For both of these definitions, the Cesàro summation method is permanent; that is, every convergent series or sequence of points is Cesàro summable and the Cesàro limit or **Cèsaro sum** equals the usual one.

cesium [sēz′ē əm] *Chemistry.* an alkali metal element having the symbol Cs, the atomic number 55, an atomic weight of 132.905, a melting point of 28°C, and a boiling point of 705°C; a soft solid that becomes liquid at just above normal room temperature; the most reactive of all elements, it decomposes water to produce hydrogen that ignites spontaneously, and reacts violently with oxygen and other substances. (From a Latin word for "blue.")

cesium-134 *Nuclear Physics.* a radioactive isotope of cesium with a half-life of 2.06 years that emits negative beta particles when it decays; commonly found in photoelectric cells.

cesium-137 *Nuclear Physics.* a radioactive isotope of cesium with a half-life of 30.2 years that emits negative beta particles during decay; currently under study for use in radiation therapy.

cesium-antimonide photocathode *Electronics.* a device, created by exposing a thin layer of antimony to cesium vapor at elevated temperatures, that generates current when exposed to blue and ultraviolet light.

cesium-beam clock *Horology.* an atomic clock in which a beam of cesium atoms is stimulated by an alternating magnetic field; when the frequency of the field is locked to the natural frequency of the atoms, it can be used for extremely precise timekeeping. The frequency of the cesium atom is 9,192,631,770 cycles per second. Also, **cesium-beam atomic clock.**

cesium-beam sputter source *Electronics.* a device that generates negative ions by dislodging them from the cesium-coated inner surface of a hollow cone.

cesium bromide *Inorganic Chemistry.* CsBr, a colorless crystalline powder; soluble in water and slightly soluble in alcohol; melts at 636°C and boils at 1300°C; used in medicine and in infrared spectroscopy.

cesium carbonate *Inorganic Chemistry.* Cs_2CO_3, a colorless to white hygroscopic crystalline powder that is very soluble in hot water; very stable and can be heated to high temperatures without loss of carbon dioxide; used in specialty glasses, brewing, mineral waters, and as a polymerization catalyst.

cesium chloride *Inorganic Chemistry.* CsCl, colorless crystals; soluble in water and alcohol and insoluble in acetone; melts at 645°C and boils at 1290°C; used in brewing, mineral waters, radio tubes, and photoelectric cells, and as a density gradient in ultracentrifuge separations.

cesium chloride structure *Materials Science.* a crystal structure possessed among others by the cesium chloride (CsCl); its structure is principally bonded ionically and therefore has an equal number of positive and negative ions. It has cubic coordination number 8, so that eight chloride ions surround a central cesium cation in the face-centered cubic unit cell; the cesium ions are at the lattice points and the chloride ions translated by a (1/2 1/2 1/2) vector.

cesium electron tube *Electronics.* a tube that generates electromagnetic energy at highly stable frequency for use in atomic clocks.

cesium fluoride *Inorganic Chemistry.* CsF, deliquescent poisonous crystals; soluble in water and slightly soluble in alcohol; melts at 682°C and boils at 1251°C; used in optics, specialty glasses, and as a catalyst.

cesium hollow cathode *Electronics.* an element within an electron tube that heats cesium at the bottom of a cylinder, producing current densities as high as 800 amperes per square centimeter.

cesium hydroxide *Inorganic Chemistry.* CsOH, a poisonous, colorless to light yellow crystalline mass, the strongest base known; very soluble in water; melts at 272.3°C; used as an electrolyte in alkaline storage batteries at subzero temperatures. Also, **cesium hydrate.**

cesium iodide *Inorganic Chemistry.* CsI, a colorless powder whose deliquescent crystals are soluble in water and alcohol; melts at 626°C and boils at 1280°C; used for infrared spectroscopy and fluorescent screens.

cesium nitrate *Inorganic Chemistry.* $CsNO_3$, colorless hexagons or cubes; soluble in water and slightly soluble in alcohol; melts at 414°C and decomposes on boiling; used in the preparation of cesium salts.

cesium oxide *Inorganic Chemistry.* Cs_2O, orange-red needles that are very soluble in water and soluble in acids; decomposes at 400°C; used in the preparation of cesium salts.

cesium perchlorate *Inorganic Chemistry.* $CsClO_4$, a crystalline solid that is a strong oxidizing agent; soluble in cold water, very soluble in hot water, and slightly soluble in alcohol; decomposes at 250°C; used in optics and specialty glasses, and as a catalyst.

cesium peroxide *Inorganic Chemistry.* Cs_2O_2, pale yellow needles; soluble in water and acids; melts at 400°C and loses its oxygen at 650°C; used in the preparation of cesium salts.

cesium phototube *Electronics.* a tube in which a cesium-coated cathode turns infrared light into electrical energy.

cesium sulfate *Inorganic Chemistry.* Cs_2SO_4, colorless crystals that are soluble in water and insoluble in alcohol; melts at 1010°C; used in brewing and mineral waters, and as a density gradient in ultracentrifuge separations.

cesium-vapor lamp *Electronics.* a low-voltage arc lamp that produces light when a current passes between two electrodes in ionized cesium vapor.

cesium-vapor rectifier *Electronics.* a device in which an alternating current is passed through a cesium vapor and onto a cesium-coated cathode in order to convert it to pulsating direct current.

cespitose *Botany.* growing or forming tufts. Also, CAESPITOSE.

Cestida *Invertebrate Zoology.* an order of comb jellies with ribbonlike bodies that have a very short tentacular axis and an elongated pharyngeal axis.

Cestoda *Invertebrate Zoology.* a subclass of tapeworms in the class Cestoidea; endoparasites of vertebrates.

Cestodaria *Invertebrate Zoology.* a subclass of tapeworms in the class Cestoidea; endoparasites of fish.

Cestoidea *Invertebrate Zoology.* the tapeworms, a class of endoparasites in the phylum Platyhelminthes, typically consisting of a differentiated scolex and a chain of proglottides.

Cetacea *Vertebrate Zoology.* a superorder composed of the orders Odentoceta and Mysticeta, including the dolphins, porpoises, and whales.

cetaceum see SPERMACETI.

cetane *Organic Chemistry.* a colorless liquid hydrocarbon of the alkane series, $C_{16}H_{34}$, used in cetane number determinations and as a solvent.

cetane index *Chemical Engineering.* an empirical method for finding the volume of cetane in a fuel, based upon API gravity and the mid boiling point.

cetane number *Chemical Engineering.* the percentage of cetane in a mixture of cetane (cetane number 100) and 1-methylnaphthalene (cetane number 0) that has the same ignition factor as the fuel being tested.

cetane number improver *Chemical Engineering.* any chemical that increases the percentage of cetane in a diesel fuel, thus increasing its ignition value.

cetavitaminic acid see ASCORBIC ACID.

cetin *Organic Chemistry.* $C_{15}H_{31}CO_2C_{16}H_{33}$, a white, waxy, crystalline solid; insoluble in water and soluble in alcohol; melts at 50°C; used in making soaps, candles, ointments, salves, and emulsions. Also, CETYL PALMITATE.

cetology *Zoology.* the scientific study of whales.

Cetomimiformes *Vertebrate Zoology.* an order of rare deep-water marine fishes with soft rays and much structural diversity.

Cetomimoidei *Vertebrate Zoology.* the whalefishes, a rare suborder of deep-sea fishes in the order Beryciformes, having a flabby body and large mouth; some species are luminous.

Cetorhinidae *Vertebrate Zoology.* the basking sharks, a family of pelagic, slow-moving fishes in the order Lamniformes, often found in large schools in coastal waters at or near the surface.

Cetraria *Botany.* a genus of foliose lichens in the family Parmeliaceae that are found chiefly in northern latitudes. Also, ICELAND MOSS.

cetrimonium bromide *Organic Chemistry.* $[CH_3(CH_2)_{15}N(CH_3)_3]Br$, a crystalline solid that is soluble in water and alcohol; melts at 237°C; used as a cationic detergent, antiseptic, and precipitant for nucleic acids and mucopolysaccharides.

Cetus [sēt´ əs] *Astronomy.* the Whale, a large and relatively dim constellation lying near the celestial equator and visible on northern autumn evenings.

cetyl *Organic Chemistry.* the group $CH_3(CH_2)_{15}-$.

cetyl alcohol *Organic Chemistry.* $CH_3(CH_2)_{14}CH_2OH$, white flakes that are insoluble in water and slightly soluble in alcohol; melts at 50°C and boils at 178–182°C (12 torr); used in perfumes, cosmetics, and pharmaceuticals. Also, 1-HEXADECANOL, PALMITYL ALCOHOL.

cetyl palmitate see CETIN.

cetyl vinyl ether *Organic Chemistry.* $C_{16}H_{33}OCH{=}CH_2$, a combustible, colorless liquid that melts at 16°C and boils at 142°C (about 5 torr); used as a copolymer with unsaturated monomers to produce internally plasticized resins.

Ceva's theorem *Mathematics.* if A, B, and C are the vertices of an arbitrary triangle, and if D, E, and F are points on the sides BC, CA, and AB, respectively, or on the extensions of those sides, but not at vertices of the triangle, then the lines AD, BE, and CF, called **Cevians,** are concurrent if and only if (BD)(CE)(AF) = (DC)(EA)(FB), where (BD), etc., denotes the length of a line segment, BD. The trigonometric form of Ceva's theorem states that the three Cevians are concurrent if and only if (sin BAD)(sin CBE) (sin ACF) = (sin DAC)(sin EBA)(sin FCB), where (sin BAD), etc., is the directed measure of angle BAD.

Cevian line *Mathematics.* a line that passes through a vertex of a triangle or tetrahedron and intersects the opposite side or face, but not another vertex.

cevitamic acid see ASCORBIC ACID.

Ceylon *Geography.* an island off the southeastern coast of India (area: 25,332 sq mi), occupied by the nation of Sri Lanka.

Ceylonese subregion *Ecology.* a distinct zoogeographical region that includes the island of Sri Lanka (Ceylon) and adjacent areas on the southern part of the Indian subcontinent.

ceylonite *Mineralogy.* a green, black, or brown variety of spinel that contains iron.

CF **1.** certainty factor. **2.** *Ordnance.* center fire.

CF or **cf** centripetal force; centrifugal force.

Cf the chemical symbol for californium.

CFC **1.** chlorofluorocarbon. **2.** *Hematology.* colony-forming cell.

cfm cubic feet per minute.

cfs cubic feet per second.

CFU *Hematology.* colony-forming unit.

CFU-S *Hematology.* colony-forming unit spleen.

CG center of gravity.

cg or **cgm** centigram.

CGA color/graphics adapter.

c-GMP *Biochemistry.* 3',5'-guanosine monophosphate, a cyclic nucleotide that promotes the intracellular reactions that generate a visual signal to the brain. Also, CYCLIC GMP.

CGN *Aviation.* the airport code for Cologne, Germany.

CGRP calcitonin gene-related peptide.

CGS or **cgs** centimeter-gram-second.

CGS system *Metrology.* centimeter-gram-second system; an international system of measurement in which the centimeter is the basic unit of length, the gram is the basic unit of mass, and the second is the basic unit of time.

CHA *Aviation.* the airport code for Chattanooga, Tennessee.

chabazite *Mineralogy.* $CaAl_2Si_4O_{12}{\cdot}6H_2O$, a colorless, white, pink, or yellow transparent to translucent trigonal mineral of the zeolite group, occurring as tabular, twinned, or cubelike crystals, having a specific gravity of 2.1 and a hardness of 4 to 5 on the Mohs scale; found in cavities in basalt, andesite, and other igneous rocks, and as a deposit in hot springs.

Chablis [shä blē´] *Food Technology.* **1.** a dry white wine made with Chardonnay grapes, produced in the Chablis region of Burgundy, France. **2.** also, **chablis.** a descriptive name for certain white wines produced elsewhere, especially in California.

chachalaca *Vertebrate Zoology.* a large gallinaceous bird of the genus *Ortalis* in the family Cracidae; a long-tailed, gray-green, tree-dweller of New World forests. (From the Aztec name for this bird, which is descriptive of its cry.)

Chadronian *Geology.* a North American geologic stage of the Lower Oligocene epoch, after the Duchesnean and before the Orellan.

Chadwick, Sir James 1891–1974, British physicist; Nobel Prize for discovery of the neutron.

Chaenichthyidae *Vertebrate Zoology.* an equivalent name for the freshwater fish family Salangidae.

chaeta *Invertebrate Zoology.* a bristle or seta on certain worms.

Chaetangiaceae *Botany.* a family of marine red algae of the order Nemaliales, characterized by gametangial and multiaxial thalli that are erect, terete, or flattened and dichotomously or adventitiously branched.

Chaetetidae *Paleontology.* a family of generally massive sclerosponges in the extinct order Chaetetida; the skeleton often breaks apart along tabular growth zones; Devonian to Permian.

chaetitid *Paleontology.* of or relating to the minutely tubular calcareous organisms formerly classified as tabulate corals but now generally seen as a type of sclerosponge, in the extinct order Chaetitida; sometimes quite large, the chaetitids were widespread in the Carboniferous and Permian.

Chaetoceraceae *Botany.* a family of marine and freshwater planktonic diatoms of the order Centrales, characterized by siliceous setae that grow as new valves are formed and that link individual organisms together in chains.

Chaetochloridaceae *Botany.* a family of the order Tetrasporales containing genera in which vegetative cells, either solitary or in gelatinous groups, attach to an aquatic plant and produce 2, 4, or 16 very long pseudoflagella.

Chaetodontidae *Vertebrate Zoology.* the butterfly fishes, a family of colorful marine fishes of the suborder Percoidei, often seen in coral reefs.

Chaetognatha *Invertebrate Zoology.* a phylum of torpedo-shaped planktonic arrowworms, predators of small crustacea.

Chaetomiaceae *Mycology.* the only family of fungi belonging to the order Chaetomiales, which is composed of fast-growing, cellulose-decomposing species that occur on such substances as soil, dung, wood, paper, or straw.

Chaetomiales *Mycology.* an order of fungi belonging to the class Pyrenomycetes, characterized by a dark ascocarp having a dense covering of hair.

Chaetomium *Mycology.* a genus of fungi belonging to the order Chaetomiales, that cause damage to cellulose and are found in soil, paper, and textiles.

Chaetonotoidae *Invertebrate Zoology.* an order of tiny unsegmented worms in the class Gastrotricha, with either no adhesive tubes or two tubes connected with the two posterior tail forks.

Chaetophoraceae *Botany.* a family of algae of the order Ulotrichales having branching filaments that may end in a fine elongated hair.

Chaetophyllopsaceae *Botany.* a family of large, prostrate, brownish and reddish liverworts of the order Jungermanniales, occurring only in New Zealand and southeast Australia; distinguished by succubous leaves and stiff, unicellular setae on leaf margins.

Chaetopteridae *Invertebrate Zoology.* a family of tube-dwelling polychaete annelid worms in the Sedentaria.

Chaetosphaeridiaceae *Botany.* a family of epiphytic aquatic algae of the order Coleochaetales, characterized by a thallus made up of clustered cells or cells united in branched filaments and a cytoplasmic seta borne by each mature cell and having a basal gelatinous sheath.

chachalaca

Chaetothyriaceae *Mycology.* a family of fungi belonging to the order Chaetothyriales, occurring on the surfaces of plant leaves, primarily in tropical regions.

Chaetothyriales *Mycology.* an order of fungi belonging to the class Loculoascomycetes, composed of plant parasites and members that live off decaying vegetable matter.

chaff *Agriculture.* **1.** the outer covering of grain, which is separated from the seed by threshing. **2.** short lengths of hay or straw. *Ordnance.* a reflector of electromagnetic radiations designed to create echoes on enemy radar; it usually consists of metal foil strips cut into lengths that can be evenly divided into the wavelength of the radar; it may be dropped from aircraft or shot into the air in projectiles. Also, CONFUSION REFLECTOR.

chaffer *Agriculture.* any device that serves to separate grain from chaff.

chaffinch *Vertebrate Zoology.* a European songbird of the species *Fringilla coelebs* in the family Fringillidae, having a blue crown and brown body.

chafing fatigue *Metallurgy.* fatigue that begins in areas previously damaged by rubbing.

chaga fungus *Mycology.* the sterile fruiting body of the Siberian fungal species *Inonotus obliquus.* Also, BIRCH CANKER.

Chagas' disease *Medicine.* a disease, common in Central and South America, that is caused by trypanosome organisms and is marked by a swollen face on one side only, fever, and swollen spleen and liver.

chain *Mechanical Devices.* a flexible length formed of metal links, rods, or plates, used for tension loads, to drag, lift, support, or restrain objects, or, in an endless loop, as a conveyor or power transmitter. *Chemistry.* a series of atoms, often of a single element such as carbon, that are joined in a line, branched line, or ring to form a molecule. *Computer Programming.* any set of data items or records that are linked together. *Geology.* **1.** in mineral deposits, describing a texture or structure in which a series of connected crystals form a chainlike pattern. **2.** any series or sequence of related, interconnected, or similar natural features, such as mountains, lakes, islands, or reefs, that are arranged more or less lengthwise. *Telecommunications.* **1.** any set of operations that are performed in sequence. **2.** a series of radio links and stations designed to operate as a group. *Cartography.* the unit of length prescribed by law for the survey of public lands in the United States; equal to 66 feet or 100 links.

chain *Mathematics.* **1.** a (nonempty) linearly ordered set; i.e., a partial ordering $\leq$ on a set C in which either $x \leq y$ or $y \leq x$ for every x and y in C. One chain is said to be a refinement of a second chain if it contains all elements of the second chain. Two chains are said to be equivalent if there exists a one-to-one, order-preserving correspondence between their elements. Also, TOTALLY ORDERED SET. A chain may consist of numerical quantities, algebraic elements such as lattice points, algebraic objects such as submodules, etc. If the linear ordering is $\leq$, the chain is said to be an **ascending chain;** if it is $\geq$, it is a **descending chain.** **2.** in integration theory, a (finite) formal sum of singular n-cubes in R^n is called an n-chain. The integral of a 1-form over a 1-chain is a *line integral;* an integral of a 2-form over a 2-chain is a *surface integral.* **3.** the collection $\{D_i\}$, $0 \leq i \leq n$, of open disks is called a **chain of disks** if

$$D_{i-1} \cap D_i \neq \varnothing,\ 1 \leq i \leq n.$$

Used in analytic continuation. Also, **circle chain.**

Chain, Ernst Boris 1906–1979, German-born British biochemist; shared Nobel Prize for research on penicillin.

chain belt *Mechanical Devices.* a conveyor or chain gear made up of parallel chains with grab hooks that engage the load to be carried. Also, **chainbelt.**

chain block *Mechanical Engineering.* a tackle used to hoist heavy loads by means of a continuous chain riding on an overhead track.

chain branching *Materials Science.* a linear polymer chain that branches off into two or more individual units, which may grow at different rates.

chain cable *Naval Architecture.* an anchor cable or other heavy cable made of chain.

chain coal cutter *Mining Engineering.* a coal cutter utilizing an endless chain that moves a jib (flat plate) to create a groove.

chain code *Computer Programming.* a binary code that is derived by arranging in a cyclical sequence some or all M-bit words constructed by cycling through a chain of bits; no word may recur before the cycle is complete. For example, the chain 010110 would yield the 3-bit code words: 010, 101, 011, and 110. *Computer Technology.* a method of describing the boundary of a region by means of successive displacements of each point on the boundary from the previous point.

chain complex *Mathematics.* a sequence $\{C_n: -\infty < n < \infty\}$ of algebraic objects such as rings, modules, or Abelian groups, together with a collection $\{\partial_n: C_n \rightarrow C_{n-1}\}$ of morphisms called boundary homomorphisms (or boundary operators), such that $\partial_{n-1} \cdot \partial_n = 0$ for each n. If C denotes a chain complex, then

$$H_n(C) = \text{Ker}\ \partial_n / \text{Im}\ \partial_{n+1}$$

is called the nth homology group of C. Also, GRADED DIFFERENTIAL COMPLEX.

chain condition *Mathematics.* the ascending or descending chain condition on modules over a ring.

chain configuration *Materials Science.* **1.** the specific spatial arrangement of the atoms and groups of atoms in a polymer molecule. **2.** the overall shape of a polymer molecule.

chain conformation *Materials Science.* the different arrangements of the subunits of the polymer chain brought about by rotations around single bonds; these include helical, folded chain, and random coil conformations.

chain conveyor *Mechanical Engineering.* a manufacturing machine that carries workpieces on one or two continuous crossbarred chains at low speed, allowing work to continue as the items move along.

chain data flag *Computer Programming.* in a channel command word, a flag bit that is set to the value of 1 to indicate a scatter read or write operation.

chain drive *Mechanical Engineering.* a power-driven conveying device consisting of a continuous-link chain that rides on toothed wheels attached to the driving shaft; used to move materials either horizontally or vertically.

chained list *Computer Programming.* a list of items or physical records stored randomly in memory, each item of which contains a pointer to the next physical record in the list.

chain feed *Mechanical Devices.* an endless chain used on machine tools or in conjunction with sheave wheels for timber feeding.

chain field *Computer Programming.* in a linked list, the part of a record that contains a pointer to the next record in the list. Also, **chain pointer.**

chain fission yield *Nuclear Physics.* the sum of various fission yields from nuclides that have the same number of neutrons, but a different number of protons in their nucleus.

chain gear *Mechanical Engineering.* a gear that consists of a series of wheels linked by chain that transmit motion from one wheel to the next.

chain harrow *Agriculture.* a type of harrow consisting of spikes attached to a network of flexible chain links, used for plowing on uneven ground.

chain homomorphism *Mathematics.* if A and C are chain complexes with boundary homomorphisms $\{\partial_n^A: A_n \rightarrow A_{n-1}\}$ and $\{\partial_n^C: C_n \rightarrow C_{n-1}\}$, respectively, a chain homomorphism $f: A \rightarrow C$ is a collection of homomorphisms $\{f_n: A_n \rightarrow C_n\}$ such that, for each n, $f_{n-1}\,\partial_n^A = \partial_n^C f_n$.

chaining *Artificial Intelligence.* an inference technique that links together related information from a variety of sources, such as another rule or the user. *Computer Programming.* **1.** a technique for storing lists of records randomly in which each record has a pointer to the next record. **2.** in early FORTRAN programming, a simple form of overlaying in which one program or link could, upon completion, call up another program for execution. **3.** in inference engines of a rule-based knowledge system, a process whereby the inference engine chains together a series of information sources to obtain the answer to a question rather than ask the user directly.

chaining search *Computer Programming.* a method of searching a linked list in which the key value produces the initial record address; if the record found at that address does not contain the desired key value, the pointer in that record is used to go to the next record in the list. The process stops when the desired record is found or the list is exhausted.

chain isomerism *Organic Chemistry.* an isomerism found in hydrocarbons involving differences in the geometry of the straight or branched skeletal chains of carbon atoms in the isomers.

chain lightning *Geophysics.* any lightning appearing as a long unbroken or zig-zag line. Also, BEADED LIGHTNING, PEARL-NECKLACE LIGHTNING.

chain locker *Naval Architecture.* the space below decks in the bow in which the anchor chain is stored when the anchor is raised.

chain loom *Textiles.* a knitting machine designed to produce flat-knit fabric.

chain mesh dredge *Oceanography.* a box-shaped sampler with a bottom of strong metal mesh, which retains sedimentary material as the dredge is dragged along the ocean floor.

chain of simplices *Mathematics.* an element of a free Abelian group generated as follows: Let G be an Abelian group with operation +, and S_i^r (where $i = 1, \ldots, n$) be the r-dimensional oriented simplices of a simplicial complex K. The set of ***r*-chains** (chains of simplices) of the form

$$x = g_1 S_1^r + \cdots + g_n S_n^r$$

is a free Abelian group if chains are added in the natural way, i.e., by adding coefficients of each oriented simplex. (It is understood that if $*S_i^r$ is the simplex S_i^r with its orientation altered, then $g(*S_i^r) = (-g)S_i^r$ for all g in G). Also, SIMPLICIAL CHAIN.

chain pipe *Naval Architecture.* a pipe through which a chain cable passes through the deck into the chain locker. Also, NAVEL, SPURLING GATE.

chain-pipe wrench *Mechanical Devices.* a wrench that utilizes a heavy chain to turn or secure a pipe. Similarly, **chain-pipe vise.**

chain plate *Naval Architecture.* one of several fittings on the side of a ship to which the shrouds are secured.

chain polymerization *Materials Science.* a reaction in which monomers, such as ethylene, change into linear polymers, such as polyethylene; the reaction steps are initiation, propagation, and termination.

chain printer *Computer Technology.* an impact printer in which the characters for printing are linked together in a chain or in a train of print slugs. Also, TRAIN PRINTER.

chain pump *Engineering.* a pump in which water is raised through small lifts by means of disks attached to an endless chain passing upward through a tube.

chain radar beacon *Telecommunications.* a radar beacon having a fast recovery time, which allows simultaneous interrogation and tracking of the beacon by multiple radars.

chain radar system *Engineering.* a system of radars used in conjunction for more comprehensive coverage of a missile in flight.

chain reaction *Nucleonics.* a self-sustaining sequence in which neutrons are produced in a fission reaction, during which a large amount of energy and several neutrons are emitted, leaving behind fission fragments; these neutrons induce subsequent fission reactions in neighboring nuclei, which in turn emit more neutrons and energy, causing the next generation or chain of fissions. *Chemistry.* a chemical reaction in which a change in a single molecule causes many molecules to change, until a stable compound is formed.

chain-reaction polymerization *Materials Science.* one process for the rapid, low-cost production of polymers, wherein an initiator which has either a free electron or an ionized group is added to the polymer and attracts one of the electrons of the carbon double bond, leaving the other electron of the broken bond unsatisfied. It attracts an electron from another monomer, continuing the process until the chain is stopped as a result of meeting either another growing segment or a terminator.

chain reflex see ACTION CHAIN.

chain rule *Mathematics.* a rule for ordinary or partial differentiation of composite functions. If y is a differentiable function of u and u is a differentiable function of x, then $dy/dx = (dy/du)(du/dx)$. Also written as

$$(f \bullet g)'(x) = f''(g(x)) \bullet g'(x).$$

If F is a differentiable function of one or more variables $u_1, \ldots, u_n$, each of which in turn is a differentiable function of one or more variables $x_1, x_2, \ldots$, then

$$\partial F/\partial x_r = \sum_{i=1}^{n} \partial F/\partial u_i \, \partial ui/\partial x_r.$$

Similar rules often hold for more complicated differentiable objects that have a composition law.

chainsaw or **chain saw** *Mechanical Engineering.* a portable power saw with cutting teeth linked together to form a continuous chain; used to cut timber.

chain scission *Materials Science.* a polymer degradation process consisting of the breaking of the backbone of the main polymer chain into fragments; this can be induced thermally, photochemically, or by ultrasonic radiation or mechanical agitation. Also, DEPOLYMERIZATION.

chain silicate see INOSILICATE.

chain stiffening *Materials Science.* the inclusion of monomers other than those of the main chain in the carbon chain or branches of polymers such as nylons, acrylics, acetals, cellulosics, and polycarbonates, resulting in stronger bonding and higher moduli.

chain termination method *Molecular Biology.* a method of stopping the extension of the DNA chain during replication by incorporating a nucleotide analog that cannot form a subsequent 3'-phosphodiester bond.

chain tongs *Mechanical Devices.* a type of wrench used to turn or hold a pipe by encircling it.

chain transfer reaction *Materials Science.* the termination of a growing polymer chain and the start of a new one; the process is usually mediated by a chain transfer agent that can be a monomer, solvent, polymer, or initiator.

chain turpentine see TURPENTINE, def. 2.

chainwall *Mining Engineering.* a method of mining in which the coal is mined from between coal pillars that support a mine roof.

chair configuration or **chair conformation** see CHAIR FORM.

chair form *Organic Chemistry.* a configuration of atoms in a molecule with more than five atoms, in which the bond angles form an irregular chairlike pattern, as in a modification of cyclohexane where the hydrogen atoms are staggered in relation to the carbon atoms. This is normally the most stable spatial arrangement of cyclohexane (as opposed to the less stable boat form.)

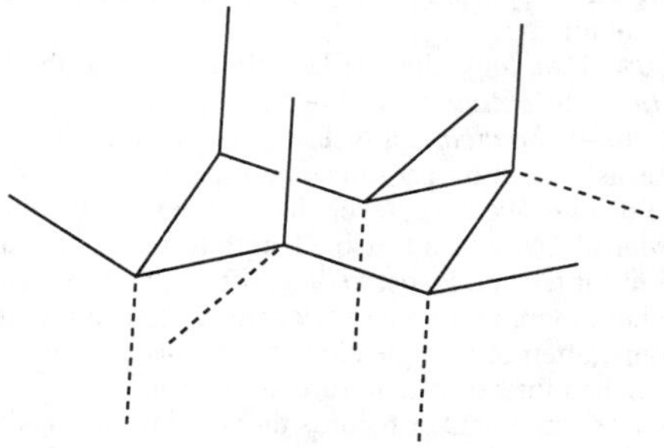

chair form

Chalastogastra *Invertebrate Zoology.* a suborder of hymenopteran insects, including sawflies and horntails, with a broad attachment of thorax to abdomen and caterpillarlike larvae.

chalaza *Botany.* **1.** the base part of an ovule where the nucleus attaches to the funiculus. **2.** the part of a seed, opposite the micropyle, where the seed coat comes together with the rest of the ovule. *Cell Biology.* an albuminous spiral band that attaches the yolk of a bird's egg to the egg membrane.

chalazion *Medicine.* a fluid-filled sac arising on the conjunctival inside surface of the eyelid.

chalazogamy *Botany.* a fertilization process in which pollen passes through the chalaza and not the micropyle.

chalcanthite *Mineralogy.* $Cu^{+2}SO_4{\cdot}5H_2O$, a pale or dark blue, transparent to translucent triclinic mineral occurring as short prismatic or thick tabular crystals with a vitreous luster, having a specific gravity of 2.28 to 2.30 and a hardness of 2.5 on the Mohs scale; found as a secondary mineral in the oxidation zone of copper-bearing sulfide ore deposits. Also, BLUE VITRIOL, BLUESTONE.

chalcedony [kal´sə dōn´ē] *Mineralogy.* a fine-grained variety of quartz.

Chalcididae *Invertebrate Zoology.* a family of hymenopteran insects in the superfamily Chalcidoidea.

Chalcidoidea *Invertebrate Zoology.* a superfamily of wasps, hymenopteran insects in the suborder Apocrita, mostly parasitic on insect larvae.

chalco- *Geology.* a prefix meaning "copper."

chalcoalumite *Mineralogy.* $Cu^{+2}Al_4(SO_4)(OH)_{12}{\cdot}3H_2O$, a turquoise-green, transparent to translucent monoclinic mineral, commonly occurring as twinned, thin, triangular platelike crystals, having a specific gravity of 2.29 and a hardness of 2.5 on the Mohs scale; found with azurite, cuprite, malachite, goethite, and quartz.

chalcocite *Mineralogy.* Cu_2S, a gray to black, opaque, brittle, monoclinic mineral occurring as twinned prisms, single tabular crystals, or in massive form, having a specific gravity of 5.5 to 5.8 and a hardness of 2.5 to 3 on the Mohs scale; found as an important ore of copper in the oxidized and enriched zones of hydrothermal sulfide vein deposits.

chalcocyanite *Mineralogy.* $Cu^{+2}SO_4$, a transparent to translucent pale green, blue, or yellowish orthorhombic mineral occurring as tabular crystals, having a specific gravity of 3.65 and a hardness of 3.5 on the Mohs scale; found in fumaroles following eruptions of Mount Vesuvius.

chalcogen [kal´kə gən] *Inorganic Chemistry.* a term for any of the elements that are part of Group VIB of the periodic table: oxygen, sulfur, selenium, tellurium, or polonium.

chalcogenide *Inorganic Chemistry.* a compound made up of a chalcogen (oxygen, sulfur, selenium, tellurium, or polonium) and a more electropositive element.

chalcogenide glass *Materials.* a glass that possesses eletrical properties similar to those of intrinsic semiconductors and containers of elements known as chalcogens, such as sulfur or tellurium.

Chalcolithic see COPPER AGE.

chalcomenite *Mineralogy.* $Cu^{+2}Se^{+4}O_3 \cdot 2H_2O$, a bright blue, transparent orthorhombic mineral occurring as minute, prismatic crystals, having a specific gravity of 3.8 to 4 and a hardness of 2 to 2.5 on the Mohs scale; found as a secondary mineral formed by the oxidation of copper and lead selenides.

chalcophanite *Mineralogy.* $(Zn,Fe^{+2},Mn^{+2})Mn_4^{+3}O_7 \cdot 3H_2O$, a blue to black, opaque, trigonal mineral occurring in drusy crusts as minute tabular crystals, having a specific gravity of 3.8 to 4 and a hardness of 2.5 on the Mohs scale; found as a secondary mineral in oxidation zones of zinc-bearing ore deposits.

chalcophile *Geochemistry.* an element having a strong affinity for sulfur, and therefore tending to be more abundant in sulfide minerals and ores than in other types of rock. Thus, **chalcophilic.**

chalcophile element see SULFOPHILE ELEMENT.

chalcophyllite *Mineralogy.* $Cu_{18}Al_2(AsO_4)(SO_4)_3(OH)_{27} \cdot 33H_2O$, a green or blue-green, transparent to translucent trigonal mineral occurring as thin tabular crystals; having a specific gravity of 2.67 to 2.69 and a hardness of 2 on the Mohs scale; found as a secondary mineral in oxidation zones of copper-bearing ore deposits.

chalcopyrite *Mineralogy.* $CuFeS_2$, a brassy yellow, brittle, opaque metallic tetragonal mineral occurring as sphenoidal crystals, having a specific gravity of 4.1 to 4.3 and a hardness of 3.5 to 4 on the Mohs scale; an important copper mineral found in medium- to high-temperature sulfide ore deposits.

chalcosiderite *Mineralogy.* $Cu^{+2}Fe_6^{+3}(PO_4)_4(OH)_8 \cdot 4H_2O$, a dark green, transparent to translucent triclinic mineral occurring as short prismatic crystals, having a specific gravity of 3.22 and a hardness of 4.5 on the Mohs scale; found as a secondary mineral in oxidation zones of copper-bearing ore deposits.

chalcostibite *Mineralogy.* $CuSbS_2$, a gray, opaque, orthorhombic mineral occurring as flattened prismatic crystals; having a specific gravity of 4.95 and a hardness of 3 to 4 on the Mohs scale; found with sulfosalts, sulfides and quartz.

chalcotrichite *Mineralogy.* a capillary variety of cuprite occurring in thin, interlacing fibrous crystals.

chaldron *Metrology.* an older unit of dry measure used in Great Britain, equal to 36 bushels.

chalicosis *Medicine.* a lung condition caused by the inhalation of fine stone particles. Also, FLINT DISEASE.

Chalicotheriidae *Paleontology.* an extensive family of perissodactyl ungulates in the extinct suborder Ancylopoda, characterized by cleft and clawlike toes, long front limbs, and a massive head; worldwide distribution of relatively small populations; Eocene to Pleistocene.

Chalicotherioidea *Paleontology.* a superfamily of perissodactyl ungulates, used in some classifications to include the Chalicotheriidae and some smaller groups in the extinct suborder Ancylopoda; extant in the Eocene to Pleistocene.

chalk *Materials.* **1.** a fine-grained limestone or a soft form of calcium carbonate composed of finely divided marine shells; it is very fine-grained, porous, and friable, and is usually white or very light-colored. It is used in putty, crayons, paints, linoleum, and polishes. **2.** a writing implement made of this substance or a similar substance. **3.** to use such a writing implement.

chalkboard *Materials.* see BLACKBOARD.

chalk-brood *Entomology.* a destructive disease of bee larvae, in which infected larvae become fluffy and swollen and later shrink, harden, and die, sometimes turning a chalk-white color.

chalking *Chemistry.* the formation of a loose, powdery substance on the surface of paint, caused by the decomposition of the binding element. *Metallurgy.* in a metallic coating, a flaw caused by undesired deposition of a powdery material. *Graphic Arts.* a fault in ink drying in which the ink rubs off a printed page even though it is dry to the touch; this is caused by a too-rapid absorption of the ink's binding vehicle into the paper.

chalkstone *Pathology.* a concretion taking the form of sodium urate deposits in the joints of the feet or hands, associated with gout.

challenge *Telecommunications.* any procedure performed by one entity in order to validate the identity of another entity.

Challenger *Space Technology.* a U.S. space shuttle that exploded after takeoff in January 1986, killing the astronauts on board.

challis [shal´ē] *Textiles.* a soft, lightweight fabric made of cotton, wool, or synthetic fiber in plain or twill weave.

chalones *Biochemistry.* a protein that blocks cell division.

chalybite see SIDERITE.

Chamaeidae *Vertebrate Zoology.* the wren-tit family, a classification under dispute in some sources; the only family of birds that is strictly North American, having one species, *Chamaea fasciata,* about six inches long, with a gray-brown back and white underbelly; its range is west of the Rockies.

Chamaeleon [kə mēl´yən] *Astronomy.* a small constellation near the south celestial pole. Also, CHAMELEON.

Chamaeleonitidae *Vertebrate Zoology.* the chameleons, a family of reptiles in the suborder Sauria of the order Squamata, characterized by their ability to change color rapidly to match their surroundings, and by very long sticky tongues, independently movable eyes, and granular skin.

chamaemyidae *Invertebrate Zoology.* a family of two-winged aphid flies in the subsection Acalyptratae.

chamaephyte *Ecology.* a plant classification for perennial plants that have winter buds at ground level or within 25 cm of ground level; usually low-growing woody or herbaceous plants common in dry or cold climates. Also, **chamaeophyte.**

chamber *Ordnance.* **1.** the part of a gun that holds the charge before firing. **2.** to insert a round of ammunition into the chamber of a gun. Thus, **chambering.** *Mining Engineering.* **1.** the place in which a miner works. **2.** a body of ore apparently filling a preexisting cavern, possessing definite boundaries.

chamber acid *Inorganic Chemistry.* a term for sulfuric acid, H_2SO_4, when prepared by the lead chamber process.

chambered nautilus *Invertebrate Zoology.* any cephalopod of the genus *Nautilus,* having a spiral, chambered shell with pearly septa. Also, NAUTILUS, PEARLY NAUTILUS.

chambered nautilus

chambering *Mining Engineering.* the action of increasing the size of a drill hole in a quarry by means of small explosive charges, in preparation for the use of a much larger one.

chamber kiln *Engineering.* a kiln in which the fire moves through a series of chambers, arranged in a circle, over a period of several days, allowing time for waste gas from the fuel to preheat objects in a chamber before the fire reaches it.

Chamberlain, Owen born 1920, American physicist; shared Nobel Prize for discovery of the antiproton.

Chamberlain, Thomas 1843–1928, American geologist; with Moulton, formulated planetesimal theory of origin of the solar system.

Chamberland filter or **Chamberland candle** *Microbiology.* a filter made of unglazed porcelain that is used to produce filtrates free of bacteria. Also, PASTEUR-CHAMBERLAND FILTER.

chamber process *Chemical Engineering.* a sulfuric acid manufacturing process in which sulfur dioxide, air, and steam are reacted in lead chambers, using nitrogen oxides as a catalyst.

chamber tomb *Archaeology.* a prehistoric tomb that contains a large burial chamber, especially such a construction used in Europe in the Megalithic period.

chambray *Textiles.* a lightweight plain-weave cotton fabric having a color warp interwoven with white filling.

chameleon [kə mēl′yən] *Vertebrate Zoology.* a lizardlike reptile of the family Chamaeleonitidae, having a long, sticky tongue, independently movable eyes, granular skin, and an ability to change color rapidly to match its environment; most species are found in sub-Saharan Africa.

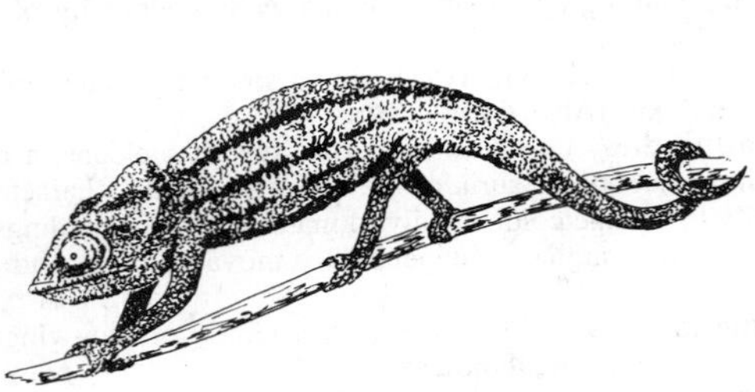

chameleon

Chameleon see CHAMAELEON.

chameolith see HUMIC COAL.

chamfer *Engineering.* **1.** a beveled edge or corner. **2.** a groove cut into wood or other materials. **3.** the angle between a beveled surface and the axis of a milling cutter.

chamfering see BEVELING.

chamfer plane *Mechanical Devices.* a plane fitted with adjustable guides, used to bevel down right-angled edges in woodwork.

chamois [sham′ē] *Vertebrate Zoology.* a goatlike mammal of the subfamily Caprinae in the family Bovidae; found in the mountains of Europe and western Asia and known for its remarkable leaping ability. *Textiles.* see CHAMOIS LEATHER.

chamois leather *Textiles.* a high-quality light-tan leather with a soft nap, originally made from the skins of the chamois of Europe but now from treated lamb, sheep, goat or split hides; it is used for clothing and as a polishing and drying material. Also, **chamois skin.**

chamosite *Mineralogy.* $(Fe^{+2},Mg,Fe^{+3})_5Al(Si_3Al)O_{10}(OH,O)_8$, a green, gray or black monoclinic mineral of the chlorite group that is dimorphous with orthochamosite and forms a series with clinochlore, having a specific gravity of 3 and a hardness of 3.1 to 3.2 on the Mohs scale; found in sedimentary ironstones with siderite and kaolinite.

Champagne *Food Technology.* **1.** a dry white sparkling wine made in the Champagne region of northeastern France. **2.** also, **champagne.** a similar, usually white sparkling wine made elsewhere.

champak *Materials.* the lustrous wood of the *Michelia champaca* timber tree of southeast Asia; used for furniture, boats, plywood, and pulp. Also, **champac.**

Champiaceae *Botany.* a family of marine red algae of the order Rhodymeniales, having cylindrical or tubular branched thalli of multiaxial construction and similar gametangial and sporangial forms.

Champlainian *Geology.* **1.** a North American provincial series of the Middle Ordovician period, occurring after the Canadian and before the Cincinnatian. **2.** formerly, the Ordovician period.

Champollion, Jean-Francois 1790–1832, French archaeologist; an early Egyptologist; deciphered the Rosetta Stone.

chan channel.

chance-constrained programming. *Mathematics.* in nonlinear mathematical programming, a process in which an optimization problem is constrained by boundaries or barrier functions.

Chance process *Mining Engineering.* a method of cleaning a coal product using a fluid mixture of sand and water, which floats off, allowing slate and other impurities to sink.

chancre *Medicine.* **1.** a lesion caused by infection with syphilis. **2.** a skin lesion symptomatic of diseases such as tularemia or sporotrichosis.

chancroid *Medicine.* a venereal disease marked by painless sores in the genital area.

Chandler, Seth Carlo 1846–1913, American astronomer; discovered a periodic wobble in the earth's motion.

Chandler motion see POLAR WANDERING.

Chandler period *Geophysics.* a period of the Chandler wobble.

Chandler wobble *Geophysics.* a constant, predictable movement in the rotation of the Earth's axis over a period of about 14 months.

Chandrasekhar, Subrahmanyan born 1910, Indian-born American astrophysicist; Nobel Prize for research on white dwarf stars.

Chandrasekhar limit *Astronomy.* the maximum mass (equal to 1.44 solar masses) that theory predicts a white dwarf star can attain without collapsing to become a neutron star or black hole.

chanduy or **chandui** *Meteorology.* a cool, descending wind that blows during the dry season from July to November in Guayaquil, Ecuador.

change chart *Meteorology.* a chart indicating the amount and direction of change of any meteorological element over a specified time interval. Also, TENDENCY CHART.

changed memory routine *Computer Programming.* a utility program that creates a change dump.

change dump *Computer Programming.* a printout or other recording of the contents of a storage device containing only those items that have changed since the last such dump.

change gear *Mechanical Engineering.* a gear that alters the speed of a driven shaft while maintaining the speed of the driving shaft.

change in storage *Hydrology.* the difference between the amount of water flowing into a reservoir or other body of water and the amount flowing out.

change of coordinates see COORDINATE TRANSFORMATION.

change of tide *Oceanography.* a shift in the direction of the tide or of a tidal current, as from high tide to low tide.

changeover switch *Electricity.* a switching device for changing an electric circuit from one set of connections to another.

change record *Computer Programming.* a record processed to alter a corresponding record in the master file. Also, TRANSACTION RECORD.

change tape *Computer Programming.* a magnetic tape containing the change records or transactions to be used in updating the master file.

changing bag *Engineering.* a lightproof bag used to enclose a camera while loading or unloading photographic film.

Chanidae *Vertebrate Zoology.* the milkfish, a family of marine fishes in the suborder Chanoidei of the order Gonorynchiformes, having no teeth and being among the most fertile of fishes.

channel a way or course along which something moves; specific uses include: *Hydrology.* **1.** a natural bed or an open, artificial conduit in which a body of water runs or may run, either continuously or periodically. **2.** a natural passageway or water course connecting two bodies of water. *Geology.* a buried or abandoned water course that is indicated by the presence of gravel or sand deposits. *Navigation.* the central or deepest part of a river or stream that carries the main flow. *Geography.* **1.** a wide strait. **2. the Channel.** the English Channel. *Agronomy.* the flat, lower part of a furrow. *Telecommunications.* the portion of a frequency spectrum that is assigned for the operations of one specific carrier, such as a television broadcasting facility to ensure clear communications. *Electronics.* **1.** a single path for transmitting an electric signal. **2.** in a field effect transistor, the conducting path between the source and the drain. *Computer Technology.* **1.** a path along which data flow between components in a computer system. **2.** a device associated with the input/output controller that transfers data between the processor and one or more peripheral devices; three types of channels are currently in use: selector, byte-multiplexer, and block-multiplexer. **3.** the portion of storage media, such as a track on a magnetic tape or disk, that is available to a single character-reading station. *Chemical Engineering.* a furrow causing excessive flow in a portion of the clay bed in the percolation filtration process, allowing a stream to pass unfiltered. *Nucleonics.* **1.** the path that carries coolant past the cladding surface to remove heat from a reactor core. **2.** a passage between fuel elements for primary coolant flow. *Building Engineering.* any reinforced concrete structural member composed of three sides forming a rectangle. *Transportation Engineering.* in a rail station, an enclosed line and its passenger platform.

channel adapter *Computer Technology.* any of a variety of devices used to allow data transfer between devices that operate at differing transfer rates. Also, INPUT/OUTPUT ADAPTER.

channel address word *Computer Programming.* **1.** in an input/output instruction, the field that contains the logical address of the channel to be activated; other fields may contain the operation code (read, write) and the desired device address. **2.** the main memory address that contains the first instruction of a channel program.

channel-attached device *Computer Technology.* an input or output device that is directly connected to the computer through data channels.

channel bank *Electronics.* the part of a carrier-multiplex terminal that converts radio frequencies into voice frequencies and vice versa.

channel bar *Geology.* an elongated deposit of sand and gravel, usually found in the course of a braided stream.

channel basin *Geology.* a former term for a long, narrow proglacial trench, valley, or channel.
channel black see GAS BLACK.
channel buoy *Navigation.* a buoy that marks the location of a navigable channel.
channel capacity *Hydrology.* the maximum flow that can be carried by a given channel without overflowing its banks. *Computer Technology.* the rate at which a channel can transfer data between main storage and an input/output device, usually measured in bytes or kilobytes per second. *Telecommunications.* the maximum signal rate that a particular channel is capable of transmitting under given conditions; usually measured in bauds or bits per second.
channel command *Computer Programming.* an instruction stored in main memory or a special register that is decoded and executed by the I/O channels; usually contains an operation code (such as read or write), flags and options, and the starting address and the word count of the data on which the operation is to be performed. Also, **channel command word, channel control command.**
channel configuration *Computer Technology.* the types, numbers, and connectivities of channel controllers, peripheral I/O devices, and the main processor to the channel. *Transportation Engineering.* the physical configuration of a highway traffic channelizing arrangement.
channel control *Hydrology.* a condition in which the stage of a stream is influenced only by discharge and the contours of the stream's bed, banks, and floodplains.
channel depth see RIDGE HEIGHT.
channel design *Computer Technology.* the specific capabilities, characteristics, schematics, and logic that are included in a particular data channel.
channel designator *Telecommunications.* a unique identification assigned to a channel for reference purposes. Also, CHANNEL SEQUENCE NUMBER.
channel director *Computer Technology.* a scanner circuit that is located between the processor and multiple data channels, polling the channels in a defined priority order and activating data transfers as they are requested.
channeled scabland *Geology.* scabland that has been deeply eroded due to scouring by glacial meltwater.
channel effect *Electronics.* the leakage that occurs over the surface path between the collector and emitter in some types of transistors.
channel-end condition *Computer Technology.* a signal that occurs when the data transmission is completed; the channel is considered busy until the CPU accepts this condition.
channeler or **channeller** see CHANNELING MACHINE.
channel erosion *Geology.* erosion caused by the repeated flow of water over the same course or in well-defined channels.
channel fill *Geology.* **1.** an alluvial deposit in a stream channel, usually where the rate at which sediment is carried by the stream is insufficient to remove the material supplied to it. **2.** the cast of a channel that is cut in shale and filled with sand.
channel flow *Hydrology.* surface runoff that moves through a long, narrow depression that is bounded by banks or valley walls. Also, CONCENTRATED FLOW.
channel frequency see STREAM FREQUENCY.
channel gradient ratio see STREAM GRADIENT RATIO.
channeling *Physics.* the flow of particles or material into a medium that contains voids or regions of lower density, so as to equalize the density throughout the medium. *Analytical Chemistry.* the establishment of flow paths in a bed of solid particles due to the passage of liquid through the particles. *Telecommunications.* the process of multiplexing two or more channels onto one band of signals by assigning each channel a discrete frequency band within the main band. Also, **channelizing.**
channeling machine *Mechanical Engineering.* a power-driven chiseling machine that consists of a series of cutting tools that move back and forth on a track; used to excavate soft rock.
channeling radiation *Physics.* the radiation emitted by charged particles that pass through a solid.
channel iron *Mechanical Devices.* a U-shaped metal fitting such as a roll bar with a web and two flanges; used in bridge and girder work.
channelization *Transportation Engineering.* an arrangement that directs the flow of highway traffic into streams, limiting or preventing movements from one stream to another.
channel-lag deposit *Geology.* the coarse residual material that accumulates as the material is removed in a channel during the normal processes of the stream.
channel light *Navigation.* a navigational light that marks the location of a navigable channel.
channel line *Hydrology.* the line of the strongest flow or fastest current of a stream.
channel marker *Navigation.* a light or buoy that marks a channel.
channel mask *Computer Technology.* in an IBM system program status word, a field that indicates which channels may interrupt the task with their completion signals.
channel miles *Telecommunications.* a measurement, in miles, multiplied by the number of communications channels between two given points.
channel morphology see RIVER MORPHOLOGY.
channel-mouth bar *Geology.* a bar resulting from a decrease in velocity as a stream enters a body of standing water.
channel net *Hydrology.* the pattern made by all stream channels within a particular drainage basin. Also, **channel network.**
channel order see STREAM ORDER.
channel pattern *Hydrology.* an aerial view showing the configuration of a limited reach of a river channel. Also, RIVER PATTERN.
channel process *Chemical Engineering.* the production of carbon black, such as channel black or lampblack, by burning natural gas in the absence of air in a closed chamber containing baffles, channels, or ducts, in which the various grades of fine carbon particles are precipitated.
channel program *Computer Technology.* a list of instructions or channel commands stored in main memory that the channel retrieves and executes sequentially to perform input and output operations.
channel protein *Biochemistry.* a protein that facilitates the diffusion of molecules or ions across lipid membranes by forming a hydrophilic pore.
channel pulse *Telecommunications.* an electrical pulse conveying channel information by modulation of that pulse, e.g., pulse-duration modulation.
channel read-backward command *Computer Programming* in a channel command, an operation code that causes the channel to transfer data from a magnetic tape device while the tape is moving backward.
channel read command *Computer Technology.* in a channel command, an operation code that causes the channel to transfer input data from a storage device into main memory.
channel reliability *Telecommunications.* the percentage of time that a channel satisfies the standards established by its operator, such as data rate, bit-error rate (BER), and channel availability.
channel roughness *Geology.* a measure of the tendency of stream channel material to resist the flow of water.
channel sample see GROOVE SAMPLE.
channel sand *Geology.* a sand or sandstone deposited in a stream bed or other channel.
channel scheduler *Computer Programming.* a program in a computer operating system that processes the list of requests for input and output operations in the appropriate order.
channel segment see STREAM SEGMENT.
channel selector *Electricity.* a switch or other control that tunes in a desired channel.
channel sense command *Computer Technology.* a signal sent to a channel requesting additional information about the status of a particular input/output device connected to that channel; the command might be sent as a result of a complex status such as a card jam, an empty card hopper, or a printer out of paper.
channel sequence number see CHANNEL DESIGNATOR.
channel shifter *Electronics.* a radiotelephone carrier circuit that reduces cross-talk between channels by moving voice-frequency channels to a higher frequency.
channel skip *Computer Technology.* a carriage-control character that instructs the printer to skip to the top of the next page or to carry out a vertical tab.
channel spacing *Telecommunications.* the frequency difference between two adjacent television or radio channels.
channel spin *Nuclear Physics.* the sum, including the magnitude and direction, of the spins of the particles active in a nuclear reaction before or after the reaction takes place.
channel splay see FLOODPLAIN SPLAY.
channel spring *Hydrology.* a depression spring issuing from the bank of a stream whose channel is below the water table.
channel status table *Computer Technology.* a table of channel status words relating to the current status of all data channels in the system.

channel status word *Computer Technology.* a computer word describing the control information that is presented to the channel by the controller and the status of the channel itself.

channel synchronizer *Computer Technology.* a device that transmits data between a computer's central processing unit and its peripheral devices.

channel-to-channel adapter *Computer Technology.* a direct-control connection device for rapid data transfer between two computers.

channel wave *Geology.* a seismic wave propagated along a surface or discontinuity within a low-velocity layer of the earth, ocean, or atmosphere.

channel width *Geology.* the distance across a channel or a stream measured from bank to bank when the stream nearly fills its channel.

channel wing *Aviation.* an airplane wing that is shaped like a half-circle (when seen from fore or aft) and that forms a trough partly surrounding the propeller, which increases lift at low speeds and facilitates short takeoff and landing.

Channidae *Vertebrate Zoology.* the snakeheads, the sole family in the suborder Chanroidei of the order Perciformes; found in fresh waters of tropical Africa and eastern Asia and characterized by an extra respiratory organ allowing respiration of atmospheric oxygen for short periods.

Chanoidei *Vertebrate Zoology.* a suborder of marine fishes in the order Gonorynchiformes, including the milkfish, kneriids, and sand eels.

Chantilly (lace) *Textiles.* a type of bobbin lace having a scroll or floral pattern outlined in heavy thread on a fine hexagonal mesh background. Also, **chantilly.** (Originally from the town of *Chantilly,* near Paris.)

Chaoboridae *Invertebrate Zoology.* the phantom midges, a family of two-winged insects in the suborder Orthorrhapha.

Chaoborous *Invertebrate Zoology.* the phantom midge; an insect closely related to the mosquitoes, but differing in being unable to suck blood; the larvae are predatory and often feed on mosquito larvae.

chaos *Chaotic Dynamics.* the dynamical evolution that is aperiodic and sensitively dependent on initial conditions. In dissipative dynamical systems this involves trajectories that move on a strange attractor, a fractal subspace of the phase space. For continuous systems chaos requires at least three dimensions in phase space. This term takes advantage of the colloquial meaning of chaos as random, unpredictable, and disorderly behavior, but the phenomena given the technical name *chaos* have an intrinsic feature of determinism and some characteristics of order. Also, CHAOTIC DYNAMICS, CHAOTIC BEHAVIOR (IN SYSTEMS), DETERMINISTIC CHAOS, NONLINEAR DYNAMICS. *Geology.* a type of breccia that consists of various sizes of irregularly shaped blocks.

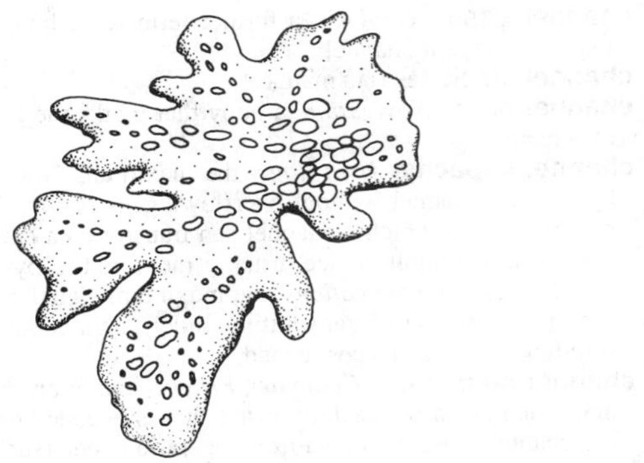

Chaos

Chaos *Invertebrate Zoology.* a genus of large, sluggish freshwater amoeboid protozoans with few to many nuclei and several short, broad pseudopods. Also, PELOMYXA.

chaotic behavior see CHAOS.

chaotic dynamics see CHAOS.

Chaoul therapy *Radiology.* a method of treatment using low-energy X-radiation and short source-to-tissue separation.

Chaoul tube *Radiology.* a low-voltage X-ray tube that can be placed 2 cm from the skin, allowing intense but superficial radiation penetration.

chaparral *Ecology.* a type of vegetation characterized by low, thickly growing evergreen shrubs or bushes with flat, broad leaves and interlacing branches; the typical natural growth of many areas with a climate of cool moist winters and long dry summers, as in much of the western United States. Also, MEDITERRANEAN SCRUB.

Chaparral *Ordnance.* a surface-to-air missile system designed for short-range, low-altitude Army air defense operations; officially designated **MIM-72.**

chapel *Architecture.* a small, usually private place of worship within a larger building such as a church, school, or hospital.

chaplet *Metallurgy.* in casting, a metallic support that holds the core.

Chaplygin equation *Fluid Mechanics.* the equation denoting two-dimensional potential flow at constant entropy without shocks.

Chapman-Enskog approximation *Physics.* a method of successive approximation used to solve the Boltzmann transport equation.

Chaotic Dynamics (Chaos)

The new discipline of chaotic dynamics, known popularly as "chaos" and listed in library catalogs as "chaotic behavior in systems," is an analytical approach to the array of real-world dynamical systems that are random, irregular, aperiodic, and unpredictable. Most natural phenomena and the mathematical equations that attempt to describe them are inherently nonlinear—so complex that they cannot be computed, predicted, or sometimes even defined. The easily integrable, continuous, well-behaved linear equations of classical dynamics exist mostly in textbooks and, in the real world, are the exception rather than the rule. By contrast, nonlinear mathematics is ubiquitous in the world of chaos, where even simple systems create difficult problems of predictability.

Chaos lays emphasis on the influence of starting conditions on the evolution of dynamical systems, and presents newly discovered laws governing the behavior of dynamical systems encountered in meteorology, physics, chemistry, engineering, ecology, and medicine, including cardiac arrhythmia and erratic nerve impulses.

Precursory intimations of chaotic behavior, dating from work as early as that of Poincaré (1892), lay dormant and were viewed mostly as curiosities until the synergistic confluence of experimental mathematics and high-speed electronic computers about twenty years ago. The analytical methods of the discipline now known as chaos permit access to quantitative numerical simulations of nonlinear dynamical systems to levels of precision previously inaccessible, with the clarifying knowledge that the complexity of Nature is not merely a consequence of randomness, but reflects some embedded order. Computers have opened windows on the world of bifurcations, period-doubling, and strange attractors—patterns poised between order and chaos—by generating images of extraordinary mathematical content and rare aesthetic appeal.

To put its importance into perspective, it is fair to say that chaotic dynamics bears the same relationship to explaining the turbulent universe of nonlinear events as relativity bears to analysis of motion at speeds near the velocity of light and that quantum mechanics bears to understanding phenomena at the atomic and molecular scale.

The entries in this dictionary labeled *Chaotic Dynamics* are representative of the vocabulary of this new discipline, which already has a large literature that is increasing beyond easy enumeration—seemingly bifurcating and period-doubling with a life of its own.

John W. McDonald
Weapons Technology Staff, retired
Los Alamos National Laboratory

Chapman-Enskog solution *Physics.* a method of solving the Boltzmann transport equation using the Chapman-Enskog theory.

Chapman-Enskog theory *Physics.* a procedure for solving the Boltzmann transport equation by successive approximations, primarily in powers of the mean free path. Also, ENSKOG THEORY.

Chapman equation *Geophysics.* a theoretical formula that equates electron density distribution in the upper atmosphere to altitude.

chapmanite *Mineralogy.* $Sb^{+3}Fe_2^{+3}(SiO_4)_2(OH)$, a green or yellow monoclinic mineral having a specific gravity of 3.69 to 3.75 and a hardness of up to 2.5 on the Mohs scale, occurring as lath-shaped crystals associated with silver and sulfide ores.

Chapman-Jouguet condition *Fluid Mechanics.* the lowest possible velocity of a detention wave.

Chapman-Jouguet plane *Thermodynamics.* a hypothetical plane behind a shock front, defining the region behind which reaction gases have reached thermal equilibrium, and the detonation velocity has become equal to the sound velocity plus the particle velocity.

Chapman region *Geophysics.* a region of the upper atmosphere in which the Chapman equation applies.

Characeae *Botany.* a family of nonvascular green plants of the order Charales, similar to the green algae but forming an intermediate group between them and the higher land plants.

Characidae *Vertebrate Zoology.* the South American tetras or characins, a large family of freshwater schooling fishes in the suborder Characoidei.

Characidiopsidaceae *Botany.* a monogeneric family of freshwater yellow-green algae of the order Heterogloeales, characterized by solitary cells with a distinct cell wall that are attached to the substrate by either a stipe or a gelatinous disk.

Characiochloridaceae *Botany.* a family of the order Tetrasporales encompassing those genera that occur almost always as solitary cells lacking pseudoflagella, live attached to planktonic organisms, and are usually enclosed in a gelatinous envelope. Also, CHLORANGIELLACEAE.

Characiopsidaceae *Botany.* a family of freshwater, brackish, and marine yellow-green algae of the order Mischococcales characterized by solitary or paired cells that are attached to the substratum directly or by a mucilaginous pad.

Characiosiphonaceae *Botany.* a monospecific family of green algae of the order Chlorococcales, composed of a cylindrical or clavate sac that attaches in clumps to substrates in shallow streams in India.

Characoidei *Vertebrate Zoology.* a suborder of freshwater fishes in the order Cypriniformes, composed of only one family, Characidae.

character *Graphic Arts.* a symbol in a font of type, such as a letter, numeral, or punctuation mark. *Computer Programming.* **1.** any of a set of basic symbols that can be grouped and ordered to express information, such as letters, numerals, punctuation marks, and special symbols. **2.** the electrical, mechanical, or magnetic representation of such a symbol that can be stored in a computer or a storage medium. **3.** the binary code that represents a character. *Psychology.* an individual's personality; all the behavior patterns and reportable inner experiences that characterize a person. *Systematics.* any feature or trait of an organism, or any difference among organisms, used to construct a classification or estimate phylogeny. *Mathematics.* **1.** the character χ of a group representation over an algebraically closed field K is a homomorphism that maps the group to the multiplicative group of K. The character of each group element is equal to the trace of the associated linear transformation or matrix. Also, **proper character.** **2.** a homomorphism from a Banach algebra into the multiplicative group of complex numbers.

character-addressable computer *Computer Technology.* a computer system in which each stored character has one unique location with its own address. Also, CHARACTER-ORIENTED COMPUTER.

character boundary *Computer Technology.* in character recognition, the real or assumed rectangle that frames a character and separates it from others, above and below as well as on either side.

character code *Telecommunications.* a unique bit pattern that is designated to represent a specific character in a character set.

character convergence *Evolution.* an evolutionary process in which two species interact with their environment in such a way that each begins to resemble the other in terms of one or more hereditary characters.

character count *Graphic Arts.* an actual or estimated count of the total number of characters and letter spaces contained in a given text or manuscript.

character data type *Computer Programming.* an interpretation applied to a string of bits that classifies it as internally representing a printable character.

character density *Computer Technology.* a measure of the number of characters that can be recorded on a given area of storage medium, such as characters per inch on magnetic tape.

character disorder see PERSONALITY DISORDER.

character displacement *Evolution.* the observed character divergence in the sympatric populations of two species as compared to allopatric populations of the same two species, presumably as a result of the selective effects of competition for common food resources.

character display terminal *Computer Technology.* an interactive video display unit that is limited to the display of printable characters.

character divergence *Evolution.* an evolutionary process in which two recently evolved species interact in such a way that each begins to resemble the other less and less in terms of one or more hereditary characters.

character emitter *Computer Technology.* an electronic circuit or electromechanical device that emits a timed pulse or series of pulses representing a character in some acceptable code.

character fill *Computer Programming.* the process of replacing all character positions in one or more storage locations with repetitions of a single character, usually blanks or zeros.

character generator *Computer Technology.* in a video display unit, the hardware or software device that causes a character to be visible on the screen, usually by brightening a dot in a matrix or by painting lines between points on the character matrix.

character group *Mathematics.* the group formed by the set of all characters in the special case of the continuous functions f from a group G (with operation $*$) to the unit circle in the complex plane, such that

$$f(x*y) = f(x)f(y)$$

for all $x, y \in G$. The product of two such characters f and g is the character h defined by $h(x) = f(x)g(x)$ for each $x \in G$. G is isomorphic to its character group if G is Abelian and locally compact. A topology on the character group can be defined as follows: N is a neighborhood of a character f if there are elements $x_1, \ldots, x_k$ of G and $\varepsilon > 0$ such that N is the set of all characters g for which

$$|f(x_i) - g(x_i)| < \varepsilon \quad \text{for } i = 1, 2, \ldots, k.$$

The character group is locally compact if G is locally compact, and discrete if G is compact.

characteristic *Statistics.* the attribute of the population that is the object of statistical study, such as family income or life span. *Mathematics.* **1.** the integral part of a logarithm. **2.** let R be a ring (or integral domain or field). If there is a least positive integer n such that $na = 0$ for all $a \in R$, then R is said to have characteristic n. If no such n exists, R is said to have characteristic 0. Denoted char $R = n$. **3.** a subgroup H of a group G is said to be a **characteristic subgroup** if $f(H)$ is a subgroup of H for every automorphism $f: G \to G$. If f is an endomorphism then H is said to be fully invariant. Every fully invariant subgroup is characteristic and every characteristic subgroup is normal. **4.** see EIGENVALUE. *Computer Programming.* the exponent part of a number expressed in floating-point representation.

characteristic actuation probability *Ordnance.* the average probability of a specific type of mine being set off during one mine-sweeping run.

characteristic actuation width *Ordnance.* the width of the path over which mines can be actuated by a single run of the mine-sweep gear. Similarly, **characteristic detection width.**

characteristic chamber length *Space Technology.* the length of a straight cylindrical tube of given diameter having the same volume or capacity as the smallest rocket engine capable of complete combustion of a given type and amount of propellant.

characteristic class *Mathematics.* a ***k*th characteristic class** is a function c that associates with each (smooth) n-dimensional bundle $\xi = \pi : E \to M$ over a manifold M an element $c(\xi)$ of the kth deRham cohomology class of M such that if $\xi' = \pi' : E' \to M'$ is another smooth n-dimensional bundle, and $(\bar{f}, f) : \xi' \to \xi$ is a smooth bundle map, then $c(\xi') = f^*(c(\xi))$, an element of $H^k(M')$. Special cases of this are the Euler (characteristic) classes, Chern class, and Pontryagin classes.

characteristic cone *Mathematics.* for a hyperbolic partial differential equation such as the wave equation, a cone that divides the domain of the partial differential equation into regions that can and cannot influence or be influenced by a disturbance at the vertex. Also called a **light cone** or **null cone** because its principal application is to the (electromagnetic) wave equation, whose characteristic curves are null geodesics in the Minkowski metric in flat space-time.

characteristic curve *Mathematics.* a curve on a given surface S, such that the direction of the tangent to the curve at any point p is a characteristic direction at p.

characteristic damage state *Materials Science.* the transition point or zone between the two stages of fatigue development in fiber-reinforced composites, characterized by complete saturation of noninteractive cracking.

characteristic detection probability *Ordnance.* the ratio of the number of mines that are actually detected during one mine-sweeping run, to the total number of mines that are present and could have been detected.

characteristic directions *Mathematics.* given a point p on a surface S, the particular pair of conjugate directions at p that are symmetric with respect to the principal directions of S at p.

characteristic distance see CHARACTERISTIC LENGTH.

characteristic distortion *Telecommunications.* signal transition displacement that is caused by previous transitions.

characteristic equation *Physics.* **1.** an analytical relationship between a set of physical variables that determines the state of a physical system. **2.** any equation that shows the relationships between a substance's volume, pressure, and temperature. Also, EQUATIONS OF STATE. *Mathematics.* **1.** given a general quasi-linear second order partial differential equation

$$\sum_{i,j=1}^{n} a_{ij}(x_1,\ldots,x_n)\partial^2 w/\partial x_i \partial x_j + F(x_1,\ldots,x_n,w,\partial w/\partial x_1,\ldots,\partial w/\partial x_n)=0,$$

linear in the second derivatives with continuous coefficients a_{ij}, the equation

$$\sum_{i,j=1}^{n} a_{ij}(x_1,\ldots,x_n)\,\partial w/\partial x_i \partial w/\partial x_j = 0$$

2. the equation $\det(\lambda I - L) = 0$; i.e., the condition for solvability of the eigenvalue equation $Lu = \lambda u$, where L is a linear operator. The roots of the characteristic equation are the eigenvalues (or **characteristic roots**) of L. **3.** in general, any equation, subject to boundary conditions and containing a parameter, that has a solution only when the parameter takes on certain values.

characteristic exhaust velocity see CHARACTERISTIC VELOCITY.

characteristic form *Mathematics.* a polynomial created by replacing each partial derivative $\partial/\partial x_i$ in the highest-order terms of an mth order quasi-linear partial differential operator by an indeterminate ξ^i. The resulting mth-degree homogeneous polynomial is the characteristic form for the operator, and its properties (such as positivity, root varieties, etc.) reveal much about the operator and its solutions; e.g., the characteristic form for the operator

$$\sum_{i,j=1}^{n} a_{ij}(x_1,\ldots,x_n)\,\partial^2 w/\partial x_i \partial x_j + F(x_1,\ldots,x_n,w,\partial w/\partial x_1,\ldots,\partial w/\partial x_n)$$

is the polynomial $\Sigma_{i,j=1}^{n} a_{ij}(x_1,\ldots,x_n)\xi^i\xi^j$. The modern term for the characteristic form is the symbol for the partial differential operator, although partial derivatives may be replaced by $(\sqrt{-1})\xi^i$ instead of just ξ^i alone.

characteristic frequency *Telecommunications.* an easily identified and easily measured frequency that is characteristic of a given transmission.

characteristic function *Physics.* a function that is the integral of some property of an optical or mechanical system over time or over the path followed by the system, and whose value for a path actually followed by the system is a maximum or a minimum value with respect to nearby paths with the same endpoints. *Statistics.* for a random variable X, the function $c(t) = E\{e^{itX}\}$, where t is any real number, and E is the expectation sign; this is important in the proof of theoretical results, including the central-limit theorem. *Mathematics.* **1.** if A is a subset of a set S, then the characteristic function of A is the function χ_A: $S \rightarrow \{0, 1\}$ such that $\chi_A(x) = 1$ if $x \in A$ and $\chi_A(x) = 0$ if $x \in S - A$. Also, INDICATOR FUNCTION. **2.** see EIGENFUNCTION.

characteristic impedance *Telecommunications.* in a coaxial cable or other transmission medium of infinite length, the ratio of the applied electromotive force to the resulting steady-state current. Also, SURGE IMPEDANCE.

characteristic length *Engineering.* any reference length that is used for convenience in engineering calculations. *Thermodynamics.* any quantity expressed in units of length that characterize the appropriate length, such as characteristic length for conduction or for diffusion. Also, CHARACTERISTIC DISTANCE.

characteristic loss spectroscopy *Spectroscopy.* the detection and analysis of the energy lost by backscattered particles when a solid surface is bombarded with monochromatic electrons.

characteristic manifold *Mathematics.* an initial manifold for a general first-order partial differential equation in n independent variables specifies a unique solution to the partial differential equation, provided that it is transverse to the characteristic directions of the partial differential equation. If an initial manifold fails to be transverse, a solution may still be possible if the manifold is a characteristic (strip) manifold; that is, if it satisfies n–1 sets of ordinary differential equations, each of which resembles the characteristic system for the partial differential equation. Uniqueness of the solution may fail, however.

characteristic overflow *Computer Programming.* in floating-point number arithmetic, the condition that occurs when the characteristic exceeds the maximum positive value allowed by the computer hardware.

characteristic polynomial *Mathematics.* **1.** the monic polynomial p_A of degree n: $p_A = \det(xI - A)$, where A is any $n \times n$ matrix over $K[x]$ and K is a commutative ring with identity. Similar matrices have the same characteristic polynomial. **2.** in general, let ϕ: $E \rightarrow E$ be an endomorphism of a free K-module E of finite rank n (such as R^n), where K is a commutative ring with identity. (ϕ has a matrix representation relative to each ordered basis, and any two such matrices are similar.) Then the characteristic polynomial p_ϕ of ϕ is p_A, where A is any matrix of ϕ relative to some ordered basis; p_ϕ is independent of the choice of basis.

characteristic radiation see CHARACTERISTIC X-RAYS.

characteristic ratio *Materials Science.* a measurement used to compare the physical properties of different polymers, equal to the ratio of the square of the average random-flight end-to-end distance to the product of a number of backbone units, times the interactions along the polymer chain.

characteristic rays see CHARACTERISTIC X-RAYS.

characteristics *Fluid Mechanics.* the space-time paths along which fluid disturbances propagate.

characteristic space see EIGENSPACE.

characteristic species *Ecology.* a species that is usually found in association with other specific types of plant and animal life and, as such, helps define the nature of a local ecology.

characteristic spectrum *Materials Science.* a series of higher-intensity peaks superimposed on the continuous white X-radiation spectrum resulting from the characteristic radiation.

characteristic strength *Materials Science.* a distribution function that best fits experimental data used in Weibull statistics.

characteristic strip *Mathematics.* a characteristic curve for the partial differential equation in n independent variables, along with a specification of an n-dimensional tangent plane at each point of the curve, consistent with the characteristic systems for the partial differential equation. It can be shown that if a characteristic strip is tangent to a solution surface at one point, then the entire characteristic curve lies in the solution surface.

characteristic system *Mathematics.* for a given partial differential equation, its characteristic system is a set of ordinary differential equations for the characteristic directions and normals to tangent planes of solutions to the original partial differential equation. With appropriate initial conditions, the characteristic system can be used to construct a solution to the original partial differential equation.

characteristic underflow *Computer Programming.* in floating-point number arithmetic, the condition that occurs when the characteristic falls below the lower bound of the allowable range.

characteristic value see EIGENVALUE.

characteristic vector see EIGENVECTOR.

characteristic velocity *Space Technology.* a performance standard for the effective combustion of propellants and production of thrust in a rocket engine; equal to the product of the combustion-chamber pressure and the nozzle throat area divided by the rate of propellant mass flow. Also, CHARACTERISTIC EXHAUST VELOCITY.

characteristic X-rays *Atomic Physics.* X-rays of definite wavelengths, characteristic of a pure substance (generally a metal) and produced when that substance is bombarded by fast electrons. Characteristic X-rays are emitted when an electron that has been displaced from an inner shell of the atom being excited (an atom in the target) is replaced by another electron that falls in from an outer shell. Also, CHARACTERISTIC RAYS, CHARACTERISTIC RADIATION.

characterization factor *Chemical Engineering.* a number that relates the variations in physical properties of paraffinic stock with a change in the amount of aromatic material present.

character macro *Artificial Intelligence.* in Lisp, a macro that is invoked by a single character in the source code.

character of the bottom *Navigation.* a description (such as muddy, soft, or sticky) of the nature of the water bottom.

character-oriented computer see CHARACTER-ADDRESSABLE COMPUTER.

character outline *Computer Technology.* the pattern created by the outermost strokes of a handwritten or printed character; one of the discriminators used in character recognition.

character pitch see PITCH.

character printer see SERIAL PRINTER.

character reader *Computer Technology.* any data-preparation device that can read, recognize, and convert written or printed characters into machine-processable form, such as an optical or magnetic reader.

character recognition *Computer Technology.* the process of converting printed or handwritten symbols into machine-readable information, especially by using an optical or magnetic scanning system and pattern-recognition technology.

character set *Computer Programming.* a set of symbols that define the characters which can be used on given hardware or in writing programs in a particular programming language, such as ASCII.

character skew *Computer Technology.* the angular displacement of a character from its normal or intended position, sometimes such that it cannot be recognized by an optical scanner.

characters per inch *Computer Technology.* the number of characters written per inch of a physical medium such as magnetic tape.

characters per pica *Graphic Arts.* for a certain typeface at a given size, the average number of characters that will fit in a space one pica wide; a measure of how wide the typeface will set in comparison with other faces.

character string *Computer Programming.* any sequence or grouping of characters, such as a variable name, construct, or code.

character string constant *Computer Programming.* a combination of letters, numbers, and other symbols specified as a constant value, analogous to a numeric constant; often bracketed by quotation marks or other delimiters to distinguish it from other strings such as variable names.

character style *Computer Technology.* the distinguishing structure of the symbols constituting a character set; a font.

character type *Psychology.* any of a number of basic personality types defined by specific character traits and patterns of behavior.

character-writing tube *Electronics.* a tube in which an electric beam produces alphanumeric and symbolic characters on a screen.

Charadrii *Vertebrate Zoology.* a suborder of shorebirds in the order Charadriiformes, including the jacanas, plovers, sandpipers, avocets, phalaropes, and curlews.

Charadriidae *Vertebrate Zoology.* the plovers and lapwings, a family of wading shorebirds in the order Charadriiformes; they are excellent flyers and often migrate in large flocks.

Charadriiformes *Vertebrate Zoology.* a diverse and cosmopolitan order of shore and aquatic birds, including plovers, gulls, grouse, snips, sandpipers, and puffins.

Charadrioidea *Vertebrate Zoology.* a superfamily of shorebirds in the suborder Charadrii of the order Charadriiformes, including the sandpipers, plovers, and phalaropes.

Charales *Botany.* the only order of the class Charophyceae; the green algae.

Charbray *Agriculture.* a large breed of white, horned beef cattle developed in Texas from the Brahman and Charolais breeds.

charcoal *Materials.* **1.** a porous carbon created by heating organic material, especially wood, in an airless environment. **2.** a writing or drawing implement made of this substance.

charcoal test *Chemical Engineering.* a test to determine the natural gasoline content of natural gas, by using activated charcoal to adsorb the gasoline fraction and then retrieving it through distillation.

Charcot-Leyden crystals *Medicine.* protein crystals found in the bronchial secretions in bronchial asthma and sometimes in the stools of patients with intestinal parasitism. Also, ASTHMA CRYSTALS, LEUKOCYTIC CRYSTALS, LEYDEN'S CRYSTALS.

Chardonnay [shar´də nā´] *Agriculture.* a green grape variety grown originally in Burgundy, France, and now also grown in California, Italy, Australia, and elsewhere. *Food Technology.* also, **chardonnay.** a dry white wine made with this grape. Also, PINOT CHARDONNAY.

Chardonnet, Hilaire 1839–1924, French chemist and physiologist; patented and manufactured rayon and other synthetic fibers.

Chargaff's rule *Molecular Biology.* the observation by Erwin Chargaff that for the DNA of any species, the number of adenine residues equals the number of thymine residues and the number of guanines equals the number of cytosines.

charge *Ordnance.* the quantity of explosive or propellant of a projectile, round, shell, bomb, or mine. *Military Science.* the act of rushing directly toward the enemy. *Engineering.* the measured amount of material used to fill a mold. *Electricity.* **1.** an electrostatic quantity, measured as a surplus or deficiency of electrons on a given object. **2.** the electrical energy stored in an insulated object, such as a battery or capacitor. **3.** the conversion of electric energy into chemical energy within a cell or battery. **4.** to direct electrical energy into a cell or battery for the purpose of such conversion. *Mechanical Engineering.* **1.** to feed refrigerant into a cooling system. **2.** the quantity of refrigerant within such a system. *Nucleonics.* **1.** in a nuclear reactor, the total amount of fuel required to attain a critical condition. **2.** to load fuel into a reactor core. *Metallurgy.* the raw material loaded into a melting or smelting furnace.

charge conjugation *Particle Physics.* the operation of changing a particle to its antiparticle or, equivalently, reversing its charge and magnetic moment. Also, **charge conjugation operation.**

charge conservation see CONSERVATION OF CHARGE.

charge-coupled device *Electronics.* a high-speed, high-density superconductor device that relies on the transfer of short-term stored charges in spatially defined depletion zones on its surface.

charge-coupled image sensor *Electronics.* a sensor in which light from various picture segments is marshalled to produce a television-type output signal. Also, SOLID-STATE IMAGE SENSOR.

charge-coupled memory *Computer Technology.* a volatile direct-access memory unit made up of charge-coupled devices; used especially when memory contents are accessed serially.

charge coupling *Electronics.* a means to store and transfer data based on the ability to store and transfer electric charges or charge carriers.

charged-current interaction *Particle Physics.* a weak interaction that changes the charges of the interacting fermions, such as beta decay.

charged demolition target *Ordnance.* a demolition target on which all charges have been placed and which is in a state of readiness.

charge-delocalized ion *Organic Chemistry.* an ion in which a single charge of a given type is spread around the molecule and not localized with any particular atom.

charge density *Materials Science.* the electrical charge per unit area, for example, on an electrode.

charge-density wave *Solid-State Physics.* a state in which an excess of static charge alternates with a region of static charge defect to form a pattern of charge oscillations merely as a sound wave in air.

charged particle *Particle Physics.* any particle containing either a positive or negative electric charge, that is, having a nonzero value for its charge; designated by 1− for a negative charge and 1+ for positive.

charged species *Chemistry.* any species having an electrical charge; a cation or anion.

charged tRNA *Molecular Biology.* a transfer RNA molecule to which an amino acid is attached. Also, ACYLATED tRNA.

charge eliminator *Electricity.* a battery charger having a low-noise, low-impedance output that can either charge a storage battery or provide a direct-current load without a storage battery in parallel.

charge establishment *Ordnance.* the process of establishing the quantity of propellant necessary to produce a desired muzzle velocity with a particular propellant in a particular weapon.

charge exchange *Physics.* a collisional process in which an electron is transferred from a neutral atom or molecule to an ion.

charge-exchange source *Electronics.* a device that generates negative ions, usually helium, by passing positive ions through a donor canal, usually containing lithium vapor, in which they collect two electrons to form negative ions.

charge independence *Nuclear Physics.* the principle that the strong force between a neutron and a proton is equal to the force between two protons or two neutrons in the same orbital and spin state. *Particle Physics.* a statement that there is no variation of the interaction force between particles if one of the particles is replaced by another particle having the same isotopic spin multiplet.

charge-injection device *Electronics.* a device that relays stored charges positioned at predetermined locations; it is used as an image sensor whose points are referenced to their horizontal and vertical coordinates.

charge-localized ion *Organic Chemistry.* an ion in which a single charge of a given type is associated with a particular atom that is more electronegative or electropositive.

charge-mass ratio *Electricity.* the ratio of a particle's electric charge to its mass. Also, SPECIFIC CHARGE.

charge mobility *Materials Science.* the movement in a semiconductor of electrons that jump the forbidden-energy gap from the filled valence band to the empty conduction band; the energized electrons are able to carry a charge toward the positive electrode, and the resulting electron holes in the valence band become available for conduction when electrons deeper in the band move up into those vacated levels.

charge neutrality *Solid-State Physics.* the ground state of matter in which there is an exact balance between all positive and negative charges in a system.

charge parity *Particle Physics.* the property of an elementary particle, either odd or even, that indicates whether or not its mirror image occurs in nature.

charge quantization *Electricity.* the fact that the electric charge of a particle must be equivalent to an integral multiple of the universal basic charge.

charger *Mining Engineering.* see BEETLE, def. 1.

charge-storage transistor *Electronics.* a transistor that produces a charge at its collector junction when a large voltage is applied at a high level to its base and at a low level to its collector.

charge-storage tube *Electronics.* a tube that stores information on its surface as electric charges.

charge-storage varactor *Electronics.* a device in which capacitance varies inversely with the voltage applied from a direct current; it is used to generate power in excess of 50 watts at ultrahigh or microwave frequencies.

charge transfer *Physical Chemistry.* the transfer of charge from one entity to another.

charge-transfer complex *Chemistry.* a species formed by the interaction of two molecules upon a transfer of charge between them; a donor/acceptor complex.

charge-transfer device *Electronics.* any device that relays stored charges positioned at predetermined locations, such as charge-coupled or charge-injection devices.

charge-weight ratio *Ordnance.* the ratio of the weight of an explosive charge to the total weight of the bomb or projectile containing the charge.

charging current *Electricity.* a current that charges an electrical storage device.

charging line *Petroleum Engineering.* a pipeline used to carry fresh charging stock of oil, gas, and crude oil to a still.

charging pump *Chemical Engineering.* a pump that supplies a pressurized flow of fluid to another unit.

charging stock *Petroleum Engineering.* any product to be treated in a particular refinery unit or still.

Charles, Jacques 1746–1823, French mathematician and physicist; formulated Charles' law on dilation of gases with heat.

Charles' law *Physics.* a law stating that the volume of a gas at constant pressure is proportional to the absolute temperature.

charm *Particle Physics.* a quantum number characterizing quarks and hadrons, which was proposed to satisfy the notion of lepton-quark symmetry. Particles carrying a nonzero charm quantum number have been definitively observed.

Charmat process *Food Technology.* a process for making champagne-style sparkling wine, in which the second fermentation takes place in a large glass-lined tank rather than in the bottle.

charmed particle *Particle Physics.* a particle whose total charm is other than zero.

charmed quark *Particle Physics.* a quark having electric charge 2/3, hypercharge 1/3, a baryon number $B = 1/3$, zero strangeness, and charm of +l. Also, *C* QUARK.

charmonium *Particle Physics.* a group of hadron particles that are composed of a charmed quark and its antiquark.

Charmouthian *Geology.* in Great Britain, a geologic stage of the Lower Jurassic period, occurring after the Sinemurian and before the Domerian.

Charnian orogeny *Geology.* a late Precambrian crustal deformation that is believed to have affected the English Midlands.

charnockite *Petrology.* any of various faintly foliated, nearly massive granites characterized by orthopyroxene and also containing quartz and feldspar.

charnockite series *Geology.* a series of plutonic rocks that are similar in composition to granitic rock but are characterized by the presence of orthopyroxene.

Charolais or **Charollais** [shä´rə lā´] *Agriculture.* a breed of large beef cattle characterized by creamy white coloration and soft hair. (From the *Charolais* region of France.)

Charon [ka´ rən] *Astronomy.* Pluto's single moon, discovered in 1978; 1190 kilometers in diameter.

charon *Molecular Biology.* any of several derivatives of the bacteriophage lambda that can serve as a cloning vector. Also, **charon vector.**

Charophyceae *Botany.* a class of green algae having an erect, branched thallus with regular nodules, and reproducing through solitary, single-celled oogonia.

Charophyta *Botany.* a group of mostly freshwater aquatic plants living totally submerged and attached to the bottom by rhizoids; they reproduce through oogonia and often become encrusted with calcium carbonate, giving them a rigid appearance.

Charpak-Massonet current distribution system *Particle Physics.* an electronic system used to determine the location of a single spark event in a spark chamber by processing data obtained from the amount of current flowing through the different paths of a multipath network.

Charpy impact test *Materials Science.* a method that is widely used to test the notch-impact strength of metals; a notched specimen of the test material is placed in a device and broken by the impact of a swinging pendulum; the amount of energy to fracture the material is calculated from the arc of the follow-through swing of the pendulum. Also, **Charpy test.**

chart *Cartography.* **1.** a map, especially one used for nautical or air navigation. **2.** to prepare such a map. *Mathematics.* let U be an open set of a topological space X and a homomorphism $h: X \rightarrow R^n$ such that $h(U)$ is an open set. Then the pair (h, U) is called an ***n*-chart**. Multiple charts (i.e., an atlas) are usually required to cover an entire manifold. The components of h are the coordinates of the chart. Charts must satisfy a compatibility condition when they overlap.

chartaceous *Botany.* having a papery appearance or texture.

chart comparison unit *Engineering.* a display that superimposes radar plan-position indicators on a navigational chart.

chart datum *Cartography.* a datum to which depths or soundings in a hydrographic survey or on a chart are referred.

chart desk *Engineering.* a rigid, flat board that is sized and designed for the inspection of charts and maps, often having storage space and a protractor on parallel arms. Also, CHART TABLE.

charted depth *Oceanography.* the vertical distance between the bottom and the tidal datum plane.

charted VFR flyway *Navigation.* a recommended flight path for aircraft on visual flight rules, which avoids areas heavily used by large, jet-powered aircraft.

charted visibility *Navigation.* the maximum distance, given on a chart, at which a navigational light can be seen under standard conditions.

charthouse *Naval Architecture.* a covered housing for the navigator's station, charts, and equipment. Also, **chart room.**

chartlet *Cartography.* a small chart.

chartometer *Cartography.* an instrument that is used for measuring distances on maps or charts.

chart reading *Navigation.* the interpretation of the various lines and symbols appearing on a chart.

chart recorder *Engineering.* a device that plots a dependent variable against an independent variable by moving a pen across a paper, or a light beam or electrode across photosensitive paper.

chart table see CHART DESK.

chase *Design Engineering.* **1.** a series of cuts, as on a screw thread, each of which follows the path of the cut before it. **2.** to groove or to cut like a screw thread. **3.** to decorate surfaces (especially metal) by embossing or engraving. *Building Engineering.* a passageway, space, or groove in a masonry wall oriented lengthwise, usually vertically, to allow ducts, pipes, or wires to be routed around, in, or through a building. *Ordnance.* the exposed part of an artillery gun, located in front of the trunnions. *Graphic Arts.* **1.** in lithography, a usually metal frame having a glass face scribed with register lines, in which film or plates are positioned for photocomposing. **2.** in letterpress printing, a frame in which type or engravings are held in place.

chase mortise *Design Engineering.* a mortise with a sloping edge that allows a tenon to be inserted when the outside clearance is small.

chase pilot *Aviation.* a pilot who escorts and advises another pilot in a separate, nearby aircraft during flight training or testing.

chaser *Engineering.* **1.** a tool used to cut threads in screws. **2.** an engraving tool. *Space Technology.* a vehicle or mission designed to retrieve or rendezvous with an object in space, such as a satellite in orbit.

chase ring *Mechanical Engineering.* a ring that restrains the blank during the hobbing process.

chasing tool *Mechanical Devices.* a steel tool resembling a punch, but with special forgings used for designing decorations on a surface.

Chasles' theorem *Mechanics.* the statement that it is always possible to represent arbitrary rigid body motions as the sum of the rotation about the center of mass of the body plus a linear translation of the center of mass. (Proven by the French mathematician Michel *Chasles,* 1793–1880.)

chasm *Geology.* a deep cleft or opening in the earth's surface, such as a fissure or a gorge, or a recess that extends below the floor of a cave. Also, ABYSS.

chasmogamy *Botany.* pollination in a flower that is fully opened. A flower in which pollination occurs in this way is **chasmogamous.**

chasmophyte *Ecology.* a plant that grows on rocks or within rock crevices. Thus, **chasmophilous, chasmophile.**

chassignite *Geology.* a stony meteorite composed almost entirely of olivine and lacking chondrules, as well as nickel-iron.

chassis *Engineering.* any major part or framework of an assembly to which other parts are attached, such as the frame upon which an automobile body is mounted or the frame on which the components of an electronic device, such as a radio, are mounted. *Electricity.* a metal frame onto which components are mounted; usually the frame is designed to be connected to ground to reduce shock hazard and electrical noise.

chassis ground *Electricity.* a connection to the metal structure (chassis) on or in which the components of a circuit are mounted.

chassis punch *Mechanical Devices.* a hand tool used to punch holes in sheet metal.

chat *Vertebrate Zoology.* any songbird known for harsh chattering notes including the true chats or chat-thrushes of the family Turdidae, the yellow-breasted chat of the family Parulidae, and the Australian chats of the family Maluridae.

Châtelet, Marquise du 1706–1749, French mathematician and physicist; translated and interpreted Newton's works.

chatoyancy *Mineralogy.* the property of certain minerals that, in reflected light, exhibit a movable sheen concentrated in a narrow band of light that changes position as the mineral is turned.

chatoyant *Mineralogy.* any mineral or gemstone having a changeable luster or color delineated by a thin band of light.

chatter *Acoustics.* vibrations, usually at a resonant frequency, of a piece of machinery, produced by continuous motion of two or more machine parts against each other, such as brushes in an alternating-current electrical motor. *Engineering.* such vibrations of a cutting tool not firmly held or of an insufficiently rigid machine, resulting in an uneven finish. *Electricity.* an undesirable rapid opening and closing of electrical contacts caused by fluctuations in the coil current. Also, CONTACT CHATTER.

chattering *Control Systems.* an uncontrolled, high-speed switching back and forth of a relay in a relay-type control system.

chattermark or **chatter mark** *Geology.* **1.** any of a series of short, densely packed curved scars or cracks made on the surface of bedrock by rock fragments carried at the base of a glacier. **2.** any mark, scratch, or scar made on the surface of a rock as a mass moves over it. **3.** a crescent-shaped mark on a pebble caused by wave action.

Chattian *Geology.* a European geologic stage of the Upper Oligocene epoch, occurring after the Rupelian and before the Aquitanian of the Lower Miocene. Also, CASSELIAN.

Chautauquan *Geology.* a North American provincial series of the Upper Devonian period, occurring after the Senecan and before the Kinderhookian.

chavicol *Organic Chemistry.* $C_9H_{10}O$, a colorless liquid that is soluble in water and miscible with alcohol; melts at 16°C and boils at 235–236°C; occurs in many essential oils. Also, BENZENE 1-ALLYL-4-HYDROXY.

Chazyan *Geology.* a North American geologic stage of the Upper Ordovician period, occurring before the Mohawkian.

Chebyshev filter *Electronics.* a filter whose pass band response exhibits two or more ripples of equal peak-to-valley amplitude.

Chebyshev polynomial *Mathematics.* any of a family of orthogonal polynomials $T_n(x)$ on $L^2[-1, 1]$ that are solutions to Chebyshev's differential equation; given by $T_0(x) = 1$ and $T_n(x) = 2^{1-n}\cos(n \arccos(x))$ for $n \geq 1$. They can be used to obtain the polynomial of a given degree that best approximates a function in $L^1[-1, 1]$.

Chebyshev's differential equation *Mathematics.* the following special case of Gauss' hypergeometric differential equation:

$$(1 - x^2)f'''(x) - xf'(x) + n^2 f(x) = 0.$$

Chebyshev's inequality *Mathematics.* given two monotone sequences of n real numbers $\{a_k\}$ and $\{b_k\}$, both either increasing or decreasing, and $r \geq 1$,

$$[(1/n)\sum_{k=1}^{n}(a_k)^r]^{1/r}[(1/n)\sum_{k=1}^{n}(b_k)^r]^{1/r} \leq [(1/n)\sum_{k=1}^{n}(a_k b_k)^r]^{1/r}.$$

If one of $\{a_k\}$ or $\{b_k\}$ is decreasing and the other increasing, the inequality is reversed.

Chebyshev's theorem see CHEBYSHEV'S INEQUALITY.

Chebyshev system *Mathematics.* a set $\{f_1(x), \ldots, f_n(x)\}$ of continuous functions defined on an open interval (a, b) such that for any set $\{c_k\}$ of real numbers, not all zero, the function

$$T(x) = \sum_{k=1}^{n} c_k f_k(x)$$

is nonzero on (a, b) except possibly at no more than $n - 1$ points of (a, b). Also, T-SYSTEM.

check *Computer Technology.* **1.** to verify the accuracy or correctness of the results of a process, data entry or transfer, or the contents of memory or registers. **2.** an instance or method of carrying out such a process.

check bit *Computer Technology.* one or more bits that, when appended to a recorded character or frame, can detect transmission errors; often a parity bit.

check character *Computer Technology.* a character, digit, or bit that is used in error detection during data transfer or transcription, often appended following a block of data.

check digit *Computer Technology.* a number that is created by a computer during the transcription of source data into machine-readable form in order to validate the data according to predefined conventions.

checkerboard *Mining Engineering.* see GRID.

checkerboard regenerator *Engineering.* an arrangement of bricks set in a regenerator in such a way as to leave passage for the movement of hot gases, absorbing and releasing heat.

checker refractory *Materials.* a furnace made of heat-resistant bricks that is used in glass making.

checkers *Engineering.* **1.** the refractory or special-shaped bricks in a checkerboard regenerator. Also, **checker bricks. 2.** engineers who perform the final check on certain stages of an engineering design.

check flight *Aviation.* **1.** a flight made to test the proficiency of a pilot or crew or to familiarize a pilot or crew with a particular flight vehicle. **2.** a flight to test the performance of a flight vehicle or component.

check-in/check-out *Transportation Engineering.* a track circuit that registers when a train enters and leaves a block.

check indicator *Computer Technology.* a computer console device such as a light, that informs the operator that an error has occurred or that a predetermined goal or condition has been reached.

checking *Materials Science.* a fault that develops on paintwork or ceramic coating to relieve minor stresses in the film, characterized by a network of fine cracks. *Metallurgy.* the formation of very fine cracks on the surface of a metallic part or of a coating.

checking program *Computer Programming.* a program designed to detect and diagnose errors in other programs or data.

check ligament *Anatomy.* any ligament that restrains the movement of a joint, especially the strong, rounded bilateral cords arising on the side of the upper part of the odontoid process and inserting in the inner side of the condyles of the occipital bone of the skull, which limit the rotation of the head.

checkline *Naval Architecture.* a line used to brake a vessel's way while it is coming alongside a pier.

checklist *Aviation.* a list carried on board an aircraft listing procedures and instructions to be followed during normal operations; usually used before takeoff.

check number *Computer Technology.* a number with one or more digits used to detect equipment malfunction during data transfer operations.

check observation *Meteorology.* an aviation weather observation that is abbreviated to include just those elements that have an important effect on aircraft operations.

checkout *Engineering.* a final, sequential test for the readiness and ability to perform of a device or a system, implemented just prior to use. *Aviation.* instruction or training to familiarize a pilot and crew with a flight vehicle, both on the ground and during a check flight; a sequence of steps taken to accomplish this instruction.

checkpoint *Navigation.* a geographical point at which an aircraft reports its position to ground control. Checkpoints may be mandatory or optional. *Military Science.* see REGISTRATION POINT. *Computer Programming.* a designated point in a program where processing is interrupted and all status information is recorded in order to be able to restart the process at that point, thus avoiding the necessity to repeat the processing from the beginning. Also, RERUN POINT, RESTART POINT.

check problem see CHECK ROUTINE.

check profile *Photogrammetry.* a profile plotted from a field survey and used to check a profile that was prepared from a topographic map; this serves to verify the accuracy of contours on the topographic map.

check protect symbol *Computer Programming.* a printed character, usually an asterisk, that is used to fill in the spaces to the left of the dollar amount of a computer-printed bank check in order to prevent the unauthorized insertion of extra digits.

check register *Computer Technology.* a register that is used to hold a copy of information being transferred for the purpose of comparison when the transfer is complete.

check routine *Computer Technology.* a program or subprogram designed to test computer components and detect any faults or malfunctions that may exist.

check study *Industrial Engineering.* the timing of an actual job or operation to confirm the validity of a standard time established for it.

check sum or **checksum** *Computer Technology.* an error detection method for a set of information such that a sum is computed modulo some number *n,* which is appended to the data; when the data is subsequently read, the sum modulo *n* is recomputed and compared with the check sum accompanying the data.

check valve *Engineering.* a pipe valve that allows fluid to flow in only one direction. Also, CLACK VALVE, NONRETURN VALVE.

checkweigher *Engineering.* a device used to establish whether the weight of a material or package is within predetermined limits.

check word *Computer Technology.* a machine word appended to a block of data that serves as a check symbol during data transfer.

cheddar or **Cheddar** *Food Technology.* a widely produced class of hard cheese, usually made with orange-colored whole milk and ripened by bacteria.

Chediak-Higashi anomaly *Pathology.* an autosomal recessive defect characterized by the aberrant nuclear structure of the leukocytes with giant malformations of the peroxidase-positive granules and cytoplasmic inclusions; it occurs in children and is characterized by recurrent skin infections, albinism, photophobia, lymphadenopathy, and hepatosplenomegaly, with death usually occurring before the eighth year.

cheek *Anatomy.* the area at the side of the face between the nose and the ear. *Metallurgy.* in casting, the interior section of a mold between the top, or cope, and the bottom, or drag, portion.

cheek cells

cheek pouch *Vertebrate Zoology.* a saclike expansion of the cheeks in some animals, usually rodents such as squirrels and hamsters, in which food is temporarily stored.

cheese *Food Technology.* a food made from milk curd that is separated from the whey, usually ripened and compressed, and often colored and salted; a common food since ancient times, made chiefly from cows' milk but also from goats, sheep, and other animals. *Textiles.* a cylindrical package of tightly wound warp yarn designed for use on a loom.

cheesecloth *Textiles.* a lightweight, open-weave cotton fabric used in making wiping cloths, gauze bandages, and tea bags.

cheese skipper *Vertebrate Zoology.* a jumping maggot-type larva of the cheese fly, *Piophila casei;* a common pest of stored cheese and other protein foods.

cheese-washer's lung *Medicine.* extrinsic allergic alveolitis due to inhalation of *Penicillium casei* spores from moldy cheese. Also, **cheese worker's lung.**

cheetah

cheetah *Vertebrate Zoology.* a large leopardlike cat of the plains of Africa and Asia, *Acinonyx jubatus,* having a small head, a slender body, and very long legs; it is unique among cats in having nonretractile claws. The cheetah is the fastest runner of any animal and can attain speeds of up to 70 miles per hour over short distances.

cheil- or **cheilo-** a combining form meaning "lip," as in *cheilosis.*

cheilitis *Medicine.* an inflammation of the lip.

cheiloplasty *Surgery.* the surgical repair of a lip deformity or surgery to change the contour of the lips.

cheilorrhaphy *Surgery.* suturing of the lip.

cheilosis *Medicine.* an irritation at the junction of the upper and lower lips.

Cheilostomata *Invertebrate Zoology.* an order of colonial bryozoans in the class Gymnolaemata; the zooids live close together but with separate walls.

cheilostomatoplasty *Surgery.* plastic surgery of the lips and mouth.

cheimaphobia *Psychology.* an irrational fear of cold or of winter.

cheir- or **cheiro-** a combining form meaning "hand," as in *cheirospasm, cheiroplasty.*

Cheiracanthidae *Paleontology.* a genus of early fishes in the extinct order Acanthodiformes; extant in the middle Devonian.

cheiromegaly *Medicine.* the abnormal enlargement of one or both hands.

cheiroplasty *Surgery.* the surgical repair of a defect or trauma to the hands.

Cheiropleuriaceae *Botany.* a monotypic family of terrestrial ferns of the order Filicales that are native to eastern Asia; characterized by netted veins, bilobed vegetative fronds, undivided fertile fronds, and a short creeping hairy rhizome.

cheirospasm *Medicine.* a spasm of muscles of the hand.

chela *plural,* **chelae.** *Invertebrate Zoology.* **1.** a claw on the limbs of certain crustaceans and arachnids. **2.** a sponge spicule.

chelate *Organic Chemistry.* a highly stable complex formed between a metal ion and more than one organic group to form a ring.

chelate laser *Optics.* a laser in which a rare-earth compound generates electromagnetic radiation by energizing the organic portion of a liquid molecule and then transferring the energy to metallic ions; it is easily pumped and has high quantum efficiency and narrow emission lines.

chelating agent *Biochemistry.* a substance that is capable of forming a chelate with a metal ion; usually used in treating heavy metal poisoning, as when ethylenediaminetetraacetic acid (EDTA) is used to promote the excretion of lead. Also, SEQUESTERING AGENT.

chelating resin *Organic Chemistry.* an ion-exchange resin with very high selectivity for specific cations.

chelation *Organic Chemistry.* the binding of a metal ion to more than one organic group to form a ring.

chelation therapy *Toxicology.* the sequestering of poisonous metals by the administration of a chelate.

chelerythrine *Organic Chemistry.* $[C_{21}H_{18}NO_4]^+$, an alkaloid from *Chelidonium majus*; melts at 213–214°C.
chelicera *Entomology.* either of the first pair of appendages in arachnids, usually modified for attacking prey and sometimes having a poison gland at the base. Also, CHELIFORE.
Chelicerata *Invertebrate Zoology.* a subphylum of the phylum Arthropoda, including spiders, ticks, mites, scorpions, and king crabs.
Chelidae *Vertebrate Zoology.* the side-necked or snake-necked turtles, a family of aquatic and semiaquatic turtles of the suborder Pleurodira, in which the neck is long and not completely hidden when retracted.
chelidonic acid *Organic Chemistry.* $C_7H_4O_6$, rose-colored, monoclinic needles; soluble in water and slightly soluble in alcohol; decomposes at 257°C; derived from the herb *Chelidonium majus.*
chelidonine *Organic Chemistry.* $C_{20}H_{19}NO_5$, monoclinic plates; insoluble in water and slightly soluble in alcohol; decomposes at 136–140°C; an isoquinoline alkaloid obtained from the root of celandine, *Chelidonium majus.*

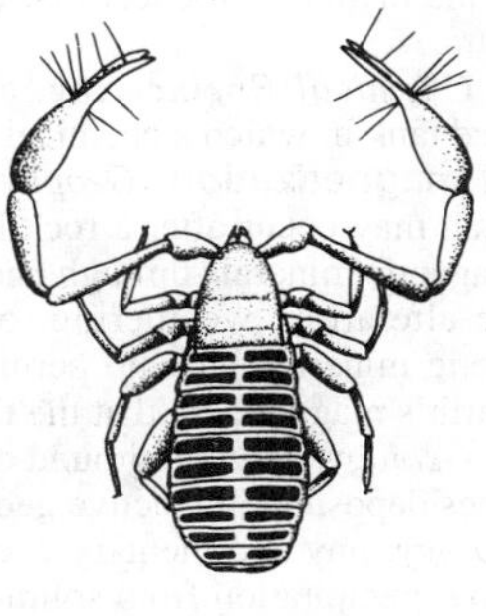

chelifer

chelifer *Invertebrate Zoology.* the book scorpion, a common pseudoscorpion found in houses.
chelifore *Invertebrate Zoology.* either of the first pair of appendages on the cephalic segment of pycnogonids. *Invertebrate Zoology.* see CHELICERA.
cheliform *Invertebrate Zoology.* having a pincerlike appendage, such as a crab's claw.
cheliped *Invertebrate Zoology.* either of the paired appendages with chelae in decapod crustaceans.
Chellean or **Chellian** see ABBEVILLIAN.
chelometry *Analytical Chemistry.* the formation of soluble chelates when a metal ion is titrated with a chelating agent.
Chelonariidae *Invertebrate Zoology.* a family of beetles, insects in the superfamily Dryopoidea.
Chelonethida see PSEUDOSCORPIONIDA.
Chelonia *Vertebrate Zoology.* an equivalent name for Testudines, the turtles, terrapins, and tortoises; an order of marine and freshwater reptiles in the subclass Anapsida, distributed in warm waters throughout the world.
Cheloniidae *Vertebrate Zoology.* the modern sea turtles, a family of giant tropical and subtropical marine turtles that lay their eggs on sandy shores and belong to the order Testudines.
Cheluridae *Invertebrate Zoology.* a family of crustaceans in the suborder Gammaridea, with no carapace and no eye stalks.
Chelydridae *Vertebrate Zoology.* the snapping turtles, a family of omnivorous or carnivorous bottom-walking freshwater turtles in the order Testudines.
chem chemical; chemist; chemistry.
chem- or **chemi-** a prefix meaning "chemical," as in *chemurgy.*
chemical *Chemistry.* **1.** of or relating to the science of chemistry or to the properties or actions of chemicals. **2.** any substance having a defined molecular composition. **3.** any substance formed by or used in a chemical reaction.
chemical agent *Ordnance.* a chemical substance used in military operations to kill or injure personnel, animals, or plants, or to damage equipment or facilities.
chemical ammunition *Ordnance.* any ammunition filled primarily with a chemical agent, such as smoke, incendiaries, or war gases. Similarly, **chemical bomb, chemical grenade, chemical mine.**
chemical antidote *Toxicology.* a chemical that counteracts the toxic action of another chemical.
chemical-biological warfare *Military Science.* the use of chemical and biological agents (such as poison gas and toxic microorganisms) to kill, injure, or seriously incapacitate personnel or to deny the use of areas, facilities, or materials. Also, CBW.
chemical bond see BOND.
chemical burn *Medicine.* a trauma to the skin caused by chemicals such as acids or alkalis.
chemical change see REACTION.
chemical composition *Petrology.* the percent by weight of the elements in a rock.
chemical compound see COMPOUND.
chemical control *Biology.* the regulation of life processes in an organism by the chemical transmission of messages, usually through hormonal action, from one part of the organism to another.
chemical conversion coating *Metallurgy.* a coating generated by a chemical reaction.
chemical crystallography *Crystallography.* the study of the internal arrangement of atoms formed from chemical compounds.
chemical dating *Analytical Chemistry.* the determination of the ages of minerals or of ancient objects by chemistry, as by radiocarbon dating.
chemical denudation *Geology.* the baring of the land surface by water carrying soluble materials into the sea.
chemical deposition *Chemistry.* the precipitation of a metal from a solution of a metallic compound, brought about by the addition of another metal.
chemical element see ELEMENT.
chemical energy *Physical Chemistry.* the energy generated when a chemical compound combusts, decomposes, or transforms to produce new compounds.
chemical engineering *Engineering.* the design and manufacture of products, processes, and operations involving basic chemical materials or chemical change.

Chemical Engineering

The term *chemical engineering* came into use in English in the late 1800s. It then described a branch of applied chemistry, called into being by the special needs for large-scale manufacturing of chemical products. Commercial production not only required larger equipment, but also became the field of study for chemical engineers and provided the first real basis for a distinct and unified discipline. This development, along with increased attention to the chemical reactor as a central element in chemical processes, played a crucial role in the creation of the modern petrochemical industry.

A further evolution occurred, beginning in the 1960s, toward a greater emphasis on the scientific and mathematical underpinnings of the field. Thus, fluid mechanics, diffusion, heat transfer, chemical reactor engineering, thermodynamics and computer-aided design became primary areas for chemical engineering research. Since then the very definition of processing has also expanded. Chemical engineers now participate in the microelectronics field, where the processes are masking and etching, and in biotechnology, where the processing elements are living organisms. The properties of polymers—the basis for the textile, rubber, plastics and paint industries—depend crucially on the forming and shaping process itself.

Chemical engineering began as and has remained the engineering discipline with strong roots in chemistry. The safe, efficient and environmentally sound operation of chemical manufacturing facilities is the primary goal of its practitioners.

W. W. Graessley
Professor of Chemical Engineering
Princeton University

chemical equilibrium *Chemistry.* a condition in which a chemical reaction and its reverse occur at the same rate, producing constant concentrations of the reacting substances and products; the position of minimum Gibbs free energy.

chemical etching *Metallurgy.* etching of the surface of a metal or alloy, caused by a chemical reaction.

chemical exchange process *Chemistry.* a method of using the differing properties of isotopes of certain lighter elements to separate them from each other by repeating a chemical reaction that involves exchange of the isotopes.

chemical film dielectric *Electricity.* a very thin film of material on one or both electrodes of an electrolytic capacitor, which conducts electricity in one direction only and thereby serves as the insulating element of the capacitor.

chemical fire extinguisher *Chemical Engineering.* any type of fire extinguisher, such as dry chemical, carbon dioxide, or vaporizing liquid, that suffocates a fire by emitting a solid, liquid, or gaseous substance.

chemical flux cutting *Metallurgy.* a high-temperature cutting process that uses a flux.

chemical focus see ACTINIC FOCUS.

chemical fog *Graphic Arts.* a blurring of all or part of an image on a photographic negative or plate, due to improper exposure of the film to chemicals, usually during developing.

chemical formula *Chemistry.* a combination of chemical symbols and numbers used to indicate the composition of a substance, such as NaCl (sodium chloride) or H_2O (water: two atoms of hydrogen, one of oxygen). Molecular formulas, which show the actual number and kinds of atoms in a chemical substance, are the most often used in chemistry; other kinds of formulas are empirical, structural, generic, and electronic.

chemical fossils *Organic Chemistry.* organic molecules, such as porphyrins and alkanes, thought to indicate that living matter existed before true fossil formation; found in strata such as the Precambrian.

chemical glassware *Materials.* borosilicate glassware containing boric oxide, B_2O_3; produced for its chemical durability as laboratory glassware in contact with chemicals.

chemical indicator *Analytical Chemistry.* a substance that shows a significant change in the chemical nature of a system by color alteration or some other visible manifestation; for example, an acid-based indicator, which indicates the end point in a titration.

chemical inhibitor see INHIBITOR.

chemical-ion pump *Chemical Engineering.* a vacuum pump using evaporated metal vapor that reacts with molecules in the gas to be removed, in order to achieve its pumping action.

chemical kinetics *Physical Chemistry.* the branch of physical chemistry concerned with the study of the rates and mechanisms of chemical reactions. Also, REACTION KINETICS.

chemically *Chemistry.* **1.** according to chemistry or its laws. **2.** by a chemical process.

chemically pumped laser *Optics.* a laser that relies on chemical reactions rather than electrical energy to achieve the pumping action necessary for the production of light pulses. Also, **chemical laser.**

chemically pure *Chemistry.* without impurities detectable by any experimental procedure.

chemical machining *Metallurgy.* the process of removing a portion of a metallic material by a controlled chemical reaction.

chemical metallurgy *Metallurgy.* the art and science of winning metal values from ores and of subsequently refining the crude metals thus obtained. Also, PROCESS METALLURGY.

chemical microscopy *Analytical Chemistry.* the utilization of microscopes to analyze the composition of a substance by observing physical structures and phases.

chemical operations see CHEMICAL WARFARE.

chemical oxygen demand *Organic Chemistry.* a measure of the quantity of oxidizable components present in water; carbon and hydrogen, but not nitrogen, are chemically oxidized in organic matter, so the consumed oxygen is only a measure of the chemically oxidizable components. Also, OXYGEN CONSUMED, DICHROMATE OXYGEN CONSUMED.

chemical pathology *Pathology.* the scientific discipline that observes and studies the chemical changes caused by diseases.

chemical peel *Medicine.* the removal of the outer layers of the skin by painting the face with a solution that burns and erodes the areas to which it is applied.

chemical polarity *Physical Chemistry.* the tendency of a molecule to be attracted or repelled by electrical charges that arise from electrons orbiting a nucleus.

chemical polishing *Metallurgy.* the process of polishing the surface of a metal or alloys by a controlled chemical reaction.

chemical precipitates *Geology.* the material formed from a solution that has precipitated.

chemical pressurization *Space Technology.* the pressurization of a rocket's propellant tanks by means of gases generated by combustion or by some other chemical reaction.

chemical pulping *Chemical Engineering.* a process that separates wood fiber from wood for paper pulp manufacture, by dissolving the lignin that fuses the fibers together through a chemical treatment.

chemical pump *Petroleum Engineering.* a small-volume, high-pressure, skid-mounted pumping unit, used for adding chemicals into the power oil that operates bottom-hole pumps, in order to limit corrosion in the system.

chemical reaction *Chemistry.* a chemical change in which the atoms or molecules of two or more substances are rearranged to form one or more additional substances, often having different properties.

chemical reactivity *Chemistry.* the tendency of a substance to react with others to form one or more products that differ from the original reactants.

chemical reactor *Chemical Engineering.* any vessel, tube, pipe, packed bed, or agitated tank in which a chemical reaction takes place.

chemical remanent magnetization *Geophysics.* that part of remanent magnetization that may occur after a rock has been formed by the production of new magnetic minerals through chemical reactions associated with diagenetic alteration, weathering, or metamorphisms; the newly formed magnetic minerals pick up permanent magnetization in the direction of the earth's magnetic field at the time.

chemical reservoir *Geology.* an underground oil or gas trap formed in limestones or dolomites deposited in inactive geologic environments.

chemical rock *Geology.* any sedimentary rock whose primary constituents are formed by precipitation from solution or suspension, or by deposition of insoluble substances.

chemical sense *Physiology.* a general sensitivity that causes residual reactions to various irritants; restricted in humans to the moist surfaces of the buccal and nasal mucous membranes, the conjunctivas of the eye, and the mucosa of the anal canal.

chemical sequencing *Molecular Biology.* a technique for determining the primary sequence of nucleotides in a DNA molecule that involves labeling the molecule with radioactive phosphorus, cleaving the molecule into fragments, and analyzing the fragments using electrophoresis.

chemical shift *Materials Science.* any variations in nuclear magnetic resonance that occur at different frequencies for each type of proton, commonly expressed in relation to the resonance of tetramethylsilane as the zero of reference. *Physical Chemistry.* a term for the slight change in a nuclear magnetic-resonance spectrum as a result of the electrons shielding the nuclei.

chemical shim *Nucleonics.* a means of providing chemical control in a reactor, supplementing mechanical control rods, by varying the concentration of a neutron absorber (such as boric acid) in the reactor's water moderator or coolant; this procedure compensates for reactivity changes during power changes and zenon transients, and for fuel depletion and fission-product buildup.

chemical similitude *Chemical Engineering.* a tested procedure to assure acceptable operation of a large-scale chemical process, the design of which is based on pilot plant data.

chemical spray *Ordnance.* the aerial release of liquid war gas intended to cause casualties, liquid smoke intended to cause aerial smoke screens, or defoliant intended to remove vegetation.

chemical sterilization *Engineering.* a process using bactericidal chemicals to sterilize air, a substance, or an object.

chemical stoneware *Materials.* a highly vitrified, glazed ceramic that is resistant to chemical attack; used for tanks and pipes.

chemical stress relaxation *Materials Science.* a relaxation in a polymer caused by a chemical phenomenon; for example, a decrease in the elasticity of silicone rubber when it is exposed to dry nitrogen.

chemical symbol *Chemistry.* a notation consisting of one or two letters that is used to represent a chemical element or one atom of such an element, such as K (potassium) or Cl (chlorine).

chemical synthesis *Chemistry.* the formation of a chemical compound by a chemical reaction involving its constituents.

chemical thermodynamics *Physical Chemistry.* the branch of physical chemistry concerned with the study of the thermodynamic properties of chemical reactions.

Chemistry

Chemistry is concerned with the synthesis, properties and reactions of molecules (combinations of atoms) and the practical application of this information (which is also in the domain of Chemical Engineering). Subdivisions of this field include Organic, Inorganic, Physical, Biological, Analytical and Nuclear Chemistry. Such hybrid names as medicinal chemistry, geochemistry, environmental chemistry, agricultural chemistry, and fuel chemistry attest to chemistry's widespread usefulness and its close ties to other fields of science and applied technology. Chemistry is perhaps the most utilitarian of all the sciences.

Chemistry lies along a spectrum between physics on the one hand and biology on the other. It overlaps and permeates both of these sister sciences to such an extent that at the interfaces there is, in fact, no clear distinction between them; physical chemistry fades into biochemistry just as biochemistry and medicinal chemistry mesh with molecular biology.

Chemistry, too, is the sole scientific discipline that has lent its name to a major segment of the manufacturing industry. Of all the sciences, it has the closest links to the world of economics and business (although some other fields, such as electronics, have found their way into this regime).

During the coming decades chemists will synthesize millions of new compounds tailored for a wide spectrum of practical uses. Equally important, or perhaps more important, will be the tremendous advances in theoretical chemistry.

Perhaps the most important progress will be made in the chemistry of life processes in biochemistry, molecular biology and related areas. Biochemical genetics should give us a great deal of control over the genetic codes leading to the potential elimination of genetic defects, and pharmaceutical chemistry should produce many new drugs to combat disease and alleviate human suffering.

Glenn T. Seaborg
University Professor of Chemistry
University of California
Nobel Laureate in Chemistry

chemical thinning *Agronomy.* a growth-regulating method to prevent overbearing.

chemical tracer *Nucleonics.* an element in which radioisotopes have been induced by neutron bombardment to be used as tracers in the study of chemical reactions and in medical applications.

chemical unconformity *Geology.* an unconformity determined by chemical analysis.

chemical vapor deposition *Materials Science.* a technique for coating the surface of a solid by exposure to a vapor, usually a metal halide vapor, under a pressure of approximately 10^{-6} mmHg, while allowing the vapor to undergo thermal decomposition or hydrogen reduction.

chemical warfare *Military Science.* the use of chemical agents (such as poison gas) to kill, injure, or seriously incapacitate personnel or to deny the use of areas, facilities, or materials.

chemical waste *Chemical Engineering.* any unusable by-product of a chemical process, especially one that is toxic or polluting.

chemical weathering *Geochemistry.* the breakdown and alteration of rocks at or near the earth's surface as a result of chemical processes, including solution, hydrolysis, ion exchange, oxidation, and biochemical reactions.

Chemiclave *Biotechnology.* a device that sterilizes surgical instruments by exposing them to the vapors of alcohol, formaldehyde, and water at high temperature and pressure.

chemigraphy *Graphic Arts.* any of various nonphotographic chemical processes used to make etchings or engravings.

chemiluminescence [kem´i loom´i nes´ens] *Physical Chemistry.* any process in which a chemical reaction produces visible light without a corresponding increase in temperature; for example, bioluminescence, such as the light generated by fireflies, is a form of chemiluminesence. Thus, **chemiluminescent.**

chemiluminescent labeling *Biotechnology.* a method used to label DNA; two different DNA probes emit light when brought together in the same region of a gene.

chemiosmotic hypothesis *Biochemistry.* the principle that a proton gradient across the inner mitochondrial membrane couples electron transport and ATP synthesis, so that electrons passing through the respiratory chain, for example, cause protons to be pumped from the matrix through to the opposite side of the inner mitochondrial membrane.

chemism *Chemistry.* **1.** chemical activity. **2.** a chemical property or relationship.

chemisorb *Chemistry.* to adsorb (a material) by chemical action.

chemisorption *Physical Chemistry.* a process by which the atoms or molecules of a gas or a liquid are chemically bonded to the surface of a solid material.

chemist *Chemistry.* a scientist who specializes in chemistry. *Pharmacology.* another term for a pharmacist, especially in British use.

chemistry the science that deals with the composition, structure, properties, interactions, and transformations of matter. The two main subdivisions of chemistry are organic chemistry and inorganic chemistry; other important branches include biochemistry, physical chemistry, and analytical chemistry.

chemo- a combining form meaning "chemical," as in *chemotherapy, chemoprevention.* Terms beginning with this form may be pronounced as either [kēm´ō-] (rhymes with "Nemo") or [kem´ō-] (rhymes with "memo").

chemoautotroph [kem´ō aut´ō trōf´] *Microbiology.* **1.** a microorganism that can derive the energy required for growth from oxidation of inorganic compounds such as hydrogen sulfide or ammonia. **2.** any autotrophic bacteria or protozoan that is not involved in the process of photosynthesis.

chemocautery *Surgery.* the cauterization of tissue by applying a caustic substance.

chemoceptor see CHEMORECEPTOR.

chemocline *Hydrology.* the boundary separating the circulating and noncirculating layers or water masses of a lake.

chemocoagulation *Surgery.* the coagulation of tissue by applying chemicals.

chemodectoma *Medicine.* an abnormal growth of the chemoreceptor system.

chemofacies *Geology.* the chemical elements that are collected or absorbed within aqueous bottom muds.

chemokinesis *Immunology.* an increased random migratory activity of cells produced by a chemical agent.

chemoluminescence see CHEMILUMINESCENCE.

chemoprevention [kēm´ō prə ven´shən] *Medicine.* the use of a chemical substance to forestall the contraction of a disease or to stop the progress of a disease.

chemoprophylaxis [kēm´ō prō fi laks´is] *Medicine.* the avoidance of the development or spread of a certain disease by administering a chemical agent.

chemoreception *Physiology.* the action of the sense organs that respond to chemical stimulation, including the organs for taste (gustation) and smell (olfaction).

chemoreceptor [kēm´ō rē sep´tər] *Physiology.* **1.** any of a group of receptors in the carotid sinus and the aorta that are sensitive to changes in blood oxygen and carbon dioxide levels and that have an effect on the rate and depth of breathing as well as on the heart rate. **2.** any neuroreceptor that responds to specific chemical stimuli. Also, CHEMOCEPTOR.

chemoresistance *Immunology.* specific resistance by cells to the action of chemicals.

chemosphere *Meteorology.* a vaguely defined region of the upper atmosphere in which photochemical reactions occur.

chemostat *Microbiology.* a continuous-flow culture system that maintains a population of microorganisms in the exponential phase of growth for an extended period of time, by limiting the concentration of an essential substrate.
chemosterilant *Entomology.* a chemical that is used to control insect plant pests by precluding their reproduction without affecting their life span or mating behaviors.
chemosynthesis *Biochemistry.* the use of inorganic substances such as carbon dioxide to synthesize carbohydrates from energy released by chemical reactions rather than by absorbed light.
chemotactic *Biology.* of or relating to chemotaxis.
chemotaxigen *Cell Biology.* any substance that facilitates the formation of an agent that is chemotactic for cells.
chemotaxis *Biology.* the movement of a cell or organism toward or away from a chemical substance. *Immunology.* the movement of granulocytes or macrophages to higher concentrations of agents known as cytotaxins.
chemotaxonomy *Botany.* the process and methods of classifying plants based on a chemical analysis of their products.
chemotherapeutic *Medicine.* of, relating to, or used in chemotherapy.
chemotherapeutic index *Pharmacology.* the relationship between the toxicity of a chemotherapeutic agent to the body and its toxicity to the cause of the disease being treated. *Virology.* the ratio between the lowest effective antiviral concentration and the highest nontoxic concentration of a compound.
chemotherapy *Medicine.* the treatment of a disease with chemicals or drugs; used especially in reference to the treatment of cancer with chemicals.
chemotroph *Biology.* any organism that creates its principal energy source by oxidizing organic or inorganic compounds.
chemotrophy *Biology.* the bending of a plant or plant organ in response to a chemical stimulus.
chemotropic *Biology.* relating to or exhibiting chemotropism; tending to move toward or away from a chemical stimulus.
chemotropism *Biology.* the tendency of an organism or part of an organism to bend toward or away from a chemical stimulus.
Chemungian *Geology.* a North American geologic stage of the Upper Devonian period, occurring after the Fingerlakesian and before the Cassadagan.
chemurgic [kə mur′jik] *Chemical Engineering.* relating to or used in chemurgy.
chemurgy [kem′ər jē] *Chemical Engineering.* an area of chemistry that is involved with the use of raw, organic, and previously unused agricultural substances to produce new, nonfood products such as varnishes and paints.
chenevixite *Mineralogy.* $Cu_2^{+2}Fe_2^{+3}(AsO_4)_2(OH)_4 \cdot H_2O$, a monoclinic greenish yellow to dark green mineral having a specific gravity of 4.38 to 4.59 and a hardness of 3.5 to 4.5 on the Mohs scale, occurring in massive form in the oxidation zone of copper deposits.
chengal *Materials.* the wood of the *Balanocarpus heimii* tree of Malaysia; used for the construction of boats and furniture.
chenier *Geology.* a continuous, narrow beach ridge built upon a plain of tidal marshes or swamps along a sea shore, often supporting evergreen oaks or pines.
chenille *Textiles.* a fabric of wool, cotton, silk, rayon, or synthetic fiber with a cut pile made from warp threads wound around soft filling threads.
Chenin Blanc [shə nēn′ blängk′; shen′in blangk′] *Botany.* a green grape grown originally in the lower Loire Valley of western France, and now also grown in California, Australia, and elsewhere. *Food Technology.* Also, **chenin blanc.** a light white wine made exclusively or mostly from this grape.
chenodeoxycholic acid *Biochemistry.* $C_{24}H_{40}O_4$, a component of bile that dissolves in alcohol, methanol, and acetic acid; found, for example, in fowl or in humans to a limited extent. Also, **chenic acid, chenodiol.**
Chenopodiaceae *Botany.* a widely distributed family of herbs and shrubs that thrive in a salty environment, having mostly simple and alternate leaves and small, mostly green flowers with no petals.
cheradophyte *Ecology.* any plant that grows on a sandbar. Thus, **cheradophilous, cheradophile.**
Cherenkov radiation *Nuclear Physics.* the visible radiation, mostly in the blue and near ultraviolet portions of the electromagnetic spectrum, emitted whenever a high-speed charged particle moves through a medium faster than the speed of light in a vacuum.
chergui *Meteorology.* a persistent, dry, dusty desert wind that blows from the east or southeast in Morocco.
Chermidae *Invertebrate Zoology.* a family of plant lice, homopteran insects in the superfamily Aphidoidea; plant pests.
Cherminae *Invertebrate Zoology.* a subfamily of beaked plant lice, homopteran insects in the family Chermidae.
Chern class *Mathematics.* a characteristic class defined for complex bundles given by various invariant polynomials in the entries of the curvature tensor. The Pfaffian is the simplest zero.
Chernozem *Geology.* a group of zonal soils developed in a temperate to cool subhumid climate and having a deep, dark to nearly black, highly organic surface horizon grading into a lighter-colored horizon over an accumulation of lime; normally occurs in subhumid climates under vegetation of medium to tall grass.
cherophobia *Psychology.* an irrational fear of happiness or joy.
cherry *Botany.* any of various trees belonging to the genus *Prunus,* especially *P. avium, P. cerasus,* and others derived from these species; cultivated for their edible fruit, which consists of a pulpy globular drupe surrounding a one-seeded smooth stone. *Materials.* the wood of various European and American cherry trees, brown, light or dark red in color, and darkening with exposure to light; used for paneling, cabinets, and fine furniture.
Cherry-Burrell process *Food Technology.* a method of making butter in which high-fat cream is pasteurized and cooled, color and salt are added, fat content is reduced, and the mixture is pumped through a device that chills, agitates, and adds texture to it.
cherry leaf spot *Plant Pathology.* a disease of cherry trees caused by the fungus *Coccomyces hiemalis,* characterized by small discolored spots on the leaves that may drop out to leave holes, and subsequently cause abnormal fruit and leaf growth.
cherry mahogany see MAKORE.
cherry picker or **cherrypicker** *Mechanical Engineering.* a tall, mobile platform, usually mounted on a truck, containing a winch, winch line, and hook that enable an operator to reach otherwise inaccessible parts of a building, light standard, treetop, or other elevated location. *Space Technology.* in the launch of a rocket vehicle carrying a manned space capsule, an electric or hydraulic crane used for the safe and speedy removal of the capsule in the event of an abort or malfunction.
cherry plum *Botany.* a small tree, *Prunus cerasifera,* that is closely related to true cherries and has a similar fruit.
cherry tomato *Botany.* a variety of tomato, *Lycopersicon lycopersicum cerasiforme,* having a red or yellow fruit that is only slightly larger than a cherry.
chersophyte *Ecology.* a plant that grows in dry wasteland habitats. Thus, **chersophilous, chersophile.**
chert *Petrology.* any hard, dense, micro- or cryptocrystalline rock composed of one or more forms of silica grains such as chalcedony and microcrystalline quartz, laced with microorganisms and occurring in nodules, lenses, and layers in shale or limestone. *Materials.* a form of this rock used in building and paving or as an abrasive.
chertification *Geology.* a process in which fine-grained quartz or chalcedony replaces minerals in other materials. Also, SILIFICATION.
Chesterian *Geology.* a North American provincial series of the uppermost Mississippian, after the Meramecian and before the Morrowan.
Chester White *Agriculture.* a breed of large hog that is white in color and has drooping ears, cross-bred from the Yorkshire, Lincolnshire, and Cheshire hogs. (From *Chester* County, Pennsylvania, where the breed was developed.)

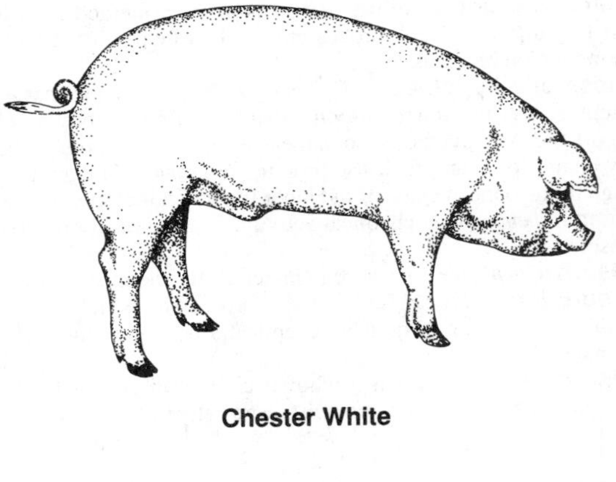

Chester White

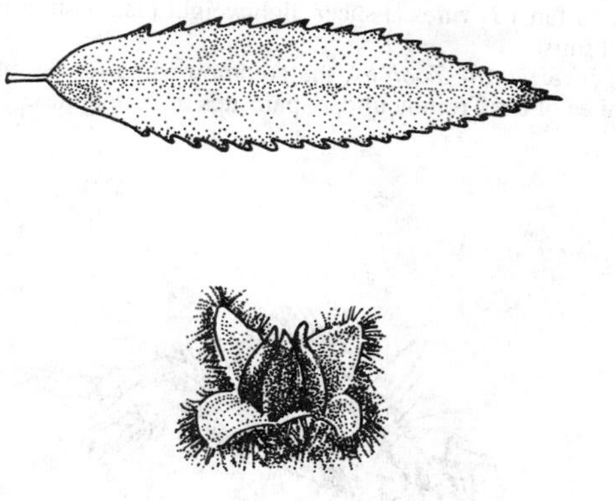

chestnut

chestnut *Botany.* any tree of the genus *Castanea,* of the beech family, characterized by toothed, elongated leaves and bearing edible nuts encased in a spiny bur. Important species include the **European chestnut,** *C. sativa,* and the **Japanese chestnut,** *C. crenata.* The **American chestnut,** *C. dentata,* has been decimated by blight.

chestnut blight *Plant Pathology.* a disease of chestnut trees caused by the fungus *Enqothia parasitica,* and characterized by swollen, cracked, girdling external cankers that spread into the trunk and kill the tree. Also, **chestnut canker.**

chestnut coal *Geology.* **1.** an anthracite coal that will pass through a 1.625-inch mesh screen, but not through a 1.1875-inch mesh screen. **2.** see NUT COAL.

chestnut oak *Botany.* any of several North American oak trees whose leaves resemble those of a chestnut tree, especially *Quercus prinus* (which is also called the **rock oak).**

Chestnut soil *Geology.* a group of zonal soils developed in temperate to cool, subhumid to semiarid climates, and having a dark-brown surface horizon grading into a lighter-colored horizon over a layer of lime accumulation.

Chevalier lens *Optics.* a lens in which an achromatic negative lens is joined with a distant collecting front lens to achieve a magnifying power of 10× with object distances of up to 3 inches.

Cheviot [shev´ē ät´; shev´ē ət] *Agriculture.* a medium-wool breed of sheep having a black nose and no wool on its face or legs. *Textiles.* **1.** a woolen fabric of plain or twill weave with a heavy rough nap, originally made of wool from Cheviot sheep. **2.** a sturdy cotton skirting fabric resembling woolen cheviot. (Named for the *Cheviot* Hills, in England and Scotland.)

chevkinite *Mineralogy.* $(Ca,Ce,Th)_4(Fe^{+2},Mg)_2(Ti,Fe^{+3})_3Si_4O_{22}$, a reddish brown to black opaque, brittle, monoclinic mineral occurring as irregular masses and prismatic crystals, having a specific gravity of 4.3 to 4.67 and a hardness of 5 to 6 on the Mohs scale; found in pumice fragments and ash from Cenezoic volcanism in the western United States; dimorphous with perrierite.

chèvre [shev´rə; shev] *Food Technology.* any cheese that is made from goat's milk. (From the French word for "goat.")

Chevrel phase *Solid-State Physics.* any of a series of ternary molybdenum compounds exhibiting superconductive properties and having the generalized chemical formula $M_xMo_6X_8$, where M represents one of many metallic elements, x takes on a value of 1 to 4, and X is either sulfur, selenium, or tellurium.

chevron *Military Science.* a badge consisting of stripes meeting at an angle, worn on the shoulder as a sign of rank, years of service time, and so on. *Architecture.* a zigzag molding commonly used in Norman architecture. *Vertebrate Zoology.* the Y-shaped arch on the ventral surface of the caudal vertebrae.

chevron cross-bedding *Geology.* cross-bedding that forms a herringbone pattern.

chevron dune *Geology.* a V-shaped dune found in vegetated areas where strong winds blow in a single direction.

chevron fold *Geology.* a fold having sharply bent crests or troughs and relatively straight limbs of equal length.

chevron mark *Geology.* a tool-mark forming rows of chevron-shaped depressions.

chevrotain *Vertebrate Zoology.* the mouse deer of West Africa and tropical Asia, a member of the family Tragulidae in the order Artiodactyla; the smallest of all ruminant mammals.

Cheyne-Stokes respiration *Medicine.* an abnormal breathing rhythm in which episodes of shallow or absent breathing are followed by episodes of rapid breathing.

Chezy formula *Fluid Mechanics.* a method of expressing the velocity of a flowing fluid in an open channel as a function of the hydraulic radius, the head loss per unit length of channel, and a constant C;

$$V = C\sqrt{(Rh*Sb)}$$

where Rh is the hydraulic radius, Sb is the head loss per unit length of channel, V is the fluid velocity, and C is a constant determined from empirical observations.

CHF congestive heart failure.

CHI *Aviation.* the airline code for the city of Chicago; Chicago's two principal airports are O'Hare International (ORD) and Midway (MDW).

chi [kī] the 22nd letter of the Greek alphabet, written as X or χ.

chi *Genetics.* an acronym for cross-over hot-spot instigators.

Chianti [kī än´tē] *Food Technology.* **1.** a light red wine that is widely produced in the Chianti region of Tuscany, central Italy, mostly from the Sangiovese and Trebbiano grapes. **2.** also, **chianti.** a similar wine made elsewhere.

chian turpentine see TURPENTINE, def. 2.

Chiari's network *Cardiology.* a network of small fibers sometimes found extending across the interior of the right atrium from the thebesian and eustachian valves to the crista terminalis. (Named for Hans von *Chiari,* 1835–1915, Austrian pathologist.)

chiasm [kī az´əm] see CHIASMA.

chiasma [kī az´mə] *plural,* **chiasmata.** [kī´əz mä´tə] *Genetics.* a point at which pairs of homologous chromatids are in contact from late prophase of meiosis to the beginning of first anaphase and at which the exchange of homologous parts between nonsister chromatids has taken place by crossing over. *Anatomy.* an X-shaped structure, such as the point at which the optic nerves cross or a point of contact between two chromosomes. Also, CHIASM.

chiasma terminalization see TERMINALIZATION.

chiasmatype theory *Cell Biology.* the hypothesis that DNA crossover, or genetic recombination, is directly related to chiasma formation.

Chiasmodontidae *Vertebrate Zoology.* the swallowers, a family of slender-bodied marine fishes in the suborder Trachinoidet of the order Perciformes, noted for capturing and consuming prey larger than themselves.

chiastobasidial *Mycology.* describing the manner in which spindles of nuclei position themselves traversely or crosswise in a developing basidium or fruiting body.

chiastolite *Mineralogy.* a variety of andalusite.

Chicago boom *Mechanical Engineering.* a boom attached to a structure in the process of construction, with an outside upright portion of the structure serving as the mast and the boom being stepped in a fixed socket clamped to the upright.

Chichibabin reaction *Organic Chemistry.* a substitution reaction in which sodium amide is used to add an amino group to pyridine.

chick *Agriculture.* a young chicken or other bird, one that is either newly hatched or about to be hatched.

chevron fold

chickadee

chickadee *Vertebrate Zoology*. any of seven bird species of American tits (or titmice) of the family Paridae and genus *Parus*; a very active and social songbird with a rounded dark head and fluffy underbelly.
Chickasawhay *Geology*. a North American, Gulf Coast geologic stage of the Oligocene epoch, occurring after the Vicksburgian and before the Anahuac.
chicken *Vertebrate Zoology*. a bird of the genus *Gallus gallus,* a common domesticated fowl belonging the the family Galliformes; raised virtually worldwide for its meat and eggs.
chicken fat clot *Hematology*. a term for blood clot that has a yellowish appearance, resembling chicken fat, because of the settling out of the erythrocytes before clotting occurred.
chickenpox *Medicine*. a communicable disease caused by a herpes virus and producing red spots on the skin that develop into fluid-filled blisters.
chickenpox virus *Virology*. a virus of the genus *Alpha* herpesvirus 3, family Herpesviridae, that causes the disease chickenpox on primary infection and zoster (or shingles) on reactivation. Also, VARICELLA ZOSTER VIRUS.
Chick-Martin test *Biotechnology*. a suspension test used to determine the effectiveness of a disinfectant on organic material; the test result is expressed as the disinfectant's phenol coefficient.
chicle *Materials*. a gummy exudate that is obtained from the bark of the evergreen tree *Achras zapota*; the main ingredient in chewing gum.
chicory *Botany*. a perennial herb, *Cichorium intybus,* of the order Campanales, cultivated for its edible roots and green leaves.

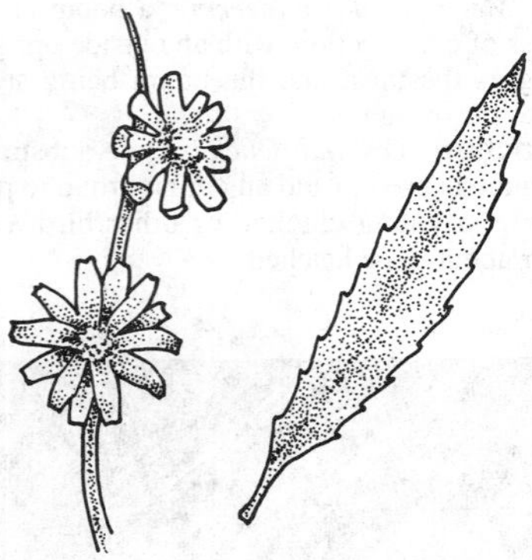

chicory

Chideruan see TATARIAN.
chief *Anthropology*. a leader, especially a hereditary leader, in a traditional or tribal society.
chief cell *Histology*. **1.** a type of cell that lines the fundic glands of the stomach and secretes pepsinogen. **2.** a type of cell in the parathyroid gland that secretes parathormone.
chiefdom *Anthropology*. a ranked society in which status is dependent upon kin; usually the office of chief is inherited and permanent, and it carries authority in economic, political, and religious activities.
chief petty officer *Military Science*. in the U.S. Navy and Coast Guard, a noncommissioned officer ranking above petty officer first class and below senior petty officer.
chiffon [shi fän′] *Textiles*. a sheer, lightweight plain cotton fabric with a soft, dull finish.
chigger *Invertebrate Zoology*. the common name for the bloodsucking larval mites of the Trombiculidae; parasites of vertebrates.

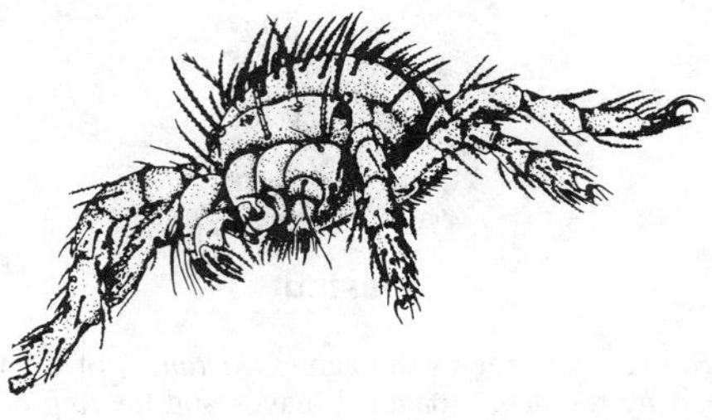

chigger

chilarium *Invertebrate Zoology*. either of a pair of processes between the bases of the fourth pair of legs on the king crab.
chilblain or **chilblains** *Medicine*. inflammation and swelling caused by exposure to cold and dampness, often accompanied by itching and burning; usually affects the hands, feet, ears, and face. Also, PERNIO.
child *Computer Programming*. in data structures, any node in a tree except the root; a direct descendant of a given node. Also, DAUGHTER, SON.
child abuse *Psychology*. the physical, mental, or emotional mistreatment of a child by a parent or other adult.
childbirth *Medicine*. the process of giving birth to a child; labor and delivery.
Childe, V. Gordon 1892–1957, British archaeologist; developed the diffusionist theory of European prehistory.
childhood schizophrenia *Psychology*. schizophrenia that has its onset before puberty, characterized by autistic thinking, extreme withdrawal, and severe developmental immaturity.
childrenite *Mineralogy*. $Fe^{+2}Al(PO_4)(OH)_2 \cdot H_2O$, a brown, transparent to translucent monoclinic mineral occurring as pyramidal, short prismatic, or thick tabular crystals, having a specific gravity of 3.2 to 3.25 and a hardness of 5 on the Mohs scale; found as fine crystals in hydrothermal vein deposits.
Child's law *Electronics*. a law stating that current in a heat-sensitive element varies directly with the three-halves power of anode voltage and inversely with the square of the distance between electrodes. Also, THREE-HALVES POWER EQUATION.
chile see CHILI.
Chilean subregion *Ecology*. a distinct zoogeographical region that includes the country of Chile and other areas either west of the Andes Mountains or south of the Amazon River basin.
Chile mill *Mechanical Engineering*. a device that crushes and grinds material between vertical rollers within a circular container.
Chile niter *Mineralogy*. a commercial name for the sodium nitrate mineral natratine that occurs naturally in caliche in northern Chile. Also, **Chile saltpeter**.
chili *plural,* **chilies.** *Agriculture*. the pungent pod of species of *Capsicum,* particularly *C. annuum longus;* widely used in cooking. *Meteorology*. a warm, dry, descending wind in Tunisia, similar to the sirocco.
chill *Metallurgy*. **1.** in casting, a high-conductivity insert that is placed in the mold to locally increase the freezing rate of the casting. **2.** white cast iron present on a gray or ductile iron.
chill-block melt spinning *Metallurgy*. a process for manufacturing rapidly solidified alloys by pouring a molten charge onto a chilled, rotating surface, such as a roll or a dish, resulting in significant structural modifications.
chill casting *Materials Science*. a method used in casting aluminum, in which the molten metal is poured from a holding furnace into stationary molds resting on elevators; when the metal at the bottom of the mold is solidified, the elevator descends while the ingots are sprayed with water to complete the solidification process.
chilled contact *Petrology*. the finer-grained portion of an igneous rock found near its contact with older rock. Also, **chilled edge, chilled margin.**

chilled iron *Metallurgy.* iron that has been cast against a thermally conductive mold.
chilled shot see HARD LEAD.
chiller *Chemical Engineering.* an oil-refining unit that cools the paraffin distillates.
chilling *Food Technology.* the process of lowering the temperature of a food product, usually by refrigeration.
chilling injury *Food Technology.* any color or texture change on the surface of a food caused by overexposure to low temperature.
chill roll *Engineering.* a roll of iron having a hard outer layer of white iron, a second layer of mottled iron, and a center of full gray iron.
chill-roll extrusion *Engineering.* a process of cooling and forming plastic film by drawing it around very smooth chill rolls containing cold water. Also, CAST-FILM EXTRUSION.
chill wind factor see WIND-CHILL FACTOR.
chill zone *Materials Science.* a narrow layer of randomly oriented and generally equiaxed grains at the surface of a casting.
Chilobolbinidae *Paleontology.* a genus of Paleozoic ostracods characterized by a straight hinge, a frill, and an oval-shaped brood pouch; Ordovician and Silurian.
Chilodonella *Invertebrate Zoology.* a large genus of fresh- or brackish-water ciliate protozoans in the subclass Holotrichia, including a number of ectoparasites, some of which are destructive to cyprinoid fish.
Chilognatha *Invertebrate Zoology.* a subclass of millipedes; arthropods in the class Diplopoda, including most typical millipedes.
Chilomastix *Invertebrate Zoology.* a genus of small, pear-shaped, four-flagellated protozoans in the order Polymastigina; parasites in the intestines of vertebrates, including humans.
Chilomonas *Invertebrate Zoology.* a genus of colorless, protozoalike algae, widely used in laboratory experiments.

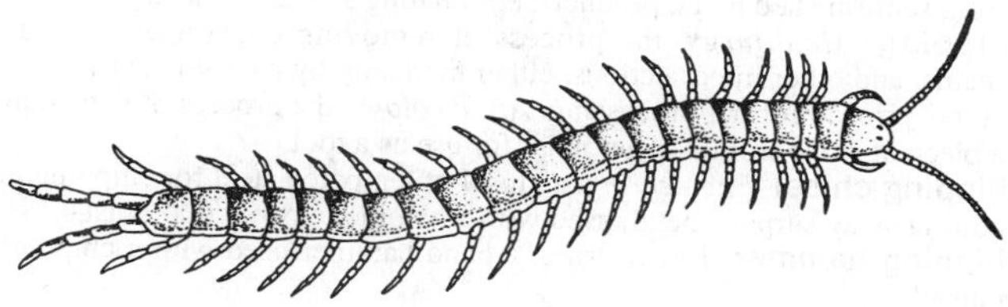

Chilopoda

Chilopoda *Invertebrate Zoology.* the centipedes, a class of many-footed carnivorous predators in the Myriapoda.
Chimaeridae *Vertebrate Zoology.* a family of marine fishes in the subclass Holocephali, having a cartilaginous skeleton but many other features more like those of the bony fishes.
Chimaeriformes *Vertebrate Zoology.* the single order in the chondrichthyan subclass Holocephali, a group of usually bottom-feeding marine fishes occurring worldwide.
chimera [kī mâr´ə] *Biology.* an organism consisting of two or more genetically distinct cell types, produced as a result of mutation, transplantation, or grafting, the fusing of different embryos, or other similar processes. (From *Chimera,* a mythological Greek monster made up of the front of a lion, the middle of a goat, and the tail of a snake.)
chimeric *Biology.* of or relating to a chimera or to chimerism.
chimeric plasmid *Genetics.* a plasmid that has received new DNA sequences enabling it to contain two connected DNA sequences, one of which represents the original plasmid molecule.
chimeric protein *Biochemistry.* a protein consisting of a mixture of sequences from different sources.
chimeric vector *Genetics.* a plasmid or bacteriophage containing new gene sequences used to connect other DNA sequences.
chimerism *Biology.* the state of containing two or more genetically distinct cell types. In humans, chimerism results from any organ transplant other than that between identical twins or from the mixture of blood cells between dizygous twins in the uterus before birth.
chi meson *Particle Physics.* a meson particle having mass 958 MeV, zero charge, zero isospin, negative parity, and (probably) zero spin.
chiming *Horology.* **1.** in a timepiece, the act of ringing a bell or melodic sequence of bells to mark each quarter-hour that passes, as opposed to striking or ringing only on the hour. **2.** of or relating to the gear train in a timepiece that controls the chiming process.
chiming train *Horology.* in a timepiece, a set of gears and pinions, usually synchronized to the going train, that causes chimes or bells to be sounded at the quarter-hours.
chimney *Building Engineering.* a vertical flue or passageway in a building, that draws up combustion by-products from a stove, furnace, or fireplace. *Electronics.* an enclosure that is placed over a heat sink to increase its dissipating ability. *Geology.* **1.** an angular, columnar projection of rock along a rough coast, that has been isolated from the main cliff by wave action. **2.** a cylindrically shaped ore body that is more or less vertical. **3.** a steep, very narrow gully in the face of a mountain. **4.** a natural vent or conduit through which magma reaches the earth's surface. **5.** in a cave, a vertical passage having rounded walls.
chimney bar *Building Engineering.* a lintel, usually of wrought iron or steel, that serves to support the masonry above a fireplace opening. Also, TURNING BAR.
chimney breast *Building Engineering.* a wall surface projecting into a room at a point where a chimney passes through it; it is typically wider than the chimney itself to allow construction of a mantle, or for aesthetic purposes.
chimney cloud *Meteorology.* a cumulus cloud having a greater vertical than horizontal extent.
chimney core *Mechanical Engineering.* the inside wall of a double-walled furnace stack, separated from the outer portion by an air space.
chimney effect *Fluid Mechanics.* the theoretical draft in a chimney produced by the difference in static head of equal columns of atmospheric air and hot flue gas when at rest.
chimney lining *Building Engineering.* a tiled flue positioned within a chimney. Also, **chimney flue.**
chimney rock *Geology.* **1.** a column of rock that rises above its surroundings or is partly isolated on the face of a slope. **2.** a chimney-shaped rocky island or mass that has been isolated from a rock cliff by wave action and weathering.
chimney shaft *Building Engineering.* the passageway through which air or smoke travels as it rises above a combustible area, such as the hearth of a fireplace.
chimney swift *Vertebrate Zoology.* a small, dark gray North American swift, *Chaetura pelagica,* that frequently builds its nest in chimneys.
chimnoclorous *Ecology.* describing plants with thin herbaceous leaves that persist throughout winter.
chimnophilous *Ecology.* thriving during winter; exhibiting maximum growth during winter. Thus, **chimnophile, chimnophily.**
chimopelagic *Ecology.* describing deep-water marine organisms that live on the surface only in winter. Also, **chimnopelagic.**

chimpanzee

chimpanzee *Vertebrate Zoology.* a tailless tropical African ape, *Pan troglodytes,* of the family Pongidae, noted for its high level of intelligence.
chin *Anatomy.* the projecting part of the lower jaw below the lower lip.
China clay see KAOLIN.

chinacrin hydrochloride see QUINACRINE HYDROCHLORIDE.

China grass see RHEA.

China oil see PERU BALSAM.

China Sea see EAST CHINA SEA; SOUTH CHINA SEA.

chinch bug *Invertebrate Zoology.* a tiny plant-bug of the order Heteroptera; a troublesome pest on cereal grains in both Europe and America.

chinchilla *Vertebrate Zoology.* a squirrel-like mammal, *Chinchilla laniger,* of the family Chinchillidae, native to the Peruvian Andes but now raised commercially all over the world for its soft gray fur.

chinchilla cloth *Textiles.* a twilled woolen fabric having a long nap that is shaped into nubs or coils; used in making coats.

Chinchillidae *Vertebrate Zoology.* the chinchillas and viscachas, a family of herbivorous South American rodents in the suborder Myomorpha, noted for their long, silky fur.

chine *Naval Architecture.* **1.** the line formed when the side of a hull meets a flat or V bottom at an angle (as opposed to a rounded bilge). **2.** the line at which the curved side of a barrel meets a flat end; casks stowed end to end are referred to as **chine and chine.**

Chinese cabbage *Botany.* a long, thick head of broad, whitish leaves of the plant *Brassica rapa pekinensis;* used in salads and Oriental cookery.

Chinese hamster ovary cells *Virology.* a line of cells that are prepared from the ovaries of Chinese hamsters and used to grow many viruses, including herpesviruses.

Chinese letter arrangement *Microbiology.* a grouping of the cells of certain bacteria, such as the *Coryneform* species, that resembles Chinese characters when observed by light microscopy.

Chinese postman problem *Mathematics.* in a graph G with weights $w(e_i)$ on the edges e_i, find a closed walk that includes every edge and minimizes the sum of the weights of the edges in the walk, an edge weight being included in the sum every time the edge is encountered in the walk.

Chinese remainder theorem *Mathematics.* **1.** if the n positive integers $\{m_k\}$ are pairwise relatively prime and $\{b_k\}$ are any n integers, then there exists an integer x satisfying all the congruences $x \equiv b_k(\text{mod } m_k)$, for $k = 1, \ldots, n$, uniquely determined (mod $m_1 m_2 \cdots m_n$). **2.** in general, let $A_1, \ldots, A_n$ be the ideals in a ring R such that $R^2 + A_i = R$ for all i and $A_i + A_j = R$ for all $i \neq j$. If $b_1, \ldots, b_n \in R$, there then exists $b \in R$ such that $x \equiv b_k(\text{mod } A_k)$, for $k = 1, \ldots, n$. Furthermore, b is uniquely determined modulo the ideal $A_1 \cap A_2 \cap \cdots \cap A_n$. (Note here that R is not assumed to possess a multiplicative identity. If R has an identity, then $R^2 = R$, so that $R^2 + A = R$ for every ideal A of R.)

Chinese white see ZINC WHITE.

Chinese windlass see DIFFERENTIAL WINDLASS.

chino *Textiles.* a cotton twill fabric often used in making military uniforms and work clothes or casual wear.

chinook [shə nůk´; chə nůk´] *Meteorology.* **1.** a warm and moist wind in the northwestern U.S. that blows from the sea to the land in spring and winter. **2.** a rapidly moving downdraft on the eastern side of the Rocky Mountains, causing a dramatic fluctuation in temperature.

chinook arch *Meteorology.* a cloud formation appearing as a flat layer of altostratus clouds over the Rocky Mountains that heralds the approach of a chinook.

chinook salmon *Invertebrate Zoology.* a large salmon, *Oncorhynchus tshawytscha,* found in the northern Pacific Ocean. It is the largest of the salmon family and is a valuable food fish. Also, KING SALMON, QUINNAT SALMON.

chintz *Textiles.* a firm, plain-woven cotton fabric, usually printed, having a glazed finish.

chiolite *Mineralogy.* $Na_5Al_3F_{14}$, a colorless or white, transparent to translucent tetragonal mineral occurring as masses and very small, rare crystals; found with cryolite in pegmatites.

chionablepsia *Medicine.* snow blindness.

Chionididae *Vertebrate Zoology.* the white sheathbills, a family of wading, fish-eating birds of the Antarctic, belonging to the order Charadriiformes. Also, CHIONIDAE.

chionophilous *Ecology.* describing a species that thrives in snow-covered habitats. Thus, **chionophile, chionophily.**

chionophobia [kī´ än ō fō´bē ə] *Psychology.* an irrational fear of snow.

chionophobous *Ecology.* describing a species that is not tolerant of snow-covered habitats. Thus, **chionophobe, chionophoby.**

Chionosphaera *Mycology.* a genus of fungi of the order Sporidiales that possess fruiting bodies shaped like stalks and form spore-producing structures called basidia.

chip *Materials.* any small piece or fragment of material. *Archaeology.* **1.** a piece of stone removed from a larger mass for use as a tool; a flake. **2.** any small piece of stone detached in the process of toolmaking. **3.** to strike off a piece of stone in toolmaking. *Electronics.* **1.** a tiny piece of superconductor material on which all the components of an integrated circuit, such as transistors or resistors, are found. Also, DIE. **2.** a popular name for an integrated circuit.

chip blower *Medicine.* a syringe of small caliber that is used to funnel a stream of air onto the surface of a tooth during the removal of decay, in order to clear away particles and remove moisture. Also, CHIP SYRINGE.

chipboard *Materials.* a low-grade, low-density cardboard usually made from recycled waste paper.

chip breaker *Design Engineering.* a groove in the face of a cutting tool that allows removed material to chip and break away.

chip cap *Design Engineering.* a plate or cap on a carpenter's plane that breaks up the wood shavings and gives the tool strength.

chip capacitor *Electronics.* a device, used to store electrical charges, that has metalized terminations so that it may bond easily onto a hybrid integrated circuit.

chip log *Engineering.* a piece of wood, in a quadrant shape and weighted around its edges, attached to a log line. *Navigation.* an older form of speed measuring device for ships. A chip of wood was streamed over the transom, and the number of knots that passed over the rail in a given time was an indication of the speed of the vessel in nautical miles per hour.

chipmunk *Vertebrate Zoology.* any of several small, striped squirrels of the genera *Tamias,* of North America, and *Eutamia,* of Asia and North America.

chippable *Archaeology.* of a stone, capable of being worked to produce a tool or other such artifact.

chipped stone see FLAKED STONE.

chipper *Engineering.* a pneumatic chisel that removes surface defects from semifinished metal products, eliminating seams and laps.

chipping *Metallurgy.* the process of removing excessive material, seams, and other imperfections, either manually by a chisel, or by a machine, prior to further processing. *Archaeology.* the process of detaching a piece of stone from a larger mass for use as a tool.

chipping chisel *Engineering.* a chisel of tempered steel for chipping or cutting away surplus metal, used with a hammer. Also, COLD CHISEL.

chipping hammer *Engineering.* a hand hammer used with a chipping chisel.

chipping station or **chipping floor** *Archaeology.* a surface or area used for the manufacture of prehistoric stone tools; identified by the presence of waste stone chips and other such evidence. Also, FLAKING STATION.

chip resistor *Electronics.* a device, used to impede current flow, that has metalized terminations so that it may bond easily onto a hybrid integrated circuit.

chip sampling *Mining Engineering.* the process of obtaining small ore or coal samples, either at random or along a line, from the width of an ore face exposure.

chip syringe see CHIP BLOWER.

chir- a combining form meaning "hand," as in *chirality.*

chiral [kī´rəl] *Chemistry.* of a molecule in a given configuration, not symmetrical with its mirror image.

chirality [kī ral´ə tē] *Chemistry.* **1.** the condition of being chiral. **2.** the direction, either left (levo, –) or right (dextro, +), in which a given chiral molecule rotates plane polarized light; "handedness." *Physics.* any characteristic associated with an object that cannot be superimposed on its mirror image due to its singleness of state properties. *Particle Physics.* the property in which a particle's spin can exist in either right- or left-handed forms, pointing along its direction of travel or against it. Also, **chiral symmetry.**

chiral symmetry group *Particle Physics.* a group of symmetry transformations that have different effects on right-handed and left-handed portions of a fermion field.

Chireix antenna *Electromagnetism.* a phase array composed of two or more coplannar square loops, connected in series.

Chiridotidae *Invertebrate Zoology.* a family of wormlike holothurian echinoderms in the order Apodida, with wheel-shaped spicules.

chiro- a combining form meaning "hand," as in *Chiroptera.*

Chirocentridae *Vertebrate Zoology.* the wolf herring, a monotypic family of large herringlike fish of the suborder Clupeoidei, characterized by voracious predatory habits and native to warm-temperate and tropical seas of the western Pacific and Indian oceans.

Chirodontidae *Paleontology.* a family of early chondrostean fish in the extinct order Palaeonisciformes and suborder Platysomoidei, characterized by a deep body, a very long dorsal fin, elongated scales, and a crushing dentition; Carboniferous to Triassic.
Chirognathidae *Paleontology.* a large family of conodonts in the order Prioniodinida; the first representatives of the order, the Chirognathidae exhibit a broad range of tooth designs, but the teeth are generally fibrous and grasp or tip the end of the jaw ramus; Ordovician.
Chiron [kī´ rän] *Astronomy.* a weakly active comet orbiting the sun between Saturn and Uranus; discovered in 1977 and formerly classified as asteroid 2060.
Chironomus *Entomology.* minute, long-legged nonbiting two-winged flies with piercing mouthparts; the aquatic larvae of various species are green, blue, yellow, colorless, or red type called bloodworms.
chiropodist [ke räp´ə dist] *Medicine.* a person who treats corns, bunions, and other conditions of the hands and feet; a podiatrist.
chiropody [ke räp´ə dē] *Medicine.* the art or profession of a chiropodist; podiatry.
chiropractic [kī´rō prak´tik] *Medicine.* **1.** a program of treatment based on the theory that the nervous system is responsible for health, and relying on spinal manipulation as a principal therapy. **2.** of or relating to such a program of treatment.
chiropractor [kī´rō prak´tər] *Medicine.* a person who practices chiropractic medicine.
Chiroptera *Vertebrate Zoology.* the bats, a large order composed of the only mammals capable of true sustained flight, having wings formed from a membrane of skin stretched between the limbs, and being mostly nocturnal.
chiropterophilous *Biology.* of flowers, being pollinated by bats.
chiropterygium *Vertebrate Zoology.* any typical limb on a vertebrate, thought to have evolved from a finlike extension.
chirospasm see CHEIROSPASM.
Chirotheuthidae *Invertebrate Zoology.* the deep-sea squids, a family of highly evolved mollusks.
chirp *Telecommunications.* a pulsed signal of a frequency that varies during a pulse time (usually from higher to lower frequency) and has an audible replica that resembles a bird's chirp. Chirp signals are employed in radar to aid in distinguishing targets from surrounding clutter.
chirp radar *Engineering.* a radar system in which each signal transmitted and reflected is compressed into an abbreviated pulse or chirp signal.
chirurgenic *Surgery.* developing or resulting from a surgical procedure.
chirurgery *Surgery.* a former term for surgery.
chisel *Mechanical Devices.* a flat, steel cutting tool having a sharp bevel on one end and a wooden handle on the other, used for dressing stone or cutting wood or metal. *Agriculture.* a plow blade that has long, penetrating points. Also, **chisel bottom.**

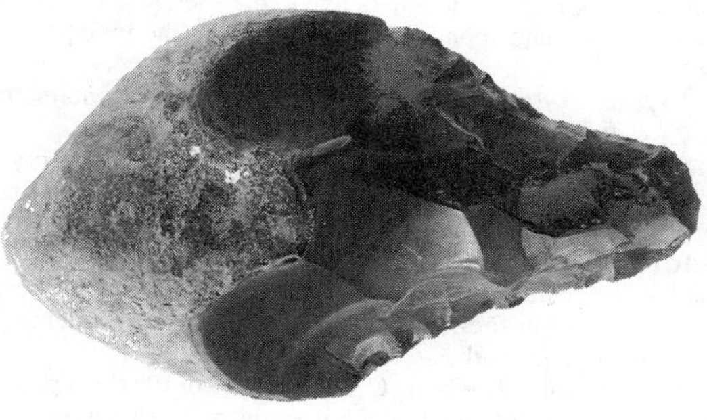

chiseling tool

chisel-edge angle *Design Engineering.* the angle formed between the chisel edge and the cutting edge of a drill, as viewed from the end of the tool.
chisel plow *Agriculture.* a type of plow that uses narrow, curved bottoms to break up the subsoil without bringing it to the surface. A **deep-tillage chisel plow** has blades that are longer and thicker to give greater penetration.
chisel point *Mechanical Devices.* a nail or spike whose point is formed by the meeting of two inclined surfaces at a sharp angle.
chisel-tooth saw *Mechanical Devices.* a circular saw having chisel-shaped teeth around its cutting edge.
chi-sites *Genetics.* extremely active short portions of DNA that stimulate crossing-over between chromosomes.
chi-square distribution *Statistics.* a special case of the gamma distribution, defined as the probability density arising as the distribution of the sum of the squares of n independent standard normal random variables; the value of $n-1$ is called the number of degrees of freedom; used in statistical estimation and goodness-of-fit tests.
chi-square test *Statistics.* the goodness-of-fit test used to determine whether a specific probability distribution function fits observed data; it is based on the sum of the standardized squared differences between the observed and the expected outcomes and has a chi-square distribution; also used in contingency table analysis.
chitin *Biochemistry.* $(C_8H_{13}NO_5)_n$, a complex carbohydrate that is derived from *N*-acetyl-D-glucosamine and forms the hard outer shell of insects, crustaceans, arthropods, fungi, and some algae.
chitinase *Enzymology.* a digestive enzyme that breaks down glycosidic chemical bonds in chitin.
chitinivorous bacterium *Microbiology.* a microorganism that is able to decompose chitinous material such as a crab shell.
chitinolytic *Biochemistry.* of or relating to an enzyme that converts chitin to chitobiose.
Chitinozoa *Paleontology.* a large group of one-celled Paleozoic microfossils similar to foraminiferids; characterized by their flask shape, they are found in high concentrations in certain strata and are important in petroleum geology; Cambrian to Devonian.
chitobiose *Biochemistry.* a disaccharide having two β-linked *N*-acetyl-D-glucosamine residues that is secreted by a number of organisms, including spiders and plant tissues.
chiton [kī´tän; kī´tən] *Invertebrate Zoology.* the common name for a member of a group of mollusks with segmented shells in the class Polyplacophora.
chitosan [kī´tə san] *Organic Chemistry.* the deacylated form of chitin ($[C_8H_{13}NO_5]_n$); used in water treatment, photographic emulsions, and for improving dyeability of synthetic fibers.
chitosome *Cell Biology.* a membranous vesicle that contains the enzyme chitin synthase, is found in certain fungi, and may either be produced by the cell or be an artifact of the isolation procedure.
Chladni figures *Mechanics.* the distinctive pattern of lines produced by putting sand or similar materials on a horizontal plate and then vibrating the plate while holding it fixed along its sides or at the center; the material forms figures along the the nodal lines of vibration. (From the 18th-century German physicist Ernst *Chladni.*)
chlamydeous *Botany.* of a plant, having a perianth.
Chlamydia *Bacteriology.* a genus of Gram-negative, coccoid bacteria of the order Chlamydiales that occur as obligate intracellular parasites in humans and other animals.
Chlamydiaceae *Bacteriology.* a family of bacteria of the order Chlamydiales, consisting of the single genus *Chlamydia.*
Chlamydiales *Bacteriology.* an order of bacteria that are parasitic for eukaryotic cells and are distinguished by a developmental cycle alternating between small infectious forms (elementary bodies) and larger noninfectious forms (reticular bodies).
Chlamydobacteriaceae *Bacteriology.* a former classification for a family of bacteria in the order Chlamydobacteriales.
Chlamydobacteriales *Bacteriology.* a former classification for an order of bacteria comprised of nonpigmented, filamentous cells.
Chlamydomonadaceae *Botany.* a family of unicellular biflagellates of the order Volvocales, characterized by a one-piece cell wall, a terete or compressed cell, and a usually cup-shaped chloroplast; common in organically enriched fresh water, soil, snow, ice, and coastal waters.
Chlamydomonadidae *Invertebrate Zoology.* a family of flagellate, colorless protozoans in the order Volvocida.
Chlamydoselachidae *Vertebrate Zoology.* the frilled sharks, a family of rare eel-like sharks in the order Hexanchiformes, found in Japanese seas and in the Atlantic and Arctic Oceans.
chlamydospore *Mycology.* a kind of resting spore that is unicellular, has thick walls, and occurs in almost all parasitic fungi.
Chlamydozoaceae *Bacteriology.* a former classification for a family of bacteria that are now classified in the family Chlamydiaceae.
chloanthite *Mineralogy.* an arsenic-deficient variety of nickel-skutterudite.

chloasma *Medicine.* a skin discoloration that sometimes occurs during pregnancy, affecting the cheeks, chin, and forehead. Also, MELASMA.

chloflurecol methyl ester *Organic Chemistry.* $C_{15}H_{11}ClO_3$, a white, crystalline solid that is soluble in water and melts at 152°C; used as a growth regulator for grass and weeds.

chlor- a combining form meaning: **1.** green, as in *chlorella.* **2.** chlorine, as in *chlordane.*

Chloracea *Bacteriology.* a family of green sulfur, photosynthetic bacteria of the order Rhodobacteriineae.

chloracne *Medicine.* a skin irritation caused by exposure to chlorine.

Chloradendrineae *Botany.* a suborder of green algae; most species are colonial, but some are capable of independent existence and movement.

chloragogen *Invertebrate Zoology.* any of the specialized yellowish cells forming the outer layer of the alimentary tract in earthworms and other annelids. Also, **chloragogue.**

chloral *Organic Chemistry.* Cl_3CCHO, a toxic, colorless, oily liquid; very soluble in water and soluble in alcohol; melts at –57.5°C and boils at 97.57°C; used in the manufacture of chloral hydrate and DDT. Also, TRICHLOROACETALDEHYDE.

chloral hydrate *Organic Chemistry.* $CCl_3CH(OH)_2$, transparent, colorless crystals; soluble in water and alcohol; melts at 57°C and boils at 97.5°C (with loss of water); used as a sedative and hypnotic drug (toxic in overdose), and in the manufacture of DDT and liniments.

chloralkane *Organic Chemistry.* a compound derived among other ways by treating an alkane with chlorine under the influence of ultraviolet light at 250–400°C.

α-chloralose *Organic Chemistry.* $C_8H_{11}O_6Cl_3$, a crystalline solid that melts at 87°C, and is soluble in acetic acid and ether; used as a bird repellent on seed grains and as a hypnotic for animals. Also, CHLORALOSANE, GLUCOCHLORAL.

chloraluminite *Mineralogy.* $AlCl_3 \cdot 6H_2O$, a colorless, white, or yellow trigonal mineral with rhombohedral crystals that occurs as crystalline crusts and stalactites; having a specific gravity of 1.67 and an undetermined hardness on the Mohs scale; found on Mount Vesuvius after the 1872 and 1906 eruptions.

chloramben *Organic Chemistry.* $C_6H_2NH_2Cl_2COOH$, a white solid, slightly soluble in water; melts at 200–201°C; used as an herbicide or plant-growth regulator. Also, 3-AMINO-2,5-DICHLOROBENZOIC ACID.

chloramine-T *Organic Chemistry.* $CH_3C_6H_4SO_2NNaCl \cdot 3H_2O$, white or slightly yellow crystals or crystalline powder, soluble in water, decomposes in alcohol and air; used as an antiseptic and an oxidizing or chlorinating reagent. Also, SODIUM *p*-TOLUENESULFOCHLORAMINE.

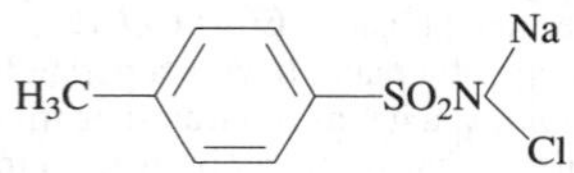

chloramine-T

Chloramoebaceae *Botany.* a family of mostly freshwater, yellow-green algae comprising the order Chloramoebales.

Chloramoebales *Botany.* a monofamilial order of yellow-green algae of the class Xanthophyceae that have solitary, free-swimming, naked cells with two unequal flagella (or sometimes only one flagellum) and reproduce by longitudinal fission.

chloramphenicol *Microbiology.* a broad-spectrum polyene antibiotic originally derived from *Streptomyces venezuelae* and now produced synthetically; it causes a reversible inhibition of bacterial protein synthesis. Also, CHLOROMYCETIN.

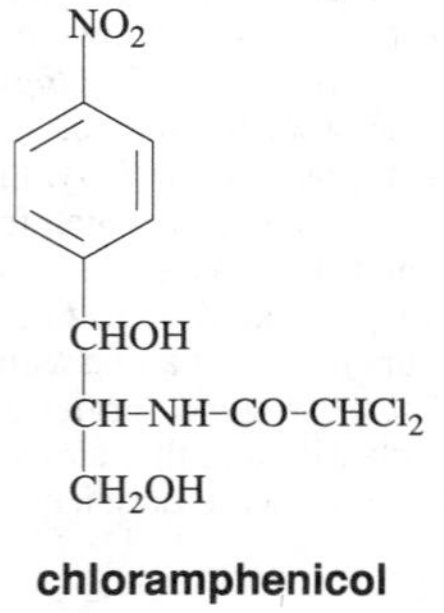

chloramphenicol

Chlorangiaceae *Botany.* a family of green algae of the order Tetrasporales having cells attaching immediately to each other.

Chlorangiellaceae see CHARACIOCHLORIDACEAE.

chloranil *Organic Chemistry.* $C_6Cl_4O_2$, yellow, monoclinic prisms; insoluble in water and slightly soluble in alcohol; melts at 290°C and sublimes at boiling point; used as an agricultural fungicide, a dye intermediate, and a reagent.

chloranilic acid *Organic Chemistry.* $C_6H_2Cl_2O_4$, red, leaflike crystals, soluble in water, that melt at 283–284°C; used in spectrophotometry and histologic staining.

Chloranthaceae *Botany.* a family of tropical and subtropical, woody or herbaceous plants of the order Piperales, characterized by scattered spherical cells containing volatile oils and seeds with much oily, starchy endosperm.

chlorapatite *Mineralogy.* $Ca_5(PO_4)_3Cl$, a pink-white or pale yellow transparent to translucent monoclinic mineral of the apatite group, occurring as prismatic crystals with a vitreous to chalky luster, and having a specific gravity of 3.1 to 3.2 and a hardness of 5 on the Mohs scale; found in calc-silicate marbles and as microgranules in meteorites.

chlorargyrite *Mineralogy.* AgCl, a transparent to translucent, gray, green, or yellow waxy cubic mineral occurring in massive form or as cubic crystals, having a specific gravity of 5.55 and a hardness of 2.5 on the Mohs scale; a secondary mineral in the oxidation zone of silver deposits, often found with native silver, jarosite, iron and manganese oxides, cerussite, and malachite. Also, CERARGYRITE.

chlorastrolite SEE PUMPELLYITE.

chlorate *Inorganic Chemistry.* any salt of chloric acid, $HClO_3$.

chlorazol black E *Microbiology.* a bacteriological triazo dye used to stain species of *Mycoplasma* and L-forms.

chlorbenside *Organic Chemistry.* $ClC_6H_4CH_2SC_6H_4Cl$, white crystals that are insoluble in water; melts at 75–76°C; used as an acaricide.

chlorbromuron *Organic Chemistry.* $C_9H_{10}ONBrCl$, a white solid that melts at 94°C; used as a broad-spectrum herbicide.

chlorcyclizine hydrochloride *Pharmacology.* $C_{18}H_{21}ClN_2 \cdot HCl$, an antihistamine occurring as a white, crystalline powder that is administered orally.

chlordane *Organic Chemistry.* $C_{10}H_6Cl_8$, a toxic, colorless, viscous liquid; insoluble in water; used as an insecticide and a fumigant.

chlordiazepoxide *Pharmacology.* $C_{16}H_{14}ClN_3O$, a benzodiazepine tranquilizer occurring as a yellow, crystalline powder that is administered orally to treat alcohol withdrawal, acute anxiety, and other conditions with symptoms consisting of anxiety, tension, and apprehension.

chlordiazepoxide hydrochloride *Pharmacology.* $C_{16}H_{15}Cl_2N_3O$, the monohydrochloride salt of chlordiazepoxide, having similar applications, occurring as a white or whitish crystalline powder that is administered orally, intravenously, or intramuscularly.

chlordimeform *Organic Chemistry.* $C_{10}H_{13}ClN_2$, colorless crystals that are soluble in water; melts at 35°C; used as an ovicide and insecticide against mites, moths, and butterflies.

Chlorella *Bacteriology.* the best-known genus of green algae in the family Oocystaceae, from which chlorellin is derived; some species are phycobionts of lichens, while others live in the tissues of freshwater ciliates, coelenterates, and sponges; used in studying the process of photosynthesis.

chlorenchyma *Botany.* parenchymatous tissue that contains chlorophyll.

chlorendic acid *Organic Chemistry.* $C_9H_4Cl_6O_4$, a white crystalline solid that is used in making dyes, insecticides, fungicides, and fire-resistant polyester resins.

chlorendic anhydride *Organic Chemistry.* $C_9H_2Cl_6O_3$, nonflammable, fine, white, free-flowing crystals, insoluble in water; melts at 239–240°C; used in fire-resistant polyester resins, in hardening epoxy resins, and in organic synthesis.

chlorfenac *Organic Chemistry.* $C_8H_5Cl_3O_2$, a colorless solid that is insoluble in water and soluble in alcohol; melts at 159°C; used as an herbicide in nonagricultural areas.

chlorfenethol *Organic Chemistry.* $C_{14}H_{12}Cl_2O$, a colorless crystalline solid; soluble in organic solvents; melts at 70°C; used to control mites.

chlorfenpropmethyl *Organic Chemistry.* $C_{10}H_{10}OC_{12}$, a brown liquid used as a weedkiller.

chlorfensulfide *Organic Chemistry.* $C_{12}H_6Cl_4N_2S$, a yellow crystalline solid that melts at 123°C; used to control mites on citrus plants.

chlorfenvinphos *Organic Chemistry.* $C_{12}H_{14}Cl_3O_4P$, a yellowish liquid that is slightly soluble in water; boils at 168°C (0.5 torr); used as an insecticide.

chloric acid *Inorganic Chemistry.* $HClO_3 \cdot 7H_2O$, a toxic compound known only in aqueous solution; very soluble in water and decomposes at 40°C; it ignites organic matter on contact and is used as a catalyst.

chloride *Chemistry.* **1.** any compound containing the chlorine atom as the negative ion (Cl^-), such as sodium chloride (NaCl). **2.** any salt or ester of hydrochloric acid (HCl).

chloride ion *Chemistry.* a negatively charged chlorine atom, Cl^-.

chloride shift *Physiology.* the exchange of chloride and bicarbonate between the plasma and red blood cells; occurs whenever bicarbonate is generated or decomposed within the red cells.

chloridization *Chemistry.* **1.** the process of treating a substance with chlorine gas or hydrochloric acid to form a chloride, especially in the extraction of metals from mineral ores. **2.** see CHLORINATION. *Metallurgy.* the chlorination of ores prior to metal extraction; for example, the formation of titanium tetrachloride from rutile prior to its reduction to titanium sponge.

chlorin *Biochemistry.* the hydroporphyrin prosthetic group found in chlorophyll; a ring structure that is related, for example, to the heme of hemoglobin.

chlorinate *Chemistry.* **1.** to introduce chlorine into a compound. **2.** to treat with chlorine.

chlorinated naphthalene *Organic Chemistry.* $C_{10}H_7Cl$, a toxic, crystalline solid that is soluble in alcohol and that melts at –2.3°C; used as a solvent and immersion liquid. Also, 1-CHLORONAPHTHALENE.

chlorinated paraffin *Organic Chemistry.* a compound produced by treating paraffin with chlorine gas at 250–400°C, with ultraviolet light.

chlorinated wool *Textiles.* wool that has been treated with chlorine in order to reduce shrinkage or improve dye absorption.

chlorination *Chemistry.* **1.** the insertion of a chlorine atom into a compound. Also, CHLORIDIZATION. **2.** the use of chlorine gas to sterilize water.

chlorinator *Chemical Engineering.* a unit used to add chlorine to organic compounds or to sterilize water with chlorine gas.

chlorine *Chemistry.* a nonmetallic halogen element having the symbol Cl, the atomic number 17, an atomic weight of 35.453, a melting point of –101°C, and a boiling point of about –35°C; a dense, pungent, greenish-yellow gas that is corrosive, toxic, and water-soluble; supports combustion and reacts violently with many substances. (Going back to a Greek word meaning "greenish" or "greenish-yellow.")

chlorine dioxide *Inorganic Chemistry.* ClO_2, a yellowish-red gas that freezes at –59.5°C and boils at 9.9°C; very reactive and unstable, exploding when heated or on reaction with organic materials; used to treat water and as a bleach.

chlorine gas *Ordnance.* a greenish-yellow, toxic gas packaged for release against enemy troops; it irritates the eyes and respiratory system, with dose-related injury or fatality. Also, **chlorine war gas.**

chlorine monofluoride *Inorganic Chemistry.* ClF, a colorless gas that freezes at about –154 °C and boils at –100.8°C; very reactive and unstable, igniting organic material on contact and reacting violently with water; used as a fluorinating agent.

chlorine monoxide *Inorganic Chemistry.* ClO, a yellowish-red gas with a strong, unpleasant odor; freezes at –20°C and boils (explodes) at 3.8°C; explodes on contact with organic materials; used in chlorination reactions.

chlorine survey *Petroleum Engineering.* the logging of a radioactivity survey taken inside the casing for determining the chlorine content of a formation.

chlorine trifluoride *Inorganic Chemistry.* ClF_3, a colorless, highly toxic gas that freezes at –83°C and boils at 11.3°C; very reactive and unstable; explodes on contact with organic material or water; used as an oxidizer and a fluorinating agent.

chlorine water *Chemistry.* a solution of chlorine gas in water; a clear yellowish liquid that deteriorates on exposure to light and air.

chlorinity *Oceanography.* a measure of the amount of chloride and other halides by mass of seawater, defined as the amount of silver required to remove the total halogen content from a 0.3285-kg sample of sea water.

chlorinolog *Petroleum Engineering.* a record of chlorine pressure and concentration found in oil reservoirs; used as an aid in locating saltwater strata. Also, **chlorine log.**

chlorion see CHLORIDE ION.

Chloriridovirus *Virology.* a genus of viruses of the family Iridoviridae that infect insects, causing a yellow-green iridescence in infected larvae.

chloriridovirus *Entomology.* a viral disease of insect larvae; infected larvae exhibit a yellow-green iridescence.

chlorite *Inorganic Chemistry.* a salt of chlorous acid, containing the univalent radical ClO_2^-. *Mineralogy.* any of the chlorite group of mostly green, platy, monoclinic or triclinic aluminosilicate minerals containing ferrous iron and magnesium; found in low-grade metamorphic rocks and as alteration products of mafic minerals.

chlorite schist *Petrology.* a metamorphic rock that contains a parallel structure of chlorite flakes, and may also contain quartz, epidote, magnetite, and garnet; formed under high-stress, low-temperature conditions and derived from igneous rocks such as dolerite or basalt.

chlorite-sericite schist *Petrology.* a low-grade, fine-grained variety of mica schist that contains no biotite.

chloritization *Inorganic Chemistry.* the introduction of a chlorite radical into a compound.

chloritoid *Mineralogy.* $(Fe^{+2},Mg,Mn)_2Al_4Si_2O_{10}(OH)_4$, a gray to black translucent triclinic and monoclinic mineral occurring in massive form or as rare, tabular crystals, having a specific gravity of 3.58 to 3.61 and a hardness of 6.5 on the Mohs scale; found as a vein mineral, as a hydrothermal alteration mineral in lavas, and in metamorphosed rocks; forms a series with carboirite.

chloritoid schist *Petrology.* a type of mica schist principally containing chloritoid.

chlormanganokalite *Mineralogy.* K_4MnCl_6, a brittle, transparent, yellow trigonal mineral having a specific gravity of 2.31 and a hardness of 2.5 on the Mohs scale; found as rhombohedral crystals lining cavities in blocks of scoria from Mount Vesuvius.

chlormephos *Organic Chemistry.* $C_5H_{12}O_2S_2ClP$, a liquid used as a soil insecticide.

chloro- *Organic Chemistry.* a combining form designating the replacement of hydrogen by chlorine.

chloroacetic acid *Organic Chemistry.* CH_2ClCO_2H, colorless to light-brownish crystals that are deliquescent and soluble in water and alcohol; melts at 61–63°C and boils at 186–191°C; used as an herbicide, preservative, and in synthetic caffeine.

chloroacetic anhydride *Organic Chemistry.* $(ClCH_2CO)_2O$, moderately toxic, colorless to slightly yellow crystals; soluble in ether; melts at 51–55°C and boils at 203°C; used as an intermediate for acetylation of amino acids.

chloroacetone *Organic Chemistry.* CH_3COCH_2Cl, a toxic, colorless liquid that is soluble in water and alcohol; boils at 119°C and melts at –44.5°C; used in color photography, insecticides, perfumes, and tear gas.

chloroacetonitrile *Organic Chemistry.* $ClCH_2CN$, a colorless liquid that is insoluble in water and soluble in alcohol; boils at 123–125°C; used as a fumigant.

chloroacetophenone *Organic Chemistry.* $C_6H_5COCH_2Cl$, white crystals that are insoluble in water; melts at 56°C and boils at 247°C; used in pharmaceuticals and riot control gas.

chloroacrolein *Organic Chemistry.* $CH_2{=}CClCHO$, a colorless liquid that boils at 29–31°C; used as a tear gas.

Chlorobacteriaceae *Bacteriology.* a former classification for a family of anaerobic, nonmotile, coccoid or rod-shaped photosynthetic bacteria found in sulfur-rich mud; it is presently classified in the family Chlorobiaceae.

***o*-chlorobenzaldehyde** *Organic Chemistry.* C_6H_4CHOCl, a combustible, colorless to yellowish liquid or powder; insoluble in water and soluble in alcohol; boils at 209–215°C and melts at 10–11.5°C; used in making dyes.

chlorobenzene *Organic Chemistry.* C_6H_5Cl, a clear, volatile liquid; insoluble in water and miscible with most organic solvents; boils at 131.6°C and melts at –45°C; used as a solvent and pesticide.

chlorobenzilate *Organic Chemistry.* $(C_6H_4Cl)_2C(OH)CO_2C_2H_5$, a carcinogenic, viscous, yellowish liquid that is slightly soluble in water; boils at 141–142°C (0.04 torr); used as a pesticide and an acaricide.

***o*-chlorobenzoic acid** *Organic Chemistry.* ClC_6H_4COOH, a nearly white, coarse powder that is soluble in hot water and alcohol; melts at 142°C; used as an intermediate and preservative.

***p*-chlorobenzoic acid** *Organic Chemistry.* $ClC_6H_4CO_2H$, a prismatic solid that is soluble in alcohol; melts at 243°C; used as an intermediate and preservative.

***o*-chlorobenzoyl chloride** *Organic Chemistry.* ClC_6H_4COCl, a colorless liquid that is insoluble in water and soluble in alcohol; boils at 227–239°C and melts at –4 to –6°C; used in pharmaceuticals and dyes.

***o*-chlorobenzyl chloride** *Organic Chemistry.* $ClC_6H_4CH_2Cl$, a colorless liquid that is insoluble in water and soluble in alcohol; melts at –17°C and boils at 216–222°C; used in dyes and pharmaceuticals.

Chlorobiaceae *Bacteriology.* a family of pigmented, photosynthetic, anaerobic bacteria of the order Pseudomonadales, occurring in aquatic habitats.

Chlorobiineae *Bacteriology.* a suborder of photosynthetic bacteria of the order Rhodospirillales, composed of the families Chlorobiaceae and Chloroflexaceae.

Chlorobium *Bacteriology.* a genus of Gram-negative, green sulfur bacteria of the family Chlorobiaceae, occurring as photosynthetic, obligately anaerobic cells in sulfide-rich aquatic environments.

chlorobium chlorophyll *Biochemistry.* the chlorophyll found in green sulfur bacteria that make use of anaerobic photosynthesis.

chlorobromide paper *Graphic Arts.* a type of emulsified contact paper that is used to make photographic enlargements and brownline proofs.

chlorobutane see BUTYL CHLORIDE.

1-chlorobutane *Organic Chemistry.* C_4H_9Cl, a toxic, flammable, colorless liquid; insoluble in water and miscible with alcohol; boils at 78.6°C and melts at –122.8°C; used in organic synthesis and as a solvent. Also, *n*-BUTYL CHLORIDE.

chlorobutanol *Organic Chemistry.* $CCl_3C(CH_3)_2OH$, combustible, colorless to white crystals; soluble in hot water and alcohol; melts at 97°C, boils at 167°C, and sublimes easily; used as an antimicrobial agent and an anesthetic. Also, ACETONE CHLOROFORM.

chlorocarbon *Organic Chemistry.* a compound consisting of carbon and chlorine.

Chlorochromatium *Bacteriology.* a complex of green phototrophic sulfur bacteria that occur in mud and stagnant waters containing hydrogen sulfide.

Chlorochromatium aggregatum *Bacteriology.* a species of rod-shaped, photosynthetic bacteria of the family Chlorobiaceae that occur as aggregates in persistent association around a large, colorless, motile bacterium.

Chlorococcaceae *Botany.* a diverse family of green algae of the order Chlorococcales, having solitary cells that are sometimes aggregated to form an extensive stratum, a uniformly or asymmetrically thickened cell wall, and isogamous or anisogamous reproduction.

Chlorococcales *Botany* nonflagellate, unicellular or colonial green algae of the order Chlorophyceae, living mostly in fresh water.

chlorocruorin *Biochemistry.* a green oxygen-transporting protein pigment that is found in the body fluids or the tissues of certain marine worms.

chlorocyclohexane *Organic Chemistry.* $C_6H_{11}Cl$, a colorless liquid with a suffocating odor; insoluble in water and soluble in alcohol; melts at –44°C and boils at 143°C.

Chlorodendrales *Botany.* an order of brackish, marine, and sometimes freshwater flagellate green algae of the class Prasinophyceae, composed of the single family Chlorodendraceae; it is characterized by cells enclosed in a lorica or wall of pectinlike material that often remain nonmotile for long periods of time, and by flagellate cells with four equal flagella.

1,1,1-chlorodiflouroethane *Organic Chemistry.* CH_3CClF_2, a flammable, colorless gas that is insoluble in water; boils at –10°C and melts at –130°C; used as a refrigerant and solvent.

chlorodifluoromethane *Organic Chemistry.* $CHClF_2$, a colorless gas that is slightly soluble in water; boils at –40.8°C and melts at –146°C; used as a refrigerant and solvent.

1-chloro-2,4-dinitrobenzene *Organic Chemistry.* $C_6H_3(NO_2)_2Cl$, toxic, combustible, pale-yellow needles; insoluble in water and soluble in hot alcohol; melts at 53°C and boils at 315°C; used in dyes and organic synthesis.

chloroethene see VINYL CHLORIDE.

2-chloroethyl alcohol see ETHYLENE CHLOROHYDRIN.

2-chloro-4-ethylamino-6-isopropyl-amine-*s*-triazine see ATRAZINE.

chloroethylene see VINYL CHLORIDE.

2-chloroethylphosphonic acid see ETHEPHON.

Chloroflexaceae *Bacteriology.* a family of photosynthetic, filamentous, gliding bacteria of the suborder Chlorobiineae.

Chloroflexus *Bacteriology.* a genus of aerobic or anaerobic bacteria of the family Chloroflexaceae; found in hot springs as filamentous cells that exhibit a gliding motility.

chlorofluorocarbon *Organic Chemistry.* a hydrocarbon in which some or all of the hydrogen atoms have been replaced by chlorine and fluorine; its use as an aerosol is prohibited because of the depleting effect on stratospheric ozone.

chloroform *Organic Chemistry.* $CHCl_3$, a colorless, heavy, volatile, toxic liquid; slightly soluble in water and miscible with alcohol; melts at –62.5°C and boils at 61.7°C; formerly used as an anesthetic; now used as a solvent, fumigant, and insecticide. Also, TRICHLOROMETHANE.

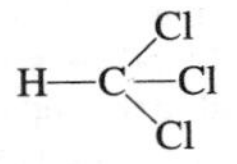

chloroform

chloroformyl *Organic Chemistry.* any carboxylic acid derivative that contains the group –C(Cl)=O.

chlorogenic acid *Biochemistry.* $C_{16}H_{18}O_9$, a substance that forms around infected tissue in some higher plants and acts as an antifungal metabolite.

Chlorogloeopsis *Bacteriology.* a genus of filamentous, sulfide-sensitive cyanobacteria, typically occurring in hot springs; the cells form trichomes that fragment into aggregates, each surrounded by a sheath.

Chlorogonium *Botany.* a genus of freshwater unicellular green algae of the family Chlamydomonadaceae, having an elongated or fusiform structure.

chloroguanide hydrochloride see PROGUANIL HYDROCHLORIDE.

Chloroherpeton *Bacteriology.* a genus of green photosynthetic bacteria that exhibit gliding motility.

chlorohydrin *Organic Chemistry.* $CH_2OHCHOHCH_2Cl$, a colorless, heavy, unstable, hygroscopic liquid; soluble in water and alcohol; melts at –40°C; used as a solvent and antifreeze agent.

chlorohydrocarbon *Organic Chemistry.* a hydrocarbon in which some of the hydrogen has been replaced by chlorine.

chlorohydroquinone *Organic Chemistry.* $ClC_6H_3(OH)_2$, white to light-tan crystals that are very soluble in water and alcohol; melts at 100°C and boils at 263°C; used as a photographic developer, organic intermediate, and bactericide.

5-chloro-8-hydroxyquinoline *Organic Chemistry.* C_9H_6ClNO, a crystalline solid that is slightly soluble in hydrochloric acid; melts at 130°C; used as a fungicide and bactericide.

Chlorokybales *Botany.* an order of sarcinoid algae of the class Chlorophyceae, composed of the monogeneric family Chlorokybaceae; characterized by a thallus made up of an aggregation of cells in a gelatinous matrix and having a subaerial habitat.

chloroma *Oncology.* a green-colored malignant tumor associated with myelogenous leukemia.

chloromethane see METHYL CHLORIDE.

chloromethoxymethane see METHOXYMETHYL CHLORIDE.

chloromethylbenzene see BENZYL CHLORIDE.

Chloromonadida *Invertebrate Zoology.* an order of plantlike flagellate protozoans, green or colorless, in the class Phytamastigophorea.

Chloromonadina or **Chloromonadophyta** see CHLOROMONADIDA.

1-chloronaphthalene see CHLORINATED NAPHTHALENE.

Chloronema *Bacteriology.* a genus of photosynthetic, gliding bacteria of the family Chloroflexaceae that are typically found in lake habitats.

4-chloro-2-oxo-3-benzothiazolin acetic acid *Organic Chemistry.* $C_8H_6O_3NSCl$, a white crystalline solid that melts at 193°C; used as an herbicide.

chloropeptide *Mycology.* a toxin produced by the fungal species *Penicillium islandicum.*

chlorophenol red see CHLORPHENOL RED.

chlorophenoxypropionic acid *Organic Chemistry.* $C_9H_9O_3Cl$, a growth regulator used to inhibit the production of fruit.

3-*p*-chlorophenyl-1,1-dimethylurea see MONURON.

***o*-chlorophenyl-*N*-methylcarbamate** *Organic Chemistry.* $ClC_6H_4OCONHCH_3$, a white, crystalline solid that melts at 90°C; used as an insecticide.

***p*-chlorophenyl-*N*-methylcarbamate** *Organic Chemistry.* $ClC_6H_4OCONHCH_3$, a white, crystalline solid that melts at 116°C; used as an herbicide.

chlorophoenicite *Mineralogy.* $(Mn,Mg)_3Zn_2(AsO_4)(OH,O)_6$, a gray-green translucent monoclinic mineral occurring as thin prismatic, elongated and striated crystals, having a specific gravity of 3.46 and a hardness of 3 to 3.5 on the Mohs scale; found in zinc-manganese-iron oxide and silicate pyrometasomatic orebodies.

Chlorophyceae *Botany.* a class of green algae containing a wide variety of types, none of which employs an apical cell for reproduction.

chlorophyll [klôr´ə fil] *Biochemistry.* a magnesium chlorin pigment, related to prophyrins, that is found in all higher plants and that gives plants their green color by absorbing red and blue-violet light and reflecting green light. Chlorophyll plays an important role in photosynthesis, and occurs in five known forms: a, b, c, d, and e.

chlorophyll a *Biochemistry.* $C_{55}H_{72}O_5N_4Mg$, a blue-green chlorophyll found in all higher plants and in algae.

chlorophyllase *Enzymology.* an enzyme that breaks down chlorophyll and occurs in all plants.

chlorophyll b *Biochemistry.* $C_{55}H_{70}O_6N_4Mg$, a yellow-green chlorophyll found in higher plants and algae.

chlorophyll c *Biochemistry.* a chlorophyll found in numerous marine algae.

chlorophyll d *Biochemistry.* a chlorophyll found in red algae. Also, 2-DESVINYL-2-FORMYL CHLOROPHYLL A.

chlorophyll e *Biochemistry.* a chlorophyll occurring only in algae belonging to the genera *Tiboneara* and *Vaucheria.*

Chlorophyta *Botany.* the green algae; those algae containing chloroplasts. *Invertebrate Zoology.* see VOLVOCIDA.

chloropia *Medicine.* an abnormal visual condition in which all objects appear green, a symptom of digitalis poisoning. Also, **chloropsia.**

chloropicrin *Inorganic Chemistry.* CCl_3NO_2, a colorless, slightly oily liquid having toxic fumes; freezes at –69.2°C and boils at 112°C; used in dyes, pesticides, and tear gas.

Chloropidae *Invertebrate Zoology.* a family of small, two-winged fruit flies, in the subsection Acalypteratae; some may bear diseases such as yaws, and the larvae of some are plant pests.

chloroplast *Cell Biology.* a large cellular organelle that is bounded by two membranes, contains a complex lammelar system of membranes (thylakoids), and functions as the site of photosynthesis in green algae and green plant cells. *Botany.* a plastid that contains chlorophyll and is a site of photosynthesis.

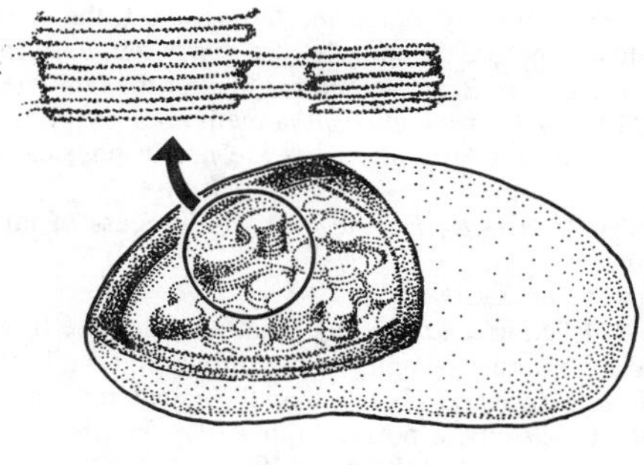

chloroplast

chloroplast DNA *Genetics.* a single circular molecule of DNA found in all green plants that contains coding sequences responsible for photosynthetic function.

chloroplatinate *Inorganic Chemistry.* any salt of chloroplatinic acid, H_2PtCl_6.

chloroplatinic acid *Inorganic Chemistry.* $H_2PtCl_6{\cdot}6H_2O$, brownish-red hygroscopic crystals; soluble in water, alcohol, and ether; melts at 60°C and decomposes above 115°C; used in platinum plating, etching, inks, and ceramics. Also, PLATINUM CHLORIC ACID.

chloroprene *Organic Chemistry.* $CH_2{=}CClCH{=}CH_2$, a flammable, toxic, colorless liquid; slightly soluble in water and miscible with acid; boils at 59.4°C; used in the manufacture of neoprene.

1-chloropropane *Organic Chemistry.* $CH_3CH_2CH_2Cl$, a liquid that is slightly soluble in water and soluble in alcohol and ether; freezes at –122.8°C and boils at 46.6°C. Also, PROPYL CHLORIDE.

2-chloropropane *Organic Chemistry.* $CH_3CCl{=}CH_2$, a liquid that is slightly soluble in water and miscible with alcohol; boils at 36°C. Also, ISOPROPYL CHLORIDE.

3-chloro-1,2-propanediol *Organic Chemistry.* $ClCH_2CH(OH)CH_2OH$, a liquid that is soluble in water and alcohol; boils at 213°C; used in making dyes and to lower the melting point of dynamite.

β-chloroproprionitrile *Organic Chemistry.* $ClCH_2CH_2CN$, a liquid that is soluble in alcohol, ether, and acetone; used in the synthesis of pharmaceuticals and polymers.

Chloropseudomonas *Bacteriology.* an ill-defined genus of aquatic bacteria of the family Chlorobiaceae, in which the type species, *C. ethylica,* now appears to be a mixture of two different bacterial species.

chloroquine *Pharmacology.* $C_{18}H_{26}ClN_3$, a synthetic substance used primarily to control malaria attacks.

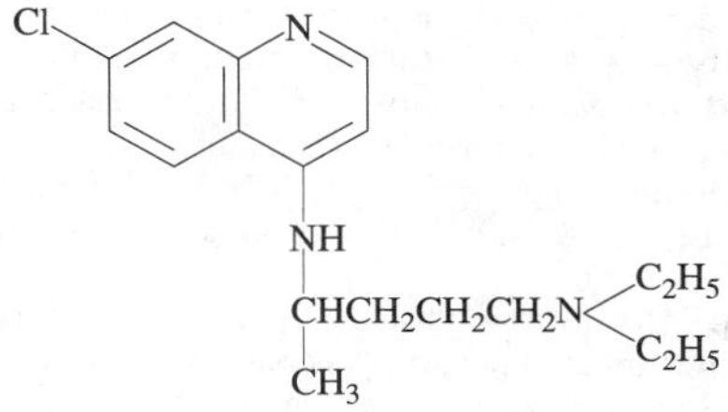

chloroquine

Chlorosarcinales *Botany.* an order of multicellular green algae in the class Chlorophyceae, composed of the family Borodinellaceae and limited to uninucleate cells that undergo desmoschisis.

chlorosis *Medicine.* a greenish-yellow skin cast, seen especially in young women, due to a chronic type of iron deficiency anemia. *Plant Pathology.* an abnormal condition that results from reduction of the chlorophyll levels in green plants and turns them yellowish.

chlorosity *Oceanography.* a property corresponding to the chlorinity of seawater but expressed as grams per 20 liters.

chlorosomes *Microbiology.* intracellular vesicles that contain the light-harvesting pigments involved in bacterial photosynthesis and are found attached to the inner membrane of members of the family Chlorobiineae.

***N*-chlorosuccinimide** *Organic Chemistry.* $C_4H_4ClNO_2$, white crystals that are soluble in water; melts at 150–151°C; used as a chlorinating agent and bactericide.

chlorosulfonic acid *Inorganic Chemistry.* $ClSO_2OH$, a fuming, colorless to pale yellow liquid with a pungent odor; decomposes in water to sulfuric acid and hydrochloric acid; melts at –80°C and boils at 158°C. It is toxic on inhalation, is a strong irritant, and can ignite flammable materials; used in detergents, pesticides, and pharmaceuticals, and for various other purposes. Also, SULFURIC CHLOROHYDRIN.

chlorothalonil *Organic Chemistry.* $C_8Cl_4N_2$, a colorless, crystalline solid that melts at 250°C; used as a fungicide.

chlorothen citrate *Pharmacology.* $C_{20}H_{26}ClN_3O_7S$, an oral antihistamine occurring as a white crystalline powder. Also, **chlorothenium citrate.**

chlorothiazide *Pharmacology.* $C_7H_6ClN_3O_4S_2$, a white crystalline powder that is slightly soluble in water; used in the treatment of hypertension and as a diuretic.

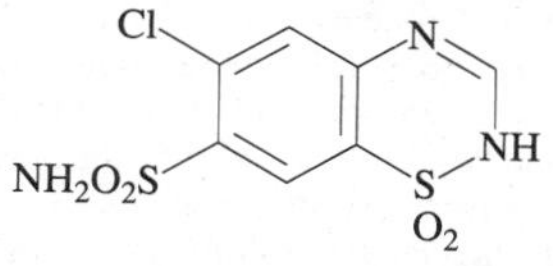

chlorothiazide

chlorothionite *Mineralogy.* $K_2Cu_2^{+}(SO_4)Cl_2$, a brilliant blue orthorhombic mineral that is soluble in water, having a specific gravity of 2.67 to 2.69 and a hardness of 2.5 on the Mohs scale; found as crystalline incrustations in fumaroles on Mount Vesuvius, Italy.

chlorothymol *Organic Chemistry.* $CH_3C_6H_2(OH)(C_3H_7)Cl$, white crystals or granular powder; soluble in water and alcohol; melts at 59–61°C; used as a bactericide.

chlorotic *Virology.* of or relating to virus-induced reduced levels of chlorophyll in plant leaves, causing yellowing.

chlorotic streak *Plant Pathology.* a virus disease of sugarcane characterized by yellow or white streaks on the leaves.

***o*-chlorotoluene** *Organic Chemistry.* $ClC_6H_4CH_3$, a toxic, colorless liquid, slightly soluble in water and miscible with alcohol; boils at 159.1°C and freezes at –36°C; used as a solvent and intermediate.

chlorotrifluoroethylene polymer *Organic Chemistry.* a heat-resistant, nonflammable, thermoplastic, colorless polymer, impervious to corrosive chemicals and resistant to most organic solvents; used in pipes, gaskets tank linings, connectors, and electronic components. Also, FLUOROTHENE.

chlorotrifluoromethane *Organic Chemistry.* $CClF_3$, a toxic, colorless, nonflammable gas that boils at –81.4°C and melts at –181°C; used in dielectrics, aerospace, and pharmaceutical processing.

chlorotrimethylsilane see TRIMETHYLCHLOROSILANE.

chlorous acid *Inorganic Chemistry.* $HClO_2$, a weak acid that is known only in solution and in the form of its salts (chlorites).

chloroxine *Organic Chemistry.* $C_9H_5Cl_2NO$, colorless crystals that are soluble in acids; melt at 180°C; used as an analytical reagent and an antibacterial.

chloroxiphite *Mineralogy.* $Pb_3Cu_2^+Cl_2(OH)_2O_2$, a brittle, translucent, green monoclinic mineral occurring as bladed crystals, having a specific gravity of 6.93 and a hardness of 2.5 on the Mohs scale; found as a secondary mineral in mendipite in Somerset, England.

4-chloro-3,5-xylenol *Organic Chemistry.* $ClC_6H_2(CH_3)_2OH$, a crystalline solid that is soluble in water and alcohol; melts at 115°C; used as a topical antiseptic and germicide.

chlorphenol red *Organic Chemistry.* $C_{19}H_{12}Cl_2O_5S$, a pH indicator that turns yellow in acid solution and red in basic solution. Also, CHLOROPHENOL RED.

chlorpromazine *Pharmacology.* $C_{17}H_{19}ClN_2S$, a phenothiazine drug occurring as a white crystalline solid that is used to treat schizophrenia and as an antiemetic and tranquilizer.

chlorpromazine hydrochloride *Pharmacology.* $C_{17}H_{20}Cl_2N_2S$, the hydrochloride of chlorpromazine, occurring as a white crystalline powder that is administered orally, intramuscularly, or intravenously as a major antipsychotic tranquilizer.

chlorpropamide *Pharmacology.* $C_{10}H_{13}ClN_2O_3S$, a white crystalline powder administered orally to lower the level of glucose in the blood, as in the treatment of diabetes mellitus.

chlortetracycline *Microbiology.* a broad-spectrum antibiotic, tradename Aureomycin, that is produced by the bacterium *Streptomyces aureofaciens* and inhibits bacterial protein synthesis; it was the first of the tetracyclines to be discovered.

chlorthiamid *Organic Chemistry.* $C_7H_5Cl_2NS$, a white crystalline solid that melts at 151°C; used as an herbicide.

chlorzoxazone *Pharmacology.* $C_7H_4ClNO_2$, a muscle relaxant that is administered orally to relieve discomfort in painful musculoskeletal disorders.

Chnoosporaceae *Botany.* a monogeneric family of brown algae of the order Dictyosiphonales, characterized by a heavily dichotomously branched thallus and axes attached by a discoid holdfast; distributed in tropical and subtropical regions throughout the world.

choana [kō´ə nə] *plural,* **choanae.** *Anatomy.* **1.** a funnel-shaped cavity. **2.** either of the paired openings between the nasal cavity and the nasopharynx. *Invertebrate Zoology.* the protoplasmic collar around the basal ends of the flagella in certain flagellates and in the flagellate cells of sponges. (From a Greek word meaning "tunnel.")

choanate fish *Paleontology.* a term sometimes used to group the Crossopterygii with the Dipneusti on the basis of their internal nares.

Choanichthyes *Vertebrate Zoology.* an equivalent name for the Sarcopterygii, a subclass of lobe-finned fishes in the class Osteichthyes.

choanocyte *Invertebrate Zoology.* any of the flagellate cells lining the cavities of a sponge. Also, COLLAR CELL.

choanoflagellates *Invertebrate Zoology.* colorless, flagellate protozoans, solitary or colonial, with stalks.

Choanoflagellida *Invertebrate Zoology.* an order of colorless flagellates, some single-celled, some colonial, in the class Zoomastigophorea.

choanosome *Invertebrate Zoology.* the inner layer of a sponge, made up of choanocytes.

CHO cells Chinese hamster ovary cells.

chock *Naval Architecture.* **1.** to secure loose articles on deck so that they will not shift with the motion of the ship. **2.** a wooden wedge that is used to secure articles in this manner. **3.** a fitting, usually consisting of paired jaws, used as a lead or guide for lines. *Mining Engineering.* a square pillar made of prop timber for supporting the roof of a mine; it is laid up in alternate cross layers, in log-cabin style, the center filled with waste.

chock-a-block *Naval Architecture.* of or relating to a tackle in which two blocks are as close together as possible. Also, BLOCK AND BLOCK, TWO BLOCKS.

chocolate *Food Technology.* a food made from cacao beans. In chocolate processing, roasted, hulled beans (or nibs) are milled to produce **chocolate liquor.** Solidified liquor (**bitter chocolate**) can be sweetened and flavored to make **sweet chocolate (milk chocolate)**.

chocolatero *Meteorology.* a moderate northerly flow of air near in the Gulf region of Mexico. Also, **chocolate gale.**

chocolate spot *Plant Pathology.* a legume disease caused by the fungus *Botrytis* and characterized by withered shoots and brown areas on the leaves and stems.

chocolate tree see CACAO.

Choeropotamidae *Paleontology.* a monogeneric family of paleodont artiodactyls in the extinct superfamily Entelodontoidea; similar to the Cebochoeridae; Eocene.

choice function *Mathematics.* suppose C is a nonempty collection of nonempty sets (such as the collection of subsets of a given set X). A choice function f is a function with domain C such that if $A \in C$, then $f(A) \in A$. Intuitively, f "chooses" an element out of each set of C. C can be regarded as an indexed family of sets, using C itself in the role of the index set with the identity mapping on C in the role of the indexing. Then f is a choice function satisfying the requirements of the axiom of choice for the Cartesian product of the sets of C; i.e., each such f can be regarded as an element of this generalized Cartesian product.

choice point *Behavior.* a position in a maze or other discrimination apparatus at which an animal is presented with two or more options and must select only one.

choice test *Behavior.* a method for investigating the preferences of animals by presenting them with simultaneous but separate options.

choke *Mechanical Engineering.* **1.** a restriction in a pipe to reduce fluid flow. **2.** see CHOKE VALVE. *Electricity.* an inductor used in a circuit to present a high impedance to frequencies above a specified frequency range without limiting the flow of direct current. Also, **choke coil.** *Electromagnetism.* a discontinuity in a waveguide surface, such as a groove, whose dimensions and shape are such that certain frequencies are impeded in their transmission through the guide. *Petroleum Engineering.* **1.** a removable nipple placed in a flow line to control the flow of gas or oil. **2.** a block or wedge that prevents a vehicle or other movable object from changing position. Also, CHUNK. *Ordnance.* **1.** the narrowing of the bore immediately before the muzzle of a gun; often used in shotguns to concentrate the pellets. **2.** an attachment to alter the bore of a shotgun.

choke coupling *Electromagnetism.* a method of coupling two parts of a waveguide system so that the parts are not in physical contact with each other.

choke crushing *Mining Engineering.* the process of taking fine ore and recrushing it.

chokedamp see BLACKDAMP.

choked flow *Fluid Mechanics.* the point at which the flow in a nozzle reaches a Mach number of unity at the throat, and the back pressure to which the nozzle discharges is less than the calculated critical pressure.

choke filter *Electricity.* a power supply filter in which a choke is the first filter element. Also, **choke input filter**.

choke flange *Electromagnetism.* the flange of a waveguide having a slot or some other discontinuity so as to impede the leakage of radiation at the joint.

choke joint *Electromagnetism.* a scheme of connecting two waveguide sections in which there is no metallic continuity at the inner surface of the guide and yet energy is permitted to pass efficiently.

choke piston *Electromagnetism.* a pistonlike post that is mounted on a waveguide such that it makes no metallic contact with the inner walls of the waveguide.

choke ring *Ordnance.* a metal ring that controls gas escape in the reaction chamber of some recoilless weapons.

choke valve *Mechanical Engineering.* a valve, usually a butterfly valve, in a carburetor intake to reduce the air supply and thus give a rich mixture for starting purposes while the engine is still cold. Also, CHOKE.

choking *Fluid Mechanics.* the modification of a nozzle to produce choked flow, such as by lowering the back pressure at the nozzle discharge below the critical pressure.

choking gas *Ordnance.* a toxic gas that causes severe, potentially fatal irritation and inflammation in the respiratory system, particularly the bronchial tubes and lungs; examples include phosgene and diphosgene.

chol- a combining form meaning "bile" or "bile duct," as in *cholate*.

cholagogic *Physiology.* relating to or causing an increase of the flow of bile into the duodenum.

cholane *Biochemistry.* $C_{24}H_{42}$, the source or parent hydrocarbon of the bile acids.

cholangiogram *Radiology.* an X-ray of the bile ducts and gallbladder.

cholangiography *Radiology.* radiographic examination of the bile ducts after injection with a contrast medium.

cholangiolitis *Medicine.* an irritated condition of the small bile cholangioles.

cholangioma *Oncology.* a tumor originating in the bile duct.

cholangiopancreatography *Radiology.* a study of the bile ducts and pancreas using X-rays, after the introduction of a radiopaque medium.

cholangitis *Medicine.* a condition in which a bile duct becomes inflamed. Also, **cholangeitis.**

cholate *Biochemistry.* a salt or an ester of cholic acid.

chole- a prefix meaning "bile" or "bile duct," as in *cholecystitis.*

Cholebrine *Radiology.* a trademark for an iocetamic acid preparation, used as a contrast medium given orally prior to gall bladder X-ray.

cholecalciferol *Endocrinology.* $C_{27}H_{44}O$, a seco-steroid formed by the actions of sunlight on its precursor molecule, the cholesterol derivative 7-dehydrocholesterol. Cholecalciferol itself is biologically inert, but it is metabolized in the kidney to the active forms, $1,25(OH)_2D_3$ and $24,25(OH)_2D_3$. *Pharmacology.* a form of this substance occurring as white, odorless crystals; used to treat rickets. Also, VITAMIN D_3.

cholecyanin see BILICYANIN.

cholecyst *Anatomy.* see GALLBLADDER.

cholecystectomy *Surgery.* the surgical excision of the gallbladder.

cholecystitis *Medicine.* an inflamed or irritated condition of the gallbladder.

cholecystogastric *Anatomy.* of or relating to a communicating channel between the gallbladder and the stomach.

cholecystography *Medicine.* an X-ray of the gallbladder.

cholecystokinin *Endocrinology.* a 33-amino-acid polypeptide hormone of the duodenum, released in response to increased amounts of free fatty acids in the intestinal lumen; release causes contraction of the gallbladder and secretion of pancreatic enzymes. Also, PANCREOZYMIN.

cholecystokinin-pancreozymin see CHOLECYSTOKININ.

cholecystostomy *Surgery.* the surgical technique of constructing a fistula in the gallbladder.

choledochal *Anatomy.* of or relating to the common bile duct.

choledochitis *Medicine.* an inflammation of the common bile duct.

choledochoduodenostomy *Surgery.* the creation of an opening between the common bile duct and the duodenum.

choledocholithiasis *Medicine.* a condition in which solidified mineral matter forms into stones in the common bile duct.

choledochoplasty *Surgery.* the surgical repair of the common bile duct.

choledochostomy *Surgery.* the surgical formation of a fistula in the common bile duct.

choleglobin *Biochemistry.* a protein or globulin that is produced by the catabolism of hemoglobin and is a chemical predecessor to the bile pigment biliverdin.

cholelith *Medicine.* a gallstone.

cholelithiasis *Medicine.* the formation of gallstones.

cholera [käl´ə rə] *Medicine.* a severe, epidemic disease that is marked by severe diarrhea leading to extreme dehydration that can result in shock, renal failure, and death. It is spread by feces-contaminated water and food.

cholera morbus *Medicine.* a severe gastroenteritis of unknown cause, occurring in hot weather and marked by vomiting, diarrhea, and colic.

cholera toxin *Toxicology.* a toxic protein that is produced by the bacterium *Vibrio cholerae* and lodges itself in the intestinal tract, preventing the mucosa from absorbing sodium and causing a severe watery diarrhea, extreme loss of fluids and electrolytes, and a state of dehydration sometimes leading to renal failure.

cholera vibrio *Bacteriology.* a biotype of the bacterial pathogen *Vibrio cholerae* that causes the human gastrointestinal disease, cholera.

choleric [kə lâr´ik] *Psychology.* **1.** one of the four basic personality types defined in ancient and medieval times, described as being hot-tempered and irritable; said to result from the predominance of yellow bile over the other bodily humors. Also, **choleric type. 2.** of or relating to this personality type.

cholerophobia *Psychology.* an irrational fear of cholera.

Cholesky decomposition *Mathematics.* an expression (decomposition) of a positive definite symmetric matrix S as a product of a lower-triangular matrix L times its transpose: $S = LL^t$; a special case of the LU decomposition.

cholestane *Endocrinology.* the C-27 hydrocarbon that is the parent structure for the synthesis of all mammalian steroids.

cholestasis *Medicine.* the suppression of the flow of bile.

cholesteatoma *Medicine.* a cystlike mass of epithelial cells and cholesterol, usually forming in the middle ear and sometimes spreading to the mastoid, causing bone loss.

cholesteric [käl´ə stâr´ik] *Materials Science.* of or relating to one of the three basic molecular orders found in liquid crystals; characterized by a sheetlike morphology or conformation.

cholesteric material *Materials.* a liquid crystal material that has a spiral shape because its molecules bend slightly from layer to layer.

cholesterol [kə les´tə rôl´] *Biochemistry.* $C_{27}H_{46}O$, the major fatty steroid alcohol of all vertebrate cells, especially animal fat, blood, liver, nerve tissue, and brain tissue; high levels of certain types have been linked to heart disease and other diseases.

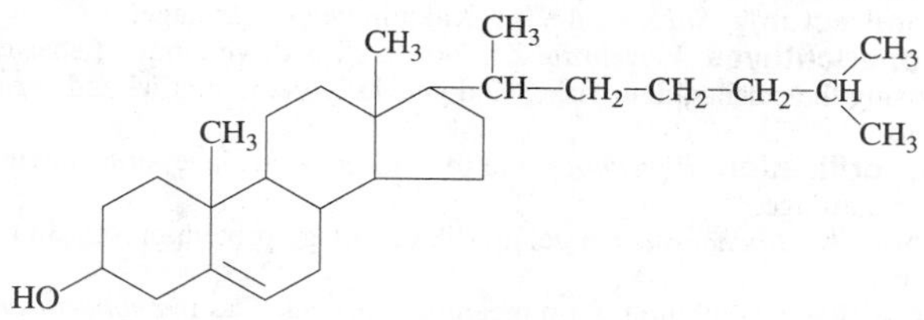

cholesterol

cholesterol oxidase *Enzymology.* an enzyme complex that catalyzes the removal of the side chain from cholesterol, converting cholesterol to pregnenolone, a precursor of the steroid hormones. Also, DESMOLASE.

cholic acid [kō´lik] *Biochemistry.* the most abundant bile acid in human bile.

choline [kō´lēn] *Biochemistry.* $C_5H_{15}O_2N$, a compound classified with the vitamin B complex that plays an important role in metabolism, inhibits the decomposition of fat in the liver, and lowers blood pressure. It is a viscous, alkaline liquid that is very soluble in water and alcohol; it is found in egg yolk, organ meat, seeds, vegetables, and legumes; used in medicine, nutrition, and as a neutralizing agent.

choline esterase I see ACETYLCHOLINESTERASE.

choline esterase II see CHOLINESTERASE.

choline hydroxide *Organic Chemistry.* $(CH_3)_3N(OH)CH_2CH_2OH$, a viscous, alkaline liquid, very soluble in water and alcohol; a member of the vitamin B complex, required for lecithin formation in humans. It is found in egg yolk, kidney, liver, heart, seeds, vegetables, and legumes; used in medicine, nutrition, and as a neutralizing agent.

cholinergic [kō´lə nur´jik] *Physiology.* a referring to nerve endings or activities that are stimulated, activated, or transmitted by choline (acetylcholine). Thus, **cholinergic nerve.**

cholinesterase *Enzymology.* an enzyme of the hydrolase class found in blood serum, the liver, and pancreas, that catalyzes the breakdown of choline esters, such as acetylcholine. Also, CHOLINE ESTERASE II, PSEUDOCHOLINESTERASE, SERUM CHOLINESTERASE.

cholla

cholla *Botany.* any of several tall, knobby, spiny cacti of the genus *Opuntia* of the southwestern U.S. and Mexico.

cholocyanin see BILICYANIN.

Cholografin *Radiology.* a trade name for preparations of iodipamide, used intravenously during cholangiograms because of its radiopaque properties.
choluria *Medicine.* the presence of bile in the urine.
choluric *Medicine.* relating to or affectd by choluria.
Chomsky, Noam born 1928, American linguist; formulated the influential theory of transformational grammar.
Chomsky hierarchy *Computer Science.* the hierarchy of formal language types: regular, context free, context sensitive, and recursively enumerable languages, each of which is a proper subset of the following class.
chondr- a combining form meaning "cartilage," as in *chondrification.*
chondral *Anatomy.* of or relating to cartilage.
chondralgia *Medicine.* a pain in the cartilage. Also, CHONDRODYNIA.
chondrectomy *Surgery.* the surgical removal of cartilage.
Chondrichthyes *Vertebrate Zoology.* a class of vertebrate fishes comprising the cartilaginous fishes and generally being marine and carnivorous.
chondrification *Physiology.* the formation of cartilage, or a conversion into cartilage.
chondrin *Biochemistry.* a gelatin-like cartilage protein present in collagen.
chondrio- a combining form meaning "cartilage," as in *chondriosome.*
chondriome *Cell Biology.* a term once used to refer to the mitochondria of a cell as a group.
chondriosome *Cell Biology.* a former term for a mitochondrion of a eukaryotic cell.
chondrite *Geology.* **1.** a stony meteorite characterized by the inclusion of small, globular aggregates, or chondrules, embedded in a fine crystalline matrix. **2.** a common trace fossil consisting of plantlike structures that radiate around a central vertical tube. *Astronomy.* a type of stony meteorite that contains chondrules and is thought to be extremely primitive chemically.
chondritis *Medicine.* an inflammation of the cartilage.
chondro- a combining form meaning "cartilage," as in *chondroskeleton.*
chondroblast *Histology.* a cell that forms the fibers and acellular matrix of cartilage. Thus, **chondroblastic.**
chondroblastoma *Oncology.* **1.** a benign tumor composed of chondroblast cellular tissue. **2.** a benign tumor originating from young chrondroblasts in the epiphysis of a bone. Also, CODMAN'S TUMOR.
Chondrobrachii *Vertebrate Zoology.* another name for Ateleopoidei, a suborder of bottom-living marine fishes in the order Lampriformes.
chondrocarcinoma *Oncology.* a carcinoma consisting of cartilaginous elements in its stroma.
chondroclast *Histology.* a cell that dissolves and absorbs cartilage.
Chondrococcus *Bacteriology.* a genus of mainly soil- and dung-inhabiting myxobacteria.
chondrocostal *Anatomy.* of or relating to the cartilages of the ribs.
chondrocranium *Developmental Biology.* the cartilaginous portions of the developing braincase of the embryo, or of the adult of cartilaginous fishes.
chondrocyte *Histology.* a cartilage cell.
chondrodite *Mineralogy.* $(Mg,Fe^{+2})_5(SiO_4)_2(F,OH)_2$, a red, brown, or yellow, transparent to translucent monoclinic mineral of the humite group occurring in massive form or as crystals of varied habit, having a specific gravity of 3.16 to 3.26 and a hardness of 6 to 6.5 on the Mohs scale; found in contact zones in limestone or dolomite.
chondrodynia see CHONDRALGIA.
chondrodysplasia see ENCHONDROMATOSIS.
chondrodystrophy *Pathology.* a morbid condition characterized by an abnormal development of cartilage.
chondrogenesis *Developmental Biology.* the development of cartilage. Also, CHONDROSIS.
chondroid *Medicine.* resembling cartilage.
chondroitin *Biochemistry.* a complex nitrogenous carbohydrate (mucopolysaccharide) that is found in cartilage.
chondrolipoma *Oncology.* a tumor containing cartilaginous and fatty tissue.
chondrology *Anatomy.* the study of the anatomy of cartilage.
chondrolysis *Medicine.* the degeneration of cartilage.
chondroma *Oncology.* a benign tumor formed from the cells that compose cartilage.
chondromalacia *Medicine.* the softening of any bone or cartilage.
chondromucin see CHONDROPROTEIN.
chondromucoid *Biochemistry.* a mucouslike substance containing a protein and chondroitin, found in cartilage.
Chondromyces *Bacteriology.* a genus of Gram-negative, gliding bacteria in the family Polyangiacaea, order Myxobacterales, that form fruiting bodies and occur in manure and on decaying vegetation.
chondropathy *Medicine.* any disease of the cartilage.
Chondrophora *Invertebrate Zoology.* a suborder of colonial free-floating coelenterates in the class Hydrozoa.
chondrophore *Invertebrate Zoology.* a structure or cavity supporting the internal hinge cartilage in bivalve mollusk shells.
chondroplasty *Surgery.* the plastic repair of cartilage.
chondroprotein *Biochemistry.* a glycoprotein that is found in cartilage. Also, CHRONDROMUCIN.
chondrosarcoma *Oncology.* a malignant growth composed of cartilage cells.
chondrosis see CHONDROGENESIS.
chondroskeleton *Anatomy.* the cartilagenous part of the skeletal system. *Zoology.* a skeleton made of cartilage rather than bone.
Chondrostei *Paleontology.* an infraclass of bony fish in the subclass Actinopterygii; characterized by a partly cartilaginous skeleton, the chondrosteans arose in the Devonian; includes several fossil orders and one extant order.
Chondrosteidae *Paleontology.* an extinct family of bony fish in the subclass Chondrostei and the extant order Acipenseriformes; characterized primarily by its very deep body; Jurassic and Cretaceous.
Chondrosteoidei *Paleontology.* a suborder of bony fish that includes the Chondrosteidae and possibly the Errolichthyidae; extant in the Triassic to Cretaceous.
chondrule *Geology.* a spherical granule or aggregate composed mainly of olivine or orthopyroxene, and occurring embedded in the matrix of certain stony meteorites.
Chondrus *Botany.* **1.** a small genus of red algae in the family Gigartinaceae with coarse branching fronds. **2.** see IRISH MOSS.
Chonecoleaceae *Botany.* a monotypic family of subtropical liverworts of the order Jungermanniales, characterized by small, delicate, prostrate plants with purplish pigmentation, many lateral branches arising from below the leaf axils, and scattered rhizoids.
Chonetes *Paleontology.* a Paleozoic genus of articulate brachiopods in the extinct order Strophomenida and superfamily Chonetacea; Chonetes had a row of short spines along its hinge line but none elsewhere.
Chonetidina *Paleontology.* an important suborder of articulate brachiopods in the extinct order Strophomenida, generally characterized by spines along the hinge-line and plano-convex shells; lower Silurian to lower Jurassic.
CHOP *Oncology.* an acronym for a cancer chemotherapy regimen that includes the drugs cyclophosphamide, hydroxydaunomycin (doxorubicin), Oncovin (vincristine), and prednisone.
chopper *Mechanical Devices.* any device, such as a knife or axe, that cuts an object into pieces. *Archaeology.* a simple stone tool used for hacking or chopping, especially one that has only a single flaked edge rather than being flaked on both sides. Also, **chopping tool.** *Aviation.* an informal name for a helicopter. *Physics.* a device that periodically interrupts a beam of light or particles so as to produce regular pulses at a detector. *Electronics.* **1.** a toothed disk that is used to interrupt a beam of light at regular intervals, or a device for interrupting electrical signals. **2.** a device, such as a knife or an axe, that cuts an object into smaller pieces.
chopper amplifier *Electronics.* a circuit used to amplify low signals by chopping the slowly varying input into a succession of higher-frequency rectangular pulses, amplifying the pulses and filtering the output to restore the slowly varying signal. Also, CONVERTER AMPLIFIER.
chopper transistor *Electronics.* a transistor that provides a continuous interruption in a circuit by converting the interrupted direct current into a square-wave alternating current whose amplitude is proportional to the strength of the current.
chopping *Physics.* the action that is performed by a chopper, producing square pulses. *Electronics.* a process that eliminates extremes in a radio signal's amplitude when they reach a predetermined level. *Photogrammetry.* the interrupting of a photographic image of a star or satellite trail, by a shutter or a similar device, to provide exact timing and orientation data for observation of aerospace vehicles against a stellar background.
chopping bit *Mechanical Engineering.* a chisel-shaped steel bit that is attached to a drill rod and used to break up hard rock.
choppy *Oceanography.* describing rough, short, irregular wave motion on the surface of the water.

chord *Acoustics.* the simultaneous playing of two or more tones in music to create a pleasing combination of overtones, which are in harmony due to a separation of at least one whole step. *Mathematics.* **1.** a line segment that intersects a curve or surface only at its two endpoints. **2.** in graph theory, given a connected graph *G* and a spanning tree *T* of *G*, the edges of *G* not in *T* are the chords relative to *T*. Also, LINKS. *Aviation.* **1.** the straight line joining the mean thickness line between an airfoil's leading edge and the trailing edge; when the airfoil has a symmetrical section, the mean line and the chord line are both straight lines and they coincide. Also, CHORD LINE. **2.** the length of such a line. **3.** broadly, the width of a wing or other surface. *Cartography.* the arc of a great circle connecting any two corners on a base line, standard parallel, or latitudinal boundary of a township.

Chordaceae *Botany.* a monogeneric family of kelps of the order Laminariales, distinguished by a lack of thallic stipe-blade differentiation and an annual sporophyte with a simple filiform axis of up to six meters in length; widely distributed throughout colder waters of the Northern Hemisphere.

chorda dorsalis see NOTOCHORD.

chordae tendineae *singular,* **chorda tendinea.** *Anatomy.* the strands of tendon that connect each cusp of the two atrioventricular valves to papillary muscles in the heart ventricles. Also, TENDINOUS CHORDS.

chordal canal see NOTOCHORDAL CANAL.

chordal thickness *Design Engineering.* on a circular gear, the thickness of each tooth as measured along the pitch circle.

chorda-mesoderm *Developmental Biology.* the tissue of the embryo of a vertebrate that develops into the notochord and mesoderm.

Chordariaceae *Botany.* a cosmopolitan family of marine brown algae of the order Chordariales, characterized by cylindrical, often slimy thalli.

Chordariales *Botany.* an order of brown algae of the class Phaeophyceae, characterized by alternating heteromorphic phases in which the filamentous sporophyte is larger than the gametophyte.

Chordariopsidaceae *Botany.* a monotypic family of brown algae of the order Chordariales, distinguished by a unique three-sided apical cell and occurring only in South Africa.

Chordata [kôr dat´ə] *Zoology.* the phylum composed of animals which at some developmental stage have a dorsal nerve cord, a notochord, and pharyngeal gill slits; includes mammals, birds, reptiles, amphibians, fish, and certain marine lower forms (such as lancelets, sea squirts, and tunicates).

Chordeuma *Invertebrate Zoology.* a genus of millipedes in the order Diplopoda.

chord length *Aviation.* the length of an airfoil section from its leading to trailing edge.

chord line *Aviation.* see CHORD, def. 1.

Chordodidae *Invertebrate Zoology.* a family of worms in the order Gordioidea.

chordoma *Oncology.* an unusual malignant tumor of the skeleton, arising from remains of the notochord.

Chordopoxvirinae *Virology.* a subfamily of the family Poxviridae, containing all the viruses that infect vertebrates.

chordotomy see CORDOTOMY.

chordwise *Aviation.* **1.** located, moving, or directed along a chord. Thus, **chordwise axis. 2.** the projected direction that is almost horizontal.

chorea [kō rē´ə] *Neurology.* any of various disorders of the nervous system that are characterized by a continuous sequence of rapid, jerky, complex movements that appear choreographed but are involuntary.

choreiform syndrome *Medicine.* a disease process mimicking chorea.

Choreocolacaceae *Botany.* a family of red algae of the order Cryptonemiales having small, cushion-shaped thalli with little or no pigmentation and believed to be somewhat parasitic on members of the family Rhodomelaceae.

chorioadenoma see CHORIONIC ADENOMA.

chorioallantoic grafting *Developmental Biology.* the transplantation of cells, tissues, or parts onto the chorioallantoic membrane of the embryonic chick.

chorioallantois *Developmental Biology.* the extraembryonic membrane that develops from the fusion of the allantois and the chorion. Also, **chorioallantoic membrane.**

choriocarcinoma *Oncology.* a cancerous growth in the chorion, or outer membrane of the fetal sac, that invades the uterine muscle and metastasizes early in the disease process.

chorioid see CHOROID.

chorioma *Oncology.* any tumor of the trophoblastic tissue.

chorion *Developmental Biology.* the outermost fetal membrane derived from extraembryonic somatic mesoderm and trophoblast; in humans, its maternal surface contains villi bathed in maternal blood, and as pregnancy progresses, a portion of it becomes the definitive placenta.

chorionic *Developmental Biology.* of or relating to the chorion.

chorionic ACTH *Endocrinology.* the placental equivalent of adrenocorticotropic hormone.

chorionic adenoma *Oncology.* a benign tumor of the outer fetal membrane, often resulting in a hydatidiform mole. Also, CHORIOADENOMA.

chorionic gonadotropin *Endocrinology.* a glycopeptide hormone produced by the fetal placenta and thought to maintain the function of the corpus luteum during the first few weeks of pregnancy; it can be detected by immunoassay in the maternal urine within days after fertilization and thus provides the basis of a commonly used pregnancy test.

chorionic somatomammotropin *Endocrinology.* a placental hormone that has some sequence homology to prolactin and growth hormone, but appears to act as a modulator of maternal glucose and fatty acid metabolism rather than as a lactogen. Also, PLACENTAL LACTOGEN, HUMAN PLACENTAL LACTOGEN, SOMATOMAMMOTROPIN.

chorioretinal *Anatomy.* of or relating to both the choroid coat of the eye and the retina.

chorioretinitis *Medicine.* irritation of the choroid and retina of the eye.

choripetalous see POLYPETALOUS.

chorisepalous see POLYSEPALOUS.

chorismate *Biochemistry.* the anionic form of chorismic acid that functions as an intermediate in the biosynthesis of the aromatic amino acids phenylalanine, tyrosine, and tryptophan; present in plants, fungi, and bacteria.

chorismite *Petrology.* a mixed rock such as vein rock, conglomerate, breccia, schist, or the like, with a macropolyschematic fabric consisting of petrologically dissimilar material of varied origins.

Choristida *Invertebrate Zoology.* an order of sponges in the class Demospongiae, having four-rayed spicules.

Choristodera *Paleontology.* an extinct monofamilial order of diapsid reptiles that were similar to crocodiles, but not closely related; Cretaceous to early Tertiary.

C horizon *Geology.* a layer of unconsolidated rock material that lies beneath the B horizon, or sometimes beneath the A horizon; it consists of weathered and disaggregated parent rock, and contains more organic matter than the horizon above it.

chorographic map *Cartography.* any map representing large areas on a small scale, as in atlases and most wall maps.

chorography *Geography.* the topography of a region or district. *Cartography.* the methodology used in describing or mapping a region or district.

choroid *Anatomy.* **1.** the vascular region of the eye between the sclera and retina. **2.** of, relating to, or resembling this region. Also, CHORIOID.

choroidal *Anatomy.* of or relating to the choroid.

choroiditis *Medicine.* an irritation of the choroid.

choroid plexus *Anatomy.* any of several masses of highly branched blood vessels in margins of the cerebral ventricles that produce and absorb cerebrospinal fluid.

chorology *Ecology.* the study of the geographical distribution of organisms; biogeography.

chorus effect *Acoustical Engineering.* a term for the effect of several different musical instruments producing tones different from any of the tones produced by the individual instruments, due to overtones and beat tones that are produced.

Chow's theorem *Mathematics.* the theorem that any compact complex subvariety of complex projective space is algebraic.

chrematophobia *Psychology.* an irrational fear of money.

Christaller, Walter 1893–1969, German geographer; studied distribution and growth of human settlements.

Christiansen effect *Analytical Chemistry.* a transparency effect that occurs when a finely powdered substance is immersed in a liquid having the same refractive index as the substance.

Christiansen filter *Optics.* a color filter that accepts only a narrow range of frequencies from a beam of light. Also, BANDPASS FILTER.

Christmas disease *Pathology.* another name for the disease hemophilia B. (From the name of the first patient with the disease who was closely studied.)

Christmas factor *Biochemistry.* a protein substance in the blood known to be involved in blood coagulation; suggested to activate factor X. Also, FACTOR IX.

Christoffel-Darboux formula *Mathematics.* suppose $\{f_n(x) : n \geq 0\}$ is a set of orthogonal polynomials such that $\int_a^b w(x)f_n(x)f_m(x)dx = 0$ for $m \neq n$. Then the Christoffel-Darboux formula is the following:

$$\sum_{m=0}^{n} f_m(x)f_m(y)^1/h_n = k_n/(k_{n+1}h_n) \cdot [f_{n+1}(x)f_n(y) - f_n(x)f_{n+1}(y)]/(x+y),$$

where $h_n = \int_a^b w(x)f_n(x)^2dx$ and k_n is the coefficient of x^n in f_n.

Christoffel symbols *Mathematics.* in differential geometry, symbols representing certain functions of the first derivatives of the (pseudo-) Riemannian metric tensor. Given a metric tensor g on an n-dimensional manifold, the **Christoffel symbols of the first kind** are

$$[ij,k] = (1/2)(\partial g_{ik}/\partial x^j + \partial g_{jk}/\partial x^i - \partial g_{ij}/\partial x^k).$$

Other notations include $[{}^{ij}_{k}]$, C^k_{ij}, or Γ_{ijk}. The **Christoffel symbols of the second kind** are

$$\{{}^{k}_{ij}\} = g^{k\sigma}[{}^{ij}_{\sigma}],$$

where g^{ij} = [cofactor of g_{ji}]/det g. In each case, summation is over repeated indices.

chrom- a combining form meaning: **1.** color, as in *chromascope*. **2.** containing chromium, as in *chromite*.

chroma *Optics.* in the Munsell color system, an attribute for judging the saturation of a color, based on how much white is present.

chroma control *Electronics.* a control that adjusts the chrominance in a color television receiver to alter the vividness of the hues.

chromadizing *Metallurgy.* the pretreatment of paints with a chromic acid solution to improve their adherence on aluminum-base products.

Chromadoria *Invertebrate Zoology.* a subclass of unsegmented nematode worms in the class Adenophorea.

Chromadorida *Invertebrate Zoology.* an order of aquatic nematode worms in the subclass Chromadoria, with a short esophagus divided into three regions.

Chromadoridae *Invertebrate Zoology.* a family of nematodes in the superfamily Chromadoroidea, found in soil or fresh water.

chromaffin *Biology.* readily stained by chromium salts.

chromaffin cell *Histology.* any of several types of cell that stain with chromium salts; found in the adrenal medulla, the paraganglia, the carotid bodies, and the suprarenal organs in lower vertebrates.

chromaffin system *Physiology.* the tissues and cells of the adrenal medulla that stain readily with chromium salts and produce adrenalin and noradrenalin.

chroma oscillator see COLOR OSCILLATOR.

chromascope *Optics.* an instrument for analyzing the optical effects of color and colored light.

chromat- a combining form meaning "color," as in *chromatin*.

chromate *Inorganic Chemistry.* any salt of chromic acid, H_2CrO_4. *Mineralogy.* any mineral containing the chromate ion, such as potassium chromate. *Metallurgy.* to perform a chromate treatment.

chromate treatment *Metallurgy.* an anticorrosion treatment based on a solution of hexavalent chromium.

Chromatiaceae *Bacteriology.* a family of photosynthetic, purple sulfur bacteria of the suborder Rhodospirillineae, typically found in sulfide-rich habitats.

chromatic *Optics.* **1.** referring to or describing something that has color. **2.** of or relating to wavelength. *Acoustics.* involving or progressing by semitones, as in the chromatic scale.

chromatic aberration *Optics.* a defect in a lens that causes it to focus the various colors in a beam of light at various points, thus producing a spectrum. *Electronics.* a defect that enlarges and blurs the focal spot on a cathode-ray tube, caused by the differences in the electron velocity distribution through the beam.

chromatic audition *Neurology.* a subjective perception of color induced by actual sound stimuli. Also, AUDITION COLORÉE, COLOR HEARING, PSEUDOCHROMESTHESIA.

chromatic dispersion see OPTICAL DISPERSION.

chromatic index *Mathematics.* in graph theory, given a finite graph G, the smallest number n such that there is an assignment of n colors to the edges of G in which no two adjacent edges have the same color. Also, **chromatic number.**

chromaticity *Optics.* the color quality of light that is determined by its hue and saturation, rather than its brightness.

chromaticity coordinates *Optics.* the numbers that represent primary colors in a color sample and meet specifications established by the International Committee on Illumination.

chromaticity diagram *Optics.* a graph that depicts specific chromaticity coordinates in a triangular pattern. Also, **chromatic diagram.**

chromatic parallax *Optics.* a parallax that is determined by position of the focal plane with respect to the wavelength.

chromatic polynomial *Mathematics.* a function $p(G, \lambda)$ whose value is the number of ways the vertices of a graph G with n vertices can be properly colored with λ or fewer distinct colors (or, analogously, the number of ways a map with n regions can be properly colored with λ or fewer distinct colors). Birkhoff's coloring theorem states that the chromatic function is actually a polynomial of the form

$$p(G, \lambda) = c_0 + c_1\lambda + c_2\lambda^2 + \cdots + c_n\lambda^n,$$

where n is the number of distinct vertices of the graph G, $c_n = 1$, $c_{n-1} = -|E(G)|$, $c_0 = 0$ if $n > 0$, and the other coefficients are integers that depend on the numbers of certain subgraphs of the graph.

chromatic resolving power *Optics.* the ability of a spectroscopic instrument to separate two barely separable wavelengths; equivalent to the ratio of their average wavelength to the difference in their wavelengths.

chromatics *Optics.* **1.** the branch of optics that studies the effects of the differing wavelengths of colors. **2.** the measurement of such properties of color as hue and saturation.

chromatic scale *Acoustics.* a musical scale based on the division of an octave into twelve notes of equal half-step intervals, from which tempered scales are derived.

chromatic sensitivity *Optics.* the least amount of change in a wavelength of light that will cause a visible color change.

chromatic vision *Physiology.* the sensation of color, based on discrimination of wavelengths of the visible spectrum, and on the presence of the color receptors (cones) in the retina.

chromatid *Cell Biology.* one of the two longitudinally adjacent strands of a chromosome that are formed as a result of chromosomal duplication; they are held together by a centromere and become visible during mitosis or meiosis.

chromatin *Biochemistry.* a stainable portion of the cell nucleus that forms a network of nuclear fibrils within the achromatin of a cell; it is deoxyribonucleic acid (DNA) attached to basic proteins called histones, and is the carrier of the genes in inheritance; during cell division it coils and folds to form the chromosomes. Also, CHROMOPLASM.

chromatin-positive *Cell Biology.* of cells, containing the X chromosome, or Barr body, such as the cells of a human female.

Chromatium *Bacteriology.* a genus of purple sulfur bacteria of the family Chromatiaceae, characterized by rod-shaped cells capable of intracellular sulfur deposition.

chromato- a combining form meaning "color," as in *chromatophore*.

chromatofocusing *Biotechnology.* a process in which macromolecules are isolated and purified through chromatography and electrophoresis.

chromatogram *Analytical Chemistry.* a plot of the elution of the components of a mixture as measured against time or volume in chromatography.

chromatographic adsorption *Analytical Chemistry.* the tendency for different substances in a mixture to be adsorbed at different rates from a moving stream of gas or liquid (mobile phase) to a stationary phase.

chromatography *Analytical Chemistry.* a technique for separating components from a mixture by placing the mixture in a mobile phase that is passed over a stationary phase.

chromatoid bodies *Invertebrate Zoology.* in certain protozoans, especially those in precystic and early cystic developmental stages, long, slender refractile bodies.

chromatophobia see CHROMOPHOBIA.

chromatophore *Biochemistry.* the chomoplast of photosynthetic bacteria, which contain bacteriochlorophyll. *Histology.* any pigmentary cell or color-producing plastid.

chromatophorotrophin *Biochemistry.* a melanocyte-stimulating hormone responsible for the movement of pigment particles within the chromatophores.

chromatoplasm *Botany.* the marginal region in the cytoplasm of blue-green algae that contains its chlorophyll and other pigments.

chromatopsia *Medicine.* a condition in which colors are imperfectly perceived or in which colorless objects appear to have color.

chromatoscope *Optics.* **1.** a telescope through which the viewer observes a star as a circle of light, not as a point. **2.** an instrument that uses light beams to mix color stimuli.

chromatosis *Medicine.* a former term indicating the pigmentation of skin, especially, abnormal pigmentation.

chromatron *Electronics.* a color picture tube that has a single electron gun and whose phosphors are laid out in parallel lines instead of dots.

chrome alum see CHROMIUM POTASSIUM SULFATE.

chrome diopside *Mineralogy.* a bright green variety of diopside containing chromium.

chrome dye *Textiles.* an acid dye containing a chromium compound that fixes the dye to the fibers.

chrome green *Inorganic Chemistry.* a dark-green powder used as a pigment, consisting of about 99% chromic oxide (Cr_2O_3), often mixed with small amounts of cobalt oxide.

chromel *Metallurgy.* **1.** an alloy used as an element of a thermocouple, consisting of about 90% nickel and 10% chromium. **2.** any of several alloys based on nickel and chromium, mainly used at high temperatures.

chrome-magnesia brick *Materials.* refractory bricks used in steel furnaces.

chrome pigment *Inorganic Chemistry.* any inorganic pigment containing chromium.

chrome plating *Metallurgy.* the process of coating a metallic component with chromium metal to improve its appearance, abrasion resistance, and corrosion resistance.

chrome red *Inorganic Chemistry.* a pigment consisting largely of basic lead chromate, $PbCrO_4 \cdot Pb(OH)_2$.

chromesthesia *Neurology.* an association of imagined sensations of color with actual sensations of hearing, taste, or smell.

chrome tanning *Chemical Engineering.* the use of chromium salts to tan animal skins.

chrome yellow *Inorganic Chemistry.* a pigment consisting largely of normal lead chromate, $PbCrO_4$.

chromia see CHROMIC OXIDE.

chromic *Chemistry.* **1.** of or relating to chromium. **2.** describing various compounds of chromium, especially those in which the element has a valence of 3.

chromic acetate *Inorganic Chemistry.* $Cr(C_2H_3O_2)_3 \cdot H_2O$, a grayish-green powder or blue-green pasty mass; soluble in water and insoluble in alcohol; toxic on ingestion; used in tanning and as a mordant and catalyst.

chromic acid *Inorganic Chemistry.* CrO_3, carcinogenic, poisonous, dark purplish red crystals; soluble in water, alcohol, and acid; melts at 196°C and decomposes at 250°C; used in chromium plating, medicine, inks, paints, and process engraving. Also, CHROMIC ANHYDRIDE, CHROMIUM TRIOXIDE.

chromic anhydride see CHROMIC ACID.

chromic bromide *Inorganic Chemistry.* $CrBr_3$, olive-green, hexagonal crystals; insoluble in cold water, soluble in hot water, and very soluble in alcohol; used as a catalyst.

chromic chloride *Inorganic Chemistry.* **1.** $CrCl_3$, toxic violet crystals, insoluble in cold water and alcohol, slightly soluble in hot water; melts at about 1150°C and sublimes at 1300°C; used in chromium plating and waterproofing, as a mordant and catalyst, and for various other purposes. **2.** $CrCl_3 \cdot 6H_2O$, the hexahydrate form, greenish-black or violet deliquescent crystals; soluble in water and alcohol and insoluble in ether; melts at 83°C. Its uses are similar to those of $CrCl_3$.

chromic fluoride *Inorganic Chemistry.* $CrF_3 \cdot 4H_2O$, fine green crystals; insoluble in water and alcohol; melts above 1000°C and sublimes at 1100–1200°C; used in dyeing and as a catalyst.

chromic hydroxide *Inorganic Chemistry.* $Cr(OH)_3$, a gelatinous green precipitate; insoluble in water and soluble in acids; decomposes to chromic oxide on heating; used as a catalyst and in tanning and dyeing. Also, **chromic hydrate.**

chromic nitrate *Inorganic Chemistry.* $Cr(NO_3)_3 \cdot 9H_2O$, toxic purple crystals; soluble in water and alcohol; melts at 60°C and decomposes at 100°C; explosive when shocked or heated and ignites organic material on contact; used as a catalyst and corrosion inhibitor.

chromic oxide *Inorganic Chemistry.* Cr_2O_3, very hard, hexagonal green crystals; insoluble in water, acids, and alkalis; melts at about 2265°C and boils at 4000°C; used as a pigment, catalyst, and abrasive, and for various other industrial purposes. Also, CHROMIUM OXIDE, CHROMIA, CHROMIUM SESQUIOXIDE, GREEN CINNABAR.

chromic phosphate *Inorganic Chemistry.* $CrPO_4$ or the hydrated form, $CrPO_4 \cdot 6H_2O$, violet triclinic crystals; insoluble in water and soluble in acids; melts at 100°C; used as a pigment and catalyst. Also, CHROMIUM PHOSPHATE.

chrominance *Optics.* the difference between a given color and a reference color with the same degree of brightness.

chrominance carrier see CHROMINANCE SUBCARRIER.

chrominance-carrier reference *Telecommunications.* **1.** a signal having the same frequency as a given chrominance subcarrier. **2.** the phase-reference of carrier-chrominance signals.

chrominance channel *Telecommunications.* a channel of a color television broadcast signal along which chrominance signals are sent.

chrominance demodulator *Electronics.* a device used in a color television receiver to extract the color elements from a video frequency wave.

chrominance frequency *Telecommunications.* the frequency of a chrominance signal; equals 3.579545 MHz.

chrominance gain control *Electronics.* a network of resistors that individually adjust the level of primary colors in a color television set.

chrominance modulator *Electronics.* a device used in color television to include the color information in the composite television signal.

chrominance primary *Telecommunications.* either of two transmission primaries, the amplitudes of which determine the chrominance of a color.

chrominance signal *Telecommunications.* in color television reception, the signal that contains the luminance and color information.

chrominance subcarrier *Telecommunications.* the subcarrier that is used in color television reception.

chrominance-subcarrier oscillator see COLOR OSCILLATOR.

chrominance video signal *Electronics.* the output voltage generated by the red, green, and blue sections of a color television camera or receiver matrix.

chromite *Mineralogy.* $Fe^{+2}Cr_2O_4$, a black, opaque, metallic, sometimes magnetic cubic mineral occurring in massive form or as octahedral crystals, having a specific gravity of 4.5 to 4.8 and a hardness of 5.5 on the Mohs scale; found in olivine-rich igneous rocks and as a common component of meteorites.

chromite sand *Materials.* black sand found primarily in South Africa and used as a nonsilica foundry sand.

chromium *Chemistry.* a metallic element having the symbol Cr, the atomic number 24, an atomic weight of 51.996, a melting point of 1900°C, and a boiling point of 2200°C; a hard, brittle, blue-white to grayish metal that forms many compounds but resists corrosion. (From a Greek word for "color," in reference to the bright colors of its compounds.)

chromium-51 *Nuclear Physics.* a radioactive isotope of chromium with a half-life of 27.7 days; created in a laboratory by bombarding naturally occurring chromium with neutrons.

chromium acetate see CHROMIC ACETATE.

chromium carbide *Inorganic Chemistry.* Cr_3C_2, very hard and resistant gray rhombic crystals; insoluble in water; melts at 1890°C and boils at 3800°C; used in coatings and components where hardness and resistivity are important.

chromium chloride **1.** see CHROMOUS CHLORIDE. **2.** see CHROMIC CHLORIDE.

chromium depletion *Materials Science.* a reduction of the amount of chromium in solid solution in stainless steels, occurring when chromium carbide is precipitated primarily at grain boundaries on heating, as in regions adjacent to a weld. The high chromium content of the carbide draws chromium from nearby regions, resulting in a galvanic action between the regions of different chromium content.

chromium dioxide *Inorganic Chemistry.* CrO_2, brownish-black crystals or powder; insoluble in water; loses its oxygen at 300°C; used in magnetic tapes and as a catalyst.

chromium dioxide tape *Electronics.* a recording tape that is coated with chromium dioxide so that it produces an audio signal with a wide dynamic range and a low noise level.

chromium fluoride **1.** see CHROMOUS FLUORIDE. **2.** see CHROMIC FLUORIDE.

chromium-gold metalizing *Electronics.* a material painted on the surface of semiconductors to prevent damage that arises when gold and aluminum come in contact.

chromium hydroxide or **chromium hydrate** see CHROMIC HYDROXIDE.

chromium nitrate see CHROMIC NITRATE.

chromium oxide see CHROMIC OXIDE.

chromium oxychloride see CHROMYL CHLORIDE.

chromium phosphate see CHROMIC PHOSPHATE.

chromium potassium sulfate *Inorganic Chemistry.* $CrK(SO_4)_2 \cdot 12H_2O$, toxic, efflorescent, violet to ruby-red crystals that melt at 89°C, lose $10H_2O$ at 100°C and $12H_2O$ at 400°C; used in tanning, dyeing, photography, and ceramics. Also, POTASSIUM CHROMIUM SULFATE, CHROME ALUM.

chromium sesquichloride see CHROMIC CHLORIDE, def. 2.

chromium sesquioxide see CHROMIC OXIDE.

chromium stearate *Organic Chemistry.* $Cr(C_{18}H_{35}O_2)_3$, a green powdery solid that melts at 95°C; used in making lubricants, plastics, and ceramics.

chromium trichloride see CHROMIC CHLORIDE.

chromium trioxide see CHROMIC ACID.

chromizing *Metallurgy.* the process of diffusing chromium into the surface layer of a metallic product.

chromo- a combining form meaning "color," as in *chromophilia.*

Chromobacterium *Bacteriology.* a genus of Gram-negative, facultatively anaerobic bacteria that occur as rod-shaped cells in both soil and water.

chromoblastomycosis *Medicine.* the invasion of a wound by fungi, producing skin eruptions that sometimes ulcerate.

chromocenter *Cell Biology.* an aggregation of heterochromatin, especially that of the polytene chromosomes of the *Drosophila* larval salivary gland cell.

chromocratic see MELANOCRATIC.

Chromodoroidea *Invertebrate Zoology.* a superfamily of nonparasitic nematodes.

chromodynamics *Particle Physics.* a theory of the interaction of quarks carrying color, in which the quarks exchange gluons in a manner that is analogous to the interaction of charged particles in electrodynamics.

chromogen *Biochemistry.* **1.** a colorless compound that can produce or become a colored compound as a result of a chemical reaction. **2.** a pigment-producing bacterium, usually belonging to the *Mycobacterium* species, that creates yellow or orange colonies.

chromogenesis *Biochemistry.* the metabolic production of pigments or colors, for example by bacteria.

chromogenic substrate *Enzymology.* a substrate that forms a colored substance when acted upon by a certain enzyme, allowing visual confirmation of enzyme activity.

chromolithography *Graphic Arts.* a once-common printing process that produces color images using a stone or metal lithographic transfer system.

chromomere *Cell Biology.* any of a series of dark-staining bands, containing condensed chromatin, along a eukaryotic chromosome.

chromometer see COLORIMETER.

chromomycin *Microbiology.* an antibiotic and antitumor agent that binds to DNA, inhibiting RNA transcription and DNA replication.

chromomycosis see CHROMOBLASTOMYCOSIS.

chromonema *Genetics.* a single coiled thread in a eukaryotic chromosome.

chromoneme *Genetics.* the DNA thread in bacterial cells or their viruses.

chromophile *Biology.* a cell, cytoplasmic structure, or tissue that is readily stained with dyes. Also, **chromophil.**

chromophobe *Biology.* a cell or cytoplasmic structure that is not readily stained.

chromophobia *Psychology.* an irrational fear or hatred of colors or of certain colors. Also, CHROMATOPHOBIA.

chromophore *Organic Chemistry.* the structural feature of a molecule responsible for its absorption of UV or visible light to give a colored compound. Thus, **chromophoric.**

Chromophycota *Botany.* **1.** a phylum of algae whose members exhibit a wide range of traits, although most members have chloroplasts. Also, **Chromophyta. 2.** in some classification systems, a very heterogenous division that extends from unicellular flagellates to enormous brown algae; characterized by the presence of chlorophylls *a* and *c* but never chlorophyll *b*, and by a periplastidial cisterna in the cells.

chromoplasm *Biochemistry.* see CHROMATIN. *Botany.* the pigmented marginal protoplasm of blue-green algae.

chromoplast *Cell Biology.* a plant cell plastid containing pigments other than chlorophyll, usually the yellow and orange carotenoid pigments that contribute color to ripe fruits and flower petals.

chromoprotein *Biochemistry.* a colored conjugated protein that contains metal pigment and has respiratory functions, such as the red hemoglobin of animals.

chromoradiometer *Engineering.* a meter that contains a substance whose color depends on the amount of X-ray radiation it has received.

chromoscope *Optics.* an instrument that is used to study various properties of color, including its value and intensity.

chromosomal RNA *Genetics.* ribonucleic acid that is associated with chromosomes, either during transcription or during DNA replication, as distinguished from RNA that is associated with protein synthesis.

chromosomal sterility *Genetics.* a situation in which sterility occurs in a hybrid due to incompatibility between the separate chromosomal lines. Also, **chromosomal hybrid sterility.**

chromosomal translocation *Immunology.* the rearrangement of DNA sequences from one chromosome to another; often associated with cancer.

chromosome *Cell Biology.* in a cell nucleus, a structure containing a molecule of DNA that transmits genetic information and is associated with RNA, histones, and nonhistone proteins. Each organism of a species normally has a characteristic number of chromosomes in its somatic cells; the normal number for humans is 46.

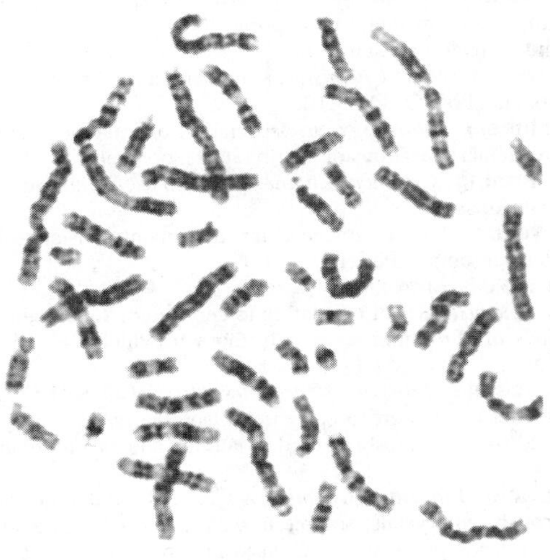

chromosome patterns

chromosome aberration *Genetics.* an abnormal chromosomal complement that results from the loss, duplication, or rearrangement of genetic material.

chromosome banding *Genetics.* **1.** the process of staining the metaphase chromosomes of cultured cells to produce characteristic patterns of bands, allowing identification of specific chromosome pairs and recognition of abnormal patterns associated with certain chromosomal defects. **2.** the pattern of alternating light and dark bands visible on the living giant chromosomes of certain fly larvae due to alternating regions of condensed and uncondensed chromatin. Also, BANDING.

chromosome complement *Genetics.* the particular group of chromosomes that are derived from a particular gametic or zygotic nucleus.

chromosome jumping *Genetics.* the process of sequencing a chromosome in noncontiguous segments; a more rapid process than chromosome walking.

chromosome puff *Cell Biology.* a region of decondensed chromatin on a polytene chromosome that is undergoing selective transcriptional activity.

chromosome walking *Genetics.* a mapping method for placing a series of overlapping restriction fragments, used mainly to determine the position of a gene on a larger DNA molecule.

chromosphere *Astronomy.* the layer of the sun, pinkish in color and composed of hydrogen, that lies between the visible surface (the photosphere) and the corona.

chromospheric network *Astronomy.* a network of strong magnetic fields in the sun's chromosphere that coincides with the supergranulation pattern of the photosphere.

chromostereopsis *Photogrammetry.* the phenomenon in which a set of different colored objects are perceived to be at different heights because of their color, even though there is no actual height difference.

chromotropic acid *Organic Chemistry.* $C_{10}H_8O_8S_2$, white needles that are soluble in water and insoluble in alcohol; used as an azo dye intermediate and analytical reagent.

chromous *Chemistry.* **1.** of or relating to chromium. **2.** describing various compounds of chromium, especially those in which the element has a valence of 2.

chromous chloride *Inorganic Chemistry.* $CrCl_2$, white deliquescent needles; very soluble in water and insoluble in alcohol and ether; melts at 824°C; used as a reagent, catalyst, and reducing agent.

chromous fluoride *Inorganic Chemistry.* CrF_2, greenish, shiny crystals; slightly soluble in water and insoluble in alcohol; melts at 1100°C and boils above 1300°C; highly toxic and irritating; used as a catalyst.

Chromulina *Botany.* a genus of green algae of the family Ochromonadaceae, distinguished by a layer of thin, organic scales and a reduced flagellum.

chromyl chloride *Inorganic Chemistry.* CrO_2Cl_2, a dark red, toxic liquid that fumes in air and reacts strongly with water; used in a variety of industrial processes.

chron *Geology.* the shortest interval into which geologic time is subdivided.

chron- a combining form meaning "time," as in *chronicity.*

chronic [krän´ik] *Medicine.* persisting over a long period of time; often said of an illness.

chronic alcoholism *Medicine.* a pathological pattern of alcohol abuse persisting over time, impairing the health and normal function of the individual.

chronic appendicitis *Medicine.* the prolonged inflammation of the appendix with the buildup of fibrous tissue in the appendiceal wall.

chronic carcinoma see SCIRRHOUS CARCINOMA.

chronic carrier *Medicine.* an individual who is host to a disease organism for a prolonged period.

chronic catarrhal enteritis *Medicine.* a former term for a prolonged or recurring intestinal inflammation producing mucus.

chronic exposure *Radiology.* a long-term exposure to radiation on a continuous or intermittent basis, generally of low intensity and associated with such symptoms as genetic damage and increased risk of cancer.

chronic glomerulonephritis *Medicine.* a progressive kidney inflammation usually resulting in permanent kidney failure.

chronic granulocytic leukemia *Oncology.* a form of leukemia characterized by abnormal, excessive, unrestrained overgrowth of granulocytes in the bone marrow, associated with a chromosomal abnormality and usually occurring between the ages of 25 and 60; symptoms include malaise, anemia, and leukocytosis. Also, **chronic myelocytic leukemia.**

chronic granulomatosis *Medicine.* the development of many granulomas, or lesions, made up of lymph and epithelial cells; characteristic of several diseases.

chronic infectious arthritis see RHEUMATOID ARTHRITIS.

chronicity index *Toxicology.* a numerical expression of the cumulative toxicity of a poison over a specified period of time.

chronic leukemia *Medicine.* a slowly progressing, usually fatal disease of the blood-cell producing organs.

chronic myeloid leukemia *Medicine.* a slowly progressing type of leukemia that affects the myeloid, or bone marrow, cells.

chronic rhinitis *Medicine.* a continued irritation of the nasal mucosa, resulting in swelling and later shriveling of the tissues.

chronic selenium poisoning see ALKALI DISEASE.

chronic tamponade *Cardiology.* the chronic compression or restricted expansion of the heart due to the accumulation of blood or other material in the pericardium or to disease of the pericardium.

chronic toxicity *Toxicology.* a poisonous effect noticed after exposure over a long period of time.

chronistor *Electronics.* a miniature elapsed-time indicator that uses electroplating principles to measure the operating time of equipment.

chrono- a combining form meaning "time," as in *chronometer.*

chronoamperometry *Analytical Chemistry.* the measurement at an electrode of the rate of change in current versus time, during a titration, upon application of a controlled potential.

chronocline *Paleontology.* a time series of fossils showing small changes in successive representatives of a taxon.

chronocyclegraph *Industrial Engineering.* a cyclographic time-and-motion study in which the light level varies to permit computation of the speed and direction of body motions.

chronogram *Horology.* the record produced by a chronograph.

chronograph *Horology.* **1.** a highly precise instrument that measures, indicates, and records the elapsed time of an event. **2.** a precise timekeeper used for navigational purposes. **3.** an instrument that records the beats of a timepiece in order to compare them to a time standard.

chronological age *Psychology.* a person's actual age as calculated from birth, as opposed to any age classification based on rate of mental development.

chronological backtracking *Artificial Intelligence.* in a tree search, a form of backtracking in which decisions that led to a failure are retracted or modified in the reverse of the order in which they were made, so that the most recently made decision in the current search tree is modified first.

chronology *Science.* **1.** the sequence of events in order of their occurrence. **2.** a written record of such a sequence of events. **3.** the scientific study of time and the sequence of events. Thus, **chronological.**

chronometer *Horology.* any exceptionally accurate timepiece, especially as defined by the Swiss Federation of Watch Manufacturers or another official rating organization. *Navigation.* an exceptionally accurate clock designed for use in celestial navigation and built to remain accurate despite harsh conditions at sea. Also, MARINE CHRONOMETER.

chronometer time *Horology.* the time of day as indicated by a chronometer set to Greenwich Mean Time.

chronometric data *Engineering.* any data focusing on the time an event occurs, the duration of the event, or the amount of time that transpires between events.

chronometric dating *Archaeology.* any method of dating that relies on chronological measurement such as calendars, radiocarbon dates, and so on.

chronometric encoder *Electronics.* an encoder that converts information into digital form by counting electrical pulses.

chronometric radiosonde *Engineering.* a radiosonde that transmits meteorological data at intervals corresponding to the magnitude of the meteorological event being evaluated.

chronometric tachometer *Engineering.* a tachometer that counts the revolutions of a shaft during a given time interval to calculate the average speed of a machine.

chronometry *Horology.* **1.** the theories and techniques of measuring time. **2.** the measurement of extremely brief periods of time.

chronopher *Electronics.* a device that generates standard time signal pulses from a clock or other timing device.

chronopotentiometry *Analytical Chemistry.* the measurement at an electrode of the rate of change in potential versus time, upon application of a controlled current.

chronoscope *Horology.* an optical or electronic instrument used for precise measurement of extremely brief periods of time.

chronospecies *Evolution.* **1.** any species for which there is paleontological evidence through a period of time. **2.** a series of species that form a continuous phyletic lineage, and whose evolutionary order is determined by the geological age of the rock in which the fossils are found. Also, PALEOSPECIES.

chronostratigraphic *Geology.* of or relating to chromostratigraphy.

chronostratigraphy *Geology.* the study of geologic history based on the age of rock strata and their time sequence. Also, TIME-STRATIGRAPHY.

chronotron *Electronics.* a device that measures the intervals between pulses on a transmission line, in order to establish the length of time between the pulses and the events that initiated them.

Chroococcales *Botany.* an order of unicellular and colonial blue-green algae that reproduce by cell division or colony fragmentation.

Chroococcidiopsis *Bacteriology.* a genus of unicellular photosynthetic cyanobacteria that divide by binary fission, forming cubical aggregates.

Chryomyidae *Invertebrate Zoology.* a family of two-winged myodarian insects in the subsection Acalypteratae.

chrys- a combining form meaning "gold," as in *chrysanthemum.*

Chrysamoebidaceae *Botany.* a family of the order Chrysamoebidales, characterized by species with a naked rhizopodial protoplast; some animal rhizopods have their origin in this family.

Chrysamoebidales *Botany.* an order of the class Chrysophyceae, containing organisms with a dominant amoeboid phase in their life history and a cell with pseudopodia or rhizopodia used for locomotion and in nutrition.

chrysanthemum *Botany.* **1.** any cultivated plant of the genus *Chrysanthemum,* such as *C. morifolium,* known for their large, diversely colored flowers. **2.** any of several composite plants, such as *C. leucanthemum.*

chrysazin see 1,8-DIHYDROXYANTHRAQUINONE.

chrysene *Organic Chemistry.* $C_{18}H_{12}$, carcinogenic, red or blue, fluorescent, rhombic plates; insoluble in water and slightly soluble in alcohol; melts at 255–256°C and boils at 448°C; used in organic synthesis.

Chryseomonas *Bacteriology.* a genus of Gram-negative, yellow-pigmented, aerobic bacteria that occur as rod-shaped cells exhibiting flagellar motility.

chrysiasis *Toxicology.* poisoning that is caused by a deposition of gold within body tissues; symptoms may include a graying of the skin and damage to the liver, kidneys, eyes, and hematopoietic system. Also, AURIASIS.

Chrysididae *Invertebrate Zoology.* the cuckoo wasps, a family of hymenopteran insects in the superfamily Bethyloidea having bright, metallic green or blue bodies.

chryso- a combining form meaning "gold," as in *chrysocarpous.*

Chrysobalanaceae *Botany.* a family of tropical dicotyledonous trees and shrubs of the order Rosales, characterized by simple, spirally arranged leaves and perfect flowers with annular nectary disks.

chrysoberyl *Mineralogy.* $BeAl_2O_4$, a variegated green to yellow, transparent to translucent orthorhombic mineral occurring as commonly twinned, thick tabular or short prismatic crystals, having a specific gravity of 3.75 and a hardness of 8.5 on the Mohs scale; found in granite pegmatites and mica schists and used as a gemstone.

Chrysocapsaceae *Botany.* a family of the order Chrysocapsales, distinguished by the lack of an organized thallus and by being embedded in mucilage.

Chrysocapsales *Botany.* an order of the class Chrysophyceae, characterized by organisms that spend most of their lives as immotile naked cells surrounded by mucilage in what is called a palmelloid organization.

chrysocarpous *Botany.* bearing yellow or golden fruit.

Chrysochloridae *Vertebrate Zoology.* the golden moles, a family of small fossorial insectivores of the suborder Lipotyphla (or Zalambdodonta, in some classifications), ranging through southeast and central Africa and characterized by a pointed head with a bare muzzle, and dense shiny fur.

Chrysococcus *Botany.* a genus of green algae belonging to the family Dinobryaceae, characterized by a secreted globular envelope around the cells that has only a very small pore through which the flagellum protrudes.

chrysocolla *Mineralogy.* $(Cu^{+2},Al)_2H_2Si_2O_5(OH)_4 \cdot nH_2O$, a brittle, transparent to opaque, blue or green monoclinic mineral usually occurring in cryptocrystalline opal-like form, having a specific gravity of 2 to 2.4 and a hardness of 2 to 4 on the Mohs scale; found in oxidized zones of copper deposits.

chrysoidine *Organic Chemistry.* $C_{12}H_{12}N_4$, free base; pale yellow needles that are slightly soluble in water and soluble in alcohol; used as an orange dye for cotton and silk.

chrysolaminarin *Biochemistry.* a storage polysaccharide of golden-brown algae.

chrysolite see OLIVINE.

Chrysomelidae *Invertebrate Zoology.* a family of leaf beetles, coleopteran insects in the superfamily Chrysomeloidea.

Chrysomeloidea *Invertebrate Zoology.* a superfamily of beetles in the suborder Polyphaga.

Chrysomonadida *Invertebrate Zoology.* an order of brown or yellow flagellate protozoans, solitary or colonial, belonging to the class Phytamastigophorea. Also, **Chrysomonadina, Chrysophyta.**

Chrysopetalidae *Invertebrate Zoology.* a family of scaled polychaete worms of the Errantia.

chrysophanic acid *Organic Chemistry.* $C_{15}H_{10}O_4$, a yellow powder obtained from rhubarb; insoluble in water and soluble in alkaline solutions; melts at 196°C; used in treating eczema and psoriasis.

Chrysophyceae *Botany.* a family of gold-brown algae belonging to the class Chrysophyta, having both freshwater and saltwater varieties, living alone or in colonies, and reproducing both asexually and sexually.

chrysoprase *Mineralogy.* an apple-green variety of chalcedony.

Chrysosphaerales *Botany.* an order of the family Chrysophyceae containing species that spend most of their lives in the coccal phase and reproduce by means of autospores or swarmers formed by the division of the mother cell.

Chrysosporium *Mycology.* a genus of fungi that belongs to the class Hyphomycetes and that causes the disease adiaspiromycosis. Also, EMMONSIA.

chrysotherapy *Medicine.* therapy with gold salts. Also, AUROTHERAPY.

chrysotile *Mineralogy.* $Mg_3Si_2O_5(OH)_4$, a subgroup name for three monoclinic and orthorhombic polymorphs of serpentine in the kaolinite-serpentine group, occurring as fine tubular fibers of silky luster and variegated green color with a hardness of 2.5 on the Mohs scale and a specific gravity of 2.5; found in cross-fiber veinlets in massive serpentine; the principal serpentine asbestos mineral.

CHS *Aviation.* the airport code for Charleston, South Carolina.

Chthamalidae *Invertebrate Zoology.* a family of barnacles in the suborder Thoracica, with sessile adults.

chub *Vertebrate Zoology.* a freshwater fish belonging to the carp family, Cyprinidae, used for gamefishing bait; larger specimens are often treated as gamefish, although not favored for food.

chubasco *Meteorology.* a severe thunderstorm characterized by strong lightning squalls occurring on the west coast of Nicaragua and Costa Rica in Central America.

chuck *Mechanical Engineering.* a device attached to the spindle of a machine tool that grips the rotating workpiece. *Mechanical Devices.* any device designed to hold a workpiece.

chucking *Mechanical Engineering.* the gripping of a workpiece in a chuck.

chucking lathe *Mechanical Engineering.* a lathe using a chuck to hold the workpiece.

chucking lug *Mechanical Engineering.* a lug that is added to a forging to allow on-center machining with one setup or chucking. *Metallurgy.* the projection of a cast or forged component, needed to control subsequent machining.

chucking machine *Mechanical Engineering.* a lathe with an attached chuck that grips the piece being worked.

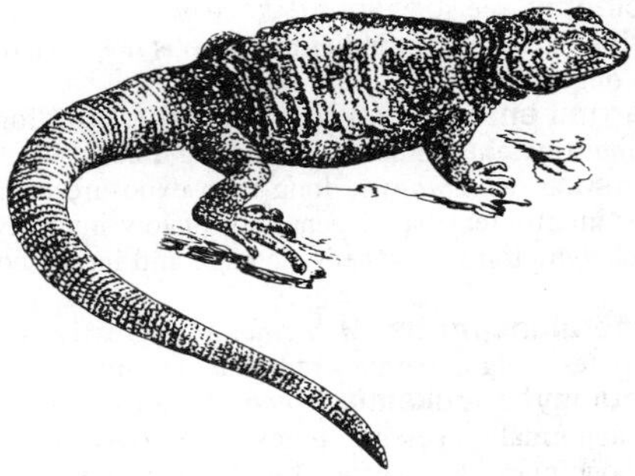

chuckwalla

chuckwalla *Vertebrate Zoology.* a large stocky lizard, *Sauromalus obesus,* of the family Iguanidae, characterized by dull, rough skin and found in arid regions of southwest North America; it eludes capture by wedging into crevices and inflating itself.

chuffing see CHUGGING.

Chugaev reaction *Organic Chemistry.* a reaction that involves the pyrolysis of a methyl xanthate to produce an olefin, COS, and methyl mercaptan.

chugging *Space Technology.* **1.** in a liquid rocket engine or other aircraft engine, a combustion instability characterized by pulsing operation and noise at low frequencies. **2.** the noise made in this type of combustion. Also, BUMPING, CHUFFING.

chukar *Vertebrate Zoology.* a rock partridge, *Alectoris graeca,* of the family Phasianidae, characterized by a red bill and feet with a dominant black mask across the eyes and down the neck; native to Eurasia but successfully introduced as a gamebird in dry parts of the western U.S.

chukar

chunk *Artificial Intelligence.* a hypothesized unit of human knowledge, such as a rule, pattern, or concept.
chunking *Psychology.* the perceptual process by which an individual collects simple units of memorized information into larger, coordinated units, or chunks, that have more significant meaning. *Artificial Intelligence.* a form of learning by forming a new chunk, whose components are existing chunks.
churada *Meteorology.* a severe rainstorm occurring during the monsoon season over the Mariana Islands in the western Pacific Ocean.
Church's thesis *Mathematics.* the generally accepted hypothesis that any function regarded naturally as computable can be computed by a suitable Turing machine; equivalent to all current general definitions of computability. Church's thesis could be overthrown only by proposing a new, generally acceptable definition of computability that Turing machines cannot execute. *Artificial Intelligence.* the thesis that computers can perform any effectively specified symbolic process. Also, **Church-Turing thesis.**
churn *Food Technology.* **1.** to agitate a liquid in order to separate out the fat globules; used to produce butter from cream. **2.** a device used to carry out this process.
chute *Engineering.* a steep conduit or trough through which materials fall or slide by gravity.
chute conveyor see JIGGING CONVEYOR.
chute system *Mining Engineering.* a method of mining that breaks ore from the surface downward into chutes, then removes it through passageways below. Also, GLORY HOLE SYSTEM, MILLING SYSTEM.
Chworinov rule *Metallurgy.* a postulate stating that the total freezing time is a function of the volume-to-surface ratio.
chyle *Physiology.* a whitish liquid, containing fat globules and other products of digestion, that is taken up by the small intestine, enters the lymphatic system, and is carried to the veins through the thoracic duct. Also, **chylus.**
chylomicron *Biochemistry.* a particle from a lipoprotein that originates in the small intestine and is found in blood or lymph; transports fats to tissues after fat ingestion.
chylothorax *Medicine.* a collection of chyle in the pleural space of the lung. Also, **chylopleura.**
chylurla *Medicine.* the discharge of chyle into the urine.
chymosin *Enzymology.* an enzyme that catalyzes the coagulation of milk, specifically the cleavage of a single bond in soluble casein K to form paracasein, which then reacts with calcium to form an insoluble curd; produced in the fourth stomach of calves and other ruminants, and used especially in cheesemaking. Also, RENNIN.
chymotrypsin *Biochemistry.* a hydrolytic enzyme secreted by the pancreas and activated in the small intestine; catalyzes the breakdown of protein.
chymotrypsinogen *Biochemistry.* the precursor, in active form, of chymotrypsin.
Chytridiaceae *Mycology.* a family of fungi of the order Chytridiales, occurring on land or in the water and composed of parasites of living or dead pollen, protozoa, aquatic fungi, or algae.
Chytridiales *Mycology.* an order of fungi of the class Chytridiomycetes, including parasites and pathogens of algae, other fungi, and plants.
Chytridiomycetes *Mycology.* a class of fungi of the subdivision Mastigomycotina, occurring on land, in fresh water, and in salt water, and including parasites of freshwater protozoa and algae.
Chytriodiniaceae *Botany.* a family of parasitic marine algae comprising the order Chytriodiniales.
Chytriodiniales *Botany.* a monofamilial order of marine algae of the class Dinophyceae that are parasitic on both plants and animals.
CI *Aviation.* the airline code for China Airlines.
CI cast iron.
Ci curie; cirrus.
cibarium *Invertebrate Zoology.* the space anterior to the mouth cavity in insects, within which food is chewed.
cibophobia *Psychology.* an irrational fear or hatred of food or eating.
Cicadellidae *Invertebrate Zoology.* a family of homopteran insects, including the leaf hoppers, in the series Auchenorrhyncha.
Cicadidae *Invertebrate Zoology.* a family of homopteran insects, including the cicadas, in the series Auchenorrhyncha.
cicatrectomy *Surgery.* the excision of a scar.
cicatricotomy *Surgery.* the incision of a scar.
cicatrix [sik′ə triks′] *Biology.* a scar or scarlike mark, especially one marking the previous attachment of an organ or structure, such as the mark left on a tree by a fallen branch. *Medicine.* see SCAR.
cicatrization *Surgery.* the process of healing by scar formation. Also, EPULOSIS.
cicatrize *Surgery.* to heal by the development of a scar.
Cichlidae *Vertebrate Zoology.* the cichlids, a family of freshwater spiny-finned fishes of the order Perciformes, found particularly in the lakes of Africa and Central America and noted as a popular aquarium fish.
Cicindelidae *Invertebrate Zoology.* the tiger beetles, a family of coleopteran insects in the suborder Adephaga.
Ciconiidae *Vertebrate Zoology.* the storks, a family of large, long-legged wading birds in the order Ciconiiformes, noted for their elaborate courtships and tight monogamous bonding.
Ciconiiformes *Vertebrate Zoology.* an order of long-necked, long-legged wading birds including the herons, bitterns, storks, ibises, and spoonbills.
cicutoxin *Toxicology.* a potent poison found in water hemlock of the species *Cicuta.*
CID *Aviation.* the airport code for Cedar Rapids-Iowa City.
CID combined immunological deficiency; charge-injection device; cubic-inch displacement.
Cidaroida *Invertebrate Zoology.* an order of sea urchins, primitive echinoderms in the subclass Perischoechinoidea.
CIE counter immunoelectrophoresis.
Cifax *Telecommunications.* **1.** the application of cryptography techniques to facsimile transmission. **2.** facsimile signals that have been enciphered to provide data security during transmission.
cigarette burning *Chemistry.* a term for a type of combustion induced in a shaped solid that is ignited at one end so that it burns toward the other end.
ciguatera *Toxicology.* poisoning due to the ingestion of a toxin normally present in some fish; symptoms may include gastrointestinal and neurological disturbances.
ciguatoxin *Biochemistry.* a toxic substance produced by the flagellate protozoan *Gambierdiscus toxicus.*
Ciidae *Invertebrate Zoology.* a family of tiny tree-fungus beetles, coleopteran insects in the superfamily Cucujoidea.
cilia [sil′ē ə] *singular,* **cilium.** *Cell Biology.* in most animal species and in some lower plant species, an array of motile, microtubular processes that cover certain cell surfaces, facilitating movement of single cells through a fluid or promoting the flow of fluid over the surface of cells. *Anatomy.* **1.** specifically, the hairlike processes that extend from certain epithelial cells, such as those lining the bronchial passages, and serve to move mucus and objects such as dust particles over the surface of the tissue. **2.** another term for the eyelashes.
ciliary body *Anatomy.* a thick, wedge-shaped body between the choroid and the iris, from which aqueous humor is secreted and in which the ciliary muscle is located.
ciliary movement *Biology.* motion of a cell accomplished by the beating of the cell's cilia.
ciliary muscle *Anatomy.* one of the muscles around the lens of the eye that aids in accommodation to changes in focal length.
ciliary process *Anatomy.* approximately 70 ridges or folds, extending from the crown of the ciliary body, which secrete aqueous humor.
Ciliatea *Invertebrate Zoology.* a class of ciliated protozoans in the subphylum Ciliophora.
ciliated *Invertebrate Zoology.* having cilia.

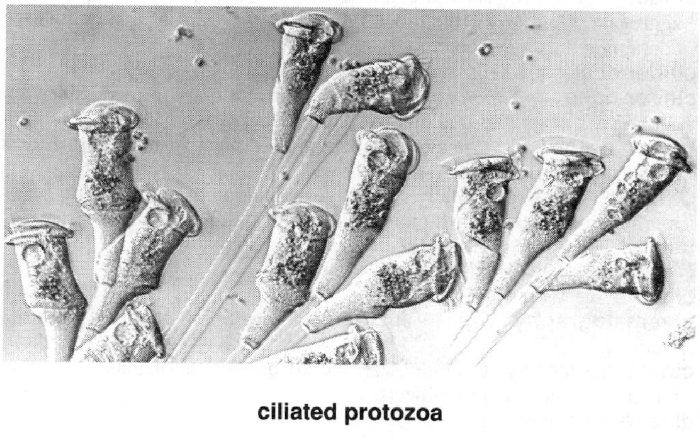

ciliated protozoa

ciliated epithelium *Histology.* a layer of epithelial tissue whose free border is covered with cilia.
ciliation *Invertebrate Zoology.* the presence of cilia in a particular location, or on a particular organism.
ciliolate *Botany.* having very fine hairs along the margin. Also, **ciliate.**
Ciliophora *Invertebrate Zoology.* a subphylum of the Protozoa, free-living ciliated single-cell animals.
cilium [sil′ē əm] *plural,* **cilia.** *Cell Biology.* a single hairlike process projecting from the free surface of some cells, composed of nine pairs of microtubules arrayed around a central pair and bounded by the cell membrane.
CIM computer-integrated manufacturing; computer input (from) microfilm.
Cimbex *Invertebrate Zoology.* a serious plant pest in the family Cimbicidae.
Cimbicidae *Invertebrate Zoology.* a family of saw flies, hymenopteran insects in the superfamily Tenthredinoidea, with leaf-eating caterpillar-like larvae.
Cimex *Invertebrate Zoology.* the common bedbug, the type genus of the family Cimicidae.
Cimicidae *Invertebrate Zoology.* a family of wingless hemipteran insects in the superfamily Cimicimorpha, blood-sucking parasites including the bed, bird, and bat bugs.
Cimicimorpha *Invertebrate Zoology.* a superfamily of land bugs, hemipteran insects in the subdivision Geocorisae.
ciminite *Petrology.* a medium-grained igneous rock consisting essentially of olivine, pyroxene, and basic calcic plagioclase rimmed with alkali feldspar.
Cimmerian orogeny *Geology.* a brief, Triassic to Jurassic crustal deformation involving Phanerozoic rock.
cimolite *Mineralogy.* a white, gray, or red hydrous aluminum silicate mineral that often appears in soft, claylike masses.
cinchona *Botany.* **1.** any of several tropical American trees of the genus *Cinchona,* of the madder family, whose dried bark is the source of quinine and other alkaloids. **2.** the medicinal bark of such a tree.
cinchonamine *Organic Chemistry.* $C_{19}H_{24}N_2O$, an alkaloid that crystallizes in rhombic needles; insoluble in water and very soluble in alcohol; melts at 186°C.
cinchonism *Toxicology.* poisoning due to excessive ingestion of cinchona or one of its derivatives, such as quinine; symptoms may include nausea, rash, tinnitus, and heart and respiratory failure.
Cincinnatian *Geology.* a North American provincial series of the Upper Ordovician period, occurring after the Champlainian and before the Alexandrian of the Silurian period.
Cinclidae *Vertebrate Zoology.* the dippers, a family of small sparrow-like aquatic birds in the suborder Oscines of the order Passeriformes, noted for their restless, bobbing motion when out of the water.
cinclides *singular,* **cinclis.** *Invertebrate Zoology.* pores in the body wall of certain sea anemones.
cinder *Materials.* **1.** a partly burned and extinguished piece of combustible material or particles of such a material. **2.** of or relating to a material that includes an aggregate of such particles. Thus, **cinder concrete.** *Metallurgy.* **1.** the scale that becomes detached during hot forging. **2.** in ironmaking, the slag formed in the blast furnace smelting process. *Volcanology.* **1.** a glassy or vesicular volcanic fragment that ranges in diameter from 4 to 32 millimeters. **2.** coarse, vesicular lava thrown out by an explosive eruption.
cinder block *Materials.* a concrete building block made with a cinder aggregate. *Metallurgy.* the block that closes the front of a blast furnace containing the cinder notch.
cinder coal see NATURAL COKE.
cinder cone *Volcanology.* a conical hill or mountain formed by the accumulation of cinders and other volcanic debris around a vent.
cinder notch *Metallurgy.* the opening in a blast furnace that allows molten slag to flow out. Also, MONKEY.
C index *Geophysics.* a daily subjective evaluation of geomagnetic activity in a given location, which is graded 0 for quiet, 1 for moderately disturbed, and 2 for very disturbed.
C indicator see C DISPLAY.
cinema see MOTION PICTURES.
cinematography *Graphic Arts.* the art or process of making motion pictures.
cinemicrography *Graphic Arts.* the art or process of making motion pictures of microscopic subjects.
cinene see DIPENTENE.
cineole or **cineol** see EUCALYPTOL.
cineraria *Botany.* a composite plant, *Senecio hybridus,* that is widely cultivated for its colorful flowers.
cinerary urn *Archaeology.* a container that holds cremated ashes.
cinerous *Biology.* **1.** being an ash-gray color. **2.** of or relating to the gray matter of nerve tissue. Also, **cineraceous, cineritious.**
cinetheodolite *Engineering.* a theodolite equipped with a motion picture camera that records precise time-correlated observations of the elevation and position of aircraft and artificial satellites.
Cingulata *Vertebrate Zoology.* the armadillos, a group of armored mammals of the suborder Xenarthra in the order Edentata.
cingulate *Biology.* having a girdle-shaped structure, usually partially or totally encircling another structure; especially used to describe the thorax or abdomen of insects.
cingulum *Anatomy.* **1.** anything that encircles a body; a band or ridge. **2.** a bundle of association fibers that partly encircles the corpus callosum near the median plane. **3.** the lingual lobe of an anterior tooth. *Invertebrate Zoology.* **1.** a colored or raised band on certain bivalve shells. **2.** the outer zone of cilia on the discs of certain rotifers. **3.** the thickened saddlelike part of an annelid's body wall. (A Latin word meaning "girdle.")
cinnabar *Mineralogy.* HgS, a red, brown, or black transparent to translucent, brittle trigonal mineral occurring in massive form and as rhombohedral, tabular, or prismatic crystals having a specific gravity of 8.09 and a hardness of 2 to 2.5 on the Mohs scale; the chief ore of mercury, usually found with pyrite, calcite, quartz, and opal in low-temperature deposits.
cinnamaldehyde see CINNAMIC ALDEHYDE.
cinnamate *Organic Chemistry.* a salt of cinnamic acid, containing the organic group $-C_9H_7O_2$.
cinnamein see BENZYL CINNAMATE.
cinnamic acid *Organic Chemistry.* $C_6H_5CH{=}CHCOOH$, a derivative of phenylalanine, produced by plants, occurring as crystals that are soluble in water and alcohol; melts at 133°C and boils 300°C; used by plants to make materials, including lignin; used in perfumes.
cinnamic alcohol *Organic Chemistry.* $C_6H_5OH{=}CHCH_2OH$, yellowish needles, soluble in water and many organic solvents; melts at 33°C and boils at 250°C; used as a deodorant and antiseptic. Also, **cinnamyl alcohol.**
cinnamic aldehyde *Organic Chemistry.* $C_6H_5CH{=}CHCHO$, a yellowish oil that is slightly soluble in water and soluble in alcohol; melts at –7.5°C and boils at 253°C; used in flavors and perfumes. Also, CINNAMALDEHYDE.
cinnamon *Botany.* any of several trees of genus *Cinnamonum* of the laurel family, especially **Ceylon cinnamon,** *C. zeylanicum,* whose inner bark is used as a spice, and **Saigon cinnamon,** *C. loureirii,* whose bark is used in preparing medicines.

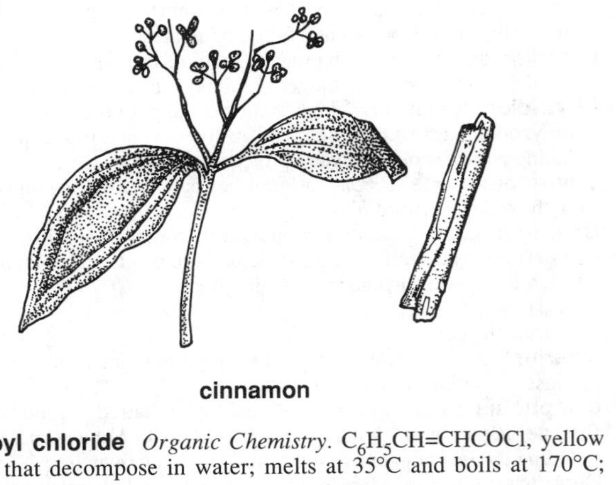

cinnamon

cinnamoyl chloride *Organic Chemistry.* $C_6H_5CH{=}CHCOCl$, yellow crystals that decompose in water; melts at 35°C and boils at 170°C; used as a reagent.
cions *Forestry.* cuttings from the branches of superior trees; used for grafting to the roots of other two-year-old trees.
CIP *Robotics.* computer-integrated production; any manufacturing process in which all aspects of the process are controlled by computer.
cipher *Telecommunications.* any cryptosystem in which plain text is encrypted to provide data security.

cipher machine *Telecommunications.* a device used for enciphering and deciphering text.

ciphertext *Telecommunications.* text or signals that are generated by cipher machines to provide data security by rendering the encrypted data unintelligible without the key.

ciphony [sī′fə nē] *Telecommunications.* a cryptography technique in which speech is converted into a series of on-off pulses and mixed with the pulses of a key generator; messages cannot be reconstructed unless an identical key generator is used.

ciphony equipment *Electronics.* any device attached to a radio transmitter, radio receiver, or telephone to scramble or unscramble voice messages.

Ciplopoda see MILLIPEDE.

CIPW classification *Petrology.* a designation for classifying igneous rocks based on the proportions of certain normative minerals in the rock. (An initialism based on the last names of the four scientists who devised it: Cross, Iddings, Pirsson, and Washington.)

Cir see CIRCINIUS.

cir. circle; circuit; circular; circumference.

circ. circular.

circa *Science.* at, about, or approximately (followed by a date). (From Latin *circum.*)

circadian [sur kād′ē ən] *Biology.* occurring or functioning in approximately 24-hour cycles.

circadian rhythm *Physiology.* a biological cycle or rhythm pattern that recurs about once every 24 hours, such as the adult sleep-wake cycle.

Circaeasteraceae *Botany.* a family of dicotyledonous herbs of the order Ranunculales, having spirally arranged leaves that arise from the rhizome or are clustered atop a stemlike hypocotyl; native to China.

circinate *Botany.* coiled from the tip toward the base, such as new fern leaves. *Biology.* having a circular configuration; ring-shaped.

circinate venation *Botany.* the arrangement of young leaves and fronds in a bud so that when it opens they uncoil like a watch spring. Also, **circinate vernation**.

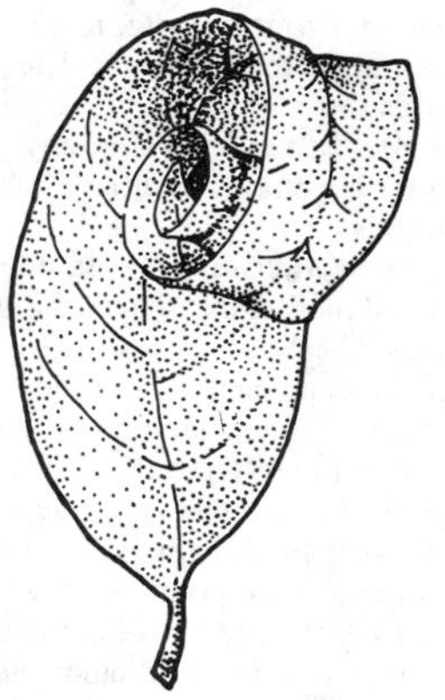

circinate venation

Circinus *Astronomy.* the Compasses, a small faint constellation of the southern celestial hemisphere adjoining Centaurus.

circle *Mathematics.* **1.** a plane curve consisting of all points a fixed distance from a given point. The equation in the plane of a circle with center (a,b) and radius c is $(x-a)^2 + (y-b)^2 = c^2$, or in polar coordinates with radius c and center (r_1, θ_1), $r^2 + r_1^2 - 2rr_1\cos(\theta - \theta_1) = c^2$. **2.** a rotation of 2π or 360°.

circle diagram *Electricity.* **1.** a circular locus describing the performance characteristic of a rotating machine, such as current or impedance. **2.** a diagram portraying a graphical solution of equations for a transmission line.

circle-dot mode *Electronics.* a method of storing binary digits in a cathode-ray tube in which one kind of digit is represented by a small circle on the screen, and the other is represented by a circle with a concentric dot.

circle of confusion *Optics.* the smallest diameter image spot that arises when aberrations in an optical system cause a point source to blur into an ever-expanding circle. Also, **circle of least confusion.**

circle of convergence *Mathematics.* the circle in the complex plane inside of which a power series converges and outside of which it diverges. That is, given a general (complex) power series $\sum_{n=0}^{\infty} c_n(z-a)^n$, if there exists a number $r > 0$ such that the series converges absolutely for all z with $|z-a| < r$ and diverges for $|z-a| > r$, then r is the radius of convergence and $|z-a| = r$ is the circle of convergence.

circle of curvature *Mathematics.* the circle that most closely fits a curve at a given point. That is, given a point p on a plane curve C with nonzero finite curvature q at p, the circle of curvature at p is the circle of radius $1/q$ tangent to C at p on the concave side; that is, the circle tangent to C at p that has the same curvature as C at p. $1/q$ is called the **radius of curvature,** and the center of the circle is called the **center of curvature.**

circle of equal altitude *Navigation.* an imaginary circle on the surface of the earth, along which a celestial body has a given altitude at a particular instant.

circle of equal probability *Space Technology.* a method of measuring the precision of the guidance system of a rocket or missile that refers to the radius of the circle at a specified distance from launch within which half of the shots land. Also, CIRCLE OF PROBABLE ERROR, CIRCULAR PROBABLE ERROR.

circle of illumination *Geology.* a circular boundary at the edge of the sunlit hemisphere that separates the earth into a light half and a dark half.

circle of inertia see INERTIAL CIRCLE.

circle of inversion *Mathematics.* the circle through which inversion with respect to a circle is to be accomplished. Inversion of a point p with respect to a given circle involves finding the point p' on the radial line through p so that the product of the distances of the two points from the center of the circle equals the square of the radius. p and p' are inverses of each other and the center of the circle is the center of inversion. The radius of the circle is the radius of inversion.

circle of latitude *Cartography.* **1.** a great circle of the celestial sphere passing through the ecliptic poles. **2.** a meridian along which latitude is measured.

circle of longitude *Cartography.* **1.** a circle of the celestial sphere parallel to the ecliptic. **2.** a circle on the surface of the earth parallel to the equator, along which longitude is measured.

circle of position *Navigation.* a line of position that is in the form of a circle. *Astronomy.* the circle of points on the earth's surface from which a given star will appear to have the same altitude above the horizon.

circle of uncertainty *Navigation.* a circle anywhere within which a craft may be located.

circle of Willis *Anatomy.* the circular union of the anterior and posterior cerebral arteries surrounding the stalk of the pituitary gland at the base of the brain.

circle shear *Mechanical Engineering.* a machine designed to cut disks from sheet metal that rolls between the cutting wheels.

circuit *Electricity.* an interconnection of electrical elements forming one or more complete paths for the flow of current. *Electromagnetism.* a network of media, usually conductors and semiconductors, that provide a closed path for electromagnetic phenomena to occur. *Mathematics.* **1.** a closed trail in a graph. **2.** a cycle in a directed graph. *Computer Programming.* a closed walk containing at least one edge. *Cartography.* a continuous line of levels, a series of lines of levels, or a combination of lines or parts of lines of levels that, together with a continuous series of measured differences of elevation, forms a loop back to the starting point of a survey.

circuit analyzer see VOLT-OHM-MILLIAMMETER.

circuit breaker *Electricity.* a current-sensitive protection switch that will automatically open a circuit when the current exceeds a certain predetermined value; unlike a fuse, a circuit breaker can be reset. Also, BREAKER.

circuit capacity *Telecommunications.* the maximum rate at which a given circuit is capable of transmitting data under given conditions, sometimes measured in channels, frequency bandwidths, or bits per second.

circuit conditioning see CONDITIONING.

circuit design *Electricity.* the process of determining the components and connections of an electrical circuit or network.

circuit diagram see SCHEMATIC DIAGRAM.

circuit efficiency *Electronics.* the efficiency that is determined by dividing the power transmitted to a system or device at a given frequency by the power provided by the electron stream at the same frequency.

circuit element see COMPONENT.

circuit grade *Telecommunications.* a designation of the type of information or signal that a given circuit can carry, such as voice or telegraph.

circuit loading *Electricity.* drawing power from a circuit.

circuit matrix *Mathematics.* let G be a finite directed graph with n enumerated edges and m enumerated circuits, with assigned orientations around the circuits. The circuit matrix B_c associated with G is an $m \times n$ matrix whose (i,j)th entry is: (i) +1, if the jth edge is in the ith circuit and its orientation agrees with the circuit orientation; (ii) −1, if the jth edge is in the ith circuit and its orientation does not agree with the circuit orientation; and (iii) 0, if the jth edge is not in the ith circuit. If G is an undirected graph, the (i,j)th entry of B_c is 1 if the jth edge is in the ith circuit and is 0 otherwise. A row of B_c is called a **circuit vector.** The submatrix of B_c corresponding to a fundamental system of cycles of G is a fundamental cycle matrix.

circuit misclosure *Cartography.* the amount by which the algebraic sum of the measured differences of elevation around a circuit fails to equal the theoretical closure of zero.

circuit noise *Telecommunications.* the noise present as a continuous background to signals in a circuit.

circuit noise level *Telecommunications.* the ratio of the circuit noise at a given point to some reference amount of circuit noise, expressed in decibels.

circuit protection *Electronics.* a feature built into a circuit that minimizes the hazard of fire or smoke when an overload or some other defect occurs.

circuit rank *Mathematics.* the number $\gamma(G)$ of edges that must be removed from an arbitrary graph G with n vertices, m edges, and k components to arrive at a spanning forest; $\gamma(G) = m - n + k$ and equals the dimension of the circuit subspace. Also, CYCLOMATIC NUMBER.

circuit reliability *Telecommunications.* the percentage of time that a circuit satisfies standards established by a user.

Circuitron *Electronics.* a network of components mounted inside a section of a device, such as the shell of an electron tube, that functions as one or more complete operating stages.

circuitry *Electricity.* the entire combination of circuits in a particular electrical system or device.

circuit shift see CYCLIC SHIFT.

circuit subspace *Mathematics.* the subspace of the vector space associated with a graph G which consists of all circuits and edge-disjoint unions of circuits of G.

circuit switching *Telecommunications.* a procedure that establishes a connection between two or more communication entities, and grants them exclusive use of the circuit until the connection is released. Also, LINE SWITCHING. *Computer Technology.* a method of network communication in which a circuit is established between the communicating nodes for the duration of the communication.

circuit theory *Electricity.* the formal analysis of conditions and relationships in electronic circuitry.

circulant matrix *Mathematics.* a matrix in which the entries consist of copies of the first row, shifted one position to the right in each successive row, with the last element moved to the first. That is, an $n \times n$ matrix in which $a_{1,j}$ is given; $j = 1, \ldots, n$; $a_{ij} = a_{i-1,j-1}$ for $i, j = 2, \ldots, n$; and $a_{i,1} = a_{i-1,n}$ for $i = 2, \ldots, n$. In particular, $a_{ii} = a_{11}$ for $i = 2, \ldots, n$.

circular *Science.* of, relating to, or having the form of a circle. *Metrology.* see CIRCULAR MIL.

circular antenna *Electromagnetism.* a folded dipole antenna that is bent into a circle.

circular birefringence *Optics.* a phenomenon in which an active medium causes right circularly polarized light to move at a different velocity than left circularly polarized light.

circular buffering *Computer Technology.* a method of temporarily storing input or output data, using a single buffer that is logically arranged in a circle with data appearing to wrap around.

circular-chart recorder *Engineering.* a device with a stylus triggered by a transmission signal that records measured values of temperature, pressure, etc., on a rotating circular chart.

circular coal see EYE COAL.

circular coil *Electromagnetism.* in eddy-current nondestructive tests, a type of test coil that surrounds an object.

circular collider *Nucleonics.* any of a family of particle accelerators, including electron-positron colliders, proton-antiproton colliders, and proton-proton colliders, in which counterrotating beams are brought into collision at experimental areas, where they generate intense bursts of energy; used to study the fundamental properties of subatomic matter.

circular cone *Mathematics.* a cone whose base curve is a circle; equivalently, a cone whose intersections with planes perpendicular to the axis are coaxial circles.

circular current *Electricity.* an often undesirable electric current traveling in a circular path.

circular cutter *Mechanical Engineering.* a shearing machine with a rotating blade, used to cut metal.

circular cylinder *Mathematics.* a cylinder whose base curve is a circle; equivalenty, a cylinder whose intersections with planes perpendicular to the axis are coaxial, congruent circles.

circular decleroism *Spectroscopy.* a differential absorption of left and right circularly polarized light by optically active materials.

circular deoxyribonucleic acid see CIRCULAR DNA.

circular dichroism *Biochemistry.* the absorption of left and right circularly polarized light; a property of molecules that are optically active, used to obtain information about the solution environment of proteins. *Optics.* a phenomenon in which planar polarization converts to elliptic polarization when plane-polarized light travels through an active medium.

circular DNA *Biochemistry.* a form of deoxyribonucleic acid (DNA) having a single-stranded or double-stranded ring; found in certain bacteria cells and in the human wart virus. Also, CIRCULAR DEOXYRIBONUCLEIC ACID, RING DNA.

circular electric wave *Electromagnetism.* a transverse electromagnetic wave having circular polarization.

circular error *Ordnance.* a bombing error measured as the radial distance between the point of impact or mean point of impact and the desired point of impact; in the case of an airburst atomic bomb, it is measured from the point of the ground directly below the explosion and the desired ground zero.

circular error average *Ordnance.* the error of a particular bombing attack; it is measured as the average radial distance between the points of impact or mean points of impact and the desired point of impact.

circular error probable *Ordnance.* an estimate of weapon system accuracy used in determining the probable damage to a target; it is the radius of a circle, centered on the mean point of impact, within which half of the missiles or projectiles aimed at the target are expected to fall.

circular file *Computer Programming.* a file structure in which a new record replaces the oldest record in the file; used mainly for storing highly volatile records.

circular form tool *Mechanical Devices.* any cutting tool whose cutting edge forms the edge of a circle, such as a circular saw; it is used in a mount to form materials.

circular horn *Electromagnetism.* a waveguide with a circular cross section that flares outward having the shape of a horn, serving as a feed for a microwave reflector.

circular inch *Metrology.* the area of a circle 1 inch in diameter.

circular linkage map *Genetics.* a circular chromosome map based on the physical structure of the DNA that comprises the chromosome.

circular list *Computer Programming.* a cyclic arrangement of data elements in which the last item contains a link to the first, allowing access to the entire list from any given point. Also, RING.

circularly polarized light *Optics.* a light beam having electric vectors that can be broken into two perpendicular elements with equal amplitudes and a phase difference of 1/4 wavelength.

circular magnetic wave *Electromagnetism.* a transverse electromagnetic wave having circular polarization.

circular measure *Metrology.* a system for expressing the measure of circles, in which 1 circle = 360 degrees or 4 quadrants, 1 quadrant = 90 degrees, 1 degree = 60 minutes, and 1 minute = 60 seconds.

circular mil *Metrology.* the area of a circle that is 1 mil in diameter. Also, CIRCULAR.

circular motion *Mechanics.* **1.** the motion of a rigid body in which each of its particles circles with equal angular velocity around an axis that is fixed with respect to the body. **2.** any motion of a particle on a circular path.

circular nomograph *Mathematics.* a special kind of nomograph in which the curves denoting the three variables are concentric circles, graded in such a way that any diameter passes through related values of the variables, usually values satisfying a given equation.

circular orbit *Astronomy.* any orbit that has an eccentricity of exactly zero, an ideal condition that is seldom or never achieved in reality.

circular paper chromatography *Analytical Chemistry.* a separation technique in which the components of a sample are eluted on a paper sheet into a series of concentric rings for identification.

circular particle accelerator *Nucleonics.* a particle accelerator that uses magnetic fields, either static or time-varying, to bend and confine the orbit of charged particles to circular "racetracks" of gradually increasing radius.

circular pitch *Design Engineering.* the distance between corresponding points of adjacent teeth on a gear wheel, measured along the pitch circle.

circular plane *Mechanical Devices.* a flexible bed plane used to plane concave or convex work surfaces.

circular polarization *Physics.* the rotation of the electric displacement vector or the electric field vector in a transverse wave propagating through a medium.

circular reference *Computer Programming.* a situation caused by a programming logic error in which two entities refer to each other, creating an endless loop.

circular saw *Mechanical Engineering.* a steel disk with teeth along its circumference, rotating on a spindle; used for cutting wood or metal.

circular screen *Graphic Arts.* a glass or film plate etched with fine concentric circles and used as a halftone screen.

circular segment *Mathematics.* one of the two portions into which a circular region is divided by a chord.

circular slide rule *Mathematics.* a slide rule arranged with its scales in a circle, so that it is never necessary to reset the slide to avoid putting the answer off the end of the rule.

circular sweep generation *Electronics.* the use of an electric current to maintain the electron beam in certain devices, such as a cathode-ray tube, moving in a circular motion at consistent speed.

circular velocity *Mechanics.* the orbital velocity necessary at a given radius to maintain a constant radius orbit.

circular vortex *Meteorology.* a flow in parallel planes in which streamlines and other isopleths are concentric circles about a common axis.

circular wait *Computer Programming.* in a multiprogramming system, the deadlock that occurs when two tasks request two or more resources in opposite order, so that each waits on the other for the remaining resource. Also, MUTUAL DEADLOCK.

circular waveguide *Electromagnetism.* a waveguide having a circular cross section.

circulate-and-weight method *Petroleum Engineering.* a technique for controlling well pressure during drilling operations by gradually increasing mud weight but immediately starting mud circulation. Also, CONCURRENT METHOD.

circulated gas-oil ratio *Petroleum Engineering.* during gas-lift operations, the ratio of the volume or cubic feet of gas put into a well to the volume (or number of barrels) of oil removed.

circulating fluid *Engineering.* a fluid pumped into a borehole through the drill stem to cool the drill bit, wash away the cuttings from the bit, and transport them out of the borehole.

circulating memory see CIRCULATING STORAGE.

circulating pump *Chemical Engineering.* a pump, usually of the centrifugal type, used to circulate liquid out of and into a process system.

circulating register *Computer Technology.* a shift register that shifts a bit from one end and reinserts it at the other end; often used in a continuously shifting mode as a delaying mechanism.

circulating scrap *Metallurgy.* in metal processing, the scrap that is produced in the operation and at least partially recycled in the same plant. Also, HOUSE SCRAP.

circulating storage *Electronics.* a circuit that uses a delay line to store pulses in a trainlike pattern; the output pulses are picked up, amplified, reshaped, and reinserted into the delay line after a certain time lag. Also, DELAY-LINE STORAGE, CIRCULATING MEMORY.

circulating system *Chemical Engineering.* a fluid system in which the process fluid is taken from and pumped back into the system.

circulation the process of moving in a circle or circuit; specific uses include: *Physiology.* movement in a regular, circuitous course, especially the movement of the blood by the rhythmic pumping of the heart through the heart and blood vessels. *Fluid Mechanics.* the circulation around the boundary of any surface element, having a given position and aspect, is ultimately proportional to the area of the element in terms of the continuity equation. *Oceanography.* the water current flow within a large area. *Mathematics.* the line integral of a fluid velocity around a closed path. *Meteorology.* the integrated line of a tangential component of the velocity field about a closed curve in an air mass. *Transportation Engineering.* the flow of traffic in a given area or system.

circulation area *Building Engineering.* that interior portion of a structure designated for personnel movement; it includes corridors, escalators, stairways, stairwells, elevators, entrances, and exits.

circulation flux *Meteorology.* the dominant fluctuation occurring as a result of mean atmospheric motion.

circulation index *Meteorology.* a measure of the magnitude of one of several aspects of large-scale atmospheric circulation patterns, such as the strength of zonal (east to west) or meridional (north to south) wind components.

circulation pattern *Meteorology.* the general geometric configuration of atmospheric circulation applied to the large-scale features of synoptic charts.

circulation rate *Cardiology.* the rate per minute at which blood flows through the vessels.

circulator *Electromagnetism.* a device having several ports for waveguides, in which energy entering the device through one of the ports is transmitted to a port that is adjacent to the first port.

circulatory *Physiology.* of or relating to circulation.

circulatory system *Anatomy.* the heart and vessels that move blood through the body.

circulin *Microbiology.* a former term for a group of peptide antibiotics produced by species of the bacterium *Bacillus circulans,* now classified as polymyxin antibiotics.

circulus *Biology.* shaped in a circle or ring; often used in reference to veins and arteries.

circum- a prefix meaning "around," as in *circumpolar.*

circumaustral *Ecology.* distributed around the high latitudes of the southern hemisphere. Thus, **circumaustral distribution.**

circumboreal *Ecology.* distributed around the high latitudes of the Northern Hemisphere. Thus, **circumboreal distribution.**

circumcenter *Mathematics.* for a polygon having a circumcircle or arbitrary triangle, the center of the circumscribed circle; i.e., the point of intersection of the perpendicular bisectors of the sides of the regular polygon or triangle.

circumcircle *Mathematics.* a circumscribed circle.

circumcision *Surgery.* the surgical removal of the foreskin of the penis.

circumduction *Physiology.* an active or passive rotation of a limb or eye, particularly the swinging of an arm or leg in a conical figure, with the joint of the limb as the base of the cone.

circumference *Mathematics.* **1.** the distance around a circle; equal to $2\pi r$ for a circle of radius r. **2.** the length of a maximal geodesic on an n-sphere. **3.** a perimeter.

circumferentor *Engineering.* a surveyor's compass with diametral projecting arms, each carrying a vertical slit sight.

circumflex coronary artery *Anatomy.* a branch of the left coronary artery that supplies blood to the left ventricle of the heart.

circumgemmal *Neurology.* a form of nerve ending in which a budlike terminal is surrounded by nerve fibrils.

circumnutation *Botany.* the rotary or elliptical movement that a stem tip evinces as it grows.

circumpharyngeal connective *Invertebrate Zoology.* either of a pair of nerve strands surrounding the esophagus in annelids and anthropods.

circumpolar *Astronomy.* of or relating to any celestial object that never sets as seen from a given latitude. Thus, **circumpolar star.**

circumradius *Mathematics.* for a regular polygon or arbitrary triangle, the radius of the circumscribed circle.

circumscissile *Botany.* opening along a transverse fissure, as a seed, fruit, or anther.

circumscribed *Mathematics.* **1.** a closed curve (or surface) is said to be circumscribed about a polygon (or polyhedron) if the polygon is inside the curve and if every vertex of the polygon is on the curve. The polygon is said to be inscribed in the curve. **2.** a polygon (or polyhedron) is said to be circumscribed about a closed curve (or surface) if the curve is inside the polygon and if every edge (or face) of the polygon is tangent to the curve. The curve is said to be inscribed in the polygon.

circumscription *Artificial Intelligence.* a form of nonmonotonic reasoning based on making explicit the assumption that the information provided is the only important information about a situation, and that it is safe to assume that unmentioned things may be considered to be as usual.

circumzenithal arc *Optics.* a rainbow that occurs when the sun's rays strike prismatic ice crystals in the atmosphere; it usually appears parallel to the horizon and lasts for only a few minutes.

ciré [sē rā′] *Textiles.* a fabric that is treated with wax, heat, and pressure in order to give it a glossy finish.

Cirolanidae *Invertebrate Zoology.* a family of crustaceans, active predators and scavengers, in the suborder Flabellifera.

cirque

cirque [surk] *Geology.* **1.** a deep, steep-walled, semicircular recess or hollow located high on a mountainside and produced by glacial erosion. **2.** any semicircular recess that resembles a cirque glacier.

cirque glacier *Hydrology.* a small mountain glacier that has not yet flowed out of the cirque.

cirque lake *Hydrology.* a small, deep, clear-water glacial lake fed by runoff from surrounding slopes and retained in a cirque. Also, TARN.

cirque niveau *Geology.* the altitude at which most cirques in an area have their floors.

cirque platform *Geology.* a level surface formed by the growing together of several cirques.

cirque stairway *Geology.* a series of cirques in the same glacial valley that formed in a row at different levels.

cirrate see CIRROSE.

Cirratulidae *Invertebrate Zoology.* a family of fringe worms in the Sedentaria; detritus burrowers and feeders in coastal waters.

Cirrhitidae *Vertebrate Zoology.* the hawkfishes, a family of shallow-water tropical marine fishes of the suborder Percoidei, characterized by voracious predatory habits and native to warm-temperate and tropical seas of the western Pacific and Indian Oceans.

cirrhosis *Medicine.* a liver disease in which parenchymal tissues die and the liver becomes filled with fibrous tissue; most commonly the result of alcohol abuse, but also caused by malnutrition or infection such as hepatitis.

cirri- a prefix used for *cirrus* in compounds such as *cirriform.*

cirriform *Meteorology.* of or relating to clouds composed of small particles, mostly ice crystals, that are widely dispersed, producing transparent or white halo phenomena. *Zoology.* having a tendril-like form.

Cirripedia *Invertebrate Zoology.* a subclass of Crustacea including the barnacles and goose barnacles; free-swimming larvae and sessile adults.

cirro- a prefix used for *cirrus* in compounds such as *cirrocumulus.*

cirrocumulus [sēr´ō kyoom´yə ləs] *Meteorology.* a principal cloud type composed of highly supercooled water droplets, ice crystals, or a mixture of both, appearing as a formation of small masses of white, fleecy clouds in a globular, wavelike pattern at heights of at least 35,000 feet.

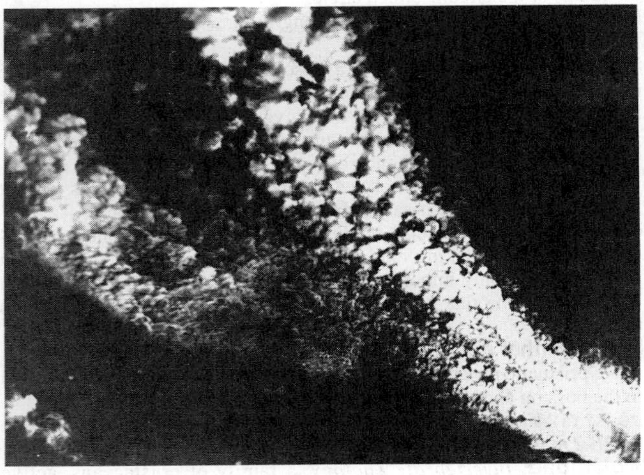

cirrocumulus clouds

Cirromorpha [sēr´ə môr´fə] *Invertebrate Zoology.* a suborder of octopuses belonging to the order Octopoda; highly evolved eight-tentacled mollusks.

cirrose *Biology.* **1.** having or resembling a long, slender appendage or filament. **2.** curly or wavy. Also, CIRRATE.

cirrostratus cloud [sēr´ō strat´əs] *Meteorology.* a principal cloud type that is composed of ice crystals and appears as a whitish veil, usually fibrous but occasionally smooth, that may cover the sky and often produces halo phenomena, generally occurring at heights of at least 35,000 feet.

cirrus [sēr´əs] *Meteorology.* a principal cloud type that is composed of ice crystals, appearing as thin, detached, featherlike white patches or narrow bands at heights of at least 35,000 feet; they are often illuminated by the sun when lower clouds are submerged in the earth's shadow. Also, **cirrus cloud.** *Vertebrate Zoology.* a hairlike tactile barbel on certain fishes. *Invertebrate Zoology.* **1.** any of the jointed thoracic appendages of barnacles. **2.** a hairlike tuft on insect appendages. **3.** the male copulatory organ in some mollusks and trematodes. **4.** the locomotory structure on some protozoa.

cirrus clouds

cirrus sac *Invertebrate Zoology.* a pouch that contains the cirrus (the male copulatory organ) in certain invertebrates.

cirrus spissatus cloud see FALSE CIRRUS CLOUD.

cis *Organic Chemistry.* a stereo descriptor referring to two groups on the same side of a double bond or of a ring.

cis- a prefix meaning "on the near side of," as in *cisalpine.*

***cis*-aconitic acid** *Biochemistry.* one of three organic acids that stimulate the oxidation of pyruvate by muscle tissue in animals.

***cis*-acting locus** *Genetics.* a genetic sequence that affects other genes on the same chromosome, usually due to positional relationships of the two loci.

cisalpine *Ecology.* of, relating to, or found on the southern side of the European Alps.

cisatlantic *Ecology.* of, relating to, or found on both sides of the Atlantic Ocean; specifically, in both Europe and North America.

CISC complex instruction set computer.

***cis* configuration** *Genetics.* a term used to designate the location of mutations in two genes on homologous chromosomes that determine the same phenotype. Two mutations occurring in two genes on the same chromosome are said to be in the *cis* configuration, as opposed to the *trans* position, which describes one mutation on each of the homologous chromosomes.

***cis* dominance** *Genetics.* the ability of a gene to affect the activity of other genes on the same chromosome.

cisele *Textiles.* a velvet fabric having a pattern of cut and uncut pile loops.

cislunar *Astronomy.* of or relating to the region of space inside the orbit of the Moon.

cissoid *Mathematics.* the graph of the cubic equation $x^3 + (x - a)y^2 = 0$; in polar form, $r = a \tan \theta \sin \theta$, with $a > 0$. Geometrically, consider a circle of radius $a/2$, centered at $(a/2, 0)$, and its tangent line $x = a$. A cissoid is the locus of points on a ray from the origin, moving from $-\pi/2$ to π/2, whose radial distance from the origin is the same as the distance along the ray from its second point of intersection with the circle to the tangent line. The curve has a cusp at the origin, is symmetric about the *x*-axis, and is asymptotic to the line $x = a$. Also, **cissoid of Diocles.**

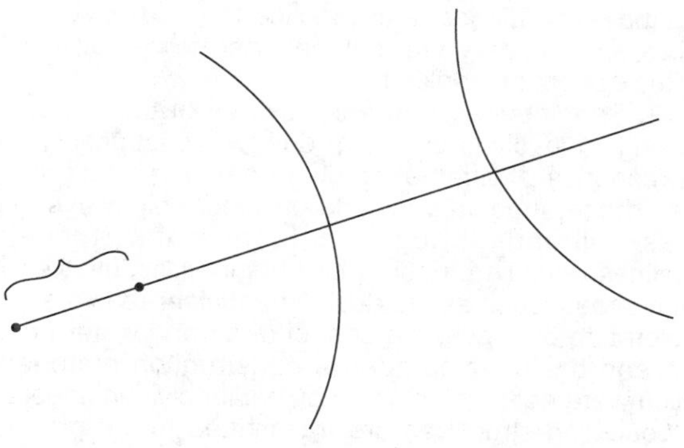

cissoid

***cis* structure** *Materials Science.* the molecular structure of certain polymers that denotes unsaturated positions on the same side of the polymer chain, as in natural rubber, in which two double-bonded carbon atoms have a CH_3 group on the same side of the chain.

Cistaceae *Botany.* a family of mainly Mediterranean dicotyledonous shrubs and herbs of the order Violales, characterized by showy flowers and hairs that are often clustered to appear stellate; includes many ornamental species and some with fragrant resins.

cistern *Anatomy.* a sac or cavity that serves as a reservoir for a natural body fluid. Also, CISTERNA. *Geology.* a natural or artificial reservoir or hollow in which water is stored.

cisterna *Anatomy.* see CISTERN.

cistern barometer *Engineering.* a barometer with a tube sealed at the top and open at the bottom that is immersed in a container of liquid.

cistern rock see LACCOLITH.

***cis-trans* isomerism** *Organic Chemistry.* of or relating to two stereoisomers that differ only in having certain atoms located on different sides of a specified plane, usually a double bond or a ring.

***cis-trans* test** *Molecular Biology.* a genetic test used to determine whether an individual mutation occurs within a single gene sequence or in distinct gene sequences on homologous chromosomes that are involved in determining the same phenotypic character.

cistron *Molecular Biology.* **1.** a part of the DNA that specifies one polypeptide chain in protein synthesis. In the theory of triple code, one cistron must have three times as many nucleotide pairs as amino acids in the chain it specifies. **2.** the nucleotide coding for a single polypeptide, excluding regulators and terminators; a structural gene.

cistron-transcription unit *Molecular Biology.* the section of the DNA molecule containing the base sequence that codes for the formation of a particular polypeptide chain.

cisvestism [siz´vest´iz əm] *Psychology.* the practice of dressing in the clothes of one's own sex, but in clothing inappropriate to one's age or status.

cisvestite [siz vest´īt] *Psychology.* a person who practices cisvestism.

Citeromyces *Mycology.* a genus of yeast fungi belonging to the family Saccharomycetaceae; its reproduction is characterized by multilateral budding of its cells.

Citheroniinae *Invertebrate Zoology.* a subfamily of moths including imperial moths and regal moths; lepidopteran insects in the family Saturniidae.

citizens' band *Telecommunications.* CB, a band of radio frequencies that is used exclusively by the citizens' radio service for signal transmission, as by motorists on the highway. Thus, **citizens'-band radio (CB radio).**

citizens' radio service *Telecommunications.* a communications service established for the use of limited-distance personal or business radio communication.

citizens' waveband see CITIZENS' BAND.

Citlaltepetl see ORIZABA.

Cit plasmid *Genetics.* an enterobacterial plasmid that encodes a citrate transport system.

citraconic acid *Organic Chemistry.* $COOHCH{=}C(CH_3)COOH$, hygroscopic crystals; soluble in water and alcohol; melts and decomposes at 90°C; used in protein chemistry for acylation of amino acids.

citral *Organic Chemistry.* $(CH_3)_2C{=}CHC_2H_4C(CH_3){=}CHCHO$, a mobile, pale-yellow liquid monoterpene found in the volatile oils of citrus plants; insoluble in water and soluble in alcohol; used as a food flavor and in perfumes.

citramalase *Enzymology.* an enzyme that catalyzes the conversion of citramalate to acetate and pyruvate.

citrate *Biochemistry.* a salt of citric acid, formed as a by-product of carbohydrate metabolism.

citrated plasma *Hematology.* blood plasma treated with sodium citrate that prevents clotting.

citrate lyase *Enzymology.* an enzyme found in cytoplasm that cleaves citrate into acetyl-*S*-CoA and oxaloacetate.

citrate synthase *Enzymology.* an enzyme that catalyzes the condensation of acetyl coenzyme A with oxaloacetic acid to produce citric acid and coenzyme A.

citrate test *Microbiology.* a standard metabolic test to determine the ability of a bacterial strain to utilize citrate as its sole carbon source.

citrene see LIMONENE.

citric acid *Biochemistry.* $C_6H_8O_7$, a tricarboxylic acid present in plants, particularly citrus fruits; commonly used as an antioxidant and flavoring agent in foods.

citric acid cycle see KREBS CYCLE.

citriculture *Botany.* the agriculture of citrus fruits.

citrine *Mineralogy.* a yellow variety of quartz. Also, **citrine quartz.**

citrine

citrinin *Mycology.* a yellow toxin produced by the fungal species *Penicillium citrinum* and a few species of *Aspergillus;* used as an antibacterial agent against Gram-positive bacteria.

Citrobacter *Bacteriology.* a genus of Gram-negative, rod-shaped bacteria of the family Enterobacteriaceae, typically motile by peritrichous flagellation; found in the intestines of humans and other vertebrates.

citron *Botany.* a thorny evergreen tree, *Citrus medica,* that is grown for its edible citrus fruit.

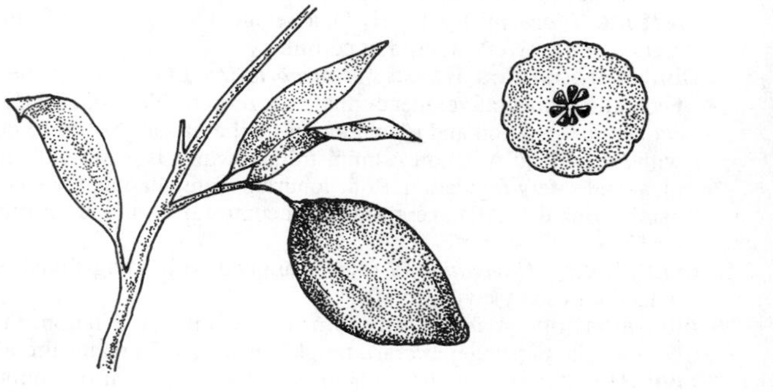

citron

citronella *Botany.* a tropical grass, *Cymbopogon nardus*, that is cultivated commercially as the source of **citronella oil,** a volatile oil used in soaps, perfumes, and insect repellents.

citronellal *Organic Chemistry.* $(CH_2)_2C{=}CH(CH_2)_2CH(CH_3)CH_2CHO$, a monounsaturated acyclic monoterpene aldehyde found in citronella oil; slightly soluble in water and soluble in alcohol; boils at 207.8°C.

citronellol *Organic Chemistry.* $(CH_3)_2C{=}CH(CH_2)_2CH(CH_3)CH_2CH_2OH$, an unsaturated acyclic monoterpene alcohol that forms the main constituent of geranium oil and the volatile oil of rose; an oily liquid that is slightly soluble in water and miscible with alcohol; boils at 224.4°C; used in perfumes.

citrulline *Biochemistry.* $C_6H_{13}O_3N_3$, an amino acid that is present in plants and formed in the liver of animals; an intermediate in the urea cycle.

citrus *Botany.* **1.** any spiny tree or shrub of the genus *Citrus* or related genera, such as the lemon, lime, orange, grapefruit, tangerine, and citron. **2.** the usually tart, sometimes sweet fruit of such trees, eaten or used to produced juices and flavorings.

citrus anthracnose *Plant Pathology.* a disease of citrus plants that is caused by the fungus *Colletotrichum gloeosporioides*. It is characterized by spotted or decayed leaves and fruit and by a withering of shoots and twigs.

citrus blast *Plant Pathology.* a citrus disease caused by the bacterium *Pseudomonas syringae* and characterized by a darkening and withering of foliage and black pitting of the fruit.

citrus canker *Plant Pathology.* a citrus disease caused by the bacterium *Xanthomonas citri* and characterized by oily, craterlike lesions on foliage, woody plant parts, and fruit.

citrus flavonoid compound see BIOFLAVONOID.

citrus scab *Plant Pathology.* a citrus disease caused by the fungus *Sphaceloma rosarum* and characterized by corky, projecting lesions or scabs on leaves, woody parts, and fruits.

city *Geography.* **1.** any large urban area. **2.** in the United States, an incorporated town. **3.** in Great Britain, a town that is the seat of a bishop or site of a cathedral.

city plan *Cartography.* a large-scale map of a city on which streets, important buildings, and significant urban features are shown.

civ civilian.

civet *Vertebrate Zoology.* **1.** any of several species of small, catlike nocturnal carnivores of the family Viverridae, found in Africa and southeast Asia and distinguished by their spotted and striped coats and their civet-secreting glands. Also, **civet cat. 2.** a brown or yellow musky-odored fatty substance secreted from a pouch near the sexual organs of Old World civet cats; used in perfumes. **3.** see CACOMISTLE.

civet

civet gland *Vertebrate Zoology.* a pouchlike scent gland near the external genitals of civet cats that secretes civet.

civetone *Biochemistry.* $C_{17}H_{30}O$, a ketone that is obtained from the secretion of the civet cat; used in perfumery.

Civil Aeronautics Board *Transportation Engineering.* the former Federal administrative agency that was responsible, until 1985, for the economic operation and regulation of civil aviation. Not to be confused with the Federal Aviation Administration, which is responsible for technical and safety regulation. Prior to airline deregulation, the Civil Aeronautics Board (CAB) exercised broad control over air carrier routes and tariffs.

civil airway *Navigation.* a route equipped with navigational aids, intended for use by civil aircraft.

civil aviation *Aviation.* any form of nonmilitary aviation, including commercial passenger aircraft, freight carriers, and private aircraft.

civil day *Astronomy.* the 24-hour period lasting from one midnight to the next.

Civil Engineering

Civil engineering is a profession applying the knowledge of mathematics and sciences gained by study, experience, and practice to develop ways to utilize economically the materials and forces of nature for the benefit of mankind.

Civil engineering is a very diverse branch of engineering. It involves planning, design, construction and maintenance of a large variety of structures and facilities. These structures include buildings, highways, airports, railroads, bridges, harbors, dams, tunnels, pipelines, etc. The modern civil engineer is involved in many areas, such as the design of nuclear reactors, the exploration of space, the control of air and water pollution and the development of construction materials, which were not much studied by earlier civil engineers.

Loads on structures are transmitted to the ground. The load-bearing qualities and the deformation characteristics of the ground play an important role in the design of many structures. This makes soil mechanics and foundation engineering important in many structural designs. In cities with rapidly expanding populations, rapid-transit systems for commuter traffic, improved highways, adequate water supply systems, improved drainage, and public health schemes are essential.

Air and water pollution has grown rapidly in the last decade. Environmental engineering, dealing with the control of pollution, has become an important branch of civil engineering. Many catastrophic failures occur during earthquakes. Better design and construction methods and better building materials are needed for structures subject to earthquake loadings. These give a few areas where researchers and developments are needed.

Tung Hua Lin
Professor of Civil Engineering, Emeritus
University of California, Los Angeles

civil engineering *Engineering.* the conception, research, design, construction, operation, and maintenance of various facilities and systems including buildings, bridges, highways, airports, water treatment plants, dams, and reservoirs.

civilization *Anthropology.* in the 19th-century evolutionary view of culture, the highest stage of social development, following savagery and barbarism.

civil time *Astronomy.* ordinary clock time, whether standard or daylight saving. *Horology.* a system of time measurement based on solar time, with a mean solar day beginning at midnight.

civil twilight *Astronomy.* the interval between sunset and the time at which the center of the sun lies 6° below the horizon.

CK creatine kinase.

Cl the chemical symbol for chlorine.

cl or **cl.** centiliter; class.

clack valve see CHECK VALVE.

cladding *Engineering.* any of various processes in which two materials are bonded together under high pressure and heat. Also, BONDING. *Nucleonics.* **1.** in nuclear reactors, a thin barrier of zirconium or stainless steel encasing fuel pins, or ceramic coating bonded to fuel pellets, to prevent corrosion of the fuel and contamination of coolant by fission fragments and fission-product gases. Also, CAN, CARTRIDGE, JACKET. **2.** the process of placing such a casing around the fuel pin or pellet before insertion into a reactor. Also, CANNING. **3.** in nuclear weapons, a plating of noble metals on fissionable weapon components to protect them from environmental corrosion and to reduce neutron exposure and accidental criticality during handling.

clade *Systematics.* a branch of the evolutionary tree; a taxon made up of an ancestor and all of its descendants.
Cladistia *Vertebrate Zoology.* the equivalent name for Polypterini, an order of bony fishes.
cladistic *Systematics.* of or relating to a clade or clades.
cladistic affinity *Evolution.* the degree of recency of common ancestry.
cladistics *Systematics.* a method of classification in which phylogenetic hypotheses are the basis for classification and the recency of common ancestry is the sole criterion for grouping taxa. *Evolution.* a method of reconstructing phylogeny that is based on the identification of derived character states. Also, **cladism**, **cladistic method, cladistic analysis**.
clad metal *Metallurgy.* a composite material, such as alclad, consisting of two or more layers bonded together.
Cladocera *Invertebrate Zoology.* an order of water fleas, small freshwater crustaceans with a transparent bivalve shell.
Cladochytriaceae *Mycology.* a family of fungi belonging to the order Chytridiales that occurs on land and in water; some species live off decaying vegetable matter, while others are parasites of microscopic animals and freshwater algae.
Cladocopa *Invertebrate Zoology.* a suborder of marine bivalves in the order Myodocopida. Also, **Cladocopina.**
cladode see CLADOPHYLL.
cladodont *Paleontology.* of or relating to the early sharks formerly known as Cladodontiformes but now generally classified in separate orders; characterized by a long, slender body, a spineless dorsal fin, and **cladodont dentition** (i.e., broad-based teeth with a single large cusp and smaller lateral cusps); Devonian and Carboniferous.
cladogenesis *Evolution.* an evolutionary change in which an ancestral lineage branches and subsequently diverges, resulting in two new lines and the loss of the ancestral line.
cladogram *Systematics.* a branching diagram showing the points of divergence or speciation from a common ancestor, and the degree of relation among species based on shared characters derived from that common ancestor.
Cladophoraceae *Botany.* a family of green algae belonging to the order Cladophorales, characterized by simple or branched filamentous thalli, often having rhizoids attached to basal cells.

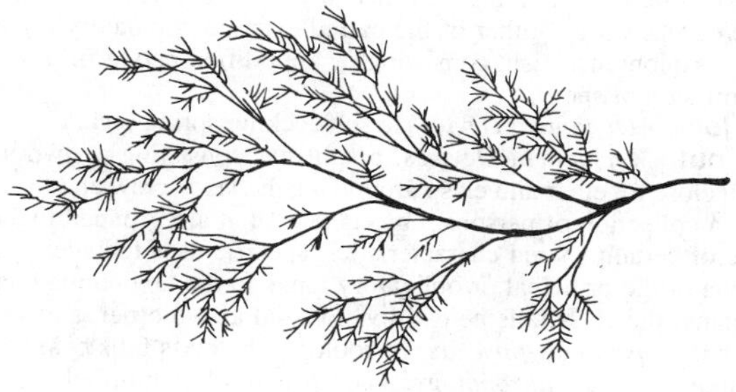

Cladophoraceae

Cladophorales *Botany.* an order of freshwater and marine algae of the class Chlorophyceae having a multinucleate thallus, chloroplasts, and simple or branched filaments; habitats range from the backs of freshwater snails, turtles, and shrimps to free flotation in fresh and marine waters.
cladophyll *Botany.* a flattened branch or stem having the function and often the appearance of a leaf. Also, CLADODE.
cladoptosis *Botany.* the shedding of branches and stems by abscission.
Cladopyxidaceae *Botany.* a family of photosynthetic marine dinoflagellates of the order Peridiniales, characterized by a spherical or slightly elongated cell body.
Cladoselachii *Paleontology.* the earliest sharks, an order of primitive elasmobranchs; characterized by large pectoral fins, dorsal fins with spines, and cladodont dentition; upper Devonian and lower Carboniferous. Also, **Cladoselachida.**
cladosporiosis *Pathology.* a common term for an infection with a cladosporium fungus, which causes rough skin, black lesions on the hands, and sometimes a brain abscess.
Cladosporium *Mycology.* a genus of fungi belonging to the class Hyphomycetes that includes species pathogenic to plants and animals.
Cladosporium resinae see HORMOCONIS RESINAE.
Cladostephaceae *Botany.* a monogeneric family of brown algae of the order Sphacelariales, characteristically reaching 25 cm in height and having a stiff, bushy, densely branched axis arising from a discoid holdfast; distributed throughout the North Atlantic and in Australia and New Zealand.
Claibornian *Geology.* a North American geologic stage of the Middle Eocene epoch, occurring after the Wilcoxian and before the Jacksonian.
Clairaut, Alexis Claude 1713–1765, French mathematician and astronomer; formulated Clairaut's theorem and differential equation.
Clairaut's differential equation *Mathematics.* a first-order differential equation of the form $y - xy' + f(y')$ for some function f.
Clairaut's theorem *Geodesy.* a theorem that relates the value of the earth's centrifugal force at the equator to the value of gravity at the equator; useful in determining the flattening of the earth.
clairvoyance *Psychology.* the power to acquire knowledge of objects and events beyond the ordinary perception of the physical senses.
clairvoyant *Psychology.* **1.** a person who is capable of clairvoyance. **2.** of or relating to clairvoyance.
Claisen condensation *Organic Chemistry.* a condensation reaction in which aldehydes or ketones react with esters in the presence of a base such as metallic sodium or sodium ethylate to form unsaturated esters. Also, **Claisen reaction.**
Claisen flask *Chemistry.* a glass flask having a branched neck, used in distillation.
Claisen-Schmidt condensation *Organic Chemistry.* a condensation of an aromatic aldehyde with an aliphatic aldehyde or ketone in the presence of a relatively strong base to form an α,β-unsaturated aldehyde or ketone.

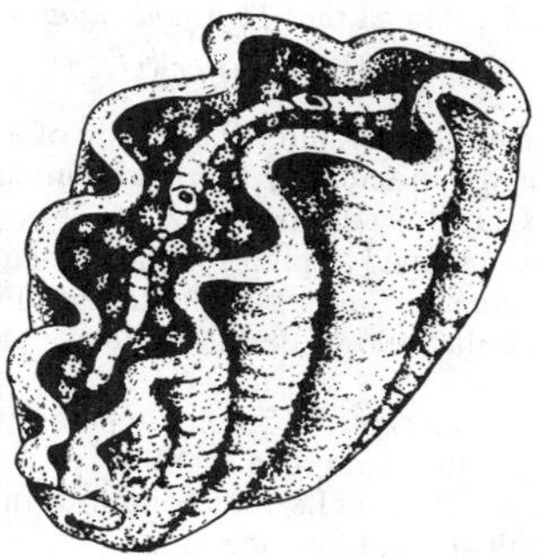

clam

clam *Invertebrate Zoology.* the common name for any of numerous species of bivalve mollusks, many of which are used as food.
Clambidae *Invertebrate Zoology.* a family of minute beetles within the superfamily Dascilloidea.
clammy *Biology.* being sticky, moist, and usually cool, such as the skin of a person in shock.
clamp *Mechanical Devices.* a usually rigid device designed to hold work in place using either opposing fixed and adjustable jaws or a slotted steel strap that can be tightened or loosened as needed. *Surgery.* any of various surgical instruments used for effecting compression.
clamp connection *Mycology.* in fungi belonging to the subdivision Basidiomycotina, a hollow growth protruding from hyphal cell walls of the hyphal network or mycelium and covering the cell walls like a clamp. Also, **clamp, clamp cell, buckle.**
clamp forceps *Surgery.* a forceps having an automatic lock; used, e.g., to compress an artery or the pedicle of a tumor. Also, PEDICLE CLAMP.
clamping the process of using a clamp; specific uses include: *Electronics.* **1.** a process that sets the operation of a device at an exact direct current level. **2.** a process that sets the level for a television picture signal each time the electron beam begins its synchronized movements across the screen. *Robotics.* the gripping, grasping, and releasing of objects on command by a robot's end effector.
clamping circuit *Electronics.* a circuit in which the voltage or current is held at a fixed level; used in a television receiver to restore the direct current component and the light value of the reproduced image.
clamping coupling *Mechanical Engineering.* a metal connecting piece that also acts as a clamp to hold two shaft ends together.

clamping diode *Electronics.* a diode that is added to a circuit to hold the voltage at a predetermined level.
clamping gripper *Robotics.* the most common type of end effector for handling components, using two-link motion, parallel-jaw motion, and combinations thereof.
clamping plate *Engineering.* a plate that attaches a mold to a machine.
clamping pressure *Engineering.* the amount of externally applied pressure necessary to prevent the fluid pressure in plastic from forcing open the mold in which it is being shaped.
clamp-on *Telecommunications.* describing a type of call-waiting system in which the caller is notified when the line is cleared.
clamp screw *Mechanical Devices.* a screw that is tightened to two parts or workpieces oriented against each other.
clamshell bucket *Mechanical Engineering.* a shovel bucket with two jaws that clamp together by their own weight when lifted by the closing line. Also, **clamshell grab.**
clamshell snapper *Mechanical Engineering.* an underwater excavating machine with a hinged-jaw bucket that automatically opens upon striking the bottom, then snaps shut on sediment when the closing line is lifted.
clam worm *Invertebrate Zoology.* the common name for any of several species of flattened polychaete annelid worms in the family Nereidae that are commonly used as bait.
clan *Anthropology.* a kin group that traces its descent from a single ancestor.
clandestine operations *Military Science.* intelligence, counterintelligence, or similar activities conducted in a manner that preserves the secrecy of the operation. Also, **clandestine ops.**
clapboard *Materials.* a long, narrow board with one edge thicker than the other edge for use as siding on a building; the thick edge of one board is laid over the thin edge of another board.
Clapeyron equation [klə pä´rän] *Thermodynamics.* an equation,

$$(dp/dT) = \Delta H/(T\Delta V),$$

that gives the pressure-temperature relationship of a phase transition, the pressure p, the change in enthalpy ΔH, and the change in volume ΔV. Also, **Clapeyron-Clausius equation.**
clapper *Electricity.* a hinged or pivoted relay armature.
clapper box *Mechanical Engineering.* a toolhead carried on a planing or shaping machine that allows the tool to clear the work's surface on the return stroke.
Clapp oscillator *Electronics.* a Colpitts-type oscillator that uses a series-resonant tank circuit to improve stability.
Clara cells *Cell Biology.* cells that contain a metabolic system for detoxifying contaminants entering the lungs.
clarain *Geology.* a coal lithotype characterized by fine banding, a semibright silky luster, and a shell-like irregular fracture. Also, **clarite.**
Clarendonian *Geology.* a North American geologic stage of the Middle Miocene epoch, after the Barstovian and before the Hemphillian.
claret [klâr´it] *Food Technology.* a term for red wine from Bordeaux or in the Bordeaux style, especially in British use.
clarification of feeling(s) *Psychology.* a therapeutic approach in which the therapist defines or summarizes the emotions that are reflected in the patient's remarks, without passing judgment.
clarifier *Telecommunications.* on an ssb transceiver, a control that enables frequency adjustment so that the frequencies of the received signal will be the same as those of the modulating signal.
clarify *Engineering.* to clear a liquid of suspended particles through filtration, centrifugation, or the addition of an enzyme.
clarifying agent *Food Technology.* any of several chemical substances used in removing cloudiness from edible liquids.
clarifying centrifuge *Mechanical Engineering.* a device that uses centrifugal force to clear liquids of solid materials or of suspended droplets of a second liquid phase.
clarifying filter *Engineering.* any filter that frees a liquid of particles and impurities, and may also remove color from the liquid.
Clariidae *Vertebrate Zoology.* the walking catfishes, a family of Asian and African catfishes in the suborder Siluroidei that are noted for their ability to respire atmospheric oxygen and even spend short periods of time on land.
clarity *Chemical Engineering.* a measure of the quantity of suspended opaque solids in a liquid, as determined by visual methods or by turbidity tests.
Clark cell *Electricity.* an early form of standard cell that had a voltage of 1.433 volts at 15°C.
Clark degree see ENGLISH DEGREE.
Clarke, Alexander 1828–1914, English geodesist; accurately determined figure of the earth; devised Clarke ellipsoids and spheroids.
Clarke, David 1938–1976, British archaeologist; founder of analytical archaeology.
clarke *Geochemistry.* a measure of the average abundance of an element in the earth's crust, expressed as a percentage or as parts per million. (Named for F. W. *Clarke,* American geochemist.)
Clarkecaris *Paleontology.* an extinct genus of malacostracan crustaceans in the order Anaspidacea; Permian.
Clarke ellipsoid of 1866 *Geodesy.* the reference ellipsoid used for charting geodetic surveys in North and Central America.
clarkeite *Mineralogy.* $(Na,Ca,Pb)_2U_2(O,OH)_7$, a dark brown, waxy, translucent mineral of unknown crystal structure occurring in massive, microcrystalline form, having a specific gravity of 6.29 to 6.39 and a hardness of 4 to 4.5 on the Mohs scale; found as an alteration product or uraninite in pegmatites.
Clarke spheroid of 1866 *Geodesy.* a reference ellipsoid having a semimajor axis of 6,378,206.4 meters, a semiminor axis of 6,356,583.8 meters, and a flattening of 1/293.46.
Clarke spheroid of 1880 *Geodesy.* a reference ellipsoid having a semimajor axis of 6,378,249.1 meters and a flattening of 1/293.46.
Clarkforkian *Geology.* a North American geologic stage of the Upper Paleocene epoch, occurring after the Tiffanian and before the Wasatchian.
Clark process *Chemical Engineering.* a process to soften water by the addition of alkaline solutions; calcium hydroxide is added to water to convert acid carbonates into normal carbonates.
clarodurain *Geology.* a transitional coal lithotype containing vitrinite, but characterized more by the presence of other macerals, such as micrinite and exinite.
clarofusain *Geology.* a transitional coal lithotype characterized by the presence of vitrinite and fusinite, together with other macerals.
clarotelain *Geology.* a transitional coal lithotype characterized by the presence of telenite, together with small amounts of other macerals.
clarovitrain *Geology.* a transitional coal lithotype characterized mainly by the presence of vitrinite, with smaller amounts of other macerals.
clasp *Mechanical Devices.* a metal device used to hold or fasten together two or more objects while allowing for an easy release.
clasper *Botany.* a twining tendril that helps support a climbing plant. *Vertebrate Zoology.* either of the paired external copulatory organs of a male elasmobranch fish, consisting of a modified pelvic fin used in the transmission of sperm.
clasp lock *Mechanical Devices.* a self-locking spring lock.
clasp nut *Mechanical Devices.* a split nut consisting of two sections which close together and clasp around a lathe lead screw and the saddle.
class a collection of persons, objects, etc., that are grouped together by virtue of certain shared characteristics; specific uses include: *Systematics.* one of the principal or obligatory ranks in the taxonomic hierarchy: in botany, the class falls below division and above order, and generally ends with *-opsida* or *-phyceae;* in zoology, the class falls below phylum and above order. *Anthropology.* a component of a stratified society that is determined by economic level. *Computer Science.* in object-oriented programming, a description of a set of similar objects; for example, Fido, an object, might be a member of the class Dogs. *Artificial Intelligence.* in a frame system or in object-oriented programming, a representation of data and methods that are shared by individual instances of the group and its subgroups. *Mathematics.* **1.** one of the three primitive (undefined) notions in the Gödel-Bernays formulation of set theory (the other two are membership and equality). Intuitively, a class is a collection of objects defined by any property that each one of the objects does or does not have. A class A is a set if and only if there exists a class B such that $A \in B$; otherwise A is a **proper class.** The axiom of class formation asserts that for any statement $P(y)$ in the first-order predicate calculus involving a variable y, there exists a class C (denoted by $C = \{x : P(x)\}$) such that $x \in C$ if and only if the statement $P(x)$ is true. **2.** a particular subset of an algebraic object, defined by some algebraic property; such as an **equivalence class,** a **quotient class,** or a **residue class.** **3.** see SET. **4.** a term for the degree of differentiability of a function; that is, if the kth partial derivatives of a function $f: R^n \to R^m$ exist and are continuous at a point c (or at every point of D) in the interior of some domain D (or at every point of D), then f is said to belong to class C^k at c (or on D). **5.** the class of an algebraic curve in the plane is the greatest number of tangents that can be drawn to it from any point in the plane not on the curve.

class A amplifier *Electronics.* an amplifier designed so that plate current flows at all times.
class AB amplifier *Electronics.* an amplifier designed so that current flows over more than half the electrical cycle, but not through the complete cycle.
class A modulator *Electronics.* a class A amplifier used specifically to supply the necessary signal power to modulate a carrier.
class B amplifier *Electronics.* an amplifier designed so that current flows only when voltage is applied and for only half the electrical cycle.
class B modulator *Electronics.* a class B amplifier used specifically to supply the necessary signal power to modulate a carrier.
class C amplifier *Electronics.* an amplifier designed so that current flows only when voltage is applied and for appreciably less than half of the electrical cycle.
class equation *Mathematics.* an equation expressing the order of a finite group G as the sum of the indices in G of the centralizers of $\{x_i\}$, where $\{x_i\}$ is a set of representatives of the distinct conjugacy classes of G; that is, $|G| = \sum_{i=1}^{n} [G : C_G(x_i)]$.
classic or **Classic** *Archaeology.* **1.** of or relating to the period of time when a culture or civilization reaches its highest point of complexity and achievement. **2.** of or relating to the civilizations of ancient Greece and Rome. **3.** in New World chronology, of or relating to the period between the Formative (Pre-Classic) and the Post-Classic; characterized by the emergence of city-states. *Architecture.* see CLASSICAL.
classical or **Classical** *Architecture.* **1.** relating to or based on the architectural style of ancient Greece and Rome; used especially to refer to public buildings, monuments, and the like of later eras that follow this style. **2.** of a style or design, simple, graceful, and well-proportioned. *Mechanics.* of or relating to Newton's laws of motion, or to applications and measurements that satisfy these laws.
classical conditioning *Behavior.* the teaching of a specific behavior by pairing a neutral stimulus with a stimulus that elicits a given response; after repeated presentations, the neutral stimulus comes to evoke the same response. For example, an animal hears a bell ring and then salivates at the sight of food; eventually the sound of the bell alone will cause salivation. Also, PAVLOVIAN CONDITIONING.
classical conductivity theory see CONDUCTIVITY THEORY.
classical dynamics see CLASSICAL MECHANICS.
classical electron radius *Electromagnetism.* a quantity expressed as $e^2/m_e c^2$, where e is the charge of the electron (in electrostatic units), m_e is the electron rest mass, and c is the speed of light; approximately equal to 2.82×10^{-13} centimeters.
classical field theory *Physics.* the study of distributions of energy, matter, and other physical quantities under conditions in which they may be regarded as continuous functions of position. Also, C-NUMBER THEORY, CONTINUUM MECHANICS.
classical groups *Mathematics.* the following families of matrix groups: general linear groups, orthogonal groups, unitary groups, and symplectic groups. With the special linear groups replacing the general linear groups, these groups also form the principal series of semisimple Lie groups.
classical mechanics *Mechanics.* a term for those aspects of mechanics based on Newton's laws of motion, in the realm where the speeds involved are small compared to the speed of light, and where the sizes of objects involved are large compared to their deBroglie wavelengths. Also, CLASSICAL DYNAMICS.
classical physics *Physics.* the discipline of physics from its early years, based on Newton's laws of motion and excluding later studies of relativity and quantum mechanics.
classical statistics *Statistics.* a term designating several different approaches to inference based on sampling properties of statistical procedures, as opposed to bayesian statistics and exploratory data analysis.
classical wave equation see WAVE EQUATION.
classic epidemic typhus *Medicine.* a severe type of typhus carried to humans by body lice.
classicism *Architecture.* principles based on classical forms.
classification the process of dividing things into classes, or the divisions arrived at by such a process; specific uses include: *Engineering.* the process of grading and sorting of particles by size, shape, or density. *Statistics.* the assignment of observations to groups; used in prediction, as in medical diagnosis, and description, as in botany. *Military Science.* **1.** the determination that official information or activities require a specific level of protection against unauthorized disclosure in order to protect the interests of national security. **2.** the designation of such determination. *Archaeology.* the process of ordering objects into classes or groups that share certain characteristics; the first step in the analysis of archaeological data.
classification paradigm *Artificial Intelligence.* the process of selecting the best alternative from a set of options.
classificatory system *Anthropology.* a kinship system in which merging of kin terms occurs, and the same term is given to a number of different persons; e.g., a mother's sister is also called "mother."
classifier *Mechanical Engineering.* any of a wide variety of devices and systems that are used to separate mixtures into constituent parts by size and/or density.
classify *Science.* to categorize into classes or other groups based on specified features.
class inclusion *Psychology.* the ability to identify an object as one example of a larger group or set, rather than to regard it simply as a unique individual item; a stage in childhood cognitive development. Also, **classification.**
class interval *Statistics.* a section of the range of a random variable into which individual observations of the random variable are grouped.
class mark *Statistics.* the value at the center of a class, found by averaging the upper and lower class limits.
class NP-complete problems *Computer Programming.* a group of computational problems, known as **nondeterministic polynomial complete problems,** for which no effective computer algorithms have been developed; the only known approach to solution requires an amount of computational time proportional to an exponential function of the size of the problem, or to a polynomial function on a nondeterministic computer that, in effect, guesses the correct answer.
classon *Particle Physics.* a massless boson that bears the quantum numbers of the two classical fields, gravitation and electromagnetism.
class switching *Immunology.* the transferring of a B lymphocyte from the synthesis of one antibody class to another.
class variable *Computer Science.* in an object-oriented language, a variable associated with a class of objects, such as the number of members of the class.
clast *Geology.* a constituent element of a clastic rock formation, such as a piece of sediment, silt, sand, or gravel.
-clast a combining form meaning "something that breaks or destroys."
clastation see WEATHERING.
clastic *Geology.* describing a rock or sediment composed mainly of broken fragments of preexisting minerals or rocks that have been transported from their places of origin. (From an ancient Greek word meaning "broken.")
clastic deformation *Geology.* a process of dynamothermal metamorphism involving the fracture, rupture, rolling out, and pulverization of mineral and rock particles.
clastic dike *Geology.* a tabular mass of clastic material that is derived from underlying and overlying beds, and that cuts across the structure or bedding of the preexisting, surrounding rock.
clastic pipe *Geology.* an almost vertically standing, irregular cylindrical or columnar mass of clastic material.
clastic ratio *Geology.* the ratio of the thickness of clastic material to the thickness of nonclastic material in a stratigraphic unit. Also, DETRITAL RATIO.
clastic reservoir *Geology.* an underground oil or gas trap formed in clastic rock, such as limestone.
clastic rock *Geology.* a sedimentary rock composed mainly of fragments that broke off from preexisting rock or that formed as a result of the chemical weathering of preexisting rock and were transported to their present site of deposition.
clastic sediment *Geology.* sediment formed by the deposition of clastic materials that have been transported by mechanical agents, such as water, wind, and gravity. Also, MECHANICAL SEDIMENT.
clastic wedge *Geology.* the clastic sediments of a geosyncline within a craton that are derived from the tectonic landmasses of the adjacent orthogeosynclinal belt.
clastogenic *Biology.* giving rise to or inducing a breakage or disruption, e.g., of chromosomes.
Clathraceae *Mycology.* the stinkhorns, a family of fungi belonging to the order Phallales.
clathrate *Biology.* like a lattice in form; having pores, holes, or cavities. *Chemistry.* a substance in which molecules of one compound are enclosed in cavities within another compound. *Petrology.* a net, web, or spongelike condition in leucite rock due to clear leucite crystals surrounded tangentially by crystals of augite. Also, ENCLOSURE COMPOUND.
clathrate compound see CAGE COMPOUND.

clathrin *Cell Biology.* a fibrous protein that forms a characteristic basketlike meshwork, or coat, surrounding coated pits and vesicles that are involved in endocytosis and exocytosis by eukaryotic cells.

clathrin-coated pit *Cell Biology.* a depression of the cell membrane that is coated with a network of clathrin protein on its cytoplasmic side and is the site of receptor-mediated endocytosis.

Clathrinida *Invertebrate Zoology.* an order of simple sponges in the subclass Calcinea.

Clathrinidae *Invertebrate Zoology.* a family of sponges in the order Clathrinida.

Clathrochloris *Bacteriology.* a genus of photosynthetic green bacteria of the family Chlorobiaceae, occurring in sulfide-rich freshwater or estuarine habitats.

Clathrulina *Invertebrate Zoology.* a genus of stalked, spheroidal amoeboid protozoans in the order Heliozoa, with the main body mass contained in a siliceous skeleton.

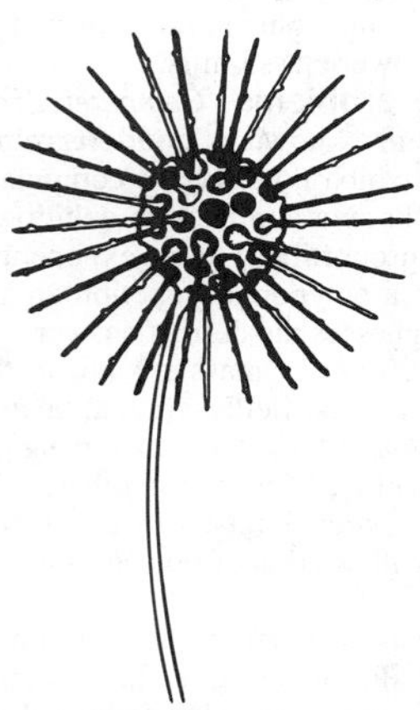

Clathrulina

Claude, Albert 1898–1983, Belgian-born American cytologist; Nobel Prize for the study of cell components with centrifuge and electron microscope.

Claude, Georges 1870–1960, French engineer; developed Claude process of liquefying air.

Claude process or **Claude method** *Physics.* a process for liquefying air and other gases, in which a gas is used to move a piston, and then moves to an expansion chamber where it cools by adiabatic expansion. *Chemical Engineering.* a process for ammonia synthesis that uses high operating pressures and a train of converters.

claudetite *Mineralogy.* As_2O_3, a coloress to white, transparent monoclinic mineral occurring as commonly twinned tabular crystals, having a specific gravity of 4.15 and a hardness of 2.5 on the Mohs scale; found as a secondary mineral formed by the oxidation of arsenic minerals.

clause *Computer Programming.* a set of consecutive COBOL words used to specify an attribute of an entry; types include data, environment, and file clauses. *Artificial Intelligence.* in resolution theorem proving, a logical formula consisting of a disjunction of literals treated as a unit.

Clausius, Rudolf 1822–1888, German physicist; a founder of thermodynamics; major advances in electrolysis and kinetic theory of gases.

clausius *Thermodynamics.* a unit of entropy increment for a closed system that receives 1000 international table calories (4186.8 joules) of heat transfer at a temperature of 1 K.

Clausius-Clapeyron equation see CLAPEYRON EQUATION.

Clausius equation *Thermodynamics.* an equation of state for gases that provides for an improvement in the van der Waals equation.

Clausius inequality *Thermodynamics.* the statement that for a cyclical thermodynamic process, the integral over a complete cycle of the amount of heat transferred to the system, divided by the absolute temperature, must be equal to or less than zero.

Clausius law *Thermodynamics.* the law that the specific heat of an ideal gas at constant volume is independent of temperature.

Clausius-Mosotti equation *Physics.* an equation representing the polarization γ of an individual molecule located in a medium having the relative dielectric constant ε and N molecules per unit volume: $\gamma = (3/4\pi N)[(\varepsilon - 1)/(\varepsilon + 2)]$.

Clausius number *Thermodynamics.* a dimensionless number used in the study of heat conduction in forced-fluid flow.

Clausius range *Physics.* a state in which the mean free path of a gas is smaller than the dimensions of its container.

Clausius statement *Thermodynamics.* an expression of the second law of thermodynamics, stating that it is not possible to construct a device to produce a heat transfer from a colder body to a hotter body without the occurrence of other changes in the surrounding environment; i.e., work must be done on the device by some outside agent.

Clausius theorem see CLAUSIUS INEQUALITY.

Clausius virial theorem *Physics.* a theorem stating that the total kinetic energy of a system of particles whose positions and velocities are bounded, averaged over a long period of time, will equal the virial of the system. Also, VIRIAL THEOREM.

Claus method *Chemical Engineering.* an industrial process that produces sulfur by a partial oxidation of gaseous hydrogen sulfide in air.

clausthalite *Mineralogy.* PbSe, a gray, opaque, metallic cubic mineral occurring in massive, sometimes foliated form; having a specific gravity of 7.8 to 8.22 and a hardness of 2.5 to 3 on the Mohs scale; found in carbonate veins in basalt and in uranium and vanadium deposits.

claustrophilia *Psychology.* an irrational desire to be enclosed in a small or confined space. Thus, **claustrophile.**

claustrophobe *Psychology.* a person who suffers from claustrophobia.

claustrophobia [klôs´trə fō´bē ə] *Psychology.* an irrational fear of being enclosed; the fear of small or confined spaces such as elevators, locked rooms, or tunnels. Also, CLEITHROPHOBIA.

claustrophobic [klôs´trə fō´bik] *Psychology.* **1.** of or relating to claustrophobia. **2.** tending to produce claustrophobia.

claustrum *Anatomy.* a thin layer of gray matter located between the external capsule and the island of Reil, and having reciprocal connections with the sensory areas of the cerebral cortex.

Claustulaceae *Mycology.* a family of fungi whose classification is in transition; commonly classified under the order Phallales and containing the single genus *Claustula,* which occurs in New Zealand.

clava *Biology.* a club-shaped structure, such as that on the end of the antennae of some insects.

clavaria see APHYLLOPHORALES.

Clavariaceae *Mycology.* a family of fungi belonging to the order Agaricales, composed of plant pathogens that cause such diseases as snowmold and root rot.

clavate *Biology.* shaped like a club; thicker at one end like a club. Also, CLAVIFORM.

Clavatoraceae *Paleontology.* an extinct family of algae known by their charophytes; several of the Clavatoraceae are important as index fossils for the period between the late Jurassic and the latest Cretaceous; detailed evolutionary studies have been made of several genera of the lower Cretaceous, showing gradual changes between very different forms at both extremes.

Clavaxinellida *Invertebrate Zoology.* an order of sponges in the class Demospongiae, with minute four-pointed spicules.

Clavibacter *Bacteriology.* a genus of Gram-positive, nonmotile, obligately aerobic bacteria that are parasitic in plants.

Claviceps *Mycology.* a genus of fungi belonging to the order Clavicipitales that are parasitic to grasses and cause plant diseases such as ergot, a rye disease.

Clavicipitaceae *Mycology.* the only family of fungi belonging to the order Clavicipitales, consisting of parasites on grass, insects, spiders, and other fungi.

Clavicipitales *Mycology.* an order of usually parasitic fungi belonging to the class Pyrenomycetes that are parasitic to other fungi, grasses, and insects.

clavicle *Anatomy.* a bone connecting the breastbone with the shoulder blade; the collarbone.

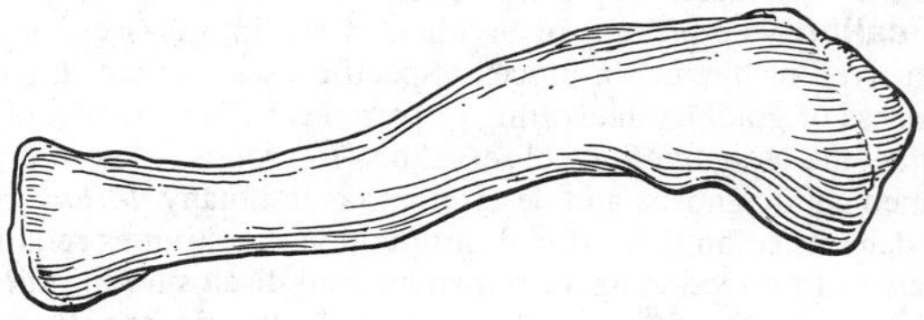

clavicle

claviculate [klə vik´yə lāt´] *Anatomy.* of or relating to the clavicle.

claviform see CLAVATE.

Clavispora *Mycology.* a genus of yeast fungi belonging to the family Saccharomycetes; its reproduction is characterized by the multilateral budding of its cells.

clavus *Invertebrate Zoology.* any rounded or fingerlike process, such as the club of an insect's antenna.

claw *Vertebrate Zoology.* a sharp, hooked nail in a vertebrate's foot. *Invertebrate Zoology.* a sharp, curved process on the tip of a limb of an insect. *Mechanical Devices.* a tool consisting of two curved prongs used to remove nails or spikes.

claw clutch *Mechanical Engineering.* a clutch consisting of two working parts that interlock when pushed together.

claw coupling *Mechanical Engineering.* a shaft coupling for instant connection or disconnection in which flanges carried by each shaft engage through teeth in corresponding recesses in their opposing faces.

clay *Geology.* **1.** a rock or mineral fragment of extremely small size, usually defined as having a diameter of less than 0.0039 millimeter, or 0.00156 inch. **2.** a soft, earthy, very fine-grained natural sediment or aggregate that is composed mainly of particles of this size, made up of hydrous silicates of aluminum mixed with various impurities. It is highly plastic and moldable when mixed with water, retains its shape on drying, and becomes rocklike and permanently hard on heating or firing. *Materials.* this material in the form of pottery, tiles, or brick. *Agronomy.* agricultural soil that consists mainly of particles of this material. *Engineering.* any plastic material consisting of particles having a diameter less than 0.074 millimeter.

clay atmometer *Engineering.* an atmometer having a porcelain container attached to a receptacle of purified water, and that measures evaporation based on the amount of water depleted from the receptacle.

clay bit *Engineering.* **1.** a bit used on clay barrels. **2.** see MUD AUGER.

clay digger *Mechanical Engineering.* a hand-held power-driven spade used for digging clay, hard soil, or soft rock.

C layer *Geology.* in the classification of the earth's interior, the seismic region equivalent to the transition zone of the upper mantle, extending from 410 to 1000 km below the surface.

clayey see ARGILLACEOUS.

clay gall *Geology.* a small, flattened, curled fragment, chip, or flake of clay that is derived from dried, cracked mud and has become embedded in sand.

clay gouge see FAULT GOUGE.

claying *Food Technology.* an ancient method of refining sugar by pouring a mixture of clay and water through a molded cone of crystallized sugar syrup.

clay ironstone *Petrology.* **1.** a clayey sedimentary rock containing large quantities of iron carbonate or iron oxide in layers of nodules or irregular beds. **2.** a clayey-looking stone that is an impure form of chalybite within carbonaceous strata, generally overlying coal seams; it contains 20–30% iron.

clay loam *Geology.* a fine-textured soil that breaks into hard lumps when dry and is plastic when moist, composed of 27–40% clay, 20–45% sand, and the remainder silt.

clay marl *Geology.* a whitish, smooth, chalky, unconsolidated deposit that is predominantly composed of clay.

clay mineral *Mineralogy.* **1.** any of a complex group of finely crystalline to amorphous, essentially hydrous aluminum silicate minerals formed by the alteration or weathering of primary silicate minerals, and found in clay deposits, soils, or shales; characterized by extremely small particle size and the ability to absorb water. **2.** in general, any mineral particle of extremely small size, less than 4 micrometers in diameter.

claymore *Military Science.* an antipersonnel mine that can be detonated by remote control.

claypan or **clay pan** *Geology.* **1.** a compact, heavy, relatively impervious subsurface layer of clay soil that is hard when dry and plastic when immersed in water. **2.** in Australia, a shallow depression that is filled with clay and silt and has a hard, sun-baked surface.

clay plug *Geology.* a term for a sedimentary deposit of silt, clay, and finely divided, well-decomposed organic material that blocks the end of an abandoned meander bend.

clay press *Engineering.* a device used to press superfluous water from slurry.

clay refining *Chemical Engineering.* a process in which a vaporized gasoline or light petroleum product flows through a bed of granular clay, where various olefins are polymerized to gums and absorbed into the clay.

clay regeneration *Chemical Engineering.* a method for rejuvenating coarse-grained absorbent clays for reuse, by removing oil from the clay with naphtha, steaming out the excess naphtha, and then removing carbonaceous material by roasting in a stream of air.

clay rock see CLAYSTONE, def. 1.

clay shale *Geology.* **1.** a shale composed primarily of clay material that becomes clay on weathering. **2.** a consolidated sediment that contains no more than 10% sand and has a silt to clay ratio of less than 1:2.

clay slate *Geology.* **1.** a low-grade slate that is less than 50% reconstituted. **2.** any slate that is derived from argillaceous rock rather than from volcanic ash.

clay soil *Geology.* a fine-grained soil containing at least 30–40% clay, which is sticky and plastic when wet and forms very hard lumps when dry. It is one of the three basic types of soil along with sand and loam.

claystone or **clay stone** *Geology.* **1.** an indurated clay that is derived chiefly from the decomposition of feldspars, and that has not been chemically altered or metamorphosed, but which is sufficiently hardened that it requires grinding to be worked. Also, CLAY ROCK. **2.** a former term for an altered feldspathic igneous rock that has been reduced to a compact mass of earthy or clayey alteration products. **3.** a flat, rounded concretionary mass often found in clay beds.

clay vein *Geology.* a roughly tabular body of clay which, like an ore vein, fills a crevice in a coal seam. Also, DIRT SLIP.

clayware see POTTERY, def. 1.

CLB technique *Genetics.* a technique used to detect sex-linked lethal and viable mutations in *Drosophila*, consisting of a crossover suppressor, a lethal allele, and the dominant marker bar eye.

Cl compound *Chemistry.* any compound of chlorine, such as ionic compounds formed from chlorine in the –1 state or covalent chloroderivative compounds.

CLE *Aviation.* the airport code for Hopkins International Airport, Cleveland, Ohio.

clean *Agronomy.* **1.** of soil, free from weeds. **2.** of or relating to clean plowing. *Aviation.* of a landing gear, flags, or other retractable device, in a retracted position.

clean and certify *Computer Technology.* the process of preparing magnetic tape for reuse by cleaning the recording surface, writing a test pattern, and checking for recording errors that indicate damage.

clean bomb *Ordnance.* a nuclear bomb that produces relatively small amounts of radioactive fallout.

clean compile *Computer Programming.* the conversion of a program from the source programming language into machine code with no significant syntax errors.

clean plowing *Agronomy.* a technique of plowing that turns over the top six to ten inches of soil, thus clearing the surface of weeds and plant debris. Also, **clean tillage, clean till, clean culture.**

clean room *Engineering.* a sterile, dust-free facility used for the assemblage of sensitive equipment, such as that used for space travel and exploration.

cleanser *Materials.* a cleaning preparation, such as a liquid or powder, for scouring sinks or bathtubs.

clean ship *Naval Architecture.* a tanker designed to carry refined petroleum products as opposed to crude or heavy oils.

clean sound *Acoustical Engineering.* a term for the type of sound reproduction provided by a high-quality amplifier, in which there is minimal distortion of the original sound.

clean track *Acoustical Engineering.* a recorded sound track that, when played back, does not pick up signals from adjacent tracks.

clean up *Graphic Arts.* to clean an offset plate to reverse or forestall the process of catching up, so that nonimage areas will repel ink.

cleanup *Aviation.* any modification to the external form of an airfoil or aircraft for the purpose of reducing its drag. *Electronics.* a process that removes residual and absorbed gases from a vacuum tube.

clear free from darkness, obscurity, or blockage; specific uses include: *Meteorology.* **1.** the state of the sky when it is cloudless or when the sky cover is less than 0.1. **2.** the general character of a day's weather when the average cloudiness has been from 0.0 to 0.3 for the 24-hour period. **3.** a popular designation of the condition of the atmosphere when it is very transparent with negligible cloudiness. *Computer Programming.* **1.** to set a register, counter, or other storage location to all zeros. **2.** to remove all display from a CRT screen. **3.** an instruction that cancels the last keyed entry or replaces the display with one or more zeros. **4.** a function key that transmits such an instruction. *Ordnance.* **1.** to unload a gun or make certain that it is empty of ammunition. **2.** to free a gun of stoppages. *Navigation.* to pass at a safe distance.

clear-air turbulence *Meteorology.* a violent disturbance in air currents, often experienced by high-speed aircraft, caused by dramatic temperature changes within a jet stream at the upper troposphere or lower stratosphere level, unaccompanied by clouds.

clearance *Mechanical Engineering.* **1.** the distance between two objects or between two mating parts when assembled together. **2.** the distance between a moving and a stationary part of a machine or between two moving parts. *Navigation.* the height of a bridge above mean high water (**vertical clearance**) or width between fenders (**horizontal clearance**). *Transportation Engineering.* in highway design, the horizontal and vertical dimension of the clear path for traffic past some potential obstruction (such as a bridge). *Aviation.* authorization for an aircraft to take off, proceed, or land, given by the airport tower or air traffic control. *Military Science.* authorization to work in a specific area, participate in a specific operation, or receive specific information that has been classified in the interests of national security. *Ordnance.* the elevation of a gun at an angle that allows the projectile to move freely between the muzzle and the desired point of impact.

clearance angle *Mechanical Engineering.* the angle between the back face or lower part of a cutting tool and the surface of the material being cut.

clearance limit *Aviation.* the location to which an aircraft is cleared by air traffic control.

clearance test *Pathology.* a gauge of the rapidity with which a substance is depleted from the blood.

clearance volume *Mechanical Engineering.* the volume enclosed by the piston and adjacent end of a reciprocating engine or compressor cylinder when the crank is at inner dead-center.

clear area *Computer Technology.* in character recognition, a designated space that must be blank and free of any print or markings unrelated to the scanning process.

clear band *Computer Technology.* a clear area between lines of print.

clear base *Graphic Arts.* any transparent material capable of carrying a photosensitive emulsion; a film base that is not coated with emulsion.

clear channel *Telecommunications.* **1.** an AM channel on which one station transmits over large areas without interference. **2.** a digital signaling format in which the 64 kilobit-per-second encoded voice information channel format is not shared with control information.

clear cutting *Forestry.* a method of harvesting timber in which all the trees are removed in a certain area of a forest, providing full sunlight in order for new seedlings to develop. Also, **clear-cutting system.**

clear-faced worsted *Textiles.* worsted fabric in which the weave is visible due to removal of the nap.

clear gasoline *Materials.* gasoline that does not contain antiknock additives.

clear ice *Hydrology.* a term for relatively transparent ice having a homogeneous structure and few air pockets.

clearstory see CLERESTORY.

clear text *Telecommunications.* text that is not in code.

clear-voice override *Telecommunications.* a feature that enables a scrambler to receive a clear message even when set to coded operation.

cleat *Mechanical Devices.* **1.** a device used to secure ropes, usually consisting of opposite-facing prongs around which the rope is wound. **2.** a wood or metal strip fixed to another strip of the same or other material, or nailed to a wall to allow fastening of objects to it. *Mining Engineering.* the main joint in a coal seam along which it breaks most easily. Also, CLEET.

cleavage *Crystallography.* the property, possessed by many crystals, of splitting readily in one or more definite directions to give smooth surfaces, always parallel to actual or possible crystal faces. *Geology.* **1.** the tendency or property of a mineral to break or split along its crystallographic planes, which are parallel to actual or possible crystal faces. **2.** the tendency or property of a rock to break into thin, parallel sheets along smooth planes. *Biochemistry.* the splitting up of a complex molecule, such as a protein, into simpler molecules by hydrolysis and with the mediation of an enzyme. *Developmental Biology.* a series of cell divisions of an ovum immediately after fertilization; the size of the embryo remains uniform as the blastomeres become smaller with each division.

cleavage cavity see BLASTOCOELE.

cleavage cell see BLASTOMERE.

cleavage fracture *Crystallography.* a fracture along the cleavage plane of a crystal.

cleavage furrow *Developmental Biology.* the groove encircling a zygote or blastomere as it undergoes cytokinesis.

cleavage nucleus *Developmental Biology.* the nucleus of a zygote after fusion of male and female pronuclei. Also, SEGMENTATION NUCLEUS.

cleavage plane *Crystallography.* a plane along which a crystal has a natural tendency to split.

cleavelandite *Mineralogy.* a variety of albite occurring in lamellar masses.

cleaver *Mechanical Devices.* a heavy knife with a broad blade, used by butchers in cutting the bones and joints of meat. *Archaeology.* a stone tool of the Paleolithic period, typically having a wide cutting edge at one end like a modern axehead.

Clebsch-Gordan coefficients see CLEBSCH-GORDAN SERIES.

Clebsch-Gordan series *Mathematics.* the series that expresses the tensor product $T^{(\mu)} \otimes T^{(\nu)}$ of two irreducible representations of a semisimple Lie group (or algebra), of degrees n_μ and n_ν, respectively, as a sum $\sum_\alpha T^{(\alpha)}$ of irreducible representations. The **Clebsch-Gordan coefficients** are the entries of the $n_\mu n_\nu \times n_\mu n_\nu$ change of basis matrix that expresses the (nonunique) basis $\{w_k^\alpha\}$ for the irreducible representation spaces (modules) in terms of the product basis $\{v_i^\mu \otimes v_j^\nu\}$. Also, VECTOR ADDITION COEFFICIENTS, WIGNER COEFFICIENTS.

cledophilous *Ecology.* describing a species that thrives on rubbish heaps or in wasteland habitats. Thus, **cledophile, cledophily.**

cleet see CLEAT.

cleft grafting *Botany.* a method of grafting in which a grafted branch (scion) is inserted directly into a cleft cut in the stock.

cleft lip *Medicine.* a fissure of the upper lip, resulting from the failure of the normal embryonic processes to form the face. Also, HARELIP.

cleft palate *Medicine.* a birth deformity of the palate in which an abnormal opening exists between the oral and nasal cavities. *Metallurgy.* in welding, a weld where the pieces to be welded contain V-shaped projections and mating V-shaped notches.

cleidocranial dysostosis *Medicine.* a rare heriditary deformity of the skeleton resulting in decreased height, sloping shoulders due to malformed or absent clavicles, malformed or redundant teeth, and late closure of the fontanel or "soft spot" in the skull.

cleistocarpous *Mycology.* of or relating to a spore-bearing structure that frees its spores by rupturing or decaying.

cleistogamy *Botany.* the production of flowers that self-fertilize prior to opening.

cleistothecium *Mycology.* a spore-bearing structure that frees its spores by rupturing or decaying. Also, **cleistocarp.**

cleithrophobia *Psychology.* see CLAUSTROPHOBIA.

cleithrum *Vertebrate Zoology.* a dermal bone on each side of the clavicle in many fishes, some amphibians, and some primitive reptiles.

Cleland's reagent see DITHIOTHREITOL.

Clemmensen reduction *Organic Chemistry.* a reaction that converts an acyl group into an alkyl group through the use of amalgamated zinc and concentrated hydrochloric acid (HCl).

Clenodiniaceae *Botany.* a family of freshwater, thecate dinoflagellates of the order Peridiniales that are photosynthetic and subglobose.

clepsydra see WATER CLOCK.

clerestory *Architecture.* an upper extension of a side wall, built above adjoining roofs and windowed to admit light into a high central room. Also, CLEARSTORY.

Cleridae *Invertebrate Zoology.* a family of checkered beetles, brightly colored predators in the superfamily Cleroidea.

Cleroidea *Invertebrate Zoology.* a superfamily of beetles in the suborder Polyphaga.

Clethraceae *Botany.* a monogeneric family of dicotyledonous tropical shrubs and small trees of the order Ericales, characterized by a stellate, hairy appearance, toothed leaves, and numerous, usually winged seeds.

cleugh see CLOUGH.

Cleveaceae *Botany.* a widespread family of typical liverworts of the order Marchantiales, characterized by dichotomous or ventral branching, simple pores in the epidermis, one to four layers of air chambers, and a sporophyte with a dehiscent capsule.

clevis *Mechanical Devices.* **1.** a U-shaped metal or iron fitting that joins a connecting rod to a plate or lever, such as an automobile brake rod to a brake lever. **2.** a U-shaped fitting with prefabricated holes to receive a bolt or pin from another attached part. *Mining Engineering.* a spring or snap hook to attach a bucket to the hoisting rope. Also, CLIVVY.

clevis pin *Mechanical Devices.* a flat-headed pin fastener used in mechanical linkages that include a clevis.

cliachite *Mineralogy.* any colloidal aluminum hydroxide that occurs as a constituent of bauxite. Also, KLIACHITE.

click a slight, sharp, quick sound; specific uses include: *Acoustical Engineering.* the sound heard on the playback of a recorded tape that has a kink or perforation in it. *Telecommunications.* **1.** a transient pulse on a transmission channel due to the momentary opening or closing of relay contacts or toy brief intermittent corruption of voice signal information. **2.** a description of audible sounds in a voice signal due to bit errors in a PCM encoded signal. *Horology.* a spring-loaded pawl that, in combination with the clickwheel, prevents the mainspring from unwinding except to power the going train. *Computer Technology.* to press on a mouse input device in order to provide an instruction to the computer.

click art *Computer Technology.* a collection of pictures, graphs, and drawings that are usually stored on a magnetic disk to be selected for use in computer-produced documents.

click filter *Electronics.* a capacitor and resistor connected across the contacts of a switch or relay in order to prevent current from surging into an adjacent circuit.

click on *Computer Technology.* to press a mouse button while the mouse pointer is at a particular location on a displayed window, thus selecting the item or command denoted by the pointer.

click track *Acoustical Engineering.* a sound track that contains a series of intentionally provided clicks; one use of such a track is to provide musicians with a sense of common tempo while recording in a studio.

clickwheel *Horology.* in the winding mechanism of a spring-driven timepiece, a ratchet wheel attached to the mainspring barrel, driven by the crown wheel to wind the mainspring and, in combination with the click, preventing the mainspring from unwinding except to power the going train. Also, RATCHET WHEEL.

client *Psychology.* a person who receives mental health care; a term preferred to *patient* in certain therapeutic settings. *Computer Technology.* **1.** a part of an operating system that requests service from another part. **2.** in data management, a user terminal requesting access to a database that resides on the host computer.

client-centered therapy *Psychology.* a therapeutic approach in which the main task of the therapist is to create a warm, accepting atmosphere in which the patient can develop insights about his inner self and gradually decide how to resolve his own conflicts, rather than becoming dependent on the therapist. Also, NONDIRECTIVE THERAPY.

client-server architecture *Computer Technology.* a configuration in which specific functions, such as database management, are performed by dedicated hardware and software.

cliff *Geography.* a high rock face that is very steep, perpendicular, or overhanging, especially near a seashore.

cliff house *Architecture.* a dwelling place built into a cliff face or canyon wall, especially those used by prehistoric Indians of the southwestern United States. Also, **cliff dwelling.**

Clifford algebra *Mathematics.* let V be a vector space over a field K and σ a symmetric form on V. A Clifford algebra for σ (sometimes called the Clifford algebra belonging to σ) is an associative algebra (and hence a vector space) $C(\sigma)$ and a linear map (depending on σ) $\rho: V \to C(\sigma)$ with the property that if $\psi: V \to L$ is a linear map of V onto a K-algebra L such that $\psi(X)^2 = \sigma(X) \cdot 1L$ ($1L$ is the unit element of L) for all $X \in V$, then there exists a unique algebra-homomorphism $\psi_*: C(\sigma) \to L$ such that the following diagram commutes:

$$\begin{array}{ccc} & \rho & \\ V & \rightarrow & C(\sigma) \\ \psi \downarrow & & \downarrow \psi_* \\ & L & \end{array}$$

$C(\sigma)$, if it exists, is uniquely determined up to isomorphism and is generated by $\sigma(V)$ as an algebra over K. A Clifford algebra can be viewed as a universal algebra over K, in which V is embedded and the square in the algebra corresponds to the value of the quadratic form in V.

Climaciaceae *Botany.* a monogeneric family of large, dull, usually dendroid mosses of the order Isobryales that form loose mats on soil along streams and in meadows and have tomentose prostrate stems that give rise to erect, secondary stems topped by a densely branched portion; found in temperate and boreal zones of the Northern Hemisphere.

Climacograptus *Paleontology.* a scandent uniserial graptolite in the order Graptoloidea and suborder Diplograptina; Ordovician.

climacophobia *Psychology.* an irrational fear of steps or stairs.

climacteric *Physiology.* a period in which the reproductive capacity of men and women decreases, culminating in women in the menopause. Also, **climacterium.** *Botany.* a period of maximum respiration in a fruit, at which time it becomes fully ripened.

Climacteridae *Vertebrate Zoology.* the monogeneric family of Australian creepers of the passerine suborder Oscines; habits are creeperlike although birds often feed on the ground.

climagram or **climagraph** see CLIMATIC DIAGRAM.

climate *Meteorology.* the statistical collective of the weather conditions of a specified area during a specified interval of time, usually several decades.

climate control *Meteorology.* the use of artificial techniques to alter or control the climate over a given area. *Mechanical Engineering.* see AIR CONDITIONING.

climatic *Meteorology.* relating to, involving, or affected by climate.

climatic change *Meteorology.* the long-term effects on the earth's climate from such features as rainfall, snowpack, air temperature, humidity, or other meteorological aspects.

climatic classification *Meteorology.* the division of the earth's climates into a worldwide system of contiguous regions, each defined by the relative homogeneity of the climatic elements.

climatic climax *Ecology.* a climax community in which climate is the main influencing factor.

climatic controls *Meteorology.* the various factors that govern the general nature of the climate over a specified area of the earth, including solar radiation, distribution of water masses, elevation, and ocean currents.

climatic cycle *Meteorology.* the alternating episodic climatic events that recur with some regularity, but are not strictly periodic. Also, CLIMATIC OSCILLATION.

climatic diagram *Meteorology.* a graphic representation of at least two simultaneous climatic events or elements, typically over an annual period. Also, CLIMAGRAM, CLIMAGRAPH, CLIMATOGRAPH, CLINOGRAM, CLINOGRAPH.

climatic divide *Meteorology.* a boundary between characteristically differing climatic regions.

climatic factor *Meteorology.* **1.** a natural or artificial element, such as the smoke or heat of a metropolitan area, that influences the climate of a region. **2.** a climatic element and its variations considered as an important influence on human activities.

climatic forecast *Meteorology.* a forecast of general weather conditions to be expected in a region over a period of years.

Climatic Optimum *Meteorology.* the period in history, from about 5000 to 2500 BC, during which temperatures were warmer than they are now in nearly all parts of the earth. Also, MEGATHERMAL PERIOD.

climatic oscillation see CLIMATIC CYCLE.

climatic province *Meteorology.* any of the thirty-four regions of the earth's surface, each of which is characterized by an essentially homogeneous climate defined mainly by temperature and rainfall and partly by wind and orography.

climatic snow line *Meteorology.* the altitude above which a flat surface fully exposed to sun, wind, and precipitation would have a net accumulation of snow over an extended period of time.

climatic zone *Meteorology.* a belt of the earth's surface within which the climate is essentially homogeneous in some respect.

Climatiidae *Paleontology.* a family of primitive fishes in the order Climatiiformes; Silurian and Devonian.

Climatiiformes *Paleontology.* an extinct order of primitive acanthodian jawed fishes, generally only a few inches long; Silurian and Devonian.

Climatioidei *Paleontology.* in some classifications, a suborder of acanthodian fishes in the order Climatiiformes.

Climatius *Paleontology.* a primitive genus of acanthodian jawed fishes in the order Climatiiformes, only 7–8 cm long; upper Silurian to lower Devonian.

climatochronology *Geology.* a method of absolute age dating of recent geologic events by using oxygen isotope ratios.

climatograph see CLIMATIC DIAGRAM.

climatography *Meteorology.* a quantitative description of the climate of a region with reference to the tables and charts that show the characteristic values of climatic elements.

climatological [klī′mə tə läj′ə kəl] *Meteorology.* relating to climate or the study of climate.

climatological forecast *Meteorology.* a weather forecast based solely on regional climate.

climatological station elevation *Meteorology.* the elevation above mean sea level used as the reference datum level for climatological records of atmospheric pressure in a given area..

climatology [klī′mə täl′ə jē] *Meteorology.* the scientific study of climate.

climatopathology *Medicine.* the study of disease as it relates to or is affected by climate.

climatophysiology *Physiology.* the study of the effects of environmental factors on human physiology and health.

climatotherapy *Medicine.* the relocation of a patient to a more amenable climate as a treatment for a medical condition.

climax *Physiology.* the period of greatest intensity, for example, in sexual excitement or in the course of a disease. *Ecology.* see CLIMAX COMMUNITY.

climax community *Ecology.* a stable community of organisms in equilibrium with existing environmental conditions that represents the final stage of an ecological succession. Thus, **climax species, climax area.**

climb *Aviation.* an increase in altitude of a flight vehicle, usually a deliberate program that may make the gain in altitude gradual or abrupt, depending on the circumstances.

climb cutting *Metallurgy.* the cutting effected when a cutter moves in the direction of the feed.

climbing bog *Geology.* a boggy area that forms on the margin of a swamp at a level higher than that of the original swamp.

climbing crane *Mechanical Engineering.* a crane designed for use on top of a multistory building, ascending with each added floor as construction progresses.

climbing dune *Geology.* a low mound, ridge, bank, or hill formed by the piling up of sand on the windward side of a cliff or mountainside. Also, RISING DUNE.

climbing fern *Botany.* any of several vinelike ferns of the genus *Lygodium,* with climbing or trailing stems; found primarily in the tropics.

climbing irons *Mechanical Devices.* a pair of spiked iron frames attached to a shoe, leg, or knee to assist in climbing trees or wooden utility poles. Also, **climbing spur.**

climbing ripple *Geology.* a cross-layered sedimentary deposit that is formed by superimposed migrating ripples, whose crests appear to be advancing upslope.

climbing stem *Botany.* any plant stem having the ability to mount over and attach to other plants or stationary structures.

cline *Biology.* a gradual variation in the inherited characteristics of an animal or plant species across different parts of its range according to varying ecological, geographic, or other factors. *Ecology.* the differences in a community structure due to changes in the slope around a mountain or ridge.

clingage *Petroleum Engineering.* oil that adheres to the wall of a measuring tank after the tank is drained.

clingfish *Vertebrate Zoology.* any fish of the family Gobiesocidae, having a sucking disc on the abdomen for clinging to stones and the like.

clinical *Medicine.* relating to or based on the actual observation and treatment of patients, rather than on experimentation or theory.

clinical chemistry *Pathology.* **1.** the chemistry of the diseases and health of humans. **2.** the chemistry used in the clinical management of patients, as in the laboratories of hospitals.

clinical depression *Psychology.* depression of sufficient severity to require intervention by a psychiatrist or psychotherapist.

clinical genetics *Genetics.* the use of genetic analysis and the application of genetic repair in the clinical treatment of human disorders.

clinical neurology *Neurology.* the branch of neurology that deals specifically with the diagnosis and treatment of neurologic disorders.

clinical pathologist *Medicine.* a physician specializing in the branch of pathology that is applied to the solution of clinical problems, especially the use of laboratory methods in clinical diagnosis.

clinical pathology *Pathology.* any part of the medical aspect of pathology as it relates to patient care, especially using laboratory methods for clinical diagnosis and solutions to clinical problems.

clinical pharmacology *Pharmacology.* the branch of pharmacology that is concerned with the origin, nature, chemistry, effects, and uses of drugs in the prevention, treatment, and control of disease in humans.

clinical psychology *Psychology.* the branch of psychology that focuses on the study, diagnosis, and treatment of mental disorders. Thus, **clinical psychologist.**

clinical thermometer *Engineering.* **1.** a thermometer with a scaled glass tube containing mercury, which expands and travels up a capillary as the temperature rises and stops at a given reading. **2.** such a thermometer having a digital readout.

clinical toxicology *Toxicology.* the branch of toxicology that deals with the diagnosis and treatment of poisoning in humans.

clinical trial *Statistics.* an evaluation of the effectiveness of medical treatments based on statistical methods.

Clinidae *Vertebrate Zoology.* the scaled blenny family, in the suborder Blennioidei of the Perciformes; found primarily in the Southern Hemisphere, this large group of small fish is characterized by long, multispined dorsal fins and often by tentacles hanging from the head and pointed nose. Also, KLIPFISH.

clinker *Geology.* **1.** a hard mass of fused ash produced in a coal furnace as a by-product of combustion. **2.** a fused rock or rough volcanic fragment, usually less than 15 centimeters in diameter, that resembles the clinker of a furnace. **3.** coal that has been altered by igneous intrusion. *Materials.* **1.** hydraulic cement in the state it issues from the rotary or shaft kiln in which it was produced. **2.** a general term for a lump of vitrified stony material.

clinker building *Design Engineering.* a method of overlapping wooden planks or steel plates, used as the outside covering of ships or boilers.

clino- a combining form meaning "slope," as in *clinoform.*

clinoamphibole *Mineralogy.* any monoclinic amphibole.

clinoaxis *Crystallography.* a diagonal or lateral axis of a monoclinic crystal, which forms an oblique angle with the vertical axis. Also, **clinodiagonal.**

clinochlore *Mineralogy.* $(Mg,Fe^{+2})_5Al(Si_3Al)O_{10}(OH)_8$, a transparent to translucent white, yellow, or green monoclinic mineral of the chlorite group, usually occurring in massive, foliated to granular form, having a specific gravity of 2.63 to 2.98 and a hardness of 2 to 2.5 on the Mohs scale; found in schists, serpentines, and other metamorphic rocks and as an alteration product of mafic minerals in igneous rocks.

clinoclase *Mineralogy.* $Cu_3^{+2}(AsO_4)(OH)_3$, a green-blue to green-black, transparent to translucent monoclinic mineral occurring as tabular and rhombohedral crystals that often form rosettes, having a specific gravity of 4.33 to 4.38 and a hardness of 2.5 to 3 on the Mohs scale; found as a secondary mineral with olivenite. Also, **clinoclasite.**

clinoenstatite *Mineralogy.* $Mg_2Si_2O_6$, a yellow, brown, or green transparent to translucent monoclinic mineral of the pyroxene group, occurring as short prismatic or tabular crystals, having a specific gravity of 3.21 and a hardness of 5 to 6 on the Mohs scale; found in porphyritic volcanic rock.

clinoferrosilite *Mineralogy.* $(Fe^{+2},Mg)_2Si_2O_6$, a transparent amber-colored monoclinic mineral occurring as tiny needlelike crystals; having a specific gravity of 3.96 and a hardness of 5 to 6 on the Mohs scale; found in lithophysae in obsidian.

clinoform *Geology.* an underwater sloping land form, such as the continental slope of the ocean.

clinogram see CLIMATIC DIAGRAM.

clinograph *Engineering.* a surveying instrument used to record deviation from the vertical; some clinographs are equipped with internal cameras or gyroscopic orientation. *Meterology.* see CLIMATIC DIAGRAM.

clinohedral class *Crystallography.* a class of crystals characterized by a single plane of symmetry parallel to the clinopinacoid and 110 axes.

clinohedrite *Mineralogy.* $CaZnSiO_4 \cdot H_2O$, a transparent, white or violet monoclinic mineral occurring as prismatic, wedge-shaped, or tabular crystals and in massive form, having a specific gravity of 3.28 to 3.33 and a hardness of 5.5 on the Mohs scale; found with willemite, axinite, and calcite at Franklin, New Jersey.

clinohumite *Mineralogy.* $(Mg,Fe^{+2})_9(SiO_4)_4(F,OH)_2$, a white, yellow, or brown transparent to translucent monoclinic mineral occurring as commonly twinned crystals, having a specific gravity of 3.17 to 3.35 and a hardness of 6 on the Mohs scale; found in veins, in serpentine and talc schist, and in contact zones in dolomite.

clinometer *Cartography.* a surveying instrument small enough to be held in one hand, combining a spirit level, a vertical circle, and a sighting device for coarse but rapid measurement of vertical angles. *Engineering.* **1.** any hand-held instrument used to measure the inclination of slopes or embankments. **2.** a device that measures the roll of a ship. **3.** see INCLINOMETER.

clinopinacoid *Crystallography.* a monoclinic crystal whose crystallographic faces are parallel to the *c* axis and the inclined axis.

clinopyroxene *Mineralogy.* any monoclinic pyroxene, such as diopside, clinoferrosilite, augite, or jadeite.

clinozoisite *Mineralogy.* $Ca_2Al_3(SiO_4)_3(OH)$, a transparent to translucent yellow, gray, pink, or green monoclinic mineral of the epidote group occurring as prismatic, striated crystals or in massive form, having a specific gravity of 3.21 to 3.28 and a hardness of 6.5 on the Mohs scale; found in metamorphosed igneous and sedimentary rocks.

clint *Geology.* **1.** any projecting ledge of hard or flinty rock, or a rocky cliff. **2.** a narrow channel between ridges in limestone, caused by the solvent action of rainwater.

Clintonian *Geology.* in New York State, a geologic stage of the Middle Silurian period, occurring after the Medinan and before the Lockportian.

clintonite *Mineralogy.* $Ca(Mg,Al)_3(Al_3Si)O_{10}(OH)_2$, a red, green, or yellow transparent to translucent monoclinic mineral of the mica group occurring as tabular pseudohexagonal crystals, having a specific gravity of 3 to 3.1 and a hardness of 3.5 to 6 on the Mohs scale; found with calcite, clinopyroxene, and phlogopite in crystalline limestone.

Clionidae *Invertebrate Zoology.* a family of marine boring sponges in the class Demospongiae.

clip *Surgery.* a metallic instrument used to bring together the edges of a wound or to prevent bleeding from small blood vessels. *Mechanical Devices.* **1.** a temporary angle iron used in building construction to hold and maintain an angle between two members or surfaces. **2.** a casting with side flanges that fits over the outside of a hose or pipe and a metal nozzle to join them.

clip art *Computer Technology.* any ready-made printed or image graphics that can be cut-and-pasted into a document.

clipboard *Computer Technology.* a "scratch pad" computer memory designed for the temporary storage of data being moved within a file or between files.

clip-dot fabric *Textiles.* a fabric having a pattern of small dots made from extra yarn of the same or another color.

clip lead *Electricity.* a short length of flexible wire with an alligator clip at one or both ends.

clipped gable see JERKINHEAD.

clipper diode *Electronics.* a diode that clips signal voltage of either polarity that surpasses a predetermined amplitude.

clipper-limiter *Electronics.* a transducer that generates an output signal whose amplitude range conforms with the input signal voltage between two predetermined points. Also, SLICER.

clipping *Computer Technology.* in graphics display, the process by which the hardware or software eliminates the part of the drawing that would fall outside the boundaries on the screen or a window. Also, SCISSORING. *Electronics.* **1.** the loss of initial or final parts of words or syllables when voice-operated devices malfunction. **2.** a flattening of the peaks of a waveform by an amplifier driven past its power capacity. *Telecommunications.* **1.** the amplitude restriction of a signal to a maximum level. **2.** a distortion of signals caused by exceeding the maximum allowable amplitude of imput signal to an amplifier.

clipping circuit see LIMITER.

clipping edge *Metallurgy.* in forging, the location where the flash is removed.

clipping level *Electronics.* the amplitude level of the output signal in a clipper-limiter.

CLIPS *Artificial Intelligence.* a widely used forward-chaining expert system shell developed at NASA.

clisere *Ecology.* an ecological succession resulting from a significant change in climatic conditions.

Clistogastra *Invertebrate Zoology.* a suborder of hymenopteran insects, including bees, wasps, and ants, having a narrow waist joining the thorax to the abdomen, and bearing legless larvae.

Clitellata *Invertebrate Zoology.* in some classifications, a major division of annelid worms containing the Oligochaeta (earthworms) and Hirudinea (leeches).

clitellum *Invertebrate Zoology.* in many leeches and earthworms, a thick, saddlelike part of the body wall that cocoons the eggs.

clitoris [klit´ə rəs; kli tôr´əs] *Anatomy.* a small, sensitive erectile organ in the female corresponding to the male penis and located at the ventral end of the vulva.

clo *Engineering.* a unit of clothing insulation that maintains normal skin temperature when the body's heat production reaches 50 kg-calories per meter squared per hour at an air temperature of 21°C and no wind.

cloaca [klō ä´kə] *Invertebrate Zoology.* the respiratory, excretory, and reproductive duct chamber in certain invertebrates. *Vertebrate Zoology.* the common body cavity into which genital, urinary, and intestinal canals discharge in monotremes, birds, reptiles, amphibians, and many fishes. *Developmental Biology.* the terminal end of the hindgut before division into the rectum, bladder, and genital primordia. *Pathology.* an opening into the involucrum of a necrosed bone.

cloacal bladder *Vertebrate Zoology.* a sac in the cloacal wall into which urine from the cloaca is forced.

cloacal gland *Vertebrate Zoology.* a sweat gland present in the cloaca of lower vertebrates, including snakes and amphibians.

clobber *Computer Programming.* in informal use, a term meaning to inadvertently destroy valid data in memory or a stored file.

clock *Horology.* **1.** a device, not conveniently portable, for measuring or recording time in units of hours, minutes, and seconds, usually by means of hands and a dial or a numerical display; it may also indicate other chronological data. **2.** to measure or record time using a clock or watch. *Electronics.* **1.** a generator used to synchronize the timing in switching circuits and the speed of the central processing unit in a computer. **2.** a device that controls timing in a system, generally by generating a steady stream of pulses at a tightly controlled constant rate.

clock control system *Control Systems.* a system that uses a timing device to generate the control function.

clock cycle *Computer Technology.* **1.** the fundamental time signal that is fixed by the hardware design of a computer and used to synchronize its operations. **2.** the length of the clock cycle.

clock drive *Engineering.* a mechanism in a telescope that allows it to move about on its axis and remain focused on the same part of the sky over time.

clocked flip-flop *Electronics.* a flip-flop circuit triggered only if trigger and clock pulses are present at the same time.

clocked logic *Electronics.* a technique in which all the flip-flops of a logic network change in accordance with logic input levels at a discrete time.

clock-face model *Cell Biology.* a theory proposed to explain the different pathways of differentiation taken by different cells in regeneration experiments using insect larval tissue.

clock frequency *Electronics.* the master frequency that controls the periodic pulses used to schedule the operation of a digital computer.

clock interrupt *Computer Technology.* an interrupt signal that is generated by a hardware clock, causing the interruption of the processor after a specified time limit has been reached.

clock method *Military Science.* a method of indicating the direction of some object of interest with respect to the orientation of the observer (usually in a vehicle); for example, "three o'clock" would be to the right of the observer. *Ordnance.* a method of ordering artillery fire by using this system.

clock motor see TIMING MOTOR.

clock mutants *Genetics.* a type of mutation in *Drosophila* that interferes with the normal 24-hour circadian activity period characteristic of the species.

clock oscillator *Electronics.* a circuit that regulates an electronic clock.

clock paradox *Physics.* a seeming inconsistency between two conclusions in relativity theory, one stating that time passes faster for an observer at rest than it does for an observer in motion, and the other postulating that all observers are equivalent.

clock pulses *Computer Technology.* a train of signals in which the separation between pulses is constant and which serves to synchronize information transfer among computer components.

clock rate *Horology.* a measurement of the accuracy of a clock based on the amount of time it regularly gains or loses each day. *Electronics.* the frequency at which a clock produces pulses, usually measured in megahertz.

clock time see CYCLE TIME.

clock track *Computer Technology.* on a magnetic tape or disk, a specific track containing a pattern of bits that provides a clock signal for synchronizing read and write operations.

clock watch *Horology.* a watch that indicates the hour with a bell, chime, or other sound; it may also indicate parts of the hour.

clockwise angle see ANGLE TO RIGHT.

clockwork *Horology.* the mechanism of a mechanical clock considered as a whole unit. Also, WORKS.

clod *Agronomy.* a lump of soil that is created by some human process of plowing or digging, as opposed to a naturally occurring aggregation of soil.

clog snow *Hydrology.* a popular term for new snow that is wet and sticky.

cloister *Architecture.* an arcaded or colonnaded courtyard, especially in a monastery.

cloister vault or **cloistered vault** see COVED VAULT.

clonal analysis *Genetics.* a genetic technique involving the study of certain alleles carried on the X chromosome; used to identify the clones within a tissue or body structure.

clonal selection theory *Immunology.* a selective theory of antibody production proposing that each B lymphocyte has the ability to respond to a specific antigen that an individual may or may not have contacted previously.

clone *Genetics.* **1.** a group of genetically identical cells produced by mitotic divisions from one original cell. **2.** a group of genetically identical organisms all descended from the same single parent by asexual processes. **3.** a group of DNA molecules derived from one original length of DNA sequences and produced by a bacterium or virus using genetic engineering techniques, often involving plasmids. *Biotechnology.* to propagate or produce a clone by transporting nuclei from body cells to enucleated eggs. *Computer Technology.* a computer, peripheral device, or software product that is an imitation in performance and appearance of a product built by a different manufacturer, usually one that sells at a higher price; for example, an **IBM clone** is a personal computer built to resemble the popular IBM PC model. (From a Greek word meaning "twig" or "shoot.")

cloned gene *Genetics.* a gene or nucleotide sequence that has been incorporated into a cloning vector.

clonic *Physiology.* of, relating to, or resembling clonus.

clonic contraction see CLONUS.

clonicity *Neurology.* the condition of being clonic.

clonicotonic *Neurology.* demonstrating or relating to both clonic and tonic muscular contractions.

cloning vector *Molecular Biology.* a plasmid used in recombinant DNA experiments as an acceptor of foreign DNA.

clonism *Neurology.* a succession of clonic spasms. Also, **clonismus.**

clonogenic *Molecular Biology.* **1.** derived from or consisting of a clone. **2.** of or relating to a population of cells or organisms derived from a single cell by asexual propagation.

clonorchiasis *Medicine.* an infection of the bile ducts caused by ingesting raw fish.

clonospasm *Neurology.* a clonic spasm.

Clonothrix *Bacteriology.* a genus of iron bacteria that grow as sheathed filaments associated with deposits of iron and manganese oxides.

clonus *Neurology.* brief muscular contraction and relaxation repeated at short, regular intervals, usually occurring in an epileptic convulsion. Also, CLONIC CONTRACTION.

clopen *Mathematics.* a set that is both open and closed. (From closed and open.)

CLOS *Artificial Intelligence.* Common Lisp Object System, a standardized object-oriented programming system that is part of Common Lisp.

close [klōz] *Computer Programming.* **1.** to perform a series of tasks to deactivate a file and store it in a state suitable for reopening at a later time. **2.** to issue a program command to carry out this process.

close [klōs] *Meteorology.* a popular description of oppressively still, warm, and humid air, commonly referring to indoor conditions.

close-boarded *Building Engineering.* of or relating to running ground wood planks that are placed adjacent to one another against the ground.

close-control radar *Engineering.* a ground control radar used in conjunction with radio transmissions to direct an aircraft toward a target until the target can be sighted.

close-coupled pump *Mechanical Engineering.* a pump containing an electric motor, with the drive and the pump mechanisms coupled on the same shaft.

close coupling *Electricity.* any degree of coupling greater than critical coupling. Also, TIGHT COUPLING.

closed *Mathematics.* **1.** a set S is closed with respect to a binary operation $\bullet$, if $x \bullet y \in S$ whenever $x \in S$ and $y \in S$. **2.** a subset of a topological space, the complement of which is open. **3.** see ALGEBRAICALLY CLOSED.

closed area *Military Science.* a designated area in or over which passage of any kind is prohibited.

closed ball *Mathematics.* a closed ball of radius R with center c in a metric space with metric μ is the set of all points x in the space such that $\mu(x, c) \leq R$. In n-dimensional Euclidean space, this is often referred to as the **closed *n*-ball**. A closed interval is a special case of a closed ball.

closed basin *Geology.* in an arid region, a body of water having no outlet, so that water escapes only by evaporation.

closed circuit *Telecommunications.* **1.** a circuit that transmits radio or television signals directly to specified customers and does not broadcast for general consumption. **2. closed-circuit.** relating to or being a circuit of this type. Thus, **closed-circuit system.**

closed-circuit communication *Telecommunications.* any communication system that is completely independent of other facilities; a self-contained operation.

closed-circuit signaling *Telecommunications.* a signaling method in which there is a certain current level in the idle state, and a signal is transmitted by increasing or decreasing the current.

closed-circuit television *Telecommunications.* a televison signal transmission system that does not broadcast to the general public.

closed-coil armature *Electricity.* an armature whose windings form a closed circuit.

closed community *Ecology.* a habitat whose niches are all so well occupied that it precludes colonization by other species. Also, CLOSED ECOLOGICAL SYSTEM.

closed contour *Geology.* on a contour map, a continuous closed loop that does not intersect the edge of any part of the map.

closed curve *Mathematics.* a curve whose endpoints coincide.

closed cycle *Thermodynamics.* an isolated thermodynamic cycle in which the working substance is used repeatedly, rather than mass being introduced to or removed from the system.

closed-cycle fuel cell *Electricity.* a fuel cell composed of reactants that are recycled by an auxiliary process, such as electrolysis.

closed-cycle reactor *Nucleonics.* a gas-cooled nuclear reactor in which the carbon dioxide coolant flows through coolant channels in the core and upper plenum to a steam generator and then through gas circulators to complete the cycle.

closed-cycle turbine *Mechanical Engineering.* a gas-driven engine that captures and recycles its heat energy.

closed die *Metallurgy.* in forging or forming, a die that restricts the material flow to the die cavity.

closed drainage *Hydrology.* surface drainage that does not reach the ocean, but collects in a sink, lake, or interior basin. Also, INTERNAL DRAINAGE, INTERIOR DRAINAGE.

closed ecological system *Space Technology.* in a spacecraft, a process that supports life by collecting urine, exhaled carbon dioxide, and other waste matter and converting them (through chemical means or photosynthesis) into oxygen, water, and food. *Ecology.* see CLOSED COMMUNITY.

closed file *Computer Programming.* a file that is no longer accessible to a program for retrieving or storing data.

closed fold *Geology.* a fold whose limbs have been so compressed that they are parallel. Also, TIGHT FOLD.

closed form *Mathematics.* a differential form ω on a manifold M is closed if its exterior derivative is zero: $d\omega = 0$.

closed fracture *Medicine.* a bone fracture that does not produce an open wound in the skin.

closed graph theorem *Mathematics.* if $\Lambda: X \to Y$ is a linear operator between Frechet spaces X and Y, and $G = \{(x, \Lambda x) : x \in X\}$ (i.e., G is the graph of Λ) is a closed set in $X \times Y$, then Λ is continuous. Also, BANACH'S CLOSED GRAPH THEOREM.

closed high *Meteorology.* a high that may be completely encircled by an isobar or contour line.

closed interval *Mathematics.* in a linearly ordered set, the subset of points x such that $a \leq x \leq b$ for some a, b; denoted $[a, b]$.

closed lake *Hydrology.* a lake having no surface outlet, so that it loses water only through evaporation or seepage and is saline or brackish.

closed loop *Computer Programming.* a segment of computer code that is repeated endlessly unless interrupted by external means. *Engineering.* see CLOSED-LOOP SYSTEM. *Neurology.* a feedback path within the nervous system in which an input is affected by its own output.

closed-loop system *Engineering.* a system that continually measures variables, compares them to a set norm, and adjusts its performance according to the feedback in order to reduce deviation from the norm.

closed-loop voltage gain *Electronics.* the voltage gain of an amplifier with feedback, given as the ratio of the output alternating current voltage to the input alternating current voltage.

closed low *Meteorology.* a low that may be completely encircled by an isobar or contour line.

closed magnetic circuit *Electromagnetism.* a magnetic circuit in which the magnetic flux is uninterrupted, such as that of a ferromagnetic core with no air gap.

closed map *Mathematics.* a function between two topological spaces that maps closed sets to closed sets.

closed nozzle *Mechanical Engineering.* a fuel nozzle that contains a shutoff valve between the fuel supply and the combustion chamber.

closed operator *Mathematics.* a linear operator T on a Hilbert space H, whose graph is a closed subspace of $H \times H$. By the closed graph theorem, a closed operator T is bounded (hence continuous) if and only if the domain of T is all of H.

closed orthonormal set see COMPLETE ORTHONORMAL SET.

closed pair *Mechanics.* two bodies whose relative motions are completely constrained from each other.

closed pores *Materials Science.* the cavities inside a material that are impermeable when the material is immersed in a liquid, in contrast to open pores such as the spaces between particles, cracks, and capillaries that extend into materials from their surfaces.

closed reading frame *Genetics.* a sequence of DNA nucleotides containing termination codons within the sequence, and therefore, not translatable into a protein.

closed reduction *Medicine.* the realignment of fractured or dislocated bones by nonsurgical manipulation.

closed respiratory gas system *Engineering.* a ventilation system used for self-contained, sealed cabins such as space capsules, supplying oxygen to inhabitants, maintaining cabin pressure, and removing exhaled carbon dioxide and water vapor.

closed ridge *Geology.* an irregularly shaped ridge of glacial material that surrounds a central depression and is produced as a stagnant ice block melts.

closed sea *Oceanography.* **1.** the opposite of open sea; a part of the ocean enclosed by headlands or within narrow straits. **2.** a part of the ocean inside the territorial limits of a country.

closed set *Mathematics.* **1.** a subset A of a given set S is said to be a closed set in S if $S - A$ is open. Finite unions and arbitrary intersections of closed sets are closed sets. **2.** in particular, a subset A of a topological space, whose complement is open, is a closed set. By theorem, A is closed if and only if it contains all its accumulation (cluster) points. Also, TOPOLOGICALLY CLOSED SET. **3.** in general, a set S is said to be closed if, given any chain C contained in S, the least upper bound of C is also in S.

closed shell *Physics.* an atomic electron shell that, by the Pauli exclusion principle, has the maximum allowable number of electrons.

closed shop *Computer Technology.* a data processing department that permits only authorized computer operators to run the equipment.

closed site *Archaeology.* a site located within a cave or rock shelter.

closed steam *Engineering.* steam that flows through a coil and heats materials indirectly.

closed subroutine *Computer Programming.* a single segment of code that is stored in memory and can be called from different places in the program. Also, OUT-OF-LINE CODE.

closed system *Thermodynamics.* a system in which no mass can cross the boundaries of the system. Also, NONFLOW SYSTEM.

closed traffic *Aviation.* air traffic that takes off, circles, and lands at a single airport without leaving that airport's traffic pattern.

closed traverse *Cartography.* a survey traverse that starts and ends on the same station, or on stations that have had their relative locations determined by other surveys.

closed universe *Astrophysics.* any model for the universe that contains enough matter to overcome by gravity its infinite expansion.

closed-world assumption *Artificial Intelligence.* in a theorem prover, logic programming language, or database system, the convention that if a question cannot definitely be proved to be true, it may be assumed to be false.

close-joint cleavage **1.** see FRACTURE CLEAVAGE. **2.** see SLIP CLEAVAGE.

closely-coupled *Computer Science.* describing a parallel computer architecture consisting of multiple central-processing units that are tightly connected; e.g., by sharing the same memory.

close-packed crystal *Crystallography.* a crystal structure consisting of a lattice of points that coincide with the centers of spheres of equal size arranged in layers so that a sphere is in contact with six other spheres in its own layer, three spheres in the layer above, and three spheres in the layer below; such a structure provides minimal interstitial volume. This is one of the two close-packed types of unit cells that form the crystal of most metals, the other being face-centered-cubic. Also, **close-packed hexagonal.**

close-packed plane *Materials Science.* a layer of atom having the highest coordination number. It is generally the plane on which slip occurs duting plastic deformation.

close-range photogrammetry *Photogrammetry.* a branch of photogrammetry that is concerned with object-to-camera distances of not more than 300 meters.

close routine *Computer Programming.* a series of statements that perform closing and storing tasks on an active file so that it will be ready for subsequent reopening.

closest approach *Astronomy.* the moment when two celestial bodies lie closest together, either positionally on the celestial sphere or physically.

close support *Military Science.* the action of a supporting force against targets or objectives so close to the force being supported that detailed integration and coordination between the supporting and supported forces is required. Thus, **close-supporting fire.**

close-talking microphone *Acoustical Engineering.* a microphone designed to be positioned close to the mouth to eliminate extraneous noise.

close-tolerance forging *Metallurgy.* forging that results in very close dimensional control, usually requiring only minor or no subsequent machining.

closing *Cartography.* the act of finishing a survey process so that its accuracy may be checked.

closing corner *Cartography.* a corner at the intersection of a surveyed boundary with a previously established boundary line.

closing line *Mechanics.* a vector used to complete a vector polygon so that the resultant, i.e., the vector sum, is zero.

closing machine *Engineering.* a device used to interweave wire into strands and to form the strands into rope. Also, STRANDING MACHINE.

closing pressure *Mechanical Engineering.* the amount of pressure on a safety valve when at rest.

closing rate see CLOSING SPEED.

closing speed *Aviation.* the speed at which two aircraft or other flight vehicles approach one another, based on the combination of their individual speeds. Also, CLOSING RATE, CLOSURE.

Closterium *Botany.* a genus of common green algae in the family Peniaceae, characterized by their crescent-shaped cells.

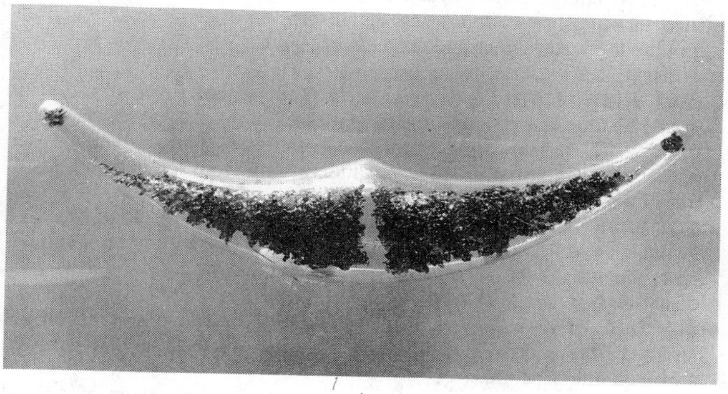

Closterium

Closterovirus group *Virology.* a genus of plant viruses with highly flexuous rod-shaped particles that infect a wide host range, primarily by aphid transmission.

Clostridium *Bacteriology.* a genus of Gram-positive, anaerobic bacteria that occur as rod-shaped cell and form endospores; found in soil and in the intestinal tracts of humans and other animals.

clostripain *Biochemistry.* a protease that functions at basic amino acyl residues to cleave amides and esters of amino acids; present in *Clostridium histolyticum.*

closure the act of closing, or something that does this; specific uses include: *Building Engineering.* **1.** the part of a brick used to complete the end of a course. **2.** an architectural screen used to form a parapet. *Aviation.* see CLOSING SPEED. *Cartography.* the act of completing a survey on the same station from which it began, thereby completing a loop; the difference in elevation between the first and last sightings should theoretically be zero. *Geology.* the vertical distance between the highest point of a fold, dome, or other structural trap and the lowest contour that encloses it. *Artificial Intelligence.* in Lisp, the combination of a function definition and a binding environment; it can be used in constructing generators or coroutines. *Psychology.* the concept that an individual has an innate tendency to fill in the missing parts of a perceived image or memory, so as to create a completed whole. *Mathematics.* **1.** the closure of a set M is the union of the set of all accumulation points of M (called the derived set of M) with M itself; equivalently, the smallest closed set that contains M; the intersection of all closed sets that contain M. **2.** given a simple undirected graph G with n vertices, the closure $c(G)$ is formed by recursively joining by an edge a pair of nonadjacent vertices whose degree sum just before the joining is at least n, until no such nonadjacent pairs of vertices exist. By theorem, $c(G)$ is independent of the order in which edges are added.

closure minefield *Ordnance.* an underwater minefield that creates such a danger that it prevents enemy ships from moving.
closure parameter *Astrophysics.* the ratio of the local density of the universe to that which is just necessary to halt its expansion; if the ratio is greater than 1, the universe is closed; if less than 1, it is open.
clot *Physiology.* **1.** a semisolid mass, such as the mass of protein fibers and blood cells that forms during the coagulation of blood. **2.** any semisolidified mass, as of blood or lymph. **3.** to form such a mass. Also, COAGULUM.
clothing monitor *Nucleonics.* a survey meter that is used to detect and measure radioactivity on outer clothing and protective garb such as coveralls, gloves, hair and shoe covers, and respirators, after possible exposure to contamination.
clot retraction time *Pathology.* the time elapsed when 50% of a coagulum recedes from the wall of its containment vessel and propels serum.
clotting factor *Hematology.* any of those substances in the blood, designated by Roman numerals I–XIII (Factor I, Factor II, and so on), whose presence in the blood is essential to the clotting process.
clotting time *Physiology.* the time required for blood coagulation as a function of certain biochemical and physical factors.
cloud *Meteorology.* **1.** an aggregate of small water droplets or ice crystals that float in the atmosphere above the earth's surface. **2.** any collection in the atmosphere of particulate matter, such as dust or smoke, that is dense enough to be perceptible to the eye. *Nuclear Physics.* a term used for the collection of orbital electrons, moving in the total electric field attributable to the nucleus of an atom, that make up the spherically symmetric charge cloud of the central-field approximation.
cloud absorption *Geophysics.* the absorption of electromagnetic radiation by the water vapor and droplets in clouds.
cloudage see CLOUD COVER.
cloud attenuator *Electromagnetism.* the attenuation of radiation caused by clouds, primarily due to scattering rather than absorption.
cloud band *Meteorology.* a broad belt of clouds, typically between ten and over one hundred miles wide, and varying from ten to hundreds of miles long.
cloud bank *Meteorology.* a fairly well-defined mass of cloud seen at a distance, covering an appreciable portion of the horizon, but not extending overhead.
cloud banner see BANNER CLOUD.
cloud bar *Meteorology.* **1.** a heavy bank of clouds appearing on the horizon at the approach of an intense tropical storm. **2.** a long, narrow, unbroken line of cloud.
cloud base *Meteorology.* the lowest level in the atmosphere of a cloud or cloud layer at which the air has a definite quantity of cloud particles.
cloudburst *Meteorology.* a popular designation for a sudden, heavy rainfall, about equal to or greater than 3.94 inches per hour.
cloud cap see CAP CLOUD.
cloud chamber *Nucleonics.* an apparatus for delineating the tracks of high-speed particles as they pass through a chamber filled with a saturated vapor that, when supercooled by adiabatic expansion, condenses to form droplets of liquid around ions left in the particle's path. Also, WILSON CLOUD CHAMBER.

cloud chamber

cloud classification *Meteorology.* any of the three systems by which clouds are grouped and distinguished; i.e., by their appearance and process of formation, their usual altitude, or their particulate composition.
cloud cover *Meteorology.* the portion of the sky cover that is attributed to clouds, usually expressed in terms of tenths of sky covered. Also, CLOUDAGE, CLOUDINESS.
cloud crest see CREST CLOUD.
cloud deck *Meteorology.* the uppermost surface of a cloud.
cloud discharge *Geophysics.* the release of lightning from an area of charge (usually negative) to an area of opposite charge within a cloud.
cloud droplet *Meteorology.* a particle of liquid water formed by the condensation of atmospheric water vapor that is suspended in the atmosphere with other drops to form a cloud. Also, **cloud drop.**
cloud-drop sampler *Engineering.* a device consisting of a sampling plate covered with a substance that, when exposed to a cloud, captures or forms impressions of cloud particles.
cloud echo *Meteorology.* a radar target signal returning from clouds alone and detected by very-short-wavelength equipment.

cloud formation

cloud formation *Meteorology.* **1.** the process by which various types of clouds are formed, usually by adiabatic cooling of ascending moist air. **2.** a particular arrangement or appearance of clouds in the sky.
cloud genus *Meteorology.* any of the ten characteristic cloud forms: cirrus, cirrocumulus, cirrostratus, altocumulus, altostratus, nimbostratus, stratocumulus, stratus, cumulus, and cumulonimbus.
cloud height *Meteorology.* the height of a cloud base above local terrain.
cloudiness see CLOUD COVER.
cloud layer *Meteorology.* a series of continuous or detached clouds, not necessarily of the same type, whose bases lie at approximately the same altitude over a given area.
cloud level *Meteorology.* **1.** the layer in the atmosphere between the bases and tops of an existing cloud form at a particular time. **2.** a layer in the atmosphere in which certain cloud genera are found, usually defined in cloud classification as high, middle, and low.
cloud mirror *Engineering.* a nephoscope that utilizes a mirror to observe cloud motion. Also, MIRROR NEPHOSCOPE.
cloud modification *Meteorology.* any artificial process that alters the natural development of a cloud, including the stimulation of precipitation or the dissipation of the cloud.
cloud particle *Meteorology.* any of the cloud droplets of liquid water or ice crystals that constitute a cloud.
cloud-phase chart *Meteorology.* a chart that indicates and distinguishes supercooled water clouds from ice-crystal clouds.
cloud physics *Meteorology.* the study of the physical, synoptic, and dynamic processes that govern the initiation, development, and dissipation of clouds and their associated precipitating effects.
cloud point *Materials Science.* the temperature and ionic strength of a solution at the point where phase separation is induced. Also, PHASE SEPARATION POINT. *Chemical Engineering.* specifically, the temperature at which wax or other solids begin to crystallize and separate from a solution of petroleum oil that is being cooled.

cloud pulse *Electronics.* the energy produced when the electron beam in a charge-storage tube is switched on and off.

cloud seeding *Meteorology.* an artificial moisture-enhancing technique by which nucleating or precipitating particles are added to a cloud to alter its natural development.

cloud shield *Meteorology.* the cloud forms found on the cold air side of the frontal system of a typical wave cyclone.

cloud symbol *Meteorology.* any one of a number of specific symbols plotted on a weather map that represent a principal cloud type observed over a defined area.

cloud system *Meteorology.* an array of clouds and precipitation that is associated with a cyclonic-scale feature of atmospheric circulation and displays typical patterns and continuity. Also, NEPHSYSTEM.

cloud test *Chemical Engineering.* an ASTM (American Society for Testing and Minerals) procedure to define the point at which solids begin to form in petroleum oil.

cloud-to-cloud discharge *Geophysics.* the release of lightning from an area of charge in one cloud to an area of opposite charge in another.

cloud-to-ground discharge *Geophysics.* the release of lightning from an area of charge in a cloud to an area of opposite charge on the ground.

cloud top *Meteorology.* the highest level in the atmosphere of a cloud or cloud layer at which the air contains a perceptible quantity of cloud particles.

cloudy *Meteorology.* the character of a day's weather when the average cloudiness has been more than 0.7 for the 24-hour period.

cloudy crystal-ball model *Nuclear Physics.* an analogy used to explain the scattering of nucleons by nuclei, in which the nucleus is presented as a sphere that partially refracts and partially absorbs the incident nucleon wave. Also, OPTICAL MODEL.

cloudy swelling *Pathology.* the primary occurrence of a toxic degenerative alteration in the cell structure after an injury or infection; cells appear enlarged and milky due to the buildup of intracellular water, but return to normal when the cause is removed.

clough [kluf] *Geography.* a steep riparian gorge or ravine. Also, CLEUGH.

clout nail *Mechanical Devices.* a construction nail having a large, flat, thin head.

clove *Botany.* any young bulb that is generated in the scales of a larger bulb. *Food Technology.* a spice derived from the unopened flowers of the *Eugenia caryophyllata,* a tree of the Myrtle (Myrtaceae) family.

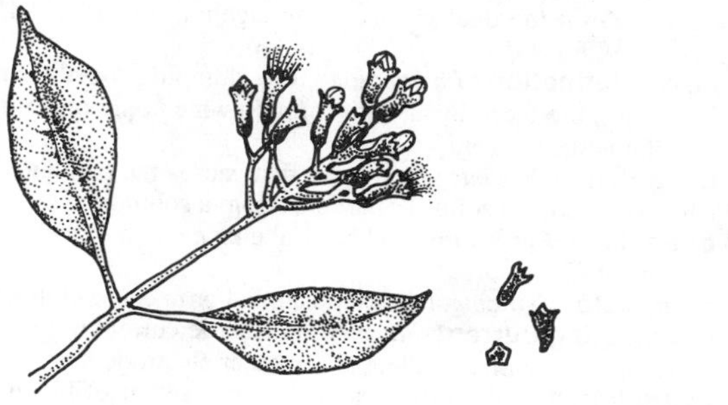

clove

Clovelly *Geology.* a North American Gulf Coast geologic stage of the Miocene epoch, occurring after the Duck Lakean and before the Foleyan.

clove oil *Materials.* an optically active, essential oil distilled from cloves; used in medicine, flavoring, perfumes, and soaps.

clover *Botany.* any herbaceous, trifoliate plant of the genus *Trifolium,* characterized by alternate leaves and small, dense, capitate flowers.

cloverleaf antenna *Electromagnetism.* an omnidirectional antenna whose radiators are arranged in the shape of a four-leaf clover.

Clovis point *Archaeology.* a long, leaf-shaped projectile point type characteristic of the early Paleo-Indian period; often found in association with mammoth bones. (From *Clovis,* New Mexico, where it was first found.)

clr clear; clearance.

CLT *Aviation.* the airport code for Charlotte, North Carolina.

clubfoot *Medicine.* a congenital deformity of the foot, which is twisted out of position or shape. Also, TALIPES.

club fungi *Mycology.* the common name for fungi belonging to the family Clavariaceae.

club moss *Botany.* any member of the genus *Lycopodium,* spore-producing, vascular green plants.

Clupeidae *Vertebrate Zoology.* a family of primitive bony fishes in the suborder Clupeoidei, including herring, shad, sardines, and menhaden.

Clupeiformes *Vertebrate Zoology.* a prolific, widespread order of herringlike fishes in the subclass Actinopterygii, generally having a scaly, compressed body and four pairs of gills.

clupeine or **clupein** *Biochemistry.* a simple protein or protamine found in the sperm of salmon or herring; forms a strong bond with deoxyribonucleic acid.

Clupoidea *Vertebrate Zoology.* a suborder of bony fishes in the order Clupeiformes, containing the herrings, sardines, and anchovies. Also, **Clupeoidei**.

cluse *Geology.* a narrow, steep-walled trench or water gap cutting crosswise through an otherwise continuous mountain range.

Clusiaceae *Botany.* a family of tropical and temperate dicotyledonous trees, shrubs, vines, and herbs of the order Theales, characterized by secretory canals or cavities with resinous juices, simple entire leaves with lateral veins and no stipules, and a berry, drupe, or capsule fruit. Also, GUTTIFERAE.

Clusiidae *Invertebrate Zoology.* a family of insects in the subsection Acalypteratae.

Clusius column *Nucleonics.* a device consisting of a long vertical cylinder with a heated wire along the axis, used for isotope separation by thermal diffusion; the temperature gradient produced by the wire causes the lighter isotope to concentrate around the wire and the heavier isotope to concentrate near the walls of the tube.

star cluster

cluster a group of similar items considered as a unit; specific uses include: *Astronomy.* any group of stars or galaxies held together by gravitation and sharing a common proper motion. *Psychology.* any group of objects, words, pictures, or events that an individual perceives as belonging together. *Ordnance.* **1.** a group of small bombs released in a single unit, usually consisting of fragmentation or incendiary bombs. **2.** a component of the pattern laid in a minefield, consisting of one to five mines; it may include antitank and antipersonnel mines, but no more than one antitank mine. *Computer Technology.* **1.** a group of disk sectors that are treated as a unit. **2.** a group of peripheral devices, tape units, or terminals that are accessed together, usually through a master controller. *Computer Programming.* **1.** in a cluster-organized file, a group of items having similar content identifiers. **2.** a group of related data types and function definitions.

cluster aggregation *Physics.* a mathematical model of a coagulation process; in a collection of particles in random motion, two particles adhere when they come within a certain fixed distance of each other.

cluster analysis *Statistics.* in multivariate analysis, the study of the relations among characters or individuals, based on grouping units that show similar patterns.

clustered file *Computer Programming.* a file in which items with similar content identifiers are automatically grouped into common classes for faster file search.
cluster expansion *Physics.* a virial expansion in which the virial coefficients are obtained from integrals of functions involving intermolecular potentials.
cluster gene *Genetics.* a gene that encodes a multifunctional protein.
clustering algorithm *Computer Programming.* in pattern recognition, a statistical method of establishing any sets of samples, among a given group of unlabeled samples, that are more similar to each other than to samples outside the set.
cluster mill *Metallurgy.* a rolling mill in which the working rolls, touching the workpiece, are supported by two or more backup rolls. Close tolerances of products, including foil, are so obtained.
cluster point *Mathematics.* an accumulation point.
cluster radioactivity *Nuclear Physics.* the emission by a nucleus of a fragment that is heavier than an alpha particle but lighter than a fission fragment, such as carbon-14.
clutch *Mechanical Engineering.* any device for engagement or disengagement of power; may be friction, electromagnetic, hydraulic, or pneumatic. *Vertebrate Zoology.* a nest of eggs or a brood of newly hatched birds.
clutch wheel *Horology.* a geared wheel on the winding shaft of a mechanical timepiece; when actively engaged with the crown wheel, it transmits the winding force from the stem to the mainspring. Also, CASTLE WHEEL.
clutter *Electromagnetism.* any interference appearing on a radar scope due to extraneous echoes.
clutter gating *Electronics.* a technique that provides switching between moving-target indicators and normal video, allowing normal video to be displayed in areas with no clutter, and moving-target indicator video to be switched in only for the clutter areas.
Clydesdale *Agriculture.* a breed of draft horse, usually bay-colored with a white face and legs, characterized by a large amount of fetlock hair. (From *Clydesdale*, Scotland, where the breed originated.)
clymenid *Paleontology.* of or relating to members of the ammonoid order Clymeniida, of the lower Devonian; clymenids are the only ammonoids that have dorsal siphuncles and are important in defining the boundary between the Devonian and the Carboniferous.
Clypeasteroida *Invertebrate Zoology.* an order of echinoderms in the subclass Euechinoidea.
clypeate *Biology.* **1.** shaped like a rounded shield. **2.** having a process thus shaped. Also, **clypeiform.**
clypeus *Invertebrate Zoology.* an anterior medial plate on the head of an insect. *Mycology.* in certain fungi belonging the subdivision Ascomycotina, a disk of tissue found around the mouth of the spore sac or perithecium.
clysis *Medicine.* the instillation of a fluid into the body by means other than the mouth.
Clytemnestra complex *Psychology.* the desire of a wife to kill her husband. (From the ancient Greek queen *Clytemnestra* of Mycenae, who with her lover Aegisthus murdered her husband Agamemnon.)
Clythiidae *Invertebrate Zoology.* a family of flat-footed flies, two-winged insects in the series Aschiza.
CM circular mil; common meter.
Cm the chemical symbol for curium.
cM centimorgan.
cm or **cm.** centimeter.
CMA certified medical assistant.
CMC cell-mediated cytotoxicity.
cmc critical micelle concentration.
cmd command. Also, **comd.**
cmdr commander. Also, **comdr.**
CMF *Oncology.* a breast-cancer chemotherapy regimen that includes the drugs cyclophosphamide, methotrexate, and 5-fluorouracil.
CMH *Aviation.* the airport code for Port Columbus International, Columbus, Ohio.
CMI cell-mediated immunity; computer-managed instruction.
C-MOPP *Oncology.* a cancer chemotherapy regimen that includes the drugs cyclophosphamide, mechlorethamine, Oncovin, procarbazine, and prednisone.
CMOS device *Electronics.* an integrated circuit that uses both PMOS and NMOS devices on the same substrate, resulting in extremely low power dissipation.
CMP capillary melting point.
CM Tauri *Astronomy.* the supernova of 1054 that gave rise to the Crab Nebula. Also, **SN 1054.**
CNC computer numerical control.
Cneoraceae *Botany.* a monogenetic family of dicotyledonous shrubs of the order Sapindales, characterized by secretory cells, small alternate leaves, and separating mericarpic drupelets; native to the Canary Islands, Cuba, and the western Mediterranean.
CNF conjunctive normal form.
Cnidaria see COELENTERATA.
cnidoblast *Invertebrate Zoology.* a nettle cell or stinging cell that produces nematocysts. Also, **cnidocyte.**
cnidocil *Invertebrate Zoology.* the part of a cnidoblast that activates discharge of the nematocyst when touched.
cnidocyst see NEMATOCYST.
cnidophore *Invertebrate Zoology.* a structure bearing nematocysts in certain coelenterates.
Cnidospora *Invertebrate Zoology.* a subphylum of spore-producing protozoans that are parasitic on invertebrates, fish, turtles, and some amphibians.
CNO cycle see CARBON-NITROGEN-OXYGEN CYCLE.
CNS central nervous system.
c-number theory see CLASSICAL FIELD THEORY.
CO *Aviation.* the airline code for Continental Airlines.
CO cardiac output; commanding officer.
Co the chemical symbol for cobalt.
co- a prefix meaning "with" or "together," as in *coaxial, cohesion.*
CoA see COENZYME A.
coacervate *Chemistry.* a mass of colloidal particles that are bound together by electrostatic attraction.
coacervation *Chemistry.* the process of forming a coacervate by the partial mixing of two or more liquids, at least one of which is in a colloidal state. *Materials Science.* the separation of a polymer solution into two or more phases.
coach screw *Mechanical Devices.* a large, square-headed wooden screw that is turned by a spanner to join two pieces of heavy timber. Also, LAG BOLT, LAG SCREW.
coadaptation *Evolution.* **1.** an evolutionary change producing mutually advantageous hereditary characteristics in two or more interacting species. **2.** a selection by which harmoniously interacting genes accumulate in the gene pool of a population, thus enabling them to survive. Also, INTERNAL BALANCE.
coadjoint representation *Mathematics.* the representation of a Lie group G on the dual g^* of its Lie algebra; given by $Ad^*(A)(\phi) = \phi \times Ad(A^{-1})$, for $A \in G$.
co-agglutination *Immunology.* the clumping together of two distinct antigens, produced by an antiserum, in which only one of the antigens is an immunizing agent.
coagulant *Chemistry.* any agent that causes the precipitation or separation of dissolved solids or particles from a solution.
coagulase *Enzymology.* a bacterial enzyme that causes blood plasma to coagulate.
coagulate *Medicine.* to become clotted or to cause clotting.
coagulating current *Surgery.* an electric current used to coagulate tissue; applied with a needle, ball, or other electrode.
coagulation *Hematology.* a clumping together of blood or lymph; the process of clot formation. *Surgery.* the disruption of tissue in order to cause formation of a clot.
coagulation factor VIII see ANTIHEMOPHILIC FACTOR.
coagulation factor XIII see FIBRINASE.
coagulation system *Immunology.* a system of twelve proteins in serum, resulting in the production of fibrin that blocks bleeding from injured vessels.
coagulator *Surgery.* a surgical device that employs electric current or light to stop bleeding.
coagulopathy *Medicine.* any disorder of blood coagulation.
coagulum *plural,* **coagula.** see BLOOD CLOT.
Coahuilan *Geology.* a North American provincial series of the Lower Cretaceous period, after the upper Jurassic and before the Comanchean.
coal *Geology.* a brown to black combustible, carbonaceous sedimentary rock formed by the compaction of partially decomposed plant material. *Materials.* this rock when burned as fuel. Coal was once the main type of fuel in all industrial countries, and it is still widely used as a source of energy for electric power plants. It is also the source of coke, which is used in the manufacture of iron and steel. (From an Old English word for this substance.)

coal ash see ASH.
coal ball *Geology.* a roundish nodule or mass of mineral and plant debris found in coal seams or adjacent rocks.
coal bank *Mining Engineering.* an exposed seam of coal.
coal barrier *Mining Engineering.* a protective pillar of coal. Also, BARRIER PILLAR.
coal basin see COALFIELD.
coal bed *Geology.* a layer of coal lying parallel to the rock stratification. Also, COAL SEAM.
coal blasting *Mining Engineering.* a method of using explosives to break coal.
coalbreaker *Mining Engineering.* see BREAKER, def. 1.
coal breccia *Geology.* naturally occurring angular fragments of coal within a coal bed.
coal clay see UNDERCLAY.
coal cutter *Mining Engineering.* a power-operated machine that draws itself by rope haulage along the coal face and cuts out a thin strip of coal from the bottom of the seam.
coal digger see FACEMAN.
coal drill *Mining Engineering.* a rotary drill, usually electric, of a light, compact design.
coal dust *Mining Engineering.* a finely divided coal; there is no established standard as to how finely coal must be divided to be termed dust.
coalesce *Science.* to join together, forming a whole.
coalesced copper *Metallurgy.* the copper produced by briquetting comminuted brittle cathode copper, sintering in a reducing atmosphere, and hot working.
coalescence *Physics.* the formation of a body by bringing together its constituent parts, as in the growth of an ice crystal by the accumulation of water molecules or the merging of two water drops to form a larger drop. *Metallurgy.* **1.** in physical metallurgy, the solid-state aggregation of small particles into larger ones. **2.** in joining, the attachment of two or more components. *Botany.* the fusion of like parts within a plant.
coalescence efficiency *Meteorology.* the fraction of all collisions between water drops of a specified size that result in coalescence.
coalescence process *Meteorology.* the growth of raindrops by the collision and coalescence of water drops and small precipitation particles.
coalescent *Chemistry.* any agent used to separate two immiscible liquids by causing the droplets of one to form a mass.
coalescent pack *Chemical Engineering.* a high-surface-area packing that consolidates liquid droplets for gravity separation from a second phase, such as immiscible liquid or gas.
coalescer *Chemical Engineering.* a process unit containing wettable, high-surface-area packing on which liquid droplets coalesce.
coal face *Mining Engineering.* see FACE, def. 4.
coalfield *Mining Engineering.* a coal-bearing area. Also, COAL BASIN.
coal gasification *Chemical Engineering.* a method of converting coal, coke, or char to a gaseous product by reaction with oxygen, steam, carbon dioxide, air, or a mixture of these.
coal getter see FACEMAN.
coalification *Geology.* the natural processes, including diagenesis and metamorphism, whereby plant material is transformed into coal. Also, CARBONIFICATION, INCARBONIZATION, INCOALATION, BITUMINIZATION.
coal liquefaction *Chemical Engineering.* the procedure for making a liquid mixture of hydrocarbons by the destructive distillation of coal.
Coal Measures *Geology.* **1.** in Europe, the sequence of rocks corresponding to the Upper Carboniferous period, broadly synchronous with the Pennsylvanian of North America. **2. coal measures.** any sequence of sedimentary rocks that includes coal beds interstratified with clays, shales, sandstones, etc.
coal mining *Mining Engineering.* the process of removing coal from the earth and preparing it for market.
coal paleobotany *Paleontology.* the division of paleobotany that deals with coal deposits, including the types of plants and other organisms associated with coal-bearing strata.
coal pebbles *Geology.* spherical masses of coal formed in sedimentary rock as a result of pressure.
coal petrology *Geology.* the study of the origin, occurrence, structure, chemical composition, and classification of coal.
coal planer *Mining Engineering.* a type of continuous coal-mining machine developed for longwall mining, in which power equipment drags a heavy steel plow back and forth across a coal face.
coal plow *Mining Engineering.* a device that pushes coal onto the face conveyor after shearing it off with steep blades.
Coalsack *Astronomy.* a dark, dense cloud of dust about 500 light-years away, lying mostly in the southern constellation Crux and obscuring the more distant stars of the Milky Way behind it.
coal seam see COAL BED.
coal-sensing probe *Mining Engineering.* a nucleonic instrument that uses a gamma-ray backscattering unit to measure the thickness of coal left on a seam floor.
coal sizes *Mining Engineering.* the sizes by which anthracite coal is marketed, determined by the diameter of the opening through or over which the coal will pass.
coal tar *Materials.* a black, thick liquid formed during the distillation of coal that yields compounds such as benzene and phenol, from which a large number of dyes, drugs, and other compounds are derived, and that finally yields a residuum, which is used in pavements.
coal-tar creosote see CREOSOTE OIL.
coal tar dye *Organic Chemistry.* a dye produced from coal tar hydrocarbons.
coal-tar light oil see LIGHT OIL, def. 1.
coal tar pitch *Materials.* a stable bituminous material used to seal and cover underground pipes.
co-altitude *Astronomy.* the angular distance between a celestial object and the zenith.
coaming *Naval Architecture.* a raised flange or low bulwark around a hatch or other deck opening, serving to protect the opening from water on deck.
Coanda effect *Fluid Mechanics.* the angular deflection of a stream of flowing water resulting when a circular cylinder is inserted across the stream.
COAP *Oncology.* a cancer chemotherapy regimen that includes the drugs cyclophosphamide, Oncovin, ara-C, and prednisone.
coapt *Surgery.* to bring together, as the edges of a wound.
coaptation suture *Surgery.* a suture that reapproximates the edges of skin, as after surgical incision.
coarctation *Medicine.* the constriction of a canal, especially of a cardiac blood vessel.
coarse *Science.* **1.** consisting of relatively large particles or elements. **2.** having a form or texture that is rough or that lacks delicacy.
coarse fragment *Geology.* a rock or mineral particle in soil that is larger than 2 millimeters in diameter.
coarse-grained *Computer Programming.* describing an arrangement of parallel computers that consists of a few large processors.
coarse-grained phenocrystalline see PHANERITIC.
coarse mine *Ordnance.* a relatively insensitive marine mine.
coarse setting *Ordnance.* the preliminary adjustment on the main scale of a gunsight.
coarse sighting *Ordnance.* the adjustment of a gunsight so that the front sight is visible through the notch in the rear sight.
coast *Engineering.* a memory feature on a radar having the ability to make the range and angle systems move in the same direction and speed as they did when following an original target. *Geography.* **1.** the edge or margin of land touching the sea; the seashore. **2.** a region of land near the sea. *Navigation.* to sail along the coast between ports.
coastal current *Oceanography.* a current running parallel to the shore and just beyond the surf zone.
coastal dune *Geology.* a sand dune that occurs along sea and lake shores.
coastal plain *Geography.* an area of lowland between the seashore and the nearest hills, usually emergent or alluvial and often sloping gently seaward.
coastal refraction *Electromagnetism.* a change in the direction of a radio wave as it crosses a coastline.
coastal sediment *Geology.* the mineral or organic deposits of formations located on or near a coast.
coast artillery *Ordnance.* **1.** artillery designed specifically for coastal defense. **2.** a military unit operating such artillery.
coaster *Naval Architecture.* a relatively small cargo vessel that is normally engaged in local coastwise service.
Coast Guard *Military Science.* **1.** a U.S. military service under the Department of Transportation; it guards the coast, protects the coastal environment, and ensures the safety, order, and effective operation of maritime traffic; in wartime, it may be placed under the Department of the Navy. **2. coast guard.** any similar service in other countries.
coast guard cutter *Naval Architecture.* any regular-service U.S. Coast Guard vessel that is larger than a boat.
coastguardsman *Military Science.* a member of a coast guard service.

Coast Guard station *Navigation.* a land base from which Coast Guard craft operate.

coasting *Navigation.* a course or movement that follows the coastline, generally near enough to be in pilot waters.

coasting flight *Space Technology.* the unpowered flight of a rocket vehicle or missile after thrust cutoff or between the burnout of one stage and the ignition of another.

coastline *Geography.* **1.** the outline of a coast, following the edge of the land including bays, but crossing rivers and minor inlets at their mouths. **2.** the shape or appearance of a coast as seen from the sea. **3.** the landward edge of a beach, usually measured as the highwater mark of spring mean tides.

coastlining *Cartography.* the process of obtaining data from which a coastline can be depicted on a chart.

coast of transverse deformation see COMPOSITE COAST.

coast piloting *Navigation.* the process of navigating in coastal waters by using terrestrial references and aids to navigation.

coast shelf see SUBMERGED COASTAL PLAIN.

coastwise navigation *Navigation.* navigation in the vicinity of a coast.

coat *Virology.* any protective shell around a viral nucleic acid, such as a capsid or layers of protein and lipids.

coated cathode *Electronics.* a cathode that is coated with a compound having a lower work function than the base metal in order to increase electron emission.

coated electrode *Metallurgy.* in consumable arc welding, a filler metal electrode that is coated with nonmetallic compounds designed to stabilize the arc.

coated filament *Electronics.* a filament that has been coated with metal oxides in order to increase electron emission.

coated lens *Optics.* a lens to which a thin transparent film of one or more layers of different refractive index has been applied in order to decrease the amount of light that is lost to reflection.

coated paper *Graphic Arts.* any paper whose surface has been covered with a smooth mineral substance, such as an adhesive, in order to improve its printing qualities.

coated pit *Cell Biology.* a depression on the surface of the plasma membrane of eukaryotic cells that is coated on its cytoplasmic surface with the protein clathrin and functions in the receptor-mediated endocytosis of low-density lipoprotein, insulin, and other ligands.

coat hanger die *Engineering.* a plastic sheet slot die having an interior that is similar in shape to a coat hanger.

coati [kō ät´ē] *Vertebrate Zoology.* any member of the genus *Nasua*, omnivorous tropical American mammals in the raccoon family Procyonidae and the suborder Caniformia, distinguished by a long, flexible snout and vertically held tail. Also, **coatimundi**.

coati

coating *Graphic Arts.* **1.** any mineral substance used to cover the surface of a printing paper, such as blanc fixe or china clay. **2.** the application of such a substance to a sheet of paper. **3.** the application of a varnish or photosensitive solution to a printing plate. *Materials Science.* a material that forms a continuous layer over a surface, or the film formed by such a material.

CoA-transferase *Enzymology.* an enzyme that transfers coenzyme A.

coaxial *Mechanics.* having a common axis.

coaxial antenna *Electromagnetism.* a vertical antenna that is fed by a coaxial line and is essentially formed by extending the inner conductor by one-quarter of a wavelength while folding the outer conductor back by about one-quarter of a wavelength.

coaxial attenuator *Electromagnetism.* an attenuator having a coaxial construction and terminations suitable for use with coaxial cable.

coaxial cable *Electromagnetism.* a transmission line that is constructed in such a way that an inner conductor is isolated from an outer conductor and both conductors share a common longitudinal axis. Also, COAXIAL LINE.

coaxial capacitor see CYLINDRICAL CAPACITOR.

coaxial cavity *Electromagnetism.* a circular cylindrical cavity resonator having a conductor along its axis that may be in contact with a tuning piston.

coaxial cavity magnetron *Electronics.* a chamber in which the central conductor makes contact with movable pistons, or other reflecting devices, in order to pick up microwaves.

coaxial circles *Mathematics.* **1.** the circles in R^3 that would be concentric except for translation along an axis perpendicular to the planes of the circles. **2.** the circles in R^2 that have the same radical axis.

coaxial connector *Electromagnetism.* an electric connector between a coaxial cable and the circuit of an electric or electronic component.

coaxial cylinder magnetron *Electronics.* a device, used to generate extremely powerful microwaves, that achieves mode separation, high efficiency, stability, and ease of mechanical tuning by coupling two different types of resonators.

coaxial cylinders *Mathematics.* cylinders sharing a common axis; that is, cylinders whose cylindrical surfaces are generated by lines passing through, and perpendicular to, concentric coplanar circles.

coaxial diode *Electronics.* a diode having the same outer diameter and terminations as the coaxial cable into which it is inserted.

coaxial filter *Electromagnetism.* a section of coaxial line having reentrant elements providing the inductance and capacitance of a filter section.

coaxial hybrid *Electromagnetism.* a hybrid junction of a coaxial transmission line.

coaxial isolator *Electromagnetism.* a device, used in-line with a coaxial transmission line, that impedes the flow of energy in one direction to a greater extent than it does in the other direction.

coaxial line see COAXIAL CABLE.

coaxial line resonator *Electromagnetism.* a resonator consisting of a length of coaxial line that is short-circuited at one or both ends.

coaxial machine gun *Ordnance.* a machine gun mounted with the main gun of a tank; it may be aimed separately in elevation, so that its line of fire crosses that of the main gun at a specific range.

coaxial relay *Electromagnetism.* a relay device that is designed to switch coaxial lines without disturbing their characteristic impedances.

coaxial speaker *Acoustical Engineering.* a moving-conductor loudspeaker housing, normally supplied by one speaker cable and consisting of two or more loudspeakers mounted such that they share a common axis for a specified sound radiation pattern.

coaxial stub *Electromagnetism.* a short section of coaxial line that serves to branch to another coaxial line or waveguide, producing some desired changes in its character.

coaxial switch *Electricity.* a multiposition switch designed to change connections between coaxial cables leading to high-frequency devices without introducing impedance mismatch.

coaxial transistor *Electronics.* a diffused-base, alloy-emitter device that conducts electricity through a thin piece of semiconductor material at medium power.

coaxial wavemeter *Engineering.* a device that measures frequencies over 100 megahertz.

coaxing *Metallurgy.* the process of applying a gradually increasing stress in order to improve the fatigue strength of a metallic sample.

cob *Mining Engineering.* to break ore using manual hammers to sort out the valuable portion.

cobalt *Chemistry.* an element having the symbol Co, the atomic number 27, an atomic weight of 58.933, a melting point of 1495°C, and a boiling point of 2900°C; a hard, ductile, and somewhat malleable shiny steel-gray metal that corrodes easily in air; it is used in the production of alloys and ceramics. (From a German word for "goblin" or "evil spirit"; a being believed by early miners to inhabit mines.)

cobalt-60 *Nuclear Physics.* a radioactive isotope of cobalt with a half-life of 5.27 years that emits gamma rays when it decays; often used as a source of gamma rays in radiation therapy and nondestructive testing.

cobalt beam therapy *Radiology.* radiation therapy using cobalt-60-supplied gamma radiation.

cobalt black see COBALTIC OXIDE.

cobalt bloom see ERYTHRITE.

cobalt blue *Inorganic Chemistry.* a green-blue pigment composed chiefly of cobalt oxide and alumina. Also, AZURE BLUE, KING'S BLUE.

cobalt bomb *Ordnance.* an atomic or hydrogen bomb with a cobalt encasement that would be transformed into lethal radioactive dust upon detonation; it is purely theoretical because its destructive properties are considered impossible to control.

cobaltic *Chemistry.* **1.** of or relating to cobalt. **2.** describing various compounds of cobalt, especially those in which it has a valence of 3.

cobaltic boride *Inorganic Chemistry.* CoB, prismatic crystals that decompose in water and are soluble in nitric acid; used in ceramics.

cobaltic fluoride *Inorganic Chemistry.* CoF_3, a brown powder that reacts readily with atmospheric moisture; insoluble in alcohol and ether; reacts with water to form a precipitate of cobaltic hydroxide; used to fluoridate hydrocarbons. Also, COBALT TRIFLUORIDE.

cobaltic hydroxide *Inorganic Chemistry.* $Co(OH)_3$, a dark brown powder that is insoluble in water and alcohol; used in cobalt salts and as a catalyst.

cobaltic oxide *Inorganic Chemistry.* Co_2O_3, a grayish-black powder that is soluble in concentrated acids and insoluble in water; decomposes at 895°C; used as a pigment and glaze.

cobaltic potassium nitrite see COBALT POTASSIUM NITRITE.

cobaltite *Mineralogy.* CoAsS, an opaque, metallic, black, gray, or white orthorhombic mineral occurring as pseudocubic or pseudopyritohedral crystals, having a specific gravity of 6.33 and a hardness of 5.5 on the Mohs scale; found in high-temperature hydrothermal vein deposits or metamorphic rocks.

cobalt-molybdate desulfurization *Chemical Engineering.* a process using cobalt-molybdate as a catalyst to desulfurize petroleum.

cobaltocalcite see SPHAEROCOBALTITE.

cobaltomenite *Mineralogy.* $CoSeO_3{\cdot}2H_2O$, a brittle, transparent, pink monoclinic mineral, having a specific gravity of 3.39 to 3.42 and a hardness of 2 to 2.5 on the Mohs scale; it occurs as crystalline crusts on sandstone.

cobaltous *Chemistry.* **1.** of or relating to cobalt. **2.** describing various compounds of cobalt, especially those in which it has a valence of 2.

cobaltous acetate *Organic Chemistry.* $Co(C_2H_3O_2)_2{\cdot}4H_2O$, reddish-violet deliquescent crystals, soluble in water, alcohol, and acid; melts at 140°C; used in inks and paint, and as a catalyst and foam stabilizer.

cobaltous acetate tetrahydrate *Organic Chemistry.* $Co(C_2H_3O_2)_2{\cdot}4H_2O$, reddish-violet deliquescent crystals that are soluble in water, alcohol, and acid; it melts at 140°C; used in inks and paint, and as a catalyst and foam stabilizer.

cobaltous bromide *Inorganic Chemistry.* **1.** $CoBr_2{\cdot}6H_2O$, red-violet deliquescent prisms that are soluble in water, alcohol, and ether; melts and loses $4H_2O$ at 47–48°C and $6H_2O$ at 130°C; used in hygrometers and as a catalyst. **2.** $CoBr_2$, the anhydrous form, bright green, hexagonal, deliquescent crystals. Also, **cobalt bromide.**

cobaltous carbonate *Inorganic Chemistry.* **1.** $CoCO_3$, red triclinic crystals that are insoluble in water and soluble in acids; decomposes at melting point; used as a pigment, catalyst, trace element, and temperature indicator. **2. basic cobaltous carbonate.** $2CoCO_3{\cdot}Co(OH)_2{\cdot}6H_2O$, a commercial form of cobaltous carbonate; red-violet prisms that are soluble in acids and insoluble in water; decomposes in hot water; used in manufacturing cobaltous oxide and in cobalt pigments and salts. Also, **cobalt carbonate.**

cobaltous chloride *Inorganic Chemistry.* **1.** $CoCl_2$, blue hexagonal hygroscopic crystals; soluble in water, alcohol, and acetone; melts at 724°C (in hydrogen chloride gas) and boils at 1049°C; used as an absorbent for ammonia, as a catalyst and reagent, and for many other industrial purposes. **2.** $CoCl_2{\cdot}6H_2O$, the hexahydrate form, red monoclinic crystals; melts at 86°C and loses its water at 110°C. Its uses are similar to those of $CoCl_2$. Also, **cobalt chloride.**

cobaltous chromate *Inorganic Chemistry.* $CoCrO_4$, brown or gray-black crystals that are insoluble in cold water and decompose at melting point; a known carcinogen; used in ceramics as a pigment. Also, **(basic) cobalt chromate.**

cobaltous fluoride *Inorganic Chemistry.* CoF_2, pink monoclinic crystals, soluble in water, slightly soluble in acids, and insoluble in alcohol and ether; melts at about 1200°C and boils at 1400°C; used as a catalyst. Also, **cobalt (di)fluoride.**

cobaltous fluorosilicate *Inorganic Chemistry.* $CoSiF_6{\cdot}H_2O$, pale red trigonal crystals that are soluble in water; an irritant to tissues; used in ceramics. Also, COBALTOUS SILICOFLUORIDE.

cobaltous hydroxide *Inorganic Chemistry.* $Co(OH)_2$, a rose-red powder that is insoluble in water and alkalis and soluble in acids; decomposes at melting point; used as a catalyst and in battery electrodes. Also, **cobalt hydroxide, cobalt hydrate.**

cobaltous iodide *Inorganic Chemistry.* $CoI_2{\cdot}6H_2O$, brownish-red, hexagonal, hygroscopic crystals that are soluble in water and alcohol; loses iodine on exposure to air, and loses its water at 27°C; used in hygrometers and as a catalyst. Also, **cobalt iodide.**

cobaltous nitrate *Inorganic Chemistry.* $Co(NO_3)_2{\cdot}6H_2O$, red crystals that are deliquescent in moist air and soluble in most organic solvents; melts at 55–56°C; used in medicine and pigments. Also, **cobalt nitrate.**

cobaltous oxide *Inorganic Chemistry.* CoO, a grayish powder or reddish crystals, soluble in acids and insoluble in water; melts at about 1795°C; used as a pigment and catalyst. Also, COBALT OXIDE.

cobaltous phosphate *Inorganic Chemistry.* $Co_3(PO_4)_2{\cdot}8H_2O$, a reddish powder that is slightly soluble in cold water and soluble in mineral acids; loses its water at 200°C; used as a pigment and animal feed supplement. Also, **cobalt phosphate.**

cobaltous silicofluoride see COBALTOUS FLUOROSILICATE.

cobaltous sulfate *Inorganic Chemistry.* **1.** $CoSO_4$, or its heptahydrate form $CoSO_4{\cdot}7H_2O$, a reddish-pink powder that is soluble in water; the anhydrous form decomposes at 735°C and the hydrated form loses its water at 420°C; used as a pigment, catalyst, and soil additive, and for various other industrial purposes. **2.** any of various other compounds of cobalt and sulfate, such as $CoSO_4{\cdot}6H_2O$. Also, **cobalt sulfate.**

cobalt oxide *Inorganic Chemistry.* **1.** also, **cobalt monoxide.** see COBALTOUS OXIDE. **2.** see COBALTIC OXIDE. **3.** any of various other compounds composed of cobalt and oxygen, such as Co_3O_4.

cobalt potassium nitrite *Inorganic Chemistry.* $K_3Co(NO_2)_6$, a yellow powder made up of tiny crystals; slightly soluble in water and insoluble in alcohol; decomposes at a melting point of 200°C; used in medicine and as a pigment. Also, POTASSIUM COBALTINITRITE, COBALT YELLOW, FISCHER'S SALT.

cobalt pyrites see LINNAEITE.

cobalt trifluoride see COBALTIC FLUORIDE.

cobalt yellow see COBALT POTASSIUM NITRITE.

cobamide see COENZYME B_{12}.

cobb see PROMONTORY.

cobber *Mining Engineering.* **1.** a device that expels waste material from ore concentrates. **2.** a worker who chips waste rocks from lumps of ore.

cobble *Geology.* **1.** a slightly elevated rounded hill. **2.** any rock fragment larger than a pebble and smaller than a boulder, ranging between 64 and 256 millimeters in diameter, and formed by abrasion in the course of transport. *Archaeology.* a stone worn smooth by the action of water, used as a core for making a stone tool. Thus, **cobble tool.**

cobble beach see SHINGLE BEACH.

Cobbs' disease *Plant Pathology.* a disease of sugarcane caused by the bacterium *Xanthomonas vascularum* that produces dwarfing, slime in the vascular bundles, linear markings on the leaves, and general breakdown of plant tissue. Also, SUGARCANE GUMMOSIS.

cobia *Vertebrate Zoology.* a voracious carnivorous fish, *Rachycentron canadum*, the sole member of the family Rachycentridae, having a sleek body, a jutting lower jaw, and a row of spines in front of the dorsal fin; found in warm seas and valued as a gamefish. Also, CRABEATER.

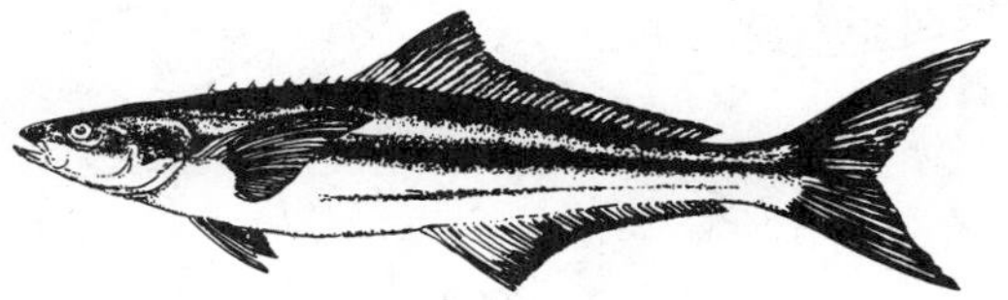

cobia

Cobitidae *Vertebrate Zoology.* the loaches, a family of eel-like freshwater fishes in the suborder Cyprinoidei and the order Cypriniformes, some of which are distinguished by their ability to utilize atmospheric oxygen. Also, **Cobitididae.**

Coblenzian or **Coblentzian** *Geology.* a European geologic stage of the Upper Lower Devonian period, occurring after the Gedinnian and before the Couvinian, and including the Siegenian and Emsian stages.

COBMAM *Oncology.* an acronym for a cancer chemotherapy regimen including the drugs cyclophosphamide, Oncovin, bleomycin, methotrexate, Adriamycin, and MeCCNU.

COBOL [kō´bôl´] *Computer Programming.* a high-level procedural language that is used for commercial data processing and other business applications. (An acronym for common business-oriented language.)

cobordism *Mathematics.* the (equivalence) relationship among mani-

cobordism *Mathematics.* the (equivalence) relationship among manifolds under which two are defined to be related if together they form the boundary of a manifold-with-boundary of one higher dimension.
coboundary *Mathematics.* in cohomology, an element of the image space of a differential d^n of a cochain complex.
coboundary operator *Mathematics.* the differentials $\{d^n: C^n \to C^{n+1}\}$ of a cochain complex, where $d^{n+1} \circ d^n = 0$ for each n.
cobra *Vertebrate Zoology.* a large, highly venomous African and Asian snake of the reptilian family Elapidae, characterized by a dilatable hood on each side of the head.

cobra

Cobra *Ordnance.* a German lightweight anti-tank missile powered by a solid-propellant rocket motor and nonjettisonable booster; the **Cobra 2000** carries a hollow 5.5-pound charge capable of penetrating over 18 inches of steel armor at a range of 1310 to 6560 feet (400 to 2000 meters). Also, **BO 810.**
cobra hemotoxin *Toxicology.* the toxic elements of cobra venom, including phospholipase A, that destroy red blood cells upon contact.
cobraism *Toxicology.* poisoning caused by cobra venom.
coca *Botany.* a South American shrub, *Erythroxylon coca,* from whose dried leaves the drug cocaine is derived.

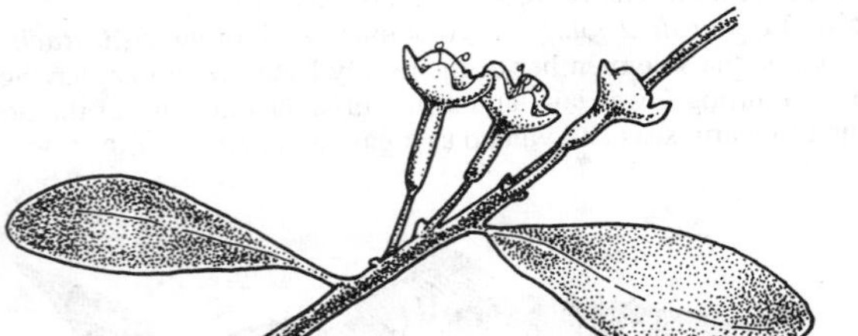

coca

cocaine *Organic Chemistry.* $C_{17}H_{21}NO_4$, an alkaloid compound; colorless to white, monoclinic prisms or powder, insoluble in water and very soluble in alcohol; melts at 98°C. Cocaine acts as a stimulant of the central nervous system and is illegal in the U.S. and many other countries; it is a drug prepared from coca leaves or by synthesis from ecgonine or its derivatives and has been used as a local anesthetic. It affects mental activity, and abuse can lead to habituation or addiction.
cocarcinogen *Oncology.* a factor that produces cancer in combination with other factors, such as a substance that potentiates the effect of a carcinogen with minimal carcinogenicity of its own.
Coccidia *Invertebrate Zoology.* a subclass of protozoans within the class Telosporea; parasitic on vertebrates and invertebrates.
Coccidiascus *Mycology.* a genus of yeast fungi belonging to the class Hemiascomycetes; *Coccidiascus legeri* is a parasite found in the intestine of the fruit fly *Drosophila.*
Coccidiniaceae *Botany.* a monogeneric family of marine dinoflagellates of the order Syndiniales that are parasitic in the nucleus and cytoplasm of several marine, neritic, armored dinoflagellates.
Coccidioides *Mycology.* a genus of fungi belonging to the family Spermophthoraceae; the only species, *C. immitis,* causes a respiratory disease in humans known as desert fever (coccidioidomycosis).
coccidioidomycosis *Medicine.* a fungal disease caused by inhaling spores and resulting in flulike symptoms such as pneumonia, fever, and infrequently, a rash. Also, DESERT FEVER, CALIFORNIA DISEASE.
coccidiosis *Medicine.* a parasitic infection caused by coccidia parasites and usually found in animals, especially domestic fowl; usually asymptomatic in humans, it causes diarrhea, tenesmus, anorexia, and nausea in animals.
coccine *Invertebrate Zoology.* a vegetative condition in protists during which reproduction does not take place.
Coccinellidae *Invertebrate Zoology.* a family of brightly colored ladybird beetles, in the superfamily Cucujoidea, that feed on aphids and other plant pests.
cocco- a combining form meaning "spherical," as in *coccosphere.*
coccobacillus *Microbiology.* a bacterial cell intermediate in morphology between a coccus and a bacillus; an extremely short bacillus.
coccoid *Microbiology.* relating to a bacterial cell that is spheroid in morphology.
Coccoidea *Invertebrate Zoology.* a superfamily of scale insects and mealy bugs in the series Sternorrhyncha.
coccolith *Botany.* interlocking calcareous plates found on the golden-brown algae. *Geology.* a fine-grained oceanic sediment composed of a mixture of undissolved minute calcite or aragonite crystals and amorphous clay-sized particles.
Coccolithophora *Invertebrate Zoology.* an order of phytoflagellate protozoans in the class Phytamastigophorea.
Coccolithophoraceae *Botany.* a family of flagellate marine algae of the order Coccosphaerales, characterized by coccolith-bearing cells with placoliths and including many fossil species.
Coccolithophorida *Botany.* golden-brown algae characterized by a pair of flagellae and a covering of coccoliths.
Coccomyxaceae *Botany.* a family of freshwater, subaerial, and marine green algae of the order Chlorococcales that form gelatinous colonies but have no contractile vacuoles in the vegetative cells.
Coccosphaerales *Botany.* an order of the flagellate algae class Prymnesiophyceae comprising the coccolith-forming species and occurring in marine phytoplankton, especially in warmer waters.
coccosphere *Botany.* a coccolithophore or its sphere-shaped skeleton.
Coccosteoidea *Paleontology.* a superfamily of arthrodire fishes in the extinct suborder Brachythoraci; Devonian.
Coccosteus *Paleontology.* a genus of arthrodire placoderms in the suborder Brachythoraci and superfamily Coccosteoidea; a joint-necked fish about one foot long, and known from European deposits of the Middle Devonian (possibly also the Upper Devonian in North America).
cocculin see PICROTOXIN.
coccus [käk´əs] *Microbiology.* a spheroid bacterial cell.
coccygeal *Anatomy.* relating to or located in the region of the coccyx.
coccygectomy *Surgery.* the surgical removal of the coccyx.
coccyx [käk´siks] *Anatomy.* the final bones in the spinal or vertebral column, formed by the combination of four incompletely formed vertebrae.
cochain complex *Mathematics.* a sequence $\{C^n: -\infty < n < \infty\}$ of algebraic objects such as rings, modules, or Abelian groups, together with a collection $\{d^n: C^n \to C^{n+1}\}$ of homomorphisms called differentials (also coboundary homomorphisms or coboundary operators), such that $d^{n+1} \circ d^n = 0$ for each n. If C denotes a chain complex, then $H^n(C) = \text{Ker}d^{n+1}/\text{Im}d^n$ is called the nth cohomology group of C. Also, **graded differential complex.**
cochannel interference *Telecommunications.* interference caused by two or more transmissions occuring simultaneously in the same channel.
cochineal [käch´ə nēl] *Invertebrate Zoology.* of, relating to, or derived from the dried bodies of female insects, *Coccus cactus,* that enclose the young larvae of the species.
cochineal solution *Organic Chemistry.* $C_{22}H_{20}O_{13}$, a red indicator dye obtained from the dried exoskeletons of female cochineal insects; used in acid-base titrations and as a source of carmine and carminic acid.
cochinilin see CARMINIC ACID.
cochlea [käk´lē ə] *Anatomy.* **1.** a spirally wound tube in the inner ear that is essential to hearing. **2.** a part of the temporal bone that contains the membranous and osseous labyrinthes.
cochlear duct see SCALA MEDIA.

Cochleariidae *Vertebrate Zoology.* the boat-billed herons, a family of broad-billed nocturnal birds in the order Ciconiiformes, usually found in the marshes and densely wooded shores of Central America.

cochlear nerve *Anatomy.* the part of the vestibulocochlear nerve concerned with hearing.

cochlear nucleus *Anatomy.* a small mass of nerve cell bodies in the medulla oblongata that receives impulses regarding sound stimuli from the cochlear nerve.

cochleate [käk´lē āt´] *Biology.* spiral-shaped, like a snail shell.

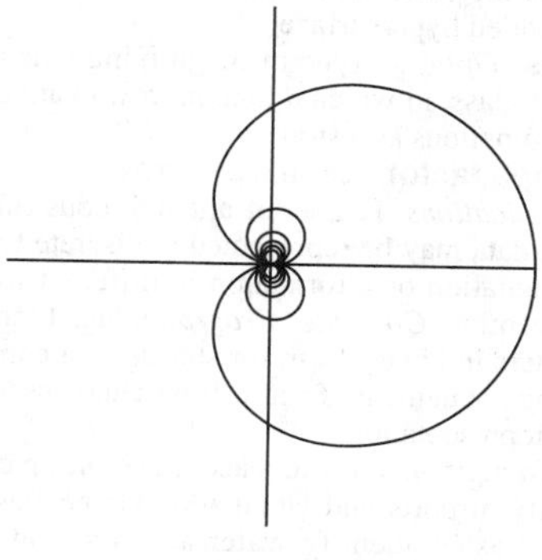

cochleoid

cochleoid *Mathematics.* the graph of the equation

$$(x^2 + y^2) \tan^{-1}(y/x) = ay$$

in polar coordinates $r\theta = a \sin\theta$. Also, OUIJA BOARD CURVE.

Cochliodontidae *Paleontology.* a family of cartilaginous fish in the extinct suborder Cochliodontoidei, having large, crushing tooth plates that exhibit a spiral, snail-like growth pattern; Devonian and Permian.

Cochliopodium *Invertebrate Zoology.* a genus of freshwater amoeboid protozoans having a hemispherical test.

Cochlonemataceae *Mycology.* a family of fungi belonging to the order Zoopagales, characterized by sticky spores and found in amoebae, nematodes, and rotifers.

cock *Vertebrate Zoology.* an adult male game bird or domesticated fowl. *Engineering.* a device for regulating or stopping the flow in a pipe, consisting of a taper plug that may be rotated.

cockatoo *Vertebrate Zoology.* a large, usually white parrot of the family Psittacidae, characterized by a large erect crest on its head, a short tail, and a worm-like tongue in a large, heavy bill used for excavating grubs and nutcracking; found in treetops in Australia and nearby islands, and often raised as pets.

cockatoo

Cockcroft, Sir John Douglas 1897–1967, English physicist; with Ernest T. S. Walton, won the Nobel Prize for being first to split the atom artificially.

Cockcroft-Walton accelerator *Nucleonics.* an electrostatic particle accelerator in which ions are accelerated through an evacuated tube by a high-potential difference provided by a voltage multiplier circuit that charges capacitors in parallel and discharges them in series between the ends of the tube; used to study the first nuclear reactions (1932) induced by artificial means.

cocked hat *Navigation.* the small triangle formed by three lines of position that do not meet at a single point. Under most conditions the fix is established inside the cocked hat at a point equidistant from the three lines of position.

cockerel *Agriculture.* a young male chicken or turkey, especially one that is less than twelve months old.

cockeyed bob *Meteorology.* a colloquial name for a thunder squall that occurs on the northwest coast of Australia during the summer.

cocking lever *Ordnance.* a lever that pulls back or lowers the striker of an automatic weapon or bomb release mechanism.

cockle *Invertebrate Zoology.* the common name for various species of marine mollusks in the class Bivalva that burrow into the bottom in shallow waters; their shells have convex radial ribs. Several species are used for food, especially in Europe. *Botany.* a term for various plants that are agricultural weeds; e.g., the **corn cockle,** *Agrostemma githago,* which is a dangerous weed around winter wheat. *Graphic Arts.* a term for bond paper having an irregular surface.

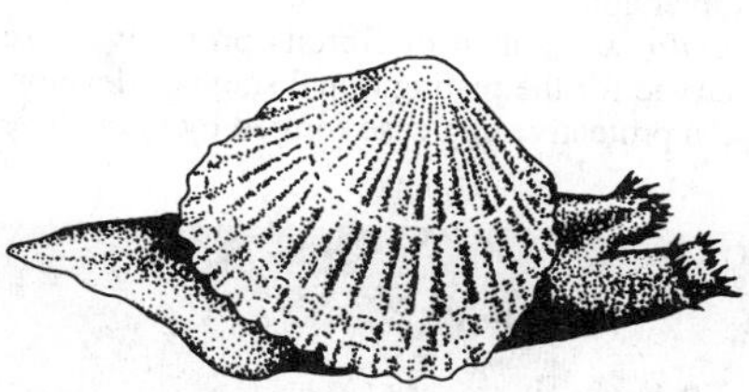

cockle

cocklebur *Botany.* the common name for several species of annual weeds that compose the genus *Xanthium*; all are native to North America and have sharp, spiny burs which cling to objects such as clothing or animal fur.

cockpit *Aviation.* a space or compartment in a flight vehicle, especially at the top and front of a small aircraft, designed to seat the pilot and sometimes the crew or passengers. *Naval Architecture.* **1.** a below-decks space in a sailing warship given over to care of the wounded in battle; ancestor of the modern sick bay. **2.** a semi-enclosed well near the stern of a sailboat, used as a conning position.

cockpit karst see CONE KARST.

cockroach *Invertebrate Zoology.* the common name for an orthopteran insect of the family Blattidae, having a flat, wide body and long, slender, segmented antennae; the front of the thorax projects over the head. It is a widespread pest in human dwellings and may carry disease; e.g., the **German cockroach,** *B. germanica.*

cocoa *Food Technology.* **1.** a powder made of cacao beans that have been fermented, dried, roasted, ground, and usually partially defatted. **2.** a beverage made by mixing this powder with hot milk or water.

cocoa butter *Food Technology.* fat that is removed from chocolate liquor.

cocoa powder *Food Technology.* a product obtained after cocoa butter is pressed from chocolate liquor; the remaining press cake is ground into cocoa powder and classified according to its fat content.

cocobola *Materials.* an extremely hard and heavy wood having orange and red bands with dark streaks, obtained from the *Dolbergia retusa* tree of Central America; it is used for canes, knife handles, and inlaying. *Botany.* the tree itself.

cocodyl oxide *Organic Chemistry.* $(CH_3)_2AsOAs(CH_3)_2$, a liquid that has an obnoxious odor and is slightly soluble in water and appreciably soluble in alcohol; boils at 150°C; used in organic synthesis. Also, ALKARSINE.

co-consciousness or **coconsciousness** *Psychology.* a mental state that is coexistent with an individual's consciousness, but outside his or her direct awareness. Thus, **co-conscious** or **coconscious.**

coconut

coconut *Botany.* the large, hard-shelled seed of the coconut palm, *Cocas nucifera,* enclosed in a thick husk and containing an edible white meat and a milky liquid known as **coconut milk.**

coconut bud rot *Plant Pathology.* a disease of the coconut palm caused by the fungus *Phytophthora palmivora* that is characterized by a rotting of the terminal bud and the leaves attached to it, eventually resulting in the death of the tree.

coconut oil *Food Technology.* a semisolid white fat or nearly colorless oil extracted from coconuts; used primarily in foods and the manufacture of soaps, cosmetics, and candles.

coconut palm *Botany.* a tall tropical palm tree, *Cocos nucifera,* that bears the coconut.

coconversion *Genetics.* the simultaneous correction of several loci during gene conversion.

cocoon *Entomology.* a silken or fibrous protective case produced by certain insect larvae for the resting pupal stage of development. *Invertebrate Zoology.* a protective egg case formed by many invertebrates.

cocoon

cocurrent line *Oceanography.* a map or chart line that connects points that have the same tidal current hour.

cocycle *Mathematics.* in cohomology, an element of the kernel space of a differential d^n of a cochain complex; called a **closed differential form** if the complex is of differential form.

cod *Vertebrate Zoology.* the common name for a large, soft-finned deep-sea food fish of the family Gadidae in the order Gadiformes; found mostly in the North Atlantic Ocean.

codan *Electronics.* an electronic circuit that activates a receiver when a signal is received. (An acronym for carrier-operated device anti-noise.)

CODAR *Acoustical Engineering.* correlation, detection, and ranging, a method of detecting submarines.

CODASYL *Computer Programming.* an organization of computer and software manufacturers and major users such as the U.S. Department of Defense. (An acronym for Conference on Data Systems Languages.)

CODASYL code *Computer Programming.* any set of program instructions that are written in a high-level language, especially COBOL, and set as a standard by CODASYL.

Codazzi-Mainardi equations *Mathematics.* for a Riemannian manifold embedded isometrically as a hypersurface (codimension = 1) in another Riemannian manifold, the Codazzi-Mainardi equations give the components of the Riemann curvature tensor which are normal to the hypersurface. These equations may be written

$$\langle R'(X, Y)Z, \nu \rangle = (\nabla_X \mathrm{II})(Y, Z) - (\nabla_Y \mathrm{II})(X, Z),$$

where X, Y, and Z are vector fields tangent to the hypersurface at a point P, ν is the unit normal to the hypersurface at P, $R'(X, Y)$ is the linear transformation representing the curvature tensor of the overlying manifold (hence the prime), and $\mathrm{II}(X, Y)$ is the (scalar) second fundamental form of the embedded hypersurface.

Coddington lens *Optics.* a type of magnifying lens that is cut from a spherical piece of glass, in which a long narrow channel around the center of the sphere functions as a stop.

Coddington shape factor see SHAPE FACTOR.

code *Telecommunications.* **1.** a set of unambiguous rules specifying the manner in which data may be represented in discrete form. **2.** a transformation or representation of information in different forms according to preassigned convention. *Computer Programming.* **1.** any system for representing characters in binary form for storing in a computer. **2.** to write computer language statements from a flow chart, pseudocode, or other program design representation.

code beacon *Navigation.* an aerobeacon, exhibiting coded flashes, that is used to identify airports and landmarks. Green flashes identify land airports, yellow flashes identify water airports, and red flashes mark hazards.

codebook *Telecommunications.* a book that contains codes arranged in systematic order and a vocabulary made up of arbitrary meanings, each accompanied by one or more groups of symbols used as equivalents to the plain-text messages.

codec *Telecommunications.* a common circuit used for both encoding for transmission and decoding for reception. Codec is in common use in PCM transmission and switching and in ISDN.

codecarboxylase *Enzymology.* a constituent of the enzyme carboxylase that causes D-amino acids to go through the process of decarboxylation.

code-check *Computer Programming.* the process of examining a coded program or routine to detect and remove errors in the coding; this can be done visually or by using a compiler or other translator that provides diagnostic capabilities.

codeclination *Cartography.* the complement of an astronomic declination; i.e., 90° minus the declination.

code converter *Computer Technology.* a device or system that automatically transforms data represented in one coding scheme into the format and structure of another coding scheme.

coded decimal *Computer Programming.* the representation of a decimal digit by four or more binary digits.

code dependence see CODE SENSITIVITY.

code dictionary *Computer Programming.* an alphabetized list of plain-language words with their corresponding coded representation. Also, DICTIONARY CODE.

code division multiple access *Telecommunications.* a form of modulation in which digital information is encoded in an expanded bandwidth format; this permits a high degree of energy dispersion in the emitted bandwidth.

code division multiplex *Telecommunications.* transmission in which a common character is divided into a number of separate, low-capacity channels or transmission paths using different codes as message labels.

coded passive reflective antenna *Electromagnetism.* an object having variable reflecting properties according to a predetermined code; used for producing an indication for a radar receiver.

coded program *Computer Programming.* a program written in a language or code that is interpretable by a specific computer.

coded stop *Computer Programming.* an instruction written into a routine such that it causes processing to stop at that point.

coded tape *Graphic Arts.* a strip of paper or other material produced from a keyboard and containing a punched or electronic code with instructions for a Monotype or phototypesetting machine.

coded track circuit *Transportation Engineering.* a track circuit that can variously detect vehicles or transmit information to them.

code element *Telecommunications.* **1.** any of a finite set of data items or signals used to compose the characters in a given code. **2.** the representation of a piece of data in accordance with a code, or the representation of a character in accordance with a coded character set.

code extension *Computer Programming.* **1.** the use of a special character to indicate that the characters that follow are to be interpreted using a different code. **2.** a layering method of increasing the number of characters that can be represented in a given code by combining characters into subgroups.

code generation *Computer Programming.* a function of a language processor, such as a compiler, by which a machine language code is created.

code generator *Computer Programming.* **1.** the part of a compiler that generates the final output code. **2.** a computer-aided software engineering tool that automatically produces a high-level language code from a program design.

code group *Telecommunications.* a set of symbols, such as letters, numbers, or other special signs in combinations assigned to represent a plain-text element. Also, **code system.**

codeine *Pharmacology.* $C_{18}H_{21}NO_3{\cdot}H_2O$, a drug obtained from opium or derived from morphine, occurring as a white crystalline powder that is administered orally as a painkiller or cough suppressant; it is addictive but less so than morphine. Also, METHYLMORPHINE.

code line *Computer Technology.* in character recognition, the area in which the character reader expects to find printed or written characters.

code optimization *Computer Programming.* a function of a language processor that examines the intermediate code or machine code and modifies it to make more efficient use of the computer resources.

co-dependent *Psychology.* a term used to describe an individual who has an ongoing close relationship with a dependent person (such as an alcoholic) and who thus may be regarded as condoning or supporting the dependent behavior in some way. Thus, **co-dependency.**

coder *Computer Programming.* a person who implements detailed program specifications and design prepared by others; sometimes used in a pejorative sense. *Telecommunications.* see ANALOG-TO-DIGITAL CONVERTER.

code ringing *Telecommunications.* **1.** a coded ringing signal sent out on a party telephone line to alert a subscriber on the line. **2.** selective calling on a multiparty line, with a combination of long and short rings.

code sensitivity *Computer Technology.* the characteristic of certain hardware or software that limits data representation to a particular code. Also, CODE DEPENDENCE.

code set *Computer Programming.* a complete list of all the allowable representations defined as a given code, such as the set of all ASCII codes. Also, **coded character set.**

codetext *Telecommunications.* a message in ciphered form.

code translation *Telecommunications.* the act of converting information from one form into another, such as the function carried out by a telegraph or data receiver whereby the text of a message is derived and recorded from the incoming digital signal.

code transparency *Computer Technology.* the characteristics of certain hardware or software that enable them to interpret data represented in any code.

code word *Military Science.* **1.** a word given a classification and a classified meaning to protect information regarding a classified plan or operation. **2.** a cryptonym used to identify sensitive intelligence data.

codex *Archaeology.* **1.** an early handwritten Christian manuscript of the Scriptures. **2.** a Mesoamerican document written in hieroglyphic or pictographic characters on bark or animal skin; contains information about pre-Columbian and post-Conquest life.

Codex Alimentarius *Food Technology.* an international code of food standards established by the United Nations Food and Agriculture Organization (FAO) and World Health Organization (WHO).

Codiaceae *Botany.* a family of green algae of the order Siphonales, having thalli made up of interwoven tubes.

codice fish see ANTARCTIC COD.

codimension *Mathematics.* in differential geometry, the difference in dimension between the ambiant space and the dimension of the embedded manifold. For example, a circle (1-dimensional) embedded in R^3 is of codimension 2.

codimer *Organic Chemistry.* **1.** a copolymer formed from the copolymerization of two different olefin molecules. **2.** a compound formed when isobutylene polymerizes with one of the two normal butylenes.

coding *Computer Programming.* the process of implementing detailed program specifications in a programming language suitable for computer processing.

coding delay *Navigation.* a modification of a Loran-A signal by the introduction of a short delay to improve readability under certain conditions.

coding strand *Molecular Biology.* the nontranscribed strand of duplex DNA having the same nucleotide sequence as the specified mRNA, except that in DNA T substitutes for U in RNA. Also, SENSE STRAND.

Codiolaceae *Botany.* a family of green algae belonging to the order Acrosiphoniales, characterized by unbranched diffuse filaments and tubular multinucleate cells; commonly found in cold, protected, marine and brackish areas.

Codium *Botany.* a genus including the green algae family Codiaceae, distinguished by its lack of leukoplasts and pyrenoids, variable thallic forms, and anchoring of the thallus by a weft of rhizoids.

codliver oil *Materials.* a pale yellow oil that is extracted from the liver of the common cod or related species; formerly in wide use in medicine as a source of vitamins A and D.

Codman's tumor see CHONDROBLASTOMA, def. 2.

codomain *Mathematics.* **1.** given a function or mapping f from a set X to a set Y, the set Y is the codomain of f. **2.** see RANGE.

codominance *Genetics.* a situation in which both of a particular pair of alleles are fully expressed in the heterozygous state.

codon *Genetics.* the basic genetic coding unit, consisting of a sequence of three nucleotides that specify a particular amino acid to be incorporated into a polypeptide chain. *Telecommunications.* a unit corresponding to one character of a coded message.

Codoniaceae *Botany.* a cosmopolitan family of liverworts of the order Metzgeriales, characterized by simple or dichotomous branching, the lack of secondary pigments, and organization into either a distinct stem and lateral leaves or a thallus with a short stalklike portion and a distal expanded portion.

codon restriction *Genetics.* the progressive inability of aging cells to translate the entire genome they carry, resulting in an inability to produce certain proteins and enzymes.

coeff. or **coef.** coefficient.

coefficient *Mathematics.* **1.** a factor in a product. **2.** specifically, a constant factor, as distinguished from a variable or variable expression.

coefficient A *Navigation.* a deviation in a magnetic compass that is constant on all headings.

coefficient B *Navigation.* a semicircular deviation in a magnetic compass that is proportional to the sine of the compass heading. It is maximum on east or west headings, zero on north or south headings.

coefficient C *Navigation.* a semicircular deviation in a magnetic compass that is proportional to the cosine of the heading. It is maximum on north or south headings, zero on east or west headings.

coefficient D *Navigation.* a quadrantal deviation in a magnetic compass that is proportional to the sine of twice the compass heading. It is maximum on intercardinal headings and zero on cardinal headings.

coefficient E *Navigation.* a quadrantal deviation in a magnetic compass that is proportional to twice the cosine of the heading. It is maximum on cardinal headings and zero on intercardinal headings.

coefficient J *Navigation.* a change in deviation of a magnetic compass for a heel of 1° while the vessel is on a north compass heading.

coefficient of coincidence *Genetics.* an experimental value that is equivalent to the observed number of double crossovers divided by the expected number.

coefficient of compressibility *Mechanics.* a measure of the fractional change in volume of a body under unit pressure; the reciprocal of the bulk modulus.

coefficient of condensation *Physics.* a ratio of the number of molecules that are condensed on the surface of a solid or liquid in equilibrium with its vapor phase, to the number of free vapor molecules striking that surface.

coefficient of correlation *Statistics.* a standardized measure of the strength of the linear dependence between two random variables, defined as the ratio between the covariance and the product of the standard deviations; a value of 1 or (–1) corresponds to a perfect positive (or negative) linear relationship, whereas a value of 0 corresponds to no correlation. Also, PRODUCT-MOMENT COEFFICIENT.

coefficient of cubical expansion *Thermodynamics.* the increase in the volume of a unit volume of a substance for each degree of increase in its temperature at constant pressure. Also, COEFFICIENT OF VOLUMETRIC EXPANSION.

coefficient of determination *Statistics.* a measure of the strength of the linear association between two variables; defined as the square of the correlation coefficient.

coefficient of expansion *Thermodynamics.* an increase in the unit length, area, or volume of a substance as the result of an increase in its temperature by one degree at constant pressure.

coefficient of form *Naval Architecture.* any of several coefficients, including the block coefficient and the prismatic coefficient, that relate the volume (and thus displacement) of a hull form to its dimensions.

coefficient of friction *Mechanics.* a constant that, when multiplied by the normal force between two bodies that are in surface contact, gives the force of friction between them necessary to start sliding. Also, FRICTION COEFFICIENT.

coefficient of kinetic friction *Mechanics.* a constant that, when multiplied by the normal force between two bodies that are in surface contact and are in motion relative to each other, gives the force of friction necessary to maintain uniform motion. Also, COEFFICIENT OF SLIDING FRICTION.

coefficient of linear expansion *Thermodynamics.* the increase in the length of a unit length of a solid for each degree of increase of its temperature at constant pressure. Also, LINEAR COEFFICIENT OF THERMAL EXPANSION, LINEAR EXPANSION COEFFICIENT.

coefficient of multiple determination *Statistics.* in multivariate regression analysis, a measure of the degree to which the variation in the dependent variable can be explained by a linear relation with one of the independent variables, when all the other independent variables are held constant.

coefficient of partial correlation *Statistics.* the square root of the coefficient of multiple determination with the sign of the corresponding regression coefficient.

coefficient of performance *Thermodynamics.* the ratio of the heat transfer to the refrigeration cycle, or supplied in a heat-pump cycle, to the amount of work done in order to operate the cycle. Unlike efficiency, it can be a number greater than 1.0.

coefficient of restitution *Mechanics.* the ratio of the relative speeds of two bodies just after and just before a collision between them along a mutual straight line; it ranges from 0, for perfectly inelastic collisions, to 1, for completely elastic ones. It is a constant at moderate speeds.

coefficient of rolling friction *Mechanics.* a constant that, when multiplied by the normal force between two bodies rolling against each other, gives the force of friction needed to maintain the uniform motion.

coefficient of sliding friction see COEFFICIENT OF KINETIC FRICTION.

coefficient of standing friction see COEFFICIENT OF STATIC FRICTION.

coefficient of static friction *Mechanics.* a constant that, when multiplied by the normal force between two stationary bodies in surface contact, gives the shear force that must be exceeded for the bodies to start sliding against each other. Also, COEFFICIENT OF STANDING FRICTION.

coefficient of strain *Mechanics.* the ratio of the distance along the strain axis between two points in a body undergoing a one-dimensional strain, to the distance between the same points before the deformation. *Mathematics.* a multiplier used to compress or expand geometrical objects along a direction parallel to an axis.

coefficient of superficial expansion *Thermodynamics.* the increase in the area of a unit area of a solid surface for each degree of increase of its temperature at constant pressure.

coefficient of thermal expansion *Thermodynamics.* the increase in size of a solid object that takes place because of an increase in temperature, either an expansion in length or an expansion in volume. See COEFFICIENT OF LINEAR EXPANSION, COEFFICIENT OF CUBICAL EXPANSION.

coefficient of variation *Statistics.* a relative measure of dispersion found by expressing the standard deviation as a percentage of the arithmetic mean.

coel- or **coelo-** a combining form meaning "cavity," as in *coelenterate, coeloblastula.*

coelacanth [sē´lə kanth´] *Vertebrate Zoology.* a common name for the bony marine fishes belonging to the order Coelacanthiformes, all of which were thought to be extinct until a single species was discovered in 1938 in the Indian Ocean, off the coast of South Africa.

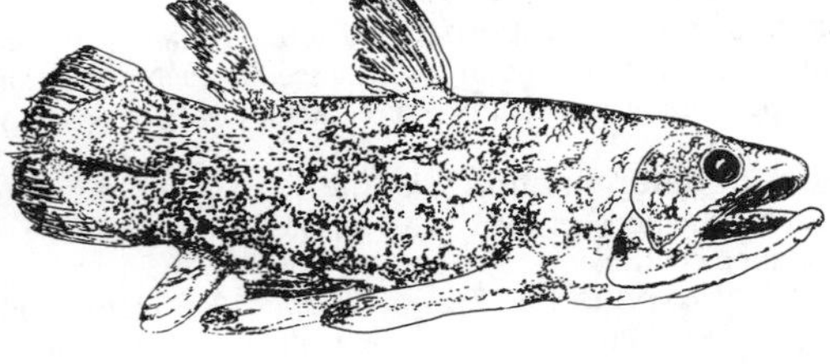

coelacanth

Coelacanthidae *Paleontology.* a family of crossopterygian fishes in the suborder Coelacanthiformes; the order was long thought to be extinct, but a single living representative, *Latimeria*, was recently discovered; characterized by fleshy lobed fins, two of them dorsal; Devonian to present.

Coelacanthiformes *Vertebrate Zoology.* the tassel-finned fish, an order of bony marine fishes comprising one existing family with one species; recently found off the coast of South Africa.

Coelacanthini *Vertebrate Zoology.* an equivalent name for Coelacanthiformes, an order of bony marine fishes.

Coelenterata *Invertebrate Zoology.* a phylum of animals with radially symmetrical bodies, a saclike internal cavity, tentacles, and nematocysts, including jellyfish, hydras, sea anemones, and corals. Also, CNIDARIA.

coelenterate [sə len´tə rāt´; sə len´tə rət] *Invertebrate Zoology.* **1.** a member of the phylum Coelenterata. **2.** of or relating to this phylum.

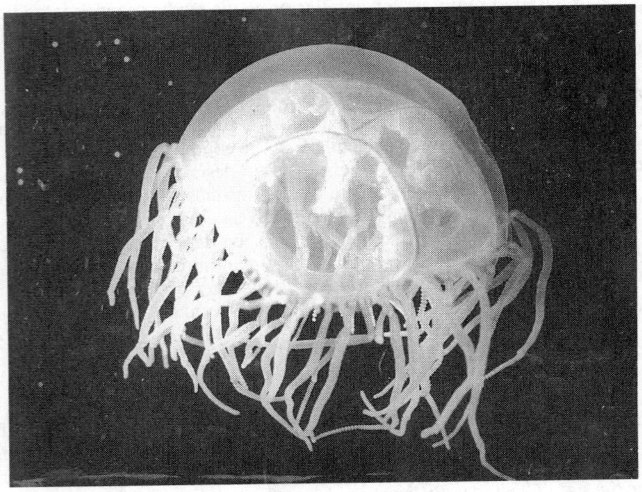

coelenterate

coelenteron *Invertebrate Zoology.* the internal cavity in coelenterates.

coeloblastula *Developmental Biology.* the normal form of blastula, which is a hollow sphere of blastomeres.

Coelodiscaceae see JAOACEAE.

Coelodonta *Paleontology.* the woolly rhino that is often depicted in Pleistocene cave drawings, a genus of large perissodactyls in the suborder Ceratomorpha and family Rhinocerotidae; Pliocene and Pleistocene.

coelom [sē´ləm] *Developmental Biology.* a secondary body cavity of all vertebrates and most invertebrates, having a membranous lining composed of specialized cells and containing the internal organs; it develops between the mesoderm layers and forms the adult pleural, peritoneal, and pericardial cavities. *Invertebrate Zoology.* a similar secondary body cavity found in members of lower invertebrate phyla, such as roundworms and rotifers; it develops within the mesoderm and serves some of the functions of a true coelem. (From a Greek word meaning "hollow.") Also, CELOM.

Coelomata *Zoology.* a division that includes all phyla having a true coelom.

coelomic [sə lōm´ik; sē´lōm´ik] *Zoology.* relating to or situated in the coelem. Thus, **coelemic fluid.**

coelomoduct *Invertebrate Zoology.* either of a pair of channels, used for excretion and reproduction, passing from the coelom to the exterior in certain invertebrates such as annelids and mollusks.

Coelomomyceteceae *Mycology.* a family of fungi that belongs to the order Blastocladiales; parasitic to insects, primarily mosquito larvae.

coelomostome *Invertebrate Zoology.* the internal opening of a coelomoduct.

Coelomycetes *Mycology.* a class of fungi belonging to the subdivision Deuteromycotina that is characterized by closed fruiting bodies of various shapes, which may open as they mature; some species are parasitic to plant material, while others live off of nonliving organic matter.

Coelopidae *Invertebrate Zoology.* a family of seaweed flies, two-winged insects in the subsection Acalypteratae; larvae feed on decomposing seaweed.

coeloplanula *Invertebrate Zoology.* a hollow planula (a free-swiming larval type) with a wall made up of two layers of cells.

Coelosphaerium *Botany.* a genus of blue-green algae in the family Chroococcaceae, characterized by its conspicuous hollow, spherical colony structure.

coelostat [sē′lə stat′] *Astronomy.* a telescope with a rotating mirror that enables it to continuously reflect the same region of the sky.

coelozoic [sē′lə zō′ik] *Biology.* living or situated within a body cavity; extracellular.

coelurosaur [sē′lür′ə sôr′] *Paleontology.* **1.** a member of the infraorder Coelurosauria, small, lightly built, carnivorous dinosaurs in the order Saurischia and suborder Theropoda. The coelurosaurs had long tails and necks and narrow skulls, and grew to about nine feet in length; they were characterized by hollow bones and a bipedal stance; and appeared in the middle Triassic, preying on insects and on the smaller therapsids (primitive mammals). They are ancestral to the later saurischians. **2.** of or relating to this infraorder. (From a Greek term meaning "hollow lizard.")

Coelurosauria *Paleontology.* an infraorder of small, carnivorous bipedal saurischian dinosaurs in the suborder Theropoda; Triassic to Cretaceous.

Coenagrionidae *Invertebrate Zoology.* a family of damsel flies, insects belonging to the order Odonata, that are brightly colored with transparent wings.

coenenchyme *Invertebrate Zoology.* the mesagloea around the polyps in compound anthozoans.

coenobiology see BIOCENOLOGY.

Coenobitidae *Invertebrate Zoology.* a family of land-dwelling decapod crustaceans in the order Anomura with lobsterlike or crablike forms.

coenobium [sə nō′bē əm′] *Invertebrate Zoology.* a colony of protozoans with constant size, shape, and cell number, but without differentiated cells.

coenocline [sēn′ə klīn′] *Ecology.* a sequence of natural communities distributed along an environmental gradient.

coenocyte [sēn′ə sīt′] *Biology.* a cytoplasmic mass bounded with a cell wall and having many nuclei; formed by growth of the cytoplasm and the division of the nucleus without the division of the cell, as in some green algae and many kinds of fungi.

coenocytic [sēn′ə sit′ik] *Mycology.* referring to filaments or hyphal masses termed mycelium that lack cell walls; occurring in molds. Also, NONSEPTATE.

Coenomyidae *Invertebrate Zoology.* a family of two-winged insects with short antennae, in the series Brachycera.

Coenopteridales *Botany.* an extinct order of ferns that had both simple and pinnate leaves. *Paleontology.* the larger of the two main groups of ferns in the order of ancestral ferns Primofilicales, the other being Cladoxylales; the Coenopteridales were widespread in the late Paleozoic but became extinct in the Triassic.

coenosis [sə nō′sis] *Ecology.* a random assemblage of organisms that have common ecological requirements.

coenosite [sēn′ə sīt′] *Ecology.* an organism that habitually shares food with another organism.

coenosium [sə nōz′ē əm] *Ecology.* a plant community.

coenosteum *Invertebrate Zoology.* the skeleton of a compound coral or bryozoan colony.

Coenothecalia *Invertebrate Zoology.* an order of colonial coelenterate anthozoans in the class Alcyonaria.

coenotype [sēn′ə tīp′] *Biology.* an organism that has the characteristic structure of its group.

coenzyme [kō′en′zīm] *Enzymologyy.* an organic nonprotein molecule, frequently a derivative of a water-soluble vitamin, that binds, as with a protein molecule (apoenzyme), to form an active enzyme (holoenzyme).

coenzyme A *Enzymologyy.* $C_{21}H_{36}O_{16}N_7P_3S$, a coenzyme that takes part in fatty acid metabolism; it is found in the cells of all living organisms.

coenzyme B$_{12}$ *Enzymology.* a form of vitamin B_{12} in which the sixth coordination position of the cobalt atom is linked covalently to the 5'-carbon of 5'-deoxyadenosine. It is the only known biomolecule to have a carbon-metal bond. Also, COBAMIDE, 5'-DEOXYADENOSYLCOBALAMINE.

coenzyme Q *Enzymology.* a compound that is structurally related to vitamin K and that functions as an electron carrier in the electron transport system of cells. Also, UBIQUINONE.

coenzyme R see BIOTIN.

coercimeter *Engineering.* an instrument that is used to measure the magnetic force of a natural magnet or an electromagnet.

coercion *Computer Programming.* in programming languages, a mechanism that automatically converts an operand from one data type to another; for example, if both floating point and integer variables appear in an arithmetic expression, the integer variables will be automatically converted to floating point.

coercive force *Electromagnetism.* the amount of applied magnetic field that is required to overcome the magnetic induction of a ferromagnetic material and bring its residual magnetism back to zero; i.e., the force required to demagnetize a material.

coercivity see COERCIVE FORCE.

Coerebidae *Vertebrate Zoology.* a former family classification strictly for the honeycreepers, now placed in the family Parulidae.

coeruleolactite *Mineralogy.* $(Ca,Cu^{+2})Al_6(PO_4)_4(OH)_8 \cdot 4\text{–}5H_2O$, a white or light blue triclinic mineral that occurs as finely crystalline or fibrous crusts and veinlets, having a specific gravity of 2.57 to 2.69 and a hardness of 5 on the Mohs scale; found with wavellite and limonite.

coesite *Mineralogy.* SiO_2, a colorless, transparent monoclinic mineral found in meteor craters as irregular grains, having a specific gravity of 2.93 to 3.01 and a hardness of 7.5 to 8 on the Mohs scale; polymorphous with quartz, tridymite, cristobalite, and stishovite.

coetaneous [kō′ə tān′ē əs] *Science.* of the same age or period; coeval.

coeval [kō ē′vəl] *Science.* existing at the same time; of the same age, period, or duration.

coexistence *Ecology.* the occurrence of two or more species in the same habitat, especially when they are potential competitors or have a predator-prey relationship.

coextrusion *Materials Science.* the process of producing continuous multilayer products in sheet, film, tubing, filament, or other forms. Separate polymer streams are fed from different extruders to a die feed block, where they are combined in the die, emerging in combined form as a continuous multilayer extrudate.

coextrusion blow molding *Materials Science.* one of the earliest technologies for the extrusion of multilayer products; extruded multilayer tube (parison) material centers and is closed within a mold, and subsequently, gas pressure expands the parison to fit the mold cavity.

cofactor *Biochemistry.* a substance or element, such as a coenzyme or a metal ion, with which another molecule must unite in order to function or to become active. *Mathematics.* given a square $n \times n$ matrix, the cofactor of the (i,j)th element is the minor of that element, multiplied by $(-1)^{i+j}$. Also, COMPLEMENTARY OR SIGNED MINOR.

cofactor recycling enzyme *Enzymology.* an enzyme that catalyzes the recycling of the nonprotein component of an enzyme required for its activity.

coffee *Food Technology.* a widely consumed beverage containing caffeine, made from the roasted and ground berries (**coffee beans**) of certain plants. *Botany.* any of the various trees and shrubs of the genus *Coffea* from which this beverage is obtained, especially *C. arabica,* the common coffee tree. (From an Arabic word; named for the region of Ethiopia where it is thought to have originated.)

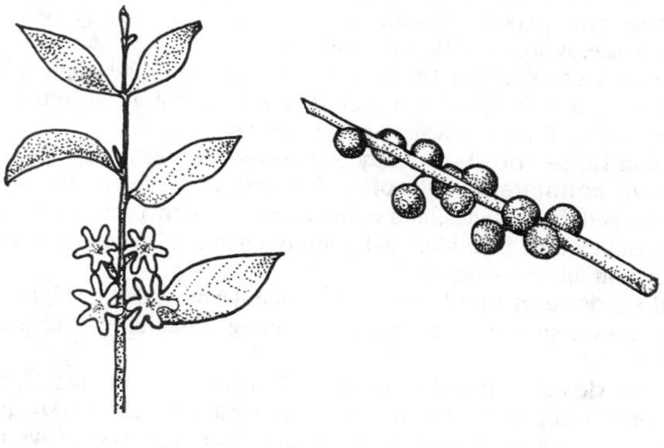

coffee

cofferdam *Naval Architecture.* **1.** a heavy double bulkhead that is used in tankers or other merchant vessels to separate oil tanks or cargo holds. **2.** a temporary watertight structure built around a leaking or damaged section of hull to allow continued operation or repair while afloat.

coffered ceiling *Building Engineering.* a particular type of ceiling constructed in such a way that its panels are sunken or recessed.

coffin *Nucleonics.* a heavily shielded container for transporting or storing reactor fuel rods containing high-activity fission products. Also, CASK.
coffinite *Mineralogy.* $U(SiO_4)_{1-x}(OH)_{4x}$, a black, brittle tetragonal mineral that appears as aggregates of fine particles, having a specific gravity of 4.64 to 5.1 and a hardness of 5 to 6 on the Mohs scale; found in black vanadium-rich ores.
cofinal [kō´fī´nəl] *Mathematics.* a subset A of a partially ordered set X (with ordering $\leq$) is said to be cofinal in X if for each element x of X there exists an element a of A such that $x \leq a$. Every totally ordered set (chain) has a cofinal well-ordered subset.
cog *Mechanical Engineering.* a wooden tooth along the edge of a gear wheel. Electromagnetism. see COGGING.
cog belt *Mechanical Engineering.* a device in which a gear or wheel engages with a flexible belt within an engine to provide accurate timing and slip-free power transmission.
cogeneration *Mechanical Engineering.* the process of supplying both electric and steam energy from the same power plant.
Cogentin *Pharmacology.* the trademark for preparations of benztropine mesylate, a compound that shows antihistamine and atropine effects; also used in the treatment of parkinsonism.
cogeoid see COMPENSATED GEOID.
cogging *Electromagnetism.* a variation in torque for motors and generators caused by varying magnetic flux due to changing alignment of the armature and rotor teeth. Also, TORQUE RIPPLE.
cogging mill see BLOOMING MILL.
Cognac or **cognac** [kōn´yak´l] *Food Technology.* **1.** a French brandy produced in or around the town of Cognac in the Burgundy region of central France. **2.** a similar brandy made elsewhere.
cognac oil see ETHYL ENANTHATE.
cognate ejecta *Volcanology.* pyroclastic material that is formed at the same time as, or during the development of, an eruption.
cognate help *Immunology.* a T lymphocyte helping a B lymphocyte, with each recognizing a different antigenic determinant on a protein.
cognate word *Linguistics.* a word or morph that is closely related in sound and meaning to one in a related language, such as "grandmere" in French and "grandmother" in English.
cognatic descent *Anthropology.* a kinship system in which the individual has the option of affiliating with the family of either parent.
cognition *Psychology.* the mental activity by which an individual is aware of and knows about his or her environment, including such processes as perceiving, remembering, reasoning, judging, and problem solving.
cognitive *Psychology.* **1.** of or relating to cognition; having to do with thoughts and ideas. **2.** of or relating to cognitive psychology.
cognitive appraisal (theory) *Psychology.* the theory that emotions evolve from people's subjective interpretation of the situation they are in; several emotions, even conflicting ones, can result from one stimulus, depending on an individual's evaluation.
cognitive arousal (theory) *Psychology.* the theory that a person experiencing some form of physical arousal, such as trembling or rapid heartbeat, will associate it with an emotional feeling that is assumed to be the cause. Also, **cognition arousal.**
cognitive behavior(al) modification *Behavior.* the application of conditioning principles and techniques to bring about changes in thought that will lead to desired behavioral changes.
cognitive behavior(al) therapy see COGNITIVE THERAPY.
cognitive consistency (theory) *Psychology.* the theory that an individual's beliefs, attitudes, and opinions will be consistent with one another, and that the individual will modify cognition or behavior in order to maintain this consistency.
cognitive development *Psychology.* the development of higher mental processes such as perceiving, reasoning, imagining, and problem solving.
cognitive development(al) theory *Psychology.* Jean Piaget's theory of a child's cognitive development, constructed around four distinct stages occurring from birth to age 11 and older. Thus, **cognitive development(al) stages.**
cognitive dissonance (theory) *Psychology.* the theory that individuals seek to avoid conflict or inconsistency in their beliefs, attitudes, and opinions, and that they will modify their cognition or behavior in order to minimize such conflict or inconsistency.
cognitive map *Behavior.* a theoretical mental representation of a place or task that allows an individual to attain a certain goal or solve a certain problem.
cognitive model *Psychology.* an approach to psychotherapy that focuses on the way individuals interpret their behavior and emotions as well as on the behavior itself.
cognitive psychology *Psychology.* the branch of psychology that attempts to explore and explain processes of sense, perception, memory, and thought, as contrasted with other approaches that focus on observable behavior or on the unconscious. Thus, **cognitive psychologist.**
cognitive restructuring *Psychology.* a therapeutic approach that systematically attempts to have the patient identify negative thought patterns and then reexamine problems in a more rational and constructive fashion.
cognitive science *Artificial Intelligence.* the study of human and animal cognition, making use of insights provided by Artificial Intelligence and neuroscience as well as psychology, linguistics, and philosophy.
cognitive style *Psychology.* the characteristic manner in which an individual approaches and solves problems and other cognitive tasks.
cognitive theory *Psychology.* the theory that the way in which individuals perceive and define themselves and their environment determines behavior and emotions; erroneous or unrealistic perceptions can thus cause depression or anxiety.
cognitive therapy *Psychology.* a form of psychotherapy that attempts to revise individuals' negative perceptions of themselves and the world around them in order to develop more productive behavior.
COGO *Computer Programming.* a problem-oriented language that is used in civil engineering to perform the geometric computations required in surveying. (An acronym for coordinate geometry.)
cogon *Botany.* any of a number of tall tropical grasses of the genus *Imperata,* used for thatching.
cogwheel *Mechanical Devices.* a gear wheel with metal teeth.
Cohen, Stanley born 1922, American biochemist; Nobel Prize with Levi-Montalcini for fundamental research on nerve growth factor.
cohenite *Mineralogy.* $(Fe,Ni,Co)_3C$, a white, opaque orthorhombic mineral that tarnishes to bronze or yellow, occurring as elongated tabular crystals, and having a specific gravity of 7.2 to 7.65 and a hardness of 5.5 to 6 on the Mohs scale; found in iron meteorites, in terrestrial irons, and in South African diamonds. Also, CEMENTITE.
coherence *Physics.* a correlation between the phases of two or more waves, so that interference effects may be produced between them, or a correlation between the phases of parts of a single wave. *Statistics.* in decision under uncertainty, avoiding acts that entail a sure loss; the foundation of subjective probability and bayesian methods.
coherence area *Optics.* the largest cross section of a beam in which light that passes through any two pinholes produces interference fringes.
coherence length *Physics.* the distance over which a wave train exhibits coherence. *Solid-State Physics.* a characteristic scale of the distance over which the order parameter describing the superconducting state can vary in a superconductive material. Also, **coherence distance.**
coherence time *Physics.* the time it takes for a wave to travel a distance equal to its coherence length.
coherency *Metallurgy.* in a solid-state transformation, the phase precipitated from a supersaturated solution that has lattice continuity with the matrix.
coherent *Geology.* describing any consolidated rock or deposit, especially one that is not easily shattered. *Geochemistry.* describing a group of elements that behave similarly in geochemical processes, such as the rare earths, or Ca-Sr.
coherent detector *Electronics.* a detector that produces an output signal whose amplitude is based on the phase of a signal rather than its strength; required for a radar display that shows only moving targets.
coherent echo *Electronics.* a radar echo that remains constant in phase and amplitude at a given range.
coherent interrupted waves *Telecommunications.* **1.** continuous waves that are interrupted at a constant audio-frequency rate. **2.** a modulation technique in which there is on-off keying of a continuous wave.
coherent light *Optics.* light waves that have the same, or nearly the same, wavelength and possess a constant phase relationship between various points in their fields, as in a laser.
coherent light communications *Telecommunications.* a form of communication that uses the optical band as a transmission medium and coherent light sources for signal generation.
coherent moving target indicator *Engineering.* a radar system that compares the doppler frequency of a target echo with the local generated frequency produced by a coherent oscillator.
coherent noise *Engineering.* any noise that disrupts all tracks of a magnetic tape at the same time and at the same intensity.

coherent optics *Optics.* a branch of optics that is concerned with utilizing coherent radiation; used to produce holograms in order to obtain three-dimensional images of objects.

coherent oscillator *Electronics.* an oscillator that provides the phase references for a radar receiver.

coherent precipitate *Physical Chemistry.* solid particles that retain the same lattice structure as the solvent from which they emerged, following a chemical or physical change in the solvent.

coherent-pulse radar *Electronics.* a radar system in which objects are located by comparing the phase between periodic changes in the frequency of pulsating and nonpulsating radio waves.

coherent pulses *Electronics.* of or relating to individual trains of high-frequency waves that are all in the same phase.

coherent radiation *Physics.* the radiation that emanates from a monochromatic source, having a definite phase relationship.

coherent reference *Electronics.* a signal, generally of stable frequency, to which other signals are phase-locked in order to establish coherency within a system.

coherent scattering *Physics.* a scattering in which the scattered radiation bears a definite phase relationship with the incident radiation, thus allowing for a stable interference pattern.

coherent signal *Electronics.* a signal having a constant phase used with an echo signal whose phase varies to measure the range of a target.

coherent source *Physics.* any source that emits coherent radiation.

coherent states *Quantum Mechanics.* the conditions of atomic systems characterized by definite quantities of energy, angular momentum, and the like, in which wave motions with different phases display interference effects.

coherent transformation *Materials Science.* a phase transformation in the production of alloys in which the crystal planes in the lattice of the precipitate are forced to be related to, or even continuous with, the crystal planes of the matrix; a distorted region is formed in the matrix so that the movement of dislocations in the matrix is impeded in the vicinity of the precipitate; a special heat treatment, such as age hardening, is usually required to produce coherent precipitation.

coherent transponder *Electronics.* a transponder whose output signal is coherent with its input signal.

coherent unit *Physics.* a unit in a system of measurement in which physical quantities, such as velocity, force, magnetic field, and strength, are defined in terms of base units.

coherer *Electricity.* a cell that has a fine granular conductor between two electrodes; formerly used as a radio-telegraph detector.

cohesion *Physics.* the attraction between the molecules of a liquid that enables drops and thin films to be formed. *Botany.* the fusion or union of similar plant parts during growth. *Geology.* in a soil, sediment, or rock, the shear strength of the cement or adsorbed water films separating individual grains at their areas of contact, independent of interparticle friction. *Computer Programming.* in structured program design, the strength of the functional relationship of processing elements within a single module; the stronger the cohesion, the better the design.

cohesional work *Physics.* the amount of energy expended in separating a single column of a liquid with a cross-sectional area of 1 cm^{-2} into two separate columns.

cohesionless *Geology.* describing a soil that does not stick together when dry or when saturated with water.

cohesive ends *Molecular Biology.* single-stranded termini on DNA molecules that are capable of base-pairing with complementary nucleotide sequences. Also, STICKY ENDS.

cohesive energy *Solid-State Physics.* the energy associated with the chemical bonding of atoms in a solid, expressed as the difference between the energy per atom of a system of free, noninteracting atoms at rest and the energy per atom of the solid.

cohesive energy density *Materials Science.* the energy per unit volume required to separate completely the atoms or molecules in a system; used to predict heats of mixing of polymeric and nonpolymeric liquids.

cohesiveness *Geology.* a property of loose, fine-grained sediments whereby the particles stick together as a result of surface forces.

cohesive soil *Geology.* a clayey or silty soil that sticks together when dry or when saturated with water.

cohesive strength *Mechanics.* **1.** strength arising from the cohesive forces between molecules of the same material. **2.** the theoretical stress required to induce fracture in a body without any accompanying plastic deformation.

cohesive terminus see COHESIVE ENDS.

Cohn, Ferdinand Julius 1828–1898, German botanist; founder of bacteriology; studied heat production in plants.

Cohnheim's theory *Oncology.* **1.** an inflammation theory in which the essential inflammatory cause is due to the emigration of leukocytes from a blood vessel to a lesion. **2.** a tumor origination theory based on tumor development from embryonic rests. Also, EMIGRATION THEORY.

cohomology *Mathematics.* the collection of homology groups associated with a cochain complex (as opposed to the homology associated with a chain complex).

cohort *Systematics.* a rank in the taxonomic hierarchy below class and above order.

coil *Electromagnetism.* a number of turns of a wire used to introduce inductance into an electric circuit. *Control Systems.* any discrete or logical result that can be transmitted as output from a controller.

coil antenna *Electromagnetism.* a loop antenna; an antenna consisting of at least one complete turn of a wire.

coil break *Metallurgy.* a defect extending across a strip or sheet, perpendicular to the rolling direction.

coiled tubular gland *Anatomy.* a gland whose secretory portion is composed of one or more coiled tubules that end in a blind extremity.

coil form *Electromagnetism.* a material, insulated at least on its surface, that is used to provide support for a coil.

coil loading *Telecommunications.* the use of a series of lumped inductors (loading coils) at regularly spaced intervals in a transmission cable to improve signal transmission.

coil neutralization see INDUCTIVE NEUTRALIZATION.

coil spring *Mechanical Devices.* a spring formed by a wire wound in helical form; applied to many devices such as the suspension system of an automobile.

coil weld *Metallurgy.* a butt weld that joins two metallic sheets at their end, in order to create a continuous product.

coil winder *Engineering.* a hand- or motor-operated apparatus that winds coils individually or in batches.

coimage *Mathematics.* given a module homomorphism $f : A \rightarrow B$, the quotient module $A/\text{Ker} f$, where $\text{Ker} f$ denotes the kernel of f.

co-immobilized enzymes *Enzymology.* enzymes that operate together in a process and are physically confined while they carry out their catalytic function.

coincidence amplifier *Electronics.* an amplifier that produces an output signal only when two input pulses are applied simultaneously.

coincidence boundary *Crystallography.* an imaginary boundary within a crystal having an exact correspondence between the points on either side.

coincidence circuit *Electronics.* **1.** a circuit that generates an output pulse only when it receives a specific number of pulses within a specified time frame. **2.** a circuit that generates an output signal only after it receives all input signals.

coincidence counting *Nucleonics.* a technique that limits or eliminates the problem of local background or false events by using radiation counters in temporal coincidence with other particle detectors, or by using two radiation detectors separated by a fixed distance to measure particles' time of flight.

coincident-current selection *Electronics.* the simultaneous application of two or more currents to select a magnetic computer storage cell for reading or writing data.

coincidence rangefinder *Optics.* a rangefinder in which two telescopes are calibrated to permit two images of a target or parts of a target to be compared or observed from different paths.

coin gold *Metallurgy.* a gold-base alloy containing 8.3% to 10% copper, suitable for making coins.

coining *Metallurgy.* **1.** in metal fabrication, a closed-die pressing operation that results in a well-defined relief on both sides of the resulting product. **2.** in powder metallurgy, the final pressing operation, designed to obtain a desired surface configuration.

coin silver *Metallurgy.* an alloy of 90% silver and 10% copper, formerly used in the manufacturing of U.S. coins.

cointegrate *Molecular Biology.* a circular molecule formed by fusing two replicons, one possessing a transposon and one without; an obligatory intermediate in the transposition process. Also, **cointegrate structure**.

co-ion *Analytical Chemistry.* any small, mobile ion in a solid ion exchanger that has the same charge as that of the fixed ions.

coitophobia *Psychology.* an irrational fear of sexual intercourse.

coitus [kō´i təs] *Physiology.* sexual intercourse between a man and a woman. *Zoology.* the process of sexual reproduction in animals.

coke *Materials.* **1.** a residue of fixed carbon and ash left after heating bituminous coal in the absence of air; used as a fuel for blast furnaces. **2.** to convert coal into such a substance. Also, PETROLEUM COKE.

coke breeze *Mechanical Engineering.* the combustion residue after making coke or charcoal, consisting of small cinders and ash that will pass through a screen with openings from 0.5 to 0.75 inch in diameter.

coke coal see COKEITE.

coke drum *Chemical Engineering.* a vessel in which coke is produced.

cokeite *Geology.* naturally occurring coal that is formed as a result of interaction with magma or by natural combustion. Also, COKE COAL, NATIVE COAL.

coke number *Chemical Engineering.* a number used to represent the amount of carbon residue in a petroleum product during a Ramsbottom carbon residue test.

coke oven *Chemical Engineering.* an oven or retort used to produce coke by the carbonization of coal.

coker *Chemical Engineering.* the unit in which coking occurs.

cokernel *Mathematics.* given a module homomorphism $f: A \rightarrow B$, the quotient module $A/\text{Im} f$, where $\text{Im} f$ denotes the image of f.

coking *Chemical Engineering.* **1.** the production of coke by heating coal for approximately 12 hours; the destructive distillation of coal to make coke. **2.** a procedure for converting heavy residual bottoms of crude oil into lower-boiling petroleum fractions and petroleum coke.

coking coal *Geology.* a variety of bituminous coal that contains about 90% carbon, suitable for the production of coke for metallurgical use.

coking still *Chemical Engineering.* a still, usually a batch still, in which the coking of crude oil residuals takes place.

col *Geology.* **1.** a narrow, sharp-edged gap in a mountain ridge or between two adjacent peaks, usually formed by the headward erosion of oppositely oriented cirques or valleys. **2.** the lowest point in the crest of a mountain ridge. Also, SADDLE. **3.** an elevated neck of land connecting two higher land masses. *Meteorology.* the intersection point between a trough and a ridge in the pressure pattern of a weather map. Also, NEUTRAL POINT, SADDLE POINT.

col- a prefix meaning "with" or "together," as in *collateral*.

cola *Botany.* the tropical tree *Cola acuminata,* cultivated for the seeds of its **cola nuts**, from which cola flavoring is obtained. *Food Technology.* a soft drink flavored with the extract of cola nuts.

Colacium *Invertebrate Zoology.* a genus of green, stalked protozoans, with a free-swimming flagellate stage and a sessile stage that often attaches to rotifers, copepods, and other tiny aquatic animals.

colanic acid *Biochemistry.* a capsular form of heteropolysaccharide that is produced by certain enterobacteria strains, such as *Escherichia coli* and *Salmonella.*

colatitude *Astronomy.* the angular distance between a celestial object and the ecliptic pole; equal to 90° minus the object's celestial latitude.

Colburn analogy *Fluid Mechanics.* an analogy that demonstrates that the resistance of the laminar sublayer to mass or heat transfer can be expressed by an appropriate modification of the Reynolds analogy.

Colburn j factor equation *Thermodynamics.* an equation showing the correspondence between heat and mass transfer in convective flows.

Colburn method *Chemical Engineering.* a graphical method using complementary equations to calculate the theoretical number of plates required for separating binary mixtures in a distillation column.

Colcemid *Organic Chemistry.* a trade name for a compound having the formula $C_{21}H_{25}NO_5$; pale yellow to white prisms that are soluble in water and alcohol; melts at 186°C; used as an antineoplastic. Also, DEMECOLCINE.

colchicine *Organic Chemistry.* $C_{22}H_{25}NO_6$, poisonous, alkaloid, yellow plates; soluble in water and alcohol; melts at 155–157°C; used to block mitosis in metaphase by preventing the synthesis of tubulin, and to treat gout and Hodgkin's disease.

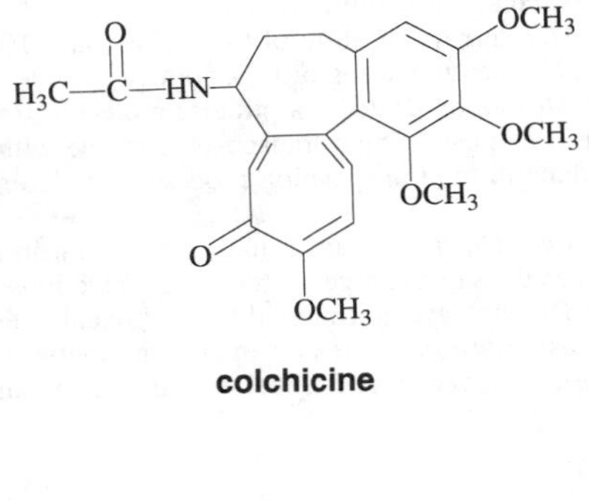

colchicine

colcothar *Inorganic Chemistry.* a form of red ferric oxide, Fe_2O_3, made by heating ferrous sulfate; formerly used as a pigment and abrasive.

cold *Medicine.* a contagious viral infection that commonly occurs in the colder seasons and climates, causing various symptoms, including nasal drip, sneezing, chills and/or fever, nasal and other mucous membrane inflammation and resultant congestion, and fatigue. Also, COMMON COLD. *Electricity.* describing a circuit that is disconnected from the power supply and at ground potential.

cold agglutination phenomenon *Immunology.* the clumping of red blood cells at 0–4°C, rather than at body temperature, which occurs during certain stages of diseases such as pneumonia.

cold agglutinin *Immunology.* any antibody that binds with an antigen more efficiently at low temperatures (below 4°C).

cold-air drop see COLD POOL.

cold-air machine *Mechanical Engineering.* a refrigeration system in which air is used as the coolant in an adiabatic cycle of compression and expansion.

cold anticyclone see COLD HIGH.

cold area *Science.* a laboratory area in which physical and chemical tests are conducted that require a continuous, specified low temperature

cold bending *Metallurgy.* bending a metallic stock at ambient temperature.

cold boot *Computer Technology.* see COLD START.

cold cathode *Electronics.* a cathode that can generate and sustain current flow without heat.

cold-cathode rectifier *Electronics.* a rectifier circuit employing a cold cathode tube. Also, GAS-FILLED RECTIFIER.

cold-cathode tube *Electronics.* a tube, such as a photoelectric cell or gas glow tube, in which current flows without generating heat.

cold cautery see CRYOCAUTERY.

cold chamber *Metallurgy.* the receptacle used in a die casting process in which hot liquified metal is poured into a cold cylinder and then forced into the die cavity. Thus, **cold chamber die casting.**

cold chisel *Mechanical Devices.* a steel chisel used in metal working to cut or chip cold metal. Also, COLD CUTTER.

cold color *Graphic Arts.* a color toward the blue end of the spectrum, or a color with a bluish tint.

cold composition see COLD-TYPE COMPOSITION.

cold-core high see COLD HIGH.

cold-core low see COLD LOW.

cold core ring *Oceanography.* a ring formed on the south side of the Gulf Stream that injects colder coastal water into the Sargasso Sea; also, CYCLONIC RING.

cold cure *Chemical Engineering.* the vulcanization of rubber at moderate temperatures with a sulfur solution.

cold cyclone see COLD LOW.

cold dome *Meteorology.* a cold-air mass considered as a three-dimensional entity.

cold drawing *Metallurgy.* the process of drawing a metal or alloy at ambient temperature. *Textiles.* the process of pulling or stretching a fabric while it is cold.

cold extrusion *Metallurgy.* **1.** the process of extruding a metal or alloy at ambient temperature. **2.** the resulting product of such process.

cold fault *Computer Technology.* a malfunction that can be observed as soon as a computer is turned on.

cold-finished steel *Metallurgy.* a steel mill product that is finished by cold rolling, drawing, or machining after a hot-working operation.

cold flow *Mechanics.* **1.** a viscous flow of a solid at ordinary temperatures. **2.** the distortion of a solid under sustained pressure.

cold-flow test *Space Technology.* a static test of a liquid rocket, designed to check a propulsion system or subsystem, including the flow of propellants.

cold forming *Metallurgy.* the process of forming a metal or alloy at ambient temperature.

cold front *Meteorology.* the leading edge of an advancing cold-air mass, often accompanied by heavy showers.

cold-front-like sea breeze *Meteorology.* a breeze that forms over the sea, moves slowly toward the coast, and then suddenly moves inland; associated with showers, a sharp wind shift, and a drop in temperature.

cold galvanizing *Metallurgy.* the process of painting iron with zinc particles held in suspension in order to maintain a zinc coating following evaporation of the solvent.

cold-gas approximation *Physics.* an approximation according to which the sound speed is less than the Alfvén speed, or the gas pressure is less than the magnetic pressure.

cold hammering *Metallurgy.* the technique of shaping certain malleable metals by beating the metal into the desired shape without heating it.
cold heading *Metallurgy.* heading a metal or alloy at ambient temperature.
cold high *Meteorology.* an anticyclone characterized by air that is colder near its center than around its periphery. Also, COLD ANTICYCLONE, COLD-CORE HIGH.
cold injury *Medicine.* any abnormal physical condition caused by exposure to reduced temperatures, such as frostbite or hypothermia.
cold inspection *Metallurgy.* in forging, a visual inspection performed at ambient temperature.
cold isostatic compression molding *Materials Science.* a forming process in which the material is encased in a rubber sheath and pressure is applied through a fluid; the formed ware is of uniform density and has lower firing shrinkage than ware produced by other methods.
cold joint *Engineering.* a joint soldered without adequate heat, so that the wire is held in place by rosin flux.
cold junction *Electronics.* the junction of two wires or strips of dissimilar metals, such as antimony and bismuth, that produce a direct current at room temperature; used in measuring instruments.
cold light *Physics.* light in the visible range of the spectrum that does not contain infrared wavelengths and therefore has little heating effect.
cold lime-soda process *Chemical Engineering.* a standard water-softening process that uses hydrated lime and soda ash to precipitate calcium and magnesium in hard water.
cold low *Meteorology.* a cyclone characterized by air that is colder near its center than around its periphery. Also, COLD-CORE CYCLONE, COLD-CORE LOW.
cold molding *Engineering.* a procedure in which a substance is shaped under pressure without heat, then hardened by subsequent baking.
cold neutron *Solid-State Physics.* a neutron with very low kinetic energy having a de Broglie wavelength on the order of a crystal lattice constant.
cold pasteurization *Food Technology.* a method of pasteurizing heat-sensitive liquids by pouring them through a microporous membrane filter; developed as a means of producing "draft beer" in cans, now also used to process wine and fruit juices.
cold plate *Mechanical Engineering.* a piece of flat metal embedded with tubes that carry liquid coolant; placed under heat-emitting electronic components to absorb heat.
cold pole *Meteorology.* the location having the lowest mean annual temperature in its hemisphere.
cold pool *Meteorology.* a region of cold air surrounded by warmer air. Also, COLD-AIR DROP.
cold pressing *Metallurgy.* in powder metallurgy, the process of forming a compact at a temperature at which sintering does not occur.
cold receptor *Physiology.* a thermoreceptor that is sensitive to cold, or to temperatures below that of the body. *Entomology.* specifically, a cell on the antenna or maxillary palps of adult insects and larvae that is sensitive to cooling temperatures.
cold rolling *Metallurgy.* a processing of metal in which a metal ingot is rolled into a slab through four-high rolling mills, either alone or in a series and below the recrystallization temperature, usually at room temperature. Thus, **cold-rolled steel.**

cold-rolled steel

cold room *Biotechnology.* a temperature-controlled room used for the preparation of enzymes and other temperature-sensitive compounds, usually maintained around 4°C.
cold saw *Mechanical Engineering.* a slow-running circular saw designed for cutting cold metal; relatively thick in proportion to its diameter and equipped with short teeth that may be either inserted or integral with the disk.
cold-sensitive mutant *Genetics.* **1.** a gene that is defective at low temperatures, but functional at normal ones. **2.** a mutant that has a higher minimum temperature of growth than the wild-type organism.
cold settling *Chemical Engineering.* a settling operation used to clarify high-viscosity products.
cold-short *Metallurgy.* the attribute of a metal or alloy that is brittle when subjected to stress below the recrystallization temperature.
cold shot *Metallurgy.* the surface defect of a casting, caused by a splash of molten metal on the mold wall. The splash freezes prematurely and does not weld to the body of the casting.
cold shut *Metallurgy.* **1.** the premature freezing of the top of an ingot before the mold is completely filled. **2.** in casting, a defect similar to a cold shot, but caused by a two separate streams of molten metal. **3.** in forging, a surface defect caused by a fold of hot metal without welding.
cold slug *Engineering.* a substance that cools below effective molding temperature as it passes through the sprue orifice of an injection mold.
cold soldering *Metallurgy.* **1.** soldering without heat. **2.** soldering with insufficient heat, thus inducing incomplete coalescence.
cold start *Mechanical Engineering.* the process of starting an internal-combustion engine after it has been turned off for an extended period of time. *Computer Technology.* the process of booting a computer from an off and empty state. Also, COLD BOOT.
cold sterilization see FOOD IRRADIATION.
cold storage *Engineering.* the storage of perishables at low temperatures maintained by refrigeration, to prolong their useful life.
cold-storage locker plant *Engineering.* a plant that provides rental steel lockers, each usually 6 cubic feet, for the cold storage of food.
cold stress *Mechanics.* stress resulting from low temperatures, which tend to deform materials.
cold stretch *Engineering.* a process performed without heat, in which plastic is pulled at the edges to improve its resistance to tension.
cold test *Chemical Engineering.* a test that lowers the temperature of a sample of oil in order to determine the temperature at which wax first forms from a lubricating oil, or the point when oil no longer flows.
cold tongue *Meteorology.* a pronounced protrusion of cold air toward the equator.
cold torpor *Physiology.* a lack of response due to prolonged exposure to cold.
cold trap *Mechanical Engineering.* a vapor-condensing tube with liquid nitrogen or other liquid coolant in its walls.
cold treatment *Metallurgy.* the thermal treatment of a metal or alloy performed at a subzero temperature.
cold trimming *Metallurgy.* in postforging operations, the process of removing flash when the product is at ambient temperature.
cold type *Graphic Arts.* **1.** any form of copy that has not been set with hot metal type. **2.** see COLD-TYPE COMPOSITION.
cold-type composition *Graphic Arts.* **1.** the production of type by any method other than the casting of molten metal. **2.** specifically, computerized typesetting as opposed to linotype. Also, COLD COMPOSITION, COLD TYPE.
cold wall *Oceanography.* a well-defined line where two water masses of different temperatures meet.
cold-water sphere *Oceanography.* the cold deep-water and bottom-water masses, where the temperature is below 8°C; one of the two major subdivisions of ocean water masses, along with the warm-water sphere.
cold wave *Meteorology.* **1.** a rapid fall in temperature within 24 hours, requiring increased protection to agriculture, industry, commerce, and social activities. **2.** a popular term for a period of very cold weather.
cold welding *Metallurgy.* the pressure welding of metals or alloys performed at ambient temperature.
cold working *Metallurgy.* any of various working processes performed at a temperature at which the material does not strain-harden.
Colebrook equation *Fluid Mechanics.* an equation giving the friction factor for the transition zone between laminar and turbulent flow in terms of the pipe diameter, roughness, and Reynolds number.
colectasia *Medicine.* the stretching of the colon.
colectomy [kə lek′tə mē] *Surgery.* the surgical removal of part or all of the colon.

colemanite *Mineralogy.* $Ca_2B_6O_{11}·5H_2O$, a brittle, transparent to translucent, colorless, white, or gray monoclinic mineral occurring as short prismatic crystals, having a specific gravity of 2.42 and a hardness of 4.5 on the Mohs scale; found in desiccated saline lake deposits in desert regions, sometimes appearing as geodes or rounded aggregates.

Coleodontidae *Paleontology.* a family of Paleozoic conodonts in the order Neurodontiformes.

Coleoidea *Invertebrate Zoology.* a subclass of mollusks including all cephalopods except Nautilus.

Coleophoridae *Invertebrate Zoology.* a family of case-bearing moths, narrow-winged lepidopterans in the suborder Heteroneura; named for the leaf-and-silk shell carried by larvae.

coleopter *Aviation.* a jet aircraft having a single annular wing with the fuselage at the center; designed for vertical takeoff and landing; known informally as a "flying barrel."

Coleoptera [kō´ lē äp´tə rə] *Invertebrate Zoology.* an order of insects composed of over 300,000 known species; members include the beetle, weevil, borer, and firefly. It is the largest order of the animal kingdom.

coleopteran [kō´ lē äp´tə rən] *Invertebrate Zoology.* **1.** a member of the order Coleoptera, especially a beetle. **2.** of or relating to this order.

coleoptile *Botany.* the primary leaf of a grass seedling.

coleorhiza *Botany.* a sheathlike structure protecting the embryonic root (radicle) of a grass seed.

Coleorrhyncha *Invertebrate Zoology.* a family of homopteran insects; the beak is at the anteroventral extremity of the face, shielded by the propleura.

Coleosporaceae *Mycology.* a former term for a family of fungi belonging to the order Uredenales which has now been classified as the family Malamsporaceae; all of its species are parasites.

coleps *Invertebrate Zoology.* a genus of small, barrel-shaped ciliated protozoans in the subclass Holotrichia, often having spines.

coleus *Botany.* a genus of showy foliage plants native to tropical Africa and Asia; several species are widely grown as houseplants for their large, colorful leaves.

col factor *Genetics.* a plasmid that contains genes for colicins.

colibacillosis *Veterinary Medicine.* diarrhea caused by neonatal infection with *Escherichia coli*; a term usually used in relation to poultry.

colic *Medicine.* **1.** any sudden attack of sharp abdominal pain, especially as accompanied by intense crying and irritability in infants, typically caused by swallowing air or overfeeding. **2.** of or relating to the colon.

colic artery *Anatomy.* any of three branches of the mesenteric artery that supply oxygen-rich blood to the colon.

colicin *Microbiology.* any of various bacteriocins produced by strains of *Escherichia coli* and *Shijella sonnei;* often lethal to other bacterial cells.

colicky *Medicine.* relating to or affected by colic.

colicoplegia see LEAD POISONING.

coliform bacteria *Bacteriology.* **1.** any fermentative, Gram-negative, rod-shaped anaerobic bacteria, typically found in the intestinal tracts of humans and other animals. **2.** specifically, any of such bacteria that ferment lactose.

Coliidae *Vertebrate Zoology.* the mousebirds or colies, a family of long-tailed African land birds in the order Coliiformes, noted for creeping around trees and bushes in a mouselike fashion.

Coliiformes *Vertebrate Zoology.* the mousebirds, a monofamilial order of long-tailed African land birds represented by the single genus *Colius.*

colinearity *Biochemistry.* the linear relationship between the amino acid sequence in a polypeptide chain and the nucleotide sequence in the DNA and mRNA coding for the chain.

colistin *Microbiology.* a former term for a group of peptide antibiotics that are now classified within the polymyxin antibiotics group.

colitis *Medicine.* an inflammation of the colon.

colitose *Biochemistry.* 3,6-dideoxy-L-galactose, a sugar present in the O-specific chains of the lipopolysaccharide of given *Salmonella* serotypes, originally extracted from *Escherichia coli.*

colitoxicosis *Toxicology.* intoxication caused by exposure to *Escherichia coli.*

colitoxin *Toxicology.* the substance in *Escherichia coli* that causes colitoxicosis.

colla *Meteorology.* a strong south to southwest wind in the Philippine Islands that is accompanied by heavy rain and severe squalls. Also, **colla tempestada.**

collaboratory see VIRTUAL LABORATORY.

collada *Meteorology.* a strong wind of 35–50 mph that blows from the north or northwest in the upper part of the Gulf of California, and from the northeast in the lower part of the gulf.

collagen *Biochemistry.* a gelatinous protein present in all multicellular organisms, particularly in the connective tissue, to which it gives strength and flexibility. Thus, **collagenous.**

collagenase *Enzymology.* an enzyme that catalyzes the breakdown of collagen into smaller compounds.

collagen disease *Medicine.* a group of diseases of the connective tissue characterized by arthritic conditions and various skin and viscera disorders.

collandria evaporator *Biotechnology.* a circulating evaporator in which steam-heated material flows up a set of short, vertically bundled tubes, and then recirculates back down a central pipe as it cools.

collapse breccia *Geology.* a breccia derived from the collapse of rock overlying a hollow area, such as a cave, or the roof above an intrusion. Also, FOUNDER BRECCIA.

collapse caldera *Volcanology.* a large, circular volcanic depression formed by the sinking of the roof of a magma chamber after withdrawal or eruption of the magma.

collapse depression *Geology.* an elongated hollow formed in the surface of a lava flow as a result of partial or complete collapse of the roof of a lava tunnel.

collapse properties *Materials Science.* those characteristics of a commercial material that allow it to resist collapse.

collapse sink *Geology.* a large depression in the ground surface resulting from the collapse of an underlying cave.

collapse structure *Geology.* any rock structure formed by the removal of support and subsequent rock slides under the influence of gravity. Also, GRAVITY-COLLAPSE STRUCTURE.

collapsing pressure *Mechanics.* the amount of external pressure that is required to collapse a body or structure.

collapsing pulse see CORRIGAN'S PULSE.

collar any of various markings or structures thought to resemble the collar of a garment; specific uses include: *Mechanical Devices.* **1.** a short, rectangular ring that is fitted or forged over a rod or shaft to locate or hold a bearing. **2.** a raised area of metal used as a reinforcement for welds. *Naval Architecture.* **1.** one of the short ropes or wires that secures the stays to the deck. **2.** a watertight fitting around a projecting beam or timber. *Invertebrate Zoology.* the clearly demarked prothorax of an insect or the choana of a choanocyte. *Virology.* a connective body part attached to the neck of some tailed phages.

collar beam *Building Engineering.* a beam positioned above and attached to the lower portion of side rafters on a sloping roof well above the wall plate, forming a roof truss.

collar bearing *Mechanical Engineering.* a bearing provided with several collars to take the thrust of a shaft or to provide adequate surfaces for lubrication of a vertical shaft.

collarbone *Anatomy.* the popular name for the clavicle, the bone that connects the breastbone (sternum) with the shoulder blade (scapula).

collard *Botany.* a large kale, *Brassica oleracea,* whose green leaves are eaten as a vegetable.

collard

collared hole *Engineering.* a small indentation that keeps a drill bit in place through the beginning of the drilling process.

collared lizard *Vertebrate Zoology.* a lizard of the genus *Crotaphytus* in the family Iguanidae, especially *C. collaris,* which is characterized by the frilled skin around its neck and head and by its ability to stand and move on two legs; found in the south central U.S. and Mexico.

collate *Computer Programming.* to merge two or more sets of data or values to form a single set in a prescribed order or sequence.

collateral bud *Botany.* an accessory bud that lies adjacent to an axial bud.
collateral bundle *Botany.* a vascular structure in which the xylem is positioned at one side of the phloem, usually the abaxial side.
collateral damage *Military Science.* damage inflicted on nonmilitary property or persons in the course of an attack on a military target.
collateral kin *Anthropology.* a group of relatives who are related indirectly through a linking relative such as a mother's brother.
collateral series *Nuclear Physics.* a decay series, nonexistent in nature and not in the direct line of the four principal heavy element radioactive series, in which radioactive nuclides disintegrate to successively low levels of energy until they become stable.
collating marks *Graphic Arts.* a set of small marks printed in designated positions on the forms of a book as a guide in collating folded signatures. Typically the first mark is near the top back edge of the first signature, and each successive mark is slightly lower; thus a straight diagonal line across the gathered signatures indicates that they are collated correctly.
collating sequence *Computer Programming.* a specific order that is assigned to a set of items, such as alphabetization.
collecting *Archaeology.* the nonscientific removal of archaeological materials from a site by residents of another area.
collecting tubule *Anatomy.* any of a number of small tubes that allow the passage of urine from the nephrons or renal tubules to the minor calices of the renal pelvis.
collection-and-distribution service *Transportation Engineering.* a transport system serving many stops, having origins and destinations spaced throughout the system.
collection trap *Analytical Chemistry.* an apparatus that collects cooled eluent, holding it for later analysis.
collective behavior *Behavior.* the behaviors that characterize a group rather than its individual members.
collective fire *Ordnance.* fire from a group of small arms aimed at a single target or area.
collective mode *Physics.* the behavior of a many-particle system dominated by an internal field set up collectively by the particle.
collective motion *Nuclear Physics.* an activity that takes place among particles in the nucleus, in which the movement of each particle corresponds with the movements of the other particles, so that their relative positions remain the same or change very slowly.
collective transition *Nuclear Physics.* a transition in which a nucleus moves from one state of collective motion to another.
collective unconscious *Psychology.* Carl Jung's term for the part of the unconscious that is common to all humanity, formed by the accumulated experience of all previous generations and affecting present human behavior and beliefs. Also, RACIAL UNCONSCIOUS.
collectivist or **collectivistic** *Anthropology.* of or relating to a culture in which the individual's loyalty to family, tribe, or ethnic group takes precedence over personal goals and rewards. Thus, **collectivism.**
collector *Electronics.* **1.** the region of a bipolar transistor into which most of the current flows. **2.** an electrode that gathers electrons or ions after they have completed their function.
collector capacitance *Electronics.* the depletion layer capacitance stored near the junction in a transistor where most of the current flows.
collector current *Electronics.* the current flowing into a transistor's collector terminal.
collector cutoff *Electronics.* the value of bias at which the current flowing across the collector fails to produce a signal.
collector junction *Electronics.* the semiconductor junction between the collector and base electrodes of a transistor.
collector modulation *Electronics.* a process for changing the shape of a signal by varying the voltage at the collector junction.
collector plate *Electricity.* a metal plate placed in the lining of an electrolytic cell in order to minimize the resistance between the cell lining and the current lead.
collector resistance *Electronics.* the resistance between the base of a transistor and the collector.
collector voltage *Electronics.* voltage generated by a direct current between the base of a transistor and the collector.
Collembola *Entomology.* the springtails, an order of tiny, primitive, wingless apterygote insects with biting mouthparts, short antennae, and abdomens fused in six segments; living in damp places, they jump by releasing an abdominal spring held in place by a hook.
collenchyma *Botany.* a collection of living, thick-walled, elongate cells functioning to support the primary growth of leaves and stems.
collenchyme *Invertebrate Zoology.* a loose mesenchyme, in many lower invertebrates such as sponges, that fills the space between the ectoderm and the endoderm in the body wall.
collencyte *Invertebrate Zoology.* any of the branched connective tissue cells constituting the collenchyme, as in sponges.
collenia *Paleontology.* a convex, turbinate Precambrian stromatolite produced by the cyanophytic (blue-green) algae of the genus *Collenia.*
Colles' fracture *Medicine.* a fracture of the radius above the wrist in which there is dorsal displacement of the distal fragment.
collet *Mechanical Engineering.* **1.** the neck section of a glass bottle that remains after it has been removed from a glass-blowing iron. **2.** a split sleeve that is used to hold work or tools during machining or grinding. **3.** specifically, a slotted sleeve with an external cone shape that fits into a similarly shaped nose of a lathe mandrel to grip circular workpieces. Also, **collet chuck.** *Horology.* in a timepiece with a balance escapement, a small circular ring that joins the inner end of the hairspring to the balance staff.
Colletidae *Invertebrate Zoology.* a family of hymenopteran insects, the colletid bees, in the superfamily Apoidea.
Colletotrichum *Mycology.* a genus of fungi belonging to the order Melanconiales and causing such plant diseases as red rot.
colliculus *plural,* **colliculi.** *Anatomy.* a small mound or elevation, as of an organ or a bone.
2,4,6-collidine *Organic Chemistry.* $C_8H_{11}N$, a colorless liquid that is soluble in water and alcohol; melts at –44.5°C and boils at 170°C; used as an intermediate and dehydrohalogenating agent.
colliding-beam accelerator *Particle Physics.* a particle accelerator in which two beams of particles are made to collide head-on, so that a greater proportion of the energy of the incident particles is available for the creation of new particles than with a fixed-target accelerator.
colliding-beam source *Electronics.* a device producing beams of polarized hydrogen or deuterium that convert to negative ions when they collide with cesium atoms.
collier *Naval Architecture.* a ship engaged in the carriage of coal.
colliery *Mining Engineering.* a complete coal-mining operation, including the mine, shops, preparation plant, and equipment.
colliery explosion see FIREDAMP EXPLOSION.
colligative *Physical Chemistry.* depending on number rather than on size or other characteristics.
colligative properties *Physical Chemistry.* those chemical properties that are dependent on the number of molecules present, rather than on the size, weight, or characteristics of those molecules; they include osmotic pressure, boiling-point elevation, freezing-point depression, and vapor-pressure lowering.
collimate *Physics.* to focus or align along a unidirectional path. *Optics.* to make rays of light parallel to a particular line or direction.
collimated beam *Physics.* a beam of particles with a narrow cross section that has little or no spreading angle.
collimating lens *Optics.* the lens in a collimator that focuses light from an object near one of its focal points onto a parallel beam.
collimating sight *Ordnance.* a sight equipped with a collimator, which produces parallel rays of light and is set parallel with the axis of the gun's bore while in horizontal position, and can be adjusted to remain focused on the target while the gun is raised or lowered.
collimation *Astronomy.* the process of adjusting a telescope to place all optical elements in precise alignment. *Radiology.* the process of eliminating peripheral portions of an X-ray beam, using radiation-absorbing tubes, cones, or diaphragms.
collimation axis *Photogrammetry.* the line through the rear nodal point of the objective lens in an optical instrument that is exactly parallel with the instrument's center line.
collimation error *Engineering.* **1.** in surveying, the failure of two nominally parallel lines of sight to have the correct angular relationship. **2.** a particular angle in a radar that causes an incorrect line of sight.
collimation method see HEIGHT-OF-INSTRUMENT METHOD.
collimator *Optics.* an instrument that creates a parallel beam of light or an infinitely distant virtual image; used in lens testing to establish focal lengths and in metrology to locate a distant object whose position is known. *Radiology.* a diaphragm or device constructed of radiation-absorbing materials, used to restrict the dimensions of a radiation beam. Also, LOCALIZER.
collinear [kə lin′ē ər] *Mathematics.* lying on the same line.
collinear heterodyning *Electronics.* an optical processing system whose correlation function is derived from an ultrasonic light modulator.

collinear transformation *Optics.* an ideal image/object relationship, in which there is a point in the object that corresponds to every unique point in the image.
collinear vectors *Mathematics.* two vectors, one of which is a scalar multiple of the other; or, in the case of a vector from the origin, two vectors having angular rotation differing by a multiple of π. Two vectors that are not collinear are linearly independent.
collineation *Mathematics.* a nonsingular linear transformation that preserves collinearity; that is, points, lines, and planes are mapped to points, lines, and planes, respectively.
collinite *Geology.* a hard coal maceral of the vitrinite group, composed of jellified and precipitated plant material.
Collins helium liquefier *Physics.* a device for liquefying helium, in which the gas is used to move a piston, and then moves to an expansion chamber where it cools by adiabatic expansion.
collinsite *Mineralogy.* $Ca_2(Mg,Fe^{+2})(PO_4)_2 \cdot 2H_2O$, a transparent to translucent, colorless, brown or white triclinic mineral usually occurring as fibrous crusts, having a specific gravity of 2.93 to 2.99 and a hardness of 3.5 to 5 on the Mohs scale; found in altered phosphate nodules from granite pegmatite.
collision *Physics.* an interaction between particles in which momentum is conserved. *Computer Programming.* in address transformation, the mapping of two or more keys to the same address. *Computer Technology.* in a broadcast network, a situation in which more than one node tries to send a message simultaneously, resulting in the garbling of both messages. *Artificial Intelligence.* a constraint identification paradigm to rule out unwanted patterns or to identify conflicting options.
collision-avoidance system *Aviation.* a radar system that warns aircraft when an airspace conflict (and therefore a potential collision hazard) occurs, directing pilots to take evasive action or taking such action automatically. Thus, **collision-avoidance radar.**
collision bearing *Navigation.* a constant bearing of an approaching craft which indicates that a risk of collision exists.
collision blasting *Engineering.* a blasting process in which different sections of the rocks are blasted out against each other.
collision broadening *Spectroscopy.* the widening of a spectral line as a result of collisions between radiation and sample particles that destroy or interrupt energy level transitions. Also, **collision line-broadening.**
collision course *Navigation.* a course that, if maintained, will cause a craft to collide with an obstacle or another craft.
collision-course homing *Ordnance.* a missile homing system in which an antenna on the missile works with built-in computers to anticipate the movement of a target and to direct the missile into the target's path, thus achieving contact in the least possible time.
collision detection *Telecommunications.* in some local-area networks, a function whereby a device trying to transmit data senses that the network is in use, waits a random time interval, and attempts to retransmit.
collision efficiency *Meteorology.* the fraction of all water drops that actually collide with other drops moving on a collision course.
collision excitation *Atomic Physics.* a process in which charged particles strike atoms in a gas or vapor and raise their energy levels.
collision frequency *Physics.* the average number of collisions that a particle makes per unit time as it moves through a gas or other medium.
collision-frequency factor *Physics.* the number of collisions per unit time and unit volume divided by a constant and then divided by the square of the concentration.
collision ionization *Atomic Physics.* a process in which highly charged particles eject electrons from atoms in a vapor or gas, causing the atoms to become ionized.
collision of the first kind *Physics.* an inelastic collision in which a portion of the kinetic energy of the incident particle is converted into energy used to excite or ionize the target particle.
collision of the second kind *Physics.* an inelastic collision in which energy from an excited particle is transferred to a moving particle, thereby increasing the kinetic energy of the moving particle.
collision-radiative recombination *Atomic Physics.* the capture of an electron by an ion in a gas that takes place along with the ejection of one or more photons.
collision theory *Physical Chemistry.* the theory that the number of new compounds formed during a chemical reaction equals the number of molecules colliding, times a factor that corrects for low-energy collisions. *Quantum Mechanics.* the analysis of interactions between particles that approach each other at relatively close ranges for brief periods.
collision vector *Space Technology.* a vector that, if maintained by a spacecraft or missile, will lead to a collision with some other object.
colloblast *Invertebrate Zoology.* an adhesive cell found on the tentacles of comb jellies.
colloclarain *Geology.* a transitional coal lithotype characterized by the presence of collinite, but containing more of other macerals.
collodion *Organic Chemistry.* a flammable, pale yellow, syrupy liquid, immiscible with water, prepared from nitrocellulose in ether and alcohol; used in coating wounds, as a dialysis membrane, and as a solvent.
collodion replication *Analytical Chemistry.* a cellulose nitrate-film mold of a specimen surface that is thin enough to be studied using an electron microscope.
colloform *Geology.* describing the spherical texture of a mineral deposit resulting from colloidal precipitation.
colloid [käl´oid] *Physical Chemistry.* **1.** a substance consisting of very tiny particles that are usually between 1 nanometer and 1000 nanometers in diameter and that are suspended in a continuous medium, such as a liquid, a solid, or a gaseous substance. **2.** the translucent, pale yellow, gelatinous substance resulting from colloid degeneration.
colloidal [kə loid´əl] *Physical Chemistry.* relating to or being a colloid or a colloidal suspension.
colloidal crystal *Physical Chemistry.* a colloidal system in which particles of identical shape and size are arranged in regular patterns analogous to crystal structure.
colloidal dispersion see COLLOIDAL SUSPENSION.
colloid goiter *Medicine.* a chronic enlargement of the thyroid gland, characterized by atrophy of the epithelium and an increase in the amount of gelatinous matter in the gland.
colloidal gold *Biotechnology.* a stable electron-dense gold colloid used to identify and quantitate material; particularly effective in electron microscopy, where it serves as a location-specific label.
colloidal graphite *Materials Science.* a colloidal suspension of very fine graphite in oil; used as a lubricant.
colloidal instability *Meteorology.* a condition of clouds in which the particles tend to aggregate into masses large enough to precipitate.
colloidal particle *Physical Chemistry.* any of the individual particles of a colloid. See COLLOID.
colloidal silica *Materials.* a colloid of fine particles of silica that are negatively charged by a small addition of alkali, forming a high-concentration solution that is used to make molded ceramics or to treat textiles.
colloidal suspension *Physical Chemistry.* a mixture of two substances, one of which (the colloid) is uniformly distributed throughout the other (the dispersion medium); either substance may be gaseous, liquid, or solid. Also, COLLOIDAL DISPERSION, COLLOIDAL SYSTEM.
colloidal system see COLLOIDAL SUSPENSION.
colloid chemistry *Physical Chemistry.* the scientific study of colloidal matter.
colloid mill *Mechanical Engineering.* a grinding machine that breaks down agglomerates into very fine particles, or shears fluid phases to produce stable emulsions containing dispersed droplets of very fine size.

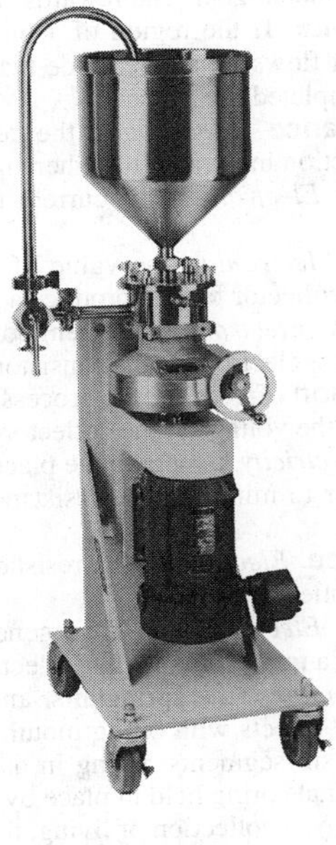

colloid mill

collophane *Mineralogy.* a massive, fine-grained member of the apatite group.

collophore *Entomology.* a thick tubular projection from the ventral surface of the abdomen of all members of the order Collembola.

Collothecacea *Invertebrate Zoology.* a suborder of rotifers in the order Monogonata; most are sessile, and some live in gelatinous tubes.

Collothecidae *Invertebrate Zoology.* the single family of rotifers in thet suborder Collothecacea.

collotype printing *Graphic Arts.* a usually direct printing process similar to lithography but using bichromated gelatin as the printing surface; used for fine art reproductions and other short-run, detailed printing jobs. Also, PHOTOGELATIN PRINTING.

colluvium *Geology.* any loose, heterogenous sediment deposited by rainwash, sheetwash, or slow continuous downslope creep, usually at the base of a cliff or slope.

Collyrites *Paleontology.* a genus of small euechinoids in the extinct order Disasteroida, characterized by their oval shape and the absence of a lantern; Jurassic and Cretaceous.

Colmol miner *Mining Engineering.* a continuous mining machine in which the coal is hewed from the solid by rotating chipping heads.

Colobognatha *Invertebrate Zoology.* an order or superorder of millipedes having 30–70 body segments, most segments bearing glands that secrete a caustic or poisonous repugnant fluid.

coloboma *Medicine.* a defect of ocular tissue, either present at birth or caused by disease or trauma.

coloclysis *Medicine.* irrigation of the colon.

colocynth *Pharmacology.* the dried pulp of the full-grown but unripe fruit of bitter cucumber, *Citrullus colocynthis.*

colocynthidism *Toxicology.* poisoning due to ingestion of colocynth; symptoms may include severe gastrointestinal disturbances.

colog cologarithm.

cologarithm [kō´lôg´ə rith´əm] *Mathematics.* the additive inverse of the logarithm; that is, if log $x = y$, then colog $x = -y = \log 1/x$.

Cologne earth see BLACK EARTH.

colominic acid *Biochemistry.* a capsular form of polysaccharide present in *Escherichia coli* K_1 strains that functions as a linear polymer of *N*-acetylneuraminic acid residues attached by α-ketosidic bonds.

colon *Anatomy.* the section of the large intestine from the cecum to the rectum.

colonel [kur´nəl] *Military Science.* in the U.S. Army, Air Force, or Marine Corps, a field-grade officer ranking between lieutenant colonel and brigadier general.

colonic [kə län´ik; kə lōn´ik] *Anatomy.* of or relating to the colon.

colonist *Ecology.* an organism that invades and colonizes a new habitat.

colonization *Ecology.* the successful occupation of a new habitat by a species not previously found in this area. *Oncology.* see INNIDATION.

colonize *Ecology.* to carry out the process of colonization.

colonnade [käl´ə näd´] *Architecture.* a row of columns supporting another member such as a beam or entablature.

colony *Ecology.* a group of organisms that have recently become established in a new habitat or that are situated together as a breeding group, such as a bird or seal colony. *Invertebrate Zoology.* a community of animals living attached or close to one another with a common organization, such as bees, ants, coral polyps, or bryozoans. *Microbiology.* a discrete group of microbial cells that are growing on a solid nutrient medium and have originated from one cell.

colony count *Microbiology.* the number of cell colonies in a given sample cultured on solid medium, with the assumption that each colony arose from a single cell.

colony-forming unit *Hematology.* any of several hematopoietic stem cells that give rise to monoclonal colonies in the spleen when transplanted into isogeneic, lethally irradiated mice.

colony hybridization *Biotechnology.* a procedure used to identify a clone containing a specific DNA sequence.

colony lift *Biotechnology.* a technique that determines the resistance of bacteria to a particular antibiotic; bacterial colonies are lifted from an agar plate and inoculated in the same position on subsequent plates containing the antibiotics being tested.

colony-stimulating factor *Immunology.* any substance that stimulates the proliferation of certain bone marrow cell populations. *Endocrinology.* any one of a group of cytokines that, when added to hematopoietic progenitor cells in culture, will cause the formation of clonal colonies of defined hematological lineages.

colopexy [käl´ə peks´ē] *Surgery.* the surgical fixation or suspension of the colon.

colophon [käl´ə fän´] *Graphic Arts.* **1.** a note at the back of a book giving information about its production, such as the typeface(s) and paper used and the designer, typesetter, and printer. **2.** a publisher's identifying emblem, often printed on book covers and title pages.

coloproctitis *Medicine.* an inflammation of the colon and rectum.

coloptosis *Medicine.* a downward displacement of the colon, a condition no longer considered to be pathological.

color *Optics.* **1.** the sensation, determined by wavelength, that is generated by light in the visible spectrum. **2.** the characteristic of light that produces specific degrees of hue, saturation, and brightness. *Particle Physics.* a property of quarks and gluons, analogous to electric charge, that describes how the strong force acts on these particles.

color additive see COLORING AGENT.

Colorado *Geography.* a river in the southwestern U.S., rising in the Rocky Mountains of northern Colorado and flowing west and south into the Gulf of California.

Coloradoan *Geology.* a geologic stage of the Middle Upper Cretaceous period.

coloradoite *Mineralogy.* HgTe, a black, metallic, opaque cubic mineral of the sphalerite group, having a specific gravity of 8.04 to 8.63 and a hardness of 2.5 on the Mohs scale; found in telluride ores.

Colorado low *Meteorology.* an area of low pressure that first appears as a cyclonic circulation in the vicinity of Colorado on the eastern slopes of the Rocky Mountains.

Colorado tick fever *Medicine.* an acute viral infection, transmitted by tick bite, occurring throughout the Rocky Mountains in the U.S. and characterized by fever, a low white blood cell count, and occasionally meningoencephalitis.

colorant *Chemistry.* any substance, such as a dye or pigment, that gives color to another substance.

color balance *Electronics.* an adjustment of electron-gun emissions to compensate for differences in light emissions of the three phosphors on the screen of a color picture tube.

color bar *Graphic Arts.* a small printed panel at the edge of a proof showing the desired density of the solid colors; used as a color check during a press run. Also, COLOR STRIP, COLOR STRIPE.

color-bar generator *Electronics.* a generator that delivers the signal needed to produce a color-bar test pattern on a color television channel.

color-bar test pattern *Telecommunications.* an arrangement of bands of colors used in television to check and correct color monitoring before broadcasting a live or taped program; also used to test and align color television receivers.

color blind or **colorblind** *Medicine.* affected with color blindness.

color blindness or **colorblindness** *Medicine.* the inability to distinguish one color from another. In the more common form, **dichromatic vision** (Daltonism), reds and greens cannot be distinguished from each other, but yellows and blues can be perceived. In the less common but more severe form, **achromatic vision,** color blindness is total and all colors appear as white, gray, or black.

color breakup *Telecommunications.* a condition whereby the primary color components of a visual display unit image separate; caused by rapid horizontal movements.

color burst *Electronics.* a short sequence of color subcarrier frequency transmitted as a reference for the chrominance signal at the beginning of each line. Also, REFERENCE BURST.

color center *Solid-State Physics.* a defect, impurity, or more complex imperfection in a lattice that absorbs light; an absorption band associated with color centers produces a colored appearance in normally transparent crystals and glasses.

color circle *Optics.* a graph in which colors are plotted along the circumference of a circle as they appear in the electromagnetic spectrum, with complementary colors facing each other across the diameter.

color code *Engineering.* a system of standard colors used as identifying conductors for polarity, and for identifying external terminals of motors and starters to facilitate making power connections between them.

color code see RESISTOR COLOR CODE.

color-color diagram *Astronomy.* a two-dimensional plot of stars, with the plot's axes being any two different-color indices.

color comparator *Analytical Chemistry.* a photoelectric instrument that compares and matches the colors of an unknown with standards for identification purposes.

color composite *Photogrammetry.* a composite photograph in which the component images are shown in different colors.

color contamination *Electronics.* an error in color reproduction caused by the incomplete separation of primaries.

color correction *Graphic Arts.* any alteration to a color negative or printing plate designed to improve the color reproduction of a given image. *Optics.* a technique that reduces chromatic aberration by allowing more than one wavelength to come to a common focus.

color-difference signal *Electronics.* a signal produced when the amplitude of a color signal is reduced by an amount equal to the amplitude of the luminance signal.

color disk *Optics.* a disk composed of red, green, and blue filters that rotates in front of the receiving screen in a color television system; used to produce color images.

colorectostomy [kō´ lə rek täs´tə mē] *Surgery.* the creation of an opening between the rectum and the colon. Also, **coloproctostomy.**

color excess *Astronomy.* the measure of reddening that a star's light has undergone before reaching the earth, as determined by the difference between the star's photometrically observed color index and the intrinsic color index of its spectral type.

color facsimile *Telecommunications.* a system for transmission of color images in which an image is scanned at a transmitter, reconstructed at the receiving station, and duplicated on some form of paper.

colorfast *Textiles.* having colors that do not fade or run when the fabric is worn, cleaned, or exposed to light.

color film *Graphic Arts.* film used in color photography, usually coated with three layers of emulsion that are sensitive to blue, red, and green light.

color filter *Graphic Arts.* in color separation and photography, a dyed sheet of plastic, glass, or gelatin used to enhance the reproduction of one color by absorbing others. *Optics.* a device that partially suppresses light waves of selected wavelengths by passing them through thin layers of semitransparent material, such as glass or film.

color force *Particle Physics.* a force acting between quarks, analogous to electromagnetism, that binds them in hadrons; thought to be the basis of all nuclear forces.

color fringing *Electronics.* a distortion of the chrominance along the boundaries of objects in a color television picture, which can cause small objects to appear fractured into various colors.

color gradients see HYPSOMETRIC TINTING.

color/graphics adapter *Computer Technology.* the first color graphics video display standard for IBM PCs, having graphics resolution of 320×200 pixels with four colors or 640×200 pixels with two colors.

color hearing see CHROMATIC AUDITION.

colorimeter [kul´ə rim´ə tər] *Analytical Chemistry.* an instrument used to measure the concentrations of solutions by comparing the relative intensities of color against standards.

colorimetric analysis see ABSORPTIOMETRIC ANALYSIS.

colorimetric photometer *Optics.* **1.** an instrument that gauges color by comparing the intensity of three primary colors against a test color. **2.** an instrument that employs a series of filters with wide spectral bands to measure color.

colorimetry [kul´ə rim´ə trē] *Optics.* the measurement and definition of unknown colors in terms of standard colors; techniques may be visual, photoelectric, or spectrophotometric.

color index *Pathology.* a blood quotient obtained by comparing the amount of hemoblobin in the red blood cells of a given patient with that of a normal individual of the patient's age and sex, and dividing the concentration of hemoglobin by the approximate number of red blood cells. *Astronomy.* the difference in magnitudes between a star's apparent brightness at any two chosen wavelengths, with the long-wavelength brightness being subtracted from the short-wavelength brightness.

coloring *Mathematics.* the assignment of a color to each vertex of a graph or each region of a map. A graph *G* is said to be **properly colored** if no two adjacent vertices are the same color. A map is properly colored if no two adjacent regions are the same color. A graph or map is ***k*-colorable** if it can be properly colored with *k* distinct colors.

coloring agent *Food Technology.* any natural or synthetic food additive that deepens or imparts color to make the food more appetizing. Also, COLOR ADDITIVE.

coloring medium *Optics.* any colored, pellucid material that colors light transmitted through it by partially absorbing the light.

color-killer circuit *Electronics.* a circuit that prevents signals in a color receiver from passing through the chrominance channel during black-and-white telecasts.

color-luminosity diagram *Astronomy.* a two-dimensional plot of stars, with the horizontal axis being photometrically-determined color and the vertical axis being intrinsic luminosity. Also, HERTZSPRUNG-RUSSELL DIAGRAM.

color-magnitude diagram *Astronomy.* a two-dimensional plot of stars, similar to a color-luminosity diagram, except that the vertical axis plots apparent brightness measured in magnitudes.

color map *Computer Technology.* a table that gives the correspondence between a color number and the values, e.g., red-green-blue strengths, that specify to the display hardware how to produce that color.

color oscillator *Electronics.* an oscillator that produces a signal in a color television at the same frequency and phase as the incoming signal. Also, CHROMINANCE SUBCARRIER OSCILLATOR, CHROMA OSCILLATOR.

color phase *Telecommunications.* the actual difference in phase between color signal components and the color reference in a color television receiver.

color-phase alternation *Telecommunications.* a color televison system in which the relative waves of the chrominance signal, sent in quadrature, are reversed in alternate lines in order to minimize phase errors and improve color performance.

color-phase detector *Electronics.* a detector in a color receiver circuit that ensures the color sections of the picture are in register with the monochromatic portions of the screen.

color photography *Graphic Arts.* any of various methods for reproducing a color image on photosensitive material.

color picture signal *Telecommunications.* **1.** in color facsimile, the signal produced by scanning. **2.** a portion of a composite television video signal that is located above the blanking signal and contains picture information.

color picture tube *Telecommunications.* a tube in which three distinct color phosphors produce a color image when scanned by an electron beam.

color printing *Graphic Arts.* any of a variety of historical and modern processes for producing a printed piece using colors other than solid black or degrees of black.

color process see PROCESS COLOR.

color-proof process *Photogrammetry.* a photomechanical method of printing that combines negative separations by successive exposures; used to produce a composite color proof on a vinylite sheet.

color purity *Electronics.* **1.** a condition in which only the desired color appears at a given point on a color television screen. **2.** the ratio of desired to undesired color components.

color saturation *Optics.* see SATURATION.

color separation *Graphic Arts.* a method for reproducing a full-color print or transparency by photographing it, usually through a treated lens to get black and through a series of three filters, each of which screens out all color except either magenta, yellow, or cyan. The separate negatives are then used to make one-color halftone plates, which can reproduce full color by overprinting.

color-separation negative *Graphic Arts.* a photographic negative reproducing only one color (magenta, yellow, cyan, or black), usually made for use in two-color or four-color printing.

color signal *Telecommunications.* an electric signal that is derived from and convertible to a visible color picture.

color solid *Optics.* a three-dimensional diagram depicting the relationships of the three general attributes of color, which are hue, saturation, and brightness.

color stability *Chemistry.* the ability of a substance to resist changes in color brought about by aging or light.

color standard *Analytical Chemistry.* a solution of known chemical composition and concentration, used in optical analysis for comparison with samples of unknown concentration.

color strip or **color stripe** see COLOR BAR.

color SU_3 see UNITARY SYMMETRY.

color-subcarrier oscillator see COLOR OSCILLATOR.

color television *Telecommunications.* **1.** a general term for the technology that is involved in broadcasting and receiving color images using chrominance signals. **2.** a television set capable of receiving chrominance signals and decoding them into color images.

color temperature *Physics.* the temperature at which a black body will emit a light of the same color as that of a given source; used to specify the color of a light source. *Astrophysics.* the temperature that a black body would need to match the energy emitted by a celestial object.

color test *Analytical Chemistry.* the analysis of a substance by comparing the intensity of the color produced in a sample with that of a standard color.

color throw *Analytical Chemistry.* discoloration of a liquid passing through the material used for the analysis bed in an ion-exchange process.

color-translating microscope *Optics.* a compound microscope that uses three wavelengths to reveal images generated by invisible forms of radiation such as ultraviolet light.
color transmission *Telecommunications.* the process of sending color images over a telecommunications system such as a television network or a color facsimile.
color transparency *Graphic Arts.* a positive color photographic image printed on transparent film.
color triangle *Optics.* a graph depicting the span of chromaticities produced by mixing three prescribed primary colors.
color vision *Physiology.* the sensation and perception of portions of the visible spectrum of light as different hues by the specialized cone receptors in the retina of the eye.
Colossendeidae *Invertebrate Zoology.* a family of long-legged deep-sea arthropods, the sea spiders, in the subphylum Pycnogonida.
colostomy [kə läs´tə mē] *Surgery.* **1.** the surgical diversion of the colon in order to bypass an area that is diseased or inflamed, forming an artificial anus by interrupting the colon at some point and joining it to an artificial opening in the abdominal wall. **2.** the opening so created.
colostrum *Physiology.* the thin, yellow, milky fluid that is secreted by the mammary glands just after parturition, characterized by a high protein and antibody content.
colotomy [kə lät´ə mē] *Surgery.* an incision of the colon. Also, LAPAROCOLOTOMY.
colp- a combining form meaning "vagina," as in *colpitis.*
colpectomy *Medicine.* the surgical removal of the vagina.
colpeurysis *Medicine.* the dilation of the vagina.

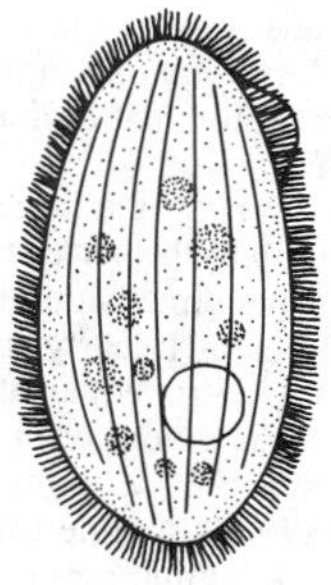

Colpidium

Colpidium *Invertebrate Zoology.* a genus of small, marine and freshwater ciliate protozoans in the order Holotrichia, widely used in biological research.
colpitis *Medicine.* an inflammation of the vagina; vaginitis.
Colpitts oscillator *Electronics.* an oscillator circuit in which the feedback is provided by a capacitor voltage divider that is part of the tuned circuit.
col plasmid see COL FACTOR.
colpo- a combining form meaning "vagina," as in *colposcope.*
colpocervical *Anatomy.* of or relating to the vagina and cervix.
Colpoda *Invertebrate Zoology.* a common freshwater genus of small, flat, kidney-shaped ciliate protozoans in the order Holotrichia.

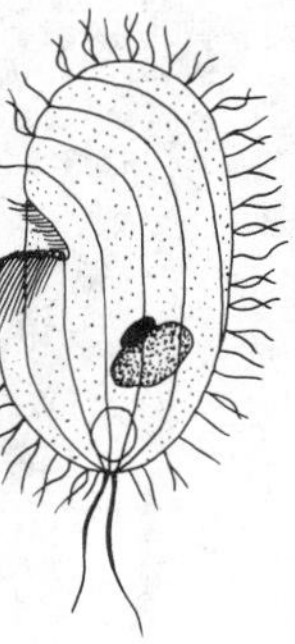

Colpoda

colpodidae *Invertebrate Zoology.* a large family of common, cosmopolitan, generally aquatic, free-living ciliate protozoans in the order Holotrichia.
colpodynia *Medicine.* a vaginal pain; vaginodynia.
colpohysterectomy *Surgery.* the surgical removal of the uterus by way of the vagina. Also, VAGINAL HYSTERECTOMY.
colpoplasty *Surgery.* plastic surgery of the vagina. Also, VAGINOPLASTY.
colporrhagia *Medicine.* a vaginal hemorrhage.
colporrhaphy *Surgery.* the suturing of the vagina, especially in order to narrow it.
colposcope *Medicine.* an instrument used to examine the cells of the vagina and cervix in vivo.
colposcopy *Medicine.* an examination of the vagina and cervix using a colposcope.
colpotomy *Medicine.* an incision of the vagina; vaginotomy.
colt *Agriculture.* a young male horse, especially one that is less than five years old.
Colt *Ordnance.* any of numerous firearms manufactured by the arms company founded by the American designer Samuel Colt; the most famous was the **Colt .45,** a single-action army revolver first manufactured in 1873.
colter *Agriculture.* the part at the front of a plow that cuts a vertical slice in the soil. Also, COULTER.
Colubridae *Vertebrate Zoology.* a very large family of terrestrial snakes in the order Squamata, containing about 75% of all known snakes.
colugo [kə loo´gō] *Vertebrate Zoology.* either of two lemurlike mammals, *Cynocephalus variegatus,* found in Asia and in the East Indies, or *C. volans,* found in the Philippines, having folds of skin on both sides of the body that aid in gliding from tree to tree. Also, FLYING LEMUR.
Columba *Astronomy.* the Dove, a small southern constellation lying southwest of Canis Major.
Columbia *Agriculture.* a large breed of long-legged sheep that produce both wool and mutton, cross-bred in the United States from the Lincoln and the Rambouillet breeds. *Space Technology.* a U.S. space shuttle, the first reusable spacecraft to orbit and return to earth (1981).

Columbia

Columbidae *Vertebrate Zoology.* the doves and pigeons, a cosmopolitan family of medium-sized birds in the order Columbiformes that are usually terrestrial or arboreal and have a dense, soft plumage.
Columbiformes *Vertebrate Zoology.* the doves and their relatives, including sandgrouses, dodos, and pigeons; a cosmopolitan order of medium-sized terrestrial or arboreal birds characterized by a small head, short beak, and swift, direct flight.
columbite *Mineralogy.* a name for oxides having the general formula $(Fe^{+2},Mn^{+2},Mg)(NbTa)_2O_6$; black, brown or greenish, submetallic, orthorhombic minerals occurring as thin to thick, tabular or short prismatic, commonly twinned crystals that may weigh up to 200 pounds, having a specific gravity of 5.1 to 5.2 and a hardness of 6 on the Mohs scale; found in granite pegmatites, and in placer deposits in areas of granitic rocks.
columbium *Chemistry.* the former name, still used by metallurgists and some others, for the element niobium.

Christopher Columbus

Columbus, Christopher 1451–1506, Italian navigator and explorer; his voyage to the West Indies in 1492 made the European community generally aware of the existence of the Americas.

columella *Anatomy.* any of several columnlike structures, such as the septum or the central axis of the cochlea. *Botany.* a structure serving as a small central axis for certain root caps, fruits, and the sporangia of liverworts, mosses, and fungi. *Invertebrate Zoology.* a columnlike part, such as the central axis of a spiral univalve shell or the central pillar in the calyx of many corals. (A Latin word meaning "small column.")

Columelliaceae *Botany.* a monogeneric family of bitter, tanniferous Andean shrubs and trees of the order Rosales, characterized by simple opposite leaves and a sympetalous corolla.

column *Architecture.* a vertical support, usually consisting of a base, shaft, and capital. *Engineering.* any vertical member used to support a compression load. *Military Science.* a formation of troops, ships, aircraft, or the like, in which the individual elements are placed and move one behind the other. *Computer Programming.* **1.** a particular character or digit position in a field or on a physical device. **2.** a vertical arrangement of data in a printed output. *Chemical Engineering.* a vertical, cylindrical container or vessel used in petroleum and chemical processing, usually to achieve a degree of separation of vapor-liquid or liquid-liquid systems by extraction, distillation, or absorption. *Nucleonics.* **1.** in nuclear testing, the hollow cylinder of water or spray thrown up by an underwater burst of a nuclear weapon, through which hot, high-pressure gases are vented to the atmosphere. **2.** a similar column of dirt formed in a shallow underground explosion. *Mathematics.* **1.** a position corresponding to a power of the number base in positional representation of numerical value. Also, PLACE. **2.** in an $m \times n$ matrix with entries a_{ij}, the m values of a_{ij} for fixed j are called, collectively, the ***j*th column**.

columnar coal *Geology.* coal having a columnar fracture structure, usually as a result of metamorphism by an igneous intrusion.

columnar epithelium *Histology.* epithelium that is composed of elongated, columnar cells.

columnar grain casting *Materials Science.* a process of casting that controls the direction of grain growth, for applications such as blades and vanes for turbines and jet engines. Directional solidification results in a structure wherein the grain boundaries run in the longitudinal direction of the part.

columnar ice *Hydrology.* ice consisting of vertical columns of ice crystals massed together.

columnaris disease *Veterinary Medicine.* a disease caused by a member of the slime bacteria group *Flexibacter columnaris* and occurring in warm-water species of fish; prevention is often achieved by lowering the water temperature and avoiding traumatic injury. Also, COTTONMOUTH DISEASE, SADDLEBACK.

columnar jointing *Geology.* in tabular bodies of igneous rock, a pattern of jointing produced as a result of contraction during cooling and characterized by the division of rocks into long, parallel prisms or pillars.

columnar resistance *Geophysics.* a measure of atmospheric electrical resistance in a column of air 1 centimeter square, measured from the earth's surface to a given altitude.

columnar section *Geology.* a scale drawing that graphically illustrates the vertical sequence and relationships of rock units occurring in a specific area.

columnar stem *Botany.* a cylinder-shaped stem having no branches and bearing leaves only at its apex.

columnar structure *Petrology.* a primary sedimentary structure with slender columns arranged perpendicular to the bedding.

column bleed *Analytical Chemistry.* the loss of carrier liquid during gas chromatography due to evaporation into the gas being analyzed.

column chromatography *Analytical Chemistry.* a separation technique employing a column packed with a stationary phase through which a mobile phase containing a mixture is poured; the components can be eluted and collected for analysis.

column collapse *Volcanology.* the collapse of a cloud column from a volcanic eruption, occurring when the density of the column exceeds that of the ambient atmosphere.

column cover *Military Science.* the protection of a column by aircraft in communication with the column; this may include reconnaissance or attack of targets threatening the column.

column crane *Mechanical Engineering.* a crane in which the projecting arm swings about a mast attached to one of the building's columns.

column drill *Mechanical Engineering.* a drilling machine supported by an upright steel column; used to excavate rock.

column gap *Military Science.* the space between two consecutive elements of a column; it may be measured in units of length or time.

column operations *Mathematics.* the rules for manipulating the columns of a matrix representing a linear transformation so that the image space of the linear transformation is unchanged.

column order *Computer Programming.* a method of defining a matrix by naming the elements one column at a time, as opposed to row order. Also, **column-major order.**

column pipe *Mining Engineering.* the large pipe through which water is conveyed from a mine pump to the surface.

column rank *Mathematics.* the dimension of the column space of a matrix; equivalently, the number of linearly independent columns of the matrix. The column rank of a matrix equals the row rank.

column space *Mathematics.* given an $m \times n$ matrix $M = (a_{ij})$ over some field F, the column space of M is the subspace of F^m (the m-dimensional vector space over F) spanned by the n columns of M regarded as vectors of F^m. The column space is the same (up to a change of basis) as the image space of the linear transformation corresponding to M.

columnar joint

column vector *Mathematics.* a matrix with only one column.

colure *Cartography.* an hour circle that passes through the equinox of the solstice.

colusite *Mineralogy.* $Cu_{26}V_2(As,Sn,Sb)_6S_{32}$, a bronze, metallic, opaque cubic mineral occurring in massive form or as tetrahedral crystals, having a specific gravity of 4.43 to 4.5 and a hardness of 3 to 4.5 on the Mohs scale; found with tetrahedrite, pyrite, and quartz.

Colydiidae *Invertebrate Zoology.* a family of beetles in the superfamily Cucujoidea.

colza oil see RAPE OIL.

com- a prefix meaning "with" or "together," as in *combination, commution.*

COM computer output on microfilm; computer output microfilmer.

coma *Medicine.* an abnormal, continuous deep unconsciousness from which a patient cannot be awakened. *Astronomy.* a bright shell of gas and dust that is given off by a comet's nucleus as it heats up under sunlight and that hides the nucleus from view. *Electronics.* a defect in a cathode-ray tube that causes the spot on the screen to appear comet-shaped when it moves from the center of the screen. *Optics.* an aberration, arising from the occurrence of zones of different magnification on the surface of a lens, in which the image of a point is deformed so that it is shaped like a comet with a small bright head; it normally occurs in the section of the vision field farthest from the principal axis.

Coma Berenices *Astronomy.* Berenice's Hair, a faint constellation in the northern sky, that lies between Leo and Bootes and is visible during the northern spring.

comagmatic region see PETROGRAPHIC PROVINCE.

coma lobe *Electromagnetism.* for a reflecting antenna, a side lobe that appears because the radiating element is not located exactly at the focal point of the dish.

Comanchean *Geology.* a North American geologic stage of the Lower and Upper Cretaceous period, occurring after the Coahuilan and before the Gulfian. Also, **Comanchian.**

Comasteridae *Invertebrate Zoology.* a family of feather stars, radially symmetrical echinoderms in the order Comatulida.

comatose [kōm´ə tōs˝] *Medicine.* in a coma or comalike state.

Comatulida *Invertebrate Zoology.* an order of echinoderms within the subclass Articulata.

comb *Vertebrate Zoology.* a fleshy crest on top of the head of chickens, pheasants, and other fowl, usually found in the males only. *Molecular Biology.* a structure in chickens whose appearance varies due to the influence of two nonallelic gene pairs; an early example of gene interactions. *Invertebrate Zoology.* **1.** a beeswax construction of hexagonal cells built by a colony of bees. **2.** a swimming plate in comb jellies. *Geology.* in a mineral vein, an aggregate that resembles a honeycomb in which individual crystals have grown perpendicular to the vein walls so that their ends project outward.

COMB *Oncology.* an acronym for a cancer chemotherapy regimen that includes the drugs cyclophosphamide, Oncovin, MeCCNU, and bleomycin.

comb antenna *Electromagnetism.* a broadband antenna consisting of half of a fishbone antenna erected vertically; used for vertically polarized signals.

combat air patrol *Military Science.* an aircraft operation intended to intercept and destroy hostile aircraft before they reach their target; it may be flown over an objective area, a protected force, an air defense area, or the critical area of a combat zone.

combat analysis *Military Science.* a theoretical analysis of a weapons system to determine its probable performance under combat conditions.

combatant ship *Naval Architecture.* a ship designed primarily for direct action with an enemy.

combat area *Military Science.* a restricted area established to prevent or minimize mutual interference between friendly forces during combat operations.

combat chart *Military Science.* a special-purpose chart that uses the characteristics of a map to represent land and those of a chart to represent sea areas, making the chart useful in military operations.

combat day of supply *Military Science.* the total quantity of supplies required to support one day of combat, calculated by multiplying a standard day of supply by the intensity factor.

combat development *Military Science.* the research, development, and testing of new concepts and materials, aimed toward maximizing their effectiveness in combat. Thus, **combat development field experiment, combat development test project,** and so on.

combat fatigue or **combat disorder** see BATTLE FATIGUE.

combat information center *Military Science.* a compartment in a warship into which information is fed and evaluated. Also, OPERATIONS ROOM.

combat intelligence *Military Science.* knowledge of the enemy, weather, and geographical conditions necessary so that a commander can plan and conduct combat operations.

combat loading *Military Science.* the arrangement of personnel and the loading of equipment and supplies in a manner designed to maximize efficiency in the anticipated tactical operation.

combat operations center *Military Science.* an organization into which information is sent for analysis and display, at which command decisions are made and orders transmitted. Also, **COC.**

combat patrol *Military Science.* a tactical unit sent out from the main military force to engage in independent fighting.

combat power *Military Science.* the total means of destructive or disruptive force that a military unit can apply against an enemy at any given time.

combat radius *Aviation.* the distance that an aircraft can fly under combat conditions, carry out its mission, and return to its base of origin with a specified fuel reserve remaining and without refueling in flight.

combat serviceable item *Ordnance.* an item of equipment that is ready to perform at its rated capacity for an extended period of time under combat conditions.

combat service support *Military Science.* noncombat assistance provided for operating combat forces; e.g., supply, maintenance, transportation, construction, engineering, administrative services, chaplain services, and health services.

combat tire *Ordnance.* a heavy-duty, pneumatic tire designed to operate without air pressure for a limited distance under combat conditions.

combat vehicle *Ordnance.* a vehicle, with or without armor, designed for a specific fighting function.

combat zone *Military Science.* the area required by combat forces to conduct operations, usually extending from the front line to a rear boundary established by the theater commander.

combed cotton *Textiles.* cotton yarn that has been cleansed of short fibers and impurities by means of wire brushes and roller cards.

comber *Oceanography.* on the open sea, a long, curling wave which spills its top as a whitecap.

combescure transformation *Mathematics.* a one-to-one correspondence between the points on two curves in space such that tangents at corresponding points are parallel.

comb filter *Electronics.* a filter network of a multiple-bandpass design that allows frequencies within a number of narrow bands to pass while blocking frequencies that fall between the bands; so named because its frequency-response curve resembles the teeth of a comb.

comb-filter distortion *Acoustical Engineering.* the loss of signals at null nodes in electroacoustical processing, caused by delays in signal processing.

comb growth unit *Biology.* a unit used for the standardization of male sex hormones.

combination *Mathematics.* an unordered selection of zero or more objects from among the members of a given finite set. In particular, the number of combinations of n objects, taken to be r at a time, equals $n!/(n-r)!r!$ and is variously denoted as $C(n, r)$, ${}_nC_r$, or $\binom{n}{r}$.

combinational circuit *Electronics.* a logic circuit in which the output is determined solely by the concurrent inputs and is independent of the previous inputs.

combination buoy *Navigation.* a buoy with more than one means of identification, such as color and sound.

combination cable *Electricity.* a cable in which the conductors are grouped in pairs, quads, or similar combinations.

combination chuck *Mechanical Devices.* a three-jaw lathe chuck used to center a workpiece automatically by independent or simultaneous jaw motion.

combination coefficient *Geophysics.* the specific rate of disappearance of small atmospheric ions, due to combination either with oppositely charged larger ions to form neutral Aitken nuclei or with the neutral Aitken nuclei themselves to form other, larger ions.

combination collar *Mechanical Devices.* a collar used to join pipe ends having left-hand and right-hand threads at opposing ends.

combination die *Metallurgy.* in die casting, a die that has two or more cavities.

combination distributing frame *Electricity.* a frame that has the functions of both a main distributing frame and an intermediate distributing frame.

combination fuse *Ordnance*. a fuse combining two or more different types of fuse mechanism, usually impact and time mechanisms.

combination influence mine *Ordnance*. a mine that is actuated by two or more influences, either simultaneously or in a predetermined order. Also, **combined influence mine.**

combination lock *Engineering*. a lock that operates by a dial that opens the lock when the dial is turned to a specific combination of numbers or symbols.

combination mill *Metallurgy*. a continuous mill that has a roughing and a finishing section.

combinatio nova [kōm´bi nä´shē ō nō´vä] *Systematics*. a new name for a species, resulting from the transfer of that species to a new genus. (A Latin term meaning "new combination.")

combination plate *Graphic Arts*. a printing plate containing both halftones and line drawings, often combined.

combination pliers *Mechanical Devices*. a slip-joint pliers in which both flat and notched sections of its jaws perform multiple work tasks, such as holding objects or cutting wire.

combination principle see RITZ'S COMBINATION PRINCIPLE.

combination rig *Petroleum Engineering*. a drilling rig equipped with machinery for both cable-tool and rotary drilling.

combination saw *Mechanical Engineering*. a saw with teeth specially designed for ripping and crosscut mitering applications.

combination spectrum *Astronomy*. the spectrum, containing two or more spectra superimposed, that is detected from a binary or multiple star system so distant that its individual stars cannot be resolved.

combination square *Mechanical Devices*. an adjustable carpenter's square fitted with a steel sliding blade and a frame edge; used for miter layout work to measure both 45° and 90° face angles.

combination system *Thermodynamics*. a system that combines characteristics of a closed system and an open system; e.g, a tank in the process of being filled or emptied.

combination trap *Geology*. a subsurface oil or gas reservoir, closure, deformation, or fault in which reservoir rock partially covers the structure.

combination unit *Chemical Engineering*. a system that combines several processes to form a functional unit.

combination vibration *Spectroscopy*. the interaction of two or more fundamental vibrational frequencies in polyatomic molecules, resulting in weak absorptions in the infrared region of a spectrum. Also, **combination modes.**

combination wrench *Mechanical Devices*. a wrench with a socket head at one end and an open head at the other.

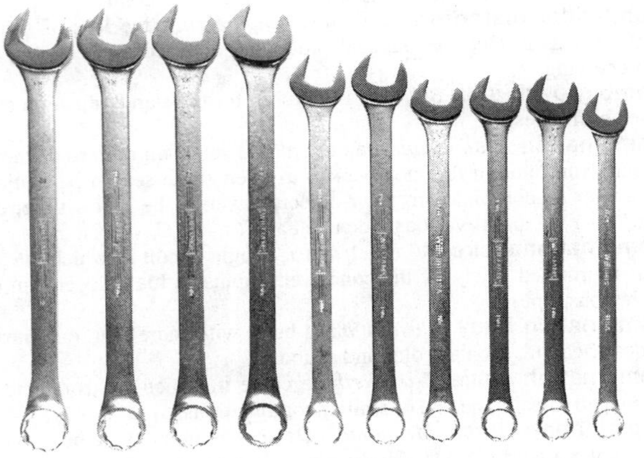

combination wrenches

combinatorial analysis *Mathematics*. the study of problems in combinatorics using analytical methods, e.g., the use of generating functions.

combinatorial topology *Mathematics*. the branch of topology used to study simple components of geometric forms, e.g., simplicial complexes, polyhedra, and so on.

combinatoric explosion *Artificial Intelligence*. the rapid growth of the number of possibilities to be explored in a search, often exponential because it is the product of the number of possible choices at each level of the search tree.

combinatorics *Mathematics*. the branch of discrete mathematics that deals with the study of arrangement of objects into sets, and in particular with two general types of problems: enumeration and existence.

combine [käm´bīn´] *Agriculture*. a farm machine that cuts, threshes, and cleans a standing crop while moving across a field; used to harvest wheat and other grains. Also, **combine-harvester.**

combine

combined carbon *Organic Chemistry*. carbon that is bound within a compound, as distinguished from free or unbound carbon.

combined cyanide *Organic Chemistry*. the cyanide portion of a complex ion that consists of the cyanide group and a metal.

combined flexure *Mechanics*. the flexure of a beam due to a combination of transverse and longitudinal loading.

combined immunological deficiency disease *Medicine*. a usually fatal disorder in which the T cells (produced by the thymus and responsible for defense against viral infection) and B cells (generated by the bone marrow and responsible for globulin and antibody production) are lacking, creating a state of lowered immunity and resistance to disease.

combined moisture *Mining Engineering*. any moisture in coal that cannot be removed by ordinary means of drying.

combined pregnancy *Medicine*. the simultaneous presence of an intrauterine and an ectopic pregnancy.

combine drill *Agriculture*. a drill that dispenses seed and fertilizer at the same time.

combined stresses *Mechanics*. bending or twisting stresses in a body, combined with tensile or compressive stresses.

combined water *Geochemistry*. molecules of water that either are bound to ions in a hydrate or are found in a homogeneous crystalline phase in a mineral.

combiner circuit *Electronics*. a circuit that mixes the luminance and chrominance signals with the synchronizing signals in a color television camera.

combing *Textiles*. a machine process to straighten and separate fibers before spinning. *Engineering*. **1.** a process in which a comb or stiff brush is run over a freshly painted surface to create a pattern. **2.** a process in which a soft stone surface is abraded or smoothed. *Building Engineering*. the uppermost row of shingles that project above the ridge line of a sloping roof on a building.

combining glass *Optics*. a glass screen that reflects display imagery to a viewer, normally at preferred wavelengths, and at the same time transmits light from the scene beyond.

combining site *Immunology*. an area on an antibody at which it joins with the determinant of an antigen or partial antigen.

combining volumes principle *Chemistry*. a law stating that at constant temperature and pressure, the volumes of gases involved in a chemical reaction are in a ratio of small whole numbers to each other, even when solids or liquids are involved in the reaction.

combining weight see EQUIVALENT WEIGHT.

comb neophoscope *Engineering*. an instrument that measures the altitude and movement of a cloud, with a turning comb mounted on the instrument so that the cloud appears to move in a direction parallel to the ends of the comb's vertical rods.

comb. nov. combinatio nova.

Combretaceae *Botany.* a family of tropical and subtropical trees and shrubs of the order Myrtales that are often climbing and are characterized by unicellular hairs, simple alternate leaves, and an indehiscent fruit adapted to distribution by water.
combustible *Materials Science.* **1.** of a material, able to burn. **2.** specifically, relatively difficult to ignite and slow to burn, as contrasted with *flammable*.
combustible loss *Engineering.* a loss of heat that occurs when fuel does not undergo complete combustion.
combustible shale see TASMANITE.
combustion *Chemistry.* the burning of a substance in an oxidizing reaction that produces heat and often light.
combustion chamber *Mechanical Engineering.* **1.** in an internal combustion engine, the space above a piston in which combustion occurs. **2.** in a furnace, any space in which combustion occurs, or the space in which combustion of gaseous products occurs. *Space Technology.* the part of a liquid rocket, ramjet, or gas turbine engine in which the combustion of propellants takes place at high pressure. Also, BLAST CHAMBER, BURNER, FIRING CHAMBER, ROCKET CHAMBER.
combustion chamber volume *Mechanical Engineering.* the volume of the combustion chamber when a piston is located at top dead center.
combustion deposit *Engineering.* the ash residue resulting from the burning of fuel, occurring on the heat-exchange surfaces of a combustion chamber.
combustion efficiency *Chemistry.* the ratio of heat produced in a combustion process to the heat that would be produced if combustion were complete.
combustion engine *Mechanical Engineering.* an engine operated by the energy released from a combustible fuel that is fed to the engine.
combustion engineering *Mechanical Engineering.* the study of heat liberated and absorbed by the combustion process as applied to furnace efficiency and design.
combustion furnace *Analytical Chemistry.* a heating device used to analyze the elemental content of organic compounds.
combustion instability *Space Technology.* the irregular combustion of fuel or propellant in a jet or rocket engine.
combustion nucleus *Meteorology.* a condensation nucleus that arises as a result of natural or industrial combustion processes.
combustion rate *Chemistry.* the rate at which a substance burns.
combustion shock *Engineering.* a sudden disturbance in an internal combustion engine that occurs when the fuel is being improperly burned, due to ignition or control errors.
combustion train *Analytical Chemistry.* the arrangement of instruments and apparatus for elemental analysis.
combustion tube *Analytical Chemistry.* a tube, resistant to high temperatures and usually composed of glass, silica, or porcelain; used to hold samples during pyrolysis for elemental analysis.
combustion wave *Chemistry.* a zone of combustion that travels along a narrow path through a burning substance.
combustor *Mechanical Engineering.* the system in a gas turbine or jet engine that contains burners, ignitors, and injection devices in addition to the combustion chamber. *Aviation.* specifically, such a device in a ramjet engine.
comdt commandant.
come-along *Mechanical Devices.* a jawed tool that grips the ends of a wire, rope, cable, or chain and is used to haul it into place or to shorten its length. Also, PULLER.
comedo [käm´ə dō] *Medicine.* a blackhead.
comedocarcinoma *Medicine.* an adenocarcinoma of the breast in which the ducts are filled with comedo-appearing cells that are actually necrotic malignant cells.
comendite *Geology.* a white alkaline rhyolite containing pyroxene or amphibole.
comes [kō´mēz´] *Astronomy.* the fainter star in a binary star system.
Comesomatidae *Invertebrate Zoology.* a family of free-living, detritus-feeding nematodes in the superfamily Chromadoroidea.
comet *Astronomy.* a small solar system body, composed mainly of ices and dust, that usually follows a highly elliptical orbit and displays a long tail of gas and dust when warmed by the sun.
cometary nebula *Astronomy.* a dusty nebula whose fan shape happens to resemble a comet.
cometary outburst *Astronomy.* a sudden brightening of a comet, typically of two to five magnitudes and lasting three to four weeks, that may be caused by gas pockets erupting, by breakup of the nucleus, by impact cratering, or by some other process not yet understood.
comet family *Astronomy.* a group of comets whose orbits have similar distances from the sun at aphelion, usually at the distance of a major planet such as Jupiter.
COM file *Computer Programming.* a command file that contains an executable object code; the file name has a COM extension.
comfort behavior *Behavior.* any of various animal behavior patterns that involve care of the body or relief of physiological tensions, such as grooming.
comfort chart *Engineering.* a graphic representation that displays relative humidity and effective temperature curves overlying rectangular coordinates of wet-bulb and dry-bulb temperatures.
comfort control *Engineering.* the regulation of heat, humidity, ventilation, and air composition in a structure or vehicle to accommodate the comfort of the inhabitants.
comfort curve *Engineering.* the line on an air temperature graph that is drawn versus a function of humidity, such as relative humidity, to show a range of conditions for which an average stationary person would feel the same degree of comfort.
comfort index see TEMPERATURE-HUMIDITY INDEX.
comfort zone *Engineering.* the range of indoor temperature, humidity, and ventilation conditions within which an average person is physically and mentally comfortable.
COMIT *Computer Programming.* a high-level language developed in the late 1950s for text processing and translation of natural languages.
Comleyan *Geology.* a geologic stage of the Lower Cambrian period.
comlognet *Telecommunications.* a computer-based data-transmission network that processes and distributes logistical information. (A term coined from <u>com</u>puter-based <u>log</u>istics <u>net</u>work.)
comm commander; commission; communication.
comma *Acoustics.* a term used in music to describe an inharmonic interval between two notes, such as C and D flat, which produces overtones less than a whole step apart.
comma-less code *Genetics.* a description of the DNA code as a continuous series of nucleotides without "punctuation marks" indicating where one codon ends and the next one begins.
command *Computer Programming.* **1.** an operation that is specified in an instruction. **2.** an instruction that is specified in a programming language. **3.** an operation given to an operating system. *Control Systems.* any control signal.
command and control system *Transportation Engineering.* the regulatory system used to govern all operations of a transportation network.
commandant *Military Science.* **1.** the chief officer of a military facility or organization. **2.** the senior officer of the U.S. Marine Corps. **3.** the head of a military school.
command character *Computer Programming.* a character code used by a control operation, such as a character used in text processing to instruct the printer to begin a new page. Also, CONTROL CHARACTER.
command control see COMMAND GUIDANCE.
command-control program *Computer Programming.* in a central computer, a program that prioritizes and processes all the requests for processing from multiple-user consoles.
command destruct *Control Systems.* a system used by a launching facility to destroy an airborne rocket or missile that it has launched, when performance indicates a potential safety hazard.
command-detonated mine *Ordnance.* a mine that is detonated by remote control.
command-driven *Computer Programming.* describing a program or system that requires the user to know and use its specified command-language statements and syntax, as opposed to selecting from plain-English, user-friendly menus. Thus, **command-driven software.**
commander *Military Science.* **1.** the chief officer of an army or subdivision of an army. **2.** in the U.S. Navy, an officer ranking above a lieutenant commander and below a captain.
commander-in-chief *Military Science.* the commander in charge of the armed forces of a nation or of several allied nations.
command guidance *Engineering.* the guiding of missiles or aircraft by signals from an external source that is operated manually or automatically. Also, COMMAND CONTROL.
command interpreter *Computer Programming.* a program that translates a physical action taken by a user (such as pressing a key or clicking a mouse) or a symbolic command into usable parameters and forwards it to the program for which the action is intended.
command key *Computer Programming.* a special key on a computer keyboard that is used to execute instructions for the computer rather than to enter a letter, number, or symbol.

command language *Computer Programming.* a means by which a user can interact with a computer to describe to the system the requirements of a job, such as the amount of storage needed or input/output devices required, and actions to be taken, such as programs to be executed. Also, JOB-CONTROL LANGUAGE.

command language interpreter *Computer Programming.* the part of an operating system that interprets the user's job-control statements.

command level *Computer Programming.* the mode of operation in which the computer operator interacts directly with the operating system using a set of direct commands, rather than through a series of programmed instructions.

command list *Computer Programming.* a sequence of operations related to input/output functions, usually generated by the central processing unit.

command mode *Computer Programming.* in a time-sharing environment, the status of a user console that is on-line but has no current task to be performed, so that the system is waiting for a command.

command module *Space Technology.* in a spacecraft, the module that carries the crew, the main communications equipment, and the reentry vehicle.

command negativism see NEGATIVISM.

command net *Military Science.* **1.** a military communication system connecting an echelon of command with some or all of its subsidiary echelons. **2.** a system of command locations controlling guided missiles.

commando *Military Science.* **1.** a small, specially trained fighting unit, originally used for surprise, hit-and-run raids during World War II; now used for similar situations, as well as for antiterrorist operations. **2.** a member of such a unit. **3.** of or relating to a raid made by such a unit.

command post *Military Science.* a unit or subunit headquarters where the commander and the staff perform their activities; if the headquarters is divided into echelons during combat, the echelon where the commander is located is the command post.

command pulses *Electronics.* the electrical representation of bit values that govern input-output devices.

command set *Telecommunications.* any radio linkup used to give or receive commands.

command speed *Transportation Engineering.* the speed set by an automatic train control system at any given time.

Commelinaceae *Botany.* a family of mostly tropical herbaceous plants of the order Commelinales, having jointed stems, sheathing leaves, and flowers with tenuous petals.

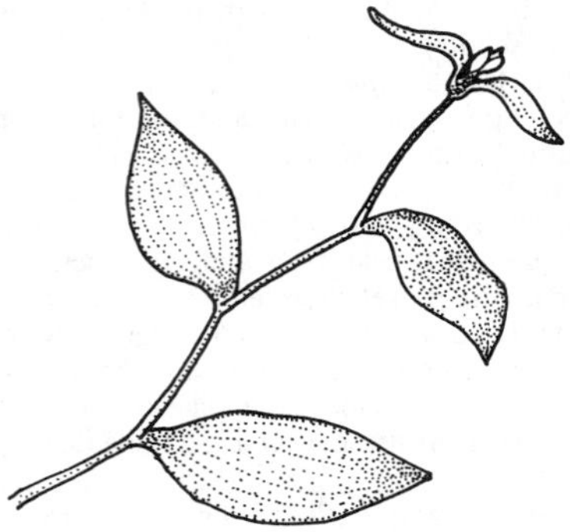

Commelinaceae

Commelinales *Botany.* an order of mostly terrestrial monocotyledonous herbs having leaves with a closed sheath and seeds with endosperm.

Commelinidae *Botany.* a subclass of mostly herbaceous monocotyledonous plants of the class Liliopsida, having mainly bisexual flowers and often a reduced perianth.

commensal *Ecology.* **1.** relating to or showing commensalism. **2.** an individual organism that participates in a relationship of commensalism.

commensalism *Ecology.* a relationship between two species in which one species benefits and the other is not affected either negatively or positively.

commensurable [kə men′sur ə bəl] *Mathematics.* describing quantities, especially the lengths of line segments, whose ratio is a rational number. The greatest common divisor and least common multiple exist for commensurable quantities.

commensurable motions *Astronomy.* a state that occurs when one of two objects orbiting a common center of mass has an orbital velocity that is an even fraction or exact multiple of the other's velocity.

commensurate orbits *Astronomy.* a condition that occurs when one of two objects orbiting a common center of mass has a period that is an even fraction or exact multiple of the other's period.

comment *Computer Programming.* a nonexecutable explanatory expression that is embedded in a computer program for documentation purposes and has no effect on program operations.

comment code *Computer Programming.* the means by which a computer recognizes part of a program as a comment, such as an asterisk or a "C" in the first character position.

comment out *Computer Programming.* to nullify an executable statement in a program by adding a comment code to the statement.

commercial aviation *Aviation.* aviation that is operated as a business for profit, as distinguished from military or private aviation.

commercial exchange see MARKET EXCHANGE.

commercial farming *Agriculture.* the production of crops and livestock primarily for sale.

commercial mine *Mining Engineering.* a mine operated to supply coal to purchasers in general, unlike a captive mine.

commercial ore *Mining Engineering.* any mineralized material that is profitable at current metal prices.

commercial speech bank see VOICE FREQUENCY.

commercial-type vehicle *Ordnance.* a term for a motor vehicle designed for civilian use that is placed into military use without significant modification.

commingling *Petroleum Engineering.* the blending of oil products with like properties, often performed to ease pipeline transportation.

comminute *Mining Engineering.* to reduce solids to minute paricles by crushing, grinding, or pulverizing.

comminuted fracture *Medicine.* a fracture in which the bone is crushed or splintered.

comminution *Engineering.* a process of material size reduction, consisting of breaking, crushing, grinding, or pulverizing. *Materials Science.* specifically, the particle size reduction of raw materials used to produce ceramics; methods used include wet or dry milling, crushing, impact, and attrition. *Metallurgy.* the process of making metallic powders by mechanical means.

comminutor *Mechanical Engineering.* a tool or machine used to break up solids into very fine particles.

commiscuum *Ecology.* an ecosystem consisting of two different species that are capable of interbreeding and producing hybrid offspring with partial fertility.

commission *Military Science.* **1.** an official document granting authority and rank to a military officer. **2.** the authority and rank granted by such a document. **3.** to put in or make ready for service or use; e.g., to *commission* an aircraft or a ship. Thus, **in commission** or **out of commission.**

commissioned officer *Military Science.* an officer who holds a commission and ranks as a second lieutenant or higher in the U.S. Army, Air Force, or Marine Corps, or as an ensign or higher in the U.S. Navy or Coast Guard.

commission ore *Mining Engineering.* any uranium-bearing ore of 0.10% U_3O_8 or higher, for which the U.S. Atomic Energy Commission has an established price.

commissure [käm′i shů r′] *Anatomy.* **1.** a joint, seam, or meeting line between two structures. **2.** a band of nerve tissue connecting corresponding parts of the brain or spinal cord.

commissurotomy [käm′i shů rät′ə mē] *Surgery.* the surgical elimination of a commissure.

committed *Ordnance.* describing a fuse that has reached a point at which the arming process will continue to completion without further external action.

committed transaction *Computer Programming.* a transaction that has completed all the required calculations and copied the results to a secure place; used to protect against system crash damage by immediately recording the results of the transaction.

common *Electricity.* **1.** shared by two or more circuits, devices, services, or users. Thus, **common branch, common user circuit. 2.** connected to a ground; grounded.

common acne see ACNE, def. 1.

common alum see ALUM.

common area *Computer Programming.* an area in storage in which to store data that is to be shared by two or more routines.

common battery *Telecommunications.* an electrical-power supply concept in which all of the direct-current energy for a telephone subsystem, such as a switching center, is supplied by one source.
common bile duct *Anatomy.* the duct formed by the union of the cystic and hepatic ducts.
common bond *Building Engineering.* see AMERICAN BOND.
common brick *Building Engineering.* a brick that is used for filling or backing in the rough work phase of building construction.
common business-oriented language see COBOL.
common carrier *Transportation Engineering.* a transport carrier that offers its services to the general public in the marketplace, rather than by individual contract, and that is therefore subject to common carrier regulations regarding rates and services.
common cold see COLD.
common cold virus *Virology.* any of a group of viruses, most of which are in the genus *Rhinovirus,* family Picornaviridae, that infect humans. Also, HUMAN RHINOVIRUSES.
common-data environment *Computer Programming.* a means of describing any storage system in which data can be accessed by any module in the system.
common declaration statement *Computer Programming.* a nonexecutable statement in FORTRAN that is used to control storage allocation by allowing two or more routines to share a common storage area.
common denominator *Mathematics.* any common multiple of the denominators of two or more fractions.
common-drain amplifier *Electronics.* a field-effect transistor amplifier in which the drain terminal is common to the input and the output. Also, SOURCE-FOLLOWER AMPLIFIER.
common envelope star *Astronomy.* one of two stars in a binary system whose outer atmosphere is so extended that it has engulfed the other star.
common fraction *Mathematics.* a fraction whose numerator and denominator are both integers.
common-gate amplifier *Electronics.* a field-effect transistor amplifier in which the gate terminal is common to the input and the output.
common-impedance coupling *Electromagnetism.* the intentional interaction between two or more circuits by means of inductance or capacitance.
common-impression system *Graphic Arts.* on a rotary press, a roller configuration having two to five plate cylinders around a single impression cylinder.
common-ion effect *Chemistry.* a phenomenon whereby an ionic compound is rendered less soluble when added to a solution containing another compound with which it has an ion in common.
common item *Industrial Engineering.* **1.** any item used in more than one activity or used or procured by more than one operation or department, including items of similar manufacture that vary in color or shape. **2.** any part or component required in the assembly of two or more complete end items.
common joist *Building Engineering.* a building beam used in a building to which floor boards are attached.
common language *Computer Programming.* a programming language or macro code that can be read and understood by different computers; such languages include FORTRAN and COBOL.
Common Lisp *Artificial Intelligence.* a large, standardized dialect of Lisp.
common logarithm *Mathematics.* a logarithm having ten as the base.
common mica see MUSCOVITE.
common mode *Electronics.* in a system processing a differential signal, the part of the signal that appears on both conductors and does not contribute to the differential component. The common mode signal is one-half of the sum of the signals measured on the two differential leads with respect to the ground reference.
common-mode error *Electronics.* a difference in voltage that appears at the output terminal of an operational amplifier, caused by the appearance of voltage at two separate input locations.
common-mode gain *Electronics.* the ratio of the common mode of the output voltage of a differential amplifier to the common mode of the input voltage; the gain of an ideal differential amplifier is zero.
common-mode rejection *Electronics.* the ratio of the common mode signals at the input and the output, respectively, of a differential amplifier. Also, IN-PHASE REJECTION.
common-mode signal *Electronics.* the algebraic average of two signals applied simultaneously to both ends of a balanced circuit, such as a differential amplifier. Also, IN-PHASE SIGNAL.
common-mode voltage *Electronics.* **1.** the amount of voltage common to both lines of a balanced amplifier. **2.** a signal carried along a transmission line in both wires of its circuit.
common multiple *Mathematics.* a number or quantity that is divisible by two or more given numbers or quantities.
common projectile *Ordnance.* a projectile designed to explode after piercing the outer armor of a tank or ship; it carries a high-explosive bursting charge.
common proper motion stars *Astronomy.* stars that share the same three-dimensional motion on the sky and were probably formed together at about the same time.
common rafter *Building Engineering.* rafters that are common to any portion of a roof and extend continuously from edge to edge without a break. Also, PRINCIPAL RAFTER.
common-rail injection *Mechanical Engineering.* a fuel injection system applied to multicylinder compression-ignition engines; an untimed pump maintains constant pressure in a pipe line, called a rail, from which branches deliver oil to mechanically operated injection valves.
common return *Electronics.* a single conductor that forms the return circuit for two or more otherwise separate circuits.
common salt see HALITE.
commonsense reasoning *Artificial Intelligence.* the ability to reason about a new and unforeseen situation using accumulated knowledge of existing conditions and previous experiences.
common-source amplifier *Electronics.* a field-effect transistor amplifier in which the source terminal is common to both the input and the output.
common storage *Computer Programming.* a provision that enables chained programs to store intermediate data results in a common area and effectively pass the data to the next program.
common trait *Psychology.* a personality trait that is common to a group of individuals and that can be used to characterize this group, as opposed to a personal trait that characterizes one individual.
common-user channel *Telecommunications.* a channel that is designated to furnish communication facilities and services to a number of users on a common basis.
common wall *Building Engineering.* a building wall that separates two dwelling units but is shared by both.
Common Water *Oceanography.* a deep water mass produced by the mixing of North Atlantic Deep and Bottom Water with Antarctic Circumpolar Water.
commotio [kə mō´shē ō] *Neurology.* **1.** a concussion or violent shaking. **2.** the shock that results from such an occurrence.
communal *Ecology.* relating to or living within a group.
communalism *Behavior.* the behavior pattern of living within a group.
commune *Ecology.* a group of individuals of the same species having a social structure and a high degree of interaction and relatedness. *Anthropology.* see CONNUBIUM.
communicable disease *Medicine.* any disease that can be transmitted from one affected organism to another; e.g., the common cold, as opposed to one that cannot be transmitted in this way; e.g., asthma.
communicating hydrocephaly *Medicine.* a form of hydrocephaly in which there remains normal communication between the brain's ventricles and subarachnoid space.
communicating word processor *Computer Technology.* a text processor that can exchange information and files with other such processors, usually via a local area network.
communication *Telecommunications.* any method or means of conveying information from one person or place to another, especially over wires or radio waves and excluding only correspondence through postal agencies, or direct and unassisted conversation.
communication band *Telecommunications.* **1.** a frequency spectrum between two defined limits. **2.** a range of frequencies between an upper and lower limit.
communication bus *Telecommunications.* **1.** a circuit over which data or power is transmitted, often serving as a common connection among a number of locations. **2.** a path over which information is transferred from any of several sources to any of several destinations.
communication channel *Telecommunications.* **1.** the smallest subdivision of a circuit that provides a single type of communication service. **2.** a path along which signals can be sent. *Computer Technology.* the portion of a storage medium that is accessible to a given reading or writing station.
communication countermeasure *Telecommunications.* any electronic interference with proper communication, such as jamming.

communication disorder *Medicine.* any condition, regardless of cause, in which an individual's ability to communicate with others is disturbed.

communication engineering *Telecommunications.* a branch of engineering specializing in the design, maintenance, and operation of telecommunications facilities and equipment.

communication protocol *Computer Programming.* a strict set of rules or procedures for initiating and maintaining data communications. Also, PROTOCOL.

communications *Engineering.* the various electronic processes by which information is transmitted from an originating source to a receiver at another location.

communications control unit *Telecommunications.* any electronic device used to control the flow of messages between stations or components in a communications network. *Computer Technology.* a small computer used to monitor and control the flow of data communications to and from the communications control unit's larger host computer.

communications intelligence *Telecommunications.* technological intelligence information derived from a communication system by other than its intended users; the use of communication systems, procedures, and equipment to intercept transmissions.

communications language *Telecommunications.* a coherent set of symbols, conventions, and rules used for representing and conveying information between two participants, including persons and machines.

communications network *Telecommunications.* a group of stations that are connected or in communication contact such that messages can be transmitted among them, and such that they are capable of intercommunication, though not necessarily on the same channel.

communications package *Computer Programming.* a common program that is written for a number of communications applications so that a variety of specific problems of data organization can be met.

communication speed *Telecommunications.* the rate of transmission of information over a communication system or device.

communications relay station *Telecommunications.* a central point that facilitates communication among radio stations by routing information to different stations, in order to increase range, interconnect radio stations at different frequencies, or use different methods of modulation.

communications satellite *Space Technology.* an artificial satellite that relays radio, television, and other signals around the world; usually follows a geostationary or geosynchronous orbit. Though it can be passive, it usually fulfills a repeater role using an onboard solar-powered transponder. Also, RELAY SATELLITE.

communications satellite tracking antenna

communications traffic *Telecommunications.* **1.** the sum of all messages that are transmitted and received over a given network. **2.** the stream of messages that are transmitted within a network or among certain networks at any given time.

communication system *Telecommunications.* a system or facility that can transfer information between persons and equipment; individual components of a single system must serve a common purpose, be technically compatible, employ common procedures, and respond in a coordinated manner to some form of control and operation. *Computer Technology.* a computer system that handles on-line, real-time applications.

communication theory *Telecommunications.* **1.** a mathematical investigation of the properties of transmitted messages, which are subject to certain probabilities of transmission failure distortion and noise. **2.** a theory dealing with the probable characteristics of the transmission of data in the presence of noise.

communication zone indicator *Electronics.* a device designed to indicate whether long-distance high-frequency broadcasts are reaching their destination.

communicator bias *Psychology.* the tendency to define as inherent personality traits in others what are actually their behavioral reactions to one's own personality.

community *Ecology.* a naturally occurring aggregation of organisms belonging to a number of different species, occupying a common habitat and interacting with each other within this area. Also, ECOLOGICAL COMMUNITY.

community psychology *Psychology.* a branch of applied psychology that deals with problems of mental health and social welfare in a community setting and involves the community members in proposed solutions to those problems.

commutating capacitor *Electronics.* a capacitor that provides commutating voltage for circuit-commutated thyristors in a self-commutated converter.

commutating pole *Electromagnetism.* one or several small poles between the main poles of a direct-current generator or motor.

commutating reactance *Electronics.* the reactance that effectively opposes the transfer of current to ensure that it remains steady as it flows from one anode to the next; for convenience, the value of the commutating reactance is designated as the reactance from phase to neutral or one-half the total reactance in the commutating circuit.

commutating reactor *Electricity.* a reactor generally used to modify the rate of current transfer rate between rectifying elements.

commutating zone *Electromagnetism.* a region containing a number of armature conductors that are short-circuited by a brush of the commutator.

commutation *Electromagnetism.* the repeated reversal of current through the windings of an armature in a direct-current motor so that direct current is provided at the brushes. *Electronics.* **1.** the transfer of current from one converter switching branch to another. **2.** the switching of currents between various paths as needed for the operation of a system or device. *Telecommunications.* a sampling of two or more channels, circuits, sources, or quantities in a cyclical or repetitive manner for multiplexing transmission over a single channel. Also, MULTIPLEXING.

commutation rules *Quantum Mechanics.* the specification of commutators of operators that in quantum physics correspond to the coordinates and momenta of a system.

commutative diagram *Mathematics.* a pictorial tool from category theory; widely used in all areas of algebra to illustrate assertions involving composition of functions.

$$\begin{array}{ccc} X & \xrightarrow{f} & Y \\ {\scriptstyle h}\downarrow & & \downarrow{\scriptstyle g} \\ & Z & \end{array}$$

commutative law *Mathematics.* given a binary operation • on the members of a set S, the rule that $a \bullet b = b \bullet a$ for all $a, b \in S$. An operation that satisfies the commutative law is called Abelian, a **commutative operation,** or, simply, **commutative.**

commutative ring *Mathematics.* a ring in which the multiplication is commutative.

commutator *Electromagnetism.* a part in a direct-current motor or generator that provides electrical continuity between the rotating armature and the stationary terminal, and also permits reversal of the current in the armature windings. *Electricity.* for rotating machines, an assembly of conductors onto which brushes are pressed to allow current to flow between the rotor and stator. *Mathematics.* given two elements a, b in a group G, the element c $(= a^{-1}b^{-1}ab)$ of G such that $ab = bac$.

commutator pulse *Computer Technology.* a pulse occurring at a particular time relative to a reference pulse that is often used to mark, clock, or control a particular bit position in a computer word.

commutator subgroup *Mathematics.* the subgroup generated by all commutators of a given group. The commutator subgroup of an Abelian group consists of the identity element. If the commutator subgroup is the entire group, the group is said to be perfect.

commutator switch *Electricity.* a switch that executes a series of switching operations in sequential order. Also, SAMPLING SWITCH, SCANNING SWITCH.

commuter rail system *Transportation Engineering.* an urban-suburban passenger service run by a mainline railway.

commuting operators *Quantum Mechanics.* the pairs of operators whose commutator equals zero, and which therefore may share simultaneous eigenfunctions.

comonomer *Chemistry.* a simple molecule that is combined with both like and unlike molecules in chains to form a polymer.

comose *Botany.* bearing a tuft of hair.

Comovirus group *Virology.* a group of plant viruses that cause systemic mosaicism and stunting in a narrow range of plants; transmitted primarily by leaf-feeding beetles. Also, COWPEA MOSAIC VIRUS GROUP.

comp. or **compd.** compound.

compact *Metallurgy.* a component manufactured by pressing metallic powder.

compact colony-forming active substance *Biochemistry.* a carbohydrate found in the cell walls of specific staphylococci producing the compact structure of colonies found in serum soft agar; without this substance, colonies in such a substrate would be diffuse.

compact disk *Acoustical Engineering.* a grooveless, read-only optical disk on which music is digitally encoded for playback on a compact disk player; it offers better sound quality and greater durability than phonograph records and conventional audio tapes. *Computer Technology.* a similar disk used to encode a computer program or data. Also, **compact disc.**

compact disk player *Acoustical Engineering.* an electronic device projecting a laser beam that scans a compact disk, decodes its digitally recorded music, and transmits the resulting signal to an amplifier or other playback device.

compact disk read-only memory see CD-ROM.

compact double layer see COMPACT LAYER.

compact H II region *Astronomy.* a dense region of ionized hydrogen less than about 10 light-years across.

compactification *Mathematics.* let X be a topological space. A compactification of X, if it exists, is any compact topological space W that contains X, or that contains a subspace homeomorphic to X. If W contains only one point in addition to X, then W is a **one-point compactification** of X.

compacting garbage collection *Computer Programming.* an automatic process of identifying memory cells that contain useless data and making them available again for reuse; includes a physical rearrangement of the cells into a contiguous array.

compaction *Engineering.* any process of consolidating granular material by mechanical means. *Geology.* in a soil or sediment, the process whereby pore space, and therefore volume, is reduced in response to an increase in the weight of overlying deposits or to pressure resulting from earth movements. *Computer Programming.* **1.** any of a number of techniques used to reduce storage space or data communication costs by reducing the redundancy in data representation. Also, COMPRESSION. **2.** in data management, the process of moving data on a list so that the available space is contiguous.

compact layer *Physical Chemistry.* in a double-layer ionic structure, the inner layer of ions that are specifically adsorbed on the electrode surface, as opposed to the adjacent outer region diffused in the electrolytic liquid. Also, COMPACT DOUBLE LAYER, HELMHOLTZ LAYER.

compact-open topology *Mathematics.* let C be the space of all continuous functions $f : X \to Y$, where X and Y are topological spaces. The topology for C generated by the subbase $W(K, U) = \{f \in C : f(K) = U$, where K is compact and U is open$\}$ is called the compact-open topology for C.

compact operator *Mathematics.* let X and Y be Banach spaces, K an operator $K: X \to Y$, and $\{x_n\}$ a sequence such that $x_n \in X$ and $\|x_n\| \le 1$. Then K is said to be compact if for any such sequence $\{x_n\}$ there exists a subsequence $\{xn_k\}$ such that the sequence $\{K(xn_k)\}$ converges in Y. Equivalently, K maps the closed unit ball onto a relatively compact set in Y. Also, COMPLETELY CONTINUOUS OPERATOR.

compactor *Mechanical Devices.* a household appliance that is used to compress trash to a smaller volume for more convenient disposal. *Mechanical Engineering.* a machine that consolidates earth by vibration, impact, or other means in order to carry a larger load.

compact radio source *Astronomy.* a radio source whose emission comes mostly from a single region smaller than about 3000 light-years across.

compact set *Mathematics.* a subset A of a topological space X is said to be compact if every open cover of A contains a finite cover of A. If this is true only for countable covers of A, then A is **countably compact.** Every compact set is also closed. Also, BICOMPACT.

compact space *Mathematics.* a topological space that is a compact set.

companded single-sideband system *Telecommunications.* a microwave system in which the frequencies produced by the process of modulation on one side of the carrier are transmitted in compacted form, and those on the other side are suppressed.

compander *Electronics.* a device that increases signal/noise ratio by compressing the volume range of a transmitted signal and reexpanding it at the receiver. (An acronym for <u>comp</u>ressor ex<u>pander</u>.)

companding *Electronics.* a process in which compression is followed by expansion; it is often used to reduce noise.

companion body *Space Technology.* **1.** a nose cone or other hardware from a launch system that remains with a spacecraft or satellite throughout its orbit and final trajectory. **2.** a natural satellite that strays in synchronous orbit with another natural satellite.

companion cell *Botany.* a specialized nucleated parenchyma cell associated with a sieve tube member in the phloem of angiosperms.

companion ladder *Naval Architecture.* either of a pair of ladders leading from the quarterdeck down to the upper deck of a sailing vessel.

companion star *Astronomy.* a general term for the smaller star(s) in a double-star or multiple-star system.

companionway *Naval Architecture.* any ladder or stair, or associated platform, that allows passage between decks.

company *Military Science.* an army unit composed of several platoons, usually commanded by a captain.

comparative cover *Cartography.* aerial photographic coverage of the same area or object taken at different times to show changes in detail.

comparative embryology *Developmental Biology.* the study of embryos of various species, with the aim of discovering their evolutionary relationships.

comparative linguistics *Linguistics.* a method of developing language families by finding cognate words between languages and determining how they are related; a protolanguage is then reconstructed from these cognates.

comparative method *Anthropology.* the collection and classification of ethnographic phenomena that are then compared to discover the source of their similarities and dissimilarities; a technique of the 1800s, then held to be the only valid approach to culture. *Linguistics.* a technique that groups languages into families, and then traces a language back to its origins.

comparative pathology *Medicine.* the study of disease in various species to discover similarities and differences in a disease and its manifestation.

comparative psychology *Psychology.* the study of behavioral differences among various species, aiming at both the understanding of animal evolution and behavior and the further understanding of human behavior.

comparative rabal *Engineering.* an approximate determination of the working order of electronic tracking equipment, which is derived from a rabal observation taken in conjunction with a rawlin observation.

comparative study *Statistics.* a statistical study in which experimental conditions are not controlled, and conclusions are drawn based on historical relationships among variables. Also, OBSERVATIONAL STUDY.

comparator *Engineering.* any instrument that compares the measurement of a gauged part with a fixed standard of measurement for the part. *Cartography.* a precision optical instrument used to determine the rectangular coordinates of a point with respect to a point on any other plane surface, such as a photograph. *Control Systems.* a device that continuously compares the value of a quantity to the value desired. *Computer Technology.* **1.** a combinational electronic circuit that compares two numbers and determines their relative magnitude. **2.** a device used to detect data transcription errors by comparing two instances of the same information.

comparator circuit *Electronics.* a circuit whose output indicates whether one of its two inputs is higher than, equal to, or lower than the other input.

comparison bridge *Electronics.* a voltage-comparison circuit that resembles a four-arm electrical bridge, in which the elements are arranged so that a zero error signal is derived if a balance exists in the circuit.

comparison lamp *Optics.* an incandescent lamp that has a consistent degree of luminosity and is compared with other light sources in photometric testing.

comparison microscope *Optics.* an instrument that consists of two microscopes with one eyepiece, coupled onto a single stand, permitting two images to be projected within a single field of view.

comparison spectrum *Spectroscopy.* a known spectrum having sharp lines that is used as a standard against which the spectrum of a sample is compared.

comparison star *Astronomy.* a star that is assumed to be invariable in brightness (or other properties) and is used as a check or reference point in astronomical observations.

comparison test *Mathematics.* let $\sum_{n=0}^{\infty} a_n$ and $\sum_{n=0}^{\infty} b_n$ be two series such that $|a_n| \le b_n$ for all but a finite number of n. Then $\sum_{n=0}^{\infty} a_n$ will converge if $\sum_{n=0}^{\infty} b_n$ converges. If $a_n \ge |b_n|$ for all but a finite number of n, then $\sum_{n=0}^{\infty} a_n$ will diverge if $\sum_{n=0}^{\infty} b_n$ diverges.

comparium *Ecology.* either of two different species that are capable of interbreeding and producing hybrid offspring with full fertility.

compartment *Naval Architecture.* one of the spaces between the bulkheads of a vessel; depending upon the vessel, one or more flooded compartments may be sealed off and the vessel will remain afloat.

compass *Engineering.* an instrument that indicates direction, usually having a magnetic needle that swings freely on a pivot and points to the magnetic north. *Navigation.* relating to or indicated by this instrument; used to form many compound terms, such as **compass amplitude, compass azimuth, compass bearing, compass direction, compass heading,** and so on. *Mathematics.* a V-shaped instrument used for constructing the locus of points a fixed distance from a given point. *Graphic Arts.* **1.** this V-shaped instrument used as a drawing and measuring tool in drafting. **2.** an electronic or manual tool used in platemaking for stripping flats.

compass adjustment *Navigation.* the process of neutralizing the effects on a compass that are caused by magnetic influences originating on the craft.

compass bowl *Engineering.* the part of a compass that serves as the foundation for the compass cord.

compass card *Engineering.* the pivoted disk on a compass that carries the magnetic needle, graduated in either degrees or compass points.

compass card axis *Engineering.* the line joining north and south, or 0 and 180 degrees, on a compass card.

compass compensation *Navigation.* the process of neutralizing the effects of degaussing currents on a magnetic compass.

compass course *Navigation.* the course of a craft relative to compass north. Similarly, **compass track.**

compass declinometer *Engineering.* a device used to determine the magnetic distribution of an area, composed of a thin compass needle that sits atop a sapphire bearing and steel pivot.

compass deviation *Navigation.* the difference between magnetic north and north as indicated on a specific compass installed in a particular craft; such deviations also vary with heading.

compass error *Navigation.* the difference between true north and compass north.

compass locator *Navigation.* a low-power radio beacon, installed as part of an instrument landing system, that can be used for navigation at close distance.

compass meridian *Navigation.* the north-south line passing through a craft, as determined by its compass.

compass north *Geodesy.* the direction north as indicated by a magnetic compass, which may be influenced by forces other than the earth's magnetic field.

compass plane *Mechanical Devices.* a plane with an adjustable sole used for smoothing concave or convex surfaces.

compass point *Cartography.* one of thirty-two divisions of a circle on a compass, equaling 11.25 degrees and indicating a direction relative to the direction given by the compass needle.

compass rafter *Building Engineering.* a rafter for special uses that is cut to a curve on one or both edges.

compass repeater *Navigation.* an instrument that gives compass readings at a location other than that of a craft's master compass.

compass roof *Building Engineering.* a roof in which each truss is applied similarly to an arch.

compass rose *Navigation.* a printed circular device showing compass directions. It may be marked in degrees, in "points" (the older system), or both.

compass saw *Mechanical Devices.* a handsaw having a number of long, narrow, tapering blades for cutouts and irregular or circular cuts.

compass transmitter *Navigation.* a device that transmits compass readings to a remote location.

compatibility the fact or condition of existing or functioning together; specific uses include: *Engineering.* the ability of two or more items or components to exist or function in the same system or environment without mutual interference. *Computer Technology.* the ability of a computer program or device to operate successfully with a given computer. *Immunology.* the ability of two different tissues or blood types to fit with each other without causing rejection. *Pharmacology.* of two substances, the capability of being administered simultaneously or mixed together in solution, so that neither one nullifies or inordinately affects the other's action.

compatibility conditions *Mechanics.* a set of six kinematic relations among the six independent components of any strain tensor that ensures the existence of a single-valued continuous displacement field. Also, ST. VENANT'S COMPATIBILITY EQUATIONS.

compatible *Engineering.* of a device or system, capable of performing with a given piece of equipment without the need for special conversion or modification. *Computer Technology.* **1.** of a program or peripheral device, capable of functioning successfully with a given computer or computer system. **2.** specifically, capable of operating with a computer or system produced by another manufacturer without the need for special linking software; for example, a Hewlett-Packard laser printer that is *compatible* with an IBM or Apple computer. Thus, **IBM-compatible, Apple-compatible.** *Botany.* of two or more plants, capable of cross-fertilization. *Immunology.* of two different tissues or blood types, able to fit with each other without causing rejection.

compatible single-sideband system *Telecommunications.* a method of independent sideband transmission in which the carrier is reinserted at a lower level after its normal suppression to permit reception by conventional AM receivers.

compensated amplifier *Electronics.* a broadband amplifier whose frequency range has been extended by altering circuit characteristics.

compensated geoid *Geodesy.* a surface derived from the geoid by application of computed values of the deflection of the vertical, which depend on the topographic and isostatic compensation. Also, COGEOID.

compensated-loop direction finder *Electronics.* a direction finder that uses two antenna systems to compensate for polarization error.

compensated pendulum *Horology.* a pendulum that is composed of two materials, zinc and steel, with each having different expansion characteristics so that the pendulum's effective length is not affected by temperature changes; used in the "Big Ben" clock of London, England.

compensated semiconductor *Electronics.* a semiconductor in which a donor imperfection partially cancels the electrical effects of an acceptor imperfection.

compensated volume control see LOUDNESS CONTROL.

compensating balance see COMPENSATION BALANCE.

compensating eyepiece *Optics.* an eyepiece that corrects for lateral color aberrations; used in microscopes with apochromatic objectives.

compensating glass *Optics.* a clear glass plate that converges or diverges rays of light in a manner similar to that of a filter, in order to maintain a focus.

compensating impurity *Solid-State Physics.* an impurity that is deliberately introduced into a crystal in order to counteract the effect of an unwanted existing impurity.

compensating network *Control Systems.* a network designed to suppress excessive vibration in a control system.

compensation *Biology.* the increased activity or relative size of a part of an organism to make up for a weakness in or loss of another part. *Psychology.* a process, either conscious or more generally unconscious, by which a person attempts to make up for some real or imagined physical or psychological shortcoming. *Cardiology.* the maintenance of adequate blood flow in response to increased demand accomplished by heart and circulatory adjustments, such as tachycardia. *Control Systems.* a method of changing a control system by reprogramming or adding new equipment in order to improve system performance. Also, STABILIZATION. *Electronics.* a modifying of supplementary action, or the effect of such action, to improve performance of a specified characteristic in a system; this characteristic is usually the system deviation.

compensation balance or **compensated balance** *Horology.* in a timepiece, a bimetallic balance wheel that compensates for the effect of temperature variations on the hairspring. Also, BALANCE BIMETALLIC, COMPENSATING BALANCE, CUT BALANCE.

compensation depth *Oceanography.* the depth at which there is just enough light to allow the amount of oxygen produced by phytoplankton through photosynthesis to balance the amount consumed by their respiration. *Geology.* see CALCITE COMPENSATION DEPTH.

compensation plate *Cartography.* a glass plate having its surface ground to a predetermined shape for insertion in the optical system of a diapositive printer or plotting instrument in order to compensate for radial distortion introduced by the camera lens.

compensation point *Botany.* in photosynthesis, the light intensity at which the amount of carbon dioxide absorbed equals the amount of oxygen released.

compensation sac *Invertebrate Zoology.* in bryozoans, a membranous sac that fills with water when tentacles are ejected and empties when they are withdrawn.

compensation signal *Engineering.* a radio signal that is recorded on a tape along with data and is used during data playback to electrically correct the results of tape-speed errors.

compensator *Control Systems.* in a feedback control system, an operating element or device used to achieve stability and thus improve system performance. Also, FILTER. *Electronics.* **1.** the portion of a direction finder that automatically applies all or part of a correction for a deviation to the direction indication. **2.** an electronic circuit that alters the frequency response of an amplifier system to achieve a desired result. *Optics.* a device that measures the phase difference between two components of elliptically polarized light to correct for displacement.

compensator gene *Genetics.* a gene located on the X chromosome in *Drosophila* that inhibits the activity of other sex-linked homozygous alleles in females.

compensatory emphysema *Medicine.* the hyperdistension of part of the lung that results when the volume of another part of the lung is reduced due to removal, contraction, or other cause.

compensatory hypertrophy *Medicine.* an increase of the size of an organ as compensation for the loss of function of an opposite, paired organ, or of the function of a portion of the same organ.

compensatory planting *Forestry.* the process of creating a forest in one area in order to restore (in part or whole) a loss in growing trees elsewhere; usually done when an area is being cleared permanently.

compensatory polycythemia *Hematology.* see SECONDARY POLYCYTHEMIA.

competence *Developmental Biology.* the ability of embryonic cells to differentiate into cell types determined by inductors. *Geology.* the ability of a wind or water current to transport particles, as determined by the diameter of the largest particle transported.

competent *Geology.* describing a bed or other geologic structure that is strong enough to transmit pressure and thus able to withstand folding pressure without flowage or alteration of thickness. *Molecular Biology.* **1.** able to react to a given morphogenetic stimulus by differentiation in a particular developmental direction. **2.** of bacterial cells, capable of incorporating foreign DNA molecules.

competent cells *Biotechnology.* a group of embryonic cells that retain the capability to differentiate themselves into other types of cells when exposed to a particular stimulus.

competing equilibria condition *Chemistry.* the competition for a particular reactant among several simultaneous chemical reactions in a complex chemical system.

competition *Ecology.* the simultaneous demand by two or more organisms or species for a necessary common resource that is in limited or potentially limited supply, resulting in a nonconfrontational struggle among those organisms or species for continued survival.

competitive *Ecology.* relating to or characterized by competition for resources.

competitive binding *Molecular Biology.* a situation in which one molecule competes with another molecule for a bonding site on a third molecule.

competitive enzyme inhibition *Enzymology.* the blocking of enzyme activity that occurs when a substance reversibly binds to the enzyme and competes with the substrate.

competitive exclusion *Ecology.* the exclusion of one species by another species when they compete for the same resources within the same area.

competitive exclusion principle *Evolution.* the principle that if two species try to occupy the same ecological niche, a superior species will eventually emerge to replace the inferior species; for two species to coexist in a stable equilibrium within a limited geographical area, they must occupy distinct ecological niches. Also, GAUSE'S LAW, GAUSE'S PRINCIPLE, LOTKA-VOLTERRA PRINCIPLE.

competitive inhibition *Biochemistry.* inhibition of an enzyme due to the reversible binding of an inhibitor molecule to the substrate binding site of the enzyme.

competitive release *Ecology.* a lessening of the intensity of competition between species within an area, as may result from a change in habitat range or resource needs by one species.

compilation *Cartography.* the production of a new or revised map or chart from existing maps, aerial photographs, surveys, new data, or other sources. *Photogrammetry.* the production of a map or chart from aerial photographs and geodetic control data by means of photogrammetric instruments. *Computer Programming.* the process of translating a high-level programming language into a machine-executable form.

compilation error *Computer Programming.* an error, often in syntax, that is detected by a compiler during compilation.

compilation history *Cartography.* the complete set of information regarding the development of a map or chart, including planning, source materials, control, compilation methods, and production techniques.

compilation manuscript *Photogrammetry.* the original drawings of a map or chart as compiled from various data, on which cartographic and related detail is depicted in colors on a stable-base medium, often a base manuscript together with its various overlays.

compilation scale *Photogrammetry.* the scale at which a map or chart is depicted on the original manuscript.

compile *Computer Programming.* to translate a source program written by a programmer in a programming language into a machine-executable form called an object program.

compile-and-go *Computer Programming.* to completely run a program written in a high-level language; this includes compiling, linking, loading, and executing in a continuous sequence.

compiled knowledge *Artificial Intelligence.* knowledge, of humans or machines, that is analogous to a compiled procedure; it can be executed, but its internal structure cannot easily be examined or understood.

compiler *Computer Programming.* a computer program that translates high-level language programs into a set of machine language instructions; most also provide error checking and diagnostic messages, code optimization, and memory usage information.

compiler-compiler *Computer Programming.* a program that produces a compiler for a language from a specification of the syntax and semantics of the language.

compiler diagnostics *Computer Programming.* a series of messages that report errors detected during compilation, ranging in severity from warnings to catastrophic error messages.

compiler-level language *Computer Programming.* any high-level programming language supported by standard compilers such that a program written for one computer may be compiled and executed on another with no modifications; e.g., FORTRAN, COBOL, or PASCAL.

compiler listing *Computer Programming.* an optional printout that is produced by a compiler and contains the source program listing, diagnostics, and other messages. Also, REFERENCE LISTING.

compiler toggle *Computer Programming.* any of a set of control commands included with the job control statements for a compilation run that invoke optional features of the compiler, such as a memory map listing.

compile time *Computer Programming.* **1.** the amount of time required for compilation. **2. compile-time.** referring to something done or doable during compilation, such as **compile-time error detection.**

compital *Botany.* of the veins of a leaf, intersecting at an angle.

complement *Food Technology.* **1.** to carry out the process of complementation. **2.** an additive used in this process. *Immunology.* a protein system in serum that combines with antibodies to form a defense against cellular antigens. *Mathematics.* **1.** for a subset A of a set S, the set $S - A$; i.e., the set of all members of S that are not in A. Thus, $S = A \cup (S - A)$. **2.** in general, an object or quantity that, when added to another object or quantity (under some type of addition such as set union, arithmetic addition, or angle addition), yields a specified result. **3.** in a bounded lattice (that is, a lattice that has both 0 and 1), a is said to be a complement of b if both $a \wedge b = 0$ and $a \vee b = 1$. **4.** the complement of a simple graph G is the graph $\bar{G}$ on the same vertex set such that any two distinct vertices are joined by an edge in $\bar{G}$ if and only if they are not joined by an edge in G. **5.** if H is a subgraph of graph G, then the complement of H in G is $G - E(H)$.

complementarity *Psychology.* a relationship in which the behavior of one person is met by contrasting behavior of the other, as when one spouse takes on a domineering role and the other is submissive. *Quantum Mechanics.* the principle that observable physical properties of a system, such as the conjugate variables of momentum and position, cannot be quantified simultaneously, thus requiring the dual aspects of particle and wave for complete description.

complementarity-determining region *Immunology.* a portion of an immunoglobin molecule that determines the binding of one specific antigen.
complementarity theory *Psychology.* the theory that two people are attracted to each other because they have contrasting needs or traits that balance one another. Also, **complementarity hypothesis.**
complementary *Electronics.* describing elements that have similar characteristics, differing by sign, such as P-type and N-type devices. *Psychology.* of or relating to complementarity; balancing or offsetting the traits of another person.
complementary angle *Mathematics.* either of a pair of angles whose sum is 90°.

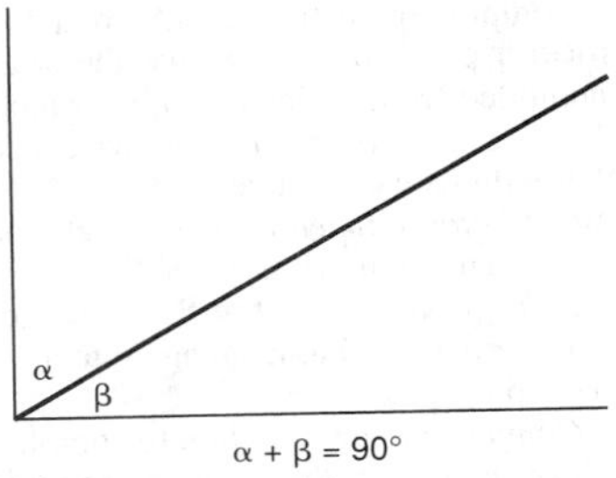

complementary angles

complementary base pairing *Genetics.* a sequence of bases in a single-stranded molecule of DNA or RNA that is paired or precisely complementary to that of another single strand and therefore capable of hydrogen bonding.
complementary bases *Genetics.* the base portion of the nucleotides that pair together by forming hydrogen bonds when DNA single strands assume a double-stranded configuration and when RNA is present in a double-stranded mode with DNA; consists of adenine-thymine and guanine-cytosine (adenine-uracil in RNA) base pairings.
complementary base sequence *Genetics.* a sequence of nucleotide bases in one strand of a DNA or RNA molecule that is exactly complementary (adenine-thymine, adenine-uracil, or guanine-cytosine) to that on another single strand.
complementary color *Optics.* either of a pair of colors that produces an achromatic color when mixed with the other.
complementary constant-current logic *Electronics.* a computer circuit design that produces extremely fast switching speeds of 3 nanoseconds.
complementary deoxyribonucleic acid see COMPLEMENTARY DNA.
complementary DNA *Genetics.* DNA that is produced in vitro using RNA as a template and RNA-dependent DNA polymerase; the resulting DNA is complementary in sequence to the RNA template. Also, COPY DNA.
complementary function *Mathematics.* the general solution of the homogeneous part of a linear differential equation of degree *n*. When added to a particular solution of the original (nonhomogeneous) differential equation, the result is a general solution of the nonhomogeneous differential equation.
complementary genes *Genetics.* two independent pairs of nonallelic genes, neither of which will produce its effect in the absence of the other. Also, RECIPROCAL GENES.
complementary logic switch *Electronics.* a device in which two transistors are paired so that one is on when the other is off.
complementary minor see COFACTOR.
complementary rocks *Petrology.* igneous rocks that are differentiated from a common magma, and whose total average composition is identical to that of the parent magma.
complementary space *Mathematics.* two closed linear subspaces U_1 and U_2 of a Banach space B are complementary spaces if every $x \in B$ can be uniquely expressed as $u_1 + u_2$, where $u_1 \in U_1$ and $u_2 \in U_2$. Such subspaces are called direct summands; denoted $B = U_1 \oplus U_2$.
complementary strand see C STRAND.
complementary symmetry *Electronics.* a circuit in which two transistors with opposite conductivity provide electricity during opposite halves of the input signal frequency cycle; commonly used in loudspeaker systems.
complementary transistors *Electronics.* two transistors with opposite conductivity that operate in the same functional unit.
complementary wave *Electromagnetism.* an electromagnetic wave that arises from a reflection in a transmission line due to a discontinuity or open end where there is a mismatch of impedance.
complementary wavelength *Optics.* a wavelength that, when mixed with a given color in proper proportions, duplicates a standard reference light. Also, **complementary dominant wavelength.**
complementation *Food Technology.* the addition of biochemical substances to a product or diet in order to improve the nutritional quality or correct a deficiency. *Nutrition.* the process of making up for the deficiency of a particular amino acid by ingesting a surplus of another at about the same time. *Computer Technology.* a method or instance of representing negative numbers by their complement value; used in computers to carry out arithmetic subtraction in adder circuits. For example, in 1's complement, each bit is replaced by its opposite. *Genetics.* the interaction of a gene or protein with a defective homologous gene or protein that is present in the same cell or organism, compensating for the defect and permitting the cell to function in spite of the presence of the defective gene. *Virology.* an intracellular interaction of two defective animal or bacterial viruses that allows the replication of both. *Mathematics.* the replacement of a set or object by its complement.
complementation group *Genetics.* a group of mutants which carry mutations that are located within the same cistron. *Virology.* a group of viruses that have similar genetic mutations such that they do not complement one another.
complementation law *Statistics.* a law stating that if A is a set and $P(A)$ is its probability, the probability of the complement of A, that is the probability of A not occurring, is $P(A^C) = 1 - P(A)$.
complementation test *Genetics.* a genetic test that indicates whether a mutation involves alleles of the same gene or of two different genes on a homologous pair of chromosomes; also used in preparing complementation maps.
complemented lattice *Mathematics.* a lattice in which every element has a complement; equivalently, a partially ordered set in which every pair of elements has both a greatest lower bound and a least upper bound.
complementer circuit *Computer Technology.* in the hardware of certain computers, an electronic circuit that assists in carrying out arithmetic operations by converting negative numbers to their complement representations.
complement fixation *Immunology.* the combining of a complement with an antigen-antibody complex, making the complement unavailable to complete a reaction in subsequent antigen-antibody complexes.
complement-fixation test *Immunology.* a test used to detect and quantitate specific antibodies or antigens by demonstrating the consummation of an active complement by antigen-antibody complexes; commonly used in the diagnosis of syphilis.
complement number system *Computer Programming.* a method of storing and performing arithmetic operations on negative numbers by representing them in their radix or diminished radix form; for example, 10's or 9's complement for negative decimal numbers, and 2's and 1's complement for negative binary numbers.
complement-on-9 see NINES COMPLEMENT.
complete *Artificial Intelligence.* **1.** of a theorem prover, able to prove any true theorem eventually. **2.** able to solve any problem in a specified class. *Mathematics.* a set of elements $\{x_\alpha\}$ is said to be complete in a normed space N if the set of all linear combinations of the elements x_α is dense in N.
complete bipartite graph *Mathematics.* a simple bipartite graph G with bipartition V_1, V_2 such that if $u \in V_1$ and $w \in V_2$, then uw is in $E(G)$. If $|V_1| = m$ and $|V_2| = n$, the graph is denoted by $K_{m,n}$ or by $K_{n,m}$ equivalently.
complete blood count *Pathology.* a quantitative analysis of the peripheral blood, including the red blood cell count, white blood cell count, erythrocyte indices, hematocrit, and differential blood count.
complete carry *Computer Technology.* in a parallel full-adder, the technique that propagates or cascades the carry bit from one bit position to the next.
complete combustion *Chemistry.* the combination of all available fuel with oxygen.
complete degeneracy *Quantum Mechanics.* a condition in which all quantum stationary states of a system have the same energy.
complete environment seeding *Agronomy.* a planting technique involving the use of a tape or layer of biodegradable material that contains both seed and plant nutrients.
complete-expansion diesel cycle see BRAYTON CYCLE.

complete fertilizer *Agronomy.* a fertilizer containing the three major plant nutrients: nitrogen, phosphorus, and potassium (potash).
complete flower *Botany.* a flower having a stamen, calyx, corolla, and carpels.
complete fusion *Metallurgy.* in welding, fusion of the entire surface of the base material.
complete graph *Mathematics.* a simple graph in which every pair of distinct vertices are adjacent. The complete graph on n vertices is usually denoted by K_n.
complete inner product space see HILBERT SPACE.
complete integral *Mathematics.* **1.** a solution of an nth-order ordinary differential equation in one variable that depends on n parameters as well as the independent variable. Also, **complete primitive. 2.** a solution of a first-order partial differential equation in m independent variables that depends on each of the independent variables as well as m parameters.
complete lattice *Mathematics.* a partially ordered set L is a complete lattice if every subset (including the empty set and any infinite subsets) H of L has a *supremum* (least upper bound) and an *infimum* (greatest lower bound). If this is true only for finite, nonvoid subsets, then L is a lattice. A complemented lattice is a special case of a complete lattice.
complete lubrication *Engineering.* the lubrication that occurs when a fluid film comes between two rubbing surfaces; slipperiness is caused only by internal fluid friction in the film. Also, VISCOUS LUBRICATION.
completely continuous operator see COMPACT OPERATOR.
completely elastic collision see ELASTIC COLLISION.
completely inelastic collision see INELASTIC COLLISION.
completely mixed bioreactor *Biotechnology.* a stirred-tank or single-stage fermenter characterized by continuous stirring of the reactor contents but no recycling of the biomass.
completely normal space *Mathematics.* a topological space X with the property that, given any subsets S_1 and S_2 of X, neither of which contains a limit point of the other, there exist two disjoint open sets in X, one containing S_1 and the other containing S_2. (In particular, if S_1 and S_2 are nonintersecting closed sets, then X is normal.)
completely randomized design *Statistics.* in analysis of variance, a design in which experimental units are randomly assigned to each factor level. Also, RANDOMIZED GROUP DESIGN.
completely regular space *Mathematics* a topological space X with the property that, for each $x_0 \in X$ and each neighborhood U of x_0, there is a continuous function $f: X \rightarrow [0, 1]$ such that $f(x_0) = 1$, $0 \leq f(x) \leq 1$ for $x \in U - \{x_0\}$, and $f(x) = 0$ for $x \in T - U$. A completely regular space is also regular.
complete metric space *Mathematics.* a metric space in which every Cauchy sequence converges to a point in the space.
complete operation *Computer Technology.* a computer operation cycle that includes fetching the instruction to be executed, decoding the operation code, obtaining the necessary operands, executing the operation, and storing the results in the correct memory location.
complete orthonormal set *Mathematics.* in a finite or infinite dimensional vector space, a set of mutually orthogonal unit vectors (i.e., an orthonormal set) that is not a proper subset of any other orthonormal set. Such a set forms an orthonormal basis for the vector space. Also, CLOSED ORTHONORMAL SET.
complete penetrance *Genetics.* a condition in which every individual having a certain genotype manifests the expected phenotype; that is, every individual who carries a certain gene exhibits the expected physical characteristics caused by that gene.
complete protein *Nutrition.* a protein food source that contains all the amino acids essential for health and growth, such as eggs and milk.
complete round *Ordnance.* a basic unit of ammunition including all explosive and nonexplosive components necessary for it to function once; the term is also applied to bombs, missiles, and rockets.
complete routine *Computer Programming.* a subprogram that performs a specific function without modification by the user; often supplied by the computer manufacturer or included in a program package.
completing the square *Mathematics.* **1.** the process of rewriting a general quadratic equation $ax^2 + bx + c = 0$ (where $a \neq 0$) in the form $(x + b/2a)^2 = (b^2 - 4ac)/4a^2$. **2.** the process of rewriting a polynomial $ax^2 + bx + c$ in the form $a(x + b/2a)^2 + (4ac - b/2)/4a$ (where $a \neq 0$). **3.** the process of writing any second-degree polynomial as a sum of a square and a number.
completion of a metric space *Mathematics.* **1.** the adjoining of limits of all Cauchy sequences to a given metric space. **2.** the metric space obtained by this process.
complex *Psychology.* a largely unconscious set of related ideas that have a common emotional tone and strongly influence a person's behavior and attitudes. *Medicine.* a group of physical symptoms or signs that appear together with some consistency. *Geology.* a large-scale, complicated, intricately related assemblage of rocks of any age or origin that cannot be easily differentiated in mapping. *Archaeology.* a patterned grouping of similar artifact assemblages from two or more sites; presumed to represent an archaeological culture. *Mathematics.* **1.** a countable set of simplices whose intersection is either a common face or the empty set such that each vertex belongs only to a finite number of the simplices. **2.** relating to or being a complex number.
complex carbohydrate *Nutrition.* any macromolecular saccharide substance, such as starch, glycogen, or cellulose polysaccharide.
complex chemical reaction *Chemistry.* a system in which several chemical reactions take place at the same time.
complex climatology *Meteorology.* the analysis of the climate of a place, or a comparison of the climates of two or more places, by the relative frequencies of various weather types or groups of such types over a specified time period.
complex compound *Chemistry.* **1.** any combination of two or more compounds or ions to give a species in which a central atom (usually metallic) is bonded to two or more usually nonmetallic atoms or ions. **2.** see COORDINATION COMPOUND.
complex conjugate *Mathematics.* **1.** for the complex number $a + bi$, the complex number $a - bi$. **2.** the matrix whose entries are the complex conjugates of a given matrix.
complex data type *Computer Programming.* numeric data that contain two real fields representing the real and imaginary parts of an imaginary number of the form $a + bi$, where i is the square root of -1.
complex declaration statement *Computer Programming.* a nonexecutable FORTRAN statement that identifies certain variables as complex data types.
complex degree of coherence *Physics.* a measure of the coherence of two waves, equal to the cross-correlation function between the normalized amplitude of one wave and the complex conjugate of the normalized amplitude of the other.
complex dune *Geology.* a dune produced by the merging of various dunes under the influence of multidirectional winds.
complex fold *Geology.* a fold that incorporates a preexisting fold having a different orientation.
complex fraction *Mathematics.* a fraction whose numerator or denominator (or both) is a fraction.
complex frequency *Engineering.* a complex number that represents exponential and damped sinusoidal motion, indicated by a constant s that represents a motion having an amplitude designated as Ae^{st}, in which A is a constant and t represents time.
complexification *Mathematics.* the process of replacing a real parameter or variable with a complex one; for example, vector components in a vector space or the parameters in a Lie group. Theorems exist that relate the properties of the original object with the properties of the complexified object.
complexing *Chemistry.* the production of a complex compound.
complexing agent see LIGAND.
complex instruction set computer *Computer Technology.* a central processing unit design featuring a large number of relatively complex instructions.
complex ion *Chemistry.* an ion having a structure in which a central atom or ion is bonded to other ions or molecules by coordinate bonds.
complexity *Computer Programming.* the level of difficulty in solving mathematically posed problems as measured by the time, number of steps or arithmetic operations, or memory space required (called time, computational, and space complexity, respectively).
complexity analysis *Computer Programming.* in structured program design, a quality-control operation that counts the number of "compares" in the logic implementing the function; a value of less than 10 is considered acceptable.
complex loci *Genetics.* loci that contain several structural genes and a common regulatory element.
complex low *Meteorology.* a region of low atmospheric pressure having more than one low-pressure center.
complex mRNA see SCARCE MRNA.
complex notation *Physics.* the representation of a physical quantity by a complex value.
complex number *Mathematics.* any number of the form $a + bi$, where a and b are real numbers and $i^2 = -1$.

complexometric titration *Analytical Chemistry.* a volumetric analysis in which a soluble, undissociated stoichiometric metal complex is formed during titration by a complexing agent.

complex ore *Geology.* any ore that yields more than one metal.

complex permeability *Electromagnetism.* a measure of the permeability of an inductor core, equal to the ratio of the magnitude of the magnetic induction of the coil to the magnitude of the field strength within the core.

complex permittivity *Electricity.* a property of a dielectric functioning as the insulating material in a capacitor; it is equal to $\varepsilon_0(C/C_0)$, where C is the complex capacitance of the capacitor when it is connected to a sinusoidal voltage source, and C_0 is the vacuum capacitance of the capacitor.

complex plane *Mathematics.* the plane of complex numbers as represented in the Argand diagram. If the single point at infinity, denoted by $z = \infty$, has been adjoined, it is sometimes called the **extended complex plane** or **inversive complex plane.** The complex plane without $z = \infty$ is called the **finite complex plane** and is designated by $|z| < \infty$. The (extended) complex plane is conformally equivalent to a sphere under the stereographic projection.

complex potential *Fluid Mechanics.* a representation of streamlines for potential flow using the complex plane for mapping.

complex-query language *Computer Programming.* a database user language designed to allow the construction of queries that may require searching multiple files and joining selected records from each.

complex reflector *Engineering.* a structure or an arrangement of structures having many surfaces, all of which reflect radar and are oriented to varying directions.

complex ripple mark see CROSS RIPPLE MARK.

complex salt *Inorganic Chemistry.* a class of salts containing two different metal atoms, of which only one is ionized in solution as the other is bound in a complex ion.

complex sensor *Robotics.* a term for any sensor used by a sophisticated robot to work within a complex environment.

complex stream *Hydrology.* a stream that has begun its second or later cycle of erosion.

complex target *Engineering.* a radar target made up of an arrangement of reflective surfaces which as a whole are smaller in size than the resolution capabilities of the radar.

complex tombolo *Geology.* a system of islands and beaches formed when several islands are connected with one another and with the mainland by a series of sand ridges or bars.

complex tone *Acoustics.* a tone that consists of a combination of pure (sinusoidal) frequencies.

complex twin *Crystallography.* a twin feldspar crystal produced as a result of both normal and parallel twinning.

complex unit *Mathematics.* a complex number whose modulus is one. Such a number has polar form $\cos\theta + i\sin\theta$.

complex valley *Geology.* a valley whose course may be partly parallel and partly transverse to the general structure of the underlying strata.

complex variable *Mathematics.* a variable that is permitted to assume complex values.

complex wave *Physics.* a wave that is composed of several frequencies.

compliance *Mechanics.* the displacement of a mechanical system under a unit applied force. *Psychology.* a situation in which an individual outwardly conforms to the wishes of an influencing source without altering his or her internal beliefs and attitudes.

compliance constant *Mechanics.* a coefficient appearing in the generalized Hooke's law (which describes the stress produced in a body as directly proportional to the strain applied).

complicate *Entomology.* of insect wings, folded lengthwise several times.

compo board see COMPOSITION BOARD.

component an individual part of an organized whole; specific uses include: *Chemistry.* **1.** an ingredient in a mixture; the part of a mixture that is analogous to a constituent of a compound. **2.** any of the smallest possible number of substances required to determine all the constituents of a chemical system. *Electricity.* a unit in an electrical system, such as a capacitor or generator, that has distinct electrical characteristics and terminals so that it can be connected to other components to form a circuit. Also, ELEMENT. *Materials Science.* a pure chemical species involved in a phase diagram; in metallurgy they are the pure metals constituting an alloy system; in ceramics, the pure oxides. *Archaeology.* a site, or a level within a site, representing the activities of one cultural group during a relatively brief period of time. *Linguistics.* one of the major subdivisions of a generative grammar. *Mathematics.* **1.** given a collection A of sets, a component of A is any connected direct summand of A other than the empty set; equivalently, a component of A is a connected set that is not contained in any other set of A. **2.** given an algebraic plane curve with the equation $h(x,y) = 0$, a component of the curve is an algebraic plane curve with equation $g(x,y) = 0$ such that $h(x,y) = g(x,y)p(x,y)$ for some polynomial $p(x,y)$. An algebraic curve with only one component is irreducible. **3.** a vector component. **4.** a tensor component.

component distillation *Chemical Engineering.* a distillation process in which a fraction that cannot be separated by normal distillation is separated by forming an azeotropic mixture.

component family *Design Engineering.* a group of parts that share certain codifiable geometric characteristics.

component rank see CUTSET RANK.

components (of a star system) see COMPANION STAR.

component substances law *Chemistry.* a law stating that each substance in a mixture retains its own specific properties, which are independent of the other substances in the mixture.

compose *Graphic Arts.* **1.** to set characters in type. **2.** to combine various film elements such as halftones and screens on one piece of film, often in preparing a printing plate.

composing room *Graphic Arts.* a facility in which type is set, especially hot metal type.

composing stick *Graphic Arts.* a usually metal frame used by a compositor to set metal type by hand.

Composita *Paleontology.* an extinct genus of smooth-shelled, articulate brachiopods in the order Spiriferida and suborder Athyrididina; extant in the upper Devonian to Permian.

Compositae *Botany.* a very large plant family that consists of more than 20,000 species and contains the most highly developed dicotyledonous plants, including food plants such as lettuce, chicory, artichokes, and endive, decorative flowers such as the aster, daisy, chrysanthemum, and marigold, and common weeds such as dandelions, ragweeds, and thistles. The family is characterized by a tightly formed head of many small epigynous flowers and sometimes by resin canals or latex ducts. Also, ASTERACEAE.

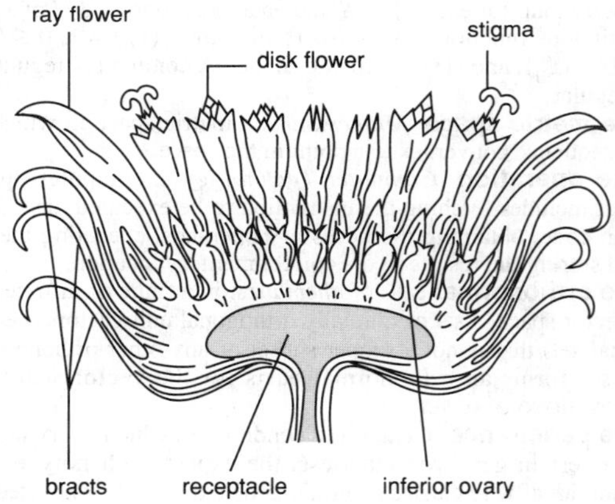

composite flower

composite an item made up of a combination of various elements; specific uses include: *Botany.* **1.** a plant that is a member of the composite family (Compositae). **2.** of or relating to this family. Thus, **composite flower.** *Materials Science.* a material or materials system composed of a mixture or combination of two or more microconstituents or macroconstituents that differ in form and chemical composition and that are essentially insoluble in each other. Thus, **composite material.** *Metallurgy.* any of several synthetically produced materials containing dissimilar components. A metal-matrix composite, which is often referred to as MMC, has a continuous metallic phase in which fibers or particles, usually nonmetallic, are embedded. *Acoustical Engineering.* the combination of multiple tracks into a master recording, representing the best possible combination of tracks.

Composite *Architecture.* a Roman order combining elements of the Corinthian and Ionic.

composite aircraft *Aviation.* **1.** an aircraft having a composite power plant. **2.** two attached aircraft, such as a large airplane carrying or towing a smaller airplane.

composite balance *Electricity.* an electric balance created by adjusting the Kelvin balance to measure amperage, voltage, or wattage.

composite band *Anthropology.* a group of unrelated families who are associated through a fictive or ceremonial kinship, or a multiple family group linked through bilateral descent.

composite cable *Electricity.* a cable composed of conductors of different gauges and types under a single sheath.

composite circuit *Electronics.* a circuit that is used simultaneously for both telephony and telegraph transmissions, separating the two by frequency discrimination.

composite coast *Geology.* a coast having alternating zones of emergence and submergence as a result of upwarping or subsidence along transverse lines. Also, COAST OF TRANSVERSE DEFORMATION.

composite color signal *Telecommunications.* a total color television signal including video, blanking, synchronizing, and chrominance component signals. Also, **composite color video signal, composite picture signal.**

composite color sync *Telecommunications.* a television signal system that encodes color and luminance information in the same signal and creates an exact alignment of sound and picture elements.

composite compact *Metallurgy.* in powder metallurgy, a compact consisting of two or more layers having different compositions.

composite cone *Volcanology.* a large volcanic cone composed of alternating layers of lava and pyroclastic material. Also, STRATOVOLCANO, COMPOSITE VOLCANO, STRATIFIED VOLCANO.

composite construction *Building Engineering.* building construction that utilizes structural materials composed of combinations of metals, alloys, or plastics as its principal foundation of building materials.

composite defense *Ordnance.* a form of antiaircraft defense integrating two or more types of artillery units.

composite dike *Geology.* a dike that results from the successive injections of different chemical and mineralogical materials into the same fissure.

composite electrode *Metallurgy.* in consumable arc welding, an electrode consisting of two or more different materials; at least one material is a filler metal.

composite family *Botany.* the popular name for the plant family Compositae, which includes such plants as the daisy, dandelion, and sunflower. See COMPOSITAE.

composite filter *Electronics.* a network composed of various types of filters that are linked together in series.

composite flash *Geophysics.* a type of lightning discharge that consists of a series of rapid (occurring at intervals of 0.05 second) lightning discharges, all following the same channel.

composite fold *Geology.* a fold having minor folds on its limbs. Also, COMPOUND FOLD.

composite gneiss *Petrology.* a banded rock typically formed by intimate association of granitic magma injected into country rocks.

composite grain *Geology.* a sedimentary particle formed from two or more distinct particles.

composite hypothesis *Statistics.* in testing, a hypothesis including more than one possible value of the parameter of interest.

composite intrusion *Geology.* an igneous intrusion composed of two or more rocks having different chemical compositions.

composite joint *Metallurgy.* a joint consisting of a mechanical and a welded attachment.

composite macromechanics *Engineering.* the analysis of composite material behavior in which the material is assumed to be homogeneous; the effects of its constituents are determined only as averaged apparent properties of the composite.

composite map *Cartography.* a map on which two or more types of information that are usually shown on separate maps are brought together for the purpose of comparison. *Mining Engineering.* a single map that shows several levels of a mine.

composite micromechanics *Engineering.* the analysis of composite material behavior, in which the constituent materials are studied microscopically to determine distinct properties; the materials' interactivity defines the properties of the composite.

composite nerve *Physiology.* a nerve that consists of both sensory fibers and motor fibers.

composite number *Mathematics.* a positive integer that has two or more prime factors; an integer that is neither prime, zero, +1, nor –1.

composite photograph *Graphic Arts.* a single photographic print that combines elements of two or more separate photographs.

composite plate *Metallurgy.* in electroplating, a multilayer coating, such as chromium onto a copper plate.

composite power plant *Aviation.* a power plant comprising two or more types of engines, such as a jet engine and a reciprocating engine.

composite profile *Cartography.* a profile derived from the highest points of an original series of profiles drawn along regularly spaced and parallel map lines.

composite propellant *Space Technology.* a solid propellant consisting of a fuel and an oxidizer, neither of which will ignite without the other.

composite pulse *Electronics.* a pulse composed of a series of overlapping pulses emanating from the same signal source, but by way of different paths.

composite sailing *Navigation.* a variation on great-circle sailing in which the vessel does not sail above a certain latitude.

composite sampler *Engineering.* an instrument that collects representative samples of oil from various levels of a storage tank.

composite seam *Geology.* a thick coal seam having two or more layers in contact wherever interrupting strata or dirt bands have wedged out.

composite sequence *Geology.* a series or succession of cyclic sediments that comprises all lithologic types in the order in which they tend to occur.

composite set *Electronics.* a network of instruments designed to function as one end of a composite circuit.

composite signal *Telecommunications.* a signaling arrangement that provides for DC signaling and dial-pulse generation beyond the range of loop signaling methods.

composite sill *Geology.* a sill formed of successive intrusions of igneous rocks which differ in chemical and mineralogical composition.

composite stream *Petroleum Engineering.* a flow of two or more fluids in one stream, such as oil and gas or different hydrocarbons.

composite symbol *Computer Programming.* a special symbol that contains two or more characters, such as ±, meaning "plus or minus."

composite topography *Geology.* a landscape whose surface features have developed during two or more cycles of erosion.

composite track *Navigation.* the track followed by a vessel in the course of composite sailing. It follows the great circle that passes through its departure and is tangent to the limiting parallel of latitude. Then it follows that latitude until it can pick up the great circle through its destination which is also tangent to the limiting parallel.

composite transposon *Genetics.* a DNA sequence flanked by insertion sequences that facilitate its transposition; a large transposon consisting of a central region that carries markers that are unconnected with transposition.

composite unconformity *Geology.* an unconformity produced by more than one occurrence of nondeposition, weathering, or erosion.

composite vein *Geology.* a large fracture zone consisting of parallel fissures and converging diagonals filled with ore and partially replaced country rock.

composite volcano see COMPOSITE CONE.

composite wall *Building Engineering.* a durable and attractive wall built with a facing material which adheres or anchors to a backing; the facing does not necessarily contribute to the strength of the wall. Also, FACED WALLS, VENEERED WALLS.

composite wave filter *Electronics.* a network of circuits that reduce the amplitude of signals operating at various frequencies.

composition *Chemistry.* **1.** the elements or compounds of which a substance is composed. **2.** the proportion by weight, or by number of atoms, of the elements making up a chemical compound. *Graphic Arts.* **1.** the process of setting characters in type. Also, TYPESETTING. **2.** the product of this process. *Materials Science.* **1.** the components making up a material or produced from it by analysis. **2.** the resultant material.

compositional maturity *Geology.* a stage of sedimentary development in which the sediment approaches the compositional end product to which it is driven by formative processes.

compositional semantics *Artificial Intelligence.* semantics in which the meaning of a phrase is composed of the meanings of its components.

composition board *Materials.* a flat sheet made of fibers processed into a pulp that is rolled and pressed. Also, COMPO BOARD.

composition diagram *Chemical Engineering.* a graphical plot that shows the solvent-solute concentration relationships during various stages of extraction operations.

composition drift *Materials Science.* in copolymerization, a phenomenon in which the likelihood of each type of monomer to add to the growing chain varies even when equimolar amounts of the monomers are initially used.

composition metal *Metallurgy.* the former name of any of several cast copper-base alloys containing tin, zinc, and lead.

composition of forces *Mechanics.* **1.** the process of establishing a single force that represents the effects of a number of coexisting forces. **2.** see RESULTANT FORCE.

composition of functions *Mathematics.* given two functions (or mappings) f and g, such that the range of f is contained in the domain of g, the composition of f and g is a new function h, given by $h(x) = g[f(x)]$, for all x in the domain of f. The usual notation is $h = g \circ f$, although sometimes $h = f \circ g$ is used (usually with $x(f)$). Function composition in general is not commutative but is associative.

composition of velocities law *Mechanics.* a law that relates the velocities of an object in two different reference frames which are moving relative to one another at a specified velocity.

composition plane *Crystallography.* the common plane between a pair of twinned crystals.

composition resistor see CARBON RESISTOR.

composition series *Mathematics.* **1.** the composition series of a group G is a subnormal series $G = G_0 > G_2 > \cdots > G_n = <e>$, where $<e>$ is the trivial group consisting of the identity element, such that each factor group G_i/G_{i+1} is simple. Every finite group has a composition series. A subnormal series is a composition series if and only if it has no proper refinements, and any refinement to a composition series is equivalent to the composition series. **2.** the composition series of a nonzero module A (over a ring R) is a normal series $A = A_0 \supset A_2 \supset \cdots \supset A_n = 0$, where 0 is the trivial module, such that each factor A_i/A_{i+1} is a nonzero module with no proper submodules. Every module over a finite ring R has a composition series. Any two composition series of A are equivalent, and A has a composition series if and only if it satisfies both the ascending and descending chain conditions on submodules.

composition surface *Crystallography.* a surface that is not necessarily planar, and that is common to a pair of twinned crystals.

compositor *Graphic Arts.* a person who sets type.

compost *Materials.* **1.** a mixture of decaying organic matter, such as rotting leaves or manure, used as a fertilizer or soil conditioner. **2.** to make or apply such a mixture.

compound *Chemistry.* a substance made up of atoms of at least two different elements held together by chemical bonds. *Biology.* made up of several similar parts that form a unified whole, such as a compound leaf or a compound stomach. *Petroleum Engineering.* a power transmission mechanism that conveys energy from the engines to the drawworks, pump, and other machines on a drilling rig.

compound acinous gland *Anatomy.* any gland made up of multiple acini.

compound alluvial fan see BAJADA.

compound angle *Engineering.* an angle that occurs when two mitered angles are joined.

compound compact *Metallurgy.* in powder metallurgy, a compact consisting of mixed particles having different compositions.

compound condition *Computer Programming.* a conditional statement that tests two or more relational expressions at the same time, combines these with operators such as "and," "or," and "not," and returns "true" or "false."

compound cryosar *Electronics.* a device composed of two cryosars whose electrical properties differ.

compound curve *Mathematics.* a curve made up of two arcs of nonequal radii, whose centers are positioned on the same side of the curve in such a way that the two curves have the same tangent at their common point.

compound die *Metallurgy.* in metal fabrication, a die that performs more than one operation.

compound elastic scattering *Nuclear Physics.* the combined effects of a collision in which kinetic energy is transferred from a projectile to a target nucleus without raising the target's internal energy to an excited state.

compound engine *Mechanical Engineering.* a displacement engine having two or more cylinders of varying sizes, designed to produce more efficient work as steam expands over two or more stages. *Aviation.* an aircraft engine consisting of a reciprocating engine from which the exhaust powers one or more turbines, which in turn provide extra thrust.

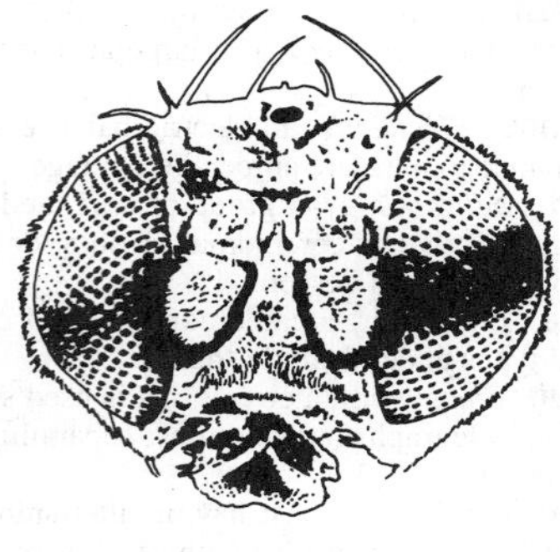

compound eye

compound eye *Invertebrate Zoology.* an eye made up of many functionally independent photoreceptor units separated by pigment cells, producing a mosaic image; such an eye is typical of insects, crustaceans, centipedes, and horseshoe crabs.

compound fault *Geology.* a zone of parallel, closely spaced fractures.

compound field winding *Electricity.* a winding composed of shunt and series coils that perform either together or against each other.

compound fold see COMPOSITE FOLD.

compound foreset bedding *Geology.* cross-bedding having a concave base and several foreset beds that dip in more than one direction.

compound fracture *Medicine.* a bone fracture in which the broken end or ends of the bone have pierced the skin. Also, OPEN FRACTURE.

compound generator *Electricity.* a DC generator having a shunt field winding outside of a series field winding on the main poles.

compound gland *Anatomy.* any gland that contains a branching system of ducts.

compounding *Mechanical Engineering.* the consequent arrangement of cylinders in an engine for the purpose of obtaining the highest degree of efficiency from an engine.

compound leaf *Botany.* a leaf whose blade is divided into leaflets.

compound lens *Optics.* a system in which two or more lenses with the same radius are used together in order to minimize or eliminate aberrations found in a single lens.

compound lever *Mechanical Engineering.* a series of levers for obtaining a large mechanical advantage, the shortest lever being connected to the longer arm of the next in the series; used in large industrial weighing and testing machines.

compound magnet *Electricity.* a group of magnets placed with like poles together so that the combined magnet operates as a single unit.

compound microscope *Optics.* **1.** a microscope that has two lenses or lens systems, one to magnify the image of the object and the other to magnify the first image. **2.** a microscope that consists of an objective and an eyepiece, as opposed to a simple microscope.

compound motor *Electricity.* a DC motor with two field windings; the primary in parallel with the armature circuit and the secondary in series with the armature circuit.

compound nucleus *Nuclear Physics.* the intermediate state of a target nucleus after absorption of an incident energetic particle, but before it decays by emitting the outgoing particle (or particles).

compound number *Mathematics.* a quantity expressed as the sum of two or more quantities of differing units; e.g., degrees, minutes, and seconds, or pounds and ounces.

compound pistil *Botany.* a pistil that is composed of two or more united carpels.

compound protein see CONJUGATED PROTEIN.

compound rest *Mechanical Engineering.* one of the main parts of a lathe, consisting of a base and an upper dovetailed portion that contains a tool post and holder.

compound ripple mark *Geology.* a cross ripple mark produced by the interference of current action with wave oscillation, thus offsetting the crests of the current ripples.

compound screw *Mechanical Devices.* a screw spindle within a nut, having threads of different (or opposite) pitch on each end of its shank to allow for high-pressure die press work. Also, DIFFERENTIAL SCREW.

compound semiconductor *Materials Science.* any of a class of compounds formed from combinations generally consisting of one element having a valence of 3 and one of valence 5, such as the compounds BP, AlSb, GaP, InP, ZnS, CdS, and HgSe.

compound shaft *Mining Engineering.* a shaft in which the upper stage is often vertical, and the lower stages may be inclined and driven into a deposit.

compound shoreline *Geology.* a shoreline whose features show at least two of the following characteristics: a submerged coast, an emerged coast, a neutral coast.

compound statement *Computer Programming.* an executable program statement that contains as parts one or more other statements.

compound tide *Oceanography.* a tidal constituent whose speed is the sum or difference of the speeds of two or more elementary constituents.

compound tubular gland *Anatomy.* a gland composed of multiple tubules, each ending in a blind extremity.

compound twins *Crystallography.* twin crystals whose individuals are united by two or more twin laws.

compound valley glacier *Hydrology.* a glacier composed of two or more ice streams that arise from tributary valleys.

compound volcano *Volcanology.* a volcano that has two or more cones, or an associated volcanic dome in its crater or on its sides.

compound winding *Electricity.* a winding that has properties of both a series and a shunt winding.

compregnate *Engineering.* the result of a process in which material is compacted into a hard mass, using pressure and heat.

compress *Science.* [kəm pres´] to cause to lose volume or to occupy less space. *Medicine.* [käm´ pres] a pad of gauze or other material applied with pressure; it may be medicated, hot or cold, wet or dry.

compressed air [kəm prest´] *Mechanics.* air that is held at a pressure higher (often many times higher) than standard atmospheric pressure, thereby increasing its density. Compressed air is used to inflate automobile tires, to operate power tools, and for various other purposes.

compressed-air blasting *Mining Engineering.* a method for breaking down coal using compressed air. Also, AIR SHOOTING.

compressed-air diving *Engineering.* a diving technique whereby the diver receives air under heavy pressure to ensure against lung collapse.

compressed-air loudspeaker *Acoustical Engineering.* a loudspeaker with an electrically actuated valve that modulates the stream of compressed air passing through it.

compressed-air power *Mechanical Engineering.* the power distributed by the pressure of compressed air as it expands; this power is used in drills, grinders, pile drivers, and locomotives.

compressed liquid see SUBCOOLED LIQUID.

compressed petroleum gas see LIQUEFIED PETROLEUM GAS.

compressibility *Mechanics.* **1.** the extent to which a material reduces its volume when it is subjected to compressive stresses. **2.** see COEFFICIENT OF COMPRESSIBILITY.

compressibility correction *Fluid Mechanics.* fluid compressibility is usually expressed in terms of a correction factor, Z, where Z is often correlated in terms of the reduced pressure and temperature and the value Z_c at the critical state.

compressibility factor *Thermodynamics.* a factor given by the relation $Z = PV/RT$, where P is the pressure, V is the volume, R is the gas constant, and T is the temperature. For 1 mole of an ideal gas, the value of the compressibility factor Z is 1.0, but for real gases, its value can be either larger or smaller than unity. Also, DEVIATION FACTOR.

compressible *Mechanics.* able to be reduced in volume.

compressible flow *Fluid Mechanics.* a flow in which density variations are not negligible.

compressible-flow principle *Fluid Mechanics.* the principle that flow is considered to be compressible if the pressure drop due to gas flow is large enough, compared with the inlet pressure, to cause a 10% or greater decrease in gas density.

compressible fluid flow *Chemical Engineering.* a gas flow in which the pressure drop, due to the flow of gas through a system as compared with the inlet pressure, is large enough to cause a 10% or greater reduction in gas density.

compression *Mechanical Engineering.* **1.** a condition in which the volume of a substance is reduced as a result of pressure changes. **2.** see COMPRESSION RATIO. *Geology.* **1.** the adjustment of the earth's crust, through folding and faulting, to the pressure of overlying sediments or to contraction stress. Also, **compressive settling, compressional movement. 2.** a system of external forces that shortens or decreases the volume of rocks. *Electronics.* a process in which the effective amplification of a signal is varied as a function of the signal magnitude, with the effective gain greater for small rather than for large signals. *Computer Technology.* a reduction of memory requirements or the load on input-output channels by eliminating redundancy in data representations.

compressional wave *Physics.* a wave in which the disturbance is parallel to the direction of propagation, as in the case of a sound wave in air. Also, PRESSURE WAVE, *P*-WAVE.

compression coupling *Mechanical Engineering.* a coupling system in which a slotted, tapered fitting is slipped over the junction of two aligned shafts and two flanges are then slipped over the fitting, thus centering the shafts and providing a pressure-sensitive surface to transmit moderate loads.

compression cup *Engineering.* a cup containing lubricant that contacts a bearing through the force of compression.

compression failure *Engineering.* a collapse or warping of a structure that results from compression.

compression forming *Materials Science.* a method of fabricating polymers: a preformed blank containing a slight excess of the material is placed in a die, the mold is closed, and under heat and pressure the material becomes plastic, while the excess is squeezed out of the mold.

compression gauge *Engineering.* an instrument that measures pressures exceeding atmospheric pressure. *Mechanical Devices.* broadly, any gauge that is used to measure compression; specifically, a gauge that is used to measure the compression ratio of an internal combustion engine in order to identify certain mechanical defects.

compression-ignition engine *Mechanical Engineering.* an internal combustion engine wherein the liquid fuel is ignited by the heat of air compressed in the cylinder into which the fuel is injected; a diesel engine is a type of compression-ignition engine.

compression lip *Materials Science.* a fracture surface feature in ceramics, used in diagnosing the cause of the fracture in question; consists of a bend or curl in the fracture as it passes through the material.

compression member *Engineering.* any component of a structure, such as a beam, that undergoes compressive stress.

compression modulus see BULK MODULUS.

compression mold *Materials Science.* a mold used in plastics manufacturing that lies open for the emplacement of material, which it then shapes by the addition of heat and the pressure of closing.

compression molding *Materials Science.* a method of molding thermosetting compounds and rubber parts: the molding powder is first placed into the open cavity of the heated mold; the mold is then closed and the plastic material is cured or hardened in the closed mold under pressure; a technique now diminishing in application.

compression plant *Petroleum Engineering.* **1.** a gas-compression facility used to supply a high-pressure gas stream to be injected into reservoir formations to augment oil yield. **2.** a pipeline installation for pumping natural gas under pressure from one site to another through a pipeline.

compression pressure *Mechanical Engineering.* in a reciprocating piston engine, the pressure developed at the end of the compression stroke without the combustion of fuel.

compression process *Chemical Engineering.* the recovery of natural gasoline fractions from gas having a high hydrocarbon content.

compression ratio *Mechanical Engineering.* in a cylinder of an internal combustion engine, the ratio of the volume above the piston at the bottom of its compression stroke to the volume above the piston at the top of its stroke. Also, COMPRESSION. *Metallurgy.* in powder metallurgy, the ratio of the volume of loose powder to that of the compact derived from such powder. *Electronics.* the ratio of the gain of a device at its lowest level to its gain at some higher level, commonly expressed in decibels.

compression refrigeration *Mechanical Engineering.* a method of refrigeration in which the refrigerant gas is compressed into liquid form, cooled by heat exchange, and revaporized upon the release of pressure.

compression release *Mechanical Engineering.* the release of compressed gas from an internal combustion engine system due to incomplete closure of an intake or exhaust valve.

compression spring *Engineering.* a spring that exerts pressure against a compressive force.

compression strength see COMPRESSIVE STRENGTH.

compression stroke *Mechanical Engineering.* the regularly occurring phase in a reciprocating engine in which fuel trapped in the cylinder is compressed by piston action.

compression test *Materials Science.* a test used for measuring compression strength, in which specimens are subjected to an increasing compressive force until failure occurs by cracking, buckling, or disintegration. *Mechanical Engineering.* a procedure in which the compression ratio within an internal combustion engine is measured in order to identify certain mechanical defects, such as worn rings. Also, BLOWDOWN.

compression waves see LONGITUDINAL VIBRATION.
compression wood *Botany.* a very dense wood found on the underside of leaning stems of conifers; a form of reaction wood.
compressive intercept receiver *Electronics.* a receiver that rapidly analyzes and sorts all signals within a broad radio-frequency spectrum by rejecting signals that differ from the ones being transmitted.
compressive strain *Mechanics.* a change in the volume of a body that occurs when it is subjected to an external force causing a change in the pressure on the body.
compressive strength *Mechanics.* the maximum compressive stress that a material can withstand without failure. Also, COMPRESSION STRENGTH.
compressive stress *Mechanics.* **1.** an external force that acts on a body to shorten it in the direction of the application of the force (generally accompanied by a certain transverse expansion according to Poisson's ratio); the body's volume is thus reduced but its shape is unchanged. **2.** a measure of this force per unit area of the body.
compressor *Mechanical Engineering.* a reciprocating or rotary pump for raising the pressure of air or another gas; this may be a single-stage or multistage unit. *Electronics.* a transducer that, for a given amplitude range of input voltages, produces a smaller range of output signals at the transmitting or recording end of a circuit.
compressor blade *Mechanical Engineering.* the primary components of centrifugal or axial flow in an air or gas compressor.
compressor station *Mechanical Engineering.* a facility in which the pressure of a gas is raised for the purpose of storing it or transporting it over transmission lines.
compressor valve *Mechanical Engineering.* an automatic valve in a compressor that operates by the pressure difference between two sides of a movable, mechanically independent, single-loaded member.
compromise network *Telecommunications.* in a telephone system, a network used with a hybrid junction to balance a connected communication circuit, such as a subscriber's loop.
compromising emanations *Telecommunications.* a term for radiation or signals that emanate from a communications circuit and that, if intercepted and analyzed, may disclose to unauthorized persons or equipment the classified or proprietary information that is being transmitted in the circuit.
compsilura *Entomology.* the larvae of the fly *Compsilura concinnata,* a caterpillar parasite; successfully introduced into the United States as a way to control gypsy moth infestations of fruit orchards.
Compsognathus *Paleontology.* a small, lightly built saurischian dinosaur in the suborder Theropoda and family Compsognathidae; it was about two feet long, weighed about eight pounds, and evidently was able to move swiftly on two legs; found mainly in lagoonal habitats of the Jurassic.
Compsopogonaceae *Botany.* a family of red algae that makes up the order Compsopogonales, being filamentous to parenchymatous thalli having two spore types.
Compsopogonales *Botany.* a family of red algae of the division Rhodophyta, composed of the single family Compsopogonaceae; widely distributed in tropical and subtropical regions.
Compton, Arthur Holly 1892–1962, American physicist; Nobel Prize for work in radiation including discovery of the Compton effect; also made important studies of cosmic rays.
Compton absorption *Quantum Mechanics.* the part of the absorption of a beam of X-rays or gamma rays associated with Compton scattering; generally greatest for medium-energy quanta and in absorbers of low atomic weight.
Compton cross section *Quantum Mechanics.* the differential cross section for the Compton scattering process.
Compton effect *Quantum Mechanics.* the elastic scattering of a photon by a massive particle (usually an electron) when the interaction is considered a collision of two otherwise free particles; this is true when the energy of the photon is comparable to or higher than the rest energy of the electron. The collision communicates kinetic energy to the electron, and the quantum energy of the photon is reduced. Also, **Compton-Debye effect.**
Compton incoherent scattering *Nuclear Physics.* the dispersion of X-rays or gamma rays by individual nucleons or electrons that takes place when the energy of the photon is strong enough to negate the effects of binding.
comptonization *Quantum Mechanics.* the process that occurs when photons scatter from particles and transfer energy to them by means of the Compton effect.
Compton recoil particle *Quantum Mechanics.* any massive particle that has undergone Compton scattering.
Compton scattering *Quantum Mechanics.* the process in which an energetic photon is scattered by a massive particle (usually an electron), and a photon with less energy emerges with a gain in the kinetic energy of the electron.
Compton shift *Quantum Mechanics.* an increase in wavelength of an emitted photon relative to that of the absorbed photon in Compton scattering.
Compton's rule *Physical Chemistry.* an empirical rule stating that if the heat generated by the fusion of an element is multiplied by the element's atomic weight, then divided by its melting point in degrees Kelvin, the quotient will equal approximately 2.
compulsion *Psychology.* **1.** an irresistible urge to perform some act, especially an irrational act, that is not motivated by the conscious will. **2.** a repetitive or ritualistic action that is engaged in for some consciously unknown reason.
compulsive *Psychology.* **1.** of or relating to compulsion: *compulsive* laughter. **2.** a person who is affected by compulsion.
compulsive behavior see COMPULSION, def. 2.
compulsive disorder *Psychology.* a neurotic disorder in which anxiety caused by internal emotional conflicts is controlled by means of obsessive and repetitive behaviors, such as repeatedly bathing or washing the hands. Also, **compulsive reaction.**
compulsive personality *Psychology.* a personality type characterized by extreme rigidity and conformity and by the tendency to be excessively organized and precise.
compulsory reporting point *Navigation.* any of a series of geographical locations that must be reported to air traffic control as an aircraft passes over them, unless air traffic control informs the aircraft that it is in radar contact.
computability *Computer Programming.* a property of a function, such that there exists an algorithm that can evaluate the function for any given inputs within the allowable domain of the function.
computable *Computer Science.* able to be determined by a computer.
computable function *Computer Programming.* a function for which there exists an algorithm such that, when given a value for the argument, it produces the value of the function.
computation *Mathematics.* **1.** the act, process, or method of calculating, especially by numerical methods. **2.** the result so obtained.
computational chemistry *Chemistry.* a branch of chemistry that uses computers to investigate chemical structures, properties, and reactions, including both numerical calculations and computer-generated graphic depictions of molecules.
computational complexity *Computer Programming.* the intrinsic difficulty of mathematical problems with respect to either the processing time or the storage space required by the algorithm.
computational numerical control see COMPUTER NUMERICAL CONTROL.
compute *Computer Science.* to carry out a computer operation or determine a value by means of a computer.
compute-bound see CPU-BOUND.
computed *Computer Science.* of a subexpression, having its value computed within a given block of code.
computed altitude *Navigation.* the angular distance of the center of a celestial body above the horizon at a given time and place, as determined by some means other than observation.
computed azimuth *Navigation.* the angular distance between north and some celestial point at a given time and place, as determined by some means other than observation.
computed azimuth angle *Navigation.* the angular distance east or west from a north or south reference direction at a given time and place, as determined by some means other than observation.
computed go-to *Computer Programming.* a conditional branch statement in which the next statement to be executed is selected from a list based on logic contained in the statement.
computed-path control *Robotics.* a robotic control process that calculates the optimal path of a robot's endpoint in order to achieve a desired result.
computed tomography see COMPUTER-ASSISTED TOMOGRAPHY.
compute mode *Computer Technology.* one of the ordinary control modes of an analog computer that produces the time-variant solution based on the input; other ordinary control modes are reset and hold. In addition, an analog computer may operate in repetititive and iterative operation control modes.

Computer Technology

Computer technology encompasses computer hardware, software, and communications. Computer hardware includes semiconductors and chips that process and store information, hardware for storage (disk and tape drives), printers and plotters, display and interactive components (CRT's, keyboards, light pens, a mouse). Types of software are operating systems (which control the operation of the hardware and software), software compilers (languages), database management systems, and application software that performs specific business or technical functions. Communications includes the equipment and software to support transmission, translation of information, routing, and communication code conversion. Computer technology is also employed in noncomputer devices such as automobiles and appliances. Computer technology has made these devices smarter, programmable, and feature rich.

Computer technology began around the turn of the century with the telephone and tabulating equipment. World War II saw the development of advanced secret military code translators and processors—some of the first computers. Computer technology advanced slowly in the 1940s and 1950s. A series of breakthroughs in transistors, integrated circuits, and large-scale integration as well as parallel advances in storage and other areas occurred in the 1960s. The decade of the 1970s was dominated by the large mainframe and minicomputers. The 1980s was the decade of microcomputers and the widespread use of computer technology in the home and office.

Computer technology continues to undergo rapid change and improvement in performance and reduced cost. Computer systems in terms of hardware double in power every three years for the same cost, and consume less space and power. The growth in power and performance of hardware has made possible and feasible more complex and capable software. Software made possible by the technology include artificial intelligence, expert systems, executive information systems, and decision support systems. Future applications will include sophisticated voice and handwriting recognition and processing (which exist today in a primitive form).

Bennet P. Lientz
Professor of Computer Technology
University of California, Los Angeles

computer *Computer Technology.* any of a variety of electronic devices that are capable of accepting data and instructions, executing the instructions to process the data, and presenting the results. A computer typically has a central processing unit (CPU), internal and external memory storage, and various input/outout devices such as a keyboard, display screen, and printer. A computer system consists of hardware (the physical components of the system) and software (the programs used by the computer to carry out its operations).

computer-aided design *Design Engineering.* an engineering tool using interactive graphics that computes design parameters and displays the design for review prior to fabrication.

computer-aided engineering *Industrial Engineering.* the use of computers to provide automatic control of an engineering operation.

computer-aided manufacturing *Industrial Engineering.* the integration of numerical control computers that provide instructions to automatic machines with other operations in manufacturing.

computer-aided production planning *Industrial Engineering.* the use of computers to create a process plan for making a specific part.

computer-aided software engineering *Computer Programming.* the systematic use of a variety of computer-based tools to improve software development and documentation; includes planning, design, implementation, and management tools.

computer analyst *Computer Science.* a person, expert in computers and information systems, who investigates a problem, derives a computer-based solution, and prepares functional and system specifications so that the solution may be implemented by programmers.

computer animation *Computer Programming.* the use of computer graphics techniques to produce animated motion pictures with three-dimensional realism and perspective.

computer architecture *Computer Technology.* the large-scale structure and hardware design of a digital computer system, including the characteristics of each component and the means by which they are interconnected.

computer art *Computer Programming.* the use of computer graphics to produce artistic images, as distinguished from the presentation of information in graphic form.

computer-assisted instruction *Artificial Intelligence.* the application of computers to present drills, exercises, and tutorial dialogue with students, sometimes in the form of an interactive dialogue.

computer-assisted tomography *Radiology.* a roentgenologic examination using a narrow radiation beam revolving about the patient, aimed at a predetermined axis, in which measured amounts of exit radiation are translated by computer into segmental images of body tissue. Also, **computer-aided tomography.**

computer code *Computer Programming.* **1.** any of a variety of binary representations of information. **2.** a set of basic pseudocommands built into the computer logic circuitry that defines the basic instruction set of a microprogrammable computer. Also, MICRO CODE.

computer composition *Graphic Arts.* any of a variety of typesetting systems in which a keyboard operator types only characters, after which typographic decisions such as hyphenation are made by a computer that produces the typeset copy.

computer control *Control Systems.* a method of control in which the variables of a process are manipulated by a computer in order to control the process.

computer control counter *Computer Technology.* a special-purpose incrementing register that holds the address of the next instruction to be fetched and executed.

computer-controlled system *Control Systems.* a control system in which a computer manipulates both the input to the system and the system's feedback.

computer crime *Computer Programming.* the misuse of a computer to commit an offense such as embezzlement or other fraud.

computer efficiency *Computer Technology.* a measure of the productive use of a computer, often the ratio of actual processing time to elapsed time.

computer graphics *Computer Technology.* **1.** the input, processing, and output of pictorial representations of objects and information. **2.** any equipment used in such graphical representation, including digitizers, scanners, and pattern recognition devices that provide input, as well as plotters, displays, laser printers, and film recorders that record output.

computer-integrated manufacturing *Industrial Engineering.* the automation and coordination by computers of all the different phases of a manufacturing process, including engineering, production, marketing, and support.

computer integration *Design Engineering.* the theoretical basis for automated production with minimal human assistance as applied to an automatic factory and manufacturing system.

computerized axial tomography see CAT SCAN.

computerized branch exchange *Telecommunications.* an automatic switchboard that, in addition to the regular routing of calls, provides extra services, such as facsimile.

computerized transverse axial tomography *Radiology.* an alternate term for computer-assisted tomography, or CAT scan.

computer literacy *Computer Science.* the knowledge and ability to make use of computers in processing data and solving problems.

computer logic *Computer Technology.* the basic principles, design, and circuitry used to implement a particular computer architecture so that it will perform a particular set of machine-code instructions.

computer-managed instruction *Computer Programming.* the use of a computer to provide course management assistance to an instructor or counselor; includes record keeping, test development and scoring, grading, and instructional material searches.

Computer Programming and Computer Science

A computer program is a representation of an algorithm in some well-defined language. Algorithms are abstract computational procedures for transforming information; programs are their concrete embodiments.

The world's first programmer was the poet Lord Byron's daughter, A. Ada Lovelace, who formulated precise instructions for the calculation of trigonometric functions and Bernoulli numbers on Charles Babbage's unfinished Analytical Engine in 1843. The total number of people who now consider programming to be part of their profession has risen to more than 5 million in the United States alone, and there are perhaps 50 million people worldwide who regularly write programs of one kind or another.

The best programs are written so that computing machines can perform them quickly and so that human beings can understand them clearly. A programmer is ideally an essayist who works with traditional aesthetic and literary forms as well as mathematical concepts, to communicate the way an algorithm works and to convince a reader that the results will be correct. Programs often need to be modified, because requirements and equipment change; programs often need to be combined with other programs. Success at these endeavors is directly linked to the effectiveness of a programmer's expository skills.

Many subtle techniques are known by which programs can be made to run considerably faster than would be possible with a naive approach. The quantitative theory of program efficiency is often called the analysis of algorithms. This field of study has many important subfields, including numerical analysis (the study of algorithms for scientific computation); complexity theory (the study of the best possible ways to solve given problems using given hardware); symbolic computation (the study of algorithms for manipulating algebraic formulas); computational geometry (the study of algorithms that deal with lines, surface, and volumes); combinatorial optimization (the study of algorithms for selecting the best of many possible alternatives); information retrieval and database theory (the study of algorithms for storing and retrieving large collections of facts); and the study of data structures (techniques of representing the relationships between discrete items of information).

Computer programming and the analysis of algorithms are, in turn, subfields of a considerably larger discipline called computer science, which deals with all of the complex phenomena surrounding computers. Computer science is known as "informatics" in French, German, and several other languages, but American researchers have been reluctant to embrace that term because it seems to place undue emphasis on the stuff that computers manipulate rather than on the processes of manipulation themselves.

Computer science answers the question "what can be automated?" Its principal subfields, besides the analysis of algorithms, presently include software engineering (the study of languages for programming, operating systems for controlling computer resources, and utility programs tailored to significant applications like accounting or desktop publishing, as well as issues of programming methodology); graphics and visualization (the development of tools for analysis and synthesis of images); computer architecture and communication (the design of machines and of networks to connect them); artificial intelligence (the development of tools for accumulating, applying, and reasoning about knowledge); human machine interaction (the study of interfaces between people and computers); robotics (the development of mobile machines with sensors); and interdisciplinary connections with virtually every other branch of science, technology, medicine, and the humanities.

Donald E. Knuth
Professor of the Art of Computer Programming
Stanford University

computer modeling *Computer Programming.* the process of providing to a computer, usually in the form of mathematical equations, a precise and unambiguous description of the system under study, including the relationships between system inputs and outputs, and using this description to simulate or model the described system.

computer network *Computer Technology.* **1.** a network connecting geographically or locally distributed computers. **2.** a resource-sharing network linking computers, file servers, and peripheral devices.

computer networking *Telecommunications.* a process by which two or more computing units are interconnected by a complex communications network, thus allowing them to communicate and work together.

computer numerical control *Control Systems.* a control system that uses a computer to generate numerical values for desired paths and tool positions. Also, COMPUTATIONAL NUMERICAL CONTROL, SOFTWIRED NUMERICAL CONTROL, STORED-PROGRAM NUMERICAL CONTROL.

computer operation *Computer Technology.* **1.** an electronic response by a computer's circuitry that occurs as a result of the interpretation of one of the computer's basic operation codes. **2.** a basic operation built into the electronic circuitry of a computer, such as add or complement.

computer-oriented language *Computer Programming.* a lower-level programming language that is developed by a computer manufacturer for use on a specific computer; sometimes an assembly language.

computer part programming *Control Systems.* the use of a computer to program a numerical control system.

computer performance evaluation *Computer Technology.* an analysis of a computer's responsiveness, throughput, and cost for the purpose of selecting a system design or improving an existing system; often uses models to represent the system and the expected workload.

computer programming *Computer Science.* the process of devising and implementing instructions that enable a computer to perform various desired operations.

computer science a branch of study that is concerned with theoretical and applied disciplines in the development and application of computers for information storage and processing, mathematics, logic, and many other areas.

computer security *Computer Technology.* a term for a set of physical, personnel, data, operational, program, and hardware safeguards and procedures designed to prevent unauthorized use of a computer or stored information.

computer system *Computer Technology.* a complete operational computer configuration including the hardware subsystem with its central processor, memory units, and input/output devices as well as the software subsystem with its programs and data structures.

computer systems architecture *Computer Technology.* a description of a total computer system configuration including its physical structure and functional behavior, the attributes of the system components, and the relationships among components.

computer technology *Computer Science.* the study and application of computer hardware and software. See COMPUTER.

computer theory *Computer Technology.* the analytic study of computer logic design, compiler design, file structures, and other basic components of computer systems.

computer utility *Computer Technology.* a commercial computer facility that provides processing to remote users, usually by means of a telephone line; considered analogous to a utility company that supplies water or electricity.

computer vision *Computer Technology.* a class of problems relating to the interpretation of images, to which computerized pattern recognition and classification techniques have been applied.
compute time see EXECUTION TIME, def. 2.
computing gunsight *Ordnance.* a gunsight with an electrical or mechanical device for calculating the angle between the line of sight to the target and the desired line of departure for the projectile.
computing power *Computer Technology.* the maximum number of operations that a specific computer can perform in a unit of time, usually measured in operations per second.
computing unit *Computer Technology.* the arithmetic-logic unit of a central processing unit (CPU).
Comstar *Space Technology.* a communications satellite that is equipped for relaying radio, telegraph, telephone, and television signals from various earth stations to other earth stations.

Comstar

Comstock refraction formula *Astrophysics.* a formula used in positional astronomy to correct for atmospheric refraction.
con *Navigation.* **1.** to control the steering of a vessel. **2.** the control so exercised, or the person who has this control. Also, **conn.**
con- a prefix meaning "with" or "together," as in *concatenation.*
conacaste *Materials.* the valuable wood of the *Enterolobium cyclocarpum* tree of the American tropics. Also, GUANA CASTE.
conalbumin *Biochemistry.* a glucoprotein present in egg whites; the noncrystalline part of egg albumin.
Conant, James Bryant 1893–1978, American chemist and educator; studied chlorophyll and hemoglobin; adviser on Manhattan Project; influential in U.S. science and mathematics education.
conarium see PINEAL BODY.
conc. concentrate; concentrated; concentration.
concanavalin A *Biochemistry.* a lectin that binds to glycoproteins with α-glucoside or α-mannoside groups; it also functions as a mitogen and agglutinates a variety of cell types; found in *Canavalin ensiformis*, or jack beans.
concatemer *Molecular Biology.* a number of DNA molecules or sequences of bases that are covalently linked in series.
concatenate *Computer Programming.* to link together or copy to adjacent positions a number of conceptually similar items, such as strings of characters. Also, CATENATE.
concatenation *Computer Programming.* the process of linking together a number of conceptually related items to form a larger, organizationally similar entity. Also, CATENATION.
concatenation of languages *Computer Science.* a language consisting of the set of sentences formed by concatenating a sentence from the first language and a sentence from the second language.
concave *Science.* curving, rounded, or hollowed inward; the opposite of convex.
concave bank *Geology.* on a curved stream, the outer bank, whose center curves toward the channel.
concave bit *Mechanical Devices.* a tungsten bit carbide drill with a concave profile for impact boring operations.
concave cross-bedding *Geology.* **1.** cross-bedding having downward arching foreset beds. **2.** cross-bedding that has been deposited on a lower, concave surface.
concave function *Mathematics.* a function f (not necessarily differentiable) is concave upward (or convex) on an interval (a,b) if $f((x_1{+}x_2)/2) \le (1/2)(f(x_1) + f(x_1))$ for all $x_1, x_2 \in (a,b)$. If the inequality is reversed, the function is said to be concave downward on (a,b). If the function is twice differentiable, then it is concave upward (resp. downward) at a point x if $f''(x)$ is > 0 (resp. < 0).
concave grating *Spectroscopy.* a reflection grating consisting of a concave mirror surface ruled with a series of narrow, closely spaced parallel grooves and used in a spectroscope to focus a given wavelength of incident light to a point. Also, ROWLAND GRATING.
concave joint *Building Engineering.* an exposed, smooth mortar joint in masonry; used for exterior walls to dispel water and reduce moisture penetration.
concave lens see DIVERGING LENS.
concave polygon *Mathematics.* **1.** a polygon that has at least one interior angle greater than 180°. **2.** a polygon with the property that there exists a straight line passing through its interior and cutting it on four or more sides.
concave spherical mirror *Optics.* a mirror that has a hollow, curved surface resembling a portion of a sphere.
concavo-concave *Science.* having two sides, both curving inward.
concavo-convex *Science.* having two sides, one curving inward and one curving outward. Also, CONVEXO-CONCAVE.
concentrate to make a substance denser, purer, or stronger; specific uses include: *Chemistry.* to increase the proportion of a dissolved substance in a solution by evaporating the solvent or by adding more of the dissolved substance. *Food Technology.* **1.** to carry out the process of concentration. Thus, **concentrated. 2.** a partially dehydrated food product. *Mining Engineering.* **1.** to separate ore or metal from its containing rock or earth. **2.** the clean product recovered in froth flotation.
concentrated fire *Ordnance.* **1.** fire from the batteries of two or more ships, directed at a single target. **2.** fire from a group of weapons directed at a single point or small area.
concentrated flow see CHANNEL FLOW.
concentrated force *Mechanics.* a force that is theoretically applied to a point or line on a body; thus, a force whose area of contact with a surface is assumed to be zero. Also, **concentrated load.**
concentration the act of making or the state of being denser, stronger, or purer; specific uses include: *Chemistry.* **1.** the process of increasing the amount of a dissolved substance in a solution by evaporating the solvent. **2.** the amount of a particular substance in a given quantity of a mixture, solution, or ore; usually stated as a percentage by weight or volume, as weight per unit volume, as molarity (a molar solution contains one gram-mole of solute per liter of solution), or as normality (a normal solution contains one gram equivalent weight of solute per liter of solution). *Hydrology.* at a specific location or over a defined area of the sea, the ratio of surface area that is covered by sea ice to the total area of the sea surface. *Food Technology.* a method of removing water from foods to make them easier to store or carry; water can be added later to bring the product to a usable state. *Mining Engineering.* the process of separating and accumulating ore from its surrounding rock or earth.
concentration cell *Physical Chemistry.* **1.** a cell in which the energy supplied by a current arises from the differences in solute concentrations at the anode and the cathode. **2.** a cell in which there is a change in energy arising from the transference of solutes from one concentration to another without a chemical reaction. **3.** an element in a potentiometer that measures the concentration of ions at a given electrode by comparing the energy supplied by a current in a known cell and an unknown cell.
concentration-dilution test *Pathology.* a method of gauging the efficiency of renal function by placing a patient under specific conditions to first concentrate and then dilute the urine, and then comparing the output with normal values.
concentration gradient *Chemistry.* the variation of concentration of a dissolved substance with distance in a solution.
concentration polarization *Physical Chemistry.* a change in the cell potential due to finite rates of mass transfer to the electrodes.
concentration scale *Physical Chemistry.* any of various numerical systems that are used to define the proportions of the constituents of a mixture.
concentration time *Hydrology.* **1.** the time required for water to flow from the farthest portion of a river basin or other watershed to the basin outlet or gauging station. **2.** in a storm, the time when the rate of runoff is equal to the rate of precipitation.

concentrator *Engineering.* **1.** a device that concentrates power into a generator. **2.** a device used to consolidate materials, or a plant where materials are consolidated. *Electronics.* **1.** an adapter system, circuit, or device that reduces a large number of inputs to a smaller number of outlets. **2.** a unit in a computer system that connects the signals from several low-speed data terminals, or other sources, into a single high-speed transmission line. **3.** see DATA CONCENTRATOR.

concentric *Science.* having a common center or axis; extending out equally in all directions from a common center. Also, **concentrical.**

concentric bundle *Botany.* any vascular bundle in which either the xylem surrounds the phloem or the phloem surrounds the xylem.

concentric cable see COAXIAL CABLE.

concentric circles *Mathematics.* circles of differing radii, lying in the same plane and having the same center.

concentric faults *Geology.* a system of faults having a common center or axis.

concentric fold *Geology.* a fold in which the parallel rock layers have not changed in thickness during deformation. Also, PARALLEL FOLD.

concentric fractures *Geology.* a system of fractures having a common center or axis.

concentric-jaw chuck *Mechanical Devices.* a small chuck whose jaws move about a center equally; used for lathes and screwing machines. Also, SELF-CENTERING CHUCK.

concentric lens *Optics.* a lens that has two spherical surfaces with the same center of curvature.

concentric locating *Design Engineering.* the process of making the axis of a tooling device coincide with the axis of the workpiece.

concentric orifice plate *Mechanical Devices.* a fluid-metering plate cross-sectionally centered in a pipe.

concentric resonator *Optics.* a type of beam resonator in which two symmetrical mirrors having the same axis of symmetry are situated so that their centers of curvature coincide on this axis.

concentric slip ring *Electricity.* a sizable slip-ring assembly composed of concentrically spaced insulators and conducting materials; used for multiple electrical connections to a rotating element.

concentric tube column *Chemical Engineering.* a carefully insulated distillation device having a high separating power, in which the outer vapor-rising annulus of the column is concentric around an inner, bottom-discharging reflux return.

concentric windings *Electricity.* transformer windings in which the low-voltage winding is in the shape of a cylinder next to the core, and the high-voltage winding, which is also cylindrical, encloses the low-voltage winding.

concept *Psychology.* **1.** a general idea that is produced by combining several separate elements into a single entity. **2.** a mental image of an idea. Also, **conception**.

conceptacle *Botany.* in brown algae, a flask-shaped cavity having an outward-opening pore in which sporangia develop.

conception *Biology.* the fertilization and implantation of an ovum, resulting in pregnancy.

concept learning *Artificial Intelligence.* a form of learning in which example data sets and classifications of the examples are given, and in which the goal of learning is to form a description of the concept implied by the examples.

conceptual dependency *Artificial Intelligence.* a theory in which the meanings of natural language sentences are represented using a small number of primitive acts that are related to other primitive acts and phrases of the sentence by deep case relations.

conceptual modeling *Computer Programming.* a technique of representing a mental image of an action or object in the form of mathematical equations and logical relationships.

conceptual schema *Computer Programming.* the logical structure of a database as a whole, as contrasted with its storage and physical schemata.

conceptus *Biology.* a fetus or embryo at any stage between conception and birth; the product of a conception.

concerted cycloaddition *Organic Chemistry.* the formation of a ring molecule by combination of two unsaturated molecules in a single step without reactive intermediates.

concerted evolution *Genetics.* the joint evolution of two or more related genes, as though constituting a single locus.

concerted reaction *Chemistry.* a reaction involving processes that simultaneously break existing bonds and form new bonds.

concertina wire *Ordnance.* a coiled barbed wire that is compressed for easier handling and extended to form a barrier for the enemy.

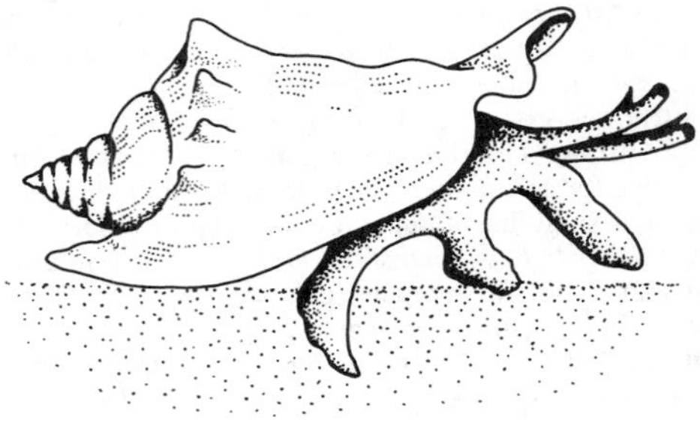

conch

conch [kängk; känch] *Invertebrate Zoology.* the common name for any of several spe- cies of gastropod mollusks in the family Strombidae, having large colorful shells.

conche or **conch** *Food Technology.* **1.** a heated mixing tank that is used to blend and knead chocolate into a smooth paste. Also, CONGE. **2.** to carry out the conching process. **3.** chocolate that has gone through this process.

conching *Food Technology.* a process whereby chocolate is blended and kneaded in a conche, then rolled, ground, and aerated. Conching improves the chocolate's viscosity, smoothness, and flavor.

conchiolin *Biochemistry.* a horny scleroprotein constituting the organic basis of mollusk shells; isomeric with ossein and closely related to keratin. Also, CONCHYOLIN.

conchoid [käng´koid´] *Mathematics.* a conchoid of a curve C, with respect to a pole point P and a line segment S of constant length in a line L, is a plane curve whose points consist of the locus of both ends of the line segment S as L rotates about P and while the midpoint of S remains on C. Conchoids of lines and of a circle through the origin are curves of order four and are called conchoids of Nicomedes and limaçons of Pascal, respectively.

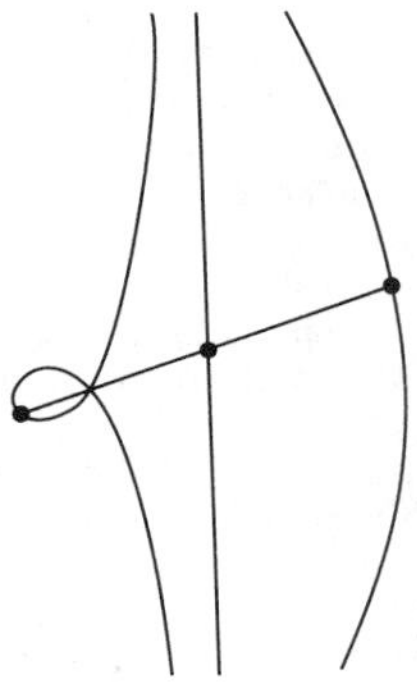

conchoid

conchoidal [käng´koid´əl] *Geology.* describing a smooth, curved, shell-shaped fracture in a glassy rock or mineral.

conchoid of Nicomedes *Mathematics.* a conchoid of the line $x = a$, with pole at the origin, given by the equation $(x - a)^2(x^2 + y^2) - b^2x^2 = 0$, where $a, b > 0$ and $a + b$ is the length of the rotating line segment. The conchoid consists of a two-part curve, asymptotic to the line $x = a$ on both sides of the line. If $b > a$, the left-hand branch of the curve forms a loop with a node at the pole. If $b \leq a$, the left-hand branch forms a cusp at the pole. Also, **conchoid of a line.**

conchology [käng´käl´ə jē] *Invertebrate Zoology.* the study of the shells of mollusks.

Conchorhagae *Invertebrate Zoology.* a suborder of tiny, deep-water wormlike animals in the class Kinorhyncha.

Conchostraca *Invertebrate Zoology.* an order of crustaceans in the subclass Branchiopoda, whose hinged carapace makes them look like mussels.

conchyolin see CONCHIOLIN.

concluded angle *Cartography.* the third angle of a triangle laid out in the process of triangulation; not a measured angle but one calculated from the other two angles.

concn. concentrate; concentrated; concentration.

concomitant *Mathematics.* let a group G act on the sets E and F. In invariant theory, a function u from E to F is a concomitant of G (i.e., of these actions of G) if it is invariant under the induced action of G; that is, if $\sigma{\cdot}(u\,(\sigma^{-1}{\cdot}x)) = u(x)$ for each $x \in E$. Also, COVARIANT, EQUIVARIANT.

concordance *Genetics.* the exhibition of similar or identical properties by genes or organisms, as in the occurrence of a particular trait in both members of a set of twins.

concordant *Geology.* describing an intrusive igneous structure having boundaries that are parallel to the foliation or bedding planes of the country rock. Thus, **concordant body, concordant injection, concordant pluton.**

concordant bedding *Geology.* an arrangement of sedimentary rocks characterized by parallel beds without angular junctions. Also, PARALLEL BEDDING.

concordant coastline *Geology.* a coastline lying generally parallel to the land structure forming the margin of an ocean basin. Also, LONGITUDINAL COASTLINE.

Concorde [kän´kôrd´] *Aviation.* a British-French supersonic aircraft designed for commercial use.

concrescence *Biology.* the union of originally separate structures by the growth of a unifying tissue beneath them; especially the convergence and fusion of the lips of a blastopore during embryogenesis.

concrete *Materials.* **1.** a hard, strong substance that is composed of cement and an aggregate such as sand and gravel which has been mixed with water and allowed to dry and harden; widely used as a building material. **2.** composed of or relating to such a material. Thus, **concrete block, concrete finish, concrete masonry.**

concrete admixture *Materials.* an addition to wet concrete that affects properties such as curing time and flow characteristics.

concrete buggy *Engineering.* a cart used to transport concrete from the mixer to the forms. Also, **concrete cart.**

concrete chute *Engineering.* an inclined metal trough through which concrete moves downward.

concrete equivalent *Radiology.* an equivalent thickness of concrete of a standard density that would shield radiation to the same degree as a material being evaluated.

concrete mixer *Mechanical Engineering.* a machine in which concrete-forming elements in fluid form are mixed mechanically.

concrete operational stage or **period** *Psychology.* the third of Jean Piaget's four stages of cognitive development, from about 7 to 12 years of age, when a child can apply logical thinking to concrete objects and events, but still lacks the ability for certain kinds of abstract or hypothetical thinking.

concrete pump *Mechanical Engineering.* a reciprocating pump that drives fluid concrete to its intended position through a pipe, generally six inches or greater in diameter.

concrete steel *Metallurgy.* steel, usually in the form of rods, used to reinforce precast concrete.

concrete vibrator *Mechanical Engineering.* a device that uses vibration to achieve a proper formation of concrete.

concretion *Geology.* a hard, dense aggregate of mineral matter formed within the pores of a sedimentary or fragmental volcanic rock by precipitation from solution, and differing in composition from the surrounding material. *Pathology.* **1.** a calculus or inorganic mass occurring in a natural cavity or in the tissues of an organism. **2.** the hardening or solidification of such a calculus or mass. **3.** an abnormal union of adjacent parts.

concretionary *Geology.* characterized by, consisting of, or tending to produce concretions.

concretioning *Geology.* the process by which rocks produce concretions, usually occurring during diagenesis or shortly after sediment deposition.

concretor *Building Engineering.* a tradesman who is skilled in spreading and leveling concrete when fabricating slabs for foundations, runways, or the like.

concurrency *Computer Technology.* a situation in which one or more operations are executing simultaneously within a computer or among multiple processors.

concurrency transparency *Computer Programming.* a situation in which multiple transactions are executed concurrently, but the environment of each transaction appears as though it were the only one being executed.

concurrent computing see CONCURRENT PROCESSING.

concurrent conversion *Computer Technology.* the transfer of stored data from one medium to another by means of a utility program while an application program continues processing.

concurrent heating *Metallurgy.* in cutting or welding, using a principal and a supplementary source of heat.

concurrent infection *Medicine.* a condition in which two or more types of infection occur simultaneously.

concurrent input/output *Computer Technology.* the reading of data from an input device or writing of data to an output device concurrently with program execution.

concurrent line *Cartography.* a line that passes through points on a map or chart having the same current hour.

concurrent operations control *Computer Programming.* the ability of an operating system to process multiple jobs at the same time, including the management of peripherals and memory.

concurrent processing *Computer Technology.* the real or logically simultaneous execution of two or more programs in the same computer system. Also, CONCURRENT COMPUTING.

concurrent programming *Computer Programming.* the development of computer programs that are specifically designed for parallel execution of several tasks or parts of the same task; used in systems with multiple processors executing simultaneously.

concurrent real-time processing *Computer Technology.* the mechanism used to control parallel execution of several tasks, some of which have real-time response requirements.

concussion *Medicine.* a violent jar or shock, or the condition that results from such an injury, especially a **brain concussion,** the result of a shocking blow to, or violent shaking of, the head. Effects of brain concussion range from temporary to prolonged loss of consciousness, with possible impairment of mental or physical functions.

concussion fracture *Geology.* one of a system of fractures in a shock-metamorphosed rock that radiate from the grain surface.

concussion fuse *Ordnance.* a bomb fuse that is set off in the air by the explosion of a previous bomb.

concussion table *Mining Engineering.* an inclined table that is agitated by a series of shocks, and that operates like a buddle. Also, PERCUSSION TABLE.

cond. condensed; condition; conductivity.

condensable vapor *Chemistry.* a gas or vapor that becomes liquid when temperature and/or pressure are changed appropriately.

condensate *Science.* a product of condensation.

condensate field *Geology.* an oil field producing both natural gas and gasoline that condenses from the gas.

condensate flash *Chemical Engineering.* the partial evaporation or flash of hot condensed liquid by decreasing the system pressure in several steps.

condensate strainer *Mechanical Engineering.* in a boiler system, a screen that removes solid particles from the steam condensate before pumping it back to the boiler.

condensate well *Mechanical Engineering.* a water-vapor collection well. *Petroleum Engineering.* a well that yields a natural gas saturated with condensable hydrocarbons of greater weight than ethane or methane.

condensation the act of making more dense or compact; specific uses include: *Meteorology.* a process by which water vapor changes to dew, fog, or cloud, brought about either by the cooling of air to its dew point or the addition of enough water vapor to bring the mixture to the point of saturation. *Chemistry.* **1.** the transformation of a gas into a liquid or solid. **2.** a reaction of two or more organic chemicals, one of the products of which is water, ammonia, or a simple alcohol. *Physics.* the region of maximum density through which compression waves (such as sound waves) travel. *Acoustics.* a measure of the instantaneous density in a medium through which a sound wave is passing, given by the expression $(r - r_o)/r$ where r is the density and r_o is the mean density at a given point in the medium. *Acoustical Engineering.* the use of electrostatic force to convert acoustic energy to electrical energy, or vice versa, whereby the capacitive charge on two electrodes separated by air is increased by sound energy, for a device such as a microphone. *Optics.* the process of focusing or collimating a beam of light. *Psychology.* a process by which one idea or image comes to represent a group of different elements.

condensation cloud *Meteorology.* a condensation mist or fog that temporarily surrounds a fireball after a volcanic explosion or the detonation of an atomic bomb.

condensation nucleus *Meteorology.* a liquid or solid particle in the atmosphere, upon which condensation of water vapor begins.

condensation polymer *Organic Chemistry.* a polymer formed by a condensation polymerization.

condensation polymerization *Organic Chemistry.* the process of forming polymers by a condensation reaction between monomers, usually with the removal of water molecules.

condensation pressure *Meteorology.* the pressure at which a parcel of moist air expanded dry adiabatically reaches saturation. Also, ADIABATIC SATURATION PRESSURE.

condensation reaction *Chemistry.* a chemical reaction in which molecules or parts of the same molecule combine, with the elimination of water or another small molecule in the process.

condensation resin *Organic Chemistry.* a resin formed by the processes of condensation polymerization.

condensation temperature *Analytical Chemistry.* the temperature at which a thin moving film of liquid coexists with the vapor from which the liquid has been condensed. *Meteorology.* the temperature at which a parcel of moist air expanded dry adiabatically reaches saturation. Also, ADIABATIC SATURATION TEMPERATURE.

condensation trail *Aviation.* a cloudlike streamer that is frequently observed behind aircraft flying in clear, cold, humid air; the moving aircraft disturbs particles of ice or water vapor, causing a pressure reduction above the wing surfaces which, when combined with water vapor in the engine exhaust gases, tends to condense, leaving a visible trail of condensed water vapor. Also, CONTRAIL, VAPOR TRAIL.

condense *Science.* to make or become more dense, more compact, or smaller; concentrate.

condensed *Science.* relating to or produced by condensation. *Graphic Arts.* describing type that is narrow in proportion to its height, especially a variation of a given typeface in which the characters are set closer together than normal; e.g., Helvetica Condensed.

condensed milk *Food Technology.* milk that is concentrated, preserved with sugar (usually sucrose), and canned.

condensed succession *Geology.* a stratigraphic sequence in which the deposits accumulated very slowly, producing a relatively thin but uninterrupted succession.

condensed system *Physical Chemistry.* a fluid system in which there is negligible vapor pressure or in which the pressure on the system greatly exceeds the vapor pressure.

condensed type *Graphic Arts.* see CONDENSED.

condenser any device or system that condenses; specific uses include: *Mechanical Engineering.* a chamber enclosing an array of tubes into which the exhaust steam from a steam engine is distributed and condensed by the circulation of cooling water through the tubes. *Electricity.* see CAPACITOR. *Optics.* a device in which a lens or series of lenses collects light from one object and projects it evenly onto another object, as in a microscope or a slide projector.

condenser box see CAPACITOR BOX.

condenser bushing see CAPACITOR BUSHING.

condenser-discharge anemometer *Engineering.* an instrument that measures the average wind velocity and operates in connection with an electrical circuit.

condenser microphone see CAPACITOR MICROPHONE.

condenser transducer see CAPACITOR TRANSDUCER.

condenser tubes *Mechanical Engineering.* the tubes through which the cooling water is circulated in a surface condenser and on whose outer surfaces the steam is condensed.

condensing capacitor see BALANCING CAPACITOR.

condensing engine *Mechanical Engineering.* a steam engine in which the steam exhaust liquefies in the vacuum space following discharge from the engine cylinder.

condensing enzyme *Enzymology.* an enzyme that catalyzes the formation of citrate from acetyl CoA and oxaloacetate.

condensing flow *Fluid Mechanics.* the flow and simultaneous condensation of vapor through a cooled pipe or other closed container.

condensing gas drive *Petroleum Engineering.* a process in which reservoir oil is displaced by gas when components of hydrocarbon of the injected gas condense in the oil that it displaces.

condis crystal *Materials Science.* a polymer mesophase in which cooperative motion between various conformational isomers occurs. (An acronym for conformationally disordered crystal.)

condition *Behavior.* to subject an organism to a particular stimulus so that it will respond in a desired (often modified) manner to subsequent stimulation; to establish the basis for a conditioned response. *Mathematics.* a mathematical premise, statement, restriction, or modifying qualification upon which a mathematical result or consequence depends. A condition is necessary if it is a logical consequence of a given statement; it is sufficient if a given statement follows as a logical consequence.

conditional *Behavior.* relating to or produced by conditioning. *Computer Programming.* of an operation, subject to a certain circumstance such as the setting of a switch or a logical comparison made during the execution of a program statement.

conditional assembly *Computer Programming.* a capability provided with some assemblers to selectively assemble or not assemble portions of a program.

conditional branching *Robotics.* a change of instruction that occurs when a previously defined condition is met.

conditional breakpoint *Computer Programming.* a position in a program at which normal execution is interrupted if a specified condition has been met, such as the violation of a programmer-asserted relationship within the program.

conditional convergence *Mathematics.* the property that an infinite series $\sum_{n=0}^{\infty} a_n$ converges but is not absolutely convergent; i.e., the series $\sum_{n=0}^{\infty} |a_n|$ diverges. Riemann's theorem states that the terms of any conditionally convergent series can be rearranged in such a way that the resulting series converges to an arbitrary given number. Convergent series whose sums are independent of the order of terms are said to be **unconditionally convergent**.

conditional distribution *Statistics.* the probability distribution of a random variable X given that a random variable Y describes the probabilities of various values of X under the additional information of the value assumed by Y; can be extended to groups of random variables.

conditional expectation *Statistics.* the expectation of a random variable X given the additional information of the value assumed by a second random variable Y; usually denoted by $E\{X|Y\}$.

conditional expression *Computer Programming.* an expression containing variables, literals, and logical operators that returns the value "true" or "false."

conditional instability *Meteorology.* the state of an air column when its lapse rate of temperature is less than the dry-adiabatic lapse rate but greater than the saturation-adiabatic lapse rate.

conditional jump *Computer Programming.* an executable instruction that transfers execution control depending on the value of a boolean or conditional expression. Also, **conditional branch, conditional goto, conditional transfer.**

conditionally compact set *Mathematics.* a set whose closure is compact. Also, RELATIVELY COMPACT SET.

conditionally periodic motion *Mechanics.* the motion of a system whose generalized coordinates undergo simple harmonic motion with differing frequencies that are not all rational fractions of each other, so that the complete motion of the system is not simply periodic. Also, PSEUDOPERIODIC MOTION.

conditionally stable circuit *Electronics.* a circuit or system that is stable when operating with specific values of input signal gain, but unstable for other values.

conditional positive regard *Psychology.* respect or affection for another person that is dependent on the individual's behavior. Also, **conditional love.**

conditional probability *Statistics.* the probability of the occurrence of an event B, given that an event A has occurred; usually indicated by $P(B/A)$. Conditional probabilities can be viewed as a special case of conditional expectations.

conditional response or **reflex** see CONDITIONED RESPONSE.

conditional statement *Computer Programming.* an executable statement that specifies alternative actions to be taken based on the outcome of the logic contained in the statement. Also, IF STATEMENT. *Mathematics.* see IMPLICATION.

conditional stimulus see CONDITIONED STIMULUS.

conditional stop instruction *Computer Programming.* a program statement that causes execution to stop if a specified condition is met, such as excessive processing time or a console switch setting.

conditional-sum *Computer Technology.* in parallel adders, the use of additional logic circuitry to reduce the total delay; related techniques include carry-lookahead and carry-complete signals.

condition code *Computer Programming.* a bit or set of bits that identifies specific conditions associated with the result of an arithmetic or logic operation.

conditioned *Behavior.* produced by or dependent on the conditioning of the individual; acquired.

conditioned lethal mutation *Genetics.* a mutation that is lethal to an organism only under certain environmental conditions (restrictive conditions) but that is not lethal under another set of conditions (permissive conditions).

conditioned reinforcer *Behavior.* a neutral stimulus that derives its reinforcing qualities from a previous conditioning experience in which it was paired with a primary reward. Thus, **conditioned reinforcement.**

conditioned response *Behavior.* in classical conditioning, a response that occurs when a neutral stimulus is presented immediately before and repeatedly with an unconditioned stimulus; eventually the neutral stimulus by itself will elicit the same response as the unconditioned stimulus. Also, **conditioned reflex.**

conditioned stimulus *Behavior.* in classical conditioning, a stimulus that evokes a response that was previously elicited by an unconditioned stimulus.

conditioned taste aversion *Behavior.* the tendency to avoid a certain food or taste because it has been paired with a negative stimulus.

condition entries *Computer Programming.* the upper right-hand portion of a decision table, containing all alternate values of the conditions listed in the condition stub.

conditioning *Behavior.* **1.** any method for teaching a specific behavior by making it contingent upon certain conditions. **2.** see CLASSICAL CONDITIONING. **3.** see INSTRUMENTAL CONDITIONING. *Engineering.* the process of performing design or installation changes to make equipment compatible for use with other equipment or to bring it in line with specific standards. *Electronics.* the process of modifying or adjusting communication transmission equipment to bring attenuation, impedance, and delay characteristics to within set limits.

condition key *Computer Programming.* in an end-user database language, a two-dimensional display into which the user may enter one or more search criteria to be applied in a query.

condition portion *Computer Programming.* the upper half of a decision table, containing a list of all the conditions to be included (the condition stubs) and all values of each condition (the condition entries).

condition stub *Computer Programming.* the upper left portion of a decision table, containing a list of all the conditions that will be considered in deciding the action to take.

condom *Medicine.* a flexible sheath or cover for the penis, worn during coitus to prevent impregnation or transmission of disease.

Condon, Edward Uhler 1902–1974, American physicist; worked on the Manhattan Project; directed Air Force study of UFOs.

Condon-Shortley-Wigner phase convention *Quantum Mechanics.* the convention that the phase of an angular momentum eigenfunction be specified such that the matrix elements of the angular momentum ladder operators J_+ and J_- are real.

condor *Vertebrate Zoology.* the largest of the American vultures, belonging to the family Cathartidae in the order Falconiformes, characterized by a bare head and neck, dull-black plumage, and a white neck ruff.

condor

Condor *Aviation.* a popular name for the B-2 bomber. *Ordnance.* an air-to-surface antiship guided missile with an optoelectronic guidance system that allows it to function beyond the range of antiaircraft guns; officially designated **AGM-53.**

Condorcet, Marquis de 1743–1794, French mathematician, philosopher, and revolutionary; treatises on integral calculus and probability.

conductance *Thermodynamics.* the heat transfer through a material divided by the temperature difference across the surfaces of the body. *Electricity.* **1.** the physical property of an element, device, branch, or system that is the factor by which the mean square voltage must be multiplied to determine the corresponding power lost by dissipation. **2.** the real part of the admittance; in a circuit with no reactance, equal to the reciprocal of the resistance; expressed in siemens and designated as G.

conduct disorder *Psychology.* a personality disorder that is identified in children who manifest severe and persistent antisocial behavior; typical patterns include stealing, lying, truancy, vandalism, and substance abuse.

conducted interference *Telecommunications.* **1.** interference resulting from radio noise or unwanted signals that enter a device by direct coupling. **2.** undesired voltage or current that is generated within a receiver, transmitter, or associated equipment and appears at the antenna terminals.

conducting glass *Materials Science.* glass that has been treated by any of various processes to increase its electrical conductivity.

conduction *Physics.* the passage of energy (heat, sound, or electricity) through a medium (the conductor) while the medium itself experiences no mass movement as a whole.

conduction band *Solid-State Physics.* an energy band in a solid in which electrons are freely mobile and can produce a net electric current.

conduction cooling *Electronics.* a process of removing undesirable heat to increase power-handling capability by mounting a device against a heatsink that transfers heat to the environment.

conduction current *Solid-State Physics.* the electric current produced by conduction electrons in motion through a solid.

conduction electron *Solid-State Physics.* an electron whose energy lies in the conduction band of a solid and is thus free to move under the influence of an electric field. Also, VALENCE ELECTRON.

conduction field *Electromagnetism.* a scalar energy field associated with a conductor through which a current is passing.

conduction pump *Engineering.* a pump that provides for the flow of molten metal or other conductive liquid through a pipe by transmitting a current across the liquid and providing a magnetic field perpendicular to the current flow.

conductive *Materials Science.* having the property of conducting heat or an electric current. Thus, **conductive elastomer, conductive rubber.**

conductive coating *Materials Science.* any coating that inhibits the accumulation of static electricity on a surface by reducing surface resistance to allow an electric current to flow through it.

conductive coupling *Electricity.* an electrical connection of two circuits by a common resistor.

conductive deafness *Medicine.* deafness resulting from impairment of the sound-conducting apparatus, the auditory canal or middle ear. Sensitivity to sound is impaired, but clarity is not changed; when sound volume is increased, the hearing is normal.

conductive equilibrium see ISOTHERMAL EQUILIBRIUM.

conductive gasket *Electricity.* a highly resilient gasket that reduces radio-frequency leakage at joints in shielding.

conductive interference *Electronics.* the unwanted extraneous power from natural or artificial sources that interferes with reception of desired signals; introduced into equipment by a power supply or power transformer.

conductive paint *Metallurgy.* a nonmetallic coating containing a dispersion of particles, such as silver or copper, that impart a reasonable electrical conductivity.

conductive paste *Medicine.* a paste or gel that lowers resistance to either electrical or sound impulses; applied to the skin for certain therapeutic and diagnostic procedures such as ultrasonic probes.

conductivity *Electricity.* a measure of the ability of a material to conduct an electrical current, equal to the reciprocal of resistivity: $s=1/r$.

conductivity cell *Electricity.* a glass vessel containing two electrodes that are set at a known distance apart and filled with a solution whose conductivity is to be determined.

conductivity modulation *Electronics.* a variation in the conductivity of a semiconductor as the density of the charge carriers within the semiconductor varies.

conductivity modulation transistor *Electronics.* a transistor whose active properties are derived from conductivity modulation caused by variations of the density of the minority carriers within the transistor.

conductivity tensor *Electricity.* a tensor that yields the current density vector when multiplied by the electric field vector.

conductivity theory *Physics.* a theory that treats the system of electrons in a metal as a gas, and uses a kinetic equation to calculate the conductivity. Also, CLASSICAL CONDUCTIVITY THEORY.

conductometer *Engineering.* an instrument that measures the rate at which materials transmit heat.

conductometric titration *Analytical Chemistry.* a titration during which the electrical conductivity of the solution is measured to determine the endpoint.

conductometry [kän′duk täm′ə trē] *Chemistry.* the measurement of the conductivity of a solution. Thus, **conductometric.** Also, **conductimetry.**

conductor *Electricity.* a material that is capable of carrying electric current, especially one that is highly suitable for this, such as copper wire. *Physics.* any material that serves or can serve as a medium for conduction, as of sound, heat, and so on. *Surgery.* a probe or sound with a groove along which a knife is passed in slitting open a sinus or fistula; a grooved director.

conductor pipe *Building Engineering.* a pipe fabricated so that rain water effectively drains off the roof of a building, usually of metal content. *Petroleum Engineering.* a short string of large-diameter casing that acts to keep the top of a well bore open and to carry drilling fluid from the well bore to the slush pit.

conduit [kän′doo it; kän′dit; kän′dwit] *Geology.* an underground channel that is completely filled with water, and is always under hydrostatic pressure. *Engineering.* any pipe through which materials may pass. *Building Engineering.* specifically, a solid or flexible tubing that houses electric wires.

conduplicate *Botany.* folded together lengthwise and in half.

condylar articulation *Anatomy.* a modified ball-and-socket joint in which a shallow socket holds the rounded end of a bone, contacting very little of the lateral surface of the "ball."

condylarth *Paleontology.* of or relating to the ancestral hoofed mammals, or ungulates, of the order Condylarthra, which arose in the late Cretaceous and disappeared in the Miocene; the condylarths appear to be transitional between insectivores and true ungulates.

Condylarthra *Paleontology.* an order of generally plantigrade mammals that were mainly herbivorous; ancestral to the perissodactyls and cetaceans; very numerous in the Paleocene and Eocene, the last of the condylarths became extinct by the end of the Oligocene.

condyle [kän′dīl] *Anatomy.* a rounded projection at the end of a bone. *Invertebrate Zoology.* a rounded process in the articulation of an arthropod appendage.

condyloid [kän′də loid] *Anatomy.* relating to or resembling a condyle or a knuckle.

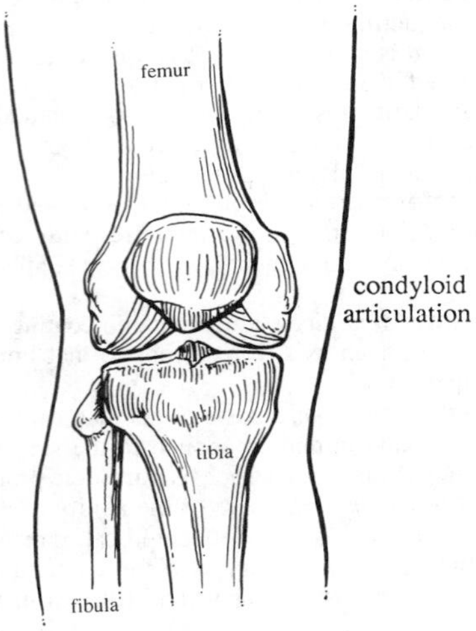

condyloid articulation

condyloma acuminatum *Medicine.* an infectious wartlike growth on the genitals, probably of viral origin. Also, VENEREAL WART.

cone *Mathematics.* **1.** a solid bounded by one section of a conical surface and a plane intersecting the section. **2.** the surface of such a solid. **3.** a closed surface consisting of a base that is bounded by a simple closed curve and a surface consisting of all line segments joining a fixed point (vertex) with the points on the boundary of the base. *Geology.* **1.** any landform having relatively steep slopes that converge to a point at the top. **2.** see VOLCANIC CONE. *Acoustical Engineering.* the cone-shaped diaphragm of a loudspeaker, made of stiff felted paper or of some similar light, durable material such as plastic. Thus, **cone loudspeaker.** *Metallurgy.* the conical part of a gas flame, near the tip orifice. *Textiles.* a cone-shaped bobbin around which yarn is spun. *Botany.* a densely packed detachable structure bearing a plant's sporangia, sporophyles, or flowers. *Histology.* a visual receptor of the vertebrate retina that can distinguish different wavelengths of the visual spectrum and is especially sensitive to bright light. *Radiology.* a cylindrical apparatus attached to an X-ray tube, used to center a radiation beam on a target field and to restrict the beam size, eliminating scattered radiation.

cone antenna see CONICAL ANTENNA.

cone biopsy *Medicine.* the excision of a cone-shaped piece of tissue from the cervix for diagnostic purposes.

cone brake *Mechanical Engineering.* a friction brake with cone-shaped rubbing parts. Similarly, **cone clutch.**

cone crater

cone crater *Geology.* **1.** a rimmed, basinlike structure at the peak of a volcanic cone, usually formed by collapse. **2.** a similar structure at the summit of a lunar cone, thought to be formed by volcanic activity.

cone crusher *Mechanical Engineering.* a device that reduces the size of materials by crushing them in the tapered space between a revolving truncated cone and an outer chamber.

cone dike see CONE SHEET.

conehead rivet *Mechanical Devices.* a rivet having a head that is shaped like a truncated cone.

cone-in-cone structure *Geology.* a sedimentary, coal, or volcanic structure characterized by the development of a series of interpenetrating cones, one within another.

cone karst *Geology.* in a tropical region, a karst characterized by a pattern of steep-walled depressions adjacent to conical hills. Also, COCKPIT KARST.

cone key *Mechanical Devices.* a tapered fitting that is placed on a shaft in order to retain a pulley wheel when its hole is larger than the portion of the shaft upon which it is keyed.

Conelrad [kän′əl rad′] *Telecommunications.* a government broadcasting system, now discontinued, that was designed to provide communications and control in the event of nuclear attack. (An acronym for Con-trol of electromagnetic radiation.)

cone mandrel *Mechanical Devices.* a centering mandrel used to center workpieces by clamping two cones with a shoulder to steady one cone and locking threads to hold the other.

Conemaughian *Geology.* a North American geologic stage of the Upper Middle Pennsylvanian.

cone of depression *Hydrology.* an inverted cone-shaped depression in the water table that develops around a well as water is withdrawn from it. Also, **cone of influence.**

cone of dispersion *Ordnance.* **1.** a cone-shaped pattern formed by a group of shots fired from a weapon with the same sight setting. Also, **cone of fire. 2.** the pattern formed by a group of pellets spreading out after the opening of a canister.

cone of escape *Geophysics.* a theoretical, cone-shaped region of the exosphere in which an atom or molecule can escape into space without collision.

cone of exhaustion *Hydrology.* a cone-shaped depression in the water table that develops around a well from which water is being withdrawn faster than it can percolate laterally.

cone of force *Archaeology.* a conical area on a stone core and its associated flake, resulting when force is applied to separate the flake.

cone of nulls *Electromagnetism.* a conical surface that is constructed from the directions of negligible radiation for an antenna.
cone of revolution *Mathematics.* the cone obtained by rotating a line about an axis intersected by the line, maintaining a constant angle of intersection. A cone of revolution is equivalent to the cone whose base curve, when taken to be in the plane perpendicular to the axis at a distance of one unit from the vertex, is a circle whose radius equals tan θ, where θ is the angle formed by the rotating line and the axis.
cone of silence *Navigation.* a cone-shaped space above a radio beacon in which the intensity of transmitted signals is sharply reduced.
cone of vision *Meteorology.* an imaginary conical surface with its apex at the eye of a given observer, and its solid angle exactly filled by the object being viewed.
cone rock bit *Mechanical Engineering.* **1.** a rotary drill with two hardened, knurled cones that cut the rock as they roll. **2.** see ROLLER BIT.
cone-roof tank *Engineering.* a tank used to store liquids, having a flattened cone-shaped top that provides a space for vapors to collect during filling operations.
cone settler *Mining Engineering.* a conical vessel fed centrally with fine ore pulp, in which the apex discharge carries the larger-sized particles, and the peripheral top overflow carries the finer fraction of the solids.
cone sheet *Geology.* one of a set of concentric curved dikes that dips or converges inward, toward the magmatic center. Also, CONE DIKE.
cone shell *Invertebrate Zoology.* the common name for any member of the mollusk family Conidae, having brightly colored, cone-shaped shells up to 4 inches long.
Conewangoan *Geology.* see BRADFORDIAN.
confabulation *Psychology.* the process of filling in memory gaps by fabricating information and details; done either consciously or, as in the case of amnesia, unconsciously.
confamilial *Systematics.* **1.** belonging to the same family. **2.** an individual belonging to the same family as another.
confection *Food Technology.* any sweetened food product, especially candy.
confectioner's coating *Food Technology.* an imitation-chocolate paste, used in place of real chocolate to coat candy, cookies, and ice-cream bars because of its lower cost and higher melting point .
confectionery *Food Technology.* **1.** a collective term for candy products. **2.** the production of candy products.
conference call *Telecommunications.* a simultaneous telephone link connecting three or more callers.
conference communications *Telecommunications.* any simultaneous transmission and reception of messages among three or more users.
Conference on Data Systems Languages see CODASYL.
confidence interval *Statistics.* a random interval having the property, before the data are observed, of covering the parameter value with at least the probability of the confidence level; as opposed to the credible interval. Also, INTERVAL ESTIMATE.
confidence level *Statistics.* the sampling probability that a confidence interval will enclose the parameter of interest before the data are observed. Also, **confidence coefficient.**
confidence limits *Statistics.* the upper and lower limits of a confidence interval.
CONFIG.SYS *Computer Technology.* a MS/DOS system configuration file that provides the operating system with certain user-specified commands, such as the number of buffers, date and time display format, and number of disk drives.
configuration the arrangement or distribution of parts of something; specific uses include: *Mechanics.* the position of all the particles of a system. *Mathematics.* a combination or arrangement of geometrical objects, such as points, lines, surfaces, and curves. *Chemistry.* the spatial arrangement of parts or elements in a molecule. *Electricity.* the arrangement of components that are interconnected to perform desired circuit functions. *Computer Technology.* the relationship of the hardware components of a computer system with each other, together with the electronic interconnectivities. *Aviation.* **1.** a certain type of flight vehicle distinguishable from others of the same model by the arrangement of its component parts or by the addition or absence of auxiliary equipment. **2.** the relative arrangement, shape, or size of key elements in a flight vehicle, such as control surfaces, wings, and engines. **3.** the external form that a flight vehicle takes as a result of the way in which the pilot has set the controls.
configurational free energy *Physics.* the unbounded energy, often associated with neighboring atoms and fields, found in a solid lattice.
configurational isomer *Chemistry.* any of a group of isomers that differ from one another in the spatial arrangement of their atoms about a chiral center. Thus, **configurational isomerism.**
configuration interaction *Physical Chemistry.* the portion of the total electronic interaction energy not described by antisymmetrized one-electron obitals.
configuration space *Mechanics.* the space spanned by a set of generalized coodinates that define the configuration of a system, with every point reperesenting a certain state of the system. *Quantum Mechanics.* the description of quantum systems in terms of spatial coordinates; wavefunctions in configuration space are Fourier transforms of those in momentum space.
confined aquifer see ARTESIAN AQUIFER.
confined charge *Ordnance.* an explosive charge loaded in a resistant container.
confined explosion *Ordnance.* an explosion occurring in a closed chamber.
confined flow *Engineering.* the passage of a liquid through a pipe or continuous container.
confined groundwater or **confined water** see ARTESIAN WATER.
confinement *Engineering.* **1.** the process of physically limiting an explosion or its effects. **2.** the degree of such a limitation.
confinement of plasma *Physics.* the maintenance of plasma in a specific location at a specific volume, often through the use of magnetic mirrors and toroidal fields.
confining bed *Geology.* a bed of relatively impermeable material that is stratigraphically adjacent to one or more aquifers.
confining liquid *Chemical Engineering.* a liquid seal that is displaced during the no-loss transfer of a gas sample from one container to another.
confining pressure *Geology.* an application of force exerted equally on all sides, such as hydrostatic pressure produced by the weight of water in the zone of saturation.
confirmation well *Petroleum Engineering.* a well drilled to prove the producing zone found by an exploratory or wildcat well; the second producing well in a new, developmental oil field.
conflict alert *Navigation.* a function of automated air traffic control systems that alerts the controllers to certain situations that require immediate attention.
conflict behavior *Behavior.* a type of behavior that results when two incompatible but equal responses are aroused by the same situation.
conflict resolution *Anthropology.* the solutions utilized by a society to settle disputes in a cohesive manner; for example, attempts to contain the damage to a group by certain destructive actions such as blood feuds. *Artificial Intelligence.* the selection of a procedure or rule from a set of conflicting procedures or rules.
conflict set *Artificial Intelligence.* in a forward-chained production system, the set of potential rule firings, that is, the set of all rules whose antecedents are satisfied by some combination of working memory elements together with the bindings of variables to those elements.
confluence *Hydrology.* **1.** the meeting or flowing together of two or more streams of glaciers. **2.** the place where such a junction occurs, or the body of water produced by this junction. *Psychology.* the fusion of several motives, responses, or instincts into one.
confluent *Hydrology.* running or flowing together.
confluent hypergeometric function *Mathematics.* a solution to the confluent hypergeometric differential equation (Kummer's equation): $zw'' + (\rho - z)w' - \alpha w = 0$. Also, KUMMER'S FUNCTION.
confocal *Mathematics.* having the same focus or foci.
confocal conics *Mathematics.* two or more conics having the same foci (and axes of symmetry); for example, a system of hyperbolas and ellipses, or a system of parabolas.
confocal coordinates *Mathematics.* coordinates of a point in the complex plane with norm greater than some constant α, in terms of the (confocal) system of ellipses and hyperbolas that have foci $(+\alpha, 0)$ and $(-\alpha, 0)$; usually $\alpha = 1$.
confocal resonator *Electromagnetism.* a wavemeter consisting of two spherical mirrors whose separation distance is adjustable; capable of measuring waves having wavelengths on the order of millimeters.
confocal scanning light microscopy *Biotechnology.* an advanced form of light microscopy in which a laser beam-illuminated subject is rapidly scanned, resulting in excellent linear resolution.
conformable *Geology.* describing a continuous sequence of rock layers formed in parallel order without interruption, or the area of contact between such layers.

conformable optical mask *Electronics.* a flexible glass substrate used as an optical mask in integrated circuit fabrication; under vacuum, the optical mask is forced into full contact with the integrated circuit substrate.

conformal array *Electronics.* an array of antennas that are electronically switched and normally used for production of special radiation beam patterns required for air navigation, radar, and missile applications.

conformal chart *Cartography.* any chart based on a conformal map projection.

conformal curvature tensor see WEYL TENSOR.

conformal group *Mathematics.* for a given (pseudo-) Riemannian manifold M, the group of transformations of the manifold that preserve angles. The conformal group for M contains the group of isometries of M.

conformality *Geodesy.* the property of transforming information from one surface to another while ensuring that an angle between any two curves on one surface is preserved in magnitude and sense by the angle between the corresponding curves on the other surface.

conformal mapping *Mathematics.* the mapping given by $w = f(z)$ is said to be conformal at $z \in D$, where D is the domain of f, if f is analytic at z and if $f'(z) \neq 0$. If two curves intersect at z, the angle of intersection at z is preserved under f. In particular, every pair of orthogonal lines is transformed into a pair of orthogonal curves under a conformal mapping.

conformal map projection *Cartography.* a map projection on which the shape of any small area of the surface map is preserved unchanged, and all angles around any point are correctly represented. Also, ORTHOMORPHIC MAP PROJECTION.

conformal metrics *Mathematics.* on a given (pseudo-) Riemannian manifold M, two Riemannian metrics g_1 and g_2 are said to be conformal if a function f exists such that $g_2(u,v) = f(x)g_1(u,v)$ for all tangent vectors u and v at all points $x \in M$. The angle between u and v is the same in either metric.

conformal reflection chart *Electromagnetism.* a mapping in the complex plane that represents the complex reflection coefficient for a waveguide junction and its image.

conformance test *Electrical Engineering.* a test performed on a product either to show conformance to certain standards or to demonstrate selected performance characteristics.

conformation *Chemistry.* the shape of a molecule, produced by the specific spatial arrangement of the units that compose it. *Crystallography.* such a representation of the shape of a molecule in which the different rotational positions are represented by torsion angles if there is a possibility for rotation about bonds.

conformational analysis *Physical Chemistry.* the study of the typical or most likely conformation of a molecule in its ground, transition, and excited states.

conformational isomers *Chemistry.* isomers of the same compound whose specific spatial arrangements correspond to energy minima for that molecule.

conformist *Psychology.* a person who tends to behave in a manner that is conventional and in agreement with prevailing social rules, codes, and customs.

conformity *Geology.* the relationships among successive sedimentary rock layers that have been deposited in orderly sequence without evidence of time lapse, erosion, or other geological disturbances. *Psychology.* the fact of being a conformist; conventional behavior.

confounded *Statistics.* of or related to the sources in an experiment that is not designed so that the effect of two or more competing sources of variation in the response can be distinguished.

confused sea *Oceanography.* a term for a disturbed water surface with no clear direction of wave travel.

confusion (of radio sources) *Astrophysics.* the combination of radio signals from two or more unresolved celestial sources; such confusion places a limit on how faint an object can be detected with a given telescope and detector.

confusion jamming *Electronics.* a military electronic countermeasure in which an incident enemy radar signal is amplified, distorted, and retransmitted, creating a false echo that will present confusing range, azimuth, and target velocity data to the operator. Also, DECEPTION JAMMING.

confusion matrix *Computer Programming.* in pattern recognition, an array in which the elements represent a statistical measure of the error in assigning classes to observed patterns.

confusion reflector *Ordnance.* see CHAFF.

conge see CONCHE.

congelifluction *Geology.* the slow, progressive lateral and downslope movement of earth materials under periglacial conditions, or of waterlogged soil and saturated surface material over frozen ground. Also, GELIFLUCTION, GELISOLIFLUCTION.

congelifract *Geology.* any angular fragment of rock broken off by frost action. Also, GELIFRACT.

congelifraction *Geology.* the mechanical fragmentation of rock or soil under pressure from water freezing within cracks or pores or along bedding planes. Also, FROST SHATTERING, FROST SPLITTING, FROST WEATHERING, FROST WEDGING, GELIFRACTION.

congeliturbate *Geology.* any unconsolidated mass of soil or earth material that has been disturbed by frost action. Also, FROST SOIL, CRYOTURBATE.

congeliturbation *Geology.* any disturbance or displacement of soil that results from the action of frost. Also, CRYOTURBATION, FROST STIRRING, GELITURBATION, FROST CHURNING.

congener *Biology.* an organism that is a member of the same genus as another animal or plant. *Chemistry.* a chemical substance that is related in some way to another, for example, as a derivative or as a member of the same group in the periodic table.

congeneric *Systematics.* **1.** belonging to the same genus. **2.** an individual belonging to the same genus as another.

congenic strain *Genetics.* a genetic variety of a species that differs from other varieties in only a small, restricted portion of its genome.

congenital [kən jen′ə təl] *Medicine.* referring to a condition or feature that is existing at, and usually before, birth.

congenital agammaglobulinemia *Medicine.* a congenital deficiency of immunoglobulins, increasing susceptibility to bacterial infection.

congenital anomaly *Medicine.* a developmental abnormality of function or structure present at the time of birth.

congenital disease *Medicine.* a disease present at birth.

congenital disorder *Genetics.* a disorder that is present at birth; may be caused by heredity or related to an error in fetal development.

congenital pathology *Medicine.* the study of diseases and abnormalities of function and structure present at birth.

congenital syphilis *Medicine.* syphilis that is acquired in utero and is manifested by any of several characteristic malformations, by neurologic changes, and by active mucocutaneous syphilis at the time of birth or shortly afterward.

conger eel *Vertebrate Zoology.* a large, scaleless, carnivorous eel of the family Congridae, found in all oceans and known as a challenging gamefish; used for food.

congestin *Biochemistry.* a toxic substance found in the tentacles of sea anemones. Also, ACTINOGESTIN.

congestion *Medicine.* the abnormal accumulation of fluid, usually blood, in a part of the body.

congestive *Medicine.* referring to or associated with abnormal accumulation of blood in a part.

congestive heart failure *Medicine.* an abnormal condition of circulatory congestion caused by heart disease, characterized by breathlessness and abnormal sodium and water retention.

conglomerate *Geology.* sedimentary rock composed of rounded fragments of rocks or minerals embedded in a fine-grained matrix of sand, silt, or cementing material.

conglomerate

conglutinant [kən gloo´ti nənt] *Surgery.* encouraging or promoting the union of the edges of a wound; adhesive.

conglutination [kən gloo´ti nā´shən] *Medicine.* the abnormal adherence of two surfaces or structures within the body. *Immunology.* a process in which a substance is added to an antibody and complement to enable them to complete, or to enhance, their agglutination process.

conglutinin *Immunology.* a protein that is present in bovine serum, and that aggregates complement-bearing immune complexes.

Congo *Geography.* a long (2760 miles) river of central Africa, flowing through Zaire and northwestern Angola to the Atlantic. Also, ZAIRE RIVER.

Congo red *Organic Chemistry.* $C_{32}H_{22}N_6Na_2O_6S_2$, a brownish-red powder, soluble in water and alcohol; used as a histological stain and a pH indicator (pH 3.0, blue; pH 5.0, red).

Congo red test *Pathology.* the intravenous injection of Congo red dye to test for the presence of amyloidosis, which is indicated if more than 60% of the dye has disappeared after one hour.

congruence *Mathematics.* **1.** the property of two geometric configurations that can be made to coincide with each other by rigid transformations. **2.** an equivalence relation ~ that satisfies the following for all a_1, a_2, b_1, b_2 on which ~ is defined: $a_1 \sim a_2$ and $b_1 \sim b_2$ imply $a_1b_1 \sim a_2b_2$. Also, **congruence relation.** The equivalence classes formed by ~ are sometimes called **congruence classes.** *Psychology.* **1.** a condition in which an individual's emotions or attitudes are consistent with the situation or experience that produced them. **2.** see CONGRUITY.

congruent *Mathematics.* **1.** of two or more numbers, having a congruence relation. **2.** of two or more figures, coinciding at all points when superimposed. Thus, **congruent triangles.** *Psychology.* relating to or showing congruence.

congruential generator *Computer Programming.* a common type of random-number generator, which produces a sequence of remainders of a large modulus by means of a linear transformation.

congruent matrices *Mathematics.* two matrices A and B are congruent if $B = TAT^T$, where T^T is the transpose of T and T is invertible.

congruent melting *Geology.* the direct transformation of a solid to a liquid having the identical chemical composition.

congruent melting point *Thermodynamics.* a point on a phase diagram for a nonstoichiometric compound in which two phases of the compound share the same melting point.

congruent numbers *Mathematics.* two integers a and b are said to be congruent modulo m if they yield the same remainder when divided by an integer m, that is, if $a - b$ is divisible by m; denoted $a \equiv b(\text{mod}/m)$.

congruent transformation *Metallurgy.* a solid-state transformation that does not result in compositional changes of the phases involved.

congruity *Psychology.* a situation in which a person's self-image is consistent with the ideal self that he or she would like to be.

Coniacian *Geology.* a European geologic stage of the Upper Cretaceous period, occurring after the Turonian and before the Santonian. Also, EMSCHERIAN.

conic [kän´ik; kōn´ik] *Mathematics.* the graph of a second-order polynomial equation in two variables. A conic may be represented as the intersection of a plane with a bipartite cone. Also, CONIC SECTION.

conical [kän´i kəl; kōn´i kəl] having the shape of a cone. Also, CONOID.

conical antenna *Electromagnetism.* a wide-band antenna whose radiating element is shaped like a cone.

conical beam *Electronics.* a cone-shaped radar beam pattern, usually produced by conical scanning.

conical buoy *Navigation.* a buoy in the shape of a truncated cone with the narrow end up; it is normally painted red and used to mark the starboard side of a channel as seen by a vessel entering from the open sea.

conical helimagnet *Solid-State Physics.* a helimagnet in which all atomic magnetic moments in a particular basal plane are aligned in a direction that forms an acute angle with a specified axis of the crystal, and the magnetic moments of successive basal planes are separated by a constant azimuthal angle.

conical horn *Acoustics.* a cone-shaped tube in which enclosed air vibrates to produce the harmonics of a fundamental frequency, which can be explained by the equation $f_n = n(10.05/L)(T_A)^{1/2}$, with f_n representing the harmonics, n representing the harmonic number, L the length of the horn, and $(T_A)^{1/2}$ the reciprocal of the fundamental.

conical horn antenna *Electromagnetism.* a horn antenna whose cross section is circular.

conical pendulum *Mechanics.* a weight attached to a cord or light rod, which is made to execute circular motion; the cord motions thus form the surface of a circular cone.

conical refiner *Mechanical Engineering.* a cone-shaped refiner with two sets of bars mounted on a rotating plug; used in paper processing.

conical refraction *Optics.* refraction in which light rays striking the surface of a biaxial crystal at a given angle split to form a cone pattern.

conical scanning *Electronics.* a sequential radar scanning in which the direction of maximum radiation creates a cone whose axis coincides with that of the radar reflector; used in missile scanning.

conical surface *Mathematics.* a surface consisting of all lines passing through a fixed point (called the vertex or apex) and an intersecting simple closed curve (called the directrix). It has two distinct sections that intersect at the vertex and are centrally symmetric at the vertex.

conichalcite *Mineralogy.* $CaCu^{+2}(AsO_4)(OH)$, a subtranslucent green orthorhombic mineral occurring as botryoidal fibrous crusts, having a specific gravity of 4.33 and a hardness of 4.5 on the Mohs scale; found as a secondary mineral in oxidized zones of copper deposits.

Coniconchia *Paleontology.* a proposed class of animals (possibly mollusks) to include the hyolithids and tentaculites; they were siphonless and had an elongate calcareous univalve, generally cone-shaped; Cambrian to Permian.

conic projection *Cartography.* a projection developed by projecting the geographic parallels and meridians onto a cone that is tangent to or intersects the sphere, and then developing the cone into a plane.

conic projection with two standard parallels *Cartography.* a conic map projection in which the surface of a sphere or spheroid, such as the earth, is conceived as developed on a cone that intersects the sphere or spheroid along two standard parallels. Also, SECANT CONIC MAP PROJECTION.

conic section see CONIC.

Conidae *Invertebrate Zoology.* a family of carnivorous mollusks in the order Neogastropoda, including the cone shells.

conidial head *Mycology.* the end part of the spore-bearing hypha in certain fungi, such as *Aspergillum.*

conidiation *Mycology.* the process by which conidia join together as if they were sexual cells or gametes.

conidioma *Mycology.* fungal tissue that bears conidia.

conidiophore *Mycology.* in fungi, a certain type of hypha that bears conidia.

conidium *Mycology. plural,* **conidia.** an asexual spore produced in certain fungi. Also, **conidiospore.**

conifer [kän´i fər; kōn´ i fər] *Botany.* any of numerous cone-bearing trees of the order Pinales, chiefly evergreen trees of the class Coniferinae including the pine, fir, and spruce, having simple, needlelike leaves.

Coniferales see PINALES.

Coniferophyta see PINACAE.

coniferous [kō nif´ər əs; kə nif´ ər əs] *Botany.* of or relating to the conifers.

coniferous forest *Ecology.* a woodland that is composed primarily of conifers, such as the Lake Superior forest of the U.S. and Canada.

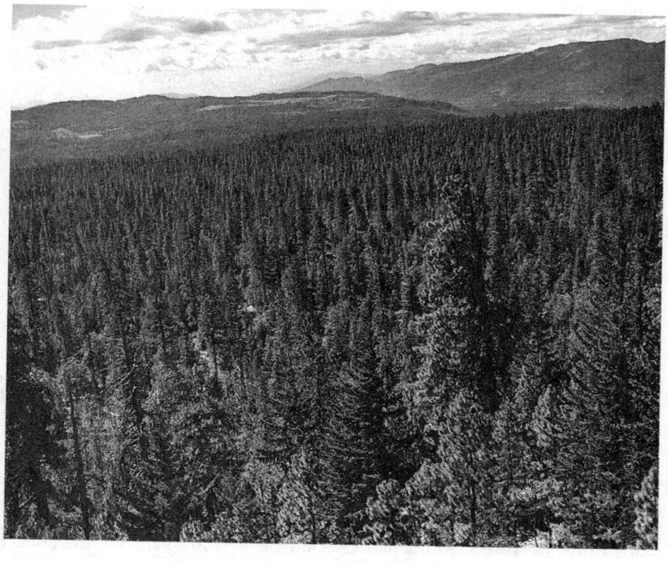

coniferous forest

coniine *Pharmacology.* $C_8H_{17}N$, a highly poisonous liquid alkaloid prepared from hemlock.

coniine hydrobromide *Pharmacology.* $C_8H_{18}BrN$, a highly poisonous derivative of coniine, formerly used in the treatment of muscle spasm.

coning *Petroleum Engineering.* the gushing of reservoir water from formation strata into an oil column, caused by excessive rates of oil production.

Coniochaetaceae *Mycology.* a family of fungi of the order Xylariales that occur on wood, dung, or soil.

Coniophora *Mycology.* a genus of fungi of the family Coniophoraceae; the species *C. puteana* ("cellar fungi") causes rot in wood.

Coniophoraceae *Mycology.* a family of fungi of the order Agaricales that occur on soil, dung, wood, and plant material or in a symbiotic relationship with plants; some species cause dry rot of wood in houses.

coniscope see KONISCOPE.

conjoined twins see SIAMESE TWINS.

conjoint tendon *Anatomy.* a tendon that is formed by a fusion of tendinous fibers at the lower ends of the interna oblique and transverse abdominis muscles.

conjugase *Enzymology.* an enzyme that catalyzes the removal of glutamates from polypteroylglutamic acid.

conjugate [kän´jə gət] joined together; paired; specific uses include: *Chemistry.* **1.** of an acid and a base, related to each other in that the loss of a proton converts the acid into the base, and the gain of a proton converts the base into an acid. Thus, **conjugate acid-base pair. 2.** of two liquids, in equilibrium. *Botany.* of a pinnate leaf, having a single pair of leaflets. *Geology.* describing faults, joints, or fissures of the same age, origin, or depositional development, or the mineral deposits occurring in such structures. *Immunology.* a substance formed by the covalent bonding of two or more types of molecules, such as fluorescein coupled with an antiglobulin molecule. *Mathematics.* **1.** given a group G with a subset H, the group element hxh^{-1} is a conjugate of x, for $h \in H$ and $x \in G$. The subgroup hKh^{-1} is the conjugate of the subgroup K. **2.** the conjugate of the complex number $a + ib$ is $a - ib$. **3.** the conjugate of the quaternion $d + ia + jb + kc$ is $d - ia - jb - kc$.

conjugate addition *Organic Chemistry.* an addition reaction in which a nucleophile attaches to the 1,4-carbons of a diene, and not to the 2,3-carbons.

conjugate binomial surds see CONJUGATE RADICALS.

conjugate branches *Electricity.* two branches of a network related so that a voltage applied to either branch produces no response in the other. Also, **conjugate conductors.**

conjugate bridge *Electronics.* a bridge in which the electrical supply circuit and the detector circuit are interchanged, as compared to a normal bridge configuration.

conjugate convex functions *Mathematics.* a pair of functions f and g on $x \geq 0$ such that $f'(0) = g'(0) = 0$, $f'(x)$ is strictly increasing for $x > 0$ (i.e., f is convex upward), and $f'(x) = 1/g'(x)$ for $x > 0$ (i.e., g is convex downward).

conjugate curve see BERTRAND CURVE.

conjugated diene *Organic Chemistry.* a compound containing two carbon-carbon double bonds alternating with single bonds.

conjugated double bonds *Organic Chemistry.* two double bonds between carbon atoms in a molecule that are separated by one single bond.

conjugate diameters *Mathematics.* for a given conic, any pair of straight lines, either of which bisects all chords parallel to the other. Conjugate diameters intersect on a line of symmetry of the conic.

conjugate directions *Mathematics.* given a point p on a surface S, the directions of a pair of conjugate diameters of the Dupin indicatrix of p on S; p must be a hyperbolic or elliptic point for conjugate directions to exist, and each such p has infinitely many pairs of conjugate directions.

conjugate distribution *Statistics.* in Bayesian inference, a family of distributions is conjugate to a given model if both the a priori and the a posteriori distribution belong to the family.

conjugate division *Mycology.* the simultaneous division of the two nuclei of a dikaryon, which occurs in certain fungi.

conjugated polyene *Organic Chemistry.* a compound containing several, usually alternate, carbon-carbon double bonds.

conjugated protein *Biochemistry.* a protein containing a nonprotein element, such as a nucleoprotein, lipoprotein, or phosphoprotein. Also, COMPOUND PROTEIN.

conjugate ellipses *Mathematics.* the pair of ellipses having the equations (in normal form) $x^2/a^2 + y^2/b^2 = 1$ and $x^2/b^2 + y^2/a^2 = 1$.

conjugate foci *Optics.* see CONJUGATE POINTS.

conjugate hyperbolas *Mathematics.* the pair of hyperbolas having the equations (in normal form) $x^2/a^2 - y^2/b^2 = 1$ and $y^2/b^2 - x^2/a^2 = 1$. They have the same asymptotes and interchanged axes.

conjugate impedances *Electricity.* impedances that have equal resistive components and reactance components that are equal in magnitude but opposite in sign.

conjugate joint system *Geology.* a system of joints of related origin consisting of two sets that intersect at right angles, but have a common dip or strike.

conjugate lines *Mathematics.* consider a conic and a line l_1 in the same plane; the line l_2 that passes through the pole of l_1 relative to the conic, and such that l_1 passes through the pole of l_2 relative to the conic, is the conjugate line of l_1 relative to the conic. l_1 and l_2 are said to be conjugate lines. The poles may be at infinity without violating the definition.

conjugate momentum *Mechanics.* the quantity p_i associated with a generalized coordinate q_i in a conservative dynamical system with Lagrangian $L(q_1, \ldots, q_n; \dot{q}_i, \ldots, \dot{q}_n; t)$, defined by $p_i = \partial L/\partial \dot{q}_i$, where $\dot{q}_i$ is the time rate of change of q_i. Also, GENERALIZED MOMENTUM, CANONICAL MOMENTUM.

conjugate particles *Particle Physics.* a pair of particles consisting of a particle and its antiparticle, which have some features in common but are opposite or inverse in other qualities.

conjugate planes *Mathematics.* consider a quadric surface and a plane P_1; the plane P_2 that passes through the pole of C_1 relative to the surface, and such that P_1 passes through the pole of C_2 relative to the surface, is the conjugate plane of P_1. P_1 and P_2 are said to be conjugate planes. The poles may be at infinity without violating the definition.

conjugate points *Optics.* **1.** two points positioned along the principal axis of a mirror or lens so that light emanating from one focuses onto the other. **2.** two points in object and image space that are imaged onto one another. Also, CONJUGATE FOCI. *Mathematics.* **1.** consider a conic, a point p_1 not on the conic, and the polar of p_1; the conjugate point of p_1 relative to the conic is a second point p_2 lying on the polar of p_1, chosen such that p_1 lies on the polar of p_2. p_1 and p_2 are said to be conjugate points. **2.** in Riemannian geometry, two points on a geodesic such that some nonzero Jacobi field on the geodesic vanishes at both points.

conjugate radicals *Mathematics.* a pair of irrational numbers of the form $a\sqrt{b} + c\sqrt{d}$ and $a\sqrt{b} - c\sqrt{d}$, where a, b, c, and d are rational but $\sqrt{b}$ and $\sqrt{d}$ are not both rational. Also, CONJUGATE BINOMIAL SURDS.

conjugate roots *Mathematics.* conjugate complex numbers that are roots of a given polynomial equation.

conjugate space see DUAL SPACE.

conjugate system of curves *Mathematics.* for a given surface S, two one-parameter families of curves on S such that through each point p of S there passes a unique curve of each family, and such that the directions of the tangents at p to these two curves are conjugate directions on S at p.

conjugate transpose *Mathematics.* given a matrix A with complex entries (a_{ij}), the conjugate transpose of A, denoted by A^*, is the transpose of the matrix whose entries are the complex conjugates of the numbers a_{ij}. If $A = A^*$, then A is said to be Hermitian.

conjugate triangles *Mathematics.* a pair of triangles such that the poles of the sides of one triangle, relative to a given conic, are the vertices of the other triangle, and vice versa.

conjugating *Botany.* reproducing by conjugation.

conjugation the act or state of being joined together; specific uses include: *Genetics.* the fusion of two gametes in sexual reproduction. *Invertebrate Zoology.* in ciliated protozoans, sexual reproduction by the temporary union of cells, with an exchange of nuclear material between the two individuals. *Botany.* in certain algae and fungi, a type of sexual reproduction in which the zygote is formed through the temporary creation of a bridgelike structure between the organisms so that a gamete can migrate from one to the other. *Bacteriology.* in bacteria, a step in the sexual cycle involving the transfer of genetic material from a donor cell to a recipient cell through direct cell-to-cell contact. *Chemistry.* the joining together of two compounds to produce another compound. *Toxicology.* the combination of a toxic product with some substance in the body to form a detoxified substance, which is then eliminated. *Mathematics.* the action of the elements of a subset H of a group G, given by $(h, x) \to hxh^{-1}$, for $h \in H$ and $x \in G$. If K is any subgroup of G, then hKh^{-1} is a subgroup of G isomorphic to K.

conjugation tube *Botany.* in conjugating green algae, the structure that allows the gametes of the one organism to migrate to the other.

conjugative plasmids see SELF-TRANSMISSIBLE PLASMIDS.

conjunction an act or state of combination or association; specific uses include: *Astronomy.* the moment when any two celestial bodies lie closest together as measured in some coordinate system; for example, right ascension or ecliptic longitude. *Mathematics.* a binary operation on logical propositions such that the resulting compound statement is true if, and only if, the two operands are true. If P_1 and P_2 are logical propositions, then their conjunction is written $P_1 \wedge P_2$ (read as "P_1 and P_2").

conjunctiva *Anatomy.* the mucous membrane that lines the eyelids and covers the exposed surface of the sclera.

conjunctival fold *Anatomy.* the fold of the conjunctiva formed at its reflection from the eyelids onto the eye. Also, PALPEBRAL FOLD.

conjunctive normal form *Artificial Intelligence.* a canonical form of writing predicate calculus formulas as a conjunction of disjuncts or clauses; used in resolution theorem proving.

conjunctive search *Computer Programming.* a search for items meeting a set of conditions combined by the logical operator AND; an item must meet all of the stated conditions.

conjunctive transformation *Mathematics.* let A be a linear operator on a finite dimensional vector space over the complex numbers. A conjunctive transformation of A is given by $B = TAT^*$, for some linear operator T, where T^* is the complex conjugate of T. A and B are said to be **conjunctive matrices.**

conjunctivitis *Medicine.* an inflammation of the conjunctiva.

conk *Mycology.* the fruiting body of certain fungi belonging to the subdivision Basidiomycotina that grow on and tend to destroy wood; it has a shelflike or bracketlike shape. Also, BRACKET FUNGUS, SHELF FUNGUS.

Connaraceae *Botany.* a family of tropical woody plants of the order Rosales, often having highly poisonous bark, fruit, and seeds and otherwise characterized by spiral leaves, small hypogynous flowers, and a dry fruit with a single aril-covered seed.

connate *Geology.* of a substance, trapped in the interstices of sedimentary or igneous rock at the time the rock was formed. *Botany.* of leaves, congenitally joined.

connate-perfoliate leaf *Botany.* any of a group of opposite sessile leaves that are connate at their bases and united around their stem.

connate water *Hydrology.* water that was trapped in the interstices of sedimentary or igneous rock; it occurs as a film of water around each grain of sand in the rocks and has existed since the rocks were formed. Also, FOSSIL WATER.

connected graph *Mathematics.* **1.** a graph in which there is a path between any pair of vertices; equivalently, a graph with exactly one component, or a graph that cannot be expressed as the union of two vertex-disjoint graphs. **2.** given a positive integer k, a graph is ***k*-connected** if it has connectivity at most k.

connected load *Electricity.* the sum of the continuous power ratings for all load-consuming equipment connected to an electric power distribution system.

connected set *Mathematics.* **1.** a set such that each pair of points can be joined by a path or curve which lies entirely in the set. **2.** a set whose only direct summands are the empty set and the set itself. **3.** in a topological space, a set that cannot be separated into two subsets, neither of which contains an accumulation point of the other; equivalently, a set whose only open subsets are the empty set and the set itself. **4.** a point set that cannot be represented as the union of two nonempty disjoint subsets. Also, ARCWISE CONNECTED SET, PATHWISE CONNECTED SET.

connected space *Mathematics.* a topological space whose only subsets that are both closed and open are the empty set and the entire space; i.e., a topological space whose points form a connected set.

connected surface *Mathematics.* a surface whose points form an arcwise connected set.

connecting bar see TOMBOLO.

connecting peptide see C PEPTIDE.

connecting rod *Mechanical Engineering.* the rod that connects the piston or crosshead to the crank in a reciprocating pump or engine.

connection the act of joining together; specific uses include: *Geodesy.* the systematic elimination of discrepancies between adjoining or overlapping triangulation networks to establish a common framework from which long-range measurements can be taken. *Electrical Engineering.* see CONNECTOR. *Mathematics.* in differential geometry, a set of rules that defines parallelism along curves on a manifold. A connection is always associated with a covariant differentiation so that, for example, a vector field Y whose covariant derivative is zero with respect to the curve's tangent vector field X is considered to be self-parallel along the curve. The manifold need not have a metric tensor to have a connection; also, several connections can coexist on the same manifold.

connection gas *Petroleum Engineering.* a gas injected into a well with the mud pump inactivated so that a connection can be made.

Connection Machine *Computer Technology.* a trade name for a massively parallel computer.

connective *Computer Programming.* a symbol written between two operands to represent an operation, such as the ampersand (&) in the expression P & V, which describes a logical AND relationship between the variables P,V.

connective equivalence *Organic Chemistry.* the relationship shared by hydrogen atoms attached to the same carbon atom.

connective tissue *Histology.* a tissue characterized by an abundance of noncellular intercellular material.

connectivity *Transportation Engineering.* in a transportation system, the ability to make connections; the quantity and quality of interchanges.

connectivity number *Mathematics.* the number n of infinite cyclic groups of a topological space X whose direct sum with the nth torsion group of X forms the n th homology group.

connector *Electrical Engineering.* **1.** a device that holds the ends of conducting wires in contact, and may be easily detached for disconnection. **2.** a coupling device between cables or between cables and a chassis or console. **3.** a plug or receptacle that can be easily joined to or separated from its mate. *Design Engineering.* any device that links or holds together objects or parts. *Industrial Engineering.* the symbol on a flow chart that indicates the flow has moved to another point.

connector block *Electrical Engineering.* an insulated base or block with terminals or connecting strips used to make semipermanent interconnections between wires or cables. Also, **connector base.**

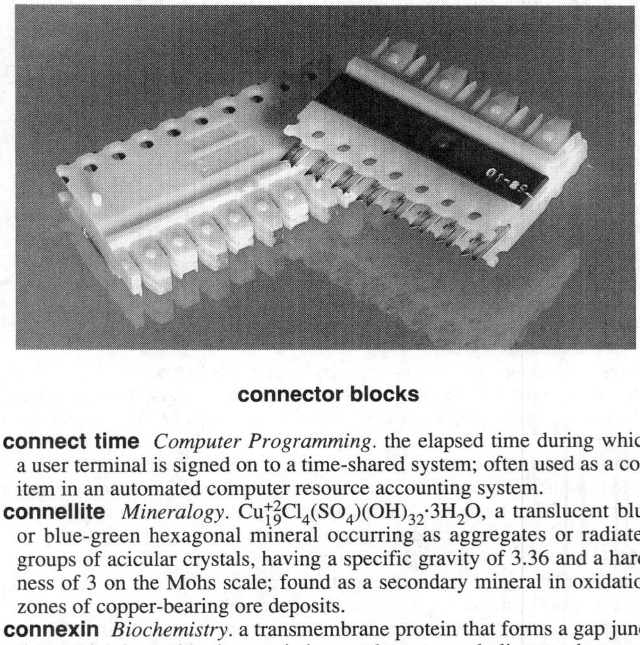

connector blocks

connect time *Computer Programming.* the elapsed time during which a user terminal is signed on to a time-shared system; often used as a cost item in an automated computer resource accounting system.

connellite *Mineralogy.* $Cu_{19}^{+2}Cl_4(SO_4)(OH)_{32}{\cdot}3H_2O$, a translucent blue or blue-green hexagonal mineral occurring as aggregates or radiated groups of acicular crystals, having a specific gravity of 3.36 and a hardness of 3 on the Mohs scale; found as a secondary mineral in oxidation zones of copper-bearing ore deposits.

connexin *Biochemistry.* a transmembrane protein that forms a gap junction, which is used by inorganic ions and most metabolites, such as sugars and amino acids, to pass through the interiors of cells.

connexon *Biochemistry.* a unit that consists of two protein particles from opposing membranes joined at a gap junction. Also, HEMICHANNEL.

conning tower *Naval Architecture.* **1.** a raised, armored position for commanding a battleship or other large combatant in action. **2.** a raised command position in the upright "sail" structure of a submarine. **3.** this structure as a whole.

connivent *Biology.* converging or coming together at the tips, but not fused together into a single part.

connotation *Psychology.* a secondary meaning that is suggested or implied by a word in addition to its direct, explicit meaning; for example, the word *July* literally means the seventh month of the year, but for many people it has the connotation of warm weather and vacation time. Also, **connotative meaning.**

connubium *Anthropology.* the largest group from which a marriage partner may be selected, often a network of neighboring groups. Also, COMMUNE.

Conocephalaceae *Botany.* a monogeneric family of large, typical liverworts of the order Marchantiales; characterized by air chambers that occur in a single layer dividing the dorsal thallus surface into coarse areolae, ventral scales with a basally constricted appendage, and simple, branching, elevated pores in the epidermis.

Conoclypidae *Paleontology.* an extinct family of euechinoids in the order Holectypoida; exocyclic with a medium to high test; extant in the Cretaceous to Eocene.

conodont *Paleontology.* a long-extinct marine animal; sometimes referred to as the **conodont animal,** since whole-body fossils have been found only in recent decades and the word *conodont* originally referred to the tooth fossils that were the main source of evidence for the animal's existence.

Conodontiformes *Paleontology.* a term formerly used to refer to a major grouping of conodont teeth.

Conodontophoridia *Paleontology.* a term formerly used in classifications that placed the conodont animal in an order of primitive fish.

conoid see CONICAL.

Conopidae *Invertebrate Zoology.* a family of wasp flies, two-winged insects in the suborder Cyclorrhapha that look like wasps and have larvae that are parasitic on wasps and bees.

conoplain see PEDIMENT.

Conopophagidae *Vertebrate Zoology.* the ant pipets, a bird family of the suborder Furnarii in some classification systems, generally incorporated into the families Formicariidae and Tyrannidae.

conoscope *Optics.* a polariscope used to examine crystals in converging rays of light.

Conrad discontinuity *Geophysics.* an area in the earth's crust, 17 to 20 km below the surface, at which compressional seismic waves experience an abrupt increase in velocity, from about 6.1 km per second to 6.4–6.7 km per second.

cons *Artificial Intelligence.* **1.** in Lisp, the function that constructs a dotted pair or basic element of list structure. **2.** to make a data structure. **3.** see CONS CELL.

consanguineal kin [kän´sang gwin´ē əl] *Anthropology.* biological relatives; those related by birth.

consanguine family [kän´sang´gwīn] *Anthropology.* a related, cohabiting group that includes a woman, her brothers, and her offspring.

consanguineous *Geology.* describing a group of naturally occurring sediments or sedimentary rocks having a common point of origin.

consanguinity [kän´sang gwin´ə tē] *Genetics.* a genetic relationship between individuals. *Petrology.* the genetic relationship between igneous rocks in a single petrographic province, presumably derived from a common parent magma; characterized by similar mineralogy, chemical composition, or texture.

cons cell *Artificial Intelligence.* the basic record from which list structures are constructed in Lisp, consisting of two pointers; it is constructed by cons, and the pointers are extracted by car and cdr. Also, CONS, DOTTED PAIR.

conscience *Psychology.* the moral and ethical quality of the mind; those values or forces that serve to inhibit an individual from carrying out harmful or unacceptable behavior, and that produce feelings of guilt if such behavior is carried out.

conscious *Psychology.* **1.** being aware of oneself and one's environment; able to perceive and react to surrounding objects and events. **2.** the part of the mind where this sense of awareness takes place.

consciousness *Neurology.* a physical and mental state of being awake, fully aware, alert, and oriented, in which the functions of the brain and the sensory stimuli are easily controlled. *Psychology.* **1.** the fact of being aware of oneself and one's surroundings. **2.** see CONSCIOUS, def. 2.

consensus *Science.* a conclusion or interpretation of data based on a general agreement or unanimity of opinion.

consensus sequence *Genetics.* a nucleic acid sequence that is shared by a number of gene sequences that are not necessarily identical; for example, the promoter regions of eukaryotic structural genes, the TATA sequence.

consequent *Geology.* describing a topographical feature that formed as a result of preexisting features or conditions. *Artificial Intelligence.* the right-hand side of an implication or rule; e.g., Q is the consequent in the rule $P \rightarrow Q$.

consequent lake *Hydrology.* a lake that occupies an existing depression or some other original inequality in a land surface. Also, NEWLAND LAKE.

consequent method see IF-NEEDED METHOD.

consequent poles *Electromagnetism.* pairs of magnetic poles in a magnetized body that are in excess of the usual single pair.

consequent reasoning see BACKWARD REASONING.

consequent stream *Hydrology.* a stream whose course or direction is dependent on the general form and slope of the existing land surface.

consequent valley *Geology.* a valley whose shape is determined either by the slope of the existing land surface or by the widening of a channel cut by a consequent stream.

conservation *Ecology.* any of various efforts to preserve or restore the earth's natural resources, including such measures as the protection of wildlife, the maintenance of forest or wilderness areas, the control of air and water pollution, and the prudent use of farmland, mineral deposits, and energy supplies.

conservation archaeology see CULTURAL RESOURCE MANAGEMENT.

conservation biology *Ecology.* the branch of the biological sciences concerned with the planning and management of natural resources, and especially with the maintenance of the balance of nature, the diversity of species and genetic material, and natural evolutionary change.

conservationist *Ecology.* **1.** a person who is professionally trained or qualified to manage natural resources. **2.** a person who advocates or supports the preservation of the earth's natural resources.

conservation law *Physics.* a general statement that a physical quantity, such as energy, momentum, or mass, is unchanged (conserved) in an interaction occurring within a closed system.

conservation of angular momentum *Mechanics.* the principle that a system under no external forces will maintain constant total angular momentum.

conservation of charge *Electricity.* a law stating that the total charge of an isolated system is fixed. Also, CHARGE CONSERVATION.

conservation of energy *Physics.* a fundamental principle stating that energy cannot be created or destroyed, but can be changed from one form to another; no violation of this principle has been found.

conservation of mass *Physics.* the principle that mass cannot be created or destroyed; this principle generally holds true in larger contexts but can be violated at the microscopic level, as in a nuclear reaction. Also, **conservation of matter.**

conservation of momentum *Mechanics.* the principle that a system under no external forces will maintain constant linear momentum.

conservation of orbital symmetry see WOODWARD-HOFFMAN RULE.

conservation of parity *Quantum Mechanics.* a principle that is obeyed by most physical interactions, according to which the parity of a wavefunction is unchanged by the interaction.

conservation of probability *Quantum Mechanics.* a requirement that the sum of probabilities of all possible final states that might issue from an initial state must equal the probability of the initial state.

conservation of quantum number *Quantum Mechanics.* a law stating that the number of quanta minus the number of antiquanta is conserved in any reaction.

conservation of vorticity *Fluid Mechanics.* the principle that the conservation equations expressing the rotation of a fluid element about an instantaneous axis contain the three angular velocity components that describe the vorticity of the fluid.

conservation task or **test** *Psychology.* a measure of a child's cognitive development, as judged by the ability to comprehend that the basic properties of a substance are not affected when superficial changes are made; e.g., eight ounces of liquid in a short, broad cup remains the same in volume when poured into a tall, thin glass.

conservation tillage or **till** *Agronomy.* any of various plowing methods that do not involve deep penetration of the soil, thus leaving plant matter on the surface.

conservative *Surgery.* of or relating to treatment by gradual, limited, or well-established procedures, as opposed to radical treatment.

conservative concentrations *Oceanography.* concentrations that undergo local changes through diffusion and advection only, except at their boundaries, two prominent examples being heat content and salinity; for instance, a conservative concentration is not affected by the presence of living organisms.

conservative dynamical system *Chaotic Dynamics.* a system that conserves energy; one for which there is an additional nontrivial function of the phase space variables that is a constant of the motion. The size of any volume element of phase space is preserved under evolution of the dynamical system, though it may change shape.

conservative elements *Oceanography.* those elements in sea water, such as chlorine, sodium, and magnesium, whose ratios to other such elements remain constant, regardless of variations in salinity.

conservative force *Mechanics.* a force for which the work done in displacing a particle from one point to another depends only on the location of those two points, and not on the path taken by the particle in moving from the initial position to the final one; e.g., the force of gravity. Thus, **conservative force field.**

conservative property *Thermodynamics.* a property of a system whose value remains constant during a process or series of processes.

conservative scattering *Electromagnetism.* the scattering of radiation without accompanying absorption.

conservative system *Mechanics.* a mechanical system in which there are no losses of energy due to dissipative processes such as friction; i.e., the sum of potential plus kinetic energy is constant. *Physics.* a system in which the amount of work expended in any physical process that transforms the system from one state to another state is independent of the process, and is instead dependent only on the initial and final states.

conserved name see NOMEN CONSERVANDUM.

conserved quantity *Mechanics.* see CONSTANT OF MOTION. *Physics.* any quantity that remains unchanged with time during the evolution of a dynamical system.

conserved vector current *Particle Physics.* a hypothesis claiming that the conserved isotopic spin current is the same as the weak hadronic vector current.

Considere construction *Materials Science.* a plot of the true stress or load against normal strain; used to determine the mechanical properties of polymers.

consistence see CONSISTENCY.

consistency *Materials Science.* **1.** the uniformity of a manufactured material. **2.** the resistance to flow of a material; viscosity. *Building Engineering.* the degree of flow or workability of concrete when analyzed by a compacting factor test. *Psychology.* see CONSONANCE.

consistency checking *Computer Programming.* in structured program design, a quality-control process that verifies the consistency and eliminates ambiguities of information from layer to layer across a family of diagrams, either data flow diagrams or structure charts. Also, BALANCING, CONSERVATION ANALYSIS.

consistency routine *Computer Programming.* a subprogram used during debugging to verify that all data elements are handled in accordance with the rules for the given data type.

consistent *Artificial Intelligence.* of a logical formula, true under some interpretation. Also, SATISFIABLE. *Mathematics.* if a mathematical theory contains no statement S such that S and $\sim S$ (the negation of S) are theorems (i.e., the theory has no contradictions), it is said to be consistent.

consistent estimator *Statistics.* an estimator that converges to the value of the population parameter as the sample size increases.

consistent system *Mathematics.* the property of a system or collection of equations that at least one set of values for the variables satisfies each equation; that is, the solution set of the system is nonempty.

consociation *Ecology.* a climax community of plants that is dominated by one particular species; such as a pine forest that is dominated by the lodgepole pine, *Pinus contorta.*

Consolan or **Consol** *Navigation.* an electronic navigation system in which the bearing of the station is determined by counting a series of dots and dashes and referring to a table or special chart (**Consol chart**).

console *Computer Technology.* the device used by a computer operator to communicate manually with the computer and to control computer operations. *Engineering.* a section of equipment designed for an operator to monitor and control a central processor and the peripheral equipment during production. *Telecommunications.* a main desk housing electronic control equipment, as in an airport tower or radar station. Also, CONTROL DESK. *Electronics.* **1.** a large cabinet containing a television set. **2.** also, **console model**. the set itself. *Space Technology.* in the ground control of a space vehicle, an array of devices used to monitor and control a particular sequence of actions such as the checkout, countdown, or launch.

console display *Computer Technology.* a presentation of computer output to the operator as a visual display, printed message, or acoustic signal.

console switch *Computer Technology.* on a computer operator's console, any of various devices that can influence the operation of a program during execution.

console terminal *Computer Technology.* an input/output device associated with a computer operator console that outputs messages and accepts keyed-in commands.

consolidant *Surgery.* an agent that encourages the healing or union of parts.

consolidated ice *Oceanography.* a solid mass of ice of various origins, consolidated by wind and currents.

consolidation a process of bringing often disparate parts together to form a whole; solidification; specific uses include: *Geology.* **1.** a process by which soft, loose, or liquid earth materials are solidified. **2.** the compaction of a soil or sediment in response to increased surface load. *Psychology.* **1.** a process by which material in short-term memory becomes fixed in long-term memory. **2.** also, **consolidation theory.** the theory that before material from short-term memory can be converted to long-term memory it must "solidify" for an interim period. *Computer Programming.* a process, usually performed by a linkage editor, combining a number of routines and subprograms that have been compiled separately into a single executable module.

consolidation test *Mining Engineering.* a test that confines a specimen laterally in a ring, then compresses it between porous plates.

Consol station *Navigation.* a radio transmitter that transmits Consolan signals.

consolute *Chemistry.* of two or more liquids, perfectly miscible under given conditions.

consolute temperature *Thermodynamics.* for a two-component liquid system of normally immiscible liquids, the upper or lower temperature limit of immiscibility, above or below which the liquids would form homogeneous solutions. Also, **consolute point.**

consonance *Acoustics.* two tones, such as a natural fifth, which when played produce overtones that are more than one step apart and therefore are pleasing to hear and are said to be harmonious. *Psychology.* a situation in which a person's behavior is consistent with his or her attitude toward that behavior; e.g., a person believes smoking is harmful and does not smoke.

consonant *Psychology.* having or showing consonance.

consortism see SYMBIOSIS.

consortium *Microbiology.* a physical association between different types of microorganisms that is beneficial to at least one of the microorganisms.

conspecific *Biology.* **1.** of or relating to organisms belonging to the same species. **2.** an individual belonging to the same species as another.

constant *Science.* a condition that remains unchanged under multiple conditions or throughout a given process; an invariable. *Mathematics.* a number, term, quantity, or mathematical object that is assumed to be fixed within the given context. *Artificial Intelligence.* in mathematical logic, a particular or abstract term that represents a single value.

constantan *Materials.* any of several nickel-copper-based alloys containing small amounts of manganese and iron, having a constant temperature coefficient of resistance; used for resistors and thermocouples.

constant angle fringes see HAIDINGER FRINGES.

constant area *Computer Programming.* a set of memory locations reserved for storing constants, static tables, and other unvarying data required for processing.

constant-bandwidth filter *Acoustics.* a frequency analyzer that uses a parallel processing of overlapping frequency filters to analyze sound frequency versus amplitude or time in separate bands that have the same bandwidth. Also, **constant-bandwidth analyzer.**

constant-conductance network see CONSTANT-RESISTANCE NETWORK.

constant-current electrolysis *Chemistry.* electrolysis in which a constant current, controlled by application of an external current, flows through an electrolytic cell.

constant-current filter *Electrical Engineering.* a filter network for connection to a source with an internal impedance so high that it is assumed to be infinite.

constant-current generator *Electricity.* a generator that outputs a constant current despite variations in load, line, or temperature.

constant-current modulation *Telecommunications.* a method of amplitude modulation in which a source of constant electrical current supplies a radio-frequency generator and a modulation amplitude in parallel. Also, HEISING MODULATION.

constant-current source *Electricity.* a source whose output current is independent of the load impedance; in practice, any source whose internal impedance is high compared to the range of load impedances used with it acts as a current source. Also, **constant-current supply.**

constant-current transformer *Electricity.* a transformer that maintains constant current in its secondary circuit under varying loads, when supplied by a constant voltage supply.

constant-delay discriminator see PULSE DEMODER.

constant-deviation fringes see HAIDINGER FRINGES.

constant element *Industrial Engineering.* an element of a job or task that does not vary in work content or performance time, regardless of changes in the process or the product.

constant-false-alarm rate *Electronics.* a constant rate of false target indications in a radar system, caused by interference or noise levels that exceed the established detection threshold.

constant folding *Computer Programming.* a compiler code generation optimization technique that computes constant expressions at compile time rather than each time the program is executed.

constant-force spring *Mechanical Engineering.* a spring having a steady restoring force no matter what the degree of displacement.

constant-gradient synchrotron *Nucleonics.* a particle accelerator in which protons, after initial acceleration by a Van de Graaff accelerator, gain energy on each revolution in an oscillating constant-gradient electric field as they pass through a series of small magnets that provide a rising magnetic field to keep the protons in a fixed orbit.

constant-head meter *Engineering.* a meter that regulates the flow of liquids by varying its opening while maintaining a constant differential pressure .

constant heat summation see HESS'S LAW.

constant-height chart *Meteorology.* a synoptic chart for a surface of constant geometric altitude above mean sea level, containing plotted data and analyses of the distribution of atmospheric pressure, wind, temperature, and humidity at a given altitude. Also, FIXED-LEVEL CHART, ISOHYPSIC CHART.

constant-height surface *Meteorology.* a surface of constant geometric or geopotential altitude measured with respect to mean sea level. Also, ISOHYPSIC SURFACE.

constant instruction *Computer Programming.* an artificial instruction that is written as a constant and is not intended to be executed as an instruction; related to a dummy or no-operation instruction.

constant-k filter *Electronics.* a highpass or lowpass filter composed of a tandem connection of a number of identical prototype L-section filters in which the product of the various series and parallel impedances is a constant value regardless of frequency within the operating range.

constant-k lens *Electromagnetism.* a solid sphere of dielectric material that is used to focus microwave plane waves onto a point in the sphere whereupon the radiation emerges from the opposite side as a parallel beam.

constant-k network *Electronics.* a ladder network of constant-k filters.

constant-level balloon *Aviation.* a balloon equipped with a feedback system, such as a pressure valve linked to its ballast supply, that allows it to float at a constant level of ambient air pressure. Also, **constant-pressure balloon.**

constant load balance *Engineering.* a weighing instrument made up of a weighted pan that is hung from a beam and that maintains a constant load of 200 grams on a counterpoised beam.

constant of aberration *Astronomy.* the greatest deflection in its position that a star can have, as viewed from the earth's surface; equal to 20.49 seconds of arc.

constant of gravitation see GRAVITATIONAL CONSTANT.

constant of motion *Mechanics.* a dynamical variable of a system that remains unchanged in the course of motion. Also, CONSERVED QUANTITY.

constant of precession *Astronomy.* the ratio of the lunisolar precession to the cosine of the ecliptic's obliquity; equal to about 54.94 arcseconds per year.

constant of the cone *Cartography.* the convergence factor of a conic map projection, ranging from infinite for the cylindrical projection to zero for the polar projection.

constant-potential electrolysis *Chemistry.* electrolysis in which the electric potential of one electrode, with respect to a stable reference electrode, is controlled by the application of a varying voltage to the cell.

constant-pressure chart *Meteorology.* a plot of constant pressure on a synoptic weather chart, including analyses of the distribution of height of the surface, wind, temperature, and humidity. Also, ISOBARIC CHART.

constant-pressure gas thermometer *Engineering.* a thermometer that measures temperature by determining the volume that a designated amount of gas takes up at a constant pressure.

constant-pressure surface *Meteorology.* a surface along which the atmospheric pressure is equal everywhere at a given instant. Also, ISOBARIC SURFACE.

constant-ratio code *Telecommunications.* a code that uses combinations of fixed ratios of ones and zeros to represent characters.

constant region *Immunology.* a portion of an immunoglobulin molecule that is characterized by the same amino acid sequence regardless of the molecule, occurring in the C terminals of the heavy and light chains. Also, C REGION.

constant-resistance network *Electronics.* a network that reflects a constant resistance to the output of a driving source, such as a power amplifier, when the load is resistive. Also, CONSTANT-CONDUCTANCE NETWORK.

constant species *Ecology.* any species that is invariably present in a given community.

constant-speed drive *Mechanical Engineering.* a mechanism that transmits motion from one shaft to another without allowing the velocity ratio of the shaft to vary.

constant-speed propeller *Aviation.* a propeller designed to maintain engine speed at a constant rpm by means of a governor that automatically adjusts the propeller's pitch.

constant-velocity recording *Acoustical Engineering.* a technique in recording in which the velocity of the recorded medium is held constant for an input signal of a given amplitude; the frequency of the signal is thus inversely proportional to the amplitude.

constant-velocity star *Astrophysics.* a star whose spectral lines correspond to those of a laboratory spectrum, indicating that the star is neither moving away from the earth nor approaching it.

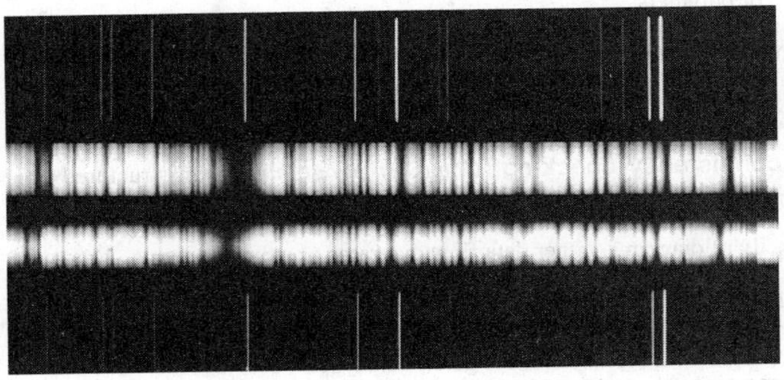

constant-velocity star

constant-velocity universal joint *Mechanical Engineering.* a joint that transmits a constant drive speed, regardless of angles; used for the drive shafts of front-wheel-drive vehicles.

constant-voltage generator *Electricity.* a generator that maintains a predetermined voltage across a load for a specified range of load resistances, line voltages, and temperature changes. Similarly, **constant-voltage transformer**.

Constellariida *Paleontology.* a suborder of stenolaemate bryozoans in the extinct order Cystoporata; one of the early cystoporates, which seem to be transitional from the Trepostomata; encrusting or marine; Ordovician and Silurian.

constellation *Astronomy.* **1.** any of the 88 officially recognized star patterns into which the celestial sphere is divided; identified and named as definite groups and usually thought of as forming certain figures or shapes in the sky. **2.** the specific region of the sky in which a particular star pattern is located.

constipated *Medicine.* affected with constipation.

constipation *Medicine.* a condition of infrequent or difficult bowel movements, with stools that are hard and dry.

constituent an element or component of something; a part; specific uses include: *Metallurgy.* a phase, or group of phases, occurring in the microstructure of an alloy and having a specific morphology. *Oceanography.* any of two or more factors that contribute to the character or force of a given oceanic phenomenon.

constituent number *Oceanography.* see ASTRONOMICAL TOTAL CONSTITUENT.

constitutional isomers *Organic Chemistry.* isomers that differ in the sequence of linkage of their constituent atoms.

constitutional theory (of personality) *Psychology.* the theory that there is a relationship between body type and personality type.

constitution diagram *Metallurgy.* in an alloy system, a diagram representing the boundaries of the phases, expressed in terms of temperatures and compositions. A constitution diagram does not necessarily represent thermodynamic equilibrium. Also, PHASE DIAGRAM.

constitutive *Genetics.* describing genes that are expressed as a function of the interaction of RNA polymerase with the promoter, without additional regulation; such genes are constantly expressed, whether or not there is demand.

constitutive enzyme *Enzymology.* an enzyme whose presence and concentration in a cell remain constant, unaffected by the presence of its substrate.

constitutive equations *Electromagnetism.* the equations $B = \mu H$, relating the magnetic induction to the magnetic field intensity, and $D = \varepsilon E$, relating the electric displacement to the electric field intensity.

constitutive expression *Genetics.* expression as a function of the interaction of RNA polymerase with the promoter, without additional regulation; sometimes used to describe expression of some functions in all cells at a low level.

constitutive heterochromatin *Genetics.* a type of heterochromatin that is condensed throughout the cell cycle, such as the chromatin of the centromere regions of chromosomes.

constitutive mutants *Molecular Biology.* mutations resulting from the increased constitutive synthesis by a bacterium of several functionally related enzymes.

constitutive property *Chemistry.* a chemical or physical property of a substance that depends on the arrangement of atoms in the molecules of the substance.

constitutive synthesis *Molecular Biology.* the continuous production of proteins and other materials that are used throughout the life of a cell.

constrained mechanism *Mechanical Engineering.* a mechanism in which all members move in a predetermined pattern.

constraint *Mechanics.* an external restriction on the motion of a system. Each constraint equation removes one effective degree of freedom from a holonomic system. Also, KINEMATIC CONSTRAINT. *Engineering.* **1.** a restriction of a solid's natural tendency to narrow when it is pulled at its ends. **2.** a restriction of the natural degrees of freedom of a system. *Artificial Intelligence.* **1.** a condition that must always be true or that must be true of any valid solution to a problem. **2.** a rule, formula, axiom, or program that expresses such a condition.

constraint-based reasoning *Artificial Intelligence.* problem solving based on reasoning about given constraints and propagating constraints to narrow the range of possible solutions.

constraint function *Mathematics.* any of a finite number of conditions on the solution to an optimization problem.

constraint matrix *Mathematics.* in mathematical linear programming, an array formed by the coefficients of the decision variables in the constraint expressions.

constraint posting *Artificial Intelligence.* the process of storing a symbolic constraint in a location where other processes working on the same problem can examine it.

constraint propagation *Artificial Intelligence.* a form of reasoning, using a network of related facts, in which a value, or range of possible values, determined for one variable constrains the possible values of the other variables to which it is related. When the range of possible values of a variable is narrowed, the constraints may allow the ranges of related variables to be narrowed, eventually resulting in consistent values or value sets for all variables in the network.

constraint satisfaction *Artificial Intelligence.* a type of problem in which the goal is to find values for a set of variables that will satisfy a given set of constraints.

constraint violation *Artificial Intelligence.* a situation in which a constraint is not true.

constriction *Medicine.* the narrowing of a passageway or channel within the body.

constricting nozzle *Metallurgy.* in plasma arc-welding or cutting, a water-cooled copper nozzle with a hole through which the arc passes.

constriction disease *Plant Pathology.* a disease of peach trees caused by a fungus of the species *Phomopsis,* resulting in the death of the plant parts of the external margin of the tree.

constrictive pericarditis *Medicine.* the inflammation, with fibrosis, of the pericardium, resulting in constriction of the heart, a restriction of the heart's pumping action, and consequent decreased blood flow.

constrictor *Aviation.* the narrowing exit end of the exhaust tube on certain ramjets.

constructible *Mathematics.* a real number α is said to be constructible, given a fundamental unit length, if by the use of straightedge and compass alone, a line segment of length α can be constructed. By theorem, if α is constructible, then α lies in some extension of the rational number field of degree 2^k for some positive integer k.

construction *Engineering.* **1.** the act of building a structure. **2.** the type of material with which a structure is built, such as wood, steel, brick, or concrete construction. *Design Engineering.* the number of strands in a wire rope and the number of wires in a strand, expressed as two numbers separated by a multiplication sign. *Textiles.* a formula for creating a specific fabric, including yarn weight and thread count.

constructional *Geology.* **1.** describing a landform that was created by any process of upbuilding, such as deposition, volcanic eruption, or diastrophism. **2.** describing an uneroded surface, such as a plain, formed by deposition. *Hydrology.* describing a stream or drainage pattern formed by runoff from a constructional landform or surface.

constructional apraxia see OPTIC APRAXIA.

construction area *Building Engineering.* the exterior and interior of building walls and the area within each partition.

construction operator *Computer Programming.* in a data structure definition, the means to create composite objects from atoms or basic objects.

construction weight *Aviation.* the weight of an airplane or unmanned rocket excluding its propellant and load. Also, STRUCTURAL WEIGHT.

construction wrench *Mechanical Devices.* an open-ended, long-handled wrench used to align holes for bolts or rivets.

constructive interference *Physics.* a situation in which the superposition of two or more waves produces an intensity greater than the summation of the intensities of the individual waves.

constructive memory *Psychology.* the utilization of one's general knowledge to provide a more complete and detailed description of some remembered event or situation.

constructivism *Psychology.* the theory that successful learning is based on collaboration and interaction between teacher and learner.

construct theory see PERSONAL CONSTRUCT THEORY.

consultation *Artificial Intelligence.* the process by which a non-expert user uses an expert system to arrive at a recommended action or solution; includes explanations of the reasoning process.

consumable electrode *Metallurgy.* an electrode designed to melt during an arc welding process.

consumable insert *Metallurgy.* in welding, a filler metal insert that is completely fused into the joint.

consumer *Ecology.* an organism that feeds on other organisms or on existing organic matter, rather than producing its own nutrients or obtaining them from inorganic sources.

consumer psychology *Psychology.* a branch of psychology that studies the effects of advertising, marketing, and packaging on the purchasing behavior of consumers.

consumer's risk *Industrial Engineering.* in quality control, the likelihood of substandard goods being missed by the quality control process and sent to the customer.

consummatory behavior [kən sum´ə tôr´ ē] *Behavior.* an action or behavior pattern that comes at the end of a series of appetitive actions, when the organism has reached some goal. Also, **consummatory act(ion), consummatory response.**

consumption *Medicine.* **1.** a general term for wasting away of the body. **2.** a former term for tuberculosis of the lungs.

consumption test *Immunology.* a test that determines the amount of antigens or antibodies removed from a system when an antigen-antibody complex is formed.

consumptive coagulopathy *Medicine.* the reduction in the blood level of one or more coagulants due to their "consumption" or depletion, because of a large amount of clotting activity in the body.

consumptive use *Hydrology.* the total amount of water that is lost from an area as a result of various factors such as evaporation and plant use. *Agronomy.* the amount of water consumed by a plant as it grows; used in determining irrigation needs.

contact a point or instance of touching or interaction; specific uses include: *Electricity.* the conducting part of a component, such as a switch or relay, that interacts with another conducting part to make or break a circuit. *Telecommunications.* the point at which an object, such as an aircraft or ship, is first detected by a radarscope or other detecting device. *Aviation.* **1.** the state of a flight vehicle as it touches a runway or other surface after being airborne. **2.** the state of an aircraft's ignition system when the circuit is closed. **3.** usually, **visual contact.** a condition permitting visual flight; used to form compounds such as **contact flight** and **contact reconnaissance.** *Medicine.* a person known to have been near enough to an infected individual to have been exposed to the transfer of infectious material. *Geology.* the boundary surface between two distinct kinds of rocks, or two rocks of different ages.

contact acid *Inorganic Chemistry.* a term for sulfuric acid, H_2SO_4, when prepared by the contact process, which is the usual production method in the United States.
contact acne *Medicine.* acne that is produced by contact with any acnegenic chemical, such as those used in industry or in cosmetic and grooming agents.
contact adsorption *Chemical Engineering.* a procedure for extracting minor constituents from fluids by stirring in direct contact with powdered or granulated adsorbents or by passing the fluid through fixed adsorbents.
contact anemometer *Engineering.* an instrument that measures wind speed by the rate at which it makes an electrical contact, the rate being dependent on wind velocity.
contact angle *Materials Science.* the angle between a solid and the tangent to a liquid surface at the contact point; useful in determining the wetting behavior of a liquid on a solid surface.
contactant *Medicine.* an allergen that can induce delayed hypersensitivity of the epidermis after one or more episodes of contact.
contact arc *Electricity.* a spark or discharge that occurs after the breaking of a current-carrying electric contact.
contact aureole see AUREOLE.
contact behavior *Behavior.* the pattern or practice of maintaining physical contact with others of the same species, as by huddling, clinging, and so on. Thus, **contact animal, contact species.**
contact binary *Astronomy.* a binary star system, one of whose components has swelled enough to fill its Roche lobe, which is transferring matter to its companion star.
contact biotrophic mycoparasite *Mycology.* a parasitic fungus that lives on the surface of other fungi without apparently damaging them.
contact bounce *Electricity.* the unwanted, uncontrolled, intermittent making and breaking of contact when relay contacts are opened or closed.
contact breccia *Petrology.* angular rock fragments resulting from the shattering of wall rocks around laccolithic and other intrusive igneous masses.
contact catalysis *Chemical Engineering.* the process of change in the structure of gas molecules when adsorbed onto solid surfaces.
contact ceiling *Building Engineering.* a ceiling in which there is direct contact between the lath and adjacent building construction; furring or building channels are not used.
contact chatter see CHATTER.
contact clip *Electricity.* in a mechanical switching device, the clip that the blade enters or embraces.
contact condenser *Mechanical Engineering.* a device in which a vapor is in direct contact with a cooling liquid; the vapor is condensed by losing its latent heat to the liquid.
contact dermatitis *Medicine.* a skin rash caused by direct contact with a primary irritant or a sensitizing antigen.
contact drop *Electricity.* the voltage drop across the terminals of a contact.
contact electricity *Electricity.* a small voltage caused by the contact of two dissimilar conductors.
contact electrification *Materials Science.* the generation of electricity by rubbing one material against another; an electronic property of polymers.
contact filtration *Chemical Engineering.* a process to extract color bodies and improve the stability of oil by mixing finely divided adsorbent clay with the oil.
contact flight *Navigation.* the condition of controlling the attitude of an aircraft and navigating it by visual references rather than by mechanical references.
contact follow *Electricity.* the distance traveled by two contacts after initial contact is made. Also, CONTACT OVERTRAVEL.
contact force *Mechanics.* the force exerted by a stationary surface on a body that is moving in contact with it. *Electricity.* the amount of force holding a pair of relay contacts together.
contact fuse see IMPACT-ACTION FUSE.
contact glass see FOCAL-PLANE PLATE.
contact goniometer *Crystallography.* a device for measuring the angles between the faces of a crystal. Direct contact is made with the crystal faces by two straight edges, and the angle between these straight edges is then measured.
contact guidance *Cell Biology.* a phenomenon in which the orientation of cells in a culture is influenced by the topography of the substratum.
contact hypothesis *Psychology.* the theory that if members of hostile groups come into increased contact with one another, the hostility between them will decrease.
contact inhibition *Cell Biology.* **1.** a temporary arrest of cell motility upon contact with another cell. **2.** a cessation of cell proliferation upon reaching a certain density, as when cultured cells reach a confluent monolayer.
contact-initiated discharge machining *Mechanical Engineering.* an electromachining processs in which the discharge is initiated when a tool and workpiece come in contact, after which the tool is withdrawn and an arc forms.
contact inspection *Engineering.* a process whereby an object is examined with an ultrasonic search unit that is put in direct contact with a thin layer of couplant, enabling information to be received from the object by the search unit.
contact lens *Optics.* a lens made of plastic that is designed to fit over the cornea and that is frequently used in place of eyeglasses to improve vision.

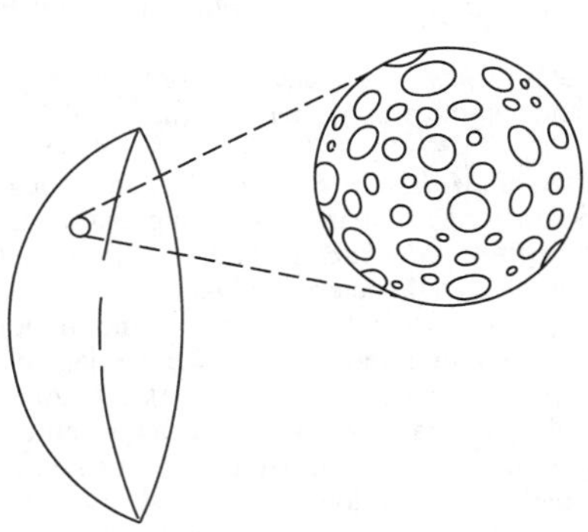

contact lens

contact log *Petroleum Engineering.* a record of electrical-resistivity data that relates to strata structures encountered along the depth of a drill hole.
contact material *Metallurgy.* any of several alloys, some based on copper or silver, that have high thermal and electrical conductivities and low contact resistance.
contact metamorphic rock *Petrology.* any rock formed by contact metamorphism.
contact metamorphism *Petrology.* a process of local thermal metamorphism in which the composition or texture of a rock is radically transformed by the contact of invading magmatic material with the country rock.
contact microphone *Acoustical Engineering.* a microphone designed to pick up sound vibrations from a solid material, such as the sounding board of a guitar.
contact mine *Ordnance.* a mine that is designed to be detonated by physical contact with a person, vehicle, or vessel.
contact mineral *Mineralogy.* any mineral formed by contact metamorphism.
contact modulation *Electricity.* the generation of square waves from a direct-current source using a fast-acting relay.
contactor *Chemical Engineering.* a column, tower, or device designed to bring two or more phases into intimate contact. *Electricity.* a device used for repeatedly closing and opening a circuit.
contact overtravel see CONTACT FOLLOW.
contact paper *Graphic Arts.* a paper coated with a photosensitive emulsion, used in making contact prints.
contact piston *Electromagnetism.* a piston or plunger in a waveguide that is allowed to make contact with the waveguide walls. Also, **contact plunger.**
contact plate see FOCAL-PLANE PLATE.
contact point *Military Science.* **1.** an easily identifiable point at which two or more military units are required to make contact. **2.** in air operations, the point at which the mission leader contacts air control.
contact-potential difference *Electricity.* the difference between the work functions of two materials in contact, divided by the electronic charge.
contact pressure *Electricity.* the amount of pressure holding a pair of contacts together.

contact print *Graphic Arts.* a same-size photographic copy made by placing a photosensitized material such as contact paper in direct contact with a positive, negative, or transparency and passing an exposing light through the original.

contact printer *Graphic Arts.* a machine used in making contact prints, usually consisting of high-intensity lights mounted on a pressurized glass frame.

contact process *Chemical Engineering.* the catalytic production of sulfuric acid from oxygen and sulfur dioxide.

contact protection *Electricity.* any technique for minimizing or completely suppressing the surge that results when an inductive circuit is interrupted; the break would otherwise cause contact arcing, leading to deterioration.

contact ratio *Design Engineering.* on a gear, the ratio of the length of the path of contact of two gears to the base pitch, equal to the average number of pairs of gear teeth in contact at any one time.

contact resistance *Electricity.* the resistance between two contacts, measured in ohms.

contact screen *Graphic Arts.* a halftone screen consisting of vignetted dots of graded density printed on a film base; designed to be placed in vacuum contact with a film or plate.

contact sensing *Control Systems.* the process of monitoring field switch contacts and converting them into digital information.

contact sensitivity *Immunology.* an epidermal cell-mediated immune reaction to chemicals placed on the skin.

contact sensor *Engineering.* an instrument that emits a signal when it detects mechanical contacts. *Robotics.* a device that senses the mechanical contact of a robot part with an external object.

contact site *Cell Biology.* either of two cell-surface proteins that enable slime mold cells to adhere to each other.

contact sparking *Electricity.* a spark occurring at contact points when two contacts break apart while carrying a current.

contact surface *Fluid Mechanics.* a surface in dynamical equilibrium across which density, temperature, or energy changes discontinuously.

contact thermography *Engineering.* a process in which the surface temperature of an object is measured by coating it with a thin layer of luminescent material and exposing it to ultraviolet rays; the surface temperature is indicated by the brightness of the coating.

contact time *Engineering.* the interval of time in which material is in direct contact with a treating agent.

contact transformation *Mathematics.* a coordinate transformation on the Cartesian product of the real line R with a cotangent bundle $T{*}M$, that maps the contact 1-form $dz - p_i\,dq^i$ to a new 1-form that is a scalar function times the original; i.e., $dz' - p_i'dq'^i = \rho(z, q^j, p_j)[dz - p_i\,dq^i]$.

contact tube *Metallurgy.* in welding, a device used to carry the current to the electrode.

contact twins *Crystallography.* two crystals growing in directions that are symmetrical about a common plane.

contact vein *Geology.* a vein developed at the boundary surface between two distinct rock formations.

contagion *Medicine.* **1.** the transmission of disease by direct or indirect contact. **2.** a disease spread in this manner. *Psychology.* a behavior pattern in which an action being performed by some members of a group is taken up by the rest of the group.

contagious *Medicine.* describing a disease or condition that can be transmitted by direct contact or close proximity.

contagious abortion *Pathology.* a condition in which reproductive organs are damaged through disease, resulting in abortion.

contagious disease *Medicine.* any disease that can be transmitted by direct contact or close proximity, such as the common cold, smallpox, tuberculosis, or AIDS.

contagious distribution *Statistics.* a distribution in which the occurrence of an event increases the probability of future events of the same kind.

contagious magic *Anthropology.* the practice of attempting to manipulate or affect the actions of another by performing a ritual on the person's possessions, nail or hair clippings, and so on.

contagious pustular dermatitis *Veterinary Medicine.* a disease that affects mainly sheep and horses, in which a parapoxvirus infects through abrasion, resulting in pustular, scab-forming lesions on the face and mouth.

container *Transportation Engineering.* a large portable compartment of standardized dimensions, used in cargo transportation. The loaded and sealed container can be transferred from one carrier vehicle to another, minimizing the need to handle its contents en route.

container car *Transportation Engineering.* a railroad car that is built to hold containers.

container ship *Naval Architecture.* a ship designed to carry cargo in standardized freight containers, in which the cargo is on-loaded and off-loaded.

container ship

containment *Nucleonics.* the combination of structures and procedures used to serve as barriers to localize radioactive materials and to keep them where they can be handled safely.

containment building *Nucleonics.* a reinforced concrete structure that encloses a reactor to confine fission products released either from the coolant system or from within the reactor itself, as well as to protect people from the hazards of an accident and to protect the reactor from exterior forces. Also, **containment structure, containment vessel.**

contaminant *Science.* any agent or action that contaminates.

contaminate *Science.* to infect or make impure by introducing foreign or undesirable material.

contamination the process of making impure, or the state of being impure; specific uses include: *Hydrology.* the pollution of water by the introduction of any substance that reduces or prevents the use of the water for ordinary purposes. *Microbiology.* the presence of an unwanted substance or microorganism in a pure, previously sterile, or defined preparation. *Nucleonics.* the presence of radioactive materials at locations where they are not wanted; the usual form is airborne particulates that can be an inhalation hazard or that adhere to surfaces of structures, objects, or persons. *Psychology.* an error in speech in which a new and invalid form is created by combining parts of two valid words. *Geology.* a process whereby country rock is incorporated into a magma, thus altering the chemical composition of the magma.

contamination monitor *Nucleonics.* a survey meter that can detect the presence of ionizing radiation from radioactive materials and measure its intensity.

contemporaneous *Geology.* describing geological features that originated, developed, or came into existence at the same time, or any rock that developed during the formation of the enclosing rocks.

content *Linguistics.* the information that is contained in or communicated by a linguistic expression. *Mathematics.* any nonnegative, finite, monotone, additive, and subadditive set function on the class of all compact subsets of a measure space.

content-addressable memory *Computer Technology.* a memory unit that can be accessed in parallel on the basis of content rather than by a specific address or location. Also, ASSOCIATIVE MEMORY.

content analysis *Computer Programming.* **1.** an analysis of text to derive attributes of the text, such as the frequency of occurrence of certain words. **2.** a process for assigning key words and content identifiers to a document and to related queries used for information retrieval.

content editing *Graphic Arts.* the process of editing that involves substative changes in the content of a manuscript, as opposed to editing for errors in spelling, grammar, usage, and the like (copy editing).

contention *Telecommunications.* a situation in which two or more devices request the use of another device, which can handle only one request at a time; e.g., when two or more stations on a multistation communication channel try to transmit at the same time or on the same channel, or when two computer terminals try to access the same printer.

contention resolver *Computer Technology.* a device or process designed to provide an orderly resolution to contention problems by implementing a priority scheme, with queues or buffers for each shareable resource; a common example is the first-in, first-out scheme.
contents *Computer Programming.* the data stored in a memory location or register.
contest competition *Ecology.* competition in which there is an unequal consumption of resources, with the more successful competitor fulfilling all its needs while the less successful obtains insufficient resources for survival or reproduction. Also, EXPLOITATION COMPETITION.
context *Linguistics.* the circumstances in which an utterance is made. *Archaeology.* see ARCHAEOLOGICAL CONTEXT. *Mycology.* in fungi belonging to the subdivision Basidiomycotina, the sterile fungal tissue making up most of the fruiting body.
context-free grammar *Computer Programming.* a grammar in which the substitution of a phrase for a component name (or metavariable) is independent of the context or surroundings of the component name.
context-sensitive grammar *Computer Programming.* a grammar in which the substitution of a phrase for a component name depends on the symbols and other component names that surround it.
context-sensitive help key *Computer Technology.* a key that when pressed causes the display of explanatory messages that are relevant to the current activity.
context switch time *Computer Technology.* in multiprogramming, the time required to save the contents of the registers of one process and load the same registers with new values for another process; an overhead cost of sharing a single processor among several activities.
contextual analysis *Computer Programming.* a process in natural language processing (in addition to syntactic and semantic analysis), in which attempts are made to relate individual sentences to one another and to the surrounding context.
contiguity *Behavior.* the fact of being closely associated in time and space, as one stimulus with another or a stimulus with a response.
contiguity theory *Behavior.* the theory that immediate reinforcement of a desired behavior is more effective than delayed reinforcement.
contiguous *Behavior.* occurring together; closely associated in time and space.
contiguous conditioning *Behavior.* conditioning that is the result of a close connection in time between the conditioned stimulus and the unconditioned stimulus.
contiguous craters *Astronomy.* small craters of similar size whose rims touch or merge and which may have formed volcanically. Also, **contiguous arc, contiguous chain.**
continent *Geography.* **1.** any of the main continuous land masses on the face of the earth, along with their neighboring islands. The continents include North and South America, Africa, Australia, Antarctica, and Europe and Asia (Eurasia). The division of the Eurasian land mass into the separate continents of Europe and Asia is traditional, even though this distinction is largely historical and cultural rather than geographical. **2.** a mainland as opposed to offshore or outlying islands, or the land as opposed to the sea.
continental accretion *Geology.* the theory that the growth of continents results from the gradual accumulation of new material around an original nucleus.
continental air *Meteorology.* an air mass whose characteristics are developed over a large land area, resulting in the basic continental characteristic of relatively low moisture content.
continental anticyclone see CONTINENTAL HIGH.
continental borderland *Geology.* the topographically complex region of the continental margin lying between the shoreline and the continental slope.
continental climate *Meteorology.* the climate typical of the interior of a land mass of continental size, characterized by ranges of temperature, low relative humidity, and moderate or light and irregular rainfall.
continental code see MORSE CODE.
continental crust *Geology.* the basement complex of crustal rocks underlying the continents and the continental shelves.
continental deposit *Geology.* any sedimentary accumulation occurring on land or in bodies of water that are not directly connected with oceans.
continental divide *Geology.* an imaginary line or ridge of high ground separating streams that flow toward opposite sides of a continent, usually into different oceans. *Geography.* also, **Continental Divide.** specifically, the high ridge of the Rocky Mountains separating streams that flow into the Pacific from those that flow into the Atlantic or the Arctic.
continental drift *Geology.* the theory that the formation of the continents was a result of the large-scale breakup and displacement of a single land mass over geologic time. Also, **continental displacement.**
continental flexure *Geology.* in an area where the seafloor and continent are in contact, the line along which warping produces an increase in the slope of the continental shelf.
continental geosyncline *Geology.* a downward flexure of the earth's crust filled with nonmarine sediments.
continental glacier *Hydrology.* a thick glacier or sheet of ice of such size that it covers a substantial portion or the entire surface of a continent. Also, **continental ice sheet.**
continental growth *Geology.* the process whereby continents expand at the expense of ocean basins.
continental heat flow *Geophysics.* the amount of heat released through the continental crust per unit time and unit area.
continental high *Meteorology.* an air mass of high atmospheric pressure seen on mean charts of sea-level pressure to overlie a continent during the winter. Also, CONTINENTIAL ANTICYCLONE.
continentality *Meteorology.* the degree to which a point on the earth's surface is subject to the influence of a land mass in all respects, usually measured by the range of temperature.
continental margin *Geology.* the total area of the seafloor lying between the shoreline and deep-ocean bottom, including the continental borderland, continental shelf, continental slope, and continental rise.
continental nucleus see CONTINENTAL SHIELD.
continental period *Geology.* the time interval during which a particular region remained above sea level, forming part of a continent.
continental plate *Geology.* one of the lithosphere plates that is composed of continental crust.
continental rise *Geology.* the smooth, gently sloping section of the continental margin lying between the continental slope and the abyssal plain.
continental shelf *Geology.* the gently sloping section of the continental margin lying between the shoreline and the continental slope. Also, **continental platform.**
continental shield *Geology.* a massive area of exposed basement rocks in a craton, theoretically indicating the point of origin of continental growth. Also, CONTINENTAL NUCLEUS.
continental slope *Geology.* the steeply inclined section of the continental margin lying between the continental shelf and the continental rise or oceanic trench.
continental terrace *Geology.* the section of the continental margin that includes the continental shelf and the continental slope.
contingency *Behavior.* the fact of being dependent on or determined by something else, such as a behavior pattern that occurs in response to a certain reinforcement or punishment.
contingency contracting see CONTRACT THERAPY.
contingency interrupt *Computer Technology.* a halt in processing caused by one of a number of conditions, such as an operator action, an arithmetic overflow, or an invalid operation code.
contingency management *Behavior.* the technique of employing presentation and withdrawal of rewards and punishments in order to shape appropriate behaviors.
contingency table *Statistics.* a table displaying categorical data from the simultaneous classification into several characteristics of the observed experimental units.
contingency theory *Psychology.* the behavioral theory that human action is determined by context and circumstances.
contingent *Behavior.* dependent on or determined by something else; occurring in repsonse to some external event or condition.
contingent punishment *Behavior.* the presentation of aversive or negative consequences for a given behavior.
continuant matrix *Mathematics.* see TRI-DIAGONAL MATRIX.
continued fraction *Mathematics.* a number represented as a number (usually a positive integer) plus a fraction whose denominator is a number plus a fraction, etc.; i.e., $a_1 + b_2/(a_2 + b_3/(a_3 + b_4/(a_4 + \cdots)))$. If the continued fraction has a finite number of terms, it is said to be terminating; otherwise it is nonterminating.
continued product *Mathematics.* a product of n factors a_k, where $n \leq \infty$; denoted by $\prod_{k=1}^{n} a_k$. If $n = \infty$, $\prod_{k=1}^{\infty} a_k$ is an infinite product.
continue statement *Computer Programming.* a FORTRAN statement that does not result in a machine operation but serves as an inert point for transfer of control; often used as the last statement in a DO loop.
continuity *Electricity.* the presence of a complete path to allow for the flow of electrical current.

continuity chart *Meteorology.* a chart of the positions of significant features of regular synoptic charts, including pressure centers, fronts, instability lines, trough lines, and ridge lines; maintained for weather analysis and forecasting.

continuity equation *Fluid Mechanics.* an equation representing the law of conservation of matter for fluid flow, stating that the mass flow into a control volume must equal the mass flow rate out of the volume.

continuous arrhythmia *Cardiology.* an irregularity in the pulse or heartbeat that goes on permanently. Also, PERPETUAL ARRHYTHMIA.

continuous at a point *Mathematics.* **1.** a function f is continuous at a point a if $\lim_{n\to\infty} f(x) = f(a)$. More generally, let f: $M_1 \to M_2$ be a function between metric spaces with metrics μ_1 and μ_2, respectively. f is said to be continuous at a point x_0 in M_1 if, given any $\varepsilon > 0$, there exists $\delta > 0$ such that whenever $\mu_1(x_0, x) < \delta$, then $\mu_2(f(x_0), f(x)) < \varepsilon$. **2.** let $f: X_1 \to X_2$ be a function between topological spaces. f is said to be continuous at a point x in X_1 if, for any sequence $\{x_n\}$ of elements in X_1 which converges to x, the sequence $\{f(x_n)\}$ converges to $f(x)$.

continuous beam *Building Engineering.* a beam that consists of three or more spans oriented in a straight line and joined together so that a load of known size and weight on the wingspan will produce a known calculated effect on the others.

continuous brake *Mechanical Engineering.* a railroad braking system in which the brakes can be applied simultaneously to all cars from a single control point.

continuous carrier *Telecommunications.* **1.** a carrier over which information is transmitted by means that do not interrupt the carrier. **2.** a signal in which the transmission of the carrier is not pulsed on and off. **3.** a radio wave of constant amplitude and constant frequency.

continuous casting *Metallurgy.* any casting process in which the product is continuously withdrawn from the mold, such as the Properzi process or the Hazlett process.

continuous cell line *Cell Biology.* any strain of cultured cells that are capable of indefinitely continued growth, such as tumor cells or transformed cells.

continuous cell recycle reactor *Biotechnology.* a high-volume fermenter that increases cell concentration and productivity by continually removing cells from the effluent and returning them to the fermentation vessel.

continuous contact coking *Chemical Engineering.* a thermal conversion process with continuous coke circulation, in which oil-wetted particles of coke travel downward through a reactor where cracking, coking, and drying take place, ultimately resulting in gas, gasoline, gas oil, and pelleted coke products.

continuous control *Control Systems.* a method of control that continuously measures a controlled quantity and uses the data to continuously correct any detected errors.

continuous cooling diagram *Materials Science.* a plot of temperature versus cooling time that can indicate when phase changes occur.

continuous cooling rate *Materials Science.* the slope of a continuous cooling diagram; a steep slope is indicative of fast cooling.

continuous cooling transformation *Materials Science.* a phase change occurring during the cooling of a material, especially austenite decomposition in steel.

continuous countercurrent leaching *Chemical Engineering.* a leaching process using continuous equipment that mechanically moves the solid and liquid through a series of leach tanks in which the flow of solvent is countercurrent to the flow of solid.

continuous culture *Microbiology.* a microbial culture that is grown in a liquid medium and maintained under constant conditions in order to prevent nutrient depletion, allowing the population to undergo balanced growth for an extended period.

continuous deformation see DEFORMATION.

continuous distillation *Chemical Engineering.* the distillation by boiling of a mixture with different component boiling points, in which a feed is supplied continuously to a fractionating column, and the product is withdrawn continuously at the top, at the bottom, and sometimes at intermediate points.

continuous distribution *Statistics.* the probability distribution of a continuous random variable; often represented by a density function. *Transportation Engineering.* such a distribution used in traffic flow analysis for vehicle acceleration distributions, trip generating functions, and other modeling purposes.

continuous dryer *Engineering.* a machine that dries material which is moving through it without interruption.

continuous-duty rating see THERMAL LIMIT.

continuous dyeing *Textiles.* a method for coloring a fabric by saturating it with a dye and then sending it without interruption through the developing, washing, and drying processes.

continuous extension *Mathematics.* a continuous function F with domain D_F is a continuous extension of another function f with domain D_f if $D_F \supseteq D_f$ and if $F = f$ when F is restricted to D_f.

continuous extension theorem see HAHN-BANACH THEOREM.

continuous fermentation *Biotechnology.* a fermentation process used to produce either a cell biomass or a secondary product; cells are kept in a state of exponential growth by the continuous feeding of a suitable medium through the fermenter.

continuous filament *Textiles.* a long, unbroken strand of synthetic fiber or raw silk.

continuous film scanner *Electronics.* a scanner containing a flying-spot kinescope that continuously scans moving motion-picture film to produce video signals for use in television broadcasting.

continuous fire *Ordnance.* **1.** firing at a normal rate without interruption for adjustment. **2.** in field artillery and naval gunfire, loading and firing at a specific rate or as rapidly as possible while maintaining accuracy and staying within the prescribed rate of fire for the weapon.

continuous-flow conveyor *Mechanical Engineering.* an enclosed belt conveyor that is pulled through a mass of granular or powdered material fed from an overhead hopper.

continuous-flow reactor *Biotechnology.* a reactor in which new material is continuously supplied and the product continuously withdrawn; used in both chemical and biological processes.

continuous form *Computer Technology.* **1.** a form that is preprinted on special paper that can be run automatically through a printer. **2.** a batch input medium, such as a register tape, that can be optically read and converted to machine-readable form.

continuous function *Mathematics.* a function that is continuous at every point in its domain; more generally, f: $X_1 \to X_2$ between topological spaces is continuous if, for every open set U contained in X_2, $f^{-1}(U)$ is an open set in X_1. That is, f is continuous if the inverse image of every open set is open. (Equivalently, f is continuous if the inverse image of every closed set is closed.). Also, CONTINUOUS OPERATOR, CONTINUOUS TRANSFORMATION.

continuous furnace *Mechanical Engineering.* a furnace in which the materials are loaded through one door and progress on a moving belt or similar means through the furnace where they are retrieved from another door. *Metallurgy.* in powder metallurgy, a furnace used for continuous sintering of compacts.

continuous group *Mathematics.* a topological group, especially a Lie group.

continuous header *Building Engineering.* in masonry, bricks that are positioned on a bed continuously with the end or cull exposed.

continuous image *Mathematics.* the image of a set under a continuous function.

continuous industry *Industrial Engineering.* a term for an industry in which successive operations are carried out in converting raw materials into finished products.

continuous kiln *Engineering.* **1.** a long oven that bakes material such as bricks or lime, with the material passing through on a moving device. **2.** a kiln in which the fire passes through progressively.

continuous loading *Electricity.* a type of loading in which the added inductance is evenly distributed along a line by wrapping a magnetic material about each conductor.

continuous loudness control see LOUDNESS CONTROL.

continuously differentiable *Mathematics.* a function f is said to be n times continuously differentiable on some subset E of its domain if it is n times differentiable at every point of E and if the nth derivative $f^{(n)}$ is continuous in E.

continuous manufacturing *Industrial Engineering.* a type of manufacturing in which a standard product such as petroleum is produced in large volume in a continuous flow.

continuous mill *Metallurgy.* a rolling mill consisting of different synchronized stations, in which the starting stock passes continuously from the first to the last operation.

continuous miner *Mining Engineering.* a machine designed to remove coal from the face, then load it into cars or conveyors without the use of cutting machines, drills, or explosives. Thus, **continuous mining.**

continuous mixer *Mechanical Engineering.* a mixer in which materials are introduced, mixed, and discharged in a steady, continuous flow; used especially for mixing concrete.

continuous operator see CONTINUOUS FUNCTION.

continuous-path robot *Robotics.* **1.** a robot that uses continuous-path control to move along a precisely defined path. **2.** a robot with a control scheme that uses computer input to specify every point along a desired path of movement. Thus, **continuous-path control.**

continuous permafrost *Geology.* an area of permanently frozen subsoil that is virtually uninterrupted by patches of unfrozen ground.

continuous phase *Chemistry.* the major component in a mixture or solution, within which the minor component is contained; for example, in fog the continuous phase is air and the dispersed phase is water or ice. Also, EXTERNAL PHASE. *Materials Science.* in the structure of a multiphase product, the phase that has continuity; that is, the matrix phase.

continuous precipitation *Metallurgy.* a solid-state transformation in which the phase that precipitates from a supersaturated solution grows without the recrystallization of the matrix.

continuous production *Industrial Engineering.* a mechanized production process that runs steadily, rather than in batches. Also, **continuous process production.**

continuous profiling *Geology.* a method of seismic exploration whereby detectors are uniformly spaced along a line in order to obtain a continuous sample.

continuous radionavigation system *Navigation.* an electronic navigation system, such as loran, that is always available; satellite systems are available only periodically.

continuous random variable *Statistics.* a random variable that can take any real number value within a given range.

continuous rating *Engineering.* a rating of a piece of equipment to indicate the constant conditions under which the equipment can operate without reduction of its service or before its power begins to dissipate; for example, a rating of a machine that gives its rated output continuously without exceeding a specified temperature rise or having any other negative effects.

continuous reinforcement *Behavior.* reinforcement in which every appropriate response is immediately reinforced or rewarded. Thus, **continuous reinforcement schedule.**

continuous reward see CONTINUOUS REINFORCEMENT.

continuous-rod warhead *Ordnance.* a type of warhead in which the explosive is encased in a continuous steel rod that shatters into thousands of fragments upon detonation.

continuous-scan thermograph *Radiology.* a device that measures variations in body temperature and displays the data as a continuous image on a cathode-ray tube.

continuous sequence *Metallurgy.* in multipass welding, the sequence consisting of passes that extend through the length of the joint.

continuous sintering *Metallurgy.* in powder metallurgy, sintering performed by moving the compact through a furnace according to a predetermined schedule.

continuous spectrum *Spectroscopy.* an uninterrupted broad band of all colors (wavelengths) that is emitted by incandescent solids; for example, by ceramics, and by liquids and gases under high pressure. *Mathematics.* let T: $X \rightarrow X$ be a linear operator on a complex Banach space X, I the identity operator on X, λ a complex number, and $T_\lambda = \lambda I - T$. Then λ belongs to the continuous spectrum of T if (a) T_λ^{-1} exists, (b) T_λ^{-1} is not continuous, and (c) the range of T_λ is dense in X. If λ does not belong to the continuous spectrum of T, then it belongs to one of the following sets: the resolvent set of T, the point spectrum of T, or the residual spectrum of T.

continuous-spectrum sound *Acoustics.* sound that consists of a wide band of frequencies, many of which are not harmonically related, so that the frequencies that are present are continuous rather than discrete.

continuous speech *Artificial Intelligence.* normal human speech, without silent pauses between words; this makes machine understanding much more difficult.

continuous spinning *Textiles.* an uninterrupted process of extruding, coagulating, washing, and winding of fiber, using a single machine.

continuous stationery *Computer Technology.* letter stock that is in a continuous form, with perforations between sheets and removable sprocket-hole edges; used for mass mailing and other correspondence.

continuous-strip camera *Photogrammetry.* an aerial camera in which the film moves continuously past a slit in the focal plane, producing a photograph in one unbroken length by virtue of the continuous motion of the aircraft.

continuous system *Control Systems.* a system in which the input signals and output signals can be changed at any time. Also, **continuous-time signal system.**

continuous telegraphy *Telecommunications.* the process of varying a carrier wave by means of dot-and-dash transmissions. Also, CONTINUOUS-WAVE MODULATION.

continuous-timing method *Industrial Engineering.* a timing technique in which a stopwatch runs continuously throughout the study and is not reset at the end of each element.

continuous titrator *Analytical Chemistry.* a titrator with a reservoir that refills the buret in order to deliver an ongoing supply of standard solution as titrant.

continuous tone *Graphic Arts.* of an image, having gradations of color density with no clear demarcation, as opposed to the distinct dots found in halftones. Continuous-tone copy can be produced only by the collotype and photogelatin processes.

continuous-tone squelch *Electronics.* a squelch system in which the receiver squelch circuit is activated by a continuous subaudible tone that is transmitted along with the desired voice modulation.

continuous transformation see CONTINUOUS FUNCTION.

continuous-tube process *Materials Science.* a process used in plastics manufacturing in which a continuous extension of plastic tubular material is connected to an arrangement of blow molds as they clamp in a specified order.

continuous variation *Genetics.* the small variations or intermediates occurring among members of the same species; the gradual additions to or diminutions of certain characteristics of a parent by generations of offspring, over a wide spectrum of conditions.

continuous wave *Electromagnetism.* a radio or radar wave with constant frequency, phase, and amplitude.

continuous-wave gas laser *Optics.* a laser in which a quartz envelope containing a mixture of helium and neon generates a beam of electromagnetic radiation.

continuous-wave jammer *Electronics.* a jammer transmitter that emits a single unmodulated frequency to interfere with enemy radar signals, appearing in the form of a picket or rail fence on an enemy radar screen. Also, RAIL-FENCE JAMMER.

continuous-wave laser *Optics.* a laser that produces a beam of coherent light within a narrow range of frequencies for a period of longer than 0.25 second. Also, CW LASER.

continuous-wave modulation see CONTINUOUS TELEGRAPHY.

continuous-wave radar *Engineering.* a radar system whereby an uninterrupted flow of radio energy is sent by a transmitter to a target that redirects a fraction of the energy to a separate antennae. Also, **continuous-wave Doppler radar.**

continuous-wave tracking system *Electronics.* a missile tracking system that keeps a continuous-wave radio beam on a target and determines the target's position from changes in the antenna that must be made to keep the beam locked on to the target.

continuous X-rays *Electromagnetism.* electromagnetic radiation having a continuous distribution rather than discrete spectral distribution, produced when high-velocity electrons strike a target.

continuum [kən tin′yoo əm] *Mechanics.* a material medium that has a continuously distributed mass, such as a bulk amount of a gas, liquid, or solid. Also, **continuous medium.** *Mathematics.* **1.** a real line. **2.** a compact, connected set, such as the set of all real numbers or any closed interval of real numbers. If the set contains more than one point, then it contains an infinite number of points. A given continuum is topologically equivalent to a closed interval of real numbers if, and only if, it has at most two noncut points.

continuum hypothesis *Mathematics.* **1.** Georg Cantor's conjecture that every infinite subset of the real numbers has the cardinal number either of the positive integers or of the real numbers. Stated symbolically, $\aleph_1 = 2^{\aleph_0}$. Both the continuum hypothesis and its negation are consistent with the usual axioms of set theory. **2.** also, **generalized continuum hypothesis.** the conjecture that $\aleph_{\alpha+1} = 2^{\aleph_\alpha}$, for each ordinal number α.

continuum index *Ecology.* the position of a community along a gradient with regard to the distribution pattern of the species in that community.

continuum mechanics or **continuum physics** see CLASSICAL FIELD THEORY.

contorted *Botany.* of proximate leaves or perianth parts, growing in such a way that they twist around one another.

contour *Cartography.* an imaginary line on the ground having all its points at the same elevation above or below a specified reference surface, usually mean sea level. *Physics.* a curve or surface composed of a locus of points that represents a boundary of constant value, such as a barometric surface of constant pressure.

contour analysis *Computer Technology.* in optical character recognition, a line-following scan that follows in turn the outline of each character in the field of view. *Computer Programming.* **1.** the development of contour maps or plots from elevation data or mathematical functions. **2.** an analysis of features such as visibility from contour map data.
contour-change line see HEIGHT-CHANGE LINE.
contour code *Meteorology.* a modification of the international analysis code in which data on the topography of constant-pressure surfaces are transmitted.
contour farming *Agronomy.* a type of cultivation that follows the natural contour of the land and thus serves as a deterrent to erosion. Also, **contouring.**
contour feather *Vertebrate Zoology.* one of the many outermost feathers that form the general covering of a bird and help establish its general body contours.
contour forming *Metallurgy.* the process of ending a sheet metal around one or more dies.
contour furrow *Agronomy.* **1.** a furrow plowed along a contour at a uniform grade. **2.** a level furrow made in a field to prevent soil erosion or run-off.
contour horizon see DATUM HORIZON.
contouring control see CONTINUOUS-PATH CONTROL.
contour integral *Mathematics.* a line integral of a complex function over a (usually closed) curve.
contour interval *Cartography.* the difference in elevation between two adjacent contours.
contour irrigation see BORDER IRRIGATION.
contourite *Oceanography.* a marine sediment deposited along contours by swift bottom currents.
contour length *Materials Science.* the total length of a freely jointed polymer chain; used to characterize the size and spatial arrangement of a polymer.
contour line *Cartography.* a line that connects points of equal elevation on a map or chart. *Meteorology.* a line of constant height on a constant-pressure chart. Also, ISOHYPSE, ISOHEIGHT. *Mathematics.* let F be a family of planes that intersect a surface S and are parallel to some plane P. The contour lines of S on P are the projections onto P of the intersections of the members of F with S.
contour map *Cartography.* a topographic map that portrays relief by the use of contour lines.
contour microclimate *Meteorology.* a portion of the microclimate that is attributed directly to the small-scale variations of ground level.
contour milling *Metallurgy.* the process of milling irregular surfaces.
contourograph *Electronics.* a device using a cathode-ray oscilloscope to produce contoured images that appear to have three dimensions.
contour plan *Mining Engineering.* a plan showing surface contours or calculated contours of coal seams to be developed.
contour plowing *Agronomy.* plowing that is done on a contour or line of equal level.

contour plowing

contour row *Agronomy.* a level row that runs at right angles to the slope regardless of the irregularities of the landscape.
contour turning *Mechanical Engineering.* a method by which a three-dimensional reproduction of a template is produced by controlling the cutting tool with a follower that moves along the template.
contour value *Cartography.* the numerical value that denotes the elevation of a contour line relative to a specified reference surface, usually mean sea level.
contra- a prefix meaning "against" or "opposite," as in *contraindication, contralateral.*
contra-angle *Surgery.* an angulation by which the working point of a surgical device is brought near to the long axis of its shaft; it may involve several angles in the shank.
contra-aperture *Surgery.* a second opening made in an abcess to facilitate the discharge of its contents. Also, COUNTEROPENING.
contraception *Medicine.* the prevention of pregnancy by artificial means, whether by physical devices or agents, or by certain practices.
contraceptive *Medicine.* **1.** describing a device or agent that prevents pregnancy while allowing intercourse. **2.** such a device or agent.
contraceptive diaphragm *Medicine.* a rubber or plastic circular device fitted over the cervix of the uterus for the prevention of conception.
contract archaeology *Archaeology.* archaeological work conducted under the direction and regulations of governments or other agencies.
contracted curvature tensor *Mathematics.* the Ricci tensor.
contracted pelvis *Medicine.* an abnormally narrow pelvis that makes childbirth difficult.
contractible *Mathematics.* suppose X is an arcwise connected topological space containing a point x_0 and let $p: X \to X$ be defined by $p(x) = x_0$ for all $x \in X$. If p is homotopic to the identity map on X, then X is said to be contractible to x_0.
contractile *Biology.* [kän trak´tīl] relating to or having the property of contracting.
contractile ring *Cell Biology.* a bandlike structure of actin and myosin that encircles an animal cell during cell division, gradually tightening to facilitate the separation of the two daughter cells.
contractile vacuole *Cell Biology.* in many protozoan cells, an organelle that maintains osmotic balance by accumulating intracellular water and periodically releasing this water extracellularly via fusion of its membrane with the plasma membrane.
contractility [kän´trak tīl´ə tē] *Physiology.* the ability to shorten in length, as in certain muscle fibers and filaments.
contracting stitching *Textiles.* the process of removing fibers within a parallel arrangement that are shorter than a desired length.
contraction *Mechanics.* a shortening in length or a decrease in volume. *Physiology.* the shortening in length of muscle fibers that occurs when they are stimulated. *Medicine.* specifically, the painful tightening of the muscles of the uterus that occurs at regular intervals during labor; as childbirth approaches the time between contractions decreases and their length and intensity increase. Contractions serve to decrease the size of the uterus and move the fetus through the birth canal. *Mathematics.* an **edge-contraction** on a graph G is the process of removing an edge e and identifying its incident vertices v and w in such a way that the resulting vertex is incident to those edges (other than e) which were originally incident to v or w. That is, after the contraction, the resulting graph H has one less vertex because $v = w$, and at least one less edge since edge e has been removed, and if there is any other vertex of G, say x, that is connected to both v and w, then those two edges are merged in H into one edge connecting x to $v = w$. A contraction of G is any graph that results from G after a series of edge-contractions.
contraction crack *Engineering.* **1.** a crack in metal that results when metal contracts inside a confining mold. **2.** a crack in exterior materials such as concrete, caused by contraction.
contraction hypothesis *Geology.* a theory that folding and thrusting are caused by the cooling and consequent shrinkage of an originally molten earth.
contraction loss *Fluid Mechanics.* the loss in mechanical energy due to friction for turbulent flow through a sudden contraction in a cross-sectional area of a duct or pipe.
contraction rule *Metallurgy.* a ruler designed to measure the dimensions that an object has after its contraction upon cooling.
contraction semigroup *Mathematics.* a semigroup of operators on a normed vector space, such that the operator norm of any member of the semigroup is less than or equal to a positive constant less than one. This means that as the semigroup parameter is allowed to become arbitrarily large, the contraction becomes arbitrarily small.

contract therapy *Behavior.* a form of behavior therapy in which the desired form of behavior is defined beforehand and reinforcements for the performance of this behavior are explicitly established. Also, CONTINGENCY CONTRACTING.

contracture *Medicine.* the permanent contraction of a muscle due to tonic spasm or to the shortening or scarring of the tissue. *Architecture.* a narrowing of the girth of a column, usually as a function of entasis.

contrail see CONDENSATION TRAIL.

contrail-formation graph *Meteorology.* a graph containing the parameters pressure, temperature, and relative humidity for critical values at which condensation trails form; used in forecasting the formation of condensation trails.

contraincision [kän´trə in sizh´yən] *Surgery.* a counterincision that is made to encourage drainage.

contraindication *Medicine.* a condition, especially any condition of disease, which renders some particular line of treatment undesirable or improper.

contralateral *Anatomy.* referring to the opposite side, usually of the body. Also, HETEROLATERAL.

contrapositive *Mathematics.* the contrapositive of the logic statement "if p, then q" is "if not q, then not p." The contrapositive has the same truth value as the original statement.

contrariety see NEGATIVISM.

contrarotating propellers *Mechanical Engineering.* a pair of propellers mounted on concentric shafts and rotating in opposite directions.

contrarotation *Engineering.* the circular path of an object that is in the opposite direction of a related object's circular path.

contra solem *Meteorology.* of or relating to motion against the sun, which is to the left in the Northern Hemisphere and to the right in the Southern Hemisphere.

contrast *Graphic Arts.* the range of light-to-dark tones within an original artwork or reproduction. *Telecommunications.* the range of light and dark values in a televised image, or the ratio between the maximum and minimum brightness values. A **high-contrast** image contains intense blacks and whites; a **low-contrast** image contains only shades of gray. *Radiology.* the difference in optical density in a radiograph, resulting from the difference in radiolucency or penetrability of the subject. *Computer Technology.* in optical character recognition, the background reflectance of a document surface within the clear band, as compared to a reference standard.

contrast control *Telecommunications.* **1.** a device, usually a variable resistor, used to adjust the range of light and dark values in a cathode-ray tube picture signal to achieve a desired level of contrast on the tube screen. **2.** a method of automatically or manually adjusting the difference between the maximum and minimum brightness values in a television picture signal by varying the amplitude of the video signal.

contrast enema see BARIUM ENEMA.

contrastes *Meteorology.* the winds that blow from opposite directions at fairly close points along the Spanish Mediterranean coast.

contrast ratio *Electronics.* the ratio of the maximum to minimum luminance values in a television picture or similar cathode-ray tube display.

contrast sharpening *Graphic Arts.* any of several techniques used to increase the delineation between light and dark areas of a photograph or other artwork, especially in platemaking.

contrast solution *Radiology.* a solution of a medium that is opaque to X-rays, used to improve visualization of a structure under study.

contrasuggestibility see NEGATIVISM.

contrasuppression *Immunology.* the action of a group of T cells that renders T helper cells resistant to the action of T suppressors.

contravane *Aviation.* a vane that inhibits or reverses the rotation of a flow. Also, COUNTERVANE.

contravariant functor *Mathematics.* let C and D be categories, and let S represent the following pair of functions: an object function that assigns each object C of C to an object $S(C)$ of D, and a morphism function that assigns each morphism f: $C \rightarrow D$ to a morphism $S(f)$: $S(C) \rightarrow S(D)$. S is called a contravariant functor if: (a) $S(1_C) = 1_{S(C)}$ for every identity morphism 1_C of C; and (b) $S(g \circ f) = S(f) \circ S(g)$ for any two morphisms f and g of C whose composite $g \circ f$ is defined. That is, the direction or sense of morphisms is reversed by the morphism function of a contravariant functor S.

contravariant index *Mathematics.* one of the superscript indices of the components of a tensor field in the Ricci calculus.

contravariant tensor *Mathematics.* a multilinear functional that accepts only covariant vectors (e.g., differential forms) as arguments. If there are r covariant vector arguments, it is a contravariant tensor of order r or contravariant r tensor. Every contravariant r tensor can be expressed as a sum of tensor products of r contravariant vectors (r in each term). The components of a contravariant r tensor are represented by symbols with r superscripts.

contravariant vector *Mathematics.* in differential geometry, a vector tangent to a manifold; i.e., a contravariant tensor of order 1.

control a device or system that regulates or directs some operation; specific uses include: *Aviation.* **1.** a switch, lever, knob, button, cable, or other device used to direct or regulate some aspect of flight. **2. controls.** the system of such devices in a cockpit, used to deflect the control surfaces of an aircraft. **3.** the control surfaces themselves. **4.** an activity or organization that regulates air traffic; air traffic control. *Science.* **1.** an experiment to evaluate another experiment to which it is identical except for one variable; in full, CONTROL EXPERIMENT. **2.** a group serving as a standard of comparison with another group to which it is similar except for one variable; in full, CONTROL GROUP. *Statistics.* to eliminate, identify, or equalize factors that could affect the outcome in designed experiments. *Control Systems.* the process or method of making a system or device operate or change according to a specific set of criteria. *Computer Technology.* the component of a computer processor that supervises operations; in full, CONTROL UNIT. *Computer Programming.* **1.** in a program, an instruction that determines the sequence of instructions to be executed. **2.** see CONTROL KEY. *Electronics.* a component such as a variable resistor, switch, or channel selector used in the operation of a piece of electronics equipment. *Cartography.* the data associated with a set of control stations and used as the basis for detailed surveys.

control accuracy *Control Systems.* the degree to which one or more variables will correspond to predetermined values in a control system.

control agent *Chemical Engineering.* in a process system under automatic control, the operation condition on which the controlled variable is dependent.

control air support *Military Science.* fighter plane support of other aircraft or of ground forces.

control algorithm *Design Engineering.* a mathematical representation of the control action to be executed in a controlled device.

control area *Navigation.* a controlled airspace extending upward from a specified limit above the earth, usually 1500 feet.

control base *Cartography.* a surface on which map projection and ground control have been plotted, and on which templates have been assembled or aerotriangulation has been accomplished, and control points thus determined have been marked.

control bit *Computer Programming.* a single-bit register used as an addressing mode bit in a microprocessor; a 1 signifies that the operand address in the microinstruction is indirect. Also, MODE BIT.

control blackboard *Artificial Intelligence.* a global process control mechanism that schedules and prioritizes tasks; analogous to the operating system in a multiprogramming environment.

control block *Computer Programming.* the area of main memory that contains control information.

control board *Engineering.* a board on which indicating instruments, key diagrams, and other accessory apparatus are mounted, and from which a system is manually operated by remote control. *Electronics.* see CONTROL PANEL.

control break *Computer Programming.* a change in the value of a key field that causes a change in processing, such as computing a subtotal after printing the last of a group of records.

control character *Computer Programming.* **1.** see COMMAND CHARACTER. **2.** in the ASCII character set, a character typed in conjunction with the control (CTRL) key.

control chart *Industrial Engineering.* a chart used to plot process outcomes in order to ensure that they lie within acceptable limits; typically used for quality control in industrial production processes.

control circuit *Electricity.* any circuit that regulates some function of a device or system. *Computer Technology.* **1.** a circuit in a computer's hardware that causes the computer to execute instructions in the proper sequence. **2.** a circuit used to transmit supervisory information to coordinate with transmission occurring on another circuit. *Electronics.* a circuit that carries the electric signals directing the performance of the controller but does not carry the main power circuit.

control code *Computer Technology.* a command sent interactively by a user to perform a particular operation on the active file or screen, often during word processing or database management. Many control codes are sent by pressing a function key or by pressing two keys simultaneously, such as *alternate (option)* and a letter key, or *control* and a function key.

control column *Aviation.* an apparatus designed to control the elevators and ailerons on an airplane; it usually consists of a lever with a wheel mounted at its upper end, the lever moving back and forth to operate the elevators and the wheel rotating to adjust the ailerons.

control computer *Computer Technology.* a special-purpose computer that is used to monitor the output and control the performance of complex devices and processes.

control counter see PROGRAM COUNTER.

control data *Computer Programming.* a data element that is used to identify, select, execute, or modify another set of data, a record, a file, or a routine.

control day *Meteorology.* a day on which the weather is supposed to provide the key for the general weather conditions of the following period, according to folklore or popular belief; for example, Groundhog Day (February 2). Also, KEY DAY.

control drive *Nucleonics.* the system of control rods that control the reaction rate in a reactor.

control echo *Engineering.* a controlled signal used in an ultrasonic inspection system, produced by consistent reflection from a surface.

control electrode *Electronics.* an electrode used to initiate or vary the current between two or more electrodes.

control element *Control Systems.* the element of a control system that actuates the machine or process being controlled.

control equipment *Design Engineering.* in machine assembly, equipment containing one or more devices that interact and manipulate a control variable.

control experiment *Science.* an experiment made to test the results of another experiment to which it is identical in all respects except for one factor or variable.

control feel *Aviation.* the degree of authority that a pilot feels over an aircraft's attitude or manueverability; such a feeling may come directly from aerodynamic forces acting on the aircraft's control surface, indirectly from devices that monitor and simulate the aerodynamic forces, or from external forces.

control field *Telecommunications.* **1.** in bit-oriented protocol data communications, a series of characters that identify the type of frame being transmitted, such as information, supervisory, or unnumbered. **2.** a fixed position in a record that contains control information.

control flight see CONTROL STRIP.

control force *Aviation.* **1.** an aerodynamic force acting on a control surface. **2.** specifically, the stick force that the pilot feels in a direct manual system.

control grid *Electronics.* **1.** a grid, usually placed between the cathode and an anode, that is used as a control electrode. **2.** the electrode of an electron tube, other than a diode, upon which a signal voltage is impressed to regulate plate current.

control-grid bias *Electronics.* the average direct-current voltage between the control grid and the cathode of an electron tube.

control group *Science.* in an experiment or survey, a group or individual that serves as a standard of comparison with another group or individual to which it is similar or even identical in all respects except for one factor or variable. *Medicine.* specifically, the individual or individuals in a controlled medical study who do not receive the treatment or medication under study.

control hierarchy *Control Systems.* a division of the control elements of a system into various levels of priority, so that the elements in higher-priority levels send control signals to elements in lower-priority levels.

control information *Computer Programming.* program information that is sent between units to control their operations.

control instruction *Computer Programming.* a program statement that controls the flow of execution, such as a jump or branch instruction.

control joint *Building Engineering.* in masonry, a joint that is utilized for expanding and contracting movements between building materials attached to it.

control key *Computer Technology.* a key on a computer keyboard that has no meaning by itself but, when pressed simultaneously with another key, gives that key a new meaning; examples are the *shift, control,* and *alternate (option)* keys.

control knowledge *Control Systems.* any information that affects the choice of an appropriate control strategy.

controllability *Aviation.* the speed and ease with which a flight vehicle responds to its controls, particularly in changing direction, attitude, or performing a complete maneuver. *Control Systems.* that property of a system by which an input signal can take the system from an initial state to a desired state within a predetermined period of time.

controllable-pitch propeller *Mechanical Engineering.* any of various aircraft or ship propellers in which the pitch of the blades can be changed while the propeller is functioning. Also, CP PROPELLER.

control lead *Computer Technology.* in data communications, a character or field that indicates that what follows is control data rather than information.

controlled access *Transportation Engineering.* a highway to which access is provided only at certain points. On non-access-controlled roads, access is permitted from any point along adjacent property. Also, LIMITED ACCESS.

controlled airport *Navigation.* an airport at which operations are carried out under ground control.

controlled airspace *Aviation.* a specified portion of airspace within which some or all aircraft may be subject to air traffic control.

controlled atmosphere *Science.* laboratory conditions involving the presence of specified gases at a specified temperature, pressure, and humidity.

controlled avalanche device *Electronics.* a semiconductor device with specific maximum and minimum avalanche-voltage characteristics that can operate and absorb momentary power surges in the avalanche region indefinitely without damage.

controlled avalanche diode *Electronics.* an avalanche diode with a well-defined avalanche voltage.

controlled avalanche transit-time triode *Electronics.* a high-speed controlled avalanche device operating at microwave frequency range.

controlled carrier modulation *Telecommunications.* a signal transmission in which the magnitude of the carrier is controlled by the signal.

controlled cooling *Metallurgy.* cooling from an elevated temperature according to a schedule of cooling rates and, at times, holding periods.

controlled-effects nuclear weapon *Ordnance.* a nuclear weapon in which the intensity of specific effects is designed to vary from the normal distribution.

controlled flooding *Agronomy.* the practice of covering an entire area with water to a depth of a few inches as a method of irrigation.

controlled fragment *Ordnance.* a piece from the casing of a warhead that is designed to break up in a particular pattern; the final shape of the piece is formed in the explosion. Also, FIRE-FORMED FRAGMENT, SELF-FORGING FRAGMENTATION.

controlled fusion *Nucleonics.* the process of combining light-element nuclei, especially of the hydrogen isotopes deuterium and tritium, in a plasma that is heated to extremely high temperatures in an appropriately shaped magnetic field or in a region defined by inertial confinement. Also, **controlled thermonuclear reaction.**

controlled-leakage system *Space Technology.* an environmental waste management system in which harmful products such as carbon dioxide and other waste are allowed to escape from a cabin and are replaced by fresh oxygen, food, and so on.

controlled map *Cartography.* a map based on precise horizontal and vertical ground control, resulting in accurate scale, azimuths, and elevations.

controlled mercury-arc rectifier *Electronics.* a rectifier tube containing mercury vapor, in which one or more electrodes control the point along the input cycle waveform at which the mercury vapor is ionized and plate current starts to flow, thus controlling output current.

controlled mine *Ordnance.* a mine, either on land or under water, that can be controlled by the user after laying, to the extent of making it safe, live, or activating the firing device.

controlled mosaic *Cartography.* a photomosaic oriented, scaled, and conforming to established horizontal ground control.

controlled parameter *Engineering.* one of several parameters that determine the value of a criterion parameter in an optimization problem.

controlled-path robot *Robotics.* a robot that uses the controlled-path system of operation.

controlled-path system *Robotics.* a computer control system that can numerically describe a path between previously programmed points.

controlled rectifier *Electronics.* a rectifier using grid-controlled devices to regulate its output current.

controlled thermonuclear reactor see FUSION REACTOR.

controlled variable *Science.* a value or condition that is manipulated, usually in order to discover its effect on another value or condition. *Control Systems.* any quantity or condition in a control system that can be changed by the system to produce a desired result. *Computer Programming.* a variable that is forced to assume a specific set or sequence of values during an iterative process; the number of these values is dependent on the number of times the loop is to be executed.

controller *Transportation Engineering.* a person who is responsible for directing the movements of aircraft in a given area; an air traffic controller. *Computer Technology.* in industrial control applications, the comparator and the control algorithms implemented within a hardware unit. *Robotics.* an information-processing device that provides instructions to a robot. *Control Systems.* a device that directs the transmission of information over the data links of a network; it is controlled by a program either within the device or in a processor to which the controller is connected.

controller hierarchy *Robotics.* the relationship between controller elements in a hierarchical control system.

controller-structure interaction *Control Systems.* the use of feedback from observation and control spillover to change a program.

control limits *Industrial Engineering.* on a control chart, the boundary limits beyond which an out-of-control or unacceptable situation is indicated. *Electronics.* a range of operating parameters established in radar performance evaluation to define the normal operation of the system.

controlling *Industrial Engineering.* the management function that involves monitoring of activities within an organization and keeping the organization on track toward its goals.

controlling depth *Navigation.* the minimum depth in a channel or approach, which limits the draft of entering vessels.

controlling element *Genetics.* either of two types of eukaryotic transposable elements (receptor and regulator elements), detectable through the abnormal activity of the genes that it affects.

controlling gene *Genetics.* a gene, such as a regulator gene, that turns cistrons on and off.

controlling obstacle *Navigation.* the highest ground or surface obstruction within a charted section of airspace, determining the minimum altitude at which aircraft can fly within the section without hazard.

control logic *Computer Programming.* the arithmetic and Boolean expressions in a program that determine the flow of control and sequence of operations.

control loop *Computer Technology.* in process control, the continuous loop formed by the sensor receiving input, the computer evaluating the input and producing a signal to an actuator that physically controls the process, and the sensor monitoring the output, thus completing the loop.

control memory *Computer Programming.* in a control unit, a writable or read-only memory that stores all the allowable microinstructions used in processing. Also, MICROCODE STORE.

control-message display *Computer Technology.* a printing or CRT device that presents processing-event messages to the operator.

control models *Design Engineering.* mathematical or computer models that describe control actions used for the purpose of simulation.

control module *Space Technology.* a module on certain rocket vehicles that receives and transmits directional signals during flight. *Computer Programming.* a subprogram within one of the control programs of an operating system. *Computer Technology.* the circuitry and registers required to perform a given function.

control operation *Computer Programming.* any action that is not directly related to obtaining the results of a computer process, such as tape rewind, memory management, or end-of-transmission signaling.

control panel *Electronics.* a panel containing an array of switches, dials, and other equipment, used to control the operation of some system or device, such as an aircraft. Also, CONTROL BOARD. *Computer Technology.* specifically, such a panel from which a computer operator monitors and controls the computer operations.

aircraft control panel

control pit *Archaeology.* a preliminary excavation done to determine the nature of a site and establish the techniques needed for actual excavation. Also, TEST PIT, SONDAGE.

control point *Aviation.* in air navigation, a point on the earth's surface over which an aircraft is scheduled to fly at a predetermined time. *Cartography.* an identifiable point on a photograph to which other data on the photograph can be related. *Transportation Engineering.* in a traffic control system, any place housing signals and switches. *Computer Technology.* **1.** in an automatic-control system, the specified value of a controlled variable such as velocity, temperature, or humidity that the controller attempts to maintain. **2.** in hardware circuitry, the point at which the output of the instruction decoder activates intercycle register-to-register communications and operations of other system resources.

control premise *Artificial Intelligence.* a clause placed in the "if" part of a rule, used as a conditional control mechanism.

control program *Computer Programming.* any of a class of programs within the operating system of a computer that control execution of other system resources, such as a scheduler program.

control read-only memory *Computer Technology.* in a computer with a microprogrammed control unit, the separate memory unit that contains all control information and microinstructions; not available to the user, in contrast with main memory.

control record *Computer Programming.* a record that contains the data used in originating, modifying, or terminating a control operation.

control register see PROGRAM COUNTER.

control reversal *Aviation.* **1.** a process in which the aeroelastic load exceeds the control surface loading of a flight-control system, causing the system to act in a direction opposite to that indicated or desired. **2.** a stick or rudder force reversal when a control surface goes hard over.

control rocket *Space Technology.* a small thruster that can be fired intermittently to alter the attitude or speed of a spacecraft.

control rod *Nucleonics.* movable pieces or assemblies of neutron-absorbing material such as cadmium, boron, and hafnium, whose movement affects the reactivity of the reactor system.

control room *Engineering.* a room from which engineers and technicians operate and control a facility such as a power plant or refinery. *Space Technology.* a room from which the flight of space vehicles is monitored and controlled. *Telecommunications.* a room that is adjacent to and communicating with a sound stage or recording studio, and from which broadcasts or recording sessions are directed.

control rule *Artificial Intelligence.* a procedure statement coded in rule form that directs the inference engine to examine the domain in a specific order.

control section *Computer Technology.* the core or governing part of a central processing unit.

control sequence *Computer Programming.* the order in which instructions are to be processed, usually in sequential order unless a branch instruction is encountered.

control signal *Telecommunications.* a signal used to convey information such as a dial tone, a busy signal, a ringing tone, and tones heard when dialing a number. *Computer Technology.* any of a set of pulses used by a microprocessor to control the transfer of data over a bus; signals include read, write, interrupt request and interrupt acknowledge, bus request and bus granted, and clock input and output.

control spillover *Control Systems.* any actions or movements in a control system that are not part of the preprogrammed process it was designed to carry out.

control spring *Mechanical Devices.* a spring often used in compasses designed so that, in any pointer deflection, its torque balances that of the instrument itself.

control statement *Computer Technology.* a statement that is interpreted by an operating system to control the processing of a job or interactive session. *Computer Programming.* see CONTROL INSTRUCTION.

control station *Cartography.* **1.** a point on the ground having a horizontal or vertical location, or both, used as a basis for obtaining the position of other points. **2.** a survey station having coordinates sufficiently accurate that the coordinates of other survey stations can be determined by reference to it.

control stick *Aviation.* a lever used to control the altitude and attitude of an airplane or to initiate a desired maneuver.

control strategy *Control Systems.* the process or technique of choosing the next action from a number of alternative problem-solving steps.

control strip *Photogrammetry.* a strip of aerial photographs taken to aid in planning and producing later photographs, or to serve as a control in assembling other strips. Also, CONTROL FLIGHT, TIE FLIGHT, TIE STRIP.

control structure *Computer Programming.* **1.** the ability of a programming language to make a necessary deviation from the normal sequential execution of instructions within a program unit; this includes facilities for selection control, exception handling, and repetition control. **2.** the manner in which control of processing in a program is organized. *Artificial Intelligence.* the method by which a program or problem-solving system determines which task to perform next.

control supervisor *Computer Programming.* in an operating system, a set of system programs that provides services and supervises the running of user programs.

control surface *Aviation.* any movable surface designed to control or maneuver a flight vehicle, especially the rudder, elevators, ailerons, dive brakes, or spoilers. *Thermodynamics.* the surface of an open thermodynamic system, which permits the flow of mass and energy into or out of the system.

control survey *Cartography.* a survey to provide horizontal and vertical positions of points to which supplementary surveys are adjusted.

control system *Engineering.* a system in which inputs and outputs are progressively altered in a well-planned way.

Control Systems

Control Systems is a field of engineering that is concerned with obtaining prespecified or desired dynamic behavior from physical systems. This desired behavior is obtained by augmenting these physical systems with controllers which exercise restraining or directing influence (or control) over the physical system, causing it to act in this prespecified way. The control is usually applied to the physical system through an actuator, a device which translates the command as physical action affecting the system.

Most control systems employ feedback and are called closed-loop systems. Closed-loop systems derive their commands from measurements of the performance attributes of the physical system being controlled. These measurements then serve as inputs to the controller which integrates the information into a control law to produce the desired commands. Most modern controllers are implemented in some form of electronics that includes a microcomputer. The measurements then serve as data to a program that runs on the computer, and the commands are the outputs of this program. This computer program is usually a mathematical algorithm based on a mathematical model of the physical system being controlled. The control program is executed repetitively at a frequency that depends on the rates at which the performance attributes change.

Control systems have a wide range of applications. They have found their way into machine tool systems, where they control everything from the motions of an individual tool to the complex operations of robotic systems and groups of machines. In the chemical process industry they are used to control quantities like levels, flow rates, and temperatures of materials in a process. They are also used in the control of both the power plants and dynamics of mobility systems from planes and trains to automobiles.

Nader M. Boustany
Section Head
Systems Analysis & Information Management Dept.
General Motors Systems Engineering

Neil Schilke
Technical Director
Systems Engineering Center
General Motors Systems Engineering

control-system feedback *Control Systems.* a signal from a control element in a control system that represents the error of the system, and can be used by the controller to change system performance by minimizing the system error.

control-systems equipment *Computer Technology.* the hardware components of a computer system that are used for process control, including sensory inputs, processor, control outputs, communications hardware, and a real-time clock.

control tape *Computer Technology.* a paper or plastic tape used to control the carriage movement on some printing output devices. Also, CARRIAGE TAPE.

control total *Computer Programming.* an editing technique designed to detect some data transmission errors; a batch control total is developed during data preparation, then used to validate the data during a subsequent input operation.

control tower *Transportation Engineering.* a raised observation platform at an airport used by air traffic controllers, providing them a general view of the airport and its surrounding airspace.

control track *Acoustical Engineering.* a supplementary audio track that is used to control the speed of the main track or enhance its effects.

control transformer *Electricity.* a synchro in which the output of the rotor is a function of the position of the shaft and the electric input to the stator. Also, **control synchro.**

control unit *Computer Technology.* the element of the control processing unit that supervises processor functions; it receives and interprets program instructions, sends control signals, and routes data to other units of the computer.

control valve *Engineering.* a valve that reduces or increases the flow of fluid through a pipe.

control vane *Aviation.* a refractory surface in the exhaust flow used to control attitude and, in turn, the trajectory of a flight vehicle when it is deflected from the neutral setting.

control variable *Control Systems.* any variable in a system that, when changed, will cause a change in the performance of the system. *Computer Programming.* a variable that assumes a specific set or sequence of values during an iterative process; the number of these values is dependent on the number of times the loop is to be executed.

control volume *Thermodynamics.* the volume enclosed by the control surface that surrounds an open thermodynamic system, which is fixed in space but which does permit the flow of mass and energy across the boundaries.

control winding *Electronics.* **1.** an excitation winding carrying a current that controls the performance of a machine. **2.** the winding used to apply a controlling magnetomotive force to the core in a saturable reactor.

control word *Computer Programming.* **1.** in channel operations, a word, often stored in a reserved memory cell, that contains the length and initial address of the block of data to be transferred. **2.** in microprogramming, a microoperational step that can be executed in a single time interval; a microinstruction.

control zone *Transportation Engineering.* the zone of airspace, defined geographically and rising upward from the surface, that is under the observation and control of a given ground control station.

contusion *Medicine.* a bruise injury without laceration, usually caused by a blow or severe compression.

Conulariida *Paleontology.* an order of problematic forms, now generally placed in the cnidarian class Scyphozoa; characterized by a pyramidal test and an often pronounced ridge pattern on the outside of the shell, which was chitinous but partly phosphatized; extant in the Middle Cambrian to Late Triassic. Also, **Conularia.**

Conulata *Invertebrate Zoology.* a subclass of free-living marine coelenterates in the class Scyphozoa, with a cone or pyramid shape and tentacles around the mouth.

Conulus *Paleontology.* an extinct genus of euechinoids belonging to the order Holectypoida; they were exocyclic and were characterized by a medium test, a flat, patterned oral side, and an almost smooth aboral dome; extant in the Upper Cretaceous.

conurbation [kän′ər bā′shən] *Geography.* a large area of unbroken urban development created by the growth and coalescence of two or more neighboring cities.

conus arteriosus *Developmental Biology.* the most anterior chamber of the heart, leading into the ventral aorta; in mammals this becomes reduced to the connection between the right ventricle and the pulmonary artery. Also, ARTERIAL CONE.

convalescence *Medicine.* the period of recovery following an attack of disease, a surgical operation, or an injury. Thus, **convalescent.**

convalescent carrier *Medicine.* a person who harbors a pathogenic agent after partially or fully recovering from its attack, and who may therefore cause the infection of others.

convalescent serum *Immunology.* a sample of serum taken from a patient who is in the recovery stage of an infectious disease.

convection *Fluid Mechanics.* a transfer of heat or mass that occurs when a fluid flows over a solid body or inside a channel while temperatures or concentrations of the fluid and the boundary are different; transfer occurs within the fluid as a consequence of the motion within the fluid relative to the flow boundary. *Geophysics.* a pattern of lateral movements of subcrustal mantle material, vertically and horizontally, due to heat variations in the earth. *Oceanography.* the movement and mixing of ocean water masses, usually caused by temperature differences between them. *Meteorology.* the predominantly vertical atmospheric motions that result in vertical transport and mixing of atmospheric properties. *Physics.* any transportation of energy or matter by currents of matter.

convection cell *Geophysics.* the lateral movement of subcrustal mantle material, due to variations of heat in the earth. *Meteorology.* a body of air in which the process of convection proceeds in an orderly fashion, giving rise to fluid motion properties.

convection coefficient see FILM COEFFICIENT.

convection cooling *Engineering.* a natural cooling process whereby hot air flows upward from the object that is being cooled.

convection current *Geophysics.* the vertical movement of air currents, upward and downward, due to temperature variation. *Meteorology.* an air current principally involved in the upward moving portion of a convective circulation such as a thermal or an updraft inherent in cumulus clouds. Also, CONVECTIVE CURRENT. *Electronics.* the time rate at which a charge in an electron current is transported through a given surface.

convection-current modulation *Electronics.* the time variation in the magnitude of the convection current passing through a surface, or the process of producing such a variation.

convection section *Engineering.* a part of a furnace in which heat from flue gases enters tubes by convection.

convection theory of cyclones *Meteorology.* a cyclone development theory proposing that the upward convection of air resulting from surface heating can be of sufficient magnitude and duration to allow the surface inflow of air to attain appreciable cyclonic rotation.

convection zone see CONVECTIVE ZONE.

convective activity *Meteorology.* the manifestations of convection in the atmosphere, including the development of convective clouds, showers, thunderstorms, squalls, hail, and tornadoes.

convective atmosphere see ADIABATIC ATMOSPHERE.

convective cloud *Meteorology.* a cloud whose vertical development, and possibly its origin, result from convection.

convective-cloud-height diagram *Meteorology.* a graph based on the dew-point formula, used in estimating the altitude of the base of convective clouds.

convective condensation level *Meteorology.* **1.** the intersecting point on a thermodynamic diagram of the vertical distribution of temperature in an atmospheric column with the saturation mixing ratio line that corresponds to the average mixing ratio in the surface layer. **2.** the level above which a lifted parcel of air becomes freely convective as a result of condensation occurring during lifting.

convective current see CONVECTION CURRENT.

convective discharge *Electronics.* the flow of a stream of particles carrying away charges from a body that has received a sufficiently high voltage charge.

convective equilibrium see ADIABATIC EQUILIBRIUM.

convective instability *Meteorology.* **1.** a column or layer of air in the atmosphere whose wet-bulb potential temperature decreases with elevation. **2.** the condition of an unsaturated parcel or layer of air that will result in instability if it is lifted bodily to saturation. Also, POTENTIAL INSTABILITY.

convective precipitation *Meteorology.* any precipitation from convective clouds.

convective region *Meteorology.* a layer in the lower atmosphere that is particularly favorable for the formation of convection, or a layer characterized by convective activity at a given time.

convective zone *Astrophysics.* a region inside a star or planet where energy is transported by hot material rising and cool material sinking.

convector *Engineering.* a heating unit that has openings for the air to enter, become warm, and then exit.

convectron *Engineering.* a device that determines an object's deviation from the vertical, using the principle that convection from a heated wire is dependent on its inclination.

convenience food *Food Technology.* a processed food product that is packaged and sold for home consumption but requires little or no preparation by the user.

conventional *Ordnance.* of or relating to nonnuclear weapons, forces, or operations; this also excludes biological weapons and most chemical agents, except existing smoke and incendiary agents and agents of the riot-control type. Thus, **conventional bomb, conventional forces, conventional warfare, conventional weapons.**

conventional current *Electricity.* the concept of current as a positive charge flow in a direction opposite to that of electrons, which are negatively charged.

conventional level *Psychology.* in Lawrence Kohlberg's analysis of moral development, the second level, showing awareness of and conformity to the rules of the group or society. This level is itself divided into stage three, which is marked by an emphasis on loyalty and commitment to other people, and stage four, marked by loyalty and commitment to the larger society. Also, **conventional morality.**

conventional milling *Metallurgy.* milling by moving the cutter in a direction opposite to that of the workpiece.

conventional programming *Computer Programming.* the use of high-level language statements to create an application program rather than starting with a software package or an applications generator.

conventional stage see CONVENTIONAL LEVEL.

converge *Science.* to meet or come together. *Mathematics.* an iterative process or infinite sequence is said to converge if a value or meaning can ultimately be ascribed to it.

converge absolutely *Mathematics.* to have the property of absolute convergence.

converge in measure *Mathematics.* a sequence $\{f_n\}$ of functions defined on a measure space M with measure μ is said to converge in measure to a function f if, given any $\varepsilon > 0$, there is an N such that $\mu(\{x : |f(x) - f_n(x)|\}) < \varepsilon$, for $n > N$.

converge in the mean *Mathematics.* a sequence $\{f_n\}$ of real-valued functions defined on a measure space M with measure μ is said to converge in the mean of order p to a function f if $\lim_{n\to\infty} \int_M |f(x) - f_n(x)|^p d\mu = 0$. In modern usage, called L^p convergence or convergence in L^p . The case $p = 2$ is known as **convergence in mean.**

convergence the process of meeting or coming together; specific uses include: *Optics.* **1.** the direction of visual lines to a near point. **2.** the tending of two objects toward a common point. *Physiology.* the process of turning the eyes inward to focus on a near or approaching point. *Evolution.* an evolutionary change producing similar hereditary characters in two or more distantly related forms as a result of their separate adaptation to common ecological conditions. *Anthropology.* the process by which similar traits are produced in different cultures by unlike circumstances. *Oceanography.* **1.** the joining of ocean currents or water masses that differ in density, temperature, or salinity, resulting in the sinking of the denser, colder, or more saline water. **2.** the zone in which such sinking occurs. *Hydrology.* the boundary separating turbid river water from clear lake water. *Meteorology.* the net flow of air into a given region. *Geology.* the gradual process in which the vertical distance between two given rock units or geologic horizons diminishes as a result of the thinning of the intervening strata. *Developmental Biology.* the movement of cells from the periphery toward the midline during gastrulation. *Anatomy.* a point where the axons of several neurons form synapses with a single or few postsynaptic neurons. *Electronics.* a condition in which the electron beams of a multibeam cathode-ray tube intersect at a specified point.

convergence circuit *Electromagnetism.* an auxiliary deflection system that maintains convergence in a color television receiver.

convergence coil *Electronics.* either one of the two coils used to ensure convergence of the electron beams in a color television receiver.

convergence constant *Cartography.* the angle at any given latitude between meridians that are 1° apart, used to determine the convergence of meridians. Also, CONVERGENCE FACTOR.

convergence control *Electronics.* a variable resistor in the high-voltage section of a color television receiver that controls the voltage applied to the three-gun picture tube.

convergence electrode *Electronics.* an electrode, in a multibeam cathode-ray tube, with an electric field that converges two or more electron beams.

convergence factor see CONVERGENCE CONSTANT.

convergence line *Meteorology.* any horizontal line along which horizontal convergence of the airflow is occurring. Also, ASYMPTOTE OF CONVERGENCE.

convergence magnet *Electronics.* a magnet assembly in a multibeam cathode-ray tube whose magnetic field converges two or more electron beams.

convergence of meridians *Cartography.* the mutual approach of the meridians on a spherical surface as they pass from the equator to the poles.

convergence phase control *Electronics.* a variable resistor that adjusts the phase of the dynamic convergence voltage in a color television receiver with a three-gun picture tube.

convergence pressure *Physical Chemistry.* the pressure at which vapor-liquid equilibrium ratios converge to unity under constant temperature conditions.

convergence zone *Oceanography.* an ocean region where a convergence of water occurs. *Acoustics.* an ocean region in which the intensity of sound is 10 to 20 decibels greater than it would be in a free field, due to various conditions of temperature and pressure that bend all the sound waves toward these areas. Occurring at depths between 750 and 4500 meters, this channel produces patterns of sound refraction that are used in sonar technology.

convergent camera *Photogrammetry.* an assembly of two cameras that take simultaneous photographs and are mounted so as to maintain a fixed angle between their optical axes; usually, this procedure increases the angular coverage in one direction along the longitudinal axis of the aircraft.

convergent die *Engineering.* a cutting tool used for molding objects, having an internal structure of grooves that converge.

convergent-divergent nozzle *Mechanical Devices.* a nozzle that first constricts the flow at the float, then expands at the exit cone area to efficiently expand gases in supersonic aircraft or steam in a steam turbine. Also, SUPERSONIC NOZZLE, PROPELLING NOZZLE.

convergent integral *Mathematics.* any integral over an unbounded region that can be assigned a finite value. Any integral that is not convergent is divergent.

convergent lens see CONVERGING LENS.

convergent position *Photogrammetry.* a split camera installation positioned so that the plane containing the camera axis is parallel to the line of flight.

convergent precipitation *Meteorology.* precipitation arising from local updrafts of moist air.

convergent sequence *Mathematics.* an infinite sequence of numbers that has a limit. Any sequence that is not convergent is divergent.

convergent series *Mathematics.* a series whose associated sequence of partial sums is convergent. Any series that is not convergent is divergent.

convergent thinking *Psychology.* the aspect of thinking that enables an individual to solve problems having a single correct answer.

convergent zone paths *Oceanography.* the property of deep sound transmission channels that produces sound-wave focusing areas at distant intervals from a source near the surface.

converging lens *Optics.* a lens that is thicker at its center than at its edge, bending light rays toward each other as they pass through it. Also, CONVERGENT LENS, CONVEX LENS, POSITIVE LENS.

converging meniscus lens *Optics.* a lens having a convex surface of greater curvature than its concave surface. Also, POSITIVE MENISCUS LENS.

conversational compiler *Computer Programming.* a language translator that transforms a program statement directly into a set of machine instructions and may display syntax error messages to the programmer.

conversational mode *Computer Processing.* a condition of real-time communications between one or more remote terminals and a time-sharing computer, in which each entry from a terminal elicits an immediate response from the computer in a manner similar to a dialogue between two persons. The remote terminal can control, interrogate, or modify a task within the computer.

conversational processing *Computer Programming.* **1.** a mode of computer operations in which each command or statement entered by the user is interpreted, executed, and answered before the next statement or command can be entered. **2.** a mode of computer operation in which there is frequent interaction between the user and the system.

conversational time-sharing *Computer Programming.* a mode of computer use in which many local and remote users may interact concurrently with a central computer.

converse [kän´vers] *Mathematics.* the converse of the logic statement "If *p*, then *q*" is "If *q*, then *p*." The converse does not necessarily have the same truth value as the original statement.

conversion the process or result of changing to a different form; specific uses include: *Metrology.* a change from one system of units to another, as from a temperature given in degrees Celsius to one given in degrees Fahrenheit. *Physics.* a change from one form of energy to another. *Computer Programming.* the transformation of data from one form, medium, or coding system to another; e.g., the *conversion* of a document created in one word-processing program to another program. *Psychology.* an unconscious process by which emotional conflict is transformed and manifested through physical symptoms. *Chemistry.* a change from one isomer to another. *Chemical Engineering.* the extent of chemical change from reactants to products in an industrial chemical process. *Nuclear Physics.* the process in which radioactive materials, principally uranium-238 and thorium-232, are bombarded with neutrons and transformed into plutonium-239 and uranium-233, respectively, for use in nuclear reactors or weapons. *Virology.* the phenomenon in which new properties, such as changes in pigmentation, colony morphology, or toxin production, are conferred on a prokaryote host by its interaction with a temperate phage.

conversion angle *Navigation.* the angular difference between the initial great-circle course to a destination and the rhumb line to that destination.

conversion coating *Metallurgy.* **1.** the process of modifying the composition of a superficial layer of a metallic object by a chemical or electrochemical reaction, in order to improve surface properties. **2.** the modified layer so produced.

conversion coefficient *Nuclear Physics.* the ratio of the number of electrons released from the nucleus per unit time to the number of photons released per unit time. Also, **conversion fraction.**

conversion disorder *Psychology.* a neurotic disorder whereby the individual suffers symptoms that seem to indicate physiological malfunctioning but have no evident organic cause. Also, **conversion hysteria, conversion reaction.**

conversion electron *Nuclear Physics.* an electron ejected from the inner shell of an atom when the excited nucleus returns to ground state and the released energy is given to the orbital electron.

conversion equipment *Computer Technology.* any device designed to convert data from one recording medium to another, such as a key-to-disk unit or an optical character reader.

conversion factor *Mathematics.* a numerical factor by which one must multiply a quantity expressed in terms of one type of unit in order to express the quantity in terms of a different type of unit; e.g., to convert a distance in meters to one in feet, one must multiply the metric distance by a conversion factor of 3.3. *Nucleonics.* either of two conversions, quality factor (QF) or relative biological effectiveness (RBE), used to convert any unit of absorbed radiation into a unit of dose equivalent.

conversion gain *Electronics.* the effective amplification of a conversion detector, measured as the ratio of voltage of output signal frequency to voltage of input signal frequency, and commonly expressed in decibels. *Nucleonics.* a quantitative measure of nuclear fuel utilization in which fissile material is formed from fertile material.

conversion program *Computer Programming.* a special program that is used to translate another program written for a certain computer into a form that will execute on a different computer or in a different programming language or dialect, so that files or documents created on one type of computer (an IBM PC, for example) may be accessed on another type (an Apple Macintosh, for example).

conversion rate *Computer Technology.* the number of complete measurements of an analog signal that an analog-to-digital converter can perform per unit time, usually expressed as cycles (conversions) per second; the inverse of conversion time.

conversion ratio *Nucleonics.* a measure of the efficiency of nuclear fuel utilization expressed as the ratio of the number of fissile nuclei produced from fertile nuclei to the number of fissile nuclei removed by fission and nonfission reactions; if the ratio exceeds 1 the reactor is a breeder, and if 1 or less the reactor is a converter.

conversion routine *Computer Programming.* **1.** a utility routine used to convert data from one recording medium to another; for example, from disk to tape. **2.** any sequence of instructions that transforms a set of data into some other form, format, or coding scheme.

conversion table *Metrology.* a list showing equivalent values in two or more measurement systems, such as miles and kilometers or Fahrenheit and Celsius temperatures.

conversion time *Computer Programming.* the time required for an analog-to-digital converter to complete the measurement of an analog signal; the inverse of conversion rate.

conversive heating *Medicine.* the conversion of radiation to heat (thermal energy) for use in thermotherapy.

convert to change the form or properties of something; specific uses include: *Chemistry.* to change one isomer to another. *Computer Programming.* **1.** to change the representation of data in either form or recording medium. **2.** to transform numbers from one number system to another, as from binary to decimal.

converted rice *Food Technology.* a type of enriched rice made by parboiling whole grains at 70°C for about 10 hours, thereby leaching vitamins and minerals into the endosperm; the rice retains much of its nutrient value even when milled and polished.

converter a device that changes something from one form to another; specific uses include: *Electronics.* **1.** a circuit or device that converts signals, frequency, voltage, current, or data from one form or mode to another. **2.** specifically, a device that changes alternating current to direct current. *Computer Technology.* a device or interface unit that transforms information from one form to another, for example, from decimal to binary, from floating point to fixed point, or from analog to digital. *Metallurgy.* a refining furnace in which air is blown through the molten bath in order to oxidize unwanted impurities. *Nucleonics.* a reactor in which a significant amount of fertile species is changed into a fissile species.

converter amplifier see CHOPPER AMPLIFIER.

converter tube *Electronics.* a multielement electrode tube that combines the mixer and local oscillator functions of a heterodyne conversion transducer.

convertiplane *Aviation.* a hybrid vehicle designed to fly as either a rotary-wing or fixed-wing aircraft. Also, **convertaplane.**

convex *Science.* curving outward; the opposite of concave.

convex combination *Mathematics.* a linear combination $\sum_{i=1}^{m} a_i v_i$ of vectors v_i, such that $a_i \geq 0$ and $\sum_{i=1}^{m} a_i = 1$.

convex curve *Mathematics.* a plane curve with the property that any straight line intersects it at no more than two points.

convex function *Mathematics.* a real function f (not necessarily differentiable on the interval) is convex (or concave upward) in the interval (a, b) if, for all $x_1, x_2 \in (a,b)$, $f((x_1 + x_2)/2) \leq (1/2)(f(x_1) + f(x_1))$. If the inequality is strictly less than, then f is said to be strictly convex. If the inequality is reversed, the function is then said to be concave downward in (a, b). If f has a second derivative on (a,b), then f is convex if and only if $f''(x) \geq 0$ for all $x \in (a, b)$.

convex hull *Mathematics.* given a collection of points in linear space, the smallest convex set containing all of the points. Also, **convex linear hull.**

convex lens see CONVERGING LENS.

convexo-concave see CONCAVO-CONVEX.

convex set *Mathematics.* a set in a linear space with the property that a line segment joining any two points of the set lies entirely within the set. The intersection of convex sets is convex.

conveyor *Mechanical Engineering.* **1.** a system or device for moving materials a short distance within a facility. **2.** see CONVEYOR BELT.

conveyor belt *Mechanical Engineering.* an endless belt, having appropriate tension and usually running on rollers, that is used to move materials in bulk from one point to another.

conveyor-belt balance *Mechanical Engineering.* a balance that weighs loose material being transported by a conveyor belt.

convivium *Ecology.* a unit consisting of two different species that would be capable of interbreeding and producing hybrid offspring but are prevented from doing so by geographical separation.

Convoluta *Invertebrate Zoology.* a genus of small, free-living marine flatworms, with green coloration caused by symbiotic algae.

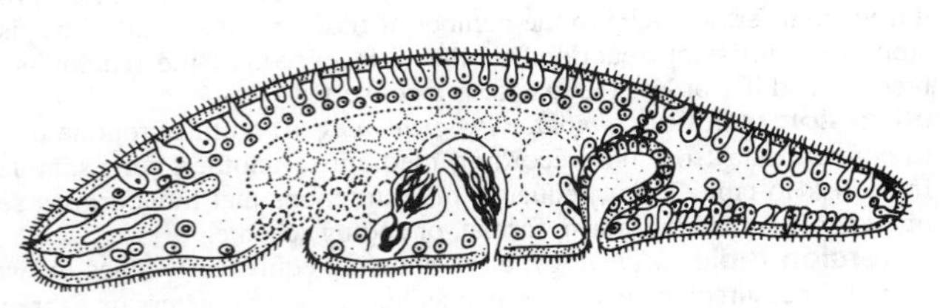

Convoluta

convolute [kän′və loot′] *Biology.* rolled up, with one part over another; coiled or twisted together.

convolute bedding *Geology.* an extreme distortion of laminae within a single layer of sedimentary rock, resulting from deformation that occurred during deposition or before cementation. Also, **convolute lamination.**

convolution *Anatomy.* a fold, twist, or coil in an organ or structure. *Textiles.* any deviation found in a cotton fiber, such as a twist. *Geology.* **1.** the process by which convoluted bedding is produced. **2.** any structure that results from such a process. *Mathematics.* **1.** multiplication in a group ring R_G, where elements of the group ring are expressed as functions from G to a field or as measures on G. For example, if G is a finite group,

$$f*g(s) = \sum_{t \in G} f(t)g(st^{-1}).$$

If G is the real numbers under addition,

$$f*g(x) = \int_{-\infty}^{\infty} f(u)g(x-u)du.$$

2. the convolution of two infinite series, $\sum_{n=-\infty}^{\infty} a_n$ and $\sum_{n=-\infty}^{\infty} b_n$, is their formal product; that is, $\sum_{n=-\infty}^{\infty} c_n$, where $c_n = \sum_{p=-\infty}^{\infty} a_p b_{n-p}$.

convolution family see FALTUNG.

convolution theorem *Mathematics.* **1.** if $f(s) = L\{F(x)\}$ and $g(s) = L\{G(x)\}$ are the Laplace transforms of $F(x)$ and $G(x)$, respectively, then $L\{F * G\} = L\{F\}L\{G\}$. That is, the Laplace transform of the convolution of two functions is the product of their Laplace transforms. **2.** if $F\{f(x)\} = F(\alpha)$ and $F\{g(x)\} = G(\alpha)$ are the Fourier transforms of $f(x)$ and $g(x)$, respectively, then $F\{f * g\} = F\{f\}F\{g\}$; that is, the Fourier transform of the convolution of two functions is equal to the product of their Fourier transforms.

convolver *Electronics.* a surface-acoustic-wave device that performs signal processing by nonlinear interactions between opposite traveling waves.

Convolvulaceae *Botany.* a family of herbs, trees, and shrubs of the order Polemoniales that have alternate leaves, regular bisexual flowers, and seeds with little or no endosperm.

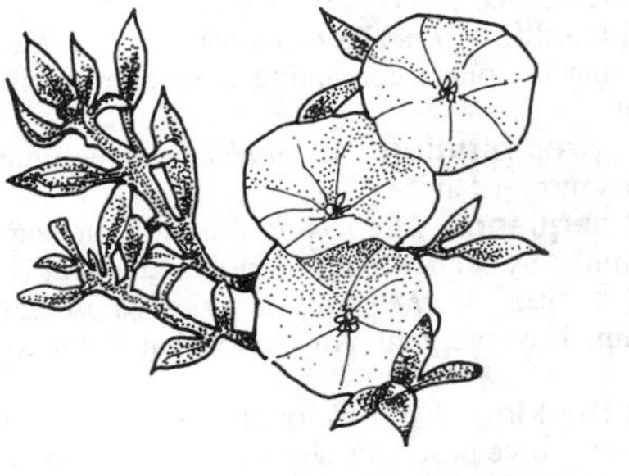

Convolvulaceae (morning glory)

convoy *Ordnance.* **1.** one or more merchant ships or naval auxiliaries escorted by warships or aircraft for the purpose of safe passage. **2.** a group of vehicles organized for the purpose of control and orderly movement, with or without escort protection.

convulsant *Neurology.* any agent that can cause violent involuntary muscle contractions.

convulsion *Medicine.* a violent and abnormal involuntary muscular contraction or series of contractions.

convulsive shock therapy see ELECTROCONVULSIVE SHOCK THERAPY.

Conwell-Weisskopf equation *Solid-State Physics.* an equation specifying the mobility of electrons in a semiconductor material containing donor or acceptor impurities, given in terms of the temperature, dielectric constant, concentration of ionized donors or acceptors, and the average distance between them.

Cook, James 1728–1779, British navigator; discovered and mapped much of the South Pacific; the first to apply Lind's cure for scurvy.

cookbook *Computer Programming.* a term for a set of step-by-step instructions for installing and using a computer or a software program.

Cookellaceae *Mycology.* a family of fungi of the order Myriangiales, characterized by reddish-brown to dark brown ascospores and occurring on plant leaves.

Cooke objective *Optics.* a lens system used in astronomical cameras consisting of a dispersive component, a biconcave lens, and two biconvex lenses.

Cooke unit *Biology.* a unit used for the standardization of pollen antigenicity.

Cooksonia *Paleontology.* a genus of tracheophytic vascular plants in the extinct division Rhyniophyta and family Rhyniaceae; probably the oldest vascular plants; specimens have been found in deposits as early as the Middle Silurian.

coolant *Materials Science.* any material, usually a liquid, that is characterized by its ability to absorb heat from its environment and transfer it away from its source. *Nucleonics.* specifically, a substance circulated through a reactor to remove or transfer heat; common coolants include light or heavy water, various gases such as air and carbon dioxide, liquid sodium, and some organic compounds.

cooled infrared detector *Electronics.* a transducer sensitive to invisible infrared radiation that must be operated at extremely low temperatures to obtain the desired sensitivity.

cooled-tube pyrometer *Engineering.* a thermometer used to measure hot flowing gases by inserting a liquid-cooled tube in the gas and measuring the heat on the outside of the tube, the mass flow rate, and the rise in temperature of the cooling liquid.

cooler nail *Mechanical Devices.* a thin wire nail that is coated with cement.

Coolidge, William David 1873–1975, American physical chemist; invented ductile tungsten and Coolidge tube.

Coolidge effect *Behavior.* the tendency of a male animal that has lost sexual interest in a female partner to regain interest upon the introduction of a new partner.

Coolidge tube *Electromagnetism.* an X-ray tube that uses a heated filament to provide electrons.

cooling channel *Engineering.* a grooved depression in a mold, through which a cooling liquid circulates.

cooling coil *Mechanical Engineering.* a tubing that transfers heat from the material or space that is being cooled to the primary refrigerant.

cooling correction *Thermodynamics.* a correction used in calorimetry to account for heat transfer between a test sample and its surroundings.

cooling curve *Thermodynamics.* a plot of temperature as a function of time, as a substance or compound is allowed to cool down; constancy of temperature indicates the point of occurrence of a first-order phase change.

cooling degree-day *Engineering.* a unit of measurement that represents the number of degrees by which the daily mean temperature exceeds a certain standard (now usually 75°F) on a given day; used in calculating the energy needed to cool the inside of a building.

cooling fin *Mechanical Engineering.* a projection or protrusion along the surface edge of a heat transfer device, designed to increase the surface area.

cooling fixture *Engineering.* a block in which a newly molded object is kept until it has cooled to a point at which its shape is solidified.

cooling load *Mechanical Engineering.* the amount of heat energy per unit of time that must be removed from a system by a cooling mechanism in order to achieve the desired temperature condition.

cooling method *Thermodynamics.* a method of determining the specific heat of a liquid, by comparing the time required for the liquid to cool down over a certain temperature range to the time required for an equal volume of water to do the same under identical conditions.

cooling pond *Chemical Engineering.* an open pond in which water, heated by an industrial process, is allowed to cool through evaporation before reuse.

cooling power *Mechanical Engineering.* the rate at which air will remove heat from a body; may be measured wet or dry.

cooling-power anemometer *Engineering.* an instrument to measure air speed, based on the principle that heat transferred to the atmosphere from an object at an elevated temperature is a function of air speed.

cooling process *Engineering.* the physical stage of an operation in which heat is removed from fluids or solids.

cooling range *Mechanical Engineering.* in a cooling tower, the difference in temperature between the entering hot water and the exiting cool water.

cooling stress *Mechanics.* stress resulting from the uneven contraction of an object upon cooling, due to an unequal temperature distribution or nonuniform material properties.

cooling tower *Engineering.* a towerlike structure in which excess heat is removed from water by convection and conduction into the air.

cooling water *Biotechnology.* water that is used to remove metabolic or process heat; the water is cycled between a heat exchanger associated with the reactor and a cooling tower or refrigeration plant.

coolometer *Engineering.* a device to measure the cooling power of air.

cool star *Astrophysics.* any star of spectral type M, whether its luminosity class is dwarf, giant, or supergiant.

cool time *Metallurgy.* in welding, a term for the time intervals between heating times.

Coombs serum *Immunology.* an antiglobulin antiserum that is used to determine the presence of Rh or other sensitivity in an organism.

Coombs Test *Immunology.* a test used to determine the presence of nonagglutinating antibodies on cells. Also, ANTIGLOBULIN TEST.

Cooper, Leon N. born 1930, American physicist; discovered Cooper pairs; shared Nobel Prize for theory of superconductivity.

cooperation *Behavior.* the interaction of two or members of the same species to produce an effect or condition that is beneficial to the participants or to others of the species.

cooperative binding *Molecular Biology.* a process in which the binding of one molecule (e.g., in a DNA strand) facilitates the binding of the next molecule.

cooperative phenomenon *Solid-State Physics.* any phenomenon resulting from the sum of contributions of individual atoms and molecules, such as ferromagnetism in which all atoms in a crystal provide aligned magnetic moments.

cooperative play *Behavior.* interactive play behavior, especially as observed in children of preschool age.

cooperative system *Engineering.* a term for a missile guidance system in which information from a remote ground station is transmitted to a missile in flight, processed by the missile, and then returned as reprocessed data to the originating station or other remote ground stations.

cooperativity *Biochemistry.* a condition in which molecules interact to make each other's action either easier (**positive cooperativity**) or more difficult (**negative cooperativity**); an effect that occurs in the binding of substrates or allosteric effectors to proteins.

Cooper-Helmstetter model *Microbiology.* a model describing the bacterial chromosome replication within the prokaryotic cell cycle.

cooperite *Mineralogy.* (Pt,Pd,Ni)S, a gray, opaque, metallic tetragonal mineral that occurs as crystal fragments or irregular grains, having a specific gravity of 9.5 and a hardness of 4 to 5 on the Mohs scale; associated with native platinum, sperrylite, and other precious-metal minerals from the bushveld complex in South Africa.

Cooper pair *Solid-State Physics.* a pair of bound electrons of opposite spin and momentum in a superconductive medium that give rise to superconductivity.

coordinate *Mathematics.* either of a pair of numbers that locate a point in a plane (or, one of three numbers that locate a point in a space). A function from an open set of a manifold to the real numbers; if a minimal but sufficient number of coordinate functions are present, so that each point in the open set can be uniquely determined by specifying the values of the coordinate functions, then the coordinate functions form a coordinate system on the open set.

coordinate addressing *Computer Technology.* **1.** a memory access method in which the cell is specified by its address (rather than by its content, as with an associative memory addressing method). **2.** a means of creating a display on a CRT screen by specifying the Cartesian coordinates of the position where illumination is desired.

coordinate axes *Mathematics.* the x, y (and z) axes in the Euclidean plane or space; on a manifold V that is a real affine or vector space with basis vectors $\{e_1, e_2, \ldots, e_n\}$, the kth coordinate axis is the set of points in V given by $\{\alpha e_k : \alpha \in R\}$. These axes exist only on such a manifold.

coordinate basis *Mathematics.* in differential geometry, the basis tangent vectors $\{\partial/\partial x^i\}$ and dual 1-forms $\{dx^i\}$ belonging to a coordinate system.

coordinate bond *Chemistry.* a chemical bond between two atoms, or an atom and a group, which share two electrons that are both supplied by one of the atoms or groups. Also, **coordinate valence.**

coordinate conversion *Cartography.* the changing of coordinate values on a map from one system of coordinates to another.

coordinate data receiver *Electronics.* a radio receiver designed to receive transmitted coordinate data and convert it to an analog or digital form for input to equipment such as radar, computers, or plot boards. Similarly, **coordinate data transmitter**.

coordinated-axis control *Robotics.* a control system that allows the axes of a robot to reach their end points simultaneously in order to give the appearance of smooth motion. Also, END-POINT CONTROL.

Coordinated Universal Time *Horology.* an international standard of time based on Greenwich Mean Time and adjusted to compensate for divergence from atomic time. Also, UNIVERSAL TIME COORDINATED.

coordinate enzymes *Enzymology.* enzymes that are produced by the action of genes controlled by a single operon and are involved in coordinate induction or coordinate repression.

coordinate geometry see COGO.

coordinate induction *Enzymology.* a process in which a single inducer molecule, usually the first substrate to be acted on in a sequence, stimulates the production of several inducible enzymes, which in turn catalyze a series of related reactions that process the substrate.

coordinate regulation *Genetics.* the correlated regulation of several genes, causing all of them to become active or inactive at the same time.

coordinate repression *Enzymology.* a process in which a single repressor molecule, usually the last end product in a sequence, inhibits the production of several repressible enzymes that catalyze a series of reactions that produce the product.

coordinates *Cartography.* any sets of values that indicate the position of a point in a given reference system.

coordinate system *Astronomy.* any system that locates objects on the celestial sphere using two parameters and a date; the most commonly used coordinate systems are horizon, equatorial, ecliptic, and galactic. *Mathematics.* a system of pairs (or triples) of numbers that locate a point in a plane (or space). On an open set U of an n-dimensional manifold, a coordinate system is a set of n coordinates that provide a one-to-one mapping of the open set U onto an open set of R^n.

coordinate transformation *Mathematics.* given two coordinate systems on an upen set U, the formulas that express the values of one set of coordinates in terms of the others. Also, CHANGE OF COORDINATES.

coordinate vector *Mathematics.* given a finite-dimensional vector space V over scalar field F with basis $B = \{b_1, b_2, \ldots, b_n\}$, the coordinate vector of a vector $v = \sum a_i b_i$ with respect to B is the column matrix $[a_1, \ldots, a_n]^T$ in F^n.

coordination *Neurology.* any combination of nerve impulses in a motor center that promotes the cooperation of appropriate muscles in a motor response.

coordination catalyst *Materials Science.* a mixture of an organometallic compound and a transition-metal compound that is used to catalyze the polymerization of olefins and dienes.

coordination compound *Chemistry.* a compound in which a group of usually nonmetallic atoms or ions (called ligands) are attached by a coordinate bond to a usually metallic central atom or ion. Also, COMPLEX COMPOUND.

coordination lattice *Crystallography.* the crystal structure of a coordination compound.

coordination number *Chemistry.* the number of points at which ligands are attached to the central atom or ion in a coordination compound. *Crystallography.* the number of nearest neighbors of a particular atom or molecule in a crystal lattice.

coordination polymer *Organic Chemistry.* a polymer that is produced through the use of catalysts that produce linear polymers with stereochemical control of the molecule.

coot

coot *Vertebrate Zoology.* a slow-flying black waterbird of the genus *Fulica* in the family Rallidae, found worldwide inhabiting inland lakes and rivers and having ducklike characteristics, including large lobed feet with an membrane acting much like webbing. Also, MUD HEN.

COP *Oncology.* an acronym for a cancer chemotherapy regimen that includes the drugs cyclophosphamide, Oncovin, and prednisone.

COP coefficient of performance.

copal *Materials Science.* **1.** any of various hard, lustrous resins obtained from tropical trees and used in making varnishes. **2.** any of various insoluble fossil resins such as kauri and manila.

COP-BLAM *Oncology.* an acronym for a cancer chemotherapy regimen that includes the drugs cyclophosphamide, Oncovin, prednisone, bleomycin, Adriamycin, and Matulane

cope *Building Engineering.* to join two molded wooden members by a profiling method, in such a way that the joint produced from the two ends resembles a miter joint; if flanges are evident, they can be cut away. *Metallurgy.* in casting, the upper section of a mold.

cope chisel *Mechanical Devices.* a chisel used to cut holes or form grooves in metal.

Copenhagen school *Quantum Mechanics.* a collective term for those physicists holding the view that the wavefunction of a system completely characterizes the system, and that therefore no other parameters or hidden variables differentiate between distinct systems having identical wavefunctions.

Copepoda *Invertebrate Zoology.* an order of crustaceans in the group Entomostraca, containing parasitic, symbiotic, and free-living forms.

cope ring *Mechanical Devices.* a cast-iron ring with logs around its periphery to which chains are attached that hold bricks and loam of a loam mold.

Copernican [kə purn´i kən] relating to or based on the theories of Nicolaus Copernicus.

Copernican principle *Astronomy.* a principle in cosmology that is based on the theories of Copernicus; it can be summarized as "the earth is not the center of the universe."

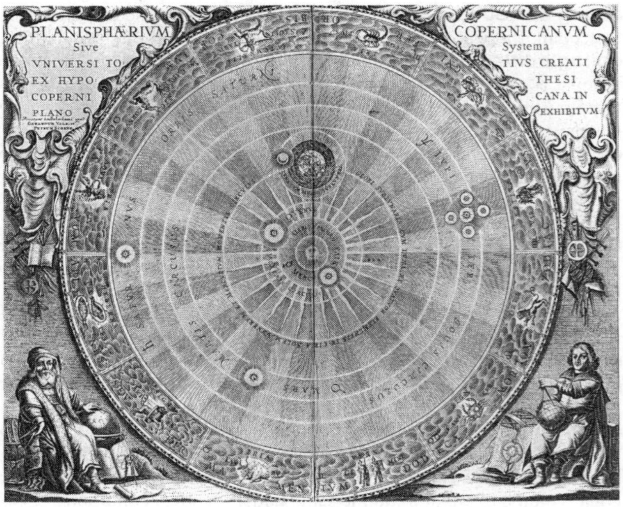

Copernican system

Copernican system *Astronomy.* the theory of the universe that has the sun at the center, the earth and other planets having circular orbits around it, and the earth rotating from west to east once a day; i.e., the modern theory formulated by Copernicus that replaced the earlier Ptolemaic theory, which held that the earth is the center of the universe.

Copernicus, Nicolaus [kə purn´i kəs] 1472–1543, Polish astronomer; formulated the Copernican theory, the foundation of modern astronomy.

Cope's rule *Evolution.* a rule stating that evolutionary change in animals tends toward an increase in body size. Also, **Cope's law**.

cop gene *Genetics.* a gene that is involved in controlling the copy number of a plasmid.

cophenetic correlation coefficient *Systematics.* a statistic that measures the degree of fit between a phenogram and a table of phenetic distances.

copia *Genetics.* a family of dispersed repetitive DNA sequences that contains a large number of closely related sequences that code for large amounts of mRNA; the copia family is a paradigm for several other types of elements whose sequences are unrelated, but whose structure and general behavior appear to be similar.

copia element *Molecular Biology.* any of the transposable DNA elements in the genetic material of *Drosophila,* existing as a family of closely related base sequences.

copiapite *Mineralogy.* $Fe^{+2}Fe_4^{+3}(SO_4)_6(OH)_2 \cdot 20H_2O$, a transparent to translucent yellow triclinic mineral occurring as thin tabular crystals and loose aggregates, having a specific gravity of 2.08 to 2.17 and a hardness of 2.5 to 3 on the Mohs scale; a secondary mineral formed by the oxidation of pyrite and other sulfides.

coping *Building Engineering.* brickwork having an end oriented and shaped to fit with a particular molding. *Architecture.* a cap or cover for a wall or chimney, often double-sloping and cut with a drip to shed water. *Mechanical Engineering.* the process of shaping or polishing stone with a grinding wheel. *Mining Engineering.* the process of splitting a stone by driving wedges into it. *Psychology.* the conscious or unconscious efforts of an individual to deal with the stresses and demands of his or her environment. Also, **coping behavior.**

coping machine *Mechanical Devices.* a device used to cut away flanges and corners of beams.

coping strategy or **strategies** *Behavior.* the behavioral and emotional adjustments that an individual makes in order to manage or alleviate a stressful situation. Also, **coping mechanism(s).**

coplanar [kō plān´är] *Mechanics.* lying or acting in the same plane.

coplanar electrodes *Electronics.* a set of two or more electrodes that are physically aligned with each other.

Copodontidae *Paleontology.* a family of primitive holocephalic fish in the extinct order Copodontiformes; may have had only one functional tooth in each jaw; Devonian and Carboniferous.

copolymer [kō päl´i mər] *Materials Science.* a polymer that is composed of polymer chains made up of two or more chemically different repeating units that can be in different sequences. *Organic Chemistry.* any polymer produced by the simultaneous polymerization of two or more dissimilar monomers.

copolymerization *Materials Science.* polymerization in which polymers are derived from more than one species of monomers.

copolymerize *Organic Chemistry.* to produce a copolymer; carry out the process of copolymerization.

COPP *Oncology.* an acronym for a chemotherapy regimen for cancer treatment that includes the drugs cyclophosphamide, Oncovin, procarbazine, and prednisone.

copper *Chemistry.* an element having the symbol Cu, the atomic number 29, an atomic weight of 63.54, a melting point of 1083°C, and a boiling point of 2595°C; a soft, reddish, ductile metal that is an excellent conductor of electricity; it has low reactivity and resists atmospheric corrosion by forming a protective cover of a green basic copper carbonate, e.g., $Cu(OH)_2 \cdot CuCO_3$. *Mineralogy.* Cu, the mineral form of this element, occurring as cubes, dodecahedra, and tetrahedra; the crystals are often elongated, flattened, or wirelike. It has a specific gravity of 8.94 to 8.95 and a hardness of 2.5 to 3 on the Mohs scale, and it is found in oxidized zones of copper-bearing sulfide ore deposits. Copper has great industrial significance in its pure state or as a base for numerous alloys such as brass and bronze. (Going back to a phrase meaning "metal of Cyprus"; from the most noted early source of this element.)

copper-64 *Nuclear Physics.* a radioactive isotope of copper with a half-life of 12.7 hours that is created by irradiating metallic copper in a reactor; commonly used as a tracer to study diffusion, corrosion, and friction in metals and alloys.

Copper Age *Archaeology.* an intermediate period between the Neolithic and the Bronze Ages, characterized by the use of copper tools. Also, CHALCOLITHIC.

copper alloy *Metallurgy.* an alloy that contains at least 50% copper.

copper amalgam *Metallurgy.* an alloy of copper and mercury.

copper arsenate see CUPRIC ARSENATE.

copper blight *Plant Pathology.* a disease of tea plants caused by the fungus *Guignardia comelliae* that is characterized by well-defined discolored spots on the leaves.

copper blue see MOUNTAIN BLUE.

copper brazing *Metallurgy.* the process of brazing or braze welding with a copper-base filler.

copper bromide **1.** see CUPRIC BROMIDE. **2.** see CUPROUS BROMIDE.

copper cable *Electricity.* a set of copper wires, used as an alternative to a single large-diameter wire for increased flexibility.

copper carbonate see CUPRIC CARBONATE.

copper chloride **1.** see CUPRIC CHLORIDE. **2.** see CUPROUS CHLORIDE.

copper chromate or **basic copper chromate** see CUPRIC CHROMATE.

copper converter *Metallurgy.* a converter used to refine impure copper.

copper cyanide see CUPRIC CYANIDE.

copper dish gum *Chemical Engineering.* a test procedure yielding the milligrams of gum in 100 mililiters of gasoline evaporated in a polished copper dish under controlled conditions.

copper fluoride **1.** see CUPRIC FLUORIDE. **2.** see CUPROUS FLUORIDE.

copper gluconate *Organic Chemistry.* $Cu[CH_2OH(CHOH)_4COO]_2$, a light-blue, fine, crystalline powder; soluble in water and insoluble in alcohol; used as a feed additive, dietary supplement, and mouth deodorant. Also, CUPRIC GLUCONATE.

copperhead

copperhead *Vertebrate Zoology.* a poisonous pit viper belonging to the family Crotalidae, found in the eastern U.S. and characterized by its copper-colored head. *Materials Science.* a pattern of reddish-brown spots that sometimes appears in the groundcoat during vitreous enameling; the spots are exposed areas of oxidized base metal.

Copperhead *Ordnance.* a gun-launched guided projectile that homes in on a target illuminated by a laser designator beam. Also, CANNON-LAUNCHED GUIDED PROJECTILE.

copper hydroxide or **copper hydrate** see CUPRIC HYDROXIDE.

coppering *Ordnance.* an accumulation of metal in the bore of a gun or firearm caused by repeated firing; it is deposited by the rotating bands or jackets of the projectiles.

copper line *Toxicology.* a green or purplish line that develops between the teeth and gums in patients with copper poisoning. Also, CORRIGAN'S SIGN.

copper loss *Electricity.* the power loss in the winding of an electromagnetic machine due to resistance.

copper monoxide see CUPRIC OXIDE.

copper nitrate see CUPRIC NITRATE.

copper number *Analytical Chemistry.* the number of milligrams of copper produced by the reduction of Benedict's or Fehling's solution by 1 gram of a carbohydrate.

copper oleate *Organic Chemistry.* $Cu(C_{17}H_{33}COO)_2$, a combustible, brown powder or greenish-blue mass, insoluble in water and slightly soluble in alcohol; used as a fungicide, insecticide, emulsifying agent, and catalyst. Also, CUPRIC OLEATE.

copper ore *Geology.* any rock from which copper can be extracted economically.

copper oxide **1.** see CUPRIC OXIDE. **2.** see CUPROUS OXIDE.

copper oxide black see CUPRIC OXIDE.

copper oxide photovoltaic cell *Electronics.* an early type of nonvacuum photovoltaic cell that generates a small voltage between the substrate and the conducting layer when exposed to light; its active element is of a layer of copper oxide in contact with a layer of metallic copper.

copper oxide rectifier or **copper oxide diode** *Electronics.* a semiconductor rectifier in which the rectifying barrier is formed by a junction of copper and cuprous oxide.

copper oxide red see CUPROUS OXIDE.

copper plate *Mining Engineering.* see APRON, def. 2.

copperplate engraving *Graphic Arts.* **1.** a printing method, common in the 18th and 19th centuries, in which characters and illustrations are directly cut into a thin copper printing plate. **2.** a print made from such a plate.

copper plating *Metallurgy.* electrochemical or electroless plating of essentially pure copper onto a substrate.

copper powder *Metallurgy.* particulate copper, which is used extensively in powder metallurgy for the fabrication of copper-base components as well as iron-base, copper-infiltered components, and also in microelectronic applications.

copper-8-quinolinolate *Organic Chemistry.* $Cu(C_9H_6ON)_2$, a toxic, yellow-green powder, insoluble in water and organic solvents and soluble in acids; used as a fungicide and in mildew-proofing of fabrics.
copper resinate *Organic Chemistry.* a combustible, green powder, derived by heating rosin oil with copper sulfate; insoluble in water and soluble in ether and oils; used as an insecticide and a metal-paint preservative. Also, CUPRIC RESINATE.
copper spot *Plant Pathology.* a disease of lawn grasses caused by the fungus *Gloeocercospora sorghi* and characterized by small, dead, light copper-colored spots on the turf.
copper steel *Metallurgy.* a steel containing a small amount of copper, intentionally added to improve corrosion resistance.
copper-strip corrosion *Engineering.* a test of the corrosive properties of a given petroleum, performed by observing its effect on a piece of copper that is immersed in it.
copper sulfate see CUPRIC SULFATE.
copper sulfide see CUPRIC SULFIDE.
copper sulfide rectifier see MAGNESIUM-COPPER-SULFIDE RECTIFIER.
copper sweetening *Chemical Engineering.* a refining process that uses cupric chloride to oxidize mercaptans in petroleum.
copperweld *Metallurgy.* a copper-clad steel.
copper wire *Metallurgy.* a wire consisting of essentially pure copper, often containing a small amount of intentionally added oxygen; widely used to carry electric current and for telecommunications.
coppice [käp´is] *Forestry.* growth consisting of small trees that are repeatedly cut down near the ground, with new growth produced on the old stumps. Also, **coppice forest.** *Botany.* any thick growth of small trees or bushes. Also, COPSE.
coppice shoot *Forestry.* any shoot arising from a bud near the base of a woody plant that has been cut back.
coppicing *Forestry.* the process of cutting broad-leaved trees close to the ground in order to produce coppice shoots.
co-precipitation or **coprecipitation** *Chemistry.* the precipitation of more than one substance at the same time. *Immunology.* the formation of an antigen-antibody complex occurring as a secondary reaction among molecules that ordinarily would not join together in conjunction with a normal antigen-antibody bonding.
Coprinaceae *Mycology.* a family of fungi belonging to the order Agaricales, which is composed of gill mushrooms and is well known for its species *Coprinus,* the ink-cap mushroom.
Coprinus *Mycology.* a genus of fungi belonging to the family Coprinaceae which have gills that are digested by the fungus after spore discharge, producing an inky fluid.
copro- or **copr-** a combining form meaning "dung," as in *coprophilous.*
coproantibody *Immunology.* an antibody occurring in the output of the intestines, generally in the feces.
coprocessor [kō prä´ses ər] *Computer Technology.* an auxiliary processing unit that works with the CPU to provide special services and speed up operations; used for mathematics, graphics, or acoustic output.
Coprococcus *Bacteriology.* a genus of Gram-positive, anaerobic bacteria that are found in the human intestinal tract as coccoid cells, growing in pairs or chains.
coprolite [käp´rə līt] *Archaeology.* ancient feces that have become fossilized; this can provide information about the human or animal activity in that particular locale, such as diet and disease.
coprophagy [käp´rə fāj´ē] *Zoology.* the act of feeding on feces.
coprophilia [käp´rə fil´ē ə] *Psychology.* an irrational interest in feces and defecation.
coprophilous [käp´rə fil´əs] *Ecology.* thriving on dung or fecal matter. Thus, **coprophile, coprophily.**
coprophobia [käp´rə fōb´ē ə] *Psychology.* an irrational fear of feces and defecation.
copse see COPPICE.
copulation [käp´yə lāl´shən] *Biology.* the act of sexual coupling by male and female.
copulatory [käp´yə lə tôr´ē] *Biology.* relating to or involved in sexual union between male and female. Thus, **copulatory organ.**
copulatory bursa *Invertebrate Zoology.* **1.** a sac that receives the sperm during mating of certain insects. **2.** the caudal expansion of male nematodes that clasps the female during mating.
copy *Graphic Arts.* **1.** edited text that is ready to be typeset or printed. **2.** any material, including artwork, that is given or ready to be given to a compositor, photographer, or printer. *Telecommunications.* **1.** to receive a message. **2.** a recorded message, or a duplicate of a recorded message. **3.** to maintain a continuous ratio receiver watch and to keep a radio log. *Computer Programming.* **1.** to prepare a duplicate version of a file or program while leaving the original intact and unaltered. **2.** to duplicate certain information at another memory or storage location, while leaving the original information intact at its source location; e.g., to *copy* text from an existing word-processing document for use in creating a new document.
copyboard *Graphic Arts.* a flat surface that is designed to hold camera copy in place; usually equipped with lights to illuminate the copy and a hinged glass cover to hold the copy in place. Also, CAMERA COPYBOARD.
copy camera *Photogrammetry.* a precision camera used in the laboratory for copying procedures. Also, PROCESS CAMERA.
copy choice hypothesis *Molecular Biology.* the hypothesis that a newly formed strand of DNA alternates between paternal and maternal copying during replication.
copy DNA see COMPLEMENTARY DNA.
copy editing *Graphic Arts.* **1.** the process of editing a manuscript to be set in type. **2.** specifically, the process of editing a manuscript to correct errors in spelling, grammar, usage, style, and format, as opposed to making substantive changes in content. Thus, **copy-edit, copy editor.**
copy error *Molecular Biology.* a mutation that results from a failure of the DNA molecule to replicate itself exactly.
copyfitting *Graphic Arts.* the process of calculating the proper type size, line width, and other adjustments needed to fit a particular piece of copy into a given space.
copying program *Computer Programming.* a utility program that copies data, a file, or a program from one input/output device to another, such as disk-to-tape.
copy-mutant *Genetics.* a mutant plasmid whose copy number is different (usually higher) from that of the wild-type plasmid. Also, **copy-number mutant, cop mutant.**
copy number *Molecular Biology.* the number of copies of a specific gene sequence in a genome or entire set of genes present in the haploid cell of a particular organism.
copy protection *Computer Programming.* any of various methods used by commercial software developers to prevent unauthorized duplication of proprietary software. Thus, **copy-protected.**
coquimbite *Mineralogy.* $Fe_2^{+3}(SO_4)_3{\cdot}9H_2O$, a colorless or purple, transparent, brittle, trigonal mineral occurring as prismatic or pyramidal crystals and granular masses; having a specific gravity of 2.11 and a hardness of 2.5 on the Mohs scale; found with sulfates such as halotrichite and voltaite.
coquina [kō kē´nə] *Petrology.* a coarse-grained, porous, easily crumbled variety of limestone composed principally of broken shells, coral fragments, and other organic debris loosely cemented together. *Invertebrate Zoology.* a small marine clam of the genus *Donax* that is a principal constituent of this material, especially *D. variabilis.*

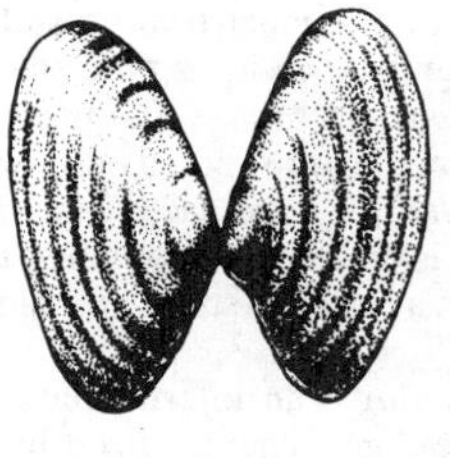

coquina

coquinoid [kō´kə noid] *Petrology.* **1.** of or relating to a mass of coquina. **2.** a lithified form of coquina. **3.** a limestone that contains whole mollusk shells similar to coquina with a substantial amount of matrix.
coracidium *Invertebrate Zoology.* the six-hooked larva hatched from a tapeworm egg, often ciliated and aquatic.
Coraciidae *Vertebrate Zoology.* the rollers, a family of Old World tropical birds in the order Coraciiformes, distinguished by their bright plumage and acrobatic flying.
Coraciiformes *Vertebrate Zoology.* a cosmopolitan order of mostly tropical Old World land birds, including the kingfishers, hoopoes, and hornbills, often characterized by their bright plumage and hole-nesting habits.
coracoid [kôr´ə koid] *Anatomy.* shaped like a crow's beak.
coracoid process *Anatomy.* a bony projection on the upper ventral surface of the scapula.

coral

coral *Invertebrate Zoology.* **1.** the calcium carbonate skeleton of certain colonial anthozoan coelenterates. **2.** the animals themselves.

coral head *Geology.* a large, knobby or dome-shaped coral growth. Also, **coral knoll.**

Corallanidae *Invertebrate Zoology.* a family of isopod crustaceans in the suborder Flabellifera; some are free-living, some parasitic.

Corallidae *Invertebrate Zoology.* a family of horny corals, coelenterates in the order Gorgonacea; forming fan- or feather-shaped colonies.

Corallimorpharia *Invertebrate Zoology.* an order of sea anemones in the subclass Zoantharia.

Corallinaceae *Botany.* a large family of red algae of the order Cryptonemiales, having thalli that are either prostrate or erect arising from a prostrate base and being adherent to a substrate or free-living in deep water; some species are used in agriculture as a source of calcium carbonate and are instrumental to the formation of coral reefs.

coralline *Invertebrate Zoology.* resembling coral.

coralline algae *Botany.* algae of the family Corallinaceae.

corallite *Invertebrate Zoology.* the skeleton of an individual polyp in a compound coral.

coralloid *Biology.* **1.** like coral in structure or appearance; branched like coral. Also, **coralliform. 2.** see CAVE CORAL.

corallum *Invertebrate Zoology.* the skeleton of a compound coral.

coral mud *Geology.* a fine-grained deposit of calcareous material derived from coral reefs.

coral pinnacle see REEF PINNACLE.

coral reef *Geology.* a wave-resistant limestone ridge found in tropical climates and composed mainly of coral fragments, usually deposited around a framework of organic remains.

coral-reef shoreline *Geology.* a shoreline formed by the buildup of large solidified ridges of coral.

coral rock see REEF LIMESTONE.

coral sand *Geology.* a coarse-grained deposit of calcareous material derived from coral reefs.

coral snake *Vertebrate Zoology.* any of several front-fanged, venomous snakes of the family Elapidae, found worldwide and characterized by bold stripes alternating red, yellow, and black.

coral snake

corange line *Cartography.* a line on a chart through points of equal tidal range.

corbel *Architecture.* a brick, block, or bracket that projects from the face of a wall and supports an overhanging member.

corbel arch or **corbeled arch** *Architecture.* an archlike brick or masonry structure that spans a gap by having successive courses project inward as they rise. Also, CANTILEVER ARCH.

Corbiculidae *Invertebrate Zoology.* a family of freshwater bivalves; mollusks in the subclass Eulamellibranchia.

Corbino disk *Electromagnetism.* a variable-resistance device utilizing the effect of a magnetic field in the flow of carriers from the center of the circumference of a semiconducting material.

corbinotron *Engineering.* a device composed of a corbino disk and a coil that creates a magnetic field at right angles to the disk.

cord *Materials.* a rope or string that is made of several strands twisted or braided together. *Metrology.* a measurement for wood, equivalent to 128 cubic feet; now used chiefly to measure stacks of wood cut for fuel. Also, CORD FOOT. *Electricity.* a flexible, insulated conductor, or several insulated conductors within one sheath.

cordage *Materials.* a collective group of ropes or cords, especially those of a ship's rigging. *Metrology.* an amount of wood, measured in cords.

Cordaitaceae *Botany.* an extinct family of Paleozoic plants of the order Cordaitales.

Cordaitales *Paleontology.* an extinct order of large gymnosperms related to the Coniferopsida; the Cordaitales commonly grew to 30 meters and were among the dominant trees from the Carboniferous to the end of the Permian, when they disappeared.

cordate *Botany.* shaped like a heart.

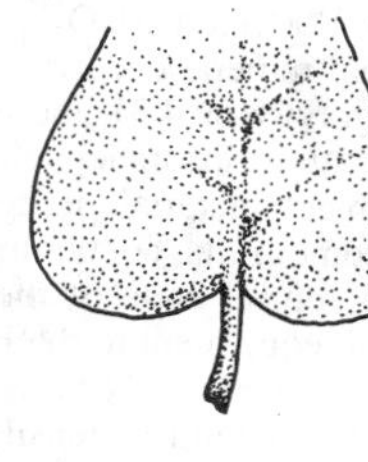

cordate leaf

cord factor *Microbiology.* a toxic bacterial cell wall lipid, composed of any mycolic acid diester of trehalose and found in species containing mycolic acids.

cord foot *Metrology.* see CORD.

cordia *Materials.* the dense, heavy wood of tropical shrubs and trees of the genus *Cordia;* used in interior decoration and construction.

cordierite *Mineralogy.* $Mg_2Al_4Si_5O_{18}$, a transparent to translucent blue orthorhombic mineral occurring in massive form and as short, prismatic crystals; having a specific gravity of 2.52 to 2.78 and a hardness of 7 to 7.5 on the Mohs scale; found in thermally altered aluminum-rich rocks and in alluvial gravels. *Materials.* a synthetically produced form of this mineral; used for refractory electronic parts and in producing refractory insulation.

cordillera [kôr´dil yâr´ə] *Geography.* **1.** a mountain system of more or less parallel ranges or ridges, as in the Andes. **2.** also, **Cordilleras, Cordeliers.** such a system in North America, usually identified as including the Rockies, Sierra Nevada, Cascades, and Coast Range.

Cordilleran Geosyncline *Geology.* the chain of western North American mountains formed during the Devonian Period.

cordite *Materials.* a smokeless, slow-burning powder that consists of nitrocellulose and nitroglycerin, with petrolatum added as a thickener and stabilizer; it has a characteristic cordlike structure. Also, PYROCELLULOSE.

cordless phone see CORDLESS TELEPHONE.

cordless switchboard *Telecommunications.* a telephone switchboard in which manually operated keys, rather than cord circuits with manually operated plug-ended cords, are used to make connections.

cordless telephone *Telecommunications.* a transportable telephone having a low-powered transmitter-receiver that communicates with a base unit.

cordon *Botany.* a plant that is trained to grow flat against a standing structure.

cordonazo [kôr′də nä′zō] *Meteorology.* a violent but brief hurricane moving northwest off the west coast of Central America and Mexico, which occurs in early October only once in several years.

cordotomy [kôr′dät′ə mē] *Surgery.* the surgical splitting of tracts of the spinal cord, either before or after a laminectomy, for relief of pain. Also, CHORDOTOMY.

cordovan [kôr′də vən] *Materials.* a tough, smooth leather made from the hides of horses. Also, **cordovan leather.**

cord tire *Mechanical Devices.* a tire built up of rubberized cord plies and cable rims with treads in parallel.

Cordulegasteridae *Invertebrate Zoology.* a family of dragonflies, insects in the order Odonata.

corduroy *Textiles.* a sturdy cotton fabric with a lengthwise cut-pile rib.

cordwood *Materials Science.* **1.** wood that is stacked and sold in cords for fuel. **2.** tree trunks or branches sawed into uniform lengths for stacking in cords. **3.** trees that are of a quality suitable only for fuel.

cordwood module *Electronics.* a circuit module configuration in which the individual components are mounted between, and at right angles to, a pair of parallel printed-circuit boards in a manner resembling stacked cordwood.

cordycepin *Organic Chemistry.* $C_{10}H_{13}N_5O_3$, needles that melt at 225–226°C; soluble in alcohol and water; a substance that slows or inhibits RNA synthesis. *Biochemistry.* this substance obtained from a variety of fungi strains, such as *Cordyceps militaris,* or scarlet caterpillar fungus.

Cordyceps *Mycology.* a genus of fungi belonging to the subdivision Ascomycotina, which occurs in soil and is parasitic of other fungi and insects, such as certain caterpillars.

Cordylidae *Vertebrate Zoology.* the girdle-tailed lizards, a family of mostly insectivorous terrestrial lizards in the suborder Scincomorpha that dwell in grasslands and rocky areas of sub-Saharan Africa and Madagascar.

cordylite-(Ce) *Mineralogy.* $Ba(Ce,La)_2(CO_3)_3F_2$, a colorless or yellow transparent, brittle, hexagonal mineral occurring as short, tiny prismatic crystals; having a specific gravity of 4.31 and a hardness of 4.5 on the Mohs scale; found in pegmatitic veins in nepheline-syenite.

core the central part of an object, material, or area; specific uses include: *Botany.* the central part of some fruits, containing the seeds, as in apples and pears. *Geology.* **1.** the central zone of the earth, originating at a depth of 2900 kilometers and equivalent to the E, F, and G layers. **2.** the central part of an anticline or syncline. *Electromagnetism.* **1.** a magnetic material placed in the center of a coil to intensify the magnetic field and increase the inductance of the coil. **2.** the part of a magnetic structure around which the magnetizing winding is placed. *Nucleonics.* the central portion of a nuclear reactor containing the fuel, coolant, and (in thermal reactors) the moderator. *Metallurgy.* **1.** in sand casting, a nonmetallic component inserted into the mold to economically produce complex castings. **2.** the portion of a case-hardened steel that has the original composition. *Engineering.* a cylindrical sample of material obtained in drilling. *Oceanography.* the area within a layer of water where its properties reach extreme values of salinity, velocity, temperature, or density. *Virology.* the portion of a virion containing nucleic acid that is enclosed by a capsid or membrane. *Archaeology.* **1.** a stone mass, often prepared or preformed, from which flakes or blades are detached. **2.** see CORE TOOL. *Graphic Arts.* the middle of a dot, especially a halftone dot.

core- a combining form meaning "pupil of the eye," as in *corectopia.*

core analysis *Geology.* a process of obtaining information about the properties and characteristics of strata by the examination of core samples taken from a borehole during drilling.

core area *Geography.* **1.** the nucleus of an ecumene. **2.** any of the more effective areas within an ecumene.

core array see CORE MEMORY.

core bank *Electronics.* a complete magnetic core memory unit consisting of a specific quantity of core arrays and related devices, such as read/write amplifiers and address registers.

core barrel *Mechanical Devices.* a hollow cylinder or pipe with a cutting bit at one end, used for extruding and holding continuous samples of drilled rocks.

core bit *Mechanical Devices.* a hollow drill bit that is used to take solid chunks of drilled material as samples.

core block *Mining Engineering.* the act of wedging core or core fragments or the impaction of cuttings inside a bit or core barrel, preventing further entry of core into the core barrel.

core blower *Metallurgy.* in casting, an apparatus for making cores by blowing and packing sand.

core-catcher case see LIFTER CASE.

cored bar *Metallurgy.* in powder metallurgy, a compact in which an interior portion has been melted by electric resistance heating.

cored electrode *Metallurgy.* a composite solder that has flux in its core.

core drill *Mechanical Engineering.* **1.** a mechanism designed to control an annular-shaped rock-cutting bit in penetrating a rock formation, producing a core sample of the formation, and lifting the sample back to the surface. **2. core-drill.** to use such a mechanism.

cored structure *Materials Science.* a structure generally occurring in the microstructures of as-cast solidified alloys, caused by a difference in chemical composition between the solid that formed first and those that formed subsequently.

core dump *Computer Programming.* a file or printed representation of the contents of main memory; used for debugging in the event of a serious error.

core enzyme *Enzymology.* the portion of RNA polymerase that possesses catalytic activity, but requires the attachment of the sigma factor before it can begin transcription.

Coregonidae see SALMONIDAE.

core-halo galaxy *Astronomy.* a radio galaxy whose emission comes from a strongly emitting core plus a weak and extended halo; about one-fifth of all extended radio sources fit this description.

Coreidae *Invertebrate Zoology.* a family of leaf-footed and squash bugs, insects in the superfamily Coreoidea.

core image *Computer Programming.* **1.** the status of an object program that is completely specified with memory addresses and is ready to be loaded and executed. **2.** a representation of the structure or contents of main memory at a particular time.

core-image library *Computer Programming.* a set of object programs that are compiled and ready to execute; usually stored on bulk storage media such as tape or disk.

core intersection *Mining Engineering.* **1.** the point in a borehole at which an ore vein or body is struck, as shown by the sample of sediment extracted from the hole. **2.** the width of the ore body, as shown by the sample of sediment extracted from the borehole. Also, **core interval.**

core iron *Metallurgy.* in founding, a strengthening iron grate in a core.

CORELAP *Industrial Engineering.* a computerized method to determine a plant layout, using numbers to represent relationships between departments. (An acronym for computerized relationship layout planning.)

coreless induction heater *Engineering.* an induction furnace having no magnetic circuit other than the charge in the furnace itself.

core logging *Geology.* a process of obtaining information about the properties and characteristics of strata by the examination of core samples taken at predetermined depth intervals as a well is drilled.

core logic *Electronics.* a digital logic circuit based on the use of magnetic cores to perform the operations.

core loss *Electromagnetism.* in an electromagnet, the energy losses associated with the changing electric field when alternating current is used, resulting in conversion of electric energy to thermal energy; this may be caused by hysteresis losses or by eddy-current losses.

core memory *Computer Technology.* a large number of magnetic cores arranged in a rectangular grid matrix to store digital data. Also, CORE ARRAY.

core-memory resident *Computer Technology.* describing an operating system, control program, or the like that is permanently stored in a computer's main memory.

coremium *Mycology.* a bundled group of spore-bearing hyphae found in certain imperfect fungi.

core molding *Metallurgy.* in casting, a molding process based on assembled cores.

corepressor *Molecular Biology.* any of a group of metabolites that when combined with repressors specifically stop the formation of enzymes involved in their metabolism.

corer *Engineering.* an device having a hollow cylindrical drill or tube to extract underground earth or marine samples.

core rod *Metallurgy.* in powder metallurgy, a rod used to produce a cavity in a compact.

core-rope storage *Computer Programming.* an early permanent storage device that stored coded information in the form of an array of magnetic cores and wires, in which the information was contained in the wires rather than the core itself; economical in the use of then-expensive core memory, since reference to a single core location would result in access to a pattern of bits or pulses.

core sample *Geology.* a piece of rock, soil, snow, or ice that is extracted by drilling and used for analysis.

core sand *Materials Science.* sand that has a binding material added to ensure cohesion and porosity after drying; used in a core in molds for casting.

core-spun yarn *Textiles.* a yarn made of staple fibers that are wrapped around a central continuous filament.

core storage *Computer Technology.* an early storage device consisting of an array of magnetic cores, each capable of storing one bit; largely replaced by semiconductor memory devices.

core test *Textiles.* the removal of a sample of wool from a bag or bale, in order to test for impurities and to grade the wool.

core tool *Archaeology.* the remainder of a core used as a tool after all flakes have been chipped off.

core yarn *Textiles.* the central fiber of a core-spun yarn.

Corey-House synthesis *Organic Chemistry.* a synthetic process for making an alkane of higher carbon number than the starting material by coupling two aklyl groups, one from lithium dialkylcopper, and the other from an alkyl halide.

Cori, Carl Ferdinand 1896–1984, and **Gerty Cori** 1896–1957, Czech-born American physicists; husband and wife who shared Nobel Prize for discovering the course of catalytic conversion of glycogen.

coriaceous *Biology.* a tough and leathery appearance and texture; used in reference to certain leaves, insects, and fungi. Also, **corious.**

coriander *Botany.* a strongly scented herb, *Coriandrum sativum,* having small white, rose, or lavender flowers and aromatic seeds that are used as a spice.

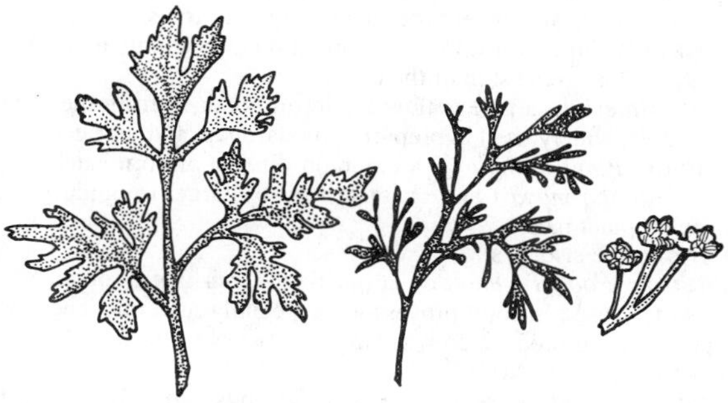

coriander

coriander oil *Materials.* an essential oil, derived from the herb species *Coriandrum sativum,* with a strong spicy taste that may be irritating when ingested; used chiefly to flavor gin.

Coriariaceae *Botany.* a monogenetic family of shrubs belonging to the order Sapindales, characterized by small, opposite leaves, terminal racemes of tiny flowers, and purplish fruit.

Cori cycle *Biochemistry.* the metabolic cycle of reactions in which glycogen is broken down and resynthesized.

coring *Materials Science.* a variation in the chemical composition of the solid phase of alloys over the freezing temperature range. The central portion of dendrites is richer in the constituent with the higher melting point, and the interdendrite species are richer in the constituent with the lower melting point. The compositional difference increases with the increasing cooling rate. *Metallurgy.* in extrusion, an axial pipe formed at the end of the extruded rod. *Petroleum Engineering.* the process of taking cylindrical samples of rock from an underground formation during the drilling operation by using a hollow drill bit; used for core analysis.

Corinthian *Architecture.* the most ornate of the classical Greek orders, characterized by its slender column and bell-shaped capital decorated with acanthus leaves.

Coriolis, Gaspard de 1792–1843, French civil engineer; analyzed the Coriolis force.

Coriolis acceleration [kôr′ē ō′lis] *Mechanics.* the apparent acceleration in a rotating reference frame due to the Coriolis force.

Coriolis coefficient *Hydrology.* a dimensionless constant, denoted by the Greek letter α, that is computed based upon the total kinetic energy at a section in a channel; α is equal to the sum of products of velocities v in elementary areas, δA cubed and δA divided by the product of the mean velocity in the section V cubed and the area of the section A. Also, ENERGY COEFFICIENT.

Coriolis correction *Navigation.* a correction to a sextant reading taken from a fast-moving craft when an artificial horizon is used. It varies with the speed, latitude, and heading of the craft.

Coriolis effect *Mechanics.* **1.** the deflection relative to the earth's surface of an object that is moving on or above the earth, due to the action of the Coriolis force. An object that is moving horizontally above the earth's surface in the Northern Hemisphere tends to show a rightward deflection, and one in the Southern Hemisphere tends to show a leftward deflection. Also, **Coriolis deflection. 2.** the effect of the Coriolis force in any rotating coordinate system. *Physiology.* the nausea, dizziness, and other physiological effects that a person feels when moving rapidly in a rotating system, as in many amusement park rides.

Coriolis force *Mechanics.* an inertial force that acts on particles moving in a reference frame that is rotating with respect to an inertial frame, given by the cross product of twice the particle's momentum in the rotating frame, with the system's angular velocity.

Coriolis operator *Spectroscopy.* an operator that gives a large contribution of energy to an axially symmetric molecule, resulting from rotation-vibration interaction when two vibrations are at equal or nearly equal frequencies.

Coriolis parameter *Geophysics.* the measurement and speed of the Coriolis force per unit mass within a given parameter.

Coriolis resonance interaction *Spectroscopy.* the perturbation of two vibrations at nearly equal frequencies in a polyatomic molecule, as a result of the energy contributed by the Coriolis operator.

Coriolis-type mass flowmeter *Engineering.* a device that measures the mass flow rate of liquids from the torque on a ribbed disk that turns at a constant speed as fluid enters its center and is accelerated radially.

Coriolus *Mycology.* a genus of fungi belonging to the order Agaricales which lives on deciduous wood, where it forms bracket-shaped spore-bearing structures or basidiocarps.

corium see DERMIS.

Corixidae *Invertebrate Zoology.* the water boatmen, the single family of hemipteran insects in the superfamily Corixoidea.

Corixoidea *Invertebrate Zoology.* a superfamily of water bugs; insects without ocelli in the subdivision Hydrocorisae.

cork *Materials.* a lightweight, spongy substance that is obtained from the outer layer of the bark of a certain Mediterranean tree; noted for its ability to be compressed into a smaller size and to resist the absorption of liquids; widely used since ancient times for bottle stoppers, fishing net floats, and various other purposes. *Botany.* **1.** the tree itself, *Quercus suber,* which is widely grown in Portugal, Spain, and elsewhere. Also, CORK OAK. **2.** in general, the outer layer of bark of any woody plant. Also, PHELLEM.

corkboard *Materials.* a material that is made of compressed cork, used for insulation and for bulletin boards.

cork cambium *Botany.* a layer of tissue beneath the cork layer, from which new cork tissue is formed. Also, PHELLOGEN.

cork oak *Botany.* see CORK, def. 1.

corkscrew *Mechanical Devices.* **1.** a device consisting of a metal spiral with a sharp point at one end and a handle or lever at the other, used to extract corks from wine bottles. **2.** any of various other devices having a spiraling shape that is suggestive of a corkscrew. *Mining Engineering.* see DRAG TWIST.

corkscrew rule *Electromagnetism.* a rule stating that a magnetic field circulates about a current filament in the direction of a corkscrew's rotation, if the current is considered in the same relative direction as the corkscrew's advancing motion.

Corliss, George Henry 1817–1888, American engineer; improved the steam engine.

Corliss valve *Mechanical Engineering.* a rotating admission and exhaust valve used on a steam engine.

corm *Botany.* the base of a stem, usually subterranean and covered with a scaly membrane, that acts as an organ of storage and vegetative propagation.

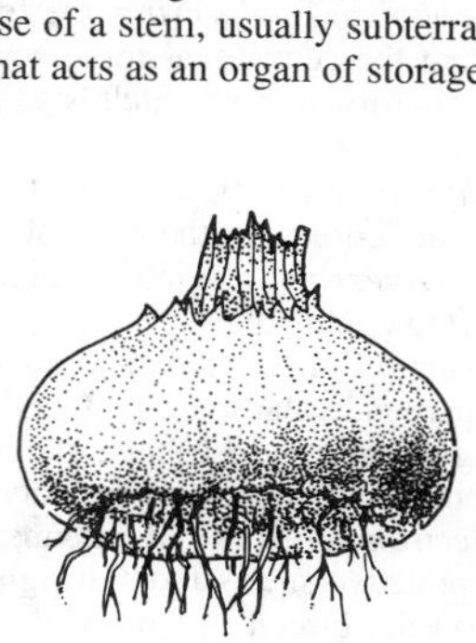

corm

Cormack, Allan MacLeod born 1924, South African-born American physicist; Nobel Prize for contributions to invention of the CAT scan.
cormatose *Botany.* any plant that is capable of forming a corm.
cormidium *Invertebrate Zoology.* the clusters of zooids dangling from the main stem of pelagic siphonophores.
cormorant *Vertebrate Zoology.* any of several seabirds of the family Phalacrocoracidae, such as *Phalacrocorax carbo,* found in America, Europe, and Asia; having a long neck and a distensible pouch under the bill in which it holds captured fish; used in China for catching fish.
corn *Botany.* **1.** a tall cereal plant, *Zea mays,* widely cultivated in many varieties in North America and elsewhere for use as a food crop, for animal feed, and as a source of oil, syrup, and other such products; it is also called **maize** (especially in British use) and **Indian corn** (because it was first grown by the American Indians and introduced to European use by them). **2.** a general term for cereal grains, such as wheat, rye, barley and so on. *Agriculture.* a term for the leading cereal crop of a particular region; e.g., wheat in England. *Medicine.* a small, painful growth on the skin, epecially the toes or feet, caused by rubbing or friction.
Cornaceae *Botany.* a family of dicotylendonous trees and shrubs having usually opposite, simple leaves and unisexual or bisexual flowers in cymes and panicles.
Cornales *Botany.* an order of woody trees, shrubs, and herbs of the subclass Rosidae.
corn borer *Invertebrate Zoology.* any of various moths whose larvae are corn pests, such as the **European corn borer,** *Ostrinisa nubilalis.*
cornea [kôr´nē ə] *Anatomy.* a transparent membrane that forms the anterior portion of the eyeball.
corneal [kôr´nē əl] *Anatomy.* relating to or affecting the cornea.
corneal reflex *Medicine.* the reflexive closing of the eyelids due to irritation of the cornea.
corned beef *Food Technology.* a brisket of beef that is cured in a brine made with salt crystals ("corn"), sugar, spices, and preservatives.
corneite *Geology.* a densely resistant fine-grained, black metamorphic rock containing quartz and mica, formed during the folding of a shale.
corner *Cartography.* **1.** a point on land where two or more boundary lines meet. **2.** a point on the earth's surface determined by surveying and marking an extremity of a boundary of a subdivision of public lands.
corner bead *Building Engineering.* **1.** a protection molding oriented vertically relative to the external angle of two intersecting surfaces. **2.** a strip of galvanized iron formed in such a way that it combines with a similarly oriented strip of metal lath; they are placed together on corners for reinforcement purposes.
corner braces *Building Engineering.* one-inch by four-inch boards that are let in to notches cut in the outer faces of the stud on wall or wood sheathing to provide lateral stability to the frame; the outer faces of the braces are then flush with the studs to which they are nailed.
corner chisel *Mechanical Devices.* a chisel with two cutting edges that meet at a right angle.
corner clamp *Mechanical Devices.* a clamp used to hold mitered joints when gluing or nailing woodwork.
corner-cube reflector *Optics.* see CORNER REFLECTOR.
corner effect *Electronics.* a rounding off of a characteristic that ideally has a sharp corner (e.g., the response at the end of the pass band). *Engineering.* the reflection of an ultrasonic beam that hits the intersection of two surfaces 90° apart at a perpendicular angle.
corner head *Building Engineering.* in plastering applications, a molding that is built into corners to prevent pieces from breaking off; it is usually made of metal.
cornering force *Mechanical Engineering.* the force produced as a vehicle turns to counteract the centrifugal force acting on it. The greater the cornering force proportional to a vehicle's weight, the more sharply it can turn.
cornering tool *Mechanical Devices.* a cutting tool with a curved edge, used in rounding off sharp corners or the internal corners of a mold.
cornerite *Building Engineering.* a reinforcement material for corners used in interior plastering applications.
corner joint *Engineering.* an L-shaped joint that results when two pieces are connected perpendicular to each other.
cornerload test *Engineering.* a test that establishes whether the load distribution has influenced the display on an analytical balance.
corner reflector *Electromagnetism.* an antenna whose reflecting surface is that of two conductive sheets fixed at right angles with a radiating element located on the bisector plane of the two sheets. *Optics.* a prism having a hypotenuse face and three mutually perpendicular surfaces, in which light passing through the hypotenuse face is reflected back to its source; used by surveyors to measure distances accurately. Also, CORNER-CUBE REFLECTOR, CUBE CORNER PRISM.
cornerstone *Building Engineering.* **1.** a stone that unites two masonry walls at a particular intersecting point. **2.** a large single stone that is the ostensible starting point of a building, having an inscription carved on the date when it was laid.
cornetfish *Vertebrate Zoology.* a tropical reef fish of the family Fistulariidae, distinguished by an extremely slender and long body and named for its long, tubular snout (about one-third of the body) resembling a cornet; further characterized by plated (scaleless) skin and a long filament extending from the central tail fin. Also, FLUTEMOUTH.
cornetite *Mineralogy.* $Cu_3^{+2}(PO_4)(OH)_3$, a blue, transparent to translucent orthorhombic mineral occurring as short prismatic crystals; having a specific gravity of 4.1 and a hardness of 4.5 on the Mohs scale; found in areas with brochantite, malachite, and chrysocolla.
Cornforth, Sir John Warcup born 1917, British chemist; Nobel Prize for research on the stereochemistry of enzyme-catalyzed reactions.
cornice *Architecture.* **1.** the upper division of a classical entablature. **2.** a projecting decorative molding crowning a building or wall.
cornice brake *Mechanical Engineering.* a device used to bend and shape sheet metal.
cornicle *Invertebrate Zoology.* the protruding dorsal tubes in aphids that secrete a waxy fluid.
corniculate *Biology.* horn-shaped or having a hornlike structure.
corniculate cartilage *Anatomy.* either of two small cartilaginous nodules in the larynx that articulate with the apices of the arytenoid cartilages and provide attachment for the aryepiglottic folds.
cornification *Physiology.* the transformation of cells into a horny layer, such as a callus on the skin of the hand.
corn oil *Materials.* a pale yellow liquid oil obtained from the germs of corn kernels, widely used in preparing foods, soap, and lubricants.
corn smut *Plant Pathology.* a common disease of corn caused by the fungus *Ustilago maydis* and characterized by large, irregularly shaped galls on all plant parts.
corn snow see SPRING SNOW.
cornstarch *Food Technology.* a purified starch that is obtained from corn; widely used in food processing, especially as a thickener, or further processed to produce corn syrup.
corn sugar see DEXTROSE.
corn syrup *Food Technology.* a clear, viscous syrup that is produced by the partial hydrolysis of starch from corn, used in confectionery as a sweetening agent.
cornu *Anatomy.* any horn-shaped process.
Cornu, Marie Alfred 1841–1902, French physicist; known for research on the speed and movement of light.
Cornu quartz prism *Optics.* a prism in which two 30° quartz prisms, one left-handed and the other right-handed, are cemented together and used along with left- and right-handed lenses to produce a rotation of polarization, in opposite directions, in each half of the prism. Also, **Cornu-double prism.**
Cornu's spiral *Mathematics.* the graph in the plane of the function $f(x,y) = \int_{-\infty}^{\infty} \exp[-y^2z^2 - z - xe^{-z}]dz$; in parametric form, the family $x = a\sqrt{\pi}\int_0^t \cos(\pi\theta^2/2)d\theta$ and $y = a\sqrt{\pi}\int_0^t \sin(\pi\theta^2/2)d\theta$. These curves have a center of symmetry at the origin and the property that the curvature at a point P on the curve is proportional to the length of an arc on the curve, measured from the origin; that is, $1/r = 1/a^2$. The curve has two arms that spiral into the asymptotic points $(a\sqrt{\pi}/2, a\sqrt{\pi}/2)$ and $(-a\sqrt{\pi}/2, -a\sqrt{\pi}/2)$.

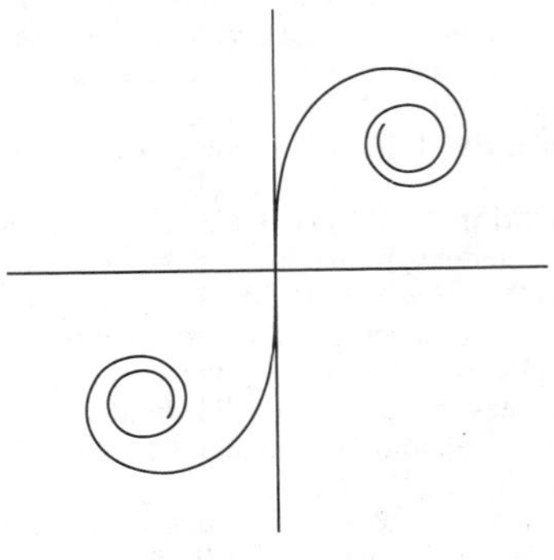

Cornu's spiral

cornwallite *Mineralogy.* $Cu_5^{+2}(AsO_4)_2(OH)_4 \cdot H_2O$, a light to dark green, subtranslucent monoclinic mineral that forms radiating botryoidal crusts; having a specific gravity of 4.52 and a hardness of 4.5 on the Mohs scale; found with olivenite, malachite, and enargite.

corn whiskey *Food Technology.* a whiskey distilled from a mash that is at least 80% corn.

coro- a combining form meaning "pupil of the eye," as in *coroscopy.*

corolla *Botany.* the petals of a flower, considered collectively.

corollate *Botany.* possessing a corolla.

corolline *Botany.* resembling or akin to a corolla.

coromell *Meteorology.* a land breeze from the south at La Paz, Mexico, prevailing from November to May, that usually begins at night and persists until midmorning.

corona [ke rō´nə] a figure or area suggesting the shape or appearance of a crown; specific uses include: *Astronomy.* the tenuous outermost layer of the sun's atmosphere that begins immediately above the chromosphere and that contains gas at temperatures of 1 to 2 million degrees kelvin. *Meteorology.* a set of prismatically colored rings, caused by diffraction of water droplets, seen around the sun or moon when viewed through a thin cloud. *Architecture.* the vertical, overhanging upper part of a cornice. *Electricity.* see CORONA DISCHARGE. *Botany.* a circular crownlike appendage of the floral perianth. *Invertebrate Zoology.* **1.** the anterior circle of cilia in rotifers. **2.** a sea urchin test. **3.** the arms and calyx of a crinoid. *Mineralogy.* a mineral zone, typically concentric with and surrounding another mineral or lying between two minerals. *Metallurgy.* in spot welding, an area around the spot that has been welded.

Corona Australis *Astronomy.* the Southern Crown, an inconspicuous constellation of dim stars lying between Sagittarius and Scorpius.

Corona Borealis *Astronomy.* the Northern Crown, a small constellation lying between Bootes and Hercules that is generally dim but contains one second-magnitude star, Gemma.

corona current *Electricity.* the current equal to the rate of charge transferred to the air from an object experiencing a corona discharge.

corona discharge *Electricity.* a luminous discharge that occurs on the surface of and adjacent to a conductor, or between two conductors on the same transmission line, when the potential gradient exceeds a critical value; caused by the ionization of the surrounding atmosphere. Also, CORONA.

coronadite *Mineralogy.* $Pb(Mn^{+4},Mn^{+2})_8O_{16}$, a black or gray, opaque, submetallic monoclinic (pseudotetragonal) mineral; having a specific gravity of 5.44 and a hardness of 4.5 to 5 on the Mohs scale; occurring as botryoidal fibrous crusts in oxidized zones of mines in California, Arizona, Australia, and Morocco.

corona failure *Electricity.* in areas of high voltage, the failure caused when corona discharge degrades a body.

coronagraph *Astronomy.* a type of telescope that artificially eclipses the sun's bright disk to reveal the lower parts of the corona and any prominences jutting from the solar limb.

coronagraph

coronal hole *Astrophysics.* a large region in the solar corona that has unusually low density and temperature and is associated with weak, diverging magnetic fields.

coronal lines *Astrophysics.* bright emission lines in the spectrum of the sun's corona.

coronal rain *Astronomy.* the hot gas in a solar prominence that is seen falling back toward the sun's surface.

coronal suture *Anatomy.* the area where the frontal and parietal bones join transversely across the vertex of the skull.

corona method *Geophysics.* a means of determining drop sizes of clouds by measuring the angular radii of the rings of the clouds' corona.

corona radiata *Histology.* the layer of cells immediately surrounding a mammalian ovum.

corona resistance *Electricity.* the ability of an insulator to withstand levels of field-intensified ionization without insulation breakdown.

coronary [kôr´ ə nâr´ē] *Anatomy.* encircling in the manner of a crown; applied to vessels, ligaments, and so on, and especially to the arteries of the heart. *Medicine.* **1.** relating to the arteries of the heart or to conditions affecting these arteries. **2.** a popular term for a coronary occlusion or coronary artery disease.

coronary angioplasty *Cardiology.* the use of a balloon catheter to surgically reconstruct atherosclerotic blood vessels of the heart in order to increase blood flow.

coronary arteriosclerosis *Cardiology.* the thickening and loss of elasticity of the walls of the coronary arteries, usually accompanied by luminal constriction.

coronary artery *Anatomy.* either of a pair of arteries that supply blood directly to the heart.

coronary artery bypass *Cardiology.* an operation in which a section of vein or other material is grafted between the aorta and a coronary artery to allow the blood to flow past an obstructive lesion in an artery. Also, **coronary bypass.**

coronary artery disease *Pathology.* any of various conditions that affect the coronary arteries and reduce blood flow to the heart, especially a narrowing or blockage of the ateries that produces the painful symptoms of angina pectoris.

coronary atherosclerosis *Cardiology.* fatty deposits on the inner surface of arteries of the heart.

coronary-care unit *Cardiology.* a specially equipped hospital unit with specially trained personnel, designed for the intensive care of patients who are suffering from severe heart disease, including myocardial infarction.

coronary circulation *Cardiology.* the bloodflow in the arteries that supply the heart.

coronary disease see CORONARY ARTERY DISEASE.

coronary failure *Medicine.* an inability of the coronary arteries to supply sufficient blood for heart function, giving rise to chronic precordial pain or discomfort.

coronary occlusion *Cardiology.* an obstruction or blockage of one of the coronary arteries, especially a complete blockage.

coronary reflex *Cardiology.* a change in the musculature of coronary arteries as a result of nervous control.

coronary sinus *Anatomy.* a vein that lies in the groove between the left atrium and ventricle on the dorsal surface of the heart; it collects blood from the heart muscle and delivers it to the right atrium.

coronary stenosis *Cardiology.* the constriction of a coronary artery due to the narrowing of the artery's lumen.

coronary sulcus *Anatomy.* the groove on the outer surface of the heart that marks the division between the atria and the ventricles. Also, ATRIOVENTRICULAR GROOVE.

coronary thrombosis *Cardiology.* a blood clot that lodges in a coronary artery, frequently causing a blockage of the blood flow and leading to myocardial infarction.

corona shield *Electricity.* a shield surrounding a point of high potential in order to redistribute the lines of electrostatic force and thus prevent corona discharge.

corona start voltage *Electricity.* the voltage at which corona discharge is initiated for a certain system.

Coronatae *Invertebrate Zoology.* an order of deep-sea coelenterates in the class Scyphozoa.

corona tube *Electricity.* a low-current gas-filled tube that utilizes corona discharge. Also, **corona-discharge tube.**

Coronaviridae *Virology.* a family of pleomorphic ssRNA-containing viruses that cause respiratory infections in humans and birds, gastroenteritis in swine, and hepatitis in mice.

Coronavirus *Virology.* the single genus of the family Coronaviridae. (So called because of their halo- or corona-shaped appearance under the microscope.)

corona voltmeter *Electricity.* a voltmeter in which the peak voltage value is indicated by the beginning of a corona at a known electrode spacing.

coroner *Medicine.* a public official whose duty is to investigate and establish the cause of a person's death, especially in the case of a death resulting from unnatural or uncertain causes.

coronizing *Metallurgy.* the process of electroplating zinc on nickel, followed by heat-treating.

coronoid *Biology.* **1.** shaped like a crow's beak; used to describe certain bony projections **2.** specifically, the membrane bone on the upper side of the lower jaw in some vertebrates.

coronoid fossa *Anatomy.* a concave space on the anterior surface of the distal end of the humerus, into which the coronoid process of the ulna rests when the elbow is flexed. Also, FOSSA CORONOIDEA.

coronoid process *Anatomy.* **1.** a flattened thin process projecting from the anterior end of the mandible, where the temporal muscle is inserted. **2.** a thin process at the proximal end of the ulna that contributes to the trochlear notch and to which the brachialis attaches.

Coronophoraceae *Mycology.* the only family of fungi belonging to the order Coronophorales which occur primarily on wood, although some are parasites of other fungi.

Coronophorales *Mycology.* a small order of fungi belonging to the class Pyrenomycetes, which is characterized by brittle ascocarps that open by disintegration of the apex and by ascospores that form a powdery mass at maturity.

coronule *Invertebrate Zoology.* the outside ring of spines on certain diatom shells.

Corophiidae *Invertebrate Zoology.* a family of amphipods, crustaceans in the suborder Gammaridea, with no carapace and no eye stalks.

corotating interaction regions *Astronomy.* regions in the solar system where abrupt changes in the velocity of the solar wind have isolated parts of the interplanetary magnetic field.

coroutine *Computer Programming.* a set of instructions that transforms inputs into outputs; unlike a subroutine, each time a coroutine is called, it begins execution at the place it left off the last time it executed rather than at the beginning.

corp. corporal.

corpora arenacea *Anatomy.* small calcareous bodies found in the pineal body of the brain. Also, BRAIN SAND.

corpora bigemina see OPTIC LOBE.

corporal *Military Science.* a noncommissioned officer, ranking above a private first class in the U.S. Army or a lance corporal in the U.S. Marines and below a sergeant.

Corporal *Ordnance.* an early U.S. mobile surface-to-surface ballistic missile powered by a liquid-propellant motor; delivering a nuclear warhead up to a maximum range of 75 nautical miles; officially designated **MGM-5.**

corpora quadrigemina *Anatomy.* four rounded eminences on the dorsal portion of the midbrain. Also, FOURFOLD BODIES, QUADRIGEMINAL BODIES, OPTIC LOBE, TECTUM.

corporate group *Anthropology.* a hierarchical society organized under a common mythological origin through which the resource base is exploited by redistribution of goods and services, regulated by hereditary leaders through ceremonial or ritual law.

cor pulmonale *Cardiology.* an abnormal cardiac condition with hypertrophy of the right ventricle of the heart, caused by hypertension of the pulmonary circulation.

corpus albicans *Histology.* the white fibrous structure in the mammalian ovary that develops from a degenerating corpus luteum.

corpus allatum *Entomology.* an endocrine gland behind the brain believed to secrete a hormone that controls moulting during the larval stages. *Invertebrate Zoology.* an endocrine structure near the brain of immature arthropods that secretes the juvenile hormone neotenin.

corpus callosum *Anatomy.* a band of white matter at the base of the longitudinal fissure connecting the two cerebral hemispheres.

corpus cavernosum *Anatomy.* the right and left cylinders of the erectile tissue, united distally but separated proximately to form the penis in the male and the clitoris in the female.

corpuscle [kôr´pus´əl] *Anatomy.* **1.** any small mass or body. **2.** specifically, a red or white blood cell. *Optics.* a particle of light in the corpuscular theory, corresponding to a photon in quantum theory.

corpuscula bulboidea see KRAUSE'S CORPUSCLES.

corpuscular radiation *Physics.* the radiation of particles such as electrons or nucleons, rather than electromagnetic radiation.

corpuscular theory of light *Optics.* the theory that light is made of a stream of particles moving at a very high velocity. Also, NEWTON'S THEORY OF LIGHT.

corpus luteum *Endocrinology.* a yellowish endocrine structure transiently formed by the granulosa cells remaining in the ovarian follicle from which an ovum has recently been discharged; these cells are lipid-rich and highly steroidogenic, producing both progesterone and estrogen, but involute unless pregnancy occurs.

corpus spongiosum *Anatomy.* the erectile tissue in a penis through which the urethra passes.

corpus striatum *Anatomy.* a subcortical mass of gray and white matter in front of and lateral to the thalamus in each cerebral hemisphere; one of the components of the basal ganglia.

corrasion *Geology.* a process of erosion in which the abrasive action of solid materials transported by wind, waves, running water, glaciers, or gravity mechanically wears away rocks or soil.

corrected altitude *Meteorology.* the indicated altitude corrected for temperature deviation from the standard atmosphere. Also, TRUE ALTITUDE.

corrected azimuth *Ordnance.* in a gun firing at a moving target, the azimuth of the bore axis, adjusted for variations in atmospheric, material, and other conditions.

correcting *Navigation.* the process of adjusting some observed phenomenon (such as a compass course or sextant altitude) to a standard representation (such as true direction).

correcting wedge *Mining Engineering.* a deflection wedge used to correct the course of a crooked borehole. Also, DEFLECTING WEDGE.

correction chamber *Engineering.* the chamber within an analytical balance that contains the material used to bring the weight to a nominal value.

correction collar *Optics.* a ring or collar around a microscope lens that adjusts the lens to work with slides and cover-glasses of various thicknesses.

correction force *Mechanics.* a force that is presumed or postulated to exist because it provides an opposing effect to correct an actual force; e.g., centrifugal force, which accounts for the fact that centripetal force does not draw rotating objects inward in a shrinking circle.

correction of soundings *Navigation.* the correction of water depths as determined by soundings to adjust for the difference between actual water surface level with the datum level shown on charts (these may differ, for example, due to tides).

correction time *Control Systems.* the time required for a controlled variable to settle within a predetermined area near the control point after any change to the operating condition or independent variables in a system.

corrective action *Control Systems.* the act of altering a process variable in order to change an operating condition in a system.

corrective maintenance *Computer Programming.* programming activities that take place after a program has been put into use, for the purpose of eliminating remaining bugs or incorrect operations. *Computer Technology.* unscheduled maintenance on hardware devices following a failure; often performed by the equipment manufacturer.

corrective network *Electricity.* an electric network that improves circuit transmission properties, impedance properties, or both. Also, SHAPING NETWORK.

corrective therapy *Medicine.* a program of progressive physical exercise and activities to improve or maintain general physical and emotional health, through individual or group participation.

corrector *Engineering.* a metal device or magnet used to adjust a magnetic compass.

corrector plate *Optics.* an element that rectifies spherical aberration on each zone of a lens, especially in telescopes.

correed relay *Electricity.* a telephone switching device made of a hermetically sealed reed capsule enclosed by a coil.

correlation *Biology.* a mutual relationship or balance existing between different functions, structures, or characteristics in a plant or an animal. *Geology.* the determination of the characteristic or chronological equivalence or relationships between geologic phenomena or events, usually in separated areas. *Statistics.* the strength of the linear dependence between two random variables. Also, PRODUCT-MOMENT CORRELATION.

correlation detection *Aviation.* a method in which moving aircraft or spacecraft are detected by comparing a signal, point to point, with an internally generated reference. Also, CROSS-CORRELATION DETECTION.

correlation, detection, and ranging *Military Science.* a system of submarine detection using sonobuoys dropped from naval aircraft; variations in time and phase of acoustic signals transmitted by radio from the sonobuoys to the aircraft indicate the location of the submerged submarines. Also, CODAR.

correlation direction finder *Engineering.* a satellite station that is part of an arrangement of stations separated from a radar facility; it receives jamming signals, and in conjunction with the other stations can help to determine the position and range of the jammers.

correlation distance *Telecommunications.* an important measure of correlation in which the line of regression that is fitted to the points on a scatter diagram is done in such a way that the sum of the squared deviation from the line is at a minimum.

correlation hole *Materials Science.* an area in a polymer melt where there is a reduced concentration of similar units.

correlation tracking system *Engineering.* a trajectory-measuring system in which various signals obtained from a single source are correlated to determine the phase difference between the signals.

correlation ultrasonic flowmeter *Engineering.* a device that measures the rate of fluid flow by recording the amount of time it takes for discontinuities in the fluid to travel between two corresponding transducers that emit and detect high-frequency sound.

correlator *Electronics.* a logic device that detects weak signals by an operation approximating the computation of a correlation function, in which a reference signal representing the expected signal is constructed and compared with the receiver input to distinguish between the noise and signal components. Also, **correlation type receiver.**

Correns, Carl 1864–1933, German biologist; rediscovered Mendel's work on genetics, in light of which he reexamined Darwin's theory.

correspondence *Cartography.* a condition that exists when corresponding images on a pair of photographs lie in the same epipolar plane; the absence of *y*-parallax.

correspondence principle *Quantum Mechanics.* the principle that the predictions of quantum and classical mechanics must correspond in the limit of very large quantum numbers.

corresponding image *Photogrammetry.* a point or line in one system of points or lines that is homologous to a point or line in another system.

corresponding image points *Photogrammetry.* the images of the same object point on two or more photographs.

corresponding image rays *Photogrammetry.* rays connecting each of a set of corresponding image points with its perspective center.

corresponding points *Physiology.* in binocular vision, the image points on the fovea of the two retinas that, when stimulated simultaneously, give the impression of a single image.

corresponding states *Physical Chemistry.* a condition in which two or more different substances have reduced pressures, volumes, and temperatures that are equal fractions of their critical pressures, temperatures, and volumes. Also, REDUCED STATES.

corridor *Ecology.* a connection between adjacent land areas that allows the passage of fauna from one area to the other. *Transportation Engineering.* **1.** a limited band of airspace connecting two general areas and containing one or more air routes. **2.** a similar band on land or water.

Corriedale *Agriculture.* a breed of hornless, white-faced sheep that produces both wool and mutton, crossbred from the Romney, Lincoln, or Leicester rams and Merino ewes. (From the *Corriedale* ranch in New Zealand, where the breed was developed.)

Corrigan's pulse *Medicine.* a pulse of abrupt and strong rise and fall found in conditions of aortic regurgitation or peripheral arterial dilation. Also, WATER-HAMMER PULSE, COLLAPSING PULSE.

Corrigan's sign see COPPER LINE.

corrode *Chemistry.* of a material, to undergo or be subjected to the process of corrosion.

Corrodentia see PSOCOPTERA.

corroding lead *Metallurgy.* a commercially pure form of lead, containing 99.94% minimum lead.

corrosion *Chemistry.* the degradation of metals or alloys due to chemical reactions with their environment, accelerated by the presence of acids or bases; for example, the rusting of metal surfaces exposed to moist air or to impure water. Also, CHEMICAL EROSION. *Materials Science.* the decay of a ceramic material, or its conversion to a nonuseful product, due to chemical action. *Geochemistry.* the partial resorption, alteration, or erosion of crystals by the solvent action of residual magma.

corrosion border *Mineralogy.* a border made up of a secondary mineral formed around an original crystal as a result of magmatic corrosion. Also, **corrosion rim.**

corrosion cell *Metallurgy.* an electrochemical cell that takes part in corrosion reactions. It can be caused, for example, by a contact between dissimilar electrodes, by differences in electrolyte concentration, or by differences in electrolyte temperature.

corrosion fatigue *Metallurgy.* the combined action of corrosion and alternative or repeated stresses, leading to fracture.

corrosion potential *Metallurgy.* the potential of a corroding surface, relative to a standard electrode.

corrosion protection *Metallurgy.* any of several methods used to alleviate corrosion, such as painting, plating, or cathodic protection.

corrosion test *Metallurgy.* any of several standard tests used to assess quantitatively the corrosion resistance of a metal or alloy in specific environments.

corrosive *Materials Science.* **1.** causing corrosion. **2.** any substance that causes corrosion, epecially one that has a greater than usual tendency to do this; e.g., hydrochloric acid, HCl.

corrosive flux *Metallurgy.* in soldering, a flux that removes oxides present on the base material.

corrosiveness *Metallurgy.* the tendency of a metal to corrode.

corrosive poison *Toxicology.* any poison that acts by dissolving or eroding tissue through chemical, physicochemical, or biologic action.

corrosive product *Chemical Engineering.* any product, usually acidic or alkaline, that corrodes or eats away surfaces of metal or other materials by electrochemical action.

corrosive sublimate *Inorganic Chemistry.* a popular name for mercuric chloride, $HgCl_2$. See MERCURIC CHLORIDE.

corrugated bar *Mechanical Devices.* a steel reinforcing bar with transverse ridges to improve its grip on concrete.

corrugated expansion joint *Mechanical Devices.* an expansion joint, usually with bellows and a corrugated diaphragm coupled to a pressure transducer.

corrugated fastener *Mechanical Devices.* a sheet-metal fastener with serrated edges, used for rough-fastening of wood objects such as crates.

corrugated lens *Optics.* a lens in which circular portions have been removed from the surface to reduce its weight while maintaining its focal power.

corrugating *Design Engineering.* the forming of parallel ridges and depressions into sheet metal, cardboard, or other material to increase stiffness and strength.

corrugation method *Agronomy.* an irrigation system used mainly for grain and hay crops, in which irrigation furrows are run 18 to 36 inches apart to wet the field between.

corrupt *Computer Programming.* to contaminate or alter information so that it is no longer usable.

Corsair *Aviation.* the popular name for the F-4U fighter.

Corsiaceae *Botany.* a family of mycotrophic herbs without chlorophyll of the order Orchidales, characterized by alternate leaves reduced to fairly large scales, solitary terminal flowers, and numerous tiny seeds with scanty endosperm; native to Chile and New Guinea.

Corsica *Geography.* a French island in the Mediterranean Sea, off the northwestern coast of Italy north of Sardinia. Also, **Corse.**

Corsiniaceae *Botany.* a widespread family of typical liverworts of the order Marchantiales, characterized by sparingly dichotomous plants, simple pores in the epidermis surrounded by thin walls, and archegonia in cavities in the thallus.

corsite *Petrology.* a spheroidal variety of gabbro composed of bytownite, hornblende hypersthene, and traces of epidote, calcite, biotite, chlorite, quartz, apatite and ores. Also, MIAGITE, NAPOLEANITE.

cortex *Botany.* **1.** the primary tissue between the vascular bundles and the epidermis of stems and roots, consisting mostly of parenchyma cells. **2.** a thickened outer layer of cells found in some algae, lichens, and fungi. *Anatomy.* the outer layer of an organ, usually surrounding an inner medula, such as the layer of gray matter covering the hemispheres of the cerebrum and cerebellum. *Invertebrate Zoology.* the outside layer of certain protozoans.

Cortez, Sea of see CALIFORNIA, GULF OF.

cortical granule *Developmental Biology.* a secretory vesicle in the cortex under the plasma membrane of animal eggs, whose contents are released during fertilization.

cortical reaction *Developmental Biology.* the changes in the plasma membrane that accompany fertilization of an ovum, one result of which is to prevent penetration of the egg by additional sperm.

cortical stimulator *Medicine.* an instrument for delivering an electric shock of gauged strength to the nervous system in nerve and mental therapy.

Corticiaceae *Mycology.* a family of fungi belonging to the order Agaricales which is representative of primitive hymenomycetes and occurs on soil, dung, wood, and plant material or in a symbiotic relationship with plants; some species are parasitic to vascular plants or other fungi.

cortico- a combining form meaning "cortex," as in *corticotropin.*

corticolous *Botany.* growing on bark, as certain lichens and fungi.

corticosteroid [kôr′ti kō ster′oid] *Endocrinology.* any one of the steroid hormones produced by the adrenal cortex, including the glucocorticoids and the mineralocorticoids. Also, ADRENOCORTICOSTEROID.

corticosteroid-binding globulin *Endocrinology.* an α globulin that binds glucocorticoids in plamsa and serves as a carrier protein for these steroids in their transport from the adrenal cortex to the target tissue. Also, TRANSCORTIN.

corticosterone [kôr′ti kō ster′ōn] *Endocrinology.* a minor mineralocorticoid hormone in humans that is secreted by the adrenal cortex; it is the main glucocorticoid of rats. Also, KENDALL'S COMPOUND B, COMPOUND B, REICHSTEIN SUBSTANCE B.

corticotropic [kôr′ti kō träp′ik] *Physiology.* exerting specific, often stimulating effects upon the cortex of the adrenal gland. Also, **corticotrophic** [kôr′ti kō träf′ik].

corticotropin see ADRENOCORTICOTROPHIC HORMONE.

corticotropin-releasing factor *Endocrinology.* a peptide hormone that is released by the hypothalamus and causes release of corticotropin (ACTH) from the anterior pituitary. Also, **corticoliberin.**

Corticoviridae *Virology.* a family of lipid-containing, icosahedral, nonenveloped bacteriophages.

Corticovirus *Virology.* the single genus of the family Corticoviridae.

corticum red thread *Plant Pathology.* a fungus disease of lawn grasses characterized by light tan to reddish-brown patches of blighted grass and pinkish to red fungus strands protruding from the leaf tips. Also, PINK PATCH.

cortina *Mycology.* **1.** the portion of a fungal fruiting body that hangs over it like a curtain. **2.** in certain mushrooms, the inner, sometimes cobwebby, tissue in the fruiting body that closes off the gill cavity.

Cortinariaceae *Mycology.* a family of fungi belonging to the order Agaricale, composed of some gill mushrooms and some species which cause wood decay; it is well known for the species *Psilocybe,* which contains hallucinogenic substances.

cortisol *Endocrinology.* the major glucocorticoid that is produced by the adrenal cortex in humans. Also, HYDROCORTISONE.

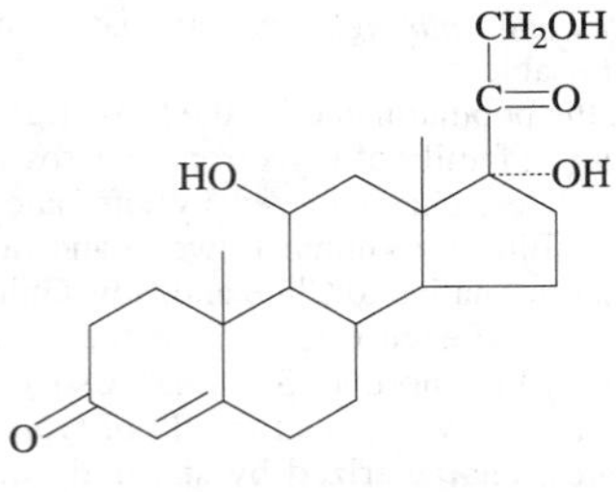

cortisol (hydrocortisone)

cortisone [kôr′ti sōn′] *Endocrinology.* a natural glucocorticoid believed to be a precursor and a metabolite of cortisol; the human adrenal cortex secretes only minute amounts, and the synthetic hormone exerts its pharmaceutical effects through its metabolic conversion to cortisol. *Organic Chemistry.* $C_{21}H_{28}O_5$, the chemical form of this substance, a white, crystalline solid that is slightly soluble in water and soluble in alcohol; melts at 220–224°C; used as an anti-inflammatory agent.

cortlandite *Petrology.* a peridotite composed essentially of olivine embedded within large anhedral crystals of hornblende. Also, HUDSONITE.

corundum *Mineralogy.* Al_2O_3, a multicolored, transparent to translucent, extremely hard, trigonal mineral occurring in granular form and as pyramidal, prismatic, tabular, and rhombohedral commonly twinned crystals; having a specific gravity of 2.82 and a hardness of 2.5 to 3 on the Mohs scale; found in limestone and dolomite, gneiss and schist, granite, and other crystalline rocks. *Materials.* a form of this mineral used as a gemstone or an abrasive.

corvée [kôr vā′] *Anthropology.* a system of enforced mandatory labor.

corvette *Naval Architecture.* **1.** originally, a sailing warship with a single gundeck, smaller than a frigate. Also, SLOOP OF WAR. **2.** in World War II, a lightly armed surface escort, smaller than a destroyer.

Corvidae *Vertebrate Zoology.* the crows, jays, and magpies; a cosmopolitan family of birds in the order Passeriformes, usually being carnivorous, aggressive, and gregarious.

Corvus *Astronomy.* the Crow, a faint and small yet distinct constellation lying west of Spica that is linked in mythology to neighboring Crater and Hydra.

corvusite *Mineralogy.* $(Na,K,Ca,Mg)_2(V^{+5},V^{+4})_8O_{26}\cdot 6–10H_2O$, a brown or black, opaque mineral of unknown crystal structure that occurs as irregular-shaped flakes; found in sandstone in uranium-vanadium mining districts.

Corylophidae see ORTHOPERIDAE.

corymb *Botany.* a nondeterminate flower formation in which the flowers open at the same level, giving the inflorescence a flat-topped appearance.

corymbose *Botany.* having the qualities or appearance of a corymb.

Corynebacteriaceae *Bacteriology.* a former classification for a family of non-spore-forming, rod-shaped bacteria in the order Eubacteriales, the genera of which have been reclassified.

corynebacteriophage *Bacteriology.* a genus of stationary, aerobic bacteriophages that are found in soil and vegetable matter, and that are parasitic and pathogenic in humans and animals.

Corynebacterium *Bacteriology.* a genus of coryneform, Gram-positive, aerobic or facultatively anaerobic, nonmotile soil bacteria of the order Actinomycetales; certain species are parasitic or pathogenic in humans and animals.

Corynebacterium diphtheriae *Bacteriology.* a species of bacteria, certain pathogenic strains of which are the causative agents of the disease diphtheria.

Corynebacterium parvum *Bacteriology.* a former term for a genus of Gram-positive, nonmotile, aerobic bacteria that are now classified as *Propionibacterium acnes.*

coryneform *Bacteriology.* **1.** of bacteria, rod-shaped with one end substantially thicker than the other. **2.** any Gram-positive, nonmotile bacteria that occur as irregularly shaped rods and resemble members of the genus *Corynebacterium.* (From a Greek term meaning "club-shaped.")

Coryneliaceae *Mycology.* the sole family of fungi belonging to the order Coryneliales, composed of species that are parasitic to conifers.

Coryneliales *Mycology.* a small order of fungi belonging to the class Pyrenomycetes which is characterized by one-walled asci and ascocarps that are erect, lobed, and split open at maturity; it is composed of parasites of conifers.

corynephage *Virology.* a bacteriophage that infects the bacterium *Corynebacterium.*

Corynocarpaceae *Botany.* a monogeneric family of dicotyledonous tanniferous trees of the order Celastrales that contain very toxic, bitter glucosides; native to New Zealand, Australia, and surrounding areas.

Corynophlaeaceae *Botany.* a family of brown algae of the order Chordariales, found worldwide on temperate shores and characterized by a dominant sporophytic phase including crustose, pulvinate, and globose forms.

Coryphaenidae *Vertebrate Zoology.* the dolphins, a family of large marine fishes in the order Perciformes, found in the Atlantic, Pacific, and Indian Oceans and characterized by a blunt nose, small teeth, and a deeply forked tail.

Coryphodontidae *Paleontology.* a family of large eutherian mammals of the extinct order Pantodonta; some genera were as large as a rhinoceros, with strong, stocky limbs and five spreading digits on the feet; they were essentially herbivorous, although they had long, saber-like canines; extant in the Paleocene to Oligocene.

Coryphodontoidea *Paleontology.* a proposed monofamilial superfamily of extinct pantodont mammals.

coryza *Medicine.* an inflammation of the nose's mucous membranes, usually accompanied by watery discharge and sneezing. Also, RHINITIS.

COS *Aviation.* the airport code for Colorado Springs, Colorado.

COS chief of staff; control air support.

cos cosine.

cosalite *Mineralogy.* $Pb_2Bi_2S_5$, a gray to white, opaque, metallic, orthorhombic mineral occurring as prismatic, often elongated crystals and in massive form; having a specific gravity of 6.76 to 6.99 and a hardness of 2.5 to 3 on the Mohs scale; found in hydrothermal or metamorphic deposits and in pegmatites.

Coscinodiscaceae *Botany.* a family of highly developed diatoms of the order Centrales, distinguished by a pseudonole on each valve and occurring principally in neritic plankton but also in the plankton of freshwater lakes.

Coscinodiscineae *Botany.* a suborder of diatoms of the order Centrales, having a marginal ring of labiate processes on each valve but no ocelli or pseudocelli.

cosec cosecant.

cosecant [kō sē´kant] *Mathematics.* **1.** in a right angle, the ratio of the hypotenuse to the side opposite a given angle. **2.** the reciprocal of the sine of a given angle or arc.

cosecant antenna *Electromagnetism.* an antenna that radiates a beam whose amplitude varies as the cosecant of the angle of depression below the horizontal.

cosecant-squared antenna *Electromagnetism.* a radar antenna designed to radiate a cosecant-squared pattern.

cosecant-squared pattern *Electromagnetism.* an antenna radiation pattern whose power varies directly with the square of the cosecant of the angle of elevation.

cosh *Mathematics.* a hyperbolic cosine.

COSIL *Robotics.* a programming language that is used for simulations of manufacturing systems.

cosine [kō´sīn´] *Mathematics.* **1.** in elementary trigonometry, the cosine of an acute angle θ is defined to be the ratio of the side adjacent to θ to the hypotenuse of a right triangle that has vertex angle θ. **2.** an even, periodic, real-valued analytic function of a real angle (rotation) with period 2π. The cosine of θ, written as cos θ, is the abscissa of the point on the unit circle obtained by moving $\theta \geq 0$ units in a counterclockwise direction along the circle from the point (1,0). (For $\theta \leq 0$, move $|\theta|$ units in a clockwise direction.) For real θ, the **cosine function** can be shown to be the real part of the function $e^{i\theta}$. When $0 < \theta < \pi/2$, the definitions are equivalent.

cosine emission law *Optics.* a law stating that the energy released from a radiating surface in any direction is proportional to the cosine of the angle between that direction and the normal. Also, LAMBERT'S COSINE LAW.

cosine pulse *Physics.* a pulse whose amplitude varies over time in proportion to the positive half-cycle of a cosine function.

cosine-squared pulse *Physics.* a pulse whose amplitude varies over time in proportion to the half-cycle of the square of a cosine function.

cosine winding *Electronics.* a coil in the deflection yoke of a cathode-ray tube with electromagnetic deflection, which reduces astigmatism over the face of the screen by dynamically focusing the beam as it is deflected.

Coslett process *Metallurgy.* a process for the phosphate coating of iron or steel, performed by immersing the workpiece in a mixture of phosphoric acid and iron filings.

cosm- a combining form meaning "world" or "universe."

Cosmarium *Botany.* a genus of common unicellular green algae of the family Oesmidiaceae.

cosmetic *Materials.* any of various substances applied to the body to improve the appearance. *Surgery.* improving the appearance, or correcting a physical defect to improve appearance.

cosmetic surgery *Surgery.* a surgical procedure performed to improve, preserve, or restore the appearance. Also, ESTHETIC SURGERY.

cosmic *Astronomy.* **1.** characteristic of or relating to the universe as a whole. **2.** relating to the portion of the universe outside the earth's atmosphere.

cosmic abundance *Astronomy.* the relative proportions of the chemical elements found in the universe as a whole, either originally, at some later time in time, or at present.

cosmic background radiation *Astronomy.* the radiation detected at microwave frequencies that corresponds to an equivalent blackbody temperature of 3 kelvins and that is seen to have an extremely smooth distribution all over the sky.

cosmic censorship hypothesis *Astronomy.* the black hole hypothesis that a singularity must always be hidden ("censored") from view by an event horizon that prevents any view of the singularity from the universe outside the horizon.

cosmic dust *Astronomy.* the particles of dust that are found anywhere in the universe, from the solar system to the intergalactic medium.

cosmic electrodynamics *Astrophysics.* the physics of the interactions of moving, charged particles and magnetic fields in planetary atmospheres, stars, and interstellar and intergalactic space.

cosmic expansion *Astronomy.* the general expansion of the universe as revealed in the steadily increasing redshifts of ever fainter and smaller, hence more distant, galaxies.

cosmic light *Astronomy.* the extremely faint and diffuse illumination of the night sky from all unresolved galaxies.

cosmic noise *Astronomy.* see COSMIC BACKGROUND RADIATION. *Telecommunications.* random noise or interference that is produced by cosmic background radiation (i.e., by radiation from outside the earth's atmosphere) and that is coupled into a communication system.

cosmic radio waves see COSMIC BACKGROUND RADIATION.

cosmic-ray composition *Astrophysics.* the makeup of cosmic rays; about 85% of cosmic rays are protons, 14% are a-particles, and the remaining 1% is made up mostly of heavier nuclei.

cosmic rays *Nuclear Physics.* a stream of high-energy charged particles consisting of nuclei, electrons, positrons, photons, and neutrinos that includes particles not only near the earth, but also in the solar atmosphere, in interplanetary space, and perhaps throughout the universe.

cosmic-ray shower *Nuclear Physics.* a cascade initiated by the interaction of a high-energy cosmic-ray proton with an oxygen or nitrogen nucleus in the earth's upper atmosphere, producing a shower of secondary radiation that consists principally of pions, muons, electrons, and gamma rays. Also, SHOWER.

cosmic-ray telescope *Engineering.* an instrument used to detect and identify the direction of cosmic rays or the material that results when cosmic rays interact with the atmosphere.

cosmic scale factor *Astronomy.* a measure of the size of the universe as a function of time.

cosmic sediment *Geology.* spherical, black magnetic particles of extraterrestrial origin, occurring in deep-ocean sediments.

cosmic string *Astrophysics.* extremely massive threads of matter, a relic from the big bang, postulated by supersymmetry cosmology theory to be major components of the early universe.

cosmic year *Astronomy.* a period of approximately 220 million years, the time it takes for the sun to make one orbit around the center of the Milky Way Galaxy.

cosmid *Genetics.* of or relating to a cosmid vector.

cosmid vector *Genetics.* a cloning vector made from a plasmid into which has been inserted the particular DNA sequences needed to package lambda phage DNA into its particle. Also, **cosmid shuttle vector.**

Cosmocercidae *Invertebrate Zoology.* a group of nematodes, in the suborder Oxyurina in some classifications, and in the suborder Ascaridina in other classifications.

cosmochemistry *Astrophysics.* the chemistry of astronomical objects and the universe as a whole.

cosmogony [käz mäg´ə nē] *Astronomy.* the study of the origin of celestial objects, usually referring to the solar system.

cosmoid scale *Vertebrate Zoology.* one of the three main types of fish scales (the others being ganoid and placoid scales), found only in lungfish and coelacanths, composed of three layers: an outer layer of cosmine (dentine), a middle layer of spongy bone, and a lower layer of laminated bone.

cosmological *Astronomy.* of or relating to the universe as a whole.

cosmological constant *Physics.* a constant in a term in the relativistic gravitational field equations representing a repulsion that just offsets the gravitational attraction which would otherwise make the matter distribution collapse.

cosmological principle *Astronomy.* a principle stating that, ignoring local irregularities, the universe looks essentially identical to every observer within it.

cosmological redshift *Astronomy.* the amount of redshift of an object that results from the expansion of the universe, as opposed to gravitational effects.

cosmology *Astronomy.* **1.** the study of the universe's origin, structure, and evolution. **2.** a theory of the origin and structure of the universe.

cosmonaut *Space Technology.* an astronaut of the USSR.

cosmopolitan *Ecology.* of an organism, having a worldwide distribution or influence wherever the habitat is suitable.

Cosmos *Space Technology.* an ongoing series of earth satellites initiated by the USSR in 1962 and used for a wide variety of missions including atmospheric research, surveillance, and communications. Also, **Kosmos.**

cosolvent *Chemical Engineering.* a nonsolvent that becomes an acceptable solvent when a small amount of active solvent is added. Also, LATENT SOLVENT, INDIRECT SOLVENT.

cospectrum *Physics.* the decomposition of the spectrum of in-phase components of the superposition of two time-dependent functions.

Cossidae *Invertebrate Zoology.* a family of lepidopterans, the carpenter or goat moths, large-bodied insects in the superfamily Cossoidea.

cos site *Molecular Biology.* a nucleotide sequence recognized for enclosing a phage DNA molecule into its capsule; a cohesive end site.

Cossoidea *Invertebrate Zoology*. a superfamily of lepidopteran insects in the suborder Heteroneura.
Cossuridae *Invertebrate Zoology*. a family of fringe worms in the group Sedentaria.
Cossyphodidae *Invertebrate Zoology*. a family of coleopteran insects, the ant guest beetles, in the superfamily Tenebrionoidea.
costa *Biology*. a rib or riblike structure, such as an anterior vein on an insect wing. *Botany*. the single midrib of a leaf, frond, or moss thallus.
costa bulb *Naval Architecture*. a streamlined ovoid bulge in a rudder, placed directly in line with a propeller to improve propulsive efficiency.
Costaceae *Botany*. a family of monocotyledonous plants of the order Zingiberales, having flowers with only one pair of pollen sacs, and spirally arranged leaves and bracts.
costal *Anatomy*. of or relating to a rib or the ribs.
costal cartilage *Anatomy*. a nonvascular cartilaginous tissue that connects the end of a rib to the sterum or to the cartilage associated with a superior rib.
costalgia [kä stal´jə] *Medicine*. pain in the ribs.
costal process *Anatomy*. a projection or outgrowth that extends laterally from the transverse process of the lumbar vertebra, homologous to the ribs. *Developmental Biology*. the embryonic mesenchymal primordium of a rib; a ventrolateral outgrowth of the caudal part of a sclerotome.
costate *Biology*. having a rib or ribs, or a similar type of connecting structure.
cost control *Industrial Engineering*. any of various systems or procedures for the management and reduction of manufacturing costs.
costectomy *Surgery*. the removal of a rib or part of a rib.
Coster, Dirk 1889–1950, Dutch physicist; discovered hafnium (with Georg von Hevesy).
Coster-Kroning transition *Atomic Physics*. a particularly intense Auger transition that shifts a hole to a higher subshell within one major shell.
costicervical *Anatomy*. of or relating to the ribs and the neck region of the spine.
costive *Medicine*. producing or suffering from constipation.
cost justification *Industrial Engineering*. a procedure used to analyze the costs, savings, and benefits to provide justification for utilization.
costoclavicular *Anatomy*. of or relating to the ribs and collarbone.
costovertebral *Anatomy*. of or relating to the ribs and vertebrae.
cost-plus *Industrial Engineering*. describing an agreement to perform contracted work in which the contractor receives payment for the actual cost of materials, equipment, and labor, plus a specific fee as profit.
cot *Molecular Biology*. the concentration of a DNA sample multiplied by the incubation time in a reaction where strand reassociation is permitted.
cot cotangent.
cot 1/2 *Molecular Biology*. the point in a reannealing sequence where half of the DNA is present as double stranded fragments. Also, HALF REACTION TIME.
cotangent *Mathematics*. **1.** in a right triangle, the ratio of the side adjacent to a given angle to the side opposite that angle. **2.** the reciprocal of the tangent.
cotangent bundle *Mathematics*. the bundle formed by taking the union of all the cotangent spaces at each point x of a manifold M; the bundle of differential 1-forms on M. The cotangent bundle is usually denoted T^*M and has a natural symplectic structure.
cotangent space *Mathematics*. the space of cotangent vectors (differential 1-forms) at a point x of a manifold M, written T_x^*M. T_x^*M is the vector space dual to the tangent space T_xM.
cotar *Engineering*. a system that determines the path of a vehicle in space through phase-comparison techniques, which determine the line of direction between a receiving antenna on the ground and a telemetering transmitter in the space vehicle. (An acronym for correlated orientation tracking and range.)
cotat *Engineering*. a trajectory-measuring system that determines an object's position in space by triangulation, based on the measurement of the object's relationship to a series of antennae base lines that are spaced at great distances from each other. (An acronyn for correlation tracking and triangulation.)
cote *Agriculture*. a pen or enclosure for small domestic animals, especially for sheep or for birds such as doves or pigeons.
cotectic *Physical Chemistry*. relating to the conditions of temperature, pressure, and composition under which two or more liquids crystallize from the same liquid.
cotectic crystallization *Physical Chemistry*. a process by which more than one solid forms from a single liquid over a limited range of falling temperatures.
coterie [kōt´ə rē] *Behavior*. a social group of animals that defends a territory against other groups of the same species.
coth *Mathematics*. a hyperbolic cotangent.
cotidal chart *Cartography*. a chart of cotidal lines that show approximate locations of high water at hourly intervals measured from a reference meridian, usually Greenwich.
cotidal line *Cartography*. a line on a chart passing through all points where high water occurs at the same time.
cotinga *Vertebrate Zoology*. any of several songbirds constituting the family Cotingidae; found chiefly in tropical regions.
Cotingidae *Vertebrate Zoology*. the cotingas, a large family of neotropical North American birds of the order Passeriformes, characterized by their brightly colored plumage and thick bills.
cot plot *Molecular Biology*. a graph showing the extent of reannealing of previously denatured DNA molecules as a function of cot value. Also, **cot curve.**
cotransduction *Genetics*. the simultaneous transduction of two or more bacterial marker genes by a single bacteriophage particle.
cotransformation *Molecular Biology*. the simultaneous transformation of two or more bacterial genes. Also, DOUBLE TRANSFORMATION.
cotransport see SYMPORT.
Cotswold *Agriculture*. a breed of white-faced, long-wool sheep characterized by a well-developed forelock. (From the *Cotswold* Hills in England.)
cottage cheese *Food Technology*. a soft, unripened, low-fat white cheese, usually coagulated with lactic acid rather than rennin and left unpressed.
cotter *Design Engineering*. a tapered rod or pin that passes through a hole in one member and rests against the surface of an encircling member, keeping both members constant to each other. Also, COTTER PIN.
cottered joint *Mechanical Engineering*. a joint in which a cotter transmits power by shear on an area perpendicular to the cotter's longitudinal axis.
cotter pin *Design Engineering*. see COTTER. *Mechanical Devices*. a half-round length of wire bent into a U shape with a split end used as a fastener to hold a nut, cotter, or hinge plates to a bolt or shaft.
Cottidae *Vertebrate Zoology*. the sculpins, a family of carnivorous fish of the suborder Cottoidei in the order Scorpaeniformes, found predominantly in cold seas.
Cottoidei *Vertebrate Zoology*. a suborder of fish in the order Scorpaeniformes, including the sculpins, oilfishes, poachers, and lumpfishes.
cotton *Botany*. any plant of the species *Gossypium*, of the mallow family, having spreading branches and broad, lobed leaves. *Materials*. the soft white fiber attached to the seeds of such plants; widely used in making textiles.
cotton anthracnose *Plant Pathology*. a cotton disease caused by the fungus *Glomerella gossypii* and characterized by nonliving, rust to light colored spots on the plant parts.
cotton ball see ULEXITE.
cotton-belt climate *Meteorology*. a warm climate characterized by dry winters and rainy summers, as is found in the region of the southeastern U.S. known as the Cotton Belt.
Cotton effect *Analytical Chemistry*. the dependence on wavelength of the optical rotary dispersion curve and the circular dichroism curve of a substance in the vicinity of an absorption band due to vibrational differences.
cotton gin *Textiles*. a machine designed to remove the seeds and hulls from cotton fibers.
cotton linters see LINTERS.
cottonmouth *Vertebrate Zoology*. a venomous snake (*Agkistrodon piscivorus*) of the viper family, having a blue-black or brown color; it inhabits swamps and low-lying branches over waters in the south and southeastern United States. (From the white interior of the mouth, which is displayed whenever the snake is threatened.)
cottonmouth disease see COLUMNARIS DISEASE.
Cotton-Mouton constant *Optics*. a constant observed in the Cotton-Mouton effect that provides the phase difference between the components of light rays parallel and perpendicular to the field.
Cotton-Mouton effect *Optics*. the phenomenon that allows certain pure liquids such as nitrobenzene or carbon disulfide, under the influence of a magnetic field, to doubly refract light in a direction that is transverse to the light rays.

cotton root rot *Plant Pathology.* a disease affecting cotton and other plants of the American Southwest, that is caused by the soil fungus Phymatotrichum and characterized by a brown rot of the lower stem and roots, a sudden wilting and death of the plant, and plant death in large circular patches.

cotton rust *Plant Pathology.* a disease of cotton caused by the rust fungus *Puccinia stakmanli* and characterized by greenish-yellow to orange bumps on the undersides of leaves.

cottonseed oil *Food Technology.* a brownish-yellow oil with a nutlike odor, obtained from the seed of the cotton plant; used in soaps, hydrogenated fats, lubricants, and cosmetics, as a cooking and salad oil, and in medicine.

cottontail *Vertebrate Zoology.* any of several small American rabbits of the genus *Sylvilagus* of the family Leporidae, native to the brush country of the Western Hemisphere and characterized by short ears, a white tail, and a brown body.

cottontail

cotton wilt *Plant Pathology.* **1.** a disease of cotton caused by the fungus *Fusarium vasinfectum,* which grows in vascular tissues and causes plants to yellow, wilt, and die. **2.** a fatal disease of cotton caused by the fungus *Verticillium alboatrum,* which produces withering.

cottonwood *Botany.* any of several species of American poplar trees (such as *Populus deltoides)* having soft white wood, alternate leaves, and canescent fruit.

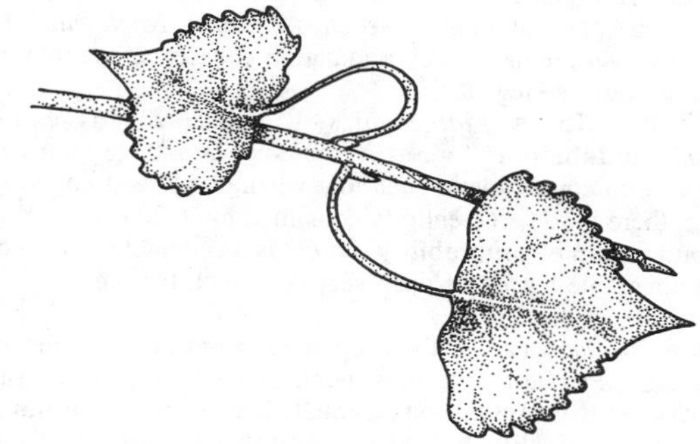

cottonwood

cotton wool *Materials.* raw, unprocessed cotton. *Medicine.* especially in British use, a soft, fluffy cotton made absorbent by removing the wax, used for surgical dressings and the like. Also, **cotton batting.**

cottony rot *Plant Pathology.* a disease of many plants, especially citrus trees, caused by the fungus *Sclerotinia sclerotiorum* and characterized by a fluffy white growth on the plant parts, root decay, and stem wilt.

Cottrell, Frederick 1877–1948, American physical chemist; invented Cottrell electrostatic precipitation process.

Cottrell atmosphere *Solid-State Physics.* one or a group of foreign atoms clustered around a dislocation in a crystal.

Cottrell hardening *Solid-State Physics.* the hardening of a solid substance by the migration and clustering of impurity atoms, which differ in size from the solvent atoms, around dislocations within a crystal.

Cottrell precipitator *Engineering.* a piece of equipment used to remove dust from gases through an electrostatic process.

cotunnite *Mineralogy.* $PbCl_2$, a colorless or white, transparent to translucent, pearly, orthorhombic mineral that commonly occurs in massive granular form; having a specific gravity of 5.8 and a hardness of 2.5 on the Mohs scale; found as veins in chalcocite, as an alteration product of galena, and as a sublimation product in fumaroles on Mount Vesuvius, Italy.

coturnism *Toxicology.* poisoning due to eating the meat of quail of the genus *Coturnix;* symptoms may include breathing difficulties, paralysis, and death.

cot value *Molecular Biology.* a measure of DNA concentration and time; used experimentally to define the conditions for the reannealing of denatured DNA.

cotyledon [kät´ə lēd´ən] *Botany.* the embryonic leaf of a seed.

cotyledonous [kät´ə lēd´ə nəs] *Botany.* of or relating to a cotyledon.

cotyledontoxin *Toxicology.* a toxic, neutral nonalkaloidal, amorphous substance found in the herbaceous plants *Cotyledon ventricosa* and *Cotyledon wallchii.*

cotylocercous cercaria *Invertebrate Zoology.* a larval form of a trematode flatworm.

Cotylosauria *Paleontology.* an archaic order that included the earliest anapsids; since it comprised both amphibians and reptiles, including several problematic genera, recent authorities prefer the term Captorhinida for the order; Carboniferous and Permian.

coua [koo´ə] *Vertebrate Zoology.* the popular term and genus name for ten species of terrestrial forest birds of the family Cuculidae inhabiting Madagascar; characterized by short wings, blue to gray plumage, and long tails; species build their own nests (unlike their relatives, the Old World cuckoos).

coucal [koo´kəl] *Vertebrate Zoology.* any of various birds of the family Cuculidae, about 27 terrestrial species of the genus *Centropus* ranging from Africa to Australia, up to 36 inches in size, having short wings, a long tail and rough, dark and white plumage; they hunt large insects and some small animals and build their own nests (unlike its relative, the Old World cuckoos).

coucal

coudé focus [koo dā´] *Optics.* the point at which light rays converge when they pass through a coudé telescope. *Astronomy.* an optical configuration used primarily with large reflectors in which the focus point remains stationary no matter where the telescope is pointing.

coudé-Newtonian-Cassegrain telescope *Optics.* a reflecting telescope designed for observation using either the coudé, the Newtonian, or the Cassegrain focus.

coudé spectrograph *Spectroscopy.* a spectrograph attached to the tube of a coudé telescope.

coudé spectroscopy *Spectroscopy.* the analysis of astronomical spectra produced using a coudé spectrograph.

coudé telescope *Optics.* a telescope that reflects light across the polar axis and then focuses it to form an image at a point on the axis where it can be seen through an eyepiece or a spectrograph.

Couette flow [koo et´] *Fluid Mechanics.* a fluid flow, dominated by viscosity, between two parallel plates in which one surface is stationary and the other is moving.

Couette-Taylor flow *Fluid Mechanics.* the motion of a fluid between coaxial cylinders experiencing relative rotation.

Couette viscometer *Engineering.* an instrument used to measure the viscosity of a liquid by recording the effect that the liquid has on the rotation of two vertical coaxial cylinders. Also, ROTATIONAL VISCOMETER.

cougar [koo´gər] *Vertebrate Zoology.* another name for the mountain lion, a large North American wild cat, *felis concolor.* See MOUNTAIN LION.

cough *Medicine.* **1.** to expel air from the lungs in a sudden, audible, and violent way, usually to keep the airways free of foreign matter or as a response to irritation. **2.** an instance of this action, or a condition that is marked by repeated coughing.

cough plate *Microbiology.* a method used to detect the bacterium *Bordetella pertussis*, the causative agent of whooping cough, by inoculation of a petri plate containing Bordet-Gengou agar with the cough droplets of a patient.

Couinae *Vertebrate Zoology.* the couas, a subfamily of ground-dwelling cuckoo birds found only in Madagascar, belonging to the family Cuculidae in the order Cuculiformes.

coulee [koo´lē] *Geology.* **1.** in the northwestern United States, a dry or intermittent stream valley, gully, or steep-walled trench. **2.** a small valley or depressed area. **3.** in the western United States, a small stream or the dry bed of such a stream.

coulisse *Engineering.* a section of wood having a notch into which another piece of wood fits or slides. Also, CULLIS.

Coulomb, Charles Augustin de [koo´läm´; koo´lōm´] 1736–1806, French engineer; crucial discoveries in electricity, friction, and magnetism; invented torsion balance; formulated Coulomb's law.

coulomb *Electricity.* the standard international unit of electric charge, equal to the charge that passes through any cross section of a conductor in one second during a constant current flow of one ampere.

Coulomb attraction *Electricity.* the electrostatic force of attraction applied by one charged particle on an oppositely charged particle. Also, ELECTROSTATIC ATTRACTION.

Coulomb barrier *Nuclear Physics.* the electrostatic forces within the high potential energy region of a nucleus that prevents positively charged particles from reaching the nucleus of an atom.

Coulomb excitation *Nuclear Physics.* the existence of a high-energy state within a nucleus, arising from the interaction of the nucleus and bombarding particles too weak to penetrate it.

Coulomb explosion *Physics.* an explosion in which the outward driving force is the electrical repulsion of charge.

Coulomb field *Electricity.* the electric field generated by a stationary-charged particle.

Coulomb force *Nuclear Physics.* the force between two charged particles, as between electrons and protons, that (by Coulomb's law) is proportional to the product of the charges and inversely proportional to the square of the distance between them and that is either repulsive or attractive, depending on the relationship of the charges.

Coulomb friction *Mechanics.* a frictional force between dry stationary surfaces; it is proportional to the normal force.

Coulomb gauge *Quantum Mechanics.* a noncovariant gauge in which there are commutation relations only for the space part of the vector potential. Also, RADIATION GAUGE.

Coulomb interactions *Electricity.* charged particle interactions due to the Coulomb forces that are exerted on one another. Also, ELECTROSTATIC INTERACTIONS.

coulombmeter *Engineering.* an instrument that measures a quantity of electricity in coulombs.

Coulomb potential *Electricity.* a scalar-point function equal to the work per unit charge used against or by the Coulomb force in moving a particle bearing an infinitely small positive charge from infinity to the field of a charged particle in a vacuum.

Coulomb repulsion *Electricity.* the electrostatic repulsion force applied by one charged particle on another charged particle of like sign. Also, ELECTROSTATIC REPULSION.

Coulomb scattering *Physics.* the scattering of charged particles through their electrical interaction.

Coulomb's law *Electricity.* a law stating that the force between two electric charges is proportional to the product of their magnitudes and inversely proportional to the square of the distance between them; the force between like charges is an attraction, and the force between unlike charges is a repulsion. Also, LAW OF ELECTROSTATIC ATTRACTION.

coulometer *Physical Chemistry.* a device that measures an electric charge in a cell by monitoring the reactions taking place when the charge or current passes through a chemical solution.

coulometric titration [koo´lə met´rik] *Analytical Chemistry.* an indirect analytical technique using a low-concentration titrant that is generated electrolytically at a constant rate; as it is generated it reacts stoichiometrically with the substance to be identified and measured.

coulometry [koo´läm´ə trē] *Analytical Chemistry.* the determination of the amount of an electrolyte released during electrolysis by measuring the number of coulombs used. Also, **coulometric analysis.**

coulostatic analysis *Physical Chemistry.* a technique that determines the amount of a given substance in a solution by measuring the amount of electricity it takes to remove the substance from the solution.

coulter see COLTER.

Coulter counter *Microbiology.* an instrument used to count the number of cells in a given volume of a suspension by monitoring decreases in electrical conductivity through a small orifice caused by the cells passing through the orifice.

Coulter pine *Botany.* a California pine tree, *Pinus coulteri*, characterized by its large, heavy cones and stout bluish-green needles.

coumachlor *Organic Chemistry.* $C_{19}H_{15}ClO_4$, a white crystalline solid; insoluble in water and soluble in alcohol; melts at 169°C; used in killing rodents.

coumarin *Organic Chemistry.* $C_9H_6O_2$, toxic, carcinogenic, colorless crystals, flakes, or powder, soluble in alcohol and slightly soluble in water; melts at 69°C and boils at 299°C; used in pharmaceuticals.

coumarin glycoside *Biochemistry.* a glycoside aromatic constituent found in numerous plants and used to create perfumes.

coumarone *Organic Chemistry.* C_8H_6O, a colorless, oily liquid that is insoluble in water and miscible with alcohol and ether; boils at 169°C; used in the manufacture of resins.

coumarone-indene resin *Organic Chemistry.* a synthetic resin polymerized from coumarone and indene; it is soluble in acetone and insoluble in water and alcohol; used in adhesives, printing inks, and friction tape.

coumatetralyl *Organic Chemistry.* $C_{19}H_{16}O_3$, a yellowish crystalline solid that is slightly soluble in water and melts at 172–176°C; used to kill rodents.

councilman body *Pathology.* an eosinophilic round body, or lesions, present in the liver of persons suffering from yellow fever, viral hepatitis, and other hepatic diseases originating from the necrosis of a single hepatic cell.

couniversal object *Mathematics.* an object T in a category C is said to be couniversal if for each object C of C there exists one and only one morphism $C \to T$. Any two couniversal objects in a given category are equivalent. Also, ATTRACTING or TERMINAL OBJECT.

counseling psychology *Psychology.* a branch of psychology that focuses on personal problems not classified as serious mental disorders, such as academic, social, or vocational difficulties of students. Thus, **counseling psychologist.**

count *Space Technology.* the progression in a countdown or plus count, usually by calling a specific number to indicate the point reached, such as "T minus 10 and counting." *Textiles.* **1.** of a textile fabric, the number of warp and filling threads per square inch. Also, THREAD COUNT. **2.** of a yarn, the mass per unit length. Also, YARN COUNT. *Nucleonics.* **1.** a single ionizing event registered by a radiation counter. **2.** the total number of ionizing events so registered.

countability axioms *Mathematics.* **1.** a space X satisfies the **first axiom of countability** if, for $x \in X$, there is a countable family of open sets $\{O_n\}$, each containing x, such that whenever O is an open set containing x, there is some O_n entirely contained in O. **2.** a space X satisfies the **second axiom of countability** if there is a countable basis; i.e., there is some countable family of open sets $\{U_n\}$ such that any open set is a union of some of the U_n's.

countable *Mathematics.* **1.** the property of a set that is either finite or can be assigned a one-to-one correspondence with the set of positive integers. Also, ENUMERABLE, DENUMERABLE. **2.** an **infinite countable set** is said to have cardinality $\aleph_0$. Also, **countably infinite.** If a set is not countable, it is *uncountable*.

countably additive *Mathematics.* **1.** let μ be a measure on a measure space M, and $\{S_i\}_{i=1}^{\infty}$ a sequence of pairwise disjoint measurable subsets of M. μ is said to be countably additive if

$$\mu\left(\bigcup_{i=1}^{\infty} S_i\right) = \sum_{i=1}^{\infty} \mu(S_i)$$

2. a real-valued set function σ is said to be countably additive if

$$\sigma\left(\bigcup_{i=1}^{\infty} S_i\right) = \sum_{i=1}^{\infty} \sigma(S_i)$$

where $\{S_i\}_{i=1}^{\infty}$ a seqence of pairwise disjoint sets on which σ is defined.

countably compact set *Mathematics.* a subset A of a topological space X is countably compact if every countable collection of open sets whose union contains A has a finite subset of open sets whose union also contains A. If A is X itself, then X is countably compact if and only if each sequence in X has an accumulation point in X.

countably subadditive *Mathematics.* **1.** let μ be a measure on a measure space M, and $\{S_i\}_{i=1}^{\infty}$ a seqence of pairwise disjoint measurable subsets of M. μ is said to be countably subadditive if

$$\mu(\bigcup_{i=1}^{\infty} S_i) \leq \sum_{i=1}^{\infty} \mu(S_i)$$

2. a real-valued set function σ is said to be countably subadditive if

$$\sigma(\bigcup_{i=1}^{\infty} S_i) \leq \sum_{i=1}^{\infty} \sigma(S_i)$$

where $\{S_i\}_{i=1}^{\infty}$ is a sequence of pairwise disjoint sets on which σ is defined.

countdown *Engineering.* a series of events leading to a climactic finish, each event in the series being in accordance with a schedule in which time is counted backward toward zero, the finishing point. *Space Technology.* specifically, the calling off of the time remaining prior to the launch or firing of a rocket vehicle; usually done in minutes, then in seconds as the countdown nears its end. *Telecommunications.* lines or marks that are inscribed on a tape or film to mark the time immediately preceding its starting point; used to assure precise cueing.

counter a device used in counting; specific uses include: *Engineering.* a device that registers and totals the number of operations performed by a machine. *Electrical Engineering.* **1.** a circuit in which an oscillator of known frequency increments a numerical output at regular intervals, so that the time between two events can be determined by indicating the number of counts that have occurred. **2.** a digital circuit, composed of a series of bistable devices such as flip-flops, that keeps track of the number of times a specific event occurs by switching from one state to the next each time a pulse is received. **3.** a device that can change from one to the next of a sequence of distinguishable states after receiving a discrete input signal. *Computer Technology.* **1.** a register whose contents go through a specified sequence of states, frequently representing consecutive binary or decimal integers. **2.** an incrementing or decrementing register that is used to control repeated sets of instructions, such as a loop, or to represent a number of occurrences.

counter *Naval Architecture.* in certain vessels, the portion of the stern that overhangs the waterline.

counter- *Military Science.* describing activities that are intended to neutralize, deny, or destroy an enemy's effectiveness; used to form many compound terms, such as **counterfire, counterforce, counterinsurgency, counterreconnaissance, countersabotage, countersubversion,** and **countersurveillance.**

counteracting selection *Evolution.* a process of natural selection operating on two or more levels of organization (as individual organisms or as a species as a whole) such that a hereditary character favored at one level may be disfavored at another.

counterattack *Military Science.* **1.** an attack intended to restore ground lost to an enemy attack, or to cut off enemy units from their line of supply. **2.** to carry out such an attack.

counterbalance see COUNTERWEIGHT.

counterbalance system see TWO-STEP GROOVING SYSTEM.

counterblow hammer *Mechanical Engineering.* a forging hammer in which the anvil and ram are forced toward each other by compressed steam or air.

counterbore *Engineering.* the process of increasing the diameter of a hole over part of its depth. *Mechanical Devices.* a drill used to accomplish this.

counterbracing see CROSS BRACING.

counter circuit *Electronics.* a circuit that provides an indication of some type after having received a predetermined number of pulses. Also, COUNTING CIRCUIT.

counterconditioning *Behavior.* a process by which an unwanted response to a stimulus is replaced by a new and incompatible response.

countercurrent *Hydrology.* a current flowing in an opposite direction to a main or comparative current. *Chemistry.* the flow, in opposite directions, of two separate phases, such as a liquid and a vapor stream or two streams of immiscible liquids, through or in contact with one another.

countercurrent distribution *Chemical Engineering.* a profile of a compound's concentration given in different ratios of two immiscible liquids.

countercurrent electrophoresis *Immunology.* a method of detecting antigens or antibodies by forcing them to move in an electrical field.

countercurrent extraction *Chemical Engineering.* a process of liquid-liquid extraction where the solvent and the process stream in contact flow in opposite directions. Also, COUNTERCURRENT SEPARATION.

countercurrent flow *Mechanical Engineering.* **1.** a heat transfer system in which two fluids flow in opposite directions. **2.** any contacting or transfer operation in which two fluid phases in either direct or indirect contact flow in opposite directions.

countercurrent leaching *Chemical Engineering.* a leaching operation in which the solvent flows through a series of leach tanks countercurrent to the flow of the solid.

countercurrent spray dryer *Engineering.* a drying machine in which the drying gases move in a direction opposite to that of the spray.

counter dead time *Nucleonics.* in an ionization chamber, an interval lasting from 100 to 400 microseconds after a counter pulse is registered, during which the detector is insensitive to further ionizing events.

counterelectromotive cell *Electricity.* a nickel-and-sodium hydroxide cell with essentially zero ampere-hour capacity; used to oppose the line voltage.

counterelectromotive force *Electromagnetism.* a voltage that is induced in an inductive circuit due to a changing current; the polarity of the voltage opposes the polarity of the voltage driving the current. Also, BACK ELECTROMOTIVE FORCE.

counterespionage *Military Science.* espionage activities or techniques intended to oppose or thwart enemy espionage.

counterevolution *Evolution.* an evolutionary change in a species resulting from detrimental interactions with another species or a population of the same species.

counterflashing *Building Engineering.* an L-shaped flashing that fits over the joint of a roof surface and a vertical wall so that one leg of the flashing runs up the wall and the other along the roof to keep rainwater from running down behind the vertical leg; it also allows expansion of the metal or differential settlement and movement of the structure without rupturing the flashing from its anchorage.

counterflow *Engineering.* the movement of a fluid in the opposite direction to a fluid flow in the same cross section of the same apparatus.

counterforce *Physics.* a contrary or opposing force or tendency.

counterfort *Building Engineering.* **1.** a cantilevered weight in the form of a pier built on the side of a material, such as a retaining wall, that is to be retained. **2.** a buttress that stengthens a basement wall against the earth's pressure.

counter frequency meter *Engineering.* an instrument having a frequency standard that measures occurrences or cycles of a periodic quantity taking place in a specific time frame or the interval between two occurrences.

counterglow *Astronomy.* another term for GEGENSCHEIN, a faint patch of light that is visible in a dark night sky, caused by sunlight reflecting from dust.

counterimmunoelectrophoresis *Immunology.* a method of immunodiffusion in which an electric current is used to produce the migration of antigen and antibody from separate wells in an agar-gel medium.

counterincision *Surgery.* a second incision next to a primary incision.

counterintelligence *Military Science.* intelligence activities or techniques that are intended to oppose or thwart enemy intelligence efforts.

counterion or **counter-ion** *Physical Chemistry.* any ion with a charge opposite to that of another given ion, especially an adjacent ion.

counterlath *Building Engineering.* **1.** a strip, usually of wood, positioned between two rafters so as to support crosswise laths. **2.** a lath positioned between timber and sheet lath. **3.** a randomly spaced lath nailed between two laths separated by an exact space. **4.** in the buildup of a partition, a lath placed on one side of it after the other side is completed.

countermeasures set *Electronics.* one or more units of military electronics equipment used to impair the operational effectiveness of enemy radar or communications systems; typically, jamming equipment.

countermine *Military Science.* **1.** a tunnel dug to destroy an enemy tunnel. **2.** an explosion device to detonate enemy mines.

counteroffensive *Military Science.* an offensive that is intended to oppose or repel an enemy offensive.

counteropening *Surgery.* a second opening across an earlier opening in an abscess or other fluid-filled cavity to enhance drainage.

counterpoise *Electricity.* a grid of wires or other conductors placed above and insulated from the ground to provide a lower system of conductors for an antenna. *Mechanical Engineering.* see COUNTERWEIGHT. *Ordnance.* a mechanism that counterbalances the weight of the breechblock on a large gun, making it easier to open.

counterpoise gun carriage *Ordnance.* a gun carriage employing a system in which the gun is returned to firing position by the action of a counterpoise that is raised by the gun's recoil.

counterpoison *Toxicology.* any poisonous substance that counteracts the action of another poison.

counterpulsation *Cardiology.* a technique for assisting the circulation and decreasing the workload of the heart by synchronizing the action of an artificial pump with the heart's contraction and relaxation cycle.

counterpuncture *Surgery.* a second opening made opposite another.

counterradiation *Geophysics.* the radiation reflected back by the atmosphere after being released from the earth; the principal action of the greenhouse effect.

counterrecoil *Ordnance.* the forward movement of a gun as it returns to firing position after the recoil.

countershaft *Mechanical Engineering.* an intermediate shaft installed between the driving and driven shafts in a belt drive, either to increase the speed ratio or because direct connection is impossible.

countersink *Mechanical Devices.* **1.** a conical recess cut in a surface to take the head of a flat-headed screw or rivet and allow it to be flush with the material to which it is attached. **2.** relating to the shaping of a hole drilled in wood or metal with a cone-shaped tool. **3.** the tapered portion of a twist drill between the pilot drill and the body.

countersinking *Engineering.* the process of making a conical enlargement at the opening of a hole to allow it to receive the head of a screw or rivet. Thus, **countersunk.**

counterstain *Biology.* a second contrasting stain applied to microscopic parts not affected by a primary stain; especially, a cytoplasmic stain used after a nuclear stain. Also, CONTRAST STAIN, AFTERSTAIN.

countertransference *Psychology.* the projection of feelings and attitudes by a psychotherapist onto a patient.

counter tube *Electronics.* **1.** an electron tube that converts an incident particle or burst of incident radiation into a discrete electric pulse that can be counted. Also, RADIATION COUNTER TUBE. **2.** a tube that produces one or more pulses after receiving a specified number of input pulses.

countervane see CONTRAVANE.

counter voltage *Electricity.* the reverse voltage that occurs across an inductor when the current through the inductor is cut off.

counterweight *Mechanical Engineering.* **1.** a nonworking weight or load that is attached to one end or side of a machine in order to balance the weight carried on the opposite end or side. **2.** a working part that is attached and positioned at least partly in order to improve the balance of a machine. **3.** a weight that is designed to counterbalance the original load carried by an elevator, hoist, or similar mechanism, so that the engine must only drive against the additional, unbalanced load. Also, COUNTERBALANCE, COUNTERPOISE.

counting chamber *Microbiology.* a specialized microscope slide that holds a known, small volume of a cell suspension for the microscopic determination of the total cell count of that sample.

counting circuit see COUNTER CIRCUIT.

counting glass *Textiles.* a magnifying lens used in determining the thread count of a fabric.

counting-rate meter *Nucleonics.* an instrument that indicates the time rate of occurrence of input pulses to a radiation counter.

counting theorem *Mathematics.* the theorem stating that each well-ordered set is similar to a unique ordinal number; that is, there exists an order-preserving one-to-one correspondence between each well-ordered set and some unique ordinal number.

country rock *Geology.* the older, preexisting rock that encloses or contains an igneous intrusion or a mineral deposit.

countwheel *Horology.* a wheel in a timepiece that advances in regular increments and determines the initiation of some other operation in the works, such as the number of blows to be struck.

couplant *Engineering.* a liquid or pasty substance used in ultrasonic examination to slow down the transmission of sound by air between a transducer and a test piece.

couple *Engineering.* to connect parts or vehicles at their ending points with a coupling device. *Chemistry.* to condense or unite two molecules. *Electricity.* to join two circuits, enabling signals to be transferred from one to the other. *Electronics.* to place two dissimilar metals or alloys in electrical contact with each other so that, when they are immersed in an electrolyte, they act as the electrodes of an electrolytic cell. *Mechanics.* a system of two equal but opposite forces that are applied at different locations, producing a torque.

coupled antenna *Electromagnetism.* an antenna that is electromagnetically coupled to another antenna.

coupled circuits *Electricity.* two or more electrical circuits configured to allow energy to flow from one circuit to the other either electrically or magnetically.

coupled columns *Architecture.* pairs of columns set close together, with wider spacing between pairs.

coupled engine *Mechanical Engineering.* a locomotive engine whose driving wheels are connected by a rod.

coupled-field vectors *Electromagnetism.* magnetic and electric field vectors that are coupled to each other in order to support the propagation of radiation according to Maxwell's equations.

coupled harmonic oscillators *Physics.* two or more linear oscillators that display an interaction, often linear or weak.

coupled modes *Acoustics.* the overlap of acoustic waves from two separate resonators, such as on the tube of a reed instrument and on the reed itself, to produce one sound.

coupled oscillators *Mechanics.* two or more systems exhibiting simple harmonic motion individually, but which interact with one another to execute more complicated motions. *Electromagnetism.* two or more circuits that are electromagnetically coupled in their alternating current operation.

coupled systems *Computer Technology.* a group of computer systems that operate in parallel, communicate with one another, and share system resources. *Physics.* two or more systems that exchange energy with one another.

coupler *Engineering.* **1.** a device that unites two separate parts. **2.** specifically, the part of a train that serves to join two railroad cars. *Electricity.* a device, usually made of capacitors and inductors, designed to transfer a surge from a generator to power equipment, while limiting energy flow from the power equipment to the generator. *Electromagnetism.* a passage that joins two cavities of waveguides, allowing them to exchange energy.

coupling *Engineering.* **1.** a fitting, usually having inside threads only, used to connect two pieces of pipe or hose. **2.** a device that joins two vehicles, particularly railroad cars. *Mechanical Engineering.* a device used to connect coaxial shafts for power transmission from one to the other. Also, SHAFT COUPLING. *Electricity.* **1.** a means of transferring power from one stage of a circuit to another stage, or from one circuit to another circuit. **2.** a hardware mechanism used to fashion a temporary connection between two wires. *Computer Programming.* the measure of the strength of the interconnection or dependence between two modules in a structured program design; loosely coupled modules have little interaction, while tightly coupled modules have a great deal of interaction.

coupling agent *Materials Science.* an agent that is added to improve the bonding of a polymer to inorganic filler materials, such as glass-reinforcing fibers; a variety of silanes and titanates are used for this purpose.

coupling aperture *Electromagnetism.* a hole in the wall of a resonator or waveguide that is used to feed or extract energy.

coupling capacitor *Electronics.* any capacitor used to couple two circuits together.

coupling coefficient *Electronics.* a value of less than 1.0 representing the degree of coupling between two circuits, expressed as the ratio of the mutual impedance divided by the square root of the product of the self impedances. *Physics.* **1.** a parameter that measures the strength of the interaction between two systems. **2.** see COUPLING CONSTANT.

coupling constant *Particle Physics.* a constant that indicates a measure of how strongly two particles interact.

coupling loop *Electromagnetism.* a small loop of wire that is inserted into a waveguide for the purpose of coupling microwave energy to or from an external circuit.

coupon *Chemical Engineering.* a polished metal strip of specified size and weight, used to determine the corrosive action of liquid or gaseous products, or to test the efficiency of corrosive-inhibiting additives.

Courant, Richard 1888–1972, German-born American mathematician; known for work in differential equations.

Courant-Friedrichs-Lowry condition *Fluid Mechanics.* in numerical hydrodynamics, the condition in which the time step in a calculation is limited to the sound-crossing line for each grid zone.

courbaril *Materials.* the hard, durable wood of the West Indian locust tree, *Hymenaea courbaril.*

Cournand, André born 1895, French-born American medical professor; with D. Richards, Nobel Prize for work in heart catheterization and circulatory pathology.

course *Navigation.* the actual or intended direction of travel of a vessel, measured clockwise through 360° from the reference direction, usually north. *Aviation.* the actual or intended line of flight of a flight vehicle. *Building Engineering.* **1.** a continuous array of bricks, shingles, or stone of uniform height oriented horizontally. **2.** in surveying, the direction and length of a survey line. *Textiles.* a row of stitches running crosswise in a knitted fabric.

course angle *Navigation.* the actual or intended direction of travel, measured clockwise or counterclockwise through 90° or 180° from the reference direction, usually north or south.

coursed masonry *Building Engineering.* blockwork laid similarly to masonry in horizontal courses of concrete blocks between 10 and 50 tons in weight. Also, **coursed blockwork.**

course error *Navigation.* the angular difference between an intended course and the course actually followed.

course indicator *Navigation.* a gyro repeater used to indicate a ship's heading. Also, SHIP'S COURSE INDICATOR.

course line *Navigation.* **1.** a line extending in the direction of the course. **2.** a line of position on or parallel to the course.

course made good *Navigation.* the course a craft actually follows in proceeding from one point to another.

course over the ground *Navigation.* the course made good as compared to the course steered.

course programmer *Control Systems.* an item used to initiate and process signals so as to set a vehicle in which it is installed along one or more projected courses.

courser *Vertebrate Zoology.* a fast-running bird of the subfamily Cursoriinae in the family Glareolidae, having three toes and short strong wings and noted for colonial nesting and hunting insects on the ground; found in Africa and southern Asia.

course recorder *Navigation.* a device that produces a graph showing headings against time. It usually derives its headings from a gyro compass.

courseware *Computer Programming.* instructional materials that use a computer as an information processing device or as a presentation medium.

courtship or **courtship behavior** *Behavior.* any behavior that leads to mating and the rearing of young.

Cousteau, Jacques [koo stō´] born 1910, French oceanographer and conservationist; co-inventor of the aqualung; known for popular underwater films.

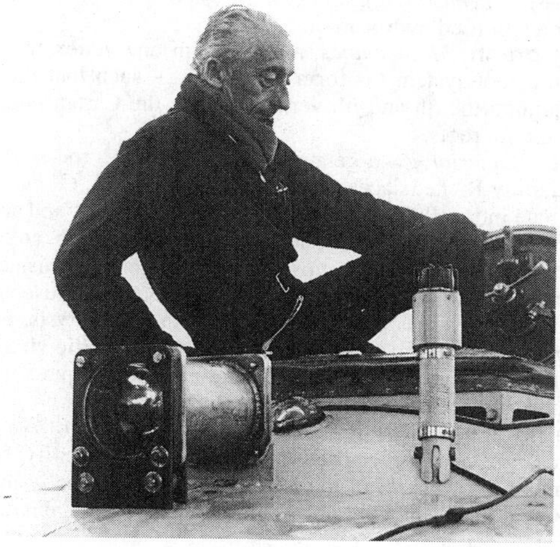

Jacques Cousteau

Coutard's method [koo tarz´; koo tardz´] *Radiology.* the administration of radiation therapy in small doses over an extended period. (From Henri *Coutard,* 1876–1950, French radiologist.)

couvade [koo väd´] *Anthropology.* the ritual practiced after birth of an offspring in some societies, in which the father assumes postpartum disability while the mother carries on normal routines.

couvillonage [koo´ vil ə näzh´] *Surgery.* the cleansing of a cavity and application of remedies utilizing either a brush or a swab.

Couvinian *Geology.* a European geologic stage of the Middle Devonian period, occurring after the Emsian and before the Givetian. Also, EIFELIAN.

covalence [kō vāl´ens] *Chemistry.* **1.** the number of shared electron-pair bonds that an atom can form. **2.** the sharing of electrons to form a chemical bond. Also, **covalency.**

covalent binding see COVALENT BOND.

covalent bond *Physical Chemistry.* a chemical bond between two atoms of the same or different elements, in which each atom contributes one electron to be shared in a pair. Also, COVALENT BONDING, ELECTRON-PAIR BOND.

covalent crystal *Crystallography.* any crystal held together by covalent bonds. Also, VALENCE CRYSTAL.

covalent hydride *Inorganic Chemistry.* a volatile compound of hydrogen and a nonmetal, e.g., ammonia, NH_3, or hydrogen sulfide, H_2S.

covalently closed circular DNA *Molecular Biology.* an uninterrupted circle of DNA that exists as a continuously covalent circular molecule.

covalent radius *Physical Chemistry.* half the distance from center to center between two atoms that are covalently bonded together.

covariance *Statistics.* a measure of linear dependence; for two random variables X and Y with expectation m_x and m_y, the covariance is $E\{(X-m_x)(Y-m_y)\}$.

covariance matrix *Statistics.* for a given set of random variables, a matrix whose element in the ith row and jth column is the covariance of X_i and X_j, and whose element in the ith row and ith column is the variance of X_i.

covariant *Physics.* having the property of remaining invariant in form under a Lorentz transformation.

covariant differential *Mathematics.* in the Ehresmann theory of principal fiber bundles, the covariant differential $D\alpha$ of a vector-valued differential k-form α is given by $D\alpha(X_1, X_2, \ldots, X_k) = d\alpha(hX_1, hX_2, \ldots, hX_k)$, where hX_j is the horizontal component of X_j.

covariant functor *Mathematics.* let C and D be categories, and let T represent the following pair of functions: an object function that assigns each object C of C to an object $T(C)$ of D, and a morphism function that assigns each morphism f: $C \to D$ to a morphism $T(f)$: $T(C) \to T(D)$. T is then called a covariant functor if: (a) $T(1_C) = 1_{T(C)}$ for every identity morphism 1_C of C; (b) $T(g \circ f) = T(g) \circ T(f)$ for any two morphisms f and g of C whose composite $g \circ f$ is defined.

covariant index *Mathematics.* one of the subscript indices of a tensor.

covariant tensor *Mathematics.* a multilinear functional that accepts only contravariant vectors as arguments; if there are s contravariant vector arguments, it is a **covariant tensor of order s** or **covariant s-tensor.** Every covariant s-tensor can be expressed as a sum of tensor products of s covariant vectors (s differential 1-forms in each term). The components of a covariant s-tensor are represented by symbols with s subscripts.

covariant theory *Physics.* a theory whose equations retain their form under a Lorentz transformation.

covariant vector *Mathematics.* a covariant tensor of order 1; i.e., a differential 1-form.

cove *Geography.* a small, usually sheltered bay or inlet. *Architecture.* a curved, concave member such as a molding.

cove beach see BAYHEAD BEACH.

coved vault *Architecture.* a vault composed of four three-sided coves that meet to form a pyramid. Also, CLOISTER VAULT; SQUARE DOME.

covellite *Mineralogy.* CuS, a blue or bright purple, opaque, submetallic, brittle, hexagonal mineral occurring in massive foliated form and as thin tabular hexagonal crystals; having a specific gravity of 4.68 to 4.7 and a hardness of 1.5 to 2 on the Mohs scale; found in oxidized zones of copper sulfide deposits. Also, **covelline.**

cover *Mining Engineering.* **1.** the depth of the rock between the mine workings and the surface. **2.** the total thickness of material overlying mine workings or a body of ore. Also, BURDEN, MANTLE.

coverage *Graphic Arts.* a quantitative and qualitative measure of the degree to which something is covered, especially a printed sheet (with ink) or the fibers of a coated paper (with emulsion). *Optics.* see COVERING POWER. *Telecommunications.* **1.** the geographic boundary of television or radio broadcast reception. **2.** the geographic area in which a jammer is capable of producing an effective jamming signal. *Computer Technology.* a measure of the effectiveness of fault-tolerance mechanisms of a computer; uses a reliability model to compute the conditional probability that the system will recover if a fault occurs.

coverboard see BINDER'S BOARD.

cover crop *Agronomy.* **1.** a crop planted to prevent erosion during a nonproductive season, such as a field planted with rye in the winter. **2.** see NURSE CROP.

covered smut *Plant Pathology.* a smut disease of grains that is caused by the fungus *Ustilago hordei* in barley and *U. avenae* in oats, in which the fungus spore masses are held together or covered for a period of time by the durable grain membrane and glumes.

cover glass *Biotechnology.* a thin rectangular glass plate that covers a mounted microscopical object or culture. Also, **coverslip.**

cover hole *Mining Engineering.* one of a group of boreholes drilled in advance of mine workings to detect water-bearing fissures or structures.

covering power *Engineering.* the degree to which an object is concealed by a coating. *Optics.* the field of view, often expressed as an angle, at which a camera lens produces its sharpest image. Also, COVERAGE. *Metallurgy.* in electroplating, the characteristic of an electrolyte that produces an acceptable plate at very low current density.

covering space *Mathematics.* a topological space X' is said to be a covering space of a topological space X if (a) both X' and X are locally arcwise connected and arcwise connected Hausdorff spaces; (b) there exists a continuous map p of X' onto such that the inverse image of any open set U in X under p is a disjoint union of open sets V_α in X'; and (c) p is a homeomorphism when restricted to each such V_α; e.g., the real line, projected by $p(x) = e^{ix}$, is a covering space for the unit circle.

cover material *Graphic Arts.* the woven cloth used to cover binder's board and forming the outermost layer of a clothbound book.

cover of a set *Mathematics.* a collection of sets whose union contains a given set; this is also known as a **covering.** If the cover has a countable number of members, it is known as a **countable cover.** It is called an **open cover** (or a **closed cover**) if all its members are open (or closed).

cover paper *Graphic Arts.* heavy paper stock used as a covering for pamphlets and the like.

cover plate *Engineering.* **1.** a plate that is fastened to the flange sections of a column or girder to increase its resistance to bending. **2.** a plate, often of glass, over an opening to a boiler or other process, for visual inspection.

coversine *Mathematics.* the coversine of θ is the quantity $1 - \sin\theta$; i.e., the difference between the radius and the sine of an angle constructed in a unit circle. Also, **coversed sine.**

cover stock *Graphic Arts.* a heavy coated paper commonly used as a covering for softbound books. Also, STOCK COVER.

covert [kuv′ərt] *Ecology.* a shelter or hiding place for animals. *Textiles.* a tight-woven twilled wool fabric containing filling threads of a single color and warp threads of two different colors.

covert [kō′ərt] *Science.* hidden, disguised, or concealed.

covert operations *Military Science.* intelligence, counterintelligence, or other military activities that are conducted so as to conceal the identity of the sponsor of the operation. Also, **covert ops.**

covert reinforcement *Behavior.* a method for increasing the frequency of a desired behavior by having the subject imagine pleasant responses to this behavior.

covert sensitization *Behavior.* a method for changing undesirable behavior through the imagination of unpleasant responses to the behavior.

covert therapy *Behavior.* a therapeutic technique in which the effects of a behavior are imagined and verbalized. Also, **covert extinction.**

covey *Vertebrate Zoology.* a small flock, hatch, or brood of birds of one kind, such as quail or partridge.

covite *Petrology.* an igneous rock composed of alkali feldspar, hornblende, sodic pyroxene, nepheline, and trace amounts of sphene, apatite, and opaque oxides; a variety of nepheline syenite.

cow *Agriculture.* **1.** a mature female bovine animal. **2.** a market term designating a female bovine animal that has produced a calf. *Vertebrate Zoology.* the mature female of certain other large mammals, such as the moose, elephant, and whale.

cowbird *Vertebrate Zoology.* a small North American blackbird of the family Icteridae, known for its symbiotic relationship with cattle, from which it picks off ticks; some lay their eggs in other birds' nests.

Cowdria *Bacteriology.* a genus of Gram-negative, nonmotile bacteria of the tribe Ehrlichieae, found within the vascular endothelial cells of ruminants.

cow-dung bomb *Volcanology.* a volcanic bomb that has a flat, rounded shape due to its striking the ground while in a molten state.

Cowell method *Space Technology.* in orbit computation, a step-by-step integration in rectangular coordinates of the orbiting body's total acceleration.

cowling *Engineering.* a metal casing that houses an engine. *Aviation.* specifically, a streamlined metal covering over all or part of an aircraft engine, usually having hinged or removable panels; designed to protect engine components and (in air-cooled engines) to promote airflow cooling. Also, **cowl.**

cowpea *Botany.* a vinelike herb, *Vigna unguiculata,* of the order Rosales having alternate, trifoliate leaves; grown for its edible seeds.

cowpea mosaic virus group see COMOVIRUS GROUP.

Cowper's gland *Zoology.* either of the two small glands that secrete a mucous substance into the male urethra. (After the English anatomist William *Cowper,* 1666–1709, who discovered the glands.) Also, BULBOURETHRAL GLAND.

cowpox *Veterinary Medicine.* a disease that erupts on the teats and udders of cows, forming small pustules that contain a virus used in the vaccination of humans against smallpox.

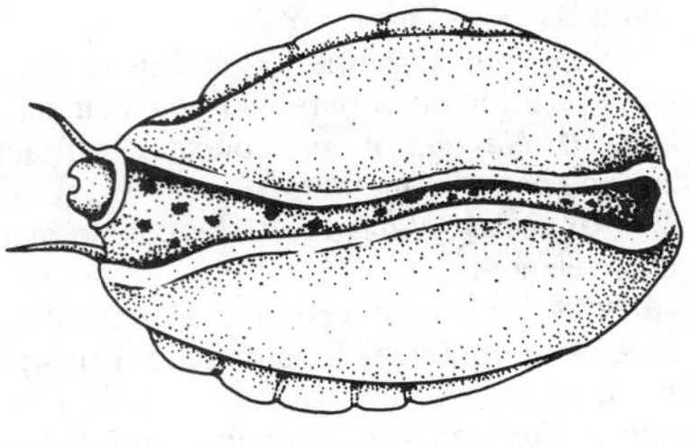

cowry

cowry *Invertebrate Zoology.* any of numerous widely distributed marine gastropod mollusks of the family Cypraeidae, with highly polished and richly colored shells, much collected and used for ornamentation and even currency.

cow shark *Vertebrate Zoology.* any of the sharks of the family Hexanchidae, having six or seven gill openings on each side of the head.

coxa *Anatomy.* **1.** the part of the body lateral to and including the hip joint. **2.** loosely, the hip joint itself. *Invertebrate Zoology.* the segment of the leg that joins to the body of certain arthropods.

coxal bone see HIP BONE.

coxal cavity *Invertebrate Zoology.* the cavity within which the leg joins to the body in certain arthropods.

coxal gland *Invertebrate Zoology.* a gland with ducts opening at the base of the leg of certain arthropods.

Cox chart *Chemistry.* a graph plotting vapor pressure against temperature for a specified hydrocarbon.

Coxeter graph *Mathematics.* a graph with one vertex for each of the roots in a root system (as for a Lie algebra), such that the number of edges joining the *i*th and *j*th vertices equals the Cartan integer relating the *i*th and *j*th roots.

Coxiella *Bacteriology.* a genus of Gram-negative, rod-shaped bacteria of the family Rickettsiaceae; it has a single species, *C. burnetti,* that is an obligate intracellular parasite of various vertebrates and arthropods.

coxitis *Medicine.* an inflammation of the hip joint. Also, **coxarthritis.**

coxopodite *Invertebrate Zoology.* the basal joint of a crustacean leg.

coxsackie virus *Virology.* any of a group of enteroviruses producing a disease that resembles poliomyelitis but without paralysis, primarily affecting children during warm weather. Also, **Coxsackie virus.**

Cox's model *Statistics.* a regression-based technique for estimating how survival depends on a set of independent variables.

coyote [kī ōt′ē] *Vertebrate Zoology.* a small wolf of western North America, noted for its keen senses, speed, and adaptability; belonging to the genus *Canis latrans* of the family Canidae; originally living mainly in the Great Plains area but now found in many other habitats.

coyote

coyote blasting *Mining Engineering.* a method of blasting in which large charges are fired in small adits or tunnels driven at the level of the floor in the face of a quarry or the slope of an open-pit mine. Also, GOPHER-HOLE BASTING, HEADING BLASTING.

COZI *Telecommunications.* a sounding system that measures the clarity with which a high-frequency broadcast reaches its intended destination. (An acryonym for Communications Zone Indicator.)

CP *Aviation.* the airline code for Canadian Airlines International Ltd.

CP candlepower; center of pressure; chemically pure.

cpd. compound.

cpd gene *Genetics.* a gene that encodes cAMP phosphodiesterase.

cpDNA chloroplast DNA.

CPE cytopathic effect.

C peptide *Biochemistry.* a metabolically inactive polypeptide chain connecting proinsulin and insulin. Also, CONNECTING PEPTIDE.

C period *Microbiology.* the time required for one round of chromosomal DNA replication in bacteria.

CPH *Aviation.* the airport code for Copenhagen, Denmark.

CPI California Personality Inventory.

cpi *Graphic Arts.* characters per inch.

CP invariance *Particle Physics.* a principle stating that the laws of physics are unchanged by any combination of space inversion (P) and charge conjugation (C) operations.

CPM see CRITICAL PATH METHOD.

CP/M *Computer Programming.* an operating system used for early eight-bit microcomputers. (An acronym for Control Program for Microprocessors.)

C polysaccharides see C SUBSTANCES.

cpp *Graphic Arts.* characters per pica.

CPS controlled-path system.

CPT theorem *Particle Physics.* the theory that, for any process, its mirror image, antiparticle, and time-reversed process will look exactly the same as the original.

CPU *Computer Technology.* central processing unit, the part of a computer that controls the interpretation and execution of machine language instructions.

CPU-bound *Computer Programming.* of a computer program, performing mostly calculations, with little input or output. Also, COMPUTE-BOUND, PROCESS-BOUND, PROCESSOR-LIMITED.

CPU time *Computer Technology.* **1.** the computational time required to execute program instructions in a central processing unit. **2.** the total amount of CPU time used by a program.

CQR anchor *Mechanical Devices.* an anchor fabricated in the shape of a plow.

CR or **C.R.** *Psychology.* conditioned response; continuous reinforcement.

crab *Invertebrate Zoology.* **1.** the common name for a wide variety of crustaceans in the order Decapoda, most with a flattened carapace and well-developed legs; many types are used as food, such as the **blue crab** and the **Dungeness crab. 2.** any of various members of the Merostomata, primitive arthropods such as the horseshoe crab. *Cartography.* the condition in which the sides of an aerial photograph are not parallel or perpendicular to the direction of flight.

crab

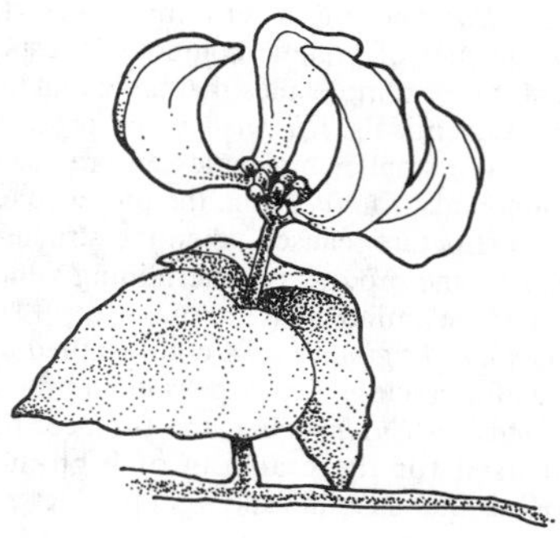

crab apple

crab apple *Botany.* **1.** a wild apple tree, *Malus sylvestris,* bearing a small, sour fruit that is used in making jellies and preserves. **2.** any of various apple trees having similar fruits. (From *crab* in the sense of "sour" or "sharp.") Also, **crabapple.**

crabbing *Textiles.* a process for setting the colors of the warp and weft threads of a fabric by winding the fabric tightly on a roller, subjecting it to a stream of boiling water, and then cooling it.

crabeater see COBIA.

crabgrass *Botany.* any of a variety of wild, weedlike grasses with creeping or decumbent rooting stems that often invade turf or cultivated areas.

crab locomotive *Mining Engineering.* a trolley locomotive fitted with a winch for hauling mine cars from workings where there is no trolley wire.

Crab Nebula *Astronomy.* the expanding cloud of turbulent gas in Taurus left by the supernova explosion of 1054.

crab plover *Vertebrate Zoology.* a wading bird of the family Dromadidae, having black and white plumage and inhabiting the northern and western shores of the Indian Ocean.

Crab pulsar *Astronomy.* the rapidly rotating neutron star that lies at the center of the Crab Nebula and which is the remnant of the star that exploded as a supernova.

crabtree see CRAB APPLE.

Crabtree effect *Mycology.* in certain yeasts, the effect of increasing fermentation by providing glucose as nourishment.

crabwood see ANDIROBA.

crachin *Meteorology.* a period of drizzle or light rain often accompanied by low stratus clouds and poor visibility, occurring in the China Sea between January and April.

Cracidae *Vertebrate Zoology.* a family of tropical American arboreal birds in the order Galliformes, including the curassows, guans, and chachalacas.

crack *Materials Science.* a fracture in a material due to an applied or internal stress, generally originating at a defect in the material. **Edge cracks** extend from the surface into the material, and **center cracks** are contained entirely within it. Cracks act as stress concentrators and may cause brittle materials to break at applied stresses that are below the nominal yield strength. *Chemistry.* to break a chemical compound into simpler compounds. *Engineering.* to slightly open something, particularly a valve.

crack arrester *Naval Architecture.* a plate attached to metal plating, or a hole bored through plating, in either case intended to prevent the onset or spread of cracks in the plating.

crack-deflection toughening *Materials Science.* the tendency for cracks to be deflected from their original direction when they reach an interface; a potent mechanism for increasing resistance to crack propagation in ceramic matrix composites.

cracked *Materials Science.* describing oils that are produced by the process of cracking, rather than by distillation.

cracked residue *Chemical Engineering.* the residue of fuel that is produced from the cracking of hydrocarbons.

cracked stem *Plant Pathology.* a disease of celery caused by a deficiency of boron, which is characterized by brown, mottled leaves, long brown streaks on the stems, and a ragged, crosswise cracking of the abnormally brittle stalks.

cracker *Computer Programming.* an informal term for a person who attempts to break into computer systems, read protected data, steal passwords, and so on.

cracking *Chemical Engineering.* a refining process that decomposes and recombines molecules of organic compounds, especially hydrocarbons, by using heat, to form molecules that are suitable for motor fuels, monomers, and petrochemicals. *Chemistry.* any process of breaking up organic compounds into smaller molecules and reassembling the products into other compounds. *Engineering.* the presence of relatively large cracks penetrating a structure, caused when the structure undergoes extreme stress. *Genetics.* the process of determining which genetic codon is specific for each of the amino acids.

cracking coil *Chemical Engineering.* a coil installed in a heated chamber or furnace, used for cracking heavy petroleum products.

cracking still *Chemical Engineering.* the furnace, reaction chamber, and fractionator used for the cracking of high-molecular-weight petroleum compounds into gasoline and light products.

crackle breccia see SHATTER BRECCIA.

Cracticidae *Vertebrate Zoology.* a bird family of the suborder Oscines that includes ten species from Australia and New Guinea, including the bell magpies, butcherbirds, and piping crows.

cradle *Engineering.* a framework that carries, supports, or restrains material or engines. *Ordnance.* the nonrecoiling structure of a gun on which the gun slides during recoil and counterrecoil. *Textiles.* in weaving jacquard fabrics, a device used to catch the punched cards that fall from a jacquard head.

cradle cap *Medicine.* a common seborrheic dermatitis of infants, giving rise to heavy, greasy crusts on the scalp.

cradle vault see BARREL VAULT.

cradlework *Industrial Engineering.* work that is performed at a bench or similar work location, rather than at a large machine or a field site.

CRAFT *Industrial Engineering.* a computerized method for determining plant layout, using letters of the alphabet to represent relationships between departments, in order to minimize the cost of materials handling.

crag *Geology.* a steep, jagged point or hill of rock, commonly found projecting from a mountainside.

Craigie's tube method *Microbiology.* a technique used for the separation of motile from nonmotile microorganisms, in which both types of cells are inoculated into an open-ended tube placed vertically into semisolid agar, allowing the motile cells to migrate out of the tube into the medium.

Crambiinae *Invertebrate Zoology.* a subfamily of lepidopteran insects, the snout moths, in the family Pyralididae.

Cramer's rule *Mathematics.* a method of solving a system $Ax = b$ of n independent linear equations in n independent unknowns, where A is the matrix of coefficients, x is the column vector whose entries are the n independent variables, and b is the column vector whose entries are the constant terms of the n linear equations. The value for the jth variable is given by the fraction whose denominator is $\det A (\det A \neq 0$ if the equations are independent), and whose numerator is the determinant of the matrix obtained from A by replacing the jth column with the entries of b.

cramp *Medicine.* a painful involuntary contraction of a muscle. *Mechanical Devices.* a wood or metal plate or bar with right-angle ends that act as a fastener for wood pieces during assembly.

crampon *Mechanical Devices.* a hook that engages a load by means of a draw link, shackle pin, cable socket, or sling chain; used for heavy loads of rock or timber.

cranberry *Botany.* **1.** any of several plants of the heath family, genus *Vaccinium,* that grow wild in bogs and are cultivated for their edible berries, such as the **American** or **large cranberry,** *V. macrocarpon,* and the **European** or **small cranberry,** *V. oxycoccus.* **2.** the acidic red fruit or berry of such plants, used for sauces and flavorings.

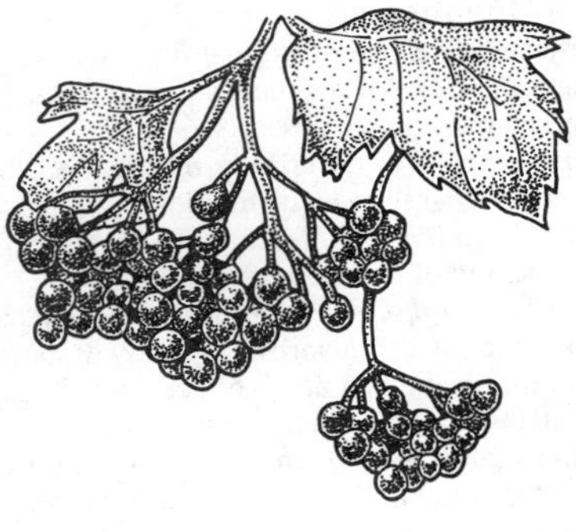

cranberry

Cranchiidae *Invertebrate Zoology.* a family of cephalopods, mollusks in the subclass Dibranchia, having two gills and an internal shell or no shell.

crandallite *Mineralogy.* $CaAl_3(PO_4)_2(OH)_5{\cdot}H_2O$, a yellow, white, or gray, transparent to subtranslucent, somewhat chalky, trigonal mineral occurring in nodular fibrous masses and as tiny prismatic crystals; having a specific gravity of 2.78 to 2.92 and a hardness of 5 on the Mohs scale; found with rare secondary phosphate minerals in variscite nodules. Also, PSEUDOWAVELLITE.

crane

crane *Vertebrate Zoology.* any of several species of long-necked, long-legged wading birds in the family Gruidae of the order Gruiformes, found in the Northern Hemisphere and Africa. *Mechanical Engineering.* **1.** a power-operated hoisting machine with lifting tackle and a pivoted boom that allows movement of loads horizontally as well as vertically. **2.** of, relating to, or equipped with such a machine. Thus, **crane barge, crane motor, crane truck.**

crane ship *Naval Architecture.* a ship designed to support a very high load lift capacity crane.

crani- a combining form meaning "cranium," as in *craniectomy.*

craniectomy [krān´ē ek´tə mē] *Surgery.* the surgical removal of parts of the cranium, frequently in strips.

craniad *Anatomy.* toward the cranium.

cranial *Anatomy.* of or relating to the cranium, or to the upper end of the body.

cranial capacity *Anatomy.* the volume of the skull; it approximates the volume of the brain, and averages 1450 cc in adult humans.

cranial deformation *Anthropology.* the cultural practice of deliberately molding the head of an infant by keeping it against a board or in some other kind of restraint.

cranial flexure *Developmental Biology.* the sharp, ventrally concave bend in the developing midbrain of the embryo. Also, CEPHALIC FLEXURE, MESENCEPHALIC FLEXURE.

cranial fossa *Anatomy.* any of three hollows in the surface of the skull at which the cerebellum and the cerebrum are lodged.

cranial index *Anthropology.* an anthropometric formula in which the cranial breadth is divided by the cranial length and multiplied by 100; used to classify skulls into cranial types.

cranial nerve *Anatomy.* any of twelve pairs of nerves containing nerve fibers that arise from particular centers in the brain.

cranial types *Anthropology.* the classification of skulls into three categories: dolichocephalic, mesocephalic, and brachiocephalic.

Craniata *Vertebrate Zoology.* a taxonomic subdivision of the phylum Chordata, including all chordates having a cranium, i.e., all vertebrates and cyclostomes but not the Protochordata.

Craniidina *Invertebrate Zoology.* a subdivision of branchiopods in the order Acrotretida, with a pedicle attached to a surface by cement.

cranio- a combining form meaning "cranium," as in *craniometry.*

craniocele [krān´ē ə sēl´] *Medicine.* a congenital protrusion of the brain through a defect in the skull. Also, ENCEPHALOCELE.
craniocerebral [krān´ē ō sə rē´brəl] *Anatomy.* of or relating to the cranium and the brain.
craniomalacia *Medicine.* an abnormal softness in the skull.
craniometer *Anthropology.* an instrument that measures the external dimensions of the skull.
craniopathy [krān´ē ə path´ē] *Medicine.* any disease of the skull. Thus, **craniopathic.**
craniopharyngeal pouch see RATHKE'S POUCH.
craniopharyngioma [krān´ē ō fə rinj´ē ō´mə] *Medicine.* a tumor of epithelial cell origin in the craniopharyngeal canal.
cranioplasty *Medicine.* the surgical correction of defects in the cranium, usually accomplished by implantation of plates of metal, bone, or other materials.
cranioscopy *Medicine.* a diagnostic examination of the cranium.
craniotome *Surgery.* an instrument used to cut the skull.
craniotomy [krān´ē ät´ə mē] *Surgery.* an operation on the skull or an incision into the skull.
craniotopography [krān´ē ō täp äg´rə fē] *Neurology.* the study of the relationship between the surface of the skull and various parts of the brain within.
cranium *Anatomy.* the skeleton of the head, especially that portion of the skull that encloses the brain.

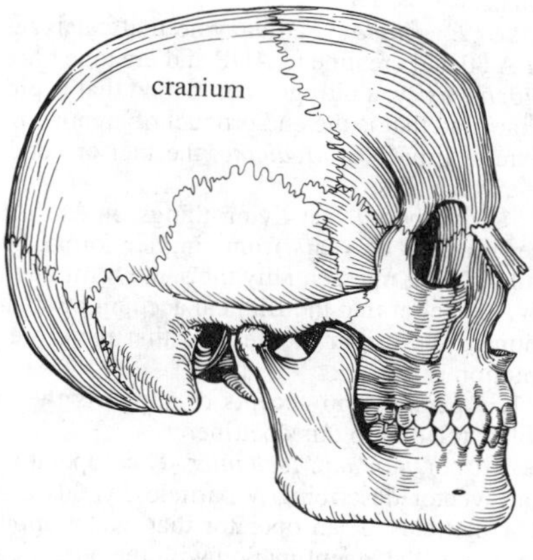

cranium

crank *Mechanical Engineering.* a device composed of an arm attached to a shaft, with a pin located at the outer edge of the arm and parallel to the shaft; it may either translate reciprocating motion to a component attached to the pin or transform such motion into rotary motion of the shaft.
crank angle *Mechanical Engineering.* **1.** the angle between the crank and a given reference direction. **2.** in a slider crank mechanism, the angle between the crank and a line from the crankshaft to the piston.
crank axle *Mechanical Engineering.* **1.** any axle that is attached to a crank. **2.** specifically, an axle bent at both ends to accommodate a large body with large wheels.
crankcase *Mechanical Engineering.* a boxlike casing for the crankshaft and connecting rods of certain engines; in an autmobile engine, the bottom is a reservoir where hot motor oil is collected and cooled before being recirculated by a pump.
crankpin *Mechanical Devices.* the machined cylindrical and movable section of a crank on a crankshaft onto which a connecting rod of an engine or linkage is fitted.
crank press *Mechanical Engineering.* a punch press in which a crank applies power to the slide.
crankshaft *Mechanical Engineering.* the shaft around which a crank rotates; in most engines or machines, it is the main shaft that transmits power from the crank to the connecting rod.
crank throw *Mechanical Engineering.* **1.** the radial distance from the main shaft to the pin of a crank, equal to one-half the stroke of a reciprocating component attached to the pin. **2.** the web and pin of a crank.
crank web *Mechanical Engineering.* the arm of a crank, usually a flat, rectangular section, connecting the pin to the shaft or connecting two adjacent pins.
crannog *Archaeology.* a small, artificially constructed island in a lake or bog, upon which a fortified structure is built; typical of prehistoric Ireland and Scotland.
crappie *Vertebrate Zoology.* a freshwater sunfish of the family Centrarchidae that is native to the eastern United States but widely introduced in other areas; a popular food and sport fish.

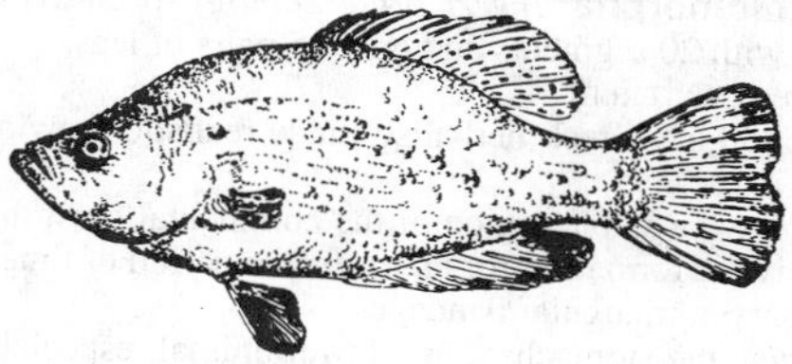

crappie

craseur [kra sûr´] *Surgery.* a wire that is looped around a part and tightened to cut it; a snare.
crash *Computer Technology.* **1.** a severe system failure that causes a system to become inoperative, usually requiring operator intervention, rebooting, or maintenance before processing can resume. **2.** to undergo such a system failure; to become inoperative.
craspedon *Invertebrate Zoology.* a coelenterate medusa stage that has a velum.
craspedote *Invertebrate Zoology.* having a velum.
Crassatella *Paleontology.* a genus of burrowing bivalves in the order Veneroida and superfamily Crassatellacea; although it appears in early Tertiary strata in Britain that were formed in relatively cold water, it is currently restricted to warmer waters; Cretaceous to the present.
Crassulaceae *Botany.* a family of succulent dicotyledonous plants in the order Rosales.
crater *Geology.* **1.** a bowl-shaped, generally steep-walled surface depression or hole formed from the impact of a falling body or by a bomb explosion. **2.** a similar depression around the vent of a volcano or geyser. *Astronomy.* a similar depression that lies on a planet or moon's surface; most of the craters in the solar system have been formed by the impact of meteorites. *Mechanical Engineering.* a depression in the face of a cutting tool caused by chip contact. *Metallurgy.* in arc welding, an undesirable depression formed at the end of the bead.

crater

Crater *Astronomy.* the Cup, a small and faint southern constellation visible in spring in the Northern Hemisphere; associated in mythology with Corvus and Hydra.
crater arc *Astronomy.* a crater chain with a curved form.
crater chain *Astronomy.* a line of craters, such as the Stadius chain near Copernicus on the moon, that is thought to result from secondary impacts of rocks ejected when a larger nearby crater was formed.
crater cone *Volcanology.* a cone formed around a volcanic vent by erupting lava.
crater crack *Metallurgy.* in arc welding, a crack, usually complex, occurring in the crater.
crater lake *Hydrology.* a deep, freshwater lake that is formed by the accumulation of rainwater and groundwater within a volcanic crater or caldera.

crater lamp *Electronics.* a neon-filled glow discharge tube in which the glow is concentrated at a crater-shaped depression at one end of the cathode; the brightness is proportional to the current through the tube and can be modulated with a signal.

craterlet *Astronomy.* any crater whose diameter is small enough (ranging from a few centimeters to a few kilometers) that it lies near the lower limit of resolution for an observer. Also, CRATER PIT.

Craterostigmomorpha *Invertebrate Zoology.* primitive centipedes of Australasia with 20 segments and only 15 pairs of legs.

crater pit see CRATERLET.

Crateuas c. 80 BC, Greek herbalist; made realistic drawings of medicinal plants.

craton *Geology.* a stable region of the continental crust that has not undergone active deformation for an extensive period of time.

cravat *Surgery.* a triangular bandage.

craw *Zoology.* the stomach of any lower animal, especially the crop of an insect or bird.

crawler *Mechanical Engineering.* **1.** either one of a pair of continuous sprocket-driven belts used instead of wheels as a means of propulsion by some tractors, bulldozers, and other heavy machinery. **2.** a piece of machinery equipped with this form of propulsion. **3.** having such a form of propulsion. Thus, **crawler drill, crawler crane, crawler tractor.**

crawler drill

crawl space *Building Engineering.* an area above or below the livable area within a building, accessible only by crawling on all fours, as the clearance is less than the height of an average human being; it allows access to plumbing and wiring for maintenance and repair functions.

crayfish *Invertebrate Zoology.* the common name for a number of lobsterlike freshwater decapods, crustaceans in the section Astacura.

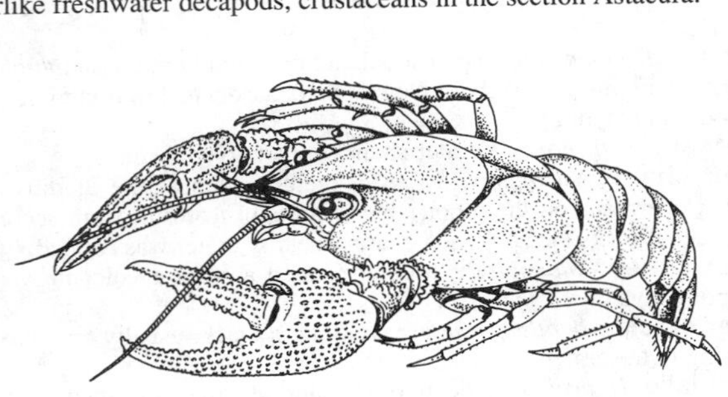

crayfish

crayoning *Graphic Arts.* the use of a crayon to correct minor defects on a printing plate.

craze *Materials Science.* to form localized regions of fine hairline cracks on a material.

crazing *Materials Science.* a pattern of fine hairline cracks occurring when a material such as plastic or enamel experiences brittle fracture.

CRC cyclic redundancy check.

CRC character *Computer Technology.* a character recorded at the end of a block of data on tape or disk to allow detection and correction of recording errors.

CRC error *Computer Technology.* an error indicated by the incorrect match of a CRC character.

C-reactive protein *Biochemistry.* a globulin that forms a precipitate with the somatic C-polysaccharide of the pneumococcus in vitro; a protein not normally detected in the serum but whose presence there is a sensitive indicator of inflammation of infectious or noninfectious origin.

cream *Food Technology.* one of the two main constituents of whole milk, along with nonfat (skim) milk. Cream is the yellowish fatty liquid that rises to the top when milk is allowed to separate; it is generally about 30 to 35% butterfat.

cream cheese *Food Technology.* a soft, unripened high-fat cheese usually lightly pressed into blocks.

cream ice see SLUDGE ICE.

cream of tartar see POTASSIUM BINOXALATE.

creatine *Biochemistry.* a nitrogenous compound that is synthesized in the body; phosphorylated creatine is an important storage form of high-energy phosphate.

creatine kinase *Enzymology.* an enzyme that catalyzes the reversible conversion of ATP and creatine to ADP and creatine phosphate.

creatinine *Biochemistry.* a nitrogen compound that is present in muscle, blood, and urine, and that is the end product of creatine metabolism.

creatinuria [krē´at i nůr´ē ə] *Medicine.* the fact or condition of having excess creatine in the urine.

creationism **1.** the theory that living things on earth did not develop gradually over millions of years from simpler forms, but were created by a supernatural being in essentially the same form that they now have. **2.** specifically, the belief that the Biblical account of creation is an accurate description of the origin of life, and that therefore the Darwinian theory of evolution is false.

creationist **1.** a person who accepts or supports the doctrine of creationism. **2.** of or relating to this doctrine.

creation operator *Quantum Mechanics.* **1.** an operator that, when applied to a state vector describing N particles, yields a state vector describing $N + 1$ particles. **2.** an operator that, when applied to a known eigenfunction, yields the eigenfunction with the next greater eigenvalue. *Computer Programming.* a function that allocates new data records at run time.

creation science a term used by adherents of creationism to describe their efforts to develop science-based arguments to support creationism or to challenge the theory of evolution; for example, the argument that there is no evidence in the fossil record of the vast number of intermediate forms that would be necessary for one-celled organisms to evolve into highly complex forms such as birds and mammals.

creativity *Psychology.* the ability to develop and produce new forms of art or science or to apply new and original ideas to problem-solving.

creatotoxism see KREOTOXISM.

creatoxin see KREOTOXIN.

creche [kresh] *Behavior.* among birds, a gathering of the young of several families, tended by one or more adults. (From French for "crib.")

credence *Electromagnetism.* a measure of confidence in a radar system's capability for target detection.

Credé's method [krə dāz´] *Medicine.* the application of silver nitrate to a newborn's eyes for the prevention of ophthalmia neonatorum. (From Karl Sigmund Franz Credé, 1819–1892, German gynecologist.)

credible interval *Statistics.* an interval computed from a posterior distribution, as opposed to the confidence interval.

credit-assignment problem *Artificial Intelligence.* the determination of which rule, heuristic, or procedure should receive credit (or blame) for the success (or failure) of a game-playing or decision-making program.

crednerite *Mineralogy.* $CuMnO_2$, a black, opaque, bright metallic monoclinic mineral that occurs as earthy coatings or thin, six-sided plates in spherulitic groupings, having a specific gravity of 5.34 to 5.38 and a hardness of 4 on the Mohs scale; found as a secondary mineral intergrown with hausmannite and psilomelane.

creedite *Mineralogy.* $Ca_3Al_2(SO_4)(F,OH)_{10}\cdot 2H_2O$, a brittle, transparent, colorless, white, or amethyst-colored monoclinic mineral occurring as short prismatic crystals in sprays or clusters, having a specific gravity of 2.71 to 2.72 and a hardness of 4 on the Mohs scale; found in cavities of banded fluorite or embedded in pyrite.

creek *Hydrology.* **1.** any natural stream of flowing water, regarded as larger than a brook but smaller than a river. **2.** a small inlet, bay, or narrow recess in the shore of a lake, river, or sea. **3.** see TIDAL CREEK. *Geology.* see ARROYO.

creep *Geology.* the slow, relatively imperceptible downward movement of soil or broken rock under the influence of gravity or sheer stress. *Mining Engineering.* any slow movement of mining ground. *Engineering.* the tendency for a rating or characteristic to change slowly, especially to deteriorate, with prolonged use or age. *Materials Science.* the deformation with time of materials under stress, usually at elevated temperatures; the deformation frequently occurs at grain boundaries. *Building Engineering.* specifically, the slow deformation and movement of concrete under stress; it is useful as a uniform load-stabilizing feature in building construction. *Mechanics.* a phenomenon in which strain in a solid increases with time when the stress producing the strain is held fixed. *Mechanical Engineering.* any undesirable play or movement in a mechanism. *Aviation.* the continued deflection of an instrument needle after an initial deflection caused by an exciting force. *Graphic Arts.* **1.** during a press run, a forward movement of an offset press blanket. **2.** in a bound book, a condition in which the middle pages of a printed signature extend slightly beyond the edges of the front and back pages. *Agriculture.* a fenced area for young animals that is not accessible to their mothers, used especially for feeding.

creepage *Electricity.* the conduction of electricity across the surface of a dielectric.

creep buckling *Mechanics.* buckling that occurs after a structural member has suffered from creep due to the application of a compressive load over a prolonged period of time.

creeper *Mechanical Engineering.* a low work platform supported by small casters; used to lie on while working beneath an automobile. *Mining Engineering.* a continuous chain that hauls mine cars up an incline by catching the axles on projecting bars. *Vertebrate Zoology.* a bird of the family Certhiidae, native to the Northern Hemisphere and known for its habit of scurrying up trees in search of insects. Also, TREE CREEPER.

creeper

creep error *Engineering.* an error that appears in a digital analytical balance when it processes, prints, or reads a value before it arrives at its last position.

creeping barrage *Ordnance.* a barrage in which the fire of all participating units remains in the same relative pattern while advancing in steps of one line at a time.

creeping disk *Invertebrate Zoology.* the smooth, adhesive undersurface of the foot or body of certain invertebrates, used for creeping.

creeping eruption *Medicine.* the presence of a growing red line on the skin, caused by the presence of larva of hookworms or certain roundworms burrowing in the dermis. Also, LARVA MIGRANS.

creeping flow *Fluid Mechanics.* flow at a very low Reynolds number, where viscous effects dominate.

creeping mine *Ordnance.* a buoyant mine that is held below the surface by a weight, thus allowing it to creep freely along the seabed with the force of a stream or current.

creep limit *Materials Science.* the maximum stress that a material can withstand at a given temperature, assuming a creep rate of 1% deformation per 1000 or 10,000 hours; it provides a means to compare the creep strength of different materials.

creep recovery *Materials Science.* the reversal of some portion of the strain occurred during a creep test after the applied stress is released.

creep rupture *Materials Science.* the failure of a material that has suffered deformation from creep; it usually follows a large local reduction in the area of the material, which results in an increase in the true stress.

creep rupture strength *Materials Science.* the stress that will lead to the rupture of a material after a given time and at a given temperature.

creep strength *Materials Science.* the maximum stress that a material can withstand without exceeding its creep limit at a given temperature.

creep test *Materials Science.* any of various methods used to determine the resistance of materials to creep, particularly by placing the material under a constant stress.

C region see CONSTANT REGION.

cremation pit *Archaeology.* a depression in which the remains of a cremation are buried.

cremnophobia *Psychology.* an irrational fear of cliffs or precipices.

cremocarp *Botany.* any two-seeded fruit that divides into a pair of mericarps upon maturity.

cremone bolt *Mechanical Devices.* a pair of vertical rods with sliding shafts controlled by a knob; used to engage both the top and the bottom of double doors or a casement window frame. Also, **cremorne bolt.**

crenate or **crenated** *Biology.* having a notched or scalloped edge, such as on many leaves and on the shells of some mollusks.

crenation *Physiology.* an abnormal shrinkage of an erythrocyte, often caused by osmotic changes in the surrounding blood plasma. Also, **crenulation.**

crenitic *Geology.* describing the raising of subsurface minerals by the action of spring water.

Crenothrix *Bacteriology.* a genus of iron bacteria that occur in water as sheathed, filamentous cells and may be associated with iron or manganese oxides.

Crenotrichaceae *Bacteriology.* a former classification for a family of bacteria in the order Chlamydobacteriales, characterized by trichomes that are differentiated at the tip and base.

crenulate or **crenulated** *Biology.* having a minutely notched or scalloped notched edge.

crenulation cleavage see SLIP CLEAVAGE.

Creodonta *Paleontology.* an extinct order of eutherian mammals that includes two families, the oxyaenids and the hyaenodontids; mostly small but some bear-sized; the creodonts are related to early Carnivora, and they offer several good examples of parallel evolution with members of that order; upper Cretaceous to Miocene.

creole [krē´ōl] *Linguistics.* **1.** a more developed language that evolves from a pidgin language and becomes the native or first language for the offspring of pidgin speakers; for example, African slaves having different dialects created a pidgin to understand one another on American plantations, and their children then developed a creole language. **2. Creole.** the version of the French language spoken by descendants of the original French settlers of Louisiana.

creosol *Organic Chemistry.* $C_8H_{10}O_2$, a monomethyl ether of guaiacol; a colorless to yellowish liquid that is slightly soluble in water and miscible with alcohol; boils at 220°C and melts at 5.5°C; a major constituent of creosote.

creosote *Materials.* an oily liquid with a burning taste, obtained by distilling coal and wood tar; used as an antiseptic and wood preservative.

creosote bush *Botany.* any of a number of evergreen shrubs of the genus *Larrea,* having resinous foliage and yellow flowers, and occurring mostly in the arid regions of the southwestern U.S. and Mexico.

creosote oil *Materials.* a yellow-to-green oily liquid that boils between 240 and 270°C, is immiscible with water and soluble in alcohol, and is derived by the fractional distillation of coal tar; used as a wood preservative, fungicide, and disinfectant. Also, COAL-TAR CREOSOTE.

crepe *Textiles.* a plain-woven fabric distinguished by its crinkled surface, which may be produced by varying the tension on warp yarns or by mixing yarns that shrink differently.

crepe paper *Materials.* a soft, lightweight, puckered paper that is stretchable and is used as a decoration or in craft projects.

crepe ring see C RING.

Crepidotus *Mycology.* a genus of fungi belonging to the order Agaricales, occurring in the shape of lobes or shells on fallen leaves and twigs in forests.
crepitation *Medicine.* the sound made by rubbing together the ends of a fractured bone, or bone and irregular cartilage.
crepuscular [krə pusk´yə lər] *Astronomy.* relating to sunrise or sunset. *Zoology.* active or primarily active at the time around dawn or dusk, as are rabbits, deer, bats, and other mammals, as well as various insects.
crepuscular rays *Astronomy.* rays of light seen shortly before sunrise or after sunset that are caused by sunlight streaming past clouds below the horizon and scattering off dust in the air.
crescent *Science.* a shape defined by a convex and concave edge, resembling a segment of a ring tapering to a point at each end. *Astronomy.* **1.** a heavenly body that appears to have this shape. **2.** specifically, the moon when it has such a shape. See CRESCENT MOON. *Invertebrate Zoology.* **1.** an organism or an anatomical structure with this shape; i.e., the gametocyte of a falciparum malaria parasite. **2.** any of a number of small butterflies with white spots on their wings.
crescent beach *Geology.* a curved beach lying concave toward the sea at a bayhead or at the mouth of a stream that enters a bay.
crescent beam *Engineering.* a beam having arches of varying sizes; the central arch is always the largest.
crescentic dune see BARCHAN.
crescent moon *Astronomy.* a name for the moon when it appears to have a shape that resembles a segment of a ring tapering to a point at each end; that is, the moon in its phases between the last quarter and first quarter, except for the new moon. The **waning crescent** follows the last quarter and the **waxing crescent** precedes the first quarter.
crescent phase *Astronomy.* the appearance of the moon or an inferior planet when its disk is less than half illuminated by the sun.
crescent-type cross-bedding see TROUGH CROSS-BEDDING.
***m*-cresol** *Organic Chemistry.* $CH_3C_6H_4OH$, a colorless, yellowish, or pinkish liquid, soluble in water and alcohol; melts at 11–35°C and boils at 191–203°C; used as a disinfectant and in the manufacture of herbicides, surfactants, and synthetic food flavors. Also, METHYLPHENOL.

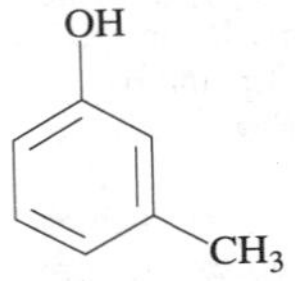

***m*-cresol**

cresol red *Organic Chemistry.* $C_{21}H_{18}O_5S$, a reddish-brown powder, soluble in water and alcohol; an acid-base indicator (pH 7.2, yellow; pH 8.8, red; pH 2–3, orange or amber); used in histological staining.
cress *Botany.* any of several plants of the mustard family (Cruciferae) that are grown for their edible leaves, especially watercress (*Rorripa*).
crest the top or upper part; specific uses include: *Design Engineering.* the top of a screw thread. *Geology.* see CREST LINE. *Anatomy.* see CRISTA.
crestal plane see CREST PLANE.
crest clearance *Design Engineering.* in a screw, the distance between the crest of a thread and the root of the thread with which it mates.
crest cloud *Meteorology.* a standing cloud that remains in the same leeward position relative to the mountain ridge over which it forms.
crest factor *Physics.* the ratio of the peak amplitude to the average amplitude of any periodically varying function.
crest length *Oceanography.* the length of a wave along its crest.
crest line *Geology.* a line connecting the highest points of a fold.
crestomycin see PARAMOMYCIN.
crest plane *Geology.* a surface connecting the highest points of the rock beds of an anticline. Also, CREST SURFACE, CRESTAL PLANE.
crest stage *Hydrology.* the highest elevation reached by the water surface at a given point during a stream rise.
crest surface see CREST PLANE.
crest value see PEAK VALUE.
crest voltmeter *Electricity.* a voltmeter that reads the peak value of the voltage applied to its terminals.
cresylic acid *Materials Science.* any of a group of acids that boil above 204°C, derived from petroleum and coal tar; used in electrical insulation, solvents, and pesticides.
Cretaceous [krə tā´shəs] *Geology.* **1.** the final period of the Mesozoic era, occurring after the Jurassic and before the Tertiary period (between 136 and 65 million years ago). **2.** the rocks formed during that time.
Crete [krēt] *Geography.* a large island in southeastern Greece, separating the Aegean Sea from the eastern Mediterranean.
cretin [krēt´ən] *Medicine.* a person affected with cretinism.
cretinism [krēt´ən iz əm] *Medicine.* a chronic condition that is caused by a congenital lack of thyroid secretion, characterized by arrested physical and mental development, dystrophy of the bones and soft parts, and lowered basal metabolism.
Creutzfeldt-Jakob disease [kroits´felt yä´käp] *Medicine.* a rare, usually fatal, transmissible subacute spongiform encephalopathy, caused by a slow virus and characterized by progressive dementia and sometimes wasting of the muscles, tremor, athetosis, ataxia, spastic dysarthria, and other neurological disturbances.

crevasses

crevasse [krə vas´] *Geology.* **1.** a deep crack or fissure in a glacier, resulting from differential ice movement. **2.** a deep vertical opening in the earth that appears after an earthquake. **3.** a wide break in a river bank or natural levee.
crevasse filling *Geology.* an accumulation of sand and gravel remaining in a crevasse as a result of the melting of glacial ice.
crevasse hoar *Hydrology.* a deposit of thin ice crystals having the shape of a leaf, plate, or cup, forming and growing in a glacial crevasse or other large, open space where water vapor can condense under calm, still conditions.
crevice corrosion *Metallurgy.* corrosion caused by concentration cells in crevices and other locations where oxygen or metal ions are depleted because of inadequate diffusion.
crew-served *Military Science.* a term for a weapon or piece of equipment that is served or operated by two or more individuals.
criador *Meteorology.* a rain-bearing west wind of northern Spain.
crib *Engineering.* **1.** a structure of horizontally, cross-piled, squared timbers used to support a structure above. **2.** a structure composed of one or more layers of timber or steel joists, upon which a load may be spread. **3.** a place to store tools. *Geology.* see ARETE.
Cribbiaceae *Mycology.* a family of fungi belonging to the order Cribbiaceae; found on wood, primarily in Australia.
crib death *Medicine.* the abrupt and unexplainable death of an apparently healthy infant, typically occurring between the ages of three weeks and five months. Also, SUDDEN INFANT DEATH SYNDROME.
cribellum *Invertebrate Zoology.* **1.** an accessory spinning organ in certain spiders. **2.** a chitinous plate with gland duct openings in insects.
Cribrariaceae *Mycology.* a family of fungi of the order Liceales, characterized by stalked sporangia and brightly colored spores; it occurs primarily on decaying wood.
cribriform *Biology.* marked by many small perforations; sievelike.

Bacteria & Viruses

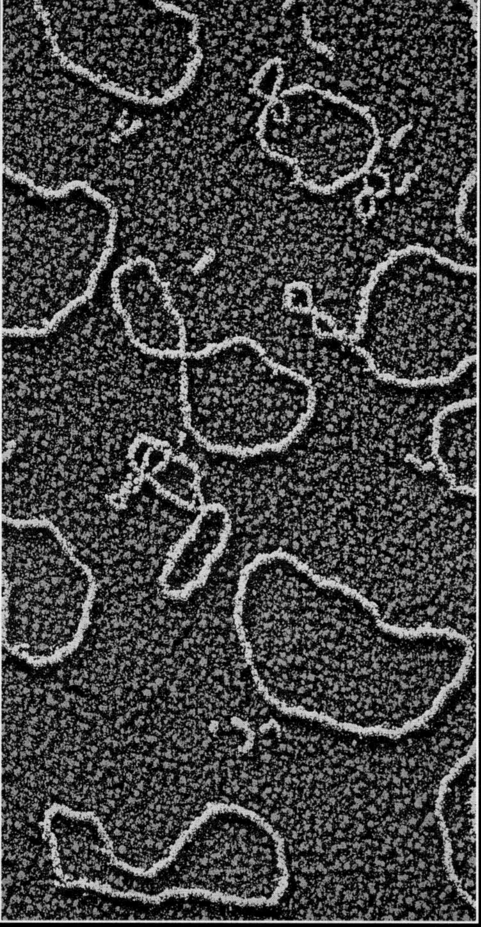

False-color transmission electron micrograph of plasmids of bacterial DNA extruded from *Escherichia coli* cell (not shown). Dr. G. Murti/SPL/Photo Researchers.

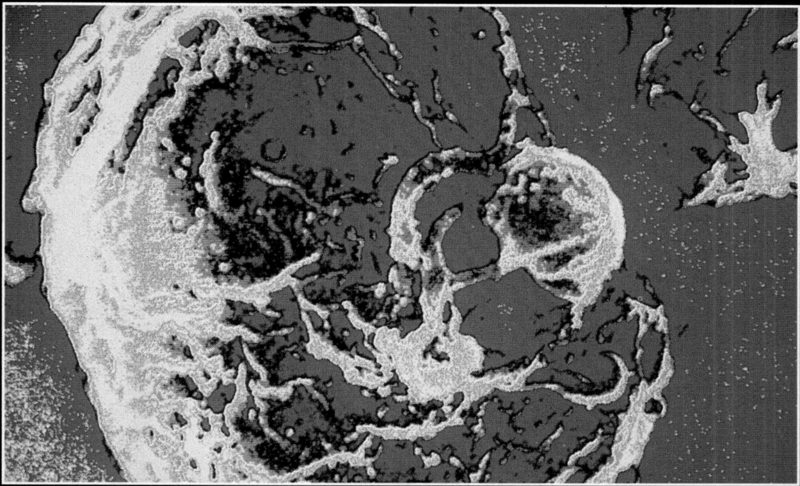

Kaposi's sarcoma: T cell and virus. Howard J. Sochurek Inc./The Stock Market.

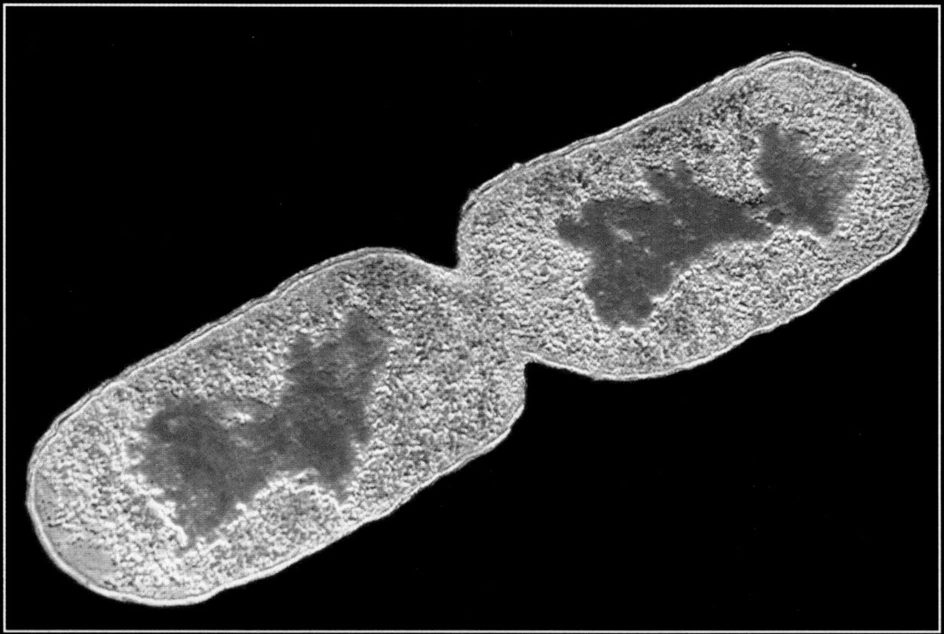

False-color transmission electron micrograph of the common bacterium *Escherichia coli*. CNRI/SPL/Photo Researchers.

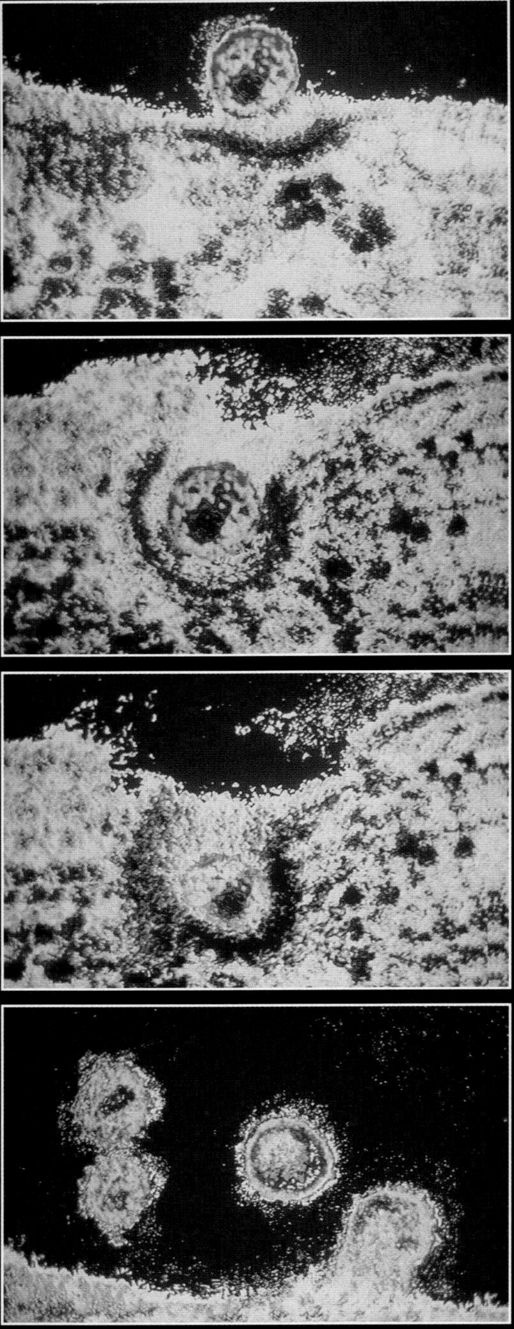

AIDS virus. (Top) AIDS virus contacting lymphocyte. (Middle panels) AIDS virus penetrating lymphocyte. (Bottom) AIDS virus being released from cell. Petit Format/Photo Researchers.

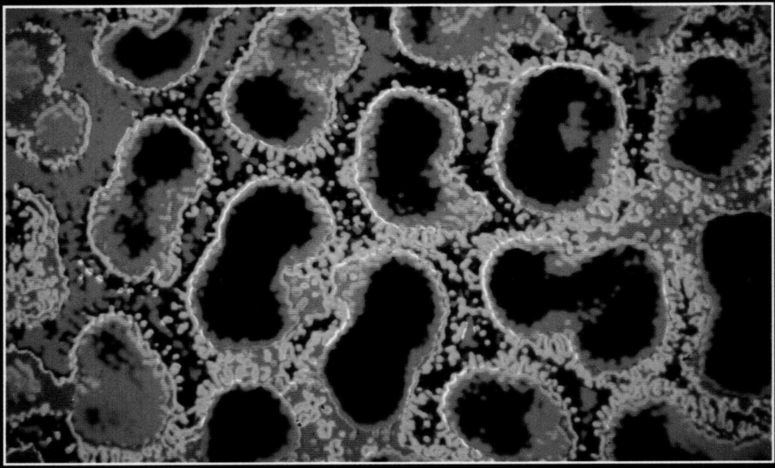

Influenza virus. Howard J. Sochurek Inc./The Stock Market.

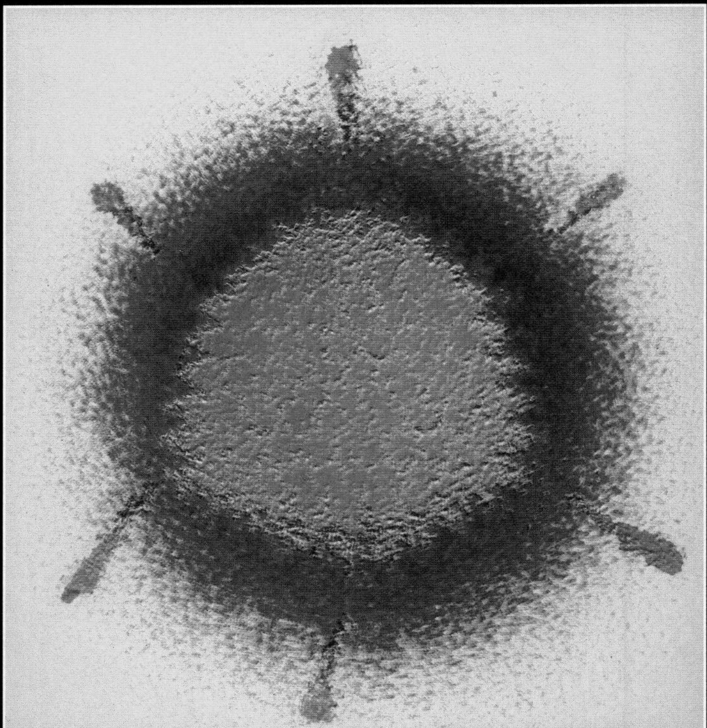

False-color transmission electron micrograph of an adenovirus, which causes an infection of the upper respiratory tract and has symptoms resembling a common cold. Biozentrum/SPL/Photo Researchers.

cribriform fascia *Anatomy.* a perforated fibrous tissue that covers the saphenous opening; part of the superficial fascia of the thigh.

cribriform plate *Anatomy.* **1.** a thin plate of ethmoid bone that fits into the frontal bone supporting the olfactory lobes to the cerebrum and contains many openings for the passage of the olfactory nerves. **2.** a perforated portion of the sclera where the fibers of the optic nerve pass.

Cricetidae *Vertebrate Zoology.* a rodent family in the suborder Myomorpha including the gerbils, hamsters, voles, lemmings, and New World rats and mice; most species are nocturnal, have no premolars, and have a naked or scaly tail.

Cricetinae *Vertebrate Zoology.* a subfamily of the family Cricetidae in the suborder Mymorpha that includes New World rats and mice, hamsters, and mole rats.

cricket *Invertebrate Zoology.* any member of the family Gryllidae (order Orthoptera), leaping insects closely related to grasshoppers but characterized by three tarsal segments, a complex ovipositor, long unsegmented anal cerci, wing venation, and a generally brown color; they have long antennae and hind legs modified for jumping. The males rub their forewings together to make a distinctive chirping noise. *Building Engineering.* a structure on the sloping roof of a building, designed to divert rainwater around an obstruction such as a chimney.

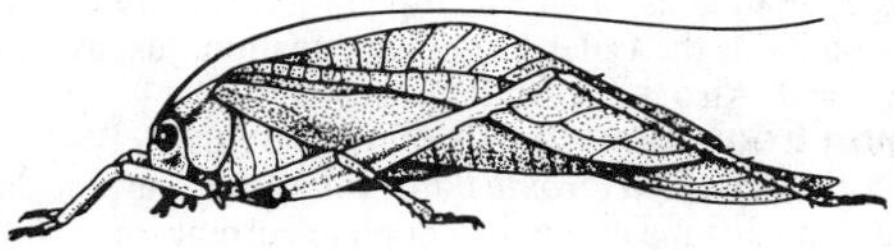

cricket

Crick, Francis born 1916, English biologist; with James Watson, discovered double helix and built Watson-Crick model of DNA molecule.

Crick strand *Genetics.* in Watson-Crick-type DNA, the strand that is not transcribed in vivo; the antisense strand. Also, C STRAND.

cricoid *Anatomy.* resembling a ring; ring-shaped.

cricoid cartilage *Anatomy.* a ring-shaped cartilage in the larynx.

cricondenbar *Physical Chemistry.* the highest pressure at which two different phases of a substance can coexist.

cricondentherm *Physical Chemistry.* the highest temperature at which two different phases of a substance, such as a liquid and a vapor, can coexist.

cri-du-chat syndrome [krē´doo shā´] *Genetics.* a genetically determined syndrome in humans that is caused by a loss of a portion of the short arm of chromosome 5. It produces several congenital malformations, and a peculiar cry in infants that sounds like the mewing of a cat. Also, CAT CRY SYNDROME. (From French; literally, "cat's cry.")

Crile, George Washington 1864–1943, American surgeon; improved treatment of shock, hemorrhage, and high blood pressure.

criminal abortion *Medicine.* any surgery for termination of pregnancy performed in violation of the law, as by an unlicensed practicioner.

crimp *Materials Science.* to make a material, such as lumber, wavy or warped. *Metallurgy.* to seal together metal parts by pressing or overlapping. *Textiles.* the curl or waviness of a fiber or wool, sometimes measured in comparison to its straightened length.

crimp contact *Electricity.* a contact with a hollow, cylindrical back portion allowing bared wire to be inserted and crimped to make contact between the metal and wire. Also, SOLDERLESS CONTACT.

crimped yarn *Textiles.* yarn that has been heat set in order to give it alternating bends.

crinal *Metrology.* a unit of force equal to 0.1 newton.

C ring *Astronomy.* the gauzy-appearing, innermost ring of the three rings of Saturn visible from earth by a small telescope; it is 19,000 kilometers wide and adjoins the inner edge of the B ring. Also, CREPE RING.

crinogenic *Medicine.* causing a gland to secrete.

crinoid *Invertebrate Zoology.* belonging to the class Crinoidea.

crinoidal limestone *Petrology.* a rock formed on the sea floor, composed predominantly of crystalline joints of encrinites with foraminiferans, corals, and mollusks.

Crinoidea *Invertebrate Zoology.* a class of radially symmetrical echinoderms, the sea lilies or feather stars; adults are flower shaped, either free living or sessile with a stem.

crinoline *Textiles.* a stiff open-weave cotton or horsehair fabric; often used for lining skirts or making heavy bandages.

crinophagy *Cell Biology.* a lysosomal degradation of secretory vesicles that occurs in secretory cells to avoid overproduction of a secretory product.

Crinozoa *Invertebrate Zoology.* a subphylum of radially symmetrical marine animals in the phylum Echinodermata.

cripple *Building Engineering.* a term for a structural member that is built to a length shorter than normal, such as a stud used beneath a window sill.

crippled mode *Computer Technology.* a term for a system that is operating at less than full capability.

crippled strain *Genetics.* a strain of bacterium that has been genetically impaired so that it will grow only in a highly specialized medium containing essential compounds that it can no longer synthesize; used in genetic experiments that could pose a danger to wild organisms. (So called because the strain cannot survive if it should accidentally escape from the laboratory.)

crippling see LOCAL BUCKLING, def. 2.

crisis *Medicine.* a sudden turn in the course of an illness, whether for better or worse; used especially to refer to the turning point at which an illness suddenly abates, and probable recovery is indicated. *Psychology.* see IDENTITY CRISIS.

crisis period *Biotechnology.* a condition in which a cell culture enters a time period during which most of the secondary cells die even though the required substances and conditions to maintain life are available.

crispation number *Physics.* a dimensionless number used frequently in convection current analysis, obtained by dividing the product of the dynamic viscosity of a fluid and its thermal diffusivity by the product of the undisturbed surface tension and layer thickness.

criss-cross inheritance *Genetics.* the passing on of sex-linked traits by a mother to her male offspring, or by a father to his female offspring.

crissum *Vertebrate Zoology.* in birds, the area surrounding the cloacal aperture, or the feathers on that area.

crista *plural,* **cristae.** *Anatomy.* a projecting structure, especially one surmounting a bone or its border. *Cell Biology.* any of numerous infoldings of the inner membrane of the mitochondria of eukaryotic cells.

cristate *Biology.* having a crest or crista.

Cristispira *Bacteriology.* a genus of Gram-negative, helical bacteria of the family Spirochaetaceae, occurring as parasites in clams and other mollusks.

cristobalite *Mineralogy.* SiO_2, a white or gray translucent to opaque tetragonal mineral occurring as fibrous masses and pseudo-octahedral, commonly twinned crystals, having a specific gravity of 2.33 and a hardness of 6.5 on the Mohs scale; found in igneous rocks or as opaline silica formed at low temperatures; polymorphous with quartz, tridymite, coesite, and stishovite.

criterion-based testing *Psychology.* a testing approach that compares an individual's test performance with a previously established standard. Also, **criterion-referenced testing.**

crith *Metrology.* a unit of measure for gases, equal to the mass of 1 liter of hydrogen at standard pressure and temperature.

crithidia *Invertebrate Zoology.* **1.** a genus of protozoans, in the family Trypanosomatidae, endoparasitic flagellates found in the digestive tract of insects. **2.** a developmental stage of some flagellate protozoans parasitic in the bloodstream.

critical *Science.* describing a state, level, or value at which a significant change takes place, such as the point at which two different phases of a substance form one phase. *Nucleonics.* describing a fissionable material that has enough mass to sustain a chain reaction. *Medicine.* **1.** relating to or being in critical condition. See CRITICAL CONDITION. **2.** of or relating to the crisis stage of a disease.

critical absorption wavelength *Spectroscopy.* the wavelength at which an absorption discontinuity occurs for an atom of a particular element.

critical activity *Industrial Engineering.* in a PERT network, an activity whose delay will cause the entire project to be delayed.

critical altitude *Aviation.* **1.** an altitude above which the performance of equipment, such as the supercharger of an engine or the propulsion system of a guided missile, falls off. **2.** an altitude above which harmful cosmic radiation effects may occur.

critical angle *Acoustics.* the angle of incidence at which acoustic waves will not penetrate a medium but will be totally reflected. *Optics.* the maximum angle of incidence for which light can be refracted from a medium with a large index of refraction to one with a smaller index. Light striking a surface at an angle greater than the critical angle is totally reflected back into the first medium.

critical angle of attack *Aviation.* on a given airfoil, the minimum angle of attack at which extensive flow separation occurs, causing a loss of lift and increase of drag. Also, ANGLE OF STALL, STALLING ANGLE OF ATTACK.

critical angle refractometer *Optics.* a refractometer in which the refraction of a given medium is gauged by observing the angle at which the entire refraction occurs with respect to another medium whose index of refraction is known.

critical anode voltage *Electronics.* the anode breakdown voltage in a gas tube.

critical assembly *Nucleonics.* the assembly of sufficient fissionable material to maintain a zero-power fission chain, in order to study the behavior of fissionable material of different isotopic enrichment and in different configurations; used to obtain fundamental physics data and in designing nuclear reactors, weapons, and safety systems.

critical bottom slope *Oceanography.* a measure of the depth distribution of an ocean in relation to its latitude.

critical compression ratio *Mechanical Engineering.* the lowest compression ratio that allows compression ignition of a given fuel.

critical condensation temperature see TRUE CONDENSING POINT.

critical condition *Medicine.* a classification in which a patient's condition is considered to be life-threatening, as indicated by unstable vital signs, unconsciousness, or other such negative conditions; the most unfavorable status in appraising a patient's condition.

critical constant *Physical Chemistry.* a maximum or minimum physical value that is characteristic of a substance, such as the value of temperature, pressure, and volume below which a gas can be liquefied.

critical cooling rate *Metallurgy.* the minimum cooling rate that prevents unwanted transformations of a metallic structure.

critical coupling *Electricity.* the degree of coupling that provides the optimum transfer of signal energy between radio-frequency resonant circuits when both are at the same frequency. Also, OPTIMUM COUPLING.

critical current *Solid-State Physics.* the current rating for a superconductive material at a specified constant temperature and in the absence of an external magnetic field, above which the material behaves normally and below which the material is still superconductive.

critical current density *Physical Chemistry.* the amount of current that is needed to cause a sudden change in one of the variables in an electrolytic process.

critical damping *Physics.* the amount of damping that is just sufficient to keep a system whose free oscillation is hindered by viscosity or friction from oscillating; the amount of damping that separates underdamping and overdamping.

critical defect *Industrial Engineering.* the most serious classification of product defects; a defect that renders the product completely unfit to serve its intended use.

critical density *Astrophysics.* an average density of matter for the entire universe that is just sufficient for gravity to halt its expansion. *Geology.* in a saturated granular material undergoing rapid deformation, the point below which the material will lose strength and above which it will gain strength.

critical depth *Hydrology.* the depth at which water flowing in a channel is at minimum energy with respect to the bottom of the channel.

critical distance *Acoustics.* the point at which the reverberant field equals the direct field.

critical elevation *Cartography.* the highest elevation on a map or chart.

critical entanglement chain length *Materials Science.* the weight-average number of chain atoms in a polymer molecule that are required to cause intermolecular entanglement; an important determining factor in viscosity.

critical equation *Nucleonics.* one of a family of multigroup mathematical expressions that establish boundary conditions for determining the dimensions that a bare reactor must have in order to be critical.

critical equatorial velocity *Astronomy.* in fast-rotating objects, the equatorial velocity at which centrifugal force equals gravity and the object is on the verge of breaking up.

critical experiment *Nucleonics.* an experimental procedure that is conducted under controlled conditions, in which masses of nuclear fuel that are initially subcritical are assembled gradually to reach a critical mass and attain a self-sustaining, near-zero-power, chain reaction.

critical exponent *Thermodynamics.* a dimensionless quantity n that characterizes the temperature dependence of a thermodynamic property of a substance when it is near its critical temperature T_c; the dependence takes the form $|T - T_c|^n$.

critical facility *Nucleonics.* a laboratory that is equipped for the safe operation and monitoring of critical experiments, including provision for acquiring fundamental physical data and conducting neutron-multiplication measurements for nuclear weapons and for power and propulsion reactors.

critical field *Electronics.* the lowest theoretical magnetic flux density, at a steady anode voltage, that would prevent an electron emitted from the cathode at zero velocity from reaching the anode in a magnetron. Also, CUTOFF FIELD.

critical flicker frequency *Optics.* the frequency at which the luminosity of a light source appears to fluctuate for half the time and remain steady for half the time.

critical flow *Fluid Mechanics.* flow for which the Froude number and the Mach number are equal to one.

critical flow prover *Petroleum Engineering.* an instrument used for measuring the velocity of gas flow while the open-flow testing of gas wells is occurring.

critical frequency *Electronics.* **1.** a frequency where a network function has a pole or a zero. **2.** see PENETRATING FREQUENCY. *Electromagnetism.* for a specified layer of the ionosphere, the limiting frequency below which radio waves would be reflected when propagating at vertical incidence.

critical function *Mathematics.* the function at which a given functional is extremized; analogous to a critical point in ordinary calculus. A critical function satisfies the Euler-Lagrange equation, just as a critical point satisfies $f'(x) = 0$. Also, EXTREMAL.

critical fusion frequency see FUSION FREQUENCY.

critical grid current *Electronics.* the amount of current drawn by the control grid in a gas tube at the instant plate current starts to flow.

critical grid voltage *Electronics.* the amount of grid voltage at which plate current starts to flow in a gas tube.

critical gun pull *Ordnance.* the maximum pulling force of the feeder mechanism of an automatic gun.

critical humidity *Chemical Engineering.* the point above the humidity in the atmosphere of a system at which a crystal of a water-soluble salt will always become damp, due to absorption of moisture from the atmosphere, and below which it will always stay dry. *Metallurgy.* the humidity level above which the corrosion rate of a metal or alloy will rapidly increase.

critical illumination see SOURCE-FOCUSED ILLUMINATION.

criticality *Science.* the fact or condition of being critical. *Nucleonics.* specifically, the condition of a nuclear system, as determined by the dimensions and composition of fissionable fuel and other components, in which the fission reaction is self-sustaining.

critical level of escape *Geophysics.* **1.** the level at which a particle that is moving rapidly upward in the atmosphere will have a $1/e$ (e is the base of natural logarithm) probability of colliding with another particle as it leaves the atmosphere. **2.** the level at which the horizontal mean free path of an atmospheric particle is equal to the scale height of the atmosphere.

critical locus *Physical Chemistry.* the line that connects the critical points of a substance's conversion from a liquid to a gas, as applied to multicomponent mixtures and plotted on a pressure versus temperature graph; for example, the critical loci for the methane-propane-pentane system.

critical Mach number *Aviation.* the freestream Mach number that will yield a local Mach number of 1.0 at a given point on an aircraft. If, for example, an airplane traveling at Mach number 0.85 attained a Mach number 1.0 about the wing, then the critical Mach number with respect to the wing would be 0.85. Often critical Mach number is erroneously considered to be equivalent to force divergence Mach number. Actually force divergence Mach number may be lower or higher than critical Mach number.

critical magnetic field *Solid-State Physics.* the magnetic field strength for a superconductive material at a specified temperature, above which temperature the material is normal and below which the material is superconductive.

critical mass *Nucleonics.* the minimum mass of a fissionable material of a specified isotopic composition and density in which, when assembled in a suitable configuration in a given location, the generation and loss of neutrons are precisely balanced to produce a self-sustaining chain reaction. Thus, **critical mass reaction.**

critical micelle concentration *Physical Chemistry.* a defined level at which a colloidal aggregate of molecules (micelle), such as a household detergent, will dissolve in a liquid.

critical moisture content *Chemical Engineering.* the average moisture content throughout a solid material that is in the process of being dried.

critical opalescence *Optics.* opalescence that is caused by strong density fluctuations in a medium near a critical point.

critical path *Industrial Engineering.* in a PERT network, the path representing the most time consumed; equivalent to the total time required for the project.

critical-path method or **technique** *Industrial Engineering.* a project-management system in which all aspects of the project are depicted in a sequence and this data is then translated into a schedule.

critical period or **phase** see SENSITIVE PERIOD.

critical phenomena *Physical Chemistry.* the physical properties, such as temperature and pressure, of liquids and gases at the point when they are about to unite. Also, CRITICAL PROPERTIES.

critical point *Physical Chemistry.* the state at which the properties of the vapor phase of a substance become indistinguishable from those of the liquid phase at the same pressure and temperature, so that the substance exists as a single phase. *Mathematics.* **1.** a critical point of a function f is a point x at which $f'(x)$ is either zero or does not exist. **2.** a critical point of an autonomous system of differential equations is a point at which each of the given differential equations vanishes.

critical point drying *Microbiology.* a procedure used in preparing a specimen for electron microscopic examination, in which damage to the specimen during its drying step is minimized by avoiding exposure to a liquid-gas boundary.

critical potential *Electricity.* a potential across a device that, when exceeded, causes the current through that device to increase sharply. *Atomic Physics.* the amount of energy that is needed to boost an electron to a higher energy level, such as the resonance potential, or to eject it from the atom, such as the ionization potential.

critical pressure *Physical Chemistry.* the pressure associated with the critical point of the liquid-vapor state of a substance; i.e., the pressure of the substance at its critical temperature. *Fluid Mechanics.* the pressure equal to the stagnation pressure multiplied by the quantity $2/(k + 1)$ raised to the $(k/(k-1))$ power, where k is ratio of the specific heat ratio for the gas.

critical pressure ratio *Fluid Mechanics.* the pressure ratio (ratio of freestream pressure to stagnation pressure) at which the flow per unit area is at the maximum.

critical properties see CRITICAL PHENOMENA.

critical radius ratio *Materials Science.* the ratio of the radii of the anions and cation when the anions just touch each other and contact the central cation.

critical range *Metallurgy.* a temperature range in which a solid-state transformation occurs. Also, TRANSFORMATION RANGE.

critical reactor *Nucleonics.* a nuclear reactor in which there is a balance between the number of neutrons produced in fission and the number lost in the reactor by absorption or by leakage from its surface.

critical region *Computer Science.* a region of a process within which it accesses, and usually modifies, a shared variable. If critical regions of multiple processes (involving the same variable) overlap, errors may occur. Also, **critical section.** *Statistics.* see REJECTION REGION.

critical resolved shear stress *Materials Science.* the applied stress resolved in the slip direction in the slip plane of a crystal to initiate plastic flow. It equals the applied stress times the cosines of the angles between the direction of stress application and the normal to the slip plane and the slip direction.

critical Reynolds number *Fluid Mechanics.* the number that corresponds to the transition from laminar flow to turbulent flow as the velocity is increased.

critical shear stress *Materials Science.* the shear stress needed to initiate slip in a given crystal plane and in a given direction.

critical size *Nucleonics.* the physical dimensions of a uranium-moderator system for which the number of neutrons produced by fission is equal to those lost by escape and capture, depending upon the isotopic composition of the uranium, the moderator configuration, and the presence of materials that cause parasitic capture of neutrons.

critical slope *Hydrology.* the slope of a channel or conduit that will produce a critical flow, or a change in the flow of water from laminar to turbulent.

critical solution temperature *Physical Chemistry.* the temperature above which a pair of liquids will dissolve into each other at all proportions; in some cases two liquids also have a lower critical solution temperature and will mix except in a medium temperature range.

critical speed *Aviation.* a speed that is considered highly significant for one reason or another, such as the lowest speed at which an aircraft can become or remain airborne, the speed at which compressibility effects are encountered, or the speed corresponding to a critical Mach number, intolerable buffet, unsatisfactory or unsafe stability, or the like. *Fluid Mechanics.* **1.** a speed for compressible flow that is numerically equal to the square root of the product of the critical temperature, the gas constant R, and the specific heat ratio of the gas k, or the equivalent expression using the stagnation temperature. Also, CRITICAL VELOCITY. **2.** for open channel flow, the speed at which the Froude number becomes numerically equal to unity. *Mechanical Engineering.* the rotational speed of a shaft at which some periodic disturbing force coincides with the natural frequency of the shaft and its attached masses, resulting in dynamic instability.

critical strain *Materials Science.* the minimum strain that is adequate for recrystallization to occur upon subsequent annealing. *Metallurgy.* the strain at the yield point of a metal or alloy.

critical stress intensity *Materials Science.* the maximum stress intensity that a material can tolerate without plastic deformation; a measure of the fracture-toughness of the material. The stress intensity factor, K_I, can be calculated according to the following formula: $K_I = sa\sqrt{pc}$, where s is the stress in the absence of a crack, a is a dimensionless geometric factor, and c is the depth of an edge crack (or half the length of an interior crack).

critical temperature *Physical Chemistry.* the temperature above which a substance has no transition from the liquid to the gaseous phase; that is, the critical temperature of a gas is the highest temperature at which it can be liquefied, regardless of the pressure applied. *Fluid Mechanics.* a temperature equal to the stagnation temperature multiplied by the quantity $2/k + 1$, where k is the specific heat ratio for the gas.

critical value *Mathematics.* a value in the range of a function f corresponding to a critical point in the domain of f. *Statistics.* the value of a test statistic that separates the acceptance and rejection regions in statistical testing.

critical velocity *Physics.* the maximum velocity at which a fluid can flow without becoming turbulent. *Fluid Mechanics.* see CRITICAL SPEED. *Space Technology.* the speed of sound under given conditions at the throat of a rocket nozzle. Also, THROAT VELOCITY.

critical vibration *Mechanical Engineering.* any vibration that may harm a structure.

critical voltage *Electronics.* the highest theoretical value of steady anode voltage, at a given steady magnetic flux density, at which electrons emitted for the cathode at zero velocity would fail to reach the anode in a magnetron. Also, CUTOFF BIAS, CUTOFF VOLTAGE.

critical volume *Physics.* the volume of one mole of a substance when the substance is at critical temperature and critical pressure.

critical weight *Engineering.* the weight added to a bit in a drilling operation that causes the bit to resonate at the angular speed of the rotating shaft. Heavy collars often perform this function.

critical zone *Fluid Mechanics.* the place on a graph where the friction factor versus the Reynolds number shows unstable flow between the transition to turbulent flow and laminar flow. *Ordnance.* the area over which a bomber aircraft must maintain straight, level flight so that the bombsight can be operated properly and the bomb dropped accurately.

crivetz [krə vetz′] *Meteorology.* a cold north-to-east wind in Romania that may occur at any season.

CRM cross-reacting material; cultural resource management.

CRO cathode-ray oscilloscope; cathode-ray oscillograph.

croaker *Vertebrate Zoology.* any of several fishes of the family Sciaenidae that are noted for making a croaking noise, particularly *Micropogonias undulatus,* found off the Atlantic coast of the southern U.S.

Crocco's equation *Fluid Mechanics.* an equation that shows how entropy varies normal to streamline flow; it equates the product of the rate of change of entropy normal to streamlines and the temperature, to the rate of change of stagnation enthalpy normal to streamlines, with the product of vorticity of flow and specific internal energy added to it.

Crocco's theorem *Fluid Mechanics.* a theorem derived from Crocco's equation, stating that if the flow is steady, adiabatic, and irrotational, then it must be isentropic throughout; conversely, if the flow is not isentropic, then it must be irrotational.

Crocco's variables *Fluid Mechanics.* a dependent variable in the momentum integral equation is the shearing stress, designated by the Greek letter τ, and for the independent variables either u/U (specific internal energy divided by internal energy) and x (the distance in the direction of flow) or simply u and x.

crochet [kró´shá´] *Textiles.* **1.** a type of needlework in which thread or yarn is loosely interwoven in a series of interconnecting loops, in a manner similar to knitting. **2.** to prepare a fabric or garment in this manner. *Mechanical Devices.* see CROCHET HOOK. *Invertebrate Zoology.* **1.** a structure thought to resemble a crochet hook; a minute hook, curved spine, or projection, as on the caterpillar proleg or the cremaster of pupae. **2.** the setae of certain earthworms.

crochet file *Mechanical Devices.* a thin file with a flat, round edge that tapers to a point.

crochet hook *Mechanical Devices.* a distinctively shaped needle with a hooked end, used for crochet work.

crocidolite *Mineralogy.* a fibrous variety of riebeckite.

Crockett magnetic separator *Mining Engineering.* an assembly for separating ore pulp, consisting of a series of flat magnets below a continuous belt submerged in a tank through which the pulp flows; magnetic solids that adhere to the underside of the belt are dragged clear.

crocking *Textiles.* the rubbing off of dye from a noncolorfast fabric.

crocodile *Vertebrate Zoology.* the genus *Crocodylus* in the family Crocodylidae; a large lizardlike aquatic reptile having long and powerful jaws that narrow toward a snout, a hard palate, socketed teeth, and a covering of heavy plates and epidermal scales.

crocodile

crocodile bird *Vertebrate Zoology.* an African plover of the family Glareolidae, noted for its symbiotic relationship with crocodiles, from which it picks off insect parasites and gives danger warnings.

crocodile clip see ALLIGATOR CLIP.

Crocodilia [kräk´ə dil´ē ə] *Vertebrate Zoology.* an order in the subclass Archosauria (which includes extinct dinosaurs) of the class Reptilia; large, lizardlike reptiles that are voracious and carnivorous; it includes alligators, crocodiles, caimans, and gavials. Also, **Crocodylia.**

crocodiling see ALLIGATORING.

Crocodylidae *Vertebrate Zoology.* a family in the order Reptilia that contains the caimans of Central and South America, the alligators of North America and China, and the true crocodiles of Africa, southeast Asia, India, Australia, and tropical America.

Crocodylinae *Vertebrate Zoology.* a subfamily of the family Crocodylidae in the order Crocodilia, comprising the true crocodiles of Africa, southeast Asia, India, Australia, and tropical America.

crocoite *Mineralogy.* $PbCrO_4$, a very brittle, transparent to translucent red-orange monoclinic mineral occurring as slender prismatic, commonly hollow crystals, having a specific gravity of 6 and a hardness of 2.5 to 3 on the Mohs scale; it is usually found as a secondary mineral in oxidized zones of ore deposits containing lead and chromium. Also, **crocoisite.**

crocus *Botany.* any of about 80 species of cormous flowering plants of the genus *Crocus,* of the iris family, grown for their brightly colored, solitary flowers. *Materials.* a fine powder consisting of iron oxide, used for polishing. Also, **crocus martis.**

crofting *Textiles.* a process of bleaching linen in the sun after treating it with an alkaline solution.

crogball see BALLSTONE.

Crohn's disease *Pathology.* a chronic inflammation of any portion of the intestine, commonly affecting the terminal ileum. Also, REGIONAL ENTERITIS, REGIONAL ILEITIS.

Croixian [kroi´ən] *Geology.* a North American provincial series of the Upper Cambrian period, occurring after the Albertan and before the Canadian of the Ordovician period. Also, **Croixan.**

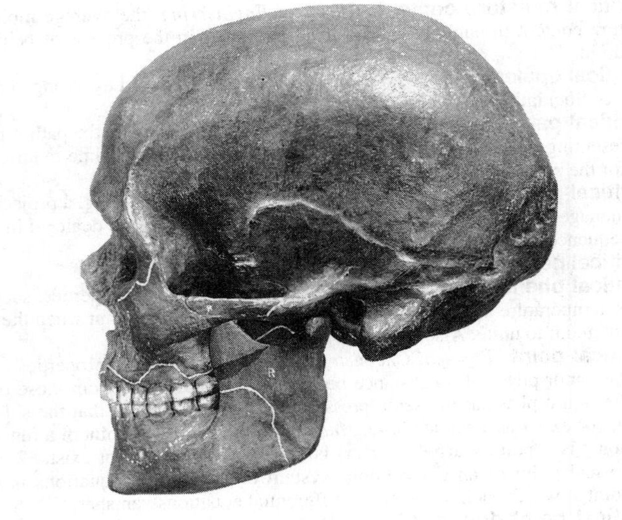

Cro-Magnon

Cro-Magnon [krō mag´nən; krō man´yən] *Anthropology.* the common term for the group of hominids, also called the **Cro-Magnon race** or **Cro-Magnon man,** that once lived in Europe and western Asia, with features including a high cranium, a broad and upright face, and cranial capacity about the same as modern humans but less than that of Neanderthals; the males were as tall as 6 feet. They appeared in Europe in the upper Pleistocene, about 40,000 years ago; their geographic origin is still unknown. Their skeletal remains show a few small differences from modern humans, but they are still generally classified as the earliest known representatives of the same subspecies, *H. sapiens sapiens.*

cromlech [käm´lek] *Archaeology.* a popular term used to refer to Megalithic tombs, especially those found in Wales.

Cromwell Current *Oceanography.* an eastward-setting equatorial undercurrent in the Pacific Ocean, extending from about 150 E to 92 W. Also, **Cromwell Undercurrent.**

Cronartium *Mycology.* a genus of rust fungi belonging to the order Uredenales which causes such pine tree diseases as blister rust.

Cronin, James Watson born 1931, American physicist; with Fitch, Nobel Prize for work on asymmetry of subatomic particles.

croning process *Metallurgy.* in casting, one of several shell molding processes.

cronism [kōn´iz əm] *Behavior.* the act or practice of a parent killing and eating its own young. (From *Cronus,* a Greek mythical figure who devoured his children.)

Cronstedt, Baron Axel 1722–1765, Swedish chemist; discovered nickel.

cronstedtite *Mineralogy.* $Fe_2^{+2}Fe^{+3}(SiFe^{+3})O_5(OH)_4$, a black, translucent to opaque monoclinic and trigonal mineral of the kaolinite-serpentine group occurring as three- and six-sided pyramidal crystals; having a specific gravity of 3.34 to 3.45 and a hardness of 3.5 on the Mohs scale; found mainly in Brazil, Czechoslovakia, and Romania.

crooked hole *Petroleum Engineering.* a borehole drilled at an angle; underground hard-rock formations lying at angles from the horizontal cause the drill bit to stray from the vertical.

crooked-hole country *Petroleum Engineering.* a term for an area or field in which drilling tends to produce a number of crooked holes, due to sharply inclined formations.

Crookes, Sir William 1832–1919, English chemist and physicist; discovered thallium and uranium X_1; invented the radiometer and the cathode-ray tube.

Crookes' dark space see CATHODE DARK SPACE.

crookesite *Mineralogy.* $Cu_7(Tl,Ag)Se_4$, a gray, opaque, metallic tetragonal mineral; massive and compact in habit, having a specific gravity of 6.9 and a hardness of 2.5 to 3 on the Mohs scale; found as veinlets with other selenides in calcite with quartz in copper mines in Sweden. (After Sir William *Crookes.*)

Crookes radiometer *Physics.* a device used to detect radiation, consisting of four vanes that are white on one side and black on the other, mounted inside an evacuated bulb in a turnstile fashion with minimal turning friction.

Crookes tube *Electronics.* an early form of cathode-ray tube.

crop *Agriculture.* **1.** any plant, either raised or naturally occurring, that is harvested for food or used for some other purpose. **2.** the yield of any such product in one season or place. **3.** a similar yield of livestock: a *crop* of lambs. *Forestry.* **1.** the standing trees on a forest area. **2.** the harvestable yield of lumber for a given forest area. *Vertebrate Zoology.* in birds, the pouched enlargement of the lower esophagus, used for storing and softening food before it passes on to the gizzard or stomach. *Invertebrate Zoology.* a similar sac in the food passage in earthworms and some insects. *Graphic Arts.* to mark off or cover an unwanted part of a photograph that is not to be printed. *Metallurgy.* **1.** the end of an ingot, which is discarded. **2.** to cut an end of an ingot, billet, or bar.

crop coal *Mining Engineering.* a term for inferior coal found near the surface of the earth.

crop-dusting or **crop dusting** *Agronomy.* the spreading of powdered insecticides, pesticides, or herbicides over an agricultural area from a low-flying airplane. Thus, **crop-duster.**

cropland *Agronomy.* the area of a farm that is actually used to grow crops, as opposed to that occupied by farm buildings, uncultivated land, and so on.

crop marks *Archaeology.* variations in the color or growth of surface vegetation that indicate the outline of buried archaeological features, such as walls, pits, or buildings; visible by aerial observation or photography.

crop out see OUTCROP.

cropping *Graphic Arts.* the process of marking off or covering an unwanted part of a photograph, usually along an edge, so that it does not appear when the photograph is printed.

cropping system *Agronomy.* a system of growing different crops in succession on the same land to conserve soil nutrients. Also, **cropping plan, cropping method.**

crop residue *Agronomy.* see RESIDUE.

crop rotation *Agronomy.* the practice of growing several different crops on the same land in successive years or seasons; done to replenish the soil, curb pests and diseases, and so on.

crop year *Agronomy.* the period of time from the planting to the harvesting of a crop.

cross *Genetics.* **1.** to crossbreed. **2.** an organism produced by crossbreeding. *Meteorology.* see SUN CROSS.

cross-addicted *Medicine.* simultaneously addicted to two or more substances.

cross antenna *Electromagnetism.* an antenna constructed of two straight horizontal antennas that bisect each other at right angles and are fed at their intersection.

cross assembler *Computer Programming.* an assembly program that runs on one computer system and assembles machine-code instructions for a different computer.

cross axle *Mechanical Engineering.* **1.** a driving axle connected to a crank or cranks at a 90° angle. **2.** a driving axle operated by levers at its ends.

crossband *Telecommunications.* a radio-telegraph network operational procedure in which calling stations contact other stations using one frequency, and then shift to another frequency to transmit their message. The called stations then respond using a third frequency.

crossbar *Building Engineering.* a horizontal bar that connects and extends across bearing bars at an angle of 90º; it is most readily used in grating applications.

crossbar micrometer *Engineering.* an instrument used to determine the positions of celestial objects, composed of two bars set perpendicular to each other in the focal plane of a telescope and tilted to the east-west path of stars by 45°.

crossbar switch *Electricity.* a switch having several vertical paths and several horizontal paths, operated electromagnetically to connect any one vertical path with any one horizontal path.

crossbar switching system *Telecommunications.* an automatic switching system that consists of crossbar switches, selecting mechanisms, and common control circuits that are used to select and test the switching paths and control the operation of the selecting mechanism. Switching information is received and stored by control mechanisms that determine the call type and treatment and then accordingly establish the appropriate circuit connection in a communication system.

crossbeam *Building Engineering.* a transverse beam in a building structure, usually as a joist in roofing applications.

cross bearing *Navigation.* bearings on two or more objects; when the bearings are plotted on a chart, the intersection gives the position at the time the bearings were taken.

cross-bedding

cross-bedding *Geology.* in a stratified rock, the internal arrangement of layers at various angles to the main depositional surface. Also, CROSS-LAMINATION, FALSE BEDDING, DIAGONAL BEDDING, CROSS-STRATIFICATION.

cross-belt drive *Mechanical Devices.* a belt drive forming a figure eight between two pulleys with connecting shafts that rotate in opposite directions.

crossbill *Vertebrate Zoology.* any of various finches of the genus *Loxia* of the family Fringillidae, characterized by crossed mandibles on the beak; native to the Northern Hemishere and American tropics.

crossbill

crossbolt *Mechanical Devices.* a locking mechanism with two bolts that move in opposite directions; used in door-frame lock hardware.

cross bond *Building Engineering.* in masonry work, a bond-brick orientation of alternating lengthwise and endwise bricks with a course of bricks laid lengthwise.

cross bracing *Building Engineering.* the buildup of a frame of a building in which two diagonal braces are nailed into each panel or stud of a truss and stabilize it to carry wind loads. Also, COUNTERBRACING.

cross-bred or **crossbred** *Agriculture.* **1.** of or relating to an animal or plant produced by parents that are of two different breeds or varieties. **2.** of or relating to an animal having two purebred parents, but of different breeds. **3.** an animal or plant produced by crossbreeding.

crossbreed *Biology.* **1.** to cross two distinctive varieties of a species to produce a new hybrid form. **2.** a hybrid organism.

cross-breeding *Genetics.* the interbreeding, either in the laboratory or in the wild, of two relatively inbred strains. *Agriculture.* the breeding or pollinating of animals or plants that belong to different species, races, breeds, or varieties, in order to combine their best characteristics in the resulting offspring. Also, **crossbreeding.**

cross-bridging *Building Engineering.* a bridge composed of criss-crossed pieces of wood for added strength. Also, HERRINGBONE BRIDGING, HERRINGBONE STRUTTING.

cross cap *Mathematics.* a nonorientable surface formed by a Möbius strip whose boundary has been deformed into a circle in such a way that the Möbius band must intersect itself. The curve of intersection is regarded as two different curves, each belonging to only one of the two parts of the surface which cross along the curve. A cross cap occurs, for example, when the boundary of a Möbius strip is identified with the boundary of a hole in a 2-manifold.

cross-color *Electronics.* the interference in the chrominance channel in a color television receiver caused by crosstalk from monochrome signals.

cross compiler *Computer Programming.* a compiler that runs on one computer and produces an object program that will run on a different computer.

cross-correlation *Acoustics.* a function that describes the dependence of the values of a function (or data) at two points, x and y, at one time on the values at the two points at another time.

cross-correlation detection see CORRELATION DETECTION.

cross-correlator *Electronics.* a circuit to detect the presence of a weak signal whose waveform is known, by integrating the product of the incoming noisy signal and a waveform of expected shape.

cross-country mill *Metallurgy.* a rolling mill with tables parallel to a connecting, crossover table that is used to transfer the material being rolled.

cross-coupling *Aviation.* a linking of two motions such as yaw and roll, particularly when the motions are abnormal or not as expected. *Telecommunications.* **1.** the coupling of a signal from one channel, circuit, or conductor to another where it becomes an unwanted signal. **2.** an unwanted transfer of interfering power from one circuit to another.

cross-cousin *Anthropology.* the child of a sibling who is the opposite sex, such as mother's brother's child or father's sister's child.

cross-cultivation *Agronomy.* a method of plowing whereby two plowings of the same area are made at right angles to each other.

cross-cultural correlates *Anthropology.* equivalent elements that are universal in several cultures, such as the customs of child-rearing that deal with basic human growth and needs; e.g., breast-feeding, toilet-training, language, and self-control.

cross-cultural research *Anthropology.* an ethnographic study in which a cultural element is examined across several societies to determine the sources of variation among cultures.

crosscurrent *Fluid Mechanics.* a flow orthogonal to the mean direction of a jet or current.

crosscurrent extraction *Analytical Chemistry.* a multistage extraction technique using repeated contacts of the feed solution with fresh solvent, resulting in a series of extract streams of diminishing concentration.

cross curve see CRUCIFORM CURVE.

crosscut *Engineering.* a cut made in wood that is perpendicular to the grain. *Mining Engineering.* **1.** a small passageway at right angles to the main entry of a mine, connecting it with a parallel entry or air course. **2.** a passageway cutting across the geological structure of a mine.

crosscut chisel *Mechanical Devices.* a narrow chipping or cold chisel used to cut grooves in heavy materials across a planed surface.

crosscut file *Mechanical Devices.* a file with a rounded edge at one end and a thin edge at the other end for sharpening straight-sided sawing teeth with round gullets.

crosscut saw *Mechanical Devices.* a saw with usually eight teeth per inch oriented in an equilateral triangle, used to cut across the grain of lumber.

crosscutting relationship *Geology.* the principle that a body of rock cutting across another rock is younger than the one it cuts across.

cross-dating *Archaeology.* the practice of dating artifacts according to their association with other items having an established date, such as dated coins or buildings, or ceramics of known manufacture.

cross-dresser see TRANSVESTITE.

cross-dressing see TRANSVESTISM.

cross drum boiler *Mechanical Engineering.* a boiler in which the axis of the horizontal drum is perpendicular to the axis of the main bank of tubes.

cross dyeing *Textiles.* the dyeing of a fabric containing two or more fibers that react differently to the dyeing process.

crossed belt *Mechanical Engineering.* a driving belt that is twisted between the driving and driven pulleys so that they operate in opposite directions.

crossed electrophoresis *Immunology.* a test used to determine the components of antigens, in which a seriological procedure separates the antigens according to an electrical field and the antigens are then electrophoresed into an antibody-containing gel at right angles to the first separation.

crossed-field amplifier *Electronics.* a high-power amplifier operating in the microwave spectrum and utilizing strong perpendicular magnetic and electric fields that interact with the electron beam emitted from the cathode to produce the power output.

crossed-field device *Electronics.* a high-vacuum electron tube in which voltage is applied to produce an electric field that is perpendicular both to a magnetic field and to the direction of propagation of a radio-frequency delay line.

crossed-field multiplier phototube *Electronics.* a sensitive optical detector in which strong, perpendicular electric and magnetic fields are used to produce output from a multiplier phototube.

crossed grid *Radiology.* a set of two radiographic grids of parallel lead strips arranged at 90° angles to each other.

crossed laterality *Neurology.* in voluntary motor acts, the preferential use of contralateral members of different pairs of organs, such as the right hand and the left foot.

crossed lens *Optics.* a lens that employs curvature radii to correct spherical aberration in incident parallel rays.

crossed-needle meter *Engineering.* a device that contains two analog meters with pointers centered at different positions, so that when they cross it displays the value of a few functions of each reading.

crossed paralysis *Medicine.* a paralysis that affects one side of the face and the opposite side of the body.

crossed prisms *Optics.* a pair of Nicol prisms that are positioned with their principal axes at right angles to each other, so that light transmitted through one is obstructed by the other.

crossed reactivity *Immunology.* the reaction of antisera with different antigens due to some shared antigenic determinants.

cross-eye *Medicine.* the popular term for a condition in which one eye turns inward toward the other (esophoria), or both eyes turn inward at the same time (esotropia). Thus, **cross-eyed.**

crossfade or **crossfading** *Acoustical Engineering.* the gradual mixing from one sound source or group of sources into another source or group of sources.

cross fault *Geology.* **1.** a minor fault that crosses a major fault. **2.** a fault whose strike is perpendicular to the general trend of the region or the associated strata.

cross-fertilization *Botany.* the fertilization of the egg of one individual with the pollen of another. *Zoology.* **1.** fertilization occurring between different species or varieties in which hybrids may occur. **2.** the mutual exchange of sperm between two hermaphrodites, resulting in fertilization.

cross fire or **crossfire** *Military Science.* the lines of gunfire from two or more positions that cross one another. *Telecommunications.* an interfering current in one signaling channel, resulting from signal currents in another channel transmitting from the same end at which the interference is measured.

cross-flight photography *Photogrammetry.* single photographic strips having stereoscopic overlap between exposures and a flight direction at right angles to that of coexistent area-coverage photography.

cross-flow *Aviation.* **1.** a flow passing over or through another flow, such as a spanwise airflow over a wing. **2.** of or relating to axial and cross-flow components for analysis purposes.

cross-flow plane *Aviation.* a plane that is perpendicular to the freestream velocity.

cross flux *Electromagnetism.* in an electric motor or rotator, the perpendicular component of magnetic flux with respect to the flux produced by the field magnets.

cross fold *Geology.* a secondary fold that intersects a preexisting fold having a different orientation.

crossfoot *Computer Programming.* a method of data validation in which various rows or columns of numbers are summed and the totals compared.

cross forging *Metallurgy.* a preliminary forging operation, performed only at selected locations of the starting stock.

cross fracture *Geology.* a small joint structure that develops between fringe joints.

cross-furring ceiling *Building Engineering.* the attachment of furring members perpendicularly to beam runners or other structural members in a ceiling.

crosshair *Optics.* a straight line that establishes the line of sight in a surveying telescope or other sighting instrument; usually one of a pair of such lines intersecting at the center of the field of vision.

crosshatch generator *Electronics.* a signal generator that produces a test pattern consisting of a grid of horizontal and vertical lines, for testing and adjusting color television receivers.

crosshaul *Mechanical Engineering.* a loading device that consists of a power-driven chain connected on opposite sides of a vehicle; it is looped around an object and rolls the object into the vehicle.

crosshead *Mechanical Engineering.* a reciprocating block that forms a junction between the piston-rod and connecting-rod of an engine; it usually slides between guides.

cross heading *Mining Engineering.* a mine passage driven for ventilation from the airway to the gangway, or from one breast through the pillar to the adjoining working. Also, **cross gateway, cross hole**.

cross hybridization *Molecular Biology.* the hybridization of a probe to a nucleotide sequence that is less than 100% complementary.

crossing angle see ANGLE OF CUT.

crossing file *Mechanical Devices.* a taper file consisting of two half-round files placed back to back.

crossing-over *Genetics.* the exchange of genetic material between homologous chromosomes during meiosis, resulting in new combinations of genes.

crossing symmetry *Particle Physics.* a statement that the amplitude associated with the process of the creation of a particle with four-momentum P_m is the same as that for the process of destruction of an antiparticle with four-momentum $-P_m$.

cross joint *Geology.* a joint whose strike lies perpendicular to the lineation of the rock bedding. Also, TRANSVERSE JOINT.

cross-lamination see CROSS-BEDDING.

cross-level *Engineering.* to level at a 90° angle in relation to the principle line of sight.

crossline screen *Graphic Arts.* a glass halftone screen covered with a grid of opaque lines.

cross-link density *Materials Science.* the density or extensiveness of chemical links that form between two polymers or between polymers and other substances.

cross-linking *Organic Chemistry.* the attachment of chains of a polymer to one another to make the polymer into a single network with increased strength and resistance to solvents.

cross-magnetizing effect *Electromagnetism.* a distortion imposed on the magnetic flux across the air gap of an electric motor due to the armature reaction.

crossmarks *Graphic Arts.* marks that are placed on copy to serve as guides for proper register during composition or printing.

cross-matching *Immunology.* a test used prior to a blood transfusion to determine donor-recipient compatibility, in which red blood cells and sera from two different individuals are mixed together to determine if the sera contain antibodies that will react with the red blood cells of the other individual.

cross modulation *Telecommunications.* **1.** intermodulation due to the modulation of a carrier of a desired signal by an undesired signal. **2.** interference that occurs when a carrier signal is modulated by an interfering and unwanted signal, as well as being modulated by its desired signal. **3.** a form of signal distortion due to modulation of the carrier of the desired signal by an unwanted wave.

cross multiplication *Mathematics.* given the equation $a/b = c/d$, the equivalent $ad = cb$, where a, b, c, and d may be any quantity or object, as long as b and d are nonzero.

cross-neutralization *Electronics.* a method of reducing distortion in a push-pull amplifier, in which each half of the amplifier applies negative feedback signals to the other half through neutralizing capacitors.

Crossocarpaceae *Botany.* a family of red algae of the order Cryptonemiales, characterized by multiaxial laminate or leaflike thalli, with gametangial and sporangial thalli similar in form and morphology.

crossopterygian [krä´säp tə rij´ē ən] *Vertebrate Zoology.* **1.** a member of the order Crossopterygii. **2.** of or relating to this order.

Crossopterygii *Vertebrate Zoology.* an order of bony, lobed-fin fishes, regarded as ancestral to the amphibians. It is composed of three suborders: Rhipidistia, Onychodontiformes, and Coelacanthiformes; the first two are extinct, but the latter is represented by the recently discovered coelocanth, *Latameria chalumnae.* Until this discovery the order, which arose in the Devonian, was thought to have been extinct since the end of the Cretaceous.

Crossosomataceae *Botany.* a North American family of shrubs of the order Dilleniales, having small, simple leaves and perigynous flowers.

crossover *Electricity.* a point at which two insulated conductors cross. *Genetics.* the end result of the exchange of genetic material between homologous chromosomes during meiosis.

crossover distortion *Electronics.* signal distortion occurring in amplifiers as input signals cross over, or pass through, their zero references.

crossover experiment *Medicine.* an experimental design composed of different treatments in which each subject or group of subjects undergoes each treatment.

crossover fixation *Genetics.* a possible consequence of an unequal crossing-over that allows a mutation in one member of a tandem cluster to spread through the whole cluster, or to be eliminated.

crossover flange *Engineering.* a projecting rim or a pipe that connects other such rims of various working pressures.

crossover frequency *Acoustical Engineering.* **1.** the frequency of a crossover network at which equal amounts of energy are supplied to the woofers and to the tweeters. **2.** see TRANSITION FREQUENCY.

crossover joint *Petroleum Engineering.* a casing length that has different threads at either end to allow changes from one thread type to another in a casing string.

crossover network *Acoustical Engineering.* an electroacoustical network designed to supply high-frequency signals to tweeters and low-frequency signals to woofers.

crossover point *Biochemistry.* the point in a multienzyme system at which the intermediates of a metabolic sequence change concentrations.

crossover region *Genetics.* a chromosomal area in which a crossover occurs.

crossover spiral see LEAD-OVER GROOVE.

crossover value *Genetics.* a numerical representation of the degree of linkage between two loci on the same chromosome, obtained by comparing the number of offspring in which meiotic crossover has occurred to the total number of offspring resulting from a specific cross.

crossover voltage *Electronics.* the voltage of a secondary-emitting surface in a charge-storage tube at which the secondary emission is unity; the crossovers are numbered in progression with increasing voltage.

cross-peen hammer *Mechanical Devices.* a hammer with a wedge-shaped edge at one end of its head running crosswise to the direction of the handle.

cross polarization *Electromagnetism.* in an electrically polarized system, a component of the electric field that is perpendicular to the desired polarization.

cross-pollination *Botany.* the transfer of pollen from one flower to the stigma of another.

cross product *Mathematics.* an anticommutative operation on the vectors of Euclidean 3-space, denoted by $a \times b = c$, where a, b, and c are vectors. The cross product is given by $a \times b = \|a\| \|b\| \sin \theta u$, where $\|\cdot\|$ denotes the norm or magnitude of a vector, θ is the angle from a to b, and u is a unit vector perpendicular to the plane of A and B such that a, b, and c form a right-handed system. The cross product distributes over vector addition, commutes with scalar multiplication, and is zero if, and only if, a and b are parallel or at least one is the null vector. $\|a \times b\|$ is the area of the parallelogram with sides a and b. Also, VECTOR MULTIPLICATION, VECTOR PRODUCT.

cross-protection *Virology.* the protection from a virus that a host has acquired through infection from a similar virus.

cross ratio *Mathematics.* let A, B, C, and D be four collinear points (or concurrent lines) in the projective plane whose parameters with respect to a given pair of base points (lines) on the line (point) they determine are (a, a'), (b, b'), (c, c'), and (d, d'), respectively. The cross ratio of A, B, C, and D is

$$R(AB, CD) = [(ac' - ca')(bd' - db')]/[(ad' - da')(bc' - cb')].$$

The cross ratio is independent of the parameterization and satisfies the following relationshisp: (a) $R(AB, CD) + R(DB, CA) = R(AB, CD) + R(AC, BD) = 1$; (b) $R(AB, CD) = R(BA, CD)^{-1}$; and (c) $R(AB, CD) = R(AB, DC)^{-1}$. The cross ratio of four concurrent lines equals the cross ratio of the four points in which the lines are intersected by an arbitrary line not passing through the point of concurrence of the lines.

cross-reacting material *Biochemistry.* a functionally inactive protein, produced by a mutant structural gene, that reacts with antibody to the normal protein.

cross-reference program *Computer Programming.* the part of a compiler or other software that produces cross-reference lists of variable names and labels or numbers of statements in which each variable name appears; usually used as program documentation and debugging aids.

cross-resistance *Toxicology.* the tolerance to a normally toxic substance that is acquired by exposure to a related substance, such as the resistance of the immune system to a pathogenic substance or of an insect population to an insecticide.

cross ripple mark *Geology.* a ripple mark produced by the interference of at least two sets of ripples, in which one set forms after or simultaneously with the other. Also, INTERFERENCE RIPPLE MARK, DIMPLED CURRENT MARK, COMPLEX RIPPLE MARK.

cross-rolling *Metallurgy.* the process of rolling a metallic plate or sheet by alternating between two perpendicular rolling directions.
cross sea *Oceanography.* a series of waves or swells that cross another series of waves at an angle.
cross section or **cross-section** *Graphic Arts.* a drawing depicting the components of an object and their relationships as if the object had been sliced open at a right angle to an axis. *Cartography.* a horizontal grid system laid out on the ground for determining contours, quantities of earthwork, and so on, by means of elevations of the grid points. *Geology.* a diagram or actual cut showing subsurface geologic features transected by a given vertical plane. *Mathematics.* **1.** the intersection of a geometric figure in Euclidean 3-space with a plane, usually perpendicular to some axis of symmetry. **2.** in general, the intersection of an n-dimensional geometric figure with a lower-dimensional hyperplane. *Nucleonics.* a measure that is proportional to the probability of a neutron reaction; some nuclides have cross sections that vary inversely to the velocity of the neutron.
cross-sectional data *Statistics.* data gathered on several individuals at the same time, as opposed to longitudinal data.
cross section per atom *Nuclear Physics.* the probability that a given nuclear reaction will occur when only one natural isotope is involved per unit area and one target is located somewhere within that unit area.
cross slide *Mechanical Engineering.* a part of a planing machine or lathe on which the toolholder is mounted so that it may be moved perpendicularly to the bed of the machine.
cross-slip *Materials Science.* the process by which a screw dislocation moving on one slip plane shifts to another intersecting slip system.
cross-slope furrow *Agronomy.* a furrow running along the contour of a slope.
cross spectrum *Physics.* the complex vector sum of the quadrature spectrum and the cospectrum.
cross-staff *Navigation.* an early navigational tool consisting of a wooden shaft along which a perpendicular cross-piece could be slid. The navigator sighted along one end of the shaft and slid the cross-piece until it exactly filled the space between the horizon and the body being observed. Marks on the shaft gave the altitude of the body.
cross-stratification *Geology.* **1.** a general term for any arrangement of layers deposited at an angle to the original dip of the formation. **2.** see CROSS-BEDDING.
crosstalk *Acoustical Engineering.* **1.** a phenomenon in which a signal on one channel of a sound system is processed by another channel, due to poor separation between the channels or improper pickup of the sound. **2.** see MAGNETIC PRINTING. *Telecommunications.* any unwanted transfer of energy from one circuit to another circuit.
crosstalk coupling *Telecommunications.* the ratio of power in one circuit to the power induced in another circuit observed at definite points in the circuit under specified terminaling conditions; usually expressed in decibels. Also, **crosstalk loss.**
crosstalk level *Telecommunications.* the effective power of a crosstalk signal, expressed in decibels below a reference level such as one milliwatt.
crosstalk unit *Telecommunications.* a measurement of the coupling of two circuits.
cross termination *Materials Science.* the tendency of monomer radicals to form paired-electron covalent bonds, thus ending the process of copolymerization.
cross-thread *Engineering.* to screw together two threaded pieces, such as a nut and a bolt, without aligning the threads properly, often with undesirable results.
crosstie *Engineering.* one of a series of timber or concrete beams that lie perpendicular to and beneath a railway track to provide support and gauge maintenance.
crosstip screwdriver *Mechanical Devices.* a screwdriver, often magnetic, with a split tip to grasp screws firmly and position them into the threads of a hole.
cross-tree *Naval Architecture.* horizontal timers at the head of the masts.
cross-tolerance *Medicine.* a decreased sensitivity to a second drug or poison that results from an acquired decrease in sensitivity to a first drug or poison.
cross valley see TRANSVERSE VALLEY, def. 1.
cross vein *Mining Engineering.* a vein that intersects an older, larger, or more productive vein.
cross-ventilation *Building Engineering.* an air flow that moves from one side of a room to another, as between two open windows.
crosswind or **cross wind** *Meteorology.* any wind with a component directed perpendicularly to the course of an exposed, moving object. *Aviation.* specifically, a wind that does not blow parallel to the runway or the course of an aircraft.
cross-wire weld *Metallurgy.* a weld performed at the intersection of crossed wires or bars.
Crotale *Ordnance.* a French surface-to-air missile designed for all-weather, close-range antiaircraft operations, powered by a solid-propellant motor and equipped with radio command guidance; delivers a 33-pound high-explosive warhead at a speed of Mach 2.3 and a range of 0.3 mile to 5.3 miles. There are land-based and shipborne versions.
Crotalidae *Vertebrate Zoology.* the rattlesnakes and pit vipers; a widely distributed family of venomous snakes in the suborder Serpentes, characterized by their paired erectile fangs and the pitlike depression between each nostril and eye.
crotalotoxin *Toxicology.* a poisonous substance found in the venom of the rattlesnake *Crotalus terrificus.*
crotonaldehyde *Organic Chemistry.* $CH_3CH{=}CHCHO$, a flammable liquid that is soluble in water; boils at 104°C and melts at –76.5°C; used in various chemical syntheses including the production of maleic acid, butyl alcohol, and resins.
crotonic acid *Organic Chemistry.* $CH_3CH{=}CHCOOH$, a white, crystalline solid, soluble in water and alcohol; melts at 72°C and boils at 185°C; used in the synthesis of resins, polymers, plasticizers, and drugs.
crotonism *Toxicology.* poisoning due to consumption of croton oil; symptoms may include severe gastrointestinal disturbances, headache, vertigo, and death from respiratory failure.
croton oil *Materials.* a brownish yellow liquid that is slightly soluble in alcohol and soluble in ether, derived from the seed of the plant species *Croton tiglium;* used as a purgative.
crotoxin *Toxicology.* a crystalline neurotoxin found in rattlesnake venom.
croup [kroop] *Medicine.* a continuous spasmodic laryngitis found especially in infants and causing periods of resonant barking cough, hoarseness, and difficult breathing.
croup-associated virus *Virology.* a parainfluenza virus that infects mainly young children and causes hoarseness, resonant cough, and respiratory problems.
croute calcaire see CALICHE.
Crout's algorithm *Mathematics.* a technique for the decomposition of an $n \times n$ matrix A into the form $A = LU$, where L is a lower triangular matrix and U is an upper triangular matrix. This method, sometimes called **Crout reduction** or ***LU* decomposition,** allows for easy numerical solutions of simultaneous linear equations.
crow *Vertebrate Zoology.* a common name for any of the large, generally glossy black birds of the family Corvidae, including the raven, carrion crow, rook, Cornish chough, and jackdaw.

crow

crowbar *Mechanical Devices.* an iron or steel bar with a generally flat profile but a bend in its shaft and a chisel-like point near its wedge-shaped head; it is used for leverage or lifting a heavy load a short distance. *Electricity.* the shunt used in a crowbar voltage protector.
crowbar voltage protector *Electricity.* a protective mechanism that quickly places a low-resistance shunt across the output terminals of a power supply whenever a preset voltage limit is exceeded.
Crowe process *Metallurgy.* an operation in gold cyaniding, following the gold separation, in which the tailings are filtered.

crown (bald eagle)

crown the uppermost part of something; specific uses include: *Anatomy.* **1.** the top of the head. **2.** the part of a tooth that is covered by enamel. **3.** the topmost part of any structure. *Medicine.* an artificial replacement for the crown of a tooth. *Botany.* **1.** the part of a plant where the root and stem meet. **2.** the upper branches forming the shape of a tree. *Civil Engineering.* **1.** the highest point of an arch or tunnel. **2.** the highest point of a cross section of a paved roadway. *Horology.* in a spring-driven timepiece, the button that extends beyond the case on a stem and is turned by hand to wind the mainspring. *Metallurgy.* **1.** in a sheet or roll, a gradual increase of thickness, or diameter, from the edges to the center. **2.** in metal fabrication, the upper part of a press, containing the working tool. *Mining Engineering.* a timber crossbar up to 16 feet long, supported at each end by two heavy legs. *Mechanical Devices.* **1.** a high spot formed on a tool joint shoulder when a drill pipe becomes unstable. **2.** the section of a drill bit inset with diamonds.

crown block *Petroleum Engineering.* a stationary pulley system joined to the top of a derrick for supporting as well as lowering and raising the drilling tools; the sheaves and supporting members are attached to the lines of the traveling block and hook.

crown canopy or **crown cover** see CANOPY.

crown-cut see BACKSAWING.

crown ether *Organic Chemistry.* a cyclic ether containing several oxygen atoms, so named because its molecular model resembles a crown.

crown fire *Forestry.* a forest fire that spreads, usually at great speeds, along the upper branches (crown) of trees and shrubs.

crown gall *Plant Pathology.* a bacterial disease of various plants caused by *Bacterium tumefaciens* and characterized by tumorous enlargements on the roots or the stem near the root crown.

crown glass *Materials.* **1.** an optical glass that has low dispersion and a low index of refraction. **2.** a soda-lime-silica glass.

crown grafting *Botany.* a method of plant propagation in which the scion is grafted directly onto the root stock.

crowning *Medicine.* the phase in the second stage of labor when the fetal head is visible and the largest diameter of the head is encircled by the stretched vulva.

crown platform *Petroleum Engineering.* a platform at the top of a derrick that allows access to the sheaves of the crown block and also creates a safe place for working with the gin pole.

crown rot *Plant Pathology.* any plant disease caused by fungi or nutritional deficiency, in which the part of the plant at or near ground level deteriorates.

crown rust *Plant Pathology.* a disease of oats and other grasses caused by the rust fungus *Puccinia coronata* and characterized by rounded light-orange fungal growths on the leaves.

crown saw *Mechanical Devices.* a rotary saw with a hollow cylinder formed with teeth around its edge, used for cutting circular or cylindrically shaped holes.

crown sheet *Mechanical Engineering.* the plate that forms the top of the furnace in a fire-tube boiler.

crown wheel *Horology.* **1.** a wheel in the winding mechanism of a spring-driven timepiece that is engaged by the winding pinion and in turn engages the clickwheel on the barrel to wind the mainspring; the crown wheel may be composed of two wheels functioning as a unit, the lower crown wheel engaging the winding pinion and the upper crown wheel engaging the clickwheel. Also, WINDING WHEEL. **2.** any wheel in a timepiece with teeth set perpendicular to its plane of rotation, such as the escape wheel in a verge escapement.

crow's nest *Naval Architecture.* in early sailing ships, a platform or structure at the top of the mainmast, used as a lookout post. *Engineering.* any similar elevated structure used for observation.

CRP *Aviation.* the airport code for Corpus Christi International.

CRP C-reactive protein.

CRT *Computer Technology.* **1.** short for cathode-ray tube. **2.** a computer monitor or other display device employing a cathode-ray tube.

Cru see CRUX.

cruciate [kroo´shē āt´] *Anatomy.* resembling a cross; cross-shaped. *Botany.* having petals or leaves arranged in the form of a cross.

crucible [kroo´sə bəl] *Metallurgy.* a refractory container used to melt metals and alloys.

crucible melt extraction *Metallurgy.* one of several rapid solidification processes in which the product is extracted from molten metal contained in a crucible.

crucible steel *Materials.* a hard steel that is made by melting wrought iron in a crucible with charcoal and ferromanganese to separate the slag and oxides, reducing the sulfur and phosphorus content.

crucifer [kroo´sə fər] *Botany.* any member of the family Cruciferae, such as the cabbage, cauliflower, turnip, and wallflower.

Cruciferae *Botany.* a large family of dicotyledonous herbs characterized by alternate leaves and by four-petaled flowers having petals arranged in the form of a cross. Also, MUSTARD FAMILY.

cruciferous [kroo sif´ə rəs] *Botany.* of, relating to, or resembling a crucifer or the family Cruciferae.

cruciform [kroo´sə fôrm´] *Science.* shaped like a cross. *Aviation.* of two surfaces, crossing one another so as to form a symmetrical X or +, as in a cruciform tail or cruciform empennage.

cruciform core *Electricity.* a transformer core in which all windings are on the center leg; four additional legs serve as return paths for the magnetic flux.

cruciform curve *Mathematics.* a plane curve that is the locus of the equation $x^2y^2 - a^2(x^2 + y^2) = 0$ in Cartesian coordinates, and that thus forms a curve that is symmetric about the origin, with four branches, one in each quadrant. The four lines $x = \pm a$, $y = \pm a$ are asymptotes. (So called because it has the shape of a cross.)

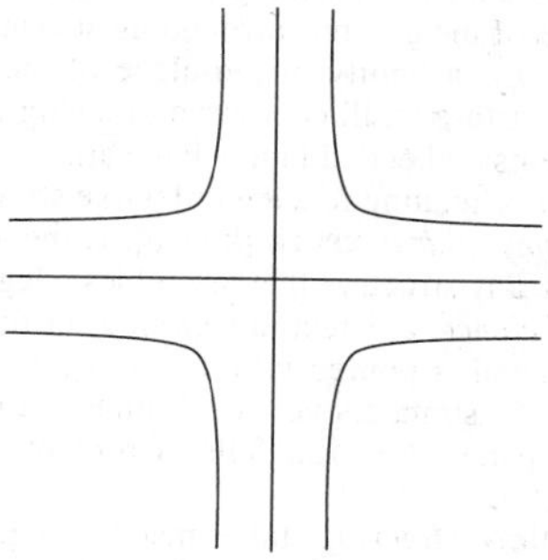

cruciform curve

cruciform structure *Molecular Biology.* a cross-shaped configuration of DNA produced by complementary inverted repeats pairing with one another on the same DNA strand, rather than normal pairing with a different strand.

crude assay *Chemical Engineering.* a method to determine the general distillation characteristics and other information concerning the quality of a crude oil.

crude cumulative gas *Petroleum Engineering.* the measurement of total gas production from a reservoir, often shown graphically in parallel relationship to the total or cumulative oil yielded from the same reservoir.

crude desalting *Chemical Engineering.* the washing of crude oil with water in one or a series of vessels, to remove materials such as dirt, silt, and water-soluble minerals.

crude drug *Pharmacology.* any raw or unrefined drug in its natural form, especially one taken directly from a plant source.

crude fiber *Nutrition.* the indigestible parts of food that serve no nutritive purpose, but do have value in the diet.
crude oil *Geology.* petroleum in its natural liquid state as it emerges from the earth, before refining.
crude ore *Mining Engineering.* unconcentrated ore as it leaves the mine.
crude protein *Nutrition.* incomplete proteins lacking essential amino acids.
crude still *Chemical Engineering.* the distillation equipment used to separate crude oil into various fractions and products.
crufomate *Organic Chemistry.* $C_{12}H_{19}ClNO_3P$, crystals that are insoluble in water and soluble in acetone; melts at 61°C and boils at 117°C (0.01 torr); used as an insecticide.
cruise *Navigation.* to travel at cruising speed. See CRUISING SPEED.
cruise control *Mechanical Engineering.* a control in an automobile that allows for travel at a single constant speed by automatically adjusting the accelerator to maintain that speed. *Aviation.* the operation of an aircraft with the object of obtaining the greatest practical efficiency, usually at cruising speed.
cruise missile *Aviation.* a long-range pilotless missile launched from a fixed or mobile platform and designed to fly a given course as dictated by the tactics of the situation or environment; flight is usually at a relatively low altitude to avoid detection by radar.
cruiser *Naval Architecture.* **1.** a large surface combat vessel combining relatively powerful armament with high speed and long range, but usually with little protective armor. **2.** in earlier use, a warship designed to raid enemy shipping, or to pursue such raiders.
cruising altitude *Aviation.* **1.** the altitude at which a given aircraft flies most efficiently over an extended distance, in terms of such factors as fuel economy in relation to speed; it is less than the highest altitude attainable by the aircraft. **2.** any constant altitude that is maintained for an extended period during level flight.
cruising range *Aviation.* the maximum distance from its base that an aircraft can safely fly and return, starting with full fuel tanks and returning with a specified amount of fuel remaining as a safety factor. Also, **cruising radius.**
cruising speed *Navigation.* a constant speed that is the most efficient for sustained travel for the craft or vehicle in question, based on such factors as fuel economy, safety, and passenger comfort; it will always be less than the maximum speed possible.
crumb *Geology.* a soft, porous, and rounded particle of soil, 1 to 5 millimeters in diameter.
crump *Geology.* a sudden movement of the earth that results from the failure under stress of the ground surrounding subsurface workings.
Cruoriaceae *Botany.* a family of red algae of the order Gigartinales, characterized by crustose thalli composed of a single layer of cells from which erect filaments embedded in mucilage arise.
crural *Anatomy.* of or relating to a leg or leglike structure.
crus [kroos] *Anatomy. plural,* **crura** [kru̇r´ə]. **1.** the lower limb, from the knee to the foot. **2.** any structure that resembles a leg.
crush *Materials Science.* a defect in a casting resulting from a defect in the mold before metal is poured into it. *Mining Engineering.* **1.** a general settlement of the strata above a coal mine due to the breakdown of coal, often accompanied by local falls of roof in mine workings. **2.** a species of fault in coal.
crushable ceramics *Materials.* tubes made of high-purity magnesia or alumina that are used primarily to insulate metal-sheathed thermocouples and heating elements.
crush breccia *Geology.* a breccia that is formed in place by mechanical fragmentation resulting from movements in the earth's crust.
crush conglomerate *Geology.* a rock made up of rounded particles, formed in place by the deformation of brittle, closely jointed rocks.
crushed stone *Materials.* commercial stone, usually granite, limestone, or fine-grained igneous rock; used especially for roads, concrete making, and railway ballast.
crusher *Mechanical Engineering.* any of a wide variety of machines used to crush rock or other materials; such machines are generally used in the early stages of size reduction.
crush fold *Geology.* a large fold presumed to be formed by movements of the earth that would create a mountain chain or an oceanic deep.
crush-forming *Engineering.* a process in which a metal roll is rotated into the face of a grinding wheel and forces it into a shape.
crushing *Mining Engineering.* **1.** the process of reducing ore or quartz by stamps, crushers, or rolls. **2.** the quantity of ore pulverized or crushed in a single operation.
crushing strain *Mechanics.* compressive strain that leads to the failure of a material.
crushing strength *Mechanics.* the compressive stress required to fracture a material.
crushing test *Engineering.* any test that applies extreme pressure to stone to determine if it is strong enough to be used for roads or building processes. *Metallurgy.* a standardized test that assesses the maximum compressive lead born by a component without failure.
crush syndrome *Medicine.* the collection of symptoms, including edema, that accompany renal failure and are brought on by the crushing of a body mass, usually a large muscle.
crush zone *Geology.* an area of crushed, ground rock material and crush breccia along a fault.
crust *Geology.* the outermost layer of the earth, lying above the Mohorovicic discontinuity and composed of sial (continental crust), or sial and sima (oceanic crust). *Hydrology.* **1.** see ICE RIND. **2.** see SNOW CRUST.
Crustacea *Invertebrate Zoology.* a class of arthropods in the subphylum Mandibulata, with two pairs of antennae, a pair of mandibles, and often many other appendages modified for various purposes. They are mostly aquatic, but a few are terrestrial.
crustacean [krus tā´shən] *Invertebrate Zoology.* **1.** any of the various members of the class Crustacea; included are such edible aquatic forms as lobsters, crabs, shrimp, and crayfish, as well as barnacles, wood lice, and water fleas. **2.** of or relating to this class.

crustacean

crustal abundance *Geochemistry.* the relative average content or mean concentration of an element in the earth's crust, including the atmosphere and the oceans.
crustal plate *Geology.* a large tabular section of the lithosphere that reacts as a unit to tectonic forces.
crustecdysone *Biochemistry.* a hormone that causes crustaceans to molt.
crustose *Botany.* having or forming a crust; commonly used to describe certain lichens having a crustlike thallus.
crutch *Medicine.* any of various devices that are used to aid in walking, typically a wooden or metal staff with a hand grip at waist height and a padded crosspiece that fits under the armpit. *Horology.* in a pendulum clock, a forked lever mounted on the pallet arbor and connecting the pallets to the pendulum rod.
Crux [kruks; kru̇ks] *Astronomy.* the Southern Cross, a small but bright and distinct constellation of the Southern Hemisphere.
Crv see CORVUS.
CRW *Aviation.* the airport code for Charleston, West Virginia.
cry- a combining form meaning "cold" or "freezing," as in *cryesthesia.*
cryalgesia *Neurology.* pain induced by the application of cold.
cryesthesia *Neurology.* abnormal sensitivity to cold.
cryo- a combining form meaning "cold" or "freezing," as in *cryogenic.*
cryoablation [krī´ō ə blā´shən] *Surgery.* the removal of a part, such as a wart, by freezing.
cryobiology [krī´ō bī äl´ə jē] *Biology.* the study of the effects of low temperatures on living organisms, especially extremely low temperatures. Thus, **cryobiologist, cryobiological.**
cryobiosis [krī´ō bīō´sis] *Physiology.* the carrying on of bodily functions at low temperatures.
cryocautery [krī´ō kôt´ə rē] *Surgery.* the destruction of tissue by application of extreme cold. Also, COLD CAUTERY.

cryochem process *Chemical Engineering.* a freeze-drying procedure that involves conduction heat transfer to the frozen solid secured on a metallic surface.

cryoconite *Geology.* a dark, finely textured dust powder carried by wind and deposited on a snow or ice surface. Also, KRYOKONITE. *Mineralogy.* a mixture of garnet, sillimanite, zircon, pyroxene, and quartz.

cryoconite hole *Geology.* a cylindrical depression containing cryoconite that absorbs solar radiation, thus causing the melting of the neighboring glacier ice.

cryoelectronics *Electronics.* a field of engineering that studies the design and functioning of electronics systems, circuits, and devices at temperatures approaching absolute zero (0 K or –270°C); particularly as applied to the phenomenon of superconductivity. Also, CRYOTRONICS.

cryofixation *Microbiology.* the fixation of cells for electron microscopic examination by freezing.

cryogen [krī´ō jən] *Chemistry.* **1.** any substance that is used to induce freezing. **2.** see CRYOGENIC FLUID.

cryogenic [krī´ō jen´ik] *Physics.* **1.** of or relating to cryogenics. **2.** relating to or occurring at extremely low temperatures.

cryogenic coil *Physics.* a coil cooled to extremely low temperatures so as to reduce its electrical resistivity.

cryogenic conductor see SUPERCONDUCTOR.

cryogenic device *Physics.* any device operating at extremely low temperatures.

cryogenic engineering *Engineering.* a branch of engineering that deals with processes that are performed at very low temperatures.

cryogenic film *Computer Technology.* a coating of superconducting material that could maintain or store a current indefinitely when kept at a temperature near absolute zero.

cryogenic fluid *Chemistry.* any fluid that changes from a liquid to a gas at less than 110 K at standard pressure. Also, CRYOGEN, CRYOGENIC LIQUID.

cryogenic freezing *Food Technology.* a method of freezing food products at extremely low temperatures by immersing them in a cryogenic gas such as liquid nitrogen or liquid carbon dioxide.

cryogenic gyroscope *Engineering.* a gyroscope that operates under very low temperatures, having a central rotating disk of superconducting niobium which spins while in levitation. Also, SUPERCONDUCTING GYROSCOPE.

cryogenic liquid see CRYOGENIC FLUID.

cryogenic period *Geology.* a period of geologic history during which large bodies of ice formed at or near the poles and the climate supported the growth of continental glaciers.

cryogenic pump *Physics.* a pump capable of producing a very low vacuum (with pressures of about 10^{-8} mm Hg) through the cooling of surfaces to about 20 K. Also, CRYOPUMP.

cryogenics *Physics.* a field of physics devoted to the study of processes that occur at extremely low temperatures.

cryogenic spectroscopy *Spectroscopy.* spectroscopic analysis carried out on a sample that is maintained at cryogenic temperatures (generally liquid nitrogen and below).

cryogenic steel *Materials.* an alloy steel used for low-temperature applications such as liquid oxygen tanks. Also, LOW-TEMPERATURE STEEL.

cryogenic temperature *Physics.* a temperature typical of cryogenic fluids; usually 77 K or lower.

cryogenic transformer *Electronics.* a superconducting transformer, especially designed to operate at temperatures approaching absolute zero.

cryoglobulin *Immunology.* aberrant plasma proteins, seen in patients with multiple myeloma, that precipitate, gel, or crystallize at low temperatures (4.4–21°C) and redissolve upon reaching body temperature.

cryohydrate *Chemistry.* **1.** a salt that contains water of crystallization at low temperatures. Also, CRYOSEL. **2.** a crystal that is produced by freezing a supersaturated solution and that has the same proportions of solute and solvent as the solution. **3.** a mixture of ice and a salt combined in a proportion designed to have the lowest possible melting point.

cryohydric point *Physical Chemistry.* the lowest point at which salt water freezes.

cryolaccolith see HYDROLACCOLITH.

cryolite *Mineralogy.* Na_3AlF_6, a white, brown, or red transparent to translucent, brittle monoclinic mineral occurring in massive form and as short prismatic or cuboidal, commonly twinned crystals, having a specific gravity of 2.5 and a hardness of 2.97 on the Mohs scale; it is found as small masses in quartz-feldspar pegmatite and veins with zircon crystals.

cryolithionite *Mineralogy.* $Na_3Li_3Al_2F_{12}$, a colorless or white transparent brittle cubic mineral occurring as dodecahedral crystals, having a specific gravity of 2.77 and a hardness of 2.5 to 3 on the Mohs scale; found with cryolite in pegmatite.

cryolithology *Hydrology.* the study of the nature, structure, and development of underground ice, especially in permafrost regions.

cryology *Hydrology.* the scientific study of snow and ice. *Mechanical Engineering.* the study of refrigeration at low temperatures ranging down to absolute zero.

cryomagnetic *Physics.* relating to the production of extremely low temperatures through the adiabatic demagnetization of paramagnetic salts.

cryometer *Engineering.* a thermometer used specifically for the measurement of low temperatures.

cryomorphology *Geology.* a branch of geomorphology that involves the study of the processes and features of cold climates.

cryonics [krī än´iks] *Medicine.* any of various techniques that involve the application or use of low temperatures for therapeutic purposes, such as to temporarily anesthetize a surface injury. Thus, **cryonic.**

cryopedology *Geology.* a branch of geology that involves the study of frost action and the occurrence of frozen ground.

cryophilous *Ecology.* thriving at low temperatures. Thus, **cryophile, cryophily.** Also, **cryophilic.**

cryophysics *Physics.* a branch of physics that is concerned with processes and phenomena at temperatures approaching absolute zero.

cryophyte *Ecology.* an organism that lives or thrives at low temperatures. *Botany.* specifically, a plant that grows on snow or ice.

cryoplanation *Geology.* land erosion that results from processes associated with intensive frost action.

cryoprecipitate *Biochemistry.* the solid matter that precipitates when a biological substance is cooled to low temperatures.

cryopreservation *Biotechnology.* the process of preventing materials from spoiling or deteriorating by maintaining them at extremely low temperatures.

cryoprobe *Surgery.* an instrument used to apply extreme cold to tissue.

cryoprotective *Hematology.* capable of protecting against injury due to freezing, as glycerol protects frozen red blood cells.

cryoprotein *Hematology.* any blood protein that precipitates on cooling, such as cryoglobulin or cryofibrinogen.

cryopump see CRYOGENIC PUMP.

cryoresistive transmission line *Electricity.* an electric power transmission line whose conducting cables are cooled to the temperature of liquid nitrogen, 77 K (–196°C), thereby causing a reduction of the conductor resistance by a factor of about 10, allowing increased transmission capability.

cryosar *Electronics.* a cryogenic semiconductor device designed for very high speed computer switching and memory applications.

cryosel see CRYOHYDRATE, def. 1.

cryoscope *Engineering.* an instrument that determines the point at which a liquid freezes.

cryoscopic constant *Analytical Chemistry.* a proportionality constant used to calculate freezing point depression, which is dependent on concentration.

cryoscopy *Analytical Chemistry.* a phase equilibrium technique used to determine the molecular weight of a solute by dissolving it in a solvent and measuring the freezing point.

cryosistor *Electronics.* a cryogenic semiconductor device that, after ionization, can act as a three-terminal switch, a pulse amplifier, an oscillator, or a unipolar transistor.

cryosorption pump *Mechanical Engineering.* a high-vacuum pump that uses an adsorbent such as activated charcoal cooled by liquid nitrogen to reduce pressure in the enclosed atmosphere.

cryosphere *Geology.* the region of the earth's surface that is frozen throughout the year.

cryostat *Engineering.* a thermostat that maintains low temperatures in an environment.

cryosurgery *Surgery.* the destruction of tissue by the application of extreme cold, as in the treatment of various skin disorders such as warts. Thus, **cryosurgical.**

cryotherapy *Medicine.* the therapeutic use of cold or refrigeration.

cryotron *Electronics.* a superconductive device in which current in one or more input circuits magnetically controls the superconducting-to-normal transition in one or more output circuits, provided that the current in each output circuit is less than its critical value.

cryotronics see CRYOELECTRONICS.

cryoturbation *Geology.* disturbance of the soil surface by the action of freezing or of alternate freezing and thawing.

Cryphaeaceae *Botany.* a tropical and subtropical family of mostly dull mosses of the order Isobryales that form loose mats on trees and rocks; characterized by creeping stems with numerous short lateral branches that sometimes form pendulous sheets, by lateral sporophytes, and by a single costa ending below the apex.

crypsis *Ecology.* the ability of an organism to camouflage and conceal itself through its natural appearance.

crypt *Anatomy.* a blind pit or tube that opens onto a free surface.

crypt- a combining form meaning "secret" or "hidden," as in *cryptanalysis.*

cryptanalysis [krip´tə nal´ə sis] *Linguistics.* an analysis of coded or secret text, including the steps, operations, and procedures required to convert an encrypted message into plain text without prior knowledge of the key employed in the encryption process.

cryptand *Organic Chemistry.* a bicyclic or cyclic compound of higher order that surrounds and shields an ion in the center of a molecule.

cryptate *Organic Chemistry.* a complex molecule consisting of a central ion surrounded and shielded by a cyclic compound.

Crypteroniaceae *Botany.* a monogeneric family of tropical dicotyledonous trees in the order Myrtales, often accumulating aluminum and characterized by opposite, simple leaves and numerous tiny flowers borne in axillary racemes, spikes, or panicles.

Crypthecodiniaceae *Botany.* a family of marine flagellates of the order Peridiniales, nonphotosynthetic saprophytes that live on rotting large brown algae; only one species is currently known, but evidence points to as yet undescribed species.

cryptic [krip´tik] *Zoology.* relating to or describing the ability of an organism to conceal itself by means of coloration and markings that resemble the surrounding environment.

cryptic coloration *Zoology.* protective coloration that allows an organism to blend in with its environment and thus remain concealed from predators or prey; often occurring in insects, including such unusual adaptations as the ability of stick insects to match their color to a changing background by moving pigment granules in their epidermal cells.

cryptic mutant *Cell Biology.* a cell possessing a mutation that cannot be detected phenotypically.

cryptic plasmid *Molecular Biology.* a plasmid having no apparent effect on the phenotype of the cell where it occurs.

cryptic satellite *Genetics.* a satellite DNA sequence not identified as such by a separate peak on a density gradient; that is, it remains present in main-band DNA.

cryptic species *Systematics.* a species so similar to another that the two are difficult to distinguish from each other. Also, SIBLING SPECIES.

cryptic virus *Virology.* any inapparent virus, especially one of the Cryptovirus group, which cause no symptoms in plants.

crypto- a combining form meaning "secret" or "hidden," as in *cryptography.*

Cryptobiidae *Invertebrate Zoology.* a family of colorless protozoans with two flagella, in the order Kinetoplastida.

cryptobiosis *Physiology.* a condition in which all external signs of metabolic activity are absent from a dormant organism.

cryptobiotic *Zoology.* describing an organism that typically hides or conceals itself.

Cryptobranchidae *Vertebrate Zoology.* the giant salamanders, a family in the suborder Cryptobranchoidea in the order Caudata; generally aquatic and having a flattened body, loose skin, and lidless eyes.

Cryptobranchoidea *Vertebrate Zoology.* a suborder in the order Caudata containing the most primitive living salamanders, including the Asiatic and giant salamanders; distinguished by external fertilization and aquatic larva.

Cryptocerata *Invertebrate Zoology.* in some classifications, a division of hemipteran insects including the water bugs or Hydrocorisae.

Cryptochaetidae *Invertebrate Zoology.* a family of two-winged insects in the subsection Acalypteratae, with small or rudimentary calypters.

cryptoclastic *Geology.* describing a clastic rock composed of microscopic finely broken or fragmented particles.

cryptoclimate *Engineering.* the temperature and moisture conditions of an enclosed space.

cryptoclimatology *Meteorology.* the study of climates of confined spaces.

Cryptococcaceae *Mycology.* a former family of fungi belonging to the class Deuteromycetes, including many species pathogenic to humans, such as the genus *Candida.*

Cryptococcales *Mycology.* an order of imperfect fungi whose classification is uncertain; it is sometimes classified as belonging to the class Hyphomycetes, was formerly classified under Blastomycetes, and has characteristics that link it to the subdivision Basidiomycotina; its perfect or sexual stage is not known.

cryptococcal meningitis *Medicine.* an inflammation of the meninges brought on by the yeastlike organism, cryptococcus.

cryptococcosis *Medicine.* a cryptococcus infection that may affect the lungs, skin, and usually the brain, nervous system, and their meninges.

Cryptococcus *Mycology.* a genus of yeast fungi belonging to the class Hyphomycetes which includes species pathogenic to humans and other animals, such as *C. neoformans;* its reproduction is characterized by the multilateral budding of cells.

cryptocrystalline *Geology.* describing the texture of an aggregate characterized by crystalline constituents too small to be distinguished under an ordinary microscope. Also, MICROCRYSTALLINE.

Cryptodira *Vertebrate Zoology.* the modern turtles, an infraorder of the reptilian order Testudines containing most living terrestrial and aquatic turtles, tortoises, and terrapins; characterized by the ability to withdraw the neck directly into the shell and a pelvis that is not fused to the shell.

Cryptodonta *Invertebrate Zoology.* a subclass (order Nuculoida) of primitive clams in the family Solemyidae, having almost no hinge.

cryptoexplosion structure *Geology.* a more or less circular structure formed by a sudden, often explosive, release of energy and showing intense rock deformation unrelated to volcanic or tectonic activity. Also, **cryptoexplosive structure.**

cryptogram [krip´tə gram´] *Linguistics.* a message that has been encrypted; i.e., a message whose content is not evident from ordinary interpretation. *Virology.* a descriptive code that is used to record certain basic properties of a virus, including its shape, type and weight of nucleic acid, host range, and vector.

cryptographic [krip´tə graf´ik] *Linguistics.* **1.** of or relating to cryptography. **2.** in an encrypted form; using a code or cipher.

cryptographic algorithm *Telecommunications.* a set of mathematically expressed rules for enciphering and deciphering data by effecting a series of conversions controlled by the use of a cryptographic key.

cryptographic bitstream *Telecommunications.* a stream of bits that is connected with a plain-text message in order to form a cryptogram, or a stream to decode plain text from a cryptogram.

cryptographic key *Telecommunications.* a cipher or code that is used to set or adjust cryptographic equipment at a transmission or receiving station so that messages can automatically be enciphered at the transmission station or deciphered at the receiving station.

cryptography [krip täg´rə fē] *Linguistics.* a branch of communications devoted to the design and use of ciphers, including principles and methods for converting plain text into unintelligible form and converting apparently unintelligible ciphered text into intelligible form, usually by means other than cryptanalysis.

Cryptograptus *Paleontology.* a genus of scandent biserial graptoloids in the suborder Glossograptina and family Cryptograptidae; Ordovician.

cryptohalite *Mineralogy.* $(NH_4)_2SiF_6$, a white or gray transparent cubic mineral occurring in massive or arborescent forms, having a specific gravity of 2 to 2.01 and a hardness of 2.5 on the Mohs scale; dimorphous with bararite; found as a sublimate in coal basins and on Mount Vesuvius.

cryptology [krip täl´ə jē] *Linguistics.* the branch of cryptography that deals with the hidden, disguised, or encrypted meanings in messages; used in communications security and intelligence.

cryptomedusa *Invertebrate Zoology.* the last stage in the reduction of a hydroid medusa to a rudiment having sex cells in the gonophore.

cryptomelane *Mineralogy.* $K(Mn^{+4},Mn^{+2})_8O_{16}$, a gray or black opaque, metallic monoclinic (pseudotetragonal) mineral occurring as massive, fine-grained, porous aggregates, having a specific gravity of 4.3 and a hardness of 1 to 6.5 on the Mohs scale; found as a secondary mineral with pyrolusite and other manganese oxides.

cryptomenorrhea *Medicine.* the occurrence of symptoms of menstruation each month without the flow of blood, as in imperforate hymen.

cryptomitosis *Invertebrate Zoology.* cell division occurring in certain protozoans, in which a modified spindle forms and chromatin assembles, but with no apparent chromosome differentiation.

Cryptomonadaceae *Botany.* a family of freshwater and marine flagellates of the order Cryptomonadales, characterized by almost equally long flagella attached to a subapical furrow, a laterally compressed cell body, and a variety of pigmentations based on specific physiological conditions.

Cryptomonadales *Botany.* an order of biflagellates of the class Cryptophyceae that occur in freshwater, brackish, and marine habitats and are characterized by a dominant motile phase.
cryptomonadid *Biology.* a member of the order Cryptomonadida.
Cryptomonadida *Biology.* a small order of plantlike flagellate protozoans, in the class Phytomastigophorea, with one or two flagella arising in a pit and yellowish-brown chloroplasts; also classified as algae in the class Cryptophyta. Also, **Cryptomonadina.**

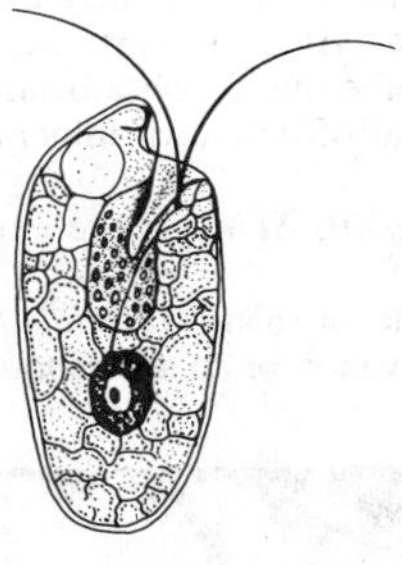

Cryptomonadida

Cryptomycetaceae *Mycology.* a family of fungi belonging to the order Phacidiales, composed of species that are parasitic and pathogenic to plants, causing such diseases as canker.
Cryptonemiaceae *Botany.* a large family of red algae of the order Cryptonemiales, characterized by erect or flattened thalli that often have proliferous branching, and by superficially formed spermatangia that sometimes produce large, colorless patches.
Cryptonemiales *Botany.* a diverse order of red algae of the division Rhodophyta, having thalli that range from crustose and discoid to erect and frondose, loose to compact filaments, and diverse reproductive structures.
cryptonephridic *Invertebrate Zoology.* of or relating to certain insects whose Malpighian tubules are independently attached to the hindgut, instead of being free.
cryptopart *Linguistics.* a portion of a cryptotext that has a specific message indicator.
Cryptophagidae *Invertebrate Zoology.* a family of coleopteran insects, the silken fungus beetles, in the superfamily Cucujoidea.
Cryptophyceae *Botany.* in some systems of classification, a type of biflagellate algae equivalent to the Bryophyta.
Cryptophyta *Botany.* a class of algae having flagella and reproducing by longitudinal fission; the same organisms are also classified as protozoans in the order Cryptomonadida.
cryptophyte [krip′tə fīt′] *Botany.* a plant that produces buds, either underground or underwater.
Cryptopidae *Invertebrate Zoology.* a family of centipedes in the order Scolopendromorpha.
cryptorchidopexy see ORCHIOPEXY.
cryptorchism [krip′tôr′kiz əm] *Medicine.* a developmental abnormality in which the testes fail to descend normally into the scrotum and remain in the abdomen. Also, **cryptorchidism.**
cryptosporidiosis [krip′tō spôr′i dī ō′sis] *Medicine.* an enteric disease caused by protozoa of the genus *Cryptosporidium* and characterized in humans by severe diarrhea.
cryptosporidium *Invertebrate Zoology.* a coccidial parasite found in the intestinal tracts of various vertebrates; it has been implicated in human intestinal disease.
Cryptostomata *Paleontology.* an extinct order of colonial bryozoans in the class Gymnolaemata; colonies consist of short individual tubes, with a septum half-closing the aperture at the bottom of a vestibule at the outer end of the tube; Ordovician to Permian.
cryptotext *Linguistics.* the text part of a message that is in coded form.
cryptotope *Immunology.* a latent antigenic determinant that becomes functional only when the protein molecule is broken or degraded.
cryptotoxin *Toxicology.* any substance whose toxic characteristics are not readily apparent or are hidden by other benign characteristics.
cryptovirogenic *Virology.* of, relating to, resembling, or having qualities of a cryptovirus.
Cryptovirus *Virology.* one of the cryptic viruses, a genus of plant viruses that consist of isometric particles, are transmitted through seeds, and cause no apparent symptoms.
cryptovolcanic *Geology.* produced by hidden volcanic activity.
cryptovolcanic structure *Geology.* a highly deformed, roughly circular structure that is thought to be produced by concealed or subterranean volcanic explosions, and that therefore shows no direct evidence of volcanic activity.
cryptoxanthin *Biochemistry.* a yellow carotenoid pigment found throughout nature; for example, in egg yolk and yellow corn; convertible by many animal livers to vitamin A.
Cryptozoic *Geology.* the geologic time interval for which the corresponding rocks show limited evidence of primitive life forms.
crypts of Lieberkühn *Anatomy.* the tubular glands of the small intestine. (Named for Johann Nathaniel *Lieberkühn,* 1711–1756, German anatomist.)
crypts of Morgagni see SINUS OF MORGAGNI.
crysophyta see SILICOFLAGELLIDA.
crystal *Crystallography.* any homogeneous solid that has a regularly repeating atomic arrangement; it may be a chemical element, a compound, or an isomorphous mixture. *Electronics.* a piezoelectric element, such as quartz, capable of transforming mechanical vibrations into electrical impulses or vice versa; generally used for precision frequency control in electronics equipment. *Horology.* the transparent covering that fits into the case of a watch to protect and in some cases magnify the dial and hands. Also, WATCH CRYSTAL.
crystal activity *Electronics.* an indication of how well a finished slab of crystal utilizes the piezoelectric effect; determined by measuring the amplitude of vibrations of the crystal slab under controlled conditions.
crystal aerugo see CUPRIC ACETATE.
crystal audio receiver *Electronics.* a crystal detector in which the nonlinear characteristic of a crystal diode is used to transform an audio-modulated radio-frequency signal directly into an audio signal.
crystal blank *Electronics.* a slab of piezoelectric or semiconductor crystal material that has been cut to approximate final size, but not yet processed into final form.
crystal calibrator *Electronics.* a crystal-controlled oscillator used as a standard against which the frequency tuning of a receiver or transmitter can be checked.
crystal carbonate see SODIUM CARBONATE MONOHYDRATE.
crystal cartridge *Acoustical Engineering.* the element in a crystal microphone or pickup containing the piezoelectric material.
crystal class *Crystallography.* one division within a crystal-classification system that is based on symmetry.
crystal clock see QUARTZ CRYSTAL CLOCK.
crystal control *Electronics.* the use of a piezoelectric crystal to control the frequency of an oscillator circuit.
crystal-controlled oscillator *Electronics.* an oscillator whose frequency is controlled by a piezoelectric crystal. Also, CRYSTAL OSCILLATOR.
crystal-controlled transmitter *Electronics.* a radio transmitter whose carrier frequency is directly controlled by a crystal oscillator.
crystal current *Electronics.* the actual current that flows through a crystal incident to the operation of the associated equipment, such as the radio-frequency current flowing through the crystal unit of a crystal-controlled oscillator.
crystal cutter *Acoustical Engineering.* a recording stylus made of a material having piezoelectric properties, such as sapphire or corundum.
crystal defect *Crystallography.* any departure of a crystal from symmetry, often caused by vacancies, disorders, voids, or stray crystal forms.
crystal detector *Electronics.* a mineral or crystalline material with rectification properties that converts an alternating current into a pulsating current.
crystal diffraction *Solid-State Physics.* the diffraction of X-rays, electrons, or cold neutrons with wavelengths comparable to the order of the atomic spacing in the crystal.
crystal diffraction spectrometer see BRAGG SPECTROMETER.
crystal dynamics see LATTICE DYNAMICS.
crystal faces *Crystallography.* the smooth, flat bounding surfaces of a crystal that intersect to form sharp edges, and that show definite symmetrical relationships, giving an indication of the internal structure of the crystal.
crystal field theory *Physical Chemistry.* the theory that an ion group bound to a metal's central atom is the source of the negative charge that causes the metal to respond to an electric field.
crystal filter *Electronics.* a filter circuit that uses a piezoelectric crystal to establish the center frequency of the passband.

crystal form *Crystallography.* **1.** the geometric shape of a crystal. **2.** crystal faces or planes in a particular crystal class that are symmetrically equivalent to a single face. *Geology.* see PRISM.

crystal glass *Materials.* a clear glass containing lead, which increases its brilliancy, strength, and clarity.

crystal gliding *Crystallography.* a combined operation in which a molecule is related to another by translation, which is half the spacing between lattice points, followed by reflection.

crystal grating *Spectroscopy.* a grating in which a crystal lattice diffracts an incident beam of X-rays or gamma rays.

crystal growth *Crystallography.* a controlled change of state, or phase change, to the solid state of a crystal, using direct or indirect techniques.

crystal habit *Crystallography.* the usual appearance of a crystal of a particular substance. It is found that the most rapidly growing faces of a crystal are the smallest and the least well developed. Crystal habit may be strongly dependent upon the conditions of crystallization, especially the solvent used.

crystal headphone *Acoustical Engineering.* a headphone that uses quartz, Rochelle salts, or a similar crystalline piezoelectric material as the active element.

crystal holder *Mechanical Devices.* a holder and housing for a quartz crystal plate, providing support and protection for connections within it.

crystal hydrophone *Acoustical Engineering.* a type of crystal used for detecting or responding to sounds under water.

crystal indices see MILLER INDICES.

crystal laser see GLASS LASER.

crystal lattice *Crystallography.* a regularly spaced array of points that represents the structure of a crystal. Crystals are composed of groups of atoms repeated at regular intervals in three dimensions with the same orientation. Each such group of atoms is replaced by a representative point; the collection of points so formed is the **space lattice** or **lattice of the crystal.** Each crystal lattice is a Bravais lattice.

crystal-lattice filter *Electronics.* a filter circuit that uses piezoelectric crystals to establish the limits of the passband.

crystalliferous bacteria *Bacteriology.* bacteria, most often *Bacillus thuringiensis,* that are characterized by the creation of a protein crystal in the sporangium during spore formation.

crystallin see CRYSTALLIN PROTEIN.

crystalline [kris′tə lin; kris′tə līn′] *Crystallography.* relating to or having a crystal structure.

crystalline anisotropy *Solid-State Physics.* the tendency of a crystal to have different properties in different directions.

crystalline carbonate *Geology.* a sedimentary rock consisting mainly of carbonate minerals in which recrystallization and replacement have rendered the depositional texture unrecognizable.

crystalline chloral see CHLORAL HYDRATE.

crystalline chondrite *Geology.* a crystalline, stony meteorite characterized by firm, round, radial chondrules that break the matrix.

crystalline double refraction *Optics.* the splitting of a wavefront that occurs when a wave disturbance extends through an anisotropic crystal.

crystalline field *Solid-State Physics.* the electric field within a crystal produced internally by localized charges such as ions.

crystalline fracture *Metallurgy.* a fracture that generally follows selected planes in the crystals of a metallic structure.

crystalline frost *Hydrology.* a hoarfrost exhibiting a simple macroscopic crystalline structure.

crystalline-granular texture *Petrology.* a primary texture of sedimentary rocks precipitated directly from water, such as rock salt and gypsum.

crystalline laser *Optics.* a laser that creates a coherent beam of light from a pure or doped crystal.

crystalline lens see LENS.

crystalline melting *Materials Science.* the disappearance of the polymer crystalline phase with accompanying changes in the physical properties of the polymer, including changes in density, refractive index, heat capacity, and transparency.

crystalline polymer *Chemistry.* a polymer chain folded back on itself in a regular pattern to form a material with the properties of a crystal.

crystalline porosity *Geology.* the property of permeability in crystalline limestone and dolomite, allowing the formation of underground oil reservoirs.

crystalline rock *Petrology.* **1.** rock that is made up of clearly crystallized minerals. **2.** any igneous or metamorphic rock.

crystalline solid *Materials Science.* a solid in which the arrangement of atoms, ions, or molecular groups repeats itself in three dimensions.

crystalline style *Invertebrate Zoology.* a long, tubular, translucent, proteinaceous, and gelatinous mass in the stomach of many bivalve mollusks and some gastropods; its abrasive function releases digestive enzymes. Also, **crystalline stylet.**

crystallinity [kris′tə lin′ə tē] *Crystallography.* a measure of the crystalline properties and qualities of a substance. *Materials Science.* in polymers, a type of molecular structure characterized by uniformity and compactness, caused by the existence of solid crystals with a definite geometric form.

crystallin protein *Biochemistry.* the major soluble protein present in the eye lens of vertebrates. Also, CRYSTALLIN.

crystallite *Geology.* in the initial crystallization of a magma, a microscopic embryonic crystal of unknown mineralogical composition that does not polarize light.

crystallizability *Materials Science.* the extent to which a material forms crystals.

crystallization [kris′tə lə zā′shən] *Materials Science.* the formation of crystals from a liquid, a vapor, or an amorphous solid.

crystallization

crystallization path *Materials Science.* a line or curve on an equilibrium phase diagram showing the composition of the solid phases in the system being studied.

crystallization regime *Materials Science.* one of three different types of polymer crystallization kinetics; each differs from the others by the rate at which chains are deposited on the surface of crystals.

crystallized intelligence *Psychology.* the sum of an individual's mental skills and abilities acquired through education and experience.

crystallizer *Chemical Engineering.* a vessel in which dissolved solids in a solution are precipitated from the solution by cooling or evaporation, and then recovered as solid crystals with a specified size range.

crystalloblast *Mineralogy.* a crystal of a mineral that is formed by the metamorphic processes.

crystalloblastic *Geology.* describing a crystalline rock texture resulting from recrystallization during metamorphism under conditions of high viscosity and directed pressure.

crystalloblastic order see CRYSTALLOBLASTIC SERIES.

crystalloblastic series *Geology.* a sequence of metamorphic minerals arranged in order of decreasing energy of crystallization, so that the crystals of any listed mineral are completely bounded by their own crystal faces wherever they border on simultaneously developed crystals of all minerals occupying lower positions. Also, IDIOBLASTIC SERIES, CRYSTALLOBLASTIC ORDER.

crystallogram *Crystallography.* an X-ray diffraction pattern revealing the arrangements of atoms in a crystal lattice.

crystallographic [kris′tə lə graf′ik] of or relating to crystallography.

crystallographic axes *Crystallography.* the lines of reference that intersect in the center of a crystal, and whose lengths serve to determine the system to which a crystal belongs.

crystallographic plane *Crystallography.* a crystal face, crystal plane, cleavage plane, or lattice plane that can be defined mathematically with respect to the directions and lengths of the crystallographic axes.

crystallographic texture *Mineralogy.* a texture of mineral deposits formed by replacement or exsolution when the distribution of the inclusions are dictated by the crystallography of the host mineral.

Crystallography

Crystallography is the science of crystals. Classical crystallography, the study of the angles between natural plane faces on crystals and of their optical properties, was of great value in chemistry, mineralogy, and other fields of science, especially in the identification of substances.

Modern crystallography started about 1890, when the 230 space groups (combinations of symmetry operations possible for a crystal) were discovered, and developed after 1913–1914, when diffraction of X-rays by crystals was discovered by Max von Laue and the Bragg equation was formulated by Lawrence Bragg. These discoveries led to much progress in atomic physics through the determination of the wavelengths of X-rays, and even in nuclear physics through the analysis of gamma-ray spectra.

Modern chemistry, based largely on knowledge about chemical bonds and the structures of molecules as determined by X-ray diffraction studies of crystals, could not have developed without X-ray crystallography. Molecular biology is also a product of X-ray crystallography. The alpha helix and the pleated sheets of protein molecules and the double helix of DNA were discovered by the aid of the X-ray diffraction patterns, even before the powerful new direct methods of crystallographic analysis of diffraction patterns were developed.

Now these new methods have been applied in the determination of the detailed atomic arrangements of hundreds of protein crystals, giving remarkable insights into the nature of living organisms. Our present great understanding of the nature of the world of atoms, molecules, minerals, and human beings can be attributed in large part to crystallography.

Linus Pauling
Research Professor
Linus Pauling Institute of Science and Medicine
Nobel Laureate in Chemistry
Nobel Laureate in Peace

crystallography [kris´tə läg´rə fē] the study of the properties and formation of crystals.

crystallomagnetic *Solid-State Physics.* referring to the magnetic properties of crystals.

crystallophobia *Psychology.* an irrational fear of glass or glass objects.

crystal loudspeaker *Acoustical Engineering.* a loudspeaker that responds to an audio signal voltage applied to a Rochelle salt or similar crystalline piezoelectric material in order to create mechanical displacement and thereby produce sound.

crystal microphone *Acoustical Engineering.* a microphone in which pressure from sound waves causes piezoelectric crystals to produce an electric signal; the sound waves either strike the crystals directly or strike a diaphragm that contacts the crystals.

crystal mixer *Electronics.* a crystal receiver that can be fed simultaneously from two input frequencies to produce a lower output frequency; as, for example, a crystal radar receiver in which the received frequency is mixed with the output of a local oscillator to produce an intermediate frequency signal.

crystal monochromator *Spectroscopy.* a spectrometer that uses a single crystal of lead or other element to isolate a narrow band of wavelengths from a collimated incident beam of slow neutrons from a reactor.

crystal morphology *Crystallography.* study of the shape or form of a material. With crystals, a description of the crystal faces and the angles between them can often be used for identification.

crystal optics *Optics.* the study of the transmission of electromagnetic radiation through crystals, especially anisotropic crystals.

crystal oscillator see CRYSTAL-CONTROLLED OSCILLATOR.

crystal oven *Engineering.* an oven having a stabilized temperature that is maintained by a crystal unit.

crystal pickup *Acoustical Engineering.* a phonograph pickup that uses piezoelectric material to convert sound energy into electromagnetic waves.

crystal plate *Electronics.* the final form of a piezoelectric crystal slab upon completion of lapping, etching, and metal coating prior to installation in the crystal holder. Also, QUARTZ PLATE.

crystal pulling *Crystallography.* the formation of crystals from a melted substance by pulling solid material from the liquid.

crystal rectifier see SUPERCONDUCTOR DIODE.

crystal resonator see QUARTZ-CRYSTAL RESONATOR.

crystal sandstone *Geology.* a silicate rock containing quartz grains that have been enlarged and transformed into crystals by deposition of silica in an orderly arrangement.

crystal set *Electronics.* a simple radio receiver consisting only of a resonant circuit, a crystal detector, and a headset or low-power speaker.

crystal settling *Geology.* a process occurring in a magma whereby crystals sink to the bottom as a result of their greater density. Also, **crystal sedimentation.**

crystal shutter *Electromagnetism.* a coaxial cable or mechanical waveguide shorting switch that, when closed, prevents radio frequency energy from reaching and damaging a crystal detector.

crystal spectrometer see BRAGG SPECTROMETER.

crystal-stabilized transmitter *Electronics.* a radio transmitter with automatic frequency control in which a crystal oscillator establishes the reference frequency.

crystal structure *Crystallography.* the mutual arrangement of the atoms, molecules, or ions that are packed together in a regular way to form a crystal.

crystal symmetry *Crystallography.* the entire assemblage of rows and patterns of atoms that have a definite arrangement in each crystal.

crystal system *Crystallography.* one of the seven configurations of crystals, classified in terms of their symmetry, and corresponding to the seven fundamental shapes for unit cells consistent with the 14 Bravais lattices: cubic, tetragonal, orthorhombic, monoclinic, triclinic, hexagonal, and rhombohedral.

crystal transducer see PIEZOELECTRIC TRANSDUCER.

crystal tuff *Geology.* compacted volcanic ash and dust consisting mainly of crystals and crystal fragments.

crystal unit *Electronics.* a unit or assembly within a piece of electronic equipment in which one or more crystal plates are installed.

crystal video receiver *Electronics.* a microwave receiver that contains a crystal detector to demodulate the received signal, and also a video amplifier.

crystal video rectifier *Electronics.* a crystal detector in which the nonlinear characteristic of a crystal diode is used to transform a video-modulated radio frequency signal directly into a video signal.

crystal violet see METHYL VIOLET.

crystal vitric-tuff *Geology.* compacted volcanic ash and dust containing crystal fragments and fragments of volcanic glass.

crystal whiskers *Crystallography.* crystals that are grown in filamentary forms.

crystal zone *Crystallography.* a series of three or more nonparallel faces of a crystal whose intersecting edges are parallel to a common line through the center of the crystal.

crystosphene *Hydrology.* a mass or sheet of clear ice buried and wedged between beds of other material, such as frozen spring water that rises and spreads laterally under swamps in a tundra region.

CS or **C.S.** conditioned stimulus.

C scan see C DISPLAY.

C scope *Electronics.* a radarscope that produces a C display.

CSF colony stimulating factor.

C shell *Computer Science.* a shell program for Unix systems, having a syntax similar to the C programming language.

CSMA/CD *Telecommunications.* an access method for multiple users of a computer network in which the user listens before transmitting. If there is a signal on the network, the user waits until it is clear; if two or more users attempt to transmit at exactly the same moment, each will detect a collision, cease attempting transmission, and wait a random interval before again attempting to access the network. (An acronym for carrier-sense multiple access with collision detection.)

C stage *Materials Science.* the final stage in a thermosetting resin reaction.

C stage thermosetting resin *Materials Science.* a resin that is insoluble and infusible and forms in the final stage in a thermosetting resin reaction. Also, RESITE.

C star see CARBON STAR.

C strand *Genetics.* **1.** the strand that is complementary to the genomic strand in the replicative form of a single-stranded DNA virus. Also, COMPLEMENTARY STRAND. **2.** see CRICK STRAND.

C substance *Biochemistry.* one of a series of serologically unique carbohydrates that are present in *Streptococcus*; used for identification. Also, C POLYSACCHARIDE.

CTD recorder see SALINITY-TEMPERATURE-DEPTH RECORDER.

ctene [tēn] see COMB.

ctenidium [tə nid´ē əm] *Invertebrate Zoology.* **1.** the feathery respiratory structure of certain mollusks. **2.** a row of spines on the head or thorax of certain fleas.

Ctenocladaceae *Botany.* a family of green algae of the order Ctenocladales, characterized by a filamentous thallus composed of prostrate and erect systems or packets of uninucleate cells.

Ctenocladales *Botany.* an order of green algae of the class Chlorophyceae, characterized by uniseriate branched filaments made up of uninucleate cells that each have a parietal laminate chloroplast; species are found in subaerial, freshwater, brackish, and marine environments.

Ctenodactylidae *Paleontology.* a family of rodents in the suborder Sciurognathi and superfamily Ctenodactyloidea; common in Asia in the Eocene and still represented by four pika-like genera in Africa.

Ctenodonta *Paleontology.* "comb-tooth," an extinct genus of burrowing pelecypod bivalves in the subclass Palaeotaxodonta; Ordovician.

Ctenodrilidae *Invertebrate Zoology.* a family of fringe worms, polychaete annelids in the group Sedentaria.

ctenoid scale [ten´oid] *Vertebrate Zoology.* a thin, bony fish scale having one edge toothed like a comb; found chiefly on the spiny-rayed fish of the superorder Acanthopterygii.

Ctenophora *Invertebrate Zoology.* a phylum containing the comb jellies, gelatinous marine organisms with eight rows of comblike locomotory cilia.

ctenophore [ten´ə fôr´] *Invertebrate Zoology.* any member of the phylum Ctenophora.

Ctenostomata *Invertebrate Zoology.* an order of colonial bryozoans in the class Gymnolaemata.

Ctenostomatida see ODONTOSTOMATIDA.

Ctenothrissidae *Paleontology.* a family of Paleozoic euteleost fishes in the extinct order Ctenothrissiformes; extant in the Upper Cretaceous.

Ctenothrissiformes *Paleontology.* a short-lived order of euteleost fishes; deep-bodied and about a foot long; may have been ancestral to the spiny teleosts; widespread but restricted to the Upper Cretaceous.

C terminus *Immunology.* the carboxyl end of a polypeptide chain of a protein molecule. Also, **C terminal.**

ctn cotangent.

CTP *Biochemistry.* cytidine 5'-triphosphate, a compound made by amination of uridine triphosphate during pyrimidine biosynthesis.

CTRL *Computer Programming.* **1.** an abbreviation for control, used in specifying control characters, as in "CTRL-D" to represent a D typed with the control key held down. **2.** the control key itself.

CT scan see CAT SCAN.

C-tube bourdon element *Engineering.* a flexible, hollow, tube shaped in a C-curve and used to measures changes in pressure; the tube flexes slightly with each change in internal gas or liquid pressure.

C-type asteroid *Astronomy.* the most numerous class of asteroid, whose reflected spectrum and color closely resemble those of carbonaceous chondrite meteorites.

C-type virus particle see TYPE C ONCOVIRUS GROUP.

Cu the chemical symbol for copper.

cu. cubic; cumulus; cumulative.

Cuba *Geography.* the largest island in the West Indies (area: about 44,000 square miles.), in the Caribbean Sea south of Florida.

cubane *Organic Chemistry.* C_8H_8, a polycyclic organic compound whose three-dimensional shape is cubic; existing in the form of rhombic crystals; melts at 130–131°C (sealed tube) and decomposes at 200°C.

cubanite *Mineralogy.* $CuFe_2S_3$, a bronze or yellow, opaque, metallic orthorhombic mineral; massive in habit or as thick tabular, commonly twinned crystals, having a specific gravity of 4.03 to 4.18 and a hardness of 3.5 on the Mohs scale; found in deposits formed at high temperatures.

cube *Mathematics.* **1.** a six-sided polyhedron (a hexahedron) whose faces are congruent squares (and thus whose face angles are right angles); a cuboid with equal edges. The cube is one of five Platonic solids. **2.** in Euclidean n-space, the ***n*-cube** is the Cartesian product of n closed intervals of equal length. **3.** in integration theory, a **singular *n*-cube** in a starlike region A is a continuous function c: $[0, 1]^n \to A$. ($[0, 1]^n$ denotes the n-fold Cartesian product of [0, 1] with itself.) Letting R^0 and $[0, 1]^0$ both denote $\{0\}$, a singular 0-cube is then a function c: $\{0\} \to A$; i.e., a point in A. A singular 1-cube is a curve. The **standard *n*-cube,** denoted I^n: $[0, 1]^n \to R^n$ is defined by $I^n(x) = x$ for $x \in [0, 1]^n$. **4.** in graph theory, the ***n*-cube** is the graph whose vertices are the 2^n sequences of ones and zeros of length n and whose edges join vertices that differ in exactly one component; usually denoted Q_n.

cubeb *Botany.* a climbing shrub, *Piper cubeba,* whose fruit is dried and used as a spice in Indian cuisine.

cubebism *Toxicology.* poisoning due to consumption of cubeb, the dried unripe fruit of *Piper cubeba*; symptoms may include severe gastrointestinal disturbances, fever, coma, and even death.

cube corner prism see CORNER REFLECTOR.

cube of a number *Mathematics.* the cube of a number or quantity x is $x \cdot x \cdot x$, denoted x^3.

cube root *Mathematics.* the cube root of a number or quantity x is that number or quantity whose cube equals x; denoted $\sqrt[3]{x}$ or $x^{1/3}$.

cube-surface coil *Electromagnetism.* five square coils arranged over the surface of a cube so as to produce a magnetic field that is approximately uniform over a large accessible volume.

cubic *Mathematics.* describing a volume unit; if x is a unit of length, cubic x refers to a unit of volume consisting of a cube of sides with length x. Thus a cubic meter, or a meter cubed, is the volume of a cube whose sides measure one meter each.

cubical antenna *Electromagnetism.* an antenna in which the radiating elements are positioned to form a cube.

cubical expansion *Physics.* a change in the volume of a material due to a change in pressure or temperature.

cubical parabola *Mathematics.* the graph of the equation $y = x^3$ in a plane, with x and y as the Cartesian coordinates.

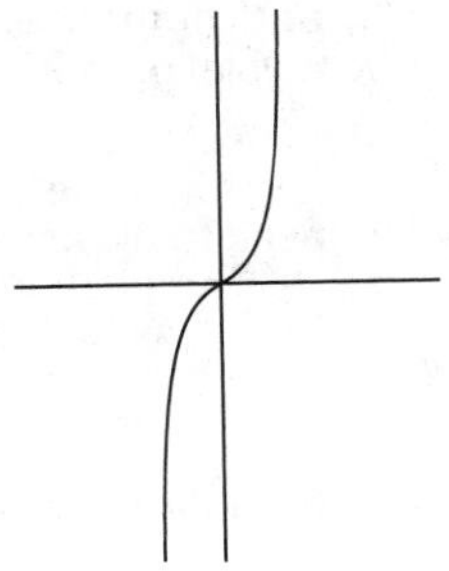

cubical parabola

cubic boron nitride *Materials.* a synthetic cutting-tool material that is second only to diamonds in hardness, made by transforming boron nitride, BN, from hexagonal to cubic form at high pressure and temperature with a suitable catalyst.

cubic centimeter *Metrology.* a unit of volume equal to 0.000001 cubic meter, or 0.06 cubic inch.

cubic crystal *Crystallography.* a crystal whose unit cell contains threefold rotation axes along all four body diagonals, and fourfold axes parallel to each crystal axis; there also are twofold axes. As a result, all axial lengths are identical by symmetry and all interaxial angles are 90° ($a = b = c$, $\alpha = \beta = \gamma = 90°$). Also, CUBIC SYSTEM, ISOMETRIC SYSTEM.

cubic curve *Mathematics.* a plane curve with the equation $f(x, y) = 0$, where $f(x, y)$ is a cubic polynomial in x and y.

cubic decimeter *Metrology.* a unit of volume equal to 1000 cubic centimeters, or 0.0353 cubic foot.

cubic dekameter *Metrology.* a unit of area equal to 1000 cubic meters, or 1308 cubic yards.

cubic foot *Metrology.* a unit of volume equal to 1728 cubic inches, or 0.037 cubic yard.

cubic foot per second see CUSEC.

cubic graph *Mathematics.* an undirected graph in which every vertex has degree 3; if, in addition, the graph is bipartite, it is called a **bicubic graph.**
cubic hectometer *Metrology.* a unit of area equal to 1,000,000 cubic meters, or 1,308,000 cubic yards.
cubic inch *Metrology.* a unit of volume equal to 0.00058 cubic foot, or 16.4 cubic centimeters.
cubic measure *Mechanics.* a unit or set of units for the measurement of volume. *Metrology.* a system using cubic units, especially one in which 1728 cubic inches equal 1 cubic foot, or 1000 cubic millimeters equal 1 cubic centimeter. Also, **cubic measurement.**
cubic meter *Metrology.* a unit of area equal to 1,000,000 cubic centimeters, or 1.3 cubic yards.
cubic meter per second see CUMEC.
cubic millimeter *Metrology.* a unit of area equal to 0.001 cubic centimeters, or 0.00006 cubic inch.
cubic packing *Crystallography.* the prearrangement of constituents so as to shape a substance into a cubic crystal.
cubic phages *Virology.* viruses with icosahedral symmetry that are isolated from prokaryotes.
cubic plane *Crystallography.* a plane that is parallel to the face of a cubic crystal.
cubic polynomial *Mathematics.* a polynomial having at least one term of degree 3 and no term of degree greater than 3.
cubic spline *Mathematics.* a spline of degree 3; used in preference to splines of higher degree because of its lack of oscillation between knots.
cubic system see CUBIC CRYSTAL.
Cubiculosporaceae *Botany.* a monospecific family of red algae of the order Gigartinales; characterized by erect, branched, multiaxial thalli arising from discoid bases and by spermatangia that occur in superficial, irregular, slightly sunken patches.
cubic yard *Metrology.* a unit of area equal to 27 cubic feet, or 0.765 cubic meter.
cubit *Metrology.* a historic measure of length, based on the length of the forearm; usually equal to 18 inches.
cuboctahedron *Mathematics.* one of the thirteen Archimedean solids, whose faces consist of six regular octahedra and eight equilateral triangles. It may be visualized as a cube with the corners sliced off. Also, **cubo-octahedron.**
cuboid *Mathematics.* a right parallelopiped with a rectangular base. *Anatomy.* a cube-shaped structure or part. *Invertebrate Zoology.* the main vein in the wing of many insects, especially flies. Also, **cuboidal.**
cuboidal epithelium *Histology.* epithelium that is composed of cube-shaped cells.
cuboid bone *Anatomy.* a bone on the lateral side of the tarsus, between the calcaneus and the fourth and fifth metatarsal bones.
Cubomedusae *Invertebrate Zoology.* an order of coelenterates in the class Scyphozoa, with a cubic umbrella.
Cuccia coupler see ELECTRON COUPLER.

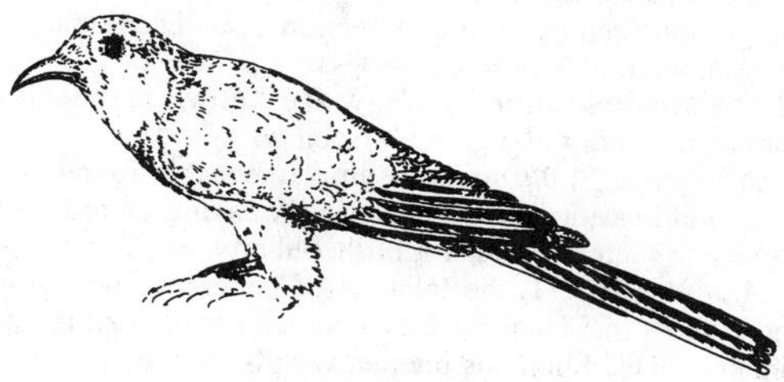

cuckoo

cuckoo *Vertebrate Zoology.* the common name for about 130 species of insectivorous arboreal birds of the order Cuculiformes; the European species is noted for laying its eggs in the nests of other birds for them to hatch.
cuckoo-shrike *Vertebrate Zoology.* an arboreal songbird of the family Campephagidae that inhabits areas throughout Africa, Asia, and the Pacific Islands; although not related to the cuckoo, it resembles it with long, pointed wings and a rounded tail; a fruit and insect eater.
Cucujidae *Invertebrate Zoology.* a family of predatory coleopteran insects, the flat-back beetles, in the superfamily Cucujoidea.
Cucujoidea *Invertebrate Zoology.* a superfamily of beetles, coleopteran insects in the suborder Polyphaga.
Cuculidae *Vertebrate Zoology.* the cuckoos, anis, and roadrunners; medium-sized birds of the order Cuculiformes, characterized by a strong, curved beak, a long tail, and slender bodies.
Cuculiformes *Vertebrate Zoology.* an order of generally arboreal birds that contains the touracos, cuckoos, anis, and roadrunners and is widely distributed in temperate areas.
cucullate *Biology.* hood-shaped or having a hood.
cucullus *Invertebrate Zoology.* the hood-shaped anterior dorsal shield of the cephalothorax in pseudoscorpions and Ricinulei arachnids.
Cucumariidae *Invertebrate Zoology.* a family of echinoderms in the order Dendrochirotida, with ten to thirty much branched tentacles.

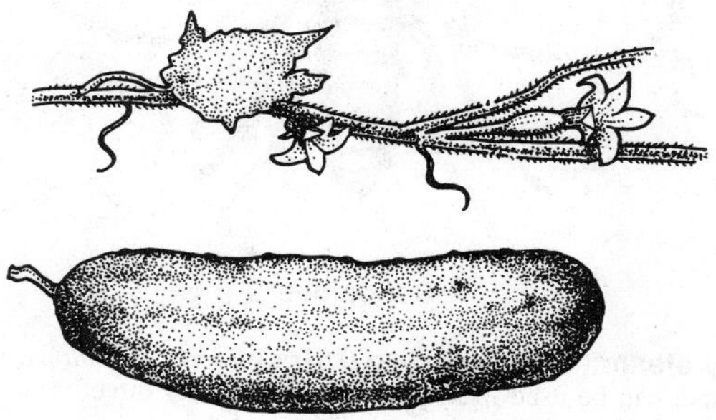

cucumber

cucumber *Botany.* **1.** a creeping vine, *Cucumis sativus,* of the gourd family, widely cultivated in many varieties for its edible fruit. **2.** the edible, fleshy fruit of this plant, having a cylindrical shape with rounded ends and a green, warty skin; eaten raw, either alone or in salads.
cucumber mildew *Plant Pathology.* **1.** a powdery whitish growth that occurs on cucumbers and melons due to the fungus *Erysiphe cichoracearum.* **2.** a downy whitish growth that occurs on cucumbers and melons due to the fungus *Peronoplasmopara cubensis.*
cucumber mosaic *Plant Pathology.* a virus disease of cucumbers, tomatoes, and related fruits that is characterized by stunted, bushy growth and yellowed or mottled foliage and fruit.
Cucumovirus group *Virology.* a group of ssRNA-containing, multicomponent plant viruses that infect cucurbits and solanaceous plants mechanically as well as through aphids and seeds. Also, **cucumber mosaic virus group.**
Cucurbitariaceae *Mycology.* a family of fungi belonging to the order Pleosporales that live off decaying plant material or as parasites on plant leaves and branches.
cucurbit wilt *Plant Pathology.* a disease of cucumbers and related plants caused by the motile bacterium *Erwinia tracheiphila* and characterized by sudden wilting and withering of the plant.
cue *Behavior.* a signal that elicits a particular behavior based on previous experience.
cue circuit *Electronics.* a unidirectional communication circuit used in radio and television for conveying program control information.
cuesta *Geology.* an asymmetrical hill or ridge characterized by one gently sloping side that generally conforms to the dip of the strata and one steeply sloping side that was formed by the outcrop of resistant rock.
Cuisian *Geology.* a European geologic stage of the lower Eocene epoch, occurring after the Ypresian and before the Lutetian.
cul-de-sac [kul′də sak′] *Civil Engineering.* a street that is closed at one end; a dead-end street. *Anatomy.* **1.** a blind pouch or sac. **2.** a tubular cavity closed at one end. (From French; literally, "bottom of the sack.")
culdocentesis *Medicine.* the removal by aspiration of pus or other fluid material from the rectouterine pouch through a vaginal puncture.
culdoscope *Medicine.* an instrument for examining the rectovaginal pouch and pelvic viscera.
Culex [kyoo′leks; koo′leks] *Invertebrate Zoology.* a large, common genus of mosquitoes that feed on human blood and transmit filariasis.
Culicidae *Invertebrate Zoology.* the mosquitoes, a family of two-winged insects in the series Nematocera, with piercing mouth parts and long legs.
Culicinae *Invertebrate Zoology.* a subfamily of mosquitos in the dipteran family Culicidae.
Culicinomyces *Mycology.* a genus of fungi belonging to the class Hyphomycetes.

Culicoides *Invertebrate Zoology.* a genus of minute biting flies, dipteran insects in the family Ceratopogonidae that feed on various warm-blooded animals and on mosquitos; the larvae live in fresh water and damp places.

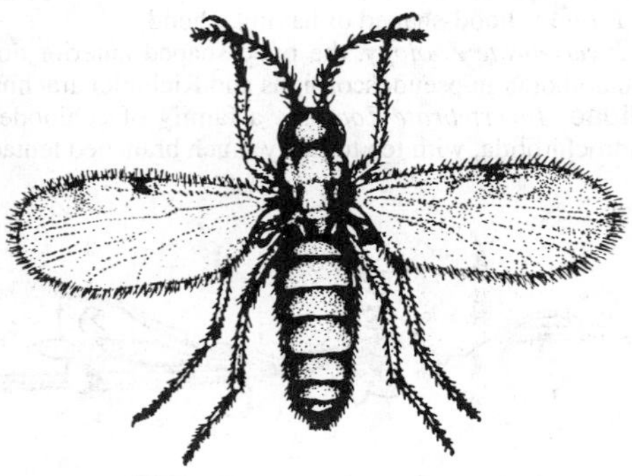

Culicoides

culinary steam [kul′i nâr′ē; kyoo′li nâr′ē] *Food Technology.* very pure steam that can be used in food preparation, either directly injected into food or indirectly used to heat and purify water.

cull *Agriculture.* **1.** a fruit, vegetable, or grain product having some defect that makes it unmarketable or less marketable. **2.** an animal that is inferior or unsuitable. **3.** to remove such worthless or inferior items from a crop, herd, or flock. *Chemical Engineering.* the material remaining in the transfer chamber after a plastics mold has been filled.

cullet *Engineering.* pieces of broken or waste glass that are suitable for remelting.

cullis see COULISSE.

culm *Botany.* **1.** the jointed, usually hollow stem of certain grass plants. **2.** the stem of sedges. *Mining Engineering.* any anthracite fine coal that will pass through a screen with 1/8-inch holes.

culmen *Vertebrate Zoology.* the longitudinal ridge along the upper section of a bird's bill.

culmination *Astronomy.* the instant when a celestial object crosses the meridian of an observer. *Geology.* the highest point of any structural feature, such as a fold.

cultigen *Biology.* **1.** a cultivated species or a variety of an organism that has no known wild ancestor, such as the cabbage. **2.** a variety produced by selective breeding. Also, **cultivar**.

cultivate *Agriculture.* **1.** to plant, tend, harvest, and improve plants. **2.** to till the soil both prior to and during the growing of crops.

cultivator *Agriculture.* a farm implement that is used to break up the surface of the soil and remove weeds near growing plants.

cultivator

cultural *Anthropology.* relating to or involving culture. *Biology.* relating to the nongenetic transmission of information from one generation to the next.

cultural adaptation *Anthropology.* the learned behavioral means by which humans adapt to and manipulate their environment.

cultural anthropology *Anthropology.* a major branch of anthropology that examines human culture and compares the variations and similarities of the cultural traits of distinct groups of people.

cultural community *Ecology.* an ecological community that is established or influenced by human intervention.

cultural context *Anthropology.* the analysis or study of a culture only within the context of its own time and place, as opposed to comparing it with other cultures that are chronologically or geographically distinct.

cultural desert *Ecology.* a term for an area that lacks flora and fauna because of the effects of human activity, as in a heavily populated urban site.

cultural diffusion *Anthropology.* the transmission or borrowing of certain culture traits from the group of origin into a foreign group; usually technological elements rather than those of social organization.

cultural ecology *Anthropology.* the study of the relationships between a culture and its natural environment.

cultural evolution *Anthropology.* a 19th-century theory in which cultures were viewed as having progressive stages of development; nomadic, agricultural, or pastoral societies were regarded as primitive in comparison with the industrial societies of Western Europe.

cultural geography *Geography.* a branch of geography dealing with the relationships between human societies and the environment.

cultural historicism *Anthropology.* the methodology used to trace cultural elements in regional societies in the reconstruction of a common historical origin.

cultural lag *Anthropology.* the failure of one element of a culture to keep pace with changes in other aspects of the culture; e.g., a situation in which rapid technological change is not accompanied by change in the nonmaterial culture.

cultural landscape *Geography.* any portion of the natural environment that has been altered by human influence.

cultural materialism *Anthropology.* the thesis that culture develops proportionately to the output of energy put forth; it evolves in direct response to technological progress.

cultural operations *Forestry.* a general term for actions that are not directly revenue-producing, but that are undertaken to assist the development and regeneration of a forest crop or to minimize the damage caused by felling and extraction.

cultural pluralism *Anthropology.* the existence of more than one culture within a society, such as that which exists in a colonial situation or in a metropolis.

cultural relativism *Anthropology.* the principle or approach that a particular culture can be fully understood only when it is analyzed within its own environmental context rather than in comparison with other cultures.

cultural resource management *Archaeology.* a professional area of archaeology that focuses on the protection of archaeological sites from urban development or natural processes.

cultural resources *Anthropology.* the known archaeological and ethnographic inventory of a group or society.

cultural universal *Anthropology.* an element that is found in all cultures; e.g., childrearing practices such as weaning or toilet-training, or rites of passage such as rituals for birth, puberty, and death.

culture *Anthropology.* **1.** the total set of beliefs, values, customs, and behavior patterns that characterizes a human population; the noninstinctive manner in which humans interact with or manipulate their environment. **2.** any specific example or stage of this: the *culture* of the ancient Egyptians. *Archaeology.* similar or related assemblages found in several sites within a defined area during the same time period, considered to represent the activities of one specific group of people; often named for a particular site or an artifact. *Biology.* a population of microorganisms (especially bacteria or fungi), tissue cells, or other living organisms cultivated on a specially prepared solid or liquid nutrient and in a controlled environment; used for scientific study or medicinal applications. *Cartography.* artificial features of the earth's surface to be included on a specific map, such as roads or buildings; contrasted with features formed exclusively by the forces of nature (natural detail).

culture-and-personality *Anthropology.* an ethnological method in which personality types found in various cultures are analyzed according to their particular cultural and environmental contexts.

culture area *Anthropology.* a geographic region in which several different societies share cultural traits and patterns of living.
culture center *Anthropology.* the point of origin for the culture traits that constitute a culture area.
culture change *Anthropology.* the set of processes by which humans adapt to new elements in their environment, including discovery or invention, and diffusion or acculturation.
culture circle see KULTURKREIS.
culture climax *Anthropology.* the greatest density or aggregation of a particular set of cultural traits, usually derived statistically.
culture collection *Bacteriology.* an organized assembly of characterized microorganisms.
culture complex *Anthropology.* a discrete group of cultural traits shared by several societies in a geographical area.
culture contact *Anthropology.* any circumstances under which a foreign culture has influenced another culture, the result of which is the appearance of one or more cultural traits, such as a new word or tool.
culture core *Anthropology.* the set of cultural traits that make up the subsistence of a society; includes the resource base, the means of production, the labor pattern, and the distribution of the product.
cultured forest *Ecology.* a wooded area established by planting tree species that are not native to the region.
culture dish see PETRI DISH.
cultured pearl *Invertebrate Zoology.* a natural pearl that is produced by artificial methods, as by introducing a seed pearl into an oyster's mantle.
cultured prairie *Ecology.* a term for a cultivated field or grassland. Similarly, **cultured forest.**
culture-free test *Behavior.* a test that is intended to be free of cultural influences so that individuals' scores will not be affected by their social background. Also, **culture-fair test.**
culture lag see CULTURAL LAG.
culture medium *Bacteriology.* a liquid or solid preparation of nutrients required for the growth of a particular microorganism.
culture shock *Anthropology.* feelings of disorientation, depression, and loss experienced by an individual who is suddenly immersed in a foreign or unfamiliar culture.
culture trait *Anthropology.* the most basic unit of culture that can be analyzed, such as a hairstyle, container, or projectile point.
culture type *Anthropology.* a particular technology and its relationship to the specific environment, which produces the slight variations among societies within a culture area.
culturology *Anthropology.* the study of cultural phenomena only in the context of culture itself, without comparisons to tenets of sociology, biology, or psychology.
culvert *Engineering.* a large pipe or covered structure that carries drainage or a watercourse underground.
Cumacea *Invertebrate Zoology.* an order of small marine crustaceans with a well-developed head carapace.
cumatophyte *Ecology.* a plant that lives or thrives in surf conditions. Also, **cumaphyte.**
cumberlandite *Petrology.* a coarse-grained, ultramafic, ultrabasic rock composed principally of ferriferous olivine, magnetite, and ilmenite with small amounts of calcic plagioclase and spinel.
Cumberland method of mining see BLOCK CAVING.
cumbraite *Petrology.* an intermediate volcanic rock containing phenocrysts of very calcic plagioclase in a groundmass of labradorite, orthopyroxene, augite, magnetite, and abundant glass.
cumec *Metrology.* a measure of volume flow equal to one cubic meter per second.
cumene *Organic Chemistry.* $C_6H_5CH(CH_3)_2$, a toxic, narcotic, colorless liquid, insoluble in water and soluble in alcohol; boils at 152.7°C; used in the production of phenol and acetone, and as a solvent.
cumene hydroperoxide *Organic Chemistry.* $C_6H_5C(CH_3)_2OOH$, a toxic, combustible, colorless to pale-yellow liquid, slightly soluble in water and soluble in alcohol; used in the production of phenol and acetone, and as a polymerization catalyst.
cumengeite *Mineralogy.* $Pb_{21}Cu^{+2}_{20}Cl_{42}(OH)_{40}$, a translucent, deep blue tetragonal mineral that occurs as octahedral or cubo-octahedral crystals; having a specific gravity of 4.67 to 4.7 and a hardness of 2.5 on the Mohs scale; it is found as a secondary mineral with boleite. Also, **cumengite.**
cumidine *Organic Chemistry.* $(CH_3)_2CHC_6H_4NH_2$, a colorless liquid that is insoluble in water; boils at 225°C and melts at –63°C; used as a reagent.

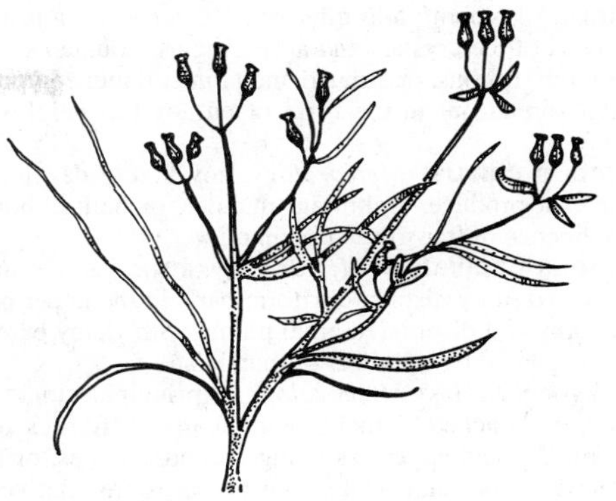

cumin

cumin [kyoo´mən] *Botany.* a small plant, *Cuminum cyminum,* of the parsley family (Umbelliferae), having seeds that are used as an aromatic spice.
cumin oil *Materials.* an essential oil derived from the plant species *Cuminum cyminum,* used in flavoring and perfumes.
cummingtonite *Mineralogy.* $(Mg,Fe^{+2})_7Si_8O_{22}(OH)_2$, a translucent to opaque, green or brown fibrous monoclinic mineral of the amphibole group that forms a series with magnesiocummingtonite and grunerite; having a specific gravity of 3.1 to 3.47 and a hardness of 5 to 6 on the Mohs scale; found in metamorphosed rocks.
cum sole [kům sō´lā] *Geophysics.* relating to motion in the same direction as the sun; i.e., clockwise in the Northern Hemisphere, counterclockwise in the Southern. (From Latin, meaning "with the sun.")
cumulants *Statistics.* coefficients in the power expansion of the logarithm of the characteristic function.
cumulate *Petrology.* any igneous rock formed from crystal accumulation near the cooling margins of a crystallizing magma.
cumulated dose see ACCUMULATED DOSE.
cumulated double bonds *Organic Chemistry.* two double bonds on the same carbon atom.
cumulative *Science.* increasing or growing by accumulation.
cumulative compound generator *Electricity.* a compound generator in which the series field is used to assist the magnetomotive force of the shunt field.
cumulative compound motor *Mechanical Engineering.* a motor that operates in a manner between that of shunt-wound (constant-speed) and series-wound (variable-speed) motors.
cumulative distribution function *Statistics.* a function describing the probability that the random variable X has a value less than or equal to a specified value x, denoted $F(x) = P(X \leq x)$; for a continuous randon variable it is the integral of the probability density function f up to x. Also, DISTRIBUTION FUNCTION.
cumulative dose see ACCUMULATED DOSE.
cumulative double bonds *Chemistry.* the bonds found in a chain of three or more carbon atoms joined by double bonds, e.g., –C=C=C–.
cumulative excitation *Atomic Physics.* a process in which the energy level of an atom is increased by bombarding it with charged particles.
cumulative feature *Archaeology.* a feature that has been formed without a plan or constraints, such as a quarry pit or a midden (refuse dump).
cumulative frequency distribution *Statistics.* a listing for each of several classes of the number of cases that are less than or equal to the upper limit of each class, or greater than or equal to the lower limit.
cumulative ionization *Atomic Physics.* a method of dislodging electrons from an atom that has remained highly energized for an extended period (metastable) by bombarding the atoms with charged particles.
cumulative radiation dose *Radiology.* the total dose resulting from a series of exposures to radiation.
cumulative sum control chart *Industrial Engineering.* a control chart in which current deviations from the norm are added to the sums of previous deviations, resulting in heightened sensitivity to abrupt changes in the process level.
cumulene *Organic Chemistry.* a compound containing two or more consecutive double-bonded carbon atoms derived from acetylene.
cumuliform *Meteorology.* of or relating to clouds characterized by vertical development in the form of rising mounds, domes, or towers.
cumulo- a combining form meaning "cumulus," as in *cumulonimbus.*

cumulonimbus [kyoom´yə lō nim´bəs] *Meteorology.* a principal cloud type composed of ice crystals and appearing as mountains or huge towers with smooth, fibrous, or striated and almost flattened upper portions, which often spread out in the form of an anvil at heights of at least 35,000 feet.

cumulonimbus calvus *Meteorology.* any highly developed cumuliform cloud that produces lightning, thunder, or hail, although the top shows no evidence of transformation into ice.

cumulonimbus capillatus *Meteorology.* a species of cumulonimbus cloud characterized by distinct cirriform parts in the upper portion, usually in the form of a disordered anvil plume with wispy extensions, and usually accompanied by a shower or thunderstorm.

cumulus [kyoom´yə ləs] *Meteorology.* a principal cloud type in the form of dense, detached elements with sharp nonfibrous outlines that develop vertically and appear as rising mounds, domes, or towers. The sunlit portions are brilliant white, and the bases are relatively dark and nearly horizontal; any precipitation usually occurs as showers. *Geochemistry.* an accumulation of mineral crystals that precipitated from a magma and settled to form layers without further modification.

cumulus congestus *Meteorology.* a species of cumulus cloud having sharp outlines and sometimes great vertical development, characterized by a cauliflower or tower aspect of large size or by high towers with tops of cloudy puffs. Also, TOWERING CUMULUS.

cumulus humulis *Meteorology.* a species of cumulus cloud characterized by a generally flattened appearance and a small vertical development due to restriction by a temperature inversion in the atmosphere. Also, FAIR-WEATHER CUMULUS.

cumulus mediocris *Meteorology.* a species of cumulus cloud with moderate vertical development and indistinct upper protuberances of a cauliflower-type aspect; it does not produce precipitation but often develops into cumulus congestus and cumulonimbus clouds.

cumulus oophorus *Histology.* a layer of follicular cells surrounding the developing oocyte in an ovarian follicle.

CUN *Aviation.* the airport code for Cancun, Mexico.

cuneate [kyoon´ē āt] *Biology.* fanning out from a pointed base; wedge-shaped, such as some leaves or insect wings. *Anatomy.* any of three wedge-shaped tarsal bones located in the instep of the foot or ankle.

cuneiform [kyoo´nē´ə fôrm] *Archaeology.* the earliest known system of writing, consisting of triangular markings pressed on a clay tablet; developed by the Sumerians in about 3000 BC. *Biology.* see CUNEATE. (From a Latin term meaning "wedge-shaped.")

cunife *Materials.* a copper-nickel-iron alloy with a high value of remnant magnetization and of coercive field, used where hard magnetic materials are required.

Cunnersdorf twin law *Crystallography.* a rarely occurring relationship between normal twin crystals in feldspar in which the twin plane is (201).

Cunninghamellaceae *Mycology.* a family of fungi belonging to the order Mucorales, which lives primarily off of nonliving organic matter in soil and dung.

cunnus *Anatomy.* the female pudendum; the vulva.

Cunoniaceae *Botany.* a family of strongly tanniferous trees and shrubs of the order Rosales that often accumulate aluminum and are characterized by opposite or sometimes whorled leaves, small and usually perfect flowers, and winged or hairy seeds.

cup any hollow, cylindrical component that is closed at one end; specific uses include: *Metrology.* a unit of capacity, equal to 8 fluid ounces or one-half pint. Also, **cupful**. *Mathematics.* a set union, denoted $\cup$, a Boolean operation. *Metallurgy.* a cylindrical shell with one open end, fabricated during the initial stages of a deep-drawing operation.

cup-and-ball joint *Geology.* see BALL-AND-SOCKET JOINT.

cup anemometer *Engineering.* a device that measures wind speed, composed of three or four vanes with cuplike structures on their ends; wind speed is determined by the rate at which the vanes revolve around a central shaft.

cup barometer *Engineering.* an instrument that measures atmospheric pressure, composed of a glass tube that sits in a cup, with both the tube and the cup containing mercury.

cup-case thermometer *Engineering.* a thermometer in which the material to be measured is held in a cup container into which the bulb of the thermometer is immersed.

cup core *Electromagnetism.* a core that serves to shield the exterior of a coil by enclosing it.

cup crystal *Hydrology.* a form of depth hoar having the shape of a hollow hexagonal cup with stepped surfaces.

Cupedidae *Invertebrate Zoology.* a family of coleopterans, the reticulated beetles, in the suborder Archostemata.

cupel *Metallurgy.* a refractory container used in cupellation.

cupellation *Metallurgy.* a refining process for gold and silver, based on oxidation of the lead containing these precious metals.

cupferron *Organic Chemistry.* $C_6H_5N(NO)ONH_4$, creamy-white crystals, soluble in water and alcohol; melts at 164–165°C; used as an analytical reagent for separating copper and iron from metals.

cup fracture *Metallurgy.* a type of fracture that has a central depression. Also, **cup-and-cone fracture**.

cup fungi *Mycology.* fungi belonging to the class Discomycetes which are shaped like cups, especially those of the family Pezizales.

cup lichens *Botany.* any of several lichens with cup-shaped fruiting bodies or stalks, such as various species of *Cladonia*. Also, **cup moss.**

cupola [kyoop´ ə lə *Architecture.* a small dome raised on a circular base, often set atop a roof. *Metallurgy.* a furnace that is similar to a blast furnace, used especially to melt cast iron. *Geology.* a dome-shaped projection of the igneous rock of a batholith.

cupola drop *Metallurgy.* the residual material that is dropped from a cupola furnace after the molten metal is poured out.

cupped pebble *Geology.* a small sedimentary particle that has been hollowed out as a result of being subjected to solution.

cup product *Mathematics.* a product of elements of the deRham cohomology groups on a manifold M, denoted $\cup$ and given by $[\omega] \cup [\eta] = [\omega \wedge \eta]$, where $[\omega] \in H^k(M)$, $[\eta] \in H^l(M)$, and $[\omega \wedge \eta] \in H^{k+l}(M)$. This gives the deRham cohomology a ring structure. A corresponding product can be defined on the cohomology groups of any topological space.

cupr- a combining form meaning "copper," as in *cuprammonium.*

cuprammonium rayon *Textiles.* a rayon fabric made from regenerated cellulose treated in a solution of copper sulfate and ammonium.

cupreine *Organic Chemistry.* $C_{19}H_{22}N_2O_2$, a colorless crystalline solid, slightly soluble in water and soluble in alcohol; melts at 202°C; used in medicine. Also, HYDROXYCINCHONIDINE.

cupreous *Chemistry.* of, relating to, or containing copper.

Cupressaceae *Botany.* a family of evergreen, resiniferous shrubs and trees of the order Pinatae, characterized by small staminate and pistillate cones with opposite or whorled scales; the pistillate cones become leathery, woody, or berrylike at maturity; includes species of junipers.

cupri- a combining form meaning "copper," as in *cuprite.*

cupric [koo´prik; kyoo´prik] *Chemistry.* **1.** of or relating to copper. **2.** describing various compounds of copper, especially those in which the element has a valence of two.

cupric acetate basic *Organic Chemistry.* $Cu(CH_3COO)_2 \cdot CuO \cdot 5H_2O$, a blue or blue-green powder; slightly soluble in water and alcohol; used as a raw material to make Paris green. Also, BLUE VERDIGRIS, CRYSTAL AERUGO.

cupric arsenate *Inorganic Chemistry.* $Cu_3(AsO_4)_2 \cdot 4H_2O$ or $Cu_5H_2(AsO_4)_4 \cdot 2H_2O$, a toxic bluish powder, insoluble in water and soluble in dilute acids; used in insecticides and fungicides.

cupric arsenite *Inorganic Chemistry.* $CuHAsO_3$ or $Cu_3(AsO_3)_2 \cdot 3H_2O$, a fine, toxic, light green powder, insoluble in water; used in pesticides, pigments, and as a wood preservative. Also, COPPER ARSENITE, COPPER ORTHOARSENITE, SCHEELE'S GREEN.

cupric bromide *Inorganic Chemistry.* $CuBr_2$, a deliquescent black powder or crystals, very soluble in water and soluble in alcohol and acetone; melts at 498°C; used in photography, as a brominating agent, in battery elecrolytes, and as a wood preservative.

cupric carbonate *Inorganic Chemistry.* $Cu_2(OH)_2CO_3$, a toxic green powder, soluble in acids and insoluble in water and alcohol; decomposes at 200°C; used in pigments, fireworks, and insecticides, and for various other industrial purposes. Also, (BASIC) COPPER CARBONATE, ARTIFICIAL MALACHITE, MINERAL GREEN.

cupric chloride *Inorganic Chemistry.* **1.** $CuCl_2$, a brownish-yellow hygroscopic powder, soluble in water; melts at 620°C and decomposes to cuprous chloride at 993°C. **2.** $CuCl_2 \cdot 2H_2O$, the dihydrate form, bluish-green deliquescent crystals; loses water at 100°C; has a wide variety of uses, as in fungicides, disinfectants, preservatives, and purifiers, and as a mordant, catalyst, and pigment.

cupric chromate *Inorganic Chemistry.* $CuCrO_4 \cdot 2CuO \cdot 2H_2O$, a yellowish-brown powder that is soluble in nitric acid and insoluble in water; loses its water at 260°C; a known carcinogen; used as a mordant, wood preservative, and fungicide.

cupric cyanide *Inorganic Chemistry.* $Cu(CN)_2$, a poisonous, unstable yellowish-green powder, soluble in acids and alkalis and insoluble in water; used in copper electroplating.

cupric fluoride *Inorganic Chemistry.* **1.** $CuF_2\cdot 2H_2O$, blue monoclinic crystals, slightly soluble in water and soluble in acids and alcohol; a poison; used in ceramics, enamels, and batteries. **2.** the anhydrous form, CuF_2, a white crystalline powder that decomposes at 950°C.

cupric fluorosilicate *Inorganic Chemistry.* $CuSiF_6\cdot 4H_2O$, blue hygroscopic crystals, soluble in water and slightly soluble in alcohol; decomposed by heat; a poison; used to color and treat marble. Also, CUPRIC SILICOFLUORIDE.

cupric gluconate see COPPER GLUCONATE.

cupric hydroxide *Inorganic Chemistry.* $Cu(OH)_2$, a toxic blue powder, soluble in acids and insoluble in water; decomposes on heating; used in pesticides and fungicides and as a mordant, pigment, and catalyst. Also, **cupric hydrate.**

cupric nitrate *Inorganic Chemistry.* $Cu(NO_3)_2\cdot 3H_2O$ or $\cdot 6H_2O$, blue hygroscopic crystals, soluble in water and alcohol; the trihydrate form melts at 114.5°C and decomposes at 170°C, and the hexahydrate form loses $3H_2O$ at 26.4°C; used as a reagent, mordant, and nitrating agent, in paints, varnishes, and enamels, and for many other industrial purposes.

cupric oleate see COPPER OLEATE.

cupric oxide *Inorganic Chemistry.* CuO, a toxic brownish-black powder, insoluble in water and soluble in acids; melts at 1326°C; used in coloring ceramics and in many chemical and industrial processes. Also, COPPER OXIDE BLACK, COPPER MONOXIDE.

cupric resinate see COPPER RESINATE.

cupric silicofluoride see CUPRIC FLUOROSILICATE.

cupric sulfate *Inorganic Chemistry.* $CuSO_4\cdot 5H_2O$, blue crystals that lose water of crystallization to the atmosphere, white when dehydrated; soluble in water and slightly soluble in alcohol; toxic and an irritant; used in agriculture, medicine, textiles, and the leather industry, and for a wide range of other purposes.

cupric sulfide *Inorganic Chemistry.* CuS, a toxic black powder, lumps, or crystals, insoluble in water and soluble in nitric acid; decomposes at 220°C; it is used in antifouling paints, dyeing, and catalyst preparation.

cuprite *Mineralogy.* $Cu_2^{+1}O$, a brittle, translucent to transparent red mineral occurring as cubic or octahedral or dodecahedral crystals and in massive form; having a specific gravity of 6.14 and a hardness of 3.5 to 4 on the Mohs scale; an important ore mineral found in oxidized zones of copper deposits, often with native copper, azurite, and malachite.

cupro- a combining form meaning "copper," as in *cupronickel.*

cuprocopiapite *Mineralogy.* $Cu^{+2}Fe_4^{+3}(SO_4)_6(OH)_2\cdot 20H_2O$, a transparent to translucent yellow triclinic mineral occurring as tabular crystals or more often as scaly aggregates or crusts; having a specific gravity of 2.08 to 2.17 and a hardness of 2.5 to 3 on the Mohs scale; found as a secondary mineral formed by the oxidation of pyrite and other sulfides.

cupronickel *Metallurgy.* any of several copper-base, nickel-bearing alloys.

cuprotungstite *Mineralogy.* $Cu_3^{+2}(WO_4)_2(OH)_2$, a waxy green mineral of unknown crystal structure; having a specific gravity of 7.06 and an undetermined hardness on the Mohs scale; it forms microcrystalline crusts replacing outer portions of contact-molybdenian scheelite in metamorphic copper deposits.

cuprous [koo´prəs; kyoo´prəs] *Chemistry.* **1.** of or relating to copper. **2.** describing various compounds of copper, especially those in which the element has a valence of 1.

cuprous bromide *Inorganic Chemistry.* CuBr, white crystals that turn green when exposed to light; very slightly soluble in cold water and decomposes in hot water; melts at 492°C and boils at 1345°C; used as a catalyst in organic reactions.

cuprous chloride *Inorganic Chemistry.* CuCl, white crystals that turn green when exposed to air and brown when exposed to light; slightly soluble in cold water and soluble in acids; melts at 430°C and boils at 1490°C; used as a catalyst, preservative, and fungicide.

cuprous fluoride *Inorganic Chemistry.* CuF, red crystals that are soluble in hydrochloric acid and insoluble in water and alcohol; melts at 908°C and sublimes at 1100°C.

cuprous oxide *Inorganic Chemistry.* Cu_2O, toxic reddish-brown crystals, insoluble in water and soluble in acids; melts at 1235°C and boils at 1800°C; used in ceramics, electroplating, and for various other purposes. Also, COPPER OXIDE RED.

cupule *Botany.* a cuplike structure made of hardened, coherent bracts, as is found on an acorn.

curare [koo rä´rā; kyoo rä´rā] *Materials.* a highly toxic mixture of various alkaloids occurring in several species of South American trees, traditionally used as an arrow poison; it acts on the central nervous system, and its derivatives are used in medicine as a muscle relaxant.

curassow

curassow [kyůr´ə sō´] *Vertebrate Zoology.* any of several large subtropical game birds of the family Cracidae, especially the genus *Crax*; native to jungles of Central and South America.

curb roof *Architecture.* a pitched roof with a double slope on each side.

Curculionidae *Invertebrate Zoology.* a family of weevils or snout beetles in the superfamily Curculionoidea; the head is in the form of a snout used for boring into grain and plants; a serious pest.

Curcurbitaceae *Botany.* a family of dicotyledonous herbs and vines, whose members bear many varieties of gourds and melons.

curd *Food Technology.* clotted protein that is formed when fresh milk is treated with an enzyme such as rennin. *Botany.* the flower top of a cauliflower or broccoli.

curdlan *Biochemistry.* a class of β glucans that form highly elastic and resilient gels.

cure *Chemistry.* to convert a raw substance into a finished product, usually by using heat or chemicals; e.g., tanning leather, vulcanizing rubber, or salting and drying food. *Materials Science.* to change the qualities or properties of a resin material by chemical action or by applying heat. *Engineering.* see CURING.

curet [ků ret´]*Surgery.* **1.** a spoonlike device for scraping material from the wall of a cavity. Also, **curette. 2.** to use a curet.

cure time see CURING TIME.

curettage [kůr´ə tazh´] *Surgery.* the use of a curet for removing material from the wall of a cavity. Also, **curettement.**

Curie, Marie [kyůr´ē] 1867–1934, Polish-born French physicist and chemist; Nobel Prizes for research in radiation (with her husband Pierre Curie) and for isolating and analyzing radium and polonium.

Curie, Pierre 1859–1906, French physicist; discovered Curie point and piezoelectric effect; shared Nobel Prize for radiation studies.

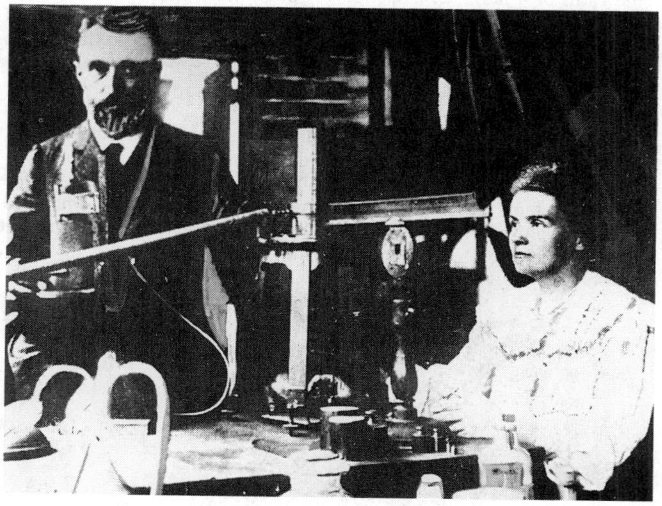

Marie and Pierre Curie

curie *Nucleonics.* **1.** a traditional unit for measuring radioactivity, defined as the activity of a radionuclide decaying at the rate of 3.7×10^{10} disintegrations per second. **2.** the quantity of a nuclide having 1 curie of radioactivity.

Curie constant *Electromagnetism.* a constant given by the magnetic susceptibility c of a normally ferromagnetic substance at a given temperature times the difference between T and T_c, where T is the absolute temperature and T_c is the Curie point; only valid for $T > T_c$.
Curie point *Physics.* the temperature above which a ferromagnetic substance becomes paramagnetic.
Curie principle *Thermodynamics.* the principle that a macroscopic cause cannot have more elements of symmetry than the effect which is produced by the cause.
Curie's law *Electromagnetism.* a law stating that the magnetic susceptibility of a paramagnetic material is proportional to the inverse of the absolute temperature.
Curie substance *Thermodynamics.* a substance whose magnetic dipole moment M is numerically equal to the Curie constant, multiplied by the magnetic intensity of the external magnetic field H, divided by the temperature T; i.e., $M = C \cdot (H/T)$
Curie temperature *Electromagnetism.* the temperature at which ferromagnetism changes to paramagnetism for a specified ferromagnetic material.
curie therapy *Radiology.* a term for any type of radiation therapy.
Curie-Weiss law *Electromagnetism.* a law stating that at temperatures above the Curie point, the magnetic susceptibility of a paramagnetic substance varies inversely with the absolute temperature between the temperature of the substance and the Curie point.
curing *Food Technology.* any of a variety of processes that employ chemicals (such as sodium chloride, nitrate, and nitrite), sugar, and spices to preserve meat while improving its taste, tenderness, and color. Corned beef, smoked bacon, and ham are cured meats. *Engineering.* a process that provides for the optimal solidification of concrete, in which the concrete is kept damp for a period extending from one week up to one month.
curing time *Materials Science.* the length of time that a substance must be kept in curing conditions in order to complete the curing process.
curiosity behavior see EXPLORATORY BEHAVIOR.
curite *Mineralogy.* $Pb_2U_5^{+6}O_{17} \cdot 4H_2O$, a transparent to translucent, deep yellow to reddish-brown orthorhombic mineral occurring in massive granules and long prismatic crystals; having a specific gravity of 7.37 to 7.4 and a hardness of 4 to 5 on the Mohs scale; found as crystal aggregates with torbernite, kasolite, and other secondary uranium minerals.
curium *Chemistry.* a synthetic radioactive element having the symbol Cm, the atomic number 96, atomic weights of isotopes 242–247.07, and melting point 1340°C; a chemically reactive, silver-white metal used for remote, small-scale power generation. (After Marie and Pierre *Curie.*)
curium-244 *Nuclear Physics.* a radioactive isotope of curium with a half-life of 18.1 years that emits alpha particles during decay.
curl *Forestry.* **1.** a long flat flake of wood cut by a knife so that it takes the form of a helix. **2.** a section of rough-hewn timber cut from the crotch of a tree and used for cutting into veneers. *Mathematics.* the curl of a vector-valued function a, denoted $\nabla \times a$, is the cross product of the del operator and a; that is,

$$\nabla \times a = (i\partial/\partial x + j\partial/\partial y + k\partial/\partial z) \times a.$$

Curl is defined only in three-dimensional space. In higher dimensions, the analogous operator is the exterior derivative.
curlew *Vertebrate Zoology.* a migratory shorebird of the genus *Numenius* in the family Scolopacidae, having a very long slender downcurved bill, streaked brown plumage, and long legs; found in the Northern Hemisphere.

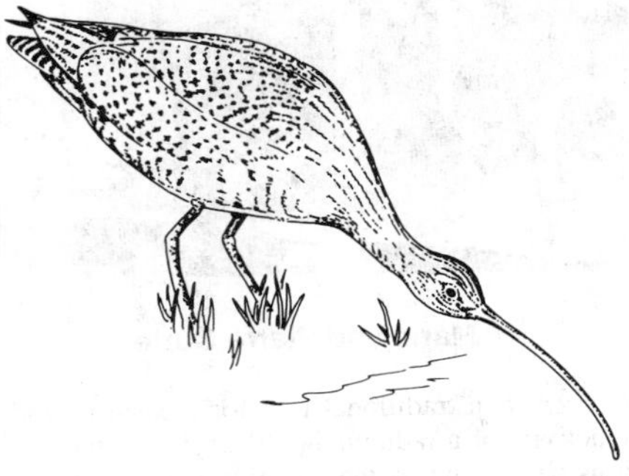

curlew

curling dies *Mechanical Engineering.* tools that shape the ends of a workpiece into a form with a circular cross section.
curling factor see GRISEOFULVIN.
curling machine *Mechanical Engineering.* a machine that uses curling dies to shape the ends of cans.
Curling's ulcer *Medicine.* a duodenal ulcer brought on after a severe burn of the skin or severe injury to the body.
curly top *Plant Pathology.* an insect-borne virus disease of plants, particularly beets, which causes stunted growth and curled and puckered leaves.
currant *Botany.* the common name for plants of some species of the genus *Ribes* or their fruits.

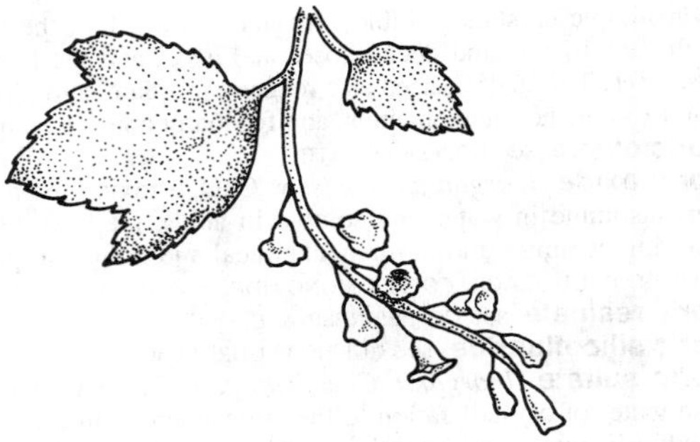

currant

currant jelly clot *Hematology.* a clot of reddish color because of the presence of erythrocytes enmeshed in it.
currant leaf spot *Plant Pathology.* **1.** a disease of currants caused by the imperfect fungus *Cercospora angulata,* and characterized by the appearance of angular discolored spots on the leaves. **2.** an anthracnose currant disease caused by the fungus *Pseudopeziza ribis* and characterized by small dark spots on the leaves and fruits.
currency pointers *Computer Programming.* a set of keys within certain records in a database, maintained automatically by the database system in order to keep track of recently accessed records.
current *Hydrology.* a horizontal movement of water in a well-defined, established pattern, as in a river or stream. *Meteorology.* the movement of a definite body of air in a certain direction. *Physics.* the amount of a quantity flowing past a reference point per unit time. *Electricity.* the rate of transfer of electrons or of positive ions, negative ions, or holes from one point to another; measured in amperes.
current amplification *Electronics.* the ratio of the output signal current in an amplifier circuit to the input signal current. Also, CURRENT GAIN.
current amplifier *Electronics.* an amplifier circuit designed to produce an output current larger than its input current.
current antinode *Electricity.* the point at which current is at a maximum along a medium that has standing waves. Also, **current loop.**
current attenuation *Electronics.* the ratio of the input current of a transducer to the output current; the inverse of current gain.
current balance *Electricity.* a device used to measure the force between current-carrying conductors.
current bedding *Geology.* any bedding structure, such as cross-bedding, produced by the action of wind or water currents.
current-carrying capacity *Electricity.* the maximum current a conductor can continuously carry without permanent damage.
current chart *Cartography.* a map of a water area depicting current speeds and directions by current roses, vectors, or other means.
current constants *Oceanography.* those characteristics of a current, at any specified locality, that remain essentially unchanging.
current-controlled switch *Electronics.* a semiconductor device in which a controlling bias current sets the resistance very high (switch off) or very low (switch on).
current cross section *Oceanography.* a graphic representation of a current shown as a vertical plane at right angles to the axis of flow; the abscissa represents the horizontal difference between surface limits, and the ordinate represents the depth.
current curve *Oceanography.* a rectangular coordinate graphic representation of the speed and duration of a tidal current; the abscissas represent the time, and the ordinates represent the speed.

current cycle *Oceanography.* a complete set of tidal current characteristics during a given period, such as a tidal day, a lunar month, or a Metonic cycle.
current decay *Metallurgy.* in resistance welding, the intentional reduction of electrical current to a level that prevents damaging stresses during cooling.
current density *Electricity.* the current flowing through a given cross-sectional area of a conductor; usually represented by a vector whose direction is in the direction of the current.
current distribution *Physical Chemistry.* the variation of current density across the surface of an electrode.
current divider *Electricity.* a mechanism employed to produce a predetermined portion of a total current to a circuit.
current drain *Electricity.* the amount of current a circuit draws from a power supply. Also, DRAIN.
current drift *Hydrology.* a shallow, broad, slow-moving current in a lake or ocean.
current efficiency *Physical Chemistry.* the ratio between the amount of electricity theoretically needed in an electrochemical process and the amount actually used.
current ellipse *Oceanography.* a graphic representation of a rotary current that uses radius vectors and vectorial angles to represent the velocity and direction of the current at different hours of the tidal cycle.
current feed *Electronics.* a method of connecting a radio-frequency transmission line to an antenna at the point at which the standing wave of the current is at maximum amplitude.
current feedback *Electronics.* a feedback voltage in amplifier circuits that is proportional to the load current.
current gain see CURRENT AMPLIFICATION.
current generator *Electronics.* a two-terminal circuit element whose terminal current is not dependent on the voltage between its terminals; ideally it would have zero internal admittance.
current gradient *Oceanography.* the rate of a current's increase or decrease in speed, relative to a given distance or period of time.
current hogging *Electronics.* a term for a condition in which one of several parallel-connected logic circuits draws more current because of improper balance among the circuits; such a condition can be cumulative and may eventually impair the operation of the associated circuits.
current hour *Oceanography.* the mean interval between the moon's transit over a reference meridian, usually Greenwich, and the time of the strength of flood current modified by the times of slack and of strength of ebb current.
current location reference *Computer Programming.* in an assembly language, a special symbol representing the current location; often used as a reference base from which to branch to another instruction that is a fixed number of statements offset from the current instruction.
current mark *Geology.* any structure resulting from the action of a water current on a sedimentary surface. Also, **current marking.**
current-mode logic *Electronics.* a logic circuit whose operation depends on switching the current from one path to another.
current node *Electricity.* a point in a transmission line, antenna, or other circuit element at which the current is zero.
current noise *Electronics.* the low-frequency electrostatic noise that is produced when an electric current flows through certain components, particularly carbon resistors. Also, EXCESS NOISE.
current profile *Oceanography.* a graphic representation of the speed of a current at different depths; the abscissa represents the speed, and the ordinate represents the depth.
current rating *Electricity.* the maximum allowable current that a device can withstand without damage; may refer to instantaneous current or continuous current for a specified period of time.
current ratio *Electromagnetism.* the ratio of the maximum to the minimum current in a waveguide.
current regulator *Electronics.* a device intended to maintain the current through a circuit at a predetermined value, usually by automatically adjusting applied voltage or circuit resistance.
current relay *Electricity.* a relay that functions at a specific current value instead of at a specific voltage value.
current ripple *Geology.* an asymmetrical ripple mark created by wind or water currents moving in a constant uniform direction over sand. Also, **current ripple mark.**
current rips *Oceanography.* turbulent movement on the surface of the water, caused by the meeting of opposing ocean currents; the waves are not progressive but rather are random vertical oscillations of the water surface.
current rose *Cartography.* a graphic depiction of currents for specified areas, using arrows at the cardinal and intercardinal compass points to show the direction toward which the prevailing current flows and the frequency with which this direction occurs for a given period of time.
current saturation see ANODE SATURATION.
current speed *Oceanography.* the rate at which a current moves, either horizontally or vertically.
current table *Oceanography.* a table giving daily predictions for the time, speed, and direction of tidal currents at selected locations.
current transformer *Electricity.* a transformer having its primary winding connected in series with the circuit to be measured or controlled; the current is measured across the secondary winding.
current velocity *Oceanography.* a rate of motion that includes current direction as well as current speed.
current-voltage dual *Electricity.* a circuit that is identical to a given circuit when quantities are replaced with dual quantities, such as current and voltage impedance and admittance.
curry *Food Technology.* a mixture of various spices such as turmeric, coriander, cayenne, and black pepper; widely used in the cuisine of India and Pakistan. Also, **curry powder.**
currying *Computer Science.* the replacement of a function of multiple arguments by a function of one argument, which returns as its value a function that can be applied to the remaining arguments. For example, (+ 1 x), the addition of 1 to x, can be replaced by (1+ x) where 1+ is a function that adds 1 to a number.
cursor [kur´sər´] *Computer Technology.* on a CRT screen, a special symbol or character, such as a flashing line or a highlighted box, that moves along as characters are entered on a keyboard, thus indicating where the next character to be entered will appear. *Mechanical Devices.* **1.** a clear cover with a radial line marked or inscribed on it, which is placed over a radar display screen in order to read bearings between 0° and 360°. **2.** the adjustable glass-lined part of a slide rule, which can be set at any point along its scale.
cursor arrow *Computer Technology.* on a terminal keyboard, any of the special-purpose keys that control the vertical and horizontal movement of the cursor.
cursorial *Vertebrate Zoology.* being adapted or specialized for running; e.g., the feet and limbs of horses.
curtain *Geology.* **1.** a wavy or folded thin sheet of dripstone hanging or projecting from the wall or ceiling of a cave. Also, DRAPERY, DRIP CURTAIN. **2.** a rock formation that connects two neighboring bastions.
curtain array *Electromagnetism.* an antenna array in which the radiating elements are half-wave dipoles suspended vertically between two suspension cables.
curtain board *Building Engineering.* a board or partition placed on a ceiling with a fire-retarding chemical agent.
curtain coating *Chemical Engineering.* a method in which a substrate is passed perpendicularly through a liquid falling sheet of coating material of low-viscosity solutions or resins.
curtain rhombic antenna *Electromagnetism.* a broadband multielement rhombic antenna in which two or more conductors join at the feed, and the terminating ends and the elements are vertically spaced at intervals of about 30–150 cm.
curtain wall *Building Engineering.* an exterior wall with no structural function in a frame building; therefore, it is not load bearing.
curtate cycloid *Mathematics.* the trochoid curve with parametric equations $x = a\phi - b \sin \phi$ and $y = a - b \cos \phi$, where $a > b$. (When $a < b$, the curve is called a prolate cycloid, and when $a = b$, a cycloid.)

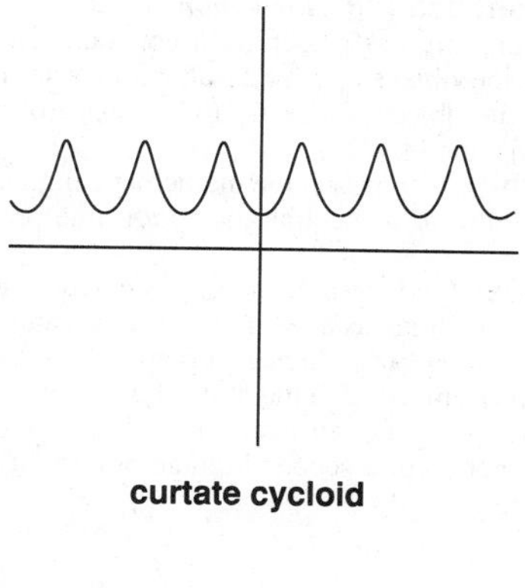

curtate cycloid

Curtius reaction *Organic Chemistry.* the rearrangement, when heated, of acyl azides, $RCON_3$, to amines, RNH_2, or isocyanates, *R*NCO. Also, **Curtius rearrangement.**

Curtobacterium cyanelles *Bacteriology.* a genus of aerobic bacteria of the order Actinomycetales, occurring as rod-shaped, coccoid cells in soil and plant litter, some species of which are plant parasites.

curvature *Mathematics.* **1.** the curvature of a circle of radius *r* at any point on the circle is defined to be $1/r$. **2.** more generally, the curvature of a curve is measured by the rate of change of the unit normal(s) to the curve with respect to arc length *s* measured along the curve. If α is the angle of normal (or tangent) to a plane curve, the curvature at a point of the curve is equal to $d\alpha/ds$. This curvature is equal to the curvature of the circle that most closely approximates the curve at the point.

curvature correction *Geodesy.* the correction applied in some geodetic procedures that takes into account the divergence of the surface of the earth, or its representing ellipsoid, from a plane.

curvature effect *Electronics.* a phenomenon in which the dielectric strength of a liquid or vacuum separating two electrodes increases as the electrode's radius of curvature decreases.

curvature form *Mathematics.* **1.** in the Cartan theory of moving frames, the curvature form is the differential 2-form Ω_j^i defined by the equation(s)

$$d\omega_j^i = -\Sigma_k\, \omega_k^i \wedge \omega_j^k + \Omega_j^i$$

where (ω_k^i) is the Cartan connection. **2.** in the Ehresmann theory of principal fiber bundles, the curvature form is the covariant differential of the Ehresmann connection; $\Omega = D\omega$.

curvature of field *Optics.* an aberration that occurs when rays from a flat object are brought onto a curved image surface by an uncorrected lens.

curvature tensor *Mathematics.* a tensor field on a differential manifold *M* formed from the first derivatives of the covariant connection that specifies the curvature of *M* at every point of *M* in every direction in the tangent space of *M*. The curvature (linear transformation) tensor $R(X, Y)$ is formed from the Koszul connection ∇ by the formula

$$R\,(X, Y)Z = \nabla_X(\nabla_Y Z) - \nabla_Y(\nabla_X Z) - \nabla_{[X,Y]}Z) = ([\nabla_X, \nabla_Y] - \nabla_{[X,Y]})\mathbf{Z}$$

$R(X, Y)$ is a derivation of degree zero on the mixed tensor algebra. On a (pseudo-) Riemannian manifold with metric tensor *g* and $\nabla_X g = 0$ for all *X*, the curvature tensor is called the **Riemann curvature tensor**, and is often exhibited with lowered indices:

$$R(W, Z, X, Y) = g(W, R(X, Y)Z)$$

curve *Mathematics.* **1.** any continuous image of a closed interval; in *n*-space, a set of *n* parametric equations $x_i = a_i(t)$. In integration theory, this definition is equivalent to a singular 1-cube. **2.** the intersection of two 2-dimensional surfaces. **3.** the image of any interval under a map of specified differentiability class; e.g., C^r curve, etc.

curved beam *Engineering.* a beam having arcs at both ends. Also, **curve beam.**

curved space-time *Physics.* a four-dimensional geometry used extensively in general relativity theory; the curvature is determined by the distribution of mass-energy.

curve fitting *Statistics.* the process of deriving the mathematical curvilinear equation that most closely describes the empirical data; often calculated by minimizing the sum of the squares of the vertical deviations about the model equation, a process known as the least-squares method.

curve follower *Computer Technology.* a peripheral device that reads the coordinates of a graph or other line for the purpose of digitizing the information into computer-readable form. Also, GRAPH FOLLOWER.

curve of constant bearing *Navigation.* the curved path that would be followed by a moving craft keeping a constant relative bearing to a fixed object. If the object is directly ahead or astern, this would be a straight line; if the object is directly to starboard or port, the resulting path would form a circle.

curve of growth *Astrophysics.* the mathematical relation that describes the equivalent width of a spectral absorption line produced by a given number of atoms.

curve resistance *Engineering.* the opposing force to the motion of a railway train along a track as a result of track curvature.

curves of form *Naval Architecture.* a graph plotting various properties of a hull form in relationship to the draft of a vessel.

curve tracer *Engineering.* an instrument that presents one voltage or current as the function of a second voltage or current, with a third serving as a parameter.

curve tracing *Mathematics.* the process of drawing the graph of a curve, using such analytic tools as symmetry, asymptotes, and derivatives (which give information about slope, critical points, convexity, etc.), as well as calculating points on the graph.

curvilinear [kur´və lin´ē ər] *Science.* consisting of, represented by, or bound by curved lines.

curvilinear coordinates *Cartography.* a term for any linear coordinates, such as polar or cylindrical coordinates, that are not a Cartesian coordinate system.

curvilinear motion *Mechanics.* motion that occurs along a curved path rather than in a straight line.

curvilinear regression *Statistics.* a method of attempting to predict values of a dependent variable from the values of one or more independent variables by determining a mathematical expression for a curve that best fits the observed data.

Cuscutaceae *Botany.* a family of leafless, parasitic herbs of the order Polemoniales, having neither chlorophyll nor internal phloem.

cusec *Metrology.* a measure of volume flow equal to one cubic foot per second.

Cushing, Harvey 1869–1939, American surgeon; pioneer in neurosurgery; conducted important research on the brain, nervous system, and pituitary gland.

Cushing's disease *Medicine.* a particular type of Cushing's syndrome affecting those patients with pituitary adenoma or hypersecretion of the pituitary.

Cushing's syndrome *Medicine.* a condition caused by increased adrenocortical secretion of cortisol resulting from adrenocortical hyperplasia or tumor and characterized by obesity, acne, amenorrhea, hypertension, abdominal pain, and weakness. Also, BASOPHILISM.

cushion plant *Ecology.* a herbaceous perennial plant growing low to the ground and forming dense masses; usually found in windy areas.

cusk eel *Vertebrate Zoology.* an eel-like marine fish of the family Ophidiidae, having a large mouth, whiskered feelers on its chin that aid in bottom-feeding, and a yellow-brown body with joined dorsal, anal, and tail fins.

cusp *Astronomy.* either tip of the moon or another heavenly body when it is observed to have a crescent shape. *Geology.* **1.** a crescent-shaped landform having curved indentations on both sides. **2.** any of a group of regularly spaced, crescent-shaped ridges or mounds of beach material formed by wave action and lying more or less perpendicular to the shoreline. *Architecture.* a pointed projection formed at the intersection of two arcs, as in Gothic tracery. *Geophysics.* a funnel-shaped area in the magnetosphere containing solar wind plasma. *Anatomy.* **1.** a leaflet or flap on a valve of the heart. **2.** one of the protuberances on or near the masticating surface of a tooth. *Mathematics.* a double point on a curve at which the two tangents are coincident; it may be visualized as a spike. A **cusp of the first kind,** or **simple cusp,** has a branch of the curve on each side of the double tangent in an arbitrarily small neighborhood of the cusp; i.e., the branches have convexities of opposite sign. A **cusp of the second kind** has both branches of the curve on the same side of the double tangent; that is, the branches have convexities of the same sign.

cuspate bar *Geology.* a crescent-shaped bar that connects to the shore at each end.

cuspate delta *Geology.* a tooth-shaped delta whose deposits have been spread by wave action uniformly on either side of the mouth of a river, thus forming two curving beaches, both concave toward the water.

cuspate foreland *Geology.* a large cape or broad triangular landform of alluvial material having crescent-shaped indentations on either side, whose apex points seaward along an open coast.

cuspate ripple mark see LINGUOID RIPPLE MARK.

cusp cap *Astronomy.* a polar region on Venus that frequently appears bright at visible wavelengths and is much more prominent when seen in the ultraviolet.

cusped magnetic field *Electromagnetism.* a magnetic field pattern that arises between two similar proximate coils that share a common axis but have oppositely directed currents.

cuspidal cubic *Mathematics.* a cubic curve having exactly one cusp, one inflection point, and no nodes.

cuspidal locus *Mathematics.* a set of points, each of which is a cusp of a member of a given family of curves.

cuspidate *Biology.* having a sharp, pointed projection or end.

cuspid valve *Anatomy.* a flap of tissue that controls the flow of blood between an atrium and ventricle of the heart.

custard winds *Meteorology.* cold easterly winds that blow along the northeast coast of England.

Custer's effect *Microbiology.* a process in which the fermentation of a substrate is stimulated by oxygen.

custom *Anthropology.* a behavior pattern, habit, or tradition that is distinctive of a particular culture.

customary system or **customary measure** see ENGLISH MEASURE.

cut any of various processes considered to be analogous to the action of a knife; specific uses include: *Geology.* **1.** to dig or hollow out a channel or other such incision by erosion. **2.** any incision produced by erosion or other natural means, as by waves. *Computer Programming.* to remove text as part of a cut-and-paste process. *Graphic Arts.* a printed illustration, especially a halftone. *Chemical Engineering.* a petroleum or chemical fraction obtained by a distillation process. *Crystallography.* a portion of a crystal that has two parallel surfaces, typically referenced to the natural crystallographic axes. *Mathematics.* **1.** a set A of arcs of a network that is a cutset in the underlying digraph and that separates the source from the sink. **2.** a subset C of a set S in a topological space such that $S - C$ is not connected.

cut-and-carry method *Metallurgy.* in stamping, a method that leaves the product attached to the original strip.

cut and fill *Geology.* **1.** a gradation or leveling process during which material eroded by waves, streams, currents, or winds from one place is deposited nearby, until the erosional and depositional surfaces become uniformly graded. **2.** a small erosional channel that has been filled in, forming a sedimentary structure.

cut and paste *Computer Programming.* **1.** a capability provided with word-processing programs that enables the user to remove a block of text and place it elsewhere in the same document or in a different document. **2.** to move a block of text from one place to another.

cut-and-patch repair *Genetics.* a method of repairing damaged DNA by replacing a defective region with a newly synthesized sequence, thereby preventing mutation or cell death. Also, EXCISION REPAIR.

cutaneous [kyoo tān´ē əs] *Anatomy.* of or relating to the skin.

cutaneous anaphylaxis *Immunology.* an acute, local inflammatory reaction or hypersensitivity caused by contact with a sensitizing agent and characterized by a skin reaction.

cutaneous anthrax see MALIGNANT PUSTULE.

cutaneous blastomycosis *Medicine.* an infection caused by *Blastomyces dermatitidis* and characterized by tumors causing suppurating lesions on the skin.

cutaneous lupus erythematosus *Medicine.* a chronic form of lupus erythematosus that causes skin lesions of the face, neck, and upper extremities.

cutaneous pain *Physiology.* the sensation caused by injury of the free nerve endings in the skin.

cutaneous reaction *Medicine.* the reaction of the skin to an irritation by contact, by an antigen, or by any substance to which the subject is sensitive.

cutaneous sensation *Physiology.* a sensation produced by stimulation of the specialized neural receptors in the skin, such as those for heat (Ruffini end organs), cold (Krause end organs), touch (Meissner corpuscles), pressure (Pacinian corpuscles), and pain (free nerve endings).

cutaneous sensitization *Physiology.* a state in which the skin becomes hyperreactive to an antigen.

cutback *Chemical Engineering.* a blending of heavier and lighter oils to bring the heavier ones to desired specifications.

cutbank *Geology.* a concave, steep slope formed by the lateral erosion of a meandering stream.

cut capacity *Mathematics.* given a network N and a cut (that divides the vertices of N into two sets C_1 and C_2), then the cut capacity is the sum of the capacities of all dipaths from C_1 to C_2.

cut edge *Mathematics.* an edge whose removal from a graph increases the number of components of the graph. Also, BRIDGE, ISTHMUS.

Cuterebridae *Invertebrate Zoology.* a family of large, hairy, two-winged insects, the botflies, in the subsection Calypteratae; larvae are parasitic under the skin of mammals.

cut film *Graphic Arts.* photographic film that is supplied in single sheets, with one sheet for each exposure, as distinguished from roll film. Also, SHEET FILM.

cut glass *Materials Science.* glass that is decorated or shaped by cutting or grinding with abrasive wheels.

cuticle *Anatomy.* **1.** the horny layer of a nail fold at the point where it attaches to the margin of the nail plate. Also, EPONYCHIUM. **2.** a layer of more or less solid substance that covers the free surfaces of epithelial cells. **3.** any fine covering. *Botany.* a waxy or fatty layer on the outer wall of epidermal cells, which is rather impervious to water.

cutie pie *Nucleonics.* a popular term for a rugged, lightweight, portable survey meter consisting of an ionization chamber for detecting and measuring both beta and gamma radiations to an accuracy of ±10%.

cutin [kyoo´tin] *Biochemistry.* the layer of waxy material that covers and protects the surface of leaves and green stems in plants.

cut-in *Control Systems.* a value that is used to close a control circuit when a predetermined temperature or pressure is reached.

cut-in angle *Electronics.* the phase angle of the input power cycle in a rectifier circuit at which conduction of a diode begins; a cut-in angle can never be exactly 0° or 180° because of the forward bias requirements of the rectifying diodes.

cutinase *Enzymology.* an enzyme that catalyzes the destruction of cutin, a waxy substance that is present in the cuticle of plants.

cutinite *Geology.* a coal maceral formed from the hardened outer cover of plant epidermal cells.

cutinization *Botany.* the deposition of cutin in cell walls, making them impermeable to water.

cutlass fish *Vertebrate Zoology.* any ribbonlike fish of the family Trichiuridae, having daggerlike teeth.

Cutler feed *Electromagnetism.* a radio-frequency antenna feeding system in which energy is transferred to a reflector through a resonator cavity at the end of a waveguide; commonly used in aircraft.

Cutleriaceae *Botany.* a family of brown algae comprising the order Cutleriales.

Cutleriales *Botany.* a monofamilial order of brown algae of the class Phaeophyceae, characterized by trichothallic growth, parenchymatous construction of both gametophytes and sporophytes, and anisogamous sexual reproduction.

cut nail *Mechanical Devices.* a flat, tapered nail cut from sheet metal rather than round wire, used for fastening applications in flooring that is subject to shifting loads.

cutoff an instance of interrupting or stopping a movement or process, or a device that does this; specific uses include: *Geology.* a boundary line indicating the areal limit of a stratigraphic unit that is not defined by natural features. *Hydrology.* the new and relatively short channel or body of water created when a stream cuts through the neck of an oxbow. *Mechanical Engineering.* the point in an engine cycle at which the fuel supply is stopped. *Physics.* a technique in which an integration limit of some variable or parameter is established because its integration contribution would otherwise produce a physically undefined quantity, usually an infinite contribution.

cutoff attenuator *Electromagnetism.* a nondissipative attenuator consisting of an adjustable-length waveguide, operating at frequencies below the cutoff frequency.

cutoff bias *Electronics.* **1.** the minimum value of negative voltage on the control grid in an electron tube that stops the flow of plate current. **2.** the value of base-to-emitter bias voltage (approximately 0 volts) in a transistor that causes collector current to stop. **3.** see CRITICAL VOLTAGE.

cutoff field see CRITICAL FIELD.

cutoff frequency *Electronics.* **1.** the upper and lower frequencies at which the gain of an amplifier is reduced to the half-power point (0.707 times the maximum gain for voltage and current amplifiers). **2.** the frequency below which radio waves are reflected by the ionosphere and above which they penetrate the ionosphere layer. **3.** the upper or lower frequency beyond which a filter or other frequency-selective device fails to respond to an input.

cutoff high *Meteorology.* a warm high that has become displaced and lies north of the basic westerly current.

cutoff lake see OXBOW LAKE.

cutoff low *Meteorology.* a cold low that has become displaced and lies south of the basic westerly current.

cutoff valve *Mechanical Engineering.* in a steam engine, a valve that stops the flow of steam to a cylinder.

cutoff voltage *Electronics.* **1.** the minimum value of negative voltage on the control grid of an electron tube that stops the flow of plate current. **2.** the value of base-to-emitter voltage (approximately 0 volts) in a transistor that causes collector current to stop. **3.** see CRITICAL VOLTAGE.

cutoff wavelength *Electromagnetism.* the wavelength that corresponds to the cutoff frequency for a particular waveguide.

cutout or **cut-out** *Electricity.* an electric device, such as a switch or circuit breaker, that is used to cut off the flow of current to a circuit. *Control Systems.* a value that is used to open a control circuit when a certain temperature or pressure is reached. *Building Engineering.* a carefully fabricated break in a beam, panel, or masonry for inserting or attaching a furnishing. *Geology.* see HORSEBACK, def. 2.

cutout angle *Electronics.* the phase angle of the input power cycle in a rectifier circuit at which conduction of a diode ceases; a cutout angle can never be exactly 0° or 180° degrees because of the forward bias requirements of the rectifying diodes.

cutout box see FUSE BOX.

cutover *Forestry.* **1.** land, usually timberland, that is cleared of trees. **2.** of or relating to an area that has been cleared of trees. *Telecommunications.* the quick transference of subscribers' lines from one exchange to another, especially from an electromagnetic to an electronic exchange.

cut plane *Robotics.* a sectional view that shows the intersection of a plane through a three-dimensional object.

cut point *Chemical Engineering.* the boiling-temperature separation between cuts of a crude oil or a base stock.

cutscore *Engineering.* a tool used in die-cutting processes that cuts material such as paper only partially, so that it may be folded.

cutset *Electricity.* a group of network branches arranged so that cutting off all the branches increases the number of separate parts, but cutting off all except one branch does not. *Mathematics.* the cutset of a graph (or digraph) is a minimal set of edges, the removal of which will increase the number of connected components in the remaining subgraph.

cutset rank *Mathematics* . the number $\xi(G)$ of edges in a spanning forest of an arbitrary graph G with n vertices, m edges, and k components. $\xi(G) = n - k$ and equals the dimension of the cutset subspace. Also, COMPONENT RANK.

cutset subspace *Mathematics.* the subspace of the vector space associated with a graph G consisting of all cutsets and edge-disjoint unions of cutsets of G.

cut-sheet printer *Computer Technology.* a printer that is capable of automatically accepting and printing on single sheets of paper rather than on continuous forms.

cut stone *Materials.* a stone or stonework dressed to a fine finish.

cutter *Engineering.* **1.** broadly, any machine, tool, or device that is used for cutting. **2.** see CUTTING TOOL. *Acoustical Engineering.* an electromagnetic or piezoelectric recording device that supports and suspends the cutting stylus. *Naval Architecture.* a term for various small, swift vessels, especially a Coast Guard vessel used to patrol coastal waters.

cutter bar *Mechanical Engineering.* **1.** in a mower, combine, or similar machine, a bar along which a knife or blade runs, protected by triangular guards. Also, SICKLE BAR. **2.** a bar that supports a cutting tool.

cutterhead *Mechanical Engineering.* a device on a machine tool that holds the cutting tool.

cutter offset *Robotics.* the amount by which a cutter has deviated from the location that was targeted by its controlling program.

cutting *Botany.* a root, stem, or leaf cut from a living plant that can produce a new plant under favorable growing conditions. *Engineering.* describing a device that cuts, or a process of cutting; used to form many compound terms, such as **cutting machine, cutting pliers, cutting torch, cutting angle, cutting time, cutting drilling,** and so on.

cutting edge *Design Engineering.* the edge of the tool that makes contact with the workpiece during a machining operation.

cutting-edge *Science.* a popular term used to describe a device or system that represents the latest and most advanced developments in its field; used especially in computer technology.

cutting fluid *Materials Science.* a liquid or gas used to cool or lubricate a cutting tool and a workpiece at their contact point. Thus, **cutting oil.**

cutting force *Design Engineering.* the force vector required to move a tool into a workpiece.

cutting-off process *Meteorology.* a sequence of events by which a warm high or a cold low originally within the westerlies becomes displaced out of the westerly current.

cutting ratio *Engineering.* the ratio between the depth of cut to the thickness of the metal chip at a specific shear angle.

cutting rule *Engineering.* the sharp steel strips that are used in a machine for cutting cardboard or other paper materials.

cuttings *Engineering.* the particles of material produced by a cutting operation. Also, BORINGS.

cutting stylus *Acoustical Engineering.* a tool that cuts grooves into a lacquer base to create a phonograph record from a master recording.

cutting tool *Mechanical Engineering.* the part of a machine tool that removes material from the workpiece. Also, CUTTER.

cuttlefish *Invertebrate Zoology.* **1.** any member of the order Sepioidea. **2.** any of various cephalopods, especially of the genus *Sepia,* squidlike fish with eight short and two long tentacles and an internal calciferous skeleton; noted for a discharge of brown ink when threatened.

cutty clay see PIPE CLAY, def. 1.

cut vertex *Mathematics.* in graph theory, the vertex of a separating set containing only one vertex. Also, ARTICULATION VERTEX.

cuvette *Chemistry.* a rectangular or cylindrical container used in various laboratory experiments; often made of quartz, with optical surfaces used to mold samples so that their light-absorbing properties can be measured. *Geology.* a basin in which sedimentation has occurred or is taking place, as opposed to one formed by the folding of preexisting rocks.

Cuvier, Baron [koo´vē ā; koov´yā} 1769–1832, French naturalist; founder of comparative anatomy and of paleontology.

Cuvieronius *Paleontology.* a bunodont gomphothere that originated in North America and spread to South America; Miocene to Pleistocene.

Cuvier's sinus see DUCT OF CUVIER.

CV cardiovascular; curriculum vitae.

Cv coxsackievirus.

CVA cerebrovascular accident.

C value *Molecular Biology.* the amount of haploid DNA that is present in the haploid genome of an organism; expressed in various ways, such as in picograms or molecular weight.

C value paradox *Genetics.* either of two paradoxes found when comparing the C values of different organisms: first, that higher organisms appear to have much more DNA than is required for a genetic code, and second, that some organisms thought to be closely related genetically have C values that differ greatly.

CVB *Oncology.* an acronym for a chemotherapy regimen for cancer treatment that includes the drugs <u>C</u>CNU, <u>v</u>inblastine, and <u>b</u>leomycin.

CVC conserved vector current.

CVD chemical vapor deposition.

CVG *Aviation.* the airport code for Greater Cincinnati.

CVP *Oncology.* an acronym for a chemotherapy regimen for cancer treatment that includes the use of the drugs <u>c</u>yclophosphamide, <u>v</u>incristine, and <u>p</u>rednisone.

CVP central venous pressure.

CW continuous wave; chemical warfare.

CW laser see CONTINUOUS-WAVE LASER.

cwt. hundredweight.

CX *Aviation.* the airline code for Cathay Pacific Airways.

Cyamidae *Invertebrate Zoology.* a family of amphipod crustaceans, the whale lice, in the suborder Caprellidea, that look like insect lice.

cyan [sī´an´] *Graphic Arts.* the blue-green color that is one of the three separation colors used in color process printing. Also, **cyan blue.**

cyan- a combining form meaning: **1.** blue, as in *cyaneous.* **2.** of, relating to, or containing cyanide.

cyanalcohol see CYANOHYDRIN.

cyanamide *Inorganic Chemistry.* **1.** H_2NCN, deliquescent crystals that are soluble in water, alcohol, and ether; melts at 42°C,; a strong irritant. **2.** see CALCIUM CYANAMIDE.

Cyanastraceae *Botany.* a monogeneric family of monocotyledonous perennial herbs of the order Liliales that are native to tropical African forests; characterized by plants arising from a tuberous-thickened base, strictly basal leaves with a closed sheath, and perfect trimerous flowers borne on a scape.

cyanate *Inorganic Chemistry.* any compound containing the isocyanate radical, NCO, derived from cyanic acid, HNCO. Also, ISOCYANATE.

cyanazine *Organic Chemistry.* $C_9H_{13}ClN_6$, a white solid that is soluble in water and alcohol; it melts at 166°C; used as an herbicide.

cyan filter *Graphic Arts.* in color separation, a screen designed to absorb red and admit blue-green (cyan).

cyanic acid *Organic Chemistry.* N≡COH, a poisonous liquid with an acrid odor, soluble in water; boils at 23.5°C; used in the formation of cyanates. Also, HYDROGEN CYANATE.

cyanide *Inorganic Chemistry.* any compound containing the CN group and derived from hydrogen cyanide, HCN; especially, the highly toxic potassium cyanide, KCN, and sodium cyanide, NaCN.

cyanide copper *Metallurgy.* copper that has been electrodeposited from a cyanide solution.

cyanide poisoning *Toxicology.* poisoning due to ingestion or inhalation of cyanides.

cyanide process *Metallurgy.* a process used to leach gold or silver from an ore for subsequent extraction.

cyanide pulp *Metallurgy.* the suspension resulting from the cyanide process, prior to separation.

cyanide slime *Metallurgy.* a metallic powder that is formed when precious metals are extracted from ore, using the cyanide process.

cyaniding *Metallurgy.* in steel, a case hardening process that uses a cyanide to supply both carbon and nitrogen to the case.

cyanine dye *Organic Chemistry.* **1.** a series of dyes consisting of two heterocyclic groups connected by a chain of conjugated double bonds containing an odd number of carbon atoms. **2.** $C_{29}H_{35}N_2I$, green metallic crystals; soluble in water; used as a sensitizer for photographic film.

cyanite see KYANITE.

cyano- a combining form meaning: **1.** blue, as in *cyanosis.* **2.** of, relating to, or containing a cyanide group, as in *cyanoacrylate.*

cyanoacetamide *Organic Chemistry.* $CNCH_2CONH_2$, toxic, combustible, white crystals, soluble in water and alcohol; melts at 119°C, decomposes at boiling point; used in organic pharmaceutical synthesis and plastics.

cyanoacetic acid *Organic Chemistry.* $N{\equiv}CCH_2COOH$, toxic, hygroscopic, white crystals that are soluble in water and alcohol; melts at 66.1–66.4°C and decomposes at 160°C; used in organic synthesis.

cyanoacrylate adhesive *Materials Science.* an adhesive based on the alkyl 2-cyanoacrylates that has excellent atomic bonding and bonding-strength properties.

Cyanobacteria *Bacteriology.* a genus of bacteria belonging to the class Oxyphotobacteria, composed of the blue-green algae; they are also classified as algae in the division Cyanophyta. See CYANOPHYTE.

cyanobacteria *Bacteriology.* a name for bacteria included in the genus *Cyanobacteria.*

Cyanobacteriales *Bacteriology.* a proposed order of bacteria that would include the genus *Cyanobacteria.*

cyanobiont *Bacteriology.* any member of the bacterial genus *Cyanobacteria* that is involved in a symbiotic association, such as with the floating freshwater fern Anabaena.

cyanocarbon *Organic Chemistry.* a hydrocarbon derivative containing the cyano (–CN) group in place of hydrogen.

cyanochroite *Mineralogy.* $K_2Cu^{+2}(SO_4)_2{\cdot}6H_2O$, a blue-green, transparent, water-soluble monoclinic mineral occurring in crusts of tabular crystals; having a specific gravity of 2.22 and an undetermined hardness on the Mohs scale; found as a fumarolic deposit formed during eruptions of Mount Vesuvius, Italy.

cyano complex *Chemistry.* any coordination compound containing a cyanide (CN) group.

cyanoderma *Medicine.* an abnormal blueness of the skin.

cyanoethylation *Organic Chemistry.* the addition of a nucleophile to a carbon-carbon multiple bond in a cyano compound such as acrylonitrile.

cyanogen *Inorganic Chemistry.* NCCN, a colorless, pungent gas that has a penetrating odor and burns with a purple flame, soluble in water, alcohol, and ether; melts at –27.9°C and boils at –21.2°C; highly toxic and flammable. It is used in organic synthesis, metallurgy, fumigants, and rocket propellants. Its hydrolysis to ammonia and oxalic acid in 1824 marked the first recorded laboratory synthesis of an organic compound from inorganic starting materials. Also, DICYAN(OGEN).

cyanogen absorption *Astrophysics.* a type of molecular absorption found in the spectra of stars of spectral type G and cooler; it is more noticeable in giant stars than in dwarfs.

cyanogen bromide *Inorganic Chemistry.* CNBr, white crystals with a penetrating odor, soluble in water, alcohol, and ether; melts at 52°C and boils at 61.4°C; highly toxic and irritating; used in organic synthesis, pesticides, and gold extraction.

cyanogen chloride *Inorganic Chemistry.* CNCl, a colorless gas or liquid, soluble in water, alcohol, and ether; melts at –6°C and boils at 12°C; toxic and an irritant; used in organic synthesis, tear gas, and pesticides.

cyanogen fluoride *Inorganic Chemistry.* CNF, a colorless gas that is insoluble in water; sublimes at –72°C; toxic and an irritant; used in organic synthesis and tear gas.

cyanogen iodide *Inorganic Chemistry.* CNI, colorless needles with a pungent odor, soluble in water, alcohol, and ether; melts at 146.7°C; highly toxic and irritating; used in taxidermy. Also, IODINE CYANIDE.

cyanoguanidine see DICYANDIAMIDE.

cyanohydrin *Chemistry.* a compound having a carbon atom attached to both a cyano group and a hydroxyl group. Also, CYANALCOHOL.

cyanometer *Optics.* an instrument that measures the amount of blueness in a light.

cyanometry *Optics.* the analysis of blue light.

Cyanomyovirus *Virology.* a genus of viruses of the family Myoviridae, marked by long contractile tails, that infect blue-green algae.

cyanophage see PHYCOVIRUS.

cyanophilous *Biology.* **1.** a cell or other entity that is readily stained with blue or green dyes. **2.** relating to or being such an entity.

cyanophose *Neurology.* a sensation or perception of blue color.

cyanophosphos *Organic Chemistry.* $C_{15}H_{14}NO_2PS$, a white crystalline solid that melts at 83°C; used as an insecticide.

Cyanophyceae *Botany.* the blue-green algae; organisms capable of photosynthesis but having a cell structure similar to bacteria.

cyanophycean starch *Biochemistry.* a storage polysaccharide present in the cells of blue-green algae as granules or rods.

cyanophycin *Biochemistry.* a food-storage peptide composed of arginine and aspartic acid; found in blue-green algae cells.

Cyanophyta *Botany.* a division containing the blue-green algae, members of which have no distinct nucleus and do not reproduce sexually.

cyanophyte [sī an´ə fīt] *Biology.* a type of blue-green algae, characterized by the possession of chlorophyll a, the absence of bacteriochlorophylls, and the ability to carry out oxygenic photosynthesis; they are the only organisms that fix both carbon dioxide and nitrogen. They are different from true algae, and are now generally regarded as more closely related to bacteria, and are thus classified as the genus *Cyanobacteria. Paleontology.* a Precambrian form of this organism, among the earliest instances of life on earth, occurring in rocks about 3 billion years old. They have been found in early Precambrian chert deposits in Canada, Australia, and South Africa.

cyanopia *Medicine.* a visual defect in which all objects appear blue.

cyanoplatinite see PLATINOCYANIDE.

Cyanopodovirus *Virology.* a genus of viruses of the family Podoviridae that infect blue-green algae.

cyanosis [sī´ə nō´sis] *Medicine.* a bluish or purplish appearance of the skin caused by insufficient oxygenation of the blood, and often due to cardiac abnormality.

cyanosomes *Bacteriology.* phycobilisomes that are found in cyanobacteria; intracellular structures attached to thylakoid membrane surfaces that transfer absorbed light energy to chlorophyll a.

cyanostylovirus *Virology.* a proposed genus of viruses of the family Styloviridae that infects blue-green algae.

cyanotic [sī´ə nät´ik] *Medicine.* relating to blueness of the skin.

cyanotrichite *Mineralogy.* $Cu_4^{+2}Al_2(SO_4)(OH)_{12}{\cdot}2H_2O$, a pale to deep blue, silky orthorhombic mineral that occurs in tufts, aggregates, or velvety coatings of fine acicular crystals, having a specific gravity of 2.74 to 2.95 and an undetermined hardness on the Mohs scale; found as a secondary mineral in oxidization zones of copper-bearing ore deposits.

Cyanthomonadaceae *Botany.* a monotypic family of freshwater flagellates belonging to the order Cryptomonadales, characterized by compressed cells with a coronate ring of trichocysts at the anterior end, no chloroplasts, and the ability to ingest food particles.

cyanuric acid *Organic Chemistry.* $HOCNC(OH)NC(OH)N{\cdot}2H_2O$, white crystals that are very soluble in hot water and acids and insoluble in alcohol; decomposes to cyanic acid at 320°C; used as an intermediate for chlorinated bleaches, herbicides, and whitening agents.

Cyatheaceae *Botany.* a family of tree-ferns having a distinct trunk, sporangia with complete annulus, and sori located in the forks of veins.

cyathium *Botany.* a type of inflorescence of the family Euphorbiaceae, in which a single female flower is surrounded by numerous, simple male flowers, the whole giving the appearance of a single blossom.

Cyathoceridae see LEPICERIDAE.

cyathozooid *Invertebrate Zoology.* the primary or stem zooid in certain tunicate colonies.

Cyathus *Mycology.* a genus of fungi belonging to the order Nidulariales, commonly known as bird's nest fungi; it is characterized by the funnel shape of its spore-bearing structure or basidiocarp.

Cybele *Astronomy.* asteroid 65, discovered in 1861 and having a diameter of about 230 kilometers; belongs to type CPF.

cybernation *Computer Science.* the application of computerized controls to an automated process, so that the control system responds to operating conditions or outputs and adjusts the process accordingly.

cybernetics [sī´bər net´iks] *Science.* the scientific study of communication and control, especially so as compare the communication and control systems of humans and other living organisms with those of complex machines. *Robotics.* this science as it relates specifically to the development and operation of automatic control equipment. (Coined by Norbert Weiner from a Greek term meaning "pilot" or "helmsman.")

cybernetist *Science.* a person who studies or specializes in cybernetics.

cyberspace *Computer Science.* an artificial world formed by the display of data as an artificial three-dimensional space, which the user can manipulate and "move through'' by issuing commands to the computer.

cyborg [sī´bôrg´] *Robotics.* a popular term for a human being having electronic or mechanical devices performing some vital physiological functions. (An acronym for cybernetic organism.)

cybotaxis *Physics.* a three-dimensional arrangement of molecules, usually applied to molecules in a liquid crystal.

cybrid *Genetics.* an organism or group of cells produced by combining the cytoplasm of one cell type with the nucleus of another cell type.

cycad *Botany.* any plant of the order Cycadales.

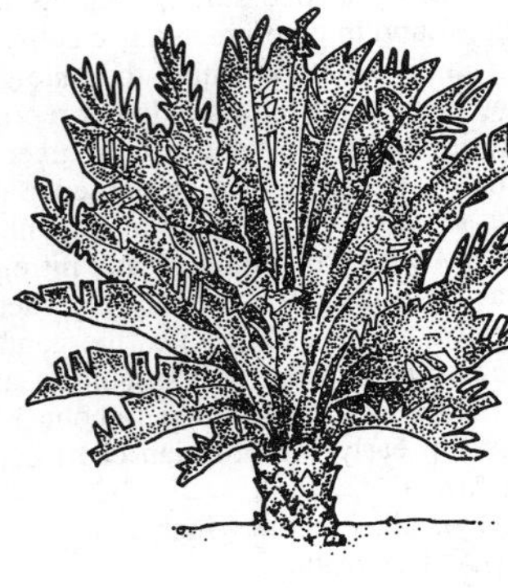

cycad

Cycadaceae *Botany.* the single family of the order Cycadales, gymnosperms having palmate leaves with a large midrib but no flanking veins.

cycadaceous *Botany.* relating to or resembling the order Cycadales.

Cycadales *Botany.* an order of gymnosperms characterized by broad, unbranched stems with large, pinnate leaves at the end.

Cycadeoidaceae *Paleontology.* one of the principal families of Mesozoic cycadeoids in the order Cycadeoidales; shorter and wider than the Williamsoniaceae and Wielandiaceae, the Cycadeoidaceae were almost globular in shape, three feet high and three feet across.

Cycadeoidales *Paleontology.* an extinct order of treelike plants that arose in the early Mesozoic and persisted to the end of the era; they are not ancestral to cycads.

Cycadofilicales see PTERIDOSPERMALES.

Cycadopsida *Botany.* a class of gymnosperms including living and extinct cycadlike plants.

cychrus *Invertebrate Zoology.* any of various large black beetles that feed on the soft body of snails by piercing the snail's shell with their long narrow head and thorax.

cycl- a combining form meaning "cycle," as in *cyclarthrosis.*

cyclamate *Organic Chemistry.* any one of the sweet, but nonnutritive salts of cyclamic acid, whose use is currently banned in the U.S.

cyclamic acid *Organic Chemistry.* $C_6H_{11}NHSO_3H$, a white, crystalline solid that is soluble in water and alcohol; melts at 170°C; a suspected carcinogen; salt used as a nonnutritive sweetener and acidulant. Also, CYCLOHEXYLSULFAMIDIC ACID.

Cyclanthaceae *Botany.* the single family of the order Cyclanthales.

Cyclanthales *Botany.* an order of tropical small shrubs or vines having deeply lobed leaves and spirally alternating, unisexual flowers.

cycle a repeating series of events; specific uses include: *Engineering.* the process of operating a machine through one production series that can repeat itself. *Industrial Engineering.* a series of elements that occur in a given order and that constitute one operation or unit of work. *Fluid Mechanics.* a process in which the initial and final states of a system are identical; i.e., the initial and final states have the identical values for all respective properties. *Mathematics.* **1.** let $i_1, i_2, \ldots, i_r$ be distinct elements of $I_n = \{1, 2, \ldots, n\}$. Let $(i_1i_2i_3 \cdots i_r)$ denote the permutation that maps i_1 to i_2, i_2 to i_3, $\ldots$, i_{r-1} to i_r, i_r to i_1, and maps every other element of I_n to itself. Then $(i_1i_2i_3 \cdots i_r)$ is called a **cycle of length *r*** or an ***r*-cycle. 2.** an elementary circuit in a graph containing at least three vertices. **3.** a chain whose boundary is identically zero; i.e., a member of the kernel of a boundary homomorphism.

cycle annealing *Metallurgy.* a process of annealing performed according to a preset schedule of time-temperature cycles.

cycle clock *Computer Technology.* in a central processor, an electronic circuit that emits pulses at regularly timed intervals to synchronize and control processing.

cycle count *Computer Programming.* the increase or decrease in a cycle index counter by 1 or by a specified step integer.

cycle criterion *Computer Programming.* the number of times that a cycle or loop is to be executed.

cyclegraph technique *Industrial Engineering.* a technique for the study of work motion, in which lights are affixed to the hands or other parts of the worker's body and a long-exposure still photograph is taken to show motion patterns.

cycle index *Computer Programming.* a means of counting how many iterations of a cycle have been executed. In an incrementing counter, the number of repetitions is equal to the index; in a decrementing counter, the number of repetitions is equal to the desired number minus the index. Thus, **cycle index counter.**

cycle length *Transportation Engineering.* the period of time needed for a traffic signal to pass through a single complete cycle.

cycle of denudation see CYCLE OF EROSION.

cycle of erosion *Geology.* **1.** the complete series of natural changes or stages, from youthful to mature to old-age, undergone by a newly uplifted or exposed land surface as it is eroded into mountains and valleys, and finally into a featureless plain or other such base level. **2.** the time period during which a newly uplifted land area is reduced to base level. Also, GEOMORPHIC CYCLE, EROSION CYCLE, CYCLE OF DENUDATION, GEOGRAPHIC CYCLE.

cycle of fluctuation see PHREATIC CYCLE.

cycle of sedimentation *Geology.* **1.** the ordered, recurring sequence of related processes and conditions recorded in sedimentary deposits. Also, SEDIMENTARY CYCLE. **2.** the deposition of sediments in a basin by the recurring changes of water levels. **3.** see CYCLOTHEM.

cycle per minute *Physics.* a unit of frequency, denoted by either cpm or min^{-1}, obtained by multiplying the number of Hertz by 60.

cycle per second see HERTZ.

cycle reset *Computer Programming.* the process of returning a cycle index to its initial value or to another starting value.

cycle skip see SKIP LOGGING.

cycle stealing *Computer Technology.* a technique for memory sharing between a CPU and an input/output device, in which the processor activity is suspended for one memory cycle to allow memory access by the input/output system.

cycle stock *Chemical Engineering.* an unfinished product removed from a unit of a refinery process, then recharged to the process at an earlier point in the operation.

cyclethrin *Organic Chemistry.* $C_{21}H_{28}O_3$, a mixture of isomers; a toxic, brown liquid; soluble in organic solvents; used as an insecticide.

cycle timer *Electronics.* a mechanical or electronic control device that opens or closes contacts according to a preprogrammed schedule. Also, PROGRAMMED TIMER.

cycle timing diagram *Computer Programming.* a diagram or table that describes the register and other activities that occur at each clock cycle during the execution of a machine instruction.

cycliacylation *Organic Chemistry.* a reaction that produces a cyclic molecule through acylation.

cyclic [sik´lik] relating to, moving in, or being a cycle. Also, **cyclical.**

3',5'-cyclic adenosine monophosphate see C-AMP.

cyclic adenylic acid *Biochemistry.* an adenylic acid isomer that regulates cellular activities in most animals, some higher plants, and bacteria.

cyclical fluctuations *Statistics.* in time series analysis, the repeated upward or downward deviations around a trend line. Also, **cyclical component.**

cyclically permuted sequence *Molecular Biology.* any DNA sequence whose constituents are all circular arrangements of the same genes in the same order, but whose linearized molecules have termini in different positions.

cyclic amide *Organic Chemistry.* a molecule in which a functional amide group is incorporated into a cyclic compound.

cyclic AMP see C-AMP.

cyclic anhydride *Organic Chemistry.* a cyclic compound having the basic formula *R*–CO–O–CO–*R'*, formed when two acid groups in the same molecule react with a loss of water.

cyclic binary code see GRAY CODE.

cyclic chronopotentiometry *Analytical Chemistry.* an electrochemical method of analysis in which current reversal is effected at the working electrode and the potential is measured against time.

cyclic code *Computer Programming.* **1.** a code generated during data transmission for the purpose of detecting and correcting errors occurring during magnetic tape, disk, and data communications. **2.** a binary code in which sequential numbers differ from one another in only one bit position.

cyclic coil see RANDOM COIL.

cyclic compound *Organic Chemistry.* an organic compound whose structure is characterized by at least one closed ring of atoms of the three major types: alicyclic, aromatic (or arene), or heterocyclic.

cyclic coordinate *Mechanics.* a generalized coordinate that does not appear in the Lagrangian of a conservative holonomic dynamical system.

cyclic currents see MESH CURRENTS.

cyclic curve *Mathematics.* **1.** a curve generated by a fixed point on a circle, as the circle rolls smoothly along a given curve. Examples include cycloids, epicycloids, hypocycloids, etc. Also, CYCLOIDAL CURVE. **2.** the intersection of a quadric surface and a sphere. Also, SPHERICAL CYCLIC CURVE. **3.** the stereographic projection of a spherical cyclic curve. Also, PLANE CYCLIC CURVE.

cyclic deformation *Materials Science.* the deformation of a material under conditions of cyclic stress.

cyclic feeding *Computer Technology.* in character-reading equipment, a process whereby the pages of an input document are fed into the reader at a predetermined and regular rate.

cyclic GMP see C-GMP.

cyclic group *Mathematics.* a group whose elements are all powers of some fixed element. Every finite group of prime order is cyclic.

cyclic identity *Mathematics.* any identity of the form $A_{ij}A_{kl} + A_{ik}A_{lj} + A_{il}A_{jk} = 0$; used in a wide variety of subjects, including tensor analysis and invariant theory; it depends on the subscripted objects exhibiting symmetry properties (usually antisymmetry).

cyclic magnetization *Electromagnetism.* a condition in which the magnetizing force varies periodically between two limits at a low enough frequency so that the magnetic induction has the same value for corresponding points in successive cycles.

cyclic module *Mathematics.* a module M over a ring R is cyclic if there exists an element x of M such that every element of the module is of the form rx for some $r \in R$.

cyclic redundancy check *Telecommunications.* a polynomial code that is widely used in data communications as a method of detecting bit errors; formed from message bits that are passed through a feedback shift register.

cyclic salt *Oceanography.* a salt that is introduced into the atmosphere from the sea surface by evaporation and bursting bubbles, blown inland, and then returned to the sea by normal drainage processes.

cyclic sedimentation *Geology.* a regularly repeated sequence of sedimentary deposition. Also, RHYTHMIC SEDIMENTATION.

cyclic shift *Computer Technology.* a register operation that circulates the bits of the register around the two ends as though the register represents a ring.

cyclic storage *Computer Technology.* a type of mass storage device that rotates so that data may be read or written only at the time the read/write head is positioned over the desired physical memory location; examples are magnetic drum or disk storage devices. Also, CIRCULATING STORAGE, REGENERATIVE STORAGE.

cyclic succession *Ecology.* an ecological succession in which the original community is ultimately restored.

cyclic train *Mechanical Engineering.* a gear system in which one or more of the gear axes rotates around a fixed axis.

cyclic twinning *Crystallography.* **1.** a twinning of crystals under the same twinning laws at regular intervals. **2.** the crystal growth occurring along a circular path, as in twins of chrysoberyl.

cyclic voltammetry *Analytical Chemistry.* an electrochemical technique in which one or more triangular potential sweeps are applied to the working electrode and the resulting current measured.

Cyclidium *Invertebrate Zoology.* a common genus of freshwater and marine ciliate protozoans, in the subclass Holotrichia.

cyclin *Cell Biology.* in certain eukaryotic cells, a protein that undergoes programmed degradation at a specific point of the cell cycle. These proteins play a role in regulation of the cell cycle by activating the $p34^{cdc2}$ kinase.

cycling the process of moving or operating in a cycle; specific uses include: *Engineering.* a series of operations in a processing plant that are repeated periodically in the same sequence. *Petroleum Engineering.* an operation in which gas flowing from a gas reservoir passes through a separation system or processing plant, after which the remaining gas is returned to the reservoir. *Control Systems.* a series or pattern of periodic changes in the value of a controlled variable.

cyclitis *Medicine.* an inflammation of the ciliary structure.

cyclization *Organic Chemistry.* a chemical reaction that causes parts of a molecule to join so that a ring of atoms is formed within the molecule.

cyclizine hydrochloride *Pharmacology.* $C_{18}H_{23}ClN_2$, an antihistamine occurring as a white crystalline powder or as small colorless crystals; administered orally to relieve or prevent the nausea and vomiting caused by motion sickness.

cyclo- a combining form meaning "cycle," as in *cyclotron.*

cycloaddition *Organic Chemistry.* a reaction in which unsaturated molecules combine to form rings.

cycloalkane *Organic Chemistry.* a saturated carbon chain forming a loop or ring.

cycloalkene *Organic Chemistry.* a cyclic unsaturated hydrocarbon having at least one carbon-carbon double bond in the ring.

cycloalkyne *Organic Chemistry.* a ringed hydrocarbon that contains a triple bond.

cycloamylose *Biochemistry.* a cyclic oligomer of glucose.

cycloate *Organic Chemistry.* $C_{11}H_{21}NOS$, a yellow liquid that boils at 145°C (10 torr); used as a weedkiller.

cyclobarbital *Pharmacology.* $C_{12}H_{16}N_2O_3$, a short-acting barbiturate that is administered orally, chiefly to induce sleep and also as a postoperative sedative; abuse can lead to addiction.

cyclobutadiene *Organic Chemistry.* C_4H_4, a highly unstable compound that has been isolated only in the form of metal complexes.

cyclobutane *Organic Chemistry.* C_4H_8, a flammable, colorless gas, insoluble in water and soluble in alcohol; boils at 13°C and melts at –80°C. Also, TETRAMETHYLENE.

cyclobutene *Organic Chemistry.* C_4H_6, a flammable gas that boils at 2°C. Also, **cyclobutylene**.

cycloconverter *Electricity.* a thyristor-based device that converts an alternating-current voltage of one frequency to another frequency; generally, the output frequency is limited to one third of the input frequency.

Cyclocystoidea *Paleontology.* a class of primitive, disk-shaped echinoderms in the subphylum Echinozoa, characterized by a marginal ring of heavy plates and by flexible radial plates on the dorsal surface; extant in the Ordovician to Devonian.

cyclodehydration *Organic Chemistry.* a dehydration reaction in which a ringed molecule is formed by the removal of water from an acyclic molecule.

cyclodextrin *Organic Chemistry.* a compound containing six or more D-glucose units joined through 1,4-α linkages in such a way as to form a ring; used as complexing agents and in enzymology.

cyclodialysis *Medicine.* a surgical procedure for the treatment of glaucoma in which a passage is created between the anterior chamber and the suprachoroidal space of the eye. Also, HEINE'S OPERATION.

cyclodiathermy *Medicine.* a therapeutic procedure for the treatment of glaucoma by destroying a portion of the ciliary body through diathermy.

cyclodiolefin *Organic Chemistry.* a class of cycloalkenes having two double bonds.

cyclododecatriene *Organic Chemistry.* $C_{12}H_{18}$, a 12-carbon isomeric cyclic compound produced by treating butadiene catalytically with a nickel complex; used to manufacture nylon.

Cycloganoidei see HALECOMORPHI.

cyclogenesis *Meteorology.* any strengthening or development of cyclonic (counterclockwise) circulation in the atmosphere.

cyclograph *Engineering.* an instrument used to test the metallurgical properties of steel, consisting of a sensing coil attached to a cathode-ray screen on which the properties are identified according to the changing patterns that appear on the screen.

1,3-cyclohexadiene *Organic Chemistry.* a liquid compound that is insoluble in water and soluble in alcohol; melts at –89°C and boils at 80.5°C. Also, DIHYDROBENZENE.

cyclohexane *Organic Chemistry.* C_6H_{12}, a flammable, moderately toxic, colorless liquid that is insoluble in water and soluble in alcohol; melts at 6.55°C and boils at 80.74°C; used in the manufacture of nylon and as a solvent. Also, HEXAMETHYLENE, HEXANAPHTHENE.

cyclohexanol *Organic Chemistry.* $C_6H_{11}OH$, a combustible, hygroscopic, colorless, needlelike, crystalline solid, soluble in water and alcohol; melts at 25.5°C and boils at 161.1°C; used in soap making and as a solvent and blending agent.

cyclohexanone *Organic Chemistry.* $C_6H_{10}O$, a toxic, combustible, colorless liquid that is slightly soluble in water and alcohol; melts at –32°C and boils at 155.65°C; used in organic synthesis. Also, PIMELIC KETONE.

cyclohexene *Organic Chemistry.* C_6H_{10}, a flammable, toxic, colorless liquid, insoluble in water and soluble in alcohol; melts at –103.5°C and boils at 82.98°C; used in organic synthesis and as a catalyst.

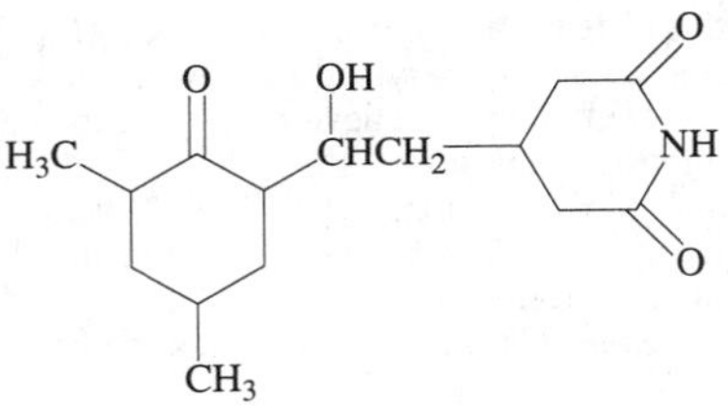

cycloheximide

cycloheximide *Molecular Biology.* an antibiotic that inhibits protein synthesis in eukaryotic cells by blocking translation of messenger RNA on the ribosome; derived from *Streptomyces griseus.*

cyclohexylsulfamidic acid see CYCLAMIC ACID.

cycloid [sī´kloid´; sik´loid] *Science.* being or resembling a circle; circular or nearly circular. *Mathematics.* a cyclic curve in which the guiding curve for the rolling circle is a straight line. Parametric representation is

$$x = a(t - \lambda \sin t), \; y = a(t - \lambda \cos t),$$

where $a > 0$ is the radius of the rolling circle, t is the angle of rotation, and $-\infty < t < \infty$. If $\lambda = 1$, then the curve is a **common cycloid.** If $\lambda > 1$, the curve is an **extended cicloid** or **prolate cycloid,** generated by a fixed point on the extension of a radius beyond the boundary of the rolling circle, at a distance λa from the center of the circle. If $\lambda < 1$, it is a **contracted cycloid** or **curtate cycloid,** generated by a fixed point on a radius inside the boundary of the rolling circle, at a distance λa from the center. Also, CYCLOIDAL.

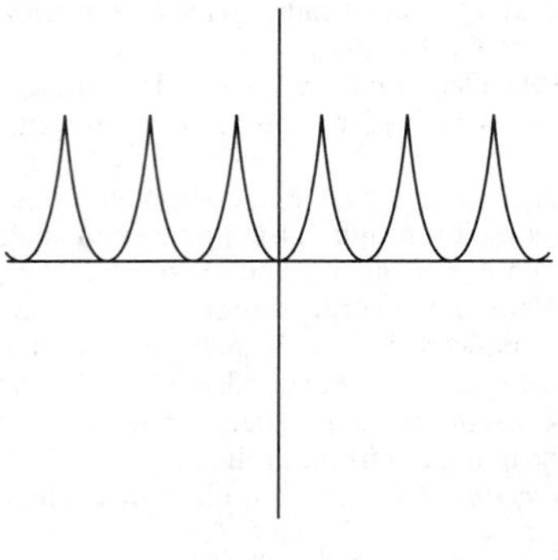

cycloid

cycloidal [sī´kloid´ əl; sik´loid əl] *Science.* resembling a circle; circular or nearly circular. *Mathematics.* see CYCLOID.

cycloidal curve see CYCLIC CURVE.

cycloidal gear teeth *Design Engineering.* gear teeth having a cycloidal profile; generally used in clockworks and timers.

cycloidal mass spectrometer *Spectroscopy.* a limited-range mass spectrometer using a cycloidal magnetic field path, rather than the more conventional circular path.

cycloidal pendulum *Mechanics.* a heavy weight suspended by a cord that hangs between two cycloid-shaped metal constraints. This type of pendulum executes motions whose period of oscillation is completely independent of the swing amplitude.

cycloidal propeller *Naval Architecture.* a propeller mounted in the horizontal plane and rotating around a vertical axis; used by some shallow draft craft.

cycloid scale *Vertebrate Zoology.* a thin, overlapping, rounded or oval-shaped fish scale with smooth edges, such as those on salmon and many other fishes.

cyclo-ligase *Enzymology.* any of several enzymes that catalyze the formation of a ring within a molecule while simultaneously hydrolyzing a nucleoside triphosphate.

cyclolysis [sī´kläl´ə sis] *Meteorology.* any weakening or dissipation of cyclonic (counterclockwise) circulation in the atmosphere.

cyclomatic number *Computer Programming.* a popular method of measuring software complexity by examining the pattern of the control flow in the program; often expressed as a directed graph.

cyclomorphosis [sī´klō môr fō´sis] *Ecology.* cyclical changes in the form of an organism, especially in response to changes in the seasons or presence of enemies.

cyclonal [sī´klōn´əl] *Meteorology.* relating to or resembling a cyclone.

cyclone [sī´klōn´] *Meteorology.* **1.** a large-scale, low-pressure atmospheric system with winds that spiral inward toward a center of lowest pressure. **2.** a popular term for various violent, small-scale circulations of air, such as tornadoes, waterspouts, or dust devils, or for any very strong wind. *Mechanical Engineering.* **1.** any of various cone-shaped devices that are used to classify fine powders or to separate fine particles and dusts from air or other gases through centrifugal action. Also, **cyclone collector, cyclone classifier. 2.** a similar apparatus that extracts suspended particles from flowing liquid; used especially to classify solids in mineral pulp.

cyclone family *Meteorology.* a series of wave cyclones that occur in the interval between two successive outbreaks of polar air and travel along the polar front in an eastwardly direction toward the pole.

cyclone furnace *Engineering.* a slow-heating water-cooled furnace having a horizontal cylindrical shape in which fuel is fired cyclonically.

cyclone wave *Meteorology.* **1.** a disturbance in the lower troposphere of wavelength 1000–2500 km; a migratory high- or low-pressure system. **2.** a frontal wave with a center of cyclonic circulation at its crest.

cyclonic [sī klän´ik] *Meteorology.* of or relating to a cyclone. *Geophysics.* relating to rotation in the same direction as the earth when viewed from the local vertical; that is, counterclockwise in the Northern Hemisphere, clockwise in the Southern Hemisphere, and undefined at the Equator.

cyclonic scale *Meteorology.* the scale of the migratory cyclone waves of the lower troposphere. Also, SYNOPTIC SCALE.

cyclonic shear *Meteorology.* a horizontal wind shear that tends to produce cyclonic rotation of individual air particles along the line of flow.

cyclonite *Organic Chemistry.* $(NO_2)CH_2N(NO_2)CH_2N(NO_2)CH_2$, toxic, white, orthorhombic crystals that are insoluble in water and slightly soluble in alcohol; melts at 205–206°C; used as an explosive, 1.5 times as powerful as TNT. Also, *sym*-TRIMETHYLENETRINITRAMINE, RDX, CYCLOTRIMETHYLENETRINITRAMINE.

1,5-cyclooctadiene *Organic Chemistry.* C_8H_{12}, a combustible liquid that is insoluble in water; it melts at –70°C and boils at 150.8°C; used as a resin intermediate.

cyclooctane *Organic Chemistry.* C_8H_{16}, a combustible, colorless liquid that melts at 14°C and boils at 148°C.

cyclooctatetraene *Organic Chemistry.* C_8H_8, a combustible, yellow or white liquid, soluble in alcohol; melts at –4.68°C and boils at 140.5°C; used in organic research.

cycloolefin see CYCLOALKENE.

cycloparaffin *Organic Chemistry.* any cycloalkane. Also, NAPHTHENE.

cyclopean [sīk´lə pē´ən] see MOSAIC.

cyclopean concrete *Materials Science.* concrete that contains large stones.

cyclopean stairs *Geology.* a glacial trough landscape created by the melting of ice and characterized by a series of variably steep and gently sloped areas and small lakes.

1,3-cyclopentadiene *Organic Chemistry.* C_5H_6, a toxic, colorless liquid, insoluble in water and soluble in alcohol; melts at –97.2°C and boils at 42°C; used as a chemical intermediate and in organic synthesis.

cyclopentadienyl anion *Organic Chemistry.* the organic group $-C_5H_5$, a derivative of cyclopentadiene.

cyclopentane *Organic Chemistry.* C_5H_{10}, a colorless liquid that is insoluble in water and soluble in alcohol; boils at 49.2°C and melts at –94°C; used as a solvent.

cyclopentanoid *Organic Chemistry.* of or relating to, containing, or similar to cyclopentane.

cyclopentanol *Organic Chemistry.* C_5H_9OH, a combustible, colorless, viscous liquid, soluble in water and alcohol; boils at 140.8°C and melts at –19°C; used in pharmaceuticals and perfume. Also, **cyclopentyl alcohol.**

cyclopentanone *Organic Chemistry.* C_5H_8O, a flammable, white liquid that is insoluble in water and soluble in alcohol; boils at 130.6°C; it is used in pharmaceuticals, biologicals, insecticides, and rubber chemicals.

cyclopentene *Organic Chemistry.* C_5H_8, a flammable, moderately narcotic, colorless liquid, soluble in alcohol; boils at 44.2°C and melts at –135.21°C; used in organic synthesis and as a cross-linking agent.

Cyclophoracea *Invertebrate Zoology.* a superfamily of aquatic gastropod mollusks in the order Prosobranchia, having two tentacles.

cyclophosphamide *Pharmacology.* a cytotoxid alkylating agent of the nitrogen mustard group; used, often in combination with other agents, as an antineoplastic in the treatment of Hodgkin's disease, lymphosarcoma, acute lymphocytic leukemia, ovarian carcinoma, and many other conditions; also used as an immunosuppressive agent to prevent transplant rejection.

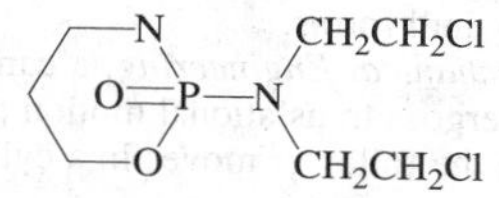

cyclophosphamide

Cyclophyllidae *Invertebrate Zoology.* an order of worms, including most tapeworms, in the subclass Cestoda; parasitic on warm-blooded vertebrates.
cyclopia [sī klō´pē ə] *Medicine.* a birth defect in which only a single eye orbit is present. (From *Cyclops,* the one-eyed giant of Greek mythology.) Also, SYNOPHTHALMIA.
cyclopiazonic acid *Toxicology.* a fungal toxin produced by species of *Penicillium* and *Aspergillus;* poisonous to certain animals.
Cyclopinidae *Invertebrate Zoology.* a family of small crustaceans in the suborder Cyclopoida.
Cyclopoida *Invertebrate Zoology.* a suborder of small crustaceans, in the order Copepoda, having one eye and long antennae.
cyclopolymerization *Materials Science.* the propagation of a polymer by the alternate addition of a monomer and the formation of a six-member (benzene) ring.
cyclopropane *Organic Chemistry.* C_3H_6, a highly flammable, moderately toxic, narcotic, colorless gas; soluble in water and alcohol; boils at –32.9°C and melts at –126.6°C; used in organic synthesis and as an anesthetic. Also, TRIMETHYLENE.
cyclopropanoid *Organic Chemistry.* a compound built around three carbon atoms arranged in a ring.
cyclops [sī´kläps´] *Invertebrate Zoology.* a very common freshwater copepod crustacean of the genus *Cyclops*; some species are intermediate hosts for human parasites. (So called because it has a single median eye in the front of the head, like the one-eyed giant *Cyclops* of Greek mythology.)

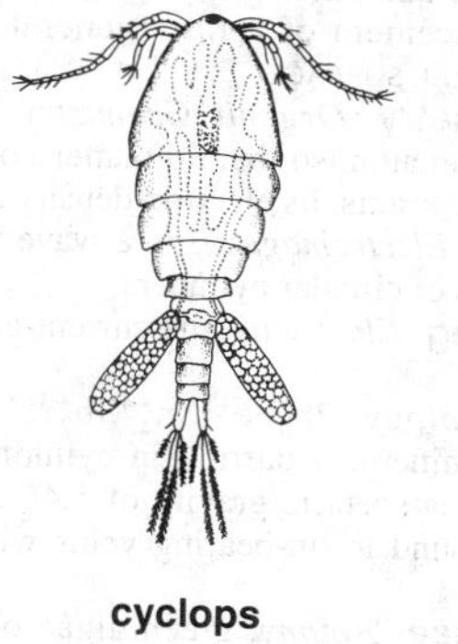

cyclops

Cyclopteridae *Vertebrate Zoology.* the lumpfishes and snail-fishes, a family of marine bottom fishes in the suborder Cottoidei, having either elongated or globular bodies and often having a sucking disc for attachment to stones.
cycloreversion *Organic Chemistry.* the reversal of a cycloaddition reaction.
Cyclorhagae *Invertebrate Zoology.* a suborder of tiny deep-sea animals in the class Kinorhyncha.
Cyclorrhapha *Invertebrate Zoology.* a suborder of true flies in the order Diptera, including the common house fly.
cycloserine *Microbiology.* a broad-spectrum antibiotic produced by the bacterium *Streptomyces orchidaceus;* it is a structural analog of the amino acid D-alanine, and inhibits the synthesis of the bacterial cell wall peptidoglycan; it is used in the treatment of tuberculosis and urinary tract infections.
cyclosis [sī klō´sis] *Cell Biology.* the rotational movement or streaming of cytoplasm around a cell, especially in plants.
Cyclosporeae *Botany.* brown algae of the Phaeophyta, characterized by a parenchymatous thallus, no alternation of generations, and sporangia contained in conceptacles.
cyclosporin A *Biochemistry.* a cyclic peptide that has a selective action on the generation of helper T cells, which do not become functional while the drug is present; produced by fungi and used as an immunosuppressive agent, particularly in the suppression of graft rejection.
Cyclosteroidea *Paleontology.* one of several little-known mid-Paleozoic classes of echinozoan echinoderms; extant in the Ordovician to Devonian.
Cyclostomata *Invertebrate Zoology.* an order of tubular bryozoans in the class Stenolaemata. *Vertebrate Zoology.* an obsolete designation for the lampreys, an order of eel-like fish now grouped in the order Petromyzoniformes.
cyclostrophic wind *Meteorology.* the horizontal wind velocity for which the centripetal acceleration exactly balances the horizontal pressure force.
cyclothem *Geology.* **1.** a recurring series of beds deposited during a single sedimentary cycle as typified by the deposits of the Pennsylvanian period. **2.** any cycle applied to sedimentary rocks. Also, CYCLE OF SEDIMENTATION.
cyclothymia [sīk´lō thī´mē ə] *Psychology.* a mood disorder characterized by fluctuations between elation and depression.
Cyclothyris *Paleontology.* an extinct genus of articulate brachiopods in the order Rhynchonellida; extant in the Cretaceous.
cyclotomic polynomial [sīk´lə täm´ik] *Mathematics.* **1.** any polynomial of the form $g(x) = x^{p-1} + x^{p-2} + \cdots + x + 1$, where p is prime. Such a polynomial is irreducible over the real numbers. Note that $g(x) = (x^p-1)/(x-1)$. **2.** in general, let n be a positive integer, K a field such that the characteristic of K does not divide n, and F a cyclotomic extension field of order n of K. Then the nth cyclotomic polynomial over K is the monic polynomial

$$g_n(x) = (x - \zeta_1)(x - \zeta_2)\cdots(x - \zeta_n),$$

where $\zeta_1, \zeta_2, \ldots, \zeta_n$ are all the distinct primitive nth roots of unity in F.
cyclotomy [sī klät´ə mē] *Mathematics.* the geometric theory of dividing a circle into equal sectors, or equivalently, of constructing regular polygons; analytically equivalent to finding nth roots of unity.
cyclotrimethylenetrinitramine see CYCLONITE.
cyclotron [sīk´lə trän´] *Nucleonics.* a particle accelerator consisting of halves of a hollow cylinder, called dees because their shape resembles a letter D, connected to a high-frequency alternating voltage source in a uniform perpendicular magnetic field; charged particles (protons, deuterons, or ions) injected near the midpoint of the gap between the dees are propelled in a spiral of increasing radius so that the pathlength increases with the particles' speed until they are deflected as a high-energy beam to a target.
cyclotron-frequency magnetron *Electronics.* a magnetron constructed to produce microwave energy through the synchronous interaction of an electron stream oscillating under the influence of a radio-frequency electrostatic field and the field itself.
cyclotron magnet *Nucleonics.* one of the powerful electromagnets that provide a constant magnetic field in a cyclotron to force ions into circular paths for acceleration.
cyclotron radiation *Electromagnetism.* the radiation emitted by charged particles as they orbit in a magnetic field. Also, **cyclotron emission.**
cyclotron resonance *Physics.* a resonance that occurs when the gyro frequency of a charged particle in a magnetic field equals the frequency of an electric field which acts perpendicular to the magnetic field.
cyclotron resonance heating *Physics.* a method of heating a plasma by compressing and expanding the plasma at a frequency approaching the cyclotron frequencies of the ions in plasma.
cyclotron wave *Electromagnetism.* the electron beam wave of a traveling wave tube.
Cydippidea *Invertebrate Zoology.* an order of comb jellies in the phylum Ctenophora, having well-developed tentacles.
Cydnidae *Invertebrate Zoology.* a family of hemipteran insects, the ground or burrower bugs, in the superfamily Pentatomorpha.
cyesis *Medicine.* another term for pregnancy.
Cyg see CYGNUS.
cygnet [sig´nət] *Vertebrate Zoology.* a popular name for a young swan.
61 Cygni [sig´nī] *Astronomy.* a visual binary star that is 9 light-years away in the constellation Cygnus; in 1838, it became the first star to have its distance measured.

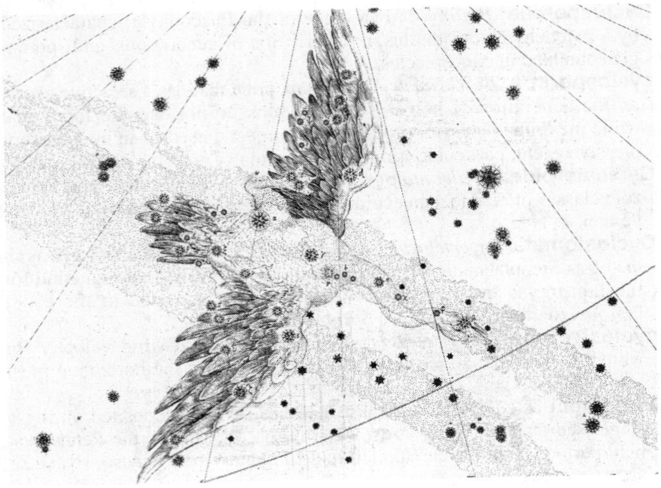

Cygnus

Cygnus [sig´nəs] *Astronomy.* the Swan, a prominent constellation of the Northern Hemisphere, lying in the Milky Way and visible on summer and autumn evenings. Also, NORTHERN CROSS.

Cygnus A *Astronomy.* a strong source of radio emission in Cygnus, emanating from a galaxy flanked by two lobes of gas ejected from the galaxy.

Cygnus Loop *Astronomy.* a supernova remnant roughly 100 light-years in diameter and about 50,000 years old. Also, VEIL NEBULA.

Cygnus X-1 *Astronomy.* a binary star system in Cygnus containing a hot supergiant and a smaller unknown star; it emits X-ray and optical radiation that is variable over times ranging from milliseconds to years.

Cygnus X-3 *Astronomy.* an X-ray binary star system in Cygnus with a period of five hours; it contains a neutron star and also emits in the infrared and radio parts of the spectrum.

cyhexatin *Organic Chemistry.* $C_{18}H_{34}OS_n$, a white solid that is insoluble in water and slightly soluble in organic solvents; it melts at 195–198°C; used to kill plant-eating mites.

cylinder *Mathematics.* **1.** a closed surface consisting of two parallel bases that are bounded by simple closed curves and a lateral surface consisting of all line segments joining corresponding points on the boundaries of the bases. **2.** a solid having this form. *Engineering.* any of various practical devices having the shape of a cylinder, such as the chambers of a revolver that hold the cartridges, or a container in which compressed gas is stored for use in pressurized operations. *Mechanical Engineering.* in an internal-combustion engine, a cylindrical chamber in which the pressure of a heated liquid fuel moves a sliding piston; the part of an engine in which the combustion energy of the fuel is converted to mechanical force. *Computer Technology.* a vertical set of tracks, with one track for each recording surface of a disk pack.

cylinder block *Mechanical Engineering.* in an internal-combustion engine, the metal casing in which the cylinders are bored; it also contains channels for cooling water.

cylinder function *Mathematics.* any solution of Bessel's equation.

cylinder gap *Graphic Arts.* a space between the cylinders of an offset press, used to house the mechanism for the grippers, plate clamps, and blanket bar.

cylinder head *Mechanical Engineering.* a removable top component of a reciprocating engine, pump, or compressor that, when bolted in place, provides a gas-tight seal for the cylinders; in an internal-combustion engine, it also contains valves, valve ports, combustion chambers, and cooling water channels.

cylinder press *Graphic Arts.* a type of letterpress in which the printing plate or type is laid on a flat surface and the paper, attached to a cylinder, is rolled over it. Also, FLAT-BED CYLINDER PRESS.

cylindrarthrosis *Anatomy.* a joint between bones whose articular surfaces are cylindrical, such as the proximal articulation between the radius and ulna.

cylindraxile *Neurology.* a process of a neuron in which impulses travel away from the cell body and are transmitted to other nerve cells or effector organs; an axon.

cylindrical *Science.* of, relating to, or having the form of a cylinder. Also, **cylindric.**

cylindrical antenna *Electromagnetism.* an antenna whose element of radiation is that of a hollow circular cylindrical conductor.

cylindrical array *Electronics.* a two-dimensional array of antenna elements whose corresponding points lie on a cylindrical surface.

cylindrical bow *Naval Architecture.* a bow that, as viewed from above, takes the form of a segment of a wide-diameter cylinder; characteristic of slow, high-capacity hull forms.

cylindrical cam *Mechanical Engineering.* a cam mechanism in which the cam follower undergoes translational motion parallel to the camshaft as a roller attached to the follower moves in a cylinder that is concentric with the camshaft.

cylindrical capacitor *Electricity.* a capacitor composed of two equal-length, concentric metal cylinders, with a dielectric located between the cylinders. Also, COAXIAL CAPACITOR.

cylindrical cavity *Electromagnetism.* a cavity resonator having the interior geometry of a right circular cylinder.

cylindrical-coordinate robot *Robotics.* a robot in which the movement of a manipulator can be defined as points within the coordinates of a cylinder.

cylindrical coordinates *Mathematics.* a curvilinear coordinate system in R^3 consisting of polar coordinates in the (x,y) plane and a rectangular distance measured perpendicular to the (x,y) plane. The coordinates (r,θ,z) of a point are determined by projecting the point onto the (x,y) plane, finding the polar coordinates r and θ of this projected point, and taking as the third coordinate z the perpendicular distance of the point to the (x,y) plane. When r is fixed and θ and z vary, a cylindrical surface is formed; hence the term cylindrical coordinates.

cylindrical-film storage *Electronics.* a magnetic core memory in which each individual core consists of a thin film of ferromagnetic material on the surface of a short glass cylinder.

cylindrical grinder *Mechanical Engineering.* a machine designed to accurately finish cylindrical shapes with a high-speed abrasive wheel; the work is rotated by the head stock of the machine and the wheel is automatically moved along it under a flow of coolant.

cylindrical helix *Mathematics.* a curve on the lateral surface of a cylinder that intersects the elements at a constant angle; for example, the wire spring on a spiral-bound notebook.

cylindrical reflector *Electromagnetism.* a reflector that has a parabolic cross section and a focal line rather than a focal point.

cylindrical surface *Mathematics.* a surface consisting of all lines parallel to a given line and passing through a simple closed curve; that is, the parallel displacement of a line (generator) along a curve (base curve). Also, LATERAL SURFACE.

cylindrical symmetry *Organic Chemistry.* a symmetry of electron density present in an atom so that no matter how the molecule is turned about the internuclear axis, its electron density appears the same.

cylindrical wave *Electromagnetism.* a wave whose phase fronts form concentric surfaces of circular cylinders.

cylindrical winding *Electricity.* the current-carrying element of a core transformer.

cylindrite *Mineralogy.* $Pb_4Fe^{+2}Sn_4^{+4}Sb_2^{+3}S_{16}$, a dark gray, opaque, metallic triclinic mineral occurring in cylindrical forms and spherical aggregates, having a specific gravity of 5.46 and a hardness of 2.5 on the Mohs scale; found in tin-bearing veins with pyrite, franckeite, and sphalerite.

Cylindrocapsaceae *Botany.* green algae of the order Ulotrichales, characterized by sexual reproduction, large chloroplasts, and thick, stratified cell walls.

Cylindrocarpon *Mycology.* a genus of fungi belonging to the class Hyphomycetes, formerly classified under the genus *Fusarium* because of similar spores; most commonly occurring in soils; some of its species are pathogenic.

cylindrodendrite *Neurology.* a collateral branch of a neuron's axon.

cylindroma *Oncology.* **1.** a usually benign tumor consisting of cylindrical epithelial masses containing small basophilic and larger pale-staining cells surrounded by pink hyaline sheaths, usually arising on the scalp early in life. **2.** an adenoid cystic carcinoma.

Cylindrospermum *Bacteriology.* a genus of filamentous cyanobacteria in which the heterocysts develop only at the termini of the trichomes; occurring in colonies attached to the leaves of certain floating plants, such as duckweed.

cyma [sī´mə] *plural,* **cymae** [sī´mē] or **cymas.** *Architecture.* a molding having a profile of two contrary curves.

cymb- or **cymbo-** a combining form meaning "boat-shaped," as in *cymbocephalic*.

Cymbellaceae *Botany*. a family of freshwater and marine diatoms of the order Pennales, distinguished by valves that are asymmetrical at the apical or transapical axis and having a polyphyletic shape.

cymbocephalic *Anthropology*. having a skull with a back-sloping forehead and a severe protrusion from the back.

cyme [sīm] *Botany*. a determinate inflorescence giving a flat-topped appearance in which each primary axis ends in a single blossom, with subsequent blossoms occurring later on smaller, lateral branches.

cymene *Organic Chemistry*. any one of the isomeric hydrocarbons composed of benzene rings carrying one methyl and one $-CH(CH_3)_2$ or isopropyl group.

Cymodoceaceae *Botany*. a family of monocotyledonous, dioecious, rhizomatous, submerged marine herbs of the order Najadales, native to tropical and subtropical seas.

cymose [sī´mōs´] *Botany*. bearing or resembling cymes.

cymose inflorescence see DETERMINATE INFLORESCENCE.

Cymothoidae *Invertebrate Zoology*. a family of isopod crustaceans in the suborder Flabellifera; parasitic on fish.

cymrite [kim´rīt´] *Mineralogy*. $BaAl_2Si_2(O,OH)_8 \cdot H_2O$, a colorless, transparent to translucent, satiny monoclinic mineral occurring as prismatic to platy crystals, having a specific gravity of 3.413 and a hardness of 2 to 3 on the Mohs scale; found in manganese mines and copper deposits.

cyn- or **cyno-** a combining form meaning "of, relating to, or resembling a dog," as in *cynanthropy, cynophobia*.

Cyniclomyces *Mycology*. a genus of fungi of the family Saccharomycetaceae, requiring high levels of carbon dioxide for growth.

Cynipidae *Invertebrate Zoology*. a family of gall wasps, hymenopteran insects in the superfamily Cynipoidea.

Cynipoidea *Invertebrate Zoology*. the gall wasps, a superfamily of hymenopteran insects in the suborder Apocrita, which produce galls on oaks.

cynodontism *Anthropology*. a condition in which the pulp cavity of the tooth is constricted, as in modern humans.

Cynoglossidae *Vertebrate Zoology*. the tongue soles, a family of tropical Asiatic fish in the suborder Soleoidei of the order Pleuronectiformes, noted for having both eyes on the left side of the head.

cynophobia *Psychology*. **1.** an irrational fear of dogs. **2.** a neurosis, sometimes precipitated by a dog bite, in which the individual reproduces the symptoms of rabies.

Cyperaceae *Botany*. a single family of reeds and sedges of the order Cyperales, having a solid, triangular culm; found in marshy areas.

Cyperales *Botany*. an order of monocotyledonous plants characterized by a pithy culm, spiked flowers, and a scaly perianth. Also, POALES.

Cypheliaceae *Botany*. a family of crustose lichens belonging to the order Caliciales.

Cyphellaceae *Mycology*. a family of fungi of the order Agaricales, characterized by small basidiocarps of various shapes and occurring on soil, dung, wood, and plant material or in a symbiotic relationship with plants; some species are parasitic to vascular plants or other fungi.

cyphonautes *Invertebrate Zoology*. the free-swimming, ciliated, conical, bivalve larva of certain bryozoans.

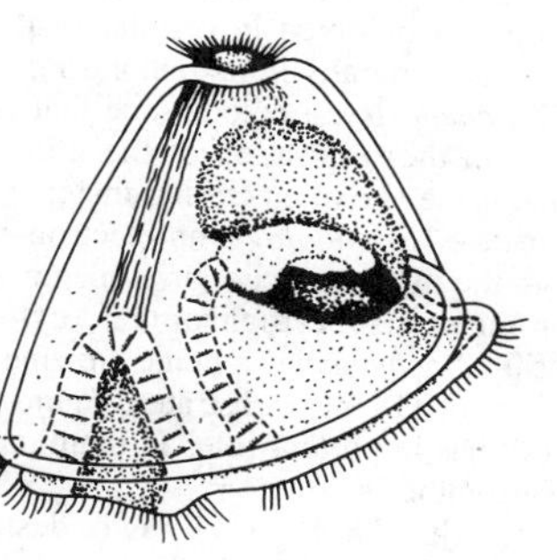

cyphonautes

Cyphophthalmi *Invertebrate Zoology*. a family of harvestmen; small, mitelike arachnids in the order Phalangida.

Cypraecea *Invertebrate Zoology*. a superfamily of gastropod mollusks in the order Prosobranchia, having two tentacles.

Cypraeidae *Invertebrate Zoology*. a family of marine snails with colorful shells, belonging to the order Pectinibranchia.

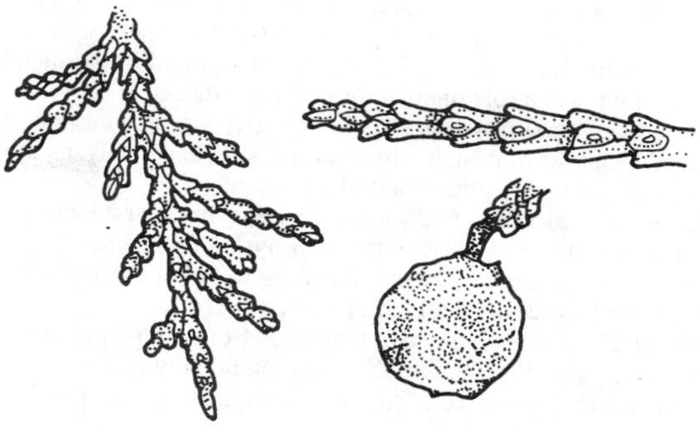

cypress

cypress *Botany*. the common name for the coniferous evergreen trees and shrubs of the genus *Cuperssus*.

Cypridacea *Invertebrate Zoology*. a superfamily of mostly freshwater bivalve ostracod crustaceans, in the suborder Podocopa.

Cypridea *Paleontology*. an extinct genus of freshwater ostracods in the superfamily Cypridacea; Middle Jurassic to Lower Cretaceous.

Cypridinacea *Invertebrate Zoology*. a superfamily of ostracod crustaceans in the suborder Myodocopa, having a round back and calcified carapace; some are bioluminescent.

cypridophobia *Psychology*. an irrational fear of being infected with a venereal disease.

Cyprinidae *Vertebrate Zoology*. a large, cosmopolitan family of freshwater fishes of the suborder Cyprinoidei of the order Cypriniformes, including the minnow, carp, and barb; characterized by convex dorsal and ventral surfaces, no adipose fin, thin lips, and a gas bladder not covered with bone. It includes more species of fish than any other family.

Cypriniformes *Vertebrate Zoology*. a very large order of usually soft-rayed freshwater fishes living on all continents except Australia; it includes the tetras, characoids, darters, piranhas, electrical eels, carps, minnows, and barbs, among others.

Cyprinodontidae *Vertebrate Zoology*. the killfishes, a family of very robust fishes in the suborder Cyprinodontoidei of the order Atheriniformes, noted for their ability to adapt to extremes of salinity and heat.

Cyprinodontoidei *Vertebrate Zoology*. a suborder of small surface-living fishes in the order Atheriniformes, including ricefishes and killfishes.

Cyprinoidei *Vertebrate Zoology*. a large suborder of carplike freshwater fishes in the order Cypriniformes, including the characins, electric eels, suckers, carps, minnows, barbs, and loaches.

Cypris *Invertebrate Zoology*. a large genus of ostracod bivalve crustaceans living in stagnant water.

cypris larva *Invertebrate Zoology*. a larval stage with a bivalve carapace in Cirripedia.

Cyrillaceae *Botany*. a family of dicotyledonous tanniferous shrubs and small trees in the order Ericales, characterized by alternate, simple, and entire leaves without stipules and by perfect flowers borne in racemes.

Cyrtophorina see GYMNOSTOMATIDA.

cyrtopia *Invertebrate Zoology*. an ostracod crustacean larva with a long first pair of antennae and lost swimming capability in the second pair.

Cyrtopodaceae *Botany*. a family of large, dull-green mosses of the order Isobryales that form loose mats on tree bark; characterized by creeping primary stems ascending to pendulous irregularly branched secondary stems and lateral sporophytes; native to Australasia, New Caledonia, Hawaii, and the Philippines.

cyrtosis *Medicine*. an abnormal curvature of an extremity or the spine.

cyst *Pathology*. an abnormal sac or pouch that contains fluid, gas, or some semisolid material. *Biology*. **1.** any closed cavity or sac, normal or abnormal, especially one containing a liquid or semisolid material. **2.** a cavity or cell in which reproductive bodies, embryos, or bacteria lie in a resting stage. **3.** this resting stage itself.

cyst- a combining form meaning "cyst" or "bladder," as in *cystectomy*.

cystadenoma see ADENOCYSTOMA.

cystamine *Organic Chemistry*. $C_4H_{12}N_2S_2$, viscous oil; soluble in water and alcohol; decomposes on distillation.

cysteamine *Organic Chemistry*. C_2H_7NS, a crystal with a disagreeable odor; soluble in water; melts at 97–98.5°C; used as a drug to treat radiation sickness.

cystectomy *Surgery*. the surgical removal of a cyst or of the bladder.

cysteine [sis´tēn´] *Biochemistry*. one of the twenty amino acids used by proteins; it contains a sulfur group that can bind to another sulfur to form protein cross-links, is the primary compound in the metabolism of sulfur, and plays a role in cofactor biotin synthesis.

cysteine carboxypeptidase *Enzymology*. an enzyme that contains a sulfhydryl group of cysteine in its active site, and catalyzes the hydrolysis of the terminal peptide bond of a protein.

cysteine proteinase *Enzymology*. an enzyme that contains a cysteine in its active site and catalyzes the hydrolysis of proteins.

cystic [sis´tik] *Anatomy*. **1.** of or relating to either the gallbladder or the urinary bladder. **2.** of or relating to a cyst.

cystic duct *Anatomy*. the duct that leads from the gall bladder and joins with the hepatic duct to form the common bile duct.

cysticercoid *Invertebrate Zoology*. a larval stage of some tapeworms, found in arthropods.

cysticercosis *Medicine*. an infestation with cysticerci, the larvae of tapeworms.

cysticercus *plural,* **cysticerci.** *Invertebrate Zoology*. the bladder worm, an inactive tapeworm larva found in an intermediate host that develops when eaten by a primary host.

cystic fibrosis *Medicine*. a hereditary disease in which the exocrine glands secrete abnormally viscid mucus that blocks intestinal and branchial ducts, causing patients to have difficulty breathing.

cystidium *Mycology*. a sterile cell found in the spore-bearing structure or basidium of certain fungi belonging to the subdivision Basidiomycotina. Also, STERILE CELL, CYSTID.

cystine *Biochemistry*. an amino acid that is produced by the digestion or acid hydrolyis of protein; sometimes found in the urine and in the kidneys in the form of small hexagonal crystals; a primary sulfur-containing compound of the protein molecule.

cystine stone *Pathology*. an obstruction of the urinary tract composed of cystine, an amino acid that normally aids in tissue repair.

cystinosis [sis´tə nō´sis] *Medicine*. a lysosomal storage disorder of unknown molecular defect, characterized by widespread deposition of cystine crystals in reticuloendothelial cells. The three forms of the disease are the infantile nephropathic type, the adolescent nephropathic type, and the benign or adult nephropathic type.

cystinuria [sis´tə nůr´ē ə] *Medicine*. excessive cystine in the urine.

cystitis [sis tīt´is] *Medicine*. an inflammation of the bladder.

cysto- a combining form meaning "cyst" or "sac," as in *cystoscope*.

Cystobacteraceae *Bacteriology*. a family of gliding, rod-shaped bacteria of the order Myxobacterales, distinguished by the production of microcysts in sporangia.

cystoblast *Entomology*. an insect ovarian germ-line cell that divides to form a cystocyte.

cystocarcinoma *Oncology*. a carcinoma characterized by cysts. Also, **cystoepithelioma.**

cystocarp *Botany*. the fruiting structure of red algae arising from the fertilization of the carpogonia.

cystocele *Medicine*. the hernial protrusion of the urinary bladder.

cystocercous cercariae *Invertebrate Zoology*. fluke larvae, which are able to retract the body into a hollow in the tail.

cystocyte *Entomology*. an insect blood cell that imitates the clotting of blood; when hemolymph flows from a wound, these cells enlarge, break down, and form threads around which plasma jells.

cystofibroma *Oncology*. a fibroma characterized by cysts.

Cystografin *Radiology*. a trade name for a preparation of diatrizoate meglumine, a contrast medium given intravenously to permit radiography of the venous supply of various organs.

cystogram *Medicine*. an X-ray of the bladder.

cystography [sis´täg´rə fē] *Medicine*. radiography of the bladder after injection of the organ with an opaque solution.

Cystoidea *Paleontology*. a former classification for the attached diploporan and rhombiferan echinoderms of the middle Paleozoic, now considered separate classes; extant in the Diploporita and Rhombifera.

cystolith *Botany*. a concentration of calcium carbonate inside the epidermal cells of certain plants, such as nettles. *Medicine*. a calculus within the bladder.

cystolithectomy *Surgery*. the removal of a bladder stone by incision of the bladder. Also, **cystolithotomy.**

cystoma *Medicine*. a tumor that contains cysts.

cystometer *Medicine*. an instrument for studying the functioning of the bladder by measuring its pressure and volume.

cystoplasty *Surgery*. plastic repair of the bladder.

cystoplegia *Medicine*. paralysis of the bladder.

Cystoporata *Paleontology*. an extinct order of colonial bryozoans, similar to the Trepostomata, that form massive calcareous zoaria; extant in the Lower Ordovician to Triassic.

cystopyelitis [sis´tō pī ə lī´tis] *Medicine*. an inflammation involving both the urinary bladder and the pelvis of the kidney.

cystopyelography [sis´tō pī´ə läg´rə fē] *Medicine*. an X-ray of the urinary bladder and pelvis of the kidney.

cystopyelonephritis [sis´tō pī´ə lō nə frī´tis] *Medicine*. the simultaneous inflammation of the bladder, the pelvis of the kidney, and the contents of the kidney.

cystorrhaphy [sis´tôr´ə fē] *Surgery*. suture of the bladder.

cystorrhea [sis´tə nůr´ē ə] *Medicine*. an abnormal discharge from the bladder.

cystorrhexis [sis´tə reks´is] *Medicine*. the rupture of the bladder.

cystosarcoma [sis´tō sär kō´mə] *Oncology*. a sarcoma that contains cellular stroma characterized by the formation of cysts or cystlike foci which sometimes metastasize.

cystoscopy [sis´täs´kə pē] *Medicine*. visual examination of the urinary tract with an instrument inserted through the urethra. Thus, **cystoscope.**

cystose [sis´tōs´] *Medicine*. resembling or containing a cyst or cysts.

Cystoseiraceae *Botany*. a common family of brown algae of the order Fucales, having a monopodial main axis with long primary branches, from which secondary laterals arise, and a discoid or conical holdfast.

cystostomy [sis´täs´tə mē] *Surgery*. the formation of an opening into the urinary bladder.

cystoureteritis [sis´tō yů rē tə rī´tis] *Medicine*. the simultaneous inflammation of the bladder and of one or both ureters.

cystourethrogram [sis´tō yů rēth´rō gram] *Medicine*. an X-ray of the bladder and the urethra. Thus, **cystourethrography.**

cyt- a combining form meaning "cell," as in *cytidine*.

Cytheracea *Invertebrate Zoology*. a superfamily of marine ostracods in the suborder Podocopa; mostly crawling and digging forms.

Cytherellidae *Invertebrate Zoology*. a family of marine ostracod bivalves in the suborder Platycopa.

cytidine *Biochemistry*. a cytosine ribonucleoside that is important in the metabolism of all living organisms. Also, CYTOSINE RIBOSIDE.

cytidine monophosphate see CYTIDYLIC ACID.

cytidine 5'-triphosphate see CTP.

cytidylic acid *Biochemistry*. a ribonucleotide that plays a role in the synthesis of glycoproteins. Also, CYTIDINE MONOPHOSPHATE.

cytisism *Toxicology*. poisoning caused by ingestion of seeds of the tree *Cytisus laburnum*; symptoms may include severe gastrointestinal and respiratory disturbances, coma, paralysis, and death.

cyto- a combining form meaning "cell," as in *cytoplasm*.

cytoarchitecture *Neurology*. the arrangement or pattern of cells in a tissue or organ, especially in the brain.

cytobiosis *Cell Biology*. a symbiotic relationship in which one of the symbionts occurs within the cell of the other.

cytochalasin *Biochemistry*. a fungal or mold metabolic product that inhibits cellular activities, such as movement, without affecting certain changes that take place in the cell nucleus before cell division.

cytochemistry *Cell Biology*. the study of the chemical composition, structure, and processes of cells.

cytochrome [sīt´ə krōm´] *Biochemistry*. an electron-carrying iron prophyrin pigment protein involved in cellular respiration; present in all plant and animal cells, generally in mitochondria.

cytochrome c *Biochemistry*. a cytochrome that plays a central role as an electron carrier in the respiration of the mitochondria; contains an iron porphyrin that is the site of electron transfer.

cytochrome oxidase *Enzymology*. an electron-carrying protein complex that occurs in the inner mitochondrial membrane, active in the oxidation of cytochrome c. Also, **cytochrome a_3, cytochrome c oxidase.**

cytochrome P450 *Biochemistry*. a heme enzyme that catalyzes the insertion of one atom of O_2 into substrates; found in animal, plant, and bacterial species, it has important roles in mammalian liver and adrenal cortex (steroid) biosynthesis.

cytocidal [sis´tə sīd´əl]*Cell Biology*. killing or destroying a cell or cells.

cytocrine gland *Cell Biology*. any cell whose secretions move directly from one cell to another, such as a melanocyte.

cytodendrite *Neurology*. any dendrite arising from a part of the cell other than the axon.

cytodiagnosis *Pathology*. the diagnosis of disease on the basis of the microscopic analysis of cells.

cytodifferentiation *Cell Biology*. the process of cellular differentiation in terms of morphological rather than chromosomal changes.

cytodistal *Neurology.* of a part of an axon or other neuronal process, located away from the cell body.

cytoduction *Mycology.* in fungi, a sexual process in which two cells fuse but do not undergo nuclear fusion, resulting in the mixing of cytoplasmic elements and the retention of two separate nuclei.

cytofluorometry *Biotechnology.* an automated separation and analysis technique for cells or chromosomes that relies on the detection of specific fluorescent markers.

cytogamy [sī täg´ə mē] *Biology.* the conjugation of cells.

cytogenetics *Genetics.* the study of genetics through microscopic analyses of chromosomes and chromosomal aberrations.

cytogenous gland *Physiology.* a cell-producing gland.

cytohet *Cell Biology.* a eukaryotic cell that is heterozygous for at least one cytoplasmic gene.

cytokeratin *Cell Biology.* a member of the intermediate filament class of proteins found in cytoskeletal filaments of epithelial cells.

cytokines *Immunology.* the soluble molecules being produced by cells that mediate reactions between cells, usually used for biological response modifiers.

cytokinesis *Cell Biology.* the division of a cell into two daughter cells following mitosis.

cytokinin *Biochemistry.* a plant hormone that influences cell division, plant metabolism, and the synthesis of ribonucleic acid and proteins.

cytological biopsy *Surgery.* a procedure in which cells are retrieved by various methods for pathological examinations, as by irrigation of a hollow viscera.

cytology [sī täl´ə jē] *Biology.* the study of the structure, function, behavior, growth, and reproduction of cells and cell components; now more often called cell biology. See CELL BIOLOGY.

cytolysis *Cell Biology.* the process of breaking open a cell by disrupting the plasma membrane. *Pathology.* cell dissolution or disintegration.

cytolysosome *Cell Biology.* a large intracellular lysosome involved in active digestion of its own cell.

cytolytic *Pathology.* referring to a process or substance that has a destructive effect on cells.

cytolytic reaction *Immunology.* the effect of an antibody on sensitized cells that results in the death of the target cell.

cytoma *Oncology.* a cell tumor, such as a sarcoma.

cytomegalic inclusion disease *Medicine.* a condition caused by viral infection and occurring in newborn infants, characterized by the presence of inclusion bodies in the cytoplasm and nuclei of enlarged cells and resulting in hepatosplenomegaly, fever, and retardation.

Cytomegalovirus group see BETAHERPESVIRINAE.

cytomere *Cell Biology.* **1.** in parasitic protozoa, a multinucleate structure that segments to form merozoites. **2.** the cytoplasmic portion of a spermatozoon.

cytomorphosis *Cell Biology.* the sequence of developmental modifications that occur throughout the life of a cell.

cyton *Cell Biology.* the cell body of a neuron.

cytopathic *Cell Biology.* causing manifestations of disease in cells.

cytopathic effect *Virology.* a change or abnormality in the microscopic appearance of cultured cells due to virus infection.

cytopathology *Pathology.* a discipline in medicine that specializes in the role of cells in disease.

cytopenia *Pathology.* a deficiency or inadequacy of cellular constituents in the bloodstream.

Cytophaga *Bacteriology.* a genus of Gram-negative, gliding bacteria of the order Cytophagales, occurring in soil or aquatic habitats, and distinguished by the ability to hydrolyze polysaccharides.

Cytophagaceae *Bacteriology.* a former classification for a family of pigmented, motile bacteria in the order Cytophagales.

Cytophagales *Bacteriology.* an order of Gram-negative, gliding bacteria that occur in soil, freshwater, marine, or estuarine habitats.

cytopharynx *Invertebrate Zoology.* a channel connecting the surface with the protoplasm in certain protozoans, generally used to ingest food.

cytophilic *Immunology.* referring to substances having the propensity to bind cells, such as cytophilic antibodies.

cytoplasm [sīt´ə plaz´əm] *Cell Biology.* the cellular region within the plasma membrane, including the cytosol and the organelles but excluding the nucleus. Thus, **cytoplasmic.**

cytoplasmic gene *Genetics.* any gene that occurs outside a bacterial chromosome or eukaryotic nucleus.

cytoplasmic inheritance *Genetics.* the inheritance of characteristics via DNA found in the extranuclear organelles of mitochondria and chloroplasts. Also, EXTRACHROMOSOMAL INHERITANCE.

cytoplasmic membrane *Cell Biology.* a protein-containing phospholipid bilayer that surrounds a cell, defining the interface between a cell and its environment, and displaying a selective permeability toward the entry of molecules into the cell.

cytoplasmic polyhedrosis virus group *Virology.* a group of entomopathogenic viruses of the family Reoviridae that infect arthropods.

cytoplasmic streaming *Cell Biology.* a directional flow of cytoplasm that occurs in some larger cells, possibly facilitating the dispersal of metabolites and other cellular components.

cytoplast *Cell Biology.* an anucleate cell produced by cytochalasin treatment.

cytoproct *Invertebrate Zoology.* a permanent pore in certain protozoan ciliates, through which indigestible food residues are discharged. Also, **cytophage.**

cytoproximal *Neurology.* of a part of an axon or other neuronal process, located near the parent cell.

cytopyge *Invertebrate Zoology.* a point of waste discharge in the body of a protozoan.

cytosegresome *Cell Biology.* a membrane-enclosed vesicle that contains cellular constituents of autophagous eukaryotic cells.

cytosine *Biochemistry.* $C_4H_5N_3O$, one of the four bases found in the RNA and DNA of all living organisms.

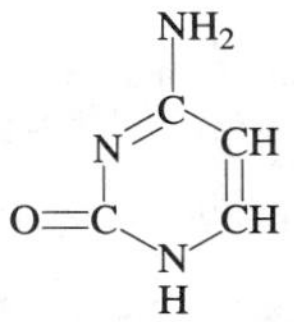

cytosine

cytoskeleton *Cell Biology.* in eukaryotes, a cytoplasmic network of protein filaments, including actin microfilaments, microtubules, and intermediate filaments, playing an essential role in cell movement, shape, and division.

cytosol *Cell Biology.* the soluble portion of the cytoplasm, including cytoskeletal fibers, molecules, and small particles such as ribosomes, but excluding the membrane-enclosed organelles.

cytosome *Cell Biology.* the cytoplasmic fraction of a cell, as distinguished from the nuclear fraction.

cytostatic *Cell Biology.* inhibiting cell growth. *Immunology.* **1.** referring to an area or process in which the movement and accumulation of blood cells is decreased. **2.** of a drug, impeding cell division.

cytostome *Invertebrate Zoology.* a pore opening through which waste is discharged in many unicellular organisms, such as Ciliophora.

cytotaxis *Cell Biology.* the movement of cells with respect to a specific source of stimulation. Thus, **cytotaxin.**

cytotoxic *Cell Biology.* having a toxic effect on cells.

cytotoxic T cell *Cell Biology.* a type of T lymphocyte that is able to recognize and lyse virus-infected cells or foreign cells.

cytotrophoblast *Developmental Biology.* the inner layer of the trophoblast.

cytotropic see CYTOPHILIC.

cytotropism *Biology.* the movement of individual cells or cell groups toward or away from one another or a stimulus.

cytozoic *Biology.* living within a cell; intracellular.

Cyttariaceae *Mycology.* the single family of fungi belonging to the order Cyttariales, which is composed of species that are exclusively parasitic to the *Nothofagus* in the Southern Hemisphere.

Cyttariales *Mycology.* an order of fungi belonging to the class Discomycetes, which is characterized by having asci that develop simultaneously with their spore cases.

Cy VADIC *Oncology.* an acronym for a cancer chemotherapy regimen that includes the drugs cyclophosphamide, vincristine, Adriamycin, and imidazole carboxamide.

Czapek's medium [chä´peks] *Microbiology.* a nutrient medium that includes sugar, water, salt, and agar, used for the culture of saprotrophic fungi and soil bacteria. Also, **Czapek's agar, Czapek-Dox medium.**

Czochalski process [chō käl´skē] *Crystallography.* a method of growing crystals by placing a seed crystal in contact with molten material and then gradually pulling it away from the melt. The result is a rod-shaped crystal.

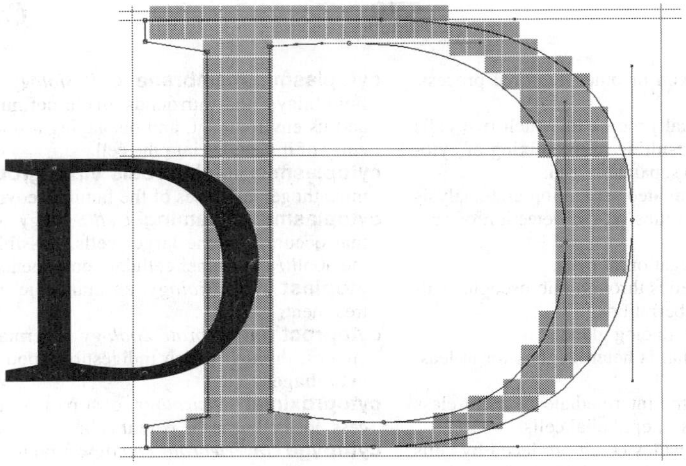

D *Science.* the fourth in a sequence or group. *Computer Programming.* in a hexadecimal numbering system, the character that represents a decimal 13.

D the chemical symbol for deuterium.

D or **D.** derivative.

d or **d.** date; degree; diameter; distance; dorsal; distal; duration.

d. an abbreviation for Latin *detur,* "let it be given"; *dies,* "day"; and *dexter,* "right."

2,4-D see 2,4-DICHLOROPHENOXYACETIC ACID.

3-D see THREE-D.

DA or **D.A.** developmental age; dental assistant.

da. day; days.

d.a. let it be given to. (From Latin *detur ad.*)

DAB *Aviation.* the airport code for Daytona Beach, Florida.

DAC or **dac** digital-to-analog converter.

dachiardite *Mineralogy.* $(Ca,Na_2,K_2)_5Al_{10}Si_{38}O_{96}\cdot 25H_2O$, a transparent, colorless monoclinic mineral of the zeolite group occurring as twinned, prismatic crystals with two perfect cleavages, having a specific gravity of 2.165 to 2.206 and a hardness of 4 to 4.5 on the Mohs scale; found with other zeolites in granite pegmatites.

Dacian *Geology.* an Eastern European stage of the Upper Pliocene epoch, occurring after the Pontian and before the Rumanian.

dacite *Geology.* a very fine-grained extrusive rock that is similar in composition to andesite, but that contains less calcic feldspar. Also, QUARTZ ANDESITE.

dacite glass *Geology.* a glass formed naturally as a result of the rapid cooling of dacite lava.

dacitoid *Geology.* an extrusive rock similar in composition to decite, but lacking modal quartz.

D/A converter see DIGITAL-TO-ANALOG CONVERTER.

Da Costa's syndrome see NEUROCIRCULATORY ASTHENIA.

Dacron *Materials.* the trade name for a type of polyester fiber, characterized by its high degree of resilience, that is widely used in the textile industry.

dacry- or **dacryo-** a combining form meaning "tears," as in *dacryocyst, dacryocystography.*

Dacrymyces *Mycology.* a genus of fungi belonging to the subclass Metabasidiomycetidae; the species *D. deliquescens,* found on decaying wood, is characterized by its gelatinous, bright orange fruiting body.

Dacrymycetaceae *Mycology.* a family of fungi belonging to the order Metatremellales, which is characterized by brightly colored basidia; occurring on decaying organic matter in tropical to temperate regions.

Dacrymycetales *Mycology.* an order of fungi of the class Hymenomycetes, occurring on wood and characterized by brightly colored, gelatinous spore-bearing structures or basidia shaped like tuning forks.

dacryocyst see LACRIMAL SAC.

dacryocystitis *Medicine.* an inflammation of the lacrimal sac.

dacryocystography *Radiology.* an examination of the lacrimal sac by X-ray after filling it with a radiopaque substance.

dacryocystorhinostomy *Surgery.* the creation of an opening between the nasal cavity and the lacrimal sac. Also, **dacryorhinocystostomy.**

dacryocystostomy *Surgery.* the surgical creation of a new opening in the lacrimal sac; incision of the lacrimal sac.

dacryolith *Medicine.* a stone in the lacrimal sac or duct.

dacryolithiasis *Medicine.* the formation of stones in the tear duct.

dacryon *Anatomy.* the junction of the lacromaxillary and the frontomaxillary sutures on the medial wall of the orbit, used as a reference point in craniometry.

dacryosinusitis *Medicine.* an inflammation of the lacrimal duct and the ethmoid sinus.

DACT dactinomycin.

dactinomycin *Pharmacology.* an antibiotic produced by *Streptomyces pavulus,* used as an antineoplastic agent for treatment of rhabdomyosarcoma and Wilms' tumor in children; it is also effective against Ewing's sarcoma, Kaposi's sarcoma, and osteogenic and soft-tissue sarcomas.

Dactylioceras *Paleontology.* a genus of planulate ammonites of the Jurassic that has figured in investigations of continental drift; it is also significant as a zone fossil because it always occurs in a stratum just above that of Amaltheus.

dactylitic *Geology.* describing a rock texture produced by the intergrowth of two different minerals whereby fingerlike projections of one mineral penetrate the other.

dactylitis *Medicine.* an inflammation of a toe or finger.

dactylo- or **dactyl-** a combining form meaning "finger" or "toe," as in *dactylospasm, dactylitic.*

Dactylochirotida *Invertebrate Zoology.* a widely distributed order of sea cucumbers found mainly in deep water.

Dactylococcopsis *Bacteriology.* a genus of unicellular bacteria of the cyanobacteria group, occurring as elongated cells that taper at the poles and sometimes form chains.

dactylognathite *Invertebrate Zoology.* the distal segment of a maxilliped in crustaceans.

dactylogram *Science.* a fingerprint taken for purposes of identification.

dactylography *Science.* the study of fingerprints.

Dactylogyroidea *Invertebrate Zoology.* a superfamily of trematodes that are parasitic on the skin, gills, and sometimes the urinary tract of fish.

dactylology *Science.* the scientific study of communication by finger or hand signs.

Dactylopiidae *Invertebrate Zoology.* a family of insects including the cochineal insect, a source of red dye.

dactylopodite *Invertebrate Zoology.* the distal segment of a leg in a crab, shrimp, lobster, or similar crustacean; it may be a claw or part of a pincer.

dactylopore *Invertebrate Zoology.* any of the small openings on the surface of corals through which the feelers of polyps are extended.

Dactylopteridae *Vertebrate Zoology.* the single family of the order Dactylopteriformes, small marine fishes with greatly elongated pectoral fins, which allow them to glide above the water for short distances, although they are primarily bottom-dwellers. Also, FLYING GURNARD.

Dactylopteriformes *Vertebrate Zoology.* an order of marine fishes comprising the family Dactylopteridae and belonging to the class Osteichthyes.

Dactylopteroidei *Vertebrate Zoology.* in some taxonomical systems, a suborder of marine shore fishes having very expansive pectoral fins in the order Perciformes.

Dactyloscopidae *Vertebrate Zoology.* a family of the order Perciformes, containing small, predominantly salt-water tropical fishes characterized by a somewhat bulky head, a large and almost vertical mouth, and eyes positioned on the top of the head. Also, SAND STARGAZER.

dactylospasm *Medicine.* a cramp or twitching of the fingers or toes.
Dactylosporangium *Bacteriology.* a genus of mycelial bacteria of the order Actinomycetales, occurring in soil and forming spores both in the sporangia and on the substrate mycelium.
dactylosternal *Vertebrate Zoology.* of turtles, having marginal processes suggesting fingers that join the ventral part of the shell to the carapace.
dactylozooid *Invertebrate Zoology.* any of the long, defensive tentacles bearing stinging cells in certain corals.
dactylus *Invertebrate Zoology.* in certain insects, the second tarsal segment, after an enlarged first tarsal segment.
DAD or **dad** data-base action diagram.
dado [dā´dō] *Architecture.* **1.** the middle part of a column pedestal, between the base and the surbase. Also, DIE. **2.** a decorative paneling on the lower part of an interior wall. **3.** the ornaments on such paneling.
dado head *Mechanical Engineering.* a power-saw tool used to cut flat bottom grooves in wood; consists of two circular saws with one or more chippers in between.
dado plane *Mechanical Devices.* a narrow plane equipped with two spurs and an adjustable depth stop; used to cut grooves in wood.
Daedalea *Mycology.* a genus of fungi belonging to the order Agaricales, occurring on wood and characterized by pores shaped like labyrinths.
daemon see DEMON.
daffodil *Botany.* a bulbous plant, *Narcissus pseudonarcissus,* of the amaryllis family, bearing a single bright-yellow flower that blooms in early spring.

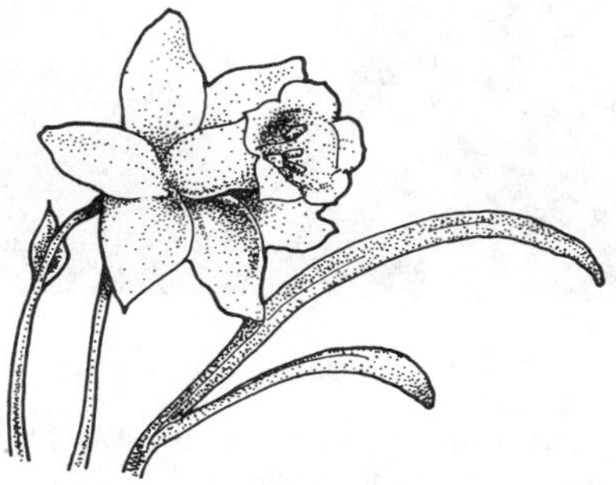

daffodil

DAG *Computer Science.* directed acyclic graph; a graph consisting of a set of nodes and directed arcs (arrows) between nodes, such that no circular paths (cycles) exist.
dag or **dag.** dekagram; dekagrams.
dagala see STEPTOE.
dagger board *Naval Architecture.* a long narrow board that slides vertically through a slot or trunk in the hull, providing additional stability and leeway resistance when down and creating a shallower draft when up; it is distinguished from a centerboard by its vertical sliding motion as opposed to the pivoting motion of a centerboard.
Dagor lens *Optics.* an astigmatic lens that is composed of two separate systems with three or more lenses that are nearly symmetrical.
Daguerre, Louis [də gâr´] 1787–1851, French painter and inventor; with Niépce, invented the daguerreotype photographic process.
daguerreotype [də gâr´ē ō tīp´] *Graphic Arts.* **1.** an early form of photography in which a silver or silver-coated plate was photosensitized with iodine and developed with mercury vapor. **2.** a photograph made by using this method. (Named for Louis *Daguerre.)*
DAH disordered action of the heart.
Dahlin's algorithm *Control Systems.* a variation of the deadbeat algorithm by which dead time is reduced in order to reduce ringing in the control system.
dahoma *Materials.* a yellowish-brown wood from the west African tree *Piptadena africana* that is used to build ships and docks. Also, AFRICAN GREENHEART.
daily keying element *Telecommunications.* a part of a cipher key that is changed every 24 hours while other parts remain unchanged.
daily permissible dose *Nutrition.* the amount of a substance that can be consumed daily without causing ill effects.
daily recommended dose *Nutrition.* the amount of a substance that, if consumed daily, is likely to aid in the maintenance of normal health.
daily retardation *Oceanography.* the amount of time by which corresponding tides occur later each day (about 50 minutes on average).
daily variation see DIURNAL VARIATION.
Daimler, Gottlieb [dīm´lər] 1834–1900, German engineer; developer of the first lightweight internal-combustion automobile engine.
dairy *Agriculture.* a place where milk products are manufactured and stored.
dais [dā´əs; dī´əs] *Architecture.* a raised platform at one end of a hall used for the seating of speakers, dignitaries, or honored guests.
daisy chain *Computer Technology.* a data communication scheme that consists of one bus connecting the CPU to all peripheral devices, over which the signal is passed serially from one device to the next. Also, BROADCAST MODE, PARTY LINE.
daisy-chain interrupt *Computer Technology.* a method of priority selection among devices requesting service from the CPU, consisting of a serial connection of all devices on a single bus with the highest priority device closest to the CPU, followed by lower priority devices in order of priority.
daisyhead colonies *Bacteriology.* colonies of the bacterium *Corynebacterium diphtheriae* that resemble daisies, with dark centers and scalloped edges, when grown on blood-tellurite medium.
daisy wheel *Computer Technology.* a disk that contains a character set on its circumference and is rotated by the printer to position the characters for printing. Also, PRINT WHEEL.
daisy-wheel printer *Computer Technology.* an impact printer whose print elements are placed at the ends of the spokes of a rotating wheel; when the proper character is in position, a hammer causes it to strike the print ribbon. Also, WHEEL PRINTER.
Dakin's solution or **Dakin solution** *Mycology.* a chemical solution containing sodium hypochlorite and sodium bicarbonate that is used to cleanse wounds. Also, ALKALINE SODIUM HYPOCHLORITE SOLUTION.
Dakota *Aviation.* a popular name for the C-47 troop transport, and by extension for its civilian counterpart, the DC-3 passenger airplane.
Dakotan *Geology.* a geologic stage of the Lower Upper Cretaceous period.
DAL *Aviation.* the airport code for Love Field, Dallas, Texas.
dal or **dal.** dekaliter; dekaliters.
dalapon *Organic Chemistry.* CH_3CCl_2COOH, a liquid that is soluble in water and alcohol; boils at 185–190°C; used as an herbicide; a major constituent of Agent Orange, a defoliant used in the Vietnam War.
Dalatiidae *Vertebrate Zoology.* the spineless dogfishes, a shark family sometimes placed in the order Squaliformes.
Daldinia *Mycology.* a genus of fungi belonging to the order Xlariales; *D. concentrica* occurs on such deadwood branches as ash.
Dale, Sir Henry Hallet 1875–1968, British physiologist; shared Nobel Prize for research on chemical transmission of nerve impulses.
d'Alembert, Jean le Rond [dä läm bâr´] 1717–1783, French mathematician; formulated d'Alembert's principle and first general theory of winds; pioneer in partial differential equations; science editor of the first modern encyclopedia.

d'Alembert

d'Alembertian [də läm bâr´ tē ən] *Mathematics.* a second-order linear partial differential operator in four-dimensional space-time (i.e., Minkowski space); denoted $\cdot = \partial^2/\partial x^2 + \partial^2/\partial y^2 + \partial^2/\partial z^2 - 1/c^2\, \partial^2/\partial t^2$, where c is the speed of light. It is the canonical operator on Minkowski space analogous to the Laplacian on Euclidean space.

d'Alembert's paradox *Fluid Mechanics.* the paradox that in inviscid idealized fluid dynamics, there is no force, either drag or lift, on an arbitrary body moving steadily without circulation; this is resolved by including viscosity.

d'Alembert's principle *Mechanics.* the principle that a system under acceleration can be treated as a static system with the addition of all the related inertial forces, and that in this frame the net force on it is zero.

d'Alembert's test *Mathematics.* let $X = (x_m)$ be a sequence of nonzero elements of n-dimensional Euclidean space (with norm $\| \ \|$); the series $\sum_{m=1}^{\infty} x_m$ converges absolutely if $\lim_{m\to\infty} \|x_{m+1}\|/\|x_m\| < 1$. The series diverges if $\lim_{m\to\infty} \|x_{m+1}\|/\|x_m\| > 1$. Known as the (Cauchy) ratio test when $n = 1$. Also, GENERALIZED RATIO TEST.

Dalén, Nils Gustaf [dä lān´] 1869–1937, Swedish physicist; Nobel Prize for the invention of the automatic lighting regulator.

Dalitz pair *Particle Physics.* the combination of electron and positron particles resulting from the decay of a pion into this pair and a photon. (From Richard Henry *Dalitz,* born 1925, British physicist.)

Dalitz plot *Particle Physics.* a drawing showing the distribution of certain three-particle configurations that result when an elementary particle decays or experiences any high-energy nuclear reaction.

Dallinger, W. H. 1842–1909, British microbiologist; with Drysdale, carried out marathon observation of single microorganism.

Dallis grass *Botany.* any of several tall, perennial forage and pasture grasses of the genus *Paspalum* in the order Cyperales, introduced from the tropics and common in the southern United States.

Dall sheep *Vertebrate Zoology.* a wild white sheep, *Ovis dalli,* of the family Bovidae; found in northwestern North America and characterized by large, ribbed, spiral horns and sharp, elastic hooves for gripping rocky mountainsides.

Dall sheep

Dall tube *Mechanical Engineering.* a device that is used to measure the flow rate of fluid through a section of pipe; similar to a venturi tube.

Dalmanites *Paleontology.* a widespread family of middle Paleozoic trilobites in the order Phacopida and superfamily Dalmanitidae, distinguished by a long tail spine and raised compound eyes with large lenses.

Dalmau plate technique *Mycology.* a technique used to study the way yeast fungi form the mass of filaments or hyphae termed the pseudomycelium or mycelium.

Dalton, John 1766–1844, English chemist and physicist; formulated the law of partial pressures of gases; proposed the atomic theory of matter; studied color-blindness (Daltonism).

dalton *Biochemistry.* a unit of mass that equals the atomic weight of a hydrogen atom or 1.657×10^{-24} g. (Named for John *Dalton.)*

Daltoniaceae *Botany.* a mostly tropical family of usually glossy and complanate mosses of the order Hookeriales that grow on humus, tree trunks, and leaves; characterized by simple to dendroid or pinnately frondose stems, spreading to erect leaves, and lateral sporophytes.

Dalton's atomic theory *Chemistry.* the atomic theory formulated by John Dalton, specifically the assumptions that: elements are made up of atoms that are identical for a given element in terms of mass, size, and other properties; atoms of a given element have properties differing from those of other elements; atoms can combine in constant relative numbers to form new compounds.

Dalton's law *Physics.* a law stating that the total pressure of a mixture of nonreacting gases in a closed container is equal to the sum of the individual pressures of the gases in the same container.

Dalton's temperature scale *Thermodynamics.* a temperature scale in which the Dalton temperature t is related to the absolute temperature T (Celsius) by the equation $T = 273.15(373.15/273.15)^{t/100}$.

dam *Civil Engineering.* any barrier designed to obstruct the flow of water. *Agriculture.* a horse that is the mother of a foal.

dam

Dam, Henrik 1895–1976, Danish biochemist; shared Nobel Prize with Doisy for discovery and analysis of vitamin K.

dam or **dam.** dekameter; dekameters.

damage *Military Science.* **1.** injury to persons, equipment, or installations that does not cause complete destruction; measured in progressive levels or categories; the definition of the level or category varies with the target and source of damage. **2.** to inflict such injury.

damage area *Military Science.* the area around a mine-sweeper within which a mine explosion is likely to interrupt operations.

damage assessment *Military Science.* **1.** a determination of the effect of an attack upon a target or targets. **2.** a determination of the effect of a compromise of classified information upon national security.

damage control *Military Science.* aboard a ship or aircraft, the action necessary to preserve and reestablish structural integrity, stability, maneuverability, and weapon power; to limit the spread of fire or contamination by toxic agents; and to care for wounded personnel.

damage potential *Military Science.* the expected damage to a specific target by a projectile or explosive.

damage radius *Military Science.* in naval mine warfare, the average distance from a ship within which a specific type of mine must detonate in order to cause a specific amount of damage. Also, RADIUS OF DAMAGE.

damaging stress *Mechanics.* the force that causes damage to a structure which will impair its function or shorten its lifespan, expressed in terms of stress per unit area.

Damascus steel *Materials.* a hard, flexible steel having a decorative pattern of wavy lines, originally used especially for sword blades. Also, **damask steel.**

damask *Textiles.* a fabric with a reversible, closely woven satin-pattern against a contrasting sateen background, made on a jacquard loom. (Named after the ancient city of *Damascus,* where such patterns were first woven in silk.)

dam gene *Genetics.* a gene that encodes a DNA adenine methylase.

damkjernite *Petrology.* a melanocratic dike rock composed of biotite and pyroxene phenocrysts in a fine groundmass of biotite and titanaugite, with trace amounts of muscovite, chlorite, epidote, orthoclase, and late magmatic calcite.

Damköhler number I [däm´kur lər] *Physics.* the ratio of the time required for a fluid to flow a specified distance to the time required for a specified process or chemical reaction to take place. Also, **Damköhler's ratio.**

Damköhler number II *Physics.* the ratio of the rate of a chemical reaction to the molecular diffusion rate.

Damköhler number III *Physics.* the ratio of the heat liberated by a chemical reaction to the bulk transport rate of heat in a fluid.

Damköhler number IV *Physics.* the ratio of the heat liberated in a chemical reaction to the conductive heat transfer rate.

dammar *Materials.* a copal-like resin derived largely from dipterocarpaceous trees of southern Asia, used for making varnish.

damp *Engineering.* the reduction of fire in a furnace when damp coals or ashes are placed on the fire bed. *Mining Engineering.* a toxic gas present in a mine. *Physics.* to gradually diminish the amplitude of a wave or oscillation.

damp course *Building Engineering.* a vertical or sloping waterproof membrane or skin composed of an impervious material such as asphalt, copper sheet, polythene film, blue brick, or slate, placed upon a wall to keep out water.

damp down *Metallurgy.* to interrupt the flow of air in a blast furnace.

damped collision see DEEP INELASTIC COLLISION.

damped harmonic oscillation *Physics.* the vibration of an oscillator which oscillates under the influence of a restoring force that is directly proportional to the displacement from the equilibrium, and a force that is proportional to the instantaneous velocity of the oscillator, usually viscosity or friction.

damped oscillation *Physics.* a free vibration whose amplitude dies away asymptotically from an initial maximum to zero amplitude, usually with an exponential envelope; for example, the note from a struck tuning fork. Also, **damped vibration.**

damped wave *Physics.* **1.** a wave emanating from a source that is subject to damped oscillation occurring at the source. **2.** a wave whose amplitude is diminished with increasing distance from the source due to losses in the medium through which it propagates.

dampener *Engineering.* a device used to lessen pulsations on reciprocating machinery. *Graphic Arts.* in offset lithography, any of a set of cloth-covered rollers designed to spread dampening solution over the printing plate.

dampening solution or **dampening etch** *Graphic Arts.* a solution of water, gum arabic, and any of various acids, designed to desensitize non-image areas of a lithographic plate. Also, FOUNTAIN SOLUTION, WETTING AGENT.

damper or **damper tube** *Electronics.* any circuit or device intended to limit unwanted oscillations, such as the damper circuit in a television receiver that ensures a stable picture by suppressing unwanted oscillations in the horizontal deflection drive circuit. Also, DAMPING DIODE.

damper loss *Engineering.* any reduction in the rate of flow or in the pressure of a gas across the damper of a stove or furnace.

damper winding *Electricity.* a special short-circuited motor winding that opposes pulsation or rotation of the magnetic field, and consists of several conducting bars on the field poles of a synchronous machine. Also, AMORTISSEUR.

damping *Engineering.* **1.** the process of quieting a vibrating motion. **2.** the reducing of reverberation by covering walls with sound-absorbing materials or using decoupling techniques.

damping capacity *Mechanics.* the ability of a material or system to dissipate the energy imparted to it by an impulse force or during impulsive forcing.

damping diode see DAMPER.

damping factor or **damping coefficient** *Physics.* the ratio of the amplitude of a vibration to that of its succeeding vibration in an underdamped vibrational system. Also, DECAY FACTOR.

damping magnet *Electromagnetism.* a permanent magnet whose field generates eddy currents in a moving conductor which subsequently creates a field that opposes the motion of the conductor and provides damping of the motion.

damping-off *Plant Pathology.* a condition of plant seeds or seedlings in which a parasitic fungus invades the plant near the ground level and causes the decay of the seeds or the wilting of the seedlings.

damping ratio *Physics.* the ratio of the amount of resistive mechanism in an underdamped vibrating system to that of the system if it were critically damped.

damping resistor *Electricity.* **1.** a shunt across a coil that prevents ringing. **2.** a resistor that provides critical damping of a galvanometer.

dampproof *Materials.* describing a material that resists the effects of dampness; for example, bituminous materials or silicon applied to a wall or other surface by brushing, spraying, or troweling. *Building Engineering.* to subject a structure or a structural element to dampproofing.

dampproofing *Building Engineering.* the process of constructing or treating a structure so that it will resist the harmful effects of dampness, as by the drainage of surface water by sloping the ground surface, the application of a dampproof coating to the exterior surface below ground level, and the use of portland cement with a water-repellent base.

damselfly or **damsel fly** *Invertebrate Zoology.* a name for the smaller and more delicate members of the dragonfly family, Zygoptera.

Danaidae *Invertebrate Zoology.* a family of large, brightly colored tropical butterflies.

danalite *Mineralogy.* $Fe_4^{+2}Be_3(SiO_4)_3S$, a red to gray cubic mineral occurring as octahedral or dodecahedral crystals, and in massive form, having a specific gravity of 3.31 to 3.46 and a hardness of 5.5 to 6 on the Mohs scale; found in granite pegmatites, gneiss, and hydrothermal veins.

dan buoy *Navigation.* a buoy consisting of a ballasted float with a staff supporting a flag or light.

danburite *Mineralogy.* $CaB_2(SiO_4)_2$, a transparent to translucent, colorless to yellowish brown orthorhombic mineral with a vitreous to greasy luster and a conchoidal fracture, having a specific gravity of 2.97 to 3.02 and a hardness of 7 to 7.5 on the Mohs scale; found with feldspar in dolomite.

dance of bees *Entomology.* the elaborate movement patterns used by worker bees to communicate to others in the hive the distance to and direction of food sources.

dancing dervish or **dancing devil** see DUST WHIRL.

Danckwerts model [dangk´vurts] *Chemical Engineering.* a theory for the calculation of mass transfer between liquid and gas in a packed absorption tower; it is assumed that the gas in the column contacts new liquid that is brought to surface by the liquid eddies from the interior of the liquid body.

D and C or **D&C** see DILATION AND CURETTAGE.

dandelion *Botany.* a weed of the genus *Taraxacum* having bright yellow flowers and edible notched leaves, especially *T. officinale.*

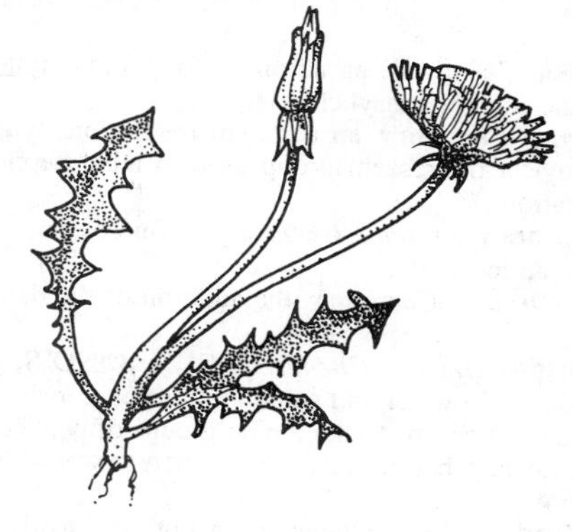

dandelion

dandruff *Medicine.* thin, dry, whitish scales cast off from the scalp.

dandy roll *Mechanical Engineering.* a roller used with a papermaking machine, consisting of a hollow cylinder covered with wire cloth; used to compress and strengthen the web of the paper.

danger angle *Navigation.* the minimum or maximum angle (either vertical or horizontal) between two points as observed from a vessel that must be maintained to clear some danger.

danger bearing *Navigation.* the maximum or minimum bearing of a point for clearing an offshore danger.

danger buoy *Navigation.* a floating aid to navigation used to mark an isolated hazard.

danger coefficient *Nucleonics.* in experimental reactors, a measure of the poisoning effect, or change in reactivity, of inserting a unit mass of neutron absorber of unknown neutron absorption cross section into a reactor core; by comparing the resultant reactivity change with that of a calibrated control rod, the numerical value of the danger coefficient can be determined as proportional to the square of the neutron flux at the location of the absorber.

danger line *Navigation.* a line drawn on a chart to indicate the limit for safe navigation of a specific area.

dangerous semicircle *Meteorology.* the half of a tropical cyclone to the right of the direction of movement of the storm in the Northern Hemisphere and to the left in the Southern Hemisphere; this half has the strongest winds and heaviest seas, so that a sailing ship tends to be carried into the path of the storm.

danger space *Military Science.* **1.** the zone between a weapon and its target where the trajectory is at or below 1.8 meters, the average height of a man. **2.** the volume around the bursting point of an antiaircraft projectile within which an aircraft will suffer damage.

dangling bond *Solid-State Physics.* an unsaturated bond found at the surface of a solid; more common in nonmetals than in metals.

dangling reference *Computer Science.* in the execution of a program in a recursive language, a situation in which a pointer to storage that is allocated on the execution stack is retained after the routine associated with that stack frame has exited, possibly resulting in errors.

Danian *Geology.* a European stage of the lowermost Paleocene epoch or of the uppermost Cretaceous period, occurring after the Maestrichtian and before the Montian.

Daniell, John Frederic 1790–1845, English chemist and physicist; invented the Daniell cell.

Daniell cell *Physical Chemistry.* a device that produces an electric current from the interactions of two electrodes, having a copper anode immersed in a copper sulfate solution and a zinc-mercury cathode in a sulfuric acid or zinc sulfate solution, separated by a porous partition.

Daniell hygrometer *Engineering.* a device used to measure the dewpoint.

Daniell integral *Mathematics.* a concept of integration that does not depend on the concept of measure. In particular, let L be a vector space of real-valued functions on some set X with the property that $|f|$ is in L for each f in L. A linear functional I on L is then called a Daniell integral if the following conditions are satisfied: (a) $I(f) \geq 0$ for each nonnegative function f in L; and (b) if ϕ_n is a sequence of functions in L such that $\lim_{n\to\infty} \phi_n(x) = 0$ for each point x in X, then $\lim_{n\to\infty} I(\phi_n) = 0$ for each point x in X. I is sometimes called a **Daniell functional.**

dannemorite *Mineralogy.* $Mn_2(Fe^{+2},Mg)_5Si_8O_{22}(OH)_2$, a yellowish-brown to greenish-gray monoclinic amphibole, having a specific gravity of 3.34 and a hardness of 5 to 6 on the Mohs scale; found in metamorphic rocks.

DANS *Organic Chemistry.* an acronym for 5-dimethylamino-1-naphthalenesulfonic acid, or dansyl chloride.

dansyl *Organic Chemistry.* an acronym for 5-dimethylamino-1-naphthalenesulfonyl, a fluorescent group used to label peptides for amino acid identification.

dansylamino acid *Organic Chemistry.* a fluorescent combination of dansyl and an amino acid.

dansylation *Organic Chemistry.* the addition of the dansyl group to a peptide.

dansyl chloride *Organic Chemistry.* $C_{12}H_{12}ClNO_2S$, yellow-orange crystals; insoluble in water and soluble in acetone; melts at 66.5–68°C; used to add dansyl groups to the amino groups of peptides and proteins and in fluorescent labeling. Also, 5-DIMETHYLAMINO-1-NAPHTHALENESULFONIC ACID.

dansyl method *Organic Chemistry.* a highly sensitive fluorescence procedure to analyze N-terminal residues at the end of a peptide chain.

danta *Materials.* the red, heavy, elastic wood of an African timber tree, *Nesogordonia papaverifera,* which is used in the construction of furniture and ballroom floors.

danthron see 1,8-DIHYDROXYANTHRAQUINONE.

Danube *Geography.* a long (1770 miles) river flowing from southwestern Germany across central Europe into the Black Sea.

Danysz phenomenon *Immunology.* the demonstration of the variations of toxicity that occur when equivalent amounts of toxin and antitoxin are mixed; the toxicity varies according to whether the toxin is added in one stage (resulting in a nontoxic mixture) or in increments (resulting in a toxic mixture).

dao see SENGKUANG.

dap *Building Engineering.* **1.** a notch that is cut in a timber in order to receive part of another timber. **2.** to cut such a notch, or to fit together with such a notch.

Daphniphyllaceae *Botany.* the monogeneric family of dioecious trees and shrubs comprising the order Oaphniphyllales; members contain a unique type of alkaloid, are tanniferous but without ellagic acid, and are native mostly to eastern Asia and Malaysia.

Daphniphyllales *Botany.* an order of dicotyledonous plants of the subclass Hamamelidae, consisting of a single family with one genus, *Daphniphyllum,* characterized by production of a unique type of alkaloid and sometimes iridoid compounds; native to eastern Asia and Malaysia.

daphnism *Toxicology.* poisoning due to plants of the genus *Daphne,* especially from contact with the berries of *Daphne mezereum;* symptoms may include blistering, internal bleeding, and kidney damage.

Daphoenidae *Paleontology.* a family of canid carnivores of the Oligocene, belonging to the superfamily Miacoidea.

dapsone *Pharmacology.* $C_{12}H_{12}N_2O_2S$, an antibacterial occurring as a white or creamy-white powder, used as a bacteriostatic for a broad range of organisms. Also, DIAMINODIPHENYLSULFONE, 4,4'-SULFONYLBISBENZENAMINE.

daraf *Electricity.* the unit of elastance, which equals the reciprocal of capacitance. (A term coined by reversing the spelling of *farad.)*

darapskite *Mineralogy.* $Na_3(SO_4)(NO_3)\cdot H_2O$, a colorless monoclinic mineral; platy to granular in habit or as long prismatic to thin tabular crystals; having a specific gravity of 2.2 and a hardness of about 2.5 on the Mohs scale; found in nitrate deposits.

Darboux's formula [där booz′] *Mathematics.* a general formula expressing a function $f(z)$, analytic in a neighborhood of a, as a sum of terms involving derivatives of f evaluated at a and z and an auxiliary polynomial ϕ of degree n evaluated at 0 and 1. In particular,

$$\phi^{(n)}(0)[f(z) - f(a)]$$

$$= \sum_{m=1}^{\infty} (-1)^{m-1}(z-a)^m[\phi^{(n-m)}(1)f^{(m)}(z) - \phi^{(n-m)}(1)f^{(m)}(a)]$$

$$+ (-1)^n(z-a)^{n+1}\int_0^1 \phi(t)f^{(n+1)}(a + t(z-a))\,dt.$$

For $\phi(t) = (t-1)^n$, Darboux's formula yields the Taylor series for f.

Darboux's theorem *Mathematics.* **1.** suppose f is a real differentiable function on $[a,b]$ with $f'(a) = A$ and $f'(b) = B$. If C is any real number between A and B, then there is a point c with $a < c < b$ such that $f'(c) = C$. **2.** let ω be a symplectic 2-form on a smooth manifold X. Then for each $x \in X$ there is a local coordinate system about x in which ω is constant; i.e., there exist coordinates such that

$$\omega = dx^1 \wedge dx^2 + dx^3 \wedge dx^4 + \cdots + dx^{n-1} \wedge dx^n.$$

darcy or **darcie** *Fluid Mechanics.* a unit of permeability defined as the passage of 1 cc of fluid, having 1 centipoise viscosity, flowing in 1 second under a pressure of 1 atmosphere, through a porous medium having a length of 1 centimeter and a cross-sectional area of 1 square centimeter. *Petroleum Engineering.* this unit used as a measure of rock permeability; because a darcy is too large to characterize many oil-producing rocks, the permeabilities used in the oil industry are expressed in units that are one-thousandth as large, such as **millidarcies;** commercial gas and oil sands show permeabilities from a few millidarcies to several thousand. (From Henri *Darcy,* 1803–1858, French engineer.)

Darcy's law *Fluid Mechanics.* a fundamental law of porous media, discovered by Henri Darcy in 1856, stating that the flow rate Q is proportional to the cross-sectional area A, inversely proportional to the length L of the sand-filter flow path, proportional to the head drop δH, and proportional to the hydraulic conductivity K that represents the constant of proportionality.

Darcy-Weisbach equation *Fluid Mechanics.* the general equation for steady, incompressible flow in conduits, derived from the mechanical energy balance.

Dardanelles [där′də nelz′] *Geography.* a narrow strait connecting the Sea of Marmara with the Aegean arm of the Mediterranean.

dark adaptation see SCOTOPIC VISION.

dark cloud see DARK NEBULA.

dark conduction *Electronics.* the conduction of current by a photosensitive material, even though the source of light has been removed.

dark current see ELECTRODE DARK CURRENT.

dark-current pulse *Electronics.* a pulse generated by a phototube in the absence of incident radiation on the phototube.

dark discharge *Electronics.* an electric discharge that does not emit light.

dark-eclipsing star *Astronomy.* the darker star of two in an eclipsing binary system.

dark-field illumination *Optics.* the technique of lighting a microscope with diffused light by passing the beam through a condenser before it reaches the specimen; used to study tiny particles or lines. The particles appear bright against a dark background. Also, **dark-ground illumination.**

dark-field imaging *Materials Science.* an imaging in which a disk or aperature allows only light or electrons diffracted, refracted, or diffused from the specimen to reach the lens, thus causing only certain details in the specimen to appear bright on a dark field.

dark halo crater *Astronomy.* a crater, possibly volcanic in origin, that is surrounded by an ejecta blanket darker than the adjacent landscape.

dark-line spectrum *Spectroscopy.* an absorption spectrum consisting of dark lines or bands superimposed on a continuous spectrum; occurs when white light passes through a sample.

darkling beetle *Invertebrate Zoology.* any of various dark, slow-moving nocturnal and generally vegetarian beetles of the family Tenebrionidae, whose mealworm larvae are bred as food for captive birds and that also are pests of various flour-based and grain-based products.

dark matter *Astronomy.* nonluminous matter in a galaxy whose exact nature is unknown, but whose presence is inferred from the observed motions of the galaxy's stars and gas clouds.

dark nebula *Astronomy.* an interstellar cloud of dust lying in the plane of the Milky Way that is visible because it lies in front of a brighter background, either a glowing gas cloud or a field of stars.

dark of the moon *Astronomy.* a general term for the lunar phases between the last quarter and first quarter, when the moon is out of the sky for most of the nighttime hours.

dark reaction *Graphic Arts.* a hardening of the photosensitive coating of a printing plate that occurs without exposure to light, usually due to excessive heat or humidity.

dark repair *Molecular Biology.* a repair of damaged DNA by special enzyme systems that act optimally in the absence of visible light.

dark resistance *Electricity.* the resistance of a selenium cell or other photoelectric device in total darkness.

darkroom *Graphic Arts.* a room that is specially designed to exclude forms of light that affect photosensitive materials; used for developing photographic negatives and positives.

darkroom filter *Optics.* a filter made of glass, gelatin, or some other material that permits the transmission of specific wavelengths of a light source so that it does not expose photographic film or paper.

dark-ruby silver or **dark-red silver ore** see PYRARGYRITE.

dark segment *Meteorology.* a bluish-gray band appearing along the horizon opposite the rising or setting sun, below the anti-twilight arch.

dark slide *Photogrammetry.* a thin metal or fiber plate that, after being inserted in a camera magazine, renders it lightproof.

dark spot *Electronics.* a characteristic of television camera tubes in which electron clouds, formed within the tube, cause a dark area to appear on the received television picture.

dark star see DARK-ECLIPSING STAR.

dark-trace tube *Electronics.* a cathode ray display tube in which the information is presented as dark traces on a very bright background and, when properly illuminated, the display can be projected onto a large screen. Also, SKIATRON.

Darlington amplifier *Electronics.* a compound emitter-follower amplifier consisting of two transistors with both collectors connected; the emitter of the input transistor is connected to the base of the output transistor, and the output signal is taken from the emitter of the second transistor; the amplifier is characterized by a high input impedance and a large current gain. Also, **Darlington pair.**

d'Arsonval galvanometer [där´sōn val´] *Engineering.* a device used to measure direct electric current, by measuring the movement of a wire coil suspended by a thin metallic ribbon in relation to a fixed magnet when current is applied to the coil. (From the French physicist Jacques-Arsène d'Arsonval, 1851–1940.) Also, LIGHT-BEAM GALVANOMETER.

dart *Invertebrate Zoology.* a projectile ejected from a dart sac in certain snails that penetrates another snail and stimulates copulation.

dart configuration *Aviation.* a configuration in which the control surfaces of a flight vehicle are located at the tail.

darter *Vertebrate Zoology.* a small, slender-bodied, brightly colored fish of the family Percidae, native to the Northern Hemisphere; named for the darting movements they make in catching prey along the bottom.

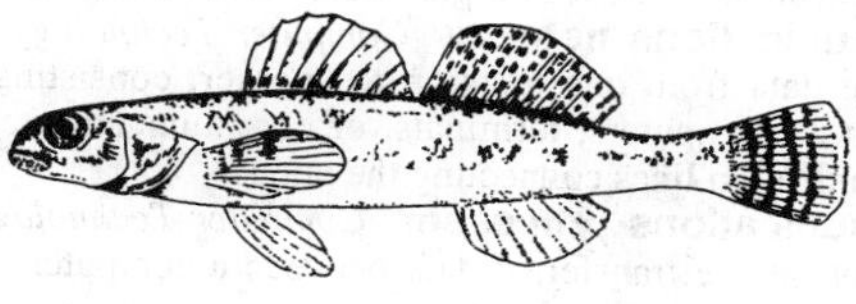

darter

dart leader *Geophysics.* the leader in chain lightning that, after the first lightning stroke, initiates the subsequent strokes.

dart sac *Invertebrate Zoology.* a pouch in certain snails that contains a projectile associated with reproduction.

Darwin, Charles 1809–1882, English naturalist; famous for highly influential theories of evolution and natural selection.

Darwin, Erasmus 1731–1802, English physician and naturalist; grandfather of Charles Darwin; proposed an earlier theory of evolution.

Darwin, George 1845–1912, English mathematician and astronmer; son of Charles Darwin; leading theorist on the moon and tides.

darwin *Evolution.* a unit used to measure the rate of evolutionary change, given as the increase or decrease in any hereditary character multiplied by a factor of 2.7 per million years.

Darwin-Doodson system *Geophysics.* a technique for the prediction of tides by depicting them as sums of the harmonic functions of time.

Darwin ellipsoids *Astronomy.* the ellipsoidal shapes that tidal effects create in homogeneous objects moving in circular orbits.

Darwin glass *Geology.* blobs, drops, or twisted shreds of silicate-rich vesicular glass found in the Mount Darwin range in western Tasmania. Also, QUEENSTOWNITE.

Darwinian [där win´ē ən] *Science.* **1.** relating to Charles Darwin or his evolutionary theories. **2.** someone who accepts or advocates Darwin's theories; specifically, a person who supports the idea that natural selection is the valid explanation for the presence of diverse species on earth.

Darwinian anthropology *Anthropology.* the study of anthropological data that analyzes culture from the perspective of Darwin's evolutionary tenets of natural selection and the survival of the fittest.

Darwinism *Evolution.* a theory of the mechanism of evolution first developed by Charles Darwin and Alfred Wallace, proposing that among the individual organisms of a given species, those with traits well-adapted to suvival are more likely to reproduce and pass on these traits to their offspring, while those with less-favorable traits are likely to die out, in a process known as *natural selection.* Ultimately, this reinforcement of favorable traits leads to the origin of a new species with these traits. Also, **Darwin's theory, Darwinian evolution.**

Darwinist see DARWINIAN, def. 2.

Darwin's finch *Vertebrate Zoology.* any of fourteen finch species of the subfamily Geospizinae that are endemic to the Galapagos Islands, but related to finches on the South American continent; distinguished by great variation in bill shape. Also, GALAPAGOS FINCH.

Darwinulacea *Invertebrate Zoology.* a superfamily of small, mostly freshwater, ostracod crustaceans. Also, **Darwinuloidea.**

Darzens' procedure *Organic Chemistry.* a reaction involving condensation of aldehydes and ketones with α-halo esters in the presence of bases to form glycidic esters. Also, **Darzens' reaction.**

DAS or **das** direct-access storage.

Dasayatidae *Vertebrate Zoology.* a family of mostly saltwater rays of the order Myliobatiformes, mostly found in temperate to warm waters, although some species are found in freshwater; characterized by one or two stout, serrated spines down the back of the whiplike tail, used to inflict injury.

Dascillidae *Invertebrate Zoology.* a family of soft-bodied plant beetles.

Dascilloidea *Invertebrate Zoology.* a superfamily of beetles.

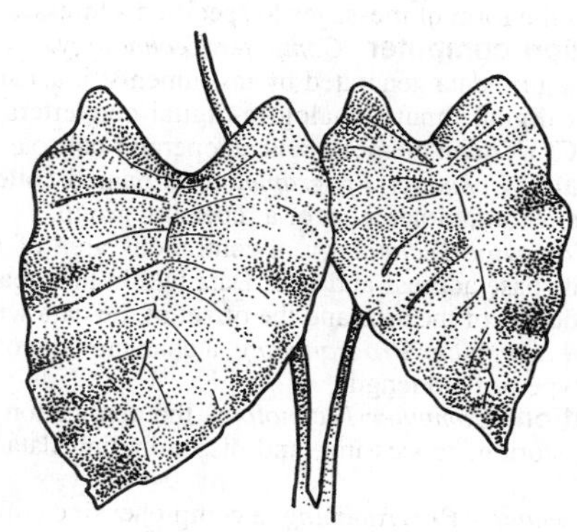

dasheen

dasheen *Botany.* the taro plant, *Colocasia esculenta,* of the order Arales; grown in the tropics for its edible starchy rootstock and used in temperate regions as an ornamental.

dashpot *Mechanical Engineering.* a device that is used to damp the vibration or control the motion of a mechanism; it consists of a piston that is connected to the component to be controlled and slides within a liquid-filled cylinder.

Dasyaceae *Botany.* a small family of marine red algae of the order Ceramiales, characterized by terete or flattened, branched uniaxial thalli and spermatangia that occur in great quantities on special axes.

Dasycladaceae *Botany.* a family of coenocytic green algae in the order Dasycladales, characterized by a central stem from which whorls of filaments develop.

Dasycladales *Botany.* an order of marine algae in the division Chlorophyta, characterized by a lime-encrusted thallus composed of nonseptate, extensively branched tubes.

dasymeter *Physics.* an instrument commonly used to determine the density of a gas.

Dasyonygidae *Invertebrate Zoology.* a family of biting lice, found only on hyraxes and similar rodents.

Dasypodidae *Vertebrate Zoology.* the armadillos, a family of mammals of the order Edentata.

Dasytidae *Invertebrate Zoology.* a large family of predatory and pollen-eating beetles. Also, MELYRIDAE.

Dasyuridae *Vertebrate Zoology.* a family of small, mainly nocturnal carnivorous marsupials of the order Marsupicarnivora, native to Australia, New Guinea, and Tasmania and including the native cats, pouched mice, banded anteater, Tasmanian devil, and related forms.

Dasyuroidea *Vertebrate Zoology.* a superfamily that is made up of the families Dasyuridae, Notoryctidae, Thylacinidae, and Myrmecobiidae; marsupials in which the marsupium, if present, opens posteriorly, the second and third digits of the hind feet are didactylous, and the tails are nonprehensile.

DAT digital audiotape.

data [da´tə; dā´tə] the plural of **datum.** *Science.* **1.** two or more individual facts or pieces of information,. **2.** a body of facts, information, or knowledge, particularly when derived from scientific observation or experimentation. *Computer Programming.* **1.** a group of one or more characters (alphanumeric, binary, or other), representing basic elements of information that can be processed or produced by a computer. **2.** the representation of facts, numbers, or concepts that can be communicated, stored, and processed to form information. Used to form many compound terms, such as **data analysis, data collection, data distribution, data preparation, data sharing, data transmission,** and so on.
Data, although originally a plural form, is now often regarded as a collective term and used with a singular verb, especially in computer fields (e.g., "Data *is* often stored on backup disks."). In certain academic contexts, however, a plural verb is still preferred (e.g., "The experimental data *are* not conclusive.").

data abstraction *Computer Programming.* **1.** a method of simplifying programming by restricting the knowledge of the data structure to the program module that creates the data type. **2.** a description of the properties of a class of related data types.

data acquisition *Computer Programming.* any accumulation of data for some later use, such as processing or analysis. *Telecommunications.* specifically, a facility or program used to gather data from a group of addressees, to assemble data within a communications system, and to deliver data in the form of messages to specified addresses.

data-acquisition computer *Computer Technology.* a computer system used to capture data generated by instruments; it usually consists of analog and/or digital inputs, analog-to-digital converters, disk or tape data storage, CPU, main memory, and an operator console.

data aggregate *Computer Programming.* a named collection of data items within a record; either a list or a repeating group.

data area *Computer Technology.* a contiguous area of main storage, specified by its base address and size. Data within the area is referenced by the base address of the area and the offset of the data within the area.

data attribute *Computer Programming.* a characteristic of a set of data, such as data type or field length.

data automation *Computer Technology.* the application of computers to gathering, storing, processing, and disseminating data and information.

data bank *Computer Programming.* a comprehensive collection of data derived from a variety of sources and stored so that it is available to a number of users.

database *Computer Programming.* an integrated collection of data that supplies information in a variety of forms or for a variety of applications. Also, **data base, data-base.**

database administrator *Computer Technology.* the person or group of people responsible for the definition, control, and use of an organization's database.

database machine *Computer Technology.* a special-purpose add-on computer system dedicated to performing database functions.

database management system *Computer Programming.* a collection of special-purpose programs that support and maintain the semipermanent storage of user-owned data; includes the software necessary for storage, retrieval, inquiry, and reporting.

database publishing *Computer Science.* the publication of information from a computer database, in the form of a directory, reference book, or the like.

database server *Computer Programming.* the software that provides access to a database for entering, updating, or deleting data.

data break *Computer Technology.* a facility that permits data input or output without interrupting program operation.

data buffering *Computer Technology.* **1.** the temporary gathering and storage of data in high-speed memory so that it will be available when needed for processing. **2.** the temporary storage of input or output data to allow transfer of data between two devices that operate at different rates.

data bus *Computer Technology.* a link that allows data to be exchanged between the CPU, main memory, and an input/output device.

data capture *Computer Programming.* the gathering or collecting of input data for processing, especially later processing for purposes other than that for which it was originally entered.

data carrier *Computer Technology.* any medium used to contain and physically transport data, such as magnetic tape reels, magnetic disks, or punched cards. Thus, **data-carrier storage.**

data cartridge *Computer Technology.* a removable cartridge that contains permanently recorded data.

data catalog *Computer Programming.* a complete list of the full names of all the data elements used by an organization.

data cell *Computer Technology.* **1.** the smallest representation of data, usually one bit. **2.** a high-capacity auxiliary storage medium in which data is recorded onto magnetic strips arranged in cells.

data cell drive *Computer Technology.* a soft-surface direct access storage device, now used only for very large on-line databases, that selects and extracts the desired strip from the magazine and wraps it around a read/write drum; after processing, the strip is automatically peeled off the drum and returned to the magazine.

data center *Computer Technology.* an area containing automatic data processing equipment and personnel who centralize and control data processing services for others in the organization; usually managed as a separate organizational entity. Also, DATA-PROCESSING CENTER.

data chain *Computer Programming.* two or more data elements, words, or codes that are linked in a way that provides meaningful information.

data chaining *Computer Programming.* the storage or retrieval of elements of a physical record in or from more that one memory location by means of individual I/O commands. One I/O buffer may be chained to the next so that when the first buffer is filled, the I/O channel will automatically switch to the next buffer without losing data.

data chamber *Photogrammetry.* the portion along the margins of an aerial photograph where ancillary data, such as the time, altitude, frame number, and other information required for identification and correlation, are recorded.

data channel *Computer Technology.* a bidirectional data path between a processor and input/output devices. Also, INPUT/OUTPUT CHANNEL, READ-WRITE CHANNEL.

data circuit *Electronics.* a radio or wire communication system specifically designed for transmission of digital data.

data code *Computer Programming.* any character or short group of characters used to represent data according to a set of rules; often used to reduce data storage requirements.

data communication control character see TRANSMISSION CONTROL CHARACTER.

data communications *Computer Technology.* the use of communication lines to handle the flow of digital data between computers.

data-communications network *Computer Technology.* a means of transmitting data from one location to another, consisting of a set of nodes, such as computers, terminals, or communication control units, and communication lines connecting the nodes.

data-communications processor *Computer Technology.* a processor that controls the transfer of data between a computer system and a network via communication lines.

data compaction *Computer Programming.* an irreversible form of data compression such as eliminating leading zeros or trailing blanks.

data compression *Computer Programming.* a technique used to reduce data storage or communication requirements by altering the form of data to reduce the number of bits required to represent a given set of information.

data concentrator *Computer Technology.* a small computer that is programmed to accept traffic from a number of lines and place it on a common line or to distribute traffic from the common line to remote terminals. Also, CONCENTRATOR.

data control *Computer Technology.* a service performed by personnel at a computing center in conjunction with quality control and computer resources scheduling.

data conversion *Computer Programming.* the process of transforming data files from one coding system to another or from one storage medium to another.

data declaration *Computer Programming.* a programming language statement that indicates the data type of a variable used in the program and allows proper interpretation of the variable when it is accessed.

data definition *Computer Programming.* a program statement that provides attributes of the data to be processed, including its size and type.

data dependence graph *Computer Programming.* a graphical representation of a program in a dataflow language, with each node representing a function and each arc carrying a value.

data descriptor *Computer Programming.* a pointer to a data segment in memory containing the address of the beginning of the array and the number of elements.

data-description language *Computer Programming.* a language associated with a database management system and a host high-level programming language such as FORTRAN, used to specify the way that data is to be stored and maintained in a database. Also, **data-definition language.**

data dictionary *Computer Programming.* a catalog that provides name, structure, and usage information for all data types stored in a database.

data directory *Computer Programming.* lists or tables that provide quick reference to relevant information about the data used by an information system, such as name and attributes; usually accompanies a data dictionary.

data display *Computer Technology.* the on-line visual representation of data, usually on a CRT screen or other video terminal under control of a computer.

data division *Computer Programming.* the section in a COBOL program that contains statements defining all the data to be processed by the program.

data-driven execution *Computer Programming.* the processing of a program in a dataflow system in which the arrival of data at an action operator triggers the execution of that action.

data-driven search *Artificial Intelligence.* a search in which the arrival of data causes conclusions to be drawn, which cause further conclusions, and so on; forward chaining.

data element *Computer Programming.* the smallest unit of named data within a record, file, or database; the smallest unit of data that has informational meaning.

data encryption *Computer Programming.* the coding of classified or sensitive data by a predetermined technique in such a way as to be uninterpretable without decryption information.

data entry *Computer Programming.* the process of converting data from human readable form, such as time cards or sales slips, into computer readable form.

data-entry terminal *Computer Technology.* a device with a keyboard and a data display used by keyboard operators to transcribe data from source documents directly into a computer, or onto a disk or tape.

data error *Computer Programming.* a common type of computer error, caused by erroneous data entry.

data-exchange system *Computer Technology.* a data communication system consisting of hardware devices and data-transmission software that accepts data from one or more sources, sorts and prioritizes the data, and transmits it to one or more destinations.

data expansion *Computer Programming.* the return of compressed data to its original length and form.

data export *Computer Programming.* the process of obtaining data from one program, such as a database, in a form that is acceptable to another program, such as a spread sheet or a word processor.

data field *Computer Programming.* a space allocated to a data element on a physical record.

dataflow or **data flow** *Computer Programming.* a method of describing a process in terms of actions that are controlled by the arrival of data at the action operator, as opposed to control flow models that are based on instruction pointers. Thus, **dataflow analysis.**

dataflow diagram *Computer Programming.* a modeling tool used to represent a system that is automated, manual, or a combination of both, in which the four components (dataflow, process, data store, and data originator/terminator) are graphically represented.

dataflow language *Computer Programming.* a modeling language used to describe the information paths in programs in which parallel or asynchronous processes occur.

dataflow system *Computer Technology.* a computer system in which basic operations are initiated by the availability of the operands instead of by the sequential flow of control. Also, NON-VON NEUMANN ARCHITECTURE.

dataflow technique *Computer Technology.* an approach to computer systems organization and design that specifies the movement of data through the system and the transformations that occur, as opposed to a traditional system design using process or control flow.

data formatting *Computer Programming.* the use of programming language statements to specify the way data to be used in the program is held in the file, including the data type and field length.

data frame *Computer Technology.* see FRAME.

data-handling system *Computer Technology.* a group of automatic or semiautomatic devices that collect, transmit, receive, and store data in digital form; often used with instrumented data collection devices.

data hierarchy *Computer Programming.* a data structure that defines succeedingly more detailed or lower-ranking subsets.

data import *Computer Programming.* the ability to read and use information produced by another program or database management system.

data independence *Computer Programming.* the abstraction of data from the processing so that a change in the logical or physical structure of the data has no effect on the application program's view of the data.

data integrity *Computer Programming.* the quality of data that is complete, accurate, and consistent.

data-intense application *Computer Programming.* the use of a computer system involving large volumes of data and repetitive transaction processing.

data interchange *Computer Programming.* the accessibility of data from one program to a separate program, using software standards for data representation.

data-interchange format *Computer Technology.* a standard data representation that permits data from one program to be accessible to another.

data item *Computer Programming.* a single unit of data; the smallest unit of named data. Also, DATUM.

data level *Computer Programming.* a reference number used to indicate to a compiler the position or rank of a data element in a hierarchical data structure.

data library *Computer Programming.* a catalog of all the disks, tapes, documentation, user's manuals, and procedures that are relevant to using data resources.

data link *Computer Technology.* the terminal installation and interconnecting circuits that permit the exchange of information directly between two different stations. *Telecommunications.* a pair of data stations and their connecting network, operating in such a manner that information can be exchanged between them.

data link layer *Computer Technology.* layer 2 of the seven-layer Open System Interconnection (OSI) model for network architectures; manages the transmission circuit, transforms data bits into data frames, transmits the frames, and processes acknowledgment frames in return.

data logging *Computer Programming.* a recording of digitized analog data, digital data, and clock data to form a record of data values over time.

data management *Computer Programming.* a general term referring to system functions that maintain and provide access to stored data, including access to storage hardware, enforcement of data storage conventions, and regulation of the use of input/output devices; usually concerns data only during a program's execution as opposed to database management that is concerned with more lasting or complex data. Thus, **data management program.**

data manipulation *Computer Programming.* a general term referring to the functions required to process data in order to make further use of it, including sorting, merging, editing, and summarizing. Thus, **data manipulation language.**

datamation *Computer Technology.* automatic data processing. (From *data automation.*)
data medium *Computer Technology.* any material on which data can be physically stored, such as a magnetic tape or disk.
data model *Computer Programming.* a graphic database design tool used to represent the properties of stored data and their relationships to each other, independent of software, hardware, or performance considerations.
data name *Computer Programming.* a symbolic name that a programmer assigns to a variable or constant used in a program.
data organization *Computer Technology.* the physical and spatial arrangement of records in storage; examples include sequential, indexed sequential, partitioned, and direct.
data packet *Computer Technology.* a relatively small unit of data transmitted over a packet switching network as part of a message that is sent from one computer to another.
data patch panel *Telecommunications.* a sequence of a patch bay, such as a board, panel, or console, in which circuits are terminated in jacks mounted on the surface; the other end of the circuit is connected to various points in a communication system to permit interconnection, monitoring, or testing via the jacks.
data plotter *Computer Technology.* an output device that provides hardcopy graphics or pictorial representations of computer-processed data, usually as line drawings.
data processing or **data-processing** *Computer Technology.* a general term referring to any operations performed on data by a computer system, in accordance with specified rules and procedures.
data-processing center see DATA CENTER.
data processor *Computer Technology.* **1.** any device used to perform operations on data. **2.** a person involved in performing such operations.
data protection *Computer Technology.* the application of procedures and safeguards to prevent deliberate or accidental data loss or damage, or the access of unauthorized persons to sensitive, classified, or private data. Also, DATA SECURITY.
data purification *Computer Programming.* an attempt to screen the maximum number of errors from a set of input data prior to submission for processing.
data rate *Computer Technology.* the speed at which circuits or devices transmit digital data, usually measured in bits per second. Also, **data transfer rate.**
data record *Computer Programming.* an organized and identifiable set of data elements that are related in some way. Also, RECORD.
data reduction *Computer Programming.* the process of obtaining only the useful data from a larger set of data through computation, summarization, aggregation, or other abstracting methods. Also, REDUCTION.
data redundancy *Computer Programming.* the replication of specific data item values in more than one file within a database.
data register *Computer Technology.* a buffer register in the CPU used to hold a word before it is written to memory or after it is read from memory.
data representation *Computer Programming.* the machine-readable code or format that is interpreted and produced by computer instructions.
data retrieval *Computer Programming.* the process of searching for and selecting data in files or in a database stored in a computer.
data rules *Computer Programming.* specific conditions, conventions, and restrictions that must be met when processing data.
data security see DATA PROTECTION.
data set *Computer Programming.* a set of related data records in a form that can be used by a computer.
data set migration *Computer Programming.* the moving of infrequently used or inactive data from on-line storage to archival or back-up storage.
data sheet *Computer Technology.* a form for collecting or recording source data that is designed for ease and convenience of transcription to machine-readable format.
data sink *Computer Technology.* the equipment at the receiving location that accepts and stores data signals after transmission via a data communications channel.
data source *Computer Technology.* the equipment at the sending location that supplies the data signals to be transmitted over a data communications channel.
data statement *Computer Programming.* a nonexecutable programming language statement that declares constant data elements to be used in the program.
data station *Computer Technology.* a remote terminal device used for communicating with a centrally located computer, as well as for off-line processing.
data stream *Computer Technology.* serial data being transmitted over a channel.
data structure *Computer Programming.* the organization of stored data or records in a regular or characteristic way, such as in an array, tree, list, queue, or stack.
data tablet see DIGITIZING PAD.
data tracks *Computer Technology.* sequences of positions where data can be recorded, either on a continuous medium, such as magnetic tape, or on a rotating medium, such as a drum or disk.
data transfer *Computer Technology.* the movement of data from one location to another without changing the information content.
data transparency *Computer Technology.* the property of transmitted data such that the data may not have any characteristics that are meaningful to the transmission protocol and the transmission may not interfere with any data pattern.
data type *Computer Programming.* a description of a particular class of data in a program, including the representation of the data, its components and their types, and operations that can be performed on it.
data under voice *Telecommunications.* a telephone service providing wideband digital signals (up to 56 kb/s) that are carried on existing microwave radio systems at the lower end of the frequency spectrum being transmitted, in addition to the usual multiplexed voice signals.
data unit *Computer Programming.* a set of characters or digits treated together as a single unit.
data verification *Computer Programming.* the process of checking input data for transcription errors.
data word *Computer Programming.* a computer word containing data to be processed.
date-time group *Telecommunications.* a set of characters that are used in a message to express the date and time in a prescribed format, such as day-month-year-hour-minute.
Datiscaceae *Botany.* a mainly temperate family of dicotyledonous perennial herbs and large trees in the order Violales, characterized by apetalous flowers, capsular fruit, and numerous tiny seeds with little or no endosperm.
datolite *Mineralogy.* $CaBSiO_4(OH)$, a brittle, varicolored monoclinic mineral with a vitreous luster, occurring in massive and crystalline forms, having a specific gravity of 2.8 to 3 and a hardness of 5 to 5.5 on the Mohs scale; found as a secondary mineral in cavities and veins in basic igneous rocks.

datolite

datum [dat´əm; dāt´əm] *Science. plural,* **data.** an individual piece of information, such as a fact or statistic. *Computer Programming.* see DATA ITEM. *Engineering. plural,* **datums.** in a horizontal control survey, a base consisting of the latitude and longitude of a point, the azimuth of a certain line from this point, and two constants used in defining the terrestrial spheroid.
datum-centered ellipsoid *Geodesy.* the ellipsoid that gives the best fit to the astrogeodetic network of a particular datum; it therefore does not necessarily have its center at the center of the earth.

datum horizon *Geology.* an easily recognizable, extensive stratigraphic bed or formation used as a reference surface in determining the positions of rock strata when making comparisons of thicknesses of strata, or from which contours are drawn in making structure-contour maps. Also, STRUCTURAL DATUM, CONTOUR HORIZON.

datum plane *Engineering.* a permanently situated plane or surface to which soundings, elevations, or other data are referred. Also, **datum level.**

datum point *Cartography.* any point of known location that serves as a reference or base for the measurement of other quantities.

datum transformation *Geodesy.* the systematic elimination of discrepancies between adjoining or overlapping triangulation networks from different datums, accomplished by moving the origins, rotating, and stretching the networks to fit each other.

daturism *Toxicology.* poisoning due to ingestion of nightshade or other plants of the genus *Datura*; symptoms may include hallucinations and delirium.

Daubentoniidae *Vertebrate Zoology.* the aye-ayes, a monotypic family of arboreal nocturnal primates found only in northern Madagascar, in which all the digits except the big toe have claws rather than nails.

daubing *Building Engineering.* the covering or coating of walls with a soft adhesive matter, such as plaster, using a spreading motion.

daubreeite *Mineralogy.* $BiO(OH,Cl)$, a dull, yellowish tetragonal mineral with a massive, earthy appearance, having a specific gravity of 6.5 to 7.56 and a hardness of 2 to 2.5 on the Mohs scale; found mixed with clay in mines in Bolivia and Utah.

dauermodification *Cell Biology.* a phenotypic modification that is acquired by a cell and may persist in the cytoplasm for a few generations but is not incorporated into the genetic material of the cell. (A term based on the German word *Dauer,* meaning "duration.")

daughter *Nuclear Physics.* a term used to denote a nuclide produced by the radioactive decay of another nuclide. Also, DECAY PRODUCT.

daughter board *Computer Technology.* an integrated circuit card that is plugged into the mother board or back plane of a computer. Also, **daughter card.**

daughter cell *Cell Biology.* either of two cells that are produced from the mitotic division of a parent cell.

daunorubicin *Oncology.* an antineoplastic antibiotic substance that is isolated from *Streptomyces peucetius* and used to treat cancers such as acute lymphocytic and granulocytic leukemia. Also, **daunomycin.**

Dauphine twin law *Crystallography.* a twin law specifying a relationship between twin crystals in quartz, in which the twin is formed by the rotation of two left-handed or right-handed crystals 180° about the *c* axis. Also, **Dauphine law.**

Dausset, Jean born 1916, French physiologist; shared Nobel Prize for showing the existence of major histocompatibility complex in humans.

Davalliaceae *Botany.* a family of mostly epiphytic ferns of the order Filicales, characterized by a simple scaly rhizome, simple to finely divided fronds, and sori with indusium on lower leaf surfaces; includes the rabbit's-foot, sword, and Boston ferns.

Davenport, Thomas 1802–1851, American inventor; he invented the early **Davenport electric motor,** which he then used to power the first model electric locomotive (1835).

Davenport electric motor

Davian *Geology.* **1.** a European subdivision of the Upper Cretaceous period. **2.** describing various types of limestone formations found in Denmark, southeastern France, Spain, and Portugal.

Davida *Astronomy.* asteroid 511, discovered in 1903 and measuring 318 kilometers in diameter; it belongs to type C.

davidite-(La) *Mineralogy.* $(La,Ce)(Y,U,Fe^{+2})(Ti,Fe^{+3})_{20}(O,OH)_{38}$, a dark-brown to black, opaque, uraniferous trigonal mineral; massive in habit or occurring as cubelike tabular or pyramidal crystals; having a specific gravity of 4.42 and a hardness of 6 on the Mohs scale.

Davidson current *Oceanography.* a Pacific Ocean countercurrent that sets northward in winter along the west coast of the U.S. from northern California to at least 48°N.

Davidsoniaceae *Botany.* a monospecific family of small, dicotyledonous, tanniferous trees of the order Rosales, distinguished by pungent red hairs that cover twigs, leaves, and fruits on young specimens, large, pinnately compound leaves, and tart, plumlike edible fruits; native to Australia.

Davidsonina *Paleontology.* a spire-bearing genus of articulate brachiopods of the Lower Carboniferous.

daviesite see HEMIMORPHITE.

davisonite *Mineralogy.* a mixture of crandallite and adatite.

Davisson, Clinton 1881–1958, American physicist; shared Nobel Prize for the discovery of electron diffraction by crystals.

Davis's quadrant *Navigation.* the backstaff or sea quadrant, invented by John Davis about 1590. When using this instrument, the navigator turned his back to the sun and used a shadow cast by the instrument to determine the zenith distance of the sun.

davit *Naval Architecture.* a pair of upright or angled cranes, often with curved tops; it is often used as a hoist and holding area for a ship's boat.

Davy, Sir Humphry 1778–1829, English chemist; the first to isolate sodium, potassium, calcium, magnesium, chlorine, barium, and strontium; invented the Davy mining lamp.

Davy Crockett *Ordnance.* the popular name for a mobile rocket launcher designed to fire a small nuclear warhead; the **M-28** version can be handled by a three-man crew and has a range of 1.5 miles; the heavier **M-29** has a range of 3 miles. (Named for *Davy Crockett,* 1786–1836, American frontiersman and folk hero.)

Dawes' limit *Optics.* the limit placed on the resolving power of a telescope by the effects of diffraction.

dawn *Astronomy.* the first light in the sky before sunrise, equivalent to morning astronomical twilight.

dawn side *Astronomy.* the hemisphere of a planet or moon that lies nearest the morning terminator.

Dawson, George Mercer 1849–1902, Canadian geologist; son of Sir John Dawson; studied surface geology and glacial phenomena.

Dawson, Sir John 1820–1899, Canadian geologist; with Sir Charles Lyell, studied strata and fossils of eastern Canada.

Dawsoniales *Botany.* an order of mosses with rigid, erect stems growing from a rhizomatous base; distinguished from the related Polytrichales by capsule form and gametophore size.

dawsonite *Mineralogy.* $NaAl(CO_3)(OH)_2$, a transparent, colorless or white orthorhombic mineral with a perfect cleavage, having a specific gravity of 2.44 and a hardness of 3 on the Mohs scale; found as a low-temperature mineral in shale.

day *Science.* in general use, a period of 24 hours. *Astronomy.* **1.** a period of time that is based on the spinning of the earth on its axis as it moves around the sun, varying slightly from 24 hours and measured, for example, from noon to noon (a solar day) or from one transit of a star to its next transit (a sidereal day). **2.** a similar measurement of the rotation of the moon or another planet. *Architecture.* a division of a window, especially of a large church window.

DAY *Aviation.* the airport code for Dayton, Ohio.

day beacon *Navigation.* an unlighted fixed aid to navigation.

day clock *Computer Technology.* a clock that represents ordinary "wall clock" time and can be read by a computer. Also, TIME-OF-DAY CLOCK.

daydream *Psychology.* dreamlike thinking while awake; a fantasy or pleasant mental image indulged in during waking hours.

dayglow *Astronomy.* a faint luminosity observed in the daytime sky, caused when charged particles from the sun strike atoms and molecules in the earth's atmosphere.

Dayia *Paleontology.* a genus of Silurian brachiopods of the extinct suborder Spiriferacea and family Atrypidae.

daylight control *Engineering.* a device, often used in outdoor lamps, that causes automatic electrical activation when it is sufficiently dark and turns off when daylight returns.

daylight lamp *Electricity.* an incandescent or fluorescent lamp that emits light with a spectral distribution approximating that of daylight.
daymark *Navigation.* any distinctive structure that can be used as an aid to navigation by daylight.
day neutral *Botany.* relating to the capability of developing and maturing independent of the amount of daylength, with the onset of flowering controlled by other factors.
day number *Astronomy.* a numerical quantity used in reducing an object's mean place on a given date to its apparent place.
day of supply see ONE DAY'S SUPPLY.
day's run *Navigation.* a term for the distance traveled by a vessel in one day, usually from noon to noon.
daytime visual range *Meteorology.* the distance in daytime at which the apparent contrast between a given target and its background becomes equal to the threshold contrast of an observer; a function of the atmospheric extinction coefficient (absorptivity). Also, VISUAL RANGE.
dazomet *Organic Chemistry.* $C_5H_{10}N_2S_2$, a white crystalline solid that is soluble in alcohol and decomposes in alcohol; melts at 106–107°C; used as a fungicide.
DB or **db** database.
dB or **db** decibel; decibels.
DBA database administrator.
dBa adjusted decibels.
dBf decibels above 1 femtowatt.
db galaxy *Astronomy.* a galaxy of the cD type that has two nuclei, giving it a "dumbbell" appearance.
dBk decibels above 1 kilowatt.
dbl. or **dbl** double.
***d*-block element** *Chemistry.* a term for a transition element that has an unfilled *d* subshell.
dBm decibels above 1 milliwatt.
DBMS or **dbms** database management system.
DBP diastolic blood pressure.
dBp decibels above 1 picowatt.
dBrn decibels above reference noise.
DBS or **dbs** direct-broadcast satellite.
dBv decibels above 1 volt.
dBw decibels above 1 watt.
dBx decibels above reference coupling.
DC or **D.C.** direct current. Also, **dc** or **d.c.**
DC or **dc** data communications.
D.C. or **d.c.** from the beginning. (From Italian *da capo.*)
DC-3 *Aviation.* a twin-engine passenger airplane designed by Douglas Aircraft; the original commercial version of the C-47 miltary transport.
DC-6 *Aviation.* a four-engine passenger airplane designed by Douglas Aircraft; the commercial version of the C-118 military transport.
DC-8 *Aviation.* one of the earliest passenger jet airliners, designed by Douglas Aircraft and first flown in 1958.
DC-9 *Aviation.* a widely used twin-engine jet airliner designed by McDonnell Douglas Corporation, carrying about 90 passengers.
DC-10 *Aviation.* a three-engine wide-body jet airliner designed by McDonnell Douglas Corporation, carrying about 270 passengers.
DCA *Aviation.* the airport code for National Airport, Washington, D.C.
D cable *Electricity.* a two-conductor cable in which each conductor resembles the shape of the capital letter D, having insulation between the conductors and between the conductors and the sheath.
DC amplifier see DIRECT-CURRENT AMPLIFIER.
DCC *Oncology.* a tumor-suppressor gene found to be altered in carcinoma of the colon.
D.Cc. or **DCc** double concave.
dc casting see DIRECT-CHILL CASTING.
DC dump see DIRECT-CURRENT DUMP.
DCE or **dce** data communications equipment.
DChE or **D.Ch.E.** Doctor of Chemical Engineering.
DCMO see CARBOXIN.
DC restorer see DIODE CLAMPING CIRCUIT.
DCTL direct-coupled transistor logic.
DC-to-DC converter *Electricity.* an electronic circuit that converts one DC voltage to another; often a transformer is used to step voltage up or down and to provide DC isolation.
DC voltage see DIRECT VOLTAGE.
dcwv direct-current working voltage.
D.Cx. or **DCx** double convex.
DD dishonorable discharge.
d.d. let it be given to. (From Latin *detur ad.*)
D-day *Military Science.* the day on which a specific operation commences or is scheduled to commence.
DDC Dewey Decimal Classification; direct digital control.
DDD direct distance dialing.
DDD see 2,2-BIS(*p*-CHLOROPHENYL)-1,1-DICHLOROETHANE.
DDL data definition language; data description language.
DDS Dewey Decimal System.
DDS or **D.D.S.** Doctor of Dental Surgery. Also, **DDSc, D.D.Sc.**
DDT *Organic Chemistry.* $(ClC_6H_4)_2CHCCl_3$, colorless needles or white to slightly off-white powder, insoluble in water and slightly soluble in alcohol; melts at 108.5°C; used as an insecticide and pesticide. (An abbreviation of dichlorodiphenyltrichloroethane.)

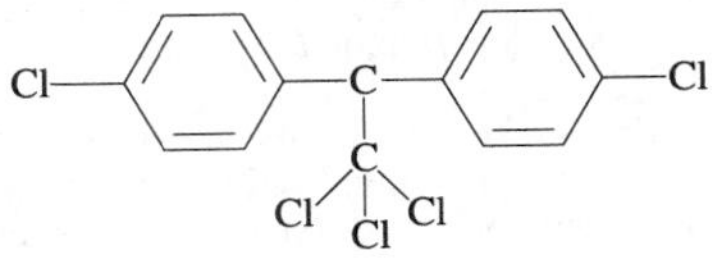

DDT

DE or **D.E.** Doctor of Engineering.
de- a prefix denoting negation, privation, cessation, or movement down or away from, as in *decompression, dehumidify, de-evolution, descent.*
deac. deaccentuator.
deaccentuator *Electronics.* a circuit that reduces the high-frequency response of audio signals in a frequency modulation receiver in order to compensate for the undesired accentuation of high-frequency audio that normally occurs at the transmitter.
deacetylate *Organic Chemistry.* to remove acetyl groups from a molecule.
deacidify *Chemistry.* to reduce the acidity of a substance.
Deacon process *Chemical Engineering.* a method of producing chlorine by the oxidation of hydrogen chloride with oxygen at 400–500°C over a copper salt catalyst.
deactivating group *Organic Chemistry.* a substituent group that when added to benzene causes the derivative molecule to react more slowly than benzene.
deactivation *Chemistry.* **1.** the process of diminishing the chemical reactivity of a substance. **2.** in photochemistry, the collision of a light-activated molecule with another molecule, resulting in a lowering of the energy of the first molecule.
dead *Biology.* without life; having experienced death. *Science.* used in various contexts to describe a situation analogous to death, as by being final, extreme, precise, minimal, and so on. *Acoustics.* describing an area that does not allow significant sound reflection, such as a recording studio equipped with sound-absorbing materials. *Electricity.* relating to a current-carrying part that is unelectrified or free from signals or fields.
dead ahead *Navigation.* bearing 0° relative to the craft's heading.
dead-air space *Building Engineering.* a sealed air space between two material structures, such as the hollow area between two walls.
dead area *Engineering.* see BLIND SPOT.
dead arm *Plant Pathology.* a disease common in grapes that is caused by the fungus *Cryptosporella viticola* and that destroys the main lateral shoots and branches (or arms) with dark, elongated cankers.
dead astern *Navigation.* bearing 180° relative to the craft's heading.
dead axle *Mechanical Engineering.* an axle that has a wheel mounted to it but does not rotate with the wheel.
dead band *Electricity.* **1.** the range of values over which a measured variable can change without affecting the output of a magnetic amplifier or automatic control system. **2.** a radio frequency band that is not used by stations.
deadbeat *Mechanics.* describing an oscillating system, such as a meter, that comes to rest in a new position without overshooting it. Thus, **deadbeat reaction, deadbeat response.**
deadbeat algorithm *Control Systems.* an algorithm used to minimize dead time following set-point changes, assuming that the process can be modeled as approximately a first-order-plus-dead-time system.
deadbeat compass see APERIODIC COMPASS.
deadbeat escapement *Horology.* an escapement mechanism in a pendulum clock in which the escape wheel and pendulum come to a complete stop, rather than recoiling slightly, when the wheel is engaged by the pallet; more accurate but less sturdy than the anchor escapement.

dead block *Engineering.* a device located on the ends of train passenger cars, designed to absorb impact shock.

dead bolt *Mechanical Devices.* a bolt that is moved directly into position with a key or knob, rather than by spring action. Also, DEADLOCK.

dead-bright *Metallurgy.* the attribute of a metal or alloy that has been thoroughly polished.

dead center *Science.* any position or point at the exact center. *Mechanical Engineering.* in a reciprocating engine, either of two positions, occurring at either end of the stroke, when the crank and connecting rod are in the same line and the force exerted on the crank by the piston and connecting rod is zero. *Mechanical Devices.* in a lathe, the stationary support that is used to hold a workpiece.

dead end *Acoustics.* the portion of a sound-recording area that produces minimal sound reflection, as opposed to the live end where considerable reverberation takes place. *Electricity.* a portion of a tapped coil through which no current is flowing at a particular bandswitch position.

dead-end effect *Electricity.* the energy absorption of unused portions of a tapped coil.

dead-end polymerization *Materials Science.* an incomplete polymerization caused by a limited amount of initiator; used experimentally to determine initiator efficiency ratios.

dead fault *Geology.* a fault along which all movement has stopped.

dead flat *Naval Architecture.* the portion of a ship's hull amidships along which the hull cross section is constant, and at which its beam is greatest; sharp lined hulls usually reach maximum beam only at one point on their length, and have little or no length of dead flat.

dead furrow *Agriculture.* a furrow that is about the width of two plow bottoms rather than one, as may be left in the middle of a strip of plowed land or between two strips.

dead glacier see STAGNANT GLACIER.

dead ground *Geology.* land that is obscured from view by intervening hills, ridges, or other obstructions.

dead halt *Computer Technology.* a system halt from which there is no recovery without operator interaction; may be deliberately programmed or caused by a catastrophic processing error. Also, DROP-DEAD HALT.

deadhead *Metallurgy.* the part of a casting that fills the opening through which the molten metal was delivered to the mold. *Transportation Engineering.* **1.** a nonpaying passenger. **2.** a nonrevenue trip by a revenue vehicle, as when it returns empty to its origin.

dead ice see STAGNANT ICE.

dead lake *Hydrology.* a dried-up lake that has been filled in by vegetation.

deadlight *Naval Architecture.* a shutter manipulated from inside a ship and designed to protect a glass porthole during heavy weather or to black out the ship's lights during engagement with the enemy.

dead line *Geology.* on a batholith, the line between the upper mineral deposits from which metals can be extracted and the lower deposits which are economically barren.

dead load see STATIC LOAD.

deadlock *Computer Technology.* a situation in a multiprogramming system in which a process cannot proceed because it is waiting for an event that will never occur, such as a resource assignment in which two tasks simultaneously require use of resources assigned to the other task. Also, **deadly embrace.** *Mechanical Devices.* see DEAD BOLT.

deadly nightshade *Botany.* a perennial plant, *Atropa belladonna,* indigenous to central and southern Europe and cultivated in North America, both as an ornamental and as a source of various anticholinergic alkaloids such as atropine and belladonnine. Also, BELLADONNA.

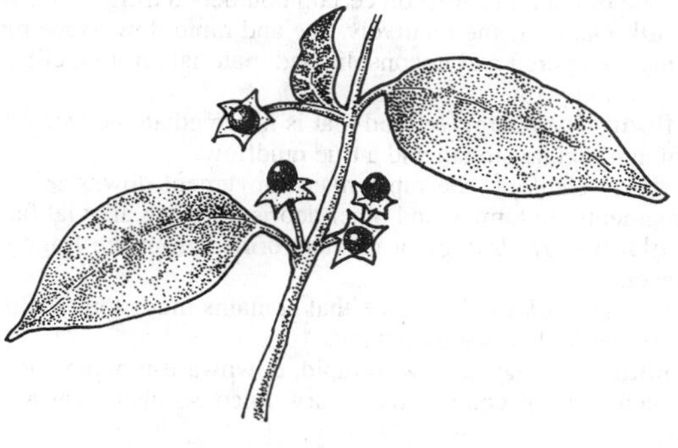

deadly nightshade (belladonna)

deadman see ANCHOR BLOCK.

deadman control *Engineering.* a safety mechanism requiring constant pressure or manipulation by a human operator; it stops a machine or vehicle automatically if the operator becomes incapacitated or inattentive. Similarly, **deadman switch.**

deadman's brake *Mechanical Engineering.* a safety device used on electric vehicles or machinery, designed so that the vehicle or machine will stop if the operator releases pressure on the control, usually a handle or foot pedal. Similarly, **deadman's handle, deadman's pedal.**

dead mine *Ordnance.* a mine that has been neutralized, sterilized, or rendered safe.

dead reckoning *Navigation.* the process of determining the new position of a ship or aircraft from a known previous position, by calculating the time, speed, and direction traveled from the old position; an estimation of position rather than an exact determination.

dead-reckoning equipment *Navigation.* a device that continuously calculates the dead-reckoning position of a craft. Also, **dead-reckoning analyzer.**

dead-reckoning plot *Navigation.* lines placed on a chart or a plotting sheet to determine position based on dead reckoning.

dead-reckoning position *Navigation.* an estimated position as determined by dead reckoning.

dead-reckoning track *Navigation.* a line on a chart or plotting sheet that connects the successive dead-reckoning positions of a craft.

dead-rise *Naval Architecture.* the angle between the horizontal and the bottom of any vessel that is not flat-bottomed.

dead room *Acoustical Engineering.* a room constructed of or lined with sound-absorbing material to minimize internal sound reflection.

Dead Sea *Geography.* a salt lake in the Great Rift Valley on the Israel-Jordan border; the lowest point on earth (about 1300 feet below sea level).

dead sea *Hydrology.* any body of water from which sedimentary salts have been or are being precipitated.

dead short *Electricity.* a short-circuit path having extremely low resistance.

dead soft *Materials Science.* of a metal or alloy, having the minimum possible hardness. Thus, **dead-soft steel.**

dead space *Anatomy.* the volume of the air passages from the nose down to the level where oxygen and carbon dioxide are interchanged. Also, **anatomical dead space.** *Physiology.* an area of the alveoli where very little gas is being exchanged. Also, **physiological dead space.** *Surgery.* the empty cavity that is left following the closure of a surgical incision or a wound, permitting the accumulation of blood or serum and resulting in a delay in healing. *Thermodynamics.* a space whose volume is occupied by a gas with a different temperature from that of the main body. *Ordnance.* an area that is within the range of a weapon, radar, or observer but cannot be covered by fire or observation because of obstacles, natural obstructions, or limitations of the device. Also, **dead zone.**

dead spot *Telecommunications.* **1.** a geographic area in which radio or television signals are weak or nonexistent, usually due to distance from the transmitter, obstructions, or ionospheric layer reflection skip distances. **2.** a band of frequencies, within the overall tuning range of a radio receiver, exhibiting low sensitivity; usually caused by improper tracking of the tuning circuits.

deadstick landing *Aviation.* a landing without engine power.

deadstock or **dead stock** *Agriculture.* the tools, machinery, and equipment used to operate a farm.

dead-stroke *Mechanical Engineering.* having little or no recoil after a stroke or upon impact; such a device may utilize a spring mechanism to reduce recoil. Thus, **dead-stroke hammer.**

dead time *Science.* any time during which a system is not operating productively; downtime. *Engineering.* the time lag in which a system having just responded to a signal or event is unable to respond to another signal or event. Also, INSENSITIVE TIME. *Computer Programming.* a delay purposely inserted between two related actions in order to prevent overlap or to permit another event, such as a control action, to take place. *Control Systems.* the time required for a process control system to respond to a change in an input signal.

dead-time compensation *Engineering.* any changes that are made in a program or system to allow for dead time. Similarly, **dead-time correction.**

dead water *Oceanography.* sea water in which a thin layer of fresher water overlies a layer of more saline water. *Navigation.* a phenomenon in which the efficiency of a ship's propulsion decreases when moving through such water; associated with internal waves near the keel.

dead weight *Transportation Engineering.* the weight of a truck, railroad car, or the like, not including its load or contents. *Mechanics.* see STATIC LOAD.

deadweight gauge *Engineering.* a calibrating apparatus that measures fluid pressure in which weight is applied to the top of a vertical piston being acted upon by the fluid.

deadweight tonnage *Naval Architecture.* a ship's maximum loaded displacement minus its empty displacement; it is equal to the maximum load that the ship can safely carry. Also, **deadweight capacity**.

dead zone unit *Computer Technology.* an analog computer device that outputs a constant signal value for all input values within a specified range.

deaeration *Chemistry.* the removal of air or oxygen from a solution, for example by bubbling with an inert gas. *Engineering.* any process of removing gases from a substance.

deaerator *Mechanical Engineering.* in high-pressure steam boilers, a vessel in which the boiler feed water is heated at lower pressure in order to remove dissolved gasses that might damage the boiler metal.

deaf *Medicine.* affected with deafness.

deafferentation *Neurology.* the process of eliminating or interrupting afferent (sensory) impulses by destroying or damaging afferent nerve fibers.

deafness *Medicine.* lack of the sense of hearing, or profound hearing loss; moderate loss of hearing is usually distinguished as *hearing loss.*

deal *Forestry.* **1.** a board of fir or pine of any length, usually having cross-sectional dimensions of 10 × 3 inches. **2.** made of fir or pine, such as a *deal table.*

dealfish *Vertebrate Zoology.* a ribbonfish of the family Trachypteridae; a long, slender fish inhabiting worldwide mid-ocean waters, having a short head, a narrow mouth, and a dorsal fin extending along the length of its back.

dealkalization *Materials Science.* the removal of alkyl groups from a compound or the reduction of alkalinity as in the neutralization process.

dealkalize *Chemistry.* to reduce the alkalinity of a substance.

dealkylate *Chemistry.* to remove attached hydrocarbon groups.

deallocation *Computer Technology.* the release of a previously assigned resource for use by another program.

dealuminization *Chemistry.* the process of removing aluminum from a compound.

deamidase see DEAMINASE.

deamidate *Organic Chemistry.* the removal of an amido group from a molecule.

deaminase *Enzymology.* an enzyme of the hydrolase class that catalyzes the removal of an amino group from compounds, releasing ammonia. Also, AMINOHYDROLASE, DEAMIDASE.

deaminate *Organic Chemistry.* the process of removing an amino group from a molecule. Thus, **deamination.**

deaminating enzyme *Enzymology.* an enzyme that removes an amino group from a molecule.

Dean number *Fluid Mechanics.* a dimensionless group that applies to flow in curved channels and adjusts the Reynolds number by the ratio of centrifugal force to inertial force.

deanol see 2-DIMETHYLAMINOETHANOL.

deashing *Chemistry.* a process by which both anions and cations of inorganic salts are removed from solution through adsorption by ion-exchange resins.

death *Biology.* the termination of life; the permanent cessation of the activity and vital functions of an organism. *Medicine.* a medical determination that this condition exists; the definition is subject to many legal, ethical, and philosophical considerations but is generally categorized in the following ways: the complete absence of spontaneous heartbeat or respiration (also called **clinical death**); the complete absence of activity at the cellular level (**cell death**); the complete lack of brain activity for an extended period (**brain death**). *Physiology.* the cessation of vital activity of an individual cell or tissue within an organism.

death fantasy *Psychology.* a fantasy in which the individual imagines himself to be dead but still able to observe reality, such as the events of one's own funeral.

death feigning *Behavior.* a behavior pattern of remaining motionless as if dead, as a means of deceiving a predator that only will attack moving prey.

death instinct *Psychology.* an unconscious drive that leads an individual toward death or self-destruction. Also, **death wish.**

death loss *Agriculture.* the rate at which cattle or other herd animals die off or are expected to die off.

death point *Physiology.* an environmental limit beyond which an organism is unable to survive.

death rate see MORTALITY RATE.

Death Valley *Geography.* a desert valley in eastern California, containing the lowest point of land in the Western Hemisphere (282 feet below sea level).

DeBakey, Michael born 1908, American surgeon; pioneer in transplant surgery; helped develop artifical heart and blood vessels.

debark *Military Science.* to carry out the process of debarkation.

debarkation *Military Science.* the unloading of troops, equipment, or supplies from a ship or aircraft. Thus, **debarkation area, debarkation unit.**

Debaryomyces *Mycology.* a former term for the fungal class Saccharomyces. Also, TORULASPORA.

debatable time *Computer Technology.* **1.** in a computer resource accounting system, the time that cannot be associated with a specific user or application. **2.** a delay in processing that has no obvious cause.

debenzylation *Organic Chemistry.* the process of removing a benzyl group from a molecule.

debitage *Archaeology.* waste chips or debris resulting from the manufacture of stone tools; found in large quantity in a tool-making area.

debit card *Computer Technology.* a coded plastic card issued by a bank and used for electronic transfers of funds to merchants' accounts.

deblocking *Computer Technology.* the process of breaking up a block of logical records into individual records for further processing.

Deborah number *Mechanics.* a dimensionless parameter equal to the relaxation time of a process, divided by the time after which the process is observed.

deboss *Metallurgy.* in forging, the removal of unwanted protrusions.

debossed *Graphic Arts.* stamped in relief, so that the image is set below the surrounding surface; the opposite of embossed.

Debot effect [də bō′] *Graphic Arts.* a manifestation of the Herschel effect in which an internal latent photographic image is converted to a surface latent image by red or infrared radiation.

Debouillet [deb′oo lā′] *Agriculture.* a breed of bare-faced sheep raised for their wool and also sometimes for mutton, crossbred in the U.S.

debouncing *Computer Technology.* a technique used to prevent the recognition of transient closures of a key or a switch, often by using a reset-set flip-flop or a time-delay mechanism.

debranching enzyme *Enzymology.* an enzyme that removes branching segments from glycogen during its catabolism.

deBranges' theorem *Mathematics.* if f is a one-to-one analytic function that maps the complex unit disk into itself and has the power series expansion $z + a_2x^2 + a_3x^3 + \cdots$, then $|a_n| \le n$ for all n. Formerly known as the **Bieberbach conjecture.**

débridement *Surgery.* the removal of foreign matter and devitalized or infected tissue from a wound or lesion in order to expose healthy tissue.

debrief *Aviation.* to carry out a systematic interrogation of a military air crew or space flight crew immediately after a mission, in order to obtain maximum useful information. Thus, **debriefing.**

debris [də brē′] *Biology.* organic waste from dead or damaged tissue that serves as food for certain scavengers. *Geology.* any accumulation of loose fragments of rock, soil, or organic matter resulting from the weathering and disintegration of rock.

debris avalanche *Geology.* the sudden and rapid sliding of unsorted mixtures of rock and soil down steep slopes.

debris cone *Hydrology.* on a glacier, a mound of ice or snow covered by a thin layer of debris that protects the underlying material from ablation. Also, GLACIAL CONE. *Geology.* **1.** see ALLUVIAL CONE. **2.** a cone-shaped pile of debris formed on certain boulders during a landslide.

debris fall *Geology.* the relatively free and rapid downslope movement of mainly weathered and unconsolidated materials from a cliff, cave, or arch. Also, SOIL FALL.

debris flood *Hydrology.* a flood that is intermediate between the turbid flood of a mountain stream and a true mudflow.

debris flow *Geology.* the rapid mass movement downslope of coarse rock fragments, soil, mud, and other debris, often on alluvial fans.

debris glacier *Hydrology.* a glacier formed from ice fragments of a taller glacier.

debris ice *Hydrology.* **1.** sea ice that contains mud, soil, shells, stones, and other material. **2.** see BRASH ICE.

debris slide *Geology.* a slow to rapid, downward-moving and forward-rolling landslide of comparatively dry, unconsolidated earth, soil, and rock debris.

debris slope see TALUS SLOPE.

de Broglie, Louis [də brōg´lē] 1892–1987, French physicist; Nobel Prize for the theory of the wave nature of electrons, a foundation of the field of quantum mechanics.

de Broglie relation *Quantum Mechanics.* a relation between the mass and velocity of a particle and the wavelength of the associated wave that postulates the wavelike behavior of electrons in which the energy of a quantum particle is proportional to the frequency of the quantum wave.

de Broglie's theory *Quantum Mechanics.* the theory that a particle may be represented by a wave, based on the coexistence of particle and wave attributes, whose frequency is given by the de Broglie relation.

de Broglie wave *Quantum Mechanics.* the wave associated with a particle by virtue of its motion.

de Broglie wavelength *Quantum Mechanics.* the wavelength of a "matter wave" for a particle of mass m given by the equation $\lambda=h/p$, where h is Planck's constant and p is the momentum, combining the corpuscular and undulatory concepts of light applicable to all particles from the subatomic to the macroscopic.

debromoaplysiatoxin *Biochemistry.* a toxin resembling aplysiatoxin, which is secreted by certain species of blue-green algae.

de Brun-van Eckstein rearrangement *Organic Chemistry.* a reaction used to prepare ketoses by mixing an aldose or ketose with aqueous calcium hydroxide, producing a mixture of monosaccharides and unfermented ketoses.

deb. spis. of the proper consistency. (From Latin *debita spissitudine.*) Also, **Deb. spis.**

debubblizer *Engineering.* a worker or device that eliminates bubbles from plastic tubing or rods.

debug *Computer Programming.* to identify, isolate, and correct errors in a program or malfunctions in computer hardware.

debugging routine *Computer Programming.* any routine used during program development to assist in locating and diagnosing programming errors, such as snapshop dump, postmortem dump, or trace routines.

debugging statement *Computer Programming.* a temporary instruction, such as a print command, strategically placed in a program to help in locating errors.

debunching *Electronics.* the tendency for a beam of electrons in a velocity-modulated tube to spread because of mutual repulsion.

deburr *Metallurgy.* to carry on the process of deburring.

deburring *Metallurgy.* a manufacturing process in which the surfaces of metal are cleaned of burrs and other imperfections.

Debye, Peter [de bī´; də bī´] 1884–1966, Dutch-born American physical chemist; Nobel Prize for study of bipole moments and diffractions of X-rays and electrons in gases.

debye *Electricity.* a unit of electric dipole moment, equivalent to 10^{-18} Franklin centimeter.

Debye attraction see INDUCTION FORCE.

Debye effect *Electromagnetism.* an effect in which radiation is selectively absorbed into a dielectric material due to molecular dipoles.

Debye equation *Solid-State Physics.* an equation specifying the lattice contribution to specific heat, which agrees with the Dulong-Petit law at higher temperatures and with the Debye T^3 law at low temperatures.

Debye equation for polarization *Physics.* a model for predicting the polarization of a nonconducting substance through calculations involving a single molecule of the substance.

Debye-Falkenhagen effect *Physical Chemistry.* a phenomenon in which the application of a high-frequency voltage increases an electrolyte's conductivity.

Debye force see INDUCTION FORCE.

Debye frequency *Solid-State Physics.* the frequency that defines the Debye temperature Q, expressed as $n=kQ/h$, where k is the Boltzmann constant and h is the Planck constant.

Debye-Huckel theory *Physical Chemistry.* the assumption that every ion in an electrolytic solution is surrounded by an ion atmosphere (a cluster of oppositely charged ions), which restricts ionic mobility when a current passes through the medium; this assumption provides an explanation for the deviation from ideal behavior of dilute solutions.

Debye-Jauncey scattering *Solid-State Physics.* the scattered X-ray radiation in diffraction analysis of a crystal, which is incoherent and reflects at angles between those of the Bragg angles.

Debye potentials *Electromagnetism.* two potentials, P_e and P_m, that define electric and magnetic fields due to radiation or scattering from a homogeneous isotropic medium.

Debye relaxation time *Physical Chemistry.* the time during which the movement of ions in a solution is retarded by the ions of the opposite charge (ion atmosphere) surrounding them.

Debye-Scherrer method see DEBYE-SCHERRER X-RAY DIFFRACTION.

Debye-Scherrer X-ray diffraction *Materials Science.* a method of diffraction analysis in which a crystal specimen in powder form is affixed to a fiber or thin-walled silica tube, then placed in the path of a monochromatic X-ray beam; the resulting reflections are recorded on a cylindrical film coaxial with the specimen. to obtain information about the structure and identity of the material.

Debye-Sears ultrasonic cell *Acoustics.* a method of creating light images of ultrasonic beams by using acoustic waves as optical grating to diffract light on either side of a central spot, and then focusing the undiffracted light to a pinhead hole which is the only source of light from the cell.

Debye shielding length *Particle Physics.* in a plasma, the distance beyond which the electric field of a charged particle is shielded by particles having the opposite charge. Also, **Debye length, Debye-Huckel screening radius.**

Debye specific heat *Solid-State Physics.* the specific heat of a solid calculated on the assumption that the atoms in the crystal lattice oscillate over a continuum of frequencies, with a maximum cutoff frequency established such that the number of vibrational modes equals the number of degrees of freedom of the lattice; this theory is a generalization of Einstein's equation for specific heat which assumes a fixed frequency of oscillation.

Debye T^3 law *Solid-State Physics.* a law stating that at temperatures which are low compared to the Debye temperature, the specific heat of a solid at constant volume varies proportionally to the cube of the absolute temperature.

Debye temperature *Solid-State Physics.* a characteristic temperature denoted by Q and defined by the relation $kQ=hn$, where k is the Boltzmann constant, h is Planck's constant, and n is the Debye frequency; used in computing the Debye specific heat of a crystal lattice.

Debye theory *Electricity.* a theory stating that the orientation polarization of polar molecules, in which the molecules possess a singular relaxation time, and the plot of the imaginary part of the complex relative permittivity against the real part of the complex relative permittivity, is a semicircle.

Debye-Waller factor *Solid-State Physics.* a reduction factor that accounts for the thermal motion of the atoms in a crystal and thus produces a lower intensity of coherently scattered X-rays, electrons, or neutrons.

dec or **dec.** decimeter; decimeters.

dec. pour off. (From Latin *decanta.*) Also, **Dec.**

deca- or **dec-** a metric combining form meaning "ten," as in *decaliter*.

decaborane *Inorganic Chemistry.* $B_{10}H_{14}$, colorless to white crystals that are slightly soluble in cold water and decompose in hot water; melts at –99.5°C and boils at 213°C; may explode or ignite in contact with heat or flame or with oxygenated solvents; used as a catalyst, corrosion inhibitor, fuel additive, and for many other purposes.

decade *Science.* a period of 10 years, especially a 10-year period beginning with the year 1 or 0, such as 1941–1950 or 1980–1989. *Electricity.* the interval between two electrical quantities, such as frequency or resistance, in which the ratio of one to the other is ten.

decade box *Electricity.* an assembly of precision resistors, coils, or capacitors whose individual values vary in submultiples and multiples of ten; it can be set to any value within its range by selector switches or jacks and is used primarily in laboratory work.

decade scaler *Electronics.* a scale-of-ten electronic counter that delivers one output pulse for each group of ten input pulses. Also, **decade counter.**

decaffeinate *Food Technology.* to remove caffeine from a substance.

decaffeinated coffee *Food Technology.* coffee from which most of the caffeine has been removed by soaking the beans in chemical solvents or water; it usually has about 3 mg caffeine per 150-ml cup, compared with 75–150 mg for regular coffee.

decagon *Mathematics.* a polygon having ten sides.

decagram see DEKAGRAM.

decahedron *Mathematics.* a polyhedron having ten faces.

decahydrate *Chemistry.* a compound that contains ten molecules of water of crystallization.

decalcification *Chemistry.* the process of removing calcium from a calcified material.

decalescence *Metallurgy.* in an iron or steel, the phenomenon occurring when the low-temperature body-centered cubic phase transforms to the high-temperature body-centered cubic phase upon heating. The transformation is accompanied by a temperature drop.

Decalin *Organic Chemistry.* $C_{10}H_{18}$, a trade name for a combustible, isomeric, colorless liquid, insoluble in water and soluble in alcohol; *cis* form boils at 195.65°C and melts at –43.2°C; *trans* form boils at 187.5°C and melts at –30.4°C; used in paints and lacquers, and as a solvent. Also, **decahydronaphthalene.**

decaliter see DEKALITER.

decamerous *Botany.* having ten parts per whorl.

decameter see DEKAMETER.

decametric wave *Electromagnetism.* an electromagnetic wave having a wavelength in the range of 10–100 meters.

***n*-decanal** *Organic Chemistry.* $CH_3(CH_2)_8CHO$, a combustible, colorless to light-yellow liquid, insoluble in water and soluble in alcohol; boils at 208°C; found in essential oils and used in flavoring and perfumes. Also, CAPRALDEHYDE, DECYL ALDEHYDE.

De Candolle, Augustin Pyramus 1778–1841, Swiss botanist; wrote a systematic classification of plants, much of which is still in use.

decane *Organic Chemistry.* $CH_3(CH_2)_8CH_3$, a combustible, narcotic, colorless liquid, insoluble in water and soluble in alcohol; melts at –29.7°C and boils at 174.1°C; used in organic synthesis and jet fuel research, and as a solvent.

decanedioic acid see SEBACIC ACID.

decanning *Nucleonics.* the removal of reactor fuel elements from their cladding, performed under controlled conditions to ensure containment of fission and activation products and actinide elements.

decanoic acid see CAPRIC ACID.

decanoic anhydride see CAPRIC ANHYDRIDE.

1-decanol *Organic Chemistry.* $CH_3(CH_2)_8CH_2OH$, a combustible, colorless or water-white liquid that is insoluble in water and soluble in alcohol; melts at 6°C and boils at 228°C; used in lubricants, detergents, and plasticizers. Also, DECYL ALCOHOL.

decantation *Engineering.* the pouring off of liquid without disturbing underlying sediment.

decanter *Engineering.* a container in which solids from a liquid sink to the bottom allowing the clear liquid on top to be withdrawn or decanted.

decapitation *Surgery.* **1.** the surgical removal of the head of a person or animal. **2.** the separation of the head of a bone from the shaft.

decapod *Invertebrate Zoology.* **1.** a member of the order Decapoda.. **2.** of or relating to this order.

Decapoda *Invertebrate Zoology.* **1.** a large order of crustaceans containing the crabs, hermit crabs, shrimp, and lobsters. **2.** the alternate name for the molluskan order Teuthoridea, containing squids.

decarbonize *Chemistry.* to remove carbon from steel. Thus, **decarbonization.**

decarboxylate *Organic Chemistry.* to remove carboxyl groups from a molecule, especially amino acids and proteins.

decarburize *Metallurgy.* to remove, or lose, carbon from a superficial layer of a steel Thus, **decarburization.**

decastere see DEKASTERE.

decatenate *Computer Technology.* to separate a data element, record, or character string into two or more parts.

decating *Textiles.* a finishing process for worsted, woolen, and other fabrics, in which they are steamed while under rollers to set the width and length and add luster and smoothness. Also, **decatizing.**

decay a decline into an inferior or weakened condition; specific uses include: *Biology.* the gradual decomposition of dead organic matter. *Physics.* the progressive diminishing of the amplitude of a quantity, either spatially or temporally. *Nuclear Physics.* see RADIOACTIVE DECAY. *Geochemistry.* see CHEMICAL WEATHERING. *Oceanography.* **1.** the weakening of a wave after it has left the fetch and passes through a calm region; the significant wave length increases and the significant wave height decreases. Also, **decay of waves. 2.** the region of lighter winds through which waves travel as swells after leaving the fetch.

decay clock *Horology.* a system that measures long periods of time (thousands of years) by determining the ratio of the amount of a radioactive element remaining in an object after radioactive decay to the amount of the element originally found in the object, based on known, steady rates of decay.

decay constant *Physics.* the positive constant associated with the variable appearing in the exponent of an exponential term, as in e^{-ct}, where c is the decay constant and t is a variable, such as time. *Nuclear Physics.* the constant of proportionality that relates the rate of disintegration of a particular radioactive species to the total number of atoms in the element under consideration; the decay constant is uniquely characteristic of the species and can identify it as precisely as qualitative chemical analysis. Also, **decay coefficient.**

decay cooling *Nucleonics.* the continued circulation of coolant for many days after reactor shutdown, in order to ensure the removal of decay heat and thus prevent damage to fuel elements and possible contamination by fission products.

decay curve *Nuclear Physics.* a graph that tracks the activity of a radioactive sample over time, showing how much radioactive material remains at any given time.

decay factor see DAMPING FACTOR.

decay family see DECAY SERIES.

decay gammas *Nuclear Physics.* the gamma rays that are emitted from radioactive isotopes during decay.

decay heat see AFTERHEAT.

decay mode *Nuclear Physics.* one of two or more alternative and competing modes by which a radioactive nuclide or particle undergoes decay, such as alpha- or beta-ray emission, spontaneous fission, or electron capture.

decay product see DAUGHTER.

decay rate *Nuclear Physics.* the continuously changing rate of disintegration of an active species, in which the number of atoms disintegrating in a unit time is proportional to the total number of remaining atoms of that species then present.

decay series *Nuclear Physics* . a sequence of nuclides that arise from and are transformed by radioactive decay until a stable isotope is produced, as in the actinium series. Also, DISINTEGRATION FAMILY, DISINTEGRATION SERIES, RADIOACTIVE CHAIN, DECAY FAMILY.

decay theory *Psychology.* the theory that certain items in long-term memory fade away with the passage of time.

decay time *Physics.* the amount of time required for the amplitude of an exponentially decaying quantity to be reduced to e^{-1} (36.8%) of its original value.

Decca *Navigation.* a British electronic navigational aid in which hyperbolic lines of position are determined by measuring the phase difference between synchronized continuous wave radio signals.

Decca chain *Navigation.* a group of radio transmitters used in the Decca system. Each group consists of one master and three slave transmitters,which ideally are equally spaced around the circumference of a circle 140 to 160 miles in diameter.

Decca chart *Navigation.* a chart overprinted with hyperbolic lines relating to Decca transmitters. Also, **Decca lattice chart.**

Deccan basalt *Geology.* a fine-grained, nonporphyritic, basaltic lava covering an extensive area of the Deccan region of southeast India. Also, **Deccan trap.**

decelerating electrode *Electronics.* an electrode to which a potential is applied to reduce the velocity of the electrons in the beam of an electron-beam vacuum tube.

deceleration *Mechanics.* **1.** a decrease in velocity with time. **2.** the rate of such a decrease.

deceleration coefficient *Mechanics.* the ratio of an object's deceleration in a fluid to the square of its velocity.

deceleration parachute *Aviation.* a parachute attached to an airplane and deployed to slow the aircraft during landing.

deceleration parameter *Astrophysics.* a number that expresses the rate of slowdown in the expansion of the universe.

deceleration time *Computer Technology.* the time it takes for a spinning magnetic tape reel to stop moving after the reading or writing of a block of data.

decelerometer *Engineering.* an instrument designed to measure the deceleration of an object or vehicle.

deceleron *Aviation.* an assembly composed of an aileron and a speed brake, essentially an aileron divided laterally in half; used on certain jet airplanes. (An acronym for decelerator and aileron.)

December solstice *Astronomy.* the moment when the sun reaches its most southerly declination on the celestial sphere; this marks the start of winter in the Northern Hemisphere and the beginning of summer in the Southern.

decentered lens *Optics.* a lens whose optical and geometrical centers do not coincide, having the effect of a lens combined with a weak prism.

decentralized data processing *Computer Technology.* the collection and maintenance of data at individual subdivisions of an organization or at various geographical locations, each of which may be connected to any or all of the other units.

decentration *Psychology.* the stage of development in which a child moves beyond the egocentric thinking that is characteristic of earlier childhood to develop an awareness of the larger world.

deception *Military Science.* a measure intended to mislead the enemy by manipulating, distorting, or falsifying evidence and information. *Electronics.* an electronic countermeasure in which false or misleading signals are transmitted to an enemy radar by radiating spurious signals synchronized to the radar or by reradiating the radar pulses from extraneous reflectors. Also, **deception means.**

deception jamming see CONFUSION JAMMING.

decerebrate *Neurology.* **1.** to eliminate an animal's higher cerebral functions by transecting the neuraxis through the midbrain. **2.** a brain-damaged person exhibiting neurologic signs similar to those of a decerebrated animal.

decerebration *Neurology.* the experimental procedure of removing cerebral function either by destroying the forebrain or by isolating lower cerebral centers.

Déchelette, Joseph [dāsh´let´] 1862–1914, French archaeologist; wrote on European prehistory.

dechlorination *Chemistry.* the process of removing chlorine from a substance.

decholesterolization *Hematology.* the process of extracting cholesterol from the blood. Also, **decholesterinization.**

deci- a metric prefix meaning "one-tenth," as in *deciliter* (0.1 liter).

deciare *Metrology.* a unit of area equal to 0.1 are.

decibar *Metrology.* a unit of pressure equal to 0.1 bar.

decibel *Metrology.* a standard unit of measurement based upon a logarithmic scale for the level of sound. *Physics.* a dimensionless quantity, used to describe a comparison of power levels, with the number of decibels equal to 10 times the logarithm of the ratio of the powers.

decibel loss *Telecommunications.* the loss, expressed in decibels, of a signal over a conductor.

decibel meter *Engineering.* an instrument for measuring the level of sound pressure, using a scale calibrated in logarithmic steps.

decibels above 1 volt *Electricity.* a power level equal to the ratio of voltage at any point in a transmission system to a reference level of 1 volt; a negative value means decibels below 1 volt.

decibels above 1 watt *Electricity.* a power level equal to the ratio of power at any point in a transmission system to a reference level of 1 watt; a negative value means decibels below 1 watt. Thus, **decibels above 1 kilowatt, decibels above 1 milliwatt,** and so on.

decibels above reference coupling *Electricity.* a measure of the coupling between two circuits, expressed in relation to a reference value of coupling that gives a specified reading on a specified noise-measuring set with a test tone of 90 dBa on one circuit.

decibels above reference noise *Electricity.* the ratio of the noise level at a selected point in a circuit to a lower reference noise level of –90 dBm.

decibels adjusted see ADJUSTED DECIBELS.

decidability *Mathematics.* the fact of being decidable; a property of problems such that there is an algorithm that can provide a definite true or false answer within a given time.

decidable *Mathematics.* **1.** referring to a class of mathematical problems for which there exists an accepted and effective algorithm for finding a desired solution. **2.** referring to a statement or formula S of an axiomatic theory with the property that exactly one of S and $\sim S$ (the negation of S) is a theorem. If S is not a decidable statement or formula, then S is said to be **undecidable** within the given axiomatic theory. *Artificial Intelligence.* of a theorem-proving or decision-making task, able to be decided one way or the other (e.g., proved or disproved) within a finite time that can be predetermined from the size of the input problem.

deciduation *Medicine.* the shedding of endometrial tissue during menstruation.

deciduoma *Oncology.* an intrauterine accumulation of decidual cells.

deciduous [di sid´joo əs] *Biology.* falling off or being shed during a certain season or at a particular stage of growth, such as leaves or antlers.

deciduous placenta *Developmental Biology.* a placenta in which the decidua or maternal parts of the placenta separate from the uterus and are cast off along with the fetal parts.

deciduous teeth *Anatomy.* the 20 temporary teeth that are the first to erupt from the gum and are replaced by permanent teeth. Also, BABY TEETH, MILK TEETH.

decigram *Metrology.* a unit of mass that is equal to 0.1 gram. Also, **decigramme.**

decile *Statistics.* any of the nine points that divide a distribution into ten equal parts; the *i*th decile is a value such that *i* times 10% of the distribution is on its right; the 5th decile is the median.

deciliter *Metrology.* a unit of capacity equal to 0.1 liter. Also, **decilitre.**

decimal *Mathematics.* referring to or based in the number 10; relating to tens or tenths.

decimal attenuator *Electronics.* an attenuator used to decrease voltage or current in discrete decimal steps.

decimal balance *Engineering.* a balance in which one arm is ten times the length of the other, thus allowing large objects to be weighed using small weights.

decimal-binary switch *Electricity.* a switch connecting a single input lead to a combination of four output leads (representing 1, 2, 4, and 8) for each of the decimal-numbered settings of the control knob; thus for position 5, output leads 1 and 4 would be connected to the input.

decimal code *Computer Programming.* **1.** a system for representing each decimal digit in a binary code, usually with four or more bits for each decimal digit. **2.** a coding scheme that makes use of ten distinct states, such as the decimal numbering system with states 0 through 9.

decimal-coded digit *Computer Programming.* a digit or other character represented by a combination or arrangement of decimal digits.

decimal number system *Mathematics.* the representation of real numbers in base ten, using the digits 0, 1, 2, 3, 4, 5, 6, 7, 8, 9 and radix notation.

decimal processor *Computer Technology.* an early high-performance computer whose stored computer words, both in main core memory units and on magnetic drums, were made up of 11 or 12 decimal digits in binary-coded decimal form.

decimal-to-binary conversion *Computer Programming.* **1.** the transformation of decimal numbers into their binary representations. **2.** the mathematical process of producing the binary (base 2) equivalent of a number in the decimal (base 10) numbering system.

decimeter *Metrology.* a unit of length that is equal to 0.1 meter. Also, **decimetre.**

decimetric wave *Electromagnetism.* an electromagnetic wave having a wavelength in the range of 0.1 to 1 meter, corresponding to a frequency range of 300 to 3000 megahertz.

decineper *Physics.* a unit of voltage and current attenuation in lines and amplifiers, equal to 0.1 neper.

decinormal *Chemistry.* a solution that contains one-tenth of the gram-equivalent weight of solute per one liter of solution.

decision *Computer Programming.* the process by which a computer selects one of the available alternative courses of action based on prescribed responses to certain conditions.

decision box *Computer Programming.* a diamond-shaped symbol in a flowchart that indicates a logical choice or a branch in the process flow.

decision element *Electronics.* a circuit that performs logic operations on digital input data and expresses the result in its output; it is usually a component of a larger logic system or subsystem.

decision height *Navigation.* a height or altitude at which, during an instrument approach, the pilot of an aircraft must decide to either continue the approach or execute a missed approach.

decision mechanism *Computer Technology.* the recognition logic system of an optical character reader that accepts or rejects the identification of a character.

decision model *Design Engineering.* a particular set of functions and conditions whereby activities are conducted through a robotic controller by the use of logical determinations, e.g., true and false tests, yes and no choices, zeros and ones, and on/off conditions.

decision rule *Statistics.* any sample-based criterion to choose among alternatives; for example, in hypothesis testing, a statement about the sample statistic that determines whether the null hypothesis is to be accepted or rejected.

decision-support system *Computer Programming.* a computer-based system that helps decision makers confront poorly structured problems through direct interaction with data and models.

decision table *Computer Programming.* a tabular representation of various actions and their associations with combinations of conditions; used as a visual aid in management decision-making or by a program.

decision theory *Statistics.* the study of the principles and methodologies for making decisions with uncertain consequences.

decision tree *Statistics.* a graphic display of forks and branches in the shape of an inverted tree, used to analyze sequential decision problems; each fork may represent a decision or event whose outcome is unknown. *Industrial Engineering.* such a diagram used for management decisions that occur in sequence; the diagram shows each possible outcome of a first decision, and its cost and payoff, along with each decision following from that outcome, from which possible outcomes and decisions branch in turn. *Artificial Intelligence.* see DISCRIMINATION NETWORK.

deck *Naval Architecture.* any of the horizontal or cambered platforms on a vessel that correspond to the floors in a house; some decks may not extend the length of the vessel, but all decks extend across the full width. Used to form various compound terms, such as **deck beam, deck chair, deck fitting, deck machinery,** and so on.

deck erection *Naval Architecture.* any superstructure rising above a ship's upper deck.

deckhead *Naval Architecture.* the underside of a deck, corresponding to the ceiling of a house.

deck height *Naval Architecture.* the vertical separation between successive decks.

deckhouse *Naval Architecture.* any houselike structure built on a ship's main or upper deck.

deckle *Engineering.* a detachable wooden frame affixed to the edges of a papermaking mold.

deckle edge *Graphic Arts.* a rough edge on a sheet of paper, such as the feathery edges on some book stock.

deckle rod *Engineering.* a small rod that, when inserted into the ends of the extrusion coating die, can adjust the die opening.

deckle strap *Engineering.* a rubber band on a papermaking machine that determines the web width.

deck light *Naval Architecture.* a glass window fitted into a ship's deck in order to permit light below.

deck line *Naval Architecture.* the vertical level at which the side of a deck meets the side of the hull.

deck log *Navigation.* a written record of the navigational movements of a vessel, along with important events.

deck roof *Building Engineering.* a type of roof design that is virtually flat but does not have a parapet wall, as opposed to a flat roof.

deck stringer *Naval Architecture.* the fore and aft stringer plating of a deck lying adjacent to the side of the hull. Also, **deck stringer plate.**

deckzelle *Invertebrate Zoology.* any of the columnar or cuboidal supporting or epithelial cells found in certain hydroids.

declaration *Computer Programming.* a term used for procedure-oriented language elements that do not call for any processing actions, but rather describe such things as data formats, record layouts, and the grouping of data items into arrays.

declarative instruction *Computer Programming.* a nonexecutable statement, usually in an assembly-language program, that defines and describes variables, arrays, files, and other items used in the program.

declarative representation *Artificial Intelligence.* a form of knowledge representation in terms of formulas, data structures, or rules of a restricted form, as opposed to a procedural representation.

declarative statement *Computer Programming.* a high-level language statement that describes variables, arrays, files, or other items used in the program, as opposed to an imperative statement that specifies an operation.

declinate *Biology.* being bent or curved downward or to one side.

declination *Geophysics.* the horizontal angle, varying with geography, between true north and magnetic north.

declination arc *Cartography.* a graduated arc, attached to the alidade of a surveyor's compass or transit, on which the magnetic declination is marked.

declination axis *Astronomy.* **1.** a celestial object's angular distance north or south of the celestial equator. **2.** in an equatorial telescope mounting, the axis that lies at right angles to the polar axis and which affords movement in declination.

declination circle *Astronomy.* **1.** in equatorial coordinates, the circles parallel to the celestial equator that mark off angular distance north or south from it. **2.** the calibrated dial on the declination axis of an equatorial mounting that indicates the declination toward which the telescope is pointing.

declination compass see DECLINOMETER.

declination variometer *Engineering.* an instrument that measures variation in declination of the earth's magnetic field.

declinatoire *Cartography.* a combined magnetic compass and straightedge, suitable for use on a plane table to mark the magnetic meridian.

declinator *Surgery.* an instrument used to retract tissue or organs during an operative procedure.

declining population *Ecology.* a term for a population in which older individuals significantly outnumber younger ones.

declinometer *Engineering.* an instrument similar to a surveyor's compass, used in determining magnetic declination, having a line of sight that can be rotated to conform to a compass needle or to any desired setting on the horizontal circle. Also, DECLINATION COMPASS.

declivity *Geology.* **1.** a steep or overhanging slope descending downward from a point of reference. **2.** any inclination or downward deviation from the horizontal.

decoct. a decoction, or extract. (From Latin *decoctum.*) Also, **Decoct.**

decode *Telecommunications.* **1.** to interpret a code. **2.** to convert encoded language into its plain-language equivalent by means of a code. *Computer Programming.* to convert data by reversing the effect of some previous encoding.

decoder *Telecommunications.* **1.** a device that decodes data or signals. **2.** a device that has a number of input lines, any of which may carry signals, and a number of output lines, of which not more than one may carry a signal at any given moment. *Electronics.* a circuit that responds to a particular coded input signal while rejecting others. Thus, **decoder circuit.** *Computer Technology.* **1.** a matrix of logic switches that selects one of 2^n outputs based on n input values. Also, LOGIC DECODER. **2.** see DECODING GATE.

decoding gate *Computer Programming.* a combinational logical circuit that converts binary information from one coded form to another; often used to select a device from its binary address. Also, DECODER, RECOGNITION GATE.

decollator *Computer Technology.* a device used to separate multiple copies of continuous computer output forms.

décollement [dā käl′mənt] *Geology.* a detachment of sedimentary strata produced by and associated with folding and overthrusting, whereby a series of beds may shear off, slide over, and fold independently of the underlying strata. Also, DETACHMENT.

décollement fold *Geology.* a fold that has developed independently of the surrounding rocks as a result of décollement.

decolorizing carbon *Organic Chemistry.* activated carbon; a highly adsorbent carbon, usually of vegetable origin, that is used for decolorizing liquids.

decommissioning *Nucleonics.* the shutdown and dismantling of a nuclear reactor at the end of its useful life, in accordance with carefully controlled procedures for recovering usable fuel and components, decontaminating the site and equipment, and implementing long-term security measures to prevent unauthorized access and exposure.

decommutation *Electronics.* the process of extracting one or more signals from a composite signal created by a commutation process.

decommutator *Electronics.* the equipment used in extracting commutated signals. Also, DEMODULATOR.

decomposable measure *Mathematics.* a measure μ on a measure space (X,B,μ) is said to be decomposable if there exists a family $\{S_\alpha\}$ of disjoint subsets of X, each of finite measure, with the property that if E is any subset of X with $\mu(E \cap S_\alpha) = 0$ for all α, then $\mu(E) = 0$. The family $\{S_\alpha\}$ is called a **decomposition** for μ.

decomposable tensor *Mathematics.* a tensor that can be expressed as the tensor product of other tensors (a general tensor is expressed as a sum of decomposable tensors). In component form, this property is easily recognized; e.g., $t^i_{jkl} = a^i_j b_k c_l$ is decomposable.

decomposer *Ecology.* any organism that feeds by breaking down organic matter from dead organisms, such as bacteria and fungi.

decomposition *Chemistry.* a process in which one or more substances break down into simpler molecular substances, as from the effects of heat, light, chemical or biological activity, and so on. *Geochemistry.* see CHEMICAL WEATHERING. *Computer Programming.* the process used to break a complex operation into smaller and simpler suboperations to aid in writing correct code. Also, TOP-DOWN DESIGN. *Robotics.* the breaking down of complex tasks into a hierarchy of simpler tasks that fulfill the objectives of the process controller.

decomposition potential *Physical Chemistry.* the minimum amount of electromotive force that is required to allow an electrolytic cell to be recharged. Also, **decomposition voltage.**

decompound *Botany.* **1.** divided into compound divisions. **2.** composed of compounds, with each segment also being compound; e.g., a bipinnate leaf.

decompression *Physics.* the act of reducing pressure or the effects of pressure.

decompression chamber *Engineering.* a chamber in which high atmospheric pressure conditions can be simulated; used to treat a scuba diver who has ascended too quickly.

decompression table *Engineering.* numerical tables used by divers to determine decompression time for a dive of a given depth.

deconcentrator *Engineering.* a device that eliminates or diminishes dissolved or suspended material in feed water.

deconditioning see COUNTERCONDITIONING.

decontamination. *Toxicology.* the removal or inactivation of a poison or harmful agent from a living or nonliving object. *Ordnance.* the process of making a person, object, or area safe by destroying, neutralizing, or removing chemical agents, biological agents, or radioactive material. Thus, **decontamination station.** *Nucleonics.* the reduction or removal of contaminating radioactive material from the surfaces of structures, objects, or persons by one or more of the following operations: treating the surface in order to remove or decrease the contamination, letting the material stand so that the radioactivity is decreased as a result of natural decay, or covering the contamination in order to immobilize and attenuate the radiation emitted.

decontamination factor *Nucleonics.* the ratio of the initial amount of contamination, expressed in terms of concentration or activity of radioactive material, to the final count resulting from a decontamination process. Also, **decontamination index.**

decoppering agent *Ordnance.* material designed to remove coppering from the inside of a gun bore; it is either manufactured into the propelling charge or loaded into the chamber with the charge.

Decorated style *Architecture.* the middle phase (from about 1250 to 1350) of English Gothic architecture, characterized by elaborate tracery and decoration, lierne vaults, and ogee arches.

decorative laminate *Materials.* fabric or paper impregnated with resin, usually phenolic and melamine; used to decorate walls, furniture, and doors. Also, **decorative thermosetting laminate.**

decorticate *Biology.* not having an outer layer or cortex; stripped or peeled.

decortication *Surgery.* the removal of the cortical substance of a structure or organ such as the lung, kidney, or brain.

decouple *Science.* to become separated or disconnected. *Engineering.* **1.** to reduce the seismic effect of nuclear or other types of explosions by detonating them in the center of an underground cavity. **2.** to separate or isolate double walls in order to decrease or eliminate noise that tends to come through a wall or partition.

decoupling *Electricity.* the process of eliminating or effectively minimizing transfer or feedback of energy from one circuit to another. *Computer Technology.* the process of separating a system into its component parts, in order to study how the individual parts function and interface with one another. *Robotics.* a coordinate system based on the positions of a robot's joints in space.

decoupling era *Astrophysics.* the time, about 30,000 years after the Big Bang, when the temperature of the universe was about 3000 K and matter and radiation decoupled, whereupon the universe became transparent.

decoupling filter *Electronics.* a filter connected at a common point between two or more circuits to reject frequencies that cause interference through interstage coupling.

decoupling network *Electricity.* a network used to prevent or effectively minimize the interaction of two electric circuits.

decoy *Ordnance.* an imitation of a person, object, or phenomenon designed to deceive enemy surveillance devices or mislead enemy evaluation; e.g., aerial reflectors designed to deceive enemy radar. Thus, **decoy guided missile, decoy airfield, decoy tank.**

decoy ship *Ordnance.* a warship camouflaged as a merchant ship, with its armament and other fighting equipment hidden in such a manner that it can be quickly unveiled and put into action. Also, Q SHIP.

decoy target *Ordnance.* a decoy designed to simulate various types of field equipment; it is made of prefabricated parts, thus allowing it to be quickly assembled under combat conditions.

decoy transponder *Electronics.* a transponder used in radar deception.

decreasing function *Mathematics.* any real-valued function f of a real variable x such that the value of $f(x)$ will decrease as x increases; that is, $f(a) < f(b)$ if $a > b$. Also, MONOTONE DECREASING FUNCTION.

decrement *Mathematics.* **1.** the amount by which a variable is decreased. **2.** to decrease the value of a variable. *Physics.* the ratio of the amplitude of an oscillation to that of its succeeding oscillation in an underdamped vibrating system. *Hydrology.* see GROUNDWATER DISCHARGE.

decrement factor *Electrical Engineering.* the ratio of the amplitude of one damped oscillation in a series to the following one.

decrement field *Computer Programming.* the part of an instruction that contains the value that is to be subtracted from the contents of a specified register.

decremeter *Engineering.* an instrument that measures the damping of an electromagnetic wave train.

decrepitation *Geophysics.* a process of breaking up mineral rock by heating, usually accompanied by a cracking noise.

decrypt *Telecommunications.* to electronically or mechanically convert encrypted text into its original form.

decryption *Telecommunications.* the process in which a coded message is deciphered and returned to its original text; the reverse of encryption.

decub. lie down; lying down. (From Latin *decubitus.*) Also, **Decub.**

decubitus position [di kyoob´ə tis] *Surgery.* the position of an individual lying on a horizontal surface; designated according to the portion of the body resting on the surface, as follows: **dorsal decubitus,** lying on the back; **left lateral decubitus,** on the left side; **right lateral decubitus,** on the right side; **ventral decubitus,** on the abdomen.

decumbent *Botany.* lying flat or running along the ground, usually with the tip being erect and growing upward.

decurrent *Botany.* **1.** extending below the point of insertion. **2.** running downward, especially a leaf or leaf base extending to form a wing or ridge along the stem.

decussate *Botany.* relating to the arrangement of leaves or branches in alternating pairs at right angles to those above and below.

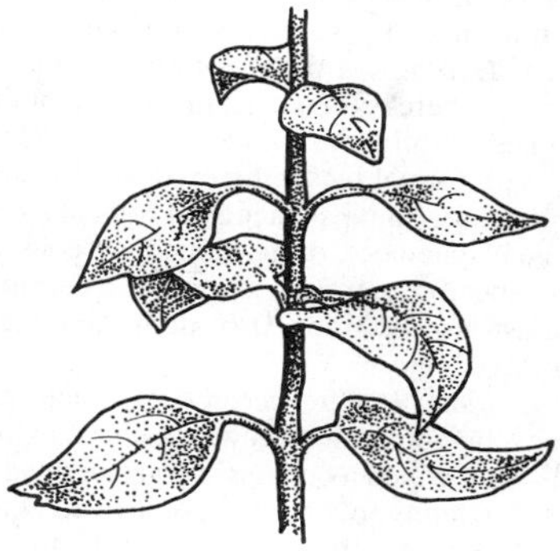

decussate

decussate structure *Mineralogy.* in thermally metamorphosed rock, a crisscross microstructure of certain minerals, especially those having a flaky or columnar habit.

decyl [des´əl] *Organic Chemistry.* a group of univalent radicals consisting of a straight chain, acyclic hydrocarbon containing ten carbon atoms, $CH_3(CH_2)_8CH_2$–.

decyl acetate *Organic Chemistry.* $CH_3(CH_2)_9OCOCH_3$, a combustible liquid with a floral odor, insoluble in water and soluble in alcohol; boils at 187–190°C; used in perfumes.

***n*-decyl alcohol** see 1-DECANOL.

decyl aldehyde see *n*-DECANAL.

decylic acid see CAPRIC ACID.

Dedekind, Richard [dā´də kint] 1831–1916, German mathematician; formulated theory of the real number system based on Dedekind cuts.

Dedekind cut *Mathematics.* a division of the elements of the number line or, more generally, an ordered field F into two nonempty disjoint subsets, A and B, such that if $x \in A$ and $y \in B$ then $x < y$. If F is the field of rational numbers, then a unique real number may be associated with each Dedekind cut; used to construct the real numbers as an extension field of the rational numbers.

Dedekind domain *Mathematics.* an integral domain in which every proper ideal is the product of a finite number of prime ideals. Every principal ideal domain is a Dedekind domain, and every Dedekind domain is Noetherian.

Dedekind sum *Mathematics.* let h and k be relatively prime integers with k positive. Then the Dedekind sum for h and k is given by $s(h, k) = \sum_{\mu=1}^{k} ((h\mu/k))\,((\mu/k))$, where the symbol $((x))$ denotes 0 if x is an integer and $x - [x] - 1/2$ if x is not an integer, and where $[x]$ is the greatest integer function of x. Using the theory of elliptic functions, it can be shown that

$$s(k, h) + s(h, k) = -1/4 + 1/12(h/k + 1/hk + k/h).$$

This is the reciprocity formula for Dedekind sums.

Dedekind test see DIRICHLET TEST FOR CONVERGENCE.

dedendum *Design Engineering.* in a gear, the difference between the radius of the pitch circle and the radius of its root circle.

dedendum circle *Design Engineering.* the circle describing the bottoms of the spaces between gear teeth.

dedicated *Engineering.* of a device, part, or system, having a single application or purpose, or functioning with a single machine. *Computer Technology.* specifically, describing a system or program that has a single specific purpose. Thus, **dedicated computer.**

dedicated function key *Computer Technology.* a key on a keyboard used exclusively for a single application or purpose, such as a cursor control key.

dedicated line *Computer Technology.* a channel or communication link that transmits data for only one subscriber or for one specific use, and that is always available to the user. Also, LEASED LINE, PRIVATE LINE.

dedicated terminal *Mechanical Devices.* a computer terminal connected to only a single data-processing system that provides a direct communication link.

dedicated word processor *Computer Technology.* a computer system that performs only the single function of word processing.

dedifferentiation *Biology.* the loss of specialized form, function, or structures and reversion to a more primitive or generalized state.

de d. in d. from day to day. (From Latin *de die in diem.*)

dedolation *Surgery.* **1.** a sensation as if the limbs had been bruised. **2.** the removal of a thin slice of skin by a slanting cut.

dedolomitization *Geology.* a process during contact metamorphism under low pressure, whereby magnesium in a dolomite is used in the formation of magnesium minerals, producing an enrichment in calcite.

deduction *Science.* a formal reasoning process in which the conclusion is reached as a result of a finite sequence of logical steps, each of which is an axiom, a given statement, or an immediate consequence of earlier statements. If the specific premises given are true, then it follows that the conclusion must also be true. Also, **deductive method, deductive reasoning, deductive logic.**

deduction tree *Artificial Intelligence.* a data structure, in the form of a tree or directed acyclic graph, that shows the facts, axioms, and intermediate conclusions from which a conclusion was derived.

deductive *Science.* relating to, based on, or arrived at by deduction.

de Duvé, Christian born 1917, Belgian biochemist; Nobel Prize for study of cell components; discovered lysosomes and peroxisomes.

dee *Nucleonics.* one of two flat semicircular electrodes that are placed in the perpendicular magnetic field of a cyclotron and subjected to alternating radio-frequency voltage to accelerate charged particles. (So called because they have the shape of the letter D.)

dee line *Nucleonics.* the gap between dees of a cyclotron across which a particle is repetitively accelerated as it crosses from dee to dee.

deemphasis *Acoustical Engineering.* the introduction of a complementary frequency response to restore the frequency spectrum to its original form, by offsetting higher frequencies that were introduced in preemphasis. Thus, **deemphasis network.**

deenergize *Electricity.* to stop the flow of current in a circuit or to remove voltage from a circuit, as by opening a switch or disconnecting from a power source.

deep *Oceanography.* an extensive portion of the ocean bottom having great depth, generally below 6 kilometers in depth.

deep case *Linguistics.* the semantic case relation between a verb and a related phrase, e.g., in "mother baked" and "the pie baked" both "mother" and "pie" are in the subject surface case, but "mother" is in agent deep case, while "pie" is in patient deep case.

deep dragonfish see DRAGONFISH.

deep-draw *Metallurgy.* to force a sheet metal deep into a die, in order to form a cup or similar shapes.

deep easterlies see EQUATORIAL EASTERLIES.

deep-etch *Graphic Arts.* to make a lithographic printing plate, especially a deep-etch plate.

deep-etch plate *Graphic Arts.* a type of printing plate made from a line or halftone film positive, rather than a negative, which is placed in contact with the plate so that image areas are etched below the plate's surface; used for fine color-process printing and for long press runs.

deep fascia *Anatomy.* a dense, firm fibrous membrane that invests the trunk and limbs, and gives off sheaths to the various muscles.

deep-focus earthquake *Geology.* an earthquake having its focus at a depth greater than 300 kilometers below the surface.

deep fording *Ordnance.* the ability of a self-propelled gun or ground vehicle to cross a water obstacle with its wheels or tracks in contact with solid ground. Also, **deep-fording capability.**

deep freezing see FLASH FREEZING.

deep hibernation *Physiology.* the behavior of certain animals that sleep through the winter, during which time the bodily functions of the animal slow down and activity stops.

deep inelastic collision *Nuclear Physics.* a type of nuclear reaction that takes place when the collision of two heavy ions lies between grazing and a direct hit, resulting in the dissipation of kinetic energy in reaction fragments. Also, **deep inelastic transfer.**

deep inguinal ring *Anatomy.* the inguinal canal opening in the abdomen.

deep knowledge *Artificial Intelligence.* an underlying, fundamental knowledge that is not as easily modeled as surface knowledge.

deep minefield *Ordnance.* an antisubmarine minefield that can be safely crossed by surface ships.

deep model *Artificial Intelligence.* a knowledge model that contains all the information elements and logic used by experts in their reasoning.

deep pain *Physiology.* a sensation of pain in the deep muscles, joints, and cutaneous tissues, described as dull, aching, and poorly localized.

deep palmar arch *Anatomy.* the deep arterial arch formed by the radial and deep branch of the ulnar artery, and located along the flexor tendons of the hand. Also, ARCUS PALMARIS PROFUNDUS, DEEP VOLAR ARCH.

deep-penetration bomb *Ordnance.* a bomb designed to penetrate its target before exploding. Also, EARTH-PENETRATING WEAPON.

deep roentgen-ray therapy *Radiology.* the application of high-energy X-radiation produced by at least 150 kilovolts, capable of penetrating deep structures within the body.

deep-scattering layer *Oceanography.* a depth range in an ocean, varying on a diurnal cycle and according to geographic area from about 200 to 1300 feet, in which small marine organisms are present in dense concentrations that cause considerable scattering of sound waves.

deep-sea basin *Geology.* a more or less equidimensional depression in the ocean floor that varies in extent.

deep-sea channel *Geology.* a trough-shaped, low-relief valley located on the deep-ocean floor beyond the continental rise.

Deep Sea Drilling Project *Oceanography.* a research project from 1967 to 1983 that used a specially built vessel, the Glomar Challenger, to retrieve drill samples of seafloor sediments to extended depths.

deep-sea plain *Geology.* a broad, almost flat area that forms most of the ocean floor.

deep-sea trench *Geology.* see TRENCH, def. 3.

deep sleep *Physiology.* the third and fourth stages of the non-REM sleep cycle, during which the sleeper is difficult to awaken and brain waves are slow and of high amplitude.

deep space *Astronomy.* **1.** originally, any area of space beyond the moon's orbit. **2.** in current use, the area outside our own solar system.

deep-space *Space Technology.* relating to or located in deep space. Thus, **deep-space probe.**

Deep Space Network *Space Technology.* a comprehensive, worldwide system for tracking and communicating with automated scientific spacecraft in deep space.

deep structure *Linguistics.* **1.** in Noam Chomsky's theory of transformational grammar, a representation of the meaning of a sentence that can be transformed into a surface structure or actual sentence; e.g., "John hit the ball" and "The ball was hit by John" have different surface structures but the same deep structure. **2.** more generally, a representation of the meaning of a sentence that contains less syntactic detail and more semantic detail than the surface representation.

deep-submergence rescue vehicle *Naval Architecture.* a vessel designed to rescue crew members from crippled submarines lying at great depth. Also, DSRV.

deep-submergence rescue vehicle

deep supporting fire *Ordnance.* fire directed at targets not in the immediate vicinity of the supported force; it is used to neutralize, destroy, or interfere with enemy reserves, weapons, command, supply, communications, and observations.

deep tank *Naval Architecture.* a tank located near midship at the lowest level of the hull; it may be used for ballast or to hold cargo.

deep trades see EQUATORIAL EASTERLIES.

deep underwater nuclear counter *Oceanography.* a submersible gamma-ray spectrometer used by ships to record ocean gamma radiation at a given locality.

deep vein thrombosis *Cardiology.* formation of a blood clot in a major vein, which occurs in the limbs or trunk of the body, leading to blockage of venous blood flow in the region.

deep volar arch see DEEP PALMAR ARCH.

deep-waisted *Naval Architecture.* of or related to a ship in which the forecastle and quarterdeck are substantially higher than the upper deck, thus creating a higher profile at the bow and stern than amidships.

deep water *Oceanography.* **1.** water deep enough that surface waves are not affected by conditions on the bottom. **2.** water masses that originated in polar regions but that are now found in the deepest regions of the ocean, generally below 3 kilometers in depth.

deep waterline *Naval Architecture.* the deepest level to which a ship may safely be loaded.

deep-well pump *Mechanical Engineering.* **1.** any pump that delivers water and other liquid from a well, shaft, or borehole. **2.** specifically, one of a series of centrifugal pump impellers mounted on a single rotating shaft operated by an electric motor at the surface.

white-tailed deer

deer *Vertebrate Zoology.* **1.** any of numerous ruminant mammals of the family Cervidae, most males of which have solid, deciduous horns or antlers and two large and two small hoofs on each foot. **2.** specifically, any of the small- or medium-sized species of this family, as distinguished from certain larger forms such as the moose or elk.

deerhorn antenna *Electromagnetism.* a dipole antenna whose ends are bent back to reduce air resistance on aircraft.

Deerparkian *Geology.* a North American geologic stage of the Lower Devonian, after the Helderbergian and before the Onesquethawan.

de-escalate *Ordnance.* **1.** to reduce the level of confrontation or conflict. **2.** specifically, to reduce the possibility of nuclear war. Thus, **de-escalation.**

de facto refuse [dā fak´tō] *Archaeology.* artifacts that are left behind when a settlement or activity area is abandoned.

de facto standard *Computer Technology.* a procedure or definition that, by virtue of common usage and acceptance, has become a standard in the industry.

default *Computer Technology.* a value, option, course of action, or attribute that is assumed by the program or system in the absence of other specification or direction by the user.

default drive *Computer Programming.* the disk drive assigned by the computer operating system when the user does not specify otherwise.

default reasoning *Artificial Intelligence.* reasoning based on things that are usually, but not necessarily, true; e.g., the premise that a given individual can drive a car because most people can drive.

default value *Artificial Intelligence.* in a frame system, a slot value that is stored in a class and inherited by instances of the class for which no value for that slot is defined.

defaunation *Ecology.* the driving off or depletion of animal life in an area. Thus, **defaunated.**

DEFCON *Ordnance.* a system for establishing the state of alert of an air defense system. The normal peacetime state is DEFCON 5. A status of DEFCON 0 would indicate that a nuclear attack is in progress. (An acronym for defense condition.)

defect *Science.* an imperfection that impairs the structure, composition, or function of an object. *Crystallography.* specifically, a lattice imperfection in a crystal that may be due to the introduction of a minute proportion of a different element into a perfect lattice.

defect chemistry *Solid-State Physics.* the study of crystal defects produced under conditions of temperature fluctuations or exposure to other conditions, such as electromagnetic radiation.

defect cluster *Crystallography.* a deviation from the internal symmetry of a crystal resulting in regrouping, due to impurities or to ions being missing or out of position.

defect conduction *Solid-State Physics.* the conduction of holes in the valence band of a semiconductor.

defective *Science.* having one or more defects; faulty. *Virology.* not able to replicate.

defective interfering particles *Virology.* virus particles that arise as a result of a mutation in a nondefective (standard) virus and interfere with the latter's replication. These particles can reproduce only in the presence of the standard virus. Also, **DI particles.**

defective number see DEFICIENT NUMBER.

defective phage *Virology.* a bacteriophage virion that cannot replicate in a host cell.

defective virus *Virology.* any virus that cannot replicate on its own.

defect lattices *Materials Science.* the defects, such as edge dislocations or vacancies, at the atomic level in the crystal lattice of a material that reduce the bonding forces in the lattice and the ability of the material to withstand shear stress.

defect motion *Crystallography.* the migration of a point defect from lattice point to lattice point within a crystal.

defeminization *Physiology.* the loss of female sexual characteristics.

defense *Military Science.* of or relating to the protection of a nation, its forces, or its allies. Thus, **defense area, defense emergency, defense readiness (conditions), defense information,** and so on. *Psychology.* see DEFENSE MECHANISM.

defense mechanism *Psychology.* any of various mental devices or attributes by which the individual unconsciously protects against guilt, anxiety, or unwanted memories. Also, DEFENSE.

defense-suppression missile *Ordnance.* an air-launched missile designed to patrol an area under its own guidance system in order to clear a path for tactical aircraft by seeking, attacking, and disabling a hostile air-defense radar system.

defense-suppression missile

defensive *Military Science.* of or relating to the protection of a nation, its forces, or its allies. Thus, **defensive coastal area, defensive fire, defensive mine countermeasures, defensive minefield, defensive sea area, defensive zone,** and so on.

defensive behavior *Behavior.* any activity by which an animal attempts to avoid falling prey to a predator.

defensive identification *Psychology.* the process of a child's identifying with the parent of the same sex as means of alleviating the guilt produced by the Oedipal conflict.

defensiveness *Psychology.* an overdeveloped tendency to defend oneself against any criticism and to justify one's actions to others.
deferent *Astronomy.* according to the Ptolemaic theory of the earth-centered universe, the large orbital circle on earth that a planet follows,
deferred entry *Computer Programming.* the passing of processing control to a subroutine in response to an asynchronous event.
deferred junction *Hydrology.* the point at which a deferred stream joins the main stream. Also, **deferred tributary junction.**
deferred processing *Computer Programming.* the postponement of noncritical batch processing until a time of low computer usage.
deferred stream *Hydrology.* a tributary stream that flows parallel to the main stream for a significant distance before joining it downstream. Also, **deferred tributary.**
defervescence see LYSIS.
defibrillation [dē fib´rə lā´shən] *Medicine.* the arrest of fibrillation (irregular, inefficient contraction) of the atrial or ventricular caridac muscles, with the restoration of normal rhythm; this is usually accomplished by electric shock.
defibrillator [dē fib´rə lā´tər] *Medicine.* a device to apply electric current for the purpose of restoring the normal rhythm of the heart.
defibrinated blood *Hematology.* whole blood from which fibrin was separated during the clotting process.
deficiency *Mathematics.* let $G = (\{V_1, V_2\}, E)$ be a bipartite graph with bipartition V_1, V_2, and edge set E. The deficiency of a subset U of V_1 is the integer $\text{def}(U) = |U| - |N(U)|$, where $N(U)$ is the set of all vertices of G adjacent to a vertex of U, and $|\ |$ denotes cardinality. The V_1 deficiency of G is the integer $\text{def}(G) = \max\{\text{def}(U) : U \text{ is a subset of } V_1\}$. The matching number of bipartite graph G is $|V_1| - \text{def}(G)$.
deficiency anemia see NUTRITIONAL ANEMIA.
deficiency disease *Medicine.* a diseae resulting from a lack of specific nutrients, such as proteins, essential amino acids, fatty acids, vitamins, or trace minerals; for example, scurvy; beriberi, rickets.
deficiency indices *Mathematics.* a pair of positive integers that express the degree to which a closed symmetric linear operator T on a Hilbert space X fails to be self-adjoint; namely, the dimensions of the two spaces $D^+(T) = \{x \in X : T^*x = +ix\}$ and $D^-(T) = \{x \in X : T^*x = -ix\}$, where T^* is the adjoint of T. The domain $D(T^*)$ of T^* contains the domain $D(T)$ of T, and it can be shown that $D(T^*) = D(T) \oplus D^+(T) \oplus D^-(T)$. If an operator has deficiency indices (0, 0), it is self-adjoint. Also, **defect indices.**
deficiency needs *Psychology.* Abraham Maslow's term for basic biological needs, such as food or air.
deficient number *Mathematics.* a positive integer that is greater than the sum of its factors, including 1 but excluding itself; for example, the factors of 10 (1, 2, and 5) total 8. A positive integer that is neither abundant nor deficient is said to be perfect. Also, DEFECTIVE NUMBER.
defilade [def´ə läd´] *Ordnance.* **1.** protection from enemy observation and fire provided by an obstacle such as a hill or ridge. **2.** a vertical distance that conceals a position from enemy observation. **3.** to shield from enemy observation or fire by using natural or artificial obstacles.
defile *Geology.* a long, narrow, steeply walled pass or passage, usually leading to a larger pass.
defined *Computer Science.* of a variable, having received a value prior to a given point in a program.
defined medium *Biotechnology.* a nutritional medium for a cell culture that is of a known composition and can be duplicated whenever required. Also, SYNTHETIC MEDIUM.
definite composition law *Chemistry.* a law stating that a given chemical compound is always composed of the same elements in the same proportions by weight. Also, **definite proportions law.**
definite integral *Mathematics.* **1.** an integral $\int_a^b f(x)\,dx$ in which limits of integration (a and b) are given. **2.** more generally, an integral over a fixed domain of integration.
definite reference *Artificial Intelligence.* in natural language, a phrase that refers to a particular object, e.g., a noun phrase using a determiner such as *the, one,* or *that.*
definition the process of establishing some value or condition with precision; specific uses include: *Graphic Arts.* the clarity of a printed or photographed piece. *Electronics.* the fidelity with which the fine detail of an image or sound is reproduced by electronic equipment. Also, RESOLUTION. *Optics.* the perceived degree of clarity or detail in an image that is reproduced by a lens. *Telecommunications.* in a facsimile system, the distinctness or clarity of detail or outline in a record sheet, object copy, or other reproduction.
definitive *Biology.* fully developed or matured.
definitive host *Biology.* a host in which a fully developed parasite is able to sexually reproduce.
deflagration *Chemistry.* a rapid combustion process that gives off heat and light.
deflagration spoon *Chemistry.* a spoon with a long handle, used in deflagration demonstrations.
deflashing *Engineering.* the finishing procedure by which the excess plastic or metal (flash) is removed from a molding.
deflation *Geology.* the process of erosion by the sweeping away of loose, dry, fine-grained particles as a result of wind action.
deflation armor see DESERT ARMOR.
deflation basin *Geology.* a topographic depression formed and maintained by wind erosion. Also, WIND-SCOURED BASIN. *Hydrology.* an erosion basin excavated by ice in front of a barrier that obstructs the path of a glacier.
deflation lake *Hydrology.* a shallow lake, especially in an arid or semi-arid region, which lies in a basin that was formed mainly by wind erosion.
deflecting torque *Mechanics.* the torque that drives an instrument, such as a meter needle, where deflection and torque are proportional.
deflecting wedge see CORRECTING WEDGE.
deflection *Science.* a movement away from a straight line or path, or from the intended course. *Electronics.* the controlled movement of the electron beam in a cathode-ray tube by varying the electromagnetic or electrostatic fields, in order to create a trace on the surface of the screen. *Engineering.* the amount of bending or twisting of a loaded structural member. *Petroleum Engineering.* an alteration in the angle of a well bore in oil well drilling. *Ordnance.* see DEFLECTION ANGLE.
deflection angle *Ordnance.* **1.** the angle between the directed course of a projectile and the actual course. **2.** the angle between the line of sight to the target and the line of sight to the aiming point. *Cartography.* **1.** a horizontal angle measured from the prolongation of the preceding line; angles to the right are positive, those to the left are negative. **2.** a vertical angle, measured in the vertical plane containing the flight line by which the datum of any model in a stereotriangulated strip departs from the datum of the preceding model.
deflection anomaly *Geodesy.* the difference between an uncorrected value of the deflection of the vertical, as determined by observation, and the value after being corrected in accordance with certain assumptions made with reference to the physical condition of the geoid.
deflection bit *Mechanical Devices.* a noncoring, conical bit designed to work with a deflection wedge to direct the cutting of a borehole.
deflection board *Ordnance.* in artillery operations, an instrument for correcting the azimuth or deflection for wind, drift, and other such factors. Also, GUN DEFLECTION BOARD.
deflection coil *Electronics.* one coil of the set that composes a cathode-ray tube deflection yoke.
deflection curve *Mechanics.* the shape assumed by a horizontal bar, supported at both ends, subject to the application of a vertical force.
deflection defocusing *Electronics.* an inherent tendency in a cathode-ray tube for the electron beam to become less focused as it is deflected toward the outer perimeter of the screen.
deflection display see DISTRACTION DISPLAY.
deflection electrode *Electronics.* one electrode of a pair in a cathode-ray tube that produces one side of the electrostatic field, creating deflection of the electron beam in one axis. Also, DEFLECTION PLATE.
deflection error *Ordnance.* **1.** in artillery, the distance, measured perpendicular to the line of fire, between the target and the bursting point of a projectile or mean bursting point of a group of projectiles. **2.** in bombing, the distance, measured perpendicular to the aircraft's approach, between the target and the point of impact.
deflection factor *Electronics.* the amount of deflection voltage or current needed to produce a given deflection on a cathode-ray tube screen.
deflection magnetometer *Engineering.* a magnetometer that measures magnetic fields by calculating their angular deflections from a small magnet moving across a horizontal plane.
deflection meter *Engineering.* a flowmeter that places differential pressure across a diaphragm or bellows to create a deflection that is proportional to the pressure.
deflection of the plumb line *Geodesy.* a term that is similar to the deflection of the vertical, except that the sign of the value is reversed, and the view is considered downward.
deflection of the vertical *Geodesy.* the angular difference, at any place, between the upward direction of a plumb line (the vertical) and the perpendicular (the normal) to the reference spheroid.

deflection pattern *Ordnance.* in naval gunfire support, the lateral interval between center of flank bursts or impacts.
deflection plate see DEFLECTION ELECTRODE.
deflection polarity *Electronics.* the relation between the polarity of the voltage applied to a pair of deflection electrodes in a cathode-ray tube and the direction in which the electron beam moves.
deflection pool *Hydrology.* a pool that occupies a depression hollowed out by a stream at a curve in its course.
deflection probable error *Ordnance.* the mean directional error of a projectile fired at a specific deflection; estimated as one-eighth of the greatest width of the dispersion pattern.
deflection scale *Ordnance.* a scale on a gunsight used to correct deflection or change the direction of the gun.
deflection sensitivity *Electronics.* the amount of deflection produced by a given deflection voltage or current applied to a cathode-ray tube; the inverse of deflection factor.
deflection ultrasonic flowmeter *Engineering.* a flowmeter that determines velocity by measuring the deflection of a high-frequency sound beam across the flow.
deflection voltage *Electronics.* the voltage applied between a pair of deflection electrodes to control the movement of the electron beam across the screen.
deflection wedge *Mechanical Devices.* a tool inserted in a borehole to direct a drill bit.
deflection yoke *Electronics.* an assembly of electromagnetic coils installed around the neck of a cathode-ray tube to control the deflection of the electron beam.
deflectometer *Engineering.* an instrument that measures the amount of bending which an object suffers during a transverse test.
deflector *Engineering.* an object that shifts the flow of a forward-moving stream.
deflexed *Biology.* bent downward at a sharp angle.
deflocculant *Chemistry.* a substance used to break up aggregated particles in solution, thus forming a colloidally stable suspension.
deflocculation *Materials Science.* the reduction in viscosity of a slip or slurry by the addition of a small amount of a suitable electrolyte, such as sodium silicate or sodium carbonate.
defloration *Medicine.* the rupturing of the hymen in sexual intercourse, in vaginal examination, or by manipulation.
defluorination *Chemistry.* the process of removing fluorine from a substance.
defluvium [də floo′vē əm] *Pathology.* **1.** a falling out, as of hair or nails. **2.** in particular, a loss or shedding of hair by the mother after the birth of a baby. **3.** the discharge or emission of fluid.
defoamer *Food Technology.* a surface-active agent, such as a liquid glyceride, used to control foam produced in the processing of baked goods, candies, jams and jellies, and fruit juices.
defoaming agent *Chemistry.* a substance that reduces the surface tension of a liquid, thus inhibiting the formation of bubbles in the liquid when it is agitated; for example, sulfonated oils, silicone fluids, or organic phosphates. Also, ANTIFOAMING AGENT.
defocus *Engineering.* to cause a beam of light or other form of electromagnetic radiation to shift from an accurate focus when it reaches the target surface.
defocus-dash mode *Electronics.* a method of writing binary data onto the screen of a storage cathode-ray tube, in which the writing beam forms a dash on the screen for one polarity of binary digit and is defocused for the opposite polarity.
defocus-focus mode *Electronics.* a method of writing binary data onto the screen of a storage cathode-ray tube, in which the writing beam forms a tightly focused spot on the screen for one polarity of binary digit and is defocused for the opposite polarity.
defogger see DEFROSTER.
defoliant *Materials Science.* a toxic chemical that causes trees, shrubs, and other plants to shed leaves prematurely or unnaturally. *Ordnance.* a chemical that removes leaves from plants to damage enemy agriculture or to prevent enemy concealment in foliage. Thus, **defoliant operation.**
defoliate *Botany.* **1.** to remove the leaves from a plant, usually prematurely. **2.** to be without leaves, particularly from insect depredation or after annual shedding.
defoliating agent see DEFOLIANT.
defoliation *Botany.* the loss, shedding, or removal of leaves from a tree or other plant, especially prematurely.
De Forest, Lee 1873–1961, American inventor; a pioneer in radio technology and broadcasting; invented the triode (audion) vacuum tube.

deforestation

deforestation *Forestry.* the process of clearing land of trees or forest.
deformability *Hematology.* the ability of certain cells, such as erythrocytes, to change shape as they pass through narrow spaces, such as capillaries. Also, **deformation.**
deformation *Mechanics.* any change in the shape and size of a body. *Mathematics.* a mapping of a space onto itself homotopic to the identity map; intuitively, a transformation that alters geometric shapes by shrinking, twisting, or stretching without breaking lines or surfaces. Also, CONTINUOUS DEFORMATION.
deformation bands *Metallurgy.* bands that are produced in a metallic crystal by deformation; caused by variable rotations of crystal portions.
deformation curve *Mechanics.* a graph for a material showing stress as the abscissa and strain as the ordinate. Also, STRESS-STRAIN RELATION.
deformation-density map *Crystallography.* the difference between the electron density at high resolution of a crystal structure and that expected for a molecule composed of isolated spherical atoms. Deformation density maps give some indication of the bonding characteristics of molecules and of the orientations of lone-pair electrons.
deformation ellipsoid see STRAIN ELLIPSOID.
deformation energy *Nuclear Physics.* the critical amount of energy represented by the elongation or deformation of a target nucleus that, with the neutron energy imparted to the nucleus, influences the ease of fission.
deformation fabric *Petrology.* a kind of rock fabric in which the orientation of the fabric elements was produced by external stress. Also, TECTONIC FABRIC.
deformation lamellae *Mineralogy.* a slipband produced during tectonic deformation by active intracrystalline slip or by shock.
deformation path or **deformation pattern** see DEFORMATION PLAN.
deformation plan *Geology.* the synthesis of the sequence of events in the deformation of a rock, based on the analysis of the rock's fabric. Also, MOVEMENT PICTURE, MOVEMENT PLAN, DEFORMATION PATH, DEFORMATION PATTERN.
deformation plane *Geology.* a plane that is perpendicular to the plane separating adjacent flow layers and parallel to the direction of movement. Also, DISPLACEMENT PLANE.
deformation potential *Solid-State Physics.* the effective electric potential set up by a local deformation in the crystal lattice of a semiconductor or conductor.
deformation thermometer *Engineering.* a thermometer that contains transducing elements which deform when heated.
deformation twin *Crystallography.* a twin crystal produced by slipping along a plane in response to plastic deformation in the original crystal.
deformed bar *Building Engineering.* a steel rod or bar used for concrete reinforcement, having surface projections or indentations to improve the mechanical bond between the steel and the concrete.
deformed nuclei *Nuclear Physics.* the nuclei found in the rare earths and actinides with nonspherical equilibrium shapes, in which even small deformations produce sizeable quadrupole moments; only in such nuclei with substantial distortions can rotational motion be observed.

deformeter *Engineering.* a gauge that is applied to models and that determines actual stresses on the model or on a real structure.

defrost *Engineering.* to remove ice or frost from an object.

defroster *Mechanical Engineering.* equipment that is designed to melt frost, ice, and condensation from the windshield of an automobile or other vehicle; it generally consists of a series of ducts that deliver hot, dry air from the engine. Also, DEFOGGER.

DEFT *Microbiology.* a technique by which microorganisms can be rapidly detected in a given liquid, such as milk, using a filtering procedure followed by fluorescent staining and microscopic examination. (An acronym for direct epifluorescent filter technique.)

defun *Artificial Intelligence.* **1.** in Lisp, the function that is used to define a new function. **2.** a function definition.

defuse *Ordnance.* to remove the fuse from a bomb, mine, or the like.

deg. degree; degrees.

degas *Electronics.* to expel or exhaust all gases from an electron tube during the manufacturing process, including gases absorbed into the internal parts of the tube.

degasifier *Metallurgy.* in liquid metal processing, an addition agent that is used to remove solute gas from the melt.

degassing *Chemistry.* the removal of a gas. *Toxicology.* the treatment of a person or object that has been subjected to the fumes of a gas. *Metallurgy.* the volatilization of foreign matter from the surface of a metal. *Electromagnetism.* see DEGAUSSING. *Engineering.* see BREATHING.

degauss [dē gous′] *Electronics.* to remove an unwanted magnetic field from a piece of electronic equipment; demagnetize. *Computer Technology.* to erase information from a magnetic tape, disk, or drum by demagnetizing the oxide surface. *Electromagnetism.* to render a ship safe from magnetic mines by neutralizing its magnetic field; this is accomplished by surrounding it with a series of electric cables to set up an equal and opposite magnetic field, thereby neutralizing it. Also, DEPERM.

degaussing *Electromagnetism.* the process of neutralizing the strength of the magnetic field of an object. Thus, **degaussing cable.**

degeneracy *Physics.* a condition in which a system has two or more eigenstates with identical eigenvalues. *Mathematics.* for a parametrized family of mathematical objects, a condition in which usually distinct properties of the family coalesce for certain values of the parameters; the opposite of bifurcation. *Genetics.* see DEGENERATE CODE.

degeneracy pressure *Physics.* the pressure exerted by a degenerate electron or neutron gas.

degenerate amphidromic system *Cartography.* a system of cotidal lines that appears to have its center or nodal (no-tide) point on land rather than on the open ocean.

degenerate amplifier see DEGENERATE PARAMETRIC AMPLIFIER.

degenerate conduction band *Solid-State Physics.* an energy band characterized by at least two quantum states having the same principal quantum number but different angular momentum states.

degenerate code *Genetics.* a genetic code in which amino acids are coded by two or more different codons. **Degeneracy** occurs because of the fact that of the 64 possible base triplets, 3 are used to code the stop signals, and the other 61 are left to code only 20 different amino acids.

degenerate electron gas *Physics.* an electron gas that, because of its very high density or very low temperature, is far below the Fermi temperature.

degenerate matter *Physics.* a dense matter in which fermions (electrons, neutrons, protons) must occupy states of high kinetic energy in order to satisfy the Pauli exclusion principle. *Astrophysics.* in white-dwarf and neutron stars, a highly dense state of matter in which atoms have all their energy levels completely filled with electrons.

degenerate parametric amplifier *Electronics.* an electron-beam amplifier device that operates from an AC source twice the frequency of the input signal; it is capable of operating at microwave frequencies and is characterized by a low noise figure. Also, DEGENERATE AMPLIFIER.

degenerate semiconductor *Solid-State Physics.* a semiconductor in which the conduction approaches that of a simple metal.

degenerate star *Astrophysics.* a star made up of degenerate matter; a white dwarf or neutron star.

degeneration *Electromagnetism.* a variation in an amplifier due to unwanted negative feedback. *Physics.* a phenomenon occurring in gases at very low temperatures, in which the molar heat capacity drops to less than 3/2 the gas constant.

degenerative *Medicine.* describing a disease in which there is an ongoing pathological change of cells or tissues to a lower or less active form.

degenerative recrystallization see DEGRADATION RECRYSTALLIZATION.

Degeneriaceae *Botany.* a family of dicotyledonous plants in the order Magnoliales represented by a single Fijian tree species, *Degeneria vitiensis*; characterized by alternate, simple, entire leaves, a primitive, solitary, bisexual flower with a single carpel and fleshy stamens, and an indehiscent, leathery fruit.

DeGennes reptation theory *Materials Science.* a theory stating that polymer molecules, which are macromolecular chains, move by twisting or "snaking" around a three-dimensional network of fixed obstacles (in the manner of a snake in a cage).

deglaciation *Hydrology.* the uncovering of a land mass from beneath glacial ice as a result of the melting and shrinking of the ice.

deglitcher *Electronics.* a special circuit that suppresses switching transients to prevent the generation of spurious digital pulses.

deglut. (taken) orally. (From Latin *deglutiatur,* "let it be swallowed.") Also, **Deglut.**

deglutition *Physiology.* the act of swallowing.

degradation a loss or reduction of the quality, integrity, or character of something; specific uses include: *Organic Chemistry.* a chemical reaction that breaks down a molecule into smaller parts. *Materials Science.* the reduction of a polymer into smaller units, usually as the result of heat, light, and oxidation. *Geology.* the wearing down of land by erosion. *Hydrology.* **1.** the vertical erosion of a stream to establish or maintain uniformity of grade. **2.** a shrinkage or disappearance of permafrost. **3.** any lowering of a stream bed. *Thermodynamics.* a concept embodied in the second law of thermodynamics, which accounts for an increase in the entropy of a system as energy is converted into forms from which it is increasingly difficult to extract useful work. *Physics.* the loss of kinetic energy of a particle solely by collision. *Computer Technology.* the operating condition of a computer system with malfunctioning components that continues to function but with diminished capability.

degradation failure *Engineering.* a failure in a device that arises from change in a parameter and causes the device to fall below minimum standards.

degradation product *Petroleum Engineering.* in the processing of petroleum, a product with low value, or a contaminant created while a reaction is taking place.

degradation recrystallization *Mineralogy.* recrystallization that results in a relative decrease in crystal size. Also, DEGENERATIVE RECRYSTALLIZATION, GRAIN DIMINUTION.

degraded illite *Mineralogy.* illite whose potassium content has been depleted due to the leaching effects of water.

degrading stream *Hydrology.* a stream that is actively downcutting its channel or valley.

degranulation *Immunology.* a process occurring in cells that contain granules of histamine, serotonin, and heparin, in which antigens combine with cytophilic antibodies, causing the cells subsequently to release histamine and other substances.

degreaser *Engineering.* a machine used to remove grease or oil from an object. *Materials Science.* a chemical, usually an organic liquid, that removes natural fats, oils, and waxes.

degree *Metrology.* **1.** any of various units of measure for physical conditions such as temperature or pressure. **2.** specifically, one increment of the Celsius, Fahrenheit, or Kelvin temperature scales. *Mathematics.* **1.** a unit of angular measure that is equal to 1/360 of a full rotation; indicated by the symbol °. **2.** the value of an exponent, usually assumed to be an integer. **3.** the degree of the monomial $ax_1^{k_1} x_2^{k_2} \cdots x_n^{k_n}$, where $a \neq 0$, is the sum $k_1 + k_2 + \cdots + k_n$. **4.** given a multinomial, the maximum of the degrees of its individual terms. Also, **degree of the multinomial.** In particular, the **degree of the polynomial** $p(x) = \sum_{i=0}^{n} a_i x^i$ is the greatest index i for which $a_i \neq 0$. If $a_i = 0$ for all i, the polynomial is said to have no degree, or sometimes degree $-\infty$. **5.** the **degree of a vertex** v, denoted deg(v), in a graph is the number of edges incident on it, each loop counting as two edges. If the graph is a directed graph, then the **in-degree** of a vertex is the number of arcs for which the vertex is a terminal point and the **out-degree** of a vertex is the number of arcs for which the vertex is an initial point.

degree-day *Metrology.* any of various units of measurement that represent one degree of variation from a given standard temperature on a given day; for example, cooling degree-day, heating degree-day, and growing degree-day; used in calculating air-conditioning, heating, and agricultural needs, respectively.

degree of crystallinity *Organic Chemistry.* the extent to which a polymer is composed of solid regions of crystallites; the fraction of a bulk polymer that consists of regions showing three-dimensional order or crystallinity.

degree of current rectification *Electronics.* the quotient of the average value of the pulsating direct current from a rectifier and the effective (root mean square) value of the alternating current into the rectifier.
degree of curve *Cartography.* the number of degrees of angular measure at the center of a circle that is subtended by a chord 100 feet in length.
degree of disorder *Robotics.* the amount of irregular or abnormal functioning in a robotic system.
degree of freedom *Physical Chemistry.* **1.** any of the variable physical quantities of a given system, such as its pressure, temperature, or composition, that must be specified in order to define the system. **2.** any of the unique ways that an individual particle absorbs energy. *Mechanics.* **1.** any of the independent modes in which a system can continuously change shape or move in space. **2.** the total number of such modes. *Statistics.* the excess of the number of data points over the number of parameters.
degree of polymerization *Organic Chemistry.* the extent to which monomers have reacted to form a polymer; the average number of monomer units per polymer molecule.
degree of risk *Military Science.* the acceptable risk to friendly forces from the detonation of a nuclear weapon in attacking a nearby enemy target; the level of risk is specified by the commander, depending upon the tactical conditions; it may be defined as **emergency risk, moderate risk,** or **negligible risk.** Also, **degree of nuclear risk.**
degree of voltage rectification *Electronics.* the quotient of the average value of the pulsating direct voltage from a rectifier and the effective (root mean square) value of the alternating voltage into the rectifier.
de Gua's rule *Mathematics.* let $a_0 + a_1x + a_2x^2 + \cdots + a_nx^n = p(x)$ be a polynomial with real coefficients, where the sequence of coefficients a_0, $a_1, a_2, \ldots, a_n$ has a group of r consecutive zero terms. Then de Gua's rule is the theorem that the number of imaginary roots (counting multiplicities) of $p(x)$ is at least N, where: (i) $N = r - 1$ if r is odd and the coefficients immediately preceding and following the group of r consecutive zero coefficients have unlike signs; (ii) $N = r$ if r is even; and (iii) $N = r + 1$ if r is odd and the coefficients immediately preceding and following the group of r consecutive zero coefficients have like signs.
degumming *Textiles.* the process of using soap to remove sericin, a natural gluelike substance, from silk fiber or fabric in order to improve the quality of the fabric.
de Haas-Van Alphen effect *Solid-State Physics.* a phenomenon that is observed in complex metals at low temperatures, in which the diamagnetic susceptibility of the conduction electrons varies periodically with the inverse of an external magnetic field applied perpendicular to a principal axis of the crystal.
dehalogenation *Organic Chemistry.* the removal of halogen from a compound.
dehiscence *Botany.* the spontaneous opening of a fruit, seed pod, anther, or sporangium at maturity to discharge its contents, usually breaking along certain lines or directions.
Dehmelt, Hans born 1922, German-born American physicist; Nobel Prize for work in trapping and observing single ions and electrons.
dehrnite see CARBONATE-FLUORAPATITE.
dehumidifier *Mechanical Engineering.* a device that removes moisture from the atmosphere.
dehydrase *Biochemistry.* an enzyme that catalyzes the removal of the elements of water from a substrate.
dehydratase see HYDRO-LYASE.
dehydrate *Chemistry.* to remove water. *Physiology.* to lose or lack fluids in the body. *Food Technology.* to carry out a process of dehydration.
dehydrated *Chemistry.* lacking water. *Medicine.* suffering from dehydration. *Food Technology.* describing a food that has been prepared by dehydration, such as a powdered soup mix.
dehydration *Chemistry.* the removal of water from a compound. *Medicine.* a condition in which there is an excessive loss of water from the body tissues, caused by such factors as prolonged diarrhea or fever, repeated vomiting, and excessive perspiration or urination. Extreme dehydration may lead to shock or even death. *Food Technology.* **1.** the process of removing water from a food product in order to preserve it or make it more portable. **2.** the drying of foods by factory methods, as opposed to natural or air-drying.
dehydrator *Engineering.* a device used to remove water.
dehydroacetic acid *Organic Chemistry.* $C_8H_8O_4$, combustible, toxic, prismatic white crystals, insoluble in water and soluble in acetone; melts at 108.5°C and boils at 270°C; used as a fungicide and bactericide, and in medicated toothpastes.
dehydroalanine *Endocrinology.* an inactive metabolic by-product of the synthesis of thyroxine or triiodothyronine.
dehydroascorbic acid *Organic Chemistry.* $C_6H_6O_6$, the oxidized form of ascorbic acid; fine needles that are soluble in water; decomposes at 225°C; used in nutrition and medicine.
dehydrobilirubin see BILIVERDIN.
7-dehydrocholesterol *Biochemistry.* $C_{27}H_{44}O{\cdot}H_2O$, a cholesterol derivative that is the metabolic precursor of vitamin D_3, forming that vitamin upon ultraviolet irradiation; found in the skin of humans and animals. Also, **dehydrocholesterol.**
dehydrocholic acid *Organic Chemistry.* $C_{24}H_{34}O_5$, a fluffy, white powder; insoluble in water and slightly soluble in alcohol; melts at 231–240°C; used in medicine and pharmaceuticals.
dehydroepiandrosterone *Organic Chemistry.* $C_{19}H_{28}O_2$, an androgenic steroid, occurring in normal human urine and synthesized from cholesterol, that crystallizes in dimorphous needles or leaflets; soluble in alcohol; needles melt at 140–141°C and leaflets at 152–153°C; are used in medicine and biochemical research. Also, **dehydroisoandrosterone, dehydroandrosterone.**
dehydrogenase *Enzymology.* any enzyme that catalyzes the transfer of hydrogen and electrons from one compound to another.
dehydrogenation *Organic Chemistry.* the removal of one or more hydrogen atoms from a molecule.
dehydrohalogenation *Chemistry.* the process of removing hydrogen and a halide from a compound.
deicer or **de-icer** *Engineering.* a device or substance used to remove unwanted ice formations, as from the engine or windshield of a motor vehicle, the wings or body of an aircraft, and so on.
deicer boot *Aviation.* an inflatable rubber strip that is attached to the leading edge of an airfoil and is activated in order to break up ice that has formed there.
deicing *Engineering.* the act of removing ice from an object by the application of chemicals, by heating, or by other physical means.
deimatic behavior or **deimatic effect** *Behavior.* a behavior pattern in which a threatened animal's appearance abruptly changes to give the impression that it is a more dangerous or less palatable species. Also, STARTLE DISPLAY.
Deimos *Astronomy.* the more distant and smaller of the two moons of Mars, measuring 15 km by 12 km by 11 km; discovered by Asaph Hall in 1877.
deindividuation *Psychology.* a situation in which the members of a group have lost their individual identities and merged anonymously into the group, adopting its values and behavior patterns.
Deinococcus *Bacteriology.* a genus of Gram-negative bacteria that occur as red-pigmented, coccoid cells.
Deinotheriidae *Paleontology.* an extinct family of proboscidean mammals in the suborder Deinotherioidea, distinguished by large, downturned tusks on the lower jaw; flourished from the Miocene to the end of the Pleistocene, when they became extinct, along with many other large mammals on all continents.
Deinotherioidea *Paleontology.* an extinct suborder of late Cenozoic mammals in the order Proboscidea; similar to but probably not directly ancestral to elephants.
deion circuit breaker *Electricity.* a circuit breaker designed so that the current-interrupting arc is magnetically blown into a stack of insulated copper plates; the resulting short arcs in series become almost instantaneously deionized when the current falls to zero in the AC cycle, so the arc cannot reform.
deionization *Chemistry.* a process in which all charged or potentially charged species are removed from a solution by an ion-exchange reaction.
deionization time *Electricity.* the time required for an ionized gas to become neutral after removal of the ionizing voltage.
Deiphon *Paleontology.* a genus of spiny Silurian trilobites in the suborder Cheirurina and family Cheiruridae.
Deister phase *Geology.* a Late Ammerian phase, a subdivision of the Jurassic period between the Kimmeridgian and Lower Portlandian.
déjà vu [dé´zhä voo´] *Psychology.* **1.** the impression that some situation has been experienced before, when actually it is being encountered for the first time. **2.** in popular use, a feeling that the past is being repeated or that something new is familiar. (From a French phrase meaning "already seen.") Similarly, **déjà pensé** ("already thought"), **déjà entendu** ("already heard"), **déjà fait** ("already done"), **déjà eprouvé** ("already experienced"), **déjà raconté** ("already told").
dejection cone see ALLUVIAL CONE.

deka- or **dek-** a metric combining form meaning "ten," as in *dekaliter* (10 liters).
dekagram *Metrology.* a unit of measure for mass, equal to 10 grams. Also, **dekagramme.**
dekaliter *Metrology.* a unit of measure for capacity, equal to 10 liters. Also, **dekalitre.**
dekameter *Metrology.* a unit of measure for length, equal to 10 meters or 32.81 feet. Also, **dekametre.**
dekapoise *Fluid Mechanics.* a unit of viscosity in the absolute metric system in which 1 dekapoise equals 10 poises; 1 poise = 0.1 kg/m sec = 1 g/cm sec.
DEL *Aviation.* the airport code for Delhi, India.
del delta.
del. delete; deletion.
Delaborne prism *Optics.* a prism in which light entering at an angle parallel to the hypotenuse face reflects at the same angle so that when the prism rotates, it rotates the image at twice that angle. Also, DOVE PRISM.
delafossite *Mineralogy.* $CuFe^{+3}O_2$, an easily fusible, black, opaque, trigonal mineral occurring in massive form or as tabular crystals, having a specific gravity of 5.41 and a hardness of 5.5 on the Mohs scale; found as a secondary mineral in copper ore deposits.
Delamareaceae *Botany.* a family of brown algae of the order Dictyosiphonales, characterized by parenchymatous sporophytes with branched or unbranched cylindrical axes and a thallus covered by enlarged cortical cells.
Delambra's analogies see GAUSS FORMULAS.
delaminate *Science.* **1.** to split a material into mutiple layers. **2.** a material that is split in this way.
delamination *Biology.* a separation into layers, especially the division of tissue cells into layers. *Developmental Biology.* the separation of blastoderm into epiblast and hypoblast during chick embryo development. *Engineering.* the separation of a laminate into its basic layers. *Materials Science.* a subcritical damage to the interfaces between the plies in a laminate composite that causes a reduction in the load-carrying capacity of the composite.
de Lange's syndrome *Medicine.* a congenital syndrome that is characterized by severe mental retardation in association with many physical abnormalities, including short stature, low-set ears and webbed neck, a flat nose bridge with tilted tip, coarse hair growing low on the forehead, on the neck, and across the base of the nose, and flat hands with short pointed fingers. A person who is affected with this condition may be referred to as an **Amsterdam dwarf**. Also, TYPUS DEGENERATIVUS AMSTELODAMENSIS.
de la Rue and Miller's law *Electronics.* a principle stating that the minimum sparking voltage in a gas-filled device is directly proportional to the gas pressure and the distance between the electrodes to which the voltage is applied.
de la Tour method *Analytical Chemistry.* a method of measuring critical temperature by sealing a sample in a tube and heating it; the critical temperature is obtained when the meniscus disappears.
delay an interruption or pause in some process or activity; specific uses include: *Industrial Engineering.* an abnormal or unwarranted event that causes a production process to temporarily halt or fall behind its schedule or normal rate of progress. *Telecommunications.* **1.** lost or waiting time introduced because a call cannot be connected immediately or because a message or signal cannot be transmitted immediately, usually due to a queue on the required line or circuit. **2.** the amount of time by which a signal is delayed or an event retarded.
delay allowance *Industrial Engineering.* in time-motion studies, time that is built into a production schedule to accommodate contingencies or interruptions that are outside the control of the worker.
delay blasting cap *Engineering.* a blasting cap whose detonation occurs after a set interval of time has elapsed. Also, **delayed-action detonator.**
delay circuit see TIME-DELAY CIRCUIT.
delay counter *Computer Technology.* a register in a CPU that can temporarily delay a program long enough to complete the execution of an operation.
delay distortion *Electronics.* a distortion in transmitted signals caused by different rates of change of phase shift over the frequency range of the equipment. Also, ENVELOPE DELAY DISTORTION.
delay Doppler mapping *Cartography.* a method for mapping a planet by illuminating it with radar and then measuring the Doppler shift resulting from the planet's rotation.
delayed-action *Ordnance.* describing an explosive device that utilizes a delay fuse, thus creating a period of time, from a fraction of a second to several days or more, between actuation of the device and explosion. Thus, **delayed-action bomb, delayed-action mine.**
delayed agc delayed automatic gain control.
delayed alpha particle *Nuclear Physics.* an alpha particle that is emitted by an excited nucleus long after beta decay has taken place.
delayed automatic gain control *Electronics.* an automatic gain control circuit that preserves weaker signals in a receiver by acting only on inputs above a predetermined amplitude. Also, **delayed automatic volume control.**
delayed classical conditioning *Behavior.* a type of classical conditioning in which an unconditioned stimulus is presented while the conditioned stimulus is still present. Also, **delayed conditioning.**
delayed combustion *Engineering.* the combustion that occurs after gas exceeds the furnace volume of a boiler.
delayed-contact fire *Ordnance.* a firing system that explodes a mine a specific period of time after physical contact.
delayed cracking *Materials Science.* cracking that is caused by flaws occurring after a material has been in service for some time, generally due to cyclic stresses or corrosive conditions, and in which small flaws grow until a critical crack size is reached, followed by rapid failure.
delayed critical *Nucleonics.* a state of criticality in which the reactivity is equal to the fraction of delayed neutrons.
delayed graft *Surgery.* **1.** a graft that has been postponed due to infection of the site, in order to await a healthy bed of granulated tissue. **2.** a skin graft that has been partially removed from the donor site and then replaced to await grafting to a new recipient bed.
delayed hypersensitivity *Immunology.* an increased inflammatory reaction to an antigen that occurs in a sensitized individual 24 to 48 hours after contact with the antigen.
delayed junction see DEFERRED JUNCTION.
delayed-line-time-compression see DELTIC METHOD.
delayed-matching-to sample see DMTS.
delayed menstruation *Medicine.* the delay of the onset of menstruation beyond the sixteenth year.
delayed neutron *Nuclear Physics.* a neutron arising from fission but not instantaneously released, sometimes delayed up to seconds; resulting from the breakdown of fission products and helpful in the control of reactors.
delayed neutron fraction *Nuclear Physics.* the ratio of the average number of delayed neutrons to the total number of neutrons per fission event, amounting to about 0.75%.
delayed-opening chaff *Ordnance.* a radar-reflecting device utilizing a parachute and time fuse to release chaff bundles behind the delivering aircraft.
delayed PPI *Electronics.* a plan-position indicator in which the start of each sweep is delayed for a predetermined time after transmission of the main pulse.
delayed repeater satellite *Space Technology.* a communications satellite that receives messages, stores them, and then retransmits them on command.
delayed runoff *Hydrology.* **1.** runoff from melted snow or ice. **2.** runoff that sinks into the ground and later discharges into streams.
delayed sweep *Electronics.* a technique in which the start of a sweep is delayed for a predetermined time after the reference pulse in order to enhance the clarity of later signals by allowing them to be seen on an expanded time scale.
delayed yield *Metallurgy.* plastic deformation that occurs with some delay after stress application.
delay element *Ordnance.* a component of a propellant-actuated device that creates a delay between the actuation of the device and the ignition of the propellant.
delay equalizer *Electronics.* a corrective circuit designed to reduce the delay distortion in a system by compensating for the different rates of change of phase shift over the frequency range of the system.
delay-frequency distortion *Telecommunications.* distortion caused by the difference between maximum and minimum propagation time delay of the electromagnetic waves with different frequencies within a signal.
delay line *Electronics.* an artificial transmission line or equivalent device designed to provide a time delay of a signal between input and output.
delay-line memory or **delay-line storage** see CIRCULATING STORAGE.

delay multivibrator *Electronics.* a monostable circuit capable of producing an output pulse at a predetermined time after receipt of a trigger input pulse.

delay relay *Electricity.* a relay that has a predetermined delay between energization and contact closing. Also, TIME-DELAY RELAY.

delay unit *Computer Technology.* a device in an analog computer that produces an output signal that is a delayed form of the input signal. Also, TRANSPORT DELAYING UNIT.

Delbrück, Max [del´brůk´] 1906–1981, American biologist; with Luria, Nobel Prize for work on mechanisms of bacterial viral infections.

Delbrück scattering *Nuclear Physics.* the diffusion of photons by the electric field created by a motionless charged particle.

deleading therapy *Toxicology.* a course of treatment that brings about the mobilization and elimination of lead from body tissues.

***d* electron** *Atomic Physics.* an electron that orbits a nucleus with the angular momentum of 2.

Delepine reaction *Organic Chemistry.* a reaction that involves the preparation of primary amines from alkyl halides by the use of hexamethylenetetramine followed by cleavage of the resulting salt with ethanolic hydrochloric acid.

Delesseriaceae *Botany.* a large family of predominantly marine red algae of the order Ceramiales, characterized by pseudoparenchymatous, uniaxial thalli, similar gametangial and sporangial forms, and superficially borne spermatangia, which often occur in numbers large enough to produce colorless areas on thalli.

delete *Computer Technology.* a key on a computer keyboard that automatically removes selected material from a file or document.

delete list *Artificial Intelligence.* a set of facts to be deleted from the world model when a STRIPS operator is applied.

deletion *Genetics.* the loss of part of a DNA molecule from a larger DNA molecule or a chromosome.

deletion mapping *Molecular Biology.* an analytical or laboratory experimental procedure in which the position of a gene on the chromosome is determined by the use of overlapping deletions.

deletion operator *Computer Programming.* a mechanism uniquely defined for a specific data structure that allows components of the structure to be deleted.

deletion record *Computer Programming.* an input record used when updating a master file, containing only the type identifier (delete) and the record key; the corresponding record is located in the master file and marked for deletion, removed directly, or not copied onto a new master.

Deleya *Bacteriology.* a genus of bacteria that was created to include marine microorganisms from the genera *Alcaligenes* and *Pseudomonas*, as well as species newly isolated from saline-rich soils.

deliberate *Military Science.* of or relating to military activities that are systematically planned in advance, usually without pressure from the enemy. Thus, **deliberate attack, deliberate breaching, deliberate crossing, deliberate defense.**

deliberate fire *Ordnance.* fire at a slower rate than normal; it is used to make adjustments between each round, to conserve ammunition, or for tactical reasons. Also, SLOW FIRE.

delimiter *Computer Programming.* a special character, such as a comma, that designates the beginning or end of a field, record, or string of data, but does not become part of the data.

deliquescence *Physical Chemistry.* the absorption of water from the air that eventually leads to a softened state. *Botany.* **1.** a type of branching in which the main trunk or stem divides repeatedly to form many smaller branches. **2.** the softening or liquefying of tissue or flowers upon maturing, especially the gills of mushrooms.

deliquescent *Physical Chemistry.* **1.** describing a substance that tends to absorb water vapor from the atmosphere and become liquid. **2.** specifically, describing a powdery substance that will dissolve in the water it absorbs from the atmosphere, unless kept enclosed or stoppered.

delirium [di lēr´ē əm] *Medicine.* an acute, reversible organic mental disorder, marked by confused thinking, incoherent speech, reduced consciousness, high levels of anxiety, fear, or excitement, and sometimes by delusions. *Psychology.* a state of wild excitement or enthusiasm.

delirium tremens *Medicine.* an acute psychotic condition, popularly known as the DTs, that is caused by excessive consumption of alcohol over an extended period of time; characterized by symptoms of delirium and specifically by vivid, frightening hallucinations; typically occurring during a period of abstinence or reduced consumption.

Delisle, Joseph [də lēl´] 1688–1768, French astronomer; devised a method for observing the transits of Venus and Mercury.

delivery *Medicine.* the process of childbirth.

delivery system *Computer Programming.* a hardware and/or software system designed for the delivery of an application to an end user. *Ordnance.* the means of delivering a missile or projectile to its target, especially a weapons system for the delivery of a nuclear warhead.

dellenite see RHYODACITE.

Dellinger fadeout *Telecommunications.* an effect that causes electromagnetic sky-wave signals to disappear rapidly as a result of greatly increased ionization in the ionosphere due to increased solar noise caused by solar storms. The effect lasts 10 minutes to several hours. An example is the fading of short-wave radio communications.

Del Mar man *Anthropology.* the fossil human remains from California that were mistakenly dated to 48,000 years ago through the amino acid racemization dating technique; the correct radiometric age is about 12,000 years.

Delmontian *Geology.* a North American geologic stage of the Uppermost Miocene, occurring after the Mohnian and before the Repettian.

delocalization of electrons *Organic Chemistry.* the ability of a given pair of electrons in certain organic molecules to bind together more than two nuclei.

delocalized bond *Chemistry.* a molecular bonding concept, often used to describe the pi bonding in a system of alternating single and multiple bonds; a bonding electron is seen as delocalized between the linked atoms instead of being associated with a particular atom.

delomorphous cell see PARIETAL CELL.

del operator *Mathematics.* a vector operator, symbolized ∇ and defined by $\nabla = i\,\partial/\partial x + j\,\partial/\partial y + k\,\partial/\partial z$, where i, j, and k are the component functions on R^3. In the space of functions f on Euclidean R^3 having continuous first partial derivatives, the del operator coincides with the differential; i.e., $\nabla f = df$. ∇f is also called the gradient of f. The formal scalar product of the del operator with a vector field $v = v_x i + v_y j + v_z k$ is called the divergence of v; i.e., $\mathrm{div}(v) = \nabla \cdot v = \partial v_x/\partial x + \partial v_y/\partial y + \partial v_z/\partial z$. The formal cross product of the del operator with a vector field V is called the curl of V. Many formal identities relating multiple applications of the gradient, divergence, and curl operators exist and form the basis for the Gibbs algebra or vector analysis. However, most of these identities hold only in R^3 and do not generalize to higher dimensions. Gradually, the del operator and vector analysis are being replaced by the algebra of differential forms and Riemannian geometry.

delorenzite see TANTEUXENITE.

delphinidin *Biochemistry.* $C_{15}H_{11}ClO_7$, a purple or brownish-red anthocyanin found in plants.

Delphinidinae *Vertebrate Zoology.* the marine dolphins, a family of moderate- to small-sized toothed whales including several types of dolphins, and the porpoise, pilot whale, and killer whale.

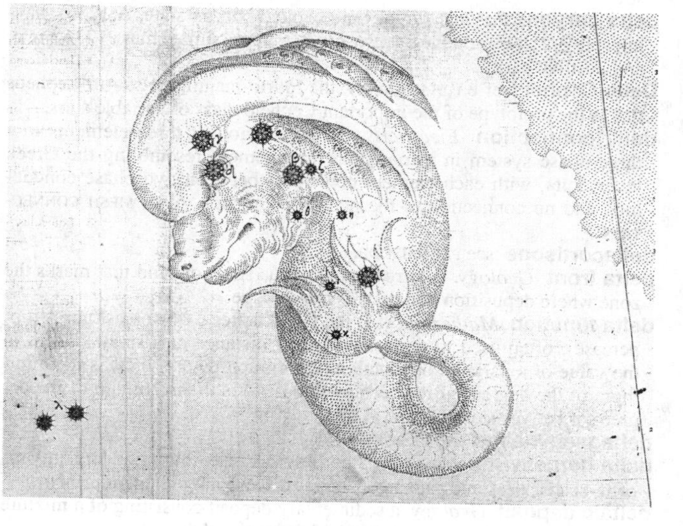

Delphinus

Delphinus *Astronomy.* the Dolphin, a small but distinct constellation lying near the celestial equator and visible on autumn evenings in the Northern Hemisphere.

Delrin *Materials.* a trade name for a highly crystalline homopolymer acetal resin that is rigid, hard, and resistant to corrosion.

delta

delta *Science.* **1.** the fourth letter of the Greek alphabet, written as Δ,δ and equivalent to D,d in the Roman alphabet. **2.** the fourth in a series. **3.** triangular; having the shape of the Greek letter Δ. *Geology.* a nearly level, often triangular alluvial plain occurring between diverging branches of the mouth of a river. *Electronics.* **1.** any electrical connection or circuit that is characterized by a triangular schematic arrangement of the main elements. **2.** the difference between the partial-select outputs of a magnetic cell in a one state and in a zero state. *Mathematics.* see DELTA FUNCTION.

Delta *Space Technology.* a three-stage launch missile used by NASA to place a wide variety of satellites in orbit. Also, THOR-DELTA.

Delta Aquarids *Astronomy.* a diffuse meteor shower that appears to come from an area near the star Delta in Aquarius, and that reaches a broad peak of activity in late July and early August.

delta bar *Geology.* a ridgelike accumulation of alluvial material formed by a tributary stream that is building a delta into the channel of the main stream.

delta baryon *Particle Physics.* an excited baryon state, Δ(1236), having spin 3/2, hypercharge +1, isospin 3/2, approximate mass 1236 MeV, and positive parity.

Delta Cephei [sef′ē ī] *Astronomy.* the fourth-magnitude star in Cepheus that is the prototype of the important Cepheid class of variable stars.

delta connection *Electricity.* the connection of three elements in a three-phase system in a triangular arrangement resembling the Greek letter delta, with each element connected between two phase conductors, and no connection to the neutral conductor. Also, MESH CONNECTION.

deltacortisone see PREDNISONE.

delta front *Geology.* a narrow, continuous sheet of sand that marks the zone where deposition in a delta is most active.

delta function *Mathematics.* **1.** the Greek letter δ (lowercase) or Δ (uppercase); often used to represent a small distance or a small change in the value of a variable or function. **2.** in graph theory, $\delta(G)$ is the minimum of the degrees of the vertices and $\Delta(G)$ is the maximum of the degrees of the vertices of graph G.

deltageosyncline see EXOGEOSYNCLINE.

delta herpesvirus *Virology.* a herpesvirus resembling the human varicella-zoster virus and causing a similar disease in nonhuman primates.

deltaic deposit *Geology.* a sedimentary deposit consisting of a mixture of sand, clay, and organic matter laid down in a delta.

delta iron *Metallurgy.* the high-temperature variety of body-centered cubic iron. Also, **delta ferrite**.

deltaite *Mineralogy.* a name for a mixture of crandallite and hydroxylapatite.

delta kame *Geology.* a hill of stratified sand and gravel that was deposited by a meltwater stream flowing into a proglacial lake. Also, KAME DELTA.

delta lake *Hydrology.* a lake that is formed along the edge of or within a delta.

delta matched antenna *Electricity.* a single-wire antenna to which the leads of an open-wire transmission line are connected in the shape of a Y. Also, Y ANTENNA.

delta matching transformer *Electricity.* the Y-shaped matching section of a delta-matched antenna.

delta meson *Particle Physics.* a scalar meson resonance having positive charge conjugation parity, isospin 1, zero hypercharge, and approximate mass 962 MeV.

delta modulation *Electronics.* a transmission system in which each sample of the signal is transmitted by the value with which it differs from the preceding sample.

delta moraine see ICE-CONTACT DELTA.

delta network *Electricity.* a group of three branches connected in series to form a mesh.

delta plain *Geology.* a plain formed by the deposition of alluvium on the landward part of a large delta.

delta pulse-code modulation *Electronics.* a transmission system in which each sample of the signal is transmitted by a value, coded as a binary number pulse sequence, showing how much it differs from the preceding sample.

delta ray *Atomic Physics.* an electron emitted when an ionizing element, such as an alpha particle, passes through matter.

delta rhythm *Physiology.* the pattern of slow waves (delta waves) of an electroencephalogram, having a frequency of less than 3 $^1/_2$ per second, and typically occurring during deep sleep, in infancy, and in serious brain disorders.

Delta Scuti star *Astronomy.* a variable star that is characterized by a period of less than 0.3 day, a small range of variability, and a type A or F spectrum.

Deltatheridium *Paleontology.* a small, probably carnivorous mammal in the subclass Theria and family Deltatheridiidae, probably ancestral to the creodonts and the Carnivora; extant in the Upper Cretaceous.

delta transformer *Electricity.* a three-phase electrical transformer whose winding ends are joined to form a triangle.

delta V *Space Technology.* ΔV, a mathematical expression for a change in velocity, especially referring to spacecraft; designates the velocity change required to transfer a spacecraft from one orbit to another.

delta virus *Virology.* a defective ssRNA-containing virus that replicates only in the presence of a helper virus of the family Hepadnaviridae, such as the hepatitis B virus; can infect humans, often exacerbating preexisting hepatitis infections.

delta wave *Physiology.* one of the high-amplitude, slow, and regular brain waves that characterize the stages of deep sleep.

delta wing *Aviation.* a triangularly shaped aircraft wing having a low aspect ratio, a sharply tapered leading edge, a straight trailing edge, and a pointed tip.

delta-Y transformation see Y-DELTA TRANSFORMATION.

deltic method *Electronics.* a method of sampling incoming waveforms and reference signals, in which the samples are compressed in time and compared by autocorrelation. (An acronym for <u>de</u>layed-<u>l</u>ine-<u>ti</u>me-<u>c</u>ompression.)

deltohedron *Crystallography.* a hemihedral isometric crystal form having 12 quadrilateral faces. Also, **deltoid dodecahedron.**

deltoid *Science.* triangular in outline. Also, **deltoidal.** *Anatomy.* of or relating to the deltoid muscle.

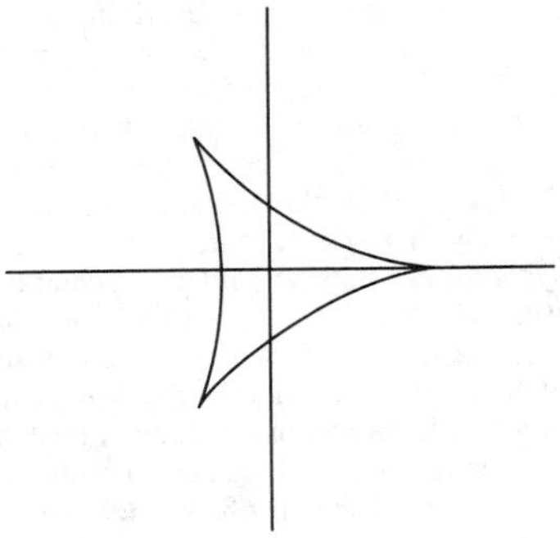

deltoid

deltoid ligament *Anatomy.* the interior or medial ligament of the ankle.
deltoid muscle *Anatomy.* the large triangular muscle covering the shoulder joint.

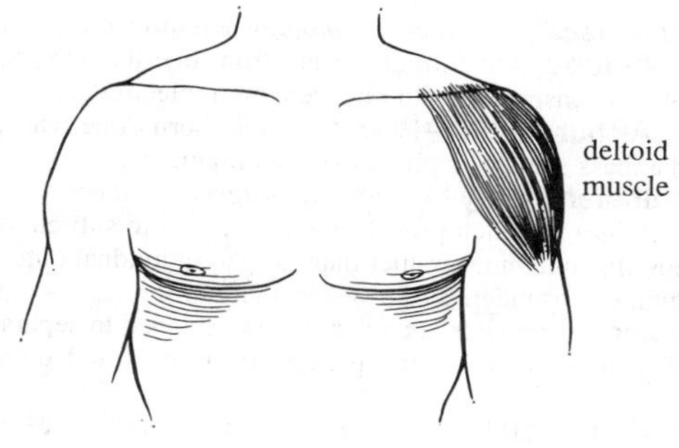

deltoid muscle

deltopectoral *Anatomy.* of or relating to the pectoral muscles and the deltoid muscles.
delubrum *Architecture.* **1.** a Roman temple, shrine, or sanctuary. **2.** a church font.
delusion *Psychology.* a false belief that is held by an individual despite obvious evidence to the contrary, and that is inconsistent with the beliefs of others of the same cultural group.
delusional (paranoid) disorder see PARANOIA.
delusion(s) of grandeur *Psychology.* the delusion that one has taken on the identity of a famous or important person.
delusion(s) of influence *Psychology.* the delusion that one's thoughts and actions are being controlled by hostile outside forces. Also, **delusion(s) of control.**
delusion(s) of observation *Psychology.* see DELUSION OF REFERENCE.
delusion(s) of persecution *Psychology.* the delusion by an individual that certain persons or groups are actively plotting to harm him.
delusion(s) of reference *Psychology.* the delusion by an individual that all the actions of people around him are specifically directed toward him and are intended to affect him negatively.
delustering *Textiles.* the process of reducing the brightness of a synthetic fabric by one of several means; for example, by adding chemicals or darker pigment to the dyes.
delvauxite *Mineralogy.* $CaFe_4^{+3}(PO_4,SO_4)_2(OH)_8 \cdot 4\text{–}6H_2O$(?), an inadequately described yellowish to reddish-brown amorphous mineral having a specific gravity of 1.85 to 2.83 and a hardness of 2.5 to 4 on the Mohs scale; found near Vise, Belgium.
dely delivery.
dem. demurrage.
demagnetization *Electromagnetism.* the complete or partial elimination of permanent magnetic fields.
demagnetize *Electromagnetism.* to remove magnetic properties from an object, especially unwanted magnetic properties.
demagnetizer *Electromagnetism.* a device used for removing undesired residual magnetism from an object, such as a magnetic tape head.
demal *Chemistry.* a solution of one gram-equivalent of solute dissolved in one cubic decimeter of solvent.
demand capacity ratio *Transportation Engineering.* the ratio between the number of trips desired by the users of a transport line and the line's maximum traffic capacity. A high demand capacity ratio produces congestion and delays; a low demand capacity ratio implies extensive unused capacity.
demand characteristic *Psychology.* a clue or bias in a psychological experiment that conveys to the subject the response that the experimenter is anticipating.
demand-driven execution *Computer Technology.* a mode of computer operation in which no computation is performed until the results are needed as input for another computation.
demand factor *Electricity.* the ratio of the maximum power demanded by a system to the actual power used. Also, **demand.**
demand limiter see CURRENT LIMITER.
demand paging *Computer Programming.* a virtual storage technique whereby segments of program instructions or data are loaded from disk to main memory only when requested by the processor; a form of swapping.
demand processing *Computer Programming.* the real-time processing of data as soon as it is available or needed, as in an airline reservation system.
demand-pull system *Industrial Engineering.* a system of inventory control in which each work station produces at a rate that is determined by the input requirements of the next station.
demand rate *Electricity.* a rate of electric power based on the maximum amount that must be kept available to a customer.
demand reading *Computer Programming.* an input operation that reads a single record from storage when it is required by the CPU, as opposed to reading blocks of data into a buffer for faster access.
demand regulator *Engineering.* an element in a open-circuit diving system that allows a scuba diver to exhale directly into the water without inhaling carbon dioxide.
demand-responsive system *Transportation Engineering.* a small-scale public transportation system (such as a jitney) whose routes and schedules may change to meet passengers' demands.
demand staging *Computer Programming.* the process of moving blocks of data, upon request from a program, from one storage device to another device having a shorter access time, as opposed to anticipatory staging.
demand stop see REQUEST STOP.
demand system *Engineering.* an automatic oxygen-dispensing system in an airplane in which the flow is regulated by the demand of a flyer's body.
demand writing *Computer Programming.* an output operation that writes a single record directly to an output device when the record is produced, as opposed to transferring data to a buffer for block outputs.
demantoid *Mineralogy.* a green, transparent, gem-quality variety of andradite with a diamantine luster.
demarcation *Anatomy.* the marking off or ascertainment of boundaries.
De Marre formula *Ordnance.* a formula expressing the relationship between certain characteristics of a projectile and its effectiveness in penetrating armor.
demasking *Chemistry.* the process of restoring normal chemical reactivity to a masked substance.
Dematiaceae *Mycology.* a family of fungi belonging to the class Hyphomycetes that are characterized by dark-colored filaments (or hyphae) and spores.
Dember effect *Electronics.* an effect in which light shining on one region of a two-region semiconductor stimulates the diffusion of hole-electron pairs, thus producing a voltage between the regions. Also, PHOTODIFFUSION EFFECT.
Dembowska [dem bäf´skə] *Astronomy.* asteroid 349, very red in color and of unique type (R); it was discovered in 1892 and measures about 160 kilometers in diameter.
deme [dēm] *Ecology.* a local population of a species viewed as a randomly interbreeding group. Also, PANMICTIC UNIT.
demecolcine see COLCEMID.
dementia [də men´shə] *Psychology.* **1.** an organic mental disorder characterized by loss of memory, impairment of judgment and abstract thinking, and changes in personality. **2.** an older term for mental illness.
dementia praecox *Psychology.* an older term for SCHIZOPHRENIA.
dementia pugilistica see PUNCHDRUNK ENCEPHALOPATHY.
dementia simplex *Psychology.* an older term for SIMPLE SCHIZOPHRENIA.
Demerec Convention *Microbiology.* a procedure established for the nomenclature of bacterial genetics.
demersal *Biology.* living near or on the bottom of the sea, or in very deep water.
demethylation *Organic Chemistry.* the removal of methyl groups from a molecule, usually replaced by hydrogen atoms.
demethylchlortetracycline *Microbiology.* a derivative of the broad-spectrum antibiotic tetracycline.
demeton-*S*-methyl *Organic Chemistry.* $C_8H_{15}O_3PS_2$, an oily liquid that is very slightly soluble in water; used as a systemic insecticide.
demeton-*S*-methyl sulfoxide *Organic Chemistry.* $C_6H_{15}O_4PS_2$, a liquid that is slightly soluble in water; used as a systemic insecticide.
demi- a prefix meaning "half," as in *demilune.*
demicyclic rusts *Mycology.* rust fungi that include most species of the genus *Gymnosporangium;* characterized by the formation of all types of spores except for the so-called urediospores or summer spores, which are able to germinate very rapidly into fungi.
demilitarized zone *Military Science.* a defined area in which military forces or installations are prohibited.

demilune *Biology.* crescent-shaped, like a new moon; especially a crescent-shaped cell in the salivary gland.
demister *Mechanical Engineering.* a British term for a defroster.
Democritus c. 460–370 BC, Greek philosopher; first to propose an atomic theory of matter.
Demodicidae *Invertebrate Zoology.* a family of arachnids that includes the pore mites, skin parasites of mammals.
demodifier *Computer Programming.* a data element that restores a modified instruction to its original state.
demodulate *Telecommunications.* to undo or reverse the effects of modulation; for example, to remove the intelligence-bearing signal from a modulation carrier.
demodulator *Telecommunications.* a circuit that recovers the signal that was used to modulate a carrier. *Electronics.* see DECOMMUTATOR.
demographic or **demographical** *Statistics.* of or relating to demography or demographics.
demographics *Statistics.* the characteristics of a specific population, such as its average age, reproductive status, and future life expectancy. *Science.* systems or procedures that are based on such characteristics.
demography *Statistics.* the systematic study of population or of a specific population, including such aspects as size, age, distribution, and birth and death rate.
Demoivre, Abraham [dŭm wäv′rə] 1667–1754, French-born English mathematician; known for hisfor theorem and work on probability.
De Moivre's theorem *Mathematics.* a rule for calculating the power of a complex number expressed in polar form; i.e., if $z = r(\cos\theta + i\sin\theta)$ then $z^n = r^n(\cos n\theta + i\sin n\theta)$.
demolition *Military Science.* **1.** the destruction of structures, facilities, or materials by use of fire, water, explosives, mechanical, or other means. **2.** describing personnel or materials involved in demolition. Thus, **demolition duty, demolition (firing) party, demolition kit, demolition target.**
demolition belt *Military Science.* an area of land that is sown with explosive charges, mines, and other obstacles to deny use to enemy troops and provide protection to friendly troops.
demolition block *Ordnance.* an explosive charge used for demolition; it is usually packaged in a nonmetallic container.
demolition bomb *Ordnance.* a bomb that penetrates the surface of the earth before exploding, thus combining the force of its blast and the underground explosion.
demon *Artificial Intelligence.* a procedure that is activated automatically when a certain condition occurs, such as an if-added method. *Computer Science.* **1.** a small process that runs within a multiprocessing system and performs a utility function, such as printing spooled print files. Also, DAEMON, PARASITE, SYMBIONT. **2.** a small program that is normally inactive, but that monitors a particular condition and becomes activated when that condition occurs.
demon of Maxwell *Thermodynamics.* an imaginary creature who is able to violate the second law of thermodynamics, which states that it is not possible for entropy to decrease in an isolated system with no heat transfer. The demon controls a tiny door in an adiabatic wall between two gas-filled vessels of the same temperature; it opens and closes the door so as to concentrate faster (and hotter) molecules in one vessel and slower (and colder) ones in the other, thus decreasing the total entropy of the system. (Named for the Scottish physicist James Clerk *Maxwell.*)
demonophobia *Psychology.* an irrational fear of demons and devils. Also, **demonia, demonomania.**
Demon Star see ALGOL.
demophobia *Psychology.* an irrational fear of crowds.
De Morgan, Augustus 1806–1871, English mathematician; wrote major theoretical works on arithmetic, algebra, trigonometry, and logic.
De Morgan's laws *Mathematics.* **1.** theorems in symbolic logic that express the interaction of the Boolean operators AND, OR, and NOT. In particular, for statements p and q, (i) NOTp AND NOTq = NOT(p OR q); and (ii) NOTp OR NOTq = NOT(p AND q). Written symbolically, (i) $\neg p \wedge \neg q = \neg(p \vee q)$; and (ii) $\neg p \vee \neg q = \neg(p \wedge q)$, respectively. Also, **De Morgan's rules. 2.** analogous theorems in set theory that express the interaction of the set theoretic operations of union, intersection, and complement. In particular, for sets A and B, (i) $A' \cap B' = (A \cup B)'$; and (ii) $A' \cup B' = (A \cap B)'$.
De Morgan's test *Mathematics.* a test for convergence of the series $\sum_{n=1}^{\infty} a_n$ used when the ratio test fails; i.e., when $\lim_{n\to\infty} |a_n/a_{n-1}| = 1$. The series converges absolutely if there exists a positive real number ε such that $\lim\sup\{n\,(|a_n/a_{n-1}| - 1)\} = -1 - \varepsilon$.
demorphism see WEATHERING.
Demospongiae *Invertebrate Zoology.* a class of sponges with a skeleton of spongin fibers alone or in combination with silicaceous spicules, including commercial sponges.
demount *Computer Technology.* to remove a tape reel or disk from its drive.
demountable pack *Computer Technology.* a removable disk pack.
demountable tube *Electronics.* an electron tube that can be partially disassembled for inspection and replacement of electrodes.
Dempster, Arthur J. 1886–1950, Canadian-born American physicist; developed a mass spectrograph; discovered uranium-235.
Dempster-Shafer theory *Artificial Intelligence.* a theory of reasoning under uncertainty in which probability is assigned to subsets of the universe of possible outcomes, rather than only to individual outcomes.
DEMS digital electronic message system (service).
demultiplexer *Electronics.* a device that is designed to separate signals previously combined by a multiplexer and transmitted over a single channel.
demultiplexing circuit *Electronics.* a circuit that acts as a demultiplexer.
demustardization *Toxicology.* the treatment of poisoning due to exposure to mustard gas.
Demyanov rearrangement *Organic Chemistry.* the expanding or shrinking of a cyclic amine by treatment with nitrous acid.
demyelination *Pathology.* the loss or obliteration of myelin, a fatty material containing cholesterol, from the covering of the peripheral nerve.
DEN *Aviation.* the airport code for Stapleton International, Denver, Colorado.
denaturant *Chemistry.* a substance used to denature a product such as alcohol. *Organic Chemistry.* a substance, such as an organic solvent, a detergent, or concentrated urea, that converts a protein into a random coil by breaking its disulfide bridges. *Nucleonics.* a nonfissile material added in low concentration to a fissile isotope of the same element in order to render a reactor fuel unsuitable for use as a weapons material, without a difficult isotope separation or other physical or chemical process.
denaturation *Chemistry.* the process of denaturing a substance. *Molecular Biology.* any process in which the molecular structure of a substance, especially that of a protein or nucleic acid, is artificially altered using heat, radiation, or an acidic solution, in order to eliminate or modify one or more of its characteristic chemical, physical, or biological properties; the process usually involves the breaking of hydrogen bonds.
denaturation of DNA see DNA MELTING.
denature *Chemistry.* to add a chemical substance to a product to render it unpleasant or unfit for human consumption, so that the product may be used exclusively for other purposes. *Biochemistry.* to change irreversibly the structure or conformation, and thus the solubility and other properties, of a protein by heating, shaking, or treating with acid, alkali, or other species.
denatured alcohol *Chemistry.* ethyl alcohol to which another substance, such as methyl alcohol, has been added to make it unfit for use as a beverage.
denatured protein *Biochemistry.* a protein having unfolded or disordered polypeptide chains, which render the molecule less soluble and usually nonfunctional.
denaturize *Chemistry.* to subject to denaturation.
DENDRAL *Artificial Intelligence.* an expert system that discovers the most likely molecular structure and atomic composition based on mass spectrometric data and built-in chemical knowledge; developed at Stanford University for use in the chemical industry. Also, HEURISTIC DENDRAL.
dendraxon *Neurology.* a nerve cell whose axon branches into terminal filaments immediately after leaving its parent cell.
dendri- or **dendr-** a combining form meaning "tree," as in *dendritic, dendroid.*
dendriceptor *Neurology.* a receptive point at an end of the branching processes of a dendrite where it can enter into contact with and be stimulated by the axon endings of other neurons.
dendrite *Neurology.* a branchlike extension of the cytoplasm of a neuron; composing most of the neuron's receptive surface, dendrites resemble axons in structure but generally extend into treelike processes, especially in multipolar neurons. Also, DENDRON. *Materials Science.* a small solid protuberance that forms at the interface between a solid and a liquid.

dendritic *Science.* branched like a tree. *Neurology.* relating to or having dendrites.

dendritic drainage pattern *Hydrology.* a pattern of natural stream courses in which the streams branch irregularly in all directions and at any angle.

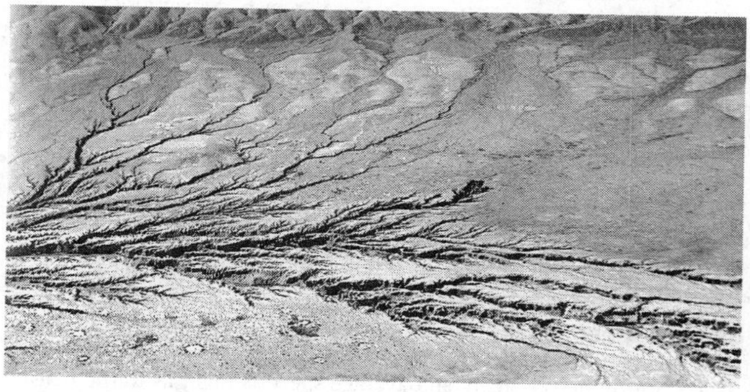

dendritic drainage

dendritic powder *Metallurgy.* a metal or alloy powder that has an acicular and branched morphology.

dendritic segregation see CORING.

dendro- a combining form meaning "tree," as in *dendrophagous.*

Dendrobatinae *Vertebrate Zoology.* the poison frogs, a family of small, diurnal, terrestrial frogs of the order Anura and native to the New World tropics; noted for their vivid warning colors and highly toxic skin, used in some cultures for poison darts and arrows.

dendrobrachiate gill *Invertebrate Zoology.* a highly branched respiratory structure found in certain decapod crustaceans.

Dendroceratida *Invertebrate Zoology.* an order of sponges with a skeleton of spongin fibers or no skeleton.

Dendrochirotacea *Invertebrate Zoology.* a subclass of sea cucumbers having retractible front ends and generally present podia and respiratory trees.

Dendrochirotida *Invertebrate Zoology.* an order of shallow-water, marine sea cucumbers having up to 30 highly branched tentacles.

dendrochronology *Archaeology.* a method of dating wooden objects by analyzing the pattern of their annual rings and comparing this pattern to an established tree-ring sequence for the region.

Dendrocolaptidae *Vertebrate Zoology.* the woodcreepers and woodhewers, a large family of tropical American birds belonging to the suborder Furnarii, characterized by a strong and compressed bill with which they dig out insects as they climb over trees.

dendrodochiotoxicosis *Medicine.* a mycotoxication found in horses and also in humans, caused by the fungus *Dendrodochium toxicum.*

dendrogram *Biology.* a branching diagram used to show relationships between members of a group; a family tree with the oldest common ancestor at the base, and branches for various divisions of lineage.

dendrohydrology *Hydrology.* the study of tree-ring configuration to determine hydrologic occurrences; variations in the width reveal variation in precipitation or water flow.

dendroid *Biology.* of or relating to a tree or its growth patterns; resembling a tree in form.

Dendroidea *Paleontology.* the earliest order of the class Graptolithina, colonial marine invertebrates extant through most of the Paleozoic; dendroids first appear in Middle Cambrian strata, while the other graptolite orders (including the most numerous, the graptoloids) first appear in the Lower Ordovician.

dendrology *Forestry.* the branch of forestry that focuses on the identification and classification of trees and shrubs.

dendrometer *Forestry.* a device for measuring the diameters and heights of trees or logs, based on principles concerning the relation of the sides of similar triangles.

Dendromurinae *Vertebrate Zoology.* the African tree mice, a subfamily of the Old World family Muridae.

dendron see DENDRITE.

dendrophagous *Zoology.* feeding mainly on trees.

dendrophobia *Psychology.* an irrational fear of trees.

dendrophysis *Mycology.* a hypha or filament that branches out on trees; found in fungi belonging to the family Cyphellaceae.

Dendropithecus *Anthropology.* an early Miocene hominoid fossil form from Kenya that had long limbs and a relatively short face, belonging to the superfamily Hominoidea and family Pliopithecidae.

Deneb *Astronomy.* Alpha (α) Cygni; at magnitude 1.3, the brightest star in the constellation Cygnus and a supergiant of spectral type A lying about 1500 light-years distant.

denervation *Physiology.* the removal, blockage, or destruction of the nerve supply to an organ.

dengue [deng´gē´] *Medicine.* a tropical or subtropical epidemic disease, transmitted to humans by the *Aedes* mosquito; charcterized by fever and severe pain in the joints, muscles, and head. Also, BREAKBONE FEVER.

dengue virus *Virology.* a flavivirus occurring as four antigenically related types that cause dengue. Also, BREAKBONE FEVER VIRUS.

denial *Psychology.* a defense mechanism in which a negative event or observation is denied or kept out of conscious awareness to avoid anxiety or pain.

denidation *Medicine.* the breakdown and expulsion of the superficial portion of the uterine mucosa in the menstrual cycle.

denier [də nēr´] *Textiles.* an international unit used to denote the fineness or coarseness of silk, nylon, and rayon yarn, designated by the weight in grams of 9000 meters of the yarn. The higher the denier number, the coarser the yarn.

denim *Textiles.* a sturdy, diagonally woven fabric of heavy, tightly twisted cotton yarns with a solid color warp and a lighter weft; used in blue jeans. (Originally called "serge de Nimes" after a town in France.)

Denisyuk hologram *Optics.* a type of hologram that can be viewed in white light, using the depth dimension of the emulsion.

denitrification *Chemistry.* **1.** the process of removing nitrogen or nitrogen compounds from a substance. **2.** the process of reducing into a gas such as nitrogen. *Biochemistry.* the liberation by bacteria of elementary nitrogen from nitrogenous compounds in the soil. Also, **denitration.**

denitrify *Chemistry.* **1.** to remove nitrogen or nitrogen compounds from. **2.** to reduce a nitrate or nitrite into a gas such as nitrogen. *Biochemistry.* to liberate nitrogen from nitrogenous soil. Also, **denitrate.**

denitrifying bacteria *Microbiology.* any strain of bacteria able to utilize nitrate or nitrite in an energy-yielding metabolic sequence which eventually produces nitrogen gas.

denitrogenate *Physiology.* to remove dissolved nitrogen from the body in order to prevent the occurrence of certain diseases or conditions.

Denjoy integral *Mathematics.* in Lebesgue integration theory, the fundamental theorem of calculus relates an absolutely continuous function f on an interval $[a,b]$ to its L^1 derivative f'. The Denjoy integral extends this to functions that fail to be absolutely continuous and whose derivatives fail to be in L^1, but which still can be viewed as the indefinite integral of their derivatives; e.g., $f(x) = x^2\sin(x^{-2})$ on $[0, 1]$.

Dennstaedtiaceae *Botany.* a family of large terrestrial ferns of the order Filicales, characterized by marginal, cuplike sori at leaf margins, creeping hairy or scaly rhizomes, and large fronds; includes the bracken, the world's most widespread fern.

denominator *Mathematics.* the (nonzero) quantity b in a fraction a/b, such as 2 in the fraction 1/2 (1 is the numerator).

denotation *Psychology.* the specific, literal meaning of a word, excluding any suggested or implied meaning that may be associated with it; for example, the denotation of the word "December" is the twelfth month of the year; this is apart from *connotations* of the word such as snow, cold weather, and the Christmas holiday. Also, **denotative meaning.**

de novo *Science.* coming from something new; over again. *Biochemistry.* denoting the synthesis of a specified molecule from very simple precursors, rather than the addition or subtraction of a side chain to an already complex molecule. (Latin for "anew" or "once again.")

dens *Anatomy.* **1.** a tooth or toothlike structure. **2.** an upwardly projecting process of the second cervical vertebra. (From the Latin word for "tooth.") Also, ODONTOID PROCESS.

dense *Mathematics.* a subset S of a topological space X is said to be **dense in** X if the closure of S is all of X; equivalently, if every point in X belongs to S or is an accumulation point of S.

dense-air refrigeration cycle see REVERSE BRAYTON CYCLE.

dense binary code *Computer Programming.* a coding system in which all possible bit patterns of the code are used to represent characters, as opposed to one in which certain combinations carry no meaning.

dense body *Cell Biology.* any intracellular region of high electron density.

dense connective tissue *Histology.* a type of connective tissue that contains a sufficient number of collagenous fibers to obscure the cells and fluid matrix.

dense-in-itself set *Mathematics.* a set S that is contained in its derived set; i.e., a set in which every point is an accumulation point. If in addition S is closed (i.e., if S contains all its accumulation points), then S is said to be perfect.

dense list *Computer Programming.* a list in which all or most of the cells are filled.

dense-media separator *Mining Engineering.* a device that causes heavier ores to sink and lighter ores to float by dispersion of a heavy mineral into water.

dense ring *Mathematics.* let $\mathrm{Hom}_D(V, V)$ denote the endomorphism ring of a vector space V of dimension $\dim(V)$ over a division ring D. A subring R of $\mathrm{Hom}_D(V, V)$ is called a dense ring of endomorphisms of V if, for every positive integer $n \leq \dim(V)$, every linearly independent subset $\{u_1, \ldots, u_n\}$ of V and every arbitrary (possibly dependent) subset $\{v_1, \ldots, v_n\}$ of V, there exists an endomorphism ϕ in R such that $\phi(u_i) = v_i$ for $i \leq n$. R is also called a **dense subring** of $\mathrm{Hom}_D(V, V)$. If $\dim(V)$ is finite, then the only dense subring of $\mathrm{Hom}_D(V, V)$ is $\mathrm{Hom}_D(V, V)$. By theorem, an arbitrary primitive ring is isomorphic to a dense ring of endomorphisms of some vector space.

densify *Engineering.* **1.** to cause a material to have greater density by impregnating it with additives under heat and pressure. **2.** to incerease density by removal of porosity through thermal processing.

densimeter *Engineering.* a device that determines the specific gravity of a substance or its exact gravity.

densitometer *Engineering.* an instrument that determines the optical density of a substance by measuring the intensity of the light it reflects or transmits. *Radiology.* a photoelectric apparatus used to determine the opacity of developed film, such as a radiograph.

densitometry *Radiology.* the measurement of the density of a material in terms of the density of another material, or a set standard.

density the concentration of items in a given system or area; specific uses include: *Mechanics.* the concentration of matter of an object, measured as the ratio of its mass to its volume; the mass per unit volume of a substance, expressed in such units as kg/m^3, g/cm^3 or lb/ft^3. *Graphic Arts.* **1.** in printing, the degree of ink coverage on a given sheet. **2.** on a photographic negative or positive, the amount of emulsion or dye per unit area. *Computer Technology.* the number of bits that can be recorded on a measurable unit of storage medium, such as an inch of magnetic tape or one track on a disk. *Ecology.* the number of individuals of the same species that live in a given area. *Optics.* see OPTICAL DENSITY. *Transportation Engineering.* the number of vehicles per mile on a given stretch of road at any moment. *Ordnance.* the average number of land mines per meter of minefield front.

density bombing *Ordnance.* bombing in which a specific tonnage of bombs is dropped on a specific area in order to ensure success in striking certain targets within the area.

density bottle see SPECIFIC-GRAVITY BOTTLE.

density correction *Engineering.* a correction applied to a measurement because of the variation of air density with respect to temperature.

density-dependent *Biology.* affected by the number of individual organisms that are present.

density-dependent factor *Ecology.* an environmental factor that affects the ability of a species to thrive as a direct result of that species' population density, tending to retard population growth (by increasing deaths or decreasing births) as density increases, or to enhance growth (by decreasing deaths or increasing births) as density decreases.

density-dependent inhibition of growth *Cell Biology.* the cessation of cell division in a population of cells due to high density and resultant cell contact of the population; an expression of contact inhibition.

density function *Mathematics.* **1.** given a discrete random variable X defined on a probability space (Ω, F, P), the function f defined by $f(x) = P\{X = x\}$. Also, **discrete density function** for X. The distribution function F of X may be defined in terms of the density function; i.e., $F(x) = P\{X \leq x\} = \Sigma_{t \leq x} f(t)$. **2.** given a continuous random variable defined on a probability space (Ω, F, P), a nonnegative function f defined by $\int_{-\infty}^{\infty} f(x)dx = 1$ and $\int_a^b f(x)\,dx = P\,\{a < X < b\}$; this is also called a **continuous density function** for X. The distribution function F of X may be defined in terms of the density function; i.e., $F(x) = P\{X \leq x\} = \int_{-\infty}^{x} f(t)dt$.

density gradient centrifugation *Analytical Chemistry.* a method of separating the components of a mixture according to the density of their particles by spinning in a centrifuge.

density-independent factor *Ecology.* a factor that influences a population but that is independent of the existing population density, such as climate.

density indicator or **density gauge** see DENSIMETER.

density modification *Crystallography.* a computational method for improving the phases used to calculate an electron density map from X-ray diffraction data. An "envelope" defining the approximate boundary of the molecule is determined from the electron density map, and all electron density outside this envelope is set to its average value. A new set of phases is then determined by Fourier inversion of this "solvent flattened" map. The usefulness of the method increases as the fraction of solvent in the unit cell increases. Also, SOLVENT-FLATTENING.

density modulation *Electronics.* a process in which the density of the electron beam in an electron tube is varied with time, resulting in a bunching of electrons.

density of states *Solid-State Physics.* a function of the energy in a solid given by the number of permitted quantum states in the energy range between E and $E + dE$, per unit volume of the material.

density parameter *Astrophysics.* the ratio of the local mean mass density of the universe to the critical density.

density rule *Forestry.* a grading system for lumber that is based on the width of annual rings.

density-speed relationship *Transportation Engineering.* the relationship between traffic density and average vehicle speed. At low densities, traffic moves freely and average speed is high; as traffic density increases and cars can no longer move freely, average speed drops.

density-speed-volume relationship *Transportation Engineering.* the relationship among traffic density, average vehicle speed, and total traffic volume. Increasing density corresponds to increasing volume for a given average speed, and at low densities the volume is nearly proportional to density. As roadway capacity is approached, however, further increases in density force a reduction in traffic speed, limiting further increases in total volume. Excessive density ("bumper to bumper traffic") may so reduce average speeds that total traffic volume is reduced.

density stratification *Hydrology.* the division of a lake or other body of water into horizontal layers as a result of differences in density; the lightest layer occurs near the top and the heaviest layer near the bottom of the lake.

density-tolerant *Ecology.* relating to or designating a species that is able to tolerate a high level of population density without territorial dispersal or a drop in birth rate. Similarly, **density-intolerant.**

density transmitter *Engineering.* an instrument that records the density of a flowing stream by measuring the buoyancy of a hollow chamber immersed in the liquid.

density-volume relationship *Transportation Engineering.* the relationship between traffic density and total traffic volume. At low densities, density and volume are nearly in proportion; as roadway capacity is approached, however, increased density no longer increases total volume, but leads to lower average speeds instead.

density wave *Physics.* a wave that causes the density of the matter through which it passes to rise alternately above and below its mean value.

density-wave theory *Astrophysics.* a theory that explains star formation and spiral arms in the disk of a galaxy by the effects of a density wave rotating with constant angular velocity.

densofacies see METAMORPHIC FACIES.

Densovirus *Virology.* a genus of viruses of the family Parvoviridae that infect a wide range of insects.

dent. dental; dentist; dentistry.

dent- a combining form meaning "tooth," as in *dentation.*

dental *Anatomy.* relating to a tooth or teeth. *Medicine.* relating to dentists or the practice of dentistry.

dental amalgam *Materials.* a soft mixture of mercury and alloys of silver, tin, and copper that becomes very hard when set and is used to fill tooth cavities.

dental arch *Anatomy.* the arch formed by the alveolar process of each jaw, with or without the attached teeth.

dental caries see CARIES.

dental cement *Materials.* a material used to attach dental implants to bone. Also, PMMA.

dental composite *Materials.* a durable, particle-reinforced polymer that is used for dental restorative material.

dental coupling *Mechanical Engineering.* a flexible coupling device that is used to join a reduction-gear pinion shaft to a steam turbine; consists of a short shaft with a set of gear teeth at each end.

dental enamel *Anatomy.* a hard biological material composed of a calcium phosphate salt that covers the crown of the tooth.

dental epithelium *Histology.* the epithelial cells marking the limit of the dental organ.

dental follicle see DENTAL SAC.

dental formula *Vertebrate Zoology.* a concise tabular statement of the dentition of a mammal; in the usual statement the numbers and kinds of teeth in the upper and lower jaw are written above and below a horizontal line. Also, **dental index.**

dentalgia *Medicine.* a medical term for toothache.

dental gold *Metallurgy.* gold that is at least 99.95% pure and is thus suitable for use in dental fillings.

Dentaliidae *Invertebrate Zoology.* a family of tooth shell mollusks of nearly worldwide distribution.

dental pad *Vertebrate Zoology.* a firm ridge replacing the incisors in the upper jaws of cud-chewing herbivores.

dental papilla *Developmental Biology.* a projection of the mesenchymal tissue of the developing jaw into the cup of the enamel organ; the outer layer develops into odontoblasts that form the dentin of the tooth. Also, DENTINAL PAPILLA.

dental plate *Vertebrate Zoology.* a flattened, often sharp-edged plate representing fused teeth in parrot fish and related forms. *Invertebrate Zoology.* a similar plate found in place of teeth in some invertebrates, such as certain worms.

dental porcelain *Materials.* a vitrified feldspar material that is used in a molded form to make denture teeth, or ground to a powder for crown and bridgework material.

dental prosthetics see PROSTHODONTICS.

dental pulp *Histology.* the vascular connective tissue found inside the pulp cavity of a tooth.

dental ridge *Developmental Biology.* the prominent border of a cusp or margin of a tooth.

dental sac *Developmental Biology.* a concentric layer of connective tissue in which the enamel organ and dental papilla are embedded, and that completely surrounds the developing tooth after disintegration of the epithelial attachment connecting the enamel organ with the dental lamina. Also, DENTAL FOLLICLE.

dental wax *Materials.* a plastic material used primarily to make an impression of an oral structure that serves as a pattern for the construction of cast and molded dental devices.

dentate *Biology.* having teeth or teethlike marginal projections.

dentate nucleus *Anatomy.* an ovoid nerve cell mass located in the center of each of the cerebellar hemispheres; the fibers arising from the cells are located in the superior cerebellar peduncle.

denti- a combining form meaning "tooth," as in *dentition.*

denticle *Zoology.* a small tooth or toothlike structure.

denticulate *Zoology.* having small teeth or teethlike structures.

dentifrice [den´tə fris] *Materials.* any substance that is used to clean and polish the teeth, such as toothpaste.

dentigerous see DENTATE.

dentil *Architecture.* any of a row of small rectangular blocks resembling teeth, used to decorate a classical cornice.

dentilingual *Anatomy.* of or relating to the teeth and tongue.

dentin *Histology.* the bonelike tissue that forms the bulk of a vertebrate tooth.

dentinal papilla see DENTAL PAPILLA.

dentinoblast *Histology.* an embryonic mesodermal cell that produces dentin.

dentinogenesis *Physiology.* the formation of dentin in the teeth. Also, **dentification, dentinification.**

dentist *Medicine.* a person who is professionally trained and licensed to practice dentistry.

dentistry *Medicine.* the practice of identifying and treating diseases, disorders, and abnormalities of the teeth and surrounding areas of the mouth, including such activites as the repair and restoration of teeth and the replacement of missing teeth.

dentition *Vertebrate Zoology.* **1.** the kind, number, and arrangement of the teeth of man and animals. **2.** the development and cutting or eruption of teeth.

dento- a combining form meaning "tooth," as in *dentosurgical.*

denture *Medicine.* an artificial structure used to replace a missing tooth or group of teeth.

denudation *Geology.* **1.** the various combined effects that result in the general wearing away of the earth's surface, including natural processes such as weathering, erosion, mass wasting, and transportation. **2.** the uncovering or exposure of bedrock by the erosion and removal of overlying material. *Surgery.* the act of laying bone, as in the removal of the epithelial layer through surgery, trauma, or pathological change.

denumerable see COUNTABLE.

deodorant *Materials.* any substance or commercial preparation used to remove or mask unpleasant odors by adsorption, replacement, neutralization, or oxidation.

deodorizing *Chemical Engineering.* any process for eliminating odor-creating substances from a product.

deoperculate *Botany.* **1.** lacking an operculum. **2.** having an operculum that does not separate spontaneously, as in mosses and liverworts.

deorbit *Space Technology.* to depart intentionally from a spacecraft orbit, usually to enter a descent phase or to change course.

deoxidation *Chemistry.* the process of removing oxygen from a substance. *Metallurgy.* specifically, a process in which aluminum, silicon, or other easily oxidized materials are added to liquid metals prior to the solidification in order to eliminate dissolved oxygen.

deoxidize *Chemistry.* **1.** to remove oxygen from a substance. **2.** to reduce a substance from an oxide state. *Metallurgy.* specifically, to remove oxygen from a molten metal or alloy bath.

deoxidizer *Chemistry.* a substance that has the ability to either remove oxygen from a substance or reduce oxide compounds. Also, **deoxidant.**

5'-deoxyadenosylcobalamine see COENZYME B_{12}.

deoxycholate *Biochemistry.* a salt or ester of deoxycholic acid.

deoxycholic acid *Biochemistry.* $C_{24}H_{40}O_4$, an unconjugated bile acid.

deoxycorticosterone *Endocrinology.* a very potent mineralocorticoid hormone, normally produced in very small amounts by the adrenal cortex. Plasma levels of this substance are vastly elevated in pregnancy, apparently as a result of the extra-adrenal conversion of progesterone via a 21-hydroxylase.

deoxygenate [dē´äk´sə jə nāt´] *Chemistry.* to remove free oxygen from a substance. *Physiology.* to deprive the blood of sufficient oxygen. Thus, **deoxygenation.**

deoxyribonuclease *Enzymology.* an enzyme that catalyzes the hydrolysis of deoxyribonucleic acid to nucleotides.

deoxyribonucleic acid [dē äk´sē rī´bō noo klā´ik] *Biochemistry.* DNA, the substance that is the carrier of genetic information, found in the chromosomes of the nucleus of a cell. See DNA.

deoxyribonucleic acid ligase see DNA LIGASE.

deoxyribonucleic acid polymerase see DNA POLYMERASE.

deoxyribonucleoprotein *Biochemistry.* a nucleoprotein in which the nucleic acid sugar is D-2-deoxyribose.

deoxyribonucleoside *Genetics.* a compound that consists of a purine or pyrimidine base attached to a 2-deoxyribose molecule.

deoxyribonucleotide *Genetics.* a compound that consists of a purine or pyrimidine base attached to a 2-deoxyribose molecule that is bonded to a phosphate group.

deoxyribose [dē äk´sē rī´bōs] *Biochemistry.* $C_5H_{10}O_4$, an aldopentose found in DNA, deoxyribonucleotides, and deoxyribonucleosides.

deoxyribovirus *Virology.* any virus that contains a DNA genome.

deoxy sugar *Biochemistry.* a sugar that has one or more hydroxyl groups replaced by hydrogen.

DEP see DIETHYLPHTHALATE.

dep. purified. (From Latin *depuratus.*) Also, **Dep.**

departure *Navigation.* **1.** the last pilotage fix by a vessel prior to its putting to sea. **2.** the distance to east or west made good by a vessel. *Cartography.* the orthographic projection of a line on an east-west axis of reference, giving the difference of the meridian distances or longitudes of the ends of the line; the departure is positive for a line having an azimuth or bearing in the northeast or southeast quadrant, and negative for a line having an azimuth or bearing in the southwest or northwest quadrant.

departure control *Transportation Engineering.* the control of the separation of aircraft departures from an airport by air traffic control.

departure point *Navigation.* the geographical position determined by a vessel's final pilotage fix on putting to sea, and at which its dead-reckoning track is begun.

depauperate *Biology.* exhibiting inferior size or development.

dependency *Medicine.* the condition of having a habitual craving or need for some chemical substance, such as alcohol or a drug. *Psychology.* the fact of relying heavily on another or others for emotional support and for advice and reassurance about the nature of one's actions. *Statistics.* any form of relationship between random variables; for example, correlation. *Computer Programming.* a relationship between jobs or processes such that one must be completed before the other can begin. Also, **dependence.**

dependency-directed backtracking *Artificial Intelligence.* a nonchronological truth maintenance method used to clean up database entries that follow a failure or dead-end path.

dependency needs *Psychology.* basic needs, originating in infancy, for mothering, comfort, security, love, protection, and shelter.
dependent *Medicine.* **1.** a person who has a craving for or addiction to some chemical substance, such as alcohol. **2.** relating to or affected by dependency.
dependent-demand inventory *Industrial Engineering.* a system of inventory control in which the demand for one item of inventory is directly related to the demand for other inventory items, such as a demand for tires and hubcaps in order to meet the demand for new automobiles.
dependent equations *Mathematics.* a set of equations, at least one of which is satisfied by a simultaneous solution of some subset of the others.
dependent events *Statistics.* events such that the occurrence of one modifies the probability of the occurrence of the other; thus, the conditional probability of event A, given that B has occurred, is different from the independent probability of A.
dependent personality disorder *Psychology.* a variation from normal personality characterized by low self-esteem, excessive reliance on the support or approval of others, and an inability to make decisions or take responsibility.
dependent rank or **ranking** *Behavior.* an animal's position in a rank order that is dependent on or influenced by the status of other group members.
dependent variable *Mathematics.* the value of a function that is determined by the function and the value or values chosen for its independent variable(s); for example, y in the equation $y = f(x)$. Also, OUTPUT VARIABLE, OUTPUT OF THE FUNCTION. *Psychology.* in psychological experimentation, the specific behavior pattern that is produced in the subject by the experimenter's varying or manipulating some condition of the experiment.
Dependovirus *Virology.* a genus of defective viruses of the family Parvoviridae that are dependent upon a coinfecting adenovirus for their replication. Also, ADENO-ASSOCIATED VIRUS.
depentanizer *Chemical Engineering.* a fractionating column to remove pentane and lighter fractions from hydrocarbon mixtures.
Deperetellidae *Paleontology.* an Eocene family of perissodactyl mammals of the superfamily Tapiroidea.
depergelation *Hydrology.* the decay or thawing of permafrost.
deperm see DEGAUSS.
depersonalization [dē pur′sən əl i zā′shən] *Psychology.* an individual's feeling that he has lost his sense of identity and become strange or unreal. Thus, **depersonalization disorder.**
dephlegmation [dē fleg mā′shən] *Chemical Engineering.* the partial condensation of vapor to form a liquid containing higher boiling constituents than the original vapor, often done in a distillation operation.
dephlegmator [dē fleg′māt ər] *Chemical Engineering.* a device used in fractional distillation to cool the vapor mixture, thus condensing less volatile boiling fractions. Also, FRACTIONATING COLUMN.
dephosphorize [dē fäs′fə riz] *Metallurgy.* to remove phosphorus from a metal or alloy.
depilation [dep′i lā′shən] *Engineering.* the process of removing hair.
depilatory [də pil′ə tôr ē] *Materials.* a substance, usually a sulfide, used to remove hair from skin.
depleted uranium *Nucleonics.* the uranium-238 isotope that remains after the fissile uranium-235 isotope has been removed from source nuclear material by isotope separation.
depletion an often harmful decrease in the supply of something; specific uses include: *Ecology.* the use of a resource at a faster rate than it can be replenished through natural processes. *Agronomy.* decreased soil productivity resulting from a loss of plant nutrients. *Electronics.* the reduction of the density of charge carriers in the region of a semiconductor P-N junction, due to a reverse biasing of the junction. *Nucleonics.* the process of removing or consuming the principal substance of raw materials, as in the concentration of ores and isotopes or by the burnup of fuel in a nuclear reactor.
depletion drive *Petroleum Engineering.* a well drive displacement mechanism for expelling hydrocarbons from porous reservoir formations; types of drives are water or gas, either injected or natural, and injected LPG. Also, SOLUTION GAS DRIVE, DISSOLVED GAS DRIVE.
depletion effect *Ecology.* a decline in the rate of immigration of nonnative species to an area as the resident species increase in population.
depletion layer *Electronics.* the region of a reverse-biased P-N junction in which there are no charge carriers, so that current does not flow and the junction acts as an insulator rather than as a semiconductor. Also, BARRIER LAYER, DEPLETION REGION, SPACE-CHARGE LAYER.
depletion-layer capacitance *Electronics.* the capacitance of the depletion layer of a semiconductor. Also, BARRIER-LAYER CAPACITANCE, JUNCTION CAPACITANCE.
depletion-layer rectification *Electronics.* the rectification that occurs at the contact between dissimilar materials, such as a semiconductor P-N junction, as the energy levels on each side of the discontinuity are readjusted.
depletion-layer transistor *Electronics.* a transistor whose operation depends directly on the flow of carriers through depletion layers.
depletion method *Molecular Biology.* a laboratory procedure for isolating mRNA specific to a particular gene; involves hybridization of total cellular mRNA to a specific segment of DNA.
depletion mode transistor *Electronics.* a field-effect transistor in which the gate is reverse-biased with respect to the channel, and which can be driven from a state of full conduction to nonconduction by varying the gate voltage.
depletion region see DEPLETION LAYER.
depletion-type reservoir *Petroleum Engineering.* an oil reservoir that is constantly in a state of equilibrium between the gas and liquid phases; oil is displaced during production by the expansion of gas set free from solution in the oil.
deployed nuclear weapons *Ordnance.* nuclear weapons that have been authorized by the Joint Chiefs of Staff for transfer to the storage facilities, transportation, or delivery units of the armed forces.
deployment *Military Science.* the act of extending, rearranging, or relocating military forces, usually in preparation for action.
depocenter *Geology.* a site of maximum deposition, especially the thickest part of a stratigraphic unit in a depositional basin.
depolarization [dē pōl′ər i zā′shən] *Electricity.* **1.** the removal, decrease, or prevention of agents that cause polarization in an electric cell. **2.** the addition of substances to depolarize a cell. *Optics.* a process in which a beam of polarized light is reflected in all directions perpendicular to its axis so that its vibrations no long occur along a single plane.
depolarizer *Physical Chemistry.* **1.** a substance that tends to reduce the polarization of an electrode. **2.** a substance that is preferentially oxidized or reduced to preclude undesired reactions.
depolymerization [dē pōl′ē mər i zā′shən] *Organic Chemistry.* the decomposition of a polymer into its constituent monomers.
deposit *Geology.* **1.** any natural accumulation or laying down of consolidated or unconsolidated material from a source. **2.** to carry out or undergo this process.
deposit attack *Metallurgy.* corrosion occurring in local areas where a deposit is present on the surface.
deposit feeder *Invertebrate Zoology.* any animal that eats detritus and other organic materials which collect at the bottom of an aquatic habitat. Also, DETRITUS FEEDER.
deposit gauge *Engineering.* an instrument that measures the quantity of particles deposited at a specified area within a specified time period; used in air pollution studies.
deposition [dep′ə zish′ən] *Geology.* the process in which material from any source is laid down or accumulates naturally, especially in beds, veins, or irregular masses of any kind. *Archaeology.* any of the various processes by which artifacts move from active use to an archaeological context, such as loss, disposal, abandonment, burial, and so on. *Toxicology.* the accumulation of a poison in body tissues.

glacial deposition

depositional [dep´ə zish´ə nəl] *Archaeology.* of or relating to archaeological deposition. A **depositional stratum** or **unit** is a separate layer of material at a site. A **depositional history** is the order in which objects are deposited at a site.
depositional dip see PRIMARY DIP.
depositional fabric *Petrology.* an arrangement of detrital particles settled from suspension or of crystals from a differentiating magma; determined by the surface plane on which they rest.
depositional mark or **depositional marking** *Geology.* an irregularity, such as a scour or tool mark, that forms on a bedding plane during deposition.
depositional strike *Geology.* the strike of the active depositional surface.
depositional topography *Geology.* the existing topography on the active depositional surface.
deposition potential *Physical Chemistry.* the minimum amount of electric potential required to cause a given electrode to begin discharging ions. Also, DISCHARGE POTENTIAL.
deposition rate *Metallurgy.* in welding or in electroplating, the rate at which the bead, or the plate, is deposited.
depot [dē´pō´] *Military Science.* **1.** an activity or facility for handling equipment and supplies, including receipt, storage, accounting, issue, maintenance, procurement, manufacture, assembly, research, salvage, and disposal. Also, SUPPLY DEPOT. **2.** an activity or facility for handling personnel, including reception, processing, training, assignment, and forwarding of replacements. Also, PERSONNEL DEPOT.
depressed *Psychology.* suffering from depression.
depressed center car *Engineering.* a railroad flat car that has a lowered center section where oversized loads are placed so they can clear tunnels.
depressed equation *Mathematics.* an equation obtained from a given equation by reducing the number of roots; for example, the depressed equation $2x^2 + 7x + 6 = 0$ is obtained from $2x^3 + 5x^2 - x - 6 = 0$ by dividing out $x - 1$.
depressed fracture *Medicine.* a fracture of the skull in which the fractured part is flattened below the normal level.
depression *Geology.* **1.** a relatively sunken or low-lying area of the earth's surface, especially one having no natural outlet for surface drainage. **2.** a lowered, sunken, or downthrusted area in the earth's crust. *Meteorology.* see LOW. *Psychology.* a mental state characterized by extreme feelings of sadness, despair, hopelessness, and low self-esteem, often accompanied by physical symptoms such as loss of appetite and energy; the negative mood is out of proportion to any actual event or condition that may have precipitated it.
depression contour *Geology.* on a topographic map, a continuous closed contour line representing a depression having no surface outlet and distinguished from normal contour lines by hachure marks on the downslope side.
depression of freezing point see FREEZING-POINT DEPRESSION.
depression spring *Hydrology.* a spring whose waters flow from permeable material onto the land surface because the land slopes down to the water table.
depression storage *Hydrology.* water from precipitation that accumulates and is retained in surface depressions, such as puddles or ditches.
depressive *Psychology.* of or relating to depression.
depressive disorder *Psychology.* see DEPRESSION.
depressor *Anatomy.* any muscle, agent, apparatus, or instrument that depresses. *Chemical Engineering.* an agent that retards or prevents a chemical process or reaction; a negative catalyst.
depressor nerve *Anatomy.* an afferent nerve whose stimulation causes a fall of blood pressure via a brainstream reflex. Also, AORTIC NERVE.
deprivation *Behavior.* a group of negative behavioral patterns that result from an animal's being isolated during its early life. *Psychology.* a group of symptoms found in young children and associated with lack of maternal contact, including withdrawal, depression, and apathy. Also, **deprivation syndrome.**
deprivation experiment see EXPERIENCE DEPRIVATION.
depropagation *Materials Science.* a chain depolymerization by successive release of monomer units from a chain end; it may be initiated by ultraviolet light, oxygen, ozone, or other similar nonpolymeric agents.
depropanize *Chemical Engineering.* the elimination of propane and sometimes lighter fractions from a petroleum fraction in oil processing.
depropanizer *Chemical Engineering.* a fractionating column that removes propane and lighter components in a gasoline or petrochemical plant.
deproteinize *Organic Chemistry.* to remove proteins from tissue or material.
depside *Organic Chemistry.* a phenolic plant product containing ester links.
depsidone *Organic Chemistry.* a member of a class of compounds consisting of depside esters that are also cyclic ethers.
Depterifores *Vertebrate Zoology.* in some classification systems, an equivalent term for Lepidosireniformes, the lungfishes.
depth a vertical distance from top to bottom; specific uses include: *Oceanography.* the vertical distance below the sea surface at any given point, especially the sea floor. *Hydrology.* the vertical measure of any body of water. *Artificial Intelligence.* the number of operator applications between the root of a search tree and a given node.
depth bomb *Ordnance.* a bomb designed to be dropped from an airplane onto underwater targets.
depth bound *Artificial Intelligence.* the maximum depth that is permitted in a bounded depth-first search.
depth charge *Ordnance.* an explosive charge used in antisubmarine warfare; it is launched from the deck of a surface ship, propelled by a mortar or missile, or dropped from a helicopter or fixed-wing aircraft, and detonated at a preset depth.
depth curve *Cartography.* a line on a map or chart connecting points of equal depth below the hydrographic datum.
depth excess *Oceanography.* the difference between the depth of the sea floor at a given point and the depth where the sound velocity equals either the surface velocity or the maximum velocity in the surface layer.
depth factor *Oceanography.* the ratio of the height of a wave to its ideal height in deep water with refraction eliminated. Also, SHOALING FACTOR, COEFFICIENT.
depthfinder *Engineering.* a device utilizing radar or ultrasonic waves to determine the depth of the sea.
depth-first search *Artificial Intelligence.* a procedure used by most backward chaining inference systems that gathers as little information as possible in the attempt to find a path from the initial question to the final goal.
depth gauge *Mechanical Devices.* **1.** a device used for measuring to tolerances of one thousandth of one inch when digging a hole, groove, or slot. **2.** a device clamped to a drill bit and used as a means of measuring the depth of a borehole. Also, **depth gage.**
depth hoar *Hydrology.* delicate stepped or layered ice crystals that are formed below the surface of snow, usually near the bottom of a thick snow cover. Also, SUGAR SNOW.
depth ice see ANCHOR ICE.
depth magnification *Optics.* the ratio of the distance separating two points on the axis on the image side of an optical system to the distance separating their conjugate points on the object side. *Mechanical Devices.* a screw-action-type depth gauge with a micrometer scale.
depth marker *Engineering.* a thin board used to distinguish snow or ice surface that has been completely covered with freshly fallen snow.
depth of compensation *Oceanography.* the depth in the ocean at which the production of oxygen and its consumption are equal. *Geophysics.* the depth (100 to 117 km below the surface) at which the slow-moving viscous material that supports the brittle material of the earth's crust adjusts to changes in the load on the surface.
depth of engagement *Design Engineering.* the depth to which mating threads intermesh, measured perpendicular to the axis.
depth of field *Optics.* the distance in object space over which a camera lens focused on a given subject can provide adequate definition or clarity.
depth of fusion *Metallurgy.* in welding, the depth of base metal, or of a previous bead, that is molten during each pass.
depth of isostatic compensation *Geodesy.* the depth below sea level at which the gravitational effect of masses extending above the surface of the geoid is approximately counterbalanced by a deficiency of density below.
depth of no motion *Oceanography.* the depth at which submarine water movement caused by surface waves practically ceases; generally at a depth equal to a surface wavelength.
depth of thread *Design Engineering.* the distance that the base of a thread is below its crest, measured perpendicular to the axis.
depthometer *Petroleum Engineering.* an instrument used for determining the depth of a particular zone or the full depth of the well.
depth perception *Physiology.* the perception of solidity of a visual object and its location in the spatial field, through the fusion in the brain of the two slightly dissimilar images from the two eyes.

depth profiling *Spectroscopy.* the determination of crystal layer separation using spectroscopic methods.

depth sounder *Engineering.* a device used to measure the depth of the sea under a ship.

depth-type filtration *Chemical Engineering.* a method to remove solids by passing the carrier fluid through a mass-filter medium having a circuitous route with many traps to catch the solids.

depth zone *Geology.* at various depths within the earth, a characteristic environment associated with different metamorphic phenomena.

depurination *Biochemistry.* a process in which an acid is used to remove purines from DNA.

dequeue [dē kyoo´] *Computer Science.* to remove items from a queue for service according to prearranged rules, usually first-in, first-out.

DER degeneration reaction.

der. derivative.

deRahn cohomology *Mathematics.* the cohomology, under the differential *d,* of the algebra of differential forms.

derail *Engineering.* to cause a railroad car or engine to go off the tracks.

derailment *Psychology.* a thought disorder characterized by abrupt shifts from one topic to another unrelated topic.

derangement *Mathematics.* a permutation having no fixed points.

derating *Electronics.* the reduction of the voltage, current, or maximum power dissipation rating of a device or component, usually to allow for operation at higher ambient temperatures.

derbesia *Botany.* a green algae genus of the family Bryopsidaceae, characteristically a dense tuft of sparsely branched filiform tubes bearing zoospore-producing sporangia.

derby *Metallurgy.* the product of the bomb reduction process; principally signifying magnesium-reduced uranium.

derbylite *Mineralogy.* $(Fe^{+3},Fe^{+2},Ti)_7Sb^{+3}O_{13}(OH)$, an opaque black to brown monoclinic mineral occurring as prismatic crystals, having a specific gravity of 4.53 and a hardness of 5 on the Mohs scale; found in cinnibar-bearing gravels in Brazil.

Derbyshire spar see FLUORITE.

Dercum's disease see ADIPOSIS DOLOROSA.

derealization [dē rē´ə li zā´shən] *Psychology.* a loss of one's sense of reality; the feeling that one's surroundings have changed and that the world is strange or unfamiliar.

dereism [di rē´iz əm] *Psychology.* thinking that is not in accordance with reality and experience; illogical, idiosyncratic reasoning.

derelict [der´ə likt] *Navigation.* a vessel abandoned at sea.

derelict land *Ecology.* a term for land that has been seriously damaged by industrial activity, such as strip mining.

derepression *Genetics.* an increase in the transcriptional activity of a gene sequence due to the removal of a repressor molecule from the sequence or from a neighboring control sequence.

Derepyxidaceae *Botany.* a family of biflagellate colorless or golden algae cells of the order Isochrysidales, characterized by equal flagella and a lorica through which they protrude; found in fresh or brackish waters.

derichment *Crystallography.* in gravimetric analysis by coprecipitation of salts, a system with a defined distribution coefficient that is less than one.

deriv. derivative.

derivation *Mathematics.* **1.** the process of constructing a formula or equation from other formulas or equations. **2.** see DIFFERENTIATION. **3.** more generally, any ring homomorphism that satisfies the Leibnitz rule. That is, a linear operator Φ defined on a ring such that $\Phi(fg) = \Phi(f)g + f\Phi(g)$, where *f* and *g* are elements of the ring. If Φ is the derivative and *f* and *g* are functions, this requirement is the ordinary product rule for differentiation. *Computer Science.* a diagram, list of steps, or tree structure that shows how a sentence in a language is derived from a grammar by the application of grammar rules.

derivative *Chemistry.* a chemical compound, usually organic, that is made from another chemical compound. *Mathematics.* **1.** the slope of the tangent line, if it exists, to a function at a point. Formally, let $f(z)$ be a single-valued continuous function of a real or complex variable *z* in a domain *D*. $f(z)$ possesses a derivative $z_0 \in D$ if: (a) $\lim_{h\to 0} [f(z_0 + h) - f(z_0)]/h$ exists and is finite; and (b) this limit is independent of how *h* approaches zero. The derivative of $f(z)$ at z_0, if it exists, is then denoted $f'(z_0)$ and *f* is said to be **differentiable** at z_0. The operation of forming a derivative is called **differentiation** and is indicated by d/dz or *D*, where $d/dz(f(z)) = df/dz = Df = f'(z)$ If *f* is a function of more than one variable, a derivative with respect to one variable x_j (while the others are held constant) is then called a **partial derivative** and denoted by $\partial f/\partial x_j$. **2.** more generally, let $f: R^n \to R^m$ be a continuously differentiable function defined on a region *D* of R^n. The derivative f' is the linear transformation given by the $m \times n$ matrix whose (i, j)th entry is $\partial f_i/\partial x_j$ (the Jacobian matrix), where f_i is the *i*th component function of *f* in R^m and x_j is the *j*th coordinate function in R^n.

derivative action *Control Systems.* a corrective action from a controller, the speed of which depends upon how fast the rate of error in the system is increasing. Also, **derivative compensation.**

derivative control *Control Systems.* a control process or method in which an actuator drive signal is proportional to the time derivative of the difference between the desired output and actual output.

derivative network *Control Systems.* a compensating network whose output is proportional to the sum of the input signal and its derivative. Also, LEAD NETWORK.

derivative polarography *Analytical Chemistry.* an electroanalytical technique in which the rate of change of current with respect to applied potential is measured as a function of the applied potential.

derivative rock see SEDIMENTARY ROCK.

derivative thermometric titration *Analytical Chemistry.* an analytical procedure during which a resistance-capacitance network is used to record first and second derivatives of the curve of temperature versus weight change upon heating to obtain a sharp end point.

derived algebra *Mathematics.* given a Lie algebra *L*, the ideal composed of all linear combinations of commutators of *L*; analogous to the concept of commutator subgroup.

derived character *Evolution.* a hereditary character or distinguishing feature altered from an earlier form of a species.

derived equation *Mathematics.* **1.** in algebra, any equation obtained from a given equation by performing the same operations to both sides of the given equation. **2.** in calculus, the equation obtained by differentiating both sides of a given equation.

derived measurement see DERIVED UNIT.

derived property *Thermodynamics.* a property in a system that cannot be measured by performing an experiment on the system, but that can be described in terms of a change in the system.

derived quantity see DERIVED UNIT.

derived set *Mathematics.* the set of all accumulation points of a given set.

derived sound system *Acoustical Engineering.* a stereophonic sound system in which each of the two channels feeds two loudspeakers, thus approximating a quadraphonic system.

derived unit *Metrology.* a unit of measure obtained from a combination of other measurements; for example, the newton is a derived unit of force obtained by measuring mass, length, and time.

derm- a combining form meaning "skin," as in *dermabrasion.*

derm. dermatology; dermatologist.

dermabrasion *Medicine.* a surgical procedure that uses an abrasive disk or other mechanical method to plane the skin.

Dermacentor *Invertebrate Zoology.* a genus of bloodsucking wood ticks that carry mammalian diseases.

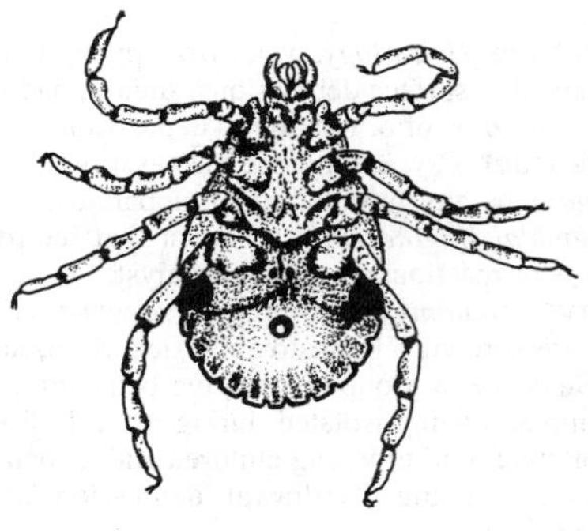

Dermacentor

Dermacentor andersoni *Invertebrate Zoology.* a species of wood tick found in the central and western United States that carries tularemia and Rocky Mountain spotted fever.

Dermacentor occidentalis *Invertebrate Zoology.* a species of wood tick found on the Pacific coast of North America.

Dermacentor variabilis *Invertebrate Zoology.* a species of tick found in the eastern United States, chiefly on dogs.

dermal *Anatomy.* of or relating to the dermis or the skin.

dermal bone *Anatomy.* any bone that ossifies directly from mesenchyme, rather than by replacement of a preceding cartilaginous structure. Also, INTRAMEMBRANOUS BONE.

dermal denticle *Vertebrate Zoology.* a toothlike scale or other conical projection on a shark's skin, composed mostly of dentine and having a pulpy central cavity.

dermalla *Invertebrate Zoology.* microscopic deposits in the bodies of sponges.

dermal pore *Invertebrate Zoology.* any one of many small openings on the surface of a sponge through which water passes into the sponge.

dermal tissue system *Botany.* the outer covering tissue of a plant, epidermis, or periderm.

Dermaptera *Invertebrate Zoology.* the earwigs; an order of thin omnivorous insects with biting mouthparts and a pair of large forceplike appendages on the hind end of the abdomen.

dermat- or **dermato-** a combining form meaning "skin," as in *dermatitis, dermatology.*

Dermateaceae *Mycology.* a family of fungi of the order Helotiales that occur on herbaceous plants or as parasites of woody plants.

Dermatemydidae *Vertebrate Zoology.* family of Central American river turtles with one living species, *Dermatemys mawi*; related to fossil species from the Late Cretaceous to Early Tertiary; has a large heavy shell with a row of plates between the carapace and plastron, much like the snapping turtle.

dermatitis [dur´mi tīt´is] *Medicine.* inflammation of the skin.

dermatitis solaris see SUNBURN.

dermatoautoplasty *Surgery.* the grafting of skin taken from another place on the patient's own body.

Dermatobia *Entomology.* a large hairy bot fly found mainly in Central America; eggs are laid on ticks and mosquitoes, which then come in contact with human skin; the eggs left behind hatch, and the larvae penetrate to feed and develop in muscle tissue.

Dermatocarpaceae *Botany.* a family of lichens in the order Pyrenulales; characterized by a squamulose or umbilicate growth form, and commonly found on calcareous or limestone soils.

dermatocranium *Anatomy.* the intramembranous bones found on the roof and sidewalls of the cranial cavity, including the bones of the face.

dermatofibroma *Oncology.* a fibrous tumorlike nodule of the skin.

dermatogen *Medicine.* the antigen of a given skin disease. *Botany.* **1.** the hollow outer layer of an apical meristem that differentiates into the epidermis. Also, PROTODERM. **2.** in the histogen theory, one of three primary meristematic tissues that differentiate into the epidermis.

dermatoglyphics *Anatomy.* the study of the patterns of ridges in the skin pattern of the soles, fingertips, and palms for identification and as a genetic indicator, especially of chromosomal abnormalities.

dermatoheteroplasty *Surgery.* the grafting of skin from a donor belonging to another species.

dermatological of or relating to the skin.

dermatologist *Medicine.* a physician whose speciality is dermatology.

dermatology [dur´mi täl´ə jē] *Medicine.* a branch of medicine that deals with the prevention, diagnosis, and treatment of diseases of the skin.

dermatoma *Oncology.* a tumor of the skin.

dermatome *Anatomy.* the area of skin that is supplied with afferent nerve fibers by a single dorsal spinal root. Also, **dermatomic area.** *Developmental Biology.* the dorsolateral part of an embryonic somite, whose cells later disperse and supply the skin with its dermal layer. *Medicine.* an instrument used to cut thin slices of skin for grafting.

dermatomycosis *Medicine.* a superficial fungus infection of the skin caused by dermatophytes, yeasts, and other fungi.

Dermatonemataceae *Botany.* a family of marine red algae of the order Nemaliales, having erect, terete, and branched gametangial thalli and multiaxial axes.

dermatoneurology *Neurology.* the study of diseases and disorders affecting the nerves of the skin.

Dermatophilaceae *Microbiology.* a family of Gram-positive, aerobic bacteria of the order Actinomycetales, characterized by a transverse division of mycelial filaments that produces groups of motile coccoid or cuboid cells.

dermatophilosis *Medicine.* an actinomycotic disease caused by *Dermatophilus congolense* and characterized by painless pustules on the hands and arms. Also, **dermatophilus infection.**

Dermatophilus *Microbiology.* a genus of Gram-positive, aerobic bacteria of the family Dermatophilaceae that grow as nonmotile branching filaments, produce motile spores, and are animal pathogens.

dermatophobia [dur´mat ə fōb´ ē ə] *Psychology.* an irrational fear of the skin or of skin disease.

dermatophyte *Mycology.* any fungus that is parasitic upon the skin, including imperfect fungi of the genera *Microsporum, Epidermophyton,* and *Trichophyton;* all thrive on substances containing keratin (such as skin, nails, and hair); various species cause such human and animal diseases as athlete's foot and ringworm. Also, **dermophyte.**

dermatoplasty [dur mat´ō plas´tē] *Surgery.* a plastic operation on the skin; the replacement of destroyed or lost skin. Also, **dermoplasty.**

dermatosclerosis see SCLERODERMA.

dermatosis [dur´mə tō´sis] *Medicine.* any disease of the skin.

Dermestidae *Invertebrate Zoology.* a family of small, destructive beetles that feed on a variety of materials including animal carcasses and grains; serious pests in museums, storehouses, and dwellings.

Dermestoidea *Invertebrate Zoology.* a superfamily of small beetles.

dermis *Anatomy.* the layer of skin just beneath the epidermis, composed of dense vascular connective tissue. Also, CORIUM, TRUE SKIN.

dermo- a combining form meaning "skin," as in *dermoplasty.*

Dermocarpa *Bacteriology.* a genus of cyanobacteria occurring as variably sized, spherical cells surrounded by an extracellular fibrous layer, and reproducing by multiple fission or by binary fission.

Dermocarpella *Bacteriology.* a genus of bacteria in the cyanobacteria group, occurring as pear-shaped aggregates of cells and producing motile endospores.

Dermochelidae *Vertebrate Zoology.* a family of leatherback marine turtles of the superfamily Chelonioidea; the largest of all living turtles (up to 1200 pounds and over 7 feet); noted for hard, leathery shell with long ridges and wide front flippers; females go to land only to nest, range in warm seas worldwide.

dermoepidermal junction *Histology.* the place where the basal layer of the epidermis meets the papillary layer of the dermis.

dermoid cyst *Medicine.* a usually benign cystic growth, the walls of which are of a dermal origin; many ovarian tumors are dermoid cysts.

dermoidectomy *Surgery.* the excision of a dermoid cyst.

dermolipectomy *Surgery.* the removal of excess fat and skin, usually from the abdomen.

Dermoptera *Vertebrate Zoology.* the colugos or flying lemurs, a small order of arboreal nocturnal mammals of southeast Asia containing the single monogeneric family Cynocephalidae; noted for their ability to glide through the air for distances up to 60 meters.

dermotoxin *Toxicology.* a poison produced by certain bacteria, especially staphylococci, that kills skin cells upon contact.

Derodontidae *Invertebrate Zoology.* a family of beetles that includes the tooth-necked fungus beetles.

derosination *Chemical Engineering.* a process in which excess resins are removed from wood by saponification with organic solvents or aqueous alkaline solutions.

derrick *Mechanical Engineering.* a crane consisting of a movable boom attached by a rotating hinge to a stationary post. *Petroleum Engineering.* the towerlike framework erected over a borehole to support the drilling tools, casing, and pipe; also includes apparatus for raising, lowering, and controlling the equipment. Also, OIL DERRICK. *Naval Architecture.* a cargo boom that is attached by a pivoting hinge to the ship's mast or kingpost. (From the name of a famous hangman of early London.)

derrick barge *Petroleum Engineering.* a type of barge used widely in offshore drilling platform construction and in oil- and gas-field development operations; fitted with at least one high capacity derrick for lifting heavy equipment onto production platforms. Also, CRANE BARGE.

derris *Botany.* the shrubs or woody lianas of the tropical genus *Derris,* in the family Leguminoseae, from which rotenone is derived. *Materials.* the root or preparation of this plant used as an insecticide.

Derxia *Bacteriology.* a genus of Gram-negative, aerobic bacteria of uncertain affiliation, occurring as rod-shaped cells, each having a single polar flagellum, and found in tropical, acidic soils.

Deryagin number *Physics.* the ratio of the thickness of a film that coats a liquid to the capillary length of the liquid.

DES data encryption standard; diethylstilbestrol.

desalinization [dē sal´ə ni zā´shən] *Chemical Engineering.* any of various methods used to remove mineral salt from ocean water or other brines, such as flash distillation, reverse osmosis, or electrodialysis. Also, **desalination, desalting.**

desander [dē´san´dər] *Engineering.* a device that clears sand from drilling fluid to protect the pumps from abrasion.

Desargues, Gérard [dā´särg´] 1591–1661, French architect and mathematician; founder of synthetic projective geometry.

Desargues' theorem *Mathematics.* a theorem of projective geometry. Let *A, B, C* and *A', B', C'* denote the vertices of two triangles. Let *P* be the intersection of the lines containing *BC* and *B'C'*, let *Q* be the intersection of the lines containing *AC* and *A'C'*, and *R* be the intersection of the lines containing *AB* and *A'B'*. The lines joining corresponding vertices of the two triangles are concurrent if and only if the points *P, Q,* and *R* are collinear. A projective plane in which Desargues' theorem holds is called a **Desargian plane.**

Desault's bandage [də sōz´] *Surgery.* a bandage binding the elbow to the side, with a pad in the axilla, for a fractured clavicle. Also, **Desault's apparatus.** (After French surgeon Pierre *Desault,* 1744–1795.)

descaling *Engineering.* any of a variety of chemical or mechanical processes for cleaning scales or metallic oxides from a metal surface.

Descartes, René [dā´kärt´] 1596–1650, French philosopher, mathematician, and natural scientist; founder of analytical geometry; formulated universal laws of motion.

Descartes

Descartes' law of refraction see SNELL'S LAW OF REFRACTION.

Descartes ray *Optics.* a type of ray that, after refraction by a sphere of transparent material such as a water droplet, returns along a route that closely parallels the original path taken by the incident ray, as in the primary rainbow.

Descartes' rule of signs *Mathematics.* the theorem that the number of positive real roots (counting multiplicities) of a polynomial $a_0 + a_1x + a_2x^2 + \cdots + a_nx^n = p(x)$ with real coefficients is equal to the number *N* of sign changes of the sequence $a_0, a_1, a_2, \ldots, a_n$ of coefficients if every root of $p(x)$ is real, or less than *N* by an even number if $p(x)$ has at least one complex root. The number of negative real roots of $p(x)$ is found by applying the rule of signs to $p(-x)$.

Descemet's membrane [des ə māz´] *Histology.* a thin layer of fibrils that forms a membrane between the posterior surface of the stroma and the anterior surface of the endothelium of the cornea.

descend *Aviation.* of a flight, to come down under control from a higher to a lower altitude.

descendant *Geology.* a topographic feature formed from a mass underlying an original feature that has been removed. *Computer Science.* a node in a tree that is a child of another node or has descended from one of its children.

descender *Graphic Arts.* in typesetting, the part of a character that extends below the baseline, such as the tail of a lowercase *p* or *q.*

descending *Anatomy.* extending downward or caudally. *Physiology.* **1.** describing any cell or organ that conducts material or impulses downward. **2.** referring to that part of the colon that carries digested food downward along the left side of the abdomen.

descending aorta *Anatomy.* the continuation of the aorta from the arch of the aorta in the thorax to the point of its division into the common iliac arteries.

descending branch *Mechanics.* the portion of an object's trajectory after its summit, along which the object's altitude decreases monotonically with time. Also, **descent trajectory, descending trajectory.**

descending chain condition *Mathematics.* a module *B* is said to satisfy the descending chain condition (DCC) on submodules if for every chain $B_1 \supset B_2 \supset B_3 \supset \cdots$ of submodules of *B* there exists an integer *m* such that $B_i = B_m$ for all $i \geq m$.

descending chromatography *Analytical Chemistry.* a separation technique using paper chromatography in which a solution is poured into the top of the developing chamber and the solute is separated from the sample as the solution moves down the paper sheet in the chamber.

descending node *Astronomy.* the point in an orbit at which an object crosses a reference plane (usually the celestial equator) from north to south. *Space Technology.* specifically, the reference point on the earth's surface that an orbiting satellite passes over as it moves south across the equator.

descending vertical angle see ANGLE OF DEPRESSION.

descensus testis *Medicine.* the downward migration of the testis from the abdomen into the scrotum shortly before the end of intrauterine life.

descensus uteri *Medicine.* the falling or sinking down of the uterus.

descent *Anthropology.* the kinship of a specific individual or family traced through time. *Aviation.* the action carried out in descending, or an instance of this.

descent guidance *Transportation Engineering.* a system that provides flights descending to land at an airport with courses, speeds, and descent rates.

Deschamps' compressor [dā´shämp´] *Surgery.* a surgical instrument used for the direct compression of the artery. (After French surgeon Joseph Francois Lois *Deschamps,* 1740–1824.)

Deschamps' needle *Surgery.* a needle with a long handle, having the eye located near its point; used for the ligation of deep arteries.

describing function *Control Systems.* a function that uses an equivalent linear transfer function to approximate a nonlinear transfer function.

descriptive anatomy *Anatomy.* the use of words and drawings to describe the patterns, types, and classes of the various bodily organs and structures, including the relationships between them.

descriptive botany *Botany.* the study of the systematic description and diagnostic characteristics of plants.

descriptive geometry *Mathematics.* the study of projecting elements of plane or space configurations onto a plane, including the determination of distance, angle, size, shape, and intersection, in geometric interpretations of physical and engineering problems. The main applications are in technical drawing techniques. Also, PROJECTION DRAWING.

descriptive kinship system *Anthropology.* a system of kin terminology in which each individual relative is designated by a distinct term.

descriptive linguistics *Linguistics.* a subdiscipline of linguistics that deals with the accurate explanation of how languages are constructed. Also, STRUCTURAL LINGUISTICS.

descriptive meteorology *Meteorology.* the study of the entire atmosphere and its varying phenomena, excluding theory. Also, AEROGRAPHY.

descriptive statistics *Statistics.* techniques that group, simplify, or summarize measurements made on a population.

descriptor *Computer Programming.* **1.** a word or phrase that is used to classify or define program elements such as a data record, a program segment, or an input/output operation. **2.** a key word that is used to identify a subject or record and to permit its storage and retrieval by computerized methods.

desensitization the process of becoming less sensitive; specific uses include: *Immunology.* the process of creating a temporary loss or suppression of hypersensitivity to an antigen in an individual, by administering a series of injections of small amounts of the antigen. *Psychology.* see SYSTEMATIC DESENSITIZATION. *Behavior.* see COUNTERCONDITIONING. *Telecommunications.* the effect on a radio or video receiver when a high-power carrier signal of nearly the same operational frequency enters the receiver, masks the latter's useful signal, and reduces its effective sensitivity.

desensitize *Neurology.* to deprive of sensation or paralyze a sensory nerve by cutting or blocking. *Psychology.* to reduce or eliminate a hypersensitive reactions to anxiety-provoking stimuli through repeated exposure during controlled relaxation exercises.

desensitizer *Graphic Arts.* **1.** a solution applied to the nonimage areas of a lithographic printing plate to make them nonreceptive to ink. **2.** a solution applied to exposed photographic film to make the emulsion less photosensitive during developing.

desert *Geography.* any region having little or no vegetation because of scant precipitation and extreme temperatures; typically, but not exclusively, found in hot climates.

desert armor *Geology.* a thin, smooth, sheetlike concentration of closely packed pebbles, gravel, and other rock fragments that covers a desert surface and protects the underlying finer-grained material from further erosion by the wind. Also, **desert pavement.**

desert devil see DUST WHIRL.

desert dome *Geology.* a dome-shaped rock surface with uniform, smooth slopes that results from the prolonged exposure of a mountain mass to desert erosion. Also, PEDIMENT DOME, GRANITE DOME.

deserticolous [dez´ur tik´ə ləs] *Ecology.* living or thriving on open land in a desert region. Thus, **deserticole.**

desertification [dez´ur tə fi kā´shən] *Ecology.* the development of desert conditions in an area that was previously not an arid environment, as a result of climatic changes or human activity.

desert peneplain or **desert plain** see PEDIPLAIN.

desert polish *Geology.* in desert regions, a smooth, glossy finish produced on the surface of rocks and other hard substances by the scouring action of windblown sand and dust. Also, WIND POLISH.

desert soil *Geology.* any of a group of zonal soils having a light-colored surface horizon over a layer of calcareous material and hardpan.

desert soil

desert varnish *Geology.* a thin, shiny, red, brown, or black stain or coating of iron or manganese oxide formed on rock surfaces after long exposure in desert regions.

desetope *Immunology.* a part of an antigen that binds to the MHC molecule; it determines the orientation of the antigen and which part will be presented to the T cell. (Derived from <u>de</u>terminant <u>se</u>lection.)

desexualization *Physiology.* the reduction of sexual development and behavior by removal of the gonads, the sources of sex hormones.

desiccant *Science.* any substance or agent that removes moisture.

desiccate *Science.* to lose or cause to lose moisture.

desiccation *Science.* any process of moisture removal. Also, EXSICCATION. *Hydrology.* a decrease or withdrawal of water from an area as a result of climatic change, erosion, or other natural or artificial processes.

desiccation breccia *Geology.* a breccia composed of fragments of a mud-cracked layer of sediment that were later deposited with other sediments. Also, MUD BRECCIA.

desiccator *Chemical Engineering.* a closed glass vessel with an airtight lid that contains a chemical to absorb moisture; used for keeping materials free from moisture.

design *Science.* **1.** a plan or sketch of a structure or process. **2.** to prepare such a plan. *Architecture.* **1.** to conceive and plan the construction of a building. **2.** a scheme for the construction and ornamentation of a building, composed of plans, elevations, renderings, and other drawings.

designation *Computer Technology.* in a computer record, an item of data that indicates the nature of the record and the manner in which it should be processed.

designator *Military Science.* a device that indicates a target for guided missiles.

design automation *Computer Technology.* the use of computers to help design logic circuitry for new computers or other kinds of circuit.

Design Engineering

Design engineering puts scientific principles to work in the creation of useful products.

Widespread study of design engineering as a formal discipline in colleges did not take place until early in the 20th century, but the real essence of design engineering—invention—is as old as man. Century after century, the products of invention—from the printing press to the telephone to the auto—have fueled human progress and improved living standards. Among notable design engineers: Thomas Edison, Alexander Graham Bell, Henry Ford, the Wright brothers.

Design engineering starts with a basic human need or problem, such as the need to go from point "A" to point "B," and solves it through the invention of a new or improved device—an airplane or an automobile, for example. In doing their work, design engineers not only rely on math, physics, chemistry, and other sciences, but they must make many choices from a fast-expanding range of new technologies vital to product design.

For example, auto design engineers must decide on the material needed for a given part. Should it be plastic, metal, or a composite? They must be acquainted with the basic principles of mechanical and fluid power transmission to decide on the size and type of motors, pumps, valves, and other components of propulsion systems. Increasingly, they must be familiar with electronic devices, such as microprocessors, that control important functions in autos. And they must know the options for attaching all the components of a product together—from a simple bolt to laser welding. Most of today's design engineers use computers to develop their products, which speeds the design process and allows them to test various options without building expensive prototypes.

A recent study of design engineering by the Simmons Market Research Bureau, New York, found that there are more than 752,000 design engineers in the U.S., and 97% of them are male. Most have engineering degrees—usually mechanical, electrical, material, or chemical engineering.

Lawrence D. Maloney
Chief Editor
Design News

design engineering *Engineering.* a field of engineering that applies to the creation of systems, devices, processes, and structures for the needs of society.

design evaluation *Design Engineering.* the process of evaluating a design relative to a set of criteria. Similarly, **design analysis.**

design factor *Engineering.* the relationship between the maximum load placed on a structure to the maximum load that a structure can safely hold.

design gross weight *Aviation.* in design calculations, the gross weight that a flight vehicle is anticipated to have if the maximum maneuverability is regarded. Thus design weight becomes the weight that the structure is designed to withstand a specified amount of, or the maximum load factor. It may be take-off weight, but is more likely to be weight with partial fuel; for example, the U.S. Air Force uses 50% internal fuel remaining; the Navy uses 60% fuel remaining.

design heating load *Engineering.* the maximum amount of heat an enclosed space will be likely to need.

design hourly volume *Transportation Engineering.* the optimum hourly traffic volume that a highway or transport system has been designed to serve.

design level of service *Transportation Engineering.* the frequency and speed of service that a transport system is designed to provide.
design load *Design Engineering.* the maximum amount of weight or other load that a structure, mechanical system, or device is designed to to be able to sustain.
design matrix *Statistics.* a matrix representing the assignment of experimental units to factor levels, typically in regression or analysis of variance.
design of experiment see EXPERIMENTAL DESIGN.
design phase *Design Engineering.* the part of a total process of bringing a product to market that involves the creation of the physical character of the product.
design pressure *Design Engineering.* the pressure used to determine the minimum thickness or other design characteristics of a pressure vessel, such as a boiler, as set by recognized code formulas.
design rules checking *Design Engineering.* a computerized system used in manufacturing to check for tolerance violations and errors.
design safety factor see FACTOR OF SAFETY.
design speed *Transportation Engineering.* the maximum safe driving speed, at light traffic densities, for which a highway is designed.
design standards *Design Engineering.* generally accepted methods, materials, units of measure, and components that are specified for a project, product, manufacturer, or industry.
design stress *Design Engineering.* the allowable maximum stress to which a machine part or member may be subjected without failure.
design thickness *Design Engineering.* the sum of the required thickness and corrosion allowances used for individual components of a pressure vessel.
design water depth *Oceanography.* in harbor design, the maximum navigable depth that the harbor is designed to provide.
design waterline *Naval Architecture.* the waterline at which a vessel floats in its normal loaded condition.
design wave *Oceanography.* in harbor design, the type of wave that the harbor is designed to protect against.
desilication *Geochemistry.* the removal of silica from magma or rock as a result of the breakdown of silicates, or from soil as a result of percolation of large quantities of rainwater.
desilter *Mechanical Engineering.* a mechanical classifier in which silt particles settle as the carrier liquid is stirred by horizontally revolving rakes.
desilverization *Metallurgy.* the removal of silver from an ore or other material.
desired effects *Ordnance.* the damage or casualties to enemy forces, facilities, and materials that a commander desires to achieve from the detonation of a nuclear weapon; damage may be classified as light, moderate, or severe; casualties may be classified as immediate, prompt, or delayed.
desired ground zero *Ordnance.* the point on the surface of the earth that is at, or vertically above or below, the center of a planned nuclear detonation.
de Sitter, Willem 1872–1934, Dutch astronomer; applied Einstein's theories to research on the age, size, and structure of the universe.
de Sitter model *Astrophysics.* an early cosmological model that was based on Einstein's field equations and that contained no matter.
de Sitter space *Physics.* a special case of Einstein's general relativity equations describing a constant-curvature vacuum.
desizing *Textiles.* the process of removing sizing materials, such as starch, from woven fabric.
desk check *Computer Programming.* a manual process of verifying correct program syntax and logic by examining the program statements without using the computer.
desktop *Computer Science.* relating to or produced by means of desktop publishing.
desktop computer *Computer Technology.* a microcomputer that is small enough to be placed conveniently on an office desk.
desktop publishing *Computer Science.* a process in which an author or publisher employs a personal computer to produce the final published product or camera-ready copy; contrasted with a traditional publishing process in which production stages such as design, artwork, typesetting, and printing are performed separately by outside contractors. (From the idea that such publications come from the author's *desktop.*)
desktop publishing program *Computer Programming.* an application package for a microcomputer with a high-quality printer; used to produce reports, books, and other publications, often permitting graphics to be embedded in the text.
deslimer *Mechanical Engineering.* any of various devices, such as solid-bowl centrifuges, that are used to remove fine particles from ore pulp, to upgrade cement fractions, or to classify abrasives and pigments.
desma *Invertebrate Zoology.* a branched spicule found in the tissues of certain sponges.
Desmanthos *Bacteriology.* a genus of gliding bacteria of the provisional family Pelonemataceae, occurring in mud as partially sheathed bundles of filamentous cells attached to a substrate.
Desmarestiaceae *Botany.* a family of brown algae of the order Desmarestiales, characterized by cylindrical, compressed, or foliose axes that are lightly or heavily branched and bushy; some species have worldwide distribution, but most are limited to Antarctica.
Desmarestiales *Botany.* an order of cold-water brown algae of the class Phaeophyceae, characterized by a very large sporophyte and a discoid or hapteroid holdfast; distributed throughout Antarctica.
Desmarest, Nicolas [dā′mä rā′] 1725–1815, French geologist; identified the role of stream erosion in the origin of valleys.
desmetryn *Organic Chemistry.* $C_9H_{17}N_5S$, a white crystalline solid that melts at 84°C; used to kill both broadleaf and grassy weeds.
desmid *Botany.* a unicellular or colonial green algae, especially of the family Desmidiaceae, characterized by green chromatophores, two semicells connected by an isthmus, lack of a siliceous skeleton, and asexual spore formation; common in fresh waters.
Desmidiaceae *Botany.* a family of unicellular or colonial desmids in the order Conjugales, comprising the placoderm or true desmids.
desmin *Biochemistry.* a muscle protein that polymerizes into filaments about 10 mm in diameter, which link the myofibrils of both smooth and skeletal muscle; particularly abundant in heart muscle.
desmine see STILBITE.
desmitis *Pathology.* an inflammation of a ligament.
desmo *Medicine.* a combining form that denotes a relationship to a band, bond, or ligament.
Desmocapsaceae *Botany.* a monotypic family of photosynthetic marine algae of the order Desmocapsales, having a motile stage in which cells produce two apically inserted flagella and an eyespot.
Desmocapsales *Botany.* an order of photosynthetic, unicellular algae belonging to the class Dinophyceae, existing either as flagellates in a gelatinous substance or as a palmelloid stage, and being both freshwater and marine.
Desmodonta *Paleontology.* a suborder of toothless bivalve burrowing mollusks, widespread from the Ordovician to the Permian.
Desmodontidae *Vertebrate Zoology.* a family of bats considered by some authorities to contain the true vampire bats.
Desmodoroidea *Invertebrate Zoology.* a superfamily of marine nematodes with a helmet-shaped structure at the head end of the body.
desmogen *Botany.* vascular merismatic tissue.
desmoid *Oncology.* **1.** an unusually firm tumor originating in the muscle sheath of the abdominal wall and consisting of unencapsulated scar tissue. Also, **desmoid tumor, desmoma. 2.** of a mass, fibrous or fibroid.
Des Moinesian or **Desmoinesian** [də moinz′ē ən] *Geology.* a North American provincial series of the Upper Middle Pennsylvanian, occurring after the Atokan and before the Missourian.
Desmokontae *Botany.* a class of solitary, motile, and predominantly marine algae in the division Pyrrophyta, with a cell wall divided into two valves. Also, DESMOPHYCEAE.
desmolase see CHOLESTEROL OXIDASE.
desmology *Surgery.* **1.** the study of the ligments, their structure and function. **2.** the art of bandaging.
Desmomastigaceae *Botany.* a monotypic family of freshwater unicellular algae of the order Desmocapsales, about which little is known.
desmoneme *Invertebrate Zoology.* a type of stinging cell that releases a long coiled tube that wraps around the prey.
desmoneoplasm *Oncology.* a tumor occurring in connective tissue.
desmopelmous *Vertebrate Zoology.* a type of bird foot characterized by united planter tendons that do not allow the hind toe to bend independent of the other toes.
Desmophyceae see DESMOKONTAE.
Desmoscolecida *Invertebrate Zoology.* an order of ringed nematodes in the family Desmoscolecidae.
Desmoscolecidae *Invertebrate Zoology.* a family of highly specialized nematodes in the superfamily Desmoscolecoidea, having a globular head, with four movable setae and a ringed pseudosegmented body.
desmose *Invertebrate Zoology.* a microfilament connecting the centrioles during mitosis in certain protozoans.

desmosine *Biochemistry.* an amino acid present in elastin and derived from four lysine chains; thought to play a critical role in the ability of elastin to return to its original shape after being stretched.

desmosome *Cell Biology.* a type of cell junction associated with intracellular intermediate filaments that firmly attaches to adjacent cells, especially to epithelial cells, where it confers strength to the tissue.

Desmostylia *Paleontology.* an extinct ungulate order of four-legged amphibians of the Tertiary; related to the proboscideans and modern manatees.

Desmostylidae *Paleontology.* a family of ungulates in the order Desmostylia.

Desmothoracida *Invertebrate Zoology.* an order of mostly sedentary, stalked, flagellate protozoans.

desorb *Physical Chemistry.* to carry out the process of desorption.

desorption *Physical Chemistry.* a physical or chemical process by which a substance that has been adsorbed or absorbed by a liquid or solid material is removed from the material.

Desor's larva *Invertebrate Zoology.* a distinctive type of larva found in certain ribbon worms that develops within the egg membrane.

despooler *Computer Programming.* a utility program that reads data from spooled output buffer and transfers it to a printer or other output device.

despun antenna *Electromagnetism.* a satellite antenna whose element is able to rotate opposite to the rotation of the satellite's body so that the antenna beam is always directed toward the same point on the earth.

desquamation [des´kwə mā´shən] *Physiology.* the shedding or scaling of the skin, and of various mucosa of organs. Also, EXFOLIATION.

dessert wine *Food Technology.* a sweet, usually fortified red or golden wine often served with dessert, such as port or Tokay.

destearinate [dē stēr´ə nāt] *Chemical Engineering.* a method of removing the lower melting point compounds from a fatty oil.

destination *Transportation Engineering.* the location at which a flight is intended to end, as opposed to any intermediate stop. For a round trip, it is identical to the origin.

destination address *Computer Programming.* the logical location to which data is to be moved, such as a memory location, a terminal, a process, or an individual user.

destination time *Computer Technology.* memory access time, including the extra clock cycles needed for indirect addressing.

destraction *Chemical Engineering.* a high-pressure process that separates high-boiling or nonvolatile material by dissolving it with the application of supercritical gases.

destressing *Mining Engineering.* the relief of pressure on abutments of excavation in deep mining by lateral drilling and blasting to loosen zones of peak stress.

Destriau effect [des´trē ou] *Solid-State Physics.* the emission of light by certain phosphor powders that are incorporated in an insulating material and subjected to an alternating electric field.

destroyer *Naval Architecture.* a multipurpose surface combat vessel with moderate size and armament, high speed, and limited protection.

destruct *Aviation.* **1.** the process of destroying a rocket vehicle, after launch but before it has completed its mission or course, because of guidance problems or other failure that makes it potentially hazardous. **2.** an instance of this process.

destruction area *Military Science.* an area in which it is planned to defeat or destroy a threat from enemy aircraft; it may be subdivided into air intercept, missile, and antiaircraft gun zones.

destruction fire *Ordnance.* fire delivered for the sole purpose of destroying material objects.

destruction system *Ordnance.* **1.** an explosive device used to destroy weapons, equipment, or supplies in order to prevent their capture by the enemy. Also, DESTRUCTOR. **2.** a system, operated by external command or preset internal means, that can be used to intentionally destroy a missile or similar vehicle in flight. Also, DESTRUCT SYSTEM.

destructive backspace *Computer Technology.* a function of a terminal keyboard in which the backward movement of the cursor on the screen automatically erases the character passed over and readies the terminal to accept another character in its place.

destructive breakdown *Electronics.* the electrical breakdown of the depletion region in a field-effect transistor causing permanent damage to the transistor.

destructive distillation *Chemistry.* a process in which a carbon-containing material such as coal or oil shale is heated in the absence of air, resulting in its decomposition into solids, liquids, and gases, with the solid end product being carbon.

destructive interference *Optics.* an interaction that occurs when two lights superimposed onto each other from two different sources produce a light having a lower combined intensity than the sum of the intensities of the original sources. Radiology. a phenomenon that results when propagated waves are out of phase so that the maximum molecular compression for one wave occurs at the same point as the maximum rarefaction for the second wave.

destructive operation *Artificial Intelligence.* in Lisp, an operation that replaces values in an existing list structure, which may change the values of other variables due to shared use of the same structure.

destructive read *Computer Technology.* the process of retrieving data from a memory location or a storage device in such a way that the data being read is destroyed.

destructive readout memory *Computer Technology.* a type of memory in which the reading process destroys the data being read; characterisitic of magnetic-core main memory.

destructive testing *Engineering.* a testing procedure in which an item is dynamically operated until failure in order to determine its limits or weaknesses.

destruct line *Aviation.* a boundary line indicating the point at which a vehicle crossing it would be destroyed; used in warfare and especially in testing of flight vehicles and missiles.

destructor see DESTRUCTION SYSTEM, def. 1.

destruct system see DESTRUCTION SYSTEM, def. 2.

Desulfobacter *Bacteriology.* a genus of Gram-negative, sulfate-reducing bacteria, occurring as rod-shaped to ellipsoidal-shaped cells in aquatic environments.

Desulfobulbus *Bacteriology.* a genus of Gram-negative bacteria that are able to reduce sulfate; occurring in aquatic habitats, usually as ellipsoidal-shaped cells bearing a single polar flagellum.

Desulfococcus *Bacteriology.* a genus of Gram-negative, coccoid bacteria that reduce sulfate, found in aquatic environments.

Desulfomonas *Bacteriology.* a genus of Gram-negative bacteria that are able to carry out dissimilatory sulfate reduction; occurring as nonmotile, rod-shaped cells in the human intestinal tract.

desulfonate [dē sul´fə nāt] *Chemistry.* to remove the sulfonate group from a molecule. Thus, **desulfonation.**

Desulfonema *Bacteriology.* a genus of Gram-positive, sulfate-reducing bacteria that grow as filamentous cells and exhibit gliding motility.

Desulfosarcina *Bacteriology.* a genus of Gram-negative, sulfate-reducing bacteria that occur in marine habitats as variably shaped cells, some of which bear a single polar flagellum.

Desulfotomaculum *Microbiology.* a genus of Gram-positive soil bacteria, characterized by the formation of endospores and the capacity for sulfate reduction during anaerobic respiration.

Desulfovibrio *Microbiology.* a genus of Gram-negative, rod-shaped bacteria that typically occur in anaerobic sediments or aquatic habitats and are capable of sulfate reduction.

desulfoviridin test *Microbiology.* an assay for the detection of the bacterial enzyme desulfoviridin, which is a sulfite reductase involved in dissimilatory sulfate reduction in certain bacteria.

desulfurize [dē sul´fə rīz] *Chemistry.* the removal or reduction of sulfur content from petroleum oil or molten metals. Thus, **desulfurization.**

desulfurization unit *Chemical Engineering.* a unit for removing sulfur or sulfur compounds used in petroleum refining, minerals processing, and other operations.

Desulfurococcus *Bacteriology.* a genus of coccoid bacteria of the archaebacteria group, typically found in Icelandic solfataric habitats.

Desulfuromonas *Bacteriology.* a genus of Gram-negative, rod-shaped bacteria that are obligately anaerobic, have a repiratory metabolism, and occur in aquatic sediments.

2-desvinyl-2-formyl chlorophyll A see CHLOROPHYLL D.

desyl *Organic Chemistry.* the organic group $C_6H_5COCH(C_6H_5)$–.

desynchronized sleep see REM SLEEP.

DET *Aviation.* the airport code for Detroit City, Michigan.

det. give. (From Latin *detur,* "let it be given.") Also, **Det.**

detachable plugboard *Computer Technology.* a removable control panel for a computer or other device that can be replaced by another panel in order to change the operation of the equipment.

detached binary *Astronomy.* a binary star system in which neither star is large enough to fill its Roche lobe.

detached core *Geology.* the inner bed of a fold that has become separated from the main body of strata as a result of extreme folding and compression.

detached-lever escapement see LEVER ESCAPEMENT.

detached meristem *Botany.* a meristematic region, originating from axillary or apical meristems, that becomes discontinuous because of the growth or differentiation of intervening tissue.

detached shock wave *Fluid Mechanics.* shock formed by a blunt body and lying ahead of the body; formed at low, but supersonic Mach number.

detached X *Genetics.* a free X chromosome in *Drosophila* that has been formed by detachment of the arms of a previously attached X chromosome.

detachment *Military Science.* **1.** a part of a unit separated from its main organization for duty elsewhere. **2.** a temporary military or naval unit formed from other units or parts of units. *Geology.* see DÉCOLLEMENT.

detail *Military Science.* the personnel assigned to a particular duty; to assign personnel to such a duty.

detail chart *Computer Programming.* a diagram that represents all the processing steps of a program. Also, **detail flowchart**.

detailed balance *Physics.* a state in which any process of energy exchange or transformation that occurs in a system in thermodynamic equilibrium is accompanied by a parallel and reverse process.

detail file *Computer Programming.* a file containing transaction data used to update a master file.

detail printing *Computer Programming.* the process of printing a line output for each record read or each transaction processed.

detection spectrometry *Spectroscopy.* the use of spectroscopic methods to detect the presence of defects, dislocations, impurities, or flaws in a substance.

detection threshold *Acoustics.* the minimum signal-to-noise ratio that a sound can have and still be detected, as measured in decibels by the equation DT = 10 log(S/N_0), with S/N_0 representing the signal-to-noise ratio.

detectivity *Electronics.* a figure representing gain over noise for solid-state detectors such as photodiodes or phototransistors.

detent [dē tent′] *Mechanical Engineering.* the catch or lever of a mechanism that locks or unlocks the movement.

detention storage or **detention** see SURFACE DETENTION.

detergent *Materials.* **1.** a soapless, water-soluble agent such as linear alkyl sulfonate that is capable of emulsifying dirt and oil, or a similar oil-soluble agent that is used in dry-cleaning solutions and lubricating oils. **2.** generally, any cleansing agent including soap. *Biochemistry.* any of a class of substances that enhance the cleansing power of solvents, having a molecular structure that features a nonpolar hydrocarbon chain with a polar end that easily ionizes and attracts water molecules.

detergent alkylate see DODECYLBENZENE.

deteriorating supplies *Military Science.* supplies that will become useless within a limited period of time, usually one or two years, whether they are utilized or not.

deterioration *Engineering.* the process by which equipment, materials, and structures lose their quality over time due to physical effects of the environment.

determinant *Mathematics.* the **determinant function** on the ring $\text{Mat}_n R$ of $n \times n$ matrices over a commutative ring R with identity I_R (or on the ring of endomorphisms of an n-dimensional R-module) is the unique alternating multilinear (scalar-valued) form det satisfying $\det(I_n) = I_R$, where I_n is the $n \times n$ identity matrix; the **determinant of a matrix** A in $\text{Mat}_n R$ is the element $\det(A)$ of R; also denoted $|A|$. Equivalently, if $A = (a_{ij})$, then $\det(A) = \sum_{\pi \in S(n)} (\text{sgn}\ \pi) a_{1\pi}(1) a2\pi(2) a_{n\pi(n)}$, where the sum is taken over all possible elements π in the symmetric group $S(n)$ of permutations on the set $\{1, 2, \ldots, n\}$, and sgn is the signum function. That is, one forms all products of n entries from A, no two of which are from the same row or column, assigns each such product the sign +1 or −1 according to specific rules, and then adds the $n!$ terms so obtained. Simplified computation rules exist for $n = 2$ and 3; the determinants of larger matrices are more easily calculated by the technique of Laplace expansion. *Control Systems.* a product of partial return differences associated with the nodes of a signal-flow graph.

determinate *Science.* having definite limits.

determinate cleavage *Developmental Biology.* any cleavage in which blastomeres are predetermined to form specific parts of the embryo. Also, MOSAIC CLEAVAGE.

determinate growth *Botany.* **1.** the growth of a stem, branch, or shoot that is limited by the development of a terminal reproductive structure. **2.** the growth that occurs only during part of the vegetative season.

determinate inflorescence *Botany.* inflorescence in which there is a flower at the top of the main stem and on each lateral branch below this, with the center flower opening first. Also, CYMOSE INFLORESCENCE.

determinate structure *Mechanics.* any structure whose stresses and strains are determined completely by the equations of statics.

determination *Analytical Chemistry.* the measurement of chemical or physical properties of compounds; for example, specific gravity determinations.

determined *Cell Biology.* of or relating to a cell that is committed to a particular pathway of differentiation but has not yet undergone any overt developmental changes.

determinism *Psychology.* the theory that all psychological and behavioral events have preceding causes. *Chaotic Dynamics.* the doctrine that all phenomena are causally determined by prior events. *Mechanics.* see CAUSALITY.

deterministic *Chaotic Dynamics.* referring to a system that is governed by definite rules of evolution leading to cause and effect relationships and predictability. There exists a set of differential equations for calculating future behavior from initial conditions.

deterministic algorithm see STATIC ALGORITHM.

deterministic chaos see CHAOS.

deterministic equation *Physics.* the equation of a determinant set equal to zero, the solution to which is used to solve a set of simultaneous equations.

deterministic finite automaton *Computer Science.* a finite automaton that has at most one transition from a state for each input symbol and no empty transitions.

deterministic model *Statistics.* a mathematical model that does not involve any uncertainties.

deterrence *Military Science.* the prevention of action by the threat of unacceptable counteraction; it is applied specifically to the threat of nuclear retaliation.

detonate *Chemistry.* to produce an explosion through a chemical change, which may be initiated by heat, friction, or mechanical impact.

detonating cord *Ordnance.* a flexible fabric tube containing a high explosive designed to transmit the detonation wave.

detonating fuse *Ordnance.* a fuse consisting of a highly explosive material contained in a waterproof textile and detonated by a blasting cap that can be fired from a great distance by way of a fuse line.

detonating net *Ordnance.* an interlaced network of detonating cord used to clear a path in a minefield; the net is placed over a mine and detonated, thus exploding the mine.

detonating rate *Mechanics.* the velocity of an explosion wave as it passes through a charge.

detonating relay *Ordnance.* a device that, when used in conjunction with a detonating fuse, can avoid short-delay blasting.

detonation *Chemistry.* an extremely rapid chemical decomposition of an explosive, accompanied by a high-pressure and high-temperature shockwave moving at a velocity of more than 1000 meters per second. *Mechanical Engineering.* in an internal combustion engine, a premature spontaneous combustion of the air-fuel mixture that occurs when the temperature of compressed air in the cylinder exceeds certain limits; accompanied by loss of power, overheating, and knocking.

detonation wave *Fluid Mechanics.* a reaction propagation rate in combustible mixtures or unstable gases, exceeding the velocity of sound and resulting in rapid pressure rises that may reach 200 to 400 times the initial pressure.

detonator *Chemistry.* a compound that is used in igniting an explosive mixture; a primer. *Engineering.* a device consisting of a sensitive primary explosive that is used to detonate a high-explosive charge.

detonator safety *Engineering.* an attribute of a fuse that occurs when detonation cannot initiate subsequent explosive train components.

detonics *Engineering.* the field of study of concerned with detonating and the performance of explosives.

detorsion *Invertebrate Zoology.* the untwisting that occurs in the digestive organs of many gastropod mollusks during postembryonic development.

detour behavior *Behavior.* a behavior pattern in which an animal finds the primary route to an objective blocked yet is still capable of reaching it by immediately taking an alternate route.

detoxification [dē täks′i fə kā′shən] *Toxicology.* the process of removing a poison or toxic substance from the body.

detoxify [dē täks′i fī] *Toxicology.* **1.** to inactivate, neutralize, or render harmless a toxin or poison. **2.** to promote the recovery of a person from the toxic effects and dependence on an addictive drug such as alcohol or heroin. Also, **detoxicate.**

detrital fan see ALLUVIAL FAN.

detrital ratio see CLASTIC RATIO.

detrital remanent magnetization *Geophysics.* the magnetization of sedimentary rock that occurs at the time of formation, and that conforms, with some depositional bias, to the earth's magnetic field or local magnetic fields if present.
detrital sediment *Geology.* a sediment formed by the deposition of loose rock and mineral material derived from preexisting rock.
detritovore *Biology.* any organism that subsists on detritus. Thus, **detritovorous.**
detritus [di trēt´əs] *Geology.* any loose rock or mineral material derived directly from older rocks by mechanical means, such as disintegration or abrasion, and transported from its place of origin. *Biology.* organic matter produced by the decay or disintegration of a substance or tissue.
detrusion see SHEAR STRAIN.
d. et s. let it be given and labeled. (From Latin *detur et signetur.*)
detune *Electronics.* to change the inductance or capacitance of a tuned circuit so that its resonant frequency differs from the incoming signal frequency.
detuning stub *Electromagnetism.* a quarter-wavelength stub used to match the impedance of a coaxial line to a sleeve stub antenna.
deutemerite *Invertebrate Zoology.* in certain protozoa, such as *Gregarina,* the posterior part of the cell that contains the nucleus. Also, **deuteromerite, deutomerite.**
deutencephalon see DIENCEPHALON.
deuter- a combining form meaning "second," as in *deuterium.*
deuteranopia *Genetics.* an inherited, sex-linked type of color-blindness in humans in which there is confusion in distinguishing red and green.
deuterate *Chemistry.* to introduce deuterium into a chemical compound. Thus, **deuteration.**
deuterated *Chemistry.* of a compound, containing deuterium.
deuteric *Geology.* describing the changes in an igneous rock during the later stages of its formation that occur as a direct result of the cooling and consolidation of the magma itself.
deuteric alteration see SYNANTEXIS.
deuteride *Chemistry.* a hydride having a deuterium atom as the hydrogen.
deuterium [doo tēr´ē əm] *Chemistry.* an isotope of hydrogen having the symbol D or ^{2}H, with one neutron and one proton in its nucleus and an atomic weight of 2.0144. Also, HEAVY HYDROGEN.
deuterium cycle see PROTON-PROTON CHAIN.
deuterium discharge tube *Electronics.* a tube that consists of a glass envelope containing deuterium gas and two metal electrodes, and that produces high-intensity ultraviolet radiation when the gas is ionized by an electric current; used primarily in spectroscopic microanalysis applications.
deuterium oxide *Inorganic Chemistry.* D_2O or heavy water; that is, water in which the hydrogen atoms are in the form of deuterium. See HEAVY WATER.
deutero- a combining form meaning "second," as in *deuterosome.*
deuterogamy [doot´ə räg´ə mē] *Botany.* **1.** any process that replaces normal fertilization. **2.** the secondary pairing of sexual cells substituting for the union of gametes, as in many fungi and algae.
Deuteromycetes *Mycology.* a term formerly used for a class of fungi belonging to the subdivision Eumycete;, its members are now classified under the subdivision Deuteromycotina.
Deuteromycotina *Mycology.* a subdivision of fungi of the division Eumycota; it includes fungi with no known sexual reproductive stages as well as fungi whose asexual (anamorphic) reproductive stage is now known. Also, IMPERFECT FUNGI, FUNGI IMPERFECTI.
deuteron [doot´ə rän´] *Nuclear Physics.* the nucleus of a deuterium atom, consisting of a proton and a neutron; because of its low binding energy, it is useful as a projectile in nuclear bombardment experiments. Also, **deuton, deuterion.**
deuteron accelerator *Nucleonics.* any of various electrostatic or time-varying-field particle accelerators designed to impart high energy to deuterons for use as projectiles in nuclear bombardment experiments, producing deuteron-proton, deuteron-neutron, and deuteron-alpha reactions.
deuteron capture *Nuclear Physics.* the absorption of a heavy hydrogen isotope by a nucleus.
Deuterophlebiidae *Invertebrate Zoology.* the mountain midges, a family of small flies having aquatic larvae.
deuteroplasm see DEUTOPLASM.
deuterosome *Cell Biology.* a region in the cytoplasm of a eukaryotic cell that may function as a microtubule organizing center for basal body assembly.
Deuterostoma or **Deuterostomata** see DEUTEROSTOMIA.
Deuterostomia *Zoology.* those phyla in which the mesoderm and coelom begin as pouches from the gut, cleavage is indistinct, and the mouth is not derived from the blastopore; a group of higher phyla that have many common features and appear to form a natural evolutionary line, this includes Chordata, Chaetognatha, Enteropneusta, Brachiopoda, and Echinodermata. Also, DEUTEROSTOMA, DEUTEROSTOMATA.
deutocerebrum *Invertebrate Zoology.* the median lobes of the brain in insects.
deutoplasm *Developmental Biology.* the nonactive material in the cytoplasm (especially material such as lipoid droplets and yolk granules) that is stored in the egg as food for the developing embryo. Also, DEUTEROPLASM.
DEV duck embryo (rabies) vaccine.
dev. development; deviation.
De Vecchis process *Mining Engineering.* a method for smelting pyrites in which roasting and magnetic concentration of the raw material is followed by reduction in a rotary kiln or electric furnace.
devel. development.
develop *Graphic Arts.* **1.** in photography, to use chemical agents in order to create a visible image from exposed film. **2.** in lithography, to remove the leftover bichromated coating from a printing plate. *Science.* to grow or progress.
developable *Cartography.* a surface that can be flattened to form a plane without compressing or stretching any part of it, such as a cone or cylinder. *Building Engineering.* describing land that is not environmentally constrained.
developable surface *Mathematics.* any surface $u(x, y) = z$ that can be isometrically mapped to a subset of the Euclidean plane. Such a surface satisfies the nonlinear second-order partial differential equation $u_{xx}u_{yy} - u_{xy}^2 = 0$. Intuitively, a developable surface is one that can be rolled out flat without distortion; e.g., a cylinder or cone.
developed *Graphic Arts.* **1.** of or relating to photosensitive material such as film that has been exposed to light and treated with chemicals, yielding a visible photographic image. **2.** of or relating to a lithographic plate that has been chemically prepared for printing.
developed blank *Metallurgy.* in stamping, a cut product that needs little or no finishing.
developed dye *Chemistry.* a colored dye that is produced from a colorless substance by means of a chemical reaction.
developed muzzle velocity *Ordnance.* the actual muzzle velocity produced by a gun.
developed ore see DEVELOPED RESERVES.
developed reserves *Mining Engineering.* **1.** the tonnage of ore that has been developed, sampled, blocked out, or exposed on at least three sides and for which the tonnage yield and quality estimates have been made. In coal mining, the tonnage of coal known to exist by development headings. Also, ASSURED MINERAL, MEASURED ORE, DEVELOPED ORE. **2.** the mineral reserves proved by underground penetration.
developed silver *Graphic Arts.* black metallic silver that precipitates from a silver-salt film emulsion by the action of a chemical developer.
developer *Organic Chemistry.* an organic compound that will combine with other organic compounds on fibers or with dyes to form a new dye. *Graphic Arts.* **1.** a reductant used to form an image in photography. **2.** a solvent for the photoresist material in lithography.
developer streaks *Graphic Arts.* unwanted white marks or lines on a developed photograph resulting from a flaw in the use or composition of a developing agent.
developing *Graphic Arts.* **1.** the process of using chemical agents to create a visible image from exposed photographic film. **2.** the process of preparing a plate for printing, especially the stage of removing the unwanted bichromated coating from the nonimage areas of a surface-type plate or the image areas of a deep-etch plate. Also, **development.**
development the growth or progress of an idea, system, or organism; specific uses include: *Geology.* the progression of changes in successive fossil groups during the deposition of the earth's strata. *Meteorology.* the process of intensification of an atmospheric disturbance, most commonly applied to cyclones and anticyclones, and implying the generation of vorticity in the atmosphere due to the addition of energy to the disturbance. *Mining Engineering.* the act of opening up a coal seam or ore body by sinking shafts and driving drifts, and installing the necessary equipment.
developmental age *Psychology.* any age equivalent that is determined by some standardized test or measurement, as distinguished from an actual chronological age. Also, **development age.**

Developmental Biology

This term has gained usage in the past 30 years for the study of the development of multicellular differentiated organisms. It overlaps the older term "embryology" (the study of embryos, starting in the 1880s).

During development, precise arrangements of different cell types arise from the relatively simple organization of the single-celled egg. In mammals, for example, at least 200 cell types (those of muscle, nerve, skin, etc.) are arranged in tissues and organs of the adult. These cell types differ in appearance and function, due largely to different proteins each synthesizes and maintains. While the various cell types usually contain the same genes, which are unchanged from those of the fertilized egg, the cell types differ in which subset of genes is used for making proteins.

As recognized by T.H. Morgan in the 1930s, development depends largely on the timely formation and specific placement of factors (RNA and proteins) that ultimately activate the use of different genes in different cell types. Two kinds of processes for forming and distributing these factors are now known: 1) Cytoplasmic localization: the oocyte (the egg precursor) and/or egg synthesizes and localizes different factors to different regions of the egg's cytoplasm, and these pass into embryonic cells divided from the egg. This process underlies the establishment of the first few regions of differing cells in the early embryo. 2) Embryonic induction: identical embryonic cells of a region form inactive factors, after which a subset of cells receives signals ("inducers") from nearby different cells and activates the factors for use. This process underlies the detailed placement of the final cell types.

Recent research, especially on the fruit fly *Drosophila,* has led to the identification of some of these factors and the genes they affect. Much work remains to be done on the intercellular signalling of induction, on the intracellular localization of materials in the egg, and on the regulatory processes by which embryos restore normal development after a loss of parts.

John C. Gerhart
Professor of Molecular and Cell Biology
University of California, Berkeley

developmental biology the study of the development of organisms.

developmental crisis *Psychology.* **1.** a brief period of stress in childhood during which the child is attempting to complete some psychosocial task, such as the establishment of identity or initiative. **2.** see DEVELOPMENTAL DISORDER.

developmental disorder *Psychology.* a delay or failure in the development of a specific ability or function, such as speech, reading, mathematical skill, and so on.

developmental genetics *Genetics.* the field of genetics that studies the relationship of hereditary patterns to embryology and the development of organisms.

developmental level *Psychology.* any period of the human life span recognized as distinct in physiological and psychological development, such as infancy, early childhood, adolescence, and so on.

developmental psychology *Psychology.* a branch of psychology concerned with physical, social, and cognitive growth and decline from birth to death. Thus, **developmental psychologist.**

developmental stage *Psychology.* **1.** a period in human life during which certain characteristic traits or behaviors become manifested. **2.** see DEVELOPMENTAL LEVEL.

development drift *Mining Engineering.* the tunnel of a mine dug from the surface or a point underground to gain access to the coal or ore.

development drilling *Mining Engineering.* the process of drilling boreholes to delineate the size, mineral content, and disposition of an ore body.

development environment *Computer Technology.* a computer system used for software development that includes aids such as compilers, debuggers, syntax checkers, and other hardware and software tools.

development index *Meteorology.* the difference in divergence between two separated, tropospheric, constant-pressure surfaces; used in forecasting cyclogenesis. Also, RELATIVE DIVERGENCE.

development rock *Mining Engineering.* any rock broken during development work, consisting of both valuable and barren rock, which is included in the tonnage sent to the reduction plant of a mine.

development system *Computer Technology.* a computer system with hardware and software that supports efficient development of applications for a particular computer; often includes design and debugging aids, emulators, editors, assemblers, and compilers.

development well *Petroleum Engineering.* a well that is drilled within the productive area as determined by appraisal wells, after proving that enough oil or gas exists for commercial production.

deviant *Psychology.* **1.** of or relating to behavior that differs sharply from what is regarded as normal, proper, or acceptable, especially in sexual conduct. **2.** a person characterized by deviant behavior. Also, DEVIATE.

deviate *Psychology.* **1.** to behave in a deviant manner. **2.** see DEVIANT.

deviation the process of departing or turning aside, as from a course, procedure, or norm; specific uses include: *Engineering.* the difference between the measured value and the expected value of a controlled variable. *Optics.* the angle that separates the incident ray striking an object or optical system from the emergent ray produced by reflection, refraction, or diffraction. *Psychology.* the fact or condition of being deviant in behavior. *Navigation.* the angular difference between magnetic and compass headings. *Statistics.* the distance of a measurement from the mean of the sample or the underlying population. *Ordnance.* the distance between the point of impact or burst and the target.

deviation absorption *Telecommunications.* distortion in a frequency-modular receiver, usually occurring as a result of inadequate amplitude-modulation rejection.

deviation factor see COMPRESSIBILITY FACTOR.

deviation hole *Petroleum Engineering.* a drilled hole that intentionally deviates from the true vertical, but by no more than 5%.

deviation IQ *Psychology.* a score on an IQ test expressed in terms of its variation from the average of the scores of the normative group.

deviation ratio *Telecommunications.* in a frequency modulation signal system, the ratio of the maximum frequency deviation to the maximum modulation frequency under given conditions.

deviation sensitivity *Navigation.* the sensitivity of a compass to deviation errors.

deviation table *Navigation.* a table of the deviations of a particular compass on a particular craft. Also, **deviation card.**

deviatoric stress *Mechanics.* the remaining stress when the spherical or mean normal stress is subtracted from the total stress.

device *Engineering.* any element, tool, or component designed for a specific use or purpose. *Electronics.* a discrete electronic element, usually an active element, such as an electron tube or diode, that cannot be divided without destroying its specific function.

device address *Computer Programming.* the binary value that identifies the desired device to the input/output controller.

device assignment *Computer Technology.* **1.** the selection of a particular input or output device by specifying its logical device number. **2.** an assignment of a device, such as a tape drive, to a particular program by the operating system.

device cluster *Computer Technology.* a group of terminals and other devices that are co-located and share a communication control unit.

device control character *Computer Technology.* a special character used to turn units on and off or to perform a specific function; often used in telecommunications.

device driver *Computer Programming.* software that interfaces with and controls the operation of an input or output unit.

device-end condition *Computer Technology.* the completion of an input or output operation, as sensed by the input/output controller.

device flag *Computer Technology.* a single-bit register that indicates the status of a particular device.

device independence *Computer Technology.* the attribute of a program for which successful execution does not depend on a particular type of physical device.

device-name assignment *Computer Technology.* the selection of a particular input or output device by specifying the symbolic name rather than the device address or logical number.

device number *Computer Technology.* a logical or physical identification number that is assigned to a particular peripheral device.

device selector *Computer Technology.* a circuit in the input/output controller that identifies and activates the desired device.

devil see DUST WHIRL.

devilline *Mineralogy.* $CaCu_4^{+2}(SO_4)_2(OH)_6{\cdot}3H_2O$, a transparent, dark green monoclinic mineral having a specific gravity of 3.13 and a hardness of 2.5 on the Mohs scale; found as crystalline crusts in copper mines. Also, **devillite.**

devil's curve *Mathematics.* the graph in the plane of the equation $y^4 - b^2y^2 = x^4 - a^2x^2$. Also, **devil on two sticks.**

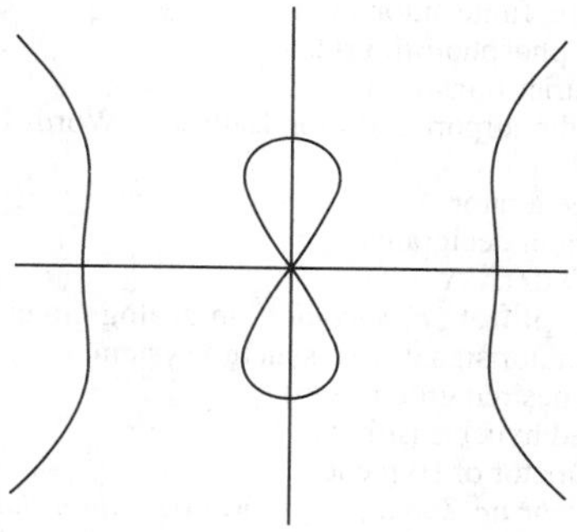

devil's curve

devil's pitchfork *Mechanical Devices.* a tool used to extract bits or other tools lost during drilling.

devitalized flour *Food Technology.* flour that has its gluten inactivated to prevent the raw flour from becoming sticky in the manufacture of bread.

devitrification *Chemistry.* the crystallization of a material upon heating, cooling, or aging. *Materials Science.* specifically, the precipitation of crystals from a glassy product, usually at high temperatures.

devolution see CATABOLISM.

Devonian [də vōn´ē ən] *Geology.* **1.** a period of the Paleozoic era, occurring after the Silurian and before the Mississippian. **2.** the rocks formed during that time. **3.** of or relating to this period or its rocks. (Associated with a notable system of rocks of this type in *Devon*shire, England.)

De Vries, Hugo [də vrēz´] 1848–1935, Dutch botanist; developed the mutation theory of evolution.

Hugo De Vries

De Vries effect *Geochemistry.* an approximate 100-year oscillation in the radiocarbon content of the atmosphere that produces variations in the apparent radiocarbon age of rock and mineral samples.

Devrinol *Organic Chemistry.* $C_{17}H_{21}NO_2$, a trade name for a brown solid that is slightly soluble in water and melts at 68°C; used as an herbicide.

dew *Hydrology.* water that forms when water vapor from saturated air cools and condenses on a surface whose temperature is above freezing but colder than the dew point of the surface air.

DEW distant early warning.

Dewar, James [doo´ər] 1842–1923, Scottish chemist; the first to liquefy hydrogen; invented the Dewar vacuum flask.

Dewar structure *Organic Chemistry.* an early bicyclic structural formula for the benzene molecule.

Dewar vacuum flask *Chemistry.* a double-walled vessel in which the space between the walls is evacuated and the surface bounding the vacuum is silvered, thus providing a high degree of thermal insulation; used frequently to store liquefied gases. Also, **Dewar flask.**

dewaterer *Mechanical Engineering.* **1.** broadly, any of various devices used to remove water from a mine or similar location, or to separate water from solid material. **2.** a wastewater treatment device that uses vacuum filtration to separate the water from the sludge.

dewatering *Engineering.* **1.** the act of removing water from a solid. **2.** the act of draining water from an enclosure or structure. **3.** the use of a chemical to remove water from a liquid such as oil or gasoline.

dewaxing *Chemical Engineering.* a process that removes solid hydrocarbon waxes from petroleum fractions, as in lube oil manufacturing. *Metallurgy.* in the lost-wax precision-casting process, the removal of the wax pattern prior to filling the mold.

dewcap or **dew cap** *Optics.* a tube that is placed at the end of a refracting telescope to prevent condensation on the objective.

dew cell *Engineering.* an instrument in which electric current passes through two wires drenched in a lithium chloride solution, causing the solution's temperature to rise until its vapor pressure equals that of the ambient air; used to determine the dew point.

dewclaw *Vertebrate Zoology.* a rudimentary, functionless inner claw or digit on the foot of some dogs and other animals that does not reach the ground, or a hoof terminating such a digit (as on the false hoof of a deer, pig, or other hoofed mammal).

dewetting *Metallurgy.* in soldering, the flow of solder away from the surface during reheating.

deweylite *Mineralogy.* a former name for a mixture of clinochrysotile lizardite with a talclike mineral.

dewindtite *Mineralogy.* $Pb_3(UO_2)_6H_2(PO_4)_4O_4{\cdot}12H_2O$, a yellow orthorhombic mineral occurring in minute tabular form or as crystals, having an undetermined hardness and a specific gravity of 5.03; found as a secondary mineral with torbernite, dumontite, and kasolite.

de Witte relation *Geophysics.* a graphical representation of the relation between electrical conductivity and distance.

dewlap *Vertebrate Zoology.* a fold of skin hanging from the neck of some bovines, reptiles, and birds. *Botany.* either of a pair of triangular areas that form a hinge at the joint of a sugarcane leaf blade.

DEW line *Military Science.* a line of radar stations located at about the 70th parallel in North America; primarily intended to provide early notice of an air attack from the Soviet Union or eastern Europe on the U.S. (An acronym for Distant Early Warning.)

dew point *Chemistry.* the temperature at which air or a given gas begins to condense to a liquid. *Meteorology.* the temperature to which a given parcel of air must be cooled at constant pressure and constant water vapor for saturation to occur. Also, **dew-point temperature.**

dew-point boundary *Chemical Engineering.* the area along which the gas-oil ratios approach zero on a phase diagram for a gas-condensate reservoir (pressure versus temperature with constant gas-oil ratios).

dew-point composition *Chemical Engineering.* the water vapor-air composition at saturation; namely, the point at which water exerts a vapor pressure equal to the partial pressure of water vapor in the air-water mixture.

dew-point curve *Chemical Engineering.* the line on a PVT phase diagram that divides the two-phase (gas-liquid) region from the one-phase (gas) region and denotes the point at a given gas temperature or pressure at which the first drop of liquid occurs.

dew-point depression *Chemical Engineering.* the process of reducing the liquid-vapor dew point of a gas by removing a fraction of the liquid from the gas. *Meteorology.* the difference in degrees between the air temperature and the dew point. Also, **dew-point spread.**

dew-point hygrometer *Chemical Engineering.* a device that is used to determine the dew point by measuring the temperature at which cooling vapor in a silver vessel begins to show condensation. Also, COLD-SPOT HYGROMETER.

dew-point pressure *Physical Chemistry.* the pressure of a system, at a fixed temperature, at which condensation or a liquid phase first appears.

dew-point recorder *Engineering.* a device that measures the dew point continuously by alternately heating and cooling the target.

dew-point reservoir *Engineering.* a hydrocarbon reservoir in a one-phase region where the temperature stays between the cricondentherm (maximum temperature and pressure in which two phases can exist) and the critical temperature. Also, RETROGRADE GAS-CONDENSATE RESERVOIR.

dexamethasone *Endocrinology.* a synthetic steroid with potent glucocorticoid activity and little mineralocorticoid activity, making it a useful therapeutic agent.

Dexaminidae *Invertebrate Zoology.* a family of sand hoppers or scuds, amphipod crustaceans in the suborder Gammeridea.

dexiocardia *Cardiology.* the location of the heart in the right side of the chest, often with the heart's orientation also rotated and the cardiac chambers transposed. Also, DEXTROCARDIA.

dexterotropic *Biology.* oriented or turning toward the right, such as the way a whorl pattern or shell formation might turn.

dextr- a combining form meaning "right," as in *dextrality.*

dextral *Invertebrate Zoology.* of a gastropod shell, turning toward the right.

dextral drag fold *Geology.* a drag fold in which the trace of a given surface bed is offset to the right; i.e., that of a Z shape.

dextral fault see RIGHT-LATERAL FAULT.

dextral fold *Geology.* an asymmetrical fold in which the longer limb appears to be offset to the right when viewed along its dip.

dextrality *Neurology.* a favoring of the right side of the body, rather than the left, in carrying out voluntary motor acts or in guiding bimanual movements; especially refers to use of the right hand.

dextran *Biochemistry.* a high-molecular-weight polymer of D-glucose, produced by enzymes on the cell surface of certain lactic acid bacteria; dextrans that are formed from sucrose by bacteria in the mouth produce dental plaque.

dextranase *Enzymology.* an enzyme that catalyzes the hydrolysis of polysaccharides into glucose.

dextran preparation *Immunology.* any of a class of polysaccharides composed of glucose residues that is not strongly antigenic and is used as a plasma extender in blood transfusions.

dextrin *Organic Chemistry.* $(C_6H_{10}O_5)_n$, any of a class of intermediate products formed from the heating or hydrolysis of starch; amorphous crystals, soluble in water and insoluble in alcohol. *Materials.* specifically, a yellow or white, amorphous, odorless powder that forms a thick liquid with strong adhesive properties; used as a paste and for sizing, and in syrups and beers. Also, STARCH GUM.

dextrinization *Organic Chemistry.* the production of a mixture of oligosaccharides and polysaccharides by the partial hydrolysis of starch or glycogen.

dextrinize *Organic Chemistry.* to treat a mixture of starches or glycogen in such a way as to produce dextrins.

dextro- a combining form meaning "right," as in *dextrorotatory.*

dextrocardia see DEXIOCARDIA.

dextromethorphan *Organic Chemistry.* $CH_3O\text{–}(C_{16}H_{24})NCH_3$, a multiringed compound used as a cough supressant.

dextromethorphan hydrobromide *Organic Chemistry.* $C_{18}H_{26}BrNO$, a multiringed compound in the form of crystals that are soluble in water; melts at 122–126°C; used as a cough suppressant.

dextropedal *Neurology.* a favoring of the right foot, rather than the left, in voluntary motor acts; right-footed.

dextrophobia *Psychology.* an irrational fear of objects on one's right.

dextropimaric acid *Organic Chemistry.* $C_{20}H_{30}O_2$, a diterpene found in the resin of the cluster pine; melts at 218–219°C.

dextrorotatory [deks´trə rō´tə tôr´ē] *Optics.* relating to or describing an optically active substance, such as a crystal, that causes a clockwise rotation of the plane of polarization when looking into the incoming light. *Organic Chemistry.* having the property, when in solution, of rotating the plane of polarized light in a clockwise direction, as measured by a polarimeter.

dextrorotatory form *Physical Chemistry.* the isomer of a compound that rotates the plane of vibration of plane-polarized light clockwise, or to the right, as distinguished from the levo form that involves counterclockwise rotation.

dextrorotatory isomer *Biochemistry.* a stereoisomer that rotates the plane of polarization of light to the right.

dextrorse *Botany.* growing toward the right as seen from outside the helix, usually to describe climbing plants.

dextrose see GLUCOSE.

dextrosinistral *Neurology.* **1.** shifting or extending from right to left. **2.** of a person, naturally left-handed but trained to use the right hand in certain motor activities.

dextrotopic cleavage *Developmental Biology.* spiral cleavage in which the third cleavage division forms a right-hand spiral.

dextroversion *Cardiology.* location of the heart in the right side of the chest, often with the heart's orientation also rotated and the cardiac chambers transposed. Also, DEXTROCARDIA.

dezincification *Chemistry.* the corrosive removal of zinc from a brass surface, leaving redeposited copper.

DF direction finder; damage-free.

DFA deterministic finite automaton.

DFP diisopropyl phosphorofluoridate.

DFT discrete Fourier transform.

DFW *Aviation.* the airport code for Dallas/Ft. Worth International Airport, Texas.

DG differential generator.

dg or **dg.** decigram; decigrams.

D galaxy see cD GALAXY.

DG synchro amplifier *Electronics.* an analog circuit used to amplify differential generator signals in a synchro system.

DGZ *Ordnance.* desired ground zero.

DH or **dh** delayed hypersensitivity.

DHg or **D.Hg.** Doctor of Hygiene.

dhole [dōl] *Vertebrate Zoology.* a wild dog, *Cuon dukhunensis,* of the family Canidae; native to India, where it hunts in packs of about thirty members and may attack mammals as large as tigers and bears.

dhole

D horizon *Geology.* a former term for the layer of soil lying beneath the B or C horizons, consisting of unweathered rock.

DHy or **D.Hy.** Doctor of Hygiene.

DI or **D.I.** diabetes insipidus; drill instructor.

di- a prefix meaning "two," "twice," or "double," as in *dichromatic, dihedral.*

di. or **dia.** diameter.

dia- a prefix meaning: **1.** through, as in *diameter.* **2.** thorough or complete, as in *diagnosis.* **3.** apart, as in *dialysis.*

diabase *Petrology.* an intrusive, medium-grained, basaltic rock consisting principally of labradorite and pyroxene, often with an ophitic texture; commonly found in dikes and sills. Also, DOLERITE.

diabase amphibolite *Petrology.* amphibole schist formed by dynamic metamorphism of diabase.

diabasic *Petrology.* of or relating to the textural quality of a basic igneous rock in which the interstices between lath-shaped feldspar crystals, usually of plagioclase, are filled with discrete crystals or grains of pyroxene, usually augite.

diabatic *Thermodynamics.* see DIATHERMAL.

diabetes [dī´ə bēt´ēz] *Medicine*. **1.** a general term for a disease that is characterized by frequent discharge of urine and excessive thirst. **2.** see DIABETES MELLITUS.

diabetes insipidus *Medicine*. a disease characterized by the excretion of large amounts of dilute urine accompanied by extreme thirst, and resulting from an inadequate output of pituitary antidiuretic hormone; this condition may be acquired, inherited, or idiopathic.

diabetes mellitus *Medicine*. a metabolic disease in which there is a deficiency or absence of insulin secretion by the pancreas; characterized by hyperglycemia, glycosuria, and alterations of protein and fat metabolism; this condition is often inherited but may be acquired. Diabetes mellitus occurs in two major forms: Type I, or **insulin-dependent diabetes mellitus,** and Type II, **non-insulin-dependent diabetes mellitus.** The condition may also be gestational (Type III), or due to impaired glucose tolerance (Type V). Type IV encompasses all other forms of diabetes, including those that are associated with pancreatic disease, hormonal changes, adverse effects of drugs, or genetic or other anomalies.

diabetic [dī´ə bet´ik] *Medicine*. **1.** of, relating to, or affected with diabetes. **2.** a person having diabetes.

diabetic amaurosis *Medicine*. a loss of vision that is associated with diabetes; characterized by capillary microaneurysms and hard or waxy exudates, and often accompanied by cataracts.

diabetic coma *Medicine*. a life-threatening condition associated with diabetic acidosis and dehydration, caused by inadequate treatment, failure to take prescribed insulin, or any stress such as trauma or surgery that increases the body's need for insulin; usually characterized by a rise and then a fall in body temperature accompanied by a drop in blood pressure. Treatment consists of administering insulin and replacing electrolytes and fluids.

diabetic gangrene *Medicine*. a moist type of gangrene associated with diabetic mellitus and often resulting from minor injuries.

diabetic ketoacidosis *Medicine*. a metabolic disorder manifested by an increase of blood concentration of hydrogen ions above normal, occurring in uncontrolled diabetes mellitus; characterized by weakness, headache, thirst, air hunger, and coma. Also, **diabetic acidosis**.

diabetic retinopathy *Medicine*. a disorder of the retinal blood vessels that is associated with diabetes. Also, **diabetic retinitis**.

diabetophobia [dī´ə bet´ə fō´bē ə] *Psychology*. an irrational fear of becoming a diabetic.

diablastic *Petrology*. of or relating to a texture in metamorphic rock that consists of intergrowth and interpenetrating multialigned, rod-shaped components.

diaboleite *Mineralogy*. $Pb_2Cu^{+2}Cl_2(OH)_4$, a bright blue tetragonal mineral occurring in tabular crystals, having a specific gravity of 5.42 and a hardness of 2.5 on the Mohs scale; found with cerussite, boleite, and linarite.

diac see THREE-LAYER DIODE.

diacetate *Organic Chemistry*. a salt or ester characterized by possessing two acetate groups (CH_3COO–).

diacetic acid see ACETOACETIC ACID.

diacetin *Organic Chemistry*. $C_7H_{12}O_5$, a combustible, colorless, hygroscopic liquid, miscible with water and alcohol; boils at 259°C; used as a solvent, plasticizer, and softening agent. Also, GLYCERYL DIACETATE.

diacetone alcohol *Organic Chemistry*. $C_6H_{12}O_2$, a flammable, colorless liquid that is soluble in water and alcohol; boils at 167.4°C and melts at –44°C; used as a solvent and stripping agent.

diacetyl *Organic Chemistry*. $C_4H_6O_2$, a flammable yellow liquid that is soluble in water and alcohol; melts at >3 to >4°C and boils at 88–89°C; used as an aroma carrier in food products. Also, DIMETHYLGLYOXAL.

diacetylmorphine see HEROIN.

diacetylurea *Organic Chemistry*. $(CH_3CONH)_2CO$, a needlelike solid with a melting point of 154°C.

diachronic *Archaeology*. denoting actions or things that occur over time, as in the study of artifacts in a region as they change across sequential periods of time. *Geology*. see DIACHRONOUS.

diachronous *Geology*. describing a lithological unit whose age varies from place to place, or a lithological unit that cuts across time planes or biostratigraphic zones. Also, TIME-TRANSGRESSIVE. *Archaeology*. see DIACHRONIC.

diacid *Chemistry*. **1.** of a base or alcohol, capable of combining with two molecules of a monobasic acid or one molecule of a dibasic acid to form a salt or ester. **2.** of an acid, having two acidic hydrogen atoms. Also, **diacidic.**

diaclinal *Geology*. describing a stream or valley that crosses a fold at right angles to the strike of the underlying strata.

Diacodectidae *Paleontology*. a primitive paleodont family of the Eocene, artiodactyl mammals of the superfamily Dichobunoidea; the type genus is *Diacodexis*.

diactine *Invertebrate Zoology*. a type of sponge spicule with two arms.

Diactophymatoidea *Invertebrate Zoology*. a superfamily of nematode worms that are parasites on earthworms and vertebrates.

diactor *Electricity*. a direct-acting regulator that controls shunt generator voltage output. (An acronym for direct-acting regulator.)

diacylglycerol *Endocrinology*. a diglyceride, present as a constituent of cellular phospholipids; when released from these phospholipids by an agonist-activated phospholipase, this molecule serves as the endogenous activator of the calcium- and calmodulin-dependent protein kinase (protein kinase C) and is part of an important intracellular signaling cascade.

diadelphous *Botany*. **1.** of a stamen having its filaments fused into two bundles. **2.** a plant having its stamens so fused.

Diadematacea *Invertebrate Zoology*. a superorder of sea urchins, echinoid echinoderms in the subclass Euechinoidea.

Diadematidae *Invertebrate Zoology*. a family of sea urchins, echinoid echinoderms in the order Diadematoida, having long spines.

Diadematoida *Invertebrate Zoology*. an order of sea urchins, round or flattened echinoid echinoderms, many having long, poisonous spines.

Diademopsis *Paleontology*. a recently discovered echinoderm of the early Jurassic; an ancestor of the large, irregular euechinoids.

diaderm *Anatomy*. a blastoderm composed of two layers, one of endoderm and the other of ectoderm.

diadochite *Mineralogy*. $Fe_2^{+3}(PO_4)(SO_4)(OH)\cdot 5H_2O$, a brown to yellow triclinic mineral having a specific gravity of 2.0 to 2.4 and a hardness of 3.5 on the Mohs scale; found as gel-like crusts in recent mine workings.

diadochy *Crystallography*. the replacement of atoms or ions by other atoms or ions at specific lattice point sites.

diadromous *Botany*. having fanlike venation; usually used to describe leaves. *Vertebrate Zoology*. migrating between sea and freshwater systems, as do some fish.

Diadumenidae *Invertebrate Zoology*. a family of small sea anemones found in shallow coastal waters.

diafocal point *Optics*. a point on a refracted ray of light that lies on a plane intersecting the axis of the refracting lens that is parallel to the ray on the opposite side of the lens.

diag. diagonal; diagram; diagnosis.

diagenesis *Geology*. in the lithification of a sediment, the chemical, physical, and biological changes that occur after its deposition but before metamorphism and consolidation.

diageotropism *Biology*. the tendency of a sessile organism or structure to grow horizontal to the ground or perpendicular to the line of gravity, such as a tree branch or root.

diagnosis *Medicine*. the identification of a disease or condition on the basis of its signs and symptoms. *Systematics*. a formal, technical description of a taxon, serving to distinguish it from other taxa. *Computer Technology*. the process of locating and identifying faults in a computer component or errors in a program. *Artificial Intelligence*. **1.** a finite classification problem involving a predetermined set of possible diseases or malfunctions. **2.** a problem in which a representation of the structure and normal operation of a mechanism is used, together with observed behavior of the mechanism, to determine what component failure(s) could account for the observed behavior.

diagnostic *Medicine*. relating to or based on a diagnosis. *Statistics*. a sample-based procedure for evaluating the correctness of an assumption underlying the inferential procedure being used.

diagnostic message *Computer Programming*. information printed by the computer that identifies improper commands, logic errors, and other faults. Also, ERROR DIAGNOSTIC.

diagnostic routine *Computer Programming*. a program that is designed to determine whether the computer system is functioning properly or to detect programming errors. Also, **diagnostic subroutine, diagnostic test.**

diagnostics *Engineering*. the information resulting from a series of automatic tests performed on a malfunctioning system to try to determine the cause of a failure.

diagonal *Mathematics*. **1.** a line connecting any two nonadjacent vertices of a polygon or a polyhedron. **2.** the main diagonal of a matrix. *Optics*. an element, such as a plane mirror or prism face, that is attached near the eyepiece of a telescope at an angle that redirects the light to improve observation or to reduce the intensity of the sun's image in order to view it directly.

diagonal bedding see CROSS-BEDDING.

diagonal bracing *Building Engineering.* a method of bracing steel industrial buildings to resist wind effects by providing additional cross bracing at the building sides.

diagonal check *Cartography.* measurements made across the opposite corners of the basic frame of a map projection to ensure the accuracy of its construction or scale.

diagonal displacement *Mechanics.* displacement along a line of applied force or torque.

diagonal fault *Geology.* a fault whose strike lies at an oblique angle to the strike of the adjacent strata, rather than parallel or perpendicular to them. Also, OBLIQUE FAULT.

diagonal horn antenna *Electromagnetism.* a horn antenna that has a square cross section and whose electric field vector is parallel to one of the diagonals.

diagonalizable matrix *Mathematics.* a (square) matrix *A* that is similar to a diagonal matrix; i. e., there exists a change of basis represented by a matrix *S* such that $S^{-1}AS$ is a diagonal matrix. Also, **diagonable matrix.** If the same similarity matrix diagonalizes another matrix *B*, then *A* and *B* are said to be **simultaneously diagonalizable matrices.**

diagonalization *Robotics.* the use of computer software to resolve the stress components of strain gauges.

diagonal joint *Geology.* a joint whose strike lies at an oblique angle to the strike of the associated sedimentary strata, or to the cleavage plane of the metamorphic rock in which it occurs. Also, OBLIQUE JOINT.

diagonally dominant matrix *Mathematics.* a square matrix with the property that the absolute value of each main diagonal element is greater than or equal to the sum of the absolute values of the off-diagonal elements in the same row or column. If the inequality is strict, then the matrix is said to be **strictly diagonally dominant.**

diagonal matrix *Mathematics.* a square matrix, all of whose entries are zero except for those on the main diagonal.

diagonal pitch *Engineering.* the distance between rivets in adjacent rows of staggered rivets.

diagonal pliers *Mechanical Devices.* pliers whose cutting jaws are at an angle to its handle; used to cut wires close to a terminal.

diagonal sheathing *Building Engineering.* an expensive method of adding rigidity to a building, using plain, matched, or shiplapped wood or other forms of sheathing boards; each board is nailed to a stud across beams in a diagonal configuration.

diagonal stay *Mechanical Engineering.* a diagonal member between the tube sheet and shell of a fire-tube boiler.

diagonal tension *Building Engineering.* the primary tensile stress for reinforced or prestressed concrete, resulting from horizontal tension and vertical shear.

diagram *Science.* a line drawing, usually accompanied by text, that is designed to explain an object or process rather than to realistically represent its details. *Mathematics.* in category theory, a pictorial representation of objects and morphisms. Vertices or letters are used to represent objects, and arrows represent morphisms from one object to another. Also, GRAPH. *Computer Programming.* a schematic representation of a process or information flow; usually less detailed than a flowchart.

diagram factor *Mechanical Engineering.* the ratio of the actual mean effective pressure of a steam engine to its ideal cycle.

diaheliotropism *Botany.* the movement of plant leaves to maintain a perpendicular position between the leaf surface and the sun's rays throughout the day.

diakinesis *Cell Biology.* the final stage of the first meiotic prophase, in which the bivalent chromosomes undergo condensation.

diakoptics *Mathematics.* a technique for solving a network, especially an electrical network, by decomposing it into components, solving the individual components, and then determining the solution of the network from the solutions of the individual components. Also, KRON'S METHOD OF TEARING.

dial *Mechanical Devices.* a graduated plate equipped with a pointer upon which various measurements are registered. *Telecommunications.* **1.** any device that generates signals that are used for selecting and establishing connections, as on a dial telephone. **2.** to initiate a direct telephone call on either a rotary type dial or a push-button or touch-tone set.

DIAL *Robotics.* a robotic control and programming language developed and used at Stanford University.

dial-a-ride *Transportation Engineering.* a radio-dispatched demand-responsive bus providing door-to-door or stop-to-stop service, usually along a regular but flexible route. Similarly, **dial-a-bus.**

dial cable *Mechanical Devices.* a braided cord or wire that provides motion for a pointer over a dial by means of a knob that is turned.

dialdehyde *Organic Chemistry.* a compound containing two aldehyde (–CHO) groups.

dialect *Computer Programming.* a version of a high-level programming language, usually containing minor variations from other versions of the same language. *Linguistics.* a variety of a language that is distinguishable from other varieties in some respect, as in pronunciation and features of vocabulary, and that is used by the people of a particular region or group; e.g., the Scottish dialect of English. A dialect is generally understandable by speakers of other dialects of the same language. *Behavior.* a variation in the sound production of a species that is found only in a particular region.

dial exchange *Telecommunications.* a telephone exchange area in which all originating calls are dialed directly.

dial feed *Mechanical Engineering.* a device that positions workpieces successively so that they can be acted upon by a machine.

dialifor *Organic Chemistry.* $C_{14}H_{17}ClNO_4PS_2$, toxic, white crystals; insoluble in water and soluble in ether; melts at 67–69°C; used as an acaricide and pesticide.

dial indicator *Mechanical Devices.* a sensitive measuring device for pressure or vacuum, calibrated to thousandths of an inch by small plunger displacements from a pointer over a circular scale. Also, **dial test indicator, dial gage.**

dialing step *Engineering.* the least amount of mass that can be placed on or lifted from a balance outfitted with dial weights.

dial key *Electricity.* the key mechanism of a subscriber's cord used to connect the dial to the line.

dialkyl *Organic Chemistry.* any molecule containing two alkyl groups.

dialkyl amine *Organic Chemistry.* an amine containg two alkyl groups attached to the amino nitrogen.

diallage *Mineralogy.* a green-to-brown or bronze-colored variety of monoclinic pyroxene (augite or diopside) occurring in lamellae or in foliated masses, having a metallic luster and a parting parallel to the front pinacoid.

dial lamp *Electricity.* a small lamp that is used to illuminate the dial of electronic equipment, sometimes also serving as a pilot lamp. Also, **dial light.**

diallyl phthalate *Organic Chemistry.* $C_6H_4(COOCH_2CH{=}CH_2)_2$, a combustible, toxic, nearly colorless, oily liquid; insoluble in water and soluble in most organic liquids; melts at –70°C and boils at 165–167°C (5 torr); used as a plasticizer and for polymerization.

dialogue *Computer Technology.* the interaction or conversation between a computer and the user. Also, **dialog.**

dialogue speaker *Acoustical Engineering.* in multi-loudspeaker home entertainment systems, a small speaker positioned near the video display to serve as the principal source for speech sounds.

dial press *Mechanical Engineering.* a punch press equipped with dial feed.

dial pulse interpreter *Electronics.* a circuit that converts the pulse strings of a dial telephone to a format that is suitable for data entry to a computer; used in areas without touch-tone capability.

dial pulsing see LOOP PULSING.

dial telegraph *Telecommunications.* a communications system that allows for transmission of written messages by code using a dialing system.

dial telephone *Telecommunications.* a telephone set having a rotatable disk that, when rotated an amount corresponding to a desired digit and released, generates signals controlling equipment at an automatic exchange.

dial tone *Telecommunications.* **1.** in telephone operation systems, an audible signal indicating to a circuit connected to an automatic exchange that the latter is ready to receive dialing signals. **2.** a steady signal used to inform the caller to start dialing.

dial train *Horology.* a gear train in a timepiece, separate from the going train and driven by the cannon pinion, that turns the hands.

dial-up *Telecommunications.* an instance of dialing a telephone call.

dial-up line *Computer Technology.* a telephone line connected to a modem that allows access to a computer or network via the telephone system.

dialuric acid *Organic Chemistry.* $C_4H_4N_2O$, a breakdown product of alloxan.

Dialypetalanthaceae *Botany.* a monospecific family of dicotyledonous trees in the order Rosales, characterized by opposite simple leaves, showy flowers borne in a terminal, branching inflorescence, and a capsular fruit containing numerous slender seeds; found primarily in Brazil.

dialysis [dī al´ə sis] *Medicine.* the diffusion of blood across a semipermeable membrane in order to remove toxic materials and maintain a proper balance of fluid and blood electrolytes; used especially in cases of improper kidney function. Also, HEMODIALYSIS. *Physical Chemistry.* the separation of substances in a solution based on the rate at which each component is diffused through a membrane; chiefly used to separate large particles (colloids) from soluble substances.
dialysis fermenter *Biotechnology.* a fermenter fitted with a selectively permeable membrane through which the substrate and other materials can penetrate but the biomass cannot, thus maintaining a high biomass concentration within the reactor.
dialyzate *Chemistry.* the material in a dialysis process that does not pass through the semipermeable membrane.
dialyzer *Chemical Engineering.* the semipermeable membrane used in dialysis, such as a collodion.
diam. diameter.
diamagnet *Electromagnetism.* a diamagnetic material, such as one of the halogens, the alkali and alkaline earth metals, or the noble gases.
diamagnetic [dī´ə mag net´ik] *Electromagnetism.* referring to or having a negative magnetic susceptibility.
diamagnetic Faraday effect *Optics.* the rotation of a polarized beam of light passing through a magnetic field at frequencies near a split absorption line.
diamagnetic resonance see CYCLOTRON RESONANCE.
diamagnetic susceptibility *Electromagnetism.* a negative magnetic susceptibility, usually a quantity on the order of -10^{-5} cm^3/mole.
diamagnetism [dī´ə mag´nə tiz əm] *Electromagnetism.* a tendency for diamagnetic materials to be repelled by magnetic fields and thus align themselves at right angles to a magnetic field.
diameter *Mathematics.* **1.** a chord passing through and bisected by the center of a circle, conic, or solid of revolution generated by a conic. For example, a diameter of a parabola (which has its center at the point at infinity on its axis) is a ray d parallel to the axis of the parabola; it bisects chords parallel to the line tangent to the parabola at the endpoint of d. **2.** the length of a diameter; the longest distance across a configuration; it may be infinite, as in the case of a parabola. **3.** more generally, given a set S in a metric space, the value $d[S] = \sup\{|s_1 - s_2|\}$, for all points s_1 and s_2 in S. A set of finite diameter is said to be bounded. If S is both bounded and closed, then S contains two points s_1 and s_2 such that $d[S] = |s_1 - s_2|$. **4.** the maximum of all of the distances between the vertices of a graph.
diameter group *Mechanical Engineering.* a parameter used in the study of flow machines such as turbines, equal to the fourth root of pressure number 2 divided by the square root of the delivery number.
diametral curve *Mathematics.* given a curve and one of its chords, the locus of midpoints of all chords parallel to the given chord.
diametral pitch *Design Engineering.* the ratio of the number of teeth on a gear to the diameter of its pitch circle, measured in inches.
diametral surface *Mathematics.* given any surface and one of its chords, any surface that passes through the midpoints of all chords parallel to the given chord. In particular, if the surface is a quadric surface, a diametral surface is always a plane and is called a **diametral plane.**
diamictite *Petrology.* a noncalcareous, unsorted or poorly sorted terrigenous sedimentary rock containing particles of many sizes. Also, MIXTITE.
diamicton *Petrology.* an unconsolidated or nonlithified diamictite. Also, SYMMICTON.
diamide *Organic Chemistry.* a compound that contains two amide ($-CONH_2$) groups.
diamidine *Organic Chemistry.* a molecule that contains two amidine ($-C{=}NHNH_2$) groups.
diamine *Organic Chemistry.* **1.** a compound containing two amino groups. **2.** see HYDRAZINE.
diamine oxidase *Enzymology.* an enzyme of the oxidoreductase class that catalyzes the aerobic oxidation of amines.
diamino- *Organic Chemistry.* a prefix indicating two amino groups.
2,7-diaminofluorene *Organic Chemistry.* $C_{13}H_{12}N_2$, a needlelike solid that is soluble in water and alcohol; melts at 164–165°C; used in chemical analysis.
2,4-diaminophenol hydrochloride see AMIDOL.
diaminophenylsulfone see DAPSONE.
diaminopimelate *Biochemistry.* a cell wall constituent present in some bacteria and blue-green algae.
diaminozide see SUCCINIC ACID 2,2-DIMETHYLHYDRAZIDE.
diammonium phosphate see AMMONIUM PHOSPHATE.

diamond

diamond *Mineralogy.* a form of carbon, a colorless, white, or occasionally tinted cubic mineral commonly occurring in octahedral crystals with a brilliant to greasy luster and a highly perfect cleavage, having a specific gravity of 3.51 and a hardness of 10 on the Mohs scale (the hardest known natural substance); found in irimberlite pipes, volcanic necks, and alluvial deposits.
diamondback rattlesnake *Vertebrate Zoology.* either of the two large, extremely venomous rattlesnakes of the genus *Crotalus,* having diamond-shaped markings on the back.

diamondback rattlesnake

diamondback terrapin *Vertebrate Zoology.* any of several edible turtles of the genus *Malaclemys,* having diamond-shaped markings on the back; found in the tidewaters of the eastern and southern U.S.
diamondbird *Vertebrate Zoology.* a tiny songbird of the family Dicaeidae, an aberrant member of the flowerpeckers from Australia; having tiny, short legs and tail, and colorful plumage with white speckles. It nests in tree holes or at ground level in sand banks.
diamond bit *Mechanical Devices.* a drill bit faced with diamonds as cutters; used for boring holes in rock. Also, BORT BIT.
diamond boring *Engineering.* high-speed drilling using either diamond or carbide boring tools.
diamond canker *Plant Pathology.* a virus disease of stone fruit trees, characterized by roughly shaped cankers on the tree bark, which cause a progressive weakening of the tree.
diamond chisel *Mechanical Devices.* a cold chisel with a diamond-shaped cutting edge, used for cutting V-shaped grooves and square corners. Also, **diamond-point chisel.**
diamond coring *Engineering.* the obtaining of rock core samples using a diamond drill.
diamond count *Design Engineering.* the number of diamonds on a diamond cut bit.
diamond crown *Mechanical Devices.* a steel drilling bit equipped with black diamonds on its face and cutting edges.
diamond cubic structure *Materials Science.* a face-centered-cubic crystal structure consisting of eight atoms per unit cell: four atoms are located at the lattice points and four are translated by 1/4, 1/4, 1/4.
diamond drill *Mechanical Devices.* a type of boring machine with bort diamonds that are rotated by steel rods about the face; used in continuous drilling for rock core samples.

diamond matrix *Geology.* the rock material in which a diamond occurs. *Design Engineering.* a metal or alloy on a grinding wheel, diamond crown, or other cutting tool in which the diamonds are embedded. *Mechanical Devices.* see BIT MATRIX.

diamond orientation *Design Engineering.* the angle or set of a diamond in a cutting tool so that the crystal face will be in contact with the material being cut.

diamond-particle bit *Mechanical Devices.* a drill bit faced with diamond fragments.

diamond pattern *Design Engineering.* the arrangement of diamonds set in a diamond drill crown.

diamond point *Mechanical Devices.* describing any cutting tool whose edge has a diamond or lozenge shape formed by two cutting edges meeting at an acute angle. Thus, **diamond-point tool.**

diamond-pyramid hardness number *Metallurgy.* a hardness measurement obtained according to the Vickers hardness test method.

diamond-pyramid hardness test see VICKERS HARDNESS TEST.

diamond reamer *Mechanical Devices.* a diamond-inset pipe device of a size larger than the drill bit and core barrel, used to enlarge or finish boreholes.

diamonds *Fluid Mechanics.* obliquely intersecting shocks formed in a supersonic jet or by reflection of a shock. Also, MACH DISKS.

diamond size *Engineering.* the number of equal-size diamonds whose total weight is one carat; used in bit setting and diamond drilling.

diamond structure *Crystallography.* a crystal structure in which an atom is located at the center of a tetrahedron whose vertices are occupied by identical atoms.

diamond stylus *Acoustical Engineering.* a high-quality stylus having a diamond tip; it has either a spherical or elliptical tip that fits into the groove on a record.

diamond tool *Mechanical Devices.* **1.** any tool with a diamond-shaped cutting edge used in drilling operations. **2.** a diamond-shaped cutter point used to make precision cuts in nonferrous metals and ceramics.

diamond-turned optics *Optics.* machined optical elements that are extremely precise.

diamond washer *Mining Engineering.* an apparatus that utilizes a vertical series of screens with 8-, 4-, 2-, and 1-mm mesh, which is operated by a shaking action that results in the separation of diamondiferous gravel.

diamond wheel *Mechanical Devices.* a circular grinding wheel with diamond dust on its abrasive surface, used to cut very hard materials such as sintered carbide or quartz.

diamyl phenol *Organic Chemistry.* $(C_5H_{11})_2C_6H_3OH$, a combustible, light yellow liquid; insoluble in water; boils at 280–295°C; used in making fungicides, detergents, plasticizers, lubricating oil additives, and synthetic resins.

diamyl sulfide *Organic Chemistry.* $(C_5H_{11})_2S$, a combustible, yellow liquid; boils at 170–180°C; used as an odorant and a flotation agent.

Diana complex *Psychology.* the desire of a female to be a male. (From *Diana,* the Greek goddess of hunting and the protector of women.)

diandrous *Botany.* having two stamens per flower.

Dianemaceae *Microbiology.* a family of slime molds in the order Trichales.

dianthovirus *Plant Pathology.* a virus that causes carnation ringspot virus and certain clover viruses; it is transmitted through the sap or soil.

Dianthovirus group *Virology.* a group of plant viruses that have a wide host range and are transmitted naturally through the soil. Also, CARNATION RINGSPOT VIRUS GROUP.

Dianulitidae *Paleontology.* an extinct family of stenolaemate bryozoans in the order Cystoporata; found in marine deposits of the Middle Paleozoic.

diapause *Physiology.* a period of delayed development or growth of an organism, associated with reduced metabolism and used as a mechanism for surviving adverse environmental conditions, as in certain insect pupae or plant seeds.

diapedesis *Hematology.* the outward passage through intact vessel walls of corpuscular elements of the blood. Also, DIAPIRESIS; EMIGRATION.

Diapensiaceae *Botany.* the sole family of the order Diapensiales, comprising alpine or arctic undershrubs with regular, bisexual flowers either solitary or in racemes, and leaves arranged in rosettes.

Diapensiales *Botany.* a monofamilial, dicotyledonous order in the subclass Dilleniidae that include low evergreen herbs and dwarf shrubs; distinguished by a tricarpellate ovary and stamens in two whorls; distributed in arctic and boreal regions.

Diaphanocephalidae *Invertebrate Zoology.* a family of roundworms; nematode parasites found in the intestines of amphibians and reptiles.

diaphanography *Medicine.* the passage of light through breast tissue for the purpose of examination by photography on infrared-sensitive film.

diaphone *Navigation.* a device that produces a fog signal by means of a slotted reciprocating piston actuated by compressed air. The signal usually consists of two separate, sequential tones.

diaphorase *Enzymology.* any of the flavoprotein enzymes that catalyze the reduction of dyes by reduced pyridine nucleotides.

diaphoresis [dī´ə fə rēt´sis] *Medicine.* sweating or perspiration, especially profuse perspiration.

diaphorite *Mineralogy.* $Pb_2Ag_3Sb_3S_8$, an opaque, metallic gray-black monoclinic mineral occurring as prismatic crystals, having a specific gravity of 5.97 to 6.04 and a hardness of 2.5 to 3 on the Mohs scale; found with galena, dolomite, and quartz.

diaphototropism *Botany.* the tendency of leaves to turn their upper surfaces to face a light source.

diaphragm [dī´ə fram´] a separating layer, sheet, or structure; specific uses include: *Anatomy.* the dome-shaped muscle that separates the thoracic and abdominal cavities and serves as the chief muscle of respiration. *Medicine.* a contraceptive device made of rubber or soft plastic material with a spring rim that is fitted over the cervix uteri to prevent the entrance of spermatozoa during intercourse. *Acoustical Engineering.* **1.** a thin, flexible disk or cone in a microphone that vibrates when contacted by sound waves, producing or varying an electric current. **2.** a similar part in a loudspeaker that vibrates in response to an electric current, thus producing sound. *Optics.* a device that prevents destructive light rays, those that would cause aberrations or glare, from entering the lens of a camera or microscope by limiting the diameter of the opening through which rays can enter the system. *Electromagnetism.* a conductive plate inserted into a waveguide to introduce and match impedance. *Radiology.* an apparatus with an adjustable opening used to limit or restrict the light or radiation that passes through it.

diaphragmatic [dī´ə frə mat´ik] of or relating to a diaphragm.

diaphragmatic hernia *Medicine.* a hernia that passes through the diaphm into the thoracic cavity.

diaphragmatic respiration *Physiology.* the movements of the diaphragm associated with inspiration and expiration during breathing.

diaphragm cell *Chemical Engineering.* a type of electrolytic cell in which anode and cathode compartments are divided by a porous membrane or diaphragm; used for the production of sodium hydroxide and chlorine from sodium chloride brine.

diaphragm compressor *Mechanical Engineering.* a compressor in which a reciprocally moving diaphragm is used to compress a small volume of gas.

diaphragm gauge *Engineering.* a pressure-sensing device that utilizes a diaphragm to measure differences in pressure between the sides of the enclosed diaphragm.

diaphragm horn *Acoustical Engineering.* a loudspeaker horn in which moving air causes a diaphragm to vibrate and produce sound.

diaphragm pump *Mechanical Engineering.* a pump in which a flexible diaphragm is set between two nonreturn valves, clamped around the edge, and attached to the center of a short-stroke reciprocating rod.

diaphragm valve *Engineering.* a fluid valve that employs a diaphragm as the open-close element.

diaphthoresis see RETROGRADE METAMORPHISM.

diaphthorite *Petrology.* a schist with minerals formed by retrograde metamorphism.

diaphysis *Anatomy.* a shaft of a long bone between the extremities. Also, SHAFT.

diapir *Geology.* an upfolded structure, such as an anticline, in which mobile, plastic core material has broken through or pierced the more brittle overlying rock. Also, DIAPIRIC FOLD, PIERCEMENT, PIERCING FOLD.

diapiresis see DIAPEDESIS.

diapiric fold see DIAPIR.

diapophysis *Anatomy.* the upper articular surface of the transverse process of a vertebra.

Diaporthaceae *Mycology.* a family of fungi belonging to the order Diaporthales that occurs on woody material and is pathogenic to higher plants.

Diaporthales *Mycology.* an order of fungi belonging to the class Pyrenomycetes, characterized by a one-walled ascus possessing a ring at its tip and a hyaline ascospore that is variable in septation; it occurs primarily on woody surfaces and is sometimes parasitic to plants.

diapositive *Cartography.* a positive photograph on a transparent medium, usually a glass plate in a plotting instrument, a projector, or a comparator.

Diapriidae *Invertebrate Zoology.* a large family of wasps whose larvae are parasitic on flies, ants, termites, and other insects.

diarch *Botany.* **1.** having two xylem and two phloem bundles or poles, as in the protoxylem of roots or the bipolar type of spindle. **2.** a tetrandrous flower in which the stamens occur in pairs, with one shorter than the other.

diarrhea [dī´ə rē´ə] *Medicine.* an intestinal disorder characterized by abnormal fluidity and frequency of fecal evacuations, generally the result of increased motility in the colon; may be an important symptom of such underlying disorders as dysenteric diseases, lactose intolerance, GI tumors, and inflammatory bowel disease. Also, **diarrhoea.**

diarsine *Organic Chemistry.* an organic compound having the general formula $(R_2As)_2$, with the arsenic atoms held together by an As–As bond.

diarthrosis *Anatomy.* a freely movable joint consisting of opposing bones covered by a layer of hyaline cartilage, which are held in apposition by ligaments that are lined by a synovial membrane around a capsule filled with synovial fluid. Also, SYNOVIAL JOINT.

diarylamine *Organic Chemistry.* an organic compound containing two aryl groups attached to the nitrogen of an amine group.

Dias, Bartolomeu [dē´əs; dē´äs] c. 1450–1500, Portuguese navigator; discovered sea route around Africa.

diaschisis *Medicine.* a loss of function in one part of the brain caused by a localized injury in another part.

diaspore *Mineralogy.* AlO(OH), a white to gray or pink orthorhombic mineral occurring as platy to acicular crystals and in massive and stalactitic forms, having a specific gravity of 3.3 to 3.5 and a hardness of 6.5 to 7 on the Mohs scale; found in metamorphic limestones and altered igneous rocks.

diasporometer *Optics.* in an optical rangefinder, the system of wedges that rotate counter to the axis of the image, used to detect deviation in the image axis.

diastalsis *Physiology.* a downward-moving wave of contraction with a preceding wave of inhibition occurring in the digestive tube.

diastase *Enzymology.* an enzyme that catalyzes the hydrolysis of starch into sugar.

diastasis *Medicine.* a simple separation of parts normally joined together, as in the separation of an epiphysis from the shaft of a long bone without a true fracture.

diastasis cordis *Physiology.* the period of time during the cardiac cycle when there is little change in the length of the atrial and ventricular muscle fibers, and the filling of the chambers with blood is slow.

diastem *Geology.* a relatively short interruption in sedimentation with little or no erosion occurring before deposition is resumed.

diastema *Anatomy.* **1.** an abnormal opening, cleft, or fissure in any part, especially when congenital. **2.** a space between two adjacent teeth in the same dental arch. **3.** a space between the maxillary lateral incisor and canine tooth, normal in many mammals. *Cell Biology.* differentiated cytoplasm found at the equatorial plane during the metaphase stage of mitosis.

diastereoisomer *Organic Chemistry.* any of the enantiomers that are not mirror images in a group of optical isomers occurring in compounds containing asymmetric carbon atoms.

diastereoptic ligand *Organic Chemistry.* a ligand whose replacement or addition produces diastereoisomers.

diastole [dī as´tə lē] *Physiology.* the dilatation, or dilatation period, of the cardiac cycle, when the atrial and ventricular muscle fibers are elongated, and the filling of the chambers with blood is rapid.

diastolic [dī´ə stäl´ik] *Physiology.* relating to the rhythmic relaxation of the muscles of the heart chambers during the time they fill with blood.

diastolic pressure *Physiology.* the arterial blood pressure taken during the diastolic period of the heart beat cycle.

diastrophism *Geology.* all of the processes whereby the earth's crust is deformed under the influence of tectonic forces, and the results of such deformation.

diastyle *Architecture.* an intercolumniation of three diameters.

diataxia *Neurology.* a muscle disorder characterized by ataxia (muscular incoordination or failure of muscular action) affecting the trunk or limbs on both sides of the body.

diathermal *Thermodynamics.* involving a heat transfer; describing a process in which there is a heat transfer in or out of the system in question. Also, **diathermic, diathermous.**

diathermal envelope *Thermodynamics.* a surface that serves to enclose and isolate a thermodynamic system, yet allows heat transfers across its boundary. Also, **diathermous envelope.**

diathermal wall *Thermodynamics.* a boundary of a system that allows heat transfer interaction between the system and its environment.

diathermanous *Physics.* of or relating to a substance that is highly transparent to infrared radiation.

diathermy [dī´ə thər´mē] *Medicine.* the production of heat in the tissues of the body because of their resistance to the passage of high-frequency currents of ultrasonic waves. In **medical diathermy** (thermopenetration), the tissues are warmed but not damaged; in **surgical diathermy** (electrocoagulation), tissues are destroyed.

diathermy interference *Telecommunications.* television interference, in the form of a herringbone pattern, that is caused by high-frequency electrical equipment.

diathesis-stress hypothesis *Psychology.* the theory that an individual with an inherent susceptibility toward mental disorder will, under stress, exhibit abnormal behavior.

diatom [dī´ə täm´] *Invertebrate Zoology.* the common name for the Bacillariophyceae, a class of unicellular microscopic algae with a symmetrical siliceous exoskeleton.

Diatomaceae *Botany.* a diverse family of freshwater and marine diatoms of the order Pennales, characterized by the lack of a raphe and only one or two labiate processes; species occur both as solitary organisms and in various types of colonies.

diatomaceous [dī´ə tə mā´shəs] relating to or consisting of diatoms or their remains.

diatomaceous earth *Geology.* a soft, fine, porou, yellow, light gray, or white s siliceous sedimentary deposit; composed mainly of the microscopic skeletons of diatoms. It is used in filtration; the particular shape of the diatoms in the deposit determine its filtering characteristics. Also, KIESELGUHR, TRIPOLITE, FOSSIL FLOUR, ROCK MEAL.

diatomaceous ooze *Geology.* a deep-sea, siliceous sediment consisting of more than 30% skeletal remains of diatoms.

diatomic [dī ə täm´ik] *Chemistry.* having two atoms per molecule.

diatomite *Geology.* the dense, consolidated equivalent of diatomaceous earth.

diatonic *Acoustics.* of or relating to a musical scale that contains five whole tones and two semitones.

diatonic scale *Acoustics.* a musical scale with separations between notes based on natural intonations; a major scale or minor scale.

diatreme *Geology.* a circular volcanic vent produced mainly by a gaseous explosion, and often filled with breccia.

diatropism *Botany.* the growth of an organ or organism that places its axis at a transverse position to the line of action of a stimulus.

Diatrymiformes *Paleontology.* an order of generally large, predatory, flightless birds in the superorder Neognathae; they were characteristic of the Early Eocene in Europe and North America.

Diatrypaceae *Mycology.* a family of fungi belonging to the order Xylariales that is composed of stromatic perithecial fungi having vegetative hyphae made up of the fungus and host tissue.

diauxic growth *Microbiology.* the phenomenon whereby a microorganism, when provided with more than one organic compound, will metabolize the preferential compound to completion before utilizing a second compound, resulting in two distinct phases of growth separated by a lag phase. Also, **diauxie.**

diaxon *Neurology.* a nerve cell having two axons or major axis-cylinder processes. Also, **diaxone**.

diazine *Organic Chemistry.* **1.** any of three isomeric, biologically important bases formed when two of the carbon atoms in the benzene ring are replaced by nitrogen atoms. **2.** a suffix indicating a ring compound having two nitrogen atoms.

diazinon *Organic Chemistry.* $C_{12}H_{21}N_2O_3PS$, a colorless liquid, slightly soluble in water and soluble in alcohol; boils at 83–84°C under reduced pressure and decomposes at 120°C; used as an insecticide.

diazoalkane *Organic Chemistry.* a saturated hydrocarbon containing the –N=N– group.

diazoamine *Organic Chemistry.* an amine containing the diazo group.

diazoaminobenzene *Organic Chemistry.* $C_6H_5N{=}NNHC_6H_5$, golden-yellow scales; insoluble in water and soluble in alcohol; melts at 98°C and explodes near 150°C; used in organic synthesis, in dyes, and as an insecticide.

diazoate *Organic Chemistry.* a salt of the type ArN=NO*M* that forms when a monovalent hydroxide such as NaOH is added to an aqueous solution of an aryldiazonium cation, ArN_2^+.

diazo coating *Graphic Arts.* a diazonium salt solution applied to lithographic plates to make them water-receptive.
diazo compound *Organic Chemistry.* any compound containing the group –N=N–.
diazo coupler *Graphic Arts.* a machine used to expose diazo material of a predetermined primary color to a color separation positive of that same color in order to produce a photomechanical color proof.
diazo group *Organic Chemistry.* the group –N=N–.
diazoic acid *Organic Chemistry.* $C_6H_5N{=}NOH$, an acid that forms when sodium hydroxide is added to an aqueous solution of a benzene diazonium salt.
diazoketone *Organic Chemistry.* a ketone of the type $RCOCHN_2$ that possesses the diazo group.
diazole *Organic Chemistry.* a monocyclic heteroaromatic compound consisting of three carbon and two nitrogen atoms.
diazo material *Graphic Arts.* a photosensitive material that will yield a positive image after exposure to a positive print.
diazomethane *Organic Chemistry.* $H_2C{=}N^+{=}N^-$, a yellow, toxic, mutagenic, carcinogenic explosive gas; soluble in alcohol and decomposes in water; melts at –145°C and boils at –23°C; used as a powerful methylating agent.
diazonium *Organic Chemistry.* the group $–N{=}N^+$.
diazonium salts *Organic Chemistry.* salts containing the group $–N{=}N^+$
diazo oxide *Organic Chemistry.* an oxide containing the group –N=N–.
diazo paper or **diazo print** *Graphic Arts.* a photographic positive print similar to a blueprint; made from another positive image, developed with ammonia fumes, and used especially for color proofing. Also, DIAZOTYPE, OZALID, WHITEPRINT, GAS PRINT, AMMONIA PRINT.
diazosulfonate *Organic Chemistry.* a salt derived from the diazosulfonic acid group.
diazosulfonic acid *Organic Chemistry.* $ArN{=}NSO_3H$, an aromatic acid containing the diazo group attached to the sulfonic acid group.
diazotization *Organic Chemistry.* a reaction using nitrous acid to transform a primary amine into a diazonium compound.
diazotroph *Microbiology.* any microorganism that is capable of nitrogen fixation.
diazotype see DIAZO PAPER.
Dibamidae *Vertebrate Zoology.* the blind skinks, a small family of limbless burrowing lizards of the order Squamata, found in Mexico and the Indo-Pacific region.
dibasic *Chemistry.* **1.** of an acid, having two hydrogen atoms that can be replaced by a monovalent metal. **2.** of a compound, having two basic monovalent atoms or groups.
dibasic acid *Chemistry.* an acid having two hydrogen atoms that can be replaced by two basic atoms or radicals to form a salt.
dibasic calcium phosphate *Inorganic Chemistry.* $CaHPO_4{\cdot}2H_2O$, white triclinic crystals that are odorless and tasteless; soluble in dilute acids, slightly soluble in water, and insoluble in alcohol; loses its water at a melting point of 109°C; used as a food supplement and in fertilizers and medicines. Also known by various other names, such as SECONDARY CALCIUM PHOSPHATE and DICALCIUM PHOSPHATE.
dibasic magnesium citrate *Organic Chemistry.* $MgHC_6H_5O_7{\cdot}5H_2O$, a white or yellow powder that is soluble in water; used in medicine.
dibasic magnesium citrate pentahydrate *Organic Chemistry.* $MgHC_6H_5O_7{\cdot}5H_2O$, a white or yellow powder; soluble in water and insoluble in alcohol; used in medicine and nutrition.
dibasic magnesium phosphate *Inorganic Chemistry.* $MgHPO_4{\cdot}3H_2O$, a white crystalline powder; slightly soluble in water and soluble in dilute acids; melts at 205°C and decomposes at 550–650°C; used as a food additive and in medicine. Also known by various other names, such as SECONDARY MAGNESIUM PHOSPHATE and MAGNESIUM HYDROGEN PHOSPHATE.
dibasic sodium phosphate *Inorganic Chemistry.* Na_2HPO_4, or its hydrated forms $Na_2HPO_4{\cdot}2H_2O$, $7H_2O$, and $12H_2O$, colorless translucent crystals or white powder with a saline taste; soluble in water and very soluble in alcohol; used in chemicals, pharmaceuticals, fertilizers, textiles, detergents, and plastics, and for a great many other purposes. Also known by various other names, such as SECONDARY SODIUM PHOSPHATE and DISODIUM PHOSPHATE.
dibenzofuran see DIPHENYLENE OXIDE.
dibenzyl see BIBENZYL.
dibenzyl disulfide *Organic Chemistry.* $C_6H_5CH_2SSCH_2C_6H_5$, a combustible, pink solid; very slightly soluble in water and soluble in alcohol; melts at 70–72°C; used as an antioxidant and antisludging agent for petroleum oils. Also, BENZYL DISULFIDE.
dibit *Computer Programming.* any one of the four arrangements of two bits: 00, 01, 10, 11.
diborane *Inorganic Chemistry.* B_2H_6, a colorless, highly reactive and toxic gas with an unpleasant odor, decomposes in cold water; it melts at –165.5°C and boils at –92.5°C; a dangerous fire risk. It has many industrial uses, including the synthesis of organic boron compounds.
Dibranchia *Invertebrate Zoology.* a subclass of Cephalopoda, including all extant cephalopod mollusks except Nautilus. Also, COLEOIDEA.
dibromide *Chemistry.* a molecule that has two bromine atoms.
dibromo- a combining form indicating the presence of two bromine atoms.
1,2-dibromo-3-chloropropane *Organic Chemistry.* $CH_2BrCHBr CH_2Cl$, a combustible, carcinogenic, brown liquid (colorless when pure); slightly soluble in water and miscible with oils; boils at 195.5°C and melts at 6.7°C; used as a pesticide, nematocide, and soil fumigant.
dibromodifluoromethane *Organic Chemistry.* CF_2Br_2, a nonflammable, colorless, heavy liquid, insoluble in water and soluble in ether; melts at –142 to –141°C and boils at 24.5°C; used in the synthesis of dyes, in pharmaceuticals, and as a fire-extinguishing agent.
dibromopropamidine isethionate *Organic Chemistry.* $C_{21}H_{30}Br_2N_4 O_{10}S_2$, prismatic needles, soluble in water and alcohol; melts at 226°C; used in cosmetics and as an antiseptic.
2,6-dibromoquinone-4-chlorimide *Organic Chemistry.* $C_6H_2Br_2{=}ClNO$, a yellow powder, slightly soluble in water and moderately soluble in hot alcohol; melts at 83°C and explodes at 120°C; used as a reagent for phenol and phosphatases.
3,5-dibromosalicylaldehyde *Organic Chemistry.* $C_7H_4Br_2O_2$; pale yellow prisms; soluble in water and alcohol; melts at 86°C; used as an antibacterial agent.
dibucaine *Organic Chemistry.* $C_{20}H_{29}N_3O_2$, a possibly allergenic, somewhat hygroscopic, colorless powder; slightly soluble in water and soluble in alcohol; melts at 62–65°C; used as an anesthetic. Also, PERCAINE, NUPERCAINE.
Dibunophyllum *Paleontology.* a variable genus of solitary Precambrian rugose coral in the extinct superfamily Zaphrenticae.
dibutyl *Organic Chemistry.* of or relating to a molecule that has two butyl groups.
dibutylamine *Organic Chemistry.* $(n\text{-}C_4H_9)_2NH$, a combustible, toxic, colorless liquid, partially soluble in water and soluble in alcohol; boils at 159.6°C and melts at –62°C; used as a corrosion inhibitor and an intermediate for dyes and insecticides.
dibutyl maleate *Organic Chemistry.* $C_4H_9OOCCH{=}CHCOOC_4H_9$, a combustible, colorless, oily liquid; insoluble in water; boils at 280.6°C; used in copolymers and plasticizers, and as an intermediate.
dibutyl oxalate *Organic Chemistry.* $(COOC_4H_9)_2$, a combustible, toxic, water-white liquid, miscible with most alcohols; boils at 234–240°C and melts at –30°C; used in organic synthesis and as a solvent.
3,4-di(*tert*-butyl)phenyl-*N*-methyl carbamate *Organic Chemistry.* $C_{16}H_{25}O_2N$, a white crystalline solid; melts at 102°C; used as an insect repellent.
dibutyl phthalate *Organic Chemistry.* $C_6H_4(COOC_4H_9)_2$, a combustible, toxic, colorless, stable, oily liquid, insoluble in water and miscible with organic solvents; boils at 340.7°C and melts at –35°C; used as a plasticizer, insecticide, and adhesive.
dibutyl succinate *Organic Chemistry.* $C_{12}H_{22}O_4$, a colorless liquid that boils at 274.5°C or 120°C (3.0 torr) and melts at –29°C; used as an insect repellent.
dibutyl tartrate *Organic Chemistry.* $C_4H_9OOC(CHOH)_2COOC_4H_9$, a combustible, light-tan liquid; miscible with organic solvents; melts at 22°C and boils at approximately 175°C (5 torr); used as a solvent and plasticizer, and in lacquers, dopes, and inks.
DIC disseminated intravascular coagulation.
Dicaeidae *Vertebrate Zoology.* a passerine songbird family of the suborder Oscines that includes the flowerpeckers and diamondbirds; centered in New Guinea and ranging to India, southeast Asia, and Australia.
dicalcium *Chemistry.* a compound that has two calcium atoms.
dicalcium phosphate see DIBASIC CALCIUM PHOSPHATE.
dicamba see 3,6-DICHLORO-*o*-ANISIC ACID.
di-cap storage *Electronics.* a device that stores information in the form of arrays of charged capacitors and utilizes diodes to control information flow.
dicarboxylic acid *Organic Chemistry.* an acid containing two carboxyl (=COOH) groups.
dicarpellate *Botany.* having two carpels.

dice crystal wafering *Materials Science.* a step in the production of transistors; after the silicon crystal has been formed and sectioned into slices (wafers), circuits are etched on the surface of the wafers, and then the wafers are notched and broken into dice.

Dicellograptus *Paleontology.* a widespread genus of Ordovician graptolites in the family Dicranograptidae and the order Graptoloidea, belonging to the Leptograptid fauna, that became extinct at the end of the Ordovician; an important index fossil assemblage.

dicentra *Botany.* any of the plants of the fumitory family, belonging to the genus *Dicentra,* having long clusters of drooping flowers, such as Dutchman's breeches or bleeding heart.

dicentric *Genetics.* describing a chromosome that has two centromeres.

dicephaly *Medicine.* the condition of having two heads.

dicerous *Invertebrate Zoology.* possessing two antennae or two tentacles.

dicetyl see DOTRIACONTANE.

Dichapetalaceae *Botany.* a monogeneric family of dicotyledonous African woody plants of the order Celastrales, species of which are highly poisonous and often have a hairy covering on all parts.

dichasium *Botany.* **1.** a cymose inflorescence in which the main axis exhibits determinate growth and forms a terminal flower, followed by the development of a pair of opposite lateral branches below, each with its own flowers, resulting in a symmetric appearance with the older flowers at the end of each unit. **2.** a cyme with a false dichotomy.

Dichelesthiidae *Invertebrate Zoology.* a family of copepod crustaceans that are ectoparasites on gills of fishes.

dichlamydeous *Botany.* having a perianth composed of both a calyx and a corolla. Also, DIPLOCHLAMYDEOUS.

dichlobenil *Organic Chemistry.* $Cl_2C_6H_3CN$, a toxic white solid that is almost insoluble in water; it melts at 144°C; used as an herbicide.

dichlofenthion *Organic Chemistry.* $C_{10}H_{13}Cl_2O_3PS$, a colorless liquid that is slightly soluble in water and miscible with organic solvents; boils at 165°C (0.1 torr); used as a pesticide.

dichlone *Organic Chemistry.* $C_{10}H_4Cl_2O_2$, toxic yellow needles that are almost insoluble in water and soluble in ether; melts at 193°C; used as a seed disinfectant, fungicide, insecticide, and organic catalyst.

dichloramine *Inorganic Chemistry.* **1.** an unstable compound not found in the pure state, formed by reaction of ammonia and chlorine. **2.** any chloramine with two chlorine atoms attached to a nitrogen atom.

dichloride *Chemistry.* a compound containing two chloride atoms.

dichloroacetic acid *Organic Chemistry.* $CHCl_2COOH$, a toxic, colorless liquid, soluble in water and alcohol; melts at 9–11°C and boils at 193–194°C; used in pharmaceuticals and medicine.

3,6-dichloro-*o*-anisic acid *Organic Chemistry.* $C_8H_6Cl_2O_3$, a brownish solid that is soluble in alcohol and slightly soluble in water; melts at 114°C; used as a broad-spectrum herbicide. Also, DICAMBA.

dichlorobenzene *Organic Chemistry.* $C_6H_4Cl_2$, any of three compounds that are derived from the chlorination of monochlorobenzene: ***m*-dichlorobenzene,** a combustible, colorless liquid that is insoluble in water and soluble in alcohol; boils at 173°C and melts at –25°C; used as a fumigant and insecticide; ***o*-dichlorobenzene,** a toxic, colorless liquid; insoluble in water and miscible with alcohol; boils at 180.5°C and melts at –17°C; used as a solvent; or ***p*-dichlorobenzene,** combustible, toxic, white crystals; insoluble in water , soluble in alcohol, and sublimes readily; melts at 53–54°C and boils at 174.1°C; used as a moth repellent, insecticide, and germicide.

1,4-dichlorobutane Organic Chemistry. $ClCH_2(CH_2)_2CH_2Cl$, a combustible, colorless liquid; insoluble in water and soluble in organic solvents; boils at 161–163°C; used in organic synthesis.

dichlorodifluoromethane *Organic Chemistry.* CCl_2F_2, a nonflammable, narcotic, colorless, noncorrosive gas; insoluble in water and soluble in alcohol; boils at –29.8°C, melts at –158°C, and stable to 550°C; used as a refrigerant and aerosol propellant. Also, FREON-12.

***p,p'*-dichlorodiphenylmethyl carbinol** *Organic Chemistry.* $(ClC_6H_4)_2C(CH_3)OH$, colorless crystals that melt at 70°C; used as an insecticide. Also, DI(*p*-CHLOROPHENYL)ETHANOL.

dichlorodiphenyltrichloroethane see DDT.

***sym*-dichloroethylene** *Organic Chemistry.* ClCH=CHCl, a flammable, toxic, narcotic, colorless liquid; slightly soluble in water and soluble in alcohol; boils at 55°C and decomposes slowly on exposure to air, light, and moisture; used as a general solvent for organic materials, perfumes, and lacquers. Also, ACETYLENE DICHLORIDE.

dichloroethyl ether *Organic Chemistry.* $[ClCH_2CH_2]_2O$, a combustible, toxic, colorless liquid; insoluble in water and soluble in organic solvents; boils at 178.5°C and melts at –50°C; used as a solvent.

dichlorofluoromethane *Organic Chemistry.* $CHCl_2F$, a nonflammable, toxic, colorless, heavy gas that is insoluble in water and soluble in alcohol; it boils at 8.9°C and melts at –135°C; used as a refrigerant and solvent. Also, FLUOROCARBON 21.

dichloroisocyanurate *Organic Chemistry.* $C_3HCl_2N_3O_3$, a compound that melts at 226.4–226.7°C; used as a topical anti-infective agent.

dichloromethane see METHYLENE CHLORIDE.

2,6-dichloro-4-nitroaniline *Organic Chemistry.* $C_6H_4Cl_2N_2O_2$, a yellow solid that is soluble in alcohol and melts at 191°C.

dichloropentane *Organic Chemistry.* $C_5H_{10}Cl_2$, a flammable, light-yellow liquid that is insoluble in water; a mixture of the dichloro derivatives of both *n*-pentane and isopentane; used as a solvent, soil fumigant, and an insecticide.

dichlorophene *Organic Chemistry.* $(C_6H_3ClOH)_2CH_2$, a toxic, light-tan powder; insoluble in water and soluble in alcohol; melts at 177°C; used as a fungicide and bactericide, and in medicine and cosmetics.

2,4-dichlorophenoxyacetic acid *Organic Chemistry.* $Cl_2C_6H_3OCH_2COOH$, a white to yellow crystalline powder; insoluble in water and soluble in alcohol; melts at 138°C and boils at 160°C (0.4 torr); used as a selective weed killer and defoliant. Also, 2,4-D.

di(*p*-chlorophenyl)ethanol see *p,p'*-DICHLORODIPHENYLMETHYL CARBINOL.

2,4-dichlorophenyl-4-nitrophenyl ether *Organic Chemistry.* $Cl_2C_6H_3OC_6H_4NO_2$, a toxic, dark-brown crystalline solid that is soluble in acetone; melts at 70°C; used as an herbicide.

dichlorotoluene *Organic Chemistry.* $ClC_6H_4CH_2Cl$, a combustible, colorless liquid that is insoluble in water and soluble in alcohol; boils at 213–214°C and melts at –13°C; used as a solvent and an intermediate.

dichlorprop *Organic Chemistry.* $C_9H_8Cl_2O_3$, colorless crystals; soluble in water; melts at 117–118°C; used as an herbicide. Also, 2,4-DP.

dichlorvos *Organic Chemistry.* $(CH_3O)_2P{=}O(O)OCH{=}CCl_2$, a poisonous liquid, slightly soluble in water and miscible with alcohol; boils at 140°C (20 torr); used as an insecticide and a fumigant.

Dichobunidae *Paleontology.* a family of artiodactyl mammals in the superfamily Dichobunoidea.

Dichobunoidea *Paleontology.* a superfamily of artiodactyl mammals of the Eocene, characterized by bunodont upper teeth; the most primitive of the paleodonts.

dichogamous [dī käg´ə məs] *Biology.* of an organism, producing male and female structures at different times. *Botany.* of a flower having pistils and stamens that mature at different times, enhancing cross-fertilization and preventing self-pollination.

dichogamy [dī käg´ə mē] *Biology.* the ability of a single organism to produce male and female structures at different times. *Botany.* the fact of having stamens and pistils maturing at different times, thus ensuring cross-fertilization.

Dichograptid fauna *Paleontology.* an important faunal assemblage of the Ordovician, widespread and including several zone fossils.

Dichograptus *Paleontology.* a genus of many-branched graptolites in the family Dichograptidae and the order Graptoloidea; one of the key members of the Ordovician dichograptid fauna.

dichoptous *Invertebrate Zoology.* having the margins of the compound eyes separate from one another.

dichotic listening *Psychology.* a testing method used to investigate the selective attention of the listener by stimulating the two ears simultaneously but with different messages.

dichotomizing search see BINARY SEARCH.

Dichotomosiphonaceae *Botany.* a monogeneric family of freshwater green algae of the order Bryopsidales, distinguished by oogamous reproduction, a tufted thallus comprised of dichotomously branched filiform tubes, and attachment to mud at water's edge by colorless rhizoids.

dichotomous [dī kät´ə məs] *Science.* relating to or showing dichotomy. *Biology.* branching by repeated divisions into two nearly equal parts.

dichotomous key *Systematics.* a device for identifying organisms based on the answers to a series of questions, with each question involving alternate choices.

dichotomy [dī kät´ə mē] *Science.* the fact of being divided into two parts or types. *Biology.* **1.** the fact of having two generally equal parts. **2.** a repeated forking or branching out. *Astronomy.* the moment when a planet or moon appears half illuminated by the sun.

dichroic mirror *Optics.* a mirror that reflects or transmits light depending on its wavelength or color.

dichroism *Optics.* the property of a material, either anisotropic or isotropic, that reveals different colors of transmitted light depending on the length of the light beam and the thickness of the material.

dichromate *Inorganic Chemistry.* any salt of dichromic acid; generally orange colored in solution.
dichromate oxygen consumed see CHEMICAL OXYGEN DEMAND.
dichromate treatment *Metallurgy.* an anticorrosion treatment that uses sodium dichromate.
dichromatic [dī krō mat´ik] *Biology.* having or exhibiting two color phases that are unrelated to sex, age, or season.
dichromatic dye *Chemistry.* a dye or indicator solution whose color varies according to its concentration.
dichromatism [dī krōm´ə tiz əm] *Medicine.* a form of defective color vision in which only two primary colors are seen and the spectrum is separated by an achromatic band; partial color blindness.
dichromic *Chemistry.* a compound containing two chromium atoms.
dichromic acid *Inorganic Chemistry.* $H_2Cr_2O_7$, an acid found only in an aqueous solution of chromium trioxide, CrO_3.
dicing *Engineering.* a process of cutting a material or object into small cubes. *Electronics.* **1.** the cutting of semiconductor material into cubes or rectangles as part of a manufacturing process for solid-state devices. **2.** the slicing of a slab of piezoelectric material into crystal blanks.
dicing cutter *Mechanical Engineering.* a machine in which sheet material is first cut into horizontal strands by blades and then fed against a rotating knife for dicing.
Dick, George Frederick 1881–1967, American bacteriologist; isolated scarlet fever streptococcus; developed scarlet fever antitoxin.
Dicke radiometer *Electronics.* an instrument that detects and measures low-level radiant energy in noise and provides a representative indication on an output display.
Dickinsoniidae *Paleontology.* a family of wormlike, flat-bodied, multisegmented coelomates of the Precambrian and Cambrian; an important member of the Ediacaran faunas.
dickinsonite *Mineralogy.* $(K,Ba)(Na,Ca)_5(Mn^{+2},Fe^{+2},Mg)_{14}Al(PO_4)_{12}(OH,F)_2$, an easily fusible, olive to grass green monoclinic mineral, occurring in tabular crystals and in foliated to micaceous masses, having a specific gravity of 3.41 and a hardness of 3.5 to 4 on the Mohs scale; found in granite pegmatites.
dickite *Mineralogy.* $Al_2Si_2O_5(OH)_4$, a white to brownish monoclinic clay mineral of the kaolinite-serpentine group, having a specific gravity of 2.6 and a hardness of 2 to 2.5 on the Mohs scale; found as a hydrothermal mineral with quartz and other sulfides.
Dicksoniaceae *Botany.* a family of tree ferns with marginal sori at the ends of veins, protected by a bivalved indusium.
Dick test *Immunology.* a skin test that determines an individual's immunity to scarlet fever, in which a scarlatinal toxin is injected and causes a reaction if an antitoxin is absent.
diclinous *Botany.* **1.** having stamens and pistils on separate flowers, either on the same or different plants. **2.** having antheridia and oogonia on separate hyphae in a fungus.
Dicnemonaceae *Botany.* a small family of large, mat-forming, usually epiphytic mosses of the order Dicranales, characterized by creeping stems with ascending branches and lateral sporophytes, and very large spores with precocious germination; occurring in Australasia and South America.
dicoccous *Botany.* composed of two one-seeded coherent carpels.
dicot *Botany.* a dicotyledonous plant.
dicotyledon [dī´kät ə lēd´ən] *Botany.* any of the great variety of plants in the Dicotyledoneae, characterized by two seed leaves (cotyledons) in the embryo, leaves with net venation, flower parts in groups of four or five, vascular tissue with a cylindrical arrangement, a cambium, and a primary root system with a taproot.
Dicotyledonae see MAGNOLIOPSIDA.
Dicotyledoneae *Botany.* one of the two subclasses of Angiospermae that includes most deciduous woody plants of temperate regions and a majority of herbaceous flowering plants; this highly developed, dicotyledonous subclass contains about 250 families.
dicotyledonous [dī´kät ə lēd´ə nəs] *Botany.* **1.** having two cotyledons. **2.** of or relating to the Dicotyledoneae.
dicovalent carbon see DIVALENT CARBON.
Dicranaceae *Botany.* a large, diverse, widespread family of ephemeral mosses of the order Dicranales, characterized by erect, little-branched stems with terminal sporophytes and often falcate-secund leaves; growing on soil, rock, humus, and trees, with the larger species forming dense cushions.
Dicranales *Botany.* a widely distributed order of mosses having dichotomous branching, lanceolate, many-ranked leaves, dense foliation, erect gametophores, and generally acrocarpous sporophytes.
Dicranemaceae *Botany.* a family of red algae of the order Gigartinales, characterized by filiform and terete or flattened and branched thalli that arise from stolonlike axes or a discoid base; all genera are restricted to southern Australia.
Dicranochaetaceae *Botany.* a monogeneric family of green algae of the order Chlorococcales, characterized by solitary or gregarious cells that are attached to the substrate and have one to four branched setae on the basal face; usually grows on aquatic angiosperms.
Dicranograptus *Paleontology.* a genus of bilateral Ordovician graptolites in the family Dicranograptidae and the order Graptoloidea; a member of the leptograptid fauna.
dicrotic [dī krät´ik] *Medicine.* of or relating to that form of the pulse in which a double beat can be felt at the wrist for each beat of the heart.
dicrotic notch *Cardiology.* a small decrease in arterial pulse or pressure marked by a notch in an arterial pulse tracing; it immediately follows the closure of the semi-lunar valves and is sometimes used as the point marking end of systole. Also, AORTIC NOTCH.
dicrotism [dī krät´iz əm] *Cardiology.* a condition in which two beats are felt in a pulse for each heartbeat.
dicrotophos *Organic Chemistry.* $(CH_3O)_2P{=}O(O)C(CH_3){=}CHC(O)N(CH_3)_2$, a toxic brown liquid that is miscible in water and boils at 400°C; used as an insecticide. Also, BIDRIN.
Dicruridae *Vertebrate Zoology.* the drongos, a family of small to medium black or gray birds of the passerine suborder Oscines; found in forests and savannas of the Old World tropics, South Africa, and Australia.
dicumarol *Biochemistry.* $C_{19}H_{12}O_6$, one of the coumarin anticoagulants, occurring as a white, crystalline powder; isolated originally from spoiled sweet clover but now produced synthetically; acts by inhibiting the hepatic synthesis of vitamin K-dependent coagulation factors. Also, **dicoumarin.**

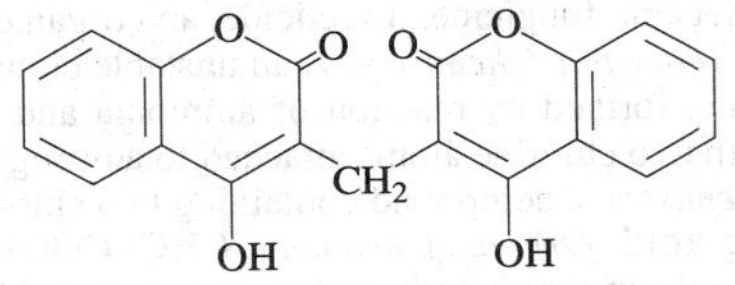

dicumarol

dictionary *Linguistics.* a book having a list of words, usually in alphabetical order, with their meaning and often with other information about them. *Computer Programming.* an alphabetized list of code names or keys used in a program, with a brief description and the meaning of each.
dictionary code see CODE DICTIONARY.
dictionary encoding *Computer Programming.* a technique for reducing data storage requirements by replacing the word with its relative position in a dictionary.
dictionary program *Computer Programming.* a program that indicates words not in its lexicon as possibly misspelled and usually offers alternate spellings; often included with a word-processing program. Also, SPELLING CHECKER PROGRAM.
Dictyoglomus *Bacteriology.* a genus of anaerobic bacteria, the sole species of which occurs in spherical aggregates comprised of varying numbers of rod-shaped cells.
dictyo- a combining form meaning "net" or "netlike," as in *dictyostele, dictyosome.*
dictyoblastospore *Mycology.* a blastospore whose septal cell walls extend both vertically and horizontally.
Dictyoceratida *Invertebrate Zoology.* an order of sponges that includes the commercial bath sponges.
Dictyochaceae *Botany.* a family of marine flagellates comprising the order Dictyochales, characterized by a siliceous skeleton made up of tubular sections, delicate pseudopodia, and a single flagellum.
Dictyochales *Botany.* the silicoflagellates, a monofamilial order in the class Chrysophyceae, considered the last survivors of a once flourishing group of organisms.
Dictyoclostus *Paleontology.* a widespread family of Middle Cenozoic vagrant brachiopods in the extinct order Strophomenida and suborder Productidina; most species have one convex valve.

Dictyonellidina *Paleontology.* a suborder of articulate brachiopods extant from the Middle Ordovician to the Permian; *incertae sedis* because of their unusual triangular umbonal plate.
Dictyonema *Paleontology.* a genus of many-branched graptolites of the Late Cambrian, in the order Dendroidea; Dictyonema is mostly sessile, but at least one species (*D. flabelliforme*) became epi-planktonic and achieved worldwide distribution at the beginning of the Ordovician; from this the planktic and very successful graptoloids developed.
Dictyosiphonaceae *Botany.* a family of membranous, saccate, tubular, and terete brown algae of the order Dictyosiphonales, characterized by generally unilocular sporangia which are borne sunken in the cortex.
Dictyosiphonales *Botany.* an order of brown algae of the class Phaeophyceae, with alternating dissimilar generations of a larger parenchymatous sporophyte and a microscopic filamentous gametophyte.
dictyosome *Cell Biology.* the Golgi apparatus of plant cells.
Dictyosphaeriaceae *Botany.* a family of green algae of the order Chlorococcales; characterized by autospores that are united in colonies by gelatinous threads or bands derived from the cell walls of previous generations; inhabiting freshwater plankton.
Dictyospongiidae *Paleontology.* a family of Paleozoic reticulate sponges in the subclass Amphidiscophora.
dictyosporal *Mycology.* describing fungi that have spores of numerous shapes.
dictyospore *Mycology.* a multicellular fungal spore having septal cell walls that extend both vertically and horizontally. Also, **dictyconidium.**
dictyostele *Botany.* a stele typically found in ferns, in which the xylem and phloem occur as separate vascular bundles encased by an epidermis, and each bundle is in a cylindrical arrangement associated with large overlapping leaf gaps.
Dictyostelia *Microbiology.* a subclass of bacteria-eating protozoa of the class Eumycetozoea, characterized by the formation of a muticellular pseudoplasmodium that differentiates into a multispored, stalked fruiting body; it includes a single order, **Dictyosteliida.**
Dictyosteliomycetes *Mycology.* in some classifications, a class of fungi of the division Myxomycota, composed of cellular slime molds that are found living in dung, humus, and soil, mainly in tropical environments.
Dictyotaceae *Botany.* a family of brown algae composing the order Dictyotales, distributed in tropical to subtropical habitats and sometimes in temperate seas.
Dictyotales *Botany.* a monofamilial order of brown algae in the class Phaeophyceae, characterized by flattened, fanlike or ribbonlike, branched or unbranched thalli, attachment by rhizoids, apical growth, and usually an alternation of isomorphic phases.
dictyotene *Cell Biology.* an extended meiotic diplotene stage that occurs in many mammalian oocytes during yolk formation.
Dictyothyris *Paleontology.* a genus of articulate brachiopods in the order Terebratulida, widespread from the Jurassic through the Quaternary.
dicyan or **dicyanogen** see CYANOGEN.
dicyanide *Chemistry.* a compound containing two cyanide groups.
dicyanodiamide *Organic Chemistry.* $NH_2C{=}NH(NHCN)$, nonflammable, pure white crystals that are soluble in hot water; melts at 207–209°C; it is used in fertilizers and organic synthesis. Also, CYANOGUANIDINE.
dicyclic *Science.* relating to or having two cycles. *Chemistry.* having a molecular structure containing two rings.
dicyclohexylamine *Organic Chemistry.* $(C_6H_{11})_2NH$, a fishy-smelling liquid; melts at –0.1°C and boils at 254–255°C; used as a solvent and corrosion inhibitor. Also, DODECAHYDRODIPHENYLAMINE.
dicyclohexylcarbodiimide *Organic Chemistry.* $C_6H_{11}N{=}C{=}NC_6H_{11}$, toxic white crystals; melts at 35–36°C and boils at 138–140°C; used as a coupling agent in peptide synthesis.
Dicyemida *Invertebrate Zoology.* an order of minute, wormlike, ciliated, mesozoan parasites found in renal organs of cephalopod mollusks.
DID direct inward dialing.
DIDA see DIISODECYL ADIPATE.
didelphic *Anatomy.* relating to or having a double uterus.
Didelphidae *Paleontology.* the opossums, a family of arboreal marsupials in the superfamily Didelphoidea; extant from the Cretaceous to the present.
Didelphoidea *Paleontology.* a superfamily of marsupials in the suborder Polyprotodonta; Upper Cretaceous to the present.
3,6-dideoxy-D-galactose see ABEQUOSE.
dideoxynucleotide *Biotechnology.* a nucleotide that lacks the 3' hydroxyl group necessary for bonding and so prevents further chain elongation on a growing polynucleotide.
dideoxy sequencing *Molecular Biology.* a method for determining the primary sequence of DNA by copying single stranded DNA by DNA polymerase I; uses electrophoresis and autoradiography to deduce the DNA sequence of the original template.
3,6-dideoxy-D-xylohexose see ABEQUOSE.
Diderot, Denis [déd´ə ró´] 1713–1784, French philosopher and critic; compiled the influential 28-volume *Encyclopedie* with Jean d'Alembert and others.
Didiereaceae *Botany.* a family of spiny, cactuslike, mostly dioecious trees of the order Caryophyllales, characterized by soft wood with a large pith, and small alternative leaves; native only to Madagascar.
Didinium *Invertebrate Zoology.* a genus of marine and freshwater protozoan ciliates that feed on other ciliates, especially *Paramecium,* which is several times larger.

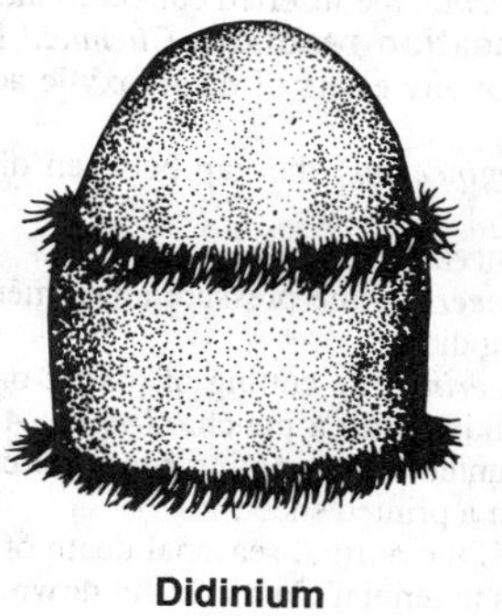

Didinium

didodecyl ether see DILAURYL ETHER.
Didolodontidae *Paleontology.* an extinct family of Early to Middle Tertiary ungulates of South America in the order Condylarthra; may be ancestral to later ungulates characteristic of South America.
Dido's problem [dí´dóz] *Mathematics.* the classical problem of maximizing the area enclosed by a curve of specified length; the solution is always a circle. (From the legend of Queen *Dido,* who was granted the right to set the boundaries of the city of Carthage; she determined that a semicircle along the coast would enclose the maximum possible area for the city.)
DIDP see DIISODECYL ADIPATE.
Didymelaceae *Botany.* a monogeneric family of dioecious evergreen trees constituting the order Didymelales.
Didymelales *Botany.* an order of dicotyledonous plants of the subclass Hamamelidae, characterized by scalariform perforations in the wood, and a pistil with one carpel; found only in Madagascar.
Didymiaceae *Mycology.* a family of fungi belonging to the order Physarales, characterized by dark spores and occurring on leaf litter, tress, dung, and plants.
didymium *Inorganic Chemistry.* a mixture of rare-earth (lanthanide) elements, mostly neodymium and praseodymium, obtained from monazite sand and used in the production of colored glass and optical filters.
Didymograptus *Paleontology.* a family of two-branched graptolites of the Lower and Middle Ordovician, in the order Graptoloidea.
didymolite see PLAGIOCLASE.
Didymosphaeriaceae *Mycology.* a family of fungi belonging to the order Melanommatales, living off decaying organic matter or as lichens on woody or herbaceous plants.
didymosporal *Mycology.* describing fungi that form spores in groups of two.
didymospore *Mycology.* a two-celled spore of imperfect fungi.
didymous *Biology.* growing or appearing in pairs; twinlike.
didynamous *Botany.* having four stamens grouped in two pairs of unequal length, as in many members of the mint family.
die *Mechanical Devices.* **1.** any of various tools or molds that are used for forming, casting, or finishing materials. **2.** a tool that is used to cut external threads. *Electronics.* **1.** a tiny diced or machined piece of semiconductor material that is used in the construction of a transistor, diode, or other semiconductor device. **2.** see CHIP. *Medicine.* a positive reproduction, in metal or another hard substance, of a tooth or its part, for use in tooth restoration.

dieback *Plant Pathology.* **1.** a progressive death of plant parts, generally starting at the outer tips and progressing toward the plant base; due to cankers, stem or root rots, nematodes, poor soil conditions, or other factors. **2. die back.** of a plant, to undergo this process.

dieb. alt. every other day; on alternate days. (From Latin *diebus alternis.*) Also, **Dieb. alt.**

die blade *Engineering.* a device attached to a die body to establish the size of the slot opening and produce plastic film of uniform thickness.

die block *Engineering.* a block that is bolted to the bed of the punch press and holds the die.

dieb. sec. every second day. (From Latin *diebus secundis.*) Also, **Dieb. sec.**

dieb. tert. every third day. (From Latin *diebus tertiis.*). Also, **Dieb. tert.**

die-cast *Metallurgy.* **1.** formed by die casting. **2.** a product formed in this way.

die casting *Metallurgy.* **1.** a process of casting in which molten metal is forced under pressure into metallic dies. **2.** a product formed in this way.

die chaser *Engineering.* the inserted cutters found in threading dies.

Dieckman condensation reaction *Chemical Engineering.* the condensation reaction of any esters of dicarboxylic acids to produce cyclic Beta-ketoesters.

die clearance *Engineering.* the gap between die members that meet during an operation.

die collar see DIE NIPPLE.

die cushion *Engineering.* the pressure attachments to a die block that are used for drawing dies.

die cutting *Engineering.* the cutting of plastic or metal sheets into desired shapes by striking with a punch. *Graphic Arts.* the use of a steel-rule die, often mounted on a printing press, to cut a designated shape such as a label from a printed sheet.

diedown *Botany.* **1.** the normal seasonal death of the aboveground portions of herbaceous perennial plants. **2. die down.** of a plant, to undergo this process.

die forming *Metallurgy.* the process of forming a metallic stock into a die in order to obtain a desired shape. Similarly, **die forging.**

die gap *Engineering.* the distance between two opposing metal faces that form the die opening.

Diego blood group *Immunology.* a group of human red blood cell antigens that is recognized by its reaction with the anti-Di(a) antibody and is commonly found in American Indians and Asiatics.

Dielasma *Paleontology.* an extinct genus of articulate brachiopods in the order Terebratulida, widespread in the Carboniferous and Permian.

dieldrin *Organic Chemistry.* $C_{12}H_{10}OCl_6$, a toxic, carcinogenic flaked solid, light tan in color; insoluble in water and soluble in organic solvents; melts at 177°C; used as an insecticide, but restricted to nonagricultural applications.

dielectric [dī´ə lek´trik] *Electricity.* describing a nonconducting material that can sustain a steady electric field and serve as an insulator. *Materials.* a nonconducting material of this type, such as glass or plastic. Also, **dielectric material.**

dielectric absorption *Electromagnetism.* the energy losses in a dielectric medium when the medium is exposed to a time-varying electric field. *Electricity.* an undesirable characteristic of certain dielectrics to retain a portion of an electric charge after removal of the electric field. Also, DIELECTRIC HYSTERESIS.

dielectric amplifier see CARRIER AMPLIFIER.

dielectric antenna *Electromagnetism.* an antenna whose radiation pattern is produced by a dielectric element.

dielectric breakdown *Electronics.* a sudden increase in electric current flow through a dielectric material when the applied electric field strength goes beyond a critical value.

dielectric constant *Electricity.* the property of a material that determines how much electrostatic energy can be stored per unit volume of the material when unit voltage is applied; it is quantified as the ratio of the capacitance of a capacitor with the material to the same capacitor with only a vacuum as a dielectric. Also, PERMITTIVITY.

dielectric current *Electricity.* the current flowing at any instant through a surface of a dielectric located in a changing electric field.

dielectric ellipsoid *Electricity.* of or relating to the locus of points r satisfying $r||K||r = 1$, for an anisotropic medium in which the dielectric constant is a tensor quantity K.

dielectric fatigue *Electronics.* a property of some dielectrics in which resistance to breakdown increases after voltage has been applied for a long period.

dielectric film *Electricity.* a film with dielectric qualities used as the central layer of a capacitor.

dielectric gas *Electricity.* a gas with a high dielectric constant, such as sulfur hexafluoride; used in laser cavities and microwave equipment.

dielectric heating *Electricity.* the heating of a nominally insulating material due to its internal losses when placed in a changing electric field.

dielectric hysteresis see DIELECTRIC ABSORPTION.

dielectric lens *Electromagnetism.* a lens made of a dielectric substance that refracts radio waves.

dielectric-lens antenna *Electromagnetism.* an aperture antenna whose beam width depends on the dimensions of a dielectric lens through which the beam passes.

dielectric loss *Electromagnetism.* the energy that is converted into heat in a dielectric material when the material is subjected to a changing electric field.

dielectric loss angle *Electricity.* an angle equal to the difference between ninety degrees and the dielectric phase angle.

dielectric loss factor *Electricity.* a factor equal to the dielectric constant multiplied by the tangent of its dielectric loss angle.

dielectric matching plate *Electromagnetism.* a plate made of a dielectric material that is used for matching purposes in waveguide technique.

dielectric phase angle *Electricity.* the angular phase difference between the sinusoidal alternating voltage applied to a dielectric and the component of the resultant alternating current having the same period.

dielectric polarization *Electricity.* **1.** a polarized effect that arises from molecules oriented in such a way that permanent dipole moments arise from an asymmetric charged distribution. **2.** see POLARIZATION.

dielectric power factor *Electricity.* the cosine of the dielectric phase angle or the sine of the dielectric loss angle.

dielectric-rod antenna *Electromagnetism.* an antenna whose radiating element is a dielectric tapered rod; surface waves propagate to the end of the rod and produce an end-fire radiation pattern.

dielectric shielding *Electricity.* the reducing of an electric field in some region by interposing a dielectric substance.

dielectric strength *Electricity.* the highest voltage a dielectric material can withstand before breakdown occurs; expressed in volts or kilovolts per millimeter of material thickness. Also, ELECTRIC STRENGTH.

dielectric test *Electricity.* a test using the application of a voltage higher than the rated value for a specified time to determine the margin of safety against the breakdown of insulating materials.

dielectric vapor detector *Analytical Chemistry.* an instrument used to measure the change in the dielectric constant of gases, especially of carrier gases in chromatography.

dielectric waveguide *Electricity.* a wave guide consisting of a solid dielectric cylinder surrounded by air.

dielectric wedge *Electromagnetism.* a wedge-shaped piece of dielectric material used in a waveguide to match the impedance to that of another waveguide.

dielectric wire *Electromagnetism.* a waveguide made of dielectric material used in circuits to transmit ultra-high-frequency signals over short distances.

dielectronic recombination *Atomic Physics.* a process in which an electron recombines with a positive ion in an ionized gas to produce a neutral atom; the process being mediated by the temporary excitation of a second electron in the ion.

dielectrophoresis *Physical Chemistry.* the property that causes an uncharged material to be activated by an electric field.

die life *Metallurgy.* the maximum number of impressions, or castings, that a die can make before the quality of the product is impaired by die erosion.

Diels, Otto 1876–1954, German chemist; shared Nobel Prize for the discovery of diene synthesis (Diels-Alder Reaction).

Diels-Alder reaction *Organic Chemistry.* an organic reaction for the synthesis of 6-membered rings in which a diene combines with a dienophile.

die match *Metallurgy.* in metal fabrication, a desirable relative alignment of the dies.

diencephalon [dī´ən sef´ə län] *Developmental Biology.* the posterior part of the prosencephalon that is made up of the thalamus, the subthalamus, and the hypothalamus. Also, INTERBRAIN.

diene see DIOLEFIN.

diene resin *Organic Chemistry.* a material containing the diene group that may easily polymerize because of the double bonds.

Dienes phenomenon *Microbiology.* a phenomenon in which two different strains of Proteus will not intermingle on an agar plate, resulting in a distinct boundary line between the two growing strains. Also, **Dienes reaction.**

diene value *Organic Chemistry.* the amount of conjugated double bonds present in a molecule of a fatty acid or fat, measured by the amount of maleic anhydride that reacts with a known weight of acid or fat.

die nipple *Petroleum Engineering.* a cylindrical fishing tool with an inside surface that is tapered upward and furnished with hardened threads; used to recover drill pipe. When placed over lost downhole drilling equipment and rotated, the inside surface cuts into the outside of the lost equipment and grips it. Also, DIE COLLAR.

dienophile *Organic Chemistry.* a compound containing a double or triple bond that seeks the diene to form an adduct in the Diels-Alder reaction.

dientamoeba *Invertebrate Zoology.* a genus of amoebas that live commensally in primate intestines.

dieresis *Surgery.* **1.** the separation or division of parts that are normally joined. **2.** the surgical separation of parts by incision, electrosurgery, or cautery.

Diesel, Rudolf [dēz´əl] 1858–1913, German automotive engineer who designed and built the diesel engine.

diesel or **Diesel** *Mechanical Engineering.* **1.** relating to or powered by a diesel engine. **2.** a vehicle or object that is powered by a diesel engine. Used to form various compounds, such as **diesel locomotive, diesel truck, diesel hammer, diesel generating station,** and so on.

Diesel cycle *Thermodynamics.* an ideal thermodynamic cycle consisting of four processes: a compression at constant entropy; a constant-pressure heat transfer to the system; an expansion at constant entropy; and a constant-volume heat transfer from the system. *Mechanical Engineering.* an actual version of this cycle used in compression-ignition (Diesel) engines: first, air is drawn into the cylinder and compressed; second, energy is added by the injection and combustion of fuel in the cylinder; third, the gases produced by this combustion expand to move the piston downward for the power stroke; fourth, the burned gases in the cylinder are expelled.

diesel-electric *Mechanical Engineering.* describing a vehicle or machine that is powered by an electric motor, which receives power from an electric generator, which in turn is powered by a diesel engine. Thus, **diesel-electric locomotive.**

diesel engine *Mechanical Engineering.* a compression-ignition engine in which a fuel pump injects measured quantities of fuel into the heated compressed-air charge of each cylinder in succession, with the fuel igniting at an almost constant temperature; earlier types sprayed the fuel with an air blast.

diesel fuel *Materials.* a combustible distillate of petroleum used as fuel for diesel engines; usually the fraction of crude oil that is distilled after kerosene. Also, **diesel oil.**

diesel index *Mechanical Engineering.* a rating system for diesel fuel based on ignition characterisitics; the higher the index, the higher the ignition quality of the fuel.

dieseling *Mechanical Engineering.* in a compresser, the explosion of the mixture of air and lubricating or fuel oil in a compression chamber or other part of the air system.

diesel knock *Mechanical Engineering.* a knock caused when the period of ignition delay in a diesel engine is so long that a large quantity of atomized fuel accumulates in the combustion chamber.

diesel squeeze *Petroleum Engineering.* a method of driving a mixture of diesel oil and cement through casing openings to mend areas that bear water without affecting the areas that bear oil.

die set *Engineering.* a complete unit consisting of a punch holder, base, and guides that can be installed as a unit on a press. Also, SUBPRESS.

die shoe *Mechanical Engineering.* a block to which a die holder is mounted in order to spread the impact over the die bed.

die stamping *Graphic Arts.* the use of a metal die to make an impression, especially on an embossed book cover.

die steel *Metallurgy.* any of several hard alloy steels suitable for fabricating dies.

diester *Organic Chemistry.* a molecule having two ester groups.

diesterase *Enzymology.* an enzyme that catalyzes the cleavage of the diester bonds holding individual nucleotides together in DNA or RNA.

diestock see SCREWSTOCK.

diestrus *Physiology.* the short periods of sexual quiescence between the estrus cycles in polyestrous animals.

diet *Nutrition.* **1.** the usual selection of food and drink consumed from day to day by a given person or group. **2.** a specific plan for the consumption of food and drink, including the proper amounts of substances containing various nutrients to maintain good health. **3.** in popular use, an eating regimen intended to bring about a weight loss through restriction of the amount and/or the type of food consumed. *Medicine.* a special program of food or drink prescribed for a given individual to correct or minimize an existing health problem, as by the elimination of certain foods. *Zoology.* the type of food normally consumed by the members of a given species. (Going back to a Greek word meaning "way of life.")

dietary fiber *Nutrition.* a general term applied to cellulose, lignin, and other complex forms of carbohydrates that are indigestible in humans but are nevertheless useful because of the laxative effect of their bulk.

dietary induced thermogenesis see SPECIFIC DYNAMIC EFFECT.

Dieterici equation of state *Thermodynamics.* an equation of state for gases, having the form $p[\exp(a/VRT)](V - b) = RT$, where p is the pressure, T is the absolute temperature, R is the gas constant, V is the molar volume, and a and b are constants characteristic of the gas.

Dieterle silver stain *Microbiology.* a bacteriological stain commonly used for spirochaetes, which appear brown or black when stained.

dietetic foods *Nutrition.* foods that are specifically processed or prepared to adhere to a particular, often restrictive diet, usually directed toward weight loss.

dietetics *Nutrition.* the science of systematic regulation of intake of food and drink for management of health and disease.

diethanolamine *Organic Chemistry.* $(HOCH_2CH_2)_2NH$, toxic, combustible, colorless crystals or liquid, very soluble in water and alcohol; melts at 28°C and boils at 269°C; used in shampoos, cleaners, polishes, and textiles.

diether *Organic Chemistry.* a molecule having two oxygen atoms with ether bonds.

1,1-diethoxyethane see ACETAL.

diethyl *Organic Chemistry.* a molecule having two ethyl groups.

diethylacetal see ACETAL.

diethyl adipate *Organic Chemistry.* $C_2H_5OCO(CH_2)_4OCOC_2H_5$, a combustible, colorless liquid, insoluble in water; boils at 251°C and freezes at –21°C; used in pigments.

diethylamine *Organic Chemistry.* $(C_2H_5)_2NH$, a highly flammable, toxic, colorless liquid, miscible with water and alcohol; melts at –49.8°C and boils at 55.5°C; used in rubber chemicals and pharmaceuticals.

diethylbenzene *Organic Chemistry.* $C_6H_4(C_2H_5)_2$, a combustible, moderately toxic, colorless liquid; insoluble in water and soluble in alcohol; boils at 179.8–184.8°C; used as an intermediate and a solvent.

diethylcarbamazine citrate *Organic Chemistry.* $C_{16}H_{29}N_3O_8$, a white, crystalline powder, very soluble in water and slightly soluble in alcohol; melts at 135–138°C; used in medicine.

diethylcarbinol see 3-PENTANOL.

diethyl carbonate *Organic Chemistry.* $(C_2H_5O)_2C{=}O$, a flammable, colorless liquid; insoluble in water and miscible with alcohol; melts at –43°C and boils at 126°C; used as a solvent and in organic synthesis.

diethylene ether see 1,4-DIOXANE.

diethylene glycol *Organic Chemistry.* $HOCH_2CH_2OCH_2CH_2OH$, a combustible, colorless, hygroscopic, noncorrosive, viscous liquid, miscible with water; boils at 244.5°C and melts at –6.5°C; used in textile manufacture, cosmetics, and paper products. Also, DIGLYCOL.

diethylene glycol distearate see DIGLYCOL STEARATE.

diethylene glycol monoethyl ether see CARBITOL.

diethylenetriamine *Organic Chemistry.* $NH_2C_2H_4NHC_2H_4NH_2$, a toxic, hygroscopic, somewhat viscous liquid, soluble in water; boils at 208°C and melts at –39°C; used as a solvent and fuel component.

diethyl ether see ETHYL ETHER.

di(2-ethylhexyl) phthalate see DIOCTYL PHTHALATE.

di(2-ethylhexyl) sebacate see DIOCTYL SEBACATE.

diethyl ketone see 3-PENTANONE.

diethyl maleate *Organic Chemistry.* $C_2H_5OOCCH{=}CHOOC_2H_5$, a combustible, water-white liquid, insoluble in water but soluble in alcohol; melts at approximately –15°C and boils at 222.7°C; used in flavorings and organic synthesis.

diethyl phosphite *Organic Chemistry.* $(C_2H_5O)_2HPO$, a combustible, water-white liquid that is soluble in water; boils at 138°C; used in paint solvents, lubricant additives, and as a reducing agent.

diethyl phthalate *Organic Chemistry.* $C_6H_4(CO_2C_2H_5)_2$, a combustible, toxic, narcotic, water-white, stable liquid, insoluble in water and miscible with alcohol; melts at –40.5°C and boils at 295°C; used in perfumes, insecticides, and plastics. Also, DEP.

diethyl pyrocarbonate *Organic Chemistry.* $C_6H_{10}O_5$, a toxic, colorless liquid with a sweet esterlike odor; soluble in alcohol and slightly soluble in water with decomposition; used as a fermentation inhibitor. Also, DEPC.

diethylstilbestrol *Biochemistry.* $C_{18}H_{20}O_2$, a nonsteroid estrogen that is used therapeutically to replace natural estrogenic hormones.

diethyl succinate *Organic Chemistry.* $(CH_2COOC_2H_5)_2$, a combustible, colorless liquid that is slightly soluble in water and miscible with alcohol; melts at –21°C and boils at 218°C; used as a chemical intermediate and plasticizer.

diethyl sulfate *Organic Chemistry.* $(C_2H_5)_2SO_4$, a combustible, toxic, noncorrosive, colorless liquid, insoluble in water and soluble in alcohol; melts at –24.4°C and boils at 208°C; used in organic synthesis.

diethyltoluamide *Organic Chemistry.* $CH_3C_6H_4CON(C_2H_5)_2$, a combustible, toxic, colorless liquid; soluble in water and alcohol; melts at 97°C and boils at 160°C (1.0 torr); used as an insect repellent and a resin solvent.

dietitian *Nutrition.* a person trained in the scientific management of meals of individuals or groups; one who arranges diet programs for health maintenance or therapeutic purposes. Also, **dietician.**

Dietl's crisis *Medicine.* paroxysmal attacks of lumbar and abdominal pain accompanied by nausea and vomiting, occurring as a result of kinking of the ureter in persons with wandering kidney.

dietrichite *Mineralogy.* $(Zn,Fe^{+2},Mn^{+2})Al_2(So_4)_4{\cdot}22H_2O$, a grayish-white monoclinic mineral occurring in fibrous masses, having a specific gravity of 1.86 and a hardness of 2 on the Mohs scale; found as a recent deposit in mine workings.

dietzeite *Mineralogy.* $Ca_2(IO_3)_2(CrO_4)$, a water-soluble dark yellow, transparent monoclinic mineral commonly occurring in fibrous or columnar masses, having a specific gravity of 3.62 and a hardness of 3.5 on the Mohs scale; found in nitrate deposits.

DIF data interchange format.

diffeomorphism *Mathematics.* a differentiable homeomorphism; equivalently, a continuously differentiable bijection with a continuously differentiable inverse.

difference *Mathematics.* **1.** the result obtained by subtracting one quantity or expression from another. The process of determining a difference is subtraction. **2.** the difference between sets A and B, denoted $A - B$ or $A\backslash B$, is the set of elements of A that are not elements of B. It is not assumed that B is contained in A, and in general, $A - B \neq B - A$. *Artificial Intelligence.* in GPS, a representation of a difference between the current state of the search and the goal; used to guide action to reduce the difference.

difference channel *Acoustical Engineering.* an audio channel that processes the difference between signals on the right and left channels of a stereo system.

difference encoding *Computer Programming.* a means of reducing data storage requirements for sequences of numeric data in which one value varies little from the previous value; successive differences from the previous value are stored instead of the actual values.

difference engine *Computer Science.* an early computer consisting of a self-controlled device that mechanized the production of the final values in a mathematical table from the pivotal values that are first computed; designed by George and Edward Sheutz in 1837, based on the analytical engine of Charles Babbage.

difference engine

difference equation *Mathematics.* **1.** any equation relating the values of a function at a set of points, each of which is a fixed interval from the previous point; in particular, any equation involving the difference operator D. **2.** a linear recursion formula.

difference-in-depth modulation *Telecommunications.* in directive systems that use overlapping lobes having modulated signals, the ratio acquired by subtracting the percentage of modulation of the smaller signal from that of the larger signal, and then dividing by 100.

difference of latitude *Cartography.* the shorter arc of any meridian between the parallels of two places, expressed in angular measure.

difference of longitude *Cartography.* the smaller angle at the pole, or the shorter arc of a parallel, between the meridians of two places, expressed in angular measure.

difference operator *Mathematics.* an operator Δ on a space of functions; defined by $\Delta f(x) = f(x + h) - f(x)$, where h is called the **difference interval** and Δf is called the **first difference** of f. Δ^0 is defined to be the identity operator; i.e., $\Delta^0 f = f$. Difference operators of higher order are defined iteratively by $\Delta^n f = \Delta(\Delta^{n-1} f)$, for $n = 1, 2, 3, \ldots$.

difference synthesis *Crystallography.* a Fourier synthesis (or map) for which the Fourier coefficients are the difference between the observed and calculated structure factor amplitudes measured for the X-ray diffraction pattern of a crystal. Such a map will have peaks where not enough electron density was included in the trial structure and troughs where too much was included. It is a valuable tool for locating missing atoms and for correcting the positions of those already present.

difference threshold see JUST-NOTICEABLE DIFFERENCE.

difference tone *Acoustics.* a tone created by the overlap of two tones, having a frequency equal to the difference of the two tones.

differentiable function *Mathematics.* a function whose derivative f' exists at every point in a given domain D is said to be differentiable on D. If, in addition, f' is continuous, then f is said to be **continuously differentiable.**

differentiable manifold *Mathematics.* let M be an n-dimensional manifold together with an atlas $\{(U_\alpha, \phi_\alpha)\}$; that is, each open set U_α of M is homeomorphic to R^n under the homeomorphism ϕ_α. M is a differentiable manifold, if whenever $x \in U_\alpha \cap U_\beta$, the function $\phi_\alpha \circ \phi_\beta^{-1}$ is a differentiable map ($\phi_\alpha \circ \phi_\beta^{-1}$ maps a subset of R^n to another subset of R^n); i.e., partial derivatives of all orders for $\phi_\alpha \circ \phi_\beta^{-1}$ exist and are continuous.

differential the fact or amount of being different; a difference; specific uses include: *Mechanical Engineering.* an epicyclic gear train that allows two shafts to rotate at different speeds while being driven by a third shaft; the sum of the rotational rates remains constant and is equal to twice the rotational speed of the driving shaft; allows the rear wheels of an automobile to rotate at different speeds when the vehicle is turning. Also, DIFFERENTIAL GEAR. *Mathematics.* **1.** the change in the value of a function of one or more variables resulting from a (usually small) change in one or more of the independent variables. **2.** given a function of several variables and a vector whose components are equal to the changes in the independent variables, a differential is a 1-form whose value is equal to the corresponding function change when applied to the vector.

differential absorption lidar *Engineering.* a device that determines the presence of atmospheric gases by comparing the return signals of two lasers with different wavelengths.

differential air thermometer *Engineering.* an instrument that measures radiant heat with a U-tube manometer containing a clear bulb at one end and a blackened one at the other end.

differential algebra *Mathematics.* **1.** a ring or algebra possessing one or more distinguished derivations. In such an algebra, a formal theory of differential equations and their solutions can be developed. **2.** the study of such algebras.

differential amplifier *Electronics.* an amplifier circuit whose output is directly proportional to the instantaneous differences between the two input waveforms.

differential analysis *Meteorology.* a synoptic analysis of change charts or vertical differential charts obtained by graphical or numerical subtraction of patterns of a meteorological element at two levels or two times.

differential analyzer *Computer Technology.* an integrating device in an analog computer that is used primarily to solve differential equations. Also, ELECTRONIC DIFFERENTIAL ANALYZER.

differential ballistic wind *Ordnance.* in bombing, a hypothetical wind equal to the difference in velocity between the ballistic wind and the actual wind at a release altitude.

differential blood count see DIFFERENTIAL LEUKOCYTE COUNT.

differential brake *Mechanical Engineering.* a brake whose operation is based on the difference between two motions.

differential calculus *Mathematics.* the branch of mathematics that studies the variation of a function resulting from changes in the independent variables, based on knowledge of the derivative(s) of the function.

differential calorimetry *Thermodynamics.* a method of measurement of the process heats: heat of reaction, heat of adsorption, heat of hydrolysis, and so on.

differential capacitance *Electronics.* the rate of change of the charge characteristic of a capacitor at a specific point on the operating curve.

differential capacitor *Electricity.* a dual-variable capacitor having one rotor and two stators arranged so that as capacitance is decreased in one section, it is increased in the other.

differential centrifugation *Cell Biology.* a fractionation procedure in which centrifugal force is used to separate subcellular particles on the basis of mass or density.

differential chemical reactor *Chemical Engineering.* a flow reactor that operates at constant temperature and very low conversions, with reactant and product concentrations close to the levels in the feed.

differential coefficient see DERIVATIVE.

differential compaction *Geology.* the uneven reduction in volume of sediments caused by uneven gravitational settling or by differences in the degrees of compactability of the earth materials.

differential comparator *Electronics.* a circuit containing one or more differential amplifiers, designed to compare the polarity and amplitude of an input voltage with respect to that of a reference voltage.

differential cross section *Physics.* an area that measures the probability that a collision will deflect the incident particle into a prescribed element of a solid angle.

differential delay *Telecommunications.* the physical difference between the maximum and minimum frequency delays that take place on any given band.

differential diagnosis *Medicine.* the determination of which of two or more diseases with similar symptoms is the one from which the patient is suffering.

differential discriminator *Electronics.* a discriminator circuit that responds only to pulse signals within a predetermined amplitude range.

differential ebuliometer *Analytical Chemistry.* a device that measures both the condensation temperature of the vapors of a boiling liquid and the boiling temperature of the liquid.

differential effects *Mechanics.* perturbations to a trajectory caused by variations from standard conditions.

differential emotions theory *Psychology.* a theory that certain basic emotions, such as fear, anger, disgust, sadness, surprise, and happiness, have evolved as part of human nature and can be found in all cultures.

differential equation *Mathematics.* an equation that relates unknown functions and derivatives (with respect to the independent variables) of these functions. If the unknown functions depend on only one independent variable, the differential equation is called an **ordinary differential equation;** otherwise it is a **partial differential equation.** To solve differential equations means to determine all functions, if any, that satisfy the differential equations. In general, such solutions are not uniquely determined by a differential equation without the introduction of additional boundary conditions.

differential erosion *Geology.* erosion that occurs at varying rates as a result of differences in the resistance and hardness of rock materials.

differential extraction *Chemical Engineering.* a theoretical limiting case of crosscurrent extraction in a single vessel. Feed is continuously processed into product and pure solvent.

differential fault see SCISSORS FAULT.

differential flotation *Mining Engineering.* a process using flotation to separate a complex ore into two or more valuable mineral components and gangue. Also, SELECTIVE FLOTATION.

differential form *Mathematics.* any completely antisymmetric homogeneous r-tensor on a differentiable n-manifold, where $0 \le r \le n$. These form a Grassmannian algebra called the **algebra of differential forms.**

differential gain control *Electronics.* a receiver-gain control circuit in which receiver gain is varied with time to keep output signal levels constant, by reducing sensitivity to close, strong signals and increasing sensitivity to distant, weak signals; used mainly in radar and loran applications.. Also, SENSITIVITY-TIME CONTROL.

differential galvanometer *Electricity.* a galvanometer having two similar but opposed coils, so that their currents tend to neutralize each other, and there is zero deflection when the currents are equal.

differential game *Control Systems.* a two-sided optimal control problem.

differential gear *Mechanical Engineering.* see DIFFERENTIAL.

differential generator *Electricity.* a generator whose shunt and series windings are in opposition to each other, in order to regulate the maximum current.

differential geometry *Mathematics.* the branch of geometry dealing with the properties of curves and surfaces.

differential heating *Metallurgy.* in a thermal treatment, the intentional creation of certain temperature gradients in the treated component.

differential heat of solution *Thermodynamics.* the partial derivative of the heat of solution with respect to the molal concentration of one of the components of the solution, under constant conditions.

differential host *Virology.* **1.** a host that serves to distinguish between various viruses because of its susceptibility to infection by one of the viruses and its exhibition of disease symptoms. **2.** a host that is used to isolate one virus from a mixture of viruses.

differential indexing *Mechanical Engineering.* in a dividing engine, a method of subdividing a circle based on the difference in motion between the index plate and the index crank.

differential input *Electronics.* an input applied to a two-input device, in a manner such that the sum of the signals on the two terminals, measured to a common reference ground, is zero. Used to form various compounds, such as **differential-input voltage, differential-input resistance, differential-input capacitance,** and so on.

differential-input measurement *Electronics.* a measurement of the signal voltage or current at the input of a differential-type circuit in which the value at each input is measured with respect to the other and not with respect to ground. Also, FLOATING-INPUT MEASUREMENT.

differential ionization chamber *Nucleonics.* a two-compartment ionization chamber in which the output ionization current is equivalent to the difference between the currents in the two compartments.

differential keying *Electronics.* a method of keying a continuous-wave transmitter to avoid chirping noises at the start and end of each keying pulse; the oscillator is keyed-on a short time before the main amplifier and keyed-off a short time after the main amplifier.

differential leukocyte count *Pathology.* a count made by observation, shown in percentages, that is used to determine the proportion of the different types of white blood cells or other cells on a stained blood smear. Also, DIFFERENTIAL BLOOD COUNT.

differential leveling *Cartography.* the process of measuring the difference of elevation between two points, in which a horizontal line of sight from a known elevation is intercepted by a vertical graduated standard.

differential manometer *Engineering.* an instrument used to measure the difference in pressure between two sources by measuring the levels of a liquid in a U-shaped tube connected to the sources.

differential-mode input *Electronics.* the voltage difference between the input terminals of a differential amplifier.

differential modulation *Telecommunications.* modulation in which the choice of the signal condition for any signal element is dependent on the choice for the previous signal element.

differential motion *Mechanical Engineering.* a mechanical movement wherein the velocity of the driven part is equal to the difference of the velocities of two parts connected to it.

differential motor *Electricity.* a DC motor whose shunt and series field windings directly oppose each other, yielding a constant speed.

differential operator *Mathematics.* let X be a space of differentiable functions on a manifold M. A differential operator T is a mapping, usually from X to itself, that is constructed out of derivatives. For example, let $M = R$; i.e., X is a space of differentiable functions of a real variable x. Then the transformation T given by $Tf(x) = f(x)f''(x) - [f'(x)]^2$ is a nonlinear differential operator.

differential permeability *Electromagnetism.* the slope of the magnetization curve for a magnetic material.

differential phase *Electronics.* the difference in phase of the high-frequency color subcarrier signal in a color television transmission system as the video signal on which it is superimposed is varied from one level to another.

differential phase shift *Telecommunications.* a type of modulation in which each signal element is a change in the phase of the carrier with respect to its previous phase angle.

differential polarography *Analytical Chemistry.* an electroanalytical technique that measures the difference in current flowing between two identical electrodes that are at the same potential but in two different solutions.

differential pressure *Physics.* a small difference between the pressures at two distinct points in a fluid system.
differential pressure gauge *Engineering.* **1.** an instrument used to measure differing pressures between two points in a system. **2.** a device used to measure differences in two fluid pressures.
differential pressure transducer *Electricity.* an instrument used to measure pressure differences between two pressure sources and to translate the difference into a change in inductance, resistance, voltage, or other electrical quality.
differential process *Chemical Engineering.* a process that causes a system to perpetually change in composition or quantity.
differential psychology *Psychology.* a field of psychology concerned with the psychological differences between groups.
differential pulley *Mechanical Engineering.* a tackle system that yields a high mechanical advantage, composed of an endless cable passing through a movable, load-bearing lower pulley and two stationary, coaxial upper pulleys of varied diameter.
differential pulse-code modulation *Electronics.* a phase-code modulation in which quantized differences between successive samples are coded into a defined number of equal-duration binary digits and used for data transmission.
differential reaction rate *Physical Chemistry.* the incremental change in reactant concentration that occurs in a differential interval of time.
differential refraction *Astronomy.* the amount by which the apparent positions of two objects at different elevations above the horizon are differently refracted.
differential reinforcement *Behavior.* the presentation of different degrees of reinforcement for two similar behaviors, in order to strengthen one and eliminate the other.
differential relay *Electricity.* a two-winding relay activated when the power difference between the windings reaches a predetermined value.
differential rotation *Astronomy.* the rotation of different parts of an object at different rates; for example, the slower rotation of the sun's polar regions as compared to its equatorial regions.
differential scanning calorimetry *Materials Science.* a technique in which an instrument measures the rate of heat evaluation or absorption of a specimen that is undergoing a programmed temperature change; generally used to indicate phase changes.
differential scatter *Engineering.* a technique in which atmospheric particles can be sensed by measuring the backscatter of laser beams of various wavelengths and comparing the results with the unique scattering signatures of known particles.
differential screw *Engineering.* a compound screw that produces motion equal to the differential motion of two component screws.
differential separation *Chemical Engineering.* a process in which gas or vapor is released from liquids by a reduction in pressure and removed from the system.
differential shrinkage *Photogrammetry.* the difference in unit contraction along the grain structure of photographic film material as compared to the unit contraction across the grain structure.
differential solution *Geochemistry.* see INTRASTRATAL SOLUTION.
differential spectrophotometry *Spectroscopy.* the photometric measurement of radiant energy absorbed by a given solution by analyzing the spectrum that represents the difference between the sample and the reference cell used.
differential splicing *Molecular Biology.* **1.** the pairing or splicing together of two particular homologous chromosomes during meiosis. **2.** alternate pathways of splicing a pre-mRNA so that the final mRNA products encode nonidentical proteins. Also, AUTOSYNDESIS.
differential stage *Electronics.* an amplifier circuit that contains a differential input, and that produces an output signal proportional to the instantaneous differences in signal voltage or current between the two inputs.
differential steam calorimeter *Engineering.* a calorimeter that compares the amount of steam condensing on one body to the amount of steam condensing on a body of known heat capacity.
differential temperature survey *Petroleum Engineering.* a technique for logging the temperature in a well that determines small temperature deviations; placed 6 ft (1.8 m) apart, two thermometers record the temperature gradient down the well, with small differential variations indicating irregularities.
differential thermal analysis *Thermodynamics.* a method used in the analysis of phase changes and the heat involved in the change, in which the temperature difference between a test sample and an inert reference sample is recorded in a uniform heating process.
differential thermogravimetric analysis *Thermodynamics.* an analytical technique that is used to measure the properties of a substance by detecting changes in its mass as it is heated; mass changes occur because of dehydration, desorption, vaporization, sublimation, pyrolysis, and chemical reactions with gases.
differential thermometer *Engineering.* a thermometer that measures temperature by using the differential thermal expansion of two metallic strips or two thermocouples.
differential thermometric titration *Analytical Chemistry.* a titration in which the titrant is added simultaneously to a mixture and a blank in identically equipped cells, and the difference in temperature is determined.
differential timing *Industrial Engineering.* a technique for timing an element of extremely short duration, in which the time of the element being studied is measured in combination with the time of the element that precedes or follows it, and is then deduced by subtraction.
differential topology *Mathematics.* the study of the topology of differentiable manifolds.
differential transducer *Electricity.* a dual-input, single-output sensor that simultaneously senses two separate sources and provides an output proportional to the difference between them.
differential transformer *Electricity.* a transformer that joins two or more signal sources to a common transmission line. Also, LINEAR VARIABLE-DIFFERENTIAL TRANSFORMER or LVDT.
differential voltage gain *Electronics.* the ratio of the output voltage of a differential amplifier to the differential input voltage.
differential voltmeter *Electricity.* a voltmeter used to measure only the difference between a known voltage and an unknown voltage.
differential wind *Ordnance.* in bombing, the vector difference between the wind at a release altitude and the wind at a given lower altitude; used to compute the differential ballistic wind.
differential winding *Electricity.* a coil winding arranged so that its magnetic field opposes that of a nearby coil.
differential windlass *Mechanical Engineering.* a windlass composed of two drums of different diameter that are rotating at the same rate; a pulley suspended below them on a line wound around the larger drum and unwound from the smaller drum is raised with mechanical advantage. Also, CHINESE WINDLASS.
differential wound field *Electricity.* a motor or generator field with series and shunt coils that oppose each other.
differentiating circuit see DIFFERENTIATOR.
differentiation *Mathematics.* the process of forming the derivative of a function *f*.
differentiation factor *Immunology.* any substance that induces maturation of cells.
differentiation theorem for Laplace transforms see INITIAL-VALUE THEOREM.
differentiator *Electronics.* a circuit that produces an output voltage which is proportional to the rate of change of the input voltage, so that the output waveform is the time derivative of the input waveform. Also, DIFFERENTIATING CIRCUIT.
diffluence *Fluid Mechanics.* the condition of tending to flow apart or of readily dissolving.
diffluent *Biology.* easily dissolved or changed into a fluid.
Difflugia *Invertebrate Zoology.* a genus of amoebas with a shell made of glued-on sand grains.

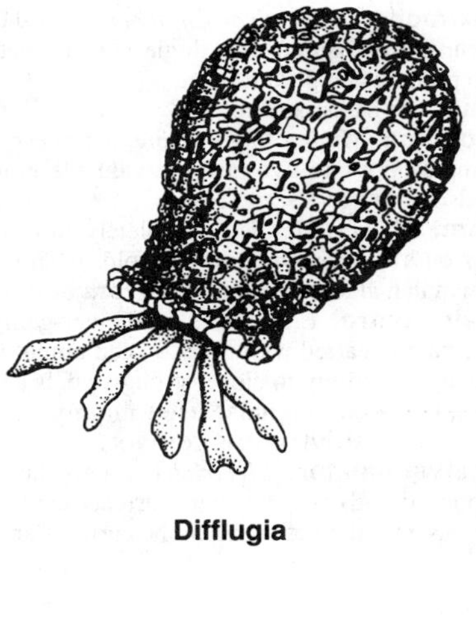

Difflugia

diffracted wave *Physics.* a wave whose propagation direction has been altered by an obstruction, rather than by reflection or refraction.
diffraction *Physics.* **1.** a phenomenon observed in the propagation of waves in which the propagation direction is changed when a wave encounters an obstruction or edge, such as an aperture. **2.** the ability of waves to spread around edges of small dimensions. *Optics.* specifically, the bending of light as it passes an obstruction.
diffractional see PULSE-HEIGHT SELECTOR.
diffraction disk see AIRY DISC.
diffraction grating *Spectroscopy.* a device consisting of a series of fine, closely spaced parallel lines, slits, or grooves scored on glass, metal, or another plane surface, used to disperse light according to wavelength. Also, GRATING.
diffraction-limited *Optics.* describing an optical system whose image quality is limited only by diffraction rather than by aberrations or manufacturing errors.
diffraction loss *Physics.* the reduction in intensity resulting from diffraction in a wave or beam.
diffraction pattern *Physics.* the interference pattern that results when diffracted waves are superposed. *Astronomy.* the image of a star formed by an optically perfect telescope; in a refractor, this consists of a small bright disk surrounded by concentric rings of light; while in a reflector, the disk will have (typically) four spikes caused by diffraction from the secondary mirror supports. *Crystallography.* the experimentally measured values of intensities, diffracting angle (direction), and order of diffraction for each diffracted beam obtained when a crystal is placed in a narrow beam of X-rays or neutrons (usually monochromatic).
diffraction rings *Optics.* circular patterns of light that appear to surround particles in the field of a microscope.
diffraction scattering *Physics.* the scattering of particles resulting from the removal of particles from a beam by inelastic processes.
diffraction spectrum *Spectroscopy.* a spectrum produced by diffraction of light at the grating surface.
diffraction symmetry *Crystallography.* the breaking up of light as it passes through a crystal so that the diffraction pattern is the same on either side of the crystal.
diffraction velocimeter see LASER VELOCIMETER.
diffraction zone *Electromagnetism.* the portion of a radio propagation path that lies outside a line-of-sight path.
diffractometer *Physics.* a device used to measure the intensities of diffracted radiation or diffracted neutrons at various angles.
diffractometry *Crystallography.* the branch of science that determines the structure of a crystal by observing the changes in amplitude or phase of an X-ray beam or other energy waves penetrating its structure.
diffuse aurora *Geophysics.* an auroral form that spreads widely and uniformally across the sky; easily seen in satellite pictures, although not often from the ground.
diffuse competition *Ecology.* the simultaneous interspecific competition among several species occupying niches that overlap marginally.
diffuse-cutting filter *Optics.* a color filter that modifies its absorption of light rays to coincide with their wavelengths.
diffused-alloy transistor see DRIFT TRANSISTOR.
diffused-base transistor *Electronics.* a transistor having a nonuniform base region and collector-base junction produced by gaseous diffusion, and an emitter-base junction that is a conventional alloy junction.
diffused junction *Electronics.* a diode or transistor junction that is produced by diffusion. Thus, **diffused-junction rectifier.**
diffused-mesa transistor *Electronics.* a diffused junction transistor in which alternating layers of P-, N-, and P-type materials are formed in the semiconductor blank by the diffusion and alloying of impurities. After emitter, base, and collector have been formed, unwanted areas are removed by an etching process, leaving a flat-topped, elevated contour.
diffused metal-oxide semiconductor *Electronics.* a semiconductor manufactured by a two-stage diffusion of impurities through a single mask opening to give precise-sized narrow channels; used on discrete field-effect transistors for ultrahigh gains and frequency performance.
diffused resistor *Electronics.* an integrated-circuit resistor that is produced by diffusion.
diffuse front *Meteorology.* a front across which wind shift and temperature change characteristics are not well defined.
diffuse galactic light *Astronomy.* starlight reflecting from interstellar dust in the Milky Way.
diffuse hypergammaglobulinemia *Medicine.* an increased concentration of immunoglobulins in the plasma, such as that frequently observed in chronic infectious diseases.
diffuse illumination *Optics.* illumination in which light emanates from a number of sources and has a high degree of scatter.
diffuse interface *Materials Science.* a method of studying dispersivity; an incident light beam is focused on a powdered sample and the resulting scattered light is collected by carefully shaped mirrors and focused on a detector.
diffuse layer *Physical Chemistry.* in a double-layer ionic structure, the outer region of ions that are distributed in a cloudlike manner in the electrolytic solution, and that contain an excess or deficiency of ions compared to this solution; as opposed to the rigid inner layer of ions (the compact layer) adsorbed on the surface of the electrode. Also, DIFFUSION LAYER.
diffuse nebula *Astronomy.* an irregular gaseous or dusty nebula.
diffuse placenta *Developmental Biology.* a placenta in which placental tissue is distributed over the chorionic membrane, as in swine.
diffuse-porous *Botany.* of or relating to wood in which the pores (vessels) are of small diameter and those formed early in the growing season have about the same diameter as those formed later; these pores are evenly distributed through the growth ring, as in such trees as the maple and birch.
diffuser *Engineering.* a chamber or duct in which a fast-moving low-pressure stream of fluid is transformed into a slow-moving high-pressure stream.
diffuse radiation *Physics.* any radiation emanating from a source and traveling in no specified direction.
diffuse reflection *Physics.* a reflection from a surface in which the reflected waves have no sharply defined direction, as in reflection off a rough surface. *Optics.* reflection of light rays that is accompanied by scattering, where excess specular reflection does not occur.
diffuse reflector *Optics.* a surface that has irregularities larger than the wavelength of the incident radiation, so that the reflected rays are scattered in a number of directions.
diffuse scattering *Crystallography.* halos or streaks that appear around intense Bragg reflections in the X-ray diffraction pattern of a crystal; they indicate the presence of disorder in the crystalline structure.
diffuse series *Spectroscopy.* for alkali elements, a set of spectral lines that represent a change in total orbital angular momentum accompanying a transition from the *p* state to the *d* state.
diffuse sound *Acoustics.* sound that travels in all directions with equal energy distribution and probability, such as an isovelocity area in the ocean.
diffuse spectrum *Spectroscopy.* a spectrum having very broad lines although there is no apparent cause for the line broadening.
diffuse transmission *Physics.* a passage of waves through a medium in which the transmitted waves propagate in all directions throughout the medium.
diffuse transmission density *Optics.* the value of the photographic transmission density, obtained by collecting and measuring the transmitted flux of light that strikes a sample.
diffusing disk see DIFFUSION DISK.
diffusing filter *Graphic Arts.* a screen used to soften the light on a subject in order to eliminate sharp contrast and detail.
diffusiometer *Physics.* a device used for monitoring the diffusion of a substance in a fluid.
diffusion a process in which something spreads out or scatters over an area; specific uses include: *Physics.* a process in which particles disperse, moving from regions of higher density to regions of lower density. *Solid-State Physics.* **1.** the thermally activated transport of charge carriers through a semiconductor material. **2.** the movement of individual atoms through a crystal lattice. *Optics.* **1.** the process by which light reflects off an irregular surface in all directions. **2.** the process by which light passes through a transparent surface. *Electronics.* a process in the manufacture of semiconductor devices in which impurity materials permeate into precisely controlled areas within a semiconductor substrate to form desired P-N junctions. *Meteorology.* an exchange of fluid parcels between regions in space in apparently random motions on a scale too small to be treated by the equations of motion. *Anthropology.* see CULTURAL DIFFUSION.
diffusion annealing *Metallurgy.* a thermal treatment that promotes solid-state diffusion.
diffusion barrier *Chemical Engineering.* a porous barrier through which gaseous mixtures pass in order to enrich the lower-molecular-weight constituent of the diffusate; employed as a multiple-stage cascade system for recovering U235F6 isotopes from a U238F6 stream.

diffusion bonding *Materials Science.* a joining technique in which two solid surfaces are contacted under high pressures and temperatures; the diffusion of atoms from one component into the other and the filling of voids produce a bond. Also, **diffusion welding.**

diffusion brazing *Metallurgy.* solid-state bonding in which a braze is used as the filler metal.

diffusion capacitance *Electronics.* the capacitance of a semiconductor junction produced by the noninstantaneous change of minority carrier charges as the voltage across the junction is changed.

diffusion coefficient *Physics.* a temperature- and composition-dependent coefficient related to the rate at which atoms or ions diffuse; it depends primarily on the temperature and activation energy. *Materials Science.* the weight of a material, in grams, diffusing across an area of one square centimeter in one second in a unit concentration gradient.

diffusion constant *Solid-State Physics.* the current density divided by the charge carrier concentration gradient in a homogeneous semiconductor.

diffusion current *Analytical Chemistry.* the limiting current reached by movement of ions in an electrolytic solution to an electrode, due to the application of a potential difference to the electrodes.

diffusion disk *Optics.* a piece of material that is marked or embossed in order to produce a softened image when attached to a camera lens. Also, DIFFUSING DISK.

diffusion equation *Physics.* an equation that expresses the time rate of change of a quantity (typically a density) in terms of the product of the diffusion coefficient and the Laplacian operating on the quantity plus any source and absorption terms. *Mathematics.* a partial differential equation, formally identical to the above equation, whose solution is the conditional probability density function f of the Wiener-Lévy process: $\partial f/\partial t = \sigma^2/2\ \partial^2 f/\partial x^2$.

diffusion gradient *Physics.* the slope of a graph of particle concentration versus position.

diffusion hygrometer *Engineering.* a hygrometer utilizing the diffusion of water vapor through a porous membrane.

diffusion layer see DIFFUSE LAYER.

diffusion length *Physics.* the mean distance that a particle travels before it is absorbed or recombines.

diffusion number *Fluid Mechanics.* a dimensionless number that is the ratio of a velocity times a length dimension divided by a diffusivity. Also, PECLET NUMBER.

diffusion of responsibility *Behavior.* the theory that people are less likely to take action in an emergency if they are part of a large group, because they feel that someone else in the group will deal with the situation. Also, **diffusion effect.**

diffusion plant *Nucleonics.* **1.** generally, any installation that uses the gaseous diffusion method for separation of isotopes, based on different rates at which gases diffuse through porous barriers. **2.** specifically, a plant that concentrates the uranium-235 isotope by diffusion through porous barriers to achieve uranium enrichment.

diffusion pump *Engineering.* a vacuum pump that utilizes heavy molecules, such as mercury vapor, to carry gas molecules out of the chamber being evacuated.

diffusion respiration *Physiology.* the exchange by diffusion of gases between the lungs and the bloodstream.

diffusion test see IMMUNODIFFUSION.

diffusion theory *Electricity.* a theory stating that for a semiconductor in which a variation of carrier concentration occurs, a motion of the carriers is produced by diffusion as well as by drift determined by mobility and electric field.

diffusion transfer *Graphic Arts.* any of several processes using a silver halide developing solution to produce, usually simultaneously, both positive and negative photographic images, as in Polaroid photography.

diffusion transistor *Electronics.* a transistor whose current depends on the diffusion of carrier, donors, or acceptors.

diffusiophoresis *Chemical Engineering.* a process in a scrubber unit in which water vapor moves toward the cold water surface carrying particles with it.

diffusivity *Physics.* a measure of the rate of heat diffusion given by the thermal conductivity of a substance divided by the product of the density and the heat capacity at constant pressure; commonly denoted by a and having MKS units of m^2/s. It varies with the nature of the involved atoms, the structure, and changes in temperature.

diffusivity analysis *Analytical Chemistry.* any process that uses diffusion effects to analyze materials in solution; for example, dialysis or polarography.

archaeological dig

dig *Archaeology.* a site that has been or is being excavated.

dig. let it be digested. (From Latin *digeratur.*) Also, **Dig.**

digallic acid see TANNIC ACID.

digastric *Anatomy.* **1.** having two bellies. **2.** of or relating to a muscle having two fleshy parts separated by tendinous tissue.

Digenea *Invertebrate Zoology.* a large subclass of trematode flatworms that are endoparasites of vertebrates.

digenesis *Biology.* the alternation of sexual and asexual reproduction in different generations, as in the asexual reproduction cycle of trematodes in the mollusk intermediate host, followed by a sexual reproductive cycle in the vertebrate host. Also, DIGENISM.

digenetic *Microbiology.* **1.** of or related to an organism that undergoes alternate asexual and sexual processes of reproduction. **2.** of or relating to a parasite that lives a portion of its life cycle in each of two different host species.

digenite *Mineralogy.* Cu_9S_5, a blue to black, opaque, trigonal mineral having a specific gravity of 5.6 and a hardness of 2.5 to 3 on the Mohs scale; found in copper deposits as intergrowths with chalcocite, chalcopyrite, and other copper minerals.

DiGeorge syndrome *Genetics.* an inherited developmental abnormality characterized by the absence of the thymus, sometimes accompanied by an abnormality in the aortic arch; often associated with the loss of a portion of chromosome 22.

digester *Chemical Engineering.* a cylindrical metal container with a cover and safety valve, used to cook or decompose substances at high temperature and pressure; used mainly to produce cellulose pulp from wood chips.

digestibility *Nutrition.* the ability of a food to be converted into chemical substances that can be absorbed and assimilated.

digestible crude protein *Nutrition.* incomplete proteins whose nutritive value is enhanced by the addition of synthetic essential amino acids.

digestion *Physiology.* the conversion of food into chemical substances that are absorbed through the epithelial lining of the small intestine and dispersed into the bloodstream to provide nourishment for the tissues of the body. *Microbiology.* the liquefaction of organic waste materials by microbiological activity. *Biotechnology.* specifically, the use of bacteria to decompose sewage material. *Chemical Engineering.* any of various chemical processes that are regarded as analogous to biological digestion, such as the removal of lignin from wood in manufacturing paper pulp and chemical cellulose.

digestive enzyme *Enzymology.* an enzyme that aids in digestion, as by hydrolyzing polysaccharides and disaccharides into monosaccharides.

digestive gland *Physiology.* any gland, such as a salivary or gastric gland or the liver or pancreas, that secretes a reactive agent (hydrochloric acid, bile, mucus, or the like) which is involved in converting food into absorbable substances to be metabolized by the body.

digestive system *Anatomy.* the organs associated with the ingestion, digestion, and absorption of food.

digestive tract *Anatomy.* the part of the digestive apparatus formed by the esophagus, stomach, and small and large intestines. Also, ALIMENTARY CANAL.

digging height see BANK HEIGHT.

diggings *Archeology.* **1.** a dig. **2.** material that is uncovered during a dig.

digicitrin *Biochemistry.* a compound present in foxglove leaves.

digicom *Electronics.* a communication system in which audio and digital signals are transmitted by wire; audio signals are converted to digital pulse trains by the delta pulse code modulation process prior to transmission and then converted back to audio signals after reception. (An acronym for digital communication.)

digit *Anatomy.* a finger or toe. Also, **digitus.** *Mathematics.* **1.** in the decimal number system, any of the numeric symbols 0, 1, 2, 3, 4, 5, 6, 7, 8, 9. **2.** In general, any numeric symbol less than the radix or base of the number system.

digit-absorbing selector *Electronics.* a dial switch that requires dialing of at least two digits to operate; the first digit dialed is absorbed by the switch, enabling it to operate when the second digit is dialed.

digital *Computer Technology.* relating to data that is represented in the form of discrete digits; as opposed to analog or continuous representation. *Engineering.* displaying data in the form of digits. Used to form various compounds, such as **digital watch, digital clock, digital thermometer, digital speedometer,** and so on.

digital audiotape *Acoustical Engineering.* a type of magnetic tape that can play or record with a high reproduction quality comparable to that of a compact disk. Also, DIGITAL TAPE, DAT.

digital circuit *Electronics.* a circuit that operates in only two states, on and off.

digital communication(s) *Telecommunications.* any communication system that utilizes digital signals in the sending and receiving of messages.

digital comparator *Electronics.* a circuit that compares two digital values and determines which is larger, or that compares a digital input value against predetermined upper and lower limits and gives pass/fail information.

digital compression *Surgery.* compression of a blood vessel by the fingers, in order to stop bleeding.

digital computer *Computer Technology.* a machine that can be programmed to accept digital data, carry out arithmetic, transfer, and logic operations on the data, and supply the results in a useful form. In modern terminology, generally called a "computer."

digital data *Computer Technology.* information represented by discrete values, as opposed to analog data that is represented in continuous form.

digital data recorder *Computer Technology.* a device that records information as discrete values on a recording medium; occasionally includes an analog-to-digital conversion capability.

digital delayer *Acoustical Engineering.* a digital device that electronically inserts a time delay in the passage of a signal from one point to another during channel processing.

digital display *Computer Technology.* a visual output consisting of character and numeric representations of digital data; especially as distinguished from analog display, such as a waveform on an oscilloscope or an indicator on a dial or scale. Also, DIGITAL READOUT.

digital filter *Electronics.* a circuit that processes signals digitally coded as a sequence of numeric values. The analog signal at the input is sampled, digitized, and coded; after filtering, the result is decoded back to analog form at the output.

digital format *Computer Programming.* the representation of variables used in a program by a number of discrete signals or by the presence or absence of signals.

digitalin *Pharmacology.* $C_{36}H_{56}O_{14}$, a cardiac glycoside extracted from the leaves or seeds of the purple foxglove, *Digitalis purpurea.*

digital integrator *Computer Technology.* a device that calculates definite integrals in which the input and output variables are represented digitally.

Digitalis *Botany.* a genus of northern, temperate herbs in the family Scrophulariaceae, having alternate leaves and bell-shaped, racemic flowers; the source of the drugs digitalis and digoxin. Also, FOXGLOVE.

digitalis *Pharmacology.* the dried leaf of the purple foxglove, *Digitalis purpurea;* a preparation of powdered digitalis is used to treat congestive heart failure and other heart disorders, its main effect being to increase the pumping action of the heart while decreasing heart rate.

digitalization *Cardiology.* a dosage schedule in which digitalis is administered for the production and maintenance of optimal therapeutic effects.

digital modulation *Telecommunications.* the process of varying one or more parameters of a carrier wave as a function of two or more finite and discrete states of a signal.

digital multiplier *Electronics.* a multiplier circuit that accepts two values in digital form and gives the product in the same digital form, usually by making repeated additions.

digital output *Electronics.* the output signal of a digital circuit representing a value or a set of conditions; the intelligence is contained in a code consisting of a sequence of one or more discrete voltage or current levels, each representing a value of zero or one.

digital plotter *Electronics.* a recorder that converts digital input data into hard copy in the form of charts or other graphic presentations.

digital printer *Computer Technology.* a narrow line printer used in industrial applications to create a permanent continuous record of numerical values, such as temperature or pressure, shown by an instrument over a period of time. Also, JOURNAL TAPE PRINTER.

digital readout see DIGITAL DISPLAY.

digital recording *Electronics.* magnetic recording in which the audio or video signal is first converted to digital form before recording on magnetic tape.

digital representation *Computer Technology.* the use of discrete values or pulses arranged in accordance with prespecified code patterns to represent data.

digital resolution *Computer Technology.* the degree of numerical precision achieved by a digital representation.

digital signal *Telecommunications.* a time-varying or spatial signal, such as sound or visual images, represented as sequences or arrays of digitized samples, as on a digital computer. Thus, **digital signal processing.**

digital simulation *Computer Programming.* the representation of physical systems and behaviors by means of computer models and discrete variables.

digital subtraction angiography *Radiology.* a computerized technique for a more clearly defined radiographic visualization and analysis of blood vessels or lymphatics.

digital synchronometer *Electronics.* a highly accurate digital time display device that compares its internal clock with time standards broadcast by the National Bureau of Standards.

digital tape see DIGITAL AUDIOTAPE.

digital television *Telecommunications.* a television receiver that uses digital amplifier circuitry, which produces improved sound and picture quality.

digital-to-analog converter *Electronics.* a converter that changes digital input data to corresponding analog output voltage values. Also, **digital-to-analog decoder.**

digital transducer *Electronics.* a transducer that generates a digital output signal, rather than an analog output, in response to input stimulus.

digital voltmeter *Electronics.* a voltmeter in which the value of the measured voltage is shown on a digital display.

digitate *Biology.* having one or more fingerlike processes or parts that radiate like fingers; palmate.

digitation *Geology.* the emergence of a minor recumbent anticline from a larger recumbent anticline.

digit-coded voice *Computer Technology.* **1.** a computer-synthesized audio output produced by generating signals and frequencies similar to human speech. **2.** an audio response constructed by assembling prerecorded words and phrases into meaningful messages.

digit compression *Computer Programming.* a method of reducing data storage requirements by storing two or more digits in a computer word.

digit emitter *Computer Technology.* a character emitter limited to numerics, excluding alphabetic and special characters.

digitigrade *Vertebrate Zoology.* walking with only the toes or digits on the ground and with the hind part of the foot more or less raised, as in most quadruped animals.

digitinervate *Botany.* having straight veins that radiate from the base or petiole, like fingers from the palm, usually with five or seven veins.

digitipinnate *Botany.* having digitate or fingerlike leaves with pinnate leaflets.

digitize *Computer Technology.* **1.** to transform graphical input data into digital form for computer processing. **2.** to assign a discrete numeric value to an analog variable by analog-to-digital conversion.

digitizer see ANALOG-TO-DIGITAL CONVERTER.

digitizing pad or **digitizing tablet** *Computer Technology.* a direct input device with a special pen or cross-hairs with which the user traces the image to be digitized; the coordinates at selected points are automatically recorded. Also, DATA TABLET, DIGITIZER, ELECTRONIC TABLET.

digitonin *Organic Chemistry.* $C_{56}H_{92}O_{29}$, a preparation made from digitalis but having no apparent effect on the heart; used as a reagent to precipitate free cholesterol in tests. Also, **digitin.**

digitoxigenin *Organic Chemistry.* $C_{23}H_{34}O_4$, a steroid aglycone that is soluble in alcohol; melts at 253°C; derived by the removal of three molecules of the sugar **digitoxose,** [$CH_3(CHOH)_3CH_2{\cdot}CHO$} from digitoxin.

digitoxin *Organic Chemistry.* $C_{41}H_{64}O_{13}$, a compound derived from digitalis leaves, usually foxglove; toxic, white plates or powder; slightly soluble in water and very soluble in alcohol; melts at 255–256°C; used in cardiac treatment.

digit period *Electronics.* the time interval between successive pulses in a series of pulses; equal to the reciprocal of the pulse repetition frequency.

digit plane *Computer Technology.* in core memory devices, a plane formed by all the bits in a particular bit position in the computer words.

digit rearrangement *Computer Programming.* a hashing scheme that selects and shifts certain digits of the original key to transform the key field into a different, often more compact, arrangement.

digitus manus [dij´ə tus man´us] *plural,* **digiti manus** [dij´ə tī]. *Medicine.* a digit of the hand; a finger.

diglossia *Anatomy.* a longitudinally split tongue.

diglucoside *Biochemistry.* any compound in which there are two glucose molecules.

diglyceride [dī glis´ə rīd] *Nutrition.* a fat composed of a glycerol molecule attached to two fatty acids; its emulsifying property is useful as an additive for shortenings and other food products.

diglycerol [dī glis´ə räl] *Organic Chemistry.* **1.** $C_6H_{14}O_5$, a combustible, viscous liquid that is soluble in water and alcohol; boils at 265–270°C; used as an emulsifier, adhesive, lubricant, and plasticizer. **2.** any mixture of ethers of glycerol.

diglycol see DIETHYLENE GLYCOL.

diglycolic acid *Organic Chemistry.* $O(CH_2COOH)_2$, a white, crystalline solid, soluble in water and alcohol; melts at 148°C; used in the manufacture of resins and plasticizers, and in organic synthesis.

diglycol laurate *Organic Chemistry.* $C_{16}H_{32}O_4$, a combustible, light straw-colored, oily, edible liquid; insoluble in water and soluble in alcohol; melts at 17–18°C and boils at 270°C; used in the manufacture of cosmetics, metal polishes, paper, leather dye, and soap.

diglycol stearate *Organic Chemistry.* $(C_{17}H_{35}COOC_2H_4)_2O$, a combustible, waxy, white solid that is soluble in hot alcohol; it melts at 54–55°C; used as an emulsifier for oils, solvents and waxes, and as a thickening agent. Also, DIETHYLENE GLYCOL DISTEARATE.

digoxin *Organic Chemistry.* $C_{41}H_{64}O_{14}$, colorless to white crystals or a white crystalline powder, insoluble in water and soluble in dilute alcohol; melts and decomposes at 235–265°C; used in the treatment of cardiac disease.

digram encoding *Computer Programming.* a data compression technique that uses infrequently occurring alphabetic characters to represent commonly occurring pairs of characters.

dihalide *Chemistry.* a compound that contains two halide ions.

dihedral [dī hē´drəld] *Mathematics.* a common line formed by two intersecting planes. Also, **dihedron**. *Aviation.* the spanwise (only upward) inclination of an airplane's wing or other support surface with respect to the horizontal; the most common design factor for providing lateral stability. Negative dihedral is usually called **anhedral.**

dihedral angle *Mathematics.* the angle formed by two intersecting planes as measured by two lines, one on each side plane, drawn perpendicular to the dihedral line from a point on the line. *Materials Science.* the boundary phase angle between two solids and a liquid under equilibrium conditions; a factor determining the processes that take place during firing of powder compacts.

dihedral group *Mathematics.* the subgroup D_n of the permutation group S_n on n objects consisting of all permutations of the form (1 2 3 $\pi(3)$ $\pi(4)$ . . . $\pi(n)$), where π is a permutation of the numbers 3, 4, . . . , n. D_n is isomorphic to (and usually identified as) the group of symmetries of a regular polygon with n sides.

dihedral reflector *Optics.* a reflector having two sides that converge at a line in order to return the beam to its source.

diheptal base *Electronics.* a 14-pin-position electron tube base, such as the base of a cathode-ray tube. Also, **diheptal socket.**

dihexagonal *Crystallography.* of or relating to a crystal structure based on a geometric figure having 12 sides separated by equal angles.

dihexagonal-dipyramidal *Crystallography.* a six-angled, six-sided geometric figure forming a double pyramid on a single base.

dihexahedron *Crystallography.* a double-figure crystal structure having six faces, as in a double cube or a double six-sided pyramid.

dihexyl see DODECANE.

dihybrid *Genetics.* a cross between two organisms that have different alleles at two gene loci.

dihydrate *Chemistry.* a compound containing two molecules of water.

dihydrazone *Organic Chemistry.* a molecule having two hydrazone (=NNH_2) groups.

dihydric *Chemistry.* containing two hydroxyl (OH) groups, such as a dihydric alcohol.

dihydro- *Chemistry.* a combining form indicating the presence of two hydrogen atoms in a compound.

dihydrochloride *Chemistry.* a compound that contains two hydrochloric acid molecules.

dihydrofolate *Enzymology.* a compound that is formed in the biosynthesis of thymidylate for DNA formation, and is essential in rapidly dividing cells.

dihydrofolate reductase *Enzymology.* an enzyme that participates in the conversion of folic acid into its active form, tetrahydrofolate, and in the biosynthesis of thymidylic acid, an essential DNA building block; important in the synthesis of purines, thymidine, and glycine.

dihydropteridine reductase *Enzymology.* an enzyme of the oxidoreductase class that catalyzes the conversion of dihydropteridine into tetrahydropteridine.

dihydrostilbestrol *Endocrinology.* a potent synthetic estrogen.

dihydrostreptomycin *Microbiology.* a derivative of the antibiotic streptomycin that is not clinically useful due to its toxicity.

dihydrotestosterone *Endocrinology.* a potent androgenic hormone, formed from testosterone in peripheral tissues by the action of 5α-reductase.

dihydroxy- *Chemistry.* a combining form indicating the presence of two hydroxyl groups.

dihydroxyacetone *Organic Chemistry.* $CH_2HOCOCH_2OH$, a colorless, hygroscopic, crystalline powder; soluble in water and alcohol; melts at 80°C; used as an intermediate and in fungicides and cosmetics.

dihydroxyacetonephosphoric acid *Biochemistry.* a phosphoric acid ester of dehydroxyacetone, which forms as an intermediate in the conversion of glycogen to lactic acid during muscle contraction.

2,4-dihydroxyacetophenone see 4-ACETYLRESORCINOL.

3,4-dihydroxyacetophenone *Organic Chemistry.* $C_6H_3(OH)_2CO$ CH_3, a needlelike white solid; melts at 115.6°C.

1,2-dihydroxyanthraquinone see ALIZARIN.

1,4-dihydroxyanthraquinone see QUINIZARIN.

1,5-dihydroxyanthraquinone *Organic Chemistry.* $C_{14}H_8O_4$, yellow to green crystals, insoluble in water and slightly soluble in alcohol and acid; melts at 280°C and sublimes at boiling point; used as an intermediate in dye manufacturing. Also, ANTHRARUFIN.

1,8-dihydroxyanthraquinone *Organic Chemistry.* $C_{14}H_6O_2(OH)_2$, orange powder or reddish brown needles, insoluble in water and soluble in alcohol and acid; melts at 193°C and sublimes at boiling point; used in dyes and medicine. Also, CHRYSAZIN, DANTHRON, ISTIZIN.

1,25-dihydroxycholecalciferol *Endocrinology.* the most active metabolic form of vitamin D_3; it is produced in the kidney, and it stimulates the absorption of dietary calcium through the intestinal wall. Also, **1,25-dihydroxyvitamin D_3.**

24,25-dihydroxychlolecalciferol *Endocrinology.* a weakly potent metabolite of vitamin D_3, produced by the kidney instead of 1,25-dihydroxycholecalciferol when plasma calcium concentrations are too high.

2,2'–dihydroxy-4,4'-dimethoxybenzophenone *Organic Chemistry.* $[CH_3OC_6H_3(OH)]_2CO$, a crystalline solid; melts at 139°C; used in paint and plastics as an ultraviolet light absorber.

dihydroxyethyl sulfide see THIODIGLYCOL.

dihydroxymaleic acid *Organic Chemistry.* $C_4H_4O_6$, a compound that occurs in *Glaucium luteum*; crystallizes as a dihydrate; soluble in water and alcohol; used as an antioxidant.

1,3-dihydroxynaphthalene *Organic Chemistry.* $C_{10}H_6(OH)_2$, a crystalline solid that is soluble in water, alcohol, and ether; melts at 124°C; used as a reagent for sugars and oils, and to identify glucuronic acid in urine. Also, NAPHTHORESORCINOL, 1,3–NAPHTHALENEDIOL.

3,4-dihydroxyphenylalanine see DOPA.

3,4-dihydroxyphenylethylamine see DOPAMINE.

dihydroxysuccinic acid see TARTARIC ACID.

diiodomethane see METHYLENE IODIDE.

3,5-diiodosalicylic acid *Organic Chemistry.* $I_2C_6H_2(OH)COOH$, a white to pale pink, crystalline powder that is soluble in water and alcohol; melts at 253°C; used in veterinary medicine.

diiodotyrosine *Endocrinology.* an amino acid and metabolic precursor of the thyroid hormones triiodothyronine and thyroxine.

diisobutylene *Organic Chemistry.* C_8H_{16}, any of a group of isomeric narcotic, colorless liquids used as intermediates, antioxidants, plasticizers, and surfactants.

diisobutyl ketone *Organic Chemistry.* $(CH_3)_2CHCH_2COCH_2CH(CH_3)_2$, a combustible, toxic, colorless liquid, immiscible with water and miscible with organic solvents; boils at 168°C and melts at –41.5°C; used as a solvent and in lacquers, stains, and organic synthesis.

diisocyanate *Organic Chemistry.* a combustible, organic compound having two isocyanate (–N=C=O) groups.

diisodecyl adipate *Organic Chemistry.* $C_{10}H_{21}OOC(CH_2)_4COOC_{10}H_{21}$, a combustible, light-colored, oily liquid; soluble in most organic solvents; boils at 239–248°C and freezes at –71°C; used as a plasticizer. Also, DIDA.

diisodecyl phthalate *Organic Chemistry.* $C_6H_4(COOC_{10}H_{21})_2$, a combustible, clear liquid; soluble in most organic solvents; boils at 250–257°C and freezes at –50°C; used as a plasticizer. ALSO, DIDP.

diisopropanolamine *Organic Chemistry.* $(CH_3CHOHCH_2)_2NH$, a combustible, white, crystalline solid, miscible with water; melts at 44–45°C and boils at 248.7°C; used as an emulsifying agent. Also, DIPA.

diisopropyl see 2,3-DIMETHYLBUTANE.

diisopropyl phosphorofluoridate *Pharmacology.* $C_6H_{14}FO_3P$, a drug used topically in glaucoma treatment to contract the pupil and thus reduce pressure in the eye. Also, ISOFLUROPHATE, DIISOPROPYL FLUOROPHOSPHATE.

dikaryon *Mycology.* **1.** a pair of nuclei found in the same cell that stem from different parent cells. **2.** of or relating to fungal cells that have two nuclei. Also, DICARYON.

dikaryophase *Mycology.* a phase in the life cycle of certain fungi, characterized by the presence of paired nuclei in the hyphal masses known as **mycelium.** Also DICARYOPHASE.

dikaryote *Cell Biology.* a cell having two haploid nuclei.

dikaryotic *Cell Biology.* of or relating to a dikaryote. *Mycology.* of or relating to a dikaryon.

dikaryotization *Mycology.* the process by which paired or dikaryotic nuclei are formed.

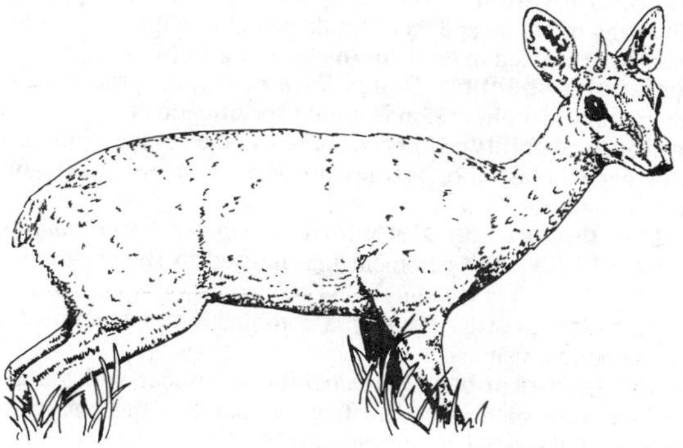

dik-dik

dik-dik *Vertebrate Zoology.* a small East African dwarf antelope of the genus *Madoqua* in the family Bovidae; it stands about 16 inches at the shoulder, weighs up to 11 pounds, and is characterized by slender legs, a pointed muzzle, and short ringed horns. (Named for the sound it makes when alarmed.)

dike *Agronomy.* a low ridge or bank of earth used to hold water in place during irrigation or to protect lowland areas against flooding. *Geology.* a tabular igneous intrusion that cuts across the beds of the adjacent rocks.

dike ridge *Geology.* any small, wall-like ridge produced by differential erosion, especially along the sides of a dike.

dike set *Geology.* a small group of closely related linear or parallel dikes.

dike swarm *Geology.* a large system of dikes that are either linear or parallel, or radiate from a single a source.

diketene *Organic Chemistry.* $C_4H_4O_2$, a flammable, colorless liquid that is soluble in common organic solvents; boils at 127.4°C and melts at –7.5°C; used in the production of pigments, pesticides, and food preservatives. Also, ACETYL KETENE.

diketone *Organic Chemistry.* a functional class name used in naming polyketones to indicate a compound that has two –C=O (carbonyl) groups.

diketopiperazine *Organic Chemistry.* any of a class of organic compounds having a dilactam ring structure formed from two α-amino acids by the joining of the α-amino group of each to the carboxyl group of the other, with the loss of two molecules of water.

diktyoma *Oncology.* a tumor of the ciliary epithelium, similar to embryonic retinal tissue.

dil. dilute; dissolve.

dilactone *Organic Chemistry.* a molecule having two lactone groups.

dilatancy *Chemistry.* the property of a viscous substance that sets solid under pressure. *Geology.* the expansion in bulk volume of a deformed mass produced by a change in structure from close packing to open packing. *Materials Science.* a rheological flow characteristic evidenced by an increase in shear viscosity with increasing rates of shear or stress.

dilatant *Chemistry.* **1.** exhibiting dilatancy. **2.** of a substance, able to increase in volume upon a change of shape.

dilatation *Physics.* any change in the volume of a substance, especially an increase in volume. *Mechanics.* specifically, an increase in volume per unit volume of a continuous substance due to deformation. *Surgery.* **1.** the condition of being stretched or dilated. **2.** the stretching of a tubular structure, cavity, or opening beyond its normal size. *Botany.* the growth of parenchyma by cell division in pith, rays, or axial system in vascular tissues; causes the increase in circumference of bark in stem and root. Also, DILATION.

dilation *Science.* **1.** the process of enlarging or expanding an opening of the body. **2.** the condition of being dilated. *Physics.* an increase in the volume of a substance, caused by thermal expansion or contraction. *Medicine.* the widening of the uterine cervix during childbirth; expressed in centimeters and progressing until **full dilation** (approximately 10 cm) at delivery. *Mathematics.* a transformation of the real or complex plane of the form $S(z) = az$, where a is positive and not equal to 1. When $a < 1$, the dilation is generally called a contraction; when $a > 1$, it is called a stretching or magnification. The origin and the point at infinity are the only fixed points of a dilation. Also, DILATATION.

dilation and curettage *Surgery.* the stretching of the cervix and scraping of the endometrium with a curette to remove contents of the uterus after incomplete abortion, to obtain specimens for diagnosis, to remove polyps, etc. Also, D AND C, DILATATION AND CURETTAGE.

dilatometer *Engineering.* an instrument used to determine the transition points of solids.

dilatometry *Physics.* the measurement of incremental changes in the dimensions of materials that are deformed by external influences.

dilaton *Particle Physics.* a hypothetical elementary particle having zero mass and zero spin; introduced in constructing a scale invariant theory of massive particles.

dilator *Physiology.* any muscle, agent, or instrument that causes the expansion of a part of the body.

dilauryl ether *Organic Chemistry.* $(C_{12}H_{25})_2O$, a combustible liquid; melts at 33°C and boils at 190–195°C; used in electrical insulators, water repellents, and lubricants. Also, DIDODECYL ETHER.

dilauryl thiodipropionate *Organic Chemistry.* $(C_{12}H_{25}OOCCH_2CH_2)_2S$, combustible white flakes that are insoluble in water and soluble in most organic solvents; extremely resistant to heat and hydrolysis; melts at 40°C; used as an antioxidant and additive for high-pressure lubricants and greases.

dilemma zone *Transportation Engineering.* the section of road leading up to a changing traffic light, on which the average driver is uncertain whether to proceed through the intersection or brake to a stop. The size and location of the dilemma zone depend on average vehicle speed.

dilepton event *Particle Physics.* a term for the scattering of a neutrino or antineutrino from a nucleus in which two leptons are among the products of the collision.

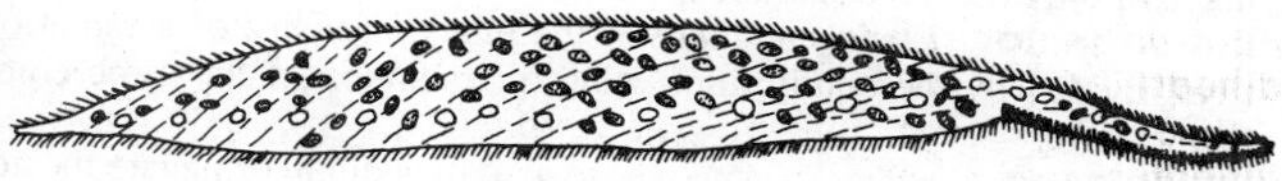

Dileptus

Dileptus *Invertebrate Zoology.* a genus of freshwater and marine ciliate protozoans, in the subclass Holotrichia, having a long necklike projection or snout.

dilinoleic acid *Organic Chemistry.* $C_{34}H_{62}(COOH)_2$, a combustible, light yellow, viscous liquid; used as a modifier for alkyd and polyamide resins, emulsify agents, and adhesives.

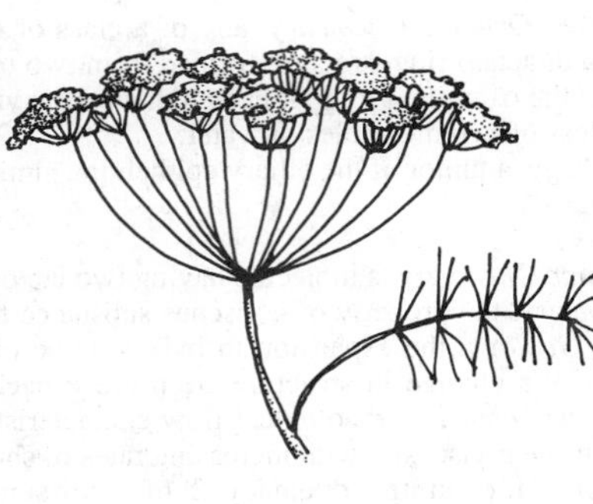

dill

dill *Botany.* a small annual or biennial herb, *Anethum graveolens,* in the family Umbelliferae; known for its aromatic leaves and seeds, which are widely used for food flavoring.

Dilleniaceae *Botany.* a family of tropical, dicotyledonous trees, woody vines, and shrubs in the order Dilleniales, characterized by alternate, entire leaves, usually regular, bisexual flowers that are solitary or cymose, hypogynous stamens, and styles that are usually free.

Dilleniales *Botany.* an order of dicotyledons in the subclass Dilleniidae that have hypogynous flowers, separate carpels, and numerous stamens.

Dilleniidae *Botany.* a subclass of dicotyledons in the class Magnoliopsida having mostly syncarpous flowers and binucleate pollen, usually bitegmic ovules, and centrifugal stamens.

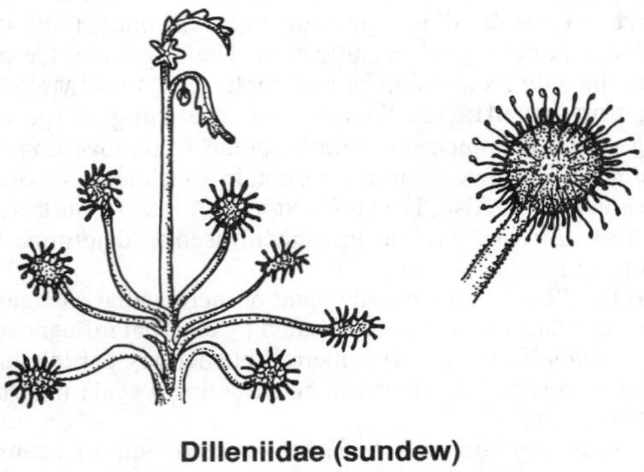

Dilleniidae (sundew)

diluc. in the early morning. (From Latin *diluculo,* "at daybreak.")

diluent *Chemistry.* a chemically inert substance added to a solution to increase the volume and reduce the concentration; a diluting agent.

dilute *Chemistry.* **1.** to reduce the concentration of a solution. **2.** describing a solution that is reduced in concentration.

dilute phase *Chemical Engineering.* the phase in liquid-liquid extraction that has a lower concentration of the material of interest.

dilution *Chemistry.* the process of reducing the concentration of solute in a solution by increasing the proportion of solvent. *Optics.* the addition of white to a color in order to reduce its saturation. *Metallurgy.* in welding, a compositional change caused by the mixing of filler and base, if of different composition.

dilution coefficient *Microbiology.* an indication of the effect that dilution of a disinfectant will have on the rate of disinfection under standard conditions.

dilution end point *Virology.* the greatest dilution of a virus preparation that still provides a measurable reaction.

dilution factor *Electromagnetism.* the energy density of a radiation field divided by the equilibrium value for radiation of the same color temperature.

dilution gene *Genetics.* a gene whose activities dilute or negate the activities of another gene.

dilution method *Microbiology.* a procedure to determine the sensitivity of a microorganism to an antibiotic by determining the lowest concentration of the antibiotic that inhibits growth.

dilution rate *Ecology.* the rate at which populations grow due to births and immigrations.

dim dimension.

dim. one half. (From Latin *dimidius.*)

dimagnesium phosphate see DIBASIC MAGNESIUM PHOSPHATE.

Dimargaritaceae *Mycology.* the only family of fungi belonging to the order Dimargaritales; most species are dry-spored and are parasitic to *Mucorales.*

Dimargaritales *Mycology.* an order of fungi belonging to the class Zygomycetes, which is characterized by cell walls (or septa) that have a central cavity containing a plug; it is parasitic to *Mucorales.*

dimedone see 5,5-DIMETHYL-1,3-CYCLOHEXANEDIONE.

dimension *Physics.* one of the units associated with a physical quantity that can always be expressed in terms of the fundamental quantities of mass, length, time, and charge. *Mathematics.* **1.** for a manifold M, the value of n such that M is homeomorphic to R^n. **2.** for a vector space V, the number of elements or vectors in a basis of V; equivalently, the number of vectors in a maximal linearly independent set. Denoted $\dim(V)$. If $\dim(V)$ is an integer n, V is said to have **finite dimension n** or to be **n-dimensional;** otherwise V is said to be **infinite-dimensional.** 3. for a simplex, one less than the number of independent points that define the simplex; i.e., an n-simplex s^n is determined by $n + 1$ vertices. **4.** for a general topological space X, any of a number of quantities (such as the Hausdorff dimension) computed for X. *Computer Programming.* the specified size of an array expressed as the number of rows by the number of columns, e.g., 2 by 3.

dimensional *Physics.* relating to or expressed as a physical quantity.

dimensional analysis *Physics.* an empirical method of calculation that is not concerned with detailed theory and magnitudes, but relies heavily on the consideration of physical units associated with the quantities involved. *Mathematics.* the techniques for judging the correctness of equations by considering the units of measurement of the variables in the equations; e.g., the units on the left side of the equation must be the same as on the right side. The goal often is, by means of change of variables, to replace physical constants by dimensionless numbers that remain invariant regardless of the units of measurement chosen.

dimensional constant *Physics.* a physical constant whose numerical value is determined by the choice of the base quantities such as those of velocity, force, or energy, but does not depend on the system of units chosen.

dimensional formula *Physics.* the representation of a physical quantity in terms of its independent fundamental quantities, such as a force that can be expressed in terms of (mass) (length)/$(\text{time})^2$.

dimensional inspection *Design Engineering.* the process of verifying whether a physical dimension is within specifications.

dimensional stability *Materials Science.* the resistance of a material, such as paper, plastic, or textiles, to changes in its dimensions under changing conditions.

dimension declaration statement *Computer Programming.* a nonexecutable FORTRAN statement that is used to specify the size of an array.

dimensionless group *Physics.* a combination of quantities having a purely numerical value.

dimensionless number *Mathematics.* a number with no associated unit of measure such as square feet or pounds; often the ratio of two numbers with the same unit of measure.

dimension line *Graphic Arts.* any of several types of lines drawn on artboards to indicate measurements of an image area, sometimes including numerical values between these lines.

dimensions *Physics.* see DIMENSION.

dimension stone *Materials.* a large, smooth block of stone that is used in construction for paving, foundation building, and curbing.

dimension theory *Mathematics.* the branch of topology that deals with various definitions of dimension.

dimer *Chemistry.* a molecule that is produced by bonding together two molecules of the same chemical structure.

dimercaprol *Pharmacology.* $C_3H_8OS_2$, a colorless liquid used as an antidote to poisoning by arsenic, gold, mercury, and other metals.

Dimeriaceae *Mycology.* a family of fungi belonging to the order Pleosporales that lives off decaying organic material or as parasites to other fungi or plants, including conifers and grasses.

dimeric water *Inorganic Chemistry.* water in which the molecules are joined in pairs by a hydrogen bond.

dimerization *Chemistry.* a reaction that forms dimers.

dimerous *Biology.* having two parts, such as a flower with two members in each whorl or an insect with two segments to each tarsus.

Dimetan *Organic Chemistry.* $C_{11}H_{17}NO_3$, a trade name for a substance consisting of toxic yellow crystals, slightly soluble in water and soluble in alcohol; melts at 43–45°C; used as an insecticide.

dimethicone *Pharmacology.* a silicone oil used as an ingredient in ointments to protect the skin against water-based irritants, especially industrial chemicals, and also against diaper rash.

dimethirimol *Plant Pathology.* a type of pesticide that is absorbed into the plant sap of ornamental plants or vegetables and enables the plant to destroy powdery mildews.

dimethoate *Organic Chemistry.* $(CH_3O)_2P{=}SSCH_2CONHCH_3$, a white solid that is slightly soluble in water but freely soluble in most organic solvents; melts at 51–52°C; used as an insecticide.

1,4-dimethoxybenzene see HYDROQUINONE DIMETHYL ETHER.

dimethoxystrychnine see BRUCINE.

dimethrin *Organic Chemistry.* $C_{19}H_{28}O_2$, a flammable, toxic, amber liquid that is soluble in water and alcohol; boils at 175°C; used as an insecticide.

dimethyl see ETHANE.

dimethylallyl pyrophosphate *Endocrinology.* an intermediate in the synthesis of terpenes.

dimethylamine *Organic Chemistry.* $(CH_3)_2NH$, a flammable, anhydrous gas, soluble in alcohol; boils at 6.88°C and melts at –92.2°C; used in the manufacture of dyes, pharmaceuticals, missile fuels, and surfactants.

2-dimethylaminoethanol *Organic Chemistry.* $(CH_3)_2NCH_2CH_2OH$, a combustible, anhydrous, colorless liquid, miscible with water and alcohol; boils at 134°C and melts at –59°C; used as an intermediate and emulsifier. Also, DEANOL.

5-dimethylamino-1-naphthalenesulfonic acid see DANSYL CHLORIDE.

4-dimethylaminopyridine *Organic Chemistry.* $(CH_3)_2NC_5H_4N$, white crystalline solid that melts at 112–114°C; used as a versatile catalyst to promote acylation. Also, DMP.

***N,N*-dimethylaniline** *Organic Chemistry.* $C_6H_5N(CH_3)_2$, a combustible, toxic, yellowish to brownish, oily liquid; insoluble in water and soluble in alcohol; melts at 2.5°C and boils at 192.5–193.5°C; used in dyes and as an intermediate, solvent, and reagent.

dimethylbenzene see XYLENE.

2,2-dimethylbutane see NEOHEXANE.

2,3-dimethylbutane *Organic Chemistry.* $(CH_3)_2CHCH(CH_3)_2$, a flammable, colorless liquid that boils at 58.0°C, and melts at –129°C; used as a high-octane fuel and in organic synthesis. Also, DIISOPROPYL.

dimethyl carbate *Organic Chemistry.* $C_{11}H_{14}O_4$, a toxic, clear, oily liquid or crystalline solid; insoluble in water and soluble in organic solvents; melts at 38°C; used as an insect repellent.

dimethyl carbonate see METHYL CARBONATE.

5,5-dimethyl-1,3-cyclohexanedione *Organic Chemistry.* $C_8H_{12}O_2$, yellowish needles or prisms that are slightly soluble in water and soluble in alcohol; melts at 150°C; used as a reagent. Also, DIMEDONE, METHONE.

2,2-dimethyl-1,3-dioxolane-4-methanol *Organic Chemistry.* $C_6H_{12}O_3$, a combustible, colorless liquid that is soluble in water and alcohol; boils at 82°C (10 torr); used as a solvent and plasticizer. Also, BLYCEROL DIMETHYLKETAL.

dimethyl ether *Organic Chemistry.* CH_3OCH_3, a highly flammable, colorless, compressed gas that is soluble in water and alcohol; boils at –24.5°C and freezes at –141.4°C; used as a refrigerant, solvent, and catalyst. Also, METHYL OXIDE, WOOD ETHER.

dimethylglyoxime *Organic Chemistry.* $CH_3C({=}NOH)C({=}NOH)CH_3$, white crystals or powder, insoluble in water and soluble in alcohol; melts at 242°C; used as a reagent and in biochemical research.

1,1-dimethylhydrazine *Organic Chemistry.* $(CH_3)_2NNH_2$, a flammable, toxic, suspected carcinogenic, colorless, hygroscopic liquid; soluble in water and alcohol; boils at 63°C and melts at –58°C; used as a component of jet and rocket fuels.

dimethylisopropanolamine *Organic Chemistry.* $(CH_3)_2NCH_2CH(OH)CH_3$, a flammable, colorless liquid that is miscible in water; boils at 125.8°C; used in methadone synthesis.

1,2-dimethyl-5-nitroimidazole see DIMETRIDAZOLE.

3,7-dimethyl-2,6-octadiene-1-ol see NEROL.

dimethylolurea *Organic Chemistry.* $CO(NHCH_2OH)_2$, colorless crystals, soluble in water; melts at 126°C; used in plywood manufacture and in permanent-press fabrics. Also, DMU.

dimethylphenol see XYLENOL.

3,4-dimethylphenyl-*N*-methyl carbamate *Organic Chemistry.* $C_{10}H_{13}NO_2$, a white, crystalline solid; melts at 79°C; used as an insecticide.

dimethylphosphoramidocyanidic acid see TABUN.

dimethyl phthalate *Organic Chemistry.* $C_6H_4(COOCH_3)_2$, a combustible, colorless, oily liquid, insoluble in water and miscible with alcohol; boils at 283.8°C and melts at 5.5°C; used as a plasticizer and in solid rocket propellants, lacquers, rubber, and insect repellents.

dimethyl sebacate *Organic Chemistry.* $[(CH_2)_4COOCH_3]_2$, a combustible, water-white liquid; melts at 26–28°C and boils at 294°C; used as a solvent and plasticizer.

dimethyl sulfate *Organic Chemistry.* $(CH_3)_2SO_4$, a combustible, carcinogenic, colorless oily liquid; soluble in water and alcohol; boils at 188°C and melts at –27°C; used as a methylating agent. Also, METHYL SULFATE.

2,4-dimethylsulfolane *Organic Chemistry.* $C_6H_{12}O_2S$, toxic, slightly yellow liquid that is slightly soluble in water; boils at 280°C; used in solvent extraction.

dimethyl sulfoxide *Organic Chemistry.* $(CH_3)_2SO$, a combustible, hygroscopic, colorless liquid, soluble in water and alcohol; melts at 18.5°C and boils at 189°C; used in medicine, veterinary medicine, pharmaceuticals, plant pathology, and nutrition.

dimethyl terephthalate *Organic Chemistry.* $C_6H_4(COOCH_3)_2$, colorless crystals that are insoluble in water and soluble in hot alcohol; melts at 141°C, boils at 288°C, and sublimes above 300°C; used to make film and polyester fiber.

dimethyl 2,3,5,6-tetrachloroterephthalate *Organic Chemistry.* $C_6Cl_4(COOCH_3)_2$, crystals that are insoluble in water and soluble in acetone; melts at 156°C; used as an herbicide. Also, DCPA.

1,3-dimethylxanthine see THEOPHYLLINE.

3,7-dimethylxanthine see THEOBROMINE.

dimetridazole *Veterinary Medicine.* an antimicrobial compound used to control various parasitic illnesses in domestic fowl and pet birds. Also, 1,2-DIMETHYL-5-NITROIMIDAZOLE.

dimiazene aceturate see BERENIL.

dimictic lake *Hydrology.* a lake that has two overturns or periods of circulation annually.

dimidiate *Mycology.* **1.** of or relating to fungi that are split into two parts. **2.** of or relating to fungal fruiting bodies that are missing one half or seeming to lack one half. **3.** of or relating to fungi in which one half is much smaller than the other.

diminished arch *Architecture.* an arch that is less than half as high as it is wide.

diminution *Botany.* the simplification or diminishing of inflorescences on successive branches.

dimity *Textiles.* a strong, sheer cotton fabric that appears to have lengthwise stripes or checks because of a bunched weaving pattern.

dimixis *Biology.* the fusion of two different nuclei in heterothallism.

dimmer *Electricity.* an electrical control device that adjusts the intensity of illumination of a lamp or other light source.

dimmerfoehn [dim´ər fān´] *Meteorology.* a foehn condition during a strong upper wind from the south in which there is a pressure difference of at least 12 millibars between the north and south sides of the Alps.

dimorphism *Science.* the fact of existing or occurring in two distinct forms. *Biology.* the existence of distinct genetically determined forms of the same species, such as distinct male and female forms or distinct young and mature forms. *Chemistry.* the fact of existing in two crystalline forms, both with the same chemical composition.

dimorphite *Mineralogy.* As_4S_3, an orange-yellow, transparent, brittle orthorhombic mineral occurring as dipyramidal crystals, having a specific gravity of 3.58 and a hardness of 1.5 on the Mohs scale; found in fumaroles.

dimorphous relating to or exhibiting dimorphism. Also, **dimorphic.**

dimple *Metallurgy.* a conical depression, such as a depression made to accommodate a rivet head.

dimpling *Metallurgy.* the formation of a depression in a sheet of metal so that rivet heads will be countersunk.

dimuon event *Particle Physics.* a collision of a neutrino or an antineutrino with a nucleus, in which two muons are produced as the result of the collision.

Dimylidae *Paleontology.* a Miocene family of mammals in the order Insectivora, distantly related to hedgehogs.

dina *Electronics.* a unit of airborne electronic countermeasures equipment used for jamming enemy radars.

Dinamoebaceae *Botany.* a monospecific family of marine, nonphotosynthetic algae that constitutes the order Dinamoebales.

Dinamoebales *Botany.* a monofamilial order of marine, nonphotosynthetic algae of the class Dinophyceae, characterized by an amoeboid state that feeds on diatoms, flagellates, and green algae.

Dinantian *Geology.* a European geologic stage of the lower Carboniferous period, occurring after the Famennian of the Devonian and before the Namurian.
Dines anemometer *Engineering.* an instrument used to measure wind speed in which the differential pressure of the pressure head and the suction head is proportional to the wind speed.
dineuric *Neurology.* having two neurons or two axons.
dineutron *Nuclear Physics.* a state, probably hypothetical, in which two neutrons are bound and take on a transitory existence during bombardment of tritium by accelerated tritons.
din gene *Genetics.* an SOS gene whose function has not been determined. (An acronym for damage-inducible genes.)
dinghy [ding´ē] *Naval Architecture.* a small rowboat or sailboat, of less than twenty feet in length, with a transom stern, especially one that is carried by a yacht or other larger boat.
dingo *Vertebrate Zoology.* a wild Australian dog, *Canis dingo,* of the family Canidae, that is the closest relative to the domestic dog; characterized by yellow to reddish-brown fur, pointed ears, and a bushy tail; eliminated in many areas due to predation on livestock.

dingo

dingot *Metallurgy.* a massive piece of primary metal produced in a bomb reaction.
Dings magnetic separator *Mechanical Engineering.* a suspended device that separates magnetic material from raw material that passes beneath it on a belt.
Dinidoridae *Invertebrate Zoology.* a family of large, heavy-bodied, plant-eating insects of the order Hemiptera.
dinitramine *Organic Chemistry.* $C_{11}H_{13}N_3O_4F_3$, a yellow solid that melts at 98–99°C; used as an herbicide.
dinitrate *Chemistry.* a compound containing two nitrate groups.
dinitrite *Chemistry.* a compound containing two nitrite groups.
2,4-dinitroaniline *Organic Chemistry.* $C_6H_3NH_2(NO_2)_2$, combustible, toxic, yellow crystals; insoluble in water and soluble in alcohol; melts at 180°C; used as an intermediate and in dye preparation.
***m*-dinitrobenzene** *Organic Chemistry.* $C_6H_4(NO_2)_2$, toxic yellow crystals; slightly soluble in water; melts at 89.9°C and boils at 302°C; used in organic synthesis.
***o*-dinitrobenzene** *Organic Chemistry.* $C_6H_4(NO_2)_2$, toxic white crystals; soluble in water; melts at 118°C and boils at 319°C; used in organic synthesis.
***p*-dinitrobenzene** *Organic Chemistry.* $C_6H_4(NO_2)_2$, toxic white crystals; slightly soluble in water; melts at 173–174°C and boils at 299°C; used in organic synthesis.
2,4-dinitrobenzenesulfenyl chloride *Organic Chemistry.* $(NO_2)_2C_6H_3SCl$, a yellow solid that is soluble in benzene, chloroform, and acetic acid; melts at 96°C; used to characterize organic compounds.
4,6-dinitro-2-*sec*-butylphenylacetate *Organic Chemistry.* $C_{12}H_{13}O_6N_2$, a brown liquid; melts at 170°C; used as an herbicide.
dinitrophenol *Organic Chemistry.* $C_6H_3OH(NO_2)_2$, toxic yellow crystals that are slightly soluble in water and soluble in alcohol; melts at 114–115°C and boils at 302°C; used in dyes, preservation of lumber, and manufacture of explosives, and as a reagent.
2,4-dinitrophenol *Biochemistry.* $C_6H_4N_2O_5$, a compound that prevents the uptake of inorganic phosphate and the production of ATP by disrupting the linkage between electron transport and ATP synthesis; useful in maintaining body temperature in hibernating mammals and in newborn mammals, including humans.

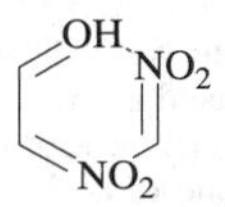

2,4-dinitrophenol

2,4-dinitrophenylhydrazine *Organic Chemistry.* $(NO_2)_2C_6H_3NHNH_2$, a bright red crystalline powder, slightly soluble in water and alcohol; melts at 200°C; used as an explosive and as a reagent to detect and characterize aldehydes and ketones.
dinitroresorcinol *Biochemistry.* $C_6H_2(NO_2)_2(OH)_2$, a green coal tar derivative; used in preparing degenerated nerve tissue for biochemical analysis.
dinitrotoluene *Organic Chemistry.* $C_6H_3CH_3(NO_2)_2$, isomeric, toxic, combustible yellow crystals or oily liquid; very slightly soluble in water and soluble in alcohol; melts at 61–92.3°C; used in organic synthesis, dyes, and explosives. Also, DNT.
dinking *Mechanical Engineering.* the process of cutting light-gauge soft metals or nonmetallic materials with a sharp, hollow punch.
Dinobryaceae *Botany.* a family of flagellates belonging to the order Ochromonadales, characterized by an envelope called a lorica surrounding each cell; species are widely distributed, mainly in freshwater phytoplankton.
dinobryon *Botany.* a vase-shaped green alga genus of the family Oinobryaceae, consisting of fibrils interwoven in a definite pattern, that is a common fouling agent of water supplies.
Dinocerata *Paleontology.* a Tertiary suborder of large, herbivorous hoofed mammals (ungulates) in the order Amblypoda; formerly called Uintatheres, from the North American locality in which they were discovered.
Dinocloniales *Botany.* a little-known order of marine algae belonging to the class Dinophyceae and composed of the single family Dinocloniaceae.
Dinoflagellata *Invertebrate Zoology.* a large class of protozoans possessing two flagella; some cause "red tides," some are bioluminescent. Also, **Dinoflagellida.**
dinoflagellate *Invertebrate Zoology.* **1.** a member of the class Dinoflagellata. **2.** of or relating to this class.
Dinophilidae *Invertebrate Zoology.* an order of polychaetous annelid marine worms.
Dinophyceae *Botany.* a unicellular, planktonic, usually solitary alga in the class Pyrrophyta, characterized by biflagellate zoospores with a transverse groove, chromatophores and possibly pyrenoids, a cellulosic cell wall, reproduction by vegetative division, and the presence of zoospores or aplanospores.
Dinophycidae *Botany.* a subclass of the algal class Dinophyceae, having a mostly unicellular and biflagellate form with a multilayered cell covering and a nucleus with permanently condensed chromosomes.
Dinophysiaceae *Botany.* a family of marine algae of the order Dinophysiales, characterized by a rounded or oval-shaped cell body and some nonphotosynthetic species.
Dinophysiales *Botany.* an order of marine planktonic algae of the class Dinophyceae, that is thecate and has a reduced epitheca.
Dinornithiformes *Paleontology.* the moas, an order of giant, wingless New Zealand ratites that became extinct in historical times before the arrival of Europeans.
dinosaur *Paleontology.* a general name for two orders of extinct reptiles, the "lizard-hipped" Saurischia and the "bird-hipped" Ornithischia, certain types of saurischian dinosaurs were the largest animals ever to live on land. Both orders were probably descended from a single thecodont ancestor of the Triassic, and both were widespread in the Jurassic and Cretaceous periods. They disappeared in the mass extinction at the end of the Mesozoic era, of causes not yet definitively established, about 60 million years before the appearance of the first nearly human apes. (From a Greek term literally meaning "terrible lizard.")
dinoseb *Organic Chemistry.* $C_{10}H_{12}N_2O_5$, a reddish brown liquid; soluble in most organic solvents; boils at 32°C; used as an herbicide. Also, **2,4-dinitro-6-*sec*-butylphenol.**

Dinosphaeraceae *Botany.* a rare monogeneric family of thecate, photosynthetic dinoflagellates belonging to the order Peridiniales.
d. in p. aeq. let it be divided in equal parts. (From Latin *dividatur in partes aequales.*) Also, **D. in p. aeq.**
DIN system *Graphic Arts.* a method of rating the exposure speed of film emulsions; used mainly in Europe. (An acronym for Deutsche Industrie Norm.)
dinucleotide *Biochemistry.* two nucleotides linked together by a 3',5'-phosphodiester bond.
Diocles of Carystus 4th century BC Greek physician; wrote on nutrition, medical botany, embryology, and zoology.
diocoel *Developmental Biology.* the cavity of the diencephalon; the third ventricle of the cerebrum.
dioctahedral *Crystallography.* **1.** a solid geometric figure having 16 faces and 16 equal angles. **2.** of or relating to an octahedral crystal structure in which two of three coordinated positions are occupied.
dioctogon *Mathematics.* a polygon with 16 sides.
dioctohedron *Mathematics.* a polyhedron with 16 faces.
Dioctophymatida *Invertebrate Zoology.* an order of nematode worms that are parasitic on earthworms and vertebrates.
Dioctophymoidea see DIACTOPHYMATOIDEA.
dioctyl *Organic Chemistry.* a molecule having two octyl groups.
dioctyl phthalate *Organic Chemistry.* $C_6H_4[COOCH_2CH(C_2H_5)C_4H_9]_2$, a pale liquid that is insoluble in water and miscible with mineral oil; boils at 231°C; used as a plasticizer and insecticide. Also, DI(2-ETHYLHEXYL) PHTHALATE.
dioctyl phthalate test *Engineering.* a technique used to test air filters in which particles of fixed size entering and emerging from the filter are counted.
dioctyl sebacate *Organic Chemistry.* $C_4H_8(COOC_8H_{17})_2$, a combustible, straw-colored liquid that is insoluble in water; boils at 248°C and melts at –55°C; it is used as a plasticizer. Also, DI(2-ETHYLHEXYL) SEBACATE.
diode *Electronics.* a two-element active electronic device containing an anode and a cathode, characterized by the ability to pass an electric current more easily from cathode to anode than from anode to cathode; a diode can be in the form of an electron tube, semiconductor device, or crystal device.
diode amplifier *Electronics.* a microwave amplifier using a specially designed diode in the tuned cavity of a microwave electron tube; its center frequencies range from about 1 to 100 GHz.
diode bridge *Electronics.* a circuit consisting of four diodes connected in a bridge configuration; the polarity of the output remains the same regardless of the polarity of the input; used primarily for converting alternating current to pulsating direct current.
diode characteristic *Electronics.* the characteristic of a multielectrode vacuum tube to form a virtual diode between the cathode and all other electrodes in the tube.
diode clamping circuit *Electronics.* **1.** a circuit that uses diodes to add a fixed voltage to a waveform in order to establish a specified reference level for the waveform. Also, DC RESTORER. **2.** a circuit that uses a diode to hold a voltage to some fixed value.
diode clipping circuit *Electronics.* a circuit that uses diodes to prevent a signal from reaching the peak amplitude it would otherwise attain; this results in a distorted waveform whose peaks appear clipped.
diode-connected transistor *Electronics.* a transistor in which the emitter and collector terminals are shorted to form a diode between the base terminal and the common emitter-collector terminal connection.
diode demodulator *Electronics.* a demodulator that uses one or more crystal or electron-tube diodes to provide a rectified output with an average value proportional to the original modulation. Also, DIODE DETECTOR.
diode detector see DIODE DEMODULATOR.
diode forward voltage *Electronics.* the voltage applied across a diode with a polarity and amplitude that cause electrical current to flow forward, from anode to cathode.
diode function generator *Electronics.* a circuit that produces specific waveforms, such as sine waves or sawtooth waves, and in which the output waveform characteristics are determined by an amplifier connected to an adjustable wave-shaping network, consisting mainly of diodes and resistances.
diode gate *Electronics.* an AND or OR logic gate circuit that uses current-steering diodes to perform the appropriate logic function; frequently used in conjunction with a digital amplifier to provide the required power gain to drive other logic circuits. Also, DIODE LOGIC.
diode laser **1.** see INJECTION LASER. **2.** see SEMICONDUCTOR LASER.
diode limiter *Electronics.* a limiter circuit that uses one or more diodes to prevent a signal from exceeding a predetermined peak amplitude; when the specified amplitude is reached, the limiter acts as a clipper circuit to prevent a further increase in amplitude.
diode logic see DIODE GATE.
diode matrix *Electronics.* any number of diodes connected in a two-dimensional array configuration, used primarily for decoding digital inputs.
diode mixer *Electronics.* a circuit that uses the nonlinear characteristics of a diode to mix two frequencies in order to produce the sum or difference of the two original frequencies; its main application is in superheterodyne radio receivers.
diode modulator *Electronics.* a circuit that uses diodes to combine a modulating signal with a carrier signal.
diode oscillator *Electronics.* **1.** an oscillator based on the breakdown or negative resistance characteristics of some diodes. **2.** an oscillator that makes use of the negative anode-to-cathode resistance of a tube diode when operated at ultrahigh frequencies.
diode pack *Electronics.* a sealed unit that contains two or more diodes.
diode peak detector *Electronics.* a circuit, usually consisting of a diode and capacitor in series, used to detect and store the peak amplitude value of a signal.
diode-pentode *Electronics.* an electron tube in which a diode and a pentode are contained within the same envelope, usually with a common cathode.
diode rectifier *Electronics.* a circuit containing one or more diodes, usually part of a power supply, that converts AC input to a pulsating DC output.
diode switch *Electronics.* a circuit containing several diodes that performs switching functions by applying forward or reverse bias voltages to the anodes of the diodes in sequence: a forward voltage allows the diode to conduct, passing the signal on its cathode; a reverse voltage causes the diode to act as an open switch.
diode theory *Electricity.* a theory stating that for a semiconductor in which the barrier thickness is comparable to or smaller than the free path of the carriers, the carriers cross the barrier without being scattered, in a manner similar to that of a vacuum-tube diode.
diode transistor logic *Electronics.* a logic circuit consisting of one or more diode gates connected to the base of a transistor, in which the AND or OR functions performed by the diode gates control the on-off status of the transistor.
diode-triode *Electronics.* an electron tube in which a diode and a triode are contained within the same envelope, usually with a common cathode.
diode voltage regulator *Electronics.* a simple voltage regulator circuit that depends for its operation on the characteristic of a Zener diode when operating in the reverse-biased breakdown region. Also, ZENER DIODE VOLTAGE REGULATOR.
diodide *Chemistry.* a compound that contains two iodine atoms.
Diodontidae *Vertebrate Zoology.* the porcupine fishes, a small family of tropical marine fishes of the order Tetraodontiformes, characterized by a body covering of sharp spines.
Diodrast *Radiology.* a trade name for a preparation of iodopyracet, a radiopaque substance that is used in radiographic studies of the urinary tract.
dioecious *Biology.* having the male and female reproductive organs in separate individuals. Most animal species are dioecious, as are some plants, such as asparagus.
dioecy *Biology.* the condition of having male and female reproductive organs on separate, distinct individuals.
dioic see DIOECIOUS.
diolefin *Organic Chemistry.* an aliphatic compound containing two double bonds. Also, DIENE.
diolefin hydrogenation *Chemical Engineering.* the hydrogenation of diolefins in C4 and C5 fractions to mono-olefins for alkylation feedstocks, using a fixed-bed catalytic reactor.
Diomedeidae *Vertebrate Zoology.* the albatrosses, a family of large sea birds of the order Procellariiformes, found in southern oceans and the Pacific.
Dioncophyllaceae *Botany.* a family of dicotyledonous vines and shrubs of the order Violales, distinguished by their method of climbing via hooked or cirrhose leaf tips, by peltate hairs on leaves and stems, and by multicellular stalked or sessile glands; occurring in tropical Africa.

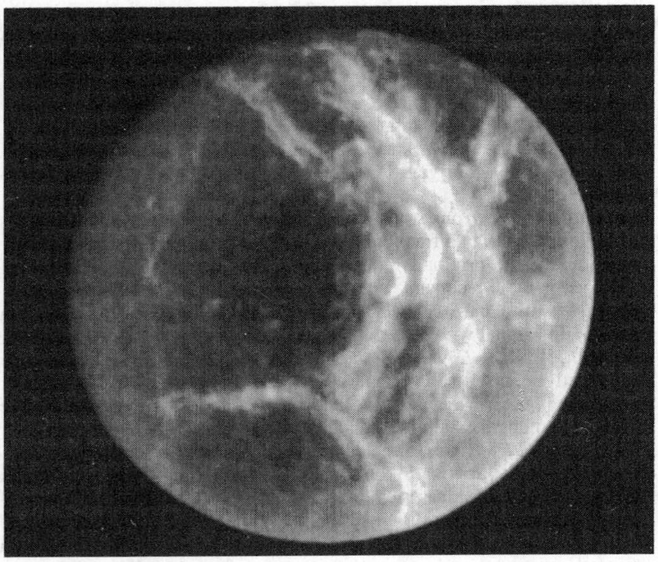

Dione

Dione [dī´ ə nē´] *Astronomy.* the fourth moon of Saturn, discovered in 1684; it measures approximately 1100 km in diameter.

-dione *Organic Chemistry.* a suffix used to indicate the presence of two keto groups in a compound.

Dionosil *Radiology.* a trade name for a preparation of propyliodone, a contrast medium used in radiographic studies of the lungs.

Dionysian personality *Anthropology.* a categorization by the anthropologist Ruth Benedict, describing a society whose members exhibit the excessive and self-involved personality exemplified by the Kwakiutl.

Dionysopithecus *Paleontology.* the earliest known catarrhine primate in Asia; similar to earlier forms in Africa, such as *Micropithecus*; extant in the Early to Middle Miocene.

Diophantine analysis *Mathematics.* the study of methods for obtaining integer solutions to algebraic equations.

Diophantine equation *Mathematics.* an algebraic equation with integer coefficients and two or more variables, for which it is desired to find a set of (usually nonnegative) integer solutions. If the equation is linear in each of the variables, it is a **linear Diophantine equation;** otherwise it is a **nonlinear Diophantine equation.**

Diophantus 3rd century AD Greek mathematician; called the father of algebra for his use of algebraic symbols and quadratic equations.

Diopsidae *Invertebrate Zoology.* the stalk-eyed flies, a family of tropical flies.

dioptase *Mineralogy.* $Cu^{+2}SiO_2(OH)_2$, a brittle, emerald-green or turquoise blue trigonal mineral occurring as prismatic crystals with a vitreous luster, having a specific gravity of 3.28 to 3.25 and a hardness of 5 on the Mohs scale; found in the oxidation zones of copper deposits.

diopter *Optics.* a unit of measure of the refractive power of a lens, equivalent to the power of a lens with a focal length of one meter; for any given lens, the refractive power is equal to the reciprocal of the focal length of the lens.

dioptometer *Optics.* an instrument that measures the refractive power of a lens or prism, used to test the ability of an eye to refract light.

dioptric *Optics.* **1.** of or relating to the refractive powers of light. **2.** of or relating to the method by which optical lenses are numbered according to their refractive powers.

dioptrics *Optics.* the branch of optics that studies the refraction of light, either by the eye or by lenses.

diorite *Petrology.* an intermediate, plutonic rock composed largely of sodic plagioclase and hornblende, but sometimes containing biotite, pyroxine, quartz, or orthoclase; used occasionally as ornamental and building stone. Also, BLACK GRANITE.

Dioscoreaceae *Botany.* a family of monocotyledonous herbs and shrubs of the order Liliales, characterized by leafy stems, tuberous roots, alternate, net-veined leaves, and regular, unisexual flowers arranged in racemes, an inferior ovary, and winged fruit and seeds; abundant in the tropics and subtropics.

1,4-dioxane *Organic Chemistry.* $C_4H_8O_2$, a flammable, toxic, carcinogenic colorless liquid; miscible with water and soluble in alcohol; boils at 101.3°C and melts at 10–12°C; used as a solvent and in lacquers, paints, dyes, cements, cosmetics, deodorants, and textile processing. Also, DIETHYLENE ETHER.

dioxide *Chemistry.* any compound that contains two oxygen atoms.

dioxin *Organic Chemistry.* a carcinogenic, teratogenic, mutagenic, white crystalline solid; melts at 295°C. It is present in defoliants such as Agent Orange, which was used in the Vietnam War; the question of its human toxicology is under continuing investigation and has been the subject of litigation. Also, 2,3,7,8-TETRACHLORODIBENZO-*p*-DIOXIN, TCDD.

1,3-dioxolane *Organic Chemistry.* $C_3H_6O_2$, a flammable, water-white liquid that is soluble in water; boils at 74°C; used as a solvent and an extractant.

dioxolone-2 see ETHYLENE CARBONATE.

dioxopurine see XANTHINE.

dioxygenase *Biochemistry.* an enzyme that catalyzes the introduction of both atoms of an oxygen molecule into its substrate, an organic compound.

dip a movement or course that goes downward; specific uses include: *Geology.* **1.** the angle made by a structural surface, such as a stratum or fault plane, with the horizontal, measured perpendicular to the strike of the structure. Also, ANGLE OF DIP, TRUE DIP. **2.** a marked depression in the land surface. *Geodesy.* the angle between the horizontal and the lines of force of the earth's magnetic field at any point. *Navigation.* the vertical angle between the visible horizon at sea and the sensible horizon, dependent on the elevation of the observer and the convexity of the earth's surface. *Mining Engineering.* the vertical angle between a vein of ore and the horizontal.

DIPA see DIISOPROPANOLAMINE.

dip angle *Navigation.* the vertical angle of the observation point between the plane of the true horizon and a sight line to the apparent horizon. *Aviation.* the vertical angle, at an air station, between the true and the apparent horizon, which is due to flight height, the earth's curvature, and refraction.

dip circle *Geodesy.* an instrument for measuring magnetic dip, consisting of a magnetic needle suspended so that it can rotate about a horizontal axis.

dipentene see LIMONENE.

dipentene glycol see TERPIN.

Dipentodontaceae *Botany.* a monospecific family of shrubs and small trees in the order Santalales about which little is known; native to southern China and Burma.

dipeptidase *Enzymology.* an enzyme that catalyzes the breakdown of dipeptides into amino acids.

dipeptide hydrolase *Enzymology.* an enzyme that catalyzes the hydrolysis of a dipeptide.

dipeptidyl I carboxypeptidase see ANGIOTENSIN-CONVERTING ENZYME.

dip equator see MAGNETIC EQUATOR.

dip fault *Geology.* a fault whose strike is parallel to the dip of the strata involved.

diphacinone see DIPHENADIONE.

diphase *Science.* having two phases.

diphase cleaning *Metallurgy.* a process of surface cleaning using an emulsion of a solvent and an aqueous phase.

diphase generator *Electricity.* a generator that delivers two alternating currents in phase quadrature.

diphead *Mining Engineering.* an inclined drift along the dip of a coal seam.

diphenadione *Organic Chemistry.* $C_{23}H_{16}O_3$, pale yellow crystals or crystalline powder; practically insoluble in water and soluble in acetone; melts at 146–147°C; used as an anticoagulant and a rodenticide. Also, DIPHACINONE.

diphenamid *Organic Chemistry.* $(C_6H_5)_2CHCON(CH_3)_2$, a toxic, white solid that is soluble in water and very soluble in alcohol; melts at 134.5–135.5°C; used in the synthesis of antispasmodics and as an herbicide and plant growth regulator. Also, *N,N*-DIMETHYL-2,2-DIPHENYLACETAMIDE.

diphenatrile see DIPHENYLACETONITRILE.

diphenoid *Crystallography.* a crystal form that has two geometric forms.

diphenol *Organic Chemistry.* a compound that has two phenol groups.

diphenyl see BIPHENYL.

diphenylacetonitrile *Organic Chemistry.* $(C_6H_5)_2CHCN$, a yellow crystalline powder; insoluble in water and very soluble in alcohol; melts at 73–73.5°C; used in the synthesis of antispasmodics and as an herbicide. Also, DIPHENATRILE.

diphenylamine *Organic Chemistry.* $(C_6H_5)_2NH$, combustible, light-sensitive, toxic, colorless to grayish crystals, insoluble in water and soluble in alcohol; melts at 52.85°C and boils at 302°C; used in solid rocket propellants, pesticides, dyes, pharmaceuticals, veterinary medicine, and rubber antioxidants.

diphenylamine

diphenylamine chloroarsine see ADAMSITE.

diphenylcarbazide *Organic Chemistry.* $(C_6H_5NHNH)_2CO$, white crystals or flakes; insoluble in water and soluble in alcohol; melts at 170°C and decomposes in light; used as a reagent and indicator for iron.

diphenylcarbodiimide *Organic Chemistry.* $C_6H_5N{=}C{=}NC_6H_5$, toxic, combustible, deliquescent crystals, slightly soluble in water and alcohol; melts at 45–46°C and boils at 331°C.

diphenyl carbonate *Organic Chemistry.* $C_{13}H_{10}O_2$, a white crystalline solid; insoluble in water and soluble in hot alcohol; melts at 78°C and boils at 302°C; used as a plasticizer and solvent.

diphenylchloroarsine *Organic Chemistry.* $(C_6H_5)_2AsCl$, toxic, colorless crystals or dark brown liquid, almost insoluble in water and soluble in carbon tetrachloride; melts at 44°C and boils at 333°C; used as a military poison gas.

diphenylene oxide *Organic Chemistry.* $C_{12}H_8O$, a crystalline solid, insoluble in water and slightly soluble in alcohol; melts at 86°C and boils at 288°C; used as an insecticide. Also, DIBENZOFURAN.

diphenylguanidine *Organic Chemistry.* $NH{=}C(NHC_6H_5)_2$, a toxic, white powder; slightly soluble in water and soluble in alcohol; melts at 148°C and decomposes above 170°C; used as a rubber accelerator. Also, MELANILINE.

diphenylhydantoin see PHENYTOIN.

diphenyl oxide *Organic Chemistry.* $(C_6H_5)_2O$, combustible, toxic, colorless crystals or liquid, insoluble in water and soluble in alcohol; melts at 28°C and boils at 259°C; used in perfumes and soaps, and in heat-transfer medium resins. Also, PHENYL ETHER.

diphenyl phthalate *Organic Chemistry.* $C_6H_4(COOC_6H_5)_2$, a combustible yellow-white powder; insoluble in water and soluble in acetone; melts at 68–70°C and boils at 405°C; used as a plasticizer.

diphosphate *Chemistry.* a compound that contains two phosphate groups.

diphosphatidyl glycerol see CARDIOLIPIN.

2,3-diphosphoglycerate *Biochemistry.* a compound found in red blood cells that binds to deoxyhemoglobin and controls the oxygen-binding affinity of hemoglobin in the red blood cells; cellular levels are based on oxygen pressure in the lungs; thus a person at sea level has less DPG than a person in the mountains.

diphosphopyridine nucleotide see NICOTINAMIDE ADENINE DINUCLEOTIDE.

diphosphoric acid see PYROPHOSPHORIC ACID.

diphosphotransferase *Enzymology.* an enzyme of the transferase class that catalyzes the transfer of diphosphate groups between molecules.

diphtheria [dif thēr′ē ə; dip thēr′ē ə] *Medicine.* an acute infectious disease caused by *Corynebacterium diphtheriae,* characterized by the formation of a tough (false) membrane lining the throat and other portions of the upper respiratory tract and by the production of a systemic toxin that may cause myocarditis and peripheral neuritis.

diphtheria toxin *Toxicology.* a poison produced by *Corynebacterium diphtheriae* that kills the cells surrounding an area infected by the bacterium. Also, **diphtherotoxin**.

diphtheritic myocarditis *Medicine.* inflammation of the muscular walls of the heart, associated with diphtheria.

diphycercal *Vertebrate Zoology.* having the tail fin divided into two equal halves by the caudal spine, as in the lungfish.

Diphyllidea *Invertebrate Zoology.* an order of tapeworms that parasitize sharks, skates, and rays.

Diphyllobothrium *Invertebrate Zoology.* a genus of medically important tapeworms, parasites of fish-eating birds and mammals; it includes the common fish tapeworm.

Diphyllobothrium latum *Invertebrate Zoology.* the fish tapeworm, a parasite reaching 15 meters in length, that infects humans and other fish-eating mammals; fish are the intermediate host. Also, **Dibothriocephalus latus.**

diphyllous *Botany.* having two leaves; bifoliate.

diphyodont *Anatomy.* having two sets of teeth, such as both a deciduous set and a permanent set.

dipicolinic acid *Biochemistry.* pyridine-2,6-dicarboxylic acid, an acid present in bacterial endospores.

dipicrylamine see HEXANITRODIPHENYLAMINE.

dip inductor see EARTH INDUCTOR.

DIPJ distal interphalangeal joint.

dip joint *Geology.* a joint whose strike is approximately perpendicular to the cleavage or bedding plane of the constituent rock.

Diplacanthidae *Paleontology.* an extinct family of toothless acanthodian fishes of uncertain order; the type genus is *Diplacanthus*; extant in the Devonian.

Diplacanthoidei *Paleontology.* a possible suborder of Devonian acanthodian fishes of uncertain order but sometimes placed in the order Climatiiformes.

diplacusis *Medicine.* a difference in perception of sound by the two ears, either in time or in pitch, so that one sound is heard as two.

diplanetism *Biology.* of or relating to the condition of having two different motile periods during a lifetime, as the zoospores of some fungi.

Diplasiocoela *Vertebrate Zoology.* in some systems, a suborder of the amphibian order Anura, characterized by a skeleton that is modified for jumping by having the anterior surface of the first seven vertebrae concave and the next eight vertebra biconcave, a firm median fusion of the shoulder girdle, and no ribs.

diplegia *Neurology.* bilateral weakness or paralysis affecting like parts on both sides of the body; most often used to describe cerebral palsy involing spastic weakness of all four limbs.

dipleurula *Invertebrate Zoology.* a bilaterally symmetrical, ciliated echinoderm larva.

diplexer *Electronics.* a coupling unit that allows two transmitters to safely share the same antenna.

diplex operation *Telecommunications.* **1.** a simultaneous one-way transmission or reception of two or more independent signals using a common element, such as a single antenna or channel. **2.** the operation of two radio transmitters into one antenna. **3.** the simultaneous utilization of different frequencies with a single antenna.

diplex reception *Telecommunications.* the simultaneous reception of two signals having a common feature, such as a single antenna or a single carrier.

diplo- a combining form meaning "double" or "in pairs," as in *diplopia, diplosome.*

Diplobathrida *Paleontology.* an order of dicyclic crinoids in the subclass Camerata, extant from the Middle Ordovician to the Upper Mississippian.

diplobiont *Biology.* an organism characterized by alternating diploid and haploid generations that are structurally dissimilar.

diploblastic *Zoology.* having a body wall composed of two layers, the endoderm and the ectoderm.

diploblastula *Invertebrate Zoology.* a flagellated larva of certain sponges that consists of two layers of tissue.

Diplocarpon *Mycology.* a genus of fungi belonging to the order Helotiales, causing such plant diseases as strawberry leaf scorch and black spot in roses.

diplochlamydeous see DICHLAMYDEOUS.
diplochromosome *Cell Biology.* a chromosome that possesses four chromatids.
Diplococcus *Microbiology.* a former classification for a genus of bacteria including species now assigned to various other genera.
diplococcus *Microbiology.* **1.** a spheroid bacterium generally ocurring in pairs due to incomplete separation after cell division. **2.** a former term for a bacterium of the genus *Diplococcus.*
diplodinium *Invertebrate Zoology.* a genus of complex protozoans in the order Spirotricha, commensal in the digestive system of ruminants.

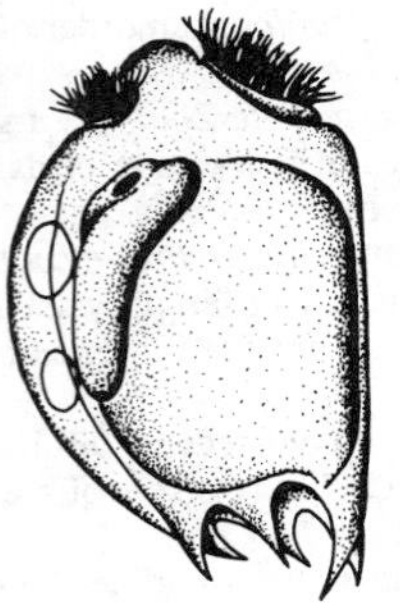

diplodinium

Diplodocus *Paleontology.* an essentially herbivorous amphibian genus of dinosaurs in the order Sauropoda. One of the largest dinosaurs, Diplodocus reached lengths of up to 90 feet; it flourished in many parts of the world from the upper Jurassic to the end of the Cretaceous. (From Greek; literally, "double-beam," referring to the Y-shaped spines on the tail vertebrae.)
dip log *Geology.* a detailed, sequential record indicating the dips of the formations traversed by boreholes.
Diplogasteroidea *Invertebrate Zoology.* a superfamily of nematode worms often found associated with insects or the fecal matter of herbivores.
diploglossate *Vertebrate Zoology.* having the end of the tongue retractile into the basal portion, as in certain lizards.
Diplograptus *Paleontology.* a family of scandent biserial graptolites of the Ordovician and Silurian, in the order Graptoloidea; widespread in the shallow seas that covered half of North America and much of other land masses in the Ordovician.
diplohaplont *Biology.* an organism characterized by alternating diploid and haploid generations that are structurally similar.
diploid *Cell Biology.* **1.** having two haploid sets of chromosomes; two copies of each chromosome are present, except (in some cases) for the sex chromosomes. **2.** a cell of this type. *Crystallography.* a 24-sided crystal form belonging to an isometric system in which the faces are arranged in pairs.
diploidization *Cell Biology.* any process that brings about the diploid condition in a cell or organism.
diploid merogony *Developmental Biology.* the development of the part of an egg containing the fused male and female pronuclei.
diploid state *Cell Biology.* the condition in which two sets of chromosomes are present. Also, **diploidy.**
diplolepidious *Botany.* double-scaled, as in the peristome of mosses with two rows of scales on the outside and one row on the inside.
Diplomonadida *Invertebrate Zoology.* an order of colorless flagellate protozoans in the class Zoomastigophorea.
Diplomystidae *Vertebrate Zoology.* a monogeneric family of small, primitive, freshwater catfish of the order Siluriformes; characterized by a smooth body and short dorsal and anal fins and native to the rivers of Chile and Argentina.
diplont *Biology.* an organism that is diploid during all of its life except during the gametic stage. All multicellular organisms belong in this category.
diplophage *Virology.* a bacterial virus that infects *Streptococcus pneumoniae.*
diplophase *Biology.* the period of any organism's life cycle during which the nuclei are diploid.
diplopia *Medicine.* a condition in which a single object is perceived as two objects; double vision.
Diplopoda *Invertebrate Zoology.* the millipedes, a class of many-legged terrestrial arthropods.
Diploporita *Paleontology.* a class of middle Paleozoic echinoderms, classified until recently with the rhombiferans in a class Cystoidea; usually pelmatozoan; characterized especially by the paired pores in most thecal plates.
Diplornaviridae *Virology.* a proposed family that would include all dsRNA-containing viruses.
diplosome *Cell Biology.* the two centrioles of a cell.
diplotene *Cell Biology.* one of the five sequential stages of the first meiotic prophase, distinguished by the initiation of desynapsis, or unpairing, of homologous chromosomes.
Diplura *Entomology.* the bristletails, an order of primitive, blind, pale insects that live in soil and feed on decaying plants and plant tissues.
dipmeter *Engineering.* a device used to determine the angle of dip and direction of geologic formations.
dipmeter advisor *Artificial Intelligence.* an expert system used in resource exploration that analyzes information from oil well logs.
dipmeter log *Geology.* a dip log based on readings of the direction and angle of dip, as measured by an instrument taken in a plane perpendicular to the borehole.
dip mold *Engineering.* a one-piece mold with an open top that is used in glassmaking to mold patterns.
dip needle *Engineering.* a device consisting of a magnetic needle on a horizontal pivot that, when set in a magnetic north-south plane, shows the angle of dip.
Dipneumonomorphae *Invertebrate Zoology.* a suborder of spiders that includes the black widows, hunting spiders, and grass spiders.
Dipneusti *Vertebrate Zoology.* in older taxonomic systems, another name for the lungfish order Dipnoi.
Dipnoi *Vertebrate Zoology.* in older taxonomic systems, an order of Choanichthyes including fossil fish and surviving lungfish, characterized by overlapping cycloid scales and dermal fin rays, a largely cartilaginous skeleton with persistent notochord, an autostylic skull, paired fins of the archipterygial type, gills covered by an overculum, and a lung or pair of lungs communicating with the ventral side of the esophagus by a short tube. *Paleontology.* fossil fish of this order, one of the oldest animal groups with living relatives; Dipneusti are represented in strata from the Devonian to the Holocene, although they become less common after the Triassic. Also, DIPTERIFORMES.
Dipodidae *Vertebrate Zoology.* the jerboas, a family of small, herbivorous, nocturnal rodents of the order Myomorpha, characterized by long hind limbs with stiff tufts of hair on the soles and a very long tail; native to semiarid Palearctic regions and northeastern Ethiopia.
dip of the horizon *Navigation.* the vertical angle between the sensible horizon and the visible horizon, measured at the eye of the observer; it is a function of height of eye.
dipolar *Chemistry.* of or relating to a dipole.
dipole *Chemistry.* a molecule containing both positively and negatively charged groups. *Electricity.* a localized positive and negative charge distribution that has no net charge, and whose mean positions of positive and negative charges do not correspond.
dipole anisotropy *Astrophysics.* an anisotropy with two directions of variance orthogonal to each other.
dipole antenna *Electromagnetism.* an antenna approximately one half-wavelength long that is split at the center for connection to a transmission line.
dipole bond *Materials Science.* the bond between two electrically or magnetically charged particles of opposite polarity, or a bond formed by coordination of two neutral halves such as noble gas elements.
dipole disk feed *Electromagnetism.* an antenna that consists of a dipole near a disk, used to reflect energy to the disk.
dipole field *Acoustics.* the sound field radiated by a dipole.
dipole moment *Electricity.* a moment resulting from an electrical dipole that forms in an asymmetrical molecule; one side of the molecule has a net negative charge and the other side of the molecule has a net positive charge.
dipole radiation *Electromagnetism.* the electromagnetic radiation produced by an oscillating electric or magnetic dipole.
dipole relaxation *Electricity.* a process in which the orientation polarization of a substance reaches equilibrium during a specified period after a change in the applied electric field.
dipole transition *Atomic Physics.* a process by which an atom absorbs or produces radiation when it changes from one energy level to another, simultaneously changing its angular momentum by at most one unit.

diporpa larva *Invertebrate Zoology.* a developmental stage in certain trematode flatworms.

dipper *Mechanical Engineering.* **1.** a digging bucket that is rigidly attached to a stick or arm of an excavating machine. **2.** a machine having such a digging bucket. **3.** of, relating to, or resembling such a machine. Thus, **dipper dredge, dipper shovel.** *Vertebrate Zoology.* any of the small, stocky diving birds of the family Cinclidae, having dense plumage and found in rapid streams and rivers.

dipping sonar *Engineering.* a sonar transducer that is dropped into the sea by a helicopter and then recovered when the search is complete.

dip pole see MAGNETIC POLE.

Diprionidae *Invertebrate Zoology.* a family of hymenopteran insects, including the conifer sawflies, in the superfamily Tenthredinoidea.

dipropyl *Organic Chemistry.* a molecule having two propyl groups.

dipropylene glycol *Organic Chemistry.* $(CH_3CHOHCH_2)_2O$, a toxic, combustible, slightly viscous, colorless liquid that is soluble in water; boils at 233°C; used as a solvent.

Diprotodontia *Vertebrate Zoology.* an order of small to large herbivorous terrestrial, arboreal marsupials, including the kangaroos, phalangers, koalas, wombats, and several extinct species; characterized by only one well-developed pair of lower incisors; native to Australia and eastern Indonesia.

Diprotodontidae *Paleontology.* an extinct family of Cenozoic marsupials of the order Diprotodonta.

diproton *Nuclear Physics.* an unstable state, probably hypothetical, consisting of two bound protons.

diprotrizoate *Radiology.* a contrast medium used in radiography of the urinary tract.

Dipsacaceae *Botany.* a family of dicotyledonous Eurasian herbs of the family Dipsacales, commonly producing iridoid compounds and sometimes alkaloids; one species, teasel, has become a widespread weed in North America.

Dipsacales *Botany.* an order of dicotyledonous herbs and shrubs in the subclass Asteridae that usually have opposite leaves, actinomorphic or weakly zygomorphic flowers, and an inferior ovary.

dip separation *Geology.* the distance between two parts of a disrupted bedding plane on either side of a fault surface, measured along the dip of the fault.

dip slip *Geology.* on a fault, the component of displacement that is parallel to the dip. Also, NORMAL DISPLACEMENT.

dip slope *Geology.* an inclined land surface whose slope is determined by and generally conforms to the direction and dip of the underlying rocks. Also, OUTFACE.

dipso- a combining form meaning "thirst," as in *dipsomania.*

Dipsocoridae *Invertebrate Zoology.* a family of true bugs that prey on insects in rotting wood, in the superfamily Dipsocoroidea.

Dipsocoroidea *Invertebrate Zoology.* a superfamily of minute true bugs, found in soil, ant nests, and leaf litter, belonging to the subdivision Geocorisae.

dipsomania *Medicine.* an irrational craving for alcoholic drink. Thus, **dipsomaniac.**

dipsophobia *Psychology.* an irrational fear of alcoholic beverages.

dipstick *Engineering.* a rod used to measure the depth of a liquid contained in a tank or container.

DIP switch *Computer Technology.* one of a set of small switches on integrated circuit boards or peripheral equipment that is used to set up or make adjustments to the interfacing equipment. (An acronym for dual in-line package.)

Diptera *Invertebrate Zoology.* the true flies, an order of insects with sucking mouth parts; nearly all possess a hind pair of wings modified to serve as stabilizers during flight.

dipteral *Architecture.* having a double peristyle.

dipteran *Invertebrate Zoology.* **1.** any member of the order Diptera. **2.** of or relating to this order.

Dipteridaceae *Botany.* a monogeneric family of eastern Asian terrestrial ferns of the order Filicales, characterized by fanlike fronds divided into deep lobes, netted veins, a bristled short-creeping rhizome, and bilateral spores.

Dipteriformes *Vertebrate Zoology.* the lungfishes, an order comprising the subclass Dipnoi in some older taxonomic systems.

Dipterocarpaceae *Botany.* a family of dicotyledonous evergreen trees in the order Theales, characterized by simple, entire, alternate leaves, regular, bisexual flowers in racemes, a superior ovary, and a nut enclosed in an enlarged calyx that forms wings; mostly tall, buttressed, and small-branched trees, dominant in tropical lowlands in Asia.

dipterous *Biology.* having two wings or winglike structures, as in certain insects and some seeds.

Dipterus *Paleontology.* a Devonian genus of bony fish in the order Dipneusti; an ancestor of the modern lungfishes.

dipulse *Electronics.* a type of binary code transmission in which 1 represents the presence of a single cycle of a sine-wave tone, and 0 represents its absence.

dipyramid *Crystallography.* a double pyramidal geometric form in which each form shares a single base.

dipyre or **dipyrite** see SCAPOLITE.

α,α'-dipyridyl *Organic Chemistry.* $(C_5H_4N)_2$, white crystals, slightly soluble in water and soluble in alcohol; melts at 69–70°C and boils at 272–273°C; used as a reagent. Also, 2,2'–BIPYRIDINE.

Dirac, Paul A. M. 1902–1984, English physicist; shared Nobel Prize for developing wave mechanics and predicting existence of positron.

Dirac covariants *Quantum Mechanics.* the components of four complex wave functions introduced in developing a first-order relativistic wave function, related to electron spin and charge.

Dirac electron theory see DIRAC THEORY.

Dirac equation *Quantum Mechanics.* a differential equation for an electron in an electromagnetic field, describing the relativistic behavior of an electron in terms of a wave equation that automatically allows for electron spin.

Dirac fields *Quantum Mechanics.* the operators in the second quantization of Dirac's theory that correspond to wavefunctions in the original theory.

Dirac hole theory *Quantum Mechanics.* the theory, based on the Pauli principle, that fermions have only occupied or unoccupied states, the symmetry of which is demonstrated by creation and annihilation operators that have the effect of holes among unoccupied fermion states.

Dirac matrix *Quantum Mechanics.* any one of four 4×4 Hermitian matrices appearing in the Dirac equation; they are formed from 2×2 Pauli spin matrices and identity matrices.

Dirac moment *Quantum Mechanics.* the magnetic moment predicted for the electron by the Dirac theory equal to *e*{hbar}/2*mc*, where *e* is the fundamental charge, *c* is the velocity of light, and *m* is the strength of a magnetic monopole.

Dirac monopole *Quantum Mechanics.* a magnetic monopole, appearing in the Dirac theory, whose strength is given by *n*{hbar}*c*/2*e*, where *n* is an integer, *c* is the velocity of light, and *e* is the fundamental charge.

Dirac notation *Quantum Mechanics.* the notation introduced by Paul Dirac employing brackets and vertical bars so that a complete bracket expression, <|>, denotes a scalar and any incomplete bracket expression, <| or |>, denotes a vector.

Dirac particle *Particle Physics.* the central core of a hadron of spin 1/2 that remains when the effects of nuclear forces are removed.

Dirac's sea *Quantum Mechanics.* the infinite quantity of electrons that, according to the Dirac hole theory, inhabit the negative energy states in empty space.

Dirac theory *Quantum Mechanics.* Dirac's relativistic theory of the electron that predicts the electron's internal angular momentum, and according to which a positron may be regarded as an empty negative energy state whose usual occupying electron has been removed. Also, DIRAC ELECTRON THEORY.

Dirac wave function *Quantum Mechanics.* a spacetime wave function consisting of a column matrix with four entries, each of which is a function of the space and time coordinates; appropriate for describing a spin 1/2 particle and its antiparticle.

diradical *Materials Science.* a molecular entity containing two unpaired electrons in atomic orbitals on different atoms. Also, BIRADICAL.

direc. prop. with proper direction. (From Latin *directione propria.*)

direct access *Computer Technology.* **1.** the ability to enter or access data directly in any location on a storage device, without reference to previously entered or accessed data. **2.** relating to or employing this capacity to locate data directly. Thus, **direct-access memory, direct-access storage device, direct-access library.**

direct-action fuse see IMPACT-ACTION FUSE.

direct address *Computer Programming.* the addressing mode in which the address portion of an instruction contains the memory location of the operand. Also, REAL ADDRESS, SINGLE-LEVEL ADDRESS.

direct-address processing *Computer Programming.* computer operations in which the operand is located at the address contained in the instruction.

direct agglutination *Immunology.* the clumping together of red cells, microorganisms, or other substances directly by a serum antibody.

direct air cycle *Aviation.* a thermodynamic propulsion cycle in which compressed air is heated in a combustion chamber and expelled through a gas turbine or ramjet to produce thrust.
direct-aperture antenna *Electromagnetism.* an antenna that has a surface or a solid for its radiating element.
direct-arc furnace *Engineering.* a furnace in which a material is rapidly heated by an electric arc running directly from electrodes to the material itself.
direct-band-gap semiconductor *Solid-State Physics.* a semiconductor in which the location of the minimum in the conduction band coincides with that of the maximum in the valence band in reciprocal space, so that radiation due to recombination of carriers occurs at high intensity. Also, DIRECT-GAP SEMICONDUCTOR.
direct-broadcasting satellite system *Telecommunications.* the broadcasting of radio and television transmissions via a geostationary satellite, directly to end-user antennas.
direct calorimetry *Nutrition.* the process of measuring heat production by immersing the body in a tank of water and measuring the temperature change in the water that is caused by body heat.
direct cell *Meteorology.* a closed thermal circulation in a vertical plane in which the rising motion occurs at a higher potential temperature than the sinking motion.
direct-chill casting *Metallurgy.* a semicontinuous casting process utilizing a short mold and water to cool the exiting casting.
direct code see ABSOLUTE CODE.
direct competition *Ecology.* the driving away of competitors from needed resources through direct aggressive behavior or the use of toxins. Also, INTERFERENCE COMPETITION, SCRAMBLE COMPETITION.
direct control *Computer Technology.* the control of one computer or peripheral device by another, with no operator interaction.
direct-coupled amplifier *Electronics.* an amplifier configuration in which the output of one stage is connected to the input of the next stage, either directly or through a resistance; used for amplifying DC voltages and slowly varying signals.
direct-coupled transistor logic *Electronics.* a logic circuit in which transistors are connected without coupling components such as capacitors or resistors.
direct coupling *Mechanical Engineering.* the direct attachment of the shaft of a motor or other driving mechanism to the shaft of a rotating mechanism. *Electricity.* the direct connection of one circuit point to another so that both direct current and alternating current can flow through the coupling path.
direct current *Electricity.* a current that flows in only one direction and has an essentially constant average value.
direct-current *Electricity.* relating to or using direct current.
direct-current amplifier *Electronics.* an amplifier circuit that can boost DC voltages. Also, DC AMPLIFIER.
direct-current component *Telecommunications.* the average luminance of a transmitted television picture.
direct-current coupling *Electronics.* a coupling used in amplifiers in which the output of one stage is connected to the input of the next stage, either directly or through a resistance; used for amplifying DC voltages and slowly varying signals.
direct-current discharge *Electronics.* the flow of direct current through a gas, which ionizes the gas and causes luminescence.
direct-current dump *Electricity.* the process of removing all DC power from a computer system or component, resulting in loss of information stored in volatile memory. Also, DC DUMP.
direct-current electrode negative *Metallurgy.* of or relating to arc welding in which the electrode is negative.
direct-current electrode positive *Metallurgy.* of or relating to arc welding in which the electrode is positive.
direct-current generator *Electricity.* a rotating electric generator that transforms mechanical power into DC power.
direct-current grid bias see GRID BIAS.
direct-current inserter *Electronics.* a circuit in a television transmitting system used for inserting a DC component into the composite video signal to be transmitted.
direct-current offset *Electronics.* a direct-current biasing level added to a signal at the input of a circuit, usually to cancel a DC component added by the circuit.
direct-current picture transmission see DIRECT-CURRENT TRANSMISSION.
direct-current power *Electricity.* the power in a DC circuit equal to voltage multiplied by the current.
direct-current power supply *Electricity.* a power unit supplying one or more DC voltages, such as a battery, transformer and rectifier, DC generator, converter, dynamotor, or photovoltaic cell.
direct-current restorer *Electronics.* a circuit that adds a fixed voltage component to a waveform in order to restore a specified reference level for the waveform; usually used to restore a DC component lost by capacitive coupling between stages.
direct-current telegraphy *Telecommunications.* a telegraph system in which an electric direct current flows during mark intervals, and no current flows during space intervals.
direct-current transducer *Electronics.* a transducer that requires a DC power source and generates a DC output signal in response to an input stimulus.
direct-current transmission *Telecommunications.* in a television picture, a signal containing a DC component that indicates the average luminance of that picture. Also, DIRECT-CURRENT PICTURE TRANSMISSION.
direct-current voltage see DIRECT VOLTAGE.
direct-current working volts *Electricity.* the maximum rated value, expressed in volts, at which a component may be operated continuously with safety and reliability.
direct-cycle reactor *Nucleonics.* a nuclear reactor design in which coolant is conducted directly to the turbines to generate power, thus eliminating separate heat transfer loops.
direct data entry *Computer Technology.* the input of data directly into a storage device from machine-readable source documents or by means of an on-line terminal, such as a key-to-disk device.
direct digital control *Control Systems.* process control that uses a digital computer to directly control petroleum, chemical, or other industrial processes.
direct distance dialing *Telecommunications.* a telephone service that enables a user to dial directly the telephones that are outside the local area without the assistance of an operator.
direct drive *Mechanical Engineering.* an arrangement in which the driving mechanism is connected directly to the driven mechanism.
direct-drive arm *Robotics.* a robot arm with joints that are linked directly to a high-torque motor.
directed angle *Mathematics.* an angle measured from one side (the initial side) to the other (the terminal side); often designating rotation in a positively or negatively defined direction; for example, $-45°$ or $+20°$.
directed distance *Mathematics.* if points A and B lie on a directed line, the directed distance AB is plus or minus the distance AB, according as the direction from A to B coincides with or is opposite that of the directed line.
directed graph *Mathematics.* a graph having a direction assigned to each edge; equivalently, a graph having an ordered pair of vertices identified with each edge. The first element of each such ordered pair is called the initial vertex of the corresponding edge, and the second element is the terminal vertex. A path in a directed graph is called a **directed path** or **dipath** if all of the edges in the path are directed in the same direction as the path. If the graph is a tree, it is called a **directed tree** or **ditree.**
directed line *Mathematics.* a line on which the positive direction is indicated.
directed number *Mathematics.* a term for a signed number.
directed set *Mathematics.* a partially ordered set D with ordering $\leq$, having the additional property that for every pair of elements a and b in S, there exists an element c of S such that $a \leq c$ and $b \leq c$. That is, every pair of elements in D has an upper bound. Also, **directed system.**
direct effect *Physical Chemistry.* a chemical reaction that occurs when ionizing radiation strikes an atom or a molecule in a material. *Behavior.* any response that is an immediate consequence of a stimulus.
direct energy conversion *Engineering.* the process of converting thermal or chemical energy into electrical energy through direct-power generators.
direct-entry terminal *Computer Technology.* a data input device connected to a computer for on-line data entry; the computer edits data as it is received and communicates with the operator, providing data validation and error-checking logic.
direct expansion *Mechanical Engineering.* in a refrigeration system, an arrangement in which the refrigerant expands in an evaporator in the airstream.
direct-expansion coil *Mechanical Engineering.* a finned refrigeration coil within which a cold fluid or evaporating refrigerant circulates.
direct field *Acoustics.* a field of sound waves that are not reflected but rather arrive at the receiver directly from the source.

direct fire *Ordnance.* fire delivered on a target, using the target itself as an aiming point. Similarly, **direct bombing, direct pointing.**

direct-fire *Engineering.* to light a furnace without first heating the air or gas.

direct-fired evaporator *Chemical Engineering.* an evaporator with a metal wall or other heating surface to separate the flame and combustion gases from the boiling liquid.

direct flight *Transportation Engineering.* a flight in which a passenger does not change planes en route, but is required to make intermediate stops before reaching the final destination.

direct-gap semiconductor see DIRECT-BAND-GAP SEMICONDUCTOR.

direct-geared *Mechanical Engineering.* describing two mechanisms that are connected in such a manner that a gear on a shaft of one mechanism meshes with a gear on a shaft of the other mechanism.

direct halftone *Graphic Arts.* a halftone negative made by photographing copy through a halftone screen rather than rephotographing a photographic negative or positive.

direct hierarchy control *Computer Technology.* **1.** the automatic manipulation and movement of data among computer memory levels, including main memory and bulk storage. **2.** in distributed systems, the control of lower level computers by supervising computers.

direct hit *Mechanics.* see HEAD-ON COLLISION.

direct-image plate *Graphic Arts.* a paper offset plate on which characters and line art are directly typed or drawn with special ribbons, pencils, and inks; a quick, inexpensive platemaking method used for short runs in which high-quality images are not needed.

direct-imaging mass analyzer *Engineering.* a secondary ion mass spectrometer in which an electrostatic immersion lens forms an image that relates the ion's place of origin on the sample surface, after which the ion crosses magnetic sectors to cause mass separation.

direct immunofluorescence *Immunology.* a process in which an unlabeled reactant (antigen) is detected with the use of a labeled reactant (antibody).

direct input/output *Computer Technology.* the transfer of data between computer main memory and input/outoput devices without the use of buffers.

direct-insert subroutine *Computer Programming.* a set of instructions that are copied into a program at each point it is to be executed, as opposed to transferring control to a single stored copy of the subroutine.

direct instruction *Computer Programming.* an instruction whose address field contains the address of the operand.

direct inward dialing *Telecommunications.* in a private branch exchange (PBX) or in central office switched business group, a feature that allows callers to dial directly from a general telephone network straight to a PBX extension without operator assistance.

direction *Science.* the position of one point in relation to another without reference to the distance between them. *Geology.* see TREND.

directional antenna *Electromagnetism.* an antenna whose characteristic pattern is more efficient in certain directions than in others.

directional beam *Electromagnetism.* an antenna signal that is concentrated in a given direction.

directional control *Engineering.* the technique of controlling motion around the vertical axis.

directional counter *Nucleonics.* a radiation monitor shielded in a manner that determines and indicates the direction of incident radiation.

directional coupler *Electronics.* a device that is used to couple radio-frequency energy to or from a wave that is traveling only in a specified direction in a transmission line or waveguide; it does not interact with waves traveling in the opposite direction. Also, DIRECTIVE FEED.

directional derivative *Mathematics.* the rate of change of a function with respect to distance in a given direction or along a specified curve. In particular, let f be a real-valued function, defined and continuous in a neighborhood of a point x_0 in R^n and let β be any point in R^n (i.e., a *direction*). Then the directional derivative of f at x_0 in the direction of β is

$$(D_\beta f)(x_0) = \lim_{t \to 0} (1/t)[f(x_0 + t\beta) - f(x_0)].$$

directional filter *Electronics.* a low-pass, bandpass, or high-pass dual-filter that separates the bands of frequencies used for transmission in opposite directions in a carrier system.

directional gyro *Mechanical Engineering.* a gyroscope on board a freely moving object that provides a reference for orienting the object with an inertial coordinate system. Also, DIRECTION INDICATOR.

directional homing *Navigation.* the act of flying or sailing a course that points the craft directly at a radio beacon (and thus at a bearing of 000 degrees relative).

directional microphone *Acoustical Engineering.* a microphone, such as a bidirectional microphone, with a well-defined directional response pattern, as opposed to one that responds equally well in all directions. Similarly, **directional hydrophone.**

directional property *Metallurgy.* any property of a metal or alloy that varies when measured in different directions.

directional relay *Electricity.* a relay that responds to the relative phase position of the current flowing through it.

directional selection *Evolution.* a natural selection process in which a single genetic variation is selected for, resulting in a unidirectional change in the overall genetic makeup of a population.

directional solidification *Metallurgy.* a process by which the solidification of a molten metal or alloy proceeds in a predetermined direction.

directional split *Transportation Engineering.* the volume of traffic in each direction on a two-way route.

directional stability *Aviation.* the ability of a flight vehicle to recover from a yawing or sideslipping condition. Also, WEATHERCOCK STABILITY.

directional structure *Geology.* any sedimentary structure, such as cross-bedding or ripple marks, that indicates the direction of the current that produced it. Also, VECTORIAL STRUCTURE.

directional well *Petroleum Engineering.* a well that is intentionally drilled to deviate from the vertical by as much as 70°, in order to bypass obstacles over the reservoir. Thus, **directional drilling.**

direction angles *Mathematics.* the angles between a line in space and the axes of a Cartesian coordinate system.

direction cosine *Mathematics.* the cosine of the angle between a vector and a coordinate axis. *Engineering.* specifically, in radar tracking, the cosine of the angle between a baseline and the line linking it to the center of the baseline target.

direction finder *Navigation.* a radio aid to navigation that uses a highly directional receiver to determine the source of an incoming signal (and thus the location of the transmitting station). Thus, **direction finding.**

direction-independent radar *Engineering.* radar that differentiates between fixed and moving targets with reflected electromagnetic waves; used as a sentry device.

direction indicator see DIRECTIONAL GYRO.

direction instrument theodolite *Photogrammetry.* a theodolite with a graduated horizontal circle that remains fixed during a series of observations, in which the telescope is pointed on a number of objects in succession, and the direction of each is read on the circle.

direction of propagation *Physics.* the direction of the time-averaged flow of energy; usually applied to the propagation of waves.

direction of relative movement *Navigation.* the direction of motion of an observed craft relative to the observing craft. As observed by a craft moving directly north, for example, a stationary object has a relative movement directly to the south. Also, **direction of relative motion.**

direction of tilt *Cartography.* the direction (azimuth) of the principal plane of a photograph, or of the principal line on a photograph.

direction rectifier *Electronics.* a rectifier circuit that converts a synchro system AC error voltage to a DC voltage; the polarity of the voltage represents the direction of the angular error and the amplitude of the voltage represents the magnitude of the error.

directive *Computer Programming.* an instruction in a source program that provides information to the compiler or assembler to control the translation process but is not translated into machine instructions.

directive coloration *Behavior.* surface markings that divert the attention of a predator to nonvital parts of the prey's body.

directive feed see DIRECTIONAL COUPLER.

directive gain *Electromagnetism.* a measure of the directional properties of an antenna, given by 4π times the radiation intensity at a given direction divided by the total power radiated by the antenna.

directive species *Behavior.* a species that is not itself prey but that attracts the attention of a predator to prey species in the area.

directive therapy *Psychology.* a treatment method in which the therapist plays an active and direct role, as by giving advice and recommending certain actions.

directivity *Electronics.* a property that causes some antennas to receive or radiate more radio-frequency energy in some directions than in others. *Electromagnetism.* **1.** the value of the directive gain of an antenna in its maximum-gain direction. **2.** the ratio of the power at the forward-wave sampling terminals of a directional coupler to the power measured at the same terminals when the direction of the forward wave is reversed.

directivity factor *Acoustics.* a factor Q that is the ratio of the square of the sound pressure, at some fixed distance and in a specified direction, to the mean squared pressure at the same distance in all directions.

directivity index *Acoustics.* the on-axis power expressed in decibels of an acoustic signal transmitted by a transducer, as given by the formula DI = $10\log Q$, where Q is the directivity factor. Also, **directivity gain.**

directivity pattern *Acoustical Engineering.* a directional sound response profile made by a transducer at a specified frequency and in a specified plane, such as the heart-shaped pattern created by a cardioid microphone.

direct keying device *Computer Technology.* computer input equipment, connected to the computer by cable or transmission lines, that enables direct data entry by means of an operator keyboard.

direct labor *Industrial Engineering.* **1.** labor that is directly required for production and contributes to individual direct unit cost. **2.** labor that advances a product toward its completion.

direct leveling *Cartography.* the determination of differences of elevation by means of a continuous series of short horizontal lines so that vertical distances from these lines to adjacent ground marks are determined by direct observation.

direct limit *Mathematics.* let $\{\boldsymbol{A}_i\}$ be a family of objects in a category A, with the property that whenever $i \leq j \leq k$, there are morphisms μ_{ij}: $\boldsymbol{A}_i \to \boldsymbol{A}_j$ such that: (a) $\mu_{jk} \circ \mu_{ij} = \mu_{ik}$, and (b) μ_{ii} is the identity morphism. The direct limit of the family $\{\boldsymbol{A}_i\}$ is a universally repelling object in the category C whose objects are the pairs $(\boldsymbol{A}, (\mu_i))$, where $\boldsymbol{A}$ is any object of A and (μ_i) is a family of morphisms μ_i: $\boldsymbol{A}_i \to \boldsymbol{A}$ such that for $i \leq j$, the following diagram commutes:

$$\begin{array}{ccccc} & & \mu_{ij} & & \\ \cdots \to & \boldsymbol{A}_i & \to & \boldsymbol{A}_j & \to \cdots \\ & \mu_i \downarrow & & \downarrow \mu_j & \\ & & \boldsymbol{A} & & \end{array}$$

direct lithography *Graphic Arts.* any method of lithographic printing by direct transfer of ink from the plate to the paper, as with a direct rotary press.

direct manipulation *Computer Technology.* the movement and altering of data by selecting and moving the displayed data with a device such as a mouse, as opposed to entering screen coordinates.

direct-memory access *Computer Technology.* an interface that provides I/O transfer of data directly to and from the memory unit and peripherals, freeing the central processing unit for other operations. Thus, **direct-memory access channel.**

direct motion *Astronomy.* **1.** an apparent eastward motion by an object in the solar system. **2.** the counterclockwise orbital motion of an object in the solar system as seen from the northern ecliptic pole. **3.** the counterclockwise rotation of a planet or moon as seen from the northern ecliptic pole.

direct nuclear reaction *Nuclear Physics.* the collision of two nuclei that takes place in the time needed for the incident particle to transverse the target nucleus and, as a result, does not mix with the nucleus as a whole but interacts only at the surface of the target nucleus.

direct numerical control *Computer Technology.* the industrial application of computers to monitor, control, and service machines; direct numerical control of machine tools by computer, as opposed to recorded control commands.

director a person or device that guides or directs; specific uses include: *Electromagnetism.* a parasitic element that is installed a fraction of a wavelength in front of a dipole receiving antenna to increase the gain of array in the direction of the major lobe. *Surgery.* a grooved probe used to guide the direction and depth of the knife during a surgical procedure. *Ordnance.* an artillery-control device that processes data from a rangefinder or radar tracks a moving target, and then transmits firing data to the gun. Also, ARTILLERY DIRECTOR.

director-sight system *Ordnance.* a system used on airborne flexible guns, in which the gunner controls the gun's aim, but the fire is controlled by a computer that computes the angle at which the gun must be fired in order to hit the target. Also, **director-type computer, director-type fire control.**

directory *Computer Programming.* **1.** a list of all data set labels within a unit of physical storage; a table of contents. **2.** the partitioning of a physical storage unit into separate files for ease in locating a specific file on a high-capacity storage medium. **3.** a collection of files and subdirectories.

direct outward dialing *Telecommunications.* a telephone system service in which a user in a private branch exchange or business group has access to the exchange network without the help of an operator.

direct phase determination *Crystallography.* a method of deriving relative phases of diffracted beams by consideration of relationships among the indices and among the structure factor amplitudes of the stronger reflections. These relationships come from the conditions that the structure is composed of atoms and that the electron density must be positive or zero everywhere. Only certain values for the phases of the Bragg reflections are consistent with these conditions. Also, **direct method.**

direct photography *Graphic Arts.* reproduction photography using direct halftones.

direct pickup *Telecommunications.* any transmission of television pictures that has not been prerecorded by photographic or magnetic means; "live" television.

direct piezoelectricity *Solid-State Physics.* the generation of electric charges on the surfaces of a particular class of asymmetric crystals subjected to certain mechanical stresses.

direct positive *Graphic Arts.* a photographic positive made without a negative by exposing photosensitive material to light.

direct-power generator *Engineering.* any device that converts thermal or chemical energy into electricity through more direct means than the conventional thermal cycle.

direct printing *Textiles.* a method of dyeing in which fabric is passed through sets of rollers, each set applying a different color or part of a pattern.

direct product *Mathematics.* let $A = \{\boldsymbol{S}_\lambda\}$ be a family of sets indexed by some set Λ. The direct product $X_\lambda S_\lambda$ is the collection of all sets $Z = \{x_\lambda\}$ with $x_\lambda \in \boldsymbol{S}_\lambda$ for each λ. x_λ is the λth coordinate or component of Z. If Λ is a finite set, then the direct product is called the Cartesian product of the sets $\boldsymbol{S}_\lambda$. If the $\boldsymbol{S}_\lambda$ are objects such as groups, rings, etc., then component-wise binary operation(s) may be defined on the elements of the direct product so that the direct product also has group (ring, etc.) structure. The direct product together with its binary operation(s) is called a complete direct sum and denoted $\Pi_\lambda \boldsymbol{S}_\lambda$. The direct product of a finite number of algebraic objects is called a **finite direct product.**

direct proof *Mathematics.* a method of proving assertion in which the desired conclusion is deduced as a logical consequence from the hypothesis and other known or given statements.

direct proportion *Mathematics.* the relationship between two variables whose ratio remains constant; e.g., $y = kx$ where k is the constant of proportionality. Also, DIRECT VARIATION.

direct quenching *Metallurgy.* **1.** the process of quenching a steel component immediately after carburizing. **2.** the process of quenching an iron casting immediately after malleablizing.

direct radial triangulation *Photogrammetry.* a graphic radial triangulation made by tracing the directions from successive radial centers directly onto a transparent plotting sheet, rather than by determining the triangulation by the template method. Also, **direct radial plot, direct radial aerotriangulation.**

direct radiator (speaker) *Acoustical Engineering.* a speaker whose radiator element directly affects the air, rather than affecting an intermediate element such as a horn.

direct rays *Radiology.* the rays emanating directly from a source without scattering or interacting with matter.

direct reading see RIGHT READING.

direct-reading gauge *Engineering.* a gauge whose pointer is actuated by direct linkage with a measuring device.

direct recording *Acoustical Engineering.* **1.** a term for any recording made from the original sound source. **2.** a recording technique in which the input signal is applied directly to the recording head with high bias and little other change.

direct-reduction process *Metallurgy.* a process that extracts iron in the solid state directly from an ore. The resulting product is called **DRI,** or **direct-reduce iron.**

direct repeat *Molecular Biology.* two or more identical or closely related DNA sequences occurring in the same orientation in the same molecule. Also, **direct repeat sequence.**

directrix *Mathematics.* the fixed line that, together with a fixed point (the focus), defines a given conic; the (constant) ratio of the distance from a point P of the conic to the focus and the distance from P to the directrix is the eccentricity.

direct runoff *Hydrology.* precipitation runoff that reaches stream channels immediately after rainfall or snowmelt.

direct scanning *Telecommunications.* in a facsimile transmitting apparatus, the scanning of the plane of the message surface along lines running from right to left commencing at the top.

direct separation *Graphic Arts.* a separation negative that is made by photographing copy through a halftone screen. Also, **direct-screen halftone.**

direct stroke *Electricity.* a lightning stroke that strikes a power or communications system.

direct sum *Mathematics.* a direct product of Abelian additive groups in which each element has all but a finite number of components equal to the additive identity; i.e., the weak direct product of additive Abelian groups. If the direct sum involves only a finite number n of components S_λ, it is often denoted $S_1 \oplus S_2 \oplus \cdots \oplus S_n$ and called a **finite direct sum.**

direct support *Military Science.* a mission requiring a force to support another force and authorizing the supporting force to directly answer the supported force's request for assistance. Thus, **direct support artillery, direct support(ing) fire.**

direct terminal repeats *Molecular Biology.* a duplication of a sequence in the same orientation at either end of a DNA molecule.

direct tide *Geophysics.* a gravitational tide (solar or lunar) of the oceans or the atmosphere that is in phase with the apparent motion of the tide-producing force.

direct transfusion see IMMEDIATE TRANSFUSION.

direct variation see DIRECT PROPORTION.

direct viewfinder *Optics.* a viewfinder in which an object is viewed directly through a lens or sight, as opposed to one in which the image is reflected.

direct-view storage tube *Electronics.* a cathode-ray tube, including one or more writing guns, a flooding gun, and selective erasing capability, in which the emission of electrons from a storage grid provides a bright display for long and controllable periods of time; used in radar and weapons-control displays on aircraft instrument panels.

direct-vision nephoscope *Optics.* an instrument into which the viewer looks directly in order to observe the movement of clouds.

direct-vision prism see AMICI PRISM.

direct-vision spectroscope *Spectroscopy.* a spectroscope using a compound prism or system of prisms arranged so that the emergent rays follow the direction of the incident rays, allowing the observer to look in the direction of the radiation source.

direct voltage *Electricity.* a voltage that does not change in polarity, such as is provided by a battery or generator; direct-current voltage.

direct wave *Telecommunications.* a wave that travels directly between its source and sink without reflecting from any object.

direct-wire circuit *Electricity.* a supervised protective communications or control line of wires connecting a transmitter and a receiver, without intermediary equipment, such as a switchboard.

Dirichlet conditions [dē ri klā′] *Mathematics.* sufficient conditions for the pointwise convergence of a Fourier series for a given function $f(x)$ defined on the interval $(-\pi, \pi)$; that is: (a) the interval $(-\pi, \pi)$ can be divided into finitely many subintervals on which $f(x)$ is continuous and monotonic, and (b) if c is a point of discontinuity of f, then except possibly for the point c, $f'(x)$ exists in some neighborhood of c.

Dirichlet function *Mathematics.* the real function f defined by $f(x) = 1$ if x is rational, and $f(x) = 0$ if x is irrational.

Dirichlet-Jordan conditions *Mathematics.* sufficient conditions for the convergence of the Fourier integral for a given real function $f(x)$; that is: (a) on any finite interval, $f(x)$ is of bounded variation and is continuous except for finitely many jump discontinuities, and (b) $f(x)$ is absolutely integrable.

Dirichlet kernel *Mathematics.* in Fourier analysis, a kernel $D_n(t)$ that, when convolved with a function $f(x)$, gives the nth partial sum $s_nf(x)$ of the Fourier series for f; i.e.,

$$s_nf(x) = (2\pi)^{-1}\int_{-\pi}^{\pi} f(x-t)D_n(t)dt,$$

where $D_n(t) = \sum_{k=-n}^{n} e^{ikt}$. It can be shown that $D_n(t) = 2n + 1$ if $e^{it} = 1$ and $D_n(t) = \sin[(n + 1/2)t]/\sin(t/2)$ otherwise, i.e., if $e^{it} \neq 1$.

Dirichlet problem *Mathematics.* **1.** the problem of determining all regions G such that for any continuous real-valued function f defined on the boundary ∂G, there exists a continuous real-valued function u defined on the closure of G that is harmonic in G and agrees with f on ∂G. **2.** equivalently, the problem of determining all regions G such that Laplace's equation is solvable in G with arbitrary boundary conditions.

Dirichlet region *Mathematics.* a region G for which the Dirichlet problem can be solved. For example, a disk is a Dirichlet region, but a punctured disk is not.

Dirichlet series *Mathematics.* any series of the form $\sum_{n=1}^{\infty} a_n n^{-s}$, where s and the coefficients a_n are real or complex numbers; used in analytic number theory; the Riemann zeta function is an important example.

Dirichlet test for convergence *Mathematics.* **1.** let $\{a_n\}$ and $\{b_n\}$ be two sequences satisfying the following conditions: (a) $\lim_{n\to\infty} a_n = 0$; (b) $\sum_{n=1}^{\infty}|a_{n+1} - a_n|$ converges; and (c) the partial sums of the series $\sum_{n=1}^{\infty} b_n$ are bounded. Then the series $\sum_{n=1}^{\infty} a_n b_n$ converges. Also, **Dirichlet test for series, Dedekind test.** **2.** let f, g, and g' be continuous real-valued functions for $x \geq c$, where g' denotes the derivative of g. Suppose that the following conditions are then satisfied: (a) $\lim_{x\to\infty} g(x) = 0$; (b) $\int_c^\infty g'(x)dx$ is absolutely convergent; and (c) $\int_c^r f(x)dx$ is bounded for $c \leq r < \infty$. Then the integral $\int_c^\infty f(x)g(x)dx$ is convergent. Also, **Dirichlet test for integrals.**

dirigible [dēr′ə jə bəl] *Aviation.* a lighter-than-air craft that is capable of being propelled and steered for controlled flight. (Going back to a Latin word meaning "to steer.")

dir. prop. with proper direction. (From Latin *directione propria.*)

dirt band *Hydrology.* **1.** in a glacier, a dark layer of debris that represents the trace of a former surface. **2.** a dark band below an ice wall. *Geology.* a thin layer of shale to soft, earthy material within a coal seam. Also, DIRT BED, DIRT PARTING.

dirt bed *Geology.* **1.** see DIRT BAND. **2.** a buried soil horizon of the geologic past that consists of partially decayed organic matter.

dirtman see GROUNDMAN.

dirt parting *Geology.* see DIRT BAND.

dirt slip see CLAY VEIN.

dirty *Computer Technology.* an informal term for a page of memory in which some value has been changed since it was read from disk.

dirty bit *Computer Technology.* in a virtual memory system, a bit that is set when a value is stored into a memory page, indicating that the page will have to be rewritten to disk before its main memory area can be reused.

dirty bomb *Ordnance.* a nuclear bomb that emits a large quantity of long-term radioactive fallout.

dirty ice *Astronomy.* grains of ice in interstellar space containing surface impurities, such as carbon and silicates.

dirty ship *Naval Architecture.* a term for an oil tanker designed to carry crude or heavy oil, as opposed to refined petroleum products.

dirty snowball *Astronomy.* a proposed model for the structure of a comet, in which the comet consists of various ices such as water or methane ("snow") mixed with dark silicate material ("dirt").

dis- a prefix meaning: **1.** apart, away; reversal, as in *discharge, disinfectant.* (From Latin *dis,* "apart.") **2.** duplicate; double, as in *dismutation.* (From Greek *dis,* "twice.")

DIS disseminated intravasculate coagulation.

dis. discharge; distance.

disable see DISARM.

disaccharide *Biochemistry.* a sugar made up of two linked monosaccharide molecules, such as sucrose, which consists of one glucose molecule and one fructose molecule.

disagreement set *Artificial Intelligence.* in unification, the set of terms in a particular argument position that have different values.

disambiguate *Computer Science.* to resolve a situation in which there are multiple possibilities, often by using an additional kind of knowledge to eliminate some possibilities. Also, RESOLVE AMBIGUITY.

disambiguating rules *Computer Science.* rules used to make decisions that would otherwise be ambiguous, e.g., rules to resolve syntactic ambiguity in a natural language sentence by applying semantic constraints.

disappearing carriage *Ordnance.* a flexible mount formerly used on heavy guns in seacoast defense; a balancing mechanism raised the gun above its protective parapet for firing and then lowered it behind the parapet for reloading.

disappearing stream see SINKING STREAM.

disappearing target *Ordnance.* a target that moves in and out of view within a short period of time.

disarm *Ordnance.* to remove the fuse or detonating device of a bomb, mine, or other explosive, thus preventing it from exploding in its usual manner. Also, DEARM, DISABLE.

disassemble *Engineering.* to break a system down into its constituent elements or parts. *Computer Programming.* to convert a machine language program into equivalent symbolic assembly language instructions.

disassembler *Computer Programming.* a program used to translate machine code into assembly language instructions.

disassortative mating *Genetics.* a system of mating in a variable population in which individuals of different phenotypes mate preferentially over individuals of the same phenotype.

disaster dump *Computer Programming.* a listing of the contents of computer storage and registers that is produced when an unrecoverable error is encountered.

Disasteridae *Paleontology.* a family of irregular burrowing euechinoids in the extinct order Disasteroida, one of the three orders of heart-shaped urchins without a lantern.

disc *Botany.* **1.** a round, fleshy, flat part growing from the receptacle of a flower, surrounding or topping the ovary and often containing nectar. **2.** the center part of a capitulum. *Science.* see DISK.

disc- a combining form meaning "relating to or resembling a disk."

discarding petal *Ordnance.* a piece of a sabot, the lightweight casing of a subcaliber projectile; the petals are peeled back by the force of the projectile and discarded in front of the muzzle.

discard rate *Archaeology.* the typical rate at which a group or society disposes of its unwanted objects; plentiful, easily replaced items tend to have a higher discard rate than rare or highly durable ones.

disc brake see DISK BRAKE.

disc diffusion test *Microbiology.* an antibiotic sensitivity assay in which a petri plate inoculated with a microbial lawn is incubated with small antibiotic-containing discs and examined for zones of growth inhibition around the antibiotics, in order to measure the sensitivity or insensitivity of the cells to antibiotics.

Disceliaceae *Botany.* a monotypic family of tiny mosses of the order Funariales, characterized by a persistent green protonema, small gametophores with long terminal sporophytes, and a single weak costa; native to temperate portions of the Northern Hemisphere.

Discellaceae *Mycology.* a family of fungi belonging to the order Sphaeropsidales; some of its species live on nonliving organic matter, while others are pathogens of plants.

disc flower *Botany.* the tubular, actinomorphic flower in the center of the inflorescence of Asteraceae. Also, DISK FLOWER.

discharge *Fluid Mechanics.* the rate at which a fluid flows out through an opening. *Electronics.* the flow of electrical current through a gas, usually causing luminescence of the gas. *Electricity.* **1.** the release of stored-up electricity from a source, such as a battery or capacitor. **2.** the conversion of chemical energy to electric energy in a battery.

discharge coefficient *Fluid Mechanics.* an empirical factor used to adjust the pressure loss-velocity of flow equation for orifices of various sizes and shapes to experimental data.

discharge efficiency see DRAINAGE RATIO.

discharge key *Electricity.* a switch for changing a capacitor suddenly from a charging circuit to a load through which it can discharge. Also, **discharge switch.**

discharge lamp *Electronics.* a lamp that produces light when the gas is ionized by an electric current passing through it at low or high pressure; e.g., fluorescent lamps, neon tubing. Also, GAS DISCHARGE LAMP.

discharge liquor *Chemical Engineering.* **1.** any liquid that has gone through a processing operation. **2.** specifically, waste liquid from a manufacturing plant.

discharge potential see DEPOSITION POTENTIAL.

discharge printing *Textiles.* the process of removing dye from a fabric with bleaching chemicals to create a pattern. Also, EXTRACT PRINTING.

discharger *Electricity.* **1.** a short-circuiting device for discharging capacitors. **2.** a device used on aircraft to reduce precipitation static.

discharge-rating curve see STAGE-DISCHARGE CURVE.

discharge tube *Mechanical Engineering.* a tube through which steam and water are released into a boiler drum. *Electronics.* a tube consisting of a glass envelope containing a low- or high-pressure gas or vapor and two metal electrodes; the gas is ionized by an electric current flowing through it when adequate voltage is applied between the electrodes.

discifloral *Botany.* having flowers in which the receptacle is enlarged and disclike.

disciform *Biology.* disk-shaped.

Discinacea *Invertebrate Zoology.* a superfamily of brachiopods whose shells are held together only by muscles and soft tissues.

discission *Surgery.* **1.** the making of an incision or cutting through some part. **2.** in cataract surgery, the process of cutting into the capsule to break up the substance of the crystalline lens.

disclimax *Ecology.* an ecological succession that is maintained below climax for an extended time, because of human or animal disturbance or because of environmental effects such as fire or climatic change. Also, DISTURBANCE CLIMAS.

disco- a combining form meaning "relating to or resembling a disk," as in *discodactylous.*

discoaster *Botany.* a star-shaped coccolith.

discoblastula *Developmental Biology.* the blastula that develops from the cleavage of a fertilized meroblastic egg, and consisting of a cellular cap separated by the blastocoel from a floor of uncleaved yolk.

discocephalous *Invertebrate Zoology.* possessing a sucker on the head.

discodactylous *Vertebrate Zoology.* having sucking disks on the toes, as in tree frogs.

discogastrula *Developmental Biology.* a modified, flattened gastrula formed from a blastoderm.

Discoglossidae *Vertebrate Zoology.* the fire-bellied toads, a family of Old World amphibians of the order Anura, including small toadlike species and large aquatic frogs; characterized by a fixed disklike tongue, and found in Europe and Asia.

discoid *Biology.* resembling a disk.

discoidal cleavage *Developmental Biology.* the cleavage that is limited to the animal pole of highly telolecithal eggs.

Discolichenes *Botany.* an order of Ascolichenes, characterized by an open discoid apothecium with a typical hymenium and hypothecium.

Discolomidae *Invertebrate Zoology.* a family of tropical coleopteran insects, the log beetles, in the superfamily Cucujoidea.

discomfort index see TEMPERATURE-HUMIDITY INDEX.

discomposition *Nucleonics.* the removal of an atom from its normal position in a crystal lattice on direct impact by a fast neutron or by a fast ion that has been previously knocked out of its lattice position.

discomposition effect *Nucleonics.* the cumulative changes in the physical and chemical properties of matter caused by discomposition.

Discomycetes *Mycology.* a class of fungi in the subdivision Ascomycotina, formerly known as Cup Fungi, so named for the cup- or disklike shape of their fruiting bodies.

discone antenna *Electromagnetism.* a broadband antenna in which the center conductor of a coaxial line terminates at a disk perpendicular to the conductor, and the outer conductor terminates at the vertex of a cone whose axis is also perpendicular to the disk.

disconformity *Geology.* an unconformity between essentially parallel beds or strata that indicates a significant interruption in the orderly sequence of deposition. Also, PARALLEL UNCONFORMITY, EROSIONAL UNCONFORMITY.

disconnect *Electricity.* to separate leads or connections to open a circuit.

disconnected set *Mathematics.* any subset of a topological space that can be expressed as the union of two nonempty disjoint open sets. A topological space that is disconnected as a subset of itself is called a **disconnected space.**

disconnecting set *Mathematics.* **1.** a set of edges of a graph G whose removal increases the number of components of G. A cutset is a minimal disconnecting set. **2.** given two distinct vertices of a graph G, a vw-disconnecting set of G is a set E of edges of G with the property that any path from v to w includes an edge of E. A vw-disconnecting set is also a disconnecting set.

discontinuity a lack of continuity; a break or interruption; specific uses include: *Physics.* **1.** the interface of an abrupt change in a medium, such as a change in density, temperature, or voltage. **2.** an interface between two different media. *Electromagnetism.* an abrupt change in the shape of a waveguide, at which a point reflection may occur. *Geology.* **1.** a surface that separates two unrelated groups of rocks or two seismic layers of the earth. **2.** any interruption in sedimentation. *Geophysics.* a seismic boundary within the earth where seismic waves abruptly change velocity. *Metallurgy.* an irregularity in a metallic component or in a metallic structure, manifested as an unexpected interruption. *Mathematics.* a point in the domain of a function f at which f is not continuous.

discontinuous amplifier *Electronics.* an amplifier that samples only a part of an entire input waveform and effectively reconstructs the output waveform by an averaging process.

discontinuous precipitation *Metallurgy.* in solid-state precipitation from a supersaturated solid solution, the discrete growth of the precipitating phase accompanied by the recrystallization of the matrix.

discontinuous reaction series *Mineralogy.* the left side of Bowen's reaction series in which the order of crystal purination occurs in abrupt phase changes.

discontinuous replication *Molecular Biology.* the synthesis of short pieces or fragments of DNA by the lagging strand during DNA replication.

discontinuous variation *Biology.* a sudden change or relatively large difference of structure, form, function, or the like, among members of an otherwise similar species, with no significant intermediate forms.

discontinuous yielding *Metallurgy.* the strain that is variable from location to location at the yield point in the plastic deformation of metals and alloys.

discopodous *Invertebrate Zoology.* having a disk-shaped foot.

Discorbacea *Invertebrate Zoology.* a superfamily of foraminiferan protozoans, in the suborder Rotaliina, with a radial calcite test.

discord see DISSONANCE.

discordance *Genetics.* the occurrence of a given trait in only one member of a pair of twins. *Geology.* **1.** an unconformity characterized by an absence of parallelism between adjacent series. **2.** see ANGULAR UNCONFORMITY.

discordant bedding *Geology.* a sedimentary structure in which parallelism of strata is lacking or in which layers are inclined with respect to the major lines of deposition.

discordant coastline *Geology.* a generally irregular coastline having many inlets, which forms where the overall structural grain of the land is transverse to the margin of the ocean basin.

discordant fold *Geology.* a fold whose axis is at an angle to the general strike of the area.

discordant junction *Hydrology.* the joining of two streams or valleys having surfaces at markedly unequal levels.

discordant pluton *Geology.* an intrusive igneous body that cuts across the bedding planes or foliation of the country rock.

discotaster *Invertebrate Zoology.* a type of eight-rayed spicule found in certain sponges.

discovery *Mining Engineering.* a legal term for the first finding of a mineral deposit in place on a mining claim; necessary before the location can be held by valid title.

discovery claim *Mining Engineering.* the first claim in which a mineral deposit is found.

discovery vein *Mining Engineering.* the mineral vein on which a mining claim is based.

discovery well *Petroleum Engineering.* an exploratory well that reveals oil in a new field or at a new level; a successful wildcat well.

discrepancy index, residual *R* *Crystallography.* an index that gives a crude measure of the correctness of a crystal structure determination. It is defined as $R = \Sigma|(|F_0| - |F_c|)|/\Sigma|F_0|$ and values of 0.06 to 0.02 are considered good. However, some partially incorrect structures have had R values below 0.10, and many basically correct but imprecise structures have higher R values. Also, R FACTOR, R VALUE.

discrete *Science.* separate or distinct.

discrete data *Computer Programming.* the representation of a variable that may assume one of a finite set of values.

discrete-event simulation *Computer Programming.* a common form of computer simulation in which the variable state changes in discrete steps.

discrete Fourier transform see FAST FOURIER TRANSFORM.

discrete-part manufacturing *Design Engineering.* a manufacturing process in which discrete parts are produced in small lots or batches between 1 and 50,000.

discrete radio source *Astrophysics.* an object that can be observationally separated at radio wavelengths from its local background; it may range from a planet to the jet from an active galaxy.

discrete random variable *Statistics.* a random variable for which the possible values constitute a countable set, as opposed to a continuous random variable.

discrete sampling *Electronics.* a method of waveform sampling that avoids appreciable negative effects on the frequency response of the channel by lengthening the duration of the individual samples.

discrete set *Mathematics.* a set with discrete topology. This topology may be induced naturally, as in, for example, the integers embedded in the real numbers.

discrete sound system *Acoustical Engineering.* a term for a multichannel sound system, such as a quadraphonic system, that possesses the ability to process each channel separately.

discrete spectrum *Spectroscopy.* a spectrum whose component wavelengths make up a separate and distinct sequence of values, as opposed to a continuum of values.

discrete system *Control Systems.* a control system that operates to control changes in incoming signals only at discrete points in time. Also, **discrete-time system.**

discrete topology *Mathematics.* the topology on a set X of points in which every subset of X is defined to be open; equivalently, the topology in which every one-point subset of X is defined to be open. X, together with such a topology, is called a **discrete topological space.**

discrete variable *Mathematics.* **1.** a variable whose domain is a discrete set. **2.** a random variable taking values in a discrete sample space; that is, a random variable X taking a countable set of values $\{x_i\}$ and probabilities $p_i = P(X = x_i)$, $i = 1, 2, 3, \ldots$. Also, DISCRETE RANDOM VARIABLE.

discretization error *Mathematics.* an inaccuracy in calculating the value of a definite integral caused by approximating (**discretizing**) the integrand.

discriminant *Mathematics.* **1.** given the quadratic equation $f(x) = ax^2 + bx + c$, the expression $b^2 - 4ac$; used to determine the nature of the roots of the equation. **2.** more generally, let $p(x)$ be a polynomial of degree n with roots $r_1, r_2, \ldots, r_n$ counting multiplicities; i.e., $p(x) = s(x - r_1) \cdots (x-r_n)$ for some number s. Then the discriminant of $p(x)$ is the product $D = s^{2n-2}\prod_{i<k}(r_i - r_k)^2$. The definitions coincide for $n = 2$.

discriminant analysis *Statistics.* a methodology for classifying an observation into two or more categories on the basis of its value.

discriminating circuit *Ordnance.* in a naval mine, the part of the operating circuit that distinguishes between the response of the detecting circuit to the passage of a ship and the response to other disturbances.

discrimination *Behavior.* the ability to respond differently to stimuli that are similar but not identical. *Telecommunications.* the ability of a receiver to distinguish, separate, identify, and detect a given signal or signal source from all other incident signals, noise, interference, or environmental effects. *Computer Programming.* the process of skipping one or more of the immediately following instructions based on the outcome of a programmed logical comparison.

discrimination learning *Behavior.* a type of learning in which the subject, when presented with two or more similar stimuli, must respond uniquely to the correct stimulus.

discrimination net or **network** *Artificial Intelligence.* a method for classifying a given data set into one of a fixed set of categories. At each interior node of the network, a data attribute is compared to several possible values or ranges, and the matching branch is taken; leaf nodes are labeled with classifications. Also, DECISION TREE.

discriminative stimulus *Behavior.* a stimulus that evokes the correct or appropriate response.

discriminator *Electronics.* a circuit that converts frequency or phase deviations from a carrier to corresponding amplitude variations.

discriminator transformer *Electronics.* a transformer used in a discriminator circuit; it is usually designed to operate at radio frequencies and contains a center-tapped secondary winding connected to a balanced detector circuit.

disdrometer *Engineering.* a device designed to measure and record the varying sizes of atmospheric raindrops.

disease *Medicine.* **1.** any abnormal condition of body functions or structure that is considered to be harmful to the affected individual; an illness or disorder. **2.** a specific illness or disorder that is identified by a characteristic set of signs and symptoms, caused by such factors as infection, toxicity, genetic or developmental defects, dietary deficiency or imbalance, or environmental effects.

disengagement *Medicine.* the emergence of the head or presenting part of the fetus from the vulva during childbirth.

disequilibrium *Genetics.* a phenomenon in which certain combinations of alleles at different loci on the same chromosome are much more common than would be expected on the basis of the chance occurrence of individual genes. Also, LINKAGE DISEQUILIBRIUM.

disharmonic fold *Geology.* a fold that varies in form or magnitude throughout the depth of the strata through which it passes.

dishing *Metallurgy.* in metal fabrication, the process of forming a depression that is not as deep as a cup.

dishpan experiment *Meteorology.* a model experiment in which fluid is differentially heated in a flat rotating pan, simulating the atmosphere and reproducing important features of general circulation and atmospheric motion.

disilane or **disilicane** *Inorganic Chemistry.* Si_2H_6, a colorless gas that slowly decomposes in cold water; soluble in alcohol and benzene; melts at –132.5°C and boils at –14.5°C; highly flammable.

disilanyl see SILANE.

disilicate *Chemistry.* a silicate compound that contains two silicon atoms.

disilicide *Chemistry.* a compound that contains two silicon atoms joined to another element.

disinclination *Crystallography.* an imperfection observed in liquid crystals in which a portion of a crystal is rotated with respect to the rest of the crystal.

disinfectant *Materials.* a chemical agent that destroys harmful bacteria or viruses, especially on inanimate objects; substances commonly used as disinfectants include: alcohols, such as isopropyl alcohol; phenols, such as cresol and carbolic acid; halogens and halogen compounds, such as chlorine, iodine, and fluorine; and mercury compounds, such as mercuric chloride. A **complete disinfectant** destroys the spores of microorganisms as well as their vegetative forms; an **incomplete disinfectant** does not destroy the spores.

disinhibition *Behavior.* a temporary increase in the strength of a conditioned response that occurs when an irrelevant stimulus is introduced. *Psychology.* the occurrence of a previously inhibited behavior after the individual has observed another or others carrying out the behavior without punishment.

disinhibitive behavior see DISINHIBITION.

disintegrate *Science.* to break or separate into parts; lose unity or solidity. *Nuclear Physics.* to undergo the process of disintegration.

disintegration *Nuclear Physics.* a change in the properties of a nucleus, whether spontaneous or induced, in which particles or photons are released. *Geology.* see MECHANICAL WEATHERING.

disintegration chain or **family** see DECAY SERIES.

disintegration energy *Nuclear Physics.* the energy released or absorbed during a nuclear or particle reaction.

disintegration rate *Nuclear Physics.* the rate of radioactive transformation, or activity, of a sample of radioactive material, measured in disintegrations per second; defined quantitatively as the product of the number of atoms originally in the sample and the decay constant.

disintegration value see Q VALUE.

disintegration voltage *Electronics.* the lowest plate voltage in a gas-filled electron tube at which destruction of the cathode-emitting surface occurs due to ion bombardment.

disintegrator *Engineering.* any of various devices used for the gradual breakup of substances, such as ore or coal, into fine powder; one type consists of two steel cages that move in opposite directions, and another type consists of a rotor with closely spaced hammers running inside a screen enclosure.

disjoint sets *Mathematics.* any two sets A and B whose intersection is empty; i.e., A and B have no members in common. Denoted $A \cap B = \varnothing$, where $\varnothing$ denotes the empty set. A collection of sets with the property that any two distinct sets in the collection are disjoint is said to be **pairwise disjoint** or a **disjoint collection.**

disjunction *Mathematics.* a binary operation on logical propositions such that the resulting statement is true if either (or both) of the two operands is true. If P_1 and P_2 are logical propositions, then their disjunction is written $P_1 \vee P_2$ (read as "P_1 or P_2"). *Artificial Intelligence.* a compound logical statement using the OR operator: true if any of its components is true. *Cell Biology.* the separation of paired chromosomes during the anaphase stage of either mitosis or meiosis.

disjunctive normal form *Artificial Intelligence.* a canonical form of writing predicate calculus formulae as a disjunction of conjuncts.

disjunctive search *Computer Programming.* a search based on the logical sum (the OR operator) of two or more criteria, as opposed to a conjunctive search using the logical product (the AND operator) of the criteria.

disjunctor *Mycology.* in certain fungi, connective tissue made of cellulose and located between the asexual spores (conidia), which are freed when the tissue eventually breaks down.

disk a circular or rounded flat plate; specific uses include: *Biology.* a round, flat part or structure, such as a red blood cell. *Anatomy.* one of the plates of fibrous material that are found between adjacent vertebrae. *Acoustical Engineering.* **1.** see COMPACT DISK. **2.** any disk-shaped recording medium, such as a phonograph record. Thus, **disk recording.** *Astronomy.* **1.** the apparent disk shape of a planet or star in a telescope. **2.** the flat part of a spiral galaxy. *Computer Technology.* a direct access medium used for high volume auxiliary memory, composed of a circular plate coated with magnetic material; the disk rotates under and is accessed by one or more read/write heads. *Mathematics.* also, **open disk.** the set of all points on the interior of a circle or 1-sphere. If the circle itself is included, the disk is called a **closed disk.** A **punctured disk** is a disk with an interior point removed. Also, DISC.

Disk and *disc* are equally correct spellings, and in some contexts they are used interchangeably. Certain automotive publications use the term *disk brakes*, while others use *disc brakes*. In terms of overall frequency of use, however, *disk* is about 50% more common than *disc*, and in certain fields, such as computers, *disk* is used almost exclusively.

disk bottom *Agriculture.* a plow blade that has the shape of a disk.

disk brake *Mechanical Engineering.* a brake system in which a disk attached to a wheel is slowed by the friction of disk pads pressed by calipers against each side of the disks. Also, DISC BRAKE.

disk cam *Mechanical Engineering.* a disk with a shaped edge that rotates about an axis perpendicular to the disk, transmitting movement to a cam follower in contact with the disk edge.

disk camera *Optics.* a camera in which the images are preserved on a photographic disk that is turned after each exposure to reveal a different film surface.

disk capacitor *Electricity.* a fixed capacitor that has a disk of dielectric material with metal-film plates deposited on its faces.

disk cartridge *Computer Technology.* a removable direct access storage medium in which the recording surface remains encased in the housing; a single platter disk pack.

disk centrifuge *Mechanical Engineering.* a centrifuge consisting of a large bowl and a set of disks that separate the solid particles in a liquid solution into thin sediment layers to create a series of shallow settling chambers that inhibit remixing of settled layers.

disk clutch *Mechanical Engineering.* a clutch that transmits power through one or more disks squeezed between a backplate and a movable pressure plate; can be disengaged by moving the plates apart.

disk colorimeter *Analytical Chemistry.* an instrument with rotating color disks that compares sample colors with a standard.

disk colter see ROLLING COLTER.

disk-core method *Archaeology.* a form of stone-tool technology in which a core is trimmed to a distinctive disk shape and flakes are then chipped off for tools.

disk crash *Computer Technology.* an event that causes damage to the drive and the disk, usually the physical contact of the read/write head with the surface of the rotating disk.

disk cultivator *Agriculture.* a farm implement that has two or more sets of disks for cultivating, used on ridges or on fields with heavy growth.

disk drive *Computer Technology.* an electromechanical unit that spins a magnetic disk and controls the read/write heads for direct access to stored data.

diskette *Computer Technology.* a flexible plastic disk coated with magnetic material and used as an inexpensive storage medium for microcomputers and minicomputers. Also, FLEXIBLE DISK, FLOPPY DISK.

disk file *Computer Technology.* a collection of related records stored on a magnetic disk.

disk filter *Engineering.* a heavy-duty vacuum filter in which the substance to be filtered is drawn through rotating disk membranes.

disk flower see DISC FLOWER.

disk formatting *Computer Programming.* the preparation of a new blank disk for recording data; this often consists of adding control data concerning sectors and tracks and establishing a directory.

disk furrower *Agriculture.* a planter attachment consisting of two disks that open a furrow for planting a seed.

disk grinder *Mechanical Engineering.* a grinding device that uses rigid abrasive disks.

disk harrow *Agriculture.* a harrow consisting of two or more groups of concave disks, used to clear weeds and debris, break up soil, cover broadcast seed, and so on.

disk harrow

disk jacket *Computer Technology.* the permanent protective sleeve that covers a diskette; the diskette is rotated within the jacket and data is read/written through a slot in the jacket.

disk loading *Aviation.* a measure of a helicopter's weight load, determined by dividing its gross weight by the disk area of the main rotor.

disk meter *Engineering.* a flow meter in which a rotating disk allows a known volume of fluid to pass through the meter.

disk mill *Engineering.* a size reduction mill in which solids, such as ore, are ground between two disks, one rotating and the other fixed.

disko- or **disk-** a combining form meaning "of, relating to, or resembling a disk," as in *diskography, diskectomy.*

disk operating system see DOS.

disk pack *Computer Technology.* a set of magnetic disks mounted on a common spindle and treated as a unit.

disk plow *Agriculture.* a plow that uses a disk rather than a share and a moldboard to cut and turn the soil.

disk population star *Astronomy.* a star belonging to a galaxy's disk.

disk sander *Mechanical Engineering.* a sander that uses a revolving abrasive disk driven by an electric motor to smooth or shape surfaces.

disk-seal tube *Electronics.* an ultra-high-frequency electron tube having closely spaced disk-shaped electrodes, providing low interelectrode capacitance along with high-power output. Also, LIGHTHOUSE TUBE.

disk sector *Computer Technology.* a pie-shaped storage surface between two adjacent radials on a magnetic disk.

disk spring *Mechanical Engineering.* a mechanical spring in which a disk or washer is supported by one force at the periphery and an opposing force at the center.

disk telescope *Optics.* a telescope through which an individual can view a solar disk.

disk thermistor *Electronics.* a disk-shaped device whose electrical resistance decreases with increases in temperature; used as a bolometer to measure temperature and, indirectly, microwave energy levels.

disk wheel *Mechanical Devices.* a spokeless vehicular wheel whose center section is solid or plated and is press-mounted on the wheel hub, giving support to the tire rim.

disk winding see PANCAKE COIL.

dislocation *Crystallography.* a discontinuity in the otherwise regularly periodic three-dimensional structure of a crystal, which can originate during crystallization or, more commonly, during deformation. If the Burger's vector is parallel to the line, it is a **screw dislocation;** if the vector is perpendicular, it is an **arc edge dislocation**. *Geology.* see DISPLACEMENT.

dislocation breccia see FAULT BRECCIA.

dislodgement *Molecular Biology.* a process (of unknown mechanism) by which a compatible plasmid, when introduced into a cell, sometimes replaces the resident plasmid.

dismicrite *Geology.* a fine-grained limestone composed mainly of micrite and sparry calcite bodies, and containing less than 1% allochems.

dismount *Ordnance.* to remove a weapon or piece of equipment from its setting, mount, or carriage.

disodium hydrogen phosphate see DIBASIC SODIUM PHOSPHATE.

disodium methylarsonate *Organic Chemistry.* $CH_3AsNa_2O_3$, a toxic, colorless, crystalline solid, soluble in water; melts above 355°C; used in pharmaceuticals and as an herbicide.

disodium phosphate see DIBASIC SODIUM PHOSPHATE.

disomaty *Cell Biology.* a division of chromosomes that results in a doubling of the chromosome number without nuclear division.

disomic *Cell Biology.* of, relating to, or having two homologous chromosomes.

Disomidae *Invertebrate Zoology.* a family of sedentary or bottom-dwelling marine polychaete annelids.

disophenol *Pharmacology.* $C_6H_3I_2NO_3$, a drug occurring as light yellow crystals; used in veterinary medicine to treat infestations of worms.

disorder *Medicine.* any abnormality or malfunction in physical or mental health. *Crystallography.* a variation from the orderly arrangement of elements of unit cells in a crystal.

disordered *Medicine.* suffering from a disorder.

disordered crystalline alloy *Solid-State Physics.* an alloy composed of two or more elements whose atoms are randomly arranged in the crystal lattice.

disorganized schizophrenia see HEBEPHRENIC SCHIZOPHRENIA.

disorientation *Medicine.* a loss of the sense of relationship to one's surroundings in terms of the ability to comprehend time, place, and people, as in organic brain syndromes.

disoriented *Medicine.* suffering from disorientation.

disparity *Artificial Intelligence.* in a binocular vision system, the displacement of a pixel between the two images due to the difference in point of observation.

dispatching *Industrial Engineering.* the selection, sequencing, and assignment of jobs to individual work centers. *Computer Technology.* the allocation of CPU processing cycles by the operating system to active programs in a multiprogramming system.

dispatching priority *Computer Programming.* in a multiprogramming system, the priority assigned to an active task by the job management scheduler function of the operating system.

dispatching rule *Industrial Engineering.* a predetermined set of steps affecting the assignment and priority of jobs to individual work centers.

dispenser *Engineering.* **1.** any person or device that dispenses a material in small amounts. **2.** a device, such as a vending machine, that releases its contents one at a time. *Ordnance.* a device that automatically releases radar chaff from an airplane.

dispenser cathode *Electronics.* an electron-tube cathode that is specially coated to continuously replenish electron-emitting material lost during operation of the tube.

dispermous *Botany.* of or relating to fruits that contain two seeds.

dispermy *Physiology.* the penetration of a single ovum by two spermatozoa.

dispersal the act or fact of dispersing; specific uses include: *Military Science.* **1.** the spreading or separating of troops, vehicles, equipment, or supplies, especially to reduce vulnerability to enemy attack. **2.** in airdrop operations, the scattering of personnel or cargo on the drop zone. *Ordnance.* **1.** a scattered pattern of hits around the mean point of impact of bombs or projectiles dropped or fired under identical conditions. **2.** in artillery fire, a scattered pattern of explosions around the mean point of burst. **3.** see DISPERSION, def. 2. *Biology.* the movement of organisms or their spores or gametes throughout the ecological niche of that particular organism. *Ecology.* the outward extension of the range of a species, usually resulting from a chance event. Also, ACCIDENTAL MIGRATION.

dispersal error *Ordnance.* the distance between the point of impact or burst of a round and the mean point of impact or burst of a group of rounds fired under identical firing conditions. Also, **dispersion error.**

disperse *Science.* to scatter individual items or particles throughout an area or substance. *Physical Chemistry.* **1.** to cause fine particles to separate throughout a bulk substance in a dispersion. **2.** also, **dispersed.** relating to or describing such particles. *Computer Programming.* to distribute or duplicate input data items into more than one output area.

dispersed elements *Geochemistry.* elements present as traces or impurities in minerals, which are either so rare or occur in such small concentrations that they seldom themselves form minerals.

dispersed phase see DISPERSE PHASE.

dispersed settlement *Geography.* a settlement pattern characterized by dwellings scattered over the land.

disperse dye *Materials.* any of various water-soluble dyes that are dispered in a water solution during the dyeing of synthetic textile fibers.

disperse phase *Physical Chemistry.* the less prevalent component (solid, liquid, or vapor) of a dispersion; for example, in paint the solid particles of coloring matter are the disperse phase and the liquid suspension is the continuous phase. Also, INTERNAL PHASE.

disperse system *Physical Chemistry.* see DISPERSION.

dispersible inhibitor *Chemistry.* any agent that easily disperses in a medium to slow down or prevent chemical activity or corrosion.

dispersing prism *Optics.* a prism that separates white light into its monochromatic elements.

dispersion a process of scattering or separating; specific uses include: *Physical Chemistry.* a two-phase system consisting of finely divided particles (the disperse phase) distributed throughout a bulk substance (the continuous phase); for example, fog is a dispersion of liquid particles in a gas; paint is a dispersion of solid particles in a liquid. Also, DISPERSE SYSTEM. *Physics.* the separation of a complex wave containing several frequencies into its individual component waves by virtue of different wave speeds. *Astrophysics.* the slowing down of a radio signal's propagation velocity as it passes through an ionized gas or plasma in the interstellar medium; short frequencies are the least affected and long ones the most affected. *Electromagnetism.* the scattering of microwave radiation by an obstruction. *Mineralogy.* the characteristic and specific manner in which a monopaque mineral will refract white light. *Space Technology.* in rocketry, a deviation from the prescribed flight path. Also, CIRCULAR DISPERSON. *Ordnance.* **1.** see DISPERSAL. **2.** in chemical and biological operations, the release of agents in liquid or aerosol form. *Military Science.* see DISPERSAL.

dispersion force or **dispersion attraction** *Physical Chemistry.* a general force of attraction between nearby atoms or molecules that arises from a temporary polarity between them, caused by the uneven distribution of electrons; independent of temperature and present in all types of matter, including the noble gases. Also, ATTRACTION FORCE.

dispersion formula *Physics.* a relationship giving the index of refraction of a particular substance as a function of the wavelength, as in the Cauchy dispersion formula or the Hartmann dispersion formula.

dispersion fuel *Nucleonics.* small particles of fissile and fertile fuel that are dispersed in a matrix in such a way that fission-fragment damage areas do not overlap, thereby providing good thermal conductivity while retaining the mechanical strength of the matrix.

dispersion interaction *Physical Chemistry.* a momentary shift in the symmetry of the electron clouds of two nearby atoms or molecules, resulting in a force of attraction between them.

dispersion ladder *Ordnance.* a diagram showing the probable dispersion of a group of shots fired under identical conditions; it indicates the percentage of shots expected to fall in each of eight zones.

dispersion measure *Astrophysics.* a quantity that expresses the amount of dispersion that a radio signal suffers; it is proportional to both the number of electrons in the path and the path's length.

dispersion medium *Physical Chemistry.* the bulk substance in which another substance is dispersed; for example, the gas in an aerosol spray mixture. Also, CONTINUOUS PHASE.

dispersion mill *Mechanical Engineering.* a size-reduction machine used to break clusters of solids in the preparation of purees, food pastes, cosmetics, pulps, paints, and the like.

dispersion of a random variable *Statistics.* the scatter of values or measurements around the arithmetic mean or other measure of central tendency.

dispersion pattern *Geochemistry.* the pattern of distribution of chemical elements in rocks or surface material, such as soil, as a result of migration away from a source, such as an ore body.

dispersion relation *Physics.* an equation relating the wavenumber k to the radian angular frequency ω, usually expressed as a derivative $dk/d\omega$ equal to some function of ω.

dispersion relations *Nuclear Physics.* the set of analogous relations between the real and imaginary parts (for example, refractivity and index of refraction, respectively) of any response function, such as the cause-and-effect pairs of force and spatial displacement, electric field and polarization, or incident and scattered waves.

dispersion strengthening *Materials Science.* a method of strengthening a material by the even distribution of particles that are small (less than 0.1 micrometer) in a matrix at 1–15% vol. The matrix is the main load-bearing component. *Metallurgy.* specifically, the strengthening of a metal or alloy by incorporation in its structure of finely dispersed particles of a stable nonmetallic compound such as aluminum oxide. Also, **dispersion hardening.**

dispersion theory see BREIT-WIGNER THEORY.

dispersion zone *Ordnance.* the area over which shots spread apart and scatter when fired with the same sight setting.

dispersive Fourier transform spectrometry *Spectroscopy.* a spectroscopic technique in which the entire range of frequencies of interest is simultaneously passed through an interferometer whose output signal is analyzed by a Fourier transform.

dispersive FT-IR *Spectroscopy.* dispersive Fourier transform in which the frequency range of interest is that of the infrared band of the electromagnetic spectrum.

dispersive lens see DIVERGING LENS.

dispersive line *Electromagnetism.* a line that delays different frequencies by different amounts of time.

dispersive medium *Electromagnetism.* a medium in which the phase velocity of electromagnetic waves is a function of the frequency.

dispersive power *Optics.* a measure of the ability of a material to separate light into different colors.

dispersivity *Materials Science.* the degree of distribution of a finely divided solid in a liquid or a solid matrix; often used to describe the distribution of macromolecules in a sample of polymers.

dispersoid *Chemistry.* the product of a dispersion.

displaced aggression *Psychology.* aggressive behavior directed at some person or object other than the original source of the hostility.

displaced ore body *Geology.* an ore body affected by displacement or disruption that occurred after its initial deposition.

displaced outcrop *Geology.* an outcrop that has moved downhill due to a landslide or soil creep.

displacement a movement away from the original position; specific uses include: *Fluid Mechanics.* **1.** the weight of fluid that is displaced by a floating body, equal to the weight of the body plus its contents. **2.** in general, the volume of fluid displaced by any floating body. *Mechanics.* **1.** any movement of a particle or body from one position in space to a new position. **2.** a description of this motion, involving the linear distance moved and the direction of the path taken. *Mechanical Engineering.* **1.** the volume through which a piston travels during a single stroke in an engine, pump, or similar mechanism. **2.** the total volume displaced by all the pistons of an engine. *Chemistry.* a chemical reaction in which one element releases another element from a compound and replaces it in the compound. *Geology.* the relative movement of rock on opposite sides of a fault, measured in any specified direction, or the magnitude of such movement. Also, DISLOCATION. *Psychology.* the shift of an emotion or impulse from the person, object, or situation toward which it was originally directed to a more acceptable substitute. *Cartography.* a shift in the position of an image on a photograph that does not alter the perspective characteristics of the photograph. *Computer Programming.* the quantity that, when added to the contents of the base register, produces the machine language address of an operand or an instruction.

displacement angle *Electricity.* the change that occurs in the phase of an alternator's terminal voltage when a load is applied.

displacement behavior *Behavior.* a behavior pattern that is seemingly irrelevant to the present situation, usually arising from conflict. Also, **displacement activity.**

displacement chromatography *Analytical Chemistry.* a separation technique using elution chromatography in which the solvent is adsorbed to a packed column (stationary phase) and the freed sample migrates down the column.

displacement compressor *Mechanical Engineering.* a compressor that relies on the displacement of a volume of air by a piston moving in a cylinder.

displacement current *Electromagnetism.* a hypothetical current used in Maxwell's equations to account for the apparent current through the space between the plates of a capacitor; it exists in the presence of a time-varying electric field.

displacement curve *Naval Architecture.* a design curve showing how a vessel's draft increases as its displacement is increased.

displacement engine see PISTON ENGINE.

displacement gyroscope *Engineering.* a gyroscope that senses, measures, and transmits angular displacement data.

displacement law see WIEN'S DISPLACEMENT LAW.

displacement length *Materials Science.* the distance between the free ends of a coiled chain polymer; an important parameter in describing the dimensions of the chain.

displacement loop see D LOOP.

displacement manometer *Engineering.* a differential manometer that measures pressure difference across a solid or liquid partition being displaced by force.

displacement meter *Engineering.* a meter that measures water flow by recording the number of times a vessel of known size and capacity is filled and emptied.

displacement parameters *Crystallography.* a description of the displacement of atoms in a crystal structure; atomic vibrations are displacements from equilibrium positions with periods that are typically smaller than 10^{-12} seconds. Because static displacements of a given atom are random from unit cell to unit cell, they will stimulate atomic vibrations.

displacement plane see DEFORMATION PLANE.

displacement pump *Mechanical Engineering.* a pump in which nonreturn valves prevent the return flow of displaced liquid during the retracting phase of the pump cycle, thus creating a pulsing action characterized by alternate filling and emptying of an enclosed volume.

displacement series see ACTIVITY SERIES.

displacement vessel *Naval Architecture.* any ship or boat that is supported in the water by displacement of its weight, as opposed to a hydrofoil or surface skimmer.

displacer-type meter *Engineering.* a specific gravity measuring device in which the liquid being measured flows continuously through a chamber containing a submerged, gas-filled cylinder whose buoyancy is recorded.

display *Electronics.* any visual presentation of the output of a unit or system, as on a cathode-ray tube or in readable characters of a digital display. *Computer Science.* in an activation record, an array of pointers to the activation records of surrounding blocks, used to access variables defined in those blocks. *Behavior.* see DISPLAY BEHAVIOR.

display behavior *Behavior.* any behavior pattern that serves to provide some kind of physical signal or communication.
display console *Computer Technology.* an input/output device that provides visual representation of data in alphanumeric or graphic form and has a keyboard for user control and data entry.
display control *Computer Technology.* routines and circuitry in the input/output control system portion of an operating system that connect the display unit to the CPU and that control data transfer and display.
display cycle *Computer Technology.* the time required to completely refresh a visual display screen.
display primary *Telecommunications.* in television broadcasting, any of the primary colors that, when carefully mixed with the other primaries, is able to produce other colors. Also, RECEIVER PRIMARY.
display processor *Computer Technology.* the part of a computer system that is dedicated to the control and display of information on a CRT screen.
display rules *Psychology.* the influences within a given society or culture that determine what is considered to be an appropriate display or expression of emotions.
display system *Computer Technology.* the combined software and hardware used to provide a visual representation of data.
display terminal *Computer Technology.* an output device that produces a visual representation of data in alphanumeric or graphic form; for example, a visual display terminal, plotter, or graphic printer.
display tube *Electronics.* a cathode-ray tube that is used to provide an electronic display.
display type *Graphic Arts.* a term for type, usually boldface and 14-point size or larger, that is used to attract attention in headlines, in advertisements, and sometimes within text.
display window *Telecommunications.* the width of the section of a frequency spectrum, usually expressed in megahertz.
disposable *Engineering.* describing a product that is intended to be discarded after use and replaced by an identical item, such as the filter unit in many systems.
disposition *Psychology.* an individual's predominant or characteristic emotional state.
dispositional attribution *Psychology.* the attribution of a person's behavior in a particular situation to internal traits of character or personality rather than to external factors of the situation.
dispositional factors *Psychology.* those influences on a person's behavior that are internal traits of character or personality, rather than external factors of the situation.
disproportionation *Chemistry.* a chemical reaction in which one compound acts as both an oxidizing and a reducing agent, thus yielding two products: a more reduced compound and a more oxidized compound.
disruptive coloration *Behavior.* surface markings that delay recognition of the whole animal by attracting the attention of the observer to certain highly distinctive elements of the color pattern.
disruptive discharge *Electricity.* a sudden, increased current flow through an insulating material due to complete failure of the material under electrostatic stress.
disruptive selection *Evolution.* a natural selection process in which extreme examples of character variation are selected for, while intermediate examples are selected against, resulting in accentuated polymorphism in the population. *Genetics.* see DIVERSIFYING SELECTION.
disruptive strength *Metallurgy.* the minimum stress level at which a metal or alloy fractures while stressed by hydrostatic tension.
dissect [dí´sekt´; di sekt´] *Biology.* to cut into an organism and separate and expose its parts for the purpose of scientific examination or study.
dissected plateau *Geology.* a plateau that has been cut by erosion into a network of hills and valleys, but that retains its original profile.
dissecting aneurysm *Medicine.* an aneurysm, usually affecting the thoracic aorta, which results from a hemorrhage that causes lengthwise splitting of the arterial wall, producing a split in the intima and establishing communication with the lumen. Also, AORTIC DISSECTION.
dissecting microscope *Optics.* a low-power microscope that magnifies the images of objects being dissected and has adequate free space beneath the objective for dissection to take place.
dissection [dí´sek´shən; di sek´shən] *Biology.* the act of dissecting; cutting apart for examination and study. *Geology.* a process of erosion in which the continuity of a level land surface is gradually cut out or destroyed by the formation of ravines or valleys. *Mathematics.* a dissection of a measure space (X, A, μ) is any finite, pairwise disjoint family of open sets $\{A_1, A_2, \ldots, A_n\}$ of A whose union is all of X; used to define the Lebesgue integral of a function on an arbitrary measure space.
dissector tube *Electronics.* a photomultiplier camera tube in which each element is scanned in a controlled sequence; the output of the tube is a series of voltage levels representing the light level of each element as it is scanned. Also, IMAGE DISSECTOR, FARNSWORTH DISSECTOR TUBE.
disseminule *Botany.* any structure in a plant that serves to disseminate or disperse a population of the same organism, such as a seed, spore, fruit, etc., that detaches and gives rise to a new plant. Also, DIASPORE.
dissepiment *Botany.* **1.** the septum or partition that divides a fruit or ovary into compartments. **2.** the trama in some fungi.
dissilient *Botany.* the process of bursting forcefully open, as the capsule or dry seed pod of some plants.
dissimulation see CATABOLISM.
dissipation *Physics.* the conversion of energy into heat by friction in a mechanical process, or by resistive joule heating in an electrical process. *Mechanics.* the loss of energy from a system as a result of this.
dissipation constant *Geophysics.* a measure of the rate at which an electrically charged particle loses its charge to the air.
dissipation factor *Materials Science.* the ratio of the loss modulus to the modulus of elasticity, expressed as a number. *Electricity.* **1.** the ratio of resistance to reactance in a capacitor or inductor. **2.** the tangent of the dielectric loss angle.
dissipation function see RAYLEIGH'S DISSIPATION FUNCTION.
dissipation line *Electromagnetism.* a length of resistive transmission line used for dissipating several kilowatts of power for a rhombic transmitting antenna; it is typically two parallel lengths of Nichrome or stainless steel wire terminated by a large, noninductive resistance.
dissipation trail *Fluid Mechanics.* a clear rift that is left behind an aircraft flying in a thin cloud layer. Also, DISTRAIL.
dissipative *Science.* characterized by or causing dissipation, or loss of energy.
dissipative dynamical system *Chaotic Dynamics.* **1.** a system that does not conserve energy but that can be maintained in a spatially or temporally complex solution by the steady infusion of energy. **2.** more generally, a system for which any volume element of phase space shrinks as the system evolves. For such systems the set of attractors is an infinitesimal subset of the phase space. The sum of the Layplunov exponents of such a system is negative.
dissipative force see NONCONSERVATIVE FORCE.
dissipative muffler *Engineering.* a device designed to absorb sound when a gas is passed through it.
dissipative system *Mechanics.* any system in which energy is not conserved, but dissipates with time, as from the effect of friction.
dissipative tunneling *Solid-State Physics.* a process of tunneling individual electrons, rather than pairs, across a thin insulator between the interface of two superconductors, resulting in the scattering of electrons at the interface.
dissociate *Physical Chemistry.* to undergo or cause to undergo the process of dissociation.
dissociation a removal from association, or the lack of association; specific uses include: *Physical Chemistry.* the separating of a molecule into fragments, such as simpler molecules, atoms, radicals, or ions, because of a change in physical conditions, as when the molecule collides with other material or absorbs electromagnetic radiation; usually reversible, as with the breakdown of weak acids in water. *Psychology.* the separating of certain ideas or emotions from the rest of a person's mental activity in order to avoid stress or anxiety. *Cardiology.* independent action of the atria and ventricles; a form of heart block. *Toxicology.* the uncoupling of a chemical from a cell or molecule to which it had become attached. *Microbiology.* a change in the phenotype of a given strain of bacteria; for example, an alteration in colony morphology from a smooth appearance to a rough, pebbled appearance. *Archaeology.* the principle that artifacts deposited together in a certain location were not necessarily used together at that location.
dissociation constant *Physical Chemistry.* a fixed quantity whose value derives from the equilibrium between the whole and fragmented forms of a molecule.
dissociation energy *Physical Chemistry.* the amount of energy needed to separate a molecule into smaller fragments, such as atoms, radicals, and ions.
dissociation pressure *Physical Chemistry.* the amount of pressure, at a specified temperature, needed to fragment a molecule.
dissociative disorder *Psychology.* a mental disorder characterized by the segregation of certain aspects of memory or personality from the individual's main consciousness and identity. Also, **dissociative reaction.**

dissociative recombination *Atomic Physics.* a process in which an electron recombines with a positive molecular ion before the molecule breaks apart; the resulting atoms from the broken molecule carry away the excess energy.
dissogeny [di säj´ə nē] *Zoology.* the fact of having two separate sexually mature periods, one during the larval stage, the other as an adult.
dissolution *Chemistry.* the dissolving of a solid or gas in a liquid.
dissolve *Chemistry.* to cause a substance to pass into a solution.
dissolved air flotation *Chemical Engineering.* a liquid-solid separation procedure in which the main mechanism for removing the suspended solid particles is the change of specific gravity of the solids in relation to the suspending liquid; separation is achieved by the occlusion of small gas bubbles formed by the release of dissolved gas to the solids. Also, AIR FLOTATION.
dissolved gas see SOLUTION GAS.
dissolved oxygen concentration *Biotechnology.* the amount of dissolved oxygen in a fermentation broth, determined by a technique using either galvanic or polarographic electrodes.
dissonance *Acoustics.* the playing together of two musical tones that are one step or less apart, so that the combination and most of the resulting overtones are inharmonic. Also, DISCORD. *Psychology.* **1.** a situation in which a person's behavior is not consistent with his attitude toward that behavior; e.g., a person believes smoking is harmful yet continues to smoke. **2.** an unpleasant feeling arising from such an inconsistency.
dissonance theory see COGNITIVE DISSONANCE THEORY.
dissonant *Psychology.* having or showing dissonance.
dissymmetrical transducer *Electronics.* a transducer with unequal input and output image impedances. Also, **dissymetrical network.**
dissymmetry [dī sim´i trē] *Science.* a lack of symmetry; asymmetry.
dissymmetry coefficient *Analytical Chemistry.* a ratio used to correct for interference occurring in light-scattering photometry of liquids, based on the intensities of scatter light at 45° and 135°.
dissymmetry factor *Optics.* a measure of circular dichroism, equivalent to the ratio of the difference in the absorption indices for left and right circularly polarized light to the absorption index for ordinary light of the same wavelength.
Distacodidae *Paleontology.* a term formerly used for a group of conodonts, now superseded by Distacodontidae.
Distacodontidae *Paleontology.* a family of simple, nonfibrous conodonts of the Ordovician and Silurian, characterized by a laminar, "cone-in-cone" internal structure of the teeth.
distaff *Textiles.* a forked stick about three feet long, used in hand spinning to hold the loose ball of carded fibers as they are spun into yarn.
distal *Biology.* **1.** away from the center or the attached end; being the outer end. **2.** far apart or widely spaced. *Invertebrate Zoology.* farthest from the base or point of attachment, as in the distal segment of a leg.
distal clot *Hematology.* a clot formed in a blood vessel distal to a ligature.
distal convoluted tubule *Anatomy.* the parts of the kidney nephrons that lie in the renal cortex of the organ between the loops of Henle and the collecting ducts.
distal hyperextensibility *Genetics.* a hereditary ability to bend the thumb sharply backward that is inherited as a simple autosomal recessive allele. Also, HITCHHIKER'S THUMB.
distance *Mechanics.* the measured length between two points. *Mathematics.* **1.** intuitively, the length of a minimal path between two given sets or geometric configurations. If A and B are single points, this is the distance function on those two points. Given subsets A and B of a metric space with metric or distance function d, the value of $\inf\{d(a, b)\}$, for all points $a \in A$ and $b \in B$. **2.** a nonsymmetric function d defined on the space of fractals; used to define a metric on the fractals. If A and B are nonempty compact subsets of R^n, then $d(A, B) = \max\{d(x, B) : x \in A\}$, where $d(x, B)$ is the distance between the sets $\{x\}$ and B. **3.** the distance between vertices v and w in an undirected graph is the length of the shortest path joining v with w if such a path exists; if no such path exists, the distance between them is ∞. **4.** in a digraph, the distance from vertex v to vertex w is the length of the shortest dipath from v to w; if no such path exists, the distance between them is ∞.
distance behavior *Behavior.* the pattern or practice of maintaining physical separation from others of the same species. Thus, **distance animal, distance species.**
distance-finding station *Navigation.* a radio beacon that sends out a sound signal at the same instant as the radio pulse. The difference in the time at which these signals are received by an observer can be converted to the distance between the station and the observer.
distance function *Mathematics.* given a set X, a symmetric, nonnegative, real-valued function d defined on $X \times X$ that satisfies the triangle inequality and has the property that $d(x, y) = 0$ if and only if $x = y$ for all x, y in X. Also, METRIC ON X.
distance mark or **marker** *Electronics.* a mark on a radar screen that indicates the distance from the radar set to a target; distance marks can be fixed at specific intervals along the radar sweep for estimating target distance, or they can be moved by turning a range knob connected to a counter for direct reading of distance. Also, RANGE MARK.
distance-measuring equipment *Navigation.* equipment that allows an airplane to determine its distance from a radio transmitter. The aircraft sends out a pulse, which triggers a pulse from the ground transmitter. By measuring the time between pulses, the equipment can calculate the distance to the station.
distance modulus *Astronomy.* the difference between the apparent magnitude of a star or galaxy *(m)* and its absolute magnitude *(M)*; this distance is given by $m - M = 5 \log(r - 5)$, where r is the distance in parsecs.
distance ratio *Mechanical Engineering.* the ratio of the distance moved by the input or effort of a machine during a given time to the distance moved by the output or load.
distance reception *Telecommunications.* the receiving of or capability of receiving signals from distant radio stations.
distance relay *Electricity.* a protective relay that removes power when a fault occurs within a predetermined distance along a circuit.
distance resolution *Engineering.* the minimum distance between two objects in order for them to be recognizable as two distinct objects on a radar screen.
distance-velocity lag *Control Systems.* a delay due to the time required for the transport of materials or the propagation of a signal from one point to another. Also, TRANSPORT(ATION) LAG.
Distant Early Warning line see DEW LINE.
distant field *Electromagnetism.* an electromagnetic field measured at a distance greater than about five wavelengths from a transmitter.
distely *Botany.* a condition in which the vascular system is divided into two separate steles, as exhibited in certain species of *Selaginella.*
distemper *Veterinary Medicine.* a highly contagious viral disease of mammals, especially of dogs and animals with unretractable claws, characterized by fever, loss of appetite, discharge from the eyes and nose, and nervous symptoms.
disthene see KYANITE.
distichous *Biology.* arranged alternately in two vertical rows on both sides of an axis, such as leaves growing on a plant stem.
distill *Chemistry.* **1.** to carry out the process of distillation. **2.** to obtain by distillation. Also, **distil.**
distillate *Chemistry.* the liquid product or condensate of a distillation process.
distillation *Chemistry.* a separation process in which a liquid is evaporated and the vapor is condensed to a liquid, usually for the purpose of purification.
distillation column see FRACTIONATING COLUMN.
distillation curve *Chemistry.* in a distillation, a plot showing temperature versus distillate volume, weight, or composition.
distillation loss *Chemistry.* the difference in volume between the starting liquid in a distillation process and the end products.
distillation range *Chemistry.* the temperature range between the initial boiling point and the end point in a distillation process.
distillation test *Chemical Engineering.* a standard method for identifying initial, intermediate, and final boiling points of petroleum products.
distilled liquor *Food Technology.* an alcoholic beverage made by distilling brewed grains (as with whiskey), wine, or fermented fruit juice (as with brandy). Also, HARD LIQUOR.
distilled mustard gas *Organic Chemistry.* mustard gas that has been distilled to reduce its odor, making it harder to detect.
distilled water *Chemistry.* water from which dissolved solids and organisms have been removed by distillation.
distillery *Chemical Engineering.* **1.** a facility where a process of distilling takes place. **2.** specifically, such a location where the distilling of alcoholic beverages takes place.
distilling flask *Chemistry.* a round-bottomed flask used for distilling liquids.
disto- a combining form meaning "distal"; used chiefly in dentistry.
distoclusion *Medicine.* malocclusion of the teeth in which those of the lower jaw articulate with the upper teeth in a position distal to normal.
distome *Invertebrate Zoology.* a type of parasitic trematode flatworm.

distorted water *Meteorology.* a multimolecular layer of water at the boundary between a mass of liquid water and the surrounding vapor, whose structure differs from that of the bulk water.

distortion *Electronics.* **1.** any undesirable change in the waveshape of a signal. **2.** any undesired difference in the proportions of a reproduced image as compared to the original image. *Optics.* an aberration in a lens or lens system in which lateral magnification varies with the distance from the optical axis, causing straight lines to appear curved. *Radiology.* a representation of a structure under study in which the image does not accurately represent its actual shape or outline.

distortion curve *Photogrammetry.* a curve that represents the linear distortion qualities of a lens, plotted with image radial distances from the lens axis as abscissas and image radial displacements as ordinates.

distortion factor *Telecommunications.* **1.** the amount by which an output waveform or pulse differs from an input waveform or pulse. **2.** the difference between the wave shapes of an original signal and the signal after it has traversed the transmission circuit.

distortion meter *Engineering.* an instrument that is designed to visibly show the harmonic content of an audio-frequency wave.

distraction or **distraction display** *Behavior.* a behavior, such as a feigned injury, that is intended to divert the attention of a predator.

distrail see DISSIPATION TRAIL.

distress frequency *Telecommunications.* a radio frequency that is used exclusively for emergency calls and messages; various frequencies are designated for use by certain mobile stations or survival craft for handling distress traffic.

distress signal *Telecommunications.* a short signal indicating that the unit sending the signal is threatened by grave or imminent danger and requests immediate assistance; often followed by a **distress message** giving the particulars of the emergency situation.

distributary *Hydrology.* **1.** a branch of a river or stream that diverges from the main flow and does not rejoin it. **2.** a channel of a braided stream. Also, **distributary channel.**

distributed amplifier *Electronics.* a wideband, multistage amplifier in which tubes are distributed along sections of an artificial delay line; the input and output capacitances of each tube form part of the total capacitance, and gain can be increased by adding more tubes.

distributed capacitance *Electricity.* any capacitance not concentrated within a capacitor, such as that between the turns in a coil or between adjacent conductors of a circuit. Also, SELF-CAPACITANCE.

distributed circuit *Electronics.* a circuit in which the electrical properties (such as resistance or capacitance) are spread throughout the circuit area and not contained in easily recognized stand-alone components.

distributed collector *Engineering.* a component in a solar heater that transports the working fluid through a network of modular focusing collectors and absorber pipes to a heat exchanger.

distributed communications *Telecommunications.* **1.** information transferred beyond a local level. **2.** information that is interchanged among several networks.

distributed constant *Electromagnetism.* a constant quantity that exists along the entire length of a transmission line. Also, **distributed parameter.**

distributed control system *Control Systems.* a system in which the control process is divided between two or more units.

distributed database *Computer Technology.* a database that is physically dispersed to separate locations, usually connected by a data network and supported by a distributed processing system.

distributed data-processing system see DISTRIBUTED PROCESSING SYSTEM.

distributed-emission photodiode *Electronics.* a photodiode that is sensitive to stimulus from a modulated laser beam; photo energy striking the device propagates a traveling wave of energy along an output transmission line.

distributed fault see FAULT ZONE.

distributed fire *Ordnance.* fire applied in a manner that will most effectively cover the target area.

distributed inductance *Electromagnetism.* an inductance distributed evenly over the entire length of a conductor, as opposed to the inductance associated with a coil or other circuit element.

distributed intelligence *Computer Technology.* the placement of processing capability in remote terminals and peripheral devices that are otherwise used for data input and output.

distributed numerical control *Control Systems.* a system in which part-classification data is distributed from a central computer to other computers that control machine tools.

distributed paramp *Electronics.* a paramagnetic amplifier that produces a high-frequency traveling wave effect by the sequential excitation of varactors spaced uniformly along a transmission line.

distributed processing system *Computer Technology.* a system of data-processing in which the processing and data storage facilities are physically dispersed and interconnected with a data communication network; it typically consists of a main computer connected with remote terminals that can access information from the main computer and carry out some data-processing operations without tying up the main computer. Also, DISTRIBUTED DATA-PROCESSING SYSTEM.

distributing frame *Electronics.* a unit of equipment that serves as a central interconnection point for the wires and cables in a telephone switching system. Also, DISTRIBUTION FRAME.

distributing point *Military Science.* a point at which supplies and ammunition obtained from a supply point by a division or other military unit are divided and distributed to subordinate units.

distributing roller *Graphic Arts.* on a printing press, a roller that picks up ink from the fountain and spreads it onto the ink plate or drum.

distributing terminal assembly *Electronics.* a unit of equipment used in a telephone-switching system for terminating and cross-connecting trunk wiring between selector switches.

distribution *Military Science.* **1.** the arrangement of troops for a battle, march, or maneuver. **2.** a planned pattern of projectiles around a target point, or of fire to cover a target area. **3.** the delivery or dispensing of orders, equipment, supplies, and services. *Archaeology.* the pattern of artifacts as they are dispersed over a site. *Mathematics.* **1.** the events of the sample space of a random variable and the frequencies with which these events are likely to occur. **2.** let U be a nonempty open subset of R^n and $D(U)$ be the topological vector space of C^∞ functions with compact support in U (i.e., test functions). A distribution on U is an element of the dual space $D(U)$ of $D(U)$; that is, a linear functional on $D(U)$. Also, GENERALIZED FUNCTION.

distribution amplifier *Acoustical Engineering.* a power distribution amplifier for audio frequencies with relatively low output impedance, so that changes in the load do not significantly affect the output voltage.

distribution cable *Electricity.* **1.** a cable extended from a feeder cable to provide service to a particular area. **2.** the transmission cable in a system from the distribution amplifier to the drop cable.

distribution center *Electricity.* **1.** the point in an AC power system at which generation, conversion, control equipment, and routing equipment are installed. **2.** a central point from which a signal is routed to points of use.

distribution coefficient *Physical Chemistry.* **1.** the ratio in which a given substance distributes itself between two or more phases. **2.** specifically, the ratio in which a given solute dissolves in two different immiscible liquids. Also, PARTITION COEFFICIENT. *Optics.* one of the tristimulus values of spectral radiations that have the same power.

distribution factor *Nucleonics.* a function of the physiological distribution of radionuclides within the critical organ of reference, such as lungs, skin, bone, kidney, thyroid, or gonads, so that localized volumes of the organ receive larger exposure than the rest.

distribution frame see DISTRIBUTING FRAME.

distribution-free method see NONPARAMETRIC METHOD.

distribution function *Mathematics.* **1.** let X be a random variable defined on a probability space (Ω, F, P). The distribution function of X is $F(x) = P(X < x)$. $F(x)$ is monotone increasing, left-continuous, and satisfies $\lim_{x\to+\infty} F(x) = 1$ and $\lim_{x\to-\infty} F(x) = 0$. **2.** let f be a finite-valued measurable function on a measure space (X, A, μ) in which $\mu(X) < \infty$. The distribution function of f is the function $g(t)$ of a real variable given by $g(t) = \mu(\{x : f(x) < t\})$. $g(t)$ is monotone increasing, left-continuous, and satisfies $\lim_{t\to+\infty} g(t) = \mu(X)$ and $\lim_{t\to-\infty} g(t) = 0$. *Statistics.* see CUMULATIVE DISTRIBUTION FUNCTION.

distribution law *Physical Chemistry.* a law stating that if a substance is dissolved in two different immiscible liquids, the ratio of the concentration of the substance in the two solutions is constant, regardless of the quantity of solute. Also, PARTITION LAW, NERNST LAW, BOLTZMANN DISTRIBUTION LAW. *Physics.* a law that describes the probability of particle distribution in a given volume of phase space or the number of particles per unit volume of phase space, through the use of a density function.

distribution of intensities *Crystallography.* the plot of intensity of Bragg reflections versus the number within that range of intensity. Intensities in the X-ray diffraction pattern of a noncentrosymmetric crystal tend to be clustered more tightly around the mean than do those from a centrosymmetric one. This forms the basis for one test for the presence or absence of a center of symmetry in the crystal.

distribution switchboard *Electricity.* **1.** a switchboard through which signals may be routed to or between various points. **2.** a switchboard for routing electric energy at a voltage common for each point of use.

distribution system *Electricity.* the circuitry associated with high-voltage switchgear, step-down transformers, voltage dividers, and other related equipment used to receive high-voltage electricity from a primary source and route it at lower voltages to substations or points of use. Also, ELECTRIC DISTRIBUTION SYSTEM.

distribution transformer *Electricity.* a transformer located near users, which steps down high-voltage power for distribution.

distributive *Mathematics.* relating to or obeying the distributive law.

distributive fault see STEP FAULT.

distributive faulting see IMBRICATE STRUCTURE.

distributive lattice *Mathematics.* a lattice in which one of the following (dual) identities holds for all elements x, y, and z of the lattice:

$$(x \wedge y) \vee (x \wedge z) = x \wedge (y \vee z) \text{ or } (x \vee y) \wedge (x \vee z) = x \vee (y \wedge z)$$

This requirement strengthens the distributive inequalities that hold in any lattice:

$$(x \wedge y) \vee (x \wedge z) \leq x \wedge (y \vee z) \text{ and } x \vee (y \wedge z) \leq (x \vee y) \wedge (x \vee z)$$

The dual of a distributive lattice is distributive, and a finite lattice is distributive if and only if it is isomorphic to a ring of sets.

distributive law *Mathematics.* the requirement that the result of an operation is independent of whether it is carried out before or after another operation. For example, in a ring, multiplication is distributive with respect to addition since $a(b + c) = ab + ac$ and $(b + c)a = ba + ca$ for all a, b, and c in the ring; in other words, multiplication distributes over addition. In general, however, addition does not distribute over multiplication. The set theoretic operations of union and intersection of sets distribute over each other; i.e., $A \cap (B \cup C) = (A \cap B) \cup (A \cap C)$ and $A \cup (B \cap C) = (A \cup B) \cap (A \cup C)$ for all sets A, B, and C.

distributor *Electricity.* in an automobile or other motor vehicle, a device with a rotating drum that directs the secondary current from the induction coil to spark plugs in the various cylinders of an internal combustion engine at correct times and in correct sequence. *Electronics.* a register that temporarily stores data transferred into or out of computer memory and functions as a distribution point for that data; information in the register can be changed without necessarily changing the contents of the memory. Also, MEMORY REGISTER.

distributor gear *Mechanical Engineering.* a gear that meshes with the camshaft gear to rotate the distributor shaft.

distributor points *Electricity.* in older ignition systems, the cam-operated contacts that create a high-voltage surge in the secondary windings of the induction coil to trigger the ignition pulse.

district forecast *Meteorology.* a general weather forecast of conditions over a specified area.

district heating *Mechanical Engineering.* the process of supplying heat from a main source to a group of buildings.

disturbance an interference with normal or existing conditions; specific uses include: *Archaeology.* **1.** the altering of an archaeological context by the effect of some unrelated activity at a later time, such as the plowing of soil in which artifacts are deposited. **2.** the nonscientific removal of an artifact from its archaeological context. *Control Systems.* the introduction of an unwanted command signal into a control system. *Geology.* a minor crustal deformation involving the bending, folding, or faulting of rock. *Meteorology.* **1.** a low-pressure area, especially one that is relatively small in size and effect. **2.** any sign of developing cyclonic circulation. **3.** an individual circulation system within the primary circulation of the atmosphere. 4. any deviation in flow or pressure associated with a weather disturbance. *Telecommunications.* a sudden increase in ionization density of the D-layer of the ionosphere, usually caused by solar flares and often resulting in greatly increased radio wave absorption (radio blackout).

disturbance climax see DISCLIMAX.

disturbed metal *Metallurgy.* a heavily deformed superficial layer of metal caused by grinding or polishing.

disturbing potential *Geodesy.* the difference between the actual value of the gravity potential at a point on the earth and the standard gravity potential.

distyle [di´stīl´; dī´stīl´] *Architecture.* having two columns across the front.

distyle in antis *Architecture.* having a pair of columns set between two antae in front.

disulfate *Chemistry.* a compound that contains two sulfate radicals.

disulfide *Chemistry.* a compound that contains two atoms of sulfur.

disulfide bridge *Genetics.* a couplet bond between two sulfur atoms that is important in linking polypeptide chains in proteins, each linkage arising as a result of the oxidation of the sulfhydryl (SH) groups on two molecules of cysteine. Also, **disulfide bond, disulfide linkage.**

disulfonate *Chemistry.* a compound that contains two sulfonate groups.

disulfonic acid *Chemistry.* a compound that contains two sulfonic acid groups.

disymmetry *Materials Science.* a phenomenon occurring in light-scattering experiments, in which large molecular particles cause a change in the scattered light intensity.

ditactic polymer *Materials Science.* a polymer that contains two sites of stereoisomerism in the main chain of the configurational base unit.

ditch *Civil Engineering.* a long, narrow, relatively shallow excavation in the earth, dug for drainage or irrigation, to bury pipes, wires, or cables, or for various other purposes. *Aviation.* **1.** to make an emergency landing on water, often involving subsequent abandonment and loss of the aircraft. **2.** an instance of this type of landing.

ditcher *Mechanical Devices.* a machine that is designed to dig a long, continuous trench of a standard width and depth. Also, **ditchdigger.**

ditching *Engineering.* the introduction of ditches around tanks to catch overflow, or along roadways to carry runoff.

diterpene *Organic Chemistry.* hydrocarbons or their derivatives that contain four isoprene units, thereby containing twenty carbon atoms and four branched methyl groups.

dither *Telecommunications.* in television broadcasting, a method of representing the entire gray scale of a picture by elements consisting of only one of two levels, black or white. *Control Systems.* see BUZZ.

dithianon *Plant Pathology.* a pesticide that is applied to fruit and vegetable plants before they come in contact with harmful fungi; it enables the plants to destroy the fungi in the case of contact.

dithiocarbamate *Organic Chemistry.* a salt of dithiocarbamic acid.

dithiocarbamic acid *Organic Chemistry.* NH_2CS_2H, colorless needles; soluble in alcohol; forms metal salts used as strong rubber accelerators. Also, AMINODITHIOFORMIC ACID.

dithiocic acid *Organic Chemistry.* the carboxy group of an organic acid in which sulfur atoms have replaced both oxygen atoms.

dithionate *Chemistry.* any salt of dithionic acid.

dithionic acid *Inorganic Chemistry.* $H_2S_2O_6$, a strong acid existing only in solution or in the form of its salts.

dithiooxamide *Organic Chemistry.* $SC(NH_2)C(NH_2)S$, a stable, orange-red powder, insoluble in water and soluble in alcohol; decomposes at 140°C; used as a chemical intermediate. Also, RUBEANIC ACID.

dithiothreitol *Organic Chemistry.* $C_4H_{10}O_2S_2$, a crystalline solid that is soluble in water and alcohol; melts at 42–43°C; used as a reducing agent for proteins and enzymes, and in biochemical research. Also, CLELAND'S REAGENT.

ditokous *Zoology.* **1.** producing two young, or laying two eggs at each birth. **2.** producing two forms of young, as in certain worms.

Ditrichaceae *Botany.* a cosmopolitan family of small, slender mosses of the order Dicranales, characterized by simple and erect stems with terminal sporophytes, undifferentiated alar cells, and a haplolepideous peristome; forming tufts on soil and rock, and containing *Ceratodon,* one of the most common mosses in the world.

Dittus-Boelter equation *Fluid Mechanics.* an empirical relation for correlating the heat transfer coefficients in forced-circulation evaporators; the dimensionless groups in the relation are the Reynolds, Prandtl, and heat transfer numbers.

ditungsten carbide *Inorganic Chemistry.* W_2C, a grayish-black powder that is insoluble in water; melts at 2860°C and boils at 6000°C; one of the hardest materials known; it has various industrial uses.

diuresis [dī´yù rē´sis] *Medicine.* the increased formation and excretion of urine.

diuretic [dī´yù ret´ik] *Medicine.* **1.** any substance or agent that promotes the increased formation and excretion of urine. **2.** increasing the excretion of urine. Coffee, tea, beer, and certain foods have a diuretic effect. *Pharmacology.* a drug or medication that is intended to have this effect; often prescribed to reduce the volume of extracellular fluid in the treatment of edema, hypertension, and other disorders.

diuretic hormone *Biochemistry.* a neurohormone that promotes water loss in insects by increasing the amount of fluid that is secreted into the Malpighian tubes.

diurnal [dī ur´nəl] *Meteorology.* relating to a complete cycle of 24 hours that recurs every 24 hours. Thus, **diurnal cycle.** *Biology.* **1.** relating to the daytime; occurring during daylight hours. **2.** occurring every day.

diurnal aberration *Astronomy.* the part of stellar aberration that comes from the rotation of the earth.

diurnal circle *Astronomy.* the apparent curving track of a celestial object relative to the horizon.

diurnal cooling *Oceanography.* the loss of heat by a body of water through surface radiation during the night.

diurnal current *Oceanography.* a tidal current that accompanies a diurnal tide.

diurnal heating *Oceanography.* solar radiation and heat absorbed by a body of water during the day.

diurnal inequality *Oceanography.* the difference between the heights of two successive high or low waters in one lunar day.

diurnal libration *Astronomy.* a very small change in the visible hemisphere of the moon that occurs because an observer, viewing from one location on earth, slightly changes position relative to the moon between moonrise and moonset.

diurnal migration *Biology.* the daily movement of oceanic organisms from deeper water to the surface in the evening and the return to deeper water at dawn.

diurnal motion *Astronomy.* an object's apparent motion from east to west across the sky, caused by the earth's rotation.

diurnal parallax *Astronomy.* the apparent angular size of the earth's radius as measured from a celestial object within the solar system. Also, GEOMETRIC PARALLAX.

diurnal range see GREAT DIURNAL RANGE.

diurnal tide *Oceanography.* a tide consisting of only one low water and one high water during a lunar day.

diurnal variation *Geophysics.* a change in the earth's magnetic field measured at a set point on a day-to-day basis. Also, DAILY VARIATION.

div. divide; division; divisor.

divalence see BIVALENCE.

divalent see BIVALENT.

divalent carbon *Organic Chemistry.* a carbon that is capable of combining with two atoms of hydrogen.

divaricate *Biology.* spread apart at a wide angle; broadly forked.

divaricator *Zoology.* the muscle in brachiopods that opens the valves of the shell.

dive *Engineering.* **1.** to plunge headfirst into water. **2.** to submerge one's body, or an object, chamber, or vessel, into an underwater environment, especially for the purpose of studying this environment or its effects. *Aviation.* **1.** a steep nose-down descent by an aircraft or missile, with or without power, in which the air speed is greater than the maximum speed in horizontal flight. **2.** to make such a descent. *Navigation.* of a vessel, to plunge down steeply into the water, as a submarine does in submerging.

dive bomber *Aviation.* **1.** an aircraft designed to release bombs during or at the end of a steep dive. **2.** specifically, a fighter-bomber aircraft of this type used in the World War II era, such as the American Dauntless (Douglas SBD 5) or German Stuka. Thus, **dive-bomb, dive bombing.**

dive bomber

dive brake *Aviation.* an air brake that is characterized by flaps or dive brakes, which can be mounted on the upper or lower surface of the wing or on the fuselage. They enable an aircraft to dive steeply at moderate speeds; common on dive bombers and sailplanes.

divergence a movement away from a central point or path; specific uses include: *Evolution.* a process by which related organisms acquire dissimilar hereditary characters as a result of natural selection or random change, and thus deviate from a common ancestral form. Also, **divergent evolution.** *Meteorology.* the net flow of air from a given layer or region of the atmosphere due to outflowing winds. *Oceanography.* a horizontal flow of water outward from a common center or zone, often associated with upwelling. *Mathematics.* **1.** the divergence of a vector-valued function $A = iA_1 + jA_2 + kA_3$, denoted div(A) or $\nabla \cdot A$, is the formal dot product of the del operator and A; i.e.,

$$\nabla \cdot A = (i\,\partial/\partial x + j\,\partial/\partial y + k\,\partial/\partial z) \cdot A = \partial A_1/\partial x + \partial A_2/\partial y + \partial A_3/\partial z$$

2. in general, on a (pseudo-) Riemannian manifold, the divergence of a covariant r tensor is a covariant $(r-1)$ tensor. *Physics.* the spatial radiation of a quantity from a source, as in electromagnetic radiation emanating from a point source. *Electronics.* the tendency of an electron stream in a cathode-ray tube to spread due to the mutual repulsion of the negatively charged electrons within the stream. *Fluid Mechanics.* the flow pattern achieved in a uniformly diverging duct with a straight axis, where the total angle between the diverging walls is not greater than about seven degrees.

divergence loss *Geophysics.* the power lost from a seismic pulse during geophysical prospecting due to the spreading effect of geologic formations within the earth.

divergence theorem see GAUSS' THEOREM.

divergent die *Engineering.* a die that has internal channels pointed to the orifice that is diverging.

divergent integral *Mathematics.* an improper definite integral that cannot be assigned a finite value.

divergent nozzle *Mechanical Devices.* a nozzle in which the cross section increases from its entry point to its exit area; used in compound impulse turbines.

divergent sequence *Mathematics.* an infinite sequence of numbers with no limit.

divergent series *Mathematics.* an infinite series whose associated sequence of partial sums is divergent.

diverging duct *Mechanical Devices.* a conduit of fluids in which the cross-section area of the interior becomes larger in the direction of flow.

diverging lens *Optics.* a lens that is thicker at its edges than at its center; spreads parallel rays of light as they pass through it. Also, CONCAVE LENS, NEGATIVE LENS, DIVERGENT LENS, DISPERSIVE LENS.

diverging meniscus lens *Optics.* a lens having a concave surface with a greater curvature than its convex surface. Also, NEGATIVE MENISCUS LENS.

diverging yaw *Space Technology.* the angle of yaw in the flight of an airplane or projectile that increases from the initial yaw, causing the airplane projectile to become directionally unstable.

diver method *Physical Chemistry.* a system to establish particle size or density by comparing how far the particle sinks in a liquid with how far a small glass plummet of a known dimension and density sinks in the same liquid.

diverse vector area *Navigation.* an area in a radar environment in which a variety of departure routes are possible.

diversified farm *Agriculture.* a farm on which a variety of crops or livestock are raised. Thus, **diversified farming.**

diversifying selection *Genetics.* a type of selection in which matings between different phenotypes are favored, resulting in an increase in heterozygosity in the population. Also, DISRUPTIVE SELECTION.

diversion *Military Science.* **1.** an attack or other action that draws the enemy's attention away from the principal operation. **2.** in naval mine warfare, a route or channel bypassing a dangerous area. **3.** any rerouting of troops or cargo.

diversion chamber *Engineering.* a chamber that can focus a stream into one or more channels.

diversion display *Behavior.* behavior that serves to attract the attention of a predator away from a more vulnerable member of a group.

diversity *Telecommunications.* in a radio communications system, a method of combatting the effects of path fading by combining two or more received signals.

diversity factor *Electricity.* the ratio of the sum of the individual maximum demands to the total maximum demand of an electrical distribution system.

diversity gain *Telecommunications.* the ratio of the signal strength that is obtained by diversity combining to the signal strength obtained by a single path or channel; expressed in decibels.

diversity radar *Engineering.* a radar system in which the antenna and video display are shared by multiple sets of transmitters and receivers, each pair operating at a different frequency.

diversity receiver *Electronics.* a radio receiver designed for space or frequency diversity reception. Also, DUAL DIVERSITY RECEIVER.

diversity reception *Telecommunications.* reception in which a resultant signal is obtained by a combination of two or more independent sources of received signal energy that carry the same modulation or information; used to minimize the effects of fading.

diverter valve see AIR BYPASS VALVE.

diverticulectomy [dī´ver tik yə lek´tə mē] *Surgery.* the excision of a diverticulum.

diverticulitis [dī´ver tik yə lī´tis] *Medicine.* inflammation of a diverticulum, most commonly when the small pockets in the wall of the colon fill with stagnant fecal material.

diverticulopexy [dī´ver tik yə lō peks´ē] *Surgery.* the surgical fixation of a diverticulum into a new position following its separation from the initial adjacent or adherent structures.

diverticulum [dī´ver tik´yə ləm] *plural,* **diverticula.** *Anatomy.* an enclosed sac of variable size that occurs normally or is created by the herniation of the lining mucous membrane through a defect in the muscular wall of a tubular organ.

Divesian see OXFORDIAN.

divide *Science.* to separate into parts. *Mathematics.* to perform an operation of division. *Geology.* the boundary between two adjacent drainage areas whose streams flow in opposite directions. Also, WATERSHED.

divide check *Computer Programming.* an overflow indicator that results during arithmetic division when the dividend has a magnitude greater that the divisor, as in an attempt to divide by zero.

divided cell *Metallurgy.* an electrochemical cell that has a diaphragm between the anodic and the cathodic compartments.

divided highway *Civil Engineering.* a highway in which the traffic lanes for opposite directions are separated by a median strip or barrier.

divided pitch *Design Engineering.* in a screw with multiple threads, the distance between corresponding points on two adjacent threads measured parallel to the axis.

divided slit scan *Computer Technology.* an electro-optical converter in an optical character reader that uses a narrow column of photoelectric cells to scan the characters and convert them into electric signals.

dividend *Mathematics.* the quantity into which another quantity, the divisor, is to be divided, producing the quotient; for example, a in the division problem $a \div b = c$.

diving bell *Engineering.* a bell-shaped watertight vessel that is supplied with compressed air and submerged in water to provide an environment for underwater excavation or other work.

diving bird *Vertebrate Zoology.* a term applied to any of various birds adapted for swimming and diving, especially ducks.

diving petrel *Vertebrate Zoology.* any of several seabirds of the family Pelecanoididae, having small, compact bodies, tubelike processes near the nostrils, and often drab plumage; found in the Southern Hemisphere.

diving rudder *Navigation.* a horizontal mounted fin used to control the operating depth of a submerged submarine. Also, **diving plane.**

diving suit *Engineering.* a weighted waterproof garment that is supplied with an air or gas mixture through a hose attached to the helmet.

divining *Mining Engineering.* a method, not scientifically accepted, that uses a hand-held stick or other device to find subsurface water or minerals. Also, DOWSING.

divining rod *Mining Engineering.* an unscientific device, usually a forked rod or tree branch, that supposedly dips when held over water, oil, gas, or minerals. Also, DOWSING ROD.

divinyl see BUTADIENE.

divinylacetylene *Organic Chemistry.* $C_2H{=}CHC{\equiv}CCH{=}CH_2$, a trimer of acetylene formed by passing it into a hydrochloric acid solution containing a metallic catalyst; used as an intermediate.

divinylbenzene *Organic Chemistry.* $C_6H_4(CH{=}CH_2)_2$, an isomeric, combustible, toxic, water-white liquid that is insoluble in water and soluble in methanol; its *meta* form boils at 199.5°C; used as a polymerization monomer. Also, VINYLSTYRENE.

divisibility *Mathematics.* the branch of elementary number theory concerned with properties of the integers, including the fundamental theorem of arithmetic, least common multiples, greatest common divisors, and the distribution of prime numbers.

divisible *Mathematics.* an integer a is said to be divisible by a nonzero integer b if the equation $a = bx$ has an integer solution. It is common usage to say that "b divides a" and write $b \mid a$.

division *Mathematics.* a noncommutative binary operation denoted by the symbol ÷; for nonzero b, $a \div b = c = ab^{-1}$, where b^{-1} is the multiplicative inverse of b. In this example, a is the dividend, b is the divisor, c is the quotient. *Military Science.* **1.** a military unit that includes all the arms and services required for sustained combat; it is larger than a regiment or brigade and smaller than a corps. **2.** a number of naval vessels or aircraft, usually of similar type, grouped together for administrative and operational purposes. *Computer Programming.* one of the four required parts of a COBOL source program: identification, environment, data, and procedure divisions. *Systematics.* a primary taxonomic rank in botany below a kingdom and above a class, corresponding to a phylum in zoology; names of this type usually end in *-phyta.*

division algorithm *Mathematics.* **1.** if a and b are integers and $b \neq 0$, there are unique integers q and r such that $a = qb + r$ and $0 \leq r < |b|$. **2.** the application of the Euclidean algorithm to polynomials; if F is a field and $f(x)$ and $g(x) \neq 0$ are two polynomials in the polynomial ring $F[x]$, then there exist polynomials $q(x)$ and $r(x)$ in $F[x]$ such that $f(x) = q(x)g(x) + r(x)$, where $r(x)$ is either the zero polynomial or has degree less than the degree of $g(x)$. Also, LONG DIVISION.

division plate *Mechanical Engineering.* **1.** in a crosshead engine, a diaphragm that surrounds the piston rod and separates the crankcase from the lower portion of the cylinder. **2.** in a machine tool, a plate used to position the plunger of an indexing head.

division ring *Mathematics.* **1.** a ring whose nonzero elements form a multiplicative group. Also, SKEW FIELD. **2.** a commutative division ring is a field. If the division ring is also an algebra, it is a **division algebra.**

division subroutine *Computer Programming.* a program that performs arithmetic division by repeated subtraction.

division wall *Building Engineering.* a principal wall from which various divisions within a building are defined.

divisor *Mathematics.* a quantity by which another quantity, the dividend, is to be divided, producing the quotient; i.e., b in the division problem $a \div b = c$.

divorced eutectic *Metallurgy.* a eutectic structure in which the component phases are not finely divided, but massive.

divulsion *Surgery.* a forcible separating, tearing, or stretching.

Dix, Dorothea 1802–1887, American social reformer; she founded institutions for the humane care of psychiatric patients.

dixenite *Mineralogy.* $Cu^{+1}Mn^{+2}_{14}Fe^{+3}(As^{+3}O_3)_5(SiO_4)_2(As^{+5}O_4)(OH)_6$, a bronze to nearly black trigonal mineral occurring in aggregates of thin folia, having a specific gravity of 4.2 and a hardness of 3 to 4 on the Mohs scale; found with adelite.

Dixidae *Invertebrate Zoology.* a family of small, frail two-winged flies with freshwater larvae, in the series Nematocera.

dizygotic twins *Genetics.* twins that develop from two separate ova fertilized by two separate sperms. Also, FRATERNAL TWINS.

dizziness *Medicine.* a condition of lightheadedness, with the feeling that one's surroundings are spinning around and one is about to fall; associated with such disorders as epilepsy, anemia, and heart disease.

djalmaite see MICROLITE.

DJD degenerative joint disease.

djerfisherite [jur fish´ə rīt] *Mineralogy.* $K_6(Cu,Fe,Ni)_{25}S_{26}Cl$, a greenish-brown mineral of undetermined hardness and specific gravity, occurring as tiny grains; found with nickel-iron, troilite, and daubreelite in meteorites.

Djulfian *Geology.* a geologic stage of the Upper Permian period.

DKA diabetic ketoacidosis.

dkg or **dkg.** dekagram; dekagrams.

dkl or **dkl.** dekaliter; dekaliters.

dkm or **dkm.** dekameter; dekameters.

DL *Oncology.* an acronym for a chemotherapy regimen for cancer that includes the drugs doxorubicin and lomustine.

DL *Aviation.* the airline code for Delta Air Lines.

dl or **dl.** deciliter; deciliters.

DM diastolic murmur; diabetes mellitus.

dm or **dm.** decimeter; decimeters.

DMA or **dma** direct memory access.

DMD or **D.M.D.** Doctor of Dental Medicine.

DME dropping mercury electrode

D meson *Particle Physics.* a chargeless pseudovector meson resonance with an approximate mass of 1285 MeV.

DMOS diffused metal-oxide semiconductor.

DMTS technique or **procedure** *Behavior.* delayed-matching-to-sample, a method of investigating animal short-term memory by presenting two alternative stimuli, one of which was previously presented.

DNA *Biochemistry.* a nucleic acid that constitutes the genetic material of all cellular organisms and the DNA viruses; DNA replicates and controls through messenger RNA the inheritable characteristics of all organisms. A molecule of DNA is made up of two parallel twisted chains of alternating units of phosphoric acid and deoxyribose, linked by crosspieces of the purine bases and the pyrimidine bases, resulting in a right-handed helical structure, that carries genetic information encoded in the sequence of the bases. Also, DEOXYRIBONUCLEIC ACID.

dnaA gene *Genetics.* an allele in *Escherichia coli* that produces a defective protein influencing DNA replication at the chromosomal origin.

DNA annealing *Molecular Biology.* the renaturation of complementary single-stranded DNA molecules into duplex DNA following their earlier denaturation; involves the formation of hydrogen bonds between pairs of bases.

DNAase see DNASE.

DNA bank see DNA LIBRARY.

dnaB gene *Genetics.* a gene in *Escherichia coli* that produces a protein influencing primer formation.

dnaC gene *Genetics.* a gene in *Escherichia coli* that produces a protein influencing primosome formation. Also, **dnaD gene.**

DNA cloning *Molecular Biology.* a technique in which DNA from one source is inserted in a cloning vector and then introduced into a host bacterium.

DNA-directed RNA polymerase see RNA POLYMERASE.

DNA-driven reaction *Molecular Biology.* a reaction in which complementary DNA greatly exceeds radioactive tracer RNA, such that the tracer will bind chiefly to highly repetitive DNA sequences.

DNA duplex *Molecular Biology.* the double helix structure of DNA, as identified by Watson and Crick, consisting of two strands of DNA wound together, usually with a right-hand twist.

dnaE gene *Genetics.* an allele in *Escherichia coli* that encodes the genetic code for subunits of DNA polymerase III, an enzyme responsible for cellular replication of DNA. Also, POLC GENE.

DNA excision *Molecular Biology.* the detachment and removal of a damaged section of DNA strand prior to repair; carried out by a number of specialized enzymes in the cell.

dnaF gene see NRDA GENE.

DNA filter assay *Molecular Biology.* the exposure of DNA to an immobilizing filter material and then to a solution that contains radioactively labeled probe DNA or RNA; used to identify complementary sequences in the original DNA sample.

DNA fingerprinting *Molecular Biology.* a laboratory technique in which the banding patterns of DNA fragments from two different individuals are compared; used in any species, including humans, to indicate relatedness.

DNA footprinting *Molecular Biology.* the use of electrophoretic gels to detect regions of DNA protected by specific proteins; a pure sample of DNA is compared to one that has been incubated with DNA-binding protein.

DNA gene *Genetics.* a gene in *Escherichia coli* that produces a protein influencing DNA replication at the chromosomal origin.

dnaG gene *Genetics.* an allele in *Escherichia coli* that encodes the genetic code for primase, an enzyme important in the cellular replication of DNA.

DNA gyrase *Enzymology.* an enzyme found in *Escherichia coli* that converts relaxed, closed-circular, duplex DNA into a negatively super-helical form prior to replication, transcription, repair, or recombination. Also, GYRASE.

DNA helicase *Molecular Biology.* an enzyme that unwinds the DNA double helix during DNA replication.

dnaH gene see DNAZ GENE.

DNA homology *Molecular Biology.* the degree of relatedness of two or more strands or portions of DNA that closely resemble one another in that they contain similar or identical sequences of bases.

DNA hybridization *Molecular Biology.* an analytical technique using radioisotope labeling to determine the degree of similarity of DNA base sequences in two species.

dnaI gene *Genetics.* a DNA gene that influences DNA replication at the replication origin.

dnaJ gene *Genetics.* a DNA gene whose function is similar to that of the closely linked dnaK gene.

dnaK gene *Genetics.* a DNA gene whose function is necessary to ensure the viability of *Escherichia coli* at high temperatures; it is also considered to be essential for phage λ DNA replication.

dnaL gene *Genetics.* an uncharacterized DNA gene.

DNA library *Molecular Biology.* a collection of cloned DNA fragments that represents at least part of the genome of an organism. Also, DNA BANK.

DNA ligase *Biochemistry.* an enzyme that connects the ends of two DNA chains by initiating the synthesis of a phosphodiester bond between the 3'-hydroxyl group on one chain and the 5' phosphate group on the other.

DNA ligation *Molecular Biology.* the joining of two DNA strands by the formation of a phosphodiester bond between their terminal nucleotides.

DNA melting *Biochemistry.* a process in which excessive heat causes the hydrogen bonds between the paired bases in the double helix to break, causing it to unravel into two separate strands.

DNA methylation *Molecular Biology.* the attachment of a methyl group to cytosine residues of eukaryotic DNA to form 5-methylcytosine.

dnaM gene *Genetics.* an uncharacterized DNA gene.

DNA modification *Molecular Biology.* any of various chemical changes that may be undergone by DNA following its initial replication.

dnaP gene *Genetics.* a DNA gene that influences DNA replication at the replication origin.

DNA polymerase *Enzymology.* an enzyme that catalyzes the formation of the DNA molecule. Also, **DNA-directed DNA polymerase, DNA nucleotidyltransferase.**

DNA polymerase I *Biochemistry.* one of three polymerases present in *Escherichia coli,* responsible for a number of DNA functions, including the sequential removal of nucleotides from the 3' end of a chain and the binding of nucleotides to the 5' end, the repair of DNA, and the removal of RNA primers.

DNA polymerase II *Biochemistry.* one of three polymerases present in *Escherichia coli,* responsible for a number of DNA functions, including a possible role in the repair of DNA damaged by ultraviolet radiation.

DNA polymerase III *Biochemistry.* one of three polymerases present in *Escherichia coli,* responsible for a number of DNA functions, including controlling the proofreading aspects of DNA replication, the removal of nucleotides sequentially from the 3' end of the strand, and the binding of nucleotides from a 5' end.

DNA polymerization *Molecular Biology.* the formation of the DNA macromolecule by an aggregate of covalently bonded repeating subunits.

DNA-protein interaction *Molecular Biology.* any molecular interaction that forms a stable complex such as a nucleosome or a transient one involving gene regulatory proteins; a crucial aspect of living systems.

DNA puff *Molecular Biology.* a localized swelling of a specific region of a chromosome due to localized synthesis of DNA or RNA.

dnaQ gene *Genetics.* an allele in *Escherichia coli* that carries the genetic code for subunits of DNA polymerase III.

DNA repair *Molecular Biology.* a process in which various enzymes cooperate to repair mutational damage incurred by the DNA; involves excision and replacement.

DNA replication *Molecular Biology.* an exact copying of a parent DNA molecule through separation of its two strands and enzymatic synthesis of two new complementary strands, each of which base pairs with one of the parental strands; this yields two identical daughter molecules.

DNA restriction endonuclease see RESTRICTION ENDONUCLEASE.

DNA-RNA hybridization *Molecular Biology.* the pairing of DNA with RNA molecules by hydrogen bonding between complementary base pairs.

DNase *Genetics.* deoxyribonuclease; an enzyme that breaks the DNA molecule into separate oligonucleotide fragments. Also, DNAASE.

DNase footprinting *Molecular Biology.* a laboratory technique utilizing a comparison of electrophoresed fragments of pure DNA and DNA that have been conjugated to proteins to determine which portions of a specific molecule are bound to and protected by a protein from attack by endonucleases.

DNA sequencing *Molecular Biology.* the process of determining the precise sequence of bases in a length of DNA, using a variety of methods including chemical sequencing and dideoxy sequencing.

DNA shearing *Molecular Biology.* the process whereby extremely long DNA molecules of high molecular weight are reduced in size by mechanical tearing; sometimes the accidental result of the preparation of a DNA sample for testing.

DNA splicing *Molecular Biology.* the ligation of DNA in vitro by the use of DNA enzymes that catalyze the formation of phosphodiester bonds.

DNA synthesizer *Biotechnology.* any of a wide variety of commercially available automated machines that produce short polynucleotide chains similar in structure to oligonucleotides.

dnaT gene *Genetics.* a DNA gene that may affect dnaC protein in vivo; its mutants are unable to destabilize the replication complex at chain termination.

DNA tumor virus *Virology.* any DNA-containing, oncogenic animal virus.

DNA twisting *Molecular Biology.* the reverse coiling or twisting of duplex DNA molecules in a direction opposite to the turns of the strands of the double helix. Also, NEGATIVE SUPERCOILING, UNDERWINDING.

DNA unwinding protein *Biochemistry.* a protein that binds to single-stranded DNA and facilitates the unwinding of the DNA during replication and recombination.

dnaW gene see ADK GENE.

DNA writhing *Molecular Biology.* a coiling of a duplex DNA molecule upon itself so that it crosses its own axis. Also, POSITIVE SUPERCOILING.

dnaX gene *Genetics.* a gene in *Escherichia coli* that carries the genetic code for certain subunits of DNA polymerase III.

dnaZ gene *Genetics.* a gene in *Escherichia coli* that carries the genetic code for certain subunits of DNA polymerase III. Also, DNAH GENE.

DNB or **D.N.B.** Diplomate of the National Board (of Medical Examiners).

DNC distributed numerical control.

Dneiper or **Dnepr** [nē´pər] *Geography.* a long (1430 miles) river that rises west of Moscow and flows southward through the Ukraine to the Black Sea.

DNF disjunctive normal form.

DNR or **D.N.R.** do not resuscitate. Also, **dnr** or **d.n.r.**

DO or **D.O.** Doctor of Osteopathy.

D_2O deuterium oxide (heavy water).

DOA or **D.O. A.** dead on arrival.

DOB or **D.O.B.** date of birth. Also, **dob** or **d.o.b.**

Dobbin's reagent *Analytical Chemistry.* a solution of mercuric chloride, potassium iodide, ammonium chloride, and sodium hydroxide, that is used to test for caustic alkalies in soap or in products containing soap.

dobby weave *Textiles.* a fabric woven on a special loom and having a small, allover design, dots, or stripes.

Dobrowolsky generator [däb´rə wäl´skē] *Electricity.* a three-wire DC generator with a balance coil connected across the armature; the midpoint of the coil produces the midpoint voltage for the system.

dobsonfly or **dobson fly** *Invertebrate Zoology.* a large, soft-bodied insect, *Corydalus cornutus*, characterized by four wings with distinct membranes, biting mouthparts, and, in the male, large mandibles jutting out from the head. (From the surname *Dobson.*)

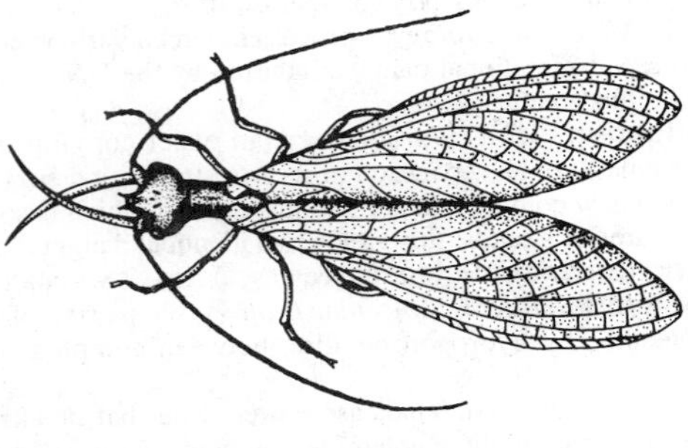

dobsonfly

Dobson prop *Mining Engineering.* a prop that is basically a self-contained hydraulic jack with an integral pump unit built into the prop, designed to yield at 25 tons. Thus, **Dobson support system.**

Dobson spectrophotometer *Spectroscopy.* a photoelectric device used to determine the ozone content of the atmosphere by comparing the solar energy at two wavelengths in the absorption band of ozone.

Dobzhansky, Theodosius [dub zhän´skē] 1900–1975, Russian-born American biologist; contributed to the synthetic theory of evolution.

dock *Civil Engineering.* **1.** a platform or other such structure built along a shore for the landing and unloading of ships. **2.** a basin or channel between two piers or wharves for the berthing of ships. **3.** see DRY DOCK. *Transportation Engineering.* a raised platform for loading and unloading trucks. *Aviation.* **1.** a place or facility for the inspection, maintenance, or repair of aircraft. **2.** an airship hangar.

docking *Navigation.* any process of guiding a craft into a dock. *Space Technology.* **1.** the mechanical coupling of two or more spacecraft, payloads, or orbiting objects. **2.** specifically, the process of sealing two manned spacecraft in orbit so that hatches can be opened between them without loss of cabin atmosphere, enabling crew members to move freely from one spacecraft to another.

docking bridge *Naval Architecture.* a raised deck and conning position at the stern of a ship.

docking keel *Naval Architecture.* one of a pair of keels fitted to the underside of a hull on each side of the main keel, designed to help support the ship when in drydock.

docking station *Robotics.* a multitask industrial robot designed to prepare, inspect, test, and package a finished product.

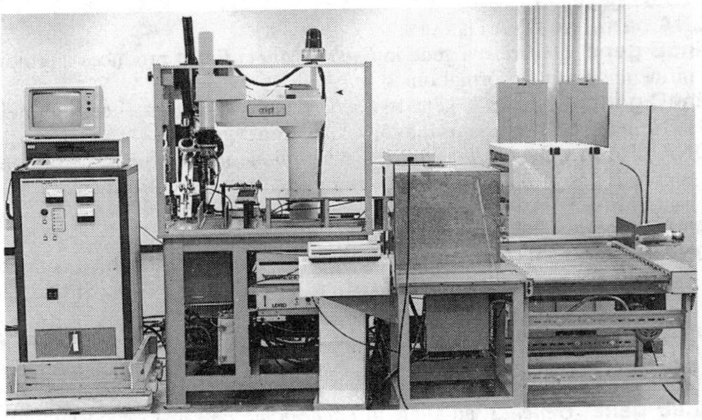

docking station

dock landing ship *Naval Architecture.* a large amphibious warfare ship designed to carry loaded landing craft aboard in a well deck, which is flooded to allow them to float out.

Docodonta *Paleontology.* one of the five early mammalian orders of the late Triassic and Jurassic, in the extinct subclass Eotheria; fossils have been found in Eurasia, North America, and Africa; the molar teeth of the docodontids were square and multicusped.

***n*-docosane** *Organic Chemistry.* $C_{22}H_{46}$, a combustible solid; melts at 45.7°C and boils at 230°C at reduced pressure; used in organic synthesis and temperature-sensing devices.

docosanoic acid see BEHENIC ACID.

1-docosanol see BEHENYL ALCOHOL.

doctor *Medicine.* **1.** a person who is educated and licensed to practice the profession of medicine as the holder of the **Doctor of Medicine** (M.D.) degree. **2.** any of various other licensed health professionals, such as a dentist, veterinarian, chiropractor, osteopath, podiatrist, or optometrist. *Science.* a person who has completed an advanced course of study in a particular field of knowledge and been awarded the highest graduate degree offered by a university, the **Doctor of Philosophy** (Ph.D.) degree. *Meteorology.* a name for a cool sea breeze that occurs in tropical and subtropical climates. (Going back to a Latin word meaning "to teach;" in medieval times it referred to a scholar or learned person.)

doctor blade *Engineering.* a metallic blade mounted near the surface of rollers to regulate the amount of liquid on the rollers. Also, **doctor knife.**

doctor-blade process *Materials Science.* **1.** a process in which a doctor blade is used to smooth a surface, as in cleaning excess color from an engraved copper plate used for printing on pottery. **2.** the parting of thin ceramic sheets or wafers of the type used in miniature condensers.

doctor roll *Chemical Engineering.* a roller instrument that removes accumulated filter cake from rotary filter drums.

doctor solution *Chemical Engineering.* a sodium plumbite solution used to rid light petroleum products of mercaptans.

doctor test *Chemical Engineering.* the use of doctor solution for the detection of sulfur compounds in light petroleum distillates.

doctor treatment *Chemical Engineering.* a refining procedure for sweetening petroleum products; sulfur and sodium plumbite convert disulfides from the odoriferous mercaptans.

document *Computer Technology.* **1.** a record containing information readable by a person or a machine. **2.** to record information about a computer program, system, or package for use during development operation or maintenance.

documentation *Computer Technology.* the organized and accessible set of recorded information needed to define and develop a computer-based system, to operate and maintain a package, program, or system, and to maintain a record of changes to an operational system.

document handling *Computer Technology.* the process of submitting a single-sheet document for character reading; it includes loading, feeding, transporting through the reader, and unloading.

document reader *Computer Technology.* a type of input device that magnetically or optically reads a limited quantity of printed characters as a document is fed through the reader.

dodec- or **dodeca-** a combining form meaning "twelve," as in *dodecyl, dodecahydron.*

dodecagon *Mathematics.* a polygon with 12 sides.

dodecahedron *Mathematics.* a polyhedron with 12 faces.

dodecahydrate *Chemistry.* of a compound, containing twelve molecules of water.

dodecahydrodiphenylamine see DICYCLOHEXYLAMINE.

dodecamerous *Botany.* having twelve pistils or twelve stamens.

dodecanal see LAURYL ALDEHYDE.

dodecane *Organic Chemistry.* $CH_3(CH_2)_{10}CH_3$, a combustible, colorless liquid, insoluble in water and very soluble in alcohol and acids; melts at –9.6°C and boils at 216.23°C; used as a solvent and in organic synthesis and jet fuel research. Also, DIHEXYL.

dodecanoic acid see LAURIC ACID.

1-dodecene Organic Chemistry. $CH_2{=}CH(CH_2) \cdot 9CH_3$, a combustible, narcotic, colorless liquid, insoluble in water and soluble in alcohol; boils at 215.9°C and melts at –33.6°C; used in dyes, perfumes, medicine, and flavor manufacture. Also, **dodecylene.**

dodecyl *Organic Chemistry.* the radical $-C_{12}H_{25}$, derived by removing one hydrogen atom from dodecane, especially $CH_3(CH_2)_{10}CH_2-$.

dodecyl alcohol see LAURYL ALCOHOL.

dodecylbenzene *Organic Chemistry.* $C_{12}H_{25}C_6H_5$, a toxic, combustible, commercial blend of isomeric, predominantly monoalkylbenzenes; used in detergents. Also, detergent alkylate.

dodecyl mercaptan see LAURYL MERCAPTAN.

Dodge crusher *Mining Engineering.* a type of jaw crusher with a hinged-bottom jaw that allows for the discharge of a uniform product.

Dodge pulverizer *Mining Engineering.* a hexagonal barrel that rotates on a horizontal axis to reduce rock and ore by means of a pulverizing action created by steel balls inside the barrel.

dodging *Photogrammetry.* a photographic process during enlargement in which light passing through parts of the negative is prevented from striking the sensitized paper by manual or electronic means.

dodine *Plant Pathology.* a pesticide that is applied to fruit trees prior to contact with harmful fungi, destroying the fungus at its source before it can attack the tree.

dodo *Vertebrate Zoology.* a clumsy, flightless, extinct bird of the genus *Raphus* or *Pezophaps* in the family Raphidae, related to the pigeons, but about the size of a turkey; formerly inhabited the islands of Mauritius, Reunion, and Rodriguez. *Engineering.* a rectangular groove carved across the grain of a board.

dodo

doe *Vertebrate Zoology.* the adult female of the deer, antelope, goat, rabbit, and certain other animals of which the male is called a buck.

Doebner-Miller synthesis *Chemical Engineering.* the heating of aniline with paraldehyde in the presence of hydrochloric acid to synthesize methylquinoline.

DOF degrees of freedom.

dog *Vertebrate Zoology.* **1.** a very common domestic animal, classified as *Canis familiaris* and bred in a great many varieties; kept as a pet and used for various purposes such as hunting and protection since early human history. **2.** any carnivorous animal belonging to the same family, Canidae, including the wolves, jackals, foxes, and so on. *Mechanical Devices.* **1.** any of various devices thought of as resembling the jaws of a dog, in that they are used for gripping, fastening, or dragging a workpiece, such as a timber, iron, or metal component. **2.** a clamp used for holding and directing work through a lathe. Also, LATHE DOG. **3.** the iron piece in a pile driver that is used to lift the hammer.

DOG degree of disorder.

dog clutch *Mechanical Devices.* a machine clutch in which the protuberances on one end fit into the hollows or recesses on the other.

dog days *Meteorology.* a popular name for the hottest part of the summer season, usually between mid-July and early September. (From Sirius, the *Dog* Star, whose rising at this time was believed to cause the hot, dry, sultry weather.)

dogfish *Vertebrate Zoology.* **1.** any of several small sharks, particularly of the genera *Mustelus* and *Squalus,* destructive to other small fishes. **2.** any of various other small fishes, such as the bowfin.

dogger *Geology.* a lumpy mass of concretionary calcareous sandstone or an irregular nodule of ironstone.

doghole *Mining Engineering.* a small opening, not as large as a breakthrough, leading from one part of a coal mine to another.

doghouse any of various shelters or structures thought to resemble a dog's house; specific uses include: *Petroleum Engineering.* **1.** a small enclosed space on a drilling rig floor used for housing various supplies. **2.** a one-room, portable shelter at a well site that is used by the drilling crew, geologist, and other workers. *Electronics.* a small weatherproof structure containing antenna-tuning components, usually located at the foot of an antenna tower.

dog iron *Mechanical Devices.* an iron bar with ends bent at right angles or with an eyelet at one end, used as a cramp or joggle to hold together stone or timber.

dog-leash technique *Archaeology.* a method of defining a recovery area by attaching a rope to a centrally located marker stake and tracing the boundary established by the circular extent of the rope.

dogleg *Navigation.* an indirect course between two points that consists of two legs of a triangle; used when the straight-line course is barred by an obstruction. *Petroleum Engineering.* an abrupt deviation in the direction of the borehole of a well which can cause tubing wear and failure; an abrupt bend in a joint of pipe. Also, ELBOW.

dog screw *Mechanical Devices.* a screw with an irregularly shaped head, used to attach a watch mechanism to a casing.

Dog Star *Astronomy.* **1.** the bright star Sirius in Canis Major. **2.** the bright star Procyon in Canis Minor.

dogtooth violet *Botany.* any of several North American lilies of the genus *Erythronium,* characterized by a hard, bulblike corm from which a pair of leaves arise in the spring and last through the summer; may bear one white blossom or, in a few species, a cluster of white flowers atop a leafless stalk; found in moist woods and thickets.

dogwood *Botany.* **1.** any tree of the genus *Cornus,* particularly *C. florida,* of America, or *C. sanguinea,* of Europe. **2.** the wood of any such tree.

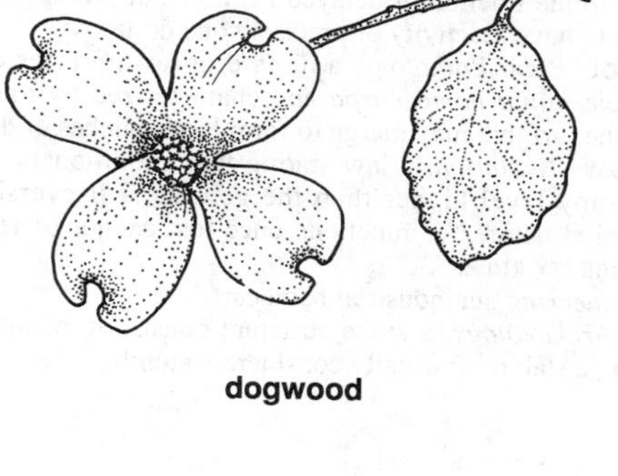

dogwood

Doherty amplifier *Electronics.* a linear radio-frequency power amplifier divided into two sections whose inputs and outputs are connected by quarter-wave networks; designed to ensure maximum undistorted transmitted power as required in amplitude-modulated transmitters.

Doisy, Edward 1893–1986, American biochemist; with Henrik Dam, he isolated vitamin K.

Doisy unit *Biology.* a unit used for the standardization of vitamin K.

Dokophyllum *Paleontology.* a genus of solitary rugose corals in the family Ketophyllidae; middle to upper Silurian.

dol *Neurology.* a unit for the measurement of pain intensity. (From Latin *dolor* for "pain.")

dolabriform *Biology.* shaped like an axe or a hatchet head. Also, **dolabrate.**

Dolan equation *Petroleum Engineering.* an empirical equation for the permeability damage factor in a reservoir by the intrusion of drilling mud or other materials.

Dolbeault complex [dōl´bō´] *Mathematics.* the cohomology of a complex manifold, analogous to the de Rham cohomology of a real manifold, but constructed with the differential (coboundary) operator $\bar{\partial}$ - ∂, the conjugate holomorphic part of the usual differential *d.*

Dolby or **dolby** *Acoustical Engineering.* **1.** relating to or denoting various devices that are used to eliminate noise from sound recording and reproduction and to improve the projection of stereophonic sound in motion-picture theaters. **2. Dolby.** a trade name for such devices. (Developed by Ray *Dolby,* born 1933, American engineer.)

Dolbyized *Acoustical Engineering.* recorded or equipped with a Dolby noise-reduction device.

Dolby noise reduction or **Dolby NR** see DOLBY.

dolchiocephalic *Anthropology.* having a relatively narrow head, in which the skull width is less than 75% of the length.

dolchiocranic see DOLCIOCEPHALIC.

doldrums *Meteorology.* a nautical name for the equatorial trough with its low pressure and light, variable winds. Also, EQUATORIAL CALMS.

dolerite see DIABASE.

dolerophanite *Mineralogy.* $Cu_2^{+2}(SO_4)O$, a brown monoclinic mineral occurring as minute, elongate crystals, having a specific gravity of 4.17 and a hardness of 3 on the Mohs scale; found as a sublimate mineral on Mount Vesuvius.

dolichocephalic *Paleontology.* having an elongated head; having a cranial index under 75.

dolichol *Biochemistry.* a substance that carries sugar residues from soluble nucleotides through membranes, such as Golgi apparatus, to polymers, such as polysaccharides.

Dolichopodidae *Invertebrate Zoology.* long-legged flies, a family of two-winged insects in the series Brachycera.

Dolichothoraci *Paleontology.* a primitive arctolepid arthrodire family of the Paleozoic, a bottom fish characterized by its long pectoral spines; later forms gradually developed shorter spines, becoming known as Brachythoraci.

doline or **dolina** *Geology.* a sinkhole or closed depression formed either by the solution of surficial limestone or by the collapse of underlying caves.

dolioform *Biology.* shaped like a barrel.

doliolaria larva *Invertebrate Zoology.* a type of ciliated pelagic larva found in crinoids.

Doliolida *Invertebrate Zoology.* an order of transparent, free-swimming tunicates in the class Thaliacea; found in warm seas.

dolipore septum *Mycology.* a kind of cell wall, in most fungi belonging to the subdivision Basidiomycotina, that has a swelling in its center shaped like a barrel.

Dolland, John 1706–1761, English instrument maker; inventor of the achromatic lens.

dollar *Nucleonics.* a unit of reactivity, defined as the ratio of a reactor's reactivity to the fraction of delayed neutrons; at prompt critical, a reactor is said to have reactivity of precisely one dollar.

dollar spot *Plant Pathology.* a disease of lawn grasses caused by the fungus *Sclerotinia homeocarpa* and characterized by round, brownish spots on the turf that may merge to form large patches of dead grass.

Dollo's law *Evolution.* a law stating that evolutionary change manifested at any level higher than the genetic is irreversible, and that anatomical structures or functions once lost cannot be regained. Also, IRREVERSIBILITY RULE.

dolly *Engineering.* an industrial hand cart.

dolmen *Archaeology.* a stone structure consisting of upright columns supporting a slab roof; usually considered a tomb.

dolocast *Geology.* the cast or impression of a dolomite crystal preserved in an insoluble residue.

dolomite *Mineralogy.* $CaMg(CO_3)_2$, a white to reddish to greenish trigonal mineral with a vitreous luster and a perfect cleavage, occurring as rhombohedral crystals or in massive form, having a specific gravity of 2.85 and a hardness of 3.5 to 4 on the Mohs scale; an important rock-forming mineral. Also, BITTER SPAR, RHOMBIC SPAR.

dolomitic limestone *Petrology.* a limestone in which the carbonate fraction consists mostly of dolomite and calcite, with calcite being more abundant; it is sometimes used as building stone. Also, **dolostone.**

dolomitization *Geology.* a process in which limestone is transformed into dolomite rock or dolomitic limestone by the replacement of the original calcite with magnesium carbonate. Also, **dolomization.**

DO loop *Computer Programming.* a set of instructions in FORTRAN that is executed repeatedly.

dolorogenic *Neurology.* producing or causing physical pain or mental anguish.

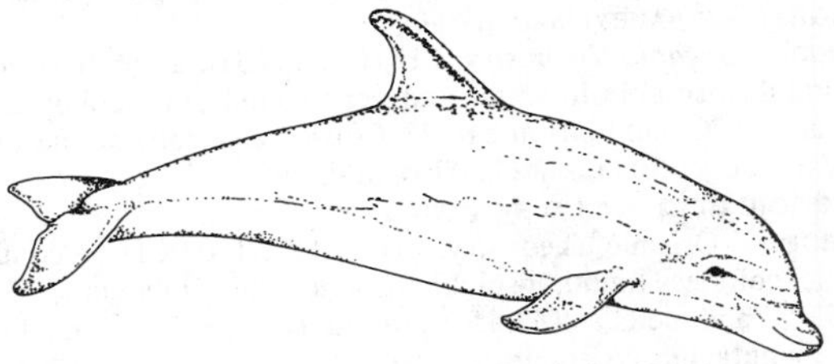

dolphin

dolphin *Vertebrate Zoology.* **1.** any of several chiefly marine, cetacean mammals of the family Delphinidae, having a fishlike body, numerous small teeth, and a head elongated into a beaklike snout, as distinguished from a porpoise. **2.** either of two large, slender fish constituting the genus *Coryphaena* that are widely distributed in tropical and temperate seas and are used as a food fish. Also, **dolphin fish.** *Civil Engineering.* an arrangement of piles to provide a mooring in the open sea or to guide ships through a narrow harbor entrance.

Dolphin see DELPHINUS.

dolphin striker *Naval Architecture.* a short spar projecting downward from the bowsprit and secured to the hull with the backropes.

dom. dominant.

Domagk, Gerhard [dō´mäk´] 1895–1964, German biochemist; discovered *Prontosil,* forerunner of the sulfonamides.

domain an area or region that is defined or identified in some way; specific uses include: *Artificial Intelligence.* an area of expertise or expert knowledge. *Biochemistry.* the region within a macromolecule that has a structure or function different from other sections of the molecule. *Immunology.* see IMMUNOGLOBULIN DOMAIN. *Crystallography.* the small regions of a crystal containing a completely oriented structure; generally used in describing ferromagnetic materials. *Computer Programming.* **1.** in the relational data model, the set of all integers, real numbers, character strings, or other types of values that an attribute may assume. **2.** the collection of all filled data fields of the same type in a relation or flat file. **3.** the set of values that a function argument can take. *Mathematics.* **1.** an open, nonempty connected set, especially of the complex plane; a region. **2.** an older term for a field. **3.** given a relation R between two sets X and Y, the set of all elements x of X such that $(x, y) \in R$ for some y in Y. **4.** given a function f, the set of all possible values for the argument of f. Also, **domain of *f*, domain of definition of *f*.**

domain expert *Artificial Intelligence.* a human expert in a particular area who provides the knowledge that is the basis for an expert system.

domain growth *Solid-State Physics.* a process in which randomly arranged atomic or molecular magnetic domains in a ferromagnet align themselves with the direction of an externally applied magnetic field.

domain rotation *Solid-State Physics.* a stage in the magnetization process in which the moments of the ferromagnetic domains are rotated against the anisotropy forces so as to align themselves with an externally applied magnetic field.

domain theory *Solid-State Physics.* a theory stating that changes in the macroscopic magnetization and polarization of a ferromagnetic, ferrimagnetic, or ferroelectric crystal arise from changes in the orientations and sizes of the atomic and molecular domains.

domatophobia *Psychology.* an irrational fear of being in a house.

dome

dome *Architecture.* a round (usually hemispherical) vaulted roof or ceiling with a circular, elliptical, or polygonal base. *Astronomy.* the hemispherical roof of an observatory building, usually rotating and with a slit through which the telescope looks out. *Acoustical Engineering.* a dome-shaped cover over the diaphragm of a loudspeaker, providing equalized sound dispersion and protection for the diaphragm. *Geology.* **1.** a circular or elliptical, almost symmetrical uplift or anticlinal structure in which the beds dip gently away in all directions from the central point of folding. **2.** a large igneous intrusion having a convex-upward surface and gently sloping sides. *Crystallography.* a form produced by a pair of corresponding planes parallel to one crystal axis, but inclined to the other axes. *Ordnance.* see SPRAY DOME.

Domerian *Geology.* in Great Britain, a geologic stage of the middle Lower Jurassic period, occurring after the Charmouthian and before the Whitbian.

domestic *Zoology.* describing an animal that has been tamed by humans for use as food, as a work animal, for a pet, and so on. *Botany.* describing a plant that is cultivated by humans for food, shade, ground cover, ornamentation, and so on. *Military Science.* of or relating to activities or events within a nation's borders; not foreign. Thus, **domestic air traffic, domestic intelligence, domestic surveillence,** and so on.

domesticate *Biology.* to control or adapt an animal or plant for human use or life with humans. Thus, **domesticated.**

domestication *Biology.* the breeding of an animal or plant to adapt it for human use or life with humans. *Anthropology.* the deliberate control of reproductive cycles of plants and animals that is the precursor for the development of agriculture and animal husbandry.

domestic induction heater *Engineering.* a cooking utensil that is heated by electrical current induced by a primary inductor.

domestic public-frequency bands *Telecommunications.* radio-frequency bands that are used exclusively for public service.

dome theory *Mining Engineering.* a theory stating that strata movements produced by underground excavations are limited by a kind of dome whose base is the area of excavation, and that the movements diminish as they extend upward from the center of the area.

domeykite *Mineralogy.* Cu_3As, a tin-white to steel-gray cubic metallic mineral occurring in reniform or botryoidal masses, having a specific gravity of 7.5 to 8.1 and a hardness of 3 to 3.5 on the Mohs scale; found in copper mines.

dominance *Ecology.* the ability of a given species, because of its size, population density, or fitness, to predominate within a community and affect or control other species there. *Behavior.* a situation in which an individual animal has higher status in a group in terms of access to food, space, or mates, so that others consistently defer or give way to this individual. *Genetics.* the tendency of certain (dominant) alleles to mask the expression of their corresponding (recessive) alleles.

dominance hierarchy *Behavior.* a social ranking system within a group of animals of the same species, in which certain forms of status and privilege are held by those ranking at the top, usually the stronger or more aggressive members of the group. Also, **dominance order.**

dominant *Behavior.* having or showing dominance; being in a superior position in a group. Thus, **dominant species.**

dominant allele *Genetics.* the allele whose phenotype is fully expressed when carried by only one of a pair of homologous chromosomes.

dominant gene *Genetics.* a gene whose phenotype is fully expressed in a heterozygote.

dominant laterality *Neurology.* a preference for using either the right (dextral) or left (sinistral) side of the body in voluntary motor activities.

dominant mode see FUNDAMENTAL MODE.

dominant wave *Electromagnetism.* an electromagnetic wave having the lowest cutoff frequency in a uniconductor waveguide.

dominant wavelength *Optics.* the wavelength of light that matches a given color when mixed with white light in the correct proportions and at the correct intensity.

dominated convergence theorem see FATOU-LEBESQUE THEOREM.

dominating integral *Mathematics.* let $f(x)$ be a positive, nonincreasing continuous function defined for $t \geq 1$, whose values at the positive integers are the terms of an infinite series $\sum_{n=1}^{\infty} a_n$; i.e., $f(n) = a_n$. Then the integral $\int_1^{\infty} f(t)dt = \lim_{n\to\infty} \int_1^n f(t)dt$ is known as the dominating integral for the series; used in the integral test for convergence of a series.

dominating series *Mathematics.* let $\sum_{n=1}^{\infty} a_n$ and $\sum_{n=1}^{\infty} b_n$ be two series. If $|a_n| \leq b_n$ for all but a finite number of n, then $\sum_{n=1}^{\infty} b_n$ is called a dominating series for $\sum_{n=1}^{\infty} a_n$; used in the comparison test for convergence of a series.

domino effect *Science.* a process in which an initial effect invariably produces a sequence of similar effects. (After the model of a row of dominoes toppling over if the first one is pushed.)

Don *Geography.* a river in the southwestern USSR, flowing southeast and then southwest through southern Russia into the Sea of Azov.

Donahue equation *Thermodynamics.* an equation that determines the heat transfer coefficient for a baffled shell-and-tube heat exchanger.

Donatiaceae *Botany.* a monogeneric family of dicotyledonous dwarfed herbs, usually placed in the order Campanulales, distinguished by often storing inulin and containing scattered tanniferous cells; one species is native to southern South America and the other to New Zealand and Tasmania.

Donati, Giovanni 1826–1873, Italian astronomer; discovered Donati's comet and five other comets; pioneer in stellar spectroscopy.

Donati's Comet *Astronomy.* comet 1858 VI, one of the brightest and most visible comets of the 19th century; discovered June 2, 1858, and last seen March 4, 1859.

Donders reduced eye *Optics.* an optical model used for calculating the dimensions and locations of images produced by the human eye.

donjon [dun´jən] *Architecture.* the inner stronghold of a castle; the keep.

donkey *Vertebrate Zoology.* the domestic ass, *Equus asinus,* of the family Equidae; used since ancient times for riding and as a work and pack animal. *Mining Engineering.* **1.** a type of winch characterized by drums that are controlled separately by clutches and brakes. **2.** see BARNEY. *Mechanical Engineering.* describing a relatively small, auxiliary machine. Used to form various compound terms, such as **donkey boiler, donkey hoist, donkey pump,** and so on.

donkey

donkey engine *Mechanical Engineering.* a small engine powered by steam or compressed air, used especially to lift cargo.

donkey power *Physics.* a unit of power equal to 250 watts or approximately one-third horsepower.

Donnan, Frederick George [dän´ən] 1870–1956, British chemist; formulated theory of membrane equilibrium.

Donnan dialysis *Physical Chemistry.* a process in which only certain types of charged particles can pass through a permeable membrane, as when smaller ions pass through but larger colloidal particles do not.

Donnan equilibrium *Physical Chemistry.* the conditions of equilibrium that exist when two solutions are separated by a membrane that is permeable to certain ions of the solutions, but not all of them; an electrical potential develops between the two sides of the membrane, and the two solutions vary in osmotic pressure. (From Frederick George *Donnan.*) Also, GIBBS-DONNAN EQUILIBRIUM.

donor one that gives or provides; specific uses include: *Medicine.* a human or other organism that provides living tissue for use in another living body; e.g., blood for transfusion or an organ for transplantation. *Microbiology.* any microorganism that gives, or donates, genetic information to a recipient cell during sexual reproduction. *Chemistry.* a molecule or the part of a molecule structure that provides an electron pair to an acceptor. Also, ELECTRON DONOR. *Solid-State Physics.* an impurity that is intentionally introduced into a pure semiconductor so as to increase the number of conduction electrons. Also, **donor impurity.**

donor-acceptor complex *Chemistry.* the relationship between a donor and an acceptor.

donor level *Solid-State Physics.* an intermediate energy level close to the conduction level in a semiconductor; filled at absolute zero, the electrons in this level can accquire energies corresponding to the conduction level at other temperatures.

Donovan bodies *Medicine.* chromatin masses at the ends of *Calymmatobacterium granulomatis* bacteria, giving them the appearance of a closed safety pin. (After Charles *Donovan,* 1863–1951, Irish physician.)

donut see DOUGHNUT.

doodlebug *Entomology.* a common name for the larva of an ant lion.

doodlebug a popular term for any of various small, squat vehicles or devices; specific uses include: *Ordnance.* **1.** a small military tank or truck. **2.** an airborne, magnetic device used to detect submarines. **3.** the German VI missile of World War II. *Mining Engineering.* any of various prospecting devices that can supposedly locate water, mineral, and oil deposits. *Mechanical Engineering.* **1.** a small tractor. **2.** a car used in railroad maintenance and repair.

Doolittle equation *Materials Science.* the mathematical relationship between polymer melt viscosity, volume occupied, and specific free volume; useful in explaining and predicting how polymers soften and flow.

door *Engineering.* a movable, solid, usually rectangular barrier (made of wood, metal, glass, or other hard materials) for opening and closing a room, hallway, building, or other enclosure. Used to form various compound terms, such as **doorstep, doorknob, doorbell, doormat,** and so on. *Architecture.* see DOORWAY.

door closer *Mechanical Devices.* **1.** also, **door check.** a door-closing regulating device composed of a piston, compression chamber, and springs. **2.** a device used in elevators to automatically open and close the doors. Also, GATE CLOSER.

doorframe or **door frame** *Architecture.* an assembly that encloses a doorway and usually supports a door; composed of two upright doorjambs and a transverse lintel.

doorhead see LINTEL.

doorjamb *Architecture.* an upright piece of wood or other material on either side of a doorway. Also, **doorpost.**

doorknob capacitor *Electricity.* a high-voltage, fixed capacitor with a plastic encasement that resembles a doorknob in size and shape.

doorknob tube *Electronics.* a UHF electron tube having a shape similar to that of a doorknob.

doornail *Building Engineering.* a large-headed nail used, especially formerly, to strengthen or ornament a door.

door rail *Building Engineering.* a horizontal member extending the full width between stiles and framing into them; it consists of a top and bottom rail and intermediate or cross rails located between them.

doorstop *Building Engineering.* **1.** a device that holds a door open, typically a metal weight or a wedge of rubber. **2.** a strip that protects the surface against which a door closes. **3.** a metal device having a rubber-covered tip to prevent a door from striking a wall or an object on a wall.

doorway *Architecture.* a passageway into a building or room.

dopa *Biochemistry.* an amino acid that is produced from the hydroxylation of tyrosine in the sympathetic nerve channels and in the adrenal gland as the first stage in the production of dopamine. Also, 3,4-DIHYDROXYPHENYLALANINE.

dopa decarboxylase *Enzymology.* an enzyme that catalyzes the removal of the carboxyl group from dihydroxyphenylalanine.

dopamine *Biochemistry.* $C_8H_{11}NO_2$, a hormonelike substance that is an important neurotransmitter in both the central and peripheral nervous systems; it occurs as an intermediate in epinephrine and norepinephrine biosynthesis. Also, 3,4-DIHYDROXYPHENYLETHYLAMINE.

dopamine hypothesis *Psychology.* the theory that schizophrenia is caused by an excess of dopamine activity in certain cell groups within the brain.

dopamine β-monooxygenase *Enzymology.* an enzyme of the oxidoreductase class, occurring in nervous tissue and in the adrenal medulla, that catalyzes a reaction between dopamine and ascorbate with molecular oxygen, producing norepinephrine and dehydroascorbate. Also, **dopamine β-hydroxylase.**

dopant or **Dopant** see DOPING AGENT.

dopa-oxidase or **dopaoxidase** *Enzymology.* an enzyme that catalyzes the oxidation of dihydroxyphenylalanine to melanin in the skin.

dope *Materials.* any of various thick liquids or pasty substances used to lubricate, absorb, prepare a surface, or produce a desired quality in another substance; specific examples include an antiknocking additive for gasoline, a varnishlike coating for aircraft wings, and an absorbent material used in the manufacture of dynamite. *Electronics.* to add a doping agent to a semiconductor material during the manufacturing process.

doped junction *Electronics.* a PN junction in a diode or transistor produced by doping.

doped solder *Metallurgy.* a solder containing an intentionally added, small amount of a material that improves the quality of the joint.

doping *Electronics.* the addition of an impurity element to a semiconductor material, such as germanium or silicon, during the manufacturing process to form P- or N-type material required for semiconductor diodes and transistors. *Engineering.* the coating of a mold or mandrel with a material that will facilitate the removal of the molded plywood part. *Metallurgy.* in powder metallurgy, the addition of a substance that accelerates sintering.

doping agent *Electronics.* a metallic impurity, such as aluminum or antimony, used in doping. Also, DOPANT.

doping compensation *Electronics.* the addition of a P- or N-type impurity to a semiconductor material to compensate for the effect of an opposite-type impurity previously added to the semiconductor.

Doppler, Christian [däp´lər] 1803–1853, Austrian physicist; formulated Doppler's principle, analyzing the Doppler effect.

Doppler broadening *Spectroscopy.* the thermally induced widening of a spectral line, especially for hot gases, as a result of differences in atomic or molecular velocities and directions in relation to the source of radiation.

Doppler current meter *Engineering.* a current meter in which the difference between a signal of known frequency projected into the water and the reverberation frequency is proportional to the speed of the water.

Doppler effect *Physics.* an apparent shift in the observed frequency of a wave due to relative motion between the source and the observer.

Doppler-free spectroscopy *Spectroscopy.* any spectroscopic technique in which laser beams are used to overcome Doppler broadening of spectral lines.

Doppler-free two-photon spectroscopy *Spectroscopy.* a type of Doppler-free spectroscopy for measuring the wavelength corresponding to a transition induced by the simultaneous absorption of two photons.

dopplerite *Geology.* a naturally occurring, brownish-black gel of humic acids found at depth in marsh and peat bog deposits.

Doppler radar *Engineering.* a radar that uses reflected electromagnetic waves to differentiate fixed and moving targets and determine target velocities.

Doppler range see DORAN.

Doppler shift *Physics.* the amount of change in the observed frequency of a wave due to the Doppler effect. Also, **Doppler frequency.**

Doppler sonar *Engineering.* a sonar technique that is based on Doppler shift measurement.

Doppler's principle *Physics.* the principle that a wave will appear to shift in frequency because of the relative motion of the observer of the wave and the wave's source; e.g., the sound of a train whistle seems to get higher as the train approaches.

Doppler tracking *Engineering.* the act of following a target with a Doppler radar system.

Doppler ultrasonic flowmeter *Engineering.* a flowmeter that utilizes the Doppler shift measurement technique in conjunction with ultrasonic waves reflected by particles in the fluid.

Doppler ultrasonography *Radiology.* the use of ultrasonic waves (usually between 1 and 10 MHz) to measure the velocity of deep body structures, such as red blood cells, in motion.

Doradidae *Vertebrate Zoology.* the thorny catfishes, a large family of South American armored catfish of the order Siluriformes, having a series of thick, overlapping bony plates along each side of the body and reputed to journey overland in search of water during dry seasons.

dorado *Vertebrate Zoology.* another name for the dolphin fish. See DOLPHIN, def. 2.

Dorado *Astronomy.* the Goldfish, a small constellation of the southern hemisphere that contains part of the Large Magellanic Cloud, a satellite galaxy of the Milky Way.

doran [dôr´an´; dôr´ən] *Engineering.* a Doppler ranging system that obtains high accuracy for the tracking of missiles; it uses phase comparison of three different modulation frequencies on the carrier wave. (An acronym for Doppler range.)

doraphobia *Psychology.* an irrational fear of touching the fur or skin of an animal.

doré [də rā´] *Metallurgy.* impure silver produced by cupellation.

Doric *Architecture.* the oldest (c. 535–430 BC) and simplest Greek architectural order, having fluted column shafts with no base, plain capitals, and a bold, simple cornice.

Dorilaidae *Invertebrate Zoology.* the big-headed flies, a family of two-winged insects with large eyes in the series Aschiza. Also, PIPUNCULIDAE.

Dorippidae *Invertebrate Zoology.* a family of decapods, the mask crabs, in the subsection Oxystomata.

dormancy *Botany.* an inactive or quiescent period in which seeds, spores, bulbs, buds, and other such vegetative and reproductive organs cease growth and development, and reduce metabolic activity, particularly respiration; it is considered a physiological response to adverse environmental conditions, and in many plants is triggered by change in photoperiod and/or temperature.

dormant *Botany.* being in a state of dormancy. *Ordnance.* the state of a mine during which a time-delay feature prevents actuation.

dormer *Architecture.* **1.** a gabled structure projecting vertically from a sloping roof. **2.** see DORMER WINDOW.

dormer window *Architecture.* a window set in a dormer.

dormouse *Vertebrate Zoology.* any of numerous small, furry-tailed, Old World rodents of the family Gliridae that resemble small squirrels, live in trees, feed on nuts and acorns, become torpid in cold weather, and yield a velvety fur used for trimming.

dormouse

Dorn effect *Physical Chemistry.* the energy generated by the movement of particles through water or another liquid.

doroid *Electromagnetism.* a coil formed in the shape of half of a toroid, and having a removable core.

Dorosomatidae see GIZZARD SHAD.

Dorothy Reed's cells see REED-STERNBERG CELLS.

Dorr, John Van Nostrand 1876–1962, American metallurgical engineer; invented the Dorr agitator, Dorr classifier, and Dorr thickener.

Dorr agitator *Mechanical Engineering.* a circular tank equipped with bottom rakes, central air lifting, and rotating top launder; used to aerate and stir pulp during cyanidation of gold ores.

Dorr classifier *Mechanical Engineering.* a horizontal-flow classifier having a rake that moves sand uphill through a rectangularly shaped tank with a sloped base; heavier materials are raked out of the tank while slime and finer sands are carried over the rear wall in suspension.

Dorr thickener *Mechanical Engineering.* a large cylindrical tank equipped with slowly circulating ploughs that move heavier materials to a central bottom discharge area while relatively clear liquid overflows along the periphery of the tank.

dors- a combining form meaning "dorsal" or "back," as in *dorsalis.*

dorsal *Anatomy.* of or lying near the back.

dorsal cavity *Anatomy.* the body cavity that is located near the posterior surface of the body, and that is further divided into the cranial cavity and the vertebral canal.

dorsal column *Anatomy.* the vertebral column.

dorsal decubitus *Surgery.* the position of an individual lying on the back on a horizontal surface.

dorsal elevated position *Surgery.* the position of an individual lying on the back with the head and shoulders elevated.

dorsal fin *Vertebrate Zoology.* the fin or finlike integumentary expansion generally developed on the back of both bony fishes and sharks.

dorsalgia *Medicine.* any pain in the back.

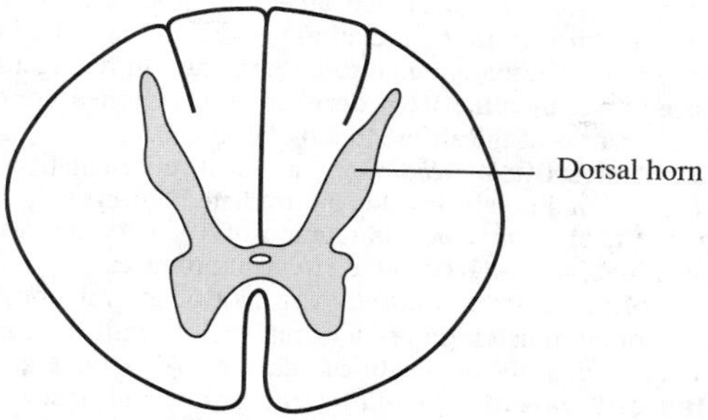

dorsal horn

dorsal horn *Anatomy.* a column of gray matter lying posteriorly in the spinal cord.

dorsalia *Vertebrate Zoology.* cartilaginous precursors of the neural arch of the vertebrae in jawless fishes.

dorsal lip *Developmental Biology.* the dorsal edge of the blastopore, in amphibian embryos; it forms the notochord and induces the neural tube and other axial structures.

dorsal position *Surgery.* the position of an individual lying on the back. Also, SUPINE POSITION.

dorsal recumbent position *Surgery.* the position of an individual lying on the back with the legs rotated outward and flexed.

Dorset *Agriculture.* a breed of white-faced short-wool sheep raised for mutton and for their lightweight wool. (From *Dorset* County, England.)

dorsi- a combining form meaning "dorsal" or "back," as in *dorsiflexion.*

dorsiferous *Botany.* in ferns, bearing the sori on the back of the fronds.

dorsiflex *Zoology.* capable of, or causing, flexing toward the back.

dorsigrade *Vertebrate Zoology.* walking on the back of the toes, as in armadillos.

dorsiventral *Botany.* having obvious dorsal and ventral sides, as most foliage leaves.

dorso- a combining form meaning "dorsal" or "back."

dorsoanterior *Anatomy.* having the back of the fetus toward the front of the mother.

dorsocephalad *Anatomy.* directed toward the back of the head.

dorsodynia *Medicine.* any pain in the back.

dorsolateral *Anatomy.* relating to the back and the side.

dorsomedial *Anatomy.* relating to the back toward the midline.

dorsomedian *Anatomy.* the median line of the back.

dorsoposterior *Anatomy.* having the back of the fetus directed toward the mother's back.

dorsoventral *Anatomy.* **1.** relating to the back and belly surfaces of the body. **2.** passing from the back to the belly surface.

dorsum *Anatomy.* **1.** the aspect of an anatomical part or structure corresponding in position to the back. **2.** see BACK.

Dorylaimoidea *Invertebrate Zoology.* a superfamily of algae-eating nematodes found in soil and freshwater environments.

Dorylinae *Invertebrate Zoology.* a subfamily of predatory ants, including the army ants.

Dorypteridae *Paleontology.* a family of advanced palaeoniscids, a widespread order of fishes in the Permian, characterized by a dorsal fin that extends almost the entire length of the body.

DOS [däs] *Computer Technology.* **1.** an operating system first developed by IBM for their small and intermediate-sized 360 model computers. **2.** a generic term now referring to microcomputer operating systems that use magnetic disks, either hard or flexible, as primary on-line storage. (An acronym for disk operating system.)

dosage *Medicine.* the giving of medicine or any other therapeutic agent in prescribed amounts for a particular patient or condition. *Genetics.* see GENE DOSAGE. *Nucleonics.* see DOSE.

dosage compensation *Genetics.* the mechanism in mammals that compensates for the dosage of genes carried on the X chromosome by inactivating one or more of the X chromosomes, which is then known as a Barr body.

dosage meter see DOSIMETER.

dose *Medicine.* the measured quantity of medicine or any other therapeutic agent to be taken at any one time or in parts over a period of time. *Nucleonics.* the total or accumulated quantity of ionizing radiation. Also, DOSAGE. The **absorbed dose** in rads represents the amount of energy absorbed from the radiation per gram of specified absorbing material and is essentially equal to the exposure in roentgens; the **biological dose** (also called the **RBE dose**) in rems is a measure of the biological effectiveness of the absorbed radiation.

dose distribution *Radiology.* a visual representation of the variation of radiation dose absorbed in an irradiated object.

dose-effect *Nutrition.* the relation of a single administered portion or quantity of a substance to the effects it produces.

dose-effect curve *Radiology.* a plot of the relationship between the level of administration of an agent, such as radiation, and its biological effect, such as the degree of cell damage. Also, DOSE-RESPONSE CURVE.

dose equivalent *Radiology.* the product of absorbed dose and the quality and distribution factors compensating for variations in biological effectiveness of different types of radiation, given in rems and sieverts.

dose fractionation *Radiology.* the practice of dividing a large dose of radiation into smaller doses administered over a period of time, used to reduce biological risks. Also, FRACTIONATION.

dose-rate meter *Nucleonics.* a portable instrument for detecting and measuring ionizing radiation and for measuring the dose rate in radiation delivered per unit time.

dose response *Nutrition.* the effects produced by a specific amount of a substance administered at a given time.

dose-response curve see DOSE-EFFECT CURVE.

dose without effect *Nutrition.* a specific amount of a substance administered at a given time that produces no perceptible reaction.

dosimeter *Nucleonics.* a radiation survey instrument, either fixed in working spaces or worn by personnel, for measuring and recording doses accumulated during possible exposure to ionizing radiation. Also, dosage meter.

dosimetric *Nucleonics.* of or relating to the measurement of radiation doses.

dosimetry *Nucleonics.* the techniques for quantitative measurement of ionizing radiation doses and dose rates, using various types of radiation survey instruments to record possible exposure of personnel working in radiation environments.

DO statement *Computer Programming.* a statement in FORTAN that specifies a DO loop by specifying the loop index variable, the values it takes, and the number of the statement that terminates the loop.

dot *Graphic Arts.* **1.** the smallest unit of ink in any given image, the distribution of which comprises halftone and color reproductions on offset presses. **2.** the units that make up a screened subject. *Electronics.* **1.** a small, round piece of metal diffused into the semiconductor material during the manufacture of alloy transistors. Also, BUTTON. **2.** a small spot of red, green, or blue phosphor on a color television screen.

dot-addressable *Computer Technology.* the capability of graphic display units or dot-matrix printers to select individual dots within a character representation.

dot angel *Electromagnetism.* a bright dot appearing on a radar screen, often on clear cloudless days, that is assumed to be due to a mass of rising air with an index of refraction different from that of the surrounding air.

dot cycle *Telecommunications.* either segment of a signal having two alternating signaling conditions.

dot etching *Graphic Arts.* in offset lithography, any of several photographic or chemical methods used on negatives to correct color or poor dot structure.

dot generator *Electronics.* a signal generator used for convergence adjustments of color television tubes, by producing a pattern of white dots or small squares on the face of the tube.

Dothideaceae *Mycology.* a family of fungi belonging to the order Dothideales that either live off nonliving organic matter or are plant parasites. Also, **Dothioraceae.**

Dothideales *Mycology.* an order of fungi belonging to the subdivision Ascomycotina; some types obtain nutrients from nonliving organic matter, while others are parasites of plants and lichenized fungi.

dot matrix *Computer Technology.* a printing technique in which characters are formed by selecting a pattern of dots from a rectangular matrix of dots.

dot matrix printer *Computer Technology.* an impact printer that forms characters as a pattern of closely spaced dots, each printed dot resulting from the actuation of a pin in the print head before impact with the print ribbon; the number of pins is a measure of the print quality. Also, **dot character printer.**

dot product see INNER PRODUCT.

dotriacontane *Organic Chemistry.* $CH_3(CH_2)_{30}CH_3$, crystals; melts at 70°C and boils at 467°C; used in research. Also, DICETYL.

dot-sequential color television *Electronics.* a color television system in which the three primary screen colors are developed in sequence on each trace.

dotted pair *Artificial Intelligence.* see CONS CELL.

dotted swiss *Textiles.* a sheer cotton fabric with a regular pattern of flocked dots; made originally in Switzerland.

dotter *Optics.* a technician who determines the optical center and optical axis of a lens with the aid of a centering device.

double *Botany.* flowers having more than one set of sepals or petals. *Engineering.* having two sides, two like parts, two simultaneous effects, and so on. Used to form a wide variety of compound terms, including the following entries and many others.

double-acting *Mechanical Engineering.* **1.** describing a machine having pistons that work in both directions, with the working fluid introduced alternately at opposite ends of the cylinder. Thus, **double-acting compressor, double-acting engine, double-acting pump,** and so on. **2.** describing a machine or device that works in both directions from a central position. Thus, **double-acting hammer, double-acting hinge, double-acting pawl,** and so on.

double-action *Engineering.* describing a device or system that carries out two or more operations in a single stroke. Thus, **double-action die, double-action forming, double-action press,** and so on. *Ordnance.* **1.** specifically, describing a firing system in which a single pull of the trigger both cocks and fires the weapon, as in a revolver. **2.** a weapon using this system.

double aging *Metallurgy.* the thermal treatment of a supersaturated solid solution that follows two consecutive, different schedules.

double altitudes see EQUAL ALTITUDES.

double arcing *Metallurgy.* in plasma arc welding or cutting, the undesirable condition that establishes a secondary arc between the workpiece and the outer part of the nozzle.

double armature *Electricity.* an armature that has two separate windings on a single core and two separate distributors.

double-armed pulley *Mechanical Devices.* a pulley of at least ten inches in width having two sets of arms that are set parallel to each other.

double-bar link *Mechanical Devices.* a type of valve link with two equal and parallel bars as opposed to a single slotted link.

double-base (junction) diode see UNIJUNCTION TRANSISTOR.

double-base junction transistor *Electronics.* a bipolar transistor in which there are two base terminals. Also, TETRODE JUNCTION TRANSISTOR.

double-beam cathode-ray tube *Electronics.* a cathode-ray tube, containing two separate and independent electron streams, that effectively functions as two separate cathode-ray tubes within the same envelope. Also, DUAL-GUN CATHODE-RAY TUBE.

double-beam spectrometer *Spectroscopy.* an instrument that measures and compares the absorption spectra of a substance having two beams of slightly different wavelengths passed through it.

double-beam spectrophotometer *Spectroscopy.* a device that measures differences in absorption of two closely related wavelengths by means of a photoelectric circuit.

double belting *Mechanical Devices.* a machine belting consisting of two lengths of leather cemented or glued together.

double bend *Mechanical Devices.* an S-shaped pipe fitting.

double beta decay *Nuclear Physics.* a possible change in the properties of a nucleus in nature, in which two protons decay into two neutrons or vice versa.

double bind *Psychology.* a situation in which an individual receives conflicting or contradictory messages as to the acceptability of certain behavior; e.g., a child is criticized for allowing a peer to bully him and then punished when he fights back.

double-bind hypothesis *Psychology.* the theory that schizophrenia is related to disordered communication between the individual and his parents, as when he is alternately praised and rebuked for similar behavior.

double-blind technique or **method** *Medicine.* a method of experimentation in which neither the subject nor the experimenter knows beforehand which condition has been assigned to the subject; e.g., whether the subject has received the drug being tested or a placebo.

double block *Mechanical Devices.* a block having two pulleys or sheaves.

double blossom *Plant Pathology.* a disease of blackberries and dewberries caused by the fungus *Fusarium rubi,* characterized by abnormal branch and twig growth and enlarged, misshapen flowers from which no fruit sets.

double bond *Chemistry.* a bond in which two atoms share two electron pairs; occurs in unsaturated compounds; represented by an equal sign, as in $H_2C=O$. Thus, **double bonding.** Also, DOUBLE COVALENT BOND.

double-bond isomerism *Physical Chemistry.* a condition in which two or more chemical compounds have the same composition, but with double bonds in different positions along the molecular chain.

double-bond shift *Organic Chemistry.* a shift occurring between a pair of valence bonds during a chemical reaction; such as butene-1, $H_2C{=}CHCH_2CH_3$, to butene-2, $H_3C{-}CH{=}CH{-}CH_3$.

double bottom *Naval Architecture.* a structural feature in a ship in which two layers of bottom plating, separated by void spaces, provides protection against flooding from grounding or other damage.

double-bottom cellular *Naval Architecture.* describing a ship's double bottom with extensive subdivision of the void spaces between the layers.

double-bounce calibration *Electronics.* the calibration of waveguide propagation delays in a radar system by using round-trip echoes; the correct range is the difference between the first and second echoes. Also, DOUBLE ECHO CHECK.

double-break switch *Electricity.* a switch that opens a connected circuit at two points simultaneously on closing.

double bridge see KELVIN BRIDGE.

double-buffered data transfer *Computer Programming.* the use of two buffer storage areas during input/output operations so that the data in one buffer is being actively processed while the other buffer is being filled or emptied.

double burn *Photogrammetry.* the intentional exposure of two or more line or halftone negatives in succession and register on the same sensitized surface. Also, DOUBLE SHOOTING.

double-button microphone *Acoustical Engineering.* a carbon microphone with an additional carbon button (a wired enclosure filled with carbon granules) to provide twice as much current variation.

double-channel simplex *Telecommunications.* any procedure that promulgates nonsimultaneous communication between two radio stations using two one-way channels.

double-click *Computer Science.* to press a mouse button twice in rapid succession, indicating a different, usually more significant, command than a single click.

Double Cluster *Astronomy.* a pair of open star clusters in the Milky Way in Perseus that are faintly visible to the naked eye.

double Compton effect *Quantum Mechanics.* a higher-order occurrence of the Compton effect, occurring with relative probability 1/137, related to the fine structure constant, in which two photons are scattered instead of one.

double-concave lens *Optics.* a diverging lens in which both surfaces are concave. Also, BICONCAVE LENS.

double-cone bit *Mechanical Devices.* a drill bit with two cone-shaped drilling elements.

double-convex lens *Optics.* a converging lens in which both surfaces are convex. Also, BICONVEX LENS.

double-coursed *Building Engineering.* in wall construction, the use of shingles that are underlaid inside sidewall applications to allow increased weather exposure and dramatic deep shadow lines at the course lines; it is made of an inferior grade of undercoursed shingles.

double covalent bond see DOUBLE BOND.

double crank *Mechanical Devices.* a versatile crank that can be connected to either side of a marine engine.

double-crank press *Mechanical Engineering.* a mechanical press with a wide slide mechanism that is operated by a crankshaft having two crank pins.

double cropping *Agronomy.* the growing of two crops in one year on the same field.

double-current generator *Electricity.* a dynamo-type generator that supplies both AC and DC from a single armature winding.

double-current signaling *Telecommunications.* **1.** a telegraph signaling method in which marks and spaces are represented by electric currents of opposite polarities rather than by a current and no current. **2.** binary transmission in which positive and negative direct current denote significant conditions.

double cusp *Mathematics.* a point on a curve that is a cusp for two or more branches of the curve. Also, TACNODE.

double-cut file *Mechanical Devices.* a file having two sets of parallel filing edges, each with teeth resembling a diamond-point cold chisel, that cross one another at angles of 45–50°.

double-cut saw *Mechanical Devices.* a saw whose teeth are set to cut on both forward and back cutting strokes. Similarly, **double-cut planer.**

double-cutting drill *Mechanical Devices.* a drill with a cutting edge design suitable for drilling in a clockwise or counterclockwise direction.

double decomposition *Chemistry.* a double displacement reaction in which two ionic compounds react to form two new ionic compounds.

double density *Computer Technology.* having twice the recording capacity of a standard disk or tape, usually because of a higher-quality recording surface.

double descent *Anthropology.* a kinship group in which a person is affiliated with his father's patrilineal kin and his mother's matrilineal kin for separate purposes. Also, **double unilineal descent.**

double-diffused transistor *Electronics.* a transistor in which two PN junctions are formed in the semiconductor wafer by gaseous diffusion of both P- and N-type impurities.

double diffusion *Immunology.* a process in which antibodies and antigens diffuse together through a gel medium to form a precipitate of homologous antigens and antibodies in optimal proportions.

double diode see DUODIODE.

double-diode limiter *Electronics.* a limiter circuit that acts as a clipper circuit to prevent the positive and negative excursions of a signal from exceeding a predetermined peak amplitude.

double-disk harrow *Agriculture.* a harrow in which two to four single-disk harrows are arranged in tandem so that the front groups throw the soil outward and the rear groups throw it inward, maintaining a level surface. Also, TANDEM HARROW.

double distribution *Chemical Engineering.* the distribution of product that results from counter double-current extraction, an arrangement in which the two liquid phases are transferred continuously and simultaneously in opposite directions through a series of contact vessels.

double-doped transistor *Electronics.* a transistor made from a grown-crystal material to which alternate P- and N-type dopants have been successively added during the crystal growth.

double-dot halftone *Graphic Arts.* a halftone produced by photographing a subject twice with different exposures, then contacting the two negatives with a photosensitive material to create a single negative or positive having a greater richness and range of tones.

double-doublet antenna *Electromagnetism.* an antenna array consisting of two half-wave dipole antennas perpendicular to each other, with one shorter than the other so as to have a broader band.

double drill *Mechanical Devices.* a drill consisting of a boring component and a countersinking component.

double ebb *Oceanography.* an ebb current that has two maxima of velocity with a smaller ebb velocity in between.

double echo check see DOUBLE-BOUNCE CALIBRATION.

double-ended bolt *Mechanical Devices.* a headless bolt threaded to receive a nut at either end. Also, STUD BOLT.

double-ended ferry *Naval Architecture.* a ferry designed so that either end can act as bow or stern; thus it is able to transit and dock with either end forward.

double expansion *Mechanical Devices.* an engine in which the working fluid is expanded in two successive cylinders, thus dividing the temperature range while decreasing condensation in each cylinder.

double exposure *Graphic Arts.* **1.** the process of exposing the same piece of photosensitive material to light twice, usually to achieve a special effect. **2.** a photographic image created in this way.

double-faced *Building Engineering.* a building constuction that has two faces from a side view, such as the interior and exterior of a masonry wall appearing as faces.

double-faced hammer *Mechanical Devices.* a hammer with two faces created by a flat head without a pane.

double fertilization *Botany.* the fertilization in most angiosperms in which one of the two sperm nuclei from the pollen tube fuses with the egg nucleus to form a diploid zygote, and the other fuses with the two polar nuclei to form the triploid primary endosperm nucleus that develops into the endosperm.

double filament lamp *Mechanical Devices.* a lamp with two filaments, usually of different resistances, only one of which is lit, allowing the lamp to emit light of varying intensities at any time.

double flood *Oceanography.* a flood current that has two maxima of velocity with a smaller flood velocity in between.

double floor *Building Engineering.* in a multistory building, the use of binding joists in a floor to support the ceiling joists below and the floor joists above.

double-frequency shift keying *Telecommunications.* a multiplex system in which two telegraph signals are combined and transmitted simultaneously by shifting among four different frequencies.

double galaxy *Astronomy.* two galaxies that appear close together in the sky, either because they are gravitationally connected or by a chance alignment.

double-glazed *Building Engineering.* describing a window or a sliding glass door that has two thicknesses of glass with an air space between them; used to provide better insulation. Thus, **double-glazing.**

double-headed nail *Building Engineering.* a nail with a round wire forming two heads which are driven into fixed concrete formwork; the second head, 12 mm higher than the first, can be used to withdraw the nail quickly with a claw hammer, if necessary.

double-headed rail *Mechanical Devices.* a reversible rail used in furniture that can be turned over when the top face becomes worn.

double headings *Mining Engineering.* a pair of coal headings driven parallel to each other and positioned side by side about 10 to 20 yards apart; used for development purposes.

double helix *Biochemistry.* a name for the characteristic spiraling shape of the two complementary strands of DNA.

double Hooke's joint *Mechanical Engineering.* a universal joint consisting of two Hooke's joints conected by a shaft; this eliminates angular displacement and angular velocity between driving and driven shaft.

double-hump fission barrier *Nuclear Physics.* the occurrence of secondary minima in potential energy, associated with decreased binding energy arising from unequal asymmetric axes in the deformed nucleus.

double-hung *Building Engineering.* describing a window that uses two vertically sliding sashes, each closing a different area of a window opening, with a counterweight on both the top and bottom.

double image see GHOST IMAGE.

double integral *Mathematics.* a definite integral on a two-dimensional manifold. By Fubini's theorem, a double integral can often be calculated as an iterated integral over one-dimensional manifolds.

double ionic layer see DOUBLE LAYER.

double jack *Mechanical Devices.* a heavy hammer, such as a sledge, that is operated with both hands. *Mining Engineering.* a double or twin-screw drill column.

double-knit or **doubleknit** *Textiles.* **1.** describing a strong, usually reversible, two-layer knitted fabric that is made using a double set of needles. **2.** a fabric or garment of this type.

double law of the mean see CAUCHY MEAN VALUE THEOREM.

double layer *Physical Chemistry.* the region of charge separation formed when an electrode meets an ionic conductor; a metal electrode in a water solution forms a specific structure consisting of the metal surface itself, an adjoining layer of adsorbed water molecules and ions, and an outer region of oppositely charged ions diffused in the liquid; this causes an electric field of considerable intensity. Also, DOUBLE IONIC LAYER, ELECTRIC(AL) DOUBLE LAYER.

double-layer capacitance *Physical Chemistry.* the capacitance at a double-layer interface, estimated by the distance that separates the oppositely charged ions from the dielectric constants of the system.

double-length number *Computer Programming.* a double-precision number containing twice the normal number of significant digits.

double limiter see CASCADE LIMITER.

double-line stream *Cartography.* a stream that is wide enough to be shown at a given scale with separate lines for each bank.

double-list sorting *Computer Programming.* an internal file sorting technique that stores the original file into memory prior to the sort process; a second, sorted file is written to another location in memory following the internal sort operation.

double load *Engineering.* a charge that has been split by inert material in a borehole.

double-loop pattern *Anatomy.* a pattern of ridges found in certain fingerprints.

double mast see A-FRAME.

double meridian distance *Cartography.* the algebraic sum of the perpendicular distances from the two ends of any line of a traverse to the initial, or reference, meridian.

double minute *Genetics.* small, paired extrachromosomal pieces of DNA that appear as small chromosomes under certain conditions.

double mirror *Optics.* a mirror consisting of two planes angled toward each other.

double-model stereotemplet *Photogrammetry.* a templet representative of the horizontal plot of two adjacent stereoscopic models that have been adjusted to a common scale.

double moding *Electronics.* the undesirable tendency of a magnetron to shift abruptly and unpredictably from one operating frequency to another.

double modulation *Telecommunications.* modulation carried out in two stages: first the subcarrier is modulated by the information-carrying modulation signal; then the modulation subcarrier is used to modulate a higher-frequency carrier.

double nickel salts see NICKEL AMMONIUM SULFATE.

double pendulum *Mechanics.* a pendulum that has been attached to the end of another pendulum.

double-pipe exchanger *Chemical Engineering.* a fluid-fluid heat exchanger composed of two concentric pipes; one fluid flows in the ring-like space between the pipes, and the other fluid flows through the inner pipe. Also, CONCENTRIC TUBE HEAT EXCHANGER.

double point *Mathematics.* a point on a curve at which the curve either is tangent to itself (a cusp) or crosses over itself (a node). Also, MULTIPLE POINT.

double-pole double-throw switch *Electricity.* a six-terminal switch used to connect one pair of terminals to either of two other pairs of terminals. Also, DPDT SWITCH.

double-pole single-throw switch *Electricity.* a four-terminal switch used to connect or disconnect two pairs of terminals simultaneously. Also, DPST SWITCH.

double-pole switch *Electricity.* a switch that functions in two separate electric circuits or in both lines of a single circuit at the same time.

double-precision *Computer Technology.* of or relating to the use of two computer words to represent a single number for increased precision. Thus, **double-precision arithmetic, double-precision number, double-precision hardware.**

double printing *Graphic Arts.* the process of photoprinting two or more halftone or line negatives in the same register on the same surface.

double-pull trigger *Ordnance.* a firing mechanism in which two distinctive mechanical stages are felt in pulling the trigger, thus providing an extra margin of safety; most military weapons use a similar trigger with longer travel in each stage.

double-pulse recording *Computer Technology.* an early magnetic memory technology in which each memory cell contained two regions that could be magnetized in opposite directions.

double pulsing *Navigation.* the transmission by one loran station of signals of two different rates. Thus, **double-pulsing station.**

double-quantum stimulated-emission device *Optics.* a laser that produces a beam from the interaction of two separate species of fluorescent ions.

double recessive *Genetics.* a dihybrid genotype in which all alleles are recessive.

double recoil *Ordnance.* of or relating to a two-carriage firing system in which the gun recoils on the top carriage, and the top carriage recoils on the bottom carriage.

double reflection *Crystallography.* a phenomenon in which X-rays that are reflected by one set of lattice planes may then be reflected by another set of planes which, by chance, are in exactly the right orientation. The resultant beam appears in the position in reciprocal space expected for a normal reflection, but is sharper in appearance on an X-ray photograph than an ordinary diffracted beam. The effect causes intensity changes in the reflections involved; it may even cause an ambiguity in the space group determination if a systematically absent reflection gains intensity by it. It can be eliminated for that particular reciprocal lattice point by reorienting the crystal. Also, RENNINGER EFFECT.

double refraction *Optics.* the separation of an electromagnetic wave in an anisotropic medium into components that have mutually perpendicular vibration directions and are propagated at different velocities. Also, BIREFRACTION.

double-rivet *Engineering.* to make a lap joint using two rows of rivets or a butt joint with four rows.

double-roll crusher *Mining Engineering.* a machine consisting of two heavy metal cylinders, usually toothed, that roll toward each other, with the crushed material discharged at the bottom; used especially in the initial crushing of coal for use in boilers.

double root *Mathematics.* a root of multiplicity two. In particular, c is a double root of a polynomial $p(x)$ if $p(x)$ can be written in the form $p(x) = (x - c)^2q(x)$ where $q(x)$ is a polynomial of which c is not a root. Also, REPEATED ROOT, EQUAL ROOTS.

double salt *Inorganic Chemistry.* a compound that is crystallized from an aqueous solution containing a mixture of ions; i.e., a molecular combination of two separate salts, such as ferrous ammonium sulfate, $FeSO_4 \cdot (NH_4)_2SO_4 \cdot 6H_2O$, which is crystallized from mixed solutions of ferrous sulfate and ammonium sulfate.

double sampling *Industrial Engineering.* a two-stage quality control procedure in which a first sampling may result in three outcomes: acceptance of the lot, rejection, or a second sampling. The second sampling will then result in either acceptance or rejection of the lot.

double screen *Electronics.* a cathode-ray tube containing two screens overlying each other, each screen having a different persistence value and color.

double sextant *Navigation.* a sextant fitted with two arms that can be used simultaneously to measure two angles from a given bearing or altitude.

double-shield enclosure *Electricity.* a cable or enclosure with two independent electromagnetic shields.

double shooting see DOUBLE BURN.

double-shot molding *Materials Science.* a technique using successive molding operations to turn out two-color parts in thermoplastic materials.

double-sideband transmission *Telecommunications.* a method of sideband transmission in which both sidebands are transmitted at full power; electromagnetic waves produced by amplitude modulation are symmetrically spaced above and below the carrier frequency and are transmitted along with the carrier wave, all at full power. Thus, **double-sideband modulation.**

double-sided disk *Computer Technology.* a magnetic disk that has recording surfaces on both sides.

double sintering *Metallurgy.* in powder metallurgy, a process consisting of two sintering operations between which a mechanical operation is performed.

double-slider coupling see SLIDER COUPLING.

double-solvent refining *Chemical Engineering.* a petroleum-refining procedure in which two solvents are used to deasphalt and solvent-treat lubricating-oil stocks simultaneously.

double-spot tuning *Telecommunications.* radio reception at two different local oscillation frequency values.

double star *Astronomy.* any two stars that appear close together; they may be physically connected by gravity (a binary star system) or by a chance alignment of stars at greatly different distances (an optical double).

double-stranded deoxyribonucleic acid *Biochemistry.* DNA that has two polynucleotide molecular chains wound round each other, with pairing between complementary bases.

double-stream amplifier *Electronics.* a microwave traveling-wave amplifier in which amplitude occurs through the interaction of two electron beams with different average velocities.

double-stub tuner *Electromagnetism.* two tuning stubs in a waveguide separated by three-eighths of a wavelength and connected in parallel with the transmission line.

doublet two items that are joined or associated in some way; a pair; specific uses include: *Physical Chemistry.* a bond between two electrons that are shared by two atoms. *Electromagnetism.* a separation of oppositely charged particles or opposite-polarity magnetic poles. *Particle Physics.* two elementary particles with the same baryon number, spin, and parity, but with slightly different masses and different charges. *Optics.* a pair of lenses having different optical qualities, combined so that distortions in one lens cancel out those that exist in the other. *Spectroscopy.* a spectral line that is split into two closely related but separate component wavelengths, indicating a transition from a single energy level to two levels having slightly different energies. *Fluid Mechanics.* a source and sink of numerically equal strengths allowed to approach each other, under the condition that as the distance between them approaches zero, their strengths increase.

double-tailed test see TWO-TAILED TEST.

doublet antenna see DIPOLE ANTENNA.

double-target leveling rod *Cartography.* any target rod with gradations on two opposite faces.

double tempering *Metallurgy.* a thermal treatment of iron and steel that follows two separate consecutive schedules, which are not always different.

double-theodolite observation *Engineering.* a computation of wind vector as a function of height by determining, at periodic intervals, the elevation and azimuth angles from two theodolites located at either end of a baseline attached to a pilot balloon.

double-throw circuit breaker *Electricity.* a circuit breaker that closes by making contact with either of two sets of contacts.

double-throw contact see MAKE-BEFORE-BREAK CONTACT.

double-throw switch *Electricity.* a switch that connects one set of two or more terminals to either of two sets of similar contacts.

double tide *Oceanography.* a double-headed tide; a high water that has two maxima of about the same height with a small depression in between, or a low water that has two similar minima with a small elevation in between.

double touch *Medicine.* a digital examination of the rectum and vagina simultaneously.

double-track tape recorder *Acoustical Engineering.* a tape recorder that records or plays back on one half of the tape when the tape is moving in one direction and on the other half when it is moving in the opposite direction

double transformation see COTRANSFORMATION.

double triode see DUOTRIODE.

doublet trigger *Electronics.* a trigger signal consisting of two pulses separated by predetermined intervals for coding.

double-tuned amplifier *Electronics.* an amplifier in which the stages are tuned to two different resonant frequencies to obtain a wider bandwidth than is possible with single-frequency tuning.

double-tuned circuit *Electronics.* a tuned circuit in which resonant components are tuned to two adjacent frequencies to achieve a wide bandwidth.

double-tuned detector *Electronics.* an FM discriminator containing two tuned circuits; the resonant frequency of each tuned circuit is adjusted to provide an overall balanced response above and below the resting frequency.

double valves *Petroleum Engineering.* two valves in series that are used as standing or subsurface traveling valves in wells; the dual valves are more reliable than one valve.

double vault *Architecture.* a dome or other vault composed of an outer skin set above an inner shell.

double-wedge cut *Mining Engineering.* a drill-hole pattern consisting of a shallow wedge within a larger, outer wedge; often used to obtain deep pull in hard rock. Also, WEDGE CUT.

double-welded joint *Metallurgy.* a joint performed by welding on both sides of a workpiece.

double-winding synchronous generator *Electricity.* a dynamo-type generator having separate armature windings in phase with each other for supplying two voltages to two independent external circuits.

double word *Computer Technology.* a storage location in memory that is two words in length and is treated as a unit.

double-work *Botany.* a technique in plant propagation in which a scion or bud is grafted to an intermediate variety that in turn is grafted to stock of another variety, often used to overcome incompatibility.

double X *Molecular Biology.* an aberrant, acrocentric, double-length X chromosome found in *Drosophila;* caused by the exposure of genetic material to radiation.

double zenith distance *Cartography.* a value of twice the zenith distance of an object, obtained by observation and not by a mathematical process.

doubling the angle on the bow *Navigation.* a method of obtaining a running fix by measuring the distance a vessel runs between two relative bearings on a fixed object, the second of which is double the first. The distance off the object at the time of the second bearing is equal to the distance run.

doubling time *Cell Biology.* the time required for a cell or a population of cells to double in number. *Nucleonics.* in a breeder reactor, the operating time interval required to produce an amount of fissile material that is equivalent to the initial mass of fissile material in the reactor; used as a measure of the reactor's performance.

doubly azimuthal map projection *Cartography.* an azimuthal map projection with two poles.

doubly linked list see SYMMETRIC LIST.

doubly linked ring *Computer Programming.* a circular data structure in which each element has backward and forward pointers; the head element is linked both to the next element and to the previous element.

doubly plunging fold *Geology.* an anticlinal or synclinal fold that reverses its direction of plunge within a particular area.

doughnut a circular object thought to resemble a ring with a hole in the middle, similar to the well-known pastry; specific uses include: *Nucleonics.* in a particle accelerator, the hollow torus between the poles of a betatron electromagnet, in which electrons are constrained to travel in a constant-radius circular path by a gradually increasing magnetic field. *Petroleum Engineering.* **1.** a device used to support a string of pipes, casing, or tubing; composed of a threaded, tapered ring of steel or a ring of wedges. **2.** a concrete structure formed around a wellhead to create a one-atmosphere chamber used for inspecting the equipment under normal atmospheric conditions.

Douglas fir *Botany.* a large, coniferous tree, *Pseudotsuga menziesii*, of the western United States, characterized by thick bark and hanging cones with bracts that extend beyond the scales. *Materials.* the yellowish brown, close-grained, soft and durable wood of this tree, used in plywood, flooring, and construction. Also, RED FIR.

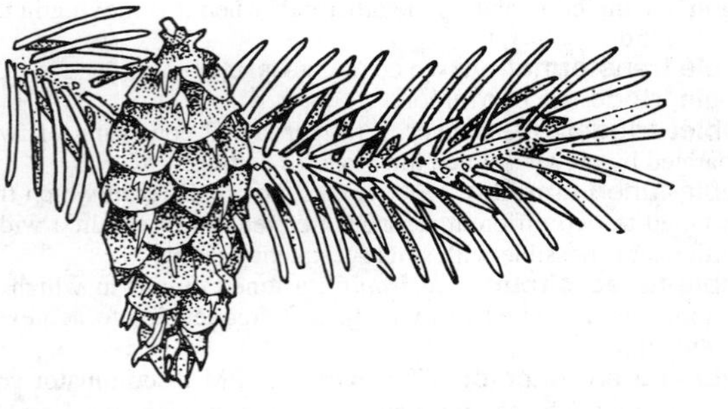

Douglas fir

douglasite *Mineralogy.* $K_2Fe^{+2}Cl_4{\cdot}2H_2O$, a light green monoclinic mineral; massive granular in habit; having an undetermined hardness and a specific gravity of 2.162; found with halite, sylvite, and carnallite.

Douglas scale *Oceanography.* one of two series of numbers from 0 to 9 that are used to define the degree of disturbance of the ocean surface.

Dounce homogenizer *Biology.* a device composed of a glass tube and pestle, used manually to break down tissue suspensions to obtain single cells or cell parts.

do-until structure *Computer Programming.* a high-level language control structure that causes a set of instructions to be executed repeatedly until a prespecified condition occurs; a test is performed after the activity is completed ensuring the loop is executed at least once.

dourine *Veterinary Medicine.* a venereal disease affecting horses, donkeys, and mules caused by *Trypanosoma equiperdum* and characterized initially by edematous inflammation of the external genitalia, glandular swelling, mucoprulent discharge from the urethra in the stallion and the vagina in the mare, and later by emaciation, paralysis, and skin plaques 2 to 10 cm in diameter.

douse *Mining Engineering.* to locate subsurface resrouces, such as water, oil, or minerals. Thus, **dousing, douser.** *Metallurgy.* to thrust a hot piece of metal into liquid during the cooling process.

DOVAP [dō´väp´] *Electronics.* a system in which the trajectory of a missile or other rapidly moving long-range object can be plotted by means of the Doppler effect, shown by radio waves bounced off the object. (An acronym for Doppler velocity and position.)

dove (mourning dove)

dove *Vertebrate Zoology.* any bird of the family Columbidae, particularly the smaller birds with pointed tails; the pure white member of this species is used as a symbol of peace.

Dove see COLUMBA.

dovekie *Vertebrate Zoology.* a small, short-billed auk of the family Alcidae that inhabits the North Atlantic, characterized by a short bill, short legs and wings, webbed feet, and a black back and white underbelly.

Dove prism see DELABORNE PRISM.

Dover, Strait of *Geography.* a narrow strait (21 miles wide) connecting the English Channel with the North Sea

dovetail *Building Engineering.* **1.** any joint formed with one or more tenons that fit tightly within corresponding mortises. **2.** to join or fit two pieces together by means of a dovetail.

dovetail anchor *Building Engineering.* a special case anchor used for stone facings in concrete walls; it is adjusted up and down to fit into the joints before the wall is poured.

dovetail brick *Mechanical Devices.* a brick with one end that is wedge-shaped or dovetailed and the other end cut to receive the dovetail or wedge of an adjacent brick.

dovetail cramp *Mechanical Devices.* a cramp used in masonry having a double dovetail to hold two building stones together.

dovetail cutter *Mechanical Devices.* a machine that performs dovetailing operations for both the inner and outer edges of wood boards.

dovetail hinge *Mechanical Devices.* a hinge with increasing width of its leaves from the hinge joint.

dovetail plane *Mechanical Devices.* a plane that is used to cut parallel-sided tongues and grooves.

dovetail saw *Mechanical Devices.* a thin-bladed saw with a fine-tooth backsaw blade; used especially for precision or detail work. Also, MITER SAW, TENON SAW.

dowel *Building Engineering.* in masonry, a wood piece drilled into a wall to receive nails for fastening woodwork or other wall fixtures. *Mechanical Devices.* **1.** any cylindrically shaped wooden or metal pin used as a guide or hole locator from one part to another. **2.** a copper, slate, nonferrous metal, or stone pin attached to each face of two stones for a strong bond between parts. Also, **dowel pin.**

dowel gauge *Mechanical Devices.* a device that can be clamped to a dowel to locate and determine the depth of a hole.

doweling jig *Mechanical Devices.* a device used to direct a bit in drilling holes to receive dowels.

dowel plate *Mechanical Devices.* a steel plate with several tapered holes that form a dowel pin pattern.

dowel screw *Mechanical Devices.* a wood dowel with threads at each end.

do-while structure *Computer Programming.* a high-level language loop control structure that causes a set of instructions to be executed repeatedly while a prespecified condition remains in effect; the test for the terminating condition precedes execution so that the loop may not be executed at all.

down *Engineering.* referring to a system that is either not working or unavailable for normal use. *Geology.* an area of high, open, grassy land, especially in the British Isles. Also, DOWNS. *Vertebrate Zoology.* the soft, fluffy feathers of a young bird, or a similar layer of inner feathers in an adult bird. *Anatomy.* see LANUGO.

downbuckle see TECTOGENE, def. 2.

down by the head *Naval Architecture.* of or relating to a ship that is floating abnormally low in the water toward the bow.

down by the stern *Naval Architecture.* of or relating to a ship that is floating abnormally low in the water toward the stern.

downcast *Mining Engineering.* the shaft through which fresh air is drawn or forced into a mine.

downcomer *Chemical Engineering.* a down-flow zone for conveying liquid from one tray in a tray column to the tray beneath.

down-converter *Electronics.* a converter whose ouput is at a frequency lower than the frequency of the input signal.

downcutting *Chemical Engineering.* a milling procedure in which the teeth of a cutting tool proceed into the work in the same direction as the feed. Also, CLIMB MILLING, CLIMB CUTTING.

downdip *Geology.* the direction downward and parallel to the dip of a stratum or bed.

down-Doppler *Acoustics.* a phenomenon in which the frequency of sound decreases due to the increase of wavelengths caused by the relative motion of the sound source, the listener, and the medium carrying the sound waves.

downdraft carburetor *Mechanical Engineering.* a carburetor in which the fuel is introduced into a downward current of air.

down feather see PLUMULE.

downflow *Chemistry.* the direction of flow of the solution in an ion-exchange process.

downhole *Petroleum Engineering.* describing equipment, tools, and instruments used in a borehole during drilling. Thus, **downhole drill.**

down-lead see LEAD-IN.

downlink *Telecommunications.* **1.** a communications path that is used to transmit signals downward to a station on the earth from a communications satellite or aircraft. **2.** to establish or utilize such a connection.

downloading *Computer Programming.* a process in which a host computer transfers a program or data in binary form to another computer or an intelligent terminal. Thus, **download, downloader.**

down lock *Aviation.* a device that locks an aircraft landing gear after it has been lowered.

down mutation *Genetics.* a promoter mutation that decreases the frequency of initiation of transcription.

downpipe see DOWNSPOUT.

down quark *Particle Physics.* a type of quark having an electrical charge of –1/3; this has been observed only paired with an up quark.

downrange *Space Technology.* the airspace along the flight course of a rocket, particularly in the direction of the target or area of impact.

downrush *Meteorology.* the strong downward-flowing air current marking the dissipating stage of a thunderstorm.

downs *Geology.* see DOWN.

Downs cell *Chemical Engineering.* a sodium-producing electrolytic cell having a steel container that is lined with brick and that has four graphite anodes projecting upward from the bottom; the cathodes are in the form of steel cylinders concentric with the anodes.

downslope time *Metallurgy.* in resistance welding, the time period during which the current is intentionally decreased.

downspout *Building Engineering.* a vertical pipe that carries rainwater from the roof or gutter to a drain. Also, DOWNPIPE, DRAINSPOUT.

Down's process *Chemical Engineering.* a procedure for producing sodium and chlorine from sodium chloride by adding fluoride and potassium chloride to the sodium chloride to reduce the melting point, then fusing and electrolyzing the mixture, so that sodium forms at the cathode and chlorine at the anode.

downstream *Hydrology.* away from the source of a stream. Similarly, **downriver.** *Meteorology.* the direction toward which a fluid is moving, implying the horizontal component of the direction of the basic stream. *Chemical Engineering.* that portion of a product stream that has passed through the system. *Molecular Biology.* **1.** the direction of nucleotide sequences extending past the 3' end of the last exon of a gene. **2.** of a nucleotide sequence in a gene, proceeding in the same direction as gene transcription.

downstream processing *Biotechnology.* any technique that separates, concentrates, and purifies the end products of fermentation, e.g., chromatography, distillation, centrifugation, and precipitation methods.

Down syndrome or **Down's syndrome** *Genetics.* a human congenital disorder caused by trisomy of chromosome 21, occurring approximately once in 1000 births and associated with advanced maternal age. It is characterized by slow physical development, mental retardation, and a number of physical abnormalities, including mongoloid facial features, and is often accompanied by heart disease, vision defects, chronic respiratory infections, and other disorders. (Named for the English physician John Langdon Haydon *Down,* 1828–1896, who first described this condition.) Also, MONGOLISM.

downthrow *Geology.* the amount of downward movement on the side of a fault.

downthrown *Geology.* describing the side of a fault that seems to have moved downward, compared to the other side. Thus, **downthrown side.**

down time *Engineering.* a period of time when a system is not in operation, either because of a failure or routine maintenance.

down-to-basin fault *Petroleum Engineering.* a deep fracture or fault in a formation that extends from an oil basin into another porous zone.

Downtonian *Geology.* in England, a sandstone series of the Lower Devonian period, formerly assigned to the uppermost Silurian.

downward compatibility *Computer Technology.* the capability of an older or smaller computer that can run the same programs as a newer or larger computer.

downward modulation *Telecommunications.* modulation in which the instantaneous amplitude of the modulated wave is less than that of the unmodulated carrier.

downwarp *Geology.* a region of the earth's crust that exhibits wide-scale downward bending.

downwash *Fluid Mechanics.* the downside of the trailing vortex system of a finite wing, which often presents a hazard to light planes following five to ten miles behind a large plane, as vortices trailing large planes have been measured in excess of 200 miles per hour.

downwind *Meteorology.* **1.** the direction toward which the wind is blowing. **2.** relating to or located in this direction.

downy mildew *Plant Pathology.* a plant disease caused by fungi of the family Peronosporaceae, which creates soft, whitish growths on diseased plant parts.

downy woodpecker *Vertebrate Zoology.* a small, black and white North American woodpecker, *Picoides pubescens,* the most common and widely distributed woodpecker of the U.S.

downy woodpecker

Dow process *Metallurgy.* a process for the electrolytic extraction of magnesium from molten magnesium chloride.

dowsing (rod) see DIVINING (ROD).

doxycycline *Pharmacology.* $C_{22}H_{24}N_2O_8 \cdot H_2O$, a semisynthetic broad-spectrum antibacterial of the tetracycline group, occurring as a yellow crystalline powder that is administered orally.

DP or **D.P.** Doctor of Podiatry; Doctor of Pharmacy.

DP or **dp** data processing; dew point; degree of polymerization.

d.p. or **dp** with proper direction. (From Latin *directione propria.)* Also, **D.P.** or **DP.**

dpdt switch see DOUBLE-POLE DOUBLE-THROW SWITCH.

D period *Microbiology.* the time between termination of a round of chromosomal DNA replication and cell division.

DPG 2,3-diphosphoglycerate.

DPH Department of Public Health.

DPH or **D.P.H.** Diploma in Public Health; Doctor of Public Health.

DPM or **D.P.M.** Diploma in Psychological Medicine; Doctor of Podiatric Medicine.

dpst switch see DOUBLE-POLE SINGLE-THROW SWITCH.

DPT diphtheria-pertussis-tetanus (vaccine).

DR dead reckoning.

DR or D.R. degeneration reaction.

Dr. Doctor.

dr. dram; drams.

Draco *Astronomy.* the Dragon, a large constellation wrapped around the northern celestial pole.

Draconematoidea *Invertebrate Zoology.* a suborder of marine nematode worms.

draconic month *Astronomy.* the time (27.2122 days) between two successive passages of the moon through its ascending node. Also, NODICAL MONTH.

Draconids *Astronomy.* two annual meteor showers that appear to radiate from Draco: the June Draconids come from Comet Pons-Winnecke and reach peak activity around June 28th, and the October Draconids are from Comet Giacobini-Zinner and peak on October 8th; the latter are also, GIACOBINIDS.

Draco system *Astronomy.* an old dwarf elliptical galaxy that belongs to the Local Group.

Dracunculoidea *Invertebrate Zoology.* a superfamily of nematode worms, parasitic on fish, reptiles, and mammals, living in the host's tissues and exiting through a skin lesion; includes the Guinea worm, up to one meter long.

Draeger breathing apparatus *Mining Engineering.* a long-service, self-contained oxygen-breathing apparatus with a lung-governed oxygen feed that allows a worker to perform normal work for up to seven hours.

Draeger escape apparatus *Mining Engineering.* a portable, self-contained oxygen-breathing apparatus; used as a defense against poisonous gases or oxygen shortages for periods as long as an hour.

draft *Fluid Mechanics.* a movement of air, especially in an enclosed space. *Mechanical Devices.* a device to control the flow of air, as in a stove or fireplace. *Naval Architecture.* the vertical distance from the waterline to the bottom of a vessel's keel. *Engineering.* **1.** the act of pulling or hauling a load, or the load itself. **2.** the area of an opening for the discharge of water. **3.** the angle of clearance in a mold that allows for easy removal of the hardened material. *Metallurgy.* the change in cross section obtained after rolling or cold drawing. *Civil Engineering.* a line of a transverse survey. Also, DRAUGHT.

draft animal *Agriculture.* any animal domesticated by humans and used to pull, haul, or carry loads; e.g., horses, cattle, mules, or elephants.

draft beer *Food Technology.* beer that is drawn from a keg or barrel, as opposed to beer in cans or bottles.

draft gauge *Engineering.* **1.** an instrument used to measure the gas density of low gas heads, or small differential gas pressures. **2.** a hydrostatic instrument used to measure the depth to which a vessel is submerged.

draft hood *Engineering.* a part of an appliance that allows for combustion products to escape from a combustion chamber.

draft horse *Agriculture.* a large, strong horse that is used for pulling heavy loads. The Percheron, Belgian, Shire, and Clydesdale are breeds of draft horses.

drafting *Graphic Arts.* the process of using lines and numbers to depict the dimensions and specifications of something to be constructed, such as a building or a mechanical device. *Textiles.* see DRAWING.

draft loss *Mechanical Engineering.* the decrease of static pressure in a boiler or furnace due to flow resistance.

draft marks *Naval Architecture.* the marks on the side of a ship's hull showing the level to which it can legally be loaded under various conditions (e.g., fresh water, summer, or winter).

draftsman *Graphic Arts.* a person who specializes in the drawing of equipment, structures, models, electrical circuits, and so on, from which prints are made for actual use. Thus, **draftsmanship.**

draft tube *Mechanical Engineering.* a flared, vertical, discharge pipe leading from a water turbine to its tailrace; it decreases outlet pressure and increases turbine efficiency.

drag to move an object along a surface with some effort; specific uses include: *Fluid Mechanics.* a force on an object parallel with the direction of the relative velocity of the flow; drag can be on an airfoil moving through the air, opposing the direction of the airfoil, or on the surface of an air duct, in the direction of the airflow. *Aviation.* the effect of this force on an aircraft or other body in flight; in combination with available thrust, drag determines the craft's speed. *Ordnance.* the force of aerodynamic resistance caused by the violent currents behind the shock front of an explosion. *Metallurgy.* in casting, the bottom portion of a mold or pattern. *Computer Science.* to move an item displayed in a window by selecting the item with the mouse, moving the item to a new position by moving the mouse, then releasing it. The displayed image of the selected item is moved on the screen as the mouse moves, giving the illusion that the item is being dragged.

drag anchor see DROGUE.

drag-body flowmeter *Engineering.* an instrument that measures the flow of liquid and determines the pressure parallel to the flow, providing a pressure reading to be used in flow equations.

drag chain *Engineering.* **1.** a chain that hangs from a truck or other motor vehicle chassis and prevents static electricity from accumulating. **2.** a chain that joins railroad cars together.

drag classifier *Mechanical Engineering.* a machine used to sort sand by weight; it consists of an endless belt equipped with transverse rakes that drag heavier, faster-settling material upward through an inclined trough while lighter, slower-settling material overflows and falls to the bottom.

drag coefficient *Fluid Mechanics.* a description of the drag force affecting a body that is moving through the air or another fluid, expressed by $F_D = C_D A_p \rho V_0^2/2$, in which C_D is the drag coefficient, A_p is the projected (silhouetted) area of the body, ρ is the density of the fluid, and V_0 is the freestream velocity. Also, COEFFICIENT OF DRAG.

drag conveyor *Mechanical Engineering.* a conveying machine used for loose material, consisting of a wide-linked endless chain with projections on either side; the chain is dragged along the bottom of a trough into which the material is fed. Also, **drag-chain conveyor.**

drag-cup generator *Engineering.* a tachometer used to measure velocity, consisting of two stationary windings (having a zero coupling) and a nonmagnetic metal cup that is revolved by the object being measured for speed. Also, **drag-cup tachometer.**

drag-cup motor *Electricity.* an induction motor having a shaft with a copper or aluminum drag cup that rotates in the magnetic field of a two-phase stator.

drag cut *Engineering.* an arrangement of drill hole cuts taken in various angles in a formation, providing for a section of rock to be removed.

drag direction *Aviation.* the direction of the relative airstream at a given moment in flight.

drag factor *Chemical Engineering.* the ratio of hindered diffusion rate to unhindered rate in a swollen dialysis membrane. Also, FAXEN DRAG FACTOR, HINDRANCE FACTOR.

drag fold *Geology.* **1.** one of a series of minor folds formed in an incompetent bed by the movement of the surrounding competent beds in opposite directions relative to each other. **2.** any fold that is a subsidiary part of a larger fold.

drag force *Physics.* a force that opposes an object moving through a fluid.

dragging of inertial frames *Physics.* a phenomenon within the context of a relativistic universe, by which a spinning mass in space will pull the space around it in the direction of its spin.

drag harrow *Agriculture.* a type of harrow having curved tines, used for deep cultivation of difficult land. Also, DUCK-FOOT HARROW.

drag-in *Metallurgy.* the transfer of a liquid into another liquid, caused by a workpiece or by an equipment component.

dragline *Mechanical Engineering.* an excavating machine consisting of a steel scoop that swings on chains from a movable jib; the scoop is cast into the material to be excavated and dragged back to the jib, where it is lifted and dumped. Also, **dragline crane, dragline excavator.**

dragline scraper *Mechanical Engineering.* a machine for moving soil, gravel, or other loose material; generally consists of a scraper attached to an endless cable or belt that is operated by a drum or sprocket wheel.

drag link *Mechanical Engineering.* a four-bar linkage designed to connect two cranks located on parallel shafts while allowing both cranks to traverse full circles.

drag-link conveyor see SCRAPER CONVEYOR.

drag mark *Geology.* a long, even striation or groove produced as a solid object is dragged across a soft sedimentary surface.

Dragon *Ordnance*. an American medium-range wire-guided antitank weapon that can be carried by individual infantry personnel, consisting of a missile, launcher, and tracker; officially designated FGM-77. *Astronomy*. see DRACO.

dragonfish *Vertebrate Zoology*. any of several marine fishes of the family Bathydraconidae, inhabiting waters along coastal Antarctica to about 2000 feet; characterized by an elongated body and shovel-like snout. Also, DEEP DRAGONFISH; ANTARCTIC DRAGONFISH.

dragonfly *Invertebrate Zoology*. a member of the insect order Odonata, suborder Anisoptera, characterized by a colorful body, long-veined translucent wings, strong eyesight, and jaws adapted for seizing prey; most of its life cycle is spent as aquatic nymph.

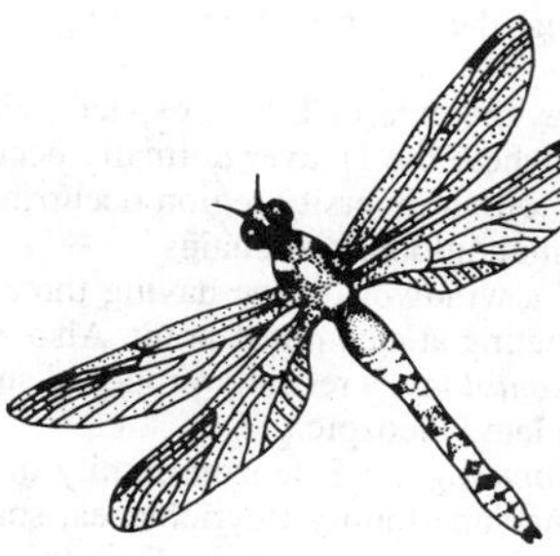

dragonfly

Dragonian *Geology*. a North American continental stage of the Lower Paleocene epoch, after the Puercan and before the Torrejonian.

drag-out or **dragout** *Metallurgy*. the transfer of a liquid out of a pool, caused by a workpiece or by an equipment component.

drag parachute *Aviation*. **1.** any of a variety of parachutes that can be deployed from high-performance aircraft under certain flight conditions to enhance control and stability or during landings to decrease speed. Also, BRAKING PARACHUTE. **2.** a drogue parachute.

drag rope *Aviation*. a long, heavy rope that can be thrown overboard from a balloon to act as a brake or variable ballast when landing. Also, TRAIL ROPE, GUIDE ROPE.

dragsaw *Mechanical Devices*. a power saw with its teeth oriented to allow cutting only on the pulling stroke; used in tree cutting.

dragshovel see BACKHOE.

drag-stone mill *Mining Engineering*. a mill in which ores are ground by dragging a heavy stone around a circular or annular stone bed.

drag technique *Metallurgy*. in arc welding, the practice of keeping the electrode in contact with the workpiece.

drag truss *Aviation*. a truss set horizontally between an airplane's wing spars to stiffen the wing structure and provide resistance to drag forces.

drag twist *Mining Engineering*. a spiral twist used to wipe a blast hole with hay before charging with black powder. Also, CORKSCREW.

drag-type tachometer see EDDY CURRENT TACHOMETER.

drag-weight ratio *Space Technology*. in a missile or rocket, the ratio of total drag at burnout to total weight.

drag wire *Aviation*. a wire designed to resist drag forces, usually running from a forward inboard point to an outboard aft point.

drain *Engineering*. **1.** any pipe, channel, or the like used to carry off excess or unwanted liquid, such as dishwater from a sink, rainwater from a paved surface, and so on. **2.** to remove liquid in this manner, such as oil from an automobile crankcase. *Surgery*. **1.** to cause the evacuation of fluid or purulent material from a cavity by a surgical incision, tapping, or manual manipulation. **2.** any device such as gauze, rubber tubing, or suture material that causes fluid to exit a wound or cavity. *Electricity*. **1.** the electrical current or power taken from a source. **2.** the load device that absorbs this current or power.

drainage *Engineering*. the process of draining, or a system for carrying out this process. Used to form various compounds, such as **drainage ditch, drainage canal.** *Surgery*. the flow or withdrawal of fluids from a wound or cavity. *Petroleum Engineering*. the movement of oil or gas in a reservoir due to reduced pressure.

drainage basin *Hydrology*. an area occupied by a closed drainage system, especially a region that collects surface runoff and contributes it to a stream channel, lake, or other body of water. Also, CATCHMENT AREA.

drainage composition *Hydrology*. a quantitative description of a given drainage basin in terms of its drainage density, stream order, bifurcation ratio, and stream-length ratio.

drainage density *Hydrology*. the average stream length within a drainage basin per unit area, calculated by dividing the total length of all streams within a drainage basin by the area of the basin.

drainage divide *Hydrology*. see DIVIDE.

drainage lake *Hydrology*. any lake having a surface outlet through which it loses water.

drainage pattern *Hydrology*. the arrangement or configuration of the natural stream courses in any given area when projected onto a flat surface. Also, **drainage network.**

drainage ratio *Hydrology*. the ratio between the amount of runoff in a given area during a given period of time and the amount of precipitation in that area during the same time period. Also, DISCHARGE EFFICIENCY.

drainage system *Hydrology*. **1.** a natural system by which a region is drained, consisting of a body of enclosed surface water, such as a stream, together with all other such bodies of water that are tributary to it. **2.** an artificial system, including the surface and subsurface conduits, that drains a region.

drainage tube *Surgery*. a tube used to evacuate fluid from a wound or cavity.

drainage wind see GRAVITY WIND.

drain casting *Materials Science*. a process used in making hollowware, in which excess slip is drained by inversion of the mold. Also, HOLLOW-CASTING.

drainpipe *Engineering*. a large pipe used to draw off excess water, such as rainwater from a roof.

drain sample *Petroleum Engineering*. a sample drawn from an oil-water, oil-brine, or oil-drilling mud separator.

drainspout see DOWNSPOUT.

drain tile *Building Engineering*. a cylindrically shaped tile made with holes, used at the base of a building foundation to carry away groundwater. Also, FIELD DRAIN.

drain wire *Electricity*. a metal conductor in contact with a foil-type signal-cable shield that provides a low-resistance ground return at any point along the shield.

Drake equation *Astronomy*. an equation that estimates the likelihood of extraterrestrial civilizations by combining factors such as the number of stars that have planets, the number of planets with life, and the average lifetime of an industrial civilization.

dram *Metrology*. a unit of weight used in the apothecaries' system of measurement, equal to 60 grains, or 0.0125 ounce (3.89 grams); in the avoirdupois system it is equivalent to 27.34 grains or 0.0625 ounce.

DRAM dynamic random-access memory.

drape forming *Materials Science*. a process in which a thermoplastic sheet is molded inside a movable frame by heating it, draping it over a male mold, and then vacuuming it.

Draper, Henry 1837–1882, American astronomer; the first to photograph a stellar spectrum (other than the sun's) and a nebula.

Draper catalogue see HENRY DRAPER CATALOGUE.

Draper classification see HENRY DRAPER SYSTEM.

Draper effect *Chemical Engineering*. an increase in volume under constant pressure when hydrogen and chlorine begin to react to create hydrogen chloride, due to an increased temperature in the reactor because of the exothermic reaction.

drapery *Geology*. see CURTAIN, def. 1.

draping *Geology*. the structural concordance of warped strata overlying a limestone reef or other hard core with the surface of the reef or core.

draught see DRAFT.

draughtsman colony *Microbiology*. a type of colony typically formed by *Streptococcus pneumoniae* incubated on blood agar, in which the colonies are steep-sided with flat tops, resembling the pieces used in the game of draughts (checkers). Also, CHECKER COLONY.

Dravidian family *Linguistics*. in historical linguistics, a language family based in central India and Pakistan.

dravite *Mineralogy*. $NaMg_3Al_6(BO_3)_3Si_6O_{18}(OH)_4$, a trigonal mineral of the tourmaline group; forms two series, with elbaite and with schorl.

draw *Engineering*. **1.** to haul a load of material. **2.** to pull an object out to a greater length; stretch. *Mining Engineering*. **1.** to raise ore, coal, rock, or the like to the surface. **2.** the distance on the surface to which the subsidence or creep extends beyond the workings. *Metallurgy*. **1.** to stretch out a metal into a long, thin shape, as to form a wire. **2.** specifically, in metal fabrication, to deform by pulling a stock through a die, or by forcing between a die and a punch. *Geology*. a gully or ravine, especially in the southwestern U.S.

drawability *Materials Science*. a measure of the relative ability of a material to be drawn out into a thin, elongated shape.

drawbar *Engineering.* **1.** a bar at the rear of a tractor that joins the tractor and a plow. **2.** a bar that joins a locomotive to a tender. **3.** a clay block in a glassmaking furnace that determines the point when sheet glass is drawn.

drawbar pull *Mechanical Engineering.* the force, measured in pounds, that is available to a locomotive or tractor to pull a load after overcoming its own tractive resistance. Similarly, **drawbar horsepower.**

draw bead *Metallurgy.* in deep drawing, an equipment component used to regulate metal flow.

drawbench *Metallurgy.* in drawing, the stationary apparatus that holds the die.

drawbridge *Civil Engineering.* a type of bridge often constructed over a waterway, in which the entire bridge or a section of it can be raised up or moved aside, as to permit passage by vessels with tall masts.

drawdown *Hydrology.* the distance by which the water level of a reservoir or other body of water is lowered as a result of the withdrawal of water. *Petroleum Engineering.* the difference between the flowing and the static bottom-hole pressure. *Graphic Arts.* a method of obtaining a rough color proof of a particular ink on a particular paper by spreading the ink into a thin film with a spatula or knife.

drawdown ratio *Engineering.* the relationship of the size of a mold to the thickness of the molded product.

draw-filing *Engineering.* a filing operation in which the file moves sideways along the work, rather than across it.

drawhead or **draw head** *Metallurgy.* in drawing, a set of rolls or dies attached to a drawbench.

drawing *Chemical Engineering.* the removal of a material after a processing operation is completed, such as metal from a casting mold, ceramic ware from a kiln, or a solid material from a furnace. *Metallurgy.* the process of reducing the cross section of a bar or wire, or the wall thickness or diameter of a tube, by pulling it through a die. *Materials Science.* the process of reheating a malleable iron in order to reduce the amount of carbon combined as cementite by spheroidizing pearlite, tempering martensite, or graphitizing both. *Textiles.* in yarn manufacture, a process of passing raw fibers through a series of rollers to lengthen and straighten them and to create greater uniformity. Also, DRAFTING.

drawing apparatus see CAMERA LUCIDA.

drawing blank *Metallurgy.* the starting stock of a drawing operation.

drawing chisel *Mechanical Devices.* a wood chisel with an oblique edge used to form tenon ends across the grain.

drawing of temper *Metallurgy.* the heating of hardened steel, followed by slow cooling.

drawing-out *Metallurgy.* in forging, a method for stretching a metal or alloy by repeated blows.

drawing pin *Graphic Arts.* a short, broad-headed pin used on drawing boards to hold down drawings and tracings.

drawknife *Mechanical Devices.* a tool, having a single long and narrow blade with two right-angled handles parallel to it, that is drawn toward the user along the face of the material to trim or shave it.

draw man see GRIZZLY WORKER.

draw mark *Metallurgy.* an undesirable surface defect that is present in a drawn bar.

drawpiece *Metallurgy.* any part that has been drawn.

drawplate *Metallurgy.* in deep drawing, the plate on which the blank rests.

drawpoint *Engineering.* a sharp steel point that is used for marking lines or piercing holes in an object.

draw slate *Mining Engineering.* a soft slate, shale, or rock, approximately 2 to 3 feet thick, lying above the coal, that falls with the coal or soon after the coal is removed.

draw tube *Biotechnology.* a coaxial tube within the body tube of a microscope that allows for adjustment of the body tube length.

drazoxolon *Organic Chemistry.* $C_{10}H_8ClN_3O_2$, toxic, yellow crystals, insoluble in water and acids, that melt at 167°C; used as a fungicide.

drd plasmid *Genetics.* a mutant conjugative plasmid that has a derepressed transfer operon.

dream *Neurology.* **1.** a mental phenomenon that occurs during sleep, in which a series of images, thoughts, and emotions resemble actual perceptions or events of the wakeful state; these experiences take place in periods of REM sleep, typically three to five times for about 20 minutes each during a full night's sleep. The source and significance of dreams have long been the subject of investigation, both scientific and speculative; Freud and others have theorized that dreams are the expression of unconscious thoughts and feelings. **2.** to have such an experience. *Psychology.* see DAYDREAM.

dream analysis *Psychology.* the interpretation of the content of dreams, in an effort to reveal their underlying content and thus the individual's unconscious mind.

dreaming sleep see REM SLEEP.

dreamless sleep see NREM SLEEP.

dream theory *Psychology.* the theory that dreams are the symbolic expression of suppressed wishes and desires, especially sexual desires.

Drechslera *Mycology.* a genus of fungi belonging to the class Hyphomycetes; its species are pathogenic to some plants.

dredge *Engineering.* a rectangular or cylindrical machine used for underwater excavation. *Mechanical Engineering.* **1.** the barge upon which such a machine is mounted. **2.** to excavate with a dredge.

dredge peat see SEDIMENTARY PEAT.

dredging *Engineering.* the process of excavating solid matter from an underwater area.

D region *Geophysics.* the area of the ionosphere, about 60 miles (100 km) above sea level, where the D layer normally occurs. *Immunology.* a segment of the gene-coding diversity region occurring in the hypervariable region of the immunoglobulin H chains.

dreikanter *Geology.* a windworn stone having three facets intersecting in three edges and meeting at a common point. Also, PYRAMID PEBBLE.

Drepanellacea *Paleontology.* a recently proposed superfamily of ostracods of the middle to late Paleozoic.

Drepanellidae *Paleontology.* a Paleozoic family of ostracods until recently classified in the superfamily Beyrichiacea; some now consider it to be the type family of a new superfamily, Drepanellacea.

Dresbachian *Geology.* a North American geologic stage of the Upper Cambrian period, occurring before the Franconian.

dress *Mechanical Engineering.* to shape a tool or material. *Mining Engineering.* **1.** to sort ore. **2.** to clean ore by breaking off fragments of the gangue from the valuable mineral. *Electronics.* the arrangement of connecting wires in a circuit to prevent undesirable coupling and feedback.

dressed and matched boards *Building Engineering.* pieces of wood or lumber of uniform size that are fabricated with a smooth finish.

dresser *Engineering.* a tool used for sharpening machinery parts so that they may be reused.

dressing *Medicine.* the material applied to a wound or infection for the purposes of providing protection and promoting healing. *Agronomy.* the process of applying fertilizer to the soil.

Dressler kiln *Mechanical Engineering.* a muffle-type tunnel kiln, originally built by the Swindell-Dressler Corporation.

Drew, Charles R. 1904–1950, American physician and hematologist; did research on plasma; established blood banks.

drewite *Geology.* a white, fine-grained, calcareous marine ooze or mud produced by the action of certain bacteria.

Drew number *Physical Chemistry.* a dimensionless group of variables used to study the diffusion of a solid into a vapor.

Dreyer's tube *Immunology.* a small glass test tube, having a flared rim and conical base, that is used in serology to observe precipitation or agglutination reactions.

Drianemaceae *Mycology.* a small family of fungi belonging to the order Trichales, found in the ground litter of Alpine regions in the spring, and on decaying wood and tree bark.

driblet cone see HORNITO.

dried gypsum or **dried calcium sulfate** see PLASTER OF PARIS.

dried milk *Food Technology.* a powdered form of milk, usually made by spraying concentrated milk into a chamber through which hot air is blown; the milk droplets dry and fall to the chamber bottom as powder.

drier *Engineering.* any device or substance used to dry some material or object. *Materials Science.* specifically, a furnace used to dehydrate ore products without changing their composition.

drift a gradual movement away from the original or intended location or setting; specific uses include: *Navigation.* the distance that a craft is moved by water or wind currents. *Oceanography.* the speed of an ocean current. *Aviation.* the lateral shift or displacement of an aircraft from its course, due to the action of the wind or other causes. *Photogrammetry.* the effect of this movement on aerial photography; it results in the edges of successive photographs being parallel but sidestepped. *Robotics.* the tendency of a robot or its manipulator to gradually move away from the targeted position. *Geology.* **1.** loose surface material transported from one place and deposited elsewhere by the action of such agents as wind, waves, currents, glaciers, or running water from a glacier. **2.** an accumulation of such material. *Mechanical Engineering.* **1.** a displacement of the gimbals of a gyroscope, typically caused by friction on its bearings. **2.** see DRIFTPIN, def. 1.

drift angle *Navigation.* the angle between the centerline of a turning vessel and a line tangent to the turning circle at the vessel's center of gravity.
drift avalanche see DRY-SNOW AVALANCHE.
driftbolt *Mechanical Engineering.* a round-shafted spike, used to fasten together heavy timbers. Also, DRIFTPIN.
drift bottle *Oceanography.* a specially designed bottle that is released into the sea to gain information on currents; it contains a card to be returned by the finder. Also, FLOATER.
drift card *Oceanography.* a card encased in a waterproof, buoyant envelope; used in the same way as a drift bottle.
drift clay see BOULDER CLAY.
drift compensation *Electrical Engineering.* the act of providing a correction to reduce the drift of the overall system. This is usually achieved by applying negative feedback to the system or by placing an element in the control circuit that drifts in the opposite direction to the system as a whole.
drift current *Oceanography.* a broad, slow-moving ocean current caused by wind. *Physics.* an electric current produced by the movement of free particles in perpendicular electric or magnetic fields.
drift dam *Geology.* an obstruction formed by the accumulation of glacial drift in a preexisting stream valley.
drift diameter *Petroleum Engineering.* **1.** the inside diameter of a wellbore joint of tubing, casing, or line pipe. **2.** in a drilling operation, the width of a hole.
drifter *Mechanical Engineering.* a mounted rack drill, used for drilling holes up to 4.5 inches in diameter.
drifter bar see DRILL COLUMN.
drift error *Computer Technology.* a type of error in the output of an analog computer caused by gradual changes in environmental conditions or slight voltage fluctuations.
drift glacier see SNOWDRIFT GLACIER.
drift ice *Oceanography.* ice that has drifted away from its place of origin.
drift indicator *Engineering.* an instrument that records any shift in direction of a moving object.
drifting *Mining Engineering.* **1.** the act of driving underground tunnels through stone. **2.** the act of tunneling along the strike of ore.
drifting mine *Ordnance.* a buoyant or neutrally buoyant mine that can move freely on or just below the surface of the water.
drifting snow *Meteorology.* snow lifted from the surface of the earth by the wind to a height of less than six feet above the surface.
drift lake *Hydrology.* a glacial lake that occupies a depression in the surface of glacial drift after the ice has all melted.
drift lead *Navigation.* a weight placed on the bottom to indicate movement of a vessel.
drift mining *Mining Engineering.* **1.** the process of using underground methods of mining to work alluvial deposits. **2.** the process of working relatively shallow coal seams by drifts from the surface.
drift mobility *Solid-State Physics.* a measure of how readily charge carriers can move under the influence of an electric field in a semiconductor or metal, given by the average drift velocity of the carrier per unit electric field. Also, MOBILITY.
driftpin *Mechanical Engineering.* **1.** a round, tapered piece of steel used for enlarging or aligning holes in metal. Also, DRIFT. **2.** see DRIFTBOLT.
drift plug *Engineering.* a wooden plug driven into a pipe to smooth out a kink or flare out an opening.
drift punch *Mechanical Engineering.* a tapered metal punch used to align rivet holes without coincidence.
drift sheet *Hydrology.* a sheetlike deposit of glacial drift laid down during a single glaciation or during a series of closely related glaciations.
drift space *Electronics.* a region within a velocity-modulated electron tube at which bunching of the electron stream occurs.
drift speed *Electricity.* the average speed at which electrons or ions flow through a medium.
drift station *Oceanography.* an ice-based scientific station in the Arctic Ocean, usually situated on an ice floe but sometimes on an ice island.
drift terrace see ALLUVIAL TERRACE.
drift transistor *Electronics.* a transistor in which high-frequency response is achieved through the nonuniform doping of the base region; this nonuniformity causes electric field gradients that, in turn, reduce transit time for the carriers. Also, DIFFUSED-ALLOY TRANSISTOR.
drift tube *Nucleonics.* in a linear accelerator, one of a series of electrodes of gradually increasing length to which alternating potential is applied for accelerating particles through gaps between the tubes.
drift velocity *Solid-State Physics.* the average velocity of charge carriers subjected to an electric field within a semiconductor, a conductor, or an evacuated region.
drift wave *Physics.* a wave observed in magnetically confined plasmas due to the presence of density gradients in the plasma.
Drilidae *Invertebrate Zoology.* the false fire beetles, a family of coleopteran insects in the superfamily Cantharoidea, whose larvae are predators of snails.
drill *Mechanical Devices.* a cutting tool with a rotating end that is used to cut or enlarge holes in metal, wood, and other solid materials. Also, **drill bit.** *Surgery.* **1.** a rotating instrument used for boring or cutting, such as into a tooth or a bone. **2.** to make or enlarge a hole in a bone or tooth. *Agriculture.* **1.** a farm implement that plants seeds by making a series of holes or furrows, dropping the seeds into them, and then covering them with soil. Also, SEED DRILL. **2.** the hole or furrow into which the seed is dropped, or a row of seeds planted in this manner. *Military Science.* **1.** repetitive exercises used to instruct, train, and discipline troops. **2.** describing personnel, objects, or inert weapons used in instruction and training. Thus, **drill ammunition, drill mine, drill sergeant.**
drill *Vertebrate Zoology.* a short-tailed, baboonlike West African monkey, *Mandrillus leucophaeus,* similar to the related mandrill but smaller.

drill

drillability *Materials Science.* the degree of ease by which a designated material may be penetrated by a drill.
drill cable *Engineering.* a strong wire rope that pulls up drilling equipment from a borehole.
drill capacity *Mechanical Engineering.* on a rotary or diamond drill, the length of a drill rod that the hoist can lift or that the brake can grip on a single line.
drill carriage *Mechanical Engineering.* a frame on which several rack drills are mounted and which moves along a track for heavy drilling needs in large tunnels. Also, JUMBO.
drill chuck *Mechanical Devices.* a device typically having adjustable jaws for securing a drill bit or other cutting tool to a spindle.
drill collar *Mechanical Devices.* a ring that holds a drill bit radially relative to a bearing. *Mineral Engineering.* a drill string-to-drill collaring device, which can weigh up to 100 tons, to alleviate torsional stresses in deep boring operations.
drill column *Mining Engineering.* a steel pipe that serves as a base on which to mount a diamond or rock drill. Also, BAR DRILL, DRIFTER BAR, JACK BAR.
drill cuttings *Engineering.* the fragmental material that surfaces when a wellhole is being drilled.
drill doctor *Mining Engineering.* a mechanic or shop that specializes in sharpening and servicing drill bits, tools, and steel.
drill drift *Engineering.* a metallic wedge used to separate two mated parts.
driller *Engineering.* a person who uses drilling tools to perform the drilling process. *Mechanical Engineering.* see DRILLING MACHINE.
driller's method *Petroleum Engineering.* a technique for killing a well by circulating high-density kill mud after the formation fluids have been circulated out of the well, in order to maintain drill pipe pressure.

drill extractor *Engineering.* an instrument used to pull broken drill pieces from a borehole.
drill extrusion ingot *Metallurgy.* an extrusion ingot that has a drilled, hollow center.
drill fittings *Engineering.* any devices that are used in a borehole as part of the drilling process. Also, DOWNHOLE EQUIPMENT.
drill floor see DRILLING PLATFORM.
drill footage *Engineering.* the depth to which a borehole is drilled, measured in lineal feet.
drill gauge *Mechanical Devices.* a thin, flat metal template with accurately sized drill holes formed and marked so that the diameters of drill bits can be checked and verified. Also, **drill gage.**
drill-grinding gauge *Mechanical Devices.* a tool that measures the angle and length of a twist drill point during grinding operations. Also, **drill-angle gauge, drillpoint gauge.**
drill hole *Engineering.* an opening that is cut or enlarged by a drill bit.
drillhole pattern *Engineering.* the specific arrangement of shot holes that form the round at the opening of a tunnel or sinking pit.
drill-in or **drilling-in** *Engineering.* **1.** the process of using a drill to make a circular hole. **2.** the operation of tunneling or stoping with a compressed-air rock drill, jackhammer, or drifter.
drilling *Engineering.* a process by which a drill forms or enlarges a hole in solid material.
drilling break *Petroleum Engineering.* a sudden gain in the penetration rate of a drill bit while drilling in a well.
drilling column *Engineering.* a round shaft of drill rods, which attaches to the drill cutting bit.
drilling machine *Mechanical Engineering.* a power machine that is used to create or enlarge a hole in a solid material. Also, DRILLER.
drilling mud *Materials.* a mixture of finely divided heavy material consisting of clay, water, and chemical additives that is pumped downhole through a drill pipe; used for such purposes as cooling the rotating bit, lubricating the drill pipe, carrying cuttings to the surface, and hindering foreign fluids from entering the wellbore. Also, **drilling fluid.**
drilling platform *Engineering.* a horizontal plank-covered area at the base of a derrick, which serves as work space and foundation for drilling equipment. Also, DRILL FLOOR.
drillings see CUTTINGS.
drill jig *Mechanical Devices.* a device used in repetitive drilling to maintain and guide a drill.
drill out *Engineering.* **1.** to finish the process of making a borehole. **2.** to drill through an obstruction in a borehole. **3.** to map out an area containing ore or petroleum by penetrating the area with a series of boreholes.
drill-over *Engineering.* a process of drilling on the outside of a borehole casing.
drill pipe *Mining Engineering.* a heavy steel pipe that is rotated to give motion to a drilling bit and is used to circulate drilling fluid.
drill plow *Agriculture.* a type of plow that has an attached apparatus for dropping seeds into the newly plowed furrow.
drill press *Mechanical Engineering.* a drilling machine that has a single vertical spindle.
drill ship *Naval Architecture.* a ship that is fitted with a rig for drilling the sea bottom in exploration for oil or gas.
drill steel *Metallurgy.* a hard alloy steel suitable for making drills.
drill-stem test *Petroleum Engineering.* a procedure for procuring a fluid sample from a formation by using a tool attached to the drillstem for testing formations; it analyzes bottom-hole pressure and aids in determining the productivity of a formation.
drill string *Mechanical Engineering.* a drill pipe assembly that links the drill bit to a mechanism imparting rotary or reciprocating motion.
D ring *Astronomy.* a faint, very tenuous ring that lies between the C ring and the cloudtops of Saturn.
drinking water *Hydrology.* any fresh water that is suitable for drinking; in general, water with less than 500 milligrams of undissolved solids per liter.
drip *Hydrology.* water that has collected on and falls from leaves, twigs, and so on. *Engineering.* any slight or intermittent flow of water or other liquid.
drip cap *Building Engineering.* a projecting outer molding designed to shed water from a door or window.
drip curtain *Geology.* see CURTAIN, def. 1.
drip-dry *Textiles.* describing a fabric that dries rapidly with few or no wrinkles when hung dripping wet.
drip edge *Building Engineering.* a strip, usually made of metal, that overhangs a building roof to deflect rainwater from the building.
drip irrigation *Agronomy.* an irrigation method in which water is applied directly to the root zone on a frequent basis, but very slowly and in small amounts. Also, TRICKLE IRRIGATION.
drip line or **dripline** *Agronomy.* an imaginary line around a tree, representing the perimeter at which water drips from the overhanging leaves; used to estimate the extent of root growth.
drive a source of power or the aplication of power; specific uses include: *Engineering.* **1.** a mechanism that imparts or transfers power to a machine or within a machine. **2.** to operate a vehicle or machine. *Behavior.* an urge to behave in a particular manner to satisfy the need for something that is physiologically necessary for survival, such as food or water. *Computer Technology.* an electromechanical device consisting of motors, read/write heads, and other components for moving magnetic tape or rotating disks and reading data from or writing data to the storage medium. *Electronics.* **1.** the application of voltage or power signals to a system, circuit, or device to cause it to perform its intended function. Also, EXCITATION. **2.** the mechanical unit in a tape recorder or a disk storage unit that moves the tape or disk.
drive chuck *Mechanical Engineering.* a mechanism that tranfers motion from a drive rod to a drill string.
drive control see HORIZONTAL DRIVE CONTROL.
drivehead *Engineering.* a fitted covering on the end of a tool that protects the tool when it is driven into a material.
drive light *Computer Technology.* a small light associated with a disk drive that indicates data read/write activity.
drive motor *Robotics.* a motor that is mounted directly on the joint of a robot or its manipulator arm, causing it to move.
driven array *Electromagnetism.* an antenna having several elements that are fed signals from a common source and are phased to produce a desired characteristic pattern.
driven element *Electromagnetism.* an antenna element that is fed directly from the transmission line.
driven gear *Mechanical Engineering.* any gear that receives transmitted motion and power from the another gear.
driven snow *Meteorology.* any snow moved by the wind and deposited in drifts.
drive pattern *Telecommunications.* an unwanted pattern of density variations in a facsimile system due to an error in the position of the recording spot.
drive pulley *Mechanical Engineering.* a pulley or drum that transfers power to a conveyor belt through contact friction.
drive pulse *Electronics.* a pulsed magnetomotive force applied to a magnetic cell from one or more sources.
driver *Engineering.* **1.** a person who operates a vehicle or machine. **2.** specifically, someone who operates a motor vehicle. *Acoustical Engineering.* an electrical circuit that drives a transducer such as a loudspeaker to produce a sound. *Electronics.* **1.** an electronic circuit that provides input for another electronic circuit. **2.** the amplifier stage that precedes the output stage in a receiver or transmitter. Also, **driver stage.** *Computer Programming.* **1.** a small program that controls external devices that interface with a computer. **2.** a program used in bottom-up software development to simulate a superordinate program when testing a subordinate module.
driver DNA *Molecular Biology.* excess complementary DNA, used in a DNA-driven reaction to ensure that radioactive tracer RNA binds chiefly to highly repetitive DNA sequences.
drive reduction theory see DRIVE THEORY.
driver element *Electromagnetism.* an antenna array that receives power directly from the transmitter.
drive rod *Mechanical Engineering.* a hollow shaft through which power is transferred from a motor to a drill string or other device. Also, **drive spindle.**
driver sweep *Electronics.* a cathode-ray tube trace that is initiated by a trigger pulse.
driver transformer *Electronics.* a transformer designed to apply signal power to a circuit or device.
drive screw *Mechanical Devices.* a fastener with a helical thread of coarse pitch screw designed to be driven into wood with a hammer and removed with a screwdriver. Also, SCREW MAIL.
driveshaft or **drive shaft** *Mechanical Engineering.* a shaft that transfers torque from a power source to a machine unit.
drive shoe *Mechanical Devices.* a sharp-edged steel sleeve attached to a drive pipe or casing to serve as its cutting edge.
drive state *Psychology.* a state in which physiological or psychological arousal directs behavior to reduce a need.

drive theory *Psychology.* the theory that behavior is motivated by the effort to satisfy or alleviate some drive. Also, DRIVE REDUCTION THEORY.

driving clock *Engineering.* an instrument that regulates the rate at which a device is being driven.

driving pinion *Mechanical Engineering.* the input gear to the differential of an automotive vehicle.

driving-point admittance *Electronics.* the ratio of resulting signal current to the signal voltage applied at the driving point of an electron tube, network, or other transducer.

driving-point function *Control Systems.* a transfer function in which input and output variables are the voltages or currents measured between a pair of terminals in an electrical network.

driving-point impedance *Electronics.* the ratio of the signal voltage applied at the driving point to the resulting signal current in an electron tube, network, or other transducer.

driving resistance *Mechanics.* the resistance against an applied force that is forcing a system to oscillate.

driving signal *Electronics.* a timing signal in a television transmitting system used to synchronize the start of a horizontal or vertical scan of the transmitted image.

driving wheel *Mechanical Engineering.* a main wheel that communicates motion to other wheels. Also, **drive wheel.**

drizzle *Meteorology.* a form of precipitation in which numerous, very small, and uniformly dispersed water drops appear to float in air currents as they fall to the ground.

drizzle drop *Meteorology.* a drop of water with a diameter between 0.2 and 0.5 mm falling through the atmosphere.

dRNA *Genetics.* a high-molecular-weight form of RNA molecule confined to the nucleus and not included in the rRNA and tRNA classes.

drogue [drōg´] any of various small objects that attach to or trail behind a vehicle; specific uses include: *Engineering.* an apparatus of wood or sailcloth that attaches to a surface buoy to measure currents. Also, DRAG ANCHOR. *Navigation.* a sea anchor or other drag-producing device used to slow a vessel. *Aviation.* **1.** a small parachute attached to an aircraft or a reentry spacecraft; used to provide stabilization and deceleration, usually preceding deployment of a main landing parachute. Also, DECELERATION PARACHUTE. **2.** a funnel-shaped device at the end of the hose of a tanker aircraft; used in flight to drag the hose out, stabilize it, and recover the probe of the receiving aircraft. *Space Technology.* the part of a connector on a spacecraft into which a docking probe fits, as on an Apollo lunar module. (A variant of *drag.*)

Dromadidae *Vertebrate Zoology.* the crab plovers, a monotypic family of white and black shorebirds of the order Charadriiformes, characterized by their stout black bills and long legs with partially webbed toes and by their diet of crabs, crustaceans, and mollusks.

dromedary *Vertebrate Zoology.* a lightweight breed of single-humped camel, *Camelus dromedarius,* that can cover up to 10 miles an hour with a swinging trot; used for travel in parts of Arabia, India, and northern Africa. Also, ARABIAN CAMEL.

dromedary

Dromiacea *Invertebrate Zoology.* a family of true crabs, decapod crustaceans in the section Brachyura. Also, DROMIIDAE.

dromo- a combining form meaning "running" or "conduction," as in *dromophobia.*

dromophobia *Psychology.* an irrational fear of running or walking about.

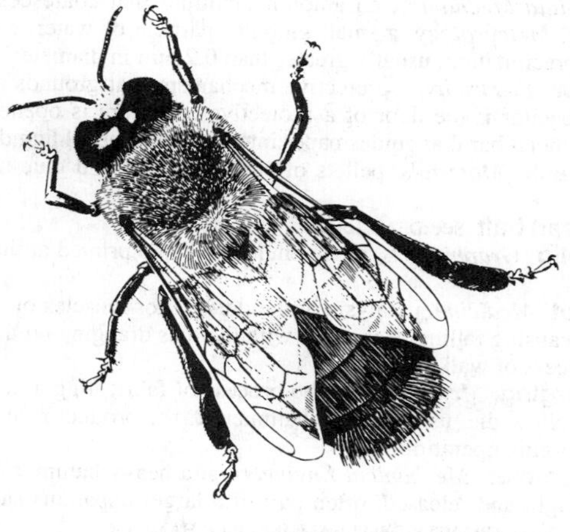

drone (bee)

drone *Invertebrate Zoology.* **1.** a male bee, produced through parthenogenesis, that serves solely in a reproductive capacity, does no work, and possesses no stinger; one of the castes of a colony. **2.** a male ant with similar attributes. *Engineering.* **1.** a land, sea, or air vehicle that is remotely or automatically controlled. **2.** the contol mechanism of such a vehicle. *Aviation.* specifically, a pilotless flight vehicle that is either preprogrammed or remotely controlled for use as an airborne target.

drone fly *Entomology.* a two-winged fly of the genus *Syrphidae* that superficially resembles a bee in appearance and as a pollen and nectar feeder, but that breeds in carrion and waste.

drongo *Vertebrate Zoology.* a small bird of the family Dicruridae in the passerine suborder Oscines, found in Africa to Australia and the Pacific Islands; characterized by glossy black plumage, a hooked beak, red eyes, and elaborate plumed tails in some species.

drongo

droop *Electronics.* a decrease in mean pulse amplitude, expressed as a percentage of maximum amplitude at a specified time after maximum amplitude has been attained.

drooped ailerons *Aviation.* any hinged trailing-edge ailerons arranged in such a way that both the right and left have a 5°–15° positive downward deflection with the control column in a neutral position; usually used when flaps are lowered to increase lift while preserving lateral control. A small amount of droop (5°) may be beneficial for high-altitude cruise.

droop governor *Mechanical Engineering.* a rotating device that controls the speed of a machine motor.

drop *Fluid Mechanics.* an amount of liquid that coalesces to form a globule. *Meteorology.* a small globular particle of water, especially of liquid precipitation, usually greater than 0.2 mm in diameter.

drop bar *Electricity.* a protective mechanism that grounds a high-voltage capacitor if the door of a protective enclosure is opened. *Graphic Arts.* a metal bar that guides paper into a printing or folding device.

drop black *Materials.* pellets of carbon, water, and glue used as pigment.

drop-dead halt see DEAD HALT.

drop folio *Graphic Arts.* a page number that is printed at the bottom of a page.

dropfoot *Medicine.* paralysis of the dorsiflexor muscles of the foot and ankle, causing falling of the foot with the toes dragging on the ground in the process of walking.

drop forging *Metallurgy.* **1.** the process of fabricating a metallic stock in a shallow die, using a drop hammer. **2.** the product resulting from a drop-forging operation.

drop hammer *Mechanical Engineering.* a heavy hammer that is lifted to a height and released; often part of a larger apparatus such as a pile driver. Also, GRAVITY DROP HAMMER, PILE HAMMER.

drop-in *Computer Technology.* an unwanted character that appears on a display screen or in a file or printout, due to one or more bits being misread or misrecorded on a tape or disk.

droplet *Meteorology.* a small globular particle of water, especially a liquid cloud particle, usually less than 0.2 mm in diameter. *Medicine.* a tiny particle of moisture, such as that which is expelled in the process of talking, coughing, or sneezing.

droplet infection *Medicine.* an infection carried from one person to another by tiny particles of moisture that contain microorganisms and that are expelled in the process of talking, coughing, or sneezing.

dropout or **drop out** *Electronics.* a momentary loss of signal in any system, causing a noticeable reduction of audio or video response. *Computer Technology.* the disappearance of a character from a display screen, or from a file or printout, due to one or more bits being misread or misrecorded on a tape or disk. *Graphic Arts.* a halftone from which some of the dots have been removed, either by retouching the negative or by etching the printing plate, in order to make certain areas (usually the background) print as completely white. Also, HIGHLIGHT HALFTONE.

dropout current *Electricity.* the maximum current at which a relay or other magnetically operated device will release to its deenergized position. Similarly, **dropout voltage.**

dropout error *Electronics.* an error that occurs in recorded magnetic tape due to foreign particles in the magnetic coating or defects in the backing.

dropout fuse *Electricity.* a fuse on a utility pole that provides rapid arc extinction upon opening and whose open or closed position is visible from the ground. Also, FLIP-OPEN CUTOUT FUSE.

dropout ratio *Electrical Engineering.* the ratio of the dropout value to the pickup value for an input quantity.

dropping angle see RANGE ANGLE.

dropping-mercury electrode *Physical Chemistry.* a device that produces mercury droplets at a constant rate, allowing electrochemical analysis to be done on a fresh mercury surface with each new drop.

dropping point *Physical Chemistry.* the temperature at which a grease liquefies under standardized conditions.

drop repeater *Electronics.* a microwave repeater station that has the capability to function as a local termination point.

drop siding *Building Engineering.* a siding composed of weatherboards with narrowed upper edges that fit into grooves or rabbets in the lower sides; the backs of the boards lie against a sheathing or studs of the wall. Also, NOVELTY SIDING.

drop-size distribution *Meteorology.* the frequency distribution of drop sizes characteristic of a given cloud or fall of rain.

dropsonde *Engineering.* an instrument attached to a parachute that is dropped from an aircraft and transmits measurements of atmospheric conditions as it descends.

dropsonde dispenser *Engineering.* a chamber in an aircraft from which dropsonde instruments are released.

dropsonde observation *Meteorology.* an evaluation of the significant radio signals received from a descending dropsonde, usually presented in terms of height, temperature, and dew point.

dropsy *Medicine.* an earlier name for edema, an abnormal accumulation of fluid in the tissues or in a body cavity. *Veterinary Medicine.* an infection of fishes caused by the bacterium *Psdudomonas punctata,* characterized by a swollen, spongelike body and protruding scales.

drop tank *Aviation.* an auxiliary fuel tank externally hooked on bomb reach hooks and designed to be discarded during flight.

drop wire *Electricity.* a wire suitable for extending an open wire or cable pair from a pole or cable terminal to a building.

Droseraceae *Botany.* a family of dicotyledonous, insectivorous herbs in the order Sarraceniales, characterized by alternate leaves, often in basal rosettes, and regular, bisexual flowers in coil-like cincinni; they capture insects using sticky leaves or traps.

drosometer *Engineering.* an instrument that measures the amount of water deposited from the atmosphere over a given place and time.

drosophila *Invertebrate Zoology.* a fruit fly of the genus *Drosophila,* especially *D. melangaster,* which is widely used in genetic research.

Drosophilidae *Invertebrate Zoology.* the fruit flies or vinegar flies, a family of small two-winged insects in the subsection Acalyptratae that feed on vinegar or decaying fruit, including the well-known fruit fly, *Drosophila melangaster.*

drought [drout] *Meteorology.* a prolonged period of abnormally dry weather that causes a serious lack of water in the affected area. Also, **drouth.**

drought

Droughtmaster *Agriculture.* a breed of beef cattle that is red in color and that is highly tick-resistant, crossbred in Australia from the Brahman and Shorthorn breeds.

drowned *Geology.* flooded or submerged.

drowned atoll *Geology.* a submerged atoll occurring at such depth that further reef growth does not reach the surface.

drowned coast *Geology.* a shoreline that underwent relative lowering as a result of inundation by the sea.

drowned stream *Hydrology.* a term for a stream that has been flooded over from the sea. Also, FLOODED STREAM.

drowned valley *Geology.* a valley whose lower end has become partly submerged as a result of a rise in sea level.

droxtal *Hydrology.* a small particle of ice that is formed by the direct freezing of supercooled water at a temperature below −30°C.

Drude, Paul [drood] 1863–1906, German physicist; formulated **Drude's theory** of metallic resistance based on electromagnetism and his theory of electrons.

Drude equation *Optics.* a formula that expresses the relationship between a given rotation for polarized light in an optically active material and the wavelength of the incident light. Also, **Drude law.**

drug *Pharmacology.* **1.** any chemical compound that is used in the prevention, diagnosis, treatment, or cure of disease, for the relief of pain, or to control or improve any physiological or pathological disorder in humans or animals. **2.** a consciousness-altering, usually habit-forming substance, especially one whose use is restricted or illegal, such as cocaine, heroin, marijuana, and amphetamines.

drug abuse see ABUSE.

drug addiction see ADDICTION.

drug dependency see DEPENDENCY.

drug idiosyncracy *Medicine.* a susceptibility to the action of drugs, peculiar to an individual person.
drug resistance *Microbiology.* the insensitivity of a bacterial strain to the action of a certain antibiotic.
drug-resistant *Microbiology.* of a microorganism, resistant to the action of drugs. Also, **drug fast.**
drug therapy *Psychology.* the use of medication to treat symptoms of psychological disorder.
drug tolerance *Medicine.* the power of taking a drug continuously or in large doses without injurious effects.
drum any of a variety of devices having the cylindrical shape of a drum; specific uses include: *Mechanical Devices.* **1.** a short cylinder revolving on an axis, used to turn other smaller wheels connected to it, such as a brake drum. **2.** a widely used shipping or storage container for liquids, capable of holding 45–416 liters (12–110 gallons). *Mechanical Engineering.* any cylindrical part of a machine. *Mining Engineering.* a horizontal cylinder or cone (or combination of the two) around which a rope or wire is wound in a hoisting mechanism. Also, HOISTING DRUM. *Chemical Engineering.* a vessel in which heated products are charged to allow volatile portions to separate. *Architecture.* **1.** one of the cylindrical blocks used to form a column. **2.** a curved or polygonal vertical wall that carries a dome. *Computer Technology.* an early direct-access storage device consisting of a rotating cylinder with data stored on tracks along the surface and a read/write head for each track.
drum armature *Electricity.* an armature that utilizes a drum winding.
drum barker *Forestry.* a long, open-ended, revolving drum in which logs are placed and rotated to loosen and remove bark.
drum brake *Mechanical Engineering.* a brake system in which a pair of brake shoes are pressed against the inner surface of a shallow metal drum rigidly attached to a wheel.
drum controller *Electricity.* an electrical controller that uses a drum switch as the main switching element.
drum feeder *Mechanical Engineering.* a rotating drum designed to distribute parts for assembly into different chute arrangements. Also, ROLL FEEDER, TUMBLER FEEDER.
drum filter *Mechanical Engineering.* a rotating cylindrical drum that is equipped with a vacuum pump and used to filter ore pulp. Also, ROTARY VACUUM FILTER.
drumlin *Geology.* a small hill composed of glacial till, characterized by a hogback outline, an oval plan, and a long axis oriented in the direction of the glacial movement that created it.

drumlin

drumlinoid see ROCK DRUMLIN.
drum mark *Computer Technology.* a special character used in magnetic drum storage to indicate the end of a variable-length block or record.
drummy *Mining Engineering.* any loose rock or coal that produces a hollow, weak sound when tapped with a bar.
drum plotter *Engineering.* a graph- or drawing-producing apparatus, such as a seismograph, that is made up of a pen that moves continuously along and across a rotating paper-covered drum.
drum printer *Computer Technology.* an older type of line printer using a character set that is engraved on the surface of a rapidly rotating solid cylinder.
drum recorder *Electronics.* a facsimile recorder consisting of a motor-driven rotating cylinder that contains the paper on which the image is to be reproduced.
drum separator *Mining Engineering.* a rotating cylindrical vessel, consisting of different and adjustable gravities, that separates run-of-mine coal into clean coal, middlings, and refuse.
drum sequencer *Robotics.* a programming device that is used to operate valves or limit switches in order to control a robot.
drumstick *Cell Biology.* a nuclear protrusion that is present in certain polymorphonuclear leucocytes of females and contains the X chromosome.
drum switch *Electricity.* a sequential switch in which the contacts are pins, segments, teeth, or surfaces on the periphery of a revolving drum; used for complex band switching or circuit changes.
drum transmitter *Electronics.* a facsimile transmitter in which the copy is mounted on a rotating cylinder.
drum winding *Electricity.* an electrical winding in which coils are on the outer surface of a cylindrical core or the inner surface of a cylindrical bore, and the two branches of a turn are under adjacent poles of opposite polarity.
drupaceous [droo pā´shəs] *Botany.* **1.** bearing or producing drupes. **2.** resembling a drupe in appearance, but not necessarily in structure.
drupe [droop] *Botany.* a simple, fleshy fruit with a skinlike exocarp, a fleshy mesocarp, and a stony or woody endocarp that contains a single seed, as in olives, peaches, and cherries. Also, STONE FRUIT.
drupelet *Botany.* an individual drupe in a fruit composed of aggregate drupes, as in raspberries and strawberries. Also, **drupel.**
druse [drooz] *Geology.* a small cavity in a rock or vein whose interior surface is encrusted with small, projecting crystals, usually of the same minerals as the enclosing rock. *Mineralogy.* a globular, compound, calcium oxalate crystal with numerous crystals projecting from its surface.
drusy *Geology.* describing a druse or a rock or vein containing many druses.
dry abrasive cutting *Mechanical Engineering.* the process of cutting a material using a rotary abrasive wheel without a liquid coolant.
dry adiabat *Meteorology.* a line of constant potential temperature on a thermodynamic diagram, representing the lifting of dry air in a dry-adiabatic process.
dry-adiabatic atmosphere see ADIABATIC ATMOSPHERE.
dry-adiabatic lapse rate *Meteorology.* the rate of decrease of temperature with height of a parcel of dry air lifted adiabatically through an atmosphere in hydrostatic equilibrium.
dry-adiabatic process *Meteorology.* an adiabatic process in a system of dry air.
dry ashing *Organic Chemistry.* to use a burner or muffle furnace to convert a compound to ash.
dry assay *Metallurgy.* an analysis of the critical elements in an ore that does not utilize a liquid means of separation..
dry avalanche see DRY-SNOW AVALANCHE.
dry-back boiler see SCOTCH BOILER.
dry battery *Electricity.* a battery composed of dry cells.
dry bed *Chemical Engineering.* a vessel filled with solid absorption materials, used to purify or recover liquid from a gas stream. *Geology.* see WASH.
dry-blast cleaning *Engineering.* a cleaning process in which a surface is blasted with abrasive material moving at a high velocity.
dry blower see DRY WASHER.
dry box *Nucleonics.* a closed, lead-lined compartment maintained at slightly negative atmospheric pressure and fitted with radiation-proof windows and rubber gloves for the safe handling of radioactive or contaminated materials during laboratory and fabrication operations. Also, GLOVE BOX. *Chemistry.* a container from which air can be withdrawn and replaced by argon or other inert gas, and which acts as an inert atmosphere for work with very reactive chemicals.
dry-box process *Chemical Engineering.* a method of removing hydrogen sulfide from industrial gases by passing the gases through boxes holding trays of wood shavings or other material coated with iron oxide.
dry-bulb thermometer *Engineering.* a thermometer that is not affected by atmospheric humidity; ordinary household thermometers are of this type. Thus, **dry-bulb temperature.**
dry cargo ship *Naval Architecture.* a ship designed to carry dry bulk cargoes as opposed to liquid cargoes.
dry cell *Electricity.* a voltage-generating primary cell in which the electrolyte is immobilized, making the battery completely portable.
dry-cell cap light *Mining Engineering.* a headlamp clipped to a belt with a bulb that, if broken, ejects automatically to prevent explosion.
dry-charged battery *Electricity.* a storage battery in which the electrolyte is drained after the plates are formed; before being placed in service, the battery is filled with electrolyte and charged for a few minutes.
dry-chemical fire extinguisher *Chemical Engineering.* a type of chemical fire extinguisher filled with a dry powder, consisting mainly of sodium or potassium bicarbonate; used mostly for extinguishing small electrical fires.

dry circuit *Electricity.* a circuit with extremely low maximum voltages and very small maximum currents, so that there is no arcing to roughen the contacts, and an insulating film develops that prevents closing of the circuit.

dry-clean *Engineering.* to clean a material using little or no liquid.

dry-cleaned coal *Mining Engineering.* any coal from which impurities have been removed without the use of liquid media.

dry-cleaning *Textiles.* a method of cleaning fabrics with organic solvents that absorb or dissolve grease, soil, etc., using little or no water.

dry climate *Meteorology.* any of several climate classifications for areas with minimal precipitation, such as desert and semidesert regions.

dry coloring *Chemical Engineering.* a process of plastics coloring by tumbling uncolored particles of plastic material with pigments and dyes.

dry contact *Electricity.* a contact through which no current flows.

dry cooling *Mechanical Engineering.* any system for cooling water in which the water is circulated through pipes that contact the cooling element, but is not directly exposed to the coolant or air.

dry corrosion *Metallurgy.* any corrosion that is not caused by moist or liquid environments.

dry course *Building Engineering.* an initial roofing bond of felt or paper material that precedes the application of tar or asphalt.

dry cow *Agriculture.* a term for a cow that does not produce milk.

dry delta see ALLUVIAL FAN.

dry-desiccant dehydration *Chemical Engineering.* a technique to remove liquids from gases by using silica gel or another solid absorbent.

dry-disk rectifier see METALLIC RECTIFIER.

dry distillation *Chemistry.* a distillation procedure carried out on dry materials.

dry dock *Naval Architecture.* **1.** a structure in which a vessel can be moored, and from which the water can then be removed, so that repairs below the water line can be carried out. **2. drydock.** to place a vessel in such a structure.

dry drilling *Engineering.* any drilling operation in which a current of air or gas is used instead of fluid.

dry electrolytic capacitor *Electricity.* an electrolytic capacitor with an electrolyte that is in a paste or a solid, rather than a liquid.

dry firn see POLAR FIRN.

dry flashover voltage *Electronics.* the minimum voltage across a clean, dry insulating material at which flashover of the surrounding air occurs.

dry fog *Meteorology.* a fog that does not moisten exposed surfaces.

dry freeze *Hydrology.* the freezing of ground without the formation of frost on exposed surfaces.

dry friction *Mechanics.* the resistance to motion of a nonlubricated system of solid bodies.

dry-fuel rocket *Space Technology.* any of a variety of rockets, particularly booster rockets, using a solid rocket propellant, usually a mixture of fuel and oxidizer. Also, SOLID-PROPELLANT ROCKET ENGINE.

dry gangrene *Medicine.* a form of gangrene in which the involved part is dry and shriveled.

dry grinding *Engineering.* a process of reducing material to fine fragments by abrasive action without the use of a liquid.

dry haze *Meteorology.* a haze produced by particles of the order of 0.1 μm that produce a bluish color against a dark background or a yellowish veil against a light background.

dry hole *Engineering.* an opening in the ground that is created without the use of water. *Petroleum Engineering.* a drill hole that has been abandoned due to a lack of production of marketable petroleum materials.

dry ice *Inorganic Chemistry.* carbon dioxide, CO_2, that has been condensed by pressure to a snowlike solid; widely used as a refrigerant. At –78.5°C carbon dioxide changes directly from a gas to a solid without an intervening liquid state.

drying *Chemistry.* **1.** the removal of most of the water from a substance (usually 92–95%), often through heat exposure. **2.** the removal of solvents other than water from a substance.

drying agent *Chemistry.* a substance having an affinity for water, used to dry fluid substances; a desiccant.

drying oil *Materials.* any of a group of organic, oily liquids, including linseed, soybean, or dehydrated castor oil, applied as a thin coating, later forming a tough, elastic layer after absorbing oxygen.

drying oven *Engineering.* an enclosed structure in which material is dried with relatively low temperatures.

Dryinidae *Invertebrate Zoology.* a family of small wasps, hymenopteran insects in the superfamily Bethyloidea, whose larvae are parasitic on leafhoppers and plant hoppers.

dry ink *Engineering.* a powdery substance that is composed of resin and pigment, used to produce images in electrophotography.

dry kiln *Engineering.* a heated chamber in which lumber is dried and seasoned.

dry labor *Medicine.* labor in which there has been a premature rupture of the amniotic sac.

dry-land farming *Agronomy.* a method of raising crops with little or no irrigation, either by growing drought-resistant crops or by allowing the soil to lie fallow every other year in order to store moisture. Also, **dry farming.**

dry limestone process *Chemical Engineering.* a method to control air pollution by exposing sulfur oxides in flue gases or the like to limestone, which changes them to disposable residues.

dry measure *Metrology.* a system of volume measurement for dry items such as grains or vegetables, in which 2 dry pints equal 1 dry quart, and 8 dry quarts equal 1 bushel.

dry mining *Mining Engineering.* a mining procedure in which efforts are made to keep all moisture out of the ventilating air.

dry mount *Graphic Arts.* to attach copy to a surface such as an artboard by the use of a wax or adhesive spray.

dry offset see LETTERSET.

Dryomyzidae *Invertebrate Zoology.* a family of small flies in the subsection Acalyptratae that frequent fungi and excrement.

Dryopidae *Invertebrate Zoology.* a family of small coleopteran insects, the long-toed water beetles, in the superfamily Dryopoidea, that feed on decaying vegetation.

dryopithecene *Anthropology.* any of the apelike fossil forms of the genus *Dryopithecus* from the Miocene and Pliocene that evolved in forest habitats of Africa and Eurasia.

Dryopithecus *Anthropology.* a widely dispersed genus of hominoids that evolved in forest habitats of Africa and Eurasia during the Miocene and into the Pliocene. (From a Greek term meaning "forest ape.").

Dryopoidea *Invertebrate Zoology.* a superfamily of small coleopteran insects in the subsection Polyphaga, water beetles that feed on decaying vegetation.

dry ore *Mining Engineering.* any ore that contains gold and silver, but insufficient lead or copper to be smelted without the addition of richer lead or copper ores.

dry permafrost *Geology.* permafrost that is loose and crumbly because it contains little or no ice or moisture.

dry permeability *Materials Science.* the degree to which dried bonded sand allows gases to travel through it during the pouring of molten material into a mold.

dry pint *Metrology.* a unit of dry measure equal to one-half dry quart or .57 liter.

dry pipe *Mechanical Engineering.* a pipe that prevents moisture from entering a steam outlet; it is positioned about the normal liquid level in the steam space of a boiler.

dry-pipe system *Engineering.* a sprinkler system used in cold climates that operates only when the air it normally contains has been vented.

dry-pit pump *Mechanical Engineering.* a pump that operates by using liquid conducted to and from the unit through the pipe.

dry placer *Mining Engineering.* a gold-bearing alluvial deposit found in arid regions.

dryplate *Graphic Arts.* a negative that is composed of a glass plate coated with a photosensitive silver-gelatin emulsion and exposed in a dry condition.

dry point or **dry point** *Analytical Chemistry.* the temperature at which the complete evaporation of a liquid occurs. *Graphic Arts.* a method of plate engraving using a hard needle instead of an acid-etching solution.

dry pressing *Engineering.* a process in which clay products are shaped by compressing moist clay powder into metal molds.

dry quart *Metrology.* a unit of dry measure that is equal to 2 dry pints or 1.1 liters.

dry quicksand *Geology.* an accumulation of sand consisting of alternating layers of firmly compacted sand and soft, loose sand that cannot support heavy loads.

dry-reed contact see REED CONTACT.

dry-reed relay *Electricity.* a relay with contacts mounted on magnetic reeds in glass tubing and an actuating coil wound around the tubing that provides the magnetic field needed to operate the relay.

dry-reed switch *Electricity.* a switch with two thin, metallic reeds in a vacuum tube; the circuit is closed when an external magnet attracts one of the reeds, which contacts the other reed; used reliably in dry circuits.

dry-relief offset see LETTERSET.

dry rot *Microbiology.* a process of timber decomposition caused by the cellulolytic action of certain microorganisms, such as the fungus *Serpula lacrymans.*

dry run a practice session used to test a process or piece of equipment for proper operation; specific uses include: *Ordnance.* simulated practice in firing or discharging weapons, especially a divebombing run without bombs. *Computer Programming.* a program-checking technique applied before using a program on a computer; the logic and coding of the program are followed manually from a flowchart and written instructions, and the results of each step of the operation are recorded.

dry sample *Mining Engineering.* an ore sample obtained by dry drilling, so that the sample is not mixed with water or other fluid.

Drysdale AC polar potentiometer *Engineering.* an instrument used to measure electrical voltage, composed of a phase-shifting transformer and a resistive voltage divider.

dry season *Meteorology.* an annually recurring period of one or more months during a year for a given region, during which precipitation is a minimum.

dry sieving *Materials Science.* a process whereby the particle sizes of powdered solids are analyzed by sifting the dry solids through a stack of mesh screens of decreasing sizes.

dry sleeve *Mechanical Engineering.* a cylinder liner that is not in contact with a refrigerant or coolant.

dry-snow avalanche *Hydrology.* one of the basic categories of snow avalanche, composed of dry, powdery snow set in motion by the wind; it is the fastest-moving type of avalanche and is estimated to be able to reach speeds of over 250 miles per hour. Also, DRIFT AVALANCHE, DUST AVALANCHE.

dry socket *Medicine.* a postoperative consequence to tooth extractions in which the blood clot in the socket disintegrates, leading to an empty socket and eventual infection.

dry spell *Meteorology.* a period, lasting not less than two weeks, during which no measurable precipitation is recorded.

dry spinning *Textiles.* **1.** a method of creating manmade fiber by forcing a polymer solution through the fine holes of a spinneret into a volatile liquid, which is then evaporated off. **2.** a method of making thick, strong yarns, such as flax, hemp, and jute, in which fibers are kept completely dry during the spinning process.

dry spot *Chemical Engineering.* **1.** an open area on laminated plastic where the surface film is incomplete. **2.** a portion of laminated glass in which there is no bonding of the plastic interlayer and surrounding glass layers.

dry-steam drum *Mechanical Engineering.* a pressurized chamber that allows steam to flow from a boiler drum.

dry-steam energy system *Engineering.* **1.** an underground energy source that emits superheated steam. **2.** a hydrothermal convective system that is controlled by vapors having temperatures of over 150°C.

dry strength *Engineering.* a measurement of the strength of an adhesive joint, taken immediately after drying under specified conditions or after a period of conditioning.

dry tabling *Mining Engineering.* a waterless process based on specific gravity differences; used to separate two or more minerals.

dry-tape fuel cell *Electricity.* a fuel cell in which the fuel is in the form of dry tape coated with fuel, oxidant, and electrolyte that converts chemical energy into electric energy.

dry test meter *Engineering.* an instrument used to measure the rate of household gas flow and to calibrate graduated measurement markings in flowmeter instruments.

dry tongue *Meteorology.* a marked protrusion of relatively dry air into a region of higher moisture content.

drywall or **dry wall** *Building Engineering.* **1.** the process of using wallboard material to cover a building wall. **2.** a wall covered with wallboard, rather than plaster.

dry wash *Geology.* see WASH, def. 3.

dry washer *Mining Engineering.* a machine for extracting gold from dry gravel. Also, DRY BLOWER.

dry well *Engineering.* any well that does not contain water or other liquid. *Civil Engineering.* a drainage pit lined with stone fragments, used to receive liquid wastes. *Nucleonics.* in boiling-water reactors, a type of primary containment consisting of a bulb-shaped steel pressure vessel within a reinforced concrete shell that forms a high-mechanical-strength enclosure and radiation shield around the reactor and coolant pumps; in a loss-of-coolant accident, steam from the wet well is released into the dry well where it is condensed, thus relieving pressure in the containment and reducing the likelihood of escape of radioactivity.

dry wire drawing *Metallurgy.* a process of wire drawing without the use of a lubricant.

DS days after sight.

ds double-stranded.

DSARC Defense Systems Acquisition Review Council.

DSC digital-to-synchro converter.

DSC or **D.S.C.** Doctor of Surgical Chiropody.

DSc or **D.Sc.** Doctor of Science.

D/S converter see DIGITAL-TO-SYNCHRO CONVERTER.

dsDNA double-stranded DNA.

DSIR Department of Scientific and Industrial Research.

D sleep see REM SLEEP.

DSM Diagnostic and Statistical Manual of Mental Disorders, the standard classification text of the American Psychiatric Association.

DSM *Aviation.* the airport code for Des Moines, Iowa.

DSP digital signal processing, dibasic sodium phosphate.

dsRNA double-stranded RNA.

DSS or dss decision support system.

DST or **D.S.T.** daylight savings time.

DSurg or **D.Surg.** Dental Surgeon.

DT diphtheria and tetanus (vaccine).

D.T. cover paper *Graphic Arts.* double-thick cover paper, consisting of two bonded sheets.

d.t.d. or **dtd** give such a dose. (From Latin *datur talis dosis.)*

DTE or dte data terminating equipment.

DTH or **dth** delayed-type hypersensitivity.

dThd or **dT** a symbol for thymidine.

DTL diode transistor logic.

DTN diphtheria toxin normal.

DTP diphtheria-tetanus-pertussis; distal tingling on percussion.

DTR or **dtr** deep tendon reflex.

D_1 trisomy syndrome see PATAU'S SYNDROME.

DTs delirium tremens.

DTS or dts digital termination service.

DTT *Aviation.* the airline code for the city of Detroit, Michigan.

DTW *Aviation.* the airport code for Detroit-Wayne County, Michigan.

D-type virus particles see TYPE D RETROVIRUS GROUP.

dual *Mathematics.* **1.** in the theory of sets, given a statement, expression, or theorem, the corresponding statement, expression, or theorem that is obtained by replacing each set by its complement, interchanging unions and intersections, and reversing all inclusions. **2.** in the theory of Boolean algebras, given a statement, expression, or theorem, the corresponding statement, expression, or theorem obtained by interchanging unions and intersections and interchanging 0 and 1. **3.** see DUAL SPACE. **4.** in category theory, a statement about objects and morphisms formed by reversing all the arrows (morphisms) in the original statement. **5.** the module of homomorphisms from a module A to its ring R of scalars; denoted $\mathrm{Hom}_R\,(A,R)$. Also, **dual module. 6.** in projective geometry, the dual of a statement about lines and points and their intersections is the statement with the terms line and point interchanged.

Dualayer distillate process *Chemical Engineering.* a process used to remove mercaptan and oxygenated compounds from fuel oil, using concentrated caustic Dualayer solution and electrical precipitation of the impurities.

Dualayer solution *Chemical Engineering.* a concentrated solution of potassium or sodium hydroxide that contains a solubilizer.

dual-beam spectroscopy *Spectroscopy.* a spectroscopic method in which two beams of light of different wavelengths are used and the resulting absorption spectra are compared.

dual-bed humidifier *Mechanical Engineering.* an air humidifier system that contains two adsorber beds in parallel, thus facilitating a continuous flow of air.

dual-capable *Ordnance.* of or relating to personnel, weapons, and equipment capable of employing or handling both nuclear and nonnuclear munitions. Thus, **dual-capable forces, dual-capable unit, dual-capable weapons.**

dual-channel amplifier *Acoustical Engineering.* an amplifier with two channels, normally referred to as channel A and channel B, used for stereophonic signal processing.

dual control *Control Systems.* a method of control that maintains an optimum balance between control errors and estimation errors in order to minimize both.

dual coordinates *Mathematics.* a line and a point in plane geometry; a line is defined by two points, and a point by the intersection of two lines.

dual-cycle boiling-water reactor *Nucleonics.* a boiling-water reactor in which steam generated in the core at high pressure drives the first stages of the turbine and is returned as high-temperature condensate to form low-pressure steam in a secondary generator that is admitted to lower stages of the turbine; by varying the flow of secondary steam to the turbine, cooler water in the core changes the void fraction and moderating effectiveness of the water, allowing the reactor to respond to varying loads without requiring motion of control rods.

dual density see DOUBLE DENSITY.

dual diode see DUODIODE.

dual disk drive *Computer Technology.* a disk-drive system that can handle two diskettes, with the capability of reading from a diskette in one drive and writing onto a diskette in the other drive.

dual diversity receiver see DIVERSITY RECEIVER.

dual-emitter transistor *Electronics.* a special transistor with two emitters; generally used in chopper circuits for the amplification of DC or very low-frequency AC signals.

dual-fuel engine *Mechanical Engineering.* an internal combustion engine that is adapted to run on either gasoline or natural gas.

dual graph *Mathematics.* given a planar graph G, the graph G^* obtained from G as follows: (a) the vertices v^* of G^* are in one-to-one correspondence with the faces of G; (b) the edges e^* of G^* join vertices of G^* that correspond to faces separated by a common edge in G. Nonplanar graphs do not have duals; matroid theory was developed to address this problem.

dual-gravity valve *Chemical Engineering.* a valve that is float-operated and functions on the interface between two immiscible liquids of different specific gravities.

dual group *Mathematics.* **1.** for a locally compact Abelian group G, the group $G^\wedge$ of complex characters of G. **2.** for an Abelian group A and the least integer m such that every element of A is of order $\leq m$, the group $\mathrm{Hom}(A,Z_m)$ of group homomorphisms from A to the group Z_m (integers modulo m). If A is finite, then it is isomorphic to its dual.

dual-gun cathode-ray tube see DOUBLE-BEAM CATHODE-RAY TUBE.

dual in-line package *Electronics.* a standard form of housing for integrated circuits consisting of a rectangular body with two rows of terminal pins bent at right angles to the body and uniformly spaced for insertion in connectors or mating holes in printed circuit boards.

dualistic theory *Hematology.* the theory that the blood cells arise from two distinct types of primitive cells, the myeloblasts and the lymphoblasts.

duality principle *Mathematics.* the general principle that a theorem is true if and only if its dual statement is true; that is, if and only if the statement obtained by replacing each object in the original theorem with its dual is true; used extensively in projective geometry, set theory, and Boolean algebra. *Electronics.* the principle that every circuit has a dual and analogous counterpart containing dual (and sometimes complementary) parameters; used in the analysis and design of electrical circuits. *Quantum Mechanics.* see WAVE-PARTICLE DUALITY.

dual laser *Optics.* a gas laser in which Brewster windows and concave mirrors mounted at opposite ends of the tube produce two distinct wavelengths at the same time from a helium-neon laser beam.

dual linear programming *Mathematics.* in mathematical programming, the process of reaching the optimal solution by solving the dual problem; e.g., in a profit maximization problem, minimizing the shadow price or opportunity cost of the limited resources.

dual memory theory *Psychology.* the theory that memory is composed of two independent systems; one element, the short-term memory, is of limited capacity, and the other, the long-term memory, is of larger capacity.

dual meter *Engineering.* an instrument that records two aspects of an electrical circuit simultaneously; for example, wattage and voltage.

dual-mode control *Control Systems.* a method of control that uses two different modes of operation, usually a linear feedback mode and a fixed-stop mode.

dual modulation *Telecommunications.* the process of varying two transmission frequencies at the same time for different purposes.

dual-phase steel *Materials.* a high-strength, low-alloy steel, with a high degree of formability, produced through a special heat-treatment process to form a ferrite and martensite microstructure.

dual-processor system *Computer Technology.* a computer system with two central processing units that can operate simultaneously and cooperatively.

dual-purpose cattle *Agriculture.* any of various breeds of cattle raised for both meat and milk.

dual-purpose reactor *Nucleonics.* a commercial fast breeder reactor cooled by liquid sodium that is designed to breed fissile plutonium-239 from fertile uranium-238 while generating useful power.

dual-purpose weapon *Ordnance.* a weapon designed to deliver effective fire against air or surface targets. Also, **dual-purpose gun.**

dual radioactive decay *Nuclear Physics.* the characteristics of a nucleus that has two or more separate and alternative decay methods.

dual-sided disk drive *Computer Technology.* a disk drive that has two read/write heads to record and access data on both sides of a diskette.

dual space *Mathematics.* **1.** the dual space of a vector space V over a field F is the vector space of all scalar-valued linear transformations (i.e., linear functionals) on V; denoted $L(V,F)$ or V^*. Also, CONJUGATE SPACE, ALGEBRAIC DUAL. **2.** the dual space of a topological vector space X over a field K is the space of all scalar-valued linear continuous functions (continuous linear forms) on X; denoted $L(X,K)$ or X^*. Also, TOPOLOGICAL DUAL.

dual thrust *Space Technology.* a rocket thrust derived from two propellant grains using the same missile propulsion section; it serves as a two-stage propulsion system that does not have the drawbacks associated with jettisoning the booster unit or sequential ignition of the sustainer grain, while offering lower cost, shorter length, and lower weight.

dual-thrust motor *Space Technology.* a solid-propellant rocket engine that produces two levels of thrust by the use of two propellant grains.

dual-tone multifrequency *Telecommunications.* a signaling method that employs fixed combinations of two specific voice frequencies, one of which is selected from a group of four low frequencies, the other from a group of either three or four relatively high frequencies.

dual-trace amplifier *Electronics.* a two-channel amplifier unit in which each channel is independently adjustable; used for amplifying the signal voltages that are to appear as vertical deflections on the associated timebase, in a dual-trace oscilloscope.

dual-trace hypothesis *Psychology.* the theory that similar information can be stored either visually or verbally.

dual-trace oscilloscope *Electronics.* an oscilloscope capable of displaying two separate and independently adjustable timebase sweeps.

dual triode see DUOTRIODE.

dualtropic *Virology.* describing a recombinant retrovirus that is formed when an ecotropic virus combines with a xenotropic virus and that is able to replicate in homologous or heterologous cells.

dual-use line *Telecommunications.* a user-access line that is normally used for voice communication but that also has special equipment for use as a digital transmission circuit or channel.

dual variables *Mathematics.* the main variables in a dual linear programming problem; e.g., the unit shadow prices of the limited resources.

dual vector space *Mathematics.* the vector space of linear functionals defined on a given vector space. Also, ADJOINT VECTOR SPACE.

Duane, William 1872–1935, American physicist; developed X-ray techniques and apparatus for cancer treatment and the Duane-Hunt law.

Duane-Hunt law *Quantum Mechanics.* a principle stating that the frequency of X rays produced in electron scattering is always less than the **Duane-Hunt limit**, which is given by eV/h, where e is the fundamental charge, V the applied voltage, and h is Planck's constant.

dub *Acoustical Engineering.* to carry out the process of dubbing.

dubbing *Acoustical Engineering.* **1.** the process of adding a new sound track or new sounds to an existing video film or recording, as by replacing the original dialogue with dialogue in another language. **2.** the process of combining two or more sound sources into a single recording, as by adding background music to narrative or dialogue. **3.** the process of transferring recorded material from one recording to another, as in copying a recorded audio tape onto a blank tape.

Dubhe *Astronomy.* Alpha (α) Ursae Majoris, an orange-yellow second-magnitude giant star in the Big Dipper; of the two pointer stars in the bowl, it is the one closer to the pole.

DuBois-Reymond, Emil [dü bwä´ rā mōn´] 1818–1896, German physiologist; performed research on animal electricity.

DuBois-Reymond, Paul 1831–1889, German mathematician; research on continuous functions and tests for convergence of series.

DuBois-Reymond theorem *Mathematics.* **1.** suppose that: (a) $y(x)$ is continuous with piecewise continuous derivatives of the first order and vanishing first variation, and (b) $f(x, y, y')$ is twice continuously differentiable with respect to all arguments, with $f_{y'y'} \neq 0$. Then $y(x)$ has continuous derivatives of the second order and $f(x,y,y')$ satisfies the Euler-Lagrange differential equation. **2.** suppose that: (a) $f(x)$ is continuous on the real interval $[a,b]$; and (b) $\int_a^b f(x)g(x)dx = 0$ for every continuous $g(x)$ satisfying $\int_a^b g(x)dx = 0$. Then $f(x)$ is constant on $[a,b]$.

René Dubos

Dubos, René [dü bōz´] 1901–1982, French-born American microbiologist; a pioneer in antibiotics.

Duboscquellaceae *Botany*. a family of nonthecate marine dinoflagellates of the order Syndiniales that are nonphotosynthetic and parasitic on the cytoplasm or nucleus of various unicellular marine organisms.

Duchemin's formula *Physics*. a formula giving the force per unit area of wind striking an inclined surface: $N=F[(2 \sin a)/(1 + \sin^2 a)]$, where F is the wind pressure on a vertical surface and a is the inclination angle.

Duchenne muscular dystrophy [dü shän´] *Medicine*. a congenital, inherited type of muscular dystrophy, believed to be caused by a deficiency of dystrophin and characterized by the degeneration and necrosis of skeletal muscle fibers and their replacement with fat and fibrous tissue; it primarily affects male children and usually leads to death before maturity due to respiratory weakness or heart failure. (Named for the French neurologist Guillaume Benjamin Amand *Duchenne*.) Also, PSEUDOHYPERTROPHIC MUSCULAR DYSTROPHY.

Duchesnian *Geology*. a North American continental stage of the Upper Eocene epoch, occurring after the Uintan and before the Chadronian.

duck *Vertebrate Zoology*. **1.** any of various wild and domesticated webfooted swimming birds of the family Anatidae, characterized by short legs and neck, a depressed body, a broad and flat bill, and a differentiation in plumage between the sexes. **2.** a female duck, as distinguished from a male drake. *Textiles*. a durable, heavy, close-woven cotton or linen cloth; lighter weights are used for suiting and heavier weights for tents, sails, and conveyor belts. *Ordnance*. **1.** see DUKW. **2.** an air intercept code meaning "trouble headed your way."

duck-billed dinosaur see HADROSAUR.

duckbill platypus see PLATYPUS.

duckbill pliers *Mechanical Devices*. a type of pliers having long, flat jaws thought to resemble a duck's bill.

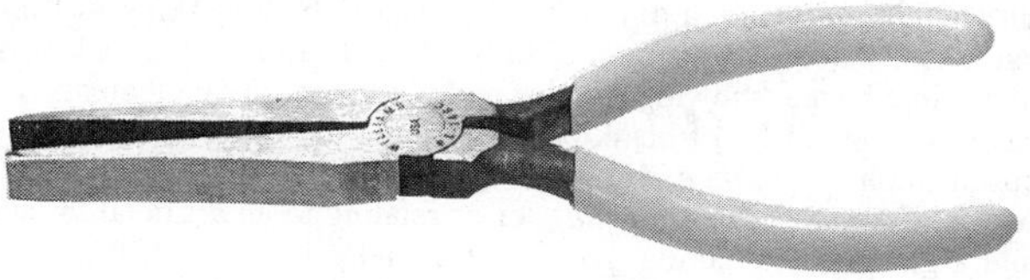

duckbill pliers

Duckeodendraceae *Botany*. a monospecific family of dicotyledonous tall trees in the order Solanales, characterized by a large red fruit with a bony stone and fibrous exocarp; native to the Amazon Basin in Brazil.

duck-foot harrow see DRAG HARROW.

Duck Lakean *Geology*. a North American Gulf Coast stage of the Miocene epoch, after the Napoleonville and before the Clovelly.

duckwheat see TARTARY BUCKWHEAT.

Ducrey test *Immunology*. a skin test used to determine whether an individual is or has been infected with the bacillus *Hemophilus ducreyi*.

duct *Mechanical Engineering*. a metal tube or casing through which air is passed to ventilate, heat, or cool a machine or part. *Anatomy*. a tubular structure that conducts fluids or secretions from their point of orgin to a point of exit. Also, **ductus.** *Geophysics*. a shallow, nearly horizontal layer in the atmosphere in which microwaves can be trapped by ducting.

ducted-fan engine *Aviation*. an aircraft engine in which a fan or propeller is enclosed in a duct. In a jet engine the enclosed fan (**ducted fan**) is used to draw in ambient air to promote jet propulsion, generating increased mass flow. Depending on the type of engine, air may be ingested either at the front of the engine (and then circulated around the combustion section) or aft of the combustion chamber. Also, FANJET.

ducted rocket see ROCKET RAMJET.

duct effect *Electronics*. a duct created by temperature inversions in the atmosphere through which radio waves appear to travel. Also, **ducting effect.**

ductile [duk´til; duk´təl] *Metallurgy*. **1.** able to be hammered out thin. **2.** able to be drawn out into wire or threads. *Materials Science*. **1.** able to be molded or shaped. **2.** able to change in form without breaking.

ductile cast iron or **ductile iron** see NODULAR CAST IRON.

ductile fracture *Materials Science*. a fracture that occurs in metals after extensive plastic deformation and is characterized by slow crack propagation.

ductility *Metallurgy*. the relative ability of a metal to be hammered thin or drawn out into wire. *Materials Science*. the relative ability of a material to be molded or shaped.

ducting *Geophysics*. an atmospheric phenomenon in which microwave signals travel more than ten times farther than the line-of-sight limit and are refracted and retransmitted by an atmospheric duct.

ductless gland see ENDOCRINE GLAND.

duct of Arantius see DUCTUS VENOSUS.

duct of Cuvier *Developmental Biology*. one of two short venous trunks (common cardinal veins) in the fetus that open into the atrium of the heart, with the right one developing into the superior vena cava; in fish, the veins leading from the cardinal veins into the sinus venosus of the heart. Also, CUVIER'S SINUSES, DUCTUS CUVIERI.

duct of Santorini *Anatomy*. an accessory duct draining the head of the pancreas, with one branch joining the pancreatic duct and the other one opening independently into the duodenum. Also, BERNARD'S CANAL.

duct of Wirsung *Anatomy*. the main secretory duct of the pancreas, which runs through the length of the gland, receiving branches along its course, and empties into the duodenum.

duct propulsion *Space Technology*. a method of propulsion achieved by ducting a surrounding fluid through an engine, applying mechanical or thermal means to add momentum, and ejecting the fluid in order to obtain a reactive force.

ductus Arantii see DUCTUS VENOSUS.

ductus arteriosus *Anatomy*. a fetal blood vessel connecting the left pulmonary artery directly to the descending aorta; after birth, it develops into a fibrous cord, the ligamentum arteriosum. Also, ARTERIAL CANAL, BOTALLO'S DUCT, PULMOAORTIC CANAL.

ductus Cuvieri see DUCT OF CUVIER.

ductus deferens see VAS DEFERENS.

ductus mesonephricus see MESONEPHRIC DUCT.

ductus venosus (Arantii) *Anatomy*. a major blood channel that develops through the embryonic liver from the left umbilical vein to the inferior vena cava. Also, DUCT OF ARANTIUS, CANAL OF CUVIER.

dud *Ordnance*. an explosive munition that has not been armed as intended or has failed to explode after being armed.

Duddell oscillograph *Electromagnetism*. an oscillograph having a mirror mounted on a moving coil that is placed in a magnetic field; the current to be observed passes through the coil, moving a beam of light that reflects off the mirror.

due *Navigation*. directly toward a cardinal point, e.g., due north.

Du Fay, Charles 1698–1739, French chemist; distinguished positive and negative electricity.

duff see ANTHRACITE FINES.

Duffing's equation [dü fā´] *Chaotic Dynamics*. an ordinary differential equation for a periodically forced, damped nonlinear oscillator; that is, damped motion in a quadratic potential with a quartic nonlinearity so that the frequency of oscillation depends on the amplitude.

Duffy alleles *Genetics*. alleles located on human chromosome 1 that control a red blood cell antigen; used frequently as a genetic marker in population studies.

Duffy blood group *Immunology.* a group of red blood cells that are identified by their distinct reaction with the immune serum anti-Fy^a.

Dufour number *Thermodynamics.* a dimensionless quantity given by the ratio of the increase in enthalpy of a unit mass during the process of isothermal mass transfer to that of a unit mass of the mixture; used in thermodiffusion analysis.

dufrenite *Mineralogy.* $Fe^{+2}Fe_4^{+3}(PO_4)_3(OH)_5{\cdot}2H_2O$, a green monoclinic mineral occurring as fibrous botryoidal crusts and rare crystals with a perfect cleavage and a weak silky luster, having a specific gravity of 3.1 to 3.34 and a hardness of 3.5 to 4.5 on the Mohs scale.

dufrenoysite *Mineralogy.* $Pb_2As_2S_5$, a blackish to lead-gray monoclinic mineral occurring in tabular, slightly elongated crystals with a perfect cleavage, having a specific gravity of 5.53 and a hardness of 3 on the Mohs scale.

duftite *Mineralogy.* $PbCu(AsO_4)(OH)$, a bright olive to gray-green orthorhombic mineral of the adelite group occurring in aggregates of minute crystals, having a specific gravity of 6.40 and a hardness of 3 on the Mohs scale; found with wulfenite, malachite, and azurite.

dugong *Vertebrate Zoology.* an herbivorous, aquatic mammal, *Dugong dugon,* characterized by a barrel-shaped body, flipperlike forelimbs, and a triangular tail; formerly widespread, but now rare.

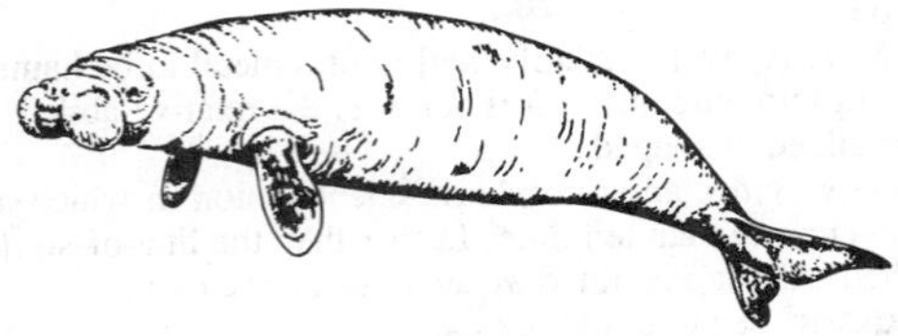

dugong

Dugongidae *Vertebrate Zoology.* the dugong and sea cow, a family of large, herbivorous, nocturnal marine mammals with notched tails and moderately cleft upper lips; found along tropical coasts of the Indian Ocean and western Pacific.

Dugonginae *Vertebrate Zoology.* the dugongs, a subfamily of aquatic mammals in the family Dugongidae and order Sirenia; characterized by the absence of nasal bones and ventrally angled dentaries.

dugout *Ordnance.* an underground shelter used to protect troops.

Duhem-Margules equation *Termodynamics.* an equation that relates the partial pressures p_A and p_B and mole fractions x_A and x_B of a liquid-vapor system of A and B: $d(\ln p_A)/d(\ln x_A) = d(\ln p_B)/d(\ln x_B)$.

Duhem's equation see GIBBS-DUHEM EQUATION.

Duhring's rule *Physical Chemistry.* a rule for estimating vapor pressure, which states that a straight line can be drawn between the temperature at which one liquid exerts a certain pressure and the temperature at which a chemically similar liquid exerts the same pressure.

duiker *Vertebrate Zoology.* a small, elusive African antelope of the family Bovidae that varies from slate blue to red with striped hindquarters. (Afrikaans for "diver," from its practice of darting for cover.)

duiker

Dukler theory *Chemical Engineering.* a theory giving the relationship of velocity and temperature distribution in thin films on vertical walls; used to compute effective thermal conductivity and eddy viscosity near the solid boundary.

DUKW *Ordnance.* an amphibious truck used by the U.S. Army to transport troops, equipment, and supplies; it is 2.5 tons and 36 feet long, with a cargo capacity of 5175 pounds, a range of 400 miles, and a speed of 50 mph. Also, DUCK.

Dulbecco, Renato born 1914, American virologist; shared Nobel Prize for research on tumor viruses.

dulcitol *Organic Chemistry.* $C_6H_8(OH)_6$, a white, crystalline powder, soluble in hot water and slightly soluble in cold water and alcohol; it melts at 188.5°C; used in bacteriology and medicine. Also, GALACTITOL.

Dulidae see PALMCHAT.

dull coal *Geology.* banded coal having a grayish color and dull appearance, and consisting mainly of clarodurain and durain.

dull emitter *Electronics.* the cathode of an electron tube that has been coated with a special material that increases the electron-emitting capability allowing the cathode to operate at lower temperatures than otherwise possible.

Dulong, Pierre Louis 1785–1838, French chemist and physicist; discovered nitrogen trichloride; with Petit, formulated a well-known law of specific heats.

Dulong-Petit law (of specific heats) *Thermodynamics.* the observation that for most solid elements at room temperature, the product of the atomic weight and the specific heat per gram is approximately the same value, about 6.2 calories per degree Celsius; for example, copper is 5.9, iron is 6.2, and silver is 6.0.

Dulong's formula *Engineering.* a formula that determines the gross heating value of coal through the analysis of the component amounts of carbon, hydrogen, oxygen, and sulfur.

dulse *Botany.* the coarse red algae in the genera *Dilsea* and *Rhodymenia,* found at northern latitudes in the intertidal zone and used as an important food source.

Dumas, Jean Baptiste André 1800–1884, French chemist; determined atomic weights of 30 elements; made important contributions in classification and quantitative analysis of organic compounds.

Dumas method *Analytical Chemistry.* **1.** a test for the presence of nitrogen in organic compounds, by heating the substance with copper oxide to convert the nitrogen to nitrogen oxides, reducing the oxides, and measuring the volume of nitrogen. **2.** a method to determine the vapor density of a volatile liquid.

Dumbbell Nebula *Astronomy.* a bright planetary nebula in Vulpecula. Also, M 27.

dumb iron *Engineering.* **1.** a tubular instrument that opens seams to prepare them for caulking. **2.** a connector between the frame of an automobile and the spring shackle.

dumb terminal *Computer Technology.* a user interface device consisting of a display screen and a keyboard with limited input/output functions and no processing capability.

dumb tool *Robotics.* an open-loop machine tool that has no sensors.

dumbwaiter *Mechanical Engineering.* a small electric or hand-operated elevator designed for quickly transporting loads between floors; used, for example, to move equipment in a construction project.

dumdum *Ordnance.* a hollow or soft-nosed bullet that flattens excessively or breaks up on contact, thus increasing its lethal anti-personnel properties; its use is forbidden by international law.

dummy *Engineering.* a device that appears to be a working instrument but actually does not operate. *Graphic Arts.* **1.** a full-size mock version of a printed page, showing the size and placement of all elements on the page. **2.** a bound set of blank pages showing the physical size of a proposed book. *Ordnance.* **1.** a nonexplosive bomb or projectile. **2.** see DECOY. *Computer Programming.* of or relating to an artificial or substitute argument, instruction, record, or address used to fulfill a requirement without affecting operations. *Telecommunications.* an item made in the same shape as a circuit element but with no operational role. *Metallurgy.* in electroplating, a cathode used to adjust the operating parameters or to preferentially plate out unwanted impurities.

dummy antenna *Electronics.* a resistive device that simulates the electrical characteristics of a transmitting or receiving antenna but is incapable of receiving or radiating radio-frequency energy; used chiefly for testing transmitters. Also, ARTIFICIAL ANTENNA.

dummy argument see DUMMY PARAMETER.

dummy block *Metallurgy.* in extrusion, a device placed between the stock and the ram.

dummy file *Computer Programming.* a nonexistent file that the program treats as actual; used to suppress the creation of files that are rarely needed.

dummy index *Mathematics.* an indexing symbol that is substituted for another in a given expression so that the expression is more convenient but its value is not changed. For example, *i* and *j* are dummy indices in the equivalent expressions $\sum_{i=5}^{9} a_i$ and $\sum_{j=0}^{4} a_{j+5}$.

dummy instruction *Computer Programming.* **1.** an artificial instruction placed in a program for a purpose other than execution as an instruction. **2.** an artificial instruction used for a specific purpose in a program such as a do-nothing instruction or a continue statement used to terminate a loop.

dummy joint *Engineering.* in a concrete structure, a groove that is cut into the slab to allow the slab to crack with minimal damage.

dummy load *Electronics.* a power-dissipating device used at the end of a transmission line or waveguide to convert transmitted energy into heat so that no energy is radiated outward or reflected back to its source. Also, ARTIFICIAL LOAD.

dummy message *Telecommunications.* a communication that generally has a meaningless content and is sent for purposes other than understanding; for example, to test whether a communications system is operating properly.

dummy module *Computer Programming.* a small program, often with only an entry and an exit, used to substitute for a subordinate module during top-down software testing.

dummy parameter *Computer Programming.* **1.** any parameter whose value is unimportant and is included only to fulfill a system requirement. Also, DUMMY ARGUMENT. **2.** another term for a formal parameter.

dummy record *Computer Programming.* a meaningless record stored only to fulfill a system requirement such as record count or file length.

dummy variable see INDICATOR VARIABLE.

Dumontiaceae *Botany.* a diverse family of red algae belonging to the order Cryptonemiales, some species of which are used in sizing paper and textiles.

dumontite *Mineralogy.* $Pb_2(UO_2)_3O_2(PO_4)_2 \cdot 5H_2O$, a yellow monoclinic mineral occurring as minute, elongated crystals, having a specific gravity of 5.65 and an undetermined hardness; found as a secondary mineral with uranophane, kasolite, and autunite.

dumortierite *Mineralogy.* $Al_7(BO_3)(SiO_4)_3O_3$, a blue, violet, or pink orthorhombic mineral usually occurring in fibrous to columnar masses, rarely in prismatic crystals; having a specific gravity of 3.26 to 3.41 and a hardness of 7 to 8.5 on the Mohs scale; found in aluminum-rich metamorphic rocks.

dump *Computer Programming.* **1.** a copy of the raw, uninterpreted contents of a computer storage device, usually in octal or hexadecimal representation; used in transferring data from one program to another. **2.** a utility program that prints the contents of memory. **3.** to transfer data from one program to another by means of a dump.

dump bailer *Engineering.* a bucket designed to carry cement or water into a well without disturbing the well's structure.

dumpcar *Mechanical Engineering.* a dumpcart that is fitted to run on a narrow-gauge railway; used especially in mining.

dumpcart *Mechanical Engineering.* a cart having either a bottom that opens downward or a body that tilts so that its contents can be discharged without handling.

dump check *Computer Programming.* a checking process in which all the digits are added during dumping and then added again when retransferring.

dumped deposit *Oceanography.* an accumulation of sediment that is deposited faster than currents and waves can redistribute it.

dumping syndrome *Medicine.* a syndrome that occurs after eating in patients with shunts of the upper alimentary canal, marked by flushing, sweating, giddiness, weakness, and vasomotor collapse.

dump power *Electricity.* any excess electric power that cannot be stored or conserved.

dump routine *Computer Programming.* a small utility program that prints the raw contents of a computer storage device.

dump scow *Naval Architecture.* a barge used to carry refuse or filler material, fitted with bottom doors that can be opened to release its load.

Dumpster *Mechanical Engineering.* a trade name for a large metal refuse bin designed to be left and filled at a remote site, then picked up and hauled away by a specially equipped truck.

dump truck *Engineering.* a truck designed to transport and dump loose materials, having a body that tilts and opens at one end to allow for unloading.

dump valve *Engineering.* a large automatic safety valve used in a tank or container to release pressure or fuel quickly in an emergency situation.

dump wagon see BOTTOM DUMP TRUCK.

dumpy level *Engineering.* a surveyor's level having the telescope with its level tube rigidly attached to a vertical spindle; capable only of horizontal rotary movement.

dunaliella *Botany.* a genus of unicellular green algae in the family Dunaliellaceae, characterized by oval, biflagellate cells and often having an eyespot; almost always present in salterns, brine lakes, and tidal pools.

Dunaliellaceae *Botany.* a family of naked biflagellate unicells of the order Volvocales, characterized by longitudinal division, a cup-shaped chloroplast surrounding a pyrenoid, and an eyespot in some species; present in salterns, brine lakes, and supralittoral pools.

Dunbarella *Paleontology.* a thin-shelled pectinoid bivalve of the Carboniferous; one of the few fossils found in association with the goniatitic ammonoids.

dundasite *Mineralogy.* $PbAl_2(CO_3)_2(OH)_4 \cdot H_2O$, a white, transparent, orthorhombic mineral occurring in spherical aggregates of radiating acicular crystals, having a specific gravity of 3.25 to 3.55 and a hardness of 2 on the Mohs scale; found as a secondary mineral in the oxidized zones of lead deposits.

dune *Geology.* a low mound or ridge of unconsolidated, windblown granular material, usually sand, that retains its characteristic shape despite the fact that it migrates from one place to another.

dune rock see EOLIANITE.

dung beetle *Invertebrate Zoology.* the common name for various beetles whose larvae feed on dung (solid animal waste); some species take balls of dung into their underground tunnels and lay their eggs on it.

dunite *Petrology.* a monomineralic rock named for the Duns Mountains of New Zealand, consisting almost solely of a magnesium-rich, nearly pure olivine with some chromite and picotite; an important source of chromium.

Dunkardian *Geology.* a North American provincial series of the uppermost Pennsylvanian-Lower Permian period, after the Monongahelan.

dunking sonar see DIPPING SONAR.

Dunning, John Ray 1907–1975, American physicist; worked on the Manhattan Project.

duodecimal number system *Mathematics.* the representation of numbers in base 12.

duodenal [doo äd´ ə nəl] *Anatomy.* relating to or situated in the duodenum.

duodenal papilla *Anatomy.* a small elevation in the wall of the duodenum through which the common bile duct and pancreatic duct empty their contents into the duodenum.

duodenal ulcer *Medicine.* a peptic ulcer or lesion, usually with inflammation, situated in the duodenum.

duodenitis *Medicine.* an inflammation of the duodenum.

duodenogram *Radiology.* a radiograph of the duodenum.

duodenography *Radiology.* an X-ray study of the duodenum after opacification with a contrast medium.

duodenoscope *Medicine.* an endoscope for examination of the duodenum.

duodenoscopy *Medicine.* the endoscopic examination of the duodenum.

duodenotomy *Surgery.* an incision into the duodenum.

duodenum [doo äd´ ə nəm] *Anatomy.* the first part of the small intestine extending from the pylorus to the junction of the jejunum at the level of the first or second lumbar vertebra on the left.

duodiode *Electronics.* an electron tube in which two diodes are contained within the same envelope. Also, DUAL DIODE, DOUBLE DIODE.

duodiode-pentode *Electronics.* an electron tube in which two diodes and a pentode are contained within the same envelope.

duodiode-triode *Electronics.* an electron tube in which two diodes and a triode are contained within the same envelope.

duomo *Architecture.* an Italian cathedral.

duoplasmatron *Electronics.* a gas-filled electron tube that generates ion beams for use in mass spectrometer applications.

duotone *Graphic Arts.* a method of two-color reproduction using two halftone plates, one black and the other gray or another light color, which are printed in register.

duotriode *Electronics.* an electron tube in which two triodes are contained within the same glass envelope. Also, DOUBLE TRIODE, DUAL TRIODE.

duplex *Engineering.* two separate systems or parts of a machine that share the same environment or operating procedure.
duplex artificial line *Electricity.* a balancing network, simulating the impedance of real line and distant terminal apparatus, that is used in a duplex circuit to make the receiving circuit unresponsive to outgoing signal currents.
duplex cable *Electricity.* a pair of twisted, insulated stranded-wire conductors that may have a common insulating covering.
duplex channel *Telecommunications.* simultaneous two-way and independent transmission across a channel in both directions.
duplex compressor *Mechanical Engineering.* an apparatus consisting of two types of compressors, such as a simple steam and a simple air cylinder, set side by side and working in combination.
duplex computer *Computer Technology.* a system consisting of duplicate computers with peripherals and circuitry that provide redundancy and ensure continued operations in case one computer fails.
duplex DNA *Molecular Biology.* DNA in the double-stranded form, the configuration of the double helix described in the Watson-Crick model.
duplexed system *Engineering.* a system having two separate sets of equipment, each having the ability to interchange with the other in case of system failure.
duplexer see TRANSMIT-RECEIVE TUBE.
duplexing *Metallurgy.* see DUPLEX PROCESS.
duplexite see BAVENITE.
duplex lathe *Mechanical Devices.* a lathe that utilizes two cutting tools, one on each side of the work, for safe, efficient operation.
duplex lock *Mechanical Devices.* a cylinder-type lock having two independent pin tumblers that act on the same bolt, requiring both a master key and a change key to lock and unlock it.
duplex material *Materials Science.* a combination of two materials bonded together in order to attain properties superior to those produced with a single melt or starting material.
duplex operation *Engineering.* a practice used in a radar operation in which two identical instruments are provided, one of which is in operation, and the other of which can be put into operation if the first one fails. *Telecommunications.* the operation of transmission equipment in such a manner that two-way conversations may be held or messages may be passed between two given points. Also, **duplexing, duplex transmission.**
duplex paper *Graphic Arts.* paper or paperboard having different colors or textures on its two sides; commonly used in direct-mail advertising.
duplex process *Materials Science.* a process that combines two methods of performing one operation. *Metallurgy.* a melting or refining process that uses two furnaces. Also, DUPLEXING.
duplex pump *Mechanical Engineering.* a positive-displacement pump in which two liquid cylinders are set side by side and geared so that their piston strokes alternate, thus producing continuous action; used especially to move a low volume of liquid at high pressure.
duplex stainless steel *Materials.* a type of austenite-ferrite stainless steel that is resistant to corrosion.
duplex tube *Electronics.* an electron tube in which two electron tubes are contained within the same envelope.
duplex uterus *Anatomy.* a uterus having two parts.
duplicate cavity plate *Engineering.* a plate used in plastics molding that is removable and holds the shape of a molding cavity.
duplicate record *Computer Programming.* a record with the same key value as another record in the same file; usually considered to be an error.
duplicating film *Graphic Arts.* a photosensitive material that yields a direct copy of a negative from a negative original (for example, in copying color transparencies) or a positive from a positive original (for example, in photocopying).
duplicating the cube *Mathematics.* the problem of constructing, with compass and straightedge alone, the side of a cube having volume exactly twice that of a given cube. It is one of a collection of impossible constructions attempted by the ancient Greeks; others include squaring the circle and trisecting the angle.
duplication *Molecular Biology.* **1.** a chromosome aberration in which a cytologically distinct portion of a chromosome occurs in more than one location in the karyotype. **2.** an additional copy of an entire chromosome.
duplication check *Computer Programming.* a performance check to verify that two independent runs of the same operation produce identical results.
duplicatus *Meteorology.* a cloud variety consisting of layers, sheets, or patches, sometimes partly merged, and occurring at slightly different levels.
dupp *Cardiology.* a term for the second cardiac sound heard by auscultation; it marks the closure of the cortic and pulsivic valves.
Dupré equation [dü prā′] *Thermodynamics.* an equation specifying the adhesional work for a gas-liquid interface: $W_{GL} = g_{GS} + g_{GL} - g_{LS}$, where g is the surface tension between the appropriate phase interface indicated by the subscripts.
Dupré's disease or **Dupré's syndrome** see MENINGISM.
durability *Engineering.* the degree to which equipment or material can withstand usage over an extended period of time.
durable-press see PERMANENT-PRESS.
durain *Geology.* a coal lithotype with a gray to brownish-black color and a dull, matte luster that occurs in bands or lenticels of variable thickness.
dura mater *Anatomy.* the outermost, toughest, and most fibrous of the three meninges covering the brain and spinal cord. Also, PACHYMENINX.
duramen see HEARTWOOD.
Durandal *Ordnance.* a French air-to-surface missile that is designed to destroy concrete runways and other hard targets; it is braked by a parachute until the solid motor fires, causing high acceleration toward ground impact; it can penetrate up to 15.75 inches of reinforced concrete before detonating a 220-pound conventional warhead with a one-second delay fuse.
durangite *Mineralogy.* $NaAl(AsO_4)F$, a translucent, orange-red monoclinic mineral occurring in prismatic and pyramidal crystals, having a specific gravity of 3.9 to 4 and a hardness of 5 to 5.55 on the Mohs scale; found in tin mines in Durango, Mexico.
Durangoan *Geology.* a North American Gulf Coast geologic stage of the Lower Cretaceous period, occurring after the LaCasitan of the Jurassic and before the Nuevoleonian.
duranickel *Materials.* a strong and resistant nickel-aluminum alloy suitable for forming by hot or cold weaking.
Durargid *Geology.* a great soil group constituting a subdivision of the Argids, and including those soils with a hardpan cemented by silica.
duration *Science.* the amount of time that a given event lasts. *Oceanography.* **1.** in wave forecasting, the length of time the wind blows from one direction over the fetch. **2.** the length of time the tide rises or falls, or the period of flow of a flood or ebb current.
duration control *Electronics.* a potentiometer control used to adjust the duration of the reduced gain time in a sensitivity time-control circuit.
Durbin-Watson statistic *Statistics.* a test statistic for detecting autocorrelation of the residuals in regression analysis. Also, **Durbin-Watson test.**
dur. dolor. while the pain lasts. (From Latin *durante dolore.*) Also, **Dur. dolor.**
durene *Organic Chemistry.* $C_6H_2(CH_3)_4$, combustible, colorless crystals that are insoluble in water and soluble in alcohol; melts at 79.3°C and boils at 193°C, and sublimes and is volatile with steam; it is used as a curing agent, rubber accelerator, and plasticizer. Also, *sym*-TETRAMETHYLBENZENE.
Dürer, Albrecht [dür′ər] 1471–1528, German painter, printmaker, and mathematician; research in anatomy, perspective, optics, and acoustics.
Dürer's conchoid *Mathematics.* the loci of points at a fixed distance a from point $Q = (q,0)$ of an extended line QR, where $R = (0,r)$ and $q + r = b$. It is represented by the equation

$$2y^2(x^2+y^2)-2by^2(x+y)+(b^2-3a^2)y^2-a^2x^2+2a^2b(x+y)+(a^2-b^2)=0.$$

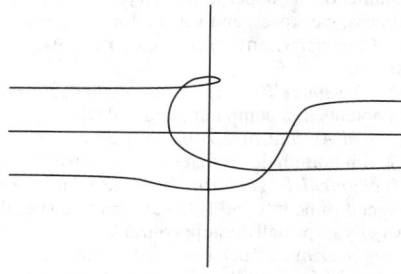

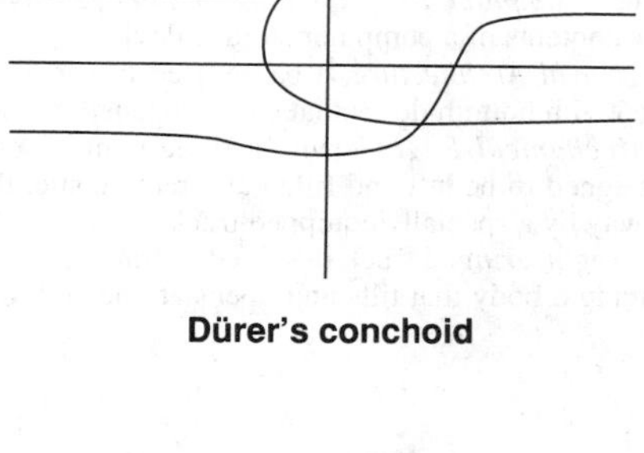

Dürer's conchoid

duress [dů res′] *Medicine.* a term for any of various reactions of the body to deleterious forces, infections, and abnormal states that tend to disturb its normal physiologic equilibrium (homeostasis).

Durfee square *Mathematics.* the largest square of dots in the upper left-hand corner of a Ferrer's graph; used in proving results involving generating functions of partitions.

Durham see SHORTHORN.

Durham (fermentation) tube *Microbiology.* a glass tube that is placed inverted into a vessel of liquid nutrient medium for the detection and collection of gas produced by a growing culture of microorganisms; used to test for indications of coliform bacteria.

durian *Materials.* the pale, easily worked, nondurable wood of the *Durio zibethinus* tree of the East Indian islands; used to make such objects as furniture, cabinets, ceiling boards, and plywood.

duricrust *Geology.* in a semiarid climate, a hard crust that forms on the surface of a soil as a result of mineral precipitation.

durinite *Geology.* a former term for the principal maceral of durain.

duripan *Geology.* in a mineral soil, a diagnostic subsurface horizon characterized by cementation by silica.

Duroc *Agriculture.* a breed of hog with red coloration and drooping ears. Also, **Duroc-Jersey**. (From *Duroc,* a stallion that lived on the New York farm where this breed was developed.)

durometer *Engineering.* an instrument that determines the hardness of a material, with the use of pressure and a small drill or blunt point.

durometer hardness *Engineering.* a measurement of a material's hardness, as determined by a durometer.

durum wheat *Botany.* a wheat, *Triticum turgidum,* whose grain is used in making pasta. Also, **macaroni wheat.**

durum wheat

Durvillaeales *Botany.* a monotypic order of kelplike algae of the class Phaeophyceae, composed of the family Durvillaeaceae and characterized by a fibrous, discoid, or bulbous holdfast, a thick woody stipe, and a leathery blade; plants grow up to 10 meters in length and are restricted to Australia, New Zealand, and subarctic regions.

Durville process *Metallurgy.* a casting process in which the mold is attached to the crucible in order to promote laminar flow of liquid metal during pouring.

DUS *Aviation.* the airport code for Dusseldorf, Germany.

düsenwind *Meteorology.* a strong east-northeast wind that blows out of the Dardanelles into the Aegean Sea, caused by a high-pressure ridge over the Black Sea.

dusk *Science.* the period of partial darkness just before sunset; the dark part of twilight.

dusk side *Astronomy.* the hemisphere of a planet or moon lying closest to the evening terminator.

dussertite *Mineralogy.* $BaFe_3^{+3}(AsO_4)_2(OH)_5$, a green trigonal mineral occurring as crusts of small tabular crystals, having a specific gravity of 3.75 and a hardness of 3.5 on the Mohs scale; found with carminite and limonite.

Dussumieridae *Vertebrate Zoology.* a group of small, round, tropical marine herrings, in some taxonomic systems included in the family Clupeidae; distinguished from other herrings by their rounded abdomen and by the fact that they do not travel in immense schools.

dust fine particles of matter; specific uses include: *Geology.* **1.** finely divided, dry, solid matter of silt- and clay-sized earthy particles, less than 0.0625 millimeter in diameter. **2.** see ASH, def. 1. *Metallurgy.* in powder metallurgy, a superfine powder, usually consisting of particles smaller than one micrometer in diameter.

dust and fume monitor *Mining Engineering.* an instrument used in mine work environments to measure and record concentrations of dust, fume, and gas over an extended period of time.

dust avalanche see DRY-SNOW AVALANCHE.

Dust Bowl *Meteorology.* **1.** a term for the region in the south-central United States that was seriously affected by drought and dust storms in 1935 and afterward. **2. dust bowl.** any of various similar regions in other parts of the world.

dust cell see ALVEOLAR MACROPHAGE.

dust chamber *Engineering.* an enclosure in which solid materials are deposited from gases as they travel through it. Also, **dust collector.**

dust-control system *Engineering.* a safety system that controls the production and transmission of dust during production operations.

dust counter *Engineering.* an instrument that measures the amount and size of dust particles per unit volume of air. Also, KERN COUNTER.

dustcover or **dustjacket** *Graphic Arts.* see JACKET.

dust devil *Meteorology.* a well-developed, small, vigorous whirlwind, usually of short duration, that is made visible by dust, sand, and debris picked up from the ground; usually ranging from 10 to more than 100 feet in diameter, and rising to an average height of about 600 feet.

dust-devil effect *Geophysics.* an abrupt and short-lived change in the vertical component of an atmospheric electric field, caused by the passage of a dust devil that is near an instrument sensitive to the vertical gradient.

duster see DUST STORM.

Duster *Ordnance.* a self-propelled, antiaircraft weapon with twin 40-mm guns, designed for use against low-flying aircraft; officially designated **M42.**

dust filter *Engineering.* a dry or sticky fabric used in the cleaning of gas to separate particles.

dust horizon *Meteorology.* a layer of dust confined by a low-level temperature inversion, appearing as a horizon when seen from above against the sky that obscures the true horizon.

dusting *Metallurgy.* the process of applying a powdered material during a metallurgical operation.

dust separator *Engineering.* any equipment or system that cleans dust from gas through precipitation or filtering.

dust storm *Meteorology.* a condition of strong winds and dust-filled air over an extensive area, following a period of drought over normally arable land; usually in the form of an advancing dust wall miles long and several thousand feet high. Also, DUSTER, BLACK BLIZZARD.

dust tail *Astronomy.* the tail of a comet made up of dust ejected from the nucleus; it is yellowish in color, shines by reflecting sunlight, and is often curved.

dust well *Hydrology.* a small pit on the surface of a glacier or sea ice produced by the differential melting of the ice beneath a dark patch of dust or other particles.

dust whirl *Meteorology.* a whirlwind over a dry and dusty or sandy area that carries dust, leaves, and other light material picked up from the ground; it typically forms as the result of strong convection during sunny, hot, calm summer afternoons. Also, DANCING DERVISH, DEVIL, DESERT DEVIL, SAND AUGER, SATAN.

Dutch Belted *Agriculture.* a breed of dairy cattle that originated in The Netherlands, characterized by a black coat with a white belt around the middle.

Dutch elm disease *Plant Pathology.* an insect-transmitted disease of elm trees caused by the fungus *Graphium ulmi,* which releases a lethal substance that rapidly destroys vascular tissue.

Dutch liquid *Organic Chemistry.* a popular name for ethylene dichloride. Also, **Dutch oil.**

dutchman *Engineering.* a piece that fills a gap between two pipes, or between a pipe and a piece of equipment, to provide for closure or proper alignment.

Dutchman's log *Navigation.* a simple speed-measuring device for vessels in which speed is calculated from the time it takes for the vessel to pass a floating object thrown off the bow.

Dutch metal *Metallurgy.* a copper-base alloy containing 15% zinc, rolled to foil; used as an inexpensive substitute for gold.

Dutch process *Chemical Engineering.* a procedure for creating white lead, in which metallic lead in containers corrodes due to the action of acetic acid and carbon dioxide, which are produced by the fermenting tanbark and manure into which the containers have been placed.

duty *Hydrology.* see DUTY OF WATER.

duty cycle *Engineering.* **1.** the amount of time it takes to start, operate, stop, and idle a machine when it is being used for intermittent duty. **2.** a percentage that expresses the amount of working time as compared to the total operating time of an intermittently working piece of equipment. *Electronics.* the ratio of the "on" period of a pulse to the total pulse period. *Nucleonics.* the fraction of time that a useful beam is available at full power in a particle accelerator. *Telecommunications.* the daily schedule of a radio transmission station.

duty cyclometer *Engineering.* an instrument that records the operation of a duty cycle.

duty of water *Agriculture.* the relationship between the quantity of water used in irrigating an area and the size of the area or the amount of crops produced there.

duty ratio *Electronics.* **1.** the ratio of the time an intermittently operating device is working to the total time available. **2.** the ratio of average to peak power in a pulsed system.

Du Vigneaud, Vincent [dü vēn´yō´] 1901–1978, American biochemist; Nobel Prize for research on pituitary hormones.

DV or **D.V.** double vibrations. Also, **dv** or **d.v.**

D value *Microbiology.* the time required to decrease the number of viable microorganisms at a given temperature to a given percentage, usually 10%, of the original number.

D variometer declination variometer.

DVM or **dvm** digital voltmeter.

DVM or **D.V.M.** Doctor of Veterinary Medicine.

DVS or **D.V.S.** Doctor of Veterinary Science.

DVST or **dvst** direct-view storage tube.

DW or **dw** deadweight; distilled water.

dwarf *Biology.* an animal or plant that is significantly shorter or smaller than others of its species, often having abnormal proportions as well, and for which there is no expectation of growth to the normal size. *Medicine.* specifically, a markedly undersized person; especially one whose bodily proportions are abnormal. *Astronomy.* see DWARF STAR.

dwarf Cepheid *Astronomy.* a pulsating variable star having a period of less than 6 hours and a type A or F spectrum.

dwarf disease *Plant Pathology.* a virus disease of blackberries, plum trees, and other plants that causes a reduction in growth and fruit production.

dwarf dud *Ordnance.* any nuclear weapon that fails to provide a yield within reasonable range of that expected with normal operation.

dwarf galaxy *Astronomy.* a galaxy having a small mass and low luminosity.

dwarfing *Plant Pathology.* an underdevelopment of a plant or plant part, caused by various types of disease agents or by faulty nutrition.

dwarfism *Medicine.* abnormal underdevelopment of the body; the condition of being undersized, especially with a lack of normal proportion.

dwarf nova see CATACLYSMIC VARIABLE.

dwarf star *Astronomy.* a star on the main sequence of the Hertzsprung-Russell diagram.

dwell *Design Engineering.* the part of a cam that allows the cam follower to remain at maximum lift for an extended period of time. *Robotics.* a programmed delay in the working cycle of a robot.

dwell time *Transportation Engineering.* the duration of a station stop, or of all station stops during a given trip. *Metallurgy.* in powder metallurgy, the time elapsed during the application of the maximum pressure.

dwigh [dwī] *Meteorology.* **1.** a sudden shower or snow flurry along the coast of Newfoundland. **2.** any coastal squall. Also, **dwy, dwey, dwoy.**

DWT or **dwt** deadweight ton; deadweight tonnage.

dwt. pennyweight.

DX *Aviation.* the airline code for Danair.

DX distance; distance reception; direct expansion.

Dx diagnosis.

Dy the chemical symbol for dysprosium.

dyad *Cell Biology.* a double chromosome arising from the halving of a tetrad during the first meiotic division. *Mathematics.* a decomposable, contravariant 2-tensor; i.e., a contravariant 2-tensor whose (i,j)th component is the product of the ith component of the first vector and the jth component of the second vector.

dyadic *Cell Biology.* of or relating to a dyad. *Mathematics.* an addition of two or more dyads.

dyadic expansion *Mathematics.* the representation of a number in base two. Also, BINARY REPRESENTATION.

dyadic operation *Mathematics.* any operation that requires two operands, such as addition or subtraction; a binary operation. *Computer Programming.* also, **dyadic Boolean operation.** a logic operation that uses two operands.

dyadic processor *Computer Technology.* a multiprocessor system in which two processors are under the control of the same operating system.

dyadic rational *Mathematics.* a rational number of the form $a/2^n$, where a and n are integers and n is nonnegative.

Dyar's rule *Entomology.* a law based on the observation that certain caterpillar parts grow in geometric progression, increasing in size by a constant ratio at each moult; used to deduce facts about the life history of an insect for which other information is lacking.

Dyazide *Pharmacology.* a trade name for a preparation of triamterene with hydrochlorothiazide, used in the treatment of hypertension and edema.

dye *Chemistry.* a natural or synthetic substance that is used in solution to impart color to another substance; distinguished from a pigment, which is used in suspension.

dyecrete process *Engineering.* a process in which organic dyes are used to permanently color concrete.

dyeing *Chemical Engineering.* the process of applying a color-producing agent to a material.

dyeing assistant *Chemistry.* any material that is used in a dyeing procedure to promote or control the actions of the dye.

dye penetrant *Metallurgy.* a liquid used to reveal fine cracks in a metallic component.

dyestuff *Materials.* any substance that yields or is used as a dye.

dyn. dynamics; dyne; dynes.

dyna- a combining form meaning "power," as in *dynatron.*

dynam- a combining form meaning "power," as in *dynamite.*

dynam. dynamics.

dynamic *Mechanics.* of or relating to the science of dynamics. *Physics.* **1.** relating to or manifesting force or power. **2.** of or relating to bodies in motion or to motion in general. **3.** describing any system that changes over time. *Acoustical Engineering.* **1.** of or relating to the range of volume of musical sound. **2.** describing an acoustical device, such as a loudspeaker or microphone, that derives electroacoustic energy from some form of motion. *Computer Technology.* of programming, processing, memory, or the like, affected by the passage of time or by the variations in power input. Also, **dynamical.**

dynamic accuracy *Control Systems.* in an automatic control system, the difference between the actual position and the position desired or commanded.

dynamic address translator *Computer Technology.* in a virtual memory system, a hardware device that uses a memory mapping table to specify the correspondence between the virtual address and the real address. Also, RELOCATION HARDWARE.

dynamical diffraction *Crystallography.* diffraction theory in which the modification of the primary beam on passage through the crystal is important. The mutual interactions of the incident and scattered beams are taken into account; this is important for perfect crystals and for electron diffraction by crystals.

dynamical equinox *Astronomy.* the intersection of the ecliptic with the celestial equator where the Sun is moving from south to north.

dynamical friction *Physics.* **1.** the friction between two materials rubbing against each other. **2.** the drag force between electrons and ions drifting with respect to one another.

dynamic algorithm *Computer Programming.* an algorithm including parameters that are determined by computations and logical comparisons during program execution.

dynamical halo model *Astrophysics.* a model to explain the propagation of cosmic rays, postulating that they originate in the galactic disk and diffuse outward and inward in a dynamical galactic halo.

dynamical mean sun *Navigation.* a fictitious sun that moves at a constant rate; used in calculating the equation of time.

dynamical system *Mathematics.* in the calculus of variations, an optimization problem in which the state of an object is described by a system of ordinary differential equations.

dynamical time *Astronomy.* the system of uniform time scales now used in almanacs in place of ephemeris time.

dynamic analogies *Physics.* the similarities between the forms of differential equations governing mechanical and acoustical systems and those of electrical systems, allowing the solving of mechanical and acoustical equations by electrical circuit analysis.

dynamic balance *Mechanics.* a running balance in a rotating system, such that the system rotates with no tendency to wobble; it occurs when the axis of forced rotation coincides with a principal axis.

dynamic balancing *Design Engineering.* see BALANCING.

dynamic behavior *Engineering.* an evaluation of equipment or system operation with respect to time.

dynamic boundary condition *Fluid Mechanics.* a boundary condition that depends on forces exerted on a boundary.

dynamic braking *Mechanics.* the slowing of a mechanism by using its motor as a generator to convert kinetic energy into electricity or heat. Unlike that in regenerative braking, energy is dissipated without reuse.

dynamic characteristic see LOAD CHARACTERISTIC.

dynamic check *Engineering.* a test that checks for the correct performance of equipment or a system during its operation.

dynamic circuit *Electronics.* a high-speed metal-oxide semiconductor circuit whose high input impedance contributes to its fast operation.

dynamic climatology *Meteorology.* a clinical approach to the study of atmospheric circulation.

dynamic-coil loudspeaker see MOVING-COIL LOUDSPEAKER.

dynamic-coil microphone see MOVING-COIL MICROPHONE.

dynamic control *Design Engineering.* real-time manipulation of a process or machine to achieve a desired output or response.

dynamic convergence *Electronics.* a condition in which the three electron beams of a color picture tube come together as they are swept vertically and horizontally over the face of the screen by the deflection circuits.

dynamic correction *Geodesy.* the quantity that must be added to the orthometric elevation of a point to obtain its dynamic number.

dynamic creep *Mechanics.* creep produced by fluctuating stress in a material.

dynamic debugging routine *Computer Programming.* a set of diagnostic instructions that execute in conjunction and interactively with the program under development.

dynamic dispatching *Design Engineering.* the use of real-time, online communication of operating orders.

dynamic dump *Computer Programming.* a memory dump produced during program execution.

dynamic elevation *Cartography.* an elevation expressed in length units, but determined by dynamic number.

dynamic equilibrium *Mechanics.* the condition of a system in which it is either at rest or moving at a constant velocity with no linear or rotational accleration; thus, the net force on the system is equal to zero. Also, KINETIC EQUILIBRIUM.

dynamic error *Electronics.* an error in the varying output signal of a transducer caused by its inadequate dynamic response.

dynamic factor *Aviation.* a ratio derived from the relationship of the load carried by a structural part of an aircraft in accelerated flight or subjected to abnormal conditions to the corresponding basic load during normal flight conditions.

dynamic fluidity *Fluid Mechanics.* the reciprocal of the absolute viscosity.

dynamic focusing *Electronics.* the automatic and continuous varying of the focus signal for a color picture tube to compensate for the tendency of the electron beams to defocus as they sweep over the flat surface of the screen.

dynamic forecasting see NUMERICAL FORECASTING.

dynamic geomorphology *Geology.* the quantitative analysis of steady-state geomorphic processes that are, to a large extent, self-regulatory. Also, ANALYTICAL GEOMORPHOLOGY.

dynamic height *Geophysics.* the distance above the geoid of points on the same equipotential surface, as measured from sea level in terms of linear units measured along a plumb line at a given latitude. *Geodesy.* a height derived by dividing the geopotential number by a constant, usually the value of normal gravity at 45° latitude.

dynamic-height anomaly see ANOMALY OF DYNAMIC HEIGHT.

dynamic holdup *Chemical Engineering.* the liquid held in a tank or process vessel, in which a constant flow of new material offsets withdrawal of the held material to maintain a constant liquid level.

dynamic ileus see SPASTIC ILEUS.

dynamic impedance *Electricity.* the electrical impedance of a device that is in operation.

dynamic leak test *Engineering.* a test to check an empty container for leaks by the use of an internal leak detector, which responds if gas escapes from the container.

dynamic level *Petroleum Engineering.* the location of the liquid surface when a well is producing oil. Also, PUMPING LEVEL.

dynamic load *Civil Engineering.* a moving, not constant load on a structural system. *Aviation.* a load imposed on a flight vehicle by acceleration, especially in response to conditions such as gusts, maneuvers, or landing; often distinguished from the aircraft's static load.

dynamic loudspeaker see MOVING-COIL LOUDSPEAKER.

dynamic map *Cartography.* a map designed to show motion, action, or change through the use of graphic symbols such as arrows or lines of greatest extent.

dynamic memory see DYNAMIC STORAGE.

dynamic metamorphism *Geology.* a regional metamorphism produced in rocks as a result of the processes and effects of crustal deformation and differential stresses. Also, DYNAMOMETAMORPHISM.

dynamic meteorology *Meteorology.* the study of atmospheric motions and their relationship to other forces as solutions to fundamental equations of hydrodynamics or other systems of equations.

dynamic microphone see MOVING-COIL MICROPHONE.

dynamic model *Engineering.* a model of an aircraft or other object having the dimensions, weight, and moments of inertia correctly proportioned so that it duplicates actual full-scale behavior. Such models may include other systems such as rivers.

dynamic noise suppressor *Acoustical Engineering.* an audio-frequency filter circuit that suppresses noise dynamically by reducing the low- and high-frequency bands when noise becomes significant.

dynamic nuclear polarization *Nuclear Physics.* the arrangement of groups of nuclei so as to align their (normally random) spin axes in the same direction.

dynamic number *Geodesy.* the expression in absolute units of the amount of work required to lift a unit of mass from sea level to a given point.

dynamic oceanography *Oceanography.* the study of ocean phenomena of force, such as waves, tides, and currents.

dynamic packing *Engineering.* a packing that functions on surfaces that are in motion and acts much like a bearing.

dynamic pickup *Electronics.* a phonograph transducer in which the motion of a stylus in the groove of the phonograph record causes a conductor to vibrate within a magnetic field, producing an electrical output signal. Also, **dynamic reproducer.**

dynamic plate impedance or **dynamic plate resistance** *Electronics.* the dynamic-state internal AC resistance between cathode and plate in a vacuum tube; computed by dividing the change in plate voltage by the resulting change in plate current, with other electrode voltages held constant. Also, AC PLATE RESISTANCE.

dynamic-pool model *Ecology.* a model for predicting optimum yields that factors in statistics of population growth and population decline.

dynamic positioning *Petroleum Engineering.* a system that uses propulsion units controlled by computers to keep a floating offshore-drilling rig in the correct position over the well.

dynamic pressure *Fluid Mechanics.* an expression of the pressure of a flowing fluid, equal to one-half the fluid density, multiplied by the fluid velocity squared.

dynamic printout *Computer Programming.* a printout produced by a program as part of its execution.

dynamic programming *Mathematics.* a quantitative method similar to linear programming, used to optimize the outcome of a sequence of decisions. *Industrial Engineering.* an operations research technique involving multistage decision-making.

dynamic random-access memory *Computer Technology.* describing direct-access storage devices whose semiconductor memory cells lose information content after a short time and must be refreshed. Also, **dynamic RAM.**

dynamic range *Electronics.* the range in signal amplitude over which a communication receiver or audio amplifier is capable of operating while producing an acceptable output; usually expressed in decibels.

dynamic recovery *Computer Technology.* in fault-tolerant computer systems, an automated self-repair capability using a special mechanism to detect a fault, switch in a spare module, and initiate the software actions that allow computing to resume.

dynamic regulator *Electronics.* a transmission regulator that continuously monitors output voltage or current and immediately responds to a change in output voltage by producing an equal and opposite correction.

dynamic relocation *Computer Programming.* the movement of all or part of an active program from one storage area to another while maintaining all necessary address references so that the program can successfully complete execution. Also, **dynamic program relocation.**

dynamic resistance *Electricity.* the electrical resistance of a device when it is in operation.

dynamic roughness *Oceanography.* the degree of roughness at a given surface location, used in calculations of the wind's force. Also, ROUGHNESS LENGTH.

dynamic routing *Transportation Engineering.* any process or policy by which a route may change in mid-trip, usually in response to traffic conditions or passenger demands.

dynamics *Mechanics.* the field of mechanics that deals with the study of motion and of the forces that bring about this motion. *Ecology.* the changes and causes of such changes in population size. Also, POPULATION DYNAMICS.

dynamic scheduling *Design Engineering.* the process of providing a changing schedule in real-time to meet operating objectives. *Transportation Engineering.* a process or policy of adjusting departures at transfer stations to increase connectivity. Also, **dynamic schedule synchronization.**

dynamic scoping *Computer Science.* a convention in a language, such as Lisp, that a variable can be referenced by any procedure that is executed after it has become bound and before it becomes unbound; thus, the scope of the variable can depend on the execution sequence.

dynamic sensitivity *Engineering.* the minimum measurement of an instrument's ability to detect the rate of a leak.

dynamic sequential control *Computer Programming.* a method of computer operation in which the sequence of operations can be altered during program execution by computing the address of the next instruction or by performing a conditional branch operation.

dynamic shift register *Computer Programming.* a shift register in the arithmetic-logic unit that is composed of volatile semiconductor memory cells.

dynamic similarity *Mechanics.* the condition that two systems, such as a real system and a model of it, behave in a similar manner, except for a difference of scale; this requires that all forces and masses be scaled appropriately. If a scale model operates at a speed exactly proportional to that of a full-size prototype, then resistance, density, length, and velocity are related as follows: $R_1/R_2 = d_1l_1^2v_1^2/d_2l_2^2v_2^2$.

dynamic speaker see MOVING-COIL LOUDSPEAKER.

dynamic stability *Mechanics.* the property of a system such that, when it is slightly perturbed from a state of steady motion, it will return to that state.

dynamic stop *Computer Programming.* a loop in a program that consists of a single jump instruction transferring control to itself; often used as a trap to indicate that an error condition has occurred.

dynamic storage *Computer Technology.* a volatile type of semiconductor memory in which the binary digit is stored as a charge on the ground capacitance and must be refreshed at regular intervals. Also, DYNAMIC MEMORY.

dynamic storage allocation *Computer Programming.* the assignment and reassignment of memory as required during computation.

dynamic subroutine *Computer Programming.* a subroutine whose coded instructions are created or modified during the execution of the calling program in accordance with specified parametric values.

dynamic symmetry *Physics.* a symmetry law related to the laws that govern the dynamic behavior of the constituents of matter.

dynamic test *Engineering.* a test that takes place when a piece of equipment is carrying an actual load or when the carrying of the load is simulated.

dynamic topography *Cartography.* a topographic map of a surface (usually isobaric) given in terms of the dynamic height of that surface.

dynamic trough *Meteorology.* a pressure trough formed on the lee side of a mountain range across which the wind blows almost at right angles. Also, LEE TROUGH.

dynamic type checking *Computer Science.* testing of the types of the values of variables at runtime, as is done in Lisp and object-oriented languages.

dynamic unbalance *Mechanical Engineering.* in a rotating machine, an imbalance that occurs along a single axial plane and on opposite sides of a rotating axis, or in two axial planes, arising from inertia on the rotating plane and forces in a single axial or differing axial planes; coincides with the value of a rotation axis, which may or may not coincide with the center of gravity.

dynamic variable *Mechanics.* a variable describing the state of a mechanical system that changes with time.

dynamic vehicle envelope see VEHICLE ENVELOPE.

dynamic vertical see APPARENT VERTICAL.

dynamic viscosity see ABSOLUTE VISCOSITY.

dynamite *Materials.* a powerful blasting explosive that was originally manufactured by the absorption of nitroglycerine into a porous base material such as charcoal or wood pulp; now generally manufactured with ammonium nitrate or cellulose nitrate rather than nitroglycerine. (From the Greek word for "power"; coined by its inventor, Alfred Nobel.)

Dynamitron accelerator *Nucleonics.* a trade name for a particle accelerator in which a radio-frequency (100 kHz) power supply is coupled to a low-capacitance direct-current rectification system to produce a high-energy beam of charged particles for research, industrial, and medical applications.

dynamo *Electricity.* an electric generator, especially one designed for direct current.

dynamo- a combining form meaning "power," as in *dynamometer.*

dynamo effect *Geophysics.* a process in which an induced current is produced in the geomagnetic field by the wind moving ions in the atmosphere.

dynamoelectric *Physics.* relating to the exchange of energy between mechanical processes and electrical processes.

dynamoelectric amplifier *Electricity.* a generator that serves as a power amplifier at low frequencies or direct current.

dynamogenesis *Neurology.* a generation of force or energy, as in muscles or nerves, to bring about movement or activity. Also, **dynamogeny.**

dynamogenic *Neurology.* of or relating to the generation of force or energy, as in muscles or nerves.

dynamometamorphism see DYNAMIC METAMORPHISM.

dynamometer *Engineering.* **1.** any of various devices used in testing a motor or engine for such characteristics as efficiency and torque, especially an instrument that measures current or the power of a motor by calculating the force between a fixed coil and a moving coil. **2.** broadly, any device used to measure mechanical force.

dynamometer multiplier *Electricity.* a multiplier with a fixed coil and a moving coil arranged so that deflection of the moving coil is in proportion to the product of the currents flowing in both coils.

dynamostatic *Electricity.* of or relating to a machine driven by AC or DC to generate static electricity.

dynamo theory *Geophysics.* a theory stating that daily variations in the earth's magnetic field are caused by the tidal movement of ionized air across the magnetic field, producing electrical currents in the lower ionosphere.

dynamotor *Electricity.* a rotating electrical machine with two or more windings on a single armature containing a commutator for DC operation and slip rings for AC operation; used primarily as an inverter to convert DC to AC.

Dynastidae *Invertebrate Zoology.* a family of large beetles, some of which possess large horns or protuberances extending from their thorax; includes the unicorn beetle of North America and the rhinoceros scarab of Europe.

dynatron *Electronics.* an electron tube, typically a tetrode, that exhibits a negative resistance effect between cathode and plate such that, within a certain operating range, plate current decreases as plate voltage is increased. Also, NEGATRON.

dynatron oscillator *Electronics.* an oscillator circuit that depends for its operation on the negative resistance effect of a dynatron.

dyne *Metrology.* a unit equal to the amount of force that will cause a mass of one gram to accelerate at a rate of one centimeter per second per second; the basic unit of force in the centimeter-gram-second system.

dyne-centimeter see ERG.

dynein *Biochemistry.* a protein that uses the hydrolysis of ATP to perform the movement in eukaryotic cilia and flagella.

Dynkin diagram *Mathematics.* a Coxeter graph of a root system with the multiple edges directed from longer to shorter roots. Single edges are undirected, since their Cartan integers equal 1.

dynode *Electronics.* an electrode that is coated with an electron-emitting surface and positioned to maximize the effects of secondary emission.

dypnone *Organic Chemistry.* $C_6H_5COCH{=}C(CH_3)C_6H_5$, a stable, combustible, light-colored liquid, insoluble in water, that boils at 340°C; used as a softening agent, plasticizer, and perfume base.

dys- a combining form meaning: **1.** bad or abnormal, as in *dysplasia.* **2.** difficult or painful, as in *dyspepsia.*

dysacusis *Medicine.* **1.** a hearing impairment that is characterized by a distortion of frequency or intensity. **2.** a condition in which certain sounds produce pain or discomfort. Also, **dysacousia, dysacousis, dysacousma.**

dysanalyte see PEROVSKITE.

dysaphia *Neurology.* an impairment or distortion of the sense of touch.

dysarthria *Neurology.* impaired articulation caused by a disturbance of muscular control, typically incurred through damage to the central or peripheral nervous system.

dysarthrosis *Medicine.* deformity or disease of a joint.

dysbarism *Medicine.* the physiologic effects resulting from changes in barometric pressure with the exception of hypoxia; decompression sickness.

dysbasia *Neurology.* a difficulty in walking, such as taking excessively fast or very small steps, generally due to a nervous system disorder.

dyscephaly *Medicine.* malformation of the head and face.

dyschondroplasia see ENCHONDROMATOSIS.

dyschromatopsia *Medicine.* any disorder or defect of color vision. Also, **dyschromasia.**

dyscinesia see DYSKINESIA.

dyscrasia *Medicine.* **1.** a morbid general state resulting from the presence of abnormal material in the blood. **2.** a term formerly used to denote disease.

dyscrasite *Mineralogy.* Ag_3Sb, a silver to tin-white orthorhombic mineral occurring in crystals and, more commonly, massive forms, having a metallic luster, a specific gravity of 9.71 to 9.74, and a hardness of 3.5 to 4 on the Mohs scale; found as a primary mineral in silver ore deposits.

dysentery [dis´in ter´ē] *Medicine.* any of various disorders marked by an inflammation of the intestines (especially of the colon) and attended by pain in the abdomen, tenesmus (straining at defecation), and frequent stools containing blood and mucus; causes include bacteria, protozoa, parasitic worms, and chemical irritants.

dysergia *Neurology.* motor incoordination due to defective nerve impulses.

dysesthesia *Neurology.* **1.** an impairment or distortion of any sense, especially that of touch. **2.** an abnormal sensation in which normal stimuli may produce an unpleasant sensation, such as itching or burning.

dysfunction *Medicine.* any abnormality of the functioning of a body part or organ. Thus, **dysfunctional.**

dysgalactia *Medicine.* abnormal or impaired secretion of milk.

dysgammaglobulinemia *Medicine.* any abnormality in the quantity or quality of serum globulins.

dysgenesia *Medicine.* **1.** the defective development of an organ or part during embryonic development. Also, **dysgenesis. 2.** an impairment of the ability to procreate.

dysgenic *Genetics.* describing a system of breeding that is detrimental to the species or tends to counteract species improvement.

dysgerminoma *Oncology.* a malignant ovarian neoplasm that is thought to be derived from primordial germ cells of the sexually undifferentiated embryonic gonad; the female counterpart of seminoma of the testis. Also, OVARIAN SEMINOMA.

dyshidrosis *Medicine.* any disturbance in sweat production or excretion.

dyskaryosis *Pathology.* an irregular alteration or maturation in the nucleus of a cell, normally observed in exfoliated cells having normal cytoplasm but hyperchromatic nuclei or deviant chromatin arrangement.

dyskinesia *Neurology.* **1.** any abnormality of voluntary movement, such as spasms or irregular and incomplete movements. **2.** bizarre spontaneous movements, such as the intermittent protrusion of the tongue and lips (facial dyskinesia). Also, DYSCINESIA.

dyskinetoplasty *Microbiology.* the absence of a normal kinetoplast.

dyslexia [dis leks´ē ə] *Medicine.* **1.** a genetic disorder, occurring more frequently in males, that is characterized by the inability to read, spell, and write words despite the ability to see and recognize letters; a typical manifestation is the reversal of letters within a word; e.g., *saw* for *was.* **2.** in popular use, a general term for reading difficulties or an inability to read. (From Greek; literally meaning "bad words.")

dyslexic [dis leks´ik] *Medicine.* **1.** a person who is affected by dyslexia. **2.** of or relating to this condition.

dyslogia *Medicine.* an impairment of reasoning powers marked by an inability to express ideas in speech.

dysmenorrhea see MENORRHALGIA.

dysmetria *Neurology.* a condition characterized by an inaccurate sense of distance or an inability to gauge distance for bodily movements.

dysmnesia *Psychology.* a condition of impaired memory. Thus, **dysmnesic.**

dysmorphism *Genetics.* **1.** a congenital malformation. **2.** see ALLOMORPHISM.

dysmorphology *Genetics.* a branch of clinical genetics concerned with the three types of structural defects: malformation, disruption, and deformation.

dysmorphophobia *Psychology.* **1.** a mental disorder in which a physically normal person is preoccupied or obsessed with some imagined defect in appearance or body shape. **2.** an irrational fear of physical deformity or of becoming deformed. Also, **dysmorphic disorder.**

dysnomia *Neurology.* a difficulty in recalling and accurately using words, especially the names of objects; partial nominal aphasia.

Dysodonta *Paleontology.* a suborder of pelecypod mollusks with an almost toothless hinge; related to the Prionodesmacea and widespread in the Paleozoic.

Dyson notation *Organic Chemistry.* a notation system that uses a single line with symbols for chemical elements, functional groups, and various ring formations.

dysopia *Medicine.* defective vision. Also, **dysopsia.**

dysostosis *Medicine.* defective bone formation.

dyspepsia *Medicine.* disturbed digestion; upset stomach.

dysphasia *Medicine.* lack of coordination in speech or inability to understand language, caused by a lesion of the central nervous system.

dysphonia *Medicine.* difficulty or pain in speaking; hoarseness.

dysphoria *Medicine.* a feeling of being ill-at-ease; restlessness.

dysphotic *Biology.* occurring or living where light is very limited, such as in marine depths.

dysplasia *Pathology.* **1.** an abnormality in tissue evolution. **2.** aberrant changes in the size, shape, and organization of adult cells.

dyspnea *Medicine.* difficult or labored breathing, usually associated with serious disease of the heart or lungs. Thus, **dyspneic.**

dyspraxia *Neurology.* a partial loss or impairment of the ability to perform voluntary coordinated movements, with no associated defect in the motor apparatus.

dysprosium [dis prōz´ē əm] *Chemistry.* a metallic element having the symbol Dy, the atomic number 66, an atomic weight of 162.50, a melting point of 1410°C, and a boiling point of 2580°C; a highly magnetic rare-earth (lanthanide) metal used to measure neutron fluxes and in lasers, phosphors, and reactor fuels. (From a Greek word meaning "hard to get at"; so named in 1886 by its discoverer, the French chemist Paul Emile Lecoq de Boisbaudran.)

dysreflexia *Neurology.* an abnormal reflex in response to stimuli.

dysrhythmia *Medicine.* any disturbance of rhythm, e.g., in breathing, speech, or brain waves.

dyssebacia *Medicine.* a scaly macular eruption that occurs primarily on the face, on the scalp, in the interscapular area, in the pubic area, and about the anus.

dyssocial behavior *Psychology.* a behavior pattern that is in conflict with the social or ethical code of the cultural group.

dyssynergia *Neurology.* any of various disturbances of muscle coordination, such as unsteadiness or disorganization of movements.

dystasia *Neurology.* difficulty in standing. Also, **dysstasia.**

dystaxia *Neurology.* difficulty in controlling voluntary movements.

dystetic mixture *Physical Chemistry.* the type of mixture that results when different substances are mixed in such a way as to produce the highest melting point possible.

dysthymia *Psychology.* a mood disorder characterized by chronic despondency and loss of energy and interest.

dystocia *Medicine.* a difficult labor or childbirth.

dystonia *Neurology.* any abnormality of muscle tone or tension.

dystrophic [dis trōf´ik] *Biology.* of or relating to defective, improper, or inadequate nutrition. *Medicine.* relating to or affected by a condition of dystrophy. *Ecology.* **1.** describing a body of fresh water that is deficient in calcium and low in plant nutrients, and that has a bottom covered with undecomposed plant fragments; typically found in acid peat areas. Thus, **dystrophic lake. 2.** describing species that are characteristically found in such waters.

dystrophy *Medicine.* any disorder arising from defective or inadequate nutrition, such as muscular dystrophy. Also, **dystrophia.**

dysuria *Medicine.* difficulty or pain in urination.

Dyticidae *Invertebrate Zoology.* a widespread family of aquatic beetles, predators on insects, fish, and mollusks.

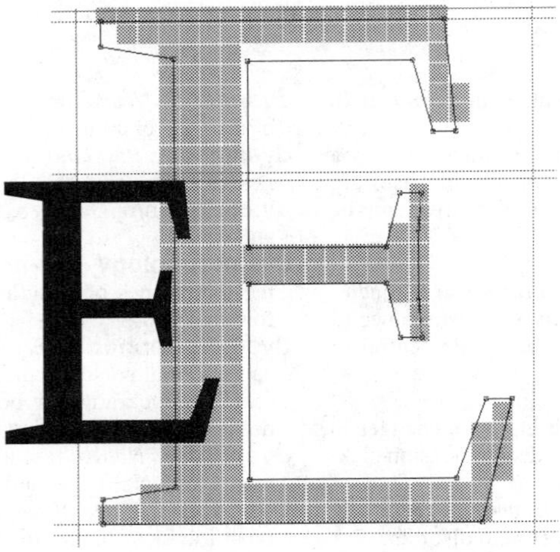

E *Immunology.* a symbol used in complement studies to show the functional lesion at the place where the complement fixation took place. *Computer Programming.* the character that represents the hexadecimal number 14.

E or **E**. eddy diffusivity; einsteinium; electric field vector; electromotive force; energy.

e a symbol for: linear strain; molar extinction coefficient; permittivity.

e *Physics.* a symbol representing the elementary charge: 1.6022×10^{-9} coulomb.

e *Mathematics.* a constant, transcendental real number that is the base of the natural (Napierian) logarithm system; it is approximately equal to 2.71828 and defined uniquely by $\int_1^e x^{-1}dx = 1$ or equivalently, by $d(e^x)/dx = e^x$. It can be shown that

$$e = \lim_{n\to\infty} (1 + 1/n)^n = \sum_{n=0}^{\infty} 1/n!$$

and that $e^{\pi i} = -1$ (Euler's formula), where $i = \sqrt{-1}$.

e. or **e** earth; east; eccentricity; electron; energy; erg; emissivity.

e- a prefix meaning "out of," as in *emission.*

e- *Chemistry.* a symbol meaning: **1.** substituted on the fifth carbon atom of an organic compound. **2.** containing a condensed double aromatic nucleus substituted in the 1.6 positions. **3.** containing an intramolecular bridge.

e^- a symbol for electron.

e^+ a symbol for positron.

EA *Immunology.* a red blood cell with an antibody attached to its surface, used to test the activity of Fc receptors.

EA *Aviation.* the airline code for Eastern Air Lines.

EAC *Immunology.* a symbol used in complement studies in which E represents an erythrocyte, A represents an anti-sheep red blood cell antibody bound to the surface membrane, and C represents a complement.

EADI electronic altitude direction indicator.

eager *Oceanography.* see BORE.

eagle *Vertebrate Zoology.* any of several large diurnal birds of prey of the family Accipitridae, noted for their strength, size, gracefulness, keenness of sight, and powers of flight.

eagle

Eagle *Aviation.* a popular name for the F-15 interceptor fighter aircraft.

Eaglefordian *Geology.* a North American Gulf Coast geologic stage of the Upper Cretaceous period, occurring after the Woodbinian and before the Austinian.

Eagle Nebula *Astronomy.* a cloud of gas and dust associated with a large open star cluster in the constellation Serpens; also designated **M 16, NGC 6611.**

Eagle's medium *Microbiology.* a nutrient medium used in tissue culture, consisting of Earle's or Hanks' balanced salt solution and additional required supplements such as amino acids, vitamins, and serum.

EAI or **E.A.I.** Engineers and Architects Institute.

E and M lead signaling *Telecommunications.* the communication between a trunk circuit and a signaling unit over two leads: an E lead that receives open or ground signals from the signaling unit and an M lead that transmits battery or ground signals to the signaling equipment. (Derived from E for rec<u>e</u>ive and M for trans<u>m</u>it.)

E antigen *Immunology.* a substance on an animal cell which stimulates an immune response to grafts by interaction with an antibody. Also, ENHANCEMENT ANTIGEN.

ear *Anatomy.* the organ of hearing and equilibrium in vertebrates, which in humans consists of the skin-covered cartilaginous external ear on each side of the head that collects sound vibrations, the middle ear in which the vibrations resonate against the tympanic membrane, and the fluid-filled internal ear that maintains balance and conducts the tympanic vibrations to the auditory nerve for transmission to the brain. Also, AURIS.

earache *Medicine.* a pain in the ear that may be sharp, dull, intermittent, or constant; it may be caused by ear disease as well as by infections and by other disorders of the nose, oral cavity, or larynx. Also, OTALGIA, OTODYNIA.

ear drops *Medicine.* a topical liquid medication for the local treatment of certain ear conditions, such as inflammation or infection of the lining of the external auditory canal or impacted cerumen.

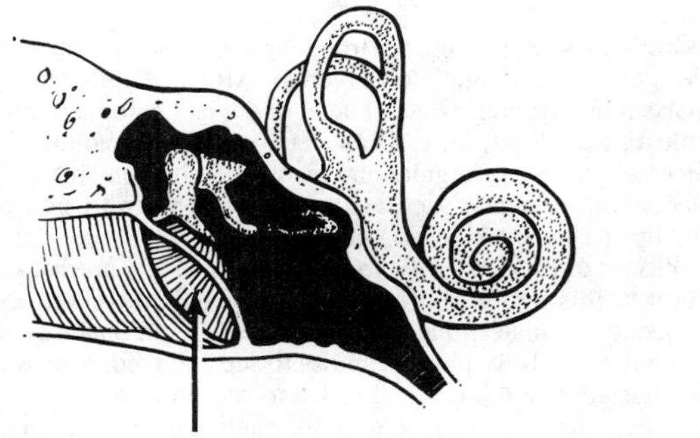

eardrum

eardrum *Anatomy.* the thin membrane that separates the external acoustic meatus from the tympanic cavity of the middle ear. Also, TYMPANIC MEMBRANE, TYMPANUM.

eared seal *Vertebrate Zoology.* any seal of the family Otariidae, which includes sea lions and fur seals, characterized by external ears and flexible hind flippers that are used to move on land.

earlandite *Mineralogy.* $Ca_3(C_6H_5O_7)_2 \cdot 4H_2O$, a white to pale-yellow monoclinic mineral occurring in fine-grained nodules, having a specific gravity of 1.95 and an undetermined hardness; found at the bottom of the Weddell Sea, Antarctica.

earless lizard *Vertebrate Zoology.* any lizard of the family Iguanidae and the genus *Holbrookia,* having no external ear opening or eardrum, thus detecting movement through excellent eyesight and sound vibrations; found in the western U.S. and Mexico.

earless seal *Vertebrate Zoology.* any seal of the family Phocidae, which includes seals that have no external ears and use their hind flippers for swimming; they move on land by wriggling and propelling themselves with their front flippers. Also, TRUE SEAL.

earliest due date *Industrial Engineering.* a dispatching rule that sequences jobs waiting in a queue, giving the highest priority to the job with the lowest order due date.

earliest finish time *Industrial Engineering.* the earliest possible time that a project, or a specific activity of a project, can be completed. Similarly, **earliest start time.**

earliest operation due date *Industrial Engineering.* a dispatching rule that sequences jobs waiting in a queue, giving the highest priority to the job with the lowest operation due date.

earlobe or **ear lobe** *Anatomy.* the fleshy lower portion of the external ear. Also, **earlap.**

early or **Early** *Geology.* occurring near the beginning of an interval of geologic time and corresponding to *lower* as applied to stratigraphic-time units.

early blight *Plant Pathology.* any plant disease, resulting in stunted growth, withering, or death of plant parts, in which the disease symptoms are visible early in the growing season.

Early English style *Architecture.* the earliest phase (c. 1180–1250) of English Gothic architecture.

early genes *Virology.* the first viral genes to be expressed in the replication cycle; they encode enzymes necessary for genome replication and subsequent gene expression.

early infantile autism see AUTISM.

Early Spring *Ordnance.* a satellite weapon system used against reconnaissance satellites.

early-type spiral *Astronomy.* a now generally disused term that stems from the idea that type S0 and SB0 galaxies might have evolved from E0 and other ellipticals and might be evolutionary precursors to other spiral and barred spiral galaxies.

early-type star *Astronomy.* a term referring to stars of spectral types O, B, A, and F0 to F5; the name derives from a former notion that hot stars such as these are at an early stage of their evolution.

early warning *Military Science.* **1.** early notification of the launch or approach of unknown weapons or weapon carriers, especially missiles and aircraft. **2.** of or relating to a system or facility designed to provide and process such information. Thus, **early warning radar, early warning control and reporting post.**

earlywood *Botany.* the first subring in an annual wood-growth ring; the part of the growth ring having larger, thinner-walled cells. Also, SPRING WOOD.

Earnshaw's elementary reaction *Organic Chemistry.* a reaction that has only one transition state, with no intermediates.

EAROM [ē´räm´] *Computer Technology.* a read-only memory in which selected locations can be reprogrammed for a limited number of times by the application of an electric field; used on IBM computers. (An acronym for electrically alterable read-only memory.)

earphone *Acoustical Engineering.* **1.** see HEADPHONE. **2.** any sound receiver that fits on, over, or in the ear, as in a telephone, hearing aid, or portable radio.

earpiece *Acoustical Engineering.* a small earphone that fits into the outer ear.

earplug *Engineering.* an object that is designed to fit into a person's ear opening and to protect the ear from such effects as loud noise, unwanted noise, or water.

ear-protector *Engineering.* a piece that fits into or onto a person's ear to protect it from loud noises that may adversely affect hearing. Thus, **ear-protective.**

ear rot *Plant Pathology.* any disease of corn caused by a variety of fungi and characterized by the deterioration and molding of the ears of the corn.

Earth

earth *Astronomy.* also, **Earth.** the planet inhabited by humans; the third planet from the sun, having an equatorial diameter of 7926 miles, a polar diameter of 7900 miles, and one satellite; its mean distance from the sun is 92.9 million miles and its period of revolution is 365.26 days. *Geology.* the loose, fragmented material that composes part of the surface of this planet. *Agronomy.* any soil that can be cultivated.

earth-centered ellipsoid *Geodesy.* a reference ellipsoid having a geometric center coinciding with the earth's center of gravity, and a semiminor axis coinciding with the earth's rotational axis.

earth-crossing asteroid *Astronomy.* an asteroid whose orbit currently crosses that of the earth, or will do so one day under the gravitational influence of the planets.

earth-current storm *Geophysics.* any variable fluctuations in an earth-current in the earth's crust; they can be as large as several volts per km.

earthenware *Materials.* a slightly porous, opaque container or decorative piece made of low-fired clay and covered with a nonporous glaze.

earthfill dam *Hydrology.* a dam constructed primarily of earth, such as the Fort Peck Dam on the Missouri River, the world's largest such dam.

earthfill dam (Fort Peck Dam)

earth-fixed coordinate system *Cartography.* any coordinate system in which the axes are stationary with respect to the earth.

earthflow *Geology.* a combination of slump and mudflow.

earth hummock *Geology.* a small, dome-shaped, uplifted mound of fine-textured earth material in a permafrost region, covered by vegetation and produced by groundwater pressure. Also, **earth mound.**

earth inductor *Engineering.* an inclinometer that measures the dip angle of the earth's magnetic field. Also, DIP INDUCTOR, INDUCTION INCLINOMETER.

earth inductor compass *Geodesy.* a compass depending for its indications on the current generated in a coil revolving in the earth's magnetic field.

earth-layer propagation *Geophysics.* the transmission of electromagnetic waves through the layers of the earth or its atmosphere.

earthlight or **earthshine** *Astronomy.* a glow on the unilluminated part of the Moon caused by sunlight reflecting off the earth.

earth movement *Geophysics.* any motion of the earth, including revolutions around the sun, rotation on its axis, and movement of the crust in relation to the mantle.

earthmover *Mechanical Engineering.* any of various large vehicles, such as a bulldozer, used to excavate and transport earth.

earthnut *Botany.* any of various roots, tubers, or underground growths, such as the peanut and the truffle, or the plants producing them.

earth orbit *Astronomy.* the elliptical path of the earth around the sun.

earth oscillation *Geophysics.* any periodic fluctuation of the earth, its oceans, or its atmosphere.

earth pig see AARDVARK.

earth pillar *Geology.* a tall column of unconsolidated or semiconsolidated earth, developed by erosion of horizontal strata in regions where most rainfall is concentrated during a short period of time. Also, FAIRY CHIMNEY, HOODOO, PENITENT.

earthquake *Geophysics.* a movement or trembling of the earth caused by a sudden release of stresses within, usually less than 25 miles below the surface; major earthquakes cause significant damage.

earthquake focus *Geophysics.* the specific point within the earth from which an earthquake originates.

earthquake intensity *Geology.* a relative measure of an earthquake's effect on people and/or structures at a particular location, based on the magnitude of the earthquake, the distance from the epicenter, and the local geology.

earthquake magnitude *Geophysics.* a measure of an earthquake's strength or the quantity of strain energy it releases, based on seismographic data.

earthquake swarm *Geophysics.* a series of minor earthquakes that occur in a limited area and time, and that are not identified with a major earthquake.

earthquake zone *Geology.* a well-defined region of the earth's crust in which movements of the crust occur, sometimes coinciding with volcanic activity. Also, SEISMIC AREA.

earth-rate correction *Navigation.* an adjustment made in an inertial guidance system for the effect of the earth's rotation on the platform leveling gyroscopes.

Earth Resources Technology Satellite *Space Technology.* the former name of **Landsat,** an artificial satellite used to gather data concerning agriculture, land use, and natural resources.

earth-rotation synthesis *Astronomy.* an observational technique in radio astronomy that uses the displacement of an antenna (caused by the earth's rotation) to simulate the angular resolving power of a much larger antenna.

earth satellite *Astronomy.* any natural body, such as the moon, that revolves about the earth. *Space Technology.* an artificial satellite launched into orbit about the earth.

earth science *Science.* any of the sciences concerned with the earth's origin, composition, and physical features, including geology, geography, meteorology, mineralogy, and oceanography.

earth shadow *Meteorology.* **1.** any shadow projecting from mountain peaks into a hazy atmosphere at sunrise or sunset. **2.** the shadow of the earth in interplanetary space.

earth-spherop *Geodesy.* the equipotential surface obtained by assigning to the spheropotential function a value such that the volume enclosed by the resulting surface is the same as the volume enclosed by the geoid.

earthstar *Mycology.* a fungus belonging to the family Geastracea that occurs on soil and has an outer covering that splits into the shape of a star.

earth station *Telecommunications.* a space communications station that is located either on the earth's surface (land or sea) or in the air; maintains communication with spacecraft and contains all radio and support equipment for communication with, and control of, a satellite. Also, GROUND STATION.

earth thermometer see SOIL THERMOMETER.

earth tide *Geophysics.* the periodic movements of the earth's crust caused by the gravitational pull of either the sun or the moon.

earth tongue *Mycology.* any of a group of fungi belonging to the genera *Geoglossum* and *Trichoglossum,* characterized by tongue-shaped fruiting bodies; found on damp soil and decaying logs.

earth wax see OZOCERITE.

earthwork or **earthworks** *Archaeology.* an early structure built from a mound or bank of earth. *Ordnance.* a similar structure that is constructed for use as a temporary or permanent fortification.

earthworm *Invertebrate Zoology.* the common name for various segmented, burrowing members of the class Oligochaeta, primarily of the genus *Lumbricus,* that burrow in soil and feed on soil nutrients and decaying organic matter.

earthy *Geology.* composed of or resembling soil. Also, **earthen.** *Mineralogy.* having a dull luster and sometimes rough to the touch.

earthy calamine see HYDROZINCITE.

ear wax or **earwax** *Physiology.* a waxy, yellowish or brownish secretion produced by certain glands of the external ear canal; cerumen.

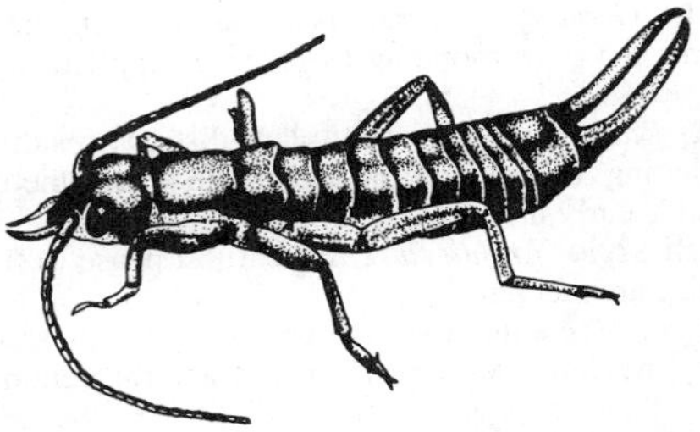

earwig

earwig *Invertebrate Zoology.* the common name for a member of the insect order Dermaptera that is characterized by a large pair of pincers at the rear of its body and is found under stones, in decayed tree bark, and in moist places; so named because of the mistaken belief that it enters a sleeping person's ear.

EAS extended area service; equivalent airspeed.

easel *Graphic Arts.* **1.** an upright stand used to display an object or to support a material being used by an artist. **2.** a form of looseleaf binding that allows the book to be stood upright for display.

easement *Civil Engineering.* a limited property right held by someone other than the principal owner of the property; for example, the right of a landowner to pass through a neighbor's property to gain access to a public road from his land, established by his historic use of this passage.

easin *Organic Chemistry.* $C_{20}H_6I_4Na_2O_5$, a brown powder that is insoluble in water; used as a dye and as a strong acid pH indicator. Also, SODIUM TETRAFLUORESCEIN, ERYTHROSINE MONOHYDRATE.

East Africa Coast Current *Oceanography.* a seasonal current off the Somali coast that is fed during the Northern Hemisphere summer by part of the South Equatorial Current and flows northeastward as a strong, narrow current to feed the Monsoon Current in the northwest Indian Ocean; it almost disappears in winter, reversing to a broad, weak southwestward flow down the Somali coast.

East African sleeping sickness see RHODESIAN TRYPANOSOMIASIS.

East African subregion *Ecology.* a distinct zoogeographical region that includes the equatorial region of Africa and also all of East Africa north of the South African subregion.

East Australia Current *Oceanography.* a broad, strong current that flows south and southeastward along the east coast of Australia between 20° and 45°S, then northward along the west coast of New Zealand.

East China Sea *Geography.* an arm of the western Pacific, bounded by China, Japan, the Ryukyu Islands, and Taiwan.

East Coast fever *Veterinary Medicine.* an acute disease of cattle that is transmitted by ticks and is caused by *Theileria parva*; characterized by fever, swelling of the lymph nodes, respiratory distress, diarrhea, emaciation, and high mortality. Also, THEILERIASIS.

Easter daisy *Botany.* a composite plant, *Townsendia exscapa,* having stalkless purplish or white flowers in a rosette of narrow leaves and found in regions of the Rocky Mountains; so named because its flowers appear at Easter time.

Easter-egging *Electronics.* a method for inspecting electronic equipment in which no particular system or procedure is followed, not unlike the approach taken by children on an Easter egg hunt.

Easter lily *Botany.* the white lily, *Lilium longiflorum,* that has become a symbol of the Easter holiday; native to Taiwan and now widely cultivated elsewhere; this tall plant with pointed leaves is brought artificially into bloom in early spring.

eastern equine encephalomyelitis *Medicine.* a viral disease occurring in horses but communicable to humans, in whom it produces flulike symptoms, with little effect on the medulla or spinal cord. Occurs along the eastern coast of the United States.

Eastern Hemisphere *Geography.* the half of the earth lying east of the Atlantic, including Europe, Asia, Africa, and Australia; usually charted eastward from 20°W to 160°E.

eastern hemlock *Botany.* a hemlock tree, *Tsuga canadensis,* characterized by horizontal branches that often droop to the ground; found in eastern North America, it is the state tree of Pennsylvania.

Eastern mixed forest

Eastern mixed forest *Ecology.* a North American wildlife community in which spruces and other northern conifers are mingled with maples and beeches; found from the Midwestern prairie east to Nova Scotia in pre-European times and now typified by forest preserves of the Catskill, Green, or Allegheny Mountains.

East Greenland Current *Oceanography.* a cold current that carries water of low salinity south and southwestward through the Denmark Strait between Iceland and Greenland.

easting *Cartography.* a linear distance eastward from the vertical grid line that passes through the origin of a grid system. *Navigation.* specifically, the distance a craft makes good to the east.

east point *Cartography.* the point on the horizon, to the right of an observer facing north, that is intersected by the prime vertical plane.

east-west effect *Astrophysics.* an effect in which fewer cosmic rays arrive at Earth from an easterly direction than from a westerly direction because of their predominantly positive electric charge and bending in the Earth's magnetic field.

Eaton agent pneumonia *Medicine.* a contagious disease of children and young adults caused by the bacterium *Mycoplasma pneumoniae* and characterized by dry cough and fever. Also, MYCOPLASMAL PNEUMONIA.

Eaton-Lambert syndrome. *Medicine.* a form of myasthenia that tends to be associated with lung cancer.

Eaton's agent *Microbiology.* the original term for the microbe *Mycoplasma pneumoniae.*

eave-lead *Building Engineering.* a lead gutter that lies behind a parapet around the edge of a building roof.

eaves *Building Engineering.* the lower projection of a roof beyond the face of constructed walls of a building or structure.

eaves board *Building Engineering.* a wood strip that is laid below a doubling course to tilt it up so that slates or tiles rest properly in the roof frame. Also, TILTING FILLET, SKEW FILLET.

eaves course see DOUBLED-COURSE.

eaves fascia see FASCIA.

eaves gutter *Building Engineering.* a trough that lies below an eave and functions as a catch basin to direct the flow of rainwater away from the roof. Also, SHUTING.

eaves molding *Building Engineering.* a cornice-shaped molding that lies beneath the eaves of a building.

eaves soffit *Building Engineering.* the horizontal surface beneath an eave.

eave trough *Building Engineering.* see GUTTER.

E.B. elementary body.

ebb *Oceanography.* the flowing back of the tide as the water returns to the sea.

ebb-and-flow structure *Geology.* a sedimentary rock structure that is characterized by alternating layers of horizontal bedding and cross-bedding and is thought to have been produced by the ebb and flow of tides.

ebb current *Oceanography.* the tidal current caused by the falling of the tide, usually settling away from shore or down a tidal stream.

ebb interval *Oceanography.* the amount of time between the transit (upper or lower) of the moon over the local or Greenwich meridian and the strength of the next ebb current.

ebb strength *Oceanography.* the ebb tidal current at the time of its greatest speed.

ebb tide *Oceanography.* the phase of the tide between a high water and the following low water.

EBCDIC *Computer Programming.* a binary code in which each character is represented by eight bits, allowing a maximum of 256 distinct characters. (An acronym for extended binary coded decimal interchange code.)

Ebenaceae *Botany.* a family of dicotyledonous, mainly tropical trees and shrubs in the order Ebenales, characterized by the absence of a latex system; a monopodial crown; simple, alternate, coriaceous leaves; usually regular, unisexual flowers in a solitary, axillary, or cymose arrangement; a superior ovary; and a calyx with fused lobes. The sole species in the eastern United States is the persimmon.

Ebenales *Botany.* an order of dicotyledonous trees and shrubs in the subclass Dilleniidae, characterized by simple, usually alternate leaves, sympetalous flowers that are regular and unisexual or bisexual, and a superior ovary.

E bend *Electromagnetism.* a smooth waveguide bend in which the axis of the waveguide remains in the plane that contains the polarization.

Eberhard effect *Graphic Arts.* an image-border defect in photography caused by insufficient agitation of developing solution and manifested as a light line along the edges of low density and a dark line along the edges of high density.

Eberth, Carl [ā´burt] 1835–1926, German virologist; discovered the typhoid (**Eberth-Gaffky**) bacillus.

Ebert ion counter *Engineering.* an aspiration condenser-type ion counter that measures the movement and concentration of small ions in the atmosphere.

EBIS electron-beam ion source.

EBL explanation-based learning.

EBM electron beam machining.

Ebola virus *Virology.* an RNA virus that occurs in the Sudan and adjacent areas of Zaire, causing an acute, highly fatal hemorrhagic fever in humans.

ebonation *Surgery.* the removal of bone fragments from a wound.

ebonite see VULCANITE.

ebony *Botany.* a tree of the genus *Diospyros* in the family Ebenaceae, found in tropical Asia and Africa. *Materials.* the hard, dark wood of this tree, or similar dark woods in other families, used in making furniture, sculpture, and jewelry.

EBR electron beam recording.

ebracteate *Botany.* lacking bracts or having reduced leaves.

ebracteolate *Botany.* lacking bracteoles or bractlets.

ébranlement *Surgery.* the removal of a polyp by twisting it until its pedicle ruptures.

Ebriaceae *Botany.* a monospecific family of marine biflagellates of the order Ebriales that are unarmored and have a complex three-dimensional internal skeleton, with three arms radiating at equal angles.

Ebriales *Botany.* the order of marine biflagellates that makes up the subclass Ebriophycidae.

Ebriida *Invertebrate Zoology.* an order of flagellated protozoans in the class Phytamastigophora, having solid, siliceous skeletons.

Ebriophycidae *Botany.* a subclass of marine biflagellates of the class Dinophyceae, characterized by a three-dimensional internal skeleton composed of silica and nonphotosynthetic phagotrophic cells, that have been observed to eat diatoms.

Ebstein's anomaly *Cardiology.* a congenital lesion of the tricuspid valve such that part of the inflow portion of the right ventricle is common with the right atrium.

ebullating-bed reactor *Chemical Engineering.* a kind of fluidized bed in which catalyst particles are suspended by the upward movement of the liquid.

ebulliometer *Physical Chemistry.* an instrument that measures the precise boiling points of solutions. Also, **ebullioscope.**

ebulliometry *Physical Chemistry.* a method for the precise measurement of a solution's boiling point. Also, **ebullioscopy.**

ebullition *Physics.* the rapid conversion of liquid to vapor by the violent eruption of bubbles, occurring when the temperature is such that the saturated vapor pressure of the liquid equals the pressure of the atmosphere.

eburnation *Pathology.* an abnormal condition in which bone becomes dense and hard.

EBV or **EB v.** Epstein-Barr virus.

EBW electron beam welding.

ec- a combining form meaning "out of," as in *eccentric.*

EC a term for blank powder. (From the phrase "empty cartridge," in which the blank or smokeless powder is chiefly used.)

ecad *Ecology.* an organism that is specifically, and occasionally uniquely, adapted to its habitat, with the characteristic adaptations not being inheritable.

ECC earth continuity conductor.

eccentric *Science.* **1.** located away from a center. **2.** of two circles, having different centers. **3.** of a body in motion, deviating from an orbit or path. *Mechanical Devices.* see ECCENTRIC ROD.

eccentric angle *Mathematics.* the angle measured at the center of an ellipse (or hyperbola) from the line passing through the foci to the line from the center to a point on the ellipse (or hyperbola). Equivalently the angle ϕ that appears in the parametric equations of the ellipse or hyperbola in normal form; namely, $x = a \cos \phi$, $y = b \sin \phi$ or $x = a \sec \phi$, $y = b \tan \phi$, respectively.

eccentric anomaly *Astronomy.* for an object moving in an unperturbed elliptic orbit, the angle (measured at the center of the ellipse) between the periapsis point and the point on a circumscribed circle where a line perpendicular to the orbit's major axis would pass through the orbiting body.

eccentric bit *Mechanical Devices.* a chisel-like drill bit in which the cutting edge extends further on one side than the other.

eccentric cam *Mechanical Devices.* a revolving disk and shaft assembly with the axis of rotation displaced from its geometric center.

eccentric contraction *Medicine.* a contraction in which the muscle fibers lengthen as the muscle responds to a resistance, allowing itself to be stretched.

eccentric gear *Mechanical Devices.* a gear with the axis of rotation offset from its geometric center.

eccentric implantation *Cell Biology.* the embedding of the blastocyst within a fold of the uterine wall, which then seals off from the main cavity.

eccentricity *Astronomy.* a number that defines how much an ellipse departs from a circle, given by $c/2a$, where a is the length of the ellipse's major axis and c is the distance between the two foci. *Mathematics.* in analytic geometry, the constant positive number that is the ratio of the distance of a point on a given conic from the focus and the distance from the directrix. The type of conic depends on the eccentricity; an ellipse has eccentricity less than 1; a parabola has eccentricity equal to 1; and a hyperbola has eccentricity greater than 1.

eccentric load *Engineering.* a load on a structure or machine part that is applied away from the central point of the structure.

eccentric orbit *Astronomy.* a closed orbit that is not a circle; for planets, the sun is at one of the two foci.

eccentric reduction *Cartography.* the correction that must be applied to a direction observed by an eccentric instrument or signal to reduce the observed value to what it should have been with no eccentricity.

eccentric ring structure *Astronomy.* the planetary rings exhibiting noncircular orbits, discovered by the Voyager spacecraft at Saturn and Neptune.

eccentric rod *Mechanical Devices.* a rod or bar that connects to an eccentric strap to transmit the motion of the strap to a target device such as a valve or pump.

eccentric-shaft press *Mechanical Devices.* a punch press with an eccentric shaft that operates by applied pressure to the slide.

eccentric sheave *Mechanical Devices.* the two-halved, off-center disk of a cam integrated with a crankshaft or axle from which the desired eccentric action is generated.

eccentric station *Cartography.* a survey point that is not in the same vertical line with the station it represents and for which the observations will be adjusted before they are combined with observations at other stations. Thus, **eccentric signal.**

eccentric strap *Mechanical Devices.* a narrow, split bearing or hoop placed around an eccentric sheave belted to the large end of a connecting rod to transmit motion through the valve rods to the valve gearing.

eccentric valve *Engineering.* a valve having a rotating cut-off body attached to one side away from the center, which serves to reduce the wear on the valve's moving parts.

ecchymosis *Medicine.* a technical term for a bruise. See BRUISE.

Eccles, Sir John Carew [ek´əlz] born 1903, Australian physiologist; shared Nobel Prize for study of nerve cells and synapses.

ECCM electronic counter-countermeasure.

Eccrinaceae *Mycology.* a family of fungi belonging to the order Eccrinales, characterized by multinucleate and nucleate sporangiospores.

Eccrinales *Mycology.* an order of fungi belonging to the class Trichomycetes, characterized by unbranched, coenocytic thalli which produce sporangiospores; it occurs in the guts of crustaceans and insects.

eccrine gland *Physiology.* a type of sweat gland in the corium of the skin that responds to elevated body temperatures by secreting a clear fluid that evaporates and cools the body.

ECCS emergency core cooling system.

eccyesis *Medicine.* an ectopic or extrauterine pregnancy.

ecdemic *Pathology.* of or relating to a disease that is observed far away from the area in which it originates.

ecdemite *Mineralogy.* $Pb_6As_2^{+3}O_7Cl_4$, a bright yellow to green, easily fusible tetragonal mineral occurring as crystals and as foliated masses and incrustations, having a specific gravity of 7.4 and a hardness of 2.5 to 3 on the Mohs scale. Also, EKDEMITE.

ecdysis *Invertebrate Zoology.* the shedding of the outer cuticle or skin during molting in insects and crustaceans. Also, MOLTING.

ecdysone *Entomology.* a steroid hormone secreted by the insect prothoracic gland that acts upon the epidermis and stimulates growth and molting.

ecdysteroid *Biochemistry.* any of various polyhydroxylated sterols regulating insect molting; also obtained from certain ferns and used as an anabolic hormone to increase vigor, especially after chemotherapy.

E cell *Electricity.* a timing mechanism that converts the current-time integral of an electrical function into an equivalent mass integral up to a maximum of several thousand microampere-hours.

ecesis *Ecology.* the successful invasion and colonization of a new habitat by migrating plants.

ECF extended care facility; extracellular fluid.

ECG electrocardiogram; electrocardiograph.

ecgonine *Organic Chemistry.* $C_9H_{15}NO_3$, plates obtained from the hydrolysis of cocaine; very soluble in water and soluble in alcohol; melts at 203°C; used as a topical anesthetic.

echelette grating *Spectroscopy.* a diffraction grating designed for the infrared region, consisting of a series of ridges or grooves on a plane surface. Also, **echellette.**

echelle grating *Spectroscopy.* a diffraction grating used to obtain high resolution and intensity at illumination angles greater than 45°.

echelon [esh´län´; esh´ə län´] *Military Science.* **1.** a military unit or level, such as a subdivision of a headquarters or a level of command. **2.** a steplike formation in which subdivisions are arranged behind each other with equal spacing.

echelon faults *Geology.* separate faults having parallel, steplike trends in a more-or-less general direction, with the individual faults parallel to one another and at an angle to that direction; believed to result from torsion in a region of differential diastrophism.

echelon grating *Spectroscopy.* a high-resolution diffraction grating consisting of approximately 20 plane-parallel plates, each extending slightly beyond the next.

Echeneidae *Vertebrate Zoology.* the remoras, a family of marine fish of the order Perciformes, characterized by a flat disk on top of the head, used to suck on to the bodies of larger fish such as sharks or marlins.

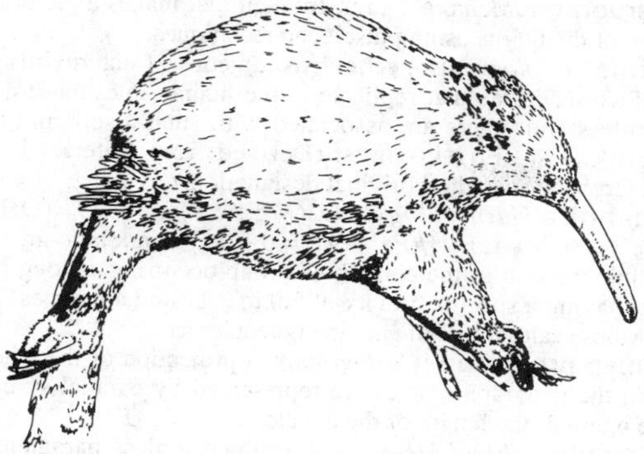

echidna

echidna [i kid´nə] *Vertebrate Zoology.* a genus of toothless burrowing monotremate mammals of the family Echidnidae, which have claws and a slender snout, eat primarily ants, and are covered with coarse hair, similar to but slightly larger than a hedgehog; found in Australia, Tasmania, and New Guinea. Also, SPINY ANTEATER.

echidnotoxin *Toxicology.* a poison that is found in viper venom.

echin-. a combining form meaning "spine," as in *echinate.*

Echinacea *Invertebrate Zoology.* a superorder of echinoids in the order Euechinoidea, including sea urchins.

echinate *Zoology.* covered with spines or prickles.

Echinidae *Invertebrate Zoology.* a family of sea urchins in the order Echinoida, with worldwide distribution.

Echiniscoidea *Invertebrate Zoology.* a suborder of the phylum Tardigrada, the water bears.

echino- a combining form meaning "spine," as in *Echinodermata.*

echinococcosis *Medicine.* the infection of a human, usually in the liver, with the parasitic tapeworm *Echinococcus granulosus* in its larval stage; dogs are the principal hosts of the adult worm.

Echinococcus *Invertebrate Zoology.* a genus of tapeworms that includes the medically important cattle, sheep, and dog tapeworms that often infect humans.

echinococcus cyst *Invertebrate Zoology.* a cyst formed in host tissue by the larva of the tapeworm *Echinococcus.*

Echinocorys *Paleontology.* a genus of irregular burrowing holasteroidal echinoids of the subclass Euechinoidea; Upper Cretaceous.

Echinocystitoida *Paleontology.* an order of echinoids in the subclass Perischoechinoidea, widespread from the Ordovician to the Permian and notable for the many genera with flattened flexible tests.

Echinodera *Invertebrate Zoology.* a group of minute, free-living marine animals. Also, KINORHYNCHA.

echinoderm [i kīn´ə durm] *Invertebrate Zoology.* any member of the phylum Echinodermata.

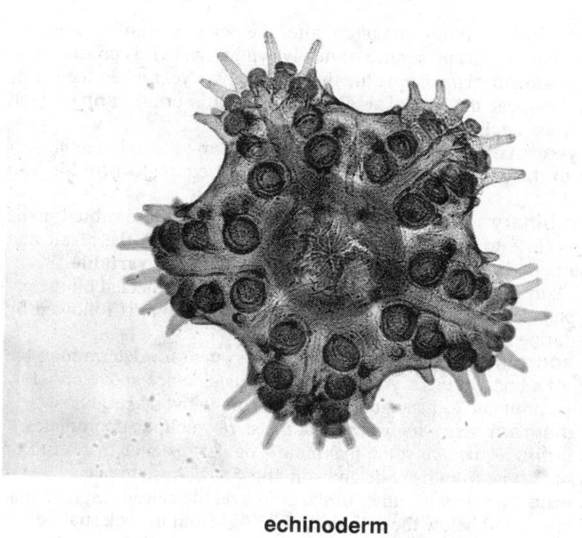

echinoderm

Echinodermata *Invertebrate Zoology.* a phylum of usually radially symmetrical marine animals with a calcareous exoskeleton, including starfish, sea urchins, sea lilies, and sea cucumbers.

Echinodiaceae *Botany.* a monogeneric family of robust, dark-green mosses of the order Hypnobryales that form loose trailing mats on aboveground roots, tree bases, soil, and wet rocks; characterized by creeping stems with branching and ascending ends that are coarse and rigid.

Echinodontiaceae *Mycology.* a family of fungi belonging to the order Agaricales, occurring primarily on woods; some species are ground and used to make dyes.

echinoid *Invertebrate Zoology.* any member of the class Echinoidea.

Echinoida *Invertebrate Zoology.* an order of sea urchins with a smooth shell, mainly sea urchins and sand dollars.

Echinoidea *Invertebrate Zoology.* the sea urchins, a class of echinoderms with a rigid, globular or disk-shaped shell.

Echinometridae *Invertebrate Zoology.* a family of sea urchins, including some rock-boring species; widely distributed on coral reefs.

echinomycin *Microbiology.* an antimicrobial and antitumor agent that inhibits DNA-directed RNA transcription. Also, QUINOMYCIN A.

echinopluteus *Invertebrate Zoology.* the bilaterally symmetrical, free-swimming larva of sea urchins.

echinopsine *Organic Chemistry.* $C_{10}H_9NO$, an alkaloid obtained from *Echinops*; needlelike crystals that are slightly soluble in water and very soluble in alcohol; melts at 152°C; it has a physiological action similar to that of strychnine.

Echinosteliaceae *Mycology.* a family of slime mold fungi belonging to the order Echinosteliales.

Echinosteliales *Mycology.* an order of slime mold fungi belonging to the subclass Myxogastromycetidae, which are characterized by the formation of tiny spores.

echinostome cercaria *Invertebrate Zoology.* the larval form of certain parasitic trematode flatworms.

Echinostomida *Invertebrate Zoology.* an order of parasitic spiny flat worms in the phylum Platyhelminthes, trematodes with a life cycle involving two or three hosts.

Echinothuriidae *Invertebrate Zoology.* a family of deepsea echinoderms with a flexible test that collapses to a disk at normal atmospheric pressure.

Echinothuroida *Invertebrate Zoology.* an order of sea urchins in the superorder Diadematacea.

Echinozoa *Invertebrate Zoology.* a subphylum of free-living echinoderms including sea urchins, sand dollars, and sea cucumbers.

Echiurida *Invertebrate Zoology.* a phylum of unsegmented marine worms with a proboscis over the mouth, found in muddy and sandy sea bottoms. Also, **Echiuroidea.**

echo *Acoustics.* a sound wave that is returned by reflection from a body or other obstacle, with sufficient time delay to distinguish it from the original direct signal. *Electronics.* **1.** a signal that reflects back to its source. Also, RETURN. **2.** the interference on a telephone line that arises from poor impedance matching or from two callers using the system at the same time. **3.** see RADAR ECHO. *Physics.* a reflected signal that is detected with sufficient delay to distinguish it from a directly received signal. *Computer Programming.* **1.** to display characters on a printer or display screen as they are typed on a keyboard. **2.** to print or display the inputs to a program. Also, ECHO-PRINT.

Echo *Space Technology.* any of a series of passive communications satellites consisting of large (100–135 ft diameter) inflated balloons made of aluminum-coated Mylar; first launched in 1960 to reflect microwave signals from point to point on earth.

echo amplitude *Electronics.* a measurement of the strength of a radar target signal by measuring the amplitude of the echo waveform.

echo area *Electromagnetism.* the effective area of a target reflecting the same amount of energy back to the radar as the actual target.

echo attenuation *Electronics.* the amount of power produced by an output terminal in a transmission line divided by the power that is reflected back to the output terminal.

echo box *Electronics.* a device that simulates radar echo by slowly releasing the energy received from a radar pulse signal into the radar's receiving system in order to test or tune the equipment.

echocardiogram *Medicine.* the record produced by echocardiography.

echocardiography *Medicine.* the delineation of heart structure and motion by means of ultrasonic waves passed through the heart and reflected backward, or echoed, when they pass from one type of tissue to another. Also, ULTRASONIC CARDIOGRAPHY.

echo chamber *Acoustics.* a specially designed small room with characteristics that provide controlled sound reflection, used to create special echo effects in recordings and broadcasts.
echo check *Computer Programming.* a method of verifying the correctness of transmitted data by returning the data to the source and comparing it with the original data.
echo contour *Electronics.* a pattern formed on a radar screen that indicates the intensity of the echo signal.
echo eliminator see ECHO SUPPRESSOR.
echoencephalogram *Medicine.* the record produced by echoencephalography.
echoencephalograph *Medicine.* the instrument employed in echoencephalography to study brain tissue.
echoencephalography *Medicine.* a diagnostic technique in which ultrasonic waves are beamed through the head from both sides and echoes are recorded as graphic tracings.
echo frequency *Electronics.* the number of oscillations in a radar target's signal that occurs within a given time frame.
echogenic *Radiology.* producing reflections of ultrasound waves.
echogram *Oceanography.* a graphic representation of a series of echo soundings, producing a continuous profile of the bottom.
echograph *Engineering.* a device that creates a graphic representation of echo soundings that originate on the sea floor.
echographia *Neurology.* a type of aphasia characterized by an inability to express one's thoughts in writing, though able to copy printed words.
echography *Medicine.* a diagnostic aid in which ultrasonic waves are directed at the tissues and a record is made of the sound waves reflected back through the tissues. Also, ULTRASONOGRAPHY.
echoic memory *Psychology.* a memory system that retains a sensory representation of a sound for a short time.
echo intensity *Electronics.* the degree of brightness displayed by a radar echo signal; generally proportional to the voltage of signal received by reflection from the target.
echolalia *Medicine.* the automatic and meaningless repetition of another's words or phrases, especially seen in schizophrenia.
echo location see ECHO RANGING.
echolucent *Radiology.* allowing passage of ultrasonic waves without generating echoes.
echo matching *Engineering.* the positioning of an antenna to create equal pulse-indications of an echo-splitting radar.
echoplex *Computer Programming.* a technique in which characters are automatically echoed to the transmitting terminal and are displayed on the screen for verification.
echo power *Electronics.* the electrical strength of a radar signal received by reflection from the target, generally measured in watts or dBm (decibels referred to 1 milliwatt).
echopraxia *Medicine.* the abnormal imitation of the actions of another person, a behavior exhibited by some schizophrenic patients.
echo-print *Computer Programming.* see ECHO, def. 2.
echo pulse *Electronics.* the energy that reaches a radar system after the signal is reflected from a target.
echo ranging *Engineering.* the determination of the distance and orientation of underwater objects by the same process that is used in sonar. Also, ECHO LOCATION.
echo-ranging sonar *Engineering.* the process whereby the distance and direction of objects is determined by the reception of the reflection of an ultrasonic pulse under water.
echo recognition *Engineering.* the determination of a sonar reflection from a target, rather than from other reflectors.
echo repeater *Acoustical Engineering.* an electronic device that can create the effect of multiple echoes by means of delay and feedback.
echosonogram *Engineering.* a graphic representation or display created with ultrasound pulse-reflection techniques.
echo sounder see SONIC DEPTH FINDER.
echo sounding *Oceanography.* a process of measuring water depth by emitting a sonic or ultrasonic signal and measuring the time it takes for the signal's echo to return from the sea floor.
echo-splitting radar *Engineering.* a type of radar that produces two echo indications on the radarscope screen by splitting the echo with special circuits.
echo suppressor *Electronics.* a device that temporarily quiets navigational equipment after a pulse has been received, to prevent the reception of delayed pulses. Also, ECHO ELIMINATOR.
echo talk *Computer Technology.* the interference created when data is echoed back to its source while the source is still transmitting.
echouterograph *Medicine.* an instrument that makes a pictorial representation of the uterus using pulse-echo techniques.
ECHOvirus *Virology.* also, **echovirus.** a genus of enteroviruses of the family Picornaviridae that replicate in the human intestinal tract; some are harmless and others are associated with such disorders as aseptic meningitis. Also, ENTEROVIRUS. (Derived from Enteric Cytopathic Human Orphan virus, their original designation.)
eckermannite *Mineralogy.* $Na_3(Mg,Fe^{+2})_4AlSi_8O_{22}(OH)_2$, $Mg/(Mg+Fe^{+2}) = 0.5 - 1$, $Fe^{+3}/(Fe^{+3}+Al) = 0 - 0.5$, a dark blue-green, brittle monoclinic mineral of the amphibole group occurring as long prismatic crystals, having a specific gravity of 3.0 to 3.17 and a hardness of 5 to 6 on the Mohs scale; found in alkaline igneous rocks.
Eckert map projection *Cartography.* a projection of the whole earth in which the geographic poles are represented by parallel straight lines that are one-half the length of the equator.
Eckert number *Fluid Mechanics.* a dimensionless parameter that is used in the study of compressible fluid flow around an obstacle, given by the square of the steady-state fluid velocity divided by the product of the specific heat at constant temperature and the difference between the temperature of the obstacle and the temperature of the fluid.
ECL emitter-coupled logic.
eclampsia *Medicine.* a disease, usually occurring during the latter half of pregnancy, marked by grand mal convulsions, coma, high blood pressure, water retention, and protein in the urine.
eclectic [i klek´tik] *Psychology.* of or relating to eclecticism.
eclecticism [i klek´tə siz´əm] *Psychology.* an approach to psychotherapy that uses techniques and perspectives drawn from various theories to suit the treatment to the specific individual. Also, **eclectic method.**
eclipse *Astronomy.* any obscuring of one celestial body as it passes into the shadow of another, such as the moon's blocking the sun from earth.

annular eclipse

eclipsed confirmation *Physical Chemistry.* a high-energy confirmation of atoms that results from rotation around a single bond in a molecule.
eclipse period *Virology.* the phase after the penetration of a virus into a cell and before the appearance of newly synthesized virus products.
eclipse seasons *Astronomy.* the times in a given year when the earth-sun line lies near the line of nodes of the moon's orbit, during which lunar and solar eclipses can occur.
eclipse year *Astronomy.* the interval between two successive returns of the sun to the same node of the moon's orbit; equivalent to 346.620 days.
eclipsing binary *Astronomy.* a binary star system whose orbital plane lies in or close to the line of sight from the earth, with the result that each star periodically eclipses the other. Also, **eclipsing variable.**
ecliptic *Astronomy.* **1.** the plane of the earth's orbit projected on the celestial sphere. **2.** the apparent path through the stars that is followed by the sun during the year.
ecliptic coordinate system *Astronomy.* a system for determining the position of a body in the sky by reference to its distance above or below the earth's orbit and to its angular distance from the vernal equinox.
ecliptic diagram *Astronomy.* a plot of the sky by ecliptic coordinates.
ecliptic limits *Astronomy.* the maximum angular distance that a full or new moon can be from its node and still allow an eclipse to occur.
ecliptic pole *Astronomy.* either of the points on the celestial sphere that are 90° above and below the ecliptic and 23.45° from the celestial poles.

eclogite *Petrology.* a dense granulose rock formed under high pressure and temperature, composed essentially of pyroxine and garnet; may also contain rutile, kyanite, and quartz.
eclogite facies *Petrology.* a metamorphic facies characterized by extremely high temperature and pressure.
eclosion *Invertebrate Zoology.* the hatching or escape of an insect larva from its egg or an adult from the pupa.
ECM electrochemical machining; electronic countermeasure.
ecmnesia *Medicine.* forgetfulness of recent events and loss of ability to learn while retaining memory for more remote events.
ecnephias *Meteorology.* a thunderstorm or squall in the Mediterranean.
ECO electron-coupled oscillator.
eco- a prefix meaning "environment" or "ecology," as in *ecosystem.*
ecodeme *Ecology.* a sub-specific group capable of interbreeding within a population.
ecofact *Archaeology.* a natural object or substance that has not been technologically altered but that has cultural signifance, such as a shell carried from the ocean to an inland settlement.
ecohazard *Ecology.* any activity or substance that may constitute a threat to a habitat or environment.
ecoline *Ecology.* the rate of genetic change that occurs in an environment due to the merging of different varieties of a plant species.
ecological *Ecology.* **1.** of or relating to the environment or to the science of ecology. **2.** relating to the prudent use or beneficial management of natural resources and the natural environment.
ecological age *Ecology.* an organism's reproductive status, which may be prereproductive, reproductive, or postreproductive.
ecological amplitude *Ecology.* the range of tolerance within a given environmental factor in which an organism can function.
ecological anthropology *Anthropology.* a branch of anthropology that specializes in the study of humans' interaction with and effect on their environment and the effect of the environment on human culture.
ecological balance see BALANCE OF NATURE.
ecological bonitation *Ecology.* a measurement of the number of individuals of a certain organism present at a given place or time.
ecological capacity see CARRYING CAPACITY.
ecological climatology *Biology.* a branch of bioclimatology concerned with the influence of climate on the physiological adaptation and geographical distribution of plants and animals.
ecological community see COMMUNITY.
ecological efficiency *Ecology.* **1.** the rate of transfer of energy from one trophic level to the next. Also, LINDEMAN'S EFFICIENCY. **2.** also, **coefficient of ecological efficiency.** the ratio of energy acquired at one trophic level to that acquired at the previous level. Also, GROSS ECOLOGICAL EFFICIENCY.
ecological equivalence *Ecology.* a situation in which two or more species with similar ecological amplitudes exist, and could, in theory, replace each other in the same ecological niche.
ecological equivalents *Ecology.* unrelated or distantly related species that fulfill the same ecological roles in different communities or geographical areas.
ecological evaluation *Ecology.* the process of assessing the environmental value of a habitat to determine its best use.
ecological indicator *Ecology.* an organism or group of organisms that reflect environmental conditions in a habitat.
ecological interaction *Ecology.* the effect that an individual of one species has on an individual of another species in the same habitat.
ecological isolation *Ecology.* the absence of interbreeding due to ecological obstacles. Also, ECOLOGICAL SEGREGATION.
ecological longevity *Ecology.* the average life span of an individual in a given population under given conditions.
ecological niche *Ecology.* see NICHE.
ecological pressure *Ecology.* the sum of all environmental factors that act as agents of natural selection.
ecological pyramid see FOOD PYRAMID.
ecological release *Ecology.* the expansion of a population's niche due to the elimination of a competitor.
ecological segregation see ECOLOGICAL ISOLATION.
ecological succession *Ecology.* a predictable and gradual process of community change and replacement, often leading toward a climax community.
ecological system see ECOSYSTEM.
ecologic fallacy *Medicine.* a false assumption that a pathogenic factor and a disease present in a population is acceptable proof that the agent is the cause of the disease in a particular individual.

Ecology

Ecology is the science concerned with interactions between organisms and the environment, on spatial scales ranging from parts of individuals to the biosphere as a whole. The range of research interests in ecology is therefore extremely broad, from theoretical population genetics to the integration of biological processes into global circulation models.

Often, ecology is divided into four major subfields: physiological ecology (concerned with interactions between individual organisms and the environment); population biology (the regulation of population growth and population size, and interactions among populations); community ecology (characteristics of the collective properties of the organisms in an area); and ecosystem ecology (regulation of the flows of energy and material in terrestrial and aquatic ecosystems). However, like any expanding field of science, ecology is developing new subfields and combining established ones at a rate that defies codification.

In the nature of their research, ecologists often study issues that relate to the effects of human activities upon populations and ecosystems. As a consequence, ecological research has identified and characterized many of the most significant problems threatening the future of humanity, including acid rain, the ongoing loss of biological diversity, and biological feedbacks to global climate change.

These contributions have led to a popular identification of "ecology" with "environmental activist," to the point that the word ecology can be embraced or rejected for reasons unrelated to the science of ecology. Many ecologists are environmental activists, and their activism often grows out of their research. However, ecology is no more defined by environmental activism than molecular biology is defined by genetic medicine.

Peter M. Vitousek
Professor of Biological Sciences
Stanford University

ecology *Biology.* the branch of the biological sciences that deals with the relationship between organisms and their environment, including their relationship with other organisms.
ecomania *Psychology.* a mental attitude whereby one is hostile and domineering toward one's own family but submissive to those in outside authority.
ecomorphology *Ecology.* the study of the relationship between the ecological relations of an individual and its morphology.
econ *Ecology.* a local vegetation type.
economic action *Military Science.* the planned use of economic measures to impair the military potential of a hostile power or to generate economic strength and stability within a friendly power. Also, ECONOMIC WARFARE.
economic analysis *Industrial Engineering.* a methodology used in resolving cost considerations in a manufacturing or production facility, whereby alternatives and objectives are analyzed and compared from the standpoint of costs and benefits.
economic anthropology *Anthropology.* a branch of anthropology that specializes in the study of subsistence spheres, exchange systems, and general economies in the context of a particular culture, or of cultures over time.
economic botany *Botany.* the study of botanical processes and products that can be used profitably.
economic determinism *Anthropology.* the theory that the evolution of cultures from simple to complex societies is based on changes in and control of the means of production.

economic entomology *Biology*. a branch of entomology concerned with the effects of economic losses that are caused by insect predation on commercially important plants and animals.

economic geography *Geography*. a branch of geography dealing with the interaction between the earth's landscape and human economic activity. Also, GEOECONOMICS.

economic geology *Geology*. the application of geologic knowledge, including the study and analysis of geologic bodies and materials, for useful or profitable purposes.

economic lot size *Industrial Engineering*. the production lot size at which production economies are maximized.

economic mineral *Mineralogy*. a term for any mineral that has commercial value.

economic mobilization *Military Science*. the preparation for and execution of changes in the national economy in order to maximize the use of resources in a national emergency.

economic order quantity *Industrial Engineering*. a system of inventory control that seeks to minimize inventory costs by ordering quantities based on previous consumption by the organization. Similarly, **economic purchase quantity.**

economic potential for war *Military Science*. the portion of the total economic capacity of a nation that can be used for warfare. Also, **economic war potential.**

economic warfare see ECONOMIC ACTION.

economizer *Engineering*. a compartment in a continuous-flow oxygen system which collects oxygen that has been exhaled by the user in order to be reused. *Mechanical Engineering*. an apparatus that uses warm flue gases exiting a steam boiler to preheat feedwater entering the boiler, thus improving boiler efficiency and economy.

economy *Computer Programming*. the ratio of the actual number of characters to be coded to the maximum number of characters that it would be possible to code.

economy of scale *Industrial Engineering*. an approach to production that is predicated on the theory that mass production of identical products tends to lower per-unit production costs.

economy of scope *Industrial Engineering*. an approach to production that is predicated on the theory that the same equipment can be used flexibly to produce small batches of a variety of products at a per-unit production cost that is close to that of mass production.

eco-organ *Ecology*. **1.** a vegetation type based on features that mirror conditions in the environment. **2.** any feature that mirrors conditions in the environment.

ecophene *Ecology*. the various phenotypes observable within a given habitat produced by a single genotype.

ecophenotype *Genetics*. a phenotype that is due to ecological conditions rather than to genetic expression.

E core *Electromagnetism*. a transformer core in the shape of an E whose windings may be wrapped on the arms of the figure.

EcoRI restriction enzyme *Enzymology*. a restriction endonuclease from *Escherichia coli* that cleaves both strands of DNA at a specific site between adenine and guanine nucleotides in the sequence 5'-GAATTC-3' on complementary strands of duplex DNA.

ecospecies *Ecology*. a taxon composed of one or more interbreeding ecotypes.

ecosphere *Ecology*. the earth and the living organisms that inhabit it, along with all the environmental factors that operate on these organisms. *Astronomy*. the region of space around a star that is considered to be capable of supporting life.

ecosystem *Ecology*. a local biological community and its pattern of interaction with its environment. Also, ECOLOGICAL SYSTEM, HOLOCOEN.

ecosystem mapping *Ecology*. the art of constructing maps of the various plants that grow in a geographical area.

ecotone *Ecology*. a transition zone between two distinct habitats that contains species from each area, as well as organisms unique to it. *Anthropology*. such an area of transition in which certain game or vegetation overlap; a region of primary importance for human subsistence.

ecotropic *Virology*. a retrovirus that can replicate only in the host of the species in which it originated.

ecotype *Ecology*. an organism that has adapted to its local environment through minor, genetically induced changes in its physiology, yet can still reproduce with other members of its species from other areas that have not undergone these changes.

ECR electron cyclotron resonance source.

ecstasy *Psychology*. a mental state of rapture and trancelike elation.

ECT electroconvulsive therapy.

ect- a combining form meaning "outer" or "external," as in *ectinites*.

ectad *Biology*. outward.

ectal *Biology*. external; outer; on the surface of.

ectasia *Medicine*. a distension or dilation, usually of a hollow structure.

-ectasia or **-ectasis** a combining form meaning: **1.** stretching; dilation. **2.** expansion; enlargement.

Ecterocoelia *Invertebrate Zoology*. a major grouping of bilateral animals that includes arthropods, mollusks, and most worms. Also, PROTOSTOMIA.

ectethmoid *Anatomy*. the paired areas of air cells contained within thin bony walls and located on the lateral aspects of the ethmoid bone.

ecthyma *Medicine*. a form of impetigo characterized by large flat pustules that ulcerate and are surrounded by inflammatory areolas.

ecto- a combining form meaning "outer" or "external," as in *ectogenesis, ectophageous*.

ectoblast see ECTODERM.

ectocardia *Medicine*. congenital displacement or malplacement of the heart, either outside or inside the thorax.

Ectocarpaceae *Botany*. a family of small to moderately sized brown algae of the order Ectocarpales that are usually loosely branched and filamentous and consist of an erect and a prostrate system of branches.

Ectocarpales *Botany*. an order of relatively simple marine brown algae of the class Phaeophyceae, usually occurring in loose tufts of branched, uniseriate filaments and ranging in size from 1 to 20 centimeters.

ectocommensal *Ecology*. an organism that benefits from living externally on another organism, without affecting its host.

ectocornea *Anatomy*. the outer layer of the cornea.

ectocrine *Ecology*. a chemical emitted by an organism when it decomposes, which affects the activity of another organism.

ectocyst *Invertebrate Zoology*. the outer layer of tissue in the walls of a bryozoan.

ectoderm *Developmental Biology*. the outermost layer of a blastodisc that develops into the epidermis and epidermal tissues, the nervous system, the external sense organs, and the mucous membranes lining the mouth and anus. Also, ECTOBLAST, EPIBLAST.

ectoenzyme *Enzymology*. an enzyme that is secreted from a cell into the surrounding medium; an extracellular enzyme.

ectogenesis *Developmental Biology*. the development of embryonic tissue in an artificial environment.

ectogenous *Medicine*. arising from causes outside the organism; exogenous.

ectognathous *Entomology*. the most common type of insect mouthparts, in which the mandibles and maxillae are well developed, visible, and outwardly projecting.

ectogony *Botany*. the influence of pollen and fertilization on the structures of the female organs of the parent, especially the effects on color, chemical composition, ripening, and abscission. Also, **ectogeny**.

ectohumus see MOR.

ectolecithal *Invertebrate Zoology*. of arthropod eggs, having the yolk situated on the periphery.

ectomere *Developmental Biology*. any of the blastomeres that takes part in the formation of the ectoderm.

ectomesoblast *Developmental Biology*. the undifferentiated cell layer that exists prior to the development of the ectoblast and mesoblast.

ectomesoderm *Developmental Biology*. a cell or cell layer that eventually develops into the mesoderm, but that has not yet differentiated. Also, MESECTOBLAST.

-ectomize a combining form meaning "deprive by excision," as in *thyroidectomize*.

ectomorph *Psychology*. a person of the ectomorphic body type.

ectomorphic *Psychology*. of or relating to a body type classification for a person having long, thin limbs and relatively little weight.

ectomorphy *Psychology*. the extent to which a person conforms to the ectomorphic body type, said to be associated with a sensitive, cautious, and inhibited personality.

-ectomy a combining form meaning "excision," as in *appendectomy*.

ectoneural system *Invertebrate Zoology*. the outer and major part of the nervous system of echinoderms; includes a nerve ring around the mouth and radial nerve extensions in the arms.

ectoparasite *Ecology*. a type of parasite that lives externally on its host; fleas and lice are **obligate ectoparasites** of man and animals.

ectophagous *Invertebrate Zoology*. of a parasite, developing and feeding on the surface of the host.

ectophloic siphonostele *Botany*. a stele in which the phloem occurs only eternal to the xylem cylinder, usually occurring in angiosperms.

ectophyte *Ecology.* a plant that lives on the surface of another organism. Thus, **ectophytic.** Also, EPIPHYTE.
ectopia *Medicine.* any abnormality of the position of an organ or a part of the body, usually congenital.
ectopic [ek täp′ik] *Anatomy.* out of place; in an abnormal location or position, as an *ectopic* pregnancy.
ectopic focus *Cardiology.* any cardiac arrhythmia that originates from an excitation impulse at a site other than the sinoatrial node; may occur in both healthy and diseased hearts, often resulting from anxiety, excitement, fatigue, or use of alcohol or tobacco.
ectopic focus theory *Cardiology.* the theory that atrial fibrillation is caused by rapid emissions from an ectopic focus.
ectopic heartbeats *Cardiology.* heartbeats that have their origin in some focus other than the sinoatrial node.
ectopic pacemaker *Cardiology.* any biological pacemaker other than the sinoatrial node.
ectopic pairing *Cell Biology.* a nonspecific pairing of nonhomologous segments of heterochromatin, specifically in *Drosophila* salivary gland chromosomes.
ectopic pregnancy *Medicine.* any development of the impregnated ovum in an abnormal position or place.
ectopic teratism *Medicine.* a congenital anomaly in which one or more parts are misplaced, such as transposition of the great vessels.
ectoplasm [ek′tə plaz′əm] *Cell Biology.* the superficial cytoplasm of ciliates, located immediately below the cell membrane, generally rather homogeneous, lacking in organelles, and relatively rigid due to the presence of microfilaments.
Ectoprocta *Invertebrate Zoology.* a subphylum of colonial bryozoans with retractable tentacles around the mouth and the anus situated just outside the ring of tentacles.
ectosome *Invertebrate Zoology.* the outer layer of a sponge.
ectostosis *Physiology.* the formation of bone beneath the perichondrium of a cartilage or the periosteum of a bone.
ectosymbiont *Ecology.* an organism that participates in a relationship of ectosymbiosis.
ectosymbiosis *Ecology.* a situation in which an organism has a mutually beneficial relationship with a host organism while remaining physically separated from the host.
ectosymbiotic commensals *Entomology.* the insects that feed on feathers, sloughed-off flakes of skin, or waxy epidermal exudates; for example, fleas and louse flies.
ectotherm *Ecology.* an organism that relies on sources of heat outside itself.
ectothermic *Ecology.* capable of maintaining body temperature by gaining heat outside itself.
Ectothiorhodospira *Microbiology.* a genus of purple sulfur bacteria with spiral, vibrioid, or rod-shaped cells, motile by means of either a single polar flagellum or a tuft of flagella; originally classified in the family Chromatiaceae, but recently proposed as the type genus of a new family, **Ectothiorhodospiracea.**
ectotrophic *Biology.* obtaining nutrients from the outside or from the surface of a host, as do certain parasitic fungi that live on the outside surface of the roots of a host plant.
ectozoon plural, **ectozoa.** *Ecology.* an animal that lives on the surface of another organism.
Ectrephidae see PTINIDAE.
ectro- a combining form meaning "congenital absence of," as in *ectrodactyly.*
ectrodactylia *Medicine.* a birth defect characterized by the lack of one or more fingers or toes or parts of them.
Ectrogellaceae *Mycology.* a family of fungi belonging to the order Saprolegniales that is parasitic to marine and freshwater algae.
ectrogeny *Medicine.* the congenital absence or marked imperfection of any organ or part of the body.
ectromelia *Medicine.* a congenital absence or marked imperfection in one or more of the limbs.
ectrosyndactyly *Medicine.* a congenital absence of some but not all of the digits, with webbing between those remaining.
ecumene [ek′yə men] *Geography.* **1.** the vital part of a nation or state, which furnishes most of the important socioeconomic elements allowing it to function. **2.** originally, the inhabitable world as known to the ancient Greeks or the entire inhabitable world. Also, ŒKUMENE, ŒCUMENE.
eczema [ek′sə mə; eg′sə mə; ik zē′mə] *Medicine.* an acute or chronic noncontagious dermatitis, usually marked by itchy red blisters followed by crusted and thickened skin.
eczematoid reaction *Medicine.* a dermal and epidermal inflammatory response characterized by edema, vesiculation, and thickening of the skin.
ED electronic dummy; effective dose.
ED_{50} effective dose 50 (median effective dose).
edaphic *Agronomy.* of or relating to the soil. *Ecology.* specifically, relating to the influence of soil characteristics on living organisms.
edaphic climax *Ecology.* a climax community whose structure and composition are largely determined by the nature of the soil.
edaphic community *Ecology.* a plant community that results from soil factors, such as marsh areas where plants are uniquely adapted to cope with shifting sediments and range of saline levels.
edaphic factors *Ecology.* the various physical, chemical, and biological properties of the soil that influence living organisms which are in association with it.
edaphology *Ecology.* the scientific study of the soil, particularly with reference to its effect on living organisms and their effect on it. *Agronomy.* this study including human land use.
edaphon *Biology.* an organism that lives in the soil, such as a fungus, a bacterium, an insect, or a worm.
Edaphosauria *Paleontology.* a suborder of primitive, mammal-like, herbivorous reptiles in the subclass Synapsida and the order Pelycosauria; they were the characteristic North American pelycosaur of the Permian and were distinguished by a spectacular "sail" that extended upward from the backbone. (Literally, "sail lizard.")
E-day *Military Science.* the day on which a NATO exercise commences.
EDC expected date of confinement.
EDD earliest due date.
Eddington, Sir Arthur 1882–1944, English astronomer; demonstrated role of radiation pressure in stellar equilibrium and relationship between a star's mass and density.

Sir Arthur Eddington

Eddington limit *Astrophysics.* a state in which radiation pressure pushing from within a star balances the inward pull of the star's self-gravitation.
Eddington's model *Astrophysics.* a model for a star in which all of its energy is transported by radiation, with no contribution from convection.
eddy *Meteorology.* a whirl or circular current of air running contrary to the general flow. *Fluid Mechanics.* a deviation in the steady flow of a fluid, causing a vortex.
eddy conductivity *Thermodynamics.* a fluid-flow quantity describing the heat transfer due to eddies in a fluid, caused by a temperature difference between the fluid near the boundary and the fluid in the center of the stream, so that hot fluid is carried to the colder regions and cold fluid to the hotter regions.
eddy correlation *Meteorology.* a means of determining the effect of the surface of the sea on the air above it by simultaneous measuring of fluctuations in the horizontal and vertical air flow from the mean.

eddy current *Physics.* a current produced in a conductor as it moves through a magnetic field. *Materials Science.* a current that is induced into a ferromagnetic material by the reversal of the electric field.

eddy-current brake *Mechanical Engineering.* a brake consisting of a mass of metal rotating in relation to a magnetic field, which induces the internal currents used as a retarding force; used, for example, in electric tramways and in loading motors for testing.

eddy-current damping *Materials Science.* the process of automatic damping by eddy currents produced by a moving conductor.

eddy-current loss *Materials Science.* in ferromagnetic metals, one of the sources of energy loss associated with the changing electric field when alternating current is used, which results in the conversion of electric energy to thermal energy.

eddy-current nondestructive evaluation *Materials Science.* a nondestructive testing technique used to evaluate structure and properties by observing the reaction between the material and an electric field.

eddy-current tachometer *Engineering.* an instrument that measures speed by the use of a rotating permanent magnet creating currents in a metal cylinder; the cylinder rotates and is attached to a pointer, which indicates the speed in relation to the rotation of the cylinder. Also, DRAG-TYPE TACHOMETER.

eddy-current test *Electromagnetism.* a test in which the change of impedance of a test coil placed close to a conducting specimen indicates the eddy current induced by the coil. *Metallurgy.* one of several nondestructive tests for detecting cracks and other flaws.

eddy diffusion *Fluid Mechanics.* the turbulent diffusion of heat, mass, or momentum caused by random eddy motions along the axis or radius of the flow channel.

eddy diffusivity of heat *Fluid Mechanics.* a constant that describes the heat flux in turbulent flow in a fluid.

eddy diffusivity of momentum *Fluid Mechanics.* a constant that describes the shear stress in turbulent flow in a fluid.

eddy mark *Geology.* any of the many overlapping spiral loops impressed on a sedimentary surface, especially on sandstone.

eddy resistance *Fluid Mechanics.* the drag of a ship caused by eddies that are cast from the hull and remove energy.

eddy velocity *Fluid Mechanics.* the variation between the instantaneous speed and the mean speed of fluid flow at a point.

eddy viscosity *Fluid Mechanics.* a property, largely dependent on flow, that is multiplied by the fluid's density and then added to the dynamic viscosity to correct for turbulence in the shear stress-viscosity equation.

Edeleanu process *Chemical Engineering.* a solvent extraction process using liquid sulfur dioxide or liquid sulfur dioxide and benzene as solvent to remove undesirable sulfur compounds from heavy petroleum products.

Edelman, Gerald born 1929, American biochemist; shared Nobel Prize for research on the chemical structure of antibodies.

edelweiss [ā´dəl wīs´; ā´dəl vīs´] *Botany.* a small composite plant, *Leontopodium alpinum,* characterized by woolly, white leaves and flowers; found in the high altitudes of the Alps and in other mountainous regions.

edema *Medicine.* a swelling of cells, tissues, or body cavities, caused by an abnormal accumulation of fluid.

-edema *Medicine.* a combining form meaning "swelling."

Edenian *Geology.* a North American geologic stage of the Upper Ordovician period, occurring after the Mohawkian and before the Maysvillian, and equivalent to the lower Cincinnatian.

Edentata *Vertebrate Zoology.* a mammalian order including sloths, armadillos, and anteaters; characterized by a small unconvoluted cerebrum, skin that is hairy or covered with bony plates, curved sharp claws, and a mouth with few, if any, small teeth.

edentate *Zoology.* lacking teeth. *Vertebrate Zoology.* **1.** belonging to the order Edentata. **2.** an animal belonging to this order.

edetate calcium disodium *Pharmacology.* $C_{10}H_{12}CaN_2Na_2O_8$, a chelating agent that is used to treat lead poisoning.

edge *Robotics.* a line that marks the boundary of an object's image in machine vision. *Mathematics.* **1.** in graph theory, a member of one of two (usually finite) sets of elements that determine a graph; i.e., an element of the **edge set.** The other set is called the **vertex set;** each element of the edge set is determined by a pair of elements of the vertex set. Denoted uv if the edge is undirected and joins vertices u and v and denoted (u, v) if the edge is directed from vertex u to vertex v. **2.** a straight line that is the intersection of two faces of a solid figure. **3.** a boundary of a plane geometric figure.

edge angle *Archaeology.* the angle of the cutting edge of a stone tool; assumed to indicate the purpose for which the tool was used.

edgeboard connector see CARD-EDGE CONNECTOR.

edge-bridging ligand *Organic Chemistry.* a ligand that forms a link from one end or side of a polyhedric metal cluster.

edge color or **edge stain** *Graphic Arts.* a gold leaf or colored stain that is applied to one or more edges of a book's pages, most often at the top.

edge-coloring *Mathematics.* the assignment of a color to each of the edges of a graph. If G is a graph with no loops, G is said to be **properly edge-colored** if no two edges meeting the same vertex are the same color. A graph is ***k*-edge-colorable** if it can be properly edge-colored with k distinct colors, and the edge-chromatic number of a graph is the smallest integer k such that the graph is k-edge-colorable.

edge detector *Artificial Intelligence.* in natural or machine vision systems, a neuron or operator that responds to the abrupt change in stimulus across a locally straight edge between two regions of approximately uniform intensity or color.

edge dislocation *Crystallography.* a dislocation in a crystal form caused by the insertion of a row of unit cells partway into the crystal and perpendicular to its edge.

edge effect *Ecology.* the effect exerted on the population structure of an area by adjoining communities. *Electricity.* an outward-curving distortion of force lines that appear close to the edges of the parallel metal plates of a capacitor.

edge focusing *Electromagnetism.* the focusing of an ion beam after the beam crosses a fringed magnetic field.

edge joint *Metallurgy.* a joint between the edges of parallel parts.

edge line *Cartography.* on a relief map, an extra-heavy line that indicates a rapid change in slope.

edger *Mechanical Devices.* **1.** a tool used to smooth, finish, or flatten the edges of a workpiece such as wood. **2.** a gardening tool used to trim vegetation. *Metallurgy.* in forging, a die portion that facilitates appropriate distribution of material during the operation.

Edgerton, Harold born 1903, American electrical engineer; invented the electronic flash.

edge species *Ecology.* a species characteristically found in marginal areas of a community.

edge stability *Metallurgy.* an index of the green strength of a powder compact.

edge water or **edgewater** *Hydrology.* the water that surrounds, borders on, or lies under a gas pool or oil-bearing formation.

edge wave *Oceanography.* a wave that travels parallel to the shoreline, its height diminishing rapidly seaward.

edge-wear analysis see USE-WEAR ANALYSIS.

edge well *Petroleum Engineering.* **1.** a well that is placed at the edge of a leased reservoir or at the edge of a gas or oil reservoir. **2.** a well that is positioned far down the side of an oil pool so as to touch the water line where water and oil meet.

edgewise structure *Geology.* a sedimentary structure consisting of flat or disk-shaped fragments whose long axes are arranged at varying and steep angles to the bedding; some of these structures have been attributed to sliding.

edging *Building Engineering.* small, solid squares that protect veneer used in cabinetry and other wood-based furniture; usually inset on the top edge face of the veneer. *Metallurgy.* in metal forming, one of several operations that improve product quality.

Ediacaran fauna *Paleontology.* one of many similar faunas first found in the Ediacara Hills of Australia and later worldwide; they represent the worldwide proliferation of the first true multicellular animals, which took place at the end of the Pre-Cambrian era; most of the Ediacara fossils are jellyfishes and soft corals of the phylum Cnidaria, and there are also worms and other animals of uncertain affinities.

edible *Nutrition.* describing any food that is fit for human consumption; able to be eaten or good to eat.

edible oil *Food Technology.* any liquid fat that is safe for eating, especially vegetable fats such as corn oil or peanut oil, used in food preparation.

edible portion *Nutrition.* the portion or parts of a food item that may be eaten after the inedible parts have been removed.

edingtonite *Mineralogy.* $BaAl_2Si_3O_{10}\cdot 4H_2O$, a white to pink orthorhombic and tetragonal mineral of the zeolite group, occurring as prismatic crystals or in massive form, having a specific gravity of 2.694 and a hardness of 4 on the Mohs scale; found with other zeolites in igneous rocks.

Thomas Edison

Edison, Thomas Alva 1847–1900, American inventor; his many inventions included the incandescent electric light, phonograph, motion picture, and carbon microphone.

Edison battery *Electricity.* a storage battery containing cells of nickel and iron in an alkaline solution. Also, NICKEL-IRON BATTERY.

E display *Electronics.* a rectangular radar display in which targets appear as intensity-modulated spots, with their range as the horizontal coordinate and their elevation as the vertical coordinate. Also, E SCAN, E INDICATOR.

edit *Graphic Arts.* to prepare writing for publication, especially something written by another person, by correcting errors in grammar, spelling, and style, by revising existing content or writing new content, and so on. *Computer Programming.* **1.** to add, modify, delete, or rearrange data. **2.** to modify a text or program file, producing a new version of the file.

editing enzyme *Enzymology.* an enzyme that proofreads a DNA molecule and corrects errors in replication or in mutational changes.

edit mask see MASK.

editor *Graphic Arts.* someone who edits a manuscript or a publication. *Computer Programming.* see EDITOR PROGRAM.

editor program *Computer Programming.* a computer program that enables a user to enter, revise, or delete data in an existing program or file.

Edman degradation *Biochemistry.* a technique for analyzing an amino acid sequence in which an amino-terminal residue is tagged and removed from a polypeptide chain without disturbing the other amino acid bonds.

edominant *Ecology.* describing a secondary species that is not dominant within a community.

EDP electronic data processing.

Edrioasteroidea *Paleontology.* a small class of echinoderms in the subphylum Echinozoa that is represented in strata from the lower Cambrian to the lower Carboniferous; most edrioasteroids had flexible tests and were stemless.

EDT Eastern Daylight Time.

EDTA ethylenediaminetetraacetic acid.

educational age see ACHIEVEMENT AGE.

educational psychology *Psychology.* a branch of psychology that is concerned with the psychological aspects of learning and with behavior in educational settings. Thus, **educational psychologist.**

eductor *Engineering.* **1.** an instrument that mixes liquids with the use of a jet pump. **2.** see EJECTOR.

eductor pump *Mining Engineering.* a pump, using water as an operating medium, for utilizing cavitation produced at a pipe construction.

edulcorate *Computer Programming.* to improve a data file by eliminating irrelevant data. *Chemistry.* to free a substance from acids, salts, or other impurities by washing.

EDVAC [ed´vak´] *Computer Technology.* the first stored-program computer; it became operational in 1952. (An acronym for electronic discrete variable automatic computer.)

Edwardsiella *Microbiology.* a genus of Gram-negative, bacillary bacteria of the family Enterobacteriaceae that exhibit flagellar motility and a facultatively anaerobic mode of respiration.

Edward's syndrome or **Edwards' syndrome** *Genetics.* a group of congenital abnormalities due to trisomy of chromosome 18, usually resulting in death in infancy. Also, TRISOMY 18 SYNDROME.

EEB or **E.E.B.** European Environmental Bureau.

EEG electroencephalogram; electroencephalograph.

EEIB or **E.E.I.B.** Environmental Engineering Intersociety Board, Inc.

eel *Vertebrate Zoology.* **1.** any of several elongated, snakelike, marine or freshwater fish of the order Apodes, having a smooth slimy skin often without scales; they are destitute of pelvic and sometimes of pectoral fins, and have the median fins confluent around the tail. **2.** any of several similar but unrelated fish, such as the lamprey.

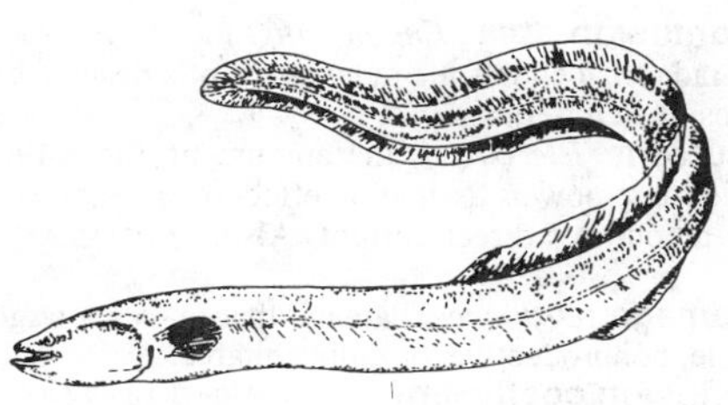

eel

eelgrass *Botany.* a grasslike marine plant, *Zostera marina,* characterized by ribbonlike leaves.

eelpout *Vertebrate Zoology.* any of various species of marine fish of the family Zoarcidae, native to cold parts of the Atlantic and Pacific Oceans and to the Arctic and Antarctic regions where they are shallow bottom feeders; characterized by thick lips with a jutting upper jaw and dorsal and anal fins that extend around a pointed tail.

EELS electron energy loss spectroscopy.

eelworm *Invertebrate Zoology.* any small nematode worm of the family Anguillulidae.

E.E.N.T. eye-ear-nose-throat.

EEPROM [ē´präm´] *Computer Technology.* a form of read-only memory that can be electrically erased and reprogrammed. (An acronym for electrically erasable programmable read-only memory.)

EER energy efficiency ratio.

EEROM [ē´räm´] *Computer Technology.* a form of read-only memory in which the entire contents can be erased electrically and reprogrammed hundreds of times without damaging the device. (An acronym for electrically erasable read-only memory.)

EF or **E.F.** Engineering Foundation.

ef- a prefix meaning "out of," as in *effluent.*

EFA essential fatty acids.

eff efficiency.

effacement *Medicine.* the slow flattening of the uterine cervix and thinning of its walls during labor.

effective action *Mathematics.* the action of a (transformation) group on a set is said to be effective only if the identity element of the group leaves every element of the set invariant. Such a group is sometimes referred to as an **effective transformation group.**

effective address *Computer Programming.* the resultant instruction or operand address after any indexing operation or other modification is performed.

effective allele *Genetics.* a polygene that produces an increase in a quantitative trait.

effective ampere *Electronics.* a time-varying or AC current that, when flowing through a resistor, produces heat at the same average rate as that produced by a steady DC current of one ampere.

effective angle of attack *Aviation.* the part of a particular angle of attack between the chord of an airfoil and an imaginary line representing the resultant velocity of the disturbed airflow; expressed as the difference between the reference and the induced angle of attack.
effective antenna length *Electromagnetism.* the electrical length of an antenna, as distinguished from its physical length.
effective area *Electromagnetism.* the ratio of the available power at the terminals of an antenna in a given direction to the power flux density of a plane wave incident on the antenna from that direction, expressed by the product of the directive gain of the antenna and the square of the wavelength, divided by 4π. *Chemical Engineering.* the area of process media that is actually involved in the process.
effective atmosphere *Geophysics.* the part of the atmosphere in which the scattered light of twilight can still be seen. Also, OPTICALLY EFFECTIVE ATMOSPHERE.
effective bandwidth *Electronics.* the bandwidth of an ideal filter, with a flat response, that would pass the same amount of energy as the given filter, when processing the same wideband input signal. Also, SPECTRAL BANDWIDTH.
effective bore line *Mechanics.* the path of a projectile in the bore of a moving launcher, as visualized in a stationary reference frame.
effective capacitance *Electricity.* the total capacitance existing between any two points of an electric circuit.
effective center *Acoustical Engineering.* a point from which sound appears to be originating; not necessarily the actual center of the sound source.
effective confusion area *Engineering.* the space occupied by countermeasure radar that equals the radar space of a designated aircraft traveling at a designated frequency.
effective current *Electricity.* the amount of alternating current that produces the same power dissipation effect on a load resistor as the corresponding amount of direct current. Also, ROOT-MEAN-SQUARE (RMS) CURRENT.
effective damage *Ordnance.* damage that makes a target inoperative, unserviceable, nonproductive, or uninhabitable.
effective diffusion coefficient see APPARENT DIFFUSION COEFFICIENT.
effective discharge area *Design Engineering.* the area of flow through a valve; used to calculate valve capacity.
effective dose *Medicine.* the amount of a drug, or the level of radiation exposure, sufficient to achieve a desired clinical improvement.
effective dose 50 see MEDIAN EFFECTIVE DOSE.
effective earth radius see EFFECTIVE RADIUS OF THE EARTH.
effective exhaust velocity *Space Technology.* **1.** the velocity of a rocket's exhaust stream, taking into account friction, heat transfer, nonaxially directed flow, and other forces. **2.** a fictitious exhaust velocity that would yield the same jet thrust as that actually observed.
effective facsimile band *Telecommunications.* the frequency band of a facsimile signal wave, equal to the width between zero frequency and maximum keying frequency.
effective field intensity *Electromagnetism.* the root-mean-square value of the inverse-distance fields when measured one mile from a transmitting antenna, covering all directions in a horizontal plane.
effective firing time *Ordnance.* the period during which an aircraft can effectively fire upon a target.
effective fragment *Ordnance.* a fragment with the proper mass, form, and velocity to achieve the desired effect upon impact with its target.
effective gust velocity *Meteorology.* the vertical component of the velocity of a sharp-edged gust that would produce a given acceleration on an aircraft flown at level flight at the design cruising speed and at a given air density.
effective half-life *Nucleonics.* the time it takes for both radioactive decay and the natural elimination processes to reduce the presence of a radioisotope in a natural organism by fifty percent.
effective height *Electromagnetism.* the vertical height above the effective ground level at which the radiation center of a transmitting or receiving antenna is located.
effective horizon *Telecommunications.* in the estimation of transmission range, the horizon based on the effective radius of the earth.
effective horsepower *Naval Architecture.* the horsepower required to tow a bare hull, without underwater fittings such as rudders and propeller shafts, at a given speed in still water and with no wind. Also, TOWROPE HORSEPOWER.
effective instruction *Computer Programming.* **1.** any instruction that does not require modification prior to execution. **2.** the resultant instruction after the basic instruction has been modified.
effective launcher line *Mechanics.* the direction in which a rocket would be traveling at the end of its burn period in the absence of gravity, but considering all aerodynamic effects.
effective length *Naval Architecture.* the mean length of a vessel's underwater hull.
effectively grounded *Electricity.* grounded through a sufficiently low impedance ground connection to prevent buildup of voltages hazardous to connected equipment.
effective mass *Solid-State Physics.* a parameter for electrons or holes in a semiconductor that differs from the mass of a free electron and depends to some extent on the particle's position in the energy band, affecting the mobility and the resulting current.
effective molecular diameter *Physical Chemistry.* a term for the general range of the electrons rotating around a molecule.
effective molecular weight *Petroleum Engineering.* the empirical relationship of oil plotted against API gravity to give the pseudoaverage molecular weight of oil for use in reservoir calculations.
effective multiplication factor *Nucleonics.* the ratio between the number of neutrons present in the current generation and that in the preceding generation, in a reactor experiencing neutron leakage.
effectiveness level *Computer Programming.* the ratio of the use time of data-processing equipment to the entire performance period of the system.
effective osmotic pressure *Physical Chemistry.* that part of the total osmotic pressure of a solution which governs the tendency of its pure solvent to pass through a semipermeable membrane.
effective percentage modulation *Telecommunications.* for a single sinusoidal input component, the ratio, expressed as a percentage, of the peak value of the fundamental component of the envelope to the average amplitude of the modulated wave.
effective permeability *Physical Chemistry.* the ability of a porous medium to absorb one liquid phase when it is in the presence of other liquid phases.
effective pitch *Aviation.* the distance a propeller advances along its flight path perpendicular to the axis of rotation during one complete revolution of the propeller.
effective population size *Genetics.* the number of reproducing individuals in the population of an organism.
effective porosity *Geology.* the percentage of the total bulk volume of a given mass of soil or rock that contains interconnected pore spaces. Also, **effective drainage porosity**.
effective potential energy *Mechanics.* the potential energy as determined in an accelerated coordinate system.
effective precipitable water *Meteorology.* the part of the precipitable water that can, in theory, actually fall as precipitation.
effective precipitation *Hydrology.* **1.** the portion of precipitation that becomes runoff. **2.** in irrigation, the portion of precipitation that is available in soil for consumptive use.
effective pressure see EFFECTIVE STRESS.
effective radiated power *Electromagnetism.* the product of the antenna power gain and the power input to the antenna.
effective radius *Astronomy.* the radius of a star calculated from its effective temperature.
effective radius of the earth *Telecommunications.* a value that is used in place of the geometric radius to correct for atmospheric refraction when the index of refraction changes linearly with altitude; under standard refraction conditions, it is equal to 4/3 the geometrical radius or 8.5×10^6 meters.
effective rake *Mechanical Engineering.* in cutting wood or other materials, the angular relationship between the cutter plane and the line made through a tooth point oriented in the direction of chip flow.
effective range *Ordnance.* the maximum distance at which a weapon may be expected to inflict casualties or damage.
effective resistance see HIGH-FREQUENCY RESISTANCE.
effective snowmelt *Hydrology.* the portion of snowmelt that becomes runoff.
effective sound pressure *Acoustics.* the root-mean-square value of sound pressure, obtained by squaring instantaneous sound pressure level measurements, averaging these over the time of one cycle, and taking the square root of this average. Also, ROOT-MEAN-SQUARE (RMS) SOUND PRESSURE.
effective speed *Computer Technology.* the speed at which a computer or device can operate over an extended period of time and that, unlike the rated speed, reflects the slowing effects of control codes, error detection, and other nonproductive activities.

effective stress *Geology.* the mean normal force per unit area that is transmitted directly from particle to particle in a mass of soil or rock. Also, EFFECTIVE PRESSURE, INTERGRANULAR PRESSURE.

effective temperature *Meteorology.* the temperature at which motionless, saturated air would create the same sensation of comfort in a sedentary worker wearing ordinary indoor clothing as that created by actual conditions of temperature, humidity, and air movement. *Astrophysics.* the surface temperature of a black body that has the same size as a star and radiates the same amount of energy per unit area.

effective terrestrial radiation *Geophysics.* the amount of infrared energy released by the earth's surface that exceeds the counterradiation reflected back by the atmosphere.

effective thermal resistance *Electronics.* the effective temperature rise of a semiconductor device per unit power dissipation of a designated junction that is higher than the temperature of a specified external reference point under conditions of thermal equilibrium.

effective thrust *Space Technology.* the theoretical thrust in the nozzle of a rocket motor or engine, without the effects of incomplete combustion or friction flow.

effective time *Computer Technology.* the specific amount of time that a computer system is in actual use and producing useful results.

effective velocity *Hydrology.* the actual velocity of groundwater as it percolates through a water-bearing substance, equal to the volume of water passing through a unit cross-sectional area divided by the effective porosity.

effector *Physiology.* a structure that produces an effect, such as contraction or secretion, in response to nerve stimulation. *Control Systems.* a mechanical device, such as a motor, solenoid, or hydraulic piston, that is used to manipulate a workpiece. *Biochemistry.* a metabolite that binds to the allosteric site on a regulatory enzyme and changes its kinetic properties. Also, MODULATOR.

effector cell *Cell Biology.* a cell that acts in response to a certain stimulus or is capable of mediating a function.

effector molecule *Molecular Biology.* a small molecule having a role in bacterial gene regulation by altering the ability of a regulatory molecule to bind to the DNA site.

effector system *Physiology.* a neuronal system composed of a motor, or efferent, neuron that carries a stimulating impulse from the central nervous system to an effector organ, causing it to react.

efferent *Physiology.* **1.** referring to a neuron that carries motor impulses away from the central nervous system. **2.** of or relating to a blood vessel through which blood flows away from the heart.

efferent ductules *Anatomy.* twelve to fourteen small tubes that carry spermatozoa from the testis to the head of the epididymis.

effervescence [ef´ər ves´əns] *Chemistry.* a bubbling of a liquid that is caused by a rapid release of gases rather than by boiling.

effervescent [ef´ər ves´ənt] *Chemistry.* **1.** of, relating to, or having the property of effervescence. **2.** of a mixture, forming a gas when hydrated.

efficiency the relative effectiveness of a system or device, especially as determined by comparing input and output; specific meanings include: *Mechanical Engineering.* the actual mechanical advantage of a machine (the ratio of force supplied to work done), divided by the ideal mechanical advantage (the distance ratio of the output and input). It is expressed as a percentage and is never greater than 100% for actual machines. Thus if a machine supplies force to a load at a ratio of 4:1, it must move at least 4:1 in relation to the movement of the load; if it actually moves 5:1, then its efficiency is 80%. *Thermodynamics.* a dimensionless quantity that characterizes the performance of a heat engine or system; given as the ratio of the work done by the engine to the amount of heat energy supplied to the engine. *Industrial Engineering.* the use of the least amount of resources to attain a desired amount of output. *Engineering.* a measurement of the amount of heat produced per fuel unit when all of the fuel has been burned. *Chemistry.* a measurement of the effectiveness of an ion-exchange system, expressed in terms of the amount of regenerant needed to remove a specific amount of adsorbed material. *Nucleonics.* the probability that a given particle or photon incident will trigger a response in a radiation detector.

efficiency expert *Industrial Engineering.* a person who is an expert in increasing the efficiency of a production process, as by the elimination of wasted motions or the improvement of technology or labor methods.

efficiency of plating *Virology.* the relative efficiency of infection in susceptible cells expressed as a proportion of the plaque count to the total number of virus particles per unit volume.

efficient estimator *Statistics.* an unbiased estimator that has the smallest variance among all such estimators using the same sample size.

effigy [ef´ə jē´] *Anthropology.* an object formed to represent a person, animal, or other animate object.

effleurage [ef´flůr äzh´] *Medicine.* a massage technique in which long, light, or firm strokes are made with the tips of the fingers, usually over the spine and back; also often used on the abdomen in the Lamaze method of childbirth.

efflorescence [ef´flə res´əns] *Botany.* **1.** the production of flowers. **2.** the season of flowering. **3.** the actual flower bloom. *Chemistry.* the loss of combined water molecules by a hydrated salt when exposed to air, resulting in a lower hydrate or anhydrous salt. *Geology.* a whitish, mealy, fluffy, or crystalline powder forming a crust or hard coating of minerals on the surface of a rock in an arid region as a result of this process. Also, BLOOM.

efflorescent [ef´flə res´ənt] *Chemistry.* of a hydrate, tending to undergo efflorescence.

effluent [ef´floo ənt; ə floo´ənt] *Hydrology.* **1.** flowing out or flowing away. **2.** a stream that flows out of a larger stream, a lake, or another body of water. *Civil Engineering.* **1.** liquid waste matter that results from sewage treatment or industrial processing. **2.** such waste liquid released into waterways. *Chemical Engineering.* any liquid or gas that is discharged from a processing operation. Also, DISCHARGE LIQUOR.

effluent cave see OUTFLOW CAVE.

effluent flow *Hydrology.* the flow of water from the ground into a body of surface water.

effluent seepage *Hydrology.* the movement of water from the land surface into the ground.

effluent stream *Hydrology.* **1.** a surface stream that flows out of a lake or any branch of a larger stream. **2.** a stream that intersects and receives its water from the water table.

efflux time *Materials Science.* the time required for a specific volume of a polymer solution to flow through a capillary tube; used to determine solution viscosity.

effort *Mechanical Engineering.* the energy that is applied to a machine to accomplish useful work.

effort-controlled cycle *Industrial Engineering.* **1.** a work cycle performed by manual work. **2.** a work cycle for which the cycle time is determined by the manual work required.

effort rating see PERFORMANCE RATING.

effort time *Industrial Engineering.* the part of a work cycle that is dependent upon the worker's skill or effort.

effuse *Botany.* spreading and diffuse, often with a poorly defined edge, as in bacterial or fungal growth. *Zoology.* having a valve edge separated by a gap.

effused-reflexed *Mycology.* describing a kind of fungal fruiting body shaped like a sheet with turned-up edges.

effusive see EXTRUSIVE.

EFL error frequency limit.

EFM electronic fetal monitor.

e-folding time *Physics.* the amount of time required for a quantity Q to decay or grow exponentially to an amount Q/e or eQ, respectively, where $e = 2.71828...$ is the base of the natural logarithm.

e format see E NOTATION.

efrapeptin *Biochemistry.* a polypeptide that binds to and blocks (F_0F_1)-type proton ATPases.

EGA *Computer Technology.* a second-generation color video display standard for IBM PCs, with graphics resolution of 640 × 350 pixels in 16 colors. (Acronym for enhanced graphics adapter.)

E galaxy see ELLIPTICAL GALAXY.

Egeria *Astronomy.* asteroid No. 13, discovered in 1850 and measuring 215 kilometers in diameter; it belongs to type C.

egest *Physiology.* **1.** to discharge a substance from the body. **2.** to eliminate undigested or unabsorbed food residues from the intestines.

egestion *Physiology.* the process of eliminating fecal material at the inferior end of the digestive tract.

EGF epidermal growth factor.

egg *Cell Biology.* a female gamete involved in the sexual reproduction of animals; an ovum. Also, **egg cell.** *Vertebrate Zoology.* in certain animals such as birds, reptiles, and fish, an external structure expelled from the female's body, consisting of this reproductive cell along with protective and nutritive material.

egg-and-dart *Architecture.* a design for an ovolo molding, consisting of alternating oval and pointed forms. Also, **egg-and-tongue.**

egg apparatus *Botany.* a group of three primordial cells (an egg and two synergids) located at the micropylar end of the embryo sac in angiosperms.

egg candling *Food Technology.* a common method of grading egg quality, performed by holding each egg up to a light source (originally, a candle) and visually inspecting the yolk.

egg case *Invertebrate Zoology.* a protective case that contains eggs outside the body, such as the silken cocoon where spiders enclose the tubes in which females lay their eggs. Also, **egg capsule**.

egger *Entomology.* the larvae of several moths of the genus *Malacosoma* that feed on the leaves of orchard and shade trees and live colonially in a dense tentlike web. (So called because of its *egg*-shaped cocoon). Also, TENT CATERPILLAR.

egg membrane *Developmental Biology.* any of several investing envelopes of the ovum. If the membrane comes from the ovum itself, it is known as primary; from the ovarian follicle, secondary; and from the oviduct, tertiary. Also, **egg envelope.**

Egg Nebula *Astronomy.* a two-component reflection nebula with a source of infrared and molecular millimeter-wave radiation lying between them.

eggplant *Botany.* **1.** a solanaceous perennial herb, *Solanum melongena*, of East Indian origin, cultivated for its fleshy, edible, purple or occasionally white or yellow fruit. **2.** the fruit of this plant used as a vegetable.

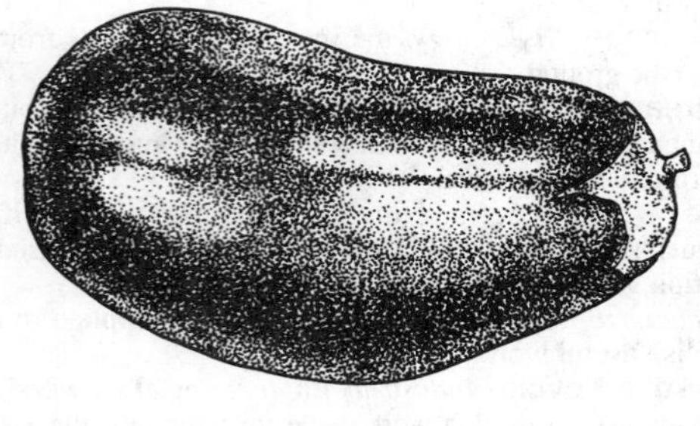

eggplant

egg raft *Zoology.* the eggs produced by certain aquatic animals that are borne on the surface of the water.

eggstone see OOLITE.

egg tooth *Vertebrate Zoology.* a calcareous prominence at the tip of the beak or upper jaw of embryo birds and oviparous reptiles, with which they break through the egg shell at hatching.

eglandular *Biology.* lacking glands.

eglestonite *Mineralogy.* $Hg_6^{+1}Cl_3O(OH)$, a brownish-yellow cubic mineral occurring in crusts or as dodecahedral crystals with an adamantine to resinous luster, having a specific gravity of 8.3 to 8.4 and a hardness of 2.5 on the Mohs scale.

Egnell's law *Meteorology.* a law stating that the velocity of nearly straight winds in the upper troposphere increases with height at approximately the same rate that the air density decreases in any one fixed place above this level.

ego *Psychology.* in Freudian theory, the part of the mind that refers to the self and that is most closely in touch with reality, enabling the individual to perceive, to reason, and to control impulses (the id) by way of inhibition and conscience (the superego). *Anthropology.* in kinship terminology, the focal person from whom all relationships in a kin study are determined.

ego boundary *Psychology.* an awareness that there is a distinction between the self and others; often lacking in persons with certain psychoses.

egocentric *Psychology.* occupied and absorbed with the self; focused on one's own interests and feelings and not concerned with others.

egocentric speech *Psychology.* a childhood speech pattern involving extended conversation that is not directed toward other people.

egocentrism *Psychology.* the fact of being egocentric; absorption with oneself and one's own interests. Also, EGOISM, EGOTISM.

ego-dystonic *Psychology.* of or relating to ideas, behaviors, or impulses that are unacceptable to the ego or self.

ego ideal *Psychology.* a more or less conscious image of the self to which a person aspires, based on a positive identification with significant figures, such as the parents, of the early childhood years.

egoism see EGOCENTRISM.

egomania *Psychology.* extreme self-centeredness or preoccupation with the self. Thus, **egomaniac.**

egophony *Medicine.* a change in the voice sound as heard through a stethoscope, resembling the bleating of a goat, often occurring in cases of pleurisy with effusion.

ego psychology *Psychology.* a psychoanalytic theory or method that focuses on the role of the ego as an independent force shaping personality and behavior. Thus, **ego psychologist.**

Egorov's theorem *Mathematics.* suppose that (f_n) is a sequence of measurable functions that converge to a real-valued function f almost everywhere on a measurable set E of finite measure. Then, given $\varepsilon > 0$, there exists a subset A of E having measure less than ε such that (f_n) converges to f uniformly on $E - A$.

ego strength *Psychology.* a trait of the ego demonstrated by emotional stability and the ability to deal with stress. Also, **ego resiliency.**

ego-syntonic *Psychology.* of or relating to ideas, behaviors, or impulses that are acceptable to the ego or self.

egotism see EGOCENTRISM.

egress *Astronomy.* the exit of a planet or moon from an eclipse, occultation, or transit.

egret [ēg´rit; ē gret´] *Vertebrate Zoology.* a mid-sized heron of the family Ardeidae and the genus *Egretta*, noted for its elaborate mating rituals utilizing long, white, lacy plumage developed during breeding; the plumes, known as **aigrettes,** were once widely used for millinery ornaments, nearly causing the extinction of several species.

egret

Egyptian asphalt *Geology.* a glance pitch, similar to asphalt, that is found in the desert between the Nile and the Red Sea, containing over 99% nonmineral content.

Egyptian cotton *Textiles.* a fine, long-staple, brownish cotton that is grown primarily in Egypt.

Egyptian jasper *Mineralogy.* a brown or banded jasper that occurs as small boulders or pebbles scattered over the surface of the desert between Cairo and the Red Sea. Also, **Egyptian pebble.**

EHF or **ehf** extremely high frequency.

Ehlers-Danlos syndrome *Genetics.* a group of hereditary abnormalities linked to deficiencies in the production of collagen and caused by both autosomal and sex-linked genes, characterized by joint hypermotility, tissue fragility, and abnormally stretchable skin due to the lack of sufficient or healthy collagen.

EHP or **ehp** effective horse power; electric horse power.

Ehrenfest, Paul 1880–1933, Austrian physicist; predicted quasi-ergodic systems; formulated Ehrenfest's equations.

Ehrenfest's adiabatic law *Quantum Mechanics.* a law stating that if the Hamiltonian of a system is varied at infinite slowness to an alternate Hamiltonian, a particle in some eigenstate of the original Hamiltonian will finish in an eigenstate of the new Hamiltonian.

Ehrenfest's equations *Thermodynamics.* a set of equations stating that for a second-order transition of phase, the derivative of the pressure P, with respect to the temperature T, is given by the equation $dP/dT = (C_P{}^f - C_P{}^i)/[TV(\gamma^f - \gamma^i)]$, where i and f refer to two different phases, γ is the coefficient of volume expansion, V is the volume, and C_P is the specific heat at constant pressure.

Ehrenfest's theorem *Quantum Mechanics.* a theorem stating that the expectation values of observables in quantum mechanics evolve in the manner predicted by classical physics.

Ehrenhaft effect *Electromagnetism.* the helical motion of fine particles that circulate about magnetic field lines exposed to light; their motion is due to radiometer effects.

Ehresmann connection *Mathematics.* a g-valued 1-form ω on a principal bundle P over base manifold M with group G such that: (i) $\omega(\sigma(X)) = X$ for all $X \in g$, and (ii) $\omega((R_a)_*Y) = \mathrm{Ad}(a^{-1})\omega(Y)$ for all $a \in G$ and all Y tangent to P. σ is the fundamental mapping of the Lie algebra g of G to vector fields on P, and $(R_a)_*$ is the derivative of the map given by multiplication on the right by $a \in G$.

Ehrlich, Paul 1854–1915, German bacteriologist; discovered cures for syphilis; Nobel Prize for research on immunity.

Ehrlichia *Microbiology.* a genus of Gram-negative, pathogenic bacteria of the tribe Ehrlichieae, family Rickettsiaceae, characterized by nonmotile, coccoid cells that grow intracellularly in the white blood cells of certain mammals.

Ehrlichieae *Microbiology.* a group of bacteria in the family Rickettsiaceae that are pathogenic for certain mammals; one species, *Ehrlichia sennetsu,* is pathogenic in humans.

ehrlichiosis *Medicine.* an infection with parasitic leukocytic rickettsiae of the genus *Ehrlichia,* producing chills, fever, cutaneous rash, severe headache, and prostration.

Ehrlich's reagent *Organic Chemistry.* $(CH_3)_2NC_6H_4CHO$, granular or leaflike crystals; slightly soluble in water and soluble in most organic solvents; melts at 74°C; used as a reagent and in dye preparation.

E-H T junction *Electromagnetism.* a junction in a waveguide system in which an E-plane T junction and an H-plane T junction intersect the main waveguide at a common point. Thus, **E-H T tuner.**

EHV or **ehv** extra high voltage.

EI *Aviation.* the airline code for Aer Lingus.

EI exposure index.

eicosanoic acid *Organic Chemistry.* $CH_3(CH_2)_{18}COOH$, crystals that are insoluble in water and soluble in alcohol; melts at 75.5°C and boils at 328°C.

eicosanoid *Endocrinology.* any of the autacoid hormones produced by the metabolism of arachidonic acid, including the leukotrienes, lipoxins, prostaglandins, and thromboxanes.

eider

eider *Vertebrate Zoology.* any of several species of heavy-bodied sea ducks of *Somateria* and related genera of the family Anatidae, native to cold areas of the Northern Hemisphere; known for soft breast feathers called **eiderdown,** used for pillows, jackets, and other warm coverings.

eidetic *Psychology.* **1.** of or relating to the exact visualization of objects or events previously seen. **2.** a person having this ability.

eidetic image or **imagery** *Psychology.* an extremely vivid and detailed visual image retained by the memory.

eidetic memory *Psychology.* the ability of the memory to retain an accurate, detailed visual image of previously seen material.

EID_{50}/HA ratio *Virology.* a method of calculating the proportion of defective particles present in a given virus, specifically the ratio between infectivity titer and the hemagglutinating titer of the prepared virus.

Eifelian see COUVINIAN.

Eiffel, Alexandre Gustave [īf´əl] 1832–1923, French architect and engineer, designed Eiffel Tower.

Eigen, Manfred [ī´gən] born 1927, German biological physicist; shared Nobel Prize for advances in studying high-speed chemical reactions.

eigenfrequency *Physics.* a characteristic frequency at which a vibrational system will oscillate.

eigenfunction *Mathematics.* a solution u of the equation $Lu = \lambda u$, where L is a linear operator (such as a partial differential operator) on a space of functions, and λ is an eigenvalue. Such a function u is of interest because the action of L leaves u essentially unchanged. *Quantum Mechanics.* one of several different characteristic functions obtained as a physically allowed solution to the Schrödinger wave equation.

eigenmatrix *Mathematics.* a square matrix, all of whose entries are zero except on the main diagonal, containing the roots of the characteristic equation of the matrix (the eigenvalues). Also, MATRIX OF EIGENVALUES.

eigenspace *Mathematics.* given a linear transformation $T: E \to E$ on a vector space E, a nonzero subspace of E composed of the zero vector and the set of elements of E that are eigenvectors corresponding to a fixed eigenvalue. Also, CHARACTERISTIC SPACE.

eigenstate *Quantum Mechanics.* a state whose state vector is an eigenvector of the matrix corresponding to some operator.

eigentones *Acoustics.* the modes of natural vibration in a specific medium; e.g., the resonant vibrations in the air columns of a pipe organ.

eigenvalue *Mathematics.* **1.** given a linear transformation $T: E \to E$ on a vector space E over a field K, an element λ of K such that $T(v) = \lambda v$ for some nonzero v in E. The eigenvalues of T, if they exist, are the roots in K of the characteristic polynomial of T. Also, CHARACTERISTIC VALUE. **2.** the eigenvalues of a graph are the eigenvalues of the adjacency matrix of the graph. *Quantum Mechanics.* an allowed characteristic value of a linear operator or corresponding dynamical variable that corresponds to a quantum state in a given force field.

eigenvector *Mathematics.* given a linear transformation $T: E \to E$ on a vector space V over a field K, a nonzero element v of V such that $T(v) = \lambda v$ for some element λ of K. Also, CHARACTERISTIC VECTOR. The set of eigenvectors corresponding to a fixed eigenvalue is a subspace of E, called an eigenspace. *Quantum Mechanics.* a nonzero vector mapped by the column representation of its corresponding eigenfunction.

eightfold way *Particle Physics.* a description of the baryon and meson states that group themselves into families of eight, possessing the property of three units of strangeness and one variety of charge, negative. (From the Eightfold Way or Path, a fundamental concept of Buddhism.)

eight-track tape *Acoustical Engineering.* a magnetic tape cartridge having eight tracks (four right and four left stereo tracks) of recorded material; used before the introduction of the four-track cassette tape.

Eijkman, Christiaan [īk´män´] 1858–1930, Dutch physician and nutritionist; Nobel Prize for research in vitamin-deficiency diseases.

Eijkman's test *Microbiology.* a test to determine the ability of enterobacteria to produce gas from lactose; a positive test indicates the presence of *Escherichia coli.*

Eikenella *Bacteriology.* a genus of Gram-negative bacteria that occur in the human oral cavity and upper respiratory tract as nonmotile, rod-shaped cells or filaments.

eikonal equation *Physics.* an equation that determines wave propagation in a nonhomogeneous medium, provided that the medium does not possess discontinuities over distances comparable to the wavelength.

Eimeria *Invertebrate Zoology.* a genus of protozoans in the order Eucoccidiida, parasitic in vertebrates; includes serious pathogenic forms.

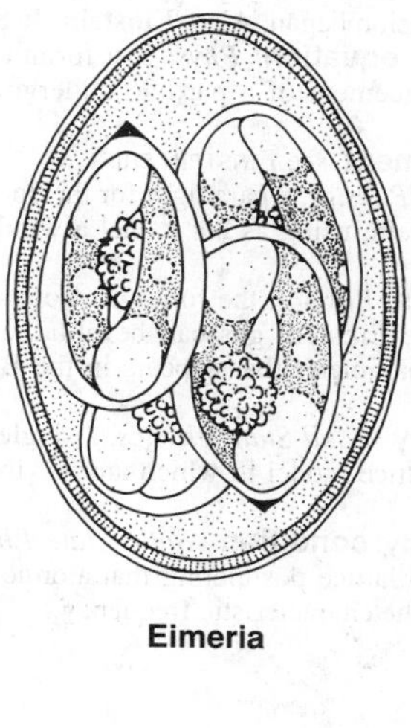

Eimeria

Eimeriina *Invertebrate Zoology.* a suborder of coccidian protozoans, parasites on arthropods and vertebrates.

Einchluss thermometer *Analytical Chemistry.* a liquid-filled laboratory thermometer capable of measuring temperatures from -201°C to 360°C.

E indicator see E DISPLAY.

einkanter *Geology.* a windworn stone having only one facet or only one sharp edge, with the face formed at right angles to the wind.

Einstein, Albert 1879–1955, German physicist who later lived in the U.S.; explained the quantum theory; formulated the theory of relativity; discovered Brownian motion.

Albert Einstein

einstein *Physics.* a unit of energy of light, given by the product of the energy of one photon at a particular frequency and Avogadro's number.

Einstein-Bohr equation *Quantum Mechanics.* the equation $\nu = \Delta E/h$, which gives the frequency of radiation during a transition between two atomic states whose energies differ by ΔE; ν is the frequency and h is Planck's constant.

Einstein coefficients *Quantum Mechanics.* a set of probability coefficients that express the probabilities of stimulated and spontaneous radiative transitions between stationary energy levels.

Einstein-de Haas effect *Physics.* the rotation observed in a freely hanging ferromagnetic mass when it becomes magnetized.

Einstein-de Haas method *Electromagnetism.* a method of determining the gyromagnetic ratio of a ferromagnetic substance by measuring the angular displacement of a cylinder suspended by a torsion fiber while reversing its magnetization.

Einstein-de Sitter model *Physics.* a simple, relativistic model that describes a universe expanding from an infinitely condensed source, and in which the laws of Euclidean geometry hold everywhere and the density of the universe is inversely proportional to the square of the time elapsed since expansion began. Also, **Einstein-de Sitter universe.**

Einstein diffusion equation *Physics.* a formula for determining the mean square displacement of a particle undergoing Brownian movement.

Einstein displacement see EINSTEIN SHIFT.

Einstein elevator *Physics.* a paradigm for the conditions of deep space developed by Einstein, using as his model a windowless elevator free-falling in its shaft.

Einstein equations *Physics.* the formulas from which the density and pressure of a Bose-Einstein gas can be obtained; given in terms of power series in a parameter that appears in the Bose-Einstein distribution.

Einstein frequency *Solid-State Physics.* a single frequency characteristic of a crystal lattice model in which atoms vibrate independently of one another.

Einstein frequency condition *Solid-State Physics.* an assumption applied to a crystal lattice postulating that atomic vibrations within the lattice are fixed at the characteristic frequency.

einsteinium *Chemistry.* a synthetic radioactive chemical element having the symbol Es, the atomic number 99, and atomic weight (of the most stable isotope) 252.08; produced in the cyclotron at the University of California, Berkeley, by bombarding berkelium and californium with helium ions and deuterons. (Named after Albert *Einstein.*)

Einstein mass-energy relation *Physics.* a fundamental formula in modern physics, $E = mc^2$, that relates the energy of a particle or system to its mass.

Einstein number *Physics.* a dimensionless number derived from the ratio of the velocity of a fluid to the speed of light, used in magnetofluid dynamics.

Einstein partition function *Physics.* the partition function for a solid, based on the Einstein frequency condition.

Einstein photoelectric law *Quantum Mechanics.* the relationship between the energy of emitted electrons and incident radiation in the photoelectric effect, expressed by the equation $E = h\nu - W$, where E is the electron energy, h is Planck's constant, ν is the frequency of the incident radiation, and W is the work function of the material.

Einstein-Planck law *Quantum Mechanics.* a law stating that the energy of a photon is given by the relationship $E = h\nu$, where E is the energy of a photon, h is Planck's constant, and ν is the frequency of the photon.

Einstein-Podolsky-Rosen experiment *Quantum Mechanics.* a thought experiment proposed in 1935 involving the simultaneous measurement of two correlated photons at distant locations, with the conclusion that quantum mechanical description of some physical systems fails to satisfy realistic criteria of completeness; paradoxically, quantum theory dictates that the particles communicate over spacelike separations. This paradox inspired the hidden-variables theories.

Einstein relation *Physics.* an equation relating the mobility of charges in an ion solution or a semiconductor to the diffusion coefficient: $m = eD/kT$, where m is mobility, e is electron charge, D is the diffusion coefficient, k is the Boltzmann constant, and T is absolute temperature.

Einstein-Rosen waves *Physics.* the gravitational waves produced by oscillating matter along an infinitely long cylindrical axis.

Einstein's absorption coefficient *Atomic Physics.* a constant of proportionality that relates the radiation-absorption rate of a sample of atoms to the radiation-energy density per unit wave number and the number of atoms in their ground state.

Einstein's equation for specific heat *Solid-State Physics.* an equation giving the specific heat of a solid crystalline substance, based on the assumption that the atomic vibrations occur at a single fixed frequency; later modifications, such as the Debye equation, allow for a continuum of possible frequencies.

Einstein shift *Physics.* the lengthening of the wavelengths of light that are emitted by bodies with strong gravitational fields, causing a displacement of spectral lines toward the red part of the spectrum; predicted in the general theory of relativity. Also, EINSTEIN DISPLACEMENT, GRAVITATIONAL REDSHIFT.

Einstein space *Mathematics.* a (pseudo-) Riemannian space whose metric tensor is proportional to its Ricci curvature tensor.

Einstein's principle of relativity *Physics.* the basic postulate of Einstein's special relativity theory, stating that the laws of nature have the same form in all inertial frames of reference.

Einstein's relativity theory see RELATIVITY THEORY.

Einstein's special relativity theory see SPECIAL RELATIVITY THEORY.

Einstein static universe *Physics.* a model, derived from Einstein's general relativity theory, that postulates a fixed universe neither expanding nor contracting.

Einstein's unified field theory see UNIFIED FIELD THEORY.

Einstein viscosity equation *Physical Chemistry.* an equation that gives the viscosity of a colloidal solution as $\eta = \eta_0(1 + k\phi)$, where η is the viscosity of the solution, η_0 is the viscosity of a pure suspending medium, ϕ is the fraction of the total volume occupied by the colloidal particles, and k is a constant coefficient based on the configuration of the particles.

Einthoven, Willem [īn´tō vən] 1860–1927, Dutch physiologist; invented the string galvanometer, forerunner of the electrocardiograph.

Einthoven's law *Cardiology.* a law stating that, in electrocardiography, if electrocardiograms are taken with three leads, at any given instant the potential of any wave in lead 2 is equal to the sum of the potentials in the two other leads.

Eisenstein, Ferdinand Gottfried Max [īz´in stīn´] 1823–1852, German mathematician; known for his work on theory of functions and theory of numbers.

Eisenstein's criterion *Mathematics.* let D be a unique factorization domain with quotient field F, and let $f(x) = \sum_{i=0}^{n} a_i x^i$ be a polynomial of degree n in $D[x]$. Then $f(x)$ is irreducible in $F[x]$ if there exists an irreducible element p of D such that p divides a_i for $i < n$, p does not divide a_n, and p^2 does not divide a_0.
Eisenstein series *Mathematics.* any series of the form

$$G_r(\omega_1, \omega_2) = \sum_{m,n} (\omega_1 m + \omega_2 n)^{-r},$$

where ω_1 and ω_2 are distinct nonzero complex numbers, and the summation is taken over all pairs of integers m and n except $m = 0 = n$. The points $\omega_1 m + \omega_2 n$ form a point lattice in the complex plane. The Eisenstein series vanishes identically for odd integers r and converges absolutely for integers $r \geq 3$.
ejaculate *Physiology.* **1.** to eject semen. **2.** the semen discharged in a single emission.
ejaculation *Physiology.* the expulsion during sexual intercourse of seminal fluid from the urethra of the penis.
ejaculatory duct *Anatomy.* the continuation of the ductus deferens from the entrance of the seminal vesicles to the prostatic urethra.
eject *Computer Programming.* the action, by a printer, of releasing a page and moving to the next. Also, PAGE EJECT.
ejecta *Science.* solid material that is thrown off or out from something. *Physiology.* waste materials that have been cast out from the body. *Volcanology.* also, **ejectamenta.** any material, including volcanic ash, pyroclasts, and bombs, that is discharged by a volcano. *Astronomy.* such fractured rocky debris excavated from a crater, either by the force of meteorite impact or by a volcanic explosion.
ejecta blanket *Astronomy.* a generally symmetrical apron of ejecta surrounding a crater; it is layered thickly at the crater's rim and thin to discontinuous at the blanket's outer edge.
ejection a process of throwing off or out; specific uses include: *Metallurgy.* the removal of a fabricated product from the mold in which it was manufactured. *Ordnance.* **1.** the expulsion of an empty cartridge case from small arms and rapid-fire guns. **2.** in air armament, the process of forcefully separating a missile, bomb, or other aircraft store from the aircraft, including the departure of air crew in an emergency.
ejection capsule *Space Technology.* a detachable compartment on an aircraft or spacecraft that may be ejected as a unit and parachuted to the ground for the recovery of personnel, instruments, and recorded data.
ejection clicks *Medicine.* sharp clicking sounds from the heart, which may be caused by the sudden swelling of a pulmonary artery or the abrupt dilatation of the aorta; often heard during examinations of individuals with septal defects or hypertension, but common and of no clinical significance in pregnant women and many other healthy individuals.
ejection port *Ordnance.* the opening through which empty cartridge cases are expelled from a firearm after firing.
ejection seat *Aviation.* an emergency device containing a detachable seat, designed to eject a pilot and his equipment safely from an aircraft.
ejector something that ejects; specific uses include: *Engineering.* any device that withdraws fluid material from an area by a steam or air jet. Also, EDUCTOR. *Metallurgy.* an instrument that removes an object from a mold when the casting process is completed. *Ordnance.* a device in the breech mechanism of a firearm that automatically expels the empty cartridge case or unfired cartridge.
ejectosome *Cell Biology.* a coiled structure of unknown function that is found just below the cell membrane in unicellular *Cryptophytes* and is discharged upon appropriate stimulation.
eka- a combining form meaning "next (after)," as in *eka-iodine*; used with the name of a known chemical element to designate an unknown element that should occur next in the same group of the periodic system.
ekdemite see ECDEMITE.
EKG electrocardiogram; electrocardiograph.
ekistics *Geography.* the scientific study of human settlements, especially the study of urban development. Thus, **ekistic, ekistician.**
ekki *Materials.* the hard and durable rust-colored wood of the tropical African trees *Lophira alata* and *Lophira procera*; used especially for flooring and railway ties.
Ekman, Vagn Wilfrid 1874–1954, Swedish oceanographer.
Ekman convergence *Oceanography.* a zone of convergence between warm and cold water masses, caused by the Ekman transport of surface water.
Ekman current meter *Oceanography.* a measuring instrument of the type designed by V.W. Ekman, consisting of a propeller and a magnetic compass hung from an immobile vessel; lowered to a desired depth and activated by messengers to measure the speed of water currents.
Ekman layer *Meteorology.* the transition boundary between the free atmosphere and the surface boundary layer in which the shearing stress is constant. Also, SPIRAL LAYER.
Ekman spiral *Meteorology.* an idealized mathematical model of the wind distribution in the planetary boundary layer of the atmosphere, within which the earth's surface has a sizable effect on the air motion.
Ekman transport *Oceanography.* the net mass displacement of water from one place to another, caused by wind blowing steadily over the surface; the net mass transport is 90° to the right (in the Northern Hemisphere) of the wind's direction.
Ekman water bottle *Oceanography.* a cylindrical tube used for deep-water sampling; its reversing mechanism closes plates at both ends when tripped by a messenger, trapping the water sample.
elaboration *Psychology.* the process or technique of relating new information to a mental image or to an already-known concept in order to retain it in long-term memory. Also, **elaborative rehearsal.**
Elachistaceae *Botany.* a family of tufted or pulvinate brown algae of the order Chordariales that range in length from microscopic to about 1 centimeter. They are completely autotrophic, although they are sometimes believed to be parasitic and agents of infection.
Elaeagnaceae *Botany.* a family of dicotyledonous, many-branched thorny shrubs in the order Proteales that are usually dioecious, characterized by leathery, silverish leaves and unisexual or bisexual flowers that occur either singly or in racemes; grown as ornamentals, especially the oleaster and sea buckthorn.
elaenia *Botany.* any of the tropical American flycatchers of the genus *Elaenia,* characterized by short crests and small bills.
Elaeocarpaceae *Botany.* a family of tropical and subtropical dicotyledonous trees and shrubs of the order Malvales that usually produce alkaloids and are sometimes tanniferous.
elaidic acid *Organic Chemistry.* $C_{18}H_{34}O_2$, white leaflets that melt at 44–45°C. Also, OCTADENSIC ACID.
elaidinization *Organic Chemistry.* the conversion of an unsaturated fatty compound from the *cis* form to the *trans* form; the resulting *trans* acid is more oxidation-resistant.
elaidin reaction *Analytical Chemistry.* an analysis using nitrous acid to differentiate between types of oils; nondrying oils form a solid isomer, semidrying oils slowly thicken, and drying oils become resinous.
elaioplast *Histology.* a leukoplast that secretes oil.
Elakatotrichaceae *Botany.* a family of unbranched green algae of the order Klebsormidiales, characterized by cells that undergo desmoschisis but lack coherence, and occurring as individuals, diads, or very short filaments.
eland *Vertebrate Zoology.* a large oxlike antelope of the family Bovidae, native to the plains and woodlands of Africa; characterized by a pale brown coat with a dark mane and tail tip, long spiraled horns, and a large dewlap from the throat.

eland

Elaphomycetaceae *Mycology.* a family of fungi belonging to the order Elaphomycetales, occurring underground and possessing spore sacs or ascocarps that resemble wood.
Elaphomycetales *Mycology.* an order of fungi belonging to the class Plectomycetes that grow underground in a symbiotic relationship with trees.

Elapidae *Vertebrate Zoology.* a family of poisonous snakes of the suborder Serpentes, having permanently erect fangs in the front of the upper jaw and found in the warmer parts of both hemispheres; includes the cobras and mambas, the coral snakes of the New World, and the majority of Australian snakes.

elapsed time *Computer Technology.* the total amount of apparent time taken by a process, measured between the apparent beginning and end of the process; it does not necessarily correspond to the actual time taken by the process itself.

Elara *Astronomy.* the seventh moon of Jupiter, discovered in 1905 and measuring 76 km in diameter.

Elasipodida *Invertebrate Zoology.* an order of deep-sea sea cucumbers in the class Holothuroidea.

Elasmidae *Invertebrate Zoology.* a family of small wasps in the hymenopteran superfamily Chalcidoidea.

elasmobranch *Vertebrate Zoology.* a cartilaginous fish of the subclass Elasmobranchii, a shark or ray.

Elasmobranchii *Vertebrate Zoology.* a subclass of cartilaginous fishes of the class Chrondrichtyes, including the sharks and rays and characterized by an upper jaw not fused to the cranium, amphistyoic or hyostylic jaw suspension, numerous teeth, and a covering of placoid scales.

Elassomatidae *Vertebrate Zoology.* the pigmy sunfishes, a subfamily of freshwater North American fish of the family Centrarchidae in the order Perciformes.

elastance *Electricity.* an electrical quantity that is equal to the reciprocal of the capacitance of a capacitor. *Medicine.* the degree to which an air- or fluid-filled organ, such as the lung or bladder, can return to its original size when a distending or compressing force is removed.

elastase *Enzymology.* a proteolytic enzyme secreted by the pancreas that digests elastin and renders it soluble.

elastic *Mechanics.* having the property of elasticity; able to return to its original shape after experiencing strain and removal of deforming stress.

elastica *Mechanics.* a plane curve made by a thin rod when forces and couples are applied only at its ends.

elastic aftereffect *Mechanics.* the lag that some materials show in recovering from elastic deformation.

elastic axis *Mechanics.* a line lengthwise through a beam; transverse surface forces must be directed toward this line in order not to produce torsion.

elastic bitumen see ELATERITE.

elastic body *Mechanics.* a solid body possessing elasticity.

elastic buckling *Mechanics.* the buckling of a member or structure under a compressive load within its elastic range.

elastic cartilage *Histology.* a type of cartilage containing an abundance of extracellular elastic fibers.

elastic center *Mechanics.* the midpoint in the cross section of a beam between the shear center and the center of twist (the three are often at the same point).

elastic collision *Mechanics.* a collision between ideally elastic bodies, such that the final kinetic energy is the same as the initial kinetic energy.

elastic constant see COMPLIANCE CONSTANT.

elastic cross section *Physics.* an area which measures the probability that an elastic collision will occur between two particles.

elastic curve *Mechanics.* the shape of the neutral surface in a beam deflected by a load.

elastic deformation *Materials Science.* a temporary deformation in a solid material that has been subjected to a load, wherein the material returns to its original shape after the load is removed.

elastic design *Building Engineering.* a former construction design in which a structure is configured to allow working stresses between 0.5 and 0.66 of the elastic limit in the framing material used.

elastic energy see STRAIN ENERGY.

elastic failure *Mechanics.* the deformation of a solid body in excess of its elastic limit, resulting in permanent deformation or fracture.

elastic fiber *Histology.* a yellow extracellular fiber that imparts elasticity to connective tissues.

elastic flow *Mechanics.* the elastic deformation of a material, or recovery from such deformation.

elastic force *Mechanics.* the resisting internal force resulting from the elastic deformation of an object, which restores the original shape when the external force is removed.

elastic hysteresis *Mechanics.* the dependence upon strain history of the relationship between stress and strain in a nearly elastic solid. Also, MECHANICAL HYSTERESIS.

elasticity *Materials Science.* a property of materials in response to stress, indicating the degree to which strain disappears from a material when the stress has been removed. *Mathematics.* the mathematical study of the behavior of elastic bodies, especially using the methods of partial differential equations.

elasticity number 1 *Fluid Mechanics.* in viscoelastic flow, a dimensionless parameter indicating the ratio of elastic force to inertial force.

elastic limit *Mechanics.* the critical stress or strain beyond which a body will yield to enduring deformation. Also, YIELD POINT LOAD.

elastic modulus *Materials Science.* the ratio between the stress placed on a material and the resulting elastic strain. The most commonly used modulus of elasticity is Young's modulus, which is usually measured in a simple tensile test and is the ratio of the applied load/unit area and the length change/unit length. Also, MODULUS OF ELASTICITY.

elasticotaxis *Microbiology.* the phenomenon whereby certain species of gliding bacteria become oriented parallel to lines of stress that are induced in a solid medium.

elastic potential energy *Mechanics.* the energy made available for use by the return of an elastic body to its original configuration.

elastic proportionality *Materials Science.* see ELASTICITY.

elastic ratio *Mechanics.* for the component of a specified stress, the elastic limit divided by the ultimate strength of a material.

elastic rebound theory *Geology.* a theory that explains faulting as an abrupt release of a progressively increasing elastic strain between rock masses on either side of a fault. Also, REID MECHANISM.

elastic recovery *Mechanics.* the portion of deformation that reverses after the causative force is removed.

elastic scattering *Materials Science.* a process in which particles collide and disperse without losing energy.

elastic strain *Materials Science.* a reversible dimensional response to stress; that is, when the stress is removed, the strain disappears.

elastic strain energy *Mechanics.* the work required to elastically deform a body.

elastic theory *Mechanics.* the theory of the elastic relationships among loads, stresses, deformations, and strains. Also, THEORY OF ELASTICITY.

elastic tissue *Histology.* a fibrous connective tissue containing an abundance of elastic fibers.

elastic vibration *Mechanics.* the oscillatory motion in an elastic solid.

elastic wave *Physics.* a mechanical wave that moves through a solid as a progressive oscillation of matter about fixed equilibrium positions.

elastin *Biochemistry.* a protein, present in artery walls, whose chains are crosslinked with desmosine and residues of lysine. *Materials.* such a protein constituting the major component of elastic fibers.

elastodynamics *Mechanics.* the field of dynamics that deals with the propagation and properties of elastic waves.

elastoid *Medicine.* a substance formed in the vessels of the uterus after delivery as a result of the hyaline degeneration of blood vessels.

elastomer *Materials.* any of numerous natural or synthetic materials having elastic properties similar to rubber.

elastomeric fiber *Textiles.* a nontextured yarn or fabric made from polymer filaments that can be stretched repeatedly up to twice its usual length and still snap back when released.

elastoplastic *Materials Science.* an elastomer that contains transient crosslinks, unlike most elastomers; it softens when heated for processing, and becomes solid yet maintains its elastic behavior upon cooling. Elastoplastics are derived from polyolefins, polyurethanes, polyesters, and styrene copolymers, and are widely used in making machine parts, housewares, automobile accessories, sporting equipment, adhesives, and toys. Also, THERMOPLASTIC RUBBER.

elastoplasticity *Mechanics.* the state exhibited by a material that has deformed both elastically and plastically.

elastoresistance *Electricity.* a variation in the electrical resistance of a material while it is subjected to a stress within its elastic limit.

elastosis *Medicine.* **1.** the degeneration of elastic tissue. **2.** a process of degenerative changes in dermal connective tissue with increased amounts of elastic material.

elater *Botany.* **1.** an elongated, filamentous, hygroscopic structure in the sporangia of liverwort sporophytes that triggers dehiscence and spore dispersal. **2.** one of the four club-shaped, hygroscopic bands that spiral around spores in horsetail (*Equisetum*). **3.** a filament in the capillitium of slime molds. *Invertebrate Zoology.* a click beetle belonging to the family Elateridae. *Entomology.* the furcula or abdominal springing organ in the wingless springtails in the order Collembola.

Elateridae *Invertebrate Zoology.* a large family of coleopteran insects, the click beetles, in the superfamily Elateroidea.

elaterite *Geology.* a brown to black, asphaltic pyrobitumen that is soft and elastic when fresh, but becomes hard and brittle on exposure to air. Also, ELASTIC BITUMEN, MINERAL CAOUTCHOUC.

Elateroidea *Invertebrate Zoology.* a superfamily of beetles in the suborder Polyphaga.

elaterophore *Botany.* **1.** a tissue bearing elaters, as in some liverworts. **2.** a column of sterile tissue in the sporangium of leafy liverworts of the order Jungermanniales.

Elatinaceae *Botany.* a family of cosmopolitan dicotyledonous herbs and half-shrubs of the order Theales that grow in shallow water or wet places and often creep or root at the nodes.

E layer *Geology.* the seismic region, in the classification of the earth's interior, equivalent to the outer core, extending from 2900 to 4710 km below the surface. *Geophysics.* a layer in the earth's atmosphere, occurring at 100 to 120 km above sea level, in which short-wave radio waves are reflected and temperature increases with increasing altitude.

elbaite *Mineralogy.* $Na(Li,Al)_3Al_6(BO_3)_3Si_6O_{18}(OH)_4$, a pink, blue, or green trigonal mineral of the tourmaline group, having a specific gravity of 3.03 to 3.10 and a hardness of 7 on the Mohs scale; occurs as gemstone crystals in granite pegmatite on the island of Elba; forms a series with dravite.

elbow *Anatomy.* the angular joint between the upper arm and the forearm; a hinge joint at which the distal end of the humerus and the proximal ends of the ulna and the radius articulate and are held in place by various tendons and ligaments.

elbow any of various features or devices having the right-angle shape of a bent elbow; specific uses include: *Geography.* a sharp bend in any geographic feature, especially a river. *Mechanical Devices.* **1.** a pipe or pipe fitting that bends approximately 90°. **2.** a two-bed arch stone whose lower bed is horizontal while the upper bed is inclined toward the arch center. Also, UNION ELBOW. *Electromagnetism.* a 90° waveguide bend of relatively short radius.

elbow catch *Building Engineering.* an L-shaped catch with a hooked end on one arm that engages with a hooked plate on the other.

Elbs reaction *Organic Chemistry.* the formation of anthracenes by dehydration and cyclization of diaryl ketones that contain a methyl or methylene group ortho to the carbonyl group.

elder *Botany.* any tree or shrub of the genus *Sambucus,* having clusters of small white flowers and a red or black berrylike fruit (**elderberry**) that is used in making jelly and wine.

elec. electric, electrical, electricity.

elecampane *Botany.* a composite weed, *Inula helenium,* characterized by large yellow flowers and aromatic leaves and root; naturalized in North America.

elective *Surgery.* of or relating to a surgical procedure that is considered to be neither mandatory nor an emergency. *Chemistry.* describing the tendency to combine with certain substances in preference to others. Thus, **elective attraction.**

elective culture *Microbiology.* a microorganism grown from a mixed culture under conditions that are selective for one kind of organism.

electr- a combining form meaning "electric" or "electricity."

Electra *Astronomy.* one of the six visible stars in the Pleiades.

Electra complex *Psychology.* the female counterpart of the Oedipus complex, involving a daughter's unconscious sexual attraction to her father, and jealousy and hostility toward her mother. (From the mythical Greek character *Electra,* who incited her brother to kill their mother.)

electret *Electricity.* a solid dielectric with permanent electric polarization; the electric equivalent of a permanent magnet.

electret headphone *Acoustical Engineering.* an earphone, such as those used in sound-powered phones, that has a diaphragm of dielectrically charged material, such as polarized plastic, thereby eliminating the need for an external power source.

electret microphone *Acoustical Engineering.* a microphone that has an electret transducer capable of converting audio waves to electromagnetic waves.

electret transducer *Electronics.* a device that produces voltage by utilizing the motion of a permanently polarized foil placed next to a metal-coated plate.

electric *Electricity.* **1.** containing, producing, arising from, or actuated by electricity. **2.** carrying electricity, or designed to carry electricity.

electrical *Electricity.* of or relating to electricity without having its properties or characteristics. Thus, **electrical engineer, electrical equipment, electrical insulator, electrical rating.**

electrical ablation *Surgery.* the removal of some organ or part using electrocautery.

electrical angle *Electricity.* an angle denoting a certain instant in an AC cycle or the phase difference between two alternating quantities.

electrical axis *Solid-State Physics.* the direction in a crystal in which electrical resistance is at a minimum.

electrical center *Electricity.* a point in the middle of an adjustable inductor or resistor that divides it into two equivalent electrical values.

electrical code *Electricity.* a set of rules directing the practical installation and application of electrically operated equipment.

electrical degree *Electricity.* the unit of measurement for an electrical angle; equal to 1/360 cycle of an alternating quantity.

electrical disintegration *Metallurgy.* the selective removal of metal or alloy by the action of an electric spark.

electrical distance *Electromagnetism.* the distance between two points in free space, expressed in terms of the time required for an electromagnetic wave to travel between the points.

electrical engineer *Engineering.* an engineer who engages in research, design, production, operation, and maintenance of electric power production and transmission facilities, telecommunication systems, or microelectronic devices and systems.

electrical engineering a branch of engineering that focuses on the design, construction, and operation of electrical systems and equipment.

Electrical Engineering

Electrical engineering is the design, analysis, and operation of electrical devices and systems. The term electrical engineering came into general use at the end of the 19th century when progress in physics and electrical science resulted in widespread practical applications of electricity. The founders of electrical science were physicists and mathematicians such as Ampere, Faraday, Gauss, and Maxwell, whose theories eventually led to the electric motor and the incandescent lamp. The practical problem of generating and distributing electricity throughout a building or a city became the province of the electrical engineer and the term well describes the task of tending the "engine," in this case an electrical engine such as the dynamo and the motor. The availability of local motive power without steam or waterwheels and light without flame created a new industry as well as a new profession.

The world of electrical power and light, that of Westinghouse and Edison, was soon joined by electrical communication. The telegraph and telephone and then the "wireless" telegraph and radio grew from the works of Morse, Bell, Hertz and Marconi. Although today we quickly distinguish between electrical power systems and radio, it should be noted that one of the first successful radio transmitters was a 5 kilowatt rotating-machine alternating-current generator, operating in the vicinity of 100,000 cycles per second or 100 kHz.

With the coming of the vacuum-tube and transistor, electronics, the behavior of the electron in vacuum and in solids, joined the field as electronic engineering, and the pertinent U.S. professional society is known as the Institute of Electrical and Electronics Engineers (IEEE). Albeit, the discipline of electrical engineering at college or university encompasses the full range from power systems through communications, computers, and, most recently, optical devices such as the laser and the camcorder. The common theme is electricity, the electron, and Maxwell's wave equation.

Robert H. Kingston
Senior Lecturer
Electrical Engineering and Computer Science
Massachusetts Institute of Technology

electrical equivalent *Analytical Chemistry.* an outside calibrated current source that is compared to the current passing through an electrolyte solution.
electrical fault see FAULT.
electrical impedance *Electricity.* the total opposition to the flow of alternating current in a circuit due to resistance and reactance.
electrical length *Electromagnetism.* the length of a conductor when expressed in terms of wavelengths, radians, or degrees.
electrically alterable read-only memory see EAROM.
electrically connected *Electricity.* joined through an electrical conducting path, such as a resistor, inductor, capacitor, or wire.
electrically erasable read-only memory see EEROM.
electrically stimulated osteogenesis *Medicine.* a bone regeneration process in which surgically implanted electrodes convey electric current, particularly at nonunion fracture sites; effective because of the different electrical potentials within bone tissue.
electrically suspended gyro *Engineeering.* a gyroscope characterized by a central rotating element that is kept in place by an electromagnetic or electrostatic field.
electrical noise *Electricity.* any unwanted electrical signals in a circuit.
electrical paper *Materials.* paper formed from wood pulp fibers that are modified with binders; used for insulation in machine parts or in the slots of small motors.
electrical potential energy *Electricity.* the ability to move an electrical charge from one point to another.
electrical pressure see PRESSURE TRANSDUCER.
electrical properties *Electricity.* the measurable characteristics of electrical circuits, such as power dissipation, current, voltage, and resistance.
electrical scanning *Electronics.* a technique in which the area covered by a transmitting or receiving antenna can be varied electrically without moving the antenna. Also, ELECTRONIC SCANNING. *Telecommunications.* the technique of scanning a surface to reproduce or transmit a picture.
electrical steel *Metallurgy.* a low carbon-iron alloy containing 0.5–5% silicon; used primarily for the cores of transformers and alternators.
electrical storm *Meteorology.* **1.** a popular name for a thunderstorm. **2.** an infrequent disturbance of the electric field in the lower atmosphere caused by strong winds and blowing dust, but without thunderstorm activity. Also, ELECTRIC STORM.
electrical thickness *Oceanography.* the vertical distance between the surface of an ocean current and an isokinetic point at which the current's speed is about 10% of its speed at the surface.
electrical transcription see TRANSCRIPTION.
electrical transport *Biochemistry.* any transport across an energy-transducing membrane.
electrical weighing system *Engineering.* an instrument that weighs an object by measuring the change in resistance caused by the deformation of a mechanical element loaded with the object.
electrical zero *Electricity.* the equivalent of neither a positive nor negative voltage, exactly at ground level.
electric anesthesia *Medicine.* a temporary anesthesia caused by the passage of an electric current through part of the body.
electric arc see ARC.
electric-arc welding *Metallurgy.* a welding process that uses an electric arc as the source of heat.
electric blood warmer *Mechanical Devices.* an apparatus for heating blood before infusions, especially in cases of massive transfusions in which cold blood might put a patient in shock.
electric boiler *Mechanical Engineering.* a steam generator that uses electrical energy as a heat source.
electric brake *Mechanical Engineering.* **1.** a braking system whose force is supplied by an adjustable spring counteracted by a solenoid, a centrifugal thruster, and an actuator, in which the actuating force is supplied by current flowing through a solenoid or an electromagnet. Also, ELECTROMAGNETIC BRAKE. **2.** an emergency braking system that is automatically applied to an electric-powered apparatus when a power failure occurs. **3.** the contact component of an electric braking system.
electric braking *Mechanical Engineering.* **1.** in an electrically driven vehicle, a system in which the motor acts as a generator, returning energy to the contact braking element. **2.** the process of applying any type of electric brake.
electric burn *Medicine.* the tissue damage resulting from heat given off by an electric current.
electric cauterization see ELECTROCAUTERIZATION.
electric cell see CELL.
electric charge *Electricity.* an accumulation of electricity in a storage battery, condenser, or other device, which may then be discharged. *Physics.* one of the basic properties of elementary particles that give rise to all electric and magnetic forces and interactions; these properties are given negative and positive algebraic signs and measured in coulombs.
electric chopper *Electromagnetism.* a chopper apparatus that uses an electromagnet driven by an AC source to vibrate a reed that periodically interrupts an electrical contact.
electric circuit see CIRCUIT.
electric clock *Horology.* **1.** a mechanical clock with a pendulum or balance wheel kept in motion by an electromagnetic switch turned on and off regularly by contacts or by an electronic circuit. **2.** a clock in which the first wheel of the going train is the rotor of a synchronous electric motor whose speed is entirely controlled by the frequency of the AC current.
electric comparator *Electricity.* a circuit that compares two input signals, which may be digital or analog, but that always results in a digital-signal output.
electric constant *Electricity.* the permittivity of empty space, equivalent to 8.854×10^{-12} farad/meter.
electric contact see CONTACT.
electric control *Electricity.* **1.** any electrical device, such as a switch or a potentiometer, that is used to vary a circuit parameter. **2.** the control of a machine by electric switches, relays, or rheostats.
electric controller *Electricity.* a device regulating the amount of electric power that is delivered to an apparatus. *Control Systems.* a component of a control system that compares the measurement of a controlled variable with a set point.
electric current *Electricity.* see CURRENT.
electric delay line *Electronics.* a device that postpones a signal's arrival time in a circuit by employing capacitive and inductive properties.
electric desalting *Chemical Engineering.* a method for removing impurities from crude oil by settling out in an electrostatic field.
electric detonator *Engineering.* an explosive device activated by a fuse wire that initiates the operation of a primer.
electric dipole see DIPOLE.
electric dipole transition *Atomic Physics.* the dominent process by which an atom produces or absorbs radiation when it changes from one energy level to another.
electric displacement *Electricity.* electric field intensity multiplied by permittivity. Also, ELECTRIC INDUCTION.
electric distribution system see DISTRIBUTION SYSTEM.
electric double layer see DOUBLE LAYER.
electric drive *Mechanical Engineering.* an electromechanical device that transmits motion from one shaft to another while controlling the velocity ratio of the shafts.
electric eel *Vertebrate Zoology.* an eel-like, freshwater fish that oftens attains a length of six feet and is capable of emitting strong electric discharges produced by electric organs consisting of modified muscle tissue situated along the ventral part of the body; found in the Amazon and Orinoco rivers and tributaries.

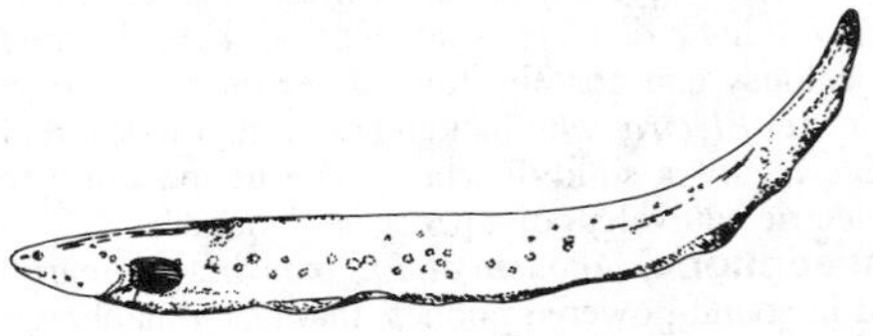

electric eel

electric energy *Electromagnetism.* **1.** the energy inherent in an array of charged particles because of their relative positions. **2.** the energy inherent in a circuit because of its position in relation to a magnetic field.
electric engine *Space Technology.* in rocketry, a reaction engine in which the propellant is accelerated by an electrical device.
electric eye see PHOTOCELL.
electric fence *Engineering.* a wire fence containing high-voltage, low-current intermittent pulses that are activated when the fence is touched.
electric field *Electricity.* **1.** a region in space in which a stationary electric charge experiences a force due to its charge. **2.** the electric force per unit test charge.
electric field intensity or **strength** see ELECTRIC INTENSITY.

electric field vector *Electromagnetism.* a vector that represents the force per unit charge acting on a positive charge in an electric field; symbolized by **E** and having units of volts per meter in the mks system.

electric filter *Electronics.* a circuit that passes selected frequencies of alternating currents while weakening other frequencies. Also, ELECTRIC WAVE FILTER.

electric fish *Vertebrate Zoology.* any of several fish that produce an electric shock by means of special organs, such as the electric eel, electric ray, or the **electric catfish,** *Malapterurus electrus.*

electric flux *Electricity.* **1.** the integral of the electric displacement that is perpendicular to a component surface. **2.** an electric line of force in a given area. Also, ELECTROSTATIC FLUX.

electric flux line see ELECTRIC LINE OF FORCE.

electric forming *Electronics.* a process in which electric energy is applied to a device, such as a semiconductor, so as to permanently change its electrical characteristics.

electric fuse see FUSE.

electric gathering locomotive see GATHERING MOTOR.

electric guitar *Acoustical Engineering.* a guitar with a built-in contact microphone, which amplifies sound produced by the guitar strings and allows various adjustments of characteristics of the sound.

electric heating *Engineering.* a process in which electric energy becomes heat energy by resisting the free flow of electric current.

electric hygrometer *Engineering.* an electrical instrument that measures the humidity of the atmosphere.

electric hysteresis see FERROELECTRIC HYSTERESIS.

electrician *Engineering.* a technician who specializes in the installation, repair, or operation of electrical equipment.

electric ignition *Mechanical Engineering.* the ignition system in an automobile or internal-combustion engine that ignites the combustible mixture in the engine cylinders with a high-voltage, high-tension spark produced between metal points in a spark plug.

electric image see IMAGE.

electric induction see ELECTRIC DISPLACEMENT.

electric intensity *Physics.* the magnitude of an electric field at a point in the field that is equal to the force that would be exerted on a small unit charge placed at that point. Also, ELECTRIC FIELD INTENSITY, ELECTRIC FIELD STRENGTH.

electricity **1.** one of the most important forms of energy, consisting of oppositely charged electrons and protons that produce light, heat, magnetic force, and chemical changes. **2.** the flow of this energy; electric current. **3.** the general phenomenon of charges at rest and in motion.

Electricity

It is hard to imagine life today without electricity. From the moment an electric alarm clock wakes us in the morning, to the time we turn off the lights at night, we are surrounded by gadgets, appliances, and machines run by electricity. At home, in the office or workplace, and in places of entertainment, indoors or outdoors, applications of electricity are all pervasive; almost every facet of human endeavor seems to depend on them critically.

Yet, very little was known about electricity until just 200 years ago. Although a Greek philosopher noted in 600 B.C. that when amber ("elektron" in Greek) is rubbed with cloth, it attracts light particles, the phenomenon remained a curiosity for 22 centuries. In 1600 A.D. a British physician, William Gilbert, published a study of "electric" behavior in several materials. Another 150 years passed before Benjamin Franklin introduced the concept of "positive" and "negative" electricity, and in 1752 showed that lightning and electricity were the same. Later in the century, Coulomb formulated his laws for forces between charged bodies.

Knowledge and uses of electricity blossomed in the 19th century, thanks to the work of several intellectual giants. Volta invented the electric pile (battery), Oersted, Ampere, Faraday, and Henry discovered electromagnetism and related phenomena, and Morse invented the telegraph, all in the first half of the century. These were followed by the invention of the telephone by Bell, electric arc lights and incandescent lamps by Edison and Swann, and electric motors and generators by Edison, Westinghouse, and many others near the end of the century. A whole host of industrial applications were introduced, based on magnetic, chemical, thermal, optical, and other properties of electricity. Fundamental studies of electromagnetic wave radiation by Maxwell and Hertz led to radio experiments by Marconi and other pioneers at the turn of the century.

New models of the atom by a number of scientists in the early years of the 20th century increased the understanding of electricity, leading to the development of vacuum tubes and electronics, and making radio a household word. World War II and the post-war period produced a surge of unprecedented creativity. Revolutionary discoveries and inventions were made by many researchers—too numerous to be named in a short article—resulting in the development and widespread use of modern-day marvels, such as television, radar, space communications, computers, and low-cost, miniaturized consumer electronics. Innumerable electrical products (many of them based on transistors, integrated circuits, and microelectronic chips) are used in almost every field of activity today.

Due primarily to electricity and electronics, the world has become a "global village"—anything happening in the remotest part of the globe can be known to the whole world community instantly, and people from all corners of the earth can interact and discuss matters of common interest in "real time." But of course not all uses of electricity are benign: it also has been used to develop highly refined weapons systems, which can seek and destroy specific localized targets or cause havoc over vast areas, with great accuracy and reliability.

What does the future hold for applications of electricity? As before, it will depend on the vision of the men leading the work. Computers and microelectronics will play a vital role; for example, combined with land and space communications, they will lead to wrist-watch sized "Dick Tracy" radios for two-way conversations anywhere world-wide. Similarly, computer-based expert systems (using neural networks, fuzzy logic, and artificial intelligence), combined with control systems and robotics, will lead to intelligent homes, offices, hospitals, and factories, where machines not only perform all ordinary chores but also make highly complex decisions and carry out difficult tasks reserved for humans today. On the other hand, their combination with missiles and sensors and sophisticated weapons (all based on electricity) can result in mass destruction, a la "Star Wars."

It is as if electricity had created a Frankenstein's monster, which can be harnessed for the good of mankind or can be allowed to run amok and destroy it. It all depends on how man adjusts to this "brave new world," and how he uses the enormous power placed at his disposal by electricity and its related technology.

S.H. Durrani
Program Manager, Advanced Systems
NASA Office of Space Communications

electric knife *Surgery.* a surgical instrument that uses electrical current to produce heat, which incises tissue by cauterization.
electric lamp *Electricity.* an electric device, such as an incandescent lamp, arc lamp, glow lamp, or fluorescent lamp, that projects light when voltage is applied across the terminals.
electric light *Electricity.* **1.** an incandescent lamp. **2.** the light produced by such a lamp.
electric line of force *Electricity.* an imaginary line in which each line segment is parallel to the direction of the electric field or the direction of the electric displacement at that point, and the density of the collection of lines is relative to the electric field or the electrical displacement. Also, ELECTRIC FLUX LINE.
electric locomotive *Mechanical Engineering.* a locomotive operated by electric power supplied from a third rail alongside or between the two track guide rails or from an overhead wire system feeding a trolley.
electric main see POWER-TRANSMISSION LINE.
electric meter *Engineering.* an instrument, such as an ampere-hour meter, that measures electrical power and totals its measurement with time. Also, POWER METER.
electric moment *Electricity.* a vector equal to the product of the magnitude of either of two charges of equal magnitude but opposite polarity and the distance between their centers.
electric monopole *Electricity.* an electric charge distribution that is centered about a point or is spherically symmetrical.
electric motor *Mechanical Engineering.* a device that converts electrical power into mechanical torque.
electric multipole *Electromagnetism.* a charge distribution from a series of common distributions, such as the electric monopole (a single charge), dipole (two opposite charges, separated by a short distance), and quadrupole (two dipoles, separated by a short distance).
electric multipole field *Electromagnetism.* the electromagnetic field produced by static or oscillating electric multipoles.
electric octupole moment *Electricity.* a quantity that describes an electric charge distribution, determined by integrating the product of the charge density, the third power of the distance from the origin, and a spherical harmonic over the charge distribution.
electric organ *Vertebrate Zoology.* the generating and sensory organs of the electric field in the electric fish, composed of electroplaques and located at the base of long, low-resistance, jelly-filled canals that radiate through the body from the head and monitor the electric field at all points over the body.
electric outlet see OUTLET.
electric polarizability *Electricity.* the induced dipole moment of an isotropic medium for which the electric polarization direction and electric field strength are equal anywhere in the medium.
electric polarization see POLARIZATION, def. 4.
electric potential *Electricity.* the potential measured by the energy of a unit positive charge at a point, expressed relative to an equipotential surface, generally the surface of the earth, that has zero potential.
electric potential gradient *Medicine.* the net difference in electric charge across the membrane of a cell.
electric power *Electricity.* electric energy per unit time, equivalent to the current flowing through an object multiplied by the voltage across it.
electric power generation *Electricity.* the large-scale production of electric power for various commercial and residential uses. Similarly, **electric power transmission.**
electric power line see POWER LINE.
electric power plant *Mechanical Engineering.* a power plant that converts raw energy into electricity, such as a hydrosteam, diesel, or nuclear generating station.
electric power station *Electricity.* a facility that generates electrical energy using generators. Similarly, **electric power substation.**
electric power system *Mechanical Engineering.* the circuitry applied to many electrical devices, in which electric energy is generated, transmitted, transformed, and distributed in the form of heat or as a driving force to other motor-controlled systems.
electric precipitation *Chemical Engineering.* a procedure using an electric field to enhance the separation of hydrocarbon reagent dispersions; often such dispersions are too fine to separate efficiently by other means.
electric propeller *Aviation.* a propeller whose pitch is regulated by an electric motor.
electric propulsion *Space Technology.* any form of rocket propulsion in which the propellant is either composed of charged electrical particles or accelerated by an electric or magnetic field.
electric protective device *Electricity.* any component in a circuit configuration used explicitly to protect other circuit components from excessive heat or current.
electric quadrupole *Electricity.* a charge distribution that generates an electric field equal to that produced by two electric dipoles whose moments have the same magnitude, but extend in opposite directions a small distance apart.
electric quadrupole lens *Electromagnetism.* a device, consisting of four electrodes arranged in alternating polarity, that focuses a beam of charged particles.
electric quadrupole moment *Electricity.* a quantity that describes an electric charge distribution, determined by integrating the product of the charge density, the second power of the distance from the origin, and a spherical harmonic over the charge distribution.
electric quadrupole transition *Atomic Physics.* a process in which an atom produces or absorbs quadrupole radiation when it changes from one energy level to another.
electric railroad *Mechanical Engineering.* a railroad having a continuous system of overhead wires or a third rail mounted alongside or between the guide rails throughout the track's length to supply electric power to a locomotive transportation system.
electric ray *Vertebrate Zoology.* any ray of the family Torpedinidae, capable of emitting strong electric discharges.
electric reactor *Electricity.* an electric circuit component that stores energy and introduces it into a circuit as an inductive or capacitive reactance.
electric scanning *Electronics.* a target-seeking method in which an operator directs the radar beam by varying the phase or amplitude of the currents flowing into various components of its antenna.
electric shock *Physiology.* a method of stimulating an excitable tissue, such as a nerve or muscle, by applying a brief electrical current. *Neurology.* the immediate effects resulting from such stimulation.
electric shock tube *Plasma Physics.* an instrument used to quickly ionize a gas, consisting of a gas-filled tube through which a high voltage is sent, allowing study of the path of the electric shock as it moves through the tube.
electric steel *Metallurgy.* any of several steels produced in an electric furnace. Also, **electric-furnace steel.**
electric storm see ELECTRICAL STORM.
electric strength see DIELECTRIC STRENGTH.
electric susceptibility *Electricity.* a measure of the polarization ease of a dielectric, equivalent to the ratio of the polarization to the product of the electric field strength and the vacuum permittivity.
electric switch see SWITCH.
electric switchboard see SWITCHBOARD.
electric tachometer *Engineering.* an instrument that measures the output voltage of a generator to determine the speed of a rotating unit.
electric transducer *Electronics.* a two-port device that processes electrical signals.
electric transient *Electronics.* a current or voltage that appears temporarily in an electric circuit whose steady-state condition has been disrupted.
electric tuning *Electronics.* the process of selecting a desired frequency on a component, such as a receiver, transmitter, or oscillator, without using mechanical devices.
electric twinning *Solid-State Physics.* a defect in crystals in which two neighboring regions have antiparallel electric moments.
electric typewriter *Mechanical Engineering.* a typewriter containing an electric motor that provides power for all keyboard operations.
electric wave see ELECTROMAGNETIC WAVE.
electric wave filter see ELECTRIC FILTER.
electride *Inorganic Chemistry.* an experimental compound made up of a positively charged ion of an alkali metal (periodic table Group Ia) and an electron; the electron functions as a chemical element in the formation of salts.
electrification *Electricity.* the process of applying an electric charge to a component or device.
electrification ice nucleus *Meteorology.* an ice nucleus formed by the fragmentation of dendritic crystals exposed to an electric field strength of several hundred volts per centimeter.
electro see ELECTROTYPE.
electro- a combining form meaning "electric" or "electricity."
The first syllable in electro- words can be variously pronounced as [i], [ē], or [ə], depending on the dialect of the speaker and on the context in which the word is used.

electroacoustic *Acoustical Engineering.* relating to or involving electroacoustics.

electroacoustic locator *Surgery.* a device for locating foreign objects in the body by amplifying the sound made when the object is touched by a probe.

electroacoustics *Acoustical Engineering.* the science or practice of converting acoustic energy to electromagnetic energy, and vice versa.

electroanalgesia *Medicine.* the application of an electric current to the spinal cord or a peripheral nerve to relieve pain.

electroanalytic chemistry *Chemistry.* the chemical analysis of compounds using electric current to produce characteristic, observable change in the substance being studied.

electroblotting *Molecular Biology.* the transfer of molecules by electrophoresis for the purpose of transfixing RNA or DNA to an immobilizing medium such as a nitrocellulose or nylon membrane.

electrocapillarity *Physics.* a change in the surface tension of a liquid that is induced by the presence of an electric field at the surface of the liquid.

electrocardiogram [i lek´trō kär´dē ə gram] *Medicine.* a graphic record made by an electrocardiograph; used to diagnose cardiac abnormalities.

electrocardiograph [i lek´trō kär´dē ə graf] *Medicine.* an instrument that records the electrical activity of the heart; used to detect abnormal electric impulses through the muscle.

electrocardiophonograph *Medicine.* an instrument that performs both phonocardiography and electrocardiography simultaneously.

electrocauterization *Medicine.* a method of removing warts or polyps by placing a needle or wire loop heated by a direct galvanic current on the tissue to be removed.

electrochemical [i lek´trō kem´ i kəl] *Physical Chemistry.* of or relating to electrochemistry; having to do with the relationship of chemical change and electric force.

electrochemical cell *Physical Chemistry.* a vessel containing two electrodes separated by an electrolyte phase. Also, CELL.

electrochemical constant see FARADAY CONSTANT.

electrochemical corrosion *Materials Science.* the removal of material from the surface of a metal by an electrochemical reaction, occurring when two metals or portions of the same material are initially at different potentials in an electrolyte.

electrochemical effect *Physical Chemistry.* a term for the process by which chemical energy is converted into electrical energy, or electrical energy into chemical energy.

electrochemical emf *Physical Chemistry.* the electrical force that results from a chemical reaction, as produced in dry batteries or fuel cells, or as occurs in nature in a galvanic reaction. Also, EMF.

electrochemical equivalent *Physical Chemistry.* the mass, in number of grams, of any substance generated or depleted when an electric current of one coloumb passes through an electrolytic solution.

electrochemical gradient *Biochemistry.* the sum of the gradients of mass and electric charge in an ion traversing a membrane.

electrochemical initiation *Materials Science.* the generation of free radicals by electrode reactions; used to begin the process of polymerization.

electrochemical machining *Metallurgy.* a process for the selective removal of a metal or alloy by anodic dissolution. Also, ECM.

electrochemical potential *Physical Chemistry.* a thermodynamic construct that conceptually separates the chemical potential into the familiar chemical potential and a term which depends on the electrical environment.

electrochemical power generation *Engineering.* a process by which chemical energy is directly converted into electrical energy, as in a battery cell.

electrochemical process *Physical Chemistry.* **1.** the chemical change that occurs in an electrolytic solution when an electric current passes through it. **2.** the chemical change that gives rise to an electric current, such as occurs in fuel cells or batteries.

electrochemical recording *Electronics.* a facsimile technique in which information is recorded by passing an electric current through a chemically sensitized area of the paper.

electrochemical series *Physical Chemistry.* a chart on which various substances, such as metals, are listed according to their chemical reactivity or electrode potential, with the most reactive ranked at the top and the others listed in descending order. Also, ELECTROMOTIVE SERIES.

electrochemical transducer *Engineering.* an instrument used to measure input parameters through the use of a chemical change, producing the measurement in terms of an electrical signal.

electrochemical valve *Electricity.* an electric valve composed of a metal in contact with a solution or compound across whose boundary current travels more readily in one direction than the other, and in which the valve action is accompanied by chemical variations.

electrochemiluminescence *Physical Chemistry.* the production of light during an electrochemical reaction. Also, ELECTROGENERATED CHEMILUMINESCENCE.

electrochemistry *Physical Chemistry.* the branch of chemistry concerned with the chemical changes that occur when a chemical reaction produces an electric current or when an electric current produces a chemical reaction.

electrochromic *Physical Chemistry.* undergoing change in color upon the passage of an electric current.

electrochromic display *Electronics.* a passive solid-state display in which an electric field controls the characteristics of light transmission and light reflection.

electrocoagulation *Medicine.* a form of surgery in which tissue is hardened or destroyed by passing high-frequency current from an electric cautery device through it.

electroconvulsive therapy *Medicine.* a treatment in which convulsions are induced by passing a low-voltage alternating electric current through the brain. *Psychology.* the use of such a technique to treat severe psychiatric disorders. Also, ECT, ELECTROSHOCK THERAPY.

electrocorticography *Medicine.* a technique in which the electrical activity of the cerebral cortex is surveyed using an electroencephalograph and electrodes attached to the brain.

electrocratic *Chemistry.* of or relating to a liquid colloidal suspension, maintained in equilibrium by the repulsion between insoluble solid particles that are positively or negatively charged.

electrode *Electricity.* a conductor by which a current enters or leaves a nonmetallic medium, such as an electrolytic cell, arc generator, or vacuum tube; the anode and cathode of an electric cell are electrodes. *Medicine.* a medium by which an electric current is conducted from the body to physiologic monitoring equipment.

electrode admittance *Electronics.* the ratio of the alternating component of the current flowing through an electrode to the alternating component of electrode voltage.

electrodecantation *Physical Chemistry.* a form of electrodialysis in which a three-part container is divided by two semipermeable membranes, with a positive electrode in one end chamber and a negative one in the other; ionic materials in the center chamber move to their respective electrodes through the membranes.

electrode capacitance *Electronics.* the capacitance between one electrode and a reference point, such as ground, another electrode, or all the other electrodes connected together.

electrode characteristic *Electronics.* the relationship between the current flowing through an electrode and the electrode voltage; usually indicated by a graph.

electrode conductance *Electronics.* the ratio of the DC current through an electrode to the DC voltage between that electrode and a reference terminal.

electrode couple *Electricity.* a pair of electrodes in an electric cell with a potential difference existing between them.

electrode dark current *Electronics.* the generally weak current that flows through a photoconductive cell, such as a camera tube, when it is shut off. Also, DARK CURRENT.

electrode dissipation *Electronics.* the power lost by an electrode in the form of heat when it is bombarded by electrons or ions.

electrode drop *Electronics.* the decrease in voltage that occurs as a result of resistance in an electrode.

electrode impedance *Electronics.* the ratio of the phasor representing the AC voltage of an electrode (relative to a reference terminal) to the phasor representing the AC current through that electrode.

electrode inverse current *Electronics.* current that flows through an electrode in the direction opposite to that for which it was designed.

electrodeposition *Physical Chemistry.* the deposit of a material, usually a metal or an alloy, at or on an electrode as a result of the passage of an electric current through a solution or suspension of the material.

electrode potential *Physical Chemistry.* the potential developed by a metal or other electrode material immersed in an electrolytic solution; usually related to the standard potential of the hydrogen electrode, which is established at zero. Also, ELECTRODE VOLTAGE.

electrode resistance *Electronics.* the ratio of the DC voltage of an electrode (relative to a reference terminal) to the DC current through that electrode.

electrodesiccation *Medicine.* an electrosurgery technique in which tissue is destroyed by burning with an electric spark; used primarily for eliminating superficial growths under local anesthesia.
electrode voltage *Electronics.* the voltage between an electrode and a given point, such as the cathode of an electron tube or ground. Also, **electrode potential.**
electrode wear control *Design Engineering.* the process of minimizing the loss of electrode material during processing.
electrodiagnosis *Medicine.* the diagnosis of a disease by recording the electrical activity of a tissue or organ.
electrodialysis *Physical Chemistry.* a process in which an electric field transports ionized material through a membrane to separate it from other liquids or ions of opposite charge.
electrodialyzer *Physical Chemistry.* a device by which electrodialysis is performed; salt can be removed from seawater by means of such a device.
electrodischarge machining *Design Engineering.* a shaping technique for the surface of a metal object, in which the object is immersed in a dielectric liquid along with a tool that emits an electric discharge of high current density and short duration.
electrodisintegration *Nuclear Physics.* a process in which a nucleus is split into two or more parts by bombarding it with electrons.
electrodynamic *Acoustical Engineering.* describing an acoustical device, such as a loudspeaker or microphone, that derives electroacoustic energy from the action of a current or the motion of a conductor in a magnetic field.
electrodynamic ammeter *Engineering.* an instrument used to measure a current traveling through an immovable coil and a movable coil that are connected in series.
electrodynamic drift *Geophysics.* the motion of charged particles in the upper atmosphere, caused by the combined force of electric and magnetic fields.
electrodynamic instrument *Engineering.* an instrument that is activated by the reaction between the current in one or more moving coils and the current in one or more fixed coils. Also, **electrodynamometer.**
electrodynamic loudspeaker see MOVING-COIL LOUDSPEAKER.
electrodynamic microphone see MOVING-COIL MICROPHONE.
electrodynamics *Electromagnetism.* the study of the relationships between electromagnetic and mechanical phenomena.
electrodynamic wattmeter *Engineering.* an instrument that is designed to measure low-frequency electricity in watts, and is activated by the torque exerted between currents carried by fixed and movable coils.
electroelution *Biotechnology.* the separation and removal of a substance by means of an electrical charge, such as in the freeing of an enzyme from an absorbent.
electroencephalogram *Medicine.* a graphic record of minute changes in the electric potential associated with the activity of the cerebral cortex, as picked up by electrodes placed on the scalp.
electroencephalography *Medicine.* a method of graphically recording brain wave activity, used to diagnose seizure disorders, brainstem disorders, tumors, or clots.
electroencephaloscope *Medicine.* an instrument used for detecting brain potentials of different sections of the brain and revealing these potentials on a cathode-ray tube.
electroendosmosis *Physical Chemistry.* the migration, under the influence of an electric field, of the liquid phase of a colloidal solution toward an electrode. Also, ELECTROOSMOSIS.
electroexplosive *Engineering.* a device or system that carries an electric impulse that activates an explosive to detonation or deflagration.
electroextraction *Physical Chemistry.* the process of recovering metal from metallic salts using electrolysis.
electrofiltration *Geology.* an electromotive force that is set up between two sides of a sheet when an electrolyte is forced through a sheet of some pervious solid dielectric.
electrofluid *Fluid Mechanics.* shear-thinning fluid whose flow properties are changed into those of a viscoplastic nature by the addition of electric-field modulation.
electrofocusing see ISOELECTRIC FOCUSING.
electroform see FORMING.
electroformed mold *Metallurgy.* a mold fabricated by electrodeposition.
electroforming *Metallurgy.* the process of fabricating a component by electrodeposition.
electrofusion *Microbiology.* an electrical technique for fusing cells or protoplasts together.
electrogalvanizing *Metallurgy.* the process of electroplating zinc on a ferrous base.
electrogas dynamics *Physics.* the extraction of kinetic energy from a moving gas to convert to electric energy.
electrogas flux-cored welding *Metallurgy.* in arc welding, the process in which the arc is struck between a flux-core filler electrode and the workpiece.
electrogenerated chemiluminescence see ELECTROCHEMILUMINESCENCE.
electrogenesis *Biology.* the production of electricity by a living plant or animal.
electrogenic transport *Biochemistry.* a transfer of charge across a membrane that generates a potential difference across the membrane.
electrogram *Meteorology.* an automatically produced record of time variations in the atmospheric electric field for a given point. *Medicine.* a record of electric activity of the heart as recorded from electrodes within the cardiac chambers or on the epicardium.
electrograph *Engineering.* a graphic representation created by the movement of an electric current on sensitized paper or an electrically controlled writing instrument. *Telecommunications.* such equipment used in facsimile transmission.
electrographic pencil *Electronics.* a device that produces conductive marks on paper; used for detection by a conductive-mark sensing device.
electrogravimetry *Analytical Chemistry.* a technique in which a suitable electrode is weighed before and after electrolysis to determine the amount of metal deposited on the electrode from the sample solution. Also, **electrodeposition.**
electrohemodynamics *Medicine.* a noninvasive technique for measuring the mechanical properties and hemodynamic characteristics of the vascular system.
electrohydraulic heart *Medicine.* an artificial heart whose ventricles are driven by the alternate pumping of a fluid rather than by compressed air, and is powered by a compact electric motor.
electrohydrodynamic ionization mass spectroscopy *Spectroscopy.* a mass spectroscopic technique for the analysis of nonvolatile molecules that involves the use of a volatile solvent with a high dielectric constant, together with high electric-field gradients to induce ion emission.
electrojet *Geophysics.* a narrow belt of intense electric current flowing through the lower ionosphere in the equatorial and polar regions that creates auroras.
electrokinetic [i lek´trō ki net´ik] *Chemistry.* relating to electrokinetics; having to do with electricity in motion. *Physical Chemistry.* relating to one of the four types of electrokinetic effects, which are electroosmosis, electrophoresis, streaming potential, and sedimentation potential.
electrokinetic effects *Physical Chemistry.* the conditions associated with relative movement between a charged surface and an electrolytic solution; the surface is ordinarily either a solid particle suspended in the solution or the medium through which the solution flows. Also, **electrokinetic phenomena.**
electrokinetic potential see ZETA POTENTIAL.
electrokinetics *Electromagnetism.* the study of the motion of electric charges.
electrokinetic transducer *Electricity.* a device that converts dynamic physical forces into electric signals.
electrokinetograph *Engineering.* a device designed to measure the speed of ocean currents, based on their electrical effects in the earth's magnetic field.
electrokymograph *Medicine.* an apparatus that uses both a photoelectric recording system and a fluoroscope to make a record of the movements of a shadow within the fluoroscopic field; once commonly used to study the heart.
electrolepsy *Neurology.* a type of chorea characterized by a continuous sequence of sudden, violent, rapid, jerky movements that appear synchronized but are involuntary.
electroless plating *Metallurgy.* plating from an aqueous solution on any surface, caused by an autocatalytic chemical reduction.
electroluminescence [i lek´trō loom´i nes´əns] *Electronics.* **1.** the emission of light by a phosphor or semiconductor that is excited by an electromagnetic field. **2.** the direct, nonthermal conversion of electrical energy into light energy in a liquid or solid.
electroluminescent display *Electronics.* a display device that presents data, usually in the form of numbers or letters, when alternating current is applied to its electroluminescent segments.

electroluminescent panel *Electronics.* a panel that produces a low-intensity light when an alternating current is applied to the phosphor centered between its two electrodes.

electrolysis [i lek´träl´ ə sis] *Physical Chemistry.* a process in which the passage of an electric current through an electrolytic solution or other suitable medium produces a chemical reaction, such as that which occurs in a battery. Also, ELECTROLYTIC PROCESS. *Medicine.* the destruction of tumors by an electric current.

electrolyte [i lek´trə līt´] *Physical Chemistry.* any liquid or solid substance that while in solution or in its pure state will conduct an electric current by means of the movement of ions; usually it is a solution of water and acids or metal salts. Also, BATTERY ELECTROLYTE. *Physiology.* any of certain inorganic compounds, such as sodium, potassium, magnesium, and calcium, which dissociate fluids into ions that conduct electric currents and that constitute the major force in controlling fluid balance within the body. The careful monitoring of electrolytes and replacement of fluid and electrolytes are important in the care of many illnesses.

electrolyte acid see BATTERY ACID.

electrolyte-activated battery *Electricity.* a reserve battery with an aqueous electrolyte kept in a separate chamber, from which a mechanism forces it into the cells of the battery for activation.

electrolyte balance *Medicine.* the equilibrium between electrolytes in the body.

electrolytic [i lek´trə lit´ik] *Physical Chemistry.* relating to, causing, or involving electrolysis.

electrolytic analysis *Analytical Chemistry.* a method of analysis based on determining the amount of chemical change that takes place at an electrode.

electrolytic capacitor *Electricity.* a capacitor in which an electrolyte serves as a plate; the other plate is wound aluminum foil. A thin layer of oxidation on the foil is the dielectric. Electrolytic capacitors have comparatively high values of capacitance for their size; however, they also have high leakage current and carry the requirement to observe polarity. Also, POLARIZED CAPACITOR.

electrolytic cell *Physical Chemistry.* an electrochemical cell in which the reactions are driven by the use of an external potential greater than the thermodynamic or reversible potential of the cell.

electrolytic conductance *Physical Chemistry.* the movement of charged particles through a medium.

electrolytic conductivity *Physical Chemistry.* the ability of charged particles to move through a medium.

electrolytic copper *Metallurgy.* refined copper produced by electrodeposition.

electrolytic dissociation *Chemistry.* the ionization of a solute in solution. *Physical Chemistry.* the breaking up of a neutral ionic compound into two or more oppositely charged ions, usually by the effect of dissolution.

electrolytic grinding *Mechanical Engineering.* a combined grinding and machining operation in which the abrasive cathodic grinding wheel is in contact with the anodic surface of an electrolyte; used with a metal-bonded and diamond-impregnated grinding wheel.

electrolytic interrupter *Electricity.* an interrupter component that is tilted to change the current through it.

electrolytic machining *Mechanical Engineering.* an electrochemical process, similar to electroplating, in which the workpiece acts as an anode and the tool as a cathode.

electrolytic mercaptan process *Chemical Engineering.* a procedure used to remove mercaptans from refinery streams by using an electrolytic solution.

electrolytic model *Petroleum Engineering.* a laboratory-model simulation of steady-state fluid flow through porous media found in a reservoir. Also, POTENTIOMETRIC MODEL, OIL-FIELD MODEL, GELATIN MODEL.

electrolytic pickling *Metallurgy.* the process of removing surface oxides from a metallic product by electrochemical means.

electrolytic plating see ELECTROPLATING.

electrolytic potential *Physical Chemistry.* the smallest difference in electronic energy between an electrolytic solution and an electrode that is immersed in the solution that will lead to the acceptance or donation of electrons.

electrolytic powder *Metallurgy.* the metallic powder produced by electrodeposition, either directly or by comminuting a bulk electrodeposit.

electrolytic process *Physical Chemistry.* see ELECTROLYSIS.

electrolytic recording *Electronics.* a document created by passing an electric current through a stylus onto chemically treated paper.

electrolytic rectifier *Electricity.* a rectifier with an electrolytic conductor between its electrodes.

electrolytic rheostat *Electricity.* a variable rheostat whose electrodes are submerged in a conducting liquid.

electrolytic separation *Physical Chemistry.* a process in which a charged electrode collects ions from a solution.

electrolytic solution *Physical Chemistry.* a liquid that contains a solvent, usually water, and an associated ionic compound, such as an acid.

electrolytic switch *Electricity.* a switch whose terminals are submerged in an electrolytic substance.

electrolytic tank *Engineering.* a structure that serves as a model for an electron-tube system or an aerodynamic system, in which voltages are applied to test electron-tube design or to aid in computing ideal fluid flow.

electrolytic tough pitch *Metallurgy.* the electrolytic copper intentionally alloyed with a controlled amount of oxygen which internally oxidizes harmful impurities and improves electrical conductivity.

electromachining *Mechanical Engineering.* the use of electric power to shape an object or remove material from a workpiece.

electromagnet [i lek´trō mag´nət] *Electromagnetism.* a magnet, consisting of a coil wrapped about a soft iron or steel core, which becomes strongly magnetized when current flows through the coil.

electromagnetic [i lek´trō mag net´ik] *Physics.* **1.** of or relating to the interaction of electric and magnetic fields, both static and dynamic. **2.** of or relating to the science of electromagnetism.

electromagnetic amplifying lens *Electromagnetism.* a system composed of several waveguides arranged symmetrically about an excitation medium so that they are excited with equal amplitude and phase to provide an effective energy gain.

electromagnetic brake see ELECTRIC BRAKE.

electromagnetic cathode ray tube *Electronics.* a cathode ray tube in which the beam of electrons produced in the tube's filament is deflected by a magnetic field.

electromagnetic clutch *Mechanical Engineering.* a friction clutch that operates by the action of electric power from a dynamo on a magnetic coupling between conductors.

electromagnetic compatibility *Electronics.* the ability of radio equipment to operate efficiently in a given electromagnetic environment.

electromagnetic complex *Electromagnetism.* an electromagnetic configuration of an installation that includes all radiators of significant amounts of energy.

electromagnetic coupling *Electromagnetism.* a coupling between circuits or conductors that are mutually affected by the same electromagnetic field.

electromagnetic current *Electronics.* the movement of charged particles in the atmosphere that give rise to electric and magnetic fields, such as those in the ionosphere that transmit radio signals.

electromagnetic deflection *Electronics.* the use of a magnetic field or external horizontal- and vertical-deflection coils to deflect an electron beam in a television picture tube or an oscilloscope.

electromagnetic disturbance *Electricity.* a random or periodic electromagnetic phenomenon that is superimposed on a desired signal.

electromagnetic field *Electromagnetism.* an oscillating electric field and its associated magnetic field acting at right angles to each other and at right angles to their direction of motion.

electromagnetic field tensor *Electromagnetism.* an antisymmetric second-rank tensor whose elements are proportional to the electric and magnetic fields.

electromagnetic flowmeter *Engineering.* an instrument that measures the rate of flow, without interruption of the flow, by producing an electromagnetic field in a liquid that creates an interior current (proportional to the actual flow rate), which is detected by two electrodes.

electromagnetic focusing *Electronics.* a method of adjusting the electron beam in a television picture tube by varying the direct current flowing through the coils attached to the tube, thus altering the surrounding magnetic field.

electromagnetic induction *Electromagnetism.* the generation of an electromotive force by changing the magnetic flux through a closed loop circuit or by moving a conductor across the magnetic field.

electromagnetic inertia *Electromagnetism.* the characteristic delay of response in an electric circuit as it reaches its maximum or zero value after the source voltage is applied or removed.

electromagnetic interaction *Particle Physics.* a long-range force involving the electric and magnetic properties of elementary particles.

electromagnetic interference *Electricity.* the harmful impairment of a desired electromagnetic signal by an electromagnetic disturbance.

electromagnetic lens *Electronics.* a device in which the magnetic field surrounding a coil forces an electron beam passing through the coil to refract in a manner similar to light passing through an optical lens.

electromagnetic log *Engineering.* an instrument used to measure and record the movement rate of a vessel through water, having an electromagnetic sensing element that hangs from the vessel and produces a voltage which is directly proportional to the vessel's speed.

electromagnetic logging *Engineering.* a system that measures and records electromagnetic currents within a formation that is being drilled.

electromagnetic mass *Electromagnetism.* the mass of a moving charge whose kinetic energy appears to account for the discrepancy between the total field energy and the field energy of a purely static electric field.

electromagnetic mirror *Electromagnetism.* a surface or region that is capable of reflecting electromagnetic waves.

electromagnetic moment *Electromagnetism.* the vector magnetic moment of a current-carrying coil, equal to the product of the current, the number of turns, and the area of the coil. The direction is given by the right hand rule.

electromagnetic momentum *Electromagnetism.* the momentum associated with electromagnetic radiation; its volume density is given by the Poynting vector divided by the speed of light squared.

electromagnetic mixing *Metallurgy.* the process of mixing molten metals or alloys by applying an electromagnetic field to produce eddies.

electromagnetic noise *Electricity.* any undesired electromagnetic disturbance.

electromagnetic oscillograph *Electromagnetism.* an oscillograph that records signal waveforms by means of a moving-coil galvanometer.

electromagnetic potential *Electromagnetism.* the electrostatic scalar potential plus the magnetic vector potential.

electromagnetic propulsion *Space Technology.* any form of rocket propulsion possessing both magnetic and electric properties, especially one in which the propellant is accelerated by an electromagnetic field. *Aviation.* specifically, such power in a flight vehicle generated by the electromagnetic acceleration of a plasma fluid.

electromagnetic pulse *Electromagnetism.* an electromagnetic reaction of large magnitude resulting from a thermonuclear explosion.

electromagnetic pump *Electricity.* a device that moves conductive liquid through a pipe by sending current through the liquid; used in a nuclear reactor cooling system for liquid alkali metal circulation.

electromagnetic radiation *Electromagnetism.* the emission and propagation of radiation associated with a periodically varying electric and magnetic field traveling at the speed of light, including gamma rays, X-rays, and ultraviolet, light, infrared, and radio waves.

electromagnetic reconnaissance *Electronics.* **1.** the process of locating electromagnetic radiation devices, such as radar or missile-guidance systems, operated by hostile forces. **2.** the use of electromagnetic equipment to locate enemy activity in a particular region.

electromagnetic relay *Electromagnetism.* a switch that opens and closes contacts by the action of an electromagnetic element.

electromagnetic rocket see PLASMA ROCKET.

electromagnetic separator *Electromagnetism.* a device that separates ions of different mass by passing them through a combination of electric and magnetic fields.

electromagnetic shielding *Electromagnetism.* a means of preventing two circuits from electromagnetically coupling by placing at least one of the circuits in a grounded enclosure of magnetic conductive material.

electromagnetic shock wave *Electromagnetism.* an intense electromagnetic wave resulting from the coincidence of several waves of different velocities in a nonlinear dispersive medium.

electromagnetic spectrum *Electromagnetism.* the total range of the frequencies or wavelengths of electromagnetic radiation, ranging from the longest radio waves to the shortest cosmic rays.

electromagnetic surveying *Engineering.* a method of surveying a subsurface area in which electromagnetic waves at the surface penetrate the earth and produce new waves according to the ore bodies they contact; the waves are detected and recorded by instruments.

electromagnetic susceptibility *Electricity.* a measure of the tolerance of a circuit to undesired electromagnetic energy.

electromagnetic system of units *Electromagnetism.* a unit system, based on the cgs system, in which the unit of current is that which produces a force of 2 dyn per cm of length between two parallel, infinitely long conductors separated by 1 cm.

electromagnetic theory of light *Electromagnetism.* a theory stating that light consists of electromagnetic radiation and therefore obeys Maxwell's equations.

electromagnetic unit *Electromagnetism.* a unit based primarily on the magnetic effect of an electric current.

electromagnetic wave *Electromagnetism.* a wave generated by the oscillation of a charged particle and characterized by periodic variations of electric and magnetic fields.

electromagnetic wave filter *Electromagnetism.* a device that allows electromagnetic waves of certain frequencies to pass while effectively attenuating others.

electromagnetism [i lek´trō mag´nə tiz əm] *Physics.* **1.** the interaction between magnetism and electricity, and the phenomena produced by the interaction. **2.** the study and application of electromagnetic phenomena.

Electromagnetism

Electromagnetism underlies such a wide range of phenomena—from the wiggle of a compass needle to the X-rays taken by your family dentist—that we employ several vocabularies to describe our encounters with it. Electromagnetism (EM) effects include the "static" forces that electric charges and currents exert on one another, the radio waves we depend upon for much of worldwide communication, the light we see by, and, at the highest energies, the gamma rays generated in stars and particle accelerators.

The first great step toward a unified understanding of EM came in 1864, when James Maxwell conceived his electromagnetic theory of light, a phenomenon then imagined as mechanical vibrations of a pervasive medium called the "luminiferous ether." Maxwell's theory, as described in the set of elegantly powerful equations which bears his name, not only accounted for all the known phenomena, but also predicted that oscillating electric currents would produce electromagnetic waves traveling at the speed of light. Heinrich Hertz's 1887 experiments verified this prediction and seemed to settle the issue.

But twentieth-century scientists took these matters further. Most notably, Albert Einstein's explanation of the photoelectric effect showed that the generation and absorption of light is best described in terms of streams of wave-packets, or photons, each of which contains an amount of energy precisely related to its frequency. These photons, which constitute the building blocks of electromagnetic phenomena, exhibit a duality of form, combining the classical behavior of both waves and particles. At long wavelengths, their wavelike behavior manifests itself in resonant microwave cavities, much like organ pipes, while at the shortest wavelengths, a single photon can slam an electron right out of its orbit, like a well-executed billiard shot.

Most recently, quantum electrodynamics has provided us with a formulation which unites classical electromagnetic theory with quantum mechanics. Today, for example, physics students are taught that charged particles repel (and attract) one another by the constant exchange of virtual photons, like children in rowboats playing "catch," albeit with very tiny "baseballs" (and "boomerangs").

Arno Penzias
Vice President, Research
AT&T Bell Laboratories
Nobel Laureate in Physics

electromanometer *Engineering.* an electronic instrument that is used to gauge the pressure of liquids or gases.

electromechanical [i lek´trō mə kan´i kəl] *Mechanical Engineering.* of a mechanical device, system, or process, actuated or controlled by electromagnetism or electrostatics.

electromechanical brake *Mechanical Engineering.* a brake whose force is obtained partly as a result of the attraction of two magnetized surfaces and partly by mechanical means.

electromechanical circuit *Electricity.* a circuit that consists of one or more electromechanical components.

electromechanical dialer *Electronics.* a device on a telephone that activates a group of precoded numbers when a user presses a button.

electromechanical plotter see PLOTTER.

electromechanical recording *Electronics.* a document created by varying the signal of electric current flowing through a stylus moving across a sheet of paper.

electromechanical relay *Electromagnetism.* a device consisting of a coil and an armature that moves under the influence of a magnetic field produced by the coil to close or open a contacting switch.

electromechanical transducer *Electronics.* a device that converts mechanical energy into electrical energy or vice versa.

electromechanics *Mechanical Engineering.* the science and technology of electromechanical devices, systems, or processes.

electrometallurgy *Metallurgy.* the art and science of extracting and refining metal values from ores by electrical or electrochemical processes.

electrometer *Engineering.* an instrument used to measure voltage variation without drawing current from the source. *Radiology.* an instrument used to determine fluctuations in electrostatic potential difference between charged electrodes due to radiation.

electrometer amplifier *Electronics.* an amplifier circuit characterized by low-current drift and input-current offsets, and adequate power and current sensitivities, such that it is capable of measuring extremely low current variations in a circuit.

electrometer tube *Electronics.* a component in an electrometer that has a high input impedance and low control-electrode conductance, facilitating the measurement of extremely low direct current or voltage.

electromodulation *Spectroscopy.* in absorption spectroscopy, the measurement of changes in the transmittance or reflectance of a sample solution induced by an externally applied electric field.

electromotion *Physical Chemistry.* the mechanical force produced by an electric current.

electromotive *Physical Chemistry.* relating to or producing an electric current.

electromotive force *Physical Chemistry.* **1.** the amount of energy supplied by an electric current passing through a given source, as measured in volts. **2.** the potential energy difference existing between an anode and cathode that are immersed in the same electrolytic solution, or that otherwise adjoin each other.

electromotive series *Chemistry.* see ACTIVITY SERIES. *Physical Chemistry.* see ELECTROCHEMICAL SERIES.

electromyogram [i lek´trō mī´ə gram] *Medicine.* a graphic record of the electric impulses or activity in a muscle.

electromyograph [i lek´trō mī´ə graf] *Medicine.* an instrument that is used to create a graphic record of the spontaneous electrical activity of a muscle.

electromyography [i lek´trō mī äg´rə fē] *Medicine.* the recording and study of the electrical properties of skeletal muscles by means of surface or needle electrodes; useful in kinesiology and the study of neuromuscular function, extent of nerve lesion, and reflex response. Also, ELECTRONEUROMYOGRAPHY.

electron *Physics.* a stable elementary particle that is a primary constituent of ordinary matter, contained in the atoms of all elements and described as having a charge of -1.602×10^{-19} coulombs, a rest mass of 9.11×10^{-31} kg, and a spin of 1/2. Electrons flowing in a conductor constitute an electric current. *Electricity.* a unit of negative charge equal to a charge of one electron; the smallest known electrical charge.

electron accelerator *Nucleonics.* a machine that increases the energy levels of electrons in order to produce a beam of highly charged particles.

electron acceptor *Solid-State Physics.* an impurity element that is introduced into a pure semiconductor material to produce holes as charge carriers. *Physical Chemistry.* see ACCEPTOR.

electron acoustic microscopy *Physics.* a technique for producing images showing the thermal and elastic variations in the properties of an object.

electron affinity *Atomic Physics.* the energy needed to remove an electron from a negative ion in order to form a neutral atom or molecule.

electronarcosis *Medicine.* anesthesia produced by passing electric currents to the brain without causing convulsions; sometimes used in treating psychiatric disorders.

electron beam *Electronics.* a stream of electrons emitted from the same source and traveling under the influence of an electric or magnetic field in the same direction and at approximately the same speed.

electron-beam channeling *Electronics.* the process by which high-energy, high-current electron beams are transported from an accelerator through a medium of high-pressure gas directly to an intended target.

electron-beam drilling *Electronics.* the process of boring tiny holes into a material, such as a ferrite or semiconductor, with a tightly focused electron beam.

electron-beam fusion *Nucleonics.* a process in which strong electron beams implode tiny pellets of deuterium and tritium, causing them to attain the temperature and density needed to initiate a fusion reaction.

electron-beam generator *Electronics.* a device, such as a klystron, in which the velocity of an electron beam is kept at a constant level in order to produce exceedingly high radio frequencies.

electron-beam ion source *Electronics.* a source of multiply charged heavy ions used by a highly energized electron beam to ionize injected gas.

electron-beam lithography *Electronics.* lithography in which radiation-sensitive film is exposed to an electron beam.

electron-beam magnetometer *Engineering.* an instrument that measures the magnetic intensity of a magnetic field according to the movement and intensity of an electron beam, which passes through the field.

electron-beam parametric amplifier *Electronics.* a device that boosts a signal by varying the energy pumped from an electrostatic field into an electron beam traveling down the length of a tube, and then manipulating the beam at either end of the tube.

electron-beam pumping *Electronics.* a process in which an electron beam provides the energy necessary to move the majority of electrons in a semiconductor out of a ground state.

electron-beam recorder *Electronics.* **1.** a device in which an electron beam places signals or data onto film in a vacuum chamber. **2.** a device that transfers computer data onto microfilm using an electron beam.

electron-beam tube *Electronics.* a device, such as a klystron, oscilloscope tube, or television picture tube, that functions through the generation of one or more electron beams.

electron binding energy *Physics.* the minimum amount of energy required to extract an electron from an atom or molecule.

electron-bombardment-induced conductivity *Electronics.* in a multimode display-storage tube, a process in which an electron gun is used to erase the image on a cathode-ray tube interface.

electron capture *Atomic Physics.* a process in which an innershell electron is captured by the nucleus of its own atom, decreasing the atomic number of the atom by one unit; the process is accompanied by the emission of a neutrino. Also, **electron attachment.**

electron capture detector *Analytical Chemistry.* a device used in gas chromatography, in which carrier gas molecules flowing through the ionization chamber are radiated and low-energy electrons are formed; certain compounds entering the chamber have an affinity for these electrons, and this decrease in electrons is recorded for component identification.

electron carrier *Biochemistry.* a protein, such as flavoprotein or cytochrome, that can gain and lose electrons in either direction and can transport electrons from one compound to another or to oxygen.

electron charge *Physics.* the charge of one electron, equal to -1.602×10^{-19} coulombs or -4.803×10^{-10} statcoulombs (approximate values).

electron cloud *Atomic Physics.* an average region around the nucleus of an atom, in which the electrons are predicted to be at certain states of excitation.

electron compound *Metallurgy.* in a phase diagram, one of several homogeneous phases that has a specific crystal structure and a specific valence electron-to-atom ratio.

electron configuration *Atomic Physics.* a configuration that shows the way in which the electrons in an atom occupy, in order of increasing energy, the available orbitals and spin states.

electron-coupled oscillator *Electronics.* a device that generates alternating current; characterized by circuitry that feeds a portion of the generated energy back into the system to sustain its operation, and by an electron stream that is coupled between the screen and the plate to reduce the effects of the load. Also, DOW OSCILLATOR.

electron coupler *Electronics.* a device that increases the power of a microwave tube by subjecting its electron beam to alternating periods of acceleration and retardation. Also, CUCCIA COUPLER.

electron coupling *Electronics.* a process by which two circuits within an electron tube transfer energy generated by the electron stream passing between the electrodes of one of the circuits.

electron cyclotron resonance source *Electronics.* a source of multiply charged heavy ions that uses microwave power to increase electron energy to extremely high levels in two magnetic-mirror confinement chambers connected in series.

electron cyclotron wave *Physics.* a circularly polarized wave found in a plasma that runs parallel to the magnetic field produced by currents outside the plasma.

electron density *Physics.* **1.** the number of electrons per unit volume. **2.** the quantum mechanical probability density for an electron.

electron-density map *Crystallography.* a contoured representation of electron density at various points in a crystal structure, expressed in electrons per cubic angstrom and highest near atomic centers. The map is calculated using a Fourier synthesis (a summation of waves of known amplitude, frequency, and phase).

electron device *Electronics.* a device whose operation involves the motion of electrical charge carriers in a vacuum, gas, or semiconductor.

electron diffraction *Physics.* an interference phenomenon among electrons observed when they are scattered off atoms or molecules in a crystal to form a diffraction pattern.

electron-diffraction analysis *Physics.* the study of crystal structure by electron-diffraction methods.

electron diffractograph *Physics.* a device that produces an electron-diffraction pattern by focusing an electron beam onto a crystal specimen, allowing the user to study the crystal structure of the specimen.

electron-discharge machining *Metallurgy.* the process of machining performed by electric sparks in a nonconductive liquid medium.

electron distribution *Physics.* a function that describes the electron density in phase space.

electron-distribution curve *Physical Chemistry.* a line on a graph that shows the electron distribution in a solid at different energy levels.

electron donor *Physical Chemistry.* see DONOR.

electronegative *Electricity.* **1.** having a negative electric charge. **2.** having the ability to act as the negative electrode in an electric cell. *Physical Chemistry.* relating to an atom or molecule that tends to draw in electrons from outside the system; nonmetals are generally electronegative.

electronegative atom *Physical Chemistry.* an atom that readily accepts electrons and thus has a tendency to acquire a partial negative charge in a covalent bond, or to form a negative ion.

electronegative potential *Physical Chemistry.* an electrode's potential stated as negative in relation to the standard potential of the hydrogen electrode, which is established at zero.

electronegativity *Physical Chemistry.* the relative power of an atom or molecule to attract electrons to itself; the tendency to become negatively charged.

electronegativity scale *Physical Chemistry.* a numerical expression of the relative negativity of various substances, based on calculations using bond energies and assigning a value of 4 to fluorine, the most electronegative element; devised by Linus Pauling. Also, PAULING SCALE.

electron emission *Electronics.* the release of electrons from a given material into the surrounding space; may be caused by heat, light, impact, chemical disintegration, or an electric field.

electron emitter *Electronics.* a material from which electrons are released.

electron energy level *Atomic Physics.* the quantum-mechanical representation of the energy level of an electron in an atom, which determines the orbit of the electron about the nucleus.

electron-energy-loss spectroscopy *Spectroscopy.* photoelectron spectroscopy; the use of electron beams to induce transitions between electronic energy levels; the study of the distribution of energy lost by scattered electrons when a substance is bombarded with monochromatic electrons. Also, ELECTRON IMPACT SPECTROSCOPY.

electroneural transport *Biochemistry.* the movement of charged or uncharged species across an energy-transducing membrane that does not result in a net transfer of charge.

electroneuromyography see ELECTROMYOGRAPHY.

electroneutral *Physical Chemistry.* describing a substance that has no net electric charge; neither positive nor negative.

electroneutrality *Physical Chemistry.* the condition of a medium that is electroneutral, such as an electrolytic solution in which the negatively and positively charged ions are distributed so that the solution as a whole is neutral.

electron flow *Electricity.* the movement of electrons through a conductive material.

electron gun *Electronics.* a device, within an electron tube or cathode-ray tube, that produces an electron beam and directs its movement.

electron-gun-density multiplication *Electronics.* the ratio of the average amount of current density at a given aperture to the average current density at the cathode surface.

electron hole *Solid-State Physics.* in a semiconductor, the electron vacancy in the valence band that occurs when an electron jumps the gap from the filled valence band to the empty conduction band; serves as a positive charge carrier, allowing electrons deeper in the band to move into the vacated area.

electron-hole droplets *Solid-State Physics.* an electronic excitation, observed at low cryogenic temperatures in silicon and germanium, in which an electron-hole Fermi liquid is formed in an unstable state.

electron-hole pair *Solid-State Physics.* a conduction electron in the conduction band and an accompanying electron hole in the valence band, which result when an electron jumps the gap in an intrinsic semiconductor.

electron-hole recombination *Solid-State Physics.* a process in which a hole is recombined with an electron within a doped semiconductor, accompanied by a release of energy, typically in the form of radiation.

electronic [i lek´trän´ik] *Electronics.* **1.** relating to a device whose operation involves the motion of electrical charge carriers in a vacuum, gas, or semiconductor. **2.** relating to the study of such devices.

electronic absorption spectrum *Spectroscopy.* any spectrum produced by the absorption of electromagnetic radiation by atoms, ions, or molecules as a result of electron excitation.

electronically agile radar *Engineering.* an airborne radar in which the beam from phased array antennae changes shape and direction according to electronic speeds.

electronic alternating-current voltmeter *Electronics.* a device that measures voltage in amplifier-rectifier circuits.

electronic angular momentum *Atomic Physics.* the entire angular momentum including contributions from the spins of all the electrons in an atom.

electronic azimuth marker *Electronics.* a line on a radar screen that indicates the bearing of an airborne target.

electronic balance *Analytical Chemistry.* a microbalance in which the sample weight is obtained automatically, based on the force produced by current in a coil in a magnetic field.

electronic band spectrum *Spectroscopy.* a spectrum characteristic of molecules that consists of bands of spectral lines corresponding to electron transitions accompanied by vibrational or rotational transitions.

electronic bearing cursor *Electronics.* a line on a radar screen that indicates the bearing of a marine target.

electronic calculator *Electronics.* a device in which transistors perform mathematical calculations and utilize light-emitting diodes or liquid crystals to display the results.

electronic camouflage *Electronics.* the use of electronic properties to minimize or negate the presence of echoes in a radar system.

electronic cash register *Engineering.* a register with a component that scans the symbols on a package label, then converts them to digital form to indicate the item price and, in some cases, maintain a record of sales and inventories; used in retail stores.

electronic chart reader *Computer Technology.* an input device that can scan curves from a continuous paper feed and convert them to digital data.

electronic circuit *Electronics.* **1.** a circuit that contains active components, such as tubes and transistors, as opposed to one that contains only passive components, such as resistors and switches. **2.** a circuit in which the balance of electrons in a given component, such as a tube, transistor, or amplifier, is disturbed by something other than an applied voltage.

electronic clock *Horology.* **1.** a clock that uses electronic circuits to count the number of oscillations in quartz crystal to determine the timekeeping impulses that activate a digital display. **2.** a clock in which the timekeeping impulse is provided by the oscillations of a tiny tuning fork attached to an electronic circuit. Also, ELECTROMAGNETIC CLOCK.

electronic commutator *Electronics.* a device that reverses or exchanges the external connections in a transducer at a high rate of speed, thus eliminating noise and wear.

electronic component *Electronics.* a device, such as an electron tube or a transistor, that does not use mechanical means to control current and voltage in a circuit.

electronic computing units *Electronics.* in cardpunch technology, the section of a tabulating device designed to ensure that it will process the data on punch cards in a prescribed manner.

electronic confusion area *Electromagnetism.* the area on a radar screen that a target appears to occupy according to a particular radar beam.

electronic cottage *Computer Technology.* a situation in which workers use home computer terminals and communicate with a central office and other workers.

electronic counter *Electronics.* a device that uses electronic elements to count the number of pulses applied to it.

electronic countermeasures *Electronics.* a division of electronic warfare in which actions are taken to interfere with an enemy's use of the electromagnetic spectrum, used in air defense and interception.

electronic counter-countermeasures *Electronics.* a division of electronic warfare in which actions are taken to ensure friendly, effective use of the electromagnetic spectrum in spite of an enemy's use of electronic countermeasures.

electronic data processing *Computer Programming.* broadly, any data processing performed primarily on electronic equipment; usually refers to data processing that is performed on digital computers.

electronic data-processing center *Computer Technology.* a specified area used to contain a computer, its peripherals, and the personnel operating the data-processing system.

electronic data-processing management science *Computer Science.* the discipline consisting of the class of management problems that can be handled by a computer system.

electronic data-processing system *Computer Technology.* a system composed of electronic and other equipment used for high-speed data processing.

electronic defense evaluation *Electronics.* a procedure in which a pilot uses electronic countermeasures to penetrate an area monitored by radar; designed to determine the effectiveness of both radar and aircraft.

electronic differential analyzer see DIFFERENTIAL ANALYZER.

electronic display *Electronics.* an electronic component that converts electromagnetic signals into a visual display.

electronic dummy *Acoustical Engineering.* a device that is used for simulating impedance, frequency response, and other characteristics of the human body to provide approximately equivalent features in vocal and hearing processes.

electronic efficiency *Electronics.* the ratio of the power at a given frequency that is delivered by an electron stream to an oscillator or amplifier to the average power supplied to the stream.

electronic emission spectrum *Spectroscopy.* any spectrum produced by the emission of electromagnetic radiation by atoms, ions, or molecules as a result of electron excitation.

electronic energy curve *Physical Chemistry.* a graph that shows the range of energy levels in a diatomic molecule, based on the distance between the nuclei of its two atoms.

electronic engineering *Engineering.* a branch of engineering that deals with the design, fabrication, and operation of electronic devices and systems. Thus, **electronic engineer.**

electronic fetal monitor *Medicine.* a device that monitors fetal heartbeat and uterine contractions during labor. Also, EFM.

electronic fix *Navigation.* a navigational fix that is determined by electronic means, such as by the use of LORAN.

electronic flame safeguard *Mechanical Engineering.* in a burner system, an electrode that acts as a safety valve by interrupting fuel flow from the main burner when its flame is not detected.

electronic formula *Chemistry.* a structural formula in which the bonds are replaced by dots that indicate electron pairs; a single bond is equivalent to one pair of electrons shared by two atoms.

electronic fuse *Engineering.* a fuse that is ignited by a self-contained electronic element; for example, a proximity fuse.

electronic heating *Engineering.* heating that is generated by an oscillator or radio-frequency power source, which produces a radio-frequency current. Also, RADIO-FREQUENCY HEATING.

electronic horizontal-situation indicator see HORIZONTAL-SITUATION INDICATOR.

electronic humidistat *Engineering.* an instrument, used to regulate the degree of humidity, composed of two sets of alternate metal conductors that transmit any variation in humidity to a relay amplifier.

electronic imaging *Telecommunications.* a photographic system in which a sensor is placed behind a camera lens to convert an image into an electronic signal, which can be stored for later playback on a television screen.

electronic interference *Electronics.* a disturbance from nearby electrical or electromagnetic activity that causes an electronic device to malfunction.

electronic line scanning *Electronics.* **1.** in a television system, a method by which a spot of light or another energy source moves along a given path by electronic means. **2.** in facsimile copying, a method by which a spot on a cathode-ray tube moves across the copy by electronic means.

electronic listening device *Electronics.* an instrument that picks up sound waves from an ostensibly private conversation and reproduces them in a form, generally on magnetic tape, that can be used as evidence.

electronic locking *Electronics.* a method used to prevent the activation of a switch until a special sequence of signals is received by the circuit.

electronic magnetic moment *Atomic Physics.* the total amount of polarization (dipole moment) caused by the movement of electrons within an atom.

electronic mail *Telecommunications.* a computer-system communications service in which text messages are sent to a central computer or over a network and retrieved by the addressee. Also, E-MAIL.

electronic microradiography *Electronics.* a procedure in which electrons released from microscopic irradiated objects are used to produce a photographic image.

electronic motor control *Electronics.* an electronic device that varies the speed of a DC motor when it is driven by an AC power line.

electronic multimeter *Electronics.* a device that uses semiconductors or electron-tube circuits to measure resistance, current, and voltage.

electronic music *Acoustical Engineering.* music created by arranging electronically synthesized sounds into a formal pattern with musical qualities.

electronic musical instrument *Acoustical Engineering.* an instrument, such as an electronic organ or electric guitar, that uses electronic amplification.

electronic navigation *Navigation.* the use of electronic aids to determine the position and direct the course of a craft.

electronic noise jammer *Electronics.* a device that transmits a signal with a white noise component in order to prevent the functioning of a radar system.

electronic organ *Acoustical Engineering.* an instrument in which musical tones generated by electronically driven reeds are processed, amplified, and passed through a set of loud speakers.

electronic packaging *Engineering.* the technical process of assembling electronic equipment, in which components are inserted above specific holes on multilayered circuit boards and then soldered to the printed wiring, which is often on the opposite side of the board.

electronic phase-angle meter *Electronics.* an instrument that uses electronic devices, such as amplifiers and limiters, to change an AC voltage into square waves before measuring its phase angle.

electronic piano *Acoustical Engineering.* an instrument in which musical tones generated by electronically driven strings are processed, amplified, and passed through a set of loud speakers.

electronic polarization *Electricity.* the ionic energy that an electron exhibits in the presence of an electric field.

electronic position indicator *Navigation.* an electronic device that displays the navigational position of a craft.

electronic radiography *Electronics.* a procedure in which an image generated by an irradiated object is converted into a signal for television viewing.

electronic raster scanning *Electronics.* in facsimile, a method of scanning in which the motion of the scanning spot is completely controlled by electronic means. Also, **electronic scanning.**

electronic reconnaissance *Electronics.* the detection, identification, evaluation, and location of foreign electromagnetic radiations emanating from points of origin other than nuclear detonations or radioactive sources.

electronic recording *Electronics.* a method of producing a graphical record of a varying quantity or signal by controlling an electron beam with an electromagnetic field, as in a cathode-ray oscillograph.

electronic robot *Robotics.* a robot operated by electronic components, such as DC stepping motors.

Electronics

The term "electronics" refers to a large number of different phenomena and devices in which useful electrical effects are achieved through control of the motion of electrons. The fact that the time in which we live is sometimes referred to as "the electronic age" indicates what widespread consequences are involved.

In the early part of this century this control of electron motion was achieved primarily through the development of vacuum tubes. A vacuum tube consists essentially of a thermionic cathode that emits electrons into vacuum that are collected by a positively charged plate. The flow of electrons is by the potential on a third type of electrode, a grid. Small changes in the voltage on the grid result in large changes in the electrical current. Many sophisticated variations of this basic principle were put into practice and made possible the beginning of "the electronic age."

What caused this electronic age to explode, however, was the discovery by Shockley in 1948 that a similar type of control of electron flow could be achieved in solid materials rather than in vacuum. The first such solid-state amplifiers were made from the semiconductor germanium, but soon it was replaced by devices made from silicon, which has dominated the field to date ever since. One of the major advantages afforded by the use of solid-state electronic control rather than vacuum control is the possibility of an enormous reduction in the size of the devices, making possible the high packing density of devices that characterizes the most sophisticated of today's electronics.

A representative partial list of electronic devices today includes such diverse applications as rectifiers, amplifiers, integrated circuits, memories, microwave sources and receivers, light-emitting devices, light-detecting devices, and solar cells to convert solar energy into electricity.

Richard H. Bube
Professor of Materials Science & Electrical Engineering
Stanford University

electronics *Physics.* the study and application of the conduction of electric charges in various media, including vacuums, gaseous media, and semiconductors.

electronic sculpturing *Computer Programming.* a procedure by which a system can be modeled on an analog computer by interconnecting analogous system components; dynamic behavior can be simulated by altering circuit gains and reference voltages.

electronic security *Electronics.* **1.** a security system consisting of electronic circuit components, used to monitor circuit variables. **2.** a security system used to restrict access to valuable information.

electronic skyscreen equipment *Ordnance.* a device that signals the exit of a missile from a predetermined trajectory.

electronic smog *Ecology.* nonionizing radiation, such as radar or television waves, that is emitted into the environment in such quantities as to constiue a potential health danger.

electronic specific heat *Solid-State Physics.* the specific heat contributed by the motion of conduction electrons in a conductor.

electronic spectrum *Spectroscopy.* a spectrum produced by the absorption or emission of electromagnetic radiation resulting only from electron transitions, rather than from vibrational, rotational, or other types of transitions.

electronic speedometer *Engineering.* an instrument used to measure speed by means of a transducer that carries data pulses over wires to the speed and mileage indicators.

electronic spreadsheet *Computer Programming.* a type of software that arranges data in ledgerlike rows and columns and performs user-specified computations expressed as constraints between rows and columns of data.

electronic structure *Physics.* the arrangement of the electron orbitals in an atom or molecule, often described in terms of the quantum numbers, energy levels, or wavefunctions.

electronic switching *Telecommunications.* an automatic system of switching electric circuits utilizing semiconductor devices, such as transistor combinational circuits, semiconductor diode matrices, and integrated circuits, to perform the selection and switching in telephone circuitry. Thus, **electronic switching system.**

electronic tablet see DIGITIZING PAD.

electronic thermometer *Engineering.* an instrument used to measure temperature that operates by the action of an electronic sensor, which is placed adjacent to the substance being measured.

electronic tuning *Electronics.* a method of adjusting the frequency of a device by changing the controlling voltage, rather than manually adjusting the component.

electronic typewriter *Computer Technology.* a typewriter whose input is provided by an operator, but whose output is produced by electronic components.

electronic video recording *Electronics.* the act of recording color signals onto photographic film as black and white coded images.

electronic warfare *Military Science.* a military action intended to prevent the use of electromagnetic radiations by hostile forces or to retain and exploit its use by friendly forces.

electronic work function *Solid-State Physics.* the energy required to remove an electron with the Fermi energy in a solid to the energy level of an electron at rest in vacuum outside the solid.

electron image tube see IMAGE TUBE.

electron impact spectroscopy see ELECTRON ENERGY LOSS SPECTROSCOPY.

electron injection *Electronics.* **1.** the release of electrons from one solid material into another solid material. **2.** the process of forcing a beam of electrons into any large electron accelerator, such as a betatron, with an electron gun.

electron lens *Electronics.* a device that uses an electromagnetic field to refract an electron beam in a manner similar to the refraction of light by an optical lens.

electron lepton number *Particle Physics.* a conserved quantity given by the number of electrons plus the number of electron-associated neutrinos minus the number of positrons minus the number of electron-associated antineutrinos.

electron linear accelerator *Nucleonics.* a machine in which radio frequencies guide electrons along a straight line in order to produce a beam of highly charged particles.

electron mass *Physics.* the mass of an electron, which is approximately 9.11×10^{-31} kg.

electron microprobe *Physics.* a device that focuses an accelerated beam of electrons to an extremely small point on the surface of a crystal specimen so that the specimen may be studied by the effects of the electron beam. Also, ELECTRON PROBE.

electron microscope *Electronics.* an instrument that uses electrons, generally focused by electron lenses, to magnify tiny objects onto a fluorescent screen or photographic plate. It provides much greater powers of magnification than an optical microscope (up to 1,000,000 times actual size without loss of definition).

electron mobility *Solid-State Physics.* the average electron-drift velocity in a semiconductor divided by the externally applied electric field.

electron multiplicity *Physics.* the quantity $2S + 1$, in which S is the total spin quantum number of an atom in Russell-Saunders coupling.

electron multiplier *Electronics.* a tube in which current is amplified by the production of secondary electrons that result from the collisions of electrons with special targets inside the tube.

electron-neutrino *Physics.* a neutrino that obeys a conservation law together with the electron, so that the total number of electrons and electron-neutrinos minus the total number of their antiparticles remains the same.

electron nuclear double resonance *Spectroscopy.* a spectroscopic technique in which a sample is irradiated with a range of nuclear resonance frequencies while electron spin resonance absorption is observed at a single frequency.

electron number *Atomic Physics.* the number of electrons in an atom or ion.

electronograph *Electronics.* **1.** a device consisting of an electron tube in which an accelerated beam of electrons produces images on a fine-grain photographic emulsion. Also, **electronographic tube. 2.** an image produced by this device.

electronography *Electronics.* the production of an image or images using an electronograph.

electron-opaque tracer *Molecular Biology.* a metallic salt often found in association with binding proteins such as antibodies; used to detect the presence of specific molecules or structures in an electron microscopic examination of a sample.

electron optics *Electronics.* the study of the motion of electrons in an electromagnetic field, as in laser technology, light amplification, and photoelectricity. Also, ELECTRO-OPTICS.

electron orbit *Physics.* the path of an electron as it circulates about its orbital in an atom.

electron pair *Physical Chemistry.* a pair of valence electrons that creates a nonpolar connection between two adjacent atoms.

electron-pair bond see COVALENT BOND.

electron paramagnetic resonance see ELECTRON SPIN RESONANCE.

electron paramagnetism *Physics.* paramagnetism arising from the contributions of the net magnetic moments of the unpaired electrons of the individual atoms or molecules in a paramagnetic substance, due to the tendency to align magnetic moments with an external magnetic field.

electron-poor *Physical Chemistry.* being somewhat deficient in electrons, and therefore able to accept an available pair of electrons.

electron-positron storage ring *Nucleonics.* a vacuum chamber that is encased with bending and focusing magnets, in which counterrotating beams of electrons and positrons are stored for several hours and then forced to collide with each other.

electron probe see ELECTRON MICROPROBE.

electron radius *Physics.* the classical theoretical radius of an electron, 2.82×10^{-13} cm, obtained by equating the rest mass energy of an electron mc^2 to the coulomb energy e^2/r.

electron-ray tube *Electronics.* a small cathode-ray tube with a fluorescent screen on which the pattern varies with the voltage applied to the grid; used to indicate the accuracy of tuning in a radio receiver.

electron-rich *Physical Chemistry.* having available electrons that can be contributed to an electron-poor atom or group of atoms.

electron shell *Atomic Physics.* the arrangement of electrons at various distances from the nucleus of an atom, according to the energy they have; those with the least energy are in the shell closest to the nucleus, traditionally called the **K shell,** which can hold no more than 2 electrons; the **Q shell**, farthest from the nucleus, can hold 98 electrons, but is never completely filled.

electron spectroscopy *Spectroscopy.* the study of the energy of photoelectrons or Auger electrons emitted when a substance is bombarded with electrons, ions, or electromagnetic radiation.

electron spectrum *Spectroscopy.* a diagram, graph, or other display indicating the intensity of electrons emitted by an irradiated substance with respect to the kinetic energy of the electrons.

electron spin *Quantum Mechanics.* the intrinsic half-integer angular momentum of an electron.

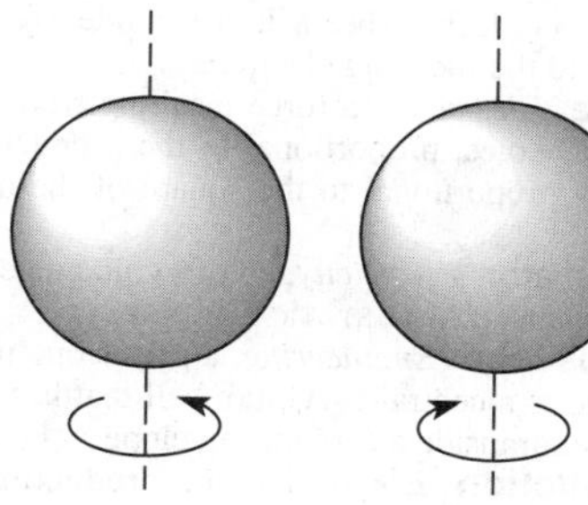

electron spin

electron spin density *Physics.* the vector sum of the spin angular momentum of electrons per unit volume.

electron spin resonance *Physics.* a resonant absorption of electromagnetic radiation by a paramagnetic substance, having unpaired electrons, when the energy levels are split by the application of a strong magnetic field. Also, ESR, ELECTRON PARAMAGNETIC RESONANCE.

electron-stream potential *Electronics.* the time average of the potential differential difference between a given point in an electron stream and the surface at which electrons are emitted.

electron synchrotron *Nucleonics.* a machine that accelerates electrons in a circular path by keeping the frequency of the accelerating stream at a constant level while increasing the strength of the magnetic field guiding the stream.

electron telescope *Electronics.* an instrument in which the infrared light of a distant object is focused onto a photocathode tube, enlarged by a series of electron lenses, and reproduced onto a fluorescent screen to form an image of the object.

electron temperature *Physics.* the kelvin temperature T that follows from setting the thermal kinetic energy $3/2\ kT$ equal to the average electron energy.

electron transfer *Physics.* the passage of an electron from one atom or molecule to another by collision or other means.

electron transport chain *Molecular Biology.* a series of electron carriers that occurs within certain membranes, such as the mitochondrial membrane, and produces energy for the cell. Also, RESPIRATORY CHAIN.

electron transport phosphorylation *Biochemistry.* the conversion of inorganic phosphate into pyrophosphate from a reaction powered by energy in a transmembrane gradient of ions generated by an electron transport chain; occurs in respiratory metabolism and in some types of fermentation.

electron transport system *Biochemistry.* a series of reduction-oxidation reactions, beginning with increased strength and ending in oxygen, that constitutes the final stage of aerobic respiration; also involved in the light reaction of photosynthesis.

electron trap *Solid-State Physics.* an impurity or defect in a semiconductor that captures mobile electrons.

electron tube *Electronics.* a device in which electrons are conducted through a vacuum or gaseous medium within a gastight chamber; used to generate, amplify, and rectify electric oscillations and AC currents.

electron-tube amplifier *Electronics.* a circuit that uses electron tubes to provide the additional power needed to amplify a signal.

electron-tube generator *Electronics.* a device that transforms direct current into a radio frequency by passing it through an electron tube in an oscillator circuit.

electron-tube static characteristic *Electronics.* the relationship among the various parameters in an electron tube when voltage and current are constant.

electron volt or **electronvolt** *Physics.* a unit of energy equal to the energy acquired by an electron accelerating through a potential difference of 1 volt; equivalent to 1.602×10^{-19} joules.

electron wave *Quantum Mechanics.* the de Broglie wave associated with an electron.

electronystagmography *Medicine.* a method in which the movements of the eye can be assessed and recorded by measuring the electric activity of the extraocular muscles.

electrooculogram *Medicine.* a record of the speed and direction of eye movements between two fixed points, determined by registering differences in electric potential; used to detect retinal dysfunction.

electro-optical imaging sensor *Robotics.* a camera or other device at the end of a robot's arm that is used to grasp or manipulate the object being worked on.

electro-optical Kerr effect *Physics.* a pattern of double refraction exhibited by certain refracting materials when exposed to an electric field.

electro-optical modulator *Telecommunications.* a device that uses the electrooptical effect to modulate a light beam.

electro-optic radar *Engineering.* a radar system that gathers information by detecting the effects of an electric field on optical phenomena.

electro-optics see ELECTRON OPTICS.

electroosmosis ELECTROENDOSMOSIS.

electroosmotic driver *Electronics.* a solution in which voltage is converted into fluid pressure by the streaming of the potential effect.

electropherotype *Virology.* **1.** a virus strain that has been distinguished by means of electrophoresis. **2.** an analysis of the proteins or nucleic acids in a virus mixture, as separated by electrophoresis.

electrophile [i lek´trə fīl´] *Physical Chemistry.* an ion or molecule that has a partial or complete positive charge, so that it can accept an electron pair or share an electron pair with another atom. *Chemistry.* see LEWIS ACID. (Literally, "electron lover.")

electrophilic [i lek´trə fil´ik] *Physical Chemistry.* **1.** relating to a process in which electrons are acquired from or shared with another atom or molecule. **2.** describing a substance with an electron deficiency.

electrophilic reagent *Physical Chemistry.* a molecule that forms a bond with another molecule by accepting an electron pair during a chemical reaction.

electrophoresis [i lek´trə fə rē´sis] *Physical Chemistry.* a process in which electrically charged particles that are suspended in a solution move through the solution under the influence of an applied electric field. Also, CATAPHORESIS.

electrophoretic *Physical Chemistry.* of or relating to electrophoresis.

electrophoretic effect *Physical Chemistry.* the tendency of an applied electromotive force to move an ionic atmosphere in a direction opposite to the motion of the central ion, thus creating a countercurrent effect that reduces the ion's velocity.

electrophoretic mobility *Biochemistry.* the tendency of cells or compounds in solution to move in an electric field toward the positive or negative electrode. *Physical Chemistry.* the velocity per unit electric field of a charged particle in a solution.

electrophoretic variants *Biochemistry.* proteins that can be divided into distinct electrophoretic components because of variations in their mobilities.

electrophoretic velocity *Physical Chemistry.* the rate at which charged particles move in a solution under the influence of an electric field.

electrophrenic respiration *Medicine.* the electrical stimulation of the phrenic nerve, used to provide respiration for patients paralyzed by acute bulbar poliomyelitis.

electrophysiology *Physiology.* the study of the electrical properties of living tissue.

electropism *Botany.* the curving movement or growth of a plant in response to slight electric currents.

electroplaque *Vertebrate Zoology.* one of a number of flattened plates forming an electric organ in certain fishes, consisting of modified muscle tissue having an electric charge. Also, **electroplax.**

electroplating *Metallurgy.* **1.** the process of plating or coating a conducting surface with a metal by a process of electrolysis. **2.** also, **electroplate.** the articles so plated. Also, ELECTROLYTIC PLATING.

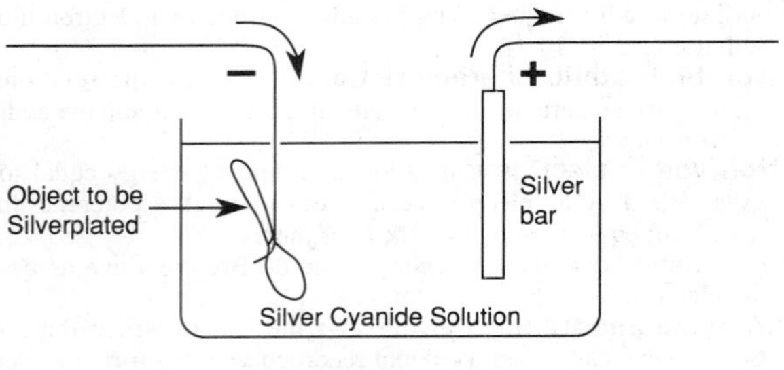

electroplating

electropolishing *Metallurgy.* the process of polishing a metallic material anodically in an electrolytic cell; often used on curved parts that cannot be polished mechanically.

electroporation *Biochemistry.* a type of osmotic transfection in which holes are produced in cell membranes by an electric current so that alien DNA molecules can enter the cells.

electropositive *Electronics.* an element that can function as the positive electrode in an electric cell. *Physical Chemistry.* relating to an atom or molecule that tends to provide electrons to an electron-acquiring substance; metals are generally electropositive.

electropositive atom *Physical Chemistry.* an atom that readily yields electrons and thus has the tendency to acquire a partial positive charge in a covalent bond, or to form a positive ion.

electropositive potential *Physical Chemistry.* an electrode's potential stated as positive in relation to the standard potential of the hydrogen electrode, which is established at zero.

electropulse engine *Aviation.* a flight vehicle engine utilizing spark discharges through which intense electric and magnetic fields are established, for periods ranging from microseconds to a few milliseconds, generating an electromagnetic force that drives plasma along the leads and away from the spark gap.

electroreceptor *Vertebrate Zoology.* any of the array of tiny, electrically sensitive ampullar organs present in sharks, electric eels, and catfish that can pick up weak electrical currents emitted by other water creatures.

electrorefining *Chemical Engineering.* a petroleum refinery procedure to aid in separating chemical treating agents from the hydrocarbon phase in light hydrocarbon streams by use of an electrostatic field. *Metallurgy.* a method of metal refining in which the metal is dissolved anodically and plated at the cathode of an electrolytic cell.

electroreflectance *Spectroscopy.* the study of reflection spectra produced by electromodulation of a sample.

electroresistive effect *Electronics.* a change in resistance against an electrical current that parallels a change in voltage applied to the current.

electroretinogram *Medicine.* a record of the changes in electric potential of the retina upon stimulation by light.

electroscope *Electricity.* a device that detects the presence of minute charges of electricity and determines their sign by means of electrostatic attraction and repulsion.

electrosensitive *Electricity.* readily affected by electric current.

electrosensitive recording *Electronics.* a document that is produced when an electric current is applied through a stylus onto a specially treated paper.

electroshock therapy see ELECTROCONVULSIVE THERAPY.

electrostatic *Electricity.* having, relating to, or utilizing electrostatic charges and their characteristics. Thus, **electrostatic instrument.** *Acoustical Engineering.* describing an acoustical device, such as a loudspeaker or microphone, that derives mechanical or electrical energy from the effect of an electrostatic field.

electrostatic accelerator *Electronics.* a device that uses an electrostatic field in a vacuum to greatly increase the velocity of charged particles. Also, ELECTROSTATIC GENERATOR.

electrostatic analyzer *Electronics.* a device that filters an electron beam by allowing only electrons within an extremely narrow velocity range to pass through it.

electrostatic attraction see COULOMB ATTRACTION.

electrostatic bond *Physical Chemistry.* a valence bond of two atoms by an electrostatic force caused by one or more electrons moving from one atom to the other.

electrostatic coalescence *Meteorology.* **1.** the coalescence of cloud drops due to the electrostatic attractions between drops of opposite charge. **2.** the coalescence of two cloud or rain drops due to polarization effects resulting from an external electric field.

electrostatic copier *Graphic Arts.* a photocopying machine designed for electrostatic printing.

electrostatic deflection *Electronics.* the movement of an electron beam due to the electrostatic field produced by electrodes on either side of the beam; primarily used in cathode-ray tubes for oscilloscopes and in old-fashioned television picture tubes.

electrostatic detection *Electronics.* a technique for locating a solid body, such as a mineral deposit, by using specialized equipment to measure the electrostatic field surrounding the body.

electrostatic energy *Electricity.* the energy contained in electricity or an electric charge at rest.

electrostatic field *Electricity.* an electric field with constant intensity, such as that produced by stationary charges.

electrostatic flux see ELECTRIC FLUX.

electrostatic focus *Electronics.* in television picture tubes, a technique for directing an electron beam in a cathode-ray tube by varying the voltage applied to the focusing electrode.

electrostatic force *Electricity.* a force resulting from the attraction of stationary, charged bodies, proportional to the product of their magnitudes and inversely proportional to the square of the distance between them.

electrostatic generator *Electricity.* a high-voltage generator operated by electrostatic induction. Also, STATIC MACHINE.

electrostatic gyroscope *Engineering.* a gyroscope in which an electrostatic field supports a central beryllium ball that is surrounded by six electrodes in a vacuum inside a ceramic envelope.

electrostatic induction *Electricity.* the production of an electric charge in an object when placed near a charged body. Also, INDUCTION.

electrostatic interactions see COULOMB INTERACTIONS.

electrostatic lens *Electronics.* a collection of plates or cylinders whose electrostatic properties cause an electron beam to refract in a manner similar to the refraction of light by an optical lens.

electrostatic loudspeaker see CAPACITATOR LOUDSPEAKER.

electrostatic memory see ELECTROSTATIC STORAGE.

electrostatic microphone see CAPACITATOR MICROPHONE.

electrostatic potential see ELECTRIC POTENTIAL.

electrostatic precipitator *Engineering.* a device that removes small foreign particles, such as dust and pollen, from the air by electrically charging and then collecting the particles on a plate that is oppositely charged.

electrostatic printing *Graphic Arts.* a process in which a positive electrostatic charge is given to a paper on which an image is projected; a bright light reverses the charge of the nonimage area so that negatively charged powdered ink adheres only to the positive image area. Pioneered by the Xerox Corporation, it is the most common dry photocopying process. Also, **electrostatic reprography.**

electrostatic repulsion see COULOMB REPULSION.

electrostatics *Electricity.* the study of the electrical phenomena associated with electric charges at rest.

electrostatic scanning *Electronics.* a scanning method in which an electrostatic field directs the electron beam.

electrostatic separation *Engineering.* a process in which an electrostatic separator sorts out a mixture of finely pulverized materials.

electrostatic separator *Engineering.* an apparatus in which a mixture of finely pulverized materials falls through a strong electric field between charged electrodes and is sorted out to fall into different receptacles.

electrostatic shielding *Electricity.* a material that inhibits the interaction of electric fields.

electrostatic storage *Electronics.* a device in which information is stored as electrostatic charges on a dielectric surface. Also, ELECTROSTATIC MEMORY.

electrostatic stress *Electricity.* an electrostatic field that acts on an insulator; the field generates polarization in the insulator and causes electrical breakdown if raised beyond a specific intensity.

electrostatic transducer see CAPACITOR TRANSDUCER.

electrostatic units *Electricity.* a cgs system of electric and magnetic units in which the basic unit of charge exerts a force of 1 dyne on another unit charge when separated from it by 1 centimeter in a vacuum.

electrostatic valence rule *Physical Chemistry.* a rule stating that in a stable state, the valence (combining power) of negatively charged atoms equals the total strength of the bonds that they have formed with nearby positively charged ions.

electrostatic voltmeter *Engineering.* an instrument that measures voltage according to the degree of attraction or repulsion between charged bodies.

electrostatic wattmeter *Engineering.* an instrument designed to measure high voltages in watts, by means of electrostatic forces.

electrostatic wave *Physics.* the wave motion of a plasma whose restoring forces are primarily electrostatic.

electrostatography *Graphic Arts.* a process in which an electrostatically charged printing plate is exposed so that the resulting positive image attracts a negatively charged resin.

electrostethophone *Medicine.* an electrically amplified stethoscope.

electrostriction *Mechanics.* the elastic deforming of a dielectric material when stressed by an electric field.

electrostrictive *Acoustical Engineering.* describing an acoustic device whose operation is based on the energy produced by the change in dimensions of an electrostrictive material, such as barium titinate. Thus, **electrostrictive microphone, electrostrictive transducer.**

electrosurgery *Medicine.* surgery that is performed using an active electrode in the form of a disk, bulb, or needle.

electrosynthesis *Chemistry.* a synthesis reaction that is induced by an electric current flow in an electrochemical cell.

electrotaxis *Biology.* the movement of cells and organisms toward or away from an electrical current stimulus. Also, GALVANOTAXIS.

electrotherapy *Medicine.* the use of low-intensity electricity to treat insomnia, anxiety, or neurotic depression.

electrothermal *Physics.* of or relating to both electricity and heat, particularly to heat produced by electrical current.

electrothermal energy conversion *Engineering.* a process in which electrical energy is directly converted into heat energy.

electrothermal process *Engineering.* a process in which an electric current is used to produce heat; used to generate higher temperatures than can be produced by combustion processes.

electrothermal propulsion *Space Technology.* a type of rocket propulsion in which the propellant is heated electrically, as an electric arc that is used to heat hydrogen gas in an arc-jet engine.

electrothermal recording *Electronics.* in facsimile, a technique in which an image is produced on a recording medium by heat generated by electronic signals.

electrothermal voltmeter *Engineering.* an instrument that measures voltage and operates as an electrothermal ammeter, using a series resistor as a multiplier.

electrotinning *Metallurgy.* the process of electrodepositing tin on a base.

electrotonus *Physiology.* the altered excitability, conductivity, or electrical state of a nerve that is produced when a constant electric current is passed through it.

electrotropism *Biology.* a curvature of sessile organisms toward or away from an electrical current stimulus. Also, ELECTROPISM.

electrotype *Graphic Arts.* a metal-plated molded replica of a relief printing plate, produced electrolytically; used for very long press runs. Also, ELECTRO.

electrovalence *Physical Chemistry.* **1.** the number of positive or negative charges that an atom acquires by transferring its electrons to another atom to form a compound. **2.** the chemical bond resulting from such a process; an ionic bond.

electrovalent *Physical Chemistry.* relating to electrovalence or to an electrovalent bond (ionic bond).

electrovalent bond see IONIC BOND.

electroweak interaction *Particle Physics.* the force that results from combining the electromagnetic force and the weak force.

electroweak theory see WEINBERG-SALAM THEORY.

electrowinning *Metallurgy.* the extraction of electrolytic metal from a solution or molten salt.

Elek plate *Microbiology.* a petri-plate method that detects the production of a diffusible toxin by a given microorganism by overlaying a culture with antitoxin and observing the formation of a toxin-antitoxin precipitate after an appropriate incubation.

Elektra *Astronomy.* asteroid 130, discovered in 1873 and measuring 189 km in diameter; it belongs to type C.

Elektrion process *Chemical Engineering.* a method of condensation and polymerization in which a light mineral oil and fatty oil mixture is exposed to an electric discharge in a hydrogen atmosphere, yielding a viscous oil.

element *Chemistry.* any of a class of substances that cannot be separated into simpler substances by chemical means; the fundamental chemical units of which all matter, at or above the atomic level, is composed. All atoms of a given element have the same nuclear charge and number of protons and electrons, but may differ in mass according to the number of neutrons in the nucleus; such varieties of an element are called isotopes. There are currently 109 known elements, although those above atomic number 103 are problematic. *Statistics.* a single individual item in a population or universe. *Mathematics.* **1.** a single member of a set. **2.** the expression following the integral sign in a definite integral. Also, INTEGRAND. **3.** an approximation to the length of a curve between two points (**element of arc length**), to the area of a region in the plane bounded by a curve (**element of plane area**), or to the volume of region in space (**element of volume**). **4.** a fundamental assumption or proposition of geometry, calculus, or other field. **5.** an entry a_{ij} of a matrix; i.e., the value that is located at the intersection of the *i*th row and *j*th column in a matrix. *Industrial Engineering.* see ELEMENTAL MOTION. *Computer Programming.* an item in a data set or array. *Computer Technology.* any circuit or device that performs a basic data-processing function. *Electromagnetism.* the part of an antenna that radiates radio waves, and that may be active (fed directly) or parasitic (no direct connection). *Electricity.* see COMPONENT.

element 104 *Chemistry.* an artificial element of atomic weight 261; the twelfth transuranium element, and first beyond the actinide series. Also, KURCHATOVIUM.

element 105 *Chemistry.* an artificial element of atomic weight 260; formed by bombarding californium-249 with nitrogen-15 ions. Also, HAHNIUM.

element 106 *Chemistry.* an artificial element of atomic weight 263, formed by bombarding californium-249 with oxygen-18 ions; its isotope 259 is created by bombarding lead-207 and lead-208 with chromium-54 ions.

element 107 *Chemistry.* an artificial element of atomic weight 262, produced by bombarding bismuth-209 with chromium-54 ions.

element 108 *Chemistry.* an artificial element of atomic weight 265, formed by bombarding lead-208 with iron-58 ions.

element 109 *Chemistry.* an artificial element of atomic weight 266, formed by bombarding bismuth-266 with iron-58 ions.

elemental analysis *Analytical Chemistry.* any of various methods used to determine the relative weight of an element in a compound.

elemental motion *Industrial Engineering.* a distinct and measurable segment of a work cycle, consisting of a single basic motion or a series of motions. Also, BASIC MOTION, ELEMENT.

elementary *Science.* relating to the earliest or most basic concepts and principles of a given field.

elementary body *Virology.* **1.** a small, round extracellular virus-associated structure, often seen under the microscope when observing scrape preparations from skin lesions of smallpox, vaccinia, or varicella zoster. **2.** an obsolete term for a virion.

elementary charge *Physics.* the fundamental unit of charge equal to the amount of charge carried by one electron or proton.

elementary event *Statistics.* a single outcome of an experiment. Also, SIMPLE EVENT.

elementary function *Mathematics.* any function of a single variable that can be represented by an arithmetical expression using only the four arithmetic operations, and exponential, trigonometric, or hyperbolic functions and their inverses. Examples of **nonelementary functions** include the gamma, Dirchlet, and greatest integer functions, as well as integrals such as $y = \int_0^x t^{-1} \sin t \, dt$.

elementary particle *Particle Physics.* one of the fundamental units of which all matter is composed, including leptons, quarks, and bosons. Elementary particles do not appear to be divisible into smaller units and so far their size is too small to measure. Also, FUNDAMENTAL PARTICLE.

elementary process *Physical Chemistry.* an event that take place at the atomic or molecular level during a larger chemical reaction.

elements of an orbit *Astronomy.* the seven quantities that describe an orbit completely by locating it in space and time: semimajor axis, eccentricity, inclination, longitude of the ascending node, longitude (or argument) of perihelion, date of perihelion passage, and period.

elements of the trajectory *Mechanics.* a trajectory's features, such as angle of fall, peak altitude, and so on.

element time *Industrial Engineering.* the time required for a single elemental motion in a work process.

elemi *Materials Science.* any of various fragrant, soft resins used in perfumery and in the making of ointments, varnishes, lacquers, and inks.

eleo- a combining form meaning "oil," as in *eleoplast.*

eleometer *Chemistry.* an instrument for determining the percentage of oil in a mixture, as well as the specific gravity of oils.

eleoptene *Chemistry.* the liquid part or hydrocarbon of a volatile oil. Also, **elaeoptene.**

eleostearic acid *Organic Chemistry.* $C_{18}H_{30}O_2$, a colorless, crystalline fatty acid that is insoluble in water and soluble in most organic solvents; its glycerol ester is a chief component of tung oil. Also, OCTADECATRIENOIC ACID.

Eleotridae *Vertebrate Zoology.* a family of small fresh-, brackish, or marine-water fishes of the suborder Gobioidei that are similar to the bogies, but have united pelvic fins; found in tropical and subtropical climates.

elephant *Vertebrate Zoology.* **1.** either of two large, four-footed, nearly hairless mammals of the family Elephantidae, the largest living land animals, having a very long prehensile trunk formed of the nose and upper lip and ivory tusks. The **African elephant,** *Loxodonta africana,* stands up to 11 feet high at the shoulder and is characterized by enormous, flapping ears and two projections at the end of the trunk. The **Indian elephant**, *Elephas maximus,* has smaller ears and a single projection at the end of the trunk. **2.** a family of bulky herbivorous mammals composing the order Proboscidea, including the modern elephants and also extinct forms such as mammoths and mastadons.

elephant (African)

elephanta *Meteorology.* a strong southeasterly wind on the Malabar coast of southwest India, occurring at the end of the monsoon season in September and October and bringing thunder squalls and heavy rain. Also, **elephant, elephanter**.

elephant bird see AEPYORNIS.

elephant ear *Aviation.* an air intake characterized by twin inlets, one on each side of the fuselage. *Space Technology.* a thick plate serving to reinforce a hatch or aperture on a rocket or missile.

elephant fish *Vertebrate Zoology.* any of the various long-snouted fish of the genus *Callorhyncus* of the family Chimaeridae that are found in the deep waters of the Southern Hemisphere.

elephant head *Botany.* a lousewort, *Pedicularis groenlandica,* characterized by spikes of crimson flowers; found in regions of arctic and western alpine North America.

elephant-hide pahoehoe *Geology.* a type of pahoehoe whose surface is covered with broad swells and pressure ridges, giving it a wrinkled and draped appearance resembling the hide of an elephant.

elephantiasis [el´ə fən tī´ə sis] *Medicine.* a chronic filarial disease characterized by enlargement and thickening of the subcutaneous and cutaneous tissues as a result of lymphatic obstruction and lymphatic edema, caused by infection with the nematode *Wuchereria bancrofti* or *Brugia malayi;* occurs primarily in the tropics.

Elephantidae *Paleontology.* a family of large proboscidean mammals that arose in the Pliocene, spread worldwide, and is still represented by two genera and two species.

elephant seal *Vertebrate Zoology.* an earless or true seal of the family Phocidae, having two species of the genus *Mirounga* that are found along the coasts near Baja California and in the Southern Hemisphere; the males weigh up to three tons; their fairly slow-moving habits almost caused their extinction by whalers in the 1800s. (So called because of the male's curved snout, resembling an elephant's trunk.)

elephant shrew *Vertebrate Zoology.* any of the African insectivores of the family Macroscelididae, characterized by a long, sensitive snout and long hind legs.

elephant's trunk *Engineering.* a long pipe, thought to resemble an elephant's trunk, that is used to remove mud, sand, and silt from the bottom of an excavation.

elephant trunks *Astronomy.* thin, dark lanes of dust that reach into glowing gaseous nebulae.

Eleutherozoa *Invertebrate Zoology.* a group name for nonsessile echinoderms.

elev. elevation.

elevate *Engineering.* to increase the angular height of an apparatus such as a gun, launcher, or optical instrument.

elevated pole *Astronomy.* the elevation in degrees of the pole as seen from a given location; equal to the site's latitude.

elevating mechanism *Ordnance.* a mechanism on a gun carriage or launcher that is used to raise or lower the weapon.

elevation *Cartography.* the vertical distance from a datum, usually mean sea level, to a point or object on the earth's surface. *Engineering.* in an engineering drawing, the front, rear, or side view(s) of a structure. *Ordnance.* in antiaircraft artillery, an alternate term for angular height.

elevation angle *Engineering.* a measurement of a line's inclination on a vertical plane between the horizontal and an ascendant line; used in astronomy and surveying. *Electromagnetism.* the angle that a beam of radiation makes with the horizontal.

elevation indicator *Ordnance.* a device that visually indicates the vertical angle between a fixed reference point and a target.

elevation of boiling point see BOILING-POINT ELEVATION.

elevation prediction correction *Ordnance.* in antiaircraft artillery, the difference between the future elevation at which a gun should be fired and the present elevation of a (moving) target.

elevation scale *Ordnance.* the scale on a gun carriage that indicates the gun's quadrant elevation. Similarly, **elevation indicator.**

elevation stop *Engineering.* a component on a gun or other apparatus that prevents its movement or elevation beyond a designated limit.

elevation table *Ordnance.* a firing table indicating the proper quadrant elevation settings for various ranges.

elevator *Mechanical Engineering.* **1.** an enclosed platform or cage used to transport people or materials vertically in a building shaft. **2.** an endless chain or belt device having buckets, scoops, arms, or trays to lift materials to a higher level. *Surgery.* an instrument used to elevate tissues or to remove osseous fragments or roots of teeth. *Aviation.* a control surface that governs an aircraft's pitch by raising or lowering the tail in flight; usually a hinged part of the horizontal surface.

elevator angle *Aviation.* the acute angle between the chord of an aircraft's elevator at a given moment and its chord in the longitudinal or neutral position; it is positive when the elevator's trailing edge is below the neutral position, and negative when it is above.

elevator dredge *Mechanical Engineering.* a dredge used to mine sand and gravel beds, using buckets mounted on a ladder.

elevator link *Petroleum Engineering.* one of the cylindrical bars that are attached and give support to a drilling rig elevator; it engages the hook of the traveling block suspended on the steel cable lines. Also, **elevator bail.**

elevator rod *Petroleum Engineering.* a steel block equipped with an opening and latching system that allows insertion of a sucker rod.

elevon *Aviation.* on the rear portion of an aircraft wing, a hinged device combining the functions of an elevator and an aileron. Usually found on delta-wing aircraft, it can be moved in the same direction on either side of the aircraft to obtain longitudinal control, or differentially to obtain lateral control.

ELF extremely low frequency.

elfin forest *Ecology.* **1.** small, stunted tree growth characteristic of forests at higher elevations in warm, moist regions. **2.** see KRUMMHOLZ. Also, **elfinwood.**

elf owl *Vertebrate Zoology.* a very small insectivorous owl, *Micropallas whitney,* of the family Strigidae; common to cactus, grass, and forest lands, and ranging from the southwestern U.S. to Mexico.

elgiloy *Materials.* a nonmagnetic, low-expansion, strong alloy with high fatigue strength and corrosion resistance, a ferrous alloy containing cobalt, chromium, nickel, molybdenum, manganese, carbon, and beryllium; used to make coils and springs.

Elgin extractor *Chemical Engineering.* a multistage extractor with spray-tower and counterflow features, in which the diameter of the base section is made larger to eliminate flow restriction at the site of light-liquid introduction.

elicited response *Behavior.* a response that consistently occurs following a particular stimulus.

eliminant *Mathematics.* the determinant of the matrix of coefficients of a system of homogeneous linear equations.

elimination *Mathematics.* **1.** any process of removing one of the unknowns from an algebraic equation. **2.** any method of solving sets of simultaneous linear equations by removing unknowns successively from all equations until one unknown is evaluated; e.g., Gaussian elimination.

elimination diet *Medicine.* a method of identifying a food allergy by sequentially omitting foods in order to determine which one or ones are responsible for symptoms.

elimination factor *Computer Programming.* in information retrieval, the ratio of the number of documents not retrieved to the total number of documents in the file.

elimination reaction *Organic Chemistry.* a reaction of one or more molecules in which a simple molecule (such as water or ammonia) is removed, with subsequent production of a new compound.

eliminator *Electronics.* **1.** any device, generally a battery, that replaces an inferior or inconvenient component. **2.** any device that diminishes or eliminates an undesirable signal or quantity.

E link *Naval Architecture.* a bracket on a compass pedestal or binnacle on which a quadrantal error corrector can be fitted.

elinvar *Metallurgy.* a ferrous alloy formerly containing 36% nickel and 12% chromium and having a nearly constant modulus of elasticity within a broad temperature range; its current composition is 33–35% nickel, 4–5% chromium, 1–3% tungsten, 0.5–2.0% manganese, 0.5–2.0% sulfur, and 0.5–2.0% carbon.

Elion, Gertrude born 1918, American biochemist; researched nucleic acid metabolism; Nobel Prize (with George Hitchings) for developing compounds used against leukemia, malaria, herpesvirus, gout, and organ transplant rejection.

ELISA *Immunology.* any of a variety of methods using an enzyme-labeled immunoreactant (antigen or antibody) and an immunosorbent (antigen or antibody bound to a solid) to identify specific serum or tissue antibodies or antigens. (An acronym for enzyme-linked immunosorbent assay.) Also, IMMUNOSORBENT ASSAY.

elixir *Pharmacology.* a clear, sweetened liquid, usually based on water and alcohol with flavoring substances, used as a vehicle to administer an oral dose of an active medication. *Chemistry.* in the practice of alchemy, a substance that was thought to be able to change base metals into gold.

ELIZA *Artificial Intelligence.* an early pattern-matching natural language program that assumed the role of therapist and carried on a realistic dialogue with the user.

elk

elk *Vertebrate Zoology.* **1.** also, **European elk.** the largest living deer, *Alces alces,* found in Europe and Asia, resembling but smaller than the related moose of North America; now rare; found in parts of Scandinavia, Germany, Russia, and Siberia. **2.** also, **American elk.** a large North American deer, *Cervus canadensis.* Also, WAPITI.

ell *Building Engineering.* **1.** a short, L-shaped connecting pipe used during the layup of utilities during building construction. **2.** a right-angle extension relative to one end of a building, forming an addition or wing to the main building.

ellagic acid *Organic Chemistry.* $C_{14}H_6O_8$, yellow crystals isolated from tannins; very slightly soluble in water and alcohol; melts at 450–480°C; used as a hemostatic. Also, GALLOGEN.

Ellesmere Island [elz´mēr´] *Geography.* a large island (area: 82,119 square miles) northwest of Greenland, in the Canadian Northwest Territories.

ellestadite *Mineralogy.* a general name for the series fluorellestadite-hydroxylellestadite, $Ca_5(SiO_4,PO_4,SO_4)_3(F,Cl,OH)$ - $Ca_5[(SiO_4),(SO_4)]_3(OH,Cl,F)$, lavender to rose hexagonal and monoclinic minerals having a specific gravity of 3.01 to 3.07 and a hardness of 4.5 to 5 on the Mohs scale; found at Crestmore, California.

Elliot's position *Surgery.* a position of the patient on an operating table, in which the lower chest is elevated by placing a support under the lower costal margin; a position that is used especially in gallbladder operations. (Named for John Wheelock *Elliot,* 1852–1925, American physician.)

ellipse *Mathematics.* a plane curve, the locus of all points such that the sum of the distance to two fixed points (foci) is constant; a conic whose eccentricity is less than 1. Equivalently, the closed conic section obtained by intersecting a cone with some planes not parallel to a generator of the cone. A general equation is $Ax^2 + 2Bxy + Cy^2 + 2Dx + 2Ey + F = 0$, where $AC - B^2 > 0$; in normal form, with foci symmetrically placed on the coordinate axes about the origin, $x^2/a^2 + y^2/b^2 = 1$. The special case $a = b$ is called a circle.

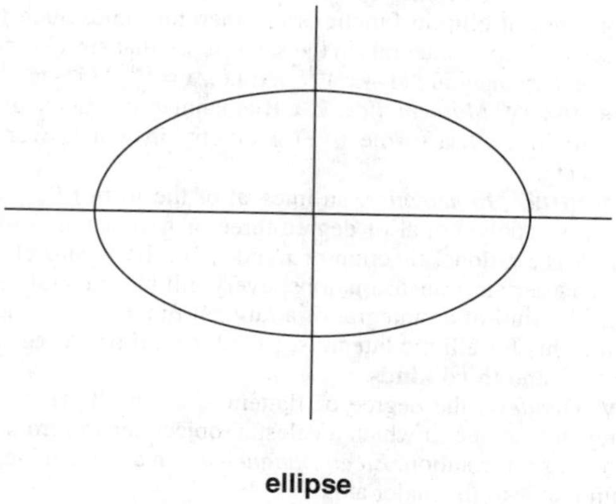

ellipse

ellipsoid *Mathematics.* **1.** a smooth closed surface, all of whose plane sections are ellipses. It is a solution surface of a general quadratic equation $Ax^2 + 2Bxy + 2Cxz + Dy^2 + 2Eyz + Fz^2 + 2Gx + 2Hy + 2Iz + J = 0$ in three dimensions whose Hessian (in this case, a matrix of constants) has a positive determinant. The simplest has the equation $x^2/a^2 + y^2/b^2 + z^2/c^2 = 1$. If at least two of a, b, and c are equal, then the surface is an **ellipsoid of revolution,** formed by rotating an ellipse about one of its axes. Plane sections perpendicular to the axis of rotation are circles. If $a = b = c$, the figure is a sphere. **2.** the solid enclosed by such a surface.

ellipsoidal *Science.* of, relating to, having the form of, or resembling an ellipsoid.

ellipsoidal coordinates *Mathematics.* the coordinates in space that are determined by three confocal quadrics.

ellipsoidal height *Geodesy.* the height above a reference ellipsoid, measured along the ellipsoidal outer normal through the point in question.

ellipsoidal lava see PILLOW LAVA.

ellipsoid(al) of wave normals see INDEX ELLIPSOID.

ellipsoidal reflector *Cartography.* a mirror surface that conforms to a portion of an ellipsoid of revolution, principally employed in several types of stereoplotter projectors.

ellipsoid of rotation *Geodesy.* the surface generated by an ellipse rotating about one of its axes.

elliptic or **elliptical** *Science.* relating to or having the form of an ellipse.

elliptical galaxy *Astronomy.* a galaxy with generally old stars that has an elliptical shape and that lacks a disk with spiral arms.

elliptical map projection *Cartography.* a projection of the surface of the earth on the inside of an ellipsoid, which is then developed into a plane.

elliptical orbit *Astronomy.* any orbit whose eccentricity is greater than zero; most closed orbits in nature are elliptical. *Mechanics.* any orbit that follows an ellipse.

elliptical polarization *Electromagnetism.* polarization in which a vector representing a wave component at any point in space describes an ellipse in a plane perpendicular to the propagation direction.

elliptical ring structure *Astronomy.* see ECCENTRIC RING STRUCTURE.

elliptical system *Engineering.* a method used in tracking or navigation whereby ellipsoids of positions are defined according to time or phase summation, relative to several fixed stations serving as the focal point for the ellipsoids.

elliptic coordinates *Mathematics.* **1.** coordinates in the plane determined by confocal hyperbolas or ellipses; given by $x = a \cosh \eta \cos \psi$ and $y = a \sinh \eta \sin \psi$, where a is constant. **2.** three-dimensional space coordinates determined by confocal conic surfaces, i.e., hyperboloids or ellipsoids; given by $x = a \cosh \eta \cos \psi$, $y = a \sinh \eta \sin \psi$, and $z = z$, where a is constant. Also, **elliptic-cylinder coordinates.**

elliptic differential equation *Mathematics.* a linear second-order partial differential equation of the form

$$0 = \sum_{i,j=1}^{n} A_{ij}(\partial^2 u/\partial x_i\, \partial x_j) + \sum_{i=1}^{n} B_i\,(\partial u/\partial x_i) + Cu + D$$

such that (a) A_{ij}, B_i, C, D are differentiable functions of $(x_1, \ldots, x_n)$, and (b) the quadratic form $\Sigma_{i,j=1}^{n} A_{ij}x_ix_j$ can be transformed to the form $\Sigma_{i=1}^{n} F_ix_i^2$, where the F_i have the same sign. The Laplace and Poisson equations are well-known examples of elliptic differential equations.

elliptic function *Mathematics.* a doubly periodic function of a single complex variable; i.e., it may be viewed as a function on a torus. By theorem, nonconstant elliptic functions are meromorphic. Such functions are inverses of elliptic integrals in the same sense that $\sin x$ is the inverse of the elementary integral $\int (1 - x^2)^{-1/2}\, dx$; i.e., $x = \int_0^{\sin x} (1 - t^2)^{-1/2}\, dt$.

elliptic geometry *Mathematics.* **1.** a Riemannian geometry of constant positive curvature. **2.** a (projective) geometry in which every pair of lines intersects.

elliptic integral *Mathematics.* an integral of the form $\int R(x, \sqrt{p(x)})\, dx$, where $p(x)$ is a polynomial of degree three or four having only simple roots, and R is a rational function of x and $\sqrt{p(x)}$. By means of reduction formulas and certain transformations, every elliptic integral can be expressed as the sum of an integral of a rational function and one of three canonical forms for elliptic integrals, called the **elliptic integrals of the first, second,** and **third kinds.**

ellipticity *Geodesy.* the degree of flattening of an ellipse or ellipsoid. *Astronomy.* the degree to which a celestial object departs from a sphere, usually because of rotation. *Electromagnetism.* in a waveguide, the ratio of the minor axis to the major axis.

ellipticity of the spheroid *Geodesy.* the ratio of the difference between the equatorial and polar radii of the earth (the semimajor and semiminor axes of the spheroid) to its equatorial radius (the semimajor axis).

elliptic point *Mathematics.* **1.** a point on a Riemannian manifold that has a neighborhood in which the Gaussian (scalar) curvature is everywhere positive. **2.** a point in the domain of a partial differential equation that has a neighborhood in which the partial differential equation is elliptic.

elliptic spring *Mechanical Devices.* a spring having graduated, laminated steel plates with convex sides oriented outward and forming an ellipse; used on locomotives and other vehicles having a rigid wheel base.

elliptocytosis *Medicine.* a mild hereditary disorder of the blood characterized by 90% or more of peripheral blood erythrocytes having an oval shape.

Ellis, Havelock 1859–1939, English psychologist; author of the seven-volume *Studies in the Psychology of Sex.*

Ellobiophycidae *Botany.* a subclass of marine dinoflagellates of the class Dinophyceae that are nonphotosynthetic and ectoparasitic of arthropods and annelids.

ellsworthite see URANPYROCHLORE.

elm *Botany.* the common name for any tall, hardwood tree of the species *Ulmus,* having simple, serrate, deciduous leaves, small, apetalous flowers, and samaroid fruits. Among the elm species widely found in North America (and greatly affected by Dutch elm disease) are the **American** or **white elm,** *U. americana,* the **slippery elm,** *U. rubra,* and the **winged elm,** *U. alata.*

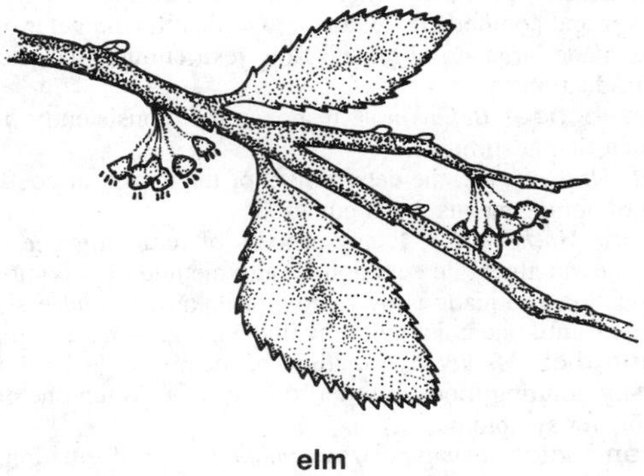

elm

elm bark beetle *Invertebrate Zoology.* either of two beetles that transmit Dutch elm disease: the primary vector, *Scolytus multistriatius,* is also called the **(smaller) European elm bark beetle;** the native North American vector, *Hyulurgopinus opaculus,* is also called the **native elm bark beetle.**

elm blight see DUTCH ELM DISEASE.

Elmidae *Invertebrate Zoology.* a family of small aquatic coleopteran insects, the drive beetles, in the superfamily Dryopoidea.

El Niño [el nēn´yō] *Oceanography.* a complex set of changes in the water temperature in the Eastern Pacific equatorial region, producing a warm current; it occurs annually to some degree between October and February, but in some years intensifies and causes unusual storms and destruction of marine life. (From Spanish for "the child;" meaning the Christ child; it typically begins at Christmas time.)

elongation *Mechanics.* the stretching of a member by tensile stress, including plastic stretching. *Radiology.* the distortion of a roentgenograph in which the image is proportionately longer than the actual structure. *Telecommunications.* the extension of a signal envelope due to a delay in the arrival of multipath components. *Astronomy.* the angular distance between two objects; most often used with reference to the sun and either Mercury or Venus.

elongation factor *Biochemistry.* a protein that forms a complex with ribosomes for the purpose of elongating polypeptide chains, then dissociates when translation is complete.

Elopidae *Vertebrate Zoology.* the tenpounders, a monogeneric family of tropical marine to estuarine fish of the order Elopiformes, characterized by a terminal mouth, small scales, and vigorous swimming.

Elopiformes *Vertebrate Zoology.* a primitive order of marine bony fish that includes tenpounders and bonefishes, characterized by a single dorsal fin made up of soft rays.

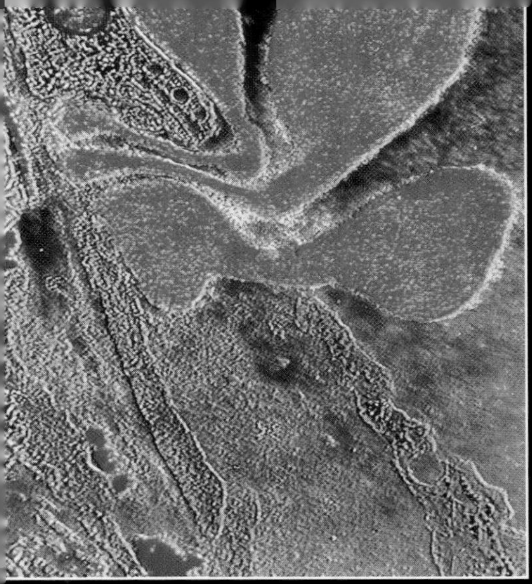

-alse-color transmission electron micrograph of red
blood cells passing through the wall of a capillary
(diapedesis). CNRI/SPL/Photo Researchers.

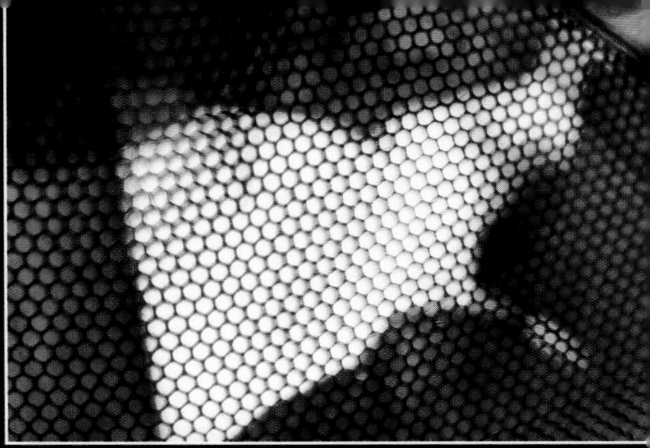

Multifaceted compound eye surface of a horsefly.
© Bruce Iverson/Science Photos.

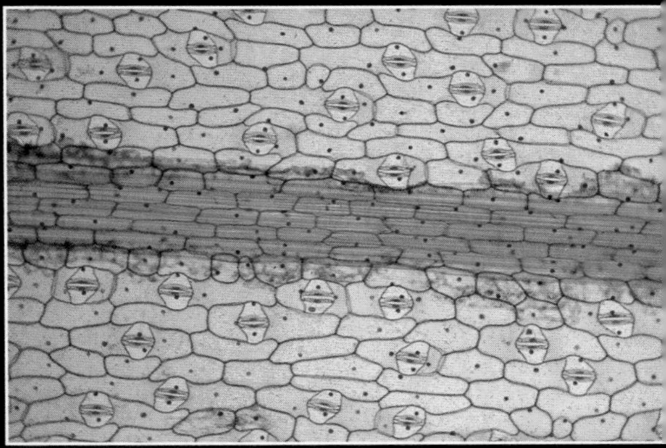

Tradescantia leaf with stomata, guard cells, nuclei, and
cell walls visible. James M. Bell/Photo Researchers.

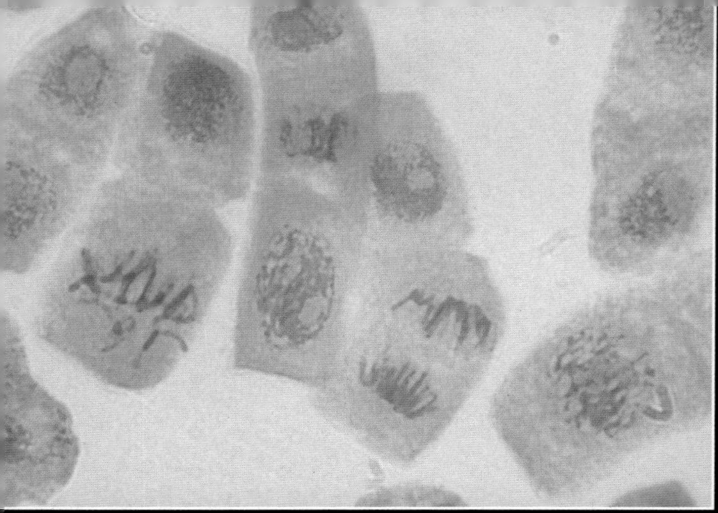

ʌitosis in the root tip of an onion. Chromosomes are visible.
ɔ Bruce Iverson/Science Photos.

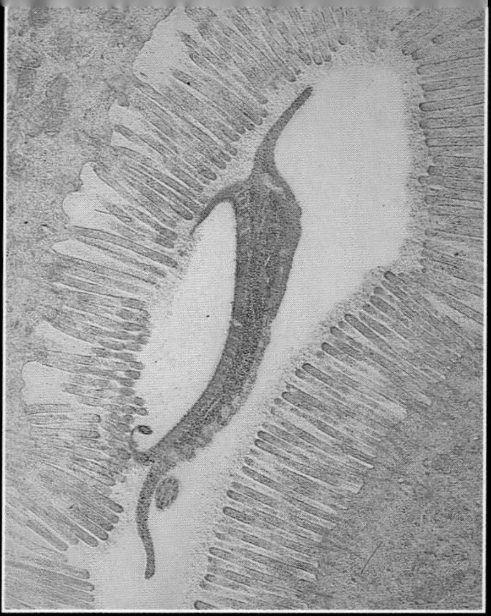

False-color transmission electron micrograph of *Giardia lamblia* (green) attached to the microvilli of the human small intestine. CNRI/SPL/Photo Researchers.

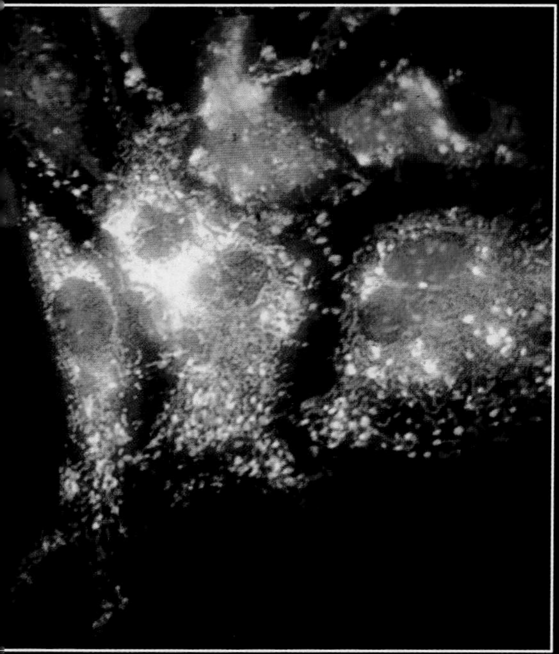

ultiple-parameter fluorescence imaging of mouse cells xhibiting wound healing. Courtesy of Center for Light icroscope Imaging and Biotechnology, an *NSF Science and* *echnology Center*, Carnegie Mellon University.

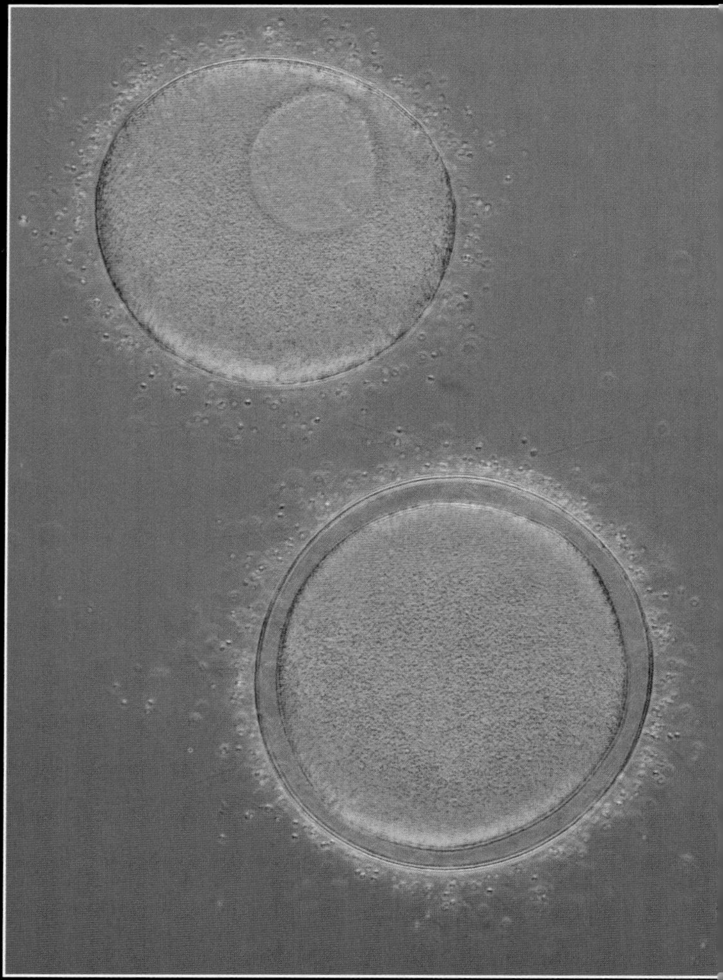

Immature sea star oocyte (upper left) after spermatozoon penetration and mature sea star oocyte (lower right) after spermatozoon penetration. Courtesy of Dr. Kazuyoshi Chiba, Tokyo Institute of Technology.

Elopoidei *Vertebrate Zoology.* the tarpons and tenpounders, a suborder of mainly marine fishes of the order Elopiformes, distinguished by a well-developed median gular plate and a mouth bordered by premaxillaries and toothed maxillaries.

ELP *Aviation.* the airport code for El Paso, Texas.

elpasolite *Mineralogy.* K_2NaAlF_6, a colorless, transparent, cubic mineral, massive in habit, having a specific gravity of 2.995 and a hardness of 2.5 on the Mohs scale.

elpidite *Mineralogy.* $Na_2ZrSi_6O_{15}\cdot 3H_2O$, a white to pale brick-red, orthorhombic mineral occurring in prismatic crystals or fibrous aggregates, having a specific gravity of 2.54 and a hardness of 5 to 7 on the Mohs scale; found in albitized nepheline-syenite pegmatites.

Elsasser's radiation chart *Meteorology.* a radiation chart used for the graphical solution of such problems as the effective terrestrial radiation, net flux of infrared radiation at a cloud base or top, and radiative cooling rates. (Named for Walter Maurice *Elsasser,* born 1904, German-born American geophysicist.)

ELSE instruction *Computer Programming.* an instruction in programming languages that denotes the actions to be taken when the conditions in the IF instruction are not met.

ELSE rule *Computer Programming.* in decision tables, a catchall rule designed to specify the action to take when conditions do not trigger one of the explicit rules.

elsinan *Biochemistry.* a glucan formed from blocks of alpha-linked maltotriose units joined by alpha bonds; it is water soluble and basically linear.

Elsinoe *Mycology.* a genus of fungi belonging to the order Dothideales, containing species that cause scab diseases in plants.

Elsinoeaceae *Mycology.* a family of fungi belonging to the order Myriangiales that occurs primarily in tropical regions and is parasitic and pathogenic to plants, causing scab diseases.

Elsonian orogeny *Geology.* a deformation affecting the Canadian Shield that occurred about 1.4 billion years ago.

Elster, Julius 1854–1920, German physicist; with Geitel, recognized instability of radioactive elements and found these to occur widely.

Elster-Geitel effect *Physics.* a phenomenon in which a heated conductor in the presence of a gas may develop an accumulation of either positive or negative charge, but in a vacuum will develop only a negative charge.

Eltonian *Geology.* a European geologic stage of the Upper Silurian period, occurring after the Wenlockian and before the Bringewoodian.

Eltonian niche *Ecology.* the ecological role of a given species in a given community.

El Tor vibrio *Bacteriology.* one of the more common biotypes of the pathogenic bacterium *Vibrio cholerae,* which causes the disease cholera.

eluant *Chemistry.* the liquid used to effect the separation of materials in an elution process.

eluate *Chemistry.* the product or substance that is separated out in an elution process.

elution *Chemistry.* a process of removing and separating substances adsorbed on a fixed bed by a stream of liquid or gas. *Materials Science.* a chromatographic separation in which an eluant is passed through the chromatographic bed after the addition of the sample. *Virology.* the dissociation of an adsorbed virus particle.

elution profile *Biotechnology.* a graph that represents the amount of a material removed from a liquid chromatography column; used to identify various compounds.

elutriation *Materials Science.* **1.** a method for removing materials from a mixture through washing, then decanting. **2.** the separation of particles according to size by submitting them to an upward current of air, water, or other fluid that will cause them to settle at different rates.

elutriator *Materials Science.* an instrument that sorts out solid particles according to size by shooting a slow stream of fluid upward through the particle mixture, so that the lighter particles float upward and the heavier particles float downward.

eluvial *Geology.* **1.** composed of or relating to eluvium, fine soil material that has been deposited by the wind. **2.** a soil horizon that has lost material through eluviation.

eluvial placer *Geology.* a mineral deposit concentrated near the decomposed outcrop of the source and deposited by rain wash, rather than by stream action.

eluviation *Agronomy.* the movement of soil material from one place to another within the soil, when there is an excess of rainfall over evaporation. It may take place downward or sidewise according to water movement.

eluvium *Geology.* **1.** fine soil material that has been moved and deposited by the wind. **2.** an accumulation of residual rock debris formed in place by weathering and disintegration. *Ecology.* a sand dune community.

ELV expendable launch vehicle.

Elysia *Invertebrate Zoology.* a genus of sea slugs, in the class Gastropoda, that sucks the contents of algae into its own digestive gland cells, where photosynthesis occurs; characterized by broad, finlike side extensions.

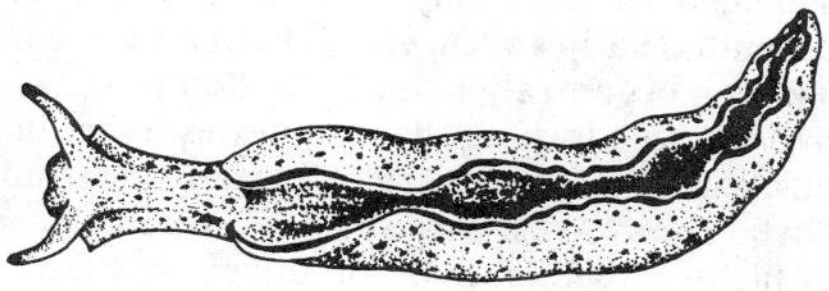

Elysia

elytra *Entomology.* the thickened horny forewings of beetles, which function to cover and protect the membranous hindwings.

elytron *Entomology.* **1.** either of the anterior pair of hard, shell-like wings in beetles that serve as protective covers for the membranous second pair of wings used for flight. **2.** any of the platelike scales covering the body of certain polychaete annelids, the scale worms.

em *Graphic Arts.* in any type font, a linear measure equal to the point size (height of the type) being used; the square of a type size. In 8-point type, for example, an em is 8 points wide. (Based on the size of a capital letter M in the given point size.) Also, **em space.**

em- a prefix meaning: **1**. to cause to be in, as in *embed.* **2.** in or within, as in *embryo.*

EM electromagnetic; electron microscope.

emaciation *Medicine.* extreme leanness or a wasted condition caused by disease or lack of nutrition. Thus, **emaciated.**

Emacs [ē´maks´] *Computer Science.* a powerful and extensible WYSIWYG screen editor. (An acronym for edit macros.)

emagram *Thermodynamics.* a plot of ln*(p)* as a function of the pressure versus temperature *T* for a substance held at constant volume.

E-mail see ELECTRONIC MAIL.

emanating power *Nucleonics.* the percentage of radon atoms that escape from the solid or solution in which they form.

emanation see RADIOACTIVE EMANATION.

emanation security *Electronics.* a form of communications security that prevents an unauthorized person from obtaining information from unintentional electronic emissions other than telecommunications.

emanometer *Engineering.* a device that measures the radon content of the atmosphere by extracting the radon through condensation or adsorption on a surface, and then recording its behavior in an ionization chamber.

emanometry *Nucleonics.* the various techniques used for determining the amount of radioactive gases escaping from the earth's surface into the lower atmosphere by counting the ions produced by the alpha particles emitted by at least one of the radioactive gases in an ionization chamber.

E-map *Crystallography.* a Fourier map, equivalent to an electron-density map for a crystal structure, with phases derived by direct methods and normalized structure amplitudes, $|E(hkl)|$ replacing $|F(hkl)|$ in the Fourier summation. Since the $|E|$ values correspond to sharpened atoms, the peaks on the resulting *E*-map are sharper than those computed with $|F|$ values.

emarginate *Science.* notched or slightly forked at the margin. *Botany.* of a petal or leaf, having a shallow notch at the apex.

emasculation *Botany.* in plant breeding, the removal of the stamens of a flower before they have shed pollen, in order to prevent self-pollination and permit cross-pollination.

embacle *Hydrology.* **1.** a pile of ice formed in a stream after a refreeze. **2.** the process whereby such a pile is formed.

EMB agar *Microbiology.* eosin-methylene blue agar, a solid culture medium used for isolation and characterization of certain bacteria, taking advantage of the addition of a colored indicator dye.

Emballonuridae *Vertebrate Zoology.* a widespread family of insectivorous bats of the order Chiroptera, characterized by an obliquely truncated face, no nose leaf, and a partially free tail; includes the sheath-tailed bats, sac-winged bats, and ghost bats.

embalm *Medicine.* to treat a cadaver with antiseptic or preservative substances for burial or dissection. Thus, **embalming.**
embankment *Civil Engineering.* **1.** a ridge constructed of earth, rock, or other materials, built to carry a highway or railroad track to a higher elevation than the surrounding terrain. **2.** a protective bank to prevent water encroachment or protect against erosion. *Geology.* a narrow depositional feature built out from the shore by the action of waves and currents. Also, BANK.
embark *Military Science.* to carry out the process of embarkation.
embarkation *Military Science.* **1.** the loading of troops with their equipment and supplies into a ship or aircraft. **2.** of or relating to such activity. Thus, **embarkation area, embarkation unit, embarkation officer, embarkation organization, embarkation team**, and so on.
embata *Meteorology.* a local onshore southwest wind in the lee of the Canary Islands, caused by the reversal of the northeast trade winds.
embatholithic *Geology.* a stage in the erosion of metalliferous batholith in which exposed areas of intruding rocks are nearly equal to the areas of invaded rock.
embayed coastal plain *Geology.* a coastal plain with many projecting headlands, bays, and outlying islands, usually the result of submergence.
embayed mountain *Geology.* a mountain that has been partially submerged, so that seawater enters its bordering valleys.
embayment *Geography.* see BAY. *Geology.* **1.** the process by which a bay is formed. **2.** a deep depression in a shoreline forming a large, open bay. **3.** a downwarped region of stratified rock that extends into an area of other rocks.
Embden *Agriculture.* a large breed of goose having white plumage, an orange bill, and deep orange shanks and toes. (From *Embden,* Germany.)
Embden-Meyerhof glycolysis *Organic Chemistry.* a process in which glucose is decomposed by enzymatic action to produce pyruvic acid; an energy source for animals and anaerobic organisms. Also, HEXOSE DIPHOSPHATE PATHWAY.
embedded *Computer Science.* describing data, a program, a language, etc., that is incorporated or implemented within another program or system, and that is usually not separately programmable by the user. Thus, **embedded language, embedded system,** and so on.
embedding *Mathematics.* see IMBEDDING.
embedment anchor *Naval Architecture.* a fixed anchor that is driven into the seabed by a ram, explosive charge, or other force.
Emberizidae *Vertebrate Zoology.* the buntings, sparrows, and cardinals, a large bird family of the passerine suborder Oscines, found primarily in the Western Hemisphere; it also includes juncos, grosbeaks, Galapagos finches, and New World sparrows.
Embioptera *Invertebrate Zoology.* the web spinners, an order of elongated, flattened insects in the order Orthoptera that look like grasshoppers and spin silk like spiders. Also, **Embiidina.**
Embiotocidae *Vertebrate Zoology.* the surfperches, a family of viviparous percoid fishes of the coastal North Pacific.
embolectomy *Surgery.* an incision made in an artery to remove a clot or embolus, often done as emergency surgery.
embolic *Medicine.* referring to an embolism or embolus.
embolism [em´bə liz´əm] *Medicine.* an obstruction caused by a clot or embolus that travels through the bloodstream and becomes lodged in a blood vessel, usually in the heart, lungs, or brain.
embolite *Mineralogy.* any of various chlorian bromargyrites or bromian chlorargyrites.
Embolomeri *Paleontology.* a suborder of primitive fish-eating amphibians in the order Anthracosauria; extant at the end of the Devonian, the embolomeres have some characteristics that are transitional to reptiles, although they are probably not ancestral to reptiles; rather, they evidently represent a failed attempt to achieve complete terrestriality, which another line of amphibians was achieving.
embolus [em´bə ləs] *Medicine.* a bit of matter that is foreign to the bloodstream, such as a large bubble of air or gas, a bit of tissue or tumor, or a piece of a blood clot, that travels through the bloodstream until it becomes lodged in a vessel and obstructs it.
emboss *Mechanical Engineering.* to raise a design in relief; carry out an embossing process.
embossing *Graphic Arts.* the process of printing or stamping in relief, with or without ink, to create a raised image area; often used on book covers. *Textiles.* the process of creating a raised design or pattern on fabric by any of several methods, including passing the fabric between hot engraved metal rollers or shearing high pile into different levels.
embossing die *Graphic Arts.* a tool, usually metal, that is designed to raise a particular image, as on book cover material.
embossing stylus *Acoustical Engineering.* a blunt-tipped recording stylus that forms a record groove by displacement of material rather than by cutting.
embouchure *Geology.* **1.** the mouth of a river. **2.** the widening of a river valley into a plain. Also, **embouchement.**
embrasure *Architecture.* **1.** an opening in a parapet between two merlons. **2.** a splayed enlargement of the interior opening of a door or window.
embrechites *Petrology.* a type of migmatite composed of permeation gneiss and injection gneiss with an augen and ribbon structure featuring crystalline shifts partly obliterated by metablastesis.
Embrithopoda *Paleontology.* an extinct order of subungulate mammals in the infraclass Eutheria, known chiefly by the single genus *Arsinoitherium,* a large, nasal-horned animal that lived in Egypt in the Oligocene; other, possibly related genera that lived in Europe and Asia were discovered in the 1970s.
embrittlement *Mechanics.* the increase in brittleness of a material because of chemical, molecular, or crystalline change, or lowered temperature.
embryectomy *Medicine.* the surgical removal of the embryo, especially in ectopic pregnancy.
embryo [em´brē ō´] *Developmental Biology.* **1.** the stage of a multicellular organism that develops from a zygote before it becomes free-living. **2.** specifically, in vertebrates, the period from after the long axis appears until all major structures are represented. In humans, this is from about two weeks after fertilization to the end of the seventh or eighth week. *Botany.* **1.** the rudimentary plant produced from fertilization and development of the zygote, usually consisting of an epicotyl, hypocotyl, radicle, and one or two cotyledons. **2.** a young sporophyte from the union of male and female sex cells when still in the seed or gametophyte.

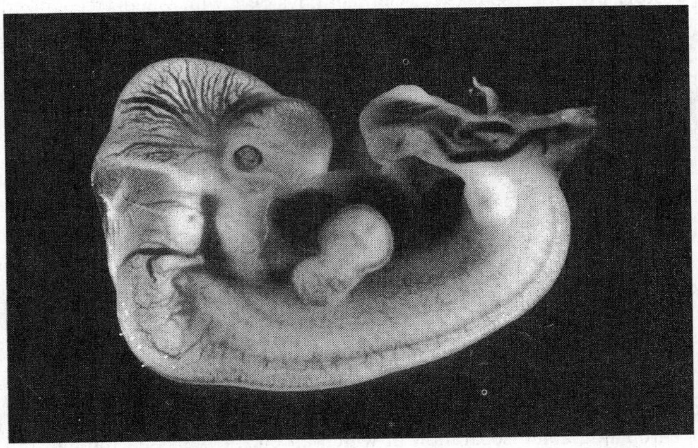

human embryo

Embryobionta see EMBRYOPHYTA.
embryo culture *Developmental Biology.* **1.** the removal of an embryo from its normal environment and its maintenance in a permissive environment. **2.** the artificial fertilization and growth of an embryo *in vitro.*
embryogenesis *Developmental Biology.* the development of an individual from a fertilized ovum; the process of embryo formation.
embryol. embryology.
embryology *Developmental Biology.* a branch of biology concerned with the development of animal and plant organisms from a fertilized egg or seed to the establishment of form and shape, including the study of processes that bring about normal and abnormal development.
embryoma *Oncology.* a tumor that contains embryonic elements.
embryonal *Developmental Biology.* of, relating to, or in the state of an embryo. Also, **embryonic.**
embryonal carcinoma cell *Oncology.* a highly malignant, undifferentiated stem cancer cell of teratocarcinoma.
embryonal-cell lipoma see LIPOSARCOMA.
embryonal leukemia see STEM CELL LEUKEMIA.
embryonate *Developmental Biology.* **1.** referring to or resembling an embryo. **2.** containing an embryo. **3.** impregnated.

embryonated egg *Virology*. an egg, usually of a hen or duck, that contains a live, developing embryo; used to identify, analyze, and culture certain viruses. Also, **embryonating egg.**

embryonic [em´brē än´ik] *Biology*. relating to or like an embryo.

embryonic differentiation *Developmental Biology*. the process in which structures become specialized during embryogenesis.

embryonic inducer *Developmental Biology*. a signal, which may consist of a soluble molecule, insoluble matrix, or other agent, by which one tissue induces an adjacent embryonic tissue to specialize.

embryonic induction *Developmental Biology*. the influence exerted by one embryonic tissue on another, causing the latter to specialize in a way that it would not do otherwise.

embryonic mass see INNER CELL MASS.

embryopathy *Pathology*. any embryonic or congenital defect resulting from faulty development.

Embryophyta *Botany*. a subkingdom of the Plantae composed of the bryophytes and vascular plants and characterized by multicellular sex organs, special conducting tissue in the sporophyte, and production of an enclosed embryo dependent on the parent for varying periods; only Myxophyta and Thallophyta are excluded. Also, EMBRYOBIONTA.

embryophyte *Botany*. any plant of the subkingdom Embryophyta.

embryo sac *Botany*. the female gametophyte of angiosperms in which fertilization of the ovum and development of the embryo occur; the most common arrangement consists of an eight-celled structure with an ovum and two synergids at the micropylar end, three antipodal cell bodies at the opposite end, and a central cell containing two polar nuclei.

embryotomy *Medicine*. **1.** the dismemberment of a fetus to remove it from the womb when natural delivery is impossible. **2.** the dissection of an embryo or fetus.

-eme *Linguistics*. a suffix meaning "a significant unit," as in *morpheme*.

emedullate *Surgery*. to extract bone marrow.

emerald *Mineralogy*. a rare, brilliant green variety of beryl, widely used as a gemstone.

emerald

emerged bog *Ecology*. a bog that grows above the water table by drawing water up through the vegetation.

emerged shoreline see SHORELINE OF EMERGENCE.

emergence *Geology*. the exposure of a land mass formerly under water, resulting from an uplift of land or a lowering of water level. *Hydrology*. the point at which an underground stream emerges to become a surface stream. *Botany*. **1.** a plant outgrowth, derived from epidermal and cortical tissues, that does not contain vascular tissue or develop into a leaf or a stem. **2.** a seedling that has germinated above ground.

emergency *Medicine*. a condition or situation that arises suddenly and usually unexpectedly, and that is potentially life-threatening to a person or a group of people.

emergency brake *Mechanical Engineering*. **1.** a hand- or foot-operated mechanical brake designed to keep a motor vehicle from rolling when it is parked. Also, PARKING BRAKE, HAND BRAKE. **2.** any brake used to stop a vehicle in an emergency.

emergency broadcast system *Telecommunications*. an interconnected network of radio transmission systems that is designed to alert and inform listeners in the event of a national crisis, such as a war or natural disaster.

emergency core-cooling system *Nucleonics*. in a nuclear reactor, a combination of independent subsystems that are actuated by the sequential operation of redundant equipment and flow paths, ensuring reliable operation of controls to introduce cooling water into the reactor core if the flow of primary coolant is decreased or lost.

emergency locator transmitter *Navigation*. a battery-operated radio transmitter carried by aircraft that broadcasts a distinctive signal on 121.5 MHz and 243.0 MHz. If "armed," it will activate automatically in the event of a crash to assist in locating the downed craft.

emergency medical technician *Medicine*. a medical specialist who is trained to provide victims of sudden, acute illness or injury with emergency care and transportation to medical facilities.

emergency medicine *Medicine*. the medical specialty that deals with acutely ill or injured patients who require immediate medical attention. Thus, **emergency medical service.**

emergency position-indicating radio beacon *Navigation*. an emergency locator transmitter used in small-craft applications, usually portable and often manually operated, such as the "Type C" unit that broadcasts on marine VHF channels 15 and 16.

emergency receiver *Telecommunications*. any receiver energized by a self-contained power supply, such as batteries.

emergency room *Medicine*. an area of a hospital that is designed, equipped, and staffed for the practice of emergency medicine. Thus, **emergency-room medicine.**

emergent *Botany*. **1.** an aquatic plant having most of its vegetative parts above water. **2.** a tree that extends above the surrounding trees.

emersion *Astronomy*. see EGRESS.

emersion zone *Ecology*. the part of a lakeshore or seashore that is submerged only at extreme high tide or high water.

emery *Materials*. **1.** a dark-brown, fine-grained variety of corundum that contains aluminum oxide in iron oxide; used as an abrasive for buffing and polishing. **2.** composed of, covered with, or relating to such a substance. Thus, **emery board, emery cake, emery cloth, emery paper, emery stone,** and so on.

emery buff *Mechanical Devices*. an emery wheel whose grinding surface consists of the finest-grade powder; used for polishing metallic surfaces.

emery rock *Petrology*. a granulose rock that contains corundum and iron ores; may be formed by magmatic segregation or metamorphism of laterite.

emery wheel *Mechanical Devices*. a high-speed grinding wheel or disk whose surface is covered with an adhesive of emery powder; used to grind metallic surfaces and to sharpen tools.

emesis *Medicine*. a technical name for vomiting.

emetic *Pharmacology*. **1.** any agent that induces vomiting. **2.** of a drug or other agent, causing vomiting.

emetine *Pharmacology*. $C_{29}H_{40}N_2O_4$, an alkaloid that is the principal active ingredient of ipecac, a common emetic used in some cases of poisoning; emetine salts are used in treating amebic disorders.

emf or **e.m.f.** *Physical Chemistry*. **1.** electromotive force. Also, **EMF, E.M.F. 2.** see ELECTROCHEMICAL EMF.

EMG electromyograph; electromyogram.

EMI or **emi** electromagnetic interference.

-emia a combining form meaning "a condition of the blood" or "a substance in the blood," as in *anemia*.

emic *Anthropology*. defining a culture from an insider's or native participant's point of view, as opposed to an outside observer's point of view.

emigrant *Ecology*. an organism that moves out of its native area.

emigration *Ecology*. the movement of an individual or group out of its native area. *Hematology*. see DIAPEDESIS.

emigration theory see COHNHEIM'S THEORY.

Emilian *Geology*. a European geologic stage of the Lower Pleistocene epoch, occurring after the Calabrian and before the Sicilian.

emiocytosis *Cell Biology*. a type of secretion, especially insulin secretion, in which cytoplasmic granules containing a secretory product first fuse with the plasma membrane and then are released to the outside. Also, REVERSE PINOCYTOSIS.

emissary sky *Meteorology*. a sky of isolated or small, separated groups of cirrus clouds that indicates the approach of a cyclonic storm.

emission the process of discharging or sending off; specific uses include: *Physics*. the release of electrons from any material due to radiation, heat, nuclear radiation, or other causes. *Electronics*. the waves radiated into space by a transmitter. *Optics*. the production of photons of wavelengths in the visible spectrum, caused by electron jumps from higher to lower energy levels.

emission characteristics *Electronics.* the relationship between the ejection of charged particles, especially electrons, and the factors, such as temperature or voltage in a filament or heater, that cause them to be ejected; often indicated graphically.

emission electron microscope *Electronics.* an instrument that projects electrons emitted from a metal surface onto a fluorescent screen, with or without focusing the electrons.

emission line *Spectroscopy.* a discrete spectral line characteristic of atomic species that corresponds to the emission of energy at a single frequency (wavelength) and results from the transition of an electron from a higher energy level to a lower energy level.

emission-line corona see E CORONA.

emission-line galaxy *Astronomy.* a galaxy, typically spiral or irregular, whose spectrum shows narrow emission lines.

emission nebula *Astronomy.* a gaseous nebula whose spectrum wholly or largely consists of emission lines from different elements.

emission security *Electronics.* the component of communications security that prevents an unauthorized person from obtaining information from unintentional telecommunications emissions.

emission spectrometer *Spectroscopy.* a device calibrated to measure the wavelengths of radiation emitted by a sample that has been vaporized by an electric discharge; used to measure the percentage composition of elements in a sample.

emission spectroscopy *Spectroscopy.* the study of spectra emitted by a specimen when it is vaporized by an electric spark or arc, arising from promotion of electrons from ground state to higher states.

emission spectrum *Spectroscopy.* a diagram, graph, or other display indicating the degree to which a substance emits radiant energy with respect to wavelength.

emission standard *Engineering.* the highest amount of pollutant allowed by law to be discharged from a vehicle, machine, or industrial process or plant.

emission tomography *Medicine.* a form of tomography in which the gamma rays of an ingested radioactive substance are recorded by detectors outside the body.

emissive power *Thermodynamics.* the total amount of energy that is radiated from a unit area surface per unit frequency emitted per unit time.

emissivity *Thermodynamics.* the ratio of the radiation emitted by a body to the radiation that would be produced by a perfect blackbody radiator of the same temperature in the same environment. Emissivity is usually denoted by the Greek letter ε (epsilon) and, for real bodies, is less than unity.

emittance *Thermodynamics.* an extensive form of emissivity.

emitted response *Behavior.* a response that is not associated with any identifiable stimulus.

emitter *Electronics.* **1.** a device that generates charged particles or waves of radiation. **2.** also, **emitter region.** in a bipolar transistor, the region that generates the charge carriers that proceed to the collector.

emitter barrier *Electronics.* an area between the emitter and base regions of a transistor where rectification takes place.

emitter bias *Electronics.* voltage applied to the emitter electrode on a semiconductor in order to establish a reference level for operation.

emitter-coupled logic *Electronics.* a logic circuit whose operation is based on pairs of transistors with their emitters connected together.

emitter follower *Electronics.* a device in which amplification is achieved by applying the input signal to the base and taking the output signal from the emitter resistor.

emitter resistance *Electronics.* in a bipolar transistor, the ratio of the emitter-to-base voltage to the emitter current.

emmenagogic *Medicine.* referring to or acting as an agent that stimulates the menstrual flow.

emmenia *Medicine.* the menses; menstruation.

emmenology *Medicine.* the study of menstruation and its disorders.

Emmenthaler see SWISS CHEESE.

emmer *Botany. Triticum dicoccum,* a species of hard red wheat having two kernels in the spikelets that remain in the glumes after threshing; grown as a forage crop in Russia, Germany, and the United States.

emmetropia *Medicine.* the state of perfect (20/20) vision in which parallel rays are focused exactly on the retina without effort of accommodation.

emmonsite *Mineralogy.* $Fe_2^{+3}Te_3^{+4}O_9 \cdot 2H_2O$, a yellow-green, triclinic mineral occurring in microcrystalline masses, having a specific gravity of 4.52 to 4.55 and a hardness of 5 on the Mohs scale; found as a secondary mineral in the oxidized zones of deposits containing tellurium.

emodin *Organic Chemistry.* $C_{15}H_{10}O_5$, orange needles that are insoluble in water and soluble in alcohol; melts at 256–257°C; used as a laxative.

emollient *Pharmacology.* **1.** a substance that softens or soothes the skin, or soothes irritated internal surfaces such as mucous membranes. **2.** soothing or softening.

emotion *Psychology.* a strong and complex feeling state that is consciously perceived, such as anger, fear, happiness, or love; emotions are accompanied by certain physiological and behavioral activities.

emotion-focused coping *Psychology.* a method of dealing with stress that reduces the negative emotions associated with a stressful situation without directly addressing the situation itself.

EMP electromagnetic pulse.

empathy *Psychology.* an ability to perceive and understand the emotions and attitudes of another or others.

Empedocles [em ped´ə klēz] c. 495–435 BC, Greek philosopher; identified four elements (earth, air, fire, and water) as the basic constituents of all matter.

empennage [am´pə näzh´] *Aviation.* a term for the assembly on the tail of an aircraft consisting of the horizontal and vertical surfaces. Also, TAIL ASSEMBLY, TAIL UNIT.

Empetraceae *Botany.* a family of small dicotyledonous evergreen shrubs belonging to the order Ericales and characterized by deeply furrowed and crowded alternate leaves with a basal pulvinus; native to the colder areas of the Northern Hemisphere and southern South America.

emphasizer see PRE-EMPHASIS NETWORK.

emphysema [em´fə sē´mə] *Medicine.* a pathological accumulation of air in tissues, especially of the lungs, causing too much inflation of the air sac and destructive changes in the alveolar walls.

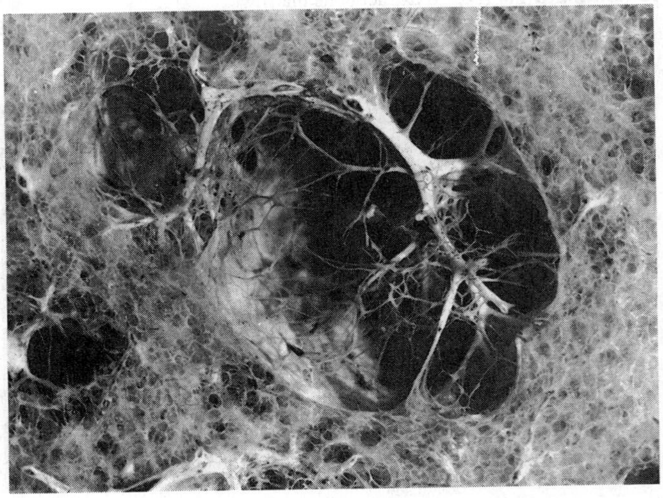

emphysema of the lung

emphysematous chest *Medicine.* the altered contour of the chest seen in pulmonary emphysema.

Empididae *Invertebrate Zoology.* dance flies and balloon flies, a family of two-winged insects in the series Nematocera.

empire cloth *Materials.* cloth that is impregnated with linseed oil and used as an insulating material.

empirical *Science.* **1.** based on actual observation or experimentation. **2.** of or relating to empiricism. Also, **empiric.**

empirical Bayes methods *Statistics.* inference procedures coupling the use of the Bayes theorem with a data-based choice of the a priori distribution.

empirical curve *Mathematics.* a curve that fits a set of observed data.

empirical distribution function *Statistics.* the cumulative distribution function of a random sample.

empirical formula *Chemistry.* a simple written representation of the chemical composition of a molecule that indicates the types and relative numbers of atoms in the molecule, but not the structure and not necessarily the number of atoms. For example, CH is the empirical formula for both acetylene (molecular formula: C_2H_2) and benzene (molecular formula: C_6H_6).

empirical orientation *Cartography.* the composite rectified adjustments of magnification, swing, easel tilt, *y*-displacement, and *x*-displacement used to re-create correctiy the exact conditions in a projected image that existed in the negative at the instant of exposure. Also, **empirical rectification.**

empirical probability *Statistics.* the estimated probability of the occurrence of an event, derived from experimentation.

empiricism *Science.* the theory that all knowledge results from experience and that any data which cannot be directly observed and verified is not valid. *Behavior.* the theory that all behavior patterns are acquired through experience.

empiricist *Science.* a person who follows or supports empiricism.

emplacement *Geology.* **1.** the process by which igneous rock is intruded into older rocks. **2.** the deposition of ore minerals by any process. *Ordnance.* **1.** a prepared position for one or more weapons or pieces of equipment. **2.** the act of fixing a gun in such a position.

emplectite *Mineralogy.* $CuBiS_2$, a grayish-white to tin-white, brittle, orthorhombic mineral with one perfect cleavage occurring in short, prismatic crystals, having a specific gravity of 6.3 to 6.5 and a hardness of 2 on the Mohs scale.

empodium *Entomology.* a small structure located between the claws of the appendages of many insects and spiders.

empressite *Mineralogy.* AgTe, a pale bronze, metallic, orthorhombic mineral occurring in compact granular masses, having a specific gravity of 7.5 and a hardness of 3 to 3.5 on the Mohs scale; found with galena and native tellurium.

empress tree see ROBINA.

empty calories *Nutrition.* a term for food items that provide calories but few important nutrients, such as candies and soft drinks.

empty-cell process *Materials Science.* a method of treating wood in which a preservative covers the wood cells without filling them.

empty-chair technique *Psychology.* a therapeutic technique in which the subject converses with an empty chair that represents another person or another aspect of the self, in order to bring out unexpressed or unacknowledged feelings. Also, **empty-chair exercise.**

empty clause *Artificial Intelligence.* in theorem proving, a clause that contains no literals and thus represents the value False or contradiction.

empty medium *Computer Technology.* a storage medium that contains only frame-of-reference data but is otherwise completely erased.

empty particles *Virology.* virus particles that do not contain any nucleic acid and that have a lower buoyant density than normal virus particles. Also, GHOST PARTICLES.

empty set *Mathematics.* a set that has no members; denoted $\varnothing$ or $\{ \}$. Also, NULL SET.

empty string see NULL STRING.

empyema *Medicine.* a collection of pus in a body cavity, particularly the space between the lung and the pleural space.

empyema tube *Surgery.* a tube for draining purulent fluid from the thoracic cavity.

Emscherian see CONIACIAN.

Emsian *Geology.* a European geologic stage of the Lower Devonian period, occurring after the Siegenian and below the Couvinian.

emu *Vertebrate Zoology.* a large, flightless but swift-running Australian ratite bird, *Dromiceius novae-hollandiae*, resembling the ostrich but smaller, with rudimentary wings and a feathered head and neck.

emu

EMU or **emu** electromagnetic unit.

emulation *Computer Technology.* **1.** the imitation of one computer system by another so that each can accept the same data or programs and produce the same results. **2.** the use of a program to simulate functions of hardware or another program.

emulation mode *Computer Technology.* the operation of a computer system to execute the instructions of a different computer, as opposed to normal mode.

emulator *Computer Technology.* hardware or software that allows a computer of one type to accept input designed for another computer and process it correctly.

emulsification *Chemistry.* a process by which one immiscible liquid is dispersed in another.

emulsification test *Chemical Engineering.* a standard laboratory technique for determining the ability of insulating oils, turbine oils, and other lubricating oils to resist emulsification.

emulsifier *Chemistry.* any agent that is used to form an emulsion. Also, **emulsifying agent.** *Food Technology.* an additive that is used to stabilize an emulsion and ensure its consistency. Emulsifiers such as lecithin and polysorbate 60 are often added to powdered beverages, puddings, and other processed food products.

emulsifying oil see SOLUBLE OIL.

emulsin *Enzymology.* an older term for β-glucosidase.

emulsion *Chemistry.* any stable mixture of two or more immiscible liquids where one liquid (in the form of fine droplets or globules) is dispersed in the other. *Food Technology.* a practical example of this used as a sauce or flavoring; mayonnaise is an emulsion of oil in egg yolks. *Pharmacology.* a preparation of one liquid distributed throughout the body of a second, usually more palatable, liquid; e.g., cod-liver oil emulsion with malt. *Graphic Arts.* a photosensitive coating for film or printing plates, usually made of silver salts suspended in a gelatin base.

emulsion breaking *Chemistry.* the process of destroying an unwanted emulsion by chemical, electrical, thermal, or physical means.

emulsion cleaner *Materials.* any cleaner that is an emulsion of an organic solvent dispersed in a water solution by an emulsifying agent.

emulsion paint *Materials.* a waterbase paint that is made from a pigmented emulsion or dispersion of a resin in water.

emulsion polymerization *Materials Science.* a process in which a monomer or mixture of monomers is emulsified in a low-viscosity aqueous medium; used for such products as polystyrene acrylics and PVC.

EMYCIN *Artificial Intelligence.* a general-purpose tool used to build expert systems; a basic inference system derived from MYCIN.

Emydidae *Vertebrate Zoology.* the terrapins, a family of mostly semiaquatic North American turtles of the order Testudines, characterized by low shells, webbed feet, and the head covered with skin or a large scale.

en *Graphic Arts.* half an em space; that is, a space equal to half the height of the type being used. Also, **en space.**

en- a prefix meaning: **1.** to cause to be in, as in *encapsulate.* **2.** in or within, as in *enarthrosis.*

enable *Computer Programming.* to activate or restore an optional feature; the opposite of disable. *Computer Technology.* **1.** an electrical signal that enables operation of another circuit. **2.** a switch on a peripheral device, such as a tape drive, that is set by an operator to indicate that it is ready for operation.

enabled instruction *Computer Programming.* a program instruction all of whose inputs are present and thus may be executed.

enabler *Psychology.* a term for a person whose acquiescence or support allows another person to continue with addictive behavior, such as alcoholism or drug abuse.

enabling pulse *Electronics.* the surge of electric current that activates a device or prepares it for a subsequent function.

enactive learning *Psychology.* learning that is based on active performance.

Enaliornithidae *Paleontology.* a family of birds in the order Hesperornithiformes and superorder Odontognathae; these birds had well-developed teeth; Cretaceous.

enamel *Medicine.* the hard substance covering the dentin of the crown of a tooth. *Materials.* **1.** a glassy, opaque ceramic coating that is fused to the surface of metal, pottery, or glass for protection and/or decoration. **2.** a paint or varnish that produces an enamel-like coating. **3.** to apply an enamel.

enamel cell see AMELOBLAST.

enameled brick *Materials.* any brick having a glazed surface.

enameled paper *Graphic Arts.* coated book paper having a high-gloss finish. Also, **enamel paper.**

enameling *Materials.* the application of an enamel coating to a surface in order to protect against corrosion and to improve appearance.

enamel kiln *Engineering.* an oven designed for the firing of porcelain enameled ware.

enamel organ *Developmental Biology.* the mass of ectodermal or endodermal cells, budded off from the dental lamina, forming the ameloblast layer of cells and eventually giving rise to the enamel cap of a developing tooth.

enamelware *Materials.* metalware that is coated with an enamel.

enanthema *Medicine.* an eruption upon the surface of a mucous membrane.

enantiomer *Chemistry.* either of two molecules of the same chemical composition that are mirror images of each other but not identical, such as one of a pair of optical isomers containing chiral carbon atoms. Also, **enantiomorph.**

enantiomeric *Chemistry.* related to another molecule as an enantiomer. Also, **enantiomorphous.**

enantiomeric excess *Chemistry.* in a synthesis that is asymmetric, a product yield in which more of a desired enantiomer is produced relative to other products.

enantiomerism *Chemistry.* the condition of being an enantiomer. Also, **enantiomorphism.**

enantiotopic ligand *Chemistry.* a ligand whose addition or replacement gives rise to an enantiomeric compound.

enantiotropy *Chemistry.* the existence of a compound in two crystalline forms, one of which is stable below a certain temperature (the transition temperature) and the other of which is stable above it.

Enantiozoa *Invertebrate Zoology.* see PARAZOA.

enargite *Mineralogy.* Cu_3AsA_4, a brittle, grayish to iron-black orthorhombic mineral occurring in massive, granular form and as prismatic crystals, having a metallic luster, a specific gravity of 4.45, and a hardness of 3 to 3.5 on the Mohs scale; found with galena and native tellurium.

enarkyochrome *Neurology.* an arkyochrome in which the chromophil is arranged in a single network.

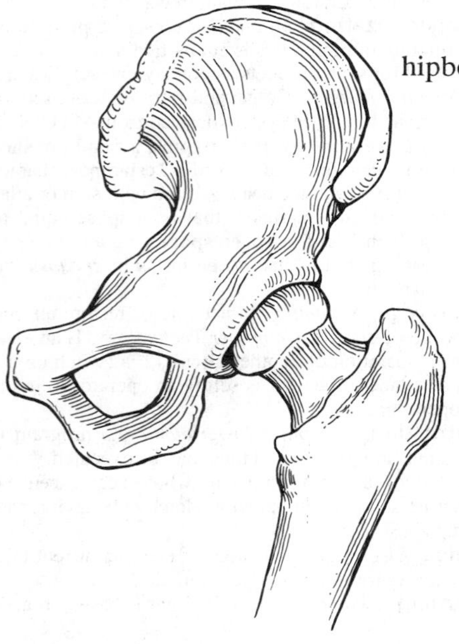

enarthrosis

enarthrosis *Anatomy.* a joint in which the globular head of one bone is received into a socket in another, as in the hip joint.

enate *Anthropology.* a person related on the mother's side.

enation *Plant Pathology.* an abnormal growth that occurs on a plant part as a result of an infection or disease.

Encalyptales *Botany.* an order of mosses of the subclass Bryidae, having broad papillose leaves and erect capsules covered with long calyptrae.

encapsidation *Virology.* a process during virion assembly in which the viral capsid incorporates the nucleic acid.

encapsulate *Science.* to enclose in or as if in a capsule. Thus, **encapsulated.**

encapsulation *Computer Science.* a method of making a software system modular by creating well-defined interface routines that deal with a particular kind of data and allowing other programs to access the data only through those routines; the interface routines encapsulate the data.

encarditis see ENDOCARDITIS.

encatarrhaphy *Surgery.* the burying of a body structure by suturing together adjacent tissues.

Enceladus *Astronomy.* the second moon of Saturn, discovered in 1789 and measuring 500 kilometers in diameter.

encephal- a combining form meaning "brain," as in *encephalitis.*

encephalitis [en sef´ə lī´təs] *Medicine.* an inflammation of the brain, typically caused by a virus transmitted by the bite of a mosquito that has previously bitten an infected individual; bacteria, environmental poisoning, or hemorrhage can also cause this condition.

encephalitis lethargica *Medicine.* an epidemic encephalitis reported in the first quarter of the 20th century, characterized by lethargy, hyperkinesia, and neurologic disability.

encephalitogen *Neurology.* any agent, normally injected, that induces encephalitis.

encephalitogenic protein *Neurology.* a myelin basic protein that induces encephalitis.

encephalization *Anthropology.* the evolutionary trend in humans in which the complexity and size of the brain have increased through time.

encephalo- a combining form meaning "brain," as in *encephalography.*

encephaloarteriography *Radiology.* the use of roentgenography to examine the blood supply of the brain, often aided by the injection of radiopaque dye.

encephalocele *Medicine.* a congenital or trauma-induced bulging of the brain through an opening in the skull.

encephalogram *Radiology.* a roentgenogram of the brain made during encephalography.

encephalography *Radiology.* an examination of the brain by radiograph after the cerebrospinal fluid has been removed and replaced with air, another gas, or contrast material.

encephalolith *Neurology.* a solid structure or concretion in the brain or ventricles.

encephalology *Neurology.* the study of the structure, functions, diseases, and disorders of the brain.

encephaloma *Medicine.* any inflammation, swelling, or tumor of the brain.

encephalomalacia *Medicine.* a softening of the brain, usually caused by infarction.

encephalomeningitis *Medicine.* inflammation of the brain and its membranes.

encephalomyelitis [en sef´ə lō mī lī´təs] *Medicine.* an inflammatory disease of the brain and spinal cord characterized by fever, headache, stiff neck, back pain, and vomiting.

encephalomyelopathy *Medicine.* any disease or disorder affecting the brain and spinal cord.

encephalomyocarditis *Medicine.* a viral disease characterized by degeneration and inflammation of the skeletal and cardiac muscle and lesions of the central nervous system.

encephalomyocarditis virus *Virology.* an enterovirus that causes encephalomyocarditis and mild aseptic meningitis; represented by four strains, occurring mostly in Africa and South America.

encephalonarcosis *Medicine.* stupor or coma caused by a brain disease or disorder.

encephalopathy *Medicine.* any defect of the structure or function of brain tissues.

encephalosclerosis *Medicine.* the hardening of the brain.

encephalosis *Medicine.* any brain disease or dysfunction resulting from an organic (structural) defect, rather than from external conditions.

encephalotomy *Surgery.* any incision of the brain.

enchondroma *Medicine.* a harmless, slow-growing tumor of cartilage cells near the ends of long bones, which may cause the bone to bulge.

enchondromatosis *Medicine.* a syndrome in which there is too much cartilage in the flared ends of many of the body's bones.

enchyma [en kī´mə] *Physiology.* the substance derived from absorbed nutritive materials; the formative juice of the tissues.

encipher *Telecommunications.* to convert plain text into an unintelligible form by means of a cipher system. Also, ENCRYPT.

Encke, Johann [eng´kə] 1791–1865, German astronomer; determined the orbit of Encke's comet; computed distance from the earth to the sun.

Encke roots *Mathematics.* given any two numbers a_1 and a_2, the (possibly complex) numbers x_1 and x_2 that satisfy the conditions $x_1 + x_2 = a_1$ and $x_1x_2 = a_2$; equivalently, $-x_1$ and $-x_2$ are roots of the quadratic equation $x^2 + a_1x + a_2$. It is customary to label x_1 and x_2 so that $|x_1| \leq |x_2|$.

Encke's comet *Astronomy.* a small, relatively inactive comet in a nearly circular orbit with a 3.3-year period; it is the parent body for the Taurid meteor shower.

Encke's division *Astronomy.* in Saturn's A ring, a dark band about 880 km across, where the ring material is very thin but not wholly absent.

enclosure compound see CLATHRATE.

encode *Linguistics.* to substitute letters, numerals, or characters to hide the meaning of a message except to certain users who know the conversion scheme. *Telecommunications.* to apply a code, usually consisting of binary digits that represent individual characters or groups of characters in a message. *Computer Programming.* **1.** to convert data into code according to a specified coding scheme. **2.** to convert program specifications or design into programming language statements.

encoded abstract *Computer Programming.* a summary that has been prepared for scanning by an automatic electronic device.

encoder *Telecommunications.* any device designed to convert messages into signals that can be transmitted by communications channels, or to change the representation of data from one form to another. *Computer Technology.* **1.** any device capable of producing machine-readable output. **2.** in character recognition, a printer designed to print forms, with certain type fonts in certain positions on the forms. **3.** a logic circuit, the opposite of a decoder, that converts a signal on one of many lines into a binary code representing the line on which the signal occurred.

encoding *Telecommunications.* the processing of messages into signals that can be transmitted by communications channels. *Psychology.* the mental process by which learned information becomes part of memory.

encoding specificity hypothesis *Psychology.* the theory that if an item is immediately associated with a memory cue, it is more likely to be remembered later.

encoding strip *Computer Technology.* in character recognition, a predefined area that has been reserved for magnetic ink characters.

encounter *Psychology.* the interaction among members of an encounter group, or sometimes between a patient and a psychotherapist.

encounter group *Psychology.* a group of people who meet together, often without a professional leader, to focus on interactions or "encounters" that are intended to promote sensitivity and self-awareness.

encounter theory *Astronomy.* a theory for the solar system's origin in which a passing star pulls material from the proto-sun, from which the planets, asteroids, and comets condense. Also, **encounter hypothesis.**

encrinal limestone *Geology.* a limestone containing 10–50% fossil crinoidal fragments.

encrinite *Paleontology.* a term usually applied to the pelmatozoan inadunate crinoids of the Triassic, especially the genus *Encrinus*, which has left some important intact fossils.

Encrinurus *Paleontology.* a common genus of trilobites in the suborder Cheirurina and the family Encrinuridae; lower Ordovician to the end of the Silurian.

encroachment a process of going beyond intended or legal limits; specific uses include: *Mining Engineering.* the act of working coal or mineral deposits beyond the boundary that divides one mine from another. Also, TRESPASS. *Petroleum Engineering.* the replacement of gas or oil by the movement of edge water or bottom water into a reservoir as the gas or oil is withdrawn, which reduces both the reservoir pressure and the remaining volume of gas or oil.

encrustation *Engineering.* the accumulation of residual materials on the interior of a furnace or kiln.

encrypt see ENCIPHER.

encryption *Computer Programming.* the conversion or encoding of information for transmission so as to prevent interpretation without the key for decryption.

enculturation *Anthropology.* the process by which culture is transmitted from one generation to the next, through daily experiences and formal education.

encyesis *Medicine.* a normal uterine pregnancy.

Encyrtidae *Invertebrate Zoology.* a family of parasitic wasps, hymenopteran insects in the superfamily Chalcidoidea.

encyst *Biology.* to enclose or become enclosed in a cyst, sac, or capsule. Thus, **encysted, encystment, encystation.**

end *Textiles.* **1.** a single warp yarn or thread. **2.** a short length of fabric; a remnant.

end- a combining form meaning "within," as in *endarterium.*

5' end *Molecular Biology.* **1.** the leading end of the mRNA molecule; once this end is inserted into a ribosome, protein synthesis begins. **2.** the end of a DNA or RNA molecule at which the 5'-carbon is not limited to another nucleotide.

Endameba *Invertebrate Zoology.* a genus of amoeboid protozoans, parasites in the intestinal tract of cockroaches and termites. Also, **Endamoeba.**

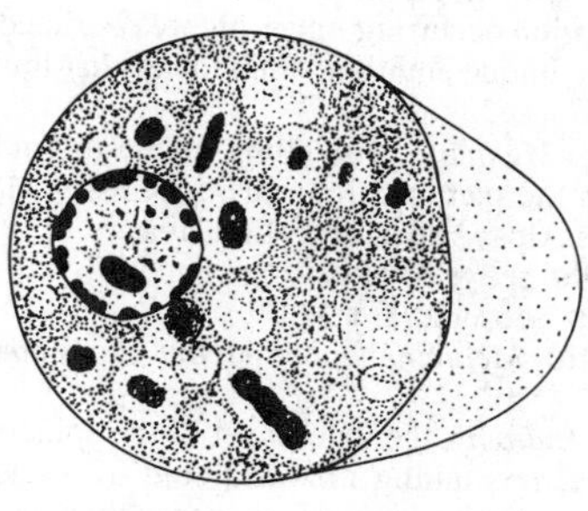

Endameba

endangered species *Ecology.* a species of animal or plant that has been identified as in danger of becoming extinct because of harmful human activity or environmental factors, and that is thus the subject of protective regulations and conservation measures.

endaortic *Anatomy.* of or relating to the interior of the aorta.

endarch *Botany.* **1.** having the protoxylem nearest to the center of the axis and younger xylem layers around it, occurring typically in angiosperms. **2.** having a single simple protoxylem around the central parenchyma.

end-around carry *Computer Technology.* the bit carried from the most significant digit place to the least significant digit place in certain kinds of computer arithmetic.

end-around shift *Computer Technology.* a register operation in which the right- and left-most positions are treated as adjacent during the shift; the bit shifted out of one end is placed in the other end. Also, CIRCULAR SHIFT.

endarterectomy *Medicine.* the excision of the diseased lining and atheromatous deposits of an artery.

endarterial *Anatomy.* within an artery.

endarteritis *Medicine.* an inflammation of the tunica intima of an artery.

endarteritis obliterans *Medicine.* a defect in which the artery walls become inflamed, blocking the opening of the vessel and blocking the smaller vessels.

end bud *Anatomy.* the remnant of the primitive knot, from which the caudal portion of the trunk arises.

end bulb *Neurology.* a spheroid or ovoid body at the terminus of a nerve fiber and dispersed in the skin, mucous membranes, muscles, joints, and connective tissue of the internal organs; such bodies take many forms, from simple end knobs to complex sensory end organs.

end bulbs of Krause see KRAUSE'S CORPUSCLES.

end cell *Electricity.* a storage battery cell that may be switched in or out of a circuit to vary the battery voltage. *Immunology.* a body cell that is no longer capable of dividing or differentiating itself.

end-cell rectifier *Electronics.* a device that provides direct current to a circuit when the voltage in the battery drops.

end-diastolic volume *Cardiology.* the volume of blood in the ventricle after the heart's relaxation (diastole). Although dependent on several factors, it is a measure of the ventricular end-diastolic fiber length and of myocardial performance. In the normal heart, the volume is 120–130 ml but can be as high as 200–250 ml.

end distortion *Telecommunications.* in start-stop systems, the shifting of the ends of all mark pulses from their proper positions in relation to the beginnings of the start signals.

end effect *Electromagnetism.* a capacitative effect at the ends of a half-wave antenna, requiring that the actual length of the dipole be cut about 5% less than a half-wavelength.

end effector *Robotics.* a tool, gripper, or other device at the end of a robot's arm that is used to grasp or manipulate an object; designed with three degrees of freedom to move in any plane, and fabricated to withstand collisions while performing on a workpiece.

Endeidae *Invertebrate Zoology.* a small family of sea spiders, arthropods in the subphylum Pycnogonida.

endellite *Mineralogy.* $Al_2Si_2O_5(OH)_4{\cdot}2H_2O$, a colorless or white monoclinic mineral of the kaolinite-serpentine group, occurring as tubular ultramicroscopic crystals, and having a specific gravity of 2.11 to 2.17 and a hardness of 2 to 2.5 on the Mohs scale. Also, HYDROHALLOYSITE.

endemia *Medicine.* any endemic disease.

endemic *Medicine.* of a disease or agent, present or usually prevalent in a geographic area or population. *Ecology.* of or relating to a native species or population occurring under highly restricted conditions due to the presence of a unique environmental factor that limits its distribution. Also, **endemial.**

endemic goiter *Medicine.* a swelling of the thyroid gland caused by lack of iodine in the diet in certain districts, especially in mountain regions such as the Alps, Andes, and Himalayas.

endemic syphilis see NONVENEREAL SYPHILIS.

endemic typhus see MURINE TYPHUS.

endemoepidemic *Medicine.* endemic, but occasionally becoming epidemic.

endepidermis *Anatomy.* the inner layer of the epidermis.

endergic *Chemistry.* taking in work; said of a chemical reaction in which the reactants have a higher free energy than the products.

endergonic *Chemistry.* characterized by the absorption of energy; said of a chemical reaction requiring energy input to proceed, such that the products have a higher free energy than the reactants; generally associated with anabolism.

endermic *Medicine.* acting through the skin by absorption, as a medication applied to the skin.

Enders, John Franklin 1897–1985, American virologist; shared Nobel Prize for isolating and cultivating polio virus.

end-fire array *Electromagnetism.* a linear antenna consisting of several parallel and coplanar elements, and whose direction of maximum radiation is along the axis of the array. Also, **end-fire antenna.**

end-group analysis *Materials Science.* a technique used to determine the molecular weight of a polymer; involves calculating the number of end-groups on a known mass of polymer molecules in solution through chemical, radiochemical, or spectroscopic analysis.

ending *Anatomy.* any termination or finish of a part, especially the peripheral termination of a nerve or nerve fiber.

endite *Invertebrate Zoology.* **1.** an appendage attached to the inner portion of an arthropod limb. **2.** a chewing ridge located on the maxillae of certain arachnids.

end item *Industrial Engineering.* the final arrangement of parts or materials in production that have been processed for their intended use.

endive [en´dīv´; än´dēv] *Botany.* **1.** a composite plant, *Cichorium endivia,* having curly-edged, slightly bitter-tasting leaves that are used in salads. **2.** see BELGIAN ENDIVE.

end-labeling *Molecular Biology.* the attachment of a radioactive label to either the 5' end or the 3' end of a DNA molecule for the purpose of analysis.

end-lap joint *Building Engineering.* a corner joint formed between the ends of two pieces of timber which intersect at an angle for a distance equal to their widths. Also, **end lap.**

endless loop *Computer Programming.* a sequence of instructions in a computer program that is executed repeatedly with no condition supplied for termination.

endless tangent screw *Navigation.* a device attached to the end of the index arm of a sextant that, by engaging a screw in teeth cut into the arc, allows very delicate adjustments to be made to the index arm. An endless tangent screw can move the full distance of the arc.

endlichite see VANADINITE (ARSENICAL).

end loss *Electromagnetism.* the difference between the physical length and the effective (electrical) length of an antenna element. *Metallurgy.* the material left over after a bar or billet is cut in the maximum number of usable segments.

end mark *Computer Programming.* a code or signal denoting the end of a unit of data.

end matter see BACK MATTER.

end member *Mineralogy.* any of the two or more chemical compounds that make up an isomorphous (solid solution) series.

end mill *Mechanical Engineering.* a milling cutter having radially disposed teeth and spiral blades; used for facing, shaping, and cutting.

end moraine *Geology.* a ridgelike mass of glacial debris that accumulates at the front, lower, or outer end of an actively flowing glacier. Also, FRONTAL MORAINE.

endo- *Science.* a combining form meaning "within," as in *endocellular. Organic Chemistry.* a combining form designating a substituent facing in toward the center of a molecule, especially one in a ring form.

endoaneurysmoplasty see ANEURYSMOPLASTY.

endobasion *Anatomy.* a point used in measuring the skull; it is located in the midline at the most anterior point of the foramen magnum.

endobatholithic *Geology.* a stage in the erosion of a metalliferous batholith in which the invaded rocks lie only as islands or roof pendants.

endobiotic *Ecology.* describing an organism that lives within the cells or tissues of a host organism.

endoblast see ENDODERM.

endobranchiate *Zoology.* describing animals that, as embryos, form gills in the hypoblast.

endobronchial tube *Surgery.* a double lumen tube that is inserted into the bronchus of one lung, allowing ventilation of one lung and deflation of the other lung; used in anesthesia and thoracic surgery.

endobyssate *Invertebrate Zoology.* a habit of substratum attachment using byssal threads common in Paleozoic bivalves; may have led to the more common present-day surface attachment (epibyssate) habit.

endocardial *Anatomy.* **1.** of or relating to the endocardium. **2.** situated or occurring within the heart.

endocardial fibroelastosis *Medicine.* a congenital defect in which the wall of the ventricle grows too much and the tissue that lines the heart becomes thick and fibrous.

endocarditis *Medicine.* an inflammation of the endocardium, often associated with a serious defect in which the lining of the heart and the heart valves become inflamed, quickly leading to death if the condition is not treated.

endocardium *Anatomy.* the endothelial membrane that lines the interior of the heart muscle; the innermost of the double membrane that surrounds the heart. Also, VISCERAL PERICARDIUM.

endocarp *Botany.* the inner, usually hard layer of the wall of the fruit or pericarp, as in the stony part of a drupe or pome.

endocast *Anthropology.* a plaster impression of the interior of a fossil cranium, used to study the brain and provide comparative information on human evolution. Also, **endocranial cast.** *Geology.* see STEINKERN.

Endoceratoidea *Paleontology.* a subclass of large mollusks in the class Cephalopoda that were extant from the Ordovician to the Silurian; the shells of some species range up to 9 meters in length and are the largest known Paleozoic fossils.

endocervicitis *Medicine.* an inflammation of the cervix of the uterus.

endocervix *Anatomy.* **1.** the mucous membrane that lines the uterine cervix. **2.** the area in which the uterine cervix opens into the uterine cavity. Thus, **endocervical.**

endochondral *Anatomy.* formed or situated within cartilage.

endochondral bone *Histology.* a bone that is formed by replacement of a cartilage model. Also, INTRACARTILAGINOUS BONE.

endochrondral ossification *Physiology.* bone formation that occurs within and replaces cartilage.

Endocladiaceae *Botany.* a small family of red algae of the order Cryptonemiales, characterized by bushy erect thalli with similar gametangial and sporangial thalli, and by spermatangia that form superficial patches.

endocoel *Invertebrate Zoology.* in some anthozoans, a portion of the gastrovascular cavity enclosed between two membranes.

endocommensal *Ecology.* an organism that benefits from living inside a host organism without affecting the host.

endocorpuscular *Cell Biology.* located within a corpuscle; for example, within an erythrocyte.

endocranial *Anatomy.* **1.** situated within the cranium. **2.** of or relating to the endocranium.

endocranitis *Neurology.* an inflammation of the inner lining of the cranium.

endocranium *Anatomy.* the endosteal outer layer of the dura mater of the brain.

endocrine [en´dō krīn´; en´də krin´] *Endocrinology.* **1.** a hormonal pathway that is characterized by the production of a biologically active substance by a ductless gland; the substance then is carried through the bloodstream to initiate a cellular response in a distal target cell or tissue. **2.** describing a pathway, organ, or structure that secretes internally. **3.** relating to internal secretions; hormonal.

endocrine gland *Endocrinology.* a ductless organ that secretes hormones directly into the circulatory system, influencing metabolism and other body processes; the **endocrine system** includes the hypothalamus, pituitary, thyroid, parathyroid, and adrenal glands, the pineal body, and the gonads. Also, DUCTLESS GLAND.

Endocrinology

Endocrinology can be defined as that branch of biologic science dealing with the actions of chemical messengers. The concept of chemical messengers dates back to the discovery of secretin in 1902 by Bayliss and Starling and to Starling's Croonian Lectures which introduced the word "hormone" and described a system that, like the nervous system, permitted action arising in one part of the body to affect a distant site. In the following decades much was learned about the multiplicity of hormone-secreting organs and the remote systems they affected. However, the bioassay systems available did not permit simple measurement of the concentrations of peptide hormones in the circulation.

In 1959 Yalow and Berson introduced the radioimmunoassay principle (RIA) and were able to measure directly circulating plasma insulin in the fasting and stimulated state. Great advances in the chemistry of peptide hormones made available other highly purified hormonal preparations which, in the next decade, led to the development of RIA's for a host of peptidal as well as non-peptidal hormones. RIA provided the sensitivity, specificity, and reliability that made possible studies of in vitro hormonal regulation that otherwise would not have been possible. The synergistic interaction between advances in the biochemistry of hormones and investigations using RIA resulted in an information explosion in endocrinology.

Endocrinology has become increasingly complicated and ever more exciting. Thus it is now known that single hormones can have many actions, many different hormones can have the same action, hormones can be found in more than one form which have or do not have biologic activity; hormones can have local actions in the cells in which they are synthesized or in adjacent cells in the same organ. What new concepts and developments will the next decades bring to this field?

Rosalyn S. Yalow
Solomon A. Berson Distinguished Professor-At-Large
Mt. Sinai School of Medicine
City University of New York
Nobel Laureate in Physiology or Medicine

endocrinology [en´dō kri näl´ə jē] *Physiology.* the study of the system of glands and tissues that produce and secrete hormones.

endocrinologist [en´dō kri näl´ə jist] *Medicine.* a physician who treats diseases and conditions of the endocrine glands. *Biology.* a scientist who specializes in the study of endocrinology.

endocrinotherapy *Endocrinology.* the treatment of disease by the use of endocrine preparations; homeotherapy.

endocuticle *Entomology.* the chitinous inner layer of the cuticle of an insect or other arthropod; during molting this layer is digested by enzymes from special glands, leaving the outer layer of the cuticle loose and ready to be cast off.

endocyclic double bond *Organic Chemistry.* a double bond that is part of a ring system molecule.

endocyst *Invertebrate Zoology.* the soft layers of tissues lining the ectocyst of a bryozoan.

endocytic vesicle *Cell Biology.* a membrane-bound vesicle that is formed by endocytosis and contains extracellular materials that are to be delivered to any of several destinations within the cell.

endocytosis *Cell Biology.* a process by which extracellular particles, fluids, or specific macromolecules are taken up by a cell via a progressive invagination and eventual pinching off of a region of the plasma membrane, forming a membrane-bound vesicle within the cytoplasm.

endodeoxyribonuclease *Enzymology.* any of several enzymes of the hydrolase class that catalyze the hydrolysis of interior bonds of deoxyribonucleotides, producing oligonucleotides or polynucleotides.

endoderm *Developmental Biology.* the innermost of the three embryonic germ layers; it develops into the epithelium of the pharynx, respiratory tract, digestive tract, bladder, and urethra. Also, ENTODERM, ENDOBLAST, ENTOBLAST, HYPOBLAST.

endodermis *Botany.* the innermost tissue forming a sheath around the vascular region in roots, rhizomes, and certain stems and characterized by a single-celled layer of cellulosic or cutinized cells and thickened Casparian strips.

endodiascope *Radiology.* an X-ray tube whose small size allows it to be placed inside a body cavity for radiography or radiotherapy.

endodontics *Medicine.* the branch of dentistry concerned with the cause, prevention, diagnosis, and treatment of conditions that affect the tooth pulp, root, and periapical tissues.

endodontist *Medicine.* a dentist who specializes in endodontics.

endodontitis *Medicine.* an inflammation of the dental pulp.

endoenzyme *Enzymology.* any intracellular enzyme that is retained and utilized by the cell that produced it; does not normally diffuse out of the cell into the surrounding medium. *Organic Chemistry.* any enzyme that forms bonds within a polymer chain.

endoergic see ENDOTHERMIC.

endoergic collision see COLLISION OF THE FIRST KIND.

end-of-arm speed *Robotics.* the time required for an end effector to arrive at its desired position.

end-of-block character *Computer Programming.* a code denoting the end of a block of data.

end-of-data mark *Computer Programming.* a code signaling the last record in a contiguous set of data records.

end-of-field mark *Computer Programming.* a data item signaling the end of a variable length field.

end of file *Computer Programming.* **1.** a termination of a quantity of data. **2.** a recorded mark on magnetic tape to indicate the end of a file of data. **3.** a signal provided to a program to indicate the reading of an end of file mark or the logical end of a file of data. **4.** an automatic procedure to handle tapes when the end of the tape is reached.

end-of-file gap *Computer Programming.* a gap of a fixed length indicating the end of a file on a tape.

end-of-file indicator *Computer Technology.* a device associated with input/output units that signals the end of a file to the system and operator.

end-of-file mark *Computer Programming.* a code signaling that the last record of a file has been read.

end-of-file routine *Computer Programming.* a routine executed at the end of file to verify that the contents of the file were read correctly.

end of message *Telecommunications.* in a tape-relay communication system, an indicator on the last format line that is used to signify the end of a message or the termination of a transmission.

end-of-record gap *Computer Programming.* a gap of fixed length denoting the end of a record on a tape.

end-of-record word *Computer Programming.* a specially formatted word used to denote or identify the end of a record.

end-of-tape mark *Computer Programming.* a reflective piece of tape denoting the end of a reel of magnetic tape.

end-of-transmission recognition *Computer Programming.* the capability of a computer to recognize the end of a data transmission, whether or not the buffer is filled.

endogamous [en däg´ə məs] relating to or practicing endogamy.

endogamy [en däg´ə mē] *Anthropology.* the practice of finding a marriage partner from within one's own restricted group or class, as in the caste system of India. *Biology.* **1.** sexual reproduction between closely related organisms. **2.** the production of a zygote by a bisexual organism.

endogastric *Anatomy.* of or relating to the interior of the stomach.

endogenetic *Geology.* describing a process that originates within the earth, including extrusive and intrusive igneous activity, crystal warping, faulting, and folding; the opposite of exogenetic. Also, **endogenic, endogenous.**

endogenote *Microbiology.* the original genome of a bacterial cell that has undergone genetic recombination and has received a fragment, or exogenote, of genetic material from a donor cell.

endogenous [en däg´ə nəs] *Biology.* originating, developing, or growing from within. *Immunology.* of an infection or disease, arising from a cause within the organism. *Psychology.* of a mental disorder or condition, attributable to internal causes rather than environmental influences.

endogenous hormones *Endocrinology.* hormones that originate within the body.
endogenous pyrogen *Biochemistry.* a protein produced by the cells of a host body that induce fever, such as those produced by white corpuscles.
endogenous retrovirus *Virology.* a retrovirus that resides in the DNA of all cells of a host species and is transmitted genetically from one generation to the next.
endogenous variable *Mathematics.* a dependent variable; a variable that is an inherent part of a mathematical model, such as the cost of labor or material in a profit model.
endogenous virus *Virology.* **1.** a virus that causes a persistent infection. **2.** a virus whose genome integrates with cellular DNA, causing vertical transmission.
endoglobular *Hematology.* situated or occurring within the blood corpuscles. Also, **endoglobar.**
endognath *Invertebrate Zoology.* the interior portion of a crustacean's mouthparts.
endognathion *Anatomy.* the inner part of the incisive bone.
Endogonaceae *Mycology.* the only family of fungi belonging to the order Endogonales, including some species used as aids in crop growth.
Endogonales *Mycology.* an order of fungi belonging to the class Zygomycetes that reproduce both sexually and asexually.
endointoxication *Toxicology.* poisoning due to toxins that are present within an organism.
endolecithal *Invertebrate Zoology.* describing a type of egg, found in certain flatworms, that contains yolk granules in the cytoplasm. Also, ENTOLECITHAL.
Endolimax *Invertebrate Zoology.* a genus of ciliated amebas that are commensal in vertebrate intestines, including one species that is found in humans (*E. nana*).

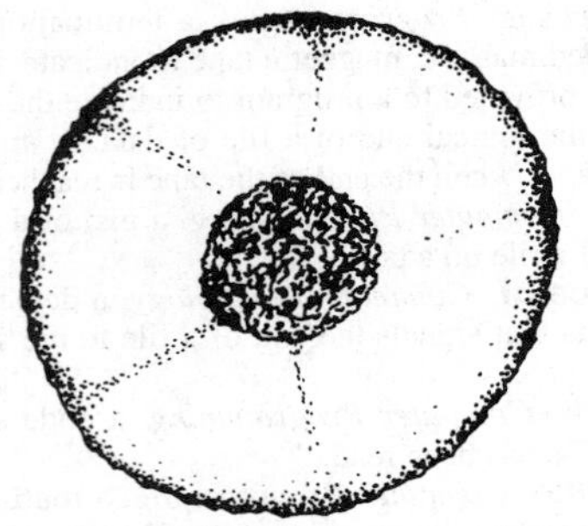

Endolimax nucleus

endolithic *Ecology.* of or related to a plant that grows within a stone, such as some lichens, algae, or fungi.
endolymph *Anatomy.* the fluid found in the membranous labyrinth of the inner ear. Also, SCARPA'S FLUID. Thus, **endolymphatic.**
endomeninx *Developmental Biology.* the mesenchymal part of the layer that covers the developing brain.
endomere *Developmental Biology.* a blastomere that serves in the formation of endoderm.
endometamorphism see ENDOMORPHISM.
endometrial *Anatomy.* of or relating to the endometrium.
endometrial carcinoma see HYSTEROCARCINOMA.
endometrioma *Medicine.* a discrete, non-neoplastic tumor or mass present in endometriosis.
endometriosis *Medicine.* a growth of endometrial tissue in various locations in the pelvic cavity that are outside the uterus.
endometriosis interna or **uterina** see ADENOMYOSIS.
endometritis *Medicine.* an inflammation of the endometrium often caused by a bacterial infection following childbirth or abortion, it may result in lesions, tubercles, or the casting off of portions of tissue.
endometrium *Anatomy.* the mucous membrane that lines the uterus; its thickness and structure vary with the phase of the menstrual cycle.
endomitosis *Cell Biology.* the replication of chromosomes within a nucleus that does not undergo subsequent nuclear division.
endomixis *Invertebrate Zoology.* a regular process of nuclear division and reorganization in certain ciliated protozoans. *Genetics.* a reproductive mechanism involving the union of a sperm and egg from one individual.
endomorph *Psychology.* a person of the endomorphic body type.
endomorphic *Psychology.* of or relating to a body type classification for a person having a round, fleshy physique.
endomorphism *Geology.* a form of contact metamorphism in which changes in an igneous rock are produced by the complete or partial assimilation of fragments of country rock, or by reaction with the country rock along the contact surface. Also, ENDOMETAMORPHISM. *Mathematics.* a homomorphism of an algebraic object (such as a group or a ring) to itself.
endomorphy *Psychology.* the extent to which a person conforms to the endomorphic body type, said to be associated with a calm, easygoing, and jovial personality.
Endomycetaceae *Mycology.* a family of fungi of the order Endomycetales occurring in soil, plant debris, slime flux, and fermenting mushrooms, and characterized by four-spored asci directly on the hyphae; some species are plant pathogens.
Endomycetales *Mycology.* an order of fungi of the class Hemiascomycetes, composed primarily of yeasts that live off decaying organic matter; used commercially in fermentation processes.
Endomychidae *Invertebrate Zoology.* a family of beetles, coleopteran insects in the superfamily Cucujoidea, that feed primarily on fungi under bark and in leaf litter.

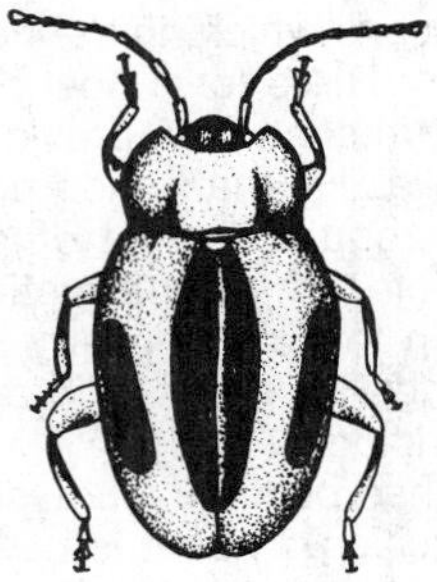

Endomychidae

endomysium *Histology.* the sheathlike layer of connective tissue that surrounds an individual muscle fiber.
endonasal *Anatomy.* located or occurring within the nose.
endoneural *Neurology.* of, relating to, or situated within a nerve.
endoneural system *Invertebrate Zoology.* the secondary nervous system in echinoderms, consisting of a marginal fiber along both sides of each arm that provides motor and muscle impulses.
endoneurial *Neurology.* of or relating to the endoneurium.
endoneuritis *Neurology.* inflammation of the connective tissue in the peripheral nerves surrounding a nerve fiber.
endoneurium *Neurology.* the layer of connective tissue that surrounds individual fibers in a nerve trunk.
endonuclear *Cell Biology.* located or occurring within a cell nucleus.
endonuclease *Enzymology.* an enzyme of the hydrolase class that catalyzes the hydrolysis of interior bonds of ribonucleotide or deoxyribnucleotide chains, producing oligonucleotides or polynucleotides.
endonuclease S_1 *Molecular Biology.* nuclease that is purified from the fungus *Aspergillus oryzae;* cleaves ssRNA and ssDNA to produce 5'-nucleoside monophosphates.
endoparasite *Entomology.* any parasite that lives inside the body of its host. Thus, **endoparasitic.**
endopeptidase *Enzymology.* any of several enzymes of the hydrolase class that catalyze the hydrolysis of peptide bonds in the interior of the peptide chain.
endoplasmic reticulum *Cell Biology.* a cytoplasmic system of membranes consisting of a series of interconnected flattened discs (cisternae). **Rough endoplasmic reticulum** is involved in the synthesis, processing, and transport to the Golgi of proteins destined for the plasma membrane, certain organelles, or secretion. **Smooth endoplasmic reticulum** is involved in the synthesis of steroids
endopod *Invertebrate Zoology.* the inner branch of a two-branched arthropod appendage. Also, **endopodite.**
endopolyploidy *Cell Biology.* a cell condition in which the chromosomes have replicated and divided repeatedly in the absence of cell division.

Endopterygota *Entomology.* a division of the insect subclass Pterygota whose members undergo complete (holometabolous) metamorphosis, with the immature stages differing from the adult in structure and habits.
endopterygote *Entomology.* any insect that undergoes complete (holometabolous) metamorphosis.
ENDOR electron nuclear double resonance.
end organ or **end-organ** *Neurology.* any of the larger, encapsulated endings of sensory nerve fibers.
endoribonuclease *Enzymology.* any of several enzymes of the hydrolase class that catalyze the hydrolysis of interior bonds of ribonucleotides, producing oligonucleotides or polynucleotides.
endorphin *Endocrinology.* any of a group of peptides produced by the pituitary gland and the brain that have pain-killing properties similar to those exhibited by morphine and are thought to regulate emotional responses. Important examples include **β-endorphin,** a peptide product of the pro-opiomelanocortin gene, and **α-endorphin**, the N-terminal 16 residue of β-endorphin.
endorsement *Artificial Intelligence.* a symbolically represented justification for belief or disbelief in a proposition.
endorser *Computer Technology.* in character recognition, a feature on magnetic ink character readers that prints a bank endorsement after a bank check has been read.
endosalpingioma see SEROUS CYSTADENOMA.
endosalpinx *Anatomy.* the mucous membrane that lines the uterine tube.
endosarc *Invertebrate Zoology.* the more fluid, interior protoplasmic region in certain unicellular organisms. Also, ENDOPLASM.
endoscope *Medicine.* an illuminated instrument used for a visual examination of the interior of a body cavity or viscus, such as the bladder.
endoscopic biopsy *Surgery.* the removal of a tissue sample by instruments passed through an endoscope.
endoscopic ultrasonography *Radiology.* a technique of visualizing body structures with ultrasound waves, using a transducer situated at the tip of a fiberoptic endoscope.
endoscopy *Medicine.* the visual inspection of any cavity in the body by means of an endoscope.
endosepsis *Plant Pathology.* a fig-tree disease caused by the fungus *Fusarium moniliforme fici* that causes the fruits to deteriorate internally.
endoskeleton *Zoology.* the cartilaginous and bony skeleton of a human or animal body, exclusive of any part of the skeleton that is of dermal origin; the opposite of *exoskeleton*. Also, NEUROSKELETON.
endosmosis *Physiology.* the transfer of a liquid across cell boundaries from outside the cell to the inside.
endosome *Cell Biology.* a cytoplasmic vesicle formed from endocytic vesicles that have shed their clathrin coats, or directly from coated pits. Also, RECEPTOSOME.
endosperm *Botany.* **1.** in angiosperms, the nutritive embryonic material that is formed by the double fertilization of the fusion of a sperm nucleus with polar nuclei and is located in the embryo sac. **2.** in gymnosperms, the food reserve derived from the megagametophyte. **3.** in ferns, the nutritive residue of the prothallus of the female gametophyte that surrounds the embryo; not homologous with the endosperm formed after fertilization.
endosperm nucleus *Botany.* **1.** in angiosperms, the triploid nucleus formed in the embryo sac by the fusion of one sperm nucleus and two polar nuclei. **2.** the male gamete that fertilizes the polar nuclei to form the endosperm.
endosteoma *Oncology.* a tumor in the internal cavity of the bone.
endostome *Botany.* **1.** the inner portion of the peristome, as in some mosses. **2.** the orifice in the inner layer of a bitegmic ovule.
endostyle *Invertebrate Zoology.* a ciliated groove used for transporting food in the pharynx of lower chordates.
endosulfan *Organic Chemistry.* $C_9H_6Cl_6O_3S$, toxic brown crystals that are insoluble in water and soluble in most organic solvents; melts between 70°C and 100°C, with the pure form melting at 106°C; used as an insecticide.
endosymbiont *Ecology.* an organism that participates in a relationship of endosymbiosis.
endosymbiosis *Ecology.* a situation in which an organism has a mutually beneficial relationship with a host organism while living inside the host's body.
endosymbiosis theory *Evolution.* a theory suggesting that the organelles of eukaryotic cells evolved from free-living prokaryotic cells that entered into symbiotic relationships with other, larger prokaryotic cells by passing through and existing inside their cell walls.
endosymbiote *Entomology.* any microbial organism that provides needed nutrients or digestive assistance for its insect host, such as the cellulose-digesting endosymbiotes living in termites.
endosymbiotic infection *Medicine.* a state achieved between a virus and its host cell in which the division of the cell is prevented, but the cell is not immediately destroyed. Also, **endosymbiosis.**
endotergite *Entomology.* a dorsal segment of the exoskeleton in insects to which muscles are attached.
endothecium *Botany.* **1.** the inner tissue of an embryonic sporophyte. **2.** the middle of three layers of an immature anther wall, usually with secondary wall thickening; it becomes the inner layer in the mature anther and aids in dehiscence.
endothelial cell *Histology.* a type of squamous epithelium cell that lines the interiors of cavities, spaces, and blood vessels.
endothelialization *Physiology.* the healing of the inner surfaces of grafts or blood vessels by endothelial cells.
endothelin *Biochemistry.* a polypeptide produced by endothelial cells that stimulates contraction of the underlying smooth muscle of blood vessel walls.
endotheliochorial placenta *Developmental Biology.* a placenta in which the syncytial trophoblast embeds maternal vessels bared to their endothelial lining.
endothelioma *Medicine.* any tumor arising from, or reassembling, endothelium; usually a benign growth.
endotheliotoxin *Toxicology.* any poison that attacks the endothelial lining of small blood vessels, causing hemorrhage.
endothelium *Histology.* the layer of squamous epithelium that lines the cavities of the heart, blood vessels, and lymph vessels.
Endotheriidae *Paleontology.* a family of insect-eating mammals in the order Protoeutheria; fossils have been found in Cretaceous strata in China.
endotherm *Physical Chemistry.* a graph of the temperature differences between a given compound and a thermally inert reference compound, such as aluminum oxide, when both are heated at a predetermined rate.
endothermic *Physical Chemistry.* **1.** relating to or describing any process in which a system absorbs heat from its surrounding environment. **2.** of a chemical process, requiring heat in order to proceed. Thus, **endothermic reaction, endothermic process.** Also, ENDOERGIC.
Endothia *Mycology.* a genus of fungi of the order Diaporthales; the species *E. parasitica* causes the plant disease chestnut blight.
Endothyracea *Paleontology.* an extinct superfamily of large, benthic foraminifera in the suborder Fusulinina; important in upper Devonian and lower Carboniferous stratigraphy; one of the three ancestral groups of the polyphyletic uniserial foraminifera.
endotoxic shock *Pathology.* septic shock due to the release of endotoxins by Gram-negative bacteria. Also, **endotoxin shock.**
endotoxin *Toxicology.* a complex bacterial toxin composed of protein, lipid, and polysaccharide, which is released only upon lysis of the cell. Also, BACTERIAL PYROGEN.
endotracheal *Anatomy.* located or occurring within the trachea. *Surgery.* performed by passage through the lumen of the trachea. Thus, **endotracheal anesthesia.**
endotracheal intubation *Surgery.* the insertion of an airway catheter through the mouth or nose into the trachea.
endotrophic *Biology.* obtaining nutrients from within; used especially in reference to certain parasitic fungi that live in the root cortex of a host plant.
endotrypanum *Invertebrate Zoology.* a digenetic (two-host) genus of *Trypanosomatidae* found in the blood of sloths and in a sandfly vector.
end piece *Horology.* in a watch, either of two flat, disk-shaped jewels into which the pivots of the balance staff are set to help reduce friction and keep out dust. Also, **end stone.**
end plate or **end-plate** *Anatomy.* a flat termination of a body part, especially of a nerve fiber that is in contact with a skeletal muscle fiber and involved in the transmission of nerve impulses to muscle.
end point *Analytical Chemistry.* a point in a titration reaction, usually indicated by a color change of the sample, at which the titrant has been added in equal proportions to the material titrated. *Chemical Engineering.* the highest thermometer reading when a determined proportion of a given liquid has boiled off in an analysis of distillation. Also, FINAL BOILING POINT. *Immunology.* the highest dilution to which an antibody or antiserum solution can be carried and still exhibit a detectable reaction with the antigen. *Mathematics.* a boundary point of an interval, line segment, or arc. *Robotics.* the point in a path of motion where a robot stops. *Industrial Engineering.* see BREAKPOINT. Also, **endpoint, end-point.**

end-point dilution assay *Virology.* a method of determining the quantity and strength of a virus in a sample by diluting the sample until it reaches a specified end point, typically the ED_{50} of the virus.
end-point rigidity *Robotics.* the amount of resistance to further movement by an end effector that has reached its end point.
end product *Engineering.* the final or resulting product, as of a manufacturing process or industry. *Physics.* the final result of a nuclear or chemical reaction, such as the stable nuclide at the end of a radioactive series.
endproduct inhibition see FEEDBACK INHIBITION.
endproduct repression see ENZYME REPRESSION.
end-pulling mechanism *Materials Science.* a model that explains how cross-link sites rearrange themselves during polymer deformation to produce a longer molecule with free ends.
end-quench (hardenability) test *Metallurgy.* a laboratory test used to assess the hardenability of a steel, in which a round bar of standard size is heated to form austenite and then end-quenched with a water stream of specified flow rate and pressure; the farther from the water-quenched end that the hardness decreases (i.e., the slower the cooling rate), the higher the steel's hardenability. Also, JOMINY TEST.
endrin *Organic Chemistry.* $C_{12}H_8Cl_6O$, a stereoisomer of dieldrin; a toxic, white crystalline powder that is soluble in water; melts at 200°C and rearranges above this point; formerly used as an insecticide.
end section see HEAD SECTION.
end-systolic volume *Cardiology.* the volume of blood in the ventricle after the heart's contraction (systole); this volume is normally 50–60 ml, but it can be as low as 10–30 ml.
endurance *Physiology.* the ability to continue an activity or function despite increasing physical or psychological stress or fatigue; stamina. *Mechanical Engineering.* the maximum amount of time that a vehicle can operate under designated conditions without refueling.
endurance limit or **endurance strength** *Materials Science.* the stress maximum that does not produce fatigue failure in a specimen subjected to a specified number of cycles (usually 10^6). Also, FATIGUE LIMIT, FATIGUE STRENGTH.
endurance ratio see FATIGUE RATIO.
end velocity *Space Technology.* in rocketry, the speed of a vehicle or stage at the moment of thrust cutoff.
endyma see EPENDYMA.
ENE east-northeast.
-ene a combining form used to denote the presence of an unsaturated hydrocarbon, as in *butylene.*
en echelon *Geology.* an overlapping or offset arrangement of geologic features. (From French, meaning "in step.")
en echelon fault blocks *Geology.* a linear zone in which the strike of the individual fault blocks is at an oblique angle to that of the entire zone.
enema *Medicine.* a procedure in which a fluid is flushed into the rectum for cleansing or treatment.
energetics *Physics.* the study of energy and its transformations from one form to another.
energetic solar particles *Astrophysics.* clouds of high-velocity electrons, protons, and nuclei emitted by solar flares with energies of 1 million to 100 million electron volts.
energized *Engineering.* describing a state of being connected to a source of energy. *Electricity.* specifically, connected to a source of electricity.
energy *Physics.* a quantity that describes the capacity to do work; commonly divided into three major classifications: kinetic (dynamic) energy, potential (static) energy, and radiant (electromagnetic) energy. *Engineering.* the use of this capacity for work to perform useful functions for humans, such as heating or cooling buildings and enclosures, powering vehicles and machinery, cooking foods, and so on.
energy absorption *Physics.* an absorption of energy into a system.
energy balance *Physics.* a mathematical statement of the law of the conservation of energy for a system in a steady state, equating the amount of input energy to the amount of output energy.
energy beam *Engineering.* a strong beam of light, electrons, or other nuclear particles that is used to cut through, shape, or process materials.
energy budget *Ecology.* a record of the flow of energy through an ecosystem population or individual.
energy coefficient *Oceanography.* the ratio of the energy in a wave per unit of crest length transmitted forward with the wave at a point in shallow water, to the energy transmitted forward with the wave in deep water.
energy conversion *Physics.* a transformation of one form of energy to another, such as the process of a falling body, in which potential energy of position is converted to kinetic energy.
energy-conversion efficiency *Mechanical Engineering.* the efficiency with which the energy of a working substance in a machine is converted into kinetic energy.
energy density *Physics.* the amount of energy per unit volume.
energy diagram *Mechanics.* a graph of potential energy versus the position of a particle, or versus one generalized coordinate of a system.
energy-dispersive X-ray analytic see ANALYTICAL ELECTRON MICROSCOPY.
energy efficiency ratio *Electricity.* the ratio of input power to output power.
energy flow *Ecology.* the passage of energy through the different trophic levels of a food chain.
energy flux *Physics.* a quantity measuring the rate of energy flow; the energy per unit time per unit area traveling across a surface element that is perpendicular to the energy flow. Also, FLUX OF ENERGY.
energy gap *Materials Science.* the energy between the top of the valence band and the bottom of the conduction band that a charge carrier must obtain before it can transfer a charge. *Solid-State Physics.* a range of forbidden energy levels between two permitted bands.
energy gradient *Physics.* the slope of a graph of energy versus position.
energy head *Fluid Mechanics.* the energy per unit weight of flowing fluid associated with its height, velocity, and pressure, usually denoted by a unit of length, as in ft-lbf/lbf≡ft, the ft-lbf being a unit of energy.
energy level *Geology.* the kinetic energy in the water of a sedimentary environment, either at the interface of deposition or several meters above it. *Quantum Mechanics.* any of the possible quantities of energies in which a system may exist.
energy-level diagram *Quantum Mechanics.* a diagram in which the possible energy levels of a system are represented by horizontal lines whose vertical positions above a zero-energy ground state correspond to their energies and show, by appropriate notation, such properties as stationary atomic states, degeneracies, shell and orbital population, and normalized eigenfunctions.
energy metabolism *Biochemistry.* the chemical reactions that produce or conserve energy within cells.
energy momentum tensor *Physics.* a 16-element tensor that gives the energy density, momentum density, and stresses for a radiation field.
energy of a charge *Electricity.* the work performed in charging a body, equivalent to $1/2\ QV$, where Q is the charge in statcoulombs and V is the electric potential in statvolts.
energy of activation *Chemistry.* the minimum quantity of energy that must be imparted to a molecule in the ground state before it can enter into a chemical reaction.
energy of rotation *Physics.* the kinetic energy associated with a rotating body having a moment of inertia I and an angular momentum ω, given by $E = I\omega^2/2$.
energy operator *Quantum Mechanics.* the Hamiltonian operator whose eigenvalues are the allowed energies of a system.
energy-product curve *Electromagnetism.* a curve derived by plotting the product of the values of magnetic induction and magnetic field strength for all points on the demagnetization curve of a ferromagnetic material.
energy profile *Physical Chemistry.* a graph that shows the changes in a system's energy levels in a chemical reaction or series of reactions.
energy spectrum *Physics.* a graphical or mathematical relation between energy and other experimental or theoretical parameters.
energy spread *Quantum Mechanics.* the statistical distribution of the possible energies of a system whose exact energy has not been established; the uncertainty in the energy.
energy state *Quantum Mechanics.* an eigenstate or stationary state of the Hamiltonian operator.
enervation *Physiology.* a lack of nervous energy; langor. *Surgery.* the removal of a nerve or a section of a nerve.
E neutrino see ELECTRON NEUTRINO.
enfilade *Military Science.* **1.** a position or fortification, such as a long trench, that makes troops vulnerable to sweeping fire. **2.** the fire directed at such a position. **3.** to maneuver into a position for such fire.
enfleurage *Chemical Engineering.* the process of removing perfume from flowers by putting them close to an odorless tallow and lard mixture that absorbs the perfume, which is then extracted from the fatty mixture by washing with alcohol.

enflurane *Pharmacology.* $C_3H_2ClF_5O$, a volatile liquid that is widely used in surgery for the induction and maintenance of general anesthesia.

eng *Materials.* a reddish-brown, hard, straight-grained wood from the timber tree *Dipterocarpus tuberculatus* of Burma.

engaged *Mechanical Engineering.* interlocked, as two wheels that are in gear with each other. *Architecture.* of a member, built so as to be or appear attached to a wall or other structure before which it stands. Thus, **engaged column.**

engagement *Medicine.* the entrance of the presenting part of the fetus into the superior pelvic strait. *Military Science.* **1.** an encounter, conflict, or battle between hostile troops, ships, or aircraft. **2.** in air defense terminology, an attack by interceptor aircraft using guns or air-to-air missiles, or the launching of ground-to-air missiles and their subsequent travel to intercept.

engagement control *Ordnance.* a ground unit that controls the launching of ground-to-air missiles to intercept enemy missiles or aircraft. Also, **engagement control station.**

engine *Mechanical Engineering.* **1.** a machine in which power is applied to do work, specifically by converting thermal energy into mechanical energy. **2.** broadly, any mechanical apparatus. **3.** the powered vehicle that pulls or drives a railroad train; a locomotive.

engine balance *Mechanical Engineering.* an assembly of moving parts designed to minimize vibrations in a reciprocating or rotating machine.

engine block see CYLINDER BLOCK.

engine car see POWER CAR.

engine cooling *Mechanical Engineering.* a cooling system in which circulating coolant, lubricating oil, and a fan are used to control the temperature of the parts of an internal-combustion engine.

engine cycle *Thermodynamics.* any series of thermodynamic processes that occur cyclically and that result in the conversion of a heat transfer into work.

engine cylinder see CYLINDER.

engine displacement *Mechanical Engineering.* the total volume displaced by an engine's cylinders, calculated by multiplying the number of pistons times the volume displaced by a cylinder moving from bottom dead center to top dead center, and expressed in cubic centimeters.

engine efficiency see EFFICIENCY.

engineer *Engineering.* a skilled professional who engages in the practice or teaching of engineering in one or more fields. *Transportation Engineering.* a person who operates a locomotive.

engineered food see FABRICATED FOOD.

engineering *Science.* the application of scientific knowledge about matter and energy for practical human uses such as construction, machinery, products, or systems.

engineering acoustics see ACOUSTICAL ENGINEERING.

engineering channel circuit *Telecommunications.* a voice or data channel or circuit that is used to operate and maintain communication services, facilities, and equipment.

engineering geology *Geology.* the application of the science of geology to the practice of engineering to promote its feasibility, safety, and profitability.

engineering materials *Materials Science.* a branch of materials science in which the relative strengths and weaknesses of many various materials are compared with respect to their use in engineering applications.

engineering psychology *Psychology.* an area of psychology that studies the relationship between human behavior and machines, especially for the purpose of designing tools and machinery. Thus, **engineering psychologist.**

engineering strain *Materials Science.* the change in length of a specimen divided by its original gauge length, $\Delta l/l_0$.

engineering stress *Materials Science.* the load applied to a specimen of a material divided by the original cross-sectional area, P/A_0.

engineering survey *Cartography.* a survey executed for the purpose of obtaining information that is essential for planning an engineering project.

engineering system of units see BRITISH GRAVITATIONAL SYSTEM OF UNITS.

engineering time see MAINTENANCE TIME.

engineer's chain *Cartography.* a measuring chain used in land surveying, consisting of 100 links that are each one foot long.

engineer's level *Cartography.* a type of precision leveling instrument for establishing a horizontal line of sight, used to determine differences in elevation.

engineer's scale *Graphic Arts.* a three-sided drafting tool, similar to a triangle, marked with incremental measurements (scale equivalents) and used for drawing plans.

engineer's strain *Mechanics.* strain calculated by dividing the change in length by the initial length; it is an approximation to true strain that is exact for infinitesimal strains.

engine friction *Mechanical Engineering.* the resistance to motion offered by contact among various moving parts of an engine.

engine intake *Mechanical Engineering.* the place or opening through which fuel enters an engine. Also, **engine inlet.**

engine knock see KNOCK.

engine nacelle *Aviation.* a self-contained compartment (nacelle) on an aircraft, housing an engine, associated devices, and sometimes landing gear.

Engineering

Engineering is the profession that creates all of the artifacts making up the modern world. The term "artifact" includes everything from bridges, housing, telephones, and television to jet aircraft, computers, nuclear reactors, medical imaging machines, and pharmaceuticals. This breadth means that engineers are of many types; from electronic, aeronautic, astronautic, and mechanical, to biomedical and chemical.

Engineering is a very old profession originating centuries before Christ, probably in Egypt. For most of its life, engineering was based on the accumulation of experience; engineers used techniques that had been found to work well. Engineering depended upon ingenuity and invention. It still does; however, a new feature has entered over the past 150 years; namely, the use of science. The great advances in modern engineering have come through discovering and using the laws of nature to accomplish goals previously impossible.

In creating new artifacts, engineers do the necessary research. Engineering research is purposeful; it is aimed at a goal, the goal of creating a new or improved product such as high-definition television, or automotive anti-lock brakes, or a service such as computer-aided teaching.

In these activities engineers often work with mathematicians and scientists who can help them understand the fundamentals involved in their artifacts. Just as important are the people who will use the results. Products and services must be useful to people without excessive effort. This feature is called "user friendly," and hinges upon thoughtful design of the interface between the human user and the system or machine. Engineering design must also aim for efficiency, beauty, safety, and environmental compatibility.

Manufacturing, too, is based upon engineering. Artifacts must be designed so they can be manufactured economically and with high quality. Automated production lines, some using robotics, can help reach these goals for large volumes. For small volumes, skilled production workers and aids for them are essential. The simultaneous design of products and suitable manufacturing techniques is known as concurrent engineering, and is essential for competing world-wide.

Excellent engineering is the key discipline for fashioning a livable world.

Dr. Edward E. David, Jr.
President; EED, Inc.

engine performance *Mechanical Engineering.* the relationship of various engine operating characteristics such as power output, revolutions per minute, fuel consumption, or ambient conditions.

engine sludge *Engineering.* the insoluble residue of oils and fuels that are produced as an internal combustion engine operates.

Engler distillation test *Chemical Engineering.* a laboratory procedure for determining the boiling range of petroleum distillates by measuring the percent of gasoline distilled at different specific temperatures.

Engler flask *Chemical Engineering.* a standardized, 100-milliliter volume flask used in the Engler distillation test.

Englerulaceae *Mycology.* a family of fungi belonging to the order Asterinales that form small blotches on the surfaces of leaves and twigs; found primarily in tropical regions.

Engler viscometer *Engineering.* an instrument designed to measure the viscosity of a substance according to the **degree Engler,** a measurement of viscosity.

English (cross) bond *Building Engineering.* a brickwork bond in which the vertical joints of the stretchers in any course line up with the centers of the stretchers above and below. Also, DUTCH BOND.

English degree *Chemistry.* a unit used to measure the hardness of water; 1 degree is equivalent to 1 part of calcium carbonate in 70,000 parts of water. Also, CLARK DEGREE.

englishite *Mineralogy.* $K_3Na_2Ca_{10}Al_{15}(PO_4)_{21}(OH)_7 \cdot 26H_2O$, a colorless or white orthorhombic mineral occurring as aggregates of curved plates, having a specific gravity of 2.68 and a hardness of 3 on the Mohs scale; found with crandallite, wardite, and other phosphate minerals.

English red *Materials.* a Venetian red pigment containing iron oxide.

English system *Metrology.* the system of measurement historically used in Great Britain, the U.S., and other English-speaking countries, in which the pound is the basic unit of mass, the yard is the basic unit of length, the gallon is the basic unit of liquid capacity, and the bushel is the basic unit of dry capacity. Also, **English measure.**

English vermillion *Inorganic Chemistry.* a bright red pigment consisting of precipitated mercury sulfide; used in paints.

engram *Physiology.* a discrete and permanent mark or trace left in nerve tissue by a stimulus. *Psychology.* a lasting trace left in the psyche by a psychic experience.

Engraulidae *Vertebrate Zoology.* the anchovies, a cosmopolitan family of mostly marine small fish of the order Clupeiformes, related to the herrings and often used for food.

engraver *Graphic Arts.* **1.** a tool used in engraving. **2.** a person whose work is engraving.

engraver's acid *Graphic Arts.* another term for nitric acid, HNO_3.

engraving *Graphic Arts.* **1.** the process of incising lines on a surface such as a stone, wood block, or printing plate. **2.** a printed work produced from such an incised surface. **3.** see PHOTOENGRAVING.

engraving

engross *Graphic Arts.* **1.** to write characters in a large, decorative hand. **2.** to produce the final handwritten or printed copy of a text, especially an official document.

engysseismology *Geophysics.* a branch of seismology that studies earthquake records from points close to the center of seismic activity.

enhanced carrier demodulation *Telecommunications.* a method of amplitude demodulation in which distortion is reduced by feeding the synchronized local carrier into the demodulator.

enhanced graphics adapter see EGA.

enhanced recovery *Petroleum Engineering.* any sophisticated recovery system for crude oil that augments the fraction of crude oil recovered from a reservoir.

enhanced spectral line *Spectroscopy.* a spectral line of a very hot source. Also, **enhanced line.**

enhancement the process of increasing, improving, or intensifying; specific uses include: *Electronics.* a process that increases current flow in a given section of a semiconductor. *Virology.* the increased yield of one virus by its joint infection with another virus.

enhancement mode *Electronics.* a characteristic of a field-effect transistor's operation in which current flows through the source-drain channel only when a voltage has been applied to the gate, and increases as the voltage increases.

enhancement antigen see E ANTIGEN.

enhancement effect *Biochemistry.* the increase in the rate of photosynthesis caused by the addition of light with a shorter wavelength.

enhancer sequence *Genetics.* a sequence whose presence helps to increase the activity of a promoter to bind RNA polymerase to initiate transcription; an enhancer can assist any promoter placed in its vicinity.

ENIAC

ENIAC [en′ ē ak′] *Computer Technology.* the first completely electronic computer, developed from 1943 to 1946. (An acronym for electronic numerical integrator and calculator .)

Enicocephalidae *Invertebrate Zoology.* a family of hemipteran insects, the gnat bugs, in the superfamily Enicocephaloidea; found in tropical and subtropical forests.

enigmatite see AENIGMATITE.

enium ion *Organic Chemistry.* a cationic portion of an ion in which the valence shell of a positively charged nonmetallic atom has two electrons less than normal, and the charged species has one covalent bond less than the associated uncharged species. Also, YLIUM ION.

enkephalin [en kef′ ə lin] *Biochemistry.* either of two simple pentapeptides **(leu-enkephalin** or **met-enkephalin)**, occurring naturally throughout the brain and spinal cord, having potent opiatelike effects and acting through opiate receptors as endogenous inhibitors of pain.

enlargement *Graphic Arts.* **1.** also, **enlarging.** the process of producing a photographic print larger than the negative. **2.** a print that is larger than the negative from which it was made.

enlargement loss *Fluid Mechanics.* energy losses in a flowing fluid occurring when the fluid passes through a sudden enlargement; such losses are usually minimized in diffusers.

enlightenment effect *Psychology.* the tendency of people who become aware of a theory of behavior to adjust their own actions accordingly, thus rendering the theory inaccurate as a predictor of behavior.

enokitake *Mycology.* the fungal species *Flammulina velutipes,* belonging to the order Agaricales, which is cultivated in Japan and Taiwan for food. Also, WINTER MUSHROOM.

enol *Organic Chemistry.* a compound that contains a hydroxyl group next to a double bond, and exhibits keto-enol tautomerism.

enolase *Enzymology.* an enzyme of the lyase class that catalyzes the dehydration of 2-phosphoglyceric acid to phosphopyruvic acid.

enolate anion *Organic Chemistry.* a delocalized anion that is formed from the removal of a proton from an enol, or from the carbonyl compound in equilibrium with the enol.

enol-keto tautomerism *Organic Chemistry.* the tautomeric migration of a hydrogen to the carbonyl group from the adjacent carbon, with subsequent movement of the double bond. The process is reversible; i.e., keto-enol tautomerism also occurs.

enophthalmos *Medicine.* the retraction of the eyeball into the socket caused by an injury or birth defect.

Enopla *Invertebrate Zoology.* a class of unsegmented ribbon worms in which the mouth is anterior to the brain.

Enoplia *Invertebrate Zoology.* a subclass of diverse nematode worms in the class Adenophora, occupying every ecological niche available to nematodes.

Enoplida *Invertebrate Zoology.* an order of free-living aquatic nematode worms that feed on diatoms, algae, or detritus.

Enoplidae *Invertebrate Zoology.* a family of mostly marine free-living nematodes in the superfamily Enoploidea.

Enoploidea *Invertebrate Zoology.* a superfamily of mostly marine nematodes.

Enoploteuthidae *Invertebrate Zoology.* a family of deep-sea squid, cephalopod mollusks found worldwide in tropical and subtropical seas.

E notation *Computer Programming.* a format used for representing floating point numbers in which a number is followed by an "E" and the power of 10 is to be applied. For example, 107.35 could be represented as 0.10735E+03. Also, E FORMAT.

enphytotic *Plant Pathology.* referring to any plant disease that occurs regularly among plants of a specific region but only in moderate severity, or an outbreak of such a disease.

enqueue [en kyoo´] *Computer Programming.* to put an item in a queue in data processing.

enquiry character *Telecommunications.* a control character denoting a request for a response from a remote station.

enrich *Food Technology.* to add nutrients to a food; carry out the process of enrichment. Also, FORTIFY.

enriched material *Nuclear Physics.* an element that has been altered to have more of a given isotope than is present in nature, such as enriched uranium, which has a higher percentage of the fissionable isotope uranium-235.

enriched milk *Food Technology.* milk that is fortified through either irradiation or the addition of vitamin D concentrate so that it contains at least 400 units of vitamin D per quart.

enriched reactor *Nucleonics.* a reactor fueled by a material that has been enriched with one or more isotopes.

enriched rice *Food Technology.* rice that is fortified with thiamine, niacin, iron, and often riboflavin, calcium, and vitamin D; rice can be enriched either by parboiling the whole grain or by soaking polished rice in a vitamin-mineral solution and then coating it with a colloid film.

enriched uranium *Nucleonics.* uranium that contains more than 0.711% by weight of uranium-235, required by many types of nuclear reactors in order to achieve criticality; it is produced from natural uranium by a process such as gaseous diffusion or centrifugation.

enriching column *Chemical Engineering.* the section of a countercurrent contactor above the feed point in which a product-rich, ascending stream from the stripping section is purified by countercurrent contact with a downward-flowing reflux stream.

enrichment *Food Technology.* the addition of vitamins and minerals to processed foods, most commonly cereal products, to replace nutrients lost in milling and processing. *Agronomy.* the addition of fertilizer, manure, or other such products to improve the soil. *Nucleonics.* a process by which the amount of a given isotope in an element is artificially increased. *Microbiology.* any process or medium that selects for and increases the proportion of a specific microorganism in a mixed population. *Virology.* the production of a defective interfering particle at the expense of its standard helper virus particle.

enrichment culture *Microbiology.* a selective nutrient medium designed to specifically promote the growth of one particular microorganism in a mixed population.

enrichment factor *Nucleonics.* the ratio between the amount of a given isotope in a material before enrichment and the amount after enrichment.

enroute chart *Cartography.* a chart of air routes in specific areas that shows the exact location of electronic aids to navigation, such as radio-direction-finder stations, radio- and radar-marker beacons, and radio-range stations. *Aviation.* an aeronautical chart that is designed to be used between terminal areas. The two classes of such charts are En Route Low-Altitude Charts and En Route High-Altitude Charts.

ENS ensign.

ensemble *Physics.* a conceptual group of an identical system used to predict the measurable properties of the system.

ensemble average *Mathematics.* an average (of a quantity related to a stochastic process) taken over an infinite number of fictitious alternate universes called an **ensemble.** In each member of the ensemble, the stochastic process is imagined to experience different random variations. The resulting average is called the expected value of the quantity, and is usually expressed as an integral over the probability density function of the stochastic process. For example, the expected value of a coin toss (heads = 1, tails = 0) is 1/2, even though such a result will never occur and can hardly be "expected." The ensemble average is conceptually distinct from the time average, and if the two averages coincide, the process is ergodic.

ensialic geosyncline *Geology.* a geosyncline whose load of sediments accumulates on sialic crust and is composed largely of clastic rock.

Ensifer *Bacteriology.* a genus of Gram-negative, aerobic bacteria that occur as flagellated, rod-shaped cells and multiply by budding; certain species of this genus prey on other bacteria.

ensiform *Biology.* long and flat with sharp edges and tapering to a point; sword-shaped.

ensilage [en´sə lij; en sī´lij] *Agriculture.* the preservation of green fodder in a silo or pit.

ensiling *Agriculture.* the process of storing green plant material in a silo to bring about fermentation.

ensimatic geosyncline *Geology.* a geosyncline whose load of sediments accumulates on a simatic crust and is composed largely of volcanic rock.

Enskog theory see CHAPMAN-ENSKOG THEORY.

enstatite *Mineralogy.* $Mg_2Si_2O_6$, a grayish-, yellowish-, or greenish-white, brittle orthorhombic mineral of the pyroxene group that is dimorphous with clinoenstatite, having a specific gravity of 3.209 to 3.431 and a hardness of 5 to 6 on the Mohs scale; found in basic igneous rocks.

ENT or **E.N.T.** ear, nose, and throat.

ent- a combining form meaning "within" or "inner," as in *entamebiasis.*

entablature *Architecture.* **1.** the upper part of a classical order, supported by a colonnade and carrying the pediment or roof plate; divided horizontally into architrave, frieze, and cornice. **2.** a similar feature in post-and-lintel construction.

Entac *Ordnance.* a French antitank missile that is powered by a two-stage solid-propellant rocket motor; it is wire-guided and carries a 9-pound high-explosive warhead capable of penetrating more than 25.5 inches of steel armor at a range of 1300 to 6600 feet; officially designated **MGM-32A.**

Entameba *Invertebrate Zoology.* a genus of amoeboid protozoans parasitic on humans and other vertebrates; some species, such as *E. histolytica,* cause amoebic dysentery. Also, **Entamoeba.**

entameba *Invertebrate Zoology.* any member of the genus *Entameba.* Also, **entamoeba.**

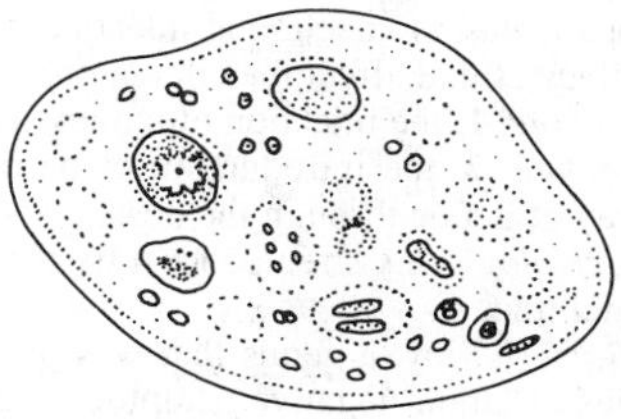

entameba

entasia *Neurology.* a spasmodic muscular contraction or constrictive spasm. Also, **entasis.**

entasis *Architecture.* the convex curving of a column, designed to counteract the optical illusion that makes straight columns appear to curve inward.

entelechy see ORTHOGENESIS.

Enteletacea *Paleontology.* a superfamily of punctate articulate brachiopods in the order Orthida and the suborder Orthidina, extant through most of the Paleozoic era.

Entelodontidae *Paleontology.* a family of primitive, even-toed, large mammals in the subclass Suina and the superfamily Entelodontoidea; the entelodonts evolved from a group of odd-toed chalicotheres in the Eocene and became extinct in the Miocene.

Entelodontoidea *Paleontology.* an extinct superfamily of artiodactyl Cenozoic mammals.

enter *Computer Programming.* **1.** to put data or records into a file or system. **2.** to confirm that a given record or an item of data is to be stored in memory by pressing the enter key.

enter- a combining form meaning "intestine," as in *enteralgia.*

enteral *Anatomy.* of, relating to, or occurring within the small intestine. Also, **enteric.**

enteralgia [en´tə ral´jə] *Medicine.* an abdominal pain that is accompanied by bowel spasms, commonly arising from the presence in the alimentary canal of some indigestible matter, but also caused by the absorption of lead into the system or by gallstones passing into the intestine. Thus, **enteralgic.**

enterectomy *Surgery.* the excision of a part of the intestine, or resection of the intestine.

enteric bacilli *Bacteriology.* any Gram-negative, rod-shaped bacteria found in the human small intestine and having a facultatively anaerobic mode of respiration.

enteric-coated *Pharmacology.* describing a capsule or tablet that is covered with a special coating so that the release of its contents is delayed until it reaches the intestines.

enteric cytopathogenic human orphan virus see ECHOVIRUS.

enteric fever *Medicine.* any of various febrile illnesses associated with enteric symptoms caused by salmonellae, especially typhoid fever and paratyphoid fever.

entericoid *Medicine.* resembling an enteric fever.

enteric plexus *Anatomy.* a network of autonomic nerve fibers located within the wall of the digestive tube.

entering angle *Mechanical Engineering.* the angle formed as the side cutting edge of a tool meets a surface of a workpiece.

enteritis *Medicine.* any inflammation of the intestine, especially of the lining of the small intestine; may be caused by a bacterial or viral infection or by certain functional disorders.

enteritis necroticans *Medicine.* an inflammation of the intestines that is caused by *Clostridium perfringens* type F; characterized by necrosis.

enter key *Computer Technology.* a special command key on a computer keyboard that, when pressed, notifies the computer that active material is now ready to be stored or processed; in some programs, it also serves other functions such as providing a page-break command.

entero- a combining form meaning "intestine," as in *enterobiasis.*

Enterobacter *Bacteriology.* a genus of Gram-negative, motile, bacillary bacteria of the family Enterobacteriaceae, characterized by lactose fermentation and found in soil, water, dairy products, and the intestinal tract of humans and other animals.

Enterobacteriaceae *Bacteriology.* a family of Gram-negative, rod-shaped, facultatively anaerobic bacteria found in soil, water, plants, and animals; they frequently occur as pathogens in vertebrates.

enterobiasis *Medicine.* an infection of the large intestine by the common pinworm *Enterobius vermicularis.*

enterocele *Medicine.* any hernia of the intestine.

enterochromaffin cells *Endocrinology.* a type of solitary hormonal cell found in the pancreas, stomach, and intestine; they produce gastrin and other peptide factors that affect the circulatory vessels.

enteroclysis *Medicine.* **1.** the injection of a nutrient or medicinal liquid directly into the bowel. **2.** the introduction of barium directly into the small bowel by way of a tube through the nose.

enterococci *Bacteriology.* any streptococcal bacteria normally found in the human intestinal tract.

Enterococcus *Bacteriology.* a genus that was proposed to accommodate certain species of Gram-negative, streptococcal bacteria that occur as facultatively anaerobic, coccoid cells in human and animal intestinal tracts.

Enterocoela *Zoology.* a group of animals, including the Echinodermata, Chaetognatha, Hemichordata, and Chordata, of which the primary distinguishing morphological characteristic is the body cavity or coelom.

enterocolitis *Medicine.* an inflammation of both the large and the small intestines, often resulting in diarrhea, ulceration, and breakdown of intestinal mucosa.

enterodynia *Medicine.* any intestinal pain.

enteroglucagon *Endocrinology.* a protein that is found in cells of the lower gastrointestinal tract, structurally homologous to glucagon but having an uncharacterized physiological function. Also, INTESTINAL GLUCAGON.

enterohepatitis *Medicine.* any inflammation of the bowel and liver.

enterohydrocoel *Invertebrate Zoology.* an anterior cavity derived from the archenteron in crinoids.

enteroinvasive *Microbiology.* of or relating to any pathogenic organism that is able to invade the intestinal mucosa. Thus, **enteroinvasion.**

enterokinase *Biochemistry.* a former term for enteropeptidase.

enterokinesia or **enterocinesia** see PERISTALSIS.

enterolith *Pathology.* an intestinal concretion created by layers of mucosa around the presence of a fruit stone or other indigestible object that was swallowed.

enterolithic *Geology.* describing a sedimentary structure that exhibits small folds resembling those produced by tectonic deformation, but that are actually produced by chemical changes affecting the volume of the rock.

enterology *Medicine.* the branch of medical science that is concerned with the intestines.

enteromorpha *Botany.* a genus of multicellular green algae in the family Ulvaceae that is able to flourish in almost all types of sea water and often grows on the bottoms of ships, characterized by a hollow tubular thallus with a wall one cell thick.

enteron *Anatomy.* the gut or alimentary canal; especially, the small intestine.

enteropathogenic *Pathology.* relating to or effective in forming a disease of the intestine.

enteropathy *Pathology.* any disease of the intestine.

enteropeptidase *Enzymology.* an enzyme of the hydrolase class that is secreted by the small intestine; it catalyzes the conversion of the inactive enzyme trypsinogen to the active form trypsin.

enteroplegia *Medicine.* a paralysis of the intestine.

Enteropneusta *Invertebrate Zoology.* the tongue worms and acorn worms, a class of free-living, wormlike, burrowing hemichordates.

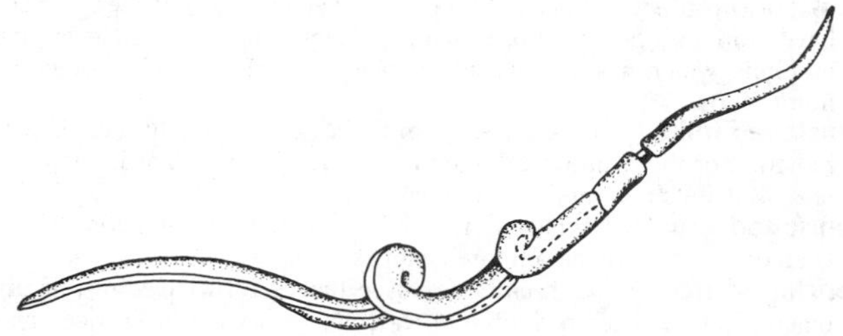

Enteropneusta

enteroptosis *Medicine.* a descent or downward displacement of the intestine in the abdominal cavity; a term based on the archaic concept that variations in the positions of abdominal organs are pathological. Thus, **enteroptotic.** Also, **enteroptosia.**

enterorrhagia *Medicine.* an intestinal hemorrhage.

enterostasis *Medicine.* the stopping of food in its passage through the intestine.

enterotoxemia see PULPY-KIDNEY DISEASE.

enterotoxin *Toxicology.* any poison that attacks the gastrointestinal tract. *Microbiology.* a toxic substance elaborated by certain bacteria, such as *Staphylococcus,* that causes the symptoms of food poisoning in vertebrates.

Enterovirus see ECHOVIRUS.

enthalpimetric analysis *Analytical Chemistry.* a thermochemical method of analysis that relies on measurement of temperature changes caused by reactions in solutions.

enthalpy [en´ thəl pē] *Thermodynamics.* heat content; a thermodynamic property of a system defined as $H = U + PV$, where H is enthalpy, U is the internal energy of the system, P is the pressure exerted on the system by its environment, and V is the volume of the system. If a steady flow process takes place at constant pressure and no work is done (other than mechanical work against the boundaries), then the change in enthalpy is equal to the heat transferred during this process.

enthalpy-entropy chart *Thermodynamics.* a plot of the enthalpy of a thermodynamic system as a function of the entropy for various values of temperature, pressure, specific volume, and so on.

enthalpy of absorption see HEAT OF ABSORPTION.

enthalpy of adsorption see HEAT OF ADSORPTION.

enthalpy of combustion see HEAT OF COMBUSTION.

enthalpy of condensation see HEAT OF CONDENSATION.

enthalpy of decomposition see HEAT OF DECOMPOSITION.

enthalpy of dilution see HEAT OF DILUTION.

enthalpy of formation *Physical Chemistry.* the change in enthalpy that takes place when a compound is formed from individual elements.

enthalpy of fusion see HEAT OF FUSION.

enthalpy of hydration see HEAT OF HYDRATION.

enthalpy of mixing *Physical Chemistry.* the difference in enthalpy between a mixture and the sum of its individual components.

enthalpy of neutralization see HEAT OF NEUTRALIZATION.

enthalpy of phase change *Physical Chemistry.* the change in enthalpy that takes place as a substance changes from one phase to another, as from a liquid to a gas. Also, **entropy of transition.**

enthalpy of reaction *Physical Chemistry.* the change in enthalpy that accompanies a chemical reaction. Also, **enthalpy of chemical reaction.**

enthalpy of solidification see HEAT OF SOLIDIFICATION.

enthalpy of solution see HEAT OF SOLUTION.

enthalpy of state change *Physical Chemistry.* the change in enthalpy that takes place as a substance changes its state, as when a gas expands.

enthalpy of sublimation see HEAT OF SUBLIMATION.

enthalpy of vaporization see HEAT OF VAPORIZATION.

enthesis *Surgery.* the use of an artificial substance to repair a deformity, defect, or injury.

entire *Biology.* having a smooth, continuous outer edge; not indented or notched.

entire function *Mathematics.* a function of a complex variable z that is defined and analytic for all finite values of z; e.g., polynomials, e^z, and $\sin z$. Also, INTEGRAL FUNCTION.

entire series *Mathematics.* a power series that converges for all values of the variables; e.g., the exponential series $e^z = \sum_{n=0}^{\infty} z^n/n!$. An entire series has an infinite radius of convergence.

entisol or **Entisol** *Agronomy.* one of the 10 major soil classifications, characterized as a mineral soil without a natural genetic horizon.

Entner-Doudoroff pathway *Biochemistry.* a series of reactions for glucose degradation that frees energy.

entoblast see ENDODERM.

Entoconchidae *Invertebrate Zoology.* wormlike gastropods, endoparasites of sea cucumbers.

entoderm see ENDODERM.

Entodiniomorphida *Invertebrate Zoology.* an order of ciliated protozoans in the subclass Spirotrichia, found widely as commensals in herbivorous mammals.

Entodinium *Invertebrate Zoology.* protozoans in the order Entodiniomorphida, with small or no cilia and adoral area of membranelles, endocommensal in the digestive tract of herbivores.

Entodontaceae *Botany.* a family of medium-sized glossy mosses of the order Hypnobryales that form loose mats on soil, humus, logs, tree bases, and sometimes rocks; characterized by the lack of a strong costa, by smooth, shiny leaves, and by freely branched creeping stems.

entognathous *Entomology.* having mouthparts with recessed mandibles and maxillae concealed by lateral folds on the head.

entolecithal see ENDOLECITHAL.

entom. or **entomol.** entomology, entomological.

entombment *Nucleonics.* a method for decommissioning a nuclear facility, such as a power plant, in which it is demolished and encased in reinforced concrete to prevent radioactive materials from entering the environment.

entomo- a combining form meaning "insect," as in *entomology.*

Entomoconchacea *Paleontology.* an extinct superfamily of large ostracods in the suborder Myodocopa; late Paleozoic.

entomogenous *Biology.* describing a fungus that lives in or on an insect body.

Entomology

Entomology is the discipline that addresses all aspects of the biology of insects, including their relation to the welfare of humanity. In the strict sense it encompasses only the "true" insects, including the 750,000 known species of the order Hexapoda or Insecta, which are characterized by the possession of three major body parts (head, thorax, abdomen) and three pairs of legs on the thorax. It is occasionally broadened to include—very loosely and inaccurately—other land-dwelling arthropods such as symphylans, mites, spiders, and centipedes.

A large part of entomology is insect systematics, and little wonder: the beetles, butterflies, and moths alone make up more than 40 percent of all known animal species. Because of their immense variety in tropical forests, insect species might easily exceed 30 million worldwide (compared with 1.4 million known species of all kinds of organisms). Their variety and huge biomass make insects major players on the land environment, as herbivores, predators, scavengers, and soil turners, the "little things that run the world."

Another key enterprise is economic entomology, the study and control of insect pests. In recent years the emphasis of this applied field has shifted from chemical insecticides, which are often injurious to health or the environment, to biological control using natural parasites and predators. Most desired is "integrated pest control," a combination of chemical and biological agents with skillful manipulation of the environment to keep populations off balance and at low densities.

Yet another division of major importance is medical entomology, dealing with poisonous insects and the insect vectors of disease. Although malaria, leishmaniasis, onchoceriasis, and a frightening array of other insect-borne diseases still cripple and kill hundreds of millions yearly, this subdiscipline is still relatively undeveloped.

Edward O. Wilson
Baird Professor of Science
Harvard University

entomology [en´tə mäl´ ə jē] *Invertebrate Zoology.* the study of insects.

Entomoneidaceae *Botany.* a family of marine, brackish, and freshwater diatoms of the order Pennales, characterized by a symmetrical valve and a raphe positioned on a high keel that descends at the valve center.

entomopathogenic *Entomology.* causing disease in insects. Thus, **entomopathogenic virus**.

entomophagous *Zoology.* feeding mainly or exclusively on insects.

entomophilic fungi *Mycology.* a species of fungi that are pathogenic to insects.

entomophilous *Ecology.* of or relating to a plant that is pollinated by insects. Thus, **entomophile, entomophily.** Also, **entomophilic.**

entomophily *Botany.* pollination by insects, as the flowers of orchids and irises. Thus, **entomophilous.**

entomophobia *Psychology.* an irrational fear of insects.

Entomophthoraceae *Mycology.* a family of fungi of the order Entomophthorales containing the genus *Entophthora,* a parasite of insects; some species are used for pest control.

Entomophthorales *Mycology.* an order of fungi of the class Zygomycetes that includes parasites and pathogens of insects, humans, and other animals.

Entomopoxvirinae *Virology.* a subfamily of viruses of the family Poxviridae that replicate in a wide range of insects; divided into three genera (A, B, and C) on the basis of morphology, genome weight, and host range.

entomopoxvirus *Virology.* any virus belonging to the subfamily Entomopoxvirinae.

Entomosigmaceae *Botany.* a monogeneric family of nonphotosynthetic flagellates of the order Gymnodiniales, characterized by a pear-shaped cell with two flagella at the anterior end and by a lack of armor.

Entomostraca *Invertebrate Zoology.* an informal name, recognized by some authorities, for a group of crustaceans including Branchiopoda, Branchiura, Cephalocarida, Copepoda, and Cirripedia.

Entoniscidae *Invertebrate Zoology.* a family of isopod crustaceans that are parasites of decapod crustaceans.

entoplastron *Vertebrate Zoology.* a median bony plate of the anterior part of the plastron of turtles that is considered homologous with the interclavicle of other reptiles.

Entoprocta *Invertebrate Zoology.* a group of stalked, sessile aquatic invertebrates with a U-shaped digestive tract so that both mouth and anus are within the circle of tentacles.

Entorrhiza *Mycology.* a genus of smut fungi of the order Ustilaginales that are parasites of rushes and other roots.

Entosiphon *Invertebrate Zoology.* a genus of colorless freshwater protozoans having two flagella.

entrail pahoehoe *Geology.* a type of pahoehoe whose surface is covered with numerous, small, intertwined toes of lava, giving it an appearance that resembles a mass of entrails.

entrained fluid *Fluid Mechanics.* the carryover of small liquid droplets or fine mists in a gas stream from a contactor or process vessel, occurring when the gas velocity is greater than the settling velocity of the droplets.

entrainer *Chemical Engineering.* an additive of organic water-insoluble liquid that forms an azeotrope with a component in a liquid mixture, to help in separations that are impossible or difficult to accomplish by normal distillation procedures.

entrainment *Hydrology.* **1.** the picking up and transporting of materials by currents. **2.** the transferring and mixing of water masses by friction between opposing currents. *Chemical Engineering.* any process in which particulates are caught up in and then carried away by a flowing gas or liquid. *Meteorology.* the integration of environmental air into a preexisting, typically upward-moving air current.

entrance *Building Engineering.* a means or place of going in, such as a gate or passageway. *Naval Architecture.* the forward portion of a ship's underbody, from the stem to the area of widest beam. *Computer Programming.* see ENTRY POINT.

entrance angle *Engineering.* the widest angle at which molten material is emplaced in a die during the molding process.

entrance cable *Electricity.* a cable that is used to transmit electric power from an outside powerline into a building.

entrance length *Fluid Mechanics.* the distance downstream in a pipe from the pipe's entrance to the location at which fully developed flow begins.

entrance loss *Fluid Mechanics.* a minor head loss encountered as a flowing fluid enters a pipe; the loss coefficient and therefore the head loss depends upon the contraction ratio and the type of entrance.

entrance region *Meteorology.* a confluence zone at the upper extremity of a jet stream.

entrance slit *Spectroscopy.* the narrow slit in a spectroscopic device through which incident radiation enters.

entrance window *Photogrammetry.* the image of the field stop formed by all the lens elements on the object side of the field stop.

entrapment *Biotechnology.* a method for immobilizing enzymes or cells by passing them through a polymeric mesh large enough for substrates and nutrients to pass, but small enough to trap the higher molecular weight biological materials.

entrenched meander *Hydrology.* an old meander that is more or less enclosed by valley walls and has been deepened vertically downward, below the surface of the valley in which it was originally formed. Also, INHERITED MEANDER.

entrenched stream *Hydrology.* a stream flowing in a narrow trench or valley that has been cut into a relatively level upland or plain.

entropy [en′ trə pē] *Physics.* a measure of the randomness or disorder of a system, expressed as $S = k \ln g$, where k is the Boltzmann constant and g is the total number of microstates admissible to the system. The total entropy of an isolated system cannot decrease when the system undergoes a change; it can remain constant for reversible processes, and it must increase for irreversible ones. Thus, since the universe can be regarded as an isolated system undergoing irreversible change, it follows that its entropy is increasing and its level of available energy decreasing. *Physical Chemistry.* **1.** for an isothermal process, the heat added or removed divided by T. **2.** more generally, the path integral of $\delta Q/T$. *Thermodynamics.* a property of a thermodynamic system, denoted by S, whose differential change is given by $dS = dQ/T$, where dQ refers to the differential amount of heat transfer introduced to the system in a reversible process, and T is the absolute temperature of the system. *Telecommunications.* a measure of the amount of information in a message, based on the number of possible equivalent messages. *Mathematics.* the entropy of a random variable X with probability density function p as the expected value of $\log(^1/p)$; i.e., the integral $-\int_{-\infty}^{\infty} p(x)\log p(x)\,dx$. Entropy represents the degree to which the values of X are dispersed.

entropy of formation *Physical Chemistry.* the change in entropy that takes place when a compound is formed from individual elements.

entropy of mixing *Physical Chemistry.* the difference in entropy between a mixture and the sum of its individual components.

entropy of phase change *Physical Chemistry.* the change in entropy that takes place as a substance changes from one phase to another, as from a liquid to a gas. Also, **entropy of transition.**

entropy of reaction *Physical Chemistry.* the change in entropy that accompanies a chemical reaction. Also, **entropy of chemical reaction.**

entropy of state change *Physical Chemistry.* the change in entropy that takes place as a substance changes its state, such as when a gas expands.

entry *Building Engineering.* an entrance, especially an open vestibule or entrance hall. *Computer Programming.* **1.** an element of a list, queue, or other data structure. **2.** the process of including a new data item in a list or other data structure. **3.** an input data item from a terminal. **4.** in a spreadsheet application, the data contained in a single cell.

entry ballistics *Mechanics.* the ballistics of a warhead or other body entering an atmosphere from outer space.

entry block *Computer Programming.* a main memory storage block assigned to each entry in a computer system; contains data associated with that entry throughout its life in the system.

entry condition *Computer Programming.* a specific condition that must be adhered to in order for a computer program or subroutine to execute correctly, such as the location of inputs or outputs, relationships between inputs, and clock signals. Also, INITIAL CONDITION.

entry corridor *Space Technology.* the region around a nominal trajectory, defining the desired landing area or indicating the design limits of a vehicle about to enter a planetary atmosphere. With too steep a trajectory, a spacecraft would burn up; with too shallow a trajectory, it would overshoot the landing area and be unable to return.

entry instruction *Computer Programming.* the first instruction executed upon transfer of control to a subroutine.

entry point *Computer Programming.* any location in a routine or subroutine to which control can be passed by another routine or subroutine; the point of execution. Also, ENTRANCE.

entry portion *Computer Programming.* the right side of a decision table consisting of the condition and action entries, the columns of which are the decision rules.

entry site *Molecular Biology.* a ribosome site that is available for initial binding of transfer RNA during genetic translation; the site where protein synthesis begins.

entry sorting *Computer Programming.* a sorting method in which records or blocks of records are placed in a buffer and then sorted into the main list prior to receiving the next record.

Entyloma *Mycology.* a genus of smut fungi of the order Ustilaginales that are parasites of certain plants and cause leaf spots.

entypy *Developmental Biology.* the covering by the endoderm of the embryonic and amniotic ectoderm; present in certain mammalian embryos.

enucleate *Medicine.* to remove an organ or a tumor whole and clean in its entirety, such as an eye from its socket or a tumor from its envelope. Thus, **enucleation.** *Cell Biology.* **1.** describing a body or form having no nucleus. Thus, **enucleated. 2.** to remove a nucleus from a cell.

enucleated cell *Cell Biology.* a cell from which the nucleus has been removed.

enumerable *Mathematics.* able to be counted; countable.

enumerable set *Mathematics.* a countable set; denumerable set.

enumerate *Computer Programming.* to generate all of the members of a set.

enuresis [en′yə rē′sis] *Medicine.* an inability to control the passing of urine; involuntary urination, especially at night.

enuretic [en′yə ret′ik] *Medicine.* **1.** relating to or causing enuresis. **2.** a person affected with enuresis.

envelope *Engineering.* a glass or metal container that houses an incandescent lamp or a vacuum tube. *Aviation.* see FLIGHT ENVELOPE. *Robotics.* the maximum range of motion available to a robot in all directions. *Telecommunications.* **1.** a curve drawn to pass through the peaks of a graph of the modulating frequency waveform of a modulated radio-frequency carrier signal. **2.** in digital data transmissions, a group of binary digits that are required for transmission system operation, such as start and stop pulses. *Virology.* a lipoprotein membrane layer that forms the outermost layer of a virus particle.

envelope delay *Telecommunications.* **1.** in a given passband of a device or transmission facility, the maximum difference of the group delay time between any two specified frequencies. **2.** a distortion that occurs when the rate of change of phase shift with frequency of a circuit or system is not constant over the frequency range that is required for transmission.

envelope delay distortion see DELAY DISTORTION.

envelope of a family of curves *Mathematics.* a curve, every point of which is tangent to a member of a family of curves.

envelope soliton *Physics.* the curve containing the solution to a differential equation, typically a wave equation, that propagates through space and time while maintaining a constant shape.

envenomation *Toxicology.* poisoning by venom from the sting or bite of an animal such as an insect, spider, or snake. *Materials Science.* the deterioration of a plastic surface, usually caused by direct contact with another material.

environment *Ecology.* **1.** usually, **the environment.** the total of all the surrounding natural conditions that affect the existence of living organisms on earth, including air, water, soil, minerals, climate, and the organisms themselves. **2.** the local complex of such conditions that affects a particular organism and ultimately determines its physiology and survival. *Physics.* the surroundings of a physical system that may have some effect on the behavior of the system. *Engineering.* the combination of all external conditions that influence the performance of a device or process. *Computer Programming.* see EXECUTION ENVIRONMENT.

environmental *Ecology.* of or relating to the environment or to an organism's particular environment. *Medicine.* caused or influenced by conditions in the environment, as in the case of certain forms of cancer.

environmental acoustics *Acoustics.* **1.** the branch of acoustics that deals with the noise and vibrations produced by motor vehicles, aircraft, machinery, industrial equipment, and other such sources of environmental sound. **2.** any of various efforts to minimize or control such noise and vibrations.

environmental archaeology *Archaeology.* the study of the relationship of human activity to the environment and environmental change, as shown by the archaeological record.

environmental biology *Biology.* the study of the effects of the environment on plant and animal biology.

environmental cab *Engineering.* the section of an earthmover or tractor in which the operator works; it is designed for the operator's comfort, having air conditioning, soundproofing, tinted safety glass, and cleaning units.

environmental control *Engineering.* the control and alteration of the environment so that it can be inhabited safely by living organisms; this includes the modification and control of soil, air, and water.

environmental control system *Engineering.* a system designed to modify the environment of a closed area, such as a space vehicle, so that its occupants can live in it and work efficiently.

environmental cracking *Mechanics.* cracking due to chemical, thermal, radioactive, or other environmental stresses.

environmental engineering *Engineering.* any technological activity that works to reduce or prevent the pollution or degradation of areas in which humans live.

environmental gradient *Ecology.* the change in an environmental factor over a given variable, such as time, temperature, or distance.

environmental impact analysis *Engineering.* an analysis of the effect on the surrounding environment of a proposed facility or structure, now commonly required by law, especially for large facilities or those producing toxic or unpleasant side effects.

environmental impact report *Engineering.* a document that reports the probable effects that a proposed construction or other land use project will have in terms of economic, aesthetic, engineering, and environmental aspects. Also, **environmental impact statement.**

environmental lapse rate *Meteorology.* the rate of temperature decrease with elevation, determined by the distribution of temperature in the vertical at a given time and place.

environmental noise see BACKGROUND NOISE.

environmental pathology *Pathology.* the study of pathology resulting from abiotic environmental agents.

environmental protection *Engineering.* the defending of living organisms and structures against adverse environmental conditions, such as the stresses of extremes in climate.

environmental psychology *Psychology.* an area of psychology that studies the relationship between human behavior and the physical environment, as for the purpose of designing work or living areas. Thus, **environmental psychologist.**

environmental range *Engineering.* the scope of the environment that a system's operation can cover efficiently.

environmental resistance *Ecology.* the factors within a particular environment that prevent an organism from achieving its full biological potential; for example, spruce trees being dwarfed by the cold weather on the tundra.

environmental test *Engineering.* a practice run of an entire system or a part of a system under simulated conditions in a laboratory, in order to determine the effects that the actual environment may have on the system's performance.

environment division *Computer Programming.* in COBOL, one of the four main program segments; describes the equipment to be used and the relationships between data files and input/output devices.

environment of sedimentation *Geology.* the aggregate of all the surrounding conditions and forces affecting a locus of sedimentation.

environment pointer *Computer Programming.* **1.** in a task descriptor, a pointer to the code and data of the task. **2.** in a stack model of block structure execution, a pointer to the current environment.

environment simulator *Engineering.* an apparatus that creates a partial or total simulation of an environment.

enzootic *Veterinary Medicine.* describing a disease that is prevalent in and restricted to animals of a certain population or locality; corresponds to the term *endemic* in human medicine.

enzootic encephalitis see BORNA DISEASE.

enzymatic browning *Food Technology.* a darkening that occurs during the freezing of fruit that has not been properly treated; caused by the oxidation at low temperatures of enzymes known as catechol-tannin substrates.

enzyme [en′ zīm′] *Enzymology.* a protein molecule in a plant or animal that catalyzes specific metabolic reactions without itself being permanently altered or destroyed. (From a Greek term meaning "in yeast;" early studies of enzymes often involved reactions in yeast enzymes.)

enzyme classification *Enzymology.* the systematic naming and classification of enzymes based on the 1972 recommendations of the Nomenclature Committee of the International Union of Biochemistry. Each enzyme is assigned a number consisting of four figures; the first figure designates one of six main divisions: oxidoreductases, transferases, hydrolases, lyases, isomerases, and ligases; the second figure denotes the subclass, the third denotes the sub-subclass, and the fourth is the serial number of the enzyme in its sub-subclass; the enzyme number is preceded by the abbreviation EC.

enzyme fermentation *Enzymology.* an enzymatically controlled series of metabolic reactions that proceed in the absence of molecular oxygen, breaking down organic compounds in an organism and yielding energy to power its metabolic activities.

enzyme induction *Enzymology.* the initiation of synthesis of a given enzyme in response to either the presence of its inducer, the substrate, or the absence of a needed product.

enzyme inhibition *Enzymology.* any interaction of a substance with an enzyme that decreases the rate of the enzymatic reaction.

enzyme kinetics *Enzymology.* the mechanics by which the rate of a reaction is controlled by the relative concentrations of an enzyme and its substrate in a specific medium.

enzyme-linked immunosorbent assay see ELISA.

enzyme repression *Enzymology.* the reduction of the rate of synthesis of an enzyme by the presence of an end product of the enzymatic reaction. Also, ENDPRODUCT REPRESSION.

enzyme separation *Enzymology.* the separation of an enzyme from a mixture of enzymes using centrifugation, electrophoresis, dialysis, filtration, or other specialized methods.

enzyme unit *Enzymology.* the amount of an enzyme that will catalyze the transformation of 10^{-6} mole of substrate per minute under specified conditions.

enzymologist *Biology.* a scientist who specializes in the study of enzymology.

Enzymology

Enzymology may be defined as the study of the properties, mechanism, and function of enzymes. Enzymes are biological catalysts that greatly accelerate the rate of the myriad of chemical reactions that make life possible. Living organisms contain thousands of different enzymes of remarkable catalytic capacity and specificity.

Enzymes are proteins—large biopolymers containing hundreds of amino acid residues selected from some 20 different amino acids. The particular amino acid present at each position of the long polymer chains is determined by the genetic code of DNA. Amino acids have side groups with different properties, such as with or without affinity for water or with negative or positive charges present. These side groups determine how one or more polymer chains fold to form a compact active enzyme. A given enzyme has a unique 3-dimensional structure that serves for the required spatial placement of atoms and groups and for the conformational adaptability required for catalytic properties.

Catalysis occurs when substrates, the molecules that undergo chemical transformation, combine with a specific site on the enzyme. Many enzymes have bound cofactors as part of their active sites. Metals, such as zinc, iron, and copper, and most vitamins function as cofactors of enzymes. The active sites of enzymes are designed so that key parts of the substrates interact with particular amino acid and cofactor groups in a manner that promotes catalytic events. After products are formed and released, the enzymes can again combine with substrates and repeat the catalytic turnover thousands of times per minute.

An understanding of enzymology underlies much of biology and medicine. For example, the replacement of a single amino acid in an enzyme by genetic error can give inheritable diseases. The effects of many drugs reflect their interaction with enzymes. Complex metabolic patterns are made possible by the catalysis and control of enzymes.

Paul Boyer,
Professor of Biochemistry
University of California, Los Angeles

enzymology [en′zə mäl′ ə jē] *Biology.* the scientific study of the nature, activities, and effects of enzymes.

enzymolysis *Enzymology.* the breakdown of any cellular structure by enzyme activity.

enzymopathy *Enzymology.* any disease condition resulting from the activity of an enzyme.

enzymosis *Enzymology.* fermentation induced by an enzyme.

Eoanthropus dawsoni see PILTDOWN MAN.

EOB end of block.

Eocambrian *Geology.* a geologic time approximately equivalent to the most recent era of the Precambrian period, according to some scientists.

Eocanthocephala *Invertebrate Zoology.* a class of spiny-headed worms that are parasitic on fish, amphibians, and reptiles.

Eocene *Geology.* an epoch of the Tertiary period, occurring after the Paleocene and before the Oligocene.

Eocrinoidea *Paleontology.* an extinct class of pelmatozoan echinoderms in the subphylum Blastozoa, extant from the middle Cambrian to the Silurian; probably ancestral to all other blastozoans.

EOF end of file.

Eogene see PALEOGENE.

Eohippus see HYRACOTHERIUM.

EOJ end of job.

eolation *Geology.* any action of the wind that affects or modifies the land surface.

eolian [ē ø′ lē ən] *Meteorology.* of or relating to the action of the wind. *Geology.* deposited or caused by wind or currents of air. Also, AEOLIAN.

eolian anemometer *Engineering.* an instrument that measures wind speed according to the pitch of eolian tones as the air moves past an obstacle.

eolian deposit *Geology.* any wind-deposited accumulations, such as loess and dune sand.

eolianite *Geology.* a consolidated sedimentary rock consisting of clastic material deposited by the wind. Also, DUNE ROCK.

eolith [ē′ ə lith] *Archaeology.* a stone showing chips or fractures that resemble those made by early peoples in tool manufacture, but that are actually caused by natural forces.

Eomoropidae *Paleontology.* a primitive family of odd-toed chalicotheres in the suborder Ancylopoda, found in early Eocene strata in Asia and North America.

Eomyidae *Paleontology.* an extinct family of rodents in the infraorder Myomorpha and the superfamily Geomyoidea; arising in the late Eocene, the Eomyidae survived until the Quaternary.

eon *Metrology.* **1.** a unit of time equal to one billion years. **2.** in popular use, a very long, indefinite period of time. Also, AEON.

eonism *Psychology.* a man's adoption of women's dress or mannerisms; transvestitism. (Named for the Chevalier *d'Eon,* 1728–1810, a French spy who adopted women's clothing on a secret mission and was later sentenced to wear such apparel for life.)

EOP or **eop** efficiency of plating.

Eophrynus *Paleontology.* a genus of spiderlike arachnids in the order Trigonotarbida; Devonian to late Carboniferous.

EOQ economic order quantity.

EOR end of record; explosive ordnance reconnaissance.

Eosentomidae *Invertebrate Zoology.* a family of primitive wingless insects of the order Protura having no eyes and no antennae.

eosere *Ecology.* an ecological succession, especially the development of a particular type of vegetation, that took place during a certain geological era or period.

eosin *Organic Chemistry.* any of various rose-colored bromine derivatives of fluorescein used as a stain or dye, especially the sodium salt of tetrabromofluorescein, $C_{20}H_6Br_4Na_2O_5$.

eosinophil *Cell Biology.* any granular leukocyte that is readily stained by eosin dye, with a nucleus usually having two lobes connected by a slender thread of chromatin, and cytoplasm containing coarse, round granules that are uniform in size. Also, **eosinocyte, eosinophilic leukocyte.** *Medicine.* any structure, cell, or tissue that is readily stained by eosin. Also, **eosinophile.**

eosinophilia *Medicine.* **1.** an increase in the number of eosinophils per unit volume of blood. **2.** the occurrence of increased numbers of eosinophilic granulocytes in certain tissues and organs.

eosinophilia granuloma *Medicine.* a benign localized growth, marked by numerous leukocytes and histiocytes, that occurs in the bones and lungs, usually in individuals between twenty and forty years of age.

eosinophilic *Cell Biology.* **1.** readily stained with eosin. **2.** of or relating to eosinophils or eosinophilia.

eosphorite *Mineralogy.* $Mn^{+2}Al(PO_4)(OH)_2 \cdot H_2O$, a rose-pink to yellowish, fusible, monoclinic mineral occurring in prismatic crystals and radial aggregates, having a specific gravity of 3.05 to 3.06 and a hardness of 5 on the Mohs scale; found in granitic pegmatites with manganese phosphates.

Eosuchia *Paleontology.* an extinct order of primitive diapsid reptiles in the subclass Lepidosauria; the eosuchians arose in the Permian and flourished until the end of the Triassic, when several families died out and the two lepidosaurian orders that still survive, the squamatans and the rhynchocephalians, diverged from the eosuchians.

EOT end of tape.

Eötvös, Baron Roland [et′vush] 1848–1919, Hungarian physicist; established the proportionality of inertial and gravitational mass; invented the Eötvös balance.

Eötvös balance *Engineering.* a highly sensitive instrument that is used to record minute gravitational differences in subsurface rocks. Also, **Eötvös torsion balance.**

Eötvös constant *Physics.* a constant used in the Eötvös equation.

Eötvös correction *Geophysics.* a correction made in gravity measurement to compensate for centripetal acceleration caused by the earth's rotation.

Eötvös equation *Physics.* an equation showing that the molecular surface energy of a substance decreases linearly with temperature, becoming zero about 60°C below the critical point.

Eötvös experiment *Physics.* a test of the equality of inertial mass and gravitational mass, performed by balancing the earth's gravitational attraction on a given body against the kinetic reaction arising from its rotation.

Eötvös rule *Thermodynamics.* the principle that the rate of change of the surface tension of a liquid with respect to temperature is a constant; in practice there are exceptions to this.

Eötvös unit *Geophysics.* a unit of gravitational gradient of acceleration, equal to 10^{-9} gallon per horizontal centimeter.

-eous a combining form meaning "composed of" or "relating to," as in *cutaneous, osseous.*

EP see EXTRA PLAY; EXTENDED-PLAY.

ep- a prefix meaning "upon," "above," or "around," as in *epaxial.*

EPA or **E.P.A.** Environmental Protection Agency.

Epacridaceae *Botany.* a mainly Australian family of dicotyledonous, heathlike trees and shrubs in the order Ericales, characterized by alternate, narrow, palmately veined leaves, regular, bisexual flowers usually in racemes or spikes, a superior ovary, and stamens and corolla lobes in equal numbers.

epaulette [ep´ə let´] *Invertebrate Zoology.* **1.** the first scale-bearing hair at the base of the costal vein in the wing of flies. **2.** any one of the branched or knobbed processes on the oral arms of many jellyfish.

epaxial *Biology.* situated on or above an axis.

epaxial muscle *Anatomy.* a muscle of the trunk that is dorsal to the transverse processes of the vertebrae and the ribs.

epeiric sea *Oceanography.* a shallow inland sea having only a restricted connection to the ocean, such as Hudson Bay.

epeirogeny *Geology.* the wide-scale, primarily vertical, crustal deformations that have affected or produced the larger features of the continents and oceans. Also, **epeirogenesis.**

epencephalon see METENCEPHALON, def. 2.

ependopathy *Neurology.* any disease or disorder of the endyma.

ependyma *Histology.* Also, **ependymal layer.** the epithelial layer lining the cavities of the brain and spinal cord. Also, ENDYMA.

ependymocyte *Neurology.* a cell associated with the ependymal layer of the central canal and ventricles.

ependymoma *Oncology.* a usually benign, slow-growing neoplasm composed of differentiated ependymal cells.

Eperythrozoon *Microbiology.* a genus of Gram-negative, parasitic bacteria of the family Anaplasmataceae that occur in the bloodstreams of a number of animal hosts.

eperythrozoonosis *Medicine.* an infection caused by any species of organisms of the genus Eperythrozoon.

ephapse *Neurology.* a point of lateral contact between nerve fibers, across which impulses are conducted directly through the nerve membranes from one fiber to another.

ephaptic transmission *Neurology.* the electrical conduction of a nerve impulse across a point of contact between neurons without the mediation of a neurotransmitter.

Ephedra *Botany.* a genus of low, nearly leafless desert shrubs belonging to the order Ephedrales, including some species from which the drug ephedrine is derived. Also, JOINT FIRS.

Ephedraceae *Botany.* the single monogeneric family composing the subclass Ephedridae.

Ephedrales *Botany.* a monogeneric order of gymnosperms in the subdivision Gneticae, characterized by dense branching, jointed stems, and scalelike leaves; considered to have advanced features that are characteristic of the angiosperms.

Ephedridae *Botany.* in some classification systems, one of three subclasses of the gymnosperm subdivision Gneticae, characterized by low-spreading, broomlike evergreen shrubs, slender green branches with scalelike leaves, and staminate and pistillate cones in clusters of two or four at the stem nodes; native to arid and desert regions of North and South America, Europe, and Asia.

ephedrine *Organic Chemistry.* $C_{10}H_{15}NO$, toxic, hygroscopic, white to colorless granules or crystals; soluble in water and alcohol; the levorotary form melts at 33–40°C and decomposes at a boiling point of 255°C; used in medicine as a bronchodilator and vasoconstrictor.

Ephemeraceae *Botany.* a family of delicate mosses of the order Funariales, characterized by minute, short-lived plants that arise from an abundant perennial protonema, simple stems with terminal sporophytes, and few leaves; plants occur on moist, bare soil.

ephemeral plant [i fēm´ ə rəl; i fem´ə rəl] *Botany.* a plant that germinates and grows rapidly, with its short life cycle often repeating several times in a growing season, such as desert plants that respond quickly following a brief rain.

ephemeral stream *Hydrology.* a stream whose channel lies above the water table and that flows only during or immediately after periods of precipitation.

Ephemerida see EPHEMEROPTERA.

ephemeris *Astronomy.* **1.** a listing of a celestial body's positions by date. **2.** a publication containing an ephemeris.

ephemeris meridian *Horology.* a postulated meridian that rotates independently of the earth's rotation at the uniform rate established by terrestrial dynamical time.

ephemeris second *Horology.* a standard unit of time established in 1956 as 1/31,556,925.9747 of the tropical year for 1900. There are 86,400 ephemeris seconds in an **ephemeris day.**

ephemeris time *Horology.* a constant-rate time scale based on the interval between two consecutive passages of the sun across the vernal equinox (the tropical year), without compensating for irregularities in the earth's rotation.

Ephemeropsidaceae *Botany.* a monogeneric family of mosses of the order Hookeriales that occur epiphytically on small branches in rain forests; characterized by a lack of vegetative leaves, a persistent, green, highly branched protonema, greatly reduced gametophytes, and unique fusiform multicellular spores.

Ephemeroptera *Invertebrate Zoology.* the mayflies, a primitive order of winged insects in the subclass Pterygota that spend most of their lives as aquatic nymphs; the fragile, membranous-winged, noneating adults live just long enough to breed.

Ephippidae *Vertebrate Zoology.* the spadefishes, a family of inshore marine fishes of the suborder Percoidei, characterized by a deep and compressed body and a protrusible jaw with rows of brushlike teeth; often found in schools in reefs of the tropical and subtropical Atlantic, Pacific, and Indian Oceans.

Ephydridae *Invertebrate Zoology.* the brine flies and shore flies, a family of two-winged insects in the subsection Acalyptratae.

ephyra *Invertebrate Zoology.* a larval, free-swimming medusoid (jellyfish) stage of scyphozoans, budded off by asexual division from the strobilia. Also, **ephyrula.**

EPI electronic position indicator.

epi- **1.** a combining form meaning "upon," "above," or "around," as in *epicardium.* **2.** *Organic Chemistry.* a combining form designating an intramolecular bridge.

epiandrum *Invertebrate Zoology.* the genital opening in male arachnids.

epibasidium *Mycology.* in certain fungi belonging to the subclass Heterobasidiomycetidae, appendages that form from cells of the spore-bearing structure (the basidium) and bear the spores.

epibenthos *Biology.* of or relating to the plants and animals living just above the sea floor, to a depth of about 100 fathoms.

epibiont *Biology.* any organism that lives on the body surface of another without feeding upon its host, as in many mosses and lichens.

epibiotic *Biology.* of or relating to an organism, such as a fungus, that lives on the surface of another organism. Also, **epibiontic.**

epiblast see ECTODERM.

epiblem *Botany.* the epidermal layer of cells of most roots and of the stems of submerged aquatic plants; it aids in water retention and protects the underlying tissues from injury and disease. Also, RHIZODERMIS.

epiboly *Developmental Biology.* a method of gastrulation by which the smaller blastomeres at the animal pole of a fertilzed ovum grow over and enclose the cells of the vegetal hemisphere.

epicadmium *Nucleonics.* the energy level above which the absorption cross section for cadmium drops very rapidly with increasing energy.

epicalyx *Botany.* a ring of fused bracts, stipules, or episepals that grows from beneath the sepals to form a secondary calyx, as in the strawberry.

epicanthus *Anatomy.* a vertical fold of skin on either side of the nose, sometimes covering the inner canthus; primarily a characteristic of certain Mongoloid peoples. Also, **epicanthic fold.**

epicardium *Anatomy.* the layer of serous pericardium on the surface of the heart. *Invertebrate Zoology.* one of two paired tubular diverticula of the ascidian (tunicate) pharynx that grow and surround the digestive viscera. Also, VISCERAL PERICARDIUM.

Epicaridea *Invertebrate Zoology.* a suborder of crustaceans in the order Isopoda, whose female members are parasitic on various marine crustaceans, while the minute males usually live attached to the female.

epicarp *Botany.* the outer layer of the pericarp or fruit wall, especially when present as a skin. Also, EXOCARP.

epicellular *Cell Biology.* located at or on the surface of a cell.

epicenter *Geology.* the point on the earth's surface directly above the focus of an earthquake.

Epichloe *Mycology.* a genus of fungi belonging to the order Clavicipitales, which includes species parasitic and pathogenic to grasses.

epichlorohydrin *Organic Chemistry.* C_3H_5ClO, a toxic, carcinogenic, flammable, highly volatile, unstable liquid; slightly soluble in water and miscible with alcohol; boils at 115.2°C; used in epoxy and phenoxy resins, in glycerol manufacture, and as a solvent for cellulose esters.

epichordal *Vertebrate Zoology.* located upon or above the notochord; usually used in reference to vertebrae or elements of vertebrae on the dorsal side of the notochord.

epiclastic *Geology.* a rock or sediment formed at the surface by mechanical deposition of detrital material from preexisting rocks.

epicolous *Ecology.* describing an organism that lives on the surface of another without any effect on the host.

epicondylalgia *Medicine.* any pain in the muscles or tendons attached to the epicondyle of the humerus.

epicondyle *Anatomy.* a protrusion on the surface of a bone, above its condyle. Also, **epicondylus.**

epicondylitis *Medicine.* an inflammation or irriation of the muscles or tendons attached to the epicondyle of the humerus; e.g., "tennis elbow."

epicone *Invertebrate Zoology.* in dinoflagellates, the part anterior to the equatorial groove.

epicontinental *Geology.* located on the continental shelf.

epicontinental marginal sea *Oceanography.* a shallow part of the ocean that overlies a continental shelf and is partially enclosed by islands or by extensions of the mainland.

epicontinental sea *Oceanography.* a general term referring to either an epicontinental marginal sea or an epeiric sea.

epicotyl *Botany.* the apical portion of the embryonic plant stem above the cotyledons and below the first true leaves, which develops into the shoot.

epicranial aponeurosis *Anatomy.* a fibrous membrane covering the cranium, serving as an attachment for the occipital and frontal scalp muscles. Also, GALEA APONEUROTICA.

epicranium *Anatomy.* the entire scalp, including the integument, aponeurosis, and muscular expansions. *Entomology.* the dorsal wall (exoskeleton) of an insect's head above and behind the frons.

epicranius *Anatomy.* a layer of muscular and fibrous tissue covering the top and sides of the skull from the occipital bone to the eyebrows.

epicuticle *Entomology.* the outermost, hardened waxy layer of the insect exoskeleton.

epicycle *Mathematics.* a circle that rolls around the circumference of another circle; a fixed point on the moving circle traces out an epicycloid. *Astronomy.* in the Ptolemaic theory of the universe, the small circle centered on the deferent in which a planet moves.

epicyclic *Mechanical Engineering.* containing two or more circular parts or elements, the smaller of which moves about inside the larger. Thus, **epicyclic gear system, epicylic gear train.**

epicycloid *Mathematics.* a particular cyclic curve; the guiding curve L is a circle of radius b, and a circle K of radius a rolls along the outside of L without slipping. The epicycloid is the path traced by a fixed point on K; for $-\infty < \phi < \infty$, the parametric representation is given by

$$x = (a + b)\cos\phi - a\cos[(a + b)\phi/a] \quad y = (a + b)\sin\phi - a\sin[(a + b)\phi/a].$$

epicycloid

epicyst *Invertebrate Zoology.* the outer layer of a cyst wall in encysted protozoans.

epideictic display *Ecology.* a behavior pattern in which the members of a group collectively show themselves to an outsider, so as to display the size and territory of their group.

epidemic *Medicine.* **1.** an extensive outbreak of a disease affecting a large number of people at the same time, usually spreading rapidly. **2.** of a disease or other health-related event, occurring suddenly in numbers clearly in excess of what is to be expected.

epidemic diarrhea of the newborn *Medicine.* a contagious diarrhea occurring in epidemics among newborn infants in hospitals, resulting in high mortality.

epidemic encephalitis *Medicine.* any of various forms of viral encephalitis occurring epidemically, such as influenza encephalitis or Japanese encephalitis.

epidemic hepatitis or **epidemic jaundice** see HEPATITIS A.

epidemic jaundice virus see HEPATITIS A VIRUS.

epidemic keratoconjunctivitis *Medicine.* a highly infectious disease caused by an adenovirus, characterized by inflammation of the cornea and conjunctiva, and sometimes resulting in permanent impairment of vision.

epidemic neuromyasthenia *Medicine.* a disease resembling poliomyelitis, with fatigue, headache, and intense muscle pain, occurring in epidemics and thought to be viral in origin. Also, **epidemic vegetative neuritis.**

epidemic parotitis *Medicine.* a technical name for mumps. See MUMPS.

epidemic pleurodynia *Medicine.* an acute infectious viral disease, generally occurring in epidemics and mainly found among children, characterized by severe pain in the stomach or lower chest, fever, headache, sore throat, fatigue, and extreme muscle aches. Also, BORNHOLM DISEASE.

epidemic roseola see RUBELLA.

epidemic tremor see AVIAN (INFECTIOUS) ENCEPHALOMYELITIS.

epidemic typhus *Medicine.* the classic, severe form of typhus that is caused by *Rickettsia prowazekii* and transmitted by the body louse, *Pediculus humanus corporis.*

epidemiology [ep´ə dēm´ ē äl´ ə jē] *Medicine.* the study of the various factors influencing the occurrence, distribution, prevention, and control of disease, injury, and other health-related events in a defined human population. Thus, **epidemiologic, epidemiologist.**

epidermal *Anatomy.* of or relating to the epidermis. Also, **epidermic.**

epidermal growth factor *Endocrinology.* a polypeptide growth factor that stimulates and sustains proliferation of many cell types of ectodermal or endodermal origin; similar or possibly identical to urogastrone.

epidermatoplasty *Surgery.* a process of skin grafting in which patches of the epidermis are used.

epidermis [ep´i dur´ mis] *Anatomy.* the outer epithelial layer of the skin, consisting of a basal layer, a prickle-cell or spinous layer, a granular layer, a clear layer, and a horny layer. Also, **epiderm.** *Vertebrate Zoology.* the surface layer of skin in vertebrates. *Invertebrate Zoology.* the single, cutinized layer of ectoderm in many invertebrates. *Botany.* the outermost parenchymatous cell layer of the primary tissues (e.g., leaves, stems, and roots) that protects against water loss, injury, and disease.

epidermization *Surgery.* **1.** the process of covering or becoming covered with epidermal cells. **2.** a process of skin grafting.

epidermodysplasia verruciformis *Medicine.* a condition, usually inherited, in which numerous flat, reddish warts appear on the feet and hands. Also, LEWANDOWSKY-LUTZ DISEASE.

epidermoid *Anatomy.* of, relating to, or resembling the epidermis or epidermal cells. *Oncology.* a brain tumor formed by the inclusion of epidermal cells.

epidermoid cyst *Medicine.* a common, noncancerous swelling under the skin.

epidermoidoma *Oncology.* a benign tumor that develops in the diploic space or between the dura mater and the inner table of the skull.

epidermolysis *Medicine.* a loosened state of the epidermis leading to the easy separation of various layers of skin, occurring spontaneously or after trauma.

epidermolysis bullosa *Medicine.* a group of rare hereditary skin diseases in which blisters develop, usually at the sites of trauma.

epididymis *Anatomy.* an elongated coiled structure along the posterior surface of the testis that merges with the ductus deferens and provides for the storage, transmission, and maturation of spermatozoa.

epididymitis *Medicine.* an inflammation of the seminal duct lying along the posterior border of the testis.

Epidinium *Invertebrate Zoology.* a genus of complex ciliate protozoans belonging to the order Entodiniomorphida, found in the stomachs of ruminant mammals.

epidiorite *Petrology.* a gabbroic or dioritic rock in which pyroxene has been transformed into a fibrous hornblende by metamorphism.

epidote *Mineralogy.* $Ca_2(Fe^{+3},Al)_3(SiO_4)_3(OH)$, a brittle, green to black monoclinic mineral, occurring as prismatic crystals, in fibrous forms, or, more often, in massive and granular forms, having a specific gravity of 3.3 to 3.5 and a hardness of 6 to 7 on the Mohs scale; found in metamorphosed igneous and sedimentary rocks.

epidote-emphibolite facies *Petrology.* metamorphic rocks that are characterized by the assemblage of hornblende-albite-epidote formed under pressures of 3000–7000 bars and temperatures of 250–450°C; these conditions lie between those that characterize greenschist and amphibolite facies.

epidotization *Geology.* a hydrothermal process in which epidote is introduced into or is formed from rocks.

epidural [ep´i dür´əl] *Anatomy.* situated on or outside the dura mater. Thus, **epidural space.** *Medicine.* see EPIDURAL ANESTHESIA.

epidural anesthesia *Medicine.* anesthesia of of the pelvic, genital, or other area by the injection of a local anesthetic into the epidural space.

epidural hematoma *Medicine.* an accumulation of blood in the epidural space.

epieugeosyncline *Geology.* a nonvolcanic geosyncline formed after an orogenic event and filled with sediments derived from an uplifted eugeosyncline. Also, BACKDEEP.

epifauna *Ecology.* those living organisms that inhabit the surfaces of water and sediment. *Paleontology.* a collective term for benthic, generally intertidal organisms that formerly lived on the substrate of the sea floor, as opposed to the infauna that burrowed into the substrate. Thus, **epifaunal.**

epigaster *Developmental Biology.* the embryonic structure that eventually develops into the large intestine. Also, HINDGUT.

epigastric *Anatomy.* relating to or affecting the epigastric region.

epigastric incision *Surgery.* a vertical incision into the midline of the upper part of the abdomen.

epigastric region *Anatomy.* the upper middle region of the abdomen; the pit of the stomach. Also, **epigastrium.**

epigeal *Botany.* **1.** describing germination in which the cotyledons rise above the ground, forming the first leaves of the plant. **2.** describing a plant or a plant part that grows above the ground. *Entomology.* describing insects that live near or on the ground, as on low vegetation. Also, **epigean, epigeous.**

epigene *Geology.* formed or taking place on or near the earth's surface.

epigenesis *Developmental Biology.* the gradual development of an embryo in which structures do not preexist but rather arise de novo.

epigenetic *Geology.* **1.** formed or originating near the earth's surface. **2.** formed after the deposition of sediment or the formation of the enclosing rock.

epigenetics *Genetics.* the study of the mechanism that produces phenotypic effects from gene activity during differentiation and development.

epigenous *Botany.* growing on the surface, especially on the upper surface of a plant or plant part, as fungus on a leaf.

epiglottal *Anatomy.* relating to or affecting the epiglottis.

epiglottis *Anatomy.* a lidlike cartilaginous structure that overhangs the entrance to the larynx and folds over to prevent food from entering the larynx and the trachea while swallowing.

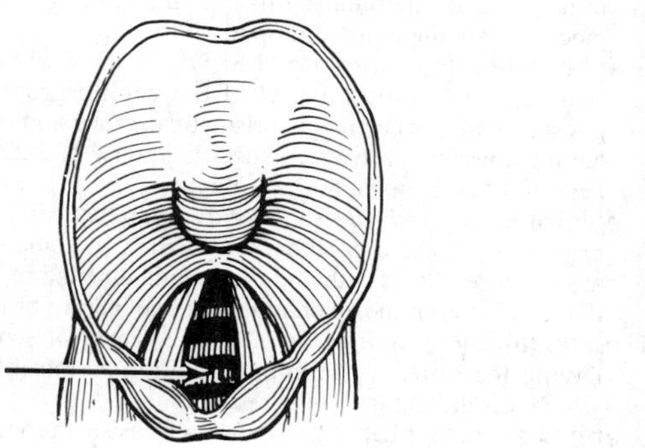

epiglottis

epigonal fold see INGUINAL FOLD.

epigraphy *Archaeology.* the study of ancient inscriptions on buildings, statuary, tablets, and other such objects. An expert in such studies is an **epigrapher** or **epighraphist.**

epigynous *Botany.* **1.** describing a flower whose calyx, corolla, and stamens grow above the inferior ovary, which is enclosed by the receptacle. **2.** having antheridia growing above the oogonium.

epigynum *Invertebrate Zoology.* a flap that covers the genital pore in female spiders. Also, **epigyne.**

epilation *Medicine.* the removal of hair roots.

epilepsy [ep´ə lep´sē] *Neurology.* any of several neurologic disorders characterized by an impairment or loss of consciousness, abnormal motor skills, and sensory disturbances; marked by recurrent, unpredictable seizures or fits and normally brought on by transient electrical disturbances of the brain. (Going back to a Greek phrase meaning "to seize upon.") Also, **epilepsia.**

epileptic [ep´ə lep´tik] *Neurology.* **1.** of or relating to the condition of epilepsy. **2.** a person who has epilepsy.

epileptic state *Neurology.* a state in which repeated epileptic convulsions are uninterrupted, resulting in prolonged unconsciousness; often severely damaging or fatal.

epileptiform *Neurology.* having characteristics of epilepsy, especially of an epileptic seizure.

epileptogenic *Neurology.* likely to produce severe or sudden spasms, such as those occurring in an epileptic seizure.

epileptoid *Neurology.* resembling epilepsy or its manifestations.

epileptologist *Neurology.* a medical practitioner who specializes in the study and treatment of epilepsy.

epileptology *Neurology.* the study of epilepsy.

epilimnion *Hydrology.* the relatively warm and uniformly mixed uppermost layer of water in a lake, especially the less dense, oxygen-rich upper layer of a thermally stratified lake.

epilithic *Ecology.* of or relating to organisms, such as lichens, that grow on stone or rock.

epiloia see TUBEROUS SCLEROSIS.

epimagma *Geology.* a relatively gas-free, vesicular, semisolid magmatic residue with a pasty consistency, usually formed by the cooling of lava.

epimastigote *Cell Biology.* a cell form occurring during certain stages of the life cycle of some parasites, such as trypanosomes.

epimer *Organic Chemistry.* either of two optical isomers that differ from each other only at the relative position of the hydrogen and the hydroxyl group on the last asymmetric carbon of the chain; e.g., D-glucose and D-mannose.

epimerase *Enzymology.* an enzyme that catalyzes a change in the distribution of hydroxyl groups on a substrate.

epimere *Anatomy.* the dorsal portion of the trunk mesoderm of chordates that forms the skeletal musculature.

epimerite *Cell Biology.* at the anterior of certain parasitic cells, a specialized region containing a septum that attaches to the host.

epimerization *Organic Chemistry.* the change of configuration in an optically active compound in which only one of several asymmetric centers is changed to an epimer.

epimeron *Invertebrate Zoology.* the posterior lateral part of a pleuron in cetain arthropods.

Epimetheus *Astronomy.* the eleventh moon of Saturn, discovered in 1980 and measuring 70 km by 60 km by 50 km; it shares its orbit with Saturn's tenth moon, Janus.

Epimorpha *Invertebrate Zoology.* a subclass of the centipede class Chilopoda in which females lay egg clusters in burrows and protect the eggs by coiling around them; the larvae have fully developed legs and segments.

epimorphosis *Physiology.* the regeneration of a part of an organism by rapid cell growth at the place of injury.

epimyocardium *Developmental Biology.* the meial mesodermal wall of the embryonic pericardial cavity that eventually surrounds the heart tube and forms the myocardium and epicardium of the heart. Also, MYOEPICARDIAL MANTLE.

epimysium *Histology.* the fibrous connective tissue sheath that surrounds an entire muscle.

epinasty *Botany.* **1.** leaf or stem growth in which the upper surface grows and causes a downward curvature of the structure, as in flowers to expose sexual organs. **2.** the downward bending of leaves caused by exposure to ethylene. **3.** the eccentric thickening of a horizontal root or branch.

epinephrine [ep´i nef´rən] *Endocrinology.* a catecholamine hormone that is secreted by the adrenal medulla and stimulates the sympathetic branch of the autonomic nervous system. *Pharmacology.* $C_9H_{13}NO_3$, a form of this substance, either prepared synthetically or extracted from the adrenal glands of sheep or cattle; used to stimulate heart and blood pressure and as a bronchodilator for acute asthma.

epineritic *Oceanography.* of or relating to the very top of the neritic zone, at and just below the surface, down to a depth of 40 meters.

epineural *Anatomy.* situated upon a neural arch of a vertebra. *Invertebrate Zoology.* **1.** in echinoderms, lying above the radial nerve. **2.** in arthropods, the nervous tissue dorsal to the ventral nerve cord.

epineurium *Histology.* the connective tissue covering a peripheral nerve.

epipelagic zone *Oceanography.* the upper zone of the ocean, down to about 200 meters, the maximum depth for photosynthesis.

epipelic *Ecology.* of or relating to organisms that inhabit the surfaces of water or sediment, or where water and sediment meet.

epipetalous *Botany.* having stamens attached to the petals, as in flowers with a tubular corolla.

epipharynx *Invertebrate Zoology.* a tasting organ attached to the roof of the mouth in many insects.

epiphenomenon *Science.* an accessory, exceptional, or accidental occurrence in the course of any event or process. *Medicine.* a secondary or additional symptom.

epiphloedal *Ecology.* of or relating to organisms that live on the surfaces of tree bark. Also, **epiphloedic.**

epiphora *Medicine.* an abnormal flow of tears down the cheek.

epiphragm *Invertebrate Zoology.* a film of calcified slime that temporarily closes the shell of hibernating land snails. *Botany.* **1.** a membrane covering the capsule aperture in mosses and bordered by the peristome teeth. **2.** any membrane covering the mouth of a deoperculate capsule. **3.** a membrane covering the sporophores in some fungi.

epiphyll *Botany.* a plant that sprouts on the surface of leaves, but is not parasitic. Thus, **epiphyllous.**

epiphyseal *Anatomy.* of, relating to, or resembling the ephiphysis. Also, **epiphysial.**

epiphyseal arch *Developmental Biology.* the embryonic structure, present in the roof of the third ventricle, from which the pineal body forms.

epiphyseal disk or **epiphyseal plate** *Anatomy.* **1.** a thin layer of cartilage lying between the epiphysis and newly forming tissue in the shaft of a growing long bone, and that is obliterated once growth is completed. Also, GROWTH DISK, GROWTH PLATE. **2.** the wide articular surface with a slightly elevated rim on either end of the centrum of a vertebra.

epiphyseal lines *Medicine.* the areas at each end of a long bone that appear when maturity of the bone is reached.

epiphysiolysis *Medicine.* the separation of an epiphysis from the shaft of a bone.

epiphysis *plural,* **epiphyses.** *Anatomy.* the end portion of a long bone that is separated from the shaft of the bone by a small area of cartilage during bone growth, and that is consolidated with the shaft when growth is complete.

epiphyte *Ecology.* **1.** a plant that uses another plant, usually a tree, for support but not for nourishment; most commonly found in tropical rain forests. **2.** a plant that grows naturally in this manner on nonliving structures, such as walls, chimneys, poles, and the like. Also, ECTOPHYTE.

epiphytotic *Plant Pathology.* of or relating to any plant disease that develops suddenly and sporadically and that causes widespread damage, in the manner of an epidemic.

epiplankton *Biology.* the part of ocean plankton occurring between the surface of the sea and a depth of about 100 fathoms (180 meters).

epiplasm *Mycology.* the cell matter or cytoplasm in fungi belonging to the subdivision Ascomycotina that is left in the spore sac (known as an ascus) after the ascospore has been formed.

epipleural *Anatomy.* located on a pleural element. *Vertebrate Zoology.* **1.** arising from or attached to a rib. **2.** a spine or bone arising from the rib of a fish and passing toward the lateral line.

epiploic foramen *Anatomy.* the opening from the larger into the smaller peritoneal cavity behind the portal fissure of the liver. Also, OMENTAL FORAMEN, FORAMEN OF WINSLOW.

epipodite *Invertebrate Zoology.* a respiratory process attached at the base of the thoracic limbs of many arthropods.

epipodium *Botany.* **1.** the apical portion of a leaf or of an embryonic phyllopodium. **2.** the stalk of the ovarian disk. *Invertebrate Zoology.* **1.** either of two paired lateral lobes of the foot in some gastropod mollusks. **2.** the elevated ring on an ambulacral plate in echinoderms.

epipolar plane *Cartography.* **1.** any plane that contains the epipoles. **2.** thus, any plane containing the air base.

epipolar ray *Cartography.* **1.** the line on the plane of a photograph joining the epipole and the image of an object. **2.** the trace of an epipolar plane on a photograph.

epipole *Cartography.* in the perspective projection of two photographic images onto a surface, either of the two points where the planes of the photographs are cut by the air base.

epiproct *Invertebrate Zoology.* a plate above the anus of certain insects representing the dorsal part of the tenth or eleventh abdominal segment.

epipubis *Vertebrate Zoology.* an unpaired cartilage or bone in front of the pubis in some amphibians and other vertebrates.

episclera *Anatomy.* the loose connective tissue that forms the external surface of the sclera.

episcleral *Anatomy.* **1.** overlying the episclera. **2.** of or relating to the episclera.

episepalous *Botany.* having stamens attached to the calyx.

episio- or **episi-** a combining form denoting a relationship to the vulva.

episiostenosis *Medicine.* the narrowing of the vulvar opening.

episiotomy [ə pēs´ē ät´ə mē] *Medicine.* a surgical procedure performed during childbirth in which the opening of the vagina is enlarged to avoid undue laceration.

episode *Geology.* **1.** a unit of geologic time during which the rocks of a substage were formed. Also, PHASE. **2.** any distinctive and significant event or series of events in the geologic history of a region or feature.

episodic memory *Psychology.* a memory of a specific experience that includes the time and location of the event as well as its nature.

episome *Genetics.* a genetic element that is found as part of the normal cellular chromosome or as a free element in bacteria.

epispadias *Medicine.* a birth defect in which the urethra opens on the dorsum of the penis, posterior to its normal opening.

epispinal *Anatomy.* situated upon the spinal cord or spinal column.

epi spiral *Mathematics.* one of the family of curves with polar coordinate equation $r = a \sec n\theta$, where n is an integer and a is real.

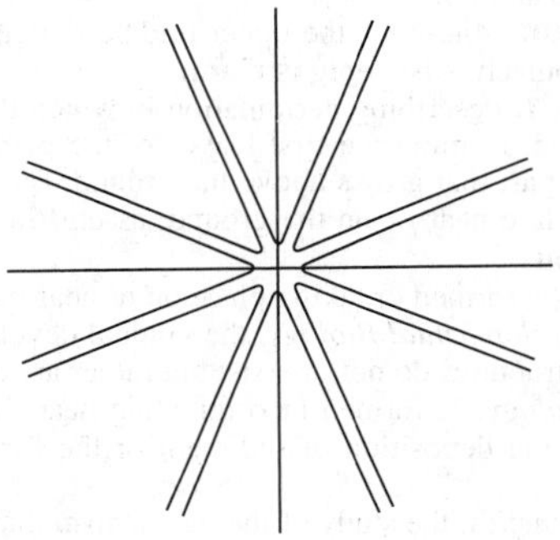

epi spiral

epistasis *Genetics.* the interaction of nonallelic genes in which an allele of one gene masks the phenotypic expression of the allelic alternatives of another gene. Thus, **epistatic.**

epistaxis *Medicine.* a nosebleed or nasal hemorrhage.

episternum *Vertebrate Zoology.* the upper part of the sternum or breastbone in mammals. *Invertebrate Zoology.* a lateral division or piece of a somite of an arthropod.

epistilbite *Mineralogy.* $CaAl_2Si_6O_{16}{\cdot}5H_2O$, a white to colorless, easily fusible, monoclinic mineral of the zeolite group, dimorphous with goosecreekite, occurring as prismatic crystals and in granular form, and having a perfect cleavage, a specific gravity of 2.22 to 2.28, and a hardness of 4 to 4.5 on the Mohs scale.

epistome *Invertebrate Zoology.* **1.** a flap that covers the mouth of some bryozoans. **2.** the area above the labrum in some dipteran flies. **3.** the area between the mouth and the second antenna in some crustaceans. **4.** the area between the epicranium and the labrum of many insects.

epitaxial layer *Solid-State Physics.* a layer of semiconductor material having the same crystalline orientation as that of the substrate upon which it is grown.

epitaxial transistor *Electronics.* a transistor produced by growing one or more layers of single-crystal material on top of a wafer substrate of the same material.

epitaxy *Crystallography.* the oriented overgrowth of one crystalline material upon the surface of another. There is often an approximate agreement in lattice spacings in the two components. Thus, **epitaxial.** Also, **epitaxis.**

epithalamus *Anatomy.* a portion of the diencephalon that contains the habenular nuclei, the stria media, and the pineal body.

epitheca *Invertebrate Zoology.* **1.** an external calcareous layer around the basal portion of the theca of many corals. **2.** the outer or upper valve of a diatom cell wall.

epithecium *Mycology.* in fungi, a layer of tissue over the spore sacs.

epithelia *Histology.* the plural form of *epithelium.*

epithelial [ep´ə thē´lē əl] *Histology.* of, relating to, or composed of epithelium or epithelial cells.

epithelial cells *Cell Biology.* cells that form the barrier between an organism and its external environment. These cells either may be involved in the secretion or uptake of liquids or nutrients, or may perform a simple barrier function.

epithelialization *Surgery.* the process of healing by the growth of epithelial cells over an area. Also, **epithelization.**

epitheliochorial placenta *Developmental Biology.* a placenta in which the uterine epithelial lining has not eroded but simply lies in apposition to the chorion.

epithelioid cell *Histology.* a macrophage that superficially resembles an epithelial cell.

epithelioma [ep´ə thē´lē ō´mə] *Medicine.* **1.** a tumor, either benign or malignant, derived from epithelium. **2.** in popular use, a skin cancer.

epitheliomuscular cell *Invertebrate Zoology.* a cell with an elongated base that contains contractile fibrils, common in the epidermis of many coelenterates and nematodes. Also, MUSCULO-EPITHELIUM.

epithelium [ep´ə thē´lē əm] *Histology.* an animal tissue composed of cells that are packed tightly together, with little intercellular matrix; it covers the external surface of the body and also internal surfaces such as the lining of tracts and vessels.

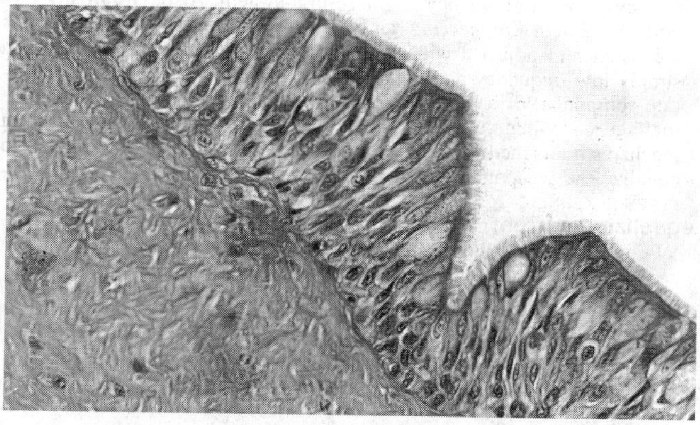

frog epithelium

epithema *Vertebrate Zoology.* a horny excresence on the bill of some birds, as in the casque of a hornbill.

Epithemiaceae *Botany.* a small family of freshwater, brackish, and terrestrial diatoms of the order Pennales, characterized by an asymmetrical apical axis and strongly developed tranverse internal costae.

epithermal *Thermodynamics.* slightly above a given thermal range.

epithermal deposit *Geology.* a hydrothermal mineral deposit formed by deposition from ascending hot solutions (at temperatures between 50°C and 200°C) in and along openings in rocks within 3000 feet of the earth's surface. Similarly, **epithermal vein.**

epithermal neutron *Nucleonics.* a neutron that has an energy level just above the thermal range, between about .02 and 100 electronvolts.

epithermal reactor *Nucleonics.* a reactor in a significant fraction of fissions that are caused by epithermal neutrons.

epithermal thorium reactor *Nucleonics.* a nuclear reactor in which epithermal neutrons, moderated by graphite or beryllium, unleash the energy contained in a uranium-thorium fuel mixture.

epithet *Systematics.* the second word in a binomen, or the second and third words in a trinomen, that identify a species or subspecies within a genus; in binomial nomenclature the epithet usually begins with a lowercase letter and is italicized.

Epithyris *Paleontology.* an extinct genus of articulate brachiopods in the order Terebratulida, widespread in the Jurassic.

epitoke *Invertebrate Zoology.* the rear portion of a marine polychaete worm in its reproductive state, swollen with eggs or sperm.

epitoky *Invertebrate Zoology.* the process by which marine polychaete worms become reproductive, shown by a seasonal modification in their gamete-bearing segments differing markedly from the usual nonsexual form.

epitonic *Neurology.* exhibiting an abnormally high degree of muscular tension or tone; tense.

epitope *Immunology.* the area of an antigenic molecule that determines the specific antibody to which the antigen binds.

epitreptic behavior *Behavior.* behavior toward another animal of the same species that tends to cause that animal to approach.

epitrichium *Developmental Biology.* the large-celled outer layer of the bilaminar fetal epidermis of mammals. Also, PERIDERM.

epitrochlear *Anatomy.* of or relating to the medial condyle of the humerus.

epitrochoid *Mathematics.* a generalization of an epicycloid in which the tracing point may be fixed anywhere on the radius of the rolling circle or its extension instead of exactly on the circumference.

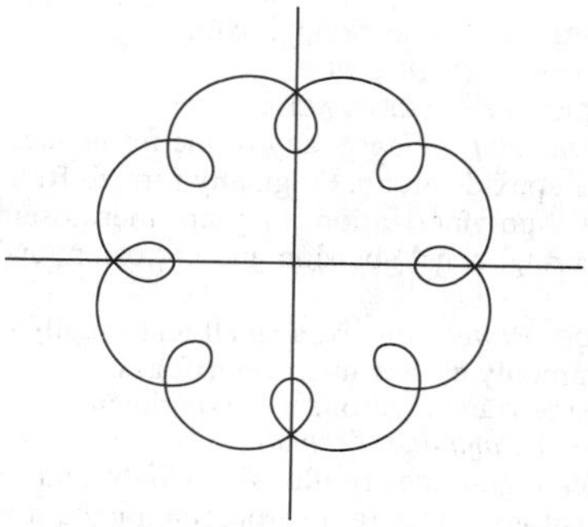

epitrochoid

epituberculosis *Medicine.* a prominent pulmonary shadow seen in X-ray films in active juvenile tuberculosis.

epitympanum *Anatomy.* the area of the middle ear above the level of the tympanic membrane that contains the head of the malleus and the body of the incus.

epitype *Immunology.* a family composed of epitopes having similar restrictive determinants.

epivalve *Invertebrate Zoology.* **1.** the upper or apical shell of certain dinoflagellates. **2.** the upper valve (epitheca) of a diatom.

epixylous *Ecology.* of or relating to an organism that grows on wood. Also, **epixylic.**

epizone *Geology.* in Grubenmann's classification of metamorphic rocks, the uppermost depth zone of metamorphism, characterized by moderate temperatures, low hydrostatic pressure, and powerful stress.

epizootic *Veterinary Medicine.* of or relating to a rapidly spreading disease affecting a large number of animals throughout a wide area; corresponds to the term *epidemic* in human medicine.

epizootic lymphangitis *Veterinary Medicine.* a chronic contagious disease of the family Equidae, also sometimes occurring in cattle but rarely in humans; characterized by thickening of a lymphatic vessel, swelling of the lymph nodes, ulcers of the mucous membranes of the nose and vulva or scrotum, and pneumonia; caused by *Histoplasma farciminosum,* which is transmitted by flies or discharges of diseased animals and enters the animal through a wound. Also, AFRICAN GLANDERS; LYMPHANIGITIS EPIZOOTICA.

epizootiology *Veterinary Medicine.* a science that deals with the frequency, distribution, cause, and control of disease in animals.

E-plane antenna *Electromagnetism.* an antenna whose radiated electric field lies in a plane parallel to the plane of the antenna.

E-plane T-junction *Electromagnetism.* a waveguide T-junction whose structure changes in the plane of the electric field. Also, SERIES T-JUNCTION.

epoch *Geology.* a unit of geologic time, longer than an age and representing a subdivision of a period during which the rocks of a particular series were formed. *Physics.* the initial set of conditions of a system, such as the initial phase angle of a simple harmonic oscillator.

eponychium *Developmental Biology.* a condensed, eleidin-rich area of the epidermis associated with the base of the nail in embryonic and adult mammals. *Anatomy.* see CUTICLE, def. 1.

eponym *Science.* **1.** a name or phrase formed from a person's name, such as *einsteinium, Addison's disease,* or the *Copernican system.* **2.** the person from whose name such a name or phrase was formed.

eponymous *Science.* **1.** named for some person. **2.** of or relating to an eponym. Also, **eponymic.**

epoöphoron [ep´ō äf´ə rän] *Anatomy.* a group of rudimentary tubules consisting of a blind longitudinal duct (Gartner's duct) and ten to fifteen transverse ductules, located in the mesosalpinx between the ovary and the oviduct; it is the homologue of the head of the epididymis in the male and the remnant of the reproductive part of the mesonephros in the female. Also, ORGAN OF ROSENMÜLLER.

epoxidation *Organic Chemistry.* the addition of an oxygen bridge across an alkene bond to yield an epoxy compound.

epoxide *Organic Chemistry.* **1.** an organic compound with a reactive group consisting of an oxygen bonded to two carbons that are bonded together. **2.** a cyclic ether of three members. Also, OXIRANE.

epoxy- *Organic Chemistry.* a prefix designating an epoxide group in a molecule.

epoxy adhesive *Materials.* an adhesive compound made from an epoxy resin.

1,2-epoxybutane see 1,2-BUTYLENE OXIDE.

epoxyethane see ETHYLENE OXIDE.

epoxy-propionic see GLYCIDIC ACID.

epoxy resin *Materials Science.* any of the resins that are based on the reactivity of the epoxide group. Originally formed from epichlorohydrin and bisphenol A polymerization, they are thermosetting, exhibit low shrinkage on curing, resist abrasion and corrosion, and are good adhesives.

Eppendorf tube *Biotechnology.* a small and usually disposable plastic tube that is commonly used in microcentrifuges.

EPR electron paramagnetic resonance; experiment.

EPROM [ē´präm´] *Computer Technology.* a read-only memory that can be erased under high-intensity ultraviolet light and reprogrammed repeatedly. (An acronym for erasable programmable read-only memory.) Also, ERASABLE READ-ONLY MEMORY, EROM.

epsilon the fifth letter of the Greek alphabet, written as E or ε.

epsilon factor *Crystallography.* in computing normalized structure factors, a factor that takes into account the fact that, depending on which of the 32 crystal classes the crystal belongs to, there will be certain groups of reflections in areas of the reciprocal lattice that will have an average intensity greater than that for the general reflections.

epsilon meson *Particle Physics.* a neutral scalar meson resonance with positive charge conjugation parity and an approximate mass of 1200 MeV.

epsilon neighorhood *Mathematics.* for a given point p in a metric space and a positive real number ε, the open ball of radius ε centered at p. On the real line it is the interval $(p - \varepsilon, p + \varepsilon)$. The term epsilon neighborhood (or **ε-neighborhood**) is often used to emphasize that ε is arbitrarily small.

epsilon structure *Solid-State Physics.* the hexagonal close-packed structure for the epsilon phase of an electron compound.

epsomite *Mineralogy.* $MgSO_4 \cdot 7H_2O$, a white or colorless, transparent, very water-soluble orthorhombic mineral usually occurring in botryoidal masses and fibrous crusts, having a vitreous to earthy luster, a bitter and saline taste, a specific gravity of 1.677, and a hardness of 2 to 2.5 on the Mohs scale; found in limestone caverns and mine workings.

Epsom salt(s) *Pharmacology.* epsomite (hydrated magnesium sulfate) crystals, used as an anticonvulsant, cathartic, and local anti-inflammatory; also used in leather tanning, fabric dyeing, and fertilizers. (Named for *Epsom,* England, where it occurs naturally in mineral waters.)

Epstein-Barr virus *Virology.* a virus of the family Herpesviridae that infects humans and is transmitted horizontally, mainly by saliva, causing infectious mononucleosis; also associated with nasopharyngeal carcinoma and Burkitt's lymphoma. (Named for Michael Anthony *Epstein,* British physician, and Y. M. *Barr,* British virologist.)

epulis *Medicine.* a generic term applied to any benign tumor or growth of the gingiva.

epulosis see CICATRIZATION.

eq equal, equation.

equal *Mathematics.* referring to two or more quantities that have the same value or that can be considered the same within a given context.

equal-area latitude see AUTHALIC LATITUDE.

equal-area projection *Cartography.* any map projection with a constant area scale; the resulting map is not conformal and therefore cannot be used for navigation.

equal-areas law *Astronomy.* the second of Johannes Kepler's three laws of planetary motion, which states that an imaginary line drawn between the sun and a planet will sweep across identical areas in equal periods of time.

equal-arm balance *Mechanics.* a simple balance with its beam supported in the center so that the pans contain exactly equal weights when balanced.

equal-energy source *Physics.* a source of acoustic or electromagnetic radiation in which all frequencies are emitted at equal energy levels.

equaling file *Mechanical Devices.* a blunt, almost parallel, slightly bulging double-cut rectangular file used for making fine tools.

equality *Mathematics.* one of the primitive or undefined notions in the Gödel-Bernays formulation of set theory, giving the relationship between two quantities that have the same value or that can be considered to be the same in the given context. Equality is an equivalence relation; i.e., it is reflexive, symmetric, and transitive.

equality gate see EXCLUSIVE-NOR GATE.

equalization *Electronics.* **1.** any of various processes by which all the frequency responses of elements in a circuit, such as those in the line or amplifier, are brought within a desired range of frequency response. Also, FREQUENCY-RESPONSE EQUALIZATION. **2.** in data transmission, a process by which frequency or phase distortion is reduced by introducing networks to compensate for the difference in attenuation and time delay at various frequencies in the transmission band.

equalizer *Mechanical Engineering.* any of various devices designed to distribute weight, pressures, strains, or other forces equally among the various parts of a machine, vehicle, or system; specific examples include the **equalizer brake** linking independent brakes on a motor vehicle or the **equalizer bar** joining axle springs on a railcar. *Electronics.* a device, usually a network of coils, capacitors, and resistors, that achieves equalization by compensating for extremes in amplitude-frequency or phase-frequency response in a system. Also, **equalizing circuit.** *Acoustical Engineering.* an element of a sound system that allows the listener to adjust a wide range of specific frequencies rather than simply low frequency and high frequency in general; it may be a separate component or a built-in feature of a receiver. *Forestry.* a machine that saws wooden stock to equal lengths. *Ordnance.* a compensating mechanism attached to the carriage of certain artillery; it is designed to equalize the weapon's weight and firing shock. *Mathematics.* see DIFFERENCE KERNEL.

equalizing current *Electricity.* a current that flows between two parallel-converted compound generators to equalize their output.

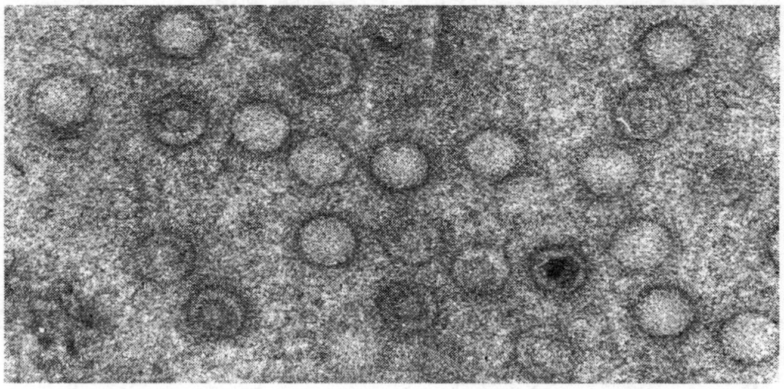

Epstein-Barr virus

equalizing pulses *Electronics.* a series of electrical pulses that precede and follow vertical synchronizing pulses in a television system, so that various elements of the picture will be aligned correctly.

equal-listener response scale *Acoustics.* a sound scale of measurement based on units such as sones, for which the loudness of a sound is estimated by averaging the threshold of response of various listeners.

equal-loudness contour see FLETCHER-MUNSON CURVE.

equally tempered scale *Acoustics.* a musical scale in which the intervals between notes are specific intervals, with a 9/8 ratio between the higher and lower tone for a whole step and a 256/243 ratio for half steps.

equal root see DOUBLE ROOT.

equal sets *Mathematics.* sets with exactly the same members. The axiom of extensionality asserts that two classes with the same elements are equal.

equal tails curve see SYMMETRIC DISTRIBUTION.

equal-zero indicator *Computer Programming.* an internal indicator that is set when an arithmetic operation's result is zero.

equate *Mathematics.* to form a mathematical statement that two expressions are equal to one another.

equation *Mathematics.* a mathematical statement of equality between two expressions T_1 and T_2; denoted $T_1 = T_2$. T_1 and T_2 are called the **left side** and **right side of the equation,** respectively, and = is called the **equal sign.** *Chemistry.* a symbolic expression of a chemical reaction that indicates the relative amounts of reactants consumed and products formed, and thus the cumulative chemical changes. In the equation $2NaOH + H_2SO_4 = Na_2SO_4 + 2H_2O$, for example, 2NaOH and H_2SO_4 are the reacting substances, and Na_2SO_4 and $2H_2O$ are the products; the = sign is read as "produces," "yields," or "gives."

equation clock *Horology.* a clock that shows the equation of time, or the difference between the mean solar time and the apparent solar time.

equation of motion *Quantum Mechanics.* one of the set of differential equations describing the evolution of a wavefunction in time. *Mechanics.* any equation that describes the motion of a body or system.

equation of piezotropy *Thermodynamics.* an equation that is applicable to certain fluids, stating that the derivative of the density with respect to time is equal to the pressure derivative with respect to time, multiplied by a function of the thermodynamic variables.

equation of state *Physical Chemistry.* an expression that describes the state of a substance in terms of the relationship of the basic physical quantities of volume, pressure, and temperature for a given mass; used for pure substances. Also applied to the relationships among other thermodynamic variables.

equation of the center *Astronomy.* the difference between an object's actual position in an elliptical orbit and the position the body would have if its angular motion were uniform.

equation of time *Horology.* the difference, due to the earth's elliptical orbit and variable orbital velocity, between apparent solar time and mean solar time, ranging from minus 14 minutes in February to over 16 minutes in November.

equation solver *Computer Technology.* a calculating device or program that solves systems of simultaneous equations, such as linear, nonlinear, or differential equations, and finds roots of polynomials.

equator *Astronomy.* an imaginary line encircling the surface of a celestial body, midway between the poles and in a plane perpendicular to the axis of rotation. *Cartography.* also, **Equator.** the great circle of the earth, midway between the North Pole and the South Pole, forming the line of 0° latitude. *Anatomy.* an imaginary line on a spherical organ, dividing its surface into two approximately equal parts.

equatorial [ēk´wə tôr´ē əl; ek´wə tôr´ē əl] *Astronomy.* of or relating to an equator. *Geography.* of, relating to, or characteristic of the regions at the earth's equator. Thus, **equatorial temperature.**

equatorial acceleration *Astrophysics.* the value of an object's gravitation acceleration as measured at its equator.

equatorial air *Meteorology.* the air of the doldrums or of the equatorial trough.

equatorial axis *Geodesy.* the diameter of the earth described between two points on the equator.

equatorial bulge *Geodesy.* the excess of the earth's equatorial diameter over its polar diameter.

equatorial calms see DOLDRUMS.

equatorial convergence zone see INTERTROPICAL CONVERGENCE ZONE.

equatorial coordinate system *Astronomy.* a commonly used celestial coordinate system that locates objects by right ascension and declination.

Equatorial Countercurrent *Oceanography.* a current flowing eastward near the equator in all oceans, generally between the westward-flowing North and South Equatorial Currents.

Equatorial Current *Oceanography.* either of two currents, the North Equatorial Current or the South Equatorial Current.

equatorial dry zone *Meteorology.* an arid region of the equatorial trough, such as the area just south of the equator in the central part of the equatorial Pacific.

equatorial easterlies *Meteorology.* the very deep trade winds in the summer hemisphere that extend to at least 8–10 km in height. Also, DEEP EASTERLIES, DEEP TRADES.

equatorial electrojet *Geophysics.* a narrow belt of intense electric current found over the magnetic equator.

equatorial front see INTERTROPICAL FRONT.

equatorial gravity value *Geodesy.* the mean acceleration of gravity at the equator, approximately equal to 978.03 cm/sec^2.

equatorial mounting *Astronomy.* the mounting of a structural base designed to support an equatorial telescope, having a polar axis that is parallel to the earth's axis and a perpendicular declination axis that is supported by the polar axis.

equatorial orbit *Astronomy.* an orbit that lies in an object's equatorial plane.

equatorial plane *Mechanics.* the horizontal symmetry plane, if one exists, through a body revolving about a vertical axis of symmetry. *Astronomy.* the plane containing a celestial object's equator. *Cell Biology.* the location of the chromosomes during the metaphase stage of mitosis, when they are aligned in the center of the mitotic spindle with their kinetochores equidistant from the spindle poles and attached to spindle fibers. Also, **equatorial plate.**

equatorial projection *Cartography.* a map projection centered on the equator.

equatorial radius *Geodesy.* the radius assigned to the great circle comprising the terrestrial equator.

equatorial telescope *Astronomy.* an astronomical telescope that rotates from an axis parallel to the earth's axis and automatically maintains a star on which it remains fixed in its field of view.

equatorial tidal currents *Oceanography.* nearly semidiurnal tidal currents that occur approximately every two weeks when the moon is over the equator.

equatorial tide *Oceanography.* a tide that occurs approximately every two weeks when the moon is over the equator, characterized by minimal diurnal inequality.

equatorial trough *Meteorology.* **1.** a zone of low pressure lying between the subtropical high-pressure belts of the Northern and Southern Hemispheres. **2.** see METEOROLOGICAL EQUATOR.

Equatorial Undercurrent see CROMWELL CURRENT.

equatorial vortex *Meteorology.* a closed cyclonic circulation that extends across the equatorial trough, developing from an equatorial wave.

equatorial wave *Meteorology.* a wavelike disturbance of the equatorial easterlies that extends across the equatorial trough; occurring frequently over the western Pacific Ocean.

equatorial westerlies *Meteorology.* the westerly winds found in the equatorial trough, separated from the midlatitude westerlies by a broad belt of easterly trade winds.

equi- a combining form meaning "equal," as in *equiangular.*

equiangular *Mathematics.* having all angles equal.

equiangular hyperbola see RECTANGULAR HYPERBOLA.

equiangular polygon *Mathematics.* a polygon, all of whose interior angles are equal.

equiangular spiral see LOGARITHMIC SPIRAL.

equiangular spiral antenna *Electromagnetism.* a frequency-independent broadband antenna that radiates a wide circularly polarized beam on either side of its surface.

equiaxed structure *Materials Science.* a grain shape having approximately equal dimensions in the three coordinate directions.

equiaxial *Mathematics.* having three axes of the same length. Also, **equiaxied.**

equicontinuous family of functions *Mathematics.* a collection F of functions (f_α) defined on a set S with the property that for every $\varepsilon > 0$ there exists a $\delta > 0$ depending only on ε such that if x_1 and x_2 are points in S with $|x_1 - x_2| < \delta$, then $|f_\alpha(x_1) - f_\alpha(x_2)| < \varepsilon$ for all functions f_α in F.

Equidae [ēk´wə dē] *Vertebrate Zoology.* a monogeneric family of perissodactyl ungulate mammals consisting of the horses, asses, zebras, and various related extinct animals such as *Hyracotherium.*

equidensity technique *Analytical Chemistry.* the measurement of photographic film emulsion density used in interference microscopy.

equidistant *Mathematics.* in a metric space, the property of being the same distance from a given reference point, line, or other object.

equigranular *Petrology.* of or relating to the texture of rocks whose essential minerals are all of the same order of size.

equi-inclinational technique *Crystallography.* any of various X-ray diffraction camera techniques, as in the equi-inclination Weisenberg camera, in which the incident beam is inclined at a selected angle to the axis of rotation of the crystal instead of being normal to it (as is usually the case).

equilateral *Mathematics.* having sides that are equal or congruent.

equilateral polygon *Mathematics.* a polygon having sides that are all of equal length.
equilateral polyhedron *Mathematics.* a polyhedron, all of whose faces are congruent; i.e., one of the five Platonic solids.
equilibrant *Mechanics.* a single force that is able to act in an equal and opposite way to a set of forces to bring a system into equilibrium.
equilibration *Psychology.* as described by Jean Piaget, the operation together of the two processes of assimilating new information and accommodating to this information.
equilibrator *Ordnance.* a mechanism that produces a balancing force around the trunnions of a gun carriage, facilitating elevation of the gun.
equilibristat *Engineering.* an instrument that measures the degree to which a railroad car deviates from equilibrium as it rounds a curve.
equilibrium a state or condition of balance or equality; specific uses include: *Physics.* the state of any system in which opposing forces balance each other. *Chemistry.* a reversible reaction in which the rates of forward and reverse reactions are equal and the concentrations of reactants and products are unaltered. *Mechanics.* the state of a system in which each constituent particle experiences a net force of zero. This requires two conditions: the system either is experiencing no linear motion at all or is moving in a straight line at a constant speed; the system either is experiencing no rotational motion at all or is rotating about a fixed axis at a constant rate. *Physiology.* the state or condition in which the amount of material taken into the body equals the amount discharged. *Psychology.* mental or emotional balance; equanimity.
equilibrium brightness *Electronics.* the type of brightness that appears on a cathode-ray tube when the pattern on the tube corresponds to the level of energy entering the tube.
equilibrium constant *Chemistry.* symbol K^0, a number expressing the relationship between the concentrations of substances in a reversible chemical reaction at a given temperature, equal to the product of the molar concentrations of the substances on the right side of the equation divided by the product of the molar concentrations of the substances on the left side; each concentration raised to the power of its coefficient in the equation for the reaction.
equilibrium diagram *Physical Chemistry.* a graph that demonstrates the relationship among temperature, pressure, and vapor and liquid compositions for a substance.
equilibrium dialysis *Analytical Chemistry.* a measurement of the degree of ion bonding of a protein; performed by calculating the diffusion of ions across a barrier impermeable to the protein.
equilibrium flash vaporization *Chemical Engineering.* a procedure in which a continuous stream of liquid feed is partially vaporized in a vessel or column, while vapor and liquid portions are continuously withdrawn, creating the vapor-liquid equilibrium. Also, SIMPLE CONTINUOUS DISTILLATION, CONTINUOUS EQUILIBRIUM VAPORIZATION.
equilibrium gas saturation *Petroleum Engineering.* in a reservoir, the state of zero relative permeability of a wetting/nonwetting phase system; the relation of the nonwetting phase, such as oil, to the wetting phase, such as water, when the saturation of the nonwetting phase is so minute that few pores contain it. Also, CRITICAL GAS SATURATION.
equilibrium isocline *Ecology.* a line indicating that the growth rate of a population is zero on a population graph showing combinations of competing populations.
equilibrium line *Hydrology.* the boundary between the accumulation area and the ablation area on a glacier.
equilibrium melting temperature *Materials Science.* the temperature at which a material melts under infinitely slow heating rates.
equilibrium moisture content *Physical Chemistry.* the point at which there is no further moisture loss when solid material is dried by air currents whose temperature and moisture content remain constant.
equilibrium orbit *Nucleonics.* a path along which particles in an accelerator oscillate, such as a ring in a synchrotron or a point moving through space in a linear accelerator. Also, STABLE ORBIT.
equilibrium point *Mechanics.* a point at which a particle will tend to remain, or to which it will tend to return, in order to achieve a condition of stable equilibrium; e.g., a ball placed inside a large rounded bowl will remain at the bottom if placed there and roll to the bottom if placed on the sides; the bottom of the bowl is the equilibrium point.
equilibrium potential *Physical Chemistry.* the reversible potential at which no faradaic processes occur.
equilibrium prism *Physical Chemistry.* a model that displays the effects of composition changes on some essential property of a multicomponent mixture, such as the temperature at which a solid begins to decompose.
equilibrium profile see PROFILE OF EQUILIBRIUM.
equilibrium ratio *Physical Chemistry.* the relationship between each phase of a substance that undergoes change in a closed system, at the point where no further change occurs.
equilibrium solar tide *Geophysics.* the form the atmosphere would take if determined solely by gravitational forces and discounting the earth's rotation relative to the sun.
equilibrium solubility *Physical Chemistry.* the maximum quantity of one substance that can be dissolved into a given mass of another substance at a specified temperature and pressure.
equilibrium spheroid *Geophysics.* the shape that the earth would take if it were entirely covered by an ocean of uniform depth, not subject to tidal forces.
equilibrium still *Analytical Chemistry.* a recirculating distillation apparatus used to determine vapor-liquid equilibrium.
equilibrium theory *Oceanography.* an ocean water model that assumes an ideal earth completely and uniformly covered by water of considerable depth, and also assumes that the water responds instantly to the tide-producing forces of the sun and moon to form a surface in equilibrium; this model does not take into account any of the other actual tidal components, such as inertia, the Coriolis effect, friction, or irregular distribution of land masses.
equilibrium tide *Oceanography.* the hypothetical tide formed under the assumptions of equilibrium theory.
equilibrium valve see BALANCED VALVE.
equilibrium vaporization ratio *Physical Chemistry.* the relationship between the vapor phase and the liquid phase of a solution undergoing change in a closed system, when the system reaches the point at which no further change occurs.
equilibrium vapor pressure *Physics.* the vapor pressure of a system at equilibrium.
equine [ē´kwīn´; ek´wīn´] *Vertebrate Zoology.* **1.** relating to or resembling a horse. **2.** relating to or being a member of the family Equidae.
equine encephalitis *Pathology.* a viral disease of horses that can be transmitted to humans. In horses, it can be fatal; symptoms include fever, an irregular gait, lack of coordination, drowsiness, and occasional convulsions. In humans, the disease causes flulike symptoms with little effect on the brain and spinal cord. Also, **equine encephalomyelitis.**
equine encephalitis virus *Virology.* any of a group of arboviruses causing encephalomyelitis in horses, mules, and humans, transmitted by mosquitoes with a reservoir of infection in birds. Also, **equine encephalomyelitis virus.**
equine ulcerative lymphangitis see PSEUDOGLANDERS.
equinoctial *Astronomy.* of or relating to an equinox, to the equality of day and night, or to the celestial equator. *Botany.* of a flower, opening regularly at or about a particular hour.
equinoctial colure *Astronomy.* the hour circle passing through 0 hours and 12 hours right ascension, the points at which the sun crosses the celestial equator going north and south, respectively.
equinoctial rains *Meteorology.* the rainy seasons occurring regularly at locations within a few degrees of the equator at or shortly after the equinoxes.
equinoctial storm *Meteorology.* a violent storm of rain and wind that is associated with the time of the equinox. Also, LINE GALE, LINE STORM.
equinoctial tide *Oceanography.* a tide that occurs when the sun is near equinox, characterized by increased spring tide ranges.
equinoctial year see TROPICAL YEAR.
equinox *Astronomy.* the moment when the center of the sun crosses the celestial equator, in either a northbound or southbound direction.

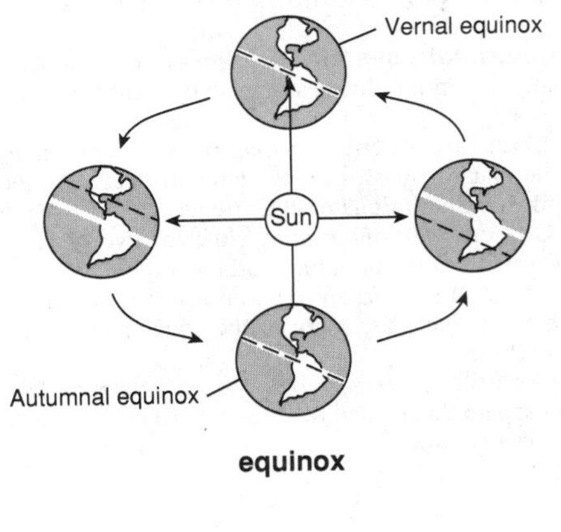

equinox

equiparte [ēk´wē pär´tā] *Meteorology.* cold and heavy rains in Mexico, lasting for several days during October to January. Also, **equipatos.**

equipartition *Chemistry.* **1**. an orderly distribution of atoms, as in a crystal structure. **2**. an equal distribution of a solute between two solvents.

equipartition law *Physics.* a law stating that the mean energy of the molecules of a gas in thermodynamic equilibrium is divided equally among the various degrees of freedom of the molecules. Also, **equipartition of energy.**

equiphase wave surface *Physics.* a surface on which the phases of the field vectors of a wave are identical at the same instant of time.

equiphase zone *Geophysics.* a region of space in which it is impossible to distinguish differences in phases between two radio signals.

equiphobia *Psychology.* an irrational fear of horses.

equipoise *Mechanics.* an equal distribution of weight.

equipollent *Mechanics.* describing two systems of forces that have equal linear and angular vector sums about each of their corresponding points.

equipotent *Science.* equally powerful or effective.

equipotential cathode see INDIRECTLY HEATED CATHODE.

equipotentiality *Science.* a state or condition of having similar and equal power; the capacity of developing in the same way and to the same extent.

equipotential line see ISOPIESTIC LINE.

equipotential surface *Geophysics.* a surface along which the gravity is constant at all points and for which the gravity vector is normal to all points. Also, LEVEL SURFACE. *Electricity.* a surface of a field of force having the characteristic that every point of the surface is at the same potential; such surfaces are normal to the flux lines.

equipotent sets *Mathematics.* equivalent sets.

equipressure contour *Petroleum Engineering.* a map or plot of the equal isopressure flow network within a reservoir that is used to find locations for water-injection wells for flood coverage of an areal pattern in a reservoir.

Equisetales *Botany.* an order of herbaceous tracheophytes in the class Equisetinae, characterized by small, scalelike leaves, longitudinal ridges along the silicaceous stem, whorls of branches and leaves at the nodes, a siphonostele perforated at the nodes, and peltate sporangiophores.

Equisetineae *Botany.* a class of the division Equisetophyta characterized by leaves and branches alternating at the nodes, longitudinally ribbed stems, an absence of leaf gaps in the siphonostele, a sporophyte with stem, leaf, and root, and sporangia on distinct sporangiophores.

Equisetophyta *Botany.* a division of the subkingdom Embryophyta, of which *Equisetum* is the sole living genus; it was formerly more extensive and diverse, reaching its height during the Upper Carboniferous.

Equisetopsida *Botany.* a class of the division Equisetophyta, whose members comprised much of the flora in swamp habitats during the Upper Carboniferous.

equisignal *Telecommunications.* of or related to two signals of equal strength.

equisignal surface *Electromagnetism.* the region surrounding a transmitting antenna formed by all points that have equal field strength during transmission.

equitability *Ecology.* a property of population distribution that refers to the numerical equality of various species populations in a community; maximum equitability is attained when all species maintain approximately equal populations.

equitant *Botany.* describing leaves that are folded to overlap transversely at the base.

equiv. equivalency, equivalent.

equivalence the condition of being equivalent; having equal value; specific uses include: *Mathematics.* a logic operator on a set of statements such that the value of the operator is true if each of the statements is true or if each is false, and false otherwise. If P and Q are equivalent statements, it is customary to say "P if and only if Q;" denoted P iff Q or $P \Leftrightarrow Q$. Also, BICONDITIONAL OPERATOR. *Cartography.* the property of having the ratio of the distances between points on a map and the differences between the corresponding points on the earth's surface be the same for all points. *Immunology.* the ratio of antigen to antibody concentration at which maximal combination occurs.

equivalence classes *Mathematics.* a mutually disjoint subset of a set A on which an equivalence relation R is defined; two elements belong to the same class if and only if they are equivalent. The class of all equivalence classes in A determined by R is often denoted A/R and called the **quotient class** of A by R.

equivalence gate or **equivalence element** see EXCLUSIVE-NOR GATE.

equivalence point *Chemistry.* the point in a titration process at which the amount of titrant is chemically equivalent to the amount of material titrated.

equivalence principle *Physics.* a principle of relativity theory stating that the mass of a body as measured by its inertia is equal to the mass of the body as measured by the force of a gravitational field on that body.

equivalence relation *Mathematics.* a binary relation between elements of a set that is reflexive, symmetric, and transitive on that set.

equivalent having the same value; specific uses include: *Mathematics.* of two sets, able to be placed in a one-to-one correspondence. *Chemistry.* having the same valence. *Medicine.* describing a symptom replacing one that is usual in a given disease. *Geology.* describing the strata or formations that correspond in geologic time or stratigraphic position, especially those contemporaneous in time of deposition or formation, or those containing the same fossil forms.

equivalent absorption area *Acoustics.* the amount of surface area of a perfectly absorbing medium required to absorb an equivalent amount of sound energy as some actual medium, given equal time and similar conditions.

equivalent airspeed *Aviation.* an aircraft's indicated or calibrated airspeed, as corrected for instrument error, position error, and compressibility.

equivalent-barotropic model *Meteorology.* a model atmosphere in which the vertical shear of horizontal wind is proportionate to the horizontal wind itself in a frictionless, adiabatic flow characterized by hydrostatic and quasi-geostrophic equilibrium.

equivalent bending moment *Mechanics.* a single bending moment whose resulting maximum stress would be equivalent to that of the actual bending and twisting moments together.

equivalent binary digits *Computer Programming.* the number of binary digits required to express one radix digit of a different base. For example, approximately $3^1/_3$ binary digits are required to express one decimal (base 10) digit.

equivalent blackbody temperature see BLACKBODY TEMPERATURE.

equivalent circuit *Electricity.* a circuit that consists of passive components and ideal current sources and that is electrically equivalent to a more complex circuit.

equivalent conductance *Physical Chemistry.* a property that equals the exact conductance of an electrolyte divided by the concentration of solutes in the solvent.

equivalent electrons *Atomic Physics.* the electrons that occupy the same orbital space within an atom and, therefore, have the same principal and orbital quantum numbers.

equivalent isotropically radiated power *Telecommunications.* a satellite footprint signal power level that is limited to 37 dB W by international agreement.

equivalent map projection see EQUAL-AREA PROJECTION.

equivalent nitrogen pressure *Physical Chemistry.* a method of standardizing pressure calculations by assuming that the actual gases present have the same molecular density as nitrogen.

equivalent noise conductance *Electronics.* the spectral density of a noise-current generator, measured at a specified frequency and expressed in conductance units.

equivalent noise pressure *Acoustical Engineering.* the noise level that a transducer introduces onto a channel with no actual electroacoustical input; normally a very low value, such as –126 decibels. Also, **equivalent input noise level.**

equivalent noise resistance *Electronics.* the spectral density of a noise-voltage generator, measured at a specified frequency and expressed in resistance units.

equivalent noise temperature *Electronics.* the temperature at which an ideal resistor, whose resistance is the same as that of the given noise-generating component, would generate the same noise energy as the given component.

equivalent nuclei *Physical Chemistry.* the nuclei in a molecule that can be transformed into each other by various processes, such as rotation or reflection, so that the molecule remains stable.

equivalent orifice *Mechanical Engineering.* a theoretical sharp-edged orifice used as a reference in measuring the resistance offered by a fan or other rotating device.

equivalent periodic line *Electricity.* a periodic line at a given frequency that has electrical behavior equivalent to that of the uniform line when measured at its terminals.

equivalent positions *Crystallography.* the complete set of atomic positions produced by the operation of the symmetry elements of the space group upon any general atomic position in a crystal.

equivalent potential temperature *Meteorology.* the equivalent temperature of an air sample that is brought adiabatically to a pressure of 1000 mbars.

equivalent reflections *Crystallography.* the principle that when a complete set of X-ray diffraction data has been collected, there are eight measurements for the intensities of each *h, k, l,* corresponding to combinations of positive and negative values of each. Some of these reflections that are equivalent by the symmetry of the crystal have (within experimental error) identical intensities. For high-symmetry crystals, other reflections may also be equivalent, e.g., *hkl, klh,* and *lhk* for cubic crystals.

equivalent resistance *Electricity.* the value that the resistance of an equivalent circuit must have so that the loss in it represents the total loss occurring in the actual circuit.

equivalent sets *Mathematics.* two sets *A* and *B* are said to be equivalent if there exists a one-to-one correspondence between their elements. Intuitively, *A* and *B* have the same number of elements.

equivalent temperature *Meteorology.* the temperature that would be reached by a sample of moist air if all the water vapor in the sample were condensed out at constant pressure, and the latent heat thus released used to raise the temperature of the sample.

equivalent twisting moment *Mechanics.* a single twisting moment whose resulting maximum shear stress would be equivalent to that of the actual bending and twisting moments together.

equivalent vapor volume *Petroleum Engineering.* the volume that would be taken up by a barrel of oil if the oil were converted to a vapor.

equivalent vertical photograph *Photogrammetry.* a theoretically vertical photograph taken at the same camera station with a camera whose focal length is equal to that of a camera taking a corresponding tilted photograph.

equivalent viscous damping *Mechanics.* a concept that simplifies calculations of vibratory systems by equating all of the actual types of damping to an assumed viscous equivalent.

equivalent weight *Chemistry.* the gram weight of a given element that will combine with or replace one mole (1.008 grams) of hydrogen atoms or one-half mole (8 grams) of oxygen atoms; a mass equal to the element's atomic weight divided by its valence. Also, COMBINING WEIGHT.

equivalent width *Physics.* the amount of total absorption of radiant energy as measured by the width of the absorption line or absorption band characteristic of the specimen.

equiviscous temperature *Chemical Engineering.* a method of expressing viscosity as measured in a tar efflux viscometer.

equivocation *Telecommunications.* the entropy of the occurrence of specific messages at a message source, given the occurrence of specific messages at a message sink connected to the message source by a specific channel; equal to the mean additional information content that must be supplied per message at the message sink to correct the received messages affected by a noisy channel.

Equoidea *Vertebrate Zoology.* in some classifications, a superfamily of perissodactylous mammals in the suborder Hippomorpha, composed of all living horses and extinct related forms.

Equuleus *Astronomy.* the Foal, a small faint constellation of the northern hemisphere that is visible on autumn evenings.

Equus [ek´wəs] *Vertebrate Zoology.* the genus composing the family Equidae, including horses, asses, zebras, and related extinct animals.

Er the chemical symbol for erbium.

ER emergency room; electroreflectance; endoplasmic reticulum.

era *Geology.* a unit of geologic time that includes two or more periods grouped together. *Science.* a distinctly identified period of history.

ERA electron ring accelerator.

eradicant *Plant Pathology.* a pesticide spray that is used to kill a parasitic organism at its source, to prevent it from infecting a plant.

erasability of storage *Computer Technology.* the ability of data locations within storage media to be erased and replaced with new data.

erasable (programmable) read-only memory see EPROM.

erasable storage *Computer Technology.* any storage device whose contents can be erased and reused, such as a magnetic tape or disk.

erase *Computer Programming.* **1.** to remove data from a storage device. **2.** to change all bits in a storage device to zeros. *Electronics.* to remove recorded signals from a magnetic tape by running it through a magnetic field (**direct-current erase**) or through a high-frequency alternating magnetic field (**alternating-current erase.**)

erase oscillator *Electronics.* a circuit in a tape recorder that generates the high-frequency signals needed to erase the tape.

erasing head *Acoustical Engineering.* the device in a tape recorder that removes previously recorded material.

erasion *Surgery.* an older term for the removal of tissue by scraping (curettage).

Erasistratus c. 300–250 BC, Greek physician and physiologist; distinguished the cerebrum from the cerebellum and sensory from motor nerves; observed lacteals.

Eratosthenes c. 276–194 BC, Greek astronomer and geographer; calculated the circumference of the earth; devised the Sieve of Eratosthenes.

erb *Genetics.* either of two oncogenes (**erbA** or **erbB**) identified as the transforming determinants in avian erythroblastosis virus.

ERBB-1 *Oncology.* a proto-oncogene that is amplified in squamous cell carcinoma and astrocytoma.

ERBB-2(NEU) *Oncology.* a proto-oncogene that is amplified in adenocarcinoma of the breast, ovary, and stomach.

erbia see ERBIUM OXIDE.

erbium *Chemistry.* a solid rare-earth (lanthanide) element of the yttrium group having the symbol Er, the atomic number 68, an atomic weight of 167.26, a melting point of 1522°C, and a boiling point of about 2500°C; a soft, malleable solid with a metallic luster and high electrical resistivity, used in lasers, alloys, and nuclear reactors. (From *Ytterby,* Sweden, where it was first found and identified.)

erbium halide *Inorganic Chemistry.* any compound consisting of the element erbium with an element from the halide ion group (periodic table Group VIIa) such as fluoride or chloride.

erbium nitrate *Inorganic Chemistry.* $Er(NO_3)_3{\cdot}5H_2O$, large, reddish crystals that are soluble in water, alcohol, and ether; loses $4H_2O$ at 130°C; may explode if heated or shocked.

erbium oxalate *Organic Chemistry.* $Er_2(C_2O_4)_3{\cdot}10H_2O$, a corrosive, reddish, microcrystalline powder, soluble in water and dilute acids; decomposes at 575°C; used to separate erbium from common metals.

erbium oxide *Inorganic Chemistry.* Er_2O_3, a pink powder that is insoluble in water and that readily absorbs water and carbon dioxide from the air; not capable of being melted; used as a phosphor activator and in infrared absorbent glass. Also, ERBIA.

erbium sulfate *Inorganic Chemistry.* $Er_2(SO_4)_3{\cdot}H_2O$, rose red, monoclinic crystals that are soluble in water; loses its water at 400°C; used to find the atomic weight of erbium.

erbon *Organic Chemistry.* $(Cl)_3C_6H_2OC_2H_4OC(O)C(Cl)_2CH_3$, a toxic, white solid that is insoluble in water; melts at 49–50°C; used as an herbicide.

Erb's atrophy see PROGRESSIVE MUSCULAR DYSTROPHY.

ERDA or **E.R.D.A.** Energy Research Development Administration.

erect *Building Engineering.* to build or raise up a structure; construct. *Botany.* of a stem, leaf, or other plant part, vertical throughout; upright. *Mathematics.* to construct a figure or draw a line upon a given base, line, and so on. *Optics.* of an image, oriented in the normal position; the opposite of inverted.

erectile [i rek´tīl] *Physiology.* capable of erection.

erectile tissue *Anatomy.* tissue that contains large venous spaces with which arteries communicate directly, as in the penis, clitoris, nasal mucosa, or smooth muscles of the nipples.

erection *Physiology.* **1.** the condition of being made rigid and elevated. **2.** specifically, the enlargement and stiffening of the penis produced by vascular engorgement following physical or psychic stimulation. *Building Engineering.* any structure that is erected, especially a building or bridge.

erection stress *Building Engineering.* the stresses that a building component is subjected to during construction.

erector *Anatomy.* a muscle that raises a body part or keeps it erect.

erect stem *Botany.* a stem that grows upright.

Eremascoideae *Botany.* a monogeneric subfamily of yeasts having septate mycelia and spherical asci with rounded ascospores.

eremeyevite see JEREMEJEVITE.

Eremolepidaceae *Botany.* a family of dicotyledonous tropical American shrublets of the order Santalales that are hemiparasitic on tree branches and have thick, superficial haustoria.

Eremomycetaceae *Mycology.* a family of fungi belonging to the order Dothideales which live off of nonliving organic matter in animal dung or soil.

eremophobia *Psychology.* an irrational fear of being alone and of uninhabited places.

eremophyte *Botany.* any plant that grows in desert conditions.

Eremosphaeraceae *Botany.* a family of noncoenobial green algae of the order Chlorococcales, characterized by spherical to boat-shaped cells containing a large central nucleus held by cytoplasmic strands, a thin or finely stratified cell wall, and oogamous sexual reproduction.

Erethizontidae *Vertebrate Zoology.* the arboreal or New World porcupines, a family of chiefly arboreal rodents of the order Hystricomorpha, characterized by a more or less prehensile tail, soles of the feet specialized for climbing, and typical barbed spines.

ereutho- see ERYTHRO-.

ERG electroretinogram.

erg *Metrology.* a unit of measure for energy, equal to the amount of work done by a force of 1 dyne acting through a distance of 1 centimeter; it is equivalent to 2.4×10^{-8} gram calories, or to 0.624×10^{12} electronvolts. *Geology.* **1.** a desert area that is covered with sand dunes. **2.** specifically, the sandy tracts of the Sahara.

erg- a combining form meaning "work," as in *ergate*.

Ergasilidae *Invertebrate Zoology.* a family of small copepod crustaceans, in the suborder Cyclopoida, with parasitic females and free-swimming males.

ergasiophobia *Psychology.* a pathological aversion to work.

ergasiophyte *Ecology.* a cultivated plant introduced into an area by humans. Also, **ergasiaphyte.**

ergastic *Cell Biology.* relating to by-product materials of cellular activity, including storage deposits such as starch and secretory products.

ergastoplasm *Cell Biology.* a term formerly used to denote ribosome-studded endoplasmic reticulum.

ergate *Invertebrate Zoology.* a worker ant.

ergo- a combining form meaning "work," as in *ergometer*.

ergocalciferol *Organic Chemistry.* $C_{28}H_{44}O$, the synthetic form of vitamin D_2, in the form of white, odorless crystals that are insoluble in water; melts at 115–118°C; used in nutrition. Also, CALCIFEROL.

ergochromes *Biochemistry.* a group of slightly acidic, yellow pigments found in *Ergot sclerotia* and in *Penicillium oxalicum*.

ergodic [ur gäd´ik] *Science* relating to a system or process in which the final distribution of states is independent of the initial state.

ergodicity *Chaotic Dynamics.* a condition in which the trajectory of a classical system uniformly covers the phase space allowed by the conservation of energy, so that time averages can be replaced by averages over the corresponding phase space.

ergodic theorem *Mathematics.* let T be a measure-preserving transformation on a probability space (Ω, F, P); let J denote the sigma field of sets A in F with the property that $T^{-1}A = A$ (i.e., the sets in F that are invariant under T); and let X be a random variable of bounded expectation (i.e., $E|X| < \infty$). Then, except possibly on a set of probability zero, the conditional expectation $E(X|J) = \lim_{n\to\infty} 1/(n+1) \sum_{k=0}^{n} (T^k\omega)$.

ergodic theory *Physics.* a theory that attempts to explain and predict the observed macroscopic behavior of matter starting from microscopic particle dynamics. *Mathematics.* the study of measure-preserving transformations, including the development and application of theorems concerning limits of probabilities and weighted means.

ergodic transformation *Mathematics.* a measure-preserving transformation on a probability space (Ω, F, P) is said to be ergodic if for every set A in F with the property that $T^{-1}A = A$ (i.e., A is a member of the sigma field J of sets invariant under T), then either $P(A) = 0$ or $P(A) = 1$.

ergograph *Engineering.* an instrument that produces a graphic representation of the measurement of muscular work capacity that it records.

ergometer *Engineering.* a device that records the measurement of muscular work under controlled conditions.

ergon *Genetics.* a unit used to represent the stability of a given gene throughout a lifetime; it is a function of the ratio of the adenine-thymine to guanine-cytosine content of the gene, and is reflected in the persistence of the resultant phenotypical trait.

ergonometrics *Industrial Engineering.* the measurement of body movements, alertness, fatigue, and other human factors involved in an industrial work process.

ergonomics *Industrial Engineering.* the study of physical and mental factors that affect people in work settings; used in the design of work sites, work processes, and so on; e.g., the design of computer work stations so that users will have minimal strain on posture and vision.

ergosome *Cell Biology.* a term once used to describe a group of ribosomes, or a polysome, attached to the endoplasmic reticulum of eukaryotic cells.

ergosphere *Physics.* that region around a rotating black hole, bounded on the inside by the event horizon and on the outside by the static limit. Also, **ergoregion.**

ergosterol *Biochemistry.* $C_{28}H_{44}O$, a sterol that produces vitamin D_2 when exposed to ultraviolet light or activated by electrons; present in ergot, fungi, bacteria, algae, yeast, molds, and higher plants.

ergot [ur´gət] *Plant Pathology.* a disease of such cereals as rye, wheat, and oats caused by the fungus *Claviceps purpurea. Medicine.* any of five toxic, isomeric pairs of alkaloids derived from this fungus; used in the treatment of migraine. *Vertebrate Zoology.* a small horny mass in the tuft of hair at the flexion surface of a horse's fetlock.

ergotinine *Organic Chemistry.* the alkaloid isomer of ergotoxine, consisting of a 1=1=1 mixture of ergocornine, ergocristine, and ergocryptine; crystallizes in long needles from acetone solution and melts at 229°C; soluble in alcohol.

ergotism *Toxicology.* an acute or chronic intoxication caused by ingesting grain infected with ergot fungus, or from the misuse of drugs containing ergot; effects may include cerebrospinal symptoms, severe gastrointestinal disturbances, and a type of gangrene. Also, SAINT ANTHONY'S FIRE.

ergotoxine *Organic Chemistry.* the alkaloid isomer of ergotinine, consisting of a 1=1=1 mixture of ergocornine, ergocristine, and ergocryptine; it occurs in orthorhombic crystals, is soluble in alcohol, and melts at 190°C; it was once used to aid in contraction of the uterus during childbirth. *Toxicology.* a toxic crystalline form of this alkaloid, found in ergot and formerly used medicinally.

Erian *Geology.* a North American provincial series of the Middle Devonian period, occurring after the Ulsterian and before the Senecan.

Erian orogeny *Geology.* a brief, late Silurian crustal deformation involving Phanerozoic rock, and equivalent to the last part of the Caledonian orogenic era. Also, HIBERNIAN OROGENY.

Ericaceae *Botany.* a cosmopolitan family of dicotyledonous, shrubby plants in the order Ericales, characterized by leathery, evergreen leaves, regular, bisexual flowers occurring either singly or in racemes, and twice as many stamens as corolla lobes; usually growing in acidic habitats and dependent on nitrogen-fixing mycorrhizae. It is commonly called the heather family and includes many species of the genera *Rhododendron* and *Erica*.

ericaceous *Botany.* heatherlike; belonging to the family Ericaceae.

Ericales *Botany.* an order of dicotyledonous shrubs in the subclass Dilleniidae characterized by simple leaves, bisexual, usually regular flowers, fused petals, and a compound ovary.

ericeticolous *Botany.* growing on or native to a heath or similar habitat.

Erichsen value *Metallurgy.* a quantitative measure of deep drawability that is obtained by the **Erichsen test.**

ericophyte *Ecology.* a plant that grows in marshy, wetland areas.

Ericsson, John 1803–1889, Swedish-born American naval engineer; designed steamships, screw propellers, and the warship *Monitor*.

Ericsson cycle *Thermodynamics.* a thermodynamic cycle that consists of four processes: a constant-temperature compression, a constant-pressure heat transfer to the working fluid; a constant-temperature expansion, and a constant-pressure heat transfer out of the system. This is an ideal cycle whose efficiency could equal that of a Carnot cycle; it is used as a model for actual gas turbines.

Eridanus *Astronomy.* the River, a long dim constellation of the southern hemisphere that winds southwest from Orion.

Erie, Lake *Geography.* the fourth largest and the southernmost of the Great Lakes.

Erikson, Erik born 1902, American psychologist; studied psychosocial development.

Erinaceidae *Vertebrate Zoology.* the hedgehogs and moonrats, a family of nocturnal mammals of the order Insectivora, having acute sight and hearing, limbs with five digits, and in the more advanced species, hairs on the back modified to form spines; found throughout the Old World.

erinaceous *Vertebrate Zoology.* of, relating to, or resembling a hedgehog (the family Erinaceidae).

erineum *Plant Pathology.* an abnormal felty, hairy growth on a leaf epidermis that is caused by certain mites.

Erinnidae *Invertebrate Zoology.* a family of two-winged flies in the series Brachycera.

Erint *Ordnance.* a U.S. missile system designed to intercept and destroy enemy missiles in flight. (An acronym for extended range intercept technology.)

Eriocaulaceae *Botany.* a family of monocotyledonous, perennial herbs in the order Eriocaulales, characterized by grasslike leaves in basal rosettes, elongate or cormlike stems, and regular, unisexual flowers borne in dense heads, and growing in swampy habitats.

Eriocaulales *Botany.* an order of monocots in the subclass Commelinidae, having a reduced or absent perianth and unisexual flowers grouped on a long peduncle.

Eriococcidae *Invertebrate Zoology.* the mealybugs, a family of plant-sucking insects that are serious pests. Also, PSEUDOCOCCIDAE.

Eriocraniidae *Invertebrate Zoology.* a family of tiny diurnal moths, lepidopteran insects in the superfamily Eriocranioidea.

Eriocranioidea *Invertebrate Zoology.* a superfamily of small moths, lepidopteran insects in the suborder Homoneura having reduced, nonbiting mouthparts; larvae are leaf miners.

eriodictyol *Organic Chemistry.* $C_{15}H_{22}O_6$, a compound isolated from *Eriodictyon californicum*, occurring in needlelike crystals from dilute alcohol solution; slightly soluble in boiling water, hot alcohol, and glacial acetic acid; used medicinally as an expectorant.

erionite *Mineralogy.* $(K_2,Ca,Na_2)_2Al_4Si_{14}O_{36}\cdot 15H_2O$, a white, hexagonal mineral of the zeolite group, occurring in minute slender crystals and fine, fibrous aggregates, and having a specific gravity of 2.02 to 2.07 and an undetermined hardness; found with opal in fractures in rhyolitic tuff and with paulingite in basalt.

Eriophyidae *Invertebrate Zoology.* the bud mites or gall mites, a family of minute, wormlike mites in the suborder Trombidiformes; parasitic on deciduous plants.

eriophyllous *Botany.* having leaves covered by a woolly or cottony pubescence.

Eris *Ordnance.* a U.S. missile system designed to intercept and destroy enemy missiles in outer space. (An acronym for exoatmospheric reentry vehicle interceptor system.)

erlang *Telecommunications.* an international dimensionless unit of the average traffic intensity of a facility during a period of time, equal to the ratio of time during which a facility is occupied to the time it is available for occupancy; one erlang is the maximum intensity for a single facility.

Erlanger, Joseph 1874–1965, American physiologist; shared Nobel Prize with Gasser for work on the functions of synapses.

Erlenmeyer, Richard August 1825–1909, German chemist; invented the Erlenmeyer flask; helped develop modern structural notation.

Erlenmeyer flask *Chemistry.* a conical glass flask with a flat bottom and a short, straight-sided neck; used in applications involving liquids.

Erlenmeyer synthesis *Organic Chemistry.* the condensation of an aldehyde with α-acylamino acid in the presence of acetic anhydride and sodium acetate to give an azlactone.

ERM enteric redmouth.

ermine *Vertebrate Zoology.* a Northern European weasel of the family Mustelidae, especially the species *Mustela erminea*, known for its fine white winter pelt, which is highly prized and which was once reserved for royalty. The winter phase of many other weasels is now known as ermine in the fur trade. Also, STOAT.

ermine

erode *Geology.* **1.** to wear away the land, as by the action of wind, water, or glaciers. **2.** to produce or alter a landform by such wearing away.

erogenous [ə räj´ə nis] *Physiology.* especially sensitive to sexual stimulation or arousal. Thus, **erogenous zones.** *Psychology.* see EROTOGENIC.

Eros [âr´ōs] *Psychology.* the instinct for life or pleasure, one of Freud's two primary motives for human conduct. *Astronomy.* asteroid 433, discovered in 1898 and measuring 40 by 14 by 14 km; it has a Mars-crossing orbit and belongs to type S. (From *Eros,* the Greek god of love.)

erose *Biology.* having the outside edge irregularly notched, as if eroded; generally applied to leaves.

E rosette *Immunology.* a lymphocyte surrounded by a cluster of (usually sheep) red blood cells.

E-rosette forming cell *Immunology.* a T lymphoid cell that forms clusters of red blood cells with sheep red blood cells.

erosion *Geology.* a combination of processes in which the materials of the earth's surface are loosened, dissolved, or worn away, and transported from one place to another by natural agents. *Materials Science.* the removal of material from a component by the mechanical attack of a solid or a liquid. *Medicine.* the superficial destruction of a surface area by infection or trauma.

erosional *Geology.* relating to, characteristic of, or caused by erosion.

erosional unconformity *Geology.* **1.** a surface that separates older, eroded rocks from younger, overlying sediments. **2.** see DISCONFORMITY.

erosion-corrosion *Materials Science.* an acceleration in the rate of corrosion attack in metal due to the relative motion of a corrosive fluid and a metal surface.

erosion cycle see CYCLE OF EROSION.

erosion pavement *Geology.* a surface layer of pebbles and other rock fragments that develops after sheet or rill erosion, preventing the underlying soil from further erosion.

erosion plain *Geology.* any plain that was produced by erosion.

erosion ridge *Hydrology.* one of a series of small ridges formed on a snow surface by the abrasive action of particles carried by windblown snow.

erosion surface *Geology.* a land surface shaped by the wearing action of streams, ice, rain, winds, or other atmospheric agencies.

erosive *Geology.* causing erosion. *Agronomy.* of or relating to the tendency of a plant to allow or promote erosion, such as cotton, peanuts, or other row crops that require extensive cultivation.

erotic *Psychology.* involving or producing sexual feelings or desire.

erotogenic *Psychology.* producing erotic feelings.

erotomania *Psychology.* an irrationally strong or persistent desire for sexual gratification.

erotophobia *Psychology.* an irrational fear of sexual activity or sexual arousal.

Erotylidae *Invertebrate Zoology.* a family of fungus beetles, coleopteran insects in the superfamily Cucujoidea.

ERP effective radiated power.

Erpodiaceae *Botany.* a widespread family of small, dull mosses of the order Orthotrichales, characterized by creeping stems with multiple prostrate branches, by terminal sporophytes borne on short, lateral branches, and by the lack of a costa; plants form mats on tree trunks and sometimes on rocks.

Errantia *Invertebrate Zoology.* a superfamily of polychaete annelids, often with long, flattened bodies.

erratic *Geology.* **1.** a large rock fragment that differs in lithology from the bedrock on which it rests, especially a fragment that has been transported from a distant source by glacial or floating ice. **2.** of or relating to such a rock fragment.

errhine [er´in] *Physiology.* inducing a nasal discharge. *Pharmacology.* any medicine that promotes nasal discharge or secretion.

error *Science.* **1.** a difference between the approximate or observed value and the actual or true value of a quantity. **2.** the degree of such a difference. *Medicine.* a defect in structure or function. *Computer Programming.* the inability to carry out a requested operation.

error analysis *Statistics.* the study of the observed discrepancies between the data and the values predicted by a model, often aimed at checking assumptions and diagnosing faults in the model.

error burst *Computer Programming.* in data communication, a sequence of signals that contains one or more errors.

error catastrophe *Genetics.* a model of aging that suggests that mutations in the protein-synthesizing apparatus have a cascading effect on a widening series of cell products, causing cell malfunction.

error character *Computer Programming.* a control character denoting an error in the data being processed and usually the amount of preceding or following data to ignore.

error-checking code see ERROR-DETECTION CODE.

error code *Computer Programming.* a character or code used to indicate an error in a block of data; the error can then be corrected or ignored when the data is processed.

error coefficient *Control Systems.* a value obtained by dividing the steady-state value of a control system's output, or some portion of that output, by the steady-state actuating signal. Also, **error constant.**

error compensation *Design Engineering.* the process of correcting for the difference between a desired target or dimension and the target or dimension achieved.

error-correcting code *Computer Programming.* data representation by a code designed in such a way as to facilitate detection and correction of certain errors. Also, SELF-CORRECTING CODE.

error-correcting telegraph system *Telecommunications.* a system employing an error-correcting code and configured so that some or all signals detected as being in error are automatically corrected at the receiving terminal before delivery to the data sink. Also, **error-correcting system.**

error correction *Engineering.* any process of correcting a computed, measured, or observed value based on its deviation from the proper or expected value. *Computer Programming.* specifically, a device that automatically detects and corrects a machine error of dropping a bit or picking up an extraneous bit without causing the machine to stop or enter programmed recovery routines.

error-detection code *Computer Programming.* data representation augmented by a code, such as a parity bit or CRC word, designed to facilitate detection of errors. Also, ERROR-CHECKING CODE, SELF-CHECKING CODE.

error-detection routine see DIAGNOSTIC ROUTINE.

error diagnostic see DIAGNOSTIC MESSAGE.

error-frequency limit *Computer Technology.* a fixed number of errors that can be detected within a given time period before a machine check interrupt is initiated.

error function *Mathematics.* any of these complex functions:

$$erf(z) = 2\pi^{-1/2}\int_0^z e^{-t^2}dt \qquad erfc(z) = 2\pi^{-1/2}\int_z^\infty e^{-t^2}dt$$

$$w(z) = e^{-z^2}(1 + 2i\pi^{-1/2}\int_0^z e^{t^2}dt) = e^{-z^2}erfc(-iz)$$

Used in boundary value problems and probability theory. Detailed tables exist for various values of the error functions. Also, **error integral.**

error interrupt *Computer Programming.* a signal that causes a halt in program execution due to errors that cannot be corrected.

error list *Computer Programming.* a list of invalid or erroneous instructions in a source program, generated by the compiler. Similarly, **error report, error log.**

error message *Computer Programming.* a displayed or printed message indicating the detection of an error, especially an invalid or erroneous instruction given by the operator.

error modeling *Design Engineering.* the use of computer or mathematical relationships to describe what error might be expected.

error of closure *Engineering.* **1.** in surveying, the amount of variation from the initial measurement in the length of the azimuth of the first line of traverse, when taken after finishing the circuit or the amount of variation from 360° of the total angles measured around the horizon. **2.** the amount by which a quantity obtained by a series of related measurements differs from the true or fixed value of the same quantity.

error of perpendicularity *Navigation.* an error in the reading of a marine sextant because the index mirror is not perpendicular to the frame of the instrument.

error-prone repair *Genetics.* a gene repair system that is prone to error, thus causing mutations and leaving a template strand that contains defects, causing DNA synthesis to stall.

error range *Statistics.* a measure of dispersion of errors given by the difference between the highest and lowest values. *Computer Programming.* **1.** the range of possible values for an error. **2.** the range of values for a data item that will cause an error condition.

error ratio *Computer Programming.* the ratio of the number of erroneous bits or data items to the total number of bits or data items during a given time period. Also, **error rate.**

error routine *Computer Programming.* a routine that takes control of a program when an error occurs and reports or tries to correct the error.

error signal *Control Systems.* in an automatic control system, a signal that adjusts the alignment between the controlling and controlled elements; it represents the difference between a sensing signal and a constant reference signal.

errors of measurement *Statistics.* the part of the variation in the distribution of items that is due to faulty measurement instruments, observation errors, and other human factors.

ertor *Meteorology.* the effective temperature of the ozone layer of the atmosphere. (An acronym for effective radiational temperature of the ozone region.)

ERTS earth resources technology satellite.

erucic acid *Organic Chemistry.* $C_8H_{17}HC{=}CH(CH2)_{11}COOH$, a combustible solid that is insoluble in water; melts at 33–34°C and boils at 264°C (10 torr); derived from mustard seed and rapeseed (now canola); used in chemical preparation and water-resistant nylon.

eruciform *Entomology.* shaped like a caterpillar.

eructation *Physiology.* belching.

erumpent *Botany.* **1.** developing initially underground and then bursting above ground and spreading. **2.** describing fungi or algae that project through host tissue.

eruption *Volcanology.* the often sudden and violent ejection of solid, liquid, or gaseous materials onto the earth's surface from a volcanic vent, a fissure, or group of fissures. *Medicine.* an outbreak of visible efflorescent skin lesions due to disease, marked by redness and prominence; a rash.

eruption

eruptive prominence *Astronomy.* a prominence on the sun characterized by rapid, explosive changes.

eruptive rock *Petrology.* **1.** a rock formed from solidified magma that reached the earth's surface while molten and then cooled. **2.** rock material spewed from a volcanic cone.

eruptive variable *Astronomy.* a star whose irregular brightness changes are explosive or erratic.

ERV expiratory reserve volume.

Erwinia *Bacteriology.* a genus of Gram-negative, rod-shaped, facultatively anaerobic bacteria of the family Enterobacteriaceae, able to carry out mixed acid fermentation of sugars; it causes dry necroses, wilts, and soft rots in plants. (Named for *Erwin* F. Smith, 1854–1927, American bacteriologist.)

Erwinieae *Bacteriology.* in some classifications, a tribe of Gram-negative bacteria of the family Enterobacteriaceae, occurring as rod-shaped cells with a facultatively anaerobic mechanism of respiration.

erysipelas *Medicine.* a contagious disease of the skin and subcutaneous tissue caused by infection with *Streptococcus pyogenes.*

erysipeloid *Medicine.* an infection of the hands marked by blue-red bumps or patches and sometimes redness, caused by handling meat or fish infected with *Erysipelothrix rhusicpathiae.*

Erysipelothrix *Bacteriology.* a genus of Gram-positive, nonmotile, rod-shaped or filamentous bacteria, occurring as parasites in mammals, birds, and fish.

Erysiphaceae *Mycology.* a family of parasitic fungi known as the powdery mildews belonging to the order Erysiphales; many species are pathogenic to crop plants.

Erysiphales *Mycology.* an order of parasitic fungi belonging to the class Pyrenomycetes that cause plant diseases such as sooty mold and powdery mildew.

Erysiphe *Mycology.* a genus of parasitic powdery mildew fungi belonging to the order Erysiphales; its species are parasites of such plants as grass and cereals.

erythema [er´ə thē´mə] *Medicine.* any redness of the skin produced by congestion of the capillaries, such as that seen in blushing or sunburn.

erythema multiforme *Medicine.* a hypersensitivity syndrome caused by various known and unknown factors, characterized by inflammatory lesions occurring on the neck, face, and legs.

erythema nodusum *Medicine.* an eruption, usually on the legs, of pink to blueish nodules.
erythema threshold *Radiology.* the quantity of radiation dose required to produce redness of the skin at the point of treatment.
erythematous *Medicine.* characterized by erythema.
erythemogenic *Medicine.* causing erythema.
erythrasma *Medicine.* a chronic, superficial bacterial skin infection caused by *Corynebacterium minutissimum* and characterized by sharply demarcated dry, brown, slightly scaly, and slowly spreading patches appearing on the body folds and toe webs.
erythremia *Medicine.* an abnormal increase in the number of red blood cells, especially associated with polycythemia vera.
erythrene see BUTADIENE.
erythrite *Mineralogy.* $Co_3(AsO_4)_2 \cdot 8H_2O$, a pink to red monoclinic mineral of the vivianite group, occurring as prismatic, commonly striated crystals, or in fibrous forms, and having a specific gravity of 3.09 to 3.18 and a hardness of 1.5 to 2.5 on the Mohs scale; found as a secondary mineral in oxidized zones of cobalt-bearing ore deposits. Also, **erythrine.**
erythritol *Organic Chemistry.* $CH_2OHCHOHCHOHCH_2OH$, white crystals that are soluble in water and slightly soluble in alcohol; melts at 121–122°C; used medicinally as a vasodilator. Also, ERYTHROL.
erythrityl *Organic Chemistry.* C_4H_9, the univalent radical from erythritol.
erythrityl tetranitrate *Pharmacology.* $C_4H_6(NO_3)_4$, a synthetic compound with actions similar to those of nitroglycerin; used as a coronary vasodilator. Also, **erythritol tetranitrate.**
erythro- or **erythr-** a combining form meaning "red," or denoting a relationship to redness or red blood cells, as in *erythrocyte, erythrite.*
Erythrobacter *Bacteriology.* a genus of Gram-negative, pigmented bacteria that possess a respiratory metabolism, occurring as rod-shaped, flagellated cells in seaweeds located at high-tide levels.
erythroblast *Hematology.* a nucleated cell in bone marrow that is the earliest developmental stage of an erythrocyte.
erythroblastosis *Hematology.* the presence of erythroblasts in the peripheral blood.
erythroblastosis fetalis *Medicine.* a type of anemia sometimes occurring in the Rh-positive newborns of Rh-negative mothers, characterized by the presence of numerous erythroblasts in the circulation, edema, and enlargement of the liver and spleen.
erythrochromia *Neurology.* a red appearance of the spinal fluid resulting from the presence of red blood cells and indicating hemorrhage.
erythrocruorin *Biochemistry.* a substance that is present in the blood vessels of some invertebrates, such as marine worms and certain mollusks; it resembles hemoglobin in that it contains a heme and transports oxygen.
erythrocyte *Hematology.* the major cellular element of the peripheral blood, containing hemoglobin and specialized to carry oxygen. In humans, the mature form is normally a nonnucleated, yellowish, biconcave disk that is adapted to carry oxygen by virture of its configuration and hemoglobin content. Also, RED BLOOD CELL, CORPUSCLE.
erythrocythemia *Medicine.* an increase in the number of erythrocytes in the blood, as in erythrocytosis and polycythemia vera.
erythrocytic *Hematology.* relating to or characterized by the presence of red blood cells.
erythrocytopenia *Hematology.* a deficiency in the number of red blood cells; anemia. Also, ERYTHROPENIA.
erythrocytophagous *Hematology.* relating to or affected by erythrocytophagy.
erythrocytophagy *Hematology.* the engulfment or consumption of erythrocytes by other cells, such as the histiocytes of the reticuloendothelial system.
erythrocytopoiesis see ERYTHROPOIESIS.
erythrocytosis *Hematology.* **1.** any abnormal increase in the number of circulating red blood cells; an accompaniment of various disorders. **2.** see SECONDARY POLYCYTHEMIA.
erythroderma *Medicine.* any dermatitis linked with abnormal redness of the skin. Also, **erthrodermia.**
erythrogenesis *Hematology.* the production of erythrocytes.
erythrogenic *Hematology.* **1.** producing erythrocytes. **2.** causing erythema.
erythroidine *Organic Chemistry.* $C_{16}H_{19}NO_3$, an alkaloid existing as either α-erythroidine or β-erythroidine; the β form has a physiological effect similar to that of curare.
erythrol see ERYTHRITOL.
erythroleukemia *Hematology.* a malignant blood disorder characterized by neoplastic proliferation of erythroblastic and myeloblastic elements, with atypical erythroblasts and myeloblasts in the peripheral blood, and showing a variable clinical course.
erythromelalgia *Hematology.* a rare disease affecting chiefly the extremities, marked by paroxysmal, bilateral vasodilatation with burning pain, redness, and increased skin temperature.
erythromycin *Bacteriology.* a macrolide antibiotic elaborated by the bacterium *Streptomyces erythreus* that is most effective against Gram-positive bacteria, inhibiting bacterial protein synthesis. *Pharmacology.* $C_{37}H_{67}NO_{13}$, a preparation of this substance, occurring as a yellow crystalline powder, used in the treatment of infections caused by susceptible organisms, especially in patients who are allergic to penicillin, in infections that are resistant to penicillin, and in legionnaire's disease.

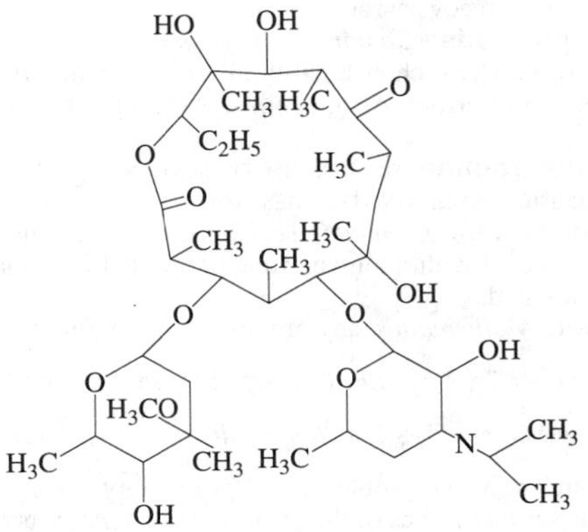

erythromycin

Erythropeltidaceae *Botany.* a small family of marine red algae belonging to the order Bangiales, characterized by discoid and pseudoparenchymatous thalli.
erythropenia see ERYTHROCYTOPENIA.
erythrophage *Hematology.* a phagocyte that takes up erythrocytes and blood pigments.
erythrophilous *Biology.* staining readily with red dyes.
erythrophleine *Organic Chemistry.* $C_{24}H_{39}NO_5$, an alkaloid that is isolated from the bark of *Erythrophleum guineense*; used medicinally.
erythrophobia *Psychology.* an irrational fear of blood or of red-colored objects in general.
erythrophore *Vertebrate Zoology.* a reddish-purple pigment-bearing cell found in the dermis of many vertebrates.
erythropoiesis *Hematology.* the production of red blood cells. Also, ERYTHROCYTOPOIESIS.
erythropoietin *Endocrinology.* a glycoprotein hormone, secreted chiefly by the kidneys in adults and by the liver in children, that acts on the stem cells of the bone marrow to stimulate the production of red blood cells.
erythropsia *Medicine.* an abnormality of vision in which all objects appear red. Also, **erythropia.**
erythrose *Organic Chemistry.* $HOCH_2(CHOH)_2CHO$, a tetrose sugar obtained from erythrol that is very soluble in water and alcohol.
erythrosiderite *Mineralogy.* $K_2Fe^{+3}Cl_5 \cdot H_2O$, a red, orthorhombic mineral occurring as tabular crystals, and having a specific gravity of 2.37 and an undetermined hardness; found as an efflorescence on rinneite.
erythrosin *Organic Chemistry.* $C_{13}H_{18}O_6N_2$, an orange-red crystalline solid, slightly soluble in water and soluble in alcohol.
erythrosin monohydrate see EASIN.
erythrosis *Medicine.* **1.** an overproliferation of erythrocytopoietic tissue. **2.** the unusual red skin color of individuals with polycytemia.
Erythroxylaceae *Botany.* a homogeneous tropical family of dicotyledonous, woody trees and shrubs in the order Linales, characterized by usually alternate, entire leaves and regular, bisexual flowers arranged in radial symmetry; includes the coca plant.
Erzgebirgian orogeny *Geology.* a brief, early Upper Carboniferous crustal deformation involving Phanerozoic rock.
Es the chemical symbol for einsteinium.
ESA or **E.S.A.** European Space Agency.

Esaki, Leo born 1925, Japanese physicist; shared Nobel Prize for research contributing to development of the tunnel (Esaki) diode.

Esaki diode see TUNNEL DIODE.

Esbach's reagent *Pathology.* an assay employed to determine albumin in urine, consisting of one part picric acid and two parts citric acid in 97 parts water.

escalate *Ordnance.* to raise the level of a confrontation, as by introduction of heavier or more lethal weapons than have been so far employed.

escalator *Mechanical Engineering.* a continuously moving stairway on an endless loop, used to carry passengers between two levels in a building.

E scan *Electronics.* see E DISPLAY.

escape *Botany.* **1.** of a cultivated plant, to grow wild. **2.** a usually cultivated plant growing wild in fields or by roadsides, generally surviving but not well naturalized. *Space Technology.* of a rocket, to achieve escape velocity. *Computer Programming.* to prefix a character with an escape character. *Architecture.* see APOPHYGE.

escape behavior *Behavior.* any behavior that removes an animal from a dangerous or threatening situation.

escape character *Computer Programming.* **1.** a special character denoting that the succeeding character or characters belong to a different code than the one currently in use or are to be treated specially. **2.** a character, ESC, in the ASCII character set and on many keyboards. **3.** a character that indicates that the succeeding character is to be taken literally as a character, rather than interpreted as punctuation.

escape conditioning *Behavior.* a type of learning in which an organism is presented with an aversive stimulus from which it may escape by making a particular response. Also, **escape training.**

escape hatch *Engineering.* a small door that provides an alternative exit from a compartment in a submarine or aircraft.

escapement *Mechanical Engineering.* a usually ratcheted mechanism that allows a component to move forward automatically to the next position in an assembly line. *Horology.* specifically, the mechanism in a mechanical timepiece that releases the power driving the timepiece in steady timekeeping impulses, thus controlling the movement of the going train and the timepiece's ability to keep accurate time.

escape orbit *Astronomy.* any orbit whose apoapsis lies at infinity; a parabolic or hyperbolic orbit.

escape rocket *Space Technology.* on a launch pad, a small rocket attached to the leading end of an escape tower; used in an emergency to accelerate and separate the capsule or payload from the booster vehicle.

escape tower *Space Technology.* on a launch pad, a trestle tower situated on top of the space capsule; used to provide thermal protection in case the escape rocket is needed during ascent; if ascent is normal, the tower is separated from the capsule.

escape trunk *Naval Architecture.* an air-lock emergency egress on the upper hull of a submarine, from which crew members can escape if the submarine is trapped under water.

escape velocity *Astronomy.* the velocity that an object needs to reach parabolic or hyperbolic orbit around its primary, which permits it to escape to infinity. *Space Technology.* the speed at which a given spacecraft can overcome the gravitational pull of a planetary atmosphere.

escape wheel or **escapewheel** *Horology.* a toothed wheel in the escapement mechanism of a mechanical timepiece, and the last wheel of the going train, that is periodically stopped and released by the pallets, thus releasing the power driving the timepiece in steady timekeeping impulse. Also, SCAPEWHEEL.

escaping tendency *Physical Chemistry.* **1.** the tendency of a solvent's molecules to pass from the solvent into the solution. **2.** in general, the tendency of molecules to escape from one phase to another.

escarole *Botany.* a broad-leafed variety of endive, *Cichorium endivia,* often used in salads.

escarpment *Geology.* **1.** a more or less continuous cliff or steep slope, generally separating two level or gently sloping areas, that is produced by erosion or faulting. **2.** the steep slope of a cuesta. *Military Science.* ground that has been cut away vertically to prevent an enemy approach to a fortification.

eschar *Medicine.* a scab or dry crust resulting from a burn, infection, or skin disease.

escharotic *Medicine.* caustic, producing a slough or an eschar.

escharotomy *Surgery.* incision of the eschar, or slough, that is formed on the skin and underlying tissues in severe burns; the eschar sometimes forms a tight band around an affected limb and cutting it allows the edges to separate and helps restore circulation to the distal tissues of the limb.

Escherichia [esh´ə rik´ē ə] *Bacteriology.* a genus of Gram-negative, rod-shaped, facultatively anaerobic bacteria of the family Enterobacteriaceae, found among the normal flora of the intestinal tract of many animals. Of important scientific value as a widely studied laboratory microorganism, its principal species is the ubiquitous **E. coli,** the predominant facultative organism of human and animal intestines. (Named for Theodor *Escherich,* 1857–1911, German physician.)

Escherichieae *Bacteriology.* in some classifications, a tribe of enteric, Gram-negative, rod-shaped, facultatively anaerobic bacteria of the family Enterobacteriaceae.

Eschka mixture *Analytical Chemistry.* a mixture of magnesium oxide and anhydrous sodium carbonate, used to detect the presence of sulfur in coal by heating the mixture with the coal sample and adding hydrochloric acid and barium chloride; a precipitate of barium sulfate indicates a positive test.

Eschrictiidae *Vertebrate Zoology.* the gray whale, a monotypic family of the baleen order Mysticeta, characterized by a relatively slender body with no dorsal fin and confined to the northern Pacific Ocean and adjacent areas; at one time, nearly hunted to extinction.

Eschweiler-Clarke modification *Organic Chemistry.* a variation of the Leuckart reaction, involving reductive methylation of primary and secondary amines or ammonia using formaldehyde and formic acid.

E scope *Electronics.* a radarscope that produces an E display.

escort *Military Science.* **1.** a combat force, ship, or aircraft that accompanies a convoy or another force, ship, or aircraft for the purpose of protection from enemy attack. Similarly, **escort forces. 2.** an armed guard that accompanies a convoy, a train, a group of prisoners, personnel, etc., either for protection or as a mark of honor. **3.** to act as an escort.

escort fighter *Aviation.* a fighter aircraft designed to accompany heavy bombers on long-range missions.

esculin *Pharmacology.* $C_{15}H_{16}O_9$, a glycoside obtained from the bark of the horse chestnut tree, *Aesculus hippocastanum,* used to treat fever and sunburn. Also, **esculoside.**

escutcheon [ə skuch´ən] *Mechanical Devices.* a protective metal plate around a panel-mounted key hole, radio dial, window, control knob, or door knob, as a flange, shield, or border. Also, **escutcheon plate.**

escutcheon pin *Mechanical Devices.* a small metal ornamental pin or nail having a half-round head, used to attach an escutcheon.

ESD external symbol dictionary.

ESE east-southeast.

eserine see PHYSOSTIGMINE.

ESF or **E.S.F.** European Science Foundation.

esia *Materials.* the heavy, reddish-brown wood of the *Combretodendron macrocarpum* timber tree of Africa; used in heavy construction.

-esis a suffix meaning "action" or "process," as in *prosthesis.*

eskebornite *Mineralogy.* $CuFeSe_2$, a brass-yellow, opaque, tetragonal mineral of the chalcopyrite group occurring as minute included grains and small tabular crystals, having an undetermined specific gravity and a hardness of 2.5 to 3.5 on the Mohs scale; found in dolomite veins with chalcopyrite.

esker *Geology.* a long, low, steep-sided, narrow, winding ridge of stratified gravel and sand deposited by subglacial streams flowing through crevasses and tunnels in stagnant or retreating glaciers; found in Maine, Canada, Sweden, and Ireland. Also, **eskar, escar, escer.**

Eskimo-Aleut *Linguistics.* a Native-American language family of the Arctic and northern regions; in some schemes, one of the three major American language families.

eso- a combining form meaning "within," as in *esophagus.*

ESOC or **E.S.O.C.** European Space Operations Center.

Esocidae *Vertebrate Zoology.* the pikes, a family of elongated voracious freshwater fish of the order Salmoniformes, coextensive with the genus *Esox* and characterized by a beaklike snout and sharp teeth.

Esocoidei *Vertebrate Zoology.* a small suborder of elongated freshwater fish in the order Salmoniformes, found in the Northern Hemisphere and including the pikes and mudminnows.

esophageal [ə säf´ə jē´əl] *Anatomy.* of or relating to the esophagus.

esophageal diverticulum *Medicine.* a herniation through the posterior pharyngeal wall, more frequently appearing on the left side of the neck.

esophageal fistula *Medicine.* an abnormal tract between the esophagus and some portion of the skin through an external opening.

esophageal glands *Anatomy.* mucous glands in the submucosa of the esophagus.

esophageal hiatus *Anatomy.* the opening in the diaphragm for the passage of the esophagus and the vagus nerves.

esophageal teeth *Vertebrate Zoology.* the series of enamel-tipped hypapophyses of the posterior cervical vertebrae of certain snakes that penetrate the esophagus and act as teeth to break the shells of eggs.
esophagectasia *Medicine.* the dilation of the esophagus.
esophagitis *Medicine.* an inflammation of the lining of the esophagus caused by infection or back flow of gastric juice from the stomach.
esophagocele *Medicine.* a hernia of the esophagus; abnormal distension of the esophagus.
esophagodynia *Medicine.* any pain in the esophagus.
esophagogastrodenoscopy *Medicine.* an endoscopic examination of the esophagus, stomach, and duodenum.
esophagogastroplasty *Surgery.* plastic surgery of the esophagus and stomach. Also, CARDIOPLASTY.
esophagogastrostomy *Medicine.* the establishment of a surgical opening between the esophagus and the stomach.
esophagogram *Radiology.* a radiograph of the esophagus. Also, **esophagram.**
esophagography *Radiology.* the radiography of the esophagus, often aided by the patient's ingestion of a radiopaque medium.
esophagomalacia *Medicine.* the softening of the esophagus.
esophagoptosis *Medicine.* a downward displacement of the esophagus.
esophagus [ə säf′ə gəs] *Anatomy.* a narrow tubular portion of the digestive tract that is approximately 10 inches or 25 centimeters in length and that connects the pharynx and the stomach.

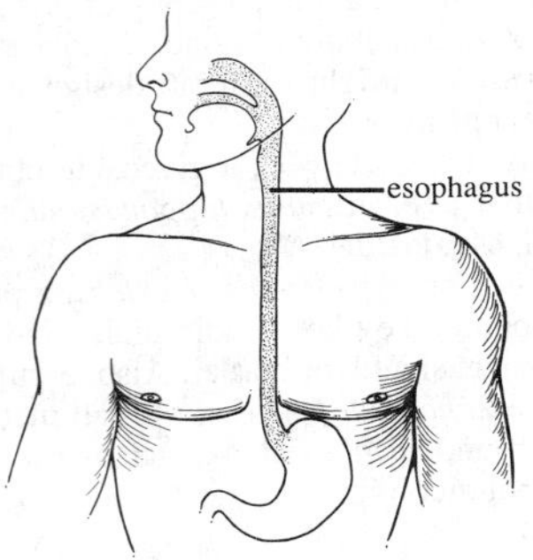

esophagus

esoteric name *Computer Programming.* a name a user assigns to a specific device in order to differentiate between devices of the same type.
esotropia *Medicine.* an inward deviation of one eye in relation to the other eye when the eyes are not focused on an object.
ESP *Psychology.* extrasensory perception; perception by some means other than the five physical senses; e.g., the precognition of future events or thought transference from one person to another.
espalier [i spal′yá; i spal′yər] *Agriculture.* **1.** a trellis or frame on which a fruit tree or shrub is trained to grow, usually in one plane against a wall. **2.** a plant so trained. **3.** to train or grow a plant on an espalier. Thus, **espaliered.**
espalier drainage pattern see TRELLIS DRAINAGE PATTERN.
esparto *Botany.* also, **esparto grass.** any of several Mediterranean grasses, especially *Stipa tenacissima,* used in making paper pulp, cordage, and wax. *Materials.* a hard vegetable wax extracted from esparto grass; often blended with varnishes, polishes, or other waxes.
ESR electron spin resonance; erythrocyte sedimentation rate.
ESRO or **E.S.R.O.** European Space Research Organization.
ESSA environmental survey satellite.
essence *Food Technology.* an extract or other derivative containing one or more distinctive qualities of a food product, such as taste or aroma. *Materials Science.* see ESSENTIAL OIL.
essence recovery *Food Technology.* a process by which volatile flavor constituents are restored to a food after being lost during some stage of processing; usually accomplished by trapping the evolved vapors, condensing them, and re-inoculating them into the food.
essential *Nutrition.* of or relating to any substance necessary for normal development and health maintenance that must be supplied by items in the diet, such as essential amino acids. *Materials Science.* of or relating to an essence or essential oil. *Volcanology.* of pyroclastic material, derived directly from magma.
essential amino acid *Biochemistry.* an amino acid, such as arginine, histidine, and lysine, that cannot be produced by vertebrates and must be supplied by the diet.
essential fatty acid *Biochemistry.* a group of polyunsaturated fatty acids, such as linoleic acid and linolenic acid, that must be included in a mammal's diet.
essential hypertension *Medicine.* high blood pressure for which no cause can be found, and which is often the only disorder.
essentially bounded function *Mathematics.* a measurable function f on a measure space (X, A, μ) is said to be essentially bounded if there exists a real number a such that $\mu(A \cap E) = 0$ for all sets E in A of finite measure, where $A = \{x \in X: |f(x)| > a\}$; i.e., A is locally μ-null.
essential oil *Materials Science.* any of a class of volatile oils extracted from certain plants and exhibiting a plant's characteristic odor; used especially in perfumes, flavorings, and pharmaceuticals.
essential singularity *Mathematics.* a singular point in the domain of a complex function that is neither a pole nor a removable singularity.
essential supremum *Mathematics.* given a measurable function f that is on a measure space (X, A, μ), the infimum of all numbers a such that $\mu(A \cap E) = 0$ for all sets E in A of finite measure, where $A = \{x \in X: |f(x)| > a\}$. Denoted $\|f\|_\infty$, ess sup$|f|$, or vrai max$|f|$.
EST eastern standard time; electroshock therapy; electric shock therapy.
established cell line *Microbiology.* a tissue culture cell population that has undergone transformation and is capable of unlimited growth.
established flow *Fluid Mechanics.* fully developed flow in which the velocity profile shape does not change with increasing distance downstream.
establishment *Oceanography.* the period of time between the transit (upper or lower) of the moon over the local or Greenwich meridian and the next high water.
ester *Organic Chemistry.* any organic compound that is formed by combining an acid with an alcohol, and eliminating water; e.g., caroboxylic acid esters, $R–O–CO–R'$.
esterase *Enzymology.* any of a subclass of enzymes of the hydrolase class that catalyze the hydrolysis of ester bonds, yielding an alcohol or phenol and an acid anion.
Estérel twin law *Crystallography.* a twin law specifying the relationship between parallel twins in feldspar in which the *a*-axis is the twin axis and the composition plane is parallel to the *a*-axis.
ester exchange *Organic Chemistry.* a reaction between an ester and another compound in which an exchange of acyl groups occurs, resulting in the formation of another ester; used in the production of polyesters, plasticizers, and polycarbonates.
ester gum *Organic Chemistry.* a hard resin formed by esterification of natural resins with polyhydric alcohol; used in paints, varnishes, and lacquers.
esterification *Organic Chemistry.* the process of ester formation by the reaction of acid with an alcohol, catalyzed by hydrogen ions.
ester interchange see INTERESTERIFICATION.
estersil *Organic Chemistry.* an ester of –SiOH and a monohydric alcohol; used as a filler in silicone rubbers and plastics, and in printing inks.
estetrol *Endocrinology.* an estrogenic steroid, 15,16-dihydroxyestradiol, that is produced from estriol by the fetal liver and crosses the placenta to appear in the maternal blood; its levels may be monitored during the last trimester of pregnancy to provide an index of fetal health.
esthacyte *Invertebrate Zoology.* a simple sensory cell found in lower animals, such as sponges.
esthematology *Neurology.* the scientific study of the senses and sense organs.
esthesia *Physiology.* the perception of sensory impressions. Thus, **esthesic.**
esthesio- a combining form meaning "feeling," as in *esthesiogenic.*
esthesiometer *Engineering.* a device used to measure sensory discrimination by calculating the distance by which two points must be separated when pressed against the skin, so that each is distinctly felt. Also, AESTHESIOMETER.
esthesioneuroblastoma *Medicine.* a radiosensitive glioma occurring in the nasal cavity.
esthesioneurosis *Neurology.* any disorder of the sensory nerves.
esthetic *Neurology.* of or relating to the mental perception of sensations.
esthetic surgery see COSMETIC SURGERY.
esthiomene *Medicine.* a chronic ulcerative lesion of the vulva, due to lymphogranuloma venereum.

estimate *Science.* **1.** a statement based on an approximate measurement of a value; a rough calculation. **2.** to make such a statement or calculation. *Statistics.* a value computed from sample data used to approximate a population parameter.
estimated position *Navigation.* the most probable position of a craft that can be determined from incomplete or doubtful information.
estimated standard deviation *Crystallography.* a measure of the precision of a quantity. If the distribution of errors follows a normal curve, there is a 99% chance that a given measurement will differ by less than 2.7 e.s.d. from the mean.
estimated time *Industrial Engineering.* the predicted time that an industrial work element or operation can be expected to require, determined without making a detailed study.
estimated time of arrival *Navigation.* an estimate, based on current conditions, of the time a craft will arrive at its destination. Also, ETA.
estimated time of departure *Navigation.* the projected time of departure of a craft, especially from a future intended intermediate stop, equal in that case to the estimated time of arrival at that point plus the estimated stay time there. Also, ETD.
estimated time of interception *Navigation.* the estimated time at which a craft or missile will intercept a target.
estimation *Statistics.* the inference about the value of unknown population parameters, based on information contained in a sample drawn from the population.
estimator *Statistics.* the rule for deriving an estimate from a sample.
estival *Astronomy.* of or relating to summer.
estivate *Biology.* to pass the summer in a certain manner or condition, often in a dormant or torpid state.
estivation *Zoology.* a dormant state of decreased metabolism in which certain animals, such as tropical amphibians, survive hot summer in a torporous condition. *Botany.* the manner in which petals and sepals are folded in the flower bud; the arrangement of the perianth or its parts in the bud. Also, AESTIVATION.
estivoautumnal *Astronomy.* of or relating to summer and autumn.
estradiol-17α *Endocrinology.* a less active isomer of the estrogenic steroid hormone estradiol-17β. Also, **α-estradiol.**
estradiol-17β *Endocrinology.* the main and most potent estrogenic steroid hormone; secreted by the ovaries, it regulates female sexual development and reproductive function. Estradiol also sustains the proliferative phase of the uterine endometrium during the menstrual cycle and acts on other tissues, for example in the breast and central nervous system. Also, **β-estradiol.**

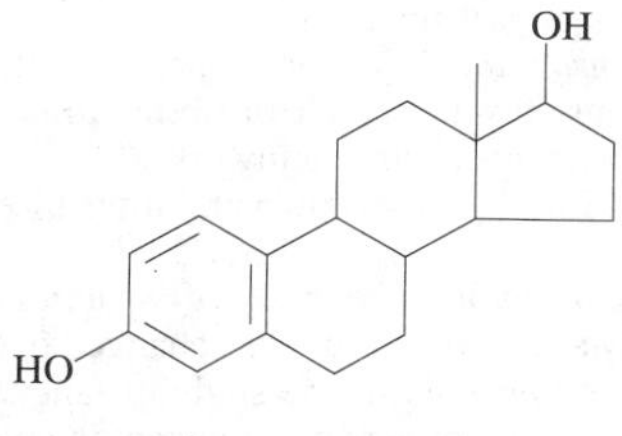

estradiol-17β

estragole *Organic Chemistry.* $C_6H_4(C_3H_5)(OCH_3)$, a colorless liquid that is soluble in alcohol; boils at 216°C; used in perfumery and flavors.
estrane *Endocrinology.* the parent steroidal precursor of the estrogens.
estriol *Endocrinology.* an estrogenic steroid hormone less potent than estradiol, formed from the conversion of estradiol or estrone in the liver, or from placental conversion of dehydroepiandrosterone from fetal sources. Also, TRIHYDROXYESTRIN.
estrogen [es´trə jin] *Endocrinology.* any of a family of steroid hormones that regulate and sustain female sexual development and reproductive function. Also, **estrogenic hormone.**
estromedins *Endocrinology.* any of several polypeptide growth factors produced by the uterus, kidney, or liver in response to estrogenic stimulation.
estrone *Endocrinology.* a minor estrogenic steroid hormone, produced chiefly from peripheral conversion of androgenic steroids.
estrous cycle *Physiology.* the recurring periods of heat, or estrus, in the adult female of most mammals and the correlated changes in the reproductive tract from one period to the next.
estrus *Physiology.* a recurrent, restricted period of sexual receptivity in female mammals other than humans, marked by intense sexual urge. Also, HEAT.
estuarine *Geology.* formed, occurring in, or characteristic of estuaries.
estuarine deposit *Geology.* a sedimentary deposit laid down at the head or floor of an estuary, and characterized by fine-grained, clayey, or silty sediments mixed with decomposed organic matter.
estuarine environment *Oceanography.* the physical conditions that characterize an estuary.
estuarine oceanography *Oceanography.* the study of the physical, geological, chemical, and biological characteristics of estuaries.
estuary *Geography.* an arm of the sea at the mouth of a river, or the drowned mouth of a river, where the tide meets the river current.
ESU or **esu** electrostatic unit.
ESV earth satellite vehicle.
ET eastern time; ephemeris time.
Et ethyl group.
ETA estimated time of arrival.
eta [ā´tə; ē´tə] the seventh letter of the Greek alphabet, written as Η or η.
Eta Aquarids *Astronomy.* a meteor shower derived from Halley's Comet that reaches peak activity around May 4; it appears to originate near the star Eta in Aquarius.
Eta Carinae Nebula *Astronomy.* a large region of ionized hydrogen surrounding the erratically variable supergiant star Eta Carinae.
eta meson *Particle Physics.* a neutral pseudoscalar meson having positive charge parity, zero isotopic spin, zero hypercharge, mass 549 MeV, and a mean lifetime of approximately 10^{-18}.
Etard reaction *Organic Chemistry.* the oxidation of methylated homologs of benzene into aromatic aldehydes by use of chromyl chloride, CrO_2Cl_2.
état [ā´tä´] *Medicine.* a physical state or condition. (French for "state.")
etch *Metallurgy.* **1.** to apply corrosive materials to a metallic surface in order to clean it or increase its roughness. **2.** the process of etching. *Graphic Arts.* **1.** to carry out any of the processes known as etching. **2.** an acid or other corrosive agent used in such a process. *Geology.* to cut by erosion.
etch crack *Metallurgy.* a crack caused by etching a stressed metal or alloy.
etched circuit *Engineering.* a printed circuit that is formed when unwanted parts of a layer of conductive material attached to an insulated base are removed.
etch figures *Crystallography.* cavities that form on the surface of a crystal when it is attacked by certain solvents. The symmetries of these cavities will give information on the symmetry of the underlying structure.
etching *Graphic Arts.* **1.** in intaglio printing, the process of incising an image on an acid-resistant plate, and then treating the plate with acid. **2.** an intaglio print. **3.** in deep-etch printing, the process of applying an acid solution to remove a hardened gum solution from nonimage areas of a plate. **4.** in lithography, the process of applying a chemical solution designed to increase water receptivity on nonimage areas of a plate.

etching

ETD estimated time of departure.

etesian *Meteorology.* any of the prevailing northerly summer winds in the eastern Mediterranean, especially the Aegean Sea, that bring clear skies and dry, cool weather.

etesian climate see MEDITERRANEAN CLIMATE.

ethanal see ACETALDEHYDE.

ethane *Organic Chemistry.* C_2H_6, a flammable, colorless gas, insoluble in water and soluble in alcohol; boils at –88.63°C; used in organic synthesis, as a fuel, and in refrigeration. Also, DIMETHYL.

1,2-ethanedithiol *Organic Chemistry.* $HSCH_2CH_2SH$, a liquid that is soluble in alcohol; boils at 144–146°C; used as a metal complexing agent.

ethanethiol *Organic Chemistry.* C_2H_5SH, a toxic, flammable, colorless liquid with a penetrating skunklike odor, slightly soluble in water and soluble in alcohol; boils at 36°C; used as a liquified natural gas odorant, stabilizer, and adhesive. Also, ETHYL MERCAPTAN, THIOETHYL ALCOHOL.

ethanoic acid see ACETIC ACID.

ethanol *Organic Chemistry.* C_2H_5OH, a flammable, colorless, volatile liquid, miscible with water and alcohol; melts at –117.3°C and boils at 78.5°C; used as a resin and fat solvent, and in antifreeze, beverages, and gasohol. Also, ETHYL ALCOHOL, GRAIN ALCOHOL.

ethanolamine *Organic Chemistry.* $HOCH_2CH_2NH_2$, a combustible, colorless, hygroscopic liquid, miscible with water; boils at 170.5°C; used in dry-cleaning fluids and paints, and as a corrosion inhibitor and rubber accelerator.

ethanolurea *Organic Chemistry.* $NH_2CONHCH_2CH_2OH$, a solid that melts at 71–74°C; its formaldehyde condensation products are water-soluble and permanently thermoplastic.

ethedine see ETHYLIDINE.

ethene see ETHYLENE.

ethenol see VINYL ALCOHOL.

ethephon *Organic Chemistry.* $C_2H_6ClO_3P$, a white solid that is very soluble in water; melts at 75.74°C; used as a growth regulator for fruit. Also, 2-CHLOROETHYLPHOSPHONIC ACID.

ether *Pharmacology.* a preparation of ethyl ether (or other ether compound, such as vinyl ether) used as an inhalational anesthetic. *Organic Chemistry.* any organic compound having an oxygen atom bonded to two carbon atoms, with the general formula *R–O–R′*. *Electromagnetism.* a hypothetical pervasive medium once thought to support the propagation of electromagnetic waves.

ether drag *Electromagnetism.* a hypothesis claiming that ether is dragged along with the motion of matter.

ether drift *Electromagnetism.* a hypothesis claiming that ether moves relative to the motion of the earth.

ethereal *Organic Chemistry.* relating to, prepared with, containing, or resembling an ether.

etherhydrolase *Enzymology.* an enzyme of the hydrolase class that catalyzes the hydrolysis of an ether.

etherification *Organic Chemistry.* the process of producing ether from an alcohol.

etherize *Medicine.* to put under the anesthetic influence of ethyl ether by inhalation.

ethernet *Computer Technology.* a high-speed local area network using coaxial cable, broadcast transmission, and variable time delays in the event of a collision.

ether thermoscope *Physics.* a simple device used for detecting radiant heat, constructed of an evacuated J-shaped tube with bulbed ends, the lower end blackened and partially filled with ether so that with exposure to radiant heat, the vapor pressure in the lower bulb causes the ether to rise up the length of the tube.

ethidene or **ethidine** see ETHYLIDENE.

ethidium bromide *Organic Chemistry.* $C_{21}H_{20}BrN_3$, dark red crystals that melt at 238–240°C; used in veterinary medicine and as an analytical dye. *Biochemistry.* a form of this dye that prevents transcription and DNA replication by binding to specific regions of the DNA molecule, such as dsDNA.

ethinyl *Organic Chemistry.* $CH_3–C≡$; a radical from acetylene.

Ethiodol *Radiology.* a trademark for a preparation of ethiodized oil, used as a radiopaque medium.

ethiolate *Organic Chemistry.* $C_7H_{15}ONS$, a yellow liquid that boils at 206°C; used as a preemergence herbicide.

ethionic acid *Organic Chemistry.* $HOSO_2CH_2CH_2SO_2OH$, an unstable diacid found only in solution. Also, ETHYLENESULFONIC ACID.

ethionine *Biochemistry.* $C_6H_{13}NO_2S$, an amino acid that is the biological antagonist and ethyl homologue of methionine.

Ethiopian region *Ecology.* a zoogeographical region extending from the mid-Sahara Desert to southern Africa that has served as a major center of evolution, producing a number of unique species such as great apes, elephants, and giraffes. It includes the East African, West African, Malagasy, and South African subregions.

ethirimol *Organic Chemistry.* $C_{11}H_{19}N_3O$, a crystalline solid that is soluble in water and acids; melts at 159–160°C; used as a fungicide.

ethmo- a combining form meaning "sieve," as in *ethmocarditis.*

ethmocarditis *Cardiology.* inflammation of the connective tissue of the heart.

ethmoid *Anatomy.* having small perforations like a sieve.

ethmoid bone *Anatomy.* a fragile cubical bone located in the anterior portion of the base of the cranium, forming the lateral wall of the nasal cavity and the medial wall of each orbit.

ethmoid sinus *Anatomy.* either of two sinuses located in the lateral portions of the ethmoid bone. They consist of several groups of air-filled sacs known as the **ethmoidal cells.**

ethmolith *Geology.* a discordant igneous intrusion having a funnel-shaped cross section.

ethmoturbinal *Anatomy.* of or relating to the superior and middle nasal conchae. Also, **ethmoturbinate.**

ethnic *Anthropology.* **1.** relating to or belonging to an ethnic group. **2.** in popular use, a member of an ethnic group.

ethnic group *Anthropology.* an identifiable group that has historically retained certain distinctive cultural traits; e.g., the Basque people of southwestern Europe.

ethnicity *Anthropology.* **1.** the fact of belonging to an ethnic group. **2.** ethnic traits in general.

ethnoarchaeology *Archaeology.* the use of archaeological techniques and data to study living cultures, especially current or recent aboriginal groups such as the Inuit or Bushmen.

ethnobiology *Anthropology.* the study of the way various cultural groups make use of or interact with the animals and plants of their environment. Thus, **ethnobiologist, ethnobiological.**

ethnobotany *Anthropology.* the study of the way plants are identified, classified, and used by various cultural groups. Thus, **ethnobotanist, ethnobotanical.**

ethnocentric relating to or characterized by ethnocentrism.

ethnocentrism *Anthropology.* **1.** the tendency to view other cultures in terms of the values or customs of one's own culture. **2.** specifically, the practice of regarding traditional, nontechnological cultures as inferior to industrialized Western cultures. *Psychology.* the belief or attitude that one's own culture, ethnic group, or society is superior to all others. Also, **ethnocentricism.**

ethnographic analogy *Anthropology.* the comparative use of studies from living societies, especially aboriginal or nontechnological societies, in the analysis of past societies.

ethnography *Anthropology.* the comprehensive, descriptive study of a particular culture, usually the result of observation and in-depth interviews with key informants during fieldwork.

ethnohistorical *Anthropology.* of or relating to ethnohistory. Also, **ethnohistoric.**

ethnohistory *Anthropology.* the study of written and oral histories in the analysis of how a specific culture has changed over time.

ethnolinguistics *Anthropology.* the study of language as an aspect of culture, particularly of the effects of language on culture and culture on language.

ethnological *Anthropology.* **1.** relating to or based on culture. **2.** of or relating to ethnology. Also, **ethnologic.**

ethnology *Anthropology.* the comparison of specific features of different cultures in an attempt to establish similarities and disparities between them. Thus, **ethnologist.**

ethnomusicology *Anthropology.* the study of the traditional or folk music of a particular culture, especially one that is not part of the European classical genre, including sounds, songs, instruments, and associated ceremonies.

ethnopsychology *Psychology.* a branch of psychology that studies the influence of cultural and ethnic factors on behavior.

ethnoscience *Anthropology.* the study of the system of knowledge of nature and the physical world held by a particular cultural group, especially a group not employing formal scientific data and analysis. Thus, **ethnoscientist, ethnoscientific.**

ethnosemantics *Anthropology.* the study of how the members of a culture use language to describe certain fundamental and universal classifications, such as color, kinship, weather, plants, and animals.

ethnozoology *Anthropology.* the study of the way various cultural groups make use of, interact with, or classify the animals of their environment. Thus, **ethnozoologist, ethnozoological.**

ethobrom see TRIBROMOETHANOL.

ethogram *Behavior.* a survey or description of all the characteristic actions and behaviors of a given species. Also, BEHAVIORAL INVENTORY.

ethohexadiol *Organic Chemistry.* $C_3H_7CH(OH)CH(C_2H_5)CH_2OH$, a combustible, colorless, slightly viscous, hygroscopic liquid that is partially soluble in water and soluble in alcohol; boils at 244°C; used in cosmetics, and as an insect repellant and chelating agent for boric acid. Also, **2-ethylhexanediol-1,3.**

ethologist *Psychology.* a person whose work or speciality is ethology.

ethology *Behavior.* the study of animal behavior, especially of animals in their natural environment rather than in a laboratory or in captivity.

ethopropazine hydrochloride *Pharmacology.* $C_{19}H_{24}N_2S{\cdot}HCl$, a phenothiazine derivative, used as an antiparkinsonian agent.

ethoxide *Organic Chemistry.* a compound formed when a monovalent metal replaces the hydrogen of the hydroxy group in ethanol.

ethoxy *Organic Chemistry.* C_2H_5O–, a radical from ethyl alcohol.

2-ethoxyethanol see CELLOSOLVE.

ethoxylation *Chemical Engineering.* a catalytic procedure in which ethylene oxide is added directly to an aliphatic alcohol or to an alkyl phenol.

ethoxyquin *Organic Chemistry.* $C_{14}H_{19}NO$, a toxic yellow liquid that melts at approximately 0°C and boils at 125°C (2 torr); used as an insecticide, antioxidant, and growth regulator for pears and apples in storage. Also, **6-ethoxy-1,2-dihydro-2,2,4-trimethylquinoline.**

ethyl *Organic Chemistry.* the hydrocarbon radical $CH_3{\cdot}CH_2$, usually written C_2H_5 or Et-.

ethyl acetate *Organic Chemistry.* $CH_3COOC_2H_5$, a toxic, flammable, colorless liquid, slightly soluble in water and soluble in alcohol; boils at 77°C and melts at –83.6°C; used as a general solvent and in organic synthesis, smokeless powders, and pharmaceuticals.

ethyl acetoacetate *Organic Chemistry.* $CH_3COCH_2COOC_2H_5$, a combustible, toxic, colorless liquid that is soluble in water; it boils at 180–181°C; used in organic synthesis, lacquers, vitamin B, antimalarials, and flavorings. Also, DIACETIC ESTER.

ethylacetylene *Organic Chemistry.* $C_2H_5C{\equiv}CH$, a flammable, colorless gas that is insoluble in water and soluble in alcohol; boils at 8.3°C; used as a specialty fuel and in organic synthesis. Also, 1-BUTYNE.

ethyl acrylate *Organic Chemistry.* $CH_2{=}CHCOOC_2H_5$, a toxic, flammable, colorless liquid that is soluble in alcohol; boils at 99.4°C; used as a monomer for acrylic resins.

ethyl alcohol see ETHANOL.

ethylamine *Organic Chemistry.* $CH_3CH_2NH_2$, a flammable, colorless, volatile liquid that is soluble in water and alcohol; boils at 16.6°C; used as a dye intermediate and in detergents, petroleum refining, and organic synthesis.

ethyl *p*-aminobenzoate hydrochloride *Organic Chemistry.* $C_6H_4NH_2CO_2C_2H_5{\cdot}HCl$, a toxic, white, crystalline powder that is very slightly soluble in water and alcohol and soluble in dilute acids; melts at 88–92°C; used in medicine and suntan preparations. Also, ANESTHESOL, BENZOCAINE, PROCAINE HYDROCHLORIDE.

ethyl amyl ketone *Organic Chemistry.* $CH_3CH_2CO(CH_2)_4CH_3$, a narcotic, combustible, colorless liquid that is insoluble in water and soluble in alcohol; boils at 157°C; used in perfumery and as a solvent for nitrocellulose.

ethylation *Organic Chemistry.* the formation of a compound by introduction of the ethyl group (C_2H_5).

ethylbenzene *Organic Chemistry.* $C_6H_5C_2H_5$, a toxic, flammable, colorless liquid that is almost insoluble in water and soluble in alcohol; boils at 136.2°C; used as a solvent and in the production of styrene. Also, PHENYLETHANE.

ethyl benzoate *Organic Chemistry.* $C_6H_5CO_2C_2H_5$, a colorless liquid, insoluble in water and soluble in alcohol; boils at 212.9°C; used in flavoring, perfumery, and as a solvent for cellulose derivatives.

ethyl borate *Organic Chemistry.* $(C_2H_5)_3BO_3$, a flammable, colorless liquid that is soluble in water; boils at 120°C; used in antiseptics, disinfectants, and antiknock compounds. Also, TRIETHYL BORATE.

ethyl bromide *Organic Chemistry.* C_2H_5Br, a toxic, flammable, colorless liquid, soluble in water and alcohol, that boils at 38.4°C; used in organic synthesis and as a refrigerant, anesthetic, and fumigant.

2-ethyl-1-butene *Organic Chemistry.* $CH_3CH_2(C_2H_5)C{=}CH_2$, a combustible, colorless liquid, soluble in alcohol and insoluble in water; boils at 64.95°C; used in organic synthesis. Also, UNS-DIETHYLETHYLENE.

2-ethylbutyl acetate *Organic Chemistry.* $C_2H_5CH(C_2H_5)CH_2O$ $OCCH_3$, a colorless liquid with a boiling range of 155–164°C; used as a nitrocellulose solvent and in flavoring.

2-ethylbutyl alcohol *Organic Chemistry.* $CH_3CH_2CH(C_2H_5)CH_2OH$, a combustible, colorless, stable liquid, slightly soluble in water and miscible in most organic solvents; boils at 148.9°C; used as a solvent and in perfumery, drug synthesis, and flavoring.

ethyl butyl ketone *Organic Chemistry.* $CH_3(CH_2)_3COCH_2CH_3$, a combustible, clear liquid that is insoluble in water and soluble in alcohol, with a boiling range of 142.8–147.8°C; used as a solvent mix.

ethyl butyrate *Organic Chemistry.* $C_3H_7CO_2C_2H_5$, a flammable, colorless liquid, almost insoluble in water and soluble in alcohol; boils at 121°C; used in flavoring extracts and perfumes, and as a solvent.

ethyl caprate *Organic Chemistry.* $C_9H_{19}COOC_2H_5$, a combustible, colorless liquid that is insoluble in water and soluble in alcohol; boils at 245°C; used in organic synthesis and as a flavoring, especially in wine. Also, ETHYL DECANOATE.

ethyl caproate *Organic Chemistry.* $C_5H_{11}COOC_2H_5$, a combustible, colorless to yellowish liquid, insoluble in water and soluble in alcohol; boils at 167°C; used in artificial fruit essences and organic synthesis. Also, ETHYL HEXOATE.

ethyl caprylate *Organic Chemistry.* $CH_3(CH_2)_6COOC_2H_5$, a combustible, colorless liquid that is insoluble in water and soluble in alcohol; boils at 207–209°C; used in flavorings and artificial fruit essences. Also, ETHYL OCTOATE.

ethyl carbamate see URETHANE.

ethylcellulose *Organic Chemistry.* a combustible, white, granular thermoplastic solid that has a softening range of 100–130°C, and that is soluble in most organic solvents and insoluble in water; used in hot-melt adhesives and coatings, paper, textiles, printing inks, and food and feed additives.

ethyl chloride *Organic Chemistry.* C_2H_5Cl, a highly flammable gas at room temperature, and a colorless, volatile liquid when compressed; slightly soluble in water and miscible with most organic solvents; boils at 12.5°C; used as an anesthetic and solvent, and in organic synthesis and insecticides.

ethyl chloroacetate *Organic Chemistry.* $ClCH_2CO_2C_2H_5$, a combustible, water-white, mobile liquid, insoluble in water and soluble in alcohol; boils at 144.2°C; used as a solvent and in organic synthesis, military poison gas, and dyes.

ethyl cinnamate *Organic Chemistry.* $C_6H_5CH{=}CHCOOC_2H_5$, a combustible, limpid, oily liquid; insoluble in water and soluble in alcohol; boils at 271°C; used in perfumery and flavoring.

ethyl crotonate *Organic Chemistry.* $CH_3CH{=}CHCOOC_2H_5$, a flammable, predominantly *trans*, water-white liquid, insoluble in water and soluble in alcohol; boils at 142–143°C; used as a solvent, a softening agent, and in lacquers and organic synthesis.

ethyl crotonic acid *Organic Chemistry.* $CH_3CH{=}C(C_2H_5)CO_2H$, colorless, monoclinic crystals that sublime at 40°C; used as a peppermint flavoring.

ethyl cyanide *Organic Chemistry.* C_2H_5CN, toxic, flammable, colorless liquid, soluble in water and alcohol, that boils at 97.4°C; used as a dielectric fluid.

ethyl decanoate see ETHYL CAPRATE.

***S*-ethyl *N,N*-dipropylthiocarbamate** *Organic Chemistry.* C_2H_5SC $(O)N(C_3H_7)_2$, an amber liquid that is slightly soluble in water; used as a preemergence herbicide.

ethyl enanthate *Organic Chemistry.* $CH_3(CH_2)_5COOC_2H_5$, a combustible, clear, colorless oil, insoluble in water and soluble in alcohol; boils at 187°C; used as an artificial flavoring in liquors and soft drinks. Also, COGNAC OIL.

ethylene *Organic Chemistry.* $H_2C{=}CH_2$, a highly flammable, colorless gas, slightly soluble in water and alcohol; boils at –103.9°C; used in plastic and resin preparation, in welding and cutting metals, and as an anesthetic, a refrigerant, and a fruit ripening accelerator. Also, BICARBURETTED HYDROGEN, ETHENE.

ethylene carbonate *Organic Chemistry.* $(CH_2O)_2CO$, a combustible, colorless, odorless solid or liquid, soluble in water and alcohol; melts at 36.4°C and boils at 248°C; used as a solvent, plasticizer, and intermediate, and in organic synthesis. Also, DIOXOLONE-2.

ethylene chloride *Organic Chemistry.* $ClCH_2CH_2Cl$, a toxic, carcinogenic, flammable, oily liquid that is slightly soluble in water and miscible with most common solvents; boils at 83.7°C; used as a solvent and fumigant, and in lead-containing anti-knock gasoline. Also, ETHYLENE DICHLORIDE, DUTCH OIL.

ethylene chlorobromide *Organic Chemistry.* CH_2BrCH_2Cl, a colorless, volatile liquid that is insoluble in water and soluble in alcohol; it boils at 107–108°C; it is used as a solvent and fumigant, and in organic synthesis.

ethylene chlorohydrin *Organic Chemistry.* $ClCH_2CH_2OH$, a colorless liquid, miscible in water and alcohol; melts at –67.5°C and boils at 128°C; used as a solvent, in organic synthesis, and in insecticides. Also, 2-CHLOROETHYL ALCOHOL.

ethylene cyanide *Organic Chemistry.* $C_2H_4(CN)_2$, a combustible, waxy, colorless solid, soluble in water and alcohol; melts at 57–57.5°C; used in organic synthesis. Also, SUCCINONITRILE.

ethylene cyanohydrin *Organic Chemistry.* $HOCH_2CH_2CN$, a combustible, toxic, straw-colored liquid that is miscible with water; it boils at 227– 228°C with slight decomposition; it is used as a solvent and intermediate.

ethylenediamine *Organic Chemistry.* $NH_2CH_2CH_2NH_2$, a toxic, colorless liquid soluble in water and alcohol; boils at 116–117°C; used in the manufacture of chelating agents, and as a fungicide, solvent, and corrosion inhibitor.

ethylenediaminetetraacetic acid *Organic Chemistry.* $(HOOCCH_2)_2NCH_2CH_2N(CH_2COOH)_2$, colorless crystals that are slightly soluble in water and insoluble in most common organic solvents; decomposes at 240–250°C; used in detergents and as a metalchelating agent, a blood anticoagulant, and a food preservative.

ethylene dibromide *Organic Chemistry.* $BrCH_2CH_2Br$, a nonflammable, toxic, carcinogenic, colorless liquid that is slightly soluble in water and miscible with most organic solvents; boils at 131°C; used as a solvent and fumigant.

ethylene dichloride see ETHYLENE CHLORIDE.

ethylene glycol bis(β-aminoethyl ether) *N,N′*-tetraacetic acid *Organic Chemistry.* $[-CH_2OC_2H_4N(CH_2COOH)_2]_4$, crystals that are soluble in water and that decompose at a melting point of 241°C; used as a chelating agent.

ethylene glycol bis(trichloroacetate) *Organic Chemistry.* $C_4H_4Cl_6O_4$, a white solid that melts at 40.3°C; used as an herbicide.

ethylene glycol diacetate *Organic Chemistry.* $CH_3COOCH_2CH_2OOCCH_3$, a combustible, colorless liquid, slightly soluble in water and soluble in alcohol; boils at 190.5°C; used as a solvent and plasticizer, and in printing inks and explosives.

ethylene glycol monophenyl ether see PHENOXYETHANOL.

ethyleneimine *Organic Chemistry.* $(CH_2)_2NH$, a colorless liquid that is soluble in water; boils at 57°C; used as an intermediate and stabilizer. Also, AZIRIDINE.

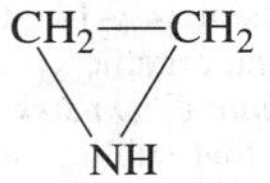

ethyleneimine

ethylene nitrate *Organic Chemistry.* $O_2NOCH_2CH_2ONO_2$, a yellow liquid that is insoluble in water. Also, GLYCOL DINITRATE.

ethylene oxide *Organic Chemistry.* $(CH_2)_2O$, a highly flammable, probably carcinogenic, colorless gas or liquid that is soluble in organic solvents and miscible with water and alcohol; boils at 10.73°C; used in ethylene glycol manufacture, surfactants, as a petroleum demulsifier, a fumigant, and a sterilizer, and for various other industrial purposes. Also, OXIRANE, EPOXYETHANE.

ethylene resin *Organic Chemistry.* a thermoplastic material that is composed of ethylene polymers derived from a synthesis achieved at elevated temperature and pressure in the presence of catalysts. Also, POLYETHYLENE.

ethylenesulfonic acid see ETHIONIC ACID.

ethyl ester see TABUN.

ethylethanolamine *Organic Chemistry.* $C_2H_5NHCH_2CH_2OH$, a combustible, colorless liquid that is soluble in water and alcohol; boils at 167–169°C; used as a solvent and intermediate.

ethyl ether *Organic Chemistry.* $(C_2H_5)_2O$, a very flammable, colorless, mobile liquid, slightly soluble in water and soluble in alcohol; it melts at –116.3°C and boils at 34.6°C; used as an anesthetic and industrial solvent, and in organic synthesis. Also, DIETHYL ETHER, SULFURIC ETHER. *Pharmacology.* see ETHER.

ethylethylene see 1-BUTENE.

ethyl formate *Organic Chemistry.* $HCOOC_2H_5$, a narcotic, highly flammable, water-white liquid, slightly soluble in water and soluble in alcohol; boils at 54.3°C and melts at –80°C; used as a solvent, fumigant, and larvicide, and in synthetic flavors.

ethyl hexanoate see ETHYL CAPROATE.

2-ethylhexanoic acid *Organic Chemistry.* $C_4H_9CH(C_2H_5)COOH$, a combustible liquid, slightly soluble in water; boils at 226.9°C and freezes at –83°C; used as a paint and varnish drier. Also, BUTYLETHYLACETIC ACID.

2-ethyl-1-hexanol *Organic Chemistry.* $CH_3(CH_2)_3CH(C_2H_5)CH_2OH$, a combustible, colorless liquid, slightly soluble in water and miscible with most organic solvents; boils at 183.5°C; used as a plasticizer, defoaming and wetting agent, and in organic synthesis, paints, textile finishing, rubber, and dry cleaning. Also, OCTYL ALCOHOL.

2-ethylhexyl acetate *Organic Chemistry.* $CH_3COOCH_2CHC_2H_5C_4H_9$, a combustible, water-white, stable liquid that is very slightly soluble in water and miscible with alcohol; boils at 198.6°C; used as a solvent.

2-ethylhexyl acrylate *Organic Chemistry.* $CH_2{=}CHCOOCH_2CH(C_2H_5)C_4H_9$, a combustible liquid that is insoluble in water and boils at 214–218°C; used as a monomer in plastics, and in paper treatment and water-based paints.

2-ethylhexylamine *Organic Chemistry.* $C_4H_9CH(C_2H_5)CH_2NH_2$, a combustible, colorless liquid, soluble in water; boils at 168–170°C; used in the synthesis of rubber chemicals, detergents, oil additives, and insecticides.

2-ethylhexyl bromide *Organic Chemistry.* $C_4H_9CH(C_2H_5)CH_2Br$, a water-white liquid that is insoluble in water and boils at 56–58°C (8 torr); used in the preparation of disinfectants and pharmaceuticals.

2-ethylhexyl chloride *Organic Chemistry.* $C_4H_9CH(C_2H_5)CH_2Cl$, a combustible, colorless liquid that is insoluble in water and boils at 172.9°C; used in the synthesis of cellulose derivatives, dyes, pharmaceuticals, textile auxiliaries, and insecticides.

ethyl-*p*-hydroxybenzoate see ETHYLPARABEN.

ethylidene or **ethylidine** *Organic Chemistry.* a $CH_3{-}CH{=}$ radical from C_2H_5, ethane. Also, ETHEDENE, ETHEDINE.

ethyl iodide *Organic Chemistry.* C_2H_5I, a combustible, toxic, narcotic, colorless liquid that turns brown on exposure to light, slightly soluble in water and soluble in alcohol; boils at 72°C and melts at –108°C; used in medicine and organic synthesis. Also, IODOETHANE.

ethylism *Toxicology.* poisoning due to excessive ingestion of ethyl alcohol.

ethyl isovalerate *Organic Chemistry.* $(CH_3)_2CHCH_2COOC_2H_5$, a combustible, colorless, oily liquid, slightly soluble in water and miscible with alcohol; boils at 135°C and melts at –99°C; used in perfumery, flavoring, and artificial fruit essences. Also, ETHYL 2-METHYLBUTYRATE.

***N*-ethylmaleimide** *Organic Chemistry.* $C_6H_7NO_2$, irritating crystals or liquid; soluble in methanol; crystals melt at 45°C; used in cancer research.

ethyl malonate *Organic Chemistry.* $CH_2(COOC_2H_5)_2$, a combustible, colorless liquid, slightly soluble in water and soluble in alcohol; boils at 198°C and melts at –50°C; used as a barbiturate synthesis intermediate and in flavoring. Also, MALONIC ESTER.

ethyl mercaptan see ETHANETHIOL.

ethyl methacrylate *Organic Chemistry.* $CH_2{=}CCH_3COOC_2H_5$, a flammable, colorless liquid that is insoluble in water; boils at 119°C and melts at –75°C; used as a polymer producer.

ethyl methyl acetylene see 2-PENTYNE.

ethyl 2-methylbutyrate see ETHYL ISOVALERATE.

ethyl nitrate *Organic Chemistry.* $C_2H_5ONO_3$, a flammable, colorless liquid that is insoluble in water and soluble in alcohol; it boils at 87.6°C; used in pharmaceuticals, organic synthesis, perfumes, dyes, and rocket propellants.

***O*-ethyl-*O*-*p*-nitrophenyl phenylphosphonothioate** *Organic Chemistry.* $C_{14}H_{14}NO_4PS$, a toxic yellow crystalline solid, insoluble in water and miscible in alcohol; melts at 36°C; used as an insecticide.

ethyl octanoate see ETHYL CAPRYLATE.

ethyl oleate *Organic Chemistry.* $C_{17}H_{33}COOC_2H_5$, a combustible, light-yellowish liquid that is insoluble in water and soluble in alcohol; it boils at 205°C; used as a solvent, plasticizer, water-resisting agent, and in flavoring.

ethyl oxalate *Organic Chemistry.* $(COOC_2H_5)_2$, a combustible, toxic, unstable, colorless liquid, slightly soluble in water and gradually decomposed by it, and miscible with alcohol; boils at 186°C and melts at –40.6°C; used as a cellulose ester and ether solvent, and in resins, pharmaceuticals, perfumes, and organic synthesis.

ethylparaben *Organic Chemistry.* $HOC_6H_4CO_2C_2H_5$, colorless crystals, insoluble in water and soluble in alcohol; melts at 115°C; used as a pharmaceutical preservative. Also, ETHYL-*p*-HYDROXYBENZOATE.

***N*-ethyl 5-phenylisoxazolium-3'-sulfonate** *Organic Chemistry.* $C_{11}H_{11}NO_4S$, crystals, soluble in water, that decompose at 207–208°C; used to form peptide bonds. Also, WOODWARD'S REAGENT K.

1-ethyl-3-piperidinol *Pharmacology.* $C_7H_{15}NO$, a drug formerly used to relieve spasm in smooth muscles such as arteries or intestines. Also, **1-ethyl-3-hydroxypiperidine.**

ethyl propionate *Organic Chemistry.* $C_2H_5COOC_2H_5$, a flammable, water-white liquid that is slightly soluble in water and soluble in alcohol; boils at 99°C; used as a solvent, flavoring agent, and in fruit syrups.

ethyl salicylate *Organic Chemistry.* $C_6H_4(OH)COOC_2H_5$, a combustible, colorless liquid that is slightly soluble in water and soluble in alcohol; boils at 231–234°C; used in perfumery and flavoring.

ethyl silicate *Organic Chemistry.* $Si(C_2H_5O)_4$, a combustible, colorless liquid that is miscible with alcohol; boils at 168.1°C and melts at –77°C; it hydrolyzes to an adhesive form of silica; used in weatherproof and acid-proof mortars, cements, refractory bricks, chemical-resistant paints, protective coatings, and as a bonding agent.

ethyl sulfide *Organic Chemistry.* $(C_2H_5)_2S$, a combustible, colorless liquid, insoluble in water and soluble in alcohol; boils at 92–93°C; used in organic synthesis, in mineral bath solvents, and in electroplating.

***O*-ethyl (*O*-2,4,5-trichlorophenyl)ethylphosphonothioate** *Organic Chemistry.* $C_{10}H_{12}OPSCl_3$, an amber liquid that is very slightly soluble in water; boils at 108°C (0.01 torr); used as an insecticide. Also, TRICHLORONATE.

ethyl urethane see URETHANE.

ethyl vanillin *Organic Chemistry.* $HOC_6H_3(OC_2H_5)CHO$, combustible, fine, white crystals, slightly soluble in water and soluble in alcohol; melts at 76.5°C. It has a strong vanilla odor and four times the flavor of natural vanilla; used in flavors and to fortify or replace vanillin.

ethynylation *Organic Chemistry.* a process in which acetylene is condensed with a reagent, such as an aldehyde, to yield an acetylenic derivative. An example is the combination of formaldehyde and acetylene to produce butynediol.

etic *Anthropology.* describing raw data about a culture from the point of view of an outside observer, without considering the cultural context.

etioallocholane see ANDROSTANE.

etiolation *Botany.* the yellowing or whitening of green plant parts due to abnormal chlorophyll development caused by a lack of light, resulting in rudimentary leaves and tall, spindly plants.

etiological *Pathology.* of or relating to etiology, or to causes of diseases. Also, **etiologic.**

etiology *Pathology.* **1.** the cause or origin of a disease or disorder. **2.** the study of the causes of diseases. *Science.* any cause or study of causes.

etioplast *Botany.* an immature chromoplast found in plants kept in the dark, containing prolamellar bodies that develop chlorophyll upon exposure to light.

etioporphyrin *Organic Chemistry.* $C_{32}H_{38}N_4$, a porphyrin with four methyl and four ethyl groups; long prismatic crystals that melt at 360–363°C; forms copper and magnesium salts.

Etna, Mount *Geography.* an active volcano (10,758 feet) in northeastern Sicily.

etoposide *Oncology.* an antineoplastic that is derived from a semisynthetic form of podophyllotoxin to treat acute monocytic leukemia, lymphoma, small-cell lung cancer, and testicular cancer. Also, VP-16.

E-transformer *Electromagnetism.* a transformer whose primary windings are wrapped on the center arm of an E-shaped core and whose secondary windings are wrapped on the outer arms.

etridiazole *Organic Chemistry.* $C_5H_5Cl_3N_2OS$, a pale-yellow solid that melts at 20°C; used as an agricultural and horticultural fungicide.

Ettingshausen, Albert von 1850–1932, Austrian physicist.

Ettingshausen coefficient *Physics.* the coefficient appearing in the equation governing the Ettingshausen effect, given by the temperature gradient divided by the product of the current density and the magnetic field strength.

Ettingshausen effect *Physics.* a difference in temperature arising across the two edges of a strip conductor when the conductor is placed so that its face is perpendicular to a magnetic field, and a current is driven through the strip longitudinally.

Ettingshausen-Nernst coefficient *Physics.* the coefficient appearing in the equation governing the Ettingshausen-Nernst effect, given by the electric field divided by the product of the temperature gradient and the magnetic field strength.

Ettingshausen-Nernst effect *Physics.* a voltage that arises at opposite edges of a strip of metal that is conducting heat in the presence of a magnetic field perpendicular to the surface of the metal. Also, NERNST EFFECT.

ettringite *Mineralogy.* $Ca_6Al_2(SO_4)_3(OH)_{12}\cdot 26H_2O$, a colorless, transparent, hexagonal mineral of the ettringite group occurring in small prismatic crystals, having a specific gravity of 1.77 and a hardness of 2 to 2.5 on the Mohs scale; found in cavities of metamorphosed limestone inclusions in lava.

etymology *Linguistics.* **1.** the origin or history of a given word. **2.** the study of the origin or history of words, especially as seen in the changing meanings of individual words. Thus, **etymologist.**

Eu the chemical symbol for europium.

eu- a prefix meaning: **1.** good, healthy, or normal, as in *eupepsia.* **2.** easy, as in *euthanasia.*

Euantennariaceae *Mycology.* a family of fungi of the order Dothideales, forming sooty mold colonies on the surfaces of leaves, stems, or branches in tropical to temperate regions.

Euascomycetes *Mycology.* in former classifications, a class of fungi including the subdivision Ascomycotina except the Hemiascomycetes.

Eubacteria *Bacteriology.* a kingdom that contains all prokaryotic organisms that are not classified in the Archaebacteria.

Eubacteriales *Bacteriology.* a former classification for an order of schizomycete bacteria composed of both coccoid and bacillary cells.

Eubacterium *Microbiology.* a genus of Gram-positive, anaerobic, rod-shaped bacteria of the family Propionibacteriacea that are normal inhabitants of the mammalian intestintal tract.

Eubasidiomycetes *Mycology.* in former classifications, a subclass of fungi of the class Basidiomycetes; its spore-bearing structure or basidium has no cross walls dividing the cells.

Eubrya *Botany.* a subclass of mosses in which the leafy gametophyte arises from the protonemata, the elevated capsule has a complex operculum, and the sporogenous tissue develops from the endothecium.

Eubryales *Botany.* an order of mosses of the subclass Eubrya characterized by perennial, erect gametophores, stems bearing leaves, and pendant capsules with a peristome of well-developed teeth.

eucairite *Mineralogy.* CuAgSe, a silver-white to lead-gray, orthorhombic metallic mineral occurring in massive and granular forms, having a specific gravity of 7.6 to 7.8 and a hardness of 2.5 on the Mohs scale.

eucalyptol *Organic Chemistry.* $C_{10}H_{18}O$, a combustible, colorless monoterpene oil, slightly soluble in water and soluble in alcohol; boils at 174–177°C; used in pharmaceuticals, flavoring, and perfumery. Also, CINEOL, CAJEPUTOL.

Eucalyptus *Botany.* a genus of evergreen trees of the family Myrtaceae (myrtles) that are native to Australia and New Guinea and have stiff, entire leaves, umbellate flowers, and woody fruit.

eucalyptus [yoo´kə lip´tis] *Botany.* any of various trees of the genus *Eucalyptus*; several types are widely grown in the U.S., especially in southern California, such as the **bluegum eucalyptus,** *E. globulus. Materials.* the hard, durable wood of this tree; used for a wide variety of construction purposes.

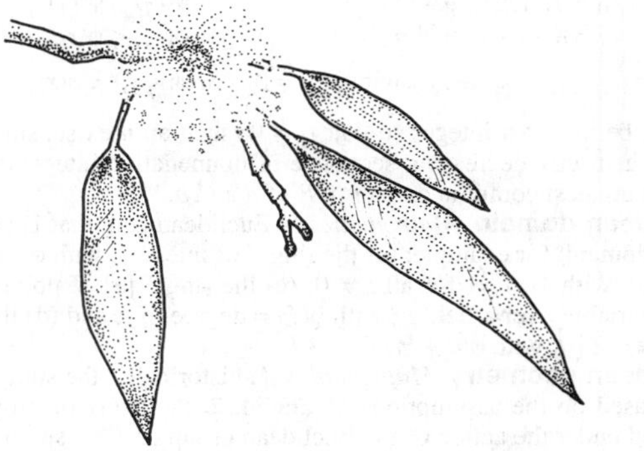

Eucalyptus

eucalyptus oil *Materials.* a volatile oil extracted from the leaves of various eucalyptus trees; widely used in the production of flavorings and pharmaceuticals.

eucapnia *Hematology.* the condition in which the carbon dioxide tension of the blood is normal.

Eucarida *Invertebrate Zoology*. a superorder of decapod crustaceans in the subclass Malacostraca, including shrimp, lobsters, and crabs, with stalked eyes and the thoracic segments fused to the carapace.

eucarpic *Mycology*. describing a fungus in which only part of the thallus converts into fruiting bodies. Also, **eucarpous.**

Eucaryotae see EUKARYOTAE.

eucaryote see EUKARYOTE.

eucatropine hydrochloride *Pharmacology*. $C_{17}H_{25}NO_3 \cdot HCl$, an anticholinergic drug occurring as a white powder and applied topically to dilate the pupil of the eye.

Euceratomycetaceae *Mycology*. a family of fungi of the suborder Laboulbeniineae which is composed of insect parasites.

Eucestoda *Invertebrate Zoology*. a subclass of true tapeworms, in the class Cestoda, including many major parasites on humans and domestic animals.

Eucharitidae *Invertebrate Zoology*. a family of wasps, hymenopteran insects in the superfamily Chalcidoidea, that parasitize ants.

Eucheuma *Botany*. a genus of red algae of the family Solieriaceae, having terete or flattened thalli, often with many spiny branchlets.

euchlorine *Mineralogy*. an emerald-green orthorhombic mineral with an approximate formula of $(K,Na)_8Cu_9^{+2}(SO_4)_{10}(OH)_6$, found as a fumarolic deposit on Mount Vesuvius.

euchroite *Mineralogy*. $Cu_2^{+2}(AsO_4)(OH) \cdot 3H_2O$, an emerald-green, rather brittle, easily fusible, orthorhombic mineral occurring in prismatic to equant crystals, having a vitreous luster, a specific gravity of 3.39 to 3.44, and a hardness of 3.5 to 4 on the Mohs scale.

euchromatin *Cell Biology*. one of the two distinct forms of chromatin, containing the majority of genetically active, transcribed DNA and distinguished from heterochromatin by remaining less condensed and staining less intensely with cytological dyes during interphase.

Eucinetidae *Invertebrate Zoology*. the plate thigh beetles, a family of coleopteran insects in the superfamily Dascilloidea, associated with fungal decay.

euclase *Mineralogy*. $BeAlSiO_4(OH)$, a colorless to pale green to blue, transparent to translucent, monoclinic mineral occurring in prismatic crystals, having a vitreous luster, a specific gravity of 3.05 to 3.1, and a hardness of 7.5 on the Mohs scale; found in griesen deposits.

Euclasterida *Invertebrate Zoology*. an order of sea stars whose members possess a distinct central, disk-shaped body.

Eucleidae *Invertebrate Zoology*. a family of lepidopteran insects in the suborder Heteroneura, large hairy moths often strikingly marked in yellow and brown. Also, LIMACODIDAE.

Euclid [yoo´klid] c. 300 BC, Greek mathematician; called "the father of geometry," author of the *Elements,* the foundation of plane geometry.

Euclidean [yoo klid´ē ən] **1.** relating to Euclid or his writings. **2.** based on the postulates or descriptions of Euclid.

Euclidean algorithm *Mathematics*. a method of finding the greatest common divisor of two elements a and b (where $b \neq 0$) of a Euclidean ring R with associated function ϕ, given as follows: Let $b = r_0$. Then by repeated use of the definition of ϕ,

$$
\begin{aligned}
a &= q_0 r_0 + r_1 \text{ with } r_1 = 0 && \text{or } \phi(r_1) < \phi(r_0)\\
r_0 &= q_1 r_1 + r_2 \text{ with } r_2 = 0 && \text{or } \phi(r_2) < \phi(r_1)\\
r_1 &= q_2 r_2 + r_3 \text{ with } r_3 = 0 && \text{or } \phi(r_3) < \phi(r_2).\\
r_k &= q_{k+1} r_{k+1} + r_{k+2} \text{ with } r_{k+2} = 0 && \text{or } \phi(r_{k+2}) < \phi(r_{k+1}).
\end{aligned}
$$

Let n be the least integer so that $r_n = 0$ (such an n exists since the $\phi(r_n)$ form a strictly decreasing sequence of nonnegative integers). Then r_{n-1} is the greatest common divisor in R of a and b.

Euclidean domain *Mathematics*. a Euclidean ring that is also an integral domain; for example: (a) the ring Z of integers, with $\phi(x) = |x|$; (b) F a field, with $\phi(x) = 1$ for all $x \neq 0$; (c) the ring $F[x]$ of polynomials f in one variable over a field F, with $\phi(f)$ = degree of f; and (d) the Gaussian integers $Z[i]$, with $\phi(a + bi) = a^2 + b^2$.

Euclidean geometry *Mathematics*. **1.** historically, the study of geometry based on the assumptions of Euclid. **2.** the study of properties preserved under the action of the Euclidean group on R^n (usually $n = 3$).

Euclidean group *Mathematics*. the group $E(n)$ of all reflections and products of reflections under transformation multiplication, which includes the group of all rotations and the group of all translations. $E(n)$ is the group of isometries of R^n. If $n = 2$ (in the Euclidean plane), then a product of two reflections is a rotation or a translation, according as the mirrors of reflection do or do not intersect, and every product of four or more reflections reduces by an even number of reflection or a product of two or three reflections.

Euclidean method (of construction) *Mathematics*. the construction of geometric figures by the use of a compass and straightedge alone.

Euclidean ring *Mathematics*. a ring R is called Euclidean if there exists a function ϕ: $R \to Z^+$ (the positive integers) such that (a) if $a,b \in R$ with $ab \neq 0$ and $a \neq 0$, then $\phi(a) \leq \phi(ab)$; and (b) if $a,b \in R$, then there exist $q,r \in R$ such that $a = qb + r$ with either $r = 0$ or $r \neq 0$ and $\phi(r) < \phi(b)$. No value is assigned to $\phi(0)$. Every Euclidean ring possesses a multiplicative identity, and every ideal of a Euclidean ring is principal. R is usually taken to be commutative. An example of a **noncommutative Euclidean ring** is the polynomial ring $P[x]$ over a skew field (division ring) P.

Euclidean space *Mathematics*. a finite-dimensional space with positive-definite inner-product. Equivalently, any Riemannian manifold isometric to the flat space R^n with chart (Id, R^n) and metric $ds^2 = \sum_{i=1}^{n} dx^i dx^i$ (where Id is the identity map); i.e., R^n with the usual distance function $d(x, y) = [\sum_{i=1}^{n} (x_i - y_i)^2]^{1/2}$.

Euclymeninae *Invertebrate Zoology*. a suborder or subfamily of segmented annelid worms in the family Maldonidae, having well-developed plaques on the body.

Eucnemidae *Invertebrate Zoology*. a family of coleopteran insects, the false click beetles, in the superfamily Elateroidea that feed on wood, fungi, and slime molds.

Eucoccidia *Invertebrate Zoology*. an order of medically important protozoa in the subclass Coccidia, parasites inside the cells of humans and other vertebrates, with alternating sexual and asexual generations. Also, **Eucocida.**

Eucoelomata *Invertebrate Zoology*. the major division of the higher invertebrates including mollusks, annelids, arthropods, echinoderms, and chordates, all having a separate mouth and anus, a true coelem, and a well-developed circulatory system.

eucoelomate *Invertebrate Zoology*. any member of Eucoelomata.

eucolloid *Physical Chemistry*. a colloid in which each dispersed particle consists of a single large molecule.

Eucommiaceae *Botany*. a monospecific family of dicotyledonous dioecious trees in the order Eucommiales, having a milky juice, storing carbohydrates as inulin, and producing an iridoid compound; native only to forests in western China.

Eucommiales *Botany*. a monospecific order of dicotyledonous trees found only in China, having simple, serrate, astipulate leaves, unisexual flowers, and latex in young plant parts.

euconodonts *Paleontology*. the later conodonts, which include all the basic types of fossil teeth, widespread and diverse from the late Cambrian to the Triassic; some authorities regard as the only true conodonts.

eucrasia *Medicine*. **1.** a state of health; a balance of factors constituting a healthy condition. **2.** a state of decreased bodily reactions to ingested or injected substances, such as drugs. (From Greek for "good mixture.")

eucrite *Mineralogy*. a type of achronditic meteorite.

Eucryphiaceae *Botany*. a monogeneric family of dicotyledonous evergreen trees and shrubs in the order Rosales, being tanniferous, producing gum or mucilage, and native to Australia, Tasmania, and Chile.

eucryptite *Mineralogy*. $LiAlSiO_4$, a colorless to white, transparent, trigonal mineral occurring in granular aggregates and single crystals, having a specific gravity of 2.66 to 2.67 and a hardness of 6.5 on the Mohs scale; found intergrown with albite in pegmatites.

Eudactylinidae *Invertebrate Zoology*. a family of copepod crustaceans in the suborder Caligoida that are external parasites on fish.

eudialyte *Mineralogy*. a pink to red to brown, trigonal mineral having a general formula of $Na_4(Ca,Ce)2(Fe^{+2},Mn^{+2},Y)ZrSi_8O_{22}(OH,Cl)_2$, occurring in tabular to prismatic crystals, and having a specific gravity of 2.74 and 2.98 and a hardness of 5 to 5.5 on the Mohs scale.

eudidymite *Mineralogy*. $NaBeSi_3O_7(OH)$, a white, glassy, monoclinic mineral, dimorphous with epididymite, occurring in tabular crystals, and having a specific gravity of 2.55 and a hardness of 6 to 7 on the Mohs scale; found in nepheline-syenite pegmatites.

eudiometer *Engineering*. a device used to measure the volume of gas during combustion, made up of a graduated glass tube closed at one end and containing wires through which electricity passes.

Eudoxus of Cnidus c. 409–355 BC, Greek astronomer; founder of empirical cosmology; accurately estimated length of solar year.

Euechinoidea *Invertebrate Zoology*. the true sea urchins, a subclass of echinoid echinoderms including most living forms

eugamy *Cell Biology*. a union of gametes, each of which contains the proper (haploid) number of chromosomes.

Eugenia *Astronomy*. asteroid 45, discovered in 1857 and measuring 214 kilometers in diameter; it belongs to type C.

eugenic [yoo jen´ik] *Genetics*. relating to the study of eugenics or to a program or policy of eugenics.

eugenics [yoo jen´iks] *Genetics*. the application of principles of heredity to the improvement of the human species. In its extreme, practical eugenics might involve encouraging the breeding of individuals designated as superior (**positive eugenics**) and discouraging or preventing the breeding of individuals designated as inferior (**negative eugenics**). Genetic counseling is a less extreme application of this science.

eugenist *Genetics*. **1.** a person who specializes in eugenics. **2.** a person who advocates a program or policy of eugenics. Also, **eugenicist.**

eugenol *Organic Chemistry*. $CH_2{=}CHCH_2C_6H_3(OH)OCH_3$, a combustible, colorless to yellowish liquid; very slightly soluble in water and soluble in alcohol and ether; it boils at 253.5°C and melts at –9°C; turns brown on exposure to air; derived by extraction of clove oil with aqueous potash and used in perfumes, in flavoring, and as an analgesic.

eugeosyncline *Geology*. the central, mobile magmatic belt of an orthogeosyncline, in which volcanic rocks are abundant. Also, PLIOMAGMATIC ZONE.

Euglena [yoo glē´nə] *Biology*. a genus of single-cell protozoa, often classified as algae, that have one or two flagella, are spindle-shaped and usually green, and are commonly found in stagnant water.

Euglenales *Botany*. an order of euglenoids with a flagellum emergent from the canal and another short, nonemergent flagellum; includes the large, freshwater genus *Euglena,* a green, mobile to rigid cell with several commercial uses.

Euglenamorphales *Botany*. an order of endozoic parasitic euglenoids having three or more emergent flagella of equal size.

Euglenida *Invertebrate Zoology*. an order of green, plantlike protozoans with one flagellum, in the class Phytomastigophorea.

Euglenidae *Invertebrate Zoology*. a family of coleopteran insects, the antlike leaf beetles, associated with rotting wood.

euglenoid *Botany*. any member of the division Euglenophyta. Also, **euglenid, euglenophyte.**

Euglenoidina see EUGLENIDA.

Euglenophyceae *Botany*. the single class of green, motile algae in the division Euglenophyta.

Euglenophyta *Botany*. a division of the Plantae composed of unicellular, flagellate organisms that have a usually naked, flattened body, colorless chloroplasts, and food stored in fat reserves; they reproduce by asexual cell division, sometimes producing resting spores, and are related to the Protozoa, possessing both plant and animal characteristics.

euglobulin *Biochemistry*. an archaic classification for a simple protein that is soluble in distilled water, but not in saline solutions.

Euglypha *Invertebrate Zoology*. a genus of mostly freshwater amebas in the class Filosa, with plated tests of siliceous scales.

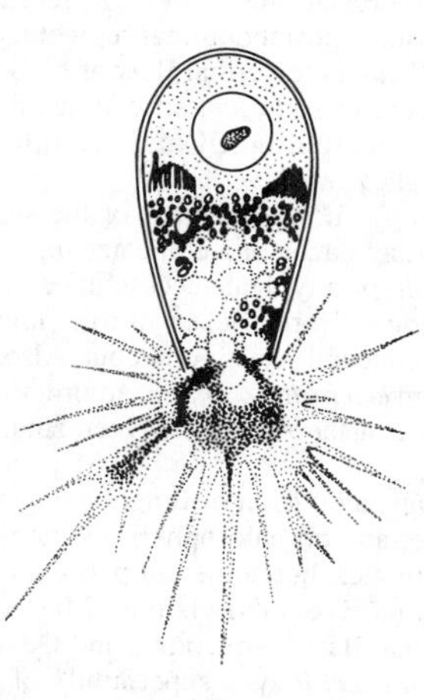

Euglypha

eugonic *Microbiology*. describing a bacterial strain that grows luxuriantly on a culture medium.

Eugregarinida *Invertebrate Zoology*. an order of protozoans in the subclass Gregarinia, parasitic in earthworms, marine annelids, and arthropods.

euhedral see AUTOMORPHIC.

Euhoplites *Paleontology*. an extinct genus of nektic ammonoids of the Cretaceous.

eukaryon *Cell Biology*. a highly organized nucleus bounded by a nuclear membrane, characteristic of cells of higher organisms. *Biology*. see EUKARYOTE.

eukaryosis *Biology*. the state of having a true nucleus, with the nuclear material surrounded by a membrane and the cytoplasm containing organelles.

Eukaryotae *Biology*. a superkingdom including all living and fossil organisms belonging to taxonomic groups above the primitive prokaryotic level; that is, all organisms containing a well-defined nucleus. Also, EUCARYOTAE.

eukaryote [yoo kâr´ē ōt´; yoo kâr´ē ət] *Biology*. an organism whose cells have a distinct nucleus, multiple chromosomes, and a mitotic cycle; this classification thus includes animals, plants, and fungi, but not bacteria or algae. Also, EUKARYON, EUCARYOTE.

eukaryotic *Biology*. relating to a eukaryon, a eukaryote, or eukaryosis.

Eulamellibranchia *Invertebrate Zoology*. a large subclass of bivalve mollusks with leaflike gill membranes and well-developed siphons, including the common freshwater mussel.

Euler, Leonhard [oi´lər] 1707–1783, Swiss mathematician; made many contributions in pure mathematics; invented calculation of sines.

Euler, Ulf Svante von 1905–1983, Swedish physiologist; shared Nobel Prize for research on role of norepinephrine in nerve impulses.

Euler angles see EULERIAN ANGLES.

Euler characteristic *Mathematics*. **1.** the Euler characteristic of an *n*-dimensional manifold *M* is the number $\chi_M = \sum_{i=1}^{n} (-1)^i \beta_i$, where β_i is the *i*th Betti number of *M*. In particular, a two-dimensional manifold has **Euler characteristic zero** if and only if it is topologically equivalent to a cylinder, torus, Möbius strip, or Klein bottle. A two-dimensional manifold has **Euler characteristic one** if and only if it is topologically equivalent to the projective plane or to a disk; it has **Euler characteristic two** if and only if it is topologically equivalent to a sphere. Sometimes called the **Euler-Poincaré characteristic**; also denoted $\varepsilon(M)$. **2.** the Euler $\varepsilon(G)$ characteristic of a graph *G* is the Euler characteristic of the surface *S* of greatest Euler characteristic, such that *G* can be topologically imbedded in *S*.

Euler-Chelpin, Hans von 1873–1964, German-born Swedish chemist; shared Nobel Prize for research on fermentation of sugars.

Euler equation *Mathematics*. **1.** in the calculus of variations, the differential equation

$$\partial L/\partial y + \sum_{k=1}^{n} (-1)^k d^k/dx^k \{\partial L/\partial y^{(k)}\} = 0,$$

in which $L = L(x, y, y', \ldots, y^{(n)})$ and $y^{(k)} = d^k y/dx^k$, and *y* is the function to be found. If *y* provides an extreme value of the functional $I[y] = \int_a^b L(x, y, y', \ldots, y^{(n)})\,dx$, then *y* satisfies the Euler equation. **2.** an ordinary differential equation of the form $f(x) = \sum_{i=0}^{n} a_{n-i} x^i d^i y/dx^i$, where the a_i are constants.

Euler force *Mechanics*. the critical axial force at which a long column will begin to buckle according to Euler's formula for long columns. Also, **Euler load.**

Eulerian [oi lâr´ē ən] relating to the Swiss mathematician Leonhard Euler. Compound entries beginning with *Eulerian* are also often written as *Euler* or *Euler's*; i.e., Eulerian/Euler/Euler's angles.

Eulerian angles *Mechanics*. a system of three angles that uniquely define the orientation of one coordinate system with reference to a second coordinate system; used to specify the orientation of body-fixed axes relative to inertial axes, so as to analyze the motion of a rotating body. Also, EULER('S) ANGLES.

Eulerian coordinates *Fluid Mechanics*. the three velocity components of the system, parallel to fixed coordinate axes.

Eulerian correlation *Fluid Mechanics*. the relationship of the flow properties at different places in space at a given moment of time.

Eulerian equation *Fluid Mechanics*. an equation of motion for frictionless fluid flow.

Eulerian graph *Mathematics*. **1.** a finite (connected) graph that has an Eulerian trail. If the restriction that the trail must be closed is removed, the graph is **semi-Eulerian.** A connected graph is Eulerian if and only if the degree of every vertex is even; a connected graph is semi-Eulerian if and only if there are no more than two vertices of odd degree. **2.** a countably infinite (connected) graph *G* that has a two-way Eulerian trail. Such a graph *G* is semi-Eulerian if there exists an infinite trail that includes every edge of *G*. A connected countable graph *G* is Eulerian only if: (a) *G* has no vertices of odd degree; (b) for every finite subgraph *H* of *G*, the infinite graph *G* – *H* (obtained from *G* by removing the edges of *H*) has at most two infinite components; (c) if every vertex of such *H* has even degree, then *G* – *H* has exactly one infinite component.

Eulerian matroid *Mathematics.* a matroid on a set S is Eulerian if S can be expressed as the union of disjoint minimal dependent subsets.

Eulerian method *Fluid Mechanics.* a framework for studying fluid mechanics in which the flow properties within an infinitesimal control volume fixed in space are considered as a function of field position and time; it is a field description as opposed to the Lagrangian method, which is a material description. Also, **Eulerian description.**

Eulerian numbers *Mathematics.* the triangular array of positive integers $A(n, r)$ such that: (a) $n, r \geq 1$; (b) initial conditions are $A(n, r) = 0$ for $n < r$ and $A(n, r) = 1$ for $r = 1$ or $r = n$; and (c) the recursion relation is $A(n + 1, r) = rA\ (n, r) + (n - r + 2)A(n, r - 1)$. Used in the study of rook polynomials, in the solution to Simon Newcomb's problem, and in various relationships with Bernoulli numbers and Stirling numbers; first derived by Euler in his 1755 work on differential calculus.

Eulerian trail *Mathematics.* **1.** also, **Euler(ian) circuit, Euler(ian) path.** a closed trail on a connected finite graph that traverses each edge exactly once. **2.** a two-way infinite trail on a connected, countably infinite graph G that traverses each edge of G exactly once; that is, an infinite edge sequence of the form $\ldots, v_{-2}v_{-1}, v_{-1}v_0, v_0v_1, v_1v_2, \ldots$ that traverses each edge exactly once. (From *Euler's* famous problem of this type on the seven bridges of Königsberg.)

Eulerian wind *Meteorology.* a wind motion only in response to the pressure force.

Euler-Lagrange equation *Mathematics.* a particular case of the Euler equation, namely $\partial(f\,(x, y, y'))/\partial y - d/dx\ (\partial(f\,(x, y, y'))/\partial y') = 0$, where $y = \partial y/\partial x$. A necessary condition in order for $y(x)$ to minimize the integral $\int_a^b f\,(x, y, y')dx$ is that y satisfy the Euler-Lagrange equation.

Euler number 1. *Fluid Mechanics.* a dimensionless parameter used in estimating the fluid friction in conduits; the ratio of the friction head to twice the velocity head. **2.** a dimensionless number that is equal to twice the Fanning friction factor. **3.** a dimensionless ratio used in the study of aerodynamic testing; the ratio of pressure forces to inertia forces. The factor of one-half is introduced into the denominator to give the dynamic pressure; expressed as $(\Delta P)/(0.5\rho V^2)$, where ΔP is the local pressure minus the freestream pressure, ρ is the fluid density, and V^2 is the square of the fluid's velocity. *Mathematics.* the coefficients E_k in the expansion $\sec z = \sum_{k=1}^{\infty} E_k/(2k)!\ z^{2k}$. The Euler numbers are all odd positive integers and satisfy the recursion relation $\sum_{k=0}^{n} \binom{2n}{2k}(-1)^k E_{n-k} = 0$ where $E_0 = 1$. They can be expressed in terms of the Bernoulli numbers. Also, SECANT COEFFICIENTS.

Euler phi function *Mathematics.* for a positive integer n, the Euler phi function of n, denoted $\phi(n)$, is equal to the number of positive integers less than or equal to n which are relatively prime to n. $\phi(n)$ is the order of the multiplicative group of integers modulo n. Also, PHI FUNCTION.

Euler product formula *Mathematics.* a formula expressed as $\sum_{n=1}^{\infty} 1/n^s = \prod_p (1-p^{-s})^{-1}$, where p ranges over all primes. Both sides of the Euler product formula converge for complex s in the halfplane Re $s > 1$, have a pole at $s = 1$, or diverge otherwise.

Euler's angles see EULERIAN ANGLES.

Euler's buckling formula see EULER'S FORMULA FOR LONG COLUMNS.

Euler's circuit see EULERIAN TRAIL.

Euler's conjecture *Mathematics.* the long-standing conjecture of Leonhard Euler, finally disproved by Bose, Shrikhande, and Parker in 1960, that there exists no pair of orthogonal Latin squares of order n for $n \equiv 2 \pmod 4$.

Euler's constant *Mathematics.* the quantity $\gamma = \lim_{n\to\infty} [\sum_{k=1}^{n} 1/k - \ln n]$, approximately equal to 0.577215. Also, MASCHERONI'S CONSTANT.

Euler's dynamical equations see EULER'S EQUATIONS OF MOTION.

Euler's equations of fluid motion *Fluid Mechanics.* a set of equations describing the unsteady motion of a nonviscous fluid, including the effects of fluid compressibility and external forces such as gravity. *Aviation.* an equation for calculating the energy removed from an airstream by a propeller or turbine.

Euler's equations of rigid body motion *Mechanics.* a system of three equations describing the motion of a single rigid body or system of connected particles in terms of a vector of linear momentum and a vector of angular momentum. Also, EULER DYNAMICAL EQUATIONS.

Euler's formula *Mathematics.* **1.** the equation $v - e + f = 2$ for the Euler characteristic of a simple polyhedron with v vertices, e edges, and f faces. Also, **Euler's theorem on polyhedra. 2.** the formula $e^{ix} = \cos x + i \sin x$; in particular, $e^{i\pi} = -1$.

Euler's formula for long columns *Mechanics.* an expression for determining the maximum axial load that a long linear elastic column can withstand without buckling. Also, EULER'S BUCKLING FORMULA.

Euler's method see EULERIAN METHOD.

Euler's theorem *Mathematics.* **1. first theorem of Euler.** the identity

$$(1+x)(1+x^3)(1+x^5)\cdots = 1 + \sum_{m=1}^{\infty} x^{m^2}/[(1-x^2)(1-x^4)(1-x^6)\cdots(1-x^{2m})]$$

2. second theorem of Euler. the identity

$$(1+x^2)(1+x^4)(1+x^6)\cdots = 1 + \sum_{m=1}^{\infty} x^{m(m+1)}/[(1-x^2)(1-x^4)(1-x^6)\cdots(1-x^{2m})]$$

The first and second theorems of Euler are useful in the study of generating functions. **3.** Euler's generalization of Fermat's theorem states that if a is relatively prime to m, then $a^{\phi(m)} \equiv 1 \pmod m$, where $\phi(n)$ is the Euler phi function of n.

Euler's theorem on homogeneous functions *Mathematics.* if $f\,(x_1, \ldots, x_n)$ is a homogeneous function of degree n, then

$$nf\,(x_1, \ldots, x_n) = \sum_{i=1}^{n} x_i\ \partial f/\partial x_i.$$

Euler's theorem on normal sections *Mathematics.* the curvature k_N of a normal section at a point P_0 on a surface S is given by the formula $k_N = k_1\cos^2\alpha + k_2\sin^2\alpha$, where k_1 and k_2 are the principal directions of the cutting plane containing the normal to S at P_0, and α is the angle made by the cutting plane with the x-axis.

Euler's trail see EULERIAN TRAIL.

Euler transformation *Mathematics.* the transformation of the series $\sum_{n=0}^{\infty} a_n$ into $\sum_{n=0}^{\infty} 2^{-n} \sum_{r=0}^{n-1} (-1)^r \binom{n-1}{r} a_r$; used to obtain a series that converges much faster than the original one or to define a type of convergence for a divergent series. The a_n are usually taken to be functions of a complex variable.

Euler triangle *Mathematics.* a spherical triangle all of whose sides (great circular arcs) and angles are less than π and no two of whose vertices A, B, and C are diametrically opposite. A **non-Euler triangle** ABC with angle $\gamma = \angle AB > \pi$ can be replaced with the Euler triangle BAC with angle $\angle BA = 2\pi - \gamma$ in the same hemisphere determined by the great circle through A and B .

eulittoral *Oceanography.* a subdivision of the littoral zone that refers to the sea floor environment between the high-water line and a depth of 40–60 meters.

Eulophidae *Invertebrate Zoology.* a large family of wasps, hymenopteran insects in the superfamily Chalcidoidea, whose larvae are parasitic on the larvae of spiders and many insects.

eulytite *Mineralogy.* $Bi_4(SiO_4)_3$, a colorless to yellow to dark-brown, cubic mineral occurring as small tetrahedral crystals and having a specific gravity of 6.1 to 6.6 and a hardness of 4.5 on the Mohs scale; found implanted on quartz.

Eumalacostraca *Invertebrate Zoology.* a grouping of crustaceans that includes shrimplike crustaceans with eight thoracic segments, six abdominal segments, and a postabdominal segment.

eumenorrhea *Medicine.* the normal flow of blood during menstruation.

Eumetazoa *Zoology.* in some classifications, a major division of the kingdom Animalia including all multicellular animals except for sponges. Also, ENTEROZOA.

eumetazoan *Zoology.* **1.** any member of the subkingdom Eumetazoa. **2.** of or relating to Eumetazoa or a eumetazoan.

eumitosis *Cell Biology.* a typical successful cell division (mitosis).

Eumycetes *Mycology.* a former class of true fungi, including the subdivisions Mastigomycota and Amstigomycota. Also, **Eumycophyta.**

Eumycetozoida *Invertebrate Zoology.* an order of protozoa recognized by some authorities, in the subclass Mycetozoia, that includes the true slime molds.

Eumycota *Mycology.* a division of fungi having well defined cell walls, filamentous hyphae, and no chlorophyll; its members primarily live off nonliving organic matter, but some are parasites of plants and animals; the division contains five subdivisions: Mastigomycotina, Zygomycotina, Ascomycotina, Basidiomycotina, and Deuteromycotina.

Eunicea *Invertebrate Zoology.* a superfamily of polychaete annelids in the group Errantia.

Eunicidae *Invertebrate Zoology.* a family of polychaete annelids, in the superfamily Eunicea, with multiple jaws.

Eunomia [yoo nō´ mē ə] *Astronomy.* asteroid 15, discovered in 1851 and measuring 272 kilometers across; it belongs to type S.

Eunotiaceae *Botany.* a family of freshwater and terrestrial diatoms comprising the suborder Raphidoidineae in the order Pennales, growing epiphitically or free-living and often found in damp sphagum moss.

eunuch [yoo´ nik] *Medicine.* a male who has undergone complete loss of testicular function from castration, inflammation, or injury.

Euomphalacea *Paleontology.* an extinct superfamily of aspidobranch marine gastropods in the order Archaeogastropoda and suborder Euomphalina; euomphalin spires are generally low.

Euomphalus *Paleontology.* a widespread genus of marine gastropods in the family Euomphalidae, found in strata from the Silurian to the Permian; shells are nearly planospiral, increasing in height only slightly in a few species.

eupantotheres *Paleontology.* an extinct order of small therian mammals in the infraclass Pantotheria, known only from the Middle Jurassic to the Lower Cretaceous; its earliest genus, Amphitherium, is probably ancestral to the later pantotheres and to the modern marsupials and placentals.

eupaverin *Pharmacology.* **1.** $C_{20}H_{21}NO_2$, the generic name of the drug Moxaverine, formerly used to prevent or relieve spasms. **2.** a former designation for 3-methyl-6,7-methylenedioxy-1-piperonylisoquinoline hydrochloride ($C_{19}H_{15}NO_4$), a smooth-muscle relaxant.

Eupelmidae *Invertebrate Zoology.* a family of slender, metallic-colored wasps, hymenopteran insects in the superfamily Chalcidoidea, that prey on and parasitize a wide variety of insects.

eupepsia *Medicine.* normal digestion.

Euphausiacea *Invertebrate Zoology.* an order of planktonic crustaceans including krill, tiny marine shrimplike animals that form a major part of the diet of most baleen whales; most are red and bioluminescent.

euphenics *Genetics.* the improvement of a genotypic problem by treatment during the lifetime of a genetically defective individual.

Eupheterochlorina *Invertebrate Zoology.* a suborder of flagellated protozoans in the order Heterochlorida.

Euphorbia *Botany.* an extensive genus of trees, shrubs, and herbs of the family Euphorbiacea that are actively poisonous, emetic, and cathartic.

Euphorbiaceae *Botany.* a family of dicotyledonous, mainly tropical herbs, shrubs, and trees in the order Euphorbiales that are often xerophytic and cactoid and are characterized by usually simple, alternate leaves, regular, unisexual flowers, and a schizocarp, dehiscent fruit; it is one of the largest flowering plant families and includes the latex-bearing *Hevea brasiliensis.* Also, PICRODENDRACEAE.

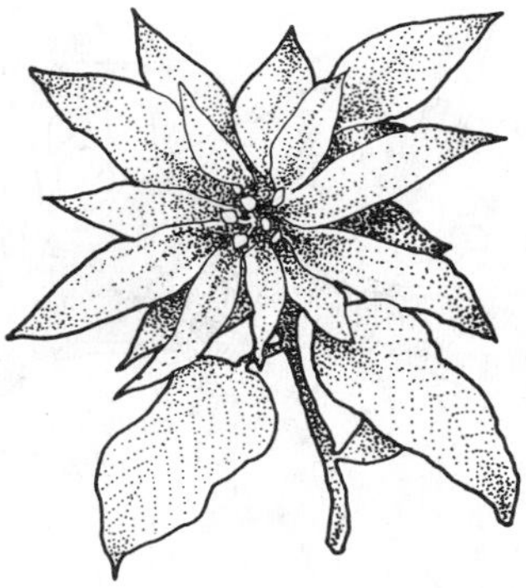

Euphorbiaceae

Euphorbiales *Botany.* the spurges, an order of dicotyledonous herbs and shrubs in the subclass Rosidae with simple leaves and unisexual, reduced flowers that are aggregated and apetalous.

euphotic zone *Ecology.* the upper waters of a lake or sea into which sufficient sunlight penetrates for photosynthesis and abundant plant growth to occur, down to about 80 meters.

Euphrates *Geography.* a river in southwestern Asia, flowing 1739 miles from eastern Turkey through Syria and Iraq to the Shatt-al-Arab near the Persian Gulf.

Euphrosinidae *Invertebrate Zoology.* a family of carnivorous polychaete annelids commonly found feeding on sponges.

Euphrosyne *Astronomy.* asteroid 31, discovered in 1854 and measuring 248 kilometers in diameter; it belongs to type C.

euplastic *Physiology.* readily organized; adapted to tissue formation. *Surgery.* healing quickly and well.

Euplexoptera see DERMAPTERA.

euploid *Genetics.* a type of polyploid possessing a chromosome number that is an exact multiple of the chromosome number found in the original nonpolyploidal form. Thus, **euploidy.**

Euplotes *Invertebrate Zoology.* a genus of large, rigid, generally egg-shaped ciliate protozoans in the order Hypotrichida, common in fresh and marine waters.

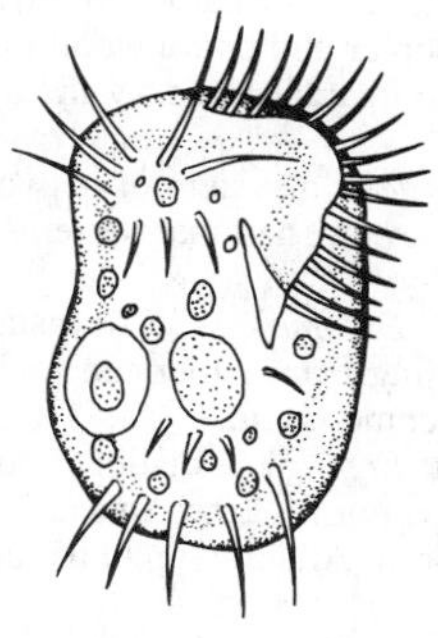

Euplotes

eupnea *Physiology.* easy or normal respiration.

Eupodidae *Invertebrate Zoology.* a family of predaceous mites, in the suborder Trombidiformes, found in soil, in leaf litter, and in caves.

Eupodiscaceae *Botany.* a large, diverse family of mostly marine diatoms of the order Centrales, characterized by valves with two or more ocelli.

Eupomatiaceae *Botany.* a monogenetic family of dicotyledonous small trees or wiry herbs of the order Magnoliales, characterized by the prescence of volatile oils and large flowers pollinated by beetles; considered morphologically isolated and confined to New Guinea and eastern Australia.

Euproopacea *Paleontology.* a superfamily of primitive, gill-breathing chelicerates in the order Xiphosurida; ancestral horseshoe crabs, the euproopacids were widespread in the Carboniferous.

Eupteleaceae *Botany.* a monogenetic family of dicotyledonous, small tanniferous trees and shrubs of the order Hamamelidales, characterized by deciduous leaves and wind- or insect-pollinated flowers; native to China, India, and Japan.

Eurasia *Geography.* the world's largest land mass, including the traditional continents of Europe and Asia.

eurhythmia or **eurhythmy** *Physiology.* a regularity of the pulse; a normal heartbeat. *Developmental Biology.* harmonious relationships in body and organ development. Thus, **eurhythmic.**

Eurocentric *Anthropology.* **1.** of or relating to the belief that the European or Western cultural tradition should be the basis for academic studies of cultural achievement. **2.** focusing on this cultural tradition to the exclusion of other non-Western cultures. Thus, **Eurocentrism.**

Europa *Astronomy.* the second Galilean moon of Jupiter, discovered in 1610 and 3130 kilometers in diameter; its composition is largely rocky.

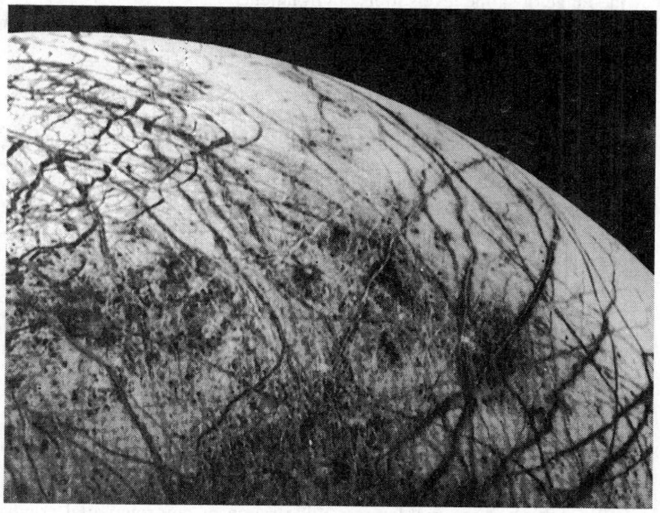

Europa

Europe *Geography.* the continent that lies east of the North Atlantic and west of the Ural Mountains and Black Sea.

European canker *Plant Pathology.* **1.** a disease of fruit and shade trees caused by the fungus *Nectria galligena* and characterized by cankers that appear as concentric rings of hardened growths on branches and the trunk. **2.** a disease of poplar and cottonwood trees caused by the fungus *Dothichiza populea* and characterized by the appearance of cankers on the limbs and trunk.

European datum *Geodesy.* a system or network of geodedic measurements compiled from several national systems, with its initial point in Potsdam, Germany.

European foulbrood *Entomology.* a devastating bacterial disease of honeybee larvae, differing from American foulbrood mainly in that affected larvae do not become viscid.

European subregion *Ecology.* a distinct zoogeographical region that includes western Europe north of the Alps and east of the Urals, and the adjacent areas of the North Atlantic up to the coast of Greenland.

European typhus see EPIDEMIC TYPHUS.

europium *Chemistry.* a rare-earth (lanthanide) element having the symbol Eu, the atomic number 63, an atomic weight of 151.96, a melting point of 826°C, and a boiling point of 1489°C; a soft gray metal, the most reactive of the rare-earth (lanthanide) elements, used as a neutron absorber in reactors and as color TV phosphors. (Named for the continent of *Europe.*)

europium halide *Inorganic Chemistry.* any compound consisting of the rare-earth (lanthanide) element europium with an element from the halogen group (periodic table Group VIIa), such as fluorine or chlorine.

europium oxide *Inorganic Chemistry.* Eu_2O_3, a white to pink powder, insoluble in water and soluble in acids to give the corresponding salt; used in phosphors and nuclear reactor control rods. Also, **europia.**

Eurotiaceae *Mycology.* a family of fungi of the order Eurotiales; most of its members belong to the genera *Penicillium* and *Aspergillus* and have spore sacs contained in closed fruiting bodies called cleistothecia.

Eurotiales *Mycology.* an order of fungi belonging to the class Plectomycetes, which includes well known genera such as the pathogenic *Asperagillus* and *Penicillium.*

eury- *Biology.* a prefix meaning "broad" or "wide."

Euryalina see BASKET STAR.

Euryapsida *Paleontology.* an extinct subclass of generally marine, Early Mesozoic reptiles, possibly ancestral to the ichthyosaurs; *incertae sedis,* the euryapsids are also sometimes called parapsids; they have an upper temporal aperture on each side of the skull, above the postorbital squamosal bones.

eurybaric *Ecology.* describing an organism that is able to tolerate a wide range of atmospheric or hydrostatic pressure in its environment.

eurybathic *Ecology.* describing an organism that lives at the bottom of a body of water.

Eurychilinidae *Paleontology.* an extinct family of dimorphic ostracods in the superfamily Hollinacea; Ordovician.

eurygamous *Invertebrate Zoology.* mating in flight, as in many insects.

euryhaline *Ecology.* describing an organism that is able to tolerate a wide range of saline levels in its environment.

euryhydric *Ecology.* describing an organism that is able to tolerate a wide range of moisture or humidity in its environment.

Eurylaimi *Vertebrate Zoology.* a suborder of the perching bird order Passeriformes, including the broadbills, asities, false sunbirds, pittas, and New Zealand wrens.

Eurylaimidae *Vertebrate Zoology.* the broadbills, a family of small to medium-sized, brightly colored birds of the order Passeriformes, most having a white dorsal patch and a large, flattened and hooked bill; native to the Old World tropics.

Eurymylidae *Paleontology.* a family of early eutherian rabbitlike mammals, now usually classed in the order Anagalida and the superfamily Eurymyloidea; the Eurymylidae are known from fossils in Upper Paleocene strata of Mongolia and are related to the rodents and the lagomorphs.

euryon *Anatomy.* the point on the right and left parietal bones that marks the greatest transverse diameter of the skull or head.

euryphagic *Ecology.* describing an organism that feeds on a wide variety of foods or food species. Also, **euryphagous.**

Euryphoridae *Invertebrate Zoology.* a family of copepod crustaceans, in the order Caligoida, ectoparasites on fish.

euryplastic *Biology.* describing an organism that has a marked ability to change, adapt to, or tolerate a wide range of environmental conditions.

Eurypterida *Paleontology.* an order of primitive, gill-breathing chelicerates in the class Merostomata; the eurypterids are found from the Ordovician to the Permian and resemble scorpions. Although very different from them in appearance, their only living relative is the horseshoe crab.

Eurypygidae *Vertebrate Zoology.* the sun bittern, a monotypic family of heronlike birds of the order Gruiformes, characterized by gray, brown, and white plumage that is often barred and mottled, a large head, a long neck, and a long bill; found in Central and South America.

eurypylous *Zoology.* having a wide opening; especially one leading from incurrent canals and to excurrent canals, as in some sponges.

eurytherm *Biology.* any eurythermic organism.

eurythermic *Biology.* describing an organism that is able to tolerate a wide range of temperatures in its environment. Also, **eurythermal**.

Eurytomidae *Invertebrate Zoology.* the chalcid flies, a large, common family of small hymenopteran insects, generally black or black and yellow; the family includes some parasites on insects and some serious plant pests.

eurytopic *Biology.* of an organism, able to tolerate a wide range of habitats or having a wide geographical distribution.

Eusiridae *Invertebrate Zoology.* a family of predaceous amphipod crustaceans in the suborder Gammaridea that feed on copepods and other planktonic crustaceans.

eusporangiate *Botany.* having sporangia derived from a group of epidermal cells, as in ferns with primitive sporangia.

eusporangium *Botany.* a sporangium of a vascular plant in which sporangia rise from a group of epidermal cells.

Eustachian tube or **eustachian tube** [yoo stā´shən] *Anatomy.* a narrow channel that connects the tympanic cavity with the nasopharynx, adjusting the air pressure in the cavity to the external pressure. (Named for Bartolommeo *Eustachio,* 1524–1574, Italian anatomist.) Also, AUDITORY TUBE.

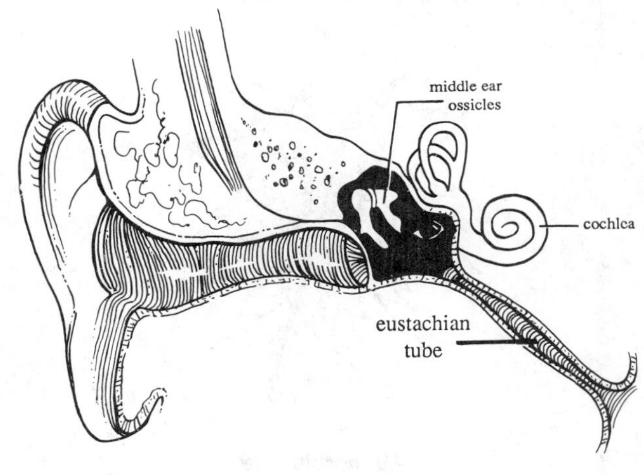

Eustachian tube

eustasy *Oceanography.* any global fluctuation in sea level caused by changes in water supply (such as ice melting) or large-scale changes in ocean basin capacity. Thus, **eustatic.**

eustele *Botany.* a modified siphonostele typical of dicotyledonous and gymnospermous plants in which the primary vascular tissues occur in a cylindrical arrangement of collateral or bicollateral vascular bundles around a pith with conjunctive tissue between.

eusternum *Invertebrate Zoology.* the anterior sternal plate in insects.

Eustichiaceae *Botany.* a monogeneric family of small, slender mosses of the order Dicranales that grow in tufts on moist cliff faces or in soil; characterized by distichous, complanate leaves, erect and sparsely branched stems, and lateral sporophytes.

Eustigmatophyceae *Botany.* a class of unicellular algae of the division Chromophycota that are mostly nonmotile and photosynthetic, live in freshwater, marine, or soil habitats, and are characterized by uniquely organized motile cells, several photosynthetic pigments, and a single parietal yellow-green chloroplast.

Eusuchia *Vertebrate Zoology.* the modern crocodiles, the sole extant suborder of the order Crocodylia, characterized by an internal nasal opening situated far back and surrounded by the pterygoid bone.

Eusyllidae *Invertebrate Zoology.* a subfamily of polychaete annelid worms.

eusynanthropic *Ecology.* describing an organism that lives or thrives in human habitats.

eusystole *Cardiology.* the regular systole of the heart, occurring normally with regard to force and time. Thus, **eusystolic.**

Eutardigrada *Invertebrate Zoology.* the water bears, an order of the phylum Tardigrada.

eutaxite *Petrology.* any rock having an eutaxitic structure.

eutaxitic *Petrology.* describing the banded, streaked, or blotched appearance observed in some eruptive rocks, resulting from alternate layering of differing texture, composition, or color; the bands were discrete features in lava or ash falls, but were subsequently drawn out while still viscous.

eutectic *Physical Chemistry.* **1.** describing a mixture in which the proportions of the substances are such that no other composition of the same materials can have a lower melting point or freezing point. **2.** an alloy or solution that has the lowest possible melting point for its components.

eutectic alloy *Metallurgy.* an alloy that has a definite minimum melting point compared with all other alloys of the same materials.

eutectic composite *Materials Science.* a two-phase composite in which one phase occurs in a fibrous form and is distributed in the matrix. Also, IN SITU COMPOSITE.

eutectic composition *Physical Chemistry.* the particular composition of materials providing the lowest temperature at which an alloy or solution will remain liquid and below which it will melt or freeze.

eutectic crystallization *Metallurgy.* the crystallization of a eutectic alloy upon cooling.

eutectic melting *Metallurgy.* in a nonhomogeneous alloy, local melting of selected volumes that have a eutectic composition.

eutectic mixture *Physical Chemistry.* the particular arrangement of a system that has reached the eutectic point. Also, **eutectic system, eutectic structure.**

eutectic point *Physical Chemistry.* the combined conditions of substance composition and temperature that produce the lowest point at which a solution or alloy will melt or freeze. *Materials Science.* in the three-plane reaction, the composition of the last remaining liquid.

eutectic temperature *Physical Chemistry.* the lowest temperature at which a solution or alloy will melt or freeze.

eutectogenic system *Physical Chemistry.* a mixture of liquids and solids undergoing transformation, in which the solids are in equilibrium with the remaining liquids at a given temperature, generally the lowest temperature.

eutectoid *Physical Chemistry.* **1.** describing a reaction in which a single solid solution decomposes at low temperature to produce two or more other solid solutions. Thus, **eutectoid reaction. 2.** an alloy or other substance formed by this process.

eutectoid alloy *Metallurgy.* a eutectic alloy formed when a substance cools in the solid state to produce two or more other solids.

eutectoid mixture *Physical Chemistry.* the particular arrangement of a system that has reached the eutectoid point. Also, **eutectoid system, eutectoid structure.**

eutectoid point *Physical Chemistry.* the point at which a solid solution is converted into two or more other solids on cooling; constant for any one alloy.

eutectoid transformation *Materials Science.* a three-phase reaction in which one solid phase transforms to two different solid phases.

eutectophyre *Petrology.* a light-colored, tufflike igneous rock characterized by a network of interlocking or intergrown quartz and orthoclase crystals. Also, **eutectofelsite.**

eutely *Invertebrate Zoology.* in many nematodes and rotifers, the peculiar feature of having a constant number of cells in the body, the number being characteristic for each species but varying among species.

Euthacanthidae *Paleontology.* a primitive acanthodian family of toothless spiny sharks in the order Climatiiformes that lived in the Devonian.

euthanasia [yoo´thə nā´zhə] *Medicine.* **1.** an easy or painless death. **2.** the practice of deliberately ending the life of a person or animal suffering from an incurable and often painful condition or disease.

euthenics *Ecology.* a science concerning changing or modifying the environment to improve future human welfare.

Eutheria *Paleontology.* the placental mammals, the subclass or infraclass that includes the most advanced mammals, with complex and well-developed brains; the eutherians arose in the latest Triassic but diversified only slightly until the end of the Cretaceous, when they began a very rapid adaptive radiation into a great diversity of habitats left open by the great extinction at the end of the Mesozoic era. *Vertebrate Zoology.* in some classifications, the contemporary placental mammals.

eutherian *Paleontology.* any member of Eutheria.

euthermic *Science.* producing or promoting warmth. *Medicine.* characterized by the proper temperature.

euthymism *Medicine.* the normal condition of thymus activity.

euthyroid *Medicine.* having normal thyroid function. Thus, **euthyroidism.**

eutocia *Medicine.* a normal labor or childbirth.

eutopic *Medicine.* **1.** situated normally. **2.** arising from the normal site or tissue.

Eutreptiales *Botany.* an order of highly mobile euglenoids characterized by two similar flagella, one directed anteriorly and the other posteriorly or laterally, and by green or colorless cells.

Eutrichosomatidae *Invertebrate Zoology.* a small family of parasitic wasps, hymenopteran insects in the superfamily Chalcidoidea.

eutrophic *Medicine.* relating to, characterized by, or promoting good nutrition. *Ecology.* of a lake or other body of water, containing a rich supply of plant nutrients and characterized by seasonal periods of oxygen deficiency as a result of excessive growth of algae.

eutrophication *Ecology.* a process that increases the amount of nutrients, especially nitrogen and phosphorus, in a marine or aquatic ecosystem; occurs naturally over geological time but may be accelerated by human activities, such as waste disposal or land drainage, leading to an increase in algae and a decrease in diversity.

eutrophy *Ecology.* the state of being eutrophic or undergoing eutrophication. *Medicine.* a state of normal, good nutrition. Also, **eutrophia.**

euxenite-(Y) *Mineralogy.* $(Y,Ca,Ce,U,Th)(Nb,Ta,Ti)_2O_6$, a brownish-black, orthorhombic mineral occurring as short, prismatic crystals and crystal aggregates, having a specific gravity of 4.3 to 5.87 and a hardness of 5.5 to 6.5 on the Mohs scale; found in granite pegmatites.

euxinic *Hydrology.* **1.** describing an environment having restricted circulation and stagnant or anaerobic conditions. **2.** of or relating to the material deposited in such environments and the process of its deposition.

EV expected value; exposure value.

eV electron volt.

EVA *Space Technology.* see EXTRAVEHICULAR ACTIVITY.

evacuate *Engineering.* to clear an area or compartment of something, specifically of gases or vapors. *Physiology.* to carry out the process of evacuation.

evacuation *Physiology.* **1.** an emptying, especially of the bowels. **2.** material emptied from the bowels; stool.

evacuator *Surgery.* a device used to empty fluid or particles from a cavity, such as the bowel or bladder.

evaginate *Biology.* to turn inside out, as a tubular organ might, or to cause a part to protrude by eversion of an inner surface. Thus, **evagination.**

eval *Artificial Intelligence.* in Lisp, the function that performs evaluation of a specified expression, allowing code that is constructed as a runtime data structure to be executed directly.

evaluation *Artificial Intelligence.* **1.** in Lisp, the process of determining the value of a symbol, number, or function applied to arguments. **2.** the process of determining the estimated value of a state during a search process. **3.** in logic, the process of determining the value of a formula given an interpretation. *Psychology.* see ASSESSMENT.

evaluation apprehension *Psychology.* the tendency of an individual's performance to be negatively affected if others are observing and evaluating his actions.

evaluation research *Psychology.* research to determine the effectiveness of implemented psychological programs or policies.

E-value see NORMALIZED STRUCTURE FACTOR.

evanescent field *Electrical Engineering.* a time-varying electromagnetic field extending into a region where the boundary conditions prevent it from propagating and where the amplitude decreases with distance. For simple cases, the decay is exponential.

Evaniidae *Invertebrate Zoology.* a family of wasps, the ensign flies, that are hymenopteran insects in the superfamily Proctotrupoidea..

Evans, Sir Arthur 1851–1941, English archaeologist; discovered and interpreted the pre-Mycenean Minoan culture at Knossos, Crete.

evansite *Mineralogy.* a colorless to white, amorphous mineral with an approximate formula of $Al_3(PO_4)(OH)_6 \cdot 6H_2O$, occurring in reniform or botryoidal masses, having a specific gravity of 1.8 to 2.2 and a hardness of 3 to 4 on the Mohs scale; found as a secondary mineral with allophane and limonite.
evaporated milk *Food Technology.* enriched whole milk that is concentrated, sterilized, and usually canned.
evaporation *Physics.* a process in which a liquid is converted to its vapor phase by adding latent heat to the liquid.
evaporation gauge see EVAPORIMETER.
evaporative *Physics.* of or relating to evaporation.
evaporative condenser *Mechanical Engineering.* an apparatus in which vapor is condensed in tubes that are cooled by the evaporation of water flowing outside of them.
evaporative cooling *Engineering.* **1.** a process that is used to lower the temperature of a large amount of liquid by putting the latent heat of vaporization of a part of the liquid to use. **2.** a process in which air temperature is lowered by evaporating water into it.
evaporative heat regulation *Physiology.* the control of body heat by evaporation of sweat.
evaporative power *Meteorology.* a measure of the degree to which the climate of a region is favorable to the process of evaporation, usually considered as the rate of evaporation from a chemically pure surface of water at the temperature of the lowest layer of the atmosphere. Also, **evaporation capacity, evaporative capacity, evaporation power, evaporativity.**
evaporator *Chemical Engineering.* equipment used to vaporize all or a portion of the solvent from a solution. *Mechanical Engineering.* **1.** the component or part in a refrigeration system where the refrigerant absorbs heat and changes from a liquid to a gas. **2.** any device in which evaporation occurs, especially one designed to concentrate a solution.
evaporimeter *Engineering.* an instrument that measures the rate at which water evaporates. Also, EVAPORATION GAUGE.
evaporite *Geology.* a nonclastic, sedimentary rock consisting of deposits of mineral salts produced by the evaporation of salt water.
evapotranspiration *Hydrology.* the total loss of water from a particular area, equal to the sum of the water lost by evaporation from the soil and other surfaces and that lost by transpiration from plants.
evapotranspirometer *Engineering.* an instrument that measures the rate at which water is discharged into the atmosphere from a designated area of the earth's surface; consists of a tank in which all water added to the tank and any water remaining can be measured.
evection *Astronomy.* an irregularity in the motion of the moon produced by the gravitational perturbation of the sun and planets.
E-vector *Electromagnetism.* the electric-field vector of an electromagnetic wave that, in free space, is perpendicular to the direction of propagation.
even-even nuclei *Nuclear Physics.* any of a group of 155 stable nuclei that have an even number of protons and an even number of neutrons; the protons and neutrons fill their allowed electronic energy levels in pairs having antiparallel spins, resulting in zero nuclear spin.
even function *Mathematics.* a function f such that $f(x) = f(-x)$ for all x in the domain of definition for f; a function f whose graph is symmetric in the y-axis.
even harmonic *Physics.* a harmonic whose frequency is an even multiple of the fundamental frequency.
evening emerald see PERIDOT.
evening gun *Ordnance.* the firing of a gun to signal the lowering of a flag in retreat. Also, RETREAT GUN.
evening star *Astronomy.* **1.** the planet Venus when it is clearly visible in the evening sky. **2.** any bright celestial object having a starlike appearance visible in the evening sky.
even keel *Naval Architecture.* a condition of floating with no list, and with an equal draft forward and aft, at a vessel's intended trim.
even number *Mathematics.* any integer that is divisible by 2.
even-odd nuclei *Nuclear Physics.* any of a group of 53 stable nuclei that have an even number of protons and an odd number of neutrons; unpaired protons and neutrons give half-integral (1/2, 3/2, 5/2, etc.) nuclear spin.
evenomation *Toxicology.* the removal or neutralization of venom from an animal sting or bite.
even parity check *Computer Programming.* a parity check in which the number of 1 bits in a word is expected to be even. An additional parity bit is appended to the word and set to the appropriate value to make the parity even.
even permutation *Mathematics.* any permutation that may be represented as the product of an even number of transpositions.
even pitch *Design Engineering.* the pitch of a screw in which the number of threads per inch is either a multiple or multiplier of the threads per inch of the lead screw of the lathe on which the screw is cut.
event something that happens; an occurrence; specific uses include: *Physics.* any specification of a position and a time that may be represented by a point in the time-space coordinate system. *Mathematics.* a measurable subset of a probability space. *Statistics.* a subset of the sample space of all possible outcomes of an experiment. *Geology.* any incident having probable tectonic significance as suggested by geologic, isotopic, or other evidence, but whose full implications are unknown. *Industrial Engineering.* the beginning or the end of a specific activity in a PERT network. *Computer Programming.* **1.** an occurrence or happening of significance to a task or program, such as the completion of an asynchronous input/output operation. **2.** a transaction or other activity that affects the records in a file.
event-driven monitor *Computer Technology.* a software monitor that performs actions upon the occurrence of certain events.
event horizon *Astronomy.* the boundary of a black hole from inside of which neither matter nor energy can escape.
Eventognathi *Vertebrate Zoology.* an equivalent name for the Cypriniformes, a large order of soft-finned freshwater fish.
eventration *Medicine.* the protrusion of the bowels from the abdomen. *Surgery.* the removal of the abdominal viscera. *Radiology.* also, **eventration treatment.** the application of radiation to internal tissues or organs through a surgical opening.
event schema *Psychology.* a series of occurrences that are related to an individual's needs and which have a definite beginning and end.
even-word boundary *Computer Programming.* a storage location whose address falls at the boundary between memory words.
Everest, Mount *Geography.* the world's highest mountain (29,028 feet) in the Himalayas on the Tibet-Nepal border.
Everest spheroid *Geodesy.* a reference ellipsoid with a semimajor axis of 6,377,276.3 meters and ellipticity of 1/300.80.
evergreen *Botany.* describing woody perennial plants that have green leaves throughout the year and that shed leaves slowly over many growing seasons; many evergreens occur in tropical regions without cold or dry seasons. In temperate and cold regions they have needlelike or scalelike leaves.
evernia *Botany.* a genus of lichens in the family Usneaceae, characterized by a fructose or pendulous thallus with a cottony medulla.
Evershed effect *Astrophysics.* the outward flow of gas in the penumbra of a sunspot.
eversion *Anatomy.* the turning outward of an organ or structure in relation to its normal position.
eversporting *Genetics.* describing a strain of organisms that continuously produce a specific mutant phenotype in successive generations rather than breed true to the parental phenotype.
everting suture *Surgery.* a method of suturing by which the edges of a wound are turned outward.
Eve's constant *Nucleonics.* a factor used in measuring the radioactivity of a substance, equivalent to the number of ions generated in air per cubic centimeter per second by one gram of the substance at a distance of one centimeter.
evisceration *Surgery.* **1.** the removal of the contents of the eye, leaving the sclera. **2.** the removal of the abdominal viscera. **3.** any extrusion of the viscera outside the body, especially through a surgical incision.
Evjen method *Solid-State Physics.* a systematic method for calculations of lattice sums, in which groups of charges whose total charge is zero are taken together so that the group contribution is small and the series for group contributions converges rapidly.
evjite *Petrology.* a gabbro of primary hornblende in which the only light-colored mineral is labradorite or bytownite.
evocation *Developmental Biology.* an embryonic induction, especially where the inducing tissue or substance is nonspecified but triggers differentiation inherent in the receptive tissue.
evoked potential *Physiology.* a sequence of electrical changes that occurs in a nerve cell membrane when evoked by a stimulus above its threshold, as seen on a brain wave tracing.
evolute *Mathematics.* the evolute of a curve γ_1 is a second curve γ_2 consisting of the centers of curvature of γ_1; equivalently, γ_2 is the envelope of the normals of γ_1. γ_1 is called an **involute** of γ_2. Each normal of the involute is a tangent of the evolute, and to a given evolute there corresponds a family of involutes.

Evolution

Charles Darwin made an important distinction in arguing that his work had included two different goals: first, to establish the fact of evolution; second, to propose a mechanism, called natural selection, as the primary cause of evolutionary change. He regarded the first goal as more important since the simple acceptance of evolution unleashed such a cascade of scientific and philosophical consequences of revolutionary import. No change in scientific thinking could be more full of wonder and implication than the shift from a world existing for only a few thousand years, with all species separately created (and subservient to humans), to an earth with a history measured in billions of years, with all creatures linked by ties of genealogical descent, and with humans arising at the latest cosmic moment and having no special or preferred status.

The theory of natural selection is equally mind-stretching in its claim that the adaptations of organisms, from the hydrodynamic efficiency of a dolphin to the beauty of a peacock's tail, arise as products of a struggle among individuals for reproductive success. Evolutionary theory, Darwin's second topic, is a ferment of exciting and useful debate, with new concepts (including neutral mutation theory, mass extinction by cometary impact, punctuated equilibrium) integrated around a Darwinian core. The fact of evolution, Darwin's first topic, is as well established as anything in science. Evolution is the central organizing principle of biology. The famous statement of our premier evolutionist, Theodosius Dobzhansky, bears repeating in closing: nothing in biology makes any sense except in the light of evolution.

Stephen Jay Gould
Professor of Geology
Harvard University

evolution *Biology.* **1.** a cumulative change in the characteristics of organisms or populations from generation to generation. **2.** the scientific study of these processes. **3.** any developmental process by which an organ or organism becomes more complex by the differentiation of its parts.

evolutionary *Biology.* **1.** of or relating to evolution or other developmental processes. **2.** of, relating to, or in accordance with a theory of evolution.

evolutionary anthropology *Anthropology.* the study of change over time in a culture or within subcultures.

evolutionary conservatism *Evolution.* the preservation of a high proportion of ancestral traits in species following diverging evolutionary pathways, resulting from the retention of a high proportion of common alleles.

evolutionary ecology *Ecology.* the integrated science of evolution, genetics, adaption, and ecology; concerned with interpretation of the structure and function of organisms, communities, and ecosystems in the context of evolutionary theory.

evolutionary method *Evolution.* a method of classification in which evolutionary history, including cladistic and morphological divergence data, is hypothetically reconstructed.

evolutionary operation *Industrial Engineering.* a process of ongoing planning involving incremental adjustments by which performance is gradually optimized.

evolutionary rate *Evolution.* the speed of evolutionary change per given unit of time.

evolutionary stable strategy *Ecology.* a strategy that, if adopted by most of a population, cannot be bettered by any other strategy, and will therefore tend to become established by natural selection.

evolutionary trend *Evolution.* the recognizable morphological changes in a species toward a given hereditary character or function over long periods of time or for many generations.

evolutionism *Anthropology.* a doctrine formulated in the late 1800s, maintaining that nontechnological societies represent an earlier or lower stage in the evolution of human culture than industrial societies. Also, **classical evolutionism.**

EVOP evolutionary operation.

eversion *Geology.* a process in which potholes are excavated in river beds as a result of the erosional action of vortices and eddies.

eversion hollow *Geology.* a pothole formed by eversion.

EVR electronic video recording.

evulsion *Surgery.* a process of forcible extraction.

Ewald-Kornfeld method *Solid-State Physics.* an extension of the Ewald method that is useful in calculating the coulomb energy associated with a dipole array.

Ewald method *Solid-State Physics.* a method for calculating lattice sums by employing mathematical techniques that make the series converge rapidly.

Ewald sphere *Solid-State Physics.* a sphere of radius $k = 2\pi/l$, where l is the X-ray wavelength centered at the k-vector origin of the incident radiation at a reciprocal lattice point of a crystal; used to determine the direction an X-ray or other beam will be reflected by a crystal lattice. *Crystallography.* see SPHERE OF REFLECTION.

Ewald-sphere construction *Materials Science.* a graphical solution to Bragg's Law in which a sphere having a radius equal to the reciprocal of the wavelength is constructed; any reciprocal-lattice point lying on the sphere obeys Bragg's Law.

ewe *Agriculture.* an adult female sheep. *Vertebrate Zoology.* the adult female of various related animals, including goats, and the smaller antelopes.

Ewing, Maurice 1906–1974, American geophysicist; established bases of plate tectonics; made early photographs and seismic studies of ocean floor.

Ewingella *Bacteriology.* a proposed genus of Gram-negative bacteria of the family Enterobacteriaceae, occurring as facultatively anaerobic, rod-shaped cells.

Ewing's sarcoma *Medicine.* a cancerous tumor that develops in the bone marrow, usually in the long bones or the pelvis.

Ewing theory of ferromagnetism *Solid-State Physics.* an early theory predicting that ferromagnetism in a permanent magnet results from the ability of an individual atom, treated as a permanent magnet, to rotate freely about its center when subjected to external fields.

EWR *Aviation.* the airport code for Newark International Airport, Newark, New Jersey.

ex- a prefix meaning **1.** out of, as in *exhale.* **2.** complete, as in *exacerbation.* **3.** former, as in *ex-member.* **4.** outer or outside, as in *exaspidean.*

exa- a combining form used to name a unit equal to one quintillion (10^{18}) times the unit with which it is combined, as in *exagram.*

exacerbation [eks as´ər bā´shən] *Medicine.* an increase in the severity of a disease or symptom.

exact chi-squared test see FISHER-YATES TEST.

exact differential equation *Mathematics.* a differential equation obtained by setting an exact differential form equal to zero.

exact differential form *Mathematics.* an alternating differential(k + 1)-form ω that is the differential of a k-form; i.e., $\omega = d\theta$ for some k-form θ.

EXAFS extended X-ray absorption fine structure.

exalate *Botany.* without expansions that resemble wings. Also, APTEROUS.

exalbuminous see EXENDOSPERMOUS.

exaltation *Psychology.* a feeling of extreme elation, often associated with delusions of grandeur.

exalted-carrier receiver *Electronics.* a device that improves reception of a fading signal by keeping the carrier component at a high level at all times.

examination *Medicine.* any inspection or investigation, especially as a means of assessing health or diagnosing a disease.

exanimation *Neurology.* a term for unconsciousness, coma, or death.

exanthem *Medicine.* **1.** a skin eruption or rash. **2.** a disease characterized by skin eruptions or rashes, such as measles, scarlet fever, or rubella. Also, **exanthema.**

exanthema subitum *Medicine.* an acute, short-lived illness of infants, similar to chickenpox or measles and probably of viral origin, marked by a high fever followed by a rash. Also, **exanthem subitum**.

exanthrope *Pathology.* any source of disease that is not situated within the human body.
exanthropic *Pathology.* of, relating to, or characteristic of an enxanthrope. *Ecology.* describing an organism that lives far from human habitats or that does not thrive in human habitats.
exarch *Botany.* having the protoxylem farthest from the center of the axis, with younger xylem nearer the axis.
exasperate *Biology.* having a rough surface covered with short, stiff bristles.
exaspidean *Vertebrate Zoology.* of birds, having a tarsal envelope that is continuous around the outer edge of the tarsus.
excavate *Archaeology.* to carry out the process of excavation.
excavation *Science.* **1.** the process of digging or hollowing out. **2.** a hollowed-out space or cavity. *Archaeology.* the scientific process of digging into the earth to discover and extract cultural remains, as a means of obtaining data about past human activity. *Medicine.* the process of hollowing out, such as the removal of decay from a tooth cavity in preparation for restoration.

archaeological excavation

excavation unit *Archaeology.* a basic area that is often the first space opened in an excavation; usually a trench, a defined feature such as a house floor, or a standard-sized square.
excavator *Mechanical Engineering.* a power-driven machine designed to dig and remove soil, gravel, or other materials. *Surgery.* **1.** a scoop-like surgical instrument used to remove the inner part of a structure or to create a cavity. **2.** a tool used in dentistry to prepare a cavity for filling.
Excellospora *Bacteriology.* a genus of thermophilic bacteria of the order Actinomycetales, occurring in soil as both substrate and aerial mycelia that develop spores.
excelsior *Materials.* a packing material usually made of fine, curled wood shavings.
except *Mathematics.* a logical operator defined as follows: the statement "P except Q" is true only when P is true and Q is false; otherwise it is false.
except gate *Electronics.* a logic device that allows a pulse to pass into the circuit only when it appears at a given set of terminals and is absent from other sets.
exceptional Jordan algebra *Mathematics.* any Jordan algebra that is not special. A Jordan algebra is exceptional if it cannot be embedded in an associative algebra with the product given by $a \cdot b = (ab + ba)/2$.
exceptional Lie group *Mathematics.* any of five Lie groups that do not belong to the four series of classical groups (the orthogonal groups on even and odd dimensional vector spaces, the symplectic groups, and the special linear groups). The exceptional groups are denoted G_2, F_4, E_6, E_7, and E_8.
exceptional space *Quantum Mechanics.* a space used in the description of a system with a finite degree of freedom in a generalization of quantum mechanics, in which observables and states are formulated in an exceptional Jordan algebra.
exception handling *Computer Programming.* the invocation of procedures or routines to deal with unexpected or abnormal situations that arise during processing.
exception-item encoding *Computer Programming.* a technique that allows uninterrupted processing of data by forcing erroneous data to an error tape for later analysis and/or correction.
exception principle *Industrial Engineering.* a theory suggesting that managers should direct their attention to abnormal values or occurrences and concentrate on returning undesirable situations to a normal level.
exception reporting *Computer Programming.* a method of examining large volumes of data and extracting only the unexpected or abnormal situations or data values. *Industrial Engineering.* in the control/feedback phase of production activity control, the process of reporting to managers any difficulties occurring on the shop floor.
excess air *Engineering.* the amount of air that is theoretically not needed in a combustion process for total oxidation.
excess coefficient *Mechanical Engineering.* the ratio $(A - R)/R$, where A is the amount of air admitted during fuel combustion and R is the amount required.
excess conduction *Solid-State Physics.* the conduction arising from excess electrons provided by donor impurities.
excess electron *Solid-State Physics.* an electron added to a semiconductor by a donor impurity and therefore available for conduction.
excess-fifty code *Computer Programming.* a numeric code in which the decimal number n is represented by the binary equivalent of $n + 50$. Also, XS-50 CODE.
excessive precipitation *Meteorology.* an unusually high rate of precipitation, generally rain, over a given period of time.
excess noise see CURRENT NOISE.
excess of arc *Navigation.* that portion of a sextant arc that indicates negative readings.
excess reactivity *Nucleonics.* reactivity that is added to a reactor to compensate for fuel burnup and the accumulation of contaminants during operation.
excess-three code *Computer Programming.* a numeric code in which the decimal number n is represented by the binary equivalent of $n + 3$. Also, XS-3 CODE.
exchange *Telecommunications.* **1.** all of the telephone equipment in a room or building equipped so that telephone lines terminating there may be interconnected as required. **2.** a city or town designated as an administrator of telephone services. *Computer Programming.* to interchange the contents of two storage locations or devices. *Quantum Mechanics.* the interchange of the states of two identical particles in proximity so as to preserve the symmetry of their respective wavefunctions.
exchangeability *Statistics.* a property of a sequence of random variables in which the order of the elements of the sequence has no effect on the joint probability distribution.
exchangeable disk storage *Computer Technology.* a storage device in which the disk pack or cartridge can be removed and replaced with another disk pack or cartridge.
exchange adsorption *Chemistry.* a type of ion exchange in which the surfaces of the adsorbent are completely saturated by two adsorbable components of the fluid phase.
exchange anisotropy *Electromagnetism.* an observation, in some mixtures of magnetic materials, that the magnetization is favored in a certain direction, rather than in the direction of a particular axis.
exchange broadening *Spectroscopy.* the widening of a spectral line as a result of frequent rapid jumps between two or more closely related energy states of the absorbing or emitting species.
exchange buffering *Computer Programming.* a data buffering technique designed to avoid the movement of data within main memory.
exchange capacity *Physical Chemistry.* a quantitative measure of the surface charge of a substance, as determined by the quantity of exchangeable ions in a given unit of the material. *Agronomy.* the capacity of a soil to retain and exchange such cations as hydrogen, calcium, manganese, and potassium.
exchange coefficient *Fluid Mechanics.* a coefficient, in turbulent flow of eddy flux, explained in correlation to coefficients of the molecular theory of gases.
exchange degeneracy *Particle Physics.* a state in which two particle systems interact and the total energy resulting from the exchange is the same as if no exchange had occurred. Also, **exchange symmetry.**
exchange diffusion *Chemistry.* a process by which atoms of the same element in two different molecules exchange places.
exchange force *Quantum Mechanics.* the nonclassical force associated with an exchange interaction generated by coupling of space and spin variables as a consequence of the Pauli exclusion principle.

exchange integral *Quantum Mechanics.* an integral over space that represents the interaction between a two-particle state and the state under exchange of particles.
exchange line *Electricity.* a line that connects a subscriber or switchboard to a commercial exchange.
exchange list *Nutrition.* a list of commonly used foods grouped according to similarities in composition, so that all those within one group may be used interchangeably in a diet.
exchange narrowing *Spectroscopy.* **1.** the narrowing of a spectral line due to the loss of hyperfine structure. **2.** for a spectral line previously broadened by some variable perturbation, a narrowing resulting from a dynamic process that exchanges different values of the perturbation.
exchange operator *Quantum Mechanics.* an operator that, acting on a two-particle system, effects an exchange of all space and spin coordinates.
exchange pairing *Genetics.* a type of pairing of homologous chromosomes during meiosis that permits crossing-over.
exchange plant *Telecommunications.* a local telephone facility.
exchange reaction *Chemistry.* a chemical reaction in which two atoms of the same element exchange places, either in two different molecules or in two different positions of the same molecule.
exchange sort see BUBBLE SORT.
exchange system *Anthropology.* the method in which resources and services are distributed in a society, including systems such as reciprocity, redistribution, or market exchange.
exchange transfusion *Medicine.* the repetitive withdrawal of small amounts of blood and replacement with donor blood, until a large proportion of the blood volume has been exchanged; used in newborn infants with erythroblastosis and in patients with severe uremia. Also, REPLACEMENT TRANSFUSION, SUBSTITUTION TRANSFUSION.
exchange velocity *Chemistry.* in an ion-exchange process, the rate of replacement of one ion by another in the exchanger.
excimer *Chemistry.* a substance formed by the joining of two atoms (or two molecules of the same chemical composition) in an excited state; an excited dimer.
excipient *Pharmacology.* a relatively inert substance, such as a syrup or elixir, that is added to a drug in order to make it easier to administer. Also, VEHICLE.
exciplex *Chemistry.* an electronically excited molecular charge complex, unstable in the ground state, that is formed from an excited acceptor molecule and a donor molecule (or vice versa).
Excipulaceae see DISCELLACEAE.
excise *Surgery.* to cut out or cut off; remove by cutting.
excision *Surgery.* a removal by cutting. *Genetics.* the enzymatic removal of a section of a DNA or RNA molecule.
excisional biopsy *Surgery.* **1.** the cutting out of tissue for examination. **2.** the removal by cutting of an entire lesion, including a margin of normal-appearing tissue, for examination.
excision repair see CUT AND PATCH REPAIR.
excitable *Biology.* describing a tissue or organism that is sensitive to or capable of responding to stimuli.
excitation the act of exciting or the state of being excited; specific uses include: *Quantum Mechanics.* the process of changing the state of a system from its ground state to a given excited state. *Atomic Physics.* a process by which the energy state of an atom or molecule is increased above the ground state, by radiation or collision. *Electricity.* the application of voltage to field coils to produce a magnetic field. *Control Systems.* the actuation of one part of a system or apparatus in order to carry out some function in another part. *Behavior.* an increase in response rate or strength.
excitation curve *Nuclear Physics.* a graph of energy states of an atom based on the energy level of incident energetic particles, showing a series of discrete quantum states.
excitation energy *Quantum Mechanics.* the energy that is necessary to change a system from a ground state to a given excited state. Also, **excitation potential.**
excitation function *Atomic Physics.* the cross section for an incident electron that raises an atom to a given excited state; expressed as a function of the electron's energy.
excitation index *Spectroscopy.* an indication of variation in excitation conditions, expressed as the ratio of the intensities of a pair of extremely nonhomologous emission lines.
excitation purity *Chemistry.* a measure of chromaticity, expressed as the ratio of a specific color to that of a reference source on a chromaticity diagram.
excitation voltage *Electricity.* the minimum voltage that is needed for the excitation of a circuit.
excitation volume *Physics.* the volume of an X-ray source used in electron-probe microanalysis.
excited state *Quantum Mechanics.* the condition of an atom at a state of higher energy than the ground state.
excited-state maser *Physics.* a maser with an amplifying transition whose terminal level is not appreciably populated at thermal equilibrium.
exciter *Electricity.* a small auxiliary generator that provides field current for an AC generator. *Electronics.* a device that furnishes a radio transmitter with carrier frequencies.
exciter lamp *Electricity.* an incandescent lamp that has a concentrated filament; used to excite a phototube or photocell.
exciter response *Electricity.* the rate of change of the main exciter voltage when resistance is changed suddenly in rotating electric machinery.
exciting current see MAGNETIZING CURRENT.
exciting line *Spectroscopy.* a spectral line from a noncontinuous source that corresponds to the frequency of electromagnetic radiation absorbed by a system in association with some particular process.
exciton *Solid-State Physics.* a bound hole-electron pair in a semiconductor, having a definite half-life during which it migrates through the crystal, and releasing its eventual recombination energy as a photon or photons.
excitor *Neurology.* a nerve whose stimulation excites greater action in the body part that it supplies.
excitron *Electronics.* a device that converts AC current to DC, characterized by the use of mercury to initiate an arc between its electrodes.
excluded middle *Mathematics.* a proposition that is either true or false is said to satisfy the law of the excluded middle. Also, **principle of the excluded middle.**
excluded volume effect *Materials Science.* any interaction between segments of a polymer chain that are moving to occupy the same space.
exclusion *Surgery.* the separation of one part of an organ from the remainder, with both parts being left in the body.
exclusion area *Nucleonics.* the section of a nuclear facility, such as a power plant, where access is limited to certain personnel.
exclusive NOR *Computer Programming.* an instruction that compares the bits of two operands and returns a 0 if one bit is a 1 and the other is a 0, and returns a 1 if both bits are 1 or both are 0.
exclusive-NOR gate *Computer Technology.* an electronic logic gate that performs the exclusive-NOR operation, returning a 0 only if one of its inputs is a 1 and the other is a 0, and returning a 1 if both of its inputs are a 1 or both are a 0. Also, XNOR GATE, EQUIVALENCE GATE.
exclusive OR *Computer Programming.* **1.** a logical operation that compares two bits and returns a 1 if one bit is a 1 and the other is a 0, and returns a 0 if both bits are 1 or both are 0. **2.** An instruction that performs an exclusive OR on corresponding bits of two operand words. *Mathematics.* a logical operator defined as follows: the statement "P exclusive or Q" is true only when exactly one of P and Q is true, and is false otherwise. Denoted $P \underline{\vee} Q$. Also, **exclusive disjunction.**
exclusive-OR gate *Computer Technology.* an electronic logic gate that performs the exclusive-OR operation. Also, XOR GATE, NONEQUIVALENCE GATE, HALF-ADDER.
exclusive segments *Computer Programming.* partitions of an overlay program structure that cannot reside in main memory at the same time.
excochleation *Surgery.* the removal of the contents of a cavity by scraping or scooping.
exconjugant *Biology.* a microorganism in its independent state just after conjugation. *Cell Biology.* a cell that is the product of conjugation.
Excorallanidae *Invertebrate Zoology.* a family of isopod crustaceans in the suborder Flabellifera; some are parasites, some free-living.
excoriation [eks kôr′ē ā′shən] *Medicine.* a scratch or abrasion of the skin or another epithelial surface.
excrement [eks′krə mənt] *Physiology.* the waste products of normal metabolism that are cast out of the body.
excrescence [eks krē′səns] *Biology.* **1.** an abnormal outgrowth, such as a wart. **2.** an abnormal increase in growth. **3.** a normal outgrowth, such as fingernails or hair.
excrete [eks krēt′] *Biology.* to eliminate waste matter from the body.
excretion [eks krē′shən] *Biology.* **1.** the elimination of metabolic waste materials by the respiratory, digestive, and integumentary systems. **2.** the material that is so discharged.
excretory [eks′krə tôr′ē] *Biology.* relating to or involved in excretion.

excretory system *Anatomy.* the system that collects and excretes nitrogenous wastes and excess water from the body in the form of urine; it includes the kidneys, ureters, urinary bladder, and urethra.

excurrent *Biology.* flowing out or providing an exit or passage out; efferent.

excurrent canal *Invertebrate Zoology.* the channel in a sponge through which water flows toward the exterior.

excursion *Physiology.* any movement occurring from a normal, or rest, position of a movable body part in performance of its normal functions. Also, **excursive movement.** *Nucleonics.* a sudden, extremely rapid rise in a reactor's power level that occurs when the production rate of fission neutrons drops.

excursion steamer *Naval Architecture.* a vessel used to take passengers on sightseeing trips or short cruises, typically of less than a day.

Exec see EXECUTIVE PROGRAM.

executable module *Computer Programming.* a program module that has been compiled and link-edited and can be run by the operating system.

executable statement *Computer Programming.* a program instruction that specifies an explicit action to be taken by the computer. Also, IMPERATIVE STATEMENT.

execute *Computer Programming.* **1.** to run a program on a computer. **2.** to perform the actions specified by a program. **3.** to execute an instruction in the CPU.

execute cycle *Computer Technology.* the part of the operation cycle during which the the CPU converts a machine-language instruction into electronic signals.

execute statement *Computer Programming.* a statement in job-control language that denotes the beginning of a job step.

execution *Computer Programming.* **1.** the process of running, or the results of having run, a program or routine. **2.** the performance of an instruction in the CPU.

execution control program see EXECUTIVE PROGRAM.

execution cycle *Computer Technology.* that part of the operation cycle during which an instruction is actually executed.

execution environment *Computer Programming.* during execution of a block-structured language, the set of local variable values and values of variables in surrounding blocks that are accessible to the running program. Also, ENVIRONMENT.

execution stack *Computer Science.* a stack of activation records or stack frames that is maintained during execution of programs in a block-structured or recursive language.

execution time *Computer Programming.* **1.** the total time required to execute an instruction. **2.** the amount of CPU time used by a program. Also, COMPUTE TIME. **3.** the portion of an instruction cycle in which work is actually performed.

executive see EXECUTIVE PROGRAM.

executive-control language *Computer Programming.* a set of instructions that allows a programmer to give commands to the executive program to control the running of a job, such as requests to mount tapes, compile and execute programs, and the like.

executive guard mode *Computer Programming.* a protection technique that prevents programs from executing instructions or accessing memory reserved for the executive program.

executive instruction *Computer Programming.* an instruction used to control the operation or execution of other routines; an instruction whose use is restricted to the executive program.

executive program *Computer Programming.* a part of the operating system that controls the execution of other programs and monitors the flow of work. Also, EXECUTIVE, EXEC.

executive routine *Computer Programming.* a computer routine designed to coordinate, direct, or control other programs or routines in the system.

executive supervisor *Computer Programming.* the portion of the executive system that controls the ordering, preparation, and execution of all jobs entering the system.

exedra *Architecture.* **1.** a semicircular or rectangular alcove with benches. **2.** in a church, an apse or niche.

exemia *Medicine.* the loss of fluid from the blood vessels, leaving behind red blood cells.

exendospermous *Botany.* lacking endosperm and thus having reserve food stored in the seed embryo. Also, EXALBUMINOUS.

exenteration *Surgery.* **1.** the removal of inner organs, especially the complete removal of the contents of a body cavity, such as the pelvis. **2.** the removal of the entire contents of the ocular orbit.

exercise *Physiology.* the performance of physical exertion in order to improve or maintain health or correct a physical deformity. *Military Science.* **1.** a military maneuver or simulated operation planned and executed for the purpose of training and evaluation. **2.** of or relating to such an activity. Thus, **exercise commander, exercise (directing) staff, exercise (planning) directive, exercise program, exercise specification.**

exeresis *Surgery.* surgical removal or excision.

exergonic *Biochemistry.* of or relating to a reaction in which the end product has less free energy than the original material, commonly associated with catabolism.

exfiltration *Military Science.* the removal of personnel units from areas under enemy control.

exfoliation

exfoliation *Science.* the process of shedding or removing outer layers; specific uses include: *Botany.* the process of shedding or falling away in scales, layers, or flakes, as the bark of some plants. *Medicine.* the peeling and flaking of tissue cells occurring in certain skin diseases, or after a severe sunburn, or in the normal process of desquamation. *Physiology.* see DESQUAMATION. *Geology.* the separation of successive thin, onion-like shells from bare surfaces of massive rock, such as granite or basalt; common in regions of moderate rainfall. Also, SPALLING, SCALING, SHEETING. *Metallurgy.* also, **exfoliation corrosion.** corrosion that causes surface layers to be lifted off.

exfoliative cytology *Pathology.* the microscopic examination of cells shed from a body surface or growth, or obtained from secretions, in order to test for malignancy or microbiological changes or to measure hormone levels.

exhalation *Physiology.* the process of breathing air outward, from the lungs to outside the body. Also, EXPIRATION. *Geophysics.* a process by which radioactive gases, formed by the decay of radioactive salts, are released from the earth's surface.

exhaust *Science.* **1.** to use up completely. **2.** the gas or vapor that leaves an enclosure. *Physiology.* to bring about exhaustion. *Chemistry.* to remove ingredients using solvents. *Mechanical Engineering.* **1.** a working fluid discharged from an engine cylinder or turbine after expansion. **2.** the parts of an engine or turbine through which exhaust escapes. Also, **exhaust system. 3**. a duct for the escape of gases, fumes, and odors from an enclosure, sometimes equipped with an arrangement of fans.

exhaust cone *Aviation.* an assembly designed to collect and direct discharge gases from a turbojet engine, consisting of an inner cone surrounded by an outer cone or casing, the open end of which forms the exhaust nozzle.

exhaust-deflecting ring *Mechanical Engineering.* a rotating ring that is mounted in the nozzle of a jet and used to deflect the exhaust stream.

exhauster *Food Technology.* a machine used to heat canned food prior to closing the can, in order to form a partial vacuum.

exhaust gas *Mechanical Engineering.* the gaseous products from an internal combustion engine or gas turbine.

exhaust head *Engineering.* an object designed to fit on the end of an exhaust pipe to reduce noise and collect oil and water.

exhaustion a state of being drained, consumed, or used up; specific uses include: *Physiology.* a state of extreme physical or mental fatigue. *Psychology.* in Hans Selye's analysis of reaction to stress, the final stage, at which the body can no longer function at an intense rate to resist the stress. Also, **exhaustion stage.** *Agronomy.* the loss of fertility by soil, to the extent that crop yields are small or no longer profitable.

exhaustion delirium *Medicine.* acute delirious reactions brought about by extreme fatigue, long wasting illness, or prolonged insomnia.
exhaustion point *Chemistry.* the point at which an adsorbent in an ion-exchange process fails to facilitate ion exchange.
exhaustion region *Electronics.* the layer in a semiconductor, located next to its metal contact, characterized by few charge carriers and almost complete ionization of atoms.
exhaustive search *Artificial Intelligence.* a search that examines all possibilities; usually not feasible due to combinatoric explosion.
exhaust manifold *Mechanical Engineering.* a branched pipe system that carries spent gases from the combustion chambers of an internal combustion engine to an exhaust system. Also, HEADER.
exhaust nozzle *Aviation.* **1.** a nozzle through which exhaust gases are ejected from the tailpipe of a jet or rocket engine. **2.** on a turbojet engine, the end of the exhaust cone leading into the tailpipe.
exhaust pipe *Mechanical Engineering.* a duct through which engine exhaust is discharged.
exhaust scrubber *Engineering.* a component of an internal combustion engine that serves to remove poisonous gases from engine exhaust.
exhaust stack *Aviation.* a pipe projecting from an aircraft engine and serving as an outlet for exhaust gases. Also, STACK.
exhaust stream *Aviation.* the stream of gaseous, atomic, or radiant matter discharged from the nozzle of a rocket or similar reaction engine.
exhaust stroke *Mechanical Engineering.* a piston stroke during which exhaust is ejected from the cylinder in a reciprocating engine. Also, SCAVENGING STROKE.
exhaust suction stroke *Mechanical Engineering.* the stroke of an engine that is characterized by simultaneous fuel exhaust and introduction of fresh fuel into the cylinder.
exhaust trail *Meteorology.* a condensation trail that forms when water vapor from an aircraft exhaust mixes with and saturates the air in the aircraft's wake.
exhaust valve *Mechanical Engineering.* in an internal combustion engine, the cylinder valve that controls the discharge of spent gas.
exhaust velocity *Fluid Mechanics.* the velocity of the fluids exiting a system. *Aviation.* see EFFECTIVE EXHAUST VELOCITY.
exhibitionism *Psychology.* **1.** a psychosexual disorder in which an individual has the impulse to publicly exhibit his genital organs to other persons. **2.** in popular use, the tendency or desire to behave in a way that will call attention to oneself. Thus, **exhibitionist.**
exhumation *Geology.* the uncovering or baring by erosion of a surface that had previously been buried by deposition.
exhumed *Geology.* see RESURRECTED.
Exiguobacterium *Bacteriology.* a genus of Gram-positive, facultatively anaerobic bacteria that grow as motile, rod-shaped or coccoid cells.
exine see EXOSPORIUM.
exinite *Geology.* a hydrogen-rich coal maceral group derived from spores, cuticle matter, resins, and waxes. Also, LIPTINITE.
existence doubtful *Navigation.* a legend on a nautical chart indicating that a danger has been reported but not confirmed.
existential anxiety or **crisis** *Psychology.* anxiety that concerns the problem of finding meaning or order in life.
existential psychology *Psychology.* an approach to psychology based on the philosophical movement of existentialism, which holds that individuals have freedom of choice and responsibility for their own actions, and that this is the source of their fear and anxiety.
existential quantifier *Mathematics.* one of the equivalent statements such as "there exists an x," "for some x," or "for at least one x"; denoted by $\exists x$.
exit *Engineering.* a passage or door that leads out of an area. *Computer Programming.* **1.** the point at which a sequence of instructions is terminated and control is passed to another sequence. **2.** the act of terminating execution of a program or subroutine. **3.** a return to the operating system when a program is finished.
exit dose *Radiology.* the amount of radiation dose absorbed at the exit site of a radiation beam passing through the body.
exite *Invertebrate Zoology.* a movable structure located on the external side of an arthropod limb.
exit pupil *Astronomy.* the image of a telescope's objective lens that is formed by an eyepiece; its size is given by dividing the aperture of the telescope by the magnification.
exit region *Meteorology.* a zone of diffluence at the downwind extremity of a jet stream.
exitus *Medicine.* death. *Anatomy.* an exit or outlet.
exit window *Photogrammetry.* the image of the field stop formed by all the lens elements on the image side of the field stop.
exmeridian altitude *Astronomy.* the altitude a celestial body has when it lies close to the observer's meridian.
exmeridian observation *Astronomy.* an altitude observation made on a celestial object lying close to the observer's meridian.
EXNOR gate see EXCLUSIVE-NOR GATE.
exo- *Science.* a combining form meaning "outer" or "outside," as in *exoskeleton, exogamy. Organic Chemistry.* a combining form denoting a substituent that is pointing toward the "outside" of a ringed molecule, or attached to a side chain rather than a ring.
Exobasidiaceae *Mycology.* the only family of fungi belonging to the order Exobasidiales, composed of plant parasites that live off the stems, leaves, and buds of flowering plants.
Exobasidiales *Mycology.* a order of fungi belonging to the class Hymenomycetes, many of its members are plant parasites that cause the host to develop abnormal growths called galls.
exobiology *Biology.* a branch of biology dealing with the search for life on other planets and elsewhere in the universe, and with the study of conditions that might give rise to extraterrestrial life, as well as the study of the effects of extraterrestrial environments. Thus, **exobiologist.**
exocardiac *Anatomy.* originating or occurring outside of the heart.
exocarp see EPICARP.
exoccipital *Anatomy.* located at the side of the foramen magnum.
Exocet [eks´ə set´] *Ordnance.* **1.** a French ship-launched surface-to-surface sea-skimming missile powered by a two-stage solid-propellant motor and equipped with inertial mid-course guidance; it carries a 220-pound conventional warhead at high subsonic speed and a maximum range of 23 miles; officially designated **MM-38,** with a later version **MM-40. 2.** a similar air-to-surface missile powered by an improved boost/sustainer; it carries a 364-pound armor-piercing warhead at high subsonic speed and a range of 31 to 43.5 miles; officially designated **AS-39. 3.** a coast defense version that uses the **MM-40** missile.
exochelin *Biochemistry.* a type of peptide produced by *Mycobacterium* in an iron-deficient substance.
exochorion *Invertebrate Zoology.* the outermost of two protective coverings of an insect egg.
exocoel *Invertebrate Zoology.* the space between pairs of adjacent mesenteries in sea anemones and coral polyps.
exocoelom see EXTRAEMBRYONIC COELOM.
Exocoetidae *Vertebrate Zoology.* the flying fishes, a family of marine fish of the order Atheriniformes, characterized by an elongate cylindrical body, a blunt snout, and an ability to leap out of the water and glide long distances; native to the Atlantic, Pacific, and Indian oceans.
exocrine *Physiology.* **1.** or or relating to external secretions. **2.** of or relating to organs or structures that secrete substances outwardly, via a duct.
exocrine cell *Cell Biology.* a secretory epithelial cell that secretes its products into the lumen of an exocrine gland.
exocrine gland *Physiology.* a gland, such as a lacrimal gland, that discharges its secretions through a duct onto an epithelial surface of the body.
exocuticle *Invertebrate Zoology.* the middle layer of the cuticle in insects and other arthropods.
exocyclic double bond *Organic Chemistry.* an external double bond connected to a ring structure.
exocytosis *Cell Biology.* a process by which a variety of secretory products are released from the cell via transport within vesicles to the cell surface and subsequent fusion with the plasma membrane, resulting in the extrusion of the vesicle contents from the cell.
exodeoxyribonuclease *Enzymology.* any of several enzymes of the hydrolase class that remove single nucleotides from the end of a DNA molecule.
exoelectron *Physics.* an electron, ejected from the surface of a metal or semiconductor in which the surface is newly formed by a fracture or abrasion.
exoenzyme *Enzymology.* an enzyme that functions outside the cell where it was synthesized.
exoergic *Physical Chemistry.* see EXOTHERMIC.
exogamous [eks äg´ə məs] *Anthropology.* relating to or practicing exogamy.
exogamy [eks äg´ə mē] *Biology.* a sexual union of gametes from two unrelated parents. *Anthropology.* the practice of finding a marriage partner outside of one's own group, as in a different kin group or another community.

exogastrula *Developmental Biology.* a malformed gastrula that cannot undergo invagination due to an overabundance of presumptive endoderm.
exogenote *Genetics.* a part of a donor genome that is transferred into an intact recipient cell during gene transfer in bacteria.
exogenous *Biology.* **1.** produced, originating, or resulting from causes outside an organism. **2.** growing by adding new layers at or near the surface. *Pathology.* of an infection or disease, caused by something outside the organism.
exogenous inclusion see XENOLITH.
exogenous retrovirus *Virology.* a retrovirus that infects a cell either from another host or from another site in the same host.
exogenous variables *Mathematics.* the predetermined, given independent variables of a mathematical model.
exogeosyncline *Geology.* a parageosyncline that accumulates its clastic sediments from the adjacent uplifted orthogeosynclinal belt that lies outside the craton. Also, DELTAGEOSYNCLINE, FOREDEEP, TRANSVERSE BASIN.
exognathite *Invertebrate Zoology.* the external branch of an oral appendage, especially in a crustacean.
Exogoninae *Invertebrate Zoology.* a subfamily of polychaete annelids, in the family Syllidae, that have short bodies with few segments.
Exogyra *Paleontology.* a distinctive extinct genus of pelecypod bivalves in the subclass Cryptodonta and the order Pterioida; Jurassic and Cretaceous.
exon *Genetics.* the portion of the DNA sequence in a gene that contains the codons that specify the sequence of amino acids in a polypeptide chain, as well as the beginning and end of the coding sequence.
exonephric *Invertebrate Zoology.* having excretory organs that empty through the body wall, as in some annelid worms.
exon shuffling *Genetics.* a process in which exons of the same gene are recombined to produce new genes.
exonuclear gene *Genetics.* any genetic sequence that is located outside the nucleus, such as in the mitochondrion or chloroplast.
exonuclease *Enzymology.* any enzyme of the hydrolase class that cleaves nucleotides from the free end of a nucleic acid; breaks down DNA sequentially, beginning at one end of the molecule.
exopathogen *Plant Pathology.* an external, nonparasitic agent, such as a virus or bacteria, that causes diseases in plants.
exopathogenesis *Plant Pathology.* a condition in which a plant becomes diseased due to a nonparasitic agent that attacks external plant parts.
exopeptidase *Enzymology.* any of several enzymes of the hydrolase class that act on the terminal peptide bonds of a protein chain.
exophoria *Medicine.* the deviation of one eye to the side, occurring when the eyes are at rest. Thus, **exophoric.**
exophthalmic *Medicine.* of, relating to, or marked by exophthalmos. Thus, **exophthalmic goiter.**
exophthalmos *Medicine.* a noticeable bulging of the eyeballs usually caused by a tumor pushing the eyeballs outward, a swelling of the brain or eyes, paralysis or injury to eye muscles, clots in the sinuses, or endocrine disorders. Also, **exophthalmus, exorbitism.**
exophytic *Biology.* growing outward. *Oncology.* proliferating on the exterior or surface epithelium of an organ or other structure in which the growth originated.
exoplasm *Cell Biology.* plasma membrane.
exopodite *Invertebrate Zoology.* the outer branch of a two-branched appendage of a crustacean.
Exopterygota *Invertebrate Zoology.* a division of winged insects in the subclass Pterygota that undergo little change between larva (nymph) and adult, having no pupal stage.
exopterygote *Invertebrate Zoology.* any member of Exopterygota.
EXOR gate see EXCLUSIVE-OR GATE.
exorheic *Geology.* relating to a region or basin characterized by external drainage. Also, **exoreic.**
Exormothecaceae *Botany.* a monogeneric family of liverworts belonging to the order Marchantiales, characterized by small and simple or bifurcate plants with distinctive volcanolike pustules in the thallus epidermis that have a single pore at the summit.
exoscopic development *Developmental Biology.* the development of the embryo following the first division of the zygote in which the outer cell gives rise to the embryo and the inner cell.
exoskeleton [eks´ō skel´ə tən] *Anatomy.* structures that develop from the ectoderm, such as hair, teeth, or nails. *Robotics.* a robotic arm having joints like those found in the arm of a human.
exosmosis *Cell Biology.* the transfer of a liquid through a cell membrane from inside the cell to the outside.
exosphere [eks´ə sfēr´] *Meteorology.* the outermost portion of the atmosphere, characterized by extremely low air density, from which atmospheric gases can escape into outer space. *Astronomy.* the part of any planet's atmosphere in which the density is low enough that the lighter atmospheric atoms can escape into space. Also, REGION OF ESCAPE.
exosporium or **exospore** *Botany.* the outer layer of a spore wall, as in pollen and bacterial spores. Also, EXINE, EXTINE.
exostome *Botany.* **1.** the outer portion of the peristome, as in some mosses. **2.** the orifice in the outer layer of a bitegmic ovule.
exostosis *Medicine.* a benign growth on the surface of the bone, characteristically capped by cartilage, and caused by chronic irritation from osteoarthritis, trauma, or infections. Thus, **exostotic.**
exoteric *Biology.* developed outside the organism.
exotheca *Invertebrate Zoology.* in corals, the tissue external to the theca.
exotherm *Chemical Engineering.* a temperature-time curve of the amount of heat produced in a chemical reaction.
exothermic *Physical Chemistry.* relating to or describing any process in which a system releases heat to its surrounding environment. Thus, **exothermic reaction, exothermic process.** Also, **exothermal.**
exotic *Ecology.* of foreign origin; not native to the region in which it is found.
exotic stream *Hydrology.* any stream that receives most of its water from the drainage system of another area.
exotoxin *Toxicology.* a potent extracellular toxin secreted by certain species of bacteria. Thus, **exotoxic.**
exotropia *Medicine.* strabismus in which the visual axis of one eye permanently deviates from that of the other, resulting in diplopia. Also, EXTERNAL STRABISMUS, WALLEYE.
exotropic *Medicine.* relating to or characterized by exotropia.
expandable space structure *Space Technology.* a space probe that can be compacted for launch and then assembled, unfolded, or otherwise expanded to its full size and form once outside the earth's atmosphere.
expanded *Materials.* of a material, processed to increase its volume. Thus, for example, **expanded brick, expanded clay, expanded plastic.**
expanded batch *Computer Programming.* in customer service systems, a higher level of processing than basic batch, in which programs perform complex computations and analyses of performance.
expanded foot *Hydrology.* a broad, bulb-shaped or fan-shaped mass of ice formed at the base of a mountain slope where a valley glacier extends onto the adjacent lowland area. Also, PIEDMONT BULB.
expanded-foot glacier *Hydrology.* a small glacier consisting of an expanded foot. Also, BULB GLACIER, FOOT GLACIER.
expanded metal *Materials.* a metal mesh or net formed by stretching a slotted piece of sheet metal; used as a reinforcement material, as lathing, and in making wire wastebaskets.
expanded scope *Electronics.* the portion of a cathode-ray tube in which the display is magnified.
expanded sweep *Electronics.* the act of accelerating the deflection of an electron beam in an oscilloscope at a selected portion of its sweep time.
expander *Electronics.* a transducer that increases the amplitude range of output voltages for a given amplitude range of input voltages.
expanding *Metallurgy.* a process by which the diameter of a cup, shell, or tube is expanded.
expanding a node *Artificial Intelligence.* in a tree search, the process of computing the set of successors of a given node.
expanding arm *Astronomy.* an arm of the Milky Way that lies on the opposite side (from the earth) of the galaxy's center and is moving outward from it.
expanding brake *Mechanical Engineering.* a brake in which an internal shoe is expanded, usually by a cam or toggle mechanism, to press against the inner surface of the brake drum. Also, INTERNAL BRAKE.
expanding population *Ecology.* a growth rate of a community characterized by the birth rate being greater than the death rate.
expanding universe *Astrophysics.* a model that postulates a universe in which all points in space are moving away from one another at speeds proportionate to their distances; a feature predicted by all cosmological theories devised since the 1930s.
expandor *Electronics.* a telephone switching system that connects a number of inlets to a greater number of outlets.
expansile *Anatomy.* capable of or prone to expansion.

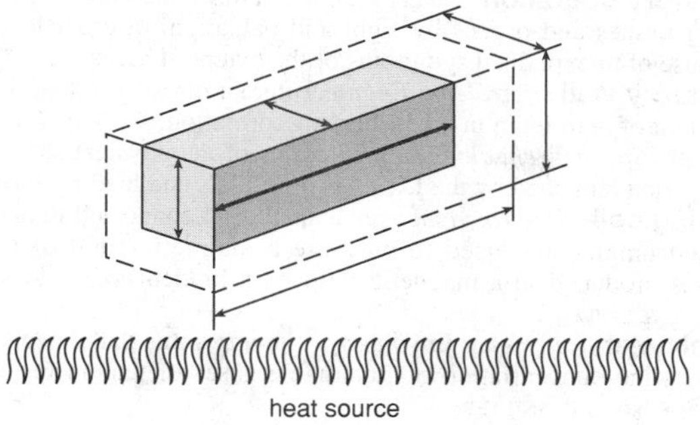

expansion by heat

expansion *Physics*. an increase in the volume of a substance while its mass remains constant. *Mechanical Engineering*. **1.** an increase in volume of the working fluid in a cylinder. **2.** a piston stroke during which such expansion occurs. *Electronics*. a process that increases the effective gain of a strong signal and decreases that of a weak signal; in an audio amplifier, this produces a greater volume range; in a facsimile system, it produces greater contrast. *Mathematics*. an expansion of a function *f* is a representation of *f* as a series, such as a power series, an asymptotic series, a series of orthogonal functions, etc. Expansions are not necessarily unique and may be divergent.

expansion bit *Mechanical Devices*. any drill bit whose cutting blade can be adjusted to various sizes. Also, EXPANSIVE BIT.

expansion board *Computer Technology*. a circuit board that can be placed in an expansion slot of a computer to increase memory or add special functions.

expansion bolt *Mechanical Devices*. a masonry or concrete bolt with a split case or sleeve that, when pulled, causes it to expand and tighten its grip like a wedge.

expansion chamber see CLOUD CHAMBER.

expansion chucking reamer *Mechanical Devices*. an adjustable machine reamer with an expansion screw for increasing its diameter within certain limits.

expansion coefficient see COEFFICIENT OF EXPANSION.

expansion cooling *Mechanical Engineering*. the cooling of a substance through adiabatic expansion.

expansion ellipsoid *Solid-State Physics*. an ellipsoid whose axes are proportional to the coefficients of linear expansion for specified directions in a crystal lattice.

expansion engine *Mechanical Engineering*. a piston-cylinder device that cools compressed air by the sudden expansion of a working fluid; used in the production of pure gaseous oxygen and an impure gaseous nitrogen stream via the Claude cycle.

expansion fissure *Petrology*. any of a system of fissures randomly radiating through feldspars and other minerals bordering olivine crystals that have been replaced by serpentine.

expansion fit *Design Engineering*. a drive or force fit obtained by inserting a chilled internal part into one at ambient temperature, whereupon the inner part warms, expanding to fit.

expansion joint *Mechanical Engineering*. **1.** a joint between two parts of a structure or machine that expands with temperature rises without distorting laterally. **2.** a pipe coupling that is resistant to temperature changes and allows motion in the piping system without creating a hazard to associated equipment. *Geology*. see SHEETING STRUCTURE.

expansion loop *Engineering*. a bend placed in a pipeline to allow for expansion and contraction of the line. It may be a partial U-shape or a complete loop. Also, **expansion bend.**

expansion opening *Engineering*. a widened area or chamber in line with a pipe or tunnel that carries liquid or air, creating a release of pressure in the pipe or tunnel by providing an area for expansion of the liquid.

expansion ratio *Mechanical Engineering*. the ratio of cylinder volume with respect to the gas pressure of a rocket chamber or a jet pipe and that at the outlet of a propelling nozzle in a reciprocating piston engine.

expansion reamer *Engineering*. a tool used to enlarge a hole, whose diameter can be adjusted by means of an expanding screw.

expansion ring *Mechanical Devices*. a U-shaped ring used for joining lengths of pipe while allowing for expansion and alleviating stress on the pipes. Also, **expansion hoop.**

expansion shield *Mechanical Devices*. a device used to secure machine bolts into a masonry wall or surface.

expansion slot *Computer Technology*. any of a series of receptacles in a personal computer that are designed to accept expansion boards or cards, thus allowing the user to upgrade functionality or increase memory with minimal effort.

expansion valve *Mechanical Engineering*. an auxiliary valve on the main slide valve of a steam engine that is designed to provide a control for the point of cutoff.

expansion wave *Fluid Mechanics*. waves that occur only in supersonic flow and that are always oblique to the surface; characterized by an increase in velocity and a decrease in pressure through the wave, and involving no loss of energy.

expansive bit see EXPANSION BIT.

expansivity see COEFFICIENT OF EXPANSION.

expectancy *Psychology*. the anticipation that a certain behavior pattern will produce a particular result. Also, **expectation.** *Medicine*. see LIFE EXPECTANCY.

expectancy theory *Behavior*. the theory that behavior is motivated by the expectation that it will produce a reward of sufficient value. Also, **expectancy-value theory.**

expectant technique *Psychology*. the practice by a psychotherapist of maintaining neutrality throughout the analysis and not evaluating the patient's behavior or stating recommendations and prohibitions.

expectation value *Quantum Mechanics*. the value of an observable that would result from measurements on a large ensemble of identical systems; the observable magnitude of a quantum mechanical operator.

expected approach clearance time *Navigation*. the time at which an aircraft is expected to be granted clearance to approach and land.

expected date of confinement *Medicine*. the predicted date of delivery for a pregnant woman, approximately 280 days from fertilization.

expected departure clearance time *Navigation*. the time assigned to an aircraft to take off in a controlled departure time program.

expected value *Mathematics*. if P is a probability measure defined on a set S, then the expected value of a function f is the integral $\int_S f\,dP$, which expresses the ensemble average value of the function. In particular, if g is a function of a discrete random variable X and probability function P, then the expected value of $g(X)$ is $E[g(X)] = \sum_n g(x_n)P(x_n)$, if the sum is convergent or finite, and where X has values $\{x_n\}$. In general, if f is the probability density function of a random variable X, then the expected value of $X = E(X) = \int_{-\infty}^{\infty} tf(t)\,dt$.

expectorant [ek spek´tə rənt] *Pharmacology*. **1.** a drug that enhances the ejection, by spitting or coughing, of mucus or other fluids from the lungs and air passages. **2.** broadly, any cough medication. **3.** of an agent, promoting ejection by spitting or coughing.

expectorate [ek spek´tə rāt] *Physiology*. to cough up and spit out substances such as mucus or phlegm from the lungs, bronchi, and trachea. Thus, **expectoration.**

expendable *Space Technology*. of a piece of equipment, designed to be consumed or used only once, such as an **expendable rocket** designed to launch a payload into orbit, or fuel tanks that are jettisoned after use.

experience deprivation *Behavior*. in raising a laboratory or domesticated animal, the deliberate withholding of certain possible experiences in order to see which capabilities develop without the benefit of those experiences. Also, ISOLATION EXPERIMENT.

experiment *Science*. **1.** a procedure that is carried out under controlled conditions in order to discover, demonstrate, or test some fact, theory, or general truth. **2.** to carry out such a process. *Statistics*. a planned data collection.

experimental *Science*. **1.** relating to, derived from, or founded upon an experiment. **2.** of the nature of an experiment; undeveloped. **3.** used for experimentation.

experimental archaeology *Archaeology*. the reproduction of past behavior to obtain or evaluate archaeological data, as by creating and using stone tools, duplicating prehistoric methods of farming, building, and travel, and so on.

experimental breeder reactor *Nucleonics*. a type of reactor in which fast neutrons sustain a chain reaction by bombarding both the core of enriched uranium-235 and the layer of natural uranium that blankets it; used for research and the breeding of more fissionable material.

experimental design *Statistics*. a plan of the relevant features of a data collection, so that the relevant inferences can be made; for example, choice of sample size, of factor levels in analysis of variance, or of questions to include in a survey. Also, DESIGN OF EXPERIMENT.

experimental error see RANDOM ERROR.

experimental psychology *Psychology.* the use of scientific experimentation to develop or evaluate psychological principles, especially the study of human or animal behavior in a laboratory setting. Thus, **experimental psychologist.**

experimental reactor *Nucleonics.* any reactor that tests new designs and systems.

experimentation *Science.* the process or practice of experimenting.

expert system *Artificial Intelligence.* an application in which problems are solved by means of an information base containing rules and data from which inferences are drawn on the basis of human experience and previously encountered problems.

expert system-building tool *Artificial Intelligence.* a nonspecific system used to construct expert systems in any domain; usually consists of assistance in forming rules and knowledge representations, and in providing an explanation facility.

expert system shell *Artificial Intelligence.* **1.** a development tool that consists of two separate software packages, a rule set manager and an inference engine. **2.** a complete expert system that contains no specific knowledge.

expiration *Physiology.* **1.** see EXHALATION. **2.** another term for death.

expiration date *Materials Science.* the date after which a material should not be sold or used; often stamped on commercial products.

expiratory *Physiology.* of, relating to, or promoting expiration.

expiratory reserve volume *Physiology.* the maximum amount of air that can be expired by the lungs from the resting expiratory level.

expiratory standstill *Physiology.* the suspension of breathing movements following an expiration.

explanation *Artificial Intelligence.* a justification of a conclusion in terms of the facts and rules that led to it.

explanation-based learning *Artificial Intelligence.* a kind of machine learning in which the result of learning from a new example establishes a general rule that explains the basis for the classification of the example.

explanation facility *Artificial Intelligence.* the component of an expert system that traces the system's reasoning process and justifies its conclusions; may include printing or displaying the relevant rules.

explant *Biology.* **1.** to remove a piece of living plant or animal tissue for experimental purposes, usually in order to start a tissue culture. **2.** tissue that is removed from its original site and transplanted, usually in an artificial medium.

explicit programming *Control Systems.* programming that uses detailed and exact descriptions of the tasks that are to be carried out.

explicit symmetry breaking *Physics.* the breaking of the exact symmetry of a system by applying a weak influence.

explode *Chemistry.* to undergo an explosion.

exploded file *Computer Programming.* a file whose records contain more information as a result of moving from one application to another.

exploder *Ordnance.* a user-controlled device that generates an electric current in a firing circuit in order to initiate an explosive charge.

exploding bridge wire *Engineering.* a system in which a bridge wire carries a high-energy electrical impulse, causing it to explode and release shock and heat energy, which in turn initiates a relatively insensitive explosive to make contact with the bridge wire.

exploitation *Engineering.* **1.** the process of extracting from the earth the oil, gas, minerals, or rocks found there as the result of exploration. **2.** the extraction and utilization of ore.

exploitation competition see CONTEST COMPETITION.

exploitative character type *Psychology.* a term used to describe an individual who attempts to satisfy his desires by coercing or manipulating others.

exploiter-victim system *Ecology.* any interaction between two species in which one benefits at the expense of another; e.g., predator and prey.

exploration *Engineering.* the process of searching for coal, minerals, or ore by means of geological surveys, geophysical prospecting, boreholes and trial pits, or surface or underground headings, drifts, or tunnels. Also, PROSPECTING. *Medicine.* an examination or investigation for diagnostic purposes.

exploratory behavior *Behavior.* any activity that helps an animal to become aware of or oriented to its surroundings.

exploratory biopsy *Surgery.* exploration combined with biopsy to determine the type and extent of neoplasms.

exploratory data analysis *Statistics.* methods for the examination of data preliminary to the formulation of probabilistic models; often based on graphics.

exploratory operation *Surgery.* an incision in one part of the body to inspect tissues and organs by sight and palpation, in order to discover the cause of unexplained symptoms or the extent of a disease.

exploratory well *Petroleum Engineering.* an oil well drilled to outline the extent of petroleum in an oil-bearing formation.

Explorer *Space Technology.* a long series of geophysical satellites and probes, first launched by the U.S. Army in 1958 and later by NASA.

exploring coil *Electromagnetism.* a small coil connected to an indicating instrument, and used to measure a magnetic field or to detect changes produced in a magnetic field by a hidden object. Also, MAGNETIC TEST COIL.

explosimeter *Petroleum Engineering.* an apparatus that detects and determines the concentration of flammable gases or vapors in the atmosphere. Also, GAS SNIFTER.

explosion *Chemistry.* an extremely rapid chemical reaction or change of state that generates heat and often gas.

explosion crater *Volcanology.* a saucer-shaped depression formed by a volcanic explosion or by the high-velocity impact of a meteorite.

explosion door *Mechanical Engineering.* a door in a furnace designed to open at a designated excess pressure.

explosion method *Thermodynamics.* a technique that is used in gas calorimetry, in which the specific heat of the gas is determined by mixing it with an explosive gas mixture with a known heat of reaction; the pressure change after the explosion is used to determine the temperature at the time of ignition.

explosion tuff *Geology.* a tuff whose constituent particles have been dropped directly in place after being ejected from a volcanic vent, rather than being washed into place.

explosion welding *Metallurgy.* solid-state welding performed by controlled explosions that exert high pressure.

explosive *Materials.* **1.** any material that has a tendency to explode, such as gunpowder, nitroglycerine, or dynamite. **2.** of or relating to such a material. Thus, **explosive filler, explosive fuel, explosive oxidizer.**

explosive bolt *Space Technology.* a bolt containing a small charge that can be exploded remotely to destroy or disengage the bolt; used to separate a satellite or stage from a launch vehicle.

explosive bonding *Engineering.* a process in which two solids are bonded under high pressures and temperatures by the detonation of a layer of explosive material spread on one of the surfaces.

explosive decompression *Aviation.* an extreme loss of air pressure in a cabin or cockpit that produces an explosion, achieving a new static condition of balance with the external pressure.

explosive disintegration *Engineering.* a process in which pressure is suddenly released on a material that contains gas or liquid, producing an explosion that blows the material apart into small fragments.

explosive echo ranging *Engineering.* a type of sonar in which a shock wave is produced underwater; a measurement of the time it takes for the return of the wave to the initiating point is made to determine distances between points underwater.

explosive eruption *Geology.* an eruption characterized by explosive phenomena and the emission of ash and cinders.

explosive evolution *Evolution.* a process of evolutionary change that splits an ancestral species or group into numerous lines of descent in a relatively short period of time. Also, **explosive radiation.**

explosive forming *Metallurgy.* metal fabrication performed by controlled explosions that exert high pressure.

explosive fracturing *Petroleum Engineering.* the use of an explosive charge in the bottom of a well to crack and fracture the formation around the borehole, in order to increase the yield of gas or oil.

explosive limits *Chemical Engineering.* the composition range of a flammable vapor, beyond which it will explode if ignited in a confined area.

explosive nucleosynthesis *Astrophysics.* the creation of elements in the shock wave immediately following detonation in a supernova.

explosive ordnance *Ordnance.* munitions that contain explosives, nuclear fission or fusion materials, or biological and chemical agents.

explosive ordnance disposal *Ordnance.* the process of identifying, neutralizing, and disposing of unexploded explosive ordnance. Thus, **explosive ordnance disposal unit, explosive ordnance disposal procedures.**

explosive rivet *Engineering.* a bolt that fits over a charge of explosive material; when the charge is activated, the bolt expands to enclose the hole securely.

explosive train *Ordnance.* a series of initiating and igniting elements arranged to cause a charge to function.

exponent [eks´pō nənt] *Mathematics.* a symbol placed to the right and above a mathematical expression (or group element or cardinal number, etc.); a shorthand way of expressing multiple multiplications of a single quantity, as in $a \times a \times a \times a = a^4$. This notation, originally defined only for positive integers, can be extended to complex numbers (α and β) that obey the following rules: (a) 0^0 is not defined; (b) $Z^\alpha Z^\beta = Z^{\alpha+\beta}$; (c) $(Z^\alpha)^\beta = Z^{\alpha\beta}$; (d) if Z has a multiplicative inverse Z^{-1}, then $Z^{-\alpha} = (Z^{-1})^\alpha$; (e) in the commutative case, $(YZ)^\alpha = Y^\alpha Z^\alpha$. Certain restrictions apply when taking roots of nonpositive numbers.

exponential [eks´pə nen´shəl] *Mathematics.* of, relating to, or expressed as an exponent or exponents.

exponential amplifier *Electricity.* an amplifier circuit whose output voltage increases exponentially with respect to its input voltage.

exponential atmosphere see ISOTHERMAL ATMOSPHERE.

exponential curve *Mathematics.* the graph in the plane of $y = a^x$, where a is a positive constant.

exponential decay *Physics.* the diminishing of a quantity Q in time or in space that follows an exponential law of the form $Q = Q_o e^{-u/c}$, where u represents either time or a spatial coordinate and c is a positive constant having identical units of u.

exponential density function *Mathematics.* the probability density $f(x) = e^{-(x/\beta)}/\beta$ with $\beta > 0$; a special case of the gamma density that is suitable for a nonnegative random variable X at $X = x$ and describes average behavior of exponential decay processes.

exponential distribution *Mathematics.* **1.** see EXPONENTIAL DENSITY FUNCTION. **2.** the distribution function that is obtained by integrating the exponential density function from 0 to x. *Statistics.* a special case of a gamma distribution where $a = 1$ in the probability density function

$$f(x) = x^{a-1}e^{-x/b} / (b^a\Gamma(a)),$$

where $x, b > 0$ and $\Gamma(a)$ is the gamma function.

exponential equation *Mathematics.* any equation that involves or includes the term e^x.

exponential function *Mathematics.* **1.** the function $f(x) = e^x$; also written $f(x) = \exp(x)$. For $z = x + iy$, the extension to the complex case is $f(z) = e^z = e^x (\cos y + i \sin y)$, by application of Euler's formula. **2.** in general, any real function g, not identically zero, that satisfies the functional equation $g(x + y) = g(x)g(y)$ for all real x and y.

exponential growth *Microbiology.* a phase of a bacterial growth curve during which cells divide at a constant rate. Also, LOGARITHMIC GROWTH. *Science.* in popular use, any rapid, regular rate of growth.

exponential horn *Acoustical Engineering.* a horn having a circular cross section whose radius increases exponentially with increasing axial distance; such a horn has desirable acoustical characteristics.

exponential integral *Mathematics.* the function $f(x) = \int_x^\infty e^{-t}t^{-1}\,dt$, where $x > 0$.

exponential law *Physics.* the principle stating that the growth or decay of a quantity is at a rate bearing an exponential relationship of the form $Q = Q_o e^{au}$, where u is a coordinate of either time or space and a is a positive constant for exponential growth or a negative constant for exponential decay.

exponential mapping *Mathematics.* **1.** suppose G is a Lie group with origin e and tangent space $T_e(G)$ at e. The exponential mapping exp: $T_e(G) \rightarrow G$ maps the line $t\gamma$ in $T_e(G)$ onto the one-parameter subgroup $g_\gamma(t)$ of G (tangent to γ at the origin); i.e., $\exp \gamma = g_\gamma(1)$. Since $T_e(G)$ is isomorphic to the Lie algebra g of G, the exponential mapping is often defined as a mapping from g to G. **2.** the mapping that sends a tangent vector to a Riemannian manifold to the endpoint of the geodesic segment or [0, 1] with tangent at 0 equal to the given vector.

exponential pulse *Physics.* any pulse whose variation in time is proportional to the variation of a critically damped harmonic oscillator.

exponential series *Mathematics.* the Maclaurin series expansion that converges for all complex numbers z, namely

$$e^z = 1 + \sum_{n=1}^{\infty} z^n/n!$$

exponential smoothing *Statistics.* a forecasting technique for the analysis of time series data.

exponential transmission line *Electricity.* a tapered transmission line whose characteristic impedance varies exponentially with distance along the line.

export *Computer Programming.* **1.** to write data in a form that can be input to another program. **2.** to send data to another program or over a network. **3.** to make symbols within a program unit accessible to other program units.

exposure the act or fact of exposing or being exposed; specific uses include: *Medicine.* the condition of being subjected to something, such as weather conditions or an infectious agent. *Radiology.* see EXPOSURE DOSE. *Graphic Arts.* **1.** the process of causing light to strike a photosensitive material, or an instance of this. **2.** the product of the duration and intensity of light striking a photosensitive material; the former is controlled by shutter speed, the latter by the aperture. **3.** a defined section of photographic film. *Meteorology.* the general surroundings of a site, particularly its orientation and openness to winds and sunshine. *Building Engineering.* the distance between shingles as measured from the butt of one shingle to that of another directly above it.

exposure dose *Radiology.* the quantity of gamma- or X-radiation at a specific site, based on a measure of ionization generated in the air at that point.

exposure factor *Nucleonics.* a quantity of radiation exposure in an individual based on the intensity of the radiation, the length of the exposure, and the distance from the source of radiation.

exposure index *Graphic Arts.* a measurement or rating of the photosensitivity of a material such as film, with higher numbers indicating greater sensitivity.

exposure interval *Photogrammetry.* the time required between successive exposures of a series of photographs for the purpose of obtaining desired forward lap.

exposure latitude *Graphic Arts.* in a given instance, the margin for error between a perfect exposure and one that will still yield an acceptable image.

exposure station see CAMERA STATION.

exposure time *Graphic Arts.* the time during which a light-sensitive material is subjected to the action of light.

expression *Mathematics.* a set of one or more symbols representing a value, function, relation, or the like. *Computer Programming.* **1.** any combination of constants, variables, or functions. **2.** a mathematical equation or formula coded in a programming language. *Chemical Engineering.* the separation of liquid from a two-phase, solid-liquid system by compression, under conditions that permit liquid to escape; the solid is confined between two compressing surfaces. Also, MECHANICAL EXPRESSION. *Genetics.* the manifestation of a heritable trait in an individual carrying the principal gene or genes that determine the trait. *Surgery.* the use of pressure to expel the contents of a vessel or cavity.

expression vector *Molecular Biology.* a cloning vehicle designed to allow and promote the expression of an inserted gene.

expressive aphasia *Medicine.* a form of aphasia in which the patient understands written and spoken words and knows what he or she wants to say, but cannot utter the words; caused by a lesion of the cortical center. Also, BROCA'S APHASIA, MOTOR APHASIA.

expressive behavior see DISPLAY.

expressiveness *Artificial Intelligence.* the ability of a knowledge representation formalism to express the features important for representing a problem domain.

expressivity *Genetics.* the degree to which a particular genotype is expressed in the phenotype, under the influence of environmental conditions.

expressway *Transportation Engineering.* a high-speed, multiple-lane highway for through traffic, having a center divider and limited access from other roads.

expulsion fuse unit *Electricity.* a vented fuse unit in which the expulsion effect of gases emitted by the arc and lining of the fuseholder extinguishes the arc.

expulsive *Physiology.* driving or forcing out; tending to expel.

exsanguinate [eks sang´gwə nāt´] *Anatomy.* **1.** to deprive of blood. **2.** anemic or bloodless.

exsecant *Mathematics.* the exsecant of θ is the quantity $\sec \theta - 1$; that is, the difference between the secant of an angle constructed in a unit circle and the radius. Denoted exsec θ.

exsect *Surgery.* to cut out any portion of tissue, bone, or organ; to excise.

exserted *Biology.* protruding beyond an enclosing structure.

exsheath *Invertebrate Zoology.* to exit from the membrane that enclosed a previous developmental stage, especially in nematodes.

exsiccation *Agronomy.* the drying up of an area from causes other than loss of rainfall. *Chemistry.* the process of depriving a crystalline substance of its water of crystallization.

exsolution *Mineralogy.* a process in which a solid solution separates during cooling into at least two distinct mineral phases without a change in bulk composition.

exsorption *Cell Biology.* the movement of substances out of cells.
exstipulate *Botany.* lacking stipules.
exstrophy *Medicine.* the often congenital turning inside out or eversion of an organ, such as the bladder.
extended area *Design Engineering.* a surface that has been extended in area without an increase in diameter, as by using pleats, rubs, or fins.
extended-area service *Telecommunications.* the extension of a toll-free area to include nearby exchange areas in return for accepting a higher tariff.
extended binary-coded decimal interchange code see EBCDIC.
extended chain *Materials Science.* in certain crystalline polymers, a morphological structure in which the size of the crystal in the chain direction is essentially equal to the extended chain length.
extended channel status word see CHANNEL STATUS WORD.
extended complex numbers *Mathematics.* a one-point compactification of the the complex numbers obtained by adjoining the point at infinity, denoted ∞; conformally equivalent to a sphere under the stereographic projection. Also, **extended complex plane.**
extended dislocation *Crystallography.* a crystal defect in a hexagonal close-packed structure consisting of a stacking fault bounded by two lines across which a fraction of a lattice constant slips into one of the alternative stacking positions.
extended-entry decision table *Computer Programming.* a decision table in which conditions, not values, are identified by the condition stubs; values are entered into the condition portion of the decision rules.
extended family *Anthropology.* a family that consists of relatives in addition to a nuclear family of two parents and their children.
extended forecast *Meteorology.* a weather forecast for a period extending beyond two days from the day of issue. Also, LONG-RANGE FORECAST, EXTENDED-RANGE FORECAST.
extended-interaction tube *Electronics.* in a microwave system, a device in which the electron stream interacts with a traveling electric field.
extended matrix *Mathematics.* given a system of m linear equations in n variables (or unknowns) written in the form $Ax = b$, where A is the $m \times n$ matrix of coefficients, x is the column vector whose entries are the n variables, and b is a column vector of m constants, the extended matrix of the system is the $m \times (n + 1)$ matrix (A, b), formed by adjoining the column b to the right side of A. The rank of (A, b) determines the behavior of the solutions, if any, to the system.
extended play *Acoustical Engineering.* **1.** see LONG-PLAYING. **2.** describing a type of record or tape having a longer playing time, especially a 45-rpm record as opposed to a 78-rpm. *Mechanical Engineering.* see EXTRA PLAY.
extended-precision word *Computer Programming.* in floating-point arithmetic, a longer or double word used to express a number for greater precision.
extended-range forecast see EXTENDED FORECAST.
extended source *Astronomy.* a source of celestial emission at any wavelength that has a measurable angular size.
extended stream *Hydrology.* a stream that is lengthened by the continuation of its course downstream across newly emerged land.
extended valley *Geology.* **1.** a valley that has become lengthened downstream as a result of the regression of the sea or the uplift of the coastal region. **2.** a valley that contains an extended stream.
extended X-ray absorption fine structure *Physics.* the observed inconsistency of absorbed radiation as a function of X-ray energy for energies just above that required to remove electrons, due to the interference between backscattered electron waves and outgoing photoelectron waves.
extender *Chemistry.* an inert substance that is added as a diluent to adjust the physical condition of a mixture, or as a filler to increase bulk weight and reduce cost per unit volume.
extend flip-flop see CARRY FLAG.
extending flow *Hydrology.* a glacial flow pattern characterized by an increase in velocity with distance downstream.
extensibility *Materials.* the degree to which a material can be stretched or distorted without breaking, often expressed as a percentage of its original size. *Robotics.* the ability to add more functions or applications to a robotic system.
extensible language *Computer Programming.* a programming language that allows definition of new language elements in terms of existing elements.
extensible system *Computer Programming.* a computer system that allows users to develop functions or capabilities to be added to the system for general use.
extension *Physiology.* the process of straightening, especially the straightening of members on either side of a flexed joint, causing an increase in the angle between them, as in straightening the leg at the knee. *Mechanical Engineering.* the movement by which the two elements of any jointed apparatus are drawn away from each other. *Mechanical Devices.* of a device, allowing for this or similar movement. *Cartography.* the act or process of extending an existing control (usually either vertical or horizontal, depending on the primary purpose of the survey) from a controlled area into an area without control. *Computer Programming.* see FILENAME EXTENSION.
extensionality *Mathematics.* the **axiom of extensionality** states that two classes A and B with the same elements are equal; that is, $x \in A$ if and only if $x \in B$ implies $A = B$.
extension bolt *Mechanical Devices.* a vertical bolt used to secure one leaf of a high double door or a standard door by using a long extended rod to slide it into position flush with the door's upper edge.
extension cord *Electricity.* a cable having one or more separately insulated conductors bundled together within an outer protective cover, a plug at one end, and an outlet at the other end; used to extend electrical power over a distance equal to the length of the cord.
extension fracture *Geology.* a fracture developing at right angles to the direction of greatest stress, parallel to the direction of compression.
extension joint *Geology.* a joint that develops parallel to the direction of compression.
extension ladder *Mechanical Devices.* a two-section ladder configured upon a sliding device that provides an extension to nearly the amount of the first section, thus doubling its length.
extension lathe *Mechanical Devices.* a lathe in which an adjustable gap is provided by sliding the headstock longitudinally as needed to accommodate a workpiece.
extension module *Space Technology.* a container mounted on a launch vehicle that allows two spacecraft (one within the module and one on top) to be carried and launched by the vehicle.
extension of a field *Mathematics.* a field F is said to be an extension of a field K (or an **extension field** of K) if K is a subfield of F. The degree of the extension, denoted $[F:K]$, is the dimension of F as a vector space with scalars in K. F is said to be a finite dimensional or infinite dimensional extension of K according as $[F:K]$ is finite or infinite. If $[F:K] = 1$, then F is a simple extension of K. If every element of F is algebraic over K, then F is an algebraic extension of K. Otherwise, F is a transcendental extension of K; i.e., there exists at least one element of F that is transcendental over K.
extension of a function *Mathematics.* if $f: Y \to Z$ is a function (or map) between the sets Y and Z, and the set X is contained in Y, then the function $g: X \to Z$ may be defined by $g(x) = f(x)$ for each $x \in X$. The function g is called the restriction of f to X, and f is called an extension of g to Y. It is customary to write $g = f \mid X$.
extension register *Computer Technology.* a register that is used as an extension of the accumulator to provide additional precision during some mathematical operations.
extension ring *Mathematics.* let S be a (commutative) ring with identity 1_S and R a subring of S containing 1_S. Then S is said to be an extension ring of R.
extension spring *Mechanical Devices.* a tightly coiled spring designed to resist a force pulling in the direction of its length.
extensive cultivation see SHIFTING CULTIVATION.
extensive farming *Agriculture.* a method of farming that uses only a relatively small amount of labor or resources per unit of land.
extensive-form analysis *Statistics.* a preposterior analysis technique that involves sequential decisions and considers all possible outcomes of sampling; similar to decision-tree analysis.
extensive property *Physical Chemistry.* a property of a system, such as its mass or total volume, that is a function of its extent or size; i.e., that changes with the quantity of material present in the system. Also, **extensive quality, extensive variable.**
extensometer *Engineering.* an instrument that measures the degree to which an object has been elongated or deformed due to stress or other disturbance.
extensor *Anatomy.* any muscle that extends a joint.
extent *Computer Technology.* contiguous physical storage locations, such as a group of disk sectors or blocks, that are allocated for a particular set of data. *Computer Programming.* the period of time in program execution during which the value of a variable is defined.
exterior *Anatomy.* situated on or near the outside; outer. *Building Engineering.* the outside surface of a building or wall.

exterior algebra *Mathematics.* let $T(V)$ be either the covariant or the contravariant tensor algebra over the vector space V. The exterior algebra $\Lambda(V)$ is $T(V)$ modulo the homogeneous ideal generated by elements of the form $x \otimes x$. If V is a vector space over a field of characteristic zero, this algebra is isomorphic to the algebra of antisymmetric tensors over V. Also, GRASSMANN ALGEBRA.

exterior angle *Mathematics.* **1.** one of four angles made by a transversal cutting two lines, and not lying between the two lines. Each exterior angle has the transversal for one of its sides and one of the two lines for the other sides. Nonadjacent exterior angles lying on opposite sides of the transversal are called alternate exterior angles. **2.** at any vertex of a convex polygon the lines containing the two sides that meet at that vertex form four angles. Each of the two vertical angles that is adjacent to the interior angle at that vertex of the polygon is called an exterior angle; the angle vertical to the interior angle is not given a special name. Each exterior angle is supplementary to the interior angle at that vertex.

exterior ballistics *Mechanics.* the branch of ballistics dealing with the behavior of a projectile after leaving the gun barrel, or the behavior of a rocket or its propellant after exiting the thrust cone.

exterior derivative *Mathematics.* the derivative d that takes p-forms (alternating covariant p-tensor fields) to $(p + 1)$-forms on a differentiable manifold; the differential (coboundary operator) in the deRham cohomology on a differentiable manifold. The exterior derivative is uniquely defined by the following properties: (a) d is linear; (b) $d(\alpha \wedge \beta) = d\alpha \wedge \beta + (-1)^p \alpha \wedge d\beta$; (c) $d^2 = 0$; (d) if f is a 0-form (i.e., a function), then df is the ordinary differential of f.

exterior differential system *Mathematics.* a set of exterior differential equations.

exterior form *Mathematics.* a differential form (an alternating covariant p-tensor field) on a differentiable manifold.

exterior of a set *Mathematics.* in a topological space X, the exterior A^e of a set A is the union of all open sets that do not intersect A; equivalently, A^e is the largest open set of X that does not intersect A. A^e may also be viewed as the interior of the complement of A. Points in A^e are called **exterior points** of A.

exterior orientation *Cartography.* the act or process of determining (analytically or in a photogrammetric instrument) the position of the camera station and the attitude of the taking camera at the instant of exposure.

exterior product *Mathematics.* the exterior product $\wedge$ of a p-form α and a q-form β (on a smooth manifold X) is a $(p + q)$-form given by

$$(\alpha \wedge \beta)(v_1, \ldots, v_{p+q}) = 1/p!q! \Sigma\{(\text{sign } \pi)[\alpha(v_{\pi(1)}, \ldots, v_{\pi(p)})\, \beta(v_{\pi(p+1)}, \ldots, v_{\pi(p+q)})]\},$$

where the v_i are tangent vectors on X and the sum is over all possible permutations π of $(1, 2, \ldots, p + q)$. The constant in front of the summation sign may vary. Also, WEDGE PRODUCT, GRASSMANN PRODUCT.

extern *Computer Programming.* a pseudoinstruction in many programming languages that informs the compiler that an operand is defined in an external module. Also, EXTERNAL DECLARATION.

external *Anatomy.* **1.** away from the center of the body. **2.** near or on the outside of the body. *Pharmacology.* of a medication, to be applied only to the outside of the body. *Psychology.* see EXTERNALIZER.

external aileron *Aviation.* an aileron that is attached to but offset from the wing.

external anal sphincter *Anatomy.* groups of striated muscle fibers that surround the anus.

external angle *Mathematics.* given an angle of measure $\theta < \pi$ (or 180°), the angle having the same sides and vertex but of measure $2\pi - \theta$ (or 360° – θ). *Building Engineering.* a vertical or horizontal surface that forms a portion of a projecting wall or other building feature. Also, ARRIS.

external armature *Electricity.* in rotating machinery, an armature that is outside the field magnets; usually, the field magnets are on the outside.

external auditory meatus *Anatomy.* the canal in the temporal bone that leads to the middle ear. Also, **external acoustic meatus.**

external brake *Mechanical Engineering.* a brake that operates by contacting the outside of a brake drum.

external buffer *Computer Technology.* a storage device outside of the computer's main storage, used to temporarily hold data.

external carotid artery *Anatomy.* an artery that originates in the common carotid and distributes blood to the anterior portions of the neck, face, scalp, ear, and dura mater.

external circuit *Physical Chemistry.* a collective term for all the connecting wires, devices, and current sources connected to an electrochemical cell.

external clot *Hematology.* a clot formed outside a blood vessel.

external-combustion engine *Mechanical Engineering.* any engine, such as a steam engine, in which fuel ignition takes place outside the place where thermal energy is converted to mechanical force. Also, HEAT ENGINE.

external declaration see EXTERN.

external delay *Computer Technology.* a time during which a computer cannot function due to external conditions, such as a power failure.

external device *Engineering.* an instrument that operates in tandem with, and is controlled by, a central system but is not an actual part of the system. *Computer Technology.* any device that is subsidiary or peripheral to a computer system, such as a remote terminal or magnetic tape drive.

external-device control *Computer Technology.* the ability of an external device to cause an interrupt during job execution.

external-device response *Computer Technology.* a response signal from an external device; e.g., indicating that the device is not in use.

external ear *Anatomy.* the portion of the ear that is external to the tympanic membrane, consisting of the external auditory meatus and the pinna or auricle.

external error *Computer Programming.* an error originating in a peripheral device, such as a disk drive, rather than in the main computer.

external fertilization *Physiology.* the union of the gametes outside the bodies of the originating organisms, as in many fish and amphibians.

external flow *Mechanics.* an unconfined flow in which there is relative motion between an object and a large mass of fluid.

external force *Mechanics.* any force exerted on a system by some agency or body outside the system; e.g., the force of gravitation of one solid body acting upon another solid body.

external galaxy *Astronomy.* an older term for a galaxy lying outside the Milky Way; all galaxies are in fact "external."

external gill *Zoology.* a specialized organ of respiration external to the body wall, found on certain aquatic insects and amphibians.

external grinding *Mechanical Engineering.* the grinding of the exterior surface of a workpiece as it is rotated.

external interrupt *Computer Programming.* an interrupt caused by a peripheral device, a time-out, or an operator action.

externalizer *Psychology.* a person who generally believes that the forces determining the outcome of events in his or her life are external and beyond the person's effective control.

external label *Computer Programming.* **1.** a label referenced in one program module and defined in another program module. **2.** a paper label affixed to the exterior of a storage medium, such as a reel of tape or diskette.

externally fired boiler *Mechanical Engineering.* a boiler having cooling tubes around its surface.

externally stored program *Computer Programming.* a program whose instructions are set up on wiring boards or plugboards and are manually inserted into the computer, as in some tabulating machines.

external memory see AUXILIARY MEMORY.

external-mix oil burner *Engineering.* a burner having a jet stream of air that strikes the liquid fuel after it leaves the burner opening.

external operation *Mathematics.* let A and X be sets. A mapping from $A \times X$ into X is called an external operation on X, and the members of A are called operators on X. If $A = X$, then the mapping is called an internal operation on X.

external phase see CONTINUOUS PHASE.

external *Q* *Electronics.* in a microwave tube, the reciprocal of the difference between the reciprocal of the loaded Q and the reciprocal of the unloaded Q.

external respiration *Physiology.* the exchange of gases between the lungs and the blood.

external sensor *Robotics.* a sensing device that measures the external environment of a robot but is not a part of the robot itself.

external-shoe brake *Mechanical Engineering.* a friction brake operated by an external element that contacts it.

external signal *Computer Programming.* a self-explanatory message to the operator, such as an on/off light.

external sorting *Computer Programming.* the use of both internal and external storage in a sorting algorithm, because the list is too large to fit in internal memory.

external storage see AUXILIARY STORAGE.

external store *Aviation.* any object that is attached to and carried on the outside of an aircraft, such as a fuel tank, rocket, or bomb.
external strabismus see EXOTROPIA.
external stress *Mechanics.* stress that is applied to a body surface from outside, as opposed to stress that exists when no external forces are being applied.
external symbol dictionary *Computer Programming.* a table of external symbols and their addresses used by the linkage editor to identify external symbol references in programs.
external table *Computer Programming.* a table that is physically located outside a program, usually in a separate file.
external thread *Design Engineering.* a screw thread cut on an outside surface.
external time *Industrial Engineering.* the time that it takes for an operator to perform an element of work when the machine or process is not in operation.
external upset casing see EXTREME LINE CASING.
external wave *Fluid Mechanics.* **1.** a wave in fluid motion that has its maximum amplitude at an external boundary. **2.** any surface wave that is found on the free surface of a homogeneous, incompressible fluid.
external work *Thermodynamics.* the work performed by a thermodynamic system as it is expanding in volume against its surrounding environment. *Industrial Engineering.* any element of work that is performed when the machine or process is not in operation, resulting in loss of machine or process time.
exteroceptor *Physiology.* a sensory structure, as in the skin, the mucous membranes, or the sense organs, that responds to stimulation from sources outside the body.
extinction the process of becoming extinct; dying out or coming to an end; specific uses include: *Evolution.* the death of every member of a given species. *Paleontology.* the failure of a taxonomic group to produce direct descendants, causing its worldwide disappearance from the record at a given point. *Behavior.* the gradual unlearning of a behavior, which occurs when the unconditioned stimulus or the instrumental reward is withheld, causing the conditioned response to diminish. *Hydrology.* the drying up of a lake, either permanently by the destruction of its basin, or temporarily by the loss of water. *Optics.* the phenomenon in which plane polarized light is almost completely absorbed by a polarizer whose axis is perpendicular to the plane of polarization. *Astronomy.* the amount by which light from a celestial body is dimmed by the atmosphere. *Crystallography.* the modification of an incident X-ray beam as it passes through a single, perfect block of a crystal. Part of the incident beam may be reflected twice so that it returns to its original direction but is out of phase with the main beam, thus reducing the intensity of the latter **(primary extinction).** When the crystal is mosaic, part of the beam will be diffracted by one mosaic block and therefore available for diffraction by a following block that is accurately aligned with the first. Thus the second block contributes less than expected to the diffracted beam **(secondary extinction).** *Physical Chemistry.* see ABSORBANCE.
extinction coefficient see ABSORPTIVITY.
extinction voltage *Electronics.* in a gas tube, the anode voltage at which the discharge ceases when the supply voltage is decreasing.
extine see EXOSPORIUM.
extirpate *Biology.* to remove, destroy, pull up, exterminate, or otherwise make extinct.
extirpation *Surgery.* the excision of an entire organ or tissue.
extra- a combining form meaning: **1.** outer or outside, as in *extrados.* **2.** beyond, as in *extragalactic.*
extracellular *Cell Biology.* located, occurring, or functioning outside a cell or cells.
extracellular fluid *Anatomy.* any body fluid outside the cells, such as the interstitial fluid, plasma, lymph, or cerebrospinal fluid; consists of ultrafiltrates of blood plasma and transcellular fluid (i.e., fluid produced by active cellular secretion), and provides a constant external environment for the cells.
extrachromosomal *Biology.* occurring or functioning outside the choromosomes.
extrachromosomal inheritance see CYTOPLASMIC INHERITANCE.
extracorporeal *Medicine.* situated or occurring outside the body.
extracorporeal circulation *Surgery.* the circulation by artificial means of body fluids outside the body, as in a heart-lung apparatus during heart operations when normal blood flow is disrupted, and in kidney dialysis to remove impurities.
extracorporeal lithotripsy *Medicine.* the breaking up of urinary calculi using ultrasonic energy from a source outside the body.
extract *Chemistry.* **1.** a substance that is obtained from a solution by extraction, often a product prepared by dissolving animal or vegetable material in a solvent and then distilling the solution (evaporating the solvent). **2.** to produce such a substance (an extract). *Pharmacology.* a concentrated drug preparation, derived from a plant or animal, prepared by separating the active ingredient in the form of a liquid, solid, or powder of known strength. *Food Technology.* a usually alcoholic solution containing concentrated food essences. Extracts made from meat, aromatic herbs, vanilla beans, and other foods are commonly used in commercial and noncommercial food preparation. *Medicine.* to perform an extraction. *Computer Programming.* **1.** to select and remove only those items from a data set that meet certain criteria. **2.** to remove certain bits or characters from specified locations in a word.
extractant *Chemistry.* in an extraction, a liquid (solvent) used to remove a solute from another liquid.
extract a root *Mathematics.* to determine the value, if it exists, of $y = x^{1/\alpha}$, given $\alpha \neq 0$. When applied to real numbers (the usual case), α is a positive integer; if x is positive, then y is positive; if α is an even positive integer, then x must be positive; if α is odd and x is negative, then y is negative.
extract instruction *Computer Programming.* an instruction that forms a new expression from selected portions of existing expressions.
extraction *Chemistry.* the use of a solvent to remove (extract) one or more components from a mixture. *Medicine.* the act of removing, especially of a tooth by means of elevators, forceps, or both.
extraction column *Chemical Engineering.* a vertical, cylindrical-shaped process vessel in which the mixtures are separated from a liquid feed by countercurrent contact with a selective solvent.
extraction fermentation *Biotechnology.* a fermentation system in which the product is continuously removed by extraction with a solvent that is soluble in water.
extraction locus *Archaeology.* a place where large amounts of material are extracted or processed, such as a quarry, clay pit, or kill site.
extraction plant *Petroleum Engineering.* a processing unit used for the separation of liquid components from wet gas or casinghead gas.
extraction rate *Food Technology.* the yield of flour obtained in the milling process expressed as a percentage of the whole grain used to produce it. White wheat flour, for example, is about 70% extracted: 100 pounds of whole wheat yield about 70 pounds of white flour.
extraction turbine *Mechanical Engineering.* a steam turbine having openings through which steam is tapped for process work at one or more stages in the expansion process.
extractive *Materials Science.* any of various oil-like hydrocarbons that can be removed from wood by an appropriate chemical solvent.
extractive metallurgy *Metallurgy.* the art and science of extracting metal values from ores, and of subsequent refining.
extractive distillation *Chemical Engineering.* a type of distillation used to separate components of eutectic mixtures, in which cooling the mixture causes one component to crystallize and the other to remain in solution.
extractor *Engineering.* any mechanism or device used to remove a material by processes such as pressure, centrifugal force, or the use of a solvent. *Chemical Engineering.* a multistage device used for removing specified components of liquid or solid feeds by contacting the feed with a selective solvent. *Ordnance.* a device in the breech mechanism of a firearm, designed to remove the empty cartridge case or unfired cartridge from the chamber. *Medicine.* an instrument used to remove a calculus or foreign body.
extractor groove *Ordnance.* in automatic weapons, a groove in the base of a cartridge case designed to accept the extractor, thus allowing the case to be removed from the chamber.
extracystic *Medicine.* situated outside a cyst or the bladder.
extracytoplasmic *Cell Biology.* either attached to or external to the outer side of the plasma membrane, within the lumen of a eukaryotic organelle or relating to the bacterial periplasmic space.
extracytoplasmic oxidation *Biochemistry.* a process in which the oxidation of a substrate on the outside of the cytoplasmic membrane or in the periplasmic region generates energy across the membrane (proton motive force) which gives rise to a number of intracellular electron and proton consuming reactions.
extrados *Architecture.* the outer curve of an arch.
extraembryonic coelom *Developmental Biology.* a part of the coelom that is external to the embryo and bordered by the chorionic mesoderm and the mesoderm of the amnion and yolk sac. Also, EXOCOELOM.
extraembryonic membrane see FETAL MEMBRANE.

extrafloral nectary *Botany.* a nectary occurring on a plant in a part other than the flower. Also, **extranuptial nectary.**

extra foresight *Cartography.* the rod reading made at an instrument station in a line of levels and on a leveling rod standing on a bench mark or another point not in the continuous line of levels.

extragalactic [eks′trə gə lak′tik] *Astronomy.* of or relating to objects lying outside the Milky Way.

extragalactic background light *Astronomy.* the diffuse light of the night sky that comes from distant but unresolved galaxies and which remains after the light from all discrete sources (e.g., stars and galaxies) has been subtracted.

extragalactic H II region *Astronomy.* a cloud of ionized hydrogen gas lying in another galaxy.

extragalactic radio source *Astrophysics.* a radio-emitting source that lies outside the Milky Way.

extragenic reversion *Genetics.* a mutation that conceals or eliminates a mutant phenotype produced by another gene.

extra-hard *Metallurgy.* describing the condition of a cold-hardened metallic material, intermediate between full-hard and extra spring.

extrahepatic *Medicine.* situated or occurring outside the liver.

extra-high voltage *Electricity.* in power distribution, a voltage greater than 240 kV. Also, **extra-high tension.**

extrajunctional receptor *Physiology.* a type of acetylcholine receptor found distributed over the surface of a muscle fiber outside the area of the neuromuscular junction.

extraneous response *Electronics.* the reception of an undesired response in a receiver or recorder that arises from the mixing of desired and undesired signals in the system.

extranuclear *Medicine.* situated or occurring outside a cell nucleus.

extraocular *Medicine.* situated outside the eye.

extraordinary wave *Geophysics.* a magnetoionic portion of a wave that has a clockwise/counterclockwise elliptical polarization when viewed below the ionosphere in the direction of propagation. Also, **extraordinary component.**

extra play *Mechanical Engineering.* a videotape recording speed that is slower than standard speed to allow for a longer recording time, usually six hours as opposed to the standard two hours for a T-120 tape.

extrapleural *Medicine.* situated outside the pleural cavity.

extrapolation [eks′trap ə la′shən] *Science.* the inference of an unknown value on the basis of a known value. *Mathematics.* the process of estimating the value of a function at a point greater or less than all points for which the value of the function is known.

extrapolation ionization chamber *Nucleonics.* a type of ionization chamber that measures variations in a single factor, such as volume or electrode separation, so that the degree of ionizing radiation can be extrapolated from the these values when they are plotted on a curve.

extrapsychic [eks′trə sī′kik] *Psychology.* occurring outside the mind, or between the mind and the external environment.

extrapulmonary *Medicine.* not connected with the lungs.

extrapyramidal system *Anatomy.* the portion of the nervous system involved in motor control, but excluding the motor cortex, pyramidal tract, and motor neurons. This includes the basal ganglia, substantia nigra, subthalamic nucleus, and part of the midbrain.

extrasensory perception [eks′trə sen′sə rē] see ESP.

extra spring *Metallurgy.* describing the strongest condition of a cold-hardened metallic material.

extrasystole [eks′trə sis′tə lē] *Medicine.* a premature contraction of the heart that is abnormal in timing or in origin of impulse.

extraterrestrial [eks′trə tə res′trē əl] *Science.* **1.** occurring or originating away from earth. **2.** a life form occurring elsewhere than on earth.

extraterrestrial life *Astronomy.* any hypothetical civilization or intelligent life forms occurring in the universe outside the earth.

extraterrestrial noise *Electromagnetism.* any electromagnetic noise caused by sources not related to the earth.

extraterrestrial radiation *Astrophysics.* electromagnetic and particle radiation that comes from space.

extratropical cyclone *Meteorology.* a migratory frontal cyclone of middle or high altitudes. Also, **extratropical low, extratropical storm.**

extrauterine [eks′trə yoot′ə rin] *Medicine.* situated or occurring outside the uterus.

extrauterine pregnancy *Medicine.* a pregnancy occurring outside the uterus.

extravasation *Medicine.* a passage into the tissues, usually of blood, serum, or lymph. *Volcanology.* the eruption of material from an opening in the earth's surface, such as lava from a vent or water from a geyser.

extravascular [eks′trə vas′kyə lər] *Biology.* **1.** outside or not contained in vessels. **2.** not having blood vessels; not vascular.

extravehicular activity *Space Technology.* an operation performed outside a spacecraft, whether in orbit or on a lunar or planetary surface.

extravehicular activity (EVA)

extraversion see EXTROVERSION.

extravert see EXTROVERT.

extraverted see EXTROVERTED.

extremal see CRITICAL FUNCTION.

extreme *Meteorology.* the highest or lowest value of a climatic element observed during a given period or during a given month or season of that period. *Mathematics.* an extreme point, extreme term, or extremum.

extreme high water *Oceanography.* the highest recorded level of the tide at a specified place.

extreme line casing *Petroleum Engineering.* a special oil or gas well casing used when there is a need for more strength and greater leak resistance than normal. Similarly, **extreme line tubing.**

extreme low water *Oceanography.* the lowest recorded level of the tide at a specified place.

extremely high frequency *Telecommunications.* a radio frequency in the range from 30 to 300 MHz.

extremely low frequency *Telecommunications.* a radio frequency in the range from 30 to 300 Hz.

extreme narrowing approximation *Spectroscopy.* in the theory of spectral-line shapes, a mathematical approximation used to indicate that the exchange narrowing of a perturbation is complete.

extreme point *Mathematics.* **1.** an end point: if S is any subset of R^n, a point z in S is said to be an extreme point of S if no distinct points x and y (not equal to z) in S and real number t with $0 < t < 1$ exist such that $z = tx + (1 - t)y$; i.e., z does not lie on the interior of the line segment connecting any two points of S. **2.** an extremum.

extreme range *Ordnance.* the maximum distance a weapon will fire or reach, whether effective or not.

extreme relativistic limit *Physics.* the limiting form of a physical property of a particle whose speed approaches the speed of light.

extreme spread *Ordnance.* the distance between the two shots farthest apart in a firing test.

extreme terms *Mathematics.* in a proportional relationship $a/b = c/d$, the quantities a and d; b and c are the means. Also, **extremes.**

extreme-value distribution *Statistics.* a distribution of record values in a set of random variables.

extreme-value problem *Mathematics.* a mathematical problem that involves finding the greatest and/or least values of a quantity.

extremity *Anatomy.* **1.** also, **extremitas.** a distal or terminal portion. **2.** a limb; an arm or leg, or sometimes specifically a hand or foot.

extremum *Mathematics.* a minimum or maximum value of a given function; it may be relative, local, or global. Also, **extreme, extreme value.**

extrinsic *Science.* coming from or originating outside.

extrinsic allergic alveolitis *Medicine.* hypersensitivity pneumonitis, a respiratory reaction to repeated inhalation of organic dust, most often in occupational environments.

extrinsic motivation *Behavior.* an incentive that stems from positive or negative reinforcements that are external to the behavior itself. Thus, **extrinsically motivated behavior.**

extrinsic photoconductivity *Electronics.* the movement of electrons that is governed by light and the impurities in the various energy bands of a given material.

extrinsic properties *Electronics.* the characteristics that arise in a semiconductor when impurities or imperfections are added to the semiconductor crystal.

extrinsic proteins *Biochemistry.* proteins originating from the outside.

extrinsic reward *Psychology.* a reward that is not perceived as being inherently connected to the behavior being rewarded.

extrinsic semiconductor *Electronics.* a semiconductor device that has had impurities added to its crystal to achieve a desired polarity or resistivity.

extrinsic variable star *Astronomy.* a star whose brightness varies due to external or geometric causes, such as being eclipsed by another star.

extro- a prefix meaning "outward" or "outside," as in *extroversion, extrogastrulation.*

extrorse *Biology.* facing or turned outward or facing away from the axis of growth; usually applied to stamens opening toward the outside of a flower.

extroversion *Biology.* a turning outward. *Psychology.* a tendency to direct one's interests and energies outward, toward other people or things, rather than inward to one's own experiences and feelings.

extrovert *Psychology.* **1.** a person who has an extroverted personality. **2.** in popular use, a person who is friendly and outgoing.

extroverted *Psychology.* characterized by extroversion; tending to direct one's attention toward other people or things.

extrudate *Engineering.* any semisoft solid material that is forced through a die to mold it into a continuous form, such as a tubing.

extrude *Engineering.* to force, thrust, or press out.

extruder *Engineering.* a device used to push a semisoft or ductile solid material through a die so that it is molded into a continuous form such as a strip or tubing.

extrusion *Engineering.* a process of forming rods, tubes, or other continuously formed pieces, by pushing hot or cold semisoft solid material through a die. *Food Technology.* a process in which starchy food is plasticized under high temperature and pressure and forced through a molder or die to a desired shape. *Volcanology.* the emission of lava or magmatic materials onto the earth's surface, and the rock formed from such an emission.

extrusion billet *Metallurgy.* the starting stock of an extrusion process.

extrusion blow molding *Materials Science.* a method of processing polymers in which the perison (extruded material) in the form of an open-end tub is surrounded by a mold, and a tube is used to increase air pressure to expand the material to fit the mold.

extrusion coating *Engineering.* a process in which a thin film of semisoft resin that has been extruded is pressed onto or into a substrate.

extrusion defect *Metallurgy.* a flaw generated in the product of the extrusion process; most commonly, a center pipe.

extrusive *Geology.* relating to any igneous material, such as a lava flow or volcanic ash, that has been ejected or poured out onto the earth's surface. Also, EFFUSIVE, VOLCANIC ERUPTIVE.

extrusive rock see VOLCANIC ROCK.

extrusome *Cell Biology.* an organelle at the cell periphery of many ciliates that is able to be discharged from the cell surface, but whose function is unclear.

extubation *Surgery.* the removal of a previously inserted tube.

exuberant granulations *Surgery.* excessive production of granulation tissue around a wound as it heals.

exudate *Medicine.* fluid, cells, or other substances that have been discharged through small pores or breaks in cell membranes, usually as a result of inflammation.

exudation *Medicine.* **1.** the escape of fluid, cells, and cellular debris from blood cells. **2.** an exudate. *Physiology.* the process of sweating. Thus, **exudative.**

exudation vein see SEGREGATED VEIN.

exumbrella *Invertebrate Zoology.* the outer convex surface of the umbrella of a jellyfish.

exuviae *Invertebrate Zoology.* the sloughed-off skin or covering of an animal, especially an arthropod. Thus, **exuvial.** Also, **exuvium.**

ex vivo [eks vē´vō] *Surgery.* outside the living body; relating to the removal of an organ from the body for surgical repair and subsequent return to the original site. (From a Latin phrase meaning "out of life.")

eye *Anatomy.* either of the pair of spherical organs serving to provide vision, each located in a bony orbit at the front of the skull and innervated by an optic nerve from the forebrain; the eye contains the following layers: sclera and cornea, choroid, and retina. Also, OCULUS. *Meteorology.* a roughly circular area of relatively light winds and fair weather found at the center of a severe tropical cyclone, varying from four miles to more than forty miles in diameter. Also, **eye of the (a) storm.** *Food Technology.* a hole formed during the bacterial ripening of cheeses such as Swiss, Emmenthal, and Gruyere.

eyeball *Anatomy.* the globe of the eye. Also, BULBUS OCULI, ORB. *Engineering.* to make an estimate or evaluation on the basis of a visual inspection.

eyeball potential *Physiology.* a series of electrical potential changes produced when the eye is stimulated by light, as shown on an electroretinogram.

eyebar *Mechanical Devices.* a metal bar with loops or eyeholes at each end.

eyebolt *Mechanical Devices.* a bolt with a looped head to receive a pin, stud, or hook.

eye coal *Geology.* coal that contains small, circular or elliptical structural disks that are parallel or normal to the bedding. Also, CIRCULAR COAL.

eye contact *Psychology.* a situation in which one person looks directly into the eyes of another; regarded as a form of nonverbal communication.

eye-ear plane *Anthropology.* the measurement of a plane extending through the top of both ear holes and the lower margin of an orbit, resulting in the maximum height of the skull above this plane. Also, ORBITAL-METAL PLANE, FRANKFURT PLANE.

eyelash *Anatomy.* any of the numerous hairs that grow in double or triple rows at the edge of an eyelid.

eyelet *Textiles.* a small hole or ring worked into a piece of fabric for receiving buttons, cords, pins, or laces.

eyeletting *Engineering.* a process that produces a hard, strengthened ring around the rim of a hole.

eyelid *Anatomy.* either of two movable folds of skin (upper and lower) that protect the front of the eyeball, having eyelashes, ciliary glands, and meibomian glands at the margin and controlled by the orbicularis oculi muscle and the facial nerve.

eye of the wind *Meteorology.* the point or direction from which a wind is blowing. Also, TEETH OF THE GALE.

eyepiece *Optics.* any lens system placed between the viewer's eye and an object that enlarges the image of the object; such systems include microscopes and telescopes. Also, OCULAR.

eye relief *Astronomy.* the distance from the eyepiece of a telescope to the location of the exit pupil.

eye screw *Mechanical Devices.* a wood or metal screw with a loop at one end.

eye socket *Anatomy.* see ORBIT.

eyespot *Invertebrate Zoology.* **1.** a simple light-sensitive organ covered by pigment or pigmented cells. **2.** a small, round spot of color on the wings of a butterfly or moth. *Botany.* **1.** a pigmented, photosensitive body lying near the base of the flagella in some motile algae and euglenids. **2.** a dark area around the hilum on some bean seeds. *Plant Pathology.* a disease found in sugarcane and other grasses that is caused by a parasitic fungus.

eyestalk *Invertebrate Zoology.* the movable stalk bearing a terminal eye in crustaceans.

eyewitness memory *Psychology.* the ability to recall a recent event that was directly witnessed; factors subsequent to the event may influence the details of the recollection.

Eyring equation *Physical Chemistry.* an equation that gives the reaction rate for a chemical reaction in terms of various constraints, including energy levels and temperature. Also, TRANSITION STATE THEORY.

E zone *Telecommunications.* one of three zones into which the earth is divided, based on longitudinal variations in the F_2 layer, to aid in predicting frequency when using the ionosphere for wave propagation; the zone roughly covers the Eastern Hemisphere.

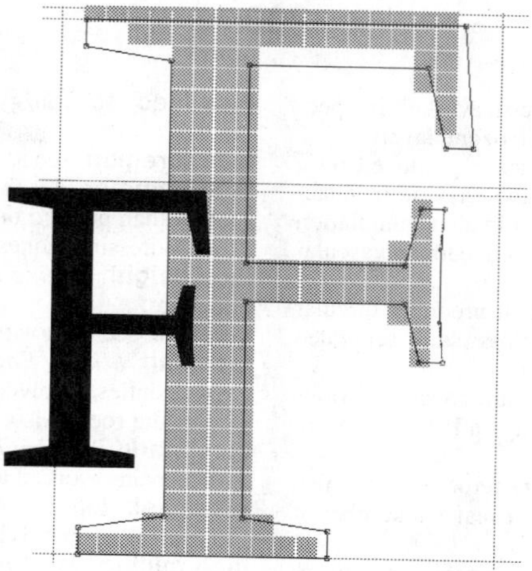

F *Computer Programming.* the character that represents the hexadecimal numeral 15.

F farad; Fahrenheit.

F the chemical symbol for fluorine.

F- or **F** *Aviation.* the U.S. military designation for fighter aircraft, as in F-84 or F-2A.

°F degrees Fahrenheit.

f farad; faraday; femto-; fluid; focal length; force; frequency.

f *Physics.* **1.** frequency in cycles per second. **2.** the Coriolis parameter.

F_1 *Genetics.* the first generation of offspring obtained from the experimental mating of two organisms. Also, FIRST FILIAL GENERATION.

F_2 *Genetics.* the second generation of offspring, obtained when two organisms from an F_1 generation mate. Also, SECOND FILIAL GENERATION.

F-2A *Aviation.* a single-engine Navy fighter used prior to and during World War II; popularly known as the Buffalo.

F-4D -1, -2 *Aviation.* modified delta-wing, carrier-based fighters.

F-4F *Aviation.* a single-engine Navy fighter used in World War II; popularly known as the Wildcat.

F-4U *Aviation.* a single-engine Navy fighter used during and (with modifications) after World War II; popularly known as the Corsair.

F-6F *Aviation.* a single-engine Navy fighter; popularly known as the Hellcat.

F-9F *Aviation.* a single-engine Navy jet fighter; popularly known as the Panther.

F-11F *Aviation.* a modified F-9F; popularly known as the Tiger.

F-14 *Aviation.* a twin-engine, two-seater, all-weather turbojet fighter interceptor designed to operate from an aircraft carrier; popularly known as the Tomcat.

F-15 *Aviation.* the Air Force's land-based counterpart to the Navy F-14; popularly known as the Eagle.

F-16 *Aviation.* a very fast and maneuverable single-engine fighter; popularly known as the Fighting Falcon.

F-51 *Aviation.* a single-engine, low-wing monoplane fighter used extensively during World War II; popularly known as the Mustang.

F-59 *Aviation.* a twin-engine jet fighter, the first such developed for the Army Air Force (1942); popularly known as the Airacomet.

F-61 *Aviation.* a twin-engine night fighter used during World War II; popularly known as the Black Widow.

F-80 *Aviation.* a single-engine jet fighter; popularly known as the Shooting Star.

F-84 *Aviation.* a variously single-jet or turboprop fighter; popularly known as the Thunderjet or Thunderstreak.

F-86 *Aviation.* a single-engine jet fighter, used in the Korean War; popularly known as the Sabre or Sabrejet.

F-89 *Aviation.* a twin-engine, two-seat, all-weather jet fighter; popularly known as the Scorpion.

F-100 *Aviation.* a sweptwing supersonic jet fighter, sometimes modified as a fighter-bomber; popularly known as the Super Sabre.

F-105 *Aviation.* a single-engine, sweptwing, supersonic jet fighter.

F-111 *Aviation.* a twin-engine, two-seater, variable sweep, all-weather attack aircraft designed for the U.S. Air Force and able to carry heavy armaments; several versions include the F-111A, F-111E, and F-111F.

FA finite automaton; field artillery.

fA femtoampere.

FAA or **F.A.A.** Federal Aviation Administration.

Fabaceae *Botany.* in some systems of classification, an equivalent name for Leguminosae.

Fabales *Botany.* an order of dicotyledonous plants having stipulate, compound leaves and fruit that is a pod; many species have symbiotic nitrogen-fixing bacteria in their roots.

Fabavirus group *Virology.* a genus of viruses that are similar in structure to Comoviruses, have a wide host range, and are transmitted mechanically and by aphids.

Faber flaw *Solid-State Physics.* a defect in a substance that serves to nucleate the growth of a superconductive region within the material.

Faber polynomials *Mathematics.* let D be a bounded, closed subset of the complex plane whose complement is simply connected, and let $f(t)$ be a function that maps the region $|t| > 1$ one-to-one and conformally onto the complement of D. Then the Faber polynomials of D are the polynomials $P_n(z)$ such that, for all z and t,

$$f'(t)/(f(t)-z) = \sum_{n=0}^{\infty} P_n(z)t^{-(n+1)}.$$

Fab fragment *Immunology.* one of a pair of identical fragments formed as an immunoglobulin molecule is broken down by the enzyme papain.

Fabian system *Mining Engineering.* a drilling system from which all other freefall drilling systems originated; no longer used in its original form.

fabric *Textiles.* a flexible cloth that consists of woven, knitted, or felted fibers. *Science.* the arrangement and physical relationship of components or constituent elements of something. *Geology.* specifically, the physical properties of a soil that depend on the spatial arrangement of its particles. *Petrology.* the spatial orientation of the constituent particles in a rock. Also, PETROFABRIC, ROCK FABRIC.

fabrication *Engineering.* the process by which individual structural or electromechanical components of a system are manufactured.

fabricator *Archaeology.* a piece of stone or bone used to chip flakes from a stone core.

fabric axis *Petrology.* one of the three orthogonal axes used as references in the orientation of fabric elements and in the description of folding and symmetry of movement in deformed rocks. Also, TECTONIC AXIS.

fabric diagram *Petrology.* an equal-area or stereographic projection of components of a rock fabric. Also, PETROFABRIC DIAGRAM.

fabric domain *Petrology.* a three-dimensional volume of more or less homogeneous rock fabric bounded by structural or compositional discontinuities.

fabric element *Petrology.* a rock fabric component ranging from an ion or atom to a mineral grain, pebble grain, lens, or layer that responds as a unit to deformative forces.

Fabriciinae *Invertebrate Zoology.* a subfamily of fanworms or featherdusters in the family Sabellidae; small, sedentary polycheate annelid worms with noncalcareous tubes.

Fabricius, Hieronymus [fa brē´sē əs] (**Girolamo Fabrizio**) c. 1533–1619, Italian anatomist; pioneer in embryology and in the physiology of venal valves.

Fabroniaceae *Botany.* a family of small, delicate, glossy mosses of the order Isobryales that form smooth mats on tree trunks, characterized by creeping stems with irregular branching and by unusual pellucid leaf cells with conspicuous cell membranes.

Fabry, Charles [fab´rē] 1867–1945, French physicist; worked in spectroscopy and interferometry; discovered atmospheric ozone layer.
Fabry disease *Genetics.* a hereditary human disease produced by a mutant form of the X-linked gene that codes for the enzyme α-galactosidase in lysosomes; the enzyme deficiency results in an accumulation of glycosphingolipids in the walls of blood vessels and leads to vascular malfunctions.
Fabry-Perot filter *Optics.* an interference filter that produces circular fringes and consists of two parallel, half-silvered glass plates separated by a thin layer of air. Also, **Fabry-Perot étalon.**
Fabry-Perot fringes *Optics.* a series of light and dark concentric rings observed when monochromatic light is passed through a Fabry-Perot interferometer.
Fabry-Perot interferometer *Optics.* a type of interferometer in which a light beam is reflected between two glass or quartz plates a number of times before it is transmitted.
Fabry-Perot method *Optics.* a method of determining the index of refraction of a prism, in which the prism is positioned so that the emergent face is perpendicular to the incident beam; the index is calculated using the angle of the prism and the angle of deviation of the beam.
facade [fə säd´] *Architecture.* the front of a building.
face *Anatomy.* **1.** the front surface of the head from the forehead to the chin. **2.** any similar anterior or presenting aspect or surface.
face any of various features or structures thought to resemble the human face; i.e., a front or outward surface; specific uses include: *Building Engineering.* the front surface or wall of a building or other structure. *Electricity.* the transparent or semitransparent glass front of a cathode-ray tube through which an image is viewed or projected. Also, FACEPLATE. *Crystallography.* see CRYSTAL FACE. *Geology.* **1.** the principal surface of a landform. **2.** the original upper surface of a rock layer. Also, FACING. *Graphic Arts.* **1.** the particular design of a set of typeset characters; a typeface. **2.** the printing surface of a piece of metal type. **3.** the front surface of a book or other printed material. *Forestry.* a cut made in a tree from which resin slowly flows. *Mining Engineering.* **1.** a working place from which coal is extracted. Also, COAL FACE. **2.** a point at which coal is being worked away in a breast or heading. Also, WORKING FACE.
face angle *Mathematics.* an angle of a polyhedron formed by two faces intersecting at an edge.
face area *Mining Engineering.* the working area by the last open crosscut in an entry or room, including the pillar being extracted or the longwall being mined.
face-bonding *Electronics.* a method of building hybrid microcircuits in which semiconductor chips, with mounting pads, are bonded face down onto the thin-film conductors of a passive substrate, such as glass or ceramic.
face boss *Mining Engineering.* a foreman in charge of operations at the working face in a bituminous coal mine. Also, FACE FOREMAN.
face brick *Building Engineering.* a high-quality decorative brick used on prominently exposed walls. Also, FACING BRICK.
face-bridging ligand *Organic Chemistry.* a ligand that forms a bridge over one of the triangular faces of the polyhedron of a metal cluster.
face-centered cubic structure *Crystallography.* a cubic unit-cell structure having one lattice point at the corners and one lattice point at the center of each face.
face-centered lattice *Crystallography.* a crystal lattice with a lattice point at the center and corners of each face. If all faces are centered, the designation is *F*; if only faces perpendicular to the *a* axis are centered, the description is *A*; and similarly for *B* and *C*.
face-centered unit cell *Crystallography.* a crystal lattice in which the centers of each face are identical in environment and orientation with the vertices. It can be considered as a crystal lattice with lattice points at the centers of faces ($x = y = 1/2$), $z = 0$, ($x = z = 1/2$), $y = 0$, and ($y = z = 1/2$), $x = 0$, as well as the corner ($x = y = z = 0$) of the unit cell.
face-discharge bit *Mechanical Engineering.* a liquid-coolant bit designed for drilling in soft formations and on a double-tube core barrel; the coolant flows through the bit and is ejected at the cutting face. Also, BOTTOM-DISCHARGE BIT, FACE-EJECTION BIT.
faced vector *Robotics.* a vector in a working envelope that can be rotated and its orientation changed.
faced wall *Building Engineering.* a wall in which the masonry composition of the facing and that of the backing are of different materials.
face edge *Building Engineering.* a trued square on the edge of a piece of wood whose working face is fabricated to assist in truing the other surfaces. Also, WORKING EDGE.
face-ejection bit see FACE-DISCHARGE BIT.
face feed *Metallurgy.* in brazing or soldering, the deposition of a filler metal.
face foreman see FACE BOSS.
face hammer *Mechanical Devices.* a hammer with a peen that is flat, rather than pointed or edged, with one blunt end and one cutting end for rough-dressing stones.
face height *Mining Engineering.* the vertical height of overburden and coal ore.
facellite see KALIOPHILITE.
faceman *Mining Engineering.* a coal miner who performs the complete set of duties involved in driving underground openings to extract coal, slate, and rock. Also, COAL DIGGER; COAL GETTER.
face mark *Building Engineering.* a mark made on the face of a piece of wood being worked to identify it as a basis for truing other surfaces.
facemask *Engineering.* a mask or helmet that fits over a worker's face to protect against flying particles and debris. Similarly, **face shield.**
face milling *Mechanical Engineering.* the process of milling a flat surface that is normal to the rotational axis of the cutting tool.
face mix *Building Engineering.* a mixture of cement and stone dust used to imitate real stone as facing for concrete blocks.
face mold *Engineering.* a template used for cutting shapes out of wood, metal, or other material.
face nailing *Engineering.* the process of securing by nails driven perpendicular to the face of the wood.
face of an *n*-simplex *Mathematics.* an *r*-face of an *n*-simplex is any simplex whose vertices are $r + 1$.
face of a plane graph *Mathematics.* the closure of any of the several connected partitions formed when the edges and vertices of a plane graph are erased
faceplate *Engineering.* a perforated plate that is mounted on the spindle of a lathe. *Electricity.* see FACE.
face sampling *Mining Engineering.* the random cutting of ore and rock samples from exposed faces of ore and waste.
facet *Geology.* a small rock face. *Geography.* any part of a surface, such as a flat or slope. *Anatomy.* **1.** a small flat area on a bone or other hard surface. **2.** also, **facette.** a spot on a tooth that has been worn smooth by chewing. *Invertebrate Zoology.* the external surface of the translucent cornea in a compound eye, usually square or hexagonal. *Artificial Intelligence.* a named component of a slot in a frame, e.g., a value facet, if-needed facet, data type facet.
facet cut *Crystallography.* the manner in which a precious stone is cut to give flat faces.
faceted pebble *Geology.* a pebble that has developed several sharp edges and flat surfaces as a result of natural processes, such as wave erosion or glacial abrasion.
faceted spur *Geology.* an angular projection having an inverted-V face as a result of faulting or the action of streams, waves, or glaciers.
face timbering *Mining Engineering.* the act of positioning safety posts at the working face to support the roof of a mine.
facet joint *Anatomy.* any of the four projections linking one spinal vertebra to the adjacent vertebra.
facework *Building Engineering.* a special material, ornamental or otherwise, on the front side or outside of a wall.
facial *Anatomy.* of or relating to the face.
facial angle *Anthropology.* in anthropometry, the angle between the line of the nasion-basion and the line of the prosthion-basion.
facial artery *Anatomy.* an artery originating from the external carotid artery with branches that supply blood to various organs and tissues in the head.

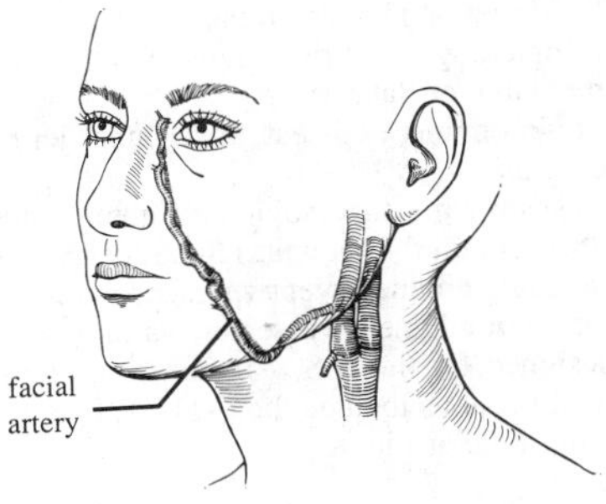

facial artery

facial bone *Anatomy.* any of the bones of the face, including the hyoid, palatine, and zygomatic bones, the mandible, and the maxilla.

facial-feedback hypothesis *Psychology.* a theory postulating that a person assuming a certain facial expression will experience the emotion associated with it.

facial index *Anthropology.* in anthropometry, the bizygomatic diameter multiplied by 100 and divided by the total facial height (the nasion-menton distance.)

facial nerve *Anatomy.* the seventh cranial nerve, consisting of a large motor root that supplies the muscles used for facial expression and a smaller root, the nervus intermedius; it attaches to the brain stem at the inferior border of the pons.

facies [fā´shēz; fā´shē ēz] *Geology.* **1.** a distinctive assemblage of characteristics within a rock unit that distinguish it from neighboring or associated rock units. **2.** the overall lithological and paleontological characteristics of a sedimentary rock, especially those that indicate its origin and environment of deposition. *Ecology.* the general appearance or character of a population or community. *Archaeology.* see PHASE. *Anatomy.* **1.** the face. **2.** the expression or appearance of the face.

facies map *Geology.* a stratigraphic map that displays the areal distribution of and variations in rock features that occur within a given geologic unit.

facies tract *Geology.* a collection of different but genetically related sedimentary facies of the same age. Also, MACROFACIES.

facilitated transport *Physiology.* the active transport of substances through cell membranes along a concentration gradient with the aid of carrier molecules. Also, **facilitated diffusion.**

facilitation *Physiology.* the process of lowering the resistance of a neural pathway by the repeated passage of an impulse along the same pathway. *Behavior.* see SOCIAL FACILITATION.

facility *Military Science.* **1.** a building or structure. **2.** any activity that provides operational assistance to military forces.

facility assignment *Computer Technology.* the allocation by the operating system of physical resources such as memory, peripheral devices, and communication equipment, as required by active programs.

facility dispersion *Telecommunications.* a distribution of circuits between two entities over more than one physical route, designed to reduce the probability of communication loss due to facility damage or other circuit failure.

facility security clearance *Military Science.* a determination granting a particular facility access to classified information of and below a particular level.

facing *Building Engineering.* an outer layer of stone on a brick wall; used as ornamentation or as protection from the elements. *Metallurgy.* **1.** in metal fabrication, the process of removing a surface layer by machining. **2.** in casting, a special sand attached to the pattern to improve the quality of the resulting casting. *Mechanical Engineering.* the use of a tool to machine the flat end of a surface as it is rotated in a spiral planar path along its axis. *Geology.* see FACE.

facing bond *Building Engineering.* a bond that consists primarily of stretchers.

facing brick see FACE BRICK.

facing paviors *Building Engineering.* a high-quality brick that is hard-burned and used as a face brick in special building applications.

facing-point lock *Civil Engineering.* a plunger that engages a rod on the switch point to lock a railroad switch track.

facing wall *Civil Engineering.* a wall or lining against the earth face of an excavation, usually made of concrete rather than wood sheeting.

facioplasty *Surgery.* plastic or reconstructive surgery of the face.

FACS fluorescence-activated cell sorter.

facsimile [fak sim´ə lē] *Telecommunications.* the full name for FAX communication. See FAX.

facsimile chart see FAX CHART.

facsimile crystallization see MIMETIC CRYSTALLIZATION.

facsimile modulation *Telecommunications.* the process of varying the frequency or amplitude of a transmitted wave in conjunction with a FAX transmission signal.

facsimile posting *Computer Technology.* the process of transferring a line of data from one set of records to another.

facsimile receiver see FAX RECEIVER

facsimile recorder see FAX RECORDER.

facsimile signal see FAX SIGNAL.

facsimile synchronizing see FAX SYNCHRONIIZING.

facsimile telegraph see FAX TELEGRAPH.

facsimile transmitter see FAX TRANSMITTER.

FACT *Transportation Engineering.* fully automatic control of trains.

-fact denoting an archaeological object; used to form words that serve to classify artifacts: *technofact, ecofact.*

factor *Artificial Intelligence.* to break a problem into relatively independent subproblems. *Mathematics.* an object or quantity a is said to be a factor of an object or quantity c if there exists another object or quantity b such that $ab = c$ (within some mathematical system such as a group); equivalently, a is any divisor of c. Expressing a given object or quantity c and the form $c = ab$ is known as **factoring**.

factor *Science.* any of several chemical substances or activities that are necessary to produce a result; often, the use of this term indicates that the chemical nature of the substance or its mechanism of action is unknown. *Biochemistry.* any of a number of substances that affect the coagulation of the blood; see the specific entries below.

factor I see FIBRINOGEN.

factor II see PHOTHROMBIN.

factor III *Biochemistry.* tissue thromboplastin; a material derived from several sources in the body and important in the formation of extrinsic prothrombin-converting principle in the extrinsic pathway of blood coagulation.

factor IV *Biochemistry.* a former term for calcium ions involved in blood coagulation.

factor V see ACCELERATION GLOBULIN.

factor VI *Biochemistry.* a factor once thought to be an activated form of factor V, but no longer considered in the scheme of hemostasis and thus has no assigned name or function at this time.

factor VII *Biochemistry.* a factor involved in the transformation of prothrombin to thrombin.

factor VIII see ANTIHEMOPHILIC FACTOR.

factor IX see CHRISTMAS FACTOR.

factor X *Biochemistry.* a storage-stable factor that participates in both the intrinsic and extrinsic pathways of blood coagulation; deficiency may cause a systemic coagulation disorder. Also, STUART FACTOR.

factor XI *Biochemistry.* a factor that links various activation factors, such as the one initiating coagulation in the blood, and the so-called hemophilic factors, whose deficiency prevents coagulation.

factor XII *Biochemistry.* a factor that initiates blood clotting and one that is deficient in hemophilia.

factor XIII see FIBRINASE.

factor XIV *Biochemistry.* a zymogen that is the inactive precursor of a serine protease. Also, PROTEIN C.

factorable polynomial *Mathematics.* a polynomial $p(x)$ is said to be factorable over a given domain, such as a ring or field, if $p(x)$ can be written in the form $a(x)b(x) = p(x)$, where $a(x)$ and $b(x)$ are both nonconstant polynomials over the same domain as $p(x)$. Factorability clearly depends on the given domain.

factor analysis *Statistics.* the study of the linear interrelationship in a set of random variables; includes the identification of linear combinations (factors) that ideally capture most of the overall variability, are close to orthogonal, and depend on only a few of the underlying variables.

factor B *Immunology.* a substance that combines with the active fragment of a complement (known as C3b) to form an alternative pathway for the initiation of complement activity.

factor comparison *Industrial Engineering.* the ranking of different jobs by several raters, according to skill, physical requirements, working conditions, and so on; evaluating each factor for its pay compensation value.

factor D *Immunology.* an active amino acid that is part of the alternative pathway for complement activation.

factor group *Mathematics.* a quotient group.

factor H *Immunology.* a glycoprotein that acts as an alternative-pathway complement inhibitor.

factor I *Immunology.* a plasma enzyme that acts in conjunction with factor H as an alternative-pathway complement inhibitor. Also, C3b INACTIVATOR.

factorial *Mathematics.* the factorial of a nonnegative integer n, denoted $n!$, is $\prod_{k=1}^{n} k$, i.e., the product of all positive integers less than or equal to n. By convention, $0! = 1$. The factorial function can be extended to complex numbers by means of the gamma function; i.e., $z! = \Gamma(z + 1)$ for z complex.

factorial design *Statistics.* an experimental design that can be used to evaluate different experimental factors simultaneously.

factoring *Mathematics.* the process of finding factors of an object or quantity, especially of an integer or polynomial. Also, SPLITTING.

factorization theorem *Mathematics.* let $\{z_n\}$ be a sequence of distinct complex numbers that has no limit point in the complex plane, and let$\{\mu_n\}$ be a sequence of positive integers. Then there exists an entire function $f(z)$ having a zero of multiplicity μ_n at z_n for each n, and having no other zeros. Then, if $z_n \neq 0$ for all n, these conditions are satisfied by $f(z) = \prod_{n=1}^{\infty}(E(z/z_n,p_n))^{\mu}n$, where $E(z,p) = (1 - z)\exp \sum_{k=1}^{p} z\, p/p$, and where the integers p_n are chosen so that the product $f(z)$ converges uniformly on compact sets. Once such a function $f(z)$ has been constructed, then the most general entire function $F(z)$ with zeros of multiplicity μ_n at z_n and multiplicity μ_0 at $z_0 = 0$ is given by $F(z) = z^{\mu}0f(z)\exp[g(z)]$, where $g(z)$ is any entire function. Also, WEIERSTRASS FACTORIZATION THEOREM.

factor of a graph *Mathematics.* given an assignment f of a nonnegative integer to each vertex of an undirected graph G, an f-factor of G is a spanning subgraph H of G such that the degree in H of each vertex v of G is $f(v)$.

factor of an integer *Mathematics.* a factor of a given positive integer n is any positive integer k that divides n. If k has no factors other than 1 and k, then k is called a prime factor of n.

factor of a polynomial *Mathematics.* one of a set of nonconstant polynomials whose product is equal to a given polynomial and whose coefficients lie in a specified domain, usually a ring or field.

factor of proportionality *Mathematics.* in the proportional relationship $A = \lambda B$, the factor of proportionality for A and B is λ.

factor of safety *Mechanics.* **1.** the ratio of the load at which a structure or material is expected to fail to the load permitted, in deference to such uncertainties as wear, consistency of manufacture, or quality of material. Also, DESIGN SAFETY FACTOR. **2.** see FACTOR OF STRESS INTENSITY.

factor of stress concentration see STRESS CONCENTRATION FACTOR.

factor of stress intensity *Mechanics.* the ratio of the ultimate stress that a structural member or material can withstand without failure to the actual or expected stress. Also, FACTOR OF SAFETY.

factory *Industrial Engineering.* an industrial production facility, especially for manufacturing. The term normally excludes extractive facilities such as mines and refineries.

factory area see VIROPLASM.

factory control *Control Systems.* in integrated computer-aided manufacturing, a module that is controlled by management personnnel and policies.

factory model *Design Engineering.* the composite of a manufacturing facility constituted by a minimum of two interconnected manufacturing centers with material-transport units such as conveyors, storage areas, and communication systems.

factory ship *Naval Architecture.* a ship equipped to process fish caught off the ship or by other ships.

facula [fak´ yə lə] *plural,* **faculae.** *Astronomy.* a bright, high-temperature patch in the solar photosphere that is visible near the limb.

facultative *Ecology.* having the ability to live and adapt to various conditions while not being restricted to those conditions or mode of life, such as bacteria that can live either with or without oxygen.

facultative factor *Ecology.* a destructive agent whose efficacy is dependent upon the density of the population it affects.

facultative heterochromatin *Genetics.* a very dense chromatin that exists as heterochromatin only in specific cells at specific times, such as the condensed, inactivated chromatin of X chromosomes in the diploid somatic cells of female mammals.

facultative parasite *Ecology.* a type of parasite organism, such as a flea, that can live separate from its host.

FAD flavin adenine dinucleotide.

fade chart *Electromagnetism.* a plot of the null areas of an air-search radar antenna.

fade-in *Telecommunications.* the gradual introduction of radio or television program material or any other received signal from zero to a normal level.

fade-out *Telecommunications.* **1.** a loss of signal due to a variation in signal cancellation from phase distortion or the simultaneous transmission of a signal at another frequency. **2.** a gradual reduction and final cessation of radio or television program material or of any other received signal from a normal level to zero.

fader *Electronics.* a device that permits an operator to gradually vary the strength of an audio or video signal.

fading *Telecommunications.* in the propagation of electromagnetic waves, the variations in received signal strength due to changes in parameters of the propagation path or to varying ionization conditions; it can impair television reception of a broadcast signal.

fading margin *Telecommunications.* the amount by which a received signal level may be reduced without causing the system or channel output to fall below a required service grade.

Fagaceae *Botany.* a family of dicotyledonous hardwood trees of the order Fagales having simple leaves, single seeded nuts in cupules, and unisexual flowers borne in small spikes or catkins; includes species of oak, beech, and chestnut.

fagaceous *Botany.* belonging to Fagaceae, the beech family.

Fagales *Botany.* an order of dicotyledonous trees of the subclass Hamamelidae, characterized by simple, alternate, stipulate leaves, unisexual flowers, and a single seeded nut having no endosperm.

Fagergren cell *Mining Engineering.* a froth-flotation cell in which a squirrel-cage rotor is driven concentrically in a vertical stator.

Fagersta cut *Mining Engineering.* a cut drilled with hand-held equipment in two steps: first the pilot hole, and second, the enlargement of the pilot hole.

fagopyrism *Toxicology.* poisoning due to toxins found in buckwheat, clover, and other plants; the symptoms may include skin irritation, photosensitivity, nausea, and vomiting.

fahlband *Geology.* a belt or belts of sulfide minerals occurring in metamorphic rock.

fahlore see TETRAHEDRITE-TENNANTITE SERIES.

Fahnestock clip *Electricity.* a flat, sheet-metal spring terminal that holds a wire in a temporary breadboard setup.

Fahr. Fahrenheit.

Fahrenheit, Gabriel [fâr´in hīt´] 1686–1736, German physicist; invented the mercury thermometer and Fahrenheit temperature scale.

Fahrenheit or **fahrenheit** *Thermodynamics.* relating to or expressed by the Fahrenheit temperature scale. Thus, **Fahrenheit thermometer.**

Fahrenheit degree *Thermodynamics.* a unit of temperature increment used in the Fahrenheit temperature scale.

Fahrenheit's hydrometer *Engineering.* a hydrometer that measures the relative density of a liquid by determining the weights necessary to sink the instrument to a scaled mark, in water and in the liquid under examination.

Fahrenheit scale *Thermodynamics.* a temperature scale that is defined so that 32 degrees corresponds to the freezing point of water and 212 degrees corresponds to the boiling point, at standard atmospheric pressure. Absolute zero on this scale occurs at –459.60°F. The scale was originally defined on the basis of 0° for the freezing point of a certain mixture of ice, water, and salt, and 96° for the normal human body temperature (later shown to be 98.6°F). Also, **Fahrenheit temperature scale.**

Fahrenholz's rule *Ecology.* the theory that there is usually a significant degree of similarity in the phylogenetic development and speciation of a parasite and its host.

failed hole *Engineering.* a term for a drill hole loaded with dynamite that failed to detonate.

fail-safe system *Engineering.* a system designed so that a component failure will not put people operating the system or other people in the vicinity at risk. *Computer Technology.* a computer system designed to avoid catastrophic results following a system malfunction or failure.

fail soft *Engineering.* a breakdown in the function of a system component that does not result in immediate or significant disruption of operations or in a degrading of output quality.

fail-soft system *Computer Technology.* a computer system that has the ability to keep some parts operating when a major element of the host system has failed; may also guard against loss of data. Also, SOFT-FAIL SYSTEM.

failure the lack of an expected or satisfactory performance. *Mechanics.* specifically, the overloading or overstraining of a structure to such an extent that it can no longer perform its required function.

failure analysis *Materials Science.* the scientific study and analysis of material failures, including the use of nondestructive testing, the study of fracture mechanics, the examination of macro- and microstructures, and the interpretation of structural and manufacturing defects.

failure distribution *Statistics.* the probability distribution of the time it takes for a randomly behaved system to fail.

failure envelope *Materials Science.* the conditions under which failure or viscoelastic rupture occurs in a highly stretched elastomer; shown by plotting the reduced stress to break against the strain to break on a log-log scale.

failure logging *Computer Technology.* an automatic recording of the status of hardware components when a system malfunction is detected; used to assist in fault diagnosis.

failure rate *Statistics.* the ratio of the failure density at a given time to the probability of surviving until that time. Also, HAZARD RATE. *Engineering.* the probability that a given operation will fail over a specified period of time or cycles of operation.

fair *Meteorology.* of or relating to pleasant weather conditions considering location and time of year, generally implying no precipitation, less than 0.4 sky cover of low clouds, and no extreme conditions of cloudiness, visibility, or wind. *Aviation.* to streamline a part or surface of a flight vehicle, especially by fitting a fairing over a nonstreamlined surface. *Computer Science.* of a scheduling algorithm, guaranteed eventually to service every request; e.g., FCFS is fair.

fairchildite *Mineralogy.* $K_2Ca(CO_3)_2$, a colorless hexagonal mineral having an undetermined hardness and a specific gravity of 2.446 that is dimorphous with Bütschliite; found as microscopic plates in the fused ashes of partially burned trees.

fair curves *Naval Architecture.* describing the smooth flow of water without sudden changes of curvature around the hull of a ship.

faired cable *Mechanical Devices.* a cable used to haul a trawling net and whose surface is designed to cut down on drag underwater.

fairfieldite *Mineralogy.* $Ca_2(Mn^{+2},Fe^{+2})(PO_4)_2 \cdot 2H_2O$, a white to pale yellow triclinic mineral occurring as prismatic crystals and foliated or fibrous masses, having a specific gravity of 3.08 and a hardness of 3.5 on the Mohs scale.

fair game *Mathematics.* **1.** a game with one player is said to be fair if its expected value is 0. It is for the player if the expected value is positive, and against the player if the expected value is negative. **2.** a game with more than one player is fair if each participant has an equal expectation of success. **3.** let $Z_1, Z_2, \ldots$ be a process in a given probability space representing a sequence of games under a given gambling system, and let S_n (a function of $Z_1, Z_2, \ldots, Z_n$) represent a player's fortune after n plays with $E|S_n| < \infty$. Then the sequence of games $Z_1, Z_2, \ldots$ under the given gambling system is fair if $E(S_{n+1}|S_n, \ldots, S_1) = S_n$ (almost surely) for all n, and unfavorable if $E(S_{n+1}|S_n, \ldots, S_1) \leq S_n$ (almost surely) for all n. Note that the sequence $S_1, S_2, \ldots$ forms a martingale.

fairing *Aviation.* **1.** a streamlined member or secondary structure designed to be fitted over a nonstreamlined surface in order to reduce the drag on a flight vehicle. **2.** a surface that has been faired.

fairlead *Mechanical Engineering.* Also, **fair lead, fairleader.** a swiveling pulley or group of pulleys that permit a rope or cable to be reeled to any direction without chafing, as on the rigging of a ship. *Aviation.* a streamlined tube through which an airplane antenna or control cable is passed; it is designed to minimize external drag and friction with other parts.

fair line *Naval Architecture.* a line formed by the intersection of an axial plane with a hull surface.

fair tide *Navigation.* a tidal current that sets in approximately the same direction as the vessel's course.

fairwater *Naval Architecture.* a device used to form fair curves around underwater hull fittings.

fairway *Navigation.* a navigable channel, natural or dredged, for vessels entering or leaving a harbor.

fairway buoy *Navigation.* any buoy marking a fairway.

fair-weather cumulus cloud see CUMULUS HUMULIS CLOUD.

fair-weather runoff see BASE RUNOFF.

fair wind *Navigation.* a wind blowing in approximately the same direction as a craft's course.

fairy bluebird *Vertebrate Zoology.* a colorful forest bird of the family Irenidae (in some classifications, the family Oriolidae) that is up to ten inches long with short legs and a slender bill, feeding on fruit, nectar, and some insects; found in southern Asia, Indonesia, and the Philippines where they are sometimes kept as caged birds. (So called because of their shimmering blue and violet colors.)

fairy chimney see EARTH PILLAR.

fairy ring *Plant Pathology.* a fungus disease of lawn grasses that appears as a fast-growing dark green ring or arc in the turf surrounding or surrounded by a ring of dead grass. (So called because it was once thought that the pattern it forms was left by dancing fairies.)

faithful functor *Mathematics.* a functor $T: C \rightarrow B$ is said to be **faithful** (or an imbedding) if, for every pair of objects $c, c' \in C$ and for every pair of parallel arrows $f_1, f_2: c \rightarrow c' \in C$ (arrows having the same domain and codomain), the equality $Tf_1 = Tf_2: Tc \rightarrow Tc'$ implies $f_1 = f_2$. The composition of faithful functors is faithful.

faithful module *Mathematics.* a (left or right) module A is faithful if its (left or right) annihilator is 0. A ring R is (left or right) primitive if there exists a simple faithful (left or right) R-module.

faithful representation *Mathematics.* a given representation of a group G is said to be faithful if the matrices corresponding to different elements of G are distinct; i.e., if G and the given representation group of G are isomorphic.

faithful species see EXCLUSIVE SPECIES.

Fajans determination *Analytical Chemistry.* a method used to determine the chloride concentration of a sample by titrating it with a standard solution of silver nitrate, using dichlorofluorescein as the indicator; the formation of a pinkish-red color indicates the end point.

fake *Naval Architecture.* **1.** a single turn of a coiled line. **2.** to lay down a line in a series of elongated bights so that it is completely exposed on the deck and will run out smoothly. Also, FAKE DOWN, FLAKE.

fake-process color *Graphic Arts.* any of several techniques for simulating four-color separation, as by overprinting a series of black-and-white negatives with different colored inks. Also, **fake-color process.**

fake set see FALSE SET, def. 1.

falcate *Biology.* curved like a sickle; hooked. *Astronomy.* having such a crescent appearance, such as the moon between New Phase and First Quarter.

falciform [fal′si fôrm] *Biology.* having the shape of a sickle or scythe.

falciform ligament *Anatomy.* a prolongation of the sacrotuberal ligament that extends along the inner border of the ramus of the ischium.

falciform ligament of the liver *Anatomy.* a sickle-shaped fold of the peritoneum on the ventral side of the liver that separates its right and left lobes and helps to attach it to the diaphragm; it extends from the coronary ligament of the liver to the umbilicus.

falciger *Invertebrate Zoology.* in polycheate worms, compound seta (bristles) with stout curved ends.

falciparum malaria *Medicine.* the most severe form of malaria, caused by the protozoa *Plasmodium falciparum* and characterized by symptoms including confusion, an enlarged spleen, stomach upset, and anemia; it is sometimes fatal.

falcon

falcon *Vertebrate Zoology.* **1.** any of several diurnal birds of prey of the family Falconidae, particularly of the genus *Falco*, having long, pointed wings and a notched bill. **2.** a female gyrfalcon or peregrine, or any of various hawks trained or adapted for use in the sport of hawking.

Falcon *Ordnance.* any of a series of air-to-air guided missiles that may be carried internally or externally on interceptor aircraft; some types of Falcon missiles were equipped with nuclear warheads.

Falconidae *Vertebrate Zoology.* a family of diurnal birds of prey of the order Falconiformes, now usually restricted to the long-winged, swift-flying falcons and the caracaras, but formerly including most hawks, eagles, buzzards, Old World vultures, and related forms.

Falconiformes *Vertebrate Zoology.* an order of chiefly diurnal flesh-eating birds having short, stout hooked bills and strong feet with four toes; the young are helpless at hatching and are fed in the nest; includes hawks, eagles, vultures, falcons, and related birds.

falculate *Zoology.* having a compressed, curved shape with a sharp point, like the claw of a cat.

Fales-Stuart windmill *Mechanical Engineering.* a windmill whose rotating component resembles a double-bladed airfoil propeller. Also, STUART WINDMILL.

Falkland Current *Oceanography.* a current flowing northward along the coast of Argentina, from 50°S to 35°S.

fall *Astronomy.* **1.** the season that begins at the autumnal equinox and ends at the winter solstice; autumn. **2.** a term for a meteorite whose fall to the ground is witnessed. *Mechanical Engineering.* in a hoisting tackle, the loose end of rope to which power is applied.

fallaway section *Space Technology*. on a launch vehicle or rocket, a structure that is designed to be separated and cast off during flight, often falling back to earth.

fallback *Geology*. any fragmental material initially ejected from an impact or explosion crater during formation, then almost immediately redeposited in, and partially filling, the crater. *Nucleonics*. the portion of material lifted into the air by an atomic explosion that ultimately returns to the detonation site. *Computer Technology*. a reserve system used in place of a host real-time system in case of failure or other emergency; often includes a backup database and software.

fallback switch *Telecommunications*. a physical switch used to reroute a communication link from a primary device to a backup device in the event of a primary device failure.

fall block *Mechanical Engineering*. a pulley block that rises and falls with a load.

fallen arch see TARSOPTOSIS.

faller *Mechanical Engineering*. a machine part whose operation depends on a falling action, as in certain spinning or stamping machines. *Textiles*. a device used to straighten worsted stock during combing.

fallfoot see TARSOPTOSIS.

falling *Cartography*. the distance measured along an established line from its intersection with a random line to a corner on which the random line was intended to close.

falling body *Mechanics*. a free body being accelerated by gravity and experiencing no other significant forces.

falling disease *Veterinary Medicine*. a terminal disease of animals caused by severe copper deficiency and manifested by collapse and heart failure.

falling-drop method *Physics*. a technique used to determine the density of a liquid by measuring the time required for a drop of the liquid to sink in a reference liquid of known density and viscosity.

falling film *Fluid Mechanics*. a liquid film that moves downward over a vertical surface, usually in laminar flow.

falling-film cooler *Engineering*. a liquid cooling system in which the cooling liquid moves down the outside of a tube, and hot process fluid flow moves upward inside the tube.

falling-film evaporator *Engineering*. a system in which the liquid evaporates as it flows down the inside surface of a tube.

falling-film still *Chemical Engineering*. a unit designed for molecular distillation, which incorporates a high evaporation rate and separation efficiency.

falling-sphere viscometer *Engineering*. a viscometer that determines the velocity of fluid from the velocity of fall of a spherical ball.

falling tide see EBB TIDE.

fall line *Geology*. **1.** a narrow zone in which a river suddenly descends, as in a waterfall. Also, FALL ZONE. **2.** a boundary zone between older, resistant rocks and younger, softer plain sediments.

fall of ground *Mining Engineering*. any rock falling from the roof into a mine opening.

Fallopian tube or **fallopian tube** [fə lō´pē ən] *Anatomy*. either of a pair of long, slender tubes that extend from the right and left cornu of the uterus to the region of the ovary of the same side, allowing for the passage of eggs from the ovaries to the uterus and of spermatozoa to the ovaries during fertilization. Also, UTERINE TUBE, SALPINX.

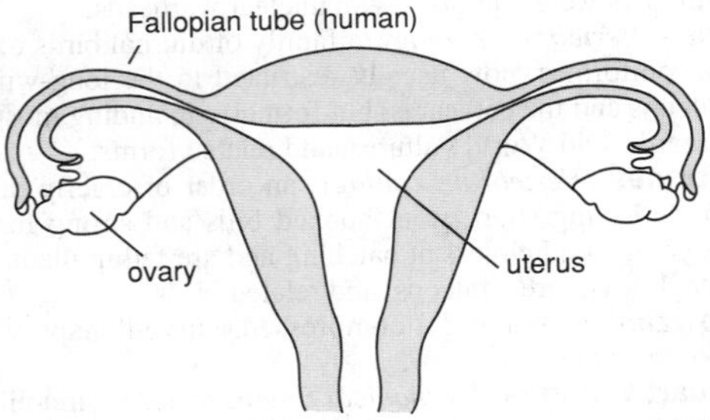

Fallopian tubes

Fallopio, Gabriele [fä lō´pē ō] 1523–1562, Italian anatomist and botanist; discovered and described the Fallopian tubes.

fallout *Nucleonics*. the particles containing radioactive materials that drop back to earth or remain suspended in the atmosphere following a nuclear explosion. Also, ATOMIC FALLOUT, RADIOACTIVE FALLOUT.

fallout area *Nucleonics*. the geographical area where radioactive materials settle following a nuclear explosion, principally determined by weather conditions.

fallout shelter *Civil Engineering*. a structure that is intended to give some protection against fallout radiation and other effects of a nuclear explosion; the maximum protection comes from shelters below ground. Also, RADIATION SHELTER.

fallout winds *Meteorology*. any winds in the troposphere that carry radioactive fallout materials.

fallow *Agriculture*. **1.** of or relating to farm land that is not in active use for the growing of crops. **2.** to remove land from active crop production.

fallow strip *Agronomy*. a strip of land that is left unplanted between rows of growing crops in order to prevent erosion.

fallow system or **fallow farming** *Agronomy*. the practice of leaving a certain area of cropland inactive during a growing season to allow it to retain nutrients and moisture for a crop the following year.

falls *Hydrology*. another name for a waterfall, especially in specific place names; e.g., Niagara Falls, Victoria Falls.

fall streak see VIRGA.

Fallstreifen see VIRGA.

fall time *Electricity*. the time required for the trailing edge of a pulse to decrease from a high to a low value; often the high and low values are the maximum amplitude and zero, and at other times they are 90% and 10% of the maximum amplitude.

fall time see PULSE DECAY TIME.

fall wind *Meteorology*. a cold, strong, downslope wind produced by an accumulation of cold air at high elevation.

fall zone see FALL LINE.

false acceptance see TYPE II ERROR.

false bearing *Cartography*. the difference between the true bearing and the back bearing caused by the convergence of meridians. *Building Engineering*. an unsupported beam, such as a window sill.

false bedding see CROSS-BEDDING.

false blossom *Plant Pathology*. **1.** a disease of cranberries caused by the fungus *Exobasidium oxycocci*, in which buds produce abnormal flowers that set no fruit. Also, ROSEBLOOM. **2.** a similar insect-transmitted viral disease. Also, WISCONSIN FALSE BLOSSOM.

false body *Physical Chemistry*. the property that causes certain colloidal substances, such as paints and inks, to solidify when left standing.

false bottom *Civil Engineering*. a temporary caisson bottom installed to add buoyancy. *Metallurgy*. in metalworking, a component inserted into a die to lengthen the die life. *Mining Engineering*. **1.** a bed of drift lying on top of other alluvial deposits, beneath which there may be a true bottom of wash resting directly on the bedrock. **2.** a flat, hexagonal or cylindrical iron die upon which ore is crushed in a stamp mill.

false branching *Microbiology*. an apparent branching of bacterial cells that normally grow in filaments, as in the cyanobacteria group, caused by the protrusion of a filament through its surrounding sheath or by the continued growth of cells separated by one or more nonviable cells.

false cirrus cloud *Meteorology*. a type of cirrus cloud whose optical thickness causes it to veil the sun and to appear grayish on the side away from the sun; often originating from the upper part of a cumulonimbus cloud. Also, CIRRUS SPISSATUS CLOUD, THUNDERSTORM CIRRUS CLOUD.

false cleavage *Geology*. **1.** a minor split in a rock, as opposed to its true or dominant cleavage. **2.** see FRACTURE CLEAVAGE.

False Cross *Astronomy*. four second-magnitude stars, Iota, Epsilon Carinae, Kappa, and Delta Velorum, that form a cross-shaped asterism sometimes confused with the constellation Crux, the Southern Cross.

false drumlin see ROCK DRUMLIN.

false easting *Cartography*. a value given to the central meridian of a coordinate system to avoid the inconvenience of using negative departs.

false form see PSEUDOMORPH.

false fruit see PSEUDOCARP.

false horizon *Navigation*. a line that lies above or below the true horizon and that can be mistaken for it.

false ice foot *Oceanography*. ice formed on a beach terrace just above the high-water line; generally caused by snow that melts further up the shore, but may also contain accretions of sea ice from spray or spring tides.

false keel *Naval Architecture*. an additional keel designed to protect the true keel or increase the vessel's draft.

false labor *Medicine*. a term for strong contractions resembling labor pains, which occur near term but are not accompanied by effacement and dilatation of the cervix.

false lapis see LAZULITE.

false ligament *Anatomy.* **1.** any suspensory ligament that is a peritoneal fold and is not of true ligamentous structure. **2.** a peritoneal connection between the vertex and sides of the bladder and the walls of the pelvis.
false-negative *Pathology.* a test result that wrongly excludes an individual from a diagnostic or other category.
false neisseriae *Bacteriology.* certain bacteria that are thought to belong to species *incertae sedis* in the genus *Neisseria* or to the subgenus *Branhamella.*
false northing *Cartography.* a value assigned to the origin of grid coordinates to avoid using negative coordinates.
false ogive *Ordnance.* a hollow cap added to the nose of a shell to improve streamlining; it may be rounded or pointed. Also, BALLISTIC CAP, WINDSHIELD.
false oolith see PSEUDO-OOLITH.
false origin *Cartography.* an arbitrary point to the south and west of a grid zone from which grid lines are numbered and grid distances are measured eastward and northward.
false-positive *Pathology.* a test result that wrongly assigns an individual to a diagnostic or other category.
false rafter *Building Engineering.* in roof construction, a short extension added to a principal rafter over an extension or cornice.
false rejection see TYPE I ERROR.
false retrieval *Computer Programming.* an unwanted or irrelevant reference retrieved during a library search.
false rib *Anatomy.* any of the lower five ribs on either side, which are not directly attached to the sternum; the first three pairs attach ventrally to ribs above, but the last two pairs are not attached at their ventral tips.
false ring *Botany.* an additional ring or partial ring of xylem formed in one growth season.
false set *Mining Engineering.* **1.** a light, temporary lagging set of timber supporting the side and roof lagging until the permanent set can be put in; the false set is then removed and used again in advance of the next permanent set. Also, FAKE SET. **2.** a temporary support for forepoles, used in driving a tunnel in soft ground. Also, HORSEHEAD.
false smut *Plant Pathology.* **1.** a palm tree disease caused by the fungus *Graphiola phoenicis,* characterized by small protruding cankers often surrounded by a yellow ring. **2.** see GREEN SMUT.
false stull *Mining Engineering.* a stull placed to provide support or reinforcement for a stull, prop, or other timber.
false target *Electronics.* the erroneous presence of a target on a radar screen that arises from a time delay.
false-target generator *Electronics.* a device that produces a delayed return signal on an enemy's radar frequency so that it will not accurately reveal a target location.
false under an interpretation *Artificial Intelligence.* of a logical formula, evaluating to the value False for a particular interpretation.
false warm sector *Meteorology.* the region, in a horizontal plane, between the occluded front and a secondary front of an occluded cyclone.
falsework *Civil Engineering.* any temporary support used to stabilize a structure until it can support itself.
falsify *Artificial Intelligence.* to cause a logical formula to have the value False.
faltung *Mathematics.* a family of functions that is closed under convolution; i.e., the convolution of any two members of the family is also a member of the family. Also, CONVOLUTION FAMILY.
falx *Anatomy.* a sickle-shaped organ or structure.
falx cerebelli *Anatomy.* a fold in the dura mater that lies in the posterior cerebellar notch.
falx cerebri *Anatomy.* a fold in the dura mater that lies in the longitudinal fissure between the two cerebral hemispheres.
falx inguinales *Anatomy.* the combined tendons of the transverse and internal oblique muscles going to the linea alba and pectineal line of the pubic bone.
famatinite *Mineralogy.* Cu_3SbS_4, a gray, opaque tetragonal mineral with a tinge of copper-red, usually massive, fine-grained in habit or as distinct equant crystals, having a specific gravity of 4.635 and a hardness of 3.5 on the Mohs scale; found in copper deposits.
FAME *Microbiology.* a fatty acid methyl ester that is prepared from the fatty acids of whole cells and is used to identify different microorganisms based on characteristic differences in cellular fatty acid content. (An acronym for fatty acid methyl ester.)
Famennian *Geology.* a European geologic stage of the uppermost Devonian period, occurring after the Frasnian and before the Tournaisian of the Carboniferous period.
familial [fə mil′yəl] *Biology.* relating to or characteristic of a family. *Systematics.* relating to a taxonomic family. *Medicine.* describing a disease or condition that occurs more often in a given family than would be expected by chance; it is usually but not necessarily inherited.
familial dysautonomia *Medicine.* a genetic disorder marked by feeding difficulties, absent corneal reflexes, indifference to pain, and other evidence of autonomic nervous system dysfunction.
familial hypercholesterolemia *Genetics.* a genetic defect inherited as a dominant allele on chromosome 14, characterized by a profusion of cholesterol in the cells and plasma of circulating blood; hyperlipoproteinemia.
familial Mediterranean fever *Medicine.* an inherited disease occurring mainly in people of Sephardic Jewish or Arabic ancestry, transmitted as an autosomal recessive trait, and characterized by recurring episodes of fever, chest pain, and rash.
familial polyposis *Medicine.* an inherited disorder characterized by multiple discrete adenomas of the colon.
family *Behavior.* a social unit in which one or more parents live with their offspring. *Systematics.* one of the principal or obligatory ranks in the taxonomic hierarchy, below the order and above the tribe or genus; the family name generally ends in *-aceae* in botany or *-idae* in zoology *Chemistry.* a group of elements having similar chemical properties, such as valency, solubility of salts, or reactivity. *Mathematics.* in a general set theoretic setting, a function x from an index set I to a set X. The range of x is called an indexed set and the value of the function x at i, called a **term of the family**, is denoted x_i. Thus the phrase "a family $\{A_i\}$ of subsets of X" is usually understood to refer to a function A from some set I of indices, into $P(X)$, the power set of X.
family formation *Design Engineering.* the process of reviewing geometry in an effort to define a family of parts.
family mold *Engineering.* a multicavity injection mold in which each cavity shapes a part of the whole product.
family practice *Medicine.* the medical specialty that is concerned with the planning and provision of the comprehensive primary health care of all members of a family, regardless of age or sex.
family practitioner *Medicine.* a doctor engaged in a family practice.
family therapy *Psychology.* therapy that involves an entire family as the subject rather than one individual; usually includes group meetings of the family with one or more therapists present.
famine fever see RELAPSING FEVER.
famphur *Organic Chemistry.* $C_{10}H_{16}NO_5PS_2$, a toxic crystalline powder; slightly soluble in water and very soluble in carbon tetrachloride and chloroform; melts at 55°C; a pesticide for lice and grubs in cattle.
fan *Mechanical Engineering.* **1.** any of various devices designed to produce a current of air by the movement of one or more broad, often rotating blades or vanes. **2.** a small vane designed to keep the blades of a windmill facing the direction of the wind. *Electromagnetism.* the volume of space periodically energized by a radar beam as it covers an established pattern. *Geology.* **1.** a low, gently sloping cone of sediments or detritus. **2.** see ALLUVIAL FAN. *Biology.* any structure that resembles an open fan, such as the tail feathers on a bird.
fan antenna *Electromagnetism.* an antenna consisting of two folded dipole elements lying in a common vertical plane with the apex at the bottom, and having a wide bandwidth in the vhf and uhf region.
fan beam *Electromagnetism.* a radio beam having an elliptical cross section with a large ratio of major to minor dimensions.
fan belt *Mechanical Engineering.* on a motor vehicle, a belt that is driven by the engine which turns a fan that blows cool air through the radiator (or, on an air-cooled engine, through the engine itself).
Fanconi's anemia *Genetics.* a rare recessive disorder characterized by hypoplasia of the bone marrow and patchy brown discoloration of the skin due to the deposition of melanin; associated with multiple congenital anomalies of the musculoskeletal and genitourinary systems. (Named for the Swiss pediatrician Guido *Fanconi.*)
Fanconi's syndrome *Genetics.* a group of acquired or inherited disorders marked by dysfunction of the proximal renal tubules, with increased excretion of amino acids, and bicarbonate and water loss.
fan cooling *Mechanical Engineering.* the use of an engine-driven fan to induce airflow through the radiator or engine of a motor vehicle.
fan cut *Engineering.* a cut in which holes of equal or greater length are drilled along a given stratum or horizontal plane, so that most of these sections will break away before the rest of the round is fired.
F and G's or **F & G's** *Graphic Arts.* unbound printed pages of a book that have been folded into signatures and arranged in the correct order for binding. (Short for folded and gathered.)

fan drift *Mining Engineering.* an enclosed airtight passage from a mine to an exhaust fan.

fan drilling *Engineering.* a drilling technique in which boreholes are drilled in several vertical and horizontal directions from a single-drill configuration.

fan efficiency *Mechanical Engineering.* the ratio of the useful power output of a fan to the power it receives, expressed as a percentage.

fan fold *Geology.* a fold in which both limbs are overturned in opposite directions.

fan-fold paper *Computer Technology.* a continuous sheet of paper that is perforated at page length intervals and is folded and stacked in a zigzag manner. Also, Z-FOLD PAPER.

fang *Vertebrate Zoology.* **1.** a long, sharp tooth with which a predator seizes, holds, or tears its prey. **2.** specifically, one of the grooved teeth of venomous snakes through which poison is injected. *Invertebrate Zoology.* a hollow toothlike structure in some spiders and centipedes with which they inject their prey with poison. *Anatomy.* **1.** a tooth that is sharp or pointed, such as a carnassial tooth. **2.** the root of a tooth.

fang bolt *Mechanical Devices.* a bolt composed of a triangularly shaped nut with sharp, protruding teeth that is used to attach metal to wood.

fanglomerate *Geology.* a conglomerate originally deposited in an alluvial fan, but cemented into solid rock since deposition.

fan-in *Electronics.* the number of inputs accepted by a logic circuit.

fanlight *Architecture.* a semicircular window over a door, with radiating sash bars resembling the ribs of a fan. Also, SUNBURST LIGHT.

fan-marker beacon *Navigation.* a directional, aeronautical radio beacon used to mark locations along an airway. The standard pattern is elliptical, though some use a kidney-shaped pattern.

fanning beam *Electromagnetism.* a narrow antenna beam that scans over a limited arc. Thus, **fanned-beam antenna.**

Fanning friction factor *Fluid Mechanics.* a dimensionless number symbolized by the product of the drop in pressure due to the friction and the diameter of the pipe, divided by the product of the kinetic energy of fluid per unit volume and the length of the pipe.

Fanning's equation *Fluid Mechanics.* a relation between the drop in pressure of a fluid in a pipe due to friction, the Reynolds number, flow rate, length and diameter of the pipe, the pipe roughness, and the gravitational acceleration.

Fanno flow *Fluid Mechanics.* an idealized gas flow in a pipe characterized by these assumptions: a perfect gas with constant specific heat and composition flows adiabatically in a pipe with constant cross section, and there are no devices in the system which may receive or deliver mechanical work.

Fano plane *Mathematics.* a system of seven lines and seven points such that each line has three points and each point lies on three lines. It may be represented by an equilateral triangle with an inscribed circle and the three angle bisectors forming the seven lines. The seven points are the three vertices of the triangle, the three midpoints of the triangle sides (which are also tangent to the circle), and the point in the center of the circle at which the bisectors meet.

fan-out *Electronics.* the number of separate input terminals (of the following stages in a circuit) that can be connected to the output terminal of a logic circuit.

fan ring *Mechanical Devices.* a metallic ring attached near the tips of a fan blade to improve the efficiency of the fan, principally used in air-cooled heat exchangers.

fan shaft *Mechanical Devices.* the axle upon which a fan is mounted. *Mining Engineering.* the ventilating shaft to which a mine fan is connected.

fan-shaped delta see ARCUATE DELTA.

fan shooting *Engineering.* the placing of seismic sensing devices in a fan-shaped array to detect irregularities in the arrival times of refracted rays; used to locate circular rock structures, such as salt domes.

fantail *Architecture.* a structure or structural member that has a number of radiating parts, such as an arch centering. *Naval Architecture.* **1.** on some vessels, the part of the stern that overhangs the water. **2.** the extreme stern of any vessel.

fantasm *Psychology.* a mental image of an absent or imaginary person or object.

fantasy *Psychology.* the experience of imagining some event or sequence of events; an imaginary situation.

Fantl unit *Biology.* a unit used for the standardization of thrombin, the enzyme that aids in blood clotting.

fan truss *Civil Engineering.* a truss whose struts are arranged like radiating lines.

fan vaulting *Architecture.* an elaborate vaulting system of the Perpendicular period in which identical ribs radiate like an unfolded fan.

FAO or **F.A.O.** Food and Agriculture Organization.

FAP *Oncology.* a tumor-suppressor gene that is found to be altered in carcinoma of the colon.

FAP fixed-action pattern.

FAR *Computer Science.* finite automaton recognizable.

farad *Electricity.* a basic unit of capacitance in the mks system; it is equivalent to the capacitance of a capacitor in which a charge of 1 coulomb produces a change of 1 volt in the potential difference between its terminals. (Named for Michael *Faraday.*)

faradaic or **Faradaic** [fâr´ə dā´ik] of or relating to Michael Faraday or his laws or theories. Also, FARADIC.

faradaic process *Physical Chemistry.* any of the fundamental processes in electrochemistry in which the passage of current at an electrode causes a chemical change. The extent of this change for a given amount of electric charge passed can be calculated by Faraday's laws.

faradaic rectification *Physical Chemistry.* the appearance of a DC component upon passage of an AC current at an electrode, because of nonlinearity of the current-potential behavior.

Faraday, Michael 1791–1867, English chemist and physicist; discovered field theory and electromagnetic induction; invented the dynamo; formulated Faraday's laws of electrolysis.

Michael Faraday

faraday *Physics.* a unit of electric charge, 9.64870×10^4 coulombs, required for 1 gram-equivalent of a substance to be liberated by electrolysis.

Faraday birefringence *Optics.* **1.** the refraction of left and right circularly polarized light that results as it passes through a medium that is parallel to an applied magnetic field. **2.** the difference between the indices of refraction under such conditions.

Faraday constant *Physical Chemistry.* a determination of the amount of electricity corresponding to one mole of electrons, given by $F = N_0 e = (6.02252 \times 10^{23})(1.6021 \times 10^{-19}) = 96{,}487$ C mole^{-1}, where N_0 is Avogadro's number and e is the charge on the electron; C is the coulombs. Also, ELECTROCHEMICAL CONSTANT.

Faraday cylinder *Electricity.* a metal cylinder that is placed around electrical equipment to protect the equipment from electromagnetic fields.

Faraday dark space *Electronics.* the relatively nonluminous portion of a glow-discharge cold-cathode tube that divides the negative glow from the positive column.

Faraday disk machine *Electromagnetism.* a device that demonstrates electromagnetic induction by rotating a copper disk between the poles of a magnet and inducing a radial voltage in the disk.

Faraday effect *Optics.* the rotation of the plane of polarization of linearly polarized light or other electromagnetic radiation when it passes through a transparent isotropic substance placed in a strong magnetic field. Also, KUNDT EFFECT, MAGNETO-OPTIC FARADAY EFFECT.

Faraday rotation *Astrophysics.* the rotation of the plane of polarization in linearly polarized light that has traversed a magnetic field.

Faraday shield *Electricity.* an electrostatic shield made of a series of parallel wires connected to a common conductor at one end that provides shielding while permitting the passage of electromagnetic waves. Also, **Faraday cage, Faraday screen.**

Faraday's law of electromagnetic induction *Electromagnetism.* a law stating that the electromotive force induced in a circuit is proportional to the time rate of magnetic flux change linked with the circuit.

Faraday's laws of electrolysis *Physical Chemistry.* **1.** the amount of chemical change produced by an electric current passing through a cell is proportional to the total quantity of the electric charge. **2.** the amount of chemical change produced in a substance by such an electric current is proportional to the equivalent weight of the substance.

faradic see FARADAIC.

faradic current *Electricity.* an intermittent and asymmetrical alternating current produced by the secondary winding of an induction coil operated by repeated interruption of a direct current in the primary winding.

faradization *Physiology.* the use of faradic current from an induction coil to stimulate muscles and nerves.

Farey sequence *Mathematics.* given positive integer N, the increasing sequence of fractions h/k with $0 \leq h \leq k$, h and k relatively prime, and $k \leq N$ is the Farey sequence of order N; used in the theory of partitions in analytic number theory.

far field *Acoustics.* **1.** a sound field in which free field spherical sound decreases 6 decibels for each doubling of radial distance. **2.** a region in the sound field that is much larger than a wavelength away from the source.

farina *Agronomy.* food product made from the middlings of hard wheat, except durum.

farinaceous *Biology.* being starchy or mealy, or appearing to have a starchy surface or texture. *Geology.* having a structure or texture that is mealy, soft, and easily pulverized. Also, MEALY.

Farinales *Botany.* a term used occasionally for an order of plants containing members that are usually grouped in the Commelinidae.

far infrared *Electromagnetism.* referring to electromagnetic radiation of wavelengths in the approximate range of 50 to 1000 micrometers.

far-infrared maser *Engineering.* a gas maser that produces a beam whose wavelength ranges from 100 to 500 micrometers.

far-infrared radiation *Electromagnetism.* electromagnetic radiation in the long-wavelength region of the infrared range: 50 to 1000×10^{-6} meters.

far-infrared spectrometer *Spectroscopy.* a spectrometer that requires a diffraction grating for spectrographic analysis.

Farinosae *Botany.* an order of monocotyledoneae, usually herbaceous, having cyclic flowers and a mealy endosperm.

farinose *Agronomy.* producing or covered with a white powdery dust, or covered with powdery hairs that can be brushed off like dust.

farm *Agriculture.* **1.** any specific and extensive area of land used to produce crops, animals, or both, along with its accompanying buildings and equipment. **2.** officially in the U.S., a place that produces $1000 or more of agricultural products annually. **3.** to cultivate a tract of land for agricultural purposes. *Food Engineering.* an area of water used for the controlled growth of seafood: a trout *farm,* a shrimp *farm.*

farmer *Agriculture.* any person who owns or manages a farm, or who earns a living by farming.

farmer's lung *Medicine.* an acute and chronic inflammatory reaction in the lungs caused by breathing dusts from moldy hay or grain.

farming *Agriculture.* the practice of agriculture.

farmstead *Agriculture.* the land and buildings of a farm.

farnesol *Biochemistry.* **1.** a plant extract that is used in perfumes. **2.** an intermediate step in the synthesis of cholesterol from mevalonic acid in vertebrates.

Farnsworth, Philo T. 1906–1971, American inventor; a pioneer in television technology; invented the image dissector.

Farnsworth dissector tube see DISSECTOR TUBE.

far point *Optics.* the farthest point on the axis of the eye at which an object can be clearly seen when the crystalline lens of the eye is relaxed. Also, PUNCTUM REMOTUM.

farringtonite *Mineralogy.* $Mg_3(PO_4)_2$, a colorless to yellow monoclinic mineral found only in meteorites, having a specific gravity of 2.74 and an undetermined hardness.

farrow *Agriculture.* **1.** to give birth to a litter of pigs. **2.** a cow that is not pregnant or that can no longer bear young.

Farr technique *Immunology.* a process that is used to measure the antigen-binding capacity of an antibody.

Farr test *Immunology.* a method of measuring antibodies in absolute amounts with the use of radioactively labeled reactants; it is based on the capacity of an antibody to combine with an antigen.

farside *Astronomy.* the hemisphere of the moon that faces away from the earth.

far-ultraviolet radiation *Electromagnetism.* electromagnetic radiation in the short-wavelength region of the ultraviolet range: 50 to 200×10^{-9} meters.

far vane *Navigation.* in a sighting instrument such as a pelorus that uses two pins or vanes, the vane farthest from the observer's eye.

FAS or **F.A.S.** Federation of American Scientists.

fascia *Anatomy.* an enveloping sheet of fibrous tissue, lying beneath the skin or enclosing muscles or groups of muscle fibers. *Building Engineering.* a flat, broad, horizontal board that covers the joint between the top of a wall and the projecting eaves. Also, **fascia board.**

fascia profunda see DEEP FASCIA.

fasciate *Botany.* describing plant parts, such as stems, that have been mashed or fused together by either an outside force or an abnormality of growth.

fasciation *Plant Pathology.* a plant malformation caused by disorganized tissue development, resulting in thin, flat, and often curved shoots and aborted or misshapen leaves.

fascicle *Botany.* a tuft or loose cluster of fibers, leaves, stems, or other plant parts.

fascicular cambium *Botany.* in a vascular bundle, the cells originating from procambium (flat strand) of cambium in between the xylem and the phloem.

fasciculate *Botany.* having bundles, tufts, or clusters of like plant parts growing out from the same area.

fasciculation potential *Physiology.* the electrical current produced by muscle twitching that involves simultaneous contraction of contiguous groups of muscle fibers.

fasciculus *Anatomy.* a small bundle of nerve, muscle, or tendon fibers, especially a bundle or group of nerve fibers that are associated because of their function.

fascine *Civil Engineering.* a cylindrical bundle of brushwood, typically 1–2 feet in diameter and 10–20 feet in length; used for facing on seawalls or riverbanks, as a dam in an estuary, or to protect a bridge, dike, or pier foundation from erosion.

Fasciola hepatica *Invertebrate Zoology.* the liver fluke, a trematode worm that lives as an adult in the liver of sheep, cattle, and sometimes humans; larvae develop in water snails.

fasciole *Invertebrate Zoology.* a band of ciliated spines on the test of certain sea urchins, believed to produce water currents and to secrete mucus for burrow maintenance.

fascioliasis *Medicine.* an infection found in many parts of the world, common in the southern and western United States; caused by the liver fluke *Fasciola hepatica,* and marked by stomach pain, fever, jaundice, hives, and diarrhea.

fasciolopsiasis *Medicine.* an infection of the intestines, common in the Far East, caused by the liver fluke *Fasciolopsis buski* and marked by constipaion, fluid pooling, and stomach pain.

Fasciolopsis buski *Invertebrate Zoology.* a trematode worm native to the Far East that parasitizes humans and pigs; its larvae develop in water snails.

fascioscapulohumeral muscular dystrophy *Medicine.* a relatively benign, autosomal dominant form of muscular dystrophy, marked by a wasting of the skeletal muscles, especially the muscles of the face, shoulders, and upper arms. Most patients have a normal life span.

fast *Graphic Arts.* of or relating to film or other material that is highly sensitive to light. *Materials Science.* **1.** of a dye or pigment, not harmed by exposure to environmental conditions such as light and air. **2.** of an explosive, detonating quickly or easily.

fast-access storage *Computer Technology.* the component of computer memory from which data is most readily accessible.

fast axis *Optics.* in the vibration of a light wave passing through a birefringent material, the vector having a low refractive index; the slow axis, which is perpendicular to the fast axis, is the vector with a high refractive index.

fast break *Metallurgy.* the interruption of the current that activates the electromagnet during magnetic particle inspection tests.

fast breeder reactor *Nucleonics.* a type of reactor in which fast moving neutrons fission highly enriched fuel and the excess neutrons are used to turn fertile material into fissionable isotopes at a breeding ratio of 1.0 or higher.

fast-burst reactor *Nucleonics.* a reactor that generates neutrons moving at microsecond pulses; used in biomedical research.
fast chemical reaction see FAST REACTION.
fast coupling *Mechanical Engineering.* a fixed or flexible coupling that permanently connects two shafts.
fast-delay detonation *Engineering.* a detonation operation that employs a blasting timer or millisecond delay caps.
fast effect *Nucleonics.* an increase in neutrons that arises from the activity of fast neutrons in a thermal reactor.
fast ejection *Astronomy.* an especially energetic part of a solar flare in which material is ejected as a single mass.
fastener *Mechanical Devices.* any of various devices or attachments, such as a snap or zipper, used to join two objects or parts that are sometimes kept separate.
fastening *Mechanical Devices.* something that fastens, such as a nail, screw, clasp, or lock. *Robotics.* specifically, the mechanical joining of parts in a robotic operation.
fast fission *Nuclear Physics.* a fission reaction, as of uranium-238, that occurs as a result of bombardment by fast neutrons with energy in excess of 1 MeV.
fast-fission factor *Nucleonics.* the ratio between the number of thermal neutrons generated by nuclear fission and the number of neutrons generated by all other types of energy in the reactor.
fast forward *Mechanical Engineering.* **1.** to wind a cassette tape forward very rapidly. **2.** the process of such winding.
fast Fourier transform *Acoustics.* the electronic processing of a signal generated by sound waves striking a transducer, whereby the signal is transformed into a digital time series for spectrum analysis. *Mathematics.* one of several algorithms for computing the discrete Fourier transform at N points. N is taken to be a power of two (by padding up the data set with zeros if necessary) and the discrete Fourier transform of length N is recursively rewritten as the sum of discrete Fourier transforms of length $N/2, N/4, \ldots, N/2^k, \ldots$, until the data have been subdivided down to transforms of length one. This vastly reduces the number of arithmetic calculations required. *Computer Science.* specifically, the algorithm, due to Cooley and Tukey, for rapidly computing the discrete Fourier transform of a set of data values; often implemented as a subroutine or in special-purpose hardware. Also, DISCRETE FOURIER TRANSFORM.
fast ice *Hydrology.* **1.** pack ice that is attached to the shore at the place where it originally formed. **2.** a term that is loosely used to denote any sea ice attached to the shore (ice foot or ice shelf), beached (shore ice), or frozen to the bottom (anchor ice).
fastigiate *Botany.* describing branches that are erect, more or less parallel to the main stem, and growing close together.
fast ion see SMALL ION.
fast-joint *Engineering.* of or relating to a joint with an immovably fastened pin.
fast neutron *Nucleonics.* a neutron with a kinetic energy that greatly exceeds some arbitrary lower limit.
fast-neutron spectrometry *Nuclear Physics.* a technique used to analyze a wide variety of nuclear reactions that involves bombarding the nucleus with fast neutrons, then observing the resonance in the reactions or the range of the radiation emitted.
fast nova *Astronomy.* a nova that declines from peak brightness over a period of weeks or months.
fast pin *Engineering.* a pin that is designed to fasten permanently, as the pin in a fast joint.
fastpin hinge *Mechanical Devices.* a hinge with a screw, pin, or nail permanently secured to a fixture such as an immovable rod that holds the two flaps of the hinge together. Also, **fast-joint hinge, fast hinge.**
fast reaction *Physical Chemistry.* a term for a chemical reaction that occurs so rapidly, under half a millisecond, that it can only be observed using a special measuring device for the experiment, such as a single-diaphragm shock tube. Also, FAST CHEMICAL REACTION.
fast reactor *Nucleonics.* a reactor in which the fission process is sustained by the absorption of fast neutrons rather than slow-moving neutrons by the nucleus; such reactors have few if any moderators such as water to slow the fission process.
fast-time constant *Electricity.* a quick response to input changes, such as resistance and capacitance, so that capacitor discharge through a resistor can occur rapidly. *Electronics.* an antijamming device used in radar video-amplifier circuits that emphasizes short-duration signals and discriminates against low-frequency components of clutter, thus suppressing rain and snow echoes. Also, **fast-time circuit.**
fast-time scale *Computer Programming.* in data processing, a time scale factor of less than one.
fast-transition theory *Paleontology.* an explanation for the absence in the fossil record of a long succession of pre-Cambrian ancestors for modern life forms, on the basis that the transition to modern forms actually took place relatively quickly (in geologic time).
fast-vibration direction *Optics.* the direction of the vector traveling with the greatest velocity in a light wave passing through a birefringent material.
fast-wave sleep see NREM SLEEP.
fat *Anatomy.* adipose tissue composed of cells containing glycerol and fatty acid; white or yellowish tissue that forms soft pads between various organs of the body, serves to smooth and round out bodily contours, and furnishes a reserve supply of energy. *Biochemistry.* one of a group of glycerides of higher fatty acids, such as palmitic acid, stearic acid, and oleic acid, that is an essential component of the human diet.
fatal *Medicine.* causing death or leading inevitably to death.
fatal error *Computer Programming.* a programming error or bug that prevents the compiler from generating the machine language program needed for execution.
Fata Morgana [fä′tə môr gä′nə] *Optics.* a complex mirage characterized by multiple distorted, and often elongated, images of a relatively flat object, as when a distant horizon appears as a mountain range or a castle. (Another version of the name *Morgan Le Fay,* a fairy associated with castles in the legends of King Arthur.)
fat body *Invertebrate Zoology.* the fatty tissue filling most of an insect body, acting as a food store. *Vertebrate Zoology.* a lobulated mass of fatty tissue attached to each genital gland in amphibians.
fat cell *Histology.* the principle cell of adipose tissue, containing one or more droplets of fat.
fat dye *Materials.* an oil-soluble dye used in coloring candles and waxy substances.
fate map *Developmental Biology.* a plan of a blastula or early gastrula stage of an embryo, showing which tissues or organs each part will normally give rise to.
fat embolus *Medicine.* a body of fat in the circulation that can block an artery; it may follow the breaking of a long bone or an injury to adipose tissue.
father-daughter complex see GRISELDA COMPLEX.
father figure *Psychology.* a person who substitutes in one's mind for one's real father and who is the focus of emotional attitudes normally directed toward one's father. Also, **father image, father substitute.**
father file *Computer Programming.* the second generation master file of the grandfather-father-son file updating concept; the current master file is termed the son, the previous version is the father, and the oldest is the grandfather which will be overwritten to become the son at the next update process.
fathogram *Oceanography.* a graphic representation of a bottom profile as measured by an echo-sounding device.
fathom *Oceanography.* a marine unit of measure, originally the distance from fingertip to fingertip of a man's outstretched arms, but now generally taken to be 6 feet or 1.8288 meters.
fathometer *Acoustics.* a device used to determine the depth of water by measuring the time interval between a transmission of an acoustic pulse and reception of its bottom echo. Also, ECHO SOUNDER, SONIC-DEPTH FINDER. *Navigation.* **Fathometer.** the trade name for a particular brand of echo sounder.
fatigue *Physiology.* **1.** a state of increased discomfort and decreased efficiency resulting from prolonged or excessive exertion. **2.** the loss of power or of the capacity to respond to stimulation, as in a muscle that is exercised until it loses its ability to contract. *Psychology.* see BATTLE FATIGUE; BEHAVIORAL FATIGUE. *Mechanics.* the progressive failure of a material due to changes in material properties resulting from repeated stress. *Electronics.* the reduction in the efficiency of light-sensitive material over time. *Materials Science.* the application of stresses which vary cyclically with time.
fatigue allowance *Industrial Engineering.* an adjustment built into production time schedules to allow for worker fatigue.
fatigue factor *Industrial Engineering.* a factor included in job timing or worker compensation adjustment to allow for worker fatigue.
fatigue failure *Metallurgy.* the failure in a metal due to a large number of applications of cyclically varying stresses, usually below the yield strength of the material.
fatigue life *Mechanics.* the number of cycles of stress a material can withstand at a specified range rate and temperature before failure.

fatigue limit *Mechanics.* the highest stress range or amplitude that a material can be expected to withstand for an infinite number of cycles without failure. Also, ENDURANCE LIMIT.

fatigue notch factor *Metallurgy.* the ratio of the fatigue limits of an unnotched and a notched test specimen.

fatigue notch sensitivity *Metallurgy.* the propensity of a material to have a high fatigue notch factor.

fatigue ratio *Mechanics.* the fatigue limit divided by the static tensile strength of a material, for a given component of stress. Also, ENDURANCE RATIO.

fatigue strength *Mechanics.* the greatest stress that a material can withstand for a particular number of cycles without failure. Also, ENDURANCE STRENGTH. *Material Science.* see ENDURANCE LIMIT.

fatigue test *Engineering.* a test that establishes the interval of alternating stress that a material can endure without breaking. Such tests are usually operated to failure.

fat liquoring agent *Materials Science.* an oil-in-water emulsion that is made soluble by a dispersing agent; made from a raw oil such as neatsfoot and used especially to replace the natural oils lost in tanning.

fat necrosis *Medicine.* a condition in which the neutral fats in the cells of adipose tissue are split into fatty acids and glycerol, usually as a result of trauma, and affecting subcutaneous fat deposits, particularly in the female breast.

Fatou-Lebesgue theorem *Mathematics.* let X be a measure space with measure μ and $\{f_n\}$ a sequence of nonnegative, extended real-valued, measurable functions defined almost everywhere on X. Suppose that there exists a Lebesgue-integrable function s defined almost everywhere in X with finite integral and such that $|f_n(x)| \le s(x)$ almost everywhere on X. Then (a) Fatou's lemma holds, and (b)

$$\int_X (\lim_{n\to\infty} \sup f_n)\, d\mu \ge \lim_{n\to\infty} \sup \int_X f_n\, d\mu.$$

If, in addition, $\lim_{n\to\infty} f_n(x)$ exists for almost all x in X, then (c) $\lim_{n\to\infty} \int_X f_n d\mu$ exists and

$$\int_X \lim_{n\to\infty} f_n d\mu = \lim_{n\to\infty} \int_X f_n d\mu.$$

Fatou's lemma *Mathematics* . let X be a measure space with measure μ and $\{f_n\}$ a sequence of nonnegative, extended real-valued, measurable functions on X. Then the following inequality holds:

$$\int_X (\lim_{n\to\infty} \inf f_n)\, d\mu \le \lim_{n\to\infty} \inf \int_X f_n\, d\mu.$$

fat-soluble vitamins *Nutrition.* the vitamins (A, D, E, and K) that are soluble in fat solvents (but not in water) and are absorbed along with dietary fats; they are not normally excreted in urine, but are stored in the body in moderate amounts.

fatty acid *Organic Chemistry.* $CH_3(CH_2)_xCOOH$, any of a variety of monobasic acids obtained or derived from animal and vegetable fats and oils used as lubricants, in cooking and food engineering, and in the production of soaps, detergents, and cosmetics.

fatty acid-derived hormones *Endocrinology.* any of the eicosanoids: prostaglandins, leukotrienes, lipoxins, thromboxanes.

fatty acid peroxidase *Enzymology.* an enzyme of the oxidoreductase class that catalyzes the oxidation of fatty acids to carbon dioxide in germinating plant seeds.

fatty acid synthetase *Enzymology.* an enzyme complex catalyzing the synthesis of long-chain fatty acids.

fatty acyl carnitine *Biochemistry.* the substance that permits fatty acids to traverse the mitochondrial membrane, produced from a reaction involving fatty acyl-coenzyme A and carnitine acyltransferase.

fatty acyl-coenzyme A *Enzymology.* an activated form of fatty acids produced by the enzyme acyl-coenzyme A synthetase. Also, ACYL-COENZYME A.

fatty alcohol *Organic Chemistry.* $CH_3(CH_2)_xOH$, any of a group of high-molecular-weight, straight-chain primary alcohols that are produced synthetically or derived from natural oils; used as a solvent in the production of pharmaceuticals, cosmetics, plastics, and detergents.

fatty amine *Organic Chemistry.* RCH_2NH_2, a normal aliphatic amine derived from natural fats and oils; used in medicine and in rubber manufacture.

fatty ester *Organic Chemistry.* $RCOOR'$, a fatty acid in which the alkyl group (R') of a monohydric alcohol replaces the active hydrogen; produced by esterifying a fatty acid with methanol or another alcohol.

fatty infiltration *Physiology.* **1.** a deposit of fat in the tissues, especially between the cells **2.** the presence of fat vacuoles in the cytoplasm of the cells of an organ.

fatty metamorphosis *Medicine.* a retrogressive process characterized by the appearance of visible fat within cells that do not ordinarily contain visible fat.

fatty nitrile *Organic Chemistry.* RCN, a high-molecular-weight ester of hydrogen cyanide; used in lube oil additives and plasticizers.

faucal *Biology.* of or relating to the throat, especially to the cavity at the back of the mouth.

fauces *Anatomy.* the passage from the mouth to the pharynx that is surrounded by the soft palate, the palatoglossal and palatopharyngeal arches, and the base of the tongue.

faucet *Engineering.* a fixture that allows water or another liquid to be drawn from a pipe or vessel.

Faugeron kiln *Engineering.* a coal-fired kiln in which a series of walls divides a tunnel into individual chambers; used to fire feldspathic porcelain.

faujasite *Mineralogy.* $(Na_2,Ca)Al_2Si_4O_{12}\cdot 8H_2O$, a colorless to white cubic mineral of the zeolite group occurring as octahedral crystals, having a specific gravity of 1.92 to 1.93 and a hardness of 5 on the Mohs scale.

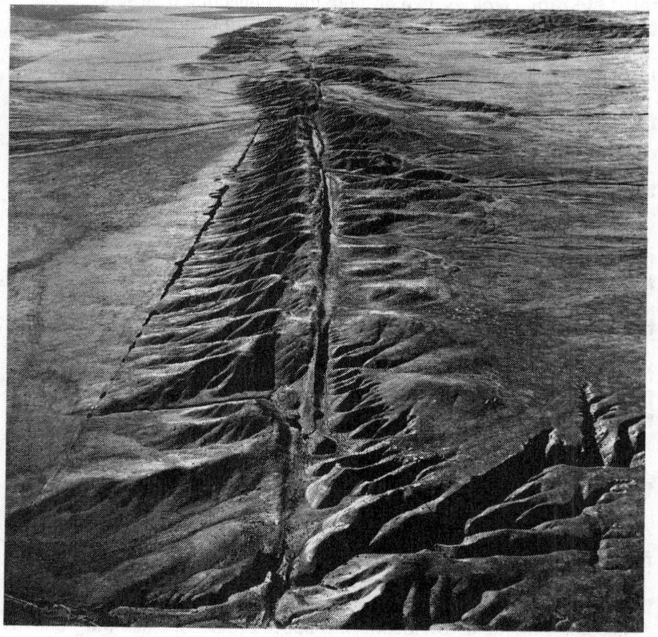

fault

fault *Geology.* a rock fracture along which movement or displacement in the plane of the fracture has taken place. *Electricity.* **1.** a defective region in a circuit or device. **2.** a failure in a circuit or device, due to an open circuit, short circuit, or ground in a circuit component or line. Also, ELECTRICAL FAULT, FAULTING.

fault basin *Geology.* a depressed area separated from the surrounding region by faults.

fault block *Geology.* a mass of rock bordered completely or partially by faults and behaving as a unit during tectonic activity or faulting.

fault-block mountain see BLOCK MOUNTAIN.

fault breccia *Geology.* angular fragments of rock formed by the shattering, shearing, crushing, or grinding together of rocks during faulting. Also, DISLOCATION BRECCIA.

fault cliff see FAULT SCARP.

fault electrode current *Electricity.* the peak current flowing through an electrode under fault conditions. Also, SURGE ELECTRODE CURRENT.

fault escarpment see FAULT SCARP.

faultfinder *Engineering.* a test set that identifies trouble spots in communications systems.

fault gouge *Geology.* any soft, uncemented, finely pulverized rock material formed by the crushing and grinding of rock material as a fault develops, filling or partly filling a fault zone. Also, CLAY GOUGE, SELVAGE.

faulting *Geology.* the action or process of fracturing and displacement that produces a fault. *Electricity.* see FAULT.

fault line *Geology.* the intersection of a fault with the ground surface, usually varying in form from straight to moderately sinuous. Also, FAULT TRACE.

fault-line scarp *Geology.* a steep cliff formed by displacement along a fault line.

fault masking *Computer Technology.* a method of fault recovery that uses a structural redundancy technique to mask faults within a set of redundant modules.

fault plane *Geology.* the surface along which rock masses on opposite sides of a fault rub against each other.

fault rock *Geology.* the crushed rock produced by the friction of the two walls of a fault rubbing against each other.

fault scarp *Geology.* an escarpment of any height produced by movement along one side of a fault. Also, FAULT ESCARPMENT, FAULT CLIFF.

fault set *Geology.* a group of parallel or near-parallel faults associated with a particular deformational event.

fault strike *Geology.* the angle, with respect to north, at which the fault plane intersects a horizontal plane.

fault surface the surface of a fracture without appreciable curvature along which displacement has taken place.

fault system *Geology.* two or more interconnecting fault sets.

fault terrace *Geology.* a step or bench on a hillside, formed by the displacement of two parallel or nearly parallel faults.

fault throw *Geology.* the amount of vertical displacement of blocks due to faulting.

fault tolerance *Engineering.* the capability of a system to perform according to design specifications regardless of changes in its internal structure or in the external environment.

fault-tolerant system *Computer Technology.* a system designed to continue successful operation in spite of hardware faults, either by automatic recovery or through backup redundant modules.

fault trace see FAULT LINE.

fault trap *Geology.* an oil or gas reservoir formed by the presence of one or more faults underground.

fault valley *Geology.* a small, narrow valley produced by the displacement of a fault.

fault vein *Geology.* a band of minerals deposited in a fault fissure.

fault wall *Geology.* the body of rock at the fault plane.

fault zone *Geology.* a generally broad area composed of many small, closely spaced rock fractures, or such an area composed of breccia or fault gouge. Also, DISTRIBUTED FAULT.

fauna [fô´nə] *Ecology.* the animal life that is present in a particular region or habitat or at a particular time. *Zoology.* animals in general, or the animals of any given region or time period.

faunal [fô´nəl] *Biology.* of or relating to fauna; having to do with animals or animal life.

faunal analysis *Archaeology.* the study of animal remains in an archaeological site, as by identifying bones or shells, examining butcher marks, and so on.

faunal extinction *Evolution.* the death of every member of a number of diverse animal groups due to global ecological circumstances that suggest a common or related cause. Also, MASS EXTINCTION.

faunal province *Ecology.* a smaller division of a faunal region, containing a distinct fauna not found in adjacent areas.

faunal region *Ecology.* a division of the animal kingdom that is determined by geographic and environmental barriers by which certain animal communities are limited, such as giraffes in Africa and beavers in North America.

faunalturbation *Archaeology.* a disturbance of the soil surface by animals, especially by the burrowing and tunneling of gophers, mice, rabbits, and the like.

faunal zone *Paleontology.* a period of time identified by a characteristic guide fauna. *Geology.* see FAUNIZONE.

faunistics *Ecology.* the study or classification of all the animals of a particular area.

faunizone *Geology.* a body of strata characterized by a distinctive assemblage of animal fossils. Also, FAUNAL ZONE.

faunula *Ecology.* the animal population of a limited area or a specific habitat.

faunule *Paleontology.* a small community of fauna, especially a group of animal fossils found only in a single stratum or in a succession of thin strata.

Faure plate [fôr] *Electricity.* a storage battery plate that consists of a lead grid containing a chemical electrolytic paste; used in automotive applications. Thus, **Faure storage battery.**

Fauveliopsida *Invertebrate Zoology.* a flabelligeridad family of polychaete annelids with a single genus and about 11 species of burrowing, sluggish nonselective marine bottom deposit feeders.

favism *Medicine.* an acute hemolytic anemia caused by eating the beans or breathing in the pollen from the *Vicia faba* plant. It results from a hereditary deficiency of glucose-6-phosphate dehydrogenase and occurs primarily in persons of southern Italian extraction.

favorable current *Navigation.* a current setting in approximately the same direction as a vessel's course.

favorable wind *Navigation.* a wind that is blowing in approximately the direction of a craft's course.

Favorskii rearrangement *Organic Chemistry.* a reaction in which α-halogenated ketones are rearranged to carboxylic acids or thin derivatures in the presence of bases.

Favosites *Paleontology.* a genus of tabulate corals in the extinct family Favositidae, extant through most of the Paleozoic but extinguished at the end of the Permian; important as guide fossils in petroleum geology.

favositid *Paleontology.* relating to or being a member of the family Favositidae.

Favositidae *Paleontology.* a family of colonial corals in the extinct order Tabulata and subclass Zoantharia; distinguished by mural pores.

favus *Medicine.* a distinctive ringworm infection, usually of the scalp, caused by fungi of the genus *Trichophyton* and marked by thick, yellow crusts, permanent scars, and loss of hair. Also, HONEYCOMB RINGWORM. *Building Engineering.* a tile or slab of marble that is cut in a hexagonal shape; produces a honeycomb pattern when used in paving.

FAX or **Fax** [faks] *Telecommunications.* **1.** a communications system for the transmission and reception of data in graphic form over channels having a bandwidth lower than is required for video signals. An electron beam scans the image of the document to be sent and composes a digital or analog signal representing the brightness of the area under the scanning beam. The signal is then transmitted to the receiver, which reproduces the image by photographic, thermal, or xerographic techniques. **2.** also, **fax.** to send a message by means of such a system. *Electronics.* also, **fax machine.** the equipment used in this process. *Graphic Arts.* the graphic material produced by this process. (Short for *facsimile.*)

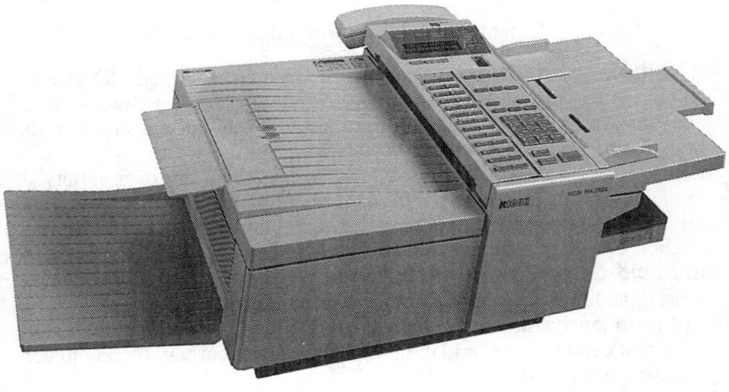

FAX machine

FAX chart *Meteorology.* a graphical representation of weather data reproduced by FAX equipment and transmitted from a central weather station to individual stations. Also, **FAX map.**

FAX receiver *Electronics.* a device that converts signals from a wire or radio channel into a picture or written material.

FAX recorder *Electronics.* the component within a FAX receiver that transfers the image onto paper.

FAX signal *Telecommunications.* a signal sent from a FAX transmitter, representing the brightness of the area under the scanning beam.

FAX signal level *Electronics.* a measure of the highest power level in a FAX system at a given point.

FAX synchronizing *Electronics.* the process by which a FAX device is able to receive and record data at a consistent rate of speed.

FAX telegraph *Telecommunications.* a telegraph system in which FAX technology is used to send documents.

FAX transmitter *Electronics.* an apparatus that converts graphic material, such as a picture or a text, into signals and sends them over a communication system, such as telephone lines.

fayalite *Mineralogy.* $Fe_2^{+2}SiO_4$, a greenish-yellow to brown orthorhombic mineral of the olivine group occurring in massive forms and as tabular crystals, having a specific gravity of 4.32 to 4.39 and a hardness of 6.5 to 7 on the Mohs scale; found in small amounts in acid and alkaline volcanic and plutonic rocks.

Faye anomaly see FREE-AIR GRAVITY ANOMALY.

faying surface *Metallurgy.* the interface between two metallic parts that are to be joined.

Fayum deposit *Anthropology.* an important site in the desert near Cairo, Egypt, that has yielded many primate fossils of the Oligocene such as *Apidium, Propliopithecus,* and *Aegyptopithecus.*

FB fixed-block.

FBA fixed-block architecture.

F band *Solid-State Physics.* an optical absorption band corresponding to excitation of an F center in an alkali-halide crystal.

fbr fast breeder reactor.

FC or **f.c.** footcandle.

FCC or **F.C.C.** Federal Communications Commission.

F center *Solid-State Physics.* an electron trapped at a negative ion vacancy in an ionic crystal; formed by the release of electrons by irradiation with X-rays or by an excess of anions in the crystal.

Fc fragment *Immunology.* the part of an immunoglobulin molecule that remains after it has been divided by the enzyme papain and has produced two identical Fab fragments.

FCFS *Computer Science.* first come, first served.

FCO *Aviation.* the airport code for Fiumicino (Leonardo da Vinci), Airport, Rome, Italy.

F corona *Astronomy.* the outer part of the sun's corona, consisting of dust that reflects sunlight.

Fc receptor *Immunology.* a binding site found on the plasma membrane of some cells that binds the Fc fragments of immunoglobulin.

FDA or **F.D.A.** Food and Drug Administation.

F display *Electronics.* a rectangular radar display with cross hairs on the scope face to aid in aiming the system; a target appears as a centralized blip when the radar antenna is aimed at it; azimuth and elevation aiming errors are indicated respectively by horizontal and vertical displacement of the blip. Also, F INDICATOR, F SCAN.

F distribution *Statistics.* any member of a family of distributions of continuous random variables formed by the ratio of two independent chi-square distributions, each divided by its degrees of freedom.

fd phage *Virology.* a virus composed of flexuous rods that is specific for bacteria containing the F plasmid; the proposed type member of the genus *Inovirus.*

F duction *Genetics.* sexduction in which the donor is mobilized by an F plasmid, or the transfer is promoted by an F' plasmid.

Fe the chemical symbol for iron. (From *ferrum,* the Latin word for this substance.)

fear *Behavior.* a motivational state in which the perception of danger produces defensive or escape behavior. *Psychology.* a strong and unpleasant emotional state brought about by the threat of danger, pain, or suffering.

fear hierarchy see ANXIETY HIERARCHY.

fear-induced aggression *Behavior.* aggression that occurs when an animal experiences fear upon being surprised or trapped.

feasibility study *Engineering.* a study conducted to determine the probability that a particular plan or system can be successfully accomplished.

feasible method see INTERACTION PREDICTION METHOD.

feasible solution *Mathematics.* any solution to a linear programming problem that does not violate any of the problem constraints.

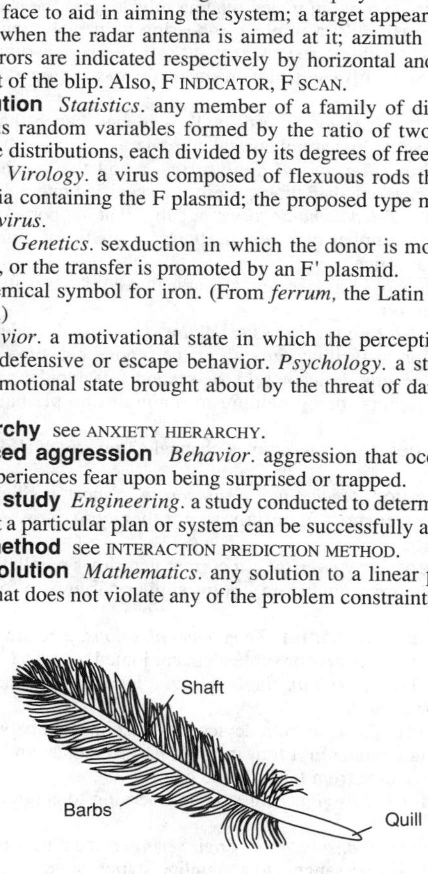

feather

feather *Vertebrate Zoology.* **1.** see BARB. **2.** a feathery tuft or fringe of hair, as in the fringe of long hair on the legs of some dogs and horses. *Aviation.* **1.** to change the pitch of a propeller so that the chords of the blades are virtually parallel to the line of flight. **2.** to turn off an engine while in flight. *Building Engineering.* a thin strip of wood used to join the edges of two adjacent butted boards. *Mechanical Engineering.* a rectangular key that connects the keyways of a shaft and a hub (e.g., of a gear) so that the hub can slide axially; used to adjust the position of the hub along the shaft. Also, **feather key.**

Feather analysis *Nucleonics.* a technique that approximates the range of the beta rays within the beta-gamma spectrum by comparing its absorption curve to the absorption curve of a pure beta emitter; used to determine the range of beta rays in aluminum.

featherback a freshwater fish of the family Notopteridae, characterized by a small, feathery dorsal fin and a very long anal fin; found in Asia and western Africa.

featheredge *Design Engineering.* a wooden tool used to give a smooth, even finish coat to plaster in corners. *Civil Engineering.* the thin edge of a gravel-surfaced road in which the thickness of the gravel is gradually increased from the edges to the centerline.

featheredging *Cartography.* a technique in which contours are progressively dropped to avoid congestion on steep slopes, and the line weight near the end of the contour is tapered. *Graphic Arts.* the thinning of overlapping edges on photographs before assembly into a mosaic, to make match lines less noticeable.

feathering *Vertebrate Zoology.* **1.** a covering of feathers. **2.** a fringe of hair, as on the legs of some dogs and horses. *Aviation.* in the event of engine failure, the process of adjusting a controllable-pitch propeller to a pitch position where the blade angle is about 90° to the plane of rotation in order to stop windmilling. Also, FULL FEATHERING. *Graphic Arts.* a printing technique, or sometimes a flaw, in which ink soaks into the paper and spreads to create a blur.

feathering propeller *Mechanical Devices.* a variable-pitch marine or airscrew propeller having an extended blade angle; used to reduce drag.

feather joint *Engineering.* a joint that is produced by cutting a mating groove into two boards, placing a feather (small spline) in each groove, and then butting them together. *Geology.* one of a series of joints that are formed in a fault zone as a result of shear and tension and resemble the barbs and shaft of a feather.

feather ore see JAMESONITE.

feather rot *Plant Pathology.* a rot that affects both living and dead tree trunks and is caused by the fungus *Poria subacida*; characterized by decaying tissue having a stringy or spongy texture.

feature *Archaeology.* any separate archaeological unit that is not recorded as a structure, a layer, or an isolated artifact; a wall, hearth, storage pit, or burial area may be designated as a feature. *Computer Technology.* in pattern recognition, a property of the image such as the number of holes or concavities, or the relative positions of line endings and beginnings; used to classify the image.

feature extraction-classification model *Computer Technology.* in pattern recognition, a method of selecting the dominant features of an image and identifying the character or object in the image by using a statistical or structural approach to consider the extracted features.

feature recognition *Robotics.* the programmed use of an object's most definitive features to identify it.

feb. dur. while fever lasts. (From Latin *febre durante.*)

febri- a combining form meaning "fever," as in *febrility.*

febrility [fə bril´ə tē] *Medicine.* having a high body temperature.

febrile [fē´brəl; feb´rəl] *Medicine.* relating to or characterized by high body temperature. Thus, **febrile disease, febrile reaction.**

febrile seizure *Neurology.* an epileptic convulsion associated with high fever, usually occurring in children and infants.

FEC forward error correction.

fecal [fē´kəl] *Physiology.* of or relating to feces.

fecal pellets *Paleontology.* petrified, fossilized excrement, mainly of invertebrates, occurring in marine deposits and in sedimentary rocks. Also, CASTINGS.

fecapentaenes *Genetics.* a group of mutagens that are produced by certain bacteria and have a highly unsaturated conjugated enol ether structure; they are believed to act as alkylating agents.

feces [fē´sēz] *Physiology.* waste materials, including undigested food and sloughed-off intestinal cells, that are expelled from the intestinal tract through the anus.

Fechner, Gustav 1801–1887, German experimental psychologist; founder of psychophysics; formulated Fechner's law.

Fechner color *Optics.* the visual sensation of color, elicited by intermittent achromatic visual stimuli capable of producing such a response.

Fechner fraction *Physiology.* a mathematical statement of Fechner's psychophysical law stating that the subjective size of a just-noticeable change in the physical stimulus is a constant.

Fechner's law *Physiology.* in sensory psychology, a law stating that the intensity of a sensation is proportional to the logarithm of the intensity of the stimulus producing it. Also, WEBER-FECHNER LAW.

fecundity *Biology* **1.** the innate capacity of an organism to form reproductive elements, such as ova or sperm, and to produce offspring or vegetation. **2.** the number of eggs produced by an organism.

fed batch system *Biotechnology.* a fermenter or anaerobic digester in which substrate is added in batches during the runs.

Federal Aviation Administration *Transportation Engineering.* the U.S. government agency that is responsible for the technical and safety regulation of aircraft, and for installing and maintaining the civilian air traffic control system.

Federal Communications Commission *Telecommunications.* a U.S. government board that is responsible for regulating all broadcasting and interstate communication by wire, radio, and television.

Federal Telecommunications System *Telecommunications.* a system of commercial telephone lines, leased by the U. S. General Services Administration, that provides services for official voice, teletypewriter, facsimile, and data transmission throughout the United States.

Fedorox stage see UNIVERSAL STAGE.

feeble-mindedness *Psychology.* in former use, a term for mental retardation or subnormal intellectual development. Thus, **feeble-minded.**

feed *Agriculture.* hay, fodder, or grains that are used as food for livestock. *Engineering.* to furnish raw material to a processing system. *Computer Technology.* **1.** the mechanical process of moving materials, such as paper or magnetic tape, through the required positions, or the device that performs this process. **2.** to provide data to a computer system by means of materials. *Electromagnetism.* a component at the end of a transmission line that radiates radio-frequency energy to or receives it from the reflector of a radar antenna. *Electronics.* **1.** to provide a device or circuit with electrical power or with a signal. **2.** a device that furnishes another device or system with power or a signal. *Mechanical Engineering.* **1.** the forward movement of a drill or cutting tool. **2.** the rate at which this occurs.

feedback *Science.* **1.** the return of information about a system or process that may effect a change in the process. **2.** the information that is returned. *Behavior.* any kind of direct information from an outside source about the effects of one's behavior. *Control Systems.* the return of part of the output of a system to the input of the system. *Robotics.* data received from sensors about a robot's position or speed. *Electronics.* a process in which a fraction of the output current or voltage is returned to the circuit, where it merges with and modifies an input signal.

feedback amplifier *Electronics.* **1.** a circuit in which a passive network returns a fraction of an amplifier output signal to its input to enhance the amplifier's performance. **2.** a device placed in the feedback path of a circuit to boost the amplitude of the feedback signal.

feedback branch *Control Systems.* a branch in a signal flow graph that is part of a feedback loop.

feedback capacitance see PLATE-GRID CAPACITANCE.

feedback circuit *Electronics.* a circuit that sends a fraction of its output signal back into its input component.

feedback compensation *Control Systems.* a method of improving the response of a feedback control system by using a compensator in the feedback path, rather than by using cascade compensation. Also, PARALLEL COMPENSATION.

feedback control *Robotics.* a feedback method to control a robot's movement and its end effector position through the use of stepping motors that control a gripper. Thus, **feedback control signal.**

feedback control system *Control Systems.* any control system in which feedback is used to compare actual performance with a standard representing the desired performance; any deviation from the standard is fed back into the control system in order to reduce the deviation. Also, CLOSED-LOOP SYSTEM.

feedback device *Robotics.* a device used to sense the position of a robot's joints or manipulator and send this information back to the robot, allowing it to adjust the position.

feedback factor *Electronics.* the fraction of an output signal that is fed back as an input signal.

feedback inhibition *Enzymology.* a negative feedback mechanism in which the end product of a metabolic pathway inhibits the activity of an enzyme that catalyzes a reaction early in the pathway. Also, END-PRODUCT INHIBITION.

feedback loop *Control Systems.* a closed transmission path that maintains a prescribed relationship between output signals and input signals in a control system. Also, **feedback control loop.** *Ecology.* a cybernetic device within a system that maintains stability via negative feedback, such as denitrifying bacteria counteracting the effects of nitrogen-fixing bacteria, or that creates significant change during positive feedback, such as an unusually cool summer hindering the winter snow melt and leading to further cooling.

feedback oscillator *Electronics.* a circuit that generates alternating current, by returning a fraction of the output signal to the input.

feedback regulator *Control Systems.* a function used to maintain a prescribed relationship between certain quantities and signals within a control system.

feedback transfer function *Control Systems.* the transfer function of a feedback path in a feedback loop.

feedback winding *Electronics.* a wire on a magnetic amplifier or saturable reactor, along which a feedback current travels.

feed chute *Ordnance.* a passage through which ammunition is guided into the breech mechanism of a machine gun.

feed-control valve *Mechanical Engineering.* a small valve used to control the advance of a hydraulic drill or cutting tool.

feed crop *Agriculture.* a crop raised to provide food for livestock rather than for human consumption, such as hay or sorghum.

feeder something that supplies food or any material thought of as analogous to food; specific uses include: *Agriculture.* **1.** a bin or hopper into which food is placed for consumption by farm animals. **2.** a young animal that is being fattened for future sale. *Mechanical Engineering.* a conveyor that controls the rate at which packages or components are assembled, separated, or delivered. *Electricity.* **1.** a transmission line that carries electric power between a transmitter and an antenna. **2.** a conductor or conductor pair that carries electric power from one point to another. *Ordnance.* a device that supplies ammunition to a weapon; it is usually controlled by an automatic or semiautomatic mechanism. *Geology.* **1.** a passage through which magma travels from the magma chamber to an intrusion. **2.** a minor ore-bearing vein that connects with a major one. *Hydrology.* see TRIBUTARY. *Metallurgy.* a runner or riser mold containing sufficient molten metal for casting operations. Also, FEEDHEAD.

feeder beach *Geology.* an artificially widened beach that nourishes downdrift beaches by natural coastal currents.

feeder-breaker *Mechanical Engineering.* a unit that crushes rock and feeds it to a materials-handling system at a specified rate.

feeder cable *Telecommunications.* a cable that supports one or more signals over a number of receiving points, usually the principal cable from a central office.

feeder canal *Civil Engineering.* a course that conducts water to a larger canal or reservoir.

feeder cell *Cell Biology.* in tissue culture, a cell that is added to another cell population to facilitate growth of the second cell type by compensating for a genetic deficiency in the second cell population, by secreting growth factors, or by helping to maintain the pH balance of the medium.

feeder channel *Oceanography.* a channel along which a feeder current flows.

feeder conveyor *Mechanical Engineering.* a short auxiliary conveyor designed to transport materials to another conveyor. Also, STAGE LOADER.

feeder current *Oceanography.* a current that flows parallel to the shore inside the breaker zone before it joins similar currents to form the neck of a rip current.

feeder distribution center *Telecommunications.* a central distributing location at which feeders or subfeeders are joined.

feeder line *Transportation Engineering.* a local line that connects with a regional or main line.

feeder reactor *Electronics.* a device that is connected between a transmitter and an antenna in a transmission line to reduce and contain disturbances that arise from faults in the line.

feeder road *Civil Engineering.* a course serving to conduct traffic to a larger road.

feeder service *Transportation Engineering.* a local transportation service that delivers passengers to a mainline station or stop.

feeder trough *Mining Engineering.* a trough connected to the conveyor pan line in a duckbill on which it rides.

feedforward *Behavior.* a behavior pattern that anticipates, rather than responds to, a change in external conditions.

feedforward control *Control Systems.* a process control that detects changes at the input level and applies a correction signal before output can be affected.

feedhead *Metallurgy.* see FEEDER.

feed holes *Computer Technology.* sprocket holes in the edges of continuous-form paper, used to maintain alignment and to advance the paper through the printer.

feeding *Building Engineering.* a chemical reaction of paint that causes it to thicken and lose opacity. *Metallurgy.* **1.** in casting, the process of delivering liquid metal to a mold. **2.** in other metallurgical operations, the process of delivering stock to a machine.

feeding mechanism *Zoology.* the various tools, techniques, and processes used by an animal to nourish itself.

feeding rod *Metallurgy.* in casting, a rod used to clear the feeder so that the liquid metal can reach the mold cavity.

feeding zone *Control Systems.* the area on the surface of a pallet or conveyor where an object that is to be manipulated is placed.

feedlot or **feed lot** *Agriculture.* a pen or enclosure in which animals are fattened for market.

feed materials *Nucleonics.* pure compounds, such as refined uranium or thorium, used as fuel elements in nuclear reactors or as fuel for the uranium enrichment process.

feed nut *Mechanical Engineering.* a threaded sleeve fitting around a feed screw.

feed pipe *Mechanical Engineering.* **1.** a pipe that carries feedwater from a pump to a boiler. **2.** any pipe that conducts a fuel or other liquid from the supply to a secondary pipe or to the point of use.

feed pitch *Design Engineering.* the distance between the midpoints of adjacent feedholes in punched paper tape.

feed preparation unit *Chemical Engineering.* any processing equipment that provides feedstock for subsequent processing.

feed pressure *Mining Engineering.* the total weight or pressure applied to a drilling stem so that a drill bit can cut and penetrate rock ore.

feed pump *Mechanical Engineering.* **1.** a pump used to supply water to a steam boiler. **2.** any pump used as a feed source to a process operation.

feed rate *Design Engineering.* the speed at which material is moved through a machine by a process.

feed ratio *Mechanical Engineering.* the number of revolutions that a drill bit and stem must turn to advance the drill bit one inch. Also, **feed speed.**

feed reel *Engineering.* the reel that supplies paper tape or magnetic tape into a system.

feed screw *Mechanical Engineering.* **1.** an externally threaded drive rod in a screw-feed or gear-feed swivel head on a diamond drill that provides motion to a feed tool. **2.** a mechanism used to carry and position workpieces for processing on a percussion drill, lathe, or the like.

feedstock *Engineering.* the raw material supplied to a machine or process. *Petroleum Engineering.* specifically, a petroleum fraction used as the raw material for a petrochemical process, as in the production of polymers, detergents, or chemicals such as ethelene and toluene.

feed tank *Engineering.* a chamber used to hold feedstock.

feedthrough *Electricity.* a conductor that connects patterns on both sides of a printed circuit board. Also, INTERFACE CONNECTION.

feedthrough capacitor *Electricity.* an insulator that delivers a desired value of capacitance between a conductor and the metal panel through which the conductor passes; used for bypass purposes in ultra-high-frequency circuits.

feed track *Computer Technology.* a track on a paper tape that contains the sprocket holes. Also, SPROCKET TRACK.

feed travel *Mechanical Engineering.* the distance a drilling machine moves its steel shank over a top-to-bottom feeding range.

feed tray *Chemical Engineering.* a tray on which untreated feedstock is introduced into the system in a tray-type separation column.

feed trough *Mechanical Engineering.* a receptacle into which feedwater overflows from a steam-boiler drum.

feedwater *Mechanical Engineering.* the treated water that is supplied to a boiler for conversion into steam.

feedwater heater *Mechanical Engineering.* an apparatus that uses steam exhaust from an engine or turbine to heat feedwater for a boiler.

feel *Aviation.* a pilot's sense or impression of his craft's operation, including factors such as its speed, direction of movement, attitude, orientation, proximity to other objects, and especially its stability and responsiveness to control.

feeler pin *Mechanical Engineering.* a thin metal sensor that allows a duplicating machine to operate only when paper is present.

feeling bottom *Navigation.* the reaction of a vessel moving in shallow water to the greater pressure and friction exerted on its hull by increasing water turbulence and velocity. *Oceanography.* the condition of a deep-water wave as it runs into shoal water and begins to be influenced by the bottom.

Fehling's reagent *Analytical Chemistry.* a test solution containing cupric sulfate, sodium potassium tartrate, and sodium hydroxide, used for detection of an aldehyde group, especially in sugars; the formation of a red precipitate indicates a positive test.

Feigenbaum constant *Chaotic Dynamics.* a universal characteristic of a sequence of period-doubling bifurcations that measure the spacings between the bifurcation points λ_n, which denote the birth of an attractor of periodicity 2^n times that of the initial periodic attractor,

$$\delta = \lim_{n\to\infty} (\lambda_n - \lambda_{n-1})/(\lambda_{n+1} - \lambda_n) = 4.669201660910299\ldots$$

and the strengths of the subharmonic peaks in the power spectrum $S(\Omega)$

$$\alpha = \lim_{n\to\infty} (S(\Omega/2^n)/S(\Omega/2^{n+1})) = 2.5029078750\ldots(\sim 8.2 \text{ dB}).$$

Feit-Thompson theorem *Mathematics.* a theorem stating that a non-Abelian simple group of finite order has an even number of elements. Equivalently, every non-Abelian group of odd order is solvable.

Fejér kernel *Mathematics.* the Cesàro sum of the Dirichlet kernels $D_k(t)$ of Fourier analysis; denoted $K_n(t)$. Specifically,

$$K_n(t) = 1/(n+1) \sum_{k=0}^{n} D_k(t).$$

It can be shown that if $\sin t/2 = 0$, then $K_n(t) = n+1$; otherwise

$$K_n(t) = 1/(n+1)\ [\sin((n+1)t/2)/\sin(t/2)]^2.$$

Fejér's theorem *Mathematics.* the Cesàro partial sums of the Fourier series of a continuous function f converge uniformly to f on compact subsets of the domain. If f is not continuous at some point x_0, the Cesàro partial sums of the Fourier series at x_0 converge to a value halfway between the left and right limits of $f(x_0)$, if they exist.

feldspar *Mineralogy.* a group of minerals having the general formula XZ_4O_8, where X = Ba,Ca,K,Na,NH_4,Sr and Z = Al,B,Si. The most common and widespread of the mineral groups, constituting 60% of the earth's crust as constituents of all types of rocks; members vary widely, but are generally white or pale tints, monoclinic or triclinic (rarely orthorhombic), having a specific gravity of 2.32 to 3.45 (mostly 2.56–2.76) and a hardness of 5.5 to 6.5 on the Mohs scale.

feldspathic graywacke *Petrology.* sandstone consisting of less than 75% quartz and chert and 15–75% detrital clay matrix, with a greater abundance of feldspar grains than rock fragments. Also, ARKOSIC WACKE, HIGH-RANK GRAYWACKE.

feldspathic quartzite see SUBARKOSE.

feldspathic sandstone *Petrology.* sandstone that is rich in feldspar, with 10–25% feldspar and less than 20% matrix material; intermediate in composition between pure quartz sandstone and arkosic sandstone.

feldspathic shale *Petrology.* a well-laminated shale featuring more than 10% feldspar in the silt size and featuring kaolinitic clay minerals in a fine matrix.

feldspathization *Geology.* a process in which a feldspar is formed in a rock, usually as a result of metamorphism.

feldspathoid or **felspathoid** *Geology.* a group of rock-forming minerals that is similar in composition to the feldspars, but deficient in silica.

f electron *Atomic Physics.* an electron that orbits a nucleus with an angular momentum of 3.

Felidae *Vertebrate Zoology.* the cats, a cosmopolitan family of carnivores of the suborder Feliforma, consisting of lithe-bodied digitigrade carnivorous mammals having soft and often strikingly patterned fur, comparatively short limbs with soft pads on the feet, usually sharp, curved retractile claws, a broad and somewhat rounded head with short but powerful jaws equipped with teeth suited to grasping, tearing, and shearing through flesh, erect ears, and typically eyes with narrow or elliptical pupils, especially adapted for seeing in dim light; includes domestic cats, *Felis domesticus*, the true cats (lion, tiger, jaguar, leopard, cougar, etc.), the cheetah, and extinct forms.

feline *Vertebrate Zoology.* belonging or relating to the genus *Felis* or the family Felidae.

Felis *Vertebrate Zoology.* a genus of mainly small cats, including the domestic cat, puma, and ocelet, that are unable to roar because of the ossification of the hyoid bone in the larynx.

fell *Forestry.* **1.** to cut down a tree or trees. **2.** the amount of timber cut down in one season.

feller-buncher

feller-buncher *Forestry.* a large caterpillar-type vehicle used to fell and harvest trees. It has hydraulic arms and fingers that grasp, lower, cut, and stack the felled tree; does not debark or delimb the tree.

fell-field *Ecology.* a community of widely scattered dwarfed vegetation that grows in the barren land above the timber line.

Fellgett's advantage *Spectroscopy.* the improvement in the signal-to-noise ratio gained by choosing the Fourier transform spectroscopy method over a diffraction grating or the Fabry-Perot method; the improvement is the result of requiring less time to scan by a factor of $(1/N)^{1/2}$ where N is the number of elements in the specimen.

fell system *Civil Engineering.* a track system for increasing traction on steep slopes, in which horizontal wheels on the locomotive grip a central rail on the tracks.

Feloidea *Vertebrate Zoology.* a superfamily of carnivorous mammals that is composed of the families Viverridae (civets and mongooses), Hyaenidae (hyenas), and Felidae (cats).

felsenmeer *Geology.* a relatively thin, flat, or gently sloping accumulation of angular blocks of rock found in place on mountaintops above the treeline. Also, BLOCK FIELD.

felsic *Mineralogy.* the light-colored silicate minerals, such as quartz and feldspar, or the rocks that contain an abundance of one or all these minerals. (An acronym of feldspar and silica.)

felsite *Petrology.* **1.** any light-colored, finely crystalline, hypabyssal igneous rock composed chiefly of quartz and feldspar, with or without phenocrysts. **2.** a rock with a felsitic texture.

felsöbanyaite *Mineralogy.* $Al_4(SO_4)(OH)_{10}\cdot 5H_2O$, a colorless or white probably orthorhombic mineral occurring as aggregates of lamellar crystals, having a specific gravity of 2.33 to 2.35 and a hardness of 1.5 on the Mohs scale. Also, **felsöbanyite.**

felsophyric see APHANIPHYRIC.

felt *Materials.* a soft woolen fabric, usually containing fur, hair, or synthetic materials, in which the fibers are matted together by heat, moisture, chemicals, and pressure, rather than by weaving.

felt paper *Materials.* a sheathing paper coated with tar and asphalt, used to protect against moisture and extreme temperatures.

felty *Geology.* of or relating to the texture of the dense, groundmass of holocrystalline igneous rocks that contain irregular arrangements of tightly and flatly pressed microlites.

Felty's syndrome *Medicine.* a spleen disorder occurring with adult rheumatoid arthritis. (Named for Augustus Roi *Felty,* American physician.)

female *Biology.* **1.** of, relating to, or being the sex that becomes pregnant and bears young. **2.** an organism of the sex or sexual phase that normally bears egg cells. *Botany.* referring to a plant that produces only an ovum or pistil. *Engineering.* describing the one of two parts into which another part (the male) fits. Thus, **female fitting.**

female connector *Electricity.* a connector that has one or several contacts in recessed openings; such as a jack, socket, or wall outlet.

female gamete *Molecular Biology.* a mature female sex cell that holds a haploid set of chromosomes; it contributes half the genetic material in sexual reproduction.

female plug *Electricity.* a plug having contacts that are separated by a recess into which one or more pins or prongs of the male plug are inserted.

female pronucleus *Developmental Biology.* the haploid nucleus of an ovum; its fusion with the nucleus of a sperm results in the formation of a fertilized egg.

female-specific phage *Virology.* a phage whose ratio of plaque titer on bacteria containing certain plasmids is much lower than that on plasmid-free strains of the same bacteria.

female-sterile mutation *Genetics.* a type of mutation common in *Drosophila* and certain other insects, in which the female is sterile because of a failure of oogenesis to proceed normally.

femic *Mineralogy.* of or relating to igneous rock having normative minerals that are rich in iron, magnesium, or calcium. (An acronym derived from iron, Fe, and magnesium, Mg.)

feminization *Physiology.* **1.** the normal induction or development of female sex characteristics. **2.** the induction or development of female secondary sex characteristics in the genotypic male.

femitrons *Electronics.* a class of field-emission microwave devices.

femoral [fem´ər əl] *Anatomy.* of or relating to the femur or to the thigh.

femoral artery *Anatomy.* an artery that extends from the external iliac artery into the thigh, originating near the inguinal ligament and branching many times to supply blood to parts of the leg and trunk.

femoral hernia *Medicine.* a hernia in which a loop of intestine descends through the femoral canal into the groin.

femoral nerve *Anatomy.* a nerve that originates from the second through fourth lumbar nerves and extends downward from the area of the inguinal ligament to the femoral triangle, branching into nerves that supply the skin of the thigh and leg, the muscles of the front of the thigh, and the hip and knee joints.

femoral ring *Anatomy.* the abdominal opening of the femoral canal.

femoral vein *Anatomy.* a major vein of the leg that follows the course of the femoral artery; it is a continuation of the popliteal vein, becoming the external iliac vein at the inguinal ligament.

femto- a combining form that is used to designate a unit equal to one quadrillionth (10^{-15}) of the unit with which it is combined, as in *femtocurie.*

femtoampere *Electricity.* a unit of low current that equals 10^{-15} A.

femtometer see FERMI.

femtoplankton *Ecology.* a plankton of less than 0.2 μm in diameter.

femtovolt *Electricity.* a unit of low voltage that equals 10^{-15} V.

femur [fē´mər] *plural,* **femora** [fem´ər ə]. *Anatomy.* the large long bone of the thigh or upper leg that articulates with the hip joint above and the tibia and patella below; the longest and largest bone in the body. Also, THIGH BONE. *Invertebrate Zoology.* **1.** the third segment from the base of an insect's leg, often greatly enlarged. **2.** the large fourth segment from the base of the legs of some crustaceans. Also, MEROPODITE.

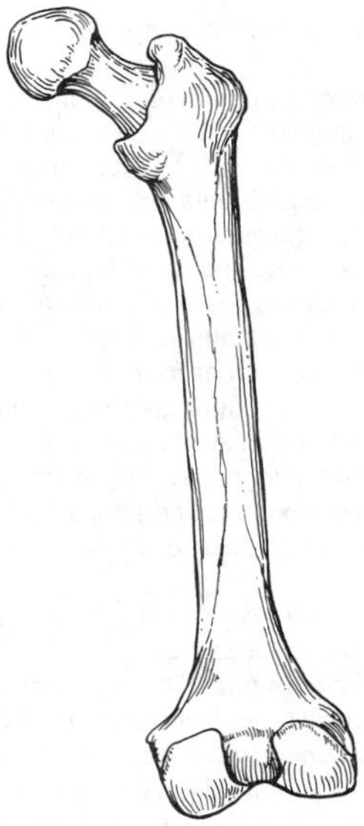

femur

fen *Geography.* a flooded peat marsh having nonacidic soil (in contrast to the acidic soil found in bogs).

fenaminosulf *Organic Chemistry.* $C_8H_{10}N_3SONa$, a yellow powder used as a fungicide.

fenazaflor *Organic Chemistry.* $C_{15}H_7Cl_2F_3N_2O_2$, greenish-yellow crystals that melt at 103°C; used as an insecticide and miticide.

fenbutatin oxide *Organic Chemistry.* $C_{60}H_{78}OSn_2$, corrosive white crystals; insoluble in water and soluble in methylene chloride; melts at 138–139°C; used to control mites in citrus fruits.

fence any of various structures or systems thought to resemble a fence; specific uses include: *Aviation.* a fixed plate or vane extending from the upper surface of a wing, other suface, or nacelle parallel to the airflow; used to prevent spanwise flow. Also, STALL FENCE. *Space Technology.* a row of radar or radio-tracking stations that monitor satellites, including early-warning radar stations. Also, RADAR FENCE. *Engineering.* a movable guide on a tool that allows for precision work. *Computer Programming.* **1.** a separation between one or more components of a processor complex and the rest of the complex, allowing maintenance to occur simultaneously with user operations. **2.** in a binary search, a criterion used as a basis for dividing a list into two parts. Also, **fence cell.**

fenchone *Organic Chemistry.* $C_{10}H_{16}O$, a colorless monoterpene oil; insoluble in water and soluble in alcohol; boils at 193°C and melts at 6°C; used as a flavoring.

fenchyl alcohol *Organic Chemistry.* $C_{10}H_{18}O$, a colorless, oily solid that melts at 39°C and boils at 201°C; produced synthetically or derived from pine oil or fennel oil; used as a solvent, flavoring, and organic intermediate. Also, ***fenchol.***

fender *Engineering.* a protective cover or enclosure over the wheel of a car or other such vehicle. *Naval Architecture.* **1.** a bag of rope or piles of timber clusters temporarily placed along a dock or pier to protect it from docking ships or other floating objects. **2.** a similar piece of protective gear on the side of a ship that protects it from a pier or an adjacent ship.

Fenestella *Paleontology.* a genus of netlike bryozoans in the extinct family Fenestellidae.

Fenestellaceae *Mycology.* a family of fungi belonging to the order Melanommatales, composed primarily of species that live off decaying woody plants and some species that are parasites of woody plants.

Fenestellidae *Paleontology.* a family of netlike Paleozoic bryozoans in the extinct order Cryptostomata; some authorities now classify this and other netlike forms as a separate order, the Fenestrata.

fenestra *Science.* a windowlike opening. *Anatomy.* an opening in an anatomical structure that is often covered by a membrane. *Medicine.* an opening that is created in a bandage or cast in order to relieve pressure or permit care of the skin. *Surgery.* **1.** an opening in the blade of a forceps. **2.** an opening in the wall of a natural or artificial tube to allow improved flow, maneuvering of instruments, and visibility. (From the Latin word for "window.")

fenestrated membrane *Histology.* any membrane with one or more openings (fenestrae), as found in the elastic membranes of the media and intima of large arteries.

fenestration *Biology.* a transparent, windowlike opening in a surface, such as one in or between bones or in a plant membrane. *Surgery.* **1.** see PERFORATION. **2.** the creation of an opening in a dressing or cast to allow inspection or treatment. **3.** the creation of an opening in the labyrinth of the ear to restore hearing in otosclerosis. *Architecture.* the arrangement and design of windows in a building.

fenitrothion *Organic Chemistry.* $C_9H_{12}NO_5PS$, a toxic yellow-brown liquid; insoluble in water and soluble in most organic solvents; boils at 118°C under reduced pressure; used as an insecticide.

fennec *Vertebrate Zoology.* a tiny nocturnal desert fox, *Fennecus zerda,* of the family Canidae, characterized by long, pointed ears, a short snout, and a light-colored coat with a blacktipped tail; found in North Africa and the Sinai and Arabian peninsulas.

fennec

fennel *Botany.* **1.** the common name for a perennial herb, *Foeniculum vulgare,* belonging to the parsley family and having feathery leaves and umbrels of small, yellow flowers. **2.** the aromatic fruits of this plant used as a spice and in medicine. Also, **fennel seed.**

fennelflower *Botany.* any of various plants of the genus *Nigella,* of the buttercup family, particularly *N. sativa,* whose seeds are used in the East as a condiment and medicine.

Fenske-Underwood equation *Chemical Engineering.* an equation in distillation-column calculations that relates the number of theoretical plates required at total reflux to the overall relative volatility, and the ratios of liquid-vapor composition on top and bottom plates.

fenson *Organic Chemistry.* $C_{12}H_9ClO_2S$, colorless crystals; soluble in organic solvents and insoluble in water; melts at 61–62°C; used as a miticide.

fenster *Geology.* see WINDOW.

fensulfothion *Organic Chemistry.* $C_{11}H_{17}O_4PS_2$, a brown liquid that boils at 138°C at reduced pressure; used as an insecticide.

fentiazon *Organic Chemistry.* $(C_6H_5)_2C_3HN_2S_2$, yellow crystals that melt at 297°C; used as a fungicide for rice.

fentin acetate *Organic Chemistry.* $C_{20}H_{18}O_2Sn$, yellow-to-brown crystals that are soluble in ether and slightly soluble in alcohol; melts at 124–125°C; used as a fungicide and algicide.

fentin hydroxide *Plant Pathology.* a fungicide that is used against such plant diseases as leaf spot, blight, and scab.

fenuron *Organic Chemistry.* $C_9H_{12}N_2O$, a white crystalline solid that is insoluble in water and slightly soluble in hydrocarbon solvents; melts at 131–133°C; used as an herbicide.

fenuron-TCA *Organic Chemistry.* $C_{10}H_{14}Cl_3N_2O_3$, white crystals that melt at 65–68°C; used to kill plants and weeds in noncrop areas. Also, **fenuron trichloroacetate.**

fepA protein *Biochemistry.* a protein in *Escherichia coli* that functions as a receptor for B and D colicins and participates in iron uptake.

feral *Biology.* **1.** having escaped from a state of domestication and reverted to the original wild or untamed state. **2.** existing naturally in nature; not cultivated or domesticated.

ferbam *Organic Chemistry.* $C_9H_{18}FeN_3S_6$, a fluffy black powder that is dispersible but only slightly soluble in water; melts and decomposes at 180°C; used as a fungicide.

ferberite *Mineralogy.* $Fe^{+2}WO_4$, a black, opaque monoclinic mineral with a submetallic luster, occurring as wedge-shaped crystals, having a specific gravity of 7.4 to 7.51 and a hardness of 4 to 4.5 on the Mohs scale; found in high-temperature hydrothermal ore veins.

Ferdinand II 1610–1670, Grand Duke of Tuscany; developed sealed alcohol thermometer; a patron of the Italian meteorological society.

fergusonite *Mineralogy.* a collective name for six minerals with the general formula (Ce,La,Nd,Y)(Nb,Ti)O_4, a brittle, brownish black monoclinic or tetragonal mineral occurring as prismatic crystals and irregular grains, having a specific gravity of 5.34 to 5.7 and a hardness of 5.5 to 6.5 on the Mohs scale; found with rare-earth minerals in granitic pegmatites and in skarns.

Fermat, Pierre de [fer´mä] 1601–1665, French mathematician; founded modern theory of numbers; cofounder of analytic geometry, infinitesimal calculus, and the theory of probabilities.

Fermat numbers *Mathematics.* numbers of the form $p_k = 2^{2^k} + 1$. Only the first five Fermat numbers (3, 5, 17, 257, and 65537) are known to be prime; about a dozen more are known to be composite. Fermat numbers arose in the attempt to construct regular polygons (*n*-gons) by Euclidean geometry (using compass and straightedge alone); algebraically this is equivalent to solving the equation

$$z^n + z^{n-1} + \cdots + z + 1 = 0.$$

Gauss showed that n must be a Fermat number, and Fermat (incorrectly) believed that all such n were prime.

Fermat's equation *Mathematics.* the Diophantine equation of the form $x^n + y^n = z^n$.

Fermat's last theorem *Mathematics.* a marginal note on Fermat's copy of the work of Diophantus stating that the equation $x^n + y^n = z^n$ "is not solvable in nonvanishing integers x, y, z for any integral $n \geq 3$," and that the proof was too long for inclusion on that page. The complete proof, if this conjecture is true, has thus far eluded mathematicians. Also, **Fermat conjecture.**

Fermat's principle *Optics.* a principle stating that an electromagnetic wave, such as a light ray, traveling from one point to another will assume the path that takes the least amount of time. Also, PRINCIPLE OF STATIONARY TIME, PRINCIPLE OF LEAST TIME.

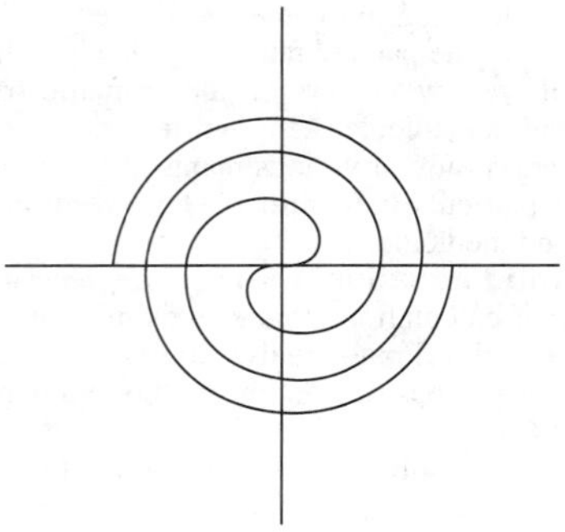

Fermat's spiral

Fermat's spiral *Mathematics.* the graph of the equation $(r - a)^2 = 4ak\theta$. Also, PARABOLIC SPIRAL.

Fermat's theorem *Mathematics.* if p is a prime and a is not divisible by p, then $a^{p-1} \equiv 1 \pmod{p}$; i.e., $a^{p-1} = np + 1$, for some integer n. Euler's generalization of this states that if a is relatively prime to m, then $a^{\phi(m)} \equiv 1 \pmod{m}$, where $\phi(m)$ is the Euler phi function of m.

ferment *Biology.* to undergo fermentation. *Biotechnology.* to decompose organic substances by microorganisms or by enzymatic activity.

fermentation *Microbiology.* the chemical decomposition of a substance, especially a carbohydrate, brought about by enzymes, bacteria, yeasts, or molds, generally in the absence of oxygen. *Biotechnology.* the process of culturing cells or other microorganisms in a container, bioreactor, or fermenter for experimental or commercial purposes.

fermentation tube *Microbiology.* a modified culture tube with an upright connecting arm closed at its end for the collection of gas produced by microorganisms growing in liquid medium.

fermenter *Biotechnology.* a fabricated vessel or bioreactor used for fermentation; these exist in a wide variety of configurations, from experimental systems of less than one liter to large commercial towers.

fermenter control *Biotechnology.* a wide variety of automatic control or monitoring facilities that regulate key processes such as foaming, automatic sequencing, and dissolved oxygen concentration in a fermenter.

Enrico Fermi

Fermi, Enrico 1901–1954, Italian-born American physicist; Nobel Prize for research in radioactivity; created first nuclear chain reaction.

fermi *Metrology.* a unit of length in the metric system equal to 10^{-15} meters; used to express nuclear measurements. Also, FEMTOMETER.

Fermi age *Nucleonics.* a value that represents the dimensions of area, not time, in which neutrons lose energy by elastic collision.

Fermi age model *Nucleonics.* a model used to study the energy lost through elastic collision between neutrons, in which it is assumed that decline takes place continuously in an infinite homogeneous medium.

Fermi beta-decay theory *Nuclear Physics.* the principle that beta decay arises from the interaction between a nucleon's source current and an electron-neutrino field.

Fermi constant *Nuclear Physics.* a universal constant, value 1.4×10^{-62} joule/meter^3, that represents the coupling between a nucleon and a lepton field; important in the Fermi beta-decay theory.

Fermi-Dirac distribution function *Physics.* a function that gives the probability that an energy state is occupied, in an ideal gas of fermions.

Fermi-Dirac statistics *Quantum Mechanics.* the mathematical formulation of the behavior of particles with completely antisymmetric wave functions and half-integer spin; particles obeying Fermi-Dirac statistics have antisymmetric eigenfunctions and are known as fermions.

Fermi distribution *Solid-State Physics.* an electron energy-state distribution in a semiconductor as given by the Fermi-Dirac distribution function, in which nearly all energy levels below the Fermi level are filled and nearly all levels above the Fermi level are empty.

Fermi energy *Physics.* the energy for which the Fermi-Dirac occupation probability equals one-half. Also, **Fermi level.**

Fermi gas *Physics.* a system of identical noninteracting fermions, particles that are subject to the Pauli exclusion principle.

Fermi hole *Solid-State Physics.* a region that surrounds an electron in a solid in which, according to band theory, the probability of finding other electrons is less than the average over the volume of the solid.

Fermi liquid *Physics.* a liquid whose particles are subject to Fermi-Dirac statistics.

fermion *Quantum Mechanics.* a particle that obeys Fermi-Dirac statistics; a half-integer spin particle.

fermion field *Quantum Mechanics.* an operator that creates or annihilates a fermion particle-antiparticle pair.

Fermi plot or **Fermi-Kurie plot** see KURIE PLOT.

Fermi resonance *Physical Chemistry.* the relationship between atoms in a polyatomic molecule that causes them to repel each other due to two vibrational levels having approximately the same energy.

Fermi's golden rule *Quantum Mechanics.* the equations relating the matrix elements of a perturbing Hamiltonian to the resulting changes in transition probabilities; the probability is proportional to the squared expectation value of the perturbing Hamiltonian.

Fermi sphere *Physics.* a sphere in momentum space whose radius equals the momentum of the most energetic fermion.

Fermi surface *Solid-State Physics.* a constant energy surface in k-space that encloses all occupied electron states at absolute zero in a crystal.

Fermi temperature *Physics.* a parameter for the Fermi-Dirac distribution function, it equals Fermi energy divided by the Boltzmann constant.

Fermi transition *Nuclear Physics.* a type of beta decay in which there is no change in the angular momentum or in the parity of the nucleus as it disintegrates into a lower state of energy.

fermium *Chemistry.* a synthetic radioactive element having the symbol Fm, the atomic number 100, and atomic weight (of the most stable isotope) 257.10; made in nuclear reactors, it has properties similar to those of erbium. (Named for Enrico *Fermi.*)

fermorite *Mineralogy.* $(Ca,Sr)_5(AsO_4,PO_4)_3(OH)$, a pinkish white hexagonal mineral of the apatite group with a massive, granular appearance, having a specific gravity of 3.52 and a hardness of 5 on the Mohs scale; found in veinlets in manganese ore.

fern *Botany.* the common name for a seedless, nonflowering vascular plant of the class Filicineae, characterized by triangular fronds that uncoil upward, true roots produced from a rhizome, and reproduction by spores located primarily on the underside of the leaves; existing in about 10,000 species and found widely in all tropical and temperate regions.

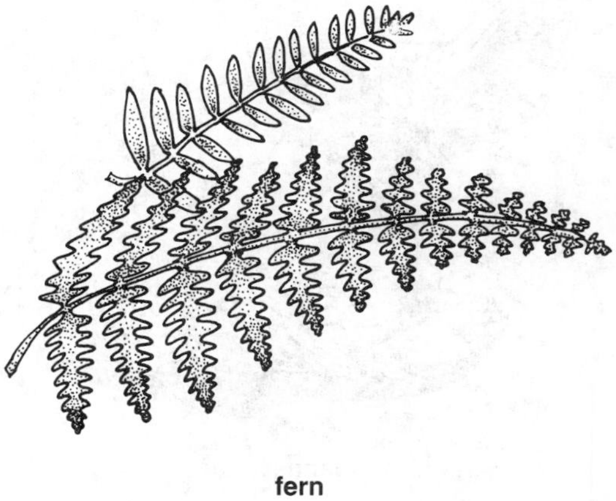

fern

Fernández-Morán particle *Cell Biology.* a particle on the inner face of the mitochondrial inner membrane that plays a role in the proton ATPase complexes.

fernandinite *Mineralogy.* $Ca_3(V^{+5},V^{+4})_{40}O_{100}\cdot 50H_2O$, a dull green monoclinic mineral of undetermined specific gravity and hardness, occurring in cryptocrystalline and fibrous forms; readily soluble in acids and partly soluble in water.

fernico *Metallurgy.* a low thermal expansion alloy consisting of about 54% iron, 28% nickel, and 18% cobalt. (From *ferrum,* the Latin word for iron.)

ferrate *Inorganic Chemistry.* a compound of ferric oxide with another oxide, such as sodium ferrite, $NaFeO_2$.

ferredoxin *Biochemistry.* an iron-containing protein that is part of the electron transport system in plants and microorganisms.

ferreed switch *Electricity.* an electromechanical switch that combines the rapid switching of bistable magnetic material with metallic contacts to maintain its position without the need for continuous current.

Ferrer's graph *Mathematics.* a configuration consisting of rows of dots arranged so that every row has at least as many dots as any row below it. Used to represent partitions of an integer; a row of k dots corresponds to the value k in the partition. Ferrer's graphs are especially useful in proving results involving partitions and generating functions. The largest square of dots at the upper left-hand corner of a Ferrer's graph is called the Durfee square.

ferret *Vertebrate Zoology.* a semidomesticated, red-eyed variety of the European polecat that was once commonly used for driving rats and rabbits from their burrows. *Ordnance.* an aircraft, ship, vehicle, or satellite that is equipped to detect, locate, record, and analyze electromagnetic radiation.

ferret

ferri- a combining form indicating the presence of a ferric ion, Fe^{3+}.

ferriamphibole *Mineralogy.* a mineral of the amphibole group containing ferric iron (Fe^{+3}).

ferric *Chemistry.* **1.** of or relating to iron. **2.** describing various compounds of iron, especially those in which the element has a valence of 3. (From *ferrum,* the Latin word for iron.)

ferric acetate *Organic Chemistry.* $Fe(C_2H_3O_2)_3$, a brown to red powder that is soluble in alcohol and insoluble in water; used as a dye mordant and wood preservative.

ferric ammonium citrate *Organic Chemistry.* $Fe(NH_4)_3(C_6H_5O_7)_2$, red granules that are used in medicine and in blueprint photography.

ferric ammonium oxalate *Organic Chemistry.* $(NH_4)_3Fe(C_2O_4)_3\cdot 3H_2O$, photosensitive green crystals that are soluble in water and alcohol; decomposes at 160°C; used in blueprint photography.

ferric ammonium sulfate *Inorganic Chemistry.* $FeNH_4(SO_4)_2\cdot 12H_2O$, lilac to violet efflorescent crystals; melts at 39–41°C and loses water at 230°C; used in medicine, analytical chemistry, and dyeing. Also, **ferric ammonium alum.**

ferric arsenate *Inorganic Chemistry.* $FeAsO_4\cdot 2H_2O$, a brown or green powder that is insoluble in water and soluble in dilute mineral acids; decomposes on heating. It is toxic and a strong irritant, and is used in insecticides.

ferric arsenite *Inorganic Chemistry.* $2FeAsO_3\cdot Fe_2O_3\cdot 5H_2O$, a brownish-yellow powder that is slightly soluble in cold water and soluble in acids and alkalis; decomposes on heating. It is nonflammable and is toxic on inhalation or ingestion.

ferric bromide *Inorganic Chemistry.* $FeBr_3$, dark red to brown deliquescent crystals; soluble in water, alcohol, and ether; sublimes and decomposes at the melting point; used in medicine and chemistry and as a bromination catalyst. Also, **ferric tribromide.**

ferric chloride *Inorganic Chemistry.* $FeCl_3$, brownish-black crystals that absorb moisture from the air; melts at 306°C and decomposes at 315°C; toxic and a strong irritant. It has many industrial uses, as in photography, sewage treatment, and water purification, and in various chemical processes. Also, **ferric trichloride.**

ferric citrate *Organic Chemistry.* $FeC_6H_5O_7\cdot 3H_2O$, photosensitive red-dish-brown scales that are soluble in water and insoluble in alcohol; used in medicine and blueprint paper.

ferric dichromate *Inorganic Chemistry.* $Fe_2(Cr_2O_7)_3$, reddish-brown granular crystals that are soluble in water and acids; toxic, highly irritating, and a fire hazard; used in pigments and dyeing.

ferric ferrocyanide *Inorganic Chemistry.* $Fe_4[Fe(CN)_6]_3$, toxic dark-blue crystals, insoluble in water; used in pigments, fertilizers, and finishes.

ferric fluoride *Inorganic Chemistry.* FeF_3, green crystals, slightly soluble in cold water and soluble in hot water and acids; melts above 1000°C. It is a strong irritant to tissues and is used in ceramics and as a catalyst.

ferrichrome *Microbiology.* an iron-based red pigment that is produced and secreted by certain microorganisms for subsequent binding and cellular uptake of extracellular iron; a siderophore.

ferric hydroxide *Inorganic Chemistry.* $Fe(OH)_3$, a brown powder that is soluble in acids and insoluble in water and alcohol; used in pigments and water purification and as a catalyst. Also, **ferric hydrate.**

ferric nitrate *Inorganic Chemistry.* $Fe(NO_3)_3\cdot 9H_2O$, colorless to pale violet crystals, soluble in water and alcohol; melts at 47.2°C and decomposes above 125°C. It is a fire hazard in contact with organic substances and a strong oxidant and irritant; used in dyeing, tanning, and chemistry.

ferric oxalate *Organic Chemistry.* $Fe_2(C_2O_2)_3$, pale-yellow amorphous scales that are soluble in water and acids; decomposes at 100°C; used in the production of pure oxygen and photographic printing paper.

ferric oxide *Inorganic Chemistry.* Fe_2O_3, a red-brown to black solid, insoluble in water and soluble in acids; melts at 1565°C. It has a wide range of industrial uses, as in metallurgy, gas purification, and magnetic tapes, and as a catalyst, pigment, and mordant. Also, **ferric trioxide, ferric oxide red.**

ferric phosphate *Inorganic Chemistry.* $FePO_4\cdot 2H_2O$, a yellow-white powder, very slightly soluble in water and soluble in acids; decomposed by heat; used as a fertilizer and food additive.

ferric resinate *Organic Chemistry.* a reddish-brown powder that is slightly soluble in alcohol and insoluble in water; used as a drying agent in paints and varnishes.

ferricrete *Geology.* a conglomerate of surface sand and gravel cemented into a mass by iron oxide.

ferric stearate *Organic Chemistry.* $Fe(C_{18}H_{35}O_2)_3$, a combustible light-brown powder that is soluble in alcohol and ether; used as a varnish drier.

ferric vanadate *Inorganic Chemistry.* $Fe(VO_3)_3$, a grayish-brown powder, insoluble in water and alcohol and soluble in dilute acids; combustible; used in metallurgy.

ferricyanic acid *Inorganic Chemistry.* $H_3[Fe(CN)_6]$, a hypothetical acid from which the ferricyanide salts are derived.

ferricyanide *Inorganic Chemistry.* any salt derived from ferricyanic acid and containing the $Fe(CN)_6^{3-}$ ion.

ferrierite *Mineralogy.* $(Na,K)_2Mg(Si,Al)_{18}O_{36}(OH)\cdot 9H_2O$, a colorless to white orthorhombic and monoclinic mineral occurring in radiating groups of tabular crystals with a perfect cleavage, having a specific gravity of 2.13 to 2.15 and a hardness of 3 to 3.5 on the Mohs scale.

ferriferous *Geology.* relating to a mineral or sedimentary rock that is rich in iron.

ferrihemoglobin *Biochemistry.* the oxidized form of hemoglobin that cannot transport oxygen and causes the blood to acquire a brownish tinge. Also, METHEMOGLOBIN.

ferrimagnetic *Solid-State Physics.* relating to or characterized by ferrimagnetism.

ferrimagnetic amplifier *Electronics.* a microwave amplifier in which the coupling inductors and transformers are made of ferrite material.

ferrimagnetic crystals *Crystallography.* magnetically ordered materials in which the anti-ferromagnetically arranged spins are not completely canceled out; as a result, there is a net magnetic moment.

ferrimagnetic material *Solid-State Physics.* a substance that exhibits ferrimagnetism; a common example is a ferrite.

ferrimagnetic resonance *Physics.* a magnetic resonance that occurs in ferrites containing two or more sublattices with different magnetizations.

ferrimagnetism *Solid-State Physics.* a phenomenon in some magnetically ordered materials in which there is incomplete cancellation of the antiferromagnetically arranged spins giving a set magnetic moment; the equivalent magnetic behavior of a nonmetal to ferromagnetism in metals.

ferrimolybdite *Mineralogy.* $Fe_2^{+3}(Mo^{+6}O_4)_3 \cdot 8H_2O$, a yellow, probably orthorhombic mineral occurring in microcrystalline fibrous forms and incrustations, having a specific gravity of 4.5 and a hardness of 1 to 2 on the Mohs scale; found as an oxidation product of sulfide ores of molybdenum.

ferrimycin *Microbiology.* an antibiotic that gains entry into a target cell by its ability to bind iron, its toxic moiety released upon subsequent intracellular hydrolysis; a siderophore analogue.

ferrinatrite *Mineralogy.* $Na_3Fe^{+3}(SO_4)_3 \cdot 3H_2O$, an easily fusible, greenish or gray to white trigonal mineral having a specific gravity of 2.55 to 2.61 and a hardness of 2.5 on the Mohs scale; found as a secondary mineral with other sulfates.

ferriporphyrin *Biochemistry.* a porphyrin that contains an iron atom in the ferric (+3) state; it cannot transport or bind oxygen.

ferrisicklerite *Mineralogy.* $Li(Fe^{+3},Mn^{+2})PO_4$, a dark brown orthorhombic mineral massive in habit and having a specific gravity of 3.41 and a hardness of 4 on the Mohs scale; forms a series with sicklerite.

ferristor *Electronics.* a small, two-winding ferrimagnetic amplifier that functions at a high-carrier frequency; may be connected as a coincidence gate, current discriminator, free-running multivibrator, oscillator, or ring counter.

ferrite *Inorganic Chemistry.* a compound consisting of the iron oxides Fe_2O_3 and FeO, with the latter being replaceable by oxides of other transition metals. The most important chemical characteristic is the magnetic moment; used in computers, tape recorders, and a wide variety of communication devices. *Solid-State Physics.* a ferrimagnetic material having a chemical formula XFe_2O_4, where X represents a bivalent metal ion. *Electricity.* this material in a powdered, pressed, and sintered form, exhibiting high electrical resistivity that minimizes eddy current losses at high frequencies. *Petrology.* an unidentifiable red, brown, or yellow iron oxide occurring as more or less transparent or amorphous grains, scales, or threads in the matrix of a porphyritic rock. *Metallurgy.* an iron-base solid solution that has a body-centered cubic crystal structure.

ferrite banding *Metallurgy.* the attribute of a steel structure that contains parallel bands of ferrite.

ferrite bead *Electronics.* **1.** a device that stores information in a magnetic environment, characterized by a ferrite powder bonded onto the signal conductors of memory matrix. **2.** a small ferrite cylinder placed over current-carrying leads to block radio frequency.

ferrite circulator *Electromagnetism.* a nonreciprocal microwave network consisting of a 45° rotator between two dual-mode transducers; used for microwave control and switching.

ferrite core *Electronics.* an element made from ferrite that maintains its polarity when charged by a pulse; commonly used in a core memory.

ferrite-core memory *Computer Technology.* an older form of memory technology that consists of a toroid of magnetic material (core) through which wires are run; current through the wire creates a magnetic field in the core in one rotation (0) or the other (1).

ferrite isolator *Electromagnetism.* an isolator that passes energy in one direction and absorbs it in the other.

ferrite limiter *Electromagnetism.* a device that regulates a linear response in an antenna circuit or receiver circuit, used to protect sensitive receivers from burnout and from blocking by a strong interfering signal.

ferrite number *Metallurgy.* in welding, a standardized measure of the ferrite content of an austenitic stainless steel weldment.

ferrite-rod antenna *Electromagnetism.* a small-reception antenna constructed from a coil wrapped about a ferrite rod; commonly used in place of a loop antenna in a radio receiver. Also, LOOPSTICK ANTENNA.

ferrite rotator *Electromagnetism.* a ferrite cylinder inside a ring-type permanent magnet placed in a waveguide to rotate the polarization plane of a wave passing through the waveguide.

ferrite switch *Electromagnetism.* a device that controls the flow of microwave power in a waveguide by rotating the electric field vector by 90°.

ferrite-tuned oscillator *Electronics.* a circuit that generates alternating current, characterized by a ferrite-loaded cavity whose resonant frequency is varied by a magnetic field.

ferritic cryogenic steel *Materials.* an alloy steel that can be used at low temperatures.

ferritic malleable iron *Metallurgy.* a white cast iron that is heat treated to form graphite nodular in ferrite matrix to make it tough, ductile, and easy to machine.

ferritic stainless steel *Metallurgy.* any of several stainless steels that have a body-centered cubic crystal structure.

ferritic steel *Metallurgy.* any of several steels that contain up to 25% chromium and about 0.1% carbon; moderately ductile, strong, and highly resistant to corrosion; used mainly for making automotive trim.

ferritin *Biochemistry.* the iron-apoferritin complex that is one of the main forms in which iron is stored in the body, it occurs at least in the gastrointestinal mucosa, liver, spleen, and bone marrow.

ferritungstite *Mineralogy.* $(K,Ca,Na)(W^{+6},Fe^{+3})(O,OH)_6 \cdot H_2O$, a bright yellow cubic mineral occurring as dipyramidal crystals and in powdery form, having a specific gravity of 5 to 5.2 and an undetermined hardness; found in cavities in limonitic gossan.

ferro- a combining form meaning "iron," as in *ferrocarbon.*

ferroacoustic storage *Electronics.* a computer data-storage device in which a conductor passes through a thin tube of magnetostritive material and into an ultrasonic driving transducer.

ferroalloy *Metallurgy.* any of several iron-base alloys containing one or more elements, added in controlled amounts to a molten ferrous bath.

ferroaluminum *Metallurgy.* a ferroalloy whose principal element is aluminum; used as an addition or to deoxidize a ferrous melt.

ferroan dolomite **1.** see ANKERITE. **2.** see KUTNOHORITE.

ferroaugite *Mineralogy.* the iron-containing variety of augite.

ferroboron *Metallurgy.* a ferroalloy whose principal element is boron.

ferrocarbon titanium *Metallurgy.* a ferroalloy containing 15–19% titanium and 6–8% carbon.

ferrocene *Organic Chemistry.* $(C_5H_5)_2Fe$, a coordination compound of ferrous iron and cyclopentadiene occurring as orange crystals that melt at 174°C; used in fuels as a combustion control additive, catalyst, and antiknock agent.

ferrocerium *Metallurgy.* a ferroalloy having cerium as its principal element.

ferrochelatase *Enzymology.* an enzyme of the lyatase class that catalyzes the addition of iron to the protoporphyrin molecule, the final step in the biosynthesis of heme.

ferrochromium *Metallurgy.* any of several ferroalloys containing 55–72% chromium and variable amounts of carbon.

ferrocolumbium *Metallurgy.* a ferroalloy containing about 50–60% columbium and up to 8% silicon.

ferrocyanic acid *Inorganic Chemistry.* $H_4[Fe(CN)_6]$, a hypothetical acid from which the ferrocyanide salts are derived.

ferrocyanide *Inorganic Chemistry.* any salt derived from ferrocyanic acid and containing the $Fe(CN)_6^{4-}$ ion.

ferrocyanide process *Chemical Engineering.* a treatment process that removes mercaptans from petroleum fuels, using sodium ferrocyanide as a solvent.

Ferrod *Geology.* a suborder of the soil order Spodosol, characterized by being well-drained and containing a large amount of elemental iron.

ferroelectric *Solid-State Physics.* referring to the property of exhibiting spontaneous electric polarization and hysteresis in a crystalline substance; analogous to the spontaneous magnetic polarization in ferromagnetism.

ferroelectric converter *Electricity.* a converter that transforms thermal energy to electrical energy by a decrease or reduction of a dielectric constant of a ferroelectric material, such as barium strontium titanate, when heated beyond its Curie temperature.

ferroelectric crystal *Solid-State Physics.* a crystal displaying ferroelectric properties.

ferroelectric domain *Solid-State Physics.* a region in the vicinity of a ferroelectric crystal over which the spontaneous electric polarization is constant.

ferroelectric hysteresis *Electricity.* the dependent polarization of ferroelectric materials under a previously existing electric field. Also, DIELECTRIC HYSTERESIS, ELECTRIC HYSTERESIS.

ferroelectric hysteresis loop *Electricity.* a graphical representation of polarization or electric displacement against an applied electric field of a material that displays ferroelectric hysteresis.

ferroelectricity *Solid-State Physics.* the alignment of domains in a material, especially a crystal, so that a net polarization remains after the electric field is removed.

ferroelectric shutter *Optics.* a camera shutter that contains a plate of ferroelectric crystal positioned between polarizers whose planes are perpendicular and that is triggered to open with pulses of up to 100 volts.

ferrogabbro *Petrology.* a gabbro rock in which the pyroxene and olivine features have an exceptionally high iron content.

ferrograph analyzer *Engineering.* an instrument that separates wear particles from a lubricating oil into gradient sizes by pumping a small sample onto a microscope slide and then generating a high-gradient magnetic field.

ferrography *Engineering.* to examine machine-bearing surface wear with a ferrograph analyzer.

ferrohornblende see BARKEVIKITE.

ferromagnet *Physics.* any ferromagnetic substance.

ferromagnetic *Solid-State Physics.* relating to or characterized by ferromagnetism; capable of spontaneous magnetic polarization.

ferromagnetic amplifier *Electronics.* an amplifier whose operations are based on time variations within a parameter, such as resistance, and on ferromagnetic properties, such as polarization, at high radio-frequency power levels.

ferromagnetic crystals *Crystallography.* aligned spin ion domains; the material has bulk magnetic moment.

ferromagnetic domain *Solid-State Physics.* the region within a ferromagnetic crystal over which the microscopic magnetic moments are aligned in a parallel. Also, MAGNETIC DOMAIN.

ferromagnetic limiter *Electromagnetism.* a power limiter that operates on the nonlinear properties of ferromagnetic materials and is used in microwave systems in place of a transmit-receive tube.

ferromagnetic resonance *Solid-State Physics.* a condition in which the apparent magnetic permeability of a ferromagnetic substance reaches a maximum value when the substance is subjected to a transverse microwave field whose frequency is equal to that of the electron precession frequency.

ferromagnetics *Electronics.* the study of properties and applications of materials with high magnetic permeability.

ferromagnetic tape *Electromagnetism.* a tape made of ferromagnetic material that is used to wind magnetic cores.

ferromagnetism *Solid-State Physics.* a phenomenon that is exhibited by certain metals and alloys (particularly those of the iron group, rare-earth, and actinide series) in which the atomic magnetic moments are capable of spontaneous magnetic polarization, resulting in drastic magnetic effects; relative magnetic permeabilities of such materials range from 1.1 to 10^6.

ferromanganese *Metallurgy.* any of several ferroalloys containing about 75–90% manganese,with variable amounts of phosphorus and carbon.

ferromanganese

ferrometer *Engineering.* an instrument designed to perform permeability and hysteresis tests on iron and steel.

ferromolybdenum *Metallurgy.* a ferroalloy containing about 55–75% molybdenum.

ferronickel *Metallurgy.* a ferroalloy whose principal element is nickel.

ferrophosphorus *Metallurgy.* a ferroalloy containing about 18% or 25% phosphorus.

ferroporphyrin *Biochemistry.* a porphyrin that contains an iron atom in the ferrous (+2) state; it is the only one that can transport oxygen.

ferroresonant circuit *Electricity.* a resonant circuit in which a saturable reactor provides nonlinear characteristics with a varying circuit, voltage, or current, providing the mechanism for tuning. *Electronics.* a circuit, having two conditions and negative resistance, that is resonant at only one quantity of AC voltage.

ferroresonant static inverter *Electricity.* a static inverter consisting of an elemental square-wave inverter system and a tuned output transformer for filtering voltage regulation and current-limiting operations.

ferroselenium *Metallurgy.* a ferroalloy containing 50–60% selenium.

ferrosilicon *Metallurgy.* any of several ferroalloys containing 10–95% silicon.

ferrosilite *Mineralogy.* $(Fe^{+2},Mg)_2Si_2O_6$, a dark green to brown orthorhombic mineral of the pyroxene group, dimorphous with clinoferrosilite, having a specific gravity of 3.96 and a hardness of 5 to 6 on the Mohs scale, and usually massive in habit; found in basic igneous rocks and in some metamorphic rocks and meteorites. Also, ORTHOFERROSILITE.

ferrospinel see HERCYNITE.

ferrotapiolite *Mineralogy.* **1.** a group of tetragonal oxides of the general formula $A^{+2}B_2^{+5}O_6$, where A=Fe,Mg,Mn,Zn and B=Nb,Sb,Ta. **2.** a rare, weakly radioactive, brown to black mineral of this group, having the formula $(Fe^{+2},Mn^{+2})(Ta,Nb)_2O_6$.

ferrotitanium *Metallurgy.* a ferroalloy containing either about 40% titanium and little carbon or about 17% titanium and 7% carbon.

ferrotungsten *Metallurgy.* a ferroalloy that contains about 70–80% tungsten.

ferrouranium *Metallurgy.* a ferroalloy whose principal element is uranium.

ferrous *Chemistry.* **1.** of or relating to iron. **2.** describing various compounds of iron, especially those in which the element has a valence of 2. (From *ferrum,* the Latin word for iron.)

ferrous acetate *Organic Chemistry.* $Fe(CH_3COO)_2 \cdot 4H_2O$, combustible greenish crystals that oxidize to brown; used as a wood preservative and in medicine and dyes.

ferrous alloy *Materials Science.* any metal alloy based primarily on iron, such as steel, stainless steel, and cast iron.

ferrous ammonium sulfate *Inorganic Chemistry.* $Fe(SO_4)_2 \cdot (NH_4)_2 SO_4 \cdot 6H_2O$, light green deliquescent crystals that are affected by light, soluble in water and acids, and insoluble in alcohol; decomposes at 100–110°C; used in analytical chemistry and metallurgy. Also, IRON AMMONIUM SULFATE, MOHR'S SALT.

ferrous arsenate *Inorganic Chemistry.* $Fe_3(AsO_4)_2 \cdot 6H_2O$, a toxic green amorphous powder that is insoluble in water and decomposes on heating; used in insecticides.

ferrous bromide *Inorganic Chemistry.* $FeBr_2 \cdot 6H_2O$, a dark green deliquescent powder, soluble in alcohol and ether and very soluble in water; melts at 27°C; used as a catalyst.

ferrous carbonate *Inorganic Chemistry.* $FeCO_3$, grayish trigonal crystals, soluble in carbonated water and decomposed by heat; used in iron deficiency anemia.

ferrous chloride *Inorganic Chemistry.* $FeCl_2$ or $FeCl_2 \cdot 4H_2O$, green deliquescent crystals, soluble in water and alcohol and readily oxidized; used in metallurgy, dyeing, medicine, and sewage treatment.

ferrous fluoride *Inorganic Chemistry.* FeF_2, green crystals that are slightly soluble in water, insoluble in alcohol, and soluble in acids and ether; melts above 1000°C; used in ceramics and as a catalyst.

ferrous hydroxide *Inorganic Chemistry.* $Fe(OH)_2$, pale green crystals or a white amorphous powder, insoluble in water and decomposed by heat; it rapidly oxidizes in air to ferric hydroxide, Fe_2O_3, turning reddish brown.

ferrous oxide *Inorganic Chemistry.* FeO, a black powder that is insoluble in water and soluble in acids; melts at about 1420°C; used as a catalyst and glass colorant and in steel manufacture. Also, IRON MONOXIDE, BLACK IRON OXIDE.

ferrous phosphate *Inorganic Chemistry.* $Fe_3(PO_4)_2 \cdot 8H_2O$, a pale blue hygroscopic powder that is insoluble in water and soluble in acids; used as a catalyst and in ceramics.

ferrous phosphide *Inorganic Chemistry.* Fe_2P, blue-gray crystals or powder, insoluble in water; melts at 1290°C; used in iron and steel manufacture.

ferrous sulfate *Inorganic Chemistry.* $FeSO_4 \cdot 7H_2O$, bluish-green crystals that absorb water from the air; soluble in water and slightly soluble in alcohol; melts at 64°C and loses $6H_2O$ at 90°C, and $7H_2O$ at 300°C; used as a pigment, catalyst, and reducing agent, in fertilizers and herbicides, in water and sewage treatment, and for various other purposes.

ferrous sulfide *Inorganic Chemistry.* FeS, blackish metallic pieces or a black precipitate, soluble in acids and insoluble in water; melts at 1193–1199°C and decomposes at boiling point; used in ceramics and pigments and to produce hydrogen sulfide.

ferrovanadium *Metallurgy.* a ferroalloy containing 55–70% vanadium.

ferroxidase see CERULOPLASMIN.

ferrucite *Mineralogy.* $NaBF_4$, a colorless orthorhombic mineral occurring as crusts of minute crystals having a specific gravity of 2.496 and a hardness of 3 on the Mohs scale; found around fumaroles on Mount Vesuvius.

ferruginous *Science.* of, relating to, containing, or resembling iron.

ferrule *Mechanical Devices.* any of various rings, bands, or other such devices used to attach, protect, or stabilize an object.

ferrum *Chemistry.* the Latin name for iron, from which the symbol Fe is derived.

ferry *Naval Architecture.* a vessel used to shuttle people or vehicles between relatively nearby points on land; most often found along rivers, or in bays or enclosed seas.

Ferry trap *Materials Science.* an entanglement or loop in a polymer caused by a physical crosslink in an existing network structure; may slide between two chemical crosslinks yielding a node for stress relaxation.

fersmanite *Mineralogy.* $(Ca,Na)_4(Ti,Nb)_2Si_2O_{11}(F,OH)_2$, a brown triclinic mineral occurring as pseudotetragonal crystals with a vitreous luster, having a specific gravity of 3.44 to 3.46 and a hardness of 5 to 5.5 on the Mohs scale; found in nepheline pegmatites.

fersmite *Mineralogy.* $(Ca,Ce,Na)(Nb,Ta,Ti)_2(O,OH,F)_6$, a dark brown to black orthorhombic mineral occurring as euhedral prismatic crystals, having a specific gravity of 4.69 to 4.79 and a hardness of 4 to 4.5 on the Mohs scale; found with columbite in marble.

fertile *Biology.* producing offspring, or having the ability to produce offspring. *Botany.* specifically, producing or having the ability to produce fruit, pollen, or abundant growth.

Fertile Crescent theory *Anthropology.* an earlier theory stating that the origin of human civilization was centered in the area known as the Fertile Crescent, between the Nile Valley and the Mesopotamian floodplain.

fertile material *Nucleonics.* a material, such as uranium-238 or thorium-232, that can be split by thermal neutrons and converted into a fissile material (plutonium-239 and uranium-233, respectively).

fertility *Biology.* the state or quality of being fertile.

fertility factor see F PLASMID.

fertilization *Biology.* **1.** the union of a male and a female gamete to form a zygote. **2.** the act of insemination, impregnation, or pollination. *Physiology.* a process marking the beginning of conception, during which the nuclei of the sperm cell and the egg cell come together and their chromosomes combine.

fertilization cone *Developmental Biology.* a swelling of the surface of an ovum at the entry site of the effective spermatozoon. Also, ATTRACTION CONE.

fertilization membrane *Cell Biology.* a membranous layer, originating from the egg coat, that surrounds the egg after entry of a sperm to prevent polyspermy.

fertilizer *Materials.* a substance that improves the plant-producing quality of the soil, such as manure or a mixture of chemicals.

fertilizer grade or **fertilizer analysis** *Agronomy.* a statement of the percentages of the three major nutrients in a mixed fertilizer, expressed as a three-part number; a bag of 10-20-10 fertilizer contains 10% nitrogen, 20% phosphorus, and 10% potassium.

fertilizin *Biochemistry.* a substance in the plasma membrane of the ovum in certain species, such as sea urchins, that is involved with sperm recognition and increased motility.

ferulic acid *Organic Chemistry.* $C_{10}H_{10}O_4$, colorless needles (*trans* form) that are soluble in water and alcohol; decomposes at 174°C; a constituent of black fir resin, with traces found in many other plants; used as a food preservative.

ferv. or **Ferv.** boiling. (From Latin *fervens.*)

fervanite *Mineralogy.* $Fe_4^{+3}(VO_4)_4 \cdot 5H_2O$, an inadequately described golden-brown, fibrous monoclinic mineral associated with highly radioactive minerals in uranium-vanadium mining districts.

fervenulin *Microbiology.* am antibiotic derived from culture filtrates of *Streptomyces fervens.* Also, PLANOMYCIN.

fescue [fes´kyoo´] *Botany.* any of the grasses of the genus Festuca.

FES oncogene *Oncology.* one of several *src* oncogenes found in feline sarcoma virus which cause sarcomas.

festinant *Neurology.* accelerating or hastening; usually refers to walking, especially as a symptom of parkinsonism, in which increasingly rapid steps are taken.

festoon *Architecture.* a carved, molded, or painted chain of flowers, fruit, or leaves suspended between two points.

Festuca *Botany.* a genus that includes more than 100 species of commonly grown pasture and lawn grasses; found worldwide, mostly in temperate or cold regions.

FET field-effect transistor.

fetal [fēt´əl] *Biology.* of or relating to a fetus. *Medicine.* describing a condition that occurs in or affects this stage of development.

fetal alcohol syndrome *Medicine.* a congenital condition involving various abnormalities and growth or developmental defects, resulting from an excessive intake of alcohol by the mother during pregancy.

fetal asphyxia *Medicine.* asphyxia of the fetus while in the uterus due to oxygen deprivation caused by interference with its blood supply, as in cord compression or premature placental separation.

fetal calf serum *Biotechnology.* serum, collected from a calf fetus, that is highly valued as a cell culture medium because it contains many growth factors.

fetal hemoglobin *Biochemistry.* the hemoglobin that is present in a fetus; distinct from adult hemoglobin in that it has unique polypeptide chains and a higher oxygen transport capacity.

fetal membrane *Developmental Biology.* a structure or tissue that forms from the fertilized egg but is not a part of the embryo itself. Also, EXTRAEMBRYONIC MEMBRANE.

fetch *Computer Programming.* to load one or more instructions or data items from main memory into the central processor. *Oceanography.* **1.** an area of the sea over which a wind of constant direction and speed blows to generate waves. **2.** the length of such an area, as measured in the direction of the wind. Also, **fetch length.**

fetch cycle *Computer Programming.* the period in a computer cycle in which the next instruction to be executed is read from memory into the arithmetic-logic unit.

fetch-execute cycle *Computer Technology.* the basic execution cycle of a computer, in which an instruction is fetched from memory, the program counter is incremented, the instruction is executed, and the cycle then repeats.

fetish [fet´ish] *Anthropology.* a material object to which supernatural or magical powers are attributed. *Psychology.* a nonsexual object or part of the body that habitually causes sexual arousal or gratification.

fetishism *Anthropology.* the worship or honoring of fetishes. *Psychology.* a condition in which sexual impulses are habitually aroused or gratified by an object or body part that is not normally associated with sexual activity; e.g., shoes, gloves, the feet. Thus, **fetishist.**

fetography *Radiology.* a roentgenogram of a fetus in utero.

fetometamorphism *Invertebrate Zoology.* a life-cycle variation in the Cntharidae (Coleoptera); the larvae hatch prematurely as legless, immature prelarvae.

α-fetoprotein *Biochemistry.* the fetal equivalent of albumin.

fetus [fēt´əs] *Developmental Biology.* **1.** the unborn young of a viviparous animal after it has taken form in the uterus. **2.** specifically, a developing human offspring in the postembryonic period, from seven or eight weeks after fertilization to the time of birth.

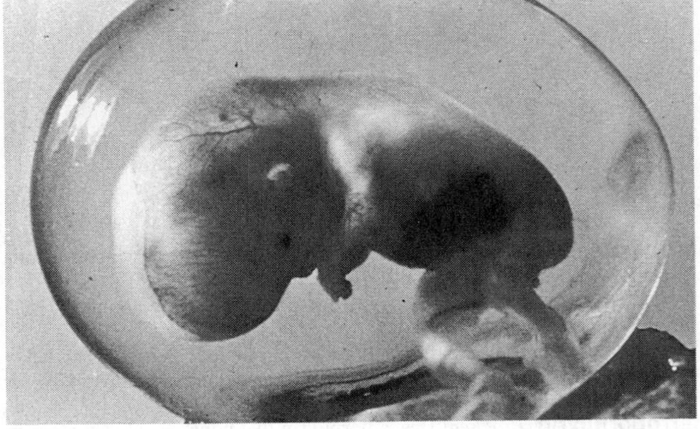

human fetus

feud or **feuding** *Anthropology.* a prolonged state of hostility or conflict between two factions within a larger cultural group involving intermittent attacks or violence, as opposed to the continuous fighting involved in warfare.

Feulgen reaction *Analytical Chemistry.* a test to distinguish aldehydes from ketones using a solution of rosaniline hydrochloride (fuchsin)-sulfurous acid; a purplish color indicates a positive test; different shades of purple are indicative of various aldehydes; a red color is produced by deoxyribonucleic acid after removal of purine bases.

fever *Medicine.* **1.** any elevation of the body temperature above the normal. **2.** any disease whose distinctive feature is elevation of body temperature.

Feyliniidae *Vertebrate Zoology.* the limbless skinks, a subfamily of reptiles in the family Scincidae, characterized by flattened heads, tiny eyes under transparent scales, and no ear opening; native to western and central Africa.

Feynman, Richard [fīn´mən] 1918–1988, American physicist; shared the Nobel Prize for developing an improved theory of quantum electrodynamics.

Feynman diagrams *Quantum Mechanics.* the schematic representations of mathematical expressions for predicting the interaction of particles, in which lines represent the path of a particle and vertices represent particle interactions.

Feynman path integral *Quantum Mechanics.* an integral over all possible paths that a particle may follow in moving from one point to another.

Feynman propagator *Quantum Mechanics.* a factor in a transition amplitude that represents a virtual particle and that corresponds to a line that connects two vertices in a Feynman diagram.

FF *Computer Programming.* the character for form feed.

F factor see F PLASMID.

FFI free from infection.

F format *Computer Programming.* fixed-length format, a data-set format in which the length of the field or record is specified; for decimal variables, the number of places to the right of the decimal point can also be specified.

FFT *Computer Science.* fast Fourier transform.

FG achromaticism see ACTINIC ACHROMATICISM.

FGF fibroblast growth factor.

F.h. let a draft be made. (From the Latin *fiat haustus.*)

FHP friction horsepower.

FHuA protein see TONA PROTEIN.

FI *Aviation.* the airline code for Icelandair.

FIA fluoroimmunoassay.

fiac *Biochemistry.* a nucleotide that combats herpes simplex viruses 1 and 2, varicella-zoster virus, and cytomegaloviruses in cell cultures.

fiber *Materials.* a thin, threadlike piece of any material. *Nutrition.* **1.** the structural part of plants and plant products consisting of carbohydrates that, when eaten, stimulate peristalsis in the intestine. **2.** food, such as whole grains, fruits, and vegetables, that contains large amounts of such carbohydrates. *Botany.* a thick-walled, narrow, elongated sclerenchyma cell, often lignified, that tapers at both ends. *Optics.* a fiber made of transparent material, such as glass, fused silica, or plastic, that is capable of conducting light signals by means of total internal reflection. *Mechanics.* an imaginary line in a solid whose linear extension describes the strain state of the solid in an intuitive way. *Mathematics.* given a bundle with base B and mapping π, the fiber at $x \in B$ is the topological space $\pi^{-1}(x)$, denoted F_x. All the fibers belonging to a given bundle are homeomorphic to a space F, which is called the **typical fiber.**

fiberboard *Materials.* a strong board made from an organic fiber, such as wood chips or bagasse, that is pressed or rolled with or without a binder; used in construction and to make paneling and containers.

fiber-bridging *Materials Science.* a phenomenon in which fibers in a ceramic-matrix composite span microcracks in the matrix and thereby strengthen the composite.

fiber bundle *Optics.* a flexible arrangement of optical fibers that is used to transmit either light signals (**incoherent fiber bundle**) or a complete optical image (**coherent fiber bundle**). *Mathematics.* a fiber bundle (E, B, π, G) is a bundle (E, B, π) together with a typical fiber F, a topological group G (called the structural group) of homeormophisms of F onto itself, and a covering of B by a countable family of open sets $\{U_j\}$ such that: (a) Locally, the bundle is trivial; that is, $\pi^{-1}(U_j)$ is homeomorphic to the topological product $U_j \times F$ for all j. The necessary homeomorphism ϕ_j: $\pi^{-1}(U_j) \rightarrow U_j \times F$ has the form $\phi_j(p) = (\pi(p), \hat{}\phi_j(p))$, and the following diagram commutes:

$$\boldsymbol{E}$$

$$\pi\downarrow \quad \downarrow\phi_j$$

$$\text{canonical projection} \cdot \phi_j = \pi$$

$$U_j \leftarrow U_j \times \boldsymbol{F}$$

$$\text{canonical projection}$$

For $x \in U_j$, $\hat{}\phi_j$ is restricted to the fiber F_x at x, or $\hat{}\phi_{j,x}$ to simplify notation, is a homeomorphism from F_x to F. (b) There is a correlation of the trivial subbundles defined on the open sets U_j covering the base B, given as follows: let $x \in U_j \cap U_k$. The homeomorphism $\hat{}\phi_{k,x} \cdot \hat{}\phi_{j,x}^{-1}$: $F \rightarrow F$ is an element of the structural group G for all such $x \in U_j \cap U_k$, and all j,k. If G has only one element, then the bundle is trivial. (c) The induced mapping $g_{jk}: U_j \cap U_k \rightarrow G$ given by $g_{jk}(x) = \hat{}\phi_{k,x} \cdot \hat{}\phi_{j,x}^{-1}$ is continuous. (E, B, π, G) is sometimes referred to as the ***E* bundle.** A fiber bundle (E, B, π, G) in which the typical fiber F and the structural group G are identical and in which G acts on F by left translation (group multiplication on the left) is called a **principal fiber bundle.**

fiber composite *Metallurgy.* a metal or alloy reinforced with fibers, including such materials as fiberglass-reinforced polymers.

fiber flax *Botany.* flax plants that are cultivated in such a way as to produce fibers suitable for processing into linen cloth.

fiberglass *Materials Science.* a glass in the form of fine, flexible fibers, widely used in the manufacturing of many industrial products such as textiles, filters, and insulators, and also used to reinforce or strengthen plastics for boat hulls, automobile bodies, aircraft parts, building panels, and other materials.

fiberglass wool *Materials Science.* a bulk form of fiberglass.

fiberizer *Engineering.* a device that converts material into fibers, especially one that beats asbestos rock into fiber or separates the fibers.

fiber metal *Metallurgy.* a feltlike material composed of metallic fibers.

fiber metallurgy *Metallurgy.* the art and science of manufacturing fiber-reinforced metallic materials.

fiber-optic *Materials Science.* **1.** relating to or composed of optical fibers. **2.** employing the technology of fiber optics.

fiber-optic current sensor *Engineering.* a test device for gauging current on high-voltage lines.

fiber-optic gyroscope *Engineering.* an instrument designed for rotation rate measurement.

fiber-optic magnetometer *Engineering.* a magnetometer in which light travels through optical fibers coiled around a distorted magnetostrictive body, producing phase changes that are measured by an interferometer.

fiber optics *Optics.* the branch of optical technology dealing with systems that transmit light signals and images over short and long distances through the use of optical fibers (transparent, hair-thin strands of glass or plastic). These fibers have a wide range of applications, as in the transmission of computer data, telephone messages, and other communications, in surgical laser instruments, in medical viewing or measuring devices, and in various other analytical instruments.

fiber-optic thermometer *Engineering.* a thermometer that measures crystal temperatures by guiding a light from a mercury lamp along an optical fiber to energize a small fluorescent crystal and causing its light to travel back along the fiber to a system, which then calculates the crystal temperature using ratios of the strengths of spectral lines in the fluorescent light.

fiber reinforcement *Materials Science.* the combination of two materials, one of which is a fiber, to produce a composite superior to either component alone; this occurs naturally in wood, and in synthetic materials such as fiberglass, boron-reinforced aluminum, graphite-epoxy, and aluminum oxide-reinforced aluminum.

fiber-saturation point *Materials Science.* the upper limit of moisture adsorption onto wood fibers; the limit of volume expansion by moisture absorption into the wood. Beyond this point, the cell cavity is filled and the wood becomes waterlogged.

fiberscope *Optics.* an optical device consisting of an eyepiece, a flexible fiber bundle, and an objective lens; the device is capable of transmitting a full-color image undisturbed when the bundle is bent, and therefore permits the viewing of objects that are inaccessible by direct sight.

fiber stress *Mechanics.* the tension or compression carried by a fiber or imaginary longitudinal element of a beam.

fiber tracheid *Botany.* a secondary xylem cell that is intermediate between a libriform fiber and a tracheid.

Fibiger, Johannes [fē′bə gər] 1867–1928, Danish pathologist; Nobel Prize for cancer research; the first to produce cancer experimentally.

Fibonacci, Leonardo [fē′bə nä′chē] c. 1170–1240, Italian mathematician; introduced Hindu-Arabic numbers to Europe; invented the Fibonacci sequence.

Fibonacci sequence *Mathematics.* **1.** a sequence a_n satisfying the recursion relation $a_n = a_{n-1} + a_{n-2}$. **2.** in particular, when the initial conditions $a_1 = a_2 = 1$ are satisfied, the resulting numbers {1, 1, 2, 3, 5, 8, 13, 21, . . . } are known as the **Fibonacci numbers.** A closed expression for the *n*th Fibonacci number is $a_n = [((1 + \sqrt{5})/2)^n - ((1 - \sqrt{5})/2)^n]/\sqrt{5}$ and the generating function is given by $f(z) = (1 - z - z^2)^{-1}$.

fibr- a combining form meaning "fiber" or "fiberlike," as in *fibratus.*

fibratus *Meteorology.* a cloud species characterized by fine hairlike striations with distinct, separated filaments whose ends are thin without tufts or hooks. Also, FILOSUS.

fibril *Biology.* a threadlike growth or tiny fiber, such as a root hair or a fine strand of a striated muscle. *Anatomy.* a fine threadlike structure either inside or outside of cells giving cellular stability. *Materials Science.* any of the small, threadlike components of a fiber; a small fiber.

fibrillar crystallization *Materials Science.* a configuration of certain polymers in which the random coil chain is interlaced with crystalline segments, creating a lumpy chain. Also, SHISH-KEBAB MORPHOLOGY.

fibrillation [fib′ri lā′shən] *Physiology.* the spontaneous contraction of muscle fibers resulting in uncoordinated muscular activity.

fibrillose *Biology.* having threadlike growths or tiny fibers.

fibrin [fī′brin] *Biochemistry.* an insoluble protein that forms a network of fibers during blood clotting.

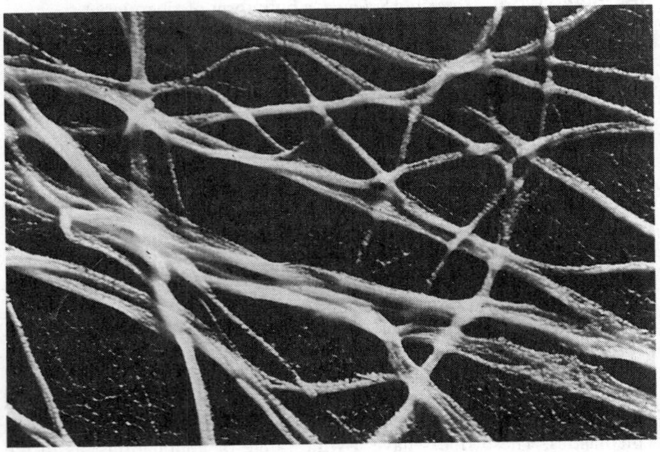

fibrin strands

fibrinase *Enzymology.* an enzyme that catalyzes the formation of covalent bonds between fibrin molecules, thus enabling fibrin to form a firm blood clot. Also, COAGULATION FACTOR XIII, FIBRIN-STABILIZING FACTOR.

fibrinogen [fī brin′ō jin] *Biochemistry.* the protein secreted by the liver into blood plasma that forms the fibrin used to clot blood.

fibrinogenopenia *Medicine.* a deficiency of fibrinogen in the blood.

fibrinoginase see THROMBIN.

fibrinoid *Biochemistry.* a substance that occurs in degenerating connective tissues.

fibrinolysis *Physiology.* the digestion of substances of a blood clot by a protein-splitting enzyme, resulting in dissolution of the clot.

fibrinous pericarditis *Medicine.* an inflammation of the pericardium marked by deposition of fibrin and leukocytes between the layers of the pericardium.

fibrin-stabilizing factor see FIBRINASE.

Fibrist *Geology.* a suborder of the soil order Histosol, composed primarily of undecomposed plant fiber and characterized by being water-saturated most of the time.

fibro- a combining form meaning "fiber" or "fiberlike," as in *fibrocartilage.*

fibroadenoma *Medicine.* a tumor of the breast that is round, movable, and firm, caused by greater than usual amounts of estrogen.

fibroblast *Histology.* the characteristic cell of fibrous tissue that produces collagenous fibers.

fibroblast growth factor *Endocrinology.* either of two members of a family of heparin-binding growth factors, **acidic** or **basic fibroblast growth factor,** which are mitogenic for a variety of cell types of mesodermal, neuroectodermal, epithelial, and endothelial origin.

fibroblastic *Petrology.* of or relating to a homeoblastic texture in metamorphic rocks due to development during crystallization of minerals with a fibrous or prismatic habit. Also, NEMATOBLASTIC.

fibroblastoma *Oncology.* a tumor that originates from a fibroblast, or connective tissue cell.

fibrocartilage *Histology.* a type of cartilage found in intervertebral disks, the pubic symphysis, and certain tendons; characterized by dense deposits of collagenous fibers. Thus, **fibrocartilaginous.**

fibrocartilaginous joint see SYMPHYSIS.

fibroepithelioma *Oncology.* a tumor that is composed of fibrous and epithelial elements.

fibroferrite *Mineralogy.* $Fe^{+3}(SO_4)(OH)\cdot 5H_2O$, a fusible, pale yellow to white, monoclinic mineral that decomposes in water, occurs in fibrous aggregates, and has a specific gravity of 1.92 to 1.95 and a hardness of about 2.5 on the Mohs scale; found as a secondary mineral formed by the oxidation of pyrite.

fibroid *Histology.* constructed of fibrous tissue.

fibroin *Biochemistry.* the protein component of silk fiber derived from silkworms and spiders, which form threads to create webs and cocoons.

fibroma *Medicine.* a noncancerous tumor largely made up of fibrous or fully developed connective tissue. Also, **fibroid, fibroid tumor.**

fibromatosis *Medicine.* the occurrence of multiple benign tumors of fibrous connective tissue.

fibromyoma *Medicine.* a benign tumor or leiomyoma of the smooth muscle.

fibromyositis *Medicine.* an inflammation of fibromuscular tissue.

fibronectin *Biochemistry.* a protein that interacts with a variety of extracellular substances in plasma and tissues (such as collagen, fibrin, and heparin) to help cells, especially fibroblasts, bind to each other; aids in tissue repair.

fibroneuroma see NEUROFIBROMA.

fibroplasia *Medicine.* a growth of fibrous connective tissue, as in the second phase of wound healing.

fibrosarcoma *Oncology.* a malignant tumor that contains connective tissue and develops suddenly from small bumps on the skin; derived from collagen-producing fibroblasts.

fibrosis [fī brō′sis] *Medicine.* **1.** a fiberlike connective tissue that occurs normally in the growth of scar tissue. **2.** the abnormal spread of fiberlike connective tissue over or replacing normal smooth muscle or other normal organ tissue.

fibrositis *Medicine.* an inflammation of fibrous connective tissue, usually characterized by pain and stiffness; often occurring in middle age. Also, MUSCULAR RHEUMATISM.

fibrous composite *Materials.* any composite material having fibers embedded in a matrix.

fibrous concrete *Materials.* concrete in which a fibrous aggregate such as asbestos or sawdust is incorporated into a sand-gravel mixture.

fibrous cortical defect see NONOSTEOGENIC FIBROMA.

fibrous dysplasia *Medicine.* an abnormal condition characterized by the fibrous displacement of bone tissue; symptoms include pain, disability, and gradually increasing deformity.

fibrous fracture *Mechanics.* the rupture of a ductile material; the dull fibrous appearance of the face is due to void coalescence.

fibrous joint *Anatomy.* any joint in which the bony elements are connected by continuous intervening fibrous tissue, making very little motion possible.

fibrous odontoma see ODONTOGENIC FIBROMA.

fibrous plaster *Materials.* a plaster slab formed by stretching canvas across a wooden frame and coating it with a layer of gypsum plaster; used as a building material.

fibrous protein *Biochemistry.* a class of highly insoluble proteins, such as α-keratin in hair and wool, fibroin in silk, and collagen in tendons, in which the polypeptide chain extends along one axis instead of folding in globular fashion; plays a structural or protective role.

fibrous skeleton *Cardiology.* the dense and mainly fibrous structure that supports the musculature of the heart.

fibrous structure *Metallurgy.* **1.** the structure of a metal or alloy that contains laminations or slag fibers. **2.** the structure of a rolled steel plate that is fine grained. Also, SILKY STRUCTURE.

fibrovascular [fī′brō vas′kyə lər] *Botany.* made up of woody fibers and ducts.

fibrovascular bundle *Botany.* a vascular bundle with associated scherenchyma, as in a leaf or a vein.

fibula [fib´yə lə] *Anatomy.* a slender bone of the lower and outer part of the leg that articulates with tibia above and the tibia and talus below.

fibular *Anatomy.* of or relating to the fibula.

ficin *Organic Chemistry.* a proteolytic enzyme occurring as a buff-colored powder that hydrolyzes fibrin, collagen, and other proteinlike material; used in food processing.

Fick, Adolph Eugen 1829–1901, German physiologist; developed fundamental laws of diffusion in living organisms.

Fickian absorption *Materials Science.* the intake of moisture by a carbon-fiber composite, resulting in a reduction of the glass transition temperature of the material.

Fick's first law *Biochemistry.* a method of describing the rate of diffusion of a substance across a surface element, as given by a formula expressing the proportionality between diffusion flux J and concentration gradient dc/dx for solute atoms in a solid solution. In this equation, $J = -D\ dC/dx$, where D is the diffusion coefficient.

Fick's principle *Cardiology.* a rule stating that the heart output is directly proportional to oxygen absorption divided by the arterial oxygen minus the mixed venous oxygen. This principle is an application of Fick's laws of diffusion.

Fick's second law *Biochemistry.* an extension of Fick's first law that is used for cases of non-steady-state diffusion in which the diffusivity is independent of time; for the one-dimensional case $dC_x/dx = (d/dx)(D(dC_x/dx))$.

ficoll *Biochemistry.* **1.** a nonionic synthetic polymer that is produced by a crosslinking reaction of epichlorohydrin and sucrose; it increases the viscosity in a medium to temper rapidly motile organisms; used in ciliar and flagellar motility studies. **2. Ficoll.** a trade name for this substance.

fictile *Science.* changing in shape or arrangement; without a permanent form or structure. Also, **fickle.**

fictitious *Cartography.* of or relating to a measurement made from an arbitrary reference line. *Navigation.* of or relating to an imaginary object that serves some useful purpose. *Mechanics.* see FICTITIOUS FORCE.

fictitious craft *Navigation.* an imaginary craft used to solve some maneuvering problem. Similarly, **fictitious ship, fictitious vehicle.**

fictitious equator *Navigation.* in a map made by the oblique cylindrical projection method, the great circle at which the cylinder is tangent to the earth.

fictitious force *Mechanics.* a force that is introduced into an accelerated or rotating coordinate system so that the system will satisfy Newton's laws; for example, the Coriolis force or the centrifugal force.

fictitious graticule *Navigation.* a network of lines on a map or chart representing fictitious parallels and fictitious meridians.

fictitious latitude *Navigation.* a coordinate based on the parallels of a fictitious graticule. Thus, **fictitious parallel.**

fictitious longitude *Navigation.* a coordinate based on the meridians of a fictitious graticule. Thus, **fictitious meridian.**

fictitious pole *Navigation.* the points on a transverse or oblique cylindrical projection that are 90° away from the great circle at which the projection is tangent to the earth.

fictitious year *Astronomy.* see TROPICAL YEAR.

Ficus [fī´kəs] *Botany.* **1.** a genus of woody, tropical or deciduous evergreen trees, shrubs, and vines of the mulberry family, characterized by a milky sap, small unisexual flowers, and fleshy, bulbous receptacles that contain their seeds; includes the fig, the banyan, and many species grown as ornamentals. **2. ficus.** any tree belonging to this genus.

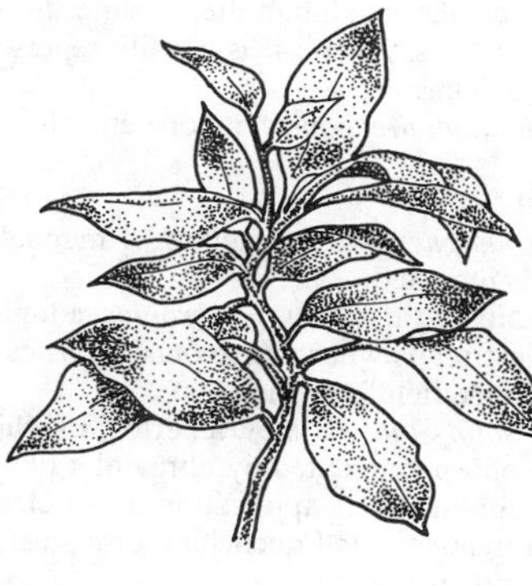

ficus

fidelity *Behavior.* the tendency of an animal to return to the same precise locale, such as its birthplace, a breeding place, winter quarters, and so on. Also, TENACITY. *Acoustics.* a measure of the ability of an electronic amplifier to accurately reproduce the characteristics of an input signal.

fido or **FIDO** *Meteorology.* an artificial fog-dissipating system in which gasoline or other fuel is burned at intervals along an airstrip. (An acronym for fog investigation dispersion operations.)

FIDO torpedo *Ordnance.* an air-launched passive acoustic homing torpedo used by the U.S. Navy and RAF Coastal Command as an antisubmarine weapon during World War II.

fiducial *Science.* accepted as a fixed basis of reference or comparison. Thus, **fiducial point, fiducial object.**

fiducial marker *Cartography.* any of a set (usually four) of small objects rigidly fastened to the interior of a camera's body so that they are photographed along with the image during the exposure.

fiducial temperature *Meteorology.* the temperature in a specified latitude at which the reading of a particular barometer requires no temperature or latitude correction.

fiedlerite *Mineralogy.* $Pb_3Cl_4(OH)_2$, a colorless to white, transparent monoclinic mineral occurring in commonly twinned tabular crystals, having a specific gravity of 5.88 and a hardness of about 3.5 on the Mohs scale; found with other secondary lead minerals in ancient lead slags that have been exposed to sea water.

Fiedler's myocarditis see INTERSTITIAL MYOCARDITIS.

field a location or area of activity; specific uses include: *Agriculture.* an open area cleared of larger vegetation and used for crops or as pasture land. *Science.* describing research or exploration that occurs outside an office or laboratory. *Military Science.* **1.** an area where a battle is fought or will be fought. **2.** any area where military or intelligence operations are carried out. **3.** of or relating to personnel, weapons, equipment, supplies, or activities used in or intended for such an area. Thus, **field exercise, field emplacement, field training, field fortification(s).** *Surgery.* the active area in an operation. *Geology.* a region characterized by having a particular mineral resource. *Hydrology.* see ICE FIELD, def. 2. *Physics.* **1.** an abstract representation of the idea that matter modifies the space around it, as when a mass sets up a gravitational force in the space around it. **2.** thus, any region, volume, or space in which a physical force is operative and influential. *Electricity.* specifically, a region in which an electrical force is active. *Electronics.* one of two or more equal parts into which a television frame is divided in interlaced scanning. *Computer Programming.* a set of one or more characters treated as a whole; a data element. *Mathematics.* **1.** a commutative division ring. **2.** see TENSOR FIELD.

field army *Military Science.* a military unit made up of two or more corps.

field artillery *Ordnance.* highly mobile artillery used or intended for field operations against ground forces; distinguished from medium, heavy, antiaircraft and antitank artillery.

field artillery observer *Ordnance.* a person who watches the effects of artillery fire, adjusts the center of impact of that fire onto the target, and reports the results to the firing agency.

field capacity *Hydrology.* the maximum amount of water held in a given soil against gravity after excess water has drained away. Also, **field-moisture capacity.**

field changes *Meteorology.* the rapid variations in the vertical component of the electric field strength of thunderstorm electricity at the earth's surface.

field classification *Photogrammetry.* field inspection and identification of features that a map compiler is unable to delineate, such as political boundary lines, place names, road classifications, and others.

field coil *Electromagnetism.* a coil of wire used to produce a magnetic field in an electromagnetic machine when current is driven through the coil.

field comparator *Cartography.* a short line measured with accuracy and precision; designed to check the lengths of apparatus used in actual field operations.

field crop *Agriculture.* a crop, such as corn or wheat, that requires a relatively large area of land to grow in relation to its profitable yield.

field data code *Telecommunications.* a standard military data transmission code consisting of seven data bits and a single parity bit.

field delimiter *Computer Programming.* a special character or other symbol used in a high-level language to provide an explicit boundary between adjacent fields; examples are spaces, slashes, parentheses, and semicolons.

field designator *Computer Programming.* a character or special symbol appearing in a field that indicates the nature of the contents of the field; for example, altitude may be expressed as feet above sea level or millimeters of mercury, and the program must be able to interpret the data properly.

field desorption *Solid-State Physics.* a method of producing a clean surface with several crystallographic orientations by applying a very strong electric field to the surface of a crystal to remove the surface atoms.

field-desorption mass spectroscopy *Spectroscopy.* a mass spectroscopic technique for the analysis of nonvolatile molecules that involves the use of high electric-field gradients and heating to cause surface desorption resulting in ion emission.

field-desorption microscope *Electronics.* a field-ion microscope in which the specimen is imaged onto a fluorescent screen using ions field-desorbed or field-evaporated from the surface of the specimen, rather than using ions supplied by an external gas.

field discharge *Electronics.* a spark that appears when a sudden change in current causes a change in voltage across a gap in a circuit.

field-discharge switch *Electricity.* a switch provided with special contacts so that a discharge resistance is connnected across the winding at the moment of breaking the field circle; used to control the field circuit of a motor or generator.

field distortion *Electromagnetism.* a distortion in the magnetic field between the poles of a generator due to the counterelectromotive force in the armature winding.

field-effect capacitor *Electronics.* a type of capacitor in which the conductive element is placed in a region of the semiconductor material depleted or inverted by an electric field.

field-effect device *Electronics.* any device whose principal characteristics are controlled by that part of the device that is under the influence of an external electric field.

field-effect diode *Electronics.* a two-terminal device in which the electrons influenced by the electric field have only one polarity.

field-effect display *Optics.* a numerical display consisting of a liquid crystal cell contained between two polarizers, the cell treated in such a manner that light is rotated 90°.

field-effect phototransistor *Electronics.* an amplification device that receives modulated light as an input signal and then modulates it further by placing it in an external electrical field.

field-effect tetrode *Electronics.* a four-terminal semiconducter device characterized by two autonomous channels whose ability to conduct current is modulated by the voltage conditions in the other.

field-effect transistor *Electronics.* a device, used to amplify a signal, in which the resistance to current flow is modulated by exposing the signal to an external electrical field.

field-effect transistor resistor *Electronics.* a field-effect transistor in which the gate is connected to the drain; used as a load for another transistor.

field-effect varistor *Electronics.* a varistor whose properties are controlled by an external electric field.

field emission *Electronics.* the excessive electron flow that occurs when an intense electric field appears between the plate and the cathode in a tube.

field-emission microscope *Electronics.* an instrument that produces a magnified image on a fluorescent screen from electrons emitted by the target.

field-emission tube *Electronics.* an electron tube in which field emission emanates from a sharp metal point.

field engineer *Engineering.* **1.** a person who supervises civil, mechanical, and electrical engineering projects in the petroleum and natural gas industries. **2.** a person who supervises a construction site crew. *Computer Technology.* a representative of a computer hardware manufacturer who is responsible for installation and maintenance of equipment at the customer's site.

field-enhanced emission *Electronics.* the increase in the number of electrons emitted from a material when an electric field nears its surface.

field flattener *Optics.* a lens that corrects for curvature of field, used to project an image onto a flat surface.

field focus *Geophysics.* an inexact term referring to the total area, discerned through observation of its effects, from which an earthquake originates.

field-free emission current *Electronics.* the current that flows from a cathode when it is not surrounded by an electric field.

field frequency *Electronics.* the number of fields transmitted per second in a television system, calculated by multiplying the frame frequency by the number of fields contained in one frame. Also, FIELD REPETITION RATE.

field galaxy *Astronomy.* a galaxy that is not member of a cluster of galaxies.

field geology *Geology.* the study of local geology by direct observation in a selected outdoor area.

field gradient *Physics.* the spatial rate of change of a field; a vector field obtained by applying the gradient operator to a scalar field.

field gun *Ordnance.* a mobile gun used or intended for field operations. Also, FIELD PIECE.

field hospital *Military Science.* a temporary hospital or medical facility set up in a combat area.

field ice *Oceanography.* a general term used to refer to all types of sea ice except new ice.

field inspection *Cartography.* the process of comparing aerial photographs with conditions as they exist on the ground, and of obtaining information to supplement or clarify that which is not readily discernible on the photographs themselves.

field intensity *Physics.* the strength of a vector field, typically characterized graphically by the concentration of field lines. *Telecommunications.* the value of an electric or magnetic field at a given point, usually in a horizontal direction; most often expressed as volts per meter for electric fields and amperes per meter for magnetic fields.

field ionization *Electronics.* a process by which gaseous atoms are ionized in an intense electric field, generally occurring near the poles of the field.

field-ion microscope *Electronics.* a microscope in which the atomic structure of the surface of a conductive material is magnified by introducing helium gas and applying a high voltage to ionize and accelerate the gas toward a fluorescent screen. Also, ION MICROSCOPE.

field length *Computer Programming.* the number of characters or bits contained in a field.

field lens *Optics.* **1.** a supplemental lens element used to increase the field of view in an optical system, adjust the imaging defects of that system, or redirect light from the subject being viewed into any subsequent lenses and toward the eyepiece. **2.** in a compound eyepiece, the lens closest to the objective.

field-line reconnection *Astrophysics.* a rearrangement of magnetic field lines that occurs when a new sunspot pair forms near an older pair: the leading spot of the first pair links its field with that of the trailing spot of the second pair to form a high-energy field, while the trailing spot of the first pair links its field with the leading spot in the second pair to create a lower-energy field.

field luminance see ADAPTATION LUMINANCE.

field magnet *Electromagnetism.* a permanent magnet that provides a strong magnetic field in an electrical device.

field map *Cartography.* a rough map that is developed in the field from direct measurement and observation, then used as a basis for a final map.

field marshal *Ordnance.* the highest army rank in many countries.

field moisture *Hydrology.* water that is present in the ground above the water table.

field officer *Military Science.* an officer with a rank of major, lieutenant colonel, or colonel.

field of fire *Ordnance.* the area that may be covered effectively by a weapon or group of weapons firing from a given position.

field of search *Electronics.* the section of space that a given radar system can effectively survey.

field of view *Optics.* the maximum area visible through a lens or the eyepiece of an optical instrument; it is usually represented by an angle and sometimes by a diameter.

field operator *Quantum Mechanics.* an operator that creates or annihilates a particle.

field piece see FIELD GUN.

field pole *Electromagnetism.* a structure of magnetic material upon which a field coil is mounted.

field-programmable logic array *Electronics.* a logic composed of an array of basic logic circuits whose interconnections can be programmed (set or changed) "in the field" (by the user).

field quenching *Solid-State Physics.* an effect in which the amount of light emitted by a phosphor excited by ultraviolet radiation or X-rays is decreased by the simultaneous application of an electric field. *Metallurgy.* the thermal treatment and quenching of a steel product at a construction site, rather than at a mill.

field rations *Military Science.* packaged or processed food rations intended to be eaten by personnel in combat areas.

field repetition rate see FIELD FREQUENCY.

field rheostat *Electricity.* a rheostat whose setting determines the current flow rate through a motor field coil.

field scan *Electronics.* a procedure in which an electron beam travels down the face of a cathode-ray tube to scan alternate lines on a television screen.

field selection *Computer Technology.* the process of isolating a specific data field within a computer word.

field-sequential color television *Telecommunications.* a color television system in which the individual primary colors are associated with sequential television fields.

field shift *Nuclear Physics.* the degree of displacement produced by alterations in the size and shape of the nuclear charge distribution that occurs when the number of neutrons in a nucleus increases. Also, VOLUME SHIFT.

Field's stain *Microbiology.* a biological stain used to detect parasites of the genera *Plasmodium* and *Trypanosoma* in blood smears.

field star *Astronomy.* a term for a star that belongs to no particular star cluster.

field stone or **fieldstone** *Materials.* any stone that is found in nature and is subsequently used in construction without being altered from its original condition.

field stop *Optics.* an aperture situated at the focal plane of an optical system that determines the boundaries of the field of view, limits the size of the image, and prevents rays of poor imaging quality from advancing to the image plane.

field strength *Physics.* the magnitude of a vector field.

field-strength meter *Engineering.* a radio receiver designed to calculate the field strength of electromagnetic energy from a radio transmitter.

field-strip *Ordnance.* to take apart a gun or other firearm for cleaning, inspection, or other purpose.

field telephone *Telecommunications.* a portable telephone designed for military use.

field theory *Physics.* any theory in which the basic quantities are fields, such as electromagnetic theory, which studies the interaction of electric and magnetic fields. *Psychology.* the theory that each person views his or her psychological environment as a whole, or field, and that the person's behavior is primarily determined by this view.

field-upgradable *Computer Technology.* of or relating to a device that is capable of being enhanced at the customer's site or at the local computer store, as opposed to being returned to the manufacturer for modification.

field weld *Metallurgy.* a weld performed at a construction site, rather than in a shop.

field wire *Electricity.* an insulated wire or cable with a fair degree of flexibility used in telephone and telegraph applications.

fieldwork *Science.* research or exploration that is carried out in an actual setting in the natural environment, rather than in a laboratory or other such facility. Similarly, **field study.**

fievre boutonneuse [fē ev′rə boo′tə nuz′] *Medicine.* a tick-borne typhus fever of characterized by headache, fever, skin rash, and a generally favorable prognosis. The disease is widely distributed along the Mediterranean, in Africa, and in the Indian subcontinent. (From French; literally, "pimply fever.")

fife rail *Naval Architecture.* a rail around the base of a mast, or along a ship's sides, with holes to hold belaying pins.

FIFO [fī′fō′] *Computer Programming.* a technique for storing and retrieving items from a list, table, or queue, in which the first item stored is the first item retrieved. *Industrial Engineering.* a management technique in which issues are dealt with in the order they arise, rather than lesser-priority issues being put aside for later decision. (An acronym for first-in, first-out.)

fifth-generation computer *Computer Technology.* a proposed knowledge-based computer system intended to combine sets of facts to create new information without explicit procedures to follow, and possibly to learn, reason, and process natural (human) languages. Such a computer system was pursued at the ICOT project in Japan.

fifth-generation language *Artificial Intelligence.* a nonprocedural artificial intelligence language in which the programmer defines a series of patterns of information to be recognized and manipulated. Examples are PROLOG, LISP, and various knowledge system shells. Also, **5GL.**

fifth nerve see TRIGEMINAL NERVE.

fifth sound *Physics.* a variation in temperature observed in helium II when it is contained in a tightly packed powder.

fifty-percent zone *Ordnance.* a term for the area centered around the mean point of dispersion or impact of a group of shots, within which half the shots fired at the same setting are expected to fall.

fig *Botany.* **1.** the common name for any of several species of trees of the genus *Ficus.* **2.** the fruit of those trees.

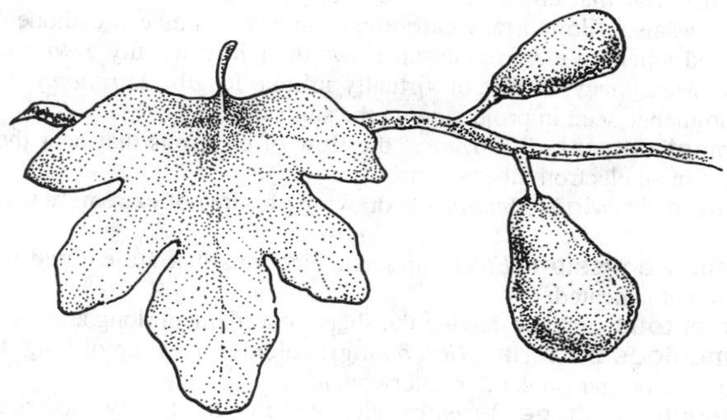

fig

fighter *Aviation.* **1.** a type of military aircraft possessing high speed, acute maneuverability, and a steep rate of climb, employed primarily to intercept and destroy missiles and/or other aircraft. **2.** in general, any relatively small, high-speed, multipurpose military aircraft designed for ground support and interdiction. Also, **fighter aircraft.**

fighter-bomber *Aviation.* a fighter aircraft that may also carry bombs or rockets. Also, **fighter ground-attack aircraft.**

fighter interceptor *Aviation.* a fighter aircraft that is designed to intercept and destroy enemy aircraft.

fighting compartment *Ordnance.* the area in a fighting vehicle where the main weapon system is fired and serviced.

Fighting Falcon *Aviation.* a popular name for the F-16 fighter airplane.

fighting load *Ordnance.* clothing, equipment, weapons, and ammunition that are carried by an individual soldier and necessary for the effective execution of a specific combat mission.

fight-or-flight reaction *Physiology.* a stage of reaction to stress in which the body is mobilized either to attack or flee from the stressful object. Also, **fight-or-flight syndrome.**

Figi disease *Plant Pathology.* a disease of sugarcane caused by a leafhopper-transmitted virus and resulting in elongated swellings on the underside of leaves and the eventual death of the plant.

Figitidae *Invertebrate Zoology.* a small family of wasplike, shining-black insects in the order Hemiptera; the larvae are parasites of insects and insect larvae.

FIGO or **F.I.G.O.** International Federation of Gynecology and Obstetrics.

figurative constant *Computer Programming.* a reserved word, such as ZERO in COBOL, that represents a numeric value, a character, or a string of repeated values or characters.

figure *Graphic Arts.* **1.** an Arabic numeral. **2.** an illustration in a printed work, especially numbered illustrations in a scholarly text.

figure-ground *Psychology.* the principle that visual information is received in the form of a separate figure or object standing out from a less-defined background. Thus, **figure-ground perception, figure-ground organization.**

figurehead *Naval Architecture.* a carving, typically of a female human figure, mounted above the cutwater of a vessel; a traditional feature of wooden sailing ships.

figure of merit *Electricity.* a performance rating that governs the choice of a device for a particular application; e.g., for a magnetic amplifier, the ratio of power amplification to control time constant. *Crystallography.* in X-ray diffraction studies of crystals, an estimate of the average precision in the selection of phase angles; used particularly in protein crystallography where phase angles are derived by isomorphous replacement methods.

figure of the earth see GEOID.

figures shift *Telecommunications.* in the five-level baudot code, a code combination that causes all following code combinations to be interpreted as uppercase lettters, numbers, and special characters.

figure stone see AGALMATOLITE.

Fijivirus *Plant Pathology.* a type subgroup of Phytorevirus that causes the sugarcane Fijivirus, a disease in which white to brown swellings occur on the undersides of the sugarcane leaves, and stunting and death of the plant follow.

filament a very slender thread or threadlike structure; specific uses include: *Biology.* a strand of cells in a long row. *Botany.* the stalk of an anther-bearing stamen. *Electricity.* a thread of tungsten, carbon, or similar material that emits light when heated by an electric current. *Electronics.* also, **filamentary cathode.** in an electron tube, a cathode that is heated when an electric current flows though it; usually a wire or ribbon. *Metallurgy.* a fiber of virtually infinite length. *Astronomy.* a solar prominence seen in projection on the sun's disk.

filament current *Electronics.* the flow of electrons that heat the filament in an electron tube.

filament drawing *Metallurgy.* drawing of wires of extremely small diameters.

filament emission *Electronics.* the electrons that are emitted when a filament is heated.

filamentous *Science.* having the shape of a filament; long and slender.

filamentous bacteria *Microbiology.* colonies made up of long threadlike microorganisms, often interwoven.

filamentous phage *Virology.* any phage of the family Inoviridae; an inovirus.

filament saturation see TEMPERATURE SATURATION.

filament transformer *Electronics.* a device that supplies the filament in an electron tube with current or voltage.

filament winding *Electronics.* in a transformer that provides several voltages for the operation of a vacuum tube circuit, the winding that provides the filament current. *Engineering.* a method of fabricating a composite structure in which continuous fiber reinforcement is wound tightly over a rotating core.

filamin *Biochemistry.* a protein that binds to F-actin, to crosslink microfilaments, thus inducing gel formation.

filaree see ALFILARIA.

Filaria *Invertebrate Zoology.* the type genus of nematode worm causing elephantiasis and filariasis in humans; adults parasitize the blood and tissue of mammals, while larvae are found in biting insects.

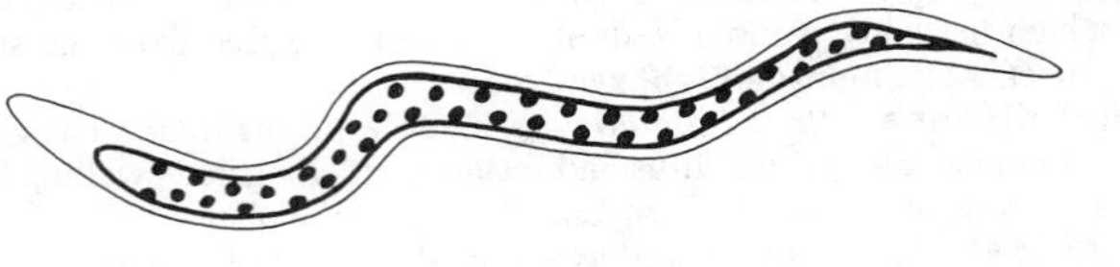

Filaria

filariasis see ELEPHANTIASIS.

filariform larva *Invertebrate Zoology.* a larval nematode resembling a filaria, with a long body and a long esophagus that is a delicate capillary tube of uniform diameter.

Filarioidea *Invertebrate Zoology.* threadlike nematode worms in the order Spirurida, parasites in birds and mammals; larvae are found in bloodsucking insects.

filar micrometer *Mechanical Devices.* an instrument mounted on a telescope or microscope consisting of a parallel set of wires, a stationary wire and a movable wire, both of which can be adjusted by a finely threaded screw to measure small distances in the focal plane.

filature *Textiles.* **1.** the process of forming a material into threads. **2.** the process of reeling raw silk from a cocoon. **3.** a reel designed for this purpose.

filbert *Botany.* **1.** the common name for plants of the genus *Corylus.* **2.** the fruit of the *Corylus.*

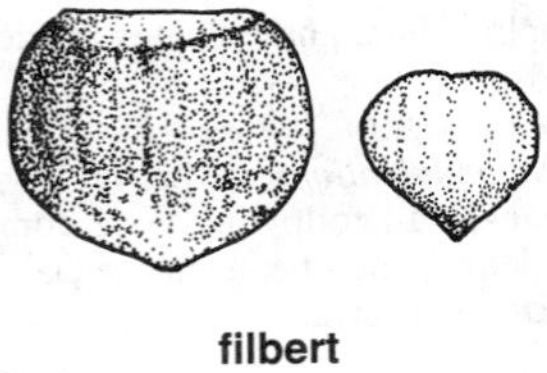

filbert

file *Mechanical Devices.* any of various steel hand tools having diagonally oriented teeth incised into its face; designed to smooth metal, wood, or plastic surfaces by means of fine, parallel cutting edges. *Computer Programming.* **1.** a collection of items with certain common aspects, organized for a specific purpose and stored or processed as a unit. **2.** any collection of data that is stored and manipulated as a named unit by a file-management system. Used to form many compound terms, such as **file backup, file catolog, file layout, file maintenance, file printout, file processing, file separator, file transfer,** and so on.

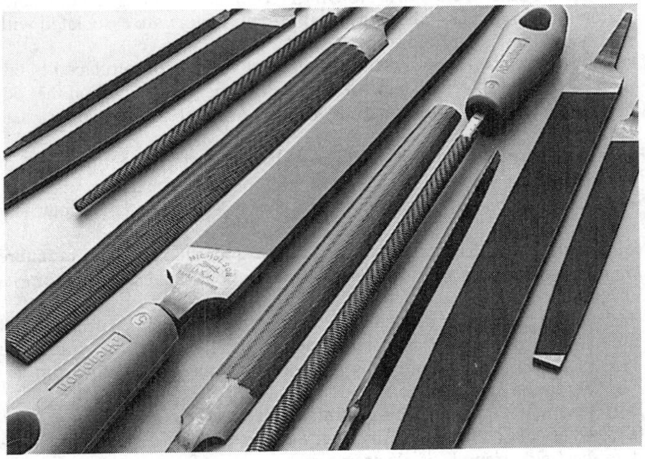

files

file band *Mechanical Devices.* an endless steel band used to hold numerous steel files in a series on a band mill or saw.

file conversion *Computer Programming.* the process of translating a file from one system of file conventions, e.g., character set, to another or of transferring a file from one medium to another.

file directory *Computer Programming.* a list of names and other relevant information relating to files associated with a particular user or application.

file-drawer problem *Psychology.* the tendency of psychological research to be biased toward confirmation of a thesis, because positive results are publicized while negative results are "filed" or disregarded.

filed waveguide *Electromagnetism.* a single insulated wire used to guide an electromagnetic wave.

file event *Computer Programming.* a single read or write access operation with a storage device.

file hardness *Engineering.* the relative hardness of a material, usually determined by attempts to cut the material with a file of standardized hardness.

file label *Computer Technology.* information found at the beginning of a disk or tape file that indicates characteristics of the file, such as name, block size, recording density, date written, and edition number. Also, HEADER LABEL.

file locking *Computer Programming.* the process of preventing access to a file by other users during file update; a method of ensuring data integrity and consistency for all users.

file-management system *Computer Programming.* system software that provides access to and allocation of storage devices, and allows users to share, create, change, and delete files.

file mark *Computer Programming.* an end-of-file mark recorded on magnetic tape.

file name *Computer Programming.* a user-supplied alphanumeric character label that uniquely identifies a set of data.

file-name extension *Computer Programming.* a code added to a file name, usually separated from the name by a period, that describes the type of file; for example, the extension .DBF may indicate a database file, or .PIC may be appended to a graphic file name. Also, EXTENSION.

file organization *Computer Programming.* the physical arrangement of records on a storage medium. Also, PHYSICAL FILE STRUCTURE.

file-oriented system *Computer Technology.* **1.** a computer system that is used primarily for processing multiple large files. **2.** a system in which secondary storage and the operating system are organized in terms of files.

file protection *Computer Technology.* **1.** a mechanism or technique used to prevent the accidental destruction of stored data by erasure or overwrite. **2.** a mechanism that allows file access only to authorized users or applications. **3.** one or more codes associated with a file to determine which users may access it.

file-protection ring *Computer Technology.* a plastic ring inserted into a groove in the inner rim of a tape reel that prevents inadvertent writing over data recorded on the tape; insertion of the ring is required in order to record data.

file recovery *Computer Programming.* the restoration of a file to an accurate condition, or to a saved previous state, following an interruption in processing due to a system failure.

file search *Computer Programming.* a process of locating and retrieving specific items of data from a file by key values or by matching data to specific criteria.

file server *Computer Technology.* hardware and software that together provide file-handling and storage functions for multiple users on a local area network.

file sharing *Computer Technology.* the use of files residing in a central storage device by multiple programs or users.

file-storage unit *Computer Technology.* a high-capacity secondary memory component of a computer system, such as a magnetic disk or tape unit.

file structure *Computer Programming.* the logical or physical method of arranging and organizing records in a file; the logical structure is the application program view of the data, while the physical structure is associated with the organization on the storage medium.

Filibacter *Bacteriology.* a genus of Gram-negative, gliding bacteria that carry out aerobic respiration and grow as filaments.

Filicales *Botany.* an order of the class Filicinae that includes those ferns whose spores originate from a single initial cell rather than from a number of initial cells.

Filicineae *Botany.* a class of the Pteridophyta; plants that reproduce by spores and are characterized by fronds and true vascular systems.

Filicophyta *Botany.* the ferns; a division of flowerless vascular plants that reproduce by spores, have creeping, erect, or trunklike stems, wiry to fleshy roots, and sporangia borne in clusters on the abaxial leaf surface.

Filicopsida *Botany.* the largest of three subdivisions of the division Filicophyta, having thin-walled sporangia often produced in sori, fronds that unroll circinately, and more strengthening tissues than other fern subdivisions.

Filicornia *Invertebrate Zoology.* a group of amphipod crustaceans in the suborder Hyperiidea, having large eyes.

filiform *Anatomy.* having a hairlike or threadlike structure. *Medicine.* of a medical or surgical instrument, having a threadlike structure or tip designed to pass through narrow openings. Thus, **filiform catheter, filiform bougie.**

filiform corrosion *Metallurgy.* corrosion of a metal or alloy that was coated with an organic substance. This corrosion proceeds preferentially along threadlike lines.

filiform lapilli see PELE'S HAIR.

filiform papilla *Anatomy.* any of the threadlike elevations that cover most of the surface of the tongue.

fill *Civil Engineering.* earth, rock, or soil used for embankments or to bring a site to a required higher elevation or level.

fill character *Computer Programming.* a nondata character or bit used to pad or fill out right- or left-justified character strings in a field that is larger than the string.

filled band *Solid-State Physics.* an energy-level band in which there are no vacancies, and the electrons do not contribute to valence or conduction processes.

filled stopes *Mining Engineering.* any stopes that are filled with barren stone, low-grade ore, or sand following the extraction of ore.

filled-system thermometer *Engineering.* a thermometer in which a pressure change from a gas, generally nitrogen or helium, or a change of liquid in the system, causes the Bourdon tube to distort.

filler *Materials Science.* **1.** an inert mineral powder or other solid that is added to a resin, paper, or other substance to add to its bulk and thereby decrease cost. **2.** a material that increases the mass of a product without affecting the properties of the product. *Building Engineering.* a material used to fill holes in a building surface, especially in preparing for painting. *Computer Programming.* in COBOL, a reserved word used in place of a data name to specify reserved but unused space in a record; often used to comply with record length requirements.

fillet [fil′it] *Architecture.* **1.** a narrow flat band between two moldings or between two flutes on an Ionic or Corinthian column. **2.** generally, a narrow, flat band of wood between two moldings or two flutes in a wood member. *Design Engineering.* a concave surface that lies between two intersecting surfaces meeting at an angle. *Neurology.* a long band or bundle of nerve fibers, such as the medial lemniscus.

fillet gauge *Mechanical Devices.* an instrument used for measuring the inside and outside radii of concave and convex surfaces such as the pins, journals, and bearings of a crankshaft. Also, RADIUS GAUGE.

fillet weld *Metallurgy.* a weld that joins two perpendicular parts.

fill factor *Mechanical Engineering.* the percentage of its rated capacity being carried at a given time by a mechanism such as a power shovel.

filling *Engineering.* any process of placing material in a cavity or opening, or the material so placed. *Medicine.* **1.** a material, such as gold or cement, inserted into a prepared dental cavity. **2.** the process of inserting and shaping such material. *Meteorology.* an increase in the central pressure of a pressure system, usually a low, on a constant-height chart, or an analogous increase in height on a constant-pressure chart.

fillowite *Mineralogy.* $Na_2Ca(Mn^{+2},Fe^{+2})_7(PO_4)_6$, an easily fusible, yellow to brown monoclinic mineral occurring in granular crystalline masses, having a specific gravity of 3.43 and a hardness of 4.5 on the Mohs scale; found in granite pegmatites.

fill terrace see ALLUVIAL TERRACE.

filly *Agriculture.* a young female horse, especially one that is less than five years old.

film *Physical Chemistry.* a very thin, continuous layer or sheet of a substance, such as a soap bubble or alcohol on water; there is no precise measure of thickness for a film, but 0.01 inch is an accepted maximum. *Materials Science.* any of various synthetic materials in the form of long, very thin sheets, such as cellophane or polyethylene, used for such purposes as wrapping and packaging foods. *Graphic Arts.* a sheet or strip of cellulose-based material covered with a light-sensitive or light-and-sound-sensitive emulsion and used for recording still or moving images. Still photographic film and radiographic film are generally composed of cellulose nitrate or cellulose acetate; motion-picture film is composed of cellulose triacetate. *Biology.* a very thin or membranelike skin or covering.

film and spots *Microbiology.* an assay used to detect species of *Mycoplasma* using an egg yolk or horse serum medium upon which this microorganism exhibits a characteristic surface film surrounded by tiny dark spots.

film badge *Nucleonics.* a device worn by individuals possibly exposed to sources of ionizing radiation, which measures the approximate amount of radiation received in a given period of time; it generally contains an emulsion and metal filters or foils contained in a metal or plastic frame. Also, BADGE METER.

film base *Graphic Arts.* the material, usually plastic or polyester, used to carry a photosensitive emulsion.

film boiling *Physical Chemistry.* the boiling that occurs when a thin, uniform film of liquid flows over a heated surface, as opposed to nucleate boiling in a pool of heated liquid.

film capacitor see PLASTIC-FILM CAPACITOR.

film-coated paper see PIGMENTED PAPER.

film coefficient *Thermodynamics.* a quantity given by the rate of heat flow per unit area out of a fluid, divided by the temperature difference between the fluid and the wall. Also, HEAT TRANSFER COEFFICIENT.

film contrast *Graphic Arts.* the contrast inherent in photographic or radiographic film.

film cooling *Thermodynamics.* a method of cooling a body by maintaining a layer of fluid over the surface of the body.

film density *Graphic Arts.* the degree of opacity of a film after exposure to light.

film-development chromatography *Analytical Chemistry.* a method of chromatography using film as the adsorbent layer instead of paper.

film integrated circuit *Electronics.* an integrated circuit in which film is substituted for an insulating substrate.

film mottle *Graphic Arts.* a flaw in a film emulsion that yields splotches, usually colored, on the exposed film or positive prints.

film optical-sensing device *Computer Technology.* an input device capable of recognizing and digitizing characters stored on microfilm.

film platen *Graphic Arts.* a unit that holds film in the focal plane during an exposure.

film pressure *Physics.* the difference in surface tension between a liquid and the same liquid with a molecular film (one molecule thick) of a particular substance.

film reader *Electronics.* a device that converts data stored on photographic film into digital form for use in a computer.

film recorder *Electronics.* a device that records data on photographic film, usually in the form of light and dark spots or transparent and opaque spots.

film resistor *Electricity.* a resistor that uses a thin layer of resistive material deposited on an insulating core in low-power applications.

film scanner *Crystallography.* a device for measuring the intensities of spots on an X-ray diffraction photograph, using a light beam to scan the photograph systematically.

film scanning *Electronics.* a procedure that converts motion picture film into electrical signals for broadcast on television.

film temperature *Thermodynamics.* the arithmetic average between the bulk temperature and the wall temperature.

film theory *Physics.* a theory concerned primarily with the flow of heat or matter across a phase interface in which at least one of the phases in question is a free-flowing fluid.

film transport *Mechanical Engineering.* the mechanism of moving photographic film through a light region when recording a motion picture soundtrack.

film vault *Graphic Arts.* a secure place to store film.

film water see PELLICULAR WATER.

FILO [fī′lō] *Computer Programming.* a technique for adding and deleting items from a stack or pushdown list, in which the most recent addition is removed first. (An acronym for first-in, last-out.)

Filobasidiaceae *Mycology.* in some classifications, a family of fungi belonging to the subdivision Basidiomycotina and characterized by long, thin spore-bearing structures lacking cell walls.

Filobasidiella *Mycology.* in some classifications, a genus of fungi belonging to the order Sporidiales; it is the perfect state of *Crytococcus neoformans,* a species occurring in soil and in pigeon excreta that causes serious diseases in humans and animals.

Filobasidium *Mycology.* a genus of fungi belonging to the order Sporidiales; the species *F. capsuligenum* occurs in sake and cider.

filoplume *Vertebrate Zoology.* a specialized hairlike feather having a slender shaft with few or no barbs.

filopodium *plural,* **filopodia.** *Invertebrate Zoology.* a filamentous pseudopodium composed chiefly of ectoplasm; found in some small amebas. *Cell Biology.* a long extension that protrudes from the cell surface of developing nerve cell axons.

filopressure *Surgery.* the compression of a blood vessel using a thread.

filoreticulopodia *Invertebrate Zoology.* filamentous psuedopodia that are branched and interconnected; found in foraminiferan protozoa.

Filosa *Invertebrate Zoology.* in some classifications, an order of protozoan amoebas with filopods in the subclass Filosia.

Filosia *Invertebrate Zoology.* a subclass of protozoans in the class Rhizopoda, having slender filopodia.

Filoviridae *Virology.* a proposed family consisting of the *Marburg* and *Ebola* viruses of Africa, which cause often-fatal viral hemorrhagic fevers in humans.

filter any device or process that serves to screen out something; specific uses include: *Engineering.* a device or porous substance through which a gas or a liquid is passed in order to remove solids or impurities. *Optics.* a device, usually consisting of a glass plate or other transparent material, used to absorb light or other forms of electromagnetic radiation at selected wavelengths. *Graphic Arts.* a colored gelatin sheet used in color separation. *Electronics.* a circuit or device that selectively passes and blocks signals entering a system in accordance with the specific requirements of that system. *Computer Technology.* a device or routine that separates signals or data in accordance with prespecified criteria. *Computer Programming.* **1.** a machine word containing a specific bit pattern used to select, test, or extract parts of other machine words. **2.** a program that selects some data from a set, while removing others. *Control Systems.* a device used by a feedback control system to improve performance or achieve stability. Also, COMPENSATOR. *Acoustical Engineering.* see ACOUSTIC FILTER. *Mathematics.* a filter on a set X is a family $F(X)$ of subsets of X such that: (a) the empty set does not belong to $F(X)$; (b) the intersection of any two subsets of X belonging to $F(X)$ is also an element of $F(X)$; and (c) any subset of X that contains an element of $F(X)$ belongs to $F(X)$; e.g., the set of neighborhoods of a point in a topological space and a (proper) dual ideal of a lattice. If every element of $F(X)$ also belongs to another filter $F'(X)$, then $F'(X)$ is said to be finer than $F(X)$. If $F'(X)$ is a filter on X such that every **finer filter** on X is identical to $F'(X)$, then $F'(X)$ is an **ultrafilter** on X. Filters and ultrafilters are used in the study of nonmetrizable topological spaces.

filterability *Engineering.* a measure of the adaptability of a liquid-solid system to filtration techniques.

filterable virus *Virology.* an older term for an infectious agent, such as a virus, that could pass through early microbiological filters.

filter bed *Civil Engineering.* a contact bed of granular material used for filtering, such as a bed of soil or sand used to filter water or sewage. Also, BACTERIA BED.

filter-cake washing *Chemical Engineering.* the removal of residual liquid impurities at the end of a filtration by washing with the flow of a solvent through the cake.

filter capacitor *Electricity.* a capacitor used in a filter, usually an electrolytic capacitor used to reduce ripple or pulsating current in the output of a power supply.

filter center *Military Science.* in an aircraft control and warning system, the location where information from observation points is interpreted for dissemination to air defense control centers and air direction centers.

filter choke *Electricity.* an iron-core coil used to reduce ripple or pulsating current in the power supply.

filter crystal *Electronics.* an element made of quartz crystal that allows only signals at specific frequencies to enter an electrical circuit.

filter discrimination *Electronics.* the ratio between the attenuations of frequencies that are suppressed by the filter and frequencies in its pass band.

filtered-particle testing *Materials science.* a technique for finding cracks in porous objects by spraying them with a liquid containing suspended particles; the particles congregate on the surface of a crack when the liquid flows through.

filter factor *Optics.* the number by which a given exposure time must be multiplied in order to compensate for the amount of light absorbed when a filter is present.

filter feeder *Invertebrate Zoology.* an aquatic animal that traps and removes food particles from a current of water passing through its body.

filter flask *Chemistry.* a heavy conical flask having a side arm and an opening to which a vacuum pump can be attached; used in vacuum filtration.

filter hybridization *Molecular Biology.* the exposure of DNA that is bound as denatured single strands to a filter material and then to a solution containing radioactively labeled probe DNA or RNA under conditions that allow reannealing; a procedure used to identify specific base sequences in the original sample.

filtering *Science.* the process of passing materials through a filter; filtration. *Transportation Engineering.* the act of interpreting reported information on movements of land, sea, or air vehicles in order to identify their probable true tracks.

filter leaf *Chemical Engineering.* the frame in a filter press that supports the filter medium.

filter medium *Materials Science.* any water-insoluble, relatively rigid material that has enough porosity to serve as a filter; natural filter media include sand, gravel, and diatomaceous earth.

filter pass band see FILTER TRANSMISSION BAND.

filter photometer *Optics.* a colorimeter in which the light length is corrected by the use of various glass filters.

filter photometry *Optics.* **1.** the analysis of solution colors with the aid of a special filter on the eyepiece of the colorimeter. **2.** the examination of Nessler tubes through a filter.

filter press *Engineering.* a device that filters water pumped between filter cloths which are placed between two iron plates that press together, forcing water out of the cloth and trapping sediments in the cloth.

filter-press cell *Physical Chemistry.* a device that generates power by passing an electric current through electrodes in series from a cathode through a porous asbestos diaphragm to an anode.

filter pump *Mechanical Engineering.* an aspirator or vacuum pump used to facilitate a filtering process, for example, by exerting negative pressure on the filtrate side of a filter drum.

filter route *Ecology.* a migration route, such as a land bridge or a mountain pass, along which some animal species pass easily, while others are barred from passing. Similarly, **filter bridge.**

filter screen *Engineering.* a screen that collects solids as a liquid passes through it; generally made from a fine metal mesh or a woven fabric.

filter section *Electricity.* an elemental RC, RL, or LC network used as a broad-band filter in a power supply, grid-bias feed, or other current-generating device.

filter slot *Electromagnetism.* a slot-shaped choke inserted into a waveguide to suppress unwanted modes.

filter spectrophotometer *Spectroscopy.* a spectrophotometer in which maximum sensitivity is obtained by using a filter (often a thin glass or other semitransparent material) to isolate a particular narrow range of wavelengths.
filter thickener *Engineering.* a device that thickens a liquid-solid mixture by filtering out some of the liquid instead of waiting for it to settle.
filter transmission band *Electronics.* a continuous range of frequencies over which the decrease in power introduced by the filter does not exceed a given value. Also, FILTER PASS BAND.
filter-type respirator *Engineering.* a device that physically traps hazardous air particles on the material in its filter.
filtrate *Science.* a liquid or gas that has passed through a filter.
filtration *Science.* the process of passing a liquid through a device or porous substance in order to remove solids or impurities. *Radiology.* the use of a solid screen of radiation-absorbing material, such as lead or aluminum, to restrict radiation wavelengths to a certain range.
filum *Anatomy.* a threadlike structure or part.
filum terminale *Anatomy.* a slender, threadlike filament of connective tissue that descends from the conus medullaris to the base of the coccyx. Also, **filum meningeale, filum spinale.**
fimbria *Biology.* **1.** any stucture forming or resembling a fringe, border, or edge. **2.** see PILUS.
fimbriate *Biology.* having fringe along the edge, such as some petals or antennae have.
fimbrin *Biochemistry.* a protein that creates bundles of filaments by longitudinally binding nearby microfilaments; present in the core of epithelial brush-border microvilli.

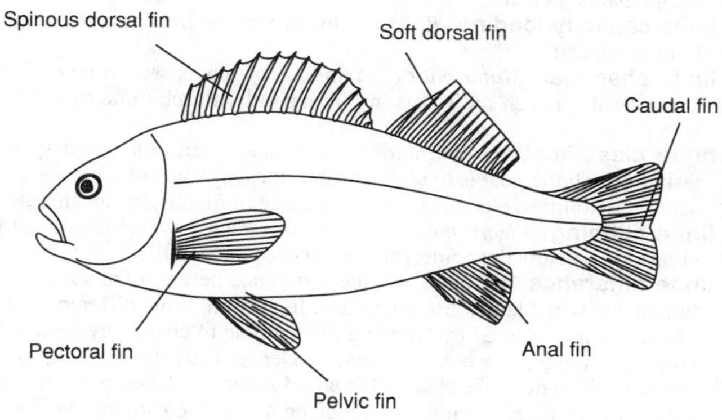

fins

fin *Vertebrate Zoology.* a membranous, winglike or paddlelike process attached to any of various fish and certain other aquatic animals that is used in propelling, balancing, or guiding the body. *Design Engineering.* a projecting flat plate or thin expansion that occurs on the side or edge of a large portion of a structure. *Aviation.* a fixed or adjustable airfoil or vane attached vertically to a flight vehicle to increase directional stability. *Mechanical Engineering.* **1.** any thin metal strip added to an air-cooled engine cylinder, gear box, or pump body to expand the cooling area. Also, COOLING FIN. **2.** any material that remains in the holes of a molded part.
final amplifier *Electronics.* in a cascade amplifier system, a circuit that feeds an amplified signal into the antenna.
final approach *Aviation.* a flight path along the runway centerline from the last turn to the runway.
final approach course *Aviation.* the proper course for a given final approach.
final approach fix *Aviation.* the specified location from which a final approach is begun.
final approach point *Aviation.* in the absence of a specified final approach fix, the point at which an aircraft is established inbound on its final approach course.
final diameter *Navigation.* the diameter of the circular path followed by a ship through 360° while maintaining a constant speed and rudder angle.
final great-circle course *Aviation.* the great-circle course at the destination.
final heading *Navigation.* the heading of a craft on arriving at its destination.
final-lock mechanism *Ordnance.* a device in a cartridge-actuated weapon that locks the striking component into final position.
final mass *Space Technology.* the mass of a rocket after burnout or cut-off.
final proof *Graphic Arts.* a fully corrected proof suitable for use in page makeup and platemaking.
final-value theorem *Mathematics.* let $f(t)$ be a function with Laplace transform $F(t)$, such that $\partial f(t)/\partial t$ also has a Laplace transform, and such that $\lim_{n\to\infty} f(t)$ exists. Then

$$\lim_{t\to\infty} f(t) = \lim_{s\to\infty} sF(s).$$

financial-planning system *Computer Technology.* a decision support computer application that enables the financial planner to examine and evaluate the probable consequences of alternative plans, based on standard financial models and prediction algorithms.
fin assembly *Ordnance.* a group of metal blades mounted on a missile, bomb, or similar projectile to provide directional stability. Similarly, **fin reinforcing assembly, fin stabilization.**
finback whale see BALAENOPTERIDAE.
finch *Vertebrate Zoology.* any of numerous small passerine songbirds of the family Fringillidae including the sparrows, grosbeaks, crossbills, goldfinches, linnets, and buntings, being of small or moderate size and rather stout build and having a short, conical bill adapted for eating seeds.
find *Archaeology.* the act of discovering archaeological remains, the remains discovered, or the location where this discovery occurs. *Astronomy.* a meteorite found on the ground whose descent was not observed.
finder *Optics.* **1.** a small telescope of low power affixed to a larger telescope so that their optical axes are parallel; used for the initial sighting and centering of objects in the field of view. **2.** the viewfinder of a camera. *Telecommunications.* in a telephone switching system, a switch or relay group that determines the route a call is to take through the system.
finder beam *Computer Technology.* a visible beam of light that is projected by some light pens onto the screen at the spot where the photo detector is focused; an aid to accurate positioning by the user.
F indicator *Electronics.* see F DISPLAY.
findspot or **find spot** *Archaeology.* the place where an archaeological find is located.
fine admixture *Geology.* the smallest particles in a mixed-size grade of a sediment.
fine ceramics *Materials Science.* fine-grained porcelain or porcelain-like compositions that are used for a variety of products including art ware and dishes.
fine delay *Navigation.* a dial on a loran indicator for making small adjustments in the position of the B trace pedestal.
fine earth *Geology.* a soil that passes smoothly through a two-millimeter sieve.
fin efficiency *Engineering.* in extended-surface heat-exchange calculations, the relationship between the mean temperature difference from surface-to-fluid to the temperature difference from fin-to-fluid at the base of the fin.
fine gold *Metallurgy.* high-purity gold, often 99.99% pure. *Mining Engineering.* gold found in extremely small particles in placer mining.
fine-grained *Geology.* **1.** describing the texture of a deposit in which the individual minerals are relatively small. **2.** describing a sedimentary rock or sediment and texture in which the individual particles are too small to distinguish without magnification. **3.** describing a soil in which silt or clay is the main component.
fine-grained architecture *Computer Technology.* a network or multiprocessor architecture consisting of many simple processors.
fine-grained habitat *Ecology.* a habitat in which entities are so small with respect to the activity patterns of an organism that the organism usually cannot usefully distinguish the quality of these entities.
fine gravel *Geology.* a gravel having particles that range from 1 to 2 mm in diameter.
fine grinder *Mechanical Engineering.* any of various machines used to pulverize a material, usually consisting of a horizontally rotating mill containing small grinding media such as balls, rods, or pebbles.
fine grinding *Mechanical Engineering.* the process of using a fine grinder to reduce a material to very fine particles, usually –100 mesh.
fine index *Computer Programming.* the secondary, or more specific, of a pair of indices used to locate particular information.
fineness *Metallurgy.* the amount of silver or gold in an alloy, often expressed in parts per thousand.

fineness modulus *Materials Science.* a number denoting the fineness of a fine aggregate or other material such as sand or paint; derived by totaling the percentages by weight of an aggregate sample retained on each of a specified set of sieves, and dividing it by 100.

fineness ratio *Aviation.* the ratio of the length of a streamlined body to its maximum diameter or other transverse dimension; applied especially to aircraft fuselages, dirigibles, missiles, and rockets.

fine papers *Graphic Arts.* any of the better grades of paper used for printing or writing, such as bond, book, or cover papers.

fines *Metallurgy.* **1.** in powder metallurgy, the portion of a metal powder that is finer than a specified size. **2.** in casting, the portion of casting sand that is much finer than the average size. See ANTHRACITE FINES.

fine sand *Geology.* **1.** a sand particle that is 0.125–0.25 mm in diameter. **2.** any soil composed largely of such sized particles.

fine-screen halftone *Graphic Arts.* a halftone of continuous-tone copy shot through a screen having 150-300 lines per inch.

fine silver *Metallurgy.* high-purity silver, usually 99.9% pure.

fine structure *Atomic Physics.* a splitting of spectral lines from an atom or a molecule; caused by the coupling of the electron spin to the orbital angular momentum.

fine-structure constant *Physics.* a dimensionless number, 7.297351×10^{-3} or approximately 1/137, that measures the strength of the electromagnetic interaction; given by the square of the electron charge divided by the product of 2 times Planck's constant, the speed of light, and the permittivity of free space.

fine stuff *Building Engineering.* a finishing coat composed of lime and plaster of Paris and applied to drywall.

fin fold *Developmental Biology.* in the embryo of a primitive vertebrate, a median fold of tissue that develops into a fin.

finger *Anatomy.* any of the five digits of the hand, each consisting of a metacarpal bone and three bony phalanges.

finger bit *Mechanical Devices.* a steel drill bit with long, thin cutting points, used for boring rock.

finger clamp *Mechanical Devices.* a flat clamp with a double-suited end that holds the workpiece while also fitting the work hole.

finger gripper *Robotics.* an end effector that uses two or more joints to grasp an object.

finger guard *Mechanical Devices.* **1.** a device that guards a worker's fingers from hazards of handling machinery. **2.** a protective device for the index or pointer of an instrument.

finger joint *Mechanical Devices.* a hinge composed of interlocking projections (fingers) that are joined by a dowel which runs through spaces between them; primarily used on the apron of a drop-leaf table.

finger lake *Hydrology.* any of a group of long, narrow lakes occupying a glacial trough or retained by a morainal dam across the lower end of a glacial valley.

Fingerlakesian *Geology.* a North American geologic stage of the lower Upper Devonian, after the Taghanican and before the Chemungian.

finger plate *Building Engineering.* a plate fixed on the side of a meeting stile of a door to prevent damage to the paint from finger marks.

fingerprint *Anatomy.* a pattern of cutaneous ridges on the bulb of the distal phalanx of a finger; the pattern of each person's fingerprints is unique to that individual. *Molecular Biology.* a typical spot pattern that is produced by electrophoresis of polypeptide fragments; obtained by the cleavage of a specific protein with a proteolytic enzyme.

fingerprinting *Biotechnology.* a technique used to determine protein structure or nucleic acid content; a two-dimensional map of the sample being tested is produced using gel electrophoresis or chromatography.

finger stop *Mechanical Devices.* an initial machine press guide with a sliding stop used to position the material to the tool for the first run.

finial *Architecture.* an ornament at the top of a spire, pinnacle, or other architectural feature.

fining *Materials Science.* a process by which molten glass becomes virtually free of all bubbles and undissolved gases.

finish *Engineering.* **1.** a surface coating applied to wood, metal, etc. **2.** to apply such a coating, or otherwise smooth a surface. *Materials.* the material used in such a coating, e.g., a varnish.

finished goods *Industrial Engineering.* items that come from a production process in their final form, ready for direct use or sale, as opposed to components or goods requiring further processing before use.

finished steel *Metallurgy.* any steel wrought product that has been further processed after completion of primary fabrication.

finisher *Civil Engineering.* a construction machine or person that aids in the smoothing of freshly placed asphalt or roadway, or the preparation of a foundation for pavement.

finish grinding *Mechanical Engineering.* the final action of a grinding operation in which the object is to obtain a smooth finish and/or precise dimensions.

finishing *Engineering.* the cleaning and polishing of metal. *Agriculture.* the fattening of animals for market.

finishing coat see SETTING COAT.

finishing die *Metallurgy.* in metalworking, the die used for the last operation.

finishing drill *Mechanical Devices.* a drill or reamer used to smooth out coarsely cut boreholes before sampling.

finishing hardware *Building Engineering.* hinges, door pulls, strike plates, and other items fabricated partly for aesthetics and visible on the exterior surfaces of a completed building or other structure.

finishing nail *Mechanical Devices.* a thin wire nail whose head can be easily countersunk and hidden with filler.

finishing temperature *Metallurgy.* in hot working, the terminal temperature.

finishing tool *Mechanical Devices.* a lathe or planer that has broad, straight cutting edges; used to remove ridges on workpieces as a final or finishing cut.

finish plate *Mechanical Devices.* a protecting plate for cylinder set screws fastened to an underplate.

finish turning *Mechanical Engineering.* a precision machining process designed to achieve a smooth finish.

finite [fī´nīt] *Science.* **1.** not infinite; having bounds or limits. **2.** able to be counted or measured.

finite automaton see FINITE STATE MACHINE.

finite automaton recognizable *Computer Science.* describing a language that is regular.

finite-capacity loading *Robotics.* the process of loading a robot to its limit or capacity.

finite character *Mathematics.* a family F of sets is said to be of finite character if for each set A, A is in F if and only if each finite subset of A is in F.

finite classification *Artificial Intelligence.* a kind of expert system task in which the goal is to classify a given case as one of a prespecified set of possibilities, e.g., diagnosing a disease or identifying an aircraft.

finite clipping *Electronics.* a process in which a waveform is cut or clipped just below the point where it exceeds an amplifier's power.

finite difference *Mathematics.* the difference between the values of a function at two distinct discrete points. In general, finite differences of a function f are obtained by allowing the variable to change by arithmetic progression, e.g., x, $x + h$, $x + 2h$, . . . Denoted $\Delta f(x) = f(x + h) - f(x)$. **Partial differences** are obtained from a function f of two or more variables by performing a finite difference on one of the variables and holding the others fixed. Used to approximate derivatives and partial derivatives. Also, ORDINARY DIFFERENCE.

finite-difference equation *Mathematics.* **1.** any equation whose terms involve finite differences. **2.** in particular, an equation in which finite-difference quotients have been substituted for derivatives.

finite-difference quotient *Mathematics.* the quotient

$$\Delta f(x)/h = [f(x + h) - f(x)]/h.$$

finite-element method *Engineering.* an approximation method for analyzing continuous physical systems, used in structural mechanics, electrical field theory, and fluid mechanics; the system to be analyzed is subdivided into small discrete elements that are interconnected at discrete node points.

finite field *Mathematics.* a field with only a finite number of elements. By theorem, the characteristic of a finite field F is a prime number p and the order or number of elements in F is $|F| = p^n$ for some integer n. Also, GALOIS FIELD.

finite free module *Mathematics.* a free module with a finite base; equivalently, a finitely generated free module. By theorem, any two bases of a finite free module have the same rank (number of elements).

finite free resolution *Mathematics.* let E be an R-module. If there exists an exact sequence

$$0 \rightarrow F_n \rightarrow F_{n-1} \rightarrow \cdots \rightarrow F_1 \rightarrow F_0 \rightarrow E \rightarrow 0$$

such that each F_i is a finite free R-module, then E is said to have a finite free resolution of length n.

finite intersection property *Mathematics.* if F is a family (collection) of sets (e.g., subsets of a topological space), F is said to have the finite intersection property if whenever $\{F_1, F_2, \ldots, F_n\}$ is in F, then

$$F_1 \cap F_2 \cap \cdots \cap F_n \neq \varnothing.$$

finitely additive *Mathematics.* **1.** let μ be a measure on a measure space M, and $\{S_i\}_{i=1}^n$ a finite sequence of pairwise disjoint measurable subsets of M. μ is said to be finitely additive if

$$\mu(\bigcup_{i=1}^{n} S_i) = \sum_{i=1}^{n} \mu(S_i).$$

2. a real-valued set function σ is said to be finitely additive if

$$\sigma(\bigcup_{i=1}^{n} S_i) = \sum_{i=1}^{n} \sigma(S_i),$$

where $\{S_i\}_{i=1}^n$ is a finite sequence of pairwise disjoint sets on which σ is defined.

finite mathematics *Mathematics.* the branch of mathematics dealing with sets, logic, and axiomatic theories, without the use of the concept of limit.

finite matrix *Mathematics.* a matrix with a finite number of rows and columns.

finite moment theorem *Mathematics.* if $f(x)$ is continuous on some interval (a, b), and if $\int_a^b f(x)x^n\,dx = 0$ for all positive integers n, then $f(x)$ is zero on (a, b).

finite precision number *Computer Programming.* a number that can be exactly expressed by a finite set of numerals, as opposed to a number such as pi that has an infinite fractional part.

finite projective plane *Mathematics.* a symmetric block design with $bn \geq 4$ and $\lambda = 1$. It is traditional to substitute the term *point* for "vertex" and *line* for "block." Equivalently, a system of at least four points and lines is a finite projective plane if and only if the following hold: (a) every pair of points is contained in exactly one line; (b) every pair of lines intersects in exactly one point; and (c) there exist three non-collinear points.

finite quantity *Mathematics.* **1.** any point on the complex plane that is not equal to infinity. **2.** any bounded quantity.

finite resource see NONRENEWABLE RESOURCE.

finite sequence *Mathematics.* an ordered list of finite length.

finite set *Mathematics.* a set with a finite number of elements. A non-empty set E is finite if it is equivalent to some natural number n; i.e., the elements of E can be put into one-to-one correspondence with the set $\{1, 2, 3, \ldots, n\}$. The empty set Ø is equivalent to $\{0\}$ and is finite by definition. A set is finite if and only if it is not equivalent to a proper subset of itself. A set that is not finite is infinite.

finite-state machine *Computer Technology.* in automata theory, a device with a finite number of distinguishable internal states or configurations.

fin keel *Naval Architecture.* a narrow keel projecting downward like a centerboard, often with a streamlined ballast weight at the bottom; it is used on racing yachts.

Fink truss *Civil Engineering.* a symmetrical steel roof truss suitable for spans of up to fifty feet, composed of a pair of braced isosceles triangles based on the sloping sides of the upper chord, with their apices joined by a horizontal tie.

Finlay, Carlos 1833–1915, Cuban biologist; identified the mosquito as the agent of yellow fever.

finned surface *Mechanical Engineering.* a tubular heat-exchange surface having finned projections that extend longitudinally along the tube length or radially around the tube surface.

finnemanite *Mineralogy.* $Pb_5(As^{+3}O_3)_3Cl$, a gray to olive green to black hexagonal mineral occurring in prismatic crystals, having a specific gravity of 7.26 and a hardness of 2.5 on the Mohs scale.

Finno-Ugric Family *Linguistics.* the language family that consists of Finnish, Hungarian, Lapppish, Estonian and other minor languages found from northern Russian to eastern Siberia.

fin rot *Veterinary Medicine.* a disease of hatchery and pet fishes resulting from bacterial infection by *Cytophaga psychrophila.* Fins and other tissues deteriorate and become necrotic, and form ulcerations with marginal hemorrhage. Also, **fin and tail rot.**

Finsen, Niels Ryberg 1860–1904, Danish physician; Nobel Prize for the development of ultraviolet therapy for *Lupus vulgaris.*

Finsen lamp *Electricity.* a multiuse lamp containing a high-temperature carbon or mercury arc resulting in a light mixture of blue, violet, and somewhat ultraviolet light; used in industrial paint testings and in treating certain skin disorders.

Finsen unit *Electromagnetism.* a unit of ultraviolet radiation intensity, equivalent to the intensity of ultraviolet radiation having a wavelength of 296.7 nanometers and an energy flux of 100,000 watts per square meter.

Finsler metric *Mathematics.* a notion of infinitesimal length on a manifold M, more general than a Riemannian metric. The Riemannian inner product on the tangent space at a point $p \in M$ is bilinear in the two vector arguments. The Finsler metric need not be bilinear. Let $\|v\|_p$ be a Banach space norm on the tangent space at $p \in M$, that is,

$$\|v\|_p > 0 \text{ for } v = (v_1, \ldots, v_n) \neq 0,$$

$$\|\lambda v\|_p = |\lambda| \cdot \|v\|_p, \text{ and}$$

$$\|v + w\|_p \leq \|v\|_p + \|w\|_p.$$

A Finsler metric is a Banach space norm $\|v\|_p$ such that the Hessian matrix $[\partial^2(\|v\|_p)^2/\partial v_i \partial v_j]$ is positive definite at all points (p, v) of the tangent bundle TM.

fin spine *Vertebrate Zoology.* a bony structure that supports the fins in some fishes.

fin waveguide *Electromagnetism.* a waveguide in which a thin metal fin is inserted longitudinally to increase the range of transmittable wavelengths.

fiord see FJORD.

fiorite see SILICEOUS SINTER.

Fior process *Metallurgy.* in steelmaking, the reduction of concentrate in a fluosolid system, often followed by briquetting and electric furnace melting.

f.i.p. finite intersection property.

fir *Botany.* the common name for various coniferous trees of the genus *Abies*, including nine species that are native to North America, such as the **balsam fir**, *A. balsamea*, the **white fir,** *A. concolor,* the **grand fir,** *A. grandis,* and the **California red fir,** *A. magnifica.*

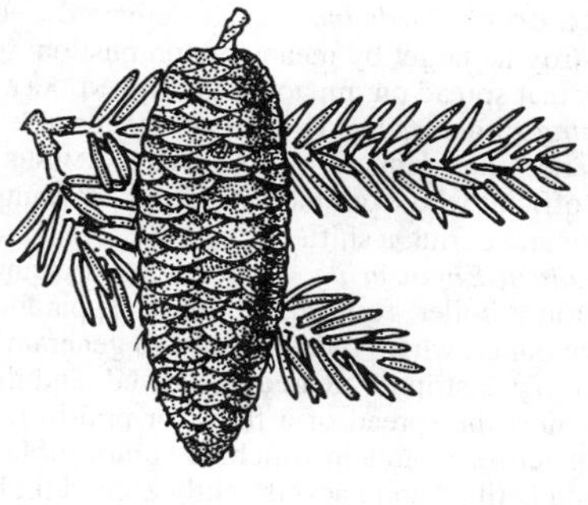

fir

fire *Chemistry.* **1.** the heat and light that is produced by combustion, in the form of a bright flame. **2.** conversely, the phenomenon of combustion as evidenced by a flame. *Engineering.* to ignite with explosives. *Ordnance.* **1.** the discharging of a weapon or the launching of a missile. **2.** the command given to discharge a weapon. **3.** relating to the effective application of fire. Thus, **fire distribution, fire effect, fire mission, fire plan, fire superiority.** *Artificial Intelligence.* to execute a rule whose antecedent is satisfied by adding its consequent, with appropriate variable substitutions, to the database of facts, or by executing a procedure associated with the rule. Also, RULE APPLICATION.

fire adjustment *Ordnance.* the correction of a weapon's aim or the explosion time of its projectile in order to ensure that the projectile will strike the desired target or burst at the desired point.

fire ant *Invertebrate Zoology.* any of various species of ants that are known for their painful, burning sting, especially the **red imported fire ant**, *Solenopsis invicta,* a major pest of the southern U.S.

firearm *Ordnance.* **1.** any weapon from which a projectile is fired using gunpowder or other solid propellant. **2.** specifically, a small weapon of this type, such as a pistol or rifle.

fire assay *Metallurgy.* a test used to analyze the purity of metals, especially gold and silver.

fireball *Astronomy.* a meteor whose brightness exceeds about –5 magnitude; many also leave glowing trains that remain faintly visible for a few seconds. *Nucleonics.* the intensely bright sphere of hot gases that forms almost simultaneously with a nuclear explosion.

Firebee *Ordnance.* a remote-control target drone used to test and train surface-to-air and air-to-air missile systems; it may be ground- or air-launched and can achieve high subsonic speeds; designated **BQM-34.**

firebelly *Vertebrate Zoology.* a toad or aquatic frog of the family Discoglossidae, characterized by a bumpy, gray-green back and limbs, bright orange markings on the belly, and skin that secretes a toxic fluid; a well-known species is the **fire-bellied toad** (*Bombina bombina*) of Europe.

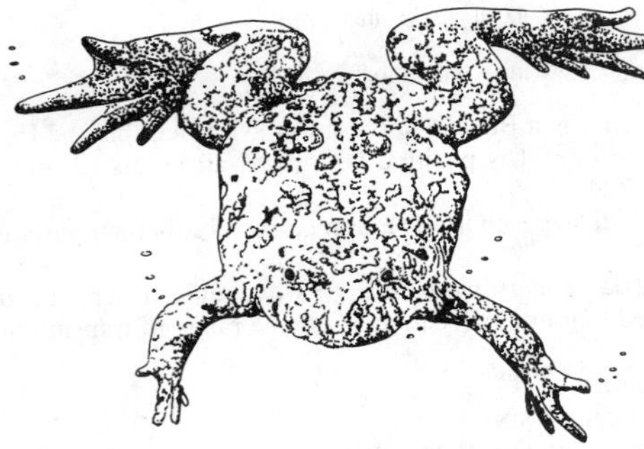

firebelly

fire blight *Plant Pathology.* a disease of apple, pear, and related trees in the rose and apple families caused by the bacteria *Ervinia amylovora*; affected areas turn black or brown as if scorched by fire, and cankers form on the bark.

fireboat *Naval Architecture.* a boat or small ship equipped with fire-fighting gear and used to fight fires aboard ships or on harborside structures.

firebomb or **fire bomb** *Ordnance.* an air-delivered weapon designed to damage or destroy its target by means of combustion; it contains incendiary materials that spread on impact. Also, INCENDIARY BOMB.

fire boss *Mining Engineering.* **1.** a person responsible for inspecting a mine for gas and other dangers. Also, MINE EXAMINER, GAS BOSS. **2.** a state-certified official who examines mines for firedamp, gas, and other dangers prior to and during a shift.

firebox *Mechanical Engineering.* **1.** an enclosed chamber where combustion occurs in a boiler, furnace, or similar apparatus. **2.** the furnace of a locomotive boiler, where fuel is burned to generate steam.

firebreak *Forestry.* a strip of cleared or plowed land that is constructed as a barrier against the spread of a forest or prairie fire. *Mining Engineering.* a strip across an area in which no combustible material is used, or in which sand is filled and packed tightly around timber supports.

firebrick or **fire brick** *Materials.* a highly heat-resistant brick made from fire clay and used to line furnaces, chimneys, and fireplaces.

fire bridge *Engineering.* a low wall that divides the hearth and the grate in a reverberatory furnace.

fire-capabilities chart *Military Science.* a chart indicating the areas that can be reached by the majority of the weapons in a unit.

fire cement *Building Engineering.* a grade of cement fabricated to withstand high temperatures. Also, REFRACTORY CEMENT.

fireclay or **fire clay** *Materials.* a clay, containing large amounts of hydrous aluminum silicates, that can resist high temperatures without disintegrating or deforming; used in manufacturing such products as crucibles and firebrick. Also, FIRESTONE.

fire climax *Ecology.* a community in which fire is the principal force maintaining stability and ecological balance, such as the long-leaf pine forest in North America.

fire construction *Building Engineering.* a type of construction that uses incombustible materials within a structure to maximize protection from fire.

fire control *Forestry.* the practice of fire protection, including methods of fire prevention, fire detection, and fire suppression. *Ordnance.* see FIRE COORDINATION.

fire-control *Ordnance.* of or relating to instruments, devices, and systems used to control the direction, volume, and timing of fire. Thus, **fire-control computer, fire-control equipment, fire-control instrument, fire-control radar, fire-control sonar, fire-control system.**

fire-control circuit *Electronics.* an electronic circuit in a fire-control system.

fire-control grid *Ordnance.* a system of squares on a military map that may be used in fire control; each square has sides of 1000 yards or meters.

fire-control quadrant *Ordnance.* a device used to measure the elevation angle of a gun in order to obtain the horizontal range of the target; it is usually attached to the gun or gun carriage.

fire coordination *Ordnance.* the use of instruments, devices, and systems to control the direction, volume, and timing of fire.

fire crack *Engineering.* a crack from thermal stress that forms on the heated side of a shell or header in a boiler or on a heat transfer surface.

firecracker *Engineering.* a cylindrical object that holds explosive material and produces noise and sparks when its fuse is lit.

fire cut *Building Engineering.* an angular cut made at the end of a joist that is configured to rest upon a brick wall.

firedamp *Mining Engineering.* **1.** a combustible gas, primarily methane, formed by the decomposition of coal or other carbonaceous matter. **2.** an airtight stopping used to isolate an underground fire and to prevent the inflow of fresh air and the outflow of foul air. Also, FIRE WALL.

firedamp alarm *Mining Engineering.* a device that gives out a signal when the methane content in the atmosphere exceeds a known value.

firedamp detector *Mining Engineering.* a portable device that detects the existence and determines the percentage of firedamp in mine air. Also, METHANOMETER.

firedamp drainage *Mining Engineering.* the collection of firedamp, most often in pipes, from coal strata. Also, METHANE DRAINAGE.

firedamp explosion *Mining Engineering.* an explosion of a flammable mixture of firedamp and air. Also, COLLIERY EXPLOSION.

firedamp fringe *Mining Engineering.* the zone of contact of the goaf gases and the ventilation air current at the face.

firedamp layer *Mining Engineering.* a sheetlike collection of firedamp under the roof of a mine roadway where the ventilation is too sluggish to remove the gas.

firedamp migration *Mining Engineering.* the movement of firedamp through the strata or goaf of a mine.

firedamp probe *Mining Engineering.* a flexible rubber tube connected to a rod, which is thrust into roof cavities to transfer a sample of air to a methanometer.

firedamp reforming process *Chemical Engineering.* an operation in which methane is combined with steam and passed through a reactor containing a nickel catalyst to obtain a mixture of carbon monoxide and hydrogen; the mixture is then blended with pure methane, resulting in a fuel of high calorific value.

fire-danger meter *Engineering.* a graph that depicts such data as seasonal changes in foliage, fuel moisture, and wind speed, to help forest rangers assess the degree of forest-fire danger.

fire detector *Engineering.* a device that senses an increase in heat and then activates fire-preventive measures, such as sounding an alarm or turning on a sprinkler system.

fire direction *Military Science.* the tactical application of firepower; it includes target selection, fire concentration or distribution, and ammunition allocation. Thus, **fire-direction center.**

fire disclimax *Ecology.* a community that is maintained through periodic fires that are followed by the emergence of new growth.

firedog *Building Engineering.* a metal support used in fireplaces because it can withstand the heat of an open fire. Also, ANDIRON.

fire door *Engineering.* the opening in a furnace or stove through which fuel is added. *Building Engineering.* a highly fire-resistant door that can be closed to provide fire protection.

fired-process equipment *Engineering.* any device that obtains heat from fuel combustion, such as a furnace, reactor, or steam generator.

fired state *Electronics.* the condition in a semiconductor switching device that arises when a triggering pulse is applied to the gate.

fire-effected *Archaeology.* of stone, showing the effects of having been heated, as for cooking activities. Also, **fire-cracked.**

fire engine *Mechanical Engineering.* a vehicle equipped with firefighting equipment including a fire pump. Also, **fire truck.**

fire escape *Building Engineering.* an outside stairway or ladder, generally made of steel, used to exit a building in the event of a fire.

fire extinguisher *Engineering.* a portable device designed to suppress fire by the use of a fire-inhibiting substance, such as water, carbon dioxide, gas, or chemical foam.

firefinder *Engineering.* a device consisting of a map and a sighting device; used in fire towers to identify forest fires.

fire flooding *Petroleum Engineering.* a technique for enhancing secondary oil recovery in a reservoir; at an injection well in the reservoir, a combustion process is begun by continually pumping gas containing oxygen downhole, and as the heat breaks down the crude, the light oil is pushed ahead through the reservoir toward the production well.

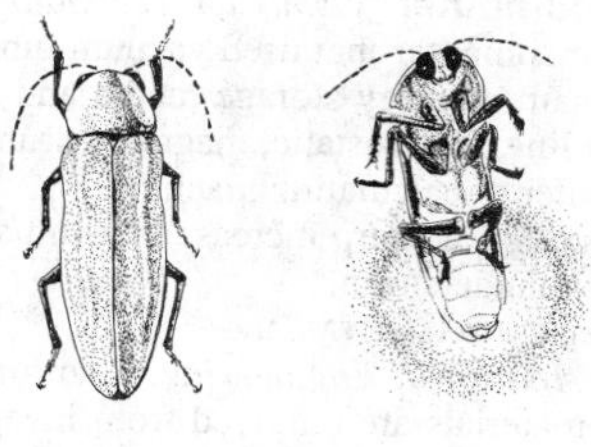

firefly

firefly *Invertebrate Zoology.* the common name for any of numerous species of nocturnal winged beetles of the family Lampyridae that emit bioluminescent light in the form of short sharp flashes or a long lingering glow, through the oxidation of luciferin.

fire for adjustment *Ordnance.* test fire intended to center the mean point of impact or burst on the target.

fire for effect *Ordnance.* fire delivered after the mean point of impact or burst has been satisfactorily established.

fire hose or **firehose** *Engineering.* a heat-resistance hose that carries water from an outlet, such as a hydrant or standpipe, to extinguish a fire.

fire hydrant *Civil Engineering.* a fixture connected to a water main provided inside buildings or outdoors to which a fire hose can be connected. Also, **fire plug.**

fire interrupter *Ordnance.* a device on aircraft guns that interrupts firing; it is usually a mechanically activated electrical switch.

fire line *Forestry.* **1.** a wide strip or trench around a fire made by clearing all logs, brush, and trees, and then scraping away the litter and soil with axes, bulldozers, and shovels. **2.** any line cleared round a fire, generally following its edge. **3.** a cleared, permanent firebreak, usually of considerable width.

fire load *Civil Engineering.* a measurement of the load of a combustible material per square foot of floor space.

fire opal *Mineralogy.* a honey-yellow to hyacinth-red, transparent to translucent opal that gives off firelike reflections.

fire partition *Building Engineering.* a wall within a building that serves to retard an advancing fire.

fireplace *Building Engineering.* the indoor extension of a chimney, in which wood or other fuel is burned.

fire point *Chemistry.* the minimum temperature at which a volatile substance will burn continuously.

firepower *Ordnance.* **1.** the amount of fire that may be delivered by a position, unit, or weapon system. **2.** the ability to deliver fire.

fireproof *Materials.* resistant to combustion. *Engineering.* to treat a surface or to build a structure with materials so as to retard or reduce combustion. *Building Engineering.* having noncombustible walls, stairways, or stress-bearing members and having all other members that could be damaged by heat protected by refractory materials.

fireproof aggregates *Materials.* building materials, including bricks, firebricks, fused clinkers, and slag, that are added to concrete for fire resistance purposes.

fire pump *Mechanical Engineering.* a pressure pump used in firefighting to propel water or chemical solutions at high pressure.

fire refining *Metallurgy.* the process of refining crude metal by oxidizing and/or reducing the melt.

fire resistance *Materials.* the extent to which a material can resist damage by fire. *Building Engineering.* the ability of a structural element to resist combustion for a specified time under conditions of standard heat intensity without burning or failing structurally. Thus, **fire resistant.**

fireroom *Mechanical Engineering.* a chamber in which the boilers of a steam-driven vessel are fired. Also, STOKEHOLD.

fire screen *Building Engineering.* a framelike wire screen used in a fireplace to protect against flying sparks and embers. Also, **fire guard.**

fire standpipe *Civil Engineering.* a safeguard to provide fire protection to upper floors of a tall structure, made of a tall vertical pipe or tank that holds water to provide a positive, relatively uniform pressure.

firestarter *Archaeology.* a wooden tool having a base with drill holes and a drill stick that is rubbed into the base to produce enough friction to give a spark. Also, **firestarter kit.**

firestone or **fire stone** see FIRECLAY.

firestop *Building Engineering.* a barrier of incombustible material, such as brick, across an open area such as a hollow wall to stop the spread of fire. Also, DRAUGHT STOP.

firestorm or **fire storm** *Ordnance.* a massive fire produced in urban areas by incendiary or fire bombs; it generates strong, inrushing winds from all sides that keep the fire from spreading while adding fresh oxygen, thus increasing intensity.

Firestreak *Ordnance.* a British close-range air-to-air missile powered by a solid-propellant rocket motor and equipped with infrared homing guidance; it carries a 50-pound conventional warhead at a speed of Mach 3 and a range of 0.75 to 5 miles.

fire support *Military Science.* fire in support of a unit in direct contact with the enemy; it may be delivered by ground, naval, or air forces and may involve close support, deep support, or direct support. Thus, **fire-support coordination.**

fire-support area *Military Science.* a maneuver area assigned to ships providing fire support for an amphibious operation. Similarly, **fire-support group, fire-support station.**

fire tower *Building Engineering.* a fireproof and smokeproof stairwell that runs the height of a building. *Forestry.* a tower, usually built on a mountain, from which a watch for forest fires is continually maintained.

fire-tube boiler *Mechanical Engineering.* a steam boiler in which hot furnace gases pass through tubes surrounded by boiler water before entering a chimney.

fire unit *Ordnance.* the complete array of elements and equipment making up a missile-launching system.

fire wall or **firewall** *Building Engineering.* a wall within a building, composed of fireproof materials and designed to retard the spread of fire. *Mining Engineering.* see FIREDAMP.

fire weather *Meteorology.* a weather condition conducive to the kindling and spreading of forest fires, characterized by low humidity, lack of precipitation during the preceding days, and high wind speed.

firing *Engineering.* **1.** the process of putting fuel and air into a furnace. **2.** the ignition of an explosive material. *Materials Science.* the process of heating a ceramic body at a high temperature to cause a ceramic bond to form. *Electronics.* **1.** the process by which gas ionizes electrons in a gas-discharge tube and initiates current flow. **2.** an action in which a pulse initiates conduction in an ionization switching device. *Ordnance.* of or relating to information or activities necessary for the proper discharge of a weapon or weapons. Thus, **firing chart, firing data, firing line, firing position, firing table, firing time. 2.** serving to discharge a weapon. Thus, **firing hammer, firing mechanism,** and so on.

firing azimuth *Ordnance.* the horizontal angle, measured clockwise from the vertical plane, of a gun in firing position.

firing circuit *Ordnance.* **1.** in land operations, an electrical circuit or pyrotechnic loop designed to detonate connected charges from a firing point. **2.** in naval mine warfare, the part of a mine circuit that completes the detonator circuit or operates a ship counter. *Electronics.* a circuit in which an ignition delivers a pulse of current to activate a cathode spot and control the time of firing.

firing interval *Ordnance.* the duration between two successive shots.

firing jack *Ordnance.* an adjustable device in certain field artillery that stabilizes and levels the gun in firing position.

firing lanyard *Ordnance.* a cord or cable that hooks onto the firing mechanism of a gun or rocket launcher, thus allowing the weapon to be discharged from a distance.

firing lock *Ordnance.* a removable component in the breech of some guns; it includes the firing pin and the mechanism that drives the pin against the primer.

firing machine *Engineering.* an electric blasting machine.

firing pin *Ordnance.* a plunger in the firing mechanism of a firearm, gun, mine, bomb, missile, or other weapon; it usually strikes and detonates a sensitive explosive, thus actuating the explosive train or propelling charge.

firing point *Ordnance.* the point in a firing circuit where the device that initiates detonation is located. *Electronics.* the degree of gas ionization that must occur before current begins to flow in an electron tube.

firing potential *Electronics.* the amount of voltage needed to achieve a self-sustaining current in a discharge tube.

firing pressure *Mechanical Engineering.* the highest pressure attained in an engine cylinder during combustion.

firing rate *Mechanical Engineering.* the rate at which fuel is fed to a burner, expressed by volume, heat units, or weight per unit time.

firing table elevation *Ordnance.* the vertical angle of the axis of the bore, measured from the horizontal plane, of a gun in position to fire at a given range under standard conditions.

firing tool *Mechanical Devices.* any hand tool used to stoke a boiler furnace, such as a shovel, rake, or slicing bar.

firmer chisel *Mechanical Devices.* a flat, thin-bladed woodworking chisel, usually 6 inches long and $^1/_8$ to 2 inches wide.
firmer gauge *Mechanical Devices.* a standard gauge having an outside bevel for cutting grooves and recesses.
Firmibacteria *Bacteriology.* a class of Gram-positive bacteria of the division Firmicutes that occur as rod-shaped or coccoid cells, some of which bear spores; includes the families Micrococcaceae, Streptococcaceae, Peptococcaceae, Bacillaceae, and Lactobacillaceae.
Firmicutes *Bacteriology.* a taxonomic division of Gram-positive bacteria of the kingdom Procaryotae in which the cell wall is composed of a thick layer of peptidoglycan containing muramic acid. Also, **Firmacutes.**
firm-joint caliper *Mechanical Devices.* an inside or outside caliper whose legs are joined at the top by a firm nut operated by hand pressure.
firmoviscosity *Mechanics.* a property of a viscoelastic substance in which the stress is proportional to a weighted sum of the strain and the strain rate.
firmware *Computer Science.* **1.** a collection of control microprograms that interpret and execute stored instructions. **2.** the implementation of software in hardware circuitry or read-only memory. (From the idea that such devices or systems represent an intermediate category between hardware and software.)
firn *Hydrology.* a loose, permeable, granular material that is over a year old, and which is transitional between snow and glacier ice.
firn basin see FIRN FIELD.
firn field *Hydrology.* any area of firn, specifically the accumulation area or upper region of a glacier, where firn is created. Also, FIRN BASIN, NÉVÉ.
firnification *Hydrology.* the process whereby snow is converted to firn, which is later transformed into glacier ice.
firn line *Hydrology.* the highest level to which winter snow remains on a glacier during the following summer. Also, **firn limit.**
firn snow see OLD SNOW.
first arrival *Engineering.* the first seismic event that appears on a seismogram.
first-arriver principle *Genetics.* a theory that proposes that the first individuals to start a new environment or adapt to a specific niche acquire a selective advantage over later arrivals.
first bottom *Geology.* the normal floodplain of a river or stream.
first category *Mathematics.* a subset of a metric space is said to be of the first category if it is the union of a countable collection of nowhere dense sets. Any set that is not of the first category is said to be of the **second category;** the Baire category theorem states that a complete metric space is of the second category. Also, MEAGER SET.
first-class current *Particle Physics.* a type of current arising from weak interaction between particles in which the properties of the charge symmetry or G-parity are identical to those of currents associated with beta decay in Fermi theory.
first come, first served *Computer Programming.* a scheduling algorithm, using a FIFO queue, in which requests are serviced in the order in which the requests were made. *Industrial Engineering.* see FIRST IN, FIRST OUT, def. 2.
first condition of equilibrium *Mechanics.* the requirement that for a system to be in equilibrium, the vector sum of all the external forces acting on the system must equal zero.
first contact *Astronomy.* the precise instant when an eclipse, occultation, or transit begins.
first countable topological space *Mathematics.* a topological space that satisfies the first axiom of countability; i.e., a topological space in which there is a countable base of neighborhoods at every point.
first-degree burn *Medicine.* in the classification of burns, the mildest injury, marked by reddening of the skin and pain.
First Family *Anthropology.* a popular name for the group of about thirteen australopithecine individuals of the species *afarensis* found in the Hadar district of Ethiopia in 1975; a rare example of distinct age groups important for studying age and sex variations within a contemporary population.
first filial generation see F_1.
first fundamental form *Mathematics.* the Riemannian metric tensor.
first-generation *Design Engineering.* of or relating to the earliest period in the development of some form of modern technology, such as television sets or telephones. *Computer Science.* **1.** specifically, of or relating to the era in computer history between the invention of the stored-program computer and the end of the use of vacuum-tube technology (i.e., up until about 1960). **2.** a computer device or system of this period.
first-generation computer *Computer Technology.* **1.** an early stored program electronic computer that used vacuum tubes as active elements and a variety of main-memory storage media and technologies, including mercury delay line, electrostatic, magnetic drum, and magnetic core. **2.** any early computer or calculating machine.
first gust *Meteorology.* a sharp increase in wind speed associated with the beginning of a thunderstorm.
first harmonic see FUNDAMENTAL.
first in, first out *Industrial Engineering.* **1.** an inventory management method in which materials are removed from inventory and used in the same order in which they were received; used especially when the material in inventory is perishable. **2.** a dispatching method in which jobs are sequenced in the same order in which they arrive. Also, FIRST COME, FIRST SERVED. *Computer Programming.* see FIFO.
first-in, last-out see FILO.
first integral *Mathematics.* the first integral of a system of ordinary differential equations is a function that is constant along solution curves of the system and that has the form

$$F(x_1, x_2, \ldots, x_n, dx_1/dt, dx_2/dt, \ldots, dx_n/dt).$$

first isomorphism theorem *Mathematics.* if $f: R \rightarrow S$ is a homomorphism of rings, then f induces an isomorphism of rings; i.e., $R/\mathrm{Ker}f$ is isomorphic to $\mathrm{Im}f$. Similar results hold for groups and modules.
first law of motion see NEWTON'S FIRST LAW.
first law of the mean *Mathematics.* the mean value theorem for derivatives or integrals.
first law of thermodynamics *Thermodynamics.* a law that is derived from the principle of conservation of energy, stating that: (a) the internal energy of an isolated system remains constant; (b) the change in the internal energy of a nonisolated system is equal to the net energy that crosses the boundaries of the system. This concept can be expressed by $\Delta E = Q - W$, where ΔE is the internal energy change, Q is the heat transfer to the system from its environment, and W is the work done by the system.
first-level addressing *Computer Programming.* an addressing mode in which the address portion of the instruction contains the location of the operand.
first-level controller *Control Systems.* a controller that is used to satisfy local objectives within a larger system. Also, LOCAL CONTROLLER.
first-level inference *Psychology.* a description of behavior that is essentially an account of observed actions, with no inference or abstraction by the observer.
first-level interrupt handler *Computer Technology.* hardware and utility software designed to service internal and external interrupt requests, such as exception conditions or traps (internal) and data transfer requests from I/O devices (external).
first lieutenant *Military Science.* in the Army, Air Force, or Marine Corps, an officer ranking above a second lieutenant and below a captain.
first light *Navigation.* the time at which morning twilight begins.
first lunar meridian *Astronomy.* the meridian of lunar longitude that passes through the point in Sinus Medii that marks 0° longitude; it is identical in function to the Greenwich meridian on the earth.
first motion *Space Technology.* the initial observable upward motion of a spacecraft at launch.
first-order *Cartography.* relating to or using surveying techniques or equipment of the highest degree of accuracy, involving strictly prescribed criteria, and usually referring to measurements obtained from direct observation of the datum.
first-order bench mark *Cartography.* a bench mark connected to the datum (usually mean sea level) by continuous first-order leveling.
first-order climatological station *Meteorology.* a meteorological station at which hourly readings of atmospheric pressure, temperature, humidity, wind, sunshine, and precipitation are made, along with regularly scheduled observations on cloud conditions and general weather.
first-order control *Cartography.* a classification of control networks meeting the following criteria for accuracy and precision: for horizontal coordinates, the maximum values of the ratio of standard deviations of distances between survey points to the distances themselves, 1:100,000; for vertical coordinates, the maximum values of the ratio of standard deviations of elevation differences in millimeters between survey points to the square root of the horizontal distance in kilometers between those points, for Class I—0.5, for Class II—0.7.
first-order kinetics *Enzymology.* a reaction in which the rate of product formation is directly proportional to the concentration of substrate.

first-order leveling *Cartography.* spirit leveling conforming to the criteria of a first-order control network; formerly known as precise leveling.

first-order logic see PROPOSITIONAL LOGIC.

first-order optics see GAUSSIAN OPTICS.

first-order predicate calculus see PREDICATE CALCULUS.

first-order reaction *Physical Chemistry.* a simple chemical reaction in which the differential rate of the reaction decreases linearly with the decrease in the reactant concentration, as in the decomposition of hydrogen peroxide or the decay of radioactive materials.

first-order spectrum *Spectroscopy.* a diffraction spectrum for which the difference in the path length traveled by radiation from adjacent slits is one wavelength.

first-order station *Meteorology.* any meteorological station staffed by National Weather Service personnel.

first-order subroutine *Computer Programming.* a subroutine called directly from the main program.

first-order transition *Thermodynamics.* a change of phase occurring at a constant temperature and pressure with a discontinuous change in volume, entropy, and enthalpy.

first-piece time *Industrial Engineering.* the time that is allowed for the production of the first piece of an order of several pieces, adjusted for factors such as setup time or a worker's unfamiliarity with the process.

first point of Aries *Astronomy.* the traditional name for the vernal equinox, so-called because when the name was given the equinox lay in the constellation of Aries (although precession has since moved it into neighboring Pisces).

first point of Libra *Astronomy.* the traditional name for the autumnal equinox, given when the equinox lay in Libra; precession has since shifted the equinox westward into Virgo.

first quarter moon *Astronomy.* the lunar phase occurring when the moon has completed the first quarter of its monthly orbit of the earth, about a week after the new moon.

first radiation constant *Physics.* a constant for the power emitted from a blackbody that is approximately equal to 3.7415×10^{-16} watt (meter)^2; used in statistical mechanics and found in the Planck radiation formula.

first selector *Electronics.* an element in a telephone that responds to the first-digit dial pulses when a call is placed.

first sergeant *Military Science.* in the U.S. Army, the highest ranking noncommissioned officer of a company, squadron, or other unit.

first strike *Military Science.* the first offensive action of a war; usually applied specifically to a nuclear war.

first-strike capability *Military Science.* the military hardware and technological resources necessary to carry out an initial nuclear attack before an enemy can respond.

first-surface mirror see SURFACE-COATED MIRROR.

first variation *Mathematics.* in the calculus of variations, Lagrange's problem is to find functions $y(x)$ that extremize a function f in the integral form $I(y) = \int_a^b f(x, y, y')dx$ and that satisfy the boundary conditions $y(x_0) = a$, $y(x_1) = b$. Assume y_0 is a solution to the problem, and let

$$F(\varepsilon) = I(y_0 + \varepsilon h) = \int_a^b f(x, y_0 + \varepsilon h, dy_0/dx + \varepsilon dh/dx)\, dx,$$

where h is a given function of x and ε is a parameter. The first variation of I with respect to h on the interval $[a, b]$ is given by $F'(0)$, where

$$F'(\varepsilon) = d/d\varepsilon \int_a^b f(x, y_0 + \varepsilon h, dy_0/dx + \varepsilon dh/dx)\, dx.$$

Fischer, Emil 1852–1919, German organic chemist; founder of the modern chemistry of sugars, proteins, and purins.

Fischer, Ernst Gottfried 1754–1831, German chemist; devised the table of equivalents used in formulating the law of definite proportions.

Fischer, Ernst Otto born 1918, German chemist; Nobel Prize for molecular studies relating to control of air pollution.

Fischer, Hans 1881–1945, German chemist; Nobel Prize for structural analysis of leaves and blood and synthesis of hemin; analyzed and synthesized biliverdin and bilirubin.

Fischerella *Bacteriology.* a genus of thermophilic bacteria of the cyanobacteria group that occur as filamentous cells, divide in more than one plane, and may be found in hot springs.

Fischer ellipsoid of 1960 *Geodesy.* a reference ellipsoid with two primary uses; in the Mercury datum it has a semimajor axis of 6,378,166.0 meters and a flattening or ellipticity of 1/298.3; in the South Asia datum the semimajor axis is 6,378,155.0 meters and the flattening or ellipticity is 1/298.3.

Fischer-Hepp rearrangement *Organic Chemistry.* the rearrangement of a nitroso derivative of a secondary aromatic amine to a *p*-nitrosoarylamine.

Fischer-Hinnen method *Electricity.* a method used to analyze a complex waveform with similar loops above and beneath a time axis.

Fischer indole synthesis *Organic Chemistry.* the formation of an indole derivative by the ring closure of an aromatic hydrazone of aldehydes or ketones in the presence of a catalyst such as zinc chloride.

fischerite see WAVELLITE.

Fischer peptide synthesis *Organic Chemistry.* a peptide synthesis whereby those peptides with a free amino group react with acid halides of α-halo acids, followed by amination.

Fischer-Riesz theorem *Mathematics.* given a measure space X with measure m, the class of functions that are defined almost everywhere and are integrable on X with respect to the measure m, together with the norm $\|f\| = \int_X |f|\, dm$, forms a complete normed space; i.e., $L^1(X)$ is a Banach space. This is not the same as the **Riesz-Fischer theorem,** which states that any sequence in $l^2(Z)$ (the square-summable doubly infinite sequences) is the sequence of Fourier series coefficients of a function in $L^2[-\pi, \pi]$.

Fischer's salt *Inorganic Chemistry.* a popular name for cobalt potassium nitrite.

fish *Vertebrate Zoology.* any of various cold-blooded, strictly aquatic, water-breathing craniate vertebrates, including the cyclostomes, elasmobranchs, and higher gilled aquatic vertebrates, with cartilaginous or bony skeletons, gills, usually fins, and typically an elongated body covered with scales. *Petroleum Engineering.* a term for an object, such as a broken tool, that is lost down a hole and that must be removed before drilling or other operations can be performed.

fishbone antenna *Electromagnetism.* a traveling-wave antenna consisting of a balanced transmission to which an array of closely spaced, coplanar dipoles is loosely coupled.

fished joint *Civil Engineering.* a structural joint using fishplates.

fisher *Vertebrate Zoology.* a large, dark brown or blackish, somewhat foxlike marten related to weasels; native to much of the forested half of North America; now very rare over much of its former range due to excessive hunting for its valuable pelt.

fisher

Fisher, Ronald 1890–1962, English geneticist; established statistical link between mutation and evolution; formulated genetic theory of natural selection.

Fisher-Behrens problem see BEHRENS-FISHER PROBLEM.

fisheries conservation *Ecology.* the steps taken to protect and preserve marine life.

Fisher information *Statistics.* a measure of the amount of information about a parameter provided by an experiment with a given probabilistic structure.

fisherman bat see NOCTILIONDAE.

Fisher's distribution see F DISTRIBUTION.

Fisher's exact test *Statistics.* a method for testing the hypothesis of independence of the counts in a 2×2 contingency table.

fishery *Ecology.* **1.** a place for harvesting or catching fish or other aquatic life. **2.** the sum of activities involved in the obtaining of fish or other aquatic life.

Fisher-Yates test *Statistics.* a nonparametric test for the equality of treatments in one-factor analysis of variance. Also, EXACT CHI-SQUARED TEST.
Fishes see PISCES.
fisheye *Metallurgy.* in any steel, the area of a fractured surface that has a whitish appearance.
fish gelatin see ISINGLASS.
fish hawk see OSPREY.
fishing space *Civil Engineering.* the space between the head and the base of a rail in which a joint bar is placed.
fishing tool *Engineering.* a device that collects objects from inaccessible locations, such as drill holes.
fish kill *Ecology.* the mass extermination of fish caused by an abrupt change in the aquatic environment, such as the absence of oxygen or the presence of water pollution.
fish ladder *Civil Engineering.* a series of stepped baffles, boxes, or stairs in a dam, which facilitates the migration of fish. Also, **fishway.**
fish lead *Engineering.* a sounding lead that is not removed from the water between soundings.
fish lice *Invertebrate Zoology.* the common name for crustacean copepods in the group Arguloida that parasitize the gills or burrow in the flesh of certain fish.
fishmouthing *Metallurgy.* see ALLIGATORING.
fish plate *Civil Engineering.* either of two steel plates that are bolted to attached lengths of a rail or beam to secure the joint between them.
fish scaling *Materials Science.* small semicircular faults on the surface of porcelain enamel, caused by pockets of hydrogen existing at the interface between the coating and the steel.
fish screen *Civil Engineering.* a screen set across a water intake or outlet to a pond, canal, pipe, etc., to prevent fish from traveling through.
fishtail *Metallurgy.* in metalworking, the irregular trailing end that is eventually discarded.
fishtail bit *Mechanical Devices.* a drill bit having a fishtail shape.
fishtail burner *Engineering.* a burner in which two jets of gas intersect with each other to produce a flame that resembles a fish's tail.
Fissidentales *Botany.* an order of about 800 species of true mosses whose members are characterized by a structure in which each leaf is attached to the ones directly above and below it by a sheathing structure.
fissile *Geology.* capable of being split easily into thin sheets along cleavage or bedding planes. *Nucleonics.* of, relating to, or describing a material that releases energy by the fission process when it absorbs thermal or slow-moving neutrons; the principal fissile materials are uranium-233, uranium-235, and plutonium-239. Also, **fissionable.**
fissiochemistry *Chemistry.* the process in which nuclear energy brings about a chemical change or reaction.
fission [fish´ən] *Nuclear Physics.* the process, either spontaneous or induced, by which a nucleus splits into two or more large fragments of comparable mass, simultaneously producing additional neutrons (on the average) and vast amounts of energy. Also, NUCLEAR FISSION. *Biology.* a method of asexual reproduction in which a parent cell divides into two or more similar and complete parts, beginning with the mitotic division of the nucleus, followed by division of the cytoplasm.
fission barrier *Nuclear Physics.* the energy difference between excited and ground states of a nucleus that must be supplied by an incident particle for fission to take place.
fission chamber *Nucleonics.* an instrument that detects and measures thermal neutrons by counting the fission fragments produced by the coating on surfaces within the detector.
fission cross section *Nuclear Physics.* the probability that an incident neutron with sufficient energy to penetrate the fission barrier will collide with and be absorbed by a nucleus, resulting in fission.
fission detector *Nucleonics.* an instrument that detects spontaneous fission with a specially treated glass upon which charged particles leave tracks, which are made visible by a chemical etching process.
fission fraction *Nucleonics.* the percentage of energy in a nuclear weapon that is caused by fission; in a thermonuclear weapon, for example, the average value is approximately 50.
fission fragments *Nucleonics.* the primary products of fission, consisting of two or more fragments of the fissioned nucleus; the nuclides first created when materials such as uranium-238 undergo fission.
fission-fusion bomb *Nucleonics.* a device that obtains its energy from the formation of a heavy nucleus from two light nuclei (fusion) and the splitting of the nucleus into two lighter nuclei (fission).
fission isomer *Nuclear Physics.* a deformed nuclear state lying in the second well of a double-hump fission barrier.
fission neutron *Nuclear Physics.* a neutron released as a result of nuclear fission, having a continuous spectrum of energy with a maximum of about 10^6 eV.
fission product *Nuclear Physics.* any nuclide, either radioactive or stable, that arises from fission, including both the primary fission fragments and their radioactive decay products. *Ordnance.* the radioactive material scattered following the explosion of a nuclear weapon.
fission-product poisoning *Nucleonics.* a condition in which fission products, such as radioactive decay particles, capture slow-moving thermal neutrons before they reach the nucleus, thus preventing fission.
fission reactor see NUCLEAR REACTOR.
fission spectrum *Nuclear Physics.* the distribution of energy carried by neutrons following fission.
fission threshold *Nuclear Physics.* the minimum amount of kinetic energy a bombarding neutron must have to cause fission in a nucleus.
fission-track dating *Geology.* a method for calculating the age in years of a geologic sample by counting the density of radiation-damage tracks formed spontaneously by the fission of uranium impurities in the sample compared with those tracks that are induced in the sample. *Archaeology.* the use of this technique to establish the age of pottery or of stones known to have been heated.
fission yield *Nucleonics.* the amount of energy generated by fission in a nuclear explosion, as opposed to the energy generated by fusion.
fissiparity *Invertebrate Zoology.* asexual reproduction involving division of an asteroid's central disk into two separate halves that each regenerate their missing parts, creating whole individuals.
fissiparous *Invertebrate Zoology.* producing new individuals through fission.
fissiped *Vertebrate Zoology.* **1.** having the toes separated to the base, as in cloven-hooved animals. **2.** of or relating to the carnivorous suborder Fissipeda, which includes cats, dogs, and bears.
Fissipeda *Vertebrate Zoology.* in older classifications, a suborder of Carnivora that includes recent land carnivores such as cats, dogs, and bears as well as extinct related forms.
fissium *Nucleonics.* a mixture of fission products added to reactor fuel to enhance the stability of the uranium-plutonium alloys when they absorb fast neutrons.
fissure [fish´yər] *Anatomy.* a deep surface groove or furrow that divides an organ into lobes. *Geology.* an extensive break or crack in rock that shows distinct separation, often filled with mineral-bearing materials. Also, JOINT FISSURE, OPEN JOINT.
fissure eruption *Volcanology.* a volcanic eruption occurring along a fracture or crack in a volcano's surface rather than from a central vent.
Fissurellidae *Invertebrate Zoology.* a family of keyhole limpets in the class Gastropoda having a hole at the apex of the conical shell.
fissure system *Geology.* a group of fissures of the same age and having parallel or nearly parallel strike and dip.
fissure vein *Geology.* a veinlike mineral deposit filling a fissure and characterized by having distinct walls.
fistula *Medicine.* an abnormal passage from an internal organ to the body surface or between two internal organs.
Fistulariidae *Vertebrate Zoology.* the cornetfishes or flutemouths, a monogeneric family of circumtropical marine fish of the order Gasterosteiformes having a long slender body, naked skin, and a tubular snout.
Fistulinaceae *Mycology.* a family of fungi of the order Agaricales, containing some edible mushrooms and some species that cause discoloration in wood.
Fistuliporidae *Paleontology.* a family of stemless marine bryozoans in the class Stenolaemata and the extinct order Cystoporata; known only from the Ordovician; characterized by exothecal pores across all possible plate sutures.
fistulous withers *Veterinary Medicine.* abscess and fistula formation on the withers of a horse, especially at the site of an injury or abrasion and subsequent infection; associated with the bacterium *Brucella abortus.*
fit *Medicine.* **1.** any episode characterized by inappropriate and involuntary motor or psychic activity, such as a paroxysm or an epileptic seizure. **2.** being in good health and capable of physical activity. **3.** the adaptation of one structure to another, such as a dental restoration to its site in the mouth. *Biology.* displaying fitness; able to survive as an individual or as a species. *Engineering.* the manner in which one part joins or aligns with another. *Design Engineering.* a dimensional relationship between mating parts in which limits of tolerances for such parts as shafts and holes result in fits of various quality, which are established by a set of uniform standards and specifications.

FITC *Biotechnology.* fluorescein isothiocyanate, a derivative of fluorescein that provides a stable fluorescent label for proteins.

fitch *Mechanical Devices.* a small, long-handled brush with chiseled or round-tapered tips, used by carpenters for finishing work.

Fitch, John 1743–1798; American inventor; developed the first practical steamboat (1786).

Fitch, Val born 1923, American physicist; with Cronin, Nobel Prize for the discovery that K-mesons violate the absolute principle of symmetry.

fitness *Medicine.* a general state of health, characterized by the ability to distribute inhaled oxygen to muscle tissue during physical activity. *Evolution.* a measurement of the ability of a given population of organisms to respond to the pressures of natural selection; the number of offspring produced, related to the number needed to maintain a constant population, sometimes measured by the intrinsic rate of natural increase of a population or subpopulations. *Ecology.* the competition within a specific community that determines which individuals survive to propagate their genes.

fitter *Engineering.* a person who assembles, maintains, and repairs machines.

Fittig's synthesis *Organic Chemistry.* the synthesis of an aromatic hydrocarbon by the sodium-catalyzed reaction of an aryl halide with an alkyl halide. Also, WURTZ-FITTIG REACTION.

fitting *Mechanical Devices.* **1.** any of various devices used to join or fit two parts together, such as pipes, engine components, or machinery. **2.** any of various small, standardized parts such as couplings, gauges, and valves, used in the construction of various types of boilers, engines, and other devices. *Electronics.* an electronic component that has a mechanical function rather than an electrical one, such as a bushing or lock nut in electrical wiring. *Mechanical Engineering.* the handwork needed to complete the assembly of a manufactured product.

fitting out *Naval Architecture.* the final stages of preparation before a vessel sets out to sea on a first voyage or on any new voyage; equipment is fitted, and gear, supplies, and crew are brought aboard.

FitzGerald, George Francis 1851–1901, Irish physicist; proposed the FitzGerald contraction, later confirmed by Lorentz.

FitzGerald-Lorentz contraction *Physics.* a relativistic contraction in the direction of its motion experienced by a body.

Fitz-Hugh–Curtis syndrome *Medicine.* gonococcal perihepatitis occurring in women, resulting from a complication of gonorrheal salpingitis; characterized by fever, upper quadrant pain, tenderness, and spasms of the abdominal wall.

five-and-ten system *Meteorology.* the most common system of representing wind speed to the nearest five knots on a synoptic chart.

five-centered arch *Building Engineering.* a building arch that has the form of a false ellipse with five apparent centers.

five-day fever see TRENCH FEVER.

five-five-five test *Microbiology.* a test to determine the effectiveness of a given disinfectant by treating a test suspension of microorganisms to a series of disinfectant dilutions and scoring the remaining viable cells.

five lemma *Mathematics.* let the following be a commutative diagram of R-modules (modules over a common ring R) and R-homomorphisms, with exact rows. Then the two results below are known as the five lemma:

$$\begin{array}{ccccccccc} A_1 & \rightarrow & A_2 & \rightarrow & A_3 & \rightarrow & A_4 & \rightarrow & A_5 \\ \downarrow_{\alpha_1} & & \downarrow_{\alpha_2} & & \downarrow_{\alpha_3} & & \downarrow_{\alpha_4} & & \downarrow_{\alpha_5} \\ B_1 & \rightarrow & B_2 & \rightarrow & B_3 & \rightarrow & B_4 & \rightarrow & B_5 \end{array}$$

(a) If α_1 is an epimorphism and α_2 and α_4 are monomorphisms, then α_3 is also a monomorphism. (b) If α_5 is a monomorphism and α_2 and α_4 are epimorphisms, then α_3 is also an epimorphism. If A_1, A_5, B_1, and B_5 are the zero modules, then the five lemma together with the following additional result is known as the **short five lemma:** (c) If α_2 and α_4 are isomorphisms, then α_3 is also an isomorphism.

five-level code *Computer Programming.* a coding system that uses five bits to represent each character.

five-level start-stop operation *Telecommunications.* a simplex mode of teletypewriter data transmission in which each code character is subdivided into five electrical units.

five-spot well pattern *Petroleum Engineering.* a network pattern of four wells located in a square with a fifth well in the center; this formation is used, for example, in reservoirs for water-injection pressure maintenance.

fix *Biology.* to kill or preserve a tissue, organ, or organism by immersion in a solution, usually for the purpose of study or examination. *Chemistry.* to reduce from volatility to a more stable state. *Biochemistry.* to change atmospheric nitrogen into another form, especially into a useful compound such as a nitrate fertilizer. *Graphic Arts.* **1.** to stabilize an image; carry out fixation. **2.** the chemical bath used in fixation. *Navigation.* an accurate position determined by crossing two or more lines of position, without reference to any former position. *Aviation.* an aircraft's position at any point along its flight path.

fixated *Psychology.* affected by a fixation.

fixation *Biology.* any process used to kill cells while preserving their structure, usually for subsequent microscopic examination. *Psychology.* a strong attachment to an object or mode of gratification of childhood, which impedes the normal formation of new attachments and patterns of behavior; associated with frustration or overindulgence at a certain stage of psychosexual development. *Graphic Arts.* in developing photographs, the use of a chemical bath (usually hypo) to convert unused halides to soluble silver, thereby stabilizing the image. *Systematics.* the determination of a type species or type specimen.

fixation of alleles *Genetics.* a condition in which only one allele in a certain gene is present in a population.

fixation probability *Genetics.* the likelihood that an allele will become fixed in a population.

fixative *Materials.* a substance that preserves, sets, or stabilizes another substance or material, such as a varnish that preserves a drawing or a chemical that prevents evaporation of a perfume.

fixator *Anatomy.* an accessory muscle that helps to steady a part of the body.

fixed-action pattern *Behavior.* a pattern of innate behavior that is characteristic of a certain species.

fixed-active tooling *Robotics.* stationary equipment in a robotic system that is activated and controlled by electronic signals.

fixed ammunition *Ordnance.* ammunition in which the cartridge case is permanently attached to the projectile, thus allowing the round to be loaded in a single unit. Also, FIXED ROUND.

fixed and flashing light *Navigation.* a steady light, varied at intervals by one or more flashes of greater intensity.

fixed arch *Civil Engineering.* a rigid structural arch with rotational and translational restraints at its supports.

fixed area *Computer Programming.* the part of main or disk memory that is assigned to specific programs, often the operating system and resident utilities; usually protected from overwriting.

fixed artillery *Ordnance.* artillery that is permanently installed for the protection of a strategic area.

fixed attenuator *Electronics.* a circuit that attenuates the level of a signal by a fixed ratio independent of frequency. Also, PAD.

fixed-bed hydroforming *Chemical Engineering.* a cyclic procedure employed in petroleum processing, in which a fixed bed of molybdenum oxide catalyst is deposited on activated alumina to form high-octane aromatic compounds.

fixed-bed operation *Chemical Engineering.* any procedure in which the additive material, such as a catalyst, remains stationary in the chemical reactor or adsorber bed.

fixed-bed reactor *Biotechnology.* a bioreactor having a fixed carrier system to which an active and growing biomass is permanently adhered or fixed.

fixed bias *Electronics.* the value of the bias voltage or current supplied from a fixed external source, such as a battery, that is independent of the signal strength of the device itself.

fixed-block *Computer Technology.* describing data transfer operations in which the blocks all contain the same number of records as specified by the program or by the computer in assigning buffers.

fixed-block architecture *Computer Technology.* a design for data storage in blocks of fixed size.

fixed-block control system *Transportation Engineering.* an automatic train control system that maintains fixed headways (usually by a block-signal system) and uses a fixed block with track circuits to detect and control vehicles.

fixed bridge *Civil Engineering.* a bridge with permanent vertical and horizontal fixity or alignment.

fixed carbon *Chemistry.* the amount of solid combustible material remaining in a sample of coal, coke, or bituminous material after the removal of moisture, volatile matter, and ash; usually expressed as a percentage of the original sample.

fixed carbon ratio see CARBON RATIO, def. 2.

fixed-coordinate system *Robotics.* a coordinate system based on fixed points in time or space.

fixed cost *Industrial Engineering.* a production cost that is independent of production level (e.g., rent or interest charges).

fixed-cycle operation *Computer Technology.* a computer process that always uses the same number of clock cycles to complete. *Transportation Engineering.* a schedule with departures at fixed intervals, such as hourly.

fixed disk *Computer Technology.* a nonremovable magnetic disk storage medium.

fixed echo *Electronics.* a blip on a radar screen that remains stationary, indicating the presence of a fixed target.

fixed-electrode method *Engineering.* a geophysical surveying technique in which one electrode remains stationary while the other is grounded at successively further distances from the stationary one; used in a self-potential prospecting system.

fixed emplacement *Ordnance.* a permanent gun emplacement, usually made of reinforced concrete.

fixed-end beam *Civil Engineering.* a structural beam supported at both ends with both rotational and translational restraints. Also, ENCASTRE BEAM. *Building Engineering.* a beam with its end secured so that the tangent to the curve taken up by the beam remains fixed when under an applied load.

fixed-end column *Civil Engineering.* a structural column supported at both ends with both rotational and translational restraints.

fixed-feed grinding *Mechanical Engineering.* the process of feeding processed material to a grinding wheel at a given rate or in predetermined increments.

fixed field *Computer Programming.* **1.** in word processing, an area that can be temporarily designated by the operator for a specific purpose. **2.** a set of fixed column locations in which certain data must appear.

fixed-field accelerator *Nucleonics.* a machine that increases the energy of charged particles while they are pushed along a circular arc by permanent magnetic fields, thus causing specific particles to travel at the same frequency despite differences in their individual speed or energy levels.

fixed-field addressing *Computer Programming.* a method for addressing a field by means of an instruction that contains the segment number, displacement, and field length.

fixed filter *Mathematics.* a filter $F(X)$ on a nonvoid set X such that the intersection of the sets of $F(X)$ is nonempty. A filter that is not fixed is free. By theorem, if X is finite, then every ultrafilter in X is fixed.

fixed fin *Aviation.* a nonadjustable vane or airfoil that is fastened longitudinally or vertically to an aerodynamic or ballistic body in order to provide stabilization.

fixed-focus *Optics.* describing a device, such as a lens or camera, having a nonadjustable focal length; commonly found in inexpensive cameras and in optical instruments designed to perform specific technical functions in dentistry and medicine.

fixed-form coding *Computer Programming.* a nonprocedural method of programming in which the fields of the instruction are fixed and assigned specific meanings with a limited set of acceptable values; often used in report generation languages such as RPG.

fixed guide see CAGE GUIDE.

fixed gun *Ordnance.* an aircraft machine gun that is rigidly mounted so that it can be aimed only by moving the aircraft.

fixed-head disk *Computer Technology.* a disk unit in which the read/write heads are permanently positioned over the recording surface, one head for each track on the disk.

fixed idea see IDÉE FIXE.

fixed-interval reinforcement *Behavior.* a reinforcement schedule in which reinforcement is given for the first correct response that occurs after a set period of time has elapsed since the last reinforcement. Also, **fixed-interval schedule.**

fixed ion *Analytical Chemistry.* a stationary ion in the matrix or lattice of a solid ion exchanger.

fixed-length field *Computer Programming.* a field in a record that always contains the same number of characters regardless of the data content.

fixed-length record *Computer Programming.* a file element that must contain an invariant number of words, characters, fields, and other components.

fixed-level chart see CONSTANT-HEIGHT CHART.

fixed light *Navigation.* a light that burns continuously at a constant intensity.

fixed logic *Computer Technology.* hardwired logic circuitry that controls computer operations.

fixed medium *Computer Technology.* **1.** data storage material that does not move during read and write operations. **2.** a data storage unit that is not removable, such as a fixed disk pack.

fixed memory see READ-ONLY MEMORY.

fixed-mooring berth *Civil Engineering.* a marine structure used to secure a ship, having a platform to support cargo-handling equipment.

fixed-needle traverse *Engineering.* a traverse with a compass equipped with a sight line that can be raised above a graduated horizontal circle, in order to obtain the azimuth angle; used in surveying.

fixed-passive tooling *Robotics.* unpowered equipment in a robotic system, such as work-holding devices and fixtures.

fixed point *Engineering.* a value derived from intrinsic properties of pure substances and used to standardize measurements. *Mathematics.* a point that is mapped to itself by a given mapping or function on a set.

fixed-point arithmetic *Computer Programming.* **1.** integer arithmetic. **2.** a computation performed on operands whose radix points are assumed to be always in the same position in the computer word. **3.** a type of arithmetic in which the operands must be shifted relative to each other or scaled, prior to calculating the result, and the result scaled prior to storing.

fixed-point attractor *Physics.* an attractor consisting of a single point in phase space and describing a stationary state of a system.

fixed-point calculation *Computer Technology.* mathematical operations involving fixed-point operands and results.

fixed-point computer *Computer Technology.* a computer limited to fixed-point arithmetic operations.

fixed-point representation *Computer Programming.* a type of numeric representation in which the radix point is always assumed to be in the same position in the word, often at the extreme left end or right end.

fixed-point theorem *Mathematics.* any of several theorems stating that a given mapping or function on a set has a fixed point under stated conditions.

fixed-position layout *Industrial Engineering.* a work flow layout in which the product remains at one location due to its weight or bulk, and equipment must be moved to that location in order to perform work on the product, as in a shipyard.

fixed-program computer *Computer Technology.* a special-purpose computer whose functions are implemented in circuitry as opposed to software. Also, WIRED PROGRAM COMPUTER.

fixed-ratio reinforcement *Behavior.* a reinforcement schedule in which reinforcement is given only after a set number of correct responses have occurred since the last reinforcement. Also, **fixed-ratio schedule.**

fixed round see FIXED AMMUNITION.

fixed-route demand service *Transportation Engineering.* a service having fixed routes but flexible departures based on passenger demand.

fixed satellite *Space Technology.* an artificial satellite that orbits from west to east at a speed that maintains its position over a given point on the earth's equator. Also, GEOSTATIONARY SATELLITE, SYNCHRONOUS SATELLITE.

fixed screen *Mining Engineering.* a stationary inclined or curved panel used to remove a large proportion of water and fines from a suspension of coal in water.

fixed-sequence robot *Robotics.* a robot whose motion is limited by a fixed object at the end of its prescribed path. Also, **fixed-stop robot.**

fixed service *Telecommunications.* a system that provides radio communication between specified fixed locations.

fixed sonar *Engineering.* a sonar in which the receiving transducer is in a fixed location, as opposed to scanning sonar.

fixed star *Astronomy.* an older term once used to distinguish the real, or "fixed," stars from the "wandering" stars, or planets.

fixed storage see PERMANENT STORAGE, def. 1.

fixed target *Ordnance.* a nonmoving or immobilized target; e.g., a building or a body of surrounded troops.

fixed transmitter *Electronics.* a transmitter that is operated in a fixed or permanent location.

fixed virus *Virology.* a strain of rabies virus that is attenuated by serial passage.

fixed-word-length computer *Computer Technology.* a computer in which all machine words contain the same number of bits.

fixing block *Building Engineering.* a block made of porous concrete material or of brick mixed with sawdust; built into the surface of a wall to provide a base for securing window frames or other joinery.

fixing moment *Mechanics*. an assumed bending moment applied at each joint and segment of a statically indeterminate structure for resolving its loads.

fixing plug *Building Engineering*. a plastic or metal plug used with screws or bolts to provide fixings to walls or concrete surfaces.

fixings *Building Engineering*. a set of supports, such as grounds and plugs, that are used to secure joinery.

fixture *Building Engineering*. an item permanently attached to a structure, such as a light or a sink. *Mechanical Engineering*. in the manufacture of interchangeable parts, a device used to hold and position a piece of work without guiding the cutting tool.

fixture models *Design Engineering*. the design of a particular work-holding tool to steady a workpiece during such manufacturing operations as robotics and inspection.

fixturing *Mechanical Engineering*. the ensemble of fixtures used in a given work process.

Fizeau, Armand [fē zō´] 1819–1896, French physicist; adapted the Doppler effect to light; measured the velocity of electricity and of light in air and water.

Fizeau fringes *Optics*. the pattern that results from the interference of the two beams of light coming from the top and bottom surfaces of a thin layer or wedge.

Fizeau interferometer *Optics*. an interferometer in which a beam of light is split, reflected a number of times, and then recombined in order to produce thin, multiple-beam interference fringes.

Fizeau toothed wheel *Optics*. a device that measures the speed of light by modifying the spin of a toothed wheel, so that light transversing one tooth reflects back through the next; it has been succeeded by the rotating mirror.

fizélyite *Mineralogy*. $Pb_{14}Ag_5Sb_{21}S_{48}$, a dark lead-gray, opaque monoclinic mineral occurring in deeply striated prismatic crystals, and having a hardness of 2 on the Mohs scale and a specific gravity of 5.56.

fjord [fē´yôrd´] *Geography*. a littoral inlet that has been gouged by glacial ice and filled by the sea, as in Norway and Alaska. Also, FIORD.

fjord

FK5 *Astronomy*. the fifth *Fundamental Katalog*, a reference catalogue giving precise positions and proper motions for about 5000 stars; its predecessor, the **FK4,** contained about 1500 stars.

fl fluid.

FL focal length.

FLA. left front anterior. (From *fronto-laeva anterior*.)

flabellate *Biology*. fan-shaped.

Flabellifera *Invertebrate Zoology*. the largest suborder in the crustacean order Isopoda, usually having a tail fan; includes the wood borers and ectoparasites of fish.

Flabelligeridae *Invertebrate Zoology*. a family of aquatic cage worms in the subclass Sedentaria; sedentary burrowers.

flabellum *Invertebrate Zoology*. a fan-shaped structure, such as the epipodite in some crustaceans or the terminal lobe of the tongue in some insects.

flaccid [flas´id; flaks´id] *Physiology*. **1.** soft and limp; lacking muscle tone or firmness. **2.** describing the normal relaxed condition of a distensible organ, such as the penis, when it is not erect. *Botany*. describing a plant that is wilted due to water deprivation; it becomes limp because the cells lack turgor.

flacherie *Invertebrate Zoology*. a fatal bacterial disease of silkworms and other caterpillars, resulting in loss of appetite, dysentery, flaccidity, and liquefaction after death.

Flacourtiaceae *Botany*. a large family of cyanogenic trees and shrubs of the order Violales, found mostly in tropical regions.

Flade potential *Metallurgy*. the potential of a passive metal or alloy before its abrupt change from the passive to the active state.

flag anything resembling a flag or banner in form or function; specific uses include: *Plant Pathology*. a withered or dead foliated branch growing from what otherwise appears to be a healthy tree; this indicates the presence of an inadequate water supply to the leaves. *Electronics*. a sheet of metal or fabric used to shield the lens of a television camera from light. *Computer Programming*. **1.** any of a variety of indicators, often a single bit, that inform of special conditions, identify members of a set, call attention to particular items of information, or mark word boundaries. Also, SENTINEL, TAG. **2.** a Boolean variable that is set to record a condition detected in a program; e.g., an error flag might be set when an error is detected. *Mathematics*. let V be an n-dimensional vector space. A flag in V is the chain of subspaces $0 = V_0 \subset V_1 \subset \cdots \subset V_n = V$, where $\dim V_i = i$. If f is an endomorphism on V such that $fV_i \subset V_i$ for all i, then f is said to stabilize (or leave invariant) this flag.

flag alarm *Engineering*. an indicator in an instrument that appears when other indicators are malfunctioning.

flagella [flə jel´ə] *Cell Biology*. the plural of flagellum.

flagellar antigen see H ANTIGEN.

Flagellariaceae *Botany*. a monogeneric family of monocotyledonous solid-stemmed, high-climbing vines of the order Restionales, lacking secondary growth but having stems arising from sympodial rhizomes.

Flagellata see MASTIGOPHORA.

flagellate [flə jel´it] *Biology*. of, relating to, or moving by means of flagella. *Invertebrate Zoology*. possessing a moveable threadlike or whip-like structure, or flagellum; specifically, those protozoa having flagella.

flagellated chamber *Invertebrate Zoology*. in sponges, an internal cavity lined with flagellated cells that assist in moving water through the sponge.

flagellation [flaj´ə lā´shən] *Psychology*. the act of whipping oneself, or of submitting to whipping by another person, for the purpose of sexual arousal or gratification. *Biology*. the arrangement of flagella on a cell or organism.

flagelliflory *Botany*. having flowers that hang freely from ropelike twigs.

flagellin *Microbiology*. the major protein subunit that makes up the filaments of bacterial flagella.

flagellum [flə jel´əm] *Cell Biology*. the long, whip-like "tail" of certain cells in both the plant and animal kingdoms, which undulates with a regular pattern; used primarily for locomotion, and also for other purposes such as attracting food particles or moving substances through a cavity. (From the Latin word for "whip.")

fla genes *Genetics*. a group of genes that are involved in the synthesis and activity of bacterial flagella.

flag flip-flop *Computer Technology*. a single-bit register that indicates the status of special conditions such as carry, overflow, sign bit, or various types of interrupts. Also, **flag bit.**

flaggy *Geology*. **1.** describing bedding having layers 1–10 cm in thickness. **2.** describing rock that tends to split into layers 1–5 cm thick and is suitable for use as flagstones.

flagman *Engineering*. any person who holds or uses a flag, as in directing traffic at a construction site.

flag officer *Military Science*. a naval officer with a rank higher than captain.

flagpole *Engineering*. any free-standing pole on which flags are displayed.

flag smut *Plant Pathology*. a smut disease of cereals and other grasses, characterized by the formation of linear sori on the stems and leaves, which burst, release black spores, and consequently cause the diseased area to fray.

flagstaff *Engineering*. a staff that is affixed to a building and used to display flags or other signals.

flag station *Transportation Engineering*. a railroad station at which trains do not stop unless signaled to do so.

flagstone *Geology.* a hard, flat, thin-bedded, fine-grained sandstone or firm shale that splits easily along bedding planes into slabs suitable for use in paving.
flag stop *Transportation Engineering.* a stop that is made only if passengers or cargo are to be taken aboard or debarked. (From the railroad practice of signaling for a stop at a station that would otherwise be bypassed.)
flail tank *Ordnance.* a tank used to detonate antitank mines, equipped with a device made of chain flails that is attached to a roller powered by the tank's engine.
flair *Civil Engineering.* the gradual widening of the flangeway of a track or rail structure near the end of a guard rail line.
flajolotite see TRIPUHYITE.
flak *Ordnance.* **1.** explosive projectiles fired from antiaircraft guns. **2.** of or relating to antiaircraft guns and their emplacement. (An acronym for a German phrase meaning "antiaircraft gun.") Thus, **flak battery, flak installation, flak tower.**
flak analysis *Military Science.* the determination of enemy antiaircraft strength, location, and effectiveness. Also, **flak intelligence.**
flake *Metallurgy.* a metal or alloy powder consisting of thin, flat particles. *Archaeology.* **1.** a piece of stone detached from a larger mass for use as a tool. Also, FLAKE TOOL. **2.** any smaller piece of stone removed during the process of toolmaking; a chip. **3.** to remove a piece of stone in toolmaking.

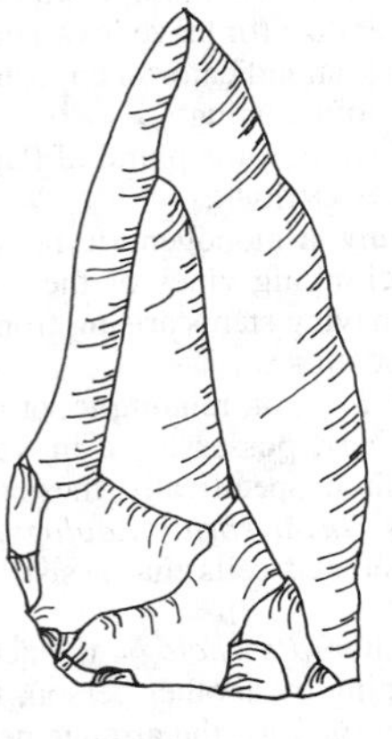

prehistoric flake tool

flaked stone *Archaeology.* any object that has been produced by one of the various percussion or pressure techniques of stone tool technology. Also, CHIPPED STONE.
flaker *Archaeology.* a pressure-flaking tool usually made from bone or antler, used to remove flakes in stone material as a point or blade is being manufactured.
flake scar *Archaeology.* a mark on a stone showing that a flake has been chipped off at that point in the making of a tool.
flake scatter *Archaeology.* a group of stone flakes discarded during tool-making; often found in a semicircle pattern where work was done.
flake tool see FLAKE, def. 1.
flaking *Engineering.* the decomposition of a bulk substance into flakes. *Archaeology.* the process of making stone tools by removing flakes from a larger mass. *Mining Engineering.* the act of breaking small chips away from a refractory face. *Chemical Engineering.* a continuous process that causes solidification of a liquid by cooling it.
flaking mill *Mechanical Engineering.* a machine for converting material to flakes.
flaking station see CHIPPING STATION.
flak jacket *Textiles.* **1.** a military jacket or vest made of heavy fabric and containing metal, nylon, or ceramic plates for protection against explosive projectiles. **2.** a modified civilian version of this garment, worn for protection by athletes. Also, **flak vest.** Similarly, **flak suit.**
flamboyant structure *Mineralogy.* the optical harmony of crystals or grains as disrupted by a divergent structure caused by slight differences in orientation.
flamboyant style *Architecture.* the last phase of French Gothic architecture, characterized by double-curved tracery resembling leaping flames.
flame *Chemistry.* the burning portion or reaction area of combustion.
flame annealing *Metallurgy.* the process of annealing performed by direct impingement of a flame.
flame arc lamp *Electricity.* an arc lamp with a high degree of volatility owing to the use of impregnated carbon electrodes such as calcium, barium, or titanium.
flame arrester *Engineering.* an assembly of screens, perforated plates, or metal-gauze packing affixed to the breather vent on a flammable-product storage tank.
flame bucket *Space Technology.* a hollow construction under a rocket launch pad, with its top open to receive hot rocket exhaust, one of its sides heavily lined and turned inward to form a flame deflector, and one or three other sides open to release heat and exhaust gases.
flame cell *Invertebrate Zoology.* in flatworms, a large, hollow cell with internal flagella or cilia that move excretory products through the body. Also, **flame bulb**.
flame cleaning *Metallurgy.* the process of cleaning a metallic surface by direct impingement of a gas flame.
flame collector *Engineering.* an instrument that removes the induction charge on an apparatus to take atmospheric electrical measurements.
flame coloration *Analytical Chemistry.* a qualitative flame test performed by fixing a substance to a platinum wire, moistening it with hydrochloric acid, placing it in a nonluminous flame, and observing the resulting color. Calcium, for example, colors the flame brick red, while sodium yields bright yellow and potassium purple.
flame cultivator *Agriculture.* a device consisting of a tank, pump, and nozzle that blows a flaming gas to destroy small weeds between rows of crops.
flame cutting *Metallurgy.* the use of an oxygen gas flame to heat metals in preparation for cutting them.
flame deflector *Space Technology.* in a vertical launch, a construction designed to deflect hot exhaust gases away from the ground or launch pad. Such devices range from small devices enclosed within the rocket launcher to heavy metal plates resembling flame buckets.
flame detector *Mechanical Engineering.* a sensing device and ignition detector that indicates whether fuel is burning and signals a controller if ignition is present.
flame-emission spectrophotometry *Analytical Chemistry.* a process of photometric analysis in which a sample solution is aspirated onto a hydrogen-oxygen or acetylene-oxygen flame and the emitted wavelength of the ions is recorded on a photographic plate. Also, ATOMIC EMISSION ANALYSIS.
flame-emission spectroscopy *Spectroscopy.* the analysis of the radiant energy emitted when the atoms of a sample solution are excited by a flame, usually oxyhydrogen or oxyacetylene.
flame excitation *Spectroscopy.* an emission technique in which the thermal energy of a high-temperature flame is used to excite the atoms of a sample.
flame gauging *Metallurgy.* a process of shallow flame cutting.
flame-hardened *Metallurgy.* describing a material produced by a process of flame hardening.
flame hardening *Metallurgy.* in the thermal treatment of steel, the process of heating a superficial layer with a flame, followed by quenching.
flameholder or **flame holder** *Aviation.* a ring, grid, or other device designed to shield a flame within a moving stream of combustible material, as in the afterburners of certain jet engines.
flame ionization detector *Analytical Chemistry.* an instrument used in gas chromatography that decomposes the neutral solute molecules in a flame into charged species and electrically measures the changes in conductivity.
flame laser *Optics.* a gas laser characterized by a self-sustaining flame and a carbon monoxide discharge, produced by the ignition of a low-pressure mixture of gases such as oxygen and carbon disulfide.
flameout *Aviation.* the termination of fuel burning, especially in a gas turbine or jet engine, for any reason other than a deliberate cutoff.
flame photometer *Spectroscopy.* a device that photoelectrically measures the characteristic wavelengths of light emitted when a sample solution is introduced into a flame.
flame photometry *Spectroscopy.* a spectroscopic technique in which flame excitation is used to produce emission spectra in order to determine the concentration of certain metals and transition elements in a sample solution.
flame plate *Engineering.* any of the plates on a boiler firebox that must withstand maximum furnace temperature.
flame plating *Metallurgy.* the process of coating an object with a metal or alloy melted in a flame.

flame polishing *Materials Science.* a post-machining procedure used on small-diameter rods or filaments; surface and subsurface flaws resulting from machining are removed by rotating the sample and passing it through a flame so that the thin surface layer melts.

flame propagation *Chemistry.* the outward spread of a flame into combustible materials.

flame retardant *Materials Science.* **1.** any chemical compound used to raise the ignition point of such materials as cloth or plastic, and thereby increase their resistance to combustion. **2.** see FLAME-RETARDED.

flame-retarded *Materials Science.* of a material, treated with chemicals in order to increase its resistance to combustion, such as a material used to make children's pajamas.

flame spectrometry *Spectroscopy.* a technique for the measurement of the wavelengths of radiation emitted when a sample is introduced into a flame.

flame spectrophotometry *Spectroscopy.* a spectroscopic technique for measuring the intensity of spectral lines in an emission spectrum produced by flame excitation.

flame spectrum *Spectroscopy.* an emission spectrum produced by vaporization of a sample in a nonluminous flame.

flame spraying *Engineering.* **1.** a technique in which powdered fragments of a plastic, along with appropriate fluxes, are projected through a cone of flame onto the surface of an object. **2.** a process in which a spray gun feeds wire into a gas flame to install a conductor on a circuit board in molten form; generally the gun has a metal mask or stencil attached. *Metallurgy.* a process of thermal spraying performed by feeding the coating material into a flame.

flame straightening *Metallurgy.* the process of improving the shape of a metallic product by local heating with a flame.

flame structure *Geology.* a sedimentary structure consisting of mud plumes that have been squeezed upward into an overlying layer at irregular intervals.

flame test *Analytical Chemistry.* any qualitative test made with a Bunsen burner, such as a bead test or flame coloration.

flamethrower or **flame thrower** *Ordnance.* a weapon that ignites and projects incendiary fuel. *Engineering.* a similar device designed for nonmilitary use, for example by gardeners to kill weeds or insects.

flame trap *Engineering.* a device that keeps a gas flame out of the supply pipe.

flame treating *Engineering.* a process of rendering inert thermoplastic objects receptive to inks, lacquers, paints, or adhesives; the object is immersed in an open flame, causing surface oxidation.

flaming *Microbiology.* the process of exposing an object, such as a bacteriological loop, to a flame for several seconds to sterilize its surface.

flamingo *Vertebrate Zoology.* any of several aquatic birds of the family Phoenicopteridae that have long legs and neck, webbed feet, a broad lamellated bill resembling a duck's but abruptly bent downward, and pinkish plumage.

flamingo

flammability [flam′ə bil′ə tē] *Chemistry.* the ease with which a material will become combustible.

flammability limit *Chemistry.* either of the limits between which a combustible gas has both enough fuel (**lower flammability limit**) and enough air (**upper flammability limit**) to support combustion.

flammable [flam′ə bəl] *Materials Science.* of a material, tending to burn; capable of supporting combustion.

flammable liquid *Materials Science.* **1.** any liquid having a flash point of less than 38°C and a vapor pressure of not over 40 psi at 38°C. **2.** any liquid that gives off combustible fumes.

Flamsteed, John 1646–1719, English astronomer; first astronomer royal; devised mural arc; compiled first accurate British star catalog.

Flamsteed number *Astronomy.* one of a series of sequential numbers assigned by John Flamsteed to naked-eye visible stars in each constellation, beginning at the westernmost star and continuing eastward in order of right ascension.

flan *Meteorology.* a term used in Scotland for a sudden gust or squall on land.

Flanders storm *Meteorology.* a term used in England for a heavy snowfall with a south wind. (Associated with the region of *Flanders,* on the North Sea opposite southern England.)

Flandrian transgression *Oceanography.* the rapid rise of the North Sea at the end of the Wurm glacial stage, between 11,000 and 3000 years ago, when the melting glaciers of Northern Europe caused the water level of the North Sea to rise from 55 meters below its present level to about 6 meters below its present level. (From *Flanders*.)

flange [flanj] *Design Engineering.* a rim or collar at the end of a pipe or tube that provides strength or a place to attach something else.

flanged coupling *Mechanical Devices.* a shaft coupling composed of two bolted, facing flanges that abut the shaft with a common bolt through them. Also, FACE-PLATE COUPLING.

flanged pipe *Mechanical Devices.* a pipe or pipe section with a flange at each end that bolts to an adjoining length of pipe.

flanged seam *Mechanical Devices.* a joint made from a furnace tube by flanging and bolting its ends together between a pair of steel rings.

flange nut *Mechanical Devices.* a broad-faced nut with a flange attached to the bottom face, used in washerless large-hole coverings.

flangeway *Civil Engineering.* an open way through a track or rail structure that furnishes a passageway for the flange of a wheel.

flanging press *Mechanical Devices.* a machine that bends the edges of plates to form a flange used to bolt the pieces together. Also, **flanging machine.**

flank the side of something; specific uses include: *Vertebrate Zoology.* the fleshy part of an animal or human between the ribs and hip. *Military Science.* **1.** the extreme right or left side of a fleet or army, or a subdivision of this. **2.** to attack or pass around the flank of the enemy. Thus, **flank guard, flank observation, flanking attack.** *Geology.* see LIMB. *Building Engineering.* the valley of a roof. *Civil Engineering.* the outer edge of a carriageway. *Mechanical Devices.* either of the inclined surfaces on a screw thread, between the root and the crest.

flank angle *Design Engineering.* the angle between the flank of a thread of a screw and a plane perpendicular to the axis when measured from the axial plane of reference.

flank hole *Mining Engineering.* **1.** a hole bored in advance of a working place, when approaching old workings. **2.** a borehole driven from the side of an underground excavation, not parallel with the center line of the excavation, to detect water, gas, or other danger.

flanking sequence *Molecular Biology.* the immediate or neighboring upstream or downstream sequence from a designated definitive structure such as a gene.

flanking window *Building Engineering.* a window adjacent to an external door.

flank wall *Building Engineering.* a side wall.

flanning *Building Engineering.* the internal splay of a window, window jamb, fireplace, or doorway from the frame to the inner face of a wall.

flap a flat, broad piece that is attached along one side to a larger body; specific uses include: *Aviation.* a moveable, usually hinged airfoil including the trailing edge of an aircraft wing, designed to increase lift or drag by changing the camber of the wing; used to slow an aircraft during landing by increasing lift. *Space Technology.* a vane affixed to a rocket and functioning as a rudder in the air or within the jet stream. *Building Engineering.* a hinge having a plate that is screwed into a door, shutter, or the like. *Surgery.* a mass of tissue that is partially or totally detached from the body for use in grafting. *Graphic Arts.* either of two parts of a book jacket that fold under the front and back covers.

flap attenuator *Electromagnetism.* a variable waveguide attenuator in which a sheet of resistive material is inserted through a nonradiating slot to provide a desired amount of power absorption.

flaperon *Aviation.* a control surface that combines the functions of a flap and an aileron.

flap gate *Civil Engineering.* a gate that uses hinges at the top of the gate to rotate around for opening and closing. Also, PIVOT LEAF GATE.

flap operation *Surgery.* any operation in which a flap or mass of tissue is raised from its normal position.

flapping *Metallurgy.* in copper fire refining, a method of accelerating the oxidation process by disturbing the slag layer.

flapping tremor see ASTERIXIS.

flap tile *Building Engineering.* a tile that is specially fabricated to fit over a hip or valley line or to catch rainwater.

flap trap *Mechanical Engineering.* an antiflooding valve consisting of an intercepting chamber fitted with a hinged metal flap that allows flow in one direction only.

flap valve *Mechanical Engineering.* **1.** a valve or sheet of flexible material fitted with a hinged flap that swings only in one direction to allow air flow. **2.** a nonreturn valve for liquids containing metal; used in low-pressure applications with O-rings.

flare a flame or burst of light, or something suggesting this; specific uses include: *Astronomy.* a sudden outburst on the sun triggered by the magnetic activity associated with sunspots, sending a stream of charged particles into space; it may cause auroras on earth. *Engineering.* a device that generates a single source of intense light for purposes of target or airfield illumination. *Optics.* undesired, non-image-forming light that passes through the lens onto the image and appears as a spot when it is concentrated, or as a haze when it is diffuse, usually the result of multiple internal reflections on the lens. *Electromagnetism.* an excessively bright spot on a radar screen. *Chemical Engineering.* a burner, usually installed outdoors in an elevated position, used to dispose of combustible waste gases from chemical or refining processes by igniting them. *Aviation.* to slow and smooth out a landing; perform a flareout. *Medicine.* **1.** a flush or area of redness on the skin, spreading out around an infective lesion or point of reaction to an irritant. **2.** a sudden exacerbation of a disease. *Design Engineering.* an expansion or thickening around an end of a cylindrical or spherical body, such as the base of a rocket. *Naval Architecture.* an outward curve of a ship's sides above the waterline, especially toward the bow, reducing the plunging of fine-lined ships; commonly associated with fast ships.

flareback *Ordnance.* the escape of flame or gas toward the rear of a gun.

flare chute *Engineering.* a flare attached to a parachute so it can remain in the air a long time.

flare dud *Ordnance.* a nuclear weapon that detonates with anticipated yield but at an altitude significantly higher than intended, thus reducing its effect on the target.

flare factor *Acoustical Engineering.* the degree of outward flare of an exponential horn, described as the constant m in the formula $S_x = S_0 \exp(mx)$, with x representing the distance from the throat and S_0 representing the throat area.

flare gas *Chemical Engineering.* a surplus or waste gas combusted by a flare.

flareout *Aviation.* a maneuver performed during the last stage of an aircraft landing, involving a slow final nose-up pitch to reduce the rate of descent toward zero at touchdown.

flare star *Astronomy.* a star that exhibits an abrupt brightening, usually of about half a magnitude, followed by a decline over several hours.

flare triangulation *Cartography.* a method of triangulation in which simultaneous observations are made on parachute flares; used for extending triangulation over lines too long to be observed by ordinary methods.

flare-type bucket *Design Engineering.* a bucket with flared sides.

flare-type burner *Engineering.* a circular burner that projects a cone-shaped flame.

flaring *Metallurgy.* the process of forming a flange.

flaser *Geology.* the streaky strands of parallel, scaly rock particles that surround the lenses of granular material in a flaser structure.

flaser gabbro *Geology.* a gabbro in which flakes of chlorite or mica swirl around eye-shaped lenses of quartz or feldspar, which formed as a result of dislocation metamorphism.

flaser structure *Geology.* a structure, in metamorphic rock, containing lenses and strands of unaltered granular minerals embedded in a matrix of highly crushed and sheared material. Also, PHACOIDAL STRUCTURE.

flash *Astrophysics.* an abrupt and short-lived increase in brightness and temperature that occurs in many kinds of stars as they evolve. *Ordnance.* the visible light created by burning propellant gasses upon the firing of a weapon or the explosion of a projectile. *Graphic Arts.* in photography, a light source designed to give very brief and bright illumination to a subject. *Materials Science.* **1.** the section of charge that oozes from the joint line of the mold cavity in plastics and rubber production, or from the joint line of the casing in metals production. **2.** the variation in color that appears on the surface of a brick. *Building Engineering.* to make a joint weather-tight by using flashing.

flash arc *Electronics.* in a large thermionic vacuum tube, an abrupt surge of current between the cathode plates at high plate voltage; may short-circuit the power supply.

flashback *Engineering.* see BACKFIRE.

flashback arrester *Engineering.* a device that prevents fire from back-passing the arrester in a torch, in order to prevent damage.

flashback voltage *Electronics.* the inverse peak voltage needed to ionize gas in a tube.

flash-bang *Ordnance.* a term for the brief interval between the time when a weapon's muzzle flash is sighted and the noise of firing is heard.

flash barrier *Electricity.* in rotating machinery, a fire-resistant structure positioned between conductors in order to minimize flashover or damage caused by it.

flashboard *Civil Engineering.* a temporary barrier at the top of a dam spillway, designed to increase storage capacity; it is usually relatively low and constructed of a series of boards.

flash boiler *Mechanical Engineering.* a steam boiler with hot tubes of small capacity designed to convert small amounts of water to superheated steam as it is fed through by a feed pump.

flash bomb *Engineering.* a bomb that lights the ground for night photography from the air.

flashbulb memory *Psychology.* a vivid memory of a single significant event in one's life, such as the death of a close relative or a celebrated public figure.

flash burn *Medicine.* a superficial but often extensive burn produced by intense heat of a brief duration, as that of explosions.

flash butt welding *Metallurgy.* resistance welding performed by heating the abutting surfaces, followed by application of pressure.

flash carbonization *Chemical Engineering.* a method of carbonization in which coal is exposed to a short residence time in the reactor in order to produce the largest possible yield of tar.

flash chamber *Chemical Engineering.* a process for converting a carbonaceous material to carbon in a higher-pressure vessel by flashing it into a low-pressure, conventional oil-and-gas separator. Also, **flash trap, flash vessel.**

flash-coat *Metallurgy.* to apply any very thin metallic coating.

flash coloring *Entomology.* normally concealed bright coloration that suddenly appears or flashes into view when an insect is disturbed, serves to frighten off potential enemies. Also, STARTLE COLORING.

flash drum *Chemical Engineering.* a drum or tower in which heavy oil or residue is flashed to a lower pressure in order to vaporize the volatile components.

flash-drying *Chemical Engineering.* the quick evaporation of liquid from a granular or porous solid by a rapid reduction in pressure or an updraft of warm air. Thus, **flash-dry.**

flasher *Electricity.* any device or circuit that flashes a light or a series of lights sequentially.

flash evaporator *Biotechnology.* an evaporator used for solvent recovery in downstream processing; the feed solution is fed into a container where the temperature is high enough that rapid evaporation of the solvent occurs.

flash exposure *Graphic Arts.* a halftoning technique in which a yellow-filtered flash is used to enhance the detail in a shadow area of a photograph.

flash extension *Metallurgy.* the remnant of a flash after trimming.

flash factor *Optics.* a number determined by the amount of light produced by a specific flashbulb and the speed of the film, used to set the lens aperture when photographing a subject at a specific distance. Also, **flash guide number.**

flash-ferm process *Biotechnology.* an inexpensive variation of the vacuum fermentation process for ethanol production in which the oxygen requirement for the yeast is met through the use of sparged air.

flash flood *Hydrology.* a sudden, localized flood of relatively great volume and short duration, generally caused by a brief, heavy rainfall in a dry area.

flash groove *Engineering.* **1.** a groove cut in a casting die so that leftover material can escape during casting. **2.** see CUTOFF.

flash hider *Ordnance.* a device attached to the muzzle of a gun to conceal or reduce the muzzle flash created upon firing. Similarly, **flash suppressor.**

flashing *Building Engineering.* a strip of tin or copper placed at the exterior junction of a building surface, such as around a chimney or window, to make joints watertight. *Chemical Engineering.* the vaporization of volatile fluids by either pressure reduction or heat. *Materials Science.* a method of burning irregularly colored face bricks by periodically stopping the air supply during firing. *Metallurgy.* in flash welding, the heating step of the process.

flashing board *Building Engineering.* a board to which pieces of flashing are attached.

flashing compound *Building Engineering.* an impermeable, elastic, nondrying aggregate used to fill the crevices between insulation blocks or other building structures.

flashing flow *Chemical Engineering.* the state of a liquid at its boiling point when it flows through a heated chamber or passage and is heated, causing partial vaporization which results in a two-phase, vapor-liquid flow.

flashing light *Navigation.* a light that burns intermittently for brief periods, the duration of light being less than the duration of darkness.

flashing-over *Electricity.* in rotating machinery, an unwanted arc that forms as a result of faulty insulation between commutator segments.

flashing ring *Building Engineering.* a ring around a pipe that seals the opening or holds it in place as it passes through a wall or floor.

flash lamp *Electronics.* a gaseous-discharge lamp that is attached to a photoflash unit which produces bursts of high-intensity light for stroboscopic photography. Also, STROBOSCOPIC LAMP.

flashless *Ordnance.* describing a propellant or propelling charge that significantly reduces or eliminates the muzzle flash upon firing. Thus, **flashless charge, flashless powder.**

flashlight *Electronics.* a portable, battery-powered device that produces a beam of light.

flash line *Engineering.* a ridged seam on a molding surface between mold faces.

flash magnetization *Electromagnetism.* the magnetization of a sample of ferromagnetic material produced by subjecting it to a current impulse of short duration, affecting only the magnetic domains near the outer layers of the material.

flash message *Military Science.* in military communications, the highest-precedence category, reserved for messages regarding initial contact with hostile forces or operational combat messages of great import; brevity is mandatory in this category.

flash mold *Engineering.* a mold that allows leftover material to escape during closing.

flashover *Electricity.* a disruptive, often luminous charge around or over the surface of an insulator, usually due to high voltage.

flashover voltage *Electricity.* the voltage needed to cause a sudden discharge of electrical energy between electrodes or conductors that are separated by air or insulating material; frequently accompanied by light. Also, SPARKOVER VOLTAGE.

flash pasteurization *Microbiology.* a process of killing bacteria in milk, in which the milk is quickly raised to a minimum temperature of 72°C for at least 15 seconds and then quickly cooled.

flash photolysis *Physical Chemistry.* a technique in which an intense flash of light is used to activate gas molecules so that they can be observed spectroscopically.

flash plate see CALIBRATION PLATE.

flash plating *Metallurgy.* the electroplating of a very thin final coating.

flash point *Chemistry.* the lowest temperature at which vapor above a volatile liquid will quickly ignite when the liquid is heated under standard conditions.

flash process *Chemical Engineering.* a liquid-vapor system in which the composition remains constant, but the ratio of gas phase to liquid phase alters as temperature or pressure changes, as in a flash drum or tank.

flash ranging *Military Science.* the process of determining the position of an enemy gun or the burst of a projectile by observing its flash. Similarly, **flash reconnaissance, flash spotting.**

flash-ranging adjustment *Ordnance.* the correction of friendly fire based on flash ranging of enemy fire.

flash ridge *Engineering.* the section on a flash mold where excess material escapes before the mold is shut.

flash roast *Mining Engineering.* a process in which finely divided sulfide mineral falls through a heated oxidizing atmosphere causing the sulfur to burn off rapidly from the ore. Also, SUSPENSION ROAST.

flash smelting *Metallurgy.* the smelting of ore concentrates by reacting them with a hot gas in a vertical furnace.

flash spectroscopy *Spectroscopy.* the study of the electron states in molecules that have absorbed energy from a brief but intense flash of light.

flash spectrum *Astrophysics.* the emission spectrum of the sun's chromosphere and lower corona, which is momentarily observable at the beginning and end of totality in a solar eclipse.

flash tank *Chemical Engineering.* the unit in a processing operation that separates the gas and liquid phases.

flash test *Electricity.* a method of insulation testing in which a higher than normal voltage is applied for a short duration.

flash vaporization *Chemical Engineering.* the fast vaporization obtained by passing a liquid through a heat source. *Petroleum Engineering.* specifically, a technique used for removing liquefied petroleum gas from storage, in which liquid is first flashed into a vapor in an intermediate pressure system, and then a second-stage regulator contributes the low pressure required to use the gas in appliances.

flash welding see FLASH BUTT WELDING.

flashy stream see FLASH FLOOD.

flask *Chemistry.* a vessel or receptacle designed to hold a substance, typically a long-necked glass vessel used to hold a liquid. *Metallurgy.* in casting, the container in which a sand mold is made.

flat something that is smooth and level; specific uses include: *Geology.* **1.** any generally smooth, horizontal, or level area or strip of land marked by little relief. **2.** see MUD FLAT. *Mining Engineering.* a lateral and horizontal ore deposit. *Mineralogy.* a term designating low-grade, uncut diamonds. *Building Engineering.* **1.** one floor of a multilevel building. **2.** any structural element on a building that is level, such as a level roof. *Naval Architecture.* a noncambered partial deck located below the main deck. *Design Engineering.* **1.** a dull or matte painted surface. **2.** a strip of iron or steel containing a rectangular cross section. *Graphic Arts.* **1.** a sheet of plastic, paper, or glass on which photographic negatives or positives are assembled in preparation for platemaking. **2.** of a photograph, having little contrast. *Acoustics.* **1.** a musical half step (100 cents) down in pitch from a specified note. **2.** the condition of being lower in musical pitch than what is intended or specified.

flat *Mathematics.* **1.** a Riemannian manifold is said to be flat if its curvature tensor vanishes identically (i.e., if it is zero everywhere). **2.** an R-module N is said to be flat if any one of the following equivalent conditions hold: (a) If $0 \to M' \to M \to M'' \to 0$ is any exact sequence of R-modules, then the tensored sequence $0 \to M' \otimes N \to M \otimes N \to M'' \otimes N \to 0$ is also exact. (b) If $f: M' \to M$ is an injective module homomorphism, then the homomorphism $f \otimes 1: M' \otimes N \to M \otimes N$ is also injective. (c) If $f: M' \to M$ is injective and M and M' are finitely generated, then $f \otimes 1: M' \otimes N \to M \otimes N$ is also injective. **3.** a coset of a subspace of a vector space; thus, if S is a k-dimensional subspace of a vector space V and if v is a vector in V, then the set $v + S$ is a k-dimensional flat in V.

flat affect *Psychology.* a lack of emotion in circumstances that would normally produce an emotional response; associated with schizophrenia. Also, FLATTENING.

flat-back stope *Mining Engineering.* an overhand stoping method in which the ore is broken in slices parallel with the levels. Also, LONG-WALL STOPE.

flat band *Building Engineering.* a square, plain impost that is made of stone.

flatbed laser scanning *Graphic Arts.* a color-separating machine in which art is placed on a flat surface as opposed to a drum for separating.

flatbed plotter *Engineering.* a device that generates graphs by directing a pen horizontally and vertically across a sheet of paper which is affixed to a flat surface; it may be connected to a computer output device.

flatbed press see CYLINDER PRESS.

flatbed truck *Engineering.* a truck having a bed that resembles a platform.

flat-blade turbine *Mechanical Engineering.* an impeller having flat blades that extend radially from the shaft.

flat cable *Electricity.* a cable consisting of parallel copper wires arranged in a plane and molded into a ribbon of flexible insulating plastic such as polyethylene.

flatcar *Engineering.* a railroad car that does not have fixed walls or a cover.

flat chisel *Mechanical Devices.* a cold chisel with a relatively wide cutting edge, used to chip a surface to a flat finish on wood or stone.

flat color *Graphic Arts.* in printing, color that is not processed and does not attempt to reproduce the natural spectrum. Also, SPOT COLOR.

flat-color printing *Graphic Arts.* printing in one or more colors other than black, using separate plates for each color but no separation or halftoning; used primarily to print line copy in solid colors.

flat-conductor cable *Electricity.* a cable consisting of wide, flat conductors arranged in a plane and molded into a protective plastic ribbon.

flat crank *Mechanical Devices.* a single-bearing crankshaft.

flat-crested weir *Civil Engineering.* a measuring weir whose crest is in the horizontal plane, the ratio of whose length to the height of the water passing over it is greater than one.

flat cut *Mining Engineering.* a way of placing the boreholes for the first shot in a tunnel starting at about 2 or 3 feet above the floor and pointed downward, so the bottom of the hole will be about level with the floor.

flat-die forging *Metallurgy.* forging between two flat dies.

flat drill *Mechanical Devices.* a rotary drill with a cutting blade in the form of two parallel beveled edges, used for drilling out cored holes.

flat-edge trimmer *Mechanical Engineering.* a machine designed to trim the notched or slotted edges of metal shells.

flat etching *Graphic Arts.* a technique for reducing the density of a photograph or printing plate by immersing the image in a chemical etching solution that reduces the exposed silver deposits.

flat-face bit *Mechanical Devices.* a diamond core drill bit with a rectangular or square cross section. Also, **flat-nose bit.**

flat fading *Telecommunications.* a type of fading in which all frequency components of a received radio signal fluctuate proportionally and in unison.

flat-field objective lens *Optics.* in photomicrography, a lens that is corrected for curvature of field and is capable of forming a flat, sharply defined intermediate image.

flat file *Mechanical Devices.* a common, double-cut file of a width four times its thickness with a tapering toward its point. *Computer Programming.* a file structure, such as a list or table, that is composed of a one- or two-dimensional array of data items.

flatfish *Vertebrate Zoology.* any fish of the marine order Heterosomata, including the halibut, sole, flounder, and turbot; distinguished by their habit of swimming on one side of a laterally compressed body and by having both eyes on the upper side.

flat jack *Civil Engineering.* a hollow steel cushion fabricated from two nearly flat disks welded around the edge and inflated with oil or cement under controlled pressure; used at arch abutments and crowns to relieve loads on the formwork.

flat line *Electromagnetism.* a radio-frequency transmission line with a standing wave ratio of one to one.

flat-lying *Geology.* describing a deposit or coal seam having a dip of 5° or less.

flat molding *Mechanical Devices.* long, thin strips of wood used for finishing surfaces.

flatness problem *Astrophysics.* the problem of explaining why the overall density of the universe has not decreased over time, as would be consistent with the concept of expansion that is inherent in the big-bang theory.

flatnose or **flat-nose** *Ordnance.* **1.** an antisubmarine projectile designed to prevent ricocheting upon impact with the water. **2.** describing a bullet with a flattened end designed to be used in cartridges for tubular magazine rifles. Thus, **flatnose bullet.**

flat of bottom *Naval Architecture.* the portion of a ship's bottom with little or no rise.

flatpack *Electronics.* an integrated circuit network in which the components are encased in a thin, rectangular package, with the connecting leads extending out from the edge of the package.

flat-plate collector *Engineering.* a collector in which a transparent lid on a shallow metal box converts sunlight into heat.

flat plowing *Agronomy.* the plowing of a field so that the furrows continously overlap each other; land plowed in this way is **flat-broken.** Also, **flat-furrow plowing.**

flat pointing *Building Engineering.* a method of pointing uncovered, internal wall surfaces to form a smooth, flat joint along the plane of the wall.

flat-position welding *Metallurgy.* see DOWNHAND WELDING.

flat rope *Mechanical Devices.* a strong and durable metal or fiber rope that resists twisting during winding by its flat, sewn or braided cross section with several strands wrapped about a central hemp core.

flats *Geography.* a tract of level land along a shore, often mud, that is flooded at high tide. Also, TIDAL FLAT. *Navigation.* a large area of shallow water, especially one in which the bottom is exposed at low tide.

flat slab *Civil Engineering.* a reinforced concrete floor construction not requiring beams and girders to transmit the floor loads to the columns.

flat sour *Food Technology.* fermentation in canned foods after sealing.

flat space *Mathematics.* any pseudo-Riemannian manifold isometric to the manifold R^n with chart (Id, R^n) and metric $ds^2 = \sum_{i=1}^{n} \varepsilon_i \, dx^i dx^i$, where $\varepsilon_i = \pm 1$ and Id is the identity map. For example, when $\varepsilon_i = +1$ for all i, the space is Euclidean; Minkowski space is flat space with $\varepsilon_1 = 1$ and $\varepsilon_i = -1$ for $i > 1$. A **locally flat space** is a space that is locally isometric to a flat space.

flat space-time *Physics.* space-time that has a completely flat geometry; the space-time postulated by special relativity.

flat spin *Mechanics.* the motion of a rotating projectile or aircraft whose yaw exceeds 45°.

flattening *Metallurgy.* **1.** a preliminary forging operation. **2.** the process of removing wavy distortions from a sheet or plate. *Geodesy.* the ratio of the difference between the earth's equatorial and polar radii (the semimajor and semiminor axes of the spheroid) to its equatorial radius (semimajor axis). Also, COMPRESSION. *Psychology.* see FLAT AFFECT.

flatter *Mechanical Devices.* **1.** a flat-faced blacksmith's hammer placed on forged work and struck by a sledge hammer. **2.** a drawplate with narrow, rectangular openings used for making flat metal or wire strips and watch springs.

flat-top antenna *Electromagnetism.* an antenna with two or more radiating wire elements that are parallel to each other and to the ground.

flat-top response see BANDPASS RESPONSE.

flat trajectory *Mechanics.* a trajectory whose gravitational deflection is negligible due to the high speed or brief duration of the flight. *Ordnance.* specifically, a trajectory with little or virtually no vertical curvature, characteristic of fire at high velocity and short range; it describes the trajectory of a pistol, rifle, or gun, as opposed to that of a howitzer or mortar. Thus, **flat-trajectory fire, flat-trajectory weapon.**

flat tuning *Electronics.* a method of adjusting a radio receiver in which a change in frequency of the received waves produces only slight changes in the current in the tuning apparatus.

flat-turret lathe *Mechanical Engineering.* a lathe having a low, flat turret on an electrically driven, cross-sliding headstock.

flatulence [flach´ə ləns] *Medicine.* an excess amount of air or gas in the stomach and intestines that may cause the organs to bloat and in some cases cause mild to moderate pain.

flat universe see EINSTEIN-DE SITTER UNIVERSE.

flatus [flāt´əs] *Medicine.* air or gas in the gastrointestinal tract that is passed through the rectum.

flat wire *Metallurgy.* a rolled or drawn (but not slit) wire that has a square or rectangular cross section.

flatworm *Invertebrate Zoology.* the common name for any of the dorsoventrally flattened worms of the phylum Platyhelminthes.

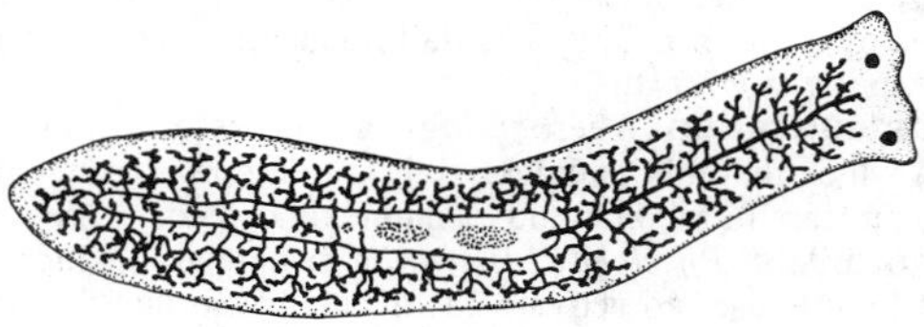

flatworm

flat yard *Civil Engineering.* a railroad switchyard where cars are moved by locomotive rather than gravity.

flaunching *Building Engineering.* a slope along the top surface of a chimney to dispel rainwater.

flav- a combining form meaning "yellow," as in *flavescence.*

flavanol *Biochemistry.* **1.** a class of flavonoid compounds that contain a hydroxyl group. **2.** a compound, without color, that gives rise to a variety of yellow plant pigments.

flavanone *Biochemistry.* any of a number of colorless crystalline compounds derived from flavone and found in higher plants; one of the major groups of flavonoids.

flavescence *Plant Pathology.* a condition in which green plant parts become yellowed or faded due to a virus disease that reduces chlorophyll.

flavin [flā´vin] *Biochemistry.* any of a group of water-soluble yellow pigments that act as coenzymes of flavoproteins; most natural flavins are fashioned from riboflavin.

flavin adenine dinucleotide *Biochemistry.* a coenzyme that acts as a hydrogen acceptor in dehydrogenation reactions in an oxidized or reduced form.

flavin mononucleotide *Biochemistry.* $C_{17}H_{21}N_4O_9P$, a phosphoric ester of riboflavin that constitutes the coenzyme of various flavoproteins.

flavin nucleotide *Biochemistry.* a flavin derivative that acts as a prosthetic group for a flavoenzyme.

Flaviviridae *Virology.* a family of enveloped ssRNA-containing viruses that are transmitted by mosquitoes and ticks and infect a wide range of animals.

Flavivirus *Virology.* the lone genus of the Flaviviridae family; members cause yellow fever, dengue, and various forms of encephalitis.

flavo- a combining form meaning "yellow," as in *flavoprotein.*

Flavobacterium *Bacteriology.* a genus of Gram-negative, aerobic, nonmotile, rod-shaped bacteria of uncertain affiliation, found in soil, water, raw meat, and milk; opportunistic pathogens in humans.

flavoenzyme *Enzymology.* any enzyme that is a flavoprotein.

flavone *Biochemistry.* a colorless crystalline flavonoid that can reverse increased capillary fragility; used to derive various yellow dyestuffs.

flavonoid *Biochemistry.* any of a large group of aromatic oxygen heterocyclic compounds that are widely distributed in higher plants; some provide pigmentation and others have physiologic properties.

flavonol *Biochemistry.* any of a class of flavone compounds that contain a hydroxyl group.

flavoprotein *Biochemistry.* a conjugated protein dehydrogenase containing flavin that carries electrons in the electron transport system.

flavor *Biology.* the quality of any substance that affects the sensation of taste. *Food Technology.* see FLAVORING. *Particle Physics.* a term used to describe the quark types (up, down, strange, charmed, top, and bottom).

flavoring *Food Technology.* any substance used to improve the taste of a food or medicine.

flavors *Artificial Intelligence.* an object-oriented programming system furnished by some Lisp dialects.

flaw *Materials Science.* a defect in a material, especially one that makes the material unacceptable for standard use. *Oceanography.* **1.** a lead between landfast ice and solid or partially broken pack ice. **2.** the seaward edge of fast ice.

flaw ice *Oceanography.* ice in widely varying shapes and sizes, resulting from shearing between pack ice and fast ice.

flax *Botany.* an annual plant of the genus *Linum* in the family Linaceae, especially *Linum usitatissimum,* having linear leaves and blue flowers and cultivated for the fibers in its stem and the oil in its seeds. *Textiles.* the fiber of this plant, used to produce linen threads or fabrics.

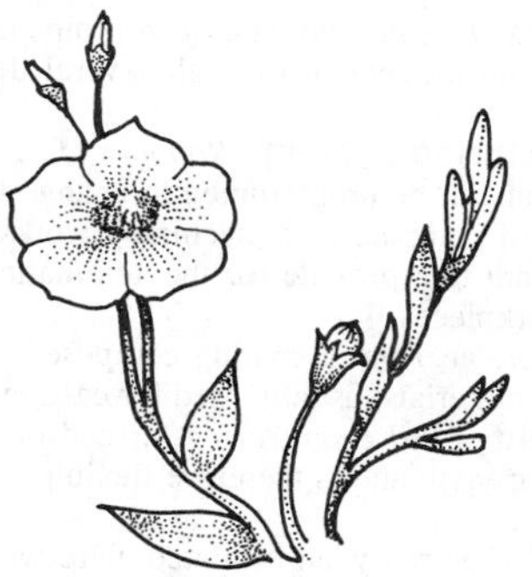

flax

flax rust *Plant Pathology.* a rust disease of flax caused by the fungus *Melamspora lini* and characterized by orange-yellow growths on leaf surfaces and stems.

flaxseed *Botany.* the seed of the flax plant, harvested for its content of linseed oil.

flaxseed ore *Geology.* an iron-bearing sedimentary deposit composed of disk-shaped oolites that have been partially flattened parallel to the bedding plane of the rock.

flax wilt *Plant Pathology.* a disease of flax caused by the fungus *Fusarium oxysporum lini*; characterized by a shriveling, drooping, yellowing, and dying of affected plants.

F layer *Geology.* the seismic region, in the classification of the earth's interior, equivalent to the transition zone between the outer and inner cores, extending from 4710 to 5160 km below the surface. *Geophysics.* a region of the high ionosphere that reflects radio waves at up to 50 MHz, thus allowing communication over the horizon; it consists of F_1 and F_2 layers during daylight and the F_2 layer only at night.

F_1 layer *Geophysics.* the part of the F layer that is below the F_2 layer during the day, has a height of 200–300 km, and is closest to the equator at noon.

F_2 layer *Geophysics.* the higher level of the F layer, having a maximum free electron density at 225 km in the polar winter to 400 km in daytime.

fl. dr. fluid dram.

flea *Invertebrate Zoology.* the common name for various small, wingless insects in the order Siphonaptera; blood-sucking ectoparasites of mammals and birds.

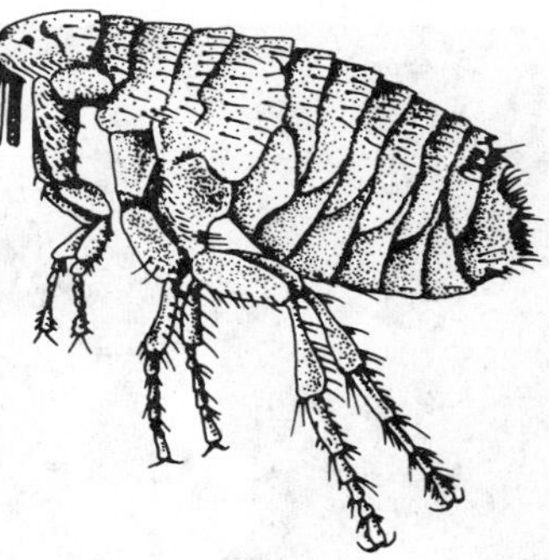

flea

fleaking *Building Engineering.* the use of reeds to thatch a roof.

flechette *Ordnance.* **1.** see AERIAL DART. **2.** a small fin-stabilized projectile designed to be loaded en masse into an artillery canister or warhead, as an anti-personnel weapon.

Flectobacillus *Bacteriology.* a genus of Gram-negative, nonmotile bacteria of the family Spirosomaceae that usually occur as curved, pigmented rod-shaped cells in aquatic habitats.

fledge *Vertebrate Zoology.* **1.** of a young bird, to acquire the feathers needed for flight. **2.** to raise a young bird to maturity. Thus, **fledged.**

fledgling *Vertebrate Zoology.* a young bird that has recently fledged.

fledgling

fleece *Vertebrate Zoology.* the coat of wool that covers a sheep or similar animal. *Textiles.* any soft fabric resembling a sheep's coat.
fleet *Military Science.* an organization of ships, aircraft, marine forces, and shore-based activities under a single commander who may exercise operational as well as administrative control. *Transportation Engineering.* the total number of vehicles or cars in a system, including active and inactive units. *Mechanical Engineering.* the sideways movement of a rope or cable when being wound on a drum.
fleet angle *Mechanical Engineering.* the maximum angle between a rope or cable and a line drawn perpendicular to the drum on which it winds.
fleet ballistic missile *Ordnance.* a submarine or shipborne strategic ballistic missile; e.g., Polaris or Poseidon. Thus, **fleet ballistic missile submarine.**
fleet broadcast *Telecommunications.* the radio broadcast received by all U.S. Navy ships and merchant vessels, consisting mostly of weather bulletins.
fleeting target *Ordnance.* a moving target that remains within observation or firing distance for such a short duration that it is difficult to take deliberate aim.
Fleming, Sir Alexander 1881–1955, Scottish bacteriologist; discovered lysozyme; shared Nobel Prize for development of penicillin.

Alexander Fleming

Fleming, John Ambrose 1849–1945, English electrical engineer; pioneer in electrical heating and lighting; invented Fleming radio rectifier.
Fleming, Sir Sandford 1827–1915, Canadian civil engineer; coordinated development of railways, telegraph, and standard time zones.
Fleming cracking process *Chemical Engineering.* a thermal cracking method for heavy petroleum fractions in a vertical shell still under pressure.
Fleming tube *Electronics.* a diode that allows negative current to flow from its heated filament to its cold electrode, but prevents it from flowing in the reverse direction.
Flemish bond *Building Engineering.* a masonry bond that consists of alternate headers and stretchers in every course, with each header centering on the stretchers in the courses above and below.
Flemish coil *Naval Architecture.* a rope coiled flat in tight concentric circles so that it resembles a circular mat.
Flemish garden wall bond *Building Engineering.* a masonry bond consisting of stretches and headers in the ratio of one to three or four in each course, with joints broken to give varied patterns.
flesh *Anatomy.* the soft muscular tissue of the body.
Flesh-Demag process *Chemical Engineering.* a procedure for making gas using a cyclic water-gas unit for feeding, and charring coal for gas generation.
flesher *Archaeology.* a long, broad-edged tool of bone, antler, or stone that is used to scrape or rub hides free of fat, sinew, hair, and other unwanted material.
fleshing machine *Mechanical Engineering.* equipment that strips flesh from hides in a tannery.
fleshy fruit *Botany.* any fruit having juicy, pulpy cellular tissue, such as peaches or melons.
Fletcher-Munson curve *Acoustics.* a graph plotting sound intensity, as expressed in decibels, in comparison with frequency, for which each point represents equal loudness over the contour. Also, **Fletcher-Munson contour.**
Fletcher radial burner *Engineering.* a burner whose gas jets are in a radial configuration.
Fleury's algorithm [flür′ēz] *Mathematics.* a method for constructing an Eulerian path in a given Eulerian graph, as follows: Start at any vertex and traverse the edges in an arbitrary manner. Erase the edges as they are traversed, and if any isolated vertices result, erase them too. At each stage, use an isthmus only if there is no alternative. The result is an Eulerian path.
flex *Mechanics.* **1.** to move from a straight course or position; bend or curve. **2.** the act of bending, or the capacity to bend.
flexed burial *Archaeology.* a method of burial in which the body is interred in a fetal position.
Flexibacter *Bacteriology.* a genus of gliding, rod-shaped bacteria of the family Cytophagaceae, occurring in soil, freshwater, or marine habitats; some species are pathogenic for fish.
Flexibacteriae *Bacteriology.* a class of gliding bacteria that includes the orders Cytophagales and Myxobacterales.
Flexibilia *Paleontology.* a small subclass of crinoids in the subphylum Crinozoa, extant from the Ordovician to the Permian; characterized by loosely joined calycal plates.
flexibility *Mechanics.* **1.** the quality or condition of offering little resistance to being bent; being able to withstand repeated bending. **2.** see ELASTICITY. *Robotics.* the ability of a robot to bend or flex repeatedly through the use of joints and links.
flexible *Mechanics.* having the property of flexibility; able to be repeatedly bent and still maintain its original shape afterward. *Design Engineering.* adaptable to a variety of parts and assembly processes. Thus, **flexible assembly, flexible automated manufacturing, flexible cells and systems, flexible fixturing, flexible flow line.**
flexible circuit *Electronics.* a device in which the electrical elements are mounted onto a pliable plastic sheet, so that it may be fitted between large components.
flexible coupling *Mechanical Engineering.* a coupling containing a resilient member such as a metal spring or rubber disk; used to connect two rigid shafts that cannot be aligned. *Electromagnetism.* a connection between two waveguides that allows a limited range of angular movement between axes.
flexible disk see DISKETTE.
flexible DNC *Robotics.* a process of DNC (distributed numerical control) in which a large number of machine tools and automated materials-handling systems are connected to a host computer.
flexible gun *Ordnance.* a gun mounted so that it can be moved both vertically and horizontally; usually applied to a machine gun mounted in an aircraft turret.
flexible-joint pipe *Engineering.* a cast-iron pipe designed to lie under water and withstand movement through several degrees without leakage.
flexible manufacturing system *Robotics.* **1.** a complex system for manufacturing that can be programmed to change from one task to another. **2.** a series of computerized machining workstations throughout a manufacturing plant that provide for the automatic production of a related group of workpieces.
flexible mold *Engineering.* a coating composed of flexible rubber or other elastomeric materials; usually used for casting plastics.
flexible pavement *Civil Engineering.* a road or runway surface that has little tensile strength and is therefore flexible; usually made of a bituminous material.
flexible resistor *Electricity.* an insulated, wire-wound resistor that can be bent, coiled, or knotted; it has the appearance of a flexible lead.
flexible route *Transportation Engineering.* a route that changes according to demand or conditions.
flexible sandstone *Geology.* a variety of itacolumite characterized by thin layers of fine-grained materials.
flexible shaft *Mechanical Engineering.* **1.** a shaft that can transmit rotary motion up to an angle of 90°. **2.** any shaft that is made of flexible material. **3.** a shaft whose bearings are designed to allow for minor misalignment.
flexible waveguide *Electromagnetism.* a waveguide that can be bent without significantly altering its electromagnetic properties.

flexion *Biology.* **1.** the motion that decreases the angle between two elements of a jointed part. **2.** the motion that bends a joint. *Robotics.* movement toward the smallest angle between two objects.

flexirubin *Organic Chemistry.* $C_{43}H_{54}O_4$, a pigment derived from *Flexibacter elegans,* occurring as violet-red needles that melt at 176°C.

Flexithrix *Bacteriology.* a genus of gliding bacteria of the family Cryptophagaceae, characterized by rod-shaped cells generally in sheathed filaments; found in marine habitats.

flex life *Materials Science.* the time to failure due to the amount of flexural fatigue imposed, an important mechanical property of polymers.

flexography *Graphic Arts.* an inexpensive web-fed rotary printing process using rubber relief plates and quick-drying inks; commonly used to print bags, labels, and wrappers.

flexometer *Engineering.* an instrument used for determing the flexibility of a material.

flexor *Anatomy.* a type of muscle that flexes or bends a limb or other body part.

flexor reflex *Physiology.* the involuntary rapid withdrawal of a body part from a painful or noxious stimulus.

flexous *Biology.* **1.** flexible. **2.** able to bend in a zigzag or wavy pattern.

flexowriter *Computer Technology.* a device resembling a typewriter that was used with early computers to accept paper tape input.

flexuous hypha *Mycology.* in rust fungi, a kind of fungal filament or hypha that extends from a certain fruiting body and acts as the female sexual structure.

flexural modulus *Mechanics.* a constant used in the calculation of bending, equal to the Young's modulus of the material times the moment of inertia (the square of the radius of gyration of a cross section perpendicular to the plane of bending). Also, **flexural rigidity.**

flexural slip see BEDDING-PLANE SLIP.

flexural strength *Mechanics.* the critical bending load that a beam can withstand without failure.

flexure *Vertebrate Zoology.* the last joint of a bird's wing. *Developmental Biology.* also, **flexura.** a bending or bent part of an organ or structure. *Geology.* any fold, curve, bend, warp, tilt, or other gentle deformation of rock strata. Also, HINGE. *Mechanics.* the deformation of a body under a load, so that points which were a straight line are now in a plane curve; the process of bending.

flexure strength *Materials Science.* the maximum tensile stress at failure for a ceramic material; determined by supporting the test specimen at both ends and applying a load.

flexure theory *Mechanics.* the application of elastic theory to long structural members deflected primarily by bending moments.

flicker *Electronics.* a variation in a regulated power supply that occurs due to sudden, brief, minimal variations in the reference voltage or the regulator. *Telecommunications.* in television, a repetitive variation in luminance of a particular area of the display that occurs when the field frequency is insufficient to synchronize the visual images. *Optics.* the visual perception stimulated in the eye by light pulsating regularly at a few cycles per second and ranging up to frequencies just below those covered by persistence of vision. *Vertebrate Zoology.* a familiar North American woodpecker of the genus *Colaptes* in the family Picidae, having a slender, downcurved bill, speckled plumage with a black breast band, and yellow or red spots on the head and wings; noted for its aberrant woodpecker behavior of hunting ants along the ground.

flicker

flicker control *Aviation.* the remote control of a flight vehicle in which only a slight motion is required to deflect the control surfaces to their maximum degree.

flicker effect *Electronics.* the noise created by the large, often erratic changes in current flow that arise from an irregularly coated cathode in an electron tube.

flicker photometer *Optics.* an instrument that compares the illumination of a test lamp to that of a standard lamp by having each one illuminate a single screen in rapid succession; variation in intensity is revealed by a flicker, while the perceived absence of flickering indicates equal luminance.

flier *Building Engineering.* a step in a straight flight of stairs.

flight *Aviation.* **1.** the movement of a vehicle or object through the atmosphere or space supported by aerodynamic, aerostatic, or reaction forces, or by orbital speed. **2.** an instance of such movement. **3.** a group of aircraft involved in a joint mission. **4.** a basic tactical unit consisting of three or four aircraft. **5.** a scheduled commercial air-carrier service, characterized by three- or four-digit identifying numbers. *Space Technology.* a NASA radio call sign meaning "Flight Director." *Building Engineering.* a series of stairs between landings or floors.

flight characteristic *Aviation.* a distinguishing feature of a flight vehicle relating to its predisposition to stall or yaw, or its ability to remain stable or controllable at a given speed.

flight control *Aviation.* **1.** the guiding and directing of airborne vehicles, especially from the ground; air traffic control. **2. flight controls.** any device or system for guiding a vehicle in flight, such as a control surface or cockpit controls.

flight-control surface see CONTROL SURFACE.

flight-control system *Aviation.* see VEHICLE CONTROL SYSTEM. *Space Technology.* in a rocket vehicle, a system of devices controlling thrust and stability; distinguished from the guidance system, which controls trajectory.

flight conveyor *Mechanical Engineering.* a conveyor in which paddles, scrapers, roller-chains, and the like are used to drag or push pulverized or granulated materials along a trough. Also, DRAG CONVEYOR.

flight deck *Aviation.* **1.** an elevated compartment occupied by the flight crew in certain large aircraft, such as the C-124 or the Boeing 747. **2.** loosely, the cockpit of any commercial or cargo aircraft.

flight distance *Behavior.* the distance to which an animal will retreat when threatened by a predator or a rival of the same species; this is predictable according to the species and environment.

flight dynamics *Aviation.* the study or monitoring of a flight vehicle's short-term motion, focusing on transient or steady maneuvers involving stability and control.

flight envelope *Aviation.* **1.** an aircraft's performance limits; specifically, the curves of speed plotted against other variables indicating the limits of speed, altitude, and acceleration that a particular aircraft can not safely exceed. **2.** a graphic representation of these limits, showing the interrelationships of operational parameters for given variables. Also, ENVELOPE.

flight feather *Vertebrate Zoology.* any of the large, stiff quills of a bird's wing or tail which are essential to flight.

flight feeder *Mechanical Engineering.* a short-length flight conveyor used to feed solid materials to a process vessel at a predetermined rate.

flight instrument *Aviation.* any of several onboard instruments used to control an aircraft's direction of flight, attitude, altitude, or speed. These include the bank-and-turn, rate-of-climb, airspeed, and flight indicators. Sometimes distinguished from navigational instruments such as the driftmeter, absolute altimeter, or directional gyro, but these may also be called flight instruments.

flight level *Aviation.* a surface of constant atmospheric pressure related to the standard pressure datum of 29.92 inches of mercury, expressed in three-digit numbers representing hundreds of feet; for example, FL 220 represents a barometric altimeter indication of 22,000 feet.

flight log *Aviation.* a flight planning document in which a pilot calculates factors such as speed, course, altitude, and fuel consumption for an upcoming flight.

flight of ideas *Psychology.* a rapid or continuous flow of speech that jumps rapidly from one topic to another without continuity of ideas; associated with manic disorders.

flight path *Aviation.* the line of movement that a flight vehicle makes or follows in the air or in space. Also, FLIGHT TRACK.

flight-path angle *Aviation.* the acute angle between a aircraft's flight path and the horizontal. This angle is positive when the craft is climbing and negative when it is descending. Also, **flight-path slope.**

flight-path computer *Computer Technology.* a special-purpose computer that is resident in an aircraft, spacecraft, or missile and provides a means to control the course and altitude in accordance with a prespecified flight plan.

flight plan *Aviation.* a pilot's intended routing and schedule for a flight, giving the pilot and aircraft identification; course, speed, and altitude to be flown; and estimated times of arrival at intermediate stops and the destination; submitted, orally or in writing, to air traffic control or a flight service station.

flight profile *Aviation.* **1.** a flight path as seen from the side. **2.** specifically, a graphic portrayal of a flight path in the vertical plane, usually plotting the vehicle's altitude against the track distance.

flight-readiness firing *Ordnance.* a brief missile system test conducted with the propulsion system operating while the missile is secured to its launcher.

flight recorder *Aviation.* any device that logs information about the flight performance of an aircraft or about conditions encountered in flight, such as weather.

flight science *Aviation.* the integrated study of the science and technology of aviation.

flight simulator *Aviation.* an electronic training device or apparatus that simulates certain flight characteristics of a specific vehicle under variable conditions; used to test and train pilots and flight crews in coping with emergencies and completing missions according to role; also used as a design and engineering aid.

flight stage *Aviation.* the portion of an aircraft's flight from takeoff to the next landing.

flight strip *Photogrammetry.* a succession of overlapping aerial photographs taken along a single course.

flight test *Aviation.* **1.** a test to determine the performance of a flight vehicle by means of actual or attempted flight. **2.** a test of a component of a flight vehicle, or of an object carried in such a vehicle, to determine its performance during actual flight. **3.** a skills test of a pilot applicant.

flight time *Aviation.* the duration of a flight from takeoff to landing.

flight-time limitation *Transportation Engineering.* the working time limitation for a flight crew, defined as beginning when crew members report to duty, and ending when the engines are turned off at the end of a flight.

flight track see FLIGHT PATH.

flinching *Industrial Engineering.* in quality control, an inspector's failure to correctly identify a borderline defect as a defect.

Flinders bar *Navigation.* one or more bars of unmagnetized soft iron placed vertically near a magnetic compass to neutralize deviation caused by induction in the vertical soft iron of a craft.

F line *Spectroscopy.* in the hydrogen spectrum and in the absorption spectrum of the sun, a green-blue line having a wavelength of 4861.33 angstroms.

flinkite *Mineralogy.* $Mn_2^{+2}Mn^{+3}(AsO_4)(OH)_4$, an easily fusible, greenish brown orthorhombic mineral occurring in aggregates of thin tabular crystals, and having a specific gravity of 3.87 and a hardness of 4.5 on the Mohs scale; found as a rare mineral on barite and other arsenates.

flint *Mineralogy.* a black to slate-gray microcrystalline granular variety of quartz, similar to chert, with a barely glistening, subvitreous luster and a conchoidal fracture; having a specific gravity of 2.65 and a hardness of 7 on the Mohs scale; ground and used in the manufacture of earthenware and porcelain in Europe.

flint

flint clay *Geology.* a gray to black, hard, smooth fireclay that exhibits conchoidal fracture and does not develop plasticity upon grinding.

flint disease see CHALICOSIS.

flintlock *Ordnance.* **1.** an early type of gunlock using a piece of flint striking against steel to produce a spark to ignite the priming powder. **2.** a rifle having this type of lock.

flint mill *Mechanical Engineering.* **1.** a ball mill that uses flint pebbles as a pulverizing agent to grind materials such as conglomerate in cement. **2.** in pottery works, a mill in which flints are ground.

flint paper *Materials.* a paper coated with crushed flint; used like sandpaper, as an abrasive.

flip chip *Electronics.* a semiconductor die in which all of the terminals are placed on one side of the substrate so that it can be flipped over for bonding onto a matching substrate, after being treated with an inert material to prevent contamination.

flip coil *Electromagnetism.* a small coil that is connected to a ballistic galvanometer or other sensitive device and used to measure magnetic fields by suddenly flipping the coil over 180°, thus reversing the magnetic flux.

flip-flop *Electronics.* **1.** a two-position relay that alternates positions with successive pulses. **2.** see BISTABLE MULTIVIBRATOR.

flip-flop model *Molecular Biology.* a model stating that a molecule that is part of a continuous lipid bilayer will move head to tail within the layer; its position relative to the two surfaces will thereby exactly reverse or flip-flop.

flipper *Vertebrate Zoology.* a broad, flat limb adapted for swimming, as on a seal, whale, or sea turtle.

flir *Ordnance.* a sensor producing visual images resulting from temperature differences. (An acronym for forward-looking infrared.)

flist *Meteorology.* a shower accompanied by a squall, in Scotland.

flitch *Civil Engineering.* either of a pair of metal plates between which another metal plate is sandwiched to add reinforcement in a beam or girder. Also, **flitch plate.**

flitch beam *Civil Engineering.* a built-up beam of flitches, between two of which a metal plate is sandwiched for reinforcement.

flitch girder *Civil Engineering.* a built-up girder of flitches, between two of which a metal plate is sandwiched for reinforcement.

FLK *Anthropology.* a site at Olduvai Gorge in East Africa, the location of many important finds of hominid remains. (An abbreviation of Frida Leakey Karongo, from the name of the wife of Louis Leakey, an early explorer of this site, and *karongo,* a Swahili word for "gully.")

FLL *Aviation.* the airport code for Ft. Lauderdale International, Florida.

float *Mechanical Devices.* **1.** a hollow, watertight structure used to mark channels or water height in a port or bay, or in a tank or boiler. Also, BUOY. **2.** a watertight device used to stabilize aircraft that make an emergency landing on water. **3.** a flat-faced trowel or file with a single set of parallel teeth for smoothing plaster or concrete. Also, **float board. 4.** a marble-polishing block or wedge. *Agriculture.* a device consisting of overlapping heavy planks that are tied together as a platform and dragged over the ground to level and compact the soil. *Geology.* a loose, isolated rock fragment or group of fragments within or on the surface of another rock, often found downslope from their source. Also, FLOATER. *Biology.* in many fishes and aquatic plants, an air cavity or spongy mass that buoys up the body of the organism. *Industrial Engineering.* **1.** extra time available to complete a task over and above the scheduled time. **2.** extra output over orders due to batch production technique. **3.** any kind of cushion or slack in a system. *Engineering.* see PLUMMET.

floatability *Materials.* the capacity of a specific material to float.

float-and-sink analysis *Mining Engineering.* a method of dividing a sample of crushed ore into fractions by passing it through a series of heavy liquids that diminish (or increase) in density in accurately controlled stages; the ore fractions either float or settle at each stage.

float barograph *Engineering.* a siphon barograph in which the motion of a float is used to record atmospheric pressure.

float chamber *Engineering.* a chamber in which a float controls the liquid level, such as the gasoline reservoir of a carburetor.

float coal *Geology.* any small, isolated masses of coal embedded in sandstone or shale, and most likely formed from eroded peat transported from its original deposit. Also, RAFT.

float collar *Petroleum Engineering.* a coupling unit placed in a casing string that holds a check valve, allowing fluid to pass downward but not upward through the casing; used on casing during cementing operations.

float control *Engineering.* a floating instrument used to send a liquid-level reading to a control apparatus, such as an on-off switch controlling liquid flow into and out of a storage tank.

float-cut file *Mechanical Devices*. a file with a single set of parallel cutting edges for use on soft materials.

floater *Medicine*. a deposit in the vitreous of the eye, usually moving about and probably composed of fine aggregates of vitreous protein, occurring as a benign degenerative change; a spot before the eyes. *Petroleum Engineering*. **1.** any drilling platform located offshore without a fixed base, such as a drill barge or drill ship. **2.** a floating-roof tank. *Oceanography*. see DRIFT BOTTLE. *Geology*. see FLOAT.

float finish *Civil Engineering*. a rough concrete finish obtained by using a wooden or steel float.

float gauge *Engineering*. an instrument that uses mechanical devices to determine the level of a liquid by measuring the height of an object floating on its surface.

float-glass process *Materials Science*. a process for producing plate glass in which glass flows from the melting furnace onto a float bath of a liquid metal such as tin, where a controlled, heated atmosphere is maintained to prevent oxidation. In the float chamber both surfaces of the glass become mirror smooth, and the sheet passes into the annealing furnace.

floating *Electronics*. of or relating to a device that is isolated from a voltage supply or isolated from the common circuit ground. *Building Engineering*. **1.** the equal spreading of plaster, stucco, or cement by means of a float board. **2.** describing the second of three coats applied with a coat board to protect the level of the screeds.

floating action *Engineering*. an action that establishes the relationship between the deviation and the speed of the final control element; generally a neutral zone, where the element remains motionless, is used.

floating address *Computer Programming*. **1.** a former term for a symbolic address that is not necessarily related to a particular base but that can be easily converted to an absolute address. **2.** see RELATIVE ADDRESS.

floating aid *Navigation*. a floating navigational aid that is anchored in position, such as a buoy.

floating axle *Mechanical Engineering*. a live axle that is used to turn the wheels of a motor vehicle.

floating battery *Electricity*. a storage battery permanently connected in parallel to another power source that keeps it in operating condition by a continuous charge at a low rate.

floating block see TRAVELING BLOCK.

floating charge *Electricity*. **1.** a continuous charge of a storage battery so as to maintain its fully charged condition; it is charged at a rate approximately equal to its losses and suitable to maintain the battery always in a fully charged condition. **2.** in a circuit, slow rates of charge suitable in compensating for internal losses and in restoring small intermittent discharges to the load circuit delivered from time to time. Also, TRICKLE CHARGE.

floating chase *Engineering*. a section of a mold that moves freely across a vertical and covers a lower member, such as a cavity or plug, in order to telescope an upper plug.

floating control *Engineering*. an element in a speed control that keeps the correction speed of the piston in a hydraulic relay proportional to the error signal.

floating crane *Civil Engineering*. a crane having a barge or scow for an undercarriage, used for waterworks and waterfront work.

floating dock *Civil Engineering*. a substitute for a dry dock, used in the repair of ships; it can be partly submerged to receive a ship and then, by pumping water out, can expose the ship's bottom.

floating dollar sign *Computer Programming*. an edit feature that eliminates leading zeros from monetary data items and prints a dollar sign immediately to the left of the most significant digit. Also, **floating currency symbol.**

floating floor *Building Engineering*. a floor that is designed to prevent impact sound, in which a layer of mineral wool or other insulating material is placed between the wearing surface and the supporting structure.

floating foundation *Civil Engineering*. a reinforced-concrete slab that distributes the load of a building across sufficient underlying soil to allow the building to float on the surface; used in soft or wet soils that provide minimal support.

floating grid *Electronics*. a type of vacuum-tube grid that remains detached from the rest of the circuit and takes on a negative potential with respect to the cathode.

floating ice see DRIFT ICE.

floating input *Electricity*. an isolated, ungrounded input circuit.

floating-input measurement see DIFFERENTIAL-INPUT MEASUREMENT.

floating lever *Mechanical Engineering*. a horizontal brake lever with a movable fulcrum, used to control railcar motion and direction.

floating mark *Cartography*. a mark seen as occupying a position above the surface of a surveyed area as the result of the stereoscopic fusion of a pair of photographs; used as a reference mark in examining or measuring a stereoscopic model.

floating neutral *Electricity*. a circuit whose neutral point can have variable potential relative to a voltage reference.

floating pan *Engineering*. a pan used for determining the amount of evaporation in a body of water, based upon the level of water in a pan floating on its surface.

floating platen *Engineering*. a platen between the main head and the press table that can be moved independently of the main head and press table; found in a multidaylight press.

floating plug *Metallurgy*. in tube drawing or extruding, a floating mandrel inside the tube.

floating-point arithmetic *Mathematics*. in computational mathematics, the representation of a numerical quantity by a sign bit s (interpreted as plus or minus), an exact integer exponent e, and a positive number mantissa M such that $1 \leq M < B$. Thus, a number is expressed as $s \times M \times B^{e-E}$, where B is the number base of the system (usually $B = 2$, but sometimes $B = 16$), and E is the bias, a fixed integer constant for any given machine and representation. The representation is said to be internal because it is machine-dependent. If $B = 10$, the representation is sometimes called floating-decimal arithmetic.

floating-point calculation *Computer Programming*. arithmetic operations that automatically account for the location of the radix point in the result.

floating-point number *Computer Programming*. an approximation of a real number using floating-point representation.

floating-point package *Computer Programming*. a subroutine or hardware device that permits floating-point arithmetic operations in a computer with no built-in floating-point capability, often involving conversions between fixed-point and floating-point representations.

floating-point processor *Computer Technology*. internal circuitry in the arithmetic-logic unit that performs floating-point arithmetic operations.

floating-point representation *Computer Programming*. a number notation scheme in which a single number is represented by two numbers: the mantissa field contains the significant digits and the exponent field the power of the radix indicating the magnitude. For example, the number 1,200,000 could be represented as the floating-point decimal number 1.2,6. meaning 1.2×10^6.

floating reserve *Military Science*. reserve troops that remain aboard ship until needed in an amphibious operation.

floating reticle *Optics*. a reticle whose image can be moved within a field of view.

floating rib *Anatomy*. the lower two ribs on either side, which are not attached at their ventral tips.

floating roof *Engineering*. a roof floating at the surface of a stored liquid that lessens vapor space and decreases evaporation potential.

floating rule *Building Engineering*. a long straightedge used to form flat surfaces during plastering or cement work.

floating sand *Petrology*. an isolated grain of quartz sand that does not appear to come in contact with surrounding grains scattered within the finer-grained matrix of a sedimentary rock, especially a limestone.

floating scraper *Mechanical Engineering*. a balanced scraper blade used to remove solids collected on a rotating drum surface.

floating zone refining *Metallurgy*. a metal purification process in which a narrow, molten zone moves along the axis of the workpiece without a mold support.

floatless level control *Engineering*. any device that measures and regulates the level of a liquid in a storage tank or process vessel without floating in it

float mineral *Geology*. any small fragment of ore transported from its bed by water or gravity.

floatoblast *Invertebrate Zoology*. in freshwater bryozoans, a free-floating statoblast (asexual reproductive body) with a float of air cells.

float switch *Engineering*. a switch whose operation is controlled by a float at the liquid surface.

float valve *Engineering*. a valve actuated directly by the float in a chamber filled with a liquid.

floc *Analytical Chemistry*. small masses of colloidal particles formed in a fluid due to the aggregation of fine suspended particles.

floccose *Botany*. covered with wool-like tufts.

flocculant *Chemistry*. a substance that induces the formation of floc in a dispersion of solids in a liquid. Also, **flocculating agent.**

floccular *Anatomy.* **1.** in the form of tufts or tresses. **2.** of or relating to the flocculus.
flocculate *Biology.* having small, hairy tufts.
flocculation *Materials Science.* the clustering of individual clay particles in aggregates or flocks; occurs when a natural clay is tempered with water. *Chemistry.* the coagulation of particle masses.
flocculent *Chemistry.* of a substance, woolly, cloudy, or flakey; not crystalline.
flocculonodular lobe *Anatomy.* a portion of the cerebellum that consists of the paired lateral flocculi and the nodulus.
flocculus *Anatomy.* any of the small paired lobules that rest on the inferior surface of the cerebellar hemisphere and form part of the flocculonodular lobe. Also, **floccule.** *Astronomy.* a bright or dark patch on the sun's surface, visible in a spectroheliogram.

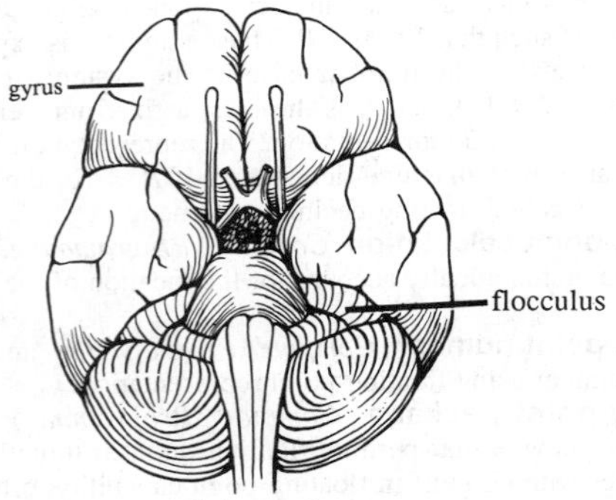

flocculus

flock *Behavior.* a group of animals, especially birds, that remain together, as for defense from predators or efficiency in locating food. *Textiles.* a velvetlike pattern on wallpaper or cloth. Also, **flocking.**
floc point *Analytical Chemistry.* the temperature at which waxes or solids form a definite floc in illuminating oils such as kerosene.
floc test *Analytical Chemistry.* a test to detect substances in illuminating oils that are rendered insoluble by heat.
floe [flō] *Oceanography.* a piece of floating sea ice of indeterminate size, other than the floating edge of an ice shelf or glacier.
floe belt *Oceanography.* a continuous narrow field of ice floes.
floeberg or **floe berg** *Oceanography.* a large mass of thick, heavily hummocked ice, formed by the piling up of many floes to heights as high as 15 meters.
floe till *Geology.* **1.** the unstratified glacial till resulting from intact deposition by a grounded iceberg in a lake adjacent to an ice sheet. Also, BERG TILL. **2.** a clay formed by deposition off boulders and stones from melting icebergs.
flogging *Building Engineering.* the process of rough-dressing an oversized timber to a desired shape.
flong *Graphic Arts.* any of a set of papier-maché sheets used to make and record a stereotype matrix.
flood *Hydrology.* the rising of a body of water so that it overflows its natural or artificial boundaries and covers adjoining land that is not usually underwater. Thus, **flood flow.** *Oceanography.* the stage of maximum height of a tide. *Agronomy.* to run water on a field to the depth of a few inches to irrigate crops. *Mechanical Engineering.* **1.** to cover or cause to be covered with fluid; for example, to cover an oil sand with water to displace the oil. **2.** to supply an excess of fuel to a carburetor, to such an extent that the engine fails to start. *Electronics.* to direct a wide range of electrons toward a storage assembly in a charge storage tube.
flood axis *Oceanography.* the average direction of a tidal current at the time of strength of flood.
flood basalt see PLATEAU BASALT.
flood basin *Geology.* **1.** a land area that is actually submerged during the highest known flood in the region. **2.** a flat, broad area between a sloping low plain and a natural levee.
flood control *Civil Engineering.* the use of canals, levees, reservoirs, floodways, retaining walls, and other means to provide protection from water overflow.
flood current *Oceanography.* the tidal current associated with a rising tide, generally setting in the direction of tidal progression.
flood dam *Civil Engineering.* a dam that protects the surrounding area from floodwater, stores floodwater, and supplies a flood of water.
flooded stream see DROWNED STREAM.
flood fringe see PONDAGE LAND.
floodgate *Civil Engineering.* a gate that controls a flow of water; the lower gate of a lock.
flood icing see ICING, def. 1.
flooding *Chemical Engineering.* a condition of a liquid-vapor counterflow contactor, such as a distillation column, in which an excessive liquid velocity causes a buildup of liquid within the unit or in the overhead gas. *Petroleum Engineering.* **1.** a secondary oil recovery method in which water is injected into a production formation or reservoir to force the oil toward the wells. Also, WATERFLOODING. **2.** the drowning out of a well due to water, caused by drilling too deeply into the sand. *Agronomy.* see CONTROLLED FLOODING. *Psychology.* a technique of treating anxiety by prolonged exposure to the feared object or situation.
flood interval *Oceanography.* the period of time between the transit (upper or lower) of the moon over the local or Greenwich meridian and the time of the following strength of flood.
flood irrigation *Agronomy.* the use of controlled flooding for irrigation, as in a rice paddy.
floodlight *Electricity.* a usually weatherproof bright light with a filament lamp or a mercury-vapor lamp and a parabolic reflector; used primarily for outdoor lighting.
floodplain *Geology.* **1.** a relatively flat strip of land bordering a river and formed from alluvial material deposited when the river overflows its banks. **2.** a relatively flat, dry lowland area that borders a stream and may become submerged during flood stages.
floodplain splay *Geology.* a small fan of clay, sand, silt, gravel, or other alluvial material produced when an overloaded stream breaks through a levee and deposits its material on the floodplain. Also, SAND SPLAY, CHANNEL SPLAY.
flood plane *Hydrology.* the position or elevation of the water surface of a stream during a specific flood.
flood plate *Microbiology.* a petri plate onto which a liquid microbial inoculum has been poured.
flood projection *Telecommunications.* in facsimile transmission, an optical method of scanning in which the subject copy is floodlighted and the scanning spot is defined by a masked portion of the illuminated area.
flood routing *Hydrology.* the determination of the progressive time and shape of a flood wave at successive points along a river.
Flood's equation *Physical Chemistry.* a formula used to determine the temperature of liquids in a binary fused salt system.
flood stage *Hydrology.* the elevation at which a stream overflows its banks and begins to cause damage along any part of its reach.
flood strength see STRENGTH OF FLOOD.
flood tide *Oceanography.* the part of the tide cycle between low water and the following high water.
flood tuff see IGNIMBRITE.
flood wall *Civil Engineering.* a levee or retaining wall that provides protection from water overflow.
floor a level, supporting surface in any structure or feature; specific uses include: *Architecture.* **1.** the bottom flat surface of a room. **2.** the horizontal structure that divides two stories of a building. *Geology.* **1.** the rock surface upon which sedimentary strata have developed. **2.** the more or less horizontal bed underlying any body of water. Also, BOTTOM. **3.** the country rock adjacent to the lower surface of an igneous intrusion. *Naval Architecture.* the lowest section of a ship's frame, usually comprised of a relatively flat area on either side of the keel. *Mining Engineering.* **1.** the upper surface of the stratum underlying a coal seam. **2.** the bottom of a coal seam or other mineral deposit.
floor beam *Building Engineering.* a beam used in framing floors during building construction. *Civil Engineering.* a beam that supports the floor of a structure.
floorboard *Building Engineering.* any of a series of planks that make up the level base of a structure.
floor cut *Mining Engineering.* **1.** a machine cut made in the floor dirt immediately underneath a coal seam. Also, BOTTOM CUT. **2.** a cut made to separate a block of stone from the quarry floor. Also, **floor break.**
floor framing *Building Engineering.* the supports, struts, and floor joists of a building.
floor function *Mathematics.* the greatest integer function.

floor guide *Building Engineering.* a groove in a floor surface into which a sliding door or partition is placed, allowing it free movement.
flooring saw *Mechanical Devices.* a pointed handsaw curved toward its toe line with teeth on both edges, used for cutting out sections, such as parts of floorboards, in order to remove them.
floor light *Building Engineering.* a window embedded in a floor that can be walked on and that allows light into a room from below.
floor line *Building Engineering.* a mark made at the lower end of a door post or other vertical member to indicate the position of the floor during construction.
floor outlet *Electricity.* an electrical outlet level with or recessed into a floor in a building or dwelling. Also, **floor plug.**
floor plate *Building Engineering.* a horizontal board on a floor that supports wall studs. *Engineering.* **1.** a piece of metal inserted on a floor to which heavy machines can be bolted. **2.** a metal plate with a roughened surface that prevents individuals from slipping.
floor sill *Mining Engineering.* a large timber laid flat, to which the drill platform boards or planking are fastened.
floor stop *Building Engineering.* a doorstop that projects from the floor to allow a door to be opened only to a certain point.
floor system *Civil Engineering.* the structural floor assembly composed of beams, girders, and floor slabs in buildings and bridges.
floor-to-floor time *Robotics.* the time required to pick up, load, machine, and unload a manufactured part.
flop gate *Mining Engineering.* an automatic gate that controls a flow of water; used in placer mining when there is a shortage of water.
flopover *Electronics.* in television reception, a flaw in which a series of frames roll up and down the screen, occurring when the vertical and horizontal sweep frequencies are not synchronized.
flopped *Graphic Arts.* of an image, inverted; may be intentional (as when an engraver's negative is inverted on a relief plate) or unintentional (as when an inverted photo appears in print).
floppy disk see DISKETTE.
FLOPS *Computer Programming.* a measure of the speed of operation of a computer. (An acronym for the number of floating point operations per second.)
Floquet theorem [flō´kā] *Mathematics.* the solution to a second-order linear differential equation, whose coefficients are periodic single-valued functions of an independent variable x, having the form $e^{ax}P(x)$, where a is constant and $P(x)$ is a periodic function.
flor- a combining form meaning "flower," as in *florula.*
flora *Ecology.* **1.** the plant life that is present in a particular region or habitat or at a particular time. **2.** plants in general; plant life as a whole. (From the name of the Roman goddess of flowers.)
floral analysis *Archaeology.* the study of plant remains in an archaeological site, including identification, association with artifacts and food processing, and so on.
floral axis *Botany.* the central axis of a flower.
floral diagram *Botany.* a cross-sectional diagram of a flower showing the various parts and their arrangement.
floralturbation *Archaeology.* disturbance of the soil surface by plants, especially by tree fall and by root growth or decay.
floration *Ecology.* the total collection of plant species in a particular area.
florencite-(Ce) *Mineralogy.* $CeAl_3(PO_4)_2(OH)_6$, a pale yellow to pink trigonal mineral occurring as small rhombohedral or pseudocubic crystals and rounded grains, having a specific gravity of 3.457 to 3.71 and a hardness of 5 to 6 on the Mohs scale; found in mica schists and placer deposits.
florentium *Nuclear Physics.* an element with the atomic number 61 and mass 147 that is produced synthetically during uranium-235 fission. Also, ILLINIUM, PROMETHIUM-147.
flores *Chemistry.* see FLOWERS.
floret *Botany.* one individual flower in the compact group of flowers of a flower cluster or of a composite plant.
Florey, Lord Howard 1898–1968, Australian-born British pathologist; shared Nobel Prize for the isolation and development of penicillin.
Florey unit *Biology.* a unit used for the standardization of penicillin.
flori- a combining form meaning "flower," as in *florigen.*
floriculture *Agriculture.* the commercial production and cultivation of flowers and ornamental plants.
Florida Current *Oceanography.* a swift current (2–5 knots) that is fed by the Yucatan Current and flows eastward through the Straits of Florida north of Cuba, where it turns northward and joins the Antilles Current north of the Bahamas to form the Gulf Stream.
floridean starch *Biochemistry.* a glucan that functions as the principle storage carbohydrate in red algae.
Florideophycideae *Botany.* the major subclass belonging to the class Rhodophyceae, a red algae having a rose red to black color and prominent cytoplasmic connections between cells. Also, **Florideae.**
floriferous *Botany.* having flowers or blooming freely, usually describing ornamental plants.
florigen *Biochemistry.* a plant hormone that causes initiation of flowering.
floristic *Botany.* of or relating to flowers or flora.
floristics *Ecology.* the study or classification of all the plants of a particular area or of the geographical distribution of particular plants.
florivorous *Zoology.* feeding mainly or exclusively on flowers.
-florous a combining form meaning "flowered."
florula *Ecology.* **1.** the plant population of a limited area or a specific habitat. **2.** a collection of plant fossils that comes from a single stratum or group.
Flory, Paul J. 1910–1985, American chemist; Nobel Prize for a method of analyzing the structures and properties of long-chain molecules.
Flory-Huggins interaction parameter *Materials Science.* a polymer solvent parameter expressing a function of the energy of mixing per unit volume; used principally in calculations involving plasticizers and miscible (amorphous) polymer blends.
Flory-Huggins theory *Materials Science.* a theory that proposed a random distribution of chain segments over the liquid lattice in dilute polymer solutions; this led to the exploration of serious discrepancies between experimental data and predictions of the theory.
Flory-Krigbaum theory *Materials Science.* an important clarification and improvement to the Flory-Huggins theory proposing that a polymer solution was made up of isolated coiled chains occupying a volume from which all other chains are excluded.
Flory-Rehner flow equation *Materials Science.* an equation relating the polymer entropy change to the heat of mixing of polymer solvent; used to determine the number of active chain segments per unit volume of rubber, or the extent of vulcanization.
Flory theta temperature *Materials Science.* the temperature level at which a second virial coefficient disappears and the coiled polymer molecules expand to their full contour length and become rod-shaped.
Floscularíacea *Invertebrate Zoology.* an order of rotifers in the class Monogononta that are sessile or free-swimming, and have a double ciliated oral corona.
flosculous *Botany.* **1.** having or composed of florets. **2.** of or relating to a tube-shaped floret.
flosculus *Botany.* a single floret. Also, **floscule**
flospinning *Metallurgy.* in metal working, forming parts by power spinning over a rotating mandrel.
floss *Metallurgy.* the slag floating on the surface of a metallic bath.
floss hole *Metallurgy.* a door on the bottom of the stack of a puddling furnace, used to remove ashes.
FLOT *Ordnance.* the front line of one's own troops; the forward line of ground troops actually engaged in combat or in a combat area. (An acronym for front line of troops.)
flot see FLOTATION TECHNIQUE.
flotation *Engineering.* a technique that separates different types of solid particles in a liquid, due to the fact that the chemical structure of some particles will cause them to absorb water while others do not. Also, FLOTATION TECHNIQUE.
flotation agent *Chemistry.* a chemical used in a flotation separation process to produce stable foam.
flotation analysis *Physics.* a technique for determining the density of a liquid by adjusting the weight of a floating body of known density until it is equal to that of the liquid.
flotation bag see FLOTATION COLLAR.
flotation cell *Mining Engineering.* an appliance in which the froth flotation of ores is performed.
flotation collar *Space Technology.* a buoyant bag on board a spacecraft that inflates and engulfs part of the outer surface if the spacecraft lands in the sea. Also, FLOTATION BAG.
flotation collector *Chemistry.* any chemical agent that increases the carrying capacity of air bubbles.
flotation technique see FLOTATION.
flotsam [flät´səm] *Navigation.* **1.** any collection of articles floating on the water. **2.** specifically, floating articles that result from a shipwreck or that are inadvertently lost overboard, as opposed to articles that are deliberately thrown overboard (jetsam).

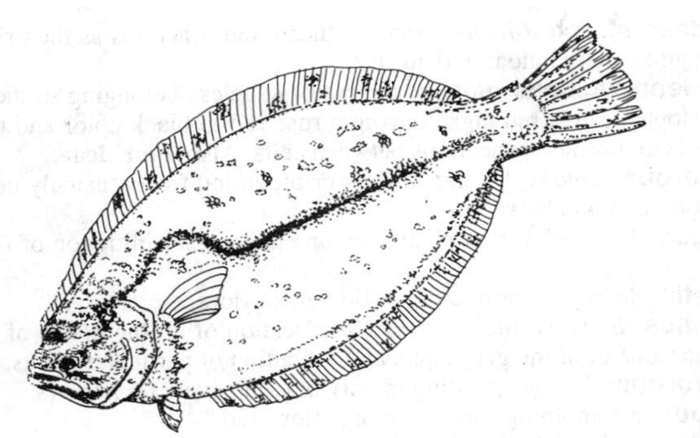

flounder

flounder *Vertebrate Zoology.* any of numerous European marine flatfishes in the families Pleuronectidae and Bothidae of the order Pleuronectiformes, characterized by eyes located on one side of the head. (From a Scandinavian name for this fish; not related to the sense of a floundering movement.)

flour *Food Technology.* a powder or fine meal ground from wheat or other cereal grains, used in baking.

flour gold *Metallurgy.* the finest size gold dust.

flouridoside *Biochemistry.* a glycoside present in red algae that regulates intracellar osmotic pressure and serves as a reserve carbohydrate.

flow to stream or well forth; specific uses include: *Fluid Mechanics.* the continuous motion of a fluid (a liquid or gas). *Physics.* the fluidlike motion of matter or charges. *Oceanography.* the total movement of water by currents in a given place, including both tidal and nontidal currents. *Geology.* **1.** any deformation of rock over time that results from stress beyond its elastic limit without loss of cohesion. Also, FLOWAGE, ROCK FLOWAGE. **2.** a continuous mass movement of unconsolidated material, usually in association with water, resembling the movement of a viscous fluid. **3.** the mass of material so moved. *Industrial Engineering.* the movement of parts or materials in a production-line operation. *Transportation Engineering.* **1.** the movement of vehicular traffic from one place to another. **2.** the movement of passengers and cargo through an airport. *Materials Science.* **1.** any plastic deformation. **2.** specifically, the large-scale deformation of polymeric material associated with the slippage of molecular chains past one another. *Civil Engineering.* a measure of the consistency of mixed concrete, mortar, or cement paste. *Computer Programming.* **1.** a general term for any sequence of events, usually as represented on a flowchart. Also, **control flow. 2.** the flow of data along paths from its definition to its uses. Also, **data flow.**

flow *Mathematics.* **1.** a flow in a network N with source set X, sink set Y, and capacity function c is a function f from the arcs of N to the nonnegative real numbers such that (a) $0 \leq f(a) \leq c(a)$ for every arc a of N, and (b) for each vertex v of $V(N) - (X \cup Y)$, the sum of the values of f on the arcs directed toward v equals the sum of the values of f on the arcs directed away from v. Ordinarily, the flow on any arc directed toward a vertex in X is zero, as is the flow on any arc directed away from a vertex in Y. **2.** a set of integral curves $x(t)$ of a vector field on a differential manifold; i.e., a solution of the system of ordinary differential equations $dx/dt = U(x)$, where U is the vector field, and x is an arbitrary point on the manifold.

flowability *Fluid Mechanics.* the ability of a body of matter to flow; liquids, gases, and particulate matter can exhibit significant flowability.

flowage *Geology.* see FLOW, def. 1.

flow analysis *Industrial Engineering.* a study of the movement of personnel or material from one place or operation to another.

flow banding *Geology.* an igneous rock structure consisting of alternating bands of layers of minerals differing in composition and texture, which are produced by a flow of lava or magma.

flow birefringence *Physical Chemistry.* the tendency of long, thin, asymmetric molecules to move in the same direction as a solution being forced through a tube or capillary.

flow bog *Ecology.* a peat bog whose surface level fluctuates according to the flow of the tides or of rainwater.

flow breccia *Geology.* a breccia formed contemporaneously with a lava flow, and composed of fragments of cooled or solidified lava that were cemented by the still fluid parts of the flow.

flow brightening *Metallurgy.* the melting of an electrodeposited metallic coating, such as tin, to improve its luster.

flow cast *Petrology.* a roll, lobate ridge, or other raised bedding feature on overlying sandstone due to the filling of depressed features produced by the flow or warp of the soft, hydroplastic underlying sediment.

flow chart or **flowchart** *Engineering.* a graph that employs a set of standard symbols to depict progressive changes in various aspects over a system.

flow cleavage *Geology.* a form of cleavage in which solid-state rock flow occurs in conjunction with the recrystallization of slaty minerals. Also, SLATY CLEAVAGE.

flow coat *Engineering.* a coating achieved by pouring a liquid material over the object and draining off the excess.

flow coefficient *Fluid Mechanics.* a constant of proportionality that relates an experimentally determined velocity of a fluid flowing in a pipe or channel to the velocity predicted by theory under specified assumptions.

flow control *Engineering.* a system that regulates the movement of gases, vapors, liquids, slurries, pastes, or solid particles in conduits or channels.

flow-control valve *Engineering.* a valve whose flow opening is controlled by the rate of flow at which liquid passes through it.

flow cytometry *Microbiology.* any technique for sorting, selecting, or counting individual cells in a suspension as they pass through a tube; applied especially to techniques involving the detection of a cell-bound fluorescent label and often used in cancer research as well as in screening for chromosomal abnormalities.

flow diagram *Industrial Engineering.* a depiction of the physical layout of a process, including the locations of all activities and operations and the directions in which personnel or material travel between these locations. *Transportation Engineering.* a map or diagram of a traffic network in which the traffic volume along each path is indicated by the width of that path in the diagram. *Computer Technology.* a variety of flowchart that emphasizes algorithms and the details of data transformation with little reference to input/output operations.

flow direction *Engineering.* the predecessor-to-successor relation between operations on a flow chart, usually indicated by arrows.

flow earth see SOLIFLUCTION MANTLE.

flow equation *Fluid Mechanics.* an equation that relates fluid properties, pipe or channel geometries, and environmental conditions.

flower *Botany.* **1.** the reproductive or seed-bearing part of a plant; a shortened branch that has modified, often colorful leaves called petals and that usually consists of a pistil, stamens, corolla, and calyx. **2.** a plant named for this part or considered in reference to this part, especially a plant that has a brightly colored, showy flower.

flowerpecker *Vertebrate Zoology.* any songbird of the family Dicaeidae that eats sticky seeds and nectar with its brush-tipped, double-tubed tongue and downcurved, serrated bill; some species have a role in seed distribution of the parasite mistletoe; native to Asia and Australia.

flowers *Chemistry.* a term for a chemical compound, such as a metallic oxide, that is formed through sublimation. Also, FLORETS.

flowers of benzene *Organic Chemistry.* a popular name for benzoic acid.

flowers of tin see STANNIC OXIDE.

flow field *Fluid Mechanics.* a spatial distribution indicating fluid velocity or density as a function of position and time.

flow graph *Mathematics.* a network with a flow defined on it.

flowing-film concentration *Materials Science.* a concentration that forms because liquid films in laminar flow have a velocity that is not the same in all depths of the film.

flowing furnace *Metallurgy.* a furnace from which molten metal can be tapped.

flowing pressure *Petroleum Engineering.* the pressure at the bottom of an oil-well bore during normal oil production. Also, BOTTOM-HOLE PRESSURE.

flowing-pressure gradient *Petroleum Engineering.* the span measured for upward liquid flow in a continuous-flow, gas-lift oil well plotted against the slope of diminishing pressure.

flowing-temperature factor *Thermodynamics.* a correction factor that is used for calculations made on flowing gases at temperatures other than 15.5°C, which is the standard temperature at which a flow equation is valid.

flowing well *Petroleum Engineering.* an oil reservoir that has enough gas-drive pressure to force the flow of oil up and out of a wellhole.

flow layer *Petrology.* in igneous rock, a layer distinguished from adjacent layers by mineral composition or structure, produced by flowage prior to complete solidification of the magma.

flow line *Hydrology.* **1.** a contour line around a body of water, indicating the water level. **2.** a line in an open channel, representing the free water surface. *Petrology.* any internal structure in igneous rock that is produced by inclusions, mineral streaks, or the subparallel orientation of tabular or prismatic crystals formed by the movement of magma or lava. Also, FLOW STRUCTURE. *Materials Science.* a mark on a molded plastic or metal article left when two input-flow fronts intersect during molding. *Petroleum Engineering.* a pipeline that takes oil from wells to a gathering center. *Computer Technology.* a path between symbols on a flowchart, representing the transfer of data or control. *Cartography.* the slope extending from the heights along the neatline to the model datum, at an angle no greater that 45°, to preclude forming the plastic sheet at a 90° angle at the neatline of a plastic relief map.

Flow-Matic *Computer Programming.* the first (1956–1958) Englishlike programming language for business data processing; the forerunner of COBOL.

flow measurement *Engineering.* the calculation of the quantity of a material that flows through a pipe, duct, or open channel.

flowmeter *Engineering.* an instrument that is designed to indicate the flow rate of a fluid flowing in a pipe.

flow microfluorometry *Microbiology.* a technique for the sorting, selection, or counting of individual cells or small particles as they pass through a small hole or tube by detecting a specific cell-conjugated fluorescent label.

flow mixer *Mechanical Engineering.* a mixing device that has jet nozzles and agitators, used to mix two or more liquids as they flow through. Also, LINE MIXER.

flow net *Fluid Mechanics.* a diagram constructed assuming fluid flow lines along which fluid particles travel and with equipotential lines that indicate surfaces of equal pressure; the diagram takes on a netlike appearance since flow lines and equipotential lines intersect at right angles.

flow noise *Acoustics.* noise produced by the dynamic flow of fluid around an object moving through the fluid; it is caused by turbulence created by the boundary surface and eddy currents due to surface roughness.

flow nozzle *Engineering.* a flowmeter in a closed conduit, composed of a short flared nozzle of reduced diameter attached to the inner diameter of a pipe that produces a brief pressure drop in flowing fluid; this process determines flow rate by measuring static pressures in front of and behind the nozzle.

flow pattern *Fluid Mechanics.* a pattern that is assumed by a two-phase (liquid-gas) combination flowing in a pipe or channel; relevant factors are the liquid-to-gas ratio, flow resistance, fluid velocities, fluid properties, and pipe dimensions.

flow plane *Geology.* the plane along which displacement occurs in both igneous and metamorphic rock.

flow process *Engineering.* a system in which fluids or solids are processed in continuous movement.

flow process chart *Industrial Engineering.* a chart on which all aspects of a production process are represented, including operations, transportations, inspections, delays, and storages.

flow rate *Fluid Mechanics.* a quantity that measures the amount of a fluid flowing across a specified unit area in a unit amount of time; it is customary to specify the amount of fluid by mass or volume. *Transportation Engineering.* the rate at which vehicles pass by a given point or through a given stretch of highway. Also, TRAFFIC VOLUME.

flow-rating pressure *Mechanical Engineering.* the value of the inlet static pressure at which a pressure-relief device is activated.

flow reactor *Chemical Engineering.* a dynamic reactor system in which reactants continuously flow into the unit and products are continuously removed.

flow regime *Hydrology.* a range of surface runoffs having similar bed forms, resistance to flow, and means of transporting sediment.

flow resistance *Fluid Mechanics.* an impedance to the flow of a fluid, including viscous effects, surface friction, bends, and constrictions.

flow shop *Industrial Engineering.* a production arrangement, such as an assembly line, in which work in progress moves in a single direction through the production process.

flow slide *Geology.* a landslide involving waterlogged material in which the slip surface is not well defined.

flow soldering *Engineering.* the soldering of printed circuit boards by passing them over a flowing wave of molten solder in a solder bath; this technique allows exact control of the immersion depth in the molten solder and minimizes heating of the board.

flowstone *Geology.* any mineral accumulation, especially of calcium carbonate, that forms on the walls or floor of a cave by the action of flowing water.

flow stress *Mechanics.* the extent of stress required to produce plastic flow. Also, YIELD STRESS.

flow string *Petroleum Engineering.* the total length of interconnected tubing or casing through which oil or gas from a well flows to the surface.

flow structure *Geology.* a primary sedimentary structure formed as a result of underwater slump or flow. *Petrology* see FLOW LINE.

flow system see OPEN SYSTEM.

flow tank *Petroleum Engineering.* a tank that receives produced oil from which the water or gas has been removed. Also, PRODUCTION TANK.

flow texture *Petrology.* a texture in fine-grained or glassy igneous rocks characterized by a wavy or swirling pattern as a result of the stream or flow lines of the once-molten material; a subparallel arrangement of prismatic or tabular crystals or microlites.

flow tract *Cardiology.* the path of bloodflow within the heart.

flow transmitter *Engineering.* an instrument designed to determine liquid flow in pipelines and convert the results into proportional electric signals to be sent to remote receivers or controllers.

flow valve *Engineering.* a valve that shuts off when the flow of a fluid reaches a certain value.

flow velocity *Geology.* a vector point function that indicates the rate and direction of water movement through soil, expressed per unit time and parallel to the direction of movement.

flow visualization *Engineering.* a method of making the disturbances that occur in fluid flow visible.

flow welding *Metallurgy.* welding performed by pouring molten filler metal.

flow work *Thermodynamics.* the work required to move mass into or out of a system, equal to the product of the pressure P and specific volume v of the system.

floxuridine *Pharmacology.* $C_9H_{11}FN_2O_5$, an antineoplastic drug used mainly to inhibit the proliferation of cancerous cells when cancer spreads to the liver from the gastrointestinal system.

fl oz or **fl. oz.** fluid ounce.

FLP. left front posterior. (From *fronto-leava posterior.*)

fl. pl. with double flowers. (From Latin *flore pleno.*)

flt. flight.

FLT. left front transverse (From *fronto-leava transversa.*)

flu *Medicine.* in popular use: **1.** a shorter term for influenza. See INFLUENZA. **2.** any viral infection having symptoms similar to those of influenza.

flu virus *Medicine.* **1.** a common term for any transitory virus that causes respiratory and gastrointestinal symptoms (**flulike symptoms**), as well as the inflammation of the mucous membranes. **2.** see INFLUENZA VIRUS.

fluctuating asymmetry *Ecology.* stochastic variations in morphology or mensuration of two sides of a body; the mean differences of the two sides center about zero.

fluctuating current *Electronics.* a condition in which direct current varies in value at an inconsistent rate.

fluctuation *Science.* a change in wave motion. *Oceanography.* a variation in the water level from mean sea level that is not caused by tide-producing forces.

fluctuation test *Microbiology.* a procedure for determining the spontaneity of mutant cell generation within a parent colony by dividing the colony into small populations and observing whether the variants occur uniformly or erratically among them.

fluctuation theory *Optics.* the theory that the diffusion of light in pure water arises from the random motion of molecules, causing variations in the density of the water and therefore variations in the refraction of the light.

flue *Engineering.* a passage for removing products of combustion from a furnace, boiler, or fireplace, or through a chimney.

flue dust *Metallurgy.* in smelting or melting, fine dust that escapes from the furnace through the stack.

flue exhauster *Engineering.* a device attached to a vent so that it furnishes a positive induced draft.

flue gas *Engineering.* the gaseous combustion product generated in a furnace.

flue-gas analyzer *Engineering.* an instrument that monitors the flue gas composition of a boiler heating unit in order to determine if the mixture of air and fuel is optimal for maximum heat output.

flue lining *Building Engineering.* a flexible stainless-steel, fireclay, or fire-resistant concrete pipe, arranged to protect chimney walls during passage of combustible materials through a flue.

fluellite *Mineralogy.* $Al_2(PO_4)F_2(OH){\cdot}7H_2O$, a colorless to white, transparent orthorhombic mineral forming dipyramidal crystals, having a specific gravity of 2.17 and a hardness of 3 on the Mohs scale; found as tiny crystals on quartz.

fluence *Physics.* a flux of particles passing across a unit area, commonly expressed in the number of particles per second.

flueric *Control Systems.* performing without the use of mechanical parts.

fluerics *Control Systems.* the technology of operating systems without the use of mechanical parts.

flufenamic acid *Pharmacology.* $C_{14}H_{10}F_3NO_2$, a drug used to treat the pain and inflammation caused by arthritis.

fluid *Science.* able to flow; moving freely. *Physics.* a nonsolid state of matter in which the atoms or molecules are free to move past each other, as in a gas or liquid.

fluid amplifier *Engineering.* a device that employs the interaction between jets of liquid rather than an electronic circuit or mechanical parts to achieve amplification.

fluid-bed polymerization *Materials Science.* a type of coordinated polymerization that uses a catalyst in the form of slurries of fine solid particles fluidized in an inert medium.

fluid-bed process *Chemical Engineering.* a process in which finely divided powders act in a fluidlike way when suspended and moved by a rising stream of gas or vapor; primarily used for catalytic cracking of petroleum distillates.

fluid catalyst *Chemical Engineering.* the finely divided solid particles used as a catalyst in a fluidized-bed process.

fluid catalytic cracking *Chemical Engineering.* a method of oil refining in which gas-oil fractions are cracked to form lower molecular weight components in a fluidized catalyst bed.

fluid coefficient *Petroleum Engineering.* during the fracturing operation, a comparison of the flow resistance to the draining of fracturing fluids into the formation.

fluid coking *Chemical Engineering.* a thermal process for continuous conversion of heavy, low-grade petroleum fractions into petroleum coke and lighter hydrocarbon products.

fluid-compressed *Metallurgy.* describing steel that has been pressed while still fluid.

fluid computer *Computer Technology.* an experimental digital computer containing air-powered fluid logic elements in place of electronic circuits.

fluid contact *Petroleum Engineering.* the location in a reservoir at which oil-water contact or gas-oil contact is found.

fluid-controlled gate valve *Mechanical Engineering.* a valve in which the valve operator is activated by fluid energy.

fluid coupling *Mechanical Engineering.* a device in which a fluid, such as oil, transmits torque from one shaft to another, producing an equal torque in the latter. Also, HYDRAULIC COUPLING.

fluid density *Fluid Mechanics.* a quantity given by the amount of mass per unit volume in a fluid.

fluid die *Mechanical Engineering.* a die that is used with a hydraulic pressure device, which makes the part to be shaped conform to the shape of a die.

fluid distributor *Engineering.* a device for the regulated distribution of fluid feed to a process unit.

fluid dram or **fluiddram** *Metrology.* **1.** a unit of liquid measure used in the U.S., equal to .125 fluid ounce or 3.697 milliliters. **2.** also, **imperial fluid dram.** a similar unit of measure used in Great Britain, equal to .125 imperial fluid ounce or 3.55 milliliters.

fluid drive *Mechanical Engineering.* a power coupling consisting of two vaned rotors in a sealed casing that is filled with oil, functioning so that one rotor, driven by the engine, moves the oil to drive the other rotor, which in turn drives the transmission; used in motor vehicles to allow a smooth start in any gear. Also, **fluid clutch**.

fluid dynamics *Physics.* the scientific study of fluids (gases and liquids) in motion.

fluid-energy mill *Engineering.* a device that grinds particles with energy generated by the collision of the particles and compressed air supplied to the grinding chamber at high speed.

fluid-film bearing *Mechanical Engineering.* an antifriction bearing used as an agent to keep rubbing surfaces apart by a film of lubricant such as oil.

fluid-flow analogy *Transportation Engineering.* a conceptual model of traffic flow based on the behavior of fluids. The intuitive power of this model is expressed in common terms such as "flow" and "bottleneck," but the fluid-flow analogy is also used to construct sophisticated models of traffic behavior.

fluid friction *Fluid Mechanics.* the resistance in fluid flow due to the interaction between fluid elements and wall surfaces; the loss in kinetic energy in such processes results in the production of heat within the fluid.

fluid fuel reactor *Nucleonics.* a nuclear reactor that is fueled by a liquid material instead of a solid.

fluid geometry *Petrology.* the distribution of fluids in the rocks of a reservoir, as influenced by the rock's composition, porosity, method of producing saturation, and wettability characteristics.

fluid hydroforming *Chemical Engineering.* a petroleum refinery catalytic cracking process used to increase the low-octane number of low-octane number stocks.

fluidic *Physics.* relating to or being a fluid; flowing freely. *Engineering.* **1.** of or relating to fluidics. **2.** describing systems or devices that use the phenomena and principles of fluidics.

fluidic flow sensor *Engineering.* a device that determines the velocity of a gas flow when a jet of air or other selected gas moves into two small openings adjacent to each other and is deflected by the flow of the gas being measured, causing the relative pressure on the two openings to become a function of gas velocity.

fluidic oscillator meter *Engineering.* a flowmeter that determines the frequency at which a liquid entering the meter will cling to one or the other of two diverging side walls.

fluidics *Engineering.* a technology that carries out sensing, control, information processing, and actuation functions with fluid dynamic phenomena rather than mechanical parts. Also, FLUIDIC TECHNOLOGY.

fluidic sensor *Engineering.* an instrument that detects the presence of an approaching or neighboring object from the back pressure generated on an air jet when the object cuts off the jet's exit area.

fluidic technology see FLUIDICS.

fluid inclusion *Petrology.* a tiny fluid- or gas-filled cavity in an igneous rock, 1–100 micrometers in diameter, formed by the entrapment of a fluid, typically that from which the rock crystallized.

fluid injection *Petroleum Engineering.* a process of forcing oil into producing wells by injecting gases or liquids under pressure into a reservoir.

fluid intelligence *Psychology.* the form of general intelligence that is largely innate and that allows the individual to adapt to unanticipated new situations.

fluidity *Fluid Mechanics.* a quantity that measures the ability of a fluid to flow; expressed as the reciprocal of the fluid viscosity.

fluidization *Chemical Engineering.* a technique in which a finely divided solid is caused to behave in the manner of a fluid by its being suspended in a moving gas or liquid; the solids treated in this way are frequently catalysts.

fluidized adsorption *Chemical Engineering.* a procedure for the fractionation or separation of vapors or gases in a fluidized bed of adsorbent material.

fluidized bed *Engineering.* a layer of hot air or gas at the bottom of a container upon which a powdered material floats; used to dry, heat, or quench.

fluidized-bed coating *Materials Science.* a technique in which a heated object is submerged into the fluidized bed of a thermoplastic resin that fuses into a continuous uniform coating over the submerged object.

fluidized bed reactor *Biotechnology.* an anaerobic digester, bioreactor, or fermenter in which the active material is kept in suspension by the upward flow of the fluid phase.

fluid level *Petroleum Engineering.* the distance from a wellhead to the surface of liquid in the casing or tubing of an oil well.

fluid logic *Engineering.* a logical operations-simulated device that uses fluid dynamic phenomena to regulate the interactions between sets of gases or liquids.

fluid-loss test *Petroleum Engineering.* the measure of fracturing fluid loss relative to the time elapsed (spurt loss) before the fluid-loss agent creates a nonpermeable layer in the pore matrix of the reservoir.

fluid lubrication *Mechanical Engineering.* a theoretical state of perfect lubrication in which the bearings and surfaces of a machine are completely covered by a viscous oil film that is sustained by the friction of the surfaces.

Fluid Mechanics

We may simplistically define fluid mechanics as the experimental and mathematical-computational study of the mechanical behavior of fluids. How, that is, does a fluid flow through a tube or a bed of sand, how does it fill a glass, what forces are required or produced, how can we calculate the velocity field, and so forth. For the most part we ignore the microstructure of the fluid, and we assume that the stresses (apart from the pressure) are proportional to the rate at which material points separate in the flow; we generally allow the density to change under the influence of pressure. These assumptions are excellent for air and water under ordinary circumstances.

It is sometimes said that, since we know the equations describing this situation, fluid mechanics must be a problem in applied mathematics. Even in this isothermal, seamless continuum without chemistry, however, physical phenomena appear that are poorly understood physically, and that are difficult to calculate, such as turbulent flow, in which the physical quantities are chaotic in space and time.

Fluid flow is responsible for most of the transport of heat and matter in the environment, in living organisms, and in technology, and hence plays a vital role in life itself and the quality of life. Fluid mechanics is thus a subject of seminal importance, permitting us ultimately to calculate these effects.

Our definition is simplistic because fluid mechanics is much more than this lapping over into many contiguous areas. It is in these overlap regions that many of the most interesting problems can be found. The microstructure must be considered very close to boundaries; it also cannot be ignored if the fluid is a rarified gas. At elevated temperatures chemistry becomes important, giving rise to other problems that are usually treated in combustion and hypersonic flow. In the presence of strong magnetic and electric fields, other forces may become important (depending on the particular fluid), giving rise to new effects; these phenomena are treated in magnetohydrodynamics, which is considered to lie in fluid mechanics.

Fluid mechanics, of course, includes the transfer of heat and matter due to the motion of the fluid, and the driving of the fluid motion due to differences in density which may be induced by temperature, as well as the effects due to temperature dependency of the materials constants, such as the viscosity.

There is another field called heat transfer, which considers some of these things in detail; the boundary between this and fluid mechanics is not precise. In materials such as polymer melts, the linearity assumption must be discarded. Some materials will hold their shape if not disturbed, but will flow when pushed; foods such as sauces and ice cream are examples. Non-Newtonian fluid mechanics and continuum mechanics deal with aspects of these nonlinear plastic materials, but some parts still lie within fluid mechanics.

John L. Lumley
Willis H. Carrier Professor of Engineering
Cornell University

fluid mechanics *Physics.* the scientific study of the mechanical properties of fluids (gases and liquids) in motion or at rest, including the observation, description, and mathematical computation of the behavior of fluids.

fluid ounce *Metrology.* **1.** a unit of liquid measure that is used in the U.S., equal to 0.0625 pint or 29.57 milliters. **2.** also, **imperial fluid ounce.** a similar unit of measure used in Great Britain, equal to 0.05 pint or 28.41 milliliters.

fluid pressure *Physics.* the pressure that is exerted by a fluid, which is directly proportional to the specific gravity at a given point and the height of the fluid above that point. *Mining Engineering.* specifically, the force that is exerted by the weight of a column of drilling fluid measured at a given point in a borehole, expressed in pounds per square inch. *Mechanical Engineering.* the force with which a stream of fluid is ejected from a pump, jet, or other opening.

fluid resistance *Fluid Mechanics.* a force that opposes the motion of a body passing through a liquid or gas.

fluid saturation *Petrology.* the gross measurement of the empty space in a reservoir rock that is occupied by a fluid.

fluid statics *Fluid Mechanics.* **1.** the study of fluids that are at rest, as opposed to those that are in motion (fluid dynamics). **2.** in particular, the study and determination of pressures in static fluids.

fluid transmission *Mechanical Engineering.* an automotive transmission having a fluid drive.

fluid viscosity ratio *Petroleum Engineering.* the ratio of the viscosity of oil in a gas-drive reservoir to that of a displacing gas, used in the calculation of unit displacement efficiency.

fluke *Invertebrate Zoology.* the common name for any parasitic flatworm belonging to the class Trematoda.

flume *Civil Engineering.* **1.** a channel of steel, reinforced concrete, or wood that carries water for industrial purposes. **2.** to divert with a flume, such as water in a stream, to expose the auriferous sand and gravel of the bed.

flumed *Engineering.* of or relating to the transportion of solids by suspension or flotation in flowing water.

fluoborite *Mineralogy.* $Mg_3(BO_3)(F,OH)_3$, a colorless or white hexagonal mineral occurring as prismatic crystals and fibrous masses, having a specific gravity of 2.89 and a hardness of 3.5 on the Mohs scale; found as crystals in thermally metamorphosed impure limestones.

fluocerite-(Ce) *Mineralogy.* $(Ce,La)F_3$, a pale-yellow to yellow to reddish-brown hexagonal mineral, granular in habit or as prismatic crystals, having a specific gravity of 5.93 to 6.14 and a hardness of 4 to 5 on the Mohs scale; found in pegmatites. Also, TYSONITE.

fluolite see PITCHSTONE.

fluometuron *Organic Chemistry.* $C_{10}H_{11}F_3N_2O$, a toxic white crystalline solid, soluble in water and alcohol, that melts at 163°C; used as an herbicide.

fluor see LUMINOPHOR.

fluor- a combining form denoting the presence of: **1.** fluorine or fluoride, as in *fluorite.* **2.** fluorescence, as in *fluorescein.*

fluoranthene *Organic Chemistry.* $C_{16}H_{10}$, a tetracyclic hydrocarbon existing as needlelike colored crystals, soluble in organic solvents; melts at 107°C and boils at 250°C; found in coal tar and petroleum.

fluorapatite *Mineralogy.* $Ca_5(PO_4)_3F$, a varicolored, hexagonal mineral occurring in granular, reniform, and fibrous forms and as prismatic crystals, having a specific gravity of 3.1 to 3.2 and a hardness of 5 on the Mohs scale; found in igneous rocks, hydrothermal veins, and bedded deposits.

fluorene *Organic Chemistry.* $C_{13}H_{10}$, a tricyclic hydrocarbon occurring as white crystalline plates; melts at 116°C and decomposes at 295°C; used in dyestuffs and resinous products.

fluorescein *Organic Chemistry.* $C_{20}H_{12}O_5$, a combustible reddish-orange crystalline powder, insoluble in water and soluble in alkali; decomposes at 290°C; used in dyes and as an indicator and diagnostic aid. *Immunology.* this powder in the form of a yellow-green dye; used to label antibodies for fluorescence.

fluorescein-conjugated antibody *Immunology.* an antibody that is formed when it is joined with fluorescein.

fluorescein isothiocyanate *Biotechnology.* a derivative of fluorescein that provides a stable fluorescent label for proteins.

fluorescence [flô res´əns; flů res´əns] *Atomic Physics.* an effect in which a substance releases electromagnetic radiation while absorbing another form of energy, but ceases to emit the radiation immediately upon the cessation of the input energy, such as the absorption of ultraviolet light by the coating in a fluorescent tube to give off visible light. *Immunology.* specifically, the light emission of a given wavelength by a substance that is activated by light of a different wavelength. *Materials Science.* a characteristic X-ray emitted by material used for chemical analysis. *Optics.* see BLOOM, def. 3.

fluorescence-activated cell sorter *Biotechnology.* a cell-separation apparatus that is able to identify and separate individual cells based on DNA content or the presence or absence of a fluorescently labeled, bound substance, such as an antibody.

fluorescence detectors *Spectroscopy.* a class of instruments that can detect light emitted by atoms or molecules excited by absorbing electromagnetic radiation.

fluorescence enhancement *Immunology.* an increased fluorescence produced by some haptens when they are bound to an antibody; the energy is absorbed from the antibody and emitted with the wavelength characteristic of the hapten.

fluorescence immunoassay *Immunology.* a sensitive technique, similar to radio immunoassay, used for measuring antigen or antibody titers, in which fluoresceinated reagents are employed so that the fluorescent reagents may be detected immediately.

fluorescence microscope *Optics.* a microscope that irradiates naturally fluorescent materials or specimens with ultraviolet, violet, and occasionally blue light, causing the materials to fluoresce and reveal structures and details that might not otherwise be visible.

fluorescence process *Graphic Arts.* **1.** a process for reproducing watercolor art using flourescent watercolor pigments in the paint and special lighting in the color separation. **2.** a process for creating highlights in color reproduction through the use of a fluorescent white pigment that is illuminated with flashing lights during photographing.

fluorescence quenching *Immunology.* the reduction of the fluorescence emitted by an antibody; produced by combining the antibody with a specific partial antigen.

fluorescence spectrum *Spectroscopy.* a spectrum produced by the emission of radiation following absorption of and excitation by radiant energy.

fluorescence yield *Nuclear Physics.* the probability that a highly energized atom will emit a photon rather than an electron.

fluorescent [flô res´ənt; flů res´ənt] *Physics.* having or exhibiting the property of fluorescence.

fluorescent antibody *Immunology.* an antibody that is labeled by a specific dye and used in the immunofluorescence technique.

fluorescent antibody test *Immunology.* a test used to detect the presence of antigens, whether they are pathogens or tissue antigens.

fluorescent dye *Chemistry.* a dye that absorbs electronic radiation and reemits it at a longer wavelength, thus giving a brilliant color.

fluorescent lamp *Electronics.* an electric light consisting of a glass tube containing a small amount of mercury and a chemically inactive gas at low pressure, usually argon; the inside of the tube is coated with phosphors that absorb ultraviolet rays and change them to visible light. Fluorescent lamps are widely used in industrial and institutional sites.

fluorescent magnetic particle inspection *Metallurgy.* a quality control method based on magnetic particles that are coated with a fluorescent dye.

fluorescent map *Cartography.* a map reproduced with fluorescent ink or printed on fluorescent paper so that the user can read the map in darkness under ultraviolet light.

fluorescent penetrant inspection *Metallurgy.* a quality control method using a fluorescent liquid that penetrates into cracks and other flaws.

fluorescent pigment *Chemistry.* a pigment that absorbs electromagnetic radiation and releases the energy in a desired wavelength.

fluorescent rays *Radiology.* secondary rays that are generated by the primary ray photon bombardment of an absorbing material, causing rearrangement within the inner electron shells of atoms.

fluorescent screen *Engineering.* a screen painted with a fluorescent substance so that it will emit visible light when bombarded with ionizing radiation such as X-rays.

fluorescent staining *Cell Biology.* the process of labeling a biological component with a fluorochrome such as acridine orange or fluorescein that emits a visible signal when exposed to illumination of a specific wavelength, allowing microscopic visualization.

fluoridation [flôr´i dā´shən; flůr´i dā´shən] *Chemical Engineering.* the practice of adding fluorine ions to municipal water supplies in concentrations of 0.8–1.6 parts per million to help prevent tooth decay.

fluoride [flôr´īd; flůr´īd] *Inorganic Chemistry.* any binary compound containing fluorine, especially a salt of hydrofluoric acid, HF.

fluorination [flôr´i nā´shən; flůr´i nā´shən] *Chemistry.* the process of introducing fluorine into a chemical compound.

fluorine [flôr´ēn; flůr´ēn] *Chemistry.* a nonmetallic element having the symbol F, the atomic number 9, an atomic weight of 18.998, a melting point of –219°C, and a boiling point of –188°C; a member of the halogen family, the most electronegative element and the strongest oxidizing agent; exists as a pungent yellow diatomic gas or liquid; used in fluoride production, and in anionic form in fluoridation compounds for drinking water and toothpaste. (From a Latin word meaning "flowing.")

fluorine dating *Archaeology.* a method for determining the related ages of Pleistocene or Holocene fossil bones from the same excavation, based on the gradual combination of fluorine in groundwater with the calcium phosphate of buried bone material.

fluorite *Mineralogy.* CaF_2, a transparent to translucent cubic mineral occurring in many colors as cubic crystals or in granular form, having a vitreous luster, a specific gravity of 3.18, and a hardness of 4 on the Mohs scale; found in various deposits and in sedimentary rocks.

fluorite objective *Optics.* an objective consisting of one or two fluorite lenses, valued for its ability to form an image with improved chromatic correction; the image quality is superior to that of achromatic lenses and is usually slightly inferior to that of apochromats.

fluoro- a combining form denoting the presence of: **1.** fluorine or fluoride, as in *fluorocarbon.* **2.** fluorescence, as in *fluoroscope.*

fluoroacetate *Organic Chemistry.* an acetate in which fluorine atoms replace the carbon-connected hydrogen atoms.

fluoroacetic acid *Organic Chemistry.* CH_2FCOOH, highly toxic colorless crystals that are soluble in water and alcohol; melts at 33°C and boils at 165°C; used to kill rodents.

fluoroalkane *Organic Chemistry.* a straight-chain, saturated hydrocarbon in which some hydrogen atoms have been replaced by fluorine atoms.

***p*-fluoroaniline** *Organic Chemistry.* $FC_6H_4NH_2$, a toxic liquid that boils at 187.4°C and melts at –1.9°C; used in herbicides and plant-growth regulators.

fluorobenzene *Organic Chemistry.* C_6H_5F, a colorless liquid, insoluble in water and miscible in alcohol; boils at 84.9°C and melts at –40°C; used as an insecticide intermediate.

fluoroborate *Inorganic Chemistry.* **1.** any salt of fluroboric acid that contains the BF_4^- ion. **2.** any of a group of compounds related to borates, in which at least one oxygen atom is replaced by a fluorine atom.

fluoroboric acid *Inorganic Chemistry.* HBF_4, a colorless liquid, strongly acidic, that decomposes at 130°C; used in producing fluoroborates and for various other purposes.

fluorocarbon [flôr´ō kär´bən; flůr´ō kär´bən] *Organic Chemistry.* **1.** any of a number of dense, inert compounds analogous to a hydrocarbon in which some or all of the hydrogen atoms have been replaced by fluorine atoms; they are widely used as solvents, refrigerants, propellants, and lubricants, and for many other purposes. **2.** loosely, a chlorofluorocarbon (a fluorocarbon that contains chlorine). In the 1970s these substances were identified as being responsible for the depletion of the ozone layer, and certain ones were banned in the U.S.

fluorocarbon-11 see TRICHLOROFLUOROMETHANE.

fluorocarbon-12 see DICHLORODIFLUOROMETHANE.

fluorocarbon-21 see DICHLOROFLUOROMETHANE.

fluorocarbon fiber *Organic Chemistry.* a fiber produced from a fluorocarbon resin.

fluorocarbon resin *Organic Chemistry.* a dense, inert polymer made up of carbon and fluorine, with or without other halogens or hydrogen.

fluorochemical *Organic Chemistry.* any organic compound in which most of the hydrogen directly attached to carbon has been replaced by fluorine; used in solvents, sealants, surfactants, and heat-transfer liquids.

fluorochromasia *Cell Biology.* the process by which a cell develops intracellular fluorescence due to the uptake of a fluorogenic substrate such as fluorescein diacetate, which enters the cell and is hydrolyzed enzymatically, producing a fluorescent molecule such as fluorescein.

fluorod *Nucleonics.* a rod that absorbs ultraviolet light and emits orange fluorescent light that can be measured with a photomultiplier to indicate its level of radiation; used in solid-state dosimeters.

fluorodifen *Organic Chemistry.* $C_{13}H_7F_3N_2O_4$, a yellow crystalline compound melting at 93°C; used as an agricultural herbicide.

fluoroform *Organic Chemistry.* CHF_3, a colorless, odorless gas that boils at –84°C and melts at –163°C; used as a refrigerant, blowing agent, and organic intermediate. Also, TRIFLUOROMETHANE.

fluorogenic substrate *Chemistry.* a material in which fluorescence is induced by another substrate through the mixing of the two.

fluorologging *Petroleum Engineering.* a well-logging method in which well cuttings are analyzed under ultraviolet light for fluorescence radiation produced by trace amounts of oil.

fluorometer *Engineering.* a measuring device that determines the fluorescent radiation produced by a sample exposed to monochromatic radiation, usually radiation from a mercury-arc lamp or a tungsten or molybdenum X-ray source that has traveled through a filter.

fluorometric analysis *Analytical Chemistry.* a technique for concentration determinations in which the sample is exposed to radiation and the amount of fluorescence emitted due to the transition of particles to a lower energy state is measured.

fluorophosphoric acid *Inorganic Chemistry.* H_2PO_3F, a viscous, colorless liquid that mixes with water and is used in metal cleaners, polishes, and protective coatings.

fluoroplastic *Materials.* a plastic composed of linear polymers with some or all of the hydrogens replaced by fluorine; fibroplastics are white, have a waxy feel, are strong, and are available in a range from rigid solids with excellent chemical resistance to rubbers; they are used commercially for such products as machine parts, wire coatings, and elastic insulators. Also, **fluorine plastic.**

fluoroscope [flôr′ə skōp′; flůr′ə skōp′] *Radiography.* **1.** a fluorescent screen that shows shadow images of an object placed between it and an X-ray device. **2.** an instrument that employs X-rays to record the inside of an opaque form for observation, such as the human body. **3.** to examine with a fluoroscope.

fluoroscopic *Radiography.* relating to or using a fluoroscope.

fluorosilicate *Inorganic Chemistry.* a salt derived from fluorosilicic acid and containing the SiF_6^{-2} radical.

fluorosilicic acid *Inorganic Chemistry.* H_2SiF_6, a colorless, fuming, extremely corrosive liquid that attacks stoneware and glass; soluble in water and decomposed by heat. It is used in water fluoridation, ceramics, disinfecting, electroplating, and for many other industrial purposes.

fluorosulfonic acid *Inorganic Chemistry.* HSO_3F, a colorless, fuming, highly irritating liquid that reacts violently with water; soluble in cold water; melts at –87.3°C and boils at 165.5°C; used as a catalyst, in electropolishing, and as a fluorinating agent.

fluorothene see CHLOROTRIFLUOROETHYLENE POLYMER.

fluorouracil *Oncology.* $C_4H_3FN_2O_2$, an antineoplastic antimetabolite that inhibits the proliferation of cancerous cells; used to relieve the discomfort of breast and gastrointestinal cancers, as a topical treatment for skin cancers, and in combination with other drugs in chemotherapy.

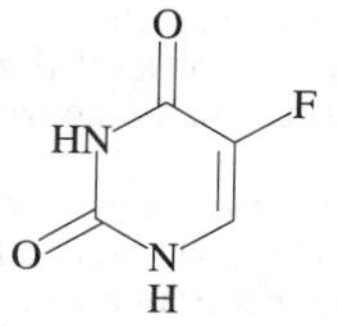

fluorouracil

fluorphor see LUMINOPHOR.

Fluosol *Hematology.* a chemical blood substitute that, when administered with oxygen, fulfills the functions of hemoglobin, but lacks clotting factors and other properties of blood.

fluosolid system *Metallurgy.* a system of ore treatment in which ore particles are kept in continuous suspension by a controlled stream of air. Also, FLUIDIZED BED.

flush *Engineering.* to remove deposits of rock fragments and other debris by flushing them with a high-velocity stream of water. *Design Engineering.* describing separate surfaces that are on the same level. *Graphic Arts.* describing a type character, image, and so on that is positioned directly against the left margin (**flush left**) or against the right margin (**flush right**). *Computer Science.* **1.** to clear a buffer by writing out or transmitting its contents. **2.** to discard the remaining data in a buffer. *Ecology.* **1.** a term for a sudden and dramatic increase in the number of plants in a community, such as the growth that occurs after a fire. **2.** any sudden, sizeable increase in the size of a population.

flush bead see QUIRK BEAD.

flush bolt *Building Engineering.* a sliding bolt that is recessed into the slider edge of a door to make it flush with the frame.

flush coat *Civil Engineering.* a coating, usually made of bituminous material, that is used to waterproof a surface.

flush cover *Graphic Arts.* a book cover, usually softbound, that is aligned with the book's pages with no overhang.

flush gate *Civil Engineering.* a gate that lies below the gate of a dam and is designed to release water for flushing a channel.

flush head *Graphic Arts.* a heading of one or more lines aligned flush left.

flushing *Civil Engineering.* the removal or reduction to a permissible level of suspended or dissolved solids or contaminants in an estuary or harbor.

flushing period *Hydrology.* the theoretical amount of time necessary for a quantity of water equal to the entire volume of a given lake to pass through its outlet.

flush joint *Engineering.* **1.** a mortar joint fabricated in flat pointing. **2.** a tongue-and-groove joint used to create a smooth, even join.

flush-joint casing *Petroleum Engineering.* a smooth joint, flush with the outer diameter of the remainder of a section length, formed by connecting lengths of casing end to end.

flushometer *Mechanical Devices.* a valve that when activated releases the exact amount of water to flush a fixture.

flush panel *Building Engineering.* a panel surface flush with the face of each stile.

flush production *Petroleum Engineering.* the first high yield from an oil well during the greatest productive period.

flush soffit *Building Engineering.* the continuous surface beneath a ceiling or stair.

flush tank *Civil Engineering.* **1.** a temporary storage tank for water and sewage for periodic release through a sewer. **2.** a small tank filled with water to flush a toilet.

flute a channel, groove, or furrow; specific uses include: *Design Engineering.* a channel or groove in a reamer, tap, or drill, especially when it is oriented parallel to the main axis of conical-shaped pieces. *Architecture.* a groove of a curved section, especially one of a series of parallel curved grooves used to decorate a column shaft. *Geology.* **1.** a shallow groove or channel formed by differential weathering and erosion, and running down the face of a relatively vertical rock surface. **2.** a small, spoon- or scoop-shaped hollow or groove with a steep end upcurrent, formed in a sedimentary structure by the scouring action of sediment-laden water currents over a muddy surface.

flute cast *Geology.* a raised, oblong or subconical bulge on the underside of a siltstone or sandstone bed, formed by the filling in of a flute. Also, SCOUR CAST, TURBOGLYPH.

fluted cutter *Mechanical Devices.* a cutter design with milling flutes for tap and reamer workpieces.

fluted point *Archaeology.* a projectile point that has a long, medial channel notched to the base of the flake.

fluted reamer *Design Engineering.* a reamer that is fluted longitudinally to cut at its sides.

flute length *Design Engineering.* on a twist drill, the length between the outside corners of the cutters to the point furthest to the back end of the flutes.

flutemouth see CORNETFISH.

fluting *Mechanical Engineering.* **1.** a set of parallel channels or grooves having a helical or longitudinal orientation. **2.** the machining process of cutting such grooves in a cylindrical or conical part such as a tap or reamer. *Building Engineering.* a pattern of long vertical grooves (flutes) cut in the surface of a column or other member.

fluting plane *Mechanical Devices.* a plane used to cut grooves on a rounded workpiece

flutter *Medicine.* a quick, irregular motion, especially one that is abnormal and that interferes with normal function, as in the heartbeat. *Acoustics.* rapid and inappropriate fluctuations in the pitch of a recorded sound, due to variations in recording or playback speed; usually caused by mechanical defects in the recording or reproduction equipment. *Electromagnetism.* a distortion or fluctuation in the strength of a signal caused by interaction with another signal or frequency. *Engineering.* an irregular motion in sections of a relief valve that arises when there is pressure but no contact between the valve disk and the seat. *Aviation.* a self-excited vibration in an aeroelastic environment wherein the vibration draws energy from the airstream and depends on the aerodynamic inertial, elastic, and dissipative properties of the system.

flutter echo *Acoustics.* repetitious rapid fluctuations in fidelity, gradually diminishing; often due to two large parallel surfaces in a recording room, or a curvature in one of the surfaces. *Electromagnetism.* a rapid succession of echoes resulting from a single transmitted pulse.

flutter valve *Engineering.* a valve that is regulated by pressure variations found in the material flowing over it.

fluvarium *Engineering.* an aquarium in which the water supply in the tanks is maintained by gravity, not by pumping action.

Fluvent *Geology.* a suborder of the soil order *Entisol,* formed in recently deposited alluvium and characterized by good drainage and a carbon content that decreases with depth.

fluvial *Hydrology.* **1.** relating to or produced by a river or the action of a river. **2.** situated in or near a river or stream. *Biology.* living or growing in or near a river or stream. (From *fluvius,* the Latin word for river.)

fluvial cycle see NORMAL CYCLE.

fluvial deposit *Geology.* an accumulation of sedimentary material carried by, suspended in, and deposited by a stream.

fluvial sand *Geology.* a sand particle ranging between 0.05 and 2.0 mm in diameter that is laid down by a stream.

fluvial soil *Geology.* any soil that has been laid down by a river or a stream.

fluviatile *Geology.* relating to, belonging to, or peculiar to rivers, especially the physical products of river action.

fluviology *Hydrology.* the scientific study of rivers.

fluviomorphology see RIVER MORPHOLOGY.

flux *Physics.* the measure of the flow of some quantity per unit area per unit time. *Chemistry.* any substance that will promote the melting of another substance to which it is added. *Materials.* a resin or similar substance that is used in soldering, welding, or brazing in order to remove oxides from the surfaces to be joined and thus promote their bonding. *Metallurgy.* in liquid metal processing, a nonmetallic material that is used to protect the metal and remove impurities. *Electromagnetism.* the electric or magnetic field lines of force that traverse a given cross-sectional area.

flux-cored welding *Metallurgy.* welding using a filler metal that contains a flux in its core.

flux density *Nucleonics.* the amount of a given type of radiation that crosses a specified area within a specified period, such as the number of photons passing through one square centimeter of a target in one second. *Physics.* a vector field that represents the differential flux of field lines per unit area.

flux-density mapping *Nucleonics.* a process by which radiation flux density is tracked within a reactor or other radiation source.

flux factor *Metallurgy.* a quality rating of silica refractories.

flux gate *Engineering.* a detector that produces an electric signal with magnitude and phase proportional to the magnetic field along its axis; used to demonstrate the direction of the terrestrial magnetic field.

flux guide *Metallurgy.* in induction heating, an equipment component that guides the magnetic flux to preferred locations.

fluxing *Metallurgy.* in several metallurgical operations, the intentional formation of a low-melting, nonmetallic material that aids processing.

fluxing ore *Metallurgy.* an ore containing constituents that promote fluxing.

flux leakage *Electromagnetism.* any magnetic flux that does not pass through the part of a magnetic circuit where it is needed.

flux line see METAL LINE.

flux linkage *Electromagnetism.* the passage of magnetic flux produced by one component or circuit through another component or closed circuit, creating a magnetic interaction between the two.

fluxmeter *Engineering.* an instrument used for the measurement of magnetic flux.

flux of energy see ENERGY FLUX.

flux path *Electromagnetism.* a path in physical space that a magnetic flux line follows.

flux pinning *Materials Science.* the prevention of flux movement and resulting quenching of a superconducter by the use of ultrathin filamentary composites that lower the risk of flux jump.

flux pump *Physics.* a cryogenic generator capable of changing small inputs of alternating current into large outputs of direct current.

flux refraction *Electromagnetism.* an abrupt change in direction of magnetic flux lines at the interface between two materials having different magnetic permeabilities.

flux sensor *Robotics.* a sensor that measures the flow of energy as a fluid.

flux unit see JANSKY.

fly *Invertebrate Zoology.* **1.** the common name for a number of species of the order Diptera, including the common housefly, blowfly, fruit fly, midge, and mosquito, characterized by one pair of wings and one pair of balancers (modified wings used for equilibrium). Found throughout the world, flies are among the most dangerous pests known, carrying germs inside their bodies or in the hair on their bodies that cause such diseases as malaria, sleeping sickness, and typhoid fever. **2.** specifically, the common housefly, *Musca domestica.* **3.** in popular use, any of various other flying insects that are not true flies, such as the mayfly or firefly. *Mechanical Engineering.* **1.** a horizontal arm that is weighted at each end and pivots about the screw of a press; when the screw is lowered, the momentum of the fly increases the force of the press. **2.** a multiblade fan used in light machinery to control speed by means of air resistance.

fly ash *Engineering.* finely divided residue, essentially noncombustible refuse, carried by the combustion gases from a furnace.

flyback *Electronics.* **1.** a sudden decrease in previously rising current or voltage. **2.** the rapid return of an electron beam to its starting position in a television picture tube or oscilloscope.

flyback power supply *Electronics.* in television equipment, a power supply that uses a flyback signal to generate high voltage.

flyby *Aviation.* a flight by an aircraft at an altitude low enough to be observed from the ground. *Space Technology.* the flight of a probe or a spacecraft close enough to a planet or other celestial object to collect scientific data or to receive a gravity assist from the planet.

flycatcher *Vertebrate Zoology.* any of several species of perching birds that dart in the air to capture insects, including the Old World songbird family Muscicapidae and the New World family Tyranidae.

flycatcher

fly cutter *Mechanical Engineering.* a single-point cutting tool, used in a milling machine to produce a flat surface. Thus, **fly cutting.**

flying angle *Aviation.* in normal level flight, the angle of attack of a wing, or the acute angle between the horizontal and an aircraft's longitudinal axis.

flying-aperture scanner *Electronics.* a component in a computer that converts text into electrical signals by flooding the text with light, then retrieving the light spot by spot from the illuminated text.

flying boat *Aviation.* a seaplane characterized by a boatlike hull or fuselage that allows the craft to float, take off, and land on water.

flying bomb *Ordnance.* **1.** originally, an explosive robot plane, designated V-1, used by Germany during World War II. **2.** in modern usage, any explosive aircraft, missile, or bomb that flies under its own power and is controlled by external or internal means.

flying bridge *Naval Architecture.* the topmost of a ship's command and conning bridges.

flying buttress *Architecture.* in Gothic architecture, a masonry brace (usually a straight bar carried on an arch) that takes the thrust of a vault to an outer buttress.

flying crane *Aviation.* **1.** a large helicopter powered by jet engines and designed to lift heavy loads; used especially to expedite the loading and unloading of military cargo ships. **2. Flying Crane.** a popular name for one such helicopter, the H-17.

flying dragon *Vertebrate Zoology.* a common name for the arboreal lizard of the Old World family Agamidae, genus *Draco,* characterized by an extensible membrane between the limbs along each side, enabling it to make long, gliding leaps through the air; found in southeastern Asia and the East Indies.

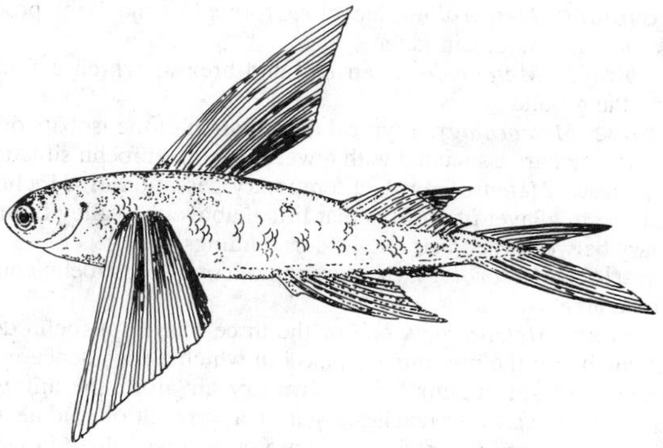

flying fish

flying fish *Vertebrate Zoology.* any of numerous fish of the family Exocoetidae, having stiff, greatly elongated pectoral fins suggesting wings that enable it to glide considerable distances through the air after leaping from the water; found chiefly in tropical and warm seas

Flying Fortress *Aviation.* a popular name for the B-17 bomber.

flying fox *Vertebrate Zoology.* any of the large fruit bats of the genus *Pteropus,* characterized by a large head, a wingspread that may be over 5 ft, and a face that resembles a fox; found in most tropical regions except South America.

flying gurnard see DACTYLOPTERIDAE.

flying head *Electronics.* a read/write head on a magnetic disk or drum that floats a microscopic distance off the moving magnetic surface.

flying level *Cartography.* a level line with relatively low accuracy, run in one direction at the close of a working day to detect large mistakes.

flying saucer *Astronomy.* a popular name for an unidentified flying object. (Early reported sightings of UFOs often decribed them as having a saucerlike shape.)

flying scaffold *Building Engineering.* a temporary support between two buildings after a structure between them has been removed.

flying shear *Metallurgy.* a shear that cuts a continuously rolled product on the flight, that is, without interrupting the rolling operation.

flying shore *Building Engineering.* a horizontal scaffolding rig used as a temporary support between two walls that face each other and are less than 9 meters apart.

flying snake *Vertebrate Zoology.* a gold and black tree-dwelling snake, *Chrysophelea ornata,* of the family Colubridae, characterized by the ability to jump from trees and glide through the air for short distances by drawing in its ventral scales; its prey includes bats, rodents, and lizards; three slender species of the genus *Chrysopelea* are found in Asia and the East Indies.

flying spot *Electronics.* the tiny point of light in a flying-spot scanner.

flying-spot microscope *Optics.* a microscope that sweeps a small, intensely bright spot of light systematically over a transparent specimen and simultaneously creates and displays a greatly magnified image of the specimen on a cathode-ray tube.

flying-spot scanner *Electronics.* a device in which a rapidly moving spot of light scans an image on a transparent screen; a phototube then absorbs the light reflected by the screen and converts it into electric signals; used for television film, slide transmission, and character recognition. Also, OPTICAL SCANNER.

flying squirrel *Vertebrate Zoology.* the common name for two groups of rodents: the subfamily Petauristinae in the family Sciuridae from North America and Eurasia, and the family Anomaluridae, a scaly-tailed group from Africa; both have a wide parachutelike membrane attached between fore- and hindlimbs and jump from trees with all limbs spread outward; may glide up to 200 feet.

flying switch *Engineering.* the occurrence of a railroad car disconnecting from the locomotive to move onto another track while both are traveling under their own momentum.

flying test bed *Aviation.* an aircraft, rocket, or other flight vehicle used to test other objects under specified conditions. Also, TEST BED.

flying veins *Geology.* a network of mineral deposits that overlap and intersect in a branchlike pattern.

fly press see SCREW PRESS.

fly rock *Engineering.* pieces of rock that are scattered when a quarry or tunnel is blasted.

Flysch [flish] *Geology.* an extensive late Cretaceous to Oligocene sedimentary formation bordering the Swiss Alps and associated with the Alpine orogeny.

flysch *Geology.* a sedimentary sequence consisting of marine deposits of dark, thinly bedded, fine-grained marls, sandstones, shales, clays, and muds.

flyway *Vertebrate Zoology.* the specific air route along which birds customarily migrate between their breeding and wintering areas.

flywheel *Mechanical Engineering.* a heavy wheel that rotates on a shaft so that its momentum imparts uniform rotational velocity to the shaft and attached machinery. Also, BALANCE WHEEL.

flywheel synchronization *Electronics.* a method of automatically synchronizing a television scanning system, in which the receiver evaluates synchronization signals from the transmitter and replaces weak signals with pulses based on the overall rate of reception.

F.M. make a mixture. (From Latin *fiat mistura.*)

FM frequency modulation; frequency modulated.

Fm fermium.

fm. fathom.

***F* martingale** *Mathematics.* let $F = \{F_t, t \geq 0\}$ be an increasing family of sigma algebras. An F martingale is an F process $\{X_t, t > 0\}$ for which the conditional expectation $E(X_t \mid F_s) = X_s$ whenever $s < t$. When all the F_t are identical, then the martingale is an ordinary (continuous parameter) martingale. The family F can be viewed as representing an increase in the amount of information over time.

f-mediated transduction *Genetics.* the induction of a genetic change in a bacterium by the introduction of new genes from another bacterium through the process of conjugation.

FMN flavin mononucleotide.

FMY *Aviation.* the airport code for Page Field, Fort Myers, Florida.

FN *Ordnance.* one of several automatic weapons and machine guns developed or manufactured by the Belgium arms company, Fabrique Nationale d'Armes de Guerre; the FN 7.62-mm light automatic rifle (FAL) is widely used.

fnp fusion point.

FNT *Aviation.* the airport code for Flint, Michigan.

***f*-number** *Optics.* an expression representing the light-passing power of a lens or lens system; the ratio of the focal length of a lens to its entrance pupil diameter, written in various forms such as $f/8, f{:}8, f8$. Also, FOCAL RATIO, STOP NUMBER.

foal *Vertebrate Zoology.* **1.** the young of an animal of the horse family, especially one that is under one year of age. **2.** to give birth to young, particularly of an animal of the horse family.

foam *Chemistry.* a dispersion of a gas in a liquid or solid, such as shaving cream or sponge rubber.

foam crust *Hydrology.* a feature of a snow surface that resembles small overlapping waves or sea foam on a beach.

foam drilling *Mining Engineering.* a method of dust suppression in which a thick foam is pumped through a drill by compressed air, and a thick sludge of foam and dust emerges from the mouth of the hole.

foaminess *Physics.* in a process of producing foam in a liquid by pumping air through it, the volume of foam produced in cubic centimeters divided by the rate of the flow of air in cubic centimeters per second.

foaming *Engineering.* any process that results in the production of foam material, whereby air or gas is mixed into a solid or liquid substance.

foam-in-place *Engineering.* a process of covering a surface with foam, whereby reactive foam ingredients are placed directly onto the surface and the foam-making process takes place on the surface to be covered.

foam line *Oceanography.* the front of a wave after it has broken and is moving shoreward.

foam mark *Geology.* a sedimentary feature characterized by a pattern of barely visible ridges and hollows, produced by the action of wind-driven sea foam on the surface of wet sand.

foam metal *Metallurgy.* an intentionally highly porous metal or alloy.

foam rubber *Materials.* a flexible foam produced by beating air into unvulcanized latex or by combining a carbonate with a strongly masticated rubber mixture; used in mattresses, cushions, and carpet pads. Also, RUBBER SPONGE.

Foamy virus *Virology.* see SPUMAVIRINAE.

focal *Science.* relating to or located at a focus.

focal curve *Mathematics.* a curve in the solution space of a first-order partial differential equation that has a characteristic direction at every point and that satisfies the strip condition.

focal distance see FOCAL LENGTH.

focal epilepsy *Neurology.* minor epileptic seizures that are predominantly one-sided or local, or that present localized features.

focal infection *Medicine.* an infection in a limited area, such as the tonsils, teeth, sinuses, or prostrate.

focal length *Optics.* the distance, usually expressed in millimeters, from the principal point of a lens or concave mirror to its focal point. Also, FOCAL DISTANCE.

focal plane *Optics.* a plane that is perpendicular to the principal axis and passes through the focal point of the axis of a mirror, lens, or lens system; the surface on which the sharpest image is formed.

focal-plane shutter *Optics.* a camera shutter consisting of one or more blinds of fabric or metal having an adjustable slot that moves rapidly across, and thus exposes, the film or plate.

focal point *Optics.* the point on the axis of a lens or mirror at which all incident parallel light rays converge or appear to diverge; a mirror has one focal point, whereas a lens has two focal points, one on each side. Also, FOCUS, PRINCIPAL FOCUS. *Mathematics.* see FOCUS.

focal power *Optics.* a measure of the ability of a symmetrical optical system to converge light beams, equivalent to the reciprocal of the focal length of that system. Also, SPEED.

focal ratio see *F*-NUMBER.

focal seizure *Medicine.* a transitory motor nerve disturbance caused by abnormal electrical activity in the brain, commonly beginning as spasms in the face, hand, or foot and spreading to other muscles.

focal spot *Radiology.* the part of the anode of an X-ray tube to which an electron stream is directed. *Metallurgy.* in electron beam or laser metal processing, the area where intensity is highest.

focus *Mathematics.* the fixed point or points in the plane that, together with the directrix or directrices, determine a conic section. Also, FOCAL POINT. *Optics.* **1.** see FOCAL POINT. **2.** to adjust the components or the position of an optical or photographic instrument so that an image appears distinct. *Geology.* the true center of an earthquake, within which the strain energy is first converted to elastic wave energy. *Electronics.* to converge or diverge electron beams by varying the voltage or current in the circuit that controls the magnetic or electric field through which the beams must pass.

focus control *Optics.* a mechanism that adjusts the focus of an image formed by a lens system. *Electronics.* a device that sharpens the image on the screen of a CRT tube by varying the current flowing through the focusing coil or by changing the position of a permanent magnet.

focused-current log *Engineering.* a resistivity log that is derived from a multielectrode arrangement.

focused grid *Radiology.* a radiographic grid composed of thin lead strips angled to focus on a line at a specific distance.

focus-forming virus *Virology.* any virus that forms foci of transformed cells in cell cultures.

focusing anode *Electronics.* an external anode used to adjust the electron beam in a cathode-ray tube.

focusing coil *Electronics.* a coil that generates a magnetic field parallel to an electron beam in order to focus the beam.

focusing collector *Engineering.* a device that gathers sunlight onto semicircular aluminum reflectors and directs the rays onto copper pipes to keep water circulating.

focusing electrode *Electronics.* the terminus to which voltage is applied in order to focus an electron beam in a cathode-ray tube.

focusing glass *Optics.* a magnifying glass that enlarges the image projected onto the viewfinder of a camera.

focusing magnet *Electronics.* a magnet that is used to create a magnetic field in order to focus an electron beam in a cathode-ray tube.

focusing scale *Optics.* a scale marked on the focusing mechanism of an optical or photographic instrument; used to set its lens-to-image distance.

focus projection and scanning *Electronics.* in a hybrid vidicon, the control of the movement of the electron beam by using a transverse electrostatic field to deflect it and an axial magnetic field to focus it.

fodder *Agriculture.* anything that is fed to livestock, especially dry roughage such as hay or straw.

foehn or **föhn** [fān] *Meteorology.* a warm, dry wind on the lee side of a mountain range, resulting from the adiabatic compression of the air descending the slopes.

foehn air *Meteorology.* the very warm, dry air characteristic of foehn winds.

foehn cloud *Meteorology.* any cloudform associated with a foehn that is of the lenticular species forming in the lee wave parallel to a mountain ridge.

foehn cyclone *Meteorology.* a cyclone formed by the foehn process on the lee side of a mountain range.

foehn island *Meteorology.* an isolated area at which a foehn has reached the ground.

foehn nose *Meteorology.* a typical deformation of the isobars on a synoptic surface chart associated with a well-developed foehn situation.

foehn pause *Meteorology.* **1.** a temporary cessation of a foehn at the ground due to a layer of cold air that lifts it above the valley floor. **2.** the boundary between foehn air and its surroundings.

foehn period *Meteorology.* the duration of continuous foehn conditions at a given area.

foehn phase *Meteorology.* one of the three phases of foehn development, which are: the preliminary phase in which a subsidence inversion separates cold surface air from warm dry air aloft; the anticyclonic phase in which warm air reaches a station as a result of cold air flowing out of the plain; and the stationary phase or cyclonic phase in which the foehn wall forms and the downslope wind becomes appreciable.

foehn storm *Meteorology.* a destructive storm during which atmospheric pressure fluctuates rapidly, frequently occurring with a high foehn in the Bavarian Alps in October.

foehn wall *Meteorology.* the steep leeward boundary of cumuliform clouds that form on the peaks and upper windward side of mountains during foehn conditions.

FOFA *Ordnance.* the use of long-range weapons to interdict a battlefield from forces following up an initial attack. (An acronym for follow on forces attach.)

fog *Meteorology.* the suspension of a visible aggregate of minute water droplets or ice crystals in the atmosphere near the earth's surface, reducing visibility to below 0.62 mile (1 kilometer). *Graphic Arts.* on a photographic negative, the veiling of an image by undeveloped silver deposits; usually caused by improper lighting or chemical imbalance in the developing solution.

fog bank *Meteorology.* a well-defined mass of fog observed in the distance, most often at sea.

fogbound *Navigation.* unable to proceed, due to dense fog.

fogbow *Meteorology.* a faintly colored circular arc formed when sunlight reflects off fog layers, consisting of drops approximately 100 micrometers or less in diameter; it resembles a rainbow except that its colors are very faint or nonexistent. Also, WHITE RAINBOW.

fog climax *Ecology.* a community whose stability is dependent on the presence of a fog bank.

fog deposit *Hydrology.* a coating of ice formed on an exposed surface by the freezing of fog.

fog drip *Hydrology.* water that drips to the ground from objects, such as leaves and branches, on which the water has collected during a fog.

fog drop *Meteorology.* an individual fog particle constituted exactly the same as a cloud drop. Also, **fog droplet.**

fog fever *Veterinary Medicine.* a colloquial name for acute respiratory distress commonly affecting adult beef cattle; it is associated with a change in grazing from dry to lush pasture (**fog pasture**).

fogged metal *Metallurgy.* a metal or alloy made dull by corrosion.

fog horizon *Meteorology.* the top of a fog layer trapped by a low-level temperature inversion so that it resembles a horizon and obscures the true horizon.

foghorn *Navigation.* a deep, loud horn used as an onshore warning signal to ships during foggy weather.

fog quenching *Metallurgy.* the process of quenching in a mist.

fog scale *Meteorology.* a classification based on the effectiveness of fog intensity in decreasing horizontal visibility.

fog signal *Navigation.* any of various sound signals used as a warning for vessels navigating during periods of fog or other reduced visibility. Thus, **fog gun, fog siren, fog whistle,** and so on.

fog track *Nucleonics.* the condensation that appears in supersaturated water vapor when charged particles pass through it; used to study the movement of particles in a cloud chamber.

fog wind *Meteorology.* a term for a humid wind that crosses the Andes east of Lake Titicaca and descends in violent squalls.

foid *Mineralogy.* the feldspathoid group of minerals.

foil *Metallurgy.* a metallic sheet or strip, thinner than a specified amount, usually 5/1000 inch or less.

foil decorating *Engineering.* a process that produces a decorative plastic object when printed paper, textile, or plastic foil is molded into the plastic object and can be seen beneath the surface of the object.

foil dosimeter *Nucleonics.* an instrument that determines radiation doses from the level of activation in a metal foil in the radiation field.

Fokker-Planck equation *Physics.* an equation describing the time dependence of a Markov process, and showing the rate of change of the distribution function of a gas whose particles undergo small-angle deflections characteristic of long-range interactions. It is expressed by

$$\sum_{i} \partial/\partial x_i (A_i \cdot f) - {}^1/_{2!}\sum_{i,j} \partial^2/\partial x_i \partial x_j (B_{ij} \cdot f) + {}^1/_{3!} \sum_{i,j,k} \partial^3/\partial x_i \partial x_j \partial x_k (C_{ijk} \cdot f) - \cdots = 0$$

where A, B, C, etc. are limiting values of the first, second, third, etc., moments of Δx as the time increment Δt of the Markov process goes to zero. This is equivalent to the Chapman-Kolmogorov integral equation. Also, KOLMOGOROV FORWARD EQUATION.

FOL first-order logic.

fold *Anatomy.* a thin, recurved margin, or doubling. *Geology.* a bend or buckle in bedded sedimentary rock or other planar structures, usually produced by deformation. *Metallurgy.* in metal working, an undesirable interface caused by the folding over of a small portion of a material without welding. Also, LAP.

foldback DNA *Molecular Biology.* a region of single-stranded DNA that has renatured by intrastrand reassociation between inverted base sequence repeats, creating a folded-back appearance.

foldbelt see OROGENIC BELT.

folded and gathered see F AND G'S.

folded cavity *Electronics.* in a klystron, an arrangement of devices that force the incoming wave to act on the electron stream at several points.

folded chain model *Materials Science.* a newer model (replacing the fringed-micelle model) that pictures sections of the molecular chains folding on themselves to form a transition from crystalline to noncrystalline regions.

folded chain structure *Materials Science.* a type of molecular order in crystalline polymers in which the chain is oriented normal or very nearly normal to the plane of the lamellae.

folded-dipole antenna *Electromagnetism.* a dipole antenna consisting of two parallel half-wave dipole elements closely spaced and joined at their ends, with one that is fed at its center.

folded horn *Acoustical Engineering.* a horn whose axis is bent or curved so as to provide a longer path over a given volume.

folded-plate roof *Building Engineering.* a roof constructed of reinforced concrete fabricated in flat plates and joined at various angles.

folded rock *Geology.* rock strata that shows a curve or bend, usually the product of deformation due to the contact between various geologic processes.

folded rock

folding *Geology.* the bending of strata, usually due to compression; of four principal types: flexure folding, flow folding, shear folding, and folding due to vertical movements. *Computer Programming.* **1.** a hashing technique in which the original key is split into two or more parts, then added together to form the more compact transformed key. **2.** the process of breaking up a large program into sections, segments, or overlays because it requires more memory space than is available. **3.** a technique for substituting one character set for another, usually in order to map a large character set into a smaller one. **4.** a computation of an operation on constants at compile time: constant folding.

folding fin *Space Technology.* **1.** a fin hinged axially to permit clearance, as on aircraft in a hangar or below deck on an aircraft carrier, or on a missile so that it can lie flat prior to a launch. **2.** a fin hinged transversely to emerge from a housing slot of a rocket or missile.

foldover *Electronics.* a distortion in which a white line frames a television screen, usually caused by a defect in the receiver's horizontal or vertical deflection circuits.

fold system *Geology.* a group or area of folds having the same tectonic origin and showing common characteristics and trends.

fold testing *Materials Science.* a flexural test that yields information on the long-term dynamic strength of a material; often used on polymers in film form.

Foleyan *Geology.* a North American Gulf Coast geologic stage of the Pliocene epoch, occurring after the Clovelly and before the Pleistocene.

Foley catheter *Surgery.* a type of balloon catheter that is used for drainage or irrigation of the bladder. (From Frederic *Foley,* 1891–1966, American urologist.)

folia *Petrology.* thin, leaflike layers or laminae occurring in gneissic or schistose rocks.

foliaceous *Botany.* of or relating to a leaf; leafy. Also, **foliate(d).**

foliage [fō′lij; fō′lē ij] *Botany.* the leaves of a plant.

foliar *Botany.* relating to, consisting of, or applied to leaves.

foliated ice *Hydrology.* a frequently wedge-shaped large mass of ice that occupies a crack formed as a result of thermal contraction in permafrost. Also, **foliated ground ice.**

foliate papilla *Vertebrate Zoology.* one of many small protuberances found on the upper surface of the tongue of many mammals, but vestigial or absent on the human tongue.

foliation *Botany.* **1.** the process of putting forth or growing leaves. **2.** of a plant, being in leaf. *Geology.* the planar arrangement of minerals, mineral bands, or other textural or structural features in metamorphic or other rocks. *Metallurgy.* the production of a metallic foil.

foliation *Mathematics.* let M be a smooth manifold of dimension n. A foliation F on M of dimension p and codimension q (i.e., $p + q = n$) is a partition $\{L_\alpha\}$ (called leaves) of M into connected subsets with the following property: For every point of M there is an open neighborhood U and a chart $(U, (x, y))$, where $(x, y) = (x_1, \ldots, x_p, y_1, \ldots, y_q)$, such that for each leaf L_α the connected components of $U \cap L_\alpha$ are defined by the equations $y =$ const, i.e., $y_i = constant$ for $i = 1, 2, \ldots, q$. The following diagrams commute:

$$\begin{array}{ccc} U & \xrightarrow{(x,Y)} & R^p \times R^q \\ {\scriptstyle x}\downarrow & & \downarrow{\scriptstyle (x,y) = (x_1,\ldots,x_p,y_1,\ldots,y_q)} \\ R^p & & \end{array} \qquad \begin{array}{ccc} U & \xrightarrow{(x,Y)} & R^p \times R^q \\ {\scriptstyle y}\downarrow & & \downarrow{\scriptstyle (x,y) = (x_1,\ldots,x_p,y_1,\ldots,y_q)} \\ R^q & & \end{array}$$

folic acid *Biochemistry.* $C_{19}H_{19}N_7O_6$, a vitamin in the B complex that is synthesized by intestinal bacteria and present in a variety of foods, including green leafy vegetables; a deficiency in its intake leads to poor growth and anemia.

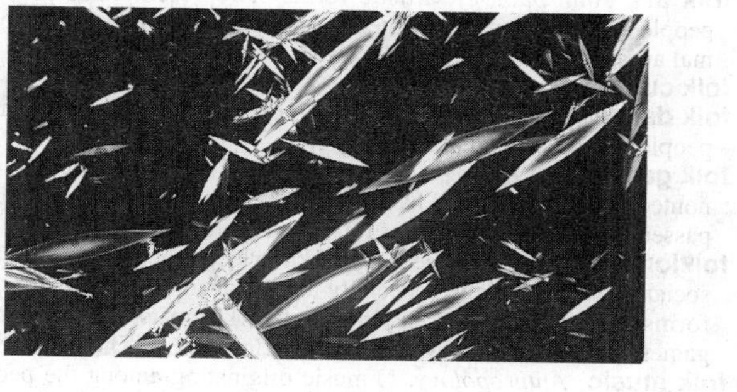

folic acid

folic acid sodium salt see SODIUM FOLATE.

folie [fä lē′] *Psychology.* the French term for insanity; often used in combination with other words to name mental disorders.

folie à deux [fä lē′ a doo′] *Psychology.* the simultaneous occurrence of the same mental disorder in two people who are closely associated, when one appears to have influenced the other. (From French; "double insanity.") Similarly, **folie à trois, folie à quatre.**

folie du doute [fä lē´ də dout´] *Psychology.* a disorder that is characterized by the inability to carry out even trivial actions without constantly verifying the result, e.g., repeatedly checking whether a door is locked or an appliance is turned off. (From French, "insanity of doubt.")

foliicolous *Biology.* growing on or parasitic on leaves, such as some fungi.

folimat *Organic Chemistry.* $C_5H_{12}NO_4PS$, an oily liquid that is soluble in water and decomposes at 135°C; used as an agricultural and horticultural insecticide.

folinic acid *Biochemistry.* $C_{20}H_{23}N_7O_7$, an important metabolite of folic acid occurring as slightly water-soluble crystals that decompose at 240–250°C; used in nutrition, medicine, and biochemical research. *Pharmacology.* this derivative of folic acid as a necessary component for the growth of *Leuconostoc citrovorum;* also used to obtain its calcium salt, leucovorin calcium, which is used to treat megaloblastic anemia. Also, LEUCOVORIN.

folio *Graphic Arts.* **1.** a page number of a book or magazine. **2.** a sheet of paper measuring 17 by 22 inches. **3.** any folded sheet of paper. **4.** a book printed on folded sheets.

foliobranchiate *Vertebrate Zoology.* having gills that resemble leaves.

foliolate *Botany.* having leaflets.

foliose *Botany.* **1.** having a leaflike appearance, especially describing lichens and algae that have a leaflike extension of the thallus. **2.** leafy.

Folist *Geology.* a suborder of the soil order Histosol, characterized by wet accumulations of organic matter less than one meter deep overlying rock or rubble.

folium of Descartes *Mathematics.* the graph in the plane of the curve with Cartesian equation $x^3 + y^3 - 3axy = 0$; the parametric representation is $x = 3at/(1 + t^3)$, $y = 3at^2/(1 + t^3)$, for $t \neq -1$. The origin (0, 0) is a double point of the curve; the vertex is $(3a/2, 3a/2)$; and the asymptote is the line $x + y + a = 0$.

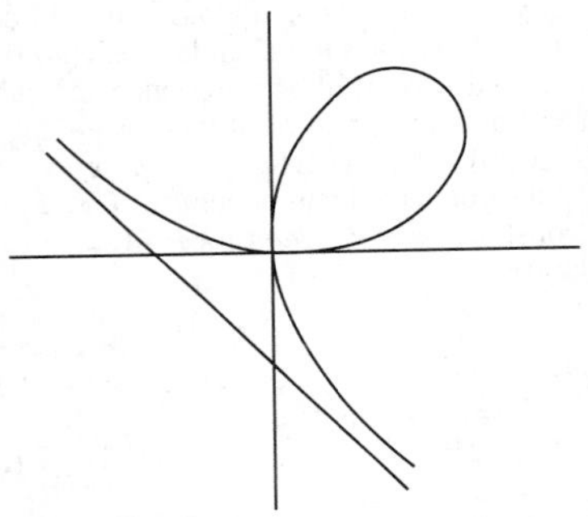

folium of Descartes

folk *Anthropology.* a body or community of people preserving a characteristic or distinctive culture from generation to generation, including its customs, legends, arts, and crafts.

folk art *Anthropology.* artistic forms or representations produced by people of a traditional or nontechnological culture who do not have formal artistic training. Thus, **folk artist.**

folk culture see FOLK SOCIETY.

folk dance or **folkdance** *Anthropology.* a dance originating among the people of a traditional or rural society.

folk game *Anthropology.* a game played by members of a traditional or nontechnological culture, having agreed-upon procedures that have been passed down over generations rather than formal rules.

folklore *Anthropology.* the body of knowledge, beliefs, and customs associated with a society, especially a rural or traditional society, in such forms as legends, tales, sayings, artwork, crafts, music, dance, and games. (A term coined by the English scholar W. J. Thoms in 1846.)

folk music *Anthropology.* **1.** music originating among the people of a traditional or rural society and passed down over successive generations. **2.** see FOLK SONG, def. 2.

folk society *Anthropology.* **1.** a small, homogeneous group of people often living in isolation who are dependent upon close contact, cooperation, and strict adherence to tradition. **2.** a traditional rural society.

folk song *Anthropology.* **1.** a song of anonymous authorship originating among the people of a traditional or rural society, passed down over successive generations and often having several versions. **2.** a modern song by a known composer that is in the style of traditional folk songs.

folk tale or **folktale** *Anthropology.* a narrative of anonymous authorship originating among the people of a traditional or rural society, passed down over successive generations and typically having an implied (but not stated) moral lesson; different cultures frequently have varying versions of the same tale.

folk taxonomy *Anthropology.* the way in which individual cultures classify the world, including the stereotypes that reflect the cultural biases of a particular group.

folkways *Anthropology.* the general pattern of social behavior of a certain group, passed down by tradition and accepted as proper, usually without being formally codified or enforced.

follicle [fäl´i kəl] *Biology.* a small, narrow cavity or sac in an organ or tissue, as those on the skin containing hair roots, or those in the ovaries containing developing eggs. *Immunology.* see LYMPHOID FOLLICLE.

follicle-stimulating hormone *Endocrinology.* a gonadotrophic glycoprotein hormone that is secreted by the gonadotrophs of the anterior pituitary gland and that stimulates spermatogenesis in the male and the growth and maturation of the ovarian follicles in the female.

follicular lymphoma *Medicine.* a variety of malignant lymphoma in which the nodules superficially resemble the follicles of normal lymph nodes.

folliculate *Biology.* having, composed of, or enclosed in follicles.

folliculitis *Medicine.* an inflammation of the hair follicles.

folliculoma *Oncology.* an ovarian tumor that is composed of granulosa or theca cells and is associated with excessive estrogen production.

follow current *Engineering.* a current at power frequency that travels through a discharge path after a high-voltage charge initiates the discharge.

follower *Behavior.* a species that practices following behavior, such as sheep or goats. *Engineering.* **1.** a drill that finishes digging a hole begun by a drill with a larger diameter, called the starter. **2.** the part of a machine that rides on a cam.

following behavior or **following response** *Behavior.* the tendency of young birds or other animals to follow closely behind the parent, or a substitute moving object, beginning shortly after birth.

following error *Control Systems.* in a contouring control system, the difference between the actual and desired or commanded positions.

following limb *Astronomy.* a former term for the eastern limb of a celestial object.

following sea *Navigation.* a sea that approaches a vessel from astern.

following spot see F-SPOT.

following wind *Meteorology.* **1.** a wind that blows in the direction of an ocean-wave advance. **2.** see TAILWIND.

follow-on echelon *Military Science.* in an amphibious operation, the echelon of assault troops, vehicles, aircraft equipment, and supplies that is required to support and sustain an assault after it has been initiated.

follow-the-pointer *Ordnance.* a scale mounted on some artillery instruments that visually indicates firing data.

follow-up *Military Science.* in an amphibious operation, the landing of reinforcements and supplies after the assault and follow-on echelons have been landed.

folpet *Organic Chemistry.* $C_9H_4Cl_3NO_2S$, a light-colored powder that is insoluble in water and slightly soluble in organic solvents; melts at 177–178°C; used as a fungicide in paints, enamels, and vinyls.

Folsom *Archaeology.* a term used to designate projectiles with Folsom points, or sites where these materials have been discovered.

Folsom point *Archaeology.* a distinctive type of projectile point used by early people of North America; found in association with extinct bison. (From *Folsom,* New Mexico, the site where these artifacts were first discovered.)

Fomalhaut [fō´mə lôt] *Astronomy.* the brightest star in Piscis Austrinus; it is a first-magnitude, blue-white star of spectral type A3, lying 22 light-years away from the earth.

Fomes *Mycology.* a genus of fungi of the order Agaricales that occurs on trees and on wood and whose fruiting bodies resemble hooves.

fomes [fo´mēz] *plural,* **fomites** [fo´mə tēz]. *Medicine.* see FOMITE.

fomite *Medicine.* any nonliving material, such as clothing or bed linen, that may harbor infectious organisms and thus serve as an agent for the transmission of disease. Also, FOMES. (From Latin for "tinder.")

fondo *Geology.* the sedimentation environment on the deep floor of a body of water.

font *Graphic Arts.* **1.** a complete set of all the characters in one size of a typeface. **2.** a particular typeface, such as Times Roman or Helvetica. Also, TYPEFONT. *Computer Programming.* a menu command used to select a desired typeface.

Fontana's stain *Microbiology.* a cytological stain that is used for spirochetes, in which ammoniacal silver nitrate renders the cells dark brown.

fontanelle or **fontanel** [fän´tə nel´] *Anatomy.* a membrane-covered space between the cranial bones in the incompletely ossified skull of a fetus or infant. *Entomology.* a small depression between the eyes that bears the opening of the frontal gland in most termite species.

Fontechevade man [fōn´tə chə väd´] *Paleontology.* an early form of *Homo sapiens,* known by the remains found at Fontechevade, France; dated from the third interglacial and probably related to Swanscombe man.

Fontinalaceae *Botany.* a mostly temperate Northern Hemisphere family of dull mosses of the order Isobryales that live submerged in slow- to fast-moving water or on wet rocks, soil, or bark near water; characterized by floating stems with wide spreading branches and by a unique endostome composed of linear papillose segments united by horizontal cell walls.

font master *Graphic Arts.* in photocompositon, a film negative containing all the characters in one typeface.

font mover *Computer Programming.* a software program that is used to implement the installation of a library of typefaces on a particular computer or to transfer a typeface from one application or location to another.

font proof *Graphic Arts.* a printed sheet or page showing all the characters in one font.

food *Biology.* **1.** a general term for any nutrient that is taken in or ingested by an organism and used by it to produce energy, build and repair tissue, and regulate body processes. **2.** such nutrients in solid form, as opposed to liquids.

food allergy *Immunology.* an abnormally sensitive reaction to food (such as vomiting or diarrhea) that occurs in a hypersensitive individual in response to a specific allergen.

food begging see BEGGING.

food-borne disease *Medicine.* a disease whose causative agents are transmitted through food.

food calorie *Nutrition.* see CALORIE.

food chain *Ecology.* the transfer of energy through various stages as a result of the feeding patterns of a series of organisms. A typical food chain begins with green plants that derive their energy from sunlight, then continues with organisms that eat these plants, then other organisms that consume the plant-eating organisms, then decomposers that break down the dead bodies of those organisms so that they can be used as soil nutrients by plants, and so on as the chain continues.

food-chain efficiency *Ecology.* the ratio of the nutritional value of the prey eaten by a predator to the nutritional value of the food consumed by the prey.

food cycle see FOOD WEB.

food engineering *Engineering.* the technology of manufacturing and preparing food.

food poisoning *Toxicology.* any of several acute illnesses, ranging in severity from mild to life-threatening, that result from eating food containing toxins or pathogenic microorganisms. Also, SITOTOXISM.

food pyramid *Ecology.* a model of the food chain with the producer organisms, generally green plants, represented at the base and primary, secondary, and tertiary consumers represented at successive, less abundant levels. Also, ECOLOGICAL PYRAMID.

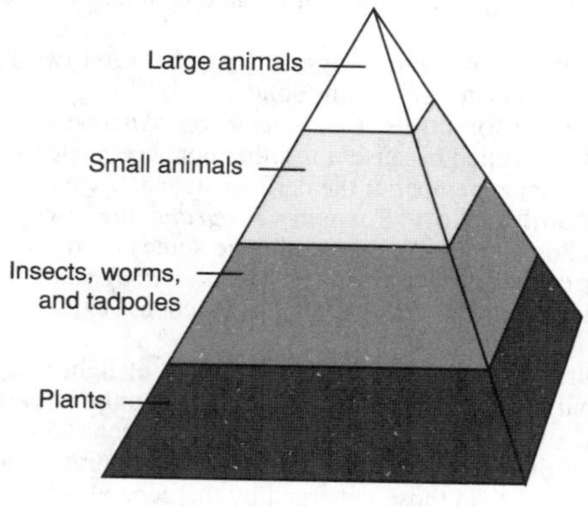

food pyramid

Food Science and Technology

Food science and technology involves the application of basic sciences (such as chemistry, biology, physics, biochemistry, microbiology, and nutrition) and engineering to the development, processing, packaging, preservation, storage, and distribution of food for the purposes of obtaining a safe, economical, and aesthetically pleasing supply of food for people worldwide. In its simplest terms, food science and technology encompasses all of the activities that occur with agricultural raw materials from the farmer's gate to the consumer's plate. Food science and food technology can be viewed as two distinct areas of scientific endeavor, but, in most academic institutions and in commerce and industry, these two areas are combined.

Food science would include the use of biology, chemistry, and physics to study the nature of foods, the causes of their deterioration, and the principles underlying the improvement of foods for the consuming public. Food technology is the application of food science to the preservation, processing, packaging, distribution, and use of food commodities.

Humans have been vitally interested in the preservation of foods since prehistoric times, when sun-drying was first used to preserve foods. Later, such practices as cooking, smoking, pickling, curing, and fermentation were developed for food preservation. Today's urban society has required the development of more sophisticated processes such as canning, freezing, and many others along with appropriate packaging technology and efficient distribution systems. Food scientists and technologists are constantly searching for improved ways to process, preserve, and package foods that will assure an adequate, nutritious, safe, and economical supply of foods for consumers. Food science and technology faces many future challenges, including concerns about the safety and nutrition of foods and the adequacy of food supplies for the burgeoning world population.

Steve L. Taylor
Professor and Head of Food Science and Technology
University of Nebraska

food science the scientific aspects of food technology.

food technology the application of scientific disciplines, such as chemistry, biology, and physics, and of practical disciplines such as engineering, to the development and distribution of food for worldwide populations.

food vacuole *Cell Biology.* a membrane-bound compartment in the cytoplasm of a protozoan cell that contains ingested food particles.

food web *Ecology.* the interconnected feeding relationships in a food chain found in a particular place and time; so called to reflect the complexity of these relationships, such as the proclivity of some organisms to feed on more than one level, or the changes that may occur from life cycle changes or the availability of food.

fool's gold *Mineralogy.* the popular name for pyrite, a mineral often mistaken for gold because of its similar appearance but having no special value. See PYRITE.

foot *Anatomy.* the lower part of the leg, having a flat undersurface and used for support in standing, walking, or running; in humans, it consists of the tarsus, metatarsus, and phalanges and the tissues encompassing them. *Invertebrate Zoology.* any of various similar appendages used for locomotion. *Mycology.* the bottom section of a fungi that attaches to the parental tissue, serving as an anchor to draw in nourishment. *Metrology.* a traditional unit of length equal to 12 inches or one-third yard; equivalent to 30.48 centimeters. *Graphic Arts.* the bottom part of a page.

foot abscess *Veterinary Medicine.* a necrotizing or purulent infection that involves the distal interphalangeal joint of the foot, caused by *Fusobacterium necrophorum* and *Coryne bacterium pyogenes.* The incidence is usually sporadic, but up to 15% of ewes in late pregnancy or rams may be affected.

footage *Engineering.* the length or quantity of a material expressed in feet. *Mining Engineering.* **1.** a measurement of the feet of borehole drilled per unit of time, or the amount necessary to complete a specific project or contract. **2.** the method of paying miners according to the running foot of work.

foot-and-mouth disease *Veterinary Medicine.* a highly infectious and widely occurring viral disease of cattle, sheep, hogs, goats, and other cloven-footed animals, characterized by fever, copious salivation, and the development of vesicles and subsequent erosions on various mucous membranes, especially in the mouth, nares, muzzle, and clefts of the feet. Also, HOOF-AND-MOUTH DISEASE.

foot block *Engineering.* a timber pad placed under props in tunneling to spread the load and distribute the weight on the underlying material.

foot bolt *Mechanical Devices.* a strong bolt located near the foot of a door in a vertical, fixed position.

foot bridge *Civil Engineering.* a bridge designed for pedestrian traffic.

footcandle *Optics.* a unit of illumination representing the light intensity over a surface of one square foot located one foot from a standard candle; equal to one lumen per square foot.

foot cell *Mycology.* **1.** a fungal hyphal cell that develops into an asexual spore. **2.** a spore cell in fungi belonging to the genus *Fusarium* that is shaped like a foot.

footer *Computer Programming.* in word processing, text that appears in the bottom margin on a printed page, often containing information such as a page number.

foot glacier see EXPANDED-FOOT GLACIER.

foot gland *Invertebrate Zoology.* **1.** in sessile bryozoans, a gland in the foot that secretes an adhesive substance. **2.** in snails, a gland in the foot the secretes a lubricant mucus to assist motion. Also, PEDAL GLAND.

foot guard *Civil Engineering.* a filler used to prevent a foot from being wedged between two rails.

footing *Building Engineering.* the lower part of a foundation that bears directly upon the earth, usually in the form of a column, in order to distribute the load over a greater area.

footlambert *Optics.* a unit of measure representing the luminance of an ideal diffuse surface reflecting or emitting 1 lumen per square foot; equal to $1/\pi$ candela per square foot.

foot-meter rod *Cartography.* a stadia rod, marked in feet and tenths on one side and meters and hundredths on the other side; used to determine distances and elevations in one unit of measurement and to check them by readings in a different unit.

footpoint *Astrophysics.* a location on the sun's photosphere where a magnetic-flux tube touches the solar surface.

foot-pound or **foot pound** *Metrology.* a unit of energy, equal to the amount of work done when a force of one pound moves an object one foot in the direction of the force.

foot-poundal *Metrology.* a unit of energy in the British absolute system, equal to the amount of work done by a force of one poundal acting through a distance of one foot in the direction of the force.

foot-pound-second *Metrology.* of or relating to a system of measurement in which the foot is the basic unit of length, the pound is the basic unit of mass, and the second is the basic unit of time.

foot-pound-second system see FPS SYSTEM.

footprint *Telecommunications.* the region of the earth's surface throughout which signals may be transmitted or received via a given communications satellite in a particular orbit. *Computer Technology.* the outline and surface area occupied by a computer and its peripheral equipment on the floor or desk top.

footprinting assay *Molecular Biology.* a technique of identifying protein binding sites on DNA by comparing a control sample of pure DNA with one that is protein-bound; a comparison of bands produced by subsequent DNA cleavage will determine the length of the sites that are protein bound or protected from cleavage.

foot rot *Plant Pathology.* any plant disease that causes the breakdown and decay of stem or trunk tissue. *Veterinary Medicine.* **1.** an infectious bacterial disease of sheep, goats, and cattle that is associated with warm, wet, unhygienic conditions, characterized by lameness due to erosion of the horny structures and inflammation of the related soft parts of the foot. Also, PODODERMATITIS. **2.** a term applied informally to a variety of degenerative foot infections of hoofed animals.

foots *Chemistry.* a mixture of soap, oil, and impurities that precipitates from an oil or wax upon standing. Also, **foot's oil.**

foot screw *Engineering.* **1.** one of three screws that joins a tribach of a platform with a plate screwed to the tripod head. **2.** a screw that serves as both a connector and a base.

foot section *Mechanical Engineering.* the part of a conveyor at the opposite end from the delivery point, consisting of a frame and a drum or sprocket upon which the chain or belt travels.

foot-section pulley see TAIL PULLEY.

footstep bearing see STEP BEARING.

footstock *Mechanical Engineering.* a device that supports a workpiece on a milling machine, usually used in connection with a dividing head.

foot survey *Archaeology.* the direct observation of a surface site by walking over it; surface features and artifacts are plotted on a site map and excavation is determined from this primary information. Also, GROUND SURVEY.

foot valve *Mechanical Engineering.* a nonreturn suction valve that is fitted in the bottom of a suction pipe of a pump or barrel to prevent the backward flow of water.

footwall *Geology.* the mass of rock underlying a fault, inclined vein, ore body, or mine working. Also, HEADING WALL, LOWER PLATE.

FOPC *Artificial Intelligence.* first-order predicate calculus.

forage *Agriculture.* **1.** any green plant material that is fed to livestock or used for silage. **2.** the search for and obtaining of such food.

foraging economy *Anthropology.* a subsistence economy based on collecting or gathering available foods; usually practiced by mobile bands and often supplemented with hunting.

foramen [fə rā´mən] *plural,* **foramina** [fə ram´i nə]. *Anatomy.* an opening that serves as a passage between two neighboring structures.

foramen magnum *Anatomy.* the large opening in the anterior and inferior part of the occipital bone, interconnecting the vertebral canal and the cranial cavity.

foramen of Luschka *Anatomy.* an opening at the end of each lateral recess of the fourth ventricle by which the ventral cavity communicates with the subarachnoid space.

foramen of Magendie *Anatomy.* a deficiency in the lower portion of the roof of the fourth ventricle through which the ventricular cavity communicates with the subarachnoid space.

foramen of Monro see INTERVENTRICULAR FORAMEN.

foramen of Winslow see EPIPLOIC FORAMEN.

foramen ovale [fə rā´mən ō vä´lā] *Developmental Biology.* an oval opening in the fetal heart of the septum secundum; the persistent part of the septum primum acts as a valve for this interatrial communication during fetal life, and after birth fuses with the septum secundum, thus closing functionally when the newborn begins to breathe and there is full circulation to the lungs, with complete closure taking several months. Also, OVAL FORAMEN.

foramen primum [fə rā´mən prē´mum] *Developmental Biology.* an orifice that exists temporarily in the anteroinferior portion of the septum primum of the embryonic heart and that permits communication between the two developing atria; it is closed early in development when the septum fuses with the atrioventricular cushions. Also, OSTIUM PRIMUM, PRIMITIVE INTERATRIAL FORAMEN.

Foraminifera or **Foraminiferida** *Invertebrate Zoology.* **1.** an order of chiefly marine protozoa in the phylum Sarcodina, whose shells are the main ingredient of chalk. **2. foraminiferan.** a member of this order.

forb *Botany.* a term for any herb plant that is not a grass.

forbesite *Mineralogy.* a mixture of cobaltoan annabergite and arsenolite.

forbidden band *Solid-State Physics.* a gap between two bands of allowed energy levels in a crystalline solid.

forbidden-character code *Computer Programming.* an illegal character or code, so called because it is either not acceptable as a valid representation or not a member of the defined alphabet.

forbidden combination *Computer Programming.* a combination of bits that has no valid meaning, according to some criteria.

forbidden-combination check *Computer Programming.* an automatic test for the occurrence of erroneous codes; examples are parity checks and redundancy checks.

forbidden line *Atomic Physics.* a wavelength of light that is not seen under normal laboratory conditions, occurring only in extremely rarefied gases, such as those found in the solar corona. *Spectroscopy.* a weak spectral line produced as a result of electron transitions having a lower probability than those governed by the general selection rules for determining the likelihood that certain transitions will take place.

forbidden structure *Evolution.* a modification in body structure that is not possible because it violates some physical law or physiological constraint.

forbidden transition *Quantum Mechanics.* a transition between two states that violates the selection rules.

Forbush, Scott born 1904, American geophysicist; discovered the Forbush decrease.

Forbush decrease *Astrophysics.* a diminishing of cosmic-ray strength as solar activity grows. Also, **Forbush effect.**

force *Military Science.* **1.** a group of military personnel, weapon systems, vehicles, and necessary support facilities. **2.** a major subdivision of a fleet. *Mechanics.* any external agent that causes a change in the motion of a free body, or that causes stress in a fixed body. *Computer Programming.* to intervene manually in the sequence of operations in a program by causing the execution of a jump instruction, often in order to bypass an error that has caused the program to halt.

force-and-torque sensor *Robotics.* a sensor in the wrist socket of a robot that measures force and torque and converts them to electronic signals that can be processed by a controller.

force-balance meter *Engineering.* a mechanism used to measure the force of the flow of liquid or air through an instrument; it uses a differential-producing primary regulator to balance the net force created by differential pressure on the sides of a diaphragm or diaphragm capsule.

force constant *Physical Chemistry.* the restoring force that tends to oppose the displacement of nuclei in a molecule, thus producing vibrations in response to a displacing force. *Mechanics.* an expression of the ratio of force to displacement in an elastic material. Also, SPRING CONSTANT, HOOKE'S CONSTANT.

force-controlled motion command(s) *Robotics.* a type of robotic control that uses computer software to provide motion information, but allows feedback or force sensing to alter the motion to adapt to the actual environment.

forced-air heating *Mechanical Engineering.* a heating system in which the positive circulation of warm air is supplied by means of a blower or fan.

forced amplification *Genetics.* a process by which gene amplification in a tissue culture is made permanent by an intense artificial selection for a certain phenotype.

forced circulation *Mechanical Engineering.* the use of a pump or other device to move liquid through pipes.

forced-circulation boiler *Mechanical Engineering.* a steam boiler in which water and steam are continuously circulated over heating surfaces by pumps for improved circulation efficiency.

forced convection *Thermodynamics.* the heat transfer by a fluid resulting from forcing the fluid by some external agency.

forced-draft *Mechanical Engineering.* describing a heating or air-conditioning system using a flow of air that is forced through pipes or ducts by a heater or blower.

forced expiratory volume *Physiology.* the volume of air that can be forced out in one second after taking a deep breath.

forced oscillation *Mechanical Engineering.* in a simple oscillator or an equivalent mechanical system, an oscillatory current produced by an external periodic driving force, so that the frequency of the current is determined by factors other than the constants of the circuit in which they are flowing. Also, **forced vibration.**

forced oscillator *Quantum Mechanics.* a system having normal modes of oscillation that is caused to vibrate with a frequency equal to that of a periodically varying force.

forced ventilation *Mechanical Engineering.* a system of ventilation in which pressurized air is forced through inlet or outlet ducts.

forced vital capacity *Physiology.* the maximum volume of air which can be expired as quickly and forcibly as possible, after a maximum inspiration.

forced wave *Fluid Mechanics.* a wave in a fluid which fits spatial and temporal irregularities at the boundary of a system and which is imposed on the system by some external agency.

force feed *Mechanical Engineering.* the lubrication under pressure of a machine, especially an internal-combustion engine.

force feedback *Robotics.* a method of measuring the amount of force exerted by an effector and feeding it back to a controller.

force field *Physical Chemistry.* the generalized region of space in which a specific force has effect and can be described by a vector.

force fit see PRESS FIT.

force gauge *Engineering.* a device that is used to measure the amount of force exerted on an object.

force main *Civil Engineering.* the discharge line of a pumping station.

force multiplier *Ordnance.* any factor that increases the combat power of a force.

force piece *Mining Engineering.* a set of timber placed diagonally across a shaft or drift to secure the ground.

force plug *Engineering.* a part of a mold that is inserted into the cavity block to exert force on a molding compound. Also, PLUNGER.

forceps [fôr´seps] *Mechanical Devices.* a pincerlike instrument with two blades, used for holding, seizing, and grasping objects. *Medicine.* such an instrument used in surgery to compress or grasp tissues and to handle sterile surgical supplies. Also, PINCERS. *Invertebrate Zoology.* a pair of modified anal cerci, curved, hard, movable appendages found in certain insects such as earwigs.

force pump *Mechanical Engineering.* a pump consisting of a barrel fitted with a solid plunger, and a valve chest with suction and delivery valves; delivers liquid under a pressure greater than its suction pressure.

force sensor *Robotics.* any sensor that can measure some kind of force and send the data back to a controller.

forcing cone *Ordnance.* a cone-shaped tapering at the beginning of the rifling in a gun tube; it eases the projectile into the rifling, thus centering it in the bore.

forcing function *Mathematics.* the function f on the right-hand side of an inhomogeneous linear differential equation, as in $Lu = f$, where L is a linear differential operator. In a second-order ordinary differential equation with constant coefficients, this function may be thought of as an actual force; thus the name.

forcipate *Biology.* shaped like or resembling forceps; forked.

forcipate trophus *Invertebrate Zoology.* a type of masticatory apparatus in certain predatory rotifers that resembles forceps and is used for grasping.

forcipressure *Surgery.* pressure applied by forceps, especially on a blood vessel to stop bleeding.

Forcipulatida *Invertebrate Zoology.* the sea stars, an order of echinoderms in the subclass Asteroidea having stalked pedicellaria of three ossicles arranged to form forceps or scissors.

ford *Hydrology.* a shallow, often narrow part of a stream or other body of water that may be crossed with relative ease.

Ford, Henry 1863–1947, American engineer and industrialist; a pioneer in automobile design and production.

Fordilla *Paleontology.* an extinct genus of primitive pelecypod isofilibranchs in the subclass Anomalodesmata; the earliest known bivalve, it occurs in widespread Cambrian deposits.

fording depth *Engineering.* the maximum depth at which a land or amphibious vehicle can be operated in water.

fore *Naval Architecture.* **1.** the front of a ship. **2.** the foremast. **3.** the forwardmost of two or more fittings or features. *Navigation.* located in, or in the direction of, the front of a craft.

fore- a prefix meaning "front" or "before" (in space or time), as in *forebrain, forecast.*

fore-and-aft *Naval Architecture.* of or related to a sail, rigging, or any other object that lies parallel to the keel of a ship. Thus, **fore-and-aft line, fore-and-aft mainsail, fore-and-aft rigged.**

forearm *Anatomy.* the portion of the arm between the wrist and the elbow. Also, ANTEBRACHIUM. *Ordnance.* a heat-insulating piece of wood or plastic that fits under the barrel of a rifle or shotgun in order to allow the shooter to steady it with his hand while firing. Also, FOREND.

forebay *Civil Engineering.* **1.** the small reservoir of a pipeline that distributes water to the consumer; it is the last free water surface of a distribution system. **2.** a reservoir feeding the penstocks of a hydroelectric power plant.

forebody *Naval Architecture.* the portion of a ship's hull forward of amidships.

forebrain *Anatomy.* the portion of the adult brain that is composed of the diencephalon and telencephalon. *Developmental Biology.* the anterior of the three primary vesicles of the embryonic neural tube of the vertebrate embryo, from which this portion of the adult brain develops. Also, PROSENCEPHALON.

forebulge *Geology.* a rise at the edge of a glacier produced by the bouyancy of the lithosphere.

forecast *Science.* a prediction of future events or conditions. *Meteorology.* **1.** specifically, a statement of expected future weather occurrences. **2.** the act of generating a weather forecast. Also, PREDICTION.

forecasting *Meteorology.* a process in which characteristics of past and current weather conditions are extrapolated and applied to predict future conditions.

forecastle [fōk´səl] *Naval Architecture.* a partial upper deck located at the front of a vessel.

forecast period *Meteorology.* the time interval for which a forecast is made.

forecast-reversal test *Meteorology.* a test in which the same verification method is applied simultaneously to a given forecast and a fabricated forecast of opposite conditions in order to evaluate the adequacy of a particular method of forecast verification.

forecast verification *Meteorology.* any process that determines the accuracy of a weather forecast by comparing the predicted weather with the observed weather of the forecast period.

foredeck *Naval Architecture.* the forward part of a ship's main deck.

foredeep *Geology.* **1.** an elongated trench along the margin of an island arc or other orogenic belt. **2.** see EXOGEOSYNCLINE.

foredune *Geology.* a coastal dune or ridge of sand oriented parallel to the shoreline along the landward limit of highest tides, and relatively stabilized by vegetation.

forefinger *Anatomy.* the second digit of the hand, located next to the thumb. Also, INDEX FINGER.

forefoot *Vertebrate Zoology.* one of the front feet of a quadruped or multiped. *Naval Architecture.* the forward end of a ship's keel, where it curves up or joins into the stem.

foregilding *Neurology.* the treatment of fresh nerve tissue and fibers with gold chloride solution, for the purpose of examining the structure and function of nerve endings.

foreground *Graphic Arts.* the portion of a photograph, illustration, or other image that is, or that appears to be, closer to the observer than the rest of the scene. *Computer Technology.* in a multiprocessing system, the environment in which execution of high-priority programs occurs.

foreground program *Computer Programming.* a high-priority program that takes precedence over concurrently running programs in allocation of system resources.

foregut *Developmental Biology.* the cephalic portion of the primitive digestive tube in the embryo; its entoderm develops into the epithelial lining of the pharynx, trachea, lungs, esophagus, stomach, and most of the small intestine.

forehand welding *Metallurgy.* a method of welding performed when the palm of the principal operating hand faces the direction of motion.

forehead *Anatomy.* the part of the face that is above the eyes.

forehearth *Metallurgy.* in a blast furnace, a hearth-level reservoir of metal or slab that is accessible through a door.

foreign-body locator *Engineering.* an instrument composed of probes that produce a magnetic field, which generates a signal when it detects the presence of foreign metallic material in tissue.

foreign element *Industrial Engineering.* a work element that is not part of a normal work cycle, such as a normal but infrequent action, an accidental interruption, or an intentional interruption.

foreign inclusion *Petrology.* a fragment or mass of country rock enclosed within an igneous rock or mass.

foreign matter *Science.* a substance that does not naturally occur in a particular place.

foreland *Geography.* a low, flat promontory. *Geology.* a stable region, usually of continental crust, bordering an orogenic belt, and toward which the rocks of the belt were thrust or overfolded.

foreland facies see SHELF FACIES.

forelimb *Anatomy.* an anterior fin, leg, or wing that is homologous to the human arm.

forellenstein see TROCTOLITE.

Forel scale *Oceanography.* a color scale designed to quantify the perception of the blue, green, and yellow colors of seawater observed against the white background of a Secchi disk.

foremast *Naval Architecture.* the forwardmost mast of a multimasted vessel; in a two-masted vessel, the foremast is usually called the main mast.

forend *Ordnance.* see FOREARM.

forensic [fə ren´sik] *Science.* of or relating to courts of law or legal proceedings, or to public discussion or debate. (From Latin for "relating to a forum or marketplace;" thus, "public.")

forensic analysis *Chemistry.* chemical analysis that is performed in connection with legal proceedings, and whose findings may be used in court.

forensic anthropology *Anthropology.* the study of skeletal and other human remains in the resolution of criminal or accidental death, requiring expertise in elements such as multiple grave analyses and ethnic or racial characteristics of the victim.

forensic chemistry *Chemistry.* the application of chemistry to matters involving civil or criminal law, as in forensic analysis.

forensic medicine *Medicine.* a branch of medicine that deals with the legal aspects of health care. Also, **forensics.**

forensic physics *Physics.* the application and discussion of physics when used to assist in argument, debate, or criminal investigation.

forensic psychiatry *Psychology.* the application of psychiatric principles in legal matters, as in dealing with issues of mental competency and responsibility. Similarly, **forensic psychology.**

forepeak *Naval Architecture.* the forwardmost compartment of a ship's hold, often a liquid tank.

forepeak bulkhead *Naval Architecture.* the bulkhead forming the afterside of the forepeak; the forwardmost transverse bulkhead.

forepoint *Cartography.* the point to which an observation is made; in surveying, for example, the point to which a foresight is directed.

forepoling *Mining Engineering.* **1.** the act of driving the poling boards beyond the last set of timbers, thus forming a roof for further advance. **2.** a method of securing loose ground by driving poles, planks, or the like ahead and on the top and sides of the timbers.

forerunner *Oceanography.* a low, long-period swell that frequently precedes the main swell from a distant storm, especially a tropical cyclonic storm.

foreset *Mining Engineering.* **1.** to place a prop under the fore or coal-face end of a bar. **2.** a timber set used at the working face for roof support.

foreset beds *Geology.* a series of gently inclined, systematically arranged layers of cross-bedded sedimentary material deposited along a steep, advancing frontal slope, such as the margin of a delta or the leeward side of a dune.

foreshaft sinking *Mining Engineering.* the first 150 feet or so of a shaft sinking from the surface; during this period the plant and services for the main shaft are installed.

foreshock *Geophysics.* an earthquake tremor that precedes the main shock of an earthquake.

foreshore *Geology.* the area of a beach or shore that is regularly covered and uncovered by the rise and fall of the tide, extending from the normal low-tide mark to the normal high-tide mark.

foresight *Engineering.* **1.** a measurement taken in a forward direction on a new survey point to calculate its elevation. **2.** a reference point made on a previously established survey point that serves to close a circuit. **3.** a measurement taken on a level rod that indicates the elevation of the point on which the rod is resting. *Cartography.* **1.** an observation of the distance and direction to the next instrument station. **2.** a point set ahead to be used for reference when resetting the transit on line or when verifying the alignment. *Ordnance.* see FRONT SIGHT.

foreskin *Anatomy.* the male prepuce; a fold of skin that covers the glans penis.

forest *Ecology.* a large area of land that is covered by trees and associated undergrowth, either growing wild or managed for the purpose of timber production. *Mathematics.* a graph that contains no cycles. A tree is a connected forest.

forestay *Naval Architecture.* **1.** on a single-masted ship, the supporting stay running forward from the mast. **2.** on a ship with more than one mast, the supporting stay running forward from the foremast to the deck near the bow.

forest climate see HUMID CLIMATE.

forest conservation *Ecology.* the total of all measures taken to protect and preserve a forest and its resources.

forest ecology *Ecology.* a field of study that examines the relationships of organisms in a forest to their environment and to each other.

forest engineering *Engineering.* a branch of engineering that addresses forestry problems in terms of long-range environmental and economic factors.

forest fire *Forestry.* any uncontrolled fire occurring in a wooded area; wildfire.

forest management *Forestry.* a professional area of forestry concerned with the overall handling and administration of forest resources to ensure maximum human benefit, continued production value, and thorough forest protection.

forest mapping *Forestry.* the process of preparing maps that show the distribution and form of various crops or stands in a forest.

forest measurement *Forestry.* the branch of forestry that focuses on measuring standing trees, cut timber, and lumber products.

forest resources *Forestry.* the natural assets of a forest such as timber, water, wildlife, grazing areas, shade trees, and recreation areas.

Forestry

Forestry applies many natural and social sciences to the management of forests for the perpetual benefit of human society. Forests are managed for many uses, i.e., for wood, water, wildlife, domestic animals, recreation, or simply natural preservation; these purposes can be integrated with agriculture or pastoral use. Because trees live a long time and are an infinitely renewable resource, planning for decades or centuries is taken more seriously in forestry than in almost any other human endeavor. Therefore, forestry tends to thrive on public or corporate ownerships that are nearly immortal or where laws require the long view.

Silviculture is the part of forestry that deals with the growing of forests and the channeling of their immense capacity to fix carbon as sugar compounds. These carbohydrates store energy to fuel the tree's vital processes and are also its chief building material; people and other organisms can also use trees for fuel and structural material. Since forest vegetation tends to fill all available growing space, forest stands can be regenerated and desirable trees favored only by deliberate creation of suitable vacancies in that growing space. This can be done only by killing trees, usually in the act of harvesting them for fuel or timber use. The patterns of such cutting operations in time and space vary widely depending on many conditions, especially the ecological adaptations of the desirable constituent species. The natural and social circumstances of each case determine the pattern or silvicultural solution that is devised.

David M. Smith
Morris K. Jessup Professor Emeritus of Silviculture
Yale University

forestry *Ecology.* **1.** the practice of growing trees for the commercial production of timber. **2.** the management of woods, forages, water, wildlife, and recreation on public lands. **3.** the study of tree growth and timber-production systems.

forest stand *Forestry.* a group of trees in a given area that consist of one species, size, or grade; used as the basic unit in forest mapping.

forest wind *Meteorology.* a light breeze that blows on calm, clear nights from forests toward open country.

forest yaws see PIAN BOIS.

fore-triangle see FORWARD TRIANGLE.

forge *Metallurgy.* **1.** a special furnace in which metal is heated before shaping. **2.** the equipment used for forging.

forgeability *Metallurgy.* a measure of the propensity of a metallic material to plastically deform without fracture when compressed at elevated temperatures.

forge delay time *Metallurgy.* in resistance welding, the time elapsed from the application of heat to the application of pressure.

forget *Artificial Intelligence.* a control blackboard command that causes the inference engine to eliminate or "forget" data concerning a particular rule or to ignore the response to a particular question.

forgetful functor *Mathematics.* a functor that "forgets" some or all of the underlying structure of an algebraic object. For example, the functor U_1: *Group* $\rightarrow$ *Set* does the following: (a) it assigns to each group G in the category *Group* the set U_1G (in the category *Set*) of all the elements of G, "forgetting" the group multiplication (and hence the group structure); and (b) it assigns to each morphism $f{:}G \rightarrow G'$ of groups in *Group* the same function f, regarded just as set function. The functor U_2: *Ring* $\rightarrow$ *Abelian* assigns to each ring R the additive Abelian group of R, and to each morphism $f{:}R \rightarrow R'$ of rings the same function f, regarded just as a morphism of addition. The multiplicative structure of the rings is "forgotten."

forging *Metallurgy.* the process of heating metallic stock and then shaping it by compression; dies may be used.

forging brass *Metallurgy.* a copper-base alloy containing about 38% zinc and 2% lead; suitable for forging.

forging hammer *Metallurgy.* a hammer with a ram that is raised mechanically and then allowed to free fall onto the forging stock.

forging plane *Metallurgy.* a plane perpendicular to the motion of the ram and containing the main face of the die.

forging press *Metallurgy.* a press that shapes hot metal by forcing it down between dies.

forging range *Metallurgy.* the temperature range within which a metallic material can be forged without fracturing.

forging stock *Metallurgy.* the metallic material that is being forged.

fork *Hydrology.* **1.** any place where two or more streams unite to form a larger waterway. **2.** the land lying within the angle made by such streams. **3.** a branch of a larger stream.

forked lightning *Geophysics.* a common form of cloud-to-ground discharge in which branching from the main lightning channel can be observed.

forklift *Mechanical Engineering.* **1.** a small vehicle equipped in front with two prongs that can slide under goods or into pallets and then be raised or lowered to move and stack goods; widely used in warehouses, shipping depots, and factories. Also, **forklift truck, fork truck. 2.** to move goods with a forklift.

fork oscillator *Electronics.* a circuit that generates alternating current, characterized by by the use of a tuning fork to determine frequency.

FOR loop *Computer Programming.* a control statement in a programming language that specifies the part of a computer program which is to be repeated a specified number of times.

form *Building Engineering.* a temporary boarding, sheeting, or molded fiberglass used to give a desired shape to poured concrete or a similar material. *Graphic Arts.* **1.** all of the pages of a printed work that are to be printed on one sheet at the same time. **2.** in letterpress printing, all of the copy for one page assembled in a chase. Also, TYPE FORM. *Systematics.* the lowest rank in the botanical taxonomic hierarchy, below subspecies and variety.

formability *Materials Science.* the capacity of a material to be bent or shaped by plastic deformation.

formal analysis *Linguistics.* the study of cognitive operations used to assemble meaningful sound sequences, and the rules by which such sequences are varied in form to generate other meanings (posited by Noam Chomsky); this methodology is extended to other classifications in kinship analysis.

formal argument see FORMAL PARAMETER.

formal charge *Physical Chemistry.* the charge assigned to a donor atom (+1) and an acceptor atom (–1) that form a covalent bond based on the sharing of electrons.

formaldehyde [fôr mal′də hīd′] *Organic Chemistry.* HCHO, a pungent polymerizable gas that is soluble in water and alcohol and boils at –19°C; commercially available as a 37–50% aqueous solution (called formalin), which is used widely in resins, disinfectants, preservatives, and fertilizers, and as a chemical intermediate.

formal derivative *Mathematics.* the formal derivative of the formal power series

$$A = \sum_{n=0}^{\infty} a_n x^n \text{ is } A' = \sum_{n=0}^{\infty} n a_n x^{n-1}.$$

The word "formal" is used to emphasize the fact that the definition of A' does not involve the concept of limits.

formal difference *Archaeology.* a method of defining variation in artifacts by differences in size and shape

formal dimension *Archaeology.* a characteristic based on a physical property of an artifact, such as its overall shape, the shape of specific parts, or its dimensions. Also, FORM ATTRIBUTE.

formal grammar *Computer Science.* see GRAMMAR.

formalism *Mathematics.* a philosophical view of mathematics in which mathematics is seen as a game; a formal language and formal rules for describing and proving theorems are introduced, a theory of the properties of the language is developed, and the goal is to show that a contradiction cannot be derived in the system. This program was shown to be impossible to achieve by Gödel's proof.

formal language *Linguistics.* language that adheres to strict standards of correct usage. *Computer Science.* a language in which abstract mathematical objects represent a vocabulary, and associated phrase-structure grammars model the syntax, of a programming or natural language.

formal logic *Mathematics.* the development of an axiomatic theory, in which provable statements or theorems are derived by logic alone from certain initially chosen axioms. A formal proof is a finite sequence of statements $S_1, S_2, \ldots S_k$, of the theory such that each S either is an axiom or comes from one or more of the preceding statements by a logical rule of inference. A theorem or provable statement is the last statement in a formal proof. (In particular, an axiom is a theorem with a one-step proof.) If S is a finite collection of statements of an axiomatic theory (called assumptions), then a statement C is deducible from S if there is a finite sequence $S_1, S_2, \ldots, S_n = C$ of statements, each of which is an axiom, a statement in S, or a statement derived from one or more preceding statements by a logical rule of inference. The sequence $S_1, S_2, \ldots, S_n = C$ is called a **formal deduction** of C from S.

formal operational stage or **period** *Psychology.* the last of Jean Piaget's four stages of cognitive development, from around the age of 12 to 14, in which the individual becomes able to carry on abstract thinking and to deal logically with issues involving the future or hypothetical situations.

formal parameter *Computer Programming.* the name that is used for a parameter in the definition of a function, procedure, macroinstruction, or subroutine, indicating that an actual parameter will be supplied by the calling program at the time of call. Also, FORMAL ARGUMENT.

formal potential *Physical Chemistry.* the measured potential of a half cell relative to a reference electrode, such as the standard hydrogen electrode when the ratio of the concentrations of the reduced and oxidized species is unity.

formal power series *Mathematics.* an expression $A = \sum_{n=0}^{\infty} a_n x^n$, in which the symbol x represents a placeholder, whose powers serve only to distinguish the coefficients a_n. The indeterminate x is never replaced by a value, and the coefficients a_n are usually taken to be elements of an integral domain (in special cases, a field). The word "formal" is used to emphasize the fact that the definition of A does not involve the concept of limits. The set of formal power series over a given integral domain with a finite number of nonzero coefficients also forms an integral domain; addition is defined term by term, multiplication is defined by the Cauchy product formula, the zero element is $0 = 0 + 0x + 0x^2 + \cdots$, and the unit element is $1 = 1 + 0x + 0x^2 + \cdots$. Formal power series are used extensively in the theory of generating functions, partitions of integers, analytic number theory, etc.

formamidase *Enzymology.* an enzyme of the hydrolase class that catalyzes the conversion of *N*-formylkynurenine to kynurenine and formate.

formamide *Organic Chemistry.* $HCONH_2$, a colorless, oily liquid that melts at 2.5°C and boils at 210.5°C with partial decomposition; used as a solvent, softener, and chemical intermediate.

formanite-(Y) *Mineralogy.* $YTaO_4$, a black, opaque, tetragonal mineral, dimorphous with yttrotantalite-(Y), having a specific gravity of 7.03 and a hardness of 5.5 to 6.5 on the Mohs scale; found as a detrital mineral in placer deposits.

forma specialis *Systematics.* in taxonomic literature, a notation used to indicate a variety of a species that develops as a result of a symbiotic relationship with a host.

format *Graphic Arts.* the general physical appearance of a printed work, including its size, shape, typeface, layout, and binding. *Computer Programming.* **1.** a specific arrangement of data for display or hardcopy output presentation. **2.** a predetermined arrangement of bits, characters, fields, lines, or the like. **3.** a programming statement that specifies the arrangement of particular input or output data elements or records.

formate *Organic Chemistry.* any compound that is derived from formic acid and contains the H·COO– radical. Also, FORMIATE.

formate dehydrogenase *Enzymology.* an enzyme of the oxidoreductase class that enables enteric bacteria to convert formic acid into carbon dioxide during fermentation. Also, **formate hydrogenlyase**.

formation *Military Science.* **1.** an ordered arrangement of troops or vehicles for a specific purpose. **2.** an ordered arrangement of two or more ships, units, or aircraft proceeding together under a commander. Thus, **formation bombing.** *Geology.* **1.** the basic rock-stratigraphic unit in the classification of local rocks, consisting of a body of rock having more or less uniform lithological characteristics, such as chemical composition, texture, and structure, which distinguish it from neighboring rock and make it useful for mapping or interpreting the geology of the region. **2.** any distinct, naturally occurring topographic feature.

formational theory *Computer Technology.* an approach to language translation by machine that assumes that symbol string formation can be defined in mathematical terms; related to transformational grammar.

formation damage *Petroleum Engineering.* the lessening of the permeability of reservoir rock due to the penetration of treating and drilling fluids into the section close to the wellbore.

formation factor *Geochemistry.* **1.** the ratio between the conductivity of an electrolyte and the conductivity of a rock saturated with that electrolyte. Also, RESISTIVITY FACTOR, FORMATION RESISTIVITY FACTOR. **2.** for a reservoir rock system, a function of the rock's porosity in terms of the degree to which precipitated minerals are bound together.

formation fracturing *Petroleum Engineering.* in a reservoir formation, a technique using hydraulic pressure to fracture rock, thus increasing oil production.

formation gas *Petroleum Engineering.* the original gas yielded underground from a reservoir.

formation resistivity *Geophysics.* the electrical resistivity of reservoir formations, used to determine their fluid content and lithography.

formation resistivity factor see FORMATION FACTOR.

formation solubility *Petroleum Engineering.* the degree of rock solubility in an oil-well acidizing solution, such as hydrochloric acid or hydrochloric-hydrofluoric acid.

formation tester *Petroleum Engineering.* an instrument used to retrieve samples of fluid from a formation in an oil reservoir.

formation-type *Ecology.* a group of geographically diverse communities with similar physiognomy and life-form and influenced by climatic and other environmental conditions.

formation-volume factor *Petroleum Engineering.* the extent or volume of an oil-bearing zone estimated to exist in an underground formation.

formation water *Hydrology.* water that is present in a formation under natural conditions, as opposed to being introduced artificially.

Formative *Archaeology.* a developmental period in New World cultural history preceding the Classic and characterized by developed food production, weaving, potterymaking, ceremonial objects and structures, and the first permanent villages. Also, PRE-CLASSIC.

formatted tape *Computer Technology.* **1.** a magnetic tape reel that has been initialized and prepared to accept data. **2.** a magnetic tape that contains a directory reference to recorded blocks of data; a desired block of data can be accessed by transporting the tape media at high speed to the proper location on the tape.

formatting *Computer Technology.* the process of preparing magnetic media for the receipt of data. *Computer Programming.* **1.** the process of arranging information and converting it to understandable form to be output to a display or hardcopy printer or plotter. **2.** specifically, the process of choosing the typeface(s), page margins, line spacing, type position, and other specifications for a document to be printed, and of executing the proper commands so that these choices will be carried out.

form attribute *Archaeology.* a characteristic of an artifact that relates to its size or shape, as opposed to its appearance or texture.

form clamp *Civil Engineering.* an adjustable clamp of steel or wood, used to hold together forms for the construction of beams, columns, and footings.

form-control buffer *Computer Technology.* a buffer that controls the vertical format of printed output.

form drag *Fluid Mechanics.* the total component of force acting in the direction opposite to the velocity of a body of particular shape moving through the fluid.

formed cutter *Mechanical Engineering.* a milling cutter shaped to the profile specified for the workpiece.

form factor *Electricity.* the ratio of the effective value of a current to its average value. Also, SHAPE FACTOR. *Mechanics.* the ratio of the modulus of rupture of a beam of given cross-sectional shape to that of a beam of equal cross-sectional area having a standard (usually square) shape. *Physics.* a function that describes the spatial variation of the matrix element for an atomic or nuclear interaction. *Quantum Mechanics.* an expression characterizing the deviation of a scattering pattern from that which would be observed from a perfect point particle, and which yields information regarding the structure of the target.

form feed *Computer Technology.* a paper movement on a printing device that brings a particular part of a form into position for printing.

form-feed character *Computer Technology.* a control character used in word-processing systems that causes the printer to move the paper so that the next line printed will appear at the top of the next page or form.

form feeding *Computer Technology.* the process of passing individual sheets or forms past the print heads or sensing devices.

form-feed printer *Computer Technology.* a hardcopy output device that prints on continuous paper.

form genus *Mycology.* a category of imperfect fungi including those fungi that have no known sexual stage of reproduction.

form grinding *Mechanical Engineering.* the grinding of cylindrical work by the use of a wheel whose cutting face is contoured to the form required and extends over the full length of the work. Also, PROFILE GRINDING, PLUNGE GRINDING.

formiate see FORMATE.

Formica *Materials.* the trade name for a laminated, heat- and chemical-resistant thermoplastic, widely used for the surface of tabletops, countertops, and so on.

formic acid *Organic Chemistry.* HCO_2H, a combustible, colorless liquid with a penetrating odor that is soluble in water, alcohol, ether, acetone, and benzene; it boils at 100.8°C and melts at 8.3°C; used in dyeing and finishing textiles, treating leather, and producing fumigants, insecticides, and refrigerants. *Biology.* this substance as it occurs in living organisms; it produces the stinging sensation that results from the bite of red ants and is also found in spiders and in pine needles and stinging nettles. (Going back to the Latin word for "ant;" it was first obtained in 1670 from the distillation of ants.) Also, METHANOIC ACID.

Formicariidae *Vertebrate Zoology.* the antbirds, a large family of the suborder Furnarii that are weak flyers and feed on ants and other insects; native to tropical America.

formication *Medicine.* an abnormal sensation, as of insects crawling in or upon the skin; a common symptom in diseases of the spinal cord and peripheral nerves, and in cocaine or amphetamine intoxication.

Formicidae *Invertebrate Zoology.* the ants, a family of social hymenopteran insects in the superfamily Formicoidea, having characteristic nodelike humps on the first or first two abdominal segments and strongly elbowed (bent) antennae.

formicivorous *Zoology.* feeding mainly or exclusively on ants.

Formicoidea *Invertebrate Zoology.* a superfamily in the order Hymenoptera containing only the ant family Formicidae.

forming *Engineering.* a bending operation that uses pressure to shape metal, plastic, glass, or other material. *Electricity.* the application of voltage to an electrolytic capacitor, electrolytic rectifier, or semiconductor during the manufacturing process to produce a permanent change in its electrical characteristics. Also, ELECTROFORM.

forming tool *Mechanical Devices.* **1.** a tool that produces its inverse shape on a workpiece. **2.** a pair of tongs with long, flat blades used to grasp and shape softened glass. Also, **form tool.**

form line *Cartography.* a dashed line that resembles a contour line, but represents no actual elevation, and has been sketched from visual observation or from inadequate or unreliable map sources to show the shape of the terrain rather than the elevation.

Formosa see TAIWAN.

form-process chart *Industrial Engineering.* a chart used to track and analyze the flow of paperwork and the type of information required by specific departments within an enterprise; used as an aid in designing forms. Also, **form-analysis chart.**

form roller *Graphic Arts.* any of the ink and dampening rollers on a press that touch the form being printed.

form stop *Computer Technology.* a sensor in a printer that stops the printer and alerts the operator with a light and/or sound when the paper supply runs out.

formula a conventional or prescribed way of doing something or expressing something; specific uses include: *Mathematics.* an equation or rule that gives a particular result. *Chemistry.* **1.** a representation of a chemical entity or relationship by symbols and figures. **2.** see CHEMICAL FORMULA. *Medicine.* a set of instructions used for preparing a particular kind of mixture, such as a medicine. *Food Technology.* a prepared mixture of milk and other nutrients used as food for a baby.

formulation *Chemistry.* **1.** the method and process of selecting the components of a mixture. **2.** the product of such a process.

formula weight *Chemistry.* the sum of the atomic weights of atoms in a compound.

formwork *Civil Engineering.* temporary boarding, sheeting, or molded fiberglass used to give a desired shape to poured concrete during its placing and curing time. Also, SHUTTERING.

formyl *Organic Chemistry.* HC(=O)–, the radical from formic acid; the aldehyde group.

***N*-formylmethionine** *Biochemistry.* the amino acid residue that catalyzes protein synthesis in bacteria, such as *Escherichia coli.*

formyltransferase *Enzymology.* an enzyme of the transferase class that catalyzes the transfer of the formyl group (HCO–) from a donor molecule to an acceptor.

Fornax *Astronomy.* the Furnace, a dim southern constellation whose brightest star is 4th magnitude.

Fornax A *Astronomy.* a 10th-magnitude elliptical galaxy that contains a double radio source at its center. Also, NGC 1316.

Fornax cluster *Astronomy.* a cluster of several dozen galaxies about 100 million light-years away.

Fornax system *Astronomy.* a 9th-magnitude dwarf elliptical galaxy that belongs to the Local Group of galaxies.

fornix *Anatomy.* a structure with an arched or vaulted shape. *Botany.* a small scale, as in the corolla tube of some plants.

fornix cerebri *Anatomy.* the efferent pathway of the hippocampus; an archlike body of nerve fibers that are beneath the corpus callosum of the cranium.

Forssmann, Werner 1904–1979, German surgeon; Nobel Prize for the development of cardiac catheterization.

Forssman antibody *Immunology.* an antibody that acts against the Forssman antigen.

Forssman antigen *Immunology.* a heterophile antigen occurring in a wide variety of animals that initiates the production of hemolysin for sheep red blood cells.

FOR statement *Computer Programming.* a program statement in an iterative structure that controls loop executions.

forsterite *Mineralogy.* Mg_2SiO_4, a white, greenish or yellowish, orthorhombic mineral, trimorphous with ringwoodite and wadsleyite; occurring in granular form or as thick tabular crystals, and having a specific gravity of 3.222 to 3.275 and a hardness of 7 on the Mohs scale; found in basic igneous rocks and thermally metamorphosed dolomitic limestones.

fort *Military Science.* **1.** a permanent military post. **2.** a fortified building or other structure. **3.** a land area containing harbor defense facilities.

FORTH *Computer Programming.* a high-level programming language that uses reverse Polish notation for computations; often used in control applications. (An acronym for <u>fourth</u>-generation language.)

fortification *Military Science.* **1.** an armed structure or position designed for defense. **2.** the act of creating such a structure or position. *Nutrition.* the process of adding nutrients to foods, particularly to foods naturally lacking in those nutrients, such as adding vitamins to milk.

Fortin barometer *Engineering.* an instrument used to measure atmospheric pressure, having a mercury cistern which increases or decreases in volume according to pressure changes, while the level of the cistern is kept at the zero of the barometer scale.

fortnightly tide *Oceanography.* a tide that occurs at intervals of a fortnight (half the oscillation period of the moon, 14.75 days).

FORTRAN [fôr´tran´] *Computer Programming.* a high-level procedure-oriented programming language that was initially designed for various mathematical, engineering, and scientific applications. (An acronym for <u>for</u>mula <u>tran</u>slation.)

Fortrat parabola *Spectroscopy.* for the lines composing the band spectrum of a molecular substance, a graph of wavenumbers with respect to serial numbers of successive lines.

fortuitous distortion *Telecommunications.* a type of random distortion that results in the intermittent shortening or lengthening of the transmitted signals; it can be caused by battery fluctuations, hits on the line, power induction, or similar accidental irregularities, and it affects transmission in any circuit coupled to the distortion source.

Fortuna *Astronomy.* asteroid 19, discovered in 1852 and measuring 219 km in diameter; it belongs to type C.

Forty Saints' storm *Meteorology.* a southerly gale over Greece occurring just before the equinox in March.

forward *Navigation.* in the direction of the front of a craft.

forward-acting regulator *Electronics.* a regulator that ensures that adjustments made to maintain the consistency of a transmission do not affect the quantity of the transmission.

forward azimuth *Cartography.* the azimuth of a previously unestablished point as measured from an established position.

forward-backward counter *Computer Technology.* a special-purpose register that can be either incremented (forward counting) or decremented (backward counting).

forward bias *Electronics.* a voltage applied to a semiconductor P-N junction to make the P side positive with respect to the N side.

forward chaining *Artificial Intelligence.* a process of reasoning in which known facts and rules are used to deduce additional facts, which are then added to the database; e.g., given the formula $A \land B \rightarrow C$, if A and B are true, C could then be deduced and added as a new fact. Also, ANTECEDENT REASONING.

forward coupler *Electronics.* a directional coupler used for sampling incident power.
forward current *Electronics.* current that flows through a P-N junction as the result of a forward bias.
forward direction *Electronics.* the direction that offers the least resistance to current flow.
forward drop *Electronics.* the voltage drop that occurs at the lowest point of resistance across a rectifier.
forward error analysis *Computer Programming.* a method for estimating the cumulative effect of errors caused by rounding and truncating during arithmetic operations on digital computers.
forward error correction *Computer Programming.* a method of error control for data transmission in which a redundant code is added to the end of the frame to be transmitted, enabling the receiving device to detect and correct errors within the received frame without retransmitting the frame.
forward extrusion see DIRECT EXTRUSION.
forward lap *Cartography.* the overlap between successive aerial photographs in the same line of flight.

forward-looking infrared system

forward-looking infrared imager *Engineering.* an infrared sensor system capable of producing a two-dimensional scan, and generating an image that represents the location and extent of infrared radiation. *Aviation.* such a system mounted on the nose of search-and-rescue helicopters; designed for night rescue work. Also, FLIR.
forward-mixing Frings aerator *Biotechnology.* a machine that precisely controls the aeration rate needed for the continuous production of vinegar; it consists of a rotating hollow-body turbine surrounded by a stator, and air is passed against the direction of the rotation.
forward mutation *Genetics.* a mutation from the wild type to a mutant condition.
forward observer *Military Science.* an observer operating with front-line troops who is trained to adhere and adjust ground or naval gunfire and pass back battlefield information; this observer may also direct close air support strikes.
forward of the beam *Navigation.* any direction between abeam and dead ahead.
forward path *Control Systems.* in a feedback control loop, the path of transmission from the loop actuating signal to the loop output signal.
forward period *Electrical Engineering.* the portion of an alternating-voltage cycle in which forward voltage appears across the rectifier circuit element.
forward perpendicular *Naval Architecture.* a vertical line extended from the intersection of the stem and the load waterline; one standard datum point for measuring a ship's length. Also, **FP.**
forward pointer *Computer Programming.* in a linked list, the part of a record that indicates the location of the next record in the list.
forward quarter *Naval Architecture.* the portion of a ship just aft of the bow and to either side.
forward reasoning *Artificial Intelligence.* a search technique in which the initial problem state is brought forward from its initial situation to one that meets the desired goal.
forward recovery time *Electronics.* the time required for the forward current or voltage to attain a specified value after a forward bias has been applied.
forward reference *Computer Programming.* during compilation, a reference to a variable that has not yet been defined.
forward resistance *Electronics.* **1.** the lowest level of resistance in a P-N junction. **2.** the resistance measured at a given forward voltage drop in a rectifier.
forward scatter *Geophysics.* the radiant energy that is scattered into space in an area bounded by a plane normal to the direction of the scatter. *Telecommunications.* a radio wave created as a result of scattering and propagating in the same direction as the wave that travels from source to load. *Physics.* the scattering of particles in which the scattering angle is between 0 and 90°.
forward transfer function *Control Systems.* the transfer function of the forward path of a feedback control loop.
forward triangle *Naval Architecture.* a triangle formed by the mast, the forestay, and the deck; in yacht racing it is used as a measure of the allowable sail area. Also, FORE-TRIANGLE.
forward wave *Electronics.* a wave in which the velocity follows the same direction as the flow of current.
FOSDIC II [fäz′dik′] *Computer Technology.* an optical mark-reading device designed in 1952 to read marks from census documents stored on microfilm and to prepare computer-readable input data. (An acronym for film optical sensing device for input to computers.)
foshagite *Mineralogy.* $Ca_4Si_3O_9(OH)_2$, a snow white, triclinic fibrous mineral having a silky luster, an undetermined hardness, and a specific gravity of 2.36; found in veins with blue calcite.
Foshay's test *Immunology.* a test in which a suspension of nonliving *Francisella tularensis* (the agent of tularemia in humans) is injected under the skin to determine if the subject has delayed hypersensitivity to the injected antigen. (From Leo *Foshay,* 1896–1961, American bacteriologist.)
foss see FOSSE, def. 1.
fossa *Anatomy.* a trench or channel. *Vertebrate Zoology.* a slender lithe mammal, *Cryptoprocta ferox,* related to the civet; the largest carnivore of Madagascar.
fossa helicis see SCAPHA.
fosse *Hydrology.* **1.** a long, narrow waterway, such as a canal or a trench. Also, FOSS. **2.** a long, narrow troughlike depression between the edge of a retreating glacier and the containing wall of its valley, or between the front of a moraine and its outwash plain.
fossil *Paleontology.* the whole or a part of a plant or animal that has been preserved, usually by mineralization and usually in a stratum of sedimentary rock; fossils are generally at least 10,000 years old. **Trace fossils,** such as footprints, result from the action of an organism. (Going back to the Latin word for "dig.")

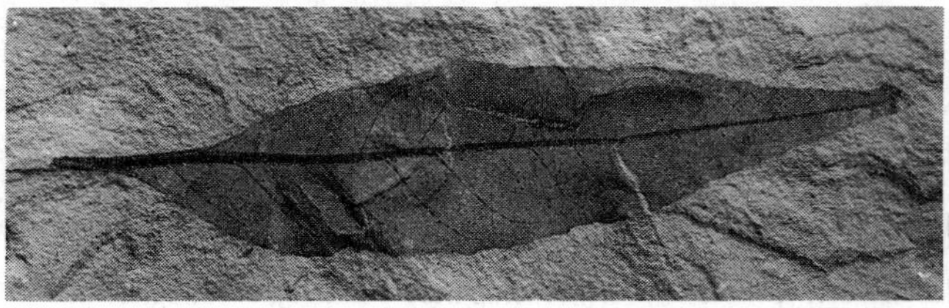

fossil

fossil directeur see TYPE FOSSIL.
fossil flour see DIATOMACEOUS EARTH.
fossil fuel *Geology.* a hydrocarbon deposit that consists of the remains of animal or vegetable life from past gelogic ages and that is now in a combustible form which is suitable for use as fuel; for example, oil, coal, or natural gas.

fossil ice *Hydrology.* **1.** ice found in frigid regions where it remains from the recent geologic past in which it was originally formed. **2.** relatively old ice found beneath the surface in a permafrost region. **3.** underground ice found in areas where present-day temperatures are not low enough to have formed it.

fossiliferous *Paleontology.* fossil-bearing; describing a stratum of rock laid down in a manner that has preserved the remains of various living organisms.

fossilization *Paleontology.* the complex process that transforms the remains of a whole or partial organism into a solid artifact, preserving, sometimes down to molecular detail, the anatomy of the original organism.

fossilized *Paleontology.* in the form of a fossil; preserved as a fossil.

fossil man *Paleontology.* the forms of the genus *Homo* that are known from fossilized skeletal remains.

fossil ore *Geology.* an iron-containing sedimentary deposit in which fragments of shells have been replaced or cemented by hematite and carbonate.

fossil permafrost see PASSIVE PERMAFROST.

fossil resin *Geology.* any naturally occurring resin, such as amber or retinite, that exuded from long-buried plant life.

fossil soil see PALEOSOL.

fossil water see CONNATE WATER.

fossil wax see OZOCERITE.

fossorial *Vertebrate Zoology.* adapted for digging or burrowing, as are the hands and feet of moles, armadillos, and aardvarks.

foster rearing *Psychology.* the raising of young by individuals other than the natural or biological parents. Also, **fostering.**

Foster's reactance theorem *Control Systems.* a function of complex variables that is analytic within its domain of definition except at a finite number of points that are poles (i.e., approach infinity).

Foucault, Jean [foo kō´] 1819–1868, French physicist; proved the relationship between velocity of light and density of media; invented the Foucault pendulum, Foucault prism, and gyroscope.

Foucault current see EDDY CURRENT.

Foucault knife-edge test *Optics.* a technique used to test for defects in the form and quality of a lens or concave mirror by moving a knife edge between the viewer and the image.

Foucault mirror *Optics.* a process in which the displacement caused by a light beam bouncing back and forth between a rapidly rotating mirror and a fixed mirror is used to calculate the velocity of the light.

Foucault's pendulum *Mechanics.* a pendulum of great length, free to swing in any direction. Over an extended time the Coriolis force will cause the plane of oscillation to rotate slowly, thus demonstrating the rotation of the earth. If the pendulum were at the North Pole, the plane would rotate exactly once each 24 hours.

Foucault test *Optics.* an optical test using an artificial star and a knife-edge that reveals the surface figure of a lens or mirror.

fougasse [foo´gas´] *Ordnance.* a type of mine that throws fragments or flammable liquid in a preset direction upon explosion.

foul *Graphic Arts.* of a galley or proof, superseded by a later, corrected version of the same material. Thus, **foul proof, foul matter.**

foulard [fə lärd´] *Textiles.* a lightweight fabric of silk or other material, having a plain or twill weave and usually printed with a small design; used for neckties, scarves, linings, and so on.

foul berth *Navigation.* **1.** an area in which a vessel at anchor is in danger of striking another vessel, the bottom, or some other obstruction. **2.** an area that has a rocky bottom providing poor holding ground for an anchor. Also, **foul bottom.**

foulbrood *Invertebrate Zoology.* any of three destructive bacterial diseases of the honeybee larvae: American foulbrood, European foulbrood, and parafoulbrood.

foul clay see PLASTIC CLAY.

Foulger's test *Analytical Chemistry.* a test using urea and stannous chloride added to an unknown solution; a blue color indicates the presence of fructose.

fouling *Naval Architecture.* **1.** the gradual accretion of barnacles, seaweed, and other materials on a ship's bottom, spoiling its performance. **2.** jamming or entangling a line. *Navigation.* a slow or drifting collision with another vessel.

fouling organism *Ecology.* **1.** one of a collection of aquatic organisms growing on the surface of floating or submerged synthetic structures and thus decreasing water flow or otherwise interfering with the function of the structure. **2.** any organism whose presence interferes with the operation of the object on which it grows.

fouling plates *Biotechnology.* metal plates that are immersed in water to attract fouling organisms; the organisms are then studied to determine the effects that environmental conditions and time have on the inhabitants of a body of water.

fouling point *Civil Engineering.* **1.** a point at a switch or turnout beyond which railroad cars must be positioned so as not to interfere with cars on the main track. **2.** the location of an insulated joint in a turnout on a signaled railroad track.

foundation *Building Engineering.* **1.** the natural or prepared ground or base on which a fabricated structure rests. **2.** the lowest division of a building wall made of masonry, partially or completely beneath the surface. *Civil Engineering.* the portion of a structure that distributes a load.

foundation coefficient *Geophysics.* a measurement indicating how much greater the effect of an earthquake would be on a given rock than on an undisturbed crystalline rock, in identical circumstances.

foundation wall *Building Engineering.* the vertical wall that lies closest to the foundation and is used to build up other horizontal sections of the structure.

founder *Geology.* the underwater submerging of crustal masses, islands, or significant sections of continents as a result of the sinking of land or a rise in sea level.

founder breccia see COLLAPSE BRECCIA.

founder effect *Evolution.* an increased diversity among a population and a step toward speciation in a colony established in isolation from the main population, since the differences in the gene pools of the colony and main population are enhanced by different evolutionary pressures resulting from different environmental conditions. *Genetics.* the genetic effect of this process; the "founder" colony's evolutionary possibilities tend to be more limited than those of the parent population.

founder principle *Genetics.* the principle that when a small sample of a larger population establishes itself as an isolated entity, its gene pool carries only a fraction of the genetic diversity represented in the parental population.

foundry *Engineering.* a workshop in which metal or glass is cast.

foundry engineering *Engineering.* a branch of engineering that involves the melting and casting of metal or glass.

foundry facing *Metallurgy.* in casting, a material applied to a sand mold to improve the surface quality of the castings.

foundry proof *Graphic Arts.* a proof pulled from a letterpress form before the form is cast into plates.

foundry type *Graphic Arts.* metal type sold in individual pieces to a compositor for handsetting.

fountain *Engineering.* a device that produces an artificial flow of water, especially one that provides water for drinking. *Graphic Arts.* a reservoir designed to hold the dampening solution on an offset printing press.

fountain effect *Fluid Mechanics.* a thermochemical effect observed in superfluid liquid helium: two vessels of helium II connected by a capillary tube with one of the vessels heated will result in a flow from the colder vessel to the hotter vessel.

fountain roller *Graphic Arts.* a roller designed to transfer dampening solution from the fountain to the press plate, usually in conjunction with a ductor roller.

fountain solution see DAMPENING SOLUTION.

fountain stop *Graphic Arts.* a movable piece of metal or other material placed on a fountain roller in order to reduce the flow of dampening solution to a given area of a plate.

fountain syringe *Surgery.* a device for injecting fluids by means of gravity.

Fouquieriaceae *Botany.* a monogeneric family of dicotyledonous, woody or fleshy, succulent shrubs and small trees of North American desert regions, belonging to the order Violales; characterized by leaves produced after rain and deciduous when the soil dries; includes the ocotillo species.

fourable *Petroleum Engineering.* a section of drill pipe, tubing, or casing composed of four joints that are connected together; every fourth joint is unscrewed in removing the pipe from a well.

four-address *Computer Programming.* a type of instruction that contains four address parts, including two operand addresses, the address for the destination of the result of the operation, and the address of the next instruction that is to be executed.

four-ball tester *Engineering.* an instrument for measuring the degree to which a lubricant is effective; it contains one ball that is driven against three immobile balls that are clamped together in a cup filled with lubricant; the efficiency of the lubricant is measured according to the wear marks on the stationary balls.

four-bar linkage *Mechanical Engineering.* a mechanism consisting of an assemblage of four links that are pinned tail to head in a closed loop and are capable of relative motion.

four-bit code *Computer Programming.* one of several weighted coding systems for representing decimal digits, in which each of the four bit positions bears a different weight; examples are 8421 code, 7421 code, and excess-3 code.

Fourcault process [fůr kō´] *Engineering.* a method in which sheet glass is made by drawing melted glass vertically upward.

four-channel *Acoustical Engineering.* of or relating to sound recording or reproduction that involves four separately processed channels to provide one unified, coherent sound. Thus, **four-channel sound.**

fourchite *Petrology.* a monchiquite without olivine or feldspar.

four-color *Graphic Arts.* relating to four-color printing or to materials printed by this method.

four-color press *Graphic Arts.* a printing press that can print four colors during one pass of the paper, thus enabling it to print in all colors simultaneously.

four-color printing *Graphic Arts.* a common printing process using four primary colors (black, cyan, magenta, and yellow) to create all colors.

four-color separation process *Graphic Arts.* a method of separating a piece of art into the four primary colors, either photographically (using filters) or electronically (using laser light).

four-color theorem *Mathematics.* the long-standing conjecture, finally proven in 1976 by K. Appel and W. Haken using reducible configurations, that every planar graph *G* is four-colorable; that is, the vertices of *G* can be assigned colors so that no adjacent vertices are the same color. Historically, the problem required the coloring of a map in the plane so that no two countries that shared a border were the same color.

four-current density *Physics.* a four-vector whose time component is the charge density.

Fourdrinier machine *Mechanical Engineering.* a machine for making paper in a continuous strip or web, characterized by a wire part whose upper forming surface is nearly horizontal.

four-element model *Materials Science.* an explanatory representation of viscoelasticity, showing how the delayed elastic deformation occurs upon application of stress and how it is followed by partial recovery upon removal of stress.

four-factor formula *Nucleonics.* a formula that gives the ratio between generation of neutrons (multiplication factor) in an infinite thermal reactor, based on the percentage of neutrons created by fission (fast fission factor), the percentage of neutrons absorbed by the reactor fuel (thermal utilization factor), and the probability that thermal neutrons will lose energy without being captured by a nucleus (resonance escape).

fourfold bodies see CORPORA QUADRIGEMINA.

four-force *Physics.* a four-vector equal to the rate of change of its four-momentum with respect to its proper time.

four-frequency diplex *Telecommunications.* a frequency-shift keying technique used in radio telegraphy in which each of the four possible signals obtainable in two telegraph channels has a different frequency.

Fourier, Jean Baptiste [fůr´ē ā´] 1768–1830, French mathematician and physicist; studied heat diffusion and conductivity.

Fourier analysis *Mathematics.* the study of the representation of functions by Fourier transforms and series and of the convergence of Fourier series. *Crystallography.* the determination of the harmonic components of a complex waveform in terms of a Fourier series.

Fourier analyzer *Engineering.* an instrument that is used to provide push-button information on the power and behavior of complex waveforms, such as underwater sounds or brain waves.

Fourier-Bessel integrals *Mathematics.* expansions of a given function $f(r)$ as an integral in Bessel functions are given by the Fourier-Bessel integrals

$$f(r) = \int_0^\infty \int_0^\infty f(\varsigma)J_n(\lambda\varsigma)J_n(\lambda r)\lambda d\lambda\varsigma d\varsigma,$$

where the J_n are Bessel functions. An expansion of $f(r)$ as a Fourier-Bessel integral exists if (a) $f(r)$ is continuous on $(0, \infty)$; (b) $f(r)$ has a finite number of maxima and minima in any finite interval; and (c) the integral $\int |f(\varsigma)|\varsigma\, d\varsigma$ converges.

Fourier-Bessel series *Mathematics.* for a function $f(x)$, the series whose n th term is

$$2J_0(j_n x)/J_1{}^2(j_n) \int_0^1 tf(t)J_0(j_n t)\, dt,$$

where the j_n are the positive zeros of the Bessel function J_0 arranged in ascending order, J_1 is a Bessel function, and $n = 0, 1, 2, \ldots$.

Fourier integral *Mathematics.* let $f(x)$ be an absolutely integrable function that satisfies the Dirichlet conditions. The Fourier integral of f is given by

$$f(x) = \int_0^\infty [a(y) \cos xy + b(y)\sin xy]\, dy,$$

where

$$a(y) = (1/\pi)\int_{-\infty}^{\infty} f(u) \cos (yu)\, d,$$

and

$$b(y) = (1/\pi)\int_{-\infty}^{\infty} f(u) \sin (yu)\, du.$$

At discontinuity points of f, define

$$f(x) = (1/2) \lim_{\varepsilon\to 0^+} [f(x+\varepsilon) + f(x-\varepsilon)].$$

$a(y)$ and $b(y)$ are sometimes called the **Fourier cosine** and **Fourier sine** transforms, respectively. Fourier integrals give expansions of nonperiodic functions. The Dirichlet-Jordan criterion is a sufficient condition for the convergence of the Fourier integral. An equivalent (complex) expression for the Fourier integral is

$$f(x) = (2\pi)^{-1}\int_{-\infty}^{\infty} dy \int_{-\infty}^{\infty} f(u)e^{iy(u-x)}\, du.$$

Fourier kernel *Mathematics.* any kernel $K(x,y)$ of an integral transform such that: (a) $K(x,y)$ is also the kernel of the inverse transform, and (b) $K(x,y)$ may be written in the form $K(x,y) = k(xy)$.

Fourier law of heat conduction *Thermodynamics.* an equation in which the rate of heat transfer dQ/dt is proportional to the quantity $A(-dT/dl)$, where A is the area normal to the direction of heat flow and dT/d is the rate of temperature change along the distance of the body in the direction of the heat flow.

Fourier-Legendre series [fůr´ē ā´ lə zhän´drə] *Mathematics.* for a function $f(x)$, the series whose nth term is

$$Pn(x)(2n+1)/2 \int_{-1}^{1} f(x)P_n(x)\, dx$$

where the $P_n(x)$ are the Legendre polynomials and $n = 0, 1, 2, \ldots$.

Fourier map *Crystallography.* a map that is computed for a periodic function by the addition of waves of known amplitude, frequency, and phase. The term is generally used for an electron density or difference electron density map obtained by X-ray diffraction techniques for a crystal structure.

Fourier number *Thermodynamics.* a dimensionless quantity symbolized by Fo_f, and given by the product of the thermal conductivity of a substance and a characteristic time divided by the product of the specific heat of the substance at constant pressure, the mass density, and the characteristic distance for the problem. It is used in the study of heat transfer under non-steady-state conditions.

Fourier series *Mathematics.* **1.** suppose a real function f is defined on the interval $-\pi < \theta < \pi$. The expansion of the form $f(\theta) = A_0/2 + \sum_{n=1}^{\infty} (A_n \cos n\theta + B_n \sin n\theta)$ is called the Fourier series of $f(\theta)$, and it converges to

$$(1/2) \lim_{\varepsilon\to 0^+} [f(\theta+\varepsilon) + f(\theta-\varepsilon)]$$

provided f is of bounded variation on $(-\pi, \pi)$. The coefficients can then be shown to equal $A_n = (1/\pi)\int_{-\pi}^{\pi} f(\theta) \cos n\theta\, d\theta$, and $B_n = (1/\pi)\int_{-\pi}^{\pi} f(\theta) \sin n\theta\, d\theta$. In actual applications, a finite number of terms are computed to give an approximation of f by a trigonometric polynomial. **2.** the **complex Fourier series** of a real function $f(x)$ of period $2L$ is given by

$$f(x) = \sum_{n=1}^{\infty} a_n e^{2\pi inx/2L}$$

The coefficients can be shown to equal

$$a_n = (1/2L)\int_{-L}^{L} f(x)e^{-2\pi inx/2L}\, dx.$$

Crystallography. the use of this theorem in representing the structure of crystal; the Fourier theorem states that any periodic function may be resolved into cosine and sine terms involving known constants. Therefore, since a crystal has a periodically repeating internal structure, this can be represented in a mathematically useful way by a three-dimensional series, to give either a three-dimensional Fourier map or an electron density map.

Fourier spectrum *Physics.* a graphical representation of the Fourier transform coefficients of a given function, indicating both the magnitude and the relative phase of each of the components.

Fourier's theorem *Mathematics.* if $f(x)$ satisfies the Dirichlet conditions in $(-\pi, \pi)$, then the Fourier series of f converges to $f(x)$ at all points of continuity of f in the interval, and approaches

$$(1/2) \lim_{\varepsilon \to 0^+} [f(x+\varepsilon) + f(x-\varepsilon)]$$

at points of discontinuity.

Fourier-Stieltjes series [für´ē ā´ stēl´yes] *Mathematics.* for a function $f(x)$ of bounded variation on $[0, 2\pi]$, the series whose nth term is $e^{inx}/2\pi\int_0^{2\pi} e^{-inx}\, df(x)$, where $n = 0, 1, 2, \ldots$.

Fourier-Stieltjes transform *Mathematics.* for a function $f(x)$ of bounded variation on $(-\infty, \infty)$ the function

$$F(x) = (2\pi)^{-1/2} \int_{-\infty}^{\infty} f(t) e^{-itx}\, df(t).$$

Fourier synthesis *Mathematics.* the determination of a periodic function from its Fourier transform.

Fourier transform *Mathematics.* the Fourier transform of a real or complex-valued function f of a real variable t is given by

$$F(x) = (2\pi)^{-1/2} \int_{-\infty}^{\infty} f(t) e^{ixt} dt$$

the transition from f to F is called the **Fourier transformation,** and the determination of f given F is called Fourier synthesis. If the Fourier transform $F(x)$ is given, the **inverse Fourier transform**

$$f(t) = (2\pi)^{-1/2} \int_{-\infty}^{\infty} F(x) e^{-ixt}\, dx$$

gives $f(t)$ back again. In general, F is complex, even for real f. *Crystallography.* in the pair of equations

$$f(x) = \sum_{-\infty}^{\infty} e^{2\pi ixy} g(y)\, dy$$

and

$$g(y) = \int_{-\infty}^{\infty} e^{-2\pi ixy} f(x)\, dx,$$

$g(y)$ is the Fourier transform of $f(x)$, and $f(y)$ is the Fourier transform of $g(x)$. In X-ray diffraction the structure factor, F, is related to the electron density, ρ, by

$$F = \int_{-\infty}^{\infty} \rho e^{i\phi}\, dV_c$$

and conversely,

$$\rho = (1/V_c) \sum_{\text{all reflections}} F e^{i\phi}$$

where $\phi = 2\pi(hx + ky + lz)$ and V_c = the volume of the unit cell. Summation replaces integration in the latter equation because the diffraction pattern of a crystal is observed only at discrete points. The intensity at a particular point of the diffraction pattern of an object is proportional to $|F|^2$ and thus to the value at that point of the square of the Fourier transform of the structure.

Fourier transform infrared spectroscopy *Analytical Chemistry.* a computerized system for obtaining and analyzing a whole single-beam infrared spectrum.

Fourier transform spectroscopy *Spectroscopy.* the study of the absorption spectrum obtained by mathematical manipulation of the Fourier transform, generated when a sample is irradiated simultaneously with all wavelengths for a short period of time; used for example in infrared and nuclear magnetic resonance spectroscopy.

four-jaw chuck *Mechanical Engineering.* a chuck having four jaws set at right angles to each other; used in lathes and other machine tools to hold rectangular or irregularly shaped workpieces.

four-layer device *Electronics.* a device, such as a silicon-controlled rectifier, that has four layers of alternating P- and N-type material to provide three P-N junctions.

four-layer diode *Electronics.* a device such as a pulse generator or memory device that has three junctions formed by terminal connections at its outer layers; it may be used to perform a number of functions in a system.

four-layer transistor *Electronics.* a device, such as a thyristor, that has several regions wherein conductivity occurs, but only three terminals.

four-level laser *Physics.* a laser that operates on four energy levels; stimulated emission results from an energy transition to an intermediate excited state instead of to the ground state.

fourmarierite *Mineralogy.* $PbU_4^{+6}O_{13}{\cdot}4H_2O$, a red orthorhombic mineral occurring in tabular crystals and compact masses and having an adamantine luster, with a specific gravity of 5.74 and a hardness of 3 to 4 on the Mohs scale.

four-phase modulation *Telecommunications.* a modulation technique in which information is encoded on a carrier frequency as a series of phase shifts, with each phase shift representing two bits of information, known as dibits.

four-piece set *Mining Engineering.* a squared timber frame used in underground driving to provide all-around support to weak ground.

four-plus-one address *Computer Programming.* a type of instruction that contains four address parts plus the address of the next instruction that is to be executed.

four-point bearing *Navigation.* a relative bearing of 045° or 315°.

four-pole double-throw *Electricity.* a twelve-terminal switch or relay contact that simultaneously connects two pairs of terminals to either of two other pairs of terminals; used to switch sets of stereo loudspeakers.

four-pole single-throw *Electricity.* an eight-terminal switch or relay contact that simultaneously opens or closes two individual pairs of circuits.

four-quadrant multiplier *Computer Technology.* a device in an early analog computer that can process high-accuracy products of two nonlinear variables at frequencies up to 1 kHz. Also, QUARTER-SQUARE MULTIPLIER.

fourré see TEMPERATE AND COLD SCRUB.

four-stroke cycle *Mechanical Engineering.* an engine cycle that is completed in four piston strokes: a suction or induction stroke, a compression stroke, an expansion or power stroke, and an exhaust stroke.

four-tape sorting *Computer Technology.* a merge sorting technique for data stored on magnetic tape that involves two input and two output tape reels.

fourth contact *Astronomy.* the precise instant at which an eclipse, occultation, or transit ends.

fourth dimension *Physics.* a dimension additional to the three rectangular coordinates of length, width, and depth; in relativity theory and in certain mathematical operations, time is used as a fourth dimension to fix an event that has variables both of space and of time.

fourth-generation *Computer Technology.* of or relating to computer devices or systems of the 1980s or later, making use of large-scale integrated circuit technology.

fourth-generation language *Artificial Intelligence.* a nonprocedural user-oriented language, including code generators, database management systems, and English-like query languages. Also, **4GL.**

fourth sound *Physics.* a pressure wave that propagates in helium II contained in a porous material, and that results solely from the motion of the superfluid component.

fourth ventricle *Anatomy.* a brain cavity that lies between the cerebellum and the brain stem.

four-track tape *Acoustical Engineering.* a type of bidirectional tape, such as cassette tape, having two tracks (right and left stereo tracks) used for one direction and a similar set of tracks for the other direction.

four-vector *Physics.* a set of four quantities that transform under a Lorentz transformation in the same way as the three space coordinates and the time coordinate of an ordinary event.

four-vector potential *Electromagnetism.* a four-vector whose components are the magnetic vector potential and whose time component is the electric scalar potential.

four-velocity *Physics.* a four-vector whose components are the rates of change of the space and time coordinates of a particle with respect to the particle's proper time.

four-vertex theorem *Mathematics.* the condition that every simple closed convex curve has at least four vertices (points of zero curvature).

four-way dip *Geophysics.* a dip in geophysical prospecting, determined by sounding devices set up in four separate directions from the shot point.

four-way reinforcing *Civil Engineering.* a method of reinforcing concrete slabs by placing rods parallel to two adjacent edges and to both diagonals of a rectangular slab.

four-way switch *Electricity.* a type of switch allowing the same light to be turned on or off in four different places, often used to control household lights.

four-way valve *Mechanical Engineering.* at the meeting point of four waterways, a valve that allows for passage between any two adjacent waterways by the use of a movable element operated by a quarter turn.

four-wheel drive *Mechanical Engineering.* an automobile drive system arrangement in which the drive shaft powers all four wheels.

four-wire circuit *Telecommunications.* a two-way circuit in which the signals are transmitted along the medium in one direction by one path, and in the other direction by another path.

four-wire line *Electricity.* a four-conductor transmission line, with three conductors that supply respective phases and a fourth conductor that is neutral.

four-wire repeater *Electronics.* a repeater used in a four-wire transmission system, comprising two amplifiers that amplify the signals propagating in the opposite directions.

four-wire subscriber line *Telecommunications.* a four-wire circuit that links a subscriber directly to a switching center.

four-wire terminating set *Electronics.* a circuit arrangement wherein four-wire circuits are terminated in a manner that permits interconnection with two-wire circuits.

fovea *plural,* **foveae** or **foveas.** *Anatomy.* a small pit or depression in the surface of a structure or organ.

fovea centralis *Anatomy.* a small depression in the center of the macula lutea of the retina.

foveal vision *Physiology.* vision that is dependent on the presence of cone receptors in the fovea centralis of the retina, characterized by maximum acuity and the perception of color.

foveate *Biology.* having small depressions or fovea; pitted.

foveola *Biology.* a small pit or depression, such as a shallow cavity in a bone; a fovea.

foveola ocularis *Anatomy.* a small depression in the floor of the fovea centralis of the retina that has no rod cells but contains rodlike elongated cones; it is the area of clearest vision, because the layers of the retina are spread aside and light can fall directly on the cones.

foveolate *Biology.* having a pattern of small depressions or pits.

fowan *Meteorology.* a scorching, dry wind of Great Britain and the Isle of Man.

fowl *plural,* **fowl** or **fowls.** *Agriculture.* **1.** any bird that is used for food, especially a domestic bird such as a chicken, duck, or turkey. **2.** the meat of such a bird.

Fowler, William Alfred born 1911, American physicist; shared Nobel Prize for studies of cosmogenic nuclear reactions.

Fowler-DuBridge theory *Solid-State Physics.* a theory based on the Sommerfeld model that takes into account thermal electron agitation in describing photoelectric emission from a metal.

Fowler function *Solid-State Physics.* a function, dependent upon frequency, photoelectric threshold frequency, and absolute temperature, that is used in the Fowler-DuBridge theory to calculate photoelectric yield.

fowlerite *Mineralogy.* (Mn, Zn)SiO_3, a variety of rhondonite containing up to 10% ZnO.

fowl pox *Veterinary Medicine.* an infectious viral disease of poultry and other birds, characterized by wartlike nodules on the skin, especially the head and face, that progress to heavy scabs, or by diphtheritic membranes in the upper digestive and respiratory tracts.

red fox

fox *Vertebrate Zoology.* any of several carnivores of the dog family, especially those of the genus *Vulpes,* related to the wolves but with shorter legs, a more pointed muzzle, large erect ears, and a long bushy tail; found in most areas of the world in various habitats.

Fox equation *Materials Science.* a mathematical relationship that explains the behavior of random copolymers, predicting the relationship between the glass transition temperature and molecular mass fraction.

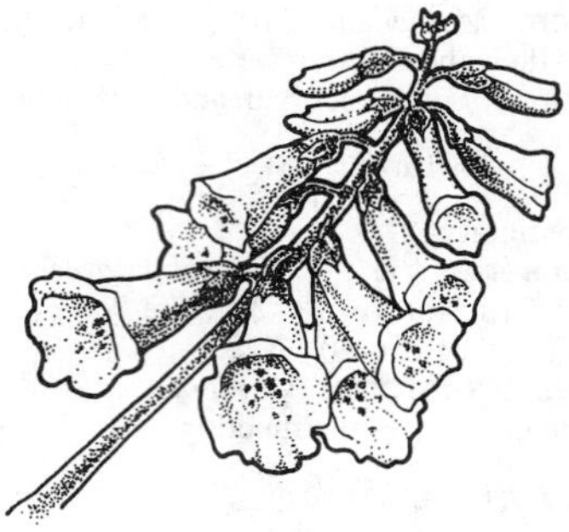

foxglove

foxglove *Botany.* any plant belonging to the genus *Digitalis,* particularly *D. purpurea,* characterized by tubular purple or white flowers on tall spikes and leaves that are the source of digitalis, a heart-stimulating medicine. (From the shape of the flowers, which are thought to suggest the fingers of a glove.)

foxhole *Ordnance.* a small pit providing protection from enemy fire, usually for one or two persons, while allowing effective fire from within.

fox lathe *Mechanical Engineering.* a lathe that has a chasing bar and leaders for cutting threads; used to turn brass.

fox message *Telecommunications.* a class of radio broadcast messages used to test a connection between transmitter and receiver; the message usually includes all the alphanumeric characters as well as many of the function characters. Also, **fox broadcast.**

foyaite *Petrology.* a variety of nepheline syenite consisting primarily of potassium feldspar with lesser amounts of dark minerals.

FP or **fp** foot-pound; freezing point. Also, **fp** or **f.p.**

FPC or **F.P.C.** Federal Power Commission.

F. pil. *Medicine.* let pills be made. (From Latin fiant pilulae.)

F pilus *Cell Biology.* a long structure that protrudes from the surface of bacterial cells and permits transfer of the fertility factor (F episome) from one cell to another during cell conjugation.

FPLA field-programmable logic array.

F plasmid *Genetics.* a circular DNA molecule present in certain bacteria that serves as a fertility factor by producing a hollow tube which allows the passage of DNA to another bacterium during conjugation. Also, FERTILITY FACTOR, F FACTOR.

fpm or **f.p.m.** feet per minute.

FPO *Aviation.* the airport code for Freeport, Bahamas.

F process *Mathematics.* let $F = \{F_t, t \geq 0\}$ be an increasing family of sigma algebras. A stochastic process $\{X_t, t > 0\}$ such that, for each t, the events $\{X_t \leq a\}$ belong to F_t for every a, is called an F process. When all the F_t are identical, the process is an ordinary (continuous parameter) stochastic process. The family F can be viewed as representing an increase in the amount of information over time.

FPS or **fps** foot-pound-second.

fps or **f.p.s.** feet per second; frames per second.

FPS system or **fps system** *Metrology.* foot-pound-second system, another name for the traditional British and American system of measurement in which the foot is the basic unit of length, the pound is the basic unit of mass, and the second is the basic unit of time.

Fr the chemical symbol for francium.

fr franklin (statacoulomb); fragment.

FRA *Aviation.* the airport code for Frankfurt, Germany.

fractal *Chaotic Dynamics.* **1.** of or relating to irregular fragmented shapes that exhibit intricate structure at all sizes so that details are reminiscent of the entire object (a property known as self-similarity). **2.** an object of fractional dimension, such as a strange attractor for chaotic behavior. *Mathematics.* a compact subset of R^n, the fractal dimension of which is not an integer; usually constructed by iterative processes that yield sets with local structure similar to smaller-scaled local structure, for example, the Cantor termary set, the Sierpenski gasket. (A term coined by the American mathematician Benoit Mandelbrot, from the Latin word *fractus,* "broken.")

fractal basin boundaries *Chaotic Dynamics.* when basin boundaries become infinitely convoluted and complexly interwoven, a region of phase space where the selection of the attractor is sensitively dependent on the initial conditions; such boundaries are often precursors of the onset of chaos as a parameter is varied.

fractal dimension *Chaotic Dynamics.* a characteristic of the strange attractors in phase space for chaotic behavior, when the dimension of an object is fractional, rather than integer. For continuous systems, chaotic strange attractors have fractional dimensions greater than two; for example, the dimension of a Lorenz chaotic attractor is about 2.04 and that of the Henon map is about 1.24. *Mathematics.* let A be a nonempty compact subset of a metric space X. For each $\varepsilon > 0$, $N(A, \varepsilon)$ denotes the minimum number of closed balls of radius ε needed to cover A. The fractal dimension of A is the number

$$D(A) = \lim_{\varepsilon \to 0} \{\ln(N(A, \varepsilon))/\ln(1/\varepsilon)\},$$

if this limit exists. The fractal dimension exists for all compact subsets of R^n and is greater than or equal to the Hausdorff-Besicovitch dimension.

fractal geometry *Chaotic Dynamics.* a generalization of Euclidean geometry that posits a noninteger dimension for describing irregular and fragmented patterns.

fractile see QUANTILE.

fraction *Science.* a distinct portion of something. *Mathematics.* an expression that is the quotient of a real or complex number and a nonzero real or complex number; written a/b, where b is restricted to be nonzero. *Chemistry.* any portion of a mixture that exhibits uniform or very similar properties, such as those obtained by fractional distillation. *Metallurgy.* the portion of a powder having particle sizes ranging between two established limits. Also, CUT. *Petroleum Engineering.* an individual, recognizable portion of crude oil, the product of a distillation or refining process.

fractional *Mathematics.* of, relating to, or constituting a fraction. *Chemistry.* of or relating to any process by which portions of a mixture are separated according to properties.

fractional arboricity *Mathematics.* given a graph G, the fractional arboricity of G is

$$\max_{H \subseteq G} |E(H)|/(|V(H)| - \omega(H))$$

where $\omega(H)$ is the number of components of H, $|S|$ is the number of elements in set S, and the maximum is taken over all subgraphs H of G for which $|V(H)|>\omega(H)$.

fractional burnup *Nucleonics.* the ratio, usually expressed as a percentage, of the number of fissions in a fuel mass to the total number of heavy atoms originally in the fuel.

fractional condensation *Chemistry.* the separation of a vaporized liquid mixture into its components through repeated condensations; separation is accomplished by differences in boiling points.

fractional coordinate *Crystallography.* an atomic coordinate expressed as a fraction of the unit cell length.

fractional crystallization *Analytical Chemistry.* a process used to separate or purify components of a mixture by lowering the temperature of the solution until one substance forms crystals and the rest of the mixture remains liquid. *Petrology.* magmatic differentiation caused by the separation of crystals and liquid prior to complete solidification; the process may produce residual liquids that are very different in composition from the original liquid. Also, FRACTIONATION, CRYSTALLIZATION DIFFERENTIATION.

fractional distillation *Chemistry.* a procedure for separating components of different boiling points from a liquid mixture by vaporizing the liquid and passing the vapor through a fractionating column.

fractional equation *Mathematic.* **1.** an equation in which one or more independent variables appear in a denominator. **2.** an equation containing fractions.

fractional horsepower motor *Electricity.* any motor that has a continuous rating of less than 1 horsepower.

fractional ideal *Mathematics.* let R be an integral domain with quotient field K. A fractional ideal of R is a nonzero R-submodule I of K such that aI is properly contained in R for some nonzero a in R.

fractional linear transformation *Mathematics.* a function of the form $f(z) = (az + b)/(cz + d)$, where the coefficients are usually taken to be complex. If the matrix $\left[\begin{smallmatrix} a & b \\ c & d \end{smallmatrix}\right]$ is used to represent such a transformation, then composition of transformations is represented by the product of the corresponding matrices. The group of fractional linear transformations is called the **Möbius group,** which is isomorphic to the projective special linear group PSL (2). Also, BILINEAR TRANSFORMATION.

fractional precipitation *Analytical Chemistry.* a series of analytical precipitations, used to improve the purity of the desired sample.

fractional sine wave *Physics.* a train of pulses whose variation in time is proportional to a truncated sine wave.

fractionating column *Chemistry.* a vertical tube containing a packing such as glass beads or metal turnings; used in fractional distillation, liquid-liquid extraction, and liquid-solid adsorption.

fractionation *Chemistry.* any of a number of methods of separating the components of a mixture into fractions, such as gel filtration, electrophoresis, or partition. *Materials Science.* any of several precipitation or phase-separation methods used to determine the molecular weight distribution of polymers, based on the tendency of polymers of high molecular weight to be less soluble than those of low weight. *Nucleonics.* a change in the isotopic composition of a substance that arises from minor differences in the physical and chemical properties of an element's isotopes; observed in nature and in waste from nuclear weapons production. *Petrology.* see FRACTIONAL CRYSTALLIZATION.

fraction defective *Industrial Engineering.* the fraction of the output of a production process that is found to be defective; a basic measure of quality control performance.

fractoconformity *Geology.* a relationship between conformable strata such that faulting of the older beds occurs simultaneously with the deposition of newer beds.

fractography *Metallurgy.* a technique used for studying and photographically recording the structural features of fractured surfaces.

fracton *Mathematics.* a local vibrational excitation of a fractal structure.

fracture *Medicine.* an injury to a bone in which the tissue of the bone is broken. *Geology.* any break in a rock that results from mechanical failure by stress, with or without displacement, including cracks, joints, and faults. *Mineralogy.* **1.** the manner in which a mineral breaks, other than along its planes of cleavage. **2.** the appearance of a mineral when broken. *Materials Science.* the separation of a material into two or more parts as a result of stress.

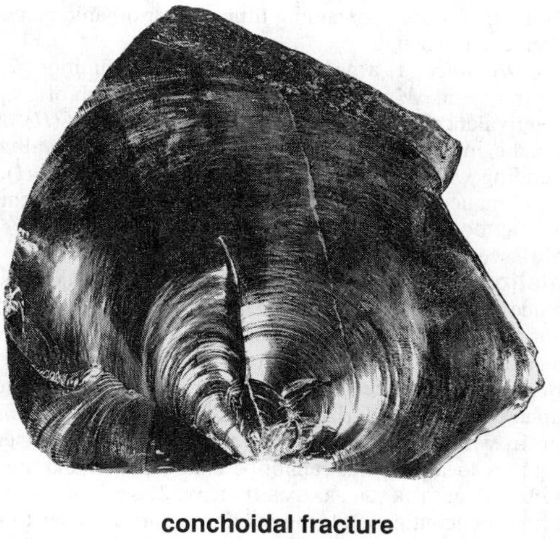

conchoidal fracture

fracture cleavage *Geology.* the splitting of a deformed but only slightly metamorphosed rock along very closely spaced parallel joints or fractures. Also, FALSE CLEAVAGE, CLOSE-JOINT CLEAVAGE.

fractured formation *Petroleum Engineering.* rock that has been broken and split by hydraulic pressure produced by injected fluids in a reservoir formation.

fracture dome *Mining Engineering.* a zone of loose or semiloose rock in the immediate hanging or footwall of a stope.

fracture initiation energy *Materials Science.* the work performed to fracture a material; required to supply the energy needed to create the fracture surfaces and to plastically deform the material if local yielding occurs prior to fracture.

fracture mechanics *Materials Science.* the analysis of structural materials and stresses in order to predict the fracture strength of materials or determine the causes of failure.

fracture mirror *Materials Science.* the smooth and flat surface of certain fractures, indicating that the crack originated at an internal flaw and traveled radially in a single plane.

fracture stress *Mechanics.* the maximum stress that a solid can withstand before fracturing.

fracture surface energy *Materials Science.* the energy required to form the new surfaces resulting from fracturing a material, especially a brittle material. Also, WORK OF RUPTURE.

fracture system *Geology.* a set of fractures formed at about the same time by similar sources of stress.

fracture test *Engineering.* a test wherein a metal part is broken and the fracture face is studied to determine information about the part including grain size, case depth, etc.

fracture toughness *Materials Science.* a property of a material, representing the material's resistance to fracture in response to high local stresses. The stress intensity factor, K_I, is used to predict the fracture toughness of most materials. For a given flawed material, fracture occurs when the stress intensity factor reaches a critical value, K_{IC}, the fracture toughness of the material.

fracture wear *Materials Science.* the wear of an abrasive due to mechanical failure of the grains.

fracture zone *Geology.* a very long, narrow, irregular area of displacement on the deep sea floor, often separating regions that differ in depth.

fracturing *Petroleum Engineering.* a method of opening up access to the pores of a hydrocarbon-bearing formation through the use of high-pressure fluid directed at the rock, causing it to crack and split.

fractus *Meteorology.* a cloud species characterized by small, irregular elements that appear ragged and shredded, as if torn, and that change ceaselessly and rapidly.

fragile *Science.* easily broken or damaged; brittle. *Artificial Intelligence.* of a program or reasoning technique, easily broken, or failing to work unless all of the inputs are exactly right.

fragile chromosome site *Genetics.* an inherited locus on chromosomes that experiences an increased incidence of acentric fragmentation and deletions.

fragility *Science.* the state or condition of being fragile.

fragility test *Pathology.* a measure of the resistance to dissolution of red blood cells in hypotonic saline solutions.

fragipan *Geology.* a hard, highly dense, slowly permeable subsurface layer of sandy, silty soil, containing little clay or organic matter, and appearing cemented when dry.

fragment *Ordnance.* **1.** a piece of an explosive munition. **2.** to break apart into fragments. *Mathematics.* let H be a subgraph of graph G. Define an equivalence relation on the edges of $E(G) - E(H)$ by placing edges e_i and e_j in the same equivalence class if there is a walk beginning with e_i, ending with e_j, and having no internal vertex in $V(H)$. The subgraphs of G induced by the equivalence classes are the fragments of H in G. Also, BRIDGES OF H, H-COMPONENTS.

fragmental see CLASTIC.

fragmentation the process of breaking up into small parts; specific uses include: *Cell Biology.* **1.** the breaking of DNA into fragments by a variety of techniques. **2.** the process by which platelets are released from megakaryocytes. **3.** the breakdown or vesiculation of a variety of organelles, especially the nuclear membrane during the prophase to prometaphase transition during mitosis. *Computer Programming.* **1.** a condition in which small areas of available memory are scattered in noncontiguous locations that cannot be amalgamated and are too small to be useful. Also, STORAGE FRAGMENTATION. **2.** the process of loading a program into noncontiguous locations in memory in order to make best use of available memory. *Metallurgy.* in a heavily cold-worked metal or alloy, the formation of variously oriented crystallite inside each single crystal grain. *Ordnance.* of or relating to munitions designed to explode into multiple small fragments. Thus, **fragmentation ammunition, fragmentation (protective) body armor, fragmentation test.**

fragmentation bomb *Ordnance.* a bomb that is designed to explode into multiple small fragments. Thus, **fragmentation bomb cluster.**

fragmentation grenade *Ordnance.* a grenade that is designed to explode into multiple small fragments.

fragmentation index *Computer Technology.* a measure of the scattered available space in a volume.

fragmentation nucleus *Meteorology.* a tiny ice particle broken from a large ice crystal, which becomes a growth center for a new ice crystal.

fragment emission *Ordnance.* the pattern of fragments created by an exploded munition, including the number of fragments and the direction, shape, and velocity of each fragment.

fragmenting *Computer Programming.* the process of breaking down a document into a set of terms, descriptors, or other components.

FRAM fleet rehabilitation and modernization.

framboid *Geology.* a microscopic aggregate of pyrite grains, often occurring in spherical clusters similar to raspberry seeds.

frame an enclosing structure or format; specific uses include: *Building Engineering.* **1.** the woodwork around windows and doors. **2.** see FRAMING. *Naval Architecture.* any of the riblike structural members of a ship's hull, curving from the keel upward to the side rail. *Graphic Arts.* **1.** a section of film that yields one positive. **2.** a vacuum device designed to hold a negative and plate in contact during exposure. *Electronics.* any of a repeating cycle of pulses. *Telecommunications.* **1.** a televison tube picture scan that mixes interleaved information. **2.** a repetitive group of signals resulting from a single sampling of all channels. **3.** in FAX technology, a rectangular area whose width is the available line and whose length varies with the service requirements. *Computer Programming.* a series of bits that includes an opening flag sequence, information bits, a closing flag sequence, and error correction bits. *Artificial Intelligence.* a technique for representing knowledge as a data structure resembling a relational data base. *Mathematics.* an ordered basis for the tangent space of a manifold at a point on the manifold. A family of frames on an open set is a **moving frame.** *Statistics.* see SAMPLING FRAME.

frame-based shell *Artificial Intelligence.* a knowledge system programming tool that represents objects and their relationships as frames, rather than in another form such as rules.

frame buffer *Computer Technology.* a temporary storage device that holds a video frame for subsequent display.

frame camera *Optics.* a camera that exposes thousands of frames of film per second, used for high-speed cinematography or special effects.

framed and braced door *Building Engineering.* a door that is bordered and secured in a frame with two stiles and three rails along the top, middle, and bottom sections, with diagonal braces along the rails.

frame frequency *Electronics.* in television, the number of times per second that the frame is scanned; equal to 30 in the United States.

frame of reference *Science.* **1.** a formal context associated with a system, presumed to exist so as to allow the system to be located and described in terms of space and time. It usually consists of an observer, a set of physical coordinates, and a means of measuring time. **2.** in popular use, any context of information that allows an individual item to be analyzed or described in more general terms. *Physics.* a coordinate system attached to a body or moving relative to a body in a specified way.

frame problem *Artificial Intelligence.* the problem of efficiently maintaining a correct and consistent world model as some aspects of the world are changed by operators (but most aspects remain the same).

framer *Electronics.* a device that adjusts FAX equipment so that, as the line progresses, the recorded elemental area has the same relationship to the record sheet as the corresponding transmitted elemental area has to the subject copy.

frameshift mutation *Genetics.* a mutation that causes the addition or deletion of a base within a coding sequence, resulting in the misreading of the code during translation because of a change in the reading frame. Also, READING FRAME SHIFT.

framework *Engineering.* the supporting skeleton of a structure, generally composed of either reinforced concrete, steel, or timber. *Geology.* **1.** an inflexible arrangement in a sediment or sedimentary rock created by particles that support one another at points of contact. **2.** a rigid, wave-resistant, calcareous structure produced by sedentary organisms, such as sponges and corals, in a high-energy environment.

framing *Building Engineering.* the skeletal structure of a building. Also, FRAME. *Electronics.* **1.** the process of synchronizing the vertical component of a video signal in order to align the top and bottom of the transmitted and received pictures. **2.** in FAX technology, the process of adjusting the picture to a desired position in the direction of line progression. *Telecommunications.* the process of aligning the characters in a digital alphanumeric transmission.

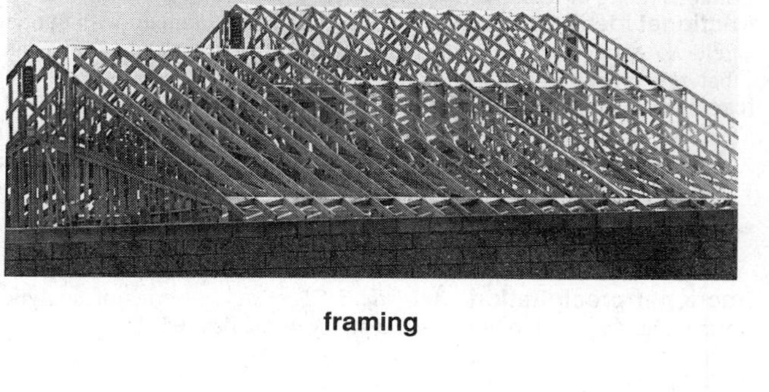

framing

framing chisel *Mechanical Devices.* a long, stout, heavy-duty chisel used by woodworkers for making mortises and other deep cuts. Also, MORTISE CHISEL.

framing control *Electronics.* a knob for centering and adjusting the height and width of a television picture.

framing square *Mechanical Devices.* a large, notched, and graduated carpenter's square with leg dimensions of 16 and 24 inches, respectively; used in building construction, particularly roof framing for rafters and stair joists. Also, STEEL SQUARE, ROOF FRAMING SQUARE.

framing table *Mining Engineering.* an inclined table that uses running water to separate ore slimes. Also, MINER'S FRAME.

Francisella *Bacteriology.* a genus of Gram-negative, aerobic bacteria of uncertain affiliation that occur as parasites and pathogens in many animals, including humans. (From Edward *Francis,* 1872–1957, American physician.)

Francisella tularensis *Bacteriology.* the type species of the genus *Francisella* and the causal agent of the disease tularemia in humans. Also, PASTEURELLA TULARENSIS.

Francis formula *Fluid Mechanics.* a formula for the rate of water flow over a rectangular weir as a function of length and head; the modified Francis formula includes two end corrections.

Francis turbine *Mechanical Engineering.* a reaction hydraulic turbine that operates on a low and medium head, with water entering radially and leaving axially; used in many large hydroelectric configurations.

francium *Chemistry.* a chemical element having the symbol Fr, the atomic number 87, and atomic weight 223; the heaviest of the alkali metals, probably existing only as radioactive isotopes, of which francium-223 is the longest-lived, with a half-life of 21 minutes. (From *France*; named by its discoverer, the French scientist Marguerite Perey.)

Franck, James 1882–1964, German physicist; Nobel Prize (with G. Hertz) for experimental confirmation of Bohr atomic model.

Franck-Condon factor *Quantum Mechanics.* see OVERLAP INTEGRAL.

Franck-Condon principle *Quantum Mechanics.* a principle regarding the vibrational wavefunctions that occur when an electronic transition takes place in a molecule; it states that the transition between two different electronic states takes place so quickly, in relation to the rate of nuclear vibration, that the nuclei do not move significantly during this transition and can thus be considered to be stationary for purposes of evaluation or calculation.

franckeite *Mineralogy.* $(Pb,Sn^{+2})_6Fe^{+2}Sn_2^{+4}Sb_2^{+3}S_{14}$, a dark gray to black, opaque, triclinic mineral occurring in foliated to radial masses and as thin tabular crystals with one perfect cleavage, having a specific gravity of 5.88 to 6.01 and a hardness of 2.5 to 3 on the Mohs scale; found in tin veins and limestone contact rocks.

Franck-Hertz experiment *Quantum Mechanics.* the first demonstration of the quantization of atomic energy levels; electrons scattered from atoms displayed energy loss corresponding to excitation of bound state transitions.

francolite see CARBONATE-FLUORAPATITE.

Franconian *Geology.* a North American geologic stage of the Upper Cambrian period, after the Dresbachian and before the Trempealeauan.

frangibility *Mechanics.* the property of a material whose grains are too weakly bonded to withstand large stress.

frangible *Mechanics.* hard but fragile; describing a material that will break or fracture from the effect of large stress.

frangible bullet *Ordnance.* a nonmetallic bullet used in firing practice; it is usually designed to leave a mark at the point of impact without penetrating the target.

frangible grenade *Ordnance.* the technical name for a Molotov cocktail, a crude incendiary grenade.

frank *Pathology.* of a condition, clinically evident; obvious.

Frank, Ilya born 1908, Soviet physicist; with Tamm, Nobel Prize for the theoretical interpretation of the Cherenkov effect.

Frank dislocation *Metallurgy.* a pinned dislocation that, under an applied stress, generates additional dislocations. This effect is at least partly responsible for strain hardening. Also, FRANK-READ SOURCE.

Frankeniaceae *Botany.* a family of dicotyledonous, halophytic, and sometimes tanniferous herbs and shrubs of the order Violales; native to mostly temperate and subtropical saline habitats in the Mediterranean region.

Frankfurt plane *Anthropology.* see EYE-EAR PLANE.

Frankiaceae *Bacteriology.* a family of bacteria of the order Actinomycetales, characterized as Gram-positive, aerobic, mycelial organisms that occur as symbionts in root nodules of nitrogen-fixing plants. The single genus is **Frankia.**

frankincense *Materials.* an aromatic gum resin obtained from various trees of the genus *Boswellia* and used in perfumery, in pharmaceutical and fumigating preparations, and as incense.

Frankland, Edward 1825–1899, English chemist; formulated the concept of valency.

Frankland's synthesis *Organic Chemistry.* the production of a zinc dialkyl from an alkyl halide and zinc. Also, **Frankland's method.**

Franklin, Benjamin 1706–1790, American statesman, inventor, printer, author, and scientist; discovered the electrical nature of lightning; invented the lightning rod; studied the movement of the Gulf Stream and the meteorological effects of volcanic eruptions.

Franklin, Rosalind Elsie 1920–1958, English chemist and molecular biologist; performed X-ray analyses of DNA molecule.

Franklin antenna *Electricity.* a base-fed vertical antenna, composed of a set of in-phase vertical half-wave elements arranged in line, that gives broadside radiation by eliminating phase reversals with loading coils or wire folds.

franklinite *Mineralogy.* $(Zn,Mn^{+2},Fe^{+2})(Fe^{+3},Mn^{+3})_2O_4$, a brittle, iron-black, cubic mineral of the spinel group with a metallic luster, occurring in octahedral crystals, and having a specific gravity of 5.07 to 5.22, and a hardness of 5.5 to 6.5 on the Mohs scale; found in zinc deposits.

Franklin oscillator *Electroncis.* a two-terminal feedback oscillator using transistors whose frequency-determining circuits are coupled to small capacitances of the input and output of a two-stage electron tube amplifier.

Frank partial dislocation *Crystallography.* a partial dislocation in which the Burgers vector is not parallel to the fault plane.

Frank-Read source *Metallurgy.* the site of a Frank dislocation.

Franz-Keldysh effect *Optics.* the lengthening in wavelengths of the optical absorption edge of a semiconductor while under the influence of a strong electric field.

Frary metal *Metallurgy.* a lead-base alloy containing one to two percent barium and calcium.

FRAS *Ordnance.* a Soviet unguided nuclear antisubmarine rocket projected through the air to drop depth charges in the vicinity of a submarine target, having a range of about 20 miles.

Frasnian *Geology.* a European geologic stage of the Upper Devonian period, occurring after the Givetian and before the Famennian.

fraternal polyandry *Anthropology.* the practice in which one woman is married to two or more brothers simultaneously, often with one joint household.

fraternal twins see DIZYGOTIC TWINS.

Frateuria *Bacteriology.* a genus of Gram-negative, obligately aerobic bacteria of the family Pseudomonadaceae, that occur on certain fruits as flagellated or nonmotile rod-shaped cells and are able to grow under acidic conditions.

fratricide *Behavior.* the killing of an animal shortly after birth by its own sibling. Also, SIBICIDE, SIBLICIDE.

Fraunhofer, Joseph von 1787–1826, German instrument maker; observed and mapped dark absorption lines in solar and stellar spectra.

fraunhofer *Spectroscopy.* in the measurement of the reduced width of a spectral line, a unit equal to 106 times the equivalent width divided by the corresponding wavelength.

Fraunhofer diffraction *Optics.* the pattern formed when a parallel beam of light falling on a narrow aperture is brought to focus and observed at an effectively infinite distance from the diffracting object. *Crystallography.* diffraction observed with parallel incident radiation; the incoming and outgoing wavefronts are planar.

Fraunhofer lines *Spectroscopy.* dark spectral lines produced as radiant energy is absorbed by cooler gases in the outer layers of the sun and other stars.

Fraunhofer region *Electromagnetism.* a distant area around a radiating antenna from which radiation appears to be coming from a single point.

Fraunhofer spectrum *Spectroscopy.* the dark-line (absorption) spectrum of the sun.

Frazer, Sir James George 1854–1941, Scottish anthropologist; studied folklore and classical history; author of *The Golden Bough.*

Frazer-Brace extraction method *Chemical Engineering.* a procedure used to draw oil from citrus fruits, in which a machine with abrasive carborundum rolls grates the peel from the fruit under a water spray; the mixture of water and peel is screened to separate the peel, allowing the oil to settle.

frazil *Hydrology.* small, disk-shaped or needle-shaped ice crystals formed in supercooled, turbulent water.

frazil ice *Hydrology.* a spongy, slushy aggregate or mass of frazil suspended in supercooled, turbulent water. Also, NEEDLE ICE.

FRC functional residual capacity.

Fréchet, Maurice-Rene [frä´she´; frä´shē´] 1878–1973, French mathematician; founded theory of abstract spaces.

Fréchet compactness see BOLZANO-WEIERSTRASS PROPERTY.

Fréchet derivative *Mathematics.* suppose X and Y are Banach spaces (where $\|\cdot\|$ denotes the norm on Y), U is an open subset of X, and $F : U \to Y$ is a bounded linear mapping of U into Y. (The collection of all such bounded linear mappings forms a Banach space $B(X, Y)$.) If, for a given point $a \in U$, there exists a mapping $(DF)_a : X \to Y$ in $B(X, Y)$ such that

$$\lim_{\varepsilon \to 0} \|F(a + x) - F(a) - (DF)_a x\| / \|x\| = 0$$

then $(DF)_a$ is called the Fréchet derivative of F at a. $(DF)_a$, if it exists, is unique. If $(DF)_a$ exists for every $a \in U$, and if the assignment $a \to (DF)_a$ is a continuous mapping of U into $B(X, Y)$, then F is said to be continuously differentiable in U.

Fréchet space *Mathematics.* **1.** a metrizable, complete topological vector space; equivalently, a (locally) convex space whose topology is induced by a complete invariant metric. **2.** a topological space X is said to be **Fréchet compact** (or to have the Bolzano-Weierstrass property) if every infinite subset of X has a limit point in X. **3.** A quasi-normed (or normed) linear space X is called a Fréchet space (or **F-space**) if it is complete.

freckle *Medicine.* a benign pigmented spot on the skin, usually resulting from exposure to sunlight, commonly on the face.

freckling *Medicine.* the process of forming freckles. *Metallurgy.* the undesirable segregation of alloy elements in a consumable-electrode remelted alloy, appearing, in a cross section, as freckles.

Fredericksburgian *Geology.* a North American Gulf Coast stage of the Lower Cretaceous period, occurring after the Trinitian and before the Washitan.

Fredholm, Eric Ivar 1866–1927, Swedish mathematician; founded modern integral equation theory; formulated Fredholm equations.

Fredholm alternative *Mathematics.* **1.** let T be a compact linear operator on a Banach space X and let $\lambda \neq 0$ be a complex number. Then exactly one of the following statements is true: (a) $Tf - \lambda f = g$ has a unique solution f for each $g \in X$; i.e., $T - \lambda I$ is an isomorphism, where I is the identity operator. (b) $Tf - \lambda f = 0$ has nonzero solutions; i.e., λ is an eigenvalue of T. Equivalently, $Tf - \lambda f = g$ has infinitely many solutions for some g and none for others. Thus, for each λ (except possibly $\lambda = 0$), the solutions span a finite dimensional subspace of X. **2.** see FREDHOLM THEOREM.

Fredholm determinant *Mathematics.* the Fredholm integral equation of the second kind can be represented as the limiting case of a system of linear equations by the following method, known as the **Fredholm method.** Divide the interval $[a,b]$ into n equal parts of size $\Delta = (a - b)/n$ and write $x_i = y_i = a + (i - 1)\Delta$. Then the following n approximate equations hold in $[a,b]$:

$$\phi(x_i) - \lambda\Delta[\sum_{j=1}^{n} K(x_i, x_j)\phi(x_j)] \approx f(x_i)$$

for $i = 1, 2, \ldots, n$. This system of equations can be viewed as an $n \times n$ matrix multiplying the $n \times 1$ column vector $(\phi(x_i))$ to yield the $n \times 1$ vector $(f(x_i))$. The determinant of the $n \times n$ matrix is the Fredholm determinant. As $n \to \infty$, the Fredholm determinant takes the form of a power series in $K(x,y)$.

Fredholm integral equations *Mathematics.* **1.** also, **Fredholm integral equation of type 1.** the Fredholm integral equation of the first kind is

$$f(x) + \int_a^b K(x, y)\phi(y)\, dy = 0$$

where the functions $K(x, y)$ (called the kernel) and $f(x)$ (called the perturbation function) are continuous in x and y on $[a, b]$ and $\phi(x)$ is to be determined. **2.** also, **Fredholm integral equation of type 2.** the Fredholm integral equation of the second kind has the unknown function $\phi(x)$ outside the integral; it is

$$f(x) + \lambda\int_a^b K(x, y)\phi(y)dy = \phi(x)$$

Fredholm operator *Mathematics.* a linear operator T between Banach spaces with closed range, such that $\alpha(T)$ and $\beta(T)$ are both finite, where $\alpha(T)$ is the dimension of the null space of T and $\beta(T)$ is the dimension of the adjoint of T. The differencex $\chi(T) = \alpha(T) - \beta(T)$ is called the index of T. Sometimes a Fredholm operator is called a Noetherian operator and the name Fredholm operator is reserved for the case $\chi(T) = 0$.

Fredholm point *Mathematics.* let X be a complex Banach space and T a linear operator on X. A complex number λ is called a Fredholm point of T if $\lambda I - T$ is a Fredholm operator on X, where I is the identity operator.

Fredholm theorem *Mathematics.* given a Fredholm integral equation of the second kind $f(x) + \lambda \int_a^b K(x, y)\phi(y)\, dy = \phi(x)$, exactly one of the following is true: (a) λ is not an eigenvalue; that is, the equation has no nontrivial solutions; or (b) the corresponding integral equation of the first kind has at least one solution, and all such solutions are mutually linearly independent.

Fredholm theory *Mathematics.* the study of Fredholm equations and their solutions.

free *Chemistry.* not combined with something else such as another element or electron; uncombined, unattached. *Mathematics.* given a graph H, a graph G such is H-free if no vertex-induced subgraph of G is isomorphic to H. $K_{1,3}$-free graphs (or claw-free graphs) have a place in the study of Hamiltonian cycles.

free Abelian group *Mathematics.* an Abelian group F that satisfies one of the four equivalent conditions: (a) F has a nonempty basis; (b) F is the (internal) direct sum of a family of infinite cyclic subgroups; (c) F is (isomorphic to) a direct sum of copies of the additive group of integers; or (d) there exists a nonempty set X and a function $\iota: X \to F$ with the following property: given an Abelian group G and a function $f: X \to G$, there exists a unique homomorphism of groups $f': F \to G$ such that $f'\iota = f$. In other words, F is a free object in the category of Abelian groups.

free admittance *Electricity.* the inverse of a blocked impedance.

free air see FREE ATMOSPHERE.

free-air anomaly *Geodosy.* the difference between observed gravity and theoretical gravity that has been figured for latitude and corrected for elevation of the station above or below the geoid, by application of the normal rate of change of gravity for change of elevation. Also, **free-air gravity anomaly.**

free-air correction *Geodesy.* the correction factor, usually expressed in milligals per meter, that is applied to observed gravity to reduce its value to the value of gravity at sea level.

free ascent *Engineering.* the act in which a diver rises to the surface in an emergency either by natural buoyancy or with the assistance of a life jacket.

free association *Psychology.* **1.** a psychoanalytic technique in which the subject says whatever comes to mind with no control or censorship; used in gaining access to repressed or unconscious material. **2.** also, **free association test.** an activity in which the subject is asked to respond to a series of cues, usually individual words, by making an immediate, spontaneous association.

free atmosphere *Geophysics.* the atmosphere at an altitude of about 600 meters, at which the influence of surface friction is assumed to be negligible and air is an ideal fluid. Also, FREE AIR.

free atom *Atomic Physics.* an unbound atom that is not influenced by the activities of adjacent ions, atoms, or molecules; assumed to exist during reactions.

free balloon *Aviation.* a usually spherical nondirigible balloon that floats without a tether, propulsion, or guidance and is able to descend by the release of gas.

freeboard *Naval Architecture.* the vertical distance from a ship's waterline to the gunwale or side of the weather deck, measured at the waist. *Civil Engineering.* the measure of height between normal water level and the crest of a dam or the top of a flume. *Analytical Chemistry.* **1.** free space in a reaction vessel. **2.** space provided above the resin bed in an ion-exchange column to allow for expansion of the bed.

freeboard deck *Naval Architecture.* the deck from which a ship's freeboard is measured for purposes of defining its margin line; it is usually the lowest open deck.

free-burning coal see NONCAKING COAL.

free carbon *Metallurgy.* in a ferrous product, the fraction of carbon that is not combined with another element.

free charge *Electricity.* an electrostatic charge that is not bound by an equal or greater charge of opposite polarity.

free convection see NATURAL CONVECTION.

free-cutting brass *Metallurgy.* any of several lead-bearing copper-zinc alloys that can be easily machined. The most widely used version contains 2.5–3.7% lead.

free cyanide *Chemistry.* cyanide that is not part of an ionic complex.

free diving *Engineering.* diving in which scuba equipment is used to assist the diver in moving freely under water.

free-drop *Engineering.* to release materials without a parachute from an airborne vehicle.

free electromagnetic field *Electromagnetism.* an electromagnetic field in space that does not interact with matter.

free electron *Physics.* an electron that has been removed from an atom and is therefore able to move about freely.

free-electron laser *Optics.* a laser in which the active medium is a beam of electrons passing through a magnetic field and yielding a powerful beam of radiation.

free-electron theory of metals *Solid-State Physics.* an early theory of metallic conduction based on the concept that outer valence electrons, which do not form crystal bonds, are free to migrate through crystal, and so form electron gas. Also, SOMMERFELD THEORY.

free-end *Building Engineering.* describing a beam that is not fixed or built in.

free energy *Thermodynamics.* **1.** a system's capacity to perform useful work; the amount of energy available for work. **2.** see GIBBS FREE ENERGY. **3.** see HELMHOLTZ FREE ENERGY.

free enthalpy *Physics.* a formula for determining the amount of energy available for work in a system after that system undergoes a change. Also, GIBBS FREE ENERGY, THERMODYNAMIC POTENTIAL.

free face *Mining Engineering.* **1.** a longwall face that does not use props between the conveyor and the coal. Also, PROP FREE FRONT. **2.** a surface near the shothole at which the rock is free to move under the force of the explosion.

free fall *Mechanics.* **1.** the ideal motion of a body when it is affected by gravity but completely unaffected by any other forces; the motion may be either descending or ascending. **2.** an actual descent that is comparable to this because air resistance is minimized. **3.** specifically, the initial stage in a parachute jump when the jumper has not yet opened the parachute. *Space Technology.* the fall or drop of a rocket or other body that is not guided, not under thrust, and not retarded by a braking device. The free fall ends when the body strikes a surface or encounters an effective braking force other than air particles.

free-falling velocity *Mechanics.* in the fall of a particle of powder through a still fluid, the speed at which the effective weight of the particle is balanced by the drag exerted on the particle by the fluid.

free fatty acid *Nutrition.* a term applied to highly reactive fragments of molecules produced during the splitting of fatty acid molecules. Free radicals may react at random with damaging effects to other fats as well as proteins, vitamins, and enzymes; they are responsible for some of the unpleasant effects of rancid fats.

free ferrite *Metallurgy.* in the solid-state transformation of a steel, the ferrite formed upon cooling that is not associated with iron carbide.

free field *Physics.* a vector field that is not affected by any other fields, such as the gravitational field of a body that is (effectively) infinitely far away from any other body. *Acoustics.* an isotropic, uniform sound field in a volume with no interacting boundaries.

free-field storage *Computer Technology.* a system of organization that permits data to be recorded on a storage medium with no constraints as to preassigned and fixed field arrangements; the information may appear anywhere in the record.

free filter *Mathematics.* a filter $F(X)$ on a nonvoid set X such that the intersection of the sets of $F(X)$ is empty. By theorem, if X is infinite, then there exists a **free ultrafilter** in X.

free fit *Design Engineering.* a liberal fit allowance in mating pieces of machine parts, especially where large variations of temperature may occur.

free flap *Surgery.* a mass of tissue that is detached from the body and reattached at a new site by connecting the blood vessels.

free flight *Mechanics.* the flight of an unattached, unguided, unpropelled body, such as a rocket descending without power or guiding direction. *Aviation.* any flight of an aircraft at less than normal power.

free-flight angle *Mechanics.* the angle between the flight path and the horizontal, at the beginning of the free flight of a rocket or projectile.

free-flight melt spinning *Metallurgy.* any of various methods for producing rapidly solidified alloys.

free-flight trajectory *Mechanics.* the path of an unpropelled object, or the portion of an object's flight path during which it is unpropelled.

free flow *Transportation Engineering.* a condition of traffic flow in which individual vehicles are free to choose their own speed, with minimal interference from bottlenecks or other flow restrictions.

free foehn see HIGH FOEHN.

freeform text *Computer Programming.* a record that contains plain, unformatted natural language text.

free gas *Physics.* any substance that naturally exists in its gaseous phase. *Petroleum Engineering.* **1.** a hydrocarbon that exists at reservoir temperature and pressure in the gaseous phase and remains so when produced under standard conditions. **2.** natural gas produced alone, not with condensate or crude oil.

free gold *Metallurgy.* a term for pure, native gold.

free graft *Surgery.* a tissue or organ from any area that is cut free and transplanted to a totally different area.

free groundwater see NONARTESIAN WATER.

free group *Mathematics.* a free group G on a set X is the set of all finite products of elements of X such that the product of two elements of X is the group identity element if and only if the given generators are inverses. Equivalently, G can be viewed as the set of all reduced words on X. The elements of X are called the generators of G.

free gyroscope *Engineering.* a gyroscope used to measure changes in the altitude of a piece of equipment through use of the property of gyroscopic rigidity. The spinning rotor is isolated from the aircraft by gimbals.

free hole *Solid-State Physics.* a hole in a semiconductor that is not bound to an exciton or an impurity.

free impedance *Electronics.* the input impedance of a transducer when its load impedance is zero. Also, NORMAL IMPEDANCE.

freeing port *Naval Architecture.* a deck level opening in a ship's bulwarks through which water on the deck can pour out.

free ion *Physical Chemistry.* an ion whose properties, such as spectrum and magnetic moment, are not significantly affected by neighboring atoms or molecules; commonly seen in ionized gases.

free joint *Robotics.* a robotic articulation having six degrees of freedom.

free-living *Biology.* **1.** living independent of a host; not parasitic. **2.** not attached; able to move freely about.

freely jointed chains *Materials Science.* a polymer chain in which there are no fixed bond angles and therefore no restriction of rotation at any bond.

free-machining steel *Metallurgy.* a steel containing any of various elements that improve machineability.

freemartin *Vertebrate Zoology.* a sexually imperfect, usually sterile female calf that is born as a twin with a male.

free meander *Hydrology.* a meander that displaces itself very easily by the abrasive action of the sediment load along the banks of the stream.

free milling *Metallurgy.* a process of reducing ore that contains free gold or silver by crushing and amalgamation, without roasting or other chemical treatment.

free-milling gold *Metallurgy.* gold that is suitable for amalgamation with mercury.

free-milling ore *Metallurgy.* any ore containing gold that can be caught with mercury.

free module *Mathematics.* a module F that is a free group with respect to its additive group; that is, a unitary module F over a ring R with identity, which satisfies one of the following equivalent conditions: (a) F has a nonempty basis; (b) F is the internal direct sum of a family of cyclic R-modules, each isomorphic as a left R-module to R; (c) F is R-module isomorphic to a direct sum of copies of R considered as a left R-module over itself; or (d) there exists a nonempty set X and a function $\iota: X \to F$ with the property that given any unitary R-module A and a function $f: X \to A$, there exists a unique R-module homomorphism $f': F \to A$ such that $f' \iota = f$. In other words, F is a free object in the category of unitary R-modules.

free moisture *Hydrology.* see FREE WATER. *Mining Engineering.* **1.** any moisture that can be removed from coal by standard air drying. **2.** the part of the total moisture lost by coal in attaining approximate equilibrium with the atmosphere to which it is exposed. **3.** any moisture not retained or absorbed by aggregate.

free molecule *Physical Chemistry.* a molecule whose properties, such as spectrum and magnetic moment, are not affected by neighboring atoms or molecules; commonly seen in ionized gases.

free molecule flow *Physics.* a flow of gas in which the mean free path of the gas particles is much greater than any characteristic dimension of the container holding the gas.

free motional impedance *Electricity.* the complex remainder of the subtraction of the blocked impedance of a transducer from the free impedance.

free object *Mathematics.* let F be an object in a concrete category C, X a nonempty set, and $i: X \to F$ a map of sets. F is said to be free on the set X if for any object A of C and any map of sets $f: X \to A$, there exists a unique morphism $f': F \to A$ of C such that $f'i = f$ (as a map of sets).

free oscillation *Physics.* a vibration in an oscillatory system that is self-sustained, having no external stimulation.
free-piston engine *Mechanical Engineering.* a piston engine in which a power piston acts directly on a compressor piston working in the opposite direction.
free port *Civil Engineering.* an isolated, enclosed, and policed port in or adjacent to a port of entry, without a resident population, and freed from normal customs duties.
free product *Mathematics.* the free product on a family of groups $\{G_i\}$ is the group $G = \prod^*_{i \in I} G_i$, i.e., the set of all reduced words on $X = \cup_{i \in I} G_i$. Equivalently, a group G with identity e is a free product of nontrivial subgroups G_i (where $G_i \cap G_k = e$ for all i, k) if each element g of G ($g \neq e$) can be uniquely expressed in the form $g = a_1 \ldots a_k$, with $a_i \in G_i$ and $a_i \neq e$.
free progressive wave *Physics.* a wave whose propagation is unaffected by boundaries or boundary conditions.
free quark *Particle Physics.* a quark that exists singly, rather than in combination with other quarks; no experimental evidence for the existence of free quarks has been found.
free radical *Chemistry.* a usually short-lived, highly reactive molecular fragment that contains one or more unpaired electrons; formed by the splitting of a molecular bond, and capable of initiating or mediating a wide variety of chemical reactions.
free-radical addition *Materials Science.* a process in which polymerization proceeds by the chain reaction addition of monomer molecules to the free-radical ends of a growing polymer.
free-range *Agriculture.* relating to or raised by free-ranging: *free-range* cattle or chickens.
free-ranging *Agriculture.* the practice of allowing grazing or feeding livestock to move freely about, with no confinement or direct control of their activity.
free recall *Psychology.* an activity in which the subject is given a list of items and later asked to recall them in any sequence. Also, **free-recall task.**
free recoil *Ordnance.* a theoretical term describing the recoil movement of a gun without recoil-absorbing mechanisms such as springs or pneumatic pressure.
free-recoil mount *Ordnance.* a gun mount used in firing tests to simulate the theoretical condition of free recoil.
free recombination *Genetics.* the principle, from Mendel's second law, that all possible zygotes of two pairs of alleles will form in new sets of gene combinations at will.
free run *Ordnance.* the movement of a projectile from the chamber to its engagement with the rifling of the bore.
free-running frequency *Electricity.* the frequency at which a normally synchronized generator operates when there is no synchronizing signal.
free-running multivibrator see ASTABLE MULTIVIBRATOR.
free-running sweep *Electricity.* a frequency sweep that is not synchronized by an applied signal and that recycles without being triggered.
free sheet *Graphic Arts.* a high-quality paper that does not contain groundwood.
Freesia *Botany.* **1.** a genus of plants of the iris family, having tubular flowers that are white or yellow; native to southern Africa. **2. freesia.** any plant of this genus. (Named after F. H. T. *Freese*.)
free space *Physics.* a theoretical concept of space devoid of all matter. *Electromagnetism.* **1.** space with no free electrons or ions. **2.** the radiation pattern of an antenna that is not affected by surrounding objects. *Materials Science.* added volume produced by a rearrangement of molecules in an amorphous polymer.
free-space field intensity *Electromagnetism.* the intensity of a radio wave measured at a point in a homogeneous medium in the absence of extraneous waves.
free-space loss *Electromagnetism.* a theoretical loss of energy in radio waves that depends only on frequency and distance while disregarding all other variable factors.
free-space propagation *Electromagnetism.* the propagation of an electromagnetic wave in free space that is observed to follow a straight-line path.
free-space radar equation *Electromagnetism.* an equation giving the characteristics of a radar signal that propagates in free space.
free-space radiation pattern *Electromagnetism.* the radiation emitted by an antenna in a pattern that would result in the absence of all reflecting, refracting, and absorbing materials.
free-space wave *Electromagnetism.* an electromagnetic wave that propagates through free space.
freestone *Botany.* **1.** describing a fruit having a pit or stone to which the fruit's flesh does not cling, making the stone easily released, as in some varieties of peaches. **2.** the pit or stone of such a fruit. *Geology.* any stone, particularly fine-grained, thick-bedded sandstone, that breaks freely and can be cut and worked easily in any direction without tending to split, thus making it suitable as building stone. *Hydrology.* see FREESTONE WATER.
freestone water *Hydrology.* water that contains little or no dissolved material.
free storage *Computer Technology.* storage that is not assigned.
free storage list *Computer Programming.* a list maintained by the data management function of an operating system or program that identifies all memory cells available for issue.
freestream capture area *Aviation.* the cross-sectional area of the column of air taken in by a ramjet engine.
free surface *Fluid Mechanics.* any boundary between two immiscible fluids.
free-swelling index *Engineering.* a method by which the free-swelling properties of coal are measured; it involves heating coal under prescribed conditions to produce a coke button, and then comparing the shape and size of the button against a series of standard profiles.
free-swimming *Zoology.* describing an aquatic organism that is not attached to a base or part of a large colony. Thus, **free swimmer.**
free-swinging meander see FREE MEANDER.
Freeth's nephroid *Mathematics.* the strophoid of a circle with respect to the center as pole, fixed point on the circumference.

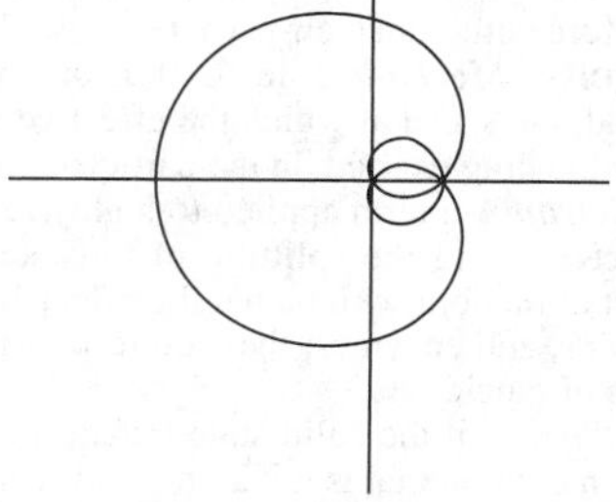

Freeth's nephroid

free tree *Mathematics.* in graph theory, a tree in which no vertex is a root.
free turbine *Mechanical Engineering.* **1.** a power-takeoff turbine that is mounted behind the main turbine/compressor assembly, drives the output shaft, and is not connected to the shaft driving the compressor. **2.** any auxiliary turbine that is a separate unit fed by a remotely produced fuel supply.
free variable *Mathematics.* in logic, a variable that is not bound; i.e., a variable whose occurrence is not within the scope of a quantifier and which can therefore be considered to be constant. *Artificial Intelligence.* in Lisp, a variable that is not bound within a function in which it appears.
free vector *Mechanics.* a vector that does not have a fixed position in space, such as a coordinate base vector.
free volume *Physics.* the volume of a containing medium holding a fluid, minus the volume occupied by the fluid molecules. *Materials Science.* the space in a solid or liquid polymer sample not occupied by polymer molecules; this allows molecular motion to take place.
free-volume theory *Materials Science.* a theory stating that for a polymer segment to move from its position in a three-dimensional framework to an adjacent site, a critical void must exist.
free vortex *Fluid Mechanics.* a two-dimensional concentric circular fluid flow in which the fluid velocity is inversely proportional to the square of the radius.
free wall *Mining Engineering.* the wall of an ore vein filling that scales off cleanly from the gouge.
freeware *Computer Programming.* a term for public-domain software programs. Also, SHAREWARE.

free water *Chemistry*. **1.** water that can be removed from another substance without changing the structure or composition of the substance. **2.** in a container holding both water and a suspension of water in another liquid, the water that is not in suspension. *Biochemistry*. the portion of water in body tissues that is not bound by macromolecules or organelles. *Hydrology*. water in the soil that is in excess of field capacity and that is thus free to move into the groundwater zone in response to gravitational force. Also, GRAVITY WATER. *Petroleum Engineering*. the water in permeable reservoir rock that usually surrounds oil and gas.

free-water content *Hydrology*. the quantity of liquid water present within a sample of wet snow.

free-water elevation or **free-water surface** see WATER TABLE.

free-water knockout *Petroleum Engineering*. a vessel or container used for piping in oil or an oil-water emulsion so that free water can settle out.

freeze *Physical Chemistry*. to undergo or cause to undergo the process of freezing. *Meteorology*. an extended period of subfreezing temperatures, typically for several days. *Engineering*. of moving parts, screws, nails, and so on, to become lodged or stuck tightly in place, as if frozen in ice. *Electronics*. **1.** to stop the projection of a motion picture film or videotape so that the action is isolated ("frozen") at a single point. **2.** a feature on a film projector or a videocassette recorder that stops the running of the film or tape in this manner.

freeze-drying *Engineering*. a process in which food or tissue is preserved, by drying it in a frozen state under high vacuum conditions; used in preparing medicines and laboratory specimens and in some forms of food preservation. Thus, **freeze-dry, freeze-dried.**

freeze etching *Physics*. the freezing of specimens and then cutting them along natural planes in order to prepare them for examination under an electron microscope.

freeze-fracture *Microbiology*. a process in which frozen samples are fractured, and carbon replicas are prepared of complementary surfaces; used to prepare cellular samples for electron microscopy, particularly in the study of surfaces or interfaces between membranes. Also, **freeze-fracturing**.

freeze-frame *Electronics*. **1.** a film technique in which a single frame of film is isolated and then reprinted as a continuous series to give the effect of the action being stopped at a single point. **2.** see FREEZE, def. 2.

freeze-out lake *Hydrology*. a term for a very shallow lake that is usually frozen over for long periods of time.

freezer *Mechanical Engineering*. **1.** a separate refrigeration unit that is held at a temperature below 32°F (typically 0–4°F), in which perishable foods can be stored in a frozen state. **2.** a compartment or unit of a refrigerator serving this same purpose.

freezer burn *Food Technology*. light-colored spots resembling burn marks, appearing on frozen food as a result of excessive moisture loss due to improper freezing or faulty packing.

freeze-sectioning *Microbiology*. the process of cutting, or sectioning, frozen-hydrated specimens with a microtome for examination by electron microscopy.

freeze sinking *Mining Engineering*. the use of circulating brine in a system of pipes to freeze waterlogged strata so that shafts can be sunk through them.

freeze-substitution *Microbiology*. a specimen preparation procedure in which a solvent replaces the ice crystals in a frozen specimen for subsequent X-ray spectrochemical analysis.

freeze-up or **freezeup** *Hydrology*. **1.** the formation of a continuous cover of ice on a body of water. **2.** the period during which a particular body of water is frozen over. *Mechanical Engineering*. **1.** the abnormal operation of a refrigeration unit due to the freezing of water over a working area. **2.** in a machine, the condition of being immobilized or inoperative due to freezing or mechanical failure.

freezing *Physical Chemistry*. the process in which a liquid is converted into a solid by the removal of heat from it; for a given pressure, this occurs at a fixed temperature in a pure substance, and over a range of temperatures in a mixture of substances. *Meteorology*. of temperatures, at or below the freezing point. *Engineering*. a process in which a part or element becomes tightly lodged or stuck in place. *Mining Engineering*. a procedure used in the drilling of water-saturated formations to prevent wall collapse, in which the drill is chilled to –34–40°C to freeze and thus consolidate the walls.

freezing level *Meteorology*. the height of the 0°C constant-temperature surface.

freezing microtome *Engineering*. an instrument for cutting sections of frozen tissue for microscopic study.

freezing mixture *Physical Chemistry*. **1.** any mixture whose freezing point is lower than the freezing points of its individual components. **2.** specifically, a mixture that is created to provide a freezing point below 0°C; for example, an ethylene glycol (antifreeze) mixture added to the water in a car's cooling system, or the application of salt to a frozen road surface.

freezing nucleus *Meteorology*. any particle within a mass of supercooled water that will initiate growth of an ice crystal about itself.

freezing point *Physical Chemistry*. the temperature at which a substance in liquid form freezes, equal to the temperature at which its solid form melts; this represents equilibrium between the liquid and solid phases. The freezing point of water at sea level is 32°F or 0°C.

freezing-point depression *Physical Chemistry*. a condition in which the freezing point of a solution is lower than the standard freezing point of the solvent in a pure state; the amount of depression is directly dependent on the amount of solute present. Also, DEPRESSION OF FREEZING POINT.

freezing precipitation *Meteorology*. any form of liquid precipitation that freezes upon impact with the ground or exposed surfaces, such as **freezing drizzle** or **freezing rain**.

freezing range *Metallurgy*. in a noneutectic alloy, the temperature range between the beginning and the completion of solidification.

Fregatidae *Vertebrate Zoology*. the frigate birds, a monogeneric family of web-footed tropical sea birds of the order Pelecaniformes, characterized by long narrow wings, a long pointed tail, a large red gular pouch in males, and rapid graceful flight.

F region *Geophysics*. the area of the ionosphere, about 120 to 400 km above sea level, in which the F_1 and F_2 layers form.

freibergite *Mineralogy*. $(Ag,Cu,Fe)_{12}(Sb,As)_4S_{13}$, a steel gray to iron black, opaque, cubic mineral containing up to 49% silver, massive in habit or as tetrahedral crystals, and having a specific gravity of 5.05 and a hardness of 3 to 4.5 on the Mohs scale; found in silver ore deposits.

freieslebenite *Mineralogy*. $AgPbSbS_3$, a steel-gray to silver-white, opaque, monoclinic mineral occurring in prismatic crystals and having a metallic luster, a specific gravity of 6.2 to 6.4, and a hardness of 2 to 2.5 on the Mohs scale; found with galena, orgenite, and siderite.

freight car *Engineering*. a railroad car that transports goods.

freighter *Engineering*. a ship or airplane that transports goods or material.

freijo [frā ē´hō] *Materials*. the durable wood of the *Cordia goeldiana* tree of the Amazon; used to make containers.

freirinite see LAVEDULAN.

fremitus [frem´i təs] *Medicine*. a sensation of vibration that is perceptible on the palpation of a body part, as in the chest.

fremontite see NATROMONT EBRASITE.

Fremy's salt see POTASSIUM BIFLUORIDE.

frenal *Anatomy*. of or relating to a frenum.

Frenatae *Invertebrate Zoology*. the suborder of butterflies and moths, lepidopteran insects, with a frenulum and reduced hindwing venation.

French chalk *Materials*. very finely ground talc, used as a dry lubricant and filler.

French coupling *Design Engineering*. a coupling that has both right-handed and left-handed threads.

French curve *Graphic Arts*. a plastic or metal drafting tool having a number of curves along one edge; used as a guide in drawing noncircular curves.

French drain *Civil Engineering*. a pipe used for subsurface drainage; may be made of earthenware, porous concrete, or plastic. Also, AGRICULTURAL DRAIN, FIELD DRAIN.

French polish *Materials*. a solution of shellac in methyl alcohol; used as a wood polish.

French press *Biotechnology*. an apparatus used to rupture cells: hydraulic pressure forces a piston into a cooled steel cylinder in which the cell sample was placed, and the cells are broken by high shear forces when they exit the cylinder due to the sudden change in pressure. Also, **French pressure cell.**

frenectomy *Surgery*. the excision of a frenum or frenulum.

Frenet-Serret formulas *Mathematics*. for a given curve in R^n, formulas for the directional derivatives of the unit vectors along the tangent, principal normal, and binormal of the curve. Also, SERRET-FRENET FORMULAS.

Frenkel defect *Solid-State Physics*. a disorder in a crystal lattice created by removing an ion from its lattice site and forcing it into an interstitial position, leaving a corresponding number of normal lattice sites vacant. Also, **Frenkel pair.**

Frenkel exciton *Solid-State Physics.* a tightly bound excited electron-hole pair that can move through a crystal.

Frenkel-Halsey-Hill isotherm equation *Physics.* an equation for the volume of a gas that is adsorbed onto a surface at a given temperature, $\ln(p/p_o)=a/v^s$, where v is the adsorbed volume, p is the gas pressure, p_o is the vapor pressure, and a and s are constants.

frenulum [fren′yə ləm] *Anatomy.* a small frenum. *Entomology.* **1.** in butterflies and moths, a bristle or group of bristles along the costal border of the hindwing, interlocking with the forewing in flight. **2.** any of the gelatinous folds or ridges supporting the umbrella in medusae.

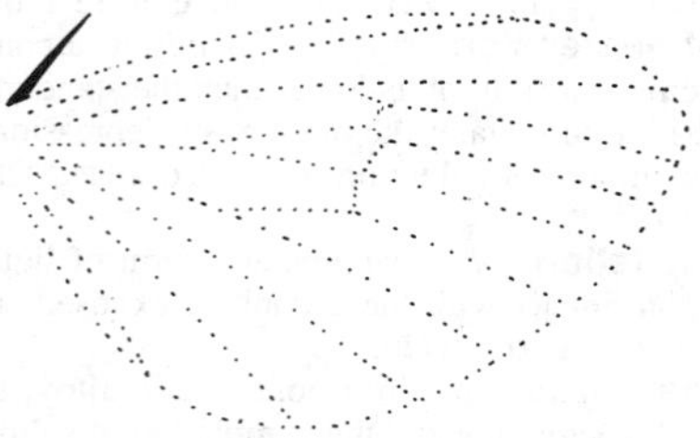

frenulum

frenum *plural,* **frena.** *Anatomy.* a fold of integument or mucous membrane that lies between two structures and limits the movement of one of those structures.

Freon [frē′än] *Organic Chemistry.* the trade name for a series of nonflammable, nonexplosive fluorocarbon or chlorofluorocarbon products that are widely used in refrigeration and air conditioning.

Freon-12 see DICHLORODIFLUOROMETHANE.

Freon-113 see TRICHLOROTRIFLUOROETHANE.

frequency *Physics.* the number of cycles or events per unit time, commonly having units of $\sec^{-1}$ (hertz). *Acoustics.* specifically, the number of sound waves per unit time that are produced by a vibrating object. *Statistics.* **1.** the number of items occurring in a given category. **2.** see RELATIVE FREQUENCY. *Mathematics.* see FREQUENCY DISTRIBUTION.

frequency agility *Military Science.* the capacity to shift radar or wireless frequency rapidly and continuously in order to avoid enemy jamming, reduce friendly interference, enhance target echoes, or carry out electronic countermeasures or counter-countermeasures.

frequency allocation *Telecommunications.* the assignment of bands of the frequency spectrum to specific categories of users; this is the frequency on which a particular transmitter has to operate, within a specified tolerance.

frequency analysis *Computer Programming.* in the analysis of algorithm complexity, the determination of the number of times parts of the algorithm are executed, for the purpose of improving the program and reducing the time required to solve the algorithm.

frequency analyzer *Electronics.* an instrument that measures the intensity of different frequency components in a particular oscillation; used to identify transmitting sources.

frequency-azimuth intensity *Electronics.* a radar display that correlates frequency, azimuth, and strobe intensity.

frequency band *Physics.* a range of frequencies that is continuous between specified limits.

frequency bridge *Electronics.* **1.** an AC bridge that can be nulled at only one frequency for a given set of bridge-arm values. **2.** a bridge used to measure unknown frequencies.

frequency changer see FREQUENCY CONVERTER.

frequency coding *Physiology.* a method of coding neuronal information using the frequency with which nerve impulses occur and can be recorded.

frequency conversion *Electronics.* the conversion of a signal from one frequency to another.

frequency converter *Electricity.* a circuit or device, such as a cycloconverter, that converts an input at one frequency to an output of a different frequency. Also, FREQUENCY CHANGER, FREQUENCY TRANSLATOR.

frequency curve *Statistics.* a curve obtained by continuous interpolation of a frequency histogram.

frequency departure *Telecommunications.* in frequency modulation, the maximum difference between the instantaneous frequency of the modulated wave and the carrier frequency. Also, **frequency deviation.**

frequency-dependent selection see APOSTATIC SELECTION.

frequency difference *Archaeology.* a method of defining variation in associated artifacts by differences in their rate of distribution at various sites.

frequency discriminator *Electronics.* a device in FM receivers that converts changes in frequency (relative to a defined center frequency) to changes in voltage level (relative to zero voltage).

frequency distortion *Electronics.* distortion that occurs when the amplification of some frequencies is different from that of others.

frequency distribution *Mathematics.* a function that measures the probability of a random variable assuming a given value; probability density function. *Statistics.* a table or graph reporting the count of the frequency with which data fall into specified categories or intervals.

frequency diversity *Telecommunications.* signal transmission in which two or more electromagnetic wave frequencies are used to simultaneously transmit the same information for greater reliability in the event of noise or other interference.

frequency divider *Electronics.* a circuit or device having an output frequency that is a fraction of its input frequency.

frequency-division multiplexing *Telecommunications.* a multiplexing technique in which the available transmission frequency band is divided into two or more narrower bands, each used as a separate communications channel. Also, FREQUENCY MULTIPLEXING.

frequency domain *Telecommunications.* a plane or two-dimensional plot on which a waveform can be represented as a function of frequency rather than time, by analyzing the effects upon individual sinusoidal components of the waveform.

frequency-domain reflectometer *Electromagnetism.* a tuned reflectometer that measures the reflection coefficients and impedance of waveguides over a wide frequency range by scanning a range of frequencies and analyzing the reflected returns.

frequency doubler *Electronics.* a circuit or device having an output frequency that is twice that of its input frequency.

frequency drift *Electronics.* a usually gradual, undesired change of a signal from its intended frequency.

frequency-exchange signaling see TWO-SOURCE FREQUENCY KEYING.

frequency factor see PRE-EXPONENTIAL FACTOR.

frequency frogging *Telecommunications.* **1.** the process of interchanging carrier frequencies in a communication channel to reduce intermodulation noise or crosstalk. **2.** the alternation of electromagnetic waves of two different frequencies at repeater sites of line-of-site microwave systems.

frequency hopping *Telecommunications.* a broadband modulation technique used for multiple access and interception-resistant transmission systems.

frequency interlace *Telecommunications.* **1.** in a television, a technique of interweaving I and J sidebands with luminance sidebands of carrier chrominance signals without causing mutual interference. **2.** in television, the intermeshing of the frequency spectrum of a modulated color subcarrier and the harmonics of the horizontal scanning frequency in order to minimize the visibility of the modulated color subcarrier.

frequency locus *Control Systems.* the path followed by the frequency transfer function or its inverse in the complex plane or on a graph representing amplitude against phase angle; used in determining the zeros of a describing function.

frequency meter *Engineering.* **1.** an instrument that measures the frequency of an alternating current against a known standard. **2.** an instrument having a graduated scale that measures the frequency of a radio current.

frequency mixing *Optics.* a process in which two or more electromagnetic waves passing as a beam through a nonlinear optical medium are combined to form another wave having a frequency equal to the sum or difference of the frequencies of the original waves. Also, **frequency multiplication.**

frequency-modulated jamming *Electronics.* jamming in which a signal is produced to cover a wide band of frequencies by varying a constant-amplitude RF signal above and below a center frequency.

frequency-modulated laser *Optics.* a gas laser in which a frequency-modulated video signal is impressed on the output beam of the laser by an ultrasonic modulation cell.

frequency-modulated radar *Engineering.* a radar whose range may be measured, due to the frequency modulation of the radiated wave which corresponds to the beat of a returning echo.

frequency modulation *Telecommunications.* FM, the instantaneous variation of the frequency of a carrier wave in reponse to changes in the amplitude of a modulating signal.

frequency-modulation broadcast band *Telecommunications.* the frequency band ranging from 88 to 108 MHz on the frequency spectrum that is used for frequency-modulated (FM) radio broadcasting.

frequency-modulation detector *Electronics.* a device that detects and demodulates a frequency-modulated wave.

frequency-modulation Doppler *Engineering.* a radar in which the radial sweep and carrier are frequency modulated.

frequency modulation-frequency modulation *Telecommunications.* a three-stage technique for multiplexing many channels onto one channel by frequency-modulating the subcarriers, then combining the modulated subcarriers, and, finally, frequency-modulating the radio carrier.

frequency-modulation laser *Optics.* a laser that contains a phase modulator inside the Fabry-Perot cavity.

frequency modulation-phase modulation *Telecommunications.* a three-stage technique for multiplexing many channels onto one channel by frequency-modulating the subcarriers, then combining the modulated subcarriers, and, finally, phase-modulating the radio carrier.

frequency-modulation receiver *Electronics.* a radio receiver that receives frequency-modulated waves and outputs corresponding sound waves.

frequency-modulation transmitter *Electronics.* a radio transmitter that transmits a frequency-modulated wave.

frequency modulator *Electronics.* **1.** a circuit or device that modulates the frequency of an oscillator. **2.** the modulator section of an FM transmitter.

frequency monitor *Electronics.* a device that is used continuously to check the frequency of a signal.

frequency multiplexing see FREQUENCY-DIVISION MULTIPLEXING.

frequency multiplier *Electronics.* a circuit or device having an output frequency that is a multiple of its input frequency.

frequency offset *Telecommunications.* a frequency shift that takes place when a signal is transmitted over an analog carrier in which the modulating and demodulating frequencies are not equivalent.

frequency-offset transponder *Electronics.* a transponder whose response frequency is different from the interrogation frequency.

frequency polygon *Statistics.* the diagrammatic form of a frequency distribution in which the variable value is represented by the abscissa and the frequency is represented by the ordinate; line segments connect the midpoint of each class at the height corresponding to the frequency of the class, forming a polygon.

frequency prediction chart *Telecommunications.* on a graph, a curve that represents the maximum usable frequency, the frequency optimum traffic, and the lowest usable frequency for transmission between two specific points for various times within a 24-hour period.

frequency pulling *Electronics.* an alteration in the frequency of a circuit, especially of a self-excited oscillator, caused by the detuning effects of an external circuit, device, or condition.

frequency recorder *Electricity.* a device that uses a frequency bridge in an alternating circuit to record known frequencies at known amplitudes for testing or measuring.

frequency regulator *Electricity.* a device that controls the frequency of an AC generator.

frequency response *Engineering.* an efficiency measurement of an instrument or system, with respect to its ability to transmit the different frequencies applied to it.

frequency-response curve *Engineering.* a graphic representation of the extent or the phase of the frequency response of an instrument or system as it relates to frequency.

frequency-response trajectory *Control Systems.* the path of a frequency-response phasor in the complex plane as the frequency is varied.

frequency run *Electronics.* a test that determines the loss characteristics of a circuit as a function of the operating frequency.

frequency-scan antenna *Electromagnetism.* a radar antenna that scans by frequency variation in one dimension.

frequency scanning *Electronics.* **1.** a controlled alternation of the transmitter frequency in a frequency-agile radar or communications system. **2.** a form of simultaneous digital monitoring of two or more channels in a programmable digital communications receiver or transceiver. **3.** the frequency-response variation in a spectrum analyzer.

frequency-separation multiplier *Electronics.* a multiplier having variables that are split into low-frequency and high-frequency components, which are then multiplied individually and the results added to obtain the required product; used to achieve broad bandwidth and high system accuracy.

frequency separator *Electronics.* in television, a circuit that separates horizontal and vertical synchronization pulses.

frequency seriation *Archaeology.* the organization of artifacts by sequence according to their frequency of appearance; based on the idea that an artifact type first steadily grows in popularity and then similarly declines.

frequency shift *Electronics.* an alteration in frequency of a radio transmitter or oscillator.

frequency-shift converter *Electronics.* a circuit that coverts the discrete frequencies in a frequency-shift signaling system to discrete voltage levels.

frequency-shift keyer *Telecommunications.* a lever that keys a transmitter for telegraph or teletype communications by shifting the carrier frequency over a range of several hundred hertz. Thus, **frequency-shift keying.**

frequency-slope modification *Telecommunications.* a method of modulation in which the carrier signal is swept periodically over the frequency band, as in chirp radar.

frequency spectrum *Physics.* a graphical display of the intensity versus frequency.

frequency splitting *Electronics.* a condition of magnetron operation in which rapid alternation occurs from one mode of operation to another, resulting in rapid change in oscillatory frequency and the consequent power loss at the desired frequency.

frequency stability *Electronics.* the extent to which a frequency remains constant during variations in temperature, current, voltage, and similar factors; usually determined on both a short-term (1-second) and long-term (24-hour) basis.

frequency stabilization *Telecommunications.* a method of modifying the center or carrier frequency so that it varies from that of a reference source by not more than a previously determined amount.

frequency standard *Electronics.* a signal source of a precise frequency used in the calibration of other signal sources.

frequency swing *Telecommunications.* the difference between the maximum and minimum frequencies above and below the carrier frequency, respectively.

frequency synthesizer *Electronics.* a generator of highly accurate signals used for test purposes, often in the form of discrete frequency steps; such signals generally originate from a single frequency source, such as a crystal oscillator.

frequency telemetering *Telecommunications.* a transmission technique involving the transmittal of an alternating-current signal from a primary element by changes in the frequency of the signal, rather than in the strength.

frequency theory *Physiology.* the assumption that the basilar membrane reacts like a telephone receiver, and that the sensation of pitch is determined by the frequency of nerve impulses reaching the auditory area of the brain.

frequency-time-intensity *Electronics.* of or relating to a radar display that correlates frequency, time, and strobe intensity.

frequency-to-voltage converter *Electronics.* a converter that provides an analog output voltage which is proportional to the frequency or repetition rate of the input signal from a flowmeter, tachometer, or other AC generating device. Also, F/V CONVERTER.

frequency transformation *Control Systems.* a transformation in which the frequency variable of a transfer function is replaced by a function of the frequency in order to synthesize a band-pass network from a low-pass network. Also, LOW-PASS/BAND-PASS TRANSFORMATION.

frequency translation *Telecommunications.* the conversion of a set of signals occupying a definite frequency band from one position in the frequency spectrum to another.

frequency translator see FREQUENCY CONVERTER.

frequency-type telemeter *Electronics.* a telemeter that translates by using the frequency of a periodically recurring electric signal.

frequency variation *Electronics.* the change over time of the deviation from a radio-frequency carrier's assigned frequency.

Frere, John 1740–1807, English archaeologist; discovered fossils and artifacts at Hoxne, England; proposed greater antiquity of humans.

fresco *Architecture.* originally, a mural painted on wet plaster; now any mural painted on plaster with water-based colors.

fresh *Geology.* describing a newly exposed, unweathered rock or rock surface. Also, UNWEATHERED. *Meteorology.* of a wind, moderately strong or brisk.

fresh breeze *Meteorology.* on the Beaufort wind scale, a wind having a speed of 19–24 miles per hour.

fresh cow *Agriculture.* a cow that has given birth to a calf and is therefore able to produce milk.

freshet *Hydrology.* **1.** a sudden rise or overflowing of a small stream as a result of heavy rains or rapidly melting snow. Also, SPATE, HIGH WATER. **2.** a small, clear, freshwater stream.

fresh gale *Meteorology.* on the Beaufort wind scale, a wind having a speed of 39–46 mph.

fresh ice *Hydrology.* **1.** see YOUNG ICE. **2.** see FRESHWATER ICE. **3.** ice that was formed on salt water but is now salt free.

fresh water or **freshwater** *Hydrology.* water with no significant amount of dissolved salts or minerals, as is normally found in streams and lakes; in general, water containing less than 1000 milligrams per liter of dissolved solids. *Biology.* also, **fresh-water.** living or found in fresh water, as opposed to ocean water.

fresh-water ecosystem *Ecology.* an area of water, either stagnant or free-flowing, characterized by low salt content.

freshwater ice *Hydrology.* ice formed by the freezing of fresh water in streams and lakes.

Fresnel, Augustin Jean [frə nel′] 1788–1827, French physicist; confirmed the wave theory of light; produced circularly polarized light.

fresnel *Physics.* a unit of frequency equal to 10^{12} cycles per second or 10^{12} hertz.

Fresnel-Arago laws *Optics.* a body of laws that describe conditions under which rays of polarized light interfere with each other.

Fresnel biprism *Optics.* a type of prism in which two prisms with very acute angles are placed base to base, used to observe interference formed by the two refracted images produced from a single source.

Fresnel diffraction *Optics.* the diffraction that occurs when the source of radiation and the point of observation are located at a finite distance from the aperture. Also, NEAR-FIELD DIFFRACTION.

Fresnel drag coefficient *Optics.* the ratio of the velocity of ether in a moving transparent medium to the velocity of the medium itself, expressed mathematically as $1-(1/n^2)$, where n is the index of diffraction of the transparent medium.

Fresnel ellipsoid *Optics.* an ellipsoid that has three perpendicular axes, used to determine the propagation velocities and vibration direction of a doubly refracting crystal. Also, RAY ELLIPSOID.

Fresnel equations *Optics.* a group of equations expressing the fraction for the two polarization components of incident radiant energy reflected or transmitted at a surface boundary between two transparent substances with different refractive indices; used to determine the specular reflection of incident radiation energy at a phase boundary. Also, **Fresnel formula.**

Fresnel fringe *Optics.* one band in a group of light and dark bands that are visible at the edge of a Fresnel diffraction shadow.

Fresnel integrals *Mathematics.* any integral of the form

$$C(x) = (2\pi)^{-1/2} \int_0^x t^{-1/2} \cos t \, dt \quad \text{or} \quad S(x) = (2\pi)^{-1/2} \int_0^x t^{-1/2} \sin t \, dt$$

Fresnel lens *Optics.* a lens characterized by a surface of stepped concentric circles, that is thinner and flatter than a standard lens of equivalent focal length; used in lighthouses, signal lenses, spotlights, traffic signals, and viewfinders.

Fresnel mirrors *Optics.* a pair of plane mirrors, tilted toward one another to form an angle just under 180°, used to produce interference of light that emanates from a single slit and reflects off both planes.

Fresnel reflection formula *Optics.* a formula that expresses the fraction of the intensity of unpolarized light partially reflected from the surface of a transparent medium.

Fresnel region *Electromagnetism.* the region of space between the near field and Fraunhofer region of an antenna, generally at a distance equal to $2d^2/\lambda$, where d is the antenna length and λ is the wavelength.

Fresnel rhomb *Optics.* a rhomb that serves as a quarter-wave retarder and can be used to obtain circularly polarized light from plane polarized light by means of total internal reflection.

Fresnel theory of double refraction *Optics.* a theory explaining birefringence in a crystal in terms of nonspherical wave surfaces.

Fresnel zones *Electromagnetism.* conical areas that exist between microwave transmitting and receiving antennas; they are usually numbered consecutively, with the first zone containing the minimum path length.

Fresnian *Geology.* a North American geologic stage of the Upper Eocene epoch, occurring after the Narizian and before the Refugian.

fret saw *Mechanical Devices.* a long, tapered handsaw with fine teeth mounted in a pressed frame, used in ornamental cutting. Also, COPING SAW.

fretting corrosion *Materials Science.* corrosion occurring at interfaces between metals that are under load and subjected to vibration and slip; appears as grooves or pits surrounded by corrosion products. Also, **fretting wear.**

Freud, Anna [froid] 1895–1982, Austrian psychiatrist, living in England; the daughter of Sigmund Freud; pioneered the psychoanalysis and therapy of children.

Freud, Sigmund [froid] 1856–1939, Austrian psychiatrist; founder of psychoanalysis; formulated a theory of personality composed of the ego, the id, and the superego.

Freudian [froid′ē ən] *Psychology.* **1.** of or relating to Sigmund Freud or his theories and doctrines. **2.** a follower or supporter of Freud, especially one who uses his methods of psychoanalysis.

Freudianism *Psychology.* the psychoanalytic theories and techniques employed by Sigmund Freud and his followers.

Freudian slip *Psychology.* according to Freud, a seemingly accidental error in speech that actually reveals some unconscious desire or motive. Also, **Freudian error**.

Freundlich isotherm equation [froind′lik] *Physics.* an equation relating the volume of a gas that is adsorbed onto a surface at a given temperature to the pressure of the gas.

Freund's adjuvant [froindz] *Immunology.* a water-in-oil solution that contains killed mycobacteria and enhances antigenicity.

Freund synthesis *Organic Chemistry.* a process in which an open-chain dihalo compound is treated with sodium to form a cycloparaffin. In the **Freund-Gustavson synthesis,** zinc is used in place of sodium.

friability *Materials Science.* the ease with which a substance can be crumbled or reduced to powder.

friable *Agronomy.* describing soils that when either wet or dry can be easily crumbled between the fingers. *Materials Science.* of a material, easily reduced to powder or crumbs.

friagem *Meteorology.* a cold-weather period in the middle and upper parts of the Amazon valley and in eastern Bolivia. Also, VRIAJEM.

fricative *Linguistics.* a sound produced by forcing air through a narrow opening in the mouth such as the sound "f."

Fricke dosimeter [frik′ə] *Nucleonics.* an instrument that measures radiation exposure by counting the number of ferrous ions transformed into ferric ions in an aerated acidic ferrous sulfate solution.

friction *Mechanics.* a force that opposes the relative motion of two material surfaces that are in contact with one another; the direction of the force on each body is opposite to the direction of its motion relative to the other body. Also, FRICTIONAL FORCE.

frictional *Mechanics.* relating to or caused by the force of friction.

frictional electricity *Electricity.* static electricity generated by rubbing one material against another.

frictional grip *Mechanical Engineering.* the friction generated by train wheels on a track.

friction bearing *Mechanical Engineering.* a solid bearing that directly contacts and supports an axle end.

friction bonding *Engineering.* a process in which a semiconductor chip is joined with a substrate by vibrating the chip under pressure and thus breaking up oxide layers, which allows the two objects to alloy.

friction brake *Mechanical Engineering.* a brake in which the stopping resistance occurs by friction.

friction breccia *Geology.* a breccia composed of angular rock fragments that were broken or crushed as a result of friction, especially between two walls of a fault that rub against each other.

friction calendaring *Engineering.* a process in which pressure is applied to an elastomeric compound to force it into a small opening in a fabric as it passes between calendar rolls.

friction clutch *Mechanical Engineering.* a clutch in which torque is transmitted by pressure of the clutch faces against each other; consists of a pair of opposed members, between which the drive is transmitted through the friction of their contact surfaces.

friction coefficient see COEFFICIENT OF FRICTION.

friction compensation *Mechanical Engineering.* a small torque, in addition to the main torque, that is added to a machine to compensate for the effects of friction of its moving parts.

friction crack *Geology.* a short, crescent-shaped break in glaciated ice or rock, occurring transverse to the direction of ice flow, and probably resulting from local increase in frictional pressure between the ice and the bedrock.

friction damping *Mechanics.* the absorption of mechanical energy of motion, vibration, or shock, and the resulting decrease in amplitude of motion or vibration, due to frictional rubbing between surfaces.

friction depth *Oceanography*. the depth of water at which the effects of wind become negligible, usually around 100 meters. Also, DEPTH OF FRICTIONAL RESISTANCE.

friction drag *Fluid Mechanics*. the drag produced when turbulent flow occurs in a boundary layer.

friction drive *Mechanical Engineering*. a power transmission system that operates by the friction forces between the contact surfaces of two wheels when one wheel rotates and is pressed against the second wheel.

friction factor *Fluid Mechanics*. a dimensionless quantity given by the Fanning friction factor multiplied by a constant; the number is used in the study of fluid friction in pipes.

friction-feed printer *Computer Technology*. a type of printer in which the paper is advanced by the movement of a roller, as with a typewriter.

friction flow *Fluid Mechanics*. a fluid flow in which a significant amount of kinetic energy is converted into heat due to a large viscosity coefficient.

friction gearing *Mechanical Engineering*. gearing in which power and motion are transmitted from one shaft to another through friction between two surfaces in rolling contact.

friction gripping *Robotics*. the use of an end effector covered with soft material to grasp an object.

friction head *Fluid Mechanics*. the amount of reduction in the head of a fluid, due to the friction between the fluid and its container and also to the intermolecular interaction.

friction horsepower *Mechanical Engineering*. the part of the gross or indicated horsepower of a machine that is dissipated through friction.

friction layer see SURFACE BOUNDARY LAYER.

friction loss *Mechanics*. the loss of energy in a system because of the force of friction; characteristic of all actual machines with moving parts.

friction pile *Civil Engineering*. a pile that is supported solely by side friction when compared to tip or bearing friction.

friction saw *Mechanical Engineering*. a toothless, high-speed circular saw used to cut metals by frictional heat, which melts the material adjacent to it; used to cut stock to length for structural parts of mild steel and stainless steel.

friction torque *Mechanics*. a moment of force that opposes rotation, produced by bearings or other surfaces in contact, by the viscosity of a lubricant, or by drag by the surrounding medium.

friction-tube viscometer *Engineering*. an instrument that measures the viscosity of a liquid by recording the pressure drop through a friction tube as the liquid is in viscous flow.

friction velocity *Meteorology*. a mathematically stated wind velocity usually applied to motion near the ground, where the shearing stress is assumed to be independent of height and approximately proportional to the square of the mean velocity.

friction welding *Engineering*. a process in which metals are mated by heat which is generated when the metal parts are rubbed together under high pressure.

Friedel-Crafts catalyst [frēd´əl] *Chemistry*. a strongly acidic metal halide such as aluminum chloride or zinc chloride, used in the polymerization of unsaturated hydrocarbons.

Friedel-Crafts reaction *Chemistry*. a reaction in which an alkyl or an acyl group replaces a hydrogen atom of an aromatic nucleus in the presence of aluminum chloride, producing a hydrocarbon or a ketone.

friedelite *Mineralogy*. $Mn_8^{+2}Si_6O_{15}(OH,Cl)_{10}$, a rose-red, monoclinic, pseudotrigonal mineral occurring most commonly in tabular crystals but also found in massive and fibrous forms, having a specific gravity of 3.04 to 3.07 and a hardness of 4 to 5 on the Mohs scale; found in manganese mines.

Friedel's law *Crystallography*. a law stating that the intensities of centrosymmetrically related Bragg reflections $(h,k,l$ and $-h,-k,-l)$ in the diffraction pattern of a crystal are equal, even for an acentric structure, provided there is no anomalously scattering atom in the structure.

Friedländer's pneumonia [frēd´len dər] *Medicine*. lobar pneumonia caused by the infecting agent **Friedländer's bacillus,** a bacterium of the species *Klebsiella pneumoniae*. (Named for Karl *Friedländer,* 1847–1887, German pathologist.)

Friedlaender synthesis *Organic Chemistry*. the formation of a quinoline derivative by the base-catalyzed condensation of 2-aminobenzaldehyde with the $-CH_2CO-$ portion of a ketone.

Friedmann, Alexander [frēd´mən] *Astrophysics*. 1888–1925, Russian mathematician and physicist; formulated an early "big bang" theory.

Friedmann equations *Astrophysics*. a set of equations devised by Alexander Friedmann in 1922, which describe the evolution of space-time and matter in the Friedmann universe model.

Friedmann universe *Astrophysics*. a widely used model for the universe based on Einstein's field equations that assumes the universe is expanding or contracting under gravity alone and that it is isotropic and homogenous. Also, **Friedmann model.**

Friedman's curve *Pathology*. a graph on which hours of labor are charted against the dilation of the cervix, which is measured in centimeters. (Named for Emanuel *Friedman,* born 1926, American obstetrician.)

Friedman's test *Pathology*. a test for pregnancy done by injecting the urine of a pregnant woman into female rabbits; a positive test results in the formation of corpora lutea in the rabbits. (Named for Maurice H. *Friedman,* born 1903, American physician.)

Friedman test *Statistics*. a distribution-free test of the hypothesis of equal effect of factor levels in two-factor analysis of variance.

Friedreich's ataxia [frēd´riks] *Medicine*. a condition of muscle weakness, loss of muscle control, weakness of the legs, and an abnormal walk, caused by a breakdown of the spinal cord. (Named after Nikolaus *Friedreich,* 1825–1882, German physician.)

Friend cell *Cell Biology*. a mouse hematopoietic stem cell that is infected with the Friend virus and that can be induced by exposure to DMSQ to undergo differentiation into red blood cells with concomitant hemoglobin production.

friendly *Computer Science*. describing computer programs or devices that are relatively easy to operate; user-friendly.

friendly fire *Ordnance*. fire from one's own forces, as opposed to enemy fire.

friendly ice *Oceanography*. an ice canopy that contains abundant large ice skylights or other openings where a submarine can surface.

friendly number see AMICABLE NUMBER.

Friend virus *Virology*. a murine leukemia genus of the family Retroviridae that causes malignant reticulopathy in mice. Also, **Friend leukemia virus.**

Fries rearrangement [frēz] *Organic Chemistry*. the conversion of a phenolic ester into the corresponding *o-* and *p*-hydroxyketone on heating with aluminum chloride or other Lewis acid catalysts.

Fries rule *Organic Chemistry*. the rule that the most stable form of a polynuclear hydrocarbon is the arrangement having the maximum number of rings in the benzenoid form; i.e., rings having three double bonds.

frieze [frēz] *Architecture*. the middle horizontal division of a classical entablature, usually ornamented with sculpted figures.

frigate [frig´it] *Naval Architecture*. **1.** a fast sailing warship, usually with a single gundeck, used for scouting and patrol. **2.** a modern surface escort warship, similar to a destroyer but usually smaller and slower.

frigate bird any of several species of seabirds in the genus *Fregata* of the family Fregatidae, having fully webbed feet.

frigid [frij´id] *Psychology*. describing a woman who is regarded as lacking in sexual feeling or desire. Thus, **frigidity.**

frigophilic *Ecology*. describing an organism that lives or thrives in cold climates. Thus, **frigophile, frigophily.**

frigorie *Thermodynamics*. a power rating unit used in refrigeration, equivalent to the extraction of 1000 calories per hour (1.16264 watts).

frigorimeter *Engineering*. an instrument designed specifically for the measurement of low temperatures.

frilled lizard *Vertebrate Zoology*. an Australian arboreal lizard of the genus *Chlamydosaurus* in the family Agamidae; known for its ability to face a predator, stand upright, and unfurl a large cape of skin until it encircles the head, making it appear much more formidable; it is also able to run on its hindlegs.

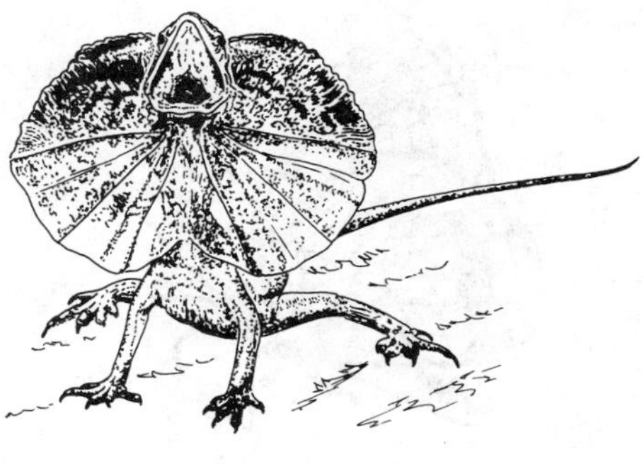

frilled lizard

F ring *Astronomy.* a very narrow, wavy ring of Saturn which lies just outside the A ring; it is confined by two small satellites and consists of at least five individual strands.
fringe *Optics.* any of the alternating light and dark bands that are produced by interference or diffraction. *Computer Programming.* the set of leaf nodes of a tree.
fringe area *Telecommunications.* the region marginally beyond the limits of radio or television transmitter range, in which signal reception is available but of poor quality.
fringed micelle model *Materials Science.* an early model of the polymer structure that involved long polymer chains of about 5000 nm wandering successively through a series of disordered and ordered regions along the length of the polymer molecule.
fringe howl *Acoustical Engineering.* a howling noise, normally below 100 hertz, due to a difference in alignment between the heads used for recording and the heads used for playback.
fringe joint *Geology.* a small-scale rock fracture formed by tension or shear peripheral to the main joint and at an angle of 5–25° to its face.
fringe magnetic field *Electromagnetism.* the portion of the magnetic field that exists outside the region between the poles of a U-shaped permanent magnet.
fringe ore *Geology.* an ore occurring at the edge of a mineralization pattern or halo. Also, HALO ORE.
fringe region *Meteorology.* the upper portion of the exosphere, at which the cone of escape equals or exceeds 180°. Also, SPRAY REGION.
fringe water see ANASTATIC WATER.
fringillid [frin´ji lid] *Vertebrate Zoology.* any member of the family Fringillidae.
Fringillidae *Vertebrate Zoology.* the finches, a cosmopolitan family of small, seed-eating passerine birds having strong short bills that are usually thick at the base, and often exhibiting well-marked sexual dimorphism with juveniles resembling the males.
fringing forest *Ecology.* a type of woodland that extends along river banks in an area otherwise devoid of trees. Also, GALLERY FOREST.
fringing groove *Ordnance.* a groove in the rotating band of a projectile, designed to collect excess metal from the band while the projectile travels through the gun bore; uncollected metal may cause a fringe at the rear of the band that can impair the flight of the projectile.
fringing reef *Geology.* an organic reef that grows directly against coastal bedrock and actually constitutes the shoreline, commonly found along tropical coasts. Also, SHORE REEF.
Frisch, Karl von 1886–1982, Austrian zoologist; Nobel Prize for discovering how bees "dance" to communicate the location of food.
Frise aileron *Aviation.* a type of aileron that has a beveled leading edge projecting beyond its inset hinges. When lowered, it forms an extension of the wing surface; when raised, its nose protrudes below the wing, increasing drag and reducing yaw. (Named for its inventor, the English engineer Leslie George *Frise.*)
frisket *Graphic Arts.* a masking device consisting of a paper sheet with cutouts for image areas.
frit *Materials.* a fused ceramic mass that is ground into a fine powder, used as the basis for glazes, ceramics, or porcelain.
Fritillaria *Invertebrate Zoology.* a genus of pelagic prochordate plankton.
fritillary *Botany.* the common name for any member of the genus *Fritillaria;* bulbous herbs with mottled or checkered flowers of the lily family. *Invertebrate Zoology.* the common name for butterflies of the genus Speyeria and others, family Nymphalidae, most with orange-brown wings.

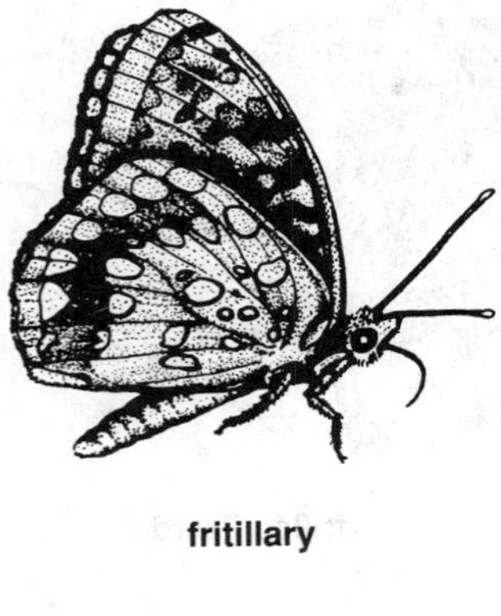

fritillary

frit seal *Engineering.* a tight closure produced when a glass binder is mixed with metallic powders.
fritting *Engineering.* a process in which glass is made with the use of heat.
Frobenius, Ferdinand Georg [frō bä´nē ůs] 1849–1917, German mathematician; made major contributions to group theory.
Frobenius reciprocity theorem *Mathematics.* let Q and T be irreducible representations of a group G and a subgroup H, respectively. Then the multiplicity of the representation Q in the representation of G induced by T is equal to the multiplicity of the representation T in the restriction of Q to H.
Frobenius theorem *Mathematics.* see MARRIAGE THEOREM.
Frobenius theorem on Bernoulli numbers *Mathematics.* a theorem demonstrating that the denominator of B_r/r contains no primes other than the denominator of B_r itself, where B_r is the rth Bernoulli number.
froe *Mechanical Devices.* **1.** a steel wedge used to split logs. **2.** a cutting tool whose handle is at right angles to the cutting edge, used to split blocks of staves and shingle.
frog *Vertebrate Zoology.* **1.** any of various stout-bodied, tailless amphibians of the order Anura, found widely throughout the world. **2.** specifically, a chiefly aquatic amphibian of the family Ranidae, with long legs, webbed feet, agile leaping and swimming abilities, and mucus-coated smooth skin. These are often called **true frogs,** while the term *toad* is reserved for short-legged, dry, warty-skinned members of Anura; both *frog* and *toad* have inconsistent common meanings. *Transportation Engineering.* a V-shaped intersection point at which a train or tram crosses from one set of rails to another. *Design Engineering.* a recessed panel on one or both of the larger faces of a brick or block so as to reduce its weight. *Agriculture.* the supporting frame of a moldboard plow.

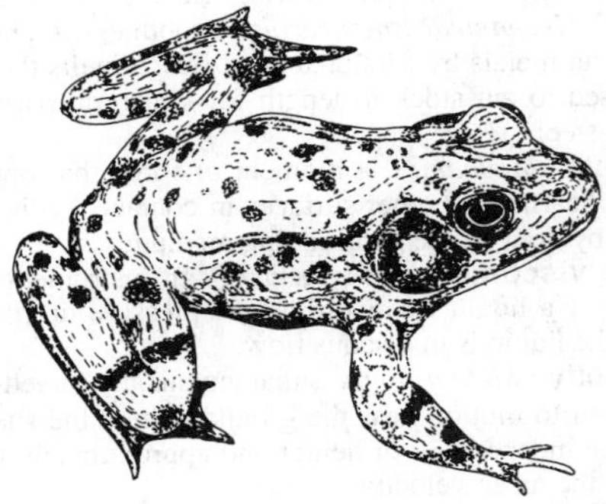

green frog

Frog or **FROG** *Ordnance.* a Soviet unguided surface-to-surface bombardment missile equipped with a nuclear or conventional warhead; **Frog 2,3,4,** and **5** are two-stage rocket-propelled weapons with a maximum range of 30 miles; **Frog 7** is a single-stage rocket-propelled weapon with a maximum range of 57 miles. (An acronym for free rocket over ground.)
frogfish *Vertebrate Zoology.* a tropical and subtropical marine fish of the family Antennariidae, being carnivorous and round-bodied and having loose, rough skin with a dangling protuberance in certain species that acts as a fishing pole to lure prey, such as that of the Sargassum angler.
frogging *Telecommunications.* a switching technique used in testing telephone equipment, by which the tip and ring leads of the test specimen are reversed relative to the driving and terminating test circuits.
frog storm *Meteorology.* the first occurrence of bad weather in the spring after a warm period. Also, WHIP-POOR-WILL STORM.
frohbergite *Mineralogy.* $FeTe_2$, an orthorhombic mineral of the marcasite group occurring as minute grains, having a specific gravity of 8.067 and a hardness of 3 to 4 on the Mohs scale; found with native gold, pyrite, and petzite.
Fröhlich's syndrome see ADIPOSOGENITAL DYSTROPHY.
Froissart bound *Particle Physics.* a limit on the rate of energy increase of the cross section of an absorptive collision between hadrons, so that the interaction radius does not increase more rapidly than the logarithm of the energy.
Fromm, Erich [främ] 1900–1980, German-born American social psychologist; described human behavior as a response to social conditions.

frond *Botany*. **1.** a compound leaf of a fern, palm, or cycad. **2.** a leaflike thallus of lichen, liverwort, or algae.

frondelite *Mineralogy*. $Mn^{+2}Fe_4^{+3}(PO_4)_3(OH)_5$, a green to greenish-black, brittle, orthorhombic mineral that appears as botryoidal crusts, isomorphous with rockbridgeite, and having a specific gravity of 3.476 and a hardness of 4.5 on the Mohs scale; found as an alteration product of manganese-iron phosphates in granite pegmatites.

frons *Entomology*. the upper or anterior surface of the head of an insect.

front *Meteorology*. the transition zone between two air masses of different density and, almost invariably, of different temperature; it may also be distinguished by a pressure trough, a change in wind direction, a moisture discontinuity, and certain cloud and precipitation forms. *Military Science*. **1.** the lateral space occupied by a military force or other element, measured from flank to flank. **2.** the direction of the enemy or, in a noncombat situation, the direction toward which the command is faced. **3.** the line of combat between two opposing forces.

front abutment pressure *Geophysics*. the energy released into the superincumbent strata.

frontal *Anatomy*. **1.** of or relating to the forehead. **2.** of or relating to the anterior portion of an organ or of the body. *Meteorology*. relating to or produced by the action of a front. Thus, **frontal precipitation.**

frontal bone *Anatomy*. a single bone that closes the front part of the cranial cavity and forms the skeleton of the forehead.

frontal contour *Meteorology*. the line of intersection of a front with a specified surface in the atmosphere, usually a constant-pressure surface.

frontal crest *Anatomy*. a ridge on the inner surface of the frontal bone that extends from the end of the sagittal suture to the point of articulation with the ethmoid bone.

frontal eminence *Anatomy*. either of two raised areas of the frontal bone on either side of the anterior surface above the eyes, forming the most prominent portions of the forehead.

frontal fire *Ordnance*. fire directed on a line perpendicular to the front of the target. Similarly, **frontal attack.**

frontal fog *Meteorology*. any fog associated with a front, usually a result of rain falling into cold stable air and raising the dew-point temperature or the sudden cooling of air over moist ground.

frontal gland *Invertebrate Zoology*. **1.** a gland at the anterior end of some worms in Platyhelminthes and Nematoda. **2.** a gland at the front of the head of certain termites from which a toxin is expelled.

frontal inversion *Meteorology*. a temperature inversion in the atmosphere that occurs during vertical ascent through a frontal zone.

frontal kame *Geology*. a mound or irregular ridge of stratified sand and gravel deposited as a steep alluvial fan against the edge of a glacier.

frontal lifting *Meteorology*. the forced ascent of warmer, less-dense air at or near a front that occurs when the relative velocities of the two air masses cause them to converge.

frontal lobe *Anatomy*. the anterior portion of the cerebral hemisphere, extending from the frontal pole to the central sulcus.

frontal moraine *Geology*. **1.** see END MORAINE. **2.** a row of hills or an elongated ridge of moraine material deposited at the front of a former glacier.

frontal nerve *Anatomy*. a branch of the ophthalmic nerve that branches into the supratrochlear and the supraorbital nerves, supplying the scalp, forehead, and upper eyelids.

frontal passage *Meteorology*. the passage of a front over a point on the earth's surface.

frontal plain see OUTWASH PLAIN.

frontal plane *Anatomy*. a longitudinal plane that is perpendicular to the sagittal plane and divides the body into anterior and posterior parts; it roughly parallels the frontal suture of the skull.

frontal plate *Anatomy*. any plane parallel with the long axis of the body. *Cardiology*. the plane of the limb lead in electrocardiography.

frontal profile *Meteorology*. the outline of a front seen on a vertical cross section oriented normal to the frontal surface.

frontal sinus *Anatomy*. either of a pair of paranasal sinuses that are located in the frontal bone and that communicate with the nasal cavities by way of the nasofrontal ducts.

frontal strip *Meteorology*. the presentation of a front on a synoptic chart as two lines drawn to represent the boundaries of the zone.

frontal suture *Anatomy*. the line of junction between the right and left halves of the frontal bone, which is normally transient, but sometimes persists in adulthood.

frontal system *Meteorology*. a system of fronts as they appear on a synoptic weather chart.

frontal terrace see OUTWASH TERRACE.

frontal thunderstorm *Meteorology*. a thunderstorm resulting from the convection induced by frontal lifting.

frontal wave *Meteorology*. a horizontal wavelike deformation of a front in the lower levels, usually associated with a maximum of cyclonic circulations in the adjacent flow.

frontal zone *Meteorology*. a three-dimensional layer of large horizontal density gradient bounded by the warmer side of a front.

front-end edit *Computer Programming*. functions of error checking and correcting that are carried out prior to passing input data to the main computer.

front-end loader *Mechanical Engineering*. a loader having a bucket or shovel at the end of an articulated arm at the front of the vehicle.

front-end processor *Computer Technology*. a small computer that receives and transmits all data passing through a computer system; it may also perform error control, data conversions, format conversions, fail-soft functions, queuing, and other input/output activities. Also, **front-end computer.**

front-end volatility *Chemical Engineering*. the volatility of lower-boiling petroleum fractions.

front focal plane *Optics*. the focal plane of a lens that lies in front of it when viewed in the same direction in which light is passing through it.

frontier of a set *Mathematics*. the boundary of the set; i.e., all points which are in the closure of a set but are not in its interior.

front loader or **front-loader** *Mechanical Engineering*. **1.** any machine or appliance that is loaded through an opening in the front, as distinguished from a top loader. Thus, **front-loading. 2.** see FRONT-END LOADER.

frontogenesis *Meteorology*. **1.** the initial formation of a front or a frontal zone. **2.** an increase in the horizontal gradient of an air mass property, such as density, and the development of the features of the accompanying wind field.

frontogenetic function *Meteorology*. a measure of the tendency of flow in an air mass to increase the horizontal gradient of a conservative property.

frontolysis *Meteorology*. **1.** the dissipation of a front or frontal zone. **2.** a decrease in the horizontal gradient of an air mass property, such as density, and dissipation of the features of the accompanying wind field.

front pinacoid *Crystallography*. see PINACOID.

front porch *Electronics*. the portion of a composite television picture signal located between the leading edge of the horizontal blanking pulse and the leading edge of the corresponding synchronizing pulse.

front sight *Ordnance*. a blade or other projection located above the barrel of a small arm near the muzzle; it is aligned with the rear sight and the target in aiming the weapon; it may be fixed, detachable, or adjustable. Also, FORESIGHT, MUZZLE SIGHT.

front slagging *Engineering*. the process of removing scum that forms on the surface of slag and molten metal as they move through a taphole.

front slope see SCARP SLOPE.

front suspension *Mechanical Engineering*. a system of springs and shock absorbers connecting the front axle to the chassis of an automobile or other vehicle; designed to reduce unwanted motion transmitted from the road.

front suspension

front-to-back ratio *Solid-State Physics.* the ratio of the electrical resistance to a current flowing through a crystal in one direction to the resistance associated with a current flowing in the opposite direction. *Electromagnetism.* a measure of the directional effectiveness of a directional antenna.

front-wheel drive *Mechanical Engineering.* an automotive system in which only the front wheels receive driving power from the engine.

frost *Hydrology.* **1.** a white, feathery deposit of interlocking ice crystals formed by the sublimation of water vapor on a surface whose temperature is below freezing. Also, HOARFROST. **2.** see PERMAFROST.

frost action *Geology.* **1.** a process in which water in pores, cracks, and other openings in the surface of a rock repeatedly or alternately freezes and thaws, causing weathering and breakup of the rock. **2.** the resulting effects of such a process.

frostbite *Medicine.* damage to tissues caused by exposure to very low temperatures in the environment, causing narrowing and damage to blood vessels resulting in oxygen starvation and tissue death. Thus, **frostbitten.**

frost boil *Geology.* **1.** an accumulation of water and mud that is released from ground ice by accelerated spring thawing. **2.** a low mound formed by local differential frost heaving at a location that is favorable for the formation of segregated ice and lacking an insulating cover of vegetation.

frost churning see CONGELITURBATION.

frost climate *Meteorology.* the climate of the regions of the earth that are perennially covered by snow and ice.

frost crack *Botany.* a usually vertical crack in tree stems that forms because wood tends to shrink more tangentially than radially in freezing.

frost day *Meteorology.* an observational day on which frost occurs.

frost flower see ICE FLOWER.

frost fog see ICE FOG.

frost hazard *Meteorology.* the risk of damage by frost that may be expressed as the probability of a killing frost on different dates during the growing season.

frost heaving *Geology.* **1.** the uneven upward expansion and distortion of the ground caused by the freezing of subsurface water, and affecting soil, rock, pavement, and other structures. **2.** any upheaval of the ground as a result of frost action.

frosting *Engineering.* a process in which a scraped metal surface is decorated with the use of a handscraper.

frostless zone *Meteorology.* the warmest part of a slope lying between the layer of cold air that forms over a valley floor on calm clear nights and the cold hill tops; it varies in level from night to night. Also, GREEN BELT, THERMAL ZONE, VERDANT ZONE.

frost line *Geology.* **1.** the maximum depth of frozen ground or soil during the winter in nonpermafrost regions. **2.** the lowest limit of permanently frozen ground in frigid regions. **3.** the altitude below which freezing does not occur in tropical regions.

frost marks *Archaeology.* variations in the amount of frost retained on the ground that indicate the presence of buried archaeological features; detected primarily by aerial photography.

frost mound *Geology.* any knoll, hill, or conical mound containing a core of ice and formed by frost heaving or hydrostatic pressure of groundwater in a permafrost region. Also, SOIL BLISTER, SOFFOSION KNOB.

frost-pattern soil see PATTERNED GROUND.

frost point *Meteorology.* the highest temperature at which moisture in the atmosphere will sublimate to form hoarfrost on a cooled polished surface.

frost ring *Botany.* an abnormal or false annual growth ring formed after frost, followed by regrowth of foliage, during the growing season.

frost smoke *Meteorology.* a rare type of ice fog that is formed at cold temperatures and composed of ice particles. Also, BARBER.

frost soil see CONGELITURBATE.

frost splitting or **frost weathering** see CONGELIFRACTION.

frost stirring see CONGELITURBATION.

frost table *Geology.* an irregular ground surface that represents the degree to which thawing has taken place in seasonally frozen ground.

frosty mildew *Plant Pathology.* a plant disease caused by fungi of the genus *Cercosporella* and characterized by whitish lesions on the plant leaves.

frost zone **1.** see ACTIVE LAYER. **2.** see SEASONALLY FROZEN GROUND.

froth *Chemistry.* a light foam.

frother *Chemistry.* a chemical compound that is used in a flotation process to stabilize foam, usually by reducing surface tension.

froth flotation *Engineering.* a method of concentrating valuable materials from lower grade ores by grinding, chemically treating, and then aerating the material so that a froth is produced, containing the valuable particles to be subsequently skimmed off.

frothing *Engineering.* a method of producing bubbles on the surface of material through chemical reaction, aeration, or other agitation.

froth promoter *Chemistry.* a substance that promotes the action of a frothing agent.

frottage [frə täzh´] *Psychology.* the act of obtaining sexual pleasure by rubbing against another person, usually a stranger in a crowd. *Medicine.* a rubbing motion, as in massage. (A French word meaning "rubbing.")

frotteur [frə tůr´] *Psychology.* a person who practices frottage.

Froude, William [frood] 1810–1879, English naval architect and engineer; invented the dynamometer.

Froude number *Fluid Mechanics.* a dimensionless constant that is significant for flows with free surface effects; the number squared may be interpreted as the ratio of inertia forces to gravity forces. (Named after William *Froude* and his son, Robert E. *Froude*.)

Froude number 1 *Fluid Mechanics.* a dimensionless number given by the square of the relative speed of a body floating on a liquid, divided by the product of a characteristic length of the body and the gravitational acceleration; the number is used in studying the motion of such a body when it generates surface waves and eddy currents.

Froude number 2 *Fluid Mechanics.* a dimensionless number given by the fluid velocity in an open channel, divided by the square root of the product of a characteristic length and the gravitational acceleration.

frozen flux *Physics.* the magnetic field lines that move with a highly conducting material such as a plasma.

frozen ground *Geology.* the soil or ground whose temperature is below freezing and usually contains water in the form of ice. Also, GELISOL.

frozen-in field *Physics.* the magnetic field trapped in a plasma.

frozen section *Biology.* a thin slice of tissue or organ from a frozen sample, used for study.

Frucht's theorem *Mathematics.* the theorem that every abstract group is isomorphic to the automorphism group of some graph.

fructan *Biochemistry.* a polysaccharide made up exclusively or primarily by fructosyl residues, hydrolyzed by a variety of microorganisms.

fructescence *Botany.* the condition or period of time in which the fruit of a plant is mature.

fructicolous *Ecology.* living on or in fruit. Thus, **fructicole.**

fructification *Botany.* **1.** the forming or bearing of fruit. **2.** the fruit or other reproductive material of a plant.

fructose *Biochemistry.* $C_6H_{12}O_6$, a sugar that is a component of a number of oligosaccharides and polysaccharides, and is generally the sweetest and most readily absorbed; in fruit juices it appears with D-glucose and in honey with glucose and sucrose. Also, **D-fructopyranose.**

frugivorous *Zoology.* feeding mainly or exclusively on fruits.

fruit *Botany.* the fully matured ovary of a seed plant, including the seed or seeds, connecting tissues, and any covering.

fruit bat *Vertebrate Zoology.* any fruit-eating bat, especially those of the family Pteropodidae found in tropical areas of the Eastern Hemisphere.

fruit bud *Botany.* a fertilized flower bud that will develop into a fruit.

fruit fly *Invertebrate Zoology.* the common name for insects in the family Tephritidae, especially of genus *Drosophila;* serious plant pests whose larvae feed on fruit or decaying vegetable matter.

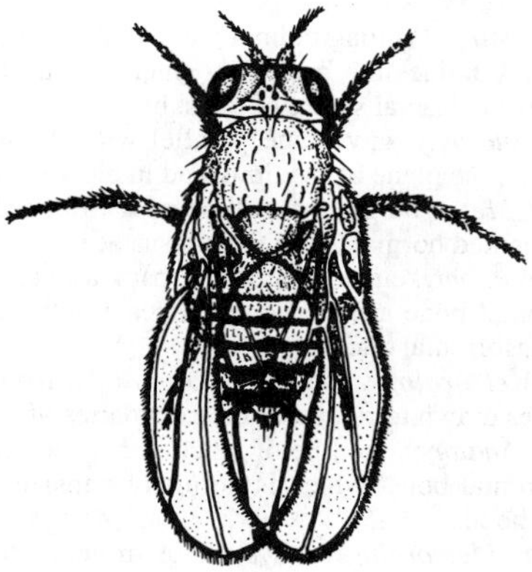

fruit fly

fruiting body *Botany.* any part of a plant that produces spores or seeds; a fructification.

Frumkin isotherm *Physical Chemistry.* a relation describing how the extent of adsorption of a substance on an electrode depends upon concentration, potential, and temperature.

frustrated internal reflectance see ATTENUATED TOTAL REFLECTANCE.

frustration *Behavior.* the failure of goal-directed behavior to achieve the goal because of some interference or obstruction. *Psychology.* an unpleasant emotional state aroused by the failure to achieve a goal. *Solid-State Physics.* a phenomenon arising from a special arrangement of antiferromagnetically interacting atoms, in which the total energy of the interaction cannot be minimized because conflicting spatial requirements make full antiferromagnetic interactions among all pairs impossible to achieve.

frustration-aggression hypothesis *Psychology.* the theory that an experience of frustration will produce aggressive behavior toward the obstacle causing this frustration; frustration is thus seen as the cause of all aggression.

frustule *Invertebrate Zoology.* **1.** the siliceous two-valved shell, sometimes including the protoplast, of a diatom. **2.** a nonciliated planulalike bud in some hydrozoans that develops into a polyp.

frustum *Mathematics.* a section of a solid between two parallel cutting planes.

frutescent *Botany.* similar to a shrub in appearance or habit.

fruticolous *Ecology.* living or growing on shrubs. Thus, **fruticole.**

fruticose *Biology.* shrublike; used especially in reference to lichens.

FRV functional residual volume.

frying noise *Electricity.* a hissing noise around transmission lines due to signal current flowing across a resistive element with multiple intermittent paths. Also, TRANSMITTER NOISE.

FS file separator.

F-scale *Psychology.* a relative measure of the level of authoritarian traits present in an individual. (Short for *Fascist* scale.) Thus, **F-score.**

F scan *Electronics.* see F DISPLAY.

F scope *Electronics.* a radarscope that produces an F display.

F_σ set *Mathematics.* any set that is the countable union of closed sets; a type of Borel set.

FSH follicle-stimulating hormone.

FSK frequency-shift keying.

f. sp. forma specialis.

f-spot or **F spot** *Astronomy.* the following or trailing of the two spots that make up a sunspot pair.

F star *Astronomy.* a star whose spectrum shows lines of hydrogen and singly ionized calcium of about equal strength; it has a surface temperature of 6000 to 7500 kelvins.

f-stop *Optics.* the aperture setting of the lens system of a camera, represented by an *f*-number.

F^+ strain *Microbiology.* a bacterial strain that contains an extrachromosomal F plasmid, or sex factor, that can be transferred to a recipient bacterial cell during cell-to-cell conjugation. Also **F^+ donor.**

F^- strain *Microbiology.* a bacterial strain that does not contain an F plasmid, or sex factor, its cells serving as recipients of genetic material from F^+ donor cells during the process of conjugation, when transfer of an F plasmid may also occur. Also, **F^- recipient.**

f-sum rule *Atomic Physics.* a rule that determines how many electrons take part in transition of energy within an atom, based on the oscillator strengths (*f*) of a given atom.

FSYS *Robotics.* an explicit robot control language that uses an interpreter and macros for monitoring.

ft feet; foot.

ftc footcandle.

F-test see VARIANCE RATIO TEST.

fth fathom.

FT-IR Fourier transform, infrared region.

ft-L foot-Lambert.

ft lb or **ft.-lb.** foot-pound.

ft. mas. div. in pil. let a mass be made and divided into pills. (From Latin *fiat massa dividenda in pilulae.*)

ft-pdl foot-poundal.

ft. pulv. let a powder be made. (From Latin *fiat pulvis.*)

fT value *Nuclear Physics.* a measure of beta decay lifetimes, or comparative half-lives, derived from the magnitude of the matrix element used to calculate the total transition rate of beta emitters.

FU finsen unit.

fu flux unit.

Fubini's theorem *Mathematics.* let X and Y be two measure spaces and f be a measurable function on $X \times Y$. Then f is integrable if and only one of the following integrals exists and is finite: (a) $\int_X du \int_Y |f|\, dv$; (b) $\int_Y dv \int_X |f|\, du$. In addition, these integrals equal $\iint_{X\times Y} |f|\, dudv$.

Fubini-Study metric *Mathematics.* the Hermitian metric on complex projective space CP^n. For example, when $n = 2$, the Fubini-Study metric is given by

$$ds^2 = [dz_1 dz_1^* + dz_2 dz_2^* + (z_2 dz_1 - z_1 dz_2)(z_2^* dz_1^* - z_1^* dz_2^*)]/Z\,,$$

where $Z = 1 + z_1 z_2^* + z_2 z_2^*$ and * denotes complex conjugation. This a real, positive definite metric on the 4-dimensional real manifold with coordinates (x_1, x_2, y_1, y_2) where $z_k = x_k + iy_k$ may be thought of as an abbreviation.

Fucaceae *Botany.* a family of branching brown algae in the order Fucales, characterized by a discoid holdfast and a flattened cylindrical or strap-shaped form.

Fucales *Botany.* in some classifications, an order of brown algae of the class Phaeophyceae.

Fuchs, Leonhard [fyooks] 1501–1566, German physician and botanist; published a large, well-illustrated flora.

fuchsia [fyoo´shə] *Botany.* any of various shrubs and trees of the family Onagraceae, widely cultivated for their bright, drooping flowers. (Named for Leonhard *Fuchs.*)

fuchsin *Organic Chemistry.* $C_{20}H_{20}ClN_3$, a synthetic rosaniline dyestuff produced by mixing rosaniline and *p*-rosaniline hydrochloride, occurring as green crystals that are soluble in water and alcohol; decomposes above 200°C; used as a red dye and as an antifungal drug.

fuchsin bodies see RUSSELL'S BODIES.

fuchsinophile *Biology.* any substance that has an affinity for the dye fuchsin.

fuchsite *Mineralogy.* a bright green variety of muscovite containing up to 4.8% Cr_2O_3.

fucoid *Botany.* of or relating to seaweeds, especially those of the genus *Fucus. Geology.* any tunnel-shaped sedimentary structure believed to be a trace fossil, but not related to a particular genus.

fucoidin *Biochemistry.* a gum found in brown algae that consists of L-fucose and sulfate acid ester groups.

Fucophyceae *Botany.* in some classifications, a class of brown algae. Also, PHAEOPHYCEAE.

fucoxanthin *Biochemistry.* one of a class of carotenoids that contain oxygen and compose the primary carotenoid pigments in diatoms and brown algae.

Fucus *Botany.* **1.** a genus of dichotomously branched brown algae in the family Fucaceae; found widely in the temperate waters of the Northern Hemisphere, usually in the intertidal zone, and harvested by the kelp industry as a source of algin. **2.** broadly, any of various brown algae.

fuel *Chemistry.* any material that evolves energy in a chemical or nuclear reaction. *Materials.* specifically, a material that can be used to provide power for an engine, power plant, or nuclear reactor.

fuel accumulator *Aviation.* see ACCUMULATOR.

fuel-air explosive *Ordnance.* a layer of liquid fuel that is delivered by bomb, shell, or rocket to an area above the target, and after mixing with air, is detonated to produce a powerful plane shock wave.

fuel assembly *Nucleonics.* a collection of fuel rods, plates, and other structural materials that are fabricated together; several such assemblies make up a reactor core. Also, **fuel element.**

fuel bed *Mechanical Engineering.* a layer of burning fuel and ash, as on a furnace grate.

fuel cell *Electricity.* a galvanic cell in which electricity is produced through the oxidation of a fuel such as methanol.

fuel-cell catalyst *Chemistry.* a substance used to make electrodes for fuel cells to increase the cell-electrode reaction rates (typically oxygen reduction or hydrogen oxidation); e.g., platinum, nickel, and silver.

fuel-cell electrolyte *Chemistry.* an ionic conductor of electricity between the electrodes of a fuel cell.

fuel-cell fuel *Chemistry.* any material that produces electricity in a fuel cell at the anode when oxygen is reduced at the cathode; hydrogen and carbon monoxide are commonly used as fuel-cell fuels.

fuel decanner *Nucleonics.* a machine that removes the metal tubes containing the enriched uranium fuel rods in a nuclear reactor by shearing the tubing into a spiral strip.

fuel filter *Engineering.* a component in a piece of equipment, such as an internal-combustion engine, that serves to remove particles from fuel.

fuel injection *Mechanical Engineering.* a system for spraying fuel directly into the cylinder or combustion chamber of a spark-ignition engine cylinder, thus eliminating the need for a carburetor; used in diesel engines and in many gasoline engines. Thus, **fuel-injection engine.**

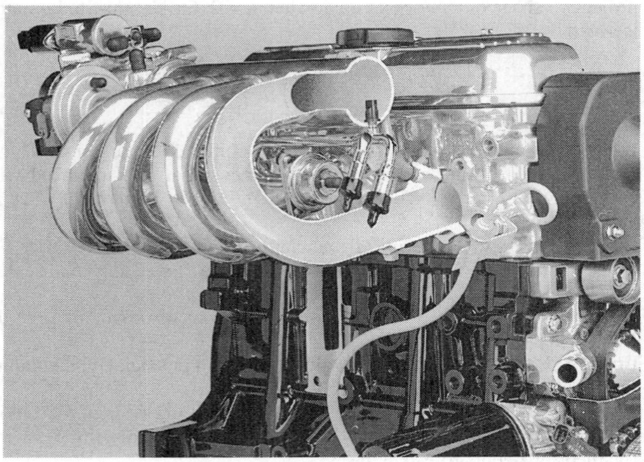

fuel-injection engine

fuel injector *Mechanical Engineering.* in the fuel injection process, the mechanism that sprays fuel into the engine cylinders or combustion chamber, consisting of a pump, valves, and nozzles.

fuel pellet *Nucleonics.* **1.** a small ball of frozen deuterium and tritium slated to be used as fuel in a futuristic laser-induced fusion power plant. **2.** a small body of fuel, often cylindrical, stacked in cans to form fuel assemblies.

fuel pump *Mechanical Engineering.* any pump used for delivering fuel from a storage tank to an engine or furnace.

fuel ratio *Petroleum Engineering.* the extent of the heating capability of a fuel in comparison to another fuel used as a standard.

fuel reprocessing *Nucleonics.* the process of retrieving unused fissionable material from a reactor's waste material.

fuel rod *Nucleonics.* a tube that contains the fuel used to generate energy in a nuclear reactor.

fuel shutoff *Space Technology.* **1.** a termination of the flow of liquid fuel into a combustion chamber or of the combustion of a solid fuel. **2.** the time or point at which this occurs. Also, SHUTOFF.

fuel system *Mechanical Engineering.* a system for the storage and delivery of fuel, including such components as a fuel tank, fuel pump, and fuel line.

fuel tank *Mechanical Engineering.* a storage component of a fuel system, often one of several composite or cellular sections.

fuel-weight ratio *Space Technology.* the ratio of the weight of a rocket's fuel to the weight of the unfueled rocket. Also, **fuel-structure ratio.**

fugacious [fyoo gā´shəs] *Botany.* falling off or fading soon after being formed, as some plant parts.

fugacity [fyoo gas´ə tē] *Thermodynamics.* a function f that is introduced as an effective substitute for pressure, to allow a real gas system to be considered by the same equations that apply to an ideal gas; the function is related to the molar Gibbs free energy by the equation $dG/d(\ln f) = RT$, where R is the gas constant and T is the temperature.

fugitive air *Mining Engineering.* air that moves through a fan but never reaches the working faces of a mine.

fugitive dye *Chemistry.* a dye that is not colorfast; used for purposes of identification.

fugue [fyoog] *Psychology.* a state of altered consciousness in which the individual may suddenly wander or travel about with no memory of his previous life; after recovery, there is no recollection of the fugue period.

fuguism see TETRODOTOXISM.

fugutoxin see TETRODOTOXIN.

Fuji, Mount *Geography.* the highest mountain in Japan (12,388 feet), in southern Honshu. Also, **Fujiyama.**

FUK *Aviation.* the airport code for Fukuoka, Japan.

Fukui, Kenichi born 1918, Japanese chemist; shared the Nobel Prize for applying quantum mechanics to the prediction of chemical reactions.

fulchronograph *Engineering.* an instrument used to record the incidence of lightning strokes, made up of a rotating aluminum disk with hundreds of steel fins on its rim which become magnetically charged when they pass between coils that have been struck by lightning.

fulcrate *Biology.* describing a structure in an organism that serves as support, such as a tendril or a stipule that stabilizes a plant.

fulcrum [fül´krəm] *Mechanics.* the pivot point about which a lever operates. *Anatomy.* any point on a body part that acts like a mechanical fulcrum, usually on a bone such as the ulna or femur.

Fulgoroidea *Invertebrate Zoology.* the lantern flies, a superfamily of insects in the order Homoptera.

fulgurator *Engineering.* an instrument used to spray salt solutions into a flame for examination.

fulgurite *Geology.* a glassy, tubular rock structure or crust formed when lightning strikes dry sand or sandy soil and causes it to fuse. Also, LIGHTNING STONE, SAND TUBE.

Fuligo *Mycology.* a genus of fungi that are slime molds belonging to the class Myxomycetes and that act as a host to fungal parasites.

full *Textiles.* **1.** to cleanse and compact cloth during manufacture by any of various methods, some using fuller's earth. **2.** describing cloth so processed.

full adder *Computer Technology.* **1.** an adder circuit in a digital computer that can handle the carry signal as well as the binary elements that are to be added. Also, THREE-INPUT ADDER. **2.** see HALF ADDER.

full annealing *Metallurgy.* a process of annealing according to a schedule that minimizes hardness.

full automatic *Ordnance.* a weapon that fires continuously as long as the trigger is depressed, as opposed to semiautomatic.

full-bore *Mechanical Engineering.* operating or moving at maximum power or speed.

full-cell process *Engineering.* a process used to preserve wood, in which the wood is placed under pressure in a vacuum to withdraw air and moisture from it before the preservative is introduced.

full charge *Ordnance.* in a gun using separate propelling charges, the largest of the charges.

full-depth avalanche see WET-SNOW AVALANCHE.

full-ended *Naval Architecture.* describing a vessel with a full, rounded bow and stern; this is associated with slow, high-capacity vessels.

fuller *Mechanical Devices.* a hammer having a half-round head, used for grooving and spreading iron. *Metallurgy.* in forging, the part of a die used for preliminary operations. *Textiles.* a person who fulls cloth.

Fuller, R(ichard) Buckminster 1895–1983, American architect, designer, and engineer; inventor of the geodesic dome.

fullerene [fül´ə rēn´] *Chemistry.* any of various cagelike molecules that consitute the third form of pure carbon (along with the forms diamond and graphite), whose prototype C_{60} (buckyball) is the roundest molecule that exists. Fullerenes are a class of discrete molecules, soccerball-shaped forms of carbon with extraordinary stability. The different forms of fullerenes are given nicknames, such as **fuzzyball, bunnyball,** and **platinum-burr ball.** (From Buckminster *Fuller*; because their configuration suggests the shape of his famous geodesic dome.)

fuller's earth *Geology.* a fine-grained, naturally occurring, highly adsorptive clay or claylike material composed primarily of montmorillonite; used as a bleaching, degreasing, or filtering agent, especially in the refining and decolorizing of oils.

full-face tunneling *Civil Engineering.* a method of excavation for a tunnel where the tunnel opening is enlarged to a desired diameter before extension of the tunnel face.

full feathering see FEATHERING.

full-force feed *Mechanical Engineering.* a system for lubricating an engine with oil under pressure.

full functor *Mathematics.* a functor $T: C \to B$ is said to be full if for every pair of objects $c, c' \in C$ and for every arrow $g: Tc \to Tc' \in B$, there is an arrow $f: c \to c' \in C$ with $g = Tf$. The composition of full functors is a full functor.

full-gear *Mechanical Engineering.* of a steam-engine valve gear, in the position giving maximum valve travel and cutoff for full power.

full-grain *Materials.* of leather, having its original grain surface intact.

full-hard *Metallurgy.* of certain alloys, at the stage in tempering just below that at which the metal cannot be formed by bending.

full jacket *Ordnance.* a lead bullet completely encased by a harder metal, usually copper. Also, **full metal jacket.**

full linear group *Mathematics.* the group of all nonsingular linear transformations on a given vector space; the group operation is composition of transformations.

full load *Electricity.* the normal rated output of an electronic device or transformer under specified conditions. Also, RATED LOAD.

full-load current *Electricity.* the output current from an electronic device or circuit when the load is at maximum under specified conditions. Also, RATED CURRENT.

full moon *Astronomy.* **1.** the instant when the moon lies directly opposite the sun in the sky, so that its illuminated side faces the earth and it appears as a round disk in the sky. **2.** the phase of the moon when this occurs.

fullness of tone *Acoustics.* the acoustic quality of a tone, based on the ratio of the loudness of a reverberant sound to the loudness of a direct sound.

full-pitch winding *Electricity.* an armature winding in which the pole pitch is equal to the span between the conductor coils.

full-pressure suit *Space Technology.* a space suit that can sustain enough gas pressure to support the wearer's normal body functions.

full-screen editing *Computer Programming.* a text processing function that allows the user to move the cursor to any position on the screen to carry out editing activities, as opposed to line editing.

full-service charge *Ordnance.* a complete round with the maximum military charge; smaller charges are measured as a percentage of a full-service charge.

full subtractor *Computer Technology.* a logic element whose inputs are a minuend, subtrahend, and borrow digit, and whose outputs are a difference digit and a new borrow digit. Also, THREE-INPUT SUBTRACTOR.

full-tracked combat tank *Ordnance.* **1.** a tank armed with one 105-mm gun, one 50-caliber machine gun, and one 7.62-mm machine gun; officially designated **M60.** Another type, officially designated **M48A3,** has a 90-mm gun. **2.** a similar tank armed with a 152-mm gun/launcher capable of firing Shillelagh missiles or conventional ammunition.

full-track vehicle *Mechanical Engineering.* a vehicle, such as a tank, that is entirely supported, driven, and steered by an endless belt or track on each side.

full-wave bridge *Electronics.* a bridge arrangement of four diode or tube rectifiers that provides full-wave rectification of the full secondary voltage of the power transformer.

full-wave control *Electronics.* a phase control that performs on both halves of each AC cycle to vary load power over the full range from zero to the full-wave maximum value.

full-wave rectification *Electronics.* a rectification process in which the negative half-cycle of the alternating input current is inverted, thus producing two positive half-cycles for each input cycle. Thus, **full-wave rectifier**.

full-wave vibrator *Electricity.* an interrupter in a vibrator-type power supply that closes contacts on both ends of its swing, thereby directing current through a transformer in alternate directions at regular intervals; used mainly in battery-operated mobile supply systems.

full width at half maximum *Physics.* a measure of the width of a resonance curve or spike-shaped function; the difference between the two frequencies for which a resonance curve has one half its maximum value.

full-width search *Artificial Intelligence.* a game-tree search technique in which all legal moves from a position are considered before any moves are discarded.

fully connected list *Computer Programming.* a list structure in which all records are connected by pointers and the last item in the list contains a pointer back to the first.

fully developed sea *Oceanography.* the maximum wave height or sea condition that can be generated by a wind of a given force blowing over sufficient fetch, regardless of duration.

fully extended chain *Materials Science.* a conformation of polymer chain in which the coil's mean square end-to-end distance is greater than the square of the radius of gyration.

fully populated board *Computer Technology.* a printed circuit board that contains the maximum number of components with no room for added logic capabilities.

fulmar *Vertebrate Zoology.* any of various seabirds of the family Procellariidae, especially *Fulmaris glacialis,* an Arctic species resembling a seagull.

fulminate [fŭl′mə nāt] *Medicine.* to rapidly worsen, such as an illness running a speedy course. Thus, **fulmination.** *Organic Chemistry.* any of various salts of fulminic acid containing the CNO– radical; used to detonate high explosives.

fulminic acid *Organic Chemistry.* C≡NOH, an unstable isomer of cyanic acid, whose salts are explosive.

fulminuric acid *Organic Chemistry.* $CNCH(NO_2)CONH_2$, colorless needles that are soluble in water; melts at 138°C and explodes at 145°C; used as a sanitizer.

fülöppite *Mineralogy.* $Pb_3Sb_8S_{15}$, a lead-gray, opaque, metallic, monoclinic mineral occurring in small prismatic or pyramidal crystals, and having a specific gravity of 5.22 to 5.23 and a hardness of 2.5 on the Mohs scale; found with zinkenite, sphalerite, and dolomite.

Fulton, Robert 1743–1815; American engineer and inventor; developed the first commercially successful steamboat (1807).

fulvene *Organic Chemistry.* C_6H_6, a yellow, readily polymerized oil boiling at 7–8°C (56 torr) that is an isomer of benzene.

Fulvia *Mycology.* a genus of fungi belonging to the class Hyphomycetes that is pathogenic to certain plants.

fumagillin *Microbiology.* a complex antibiotic that is produced by the fungus *Aspergillus fumigatus* and is an effective amebicide.

fumarase *Enzymology.* an enzyme of the lyase class that catalyzes the hydration of fumaric acid to malic acid and the reverse dehydration. Also, **fumarate hydratase.**

Fumariaceae *Botany.* a temperate family of herbaceous plants in the order Papaverales, having alternate leaves, elongated secretory cells, no latex system, and irregular flowers.

fumaric acid *Organic Chemistry.* HOOCHC=CHCOOH, colorless, odorless crystals, soluble in water and alcohol; melts at 287°C; used to make resins, paints, varnishes, and inks and as a chemical intermediate.

fumarole [fyoom′ə rōl] *Volcanology.* a vent, usually in volcanic regions, from which vapors or gases are released.

fumble *Industrial Engineering.* in work-motion studies, a sensory-motor error that is unintentional and probably not avoidable.

fume cloud *Volcanology.* a cloud of vaporous volcanic gas that rises from a body of molten lava.

fumed silica *Materials.* a calcinined ethyl silicate occurring in the form of a translucent powder; used as a substitute for carbon black in making light-colored products and to coagulate oil slicks on water so that burning off can occur.

fumes *Chemistry.* **1.** the smoky particulate matter that emanates from heated materials. **2.** vapors evolved from concentrated acids or solvents.

fumigant [fyoom′ə gənt] *Chemistry.* any vaporous toxic compound that is used as a pesticide.

fumigate [fyoom′ə gāt] *Engineering.* to use chemical compounds in a gaseous state to clear an area of insect pests or other unwanted organisms. Thus, **fumigation.**

fuming nitric acid *Inorganic Chemistry.* a term for nitric acid, HNO_3, when combined with dissolved nitrogen dioxide, NO_2, and a small amount of water; it produces visible fumes in air.

fuming sulfuric acid *Inorganic Chemistry.* a term for sulfuric acid, H_2SO_4, when combined with dissolved sulfur trioxide, SO_3; it produces visible fumes in air.

fumitremorgin *Mycology.* a toxin produced by the fungal species *Aspergillus fumigatus* that grows on foods and causes such symptoms as tremors when eaten by humans and other animals.

fumulus *Meteorology.* a delicate, almost invisible cloud veil that can occur at any of the principal cloud altitude levels.

Funariaceae *Botany.* a cosmopolitan family of small, shiny mosses of the order Funariales, characterized by simple stems with terminal sporophytes and soft leaves; one species, *Funaria hygrometrica,* is a weed that commonly grows in burned areas.

Funariales *Botany.* an order of mosses, generally small terrestrial plants with erect stems, terminal sporophytes, and broad leaves.

function *Physiology.* the special, normal, or proper activity of an organ or part. *Psychology.* a term used by Carl Jung to describe basic psychological processes such as thinking or sensing. *Chemistry.* the characteristic behavior of a chemical compound in the presence of a functional group. *Computer Programming.* **1.** a procedure that returns a single result. **2.** a precoded mathematical or logic routine. Also, FUNCTION SUBPROGRAM. *Mathematics.* **1.** if X and Y are sets, a function f from (or on) X to (or into) Y is a rule for associating with each element x of X, exactly one element $f(x)$ of Y; denoted $f\colon X \to Y$. X is the domain of f; Y is the codomain or range of f. x is called the argument, variable, or dependent variable of f; $f(x)$ is called the image of x under f. The set of all elements y of Y that are images of elements in X is a subset of Y called the image or range of f; it is sometimes denoted by $f(X)$. The terms *map, mapping, transformation, correspondence,* and *operator* can be used interchangably with *function.* **2.** a function f is sometimes represented as a collection of ordered pairs (or relation) such that if both (x, y) and (x, z) are in f, then $y = z$.

functional *Mathematics.* a generalization of the notion of function; a function whose domain and range are both sets of functions. The range may also be a set of numbers or scalars. In particular, linear mappings of a vector space V into its scalar field are called **linear functionals.** Other examples of functionals include integrals and derivatives.

functional analysis *Mathematics.* the study of spaces of functions; i.e., topological spaces in which the points of the space are functions.

functional application *Computer Programming.* a program or computer system developed for a specific activity or job, often a real-time system supporting on-going operations.

functional attribute *Archaeology.* any characteristic of an object that indicates its function, such as its form or a residue from an activity for which it was used.

functional autonomy *Psychology.* the tendency of behavior patterns to eventually become independent of the drives or motives that produced them.

functional bleeding *Medicine.* abnormal bleeding from the uterus without the presence of organic lesions.

functional bombing *Military Science.* the bombing of a specific target within a larger system or complex of targets; e.g., bombing the bridges of a transportation system or the hangars of an airfield.

functional constraint *Mathematics.* a mathematical equation representing physical conditions to be satisfied by parameters in an optimization problem.

functional decomposition *Control Systems.* the division of a large-scale control system into a set of nested generic control functions.

functional dependency *Computer Programming.* in relational databases, the relationship between two record types expressed implicitly by having a data item in each record type take its value from a common set of values.

functional diverticulum *Radiology.* the benign radiographic appearance of a saclike shadow, highlighted by a contrast medium but not apparent during subsequent direct examination.

functional electric(al) stimulation *Medicine.* the therapeutic application of controlled amounts of electric current to muscles to make them contract.

functional equation *Mathematics.* an equation between two expressions, each formed from a finite number of functions (known and unknown) and variables by a finite number of operations.

functional error recovery *Computer Programming.* a programmed method of allowing a computer to continue operations following the sensing of an error; routines for bypassing certain errors are included in the operating system.

functional fixedness *Psychology.* the tendency in problem-solving to evaluate objects or devices only in terms of their conventional use rather than in terms of all potential uses.

functional group *Organic Chemistry.* in a carbon-hydrogen molecule, an atom or group of atoms (designated as R–) replacing a hydrogen atom; a reactive group having specific properties such as a double bond.

functional group analysis *Materials Science.* an analysis of important groups on the polymer chain (such as carboxyl groups in polyesters or amino groups in polyamides), a commonly used method of determining average molecular weight.

functional independence *Mathematics.* a finite family of functions $\{f_1, \ldots, f_n\}$ defined on some open set U on a manifold is **functionally dependent** if there is a nonzero function F such that $F(f_1(u), \ldots, f_n(u))$ is identically zero as a function of u. If no such F exists, then $f_1, \ldots, f_n$ are **functionally independent.** No more than n differentiable functions can be functionally independent on an n-dimensional manifold.

functional interleaving *Computer Technology.* the process of allowing concurrent, independent input/output and computing operations with alternating access to shared memory.

functionalism *Anthropology.* the theory that culture was developed as a response to the social and psychological needs of a society and that each cultural trait has a useful function. *Psychology.* a school of psychology developed by William James and others in the late 19th century, stressing that consciousness should be analyzed in terms of its uses, or functions. Thus, **functionalist.**

functionality *Chemistry.* the extent or part of a compound that can form covalent bonds; a monofunctional compound can form one covalent bond, a bifunctional compound two.

functional language *Computer Programming.* any of a class of programming languages that are used in functional programming, typically consisting of an unordered set of equations that characterize functions and values. Also, APPLICATIVE LANGUAGE.

functional programming *Computer Programming.* a type of programming that uses function applications as the only control and data structure, rather than assignment statements as in procedure-oriented programming. Also, APPLICATIVE PROGRAMMING.

functional properties *Nutrition.* the attributes of food additives that are used for purposes not directly related to nutritional benefits, such as those that improve the color, flavor, or texture, or otherwise make a food product more attractive to the consumer.

functional residual capacity *Physiology.* the sum of expiratory reserve and residual volumes, representing the air remaining in the lungs after a normal expiration. Also, **functional residual volume.**

functional response *Ecology.* a change in the rate of exploitation of prey by an individual predator as a function of a change in prey density.

functional specification *Computer Programming.* a precise description of the effects that are intended to be achieved by a computer application, including information about required performance and available operations, provided to the user for approval during the design stage of a system. Also, **functional requirement.**

functional switching circuit *Electronics.* a circuit that implements a Boolean function and forms a basic element of a switching system; examples include AND, OR, NOT, NAND, and NOR circuits.

functional unit *Computer Technology.* any of five special-purpose computer components, including the arithmetic-logic unit, storage unit, control unit, input device, and output device.

function code *Computer Technology.* a special character or other signal carried on data storage media that controls output device actions such as carriage return, shift, or form feed.

function element *Mathematics.* an analytic function f along with its domain D; denoted (f, D). A sequence $\{(f_1, D_1), \ldots, (f_n, D_n)\}$, or chain, of function elements, where each element is an analytic continuation of the preceding one, is used in the theory of analytic continuation.

function-evaluated routine see FUNCTION SUBPROGRAM.

function generator *Electronics.* a signal generator that outputs several selectable waveforms and frequencies. *Computer Technology.* an analog computer circuit that creates a variable based on a mathematical function and one or more input variables.

function key *Computer Technology.* a special-purpose key on a keyboard that can be programmed to carry out or invoke special functions such as HELP, recalculate a spread sheet, change the form of a data display, or print a file.

function multiplier *Computer Technology.* an analog device that accepts the changing values of two functions and outputs the changing value of their product in relation to change in the independent variable.

function subprogram *Computer Programming.* see FUNCTION.

function switch *Electronics.* a switch that permits the selection of the various functions in a multifunction instrument, such as a voltohm-milliammeter.

function table *Mathematics.* a chart or table that lists the values of a function for some particular values of the variable. *Computer Technology.* an early device used to decode multiple inputs into a single output or to encode a single input into multiple outputs.

functor *Mathematics.* let C and B be categories. A functor $T: C \to B$ is a morphism of categories that consists of: (a) the object function T, which assigns to each object $c \in C$ an object $Tc \in B$; (b) the arrow function (also designated T) which assigns to each arrow $f: c \to c' \in C$ another arrow $Tf: Tc \to Tc' \in B$ so that $T(1_c) = 1_{Tc}$ and $T(g \cdot f) = Tg \cdot Tf$. That is, each identity object of C is carried to an identity object of TC and composition of arrows is preserved. *Computer Technology.* an early term for a logic circuit element.

fundamental *Physics.* the lowest frequency present in a complex signal, upon which successive harmonics are built. Also, FIRST HARMONIC.

fundamental catalog *Astronomy.* a reference catalog of precisely measured star positions and proper motions.

fundamental complex *Geology.* see BASEMENT.

fundamental constants *Physics.* the constants of physical data whose values are determined by experiment, such as the speed of light in a vacuum, the mass of an electron, and the charge of a proton.

fundamental frequency *Physics.* the lowest frequency present in a complex wave whose harmonic frequencies are integral multiple values of the fundamental.

fundamental gravity table *Geodesy.* a table giving the deformation of the geoid and its effect on gravity, computed for masses of unit density extending to various distances above and below the surface of the geoid; used as the basis for preparing special tables corresponding to particular assumptions respecting density and isostasy, for example.

fundamental interaction *Particle Physics.* one of the basic forces that act between the elementary particles of matter, including strong, weak, and electromagnetic interactions.

fundamental jelly see ULMIN.

fundamental mode *Physics.* in a vibrational system, the mode of oscillation that has the lowest frequency. Also, **fundamental mode of vibration.** *Electromagnetism.* the lowest waveguide frequency mode for a waveguide of a particular geometry.

fundamental niche *Ecology.* the largest ecological niche that an organism or species can occupy in the absence of interspecific competition or predation.

fundamental particle see ELEMENTARY PARTICLE.

fundamental quantity see BASE QUANTITY.

fundamental region *Mathematics.* any simply connected region of the complex plane that, under the action of some modular group of fractional linear transformations, tiles an open set of the complex plane. A modular function defined on a fundamental region is thus defined over the entire open set.

fundamental series *Spectroscopy.* for alkali elements, a set of spectral lines that represent a change in total orbital angular momentum accompanying a transition from the *d* state to the *f* state.

fundamental star *Astronomy.* a star whose position and proper motion are well enough known that it serves as a benchmark from which other stellar positions are measured.

fundamental strength *Geophysics.* the amount of stress, without creep, that a geologic body can withstand regardless of time.

fundamental theorem of algebra *Mathematics.* the field of complex numbers is algebraically closed; in particular, every polynomial of degree n with complex coefficients has n roots, counting multiplicities.

fundamental theorem of arithmetic *Mathematics.* any positive integer $n > 1$ may be uniquely factored into the form $n = p_1^{t_1} p_2^{t_2} \cdots p_k^{t_k}$, where $p_1 < p_2 < \cdots < p_k$ are primes and $t_i > 0$ for all i.

fundamental theorem of calculus *Mathematics.* **1.** if the integral $\int_a^b f(x)\,dx$ exists, and a function F exists such that $F'(x) = f(x)$ for all x in $[a, b]$, then $\int_a^b f(x)\,dx = F(b) - F(a)$. Also, **fundamental theorem of integral calculus. 2.** if the integral $\int_a^b f(x)\,dx$ exists, and if f is continuous at x_0 in $[a, b]$, and F is defined by $F(x) = \int_a^x f(x)\,dx$, then F is differentiable at x_0 and $F'(x_0) = f(x_0)$.

fundamental theorem of Galois theory *Mathematics.* if F is a finite dimensional Galois extension of a field K, then there is a one-to-one correspondence between the set of all intermediate fields of the extension and the set of all subgroups of the Galois group $\mathrm{Aut}_K F$ (the Galois group of F over K) such that: (a) the relative dimension of any two intermediate fields is equal to the relative index of the corresponding subgroups; in particular, $\mathrm{Aut}_K F$ has the order $[F : K]$; and (b) F is Galois over every intermediate field E, but E is Galois over K if and only if the corresponding subgroup $E' = \mathrm{Aut}_E F$ is normal in $G = \mathrm{Aut}_K F$. In that case G/E' is isomorphic to the Galois group $\mathrm{Aut}_K E$.

fundamental tone *Acoustics.* the lowest frequency in a harmonic series of frequencies.

fundamental unit see BASE UNIT.

fundamental wavelength *Physics.* the wavelength of the fundamental frequency in a complex wave or oscillating system.

F-U-N dating or **F.U.N. dating** *Archaeology.* a collective term for the techniques of fluorine, uranium, and nitrogen dating.

fundic gland see GASTRIC GLAND.

fundus *Anatomy.* **1.** the bottom or base of an organ. **2.** the part of a hollow organ that is farthest from the mouth of the organ.

fungal [fung´gəl] *Mycology.* describing a substance or process that is related to or consists of fungi.

fungate *Biology.* growing rapidly, like a fungus.

fungemia *Mycology.* the presence of fungi in the blood.

fungi [fun´jī; fung´gī] *Mycology.* the plural of FUNGUS.

Fungi *Biology.* one of five kingdoms in a commonly used system of classification; fungi are eukaryotic, incapable of spontaneous movement, contain no chlorophyll and therefore feed by ingesting other organic matter, and reproduce by means of spores; they exist in more than 100,000 species and are found in virtually all environments.

fungible *Chemical Engineering.* describing petroleum products that have similar characteristics, so that they can be blended.

fungicidin see NYSTATIN.

fungicole *Mycology.* a fungus that grows in or on other fungi. Also, **fungicolous fungus.**

fungiform *Biology.* shaped like a fungus or mushroom; i.e., having or consisting of a broad, often branched, free form on a narrower base.

fungiform papilla *Anatomy.* any of the knoblike projections on the tongue that are scattered singly among the filiform papillae.

Fungi Imperfecti see DEUTEROMYCOTINA.

fungistatic *Mycology.* referring to a substance that is able to stop fungal growth. Also, MYCOSTATIC.

fungitoxic *Mycology.* describing a substance that is poisonous to fungi.

Fungivoridae *Invertebrate Zoology.* a family of fungus gnats in the order Diptera; their larvae feed on fungae.

fungivorous *Zoology.* feeding mainly or exclusively on fungi.

fungoid *Botany.* resembling a fungus, as in having unhealthy spongy growths.

fungus [fung´gəs] *plural,* **fungi.** *Mycology.* an organism possessing cells with nuclei and rigid cell walls and lacking chlorophyll.

fungus

fungus cultivation *Entomology.* the growing of a specific fungus by ant and termite colonies for food; when a new colony is established, a piece of the fungus is often carried to the new nest and used to inoculate a growth medium composed of decaying leaves.

fungus gall *Plant Pathology.* any large, roundish, abnormal swelling on plant tissue caused by an attack of a parasitic fungus.

fungus gardens *Mycology.* fungi that are cultivated by Attine ants underground and that provide nourishment to larvae and adult ants.

funicular distribution *Chemistry.* a distribution of two phases of immiscible liquids in a porous body, in which the wetting phase is continuous over the surface of the body.

funicular polygon *Mechanics.* **1.** the two-dimensional or three-dimensional pattern formed by a cord under tension when it is acted on (or supported) by forces at various points. **2.** a figure drawn for solving equilibrium problems of rigid bodies or cables acted on in this manner.

funicular railroad *Engineering.* a railroad system used in areas of very steep gradients, in which rack-and-pinion clutches are used to grasp the sides of the rails.

funiculitis *Medicine.* **1.** an inflammation of the spermatic cord. **2.** an inflammation of the portion of a spinal nerve root lying within the intervertebral canal.

funiculus *Anatomy.* also, **funicle. 1.** a cordlike structure or part, such as the spermatic cord or the umbilical cord. **2.** one of the divisions of the white matter of the spinal cord, consisting of fasciculi or fiber tracts. *Botany.* the stalk that attaches a seed or ovule to the placenta in angiosperms. *Invertebrate Zoology.* **1.** in bryozoans, a band of mesoblastic tissue extending from the stomach to the body wall. **2.** in some insects, the slender middle part of the antennae. **3.** in certain Hymenoptera (ants, wasps, bees), the dorsal ligament connecting the petiole and propodium.

funnel *Mechanical Devices.* **1.** a wide-mouthed tool with a tapering spout, used to catch and direct poured liquids. **2.** the smokestack of a steamship, used for ventilation. *Naval Architecture.* a large engine-exhaust pipe extending above a ship's decks. Also, STACK.

funnel chest *Medicine.* a developmental deformity of the sternum, costal cartilages, and anterior portions of the ribs, in which the front of the chest is depressed.

funnel cloud *Meteorology.* see TUBA.

funneling *Astrophysics.* the common convergence on the red giant state by the evolutionary tracks of stars having a range of masses.

funoran *Biochemistry.* a galactan present in certain red algae that contains mostly D-galactose and 3,6-anhydro-L-galactose residues, with a few L-galactose residues; used as an adhesive and sizing agent.

funware *Computer Science.* a term for software programs intended for entertainment rather than for professional or educational applications.

FUO fever of undetermined origin.

fur *Vertebrate Zoology.* the thick, fine, soft, hairy coat of the skin of a mammal. *Materials.* the dressed pelt of an animal.

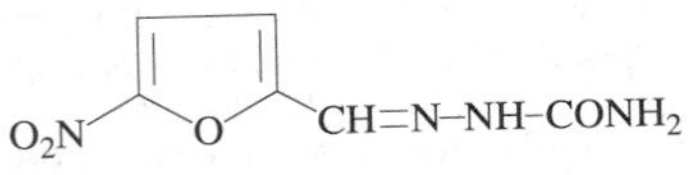

Furacin

Furacin *Pharmacology.* the trade name for a preparation of nitrofurazone, a topical anti-infective.

furan *Organic Chemistry.* C_4H_4O, a colorless monocyclic ether that melts at –86°C and boils at 31.4°C; used as an organic intermediate.

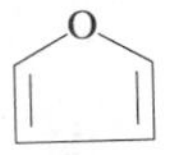

furan

furanose *Biochemistry.* a sugar that has a five-membered ring with four carbon atoms and one oxygen atom.

furanoside *Organic Chemistry.* a glycoside whose cyclic sugar component resembles that of furan.

furca *Invertebrate Zoology.* **1.** the fork-shaped posterior segment or appendages of copepods and some insects. **2.** the jumping apparatus in collembolans (springtails).

furcate *Biology.* forked, or branching like the prongs of a fork.

furcellaran *Biochemistry.* a gum obtained from a seaweed used as a gelling agent, a carrier for food preservatives, a bactericide, and in bacteriological culture media.

Furcellariaceae *Botany.* a small family of red algae of the order Gigartinales, characterized by erect gametangial thalli arising from a discoid base or stolonlike axes and sporangial thalli either similar in form or prostrate and discoid; includes one species used for agar.

furcocercous cercaria *Invertebrate Zoology.* a free-swimming, digenetic trematode larvae having a forked tail.

furcula *Entomology.* the forked tail of the *Collembola* (springtails), which when sprung from a hook under the abdomen helps propel the insect forward. *Zoology.* any forked body part or organ, such as the united clavicles of a bird.

furfuraceous *Biology.* covered with or consisting of scales or flaky particles. *Botany.* containing or resembling bran.

furfural *Organic Chemistry.* C_4H_3OCHO, a colorless liquid aldehyde that turns reddish brown on exposure to light and air, melts at –36.5°C and boils at 161.7°C; used in refining and as a solvent, intermediate, wetting agent, weed killer, and fungicide. Also, **furfurol**.

furfural extraction *Chemical Engineering.* a refining method for organic substances by contacting with furfural as a selective solvent.

furfuryl *Organic Chemistry.* C_5H_5O–, the radical from furfural.

furfuryl alcohol *Organic Chemistry.* $C_5H_6O_2$, a colorless, toxic liquid that turns brown to dark red on exposure to light and air; boils at 170°C; used in the manufacture of polymers, resins, and wetting agents.

furiotile lake *Ecology.* a subsidiary body of water that connects with the main body only at high water.

Furipteridae *Vertebrate Zoology.* the smoky bats, a small family of the order Chiroptera; found in tropical America and characterized by feet in which the first digit is much reduced or lacking.

furlong *Metrology.* a traditional measure of distance equal to 220 yards, or 1/8 mile; now mainly in nontechnical use, as in horse racing.

furnace *Engineering.* an enclosed structure in which heat is produced to create a chemical or physical change upon a substance.

furnace black *Chemistry.* a form of carbon black that is made by burning vaporized heavy oil in a furnace under oxygen-deficient conditions.

Furnarii *Vertebrate Zoology.* a suborder of neotropic birds of the order Passeriformes, most of which are sull brown and live in forests; includes the woodcreepers, ovenbirds, antbirds, and tapaculos.

Furnariidae *Vertebrate Zoology.* the ovenbirds, a family of tropical American passerine birds of the suborder Furnarii; distinguished by their huge domed nests ("ovens") built of mud and sticks and placed on the ground, in a burrow, in a tree, etc.

furniture *Graphic Arts.* **1.** a decorative printed image used to divide sections within a chapter of a book. **2.** originally, wood or metal material used to separate sections of type or fill in blank areas.

2-furoic acid *Organic Chemistry.* $C_5H_4O_3$, colorless crystals that melt at 134°C and sublime at 130°C; used as a preservative, bactericide, fumigant, and chemical intermediate.

furosemide *Pharmacology.* a high-ceiling diuretic used in the treatment of hypertension and edema.

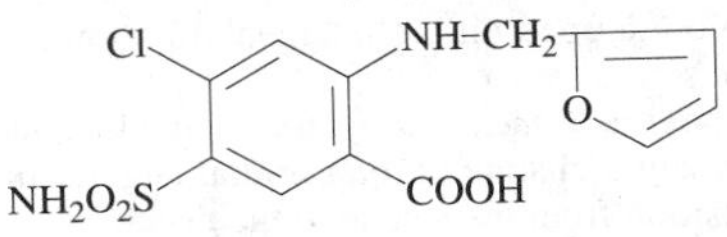

furosemide

Furovirus group *Virology.* a group of rod-shaped ssRNA-containing viruses that are naturally transmitted by fungi, causing mosaic and stunting in wheat crops. Also, SOIL-BORNE WHEAT MOSAIC VIRUS GROUP.

furring *Building Engineering.* thin strips of wood or metal applied to the joists, studs, or wall of a building, in order to level the surface, add thickness, or create an airspace. Also, **furring strips.**

furrow *Agriculture.* a long, narrow groove or trench in the earth made by a plow.

furrow irrigation *Agriculture.* an irrigation method in which the water is supplied to crops in ditches made by tilling.

furrow press see PRESS WHEEL.

furrow slice *Agriculture.* a strip of soil that is lifted and turned to one side by a plow bottom.

furrow wall *Agriculture.* the vertical side of a furrow that is dug by a plow; the turned soil is thrown away from this side.

Furry theorem *Quantum Mechanics.* a theorem stating that a closed electron-positron loop connected to an odd number of photon lines in a Feynman diagram contributes nothing to the process.

fur seal *Vertebrate Zoology.* any of various seals of the family Otariidae, such as *Callorhinus alascanus,* having luxurious underfur formerly widely used in making coats, hats, and so on.

furuncle *Medicine.* a pus-making skin infection in a gland or follicle marked by swelling, pain, and redness.

furunculosis *Medicine.* a serious skin disease marked by boils or successive crops of boils.

fusain *Geology.* a coal lithotype having a black color, silky sheen, and fibrous structure that occurs in strands and patches. Also, MINERAL CHARCOAL.

fusaric acid *Plant Pathology.* a lethal toxin derived from fungi of the genus *Fusarium,* which attacks plants and causes such disease symptoms as wilt, leaf spot, and browning of vascular tissues, and inhibition of enzyme action and production.

fusaritoxicosis *Mycology.* a disease caused by eating food contaminated by fungal molds belonging to the genus *Fusarium,* which occurs in such farm animals as chickens, pigs, horses, and cows.

Fusarium *Mycology.* a genus of imperfect fungi of the order Hypocreales, some of which are extremely pathogenic to plants and animals.

Fusarium oxysporum *Mycology.* a species of fungus that causes numerous plant diseases, such as wilts to pea plants, cotton, tomato plants, and banana plants.

Fusarium solani *Mycology.* a species of fungus that causes numerous plant diseases, such as potato and squash wilt.

fusarium wilt *Plant Pathology.* any plant disease caused by fungi of the genus *Fusarium* that results in the shrinking and drooping of the plant, following loss of turgidity.

fuse *Electricity.* a protective device based on a wire or element that melts at low temperature. When the current through the fuse exceeds the fuse rating, the wire melts and opens the circuit. *Engineering.* a combustible substance enclosed in a continuous cord, used for initiating an explosive charge by transmitting fire to it. *Ordnance.* any nonexplosive mechanical or electrical device used to set off the bursting charge of ammunition.

fusebox *Electricity.* a fireproof box that contains fuses and switches for various leads in a wiring system. Also, CUTOUT BOX.
fuse cutout *Electricity.* a disconnect switch with a fuse unit forming a portion of its flat moving conductor. Also, **fuse-disconnecting switch.**
fused aromatic ring *Organic Chemistry.* an aromatic ring having two or more carbon atoms (one or more sides) in common with another aromatic ring.
fuse diode *Electronics.* a diode that opens under particular current surge conditions.
fused junction see ALLOY JUNCTION.
fused quartz *Materials.* SiO_2, a naturally occurring fused silica glass. Also, QUARTZ GLASS.
fused-salt electrolysis *Physical Chemistry.* a process in which an electric current is passed through purified fused salts to produce a chemical reaction.
fused silica see SILICA.
fusehead *Engineering.* a section of an electric detonator made up of a pair of metal conductors connected by fine resistance wire, and bordered by a bead of igniting compound, which is set off when the firing current travels through the bridge wire.
fuselage [fyoo´sə läj´] *Aviation.* an aircraft's body structure, which houses the flight crew, passengers, and cargo and to which the wings, tail, and (in most single-engined planes) engine mount are attached.
fuse-link *Electricity.* the replaceable, current-carrying fuse element that melts.
fuse PROM *Computer Programming.* a programmable read-only memory chip that is created by selectively blowing a series of fuses to create binary patterns on the chip.
fuse wire *Electricity.* a wire made from an alloy that melts at a relatively low temperature.
fusibility *Thermodynamics.* a measure of the ability of a solid substance to be liquefied by heat transfer.
fusible *Physical Chemistry.* able to be fused or melted.
fusible alloy *Metallurgy.* any of several low melting alloys, such as those used in fire protection sprinkler systems.
fusible resistor *Electricity.* a low-value resistor that also serves as a fuse in some appliances such as televisions and receivers; it protects a system from overload and surges in power when current is first applied.
fusicoccin *Plant Pathology.* a toxin derived from the fungus *Fusicoccum amygdali* that causes disease in almond trees.
Fusicoccum *Mycology.* a genus of imperfect fungi belonging to the order Coelomycetes; the species *F. putrefaciens* causes cranberry end rot.
fusiform *Biology.* being tapered at both ends; spindle-shaped.
fusiform initial cell *Botany.* the elongated cells in a plant's cambium that give rise to the cells of the axial system in the secondary xylem and phloem.
Fusiformis see FUSOBACTERIUM.
fusigen *Biochemistry.* a compound secreted by certain fungi, such as *Aspergillus,* to facilitate synthesis and the uptake of iron.
fusillade [fyoo´sə läd´] *Ordnance.* simultaneous or continuous firing.
fusimotoneuron *Physiology.* a motor nerve fiber that innervates intrafusal fibers of the muscle spindle.
fusinite *Geology.* a principal coal maceral of fusain, composed of carbonized woody tissue.
fusinization *Geology.* the process of coalification that produces fusain.
fusion *Physical Chemistry.* the process of melting; the conversion of a solid into a liquid by using heat or pressure. *Nuclear Physics.* a nuclear process in which two light nuclei combine, at extremely high temperatures, to form a heavier nucleus and release vast amounts of energy. The explosive force of a hydrogen bomb is an example of uncontrolled fusion, and the energy of the sun and other stars is believed to derive from fusion reactions; speculated to occur at room temperature in electrochemical cells with palladium electrodes. Also, NUCLEAR FUSION.
fusion beat *Cardiology.* the phenomenon that occurs when an ectopic ventricular beat coincides with a normal beat or impulse to the ventricle; the contraction is triggered by a combination of the two influences.
fusion bomb *Ordnance.* a bomb that uses nuclear fusion for its destructive energy. Also, FUSION WEAPON, HYDROGEN BOMB, THERMONUCLEAR BOMB.
fusion casting *Materials Science.* a technique used to fabricate high-density slag-resistant refractories for the glass industry; liquid oxides melted in an arc furnace are cast into steel or sand molds.
fusion crust *Geology.* a black, thin, glassy covering formed on the surface of a meteorite by frictional heating during atmospheric flight.
fusion frequency *Physiology.* the minimum number of intermittent visual stimuli per second that produce a continuous visual sensation. Also, CRITICAL FUSION FREQUENCY.
fusion fuel *Nucleonics.* a substance that can be used to produce energy when two light nuclei form a heavier nucleus, such as deuterium and tritium, or deuterium and helium-3.
fusion nucleus *Botany.* in some seed plants, a triploid nucleus resulting from double fertilization and producing the endosperm.
fusion of cells *Biotechnology.* the process in which certain enveloped viruses, such as the *Sendai* virus, cause cells to fuse; used in the production of hybrid cells.
fusion point *Nucleonics.* the temperature above which the energy generated by nuclear fusion in a plasma surpasses the energy depleted, thus creating a self-sustaining reaction.
fusion protein *Biotechnology.* a protein molecule that brings about fusion between host cells, as between a virus envelope and a host plasma membrane in a *Sendai* virus or a *Paramyxovirus;* produced by recombinant DNA technology, usually when a gene is inserted into a plasmid vector in such a way that the terminal stop codon is deleted.
fusion reactor *Nucleonics.* a hypothetical device that can generate energy by combining two light nuclei to form a heavier nucleus; this is the the process by which stars generate energy. Also, CONTROLLED THERMONUCLEAR REACTOR.
fusion tube *Analytical Chemistry.* a device used to analyze the elements in a compound by fusing the sample with other compounds and analyzing the products.
fusion welding *Metallurgy.* any of several joining methods in which the parts to be joined are partially melted.
Fusobacterium *Bacteriology.* a genus of Gram-negative, obligately anaerobic, asporogenic bacteria of the family Bacteroidaceae that occur as rod-shaped cells or filaments in the gastrointestinal tracts of animals, including humans, and are sometimes pathogenic. Also, FUSIFORMIS.
fusula *Invertebrate Zoology.* the knob on the end of a spider's spinneret through which the silk glands open.
Fusulinacea *Paleontology.* a superfamily of large benthic foraminifers in the suborder Fusulinina that evolved from the endothyrids in the Devonian and flourished in the Carboniferous and Permian before dying out with most other fusulinids at the end of the Permian.
Fusulinidae *Paleontology.* a family of multichambered benthic foraminifera in the suborder Fusulinina, important in Late Paleozoic stratigraphy because of their rapid evolution and worldwide distribution; they disappeared in the great extinction at the end of the Permian.
Fusulinina *Paleontology.* a suborder of large foraminifera about the size of a grain of wheat; Ordovician to Triassic.
futile cycle *Enzymology.* a cycle of reactions placed under feedback controls so that conditions that activate one-half of the cycle deactivate the other half; such cycles often result only in heat production.
future label *Computer Programming.* a previously undefined label that is used in a programming instruction and that is temporarily given the address of the command in which it appears, until it becomes a permanent label having an absolute address.
future light cone *Physics.* the set of all points in space-time reached by signals traveling from a specified point at the speed of light.
fuzzy logic *Psychology.* problem-solving that involves a certain degree of inference and intuition to reach the proper conclusion; regarded as a crucial distinction between human and mechanical intelligence. *Artificial Intelligence.* a method of machine reasoning, similar to human thinking, that can process uncertain or incomplete information; characteristic of many expert systems. Also, **fuzzy thinking.**
fuzzy set *Mathematics.* a generalization of the concept of set; in particular, a subset F of a given set S for which the characteristic function $\chi_F : S \to [0, 1]$ has the entire unit interval as its range, rather than just the two values $\{0, 1\}$.
fuzzy-set theory *Artificial Intelligence.* an application of the concept of fuzzy set in expert systems for estimating the degree of certainty of conclusions.
fV femtovolt.
f value *Atomic Physics.* a quantum-mechanical quantity analogous to the number of dispersion electrons that have the same frequency within an atom. Also, OSCILLATOR STRENGTH, LADENBERG F VALUE.
F/V converter see FREQUENCY-TO-VOLTAGE CONVERTER.
FWD front-wheel drive.
fynbos [fin´bōs´] *Ecology.* a type of vegetation that is characteristic of regions with short, cool or cold, moist winters and long, hot, dry summers.

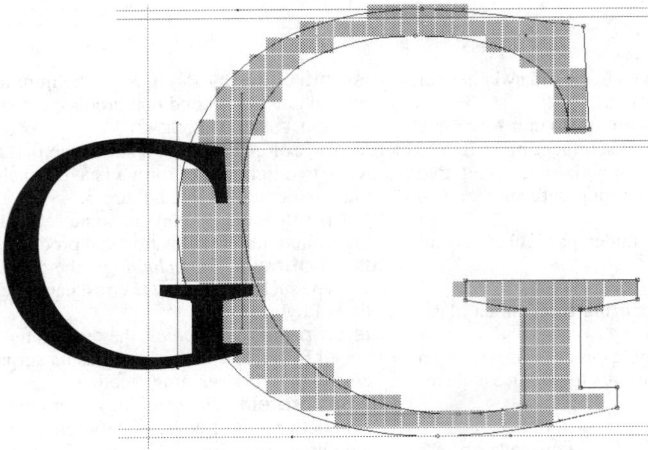

G *Electricity.* a unit of measurement indicating the capability of a circuit to conduct electricity.

G giga-; specific gravity; gravitational constant.

G or g *Metrology.* a unit of acceleration equal to the standard acceleration of gravity, or about 9.8 meters per second per second.

g an abbreviation for: gram; gravity; acceleration of gravity; gauge; giga-.

γ the Greek letter gamma.

G_0 *Cell Biology.* a resting stage in the eukaryotic cell cycle in which DNA replication and cell division cease.

G_1 *Cell Biology.* the first gap phase; a stage in the eukaryotic cell cycle that occurs after cell division and prior to DNA synthesis. During this cell cycle phase, cells monitor their growth status and do not initiate DNA replication until a minimum size and growth rate are attained.

G_2 *Cell Biology.* the second gap phase; a stage in the eukaryotic cell cycle that commences at the end of DNA synthesis and terminates at the initiation of mitosis. During this cell cycle phase, the cell monitors the condition of the DNA, so that mitosis is not initiated until replication is completed and DNA damage repaired.

ga gauge.

Ga the chemical symbol for gallium.

GA general of the army; gastric analysis.

GABA see γ-AMINOBUTRYRIC ACID.

gabbro *Petrology.* **1.** a coarse-grained intrusive igneous rock or a group of such rocks, characterized by a dark color and granular texture, consisting primarily of calcic plagioclase, such as labradorite, and clinopyroxene. **2.** of or relating to any coarse-grained, dark, igneous rock.

gabion *Engineering.* a hollow, sometimes bottomless wire cylinder that is used to hold earth materials in fieldwork or mining. Also, PANNIER. *Civil Engineering.* a cylinder of this type that is filled with stones and sunk in water to form the foundation of a jetty or dam.

gable *Architecture.* **1.** the upper, triangular portion of an end wall. **2.** the end of a double-sloping roof with the triangular piece of wall that it covers. Also, **gable roof.**

gaboon *Materials.* the light, soft-grained, reddish-brown wood of the African okume tree, *Aucoumea klaineana.* Also, **Gaboon mahogany.**

Gabor, Dennis 1900–1979, Hungarian-born English electrical engineer; Nobel Prize for the invention of holography.

Gabriel *Ordnance.* **1.** an Israeli ship-launched, surface-to-surface missile powered by a solid-propellant rocket motor and equipped with electronic guidance and terminal homing; it delivers a 330-pound conventional warhead at a maximum range of 12.6 miles. **2.** a similar air-to-surface missile that can be released in a fire-and-forget or fire-and-update mode with a maximum range of 37 miles; officially designated **Gabriel III A/S.**

Gabriel synthesis *Organic Chemistry.* a method that converts alkyl halides into primary amines by treatment with potassium phthalimide and the hydrolysis of the *N*-alkylphthalimides formed.

gad *Mining Engineering.* **1.** a small, steel wedge used for loosening seamy rock. **2.** a heavy piece of steel, 6–8 inches long, with a narrow chisel point for cutting samples, breaking out pieces of loose rock, etc.

gadder *Mining Engineering.* in quarrying, a small car or platform that carries a machine that drills a straight line of holes along its course; used to remove dimension stone. Also, **gadding car, gadding machine.**

Gadfly *Ordnance.* a Soviet land-based and shipborne surface-to-air missile. Official designation for land use **SA-11**, for naval use **SA-N-7.**

Gadidae *Vertebrate Zoology.* a family of predatory codfish and hakes of the order Gadiformes, characterized by soft rays (no spines in their fins); found primarily in cool waters in the Northern hemisphere around the continental shelf.

Gadiformes *Vertebrate Zoology.* an order of small to fairly large soft-rayed fishes that lack a swim bladder, including the eel cods, codfishes, hakes, cusk eels, and grenadiers; found almost exclusively in saltwater habitats worldwide.

gadoleic acid *Organic Chemistry.* $C_{20}H_{38}O_2$, a fatty acid obtained from glycerides in cod-liver oil, sperm oil, and herring oil; melts at 23°C; used in biochemical research.

gadolinite-(Ce) *Mineralogy.* $(Ce,La,Nd,Y)_2Fe^{+2}Be_2Si_2O_{10}$, a black monoclinic mineral of the gadolinite group occurring as irregular masses, having a vitreous luster, a specific gravity of 4.2, and a hardness of 6.5 to 7 on the Mohs scale; found in syenite pegmatites.

gadolinium *Chemistry.* a rare-earth element of the lanthanide series having the symbol Gd, the atomic number 64, an atomic weight of 157.25, a melting point of 1312°C, and a boiling point of above 3000°C; a lustrous metal having a high degree of magnetism, superconductive properties, and the highest neutron absorption cross section of any known element. (Named for the Finnish chemist Johan *Gadolin.*)

gadolinium oxide *Inorganic Chemistry.* Gd_2O_3, a white, amorphous, hygroscopic powder; very slightly soluble in water and soluble in acids to form the corresponding salts; melts at about 2330°C; used in lasers, telecommunications devices, and neutron shields.

gadolinium sulfate *Inorganic Chemistry.* $Gd_2(SO_4)_3{\cdot}8H_2O$, colorless monoclinic crystals that are slightly soluble in water; used in cryogenics.

Gaeumannomyces *Mycology.* a genus of fungi belonging to the order Diaporthales; the species *G. graminis* causes cereal diseases.

gaff *Naval Architecture.* a spar from which a square, fore-and-aft sail is hung. Similarly, **gaff-rigged, gaff sail.**

gag *Medicine.* to retch; attempt to vomit. *Surgery.* a device used to hold the mouth open.

GAG *Molecular Biology.* glycosaminoglycan, a long unbranched polysaccharide composed of repeating disaccharide subunits.

gage see GAUGE.

gageite-ITc *Mineralogy.* $(Mn^{+2},Mg,Zn)_{42}Si_{16}O_{54}(OH)_{40}$, a colorless, transparent, triclinic mineral, dimorphous with gageite-2M, occurring in radiating groups of needlelike crystals, having an undetermined hardness and a specific gravity of 3.61; found in open cavities in low-temperature hydrothermal veins at Franklin, New Jersey.

G agents *Toxicology.* a group of toxic organophosphorus compounds used as nerve gas during World War II.

gager see GAUGER.

gagger *Metallurgy.* any of several types of reinforcements of sand molds used in casting.

gaging see GAUGING.

gag reflex see PHARYNGEAL REFLEX.

gahnite *Mineralogy.* $ZnAl_2O_4$, a brittle, green to black, cubic mineral of the spinel group occurring as octahedral crystals or in granular form, having a specific gravity of 4.6 and a hardness of 7.5 to 8 on the Mohs scale; found in crystalline schists, granite pegmatites, and placer deposits.

gaign [gān] *Meteorology.* a wind of Italy that blows across mountains and causes clouds to form on their crests.

gain *Electronics.* the increase in signal power produced by an amplifier, usually expressed in decibels as the ratio of the output to the input. *Electromagnetism.* the ratio of the radiation intensity in a given direction to the radiation intensity that would be obtained if the power accepted by an antenna were radiated isotropically. Also, ANTENNA GAIN. *Control Systems.* an increase in a signal as it passes through a control system or control element. *Design Engineering.* a small hollow made in a piece of wood, into which hardware or another piece of wood is fitted.

gain asymptote *Control Systems.* an asymptote to a logarithmic graph of gain as a function of frequency.

gain control *Electronics.* **1.** the process of adjusting the gain of an amplifier. **2.** a potentiometer used to adjust the gain of an amplifier.

gain-crossover frequency *Control Systems.* the frequency at which the magnitude of a loop ratio is unity.

Gainful *Ordnance.* a Soviet surface-to-air missile designed for rapid response to aircraft flying at low and medium altitudes, powered by a solid-propellant motor and equipped with semiactive radar homing; it delivers a 175-pound warhead at a speed of Mach 2.8 and a range of 35 miles; officially designated **SA-6.**

gain margin *Control Systems.* the reciprocal of the magnitude of loop ratio at a phase crossover frequency, often expressed in decibels.

gain reduction *Electronics.* the reduction in gain of an amplifier at high- and low-frequency extremes.

gain scheduling *Control Systems.* the process of changing the parameters of a regulator in order to alleviate problems caused by variations in the process dynamics of a control system.

gain turndown *Electricity.* a device that adjusts the gain of a system or component; used to protect the transmitter from overload.

gain twist *Ordnance.* a type of rifling in which the number of turns increases from the breech end of the bore to the muzzle, thus increasing the rotation of the projectile.

Gajdusek, D. Carleton [gī´ doo shek´] born 1923, American virologist; shared Nobel Prize for discovering the kuru virus and its transmission mechanism.

gal or **Gal** *Metrology.* a unit of gravitational acceleration equal to one centimeter per second per second. (From the astronomer *Galileo.*)

galactan *Biochemistry.* any of a group of polymers of galactose that occur in plants, such as agar and carrageenan. Also, GALACTOSAN.

galactic [gə lak´tik] *Astronomy.* of or relating to the Milky Way or another galaxy. *Biology.* of or relating to milk; lactic.

galactic cannibalism *Astronomy.* the process by which a large elliptical galaxy in a cluster of galaxies attracts and engulfs smaller galaxies.

galactic center *Astronomy.* the direction toward the center of the Milky Way; lying in Sagittarius near right ascension 17h 46m and declination –29°, although the actual galactic center is hidden from view by clouds of dust and gas.

galactic cluster see OPEN-STAR CLUSTER.

galactic concentration *Astronomy.* the degree to which the light from a galaxy comes from its nucleus as opposed to its disk or other regions.

galactic disk *Astronomy.* the flat disk of a spiral galaxy that contains primarily young stars, gas, and dust clouds.

galactic equator *Astronomy.* the imaginary great circle in a galaxy that lies 90° from both poles; in our galaxy it closely parallels the middle of the visible Milky Way. Also, **galactic plane.**

galactic halo *Astronomy.* a roughly spherical collection of individual stars and globular clusters that surrounds a galaxy.

galactic latitude *Astronomy.* an object's angular distance north or south of the galactic equator as measured in degrees.

galactic light *Astronomy.* the integrated light coming from the Milky Way that illuminates the earth's night sky.

galactic longitude *Astronomy.* the angular distance between the galactic center and a point on the galactic equator at which an object's galactic meridian meets the equator, measured in degrees.

galactic magnetic field *Astrophysics.* a weak and largely disordered magnetic field, with a strength of about 5×10^{-10} tesla, that pervades the disk of the Milky Way Galaxy and controls the alignment of interstellar dust particles.

galactic meridian *Astronomy.* a great circle passing through both galactic poles and a given celestial object.

galactic nebula *Astronomy.* a general term for any kind of nebula (gas or dust) in the Milky Way Galaxy.

galactic noise *Astrophysics.* a diffuse radio signal that comes from the synchrotron radiation of electrons spiraling in the Galaxy's magnetic field. Also, **galactic radiation, galactic radio waves.**

galactic nova *Astronomy.* a nova that occurs in the Milky Way Galaxy.

galactic nucleus *Astronomy.* the central part of a galaxy, containing old stars and little dust or gas; many galaxies (including the Milky Way) also have at the center a compact, highly energetic radio source whose nature is unknown.

galactic pole *Astronomy.* the point on the sky, north or south, at which the Milky Way Galaxy's rotation axis, if extended, would meet the celestial sphere.

galactic rotation *Astronomy.* the time it takes a galaxy to spin once on its axis as measured at a given distance from its center; for the Milky Way at the sun's distance, this time is about 250 million years.

galactic windows *Astrophysics.* the regions of sky lying near the Milky Way that are unusually free of dust and gas absorption common to the galactic plane and that permit relatively unobscured views of distant galaxies.

galactic year *Astronomy.* a general term for the time it takes the sun to orbit the center of the Milky Way galaxy, about 250 million earth years.

galactitol see DULCITOL.

galactocarolose *Biochemistry.* a polysaccharide formed from the fungus *Penicillium charlesii.*

galactocele *Medicine.* a cyst caused by an obstruction in one or more of the mammary ducts.

galactocerebroside see CEREBROSIDE.

galactogen *Biochemistry.* a polysaccharide in the eggs of snails that yields galactose upon hydrolysis.

galactoglucomannan *Biochemistry.* a galactose-containing heteropolysaccharide composed of glucose and mannose residues.

galactokinase *Enzymology.* an enzyme of the transferase class that catalyzes the phosphorylation of galactose, the first in a series of reactions by which the monosaccharide galactose is brought into the glycolytic pathway.

galactolipid see CEREBROSIDE.

galactomannan *Biochemistry.* a homopolysaccharide of mannose that occurs in bacteria and various plants.

galactonic acid *Biochemistry.* $C_6H_{12}O_7$, a monobasic acid that is derived from galactose. Also, PENTAHYDROXYHEXOIC ACID.

galactophore *Anatomy.* a milk duct in the mammary gland.

galactopoiesis *Physiology.* the production of milk by the mammary glands. Also, LACTOPOIESIS.

galactorrhea *Medicine.* an excessive flow of milk or a persistent secretion of milk, irrespective of nursing.

galactosamine *Biochemistry.* an amino sugar of galactose that occurs in bacterial cell walls.

galactosan see GALACTAN.

galactose *Biochemistry.* $C_6H_{12}O_6$, a six-carbon aldohexose resembling glucose in many of its properties, but less sweet and less soluble. **D-galactose** is found in milk sugar, the raffinose of the sugar beet, many gums and seaweeds, and the cerebrosides of the brain; **L-galactose** is found in flaxseed mucilage.

galactosemia *Pathology.* an inborn error of metabolism that is due to the absence of galactose-1-phosphate uridyl transferase, resulting in an inability to convert galactose into glucose; it is often manifested by a failure to thrive in infancy and by subsequent mental retardation. Also, **galactosaemia, galactose diabetes.**

galactosidase *Enzymology.* either of two enzymes of the hydrolase class that catalyze the hydrolysis of certain galactoside residues in the formation of galactose.

galactoside *Biochemistry.* a glycoside derived by mixing galactose with an alcohol that yields galactose when hydrolyzed.

galactosuria *Medicine.* a passage of urine containing galactose, often the result of galactosemia.

galactosyltransferase *Enzymology.* any of several enzymes that catalyze the transfer of a galactized group from UDP galactose to an acceptor molecule.

galactotoxin *Toxicology.* a toxin found in stale milk.

galactotoxism *Toxicology.* poisoning due to the ingestion of galactotoxin. Also, **galactotoxismus, galactoxism, galactoxismus.**

galacturonate *Biochemistry.* a sugar acid derived from galactose by oxidation.

galacturonic acid *Biochemistry.* $C_6H_{10}O_7$, the acid produced from the oxidation of the primary alcohol group of D-galactose; a constituent in pectins and a number of plant gums.

galago see BUSH BABY.

Galapagos finch [gə lap´ə gōs´] see DARWIN'S FINCH.

Galapagos Islands *Geography.* an archipelago in the eastern Pacific, west of Ecuador.

Galapagos tortoise

Galapagos tortoise *Vertebrate Zoology.* the giant land tortoise from the Galapagos Islands, *Testuda elephantopus* of the family Testudinidae, that reaches about 400 pounds; nearly became extinct in the 1800s due to their vulnerability to predation by whaling boats because of their docile, slow-moving nature.

Galatheidea *Invertebrate Zoology.* a superfamily of crablike crustaceans in the order Decapoda, including the porcelain crabs, characterized by the abdomen curled beneath the thorax and a well-developed tail fan.

Galaxioidei *Vertebrate Zoology.* a suborder of small, freshwater fishes of the order Salmoniformes, including the icefish, smelts, salmonids, and galaxiids. Also, SALMONOIDEI.

galaxite *Mineralogy.* $(Mn^{+2},Fe^{+2},Mg)(Al,Fe^{+3})_2O_4$, a brilliant black, opaque, manganese-bearing, cubic mineral of the spinel group; granular in habit and having a specific gravity of 4.03 to 4.07 and a hardness of 7.5 to 8 on the Mohs scale; found with alleghanyite and calcite.

galaxy *Astronomy.* **1.** a gravitationally bound collection of stars, dust, and gas with a mass ranging from 100 million to 10 billion times that of the sun. **2.** also, **Galaxy.** the Milky Way Galaxy. (From the Greek word for "milk;" the Milky Way was the first galaxy known to the ancients.)

galaxy

Galaxy *Aviation.* a popular name for the C-5A military transport aircraft.

Galbulidae *Vertebrate Zoology.* the jacamars, a family of insectivorous birds of the order Piciformes, characterized by long, slender bodies and tails, and long, thin pointed bills; found in southern Mexico and Brazil.

gale *Meteorology.* **1.** a general name for an unusually strong wind. **2.** on the Beaufort wind scale, a wind whose speed measures 32–63 mph.

galea *Anatomy.* any of several structures shaped like a helmet. *Botany.* a helmet-shaped petal or upper corolla lip, as on mint plants. *Invertebrate Zoology.* **1.** in some insects, the outer of two lobes (endopodite) at the end of the maxilla. **2.** in pseudoscorpions, the spinning organ on the movable digit of the chelicera. (A Latin term meaning "helmet.")

galea aponeurotica see EPICRANIAL APONEUROSIS.

galeate *Biology.* **1.** helmet-shaped. **2.** having a galea.

Galen c. 130–200 AD, Greek physician; wrote over 400 medical texts; discovered that the arteries contain blood, not air; also studied the heart, pulse, brain, nerves, and spinal cord.

galena *Mineralogy.* PbS, a lead gray, opaque, cubic mineral having a metallic luster occurring as octahedral or cubic crystals and in massive form, having a specific gravity of 7.58 and a hardness of 2.5 to 2.75 on the Mohs scale; found in hydrothermal veins in limestones, dolomites, and other sedimentary rocks; the most important ore mineral of lead.

galenical *Pharmacology.* **1.** of or relating to the ancient system of medicine of Galen, a Greek physician and teacher of the second century. Also, **galenic. 2.** of or relating to herbal medications. **3.** any standard preparation containing one or more organic ingredients.

galenobismutite *Mineralogy.* $PbBi_2S_4$, a lead-gray to white, opaque, metallic, orthorhombic mineral occurring as crystals and columnar to compact masses, having a specific gravity of 7.04 and a hardness of 2.5 to 3.5 on the Mohs scale; found with pyrite, native gold, and quartz.

Galen's veins *Anatomy.* the small veins that pass through the pia mater of the third and fourth ventricles of the brain and empty into the great cerebral veins.

Galeomorphii *Vertebrate Zoology.* a superorder of Chrondichthyan sharks that are very small to gigantic, including bullhead sharks, carpet sharks, lamnoid sharks, and carcharhinoid sharks.

galeophobia see AILUROPHOBIA.

Galeritidae *Paleontology.* an extinct family of irregular echinoids in the suborder Echinoneina; middle to late Cretaceous.

galerne or **galerna** *Meteorology.* a cold, humid northwesterly squall that occurs behind a low pressure system over the English Channel and off the Atlantic coasts of France and northern Spain. Also, GIBOULEE.

Galilean [gal´lə lē´ən] relating to the Italian scientist Galileo or his discoveries or theories.

Galilean invariance *Mechanics.* a principle stating that Newton's laws are invariant with respect to Galilean transformations; that is, they will be the same in all inertial frames of reference.

Galilean relativity *Mechanics.* a concept based on Galileo's famous investigations into the performance of falling bodies, stating that the acceleration of a given particle will have the same value in all frames of reference that are moving at constant velocity relative to each other. This applies at ordinary speeds but not at speeds approaching the speed of light.

Galilean satellites *Astronomy.* a collective name for Io, Europa, Ganymede, and Callisto, the four largest moons of Jupiter; discovered by Galileo in 1610.

Galilean telescope *Astronomy.* a refractor telescope having a convex primary lens and a concave eyepiece lens; it has low magnification, a small field of view, and an upright image; used in opera glasses, inexpensive field glasses, and toy telescopes.

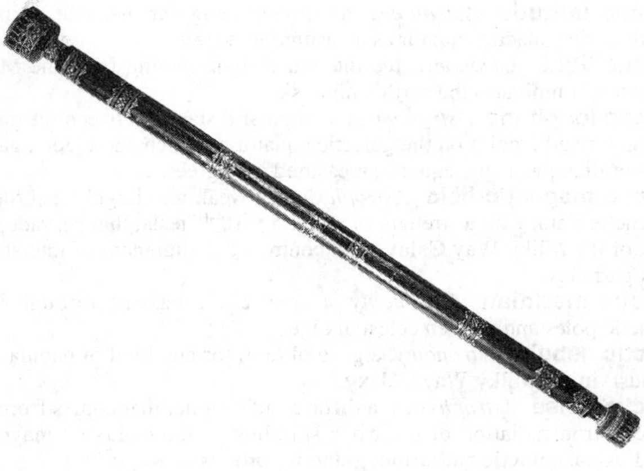

Galilean telescope

Galilean transformation *Mechanics.* a transformation of the mechanical quantities that describe a physical system from one inertial reference frame to another.

Galilean velocity addition *Mechanics.* a statement of the velocity of an object as observed from one reference frame relative to another: $v = v' + V$, where v is the velocity as measured in one frame, v' is the velocity as measured in the other, and V is the velocity of the second frame relative to the first. Thus, the velocity of an object measured by a stationary observer should equal the sum of its velocity as measured by a moving observer, plus that observer's own velocity.

galileo see GAL.

Galileo Galilei 1564–1642, Italian astronomer and physicist; discovered isochronism and the law of falling bodies; invented the hydrostatic balance, sector compass, and astronomical telescope; corroborated Copernican theory.

Galileo number *Fluid Mechanics.* a dimensionless number used in the study of circulating viscous fluids, given by the product of the gravitational acceleration, the square of the density, and the cube of a characteristic dimension divided by the square of the viscosity.

galipol *Organic Chemistry.* $C_{15}H_{26}O$, a sesquiterpene alcohol obtained from angostura bark oil; colorless crystals; melts at 89°C.

gall *Medicine.* **1.** see BILE. **2.** an excoriation or erosion. *Plant Pathology.* a large, roundish, abnormal swelling on plant tissue due to irritation by insects, mites, fungi, bacteria, viruses, or nematodes; insect oviposit and larvae are often found in galls. *Metallurgy.* in metal forming, to lose surface material because of adhesion to the die.

gallacetophenone *Organic Chemistry.* $C_8H_8O_4$, a ketone derivative of pyrogallol; a white to brownish-gray powder; soluble in water and alcohol; melts at 173°C; used as an antiseptic and thought to be a protective agent against harmful radiation. Also, 2',3',4'-TRIHYDROXYACETOPHENONE.

gallbladder or **gall bladder** *Anatomy.* a pear-shaped sac that lies under the right lobe of the liver and stores bile produced by the liver. Also, CHOLECYST.

Galle, Johann 1812–1910, German astronomer; the first to locate Neptune.

gallego *Meteorology.* a bitterly cold northerly wind of Spain and Portugal.

gallein *Organic Chemistry.* $C_{20}H_{12}O_7$, a reddish-brown crystalline powder or green scales; slightly soluble in water; decomposed by heat; used as a pH indicator and in the manufacture of dyes. Also, 4,5-DIHYDROFLUORESCEIN.

Galleriinae *Invertebrate Zoology.* a subfamily of insects in the order Lepidoptera, composed of the bee or wax moths; adults live in beehives and larvae feed on beeswax.

gallery *Architecture.* **1.** a covered corridor, arcade, or balcony. **2.** an elevated seating area in a church or auditorium. **3.** a room used to display art works. *Geology.* **1.** a horizontal passageway in a cave. **2.** any relatively horizontal underground channel. *Mining Engineering.* a subsidiary passage in a mine at a higher level than the main passage.

gallery deck *Naval Architecture.* a partial deck between the hangar deck and the flight deck of an aircraft carrier.

gallery forest or **galleria forest** see FRINGE FOREST.

gallery grave *Archaeology.* a type of European prehistoric tomb, characterized by a rectangular chamber with no separate entrance passage.

galley *Design Engineering.* a kitchen area, particularly such an area in a ship or airplane. *Graphic Arts.* **1.** a long, shallow metal tray used to store lines of composed type that will be placed into forms. **2.** see GALLEY PROOF.

galley proof *Graphic Arts.* a first proof printed from type that is set in the actual type format but not yet divided into pages; used to proofread for errors before page makeup.

gallic acid *Organic Chemistry.* $C_7H_6O_5$, obtained from nutgall tannins; colorless crystalline needles or prisms; soluble in water and alcohol; melts at 235–240°C; used in photography, tanning, ink manufacture, and pharmaceuticals. Also, 3,4,5-TRIHYDROXYBENZOIC ACID.

gallicolae *Invertebrate Zoology.* a name given to insects that produce and reside in galls, or to gall-dwelling insect development stages.

gallicolous *Plant Pathology.* living in galls,.

Galliformes *Vertebrate Zoology.* an order of fowl-like or gallinaceous terrestrial birds, mainly medium to large size, and characterized by a short, downcurved bill with a large crop used as a food reservoir, a strong gizzard that uses small stones for aid in grinding food, strong breast muscles, and short, rounded wings for rapid bursts of flight of limited distance; found worldwide except for Antarctica and Polynesia.

gallinaceous [gal'i nā'shəs] *Vertebrate Zoology.* **1.** of, belonging, or pertaining to the Galliformes, an order of fowl-like birds including chickens, turkeys, guinea fowl, grouse, pheasants, quail, curassows, guans, and megapodes. **2.** of birds, having a gizzard.

galling *Metallurgy.* a condition in which two rubbing surfaces partially weld, due to friction.

gallinule *Vertebrate Zoology.* a marshbird of the family Rallidae, characterized by long toes that enable them to walk on floating vegetation, and by a plate extending from the bill to head; found worldwide in swampy, stagnant areas.

gallinule

Gallionella *Bacteriology.* a genus of Gram-negative, iron bacteria that occur as microaerophilic, stalked cells in iron-rich, aquatic environments.

gallium *Chemistry.* a rare metallic element having the symbol Ga, the atomic number 31, an atomic weight of 69.72, a melting point of 29.78°C, and a boiling point of 2403°C; a silver-white metal some of whose compounds are used as semiconductors. (From *Gallia* or Gaul, the Latin name for what is now France.)

gallium arsenide *Inorganic Chemistry.* GaAs, dark gray crystals that are electroluminescent in infrared light; melts at 1238°C; used in electronics to transform mechanical motion into electrical impulses.

gallium arsenide laser *Optics.* a type of laser that emits light in the infrared region when an electric current is passed through its junction. Also, **gallium arsenide injection laser.**

gallium arsenide semiconductor *Solid-State Physics.* a semiconductor in which gallium and arsenic are combined in near-stoichiometric proportions and whose band gap is approximately 1.4 electron-volts.

gallium halide *Inorganic Chemistry.* any compound composed of gallium and one of the halogen elements (periodic table Group VIIa), such as chlorine or bromine.

gallium phosphide *Inorganic Chemistry.* GaP, pale orange transparent crystals or whiskers that are electroluminescent in visible light; produced at low temperature and used in semiconductor devices.

gallium phosphide semiconductor *Solid-State Physics.* a semiconductor in which gallium and phosphide are combined and whose band gap is approximately 2.3 electron-volts; can be used in high-temperature applications (up to about 870°C).

gallivorous *Vertebrate Zoology.* of or relating to insect species whose larvae devour the plant galls in which they develop.

gallnut *Plant Pathology.* any large, roundish, abnormal swelling on plant tissue that resembles a nut.

gallocyanine *Organic Chemistry.* $C_{15}H_{13}ClN_2O_5$, a green crystalline solid, insoluble in water and soluble in alcohol; used as a dye, a biological stain, and a reagent for lead.

gallogen see ELLAGIC ACID.

gallon *Metrology.* a unit of liquid measure made up of 4 quarts: (a) in the U.S., 128 fluid ounces or 231 cubic inches; equivalent to 3.785 liters. Also, STANDARD GALLON. (b) in Great Britain, 160 fluid ounces or 277.4 cubic inches; equivalent to 4.546 liters. Also, IMPERIAL GALLON.

gallop *Medicine.* a disordered rhythm of the heart with a three-sound sequence, resulting from the intensification of the normal third and fourth heart sounds. Also, **gallop rhythm.**

gallotannic acid see TANNIC ACID.

Galloway *Agriculture.* a particularly hardy breed of hornless beef cattle that is black or gray-brown in color. (From *Galloway*, Scotland, where the breed originated.)

gallstone *Pathology.* an amalgamous concretion, commonly of cholesterol crystals, bilirubin, and protein, that is formed in the gallbladder or bile duct.

gall wasp *Invertebrate Zoology.* the common name for a large group of small wasplike insects belonging to the order Hymenoptera, whose larvae produce and live in closed galls on many different plants.
Galofaro *Oceanography.* a whirlpool in the Strait of Messina between Sicily and Italy; formerly known by the classical name Charybdis.
Galois, Evariste [gal´wä´] 1811–1832, French mathematician; made major contributions in group theory and in the resolubility of algebraic equations by radicals.
Galois *Mathematics.* a field F is said to be Galois over a field K if F is a Galois extension of K.
Galois correspondence *Mathematics.* let F be Galois over K, and let E be an intermediate field; i.e., $F \supseteq E \supseteq K$. The one-to-one correspondence given in the fundamental theorem of Galois theory by assigning to each such E the Galois group $E' = \mathrm{Aut}_E F$ (of F over E) is called the Galois correspondence.
Galois extension *Mathematics.* let F be an extension field of K such that the fixed field of the Galois group $\mathrm{Aut}_K F$ is K itself. Then F is said to be a Galois extension (field) of K or is said to be Galois over K. Equivalently, an extension field F is Galois over K if and only if for any $u \in F - K$, there is a K–automorphism $\sigma \in \mathrm{Aut}_K F$ such that $\sigma(u) \neq u$.
Galois field *Mathematics.* a finite field; that is, any field with a finite number of elements.
Galois group *Mathematics.* **1.** let F be an extension field of a field K. The group of all K-automorphisms of F is called the Galois group of F over K and is denoted $\mathrm{Aut}_K F$. $\mathrm{Aut}_K F$ can be viewed as all the automorphisms of F that are automorphisms of K when restricted to K. **2.** the Galois group of a polynomial $f \in K[x]$ (a polynomial with coefficients in a field K) is the group $\mathrm{Aut}_K F$, where F is a splitting field of f over K.
Galois theory *Mathematics.* the study of the relationship of an extension field F of a field K to the group of all K-automorphisms of F. Important applications are in the theory of equations, algebraic number theory, and algebraic geometry. The motivation is the idea that knowledge of the structure of $\mathrm{Aut}_K F$ will allow one to determine whether the original problem was solvable. In particular, if $\mathrm{Aut}_K F$ is solvable, and F is the splitting field of a polynomial $p(x) \in K[x]$, then $p(x)$ can be solved by radicals. The structure of $\mathrm{Aut}_K F$ also reveals the formula for the roots, but this takes a substantial amount of extra work.
gal operon *Molecular Biology.* a DNA segment containing an operator gene and structural genes that code for proteins involved in galactose metabolism, which is transcribed as a polycistronic mRNA molecule.
Galosh *Ordnance.* a Soviet nuclear-armed missile for the exo-atmospheric interception of ICBMs. Successive models have been designated **ABM-1, SH-01,** and **SH-11.**
Galton, Sir Francis 1822–1911, English scientist; founder of eugenics; devised identification through fingerprints; discovered anticyclones.
Galtonian curve see REGRESSION LINE.
Galumnidae *Invertebrate Zoology.* a family of mites in the suborder Sarcoptiformes, scavengers that break down organic matter in soils.
galv galvanized.
Galvani, Luigi [gal vän´ē] 1737–1798, Italian anatomist; demonstrated that an electrical charge could produce muscular contractions and vice versa; an important development in physiology.
galvanic [gal van´ik] *Electricity.* of or relating to electricity flowing as a result of chemical activity. (From Luigi *Galvani.*)
galvanic battery *Electricity.* one or more galvanic cells configured so as to yield energy.
galvanic cell see VOLTAIC CELL.
galvanic corrosion *Metallurgy.* corrosion that is caused or accelerated by an electrochemical cell.
galvanic couple *Electricity.* a pair of unlike substances that generate a voltage when in contact with an electrolyte.
galvanic current *Electricity.* an essentially steady, direct current produced by galvanic action or chemical activity.
galvanic series *Chemistry.* an arrangement of metals according to their potential or ease of oxidation, with lithium at the negative (least noble) end and platinum at the positive (most noble) end.
galvanic skin response *Physiology.* a change in the electrical resistance of the skin produced by arousal or anxiety and associated with sympathetic nerve discharge. Also, ELECTRODERMAL RESPONSE.
galvanic vertigo see VOLTAIC VERTIGO.
galvanize *Metallurgy.* to coat a metallic surface, usually steel, with zinc. Thus, **galvanized, galvanization.**
galvanized steel *Materials.* a zinc-coated steel that resists corrosion.
galvanneal *Metallurgy.* to heat galvanized iron or steel until the zinc coat melts and alloys with the base.
galvanoluminescence *Physiology.* the radiation of light by tissue, caused by a galvanic electrical current.
galvanomagnetic effect *Electromagnetism.* any thermal or electrical effect that arises when a current-carrying conductor or semiconductor material is placed in a magnetic field.
galvanometer *Electricity.* an instrument that measures a small electric current by measuring the mechanical motion derived from electromagnetic or electrodynamic forces produced by the current.
galvanometer constant *Electricity.* a factor by which a certain function of a galvanometer reading must be multiplied to convert the scale reading into standard units of current.
galvanometer recorder *Acoustical Engineering.* a device that records a visual expression of an audio signal; it consists of a moving-coil galvanometer upon which a tiny mirror is mounted so that a beam of light reflected from the mirror will oscillate in accordance with the audio signal; the beam of light is made to strike a photographic plate.
galvanometer shunt *Electricity.* a resistor connected in parallel with a galvanometer to reduce sensitivity, in order to allow measurement of a larger current.
galvanotaxis *Biology.* a directed reaction of a motile organism to an electrical stimulus.
galvanotropism *Biology.* an orientation response to an electric current.
gam- a combining form meaning "married" or "united" (especially sexually), as in *gamont.*
gambrel roof *Architecture.* a peaked roof with two slopes on each side, a steeper lower slope and a flatter upper slope. Also, MANSARD ROOF.
game *Mathematics.* a mathematical model having the following three elements: (a) players, i.e., decision makers. "Nature" may take the place of one player in some applications. (b) action spaces, i.e, the set of possible actions for each player or coalition of players. The rule by which an action is chosen is called a strategy, and a sequence of actions is called a situation. (c) a payoff, i.e., a fixed function for each acting side that assigns a payoff to every situation. A solution is a set of strategies accepted by all the players. *Science.* in general, any contest between two or more players, governed by specified rules, with a particular goal.
gamebird *Biology.* **1.** any bird that is commonly hunted for food or for sport. **2.** specifically, a bird of the order Galliformes, such as the grouse, partridge, pheasant, quail, or wild turkey; mainly grain-eating, heavy-bodied, ground-nesting birds that are capable of only short, rapid flight.
game playing *Artificial Intelligence.* a development of computer programs enabling them to play games such as chess.
gamet- a combining form meaning "gamete," as in *gametangium.*
gametangial copulation *Mycology.* the process by which certain fungal gametangia join or exchange their contents.
gametangium *plural,* **gametangia.** *Biology.* any cell or organ that produces gametes.
gamete [gam´ēt] *Cell Biology.* a mature haploid reproductive cell that unites with another such cell of the opposite sex to form a diploid zygote. *Pathology.* the malarial parasite in its sexual form in the gut of the mosquito vector.
game theory *Mathematics.* the study of mathematical games, including the development of suitable concepts of solution, the investigation of their existence, and the procedures leading to solution.
gametic [gə met´ik] *Biology.* of or relating to gametes or to primitive sexual elements.
gametic copulation *Mycology.* the process by which two fungal sex cells join together.
gameto- a combining form meaning "gamete," as in *gametocyte.*
gametocyte *Histology.* an undifferentiated cell that develops into a gamete.
gametogenesis *Biology.* the formation of gametes, or reproductive cells, in reproductive organs.
gametogony *Biology.* the formation of gametes, especially in protozoans.
gametophore *Botany.* a branch of a moss or similar plant that bears the sex organs.
gametophyte [ga mēt´ə fīt´] *Botany.* an individual plant, or a haploid generation of a plant exhibiting alternating generations, that produces gametes.
game tree *Mathematics.* a graph representation of different strategic outcomes in a game. Each node or vertex represents an action space, and action spaces are connected by directed edges representing consequences of particular actions. *Artificial Intelligence.* a representation of all the possible plays or moves of two opposing players, drawn from one player's point of view.

game-tree search *Artificial Intelligence.* the search for a winning strategy in a game tree; several techniques exist.

gamma *Science.* **1.** the third letter in the Greek alphabet, written as Γ or γ. **2.** the third in a series or hierarchy, following alpha (α) and beta (β). *Chemistry.* of, relating to, or designating the third of various possible positions of atoms or groups of atoms that can be substituted in an organic compound. *Astronomy.* the third-brightest star in a given constellation. *Metrology.* a unit of measure for the strength of a magnetic field, equal to 0.00001 (10^{-5}) oersted. *Graphic Arts.* a sensometric quantity used to measure contrast in a photographic emulsion, specifically in the straight-line portion of its characteristic curve.

gamma acid *Organic Chemistry.* $C_{10}H_5NH_2OHSO_3H$, 2-amino-8-naphthol-6-sulfonic acid; white crystals; soluble in alcohol and slightly soluble in water; used as an azo dye intermediate. Also, 7-AMINO-1-NAPHTHOL-3-SULFONIC ACID.

gamma camera *Engineering.* an imaging instrument that records the spatial distribution of radioactive compounds in the human body.

gamma counter *Engineering.* an instrument that detects fast electrons created by gamma rays, and thus indicates the presence of gamma radiation.

gamma cross section *Nuclear Physics.* the probability that gamma rays will be absorbed or scattered by sample nuclei during a nuclear reaction.

gamma decay see GAMMA EMISSION.

gamma distribution *Mathematics.* the probability distribution involving two positive paramenters α and β, given by $f(x) = \beta^{\alpha}x^{\alpha-1}e^{-\beta x}/\Gamma(\alpha)$ for $x > 0$ and $f(x) = 0$ for $x \leq 0$. *Statistics.* an asymmetric, positively skewed distribution of a positive-valued random variable defined by the probability density function $f(x) = x^{a-1}e^{-x/b}/(b^a\Gamma(a))$ where x, a, $b > 0$ and $\Gamma(a)$ is the gamma function.

gamma diversity *Ecology.* the diversity of species in a range of habitats in a large geographical area.

gamma emission *Nuclear Physics.* a process in which atomic nuclei change their states to states of less energy with the simultaneous emission of electromagnetic radiations or gamma rays. Also, GAMMA RADIATION, GAMMA DECAY.

gamma field *Radiology.* a region that absorbs radiation from an unshielded or slightly shielded source of gamma radiation.

gamma flux density *Nuclear Physics.* the number of gamma rays passing through a given area in a given time.

gamma function *Mathematics.* the function Γ(α), defined by $\Gamma(\alpha) = \int_0^{\infty} x^{\alpha-1}e^{-x}\,dx$. Integration by parts gives the result $\Gamma(\alpha + 1) = \alpha\Gamma(\alpha)$ and hence $\Gamma(\alpha+1) = \alpha!$ for α a positive integer.

gamma gauge *Nucleonics.* an instrument that determines the density of a material by measuring its absorption of gamma rays.

gamma globulin *Immunology.* **1.** any serum globulin having γ electrophoretic mobility. **2.** broadly, any immunoglobulin.

gammagram *Radiology.* a graphic record of the gamma rays emitted by an object or substance.

gamma heating *Nucleonics.* the heat generated by a material that arises from the absorption of gamma rays.

Gammaherpesvirinae *Virology.* a subfamily of Herpesvirinae that infects a wide range of vertebrates, forming tumors in their hosts.

gamma iron *Metallurgy.* any iron that has a face-centered cubic crystalline structure; produced by heating alpha iron to a temperature above 900°C.

gamma irradiation *Nucleonics.* the exposure of a substance to gamma rays generated by radioactive decay.

gamma particle *Mycology.* in the fungus species *Blastocladiella emersonii* belonging to the class Chytridiomycetes, a minute organ in certain asexual spores (zoospores) that contains the genetic material deoxyribonucleic acid (DNA).

gamma prime *Materials.* a precipitated phase (Ni_3Al and/or Ni_3Ti), that imparts elevated-temperature strength and stability to nickel-base superalloys.

gamma radiation see GAMMA EMISSION.

gamma radiography *Nucleonics.* the production by gamma rays of shadowlike images on photographic film.

gamma ray *Nucleonics.* a highly energized, deeply penetrating photon that radiates from the nucleus during fission and frequently accompanies radioactive decay.

gamma-ray altimeter *Aviation.* an instrument that calculates altitude by measuring the backscatter produced when photons are transmitted to the earth from a cobalt-60 gamma source in an aircraft; used to measure altitudes below several hundred feet.

gamma-ray astronomy *Astronomy.* the branch of astronomy that studies the radiation from celestial objects at wavelengths shorter than 0.1 Angstrom; gamma rays have been detected from a few galaxies and quasars, and from certain highly evolved stars.

gamma-ray burster *Astrophysics.* an object of unknown origin emitting a short, strong flash of gamma rays with energies in the million electron-volt range.

gamma-ray detector *Engineering.* an instrument that detects and measures areas of high concentrations of gamma rays.

gamma-ray laser *Physics.* a hypothetical device that would generate coherent electromagnetic radiation of wavelengths in the range of 0.005 to 0.5 nanometer through energy transitions occurring in the nuclei of atoms. Also, GRASER.

gamma-ray source *Nucleonics.* a radioactive material that emits gamma rays in a form that can be used in radiology for medical diagnosis and therapy.

gamma-ray spectrometry *Nucleonics.* any technique for measuring the energies of gamma rays; used extensively in activation analysis.

gamma-ray telescope *Astronomy.* a telescope and detector used for observations at gamma-ray wavelengths.

gamma-ray tracking *Space Technology.* a method of triangular tracking using a cobalt transmitter in the tail of a missile.

Gammaridea *Invertebrate Zoology.* a large family of crustaceans, the scuds or sandhoppers in the order Amphipoda, with laterally compressed bodies and no carapace.

gamma scanning *Nucleonics.* a method for detecting gamma radiation in a nuclear reactor that involves counting the photons emitted from the fuel rod.

gamma structure *Solid-State Physics.* a designation of one of the Hume-Rothery rules for an electron compound having a free electron-to-atom ratio of 21 to 13, or approximately 1.62.

gamma taxonomy *Systematics.* the third stage in the development of the systematic understanding of a taxon, during which subspecies and their geographical variation are worked out.

gamma transition *Materials Science.* a reversible change that occurs in an amorphous polymer when it is heated to a certain temperature, characterized by transition from a hard, glassy, or brittle condition to a flexible or elastomeric condition.

gammil *Chemistry.* a unit of concentration equal to 1 mg of solute in 1 liter of solvent.

Gammon *Ordnance.* a large Soviet surface-to-air missile that has limited capability as an antiballistic missile weapon, delivering a conventional or nuclear warhead at a range of 100 miles; officially designated **SA-5.**

gammopathy *Immunology.* a condition of the immune system characterized by an abnormal increase in the levels of immunoglobulins in the blood.

gamo- a combining form meaning "married" or "united" (especially sexually), as in *gamogony.*

gamogony *Invertebrate Zoology.* in Sporozoa, the production of gametes for spore formation by the process of multiple fission.

gamone *Physiology.* a hypothetical fertilization-enhancing substance released by a sperm or an ovum.

gamont *Invertebrate Zoology.* in Sporozoa, the stage that will produce gametes.

gamopetalous *Botany.* of a flower, having a corolla in which the petals are united by their edges, as on the morning glory. Also, SYMPETALOUS.

gamophobia *Psychology.* an irrational fear of marriage.

gamophyllous *Botany.* having the leaves of the perianth united by their edges, with the leaves or leaflike organs more or less united one to another.

gamosepalous *Botany.* having the sepals of a plant united at their edges. Also, SYNSEPALOUS.

Gamow, George 1904–1968, Russian-born American physicist; formulated theories of radioactive decay, solar energy, stellar evolution, nuclear structure, and genetic information.

Gamow barrier *Nuclear Physics.* a nuclear potential barrier near the surface of the nucleus that inhibits the release of alpha particles.

Gamow-Condon-Gurney theory *Nuclear Physics.* a hypothesis that gives a finite but small probability for the escape of an alpha particle from an unstable nucleus by penetrating (using a tunneling process) the potential barrier near the surface of the nucleus.

Gamow-Teller selection rules *Nuclear Physics.* a principle stating that there is no parity change of the nuclear state or in spin of the nucleus during a transition, except during a transition from spin 0 to spin 0.

Gampsonychidae *Paleontology.* an extinct family of primitive Paleozoic arthropods in the order Gigantostraca (Palaeocaridacea), probably related to the eurypterids; Carboniferous.

Ganef *Ordnance.* a Soviet highly mobile surface-to-air missile that is powered by four solid-propellant boosters and a ramjet sustainer and equipped with command guidance; it delivers a conventional warhead at a range of 45 miles; officially designated **SA-4.**

gang *Psychology.* a loosely organized group, typically adolescent males of the same ethnic backgound living in one neighborhood, that unites against other groups of the same type and that is often identified with violent or criminal behavior. *Electricity.* to mechanically couple two or more variable capacitors, switches, or other components so they can be operated from the same control knob. *Agriculture.* any of the bottoms of a gangplow.

gang board see GANGPLANK.

gang chart *Industrial Engineering.* a chart depicting the simultaneous acitivies of an entire work crew and/or group of machines. Also, GANG PROCESS CHART, MULTIPLE ACTIVITY PROCESS CHART.

gang drill *Mechanical Engineering.* a set of drills operating within one machine.

ganged control *Electronics.* two or more circuit controls mounted on a common shaft to allow simultaneous control of the circuits.

Ganges *Geography.* a river in northern India, rising in the Himalayas and flowing 1560 miles southeast to the Bay of Bengal.

ganging *Graphic Arts.* **1.** in halftoning, the process of grouping and screening two or more images in order to yield a single negative that can be divided for platemaking. **2.** in printing, a process in which two or more plates or forms are combined for simultaneous printing on one large sheet.

ganglial of or relating to the ganglion.

gangliocyte *Neurology.* a nerve cell located outside the central nervous system; a ganglion cell.

gangliocytoma see GANGLIONEUROMA.

ganglioglioma *Oncology.* a tumor composed of neuroglia and mature ganglion cells.

ganglioma *Medicine.* a tumor of the central nervous system.

ganglion [gang´lē än] *plural,* **ganglia.** *Anatomy.* any knot or knotlike mass. *Neurology.* **1.** a general term for a mass of nerve cell bodies lying within or outside the brain and spinal cord. **2.** a benign cystic tumor that occurs in an aponeurosis or tendon, as in the wrist or dorsum of the foot; consists of a thin, fibrous capsule enclosing a clear mucinous fluid.

ganglionated cord *Neurology.* the principal trunk of the sympathetic nervous system, consisting of a pair of long nerve strands, one on either side of the vertebral column, and extending from the base of the skull to the coccyx.

ganglioneure *Neurology.* any cell of a sensory or autonomic ganglion.

ganglioneuroma *Oncology.* a benign tumor composed of ganglion cells. Also, GANGLIOCYTOMA.

ganglionitis *Neurology.* inflammation of a ganglion or nerve cell body.

ganglioside *Biochemistry.* any member of a class of galactose-containing cerebrosides that are found in the tissues of the central nervous system.

gang milling *Engineering.* a process in which several mill cutters on one spindle are used to mill the face and sides of material in a single operation.

gangosa *Medicine.* lesions of the nose and hard palate, considered to be a tertiary stage of yaws.

gangplank or **gang plank** *Naval Architecture.* a plank or ramp that extends from a pier or the shore to a ship's gangway. Also, GANG BOARD.

gangplow or **gang plow** *Agriculture.* a plow that has two or more bottoms; used to make parallel furrows.

gang process chart see GANG CHART.

gangrene [gang´grēn´] *Pathology.* the death of tissues due to disease or direct injury that causes a failure of blood supply, followed by bacterial invasion and putrefaction.

gangrenous [gang´grēn´əs] *Pathology.* relating to, characterized by, or resembling gangrene.

gangrenous stomatitis see NOMA.

gang saw *Mechanical Engineering.* an arrangement of parallel saws that operate simultaneously in cutting logs into strips.

gangue [gang] *Mining Engineering.* the constituents of an ore deposit that are considered valueless. *Metallurgy.* the portion of mined ore that is discarded prior to extracting valuable metals; any predominantly nonmetallic minerals associated with ore.

gangway *Naval Architecture.* **1.** the access opening on the side of a ship through which crew or passengers pass. Not properly used to refer to a gangplank or ramp leading from the gangway to a pier. **2.** a raised catwalk on either side of the waist of a sailing ship. **3.** any narrow passageway aboard ship. *Mining Engineering.* **1.** a main underground haulage road. **2.** a passageway driven into coal at a slight grade, forming the base from which the other workings of a mine are begun.

gannet *Vertebrate Zoology.* any of several large, fish-eating seabirds of the family Sulidae, characterized by a sharp bill, webbed feet, and generally white plumage with black-tipped wings; found on offshore islands in temperate seas.

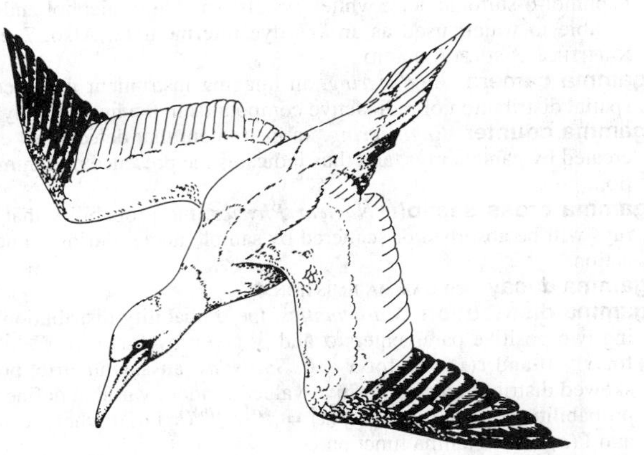

gannet

Ganodermataceae *Mycology.* a family of fungi belonging to the order Agaricales that occur exclusively on wood.

ganoid scale *Vertebrate Zoology.* a scale found on primitive ray-finned fishes such as the gar, having a rhomboidal shape and consisting of an enamel-like material packed tightly in diagonal arrangements.

ganoin *Vertebrate Zoology.* a hard, shiny, enamel-like outer layer covering the scales of certain primitive fishes (sometimes referred to as ganoid fishes). Also, **ganoine.**

ganomalite *Mineralogy.* $Pb_9Ca_5Mn^{+2}Si_9O_{33}$, an easily fusible, colorless to gray, hexagonal mineral occurring in prismatic crystals and in massive and granular forms, having a specific gravity of 5.74 and a hardness of 3 on the Mohs scale.

ganophyllite *Mineralogy.* $(K,NA)_2(Mn,Al,Mg)_8(Si,Al)_{12}O_{29}(OH)_7 \cdot 8\text{–}9H_2O$, an easily fusible, brown monoclinic mineral occurring in short prismatic crystals with a perfect cleavage and also in foliated and micaceous forms, having a specific gravity of 2.84 to 2.87 and a hardness of 4 to 4.5 on the Mohs scale; found with barite, rhodonite, and willemite.

gantlet *Civil Engineering.* a stretch of overlapping railroad track, with the rail of one track between the rails of the other track; used for narrow bridges and passes.

Gantrisin *Pharmacology.* $C_{11}H_{13}N_3O_3S$, a trade name for preparations of sulfisoxazole, a short-acting sulfonamide that is used as an antibacterial for the treatment of a wide variety of infections.

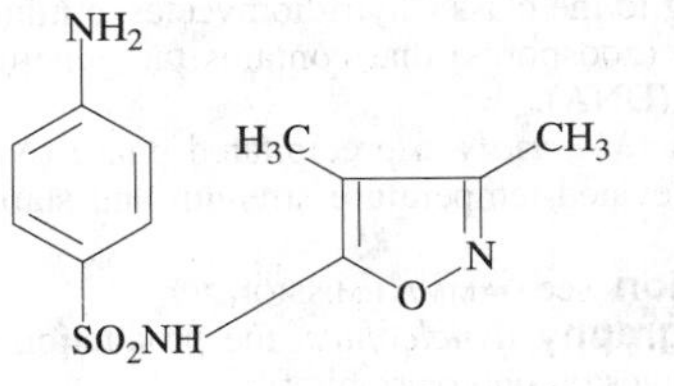

Gantrisin

gantry *Mechanical Engineering.* an adjustable hoisting machine that slides along a fixed platform or track, either raised or at ground level; used to carry and support heavy equipment or materials. Also, **gantry crane.** *Space Technology.* a similar structure used to assemble and service a large rocket on a launch pad. Also, **gantry scaffold.**

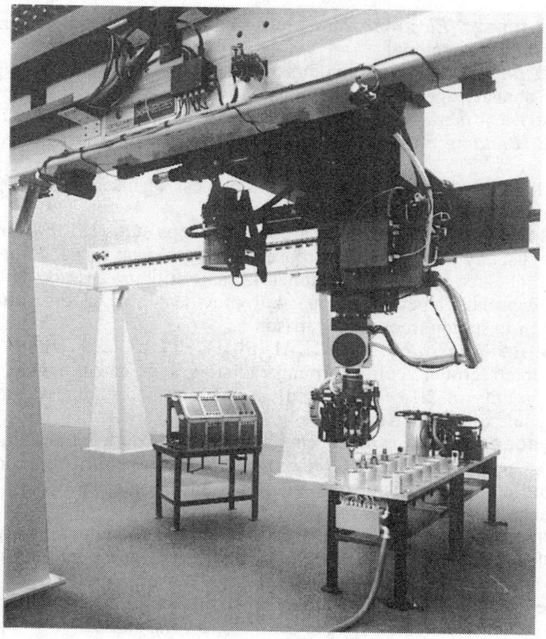

gantry robot

gantry robot *Robotics.* a continuous-path, Cartesian-coordinate, extended-reach robot that travels on a gantry. Also, **gantry-type robot.**

Gantt chart *Industrial Engineering.* a linear time chart on which scheduled performance goals are marked, allowing actual progress to be charted against the goals. (Named for Henry Laurence *Gantt*, who developed this chart in 1917.)

Ganymede [gan′i mēd′] *Astronomy.* the third and largest Galilean moon of Jupiter, 5270 km in diameter, having a composition that is about half rocky material and half water; discovered by Galileo in 1610.

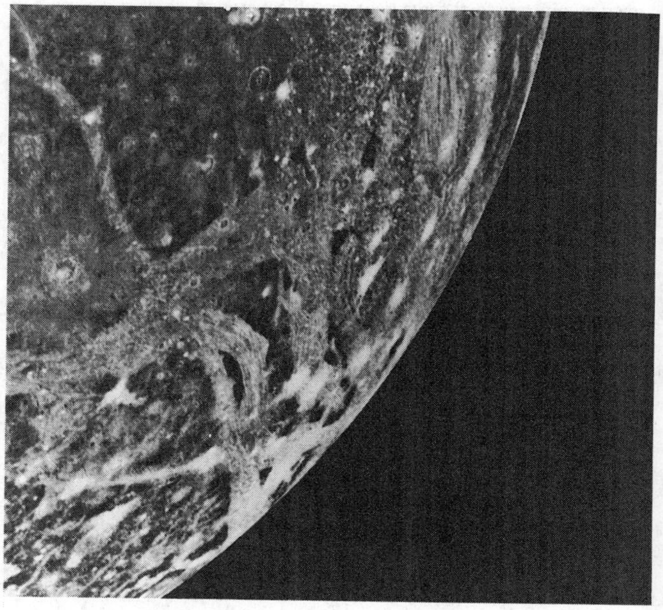

Ganymede

gap a break or opening; specific uses include: *Geography.* a narrow cut or dip in a ridge. A shallow cut is called a **wind gap;** a deep cut with water flowing through it is called a **water gap.** Also, NOTCH. *Electricity.* the distance between two electric contacts. *Electromagnetism.* an opening in a magnetic circuit filled with nonmagnetic material such as air. *Computer Technology.* a space between records or blocks of data on a tape or other magnetic medium; used to enable the medium to be stopped and restarted between reading or writing of the blocks. *Telecommunications.* an area of weakness or a lack of sensitivity in the reception pattern of an antenna. *Military Science.* an area within a minefield or obstacle belt that is free of live mines or obstacles, thus allowing a friendly force to pass through in tactical formation. Thus, **gap marker.** *Molecular Biology.* a break in one strand of a duplex DNA molecule due to the removal of one or more nucleotides.

GAPA ground-to-air pilotless aircraft.

gap bed *Mechanical Engineering.* a lathe bed having a gap near the headstock, which permits the turning of large-diameter work.

gap coding *Telecommunications.* the placement of pauses in a normally continuous transmission.

gap digit *Computer Technology.* a digit within a word used for purposes other than instructions or data, such as a parity bit or a digit used for technical reasons.

gape *Invertebrate Zoology.* in bivalve mollusks, the space between valves when the shell is closed.

gap factor *Electronics.* in a traveling-wave tube, the ratio of maximum energy gained in volts to the maximum gap voltage.

gap filling *Electromagnetism.* the process of placing auxiliary radar antennas in gap regions.

gap junctions *Cell Biology.* the common intercellular junctions creating aqueous channels between cells, through which small water-soluble molecules can pass; probably serves for intercellular communication in metabolic terms.

gap lathe *Mechanical Engineering.* a lathe having a gap bed.

gapless *Computer Technology.* describing data that is recorded in a continuous manner, without gaps between records or blocks.

gap mutation *Molecular Biology.* any heritable change within the limits of a single gene, as distinguished from changes in chromosome structure or chromosome number.

gapped tape *Computer Technology.* a magnetic tape on which blocked data is stored; includes format and all control information needed to read and use the data directly.

gap scatter *Computer Technology.* the deviation from the exact vertical alignment required of the read/write heads and the parallel tracks on the magnetic surface of a tape or disk.

gar *Vertebrate Zoology.* any one of a genus of predatory fishes characterized by an elongated body covered with ganoid scales, and having long, sharp teeth, an olive-green back, and a shiny silver belly; found primarily in freshwater habitats of North and Central America.

Garand, John C. [gə rand′] 1888–1974, American engineer and inventor; developed the M-1 rifle (**Garand rifle**), which was the standard rifle of the U.S. Army from 1936 to 1960.

garbage discarded matter; refuse; specific uses include: *Space Technology.* various objects in orbit, usually discarded or broken away from a spacecraft. Also, SPACE DEBRIS. *Computer Programming.* **1.** erroneous or irrelevant data. **2.** dynamic memory elements that are no longer in use. *Artificial Intelligence.* in Lisp, memory space that has become unused because there do not exist any pointers to it.

garbage collection *Computer Programming.* in dynamic storage allocation, a function that periodically examines memory for data that is no longer useful and restores those locations to available memory. Also, RECLAIMER.

garbage in, garbage out *Computer Programming.* a phrase meaning that the quality of the output of a computer program is dependent upon the quality of the input; commonly abbreviated GIGO.

garbin *Meteorology.* a southwesterly sea breeze in France that sets in around midmorning, reaches its maximum in midafternoon, and ceases about 5 PM.

garbled *Telecommunications.* describing an unreadable or undecipherable message caused by defective transmission or reception, faulty encoding, or deliberate tampering.

garboard strake *Naval Architecture.* the centermost strake of a hull's underbody, running directly adjacent to the keel on either side.

gardening *Astronomy.* a term used to indicate a process in which countless small impacts create a regolith on a planetary surface and then continually keep it mixed, broken up, and turned over; much as the act of cultivating a garden.

garden-path sentence *Artificial Intelligence.* a sentence that is misleading to a reader because the initially "obvious" parsing is incorrect, e.g., "The old man the boats." (From the expression "to lead down the garden path," meaning "to mislead or deceive.")

garden snail

garden snail *Invertebrate Zoology.* any of various common land snails, gastropod mollusks of the order Pulmonata, that feed on flowers and vegetables.

Gardnerella *Bacteriology.* a genus of small, anaerobic, Gram-negative, rod-shaped bacteria found in the normal female genital and urinary tracts; a major cause of bacterial vaginitis.

Gardner's syndrome *Medicine.* a familial cancer disorder with a complex of symptoms including gastrointestinal tumors, fibromas, bony tumors, and epidermal cysts. (Named for Eldon J. *Gardner,* born 1909, American geneticist.)

gargoyle [gär´goil´] *Architecture.* a projecting rooftop waterspout, often carved in the form of a grotesque human or animal figure; a common feature of Gothic cathedrals, palaces, and other buildings.

garigue *Ecology.* a low, open scrubland found only in limestone sites along the Mediterranean and characterized by evergreen shrubs and other small trees.

garland *Mining Engineering.* **1.** a channel fixed around a shaft to catch water draining down the walls and transfer it to a lower level. Also, WATER CURB, WATER RING. **2.** a frame used to heighten and increase the sides of a truck or coal tub.

garlic *Botany.* any plant of the genus *Allium,* especially *Allium sativum,* a perennial herb of the lily family cultivated for its pungent, edible bulbs. *Food Technology.* the bulb of this plant, widely used in cooking and food processing, often in a prepared form as powder or flakes.

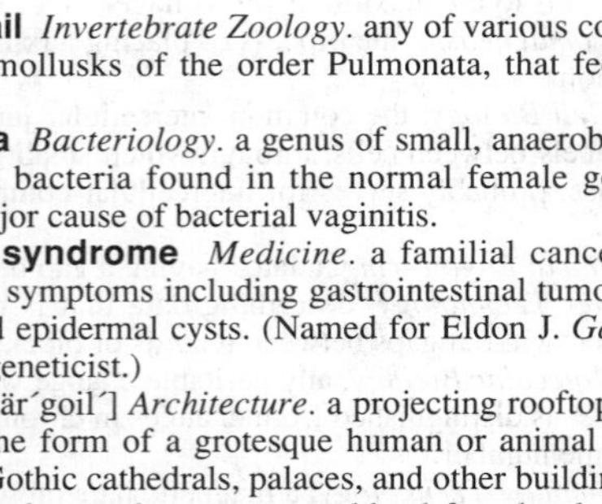

garlic

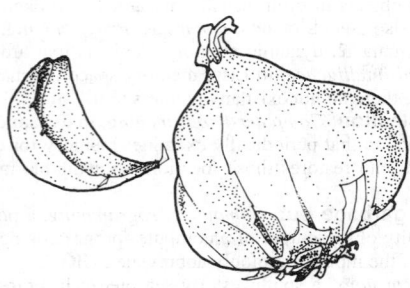

garner *Agriculture.* to gather grain for storage in a granary.

garnet *Mineralogy.* a group name for cubic silicates with the general formula $A_3B_2(SiO_4)_3$ (for hibschite and katoite, $A_3B_2(SiO_4)_{3-x}(OH)_{4x}$), where A=Ca,Fe^{+2},Mg,Mn^{+2} and B=Al,Cr^{+3},Fe^{+3},Mn^{+3},Si,Ti,V^{+3},Zr (Si is partly replaced by Al, Fe^{+3}); usually occurring as dodecahedral or trapezohedral crystals of varied color (often red), ranging from 2.8 to 4.3 (commonly 3.5 to 4.3) in specific gravity and from 6 to 7.5 in hardness on the Mohs scale; found in metamorphic rocks and as an accessory mineral in igneous rocks; used as a semiprecious stone and as an abrasive.

garnet hinge *Mechanical Devices.* a hinge consisting of an upright bar and a horizontal strap.

garnet jade see TRANSVAAL JADE.

garnet maser *Electronics.* a maser using natural or synthetic garnet as the stimulated material.

garnet paper *Materials.* an abrasive paper on which powdered garnet is the abrasive agent.

garnierite *Mineralogy.* a general term for hydrous nickel silicates.

garret *Building Engineering.* a portion of a dwelling just beneath the roof.

garreting *Building Engineering.* the process of inserting small stone splinters into joints of coarse masonry.

garrison *Military Science.* **1.** a body of troops stationed in a fortified place, usually for defensive purposes. Also, **garrison force. 2.** the place in which such troops are stationed. **3.** any military base or defensive area, especially a permanent one. **4.** to provide a town or fort with a garrison or to station troops in a garrison.

garronite *Mineralogy.* $Na_2Ca_5Al_{12}Si_{20}O_{64}\cdot 27H_2O$, an orthorhombic, pseudotetragronal zeolite forming radiating aggregates, having a specific gravity of 2.13 to 2.17 and an undetermined hardness; found as amygdule fillings in basalts.

Garryaceae *Botany.* a monogeneric family of dicotyledonous dioecious evergreen shrubs and trees in the order Cornales, commonly producing iridoid compounds and toxic alkaloids and native to western North America and the Antilles.

garter snake *Vertebrate Zoology.* **1.** a harmless, common, viviparous North American snake of the genus *Thamnophis* in the family Colubridae, characterized by a longitudinally striped pattern on the back. **2.** any of several venomous African snakes of the family Elapidae, having ringed bodies and related to the New World coral snakes.

garter snake

garter spring *Mechanical Devices.* a closed helical spring, used to exert a uniform radial force upon the object around which it is wound.

Gartner's duct *Anatomy.* a vestigial closed duct that parallels the oviduct. (Named for Hermann *Gartner,* 1785–1827, Danish anatomist.)

garúa [gä roo´ə] *Meteorology.* a dense fog or drizzle from low clouds on the west coast of South America that brings raw, cold weather and a limited amount of moisture during the winter. Also, CAMANCHACA.

Garvey-Kelson mass relations *Nuclear Physics.* a set of equations relating masses of nuclei having different numbers of neutrons and protons to predict masses of exotic nuclei that occur in stellar processes.

gas *Physics.* one of the three fundamental forms of matter, along with liquids and solids. Unlike a solid (and like a liquid), a gas has no fixed shape and will conform in shape to the space available. Unlike a liquid, it has no fixed volume and will conform in volume to the space available. In comparison with solids and liquids, gases have widely separated molecules, are light in weight, and are easily compressed. *Chemistry.* any such substance (e.g., hydrogen, oxygen) or a mixture of such substances (e.g., air, carbon dioxide). *Materials.* a shorter term for GASOLINE. *Mining Engineering.* an explosive mixture of firedamp and air. *Ordnance.* **1.** any chemical agent used as an antipersonnel weapon; its effects may range from mild irritation and incapacitation to severe injury and death; it may be delivered in liquid, solid, or vapor form. Also, WAR GAS. **2.** to use such an agent. **3.** of or relating to weapons that utilize or deliver gas. Thus, **gas attack, gas bomb, gas grenade, gas munition, gas shell, gas warfare.** *Mechanical Engineering.* **1.** powered by a combustible fuel, such as gasoline or natural gas. **2.** see GAS PEDAL. (Coined by the Flemish chemist Jan Baptista van Helmont, 1579–1644; derived from the Greek word *chaos,* meaning "space.")

GAS general adaptation syndrome.

gas-absorption operation *Chemical Engineering.* the recovery of solute gases found in gaseous mixtures of noncondensables, achieved by contacting the gas stream with a liquid solvent or a solid adsorbent.

gas adsorption *Physical Chemistry.* a process in which a high concentration of gas is held in a thin layer at the surface of a material, due to physical or chemical surface forces, such as the adsorption of ethane in porous charcoal granules.

gas amplification *Nucleonics.* the ratio of the energy absorbed to the energy released when an atom or molecule first captures or ejects an electron in a radiation-counter tube.

gas analysis *Analytical Chemistry.* the analysis of the properties and composition of a gas using techniques such as chromatography, mass spectroscopy, combustion, and chemical adsorption.

gas anchor *Petroleum Engineering.* an apparatus consisting of a 5-ft length of tubing that is used to minimize gas-in-oil froth downhole before pumping to improve the efficiency of a pump.

gas bag *Engineering.* **1.** a bag that is inflated and then inserted into a pipeline in order to inhibit the flow of gas through the line. **2.** a slang term for an airship.

gas black see CARBON BLACK.

gas boss see FIRE BOSS, def. 1.

gas-bounded nebula *Astronomy.* a hydrogen nebula kept in an ionized state by a hot, highly luminous star within it.

gas buoy *Navigation.* a lighted buoy whose light is operated by gas.

gas burner *Mechanical Engineering.* **1.** see GAS JET. **2.** any engine or appliance that burns gas as a fuel.

gas cap *Mechanical Engineering.* a metal cap designed to fit over the opening of a gas tank. *Mining Engineering.* the blue halo of ignited firedamp that shows above the yellow flame of a safety lamp; the percent of firedamp can be roughly measured by the height of the cap. *Geophysics.* the gas immediately preceding a meteoroid as it travels through the atmosphere. *Petroleum Engineering.* the section of an oil-producing reservoir filled with free gas; gas in a free condition over an oil zone.

gas capacitor *Electricity.* a capacitor composed of multiple electrodes separated by a gaseous dielectric.

gas carburizing *Metallurgy.* the process of increasing the carbon content of a superficial layer of steel by reacting the steel in a gas atmosphere.

gas cell *Electricity.* a cell whose action depends on the absorption of gases by the electrodes.

gas-cell frequency standard *Atomic Physics.* a standard using a gas cell that contains rubidium, cesium, or sodium to calibrate other visible-frequency instruments.

gas centrifuge process *Nucleonics.* a technique used to separate isotopes in a gas by spinning it in a centrifuge at high speeds, forcing the heavy isotopes toward the wall and the light isoptopes toward the center.

gas chromatography *Analytical Chemistry.* a technique for separating gas mixtures, in which the gas is passed through a long column containing a fixed absorbent phase that separates the gas mixture into its component parts.

gas-condensate liquid *Organic Chemistry.* **1.** the purified product of a distillation procedure in which the vapor of a boiling liquid is collected and recondensed to the liquid phase. **2.** the propane, butane, and pentane condensates of compressed or refrigerated natural gas.

gas-condensate reservoir *Geology.* a hydrocarbon reservoir in which the heavier constituents have condensed as a liquid phase separate from the reservoir gas.

gas-condensate well *Petroleum Engineering.* a well that produces liquefiable hydrocarbons from a gas-condensate reservoir.

gas constant *Thermodynamics.* a proportionality constant, usually symbolized by R, that appears in the equation of state for 1 mole of an ideal gas; its value is given by $R = pV/T$, where p is the pressure, V is the molar volume, and T is the absolute temperature; equal to 8.314 joules per mole-K.

gas-cooled reactor *Nucleonics.* a reactor in which a gaseous substance, such as air, helium, or carbon dioxide, is used to remove or transfer heat.

gas counter *Nucleonics.* an instrument that measures radiation by mixing it with a gas before introducing it into a counter tube, where an electric current passes through the gas converting the charged particles into electrical pulses.

gas current *Electronics.* the grid current due to the presence of gas in an electron tube.

gas cutting *Metallurgy.* the process of cutting metal with a flame produced by the combustion of a gas, such as acetylene, with oxygen.

gas cycling *Thermodynamics.* any series of thermodynamic processes involving a gaseous fluid that occur in a sequence which eventually returns the gas to its original state. *Petroleum Engineering.* specifically, a petroleum-formation recovery process in which the gas that is produced with oil is then returned to the oil sand to facilitate the production of more oil.

gas cylinder *Mechanical Engineering.* see CYLINDER.

gas dehydrator *Chemical Engineering.* a plant unit or system that removes moisture vapor from a gas stream.

gas-deviation factor see COMPRESSIBILITY FACTOR.

gas discharge *Electronics.* the conduction of electricity in a gas, resulting from the movements of ions produced by collisions between electrons and gas molecules.

gas-discharge display *Electronics.* a display in which the segments of numerical or alphanumeric characters are formed by seven or more cathode elements that are vacuum-sealed in a nonmercury gas and energized by direct current.

gas-discharge lamp see DISCHARGE LAMP.

gas doping *Electronics.* the introduction of impurity atoms into a semiconductor material by epitaxial growth, using streams of gas that are mixed prior to insertion into a reactor vessel.

gas dynamic laser *Optics.* a laser that produces continuous high levels of power by burning suitable fuel to produce carbon dioxide and nitrogen at high temperature and pressure; the thermal energy is then converted into coherent radiation.

gas dynamics *Physics.* the study of the motion of gases and of the nature and effect of such motion.

gas embolus *Medicine.* an air bubble in the cirulation, often due to rapid decompression, as when a diver ascends too rapidly.

gas engine see GASOLINE ENGINE.

gaseous *Physics.* existing in the state of a gas, as opposed to a liquid or solid state. *Chemistry.* having the characteristics of a gas.

gaseous diffusion *Chemical Engineering.* **1.** the selective transfer of gas by molecular diffusion through microporous barriers; used especially in preparing fuel for nuclear reactors. **2.** the selective solubility diffusion of gas by absorption and solution of the gas in a nonporous polymer matrix.

gaseous diffusion plant *Nucleonics.* a facility that develops fissionable fuel for nuclear power plants and armament producers by separating uranium isotopes by means of the gaseous diffusion process.

gaseous diffusion process *Nucleonics.* a process in which a gas is passed through a porous wall to separate its light isotopes from its heavy isotopes.

gaseous nebula *Astronomy.* a nebula made up primarily of gas, which is usually luminous, as opposed to dust, or dark nebula.

gaseous transfer see VOLATILE TRANSFER.

gaseous voltage regulator see GLOW-DISCHARGE VOLTAGE REGULATOR.

gas etching *Engineering.* a process by which a substance is extracted from a semiconductor unit when it combines with a gas and forms a volatile compound.

gas explosion *Mining Engineering.* an explosion of firedamp in a coal mine in which coal dust apparently did not play a significant part.

gas field *Petroleum Engineering.* a formation area that contains closely contiguous reservoirs of commercially valuable gas.

gas-filled cable *Electricity.* a power cable containing gas under pressure to serve as insulation and minimize ionization.

gas-filled photocell *Electronics.* a photocell in which the anode and photocathode are enclosed in gas at low pressure; collision of the photoelectrons with gas molecules produces positive ions, thus increasing the sensitivity of the cell.

gas-filled porosity *Geology.* a reservoir formation whose pore space is filled by gas rather than by liquid hydrocarbons.

gas-filled radiation counter *Nucleonics.* a tube that measures radiation by measuring electric current generated by the charged particles, such as electrons and photons, that accompany the gas entering the tube.

gas-filled rectifier see COLD-CATHODE RECTIFIER.

gas-filled relay *Electronics.* a thermionic tube, generally filled with mercury vapor, when it is used as a relay.

gas filter *Chemical Engineering.* an apparatus used for removal of solid or liquid particles from a flowing gas stream.

gas-fired *Engineering.* using gas as a source for heat or fuel.

gas-flame brazing *Metallurgy.* the process of brazing with a flame produced by the combustion of a gas, such as acetylene, with oxygen.

gas-flow counter tube *Nucleonics.* a counter tube that is used to measure the radioactivity of a gas passing through the tube.

gas focusing *Electronics.* a technique in which the ionization of an inert gas is used to condense an electron beam in a cathode-ray tube. Also, IONIC FOCUSING.

gas furnace *Mechanical Engineering.* **1.** a furnace that uses a gaseous fuel. **2.** a furnace that is designed to distill gases from a solid fuel, such as coal.

gas gangrene *Medicine.* a form of gangrene usually occurring in massive wounds in which there is crushing of tissue and contamination with dirt, leading to the accumulation of gas in the injured tissue due to infection from anaerobic bacteria. Also, **gaseous gangrene.**

gas generator *Mechanical Engineering.* a compressor-combustion apparatus that supplies high-energy gas flow to a gas turbine. *Chemical Engineering.* a production facility for the production of gas from coal.

gas-giant planet *Astronomy.* a planet such as Jupiter, Saturn, Uranus, or Neptune, characterized by a deep massive atmosphere surrounding a relatively small rocky core.

gas gland *Vertebrate Zoology.* a regulatory gland found in the swim bladder of certain teleost fishes, which increases or decreases the amount of gas in the bladder to maintain the body at a desired water depth.

gas governor *Mechanical Devices.* a device that is used within an appliance to maintain a constant pressure supply.

gas gun *Ordnance.* **1.** an automatic weapon in which the gas pressure released upon firing is used to actuate a piston and cylinder mechanism that extracts the used cartridge and inserts a new cartridge while the gun remains in firing position. Also, **gas-operated gun. 2.** a gun using pressurized gas, especially liquefied carbon dioxide, as a propellant.

gas heater *Mechanical Engineering.* a heating device that uses gas as a heat source and supplies heat by forced convection.

gash fracture *Geology.* a series of small open gashes or breaks that develop diagonally to a fault or fault zone.

gas hole *Engineering.* a hollow that forms during cavitation in a material that has been cast.

gash vein *Geology.* a broad mineral vein that extends vertically and narrows as it terminates within the formation it traverses.

gasification *Chemical Engineering.* the production of gaseous or liquid hydrocarbon fuels from coal.

gasifier *Chemical Engineering.* a unit for the production of gas, especially synthesis gas from coal.

gasiform see GASEOUS.

gas injection *Mechanical Engineering.* the injection of gaseous fuel into the cylinder of a gasoline engine. *Petroleum Engineering.* the injection of natural gas under high pressure into a producing reservoir through an injection well to maintain formation pressure, for the secondary recovery of oil, or for a recycling operation. Thus, **gas-injection well.**

gas ionization *Electricity.* the removal of the planetary electrons from the gas filling an electron tube, allowing the resulting ions to flow in the current through the electron tube.

gas jet *Mechanical Engineering.* a nozzle or tip from which gas issues, as in an engine or on a stove.

gasket *Engineering.* a pressuretight seal made of deformable material, typically rubber, plastic, or paper, fitting between two stationary parts; used in many applications to prevent the leaking of fluids; e.g., that of water in a plumbing system, oil in an automobile engine, and so on.

Gaskin *Ordnance.* a Soviet surface-to-air missile designed for low-altitude tactical air defense; it is solid rocket-powered, equipped with infrared homing, and is believed to deliver a larger warhead than the SA-7 at a range up to 3 miles; officially designated **SA-9.**

gas kinematics *Fluid Mechanics.* the study of the motion of a gas without regard to the causes of the motion, such as thermal effects, etc.

gas laser *Optics.* the type of laser that is used most often for practical applications, in which an electric current passes through a sealed glass or quartz tube containing a noble gas mixture, such as helium and neon, to generate a beam of monochromatic light. Also, **gas-discharge laser.**

gas law see IDEAL GAS LAW.

gas lift *Chemical Engineering.* a process for moving solids in which an upward-flowing gas stream lifts powdered or granular solid material in a closed system from one unit or vessel to another, often located at a higher elevation.

gas-liquid chromatography *Analytical Chemistry.* a separation technique in which a sample is applied to the stationary phase of a column packed with an inert solid coated with a nonvolatile liquid and the mobile phase is a gas, which is used to elute the components for detection. Also, **gas-liquid partition chromatography.**

gas magnification *Electronics.* an increase in current through a phototube, caused by the ionization of the gas within the tube.

gas making *Chemical Engineering.* the production of air gas or water gas by the action of steam and air upon hot coke.

gas manometer *Engineering.* an instrument that compares the pressure of two gases, typically a U-shaped tube in which a liquid height is measured in each leg of the tube to determine the gas-pressure difference.

gas maser *Physics.* a maser in which the interaction takes place between the microwave signal and the molecules of a gas such as ammonia.

gas mask *Engineering.* a masklike device worn by an individual to protect against exposure to poisonous or noxious gases, in which the air inhaled by the wearer is filtered through charcoal or other chemicals; used in certain industries and in warfare.

gas mechanics *Fluid Mechanics.* the study of the forces and pressures associated with a gas.

gas meter *Engineering.* an apparatus designed to measure and record the amount of gas that flows through a pipe.

gasohol [gas´ə hôl] *Chemistry.* a mixture containing approximately 90% unleaded gasoline and 10% ethyl alcohol; introduced as an energy-conservation measure and used chiefly as an alternative fuel in some automobile and truck engines. (A blend of gasoline and alcohol.)

gas-oil contact *Petroleum Engineering.* the surface in an oil reservoir at which the bottom of a gas sand contacts the top of an oil sand.

gas-oil ratio *Petroleum Engineering.* the number of cubic feet of natural gas produced with a barrel of oil; an approximation of the composition of oil and gas from a reservoir, expressed in cubic feet of gas per barrel of oil at 14.7 psia and 60°F or 15.6°C.

gasoline *Petroleum Engineering.* a volatile, liquid mixture of hydrocarbons that is obtained by refining petroleum and is used as a fuel in most internal-combustion engines.

gasoline engine *Mechanical Engineering.* an internal-combustion engine that is driven by a mixture of gasoline and air.

gasometer *Engineering.* an apparatus designed to contain and measure gas, particularly for chemical studies.

gasometric method *Analytical Chemistry.* any technique for measuring gases that involves the use of instruments or chemical reactions.

gasometry *Analytical Chemistry.* the measurement of gases.

Gaspak *Microbiology.* a gastight jar containing a mixture of chemical compounds that, after the addition of water, generates an atmosphere of hydrogen and carbon dioxide for the anaerobic culture of microorganisms on petri plates.

gas pedal *Mechanical Engineering.* a foot-operated control device that activates the accelerator of a motor vehicle. Also, GAS.

gaspéite *Mineralogy.* $(Ni,Mg,Fe^{+2})CO_3$, a light-green trigonal mineral occurring as rhombohedral crystals, belonging to the calcite group and having a specific gravity of 3.71 and a hardness of 4.5 to 5 on the Mohs scale; found in oxidized sections of nickel-sulfide deposits.

gas-permeable lens *Optics.* a contact lens that is made of a unique plastic which has the ability to permit oxygen to diffuse into and carbon dioxide to diffuse out of the cornea, allowing for increased comfort and wearing time, an increased visual field, and a wide range of vision corrections.

gas phase sequentor *Biotechnology.* a machine used to determine the amino acid sequence of proteins; the sample is treated with gas phase reagents for attachment to a carrier.

gas phototube *Electronics.* a phototube into which gas has been introduced, usually to increase its sensitivity.

gas pit *Geology.* a small circular pit surrounded by a mud mound formed by the escape of gas bubbles from the surface of a mud bar.

gas porosity *Metallurgy.* the porosity caused by gas escaping from molten metals during the solidification of a casting or a weld; usually undesirable, this phenomenon is sometimes exploited for the production of certain foam metals.

gas port *Ordnance.* **1.** a small hole that connects the bore with the cylinder of a gas-operated rifle, allowing a portion of the propellant gases to operate the breech mechanism. **2.** any opening for the passage of gas, such as the holes in a rifle receiver that release excess gas pressure. Also, **gas vent.**

gas-pressure maintenance *Petroleum Engineering.* a method of maintaining gas pressure in an oil reservoir by using gas injection, which increases hydrocarbon recovery and improves reservoir production characteristics.

gas producer *Chemical Engineering.* a device used for achieving the complete gasification of coal, using air and water-gas reactions.

gas pump *Mechanical Engineering.* a device that draws a measured amount of gasoline from a storage tank and pumps it into the carburetor of a motor vehicle.

gas reservoir *Geology.* an accumulation of natural gas found together with accumulations of crude oil in the earth's crust.

gas sand *Geology.* an oil sand or a sandstone rock that contains a large amount of natural gas.

gas scattering *Electronics.* the process of dispersing electrons or other particles in a beam by residual gas in the vacuum system.

gas scrubbing *Chemical Engineering.* a procedure that uses the action of a liquid absorbent to remove impurities from a gas.

gas seal *Engineering.* a tight closure that inhibits the flow of gas into or out of a machine.

Gassendi, Pierre 1592–1655, French philosopher, scientist, and mathematician; challenged Aristotelian dogma; revived atomic theory.

gas sendout *Petroleum Engineering.* a term for the total amount of gas that is removed from underground storage during a certain time period.

Gasser, Herbert 1888–1963, American physiologist; received Nobel Prize for research on synapses.

Gasserian ganglion *Anatomy.* a mass of nerve cell bodies at the base of the trigeminal cranial nerve. Also, SEMILUNAR GANGLION.

gassing *Engineering.* the evolution of gases during an event, especially electrolysis. *Electricity.* the production of gas in a storage battery during charging.

gas snifter see EXPLOSIMETER.

gas-solid chromatography *Analytical Chemistry.* a separation technique in which a sample is applied to the stationary phase of a column packed with a surface-active solid such as charcoal or silica gel, and the mobile phase is a gas that is used to elute the components for detection.

gas solubility *Physical Chemistry.* the amount of gas that must dissolve in a liquid before the overall system has a single phase.

gas spurt *Geology.* one of the small heaps containing organic matter that occur on the surface of certain strata; believed to be caused by the escape of gas during the early formative stages of the strata.

gas sterilization *Microbiology.* a procedure for eliminating contaminating microorganisms from a material by using sufficiently high concentrations of a lethal gas.

gas stimulation *Petroleum Engineering.* the use of nuclear explosives in the strata of a natural gas field to create greater gas flow.

gassy *Mining Engineering.* describing a situation in which a mine gives off methane or other gas in quantities that must be diluted with pure air to prevent an occurrence of explosive mixtures.

gassy tube *Electronics.* **1.** a vacuum tube that has been incompletely evacuated and thus contains a quantity of gas which degrades its performance. **2.** a vacuum tube in which gas has appeared after evacuation.

gas tail *Astronomy.* the tail of a comet formed by gas ejected from the nucleus; it is straight and bluish in color, shines by fluorescence, and points directly away from the sun.

gas tank *Petroleum Engineering.* any storage tank that is used for holding gasoline or other such combustible fuel. *Mechanical Engineering.* specifically, the fuel container in an automobile or other motor vehicle, linked by a fuel line to the engine.

Gasterellaceae *Mycology.* a family of fungi belonging to the order Hymenogastrales that grow with blue-green algae and diatoms on the surface of damp soil.

gasteroid *Mycology.* relating to fungal fruiting bodies that lack a surface layer, or that have a surface layer that is divided into chambers and is totally or partially enclosed.

Gasteromycetes *Mycology.* a class of fungi belonging to the subdivision Basidiomycotina whose spore-bearing structure or basidium is covered with an outer layer; includes stinkhorns, earthstars, and puffballs.

Gasterophilidae *Invertebrate Zoology.* a family of horse botflies, in the order Diptera, that resemble honeybees; larvae parasitize the alimentary tract of horses.

Gasterophilus *Invertebrate Zoology.* a genus of horse botflies in the family Gasterophilidae.

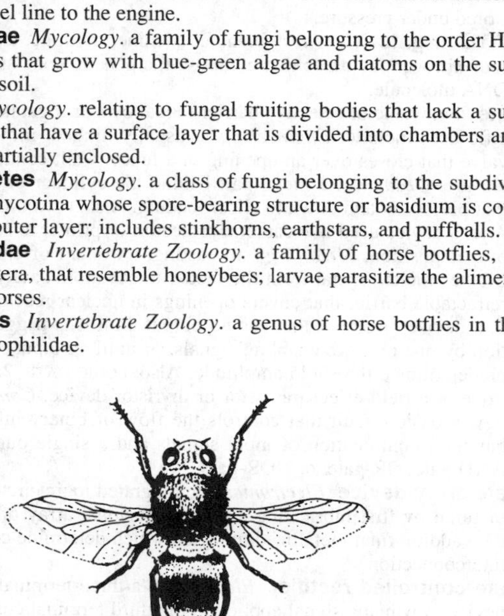

Gasterophilus

Gasterosteidae *Vertebrate Zoology.* the sticklebacks, a family of the suborder Gasterosteoidei, characterized by small, long, scaleless bodies with sharp spines along the back, and with some species covered by bony plates; found in both fresh- and saltwater habitats of the Northern Hemisphere.

Gasterosteiformes *Vertebrate Zoology.* an order of small, spiny-rayed fishes found in fresh and saltwater habitats of the Northern Hemisphere, and including the sticklebacks, trumpetfishes, seahorses, and pipefishes.

Gasterosteioidei *Vertebrate Zoology.* a suborder of marine and freshwater fishes of the order Gasterosteiformes, characterized by a protractile upper jaw with well-developed premaxillaries; includes the tube-snouts, sand eels, and sticklebacks.

Gasteruptiidae *Invertebrate Zoology.* a family of wasplike insects in the order Hymenoptera, with an ovipositor almost as long as the body.

gas thermometer *Engineering.* an instrument that measures temperature by observing the change in pressure of a gas at constant volume.

gas thermostatic switch *Electricity.* a switch in which heat increases the gas pressure in a sealed metal bellows, thus moving the bellows and closing the contacts of the switch.

gastr- a combining form meaning "stomach," as in *gastritis.*

gas tracer *Mining Engineering.* a dust cloud, chemical smoke, or gaseous or radioactive tracer that is used to detect slowly moving air currents in a mine.

gastraea *Invertebrate Zoology.* a hypothetical ancestral organism, like a gastrula larva in structure, from which primitive flatworms may have evolved. Also, **gastrea.**

gastralium *plural,* **gastralia.** *Vertebrate Zoology.* a slender, riblike, V-shaped bone found in the ventral surface of certain reptiles, which is part of the dermal structure and thus is not a true rib. *Invertebrate Zoology.* in sponges, a type of spicule located within the inner cell layer.

gas-transport laser *Optics.* a laser whose gas mixture is recycled rather than dispersed into the atmosphere, and which therefore does not require a constant supply of gas.

gas trap *Civil Engineering.* a bend or chamber in a sewer drain system that prevents gases from escaping.

gastrectomy *Surgery.* the excision of all or part of the stomach.

gastric *Anatomy.* relating to, affecting, or originating in the stomach. (Going back to the Greek word for stomach; literally meaning "gnaw.")

gastric acid *Physiology.* the strongly acid digestive secretions of the gastric glands, composed primarily of pepsin, hydrochloric acid, rennin, lipase, and mucin. Also, GASTRIC JUICE(S).

gastric antrum *Anatomy.* a dilated portion of the lower end of the stomach, between the body of the stomach and the pyloric canal. Also, ANTRUM OF WILLIS, PYLORIC ANTRUM.

gastric bypass *Surgery.* the formation of an anastomosis of all or a section of the stomach to the small intestine; performed in cases of obstruction or to treat morbid obesity.

gastric cecum *Invertebrate Zoology.* any of the elongated pouches projecting from the digestive tract of an insect.

gastric enzyme *Enzymology.* any digestive enzyme secreted by cells lining the stomach.

gastric filament *Invertebrate Zoology.* in jellyfish, a tentacle with nematocysts (stinging cells) in the gastric cavity, which kills living prey entering the stomach.

gastric gland *Anatomy.* any of numerous tubular glands in the stomach wall that secrete hydrochloric acid and pepsinogen. Also, FUNDIC GLAND.

gastric hypothermia *Medicine.* a cooling of the stomach, used sometimes for the management of bleeding ulcers.

gastric inhibitory polypeptide *Endocrinology.* a polypeptide hormone that is secreted by the small intestine and may inhibit gastric secretion and motility.

gastric juice or **gastric juices** see GASTRIC ACID.

gastric lavage *Toxicology.* the process of washing out the stomach with saline solution using a lavage tube, to remove poisons taken orally.

gastric mill *Invertebrate Zoology.* in malacostracan crabs, lobsters, and shrimp, an apparatus in the pharynx or stomach with teeth, setae, and ossicles for grinding food.

gastric ostium *Invertebrate Zoology.* in sponges, a small opening leading into a stomach pouch.

gastric pouch *Invertebrate Zoology.* in jellyfish, one of the pouchlike cavities of the stomach.

gastric secretion *Endocrinology.* the process by which substances such as hydrochloric acid or pepsin are secreted by cells in the stomach.

gastric ulcer *Medicine.* an ulcer affecting the wall of the stomach.

gastrin *Endocrinology.* any of a family of peptide hormones, differing in size and biological potency, that are secreted from the antrum of the stomach during digestion and increase the secretion of hydrochloric acid and pepsin.

gastrinoma *Oncology.* a non-beta islet cell tumor that produces gastrin and usually occurs in the pancreas; may also occur at other sites, such as the antrum of the stomach, the hilus of the spleen, and regional lymph nodes.

gastrin-releasing peptide *Endocrinology.* a gastric peptide that is highly homologous to bombesin and stimulates the secretion of gastrin.

Gastrioceras *Paleontology.* a genus of planulate ammonoids in the suborder Goniatitina, characterized by a depressed whorl section; short-ranged in time and important as an index fossil; upper Carboniferous.

gastritis [gas trīt´is] *Medicine.* an inflammation of the stomach, usually caused by gastric lesions, by the action of poison or other corrosive agents, or by radiation injury.

gastro- a combining form meaning "stomach," as in *gastroenteritis.*

gastroanastomosis *Medicine.* the formation of a surgical connection between parts of a malformed stomach, such as an hourglass stomach.

gastroblast *Invertebrate Zoology.* a feeding zooid (individual) of a tunicate colony.

gastrocnemius *Anatomy.* the large posterior superficial muscle of the lower leg that forms most of the calf.

gastrocoele see ARCHENTERON.

gastrodermis *Invertebrate Zoology.* the lining membrane of the digestive tract of coelenterates, ctenophores, and platyhelminthes; individual cells ingest and digest food particles.

gastroduodenal *Anatomy.* relating to or communicating with the stomach and duodenum.

gastroduodenitis *Medicine.* an inflammation of the stomach and duodenum.

gastroenteritis *Medicine.* an inflammation of the mucosa of the stomach and intestine; caused by viruses or bacteria, or by an intolerance to specific foods.

gastroenterologist *Medicine.* a physician who specializes in disorders of the stomach and intestines.

gastroenterology *Medicine.* the study of the stomach and the intestines and their diseases. Thus, **gastroenterological.**

gastroenterostomy *Surgery.* the establishment of a surgical communication between the stomach and small intestine.

gastroepiploic artery *Anatomy.* the large abdominal artery that delivers blood to the stomach and greater omentum.

Gastrografin *Radiology.* a trademark for a preparation of meglumine diatrizoate, a radiopaque substance used in some X-ray examinations.

gastrohepatic *Anatomy.* of, relating to, or involving the stomach and the liver.

gastrointestinal *Anatomy.* relating to or communicating with the stomach and the intestines. *Medicine.* relating to disorders of the stomach and intestines.

gastrointestinal hormone *Endocrinology.* any of a variety of hormones, such as gastrin and secretin, that are produced by the tissues of the digestive tract and regulate the physical and biochemical processes of digestion.

gastrointestinal tract *Anatomy.* the tubular organs forming a digestive pathway from the mouth to the anus, including the stomach and the intestines.

gastrojejunostomy *Surgery.* the formation of an anastomosis of the jejunum to the wall of the stomach.

gastrolith *Vertebrate Zoology.* a pebble swallowed and kept in the gizzard or stomach of certain animals, where it aids in digestion by helping to grind food.

gastrology *Medicine.* the study and sum of knowledge of the stomach, including its structure, functions, and diseases.

gastrolysis *Medicine.* the breaking up of adhesions between the stomach and adjacent organs.

Gastromyzontidae *Vertebrate Zoology.* a small family of loaches, fish of the suborder Cyrinoidei having broad, raylike bodies; found in the streams of southeastern Asia.

gastropexy *Medicine.* the replacement of a prolapsed stomach to its normal position by suturing it to other structures.

gastroplasty *Surgery.* any plastic operation on the stomach.

gastroplication *Surgery.* a surgical treatment for the relief of chronic gastric dilation by stitching a fold in the stomach.

gastropod *Invertebrate Zoology.* of or belonging to the mollusk class Gastropoda.

Gastropoda *Invertebrate Zoology.* a class of snails, slugs, limpets, and conches in the phylum Mollusca, having a well-developed head, a flattened foot, and viscera that usually undergo some degree of torsion.

gastropore *Invertebrate Zoology.* in corals, a pore or opening in the communal skeleton occupied by a gastrozooid (feeding individual).

gastroptosis *Medicine.* a downward displacement of the stomach; a term based on the outmoded concept that variation in position of abdominal organs is pathological.

gastropulmonary *Anatomy.* of or relating to the stomach and lungs.

gastroscope *Medicine.* a fiberoptic endoscope used for examining the interior of a stomach.

gastrosplenic ligament *Anatomy.* the fold of the greater omentum that attaches the greater curvature of the stomach to the spleen.

Gastrosporiaceae *Mycology.* a family of fungi belonging to the order Protogastrales that is common to grasses in Europe.

gastrostome *Invertebrate Zoology.* in corals, the orifice of a gastropore.

gastrostomy *Surgery.* **1.** the establishment of an artificial opening into the stomach. **2.** the opening so created.

gastrostyle *Invertebrate Zoology.* in corals, a spiculated projection arising from the gastropore and extending into the gastrozooid.

Gastrotricha *Invertebrate Zoology.* a small class of microscopic marine and freshwater animals related to rotifers, belonging to the phylum Aschelminthes or sometimes classified as a separate phylum.

gastrozooid *Invertebrate Zoology.* in corals, a feeding individual having a mouth, tentacles, and digestive organs.

gastrula *Developmental Biology.* the embryo in the stage of development following the blastula; this stage is formed in the process of gastrulation and contains the embryonic germ layers.

gastrulation *Developmental Biology.* the tranformation of the blastula into the gastrula, or in forms without a true blastula, the formation of the embryonic germ layers.

gas tube *Electronics.* a partially evacuated electron tube, containing a small amount of gas at low pressure, that ionizes and produces current flow during tube operation.

gas turbine *Mechanical Engineering.* a turbine that is rotated by the expansion of hot gases of combustion.

gas valve *Engineering.* a device that regulates the discharge of gas, e.g., from the extreme top of a balloon or airship.

gas viscosity *Fluid Mechanics.* the resistance of a real gas to deformation when subjected to a shear stress.

gas welding *Metallurgy.* any welding with a flame produced by the combustion of a gas, such as acetylene, with oxygen.

gas zone *Geology.* a rock formation in which natural gas is produced or stored under pressure.

GATC or **gatc** *Genetics.* guanine, adenine, thymine, and cytosine, the four nitrogenous bases that compose the genetic code stored within the DNA molecule.

gate a movable or selective barrier that allows or regulates passage; specific uses include: *Engineering.* a usually hinged, movable barrier or valve that closes over an opening in a fence, wall, conduit, or other enclosure. *Ordnance.* a metal part behind the cylinder of some revolvers that pivots out and down to allow loading. Also, LOADING GATE. *Graphic Arts.* a device used in a camera, projector, or printer to secure film in its proper position. *Metallurgy.* in casting, a system of channels that distributes molten metal from the feeder to the mold cavity. *Nucleonics.* a retractable barrier that covers openings in nuclear reactors. *Electronics.* **1.** a device or circuit that has no output until it is triggered into operation by one or more enabling signals, or until an input signal exceeds a predetermined threshold amplitude. Also, LOGIC GATE. **2.** the input electrode of a field-effect transistor or thyristor device. *Computer Technology.* a basic circuit that controls the flow of binary information, each having a combination of input signals and a single output, such as an AND-gate, OR-gate, or NOR-gate.

gate array device *Electronics.* an integrated logic circuit that is manufactured by first fabricating a two-dimensional array of logic cells and then adding final metallization layers which determine cell function and interconnection.

gate-controlled rectifier *Electronics.* a three-terminal semiconductor device in which a signal applied to the third terminal controls the unidirectional current flow between the rectifier terminals.

gated-beam tube *Electronics.* a five-element tube in which the electrons flow in a beam between the cathode and plate; the plate current will be cut off by a small increase in voltage on the limiter grid, and further increases will have a negligible effect on it.

gated sweep *Electronics.* **1.** a radar sweep whose initiation and duration are closely controlled to eliminate echoes in the image. **2.** a circuit providing such a sweep.
gate equivalent circuit *Electronics.* a unit of measurement for the relative complexity of digital circuits; equal to the number of individual interconnected logic gates necessary to perform the same function as the digital circuit under evaluation.
gate interlock *Mining Engineering.* a system designed to prevent shaft conveyances from being moved or action signals transmitted unless all shaft gates are closed.
gate pulse *Electronics.* a pulse applied to the gate electrode to actuate a gate-controlled semiconductor device.
gate theory *Neurology.* a theory suggesting that pain is determined by the interaction of elements in the spinal cord; specifically, that the entry point or "gate" for impulses transmitting pain may be blocked by the stimulation of myelinated fiber, thus inhibiting pain by preventing impulses from reaching higher levels of the central nervous system. Also, **gate hypothesis, gate-control theory.**
gate valve *Mechanical Devices.* a pipeline valve whose flat or wedge-shaped gate can be raised or lowered to control the straight-through flow of a fluid.
gate voltage *Electronics.* the voltage applied to the the gate of a field-effect transistor or thyristor.
gateway *Building Engineering.* an entrance or passageway that may be closed by a gate. *Transportation Engineering.* a station serving as the main entry point to a region, particularly an airport at which an international flight makes a final stop before or after crossing a border. *Telecommunications.* **1.** a switching system that interconnects national and international telecommunication networks. **2.** a data communication link that interconnects local area networks. **3.** a device that interfaces networks in order to allow communication between two terminals, or between a terminal and a computer.
gate winding *Electronics.* in a magnetic amplifier, a winding that produces gating action.
gather to bring together in one group or collection; specific uses include: *Graphic Arts.* in bookbinding, to assemble folded signatures in the proper order. *Mining Engineering.* **1.** to assemble and deliver loaded cars from several production points to main haulage for transport to the surface or pit bottom. **2.** to drive a heading through disturbed or faulty ground to meet the seam of coal at a convenient level. Also, EAT OUT.
gatherer *Anthropology.* a member of a society that practices gathering as a means of obtaining food.
gathering *Anthropology.* the practice of obtaining food by collecting wild plants and other naturally occurring foodstuffs, as opposed to the use of agriculture. *Building Engineering.* a tapered section in a flue duct or air duct that forms a transition between the chimney or duct passage and the flue, located almost directly above a fireplace.
gathering area *Petroleum Engineering.* the location, often down the regional dip from a hydrocarbon trap, where the gas or oil may have moved updip into the trap.
gathering line *Petroleum Engineering.* a special pipeline with a 2- to 4-in. diameter that connects the tank batteries on a lease with a booster station in the field; used to transport gas or crude oil from the field to the main pipeline.
gathering motor *Mining Engineering.* a light, electric locomotive used to haul loaded coal trucks from the work places to the main haulage system and to replace them with empties. Also, ELECTRIC GATHERING LOCOMOTIVE.
gathering ring *Engineering.* a circular clay band that is used to gather impurities from molten glass; subsequently the higher-quality glass is taken from the center of the band.
gather way *Navigation.* to begin to move; said of a ship from the time it starts to move until it has reached cruising speed.
gather write *Computer Programming.* to write a single block of data records obtained from noncontiguous areas of memory.
gating *Metallurgy.* an arrangement of connecting pipes, gates, and cavities providing a passage for molten material to enter a mold. *Electronics.* **1.** the process of using one signal to switch all or part of another signal on or off for a desired interval. **2.** the process of selecting a part of a wave that exists during a particular time interval or at a particular magnitude for purposes of observation or control. *Biochemistry.* the process of shutting off a function when the value of specific parameter attains a critical level; for example, proton ATPase activity is inhibited when the proton motive force drops below a given level.
gatism *Medicine.* rectal, vesical, or rectovesical incontinence.

Gatling gun

Gatling gun *Ordnance.* an early mechanical machine gun using from six to ten barrels fired in rotation; originally hand-cranked, but later models were motor-driven and fired up to 1200 rounds per minute. (From its inventor, American engineer Richard *Gatling,* 1818–1903.)
gatophobia see AILUROPHOBIA.
Gattermann aldehyde synthesis *Organic Chemistry.* a method that is used to convert aromatic hydrocarbons and phenols into their corresponding aldehydes by reaction with hydrogen chloride and hydrogen cyanide in the presence of metal chloride catalysts.
Gattermann-Koch reaction *Organic Chemistry.* a method that is used to convert aromatic hydrocarbons to aldehydes by reaction with carbon monoxide and hydrogen chloride in the presence of metal chloride catalysts.
Gattermann reaction *Organic Chemistry.* a modification of the Sandmeyer reaction that is used to form diaryl compounds from diazonium salts by heating the aqueous salt solution with copper bronze.
GAU-8A Avenger *Ordnance.* a 30-mm, 7-barrel revolving cannon designed to fire armor-piercing shells made of depleted uranium, and to be mounted on the front of an antitank aircraft such as the A-10A.
Gaucher's cells *Pathology.* large and atypical cells with a wrinkled appearance and one or more erratically placed nuclei containing kerasin; usually found in the spleen, lymph nodes, liver, and bone marrow of persons with Gaucher's disease. (Named for Philippe Charles Ernest *Gaucher,* 1854–1918, French physician.)
Gaucher's disease *Pathology.* a lipidosis caused by a mutation in the gene coding for glucocerebrosidase; the most common Jewish genetic disease and the most frequently seen lysosomal storage disease.
Gaudí, Antoni 1852–1926, Spanish architect whose distinctive style combined Gothic elements with Art Nouveau.
gaufrage see PLAITING.
gauge *Mechanical Devices.* any of a wide variety of instruments that are used for measuring a parameter or characteristic of an object, such as its width, weight, volume, conductivity, or pressure. *Engineering.* **1.** the thickness of a metal wire, rod, or sheet. **2.** the measurement of the sieve size that will allow the bulk of a given aggregate to pass through. *Civil Engineering.* the measured distance between the inner faces of the rails of a railroad track. *Ordnance.* the interior diameter of a shotgun's bore; expressed as the number of pure lead balls of a diameter equal to that of the bore that are required to make one pound in weight. *Building Engineering.* **1.** the part of a slate or tile that is exposed when laid in place. **2.** an amount of plaster of Paris that is mixed with mortar or common plaster to slow its setting time. *Electromagnetism.* a possible choice for an electric scalar potential or a magnetic vector potential that satisfies Maxwell's equations. Also, GAGE.
gauge block *Mechanical Devices.* a steel or chrome measuring block, accurate to a few millionths of an inch, used for precision checks of travel distances and linear measurements in manufacturing.
gauge boson *Physics.* a class of elementary particles that includes the gluon, photon, W^+, W^-, and Z^o particles, each having an integral spin.
gauge box see BATCH BOX.

gauge group *Physics.* the group of gauge transformations in a gauge theory.

gauge height *Hydrology.* see STAGE.

gauge length *Engineering.* in specimen testing, the original length of the specimen, prior to pressure applications and other testing operations.

gauge loss *Mining Engineering.* the diametrical reduction, caused by wear, in the size of a bit or reaming shell.

gauge plate *Civil Engineering.* a plate inserted between the rails of a railroad track to maintain the gauge.

gauge pressure *Mechanical Engineering.* the pressure of a fluid as shown by a pressure gauge; equal to the amount by which the pressure of the fluid exceeds the ambient atmospheric pressure. *Mechanics.* see BASE PRESSURE.

gauger *Petroleum Engineering.* an oil-field worker who collects, examines, and measures oil samples, and measures the amount of oil that is run from the producer's tank to the pipeline. Also, GAGER.

gauge theory *Particle Physics.* a comprehensive principle that is concerned with the existence of boson particles that mediate the fundamental forces of gravity, electromagnetism, and strong and weak nuclear forces.

gauge transformation *Particle Physics.* a charge of the phase of the fields of a gauge theory that does not alter the value of any measurable quantity.

gauging *Nucleonics.* a technique in which the thickness, density, or quantity of a material is determined by the amount of radiation it absorbs; the most common use of radioisotopes in industry. Also, GAGING.

Gaultheria *Botany.* a genus of shrubs belonging to the family Ericaceae, including many species that contain methyl salicylate, the basis of wintergreen flavoring.

gauntlet *Medicine.* a bandage that covers the hand and fingers like a glove.

Gause's law *Evolution.* the first part of the competitive exclusion principle, stating that two species with similar ecologies cannot coexist indefinitely in the same ecological niche. Also, **Gause's principle.**

Gauss, Karl Friedrich [gous] 1777–1855, German mathematician; proved fundamental theorem of algebra; major contributions in geometry, number theory, astronomy, and electromagnetism.

gauss *Electromagnetism.* a unit of magnetic induction in the cgs system equivalent to 1 maxwell per cm^2 or 10^{-4} weber per m^2.

Gauss-Bonnet formula *Mathematics.* let M be an oriented 2-dimensional Riemannian manifold, with Gaussian curvature K and volume element dA. Suppose that R is a polygon contained in M which is diffeomorphic to a subset of R^2, with boundary ∂R, vertices $v_1, \ldots, v_n$, and corresponding interior angles $\alpha_1, \ldots, \alpha_n$. Let ds be the volume element of ∂R and κ its (signed) geodesic curvature. Then $\int_R K\, dA = -\int_{\partial N} \kappa ds + \sum_{i=1}^{n} \alpha_i + (2-n)\pi$.

Gauss-Bonnet theorem *Mathematics.* let M be a compact, oriented 2-dimensional Riemannian manifold, with Gaussian curvature K and volume element dA. Then $\int_M K\, dA = 2\pi\chi(M)$, where $\chi(M)$ is the Euler characteristic of M.

Gauss coefficient *Mathematics.* let V be an n-dimensional vector space over a finite field of order q. The kth Gauss coefficient of V is the number of k-dimensional subspaces of V and is denoted $\left[{n \atop k}\right]_q$.

Gauss eyepiece *Optics.* an eyepiece used in spectrometers, refractometers, and telescopes to aid in positioning the axis of the optical system perpendicular to a plane-reflecting surface.

Gauss formulas *Mathematics.* formulas relating angles a, b, and c of a spherical triangle with corresponding opposite sides A, B, and C; i.e.,

$$\cos(c/2)\sin((A+B)/2) = \cos(C/2)\cos((a-b)/2)$$

$$\cos(c/2)\cos((A+B)/2) = \sin(C/2)\cos((a+b)/2)$$

$$\sin(c/2)\sin((A-B)/2) = \cos(C/2)\sin((a-b)/2)$$

$$\sin(c/2)\cos((A-B)/2) = \sin(C/2)\sin((a+b)/2)$$

Gauss hypergeometric equation *Mathematics.* the hypergeometric function ${}_2F_1(a, b, c; z)$.

Gaussian beam *Electromagnetism.* a beam of particles or radiation whose radial intensity distribution is a Gaussian curve centered about the center of the beam.

Gaussian coil *Materials Science.* a type of polymer chain conformation characterized by spherically symmetrical spatial distribution.

Gaussian curvature *Mathematics.* the scalar curvature of a manifold, equal to the product of the sectional curvatures, and equal to the twofold contraction of the Ricci tensor with the metric tensor.

Gaussian curve see NORMAL DISTRIBUTION.

Gaussian distribution *Spectroscopy.* a distribution of intensity as a function of frequency, centered about a central emission or absorption line in a given spectrum; the distribution results from the Doppler shifting of frequencies due to the motion of atomic or molecular emitters or absorbers. Infrared absorption curves are Gaussian in shape.

Gaussian elimination *Mathematics.* a method of solving a system of (possibly dependent) linear equations, written in matrix form as $Ax = b$, by performing Gaussian operations and arriving at a system having the form $A'x = b'$, where A' is an upper triangular matrix. In practice, the augmented matrix $[A, b]$ is transformed to $[A', b']$.

Gaussian error *Navigation.* a temporary error in a magnetic compass due to the lag in time it takes a steel ship to adjust to new magnetic influences, such as immediately following a turn.

Gaussian gravitational constant *Astronomy.* the constant (k) that scales an astronomical system of units using Kepler's harmonic (third) law, which is based on length measured in astronomical units, time in days, and mass in solar masses; $k = 0.01720209895$.

Gaussian integers *Mathematics.* complex numbers of form $a + bi$ where a and b are integers. That is the subset $Z[i]$ of the field of complex numbers given by $Z[i] = \{a + bi : a, b \in Z\}$, where Z is the set of integers. $Z[i]$ is a Euclidean domain with $\phi(a + bi) = a^2 + b^2$; sometimes called the ring of **Gaussian complex integers.**

Gaussian noise *Telecommunications.* the undesirable electrical disturbances whose density function follows normal bell-shaped (Gaussian) distribution.

Gaussian operations *Mathematics.* the matrix row operations used in Gaussian reduction and Gauss-Jordan elimination. These are (a) interchanging rows i and j, denoted $R_i\,'\,R_j$; (b) multiplying row i by nonzero scalar k, denoted $kR_i\,'\,R_i$; and (c) adding a scalar k times row i to row j, $i\pi j$, replacing row j with the result and leaving row i unchanged, denoted $kR_i + R_j \rightarrow R_j$. The operations have the advantages of simplicity and completeness for solving systems of equations. Further, operation (c) can be performed on a matrix without changing the determinant of that matrix. Column operations are similarly defined.

Gaussian optics *Optics.* a branch of simple optical theory that discounts aberration of lenses and deals only with paraxial rays, i.e., rays that are parallel or nearly parallel to the axis of an optical system. Also, FIRST-ORDER OPTICS, PARAXIAL OPTICS.

Gaussian polynomials *Mathematics.* given $k \geq 0$ and $l > 0$, the following rational functions $[^k l]$ of the indeterminate x may be shown to be polynomials:

$$[(1-x^k)(1-x^{k-1})\cdots(1-x^{k-l+1})]/[(1-x)(1-x^2)\cdots(1-x^l)]$$

where $[^0 l] = 0$, $[^k l] = 0$ for $l > k$, $[^k l] = 0$ for $l < 0$, and $[^k l] = [^k k-l]$. Used in the development of Schur functions.

Gaussian pulse *Physics.* a pulse whose variation in time is proportional to the variation of a Gaussian curve of the form $y = Ae^{-kx^2}$ where A and k are positive constants.

Gaussian quadrature *Mathematics.* a method of numerical integration that approximates the integral of a function over an interval by a sum of function values at a *particular set* of points, multiplied by a set of weighting coefficients. The method is exact for polynomials of degree up to $2n - 1$, where n is the number of points at which the function is to be evaluated. The particular set of points for evaluating are the (scaled) roots of the Legendre polynomials in the usual case. The roots of other orthogonal polynomials come into play if functions containing a common weighting factor are to be integrated; then the method is known as **Gauss-Chebyshev quadrature, Gauss-Laguerre quadrature,** and so on.

Gaussian reduction *Mathematics.* a collective term for the methods of Gaussian elimination or Gauss-Jordan elimination.

Gaussian system *Electromagnetism.* a system of units in which electrostatic units and electromagnetic units are combined with the appropriate conversions involving the speed of light.

Gaussian weighing method *Engineering.* a system of testing the accuracy of equal-arm balances and standard weights by placing an object in a pan and another of equal weight in another pan, then switching objects to opposite pans.

Gaussian year *Astronomy.* 365.2569 mean solar days, the year that a planet would have if it orbited the sun at a distance of 1 astronomical unit in a solar system with no other planets.

Gauss image point *Optics.* in an optical system with spherical aberration, the image point at which all paraxial rays emanating from a designated point converge.

Gauss-Jordan elimination *Mathematics.* a method of solving a system of (possibly dependent) linear equations, written in matrix form as $Ax = b$, by performing Gaussian operations to arrive at a system with the form $A'x + b'$ in which each nonzero row i of A' has a first nonzero entry $e_{i,}j_i = 1$ in column j_i, every other entry in column j_i is 0, all zero rows of A' follow all k of the nonzero rows of A', and $j_1 < j_2 < \cdots < j_k$.

Gauss law of the arithmetic mean *Mathematics.* a harmonic function is maximized only on the boundary of its domain of definition, or the function is constant.

Gauss-Legendre rule *Mathematics.* an approximation of definite integrals by the use of Gaussian quadrature.

Gauss map *Mathematics.* the (nonlinear) mapping that takes each point p of a hypersurface embedded in P^n to the point of the unit sphere in P^n corresponding to the unit normal at p.

Gauss mean value theorem *Mathematics.* let u be a regular harmonic function in a region R and let S be a sphere of area A, having center P and lying entirely within R. Then $u(P) = (1/A)\iint_S u\, dS$. In particular, if R is a plane region and C is a circle in R and perimeter c, $u(P) = (1/c) \int_C u\, ds$. That is, the value of a harmonic function at a point in a region is equal to the (surface) integral of the function on a sphere centered at the point.

gaussmeter *Engineering.* an instrument having a scale calibrated in gauss units; used to measure the intensity of a magnetic field.

Gauss points *Optics.* the focal, nodal, and principal points of a lens or lens system that are used for measurement and calculations. Also, CARDINAL POINTS.

Gauss principle *Mechanics.* the principle that a system of interconnected particles, subjected to a certain set of forces, will have nearly the same motion that the particles would have if disconnected from each other but still subjected to the same forces; i.e., the constraints on the system are minimal. Also, PRINCIPLE OF LEAST CONSTRAINT, LEAST-CONSTRAINT PRINCIPLE.

Gauss-Seidel method *Mathematics.* an iterative method for approximating the solution to a system of n linear equations in n unknowns. Each equation is written in the form $a_{ii}x_i = b_i - \sum_{j \neq i}^{n} a_{ij}x_j$. Estimates for the values of $x_2, \ldots, x_n$ are made and substituted into the first equation ($i = 1$). After solving for x_1, discard x_2, substitute the remaining values into the second equation, and solve for a new and presumably more accurate value of x_2. Continue in this manner, discarding the old value of x_k at the kth step, until a desired degree of accuracy is reached. In matrix notation, i.e., $Ax = b$, the Gauss-Seidel method is equivalent to splitting A into the matrix sum $A = L + D + U$, where D is the diagonal part of A, and L and U are the lower and upper triangular parts (with zeros on the diagonal), respectively.

Gauss test *Mathematics.* an infinite series $\sum_{n=1}^{\infty} a_n$ converges if, for all n, $a_{n+1}/a_n = 1 - x/n - f(n)/n^{\lambda}$, where f is some integer function and $\lambda > 1$. In particular, if $|a_{n+1}/a_n| = 1 - L/n + c_n/n^2$, where $|c_n| <$ constant for n greater than some fixed N, the series converges absolutely if $L > 1$ and diverges or converges conditionally if $L \leq 1$.

Gauss theorem *Mathematics.* a theorem stating that the integral over a closed surface of the inner (dot) product of any vector field and the unit normal to the surface is equal to the integral of the divergence of the vector field over the volume enclosed by the surface. A special case of the theorem stating that the integral of a differential form over the (smooth) boundary of a set is equal to the integral of the exterior derivative of the form over the interior of the set. Also, DIVERGENCE THEOREM.

Gauteriaceae *Mycology.* the only family of fungi belonging to the class Ganteriales that occur on forest soil.

Gauteriales *Mycology.* an order of fungi belonging to the class Gasteromycetes that occur on forest soil.

gavage *Medicine.* **1.** forced feeding, especially through a tube passed into the stomach. **2.** any administration of nourishment or other substances through a feeding tube. **3.** the therapeutic use of a very full diet; superalimentation. (A French word meaning "cramming.")

gavial *Vertebrate Zoology.* any member of the family Gavialidae, reptiles that are closely related to alligators and crocodiles, having a slender, elongated snout, an elongated and powerful tail, and weak legs that limit it to life in water.

gavial

Gavialidae *Vertebrate Zoology.* a family of reptiles belonging to the order Crocodilia. Only one species remains, the Indian gavial, *Gavialis gangeticus,* found only in India.

Gaviidae *Vertebrate Zoology.* a monogeneric family of aquatic predatory birds forming the order Gaviiformes, including the loons and divers; characterized by plumage that is white below and black and white checkered above and having summer habitat in lakes, taiga, and tundra, and winters generally spent at sea.

Gaviiformes *Vertebrate Zoology.* a monofamilial order of aquatic predatory birds, comprising the loon and diver family Gaviidae.

Gay-Lussac, Joseph Louis [gā'lü sak'] 1778–1850, French chemist and physicist; developed Charles's law and the law of combining volumes.

Gay-Lussac tower *Chemical Engineering.* a tower in the lead-chamber process for manufacturing sulfuric acid and nitrogen oxides.

gaylussite *Mineralogy.* $Na_2Ca(CO_3)_2 \cdot 5H_2O$, a very brittle, white to yellowish, monoclinic mineral occurring as wedge-shaped crystals having a perfect cleavage, a specific gravity of 1.99, and a hardness of 2.5 to 3 on the Mohs scale.

Gazella *Ordnance. a* Soviet high-acceleration ABM missile; officially designated **SH-08.**

gazelle *Vertebrate Zoology.* a small, swift antelope of *Gazella* and related genera of the family Bovidae, found on arid plains from Africa to India; characterized by long, ridged lyre-shaped horns, a graceful brown body with black shading and a white underbelly, and soft shiny eyes.

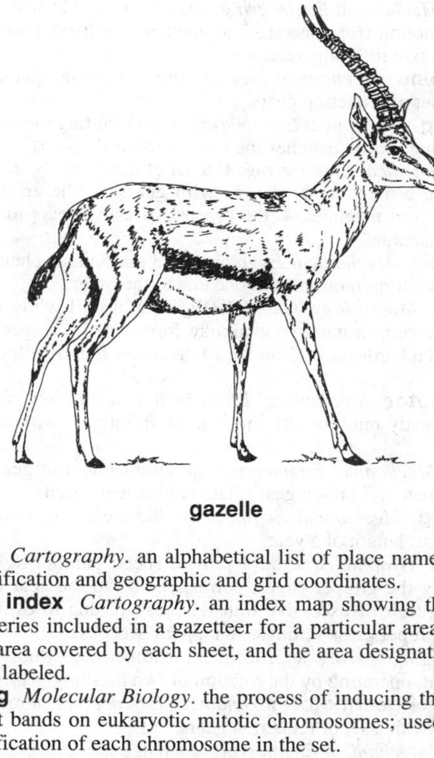

gazelle

gazetteer *Cartography.* an alphabetical list of place names giving feature identification and geographic and grid coordinates.

gazetteer index *Cartography.* an index map showing the maps of a specific series included in a gazetteer for a particular area; sheet numbers, the area covered by each sheet, and the area designation are delineated and labeled.

G banding *Molecular Biology.* the process of inducing the appearance of distinct bands on eukaryotic mitotic chromosomes; used to facilitate the identification of each chromosome in the set.

GBU-15 *Ordnance.* a television-guided U.S. glide bomb carrying a 2000-pound warhead and having a range of 5 miles.

GC *Computer Science.* **1.** garbage collection (the restoration of unused data locations to active memory). **2.** the occurrence of a garbage collection during execution. **3.** to perform garbage collection.

GC gigacycle; gonococcus.

GCA ground-controlled approach.

g-cal or **g-cal.** gram calorie.

GCD or **G.C.D.** greatest common divisor. Also, **gcd, g.c.d.**

GCF or **G.C.F.** greatest common factor.

GCM *Aviation.* the airport code for Grand Cayman, West Indies.

G.C.M. or **g.c.m.** greatest common measure.

GCR ground-controlled radar.

GCT or **G.C.T.** Greenwich Civil Time.

Gd the chemical symbol for gadolinium.

GDG generation data group.

G display *Electronics.* a rectangular radar display in which a target appears as a laterally centered blip that appears to grow wings as the target approaches; any horizontal and vertical aiming errors are indicated by horizontal and vertical displacement of the blip. Also, G INDICATOR, G SCAN.

GDL *Aviation.* the airport code for Guadalajara, Mexico.

GDP guanosine diphosphate.

Ge the chemical symbol for germanium.

geanticline *Geology.* **1.** a large, broad anticlinal structure that develops in geosynclinal sediments as a result of lateral compression. **2.** any mobile region of the earth's crust that is upwarping or has upwarped.

gear *Mechanical Engineering.* **1.** a toothed part, such as a wheel or disk, that meshes with the teeth in a similar but different-sized part in order to transmit force and motion between rotating shafts. **2.** an assembly of such parts. **3.** any of a number of configurations of such parts in a machine, such as first gear in an automobile. **4.** any system of moving parts designed to trasmit motion. **5.** any mechanism designed to perform a specific function in a machine, such as a steering gear.

gearbox *Mechanical Engineering.* **1.** the transmission of a motor vehicle. **2.** a housing for gears. Also, **gear case**.

gear cluster *Mechanical Engineering.* a set of gears that are built into or permanently attached to a shaft, designed to perform a specific function such as transmitting motion from a gearbox to the shaft.

gear cutter *Mechanical Engineering.* a circular cutter used for cutting teeth in a gear. Thus, **gear cutting.**

gear down *Mechanical Engineering.* to reduce the speed between a driving wheel and a driven wheel; shift into a lower gear.

gear drive *Mechanical Engineering.* an arrangement in which the transmission of motion from one shaft to another is achieved by direct contact between two meshing gears.

geared turbine *Mechanical Engineering.* a turbine operating with a connecting set of reduction gears.

gear forming *Mechanical Engineering.* a gear-cutting method in which a tool's cutting profile matches the desired tooth shape.

gearing *Mechanical Engineering.* **1.** a set of gear wheels. **2.** the process of equipping a machine or device with gears. **3.** the arrangement of gears on a given machine. **4.** the process of using gears to control the speed of a machine.

gearing chain *Mechanical Engineering.* a continuous chain of geared wheels transmitting motion from one end to the other.

gearksutite *Mineralogy.* $CaAl(OH)F_4 \cdot H_2O$, a dull white monoclinic mineral occurring in massive chalklike form, having a specific gravity of 2.768 and a hardness of 2 on the Mohs scale; found with cryolite and fluorite.

gearless motor *Mechanical Engineering.* a traction motor that is mounted directly onto the driving axle of an engine, without reduction gears.

gear level *Mechanical Engineering.* an arrangement of gears in which the driving gear and driven gear rotate at identical speeds.

gear loading *Mechanical Engineering.* the mechanical energy transmitted per unit length of a gear.

gear marks *Graphic Arts.* on a printed sheet, parallel light-and-dark streaks along the gripper edges, corresponding to the gear teeth of the cylinder; a result of excessive transfer pressure due to overpacking of paper or improper press settings. Also, **gear streaks.**

gear meter *Engineering.* an instrument that measures fluids by positive displacement, operating by the rotation of two meshing gear wheels.

gear motor or **gearmotor** *Mechanical Engineering.* a motor used in conjunction with a set of reduction gears.

gear pump *Mechanical Engineering.* a small pump that is used in lubrication systems, consisting of a pair of meshed gear wheels enclosed in a casing, which are contrarotated so that fluid enters the space between the teeth of each wheel and moves around the casing to the discharge port.

gear ratio *Mechanical Engineering.* the ratio of the number of teeth on a driving gear to the number of teeth on the driven gear, which determines the number of revolutions made by each gear.

gear shaper *Mechanical Engineering.* a machine that makes gear teeth by using a reciprocating cutter with a slow cutting stroke and a quick return.

gear shift *Mechanical Engineering.* a device used for changing, engaging, or disengaging gears.

gear teeth *Mechanical Engineering.* projections at evenly spaced intervals around the circumference or face of a gear wheel which, when used in conjunction with another gear or wheel, transmit force and motion.

gear train *Mechanical Engineering.* a combination of two or more engaged gears connecting driving and driven parts.

gear up *Mechanical Engineering.* to shift into a higher gear, thus raising the speed of the driven unit above that of its driver.

gear wheel *Mechanical Engineering.* a wheel having gear teeth.

gear wheel pump *Mechanical Devices.* a rotary pump consisting of two meshing gear wheels, designed to entrain fluid on one side of a casing and discharge it on the other side by contrarotation.

Geastraceae *Mycology.* a family of fungi belonging to the order Lycoperdales that comprises the earthstars and occurs in soil in temperate to tropical regions.

Geber (Jabir) 8th century AD, Persian alchemist; pioneer of medieval experimental chemistry; may have discovered sulfuric and nitric acids.

Gecarcinidae *Invertebrate Zoology.* the true land crabs, a family of crustaceans in the order Decapoda.

gecko *Vertebrate Zoology.* any one of a family of primarily nocturnal, insectivorous, and arboreal lizards of short length that have fingers and toes with an adhesive surface for climbing smooth surfaces and the ability to produce sounds such as chirping, squeaking, or barking; found in warm climates worldwide.

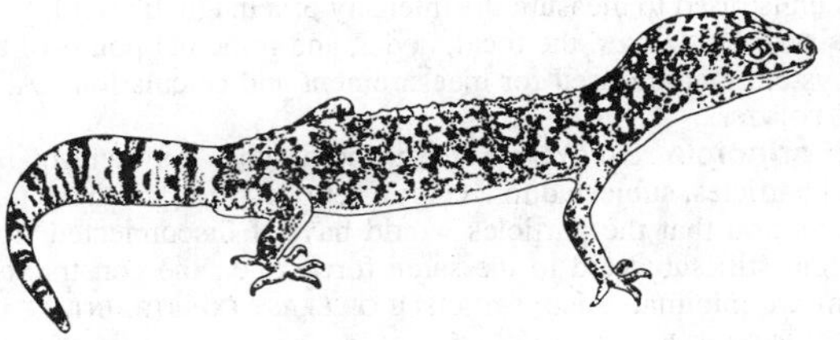

gecko

Gecko *Ordnance.* a Soviet all-weather, surface-to-air missile powered by a dual-thrust, solid-propulsion motor; it is equipped with radio-command guidance and a low-light, electro-optical tracker and has a range of 7 miles; officially designated **SA-8,** or for naval use, **SA-N-4.**

gedanite *Mineralogy.* a brittle fossil resin related to amber, having a specific gravity of 1.06 to 1.07 and a hardness of 1.5 to 2 on the Mohs scale; found near the Baltic Sea and used for beads.

Gedanken experiment *Physics.* an experiment that is conceived in thought and should work in principle but is not performed due to reasons of practicality. (From the German word for "thought.") Also, THOUGHT EXPERIMENT.

Gedinnian *Geology.* a European geologic stage of the Lower Devonian period, occurring after the Ludlovian of the Silurian period and before the Siegenian.

gedrite *Mineralogy.* $(Mg,Fe^{+2})_5Al_2(Si_6Al_2)O_{22}(OH)_2$, $Mg/(Mg+Fe^{+2})=$ 0.1–0.89, a white, green, or brown orthorhombic mineral of the amphibole group, massive to fibrous in habit, having a specific gravity of 3.15 to 3.57 and a hardness of 5.5 to 6 on the Mohs scale; found only in metamorphic rocks.

Gee *Navigation.* a British hyperbolic navigation system, similar to Loran, but limited to line-of-sight.

Gee chart *Navigation.* a chart showing lines of position and other information for use with the Gee radio navigation system.

geg *Meteorology.* a dust whirl in the deserts of China and Tibet.

GEG *Aviation.* the airport code for Spokane, Washington.

Gegenbauer differential equation *Mathematics.* the differential equation $(z^2 - 1)w'' + (2\nu + 1)zw' - \alpha(\alpha + 2\nu)w = 0$.

Gegenbauer function *Mathematics.* any of the solutions to Gegenbauer's differential equation, expressible either as hypergeometric functions or as Legendre functions of the first kind and denoted $C_\alpha^\nu(z)$. An integral expression is given by $C_\alpha^\nu(z) = -\pi^{-1}\sin(a\pi)\int_0^\infty (1 + 2tz + t^2)^{-\nu}t^{-\alpha-1}dt$.

Gegenbauer polynomial *Mathematics.* any of the solutions to the Gegenbauer differential equation for α a nonnegative integer n. In this case the Gegenbauer function reduces to a polynomial of degree n; they are the only solutions that remain finite at $z = \pm 1$. Also, ULTRASPHERICAL POLYNOMIAL.

gegenschein [gäg´ən shīn] *Astronomy.* an extremely dim patch of light about 10° across that is faintly visible to the naked eye on a dark moonless night; it lies exactly opposite the sun in the night sky and is caused by sunlight reflecting from dust.

gehlenite *Mineralogy.* $Ca_2Al(AlSi)O_7$, a grayish-green to liver-brown tetragonal mineral of the melilite group usually occurring in granular form or as short square prisms, having a specific gravity of 3.04 to 3.07 and a hardness of 5 to 6 on the Mohs scale; found in calcium-rich eruptive rocks.

Geiger, Hans [gīg´ər] 1882–1947; German physicist; invented the Geiger counter; important research on deflection of alpha particles.

Geiger-Briggs rule *Nucleonics.* a rule stating that the range of an alpha particle in dry air, at 15°C and 1 atmosphere, above 5 centimeters, is proportional to its initial velocity raised to the 3.26 power.

Geiger counter *Nucleonics.* an instrument that detects and measures the intensity of radiation by counting the number of electric pulses generated by charged particles that pass between the anode and cathode in the gas-filled tube. Also, **Geiger-Müller counter, Geiger-Müller tube.**

Geiger formula *Nucleonics.* a formula that establishes the range of an alpha particle in dry air.

Geiger-Müller probe *Engineering.* a type of radiation counter composed of a Geiger-Müller tube in a watertight enclosure; used to measure the intensity of gamma rays originating from radioactive substances inside a drilled rock. Also, **Geiger probe.**

Geiger-Nuttall rule *Nuclear Physics.* a principle that explains the inverse relationship between the half-life of alpha decay and the decay energy by establishing a (nearly) linear relationship between the logarithm of the radioactive decay constant for a number of nuclides that undergo alpha-particle decay and the logarithm of the alpha-particle range in air for each radioactive series.

geikielite *Mineralogy.* $MgTiO_3$, a bluish- or brownish-black, opaque, metallic trigonal mineral of the ilmenite group occurring as prismatic crystals or embedded grains, having a specific gravity of 4.05 and a hardness of 5 to 6 on the Mohs scale; found in metamorphosed magnesian limestones and the gem gravels of Sri Lanka.

Geissler, Heinrich [gīs´lər] 1814–1879, German physicist; invented the Geissler tube and the Geissler pump.

Geissler pump *Engineering.* a glass water pump that operates according to the Torricellian vacuum, in which a vacuum is created when mercury flows between two reservoirs, one of which is fixed and the other variable.

Geissler tube *Electronics.* a gas-filled, dual-electrode discharge tube that glows when an electric current is passed through the gas.

Geissolomataceae *Botany.* a monospecific family of dicotyledonous xeromorphic evergreen shrubs, characterized by a lack of internal phloem, unicellular pick-shaped hairs, and no nectary disk; native to the Cape Province in South Africa.

Geitel, Hans [gīt´əl] 1855–1923, German physicist; with Elster, investigated radioactive elements.

geitonogamy *Botany.* a type of fertilization in which a flower is pollinated by another flower on the same plant.

geking *Oceanography.* the process of measuring ocean movements with a geomagnetic electrokinetograph.

Gekkonidae *Vertebrate Zoology.* the geckos, a family of arboreal lizards of the order Squamata.

gel *Chemistry.* **1.** a colloidal suspension of a liquid in a solid, forming a jellylike material in a more solid form than a sol. **2.** to form a gel.

gelada *Vertebrate Zoology.* a large, terrestrial monkey, *Therophithecus gelada,* of the family Cercopithcoidae, that resembles a baboon; the males have a long shaggy mane on the neck and shoulders; native to the mountainous regions of Ethiopia.

Gelastocoridae *Invertebrate Zoology.* a family of toad bugs in the order Hemiptera that inhabit tropical and subtropical areas along the shores of ponds.

gelatin *Organic Chemistry.* a heterogeneous mixture of proteins that is obtained from the hydrolysis of collagen; strongly hydrophilic, odorless flakes or powder, soluble in hot water and insoluble in organic solvents. *Materials.* **1.** a bland, jellylike protein mixture derived by boiling the skin, bones, and other tissues of animals; used as a thickener and stabilizer in foods and manufactured products, and in photographic film, adhesives, and capsules for pharmaceuticals. **2.** any similar substance not derived from animals.

gelatinase *Enzymology.* an enzyme found in some molds and yeasts that liquefies gelatin.

gelatin duplicating *Graphic Arts.* a reprographic process in which typed or handmade copy is impressed on a paper master by a special carbon ribbon or with a hectographic pencil or ink. Also, HECTOGRAPHY.

gelatin dynamite *Ordnance.* an explosive consisting of gelatinized nitroglycerin to which cellulose nitrate has been added. Also, **gelignite.**

gelatin liquefaction *Microbiology.* a process in which gelatin protein is hydrolyzed to soluble peptides by proteolytic enzymes produced by certain bacteria and fungi.

gelatin process *Graphic Arts.* a process similar to lithography, but in which gelatin is used instead of grease and a glass plate is used instead of a flexible metal plate.

gelation *Chemistry.* the coagulation of a sol that results in a gel.

geld *Agriculture.* to remove the testes, especially of a horse.

gel diffusion see IMMUNODIFFUSION.

gel diffusion test *Immunology.* a method that is used to detect a specific antigen-antibody reaction, in which the antigen and antibody are placed in a gel and allowed to diffuse toward each other in order to form a precipitate.

gelding *Agriculture.* a castrated male animal, especially a horse.

Gelechiidae *Invertebrate Zoology.* a family of common, minute, slender-winged moths in the order Lepidoptera.

gel electrophoresis *Chemistry.* an electrophoresis (protein separation) that is carried out in a silica or acrylamide gel.

gel exclusion chromatography see GEL PERMEATION CHROMATOGRAPHY.

gel filtration *Analytical Chemistry.* a type of column chromatography that separates compounds on the basis of molecular size by using a molecular sieve, usually a zeolite; molecules larger than the largest pore size are excluded from the column.

Gelidiaceae *Botany.* a family of red algae of the order Nemaliales, having gametangial and tetrasporangial thalli of similar morphology, an inner region of colorless cells, and an outer region of chloroplast-containing cells; it includes two genera that are the principal sources of commercial agar.

Gelidiellaceae *Botany.* a monogeneric family of red algae belonging to the order Nemaliales, recently separated from the Gelidiaceae; characterized by a lack of internal rhizoids and including a species used as food and as a source of agar.

gelifluction or **gelisolifluction** see CONGELIFLUCTION.

gelifract see CONGELIFRACT.

gelifraction see CONGELIFRACTION.

gelisol see FROZEN GROUND.

geliturbation see CONGELITURBATION.

gellan gum *Biochemistry.* a polysaccharide consisting of glucose, rhamnose, and glucuronic acid residues and gels; formed by the bacteria *Pseudomonas elodea* when heated or cooled in a particular medium, such as water containing monovalent or divalent cations.

gelled cell *Electricity.* a lead-acid cell with a gelled electrolyte designed for portability.

Gell-Mann, Murray born 1929, American physicist; formulated the Law of Strangeness and the Theory of Eightfold Way; predicted the existence of omega-minus particles and quarks.

Gell-Mann–Nishijima scheme *Particle Physics.* a scheme introducing the strangeness quantum number, in which baryons, when classified according to their electric charge and their hypercharge, naturally group into octets.

Gell-Mann–Okubo mass formula *Particle Physics.* a relation that gives the masses of the particles of a unitary meson or baryon multiplet in terms of the particles' hypercharge, total isotopic spin, and other multiplet coefficients; based on SU(3) symmetry.

Gell-Mann relation *Particle Physics.* a relationship between the masses of the meson particles in a meson octet; based on the Gell-Mann-Okubo mass formula.

Gelocidae *Paleontology.* a family of ruminants in the infraorder Traguloidea, more advanced than the hypertragulids and possibly ancestral to the modern pecorans; Eocene to Pliocene.

Gelopellidaceae *Mycology.* a family of fungi whose classification is in transition, but which is commonly classified under the order Phallales containing the single genus *Gelopellis,* which occurs in South America.

gélose see ULMIN.

gel permeation chromatography *Analytical Chemistry.* a separation technique based on molecular size in which the mobile phase is a liquid and the stationary phase consists of three-dimensional networks of crosslinked polymer chains, such as beads of porous polymeric material. Also, GEL EXCLUSION CHROMATOGRAPHY, SIZE-EXCLUSION CHROMATOGRAPHY.

gel point *Physical Chemistry.* the stage at which a liquid begins to take on the semisolid characteristics of a gel.

gelsemine *Pharmacology.* $C_{20}H_{22}N_2O_2$, a poisonous alkaloid from the yellow jessamine.

gem *Mineralogy.* a cut and polished rock, mineral, or other natural material whose beauty, quality, durability, and scarcity give it intrinsic value, thus making it suitable for use in jewelry and other ornamental purposes. (From the Latin *gemma* meaning "bud.")

GEM ground-effect machine.

Gemella *Bacteriology.* a genus of Gram-positive bacteria of the family Streptococcaceae that occur as rounded cells, either singly or in pairs with adjacent sides flattened; found as parasites of mammals.

gemellary *Medicine.* of or relating to twins.

geminal *Organic Chemistry.* of or relating to two like atoms or groups attached to the same carbon atom in a molecule.

geminate *Biology.* paired; growing, occurring, or combined in pairs.

gemination *Science.* a doubling or duplication. *Physiology.* a fusion of two teeth resulting in the formation of two teeth or of a double crown formed on a single root with a single pulp canal.

Geminga *Astronomy.* a neutron star in Gemini which is the second-brightest source of high-energy gamma rays in the sky.

Gemini [jem´ə nī] *Astronomy.* the Twins, a conspicuous constellation that lies in the zodiac to the northeast of Orion. *Space Technology.* any of a series of two-person, earth-orbiting vehicles used to test rendezvous and docking in space (1965–1966).

Gemini VII

Geminids *Astronomy.* a major meteor shower that reaches peak activity around December 13th, having about 30 to 50 meteors an hour that appear to stream from Gemini.

geminiflorous *Botany.* having paired flowers.

Geminivirus group *Virology.* a group of plant viruses that contain circular ssDNA and are generally transmitted by leafhoppers and the whitefly, with a narrow host range.

gemma *plural,* **gemmae.** *Botany.* **1.** a bud. **2.** a small, budlike, asexual reproductive body of some liverworts and mosses that becomes detached from the parent plant and is able to form a new organism. *Anatomy.* any budlike or bulblike structure, such as a taste bud. Thus, **gemmaceous.**

Gemmata obscuriglobus *Bacteriology.* a species of bacteria that reproduces by budding, has a swarmer stage during the life cycle, and occurs in freshwater habitats.

gemmate *Botany.* **1.** having buds or gemmae. **2.** reproducing by budding.

gemmation *Botany.* reproduction by budding.

gemmho *Electricity.* a unit of conductance equivalent to 10^{-6} mho, that indicates the conductance of a substance having a resistance of 10^6 ohms.

gemmiform *Botany.* having the form of a gemma or bud.

Gemmiger *Bacteriology.* a genus of Gram-variable bacteria that occur as coccoid cells, reproduce by budding, and carry out anaerobic fermentation of carbohydrates.

Gemminges *Bacteriology.* a genus of bacteria that is composed of Gram-positive anaerobic cocci, and that are occasionally isolated from humans.

gemmiparous *Biology.* producing a bud, or reproducing by gemmae or buds.

gemmule *Invertebrate Zoology.* in freshwater sponges, an asexual reproductive body, resistant to freezing and drying, that germinates when conditions are right. *Biology.* any reproductive bud projecting from a cell. *Anatomy.* any of numerous small processes projecting from the dendrites of neurons. Also, DENDRITIC SPINES. *Evolution.* in the theory of pangenesis, a hypothetical living unit that bears the hereditary attributes.

Gemolite *Optics.* a binocular device that distinguishes natural gems from synthetic gems, using dark-field illumination and magnification.

gemology *Mineralogy.* the study and classification of precious and semiprecious gemstones. Thus, **gemologist.**

Gempylidae *Vertebrate Zoology.* the snake mackerels, a family of predatory marine fishes of the suborder Acanthuroidei, characterized by an eel-like body and a jutting jaw with sharp teeth; found mainly in tropics and warm waters worldwide at depths below 1200 meters.

gemsbok *Vertebrate Zoology.* an antelope of the family Bovidae, *Oryx gazella,* characterized by sharp, rapierlike horns up to four feet long, a light body with distinct black markings on the face and underbelly, and a dark, short, ridged mane that runs the length of the back to a long, tufted tail; found in eastern and southern Africa.

gemsbok

gemstone *Geology.* any naturally occurring mineral, rock, or organic material that, when cut and polished, is suitable for use as jewelry or other kind of ornament.

Gemuendina *Paleontology.* a rhenanid placoderm in the extinct family Asterosteidae; best known of the flat-bodied, raylike rhenanids; now generally thought to be nonancestral to the skates.

gen general; genus.

GEN *Computer Programming.* see GENERATE.

-gen *Medicine.* a combining form indicating an agent producing the object or state denoted by the word stem, as in *allergen* (allergy), *pathogen* (disease).

gena *plural,* **genae.** *Anatomy.* the cheek or side of the face. Thus, **genal.**

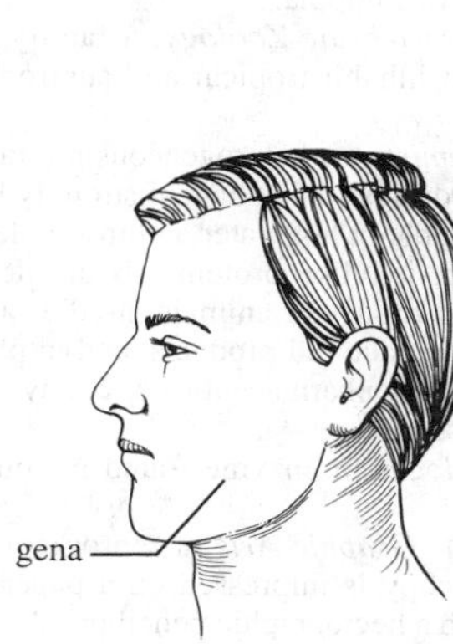

gena

gender *Biology.* sex; the category to which an individual is assigned on the basis of sex. *Electricity.* the identity of a connector as male or female.

gender identity *Psychology.* the self-awareness that one is either a male or a female.

gender role *Psychology.* the image that an individual presents to others based on culturally defined concepts of masculinity and femininity.

gender schema *Psychology.* a general conception that classifies various traits or items of information as having either a masculine or a feminine connotation.

gending *Meteorology.* a foehnlike dry wind in the northern plains of Java, caused by a wind that crosses the mountains near the south coast and pushes between the volcanoes.

gene *Genetics.* **1.** a hereditary unit that is composed of a sequence of DNA and occupies a specific position or locus. **2.** broadly, any genetic determinant.

Geneaceae *Mycology.* a family of fungi belonging to the order Tuberales, including species commonly called truffles; found in forest soil.

gene amplification *Genetics.* **1.** the synthesis of extra functional copies of genes. **2.** the duplication of the entire genome of a cell or organism. **3.** a procedure, as the treatment of a bacterial cell with an antibiotic, that increases the proportion of plasmid DNA to bacterial DNA.

gene bank see GENE LIBRARY.

gene cloning *Molecular Biology.* the process of isolating individual gene sequences in the genome of an organism, and their subsequent insertion into plasmid vectors for production in large numbers.

gene cluster *Genetics.* a set of closely linked genes that code for the same or similar products and are often grouped together on the same chromosome.

genecology *Biology.* the study of intraspecific variation and genetic composition in relation to environment.

gene conversion *Genetics.* a situation, after meiosis, in which two homologous alleles segregate in a 3:1 rather than a 2:2 ratio; may be the result of DNA polymerase's copying one allele twice and the homologous allele not at all.

gene divergence *Genetics.* the difference in nucleotide sequences between related genes that have arisen from the same ancestral gene, expressed as a percentage.

gene dosage *Genetics.* the number of times that a particular gene occurs in the nucleus of a cell or genome. Also, DOSAGE.

gene duplication *Molecular Biology.* the production of a tandem repeat of a DNA sequence by unequal crossing over or by an accident of replication.

gene expression *Genetics.* the manifestation of the genetic material of an organism as a collection of specific traits.

gene family *Genetics.* a set of genes descended by duplication and variation from an ancestral gene.

gene flow *Genetics.* the movement of genes from one population of a species to another by means of interbreeding.

gene frequency *Genetics.* the relative occurrence of a particular allele of a gene within a population, expressed as a percentage. Also, ALLELIC FREQUENCY.

gene insertion *Genetics.* the introduction or addition of one or more genes into the genome of a cell from an external source.

gene isolation *Molecular Biology.* a pair of alleles that cause a reduction in the fertility of an organism when present in the heterozygous form.

gene library *Genetics.* a collection of genes of a particular strain, including some of the unique genomic sequences of the strain, that have been cloned in a host cell. Also, GENE BANK.

gene locus see LOCUS.

gene magnification *Genetics.* a process occurring within the genome whereby organisms that are deficient in ribosomal DNA can regenerate new nucleolus organizing material.

gene manipulation see GENETIC ENGINEERING.

gene map *Genetics.* a representation of the sequence of specific genes and their relative distances from one another on a given chromosome.

gene pair *Genetics.* the two genes, represented as either identical or different alleles, that occupy the same gene locus on a homologous pair of chromosomes.

gene pool *Genetics.* all of the genes possessed by the reproductively active members of a population.

gene probe *Molecular Biology.* a biochemical that is labeled with radioactive isotopes or tagged in another way for ease of identification; used to identify or isolate a gene.

general *Military Science.* **1.** also, **general officer.** an officer of the highest rank, above a colonel. **2.** specifically, in the U.S. Army and U.S. Air Force, an officer ranking above a lieutenant general and below a general of the army or general of the air force; the insignia is four stars.

general adaptation syndrome *Behavior.* a pattern of physiological response to stress, consisting of a sequence of an alarm reaction, resistance to the stress, and exhaustion, and ultimately resulting in weakened body resistance and behavioral changes.

general anesthesia *Medicine.* the administration of anesthetic drugs resulting in loss of sensation and consciousness.

general anesthetic *Pharmacology.* any of a number of drugs, usually given by inhalation, that produce general unconsciousness, with no pain sensation over the entire body and varying degrees of muscle relaxation.

general arrangement plans *Naval Architecture.* plans showing the overall layout of a ship's structures, major fittings, and gear.

general aviation *Aviation.* a term denoting all civil (nonmilitary) aviation other than common commercial transport; includes personal flying, business flying, instructional flying, and commercial flying such as aerial photography and agricultural spraying.

general broadcast *Telecommunications.* information about weather, atmospheric conditions, map analyses, and aircraft delivered over radio waves to the U.S. Navy and merchant ships.

general chart *Cartography.* a nautical chart, of smaller scale than a coast chart but of larger scale than a sailing chart, designed for use in offshore coastwise navigation.

general circulation *Meteorology.* the complete statistical description of atmospheric motions over the earth, generated from daily flow patterns. Also, PLANETARY CIRCULATION.

general depot *Military Science.* a large supply facility serving more than one technical service.

general donor see UNIVERSAL DONOR.

general evolution *Anthropology.* the evolution of all human societies over time.

general factor *Psychology.* a general level of ability that influences an individual's test scores on various intelligence tests. Also, G FACTOR.

general farm see DIVERSIFIED FARM.

general formula see GENERIC FORMULA.

generalization *Science.* also, **generality.** a general principle or idea. *Behavior.* in conditioning, the process by which a learned response to a particular object or situation also comes to be evoked by another, similar stimulus. *Artificial Intelligence.* given a particular instance or data set, the represention of a more general case that includes the instance.

generalize *Medicine.* to spread throughout the body, as when a localized disease becomes systematic. *Psychology.* to form general ideas; carry out the process of generalization. *Cartography.* to smooth the character of features on a map or chart without destroying their visible shape. Thus, **generalized.**

generalized anxiety disorder *Psychology.* a disorder characterized by persistent tension and apprehension about one's general life circumstances; often accompanied by physiological symptoms such as headaches, sleep disturbance, fatigue, and rapid heartbeat.

generalized avoidance conditioning see AVOIDANCE CONDITIONING.

generalized coordinate *Mechanics.* one of a set of independent variables, not necessarily corresponding to the ordinates of any standard coordinate system (i.e., not necessarily having the same dimension as that of cartesian coordinates); the complete set is necessary and sufficient to specify uniquely the configuration of a system; for any holonomic system the number of independent generalized coordinates is equal to the degrees of freedom of the system so constrained.

generalized force *Mechanics.* a quantity appearing in Lagrangian mechanics corresponding to a generalized coordinate, given for a conservative system by the partial derivative of the system Lagrangian with respect to that coordinate.

generalized function *Mathematics.* see DISTRIBUTION.

generalized hydrostatic equation *Geophysics.* the vertical component of the vector equation in natural coordinates when the acceleration of gravity is replaced by the virtual gravity.

generalized hypergeometric see HYPERGEOMETRIC FUNCTION.

generalized linear models *Statistics.* a unifying approach to a variety of statistical models including regression, analysis of variance, and categorical data analysis; characterized by the additivity of the effects and the fact that the distribution of the response has to belong to a wide family known as an exponential family.

generalized momentum see CONJUGATE MOMENTUM.

generalized ratio test see D'ALEMBERT'S TEST.

generalized reciprocity *Anthropology.* a practice in which a gift is not given a specific value, nor is there a time period in which it must be repaid; often takes place in a situation in which resource availabilty is unreliable or scarce.

generalized routine *Computer Programming.* a program routine that is designed to process a wide range of jobs within a given type of application; parameters specify the details for a particular job.

generalized system *Computer Technology.* a computer system that is designed for a wide range of applications and users.

generalized theorem of the mean see CAUCHY MEAN VALUE THEOREM.

generalized transduction *Molecular Biology.* bacterial transduction in which any of the genes of the donor bacterium may be transduced; the transducing phage particles contain only donor DNA.

generalized transmission function *Geophysics.* in atmospheric-radiation theory, a set of values that varies with wavelength, with each value representing an average transmission coefficient for a small wavelength interval and a specified optical path through the absorbing gas.

generalized velocity *Mechanics.* the time rate of change of a generalized coordinate.

general manager *Industrial Engineering.* the senior management person directly responsible for several departments having different functions in an industrial facility or engineering project.

general map *Cartography.* a map of small scale used for general planning purposes.

general of the army *Military Science.* the highest rank in the U.S. Army; the insignia is five stars. Similarly, **general of the air force.**

general orders *Military Science.* **1.** a set of permanent orders that apply to an entire command, as opposed to special orders that affect individuals or small groups within the command. **2.** a set of permanent orders that govern the duties of a sentry on post.

general paresis *Medicine.* a chronic progressive form of syphilis resulting in physical and mental disturbance.

general practitioner *Medicine.* a family practice physician.

general precession *Astronomy.* see PRECESSION.

general-purpose *Computer Science.* describing a system or program that is suitable for a wide variety of applications and functions. Thus, **general-purpose language, general-purpose program, general-purpose register,** and so on.

general-purpose bomb *Ordnance.* air-delivered munition designed to destroy or damage its target by explosive effect; it is the most common type of bomb and may be used against personnel or material targets.

general-purpose computer *Computer Technology.* a computer that is designed to operate on a wide variety of applications, as opposed to one that is dedicated to a specific task, such as a word processor.

general-purpose function generator *Computer Technology.* a function generator that is designed to be changed by its operator to perform one of many different functions.

general-purpose knowledge engineering language *Artificial Intelligence.* a software package used as a tool for integrating knowledge into a computer system; for example, HEARSAY-III, ROSIE, OPS-5, or RLL.

General-Purpose Problem Solver *Artificial Intelligence.* the product of early (1957) research into developing a theory that explains how humans solve problems and into building machines that solve problems requiring intelligence.

general-purpose simulation system *Industrial Engineering.* a system originally developed to simulate the flow of information in a computer system, now used to simulate a variety of actual flow systems.

general-purpose vehicle *Military Science.* a motor vehicle used without modification for a wide variety of duties, including transportation of personnel, equipment, supplies, and ammunition.

general recombination see HOMOLOGOUS RECOMBINATION.

general relativistic collapse *Astrophysics.* a process in which a star collapsing into itself cannot release its kinetic energy and continues to collapse, creating a relativistic singularity of infinite density.

general relativity *Physics.* Einstein's theory of gravitation, which relates the gravitational field to the geometry of a curved space. See RELATIVITY THEORY.

general reserve *Military Science.* a reserve of troops under the control of the overall commander that is available for assignment to subordinate commands as required by combat conditions. Thus, **general-reserve artillery.**

general staff *Military Science.* a group of officers without command who assist high-level military commanders in planning, coordinating, and supervising operations.

general supplies *Military Science.* an intraservice classification applied to quartermaster supplies, transportation supplies, and ordnance supplies excluding ammunition.

general support *Military Science.* support provided to an operation as a whole rather than to one specific subdivision of the operation.

general-support rocket system *Ordnance.* a multiple-rocket launcher system, designed to supplement cannon artillery by delivering massive firepower in a short time against time-sensitive targets.

general term *Mathematics.* **1.** an expression involving one or more variables which, when evaluated at particular values of the corresponding index sets, give a particular member of the collection of objects being described. **2.** a symbol, subscripted by an element i of an index set I; denotes the ith member of a sequence or the ith term of a series. **3.** in general, suppose x is a function from an index set I to a set X. The range of x is called an indexed set, x itself a family, and the value of the function x at an index i a general term. A general term may involve one or more such set functions and index sets.

general transduction *Genetics.* the ability of certain bacteriophages to transduce a gene from one bacterium to another.

general war *Military Science.* armed conflict between major powers in which the total resources of the participants are employed and the independent survival of at least one of the major participants is endangered.

generate to produce or bring into being; specific uses include: *Chemistry.* to produce a gas or electric current. *Computer Programming.* **1.** to create a customized version of an operating system or software package for a particular computer system. **2.** to use a generator to produce a program or routine by selection of subsets of skeletal code under control of parameters. Also, GEN.

generate and test *Computer Programming.* a problem-solving methodology in which a sequence of candidate solutions is generated and each is tested in turn to determine if it is a correct solution.

generated address *Computer Programming.* an address calculated or determined from program logic for subsequent use by the program.

generated traffic *Transportation Engineering.* an increase in traffic arising from a new or improved transportation service or facility.

generating flow *Fluid Mechanics.* the flow of a liquid into a duct whose boundary layer flow begins at the opening of the duct and continues to grow until the duct is filled.

generating function *Mathematics.* let $a = \{a_r : r$ is a nonnegative integer$\}$ be any infinite sequence of objects, such as events, functions, numerical values, etc. Let $\{\mu_r(x)\}$ be a sequence of indicator functions. The formal power series $G(x) = \sum_{r=0}^{\infty} a_r \mu_r(x)$ is said to be a generating function for the sequence a. The indicator functions are usually chosen so that no two distinct sequences will yield the same generating function. In particular, if $\mu_r(x) = x^r$ for all r, that is, $G(x) = \sum_{r=0}^{\infty} a_r x^r$, then $G(x)$ is called the **ordinary generating function** for the sequence a. If $\mu_r(x) = x^r/r!$, that is, $G(x) = \sum_{r=0}^{\infty} a_r x^r/r!$, then $G(x) = \sum_{r=0}^{\infty} a_r x^r/r!$ is called the **exponential generating function** for the sequence a.

generating magnetometer *Engineering.* an instrument that measures the intensity of a magnetic field by rotating a coil in the field and measuring the voltage generated.

generating station *Mechanical Engineering.* a building housing the necessary equipment for the large-scale energy conversion of thermal energy (from coal, gas, etc.) to electrical energy. Also, POWER STATION.

generation *Biology.* **1.** the act or process of reproduction. **2.** a class composed of all individuals removed by the same number of successive ancestors from a common ancestor, or occupying positions on the same level in a genealogical chart. **3.** all the individuals produced within a single life cycle, having common parents and constituting a single level in a pedigree. *Computer Technology.* **1**. any of the groups used to classify computers historically, according to such features as their hardware components, logical organization, or programming techniques; thus, first-generation, second-generation, etc. **2**. any of a family of files, each one a modification of the next most recent file. **3**. in micrographics, a measure of the remoteness of the copy from the original copy, which is referred to as the first generation. *Artificial Intelligence.* the production of a statement in a language given a grammar of the language and a representation of the meaning that is to be conveyed.

generation data group *Computer Programming.* a collection of files that are maintained in chronological order; often the number of files is fixed so that when a new file is added, the oldest is deleted; each file is a **generation data set.**

generation length *Evolution.* **1.** the average length of time required for an individual of a species to develop from birth to reproducing adult. **2.** the basic time unit of natural selection for each species.

generation of waves *Oceanography.* **1.** the creation of waves by a natural or mechanical process. **2.** the creation of waves by a wind blowing over a body of water.

generation rate *Electronics.* the time rate at which electron-hole pairs are created in a semiconductor device.

generation 1 robot *Robotics.* a programmable, memory-controlled robot with two or three axes of movement, which can be equipped with a variety of attachments to work with different materials.

generation 1.5 robot *Robotics.* a sensory-controlled robot that is able to override preprogrammed control based on its own sensory data.

generation 2 robot *Robotics.* a robot that uses machine vision to perceive objects and perform manipulative operations based on the resulting input.

generation 3 robot *Robotics.* a robot that uses artificial intelligence to solve its own problems.

generation time *Microbiology.* the doubling time of a culture of microorganisms or the time elapsed from one cell division to the next. *Nucleonics.* the average time required for a neutron emitted from a disintegrating nucleus to enter another nucleus.

generative *Biology.* **1.** having the ability to produce or create. **2.** of or relating to reproduction.

generative grammar see TRANSFORMATIONAL GRAMMAR.

generative semantics *Linguistics.* a theory claiming that the meaning is the most important aspect of a sentence, and the rules of syntax and phonology are based on this semantic structure.

generativity stage *Psychology.* the seventh of Erik Erikson's eight stages of human development, during middle adulthood, when an individual is concerned with a parental role and with aiding and guiding the next generation.

generativity vs stagnation *Psychology.* the conflict or stress that can arise during the generativity stage of life, if an individual has not developed a satisfying connection with the next generation. Also, **generativity vs self-absorption.**

generator *Electrical Engineering.* a machine that converts mechanical energy into electrical energy. *Electronics.* **1.** an electronic device that converts direct-current electricity to alternating current of a desired frequency and wave shape. **2.** a source of signal, such as an oscillator or an electromechanical device, in an electronics circuit. *Computer Science.* a procedure that produces the elements of a sequence, returning the next element each time it is called; e.g., a pseudorandom number generator. *Mathematics.* a member of a set from which an algebraic object, such as a group, subgroup, ideal, module, field, etc., is constructed. In particular, let G be a group and X a subset of the elements of G. Let $\{H_\alpha\}$ be the family of all subgroups of G which contain X. Then $\cap_\alpha H_\alpha = \langle X \rangle$ is called the subgroup of G generated by the set X. If no proper subset of X generates $\langle X \rangle$, then the members of X are said to be independent. If $G = \langle X \rangle$ and X has a finite number of elements, then G is said to be finitely generated. If $a \in G$, then $\langle a \rangle$ is a cyclic subgroup of G. These definitions can easily be extended to other objects. If X is a subset of a vector space, then $\langle X \rangle$ is called the span of the set X.

generator field control *Electricity.* a system of controlling generator output by adjusting the voltage that excites the generator field.

generator set *Engineering.* a system comprising one or more generators and an energy source.

generatrix *plural,* **generatrices.** *Mathematics.* a generatrix or linear generator of a surface is a line that lies entirely in the surface and that passes through every point of the surface when moved according to a given rule; such a surface is called a ruled surface. For example, the elements of a cone represent the different positions of its generatrix. If a surface has more than one generatrix, it is called a multiply ruled surface with a system of generators. For example, the hyperboloid of one sheet $x^2/a^2 + y^2/b^2 - z^2/c^2 = 1$ has two systems of generators: (a) $x/a + z/c = u(1 + y/b)$, $u(x/a - z/c) = 1 - y/b$ and (b) $x/a + z/c = v(1 - y/b)$, $v(x/a - z/c) = 1 + y/b$ where u and v are arbitrary numbers. One generator of each system passes through each point of the surface.

gene rearrangement *Genetics.* a structural change in a chromosome that produces a change in the linear sequence of its loci.

gene redundancy *Genetics.* the presence of numerous copies of the same gene in a genome.

gene regulatory protein *Genetics.* a protein that helps to control the expression of specific genes, usually by transcriptional activities such as repression.

generic *Biology.* of or relating to a genus. *Science.* of or relating to a group or class. *Pharmacology.* of a drug name, nonproprietary; not protected by a trademark, and often descriptive of its chemical structure.

generic formula *Chemistry.* a generalized formula that applies to a series of related compounds, with a variable that stands for the number of atoms or a radical. For example, a generic formula for an alcohol is ROH, where R = a hydrocarbon radical. Also, GENERAL FORMULA.

generic point *Mathematics.* a point of a manifold that is away from any geometric singularities of the manifold, and that is surrounded by a neighborhood of nonsingular points. Typically, a proposition is said to hold at generic points, or **generically,** if it is true on some open dense subset.

genesis rock *Geology.* any rock, such as a meteorite or asteroid, that has remained unchanged since the formation of the planets.

gene splicing see SPLICING.

gene suppression *Genetics.* see SUPPRESSION.

gene switch *Genetics.* in a cell or organism, a change in expression from one gene or set of genes to another.

genet [jen´it] *Vertebrate Zoology.* any member of the genus *Genetta,* small, carnivorous, nocturnal, predatory, civetlike mammals characterized by a long body and tail, a long snout, short legs, light-colored fur with dark spots, medium-sized catlike ears, fully retractile claws, and a banded tail; found in southern Europe, Africa, and western Asia.

genet

gene therapy *Genetics.* the introduction of a gene into a cell for the purpose of correcting a hereditary disease or improving the genome.

genetic *Biology.* **1.** of, relating to, or having to do with origins or geneses. **2.** of, relating to, or produced by genes. **3.** of or relating to genetics.

genetic algorithm *Artificial Intelligence.* a class of algorithms that attempt to solve a problem by an evolutionlike process. A set of candidate solutions is evaluated in terms of quality of solution; the better candidates are allowed to reproduce, sometimes with mutations. If one candidate comes to dominate the population, it is selected as the best solution. In some versions, different candidates are combined, as in sexual reproduction.

genetic assimilation *Genetics.* a phenomenon in which a phenotypic character that originally was produced only in the presence of certain environmental conditions becomes genetically fixed in a population and no longer requires the previous environmental conditions for expression.

genetic block *Genetics.* an obstruction in a biochemical pathway caused by a mutation that interferes with or prevents the production of a needed enzyme.

genetic code *Genetics.* consecutive triplets of nucleotides in DNA or RNA that determine the sequence of amino acids in protein; nearly universal for all organisms, with minor variations in some species.

genetic colonization *Genetics.* the insertion into a host cell of genetic information from a parasite, causing the host cell to synthesize products that are useful only to the parasite.

genetic complementation *Genetics.* the restoration of wild-type function in a cell or organism containing two distinct mutations on the same chromosome.

genetic counseling *Genetics.* the genetic testing of potential parents, followed by advice from a specialist regarding the likelihood of their producing children with genetic defects or hereditary diseases.

Genetics

The science of genetics deals with the transmission and expression of hereditary characteristics and their mechanisms. In 1861, Gregor Mendel showed that the transmission of certain traits in peas was governed by discrete factors and followed definite laws.

Much later it was shown that the Mendelian laws applied to all species, including humans. The inheritance of complex traits was often shown to be due to interaction of several Mendelian factors or genes. Genes are located in a linear order on the chromosomes and can be arranged as genetic maps.

Experiments with bacteria established that deoxyribonucleic acid (DNA) is the chemical basis of the gene. James Watson and Francis Crick in 1953 proposed the now-accepted structural basis of genetic information in DNA. Later, the order of nucleotides along the DNA molecule was found to specify the linear arrangement of amino acids in proteins. The genetic code governs the translation from nucleotide order to amino acids and was found to operate in organisms as far apart as viruses and humans.

The code functions by the rule that triplets of DNA nucleotides each specify a single amino acid. Restriction enzymes cut DNA at specific sites and allow the in vitro recombination of DNA within and between organisms and between species. These and other discoveries led to the development of molecular biology.

Genetics is the science that explains biologic variation within species and between species. Genetics has become a central science underlying much of biology, evolution, and the medical sciences and is probably the most rapidly advancing field of biology. Its discoveries have not only been of great fundamental significance but are increasingly applied in agriculture, animal breeding, and medicine. As one example, genes can be placed into microorganisms or into animal cells that can be grown in culture to allow the manufacture of large quantities of gene products such as certain therapeutic agents.

There are many branches of genetics. These include formal and statistical genetics, evolutionary genetics, population genetics, developmental genetics, clinical genetics, behavioral genetics, immunogenetics, genetic epidemiology, anthropologic genetics, and others. The great variety of fields shows the high explanatory power of genetics for an understanding of biology and medicine.

Areas of ignorance remain. The behavior of movable genetic elements, the switching off and on of genes during development, and the mechanisms of gene action in the nervous system are some examples of areas where more insights are needed.

Arno G. Motulsky
Professor of Medicine and Genetics
University of Washington Medical School

genetic disease *Genetics.* any of a vast number of pathological conditions caused by mutations in the genes, such as cystic fibrosis.

genetic distance *Genetics.* a measure of the evolutionary divergence of different populations of a species, as indicated by the number of allelic substitutions that have occurred per locus in the two populations.

genetic drift *Genetics.* the random change of allelic frequencies in a population.

genetic engineering *Biotechnology.* the experimental technology developed to alter the genome of a living cell for medical or industrial use. Also, GENE MANIPULATION.

genetic equilibrium *Genetics.* a state in which the rate of an allele's forward mutation is balanced with the rate of its reverse mutation.

genetic facies *Geology.* an ancient accumulation of rocks produced by similar sedimentary processes.

genetic fine structure *Molecular Biology.* the consideration of genes at the level of nucleotide sequences, and the molecular events that are relevant to them at that level.

genetic fingerprinting *Genetics.* a comparison of the nucleotide sequences in samples of DNA to determine if they are from the same individual or genetically identical individuals, as in criminal investigations.

genetic homeostasis *Genetics.* the tendency of a population to attain equilibrium and resist changes in its genetic composition.

genetic identity *Genetics.* the relationship of two populations, as indicated by the percentage of identical genes they share.

genetic induction *Molecular Biology.* the stimulation of a specific gene by an inducer molecule acting directly or indirectly through an effect on RNA polymerase.

genetic information *Genetics.* the hereditary information contained in the nucleotide sequences of DNA or RNA.

geneticist [jə net´ə sist] *Genetics.* a specialist in the study of genetics.

genetic load *Genetics.* **1.** the deleterious genes that are carried in a population. **2.** the amount by which the average fitness of a population is lowered due to such genes, relative to the fitness that would exist if the fittest genotype in the population were to become ubiquitous.

genetic marker *Genetics.* a gene whose phenotypic expression is easily discerned; used to identify an individual or cell that carries the marker or as a probe to mark a nucleus or chromosome.

genetic material *Genetics.* DNA and RNA, the carriers of primary genetic information.

genetics *Biology.* the science of heredity, and of the mechanics by which characteristics are passed from one generation to the next.

genetic screening *Genetics.* a process of testing individuals for the presence of certain alleles, primarily those associated with human disorders such as sickle-cell anemia or Tay-Sachs disease.

genetic variance *Genetics.* the fraction of a phenotypic variance that is due to differences in the genetic constitution of individuals in a population.

gene transfer *Genetics.* the transfer of a gene to a cell by any of a variety of methods.

Geneva system *Organic Chemistry.* a system of nomenclature for organic compounds adopted in 1892, based on naming compounds as derivatives of hydrocarbons; the names correspond to the longest straight carbon chain that is present.

genic hybrid sterility see HYBRID STERILITY.

genicide see XANTHONE.

genicular *Botany.* growing on or out of a node or in the tissue of a node.

geniculate *Science.* bent at a sharp angle, like a knee or elbow.

geniculate body *Anatomy.* any of four small masses of neurons that project from the posteroinferior side of the thalamus; they relay auditory and visual sensations to the cerebral cortex.

geniculum *Anatomy.* a sharp, kneelike bend in a structure or organ.

Genie *Ordnance.* an air-to-air, unguided rocket equipped with a nuclear warhead, designed to be carried by the F-101 or F-106 aircraft; officially designated **AIR-2.**

geniocheiloplasty *Surgery.* plastic surgery of the chin and lips.

genioglossus *Anatomy.* a muscle lying beneath the tongue that depresses the tongue and protrudes it from the mouth.

Geniohyus *Paleontology.* a genus of primitive conies in the extinct family Pliohyracidae, known from deposits at Fayum, Egypt.

genioplasty *Surgery.* plastic surgery of the chin.

genital [jen´ə təl] *Anatomy.* of or relating to the genitalia. *Biology.* of or relating to reproduction or generation. *Psychology.* of or relating to the genital stage of psychosocial development.

genital atrium *Zoology.* a body cavity into which the ducts from the male and female sexual organs open.

genital coelom *Invertebrate Zoology.* in mollusks, the lumina of gonads.

genital cord *Developmental Biology.* the ridge of tissue in the caudal part of the mammalian embryo that arises through fusion of the paired genital ridges. *Invertebrate Zoology.* in crinoids, strands of primordial sex cells located in the genital canal.

genital disc *Entomology.* a set of cells in *Drosophila* and other fly larvae that are programmed to develop into the genital apparatus.

genitalia [jen ə tāl′yə] *Anatomy.* the reproductive organs, especially those on the outside of the body. Also, GENITALS.

genital pore *Invertebrate Zoology.* any pore or small opening through which eggs, sperm, or zygotes are released from the body.

genital ridge *Developmental Biology.* an elevation of thickened mesothelium and underlying mesenchyme on the ventromedial border of the embryonic mesonephros; the primordial germ cells become embedded in it, establishing it as the primordium of the testis or ovary. Also, GONADAL RIDGE.

genitals see GENITALIA.

genital scale *Invertebrate Zoology.* in Ophiuroidea (brittle stars), any of the small calcareous plates associated with the five buccal shields.

genital stage *Psychology.* in psychoanalytic theory, the final stage of psychosexual development, in which the desire for sexual gratification centers on a sexual relationship with another person. Also, **genital period, genital phase.**

genital tract *Anatomy.* the reproductive system.

genito- a combining form denoting a relation to the reproductive organs, as in *genitourinary.*

genitoplasty *Surgery.* plastic surgery of the genital organs.

genitourinary *Anatomy.* of or relating to the genital and urinary organs. Also, UROGENITAL.

genius *Psychology.* a popular term for a person who has exceptional creative abilities or intellectual powers, or who has a very high IQ.

genl general.

geno- a combining form meaning "gene," as in *genotype.*

Genoa cyclone *Meteorology.* a low-pressure system that seems to have developed around the Gulf of Genoa in the Ligurian Sea. Also, **Genoa low.**

genolectotype *Systematics.* a specimen selected after the original description of a genus to serve as the primary representative of the type species of the genus. Also, LOGOTYPE.

genome [jē′nōm′] *Genetics.* **1.** the complete gene complement of an organism, contained in a set of chromosomes (in eukaryotes), in a single chromosome (in bacteria), or in a DNA or RNA molecule (in viruses). **2.** the full set of genes in an individual, either haploid or diploid. In a human, the haploid set contains about 3 billion base pairs of DNA and 50,000–100,000 genes. Also, **genom.**

genomic [ji nōm′ik] *Genetics.* of or relating to the genome.

genomic DNA *Molecular Biology.* DNA of the genome; in bacteria it is a single circular loop, and in eukaryotes it is the entire haploid nuclear complement of DNA (chromosomes).

genomic formula *Genetics.* a method of expressing the number of sets of genetic instructions in an organism as a multiple of the haploid number, n; for example, diploid = $2n$, triploid = $3n$, tetraploid = $4n$, trisomy = $2n + 1$.

genomic library *Genetics.* a set of cloned fragments of DNA that together represent the entire genome.

genophobia *Psychology.* an irrational fear or hatred of sex.

genophore *Genetics.* the structural equivalent of chromosomes in viruses, prokaryotes, and certain organelles; contains nucleic acids but lacks the associated histones.

genospecies *Biology.* **1.** a homogenous line of descent occurring in homozygous self-fertilizing organisms, through inbreeding or cloning. **2.** all of the genotypes of a taxonomic species.

genotoxicity *Toxicology.* a measure of the potency of adverse effects of a toxin on DNA.

genotoxin *Toxicology.* any toxin that interacts with DNA, causing tumors, neoplasms, or mutations.

genotype [jen′ə tīp′; jē′nə tip′] *Genetics.* the sum total of the genetic information contained in an organism; the genetic constitution of a cell or organism. *Systematics.* the type species of a genus, or the type specimen of that genus.

gensym *Artificial Intelligence.* **1.** in Lisp, a function that generates a new symbol at runtime. **2.** a symbol so generated.

gentamycin *Microbiology.* any of a group of similar aminoglycoside antibiotics elaborated by certain species of the bacterial genus *Micromonospora* and effective against certain strains of *Pseudomonas aeruginosa.*

Gentianaceae *Botany.* a family of dicotyledonous herbaceous plants in the order Gentianales, characterized by showy flowers and a one-celled ovary; includes the gentians, buck beans, and other plants.

Gentianales *Botany.* an order of dicotyledonous herbaceous plants in the subclass Asteridae, commonly having showy flowers, an internal phloem system, and winged or tufted seeds.

gentiobiose *Biochemistry.* $C_{12}H_{22}O_{11}$, a disaccharide that yields two molecules of D-glucose upon hydrolysis; found naturally in plant glycosides such as amygdalin. Also, GLUCOPYRANOSYL-D-GLUCOSE.

gentisic acid *Organic Chemistry.* $C_7H_6O_4$, a crystalline solid, soluble in water, alcohol, and ether; melts with sublimation at 204°C; used in medicine as an analgesic and anti-inflammatory agent.

gentle breeze *Meteorology.* a term for a wind whose speed measures from 8 to 12 miles per hour on the Beaufort wind scale.

genu *plural,* **genua.** *Anatomy.* **1.** the knee; the point of articulation between the thigh and leg. **2.** broadly, any anatomical structure that is bent like the knee.

genucubital position *Surgery.* the position of an individual resting on the knees and elbows with the chest raised. Also, KNEE-ELBOW POSITION.

genufacial position *Surgery.* the position of an individual resting on the knees and face.

genupectoral position *Surgery.* the position of an individual resting on the knees and chest. Also, KNEE-CHEST POSITION.

genus [jē′nəs] *plural,* **genera.** *Systematics.* one of the principal or obligatory ranks in the taxonomic hierarchy, falling below the family and above the species; in binomial nomenclature, the genus name is always capitalized and italicized.

genus *Mathematics.* **1.** the genus of an orientable surface S (i.e., a compact 2-manifold) is the nonnegative integer $\eta(S) = 1 - \varepsilon(S)/2$, where $\varepsilon(S)$ is the Euler characteristic of S. Intuitively, a surface is of genus $p \geq 1$ if it can be distorted into (or is homeomorphic to) the surface obtained by adding p "handles" to a sphere (the plane and the sphere have genus 0). **2.** the genus $\eta(G)$ of a graph G is the genus of the orientable surface S of least genus, such that G can be topologically embedded in S (such an embedding is a minimal embedding). Clearly, $\eta(H) \leq \eta(G)$ for any subgraph H of G. Intuitively, a graph that can be drawn on a surface of genus g so that two edges of the graph intersect only at a vertex (i.e., without crossings), but that cannot be drawn on a surface of genus $g - 1$, is called a graph of genus g.

genus novum *Systematics.* in taxonomic literature, a notation made to denote a newly described genus. (A Latin term meaning "new genus.")

genu valgum *Medicine.* an inward curving of the knee, associated with greater distance than normal between the ankles.

-geny a combining form meaning "generation" or "origin."

geo *Geology.* a narrow, elongated cove or coastal inlet, walled in by steep, parallel rocky cliffs. Also, GIO.

geo- a combining form meaning "the earth," as in *geomorphology.*

geoacoustics *Acoustics.* the use of geometric echo-ranging with low-frequency seismographic waves transmitted several miles into the earth's crust to determine the geological composition and characteristics of the area. Thus, **geoacoustical.**

geoarchaeology *Archaeology.* the techniques of geology applied to archaeological issues such as dating methodology, mineral identification, soil and stratification analysis, and so on.

geobotanical prospecting *Geology.* the visual study of plant life, including distribution and morphology, to determine the possible presence of ore deposits, climatic conditions, soil composition, and so on.

geobotany *Botany.* the scientific study of the effects of various geological environments on plants. Thus, **geobotanist, geobotanical.**

Geocalycaceae *Botany.* a cosmopolitan family of medium, prostrate liverworts of the order Jungermanniales, characterized by ventral intercalary branches, scattered or ventral stem rhizoids, and large, well-developed underleaves.

geocentric *Astronomy.* of or relating to the earth as a reference point for other heavenly bodies. *Cartography.* of or relating to a measurement from the center of the earth to a specific point. *Geodesy.* the point at which the earth's axis of rotation intersects the plane of the celestial equator. *Anthropology.* describing a group or culture which holds the belief that the earth is the center of the solar system, rather than the sun.

geocentric coordinate system *Astronomy.* a system of celestial coordinates located with respect to the center of the earth. *Cartography.* a set of map coordinates that define the position of a point with respect to the center of the earth.

geocentric horizon *Cartography.* the plane through the center of the earth, parallel to the topocentric horizon.

geocentric latitude *Astronomy.* the angular distance in degrees that a celestial object lies north or south of the earth's equator. *Geodesy.* the angle formed with the major axis of the ellipse by the radius vector from the center of the ellipse to the given point.
geocentric longitude *Astronomy.* the angular distance in degrees that a celestial object lies east or west of the Greenwich meridian. *Geodesy.* see GEODETIC LONGITUDE.
geocentric parallax see DIURNAL PARALLAX.
geocentric theory *Astronomy.* **1.** the earlier theory presented by Ptolemy and others, postulating a motionless earth at the center of the solar system with the sun and the other planets revolving around it; contrasted with the modern heliocentric (sun-centered) theory. **2.** any view of cosmology in which the earth is central to the solar system.
geocentric zenith *Astronomy.* the point at which a line from the center of the earth through a point on its surface meets the celestial sphere.
geocerite *Mineralogy.* $C_{27}H_{53}O_2$ (approximately), a white, flaky, wax-like hydrocarbon occurring in brown coal.
geochemical of or relating to geochemistry.
geochemical anomaly *Geochemistry.* an unusually high concentration of one or more elements in rock, soil, vegetation, sediment, or water, or a high concentration of hydrocarbons in soil, that often indicates a nearby deposit.
geochemical balance *Geochemistry.* the study of the global distribution and migration of a particular element, mineral, or compound, including the amount liberated by weathering and transported to sediments and oceans.
geochemical cycle *Geochemistry.* the successive stages in the circulation and migration of chemical elements among the lithosphere, hydrosphere, and atmosphere during geologic changes.
geochemical evolution *Geochemistry.* **1.** over geologic time, any change in the chemical composition of some portion of the earth, such as the oceans. **2.** any alteration in the chemical composition of a rock in which the amount of a particular component exceeds the amount present in the parent rock.
geochemistry the study of the chemical elements, their isotopes, and related processes with respect to the abundance and distribution of materials within the earth's waters, crust, and atmosphere.

Geochemistry

Geochemistry is the study of the chemistry of the earth. It includes the study of the rocks and sediments that constitute the solid earth, as well as the fluids of the oceans, inland waters, and atmosphere. Geochemistry rests on a foundation of other sciences, including chemistry, geology, physics, and biology. The science includes the study of the movement, or flux, of chemical elements through the solids and fluid of the earth. Geochemistry has both a pure and an applied character; some geochemists study the materials of the earth primarily to obtain an understanding of the governing scientific principles, whereas others use their knowledge in such applied topics as the search for hidden deposits of economic minerals or the cleanup of contaminated water and soil.

Geochemistry is a relatively young discipline; the name "geochemistry" was apparently first used in 1838 by the Swiss chemist Christian F. Schonbein, the discoverer of ozone. Schonbein emphasized that the study of the chemical properties of the earth is as important as the study of the age of the geologic strata and the fossils contained therein. The foundation of modern geochemistry can probably be traced to the publication in 1908 of the classic *The Data of Geochemistry,* by Frank Wigglesworth Clark, an American mineralogist and chemist.

Donald D. Runnells
Professor and Chair of Geological Sciences
University of Colorado, Boulder

geochronological [gē´ō krän´ə läj´i kəl] *Science.* relating to or based on geochronology.
geochronology [gē´ō krə näl´ə jē] *Geology.* the study of the absolute and relative dating of structures and events in relation to the geologic history of the earth. Also, **geologic chronology.** *Archaeology.* specifically, the dating of archaeological data in association with a geological deposit or formation, such as the dating of Pleistocene human remains in the context of glacial advances and retreats.
geochronometry *Geology.* the measurement of an interval of time in relation to the history of the earth, using absolute or relative dating techniques. Thus, **geochronometric.**
Geocorisae *Invertebrate Zoology.* a suborder of land bugs in the order Hemiptera, with conspicuous antennae and an ejaculatory bulb in the male.
geocorona *Geophysics.* an envelope of ionized gases, mainly hydrogen, that encircles the earth to about 15 earth radii and emits Lyman-alpha radiation in the presence of sunlight.
geocosmology *Geology.* the study of the geologic origin and history of the earth.
geocronite *Mineralogy.* $Pb_{14}(Sb,As)_6S_{23}$, a white or greenish-white, opaque, metallic, monoclinic mineral usually occurring in massive, granular form, having a specific gravity of 6.4 and a hardness of 2.5 on the Mohs scale; found in hydrothermal veins with sulfides and other sulfosalts.
geod. geodesy; geodetic.
geode *Geology.* **1.** a rounded, more or less hollow concretion or separable nodule of rock lined with inward-projecting crystals that differ in composition from the enclosing rock. **2.** the crystal-lined cavity in such a hollowed rock.
Geodermatophilus *Bacteriology.* a genus of Gram-positive, aerobic bacteria occurring in soil, with a tendency toward mycelial growth.
geodesic [jē´ə des´ik] *Mathematics.* **1.** a curve of minimum length joining two points on a Riemannian manifold. **2.** let $_{\gamma}(t)$ be a curve in a C^{∞} manifold M, where t ranges over a real interval; the curve γ is geodesic if the tangent vector to γ is carried into itself under the parallel translation along γ defined by the connection on the manifold. *Cartography.* also, **geodesic line. 1.** the line of shortest distance between any two points on a mathematically defined surface. **2.** specifically, on the surface of the earth, a line of double curvature; usually lies between the two normal section lines determined by the two points. Also, GEODETIC LINE.
geodesically complete *Mathematics.* a manifold is said to be geodesically complete if its geodesics can all be extended to be defined for all parameter values.
geodesic coordinates *Mathematics.* any coordinate system in a neighborhood of a point P of a (pseudo-) Riemannian manifold having the property that the partial derivatives of the Riemannian metric g vanish at P with respect to the coordinates. In general, this property does not persist away from the point P unless the manifold is flat.

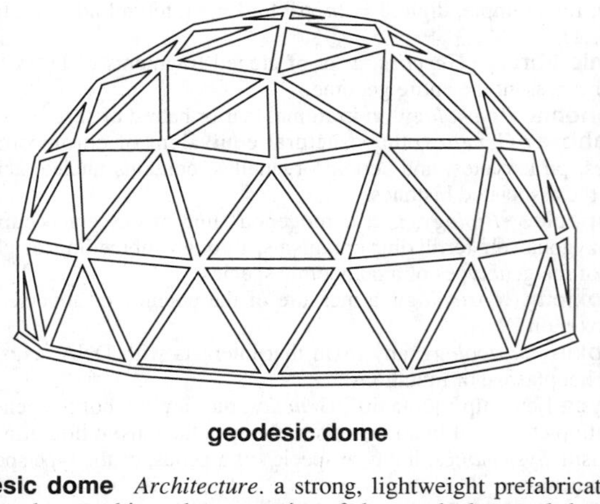

geodesic dome

geodesic dome *Architecture.* a strong, lightweight prefabricated enclosure that combines the properties of the tetrahedron and the sphere and contains no internal supports; it is made of standardized parts that allow quick assembly and dismantling. It was developed by the American engineer R. Buckminster Fuller.
geodesic motion *Physics.* the motion of a particle along a geodesic path in a four-dimensional space-time continuum.

Geodesy

Geodesy is the determination of the geometry of the earth's surface (both solid and liquid), including the time variability of this geometry.

When the surface of the earth is solid the geodetic interest is three-dimensional and involves determining the three-dimensional positions of points and monitoring changes in these positions as a function of time. Where the earth's surface is liquid the geodetic interest is in determining and monitoring the height coordinate as a function of time and horizontal position.

Because the objective of geodesy is geometric information, basic geodetic measurements are measurements of geometric quantities—distances, directions, and heights. The emphasis in geodesy is on differential measurements between points rather than absolute measurements relative to a coordinate system.

To accomplish its primary geometric objectives, geodesy has found it necessary to become involved in two other activities. These are: determination of the orientation of the earth in inertial space (e.g., earth rotation and polar motion) and determination of the earth's gravity field.

Determination of earth orientation is essential because a number of measurements of geometric quantities involve observations of extraterrestrial objects (artificial satellites, radio stars, etc.). These measurements cannot be interpreted without knowledge of earth orientation.

Knowledge of the earth's gravity field is required because a gravimetric equipotential surface which approximates near sea level, the geoid, serves as a reference surface for height determination; knowledge of the gravity field is essential to determination of the position of artificial satellites used in geodesy; and many geodetic measurements are made relative to a local coordinate system which is defined, in part, by the direction of the force of gravity at a point.

Increasingly, geodetic measurements are performed using space techniques. Global frameworks of stations are established using methods such as Very Long Baseline Interferometry, which uses radio telescopes to monitor incoming radio frequency signals from quasars, and Satellite Laser Ranging, which uses laser beams sent from ground instruments and reflected back from corner cubes on artificial earth satellites. These techniques allow worldwide positioning at the 1 to 2 cm level and determination of earth orientation.

For accurate positioning on a more local level, instruments that record the radio signals broadcast by the Global Positioning System Satellites launched by the U.S. Department of Defense are used. The vertical geometry of the ocean surface relative to the center of the earth is being determined using altimeters mounted on artificial satellites that determine the height of the satellite above the ocean surface.

William Strange
Chief Geodesist
National Geodetic Survey

geodesy [jē äd´ə sē] the study of the size and shape of the earth, the measurement of terrestrial gravitational forces, and the location of fixed points on the earth's surface. Thus, **geodesist.**

geodetic [jē´ə det´ik] *Geodesy.* of, relating to, or employing the theories, techniques, or results of geodesy.

geodetic astronomy *Geodesy.* the determination of longitudes, latitudes, azimuths, and points on the earth's surface by observation of the directions of stars, planets, the moon, and other celestial bodies.

geodetic azimuth *Geodesy.* the angle between the geodetic meridian and the tangent to the geodesic at the observer, measured in the plane perpendicular to the ellipsoidal normal of the observer, preferably clockwise from south.

geodetic constant *Geodesy.* any one of several values adopted by the International Association of Geodesy and the International Union of Geodesy and Geophysics and advocated for general use in geodetic calculations.

geodetic control *Geodesy.* a system of horizontal or vertical control stations that have been established and adjusted by geodetic methods and in which the shape and size of the earth (or the geoid) have been considered in position computations.

geodetic coordinates *Geodesy.* the quantities of latitude and longitude that define the position of a point on the surface of the earth with respect to the reference spheroid.

geodetic datum *Geodesy.* a reference surface forming the basis for the computation of horizontal-control surveys in which the curvature of the earth is considered.

geodetic equator *Geodesy.* the line of zero geodetic latitude; the great circle described by the semimajor axis of the reference ellipsoid as it is rotated about the minor axis.

geodetic graph *Mathematics.* a graph is said to be geodetic if, given any two vertices v_1 and v_2, there exists a unique path between v_1 and v_2 of minimal length.

geodetic gravimetry *Geodesy.* the science or technique of measuring gravity for the purpose of determining the size and shape of the earth.

geodetic height see ELLIPSOIDAL HEIGHT.

geodetic latitude *Geodesy.* the angle that the normal at a point on the reference spheroid makes with the plane of the geodetic equator. Also, TOPOGRAPHIC LATITUDE, GEOGRAPHIC LATITUDE.

geodetic leveling *Geodesy.* spirit leveling of a high order of accuracy, usually extended over large areas, to furnish accurate vertical control in the vertical dimension for all surveying and mapping operations; the leveling follows the geoid and its associated irregular level surfaces rather than any mathematically determined spheroid or ellipsoid and associated regular level surfaces.

geodetic longitude *Geodesy.* the angle between the plane of the geodetic meridian and the plane of an arbitrarily chosen initial meridian; it can be measured by the angle at the pole of rotation of the reference spheroid between the local and initial meridians, or by the arc of the geodetic equator intercepted by those meridians. Also, GEOCENTRIC LONGITUDE, GEOGRAPHIC LONGITUDE.

geodetic meridian *Geodesy.* a line on a reference ellipsoid that has the same geodetic longitude at every point. Also, GEOGRAPHIC MERIDIAN.

geodetic meridian plane *Geodesy.* a plane that contains the normal to the reference ellipsoid at a given point and the rotation axis of the reference ellipsoid.

geodetic parallel *Geodesy.* a line on the reference spheroid that has the same geodetic latitude at every point; a geodetic parallel, other than the equator, is not a geodesic line, but a smaller circle with its plane parallel to the plane of the geodetic equator. Also, GEOGRAPHIC PARALLEL.

geodetic position *Cartography.* the position of a point on the surface of the earth expressed in terms of geodetic latitude and geodetic longitude; an adopted geodetic datum is implied.

geodetic precession *Physics.* the movement of the axis of a spinning gyro in orbit about a heavenly body, caused by the curvature of space.

geodetic satellite *Space Technology.* an artificial satellite designed or launched especially for the purpose of gathering data regarding the size, shape, and mass distribution (gravity) of the earth as well as the precise positions of points or areas on its surface.

geodetic survey *Engineering.* a survey in which the form and size of the earth are considered; it is applicable for large areas and long lines; used to precisely locate basic points for use in controlling other surveys.

GEODSS *Ordnance.* a system used for tracking objects in space. (An acronym for ground-based electro-optical deep space surveillance.)

geodynamics *Geophysics.* the study of the processes in the earth's interior.

geoeconomics see ECONOMIC GEOGRAPHY.

geoelectric survey *Geodesy.* a survey to determine the electric potential of the earth's rocks, or resistivity.

geofact *Archaeology.* a mineral or rock resource that is found in an archaeological site; this often reveals evidence of a nearby quarry or exotic resource.

geoflex see OROCLINE.

geog geographic; geographical; geography.

Geoglossaceae *Mycology.* a family of fungi belonging to the order Helotiales composed of species commonly known as the earth tongue; found on soil, rotting wood, and plant leaves.

geoglyph [jē´ə glif´] *Archaeology.* any ground-constructed example of rock art, such as intaglios or rock alignments.

geognosy [je äg´nə sē] *Geology.* an earlier term for the science of the earth; now replaced by *geology.* (From Greek; literally "knowledge of the earth.")

geographer *Geography.* a person who studies or specializes in geography.

geographic *Geography.* of or relating to geography. *Geodesy.* of or relating to the earth considered as a general globular body, whether as the geoid, an ellipsoid, or a sphere.

geographical *Geography.* of or relating to geography. In general, the terms *geographic* and *geographical* are interchangeable, and compound terms listed here as beginning with one form can also be written with the other.

geographical botany see PLANT GEOGRAPHY.

geographical determinism *Anthropology.* a theory stating that culture can be explained by the geographical circumstances in which it is found; for example, the idea that a desert environment will produce a nomadic culture (e.g., the Taureg or the Inuit), since desert terrain facilitates movement and the lack of consistent rainfall stimulates such movement.

geographical information system *Computer Science.* a computer system specialized for storage, manipulation, and presentation of geographical information, such as topography, political subdivisions, geology, vegetation, flood plains, etc.

geographical mile *Metrology.* a distance, approximated by the nautical mile, equal to one minute of longitude at the equator of the earth (approximately 6087 feet).

geographical plot *Navigation.* a plot in which positions are displayed in terms of actual geographical position (as opposed, for example, to a relative plot).

geographical position *Astronomy.* a location on the surface of a planet, expressed in degrees of latitude and longitude. *Cartography.* such a location on the surface of the earth expressed in terms of latitude and longitude, either geodetic or astronomical.

geographic center *Geodesy.* the point on which a given area on the earth would balance, if the earth were a plate of uniform thickness.

geographic coordinates *Cartography.* a term used to designate both geodetic coordinates and astronomical coordinates. Also, TERRESTRIAL COORDINATES.

geographic cycle see CYCLE OF EROSION.

geographic latitude see GEODETIC LATITUDE.

geographic longitude see GEODETIC LONGITUDE.

geographic meridian see GEODETIC MERIDIAN.

geographic parallel see GEODETIC PARALLEL.

geographic range *Navigation.* the extreme distance at which an object or light can be seen as limited by the curvature of the earth and the heights of the object and the observer.

geographic region *Geography.* any portion of the earth's surface that has been delimited or recognized by a particular characteristic.

geographic search *Navigation.* a procedure in which search areas are assigned by geographical areas or sectors.

geographic sector search *Navigation.* a procedure in which individual search units are each assigned a radial search sector.

geographic speciation see ALLOPATRIC SPECIATION.

geographic square search *Navigation.* a procedure in which individual searching units are each assigned a square (or rectangular) search area.

geographic variation *Evolution.* any racial distinction associated with geographically distinct local populations among a large continuous population of a given species. Also, **geographical variation.**

geographic vertical see VERTICAL.

Geographos *Astronomy.* asteroid 1620, discovered in 1951 and having a generally cylindrical shape 4 km long by 1.5 km in diameter; it belongs to type S.

Geography

Central to geography is the concept of location. This is not to imply that geographers are obsessed with memorizing the coordinates of places on the face of the earth, any more than historians are consumed with memorizing lists of dates. In themselves such would be empty exercises. Not until we see locations in their global/environmental context can we understand their significance.

While Chicago is located at 42°N and 88°W, in itself this is hardly worth knowing. But in the global context that position tells us a great deal about Chicago. The wind patterns that affect its weather, its climate, and its vegetation are intimately tied to its position on earth. Its location on the shore of Lake Michigan, part of a great inland waterway, enhanced its development. Its location in the heart of an agricultural region with good connections to other places strengthened it as a distribution center. As its population and industries grew, Chicago became an industrial hub producing goods for a world market. Yet over time its population declined as businesses and people moved to the suburbs abandoning public transportation in favor of their automobiles. These kinds of dynamic interactions between the physical and human elements of places give meaning to the concept of location.

The ramifications of decisions made in one place are usually felt in others. Sometimes they are worldwide. Today we know that our discharged industrial chemicals have thinned the earth's protective ozone layer, posing serious hazards to our environment and to our very existence.

The geographer's task then is to investigate the myriad interconnections among places, the environment, and society to achieve an understanding of the integrated elements of this fragile earth. In so doing it becomes possible to predict the global consequences of the dynamic physical and human interchange.

Gilbert M. Grosvenor
President and Chairman
National Geographic Society

geography the scientific study of the surface of the earth, including such aspects as its climate, topography, vegetation, and population, as well as the effects on the earth's surface of human activity. *Science.* the topographical features of a region of the earth or another planet.

geohydrology *Hydrology.* the scientific study of subsurface water, especially the study of the geologic settings of underground water.

geoid [jē´oid] *Geodesy.* the equipotential surface in the gravity field of the earth that coincides with undisturbed mean sea level extended through the continents; this is used as the surface of reference for geodetic leveling.

geoid contour *Geodesy.* a line on the surface of the geoid that is of constant elevation with reference to the surface of the spheroid of reference.

geoid height or **geoid separation** *Geodesy.* the distance of the geoid above (positive) or below (negative) the mathematical reference spheroid. Also, UNDULATION OF THE GEOID.

geoisotherm *Geophysics.* a line connecting points of equal or constant temperatures on the earth.

geol geologic; geological; geology.

geolith see ROCK-STRATIGRAPHIC UNIT.

geologic *Geology.* of or relating to geology.

geologic age *Geology.* the absolute or relative age of a fossil, geologic event, or geologic structure in reference to the geologic time scale.

geological *Geology.* of or relating to geology. In general, compound terms listed here under *geologic* can also be written with *geological.*

geological oceanography *Geology.* the study of the composition, features, and processes of or related to the ocean floor and its margins. Also, SUBMARINE GEOLOGY, MARINE GEOLOGY.

geological survey *Geology.* **1.** a systematic study of the distribution, structure, composition, and history of the land features of a selected region. **2.** any organization that performs such a study.

geologic climate see PALEOCLIMATE.

geologic column *Geology.* **1.** a composite diagram that illustrates in columnar form, with the oldest rocks at the bottom, the sequence of stratigraphic units for a given region, or such a diagram illustrating the subdivisions of part or all of geologic time. **2.** the chronological or vertical arrangement of rock units illustrated by such a column. Also, STRATIGRAPHIC COLUMN.

geologic erosion see NORMAL EROSION.

geologic horizon *Geology.* see HORIZON.

geologic log *Geology.* a diagram of the lithological or stratigraphical units traversed by a borehole.

geologic map *Geology.* a map on which geological information, such as distribution, composition, and occurrences of bedrock, is recorded for a region.

geologic noise *Geophysics.* irregularities in observed data caused by local inhomogeneities in surrounding surface or shallow material.

geologic province *Geology.* an extensive region that is characterized by a similar geologic history, or by particular structural or physiographical features throughout.

geologic section *Geology.* a local geologic column, or any sequence of rock units found at or under the surface in a particular region. Also, STRATIGRAPHIC SECTION.

geologic structure *Geology.* see STRUCTURE.

geologic thermometer *Geology.* a mineral or aggregate of minerals whose presence defines the temperature range or limits within which the minerals were formed. Also, GEOTHERMOMETER.

geologic time *Geology.* the period of time from the end of the formation of the earth as a separate planet to the beginning of written history, as recorded and illustrated by the succession of rocks.

geologic time scale *Geology.* an arbitrary chronological arrangement of geologic events, usually in the form of a chart, used to show the absolute or relative age or duration of any portion of geologic time.

geologist *Geology.* a scientist who specializes in the study of the origin, composition, history, structure, and processes of the earth.

geology the study of the earth in terms of its development as a planet since its origin, including the history of its life forms, the materials of which it is made, the processes that affect these materials, and the products that are formed of them.

geom. geometric; geometrical; geometry.

geomagnetic *Geophysics.* of or relating to geomagnetism; having properties of geomagnetism.

geomagnetic coordinates *Geophysics.* a system of spherical coordinates based on the best fit of a centered dipole to the actual magnetic field of the earth.

geomagnetic cutoff *Geophysics.* the minimum energy needed for a cosmic ray particle to reach the top of the atmosphere at a specified geomagnetic altitude.

geomagnetic dipole *Geophysics.* the dipole created by the earth's magnetic field.

geomagnetic electrokinetograph *Engineering.* an instrument that is hung from a moving ship in order to calculate the direction and speed of ocean currents by measuring the voltage produced by the earth's magnetic field in the moving conductive seawater.

geomagnetic equator *Geophysics.* the great circle that bisects the earth's magnetic field and is perpendicular to the magnetic axis; the line of 0 geomagnetic latitude that is 90° from the geomagnetic poles.

geomagnetic field *Geophysics.* the magnetic field of the earth, with an intensity varying from about 0.25 tesla in a small region around northern Argentina to over 0.70 tesla near the south magnetic pole.

geomagnetic field reversal see GEOMAGNETIC REVERSAL.

geomagnetic interference see GEOMAGNETIC NOISE.

geomagnetic latitude *Geophysics.* the angular distance from the geomagnetic equator, measured northward or southward 90°.

geomagnetic longitude *Geophysics.* a longitudinal coordinate figured from a constant magnetic axis rather than from the earth's rotational axis.

geomagnetic meridian *Geophysics.* the meridional lines of a geomagnetic coordinate system.

geomagnetic noise *Geophysics.* any interference with radio communication caused by the earth's magnetic field. Also, GEOMAGNETIC INTERFERENCE.

geomagnetic pole *Geophysics.* either of two antipodal points marking the intersection of the earth's surface with the extended axis of a powerful bar magnet assumed to be located at the center of the earth and approximating the source of the actual magnetic field of the earth.

geomagnetic reversal *Geophysics.* a reversal of the polarity of the earth's magnetic field that has occurred in the past as indicated by the remanent magnetization found in igneous and sedimentary rock. Also, GEOMAGNETIC FIELD REVERSAL.

geomagnetic secular variation see SECULAR VARIATION.

geomagnetic storm *Geophysics.* a disruption of the geomagnetic field caused when ions unleashed by a solar disturbance reach the earth.

geomagnetic variation *Geophysics.* any change that occurs to the geomagnetic field, either short or long term.

Geology

To study the earth is to immerse oneself in the duality of history and process. This planet as an astronomical body is ancient, 4.6 billion years old, yet rocks of all ages, from a day-old lava flow to an almost 4-billion-year-old metamorphic rock, can be found in complex structural and stratigraphic arrays on the continents and sea floor. And the earth is being remade today, as it has been all through the history of the planet, by the eruptions of volcanoes, the uplifting of new mountain chains, the erosion of the land by rivers, wind, and ice, and the deposition of sediments in the sea. The constant reworking of the surface of the planet is a reflection of the dynamics of ponderous convective motions of the interior. Geology is thus the study of observable processes operating on earth today, their deeper causes, and the application of that knowledge to the reconstruction of earth history.

In the 200-year modern history of the science, geology has been transformed from its earlier primary task—the field exploration and mapping of geological formations on the continents—to a much more diverse view of the earth. Early in its development, geology became inextricably bound to paleobiology, the history of life on earth, and we have come to understand how organic evolution has profoundly affected the surface of the earth. Geology today is closely tied to geochemistry and geophysics. It now includes the geology of the sea floor as well as the continents. Geologists now study not only the solid earth, but also its interactions with the atmosphere and the oceans.

Using knowledge of modern processes we are able to reinterpret the planet's history, from its past plate tectonic movements and continental drift to the composition of ancient atmospheres and the shapes and currents of ancient oceans. From the study of the past, such as the history of recent continental glaciation, we can infer the longer-term planetary processes that have shifted earth from warm to cold climates. The present is not only the key to the past; the past helps us understand the present.

Raymond Siever
Professor of Geology
Harvard University

geomagnetism *Geophysics.* **1.** the various magnetic phenomena that are generated by the earth and its atmosphere, and by extension the magnetic phenomena in interplanetary space. **2.** the study of these phenomena.

geometric or **geometrical** *Mathematics.* of or relating to geometry; using the principles or methods of geometry. *Graphic Arts.* of a design, using or resembling the simple linear figures or forms associated with geometry.

geometrical acoustics see RAY ACOUSTICS.

geometric albedo *Optics.* the ratio of incident sunlight reflected by an object at zero phase angle to that which would be reflected by a perfectly reflecting, perfectly diffusing disk of the same size at the same distance.

geometrical dip *Cartography.* the vertical angle, at the eye of an observer, between the horizontal and a straight-line tangent to the surface of the earth; larger than dip by the amount of terrestrial refraction.

geometrical distortion *Computer Technology.* the difference in the vertical and horizontal dimension of the picture elements (pixels) in an electronic display that creates distortion in perceived objects; for example, a figure described by an equation for a circle may appear as an oval.

geometrical horizon *Cartography.* the intersection of the celestial sphere and an infinite number of straight lines tangent to the earth's surface and radiating from the eye of the observer.

geometrical isomer(ism) see GEOMETRIC ISOMER.

geometrical libration *Astronomy.* any of several oscillatory motions that allow little more than half the moon's surface to be visible at one time, including diurnal libration, libration in longitude, and libration in latitude.

geometrical optics *Optics.* the branch of physics that studies optical problems without regard to the effects of diffraction, interference, and polarization, instead assuming that light is a series of rays traveling in straight lines.

geometrical similarity *Fluid Mechanics.* a boundary-condition dependence between two types and magnitudes of fluid flow, wherein a simple change in velocity or dimensional scale renders a transformation from one type of flow to another.

geometric angle of attack *Aviation.* the normal angle of attack of a fuselage reference line, a root wing chord, or any geometric reference, as distinguished from the induced angle of attack or another angle of attack.

geometric apraxia see OPTIC APRAXIA.

geometric construction *Building Engineering.* any design or drafting process that utilizes only straightedges and circles.

geometric design *Transportation Engineering.* the clearances, curve radii, and other features of a highway, that determine its safe speed and capacity.

geometric distribution *Statistics.* the distribution of X independent and identical trials before the first successful event; if p is the probability of a success, $P(X = x) = (1 - p)p^{x-1}$.

geometric isomer *Physical Chemistry.* a compound that exists in two or more geometrically different configurations, as when atoms or groups of atoms are attached in different spatial arrangements on either side of a double bond or other rigid bond. Also, GEOMETRICAL ISOMER(ISM).

geometric mean *Mathematics.* for n quantities $a_1, \ldots, a_n$ the geometric mean is $[(a_1) \cdots (a_n)]^{1/n}$; i.e., the nth root of their product. If $a_1, a_2, \ldots, a_n$ are positive real numbers with arithmetic mean A and geometric mean G, then $A \geq G$.

geometric modeling *Design Engineering.* the design of three-dimensional objects through the use of computers which specify a workpiece in a geometrical pattern.

geometric number theory *Mathematics.* the branch of number theory that examines geometric properties of sets of n-tuples.

geometric pitch *Aviation.* the theoretical distance a propeller blade or component at a given angle would advance in one revolution were it not for distance loss due to slip; that is, if it moved along a helix having an angle equal to the blade angle.

geometric progression *Mathematics.* a sequence whose nth term has the form ar^{n-1}, where n is a positive integer, and a and r are nonzero constants; r is called the ratio or common ratio. Also, GEOMETRIC SEQUENCE.

geometric projection *Navigation.* a method of creating a map in which the points on the surface of a sphere are projected from a single point.

geometric ratio *Mathematics.* the ratio of geometric terms in a geometric progression.

geometric series *Mathematics.* any series whose individual terms form a geometric progression. The sum of the first n terms of such a series is

$$\sum_{k=1}^{n} ar^{i-1} = a(1-rn)/(1-r)$$

If $|r| < 1$, the infinite series converges and its sum is $a/(1-r)$; if $|r| \geq 1$, the series diverges.

geometric shadow *Physics.* a region which radiation would not reach, due to an obstruction, if the effects of diffraction could be neglected.

geometric structure *Design Engineering.* a particular characteristic of a robot, such as an arm, body, or end effector, which enhances its ability to perform various tasks.

geometrid *Invertebrate Zoology.* a member of the family Geometridae

Geometridae *Invertebrate Zoology.* a family of slender-bodied, broad-winged moths in the order Lepidoptera; its larvae are the caterpillars commonly called inchworms or cankerworms.

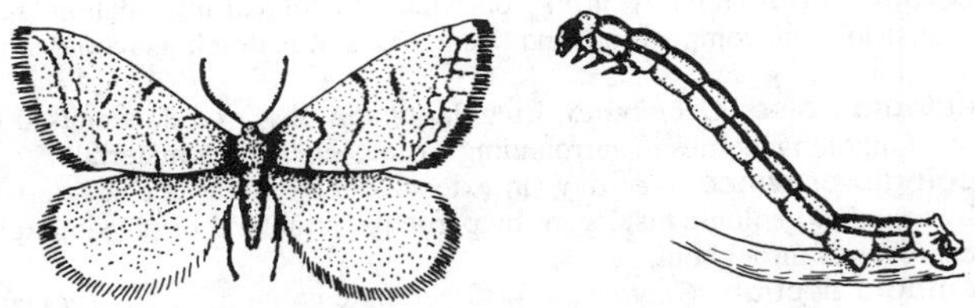

Geometridae (cankerworm)

geometrodynamics *Physics.* a theory using geometry that attempts to combine gravitational and electromagnetic theory.

Geometroidea *Invertebrate Zoology.* a superfamily of moths, lepidopteran insects in the suborder Heteroneura.

geometry *Mathematics.* the branch of mathematics dealing with those features of point configurations that are invariant under a specified group of transformations.

geomorphic cycle see CYCLE OF EROSION.

geomorphology *Geology.* the study of the surface configuration of the earth, especially the nature and evolution of present landforms, their relationships to underlying structures, and the history of geologic activity as represented by such surface features.

Geomyidae *Vertebrate Zoology.* the pocket gophers, a family of rodents of the suborder Sciuromorpha, characterized by a large flat head, a bulky body covered sparsely by stiff bristles, powerful forearms, large, curved incisors for digging extensive burrows, and fur-lined pouches extending from the face to the shoulder for carrying food; found in terrestrial and subterranean habitats from British Columbia south to Mexico and Panama, and in the southern U.S. east to Florida.

geonavigation *Navigation.* a navigational technique in which position is determined relative to terrestrial reference points, as distinguished from celestial or inertial navigation.

geonium *Atomic Physics.* a microscopic system composed of a single electron in a Penning trap that forms a synthetic atom.

geop see GEOPOTENTIAL SURFACE.

geopetal *Petrology.* of or relating to rock features that indicate top-to-bottom relations at the time of rock formation.

geopetal fabric *Petrology.* the internal structure that indicates the original orientation of the top-to-bottom strata in a stratified rock.

geophagous *Zoology.* feeding mainly or exclusively on soil.

geophilic *Ecology.* living or thriving in soil. Thus, **geophile, geophily.** *Botany.* of or relating to plants that fruit below the soil surface. Also, **geophilous.**

Geophilomorpha *Invertebrate Zoology.* centipedes, an order of arthropods in the class Chilopoda, with up to 180 pairs of legs and no eyes.

geophone *Electronics.* a seismic transducer that responds to motion of the ground at a location on or below the surface of the earth.

geophysical *Geophysics.* of or relating to geophysics.

geophysical engineering *Engineering.* a branch of engineering that utilizes physical and mathematical sciences to find mineral deposits.

geophysical prospecting *Engineering.* a method of geologic exploration that utilizes physical and mathematical sciences in measuring and identifying the composition and characteristics of a designated area of the earth's crust.

Geophysics

Geophysics as we see it in today's highly mathematical form did not take shape until the eighteenth century. However, we can date man's scientific interest in the structure of the earth and the behavior of its components back to Aristotle. There is even a twenty-five-century-old rumor that he committed suicide after his failure to explain the Mediterranean tides!

A too simple definition would be that geophysics is the application of the discipline of physics to the study of the earth, but that is not historically correct. The two subjects developed much in parallel. Both Volta and Franklin used geophysical phenomena to elucidate the understanding of electricity. Newton developed his universal law of gravitation from observations of the motions of the earth and the moon. Much of mathematical physics was a result of analysis of earth-bound problems such as the ocean tides.

It is true that today's laboratory physics is the principal route to understanding the behavior of the earth's interior. The properties of matter at high temperatures and pressures, such as viscosity, density and plasticity, are key to interpreting continental drift, sea floor spreading and the origins of the earth's magnetism. Even so, the earth remains a laboratory for fundamental physics, as the example of the several different searches being undertaken for the "fifth" force demonstrates.

The meaning of the word *geophysics* is undergoing change as well. The classical methods of geophysics are being applied to the planets now that we can reach them. Seismological techniques are being used to study the interior of the moon, and magnetic field measurements are important probes for the planets. The name will not change, however. It is a most encompassing science, ranging from petroleum exploration on earth to the understanding of the most distant planets.

William A. Nierenberg
Director Emeritus
Scripps Institution of Oceanography

geophysics [jē´ō fiz´iks] the study of the physical characteristics and properties of the solid earth, its air and waters, and its relationship to space phenomena.

geophyte [jē´ə fīt] *Botany.* a plant that survives adverse conditions by means of an underground food-storage organ, such as tubers or bulbs that sprout buds when conditions improve.

geopotential *Geodesy.* the gravity potential of the earth; the sum of its gravitational potential and the potential of its centrifugal force, in which the zero point of energy is defined to be located at sea level.

geopotential height *Geophysics.* a measure of the altitude of a point in the atmosphere expressed in terms of its potential energy per unit mass at that altitude, relative to sea level.

geopotential number *Geophysics.* a numerical value, C, given to a geopotential surface, expressed in geopotential units (gpu); nearly equivalent to height above sea level (1 gpu = 1 meter × 1 kilogal).

geopotential surface *Geophysics.* an area in which geopotential is everywhere equal; almost equivalent to height above sea level. Also, GEOP.

geopotential thickness *Geophysics.* the difference in geopotential height between two geopotential surfaces.

geopotential topography *Geophysics.* the contours of a geopotential surface as represented by lines of equal geopotential.

geopotential unit *Geophysics.* a unit of gravitational potential energy, 1 kgm, equal to the difference in potential energy of two points that are 1 m apart vertically when the strength of the gravitational field is 10 m per second squared.

geopressure *Geophysics.* a pressure of more than expected strength exerted on the earth by a subsurface formation.

GEOREF grid *Cartography.* a worldwide position-reference system that may be applied to any map or chart graduated in latitude and longitude (with Greenwich as the prime meridian), regardless of the projection. (Derived from the World Geographic Reference System.)

georgiadesite *Mineralogy.* $Pb_{16}(AsO_4)_4Cl_{14}O_2(OH)_2$, or $Pb_{16}(AsO_4)_4Cl_{14}(OH)_6$, an easily fusible, white or yellowish-brown monoclinic mineral occurring in small tabular crystals, having a resinous luster, a specific gravity of 6.3, and a hardness of 3.5 on the Mohs scale; found in altered lead slag.

georgiaite *Mineralogy.* a greenish-colored North American tektite found in the state of Georgia that is believed to be of extraterrestrial origin.

Georgian *Architecture.* English late-Renaissance architecture, the predominant style in 18th-century Britain and its North American colonies.

Georyssidae *Invertebrate Zoology.* a family of minute, mud-loving beetles, coleopteran insects in the Polyphaga.

geosensing *Botany.* the detection of or response to gravity by a plant in relation to its longitudinal axis.

geosere *Geology.* a sequence of climax communities that follow each other in geologic time as a result of changing climatic and physical conditions.

Geosiridaceae *Botany.* a monotypic mycotrophic herb family in the order Orchidales, characterized by an absence of chlorophyll, small scalelike leaves, and perfect trimerous flowers; native to Madagascar and other islands in the Indian Ocean.

geosmin *Organic Chemistry.* $C_{12}H_{22}O$, a colorless neutral oil with an earthy beet odor; boils at 270°C.

geosphere *Geology.* **1.** see LITHOSPHERE. **2.** the combination of the earth's lithosphere, hydrosphere, and atmosphere.

Geospizinae *Vertebrate Zoology.* the Darwin finches, a subfamily in some systems of New World seedeaters of the family Fringillidae.

geostatic pressure see GROUND PRESSURE.

geostationary orbit *Astronomy.* a satellite orbit in the plane of the earth's equator and 35,880 km above it, at which distance an object has an orbital period of exactly 24 hours and remains fixed in the sky as seen from a given location.

geostationary satellite see SYNCHRONOUS SATELLITE.

geostrophic [jē´ə sträf´ik; jē´ə strō´fik] *Geophysics.* of or relating to the deflective forces produced by the earth's rotation.

geostrophic approximation *Geophysics.* the use of the geostrophic wind as an approximation to the actual wind in operational forecasting or as a replacement for certain terms in the equations of motions. Also, **geostrophic assumption.**

geostrophic current *Geophysics.* a wind or ocean current in which the horizontal force is exactly balanced by the Coriolis force.

geostrophic departure *Meteorology.* the vector difference between the observed wind and the geostrophic wind speed. Also, AGEOSTROPHIC WIND.

geostrophic distance *Meteorology.* the distance along a constant-pressure surface over which the change in height is equal to the geostrophic wind speed.

geostrophic equation *Geophysics.* a formula that computes the speed of the geostrophic current.

geostrophic equilibrium *Geophysics.* the state of a moving, nonviscous liquid in which the force of the horizontal pressures exactly balances the horizontal Coriolis force at all points in the field.

geostrophic flow *Geophysics.* a flow gradient in which the horizontal pressure force exactly balances the Coriolis force.

geostrophic flux *Meteorology.* the movement of an atmospheric property by the geostrophic wind.

geostrophic vorticity *Meteorology.* the whirling effect of the geostrophic wind.

geostrophic wind *Meteorology.* the horizontal wind velocity for which the coriolis acceleration exactly balances the horizontal pressure force.

geostrophic wind level *Meteorology.* the lowest level, according to the Ekman spiral theory, in which the wind becomes geostrophic; in practice this is between 1.2 and 1.6 km, and is assumed to mark the upper limit of the frictional influence of the earth's surface. Also, GRADIENT WIND LEVEL.

geostrophic wind scale *Meteorology.* a graphical means of determining the speed of the geostrophic wind from the contour-line spacing on a synoptic chart.

geosynchronous orbit [jē´ō sin krän´əs] *Astronomy*. an orbit identical to a geostationary orbit except that the satellite's orbit does not necessarily lie in the earth's equatorial plane; thus its celestial position traces a closed figure in the sky.

geosynclinal couple see ORTHOGEOSYNCLINE.

geosynclinal facies *Geology*. uniformly deposited sedimentary facies formed in a deep-water, marine geosyncline and identified by above-average thickness, a generally argillaceous character, and a scarcity of carbonate rocks.

geosyncline [jē´ō sin´klīn] *Geology*. an extensive, basin-shaped, mobile downward subsidence of the earth's crust, caused by the deposition of considerable thicknesses of sedimentary and volcanic rocks over millions of years.

geotaxis *Physiology*. the movement of an animal in response to gravitational forces.

geotechnics *Civil Engineering*. the practical application of geological science to civil engineering problems. Thus, **geotechnical.**

geotectogene see TECTOGENE, def. 1.

geotectonic cycle see OROGENIC CYCLE.

geotectonics see TECTONICS.

geothermal *Geophysics*. relating to the internal heat of the earth.

geothermal gradient *Geophysics*. the graded rate of temperature change in soil and rock from the surface to the interior of the earth; on the average, about +10°C per kilometer.

geothermal system *Geology*. a localized geological environment in which circulating steam or hot water carries some of the earth's natural internal heat flow close enough to the surface to be gathered for use.

geothermometer see GEOLOGIC THERMOMETER.

geothermometry *Geology*. **1.** the direct or indirect measurement or approximation of the temperatures at which geologic processes take place or have taken place. **2.** the study of the earth's heat and temperatures, and their effects on geologic processes.

Geotrichum *Mycology*. a genus of fungi belonging to the class Hyphomycetes, causing such human diseases as geotrichosis, which produces lesions in the respiratory and digestive tracts and in the mouth.

geotropic [jē´ō trōp´ik] *Botany*. relating to or displaying geotropism.

geotropism [jē´ō trōp´iz əm] *Botany*. the response of a plant to gravity, as evidenced by growing patterns such as downward root growth and growth curvature. Also, GRAVITROPISM.

Gephyrea *Invertebrate Zoology*. a group of large, burrowing, marine worms including the Echiuroidea, Sipunculoidea, and Priapuloidea.

gephyrocercal *Vertebrate Zoology*. of a fish, having dorsal and anal fins smoothly joined at the aborted tail end of the vertebral column.

ger- a combining form meaning "old age," as in *geriatrics*.

Geraniaceae *Botany*. a family of dicotyledonous herbs and shrubs of the order Geraniales that includes the geranium, producing aromatic oils and having lobed or compound leaves with stipules and regular flowers.

Geraniales *Botany*. an order of dicotyledonous herbs and shrubs of the subclass Rosidae, characterized by flowers with a superior ovary and compound or lobed leaves.

geraniol *Organic Chemistry*. $C_{10}H_{18}O$, an unsaturated monoterpene alcohol found in many essential oils; a colorless or pale yellow oil with a geraniumlike odor; soluble in alcohol and insoluble in water; boils at 230°C; used in perfumes and as a constituent of synthetic fragrances.

geranium *Botany*. **1.** any of the many plants of the genus *Geranium*. characterized by regular flowers with alternating glands and petals and including the wild geranium, *G. maculatum*, of North America. **2.** any of the various plants of the allied genus *Pelargonium*, characterized by showy flowers of scarlet, pink, or white, and fragrant leaves; widely cultivated in gardens and as houseplants.

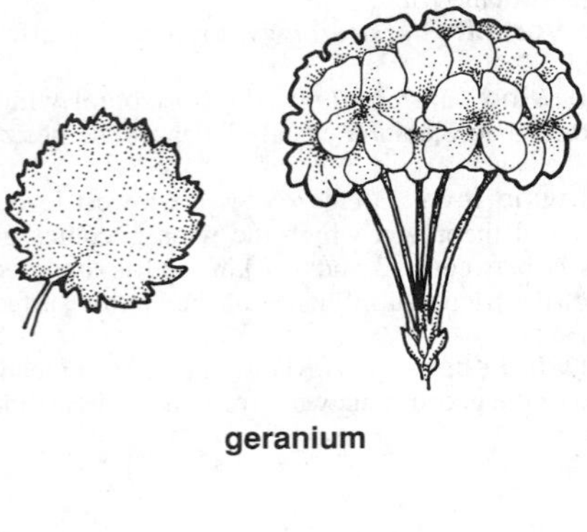

geranium

Gerardiidae *Invertebrate Zoology*. a family of small, colonial, skeletonless anthozoans in the order Zoanthidea.

Gerard reagent *Chemistry*. any of various compounds, principally quaternary ammonium compounds, that are used to separate sex hormones from urine, and ketones from fatty natural substances.

gerbil [jur´bəl] *Vertebrate Zoology*. **1.** any of numerous small burrowing rodents of the genus *Gerbillus* and related genera, having long legs used for jumping; found in Asia, Africa, and southern Russia. **2.** a jird, *Meriones unguiculatus*, used in scientific research and often kept as a pet.

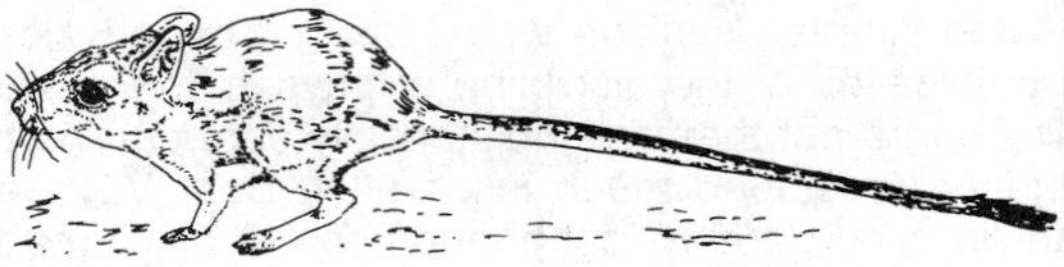

gerbil

Gerbillinae *Vertebrate Zoology*. the gerbils, a subfamily of the rodent family Cricetidae, characterized by a long, sparsely haired tail and long hindlegs.

gerhardtite *Mineralogy*. $Cu_2^{+2}(NO_3)(OH)_3$, an easily fusible, transparent, dark-green orthorhombic mineral occurring as tabular crystals, having a specific gravity of 3.4 to 3.43 and a hardness of 2 on the Mohs scale; found as a secondary mineral in oxidized zones of copper deposits.

Gerhardt's test *Pathology*. **1.** a measure of acetoacetic acid in urine, performed by adding ferric chloride, which turns it deep red, then adding sulfuric acid, after which the red disappears. **2.** a test for bile in urine, performed by shaking together urine and an equal amount of chloroform and potassium hydroxide; a yellowish-brown color is produced. (Named after Charles F. *Gerhardt*, 1816–1856, French chemist.)

geriatrics [jer´ē a´triks] *Medicine*. the branch of medical science concerned with conditions associated with aging, including the clinical problems of senescence and senility. Thus, **geriatric.**

GERL *Cell Biology*. a system of intracellular membrane-bound organelles and vesicles involved in protein synthesis, secretion, and degradation in eukaryotic cells. (An acronym of the Golgi apparatus, the endoplasmic reticulum, and the lysosomes.)

germ *Biology*. **1.** a part of an organism that is capable of developing into a new organism. **2.** an embryo, especially that of wheat. *Medicine*. any microorganism that causes disease, such as a bacterium. In technical use, the term *pathogen* is usually preferred to *germ*. *Mathematics*. a germ of differentiable functions at a point x on a manifold X is an equivalence class of functions that are differentiable in a neighborhood of x; two functions f and g belong to the same equivalence class if $f = g$ in some neighborhood of x. The set of all germs at a given point x forms an algebra and is denoted $G(x)$. (Going back to a Greek word meaning "grown" or "produced;" related to words such as *gene* and *generation*.)

germane *Inorganic Chemistry*. **1.** a compound composed of germanium and hydrogen. **2.** see GERMANIUM TETRAHYDRIDE.

germanide *Inorganic Chemistry*. a compound composed of germanium and any alkali metal or alkaline earth metal (periodic table Group I or Group IIa).

germanite *Mineralogy*. $Cu_{26}Fe_4Ge_4S_{32}$, a dark reddish-gray, opaque, cubic mineral usually occurring in massive form, having a metallic luster, a specific gravity of 4.46 to 4.59, and a hardness of 4 on the Mohs scale; found with tennanite, enargite, pyrite, galena, azurite, and malachite.

germanium *Chemistry*. a nonmetallic element having the symbol Ge, the atomic number 32, an atomic weight of 72.59, a melting point of 937.4°C, and a boiling point of 2830°C; a grayish-white solid recovered from zinc refining and used as a semiconductor in electronic devices. (Named for *Germany*, in which it was discovered.)

germanium halide *Inorganic Chemistry*. any compound composed of germanium and one of the halogen elements (periodic table Group VIIa), such as chlorine or bromine.

germanium oxide *Inorganic Chemistry*. GeO_2, a white powder used in phosphors, electronics, and infrared transmitting glass. Also, **germanium dioxide.**

germanium tetrahydride *Inorganic Chemistry*. GeH_4, a highly toxic, colorless gas; insoluble in water; it melts at –165°C, boils at –88.3°C, and decomposes at 350°C.

germanium transistor *Electronics.* a germanium semiconductor device having three or four sections or layers of N-type and P-type material; it has various uses such as amplifiers, oscillators, and switches.
German measles see RUBELLA.
German measles virus see RUBELLA VIRUS.
German R unit *Nucleonics.* a unit quantity for the dose of radiation caused by an X-ray, roughly equivalent to 1.5 roentgens per second. Also, R UNIT.
German silver *Materials.* a ductile copper-nickel-zinc alloy used to make utensils, drawing instruments, and the like. (It contains no silver, but is so named for its silvery appearance.) Also, NICKEL SILVER.
germarium *Invertebrate Zoology.* the portion of the ovary that produces eggs and the portion of the testis that produces sperm, in Rotifera and Platyhelminthes.
germ band *Invertebrate Zoology.* the band or thickening of the blastoderm of an insect egg, from which the embryo grows.
germ cell *Genetics.* in a multicellular animal, a gamete or any other cell capable of producing a gamete.
germ-free animal *Bacteriology.* **1.** an animal that is delivered, usually by caesarean section, and reared in a germ-free isolator. **2.** any animal that has no demonstrable, viable microorganisms living in close association with it.
germ-free isolator *Microbiology.* **1.** a sterile chamber in which an animal is delivered and reared in the complete absence of microorganisms. **2.** a device that provides an artificial barrier surrounding an area in which germ-free animals are housed.
germinal epithelium *Developmental Biology.* **1.** a layer of celomic epithelial cells covering the gonadial ridges as they are formed on the medial border of the mesonephroid near the root of the mesentery. **2.** the mesothelial covering of the definitive ovary.
germinal infection *Medicine.* an infection transmitted to an infant by the ovum or sperm of a parent.
germinal membrane see BLASTODERM.
germinal mutation *Genetics.* a modification of the hereditary material in cells that will develop into sperms or eggs.
germinal vesicle *Cell Biology.* the large nucleus of a primary oocyte that divides to form the female pronucleus; it is often arrested in a postsynaptic stage of meiotic prophase I.
germination *Botany.* the beginning of growth or development in a seed, spore, or zygote, especially after a period of dormancy. *Petrology.* see GRAIN GROWTH.
germ layer *Developmental Biology.* one of the three primordial cell layers (ectoderm, endoderm, mesoderm) present in an embryo, formed during gastrulation.
germ-layer theory *Developmental Biology.* the theory that the embryo forms three primary germ layers, each layer developing into specific organ derivatives.
germ line *Genetics.* in most multicellular animals, a lineage of cells that contain the complete genome of the individual and produce the reproductive cells that transfer the genome to the next generation. *Cell Biology.* **1.** the sequential development of a cell from a zygote to a germ cell. **2.** the genes as they appear in germ cells, which may be rearranged in somatic cells.
germ plasm *Biology.* the hereditary material contained in the gametes that is transmitted to the offspring.
germ sporangium *Mycology.* a fungal spore case, formed by a zygospore, that develops at the tip of a germ tube.
germ theory *Medicine.* the theory that contagious and infectious diseases are caused by microorganisms, associated with the research of Pasteur, Koch, and others.
germ tube *Mycology.* a filament in fungi that develops as a result of germination and usually becomes a hypha.
gero- a combining form meaning "old age," as in *geroderma.*
geroderma *Medicine.* the skin of old age, characterized by atrophy and loss of fat.
geronto- or **geront-** a combining form meaning "old age."
gerontology [jer´än täl´ə jē] *Physiology.* the study of the effects and problems of aging. Thus, **gerontological, gerontologist.**
gerontomorphosis *Biology.* a degree of evolutionary specialization of a species that decreases its ability to adapt and ultimately leads to its extinction.
Gerreidae *Vertebrate Zoology.* the mojarras, a family of coastal marine fishes of the order Perciformes that feed on bottom-dwelling organisms, characterized by a pointed snout with a protrusible mouth and very shiny scales on the head and body.
Gerrhosauridae *Vertebrate Zoology.* a subfamily of the girdle-tailed lizard family Cordylidae, with some species lacking front limbs; found in Africa and Madagascar.
Gerridae *Invertebrate Zoology.* the family of water striders in the order Hemiptera; long-legged insects often found on the surface of slow-moving streams and ponds.
Gerroidea *Invertebrate Zoology.* a single superfamily of hemipteran insects in the subdivision Amphibicorisae, characterized by conspicuous antennae and hydrofuge hairs covering the body.
gersdorffite *Mineralogy.* NiAsS, a brittle, opaque, silver-white to steel-gray cubic mineral of the cobaltite group usually occurring as octahedral crystals or in granular aggregates, having a metallic luster, a specific gravity of 5.9, and a hardness of 5.5 on the Mohs scale; found in uraninite and in galena, chalcopyrite, and pyrite ore.
Gerstner wave *Fluid Mechanics.* a rotational gravity wave having a finite amplitude that is produced by a set of exact theoretical equations; the practical limitations are the presumption of no irrotational flow and the assumption that the fluid is of infinite depth.
Gesell, Arnold [gə sel´] 1880–1961, American psychologist; noted for studies of child development.
Gesell development(al) schedule *Psychology.* a system of measuring childhood development that includes assessment of motor skills, cognition, language, and social behavior. (From Arnold *Gesell,* its developer.)
Gesner, Abraham 1797–1864, Canadian geologist; a pioneer in economic geology.
Gesner, Konrad von 1516–1565, Swiss physician and naturalist; author of *Historia animalium,* considered the founding work in the field of zoology.
Gesneriaceae *Botany.* a family of largely tropical dicotyledonous plants belonging to the order Scrophulariales and including the African violet and the gloxinia; characterized by perfect, often large and showy flowers and usually opposite, simple leaves.
Gestalt or **gestalt** [gə shtält´; gə stält´] *Psychology.* **1.** a basic pattern or structure that gives something its identity as a unified whole, rather than as the sum of its parts. **2.** of or relating to Gestalt psychology. *Meteorology.* a complex of weather elements that occurs in a familiar form and that may persist as an entity for a significant period of time, such as a warm front and its associated pattern of cloudiness and precipitation. (A German word meaning "form" or "structure.")
Gestalt perception *Psychology.* the capacity in an animal to recognize a stimulus combination or pattern, as opposed to recognizing only individual components.
Gestalt psychology *Psychology.* a school of psychology that originated in Germany in the early 1900s, based on the principle that perception and behavior exist as whole, unified patterns, which cannot be analyzed in terms of their individual elements nor understood as the sum of these elements. Thus, **Gestalt psychologist, Gestaltist.**
Gestalt therapy or **Gestalt psychotherapy** *Psychology.* a therapeutic approach that emphasizes the development of self-awareness and personal responsibility and a sense of unity in one's life.
gestate [jes´tāt´] *Developmental Biology.* to carry and maintain the young in the uterus from fertilization until birth.
gestation [jes tā´shən] *Developmental Biology.* the process, state, or period of gestating.
gestational [jes tā´shə nəl] *Biology.* occurring in or affecting gestation.
gestational age *Medicine.* the age of a fetus or newborn, expressed in weeks dating from the first day of the mother's last menstrual period.
gestational diabetes *Medicine.* a form of diabetes mellitus, occurring in pregnancy and disappearing after delivery, but often recurring years later; characterized by an impaired ability to metabolize carbohydrate.
gestation period *Developmental Biology.* the period of development of the young of viviparous animals from fertilization until birth; in human females, this time averages about 266 days.
gestosis *Medicine.* any toxemic manifestation of pregnancy.
getter *Physical Chemistry.* a substance that binds gases to its surface, maintaining or increasing vacuum conditions in a vacuum tube.
getter-ion pump *Engineering.* a type of high-vacuum pump in which chemically active metal deposits on the pump walls absorb active gases and ionize inert gases so that positive ions are drawn to the wall; the neutralized ions become covered by fresh deposits of metal and are either continuously or alternately deposited on the pump wall.
getter sputtering *Electronics.* the application of a vaporized metal to remove unwanted impurities remaining in a vacuum.
GeV or **Gev** giga-electron volt.

geyser *Hydrology.* a natural hot spring or fountain that periodically discharges a column of hot water and steam into the air.
geyserite see SILICEOUS SINTER.
G factor *Nucleonics.* see G VALUE. *Psychology.* see GENERAL FACTOR.
g-factor see LANDÉ G FACTOR.
g force *Physics.* a force whose magnitude equals the gravitational force acting on a body at sea level.
GG or **G.G.** gamma globulin.
GH growth hormone.
GHA Greenwich hour angle.
gharbi *Meteorology.* any fresh ocean breeze that blows in a westerly direction in Morocco.
gharra *Meteorology.* sudden, heavy squalls that pass over Libya and Africa from the northeast and are usually accompanied by heavy rain and thunder.
ghibli *Meteorology.* a hot, dust-bearing, usually south to southeasterly desert wind that blows in northern Africa.
Ghon complex *Medicine.* the combination of pulmonary and lymph node focus of tuberculosis in children.
ghost a mere shadow or trace; specific uses include: *Electronics.* on a television screen, an undesired duplicate image offset somewhat from the desired image due to a delayed, reflected signal. *Petrology.* the discernible outline of a former crystal shape, fossil, or partially obliterated rock structure bounded by inclusions, bubbles, or foreign material. Also, PHANTOM. *Ordnance.* in passive detection, an intersection of lines of direction that does not indicate a real target. *Cell Biology.* a red blood cell membrane from which the cytoplasmic components have been released by cell lysis, so that only the cell membrane is observed in microscopic examination. Also, **ghost cell.**
ghost algebraic manipulation language *Computer Programming.* a language that enables a programmer to process mathematical expressions without concern for numeric values of the mathematical symbols.
ghost dance *Anthropology.* a messianic religious movement among certain North American Indian tribes of the late 1800s, initiated by the Paiute prophet Wovoka, involving a series of ritual dances and stressing spiritual revival and return to the traditional way of life.
ghost image *Spectroscopy.* a false spectral line resulting from irregularities in the rulings on a diffraction grating.
ghost mode *Electromagnetism.* an undesirable mode of oscillation in a waveguide due to an imperfection in the wall of the waveguide.
ghost particles see EMPTY PARTICLES.
ghost signal *Electronics.* an unwanted signal on the screen of a radar indicator, such as an echo that experiences multiple reflections before reaching the receiver. Also, **ghost pulse.**
ghost spot *Plant Pathology.* a disease of tomato plants in which small white rings are formed on the surface of the fruit.
GHQ or **G.H.Q.** general headquarters.
GH-RH growth hormone-releasing hormone.
GHz gigahertz.
GI or **G.I.** galvanized iron; gastrointestinal; general issue; government issue.
Giacobinids see DRACONIDS.
Giaever, Ivar born 1929, American physicist; shared Nobel Prize for experiments in superconductivity.
giam *Materials.* the ligh-brown to dark-brown, hard, durable wood of the trees of the genus *Hopea* of Southeast Asia, used in the construction of boats.
giant see HYDRAULIC MONITOR.
giant anteater see ANTEATER.
giant cell *Microbiology.* any of the very large multinucleate cells that are present during certain diseases, such as measles.
giant cell arteritis *Medicine.* an arterial inflammatory disease manifested by giant (multinucleated) cells that affect the carotid artery branches.
giant condyloma acuminatum see BUSCHKE-LOWENSTEIN TUMOR.
giant fiber *Invertebrate Zoology.* a very large nerve fiber in many invertebrates that transmits motor impulses faster than ordinary nerves, triggering escape behavior.
giant granite see PEGMATITE.
giant nuclear resonances *Nuclear Physics.* strong excitations of atomic nuclei that appear as electric or magnetic oscillations in dipole or quadrupole modes arising from absorption of high-energy gamma radiation.
giant panda see PANDA, def 1.
giant powder *Materials.* a type of blasting powder.

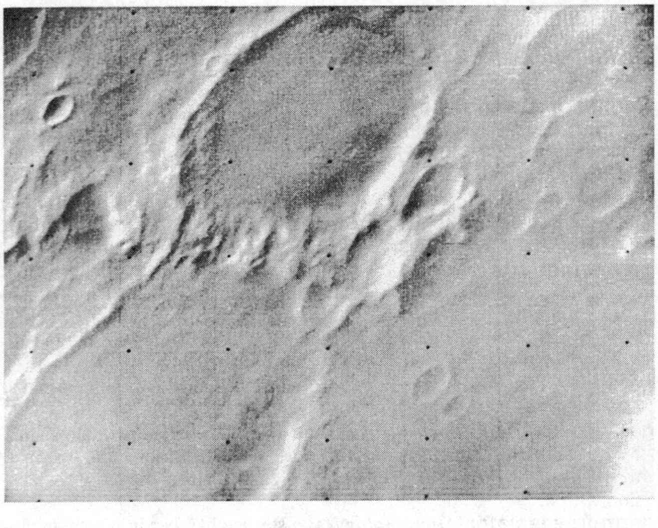

Giant's Footprint

Giant's Footprint *Astronomy.* two adjacent craters on the south polar cap of Mars, an area approximately 85 × 200 miles at 75° south latitude.
giant's kettle or **giant's cauldron** *Geology.* a cylindrical hole in the bedrock underlying a glacier bored by water that falls through a deep moulin or by boulders that rotate in the bed of a meltwater stream. Also, MOULIN POTHOLE, POTASH KETTLE.
giant squid *Invertebrate Zoology.* a cephalopod mollusk of the genus Architeuthis, the largest invertebrates, up to 55 feet in length.

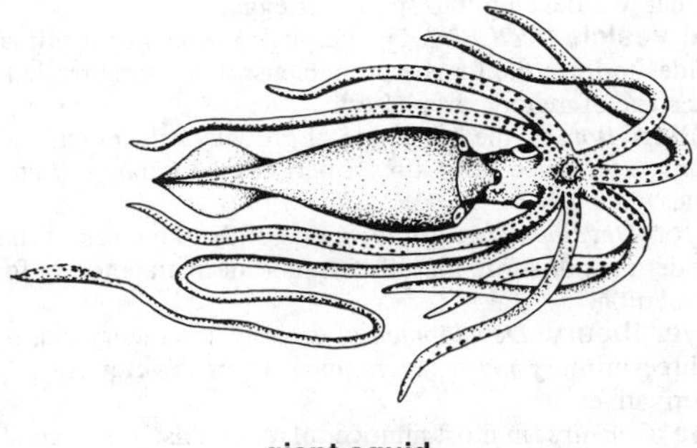

giant squid

giant star *Astronomy.* a large, luminous star with an extended tenuous photosphere around a hot core that has depleted much of its hydrogen.
Giaque, William Francis [zhyäk] 1895–1982, American chemist; with Herrick Johnston, discovered isotopes of oxygen; Nobel Prize for work in low-temperature thermodynamics.
Giaque's temperature scale *Thermodynamics.* an international temperature scale based on absolute temperature units in which the triple point of water is defined to occur at 273.16 K.
Giardia *Invertebrate Zoology.* a genus of flagellate protozoans in the order Diplomonadida.living in the intestines of mammals; may cause diarrhea in humans. (From Albert *Giard,* 1846–1908, French biologist.)

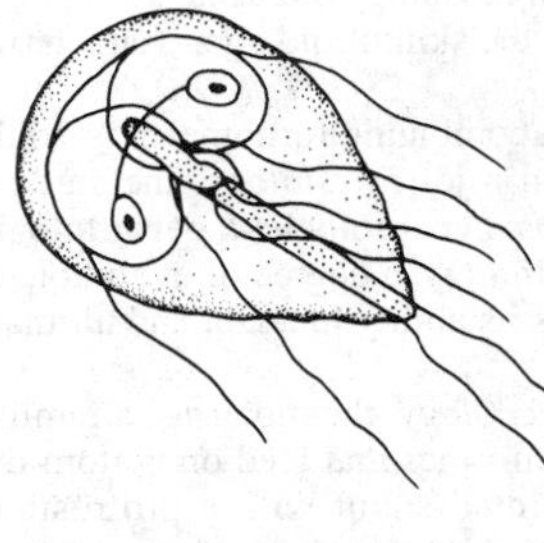

Giardia

Human Life

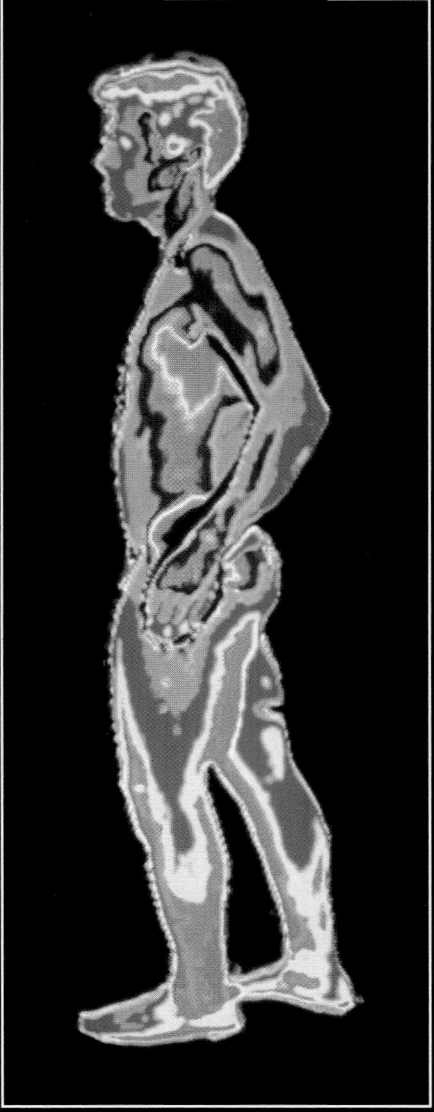

Thermogram of the human body.
Howard J. Sochurek Inc./The Stock Market.

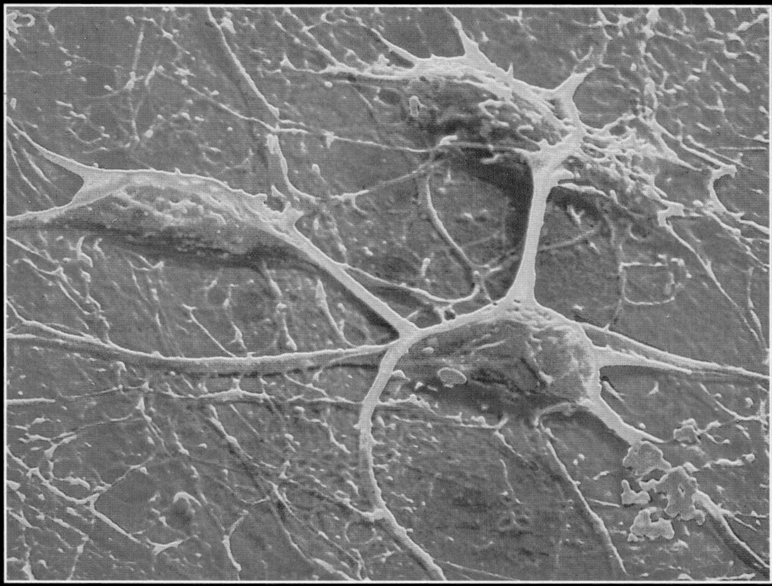

False-color scanning electron micrograph of three human neurons of the cerebral cortex, including the axons and dendrites. SPL/Photo Researchers.

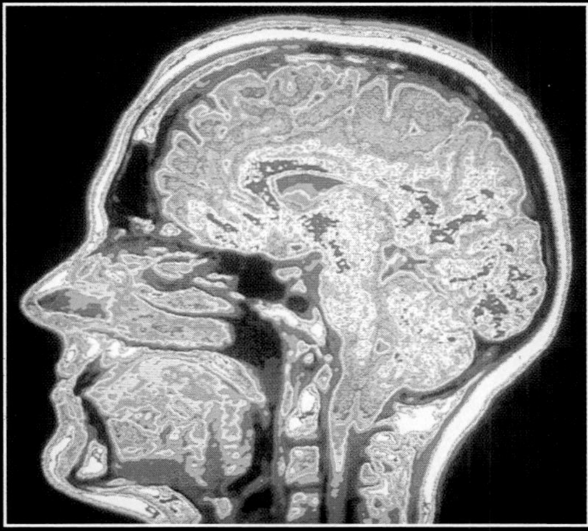

Magnetic resonance imaging scan of the human brain.
Howard J. Sochurek Inc./The Stock Market.

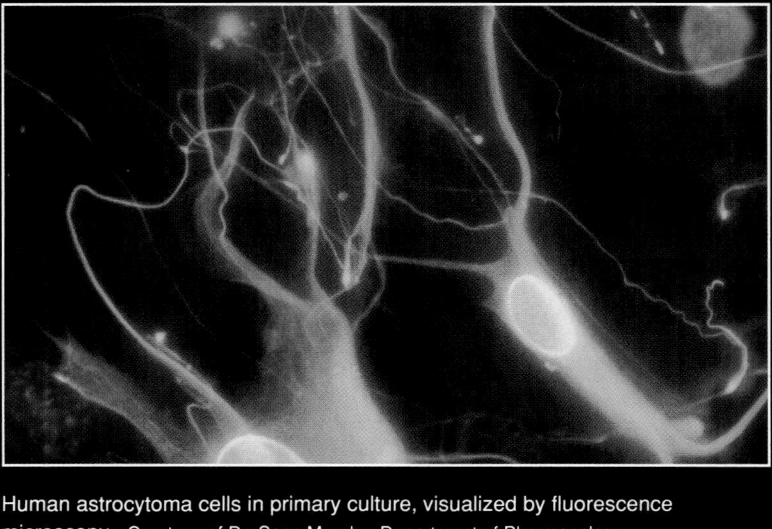

Human astrocytoma cells in primary culture, visualized by fluorescence microscopy. Courtesy of Dr. Sean Murphy, Department of Pharmacology, University of Iowa College of Medicine.

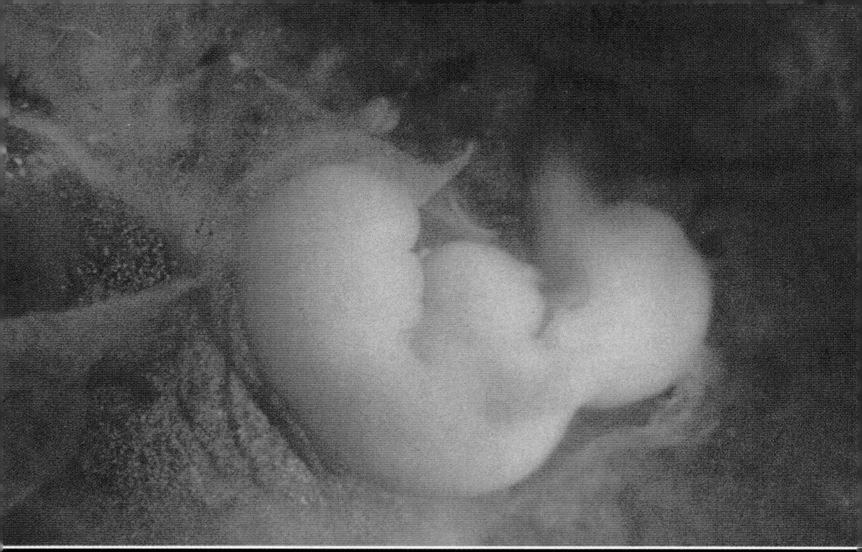

Human embryo after 28 days of development, with the head on the left.
Petit Format/Photo Researchers.

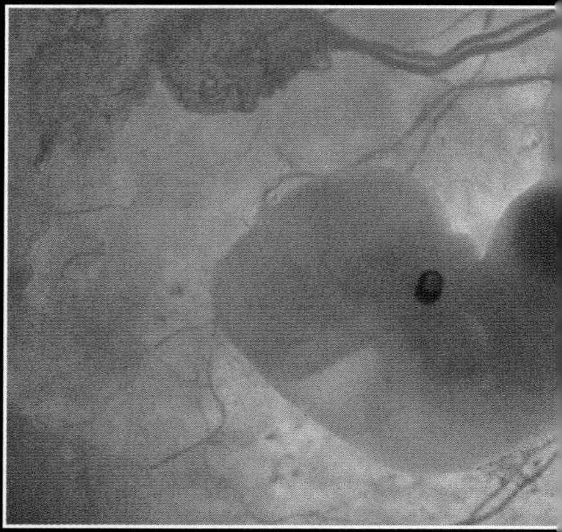

Human embryo after 7 weeks of development
Petit Format/Photo Researchers.

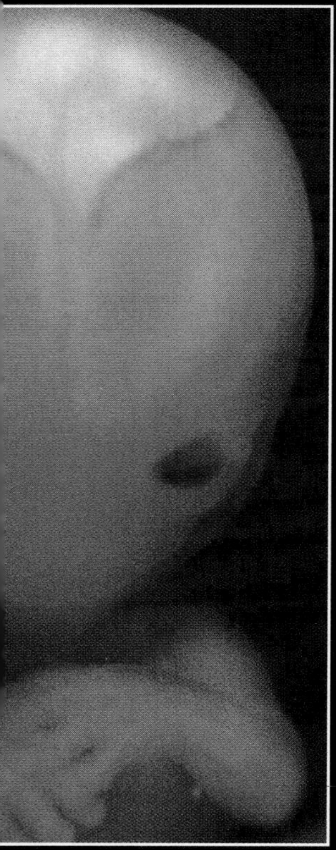

human fetus at 3 months.
rs.

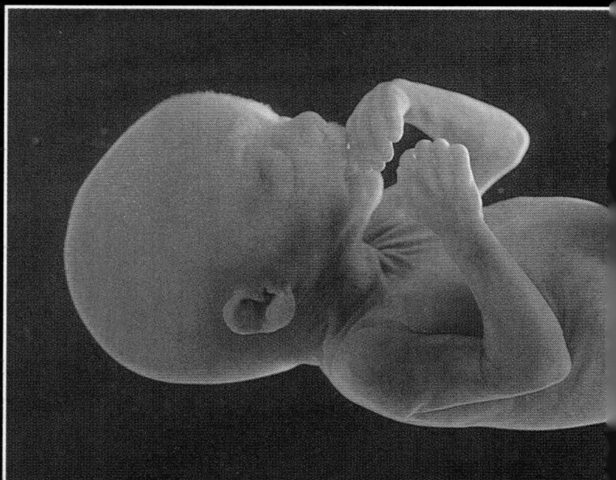

giardiasis *Medicine.* the presence of the protozoa *Giardia lamblia* in the small intestine, often without symptoms, but occasionally causing diarrhea.

gib [jib] *Engineering.* a detachable plate used to clamp parts into place and limit their range of motion. *Building Engineering.* in carpentry or ironwork, a heavy metal strap used to fasten two members together.

Gibberella *Mycology.* a genus of fungi, belonging to the order Hypocreales, that is responsible for numerous plant diseases.

Gibberella fujikuroi *Mycology.* a species of fungi, belonging to the genus *Gibberella,* that produces gibberellins, hormones that regulate plant growth and are used commercially in horticulture.

gibberellic acid *Biochemistry.* $C_{19}H_{22}O_6$, the parent compound of the gibberellins, prepared commercially by culture filtration of *G. fujikuroi.*

gibberellin *Biochemistry.* any of a group of widely occurring plant hormones that stimulate extension growth of stems and leaves.

gibbon [gib´ən] *Vertebrate Zoology.* any of various small, arboreal, acrobatic lesser apes of the genera *Hylobates* and *Symphalangus,* having extremely long arms, a small round head with wide-spaced nostrils, and a slender body with silky, dense fur; found in Southeastern Asia, Sumatra, Java, and Malaysia.

gibbon

gibbous [jib´əs] *Medicine.* a hump or swelling on a body surface.

gibbous moon *Astronomy.* the phases of the moon lying between First Quarter and Last Quarter, during which the moon looks lopsided.

Gibbs, Josiah Willard 1839–1903, American mathematical physicist; founder of chemical thermodynamics and modern physical chemistry; developed phase rule and vector analysis.

gibbs *Physics.* a unit of the amount of gas adsorbed onto a surface, equal to 10^{-6} moles per square meter.

Gibbs adsorption equation *Physical Chemistry.* a formula stating that the concentration of a solute at the surface of a liquid is greater than in the bulk liquid when added solute decreases surface tension, and less than in the bulk liquid when the added solute increases surface tension.

Gibbs adsorption isotherm *Materials Science.* an equation used to determine how surface concentration is related to surface tension in a two-component system.

Gibbs and deMarzio theory *Materials Science.* a configurational entropy model suggesting that the phenomenon of glass transition has an underlying thermodynamic basis.

Gibbs diaphragm cell *Chemical Engineering.* a cell with graphite electrodes and a cylindrical shape; used for chlorine production.

Gibbs-Donnan equilibrium see DONNAN EQUILIBRIUM.

Gibbs-Duhem equation *Physical Chemistry.* an equation, restricted to binary systems, used to calculate the change in the chemical potential of one component of a uniform solution, provided the activity of the other components is known over a range of compositions. Also, DUHEM'S EQUATION; GIBBS-DUHEM RELATION.

Gibbs elasticity *Physics.* a numerical measurement of the elasticity of a liquid film, given by (2A)*dg*/*dA,* where *A* is the surface area and *g* is the surface tension of the liquid.

Gibbs free energy *Thermodynamics.* a thermodynamic property *G* that is related to the capacity of a given system to do work, as expressed by $G = H - TS = U - TS + pV$, where *H* is the enthalpy, *T* is the absolute temperature, and *S* is the entropy of the system. Also, **Gibbs function.** *Physics.* see FREE ENTHALPY.

Gibbs-Helmholtz equation *Physical Chemistry.* an equation used to determine the effect that temperature and pressure have on the ability of a chemical reaction to perform useful work.

gibbsite *Mineralogy.* $Al(OH)_3$, a white to grayish to greenish to reddish monoclinic mineral, polymorphous with bayerite, doyleite, and nordstrandite, occurring in massive form and as tabular crystals, having a vitreous luster, a specific gravity of 2.4, and a hardness of 2.5 to 3.5 on the Mohs scale; found as an alteration product of aluminum-bearing minerals in bauxite and laterite deposits.

Gibbs paradox *Physics.* a predicted increase in the entropy of a system of two mixing gases that apparently persists even when the gases are identical, a situation in which there can be no observable mixing and thus no entropy change.

Gibbs phase rule *Physical Chemistry.* the expression $F + P = C + 2$, which demonstrates that *F,* the number of degrees of freedom or independent variables such as temperature or pressure, plus *P,* the number of coexisting phases such as ice or vapor, equals *C,* the number of components such as water or salt, plus 2; this equation applies generally to multicomponent thermodynamic systems at equilibrium. Also, PHASE RULE.

Gibbs' phenomenon *Mathematics.* let $\{f_n\}$ be a sequence of continuous functions converging to a real-valued function *f* (or a sequence of transformations of *f*), with $A = \lim_{x\to c} \inf_{n\to\infty} f_n(x)$ and $B = \lim_{x\to c} \sup_{n\to\infty} f_n(x)$. If the interval [*A*, *B*] contains points outside the interval $[\liminf_{x\to c} f(x), \limsup_{x\to c} f(x)]$, then the sequence $\{f_n\}$ is said to exhibit a Gibbs phenomenon at $x = c$. In particular, if f_n represents the *n*th partial sum of a Fourier series for *f* and *f* has a jump discontinuity at *c,* then the length of the minimal interval containing the Gibbs set at *c* is greater than the magnitude of the jump (the length of the interval [*A*, *B*]).

Gibbs sampler *Statistics.* an iterative simulation procedure used to obtain quantities that are difficult to compute in bayesian inference.

Gibbs set *Mathematics.* for a sequence $\{f_n\}$ of continuous functions converging to a real-valued function *f,* the Gibbs set at a point *c* is the set of all possible values of $\lim_{n\to\infty} f_n(x)$ as $x \to c$.

Gibbs system *Physics.* a collection of hypothetical systems forming an ensemble.

Gibbs-Thomson equation *Materials Science.* a formula used to calculate the free energy in emulsion polymerization; involving temperature, volume, and solvent.

giboulee see GALERNE.

Gibraltar, Strait of *Geography.* a narrow (8-mile) channel between Spain and Morocco, connecting the Atlantic Ocean with the Mediterranean Sea.

gid *Veterinary Medicine.* a nervous disorder of sheep, goats, and sometimes cattle, caused by infestation of the brain with the larval form of the tapeworm *Multiceps multiceps,* and characterized by forced circling movements and an unsteady (giddy) gait. Also, COENUROSIS.

Giegy-Hardisty process *Chemical Engineering.* a method of producing sebacic acid from castor oil by reaction of the acid at a high temperature with caustic alkali.

Giemsa stain *Chemistry.* a stain consisting of eosin dyes and methylene blue, used to mark hemopoietic tissues and hemoprotozoa.

GIFT *Genetics.* an in vitro procedure in which sperm and unfertilized eggs are inserted into the Fallopian tube. (An acronym for gamete intra-Fallopian transfer.)

giga- a combining form meaning "one billion" (10^9). Thus, **gigahertz, gigawatt.**

gigabit [gig´ə bit´] *Computer Technology.* one billion bits.

gigabyte [gig´ə bīt´] *Computer Technology.* a unit of storage capacity equal to 2^{30} (roughly, one billion) bytes.

gigaflop [gig´ə fläp´] *Computer Technology.* one billion floating-point operations per second; used as a measure of computer speed.

gigant- or **giganto-** a combining form meaning "huge," as in *gigantism, gigantosoma.*

gigantism [jī gan´tiz əm] *Medicine.* **1.** excessive growth, resulting in a height greater than 78–80 inches. **2.** excessive size, as of cells or nuclei.

Gigantopithecus *Paleontology.* a genus of the hominoid primate Pongidae, known from finds of the late Miocene in India and the Pleistocene in China; so-called for its large molars, which were three times larger than those of modern humans.

Giganturoidei *Vertebrate Zoology.* the giganturids, a monofamilial suborder of mesopelagic marine fishes of the order Cetomimiformes, having large, tubular eyes and a jellylike tissue beneath the skin.

Gigartina *Botany.* a type genus of red algae of the family Gigartinaceae, characterized by fleshy or cartilaginous fronds with numerous outgrowths; found mainly in the Pacific Ocean.

Gigartinaceae *Botany.* an economically important family of red algae of the order Gigartinales, characterized by branched gametangial thalli that arise from a discoid crustose base and by spermatangia formed in superficial clusters; used as a major source of carrageenans in various parts of the world.

Gigartinales *Botany.* an order belonging to the red alga division Rhodophyta, characterized by widely varying thalli, a carpogonium arising from the apical cell of an undifferentiated thallus filament, and transfer of a zygote nucleus to an auxiliary cell after fertilization.

Gigaspermaceae *Botany.* a family of very small mosses of the order Funariales, characterized by an underground creeping stem and erect bulbiform branches with soft, whitish-green leaves; plants are adapted to xeric, bare soil; found in the Southern Hemisphere.

GIGO [gī´gō´] *Computer Programming.* see GARBAGE IN, GARBAGE OUT.

gigohm *Electricity.* a unit of measurement equivalent to 10^9 ohms.

Gila monster or **gila monster** [hē´lə] *Vertebrate Zoology.* a venomous lizard, *Heloderma suspectum,* characterized by a thick body and tail, beaded orange skin with black patches, short legs, and five-toed feet with strong claws; found in desert areas of western North America, ranging from southern Nevada and Utah to Sonora in northwest Mexico. (First identified in the region of the *Gila* River of southern Arizona.)

Gila monster

Gilbert, Walter born 1932, American biochemist; shared Nobel Prize for mapping nucleotide sequence (structure of DNA).

Gilbert, William 1544–1603, English physician and scientist; author of *De Magnete,* pioneering study of electricity and magnetism.

gilbert *Electromagnetism.* a unit of magnetomotive force equal to the magnetomotive force of a closed loop of one turn in which there is a current of 1/4 (3.14) abamp.

Gilbert circuit *Electronics.* a circuit that uses the logarithmic properties of diodes and transistors to compensate for nonlinearities and instabilities in monolithic variable-transconductance circuits.

Gilbreth, Frank 1868–1924 and his wife **Lillian** 1878–1972, American industrial engineers; pioneers in techniques of scientific management.

Gilbreth's micromotion study *Industrial Engineering.* a pioneering study of basic human work motions and times, leading to the identification of elemental motions.

gilding *Graphic Arts.* the process of coating an image with a gold-colored material such as leaf.

gilding metal *Metallurgy.* a copper-base alloy containing 5% zinc.

gill *Vertebrate Zoology.* the respiratory structure of all fishes, found in sets of varying quantities and consisting of a multifolded surface membrane (for greater surface area) heavily supplied with blood vessels under a thin wall to allow for gas (i.e., oxygen) exchange. *Mycology.* in fungi belonging to the subdivision Basidiomycotina, a structure, resembling knife blades, that is located on the underside of mushroom caps and bears the spores. Also, LAMELLA. *Metrology.* **1.** a unit of liquid measure used in the U.S., equal to 4 fl oz or 118.29 ml. **2.** a similar unit of measure used in Great Britain, equal to 5 imperial fl oz or 142.07 ml. Also, IMPERIAL GILL.

gill cover *Vertebrate Zoology.* a movable, external flap following the contours of the gill chamber downward and forward beneath the jaws, behind the cheek region of most teleost fishes. Also, OPERCULUM.

Gillespie equilibrium still *Analytical Chemistry.* a recirculating distillation device used in the process of determining the azeotropic properties of liquids.

gillespite *Mineralogy.* $BaFe^{+2}Si_4O_{10}$, a red, tetragonal mineral, massive in habit, having a specific gravity of 3.3 to 3.4 and a hardness of 3 to 4 on the Mohs scale; found with quartz and celsian.

Gilliland correlation *Chemical Engineering.* a method for approximating distillation-column calculations, in which the reflux ratio and number of plates for the column are correlated and presented graphically as functions of minimum reflux and minimum number of plates.

gill net *Engineering.* a curtainlike mesh net suspended from a fishing boat to catch fish by ensnaring their gill covers in the mesh.

gill raker *Vertebrate Zoology.* a comblike skeletal structure arranged in rows, found on the gill bar of the fish gill and extending inward in the cavity of the pharynx to act as a screen against food particles and other solid substances.

gilpinite see JOHANNITE.

gilsonite *Mineralogy.* a very brittle, jet-black variety of asphaltite, having a conchoidal fracture, a specific gravity of 1.05 to 1.1, and a hardness of 2 to 2.5 on the Mohs scale; found only in the U.S. and used in waterproof coatings, lacquers, and mineral wax. Also, UINTAHITE.

gilt *Agriculture.* a female swine that has not produced a litter.

gimbal [jim´ bəl; gim´bəl] *Engineering.* **1.** a mechanical mounting frame having two mutually perpendicular axes of rotation. **2.** a support component of a gyro, which allows the spin axis to move freely. **3.** to mount an object on a gimbal

gimbaled inertial system *Navigation.* an inertial system in which the inertial sensors are mounted on gimbals and are thus unaffected by changes in the orientation of the craft.

gimbal mount *Mechanical Engineering.* a mount equipped with gimbals, commonly used to support a nautical compass or gyroscope.

Gimenez stain *Microbiology.* a method for the bacteriological staining of rickettsia cells, in which heat-fixed cells are stained green with malachite green oxalate.

gimlet *Mechanical Devices.* a small, sharp boring tool with a pointed, threaded, and fluted tip; used in woodworking.

gimmick *Electricity.* a term for a twisted, two-conductor cable that serves as a variable capacitive load, and in which capacitance is adjusted by untwisting each conductor.

gin *Agriculture.* **1.** to remove the seeds of cotton from the lint. **2.** a machine that performs this process, or a building in which the process is performed. Also, COTTON GIN. *Mechanical Engineering.* a three-legged hoisting machine equipped with pulleys, ropes, and a windlass.

gin block *Naval Architecture.* a heavy-duty metal block, usually with open metal cheeks.

G indicator see G DISPLAY.

ginger *Botany.* a plant of the species *Zingiber officinalis;* a perennial tropical herb of the family Zingiberaceae, having thick, scaly rhizomes that are ground to produce an aromatic yellow spice that is used in cooking and in medicine; native to the East Indies but now cultivated in most tropical countries.

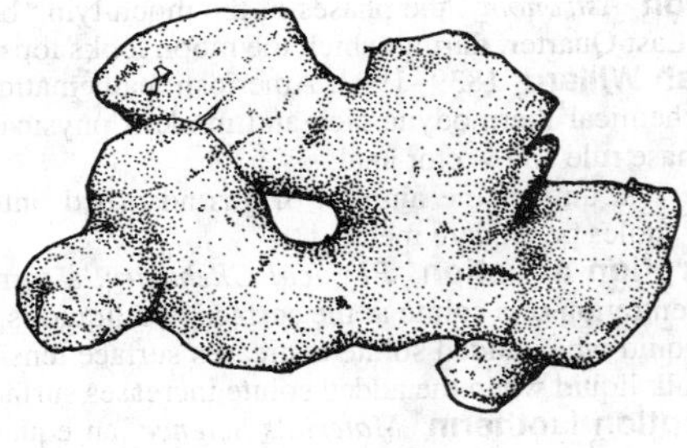

ginger

ginger blotch *Plant Pathology.* a disease of the ginger tree, possibly caused by the fungus *Pseudomonas fluorescens,* which is characterized by irregular, discolored markings on plant parts.

gingiv- or **gingivo-** a combining form meaning "gums," as in *gingivitis, gingivoplasty.*

gingiva [jin´jə və] *Anatomy.* the gums; the tissue around the necks of the teeth.

gingival [jin´jə vəl] *Anatomy.* of or relating to the gums.

gingival crevice *Anatomy.* the narrow space between the gingiva and the surface of a tooth.

gingivitis [jin´jə vī´tiəs] *Medicine.* an inflammation of the gingiva; when associated with bony changes it is referred to as periodontitis.

gingivostomatitis *Medicine.* an inflammation of the gingiva and oral mucosa.

Ginglymodi *Vertebrate Zoology.* a superorder of fishes including the gars of the order Lepisosteiformes.
ginglymoid joint *Anatomy.* a type of synovial joint that allows movement in only one plane forward and backward. ALSO, **ginglymous.**
Gini's mean difference *Statistics.* a measure of variability based on consideration of all possible pairwise differences between data points.
ginkgo [ging´kō] *Botany.* a popular, ornamental shade tree characterized by woody twigs, large fan-shaped leaves with parallel veins clustered on spur shoots, and a dicotyledonous seed with a fleshy coat; the sole surviving species of the gymnosperm family Ginkgoales, native to China and existing almost exclusively in cultivation.
Ginkgoopsida *Botany.* a class of deciduous trees in the subdivision Pinicae represented by the single order Ginkgoales. Also, **Ginkgoatae.**
ginorite *Mineralogy.* $Ca_2B_{14}O_{23}\cdot 8H_2O$, a white, monoclinic mineral occurring in rhomboid plates and pellets, having a specific gravity of 2.07 to 2.09 and a hardness of 3.5 on the Mohs scale; found as pellets in colemanite-veined basalts.
ginseng [jin´seng] *Botany.* **1.** any of the plants of the genus *Panax*, particularly the eastern Asian *P. pseudoginseng*, or North American *P. quinquefolius*, characterized by an aromatic root that is used medicinally. **2.** the root of this plant, or a preparation made from it. Also, **genseng.**

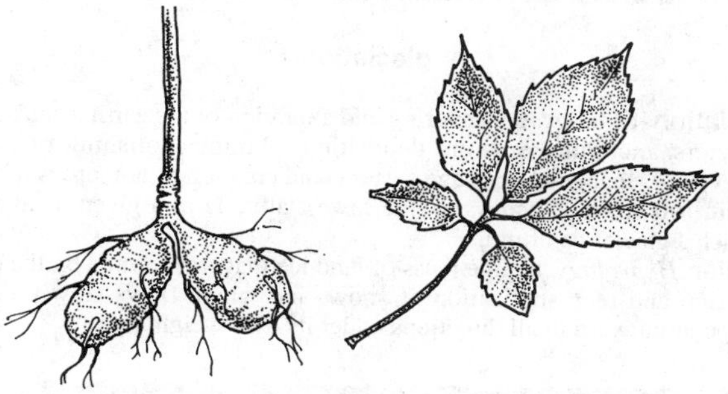

ginseng

Ginzburg-Landau superconductivity theory *Solid-State Physics.* a modification on the London theory of superconductivity that gives the spatial variation of the superfluid component of electrons; it gives the boundary energy.
Ginzburg-Landau theory *Physics.* a phenomenological theory of superconductivity that accounts for the coherence length.
GIP *Oncology.* a proto-oncogene that contains point mutations and is found in carcinoma of the ovary and adrenal gland.
gipsy see GYPSY.
giraffe *Vertebrate Zoology.* a very tall ruminant mammal, *Giraffa camelopardalis,* having a small head with a long narrow snout, a long neck, long legs and tail, and a light-colored coat patterned with dark patches; the tallest living quadruped animal, it is found in savannahs of tropical Africa where it feeds on the leaves and twigs of acacia trees.

giraffe

Giraffidae *Vertebrate Zoology.* the giraffes and okapis, a family of pecoran ruminants of the order Artiodactyla.
Giraudiaceae *Botany.* a monotypic family of densely tufted, brown algae of the order Dictyosiphonales, characterized by dense filaments attached by creeping rhizoids; found in the North Atlantic, the Mediterranean, and the Adriatic Sea.
Girbotal process *Chemical Engineering.* an absorption process for removing carbon dioxide, hydrogen sulfide, and various acid impurities from refinery gases and natural gas, using an amine as the reagent and absorbent.
girder *Building Engineering.* a large beam made of wood, metal, or reinforced concrete that is used to support joists or walls over an open area.
girdle a structure or band that encircles, limits, or confines; specific uses include: *Anatomy.* **1.** any curved or circular structure, such as the hipline. **2.** any structure that encloses or encircles another structure. *Invertebrate Zoology.* **1.** in chitons, the muscular, bristled outer edge of the mantle. **2.** in diatoms, one of the bands separating the two valves. **3.** in some dinoflagellates, the furrow circling the center of the shell. *Plant Pathology.* a plant disease in which a growth, such as a canker, surrounds a stem or branch and completely cuts off water and the nutrient supply, causing death of the plant part. *Petrology.* on a fabric diagram or equal-area projection, a common-width band of points that represent fabric element orientations.
girdle-tailed lizard *Vertebrate Zoology.* a terrestrial lizard of the Cordylidae family of Africa, having large, keeled scales encircling the heavy tail which is used in defense as a spiny club.
Girondian *Geology.* a European geologic stage of the Lower Miocene period, occurring after the Chattian and before the Langhian, and including the Burdigalian and the Aquitanian.
girt *Civil Engineering.* **1.** a timber in the second-floor corner post of a house that serves as a foundation for the roof rafters. **2.** horizontal timbers used to stiffen the frame of a building. *Engineering.* a strip of material that fits horizontally between the legs of a drill tripod or derrick, serving as a brace.
girth *Mathematics.* the girth of a graph is the length of its shortest cycle. By convention, the girth of a forest is infinite. *Science.* in general, the measure around something.
GIS *Computer Science.* geographical information system.
GI series *Medicine.* a familiar name for gastrointestinal series, an X-ray examination of the upper or lower gastrointestinal tract following an intake of barium sulfate as a contrast medium.
gismondine *Mineralogy.* $Ca_2Al_4Si_4O_{16}\cdot 9H_2O$, an easily fusible, colorless to white, monoclinic mineral of the zeolite group occurring as twinned pseudotetragonal crystals, having a specific gravity of 2.24 to 2.27 and a hardness of 4.5 on the Mohs scale. Also, **gismondite.**
githagin *Toxicology.* a poison found in the corn cockle, *Agrostemma githagin.*
githagism *Toxicology.* poisoning due to ingestion of githagin; symptoms may include severe gastrointestinal disturbances, depressed respiration, and nervous tremors.
gitogenin *Pharmacology.* $C_{27}H_{44}O_4$, a drug used as a tonic for the heart; produced by heating gitonin with hydrochloric acid.
gitonin *Biochemistry.* $C_{50}H_{82}O_{23}$, a gitogenin tetraglucoside found in the seeds of *Digitalis purpurea;* sugar composition of two galactose, one glucose, and one xylose.
gitoxigenin *Pharmacology.* $C_{23}H_{35}O_5$, a crystalline substance prepared from gitoxin and used as a tonic for the heart.
gitoxin *Pharmacology.* $C_{41}H_{64}O_{14}$, a cardiac glycoside that increases the contractile force of heart muscle and is used as a heart tonic; prepared from foxglove, *Digitalis purpurea,* but is also found in *Digitalis lanata.*
gitter cell *Pathology.* a honeycomb-shaped cell containing lipoid granules, found in the urine of those with pyelonephritis.
Givetian *Geology.* a European geologic stage of the Middle Devonian period, occurring after the Couvinian and before the Frasnian.
gizzard *Vertebrate Zoology.* a thick-walled, muscular digestive pouch in the anterior portion of the stomachs of birds and crocodilians (in birds, only fully developed in seedeaters), consisting of two large masses of muscle enclosing a flat and narrow space of tough membrane, which together grind and soften food (sometimes with the aid of pebbles) before it enters the intestine. Also, VENTRICULUS. *Invertebrate Zoology.* a similar structure in the gut of arthropods and some other invertebrates, often lined with chitin and small teeth. Also, GASTRIC MILL. *Anatomy.* a popular term for the human stomach and intestines.

gizzard shad *Vertebrate Zoology.* a forage fish of the family Clupeidae; some species of the genus *Dorosoma* have a gizzardlike stomach musculature ; found in the Atlantic, Indo-Pacific, and Australian seas.

gjá [gyä] *plural,* **gjár.** *Volcanology.* a large gaping fissure that is the source of volcanic activity in the Icelandic Rift System. (An Icelandic word meaning "chasm.")

4-GL fourth-generation (computer) language.

Gl glucinium.

glabbelo-occipital length *Anthropology.* the length described by the angle from the point above the nose and between the eyebrows (the glabella) to a point at the base of the skull (the occipital bone); used in determining cranial length.

glabella *Anatomy.* **1.** the smooth flat portion of the frontal bone between the eyebrows. **2.** the most prominent point in the midsagittal plane between the eyebrows, used as an anthopometric landmark.

glabrous *Biology.* smooth, not hairy.

glacial *Hydrology.* **1.** describing the action, features, movements, and materials produced by or derived from glaciers or ice, or a region covered by glaciers or ice. **2.** describing a geologic time period marked by an ice age. *Chemistry.* of, relating to, or tending to form into icelike crystals.

glacial abrasion *Geology.* the mechanical wearing away of the earth's surface by the action of a moving glacier and the materials carried by the glacier.

glacial acetic acid *Organic Chemistry.* CH_3COOH, a highly pure, concentrated form of acetic acid; miscible with water and alcohol; melts at 16.6°C and boils at 118°C; used as a caustic to destroy warts.

glacial advance see ADVANCE.

glacial anticyclone *Meteorology.* a semipermanent high-pressure system that is said to occur over the ice caps of Greenland and Antarctica. Also, **glacial high.**

glacial boulder *Geology.* a large rock fragment that has been transported by a glacier. Also, ICE BOULDER.

glacial cone see DEBRIS CONE.

glacial drift *Geology.* the rock material carried by glaciers or ice masses and deposited directly on the land, in the sea, or in bodies of glacial meltwater. Also, **glacial deposit, glacial debris.**

glacial epoch *Geology.* an interval of geologic time during which a large portion of the earth's surface was covered by glaciers; characterized by a significantly cold worldwide climate and widespread glacial advance toward the equator. Also, GLACIAL PERIOD.

glacial erosion *Geology.* a process in which the earth's surface is worn away by the movement of glacier ice or the materials transported by the ice, or by the action of meltwater streams. Also, ICE EROSION.

glacial flour see ROCK FLOUR.

glacial geology *Geology.* the study of the features and the effects resulting from glacial erosion and deposition by glaciers and ice sheets; also the study of the features of a specific region that has been subjected to glaciation.

glacial horn see HORN.

glacial ice see GLACIER ICE.

glacial lake *Hydrology.* **1.** a lake whose water is derived primarily from glacial melting, or such a lake occupying a basin produced by glacial erosion or deposition. **2.** see GLACIER LAKE.

glacial lobe *Hydrology.* a large, tongue-shaped mass of ice that projects from the edge of the main mass of an ice sheet or ice cap.

glacial maximum *Geology.* the time or position of the greatest glacial advance or the greatest extent of a particular period of glaciation.

glacial mill see MOULIN.

glacial outwash see OUTWASH, def. 1.

glacial period *Geology.* **1.** see GLACIAL EPOCH. **2.** an interval of geologic time characterized by one or more major advances of ice.

glacial recession or **glacial retreat** *Hydrology.* see RECESSION.

glacial scour *Geology.* a natural process in which bedrock is abraded, scratched, and polished, and surficial material removed by rock fragments is carried by or embedded in a moving glacier.

glacial stairway see CASCADE, def. 2.

glacial striation *Geology.* a series of long, fine, generally straight, parallel scratches that are inscribed on a rock surface by the continuing abrasive action of rock fragments embedded in a moving glacier. Also, **glacial scratch.**

glacial till *Geology.* see TILL.

glaciated terrain *Geology.* a land surface formerly covered and modified by a glacier or ice sheet, and exhibiting glaciation marks and glaciation features.

glaciation

glaciation *Geology.* the covering and alteration of the earth's surface by glaciers, including erosion, deposition, planing, polishing of rocks, releveling, change of drainage system, and creation of numerous lakes.

glaciation limit *Geophysics.* the lowest altitude in a given location in which a glacier can form.

glacier *Hydrology.* a large mass of land ice that was formed by the compaction and recrystallization of snow, and which flows slowly downslope or outward in all directions under its own weight.

glacier

glacier flow *Hydrology.* the slow downward or outward motion of a glacier as a result of gravitational force. Also, ICE FLOW.

glacier front *Hydrology.* the leading margin of a glacier.

glacier ice *Hydrology.* any ice that is or once was part of a glacier.

glacier lake *Hydrology.* a lake formed by the obstruction of natural drainage by the edge of a glacier or ice sheet. Also, GLACIAL LAKE.

glacier mill or **glacier well** see MOULIN.

glacier remanié *Hydrology.* a glacier formed by the regelation of ice blocks that have accumulated as a result of avalanches and icefalls from glaciers at higher levels.

glacier wind *Meteorology.* a shallow gravity wind that blows along the icy surface of a glacier as a result of temperature differences between the air in contact with the glacier and the free air at the same altitude.

glacioeustasy *Oceanography.* the state of hydrologic equilibrium between world moisture transport to the continents in the form of snow or freezing rain and to the oceans as meltwater; an example of the significance of glacioeustatic change is the drop in world sea level during the most recent ice ages: between 15,000 and 18,000 years ago, sea level was more than 100 meters lower than it is today, and 40,000 years ago it may have been more than 150 meters lower than today.

glaciofluvial *Hydrology.* relating to glacier meltwater streams or to deposits made by such streams.
glaciolacustrine *Hydrology.* derived from or related to glacial lakes or the material that is deposited into such lakes, especially deposits such as kame deltas.
glaciology *Hydrology.* the study of all processes related to any form of existing solid water, including snow and ice.
glacon *Oceanography.* a piece of sea ice larger than brash but less than 1 kilometer across.
gladiate *Botany.* shaped like a sword.
Gladiator *Ordnance.* a Soviet mobile surface-to-air missile with semiactive radar guidance; a version under development may have limited ATBM capability; officially designated **SA-12.**
gladiolus *Botany.* **1.** a perennial plant, genus *Gladiolus,* of the family Iridaceae, originating in Africa and now widely cultivated elsewhere; characterized by upright sword-shaped leaves and bright, showy flowers in many colors. **2.** the flowering spike of this plant. *Anatomy.* the middle and largest segment of the sternum.

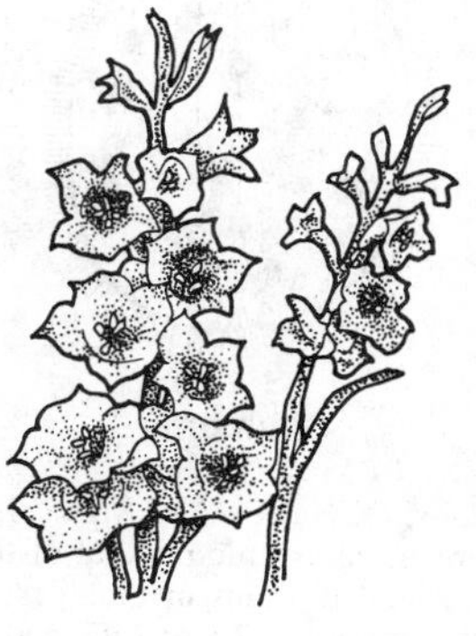

gladiolus

gladite *Mineralogy.* $PbCuBi_5S_9$, a lead-gray orthorhombic mineral occurring in prismatic crystals, having a specific gravity of 6.96 and a hardness of 2 to 3 on the Mohs scale; found with lead bismuth sulfides and quartz.
Glaisher's notation for elliptic functions *Mathematics.* the reciprocals of Jacobian elliptic functions are denoted by reversing the order of the letters that express the function; that is, $nsu = 1/snu$, $ncu = 1/cnu$, and $ndu = 1/dnu$. Quotients are denoted by writing in order the first letters of the number and denominator functions; that is, $scu = snu/cnu$, $sdu = snu/dnu$, $cdu = cnu/dnu$, etc.
glance pitch *Geology.* a variety of asphaltite characterized by a brilliant, conchoidal fracture.
glancing angle *Physics.* the complement of the angle of incidence, measured between the surface and the ray striking the surface.
gland *Anatomy.* any one of the various organs composed of specialized cells that secrete or excrete a material that is not related to ordinary metabolism, such as the pituitary gland, which produces hormones, and the spleen, which takes part in blood production. *Botany.* a secreting structure or organ. *Engineering.* an apparatus that prevents leakage at the point at which a shaft emerges from a vessel containing a fluid under pressure. (Going back to the Latin word for "acorn.")
glanders *Veterinary Medicine.* a highly fatal disease of horses and humans caused by the parasitic bacteria *Actinobacillus mallei,* affecting the mucous lining of the respiratory tract and the lungs, often characterized by a persistent nasal discharge accompanied by a chronic cough and ulcers on the skin and in the nasal mucous lining.
glans *Anatomy.* **1.** a rounded glandlike mass. **2.** the erectile tissues at the terminus of the penis and of the clitoris.
glans clitoridis *Anatomy.* the erectile tissue at the end of the clitoris, continuous with the intermediate part of the vestibular bulbs.
glans penis *Anatomy.* the cap-shaped expansion of the corpus spongiosum at the end of the penis, having the urethral opening at the center of its distal tip. Also, BALANUS.
glare *Telecommunications.* telephone interference that occurs when an outgoing call is placed simultaneously with the arrival of an incoming call on a two-way trunk, sometimes causing the trunk to become temporarily inoperative.
glare ice *Hydrology.* any smooth, shiny, highly reflective sheet of ice.
Glareolidae *Vertebrate Zoology.* a family of ternlike, insectivorous shorebirds of the order Charadriiformes, including the long-legged, short-winged coursers and the short-legged, long-winged pratincoles; found in the warm regions of Europe, Africa, Asia, and Australia.
glareous *Ecology.* describing a plant that grows in dry, gravelly soil. Also, **glareal.**
Glaser, Donald A. born 1926, American physicist; Nobel Prize for invention of "bubble chamber," used to study subatomic particles.
Glashow, Sheldon L. born 1932, American physicist; shared Nobel Prize for theory linking electromagnetism and the "weak force."
glasphalt *Materials.* a material consisting of crushed glass and asphalt that is used for surfacing roads.
glass *Materials.* **1.** a brittle, noncrystalline, usually transparent or translucent material that is generally formed by the fusion of dissolved silica and silicates with soda and lime; one of the most widely produced materials for such uses as windows, bottles and containers, automobile windshields, lenses and instruments, and many other purposes. **2.** of, relating to, containing, or resembling such a material.
glass armor *Ordnance.* a protective material composed of or containing glass.
glassblowing *Engineering.* a process of shaping glass in which air is blown through a tube into molten glass. Thus, **glassblower.**
glass-ceramic *Materials.* a predominantly crystalline product created by the controlled crystallization of glass; characterized by high thermal shock resistance and low thermal expansion.
glass cutter *Mechanical Devices.* a sharp hand tool having a diamond point or steel wheel, used to cut glass.
glass dosimeter *Nucleonics.* an instrument in which a specially treated glass rod glows with an ultraviolet light when it is irradiated by gamma rays; used to measure radiation doses.
glassed steel *Chemical Engineering.* any vessel or piping lined with glass to protect steel from corrosion caused by process streams.
glass electrode *Physical Chemistry.* **1.** a widely used chemical-measuring device, consisting of a thin membrane of specially treated glass whose two sides are in contact with different liquid solutions, one having a known pH value and the other unknown; the potential difference that develops across the membrane is due solely to the difference in hydrogen ion concentration between the two solutions and thus indicates the unknown pH. **2.** a similar device used to measure ionic concentration for substances other than hydrogen, as in monitoring the amount of fluorine in a water supply. Also, GLASS HALF CELL.
Glasser's disease *Veterinary Medicine.* a fatal bacterial disease that affects pigs (most often young pigs) with symptoms of polyarthritis, pleurisy, pericarditis, and peritonitis; meningitis occurs in some pigs, with symptoms of muscle tremor, paralysis, and convulsions.
glassfish *Vertebrate Zoology.* an Old World fish of the genus *Ambassis* and family Centropomidae, having a transparent body; found in brackish Indo-Pacific waters and popular in aquariums. Also, **glass perch.**
glass fission detector *Nucleonics.* a plate of glass on which fission fragments leave submicroscopic trails that can be seen under an ordinary microscope when chemically treated.
glass former *Materials Science.* an oxide that prevents crystallization.
glass furnace *Engineering.* an enclosure in which large amounts of glass are melted. Also, **glass tank.**
glass half cell see GLASS ELECTRODE.
glasshouse *Botany.* see GREENHOUSE.
glassine paper *Materials.* a thin, transparent, sulfite pulp paper that is used for envelope windows and sanitary wrapping paper.
glassivation *Electronics.* **1.** a transistor passivation process in which silicon semiconductor devices, complete with metal contact systems, are fully encapsulated in glass. **2.** a process in which glass is deposited on a chip to protect underlying device junctions. **3.** a process in which a dielectric material is diffused over an entire wafer to provide mechanical and environmental protection for the circuits.
glass laser *Optics.* a solid-state laser in which crystal or glass serves as the host for the active medium; the ruby or YAG laser is the most common. Also, CRYSTAL LASER, AMORPHOUS LASER.
glass-plate capacitor *Electricity.* a high-voltage capacitor in which the metal plates are separated by glass for the dielectric.
glass point see GLASS TRANSITION TEMPERATURE.
glass porphyry see VITROPHYRE.
glass pot *Engineering.* a heat-resistant vessel used in making small amounts of glass.
glass rays *Radiology.* the rays generated in an X-ray tube when the cathode rays strike the glass wall of the tube.

glass-rubber transition temperature *Materials Science.* the approximate midpoint of the temperature range over which an amorphous polymer changes from a brittle "glassy" solid condition to a highly elastic material (elastomer).
glass seal *Engineering.* an airtight closure created by molten glass, as between metal and glass.
glass snake *Vertebrate Zoology.* any of several worldwide species of limbless lizards from the genus *Ophisaurus* of the family Anguidae, characterized by snakelike movement and a very long tail that detaches easily to distract predators. Also, **glass lizard.**
glass sponge *Invertebrate Zoology.* a type of sponge with siliceous spicules, in the class Hexactinellida.
glass switch *Electronics.* a solid-state device used to control the flow of electric current, made from crystalline semiconductor materials.
glass transition *Physics.* a discontinuity in the form of a crystalline polymer, commonly brought on by a temperature gradient and typically across a region in which the form is viscous and rubbery to that in which it is hard and brittle.
glass-transition temperature *Materials Science.* the temperature above which an amorphous material becomes easily formed. Also, GLASS POINT.
glassware *Materials.* any product made of glass, especially drinking glasses.
glasswing butterfly *Invertebrate Zoology.* a common name for the butterfly *Haetera piera lepidoptera,* found in South America.

glasswing butterfly

glass wool *Materials.* a material made of threadlike glass fibers; used for insulation, packaging, and air filters.
glassy alloy see METALLIC GLASS.
Glauber, Johann Rudolf 1604–1688, German chemist; first to formulate hydrochloric acid and Glauber's salt (hydrated sodium sulfate).
glauberite *Mineralogy.* $Na_2Ca(SO_4)_2$, a pale yellow or gray monoclinic mineral occurring in tabular or prismatic crystals, having a vitreous luster, a saline taste, a specific gravity of 2.8, and a hardness of 2.5 to 3 on the Mohs scale; found in salt and nitrate deposits.
Glauber's salt *Inorganic Chemistry.* $Na_2SO_4{\cdot}10H_2O$, large transparent crystals, needles, or granular powder; soluble in water and insoluble in alcohol; melts at 32.38°C and loses its water at 100°C; used in solar heat storage, air conditioning, dyeing, and medicine. Also, SODIUM SULFATE DECAHYDRATE, SODIUM SULFATE CRYSTALS. *Mineralogy* see MIRABILITE.
glauco- a combining form meaning "gray, opaque," as in *glacophane.*
glaucocerinite *Mineralogy.* $(Zn,Cu^{+2})_{10}Al_6(SO_4)_3(OH)_{32}{\cdot}18H_2O$, a blue, hexagonal mineral occurring in radiating fibers, having a specific gravity of 2.4 and a hardness of 1 on the Mohs scale; found with malachite, gypsum, and adamite. Also, **glaucokerinite.**
glaucochroite *Mineralogy.* $CaMn^{+2}SiO_4$, a bluish-green orthorhombic mineral forming long prismatic crystals, having a specific gravity of 3.4 to 3.48 and a hardness of 6 on the Mohs scale; found with garnet, axinite, nasonite, and willemite at Franklin, New Jersey.
glaucodot *Mineralogy.* (Co,Fe)AsS, a grayish-white orthorhombic (pseudocubic) mineral of the arsenopyrite group, dimorphous with alloclasite, occurring as prismatic crystals and in massive form, having a metallic luster, a specific gravity of 6.04 to 6.05, and a hardness of 5 on the Mohs scale; found with cobaltite and pyrite.
glaucoma [glou kō´mə] *Medicine.* a group of eye diseases characterized by increased intraocular pressure and degeneration of the optic nerve head, most often caused by blockage of the channel through which aqueous humor drains (**chronic** or **open-angle glaucoma**) or by pressure of the iris against the lens (**acute** or **angle-closure glaucoma**). (From a Greek word meaning "gray" or "opaque.")
Glaucoma *Invertebrate Zoology.* a genus of freshwater ciliated protozoans in the subclass Holotrichia.

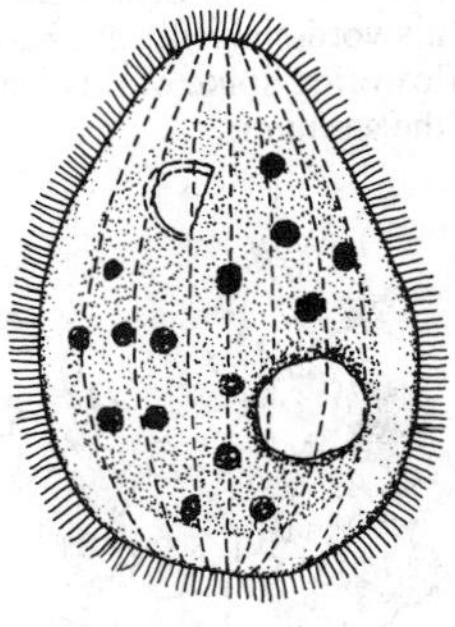

Glaucoma

glauconite *Mineralogy.* $(K,Na)(Fe^{+3},Al,Mg)_2(Si,Al)_4O_{10}(OH)_2$, a dark green to yellowish-green monoclinic mineral of the mica group usually occurring as minute, rounded grains or globules, having a dull to glistening luster, a specific gravity of 2.4 to 2.95, and a hardness of 2 on the Mohs scale; found in rocks of marine origin in deposits from most of the geological ages.
glauconitic sandstone *Petrology.* an arkosic or quartz sandstone rich in glauconite grains that impart a greenish color to the rock.
glaucophane *Mineralogy.* $Na_2(Mg,Fe^{+2})_3Al_2Si_8O_{22}(OH)_2$, Mg/(Mg+$Fe^{+2}$)=0.5–1.0, a grayish-blue to dark-blue monoclinic mineral of the amphibole group occurring in prismatic crystals and, more commonly, in fibrous, columnar or granular masses, having a vitreous to pearly luster, a specific gravity of 3.08 to 3.15, and a hardness of 6 on the Mohs scale; found in crystalline schists.
glaucophane schist *Petrology.* an amphibole schist rich in glaucophane and often containing epidote, quartz, or mica.
glaucous *Botany.* covered with a grayish- or bluish-green waxy coating that rubs off easily, as on plums and grapes.
glaves *Meteorology.* a foehnlike wind that occurs over the Faroe Islands. Also, **glave, glavis.**
G layer *Geology.* in the classification of the earth's interior, the seismic region equivalent to the inner core, extending from 5160 km below the surface.
glaze *Materials.* **1.** a smooth, glossy, glasslike coating applied to the face surface of a brick or pottery. **2.** to apply such a surface. *Hydrology.* a thin, glassy coating of ice that forms when a light rain condenses on a surface whose temperature is below freezing. Also, BLACK ICE.
glaze stain *Inorganic Chemistry.* a calcined oxide of iron, copper, manganese, or cobalt, finely ground; used as a colorant in ceramic glazes.
glazier *Engineering.* a person whose work is installing or fitting panes of glass in windows, doors, and the like.
glazier's point *Engineering.* a small triangular piece of sheet metal that is used to hold a pane of glass in place.
glazing *Materials Science.* the application to a ceramic article of a thin layer of a compound usually made from presmelted frit plus some clay, followed by firing which causes the coating to become a stable suspension. *Graphic Arts.* any of various processes for increasing the gloss on a newly developed photograph by placing the wet print on a heated metal surface; produces a deeper black than matte finish. *Building Engineering.* the process of fitting a pane of glass into a window frame. *Engineering.* the process of smoothing the abrasive edge of a wiped pipe by running a hot iron over it.
GLC gas-liquid chromatography.
GLCM *Ordnance.* an acronym for Ground-Launched Cruise Missile.

Gleason system *Design Engineering.* a standard to which all beveled gear designs are fabricated in the United States; it employs a pressure angle of 20° and its strength is increased through the use of long and short addenda comprising ratios other than 1:1, thus adding strength while avoiding undercut pinions.

gleba *Mycology.* in fungi belonging to the classes Gasteromycetes and Tuberales, the inner tissue of the fruiting body that bears the spores.

gleet *Medicine.* a slight discharge that characterizes the chronic stage of gonorrheal urethritis. *Veterinary Medicine.* an inflammation of the nasal passages of a horse, producing a thick discharge. Also, NASAL GLEET.

Gleicheniaceae *Botany.* a family of large, weedy-looking terrestrial ferns of the order Filicales that grow in masses in exposed or disturbed areas in the tropics, characterized by long, creeping, hairy or scaly rhizomes and forked fronds with dormant tips.

glenoid *Anatomy.* **1.** a depression that forms a socket in which another structure rests. **2.** resembling such a depression or socket.

glenoid cavity *Anatomy.* the oval depression on the scapula that articulates with the head of the humerus.

glessite *Geology.* a brown retinite, similar to amber, found on the shores of the Baltic Sea.

gley *Geology.* a sticky subsurface layer of soil developed under conditions of poor drainage. Also, **glei.**

gley soil *Geology.* a gray soil characterized by yellow and brown mottling produced by the partial oxidation and reduction of iron compounds under conditions of intermittent waterlogging.

glia *Neurology.* the non-neuronal cells of nerve tissue.

gliacyte *Neurology.* a peculiarly branched cell in the neuroglia. Also, **gliocyte.**

gliadin *Nutrition.* a protein present in wheat flour which, in combination with glutelin, contributes to the porous and spongy structure of bread; a protein constituent of gluten.

glial *Neurology.* of or pertaining to neuroglia or glia.

glial cell *Cell Biology.* one of the two major cell types in the nervous system, interspersed between the nerve cells to provide support and electrical insulation; types of glial cells include astrocytes, ependymal cells, oligodendrocytes, and microglia. Also, NEUROGLIAL CELL.

glide *Aviation.* **1.** the controlled descent of a heavier-than-air craft under little or no thrust, in which forward motion is maintained by gravity and vertical descent is controlled by lift forces. **2.** the flight path of such a descent. **3.** to carry out such a descent. **4.** to soar. *Linguistics.* the sounds of the letters *w* and *y* that appear before or after a vowel and must glide from the point of articulation to the vowel sound. *Crystallography.* see SLIP.

glide bomb *Ordnance.* a bomb with airfoils to provide lift, thus allowing it to "glide" toward its target when released from an airplane.

glide-bombing *Military Science.* the process of releasing bombs from a gliding airplane or of using glide bombs.

glide fold see SHEAR FOLD.

glide path or **glidepath** *Transportation Engineering.* the gradually descending path followed by an aircraft during its descent and approach to an airport. *Aviation.* **1.** the flight path of a flight vehicle as seen from the side. Also, **glide trajectory. 2.** the flight path used in a landing approach procedure and generated by an instrument-landing facility.

glide plane *Crystallography.* a symmetry element for which the symmetry operation is reflection across the plane combined with translation in a direction parallel to the plane. It is designated by a, b, or c if the translation is $a/2$, $b/2$, or $c/2$, by n if the translation is $(a + b)/2$, $(a + c)/2$, or $(b + c)/2$, i.e., halfway along one of the face diagonals, and by d if the translation is $(a + b)/4$, $(b + c)/4$, or $(c + a)/4$ or $(a + b + c)/4$.

glider *Aviation.* a fixed-wing aircraft designed to glide and sometimes to soar, usually not having any form of power plant. *Vertebrate Zoology.* any of six species of Australian marsupials of the genus *Petaurus* in the family Petaurida, having the ability to glide through the air by utilizing a membrane of skin attached to the flanks to sail as much 300 feet between trees.

glide rocket *Space Technology.* a rocket that assumes a flat, gliding attitude at cutoff, usually within the sensible atmosphere.

glide slope *Navigation.* a tightly focused radio beam transmitted from the approach end of a runway indicating the minimum approach angle that will clear all obstacles; it is one component of an instrument landing system. *Aviation.* the angle between the local horizontal and the glide path of an aircraft. Also, GLIDING ANGLE, SLOPE ANGLE.

gliding bacteria *Bacteriology.* any group of bacteria that exhibit cellular motility by means of a gliding or creeping movement that requires contact with a solid substrate.

gliding motility *Microbiology.* a relatively smooth, continuous method of cellular locomotion carried out by certain procaryotes, algae, and protazoa, and occurring only when in contact with a solid substrate.

glime *Hydrology.* a coating of ice having a consistency between that of glaze and rime.

glimmer ice *Hydrology.* ice that forms within the cracks or on the puddles of old ice.

G line *Electromagnetism.* a wire that has a circular cross section and is coated with a dielectric; used for transmitting microwave energy.

glint *Optics.* **1.** to strike and reflect off a surface at an angle. **2.** a small, brief, intense flash of light. *Ordnance.* the region that forcefully reflects light from a target. *Telecommunications.* **1.** a distorted radar-signal echo that varies in amplitude and apparent origin from pulse to pulse because the beam is being reflected from a rapidly moving object, such as a helicopter rotor. **2.** an electronic countermeasure technique in which this effect is used to degrade the tracking or seeking functions of an enemy weapons system. Also, GLITTER.

glioblastoma *Oncology.* a highly malignant brain tumor that is composed of glial tissue.

glioma *Oncology.* a tumor composed of cells and fibers representative of the support tisssue of the central nervous system.

gliopil *Neurology.* a matrix of glial cell processes, consisting principally of astrocytes.

gliosis *Medicine.* an excess of astroglia in damaged areas of the central nervous system.

gliosome *Neurology.* a small granule in the cytoplasm of neuroglial cells.

gliotoxin *Microbiology.* a toxin produced by certain fungi, such as *Trichoderma viride,* that inhibits the replication of a number of viruses and also exhibits antibacterial and antifungal activity.

Gliridae *Vertebrate Zoology.* the dormice, a family of small, squirrel-like mice of the suborder Myomorpha that are arboreal and primarily nocturnal; characterized by large eyes, rounded ears, short legs, soft and thick fur, and a generally bushy tail; found in wooded areas, hedgerows, gardens, and rocky areas of Europe, Asia, and Africa. The Gliridae arose in the early Eocene.

Glischrodermataceae *Mycology.* a family of fungi belonging to the order Sclerodermatales that occurs in charcoal, soil, and termite nests.

Glisson's capsule *Anatomy.* a connective tissue sheath surrounding the hepatic artery, portal vein, and bile ducts within the liver.

glitch *Electronics.* **1.** interference appearing as a narrow horizontal bar moving vertically in a television picture. **2.** a very brief and unwanted high-amplitude transient that occurs irregularly in an electronics system. *Engineering.* any minor malfunction in a device or system. *Astronomy.* an abrupt change in the rate at which a pulsar's spin is slowing down; thought to be caused by "star quakes" on the pulsar's surface.

glitter *Optics.* the brilliant sparkle resulting from the reflection of light by the individual facets of a surface. *Telecommunications.* see GLINT.

Glivenko-Cantelli lemma *Mathematics.* the sequence of empirical distribution functions of a random variable almost surely converges uniformly (i.e., converges uniformly in probability) to the distribution function of the random variable.

g load *Physics.* the ratio of an applied force on an object to the force due to gravity acting on the body at sea level.

global *Science.* **1.** of or relating to the entire world; worldwide; universal. **2.** relating to or using a terrestial or celestial globe. *Mathematics.* not local; a property or object is said to **hold globally** or be **defined globally** if it is valid for the entire mathematical object under discussion. *Computer Science.* **1.** of or relating to all the users in a computer network. **2.** of or relating to an entire document or file, rather than selected text within it.

global accuracy *Robotics.* a measure of the accuracy of a robot over the entire range of the working envelope.

global change *Computer Programming.* an alteration in a word-processing or pagemaking document that is made for the entire document, rather than for selected text within the document; e.g., changing the spelling "amoeba" to "ameba" wherever it occurs throughout an entire scientific manuscript. Similarly, **global correction.**

global deterioration scale *Psychology.* a list of criteria used to assess the extent of mental deterioration in an elderly person; used to determine the progression of Alzheimer's disease.

global ecology *Biology.* the study of relationships and interactions between and among organisms on a global scale.

global memory *Computer Technology.* computer memory that is accessible to an entire set of computers in a multiprocessor system.

global radiation *Geophysics.* a measure of direct and diffuse radiation received by the earth per unit horizontal surface.

global resource sharing *Computer Programming.* a feature of local-area networks that enables all of the users to share a network's resources, such as input/output and storage devices.

global sea *Oceanography.* all the seawater of the earth thought of as one ocean undergoing constant mixing.

global search *Computer Programming.* a feature of a word-processing or pagemaking program that enables the user to locate selected characters or formats wherever they occur in an entire document.

global search and replace *Computer Programming.* a feature of a word-processing or pagemaking program in which each occurrence of a given character or set of characters is found and replaced or deleted.

global stability *Ecology.* the capacity of an ecologic or taxonomic unit to withstand large-scale disturbances with little or no effect. *Chaotic Dynamics.* a characteristic of a phase-space solution in which all perturbations of the solution (regardless of their size) evolve back to the same solution.

global variable *Computer Programming.* a variable that is directly accessible by program units or blocks in addition to the one in which the varible is defined.

global warming *Meteorology.* a reported phenomenon of the 1980s and afterward, according to which the average temperature on earth is gradually increasing over its level in recent history; attributed to the increased concentration in the atmosphere of gases such as carbon dioxide that trap heat radiating upward and reradiate it toward earth.

globe *Cartography.* a spherical body on which a map of the earth (or any other body for which there is sufficient information, such as the moon) is represented, usually by cementing flat, gore-shaped segments of maps to the spherical surface, although on large globes (possibly several meters in diameter) the earth's surface may be exaggerated and shown in sculptured relief.

globe lightning see BALL LIGHTNING.

globe valve *Engineering.* a screw-down valve that regulates fluid flow in a pipeline, having a circular metal disk or ball that fits into the seating of the pipe and is moved by a threaded spindle.

Globigerina *Paleontology.* a small genus of foraminiferids, very important as an index fossil in the Cretaceous; the key organism in a Globigerina ooze.

Globigerinacea *Invertebrate Zoology.* a superfamily of protozoans in the order Foraminiferida, having a radial calcite shell.

Globigerina ooze *Oceanography.* a calcareous, deep-ocean sediment composed of at least 30% skeletal remains of foraminifera, predominantly of the genus *Globigerina.*

globin *Biochemistry.* the polypeptide (protein) chain that is associated with an iron-porphyrin group in both hemoglobin and myoglobin.

globin gene *Genetics.* any member of the related group of genes that encode one of the respiratory globin proteins produced by many animals; in humans they are clustered on chromosomes 11 and 16.

globin zinc insulin *Pharmacology.* a form of injectable insulin that provides delayed absorption and prolonged action; used in the treatment of insulin-dependent diabetes mellitus.

globoside *Biochemistry.* a ceramide oligoglycoside that contains amino sugars, simple sugars, and *N*-acetyl amino sugars; present in the red blood cells of various mammals, including hogs, humans, and sheep.

globular [gläb´yə lər; glōb´yə lər] *Science.* having the shape of a globe.

Globulariaceae *Botany.* a family of monocotyledonous, heathlike herbs and small shrubs native to Africa, Madagascar, Europe, and western Asia, belonging to the order Scrophulariales; noted for producing iridoid compounds and sometimes a bitter substance called globularin.

globular projection *Cartography.* a map projection representing a hemisphere on which the equator and a central geographic meridian are represented by straight lines intersecting at right angles, and all other meridians are represented by circular arcs connecting points of equal division on the equator with the poles; the parallels are circular arcs dividing the central and extreme outer meridians into equal parts, and the extreme outer meridian forms the limit of the projection and is a circle.

globular protein *Biochemistry.* a protein in which each polypeptide chain is coiled in three dimensions to form a globular-shaped molecule that dissolves easily in water or a water-based solvent.

globular star cluster *Astronomy.* a tightly grouped spherical cluster typically containing 100,000 to 1,000,000 stars.

globulin [gläb´yə lən; glōb´yə lən] *Biochemistry.* any of a class of simple globular proteins that are insoluble in water, soluble in dilute salt solutions, and precipitated by ammonium sulfate at 50% saturation.

globulite *Geology.* a small, usually dark, spherical crystallite found in volcanic glass.

globulomaxillary cyst *Medicine.* a cystic embryonal inclusion in the alveolar process between the lateral upper incisors and canine teeth.

glochidium *Botany.* a barbed hair or bristle, as on a cactus. *Invertebrate Zoology.* the bivalve larval stage of freshwater mussels in the family Unionidae; larvae attach to the gills or fins of fish.

glockerite see LEPIDOCROCITE.

gloea *Invertebrate Zoology.* an adhesive substance secreted by some protozoans and other lower organisms.

Gloeobacter *Bacteriology.* a genus of unicellular cyanobacteria that grow in aggregates and lack thylakoids, so that the photosynthetic and respiratory processes seem to be carried out in association with the cytoplasmic membrane.

Gloeobotrydaceae *Botany.* a family of freshwater yellow-green algae of the order Mischococcales, characterized by attached or free-floating colonies in which cells are embedded in a common mucilage.

Gloeocapsa *Bacteriology.* a genus of coccoid cyanobacteria or blue-green algae, that grow as sheath-enclosed clusters of cells in aquatic habitats.

Gloeochloris *Botany.* a genus of algae of the family Heterogloeaceae, characterized by spherical or ellipsoidal cells embedded in mucilage and forming gelatinous masses.

Gloeococcaceae *Botany.* a monogeneric family of the order Tetrasporales; planktonic gelatinous colonies of ovoid or spherical biflagellate cells.

Gloeodiniaceae *Botany.* a monospecific family of photosynthetic freshwater algae belonging to the order Gloeodiniales and characterized by nonmotile vegetative cells embedded in a gelatinous mass.

Gloeodiniales *Botany.* a monofamilial order of freshwater algae belonging to the class Dinophyceae, usually found living epiphytically in peat bogs.

Gloeothece *Bacteriology.* a genus of unicellular cyanobacteria or blue-green algae that grow as rod-shaped cells in aggregates surrounded by a sheath, possess thylakoids, and are capable of nitrogen fixation.

Gloeotrichia *Bacteriology.* a genus of freshwater, blue-green algae that grow in distinct colonies and may possess gas vacuoles.

Gloger's rule *Ecology.* a rule stating that, with regard to latitude or altitude, the pigmentation in warm-blooded animal species tends to decrease as the mean temperature decreases: individuals in colder climates are lighter colored than those in warmer climates.

Gloiosiphoniaceae *Botany.* a small family of red algae of the order Cryptonemiales, characterized by erect, gelatinous, much-branched gametangial thalli arising from a discoid base, and occasionally a discoid and crustose tetrasporangial thallus.

Glomerida *Invertebrate Zoology.* an order of millipedes in the class Diplopoda, called pill millipedes for their ability to roll into a protective ball. Also, ONISCOMORPHA.

Glomeridesmida *Invertebrate Zoology.* an order of small, eyeless tropical millipedes in the class Diplopoda, having 22 segments.

glomerulal *Anatomy.* relating to a glomerulus, especially the glomeruli of the kidneys.

glomerular filtration rate *Biochemistry.* the rate at which waste products and water pass through the glomeruli into the Bowman's capsule and down the nephron.

glomerule *Botany.* a cyme condensed into a tight cluster and resembling the flower head of a composite, as on the flowering dogwood.

glomerulonephritis *Medicine.* an acute or chronic kidney desease primarily affecting the capillary loops projecting into the lumen of a renal corpuscle.

glomerulosclerosis *Medicine.* the fibrosis and scarring that result in the senescence of renal glomeruli.

glomerulus *plural,* **glomeruli.** *Anatomy.* **1.** any small spherical cluster. **2.** specifically, a cluster of capillaries enclosed within a Bowman's capsule in the kidney; the site at which plasma is filtered into renal tubules. Also, MALPIGHIAN CORPUSCLE.

gloom *Meteorology.* a weather condition in which daylight is dramatically reduced as a result of dense cloud or smoke accumulation above the earth's surface; it does not affect surface visibility.

G loop *Molecular Biology.* a loop of unpaired strands appearing in the G region of DNA that has been extracted from the bacteriophage μ, then denatured and reannealed.

gloriosa lily *Botany.* a climbing lily of the genus *Gloriosa* characterized by brilliant red and yellow flowers; native to Africa and Asia.

glory *Optics.* see ANTICORONA.

glory hole an informal name for various types of holes or compartments; specific uses include: *Naval Architecture*. quarters on a ship for the stewards or stokers. *Engineering*. a type of furnace in which articles are reheated after manufacture. *Civil Engineering*. a conical-shaped, fixed-crest spillway. *Mining Engineering*. a vertical pit from which material is fed by gravity to hauling units. *Nucleonics*. an opening in the shield and frequently the reflector of a nuclear reactor through which a beam of charged particles, such as neutrons, is allowed to escape for research purposes. (Of unknown origin; thought to derive from an earlier word *glory* meaning "mud" or "slime.")

glory hole system see CHUTE SYSTEM.

glorypea *Botany*. either one of the two grayish-green trailing plants, *Clianthus formosus* or *C. puniceus*, characterized by showy scarlet flowers; native to Australia and New Zealand.

gloss *Optics*. **1.** the ratio between the light reflected specularly from a surface and the total amount of light reflected from the surface. **2.** the property of a surface by which reflected highlights or the images of objects appear superimposed on the surface, due to directional reflection.

gloss- a combining form meaning: **1.** tongue, as in *glossitis*. **2.** speech, as in *glossary*.

glossa *Invertebrate Zoology*. a tongue, especially one of a pair of lobes at the center of the lower lip of insects.

glossal *Anatomy*. of or relating to the tongue.

glossalgia *Medicine*. any pain or abnormal tenderness of the tongue.

glossary *Computer Programming*. in word processing, a file of commonly used words or phrases that can be retrieved in a convenient manner, as by means of a function key and a keyword.

glossate *Invertebrate Zoology*. having a glossa, a tongue, or a tongue-like structure.

glossectomy *Surgery*. the surgical removal of the tongue or of part of the tongue.

Glossinidae *Invertebrate Zoology*. a subfamily of two-winged insects, the tsetse flies, in the group Pupipara; bloodsuckers that transmit sleeping sickness.

Glossiphoniidae *Invertebrate Zoology*. a family of leeches in the order Rhynchobdellida, with a flattened body and an extendable proboscis.

glossitis *Medicine*. an inflammation of the tongue, sometimes caused by pernicious anemia.

glosso- a combining form meaning: "tongue," as in *glossopharyngeal*.

glossolalia *Medicine*. unintelligible speech in an imaginary language, as in a trance, schizophrenia, or "speaking in tongues" in a state of religious ecstasy.

glossopalatine *Anatomy*. of or relating to the tongue and the palate.

glossopharyngeal *Anatomy*. of or relating to the tongue and the throat.

glossopharyngeal nerve *Anatomy*. the ninth cranial nerve, originating in the medulla oblongata and supplying the tongue, pharynx, and parotid gland; it is essential to the sense of taste.

glossoplasty *Surgery*. plastic surgery of the tongue.

glossopterid flora *Paleontology*. a fossil plant assemblage found in Permian and lower Triassic strata in Australia, southern Africa, India, and South America (Gondwanaland); the dominant plants in the flora are a large group of seed ferns called Glossopteris, which is not yet completely systematized but includes at least 80 species.

glossopyrosis *Medicine*. a burning sensation of the tongue, sometimes the result of Moeller's glossitis.

glossospasm *Neurology*. a spasm of the tongue muscles.

glossy *Optics*. referring to or describing a surface that reflects a greater amount of incident light specularly than is reflected diffusely. *Graphic Arts*. a photograph printed on paper that has been subject to a glazing process, giving it a shiny appearance. Thus, **glossy paper, glossy print.**

glossy ibis *Vertebrate Zoology*. a long-legged wading bird having a chestnut plumage with a purplish or greenish sheen and a decurved bill; found in tropical areas of the Old World and in the southern U.S.

glossy snake *Vertebrate Zoology*. a nocturnal snake, *Arizona elegans*, characterized by smooth, shiny scales of tan with brown spots; found in the western U.S. and northern Mexico.

glost firing *Chemical Engineering*. the process of firing and glazing ceramic ware that has already been fired at a higher temperature.

glottal stop *Linguistics*. a sound produced in some languages in which the air is stopped at the glottis (an area between the larynx and vocal cords); the sound resembles that produced in the English word "little."

glottis *Anatomy*. the sound-producing region of the larynx, consisting of the true vocal cords and the vocal muscles on each side.

glotto- a combining form meaning "speech," as in *glottochronology*.

glottochronology *Linguistics*. the study of the manner in which two separate but related languages or language families have developed differently from each other, in order to estimate the time at which they diverged. *Anthropology*. the use of this technique as a means of dating cultural developments; e.g., British and American English use different words for some parts of a car, which indicates that the separation of these two groups took place before cars were invented.

gloup *Geology*. a vent in the roof of a sea cave.

glove-and-stocking anesthesia *Medicine*. a numbness of the hands and feet, sometimes caused by a disease of the peripheral nerves.

glove anesthesia *Medicine*. the loss or diminishing of sensation in the hands that may accompany inflammatory disease affecting the peripheral nerves.

glove box *Engineering*. an airtight enclosure with openings to which special long gloves are attached, allowing for handling of the contents without injury, leakage, or contamination. *Nucleonics*. see DRY BOX.

glove juice test *Microbiology*. a test to establish the disinfecting effectiveness of a given antimicrobial, surgical hand scrub.

glow discharge *Electronics*. a glow inside an electron tube due to the ionization of gas caused by a discharge of electricity. Thus, **glow-discharge tube.**

glow-discharge microphone *Acoustical Engineering*. a microphone in which the action of sound waves gives rise to variations in the current between two electrodes in a glow-discharge tube.

glow-discharge voltage regulator *Electronics*. a gas tube that regulates voltage by utilizing the variable resistance characteristics of the ionized gas within the tube. Also, GASEOUS VOLTAGE REGULATOR.

glowing avalanche see ASH FLOW.

glow lamp *Electronics*. a two-electrode electron tube in which light is produced by a glow close to the negative electrode when voltage is applied between the two electrodes.

glow plug *Mechanical Engineering*. a small, electric heater used in diesel engines to preheat the air in the cylinder to facilitate starting the engine. *Aviation*. a similar device used in gas turbines to ensure relighting when the flame is unstable, as under icing conditions.

glow potential see GLOW VOLTAGE.

glow-switch *Electronics*. in fluorescent light circuits, one of two bimetal strips in an electron tube that make contact when heated by the glow discharge.

glow-tube oscillator *Electronics*. a circuit that uses a glow-discharge tube to function as a relaxation oscillator generating fixed-amplitude periodic sawtooth waveforms.

glow voltage *Electronics*. the voltage at which a glow discharge begins in a gas tube. Also, GLOW POTENTIAL.

glowworm *Invertebrate Zoology*. the popular name for a wingless female firefly of the family Lampyridae; bioluminescence created by organs on the underside of the abdomen helps attract males for mating.

gluc- a combining form meaning "sugar" or specifically "glucose," as in *glucagon*.

glucagon *Endocrinology*. a polypeptide hormone that is produced in the alpha cells of the endocrine pancreas and released in response to decreases in blood glucose; it acts to increase blood glucose by inducing the breakdown of glycogen and by stimulating de novo synthesis of glucose.

glucan *Biochemistry*. a polysaccharide composed of D-glucose in either straight or branched chains with glycosidic linkages; members of the group include dextran, laminarin, and lichenin.

1,4-glucan branching enzyme *Enzymology*. an enzyme of the transferase class that catalyzes the transfer of a 1,4-D-glucan chain to a primary hydroxyl group in the same molecule, resulting in a new branch in a glycogen molecule or an amylopectin molecule. Also, BRANCHER, BRANCHING ENZYME, Q ENZYME.

glucan-1,4-glucosidase *Enzymology*. an enzyme of the hydrolase class that catalyzes hydrolysis of terminal 1,4-linked α-D-glucose residues from nonreducing ends of the chain with release of β-D-glucose.

glucinium *Chemistry*. a former name for beryllium.

gluco- a combining form meaning "sugar" or specifically "glucose," as in *glucogenesis*.

glucocerebroside *Biochemistry*. a glucose-containing cerebroside or acid amide of a fatty acid with glucosidic linkages.

glucocorticoid *Endocrinology*. any of a group of adrenocortical steroid hormones whose metabolic effects include stimulation of gluconeogenesis, increased catabolism of proteins, and mobilization of free fatty acids; they are also potent inhibitors of the inflammatory response (allergic response).

glucogenesis *Biochemistry.* the synthesis in living cells of the sugar glucose, occurring from precursors other than glycogen.
glucogenic amino acid *Biochemistry.* an amino acid (such as glycine, alanine, arginine, or ornithine) that degrades to a tricarboxylic acid cycle intermediate, then converts to glucose and thus enters the normal carbohydrate metabolism of the body.
glucokinase *Enzymology.* an enzyme of the transferase class that is secreted in the liver and catalyzes the phosphorylation of D-glucose; highly specific for glucose.
glucolipid *Biochemistry.* a glucose-containing compound that is synthesized by living cells and has long hydrocarbon chains as a major part of its structure.
glucomannan *Biochemistry.* a heteropolysaccharide composed of glucose and mannose residues; a major component of coniferous trees.
gluconate *Organic Chemistry.* a salt of gluconic acid containing the radical $HOCH_2(CHOH)_4COO-$.
gluconeogenesis *Biochemistry.* the formation or synthesis of glucose in the liver from nonglucose precursors such as amino acids, intermediates of glycolysis, or intermediates of the citric acid cycle.
gluconic acid *Organic Chemistry.* $C_6H_{12}O_7$, derived from glucose through oxidation; colorless crystals; soluble in water and alcohol; melts at 131°C; derived from glucose through oxidation; used in pharmaceuticals and food products, as a catalyst in textile printing, and as a cleanser.
Gluconobacter *Bacteriology.* a genus of Gram-negative, rod-shaped bacteria of the family Acetobacteraceae, distinguished by the incomplete oxidation of organic compounds and occurring in soil, plants, fruits, and flowers.
D-glucopyranose see GLUCOSE.
glucopyranoside *Biochemistry.* a cyclic form of glucose having a pyranose structure in which the carbon atoms 1 and 5 are bridged by an oxygen atom.
glucopyranosol-D-glucose see GENTIOBIOSE.
glucosamine *Biochemistry.* $C_6H_{13}NO_5$, a widely occurring amino sugar of glucose that is a component of chitin and occurs in vertebrate tissues.
glucose [gloo´kōs] *Biochemistry.* $C_6H_{12}O_6$, a six-carbon aldose that is the major sugar in the blood and a key intermediate in metabolism; used as a fluid and nutrient replenisher, usually given intravenously. Also, DEXTROSE, CERELOSE, D-GLUCOPYRANOSE.

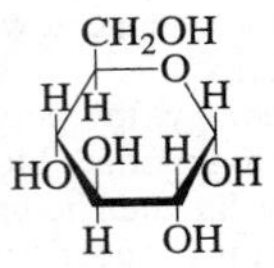

glucose

glucose effect *Enzymology.* the ability of glucose to inhibit the activity of microbial genes in producing enzymes that function in the early steps of catabolic pathways.
glucose oxidase *Enzymology.* an enzyme of the oxidoreductase class that catalyzes the oxidation of β-D-glucose by O_2 to a gluconolactone; used to estimate glucose concentration in blood or urine samples.
glucose-6-phosphatase *Enzymology.* an enzyme of the hydrolase class that catalyzes the breakdown of D-glucose 6-phosphate to free glucose and inorganic phosphate; occurs in the liver, kidneys, and intestines.
glucose phosphate *Biochemistry.* $C_6H_{13}O_9P$, a glucose derivative that is an important intermediate in glycolysis.
glucose 1-phosphate *Biochemistry.* $C_6H_{13}O_9P$, an ester formed by phosphoglucomutase that is an intermediate in the metabolism of glycogen and starch.
glucose-6-phosphate *Biochemistry.* $C_6H_{13}O_9P$, a key intermediate formed in the biochemical interactions of gluconeogenesis or carbohydrate metabolism.
glucose-6-phosphate dehydrogenase *Enzymology.* an enzyme of the oxidoreductase class that catalyzes the oxidation of glucose 6-phosphate to 6-phosphogluconic acid.
glucose phosphate isomerase *Enzymology.* an enzyme that catalyzes the isomerization of glucose 6-phosphate to fructose 6-phosphate.
glucose tolerance test *Pathology.* an oral or intravenous method of assessing the efficiency of a normal liver to absorb and store excessive quantities of glucose.
α-glucosidase *Enzymology.* an enzyme of the hydrolase class that catalyzes the hydrolysis of terminal, nonreducing 1,4-linked α-D-glucose residues with release of α-D-glucose.
glucoside *Biochemistry.* **1.** any member of a series of compounds, usually of plant origin, that may be hydrolyzed into dextrose. **2.** a plant principle that on hydrolysis yields a sugar and another principle.
glucosin *Toxicology.* a group of substances formed by the action of ammonia on glucose, some of which are potent toxins.
glucosulfone sodium *Pharmacology.* $C_{24}H_{34}N_2Na_2O_{18}S_3$, an antibacterial drug occurring as a white to faintly yellow solid, injected intravenously in the treatment of leprosy.
glucosyltransferase *Enzymology.* an enzyme that catalyzes the tranfer of a glucosyl residue to an acceptor molecule.
glucuronic acid *Biochemistry.* $C_6H_{10}O_7$, a derivative of glucose, a six-carbon sugar molecule in which the terminal carbon atom has been oxidized. Also, GLYCURONIC ACID.
β-glucuronidase *Enzymology.* an enzyme of the hydrolase class that catalyzes the hydrolysis of β-D-glucuronide.
glucuronide *Biochemistry.* a compound formed by linking glucuronic acid to another compound by means of a glycosidic bond; many toxic compounds are detoxified by being changed to a glucuronide and are then excreted in this form. Also, GLYCURONIDE.
D-glucuronolactone *Biochemistry.* $C_6H_8O_6$, a reacting agent that transfers the amino group in amino acid metabolism; a major component of practically all fibrous connective tissues in mammals; used as an antiarthritic.
glue *Materials.* **1.** a gelatin derived by boiling animal parts in water; it is hard and brittle when cooled but when heated and diluted it becomes a viscous liquid adhesive. **2.** any similar adhesive substance. **3.** to use a glue or other adhesive to stick two materials together.
glueball *Particle Physics.* a term for a particle made only of gluons with no quarks, resulting from interactions between gluons. Also, BOUND GLUE STATE.
glue block see ANGLE BLOCK.
glue-sniffing *Pathology.* the practice of inhaling the vapors of toluene, a volatile organic compound used as a solvent in certain glues, for the hallucinogenic or euphoric effect; prolonged accidental or occupational exposure or repeated use may damage a variety of organ systems.
Glugea *Invertebrate Zoology.* a large genus of sporozoans in the class Microsporidea, having a single valve and existing as intracellular parasites in various insect larvae and fish.
glumaceous *Botany.* consisting of or having glumes.
glume *Botany.* one of two basal bracts of a spikelet of grass.
glume blotch *Plant Pathology.* a disease of wheat caused by the fungus *Leptosphaeria nodurum,* characterized by necrotic lesions on the leaves and glumes of the plant.
glumiferous *Botany.* having glumes.
Glumiflorae *Botany.* in some classification systems, an equivalent name for the order Cyperales.
gluon *Particle Physics.* a massless particle that carries the strong force from one quark to another. Gluons can also interact among themselves and form particles consisting only of gluons bound together (glueballs).
glutamate *Biochemistry.* an amino acid that is an intermediate in amino acid metabolism; formed by the addition of NH_3 to α-ketoglutamate.
glutamic acid *Biochemistry.* $C_5H_9NO_4$, an aliphatic amino acid with two carboxyl groups, a product of the breakdown of proteins.
glutamic-oxaloacetic transaminase see ASPARTATE AMINOTRANSFERASE.
glutaminase *Enzymology.* an enzyme of the hydrolase class that catalyzes hydrolysis of glutamine to form ammonia and glutamic acid.
glutamine *Biochemistry.* $C_5H_{10}N_2O_3$, a proteogenic amino acid or amide of glutamic acid that promotes the regeneration of mucoproteins and the intestinal epithelium.
glutamine synthetase *Enzymology.* an enzyme that catalyzes the combination of ammonia and glutamic acid to form glutamine.
glutaraldehyde *Organic Chemistry.* $OCH(CH_2)_3CHO$, a liquid dialdehyde; soluble in water and alcohol; boils at 188°C; used as an agent for crosslinking proteins and polyhydroxy resins, and in tanning.
glutarate *Biochemistry.* the salt or ester produced by glutaric acid.
glutaric acid *Biochemistry.* $C_5H_8O_4$, pentanedioic acid, an intermediate in the metabolism of tryptophan and lysine, present in green sugar beets and water extracted from wool.

glutathione *Biochemistry.* $C_{10}H_{17}O_6N_3S$, a naturally occurring peptide, serving as a biological redox agent, a coenzyme, or a cofactor or substrate for certain coupling reactions catalyzed by ligandin; an important substance in tissue oxidation.

glutathione peroxidase *Biochemistry.* an enzyme that initiates the oxidation of glutathione with hydrogen peroxide.

glutelin *Biochemistry.* a simple protein derived from the seeds of cereals such as wheat, rice, and barley.

gluten *Biochemistry.* the protein of wheat and other grains that gives dough its elastic character; a mixture of glutelins and prolamins.

glutenin *Biochemistry.* a glutelin that is derived from wheat.

glutethimide *Pharmacology.* $C_{13}H_{15}NO_2$, a nonbarbituate drug that is related to phenobarbital, administered orally as a sedative and to induce sleep.

gluteus maximus *Anatomy.* the large buttocks muscle that extends the thigh.

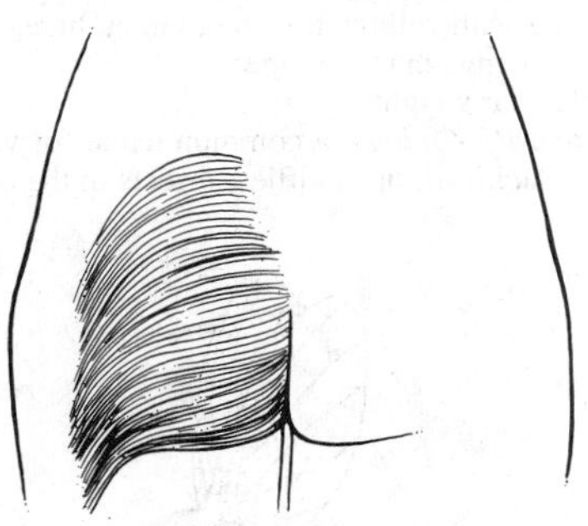

gluteus maximus

gluteus medius *Anatomy.* the middle buttocks muscle that abducts and rotates the thigh.

gluteus minimus *Anatomy.* the innermost buttocks muscle that adducts the thigh.

glutinant nematocyst *Invertebrate Zoology.* a type of nematocyst (stinging cell) with an open, sticky tube for anchoring the coelenterate when walking on tentacles.

glutinous *Botany.* having a sticky or gummy surface.

glyc- a combining form meaning "sugar" or specifically "glucose," as in *glycemia.*

glycemia *Physiology.* the presence of glucose in the blood.

glyceraldehyde *Biochemistry.* $C_3H_6O_3$, a three-carbon aldose that serves as the reference compound for the assignment of D and L configurations to amino acids, carbohydrates, and related compounds.

glycerate *Biochemistry.* the salt or ester produced by glycetic acid.

glyceric acid *Biochemistry.* $C_3H_6O_4$, a hydroxypropanoic acid that is metabolically important as an intermediate in photosynthesis, alcohol fermentation, and glycolysis.

Glyceridae *Invertebrate Zoology.* a family of polychaete worms in the subclass Errantia; burrowers with a conical first segment and a long proboscis with four jaws.

glyceride *Biochemistry.* an organic acid ester of glycerol; the natural fats are the glycerides of fatty acids.

glycerin see GLYCEROL.

glycerokinase *Enzymology.* an enzyme that catalyzes the phosphorylation of glycerol into glycerophosphate.

glycerol *Organic Chemistry.* $CH_2OHCHOHCH_2OH$, a thick, odorless, colorless, syrupy liquid with a sweet taste, prepared by hydrolysis of animal and vegetable fats and oils; soluble in water and alcohol; melts at 18°C and boils at 290°C with decomposition; used in pharmaceuticals and food, and as a solvent. Also, GLYCERIN.

glycerophosphate *Biochemistry.* the first step in the phosphatidic acid synthesis pathway. Also, ***sn*-glycerol 3-phosphate.**

glycerophosphoric acid *Biochemistry.* $C_3H_9O_6P$, a phosphate ester of glyceric acid, various forms of which are intermediates in glycolysis.

glyceryl *Organic Chemistry.* the functional 1,2,3-propanetriyl radical from glycerol.

glyceryl diacetate see DIACETIN.

glyceryl triacetate see TRIACETIN.

glyceryl tristearate see STEARIN.

glycidic acid *Organic Chemistry.* $C_2H_3OCO_2H$, a volatile liquid. Also, EPOXYPROPIONIC ACID.

glycidol *Organic Chemistry.* $C_3H_6O_2$, a combustible, colorless liquid; soluble in water, alcohol, and ether; boils at 162°C with decomposition; used as a stabilizer, and demulsifier, and in organic synthesis.

glycin *Organic Chemistry.* $C_8H_9NO_3$, crystals that are soluble in water and alcohol; melts at 245–247°C with decomposition; used as a photographic developer and acid indicator in bacteriology.

glycine [glī´sēn] *Organic Chemistry.* NH_2CH_2COOH, white, odorless crystals that are soluble in water and slightly soluble in alcohol; melts at 232–236°C; used as a nutrient, a buffering agent, a retardant to rancidity in fats, and in organic synthesis. *Biochemistry.* the natural form of this substance, a constituent in most proteins; it is the simplest proteagenic acid and is not essential in the diet, since it can be metabolized from glyoxylate. Also, AMINOACETIC ACID.

glyco- a combining form meaning "sugar" or specifically "glucose," as in *glycolysis.*

glycocalyx *Cell Biology.* a carbohydrate-rich cell coat on the extracellular side of the plasma membrane of most eukaryotic cells that may be involved in cellular recognition and confers a unique identity upon the cell.

glycocholic acid *Biochemistry.* $C_{26}H_{43}NO_6$, a carboxylic acid that is found in human bile and serves as an emulsifying agent for fat.

glycocyamine *Biochemistry.* $C_3H_7N_3O_2$, a polysaccharide that is intermediate in the biosynthesis of creatine; formed by the transfer of the guanido group from arginine to glycine.

glycogen [glī´kə jin] *Biochemistry.* $(C_6H_{10}O_5)_n$, a highly branched polysaccharide serving as a short-term storage substance, and, as such, subject to continuous synthesis and degradation.

glycogenesis *Biochemistry.* **1.** the synthesis or formation of glycogen. **2.** the production of sugar.

glycogenolysis *Biochemistry.* the catabolism or breakdown of glycogen, yielding about 10% free glucose and 90% glucose 1-phosphate; controlled by enzymes acting in response to hormones and nerve impulses. Also, **glycogeny.**

glycogenosis *Medicine.* any of at least 14 types of rare inborn errors of metabolism, due to a defect in a specific enzyme involved in glycogen catabolism. Also, **glycogen storage disease.**

glycogen synthetase *Enzymology.* an enzyme of the transferase class that catalyzes the synthesis of the amylose chain of glycogen.

glycol [glī´kôl] *Organic Chemistry.* **1.** $C_nH_{2n}(OH)_2$, a general term for a dihydric alcohol. **2.** CH_2OHCH_2OH, a sweet, poisonous, slightly viscous liquid; miscible in water and alcohol; boils at 197.6°C; used as an antifreeze. Also, ETHYLENE GLYCOL.

glycolate *Organic Chemistry.* a salt or ester of glycolic acid.

glycol dehydrator *Chemical Engineering.* the equipment used to remove water from a wet gas by contacting with glycol.

glycol ester *Organic Chemistry.* a compound formed by reaction of a glycol and an organic acid.

glycol ether *Organic Chemistry.* a colorless liquid used in detergents, as a solvent, and as a dilutent.

glycolic *Organic Chemistry.* a salt or ester of glycolic acid.

glycolic acid *Organic Chemistry.* $CH_2OHCOOH$, colorless, hygroscopic leaflets; soluble in water and alcohol; melts and decomposes at 78°C; used as a chemical intermediate in fabric dyeing.

glycolipid *Biochemistry.* any of a class of compounds in which one or more monosaccharides are glycosidically linked to a lipid; occurring in the brain and other nervous tissue.

glycolysis *Biochemistry.* **1.** the anaerobic breakdown of carbohydrates, in which a molecule of glucose is converted by a series of steps to two molecules of lactic acid, yielding energy in the form of ATP. **2.** the sequence of reactions from glucose to pyruvic acid that is common to both aerobic and anaerobic carbohydrate catabolism.

glycolytic pathway *Biochemistry.* the lengthy series of reactions followed in the anaerobic degradation of carbohydrates to produce energy, usually divided into four major steps or parts. Also, EMBDEN-MEYERHOF PATHWAY.

glyconeogenesis *Biochemistry.* the synthesis of glycogen from noncarbohydrate precursors such as fat or protein.

glyconic acid see GLUCONIC ACID.

glycophorin *Biochemistry.* a carbohydrate-rich protein that spans the red blood cell membrane.

glycophyte *Botany.* a plant that will grow only in soil that has more than a 0.5% sodium chloride solution. Most plants are glycophytes. Thus, **glycophytic.**

glycoprotein *Biochemistry.* a class of conjugated protein in which the nonprotein portion is a carbohydrate that is linked covalently to the protein and contains less than 4% hexosamine.
glycosaminoglycan *Biochemistry.* any of a group of polysaccharides consisting of repeating disaccharide units of amino sugar derivatives, found in the proteoglycans of connective tissue. Also, MUCOPOLYSACCHARIDE.
glycose *Biochemistry.* a simple sugar that is shaped like an open-chain ketone or a cyclic hemiacetal.
glycosidase *Enzymology.* any of a large group of enzymes of the hydrolase class that break the hemiacetal bonds of glycosides.
glycoside *Biochemistry.* any substance with an alcohol component in which a glycosyl group has replaced the hydrogen in the hydroxyl group.
glycosidic bond *Organic Chemistry.* a linkage between monosaccharides to form di- and polysaccharides.
glycospingolipid *Biochemistry.* any ceramide derivative that contains more than one sugar residue.
glycosuria *Medicine.* the excretion of an abnormally large amount of glucose in the urine, more than 1 g in 24 hours.
glycosyl *Biochemistry.* a radical, derived from a sugar or starch by removal of an anomeric hydroxyl group, that is linked to another molecule by means of a glycosidic bond.
glycosylation *Biochemistry.* a process in which an irreversible bond is formed between glucose and N-terminal valine in the hemoglobin beta chain to yield hemoglobin A.
glycosyltransferase *Enzymology.* any of several enzymes that transfer glycosyl groups from one molecule to another. Also, TRANSGLYCOSYLASE.
glycotropic *Biochemistry.* **1.** attracting or having an affinity for sugar. **2.** mediated by sugar. **3.** causing hyperglycemia.
glycuresis *Physiology.* the increase of glucose in the urine after an ordinary carbohydrate meal.
glycuronic acid see GLUCURONIC ACID.
glycuronide see GLUCURONIDE.
glycyl *Organic Chemistry.* NH_2CH_2COO–, the radical form of glycine, found in peptides.
glyodin *Plant Pathology.* a liquid fungicide used against apple scab and other foliage diseases of fruit trees and ornamentals.
glyoxal *Organic Chemistry.* $(CHO)_2$, yellow crystals or a light yellow liquid with a milk odor; soluble in anhydrous solvents; melts at 15°C and boils at 51°C; used in embalming fluids, permanent-press fabrics, as an insolubilizing agent, and for leather tanning.
glyoxalase *Enzymology.* an enzyme that converts methylglyoxal into lactic acid; composed of two enzymes, **glyoxalase I** (lactoylglutathione lyase) and **glyoxalase II** (hydroxyacylglutathione hydrolase).
glyoxalic acid *Organic Chemistry.* CHOCOOH, colorless crystals; soluble in water and slightly soluble in alcohol; anhydrous form melts at 98°C; aqueous solution highly corrosive.
glyoxaline or **glyoxalin** see IMIDAZOLE.
glyoxylate cycle *Biochemistry.* the series of reactions resulting in the formation of succinate from acetyl CoA, enabling carbohydrates to be made from fatty acids.
glyoxylic acid *Biochemistry.* $C_2H_2O_3$, a carboxylic acid found in green fruits, seedlings, and young leaves; the starting point for the glyoxylate cycle.
glyoxysome *Botany.* cytoplasmic organelles in plant cells, especially in the endosperm and cotyledons of oil-rich seeds, that are similar to peroxisomes, but also contain enzymes of the glyoxylate cycle that generates succinate from acetate.
glyph [glif] *Archaeology.* an individual image or design element carved in or painted on stone, such as those in hieroglyphics, pictographs, or petroglyphs.
Glyphocyphidae *Paleontology.* an extinct family of irregular echinoids in the superorder Camarodonta; some genera have echinoid plates and some diadematoid; lower Jurassic to Eocene.
glyphosate *Organic Chemistry.* $C_3H_8NO_5P$, a white solid; soluble in water; melts at 200°C and decomposes at 230°C; used as an herbicide.
glyphosine *Organic Chemistry.* $C_4H_{11}NO_8P_2$, a white solid; soluble in water; used as a growth regulator in sugarcane.
glyptal resin *Organic Chemistry.* **1.** a trademark for a group of alkyd-type polymers and plasticizers. **2.** a phthalic anhydride-glycerol polymer used in lacquers and in insulation.
Glyptocrinina *Paleontology.* a suborder of dicyclic camerate crinoids in the extinct order Monobathrida; middle Ordovician to Permian.
Glyptodontidae *Paleontology.* an extinct family of toothless, loricate, ant-eating mammals in the order Xenarthra; formerly classified in the order Edentata; extant in the Cenozoic. Also, HOPLOPHORIDAE.
glyptolith see VENTIFACT.
GM guided missile.
Gm gigameter.
gm gram or grams.
GMAT or **G.M.A.T.** Greenwich mean astronomical time.
Gmelin, Leopold [gə mā′lin] 1788–1853, German chemist and physiologist; discovered potassium ferricyanide (also called **Gmelin's salt**); wrote the first comprehensive handbook of chemistry.
gmelinite *Mineralogy.* $(Na_2,Ca)Al_2Si_4O_{12}·6H_2O$, a colorless to flesh-red hexagonal mineral of the zeolite group, forming tabular, pyramidal, or rhombohedral crystals with a vitreous luster, having a specific gravity of 2.01 to 2.17 and a hardness of 4.5 on the Mohs scale; found in basaltic lavas.
Gmelin's test [mā′linz; gə mā′linz] *Pathology.* a method of detecting and quantifying bile in bodily fluids, especially urine, by the addition of nitric acid mixed with nitrous acid; color layers appear where the solutions meet, starting with yellow, then red, violet, blue, and green.
GMT or **G.M.T.** Greenwich mean time.
GMW gram-molecular weight.
gnat [nat] *Invertebrate Zoology.* a common name for various biting flies such as midges, blackflies, or sandflies; insects in the order Diptera.

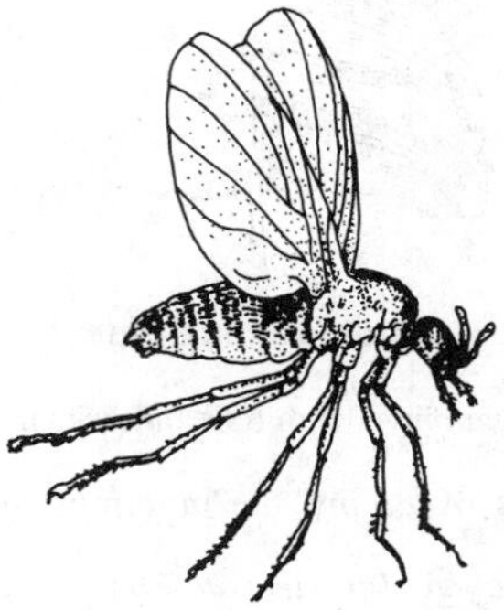

gnat

gnatcatcher *Vertebrate Zoology.* any of various tiny North and South American songbirds of the genus *Polioptila* in the family Sylviidae, that scour leaves for insects; characterized by a long tail and soft plumage.
gnateater see ANT PIPIT.
gnath- or **gnatho-** a combining form meaning "jaw."
gnathic index [nath′ik] *Anthropology.* in anthropometry, the degree of prominence of the upper jaw, expressed as a percentage of the distance from basion to nasion.
Gnathiidea *Invertebrate Zoology.* a suborder of isopod crustaceans; larvae are ectoparasitic on marine fish.
gnathion *Vertebrate Zoology.* the anterior premaxillae on or near the middle line in certain lower mammals.
gnathite *Invertebrate Zoology.* in arthropods, a mouth appendage or part.
gnathobase *Invertebrate Zoology.* in some crustaceans and arachnoids, a process on certain appendages modified for carrying or chewing food.
Gnathobdellae *Invertebrate Zoology.* a family of jawed, bloodsucking leeches, in the order Arhynchobdellida, including many important parasites of humans and other warm-blooded animals.
Gnathobelodontinae *Paleontology.* a subfamily of proboscideans in the extinct family Gomphotheriidae, characterized by a projecting, scoop-shaped jawbone; Pliocene.
gnathocephalon *Invertebrate Zoology.* on an insect's head, the several fused segments bearing the mandibles and maxillae.
gnathochilarium *Invertebrate Zoology.* in some arthropods, especially millipedes, a platelike mouth structure formed from the fusion of maxillae and labium.
Gnathodontidae *Paleontology.* a family of conodonts in the order Ozarkodinida that arose in the late Devonian and diversified widely in the Carboniferous, but became extinct in the early Permian.
gnathopod *Invertebrate Zoology.* in some crustaceans, a thoracic appendage modified for gripping.

gnathopodite *Invertebrate Zoology.* in arthropods, a modified appendage serving as a jaw, such as the maxilliped.

gnathos [nā´thōs] *Invertebrate Zoology.* in male butterflies and moths, a midventral plate on the ninth body segment.

gnathostegite *Invertebrate Zoology.* in some crustaceans, one of a pair of plates formed from the outer maxillipeds that cover associated mouth parts.

Gnathostomata *Vertebrate Zoology.* the "jaw-mouthed" vertebrates, a classification characterized by a jaw supported by bone or cartilage attached to muscle, and usually paired lateral limbs; it includes birds, mammals, reptiles, amphibians, and most fish. *Invertebrate Zoology.* a group of sea urchins in the class Echinoidea, with a lantern jaw.

Gnathostomidae *Invertebrate Zoology.* a family of parasitic nematode roundworms in the order Spirurida.

Gnathostomulida *Invertebrate Zoology.* a group of minute marine worms.

gnathothorax *Invertebrate Zoology.* in arthropods, a primary body region, including the thorax and the part of the head with the feeding organs.

gneiss [nīs] *Petrology.* a variety of foliated metamorphic rock characterized by alternating bands of dark and light minerals; the dark bands are foliated while the light are granulitic.

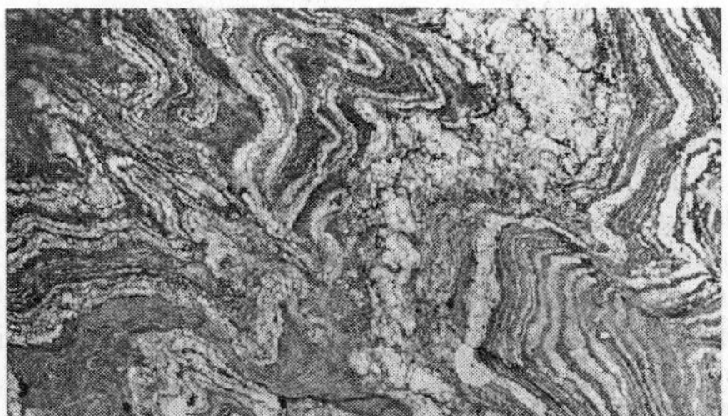

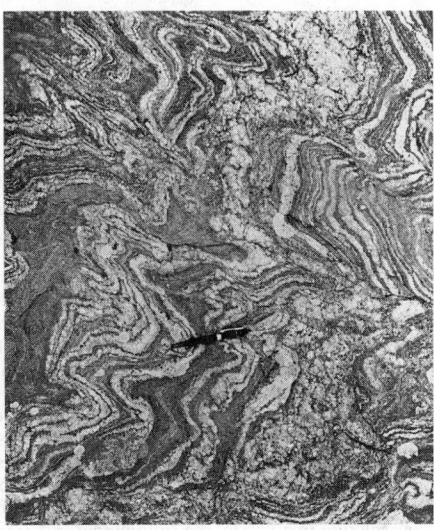

gneiss

gneissic [nī´sik] *Petrology.* relating to, resembling, or composed of gneiss.

gneissic granodiorites *Petrology.* granodiorite rocks that manifest the characteristics of gneiss.

Gnetales *Botany.* a monogeneric order of the subdivision Gneticae, characterized by vessels in the wood and generally found in warm temperate or tropical regions.

Gneticae *Botany.* a subdivision of dicotyledonous plants of the division Pinophyta, characterized by opposite leaves that range from scalelike to broad and flat, and vessels in the wood.

Gnetidae *Botany.* in some classifaction systems, one of three subclasses of the gymnosperm subdivision Gneticae, composed of the single monogeneric family Gnetaceae; characterized by climbing vines to small trees, evergreen leaves with reticulate veins, and long, slender pistillate and staminate cones; native to tropical and humid regions of South America, Africa, India, China, and New Guinea.

Gnetopsida *Botany.* a class of gymnosperms composing the subdivision Gneticae. Also, **Gnetophyta.**

gnomon [nō´män; nō´mən] *Horology.* the slanted plate or pin of a sundial that casts a shadow on the graduated disk or other such time-keeping surface. Also, STYLE. *Mathematics.* a geometric figure obtained from a parallelogram by removing a smaller similar parallelogram from a corner of the original parallelogram.

gnomonic *Horology.* of or relating to a gnomon or to a sundial.

gnomonic map *Cartography.* any chart that is based on a gnomonic projection. Similarly, **gnomonic chart.**

gnomonic map projection see GNOMONIC PROJECTION.

gnomonic projection *Cartography.* a method of creating a map in which the points on the surface of a sphere are projected from the center of the sphere onto a plane that is tangent to the sphere. This kind of map produces great distortion except near the point of tangency, but any great circle is represented as a straight line; thus it is sometimes referred to as a "great-circle chart." Also, ORTHODROMIC MAP PROJECTION.

Gnostidae see PTINIDAE.

gnotobiology *Biotechnology.* an experimental biological-culture system to which only preselected components (organisms and nutrients) are introduced; used for studying the dynamics of feeding in single or multi-species systems.

gnotobiote *Microbiology.* an environment or organism that has been specifically reared or monitored such that its entire complement of microorganisms has been identified. Thus, **gnotobiotic.**

GnRH gonadotropin-releasing hormone.

gnu [noo] *Vertebrate Zoology.* either of two species of stocky, oxlike antelopes of the genus *Connochaetes*; the **brindled gnu**, *C. taurinus,* is found on the eastern African plains and has a bull-like head with horns in both sexes and tufts of hair growing from the muzzle, throat, and chest, an upright mane, and a brown to black coat; the nearly extinct **white-tailed gnu** of southern Africa, *C. gnou,* is now protected. Also, WILDEBEEST.

gnu

GNU *Computer Science.* an initialism used to name or describe software programs that are distributed by the Free Software Foundation, such as GNU Emacs.

GO general order.

goa *Vertebrate Zoology.* a type of gazelle found in the Tibetan plateau, *Procapra picticaudata.*

Goa *Ordnance.* a Soviet two-stage, close-range, surface-to-air missile powered by a rocket motor and equipped with radio command guidance; it delivers a conventional warhead at a range of 20 miles; officially designated **SA-3**; a later ship-based version is designated **SA-N-1.**

goaf *plural,* **goaves.** *Mining Engineering.* **1.** the part of a mine from which the coal has been extracted and the space filled up. **2.** the refuse or waste left in a mine. Also, GOB.

goal *Behavior.* the result or reward that a given behavior is directed to achieve. *Artificial Intelligence.* **1.** something that is to be achieved by a program. **2.** a representation in a formal language or data structure of this objective, or of a set of conditions that, if satisfied, will accomplish the objective.

goal-coordination method *Control Systems.* a method of coordinating subproblem solutions in plant decomposition. Also, INTERACTION BALANCE METHOD OR NONFEASIBLE METHOD.

goal-directed behavior *Behavior.* behavior that is motivated by an organism's intention to approach or avoid a particular place, object, or situation.

goal-directed reasoning see BACKWARD REASONING.

goal gradient *Behavior.* the principle that closer proximity to a goal will increase the motivation to reach a positive goal or to avoid a negative one.

goal-seeking *Artificial Intelligence.* the ability of some expert systems and other decision aids to determine what values certain variables must assume in order to reach the desired goal.

goal state *Artificial Intelligence.* a state of a search that represents or satisfies the goal.

goat *Vertebrate Zoology.* **1.** an herbivorous ruminant of the genus *Capra*, related to the sheep and characterized by upward-curving, hollow horns, usually a beard, and physical agility; found in rocky and mountainous regions of the Old World and raised worldwide for their milk, flesh, hair, and hides. **2.** any of certain related varieties, including the mountain goat. *Astronomy.* the constellation Capricorn.

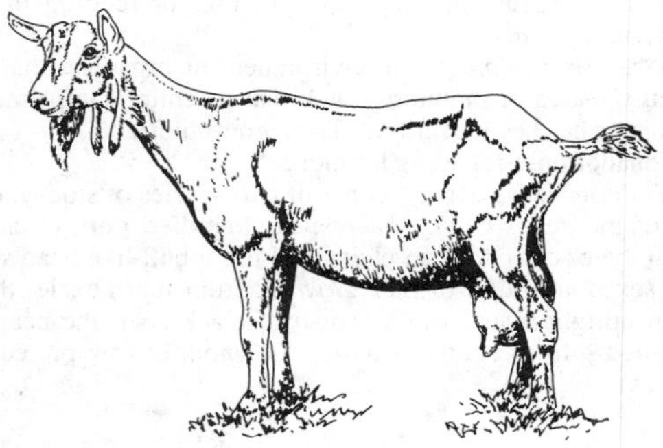

domestic goat

goatfish *Vertebrate Zoology.* any elongated marine fish of the family Mullidae having spiny rays, soft fins with a forked tail, and barbels under the jaw used as feelers to probe the bottom for invertebrates; found in tropical and temperate waters worldwide and valued as a food-fish. Also, MULLET.

goatsucker *Vertebrate Zoology.* any of various insectivorous, nocturnal birds of the family Caprimulgidae, once thought to milk goats with their extremely wide mouths. Also, NIGHTJAR.

gob see GOAF.

Gobi *Geography.* a large desert in Mongolia and northern China.

gobi *Geology.* a small, exposed, level-surfaced synclinal basin, or an accumulation of sediments deposited in such a basin.

Gobiatheriidae *Paleontology.* a single-genus family of large ungulates in the extinct suborder Dinocerata; a Chinese relative of the Uintatheriidae, the Gobiatheriidae were characterized by a long, low skull without tusks or other protuberances; middle Eocene.

Gobiesocidae *Vertebrate Zoology.* the clingfishes, a family in the order Gobiesociformes, characterized by slender, scaleless bodies, a large, broad head, and a large two-part suction disc formed from anterior ventral fins used to cling to rocks; found in shallow waters of tidal zones and temperate seas worldwide.

Gobiesociformes *Vertebrate Zoology.* an order of mostly bottom-living bonyfishes characterized by a scaleless head and body; found in shallow waters.

Gobiidae *Vertebrate Zoology.* the gobies, a family of small, benthic fishes of the suborder Gobioidei, characterized by a two-part dorsal fin and united pelvic fins forming a suction disc for clinging; some species can live out of water for extended periods of time; found in inshore, coastal, and tidal areas in all tropical and temperate zones.

Gobioidei *Vertebrate Zoology.* a suborder of fishes of the order Perciformes, including the gobies, sleepers, and wormfishes; found in tropical and temperate zones worldwide.

Goblet *Ordnance.* a Soviet ship-based anti-aircraft missile; officially designated **SA-N-3.**

goblet cell *Histology.* a mucus-secreting, glandular epithelial cell characteristic of mucous membranes such as the lining of the gut. *Invertebrate Zoology.* any of the unicellular, freshwater choanoflagelates, in the genus *Monosiga*, that attach by a thin stalk to water plants.

gobo *Acoustical Engineering.* a shield that is placed in front of a microphone to absorb certain unwanted frequencies before they are picked up by the microphone. *Graphic Arts.* a strip of dark material that is used to shield a camera lens from direct light or sometimes to create a shadow effect.

gob stink *Mining Engineering.* **1.** the odor of burning coal given off by underground fire. **2.** the odor emitted by the spontaneous heating of coal, not necessarily in the gob. Also, STINK.

goby *plural,* **gobies** or **goby.** *Vertebrate Zoology.* any fish of the family Gobiidae.

Robert H. Goddard

Goddard, Robert H. 1882–1945, American physicist; pioneer in both the theory of rocket propulsion and the design of rockets.

Gödel, Kurt [gö´dəl; gur´dəl] 1882–1945, American mathematician and logician, born in Czechoslovakia.

Gödel-Bernays set theory *Mathematics.* the formulation of axiomatic set theory that includes the primitive undefined notions of class, membership, and equality.

Gödel incompleteness theorem see GÖDEL'S PROOF.

Gödel number *Mathematics.* **1.** a unique positive integer that can be assigned to every mathematical statement (of finite length), using the following construction: (a) Assign every mathematical symbol, numerical digit, letter, etc. (i.e., every typographical symbol including blank space) to a unique positive odd integer. (b) If the statement in question has *n* typographical symbols written in a standard order, multiply the first *n* consecutive prime numbers (one for each symbol in the mathematical statement), first raising each prime number in the list to the integral power corresponding to the symbol whose spot is marked by the prime number. The resulting integer is the Gödel number of the mathematical statement, and the process of obtaining it is **Gödelization. 2.** an alternative construction is given by the following: (a) Associate the digits 1 through 9 with themselves and the digit 0 with 11. (b) Associate the letters of the alphabet with the numbers 12 though 39 consecutively, skipping multiples of ten. (c) Associate any mathematical symbols needed with succeeding numbers, skipping multiples of ten. (d) Now symbolize any axiom, theorem, formula, proof, etc., using this "dictionary," separating individual symbols with 0 and words or lines with 00. This construction gives Gödel numbers of smaller magnitude.

Gödel's proof *Mathematics.* the name given to the result that the consistency of a logical system cannot be proved within that system; that is, there exists an unprovable statement within every sufficiently complicated logical system. Also, GÖDEL INCOMPLETENESS THEOREM.

go-devil an informal term for various simple tools and devices; specific uses include: *Engineering.* a cylindrical plug made of available material, such as paper or sacking, that is put through the pump end of a pipeline and moves under pressure through the pipeline to clean it after use. Also, PIG, RABBIT. *Transportation Engineering.* **1.** a sled for transporting logs and for cultivation. **2.** a railroad car that hauls materials and transports personnel. *Agriculture.* a wide rake that is used to collect hay.

Godwin-Austen *Geography.* another name for K2, the mountain in the Karakorum range of northern Kashmir that is the second-highest (now thought to be possibly the highest) peak in the world.

Goebeliellaceae *Botany.* a monotypic family of liverworts belonging to the order Jungermanniales, noted for producing attractive festoons and characterized by wiry, rigid stems, purple rhizoids arising in fascicles from underleaf bases, and very large, single-celled spores.

Goertler parameter *Fluid Mechanics.* a dimensionless quantity given by the product of the Reynolds number and the square root of the thickness of a boundary-layer flow over a curved surface, divided by the square root of the radius of curvature of the surface.

Goethals, George Washington [gō´thəlz] 1858–1928, American civil engineer and military officer; chief engineer for the construction of the Panama Canal.

Goethe, Johann Wolfgang von [gur´tə] 1749–1832, German author, philosopher, and scientist; discovered the intermaxillary bone; wrote on plant and animal classification.

goethite [gō´thīt; gů´thīt; gur´tīt] *Mineralogy.* α-$Fe^{+3}O(OH)$, a brittle, yellowish, reddish, and blackish-brown orthorhombic mineral, polymorphous with akaganeite, feroxyhyte, and lepidocrocite; usually earthy to compact or in fibrous colloform masses, having a specific gravity of 3.3 to 4.3 and a hardness of 5 to 5.5 on the Mohs scale; widely found as an alteration product of iron-bearing minerals. (Named after *Goethe.*)

go gauge *Mechanical Devices.* a gauge that fits a given part; used in conjunction with a no-go gauge to establish maximum and minimum dimension limits for that part.

going-up theorem *Mathematics.* suppose S is an integral extension ring of R, and P_1 and P are prime ideals in R such that P_1 is properly contained in P. Further, suppose that Q_1 is a prime ideal of S such that $Q_1 \cap R = P_1$. Then there exists a prime ideal Q of S such that Q_1 is properly contained in Q and $Q \cap R = P$.

goiter *Medicine.* an enlargement of the thyroid gland, causing a swelling in the front part of the neck; may be associated with hyperthyroidism, hypothyroidism, or normal levels of thyroid function, and may be treated by surgical removal or the administration of antithyroid drugs, radioiodine, or thyroid hormone.

gold *Chemistry.* a metallic element having the symbol Au, the atomic number 79, an atomic weight of 196.967, melting point 1063°C, and a boiling point of 2800°C; it is chemically nonreactive and soluble in aqua regia. *Metallurgy.* this precious, highly malleable, high-density metal, widely used for jewelry and formerly for making coins; it also has many industrial applications, as in dentistry. (An Old English word; ultimately derived from a word for "yellow" or "shining.")

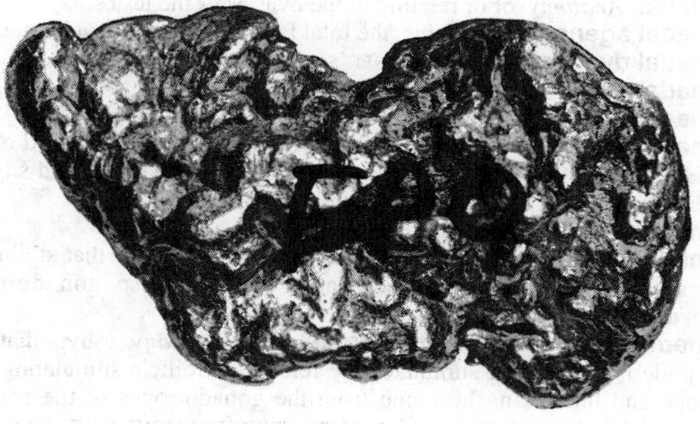

gold

gold-198 *Nuclear Physics.* a radioisotope of gold with a half-life of 2.7 days, used in medicine to treat tumors.

gold alloy *Metallurgy.* an alloy containing at least 10% gold, extensively used in dentistry, and in electrical and jewelry applications.

Goldbach, Christian 1690–1764, Prussian mathematician; formulated the **Goldbach Conjecture**, the (unproven) assertion that every even number greater than 2 is the sum of two primes.

goldbeating *Metallurgy.* the process of producing gold leaves.

Goldberger, Joseph 1874–1929, Hungarian-born American physician; identified the cause of and cure for pellagra.

gold bronze *Metallurgy.* a copper-base alloy containing 5% zinc, 3% lead, and 2% tin.

gold chloride *Inorganic Chemistry.* $AuCl_3$, red crystals; soluble in water and alcohol; decomposes at 254°C and sublimes at 265°C; used in photography, gold plating, medicine, ceramics, and glassmaking. Also, AURIC CHLORIDE, GOLD TRICHLORIDE.

golden algae *Botany.* a common name for freshwater plants of the class Chrysophyceae.

golden-brown algae *Botany.* a common name for plants of the division Chrysophyta.

golden current *Botany.* a shrub, *Ribes aureum,* of the saxifrage family, characterized by purplish fruit and drooping clusters of yellow flowers that turn reddish; found in western North America.

golden eagle

golden eagle *Vertebrate Zoology.* a diurnal bird of prey, *Aquila chrysaetos,* of the family Accipitridae that inhabits mountainous regions of the Northern Hemisphere; characterized by a great wingspan (up to 8 feet), a sharply hooked beak, a brown body and golden nape, large talons on yellow feet, and feathered legs; the golden eagle is the national bird of Mexico.

goldeneye *Vertebrate Zoology.* one of two diving ducks, *Bucephala clangula* of Eurasia and North America and *B. islandica* of North America, characterized by black and white feathers, bright yellow eyes, and white cheek patches; found primarily in northern areas.

golden mole *Vertebrate Zoology.* a small, fossorial African insectivore of the family Chrysochloridae, characterized by irridescently gold to green hair, blind eyes, long, pointed snouts, and picklike foreclaws used for burrowing; feeds mainly on earthworms.

golden ratio *Mathematics.* the ratio of two adjacent sides of a golden rectangle; equal to $(1 + \sqrt{5})/2$.

golden rectangle *Mathematics.* a rectangle that can be subdivided into a square and a smaller rectangle similar to the original rectangle.

goldenrod *Botany.* a composite plant of the genus *Solidago* characterized by long stalks with small, yellow flowers that bloom in late summer or early fall; the blossom of various species is the state flower of Kentucky and Nebraska.

goldenrod paper *Graphic Arts.* a golden-colored paper often used in platemaking to mask nonimage areas on photographic negatives. A flat masked with such paper is called a **goldenrod flat.**

golden seal *Botany.* **1.** a plant of the genus *Hydrastis canadensis* of the buttercup family, characterized by a thick yellow rootstock. **2.** the root of this plant, once used in medicine and to stop bleeding.

golden section *Mathematics.* the partition of a line segment AB by a point P, so that $AB/AP = AP/BP$. If $AP/PB = x$, then $x^2 - x - 1 = 0$ and $x = (1 + \sqrt{5})/2$. x is also called the golden ratio.

golden shiner *Vertebrate Zoology.* a small, silver freshwater minnow of the genus *Notemigonus soleucas* often used as live bait in sport fishing; native to eastern North America and introduced in western North America.

golden stars *Botany.* a plant of the amaryllis family, genus *Bloomeria crocea,* characterized by clusters of golden-orange, star-shaped flowers; native to southern California.

goldeye *Vertebrate Zoology.* a herringlike game fish of the genus *Hiodon alosoides,* found in fresh waters of central North America.

gold-filled *Metallurgy.* describing an article that has been coated or cladded with gold or a gold alloy to a specified thickness relative to the thickness of the base.

goldfinch *Vertebrate Zoology.* a small, red, black, and yellow finch of the family Carduelidae, having a delicate bill and a short, notched tail; found in Europe (genus *Carduelis*) and America (genus *Spinus*).

goldfish *Vertebrate Zoology.* a hardy orange or yellow cyprinid fish of the species *Carassius auratus,* having strongly serrated front fin spines on both the dorsal and anal fins; found in marshy pools and in standing and slow-moving waters; widely used as an aquarium and pond fish.

gold foil *Metallurgy.* a foil consisting essentially of pure gold.

Goldhaber triangle *Particle Physics.* a phase-space triangle in which the plotted points correspond to the boundaries for a high-energy reaction leading to four or more particles.

gold hydroxide *Inorganic Chemistry.* $Au(OH)_3$, a brown light-sensitive powder that is insoluble in water; it loses water easily to form the oxide; used in gilding and gold plating. Also, AURIC HYDROXIDE.

Goldie's fern *Botany.* a wood fern, *Dryopteris goldiana,* characterized by large, golden green fronds with blades that tilt backward; found in northeastern North America. (Named for its discoverer, John *Goldie,* 1793–1886, a Scottish plant collector.)

gold leaf *Metallurgy.* an extremely thin, pure gold foil, often translucent, mainly used for gilding.

gold number *Analytical Chemistry.* the number of milligrams of a protective colloid that prevents a standard gold sol in a solution of sodium chloride from coagulating.

gold oxide *Inorganic Chemistry.* Au_2O_3, a brownish-black light-sensitive powder, insoluble in water and decomposed by heat; used in gold plating. Also, AURIC OXIDE, GOLD TRIOXIDE.

gold plate *Metallurgy.* any gold or gold-alloy coating.

gold point *Thermodynamics.* the normal freezing point of gold at standard atmospheric pressure (1064.43°C); a fixed-point temperature of the International Practical Temperature Scale of 1968.

gold salt see SODIUM GOLD CHLORIDE.

Goldschmidt's law *Solid-State Physics.* a law stating that crystal structure is determined by the ratio of the numbers of constituent atoms, the ratios of their sizes, and their polarizations.

gold size *Chemistry.* a medium for affixing metallic powders to surfaces and sealing microscopic preparations permanently.

gold sodium thiomalate *Pharmacology.* either of two gold salts, $C_4H_4AuNaO_4S$ or $C_4H_3AuNa_2O_4S$, used in the treatment of the early stages of active rheumatoid arthritis. Also, SODIUM AUROTHIOMALATE.

gold sodium thiosulfate *Pharmacology.* $AuNa_3O_6S_4$, a drug used in the treatment of the early stages of active rheumatoid arthritis; also used to treat lupus erythmatosus. Also, SODIUM AUROTHIOSULFATE.

gold solder *Metallurgy.* a solder consisting of 45–80% gold and variable amounts of alloying elements, usually silver and copper.

gold stannate *Inorganic Chemistry.* a brown powder made up of a mixture of gold chloride and brown tin oxide; insoluble in water; used in coloring enamels, manufacturing ruby glass, and painting porcelain. Also, GOLD TIN PURPLE OF CASSIUS.

Goldstein, Eugen 1850–1930, German physicist; studied and named cathode rays; discovered and analyzed canal rays.

Goldstein, Joseph L. born 1940, American biochemist; shared Nobel Prize for discovering low-density lipoprotein receptors.

Goldstein's rays *Radiology.* the rays produced when X-rays interact with a radiolucent material. Also, S RAYS.

Goldstone bosons *Physics.* the bosons having zero mass and zero spin that appear in the process of spontaneous symmetry breaking.

gold tin purple see GOLD STANNATE.

gold trichloride see GOLD CHLORIDE.

gold trioxide see GOLD OXIDE.

golfada *Meteorology.* a heavy gale that occurs in the area of the Mediterranean Sea.

golf ball *Engineering.* the printing component of certain typewriters or serial printers, consisting of a removable type ball that moves across the printing surface and revolves to offer the desired character for impact.

Golgi, Camillo [gól´jé] 1844–1926, Italian physician; Nobel Prize for studies on structure of nervous system.

Golgi apparatus *Cell Biology.* a complex cytoplasmic organelle consisting of a series of layered cisternae and associated small vesicles and involved in terminal glycosylation, membrane flow, secretion, and delivery of cellular products either to the cell surface or to the appropriate intracellular destination.

Golgi cell *Anatomy.* a type of multipolar neuron found in the cerebral cortex and dorsal horns of the spinal cord.

Golgi-Mazzoni corpuscle *Anatomy.* a sensory nerve ending found in the subcutaneous tissue of the fingertips.

Golgi's corpuscle *Anatomy.* an encapsulated sensory nerve ending found in a tendon.

Golgi's theory *Neurology.* the theory that the axons of Golgi's cells and the collaterals of the axons of Deiters' cells interconnect, and that such connections are important in the neurotransmission of information.

Golgi tendon organ *Physiology.* an afferent nerve fiber having terminals in a tendon that joins muscle and bone, and reacts to an increase in muscle tension produced by stretch or contraction.

Gomberg-Bachmann-Hay reaction *Organic Chemistry.* a reaction in which diaryl compounds are formed from aryldiazonium salts and aromatic compounds in the presence of alkali.

gomitoli *Neurology.* a network of specialized capillaries in the upper region of the hypothalamus that surround the terminal arterioles of the superior hypophyseal arteries and that lead through portal veins to the andenohypophysis.

Gomontiaceae *Botany.* a monogeneric family of green algae of the order Ulvales, having irregularly shaped cells with branched rhizoids that grow into mollusk shells and irregularly branched filaments that house a nucleus and a parietal chloroplast.

Gomortegaceae *Botany.* a monospecific family of aromatic evergreen trees of the family Laurales, characterized by volatile oils contained in scattered spherical cells and an edible yellow fruit; native to Chile.

Gomphaceae *Mycology.* a family of fungi belonging to the order Agaricales that is composed of some coral fungi and characterized by pigmented spores.

Gomphidae *Invertebrate Zoology.* a family of cubtail dragonflies in the order Odonata, having well-separated eyes and an unnotched labium.

Gomphidiaceae *Mycology.* a family of fungi belonging to the order Agaricales that is composed of gill mushrooms which live primarily in symbiotic relationships with conifer trees.

gomphosis *Anatomy.* an immovable skeletal articulation in which one peglike bone is received in the cavity of another, such as a tooth in its socket.

Gomphotheriidae *Paleontology.* an extinct family of long-jawed proboscidean mammals in the suborder Euelephantoidea; standing about 9 feet at the shoulder; the **gomphotheres** were the typical bunodont mastodons of the Miocene, and were found everywhere except Australia and Antarctica; extant in the Late Oligocene to Early Pleistocene.

Gomphotheriinae *Paleontology.* a subfamily of mastodons in the extinct family Gomphotheriidae; found in Asia, Europe, and North America; Oligocene to Pleistocene.

gon- a combining form meaning: **1.** reproductive or sexual, as in *gonaduct.* **2.** of or relating to the knee, as in *gonarthritis.*

gonad [gō´nad] *Anatomy.* a gamete-producing gland; an ovary or a testis, either in the embryo or adult.

gonadal *Anatomy.* of or relating to the ovaries or the testes.

gonadal agenesis *Medicine.* the total failure of gonad development.

gonadal dysgenesis see TURNER'S SYNDROME.

gonadal ridge see GENITAL RIDGE.

gonadectomy *Medicine.* the removal of a sex gland or gonad.

gonadoblastoma *Oncology.* a dysgerminoma that is composed of gonadal elements, including germ cells, sex cord derivatives, and stromal derivatives.

gonadopium see GONOPODIUM.

gonadotropin *Endocrinology.* any hormonal substance that stimulates or sustains the functions of the ovaries or testes. Also, **gonadotropic hormone.**

gonadotropin-releasing hormone *Endocrinology.* a hypothalamic peptide hormone that stimulates the release of follicle-stimulating hormone and luteinizing hormone from the gonadotrophs of the anterior lobe of the pituitary gland. Also, LUTEINIZING HORMONE-RELEASING HORMONE.

gonangium *Invertebrate Zoology.* an asexual reproductive polyp of a hydrozoan colony producing medusa buds.

Gonapodyaceae *Mycology.* a family of fungi belonging to the order Monoblepharidales that is microscopic and occurs on land and in water, living off organic material, such as fruits or woody twigs.

gonapophysis *Invertebrate Zoology.* in insects, a tubular posterior appendage modified to serve in copulation, egg laying, or stinging.

goncalo alves *Materials.* the durable, hard, heavy wood of the *Astronium fraxinifolium* tree of the American tropics; used for furniture and building construction.

gondola [gän´də lə; gän dō´lə] *Naval Architecture.* a long, narrow, flat-bottomed boat with a raised, decorated stem and stern; used especially on the canals of Venice, Italy. *Navigation.* the passenger compartment of an airship or balloon. *Engineering.* **1.** the passenger car of an elevated cable car or tramway system. **2.** see GONDOLA CAR.

gondola car *Engineering.* an open, level-bottomed car, often having removable ends, used to transport heavy materials such as rock or steel.

Gondwanaland *Geology.* the supercontinent or landmass that supposedly fragmented millions of years ago to form the present-day continents of the Southern Hemisphere. Also, **Gondwana.**

gonglylidia *Entomology.* hyphal swellings or nodes found in fungi colonies grown by termites or ants.

Goniadidae *Invertebrate Zoology.* a family of marine polychaete worms in the subclass Errantia that live in sandy mud sediments.

goniatites *Paleontology.* the simplest of the three large groups of ammonoids, goniatitic sutures characterized by a simple pattern of lobes and saddles.

goniatitid *Paleontology.* of or relating to ammonoids in the large order Goniatitida, widespread from the Devonian to the end of the Permian.

gonidium *Biology.* an asexual reproductive cell.

gonimoblast *Botany.* a thin strand arising from the fertilized carpogonium of most red algae.

gonio- a combining form meaning "angle," as in *goniometer.*

goniometer *Engineering.* an instrument used for measuring angles or to find directions. *Electromagnetism.* specifically, a device used to determine the direction of maximum response of a received radio signal or the direction of maximum radiation for a transmitted signal.

goniometer head *Crystallography.* a device for orienting a crystal by means of translational motions and, in some models, movable arcs.

goniophotometer *Optics.* a device that measures the intensity of light reflected from a surface at different angles.

Goniopora *Paleontology.* a still-living genus of colonial reef-building corals in the order Scleractinia; massive or branching; extant from the middle Cretaceous to the present.

gonioscope *Medicine.* an optical instrument used for studying the angle of the anterior chamber of the eye.

Goniotrichaceae *Botany.* a little-known family of marine, brackish, and freshwater red algae of the order Porphyridiales, characterized by thalli consisting of cell aggregations.

gonitis *Medicine.* an inflammation of the knee joint.

gonnardite *Mineralogy.* $Na_2CaAl_4Si_6O_{20}\cdot 7H_2O$, an easily fusible, white, orthorhombic mineral of the zeolite group occurring in fine fibrous spherules, having a specific gravity of 2.26 to 2.69 and a hardness of 4 to 5 on the Mohs scale; found in cavities in basalt.

gono- a combining form meaning "reproductive" or "sexual."

gonococcal arthritis *Medicine.* arthritis caused by *Neisseria gonorrhoeae,* the same organism that causes gonorrhea.

gonococcus *Bacteriology.* a Gram-negative, intracellular bacterium *Neisseria gonorrhoeae,* which causes gonorrhea.

Gonodactylidae *Invertebrate Zoology.* the mantis shrimp, a family of crustaceans in the order Stomatopoda.

gonopalpon *Invertebrate Zoology.* a tentaclelike, sensitive structure associated with cnidarian gonophores.

gonophore *Botany.* any plant structure bearing reproductive cells, especially the prolongated axis of a flower that bears the stamens and pistils. *Invertebrate Zoology.* see GONOZOOID.

gonopodium *Vertebrate Zoology.* a copulatory organ developed from the anal fin of live-bearing fishes that is fully developed at sexual maturity; it is enlarged and rigid and, with the assistance of an internal skeletal stucture, movable to allow contact with the female genital organ, which receives the sperm through rays that form a groove for sperm passage. Also, GONADOPIUM.

gonopore *Invertebrate Zoology.* **1.** the reproductive opening through which hydrozoans release medusae to the outside. **2.** a small genital opening, as in insects and worms.

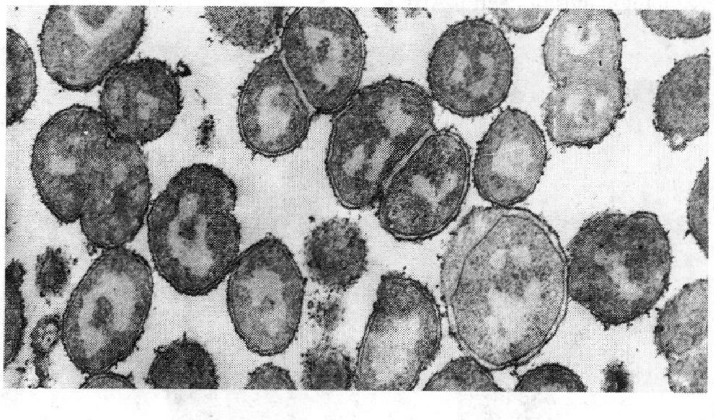

gonorrhea virus

gonorrhea [gän′ə rē′ə] *Medicine.* a venereal disease that is caused by *Neisseria gonorrhoeae,* characterized by inflammation of the mucosa of the genital tract and the discharge of pus; infection results from contact with an infected person or contact with secretions that contain the organism. Thus, **gonorrheal.**

gonorrheal vulvovaginitis *Medicine.* an infection of women with gonococci, characterized by vulvovaginitis.

gonorrhea urethritis *Medicine.* urethritis caused by infection with gonococci.

Gonorynchiformes *Vertebrate Zoology.* a small order of soft-rayed fishes characterized by a small, toothless mouth, a protractile upper jaw, and a forked caudal fin; includes the milkfishes, kneriids, cromeriids, phractolaemids, and sand eels.

gonosome *Invertebrate Zoology.* all the reproductive zooids (individuals) of a hydroid colony, including the asexual blastostyles and the sexual gonozooids.

gonosomite *Invertebrate Zoology.* in insects, the ninth abdominal segment, containing the genital structures.

gonostyle *Invertebrate Zoology.* **1.** in some Siphonophora, a reproductive individual. **2.** in some male insects, a clasping organ.

gonotheca *Invertebrate Zoology.* in certain hydrozoans, a thin, translucent exoskeleton surrounding reproductive polyps.

gonotome *Developmental Biology.* the portion of the mesoderm that eventually becomes the reproductive organs of the embryo.

gonozooid *Invertebrate Zoology.* any of the reproductive individuals of tunicate, bryozoan, or hydrozoan colonies that produce gametes.

gony- or **gonyo-** a combining form meaning "knee."

Gonyaulacaceae *Botany.* a family of marine and freshwater flagellates of the order Peridiniales, having some species that are abundant enough to discolor the water, some bioluminescent species, and some species that produce paralytic shellfish poison.

gonyautoxin *Biochemistry.* a compound, formed by dinoflagellates, that resembles saxitoxin.

gonys *Vertebrate Zoology.* a ridge along the midventral line of the lower jaw or chin of the avian beak or bill.

Goodall, Jane born 1934, English zoologist; noted for her studies of primate behavior.

Goodeidae *Vertebrate Zoology.* a family of small, freshwater, vivaparous fishes native to the highlands of Mexico and nearby areas.

Goodeniaceae *Botany.* a family of tropical and subtropical dicotyledonous, mostly perennial herbs and woody, treelike plants of the family Campnulales; characterized by a complex pollen presentation mechanism and a copious, oily endosperm.

goodness of fit *Statistics.* the degree of agreement between observed values and values predicted by a model.

Goodpasture's syndrome *Medicine.* a rapidly progressing kidney disease associated with pulmonary hemorrhage.

Goodyear, Charles 1800–1860, American inventor; discovered a process for vulcanizing rubber.

googol *Mathematics.* a term for the number 10^{100}. (Supposedly based on a child's nickname for this number.)

googolplex *Mathematics.* the number 10 to the googol power.

goongarrite see HEYROVSKYITE.

goosander see MERGANSER.

goose *Vertebrate Zoology.* any of numerous wild or domesticated waterfowl of the family Anatidae, especially of the genera *Anser* and *Branta,* that are very closely related to swans and ducks; characterized by a broad body with a flattened underside, webbed feet, and a flat, broad bill; found worldwide except in Antarctica.

Canada goose

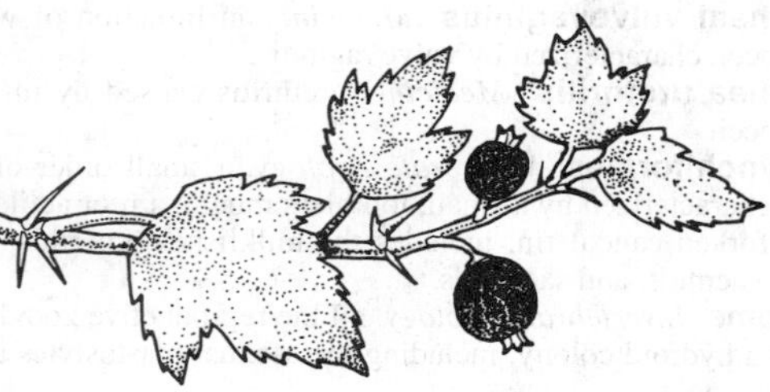

gooseberry

gooseberry *Botany*. **1.** a common name for several species of prickly, spreading bushes of the genus *Ribes* in the family Grossulariaceae, noted for producing small, acidic, edible fruit. **2.** the fruit of this species.
goosefish *Vertebrate Zoology*. a marine fish of the family Lophiidae that lures prey with a "fishing pole" hanging from a flabby, wide, flat head; characterized by large sharp teeth in a large mouth and a slender, tapered body with well-camouflaged skin; found in warm and temperate waters worldwide.
gooseflesh *Physiology*. a common name for an erection of the hairs of the skin caused by the pilomotor reflex in response to certain stimuli or changes in the environment.
goosefoot *Botany*. any plant of the genus *Chenopodium,* characterized by small, inconspicuous green flowers and weedy plants.
goose grass *Botany*. any of various types of weedlike grasses that have curved prickles on the stems or jointed stems, including cleavers, knotgrass, silverweed, arrow grass, and yardgrass.
gooseneck *Design Engineering*. a curved or bent tube, bar, or pipe. *Naval Architecture*. a swivel fitting designed to connect a boom or gaff to a mast.
gooseneck barnacle *Invertebrate Zoology*. any of the stalked barnacles in the crustacean class Cirripedia, such as those of the genus *Lepas*. Also, **goose barnacle**.
GOOVOO *Computer Programming*. a file within a generation data group, so called because of the notation used on some systems, such as G005 V002 for generation 5, volume 2.

gopher

gopher *Vertebrate Zoology*. **1.** any of various small rodents of the family Geomyidae, characterized by a large, flat head and a bulky body covered with short, stiff hair, and having powerful forearms and long, curved incisors for digging burrows; found in North American terrestrial and subterranean habitats. **2.** any of several species of small ground squirrels of the genus *Citellus* related to the chipmunks and native to prairie regions of North America.
Gopher *Ordnance*. a Soviet mobile suface-to-air missile for low-altitude defense; passive IR homing guidance; officially designated **SA-13.**
gopher hole *Engineering*. a level, T-shaped hole that is cut into rock before blasting.
gopher-hole blasting see COYOTE BLASTING.
gophering *Mining Engineering*. a method of breaking up a sandy, medium-hard overburden where usual blastholes tend to cave in.
gopher plant *Botany*. a plant, *Euphorbia lathyris,* that produces latex; considered as a possible source of crude oil and gasoline. Also, **gopherweed.**
gopher snake *Vertebrate Zoology*. the bullsnake *Pituophis melanoleucus,* which preys on rodents; found in western North America.
gopher tortoise *Vertebrate Zoology*. a burrowing land tortoise of the family Testudininae, genus *Gopherus,* characterized by a high, domed shell, vegetarian diet, and slow gait; includes the desert tortoise, a protected species; found in the southern U.S. and Mexico.
gorceixite *Mineralogy*. $BaAl_3(PO_4)(PO_3OH)(OH)_6$, a brown or white, monoclinic, pseudotrigonal mineral of the crandallite group occurring in botryoidal aggregates and pebbles, having a specific gravity of 3.1 to 3.32 and a hardness of 6 on the Mohs scale; found as spheroids in novaculite.
Gordan, Paul Albert 1837–1912, German mathematician; proved that every binary form has a finite system of invariants and covariants.
Gordiidae *Invertebrate Zoology*. a family of smooth-skinned aquatic worms in the class Gordioidae.
Gordioidea *Invertebrate Zoology*. the hairworms, a class of smooth-skinned freshwater worms in the phylum Nematomorpha.
gordioid larva *Invertebrate Zoology*. a larval stage of hairworms in the class Gordioidae; after hatching, the larvae parasitize an arthropod host.
Gordona *Bacteriology*. a former genus of aerobic, nonmotile bacteria of the order Actinomycetales.
gordonite *Mineralogy*. $MgAl_2(PO_4)_2(OH)_2{\cdot}8H_2O$, a clear to smoky-white or pale-pink, glassy, triclinic mineral of the paravauxite group occurring in sheaflike aggregates, having one perfect cleavage, a specific gravity of 2.23, and a hardness of 3.5 on the Mohs scale; found with crandallite and wardite in variscite nodules.
gore *Civil Engineering*. an irregularly shaped tract of land, generally triangular, left between two adjoining surveyed tracts because of inaccuracies in the boundary surveys or as a remnant of a systematic survey. *Cartography*. a lune-shaped map that may be fitted to the surface of a globe with a negligible amount of distortion.
Gorgas, William Crawford [gôr´gəs] 1854–1920, American physician and medical officer; coordinated the eradication of malaria and yellow fever.
gorge *Geography*. a deep, usually narrow canyon. *Oceanography*. an obstruction in a channel, river, or sea canyon, consisting of ice, rock, sediment, or plant material. *Architecture*. see CAVETTO.
gorgerin *Architecture*. the neckline portion of a capital of a column, or a feature that forms the junction between a shaft and its capital. Also, NECKING.
gorge wind see CANYON WIND.
Gorgonacea *Invertebrate Zoology*. an order of horny corals belonging to the class Anthozoa, including the sea fans, which form upright, branching colonies.
gorgonin *Biochemistry*. a protein that constitutes the horny skeleton in corals of the family Gorgonacea; generally contains iodine and bromine.
Gorgonocephalidae *Invertebrate Zoology*. a family of sea stars in the order Phrynophiurida, having simple or branched arms that can coil vertically.
gorilla *Vertebrate Zoology*. the largest living primate, *Gorilla gorilla,* including the subspecies *G. g. gorilla, G. g. graueri,* and the now-rare *G. g. beringei,* characterized by a large flattened face with wide, flaring nostrils, small eyes and ears, and a stocky body with long arms, jet black skin, and dense fur; native to western equatorial Africa and the Kivu highlands of central Africa.

gorilla

gorlic acid *Organic Chemistry.* $C_5H_7(C_{12}H_{22})COOH$, an unsaturated acid derived from sapucainha oil, found in the seeds of an Amazon Valley tree.

goshawk *Vertebrate Zoology.* a powerful short-winged, goose-sized hawk of the genus *Accipter* in the family Accipitridae that ranges in forested areas, primarily in the Northern Hemisphere; known as a fierce hunter (once used in falconry) that will take prey as large as a fox.

goshawk

goslarite *Mineralogy.* $ZnSO_4 \cdot 7H_2O$, a white to greenish to bluish orthorhombic mineral occurring in granular or fibrous forms and as efflorescent crusts, having a perfect cleavage, a specific gravity of 1.978 and a hardness of 2 to 2.5 on the Mohs scale; found in mines as a postmining alteration product of other zinc minerals.

gosling *Vertebrate Zoology.* a young goose.

gosling blast *Meteorology.* a sudden squall-like condition characterized by rain or sleet that occurs in England. Also, **gosling storm.**

gossan *Geology.* an oxidized, iron-bearing deposit overlying an ore body that contains sulfides. Also, IRON HAT, CAPPING.

GOT glutamic-oxaloacetic transaminase.

Gothic *Architecture.* the predominant Western European architecture of the late 12th to early 16th centuries, characterized by pointed arches, flying buttresses, and large colored windows.

Gothic revival *Architecture.* in the 18th and 19th centuries, a European and American movement to revive the forms of Gothic architecture.

Gotlandian *Geology.* a European geologic time period equivalent to the late Silurian. Also, **Gothlandian.**

go-to or **GO-TO** *Computer Programming.* **1.** in a high-level programming language, a branching instruction that directs the computer to another part of the program. **2.** in word processing, an instruction that directs the computer to a specific page number. Thus, **go-to statement, go-to instruction.**

Goto pair [gät´ō] *Electronics.* a circuit capable of sensing the direction of electric current, consisting of two tunnel diodes connected in series such that one is in the reverse-tunneling region when the other is in the forward-conduction region; used in high-speed gate circuits.

Gotte's larva *Invertebrate Zoology.* the ciliated free-swimming larva characteristic of many marine flatworms.

Goudsmit, Samuel Abraham 1902–1978, Dutch-born American physicist; with Uhlenbeck, discovered electron spin.

gouge *Mechanical Devices.* a chisel with a curved blade, either concave or convex, used for scooping or cutting grooves in wood or stone. *Geology.* see FAULT GOUGE.

gouge bit *Mechanical Devices.* a drill bit having a rounded cutting point whose cross section is similar to that of a gouge; used for scooping or cutting grooves in wood.

gouging *Engineering.* the process of scooping out material to create a cavity or groove. *Mining Engineering.* an operation in placer mining similar to ground sluicing. Also, BOOMING.

Gould, Stephen Jay born 1941, American paleontologist and science author; proposed theory of punctured equilibrium in evolution.

Gould's Belt *Astronomy.* a group of hot young stars, open clusters, and associated gas clouds within about 1000 light-years of the sun and tilted about 10° to 20° to the Milky Way's equator.

gourami *Vertebrate Zoology.* a general name for any of several species of freshwater fish in the suborder Anabantoidei including the families Belontiidae (several varieties), Helostamatidae (**kissing gourami**), Osphronemidae (**true gourami**), and Anabantidae (**climbing gouramis**); native to southern and central Africa, southeast Asia, and India.

gourd *Botany.* **1.** a hard-shelled fruit from various plants, particularly *Lagenaria siceraria,* whose dried shell is used for utensils, and *Cucurbita pepo,* which is used as an ornament. **2.** a plant bearing such fruit.

gout *Medicine.* a disease that results from an inborn error of uric acid metabolism, in which excess uric acid is converted to sodium urate crystals and deposited in joints; it is characterized by recurrent episodes of acute arthritis, most often of the big toe.

gouy *Physical Chemistry.* a unit that reflects the movement of electrically charged particles through a continuous medium, such as a liquid.

Gouy balance *Analytical Chemistry.* an instrument used to measure the magnetic susceptibilities of a solid or liquid.

governor *Mechanical Engineering.* a device that automatically regulates the speed of an engine or machine by varying the supply of fuel or steam according to the power demand.

gowk storm *Meteorology.* a storm or gale that occurs in England around the end of April or beginning of May.

goyazite *Mineralogy.* $SrAl_3(PO_4)_2(OH)_5 \cdot H_2O$, a colorless, yellow or pink, transparent, trigonal mineral of the crandallite group occurring in small rhombohedral crystals or rounded grains, having a perfect basal cleavage, a specific gravity of 3.22 to 3.26, and a hardness of 4.5 to 5 on the Mohs scale; found with ferberite.

gp gene product; group.

GP or **G.P.** general practice; general practitioner; geometic progression.

g parameter *Electronics.* any of four conductance parameters obtained for the equivalent-pi models of a transistor: g_{be}, g_{gc}, g_{ce}, g_m.

G parity *Particle Physics.* a quantum number associated with elementary particles that have zero baryon number and strangeness; conserved in strong interactions only. Also, ISOTOPIC PARITY.

GPD gallons per day.

G6PD glucose-6-phosphate dehydrogenase.

GPH or **gph** gallons per hour. Also, **g.p.h.**

G1 phase *Genetics.* a period of the eukaryotic cell cycle, occurring between the last mitosis and the start of DNA replication.

G2 phase *Genetics.* a period of the eukaryotic cell cycle, occurring between the end of DNA replication and the start of the next mitosis.

GPM or **gpm** gallons per minute. Also, **g.p.m.**

G protein *Biochemistry.* any of a class of guanine-nucleotide-binding proteins associated with the cytoplasmic face of the plasma membrane of mammalian cells; involved in transmitting signals from certain types of hormone and neurotransmitter receptors to intracellular pathways.

GPS gallons per second; General-Purpose Problem Solver.

GPSS *Computer Programming.* General-Purpose Systems Simulator, a programming language used for building a simulation model. *Industrial Engineering.* see GENERAL-PURPOSE SIMULATION SYSTEM.

gr or **gr.** gram; grain; grade; gravity; gross.

Graafian follicle see OVARIAN FOLLICLE.

grab *Mechanical Engineering.* any of various machines or devices used for grasping objects. *Computer Science.* to take sole control of a device or computing resource in a multiprocessing system.

grab bucket *Mechanical Engineering.* a steel bucket or cage consisting of two hinged halves that, in closing together, bite into the material on which they rest; usually attached to a crane, dredge, or hoist. Thus, **grab(bing) crane, grab(bing) dredger.**

graben *Geology.* an elongated, relatively depressed crustal block lying between two more or less parallel faults. (A German word meaning "ditch.")

grabhook *Mechanical Engineering.* a large hook used to lift stone; usually used in pairs that are chained together under tension to hold the load securely.

grab sample *Mining Engineering.* a method of sampling in which samples may be taken from the pile broken in the process of mining; often used to estimate the approximate value of material lying broken in stopes or headings.

graceful degradation *Computer Programming.* a programming methodology designed to maintain an acceptable level of performance in a computer system despite the failure of one or more of its components. Also, SOFT CRASH.

graceful graph *Mathematics.* a simple graph with e edges is graceful if there is one-to-one function $l{:}V(G)\ \{0, 1, 2, \ldots, e\}$ with the property that the set $\{|l(u) - l(v)| : uv \in E(G)\}$ includes e different numbers. It is conjectured that every finite tree is graceful.

Gracilariaceae *Botany.* a red alga family belonging to the order Gigartinales, characterized by terete, compressed, or flat thalli that are branched and multiaxial, and by densely massed spermatangia; contains many species used as a source of agar or food.

Gracilariidae *Invertebrate Zoology.* a family of moths in the order Lepidoptera; larvae mine in the leaves of plants.
gracile *Anthropology.* gracefully slender; a term used to describe a hominid having a build of this type.
Gracilicutes *Bacteriology.* a taxonomic division of bacteria proposed to include certain Gram-negative bacteria such as *Oxyphotobacteria.*
gracilis *Anatomy.* the slender muscle on the inner surface of the thigh that adducts the thigh and flexes the knee joint.
grackle *Vertebrate Zoology.* any of several species of large blackbirds of the family Icteridae that inhabit the New World, especially those of the genus *Quiscalus*; characterized by sleek, shiny, black plumage, a long, keeled tail, and a heavy pointed bill.

grackle

gradability *Mechanical Engineering.* the ability of a vehicle or machine to move a load uphill, often measured in the percent grade up which it can haul a given load. Also, **gradeability.**
gradation *Geology.* a process by which land is leveled or built up to a generally uniform grade or slope through erosion, transportation, or deposition.
gradation period *Geology.* an interval of time during which the sea level remains constant.
grade a degree or rank in a scale; specific uses include: *Food Engineering.* the classification of a food by quality, size, and so on (e.g., meat, milk). *Civil Engineering.* the slope of a roadway, usually the inclination expressed as a percentage of the horizontal distance. *Building Engineering.* the angle at which the ground meets the foundation of a building. *Hydrology.* the longitudinal profile of a stream. *Engineering.* a classification of products or materials such as pipe, explosives, or wood. *Geology.* a diameter or range of diameters of particles in a soil, sediment, or rock. *Mining Engineering.* a rating of ore based on the recoverable amount of valuable metal. *Graphic Arts.* the degree of contrast on a photographic paper; usually expressed on a scale of 0 (soft) to 5 (very hard). *Mathematics.* one-hundredth of a right angle; i.e., $\pi/200$ radian or 0.9 degree. Also, **grad.** *Telecommunications.* either of two kinds of television service, each with its own signal strength.
grade beam *Civil Engineering.* a reinforced concrete beam used for a structural foundation in the resistance of column base moments.
grade correction *Cartography.* a correction applied to a distance measured on a slope to reduce it to a horizontal distance between the vertical lines through its end points. Also, SLOPE CORRECTION.
grade crossing *Civil Engineering.* the intersection at grade of roadways, railroads, pedestrian paths, or any combination of them.
graded *Geology.* **1.** relating to a land surface or feature on which deposition and erosion are so balanced that a general slope of equilibrium is maintained. Also, AT GRADE. **2.** relating to a sediment or rock whose particles are generally the same size or within the range of a single grade. Also, SORTED.
graded bedding *Geology.* a sedimentary bedding in which each layer exhibits a gradual and progressive variation in particle size, with the coarser particles at the base and the finer particles at the top.
graded differential complex see COCHAIN COMPLEX.
graded periodicity technique *Electronics.* a method of modifying the response of a surface acoustic wave filter by varying the spacing between successive electrodes of the interdigital transducer.
graded profile see PROFILE OF EQUILIBRIUM.
graded reach *Hydrology.* **1.** any part of a stream where erosion and deposition are in equilibrium. **2.** any reach of a graded stream.
graded refractive index rod lens *Optics.* a gradient index lens used in optical fiber components, having a core refractive index that decreases with the square of the distance from the axis.
graded response *Behavior.* a response that varies in strength in proportion to the strength of the stimulus that evokes it.
graded ring *Mathematics.* let G be a commutative monoid, written additively. A G-graded ring is a ring R that can be written as a direct sum $R = \sum_{r \in G} A_r$ and such that ring multiplication maps $A_r \times A_s$ into A_{r+s}.
graded-step reinforcement *Psychology.* a therapeutic technique in which the individual is rewarded for accomplishing a series of increasingly difficult tasks.
graded stream *Hydrology.* a stream that has adjusted its slope and velocity over a period of years, so that it maintains a balance between the amount of material it can transport and the amount of material supplied to it from the drainage basin.
grade level *Hydrology.* the level attained by a stream when its longitudinal profile is a straight line, indicating its entire course has been reduced to a uniform gradient.
grade line *Civil Engineering.* a reference line or slope used in measuring the grade of a railroad or highway.
grade measurement see ARC MEASUREMENT.
grade scale *Geology.* the division of a continuous range of particle sizes into a series of classes in order to provide a systematic means for standardizing terms and for use in statistical analysis.
grade separation *Civil Engineering.* the separation of the levels at which different roads, railroad tracks, or the like cross one another, in order to avoid conflicting traffic.
gradient *Engineering.* an inclined surface, such as a roadway or ramp. *Metrology.* the rate of change of any physical quantity with respect to distance, as it progresses toward a maximum quantity. *Cartography.* a specific rate of rise or fall of a quantity against horizontal distance, expressed as a ratio, decimal, fraction, percentage or as the tangent of the angle of inclination. Also, PERCENT OF SLOPE, SLOPE. *Mathematics.* in an inner product space, the vector associated with the differential df of a function; denoted grad f or ∇f. In particular, ∇f is the vector satisfying $df(v) = (\nabla f, v)$ for all vectors v in the space. In P^n with the usual inner (dot) product, the components of ∇f and df coincide and are equal to the partial derivatives of f with respect to the coordinate functions. In Euclidean space, $\nabla f = (\partial f/\partial x)i + (\partial f/\partial y)j + (\partial f/\partial z)k$.
gradient circuit *Ordnance.* in an influence mine, a circuit that is actuated when the rate of change of the magnitude of the influence is within predetermined limits.
gradient elution *Materials Science.* an elution process in which two or more eluents of different composition are used in succession to separate the components in a single chromatographic run.
gradienter *Cartography.* an attachment to an engineer's transit with which an angle of inclination is measured in terms of the tangent of the angle rather than in degrees and minutes; may also be used as a telemeter in measuring horizontal distances.
gradient of motivation *Behavior.* the principle that the closer a goal is, the stronger the tendency to approach or withdraw.
gradient of reinforcement *Behavior.* the principle that the closer a response is to the reinforcement, the stronger it will be.
gradient plate *Microbiology.* an agar petri plate containing a substance, such as an antibiotic, the concentration of which varies gradually from one side of the plate to the other.
gradient wind level see GEOSTROPIC WIND LEVEL.
gradiometer *Geodesy.* an instrument that is used to measure gravity gradients.
gradocoen *Ecology.* the sum of all factors that impinge or encroach upon a population.
gradostat *Biotechnology.* a device that allows for a special form of continuous culture in which different solutions flow simultaneously, but in opposing directions; flow occurs through serially linked chemostats.
gradualism see PHYLETIC GRADUALISM.
graduated cylinder *Chemistry.* a cylindrical container that is marked with divisions or units of measurement.
graduated extinction *Behavior.* a technique for eliminating avoidance and fearful behaviors, in which an individual is gradually re-exposed to the fear-producing stimulus.
Graduate Record Examination *Psychology.* a combination of verbal and mathematical tests widely given to candidates for admission to graduate school.
Gräffe's method *Mathematics.* an iterative technique of approximating the roots of an algebraic equation with real coefficients.
Graf sea gravimeter *Geodesy.* a balance-type gravity meter, designed for ocean surveys, consisting of a mass at the end of a horizontal arm that is supported by a torsion-spring rotational axis.

graft *Botany.* **1.** a plant bud, shoot, or scion that is inserted in a slit or groove in the stem or stock of another plant, where it continues to grow. **2.** the place where this occurs. **3.** the plant formed by such a process. **4.** to join two plants in this manner. *Surgery.* **1.** any organ or tissue for implantation or transplantation. **2.** to implant or transplant such tissues.

graft copolymer *Materials Science.* a polymer in which the main backbone chain of molecular units has side chains attached at various points; the side chains contain different molecular units than the main chain.

graft hybrid *Botany.* a plant produced by grafting in which the skin and core derive from different species.

grafting *Surgery.* the transplantation or implantation of a tissue or organ. *Botany.* the process of creating a graft or graft hybrid.

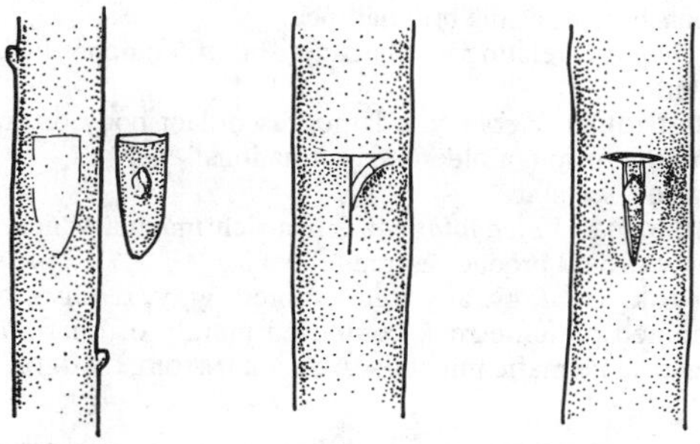

grafting

graft rejection *Immunology.* an immune-mediated destruction of tissue that has been transplanted from one individual to another.

graft-versus-host reaction *Immunology.* an immunological response in an organism that receives transplanted lymphoid cells which react against the host cells. Also, **graft-versus-host disease.**

Graham, Thomas 1805–1869, British chemist; the founder of colloid chemistry; formulated Graham's law; discovered dialysis.

Graham's law *Chemistry.* a law stating that the rate of diffusion of a gas through a liquid or porous membrane is inversely proportionate to the square root of its density and directly proportionate to its solubility coefficient.

grahamite see MESOSIDERITE.

Grail *Ordnance.* a Soviet tube-launched, shoulder-fired surface-to-air missile that is also used as an air-to-air helicopter self-defense weapon, delivering a 5.5-pound fragmentation warhead at a speed of Mach 1.5 and a range of around 2 miles; the ground-based version is officially designated **SA-7,** the sea-based version is **SA-N-5.**

grain *Botany.* **1.** the seed or seedlike fruits of wheat, oats, corn, or other cereal grasses. **2.** any of the plants on which these seeds grow. *Forestry.* the seasonal growth pattern of wood, alternating between earlywood and latewood. *Metrology.* **1.** the smallest unit of avoirdupois weight, originally based on the weight of a grain of wheat. It is equivalent to 64.8 milligrams. One pound avoirdupois equals 7000 grains. **2.** a similar unit of troy or apothecaries' weight. One pound troy equals 5760 grains. *Materials Science.* **1.** a portion of a solid material having the same crystalline structure without orientation change. **2.** an individual crystal in a polycrystalline structure. **3.** an aggregate similar in size to a crumb, but more dense and therefore less porous. *Graphic Arts.* see GRAINS.

grain boundary *Materials Science.* the interface between grains; sometimes a surface defect that serves as the boundary between two neighboring grains.

grain-boundary diffusion *Materials Science.* the diffusion of atoms along grain boundaries, which occurs more rapidly than through grains because the atoms are less closely packed.

Grain convention *Cell Biology.* a system of numbering devised for the different sets of microtubular triplets in a kinetosome.

grain diminution see DEGRADATION RECRYSTALLIZATION.

grain germ *Nutrition.* the embryo portion of the grain kernel, which is removed during milling because its richness in oils tends to give the oil a rancid taste; a basic part of a cereal grain which contains the bulk of all the vitamins and other nutrients found in the grain.

grain growth *Materials Science.* a process usually involving the movement of grain boundaries, in which some grains grow at the expense of others. *Metallurgy.* specifically, the increase of metal grain size during thermal or thermomechanical treatments. *Petrology.* the process by which carbonate sediments become a coarser-textured calcite mosaic. Also, GERMINATION.

grains *Graphic Arts.* in a photosensitive emulsion, the exposed and developed pieces of silver halide that form the visible photographic image. *Mineralogy.* see ANTHRACITE FINES.

grain size *Materials Science.* the size, or range of sizes, of grains in a polycrystalline structure. *Geology.* see PARTICLE SIZE.

grain-size strengthening *Materials Science.* an increase in strength of a material due to a decrease in the grain size. The larger grain-boundary area more effectively blocks dislocation movement.

Grallinidae *Vertebrate Zoology.* the mud-nest builders, a bird family that includes the apostle bird, the white-winged chough, and the mudlark; known for building communal nests of mud; found in forests and savannahs of Australia and New Guinea.

Gram, Hans Christian Joachim 1853–1938, Danish physician and bacteriologist; discovered Gram's method of staining bacteria.

gram *Metrology.* a basic unit of mass in the metric system, equal to the weight of one cubic centimeter of water at 4°C; equivalent to 0.035 avoirdupois ounce. Also, **gramme.**

-gram a combining form denoting something that is drawn, written, or recorded.

gram equivalent see EQUIVALENT WEIGHT.

Gramineae *Botany.* the grasses; a family of about 9000 species of monocotyledonous plants in the order Cyperales, represented in most habitats of the earth. Also, POACEAE.

graminicolous *Ecology.* **1.** living on grass. **2.** living in a grassland habitat. Thus, **gramincole.**

graminivorous *Ecology.* feeding on grass. Thus, **graminivore.**

grammar *Linguistics.* **1.** an inherent but not necessarily articulated system of rules that determines the manner in which the users of a language arrange words into meaningful sentences; e.g., the English sequence noun-verb-object, as in "Cows eat grass." **2.** a set of generally accepted principles determining what is regarded as the correct or standard usage of a language; e.g., "The cow *ate* the grass" rather than "The cow *et* the grass." *Artificial Intelligence.* a specification of a formal language in terms of a set of productions that transform nonterminal symbols (phrase names) into strings of other nonterminal symbols or terminal symbols (words or characters).

Gram matrix *Mathematics.* the Gram matrix of n real-valued, continuous functions $f_1, f_2, \ldots, f_n$ on a "well-behaved" set S is the $n \times n$ matrix whose (i,j)th entry is the real number $\int_S f_i f_j$. The Gram matrix is used to test linear independence of functions.

Grammitidaceae *Botany.* a large family of small, mostly epiphytic tropical ferns of the order Filicales, characterized by a short-creeping scaly rhizome, fronds that often have long straight hairs, sori on lower leaf surfaces, and sporangia with an erect annulus and an unusual one-rowed stalk.

gram-molecular *Chemistry.* relating to or expressed in terms of gram molecules. Thus, **gram-molecular weight, gram-molecular volume.**

gram molecule *Chemistry.* the quantity of a substance consisting of as many grams of the substance as are numerically equal to its molecular weight. Also, **grammole.**

Gram-negative *Microbiology.* of a bacterium that loses the stain or is easily decolorized by alcohol but stains with the counterstain in Gram's method; indicative of a cell wall composed of a thin layer of peptidoglycan covered by an outer membrane of lipoprotein and lipopolysaccharide. Also, **gram-negative.**

Gram-positive *Microbiology.* of a bacterium that retains the stain or resists decolorization by alcohol in Gram's method; indicative of a cell wall that is composed of a thick layer of peptidoglycan with attached teichoic acids. Also, **gram-positive.**

Gram reaction *Microbiology.* the result of testing with Gram's stain.

Gram-Schmidt orthogonalization process *Mathematics.* suppose $\{x_k\}$ is an independent sequence of members of an inner product space, with inner product (,) and norm || ||. An orthonormal sequence $\{u_k\}$ can be derived from the $\{x_k\}$ by the following inductive construction:

$$y_1 = x_1;\; y_k = x_k - \sum_{i=1}^{k-1} (x_k, y_i)/\|y_i\|^2;\; \text{and } u_k = y_k/\|y_k\|.$$

Gram's method or **Gram method** *Microbiology.* an empirical method of staining and classifying bacteria in which a fixed bacterial smear is stained with crystal violet, treated with Gram's solution, decolorized with ethanol or ethanol-acetone, counterstained with a contrasting dye such as safranin, and rinsed with water; those bacteria that retain the crystal violet stain are termed Gram-positive, while those that lose the crystal-violet stain but stain with the counterstain are termed Gram-negative.

Gram's solution or **Gram solution** *Microbiology.* a 1:15 dilution of Lugol's iodine (5 grams iodine, 10 grams potassium iodine per 100 ml purified water), used in Gram's method of staining bacteria.

Gram's stain or **Gram stain** *Microbiology.* **1.** see GRAM'S METHOD. **2.** see GRAM'S SOLUTION.

Gram type *Microbiology.* a positive/negative designation that is derived from Gram's taxonomy but is based on anatomical and biochemical properties of a bacterial cell rather than on a Gram reaction.

Gram-variable *Microbiology.* describing a bacterium that is inconsistent in its reaction to Gram's method of staining; this may reflect variations in staining technique.

grana *Cell Biology.* a series of stacks of thylakoid membranes that are found in the chloroplasts of plant cells and are rich in chlorophyll.

granary *Agriculture.* a building where grain is stored and processed.

Grand Banks *Geography.* a group of shoals off the southeastern coast of Newfoundland, famous as rich fishing grounds.

grand canonical ensemble *Physics.* an ensemble that represents a system of particles that is capable of exchanging energy and particles with its environment.

Grand Canyon

Grand Canyon *Geography.* a mile-deep gorge cut by the Colorado River in northern Arizona.

grandfather *Computer Programming.* a data set that precedes the current data set by two generations.

grandfather cycle *Computer Programming.* the period before records are replaced, during which they are retained so that other records that are accidentally lost may be reconstructed.

grandiosity *Psychology.* a delusion in which an individual has an exaggerated sense of his own importance or power. Thus, **grandiose.**

grandite *Mineralogy.* a type of garnet intermediate in composition between grossular and andradite; i.e., the grossular-andradite series.

grand mal *Neurology.* a type of epilepsy in which a sudden loss of consciousness is followed by generalized convulsions; the seizure is often preceded by an aura. Also, HAUT MAL.

grand unified field theory *Particle Physics.* a theory of elementary particles that unites three forces (weak force, strong force, and electromagnetic force) into one large field theory, stating that these three forces act as a unified force with a corresponding multiplication of energy and interactions among particles.

granellae *Invertebrate Zoology.* in some Sarcodina, small, oval, strongly refracting particles composed mainly of barium sulfate.

granide see GRANITIC ROCK.

Granit, Ragnar born 1900, Finnish-born Swedish physiologist; Nobel Prize for research on electrical properties of vision.

granite *Petrology.* a coarse-grained plutonic rock composed chiefly of quartz and feldspar with lesser quantities of mica or other colored minerals, such as hornblende, biotite, or muscovite.

granite dome see DESERT DOME.

granite-gneiss *Petrology.* **1.** a coarsely crystalline, banded metamorphic rock of granitic composition derived from a sedimentary or igneous rock. **2.** a metamorphosed granite.

granite moss *Botany.* the common name for a group of mosses of the class Bryatae, composed of two Arctic genera and characterized by a longitudinal split of the mature capsule into four valves.

granite porphyry see QUARTZ PORPHYRY.

granite series *Geology.* a sequence of granite products that evolved continuously during crustal fusion from deep-seated, syntectonic, and granodioritic products to shallower, late syntectonic (or post-tectonic) and more potassic products.

granite wash *Geology.* any material that was eroded from granitic rocks and redeposited, forming a rock having about the same major mineral constituents as the original rock.

granitic *Geology.* relating to, characteristic of, composed of, or resembling granite.

granitic batholith *Geology.* a large, discordant body of granitic rock intruded as the fusion of older land formations.

granitic layer see SIAL.

granitic magma *Volcanology.* a silica-rich magma which, if crystallized slowly, would produce a granite rock.

granitic rock *Geology.* any light-colored, gray, reddish, or greenish coarse-grained plutonic rock composed mainly of quartz along with feldspar and some mafic minerals. Also, GRANITOID, GRANIDE.

granitic rock

granitization *Geology.* a metamorphic process by which granitic rock, or a character and likeness of it, is formed from sediments.

granitoid see GRANITIC ROCK.

granivorous *Ecology.* feeding on seeds and grain. Thus, **granivore.**

granoblastic fabric *Petrology.* **1.** the texture of nonschistase metamorphic rocks in which recrystallization has formed equidimensional minerals. **2.** the secondary texture of equally sized grains due to diagenetic crystallization or recrystallization processes.

granodiorite *Petrology.* a coarse-grained plutonic rock consisting primarily of quartz, plagioclase, and potassium feldspar, with biotite, hornblende, or pyroxene as mafic constituents; it is intermediate in composition between granite and tanalite, with roughly twice as much plagioclase as orthoclase.

granogabbro *Petrology.* a plutonic rock similar to granodiorite in mineral composition but with calcic rather than sodic plagioclase.

granophyre *Petrology.* a quartz porphyry or fine-grained porphyritic granite with a groundmass of intergrown quartz and alkali feldspar.

Grantiidae *Invertebrate Zoology.* a family of calcareous sponges in the order Sycettida, having a syconoid body and triradite spicules projecting into the body cavity.

granular [gran´yə lər] *Science.* having or consisting of grains or granules. *Materials Science.* of the particles of a material, being similar in size but irregular in shape; having a grainy texture.

granular fracture *Metallurgy.* the type of metallic fracture that appears grainlike.

granular gland *Physiology.* a gland, such as one that operates during digestion, that secretes granular material into the ducts of the gland.

granular ice *Hydrology.* ice composed of small, opaque, irregularly shaped crystals that resemble grains of sand.

granularity *Science.* the degree or condition of being granular. *Computer Science.* the size of problem or data that is handled.

granular polymerization *Materials Science.* the suspension of droplets in solution by continuous agitation during polymerization; the resulting product is in the shape of irregular granules.

granular powder *Materials Science.* a powder that is essentially equiaxed but not spherical.

granulate *Chemistry.* to form into granules or small masses.

granulated metal *Metallurgy.* a metal or alloy that has been subjected to melting and separating into droplets prior to freezing.

granulation *Science.* **1.** the formation of grains or granules. **2.** a roughened surface condition. *Plant Pathology.* the dry, woody, tasteless condition of citrus fruits caused by the hardening and thickening of the juice sacs when the fruit has been picked too late in the season.

granule [gran´yůl] a small grain; specific uses include: *Medicine.* the small red grains visible in a wound during healing, made up of loops of newly formed capillaries and fibroblasts. *Virology.* the occlusion body produced by infection with granulosis viruses, a proteinaceous layer of external material that envelops the cell wall. Also, CAPSULE. *Astronomy.* a convection cell in the sun's photosphere roughly 1000 kilometers across that forms and dissipates over several minutes.

granulin *Virology.* the protein that encapsulates the occlusion body of granulosis viruses during virus replication.

granulite *Petrology.* **1.** a metamorphic rock composed of even-sized, interlocking mineral grains. **2.** a coarse, granuloblastic metamorphic rock formed at the extreme pressures and temperatures of the granulite facies and characterized by the presence of plagioclase and pyroxenes. **3.** a sedimentary rock with sand-sized aggregates of nonclastic origin, having a texture similar to that of an arenite of clastic origin. **4.** a granite with muscovite.

granulite facies *Petrology.* the set of metamorphic mineral assemblages (facies) constituting gneissic rocks with a characteristic granoblastic fabric; typical of regional dynamothermal metamorphism at temperatures above 650°C and pressures of 3000–12,000 bars.

granulitic *Geology.* **1.** relating to a rock structure that results from the production of granular or flattened fragments in a rock by crushing. **2.** a metamorphic rock texture exhibiting xenoblastic crystal development. **3.** a granular igneous rock texture in which all or nearly all the components are xenomorphic. *Petrology.* relating to any structure that is composed of granulite.

granuloblast see MYELOBLAST.

granulocyte *Histology.* a leukocyte that contains granules in the cytoplasm and possesses a multilobed nucleus. Also, POLYMORPH.

granulocytic leukemia *Medicine.* leukemia marked by a proliferation of myelopoietic cells in the bone marrow and by an unusually large amount of granulocytic cells in tissues, organs, and blood. Also, NEUTROPHILIC LEUKEMIA.

granulocytopenia *Medicine.* a deficiency of granular leukocytes in the blood.

granulocytosis *Medicine.* a condition characterized by more than the normal number of granulocytes.

granuloma *Pathology.* a general term for small growths (of nodular granulation tissue) typically formed at sites of chronic inflammation; granulomas may contain epithelioid cells, lymphocytes, giant cells with many nuclei, eosinophils, and plasma cells, and are usually surrounded by an area of fibrosis.

granuloma inguinale *Pathology.* a chronic, slowly progressive ulceration of external genitalia that is caused by *Calymmato* bacteria; assumed to be sexually transmitted, and commonly concurrent with syphilis infection.

granuloma pyrogenicum *Pathology.* a tumor with superimposed inflammation affecting the skin or other epithelial surfaces.

granulomatosis *Pathology.* any condition characterized by the formation of multiple granulomas.

granulometry *Petrology.* the measurement of grain sizes in sedimentary rock.

Granuloreticulosia *Invertebrate Zoology.* a subclass of protozoans in the class Rhizopodea, with delicate granular reticulopodia that often fuse into networks; includes the foraminiferans.

granulose *Biochemistry.* **1.** a glucan that is present in the cells of bacteria from the genus *Clostridium* just before sporulation and absent following it; this is believed to serve as a reserve source of carbon and energy for sporulation. **2.** having the surface roughened with granules. *Geology.* the structure or texture of metamorphic rock composed of essentially even-sized, interlocking mineral grains.

granulosis *Medicine.* the formation of a mass of granules. *Invertebrate Zoology.* a viral infection of insect larvae showing minute granular inclusions in the infected cells.

granulosis virus *Virology.* any of several viruses of the subgroup B of the genus *Baculovirus* that replicate, with a high host specificity, mainly in the nuclei of insects.

granulosity *Medicine.* a mass of granulations.

Granville wilt *Plant Pathology.* a disease of tobacco plants caused by the bacterium *Pseudomonas solanacearum* and characterized by drooping and shriveling of the plant, following loss of turgidity.

grape *Botany.* **1.** the common name for plants of the genus *Vitis* in the family Vitaceae; a climbing vine cultivated for its clustered, edible pulpy berries. **2.** the fruit of this plant.

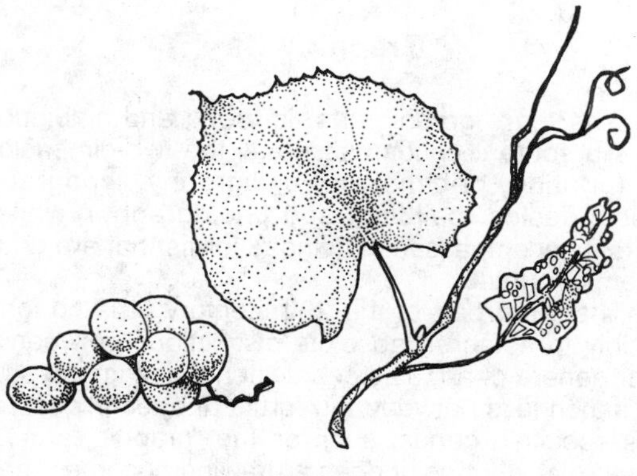

grape

grapefruit *Botany.* **1.** the common name for the species *Citrus paradisi*, an evergreen tree cultivated for its large, round edible citrus fruit and characterized by a bitter yellow rind and juicy pink or white pulp. **2.** the fruit of this tree.

grapestone *Geology.* a cluster of sand-sized grains or calcareous pellets that were cemented shortly after deposition, and resemble a bunch of grapes.

grapevine drainage pattern see TRELLIS DRAINAGE PATTERN.

grapevine stopper *Naval Architecture.* a gripper consisting of a webbed sleeve that tightens when under tension; used for temporary holding of large lines under heavy loads.

graph *Science.* a diagram or curve representing varying relationships between sets of data. *Mathematics.* **1.** a configuration consisting of vertices and edges, such that each edge connects exactly two vertices. Multiple edges between the same vertices are allowed, and an edge may join a vertex to itself (resulting in a loop). A graph that can be drawn in the plane so that two edges intersect only at a vertex (that is, without crossings) is called a **planar graph**. If a direction is assigned to each edge, then the graph is called a **directed graph** or **digraph.** **2.** the graph of an operator $T: X \to Y$ with domain $D(T)$ is the subset of $X \times Y$ with the elements (x, Tx), where $x \in D(T)$. In particular, if X and Y are the real numbers, then the graph of T is often defined to be some pictorial representation of $\{(x, Tx): x \in D(T)\}$. **3.** see DIAGRAM.

graph- a combining form meaning: **1.** writing, as in *grapheme.* **2.** containing graphite, as in *graphalloy.*

-graph a combining form denoting an instrument for writing or recording, or a record made by such an instrument.

graph component *Mathematics.* a connected subgraph of a graph G that is not properly contained in any connected subgraph of G except possibly G itself.

graphechon *Electronics.* a double-ended electron tube designed for integrating, storing, and translating information in the form of electrical signals; employed in certain computer and radar devices.

grapheme *Linguistics.* the basic unit of a writing system; a letter or symbol. *Computer Science.* a pictorial representation of a data element.

graph follower see CURVE FOLLOWER.

graphic *Science.* **1.** written or drawn. **2.** relating to, using, or represented on a graph. **3.** of or relating to the graphic arts. *Telecommunications.* see GRAPHIC. *Petrology.* of a rock, having a surface resembling written characters, due to the intergrowth of minerals. Also, **graphical.**

graphical analysis *Mathematics.* the study of mathematical objects by analysis of graphical representations; includes the obtaining of approximate solutions by graphical methods.

graphical radial triangulation *Cartography.* a radial triangulation performed by some means other than an analytical one, with the principal points as radial centers unless otherwise stated.

graphical representation *Mathematics.* any pictorial representation of ordered pairs of numerical quantities; a drawing depicting a functional relation.

graphical statics *Mechanics.* the solution of problems in statics using a scale drawing of the forces and reactions known as a free body diagram.

graphic arts **1.** the various arts and techniques used to produce reproductions of original words and images, such as engraving or lithography. **2.** the visual arts in general, including painting and photography.

Graphic Arts

The term "graphic arts" is broad and ambiguous. To some it encompasses all the two-dimensional arts (printing, printmaking, calligraphy, typography, graphic design, painting, and photography), while to others it encompasses only those forms that are drawn, written, or printed.

In the latter part of the 20th century, it is no longer possibly to establish absolute distinctions between the major genera of art, as say between painting and sculpture, much less between the forms and techniques of a more specified genus, such as the "graphic" arts. For instance, at what point does a drawing become a painting, or what is the distinction between a "fine" book and a well-produced "trade" book.

Indeed "graphic" does come from the Greek work *graphikos,* meaning written or drawn, and this has formed the basis of the traditional definition of the "graphic arts," to wit: the graphic arts are those arts—both fine and applied—that employ or utilize techniques by which written, drawn, or printed images are married to surfaces that are flat, e.g., paper, canvas, panels, walls. By extension this would also include the design or arrangement of the graphic elements (line, shape, contrast, rhythm, texture, color) into books, advertising, film titles, and the like.

Over the last four decades the techniques of art have been emancipated from the rigidly traditional, and the definition of what constitutes art has expanded to the point where the lines that theretofore separated form and technique, fine and commercial art, graphic arts and plastic arts have eroded. We are left with perhaps a single common denominator: the flatness of the surface. Yet that too is subject to question when one considers, for instance, that the pages of most books are curved surfaces which are printed on cylinders, and that many illustrations for books and publications are done in three-dimensional media.

I might further observe that in the graphic arts there is a tendency to use color flatly and decoratively, as in typography, rather than structurally, as in painting, and that, perhaps more so than the other plastic arts, there is a tendency to communicate, be it to sell a product, identify a character in a movie, locate the parts of a gastropod, tell a story, or throw a petard into the political arena.

Barry Moser
Artist and Designer
Winner of the American Book Award

graphic display *Electronics.* the display of data in a graphical form on the screen of a cathode-ray tube.

graphic granite *Petrology.* a pegmatite characterized by the intergrowth of quartz and feldspar in parallel orientation resembling Hebrew or cuneiform writing. Also, HEBRAIC GRANITE, RUNITE.

graphic intergrowth *Petrology.* an intergrowth of crystals, typically quartz and feldspar, producing a poikilitic texture in which the smaller crystals have a somewhat regular geometric outline and orientation that gives the effect of cuneiform writing.

graphic panel *Control Systems.* a master control panel that shows the status of and relationships among the equipment and operations in a system.

graphic radial aerotriangulation *Cartography.* a radial aerotriangulation done by other than analytical means.

graphic recording instrument *Engineering.* a recorder that produces a graphic representation of one or more quantities in relation to time or another variable.

graphics *Telecommunications.* titles, messages, credits, and other material in the form of words, appearing on the screen during a film or a television broadcast. *Science.* see GRAPHIC ARTS.

graphic scale *Cartography.* a graduated line on a map or chart by means of which distances may be measured in terms of ground distances. Also, BAR SCALE.

graphics interchange format *Computer Science.* a set of conventions for representing graphical images as files of characters so that they can be saved as files and transmitted across existing networks.

graphic timetable *Transportation Engineering.* a schedule shown on a line graph, with distances and stops on the vertical axis and times on the horizontal.

Graphidales *Mycology.* in some classifications, an order of fungi belonging to the subdivision Ascomycotina which includes certain lichens found on tree bark.

graphis *Botany.* the type species of the family Graphidaceae, grayish-white crustaceous lichens that are commonly found growing on bark.

graphite [graf´īt] *Mineralogy.* a black to gray, hexagonal and trigonal mineral occurring as thin tabular crystals with a perfect basal cleavage and in foliated masses, having a greasy feel, a specific gravity of 2.09 to 2.23, and a hardness of 1 to 2 on the Mohs scale; found in intensely metamorphosed rocks; widely used as a lubricant and in lead pencils and paints. (Going back to the Greek word for writing or drawing, because of its early use in pencils.)

graphitic [grə fit´ik] *Mineralogy.* relating to, containing, or consisting of graphite.

graphitic corrosion *Metallurgy.* a chemical corrosion process by which iron is leached from cast iron, leaving behind a weak, spongy mass of graphite.

graphitic steel *Metallurgy.* any of several steels that contain free carbon present as graphite.

graphitization *Organic Chemistry.* the formation of graphitelike material from organic compounds.

graphitizing *Metallurgy.* subjecting a steel to a thermal treatment that makes it graphitic.

Graphium *Mycology.* a genus of fungi belonging to the class Hyphomycetes that forms filaments or hyphae termed synnema and is parisitic to flowering plants, causing wilt diseases.

grapho- a combining form meaning "writing," as in *graphomotor.*

graphology *Linguistics.* the study of systems of writing. *Psychology.* the analysis of handwriting as an indication of personality traits.

graphomania *Psychology.* an irrational desire or impulse to write.

graphophobia *Psychology.* an irrational fear of writing.

graph theory *Mathematics.* **1.** the study of graphs and networks. **2.** the body of techniques used for drawing the graph of a function in the plane.

-graphy a combining form denoting: **1.** a process of producing graphic records or representations, as in *lithography, photography.* **2.** an art or science concerned with such records or representations, as in *biography, geography.*

grapple or **grapnel** *Mechanical Devices.* a jawed device consisting of one or more hooks or clamps; used for grasping and holding. Also, **grappling hook, grappling iron.** *Naval Architecture.* a light, four-armed or five-armed hook made fast to a line or chain, and thrown or cast to snag and hold a target, usually another vessel; it may also be used for anchoring a small boat or picking up objects from the bottom. Also, **grapline, graplin.** *Ordnance.* a device attached to the mooring of a naval mine that is designed to engage the sweep wire when the mooring is cut.

grapple skidder

grapple skidder *Forestry.* a four-wheeled tractor equipped with a grapple and used to haul logs or timber, especially over rough terrain.

Grapsidae *Invertebrate Zoology.* the square-backed crabs, a family of brachyuran shore crabs in the order Decapoda.

graptolite *Paleontology.* **1.** of or relating to the class Graptolithina. **2.** a member of this class.

Graptolithina *Paleontology.* an extinct class of colonial marine invertebrates, generally considered protochordates; the earliest were bottom-encrusting, but several sessile groups developed during the Cambrian, and then most of the graptoloids (the largest order) became planktic; the diversity and wide geographic occurrence of the Graptolithina during most of the Paleozoic make them a very important groups of index fossils; they became extinct in the middle Carboniferous. Also, **Graptozoa.**

graptoloid *Paleontology.* **1.** of or relating to the graptolites of the order Graptoloidea. **2.** a member of this order.

Graptoloidea *Paleontology.* the principal order of graptolites; generally planktic, the graptoloids arose in the latest Cambrian and persisted into the lower Devonian.

GRAS *Biotechnology.* a designation under the Federal Food, Drug and Cosmetic Act; it requires either that compounds (or additives) be closely related to others already being used or that clinical trials be performed to determine safety. (An acronym for generally recognized as safe.)

graser see GAMMA-RAY LASER.

Grashof formula *Fluid Mechanics.* an empirical formula for the amount of discharge m through an orifice of area A (square inches) when the fluid is driven by a reservoir pressure p (pounds per square inch): $m = 0.0165Ap^{0.97}$.

Grashof number *Fluid Mechanics.* a dimensionless number G used in studying natural convection in a fluid due to thermal effects: $G = bqgd^3/n^2$ where b is the thermal coefficient of volumetric thermal expansion of the fluid, q is the temperature difference responsible for the convection, g is the gravitational acceleration, d is a characteristic dimension of the system, and n is the kinematic viscosity.

grass *Botany.* any plant of the large family Gramineae, having jointed stems, long, narrow leaves, and usually a distichously arranged flower on the axis of a spikelet; includes wheat, corn, rye, and bamboo. *Electronics.* in radar, the random noise patterns on an A-scope.

Grass *Robotics.* a computer programming language that can simulate the shapes and motions of a robot's manipulator on a workstation.

grasserie *Invertebrate Zoology.* a disease of silkworms that causes spotty yellowing of the skin and internal liquefaction.

grasshopper *Invertebrate Zoology.* any of the jumping members of the insect order Orthoptera; generally straight-winged insects with stridulatory and auditory mechanisms, biting mouthparts, and simple antennae.

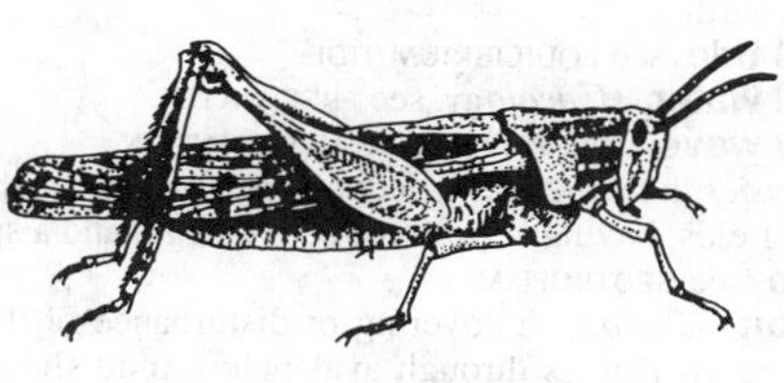

grasshopper

grasshopper engine *Mechanical Engineering.* an early steam engine in which a piston was attached to a beam, which in turn was hinged to an upright at the other end by a connecting rod suspended from its center.

grasshopper fuse *Electricity.* a small fuse with a spring attachment that sets off an alarm if the fusing wire releases the spring.

grassland *Ecology.* an area of natural vegetation dominated by grasses; an area where there is enough moisture to support grass, but not enough to support trees. Such areas are often called steppes or prairies in temperate regions and savannahs in tropical regions. *Agronomy.* a cultivated area of grass, such as a pasture.

grassland climate see SUBHUMID CLIMATE.

Grassmann, Hermann Gunther [gräs´mən] 1809–1877, German mathematician and linguist; laid the foundation of modern vector analysis; made notable studies in ancient languages.

Grassmann algebra see EXTERIOR ALGEBRA.

Grassmann manifold *Mathematics.* let V be an n-dimensional vector space. The Grassmann manifold $G_k(V)$ is the space whose points are k-dimensional subspaces of V. $G_k(V)$ is a differentiable manifold. Also, **Grassmannian manifold.**

Grassmann product see EXTERIOR PRODUCT.

Grassmann's laws *Analytical Chemistry.* a series of seven laws dealing with color identification and mixing; used as the basis of modern analytical colorimetry.

Grassot fluxmeter *Engineering.* a device used to measure the total quantity of magnetic flux that is linked with a circuit.

grass-roots deposit *Mining Engineering.* a deposit discovered in surface croppings that is easy to exploit and able to finance its own continued development.

grass sickness *Veterinary Medicine.* a noninfectious fatal disease of unknown cause affecting horses, typically those animals at pasture; the disease is afebrile and characterized by obstruction of the large intestine, reduction of gastrointestinal motility, and widespread degeneration of the autonomic nervous system. Also, EQUINE GRASS SICKNESS.

grass staggers see GRASS TETANY.

grass temperature *Meteorology.* the temperature indicated on a thermometer whose bulb is positioned at the level of the tops of short grass blades. The **grass minimum** is the lowest temperature shown on such a thermometer.

grass tetany *Veterinary Medicine.* a metabolic disturbance characterized by hypomagnesemia, occurring most commonly in adust cows and ewes, especially in heavy lactation, and undernourished beef cattle. The disease is manifested by staggering, tetany, convulsions, and death. Also, GRASS STAGGERS, HYPOMAGNESEMIA.

grate *Engineering.* **1.** a framework of parallel or crossed bars. **2.** the section of a furnace composed of fire bars or bricks that serves as a base for the fuel.

graticule *Cartography.* a network of lines on a map representing the earth's parallels of latitude and meridians of longitude. *Navigation.* the grid of reference lines on a chart, usually representing latitude and longitude. *Optics.* see RETICLE.

gratification *Psychology.* the decrease of emotional tension that follows the satisfaction of a need or desire.

grating *Engineering.* see GRATE. *Electromagnetism.* **1.** fine, parallel wires in a waveguide that allow only certain types of waves to pass. **2.** a reflective antenna element constructed of a crosshatch array of metal ribs or wires. *Spectroscopy.* see DIFFRACTION GRATING.

grating spacing *Spectroscopy.* in a diffraction grating, the distance from one diffraction center to the next in an ultrasonic wave that is producing a light diffraction spectrum. Also, **grating constant, grating interval.**

grating spectrograph *Spectroscopy.* a grating spectroscope designed to record a spectrum on photographic film.

grating spectroscope *Spectroscopy.* a spectroscope in which a diffraction grating, rather than a prism, is used to disperse incident light in order to produce a spectrum.

gratonite *Mineralogy.* $Pb_9As_4S_{15}$, a dark gray, brittle, opaque, metallic trigonal mineral forming large prismatic crystals, having a specific gravity of 6.22 and a hardness of 2.5 on the Mohs scale; found in small amounts in pyritic ore.

grattage [grə täj´] *Surgery.* the scraping or brushing of granulations in order to stimulate the healing process.

grattoir [gra twä´; gwa twär´] *Archaeology.* a flaked stone scraper, usually of the Upper Paleolithic, probably used to work wood and clean hides. (From French for "grate" or "scrape.")

grave goods *Archaeology.* artifacts associated with a burial or cremation; usually goods meant to be helpful in the afterlife or regarded as taboo for survivors to use.
gravel *Geology.* an accumulation of loose or unconsolidated, rounded rock fragments larger than sand, such as pebbles, granules, and boulders. *Materials.* rounded pebbles between 0.25 inch and 3 inches in diameter, used in concrete and paving materials. *Medicine.* a fairly coarse concretion of mineral salts, as from the kidney or bladder, of a smaller size than stones.
gravel bank *Geology.* a natural hill or exposed surface of gravel, particularly such an area from which gravel is dug.
gravel desert see REG.
gravel mine see PLACER MINE.
gravel pump *Mechanical Engineering.* a centrifugal pump used to pump a mixture of gravel and water; often lined with rubber to resist abrasion.
graver *Archaeology.* a stone tool, usually pressure-flaked, having a very sharp point that is used to cut or score soft materials such as bone and shell.
Graves' disease *Medicine.* a disorder characterized by excessive thyroid activity and usually by an enlarged thyroid gland and a marked protrusion of the eyeballs, occurring far more often in women than in men; its symptoms include weight loss, fatigue, nervousness, heat intolerance, palpitation, and a tremor of the hands. It is of unknown origin but is familial and probably autoimmune. (From Robert J. *Graves,* 1796–1853, Irish physician.)
gravid *Zoology.* describing a female that is carrying developing offspring within its body.
Gravigrada *Paleontology.* a group of giant sloths that had separated in the Oligocene (40 million years ago) into the distinct families Mylontidae, Megalonychidae, and Megatheriidae (all of which are now extinct); all inhabited shrubby grasslands, and died off during the Lower Holocene (10–12 thousand years ago).
gravimeter *Engineering.* a highly sensitive weighing instrument that is used to measure variations in the magnitude of a gravitational field or the specific gravity of a substance. Also, GRAVITY METER.
gravimetric *Engineering.* of, relating to, or based upon measurement with a gravimeter. *Geodesy.* describing the characteristics of the earth's gravity field or the techniques used to measure that field.
gravimetric absorption method *Analytical Chemistry.* a technique for measuring the moisture content of a gas by passing it through a desiccant and noting the change in weight of the desiccant.
gravimetric analysis *Analytical Chemistry.* a procedure during which a sample is converted by precipitation or combustion to a pure compound or element that is weighed.
gravimetric datum orientation *Geodesy.* adjustment of the ellipsoid of reference for a particular geodetic datum so that the differences between the gravimetric and astrogeodetic deflection components and geoid undulations are minimized.
gravimetric deflection *Geodesy.* a deflection of the vertical determined by methods of gravimetric geodesy.
gravimetric geodesy *Geodesy.* the branch of geodesy that utilizes measurements and characteristics of the earth's gravity field as well as theories regarding this field to deduce the shape of the earth and, in combination with arc measurements, the earth's size.
gravimetric geoid *Geodesy.* an approximation to the geoid as determined from gravity observation.
gravimetric projection *Cartography.* a map or chart on which contour lines represent points at which the acceleration of gravity is equal.
gravimetric survey *Geodesy.* a survey made to determine the acceleration of gravity at various places on the earth's surface.
gravimetric undulations *Geodesy.* separations between a gravimetrically determined geoid and a reference ellipsoid of specified flattening.
gravimetry *Engineering.* the scientific measurement of gravitational force.
gravitation *Physics.* a force of attraction between any two bodies having mass, the magnitude of which is dependent on the product of the two masses and the inverse square of the distance between them.
gravitational *Physics.* relating to or caused by gravitation.
gravitational acceleration *Physics.* the acceleration due to gravity; at the earth's surface, denoted by g and approximately equal to 9.8 m/s^2 or 32.2 ft/s^2 (the value varies slightly with latitude and elevation).
gravitational bremsstrahlung *Physics.* the emission of gravitational radiation by two massive objects that pass each other at a high relative velocity and deflect each other slightly.
gravitational collapse *Astrophysics.* **1.** the abrupt collapse of the core of a massive star at the end of nuclear burning, produced when the outflow of energy is no longer sufficient to balance the inward pull of the star's own gravity. **2.** the comparatively rapid contraction of an interstellar cloud of dust and gas when the gravity of the cloud is stronger than the outward pressure of its internal radiation.
gravitational constant *Mechanics.* the fundamental constant that, when multiplied by the product of the masses of two bodies and divided by the square of their separation distance, will give the gravitational attraction between them.
gravitational convection see THERMAL CONVECTION.
gravitational displacement *Mechanics.* the product of the force of gravity per unit mass and the gravitational constant. Also, **gravitational flux density.**
gravitational encounter *Astrophysics.* an encounter in which two moving bodies alter each other's direction and velocity by gravity.
gravitational equilibrium *Astrophysics.* the state in a celestial body when gravitational forces pulling inward on a particle are balanced by some outward pressure, such as radiation or electron degeneracy, so that no vertical motion results.
gravitational field *Mechanics.* **1.** the gravitational effect of a massive body, equal to the vector force that a test particle of unit mass would experience. **2.** the region of space in which gravitational attraction exists.
gravitational-field theory *Physics.* a model that treats gravity as a field rather than as a force acting at a distance.
gravitational flattening *Geodesy.* the ratio of the difference between the polar and equatorial normal gravities.
gravitational force *Mechanics.* the force of attraction between massive bodies due to gravitation. The measure of the gravitational force of a given body on earth is the weight of that body.
gravitational harmonics *Geodesy.* the spherical harmonics used in approximating the gravitational field of the earth.
gravitational instability *Astrophysics.* the unchecked collapse under gravity of a star or cloud of gas or dust; a runaway collapse. *Mechanics.* any process in which gravitation tends to pull a system that is slightly out of equilibrium farther away from equilibrium.
gravitational lens *Astrophysics.* the focusing effect of a massive and usually invisible object, such as a galaxy or cluster of galaxies, on light coming from a more distant object lying directly behind it; gravitational lenses often make single background sources appear multiple.
gravitational mass *Mechanics.* the property of matter that makes it gravitationally attractive; empirically, it is equal to inertial mass, but the fundamental reasons for this are not understood. Also, HEAVY MASS. *Physics.* the mass of a body that is used to determine the force it experiences in a gravitational field.
gravitational potential *Mechanics.* a scalar function of position equal to the work required to move a particle of unit mass to the specified position from a reference point, usually at infinity.
gravitational potential energy *Mechanics.* the potential energy stored in the gravitational fields of interacting bodies; the energy required to assemble the system from a reference configuration.
gravitational radiation *Astrophysics.* an as-yet undetected type of radiation created by some massive disturbance in matter; it travels at the speed of light and is capable of exerting a minute gravitational force on objects as it passes; predicted by the general relativity theory. Also, GRAVITATIONAL WAVE, GRAVITY WAVE.
gravitational redshift see EINSTEIN SHIFT.
gravitational repulsion see ANTIGRAVITY.
gravitational sliding *Geology.* the extensive movement of rock masses down a slope under the influence of gravity. Also, GRAVITY SLIDING, GRAVITY GLIDING.
gravitational system of units *Mechanics.* a system of physical units in which length, time, and force are regarded as fundamental, especially as opposed to one using mass instead of force. The unit of force is established as the gravitational force on a standard body at a standard location on the earth.
gravitational tide see EQUILIBRIUM TIDE.
gravitational water *Hydrology.* see FREE WATER.
gravitational wave see GRAVITATIONAL RADIATION.
graviton *Physics.* a theoretically deduced particle that is the quantum of gravitational fields, having zero mass, zero charge, and a spin of 2.
gravitropism see GEOTROPISM.
graviturbation *Geology.* a covering or disturbance of the soil surface by the force of gravity, as through avalanches, mud slides, rock slides, and the like.

gravity *Mechanics.* **1.** the phenomenon by which massive bodies are attracted to one another. In popular usage the word *gravity* is often used to describe the force of attraction itself, but technically *gravitation* is the force and *gravity* is the observed effect of this force. **2.** the gravitational attraction experienced near large massive bodies, especially on or near the surface of a planet or a star.

gravity anomaly *Geophysics.* the discrepancy between the theoretical and observed gravity in a given region.

gravity anomaly map *Cartography.* **1.** a map showing the positions and magnitudes of gravity anomalies. **2.** a map on which contour lines are used to represent points at which gravity anomalies are equal.

gravity base *Geodesy.* a point at which gravity has been measured with sufficient accuracy for use in calibrating gravimeters or as a reference in gravimetric surveys.

gravity bed *Engineering.* a moving mass composed of solid fragments that flow through a container to the bottom by gravity while process fluids move to the top; used in various devices such as coolers.

gravity cell *Physical Chemistry.* an electrolytic cell in which two ionic solutions are kept separate by the differences in their specific gravity.

gravity chute *Engineering.* any structure, such as a trough, that tilts downward to provide for the movement of material; the speed of movement is controlled by sliding friction and the angle of declination.

gravity classification see GRAVITY SEPARATION.

gravity-collapse structure see COLLAPSE STRUCTURE.

gravity concentration *Engineering.* a process in which a mixture is separated by gravity according to the various densities of its components or their various reactions to the separating medium.

gravity conveyor *Engineering.* **1.** any conveyor in which the weight of the objects being transported effects their movement from a higher to a lower place without the application of power. **2.** broadly, any structure that relies on the force of gravity to slide materials from a higher to a lower point, such as a gravity chute.

gravity corer *Engineering.* a coring instrument that penetrates downward due to the force of gravity acting upon its mass.

gravity dam *Civil Engineering.* a dam that depends on its own mass to resist overturning and sliding forces.

gravity disturbance *Geodesy.* the difference between the observed gravity and the normal gravity at the same point; distinguished from a gravity anomaly, which uses corresponding points on two different surfaces; occurs because the centrifugal force is the same when both are taken at the same point. Also, GRAVITATIONAL DISTURBANCE.

gravity drainage *Hydrology.* the removal or withdrawal of water from soil layers as a result of gravitational force.

gravity drainage reservoir *Geology.* a reservoir in which production is affected as a result of the separation of gas, oil, and water under the influence of gravity.

gravity drop hammer see DROP HAMMER.

gravity erosion see MASS EROSION.

gravity fault see NORMAL FAULT.

gravity feed *Engineering.* any device or process by which materials are moved from one point to another as a result of gravitational force; commonly used in food engineering and other industrial processes.

gravity flow *Hydrology.* **1.** the movement of ice in a glacier resulting from the steepness of the slope on which the glacier rests. **2.** see GLACIER FLOW.

gravity gliding see GRAVITATIONAL SLIDING.

gravity-gradient attitude control *Aviation.* a mechanism that adjusts a flight vehicle's position or orientation in response to external gravitational changes.

gravity haulage *Mining Engineering.* a type of haulage system in which a set of full cars is lowered at the end of a rope, and gravity force pulls up the empty cars. Also, SELF-ACTING INCLINE.

gravity incline *Mining Engineering.* an opening made in the dip of the deposit through which the ore mined is transported, usually to the next lower level.

gravity map *Geophysics.* a graphic representation of the high and low gravity variations in a given locale.

gravity meter *Engineering.* **1.** an instrument that indicates the pressure of solution specific gravities in semimicro amounts. **2.** see GRAVIMETER. *Mining Engineering.* in mineral prospecting, an electrical instrument that indicates variations in gravitation across distinct geologic areas.

gravity model *Transportation Engineering.* a mathematical formula used to calculate trip production and attraction among zones.

gravity network *Geodesy.* a network of stations at which observations are made to determine the value of gravity.

gravity prospecting *Mining Engineering.* a process in which a gravity meter is used to locate and chart the distribution of rock masses of distinct specific gravities.

gravity railroad *Engineering.* a cable railroad in which cars move down an incline by the force of gravity and are pulled up by a stationary engine, which is sometimes boosted by the force generated by a downward-moving car.

gravity reduction *Geodesy.* a combination of gravity corrections used to obtain a reduced value of gravity on the geoid.

gravity reference station *Geodesy.* a station that serves as a reference value for a gravity survey, and to which the differences at other stations are referred in a relative survey.

gravity separation *Materials Science.* a process in which the components of nonhomogeneous mixtures separate themselves by the force of gravity in accordance with their individual densities; used in ore dressing and in various industrial chemical practices. Also, GRAVITY CLASSIFICATION.

gravity settling chamber see SETTLING TANK.

gravity sliding see GRAVITATIONAL SLIDING.

gravity spring *Hydrology.* a spring that flows in response to the pull of gravity from a point where the land intersects the water table.

gravity station *Geodesy.* a station at which observations are made to determine the value of gravity.

gravity survey *Engineering.* a surveying process in which differences in gravitational force are measured at two or more points.

gravity tectonics *Geology.* the study of structures of the earth's crust produced by the downslope gliding movement of rocks primarily under the influence of gravity.

gravity tide *Geophysics.* an oceanic or atmospheric tide caused by the gravitational influence of the sun or moon.

gravity vector *Mechanics.* the direction and force of gravitation per unit mass at a specified point.

gravity wall *Building Engineering.* a retaining wall that stands upright by the force of its own weight.

gravity water *Hydrology.* **1.** see FREE WATER. **2.** water in a canal or pipeline that is delivered by gravity, rather than by pumping.

gravity wave *Fluid Mechanics.* a wave in a fluid medium whose principal restoring force is due to gravitational effects rather than compressional effects. *Oceanography.* a water wave form whose velocity of propagation is controlled primarily by gravity; all waves over 2 cm in length are gravity waves, while those under 2 cm are capillary waves. *Astrophysics.* see GRAVITATIONAL RADIATION.

gravity well *Space Technology.* an analogy in which a planet's gravitational field is represented as an energy depression out of which a spacecraft must climb.

gravity wheel conveyor see WHEEL CONVEYOR.

gravity wind *Meteorology.* a light downslope wind that usually occurs at night, caused by the cooling and increased density of air near the ground. Also, DRAINAGE WIND, KATABATIC WIND.

gravure [grə vyůr´] *Graphic Arts.* a commercial printing process that uses resin- or bitumen-coated plates or cylinders that are photographically etched or engraved in intaglio.

Grawitz's tumor see RENAL CARCINOMA.

Gray, Asa 1810–1888, American botanist; pioneer in paleobotany; wrote the *Manual of Botany,* the standard flora for the eastern U.S.

Gray, Elisha 1835–1901, American inventor; disputed Alexander Graham Bell's telephone patent; invented the harmonic telegraph.

Gray, Henry 1825?–1861, English anatomist; wrote the landmark work *Anatomy, Descriptive and Surgical* (**Gray's Anatomy**).

Gray, Stephen 1666–1736, English physicist; discovered electrical induction.

gray *Optics.* a hue between black and white. *Histology.* see GRAY MATTER. *Radiology.* the SI unit of absorbed radiation dose of one joule per kilogram and equivalent to 100 rads. Also, GREY.

gray antimony **1.** see JAMESONITE. **2.** see STIBNITE.

gray blight *Plant Pathology.* a disease of tea plants caused by the fungus *Pestalotia theae,* characterized by black dots on the leaves.

gray body or **graybody** *Thermodynamics.* a body that has the spectral distribution of a blackbody, except that its spectral emission intensity is a constant fraction less than unity of the blackbody value.

gray cast iron *Materials.* cast iron that, during solidification, permits graphite flakes to grow, which reduces the strength and ductility.

Gray clay treating *Chemical Engineering.* the use of a fixed-bed, vapor-phase treating process to selectively polymerize unsaturated gum-forming constituents such as diolefins.

gray code *Computer Programming.* a binary block code in which there are 2^n codewords, with each codeword having a length of *n* bits and differing from the preceding codeword by the complementation of a single bit. Also, REFLECTED BINARY CODE, REFLECTIVE CODE.

gray copper ore see TETRAHEDRITE-TENNANTITE SERIES.

gray earth or **gray desert soil** see SIEROZEM.

gray filter see NEUTRAL-DENSITY FILTER.

gray fox *Vertebrate Zoology.* a New World fox, *Uracyon cinereoargenteus* and *U. littoralis* of the family Canidae, having a short-haired grizzled coat, long legs, and a long bushy tail; it inhabits woodlands, open country, desert, and chaparral and is noted for its ability to climb trees.

gray iron *Metallurgy.* a cast iron in which the graphite has a flaky morphology.

gray leaf spot *Plant Pathology.* a disease of tomatoes, caused by the fungus *Stemphylium solani* and characterized by watery dark brown spots on the leaves that become gray with age.

grayling *Vertebrate Zoology.* any of several troutlike species of the genus *Thymallus,* of the family Salmonidae, inhabiting cold, northern freshwaters of North America and Eurasia; having a small mouth, silvery purple scales, and a long, flaglike dorsal fin; valued as a food and game fish.

gray managanese ore see MANGANITE.

gray matter *Histology.* the dark-colored portion of the central nervous system, composed principally of nerve cell bodies but also containing dendrites and the proximal and distal extremities of efferent and afferent neurons, respectively. *Psychology.* an informal term for the brain or intellect.

gray mold *Plant Pathology.* **1.** any fungus disease of plants that causes a graying of the diseased area. **2.** specifically, an early stage in the downy mildew disease of young grapes, in which a gray, wooly fungus covers the fruit.

gray mold blight *Plant Pathology.* a plant disease caused by species of *Botrytis* and characterized by a curling and withering of new growth, brown spots on affected plant parts, and often a mold covering affected parts during cool, humid weather.

gray scab *Plant Pathology.* a disease of willow trees caused by the fungus *Sphaceloma murrayae* and characterized by raised spots on the leaves that have gray centers and dark-brown borders.

gray scale *Graphic Arts.* the series of achromatic shades, usually divided into ten steps ranging from white through shades of gray to black, that are photographed along with a subject to ensure correct color balance and accuracy in reproduction processes. *Radiology.* a series of transparencies that represent progressive steps in the amount of transmitted radiation, used to administer predetermined amounts of radiation to a radiation-sensitive substance; the steps are calibrated in terms of transmissity, opacity, and density.

gray-scale ultrasonography *Radiology.* the use of ultrasonic waves to study deep body structures by means of a gray scale.

gray speck *Plant Pathology.* a disease of oats caused by a deficiency of manganese and characterized by grayish leaf spots, followed later by a light brownish discoloration of the blades.

graystone *Geology.* a dense, grayish-green rock of feldspar and augite that resembles basalt.

graywacke *Petrology.* a dark-gray, firmly indurated, coarse-grained sedimentary rock containing unstable mineral and rock fragments and a fine-grained clay matrix that binds the larger, sand-size detrital fragments.

graywall *Plant Pathology.* a disease of tomatoes believed to be caused by excessive exposure to sunlight and characterized by translucent gray markings on the fruit and a browning of the vascular strands.

gray whale *Vertebrate Zoology.* the most primitive baleen whale, *Eschrichtius robustus* of the monospecific family Eschrichtiidae, characterized by gray to black skin with white speckles and a slender body with low humps on its back; inhabits the northern Pacific Ocean and adjacent seas.

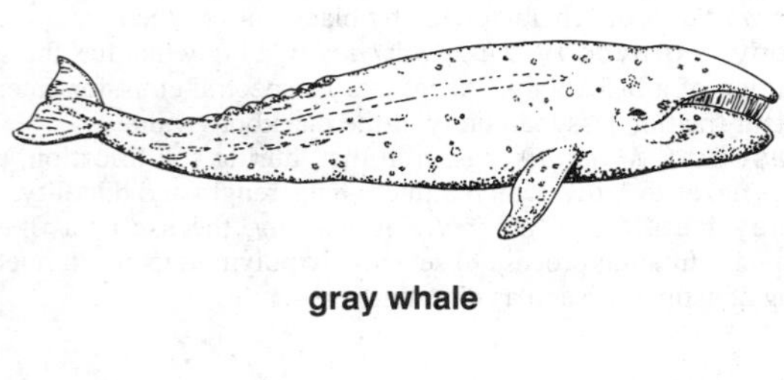

gray whale

graze *Agriculture.* of livestock, to feed on low grass or vegetation on pastures and ranges. *Ordnance.* **1.** to follow a trajectory nearly parallel to the ground and below average human height. Thus, **grazing fire. 2.** the burst of a projectile on impact with the ground. Also, **graze burst. 3.** an observation by a spotter or observer that all projectile bursts occurred on impact. **4.** in time fire, the burst of a projectile upon impact with an object at or below the level of the target.

grazer *Vertebrate Zoology.* any animal that feeds by grazing. *Ecology.* a term for a plant-eating animal that feeds on the entire plant, such as cattle or whales.

grazier *Agriculture.* a person who manages grazing animals.

grazing *Agriculture.* the act of feeding by livestock on low grass or vegetation on pastures and ranges.

grazing angle *Physics.* a very small glancing angle.

grazing incidence *Physics.* the incidence at a grazing angle in which the angle of incidence is almost 90°.

grazing-incidence telescope *Astronomy.* a telescope that brings X-rays or gamma-rays to a focus by means of a "lens" consisting of nested concentric rings, upon which the rays strike at very shallow angles and reflect to a focus.

grazing occultation *Astronomy.* an occultation in which the star or planet being affected goes in and out of view behind mountains on the limb of the Moon.

GRB *Aviation.* the airport code for Green Bay, Wisconsin.

GRE Graduate Record Examination.

grease *Materials.* **1.** a lubricant that is derived either from melted, liquid animal fat or from petroleum. **2.** any thick, oily substance that is similar to this. **3.** to apply any such substance. *Veterinary Medicine.* an inflammatory swelling in the fetlocks or pasterns of a horse's leg, accompanied by the formation of cracks in the skin and the excretion of oily matter.

grease cup *Engineering.* an internally threaded container that is filled with grease, and that screws onto a bearing to apply lubrication and relubrication.

grease gun *Engineering.* a hand-operated instrument that forces grease into bearings under high pressure.

grease ice *Hydrology.* a slushy layer of ice recently formed by the thickening of frazil on the surface of a body of water, especially on the ocean. Also, ICE SLUSH.

grease seal *Engineering.* a tight closure that prevents leakage of grease from a piston or other moving part.

grease spot *Plant Pathology.* a disease of turf grasses caused by the fungus *Pythium aphanidermatum,* characterized by patches with greasy borders of darkened leaves mixed with growths of cottony fungal filaments. Also, SPOT BLIGHT.

grease spot photometer *Optics.* a device that compares the luminous intensities of a standard light source and a test source by separating them with an opaque screen with a single translucent grease spot. Also, BUNSEN PHOTOMETER, BUNSEN SCREEN.

grease table *Mining Engineering.* an apparatus for concentrating minerals, such as diamonds, that adhere to grease.

grease trap *Civil Engineering.* a fixture in a drainage system to catch grease so it does not enter the sewer system; the trap can be removed and cleaned.

greasewood *Botany.* any plant of the genus *Sarcobatus* in the family Chenopodiaceae, especially *S. vermiculatus,* a stiff, low-growing shrub of the goosefoot family, found in the western American desert.

greasy quartz see MILKY QUARTZ.

great anteater see ANTEATER.

great ape *Vertebrate Zoology.* any of several apes of the family Pongidae, such as the gorilla, chimpanzee, and orangutan; characterized by a relatively hairless face with protruding lips and by hands having flat nails and complex fingerprints.

Great Arabian Desert *Geography.* one of the world's driest deserts (annual precipitation less than 4 inches), on the Arabian Peninsula.

Great Attractor *Astrophysics.* a proposed huge mass in space, postulated to explain the apparent movement of galaxies toward a particular point.

Great Barrier Reef *Geography.* the world's largest chain of coral formations (1260 miles long), along the northeastern coast of Australia.

Great Basin high *Meteorology.* a high-pressure system that is centered over the Great Basin region of the western United States, occurring frequently in wintertime.

Great Bear Lake *Geography.* a large freshwater lake (area: about 12,000 square miles) in the Canadian Northwest Territories.

great blue heron

great blue heron *Vertebrate Zoology.* a large American heron, *Ardea herodias,* standing 4–5 feet high and having blue-gray plumage.
great circle *Navigation.* a circle containing a segment that represents the shortest distance between two given points on the earth's surface. Thus, **great-circle route.** *Mathematics.* the circle formed by the intersection of a (two-) sphere and a plane passing through the center of the sphere. All geodesics on the sphere lie on great circles.
great-circle bearing *Navigation.* the initial direction of a great circle through any two points on the surface of the earth.
great-circle chart *Cartography.* a chart on the gnomonic map projection, in which a great circle appears as a straight line.
great-circle course *Navigation.* the direction of the great circle from the point of departure to the destination.
great-circle direction *Cartography.* the horizontal direction of a great-circle, expressed as the angular distance from a reference direction.
great-circle distance *Navigation.* the length of the shorter arc of the great circle joining two points, usually expressed in nautical miles.
great-circle sailing *Navigation.* any method used to solve the problems (such as courses and distances) involved with following a great-circle track.
great-circle track *Navigation.* the track of a vessel following, or intending to follow, an approximate great circle between departure and destination.
great diurnal range *Oceanography.* the vertical difference between mean higher high water and mean lower low water.
Great Divide see CONTINENTAL DIVIDE.
Greater Dog see CANIS MAJOR.
greater ebb *Oceanography.* the stronger of two ebb currents of a tidal day.
greater flood *Oceanography.* the stronger of two flood currents of a tidal day.
greater omentum *Anatomy.* the double fold of the peritoneum that attaches to the stomach and duodenum and drapes over the ventral surface of the intestines like an apron.
greatest common divisor *Mathematics.* **1.** the largest integer that divides each member of a finite set of integers. **2.** more generally, let X be a nonempty subset of a commutative ring R. An element d of R is then the greatest common divisor of X if: (a) d divides every element of X, and (b) if an element c of R divides every element of X, then c also divides d.
greatest elongation *Astronomy.* the maximum angular separation that an inferior planet has from the sun in a given apparition.
greatest integer function *Mathematics.* the function of a real number x whose value at x is the greatest integer less than or equal to x.
greatest lower bound *Mathematics.* the largest of the lower bounds of a set that is bounded below. Let A be a partially ordered set with partial order $\leq$, and B a nonempty subset of A. A lower bound of B (if it exists) is an element d of A such that $d \leq b$ for every $b \in B$, and a greatest lower bound (if it exists) is a lower bound d_0 of B such that $d \leq d_0$ for every other lower bound d of B.
great galago see THICK-TAILED BUSHBABY.
Great Ice Age see ICE AGE.
Great Lakes *Geography.* a chain of five large lakes in east central North America, on the border between the U.S. and Canada. They are Lakes Superior, Michigan, Huron, Erie, and Ontario.
Great Plains *Geography.* a sloping plateau in central North America extending about 400 miles eastward from the Rocky Mountains and about 2000 miles southward from Canada to Texas.
Great Red Spot *Astronomy.* a reddish elliptical region in Jupiter's South Equatorial Belt that is approximately 14,000 km in a north-south direction and has a length varying from 25,000 to 40,000 km. Its color varies in intensity from bright to dull red over time scales of years. It is now thought to be a hurricane-like atmospheric disturbance.

Great Red Spot

Great Rift *Astronomy.* a dark divide in the Milky Way between Cygnus and Ophiuchus that is caused by large clouds of interstellar dust lying between the Sun and the Milky Way's background stars.
Great Rift Valley *Geography.* a series of rift valleys running from the Jordan River Valley in southwestern Asia to Mozambique in southeastern Africa.
Great Salt Lake *Geography.* a large salty lake in northwestern Utah. With no outlet, its size and salinity vary seasonally.
great soil group *Geology.* a subdivision of a soil order consisting of a group of soils with common internal characteristics.
Great Square *Astronomy.* the Alpha, Beta, and Gamma Pegasi and Alpha Andromedae stars, forming the body of the constellation Pegasus.
great tropic range *Oceanography.* the vertical difference between tropic higher high water and tropic lower low water.
great white shark *Vertebrate Zoology.* a large, voracious, predatory mackeral shark, *Carcharodon carcharias* of the family Lamnidae, characterized by a gray back and white belly, a long pointed snout, and huge serrated, triangular teeth; found worldwide in warm and tropical seas and responsible for numerous attacks on humans.
great year *Astronomy.* a period of about 25,800 years, the time it takes precession to move the equinoxes once around the ecliptic.
grebe *Vertebrate Zoology.* any of the family Podicipitidae of carnivorous large aquatic birds with dense plumage and lobate toes; its habitat is saltwater and freshwater lakes, marshes, and coastal waters in winter; found in North and South America, Eurasia, Africa, and Australia.

grebe

Greco-Latin square *Statistics.* a pair of superimposed and orthogonal Latin squares used as an experimental design.

greedy algorithm *Mathematics.* **1.** an algorithm for solving the minimal connector problem on a connected graph G with n vertices and weighted edges. The following gives a solution: (a) Let e_1 be an edge of G with smallest weight. (b) Define $e_2, e_3, \ldots, e_{n-1}$ by choosing e_k to be an edge of smallest weight not previously chosen, so that e_k forms no circuit with the previously chosen edges $e_1, e_2, \ldots, e_{k-1}$. The subgraph of G whose edges are $e_1, e_2, \ldots, e_{n-1}$ is a spanning tree of G that satisfies the minimal connector problem. **2.** the procedure that picks the largest permissible element at each step when choosing elements of a certain set to fill a given need. For many applications, the greedy algorithm is the optimal procedure. **3.** any algorithm characterized by a step that selects an extreme element (e.g., a longest, shortest, or heaviest element) and that uses no other criterion to modify this selection.

Greeffiellidae *Invertebrate Zoology.* a superfamily of free-living nematode roundworms in the order Desmoscolecoidea.

Greek key motif *Crystallography.* a mode of folding of the polypeptide backbone of a protein that bears a resemblance to a Greek key frieze design.

Greek revival *Architecture.* a 19th-century European and American movement to revive the spirit and forms of classical Greek architecture.

green *Optics.* the color sensation that corresponds to radiation in the wavelength region 492 to 577 nanometers of the visible spectrum, and lies between blue and yellow on the color spectrum. *Metallurgy.* in powder metallurgy, the attribute of a compact that has not been sintered.

Green, George 1793–1841, English mathematician; pioneer in mathematical analysis of electromagnetism; formulated Green's theorem.

green algae *Botany.* a common name for the members of the plant division Chlorophyta, so called because their chlorophyll is not masked by other pigments.

green belt see FROSTLESS ZONE.

green chop *Agriculture.* mechanically harvested forage fed to livestock while it is still fresh. Also, **green soil.**

green cinnabar see CHROMIC OXIDE.

green copy *Graphic Arts.* during the printing of a form, the first few sheets to come off the press; used as a guide in correcting the press settings, paper alignment, or inking.

green earth *Geology.* any naturally occurring silicate, especially one bearing iron; used mainly in making green dyes and pigments.

green flash *Optics.* an apparent green coloration of the upper portion of the sun, or of the evening sky just above the setting sun, caused by atmospheric refraction.

Green function *Quantum Mechanics.* the representation of a propagator.

green gland see ANTENNAL GLAND.

green gold *Metallurgy.* a gold-base alloy containing either silver and cadmium or silver and copper.

greenheart *Materials.* the strong, hard wood of the tree *Octotea rodivei,* used to build marine structures such as ships and docks.

greenhouse *Botany.* a glass-enclosed, climate-controlled structure in which young or tender plants are cultivated, often out of season. Also, GLASSHOUSE.

greenhouse effect *Meteorology.* **1.** the warming of the earth's surface and lower atmosphere as a result of carbon dioxide and water vapor in the atmosphere, which absorb and reradiate infrared radiation. **2.** an intensification of this warming effect brought about by increased levels of carbon dioxide in the atmosphere, resulting from the burning of fossil fuels.

greenhouse gas *Ecology.* a term for a gas such as carbon dioxide or methanol that increases global temperatures by trapping solar elecromagnetic radiation.

Greenland *Geography.* the world's largest island (area: 840,000 square miles), in the North Atlantic northeast of Canada.

Greenland anticyclone *Meteorology.* the glacial anticyclone that is said to overlie Greenland; similar to the Antarctic anticyclone.

Greenland currents see EAST GREENLAND CURRENT, WEST GREENLAND CURRENT.

green laser *Optics.* a laser that utilizes a mixture of mercury and argon to produce a discharge of green laser light at 5225 angstroms.

green lead ore see PYROMORPHITE.

green machining *Materials Science.* the processes of forming and shaping that are carried out prior to the final densification of a ceramic part while the material is softer and can thus be machined more economically.

green manure *Agronomy.* plant material introduced into the soil while it is green to improve soil quality. This process is **green-manuring.**

green mold *Mycology.* any fungi of the genera *Aspergillus* and *Penicillium* that are green or form green spores.

green mud *Oceanography.* a greenish, fine-grained, oceanic mud containing chlorite and glauconite minerals that are responsible for its coloration.

green nickel oxide see NICKEL OXIDE.

greenockite *Mineralogy.* CdS, an orange to yellow, hexagonal mineral, dimorphous with hawleyite; occurring in hemimorphic pyramidal crystals and earthy coatings, and having an adamantine to resinous luster, a specific gravity of 4.82 to 4.9, and a hardness of 3 to 3.5 on the Mohs scale; found as small crystals with zeolites and coatings on sphalerite.

Greenough, George [grē´nō] 1778–1855, English geographer; pioneer in geological mapping; founder of London Geological Society.

Green Revolution *Agronomy.* a popular term for the recent great advances in agricultural productivity in certain underdeveloped countries, brought about by such factors as the introduction of superior strains of wheat and other grains, the use of pesticides, and improvements in crop management.

green rosette *Plant Pathology.* a virus disease of peanut plants characterized by severe retardation of plant development and a yellowing and bunching of the leaves.

green rot *Plant Pathology.* a deterioration that occurs on fallen deciduous trees due to the fungus *Peziza aeruginosa,* which covers the wood with deep green fuzzy growths. *Metallurgy.* a kind of high-temperature corrosion of chromium-bearing alloys, caused by oxidation and carburization.

green salt *Inorganic Chemistry.* a popular name for uranium tetrafluoride, UF_4.

greensand *Geology.* an unconsolidated marine sediment consisting principally of dark green grains of glauconite. *Petrology.* sandstone that contains green glauconite grains mixed with clay or with sand, with little or no cement.

greenschist *Petrology.* a schistose metamorphic rock, whose green color is caused by an abundance of chlorite, epidote, or actinolite.

greenschist facies *Petrology.* the set of metamorphic mineral assemblages (facies) constituting schistose rocks with a characteristic abundance of green minerals formed at low to intermediate temperatures (300–500°C) and pressures (3000–8000 bars).

Green's function *Mathematics.* let L be a linear differential operator in the variables x with homogeneous boundary conditions. Then the Green's function $G(x, x')$ for L, the parameter λ, and the given boundary conditions is the solution of $LG(x, x') - \lambda G(x, x') = \delta(x - x')$, subject to the boundary conditions. The Green's function represents the response of the system to a unit impulse, and a general solution for the parameter λ is given by multiplying the Green's function by the forcing function and integrating over the domain of the forcing function.

Green's identities *Mathematics.* formulas relating the volume integral of a function f and its gradient to a surface integral of f and its partial derivatives; for example, $\int_V u\ \nabla^2 u\ dV + \int_V (\nabla u) \cdot (\nabla v)\ dV = \int_S u(\nabla u) \cdot n\ d\sigma$, where n is the unit exterior normal. This identity and others like it are obtained from Green's theorem.

green sky *Meteorology.* a greenish tinge in part of the sky that is believed by some to be the precursor of wind, rain, or even a tropical cyclone.

green smut *Plant Pathology.* a rice disease caused by the fungus *Ustilaginoidea virens* and characterized by enlarged grains covered with a greenish powder of conidia. Also, FALSE SMUT.

green snow *Hydrology.* a snow surface having a greenish tint, caused by the growth of microscopic green algae.

Green's theorem *Quantum Mechanics.* a theorem stating that for any two functions of discrete positions, the surface integral over any volume is equal to the surface integral over the volume's boundary. *Mathematics.* the theorem that $\int_C (P\ dx + Q\ dy) = \iint_D (\partial Q/\partial x - \partial P/\partial y)\ dxdy$, where D is a bounded open set enclosed by a rectifiable simple closed curve C, the functions P and Q are continuous on $D \cup C$, and the indicated partial derivatives are bounded and continuous on D. Green's theorem is the special case of Stokes' theorem when the surface lies in the (x, y) plane; i.e., a special case of the theorem that the integral of a differential form over the (smooth) boundary of a set is equal to the integral of the exterior derivative of the form over the interior of the set.

greenstick fracture *Medicine.* an incomplete fracture of a long bone in which the bone is bent but fractured only on the outer arc of the bone; often seen in children.

greenstone *Mineralogy.* a familiar name for nephrite, a compact fine-grained variety of actinolite. *Petrology.* any basic igneous rock that has been metamorphosed or otherwise altered so as to take on a greenish color due to the presence of chlorite, actinolite, or epidote.

greenstone belts *Geology.* the elongated or arcuate landmasses within Precambrian shields, characterized by large amounts of greenstone, especially those landmasses similar to, and running north and south of, the Swaziland System.

greenstone schist *Petrology.* greenstone whose structure is foliated.

green strength *Metallurgy.* in powder metallurgy, the strength of a compact before sintering.

green sulfur bacteria *Bacteriology.* a group of anaerobic, photosynthetic bacteria that are able to utilize sulfur compounds as electron donors during metabolism; typically found in sulfide-rich aquatic environments.

green vitriol see FERROUS SULFATE.

Greenwich meridian [gren´ich] *Cartography.* the meridian through Greenwich, England, that serves as the reference for Greenwich Time; in constrast with local meridians, it is accepted almost universally as the prime meridian, or the origin of measurement of longitude.

Greenwich Time *Horology.* time as measured along the Greenwich meridian. Also, **Greenwich Mean Time, Universal Time.**

gregale *Meteorology.* a term used in the central and western Mediterranean region to denote a strong northeast wind resulting either from the passage of a depression to the south or southeast or from high pressure over central Europe or the Balkans and low pressure over Libya; it occurs most frequently in winter and may last as long as 5 days.

Gregarina *Invertebrate Zoology.* a genus of parasitic protozoans in the order Gregarinida; found in the alimentary canals of arthropods.

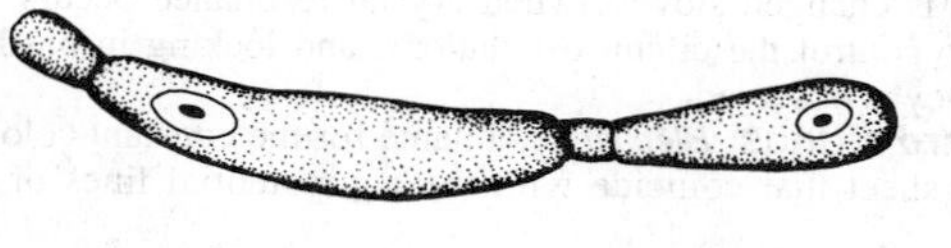

Gregarina

gregarine *Invertebrate Zoology.* **1.** any member of the protozoan order Gregarinida. **2.** of or relating to this order.

Gregarinida *Invertebrate Zoology.* an order of protozoa in the class Sporozoa, parasitic in the digestive tract of invertebrates.

gregarious [grə gâr´ē əs] *Behavior.* living in groups rather than in isolation; social.

Gregorian calendar see CALENDAR.

Gregorian telescope *Optics.* a telescope in which a concave primary mirror with a hole at the center and a secondary ellipsoidal mirror produce an erect image from a small field of view; ideal for terrestrial observations.

Gregory, James 1638–1675, Scottish mathematician; proposed theory of reflecting telescope; formulated Gregory's series, binomial series, and Taylor series.

Gregory's series *Mathematics.* the series $\pi/4 = \sum_{n=0}^{\infty} (-1)^n/(2n+1)$.

greisen *Petrology.* a pneumatolytically altered granite composed primarily of quartz, topaz, and a light-green mica.

Gremlin *Ordnance.* a human-portable Soviet surface-to-air missile. The land-based version is officially designated **SA-14**; the seabed version is **SA-N-8**.

grenade *Ordnance.* a small explosive or chemical bomb detonated by a fuse; originally thrown by hand, it is now usually launched from a special device attached to a rifle or carbine. Thus, **grenade carrier, grenade cartridge extractor, grenade launcher,** and so on.

grenadier [gren´ə dēr´] *Ordnance.* in earlier use, a soldier armed with grenades. *Vertebrate Zoology.* any of several deep-sea fishes of the family Macrouridae having a long, tapered tail.

grenatite **1.** see LEUCITE. **2.** see STAUROLITE.

Grenfell, Sir Wilfred 1865–1940, British physician; established Arctic medical services.

Grenville orogeny *Geology.* a major Precambrian deformation of the earth's crust that affected the rocks of the southeastern border of the Canadian Shield.

Grenz ray *Nucleonics.* an X-ray with little penetrating power, produced from electron beams accelerated through potentials ranging from 5000 to 15,000 volts.

grep *Computer Programming.* to search a file for text by entering one or more characters together with "wild cards," which stand for any digit, character, or string. (So called from a search command of this name.)

gressorial *Vertebrate Zoology.* being adapted for walking, as are certain birds' feet.

grey another spelling of GRAY, especially in British use. See GRAY.

greyhound *Vertebrate Zoology.* a breed of tall, short-haired dog having a pointed muzzle, arched back, slender waist, and long legs; known for its keen vision and great speed.

Greyiaceae *Botany.* a monogeneric family of dicotyledonous soft-wooded tanniferous shrubs and small trees native to South Africa and usually placed in the order Rosales; characterized by a well-developed, cupulate, extrastaminal nectary disk and numerous small seeds with copious endosperm.

GRF growth hormone-releasing factor.

grid a network of straight lines or parts; specific uses include: *Navigation.* an arbitrary system of parallel and perpendicular lines applied to a chart for reference purposes if latitude and longitude are not used. *Cartography.* a rectangular Cartesian coordinate system that is superimposed on maps, charts, and other representations of the earth's surface in an accurate and consistent manner so that identification of ground locations with respect to other locations and the computation of direction and distance to other points can be made. *Engineering.* **1.** a grating made of crossed bars. **2.** a template of equally spaced squares used as a guide for hole spacing on a chassis or printed circuit board. *Naval Architecture.* a heavy framing of timbers, used to support ships in a dock. *Computer Technology.* two mutually orthogonal sets of parallel lines in the same plane, used to specify or measure character images in OCR equipment. *Radiology.* an arrangement of thin lead strips separated by a radiolucent material used to absorb scattered radiation produced during an X-ray examination. *Electricity.* a metal framework in a storage cell or battery, used as a conductor and support for the active material. *Electronics.* in an electron tube, an electrode that controls the passage of electrons between the cathode and anode, having one or more openings through which electrons can pass under specific conditions.

grid amplitude *Archaeology.* a method of defining the location of features and artifacts on a site by plotting from a reference point oriented to magnetic north or some other known point; meridian lines run north-south and baselines east-west on a **grid square.** *Navigation.* the angle, measured to north or south, between the grid prime vertical and some other point on the celestial sphere.

grid azimuth *Navigation.* the horizontal measurement of a celestial point from a terrestrial point, measured clockwise from grid north.

grid battery see C BATTERY.

grid bearing *Navigation.* a bearing relative to grid north.

grid bias *Electronics.* a steady direct voltage between the grid and cathode of an electron tube that sets the operating point of the tube.

grid blocking *Electronics.* **1.** in an electron tube, a keying method in which a high grid voltage produces a blocking bias, which is then removed for the beginning of normal operations. **2.** in an amplifier, the blocking of capacitance-coupled stages due to an accumulated charge on the coupling capacitor during the reception of large signals.

grid capacitor *Electronics.* **1.** a capacitor in series with the grid of a tube, used for blocking purposes. **2.** a bypass capacitor in a grounded-grid tube-type amplifier. **3.** a capacitor in the grid tank circuit of a tube-type oscillator or amplifier.

grid cathode capacitance *Electronics.* the internal capacitance between the cathode and the control grid of an electron tube.

grid characteristic *Electronics.* the performance curve of grid current versus grid voltage in an electron tube.

grid circuit *Electronics.* the external circuit associated with an electron tube's control grid.

grid control *Electronics.* the control of plate (anode) current in an electron tube by means of grid voltage. Thus, **grid control tube.**

grid coordinates *Cartography.* a set of numbers or letters of a coordinate system that designate a point on a map, photograph, or chart supplied with a grid.

grid coordinate system *Cartography.* a plane-rectangular coordinate system usually based on, and mathematically adjusted to, a particular map projection, so that geographic positions may be readily transformed into plane coordinates and the computations relating to them may be made by the ordinary methods of plane surveying.

grid course *Navigation.* a course relative to grid north.

grid current *Electronics.* the current flowing between the control grid and cathode in an electron tube.

grid declination *Navigation.* the angular difference between true north and grid north.

grid drive *Electronics.* a signal that is applied to an electron tube's control grid in an amplifier, detector, or control circuit. Also, **grid excitation.**

grid drive characteristic *Electronics.* the relation between electric or light output and control-electrode voltage measured from cutoff; usually indicated graphically.

grid driving power *Electronics.* the power supplied to the biasing device and grid of an electron tube; equal to the average product of the grid voltage and the instantaneous values of the alternating components of the grid current over a complete cycle.

grid element *Electricity.* a sinuous resistor composed of wire, strap, or casting, and which is designed to heat a furnace.

grid-eye fish *Vertebrate Zoology.* any of various fishes of the the Family Ipnopidae, known for its unusual eye adaptation to deep ocean darkness: a bony plate covers the upper head to shield groups of retinal cells that perceive faint lumination.

grid-glow tube *Electronics.* a gas-discharge tube in which control electrodes initiate but do not limit plate current.

grid heading *Navigation.* the heading relative to grid north.

gridistor *Electronics.* a field-effect transistor with multiple channels of embedded grids that are produced by integrated techniques; combines characteristics of field-effect and minority-carrier injection transistors.

grid latitude *Navigation.* latitude (fictitious) on a navigational grid.

grid leak *Electronics.* a resistor that provides a direct current path in order to limit the accumulation of a charge in the grid; connected across the grid capacitor or between the grid and cathode. Also, GRID RESISTOR.

Gridley fungus stain *Microbiology.* a fungal cell stain involving the oxidation of cell wall polysaccharides and their subsequent interaction with Schiff's reagent.

grid limiter *Electronics.* a circuit that limits positive grid voltages by means of a large ohmic value resistor.

grid locking *Electronics.* a tube defect in which the grid potential becomes permanently positive due to excessive grid electron emission.

grid longitude *Navigation.* longitude (fictitious) on a navigational grid.

grid magnetic angle see GRIVATION.

grid map *Cartography.* see GRID.

grid meridian *Navigation.* a meridian (fictitious) on a navigational grid.

grid metal *Metallurgy.* any of several lead-base alloys used in lead-acid storage batteries.

grid method *Photogrammetry.* a method of plotting detail from oblique photographs by superimposing a map grid on a photograph and transferring the detail by eye; uses the corresponding lines of the map grid and its perspective as placement guides.

grid modulation *Electronics.* amplitude modulation produced by varying the direct current control-grid bias of an RF amplifier at an audio-frequency rate.

grid navigation *Navigation.* the process of navigating a craft using the directions on a grid.

grid neutralization *Electronics.* the neutralization of an amplifier produced by shifting a portion of the grid-to-cathode alternating current voltage 180° and applying it to the plate-to-cathode circuit through a neutralizing capacitor.

grid north *Navigation.* an arbitrary reference direction that is based on a grid.

grid parallel *Navigation.* a parallel (fictitious) on a navigational grid.

grid plate *Photogrammetry.* a glass plate on which an accurately ruled grid is etched; used for calibration of plotting instruments and film distortion. Also, RESEAU.

grid-plate capacitance *Electronics.* the inner capacitance between the control grid and plate of an electron tube.

grid-pool tube *Electronics.* a gas-discharge tube whose cathode is a pool of mercury and which has a control electrode or grid that begins the current flow in each cycle.

grid prime vertical *Navigation.* a prime vertical (fictitious) based on a navigational grid.

grid pulse modulation *Electronics.* modulation created by applying one or more pulses to a grid circuit in an amplifier or oscillator.

grid pulsing *Electronics.* the process of pulsing an RF oscillator by producing a negative grid bias that is sufficient to block oscillation and then removing this bias by applying a positive pulse.

grid ratio *Radiology.* the ratio of the length of the lead strips to the interval between them on a radiographic grid.

grid resistor *Electronics.* **1.** in an electron tube, a high-voltage resistor connected between the control grid and ground. **2.** see GRID LEAK.

grid return *Electronics.* the circuit path by which the control grid on an electron tube returns to either ground or B-minus.

grid rhumb line *Navigation.* a rhumb line drawn on a grid and measured from grid north.

grid spectrometer *Spectroscopy.* a spectrometer in which the entrance and exit slits have been replaced by grids patterned with opaque and transparent areas, to enable a large increase in light flux without loss of resolution.

grid suppressor *Electronics.* a fixed resistor linked in series with the control grid of a tube or the control electrode of a transistor to prevent parasitic oscillations. Also, **grid stopper.**

grid swing *Electronics.* the peak-to-peak variation of a grid excitation signal.

grid therapy *Radiology.* a method of radiation treatment using a screen with small, equally spaced perforations placed directly on the skin.

grid track *Navigation.* the track of a craft with respect to a grid reference system.

grid transformer *Electronics.* a transformer that supplies an alternating voltage to one or more grid circuits.

grid unit *Archaeology.* an individual section that has been plotted out on a site by intersecting north-south and east-west lines; usually the basic excavation unit shown on a grid map.

grid variation *Navigation.* the angular difference between magnetic north and grid north.

grid voltage *Electronics.* the DC bias voltage applied to the control grid of an electron tube.

Griebe-Schiebe method *Solid-State Physics.* a technique for observing the piezoelectric effect in small crystals in which a crystal is placed between two electrodes in an oscillator circuit and the oscillator frequency is changed slowly; when crystal resonance occurs, the crystal tends to control the circuit oscillations, and locking in on the resonant frequency is observed.

Griebhard's rings *Electricity.* a name for the constant color lines on a copper sheet that coincide with the equipotential lines of an electric field.

grief *Psychology.* the normal emotional response to a consciously recognized external loss such as bereavement, separation, or loss; its expression and duration vary greatly among cultures and individuals. Thus, **grieving.**

Griesbachian *Geology.* a European geologic stage of the Lowermost Triassic period, occurring after the Permian period.

Griess reagent *Analytical Chemistry.* a solution composed of α-naphthylamine, glacial acetic acid, and sulfanilic acid; used for detecting nitrous acid.

Griffith crack *Metallurgy.* a small crack that promotes brittle fracture.

griffithite see SAPONITE.

Griffith's criterion *Mechanics.* a theoretical threshhold of fracture in brittle materials, based on surface energy.

Griffith's tube *Biotechnology.* a stout tube, closed at the bottom and with a rough interior surface, that is used to grind tissue; grinding takes place by means of a close fitting pestle such as a glass rod.

Griffith's typing *Immunology.* a technique that is used to subdivide the Lancefield group A streptococci by means of an agglutination test.

Griffith theory *Materials Science.* a theory that fracture occurs in brittle materials because of preexisting flaws. The expression of the theory relates fracture stress to flaw size.

Griffon see GAMMON.

Grignard, Victor [grēn´yär] 1871–1935, French chemist; won Nobel Prize for the discovery of Grignard's reagents.

Grignard reaction *Organic Chemistry.* a reaction in which the addition of organomagnesium compounds to carbonyl groups or other unsaturated groups produces alcohols or ketones.

Grignard reagent *Organic Chemistry.* a class of reagents with the general formula of RMgX, where R is an alkyl or aryl group (or other such organic group), Mg is a metallic magnesium, and X is a halogen formed in the Grignard reaction; used in synthetic organic chemistry.

Grignard synthesis *Organic Chemistry.* synthesis using the Grignard reagent to form a hydrocarbon, acid, ketone, or secondary or tertiary alcohol.

grillage *Civil Engineering.* a foundation footing consisting of two or more tiers of closely spaced structural steel beams resting on a concrete block, each tier being at right angles to the one below; commonly a foundation used in marshy, treacherous soil.

grille *Building Engineering*. a usually wrought-iron, often decorative screening used to enclose an area or to provide security, as on the outside of windows or exterior doors. *Mechanical Engineering*. an opening, usually covered by grillwork, designed to admit air to cool the engine of a vehicle. *Acoustical Engineering*. a covering on the front of a loudspeaker or other such device that serves to protect the unit and often to enhance its acoustic characteristics. Thus, **grille cloth.**

grillwork *Engineering*. any material, usually metal, that is fashioned to function as or resemble a grille.

Grimm, Jakob 1785–1863, German philologist and folklorist; studied Indo-European language relationships and influenced development of comparative linguistics.

Grimm, Wilhelm 1786–1859, German folklorist; compiled noted collection of folk tales with his brother Jakob.

Grimmiaceae *Botany*. a family of dull, dark mosses of the order Grimmiales that form mats and cushions on rocks or sandy soil in northern temperate and arctic regions; characterized by creeping to erect-ascending stems with numerous branches and terminal or lateral sporophytes and by peristome reduction.

Grimmiales *Botany*. an order of small, dark mosses in the class Bryopsida that form dense mats or cushions, mostly on rock surfaces, and have terminal sporophytes and a peristome shape.

grindability *Materials Science*. the ease with which a given material can be pulverized.

grinder *Mechanical Engineering*. any of various devices used for reducing materials, finishing surfaces, sharpening tools, or grinding food, typically through the action of abrasives bonded to rotating disks or other instruments.

grinding *Mechanical Engineering*. **1.** the reduction of a material, often to a fine powder. **2.** the machining of a surface in order to remove excess materials and produce a smooth finish. *Electronics*. **1.** an operation performed on semiconductor substrates to provide a smooth surface for subsequent epitaxial deposition or diffusion of impurities. **2.** an operation performed on quartz crystals to alter their size and frequencies.

grinding aid *Mechanical Engineering*. material that serves to speed up the grinding process, as in a ball mill or rod mill.

grinding medium *Mechanical Engineering*. solids that are used in suitable mills to grind materials to a powder.

grinding mill *Mechanical Engineering*. a usually cylindrical ball, rod, or tube mill designed for crushing ore or other materials.

grinding pebbles *Mechanical Engineering*. fragments of chert or quartz that are used in grinding mills to avoid contamination with iron.

grinding relief *Mechanical Engineering*. a groove designed to facilitate certain grinding operations, especially in metallurgy.

grinding stone see MILLING STONE.

grinding stress *Mechanical Engineering*. surface compressive stress remaining on a material after grinding, which strengthens against cracking or tensile stress.

grinding-type resin *Organic Chemistry*. a resin, such as vinyl, that requires grinding before dispersal into plastisols or organosols.

grinding wheel *Mechanical Engineering*. a wheel or disk made of abrasive material and used for cutting and finishing metal and other materials.

grindstone *Mechanical Engineering*. a circular disk that revolves on an axle and is used for grinding, smoothing, or shaping materials.

grindstone truer *Mechanical Engineering*. a rotating pointed steel bar consisting of a threaded roller of steel clamped in a frame that rotates against the surface of a workpiece or grindstone; used for rectifying the rotation of a grindstone.

Grinellian niche *Ecology*. the range of values of environmental factors that are necessary to allow a species to carry out its life history; the niche can be measured by correlating the geographic distribution and associated abundance to the environmental conditions.

grip *Physiology*. a grasping or seizing; to grasp or seize. *Ordnance*. the part of a weapon designed to be held in the hand; it is usually made of two wooden or plastic pieces that are fastened to the frame of the weapon. *Medicine*. see GRIPPE.

grip and ungrip *Robotics*. the axis of motion for a pick-and-place robot.

griphite *Mineralogy*. $Na_4Ca_6(Mn,Fe^{+2},Mg)_{19}Li_2Al_8(PO_4)_{24}(F,OH)_8$, an easily fusible, brown cubic mineral occurring as nodular masses that may weigh up to 50 pounds and be up to 6 feet in diameter, and having a specific gravity of 3.64 and a hardness of 5.5 on the Mohs scale; found in granitic pegmatite.

grippe [grip] *Medicine*. an older term for influenza. Also, GRIP.

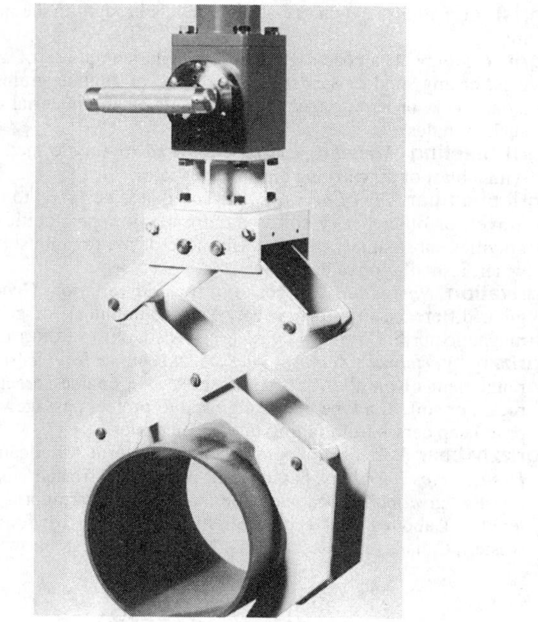

gripper

gripper *Graphic Arts*. any of a set of fingerlike metal clamps that hold the paper to the cylinder on a printing press. *Robotics*. a device on the end of a robot's arm that can grasp an object using suction cups, magnets, or an articulated end effector.

gripper edge *Graphic Arts*. on a sheet being printed, the edge that is clamped by the press gripper.

gripper margin *Graphic Arts*. on a sheet being printed, an area (typically 3/8 inch) that is left blank to allow for the gripper edge.

gripping zone *Robotics*. the area where the center of an object must be placed in order for the gripper to handle it properly.

grip safety *Ordnance*. a device on some semiautomatic pistols that prevents firing unless the grip is held firmly while the trigger is squeezed.

grip vector *Robotics*. a vector that describes the orientation of an object from a point on the wrist socket of a robot to the point where its end effector grasps the object.

griquaite *Petrology*. a coarse-grained rock consisting of garnet and diopside, sometimes containing olivine or phlogopite; found as nodular xenoliths in kimberlite pipes and dikes.

Griselda complex *Psychology*. a father's unconscious desire to keep his daughter to himself rather than allowing her to marry. (From the medieval legends of *Griselda,* a woman of great patience and self-sacrifice who represents the ideal wife.)

griseofulvin *Microbiology*. an antifungal antibiotic synthesized by the mold *Penicillium griseofulvum*. Also, CURLING FACTOR.

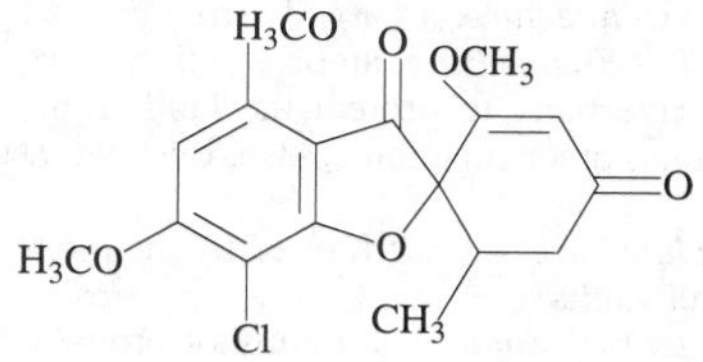

griseofulvin

griseolutein *Microbiology*. either the A or B fraction of the broad-spectrum antibiotics that are produced by *Streptomyces griseoluteus*; effective against Gram-positive microorganisms.

griseomycin *Microbiology*. an antibiotic that is produced by a bacterium similar to *Streptomyces griseolus* and is active against Gram-positive bacteria.

grist *Agriculture.* a quantity of grain that is brought to the mill for grinding.

grit *Geology.* **1.** a coarse-grained sandstone composed of angular particles, or any sand or sandstone composed of angular grains. **2.** a small stone, rock, or hard, angular sand granule or any material composed of such granules.

grit blasting *Metallurgy.* a process used to abrade metallic surfaces with a blast of air carrying abrasive particles.

grit chamber *Civil Engineering.* a cell that is designed to remove sand, gravel, or other heavy solids having declining velocities or specific gravities substantially greater than those of the organic solids in wastewater; a small type of detritus tank.

grivation *Navigation.* another term for grid variation. *Geodesy.* the angular difference in direction between geographical (or grid) north and magnetic north. (From grid variation.) Also, GRID MAGNETIC ANGLE.

grizzly *Vertebrate Zoology.* see GRIZZLY BEAR. *Engineering.* **1.** an arrangement of parallel bars that serves as a coarse screening for ore, rocks, or soil. **2.** a type of grating used to protect passageways in mines or to keep debris from falling into a water inlet.

grizzly bear *Vertebrate Zoology.* a type of North American brown bear, *Ursus arctos horribilis,* having dark brown hair frosted with white, giving it a "grizzled" appearance; the largest of all carnivores, with a body length of about 8 feet and a weight of 500–700 lbs; found mainly in western Canada and Alaska. Also, NORTH AMERICAN BROWN BEAR.

grizzly bear

grizzly chute *Mining Engineering.* a chute with a bar grizzly through which the fine material is separated from the coarse material.

grizzly worker *Mining Engineering.* in metal mining, an underground laborer who works at a grizzly. Also, DRAW MAN.

gro. or **gro.** gross.

Groeberiidae *Paleontology.* a small, extinct family of South American marsupials in the proposed suborder Groeberioidea, characterized by a superficially rodent-like dentition; extant in the Eocene.

grog *Materials.* coarse oxide particles that are bonded by finer materials to produce refractory products.

groin *Anatomy.* **1.** the area between the abdomen and thigh. **2.** see INGUINAL REGION. *Architecture.* a curved arris formed by the intersection of two vaults. *Civil Engineering.* an obstruction built out from land into the ocean or a riverbank to protect the land from erosion and sand movements, among other functions. Also, GROYNE, JETTY BREAKWATER, WING DAM.

groin vault *Architecture.* a vault formed by the intersection at right angles of two barrel vaults.

Gromiida *Invertebrate Zoology.* a family of protozoans in the subclass Filosia; widely distributed in fresh and salt water and in moist soils.

grommet *Mechanical Devices.* **1.** a ring of strong material used to strengthen an opening through which a cable or similar material passes. **2.** an insulating washer of strengthened fiber, plastic, or rubber used under nut and bolt heads to ensure tightness. Also, GRUMMET WASHER. *Ordnance.* a device that protects the rotating band of a projectile; it may be made of rope, plastic, rubber, or metal.

grommet nut *Mechanical Devices.* a type of nut having a screw thread, a rounded head, and a blind hole, often used to secure a hinge to a door.

grooming *Behavior.* the act of maintaining and caring for any part of the body surface.

groove a long, narrow indentation along a surface; specific uses include: *Geology.* a long, straight, narrow depression on the surface of a sedimentary rock. *Building Engineering.* the cut in a board made to attach to the tongue of another board. *Ordnance.* any of the shallow, spiral channels in the rifling of a gun bore, designed to transmit a spinning motion to the projectile, thus stabilizing it in flight. *Biochemistry.* in the model of Watson-Crick DNA, a channel that varies in size from 12 to 22 angstroms in width. Thus, **grooved.**

groove cast *Geology.* a round or sharp-crested rectilinear ridge formed on the underside of a sandstone bed by the filling in of a groove on the surface of underlying mudstone.

grooved bit *Mechanical Devices.* a drill bit with a cylindrical shank and a cut helical groove; used in woodworking.

grooved drum *Mechanical Devices.* a drum having a fluted surface, used to bolster and direct a rope.

groove diameter *Ordnance.* the diameter of a gun bore, as measured from the bottom of one groove to the bottom of the diametrically opposite groove.

groover *Mechanical Devices.* a cutting or punching instrument used to make grooves.

groove sample *Mining Engineering.* a coal or ore sampling that is obtained by cutting grooves along or across the road exposures. Also, CHANNEL SAMPLE.

groove weld *Metallurgy.* a weld performed in a groove between the parts to be joined.

grooving saw *Mechanical Engineering.* a circular saw designed for cutting grooves.

Gropius, Walter [grōp´ē əs] 1883–1969, German-born American architect; founder of the Bauhaus school of design.

grosbeak *Vertebrate Zoology.* any of several species of finches belonging to the family Fringillidae and subfamilies Cardinalinae and Carduelinae, characterized by large conical bills and ranging primarily in North America with a few in Eurasia.

Grosch's law *Computer Technology.* an empirical formula concerning economies of scale in computers, stating that computing power increases in proportion to the square of the cost of the system; subscribed to by the computer industry in the late 1940s.

gross anatomy *Anatomy.* the study of the structure and morphology of body organs that are visible to the naked eye.

gross area *Building Engineering.* the sum of the areas of all floors of a multistory building as measured from exterior faces of exterior walls. Also, **gross floor area.**

gross-austausch *Meteorology.* the exchange of air-mass properties and the associated momentum and energy transports produced around the world by the migration of large-scale weather disturbances in the middle latitudes.

gross ecological efficiency see ECOLOGICAL EFFICIENCY, def. 2.

gross energy *Nutrition.* the total flow of energy on which the biological processes work.

Grosseteste, Robert c. 1175–1253, English bishop, statesman, and scientist; a founder of optics; proposed theory of sound waves.

gross index see PRIMARY INDEX.

gross information content *Telecommunications.* the amount of total information, whether redundant or not, that is contained in a message.

gross model *Cartography.* the total overlap area of a pair of aerial photographs.

gross porosity *Metallurgy.* the undesirable presence of coarse porosity in a casting or weld.

gross primary production *Ecology.* the total assimilation of organic matter by an autotrophic organism or population during a given period of time over a given area.

gross production *Ecology.* the total assimilation of organic matter by an individual or population during a given time over a given area.

gross production rate *Ecology.* the rate of energy assimilation by an organism of a particular trophic level.

gross recoverable value *Mining Engineering.* the value obtained by multiplying the part of the total metal recovered by the price.

gross rubber *Chemical Engineering.* the total weight of marketable product in rubber manufacturing, including all materials used in compounding the rubber.

gross secondary production *Ecology.* the total amount of energy assimilated by a primary consumer, such as a rabbit eating plants.

gross ton see LONG TON.

gross tonnage *Naval Architecture.* the total usable enclosed space within a vessel, except for certain exempted spaces; not a measure of weight.

grossular *Mineralogy.* $Ca_3Al_2(SiO_4)_3$, a colorless to yellow, green, or brown cubic mineral, the calcium-aluminum end member of the garnet group, occurring as dodecahedral crystals and in granular form, and having a specific gravity of 3.4 to 3.6 and a hardness of 6.5 to 7.5 on the Mohs scale; found in metamorphosed calcareous rocks. Also, **grossularite.**

Grossulariaceae *Botany.* a cosmopolitan family of dicotyledonous shrubs and trees in the order Rosales that are sometimes spiny and often tanniferous and includes the currants and gooseberries; some botanists divide this large and diverse family into as many as eight smaller families.

Grothendieck group *Mathematics.* let $F(A)$ denote the set of all isomorphism classes of finitely generated A-modules, where A is a Noetherian ring, and let C be the free Abelian group generated by $F(A)$. With each short exact sequence $0 \to M' \to M \to M'' \to 0$ of finitely generated A-modules, associate the element $(M') - (M) + (M'')$ of C, where (M) is the isomorphism class of M. Let D denote the subgroup of C generated by all such short exact sequences. Then the quotient group C/D is called the Grothendieck group of A, and is denoted $K(A)$. Let $\gamma(M)$ denote the image of (M) in $K(A)$. The Grothendieck group $K(A)$ is universal in the sense that for each additive function λ from the class of finitely generated A-modules to an Abelian group G, there exists a unique homomorphism λ_0: $K(A) \to G$ such that $\lambda(M) = \lambda_0(\gamma(M))$ for all finitely generated M.

Grotthus mechanism *Physical Chemistry.* **1.** an early explanation for conductivity in electrochemical devices, assuming that an electrolyte consists of a chain of polar molecules stretching from the anode to the cathode, with current passing along the chain as a result of the two end molecules breaking up to leave a negative fragment at the anode and a positive fragment at the cathode. Also, **Grotthus chain (theory). 2.** a similar modern model used to describe the motion of hydrogen ions.

ground *Geology.* **1.** any rock, rock material, or mineralized deposit in which mineral deposition takes place. **2.** the top layer of the earth's surface. *Navigation.* **1.** the sea floor. **2.** to strike the bottom. *Aviation.* **1.** to keep on the ground; restrict or prevent a vehicle or pilot from flying. **2.** describing nonflight operations normally occurring on the earth's surface; used chiefly to form compounds such as **ground control, ground systems.** *Military Science.* operating or occurring on the ground. Thus, **ground forces, ground war,** etc. *Electricity.* **1.** a connection to the earth that conducts electrical current to and from the earth. **2.** the voltage reference point in a circuit. **3.** in an electrical system, a point with zero voltage. **4.** to carry out the process of grounding. *Materials Science.* of a material, reduced to a fine powder or dust by grinding.

ground absorption *Electromagnetism.* a loss of radio-frequency energy due to dissipation of the waves into the ground.

ground-air *Military Science.* **1.** of or related to weapons or signals that are projected from the ground to the air. Thus, **ground-air missile. 2.** of or related to military actions that involve both ground and air units.

ground area *Building Engineering.* the area of a building at ground level, as measured from exterior faces of exterior walls.

ground beetle *Invertebrate Zoology.* any of various terrestrial beetles of the family Carabidae; most are nocturnal and prey on other insects.

ground beetle

ground block *Civil Engineering.* a pulley fastened to an anchor log, used to change the horizontal tension to a vertical tension on a wire line.

ground cable *Electricity.* a heavy-duty cable connected to the earth, or some conducting body serving in place of the earth, to ground electric equipment.

ground chain *Naval Architecture.* a chain attached to an anchor as it is brought up, and used to hold the anchor clear of the ship's side.

ground check *Engineering.* **1.** a procedure in which instruments are checked before the ground launch of an airborne experiment. **2.** a procedure used to determine temperature and humidity corrections for a radiosonde system prior to the release of the radiosonde.

ground-check chamber *Engineering.* an enclosed, environmentally controlled area having instruments that check the sensing elements of a radiosonde.

ground clutter *Electromagnetism.* the interference observed on an aircraft radar caused by signal reflections from the ground.

ground coal *Mining Engineering.* the bottom of a coal seam. Also, GROUNDS.

ground conductor *Electricity.* a conductor to which electrical equipment or devices are connected in order to establish a ground connection.

ground control *Aviation.* the control of aircraft ground movements at an airport by the tower or air traffic control. *Cartography.* a specific system of points on the earth's surface with a position that has been established by ground survey, and has been referenced to the celestial sphere, the geoid, a given ellipsoid of reference, or an assumed origin.

ground-controlled approach *Aviation.* an airport approach system in which air traffic controllers transmit instructions to the approaching pilot by radio. Also, **ground-control approach.**

ground-controlled approach minimums *Aviation.* the minimum conditions of ceiling and horizontal visibility considered safe for aircraft making a ground-controlled approach to an airport.

ground-controlled approach radar *Engineering.* an aircraft landing system that communicates information from a ground radar installation to a pilot intending to land.

ground-controlled interception *Military Science.* an air-defense technique in which interceptor aircraft or guided missiles are directed toward their target(s) by instructions transmitted from the ground. Also, **ground-control interception.**

ground-controlled intercept radar *Military Science.* the radar system used to direct a ground-controlled interception system.

ground controller *Aviation.* an operator who controls aircraft activity from a station on the ground.

groundcover or **ground cover** *Botany.* any plant that grows close to the ground to provide cover or protection for the soil, especially one that is not a grass plant.

ground data equipment *Space Technology.* instruments that are located in a ground station and serve to obtain, receive, or transmit information from satellite or space vehicles.

ground detector *Electricity.* an instrument that indicates the presence of a ground.

ground dielectric constant *Electricity.* the earth's dielectric constant at a given point.

grounded-base amplifier *Electronics.* an amplifier having a transistor in a grounded-base connection.

grounded-base connection *Electronics.* a transistor circuit in which the input and output circuits share a common base electrode.

grounded-cathode amplifier *Electronics.* a tube amplifier in which the cathode is at ground potential at the operating frequency; the input is applied between the control grid and ground, and the output load is between the plate and ground.

grounded-collector amplifier *Electronics.* an amplifier having a transistor in a grounded-collector connection.

grounded-collector connection *Electronics.* a transistor circuit in which the input and output circuits share a common collector electrode, which may or may not be directly connected to circuit ground. Also, COMMON-COLLECTOR CONNECTION.

grounded-emitter amplifier *Electronics.* an amplifier having a transistor in a grounded-emitter connection.

grounded-emitter connection *Electronics.* a transistor circuit in which the input and output circuits share a common emitter electrode, which may or may not be directly connected to circuit ground.

grounded-gate amplifier *Electronics.* an amplifier having field-effect transistors in which the gate electrode is connected to ground, with input applied to the source electrode and output obtained from the drain electrode.

grounded-grid amplifier *Electronics.* an electron-tube amplifier having the control grid at ground potential at the operating frequency, with input applied between the cathode and ground and output between the plate and ground.

grounded-grid-triode circuit *Electronics.* a circuit in which the input signal is applied to the cathode and the output is obtained from the plate; the grid is at RF ground and functions as a screen between the input and output circuits.

grounded-plate amplifier see CATHODE FOLLOWER.

grounded system *Electricity.* any conducting apparatus in which at least one conductor or point is grounded.

ground effect *Aviation.* an increased lift generated by the interaction between a lift system and the ground when an aircraft is within a wingspan distance above the ground. It affects low-winged aircraft more than mid- or high-winged aircraft, because the former's wings are closer to the ground. Also, **ground cushion.** *Telecommunications.* any undesired effect that ground interference imposes on radar, radio navigation aids, or other communication systems.

ground-effect machine see AIR-CUSHION VEHICLE.

ground electrode *Electricity.* a conductor or group of conductors that are buried in the ground to provide a ground connection.

ground environment *Engineering.* the aggregate of physical, chemical, and biological factors in which equipment or a system operates on the ground. *Military Science.* in an air defense system, the facilities that provide detection, surveillance, and control of airborne weapons and vehicles; although most such facilities are located on the ground, the term may also include facilities on ships or on aircraft other than interceptors.

ground equalizer inductors *Electromagnetism.* the low inductance coils that are placed in a circuit and connected to grounding points of an antenna to distribute current appropriately.

ground equipment *Aviation.* any land-based equipment used to make an aircraft or aerospace vehicle or device fully operational. Also, AEROSPACE GROUND EQUIPMENT.

ground fallout plot *Nucleonics.* the lines on a radioactive fallout plot that indicate approximately how far the fallout will have spread after a certain amount of time has elapsed.

ground fault *Electricity.* **1.** an undesirable connection to ground. **2.** an undesirable loss of ground in a grounded system.

ground-fault interrupter *Electricity.* a fast-acting circuit breaker that senses ground-fault currents and limits the time the current can flow through the fault by opening the circuit breaker.

ground fire *Ordnance.* ground-based small arms fire directed at flying aircraft. *Forestry.* a forest fire that burns only the organic material in the soil layer and also may appear at the surface level of vegetation.

ground floor *Building Engineering.* the floor within a building that is at the same level as the ground outside the structure.

ground fog *Meteorology.* a fog condition in which 0.6 or less of the sky is covered and the fog system is not associated with the clouds above it.

ground frost *Meteorology.* a freezing condition that is damaging to vegetation, indicated when a thermometer with its bulb positioned just above a grass surface reads a grass temperature of 30.4°F or less.

ground gained forward *Photogrammetry.* the net gain per photograph in the direction of flight for a specified overlap; used to compute the number of exposures in a strip of aerial photography.

ground gained sideways *Photogrammetry.* the net lateral gain per flight for a specified sidelap; used to compute the necessary number of flight lines for an area to be photographed.

ground glass *Optics.* a glass plate that has one matte surface capable of diffusing light; used as a translucent surface in cameras to aid in focusing.

ground handling equipment see GROUND-SUPPORT EQUIPMENT.

groundhog *Vertebrate Zoology.* another name for the woodchuck, *Marmota monax.* See WOODCHUCK. *Mining Engineering.* see BARNEY.

ground ice *Hydrology.* **1.** ice found beneath the ground surface, especially a subsurface mass of clear, nonglacial ice within permafrost or seasonally frozen ground. **2.** ice formed on the ground by the compaction or freezing of snow or the freezing of rain. **3.** see ANCHOR ICE.

ground ice mound *Hydrology.* a small frost mound that contains masses of ice.

grounding *Electricity.* **1.** the process of connecting to ground or to a conductor that is grounded. **2.** the referencing of electrical circuits to the well-bonded equipotential surface. *Navigation.* the contact of a vessel with the bottom.

grounding outlet see GROUND OUTLET.

grounding plate *Electricity.* an electrically grounded metal plate on which a person stands to discharge static electricity.

grounding transformer *Electricity.* a transformer designed to provide a neutral point for grounding purposes.

ground instance *Artificial Intelligence.* a logical formula that contains no variables, or one in which constants or functions of constants have been substituted for all variables.

ground instrumentation see SPACECRAFT GROUND INSTRUMENTATION.

ground inversion see SURFACE INVERSION.

ground joist *Building Engineering.* a joist that has been blocked from the ground level of a structure.

ground layer see SURFACE BOUNDARY LAYER.

ground level see GROUND STATE.

ground log *Navigation.* a primitive means of measuring the speed of a vessel by lowering a weight to the bottom and observing the rate at which the line goes overboard.

ground loop *Electricity.* a possibly harmful loop formed when two or more points in an electric system that are nominally at ground potential are connected by a conducting path.

ground loop current *Telecommunications.* the currents from high-powered components that circulate in ground loops and generate undesired noise signals.

ground lug *Electricity.* a lug that connects a grounding conductor to a grounding electrode.

ground magnetic survey *Engineering.* a survey in which ground-based instruments are used to give a reading of the magnetic field at the earth's surface.

groundman *Mining Engineering.* **1.** a miner assigned to work on the surface. **2.** a worker who uses a pick and shovel to move overburden and coal within reach of the power shovel. Also, DIRTMAN.

groundmass *Petrology.* **1.** the relatively fine-grained material between the phenocrysts of a porphyritic igneous rock; may be crystalline and/or glassy. Also, MATRIX. **2.** see MATRIX, def. 1.

ground mine *Ordnance.* see BOTTOM MINE.

ground moraine *Geology.* any rock debris that was dragged along by, and deposited from, the base of a glacier or ice sheet. Also, BOTTOM MORAINE, SUBGLACIAL MORAINE.

ground nadir *Cartography.* the point on the ground vertically beneath the perspective center of the camera lens during aerial photography.

ground noise *Acoustical Engineering.* unwanted noise in a sound recording or reproduction that is introduced by an improperly grounded element of the system. *Geophysics.* in geophysical prospecting, a movement of the ground that was not caused by the shot.

groundnut *Botany.* any of various plants having edible tubers or other underground parts, such as the peanut.

groundnut oil see PEANUT OIL.

ground observer *Military Science.* in an air defense system, a ground-based observer who provides visual and aural information of aircraft movements. Thus, **ground observer center, ground observer organization, ground observer team.**

ground outlet *Electricity.* an electrical outlet having a receptacle equipped with a grounded contact that can be used for the connection of an equipment-grounding conductor.

ground parallel *Cartography.* the intersection of the plane of the photograph with the plane of reference of the ground during aerial photography.

ground plane *Electricity.* a conducting surface used as a common reference point for circuit returns and electric potentials. *Cartography.* the horizontal plane passing through the ground nadir of a camera station during aerial photography.

ground-plane antenna *Electromagnetism.* an antenna, consisting of a horizontal crossed dipole with a vertical element at the intersection, that produces a nondirectional horizontal pattern.

ground plate *Electricity.* a conducting plate buried in the ground to serve as a grounding electrode.

ground point of intercept *Aviation.* the point on the ground directly below the actual point of intercept of two aircraft or a missile and an aircraft.

ground position *Aviation.* the point on the ground directly below the actual position of an aircraft. Thus, **ground-position indicator.** Also, GEOGRAPHICAL POSITION.

ground potential *Electricity.* zero potential with respect to the earth.

ground pressure *Geophysics.* the pressure exerted on a rock formation by movements of the earth. Also, GEOSTATIC PRESSURE, LITHOSTATIC PRESSURE, ROCK PRESSURE.

ground protection *Electricity.* the protection of a circuit with a device that opens the circuit when a fault to ground occurs.

ground proximity warning system *Navigation.* an onboard warning system (essentially a downward-looking radar) that warns an aircrew of excessively low altitude above ground level.

ground pyramid *Photogrammetry.* a set of two components of an analytical method for determining the precise degree of photographic tilt, representing a specific spatial configuration from three ground control points (forming a triangle) to the exposure station of the photograph, which contains the identical points; when used with the photo-pyramid, the ground pyramid yields the exact analytical determination of tilt in the photograph.

ground-reflected wave *Electromagnetism.* a radio wave that is reflected off the ground at some point along its transmission.

ground resistance *Electricity.* the resistance of a connection to ground, or between two connections to ground, that depends on soil conditions and on the surface area of the ground connection.

ground return *Electricity.* **1.** the radar echoes that are reflected from the earth's surface. **2.** a lead from an electronic circuit, antenna, or power line to ground.

ground-return circuit *Electricity.* a circuit in which the earth is used to complete the circuit.

ground rod *Electricity.* a copper rod driven into the earth to serve as the ground connection.

grounds *Building Engineering.* wood pieces embedded in wall plasterings to which skirting and other joining work is attached, or as ends to plasterings around doors and windows; used to guide the thickness of a plaster coat. *Mining Engineering.* see GROUND COAL.

groundscatter propagation *Telecommunications.* multihop ionospheric propagation of radio waves traveling between the transmitting and receiving points on some route other than the great circle path; the waves bounce from the ionosphere to the earth's surface and scatter in many directions.

ground speed *Aviation.* the actual speed that an aircraft travels over the ground; it combines the craft's air speed and the wind speed relative to the craft's direction of motion. Also, SPEED OVER THE GROUND.

ground squirrel *Vertebrate Zoology.* any of several rodents of the family Sciuridae, such as squirrels of the genus *Citellus* and chipmunks of the genus *Tamias,* that live in complex underground burrows and sometimes form colonies; they are often destructive pests of cultivated land and serve as vectors of plague.

ground squirrel

ground start *Space Technology.* the generation of propellants into a large rocket from the ground during ignition and holddown so that the liftoff main-stage tanks do not need to be utilized.

ground state *Quantum Mechanics.* the lowest energy state of a quantized system. Also, GROUND LEVEL.

ground state maser *Physics.* a maser whose amplifying transition terminates on the ground state energy level.

ground station see EARTH STATION.

groundstone *Archaeology.* any rock or mineral material that can be shaped by abrasion and formed into a tool or vessel, such as granite, pumice, or steatite.

ground strafing *Ordnance.* an attack upon ground troops by low-flying aircraft using machine guns, cannons, or bombs.

ground streamer *Meteorology.* an upwardly moving column of high ion density that ascends from a point on the earth's surface toward which a stepped leader descends at the onset of a lightning discharge; it joins the stepped leader at about fifty feet above the ground.

ground substance *Biochemistry.* the nonliving matrix of connective tissue.

ground substitution *Artificial Intelligence.* a substitution of constants or functions of constants for variables.

ground-support equipment *Aviation.* any mobile or fixed land-based equipment used to support the research and development or operational activities of military aircraft or missiles, including all tools and devices in such operations. Also, GROUND HANDLING EQUIPMENT.

ground-surveillance radar *Engineering.* surveillance radar that observes and controls a vehicle's movement and location from a station on the earth's surface.

ground survey see FOOT SURVEY.

groundswell or **ground swell** *Oceanography.* a long, high ocean swell whose height rises dramatically as it passes through water that is shallower than one-tenth wavelength.

ground swing *Cartography.* an error-causing condition in electronic distance measuring that results from a reflection of the microwave beam from the ground or water surface mixing with the direct beam at the receiving antenna, thereby changing the phase of the direct beam and resulting in error in the distance measured.

ground switch see LIGHTNING SWITCH.

ground system *Electromagnetism.* a portion of an antenna system connected to a ground through a large conductive surface, such as the earth.

ground tackle *Navigation.* anchors, anchor chain, warp, and other fittings used to anchor a vessel.

ground-to-air missile see SURFACE-TO-AIR MISSILE.

ground-to-cloud discharge *Geophysics.* a type of discharge in which the primary electrical emission originates from a structure on the Earth and proceeds upward to a cloud.

ground-to-ground missile see SURFACE-TO-SURFACE MISSILE.

ground trace *Engineering.* a theoretical vertical mark that traces the movement of an airborne object passing over the earth's surface.

ground truth measurements *Geophysics.* the readings of various properties taken on the ground to verify or calibrate those taken by remote sensors, such as satellites.

ground-up read-only memory *Computer Technology.* a read-only memory that is designed from the bottom up; all of its fabrication masks are custom generated.

ground visibility *Aviation.* the horizontal visibility as observed at ground level.

groundwater or **ground water** *Hydrology.* the portion of subsurface water that is below the water table, in the zone of saturation. Also, PHREATIC WATER, UNDERGROUND WATER.

groundwater artery *Hydrology.* a tube-shaped body of permeable material encased within a matrix of impermeable or less permeable material and saturated with water under artesian pressure.

groundwater basin *Hydrology.* **1.** a subsurface basin in which groundwater collects or is retained, or from which groundwater may flow. **2.** an aquifer having more or less well-defined boundaries, as well as areas of recharge and discharge.

groundwater budget *Hydrology.* a quantitative description of the recharge, discharge, and changes in storage of an aquifer or system of aquifers. Also, **groundwater accounting, groundwater balance.**

groundwater cement *Geology.* a concentration of calcium carbonate or calcium sulfate resulting from the evaporation of groundwater near the surface or in shallow soil. Also, WATER-TABLE CEMENT.

groundwater depletion curve *Hydrology.* a graph of stream flow over a period of time, indicating the decrease in groundwater runoff after surface runoff to the stream channel has ceased.

groundwater discharge *Hydrology.* **1.** the freeing of water from the zone of saturation. **2.** the water or quantity of water thus freed. Also, **groundwater decrement.**

groundwater equation *Hydrology.* **1.** a mathematical statement of the groundwater losses and gains in a given area. **2.** specifically, the equation that will balance the groundwater budget, $R = E + S + I$, in which R represents total rainfall, E is loss through evaporation, S is stream flow out of the system, and I is the amount of replenishment.

groundwater flow *Hydrology.* **1.** the movement of water in the zone of saturation by natural or artificial means. Also, **GROUNDWATER MOVEMENT. 2.** water that has been absorbed by the ground and has become groundwater. Also, GROUNDWATER RUNOFF.

groundwater hydrology see GEOHYDROLOGY.

groundwater lake *Hydrology.* a lake whose water represents the exposure of the upper surface of the zone of saturation.

groundwater level *Hydrology.* **1.** see WATER TABLE. **2.** the height of the water table at or in a particular location or area, as determined by the level of water in wells or other depressions that intersect the zone of saturation.

groundwater mound *Geology.* a hill-shaped elevation in a water table, usually produced by the downward percolation of surface water to the water table. Also, WATER-TABLE MOUND.
groundwater movement see GROUNDWATER FLOW, def. 1.
groundwater outflow *Hydrology.* any discharge from an area that occurs as groundwater.
groundwater recession curve see GROUNDWATER DEPLETION CURVE.
groundwater recovery *Hydrology.* see RECOVERY, def. 1.
groundwater reservoir see AQUIFER.
groundwater runoff see GROUNDWATER FLOW, def. 2.
groundwater table see WATER TABLE.
groundwater withdrawal *Hydrology.* see RECOVERY, def. 2.
ground wave *Ordnance.* a disturbance in the ground created by the explosion of a mine, bomb, or other weapon. *Telecommunications.* a low-frequency radio wave that bends along the earth's surface rather than traveling through the atmosphere. Also, SURFACE WAVE.
ground ways *Civil Engineering.* the foundation type of supports made of heavy timber, used on the ground on either side of the keel of a ship under construction, providing a track for launching a ship. Also, STANDING WAYS.
ground wire *Civil Engineering.* a high strength steel wire usually of small gage used to establish a line or grade for air-blown mortar or concrete. Also, ALIGNMENT WIRE, SCREED WIRE. *Electricity.* a conducting wire used to connect electric equipment to a grounded object.
groundwood *Graphic Arts.* uncooked wood pulp left over from paper milling that is finely ground and incorporated into the rough **groundwood papers** such as newsprint.
ground zero *Ordnance.* the point on the surface of the earth at, or vertically above or below, the center of a planned or actual nuclear detonation.
group any collection or set of persons or things; specific uses include: *Chemistry.* **1.** a vertical column in the periodic table, representing a series of elements that share chemical properties, such as group I-A, which contains the alkali metals. **2.** a combination of two or more elements that tend to act as a unit in reactions, such as the hydroxyl group (–OH). *Hematology.* see BLOOD GROUP. *Military Science.* **1.** an administrative and tactical unit composed of two or more battalions, squadrons, or air wings. **2.** two or more ships or aircraft assigned for a specific purpose. *Telecommunications.* a number of voice channels (usually twelve) treated as one unit and multiplexed using single-side-band techniques and occupying the band from 40 to 180 kHz. *Mathematics.* a nonempty set G together with a binary operation $*$ defined on all of G, such that: (a) G is closed under $*$; (b) $*$ is associative; (c) an element e of G exists that is a two-sided identity element, i.e., $a * e = e * a = a$ for every $a \in G$; and (d) for every $a \in G$, there exists an element $a^{-1} \in G$ that is a two-sided inverse element of a, i.e., $a * a^{-1} = a^{-1} * a = e$. If G satisfies only the first requirement, it is a **groupoid**; if G satisfies the first two requirements, it is a **semigroup**; and if G satisfies the first three requirements, it is a monoid. If $*$ is a commutative operation, then G is commutative or Abelian. The symbol for the binary operation is often omitted and the operation indicated by juxtaposition of group elements. It if is desirable to view $*$ as either addition or multiplication, G is said to be an **additive group** or a **multiplicative group,** respectively.
group algebra *Mathematics.* **1.** a group ring $K(G)$, where K is a commutative ring. $K(G)$ is actually a K-algebra with K-module structure given by $k(\sum r_i g_i) = \sum (kr_i) g_i$, where $k, r_i \in K$ and $g_i \in G$. Equivalently, $K(G)$ is the free algebra of G over K. **2.** let G be a Lie group and dg a left invariant measure on G. The group algebra (over the complex numbers) of complex-valued continuous functions on G having compact support is defined as follows: addition is pointwise addition of functions, and multiplication $*$ is the convolution product: $f_1 * f_2(x) = \int_G f_1(g) f_2(g^{-1}x)\, dg$, where $x, g \in G$ and the Lie group operation is written multiplicatively.
group analysis *Design Engineering.* an analysis for the application of group technology.
group B rotavirus see PARAROTAVIRUS.
group bus *Electricity.* the electrical connections for a generating station in which three or more feeder lines are supplied by two bus-selector circuit breakers leading to a main and auxiliary bus.
group busy tone *Telecommunications.* a high tone fed to an outgoing trunk group to indicate to operators that all trunks in a group are busy.
group-coded record *Computer Technology.* a method for recording data on magnetic tape to detect and correct all single-bit errors; on a tape with eight data tracks and one parity track, every eighth byte is used for this purpose.
group diffusion method *Nucleonics.* a method for approximating the way that neutrons diffuse in a reactor by dividing them into groups based on their energy levels and assuming that these energy levels are maintained until they have experienced the average number of collisions needed to reduce their energy to the next lowest group.
group dynamics *Psychology.* the interactions and relationships that take place among members of a group in a social setting, such as children at play or workers in an office. Also, **group climate.**
grouped data *Statistics.* large amounts of data arranged into categories or class intervals.
grouped-frequency operation *Telecommunications.* in a two-wire carrier system, a method of grouping signals traveling in opposite directions into different frequency bands.
grouped records *Computer Programming.* two or more records associated with the same key that are combined into a single unit to conserve storage space or reduce access time.
grouper *Vertebrate Zoology.* **1.** any of several hundred species of heavy-bodied fish of the family Serranidae found in the warm oceans of the world; some grow quite large (up to six feet long and 1000 pounds) while a few are mature at one inch; characterized by a large mouth and dull green or brown scales; among the world's most important food fish. **2.** any of various rockfishes of the family Scorpaenidae.

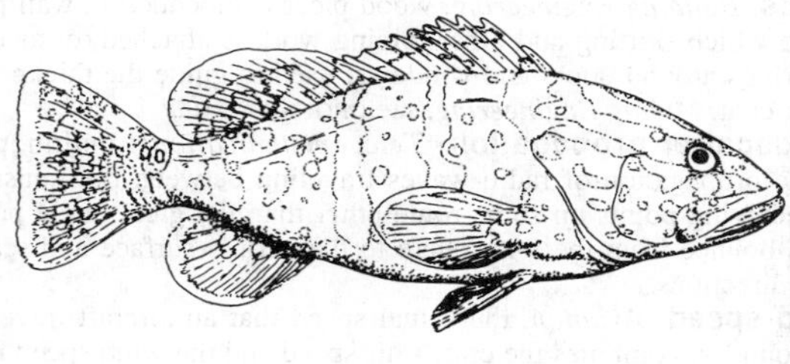

grouper

group flashing light *Navigation.* a navigational light that shows two or more flashes in a group, with the duration of light being less than the duration of darkness.
group frequency *Electromagnetism.* the frequency associated with the group velocity of waves in a waveguide or transmission line.
group incentive *Industrial Engineering.* a material incentive provided to all members of a work group to reward the group's collective performance.
group indicate *Computer Programming.* to print indicative information from only the first record of a group.
grouping *Science.* **1.** the classification of individual entities according to certain common characteristics. **2.** the process of forming a group. *Hematology.* see BLOOD GROUP. *Telecommunications.* a recurrent error in the spacing of recorded lines in a facsimile.
grouping circuits *Telecommunications.* circuits used to interconnect two or more switchboard positions to facilitate the handling of all the positions from one operator's set.
group mark *Computer Programming.* a symbol that identifies the beginning or end of a word or other unit of data.
group marriage *Anthropology.* a practice in which two or more men are married to two or more women as a single conjugal unit.
group modulation *Telecommunications.* the process of modulating a group of channels previously modulated to a specific frequency range, to a different frequency range as a group.
group molar attraction constant *Materials Science.* a mathematical value that is used to determine the solubility parameters of a polymer; employs experimental values of the heat of evaporation.
group occulting light *Navigation.* a steady navigational light with two or more brief periods of darkness in a group, with the duration of light equal to or greater than that of darkness.
groupoid *Mathematics.* a set of elements G that is closed under a binary operation whose domain is all of G.
group printing *Computer Programming.* the printing of information summarizing data held in a group of records with the same key.
group rapid transit *Transportation Engineering.* an automated guideway transit system with vehicles that carry from six to one hundred passengers and may combine to form a train; service may be scheduled or demand responsive.

group ring *Mathematics.* let G be a (multiplicative) group and R a ring. Consider the additive abelian group $R(G) = \sum_{g \in G} R$ (one copy of R for each element of G). An element $x = \{r_g\}_{g \in G}$ of $R(G)$ is defined to be a formal sum $\sum r_i g_i$ with only finitely many nonzero coordinates r_i. (Note that an element of $R(G)$ may be written in more than one way.) Addition and multiplication ($*$) are defined in the natural way, with $\sum r_i g_i * \sum s_j g_j = \sum\sum r_i s_j g_i g_j$, where $g_i g_j$ is a group element. When the product is rewritten as a formal sum, it is known as convolution. $R(G)$ is called the group ring of G over R.

group selection *Ecology.* the theory that certain behaviors, such as the decline of reproductive rates, arise from group needs and benefit the entire population rather than the individual organism; may include the controversial hypothesis that natural selection favors characteristics and behaviors (such as altruism) that enhance survival of the entire population even at the expense of certain individuals. Also, KIN SELECTION.

group-specific antigen *Virology.* an antigen specific to a particular group of viruses. Also, TYPE-SPECIFIC ANTIGEN.

group technology *Industrial Engineering.* a methodology in which production processes are grouped and manufacturing cells are established in order to produce similar components economically. Also, CELLULAR MANUFACTURING.

group theory *Mathematics.* the study of groups, including finite, discrete, topological, and Lie groups.

group therapy *Psychology.* a therapeutic method in which several individuals are treated together; often the interactions of the group are a major part of the treatment. Also, **group psychotherapy.**

group velocity *Physics.* the velocity of a packet of waves having different frequencies and phase velocities. *Oceanography.* the overall velocity of a group of component simple harmonic ocean waves; in deep water half the velocity of an individual wave but in shallow water the same as the velocity of an individual wave.

groupware *Computer Science.* software that facilitates the cooperative work of a group of people on a single problem concurrently; the people may be at different locations and communicate via networks.

grouse *Vertebrate Zoology.* any member of the family Tetraonidae of fowl-like game birds, having short, curved, strong bills; short, rounded wings; feathered legs; and plump bodies. Its habitat is in coniferous and deciduous forests and tundra throughout North and South America, northern Europe, and Asia; species include many important game birds.

grouse

grouser *Engineering.* a wooden or iron pole that is driven into a stream bottom to assist in stabilizing the position of a floating or temporarily fixed object. Also, SPUD.

grout *Materials.* **1.** a thin, coarse mortar that is used to fill the spaces between tiles, rocks, or similar adjoining objects. **2.** a fine coat of plaster used to finish interior surfaces. *Engineering.* to apply either of these materials.

grout curtain *Civil Engineering.* a row of vertical holes (**grout holes**) that are filled with grout to form a cutoff wall beneath a dam or a barrier around an excavation.

grouting *Civil Engineering.* the process of injecting grout into holes or rock formations for strengthening purposes.

groutite *Mineralogy.* $Mn^{+3}O(OH)$, a black, opaque, orthorhombic mineral forming wedge-shaped crystals, trimorphous with manganite and feitknechtite, and having a specific gravity of 4.14 and a hardness of about 5.5 on the Mohs scale; found as small crystals coating calcite.

Grove cell *Electricity.* a primary cell containing a platinum electrode submerged in a nitric acid electrolyte that is surrounded by a zinc electrode in a sulfuric acid electrolyte; it opertes on a closed circuit.

Grove's synthesis *Organic Chemistry.* the synthesis of alkyl chlorides by passing hydrochloric acid into an alcohol in the presence of anhydrous zinc chloride.

grower's year *Meteorology.* in Great Britain, a twelve-month cycle of seasonal changes that begins on November 6th.

growing season *Agronomy.* the number of days from the average date of the last killing frost in the spring to the average date of the first killing frost in the fall.

growler *Electricity.* a device designed to locate short-circuited coils in the armature of a generator or a motor, and to magnetize and demagnetize objects. *Oceanography.* a small piece of floating ice, less than 10 meters across and barely showing above water.

grown-diffused transistor *Electronics.* a transistor created by first growing the emitter and collector regions as a crystal, and then diffusing the base region while the crystal is being pulled.

grown junction *Electronics.* a P-N junction created when impurities are added in various amounts to a crystal, while it is being pulled from molten semiconductor material.

grown-junction photocell *Electronics.* a grown-junction diode used as a photoconductive cell.

grown-junction transistor *Electronics.* a transistor created by adding N- and P-type impurities to a crystal in its molten state and slicing the resultant P-N and N-P junctions from the finished crystal.

growth *Biology.* a normal increase in the size of an organism. *Physiology.* the normal progression of human development from infancy to adulthood, involving anatomic, psychological, social, intellectual, and cultural changes. *Pathology.* an abnormal formation, such as a tumor. *Microbiology.* a proliferation of cells, as in a bacterial culture.

growth curve *Microbiology.* a graphic represention of the growth of a population of microorganisms in which the density of a cell culture is plotted as a function of time. *Nucleonics.* a line that depicts an increase in the artificial radioactivity of an irradiated material over time.

growth curve analysis *Statistics.* a data analysis process consisting of repeated measurements over time of a certain characteristic.

growth disk or **growth plate** see EPIPHYSEAL DISK.

growth fabric *Petrology.* in a rock, the orientation of fabric elements due to crystal arrangement from a plane surface, such as the wall of a vein, and independent of movements or deformations due to stress. *Space Technology.* the additional weight of fuel structures, engine weight, and other materials needed to support each pound of payload added to an original payload.

growth factor *Biochemistry.* the specific substance that must be present in a growth medium to permit cell multiplication. *Oncology.* any of the polypeptides such as c-sis (PDGF gene), IL-2 gene, and EGF gene, that are encoded by a family of genes; some are proto-oncogenes.

growth-factor model *Transportation Engineering.* a method of planning trip-development patterns by grossing up from previous patterns.

growth form *Ecology.* the characteristic appearance of a plant that reflects its adaptation to the environment.

growth hormone *Endocrinology.* a protein hormone that is produced by the somatotrophs of the anterior pituitary gland, and promotes the growth of most tissues through its effects on cellular metabolism. Growth hormone induces increased protein synthesis, increased mobilization of free fatty acids, and decreased glucose utilization. Also, SOMATOTROPIN.

growth hormone release-inhibiting factor see SOMATOSTATIN.

growth hormone-releasing hormone *Endocrinology.* a hypothalamic peptide hormone that stimulates the release of growth hormone from the somatotrophs of the anterior pituitary gland. Also, SOMATOLIBERIN, SOMATOTROPIN-RELEASING FACTOR.

growth inhibitor *Molecular Biology.* any factor or specific substance that acts to inhibit or retard cell growth or cell multiplication.

growth lattice *Geology.* the inflexible, reef-building framework of an organic reef, formed in place from the calcareous skeletal remains of sessile organisms. Also, ORGANIC LATTICE.

growth precursor cell *Cell Biology.* a cell that will proliferate only during favorable environmental conditions, maintaining a low level of cellular metabolism at any other time.

growth rate *Biology.* an expression of the increase in size of an organic object per unit time, usually expressed in both absolute and relative increments. *Microbiology.* specifically, the rate of increase in the cell number or biomass of a population of microorganisms.

growth regulator *Biochemistry.* any organic compound that inhibits or accelerates the physiological processes in plants.
growth spiral *Crystallography.* a circular or spiral pattern formed by a spiral dislocation of the unit cells of a crystal.
growth step *Crystallography.* a ledge formed by one or more unit cells located on the surface of a crystal, providing the potential for crystal growth.
growth-suppressor gene see TUMOR-SUPPRESSOR GENE.
growth water *Hydrology.* the portion of total soil water available to plants for growth.
groyne *Civil Engineering.* see GROIN.
GRR *Aviation.* the airport code for Grand Rapids, Michigan.
GR-S rubber *Organic Chemistry.* a synthetic rubber, formed from styrene and butadiene; used in tires and other rubber products.
GRT group rapid transit.
grub *Entomology.* a name for the short, stout, sluggish larvae of various insects, including many beetles, certain moths, and some flies, wasps, and bees.
grub axe *Agriculture.* a farm or gardening tool with a sharp point on one end and a broad chipping blade on the other; used for digging roots or underground vegetables.
Grubbiaceae *Botany.* a monogeneric family of shrubs belonging to the order Ericales and limited to the Cape Province of South Africa; characterized by opposite, narrow, leathery leaves and an inferior ovary capped by a hairy nectary disk.
grubby *Vertebrate Zoology.* a small sculpin, *Myxocephalus aenaeus,* of coastal New England.
grub saw *Mechanical Devices.* a handsaw whose steel blade is supported by wooden slats along its back; used for cutting stone.
grub screw *Mechanical Devices.* a slotted, headless screw that prevents lateral movement when placed in a continuous thread hole between two adjacent pieces; its tension is adjusted with a screwdriver.
Gruidae *Vertebrate Zoology.* the cranes, a family of large wading birds with long legs and neck of the order Gruiformes, with distribution throughout North America, Europe, Asia, Australia, and Africa.
Gruiformes *Vertebrate Zoology.* an order of mainly wading aquatic and terrestrial birds, including cranes, rails, coots, bustards, and their allies, with distribution worldwide except Antarctica.
grummet washer *Mechanical Devices.* see GROMMET, def. 2.
Grüneisen constant *Physics.* a constant for the linear expansion of a metal, being the ratio of its coefficient of linear expansion to its specific heat. Also, **Grüneisen gamma.** *Crystallography.* in most cubic crystals, a quantity that is relatively constant given by the product of three times the coefficient of linear expansion divided by the product of the compressibility with the specific heat (at constant volume) per unit volume.
Grüneisen relation *Solid-State Physics.* an empirical relation stating that the electrical resistivity of a very pure metal is proportional to the product of the absolute temperature and a function of the ratio of some characteristic temperature and the absolute temperature.
grunerite *Mineralogy.* $(Fe^{+2},Mg)_7Si_8O_{22}(OH)_2$, $Mg/(Mg+Fe^{+2})=0–0.3$, a gray, dark-green or brown, monoclinic mineral of the amphibole group, commonly occurring in fibrous or lamellar masses, and having a silky luster, a specific gravity of 3.44 to 3.71, and a hardness of 5 to 6 on the Mohs scale; found in iron-rich metamorphosed rocks.
grünlingite *Mineralogy.* a mixture of joseite and bismuthinite.
grunt *Vertebrate Zoology.* any of various tropical and subtropical ocean fish of the family Pomadasyidae that emits a deep, guttural sound.
Grus *Astronomy.* the Crane, a small constellation of the southern hemisphere whose brightest star is second magnitude.
gruss or **grus** *Geology.* an accumulation of coarse, angular fragments derived from the granular breakdown of granite or other crystalline rock.
gr. wt. gross weight.
Gryllidae *Invertebrate Zoology.* a family of insects, the true crickets, in the order Orthoptera, having long antennae, a dark and chunky body, and a long, cylindrical ovipositor.
Grylloblattidae *Invertebrate Zoology.* the rock crawlers, an order of wingless, yellowish-brown to gray primitive insects in the order Orthoptera; found at high elevations under rocks.
Gryllotalpidae *Invertebrate Zoology.* a family of mole crickets, large insects in the order Orthoptera, that burrow in sand and mud with shovel-shaped forelimbs and eat the roots of seedlings.
Gryostemonaceae *Botany.* a family of dicotyledonous Australian trees or shrubs of the order Batales, noted for bearing mustard oil and characterized by a dry indehiscent fruit and a copious oily endosperm.
Gryphaea *Paleontology.* a genus of oysters once thought to have repeatedly evolved similar forms independently of each other, but now known to have a continuous history; especially widespread in Jurassic; upper Triassic to upper Jurassic.
GS general staff; ground speed.
GSA Geological Society of America.
GSC gas-solid chromatography; general staff corps.
G scan *Electronics.* see G DISPLAY.
G scope *Electronics.* a cathode-ray oscilloscope that produces a radar G display.
GSE ground support equipment.
G_δ set *Mathematics.* any set that is the countable intersection of open sets; a type of Borel set.
GSO general staff officer.
GSP *Oncology.* a proto-oncogene that contains point mutations and is found in adenoma of the pituitary gland and carcinoma of the thyroid.
GSR galvanic skin response.
GST Greenwich sidereal time.
G star *Astronomy.* a star whose spectrum is dominated by the H and K lines of single-ionized calcium; its surface temperature is 5000–6000 K.
g suit *Engineering.* a close-fitting, inflatable garment that exerts pressure on the lower part of the body, including the abdomen; worn by pilots and astronauts to prevent the accumulation of blood below the chest during acceleration.
GSV guided space vehicle.
GT gross ton.
gt. drop. (From Latin *gutta.*)
GT-AG rule *Molecular Biology.* the observation that the GT and AG dinucleotides occur in the first and last positions of intron sequences in eukaryotic DNA.
GTP guanosine triphosphate.
gtt. drops.
Guadalupian *Geology.* a North American provincial series of the Lower and Upper Permian period, occurring after the Leonardian and before the Ochoan.
guaiacol *Organic Chemistry.* $C_6H_4(OH)OCH_3$, colorless crystals (when pure) with aromatic odor; soluble in water and miscible with alcohol; melts at 27.9°C and boils at 205°C; toxic by ingestion and skin absorption; used as a chemical reagent and in medicine as an expectorant. Also, METHYLCATECHOL.
guaiazulene *Pharmacology.* $C_{15}H_{18}$, an anti-inflammatory agent occurring as a blue oil; derived from either chamomile oil or guaiac wood oil.
guan *Vertebrate Zoology.* a large game bird resembling a turkey, of the family Cracidae, found in tropical forests from Mexico to Central America.
guana caste see CONACASTE.
guanaco *Vertebrate Zoology.* a small, wild ancestor of the domesticated llama and alpaca, belonging to the family Camelidae; characterized by a thick brown and white coat and a slender body with a long muzzle, ears, neck, and legs and a short tail; inhabits the pampas from the Peruvian Andes to the tip of South America.

guanaco

guanajuatite *Mineralogy.* Bi_2Se_3, a bluish-gray, opaque, orthorhombic mineral, dimorphous with paraguanajuatite; occurring in acicular crystals and massive forms, and having a metallic luster, a specific gravity of 6.25 to 6.98, and a hardness of 2.5 to 3.5 on the Mohs scale; found in quartz.

guanidine *Biochemistry.* CH_5N_3, a compound formed by protein metabolism and present in urine; solutions of its hydrochloride form denature and dissolve proteins.

guanine [gwä´nēn] *Biochemistry.* $C_5H_5N_5O$, the purine 2-amino-6-oxypurine, one of the four nucleic acid bases; also a component of coenzymes and the starting material for the biosynthesis of folic acid and riboflavin.

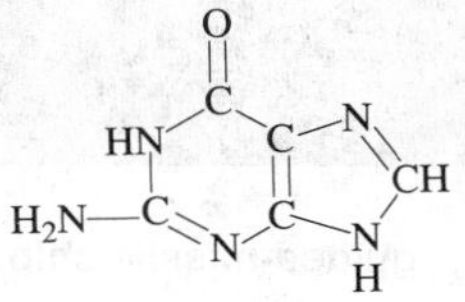

guanine

guano [gwä´nō] *Materials.* **1.** a natural manure composed chiefly of bird excrement. **2.** any similar fertilizer.

guanosine *Biochemistry.* $C_{10}H_{13}N_5O_5$, the ribonucleoside of guanine, including guanosine mono-, di-, and triphosphate; a component of many nucleotides, including nucleic acids.

3',5'-guanosine monophosphate see c-GMP.

guanosine tetraphosphage *Biochemistry.* a phosphoric acid ester of guanosine that plays an important role in the metabolism of many organisms.

guanosine triphosphate *Biochemistry.* an energy-rich molecule that is required for the synthesis of peptide bonds during translation.

guanylate cyclase *Biochemistry.* an enzyme that catalyzes the synthesis of guanosine 3', 5'-cyclic monophosphate from guanosine 5'-triphosphate; functions as a metabolic regulator in bacteria and higher cells. Also, **guanyl cyclase.**

guanylic acid *Biochemistry.* $C_{10}H_{14}N_5O_8P$, the ribonucleotide of guanine synthesized in the purine pathway, used as a flavoring and aroma; extracted from yeast nucleic acid or produced by certain microorganisms. Also, **guanosine monophospate, guanosine phosphoric acid.**

guard *Military Science.* **1.** an individual or group assigned to watch prisoners or protect an area or facility. **2.** a unit that protects the main force by fighting to gain time, while also observing and reporting information. *Engineering.* a shield, such as a screen or plate, that is used with certain machinery to prevent injury to personnel. *Medicine.* a protective device, such as one worn in the mouth during contact sports.

guard ammunition *Ordnance.* ammunition with a reduced charge designed to be used by military guards.

guard arm *Electricity.* a crossarm placed over wires to prevent other wires from falling into them.

guard band *Electronics.* a narrow, unoccupied frequency band at the upper and lower limits of an assigned channel, preventing adjacent-channel interference by preserving channel separation.

guard cell *Botany.* either of two kidney-shaped cells that surround each stoma in the epidermis of a plant and regulate the passage of gases through the stomatal pore by expanding and contracting.

guard circle *Acoustical Engineering.* the inner groove on a disk recording surface in which the stylus is prevented from being carried to the center of the turntable of a record player.

guarded command *Computer Programming.* a command that is a combination of a condition, called the guard, and a (possibly compound) statement whose execution is controlled by the condition.

guard-electrode system *Petroleum Engineering.* during electrical charting, a system of extra electrodes used to restrict the surveying current from the measuring electrode to a mainly horizontal path.

guarding or **guarding behavior** *Behavior.* a behavior pattern in which a male animal maintains continued contact or proximity with a female mate and repels other males.

guard lock *Civil Engineering.* a lock at the mouth of a dock or basin separating tidal waters from the waters in the dock or basin.

guard magnet *Mining Engineering.* an electromagnet used in a crushing system to arrest or remove tramp iron ahead of the crusher.

guard rail or **guardrail** *Civil Engineering.* **1.** a rail used as a divider between lines of traffic in opposite directions or as a safety rail on curves. **2.** a rail placed close to the outside of a railroad's inner rail on curves to keep the inner wheels of a railroad car on the track.

guard relay *Electricity.* a relay used in a linefinder circuit to ensure that only one linefinder can be connected to a line circuit at a given time.

guard ring *Electricity.* a ring-shaped electrode that surrounds one plate of a parallel-plate capacitor to decrease edge effects; used to evenly distribute the electric charge over terminal surfaces. *Thermodynamics.* a ring of material that surrounds a specimen; used in transfer experiments to ensure even, controlled heat transfer distribution.

guard shield *Electronics.* a shield enclosing the input circuit of an amplifier or instrument.

guard signal *Computer Technology.* in digitizers and converters, a signal permitting values to be read or converted only when the values are not in changing state.

guard wire *Electricity.* a grounded wire placed beneath an overhead transmission line to catch and ground the line, should it break.

guarea *Materials.* the pale, pink to reddish wood of the Nigerian pearwood.

Guarnieri body see B-TYPE INCLUSION BODY.

guava [gwä´və] *Botany.* **1.** a plant of the genus *Psidium* in the family Myrtaceae, especially *P. guajava*, a tropical American tree or shrub that produces a sweet, aromatic, juicy yellow fruit. **2.** the fruit of such a tree, used for making jam and jelly.

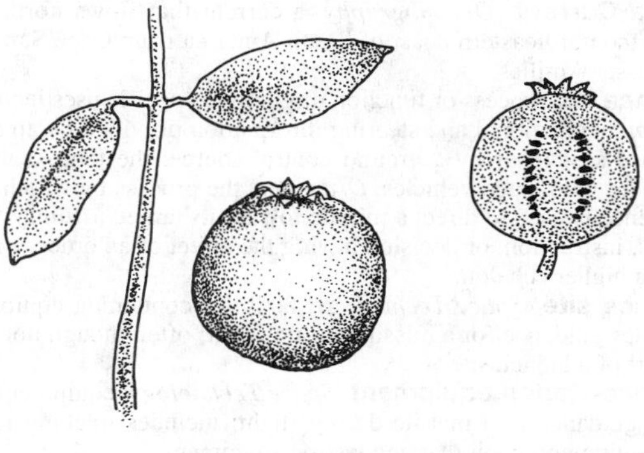

guava

guayule [gwī´yool] *Botany.* a shrub, *Parthenium argentatum*, of the family Compositae that is native to Mexico and the southwestern United States and has been cultivated as a source of rubber.

guba *Meteorology.* a term used in New Guinea to denote a sea squall.

gubernacular fold see INGUINAL FOLD.

gubernaculum *Anatomy.* any structure that guides another one. *Invertebrate Zoology.* **1.** in certain protozoans, a rear flagellum used for steering. **2.** in hydrozoans and certain nematode worms, a support structure for copulatory mechanisms.

Gubler's sign *Toxicology.* a swelling on the back of the wrist due to lead poisoning. Also, **Gubler's tumor.** (Named for Adophe Marie *Gubler*, 1821–1879, French physician.)

Gudden-Pohl effect *Electronics.* the inclination of an ultraviolet-irradiated phosphor to momentarily glow when exposed to an electric field.

Gudermannian *Mathematics.* the function y of the variable x satisfying the functional equation $\tan y = \sinh x$ (or, equivalently, $\cos y = \operatorname{sech} x$, or $\sin y = \tanh x$); denoted $y = \operatorname{gd} x$. The Gudermannian function allows one to perform many trigonometric calculations without resorting to complex arithmetic.

gudgeon *Naval Architecture.* a hingelike fitting on a rudder post, into which a rudder pintle fits. *Engineering.* **1.** any of a set of metal wheels designed to roll on tracks and rotate the roller stocks on a job press. **2.** see PIVOT.

gudgeon pin *Engineering.* in an internal-combustion engine, a projecting pin used to link the piston to the bearing of the smaller end of the connecting rod. Also, PISTON PIN, WRIST PIN.

gudmundite *Mineralogy.* FeSbS, a silver-white to steel-gray, opaque, metallic monoclinic mineral of the arsenopyrite group, occurring in commonly twinned prismatic crystals, and having a specific gravity of 6.72 and a hardness of about 6 on the Mohs scale; found in sulfide deposits.

Guerbet reaction *Organic Chemistry.* a high-temperature condensation reaction of alcohols, using sodium alkoxide or copper in a dehydrogenation, aldol condensation, and hydrogenation sequence.

Guericke, Otto von 1602–1686, German physicist; invented the air pump and used it for experiments with vacuums and air pressure.

Guernsey *Agriculture.* a breed of dairy cattle having a fawn-colored coat with white markings. (From the British island of *Guernsey.*)

guerrilla or **guerilla** [gə rilˊə] *Military Science.* **1.** a participant in guerrilla warfare. **2.** of or relating to guerrilla warfare. Thus, **guerrilla tactics.** (From Spanish, literally "little war.")

guerrilla warfare *Military Science.* military operations conducted in hostile or enemy-occupied territory by irregular forces acting independently or semi-independently of regular forces; the operations utilize such tactics as sabotage, espionage, and assassination; although guerrilla forces are usually composed of indigenous troops, they may also be composed partly or completely of professional troops.

guess-warp *Naval Architecture.* an archaic term for a line running from a vessel to the shore, and used for shifting the vessel's position; in modern usage it is called a running line.

guest computer *Computer Technology.* a computer that functions under control of another computer (the host).

guest element see TRACE ELEMENT.

guest-warp *Naval Architecture.* a line suspended from a boat boom, used for boats to tie up.

Guggenheim process *Civil Engineering.* a technique using chemical precipitation to prepare sludge for filtration; it employs ferric chloride and aeration.

Guiana Current *Oceanography.* a current that flows northwestward along the northeastern coast of South America from Cape San Roque to the Lesser Antilles.

guidance the process or function of guiding; specific uses include: *Navigation.* navigational and steering information provided by an automatic onboard system or by a ground control source; the term is used especially for unmanned vehicles. *Ordnance.* the process by which target intelligence is used to direct a missile toward its target. *Military Science.* a policy, instruction, or decision having the effect of an order when issued from a higher echelon.

guidance site *Space Technology.* facilities containing equipment that provides guidance for a missile during flight; often though not necessarily part of a launch site.

guidance-station equipment *Space Technology.* equipment that provides guidance to a missile during flight; includes tracking radar, data link equipment, computerized testing equipment.

guidance system *Navigation.* any automatic or remote system that provides navigation and steering information. *Space Technology.* specifically, in a rocket vehicle or missile, a system that processes targeting data in order to maintain the desired flight path or trajectory; operates either in conjunction with ground control or automatically within the vehicle.

guidance tape *Computer Technology.* a magnetic or paper tape containing program information to guide and control the flight of a missile or pilotless aircraft.

guide *Mechanical Devices.* any device, such as a track or bracket, by which another object is led in its proper course. *Mechanical Engineering.* see GUIDE PIN. *Mining Engineering.* **1.** any wood, steel, or steel wire rope conductor in a mine shaft to direct the movement of the cages. **2.** a pulley to lead a driving belt in a new direction or to keep it from leaving its desired direction.

guide bracket *Mining Engineering.* a steel or iron bracket fixed to a bunton to secure rigid guides in a shaft.

guide coupling *Mining Engineering.* a short coupling with a projecting reamer guide or pup to which a reaming bit is attached, used to couple a reaming bit to a reaming barrel.

guided bend test *Metallurgy.* a bend test in which the specimen is bent to a specified shape with the aid of a guide.

guided bomb *Ordnance.* an aerial bomb whose aim can be directed during its drop.

guided missile *Space Technology.* **1.** usually, a missile or missile vehicle given continuance guidance all the way to its target, as distinguished from an unguided missile (which is guided only at launch) or a ballistic missile (guided only in its upward trajectory). **2.** broadly, any missile directed during any part of its flight path or trajectory. Thus, **guided-missile control, guided-missile equipment carrier.**

guided-missile exercise head *Ordnance.* a simulated guided-missile warhead.

guided-missile ship

guided-missile ship *Naval Architecture.* a warship armed with guided missiles, usually for use against other ships, aircraft, or submarines. Thus, **guided-missile destroyer, guided-missile submarine, guided-missile cruiser, guided-missile frigate,** and so on.

guided propagation see TRAPPING.

guided rocket *Ordnance.* a guided missile powered by rocket propulsion.

guided space vehicle *Space Technology.* a space vehicle under some degree of guidance either by a pilot or through radio command, preset mechanisms, inertial guidance, or celestial guidance.

guided wave *Electromagnetism.* any wave whose propagation is confined within and parallel to boundaries of materials having different electromagnetic properties from the medium through which the wave propagates.

guide edge *Computer Technology.* the edge of a punched card, paper tape, printed sheet, or other medium, used for alignment purposes.

guide fossil *Petrology.* any fossil that may help indicate the age of a subject stratum, or help to identify correlations between rocks.

Guideline *Ordnance.* a Soviet two-stage surface-to-air missile that is powered by a solid-propellant booster and a liquid-propellant sustainer and is equipped with radio-command guidance; it delivers a 300-pound conventional warhead at a speed of Mach 3.5 and a range of 25 miles; officially designated **SA-2;** a later ship-based version is designated **SA-N-2.**

guidelines *Graphic Arts.* a set of printed or drawn lines used to orient placement of copy, as on a page layout or dummy.

guide meridian *Cartography.* in surveying, a line on the earth's surface corresponding to an astronomical meridian that intersects a point on an established base line or standard parallel, usually at intervals of 24 miles east and west of the principal meridian, and along which township comers are established.

guide mill *Metallurgy.* a hand mill that has guides at its entrance end.

guide number *Graphic Arts.* a factor used to determine the correct exposure when using a particular flashbulb; usually provided by the manufacturer.

guide pin *Mechanical Engineering.* an attachment that controls the movement of a mechanism and aligns it with its work; used in tool and die work and in printing, often in sets of two or more. Also, GUIDE.

guide post *Civil Engineering.* a post bearing signs or guide boards used to show directions along a road.

guide pulley see IDLER PULLEY.

guide rail *Engineering.* a track or rail used to steer a sliding door, window, or similar element.

guide rope see DRAG ROPE.

guide telescope *Optics.* a telescope that is attached to a larger photographic telescope to aid in the tracking of objects and in the placement of an image on a photographic plate.

guide wavelength *Electromagnetism.* the wavelength of an electromagnetic wave in a waveguide, which is always longer in an air-filled guide than in free space or an evacuated waveguide.

guideway *Transportation Engineering.* a track or channel and its supporting structures; it provides a surface on which vehicles travel and a restraint on their lateral movement.

guiding center *Electromagnetism.* a slowly moving point about which a charged particle orbits, so that the resulting motion has both a circulating component and a translational component.

guild *Ecology.* a group of closely related organisms that use similar resources and are thus potential competitors regardless of differences in taxonomy and tactics of resource acquisition.

Guild *Ordnance.* an early Soviet surface-to-air missile powered by a dual-thrust solid-propulsion motor and equipped with radio-command guidance; officially designated **SA-1.**

guildite *Mineralogy.* $CuFe^{+3}(SO_4)_2(OH)\cdot 4H_2O$, a yellow to chestnut-brown, monoclinic or triclinic mineral occurring as short, prismatic crystals with two perfect cleavages, and having a specific gravity of 2.695 and a hardness of 2.5 on the Mohs scale; found with coquimbite and other sulfates.

Guillaume, Charles Edouard [gē´yōm] 1861–1938, Swiss physicist; discovered Invar; received the Nobel Prize for improving standard measurements through work in nickel steel alloys.

Guillemin, Roger [gē´yə man] born 1924, French-born American physiologist; shared Nobel Prize for research in pituitary hormone secretion.

Guillemin effect *Electromagnetism.* an effect in which a bent rod subjected to magnetostriction tends to straighten when its length is parallel to the magnetic field.

Guillemin line *Electronics.* a network or artificial line that generates a nearly square pulse with steep rise and fall; used in radar to control pulse width during high-level pulse modulation.

guillemot [gē´yə mät] *Vertebrate Zoology.* any of several long-billed auks of the genera *Uria* and *Cepphus* in the family Alcidae, found near the Arctic Circle to Britain and the North Pacific to Southern California; a deep diver that feeds on the ocean bottom.

Guilliermondella *Mycology.* a genus of yeast fungi of the family Saccharomycetaceae that reproduce by budding.

guilt *Psychology.* a negative feeling arising from the belief that one has done something wrong or harmful; characterized by low self-esteem and the attitude that one should atone or be punished for the offense.

Guinea, Gulf of *Geography.* a wide inlet of the Atlantic formed by the bend in the western coast of Africa.

Guinea Current *Oceanography.* a current that flows eastward along the entire southern coast of West Africa, shifting north and south seasonally but always remaining above the equator.

guinea fowl

guinea fowl *Vertebrate Zoology.* any one of the subfamily Numidinae of ground-dwelling fowl-like game birds of medium size, having a richly colored naked head, slate-colored plumage with white speckles, and a short down-pointing tail; its native habitat is in sub-Saharan Africa and Madagascar, but these birds are raised for food worldwide.

guinea pig *Vertebrate Zoology.* a small, burrowing rodent of the Caviidae family, having a stocky build, short legs and ears, and a coat with black, brown, white, yellow, or gray hairs (or a combination thereof); found mainly in grasslands, forests, swamps, and rocky areas of South America, and also a popular pet and research animal.

guinea worm *Invertebrate Zoology.* the parasitic nematode roundworm *Dracunculus medinensis* that lives as an adult in the subcutaneous tissues of humans and other mammals.

Guinier-Preston zones *Metallurgy.* in the initial stage of precipitation from supersaturated solid solutions, clusters of precipitate that do not have a definitive crystalline structure. Also, G-P ZONES.

guitarfish *Vertebrate Zoology.* **1.** a bottom-dwelling fish of coastal, tropical, and subtropical waters, *Rhinobatus rhinobatus,* having a guitar-shaped body with a raylike forebody and an elongated hindbody; sandy brown on top and white underneath. Also, SPOTTED GUITARFISH, ATLANTIC GUITARFISH. **2.** any of various raylike fishes of the suborder Rhinobatoidei, including the shovelnose, sandshark, and Chinese guitarfish.

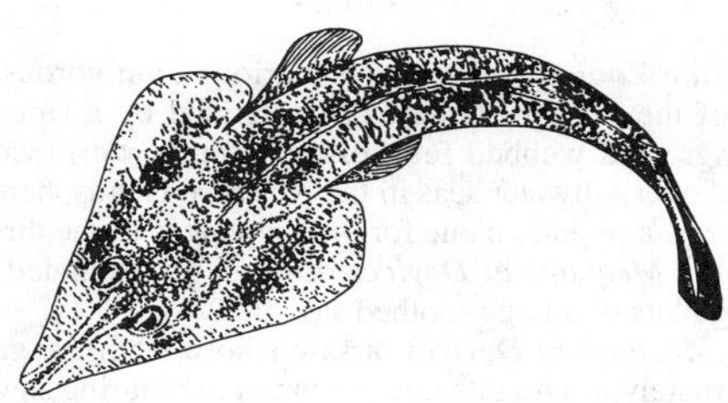

guitarfish

Gukhman number *Thermodynamics.* a dimensionless quantity given by $(T - T_m)/T$, where T is the absolute temperature of a hot gas stream flowing over a moist surface of absolute temperature T_m; used in studying the convection of heat in evaporation.

gula *Anatomy.* **1.** the upper portion of the throat or the gullet. **2.** the front of the neck. *Vertebrate Zoology.* in pelicans, a distensible pouch for holding food and positioning it for swallowing and for holding regurgitated food for feeding young; at times, used in courtship and sexual display by being inflated. Also, **gular pouch, gular sac.** *Architecture.* any molding having a deep hollow, such as a cavetto.

gular *Anatomy.* of or relating to the throat.

gulch *Geology.* an informal term for a deep, narrow ravine.

Guldberg and Waage law see MASS ACTION LAW.

Guldberg-Waage group *Chemical Engineering.* a dimensionless number used in the analysis of chemical reactions in blast furnaces, expressed in an equation that relates volumes of reacting gases and reacting products.

gulder *Oceanography.* a double low water that occurs during a double tide on the south coast of England.

gulf *Geography.* an arm of the sea, generally of deeper penetration than a bay.

Gulfian *Geology.* a North American provincial series of the Upper Cretaceous period, occurring after the Comanchean and before the Paleocene epoch.

Gulfining *Chemical Engineering.* a catalytic hydrogen-treating method used for the removal of sulfur from fuel oils and distillates to improve color, carbon residue, and stability.

Gulf Stream *Oceanography.* the well-defined western boundary current of the North Atlantic, which carries warm, saline tropical water north and northeastward along the eastern coast of the United States, joining the Labrador Current at the Grand Banks, about 40°N and 50°W, to become the North Atlantic Current; generally swift and deep, it transports a very large volume of water.

Gulf Stream Countercurrent *Oceanography.* **1.** any of several currents running counter to the Gulf Stream, on either side of it at different locations. **2.** a predicted large undercurrent beneath the Gulf Stream, observations of which are still inconclusive.

Gulf Stream eddy *Oceanography.* a detached body of water that has broken off from a meander of the Gulf Stream, noticeable because of its differences from the water surrounding it.

Gulf Stream front *Oceanography.* the sharp horizontal gradient of temperature characteristic of the Gulf Stream, frequently displaying a change of 10°C over distances of only a few hundred meters.

Gulf Stream meander *Oceanography.* one of the many variable windings that the Gulf Stream makes, especially as it becomes the North Atlantic Current, persisting a few days or weeks.

Gulf Stream system *Oceanography.* a collective term for the Florida Current, Gulf Stream, and North Atlantic Current.

gulfweed *Botany.* any seaweed of the genus *Sargassum* in the family Sargassaceae, especially *S. bacciferum,* a genus of branching brown seaweed found mainly in the tropics.

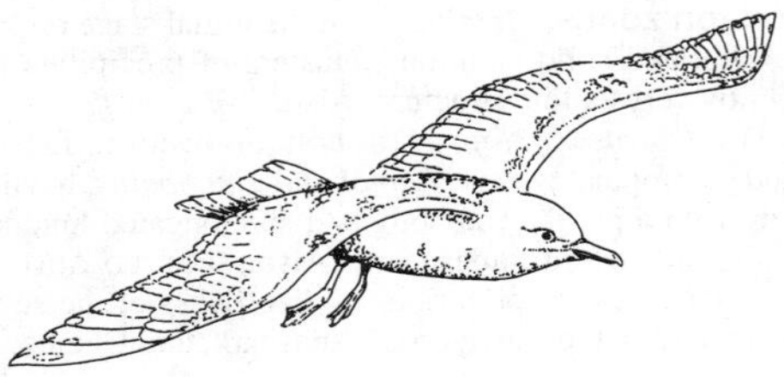

herring gull

gull *Vertebrate Zoology.* any one of various omnivorous scavenging shorebirds of the family Laridae, characterized by a stout build, long pointed wings, and webbed feet; distribution is worldwide, although concentrated over saltwater seas in the Northern Hemisphere.

gullet *Anatomy.* a popular name for the esophagus or the throat.

gulleting file *Mechanical Devices.* a blunt and rounded file used to deepen the gullets of a large-toothed saw.

gullet saw *Mechanical Devices.* a saw whose teeth are gapped or depressed alternately to mitigate dust accumulation during sawing.

Gullstrand, Allvar 1862–1930, Swedish ophthalmologist; improved the ophthalmoscope; Nobel Prize for study of visual accommodation.

gully *Geography.* a small valley cut by a stream, especially on a hillslope.

gully erosion *Geology.* the erosion of soil by running water that forms clearly defined, narrow channels which generally carry water only during or after heavy precipitation. Also, RAVINEMENT.

gully erosion

gully squall *Meteorology.* a nautical term for a violent wind squall arising from mountain ravines on the Pacific coast of Central America.

gulp *Computer Programming.* a small group of bytes treated as a unit.

gum *Materials.* **1.** a thick, viscous excretion from certain trees and plants that is hard and brittle when dry but becomes gelatinous and sticky when mixed with water. **2.** a product that is made from, contains, or resembles such a substance.

gum arabic *Materials.* the thicky, sticky exudate of certain acacia trees, especially *Acacia senegal*; used as an adhesive and emulsifier and in the manufacture of inks and pharmaceuticals. Also, **gum acacia, gum senegal.**

gum benzoin see BENZOIN.

gumbo *Geology.* a variety of fine-grained clay soil that forms a sticky, impervious, plastic mud when wet. *Botany.* see OKRA.

gumboil see PARULIS.

gumbotil *Geology.* a deoxidized, gray to dark-colored, siliceous, leached clay that develops under conditions of poor drainage from weathered till, and that represents the B horizon of fully mature soils.

gumma *Pathology.* an aggregate of rubbery dead tissue found in various organs and body sites in the third stage of syphilis. Also, **gumme.**

gummite *Mineralogy.* any of a number of yellow to orange, gumlike secondary uranium oxides.

gummosis *Plant Pathology.* in a plant, the production and exuding of gum, sap, or latex as a result of cell degeneration caused by a parasite within the plant, unfavorable growing conditions, or other environmental factors.

Gum Nebula *Astronomy.* a large supernova remnant in the south celestial hemisphere that is about 100,000 years old; it lies in Vela and is named for its discoverer, Colin *Gum.*

gum turpentine see TURPENTINE, def. 1.

gum vein *Botany.* an accumulation of resin appearing as a wide streak in some hardwoods.

gun *Ordnance.* **1.** in general usage, any weapon with a tube, barrel, or rails designed to fire a projectile through explosive or other means. **2.** in military usage, a cannon with a relatively long barrel, a relatively low angle of fire, and a high muzzle velocity; as opposed to a howitzer or mortar. *Mining Engineering.* see BOOTLEG.

gunbarrel *Chemical Engineering.* an atmospheric vessel used in the treatment of waste water from waterfloods.

gunboat *Naval Architecture.* a small armed naval vessel used for general patrol work; typically a small ship rather than a true boat.

gun breech *Ordnance.* the metal part of a cannon from the front slope to the rear face, excluding the breech mechanism.

gun burner *Engineering.* a nozzle or similar device that sprays liquid fuel into a combustion furnace. Also, **gun-type burner.**

gun camera *Optics.* a camera that is triggered by the firing mechanism of the weapon or a release button; used to assess accuracy of aim or record the results of combat.

gun carriage *Ordnance.* a mobile or fixed support for a gun; it may include the elevating and traversing mechanisms. Also, CARRIAGE.

gun charger *Ordnance.* a mechanism that retracts the breech mechanism or bolt and inserts a cartridge into the chamber of a gun. Also, CHARGER.

guncotton *Organic Chemistry.* a highly explosive nitrated cellulose made by digesting clean cotton in nitric acid and sulfuric acid; used to make smokeless gunpowder. Also, PYROCELLULOSE, CORDITE.

gundeck *Naval Architecture.* a fully armed deck on a sailing warship. Men of war had up to three complete gundecks, with additional guns on upper decks at bow and stern.

gun displacement *Ordnance.* the distance between a gun and the point from which fire is directed, either the base piece of the battery or other directing point.

gun emplacement *Ordnance.* the position from which a gun is fired, including protective structures and materials, necessary supplies, and ammunition.

gunfire *Ordnance.* the use of artillery or firearms, as opposed to bombs, grenades, torpedoes, mines, and nonexplosive weapons.

gun fore-end *Ordnance.* see FOREARM.

gun group *Ordnance.* the major components of a gun, as distinct from the gun mount.

gun handguard see HANDGUARD.

gun hoist *Ordnance.* a device to lift ammunition to the breech of a gun.

gun jack *Ordnance.* a jack used to force the tube of a gun out of firing position.

gun launcher *Ordnance.* a gun that has been adapted to launch guided missiles or rockets.

gun line *Ordnance.* the extended bore axis of an aircraft gun; for a group of aircraft guns, the mean extended axis.

gunlock *Ordnance.* **1.** originally, the ignition system that produced a spark or flame to ignite the priming powder in a firearm; gunlock types included the flintlock and matchlock. **2.** in modern usage, the assembly in a gun that includes the firing pin, hammer, spring, sear, and trigger.

gunmetal *Metallurgy.* **1.** any of several alloys having a dark gray finish. **2.** a bronze formerly used to make guns.

gun mount *Ordnance.* a support for a gun, such as a bipod, tripod, carriage, or turret.

Gunn amplifier *Electronics.* a microwave amplifier in which a Gunn diode, when placed across the terminals of a microwave source, serves as a negative-resistance amplifier.

Gunn diode *Electronics.* a two-terminal semiconductor that utilizes the Gunn effect to produce microwave oscillation or amplify an applied microwave signal.

Gunn effect *Electronics.* the formation of microwave oscillation in a DC-biased slab of *n*-type gallium arsenide in a 3-kV electric field.

gunnel *Navigation.* see GUNWALE.*Vertebrate Zoology.* a long slender, eel-like fish with short, body-length dorsal fin, belonging to the family Pholidae and found in the northern Atlantic; the **rock gunnel,** *Pholis gunnellus*, has distinctive dark spots along the upper part of its body.

Gunneraceae *Botany.* a family of dicotyledonous perennial herbs of the order Haloragales, consisting of the single genus *Gunnera* and occurring mainly in the Southern Hemisphere.

gunnery *Military Science.* the art and practice of constructing, handling, or firing guns; it is applied more specifically to the use of machine guns, aircraft guns, or artillery.

Gunn oscillator *Electronics.* an oscillator that utilizes a Gunn diode to generate a microwave frequency.

gun parallax *Ordnance.* in artillery control, the angular difference between the line from the directing point to the target and the line from the gun to the target.

gun pendulum *Engineering.* a device that determines the initial velocity of a bullet fired from a gun by suspending the gun as a pendulum and measuring its displacement upon firing.

gun reaction see RECOIL.

gunship *Ordnance.* an aircraft (usually a helicopter) used as a platform for weapons to be used against ground targets.

gunsight *Ordnance.* a device used for aiming a gun, firearm, or similar weapon; it is aligned with the shooter's eye and the target or aiming point.

gunsight computer *Ordnance.* a computer that controls a computing gunsight, calculating the angle between the line of sight to the target and the line of departure for the projectile, called the prediction angle.

gunstock *Ordnance.* the wooden or plastic piece to which the barrel and mechanism of a firearm are attached; in rifles, carbines, and shotguns, it allows the shooter to rest the weapon against his shoulder.

gun-target line *Ordnance.* an imaginary straight line from a gun to the target.

Gunter, Edmund 1581–1626, English mathematician; inventor of Gunter's chain; wrote treatise on logarithms.

Gunter's chain *Engineering.* a 66-foot, 100-link chain used as a measuring standard in surveying; 1 square chain is equal to 1/10 of an acre.

gun turret *Ordnance.* **1.** any dome-shaped or cylindrical protective structure into which one or more guns have been mounted; it is usually capable of revolving and may be located in a fort, ship, tank, airplane, or other vehicle. Also, TURRET. **2.** specifically, a dome-shaped gun mount on an airplane, especially a bomber.

gun-type weapon *Ordnance.* a nuclear device in which two or more pieces of fissionable material, each less than a critical mass, are brought together very rapidly, as by firing one piece into the other(s) by a gun in order to form a supercritical mass that will explode as the result of a rapidly expanding fission chain.

gunwale [gun′əl] *Naval Architecture.* **1.** originally, the uppermost wale, or external longitudinal stiffener on a wooden ship. **2.** the upper edge of a vessel's side. Also, GUNNEL.

Günz *Geology.* **1.** a European geologic stage of the Pleistocene epoch, after the Astian of the Pliocene and before the Mindel. **2.** the stage often considered to be the initial glacial stage of the Pleistocene in the Alps.

Günzberg reagent *Analytical Chemistry.* a solution of vanillin, phloroglucinol, and alcohol, used to test for hydrochloric acid in gastric juice; the formation of red crystals indicates a positive test.

Günz-Mindel *Geology.* the first interglacial stage of the Pleistocene epoch in the Alps, occurring after the Günz and before the Mindel.

guppy *Vertebrate Zoology.* a tiny freshwater tropical fish, *Poecilia reticulata,* that is native to northern South America and Trinidad and is widely raised in aquariums.

guppy

Gurevich effect *Solid-State Physics.* an effect occurring in conductors in which a temperature gradient is present and in which phonon-electron collisions take place; the phonons, carrying a thermal current, tend to carry the electrons with them from the hotter region to the colder region.

gurney *Medicine.* a wheeled cot or stretcher used to transport patients.

Gurney formulas *Ordnance.* a series of formulas expressing the initial velocity of a fragment as a function of the type of explosive and the ratio of explosive charge to metal weight.

gusset *Civil Engineering.* a truss joint connection usually made of steel plates. *Mining Engineering.* a V-shaped cut in the face of a heading.

gusset plate *Civil Engineering.* a steel plate that connects truss joints.

gust *Meteorology.* a sudden increase in windspeed that usually lasts no longer than 20 seconds, after which there is a lull or a slackening in the windspeed; in the United States, gusts are reported when the peak windspeed reaches at least 16 knots and the variation between peaks and lulls is at least 9 knots.

gustation *Physiology.* the sensation of taste produced by chemical stimulation of the taste buds.

gustatoreceptor *Anatomy.* a sensory cell in a taste bud.

gustatory audition *Neurology.* a subjective sensation of taste induced by actual sound stimuli.

gust-gradient distance *Aviation.* the horizontal distance from the edge of a gust along a flight path to the point at which the gust attains maximum speed.

gustiness *Meteorology.* an airflow condition characterized by gusts.

gustiness components *Meteorology.* **1.** the ratios of the average magnitudes of the component fluctuations of the wind along three perpendicular axes to the average windspeed. **2.** the ratios of the root-mean-squares of eddy velocities to the average windspeed. Also, INTENSITY OF TURBULENCE.

gustiness factor *Meteorology.* a measure of the intensity of wind gusts, expressed as a ratio of the total range of windspeeds between gusts and intermediate periods of lighter wind to the overall mean windspeed, averaged over both gusts and lulls.

gust load *Mechanics.* the transient aerodynamic load due to sudden variations of wind velocity.

gustsonde *Engineering.* an instrument that is held aloft by a parachute in order to measure turbulence; consists of an accelerometer and radio telemetering equipment.

gust tunnel *Aviation.* a wind tunnel in which models are passed over vertical jets that simulate gusts of wind.

gut *Anatomy.* a popular term for the intestines. *Developmental Biology.* the embryonic digestive tube. *Hydrology.* **1.** a narrow water passage, channel, or tidal stream connecting two bodies of water. **2.** a channel created by water in motion that is deeper than the surrounding, shallower water.

GUT grand unified (field) theory.

Gutenberg, Johann [goot′in burg] c. 1397–1468, German printer; invented the type mold and movable-type printing.

Gutenberg discontinuity *Geology.* the discontinuity at 2900 kilometers below the earth's surface that marks the boundary between the mantle and the core.

Guthrie test *Pathology.* a method for the detection of phenylketonuria in newborns using a bacterial inhibitor.

Guti weather *Meteorology.* in Rhodesia, a weather condition that occurs mainly in the summer, characterized by dense stratocumulus clouds that are frequently accompanied by drizzle; it is associated with easterly winds that bring cool and stable marine air when a high-pressure system moves in an easterly direction to the south of Africa.

gutta-percha *Materials.* a rubberlike gum that is derived from the sap of certain Malaysian trees of the sapodilla family; used in the manufacture of dental cement, insulation, and golf balls.

guttation *Botany.* the excretion of excess water as droplets from glands (hydathodes) on the leaves of many plants, commonly occurring during times of high humidity.

gutter *Building Engineering.* a channel that runs along the eaves of a building, designed to drain rainwater from the roof. *Civil Engineering.* a small channel used on the sides of roads, canals, highways, and other such structures for surface drainage. *Mining Engineering.* **1.** a drainage channel. **2.** a gob heading. **3.** the lowest and often richest portion of an alluvial deposit. *Metallurgy.* in forging, a die design feature that minimizes the formation of folds caused by flash. *Graphic Arts.* the combined inner (binding) margins of two facing pages in a book or other publication. The inner margin of one page is called the **gutter margin.** *Invertebrate Zoology.* a depression or furrow between body parts.

guttering *Engineering.* the cutting of drainage channels or gutters, as in a mine shaft or quarry.
Guttiferae *Botany.* a family of dicotyledonous trees, shrubs, lianas, or herbs in the order Theales, characterized by a resinous juice and conspicuous secretory canals or cavities in all or most of the tissues. Also, CLUSIACEAE.
guttra *Meteorology.* a term used in Iran to denote sudden squalls that occur in May.
guttulate *Mycology.* referring to fungal spores that contain small drops of oil that resemble nuclei.
guttule *Mycology.* a small drop or particle in a spore that resembles a nucleus. Also, **guttula, guttulae**.
Guttulinaceae *Bacteriology.* in some classification systems, a family of microorganisms in the Acrasiales group characterized by simple fruiting structures and component cells containing little cellulose.
guttural *Anatomy.* of or relating to the throat. *Linguistics.* **1.** of or relating to a sound made in the back of the throat. **2.** loosely, any harsh, throaty sound.
Gutzeit test *Analytical Chemistry.* a test for arsenic in which zinc and sulfuric acid are added to the sample; when filter paper treated with mercuric chloride is placed over the sample a yellow spot forms if arsenic is present.
guxen *Meteorology.* a cold wind that occurs in the Swiss Alps.
guy *Engineering.* a wire, rope, or chain that is used to secure a vertical and often temporary structure such as a mast, tower, derrick, or chimney. Also, **guy rope, guy wire.**
guy derrick *Mechanical Engineering.* a crane or derrick operating on a mast that is held upright and secured by guy ropes.
guyot *Geology.* a relatively smooth, flat-topped seamount formed in the deep oceans.
guzzle *Meteorology.* a violent blast of dry, parching wind that occurs in the Shetland Islands.
GV granulosis virus.
GVA *Aviation.* the airport code for Geneva, Switzerland.
G value *Nucleonics.* the number of molecules created or destroyed with each 100 electron volts of ionizing radiation absorbed by a substance. Also, G FACTOR.
GVH graft-versus-host.
GW gigawatt.
g-waves *Astrophysics.* see GRAVITATIONAL RADIATION.
GWEN *Telecommunications.* a communications system that is able to operate in and after a nuclear attack. (An acronym for ground-wave emergency network.)
gy gray.
Gyalectales *Mycology.* in some classifications, an order of fungi of the subdivision Ascomycotina that primarily form lichen and live symbiotically with photosynthesizing plants in humid or tropical environments.
gymna- or **gymn-** a combining form denoting nakedness, as in *gymnameba.*
gymnameba *Invertebrate Zoology.* any shell-less ameba.
Gymnamoebia *Invertebrate Zoology.* a class of free-living ameboid protozoa lacking a test.
Gymnarchidae *Vertebrate Zoology.* a monotypic family of freshwater fish of the order Mormyriformes, having an eel-like body with electrical properties, and found in the Nile, Chari, and Niger Rivers of Africa.
Gymnarthridae *Paleontology.* a family of amphibians in the subclass Lepospondyli and order Microsauria; Carboniferous and Permian.
Gymnascales *Mycology.* an order of fungi belonging to the subdivision Ascomycotina that grow primarily on nonliving organic matter and are sometimes parasitic, causing animal diseases.
gymnasium *Architecture.* a building or room designed for indoor exercise.
gymno- a combining form denoting nakedness, as in *gymnocyte, gymnosperm.*
Gymnoascaceae *Mycology.* a family of fungi belonging to the order Gymnascales which thrive on skin, hair, nails, and feathers, and also in soil and dung.
Gymnoblastea see ANTHOMEDUSAE.
gymnoblastic *Invertebrate Zoology.* in the hydrozoan suborder Anthomedusae, having unprotected medusa buds.
gymnocarpous [jim′nō kär′pəs] *Botany.* referring to fungi or lichens that have no covering on the outer, spore-bearing layer or on their fruiting bodies.
gymnocephalous cercaria *Invertebrate Zoology.* a type of liver fluke larva.
Gymnocerata *Invertebrate Zoology.* a suborder of long-horned bugs, such as water striders and stilt bugs, in the order Hemiptera, with exposed antennae longer than the head.
Gymnocodiaceae *Paleontology.* a family of small algal fossils that are preserved as irregular tubes with perforate walls; sometimes included in the Rhodophyta and sometimes in the Chlorophyta; Permian to Cretaceous.
gymnocyte *Cell Biology.* a cell that has no cell wall.
Gymnodinia *Invertebrate Zoology.* a suborder of plantlike protozoa with single flagella, in the order Dinoflagellida.
Gymnodiniaceae *Botany.* a family of free-living dinoflagellates of the order Gymnodiniales, characterized by transverse and longitudinal flagellar grooves and found in all aquatic environments.
Gymnodiniales *Botany.* an order of freshwater and marine nonthecate dinoflagellates belonging to the class Dinophyceae and distinguished by cells having well-developed transverse and sulcal grooves.
gymnogynous *Botany.* having an uncovered ovary, as some seeds that lack a pericarp.
Gymnomitriaceae *Botany.* an arctic and arctic-alpine family of usually erect liverworts of the order Jungermanniales, characterized by reddish to purplish pigments, mostly intercalary branches usually confined to shoot bases, and rhizoids scattered on the stem; further distinguished by the lack of any kind of asexual propagation.
Gymnonoti *Vertebrate Zoology.* an equivalent name for Cypriniformes, a large order of soft-rayed freshwater fish.
Gymnophiona *Vertebrate Zoology.* an order of subterranean limbless amphibians that resemble earthworms and are found in tropics and subtropics.
Gymnophlaeaceae *Botany.* a red algae family belonging to the order Gigartinales, characterized by terete or flattened multiaxial thalli that are leaflike and undivided or irregularly dichotomously divided.
Gymnopleura *Invertebrate Zoology.* a subsection of brachyuran decapod crustaceans, primitive crabs with an elongated carapace and flattened legs.
Gymnosomata *Invertebrate Zoology.* an order of small, shell-less sea slugs in the suborder Pteropoda, with elongated bodies and feet modified to form fins.
gymnosperm [jim′nō spurm′] *Botany.* a common name for any woody vascular seed plant of the division Pinophyta that has seeds that are not enclosed in an ovary or fruit, but are usually borne in cones (conifers are the most familiar group); one of the two groups of seed-bearing plants, the other being the angiosperms.
Gymnospermae see PINOPHYTA.
Gymnosporangium *Mycology.* a genus of parasitic rust fungi belonging to the class Uredenales which infect such plants as junipers and cedars.
gymnospore *Botany.* a spore that has no protective envelope.
Gymnostomatida *Invertebrate Zoology.* an order of primitive ciliate protozoans in the subclass Holotrichia, with no cilia in the oral area.
Gymnotidae *Vertebrate Zoology.* the family of predatory electrical eel-shaped fishes known as Gymnotids and belonging to the order Cypriniformes; characterized by a lack of dorsal fin rays and no adipose or pelvic fins; found in freshwater habitats of South and Central America.
Gymnotoidei *Vertebrate Zoology.* a monofamilial suborder including Gymnotid eels of the order Cypriniformes.
Gymnuridae *Vertebrate Zoology.* the butterfly rays, a family of marine fishes of the order Myliobatiformes, considered the most graceful and beautiful of the rays; having wings as broad as the body length, a short tail, and changing camouflage coloring.
GYN or **gyn** gynecology.
gyn- or **gyne-** a combining form meaning "woman" or "female," as in *gynander, gynephobia.*
gynaecandrous *Botany.* having both a flower with stamens and one with pistils on a single spike.
gynandromorph *Biology.* an organism having male and female characteristics on different parts of the body. Also, **gynander.**
gynandrous *Botany.* of certain flowers such as orchids, having stamens and pistils united in a column.
gynandry see HERMAPHRODITISM.
gynarchy *Anthropology.* a society or group that is governed or dominated by women.
gyneco- or **gynec-** a combining form meaning "woman" or "female," as in *gynecology.*
gynecography *Radiology.* the radiography of the organs or structures of the female reproductive tract.

Gynecology

Gynecology, a term that is derived from the Greek word *gynaikos* (woman), is the branch of medicine that treats diseases of the female genital tract. Historically, the modern era of gynecology began in the early nineteenth century when Dr. Ephraim McDowell successfully removed a large ovarian cyst from Jane Crawford in rural Kentucky. At about the same time, the first medical school Lecturer in Gynecology was appointed in the United States.

By the end of the nineteenth century, many medical school faculties in this country included professors of gynecology. By the early twentieth century, gynecology had disengaged from the disciplines of pediatrics and general surgery and had merged with obstetrics. This evolution was cemented by the establishment of the American Board of Obstetrics and Gynecology in 1930.

In its infancy, gynecology was principally concerned with extirpative surgery and repair of obstetrical injuries. Over the past few decades, its scope has broadened to include other areas of reproduction, such as contraception, female endocrine disorders, disturbances of menstruation, infertility, population control, sex counseling, pelvic infections, and pelvic malignancies.

In the 1970s, two subspecialties of gynecology were created by the American Board—reproductive endocrinology and gynecologic oncology. Fellowship training programs emerged and research in these subdisciplines seemed to accelerate. Two decades later, these subspecialties are flourishing. Yet, the majority of graduates of residency training programs still choose to practice general gynecology and obstetrics. Indeed, the gynecologist has become the primary physician of woman. In addition to the time-honored annual pelvic examination and Papanicolaou smear, gynecologists may provide such diverse services as diagnosis of nongynecologic disorders (breast cancer, hypertension, colon cancer, etc.), preventive medicine and lifestyle counseling (exercise, lipid profile, smoking cessation, diet, etc.), marital and sex counseling, and information about menopause and the aging process. As the American female population ages, the latter will consume a greater proportion of the general gynecologist's time, particularly with regard to information concerning the risks and benefits of estrogen replacement therapy.

Gynecology has emerged as one of the most exciting specialties, attracting the brightest of medical students. Women currently occupy half of all residency positions. Recent advances in this field include in vitro fertilization technology, improved medical therapies for endometriosis and ectopic pregnancy, outpatient treatment of preinvasive diseases of the lower genital tract, improved laparoscopic and ultrasound technology, and a much better understanding of the menstrual cycle and the etiology and treatment of infertility.

David M. Gershenson, M.D.
Professor and Deputy Chairman of Gynecology
M. D. Anderson Cancer Center
The University of Texas

gynecology [gīn´ə käl´i jē] *Medicine.* the branch of medical science that studies the diseases of women, especially those that affect the sexual organs. Thus, **gynecologist, gynecological.**

gynecomastia *Medicine.* the excessive development of mammary glands in the male.

gynephobia [gīn´ə fō´bē ə] *Psychology.* an irrational fear or hatred of women. Thus, **gynephobic**.

gyno- a combining form meaning "woman" or "female," as in *gynogamon.*

gynobase *Botany.* an elongation of the receptacle of certain flowering plants that bears the female organs of the flower.

gynodioecious *Botany.* describing a species that has some plants with only hermaphrodite flowers and others with only female flowers.

gynoecium *Botany.* the pistil or female organ of a flower, consisting of one or more carpels.

gynogenesis *Developmental Biology.* the development of an egg initiated by a spermatozoon, but without the participation of the sperm nucleus.

gynomerogony *Developmental Biology.* the development of a part of a fertilized egg containing only the female pronucleus.

gynomonoecious *Botany.* of or relating to a species in which all plants have both female and hermaphrodite flowers.

gynophore *Botany.* **1.** the stem or stalk bearing a pistil. **2.** an elongation of the receptacle of a flower between the stamens and the pistil.

gynostemium *Botany.* the column made up of the pistils and stamens and the receptacle between them.

-gynous *Botany.* a combining form meaning "having pistils (or similar organs)."

gypsite *Mineralogy.* a variety of gypsum containing dirt and sand that is found in arid regions as an efflorescent deposit over an outcrop of gypsum. Also, GYPSUM EARTH.

Gypsophila *Botany.* a genus of flowers of the pink family, having small, panicled, pink or white flowers; native to the Mediterranean and widely cultivated as an ornamental.

gypsophilous *Botany.* of a plant living or thriving in a gypsum-rich soil. Thus, **gypsophile, gypsophily.**

gypsum [jip´səm] *Mineralogy.* $CaSO_4 \cdot 2H_2O$, a colorless, white, gray, red, yellow, or brown, transparent, monoclinic mineral, occurring in tabular to prismatic crystals up to 10 feet long, and having a subvitreous to pearly luster, a specific gravity of 2.32, and a hardness of 2 on the Mohs scale; it is the most common sulfate mineral and is found in sedimentary evaporite deposits, saline lakes, and deposits associated with vulcanism; used in making cement and plaster of Paris. (Going back to the Latin word for "chalk.")

gypsum

gypsum board *Materials.* a wallboard composed mainly of gypsum.

gypsum earth see GYPSITE.

gypsum plaster *Materials.* a form of plaster composed mainly of gypsum. Also, **gypsum cement.**

gypsy or **gypsy head** *Naval Architecture.* an attachment to the windlass designed to take the anchor chain. Also, GIPSY.

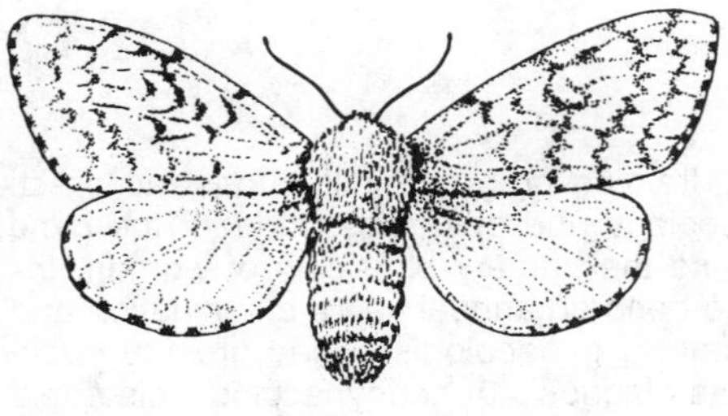

gypsy moth

gypsy moth *Invertebrate Zoology.* a tussock moth, *Lymantria dispar* or *Porthetria dispar,* of the family Lymantriidae, whose caterpillars cause serious damage to forest trees.

gyr- a combining form meaning "round" or "spiral," as in *gyration, gyrase.*

Gyracanthididae *Paleontology.* a family of acanthodian fishes, *incertae sedis* but sometimes classified in the order Climatiiformes, characterized especially by fin spines up to 40 cm long; known only from the Carboniferous of North America.

gyrase see DNA GYRASE

gyrate *Science.* twisted in a ring or spiral shape.

gyration *Mechanics.* a revolution in a circle or circles; circular or spiral movement.

gyration tensor *Solid-State Physics.* a tensor associated with an optically active crystal that, when multiplied with a unit vector whose direction coincides with the propagation of a light ray, specifies the gyration vector.

gyrator *Electromagnetism.* a waveguide device that causes no phase shift for the propagation of waves in one direction but a 180° phase shift in the other direction.

gyrator filter *Electronics.* a highly selective filter consisting of a gyrator that is terminated in a capacitor in order to have an inductive input impedance; it may be tuned with another capacitor.

gyratory breaker *Mechanical Engineering.* a rock-crushing device in which an inner, solid cone moves eccentrically within a fixed, hollow outer cone. Also, **gyratory crusher, gyratory.**

gyratory screen *Mechanical Engineering.* a sorting machine consisting of a vertical pile of horizontally oriented screens, each having a finer mesh than the one above it; eccentric motion of the machine sends material down through the series of screens.

gyre [jīr; jī´ər] *Mechanics.* a circular motion or course. *Oceanography.* **1.** one of the six great circular current systems found above and below the equator in the Indian, Pacific, and Atlantic Oceans. **2.** a closed circulation system, ranging in size from a few hundred meters to thousands of kilometers.

gyrectomy *Surgery.* an excision or resection of a cerebral gyrus or of a portion of the cerebral cortex.

Gyrinidae *Invertebrate Zoology.* a family of whirligig beetles in the order Coleoptera that swim with a gyrating movement and have flattened, oarlike legs.

Gyrinocheilidae *Vertebrate Zoology.* a monogeneric family of freshwater fishes of the suborder Cyprinoidei; found in southeast Asia and having fleshy lips on an inferior mouth which is used as a suction disc for clinging to rocks where it eats algae; respiration is through gill slits above and below gills.

gyro- a combining form meaning "round" or "spiral," as in *gyrocopter, gyroscope.*

gyrochrome *Neurology.* a neuron in which the Nissl bodies occur in the cytoplasm in a ring-shaped pattern.

gyrocompass *Navigation.* a gyroscope mounted with two degrees of freedom, which will always point to true north in order to minimize its potential energy on the rotating earth.

gyrocompass alignment *Navigation.* the adjustment of a gyrocompass so that indicated north is equal to true north.

Gyrocotylidae *Invertebrate Zoology.* an order of tapeworms in the subclass Cestodaria, fish parasites with an eversible proboscis and a posterior, ruffled, adhesive organ.

Gyrocotyloidea *Invertebrate Zoology.* in some systems of classification, a class of trematode worms.

Gyrodactyloidea *Invertebrate Zoology.* a superfamily of parasitic trematode flukes in the order Monogenea, with a posterior hooked organ; fish parasites, sometimes a serious pest in hatcheries.

gyro error *Navigation.* the total combined error in the reading of a gyrocompass, expressed in degrees east or west.

gyro flux-gate compass *Navigation.* an aeronautical magnetic compass that is stabilized by means of a gyroscope.

gyrofrequency *Astrophysics.* the frequency of rotation for an electron (or other charged particle) as it spirals in a magnetic field.

gyrogonite *Paleontology.* the calcified female reproductive body of the charophytes, a phylum of algae that has been widespread since the Silurian and is still represented by one family, the Characeae.

gyromagnetic coupler *Electronics.* a coupler in which a single-crystal YIG resonator couples two crossed stripline resonant circuits at the required low signal levels; may be used as a signal limiter or electronically tunable filter.

gyromagnetic effect *Electromagnetism.* the mechanical rotation induced in a body due to a change in its magnetization.

gyromagnetic ratio *Physics.* the magnetic dipole moment of a classical atomic or nuclear system divided by its angular momentum.

gyromagnetics *Electromagnetism.* the study of relationships between the magnetization of bodies and their angular momentum, as in the gyromagnetic effect.

gyropendulum or **gyro-pendulum** see PENDULOUS GYROSCOPE.

Gyropidae *Invertebrate Zoology.* a family of lice in the order Mallophaga, parasites of guinea pigs and other rodents in Central and South America.

gyropilot *Navigation.* a mechanical device for steering a craft that uses a gyro compass for directional information.

gyroplane *Aviation.* a rotorcraft whose chief means of propulsion is independent of the rotor system. Such craft usually use an engine to drive conventional propellers and to start their rotors, then acquire additional rotor power by the action of air when the craft is moving.

gyrorepeater *Navigation.* the part of a remote-reading gyrocompass system that shows at a distance the reading on the master gyrocompass.

gyroscope [jī´rə skōp´] *Navigation.* a small, heavy flywheel containing antifriction bearings and mounted inside supporting rings with pivots that allow rotation in any direction; when set in motion, the axle of the device maintains a constant direction; used to stabilize ships and steer aircraft.

gyroscopic drift [jī´rə skäp´ik] *Navigation.* a gradual change in the attitude of a gyroscope due to mechanical imperfections.

gyroscopic horizon *Navigation.* an artificial horizon generated by gyroscopic input.

gyroscopic mass flowmeter *Engineering.* a device used to measure the mass flow of fluid through a pipe by measuring the torque of the pipe while it is rotating.

gyroscopic precession *Mechanics.* the reaction of a free spinning body when a torque is applied to its spin axis; the axis moves steadily toward the direction of the applied torque, at right angles to the direction in which it would accelerate if the body were not rotating.

gyroscopics *Mechanics.* the branch of mechanics concerned with the study of gyroscopes and their practical application.

gyroscopic stability *Ordnance.* a mechanism that compensates for the effect of a tank's motion on the vertical movement of its main gun. Also, GYROSTABILIZATION.

gyrose *Biology.* **1.** waved or folded alternately forward and backward. **2.** marked by curved lines or circles. Also, **gyrous.**

gyrosextant *Navigation.* a sextant that has a gyroscope to indicate the horizontal.

gyrospasm *Neurology.* a rotatory spasm of the head.

gyrostabilizer *Navigation.* a motorized gyroscope used on ships and aircraft to reduce the swaying caused by wind or water movement.

gyrotheodolite *Cartography.* a theodolite with a gyrocompass attached or built in; a true azimuth reference can thereby be established in any weather, day or night, without the aid of stars, landmarks, or other visible stations.

Gyrothyraceae *Botany.* a monotypic family of creeping, purplish liverworts of the order Jungermanniales, characterized by sparse lateral-intercalary branches and a spongy stem with rhizoid pads or elevated cushions between the underleaves; limited to the Northern Hemisphere.

gyrotron *Electronics.* a device that detects system motion by comparing the frequency of a moving, vibrating tuning fork with a reference frequency that was in phase with the original frequency of the tuning fork.

gyro wheel *Navigation.* the wheel in a gyroscope that spins to resist displacement of the axis.

gyrus *Anatomy.* any of numerous ridgelike folds of the cerebral cortex.

gyttja *Geology.* an anaerobic, freshwater sediment, characterized by abundant organic debris, that is able to support aerobic life.

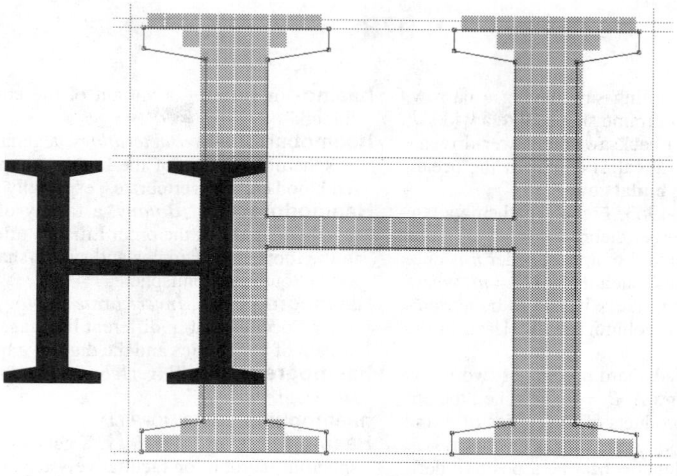

H the eighth in a sequence or group.

H the chemical symbol for hydrogen.

H or **H.** hour or hours (duration); hard; henry; henries. Also, **h** or **h**.

h coefficient of viscosity; electrolytic polarization.

h the symbol for Planck's constant.

h or **h.** hour or hours (o'clock); height; horizontal; humidity; hundred.

H- *Aviation.* the U.S. military and commercial designation for helicopter aircraft, as in *H-17*.

H-17 *Aviation.* a large jet-powered helicopter designed to lift heavy loads; popularly known as the *Flying Crane*.

ha or **ha.** hectare; hectares.

h.a. in this year. (From Latin *hoc anno.*)

HA or **H.A.** hour angle.

Haanel depth rule *Geophysics.* a rule for estimating the depth of a magnetic body provided the body can be considered magnetically equivalent to a single pole.

haar *Meteorology.* a thick, wet sea fog characterized by fine drizzle that occurs in the coastal districts of eastern Scotland and northeastern England, usually in the summertime.

Haar measure *Mathematics.* a measure μ on a locally compact topological group X such that every countable union of compact subsets of X is measurable, and for every measurable set E: (a) $\mu(E) \geq 0$, (b) there is a countable union of compact sets E such that $\mu(E) > 0$, and (c) either $\mu(xE) = \mu(E)$, for any measurable set E and any $x \in X$, or $\mu(Ex) = \mu(E)$, for any measurable set E and any $x \in X$; that is, μ is invariant under multiplication by fixed elements of X.

habenula *Anatomy.* **1.** a fold of membrane that connects two structures. **2.** a mass of cells in the dorsomedial portion of the thalamus of the brain.

habenular nuclei *Anatomy.* the gray matter of the habenula in the thalamus. Also, **habenular ganglia.**

Haber, Fritz 1868–1934, German chemist; awarded the Nobel Prize for synthesis of ammonia.

Haber-Bosch process *Chemical Engineering.* a process to produce ammonia by direct combination of nitrogen and hydrogen in the presence of iron catalysts. Also, **Haber process.**

habiline *Anthropology.* any of the hominids of the taxon *Homo habilis*, which appear in the fossil record about 1.8 million years ago, after the australopithecines and before *Homo erectus*, having a cranial capacity of about 640 cc.

habit *Behavior.* an acquired action or behavior pattern that is practiced regularly and with a minimum of voluntary control. *Botany.* the general appearance or behavior of a plant, such as erect, climbing, or prostrate. *Crystallography.* the characteristic form of any crystalline substance.

habitat *Ecology.* the place where an organism normally lives.

habitation site *Archaeology.* a general term for any area that has evidence of a domestic activity, such as food preparation.

habit plane *Crystallography.* a crystallographic plane on which a second solid phase forms, thus being common to both phases; the characteristic plane of a crystal upon which twinning, dislocations, or other phenomena occur.

habit strength *Behavior.* the strength of a habitual response, which depends on the number of reinforcements, the amount of reinforcement, the time between the stimulus and the response, and the time between the response and the reinforcement.

habitual *Behavior.* done by habit or relating to a habit.

habitual abortion *Medicine.* a condition in which there is repeated, consecutive, and spontaneous arrest of the birth process.

habituated culture *Botany.* an isolated tissue culture that develops an ability to synthesize auxin and therefore no longer needs a growth regulator. Also, ANERGIZED CULTURE.

habituation *Behavior.* the lessening or disappearance of a response to a given stimulus due to repeated exposure to this stimulus. *Medicine.* the act or process of forming a habit, as in psychological dependence on a drug or tolerance to the effects of a drug after a period of use. *Biotechnology.* the acquired or evolved ability of plant cells to grow and divide independently of any exogenous plant-growth substances.

habitus *Biology.* the external appearance, aspect, or growth form characteristic of an organism.

haboob *Meteorology.* a strong sandstorm or duststorm that blows in the deserts of the northern or central Sudan, particularly around Khartum or onto the plains of India.

hachure *Cartography.* a technique for showing sharp changes in elevation on a map by using short, wedge-shaped marks that radiate from the higher elevations and follow the direction of slope to the lower elevations.

H-acid *Organic Chemistry.* $H_2NC_{10}H_4(OH)(SO_3H)_2$, a gray powder, soluble in water, alcohol, and ether; used as an azo dye intermediate.

hacienda *Agriculture.* in Spanish-speaking countries, a large agricultural estate. *Architecture.* **1.** the main house of such an estate. **2.** in the southwestern U.S., a low, sprawling house with wide porches.

hackberry *Botany.* **1.** any tree of the genus *Celtis* in the family Ulmaceae, especially a woody tree of the eastern United States that produces small, sweet, often edible berries. **2.** the fruit of this tree.

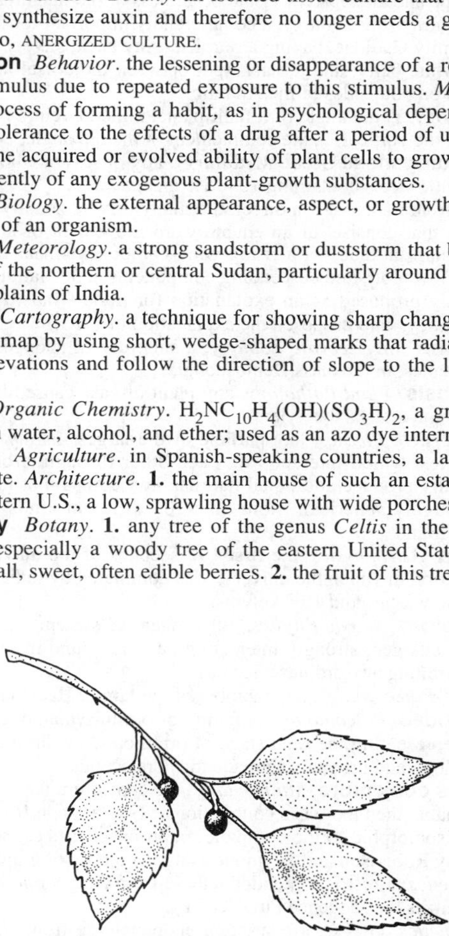

hackberry

hacker *Computer Technology.* **1.** originally, any person having highly technical computer skills, such as an assembly language programmer or a systems programmer. **2.** a popular term for a person who experiments with computer-system hardware, software, and communications, sometimes without proper authorization.

hackmanite *Mineralogy.* a variety of sodalite containing a small amount of sulfur and fluorescing orange or red on exposure to ultraviolet light.

hacksaw *Mechanical Devices.* a metal-cutting saw having a narrow, fine-toothed blade that is held within a steel frame shaped like a wide U; small hacksaws are hand-held, while larger hacksaws are power-driven.

hadal *Oceanography.* of or relating to the greatest depths of the ocean, the trenches over 6 kilometers deep. Thus, **hadal zone.**

Hadamard, Jacques [ad ə mär´] 1865–1963, French mathematician; with Vallée-Poussin, proved the prime number theorem.

Hadamard matrix *Mathematics.* a Hadamard matrix of order n is an $n \times n$ matrix H with all entries equal to ±1 and such that $HH^T = nI$, where I is the $n \times n$ identity matrix. A Hadamard matrix is said to be normalized if all entries in the first row and first column are +1. Used in the study of block designs.

Hadamard product *Mathematics.* the Hadamard product of two $n \times m$ matrices $A = (a_{ij})$ and $B = (a_{ij})$ is the matrix $A \cdot B = (a_{ij}\, b_{ij})$; i.e., the entries of $A \cdot B$ consist of the elementwise products of the entries of A and B; arises in problems with integral operators. Also, SCHUR PRODUCT.

Hadamard's inequality *Mathematics.* let $A = (a_{ij})$ be a positive definite $n \times n$ matrix with complex entries, i.e., a matrix representing a positive definite quadratic form. Then $\det A \leq \prod_{i=1}^{n} a_{ii}$.

Hadar *Anthropology.* a site in the Afar depression of Ethiopia that is famed for its excellent preservation of fossil hominids, particularly the specimen of *Australopithicus africanus* called "Lucy." *Astronomy.* Beta (β) Centauri, a blue-white giant star 390 light-years away that has a B1 spectral type and an apparent visual brightness of 0.6 magnitude.

haddock *Vertebrate Zoology.* a commercially important food fish of the codfish family Gadidae, having a purplish-gray back, silvery-gray sides, a silvery-white belly, and a black spot between the dorsal and pectoral fins; found on both sides of the Atlantic.

hade *Geology.* **1.** the angle of a fault plane, mineral vein, or lode, measured from the vertical. **2.** the complement of the dip angle.

Hadfield manganese steel *Metallurgy.* any of several abrasion-resistant austenitic steels containing 10–15% manganese.

Hadley cell *Meteorology.* a direct, thermally driven, zonally symmetric circulation that consists of an equatorward movement of air from 30° north or south latitude, with rising wind components near the equator, poleward flow aloft, and descending components at 30° latitude again; it was initially proposed as an explanation for trade winds. (Named for George *Hadley,* who proposed the explanation in 1735.)

Hadromerina *Invertebrate Zoology.* a suborder of sponges with pin-shaped spicules, in the class Demospongiae.

hadromycosis *Plant Pathology.* any plant disease caused by a fungus that enters and attacks the xylem.

hadron *Particle Physics.* any particle of the largest family of elementary particles, which interact with each other through strong interactions, usually produce additional hadrons in a collision at high energy, and are roughly spherical. More than 100 different kinds of hadrons have been found.

hadron era *Astrophysics.* the first 10^{-4} seconds of the universe's existence, a period in which the universe was dominated by radiation and its temperature was around 10^{15} kelvins.

hadronic atom *Atomic Physics.* a hydrogenlike system consisting of a negatively charged, strongly interacting particle bound in the Coulomb field and orbiting any ordinary nucleus.

hadrosaur *Paleontology.* any member of the family Hadrosauridae.

Hadrosauridae *Paleontology.* a family of ornithischian duck-billed dinosaurs, representing the final stage of ornithopod evolution; they were 4–16 mm long and probably semiaquatic; Cretaceous.

Hadwiger's conjecture *Mathematics.* the conjecture that, if a graph G is k-chromatic, then there is a contraction of G to a graph H containing a subgraph isomorphic to the complete graph on k vertices; settled affirmatively by Robertson and Seymour using the theory of graph minors.

HaeII *Genetics.* a restriction endonuclease from *Haemophilus aegyptius* that recognizes the sequence PuGCGC/Py.

HaeIII *Genetics.* GGCC, a restriction endonuclease from *Haemophilus aegyptius.*

Haeckel, Ernst [hek´əl] 1834–1919, German biologist; classified all animals as Protozoa or Metazoa; formulated the theory of recapitulation.

haemato- or **haemat-** a variant of the combining form *hemato-,* meaning "blood."

Haematococcaceae *Botany.* a family of biflagellate algal cells in the order Volvocales, distinguished by protoplasts with numerous cytoplasmic processes extending to the cell wall or into the gelatinous coenobial envelope; often associated with rainwater and particularly common in freshwater rock pools.

haematodocha see HEMATODOCHA.

haemo- or **haem-** a variant of the combining form *hemo-,* meaning "blood."

Haemobartonella *Bacteriology.* a genus of Gram-negative, rod-shaped or spheroid bacteria of the family Anaplasmataceae; parasites in or on red blood cells of vertebrates, especially cattle, dogs, cats, and rodents.

Haemodoraceae *Botany.* a family of monocotyledonous perennial geophytic herbs of the order Liliales; often characterized by red pigment in the roots and rhizomes and sword-shaped basal leaves; native mostly to the Southern Hemisphere.

Haemogregarina *Invertebrate Zoology.* a genus of protozoans in the order Coccidia that in different life stages are parasitic in the circulatory system of vertebrates and the digestive tract of invertebrates.

haemogregarine *Invertebrate Zoology.* any member of the genus *Haemogregarinae.*

haemophilia see HEMOPHILIA.

Haemophilus *Bacteriology.* a genus of Gram-negative, rod-shaped or spheroid, aerobic or facultatively anaerobic, nonmotile bacteria of the family Pasteurellaceae; they are parasites usually found in the upper respiratory tract and require one or two blood factors for survival. Also, HEMOPHILUS.

Haemoproteus *Invertebrate Zoology.* a genus of protozoan parasites in the blood of certain reptiles and birds, transmitted by bloodsucking flies.

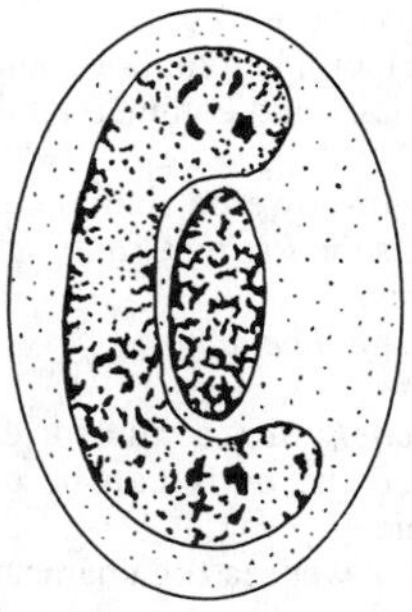

Haemoproteus

Haemosporidia *Invertebrate Zoology.* an order of minute protozoans; larvae are found in bloodsucking flies, adults as parasites in vertebrates.

Haemosporina *Invertebrate Zoology.* a suborder of parasitic sporozoan protozoans in the subclass Coccidia, including the human malarial organism.

haemotropic see HEMOTROPIC.

haemozoin see HEMOZOIN.

Haemulidae *Vertebrate Zoology.* in classification systems, an equivalent name for a Pomadasyidae, large family of marine fishes commonly known as grunts.

haff *plural,* **haffe.** *Geography.* a freshwater lagoon separated from the sea by a sandspit; characteristic of the south coast of the Baltic.

Haff disease *Toxicology.* a type of poisoning caused by arsenic found in an inlet of the Baltic Sea, believed to be present due to the introduction of cellulose-manufacturing waste water; symptoms may include severe pain in the limbs, weariness, and myglobinuria.

Hafnia *Bacteriology.* a genus of Gram-negative, facultatively anaerobic, rod-shaped bacteria of the family Enterobacteriaceae; found in birds, animals, and humans, as well as in water, soil, and sewage; becomes pathogenic when the immune system becomes sufficiently weak. It includes a single species, *H. alvei,* with a number of biotypes.

hafnium *Chemistry.* a rare chemical element having the symbol Hf, the atomic number 72, and an atomic weight of 178.49; melts at about 2100°C and boils above 5400°C; it occurs with the similar and much more common element zirconium; it is used in nuclear reactors, light-bulb filaments, and electrodes. (From *Hafnia,* the New Latin word for Copenhagen, Denmark, where it was discovered in 1923.)

hafnium carbide *Inorganic Chemistry.* HfC, a gray powder with a very high melting point, about 3890°C; used in nuclear reactor control rods.

haftplatte *Invertebrate Zoology.* an adhesive disk found in some flatworms.

Hagedorn needle *Surgery.* a curved surgical needle with flat sides, a large eye, and a straight cutting edge near the point. (Named for Werner *Hagedorn,* 1831–1894, German surgeon.)

Hagen-Poiseuille law *Fluid Mechanics.* a law stating that for fluid flowing through a circular pipe, the loss of head due to frictional effects is given by the product of 32 times the pipe length, fluid viscosity, and fluid velocity divided by the product of the gravitational acceleration, fluid density, and the square of the pipe diameter.

Hagen-Rubens relation *Optics.* an equation that relates the reflectivity of a solid surface to the frequency of radiation of its conductivity.

hagfish *Vertebrate Zoology.* the jawless fishes, comprising the order Myxiniformes and generally resembling eels but with a round mouth surrounded by tentacles, which are used to bore into the bodies of other fishes for food.

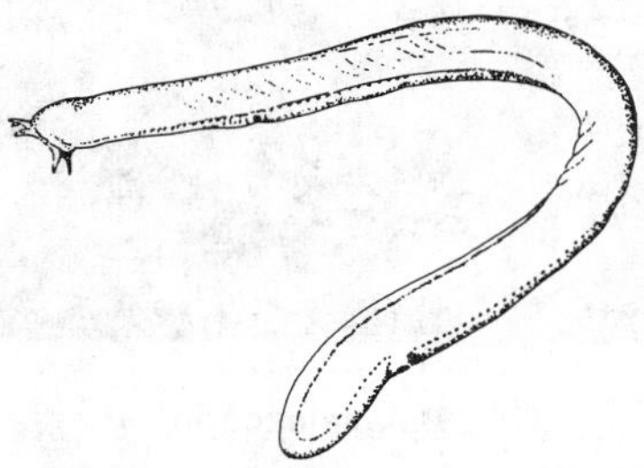

hagfish

hag gene *Genetics.* the structural gene for flagellin in enterobacteria.

H agglutinin *Immunology.* an antibody that reacts with the flagella, rather than the body, of flagellated bacteria in an agglutination reaction.

hagiophobia *Psychology.* see HIEROPHOBIA.

Hahn, Otto 1879–1968, German physical chemist; with Meitner, discovered protactinium; awarded Nobel Prize for work in atomic fission.

Hahn-Banach theorem *Mathematics.* let Y be a linear manifold in a real or complex normed vector space X and let f be a continuous linear functional on Y with norm $\|f\|$. Then f can be extended to a continuous linear functional F on X whose norm is also $\|f\|$. Also, CONTINUOUS EXTENSION THEOREM.

Hahn decomposition *Mathematics.* let μ be a signed measure defined on the class of all measurable sets of a measurable space X. By the theorem, there exist two disjoint sets A and B such that (a) $A \cup B = X$; (b) for every measurable set F, $A \cap F$ and $B \cap F$ are measurable; and (c) $\mu(A \cap F) \geq 0$ and $\mu(B \cap F) \leq 0$. A and B are said to form a Hahn decomposition of X with respect to μ. A is said to be positive with respect to μ and B is said to be negative with respect to μ.

Hahnemann, Samuel Friedrich 1755–1843, German physician; the founder of homeopathy.

hahnium see ELEMENT 105.

Hahn technique *Solid-State Physics.* a technique used to study solids by incorporating trace amounts of a radioactive element into the solid and observing the emission of radiation.

Haidinger brushes *Optics.* the faint yellow, brushlike design seen when viewing a bright surface through a Nicol prism, arising from birefringence occurring at the fovea of the eye.

Haidinger fringes *Optics.* interference fringes observed with thick, flat plates near normal incidence, such as those fringes produced by the Fabry-Perot interferometer. Also, CONSTANT ANGLE FRINGES, CONSTANT DEVIATION FRINGES.

haidingerite *Mineralogy.* $CaHAsO_4 \cdot H_2O$, a white or colorless, orthorhombic mineral occurring as fibrous or granular botryoidal coatings, having a specific gravity of 2.95 and a hardness of 2 to 2.5 on the Mohs scale; found as a secondary mineral in arsenic-bearing ore deposits.

hail *Meteorology.* any precipitation that freezes in concentric layers about a core as vertical air currents drop and lift it; its diameter is always greater than 5 mm, larger than ice or snow pellets. Hail is formed by convective clouds that are usually cumulonimbus, and is often associated with thunderstorms. Thus, **hailstorm.**

hail stage *Meteorology.* a thermodynamic process of freezing suspended water drops in an adiabatic atmosphere in which rising air is cooled below the freezing point, with the assumption that the latent heat of fusion maintains a constant temperature until all available moisture is frozen; formerly used as a model for the formation of hail.

hailstone *Meteorology.* a pellet of hail, typically 1/4 inch to 6 inches in diameter, having a shape that is spheroidal, conical, or generally irregular.

hair *Anatomy.* **1.** one of the many long, thin, threadlike structures that grow out from the skin of humans and other mammals. **2.** a dense profusion of such structures on the head of a human or forming the coat of an animal. *Biology.* a similar threadlike outgrowth in insects, spiders, and so on, or on the surface of a plant.

hairball *Veterinary Medicine.* a concretion of undigested hair within the stomach or intestines of a cat or other animal, formed as a result of the animal's licking its own coat. Also, TRICHOBEZOAR.

hair bulb *Anatomy.* an enlarged, bulbous structure at the end of a hair root, from which cells develop to form the root and then the hair shaft.

hair cell *Histology.* a specialized, ciliated sensory cell found in the inner ear of vertebrates.

haircoat *Agriculture.* the outer hair of cattle, especially the heavier winter coat.

hair cycle *Physiology.* the normal sequence of hair growth in which a new hair develops at the base of the follicle, then grows for a period (**anagen**), experiences a brief interlude between growth and resting (**catagen**), rests for a period (**telogen**), and finally drops off.

hair follicle *Anatomy.* a flasklike depression in the skin that encloses a hair and its root.

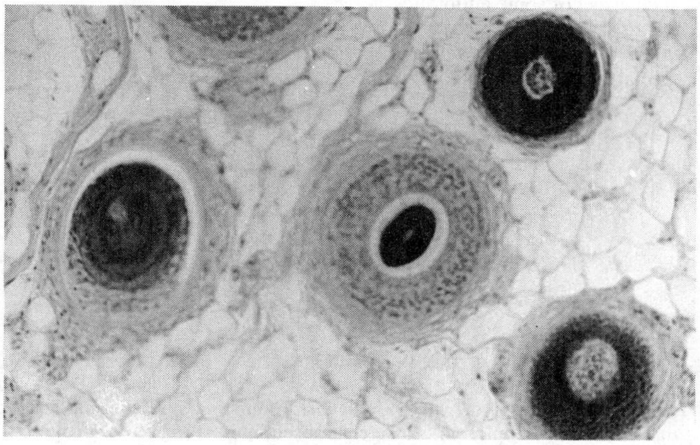

hair follicles

hair gland see SEBACEOUS GLAND.

hair hygrometer *Engineering.* a hygrometer that uses a bundle of human hairs to detect and measure relative humidity; the length of the hairs vary as they absorb moisture from, or lose moisature to, the surrounding air (or another gas), and this variance is displayed by a moving pointer or other such indicator.

hairline *Anatomy.* the lower edge of the hair along the upper forehead. *Engineering.* **1.** a line having almost no width. **2.** a fine, straight fault or bubble in glass. *Building Engineering.* a narrow crack in paint, varnish, or other surface coating. *Graphic Arts.* **1.** a very thin line on the face of type, or a typeface characterized by such lines. **2.** a fine scratch on a film negative.

hairline space see HAIRSPACE.

hair matrix carcinoma see BASAL CELL CARCINOMA.

hair papilla *Anatomy.* a structure that extends into an indentation at the base of the hair bulb and that contains a network of nerves and vessels supplying the hair root.

hairpin loop *Molecular Biology.* a loop formed by a single-stranded nucleic acid molecule when complementary sequences on the strand base pair with each other, causing the unpaired portion of the strand to loop out.

hairpin tube *Mechanical Devices.* a boiler tube that bends back upon itself to form a U shape.

hair root *Anatomy.* a general term for the lower portion of a hair, below the shaft.

hair seal *Vertebrate Zoology.* any of various seals of the family Phocidae having coarse hair and no soft undercoat.

hair shaft *Anatomy.* the thin cylindrical structure that extends through the skin to form the external portion of a hair, consisting of dead cells and containing pigment and the protein keratin.

hairspace *Graphic Arts.* in typesetting, a very thin space inserted between letters (especially capitals) to increase legibility or improve line justification. Also, HAIRLINE SPACE.

hairstone *Mineralogy.* a type of clear crystalline quartz thickly embedded with fibrous, threadlike, or needlelike inclusions of other minerals, such as rutile or actinolite. Also, NEEDLE STONE.

hair transplant *Surgery.* the surgical process of transferring clumps of hair from one site on the body to another, usually to move hair from a place where it grows more thickly to a bald spot.

hair trigger *Ordnance.* a finely balanced trigger that requires minimal force to fire the gun. (From the idea that a trigger movement of just a hair's breadth will cause the gun to fire.)

hairworm *Invertebrate Zoology.* the common name for members of the class Nematomorpha.

hairy root *Plant Pathology.* an abnormal plant condition in which the roots become fine and fibrous; it is a common phase in the crown-gall disease of the apple tree.

hairy shaker disease see BORDER DISEASE.

hairy-sphere theorem *Mathematics.* a popular name given to the theorem that it is impossible to have an everywhere nonvanishing continuous vector field on the 2-sphere. If the vector field is visualized as "hair," the theorem states that smoothly combed hair must have a whorl.

hairy tongue *Medicine.* a benign condition in which the tongue develops black hairlike elongations of the papillae, sometimes occurring as a side effect of some antibiotics.

HAI test see HEMAGGLUTINATION-INHIBITION TEST.

hake *Vertebrate Zoology.* any marine fish of the genus *Merluccius,* especially the **silver hake,** *M. bilinearis*; closely related to cod.

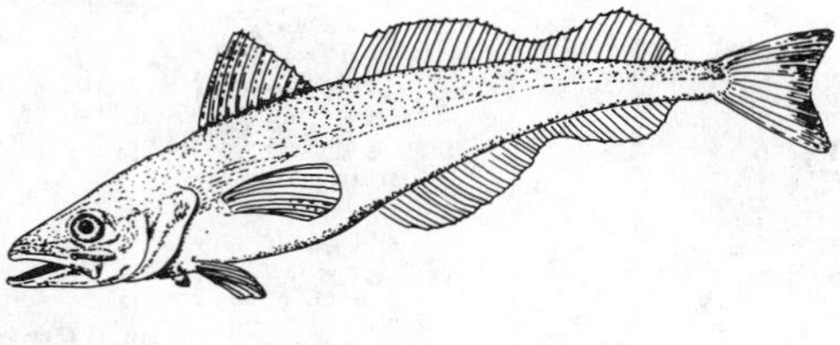

hake

Hakluyt, Richard [hak´lüt] 1552–1616, English geographer and author; his accounts of explorers' voyages aroused interest in colonization.

hal- a combining form meaning "salt," as in *halite.*

Halacaridae *Invertebrate Zoology.* a family of mostly marine parasitic mites, in the order Acarina, found in algae and bottom debris.

halation *Electronics.* the blurring of an image on a photosensitive surface caused by light diffusion.

halazone *Organic Chemistry.* $HO_2C–C_6H_4SO_2NCl_2$, white crystalline powder with strong chlorine odor; slightly soluble in water, melts at 195°C; used as a water disinfectant.

halcyon days *Meteorology.* a period of fine, tranquil weather. (From the Greek myth of the lovers Ceyx and *Alcyone,* in which the winds are suspended for seven days before and after the winter solstice.)

Haldane, J. B. S. 1892–1964, British geneticist and author; the son of John Scott Haldane; wrote many works of popular science.

Haldane, John Scott 1860–1936, Scottish physiologist; studied environmental effects on respiration; analyzed toxicity of carbon dioxide.

Haldane's evolutionary unit *Evolution.* a unit measure of increasing body size over time on an evolutionary scale, given in darwins.

Haldane's rule *Genetics.* a rule stating that if first-generation hybrids are produced between two species but with one sex absent, rare, or sterile, then that sex is the heterogametic sex. Also, **Haldane's law.**

Hale, George Ellery 1868–1938, American astronomer; invented spectroheliosocope; found Zeeman effect in sunspots; developed the Hale Reflecting Telescope at Palomar Observatory.

Halecomorphi *Vertebrate Zoology.* a superorder of fishes which includes the primitive bowfins of the order Amiiformes. Also, CYCLOGANOIDEI.

Halecostomi *Vertebrate Zoology.* a superorder of teleost fishes that includes the order Semionotiformes.

Hale cycle *Astronomy.* the approximate 22-year cycle in which the magnetic polarity of sunspot pairs reverses and then returns to its original state; during half the cycle, for example, the leading spot in every pair will have a positive polarity but during the other half, the leading spot will be negative.

Hales, Stephen 1677–1761, English clergyman; a founder of experimental physiology; made discoveries in transpiration and blood pressure; invented the pneumatic trough.

Hale telescope

Hale telescope *Optics.* the 200-inch reflecting telescope at Palomar Observatory, near San Diego, California.

half adder *Computer Technology.* an adder circuit in digital systems that can manage the two binary bits to be added, but cannot accommodate a carry signal. *Computer Programming.* see EXCLUSIVE-OR, def. 2.

half-adjust *Computer Programming.* a rounding process in which the least significant digit of a number is dropped and its value determines whether one is to be added to the next more significant digit; if it is one-half or more than one-half of the number base, one is added.

half-a-Gram stain *Microbiology.* a bacteriological staining procedure used to identify Legionellaceae cells in sputum specimens, utilizing Gram's iodine solution.

half-and-half solder *Metallurgy.* a solder that contains 50% tin and 50% lead.

half-angle formulas *Mathematics.* let a triangle T have the sides a, b, and c with the corresponding opposite angles α, β, and γ, and let $s = (a + b + c)/2$. Then the half-angle formulas for T are:

$$\tan(\gamma/2) = [(s-a)(s-b)/s\,(s-c)]^{1/2} = \sin\gamma/(1+\cos\gamma) = (1-\cos\gamma)/\sin\gamma$$

$$\sin(\gamma/2) = [(s-a)(s-b)/(ab)]^{1/2} = \pm[(1-\cos\gamma)/2]^{1/2}$$

$$\cos(\gamma/2) = [s(s-c)/(ab)]^{1/2} = \pm[(1+\cos\gamma)/2]^{1/2}.$$

half-arc angle *Meteorology.* the elevation angle of a point that bisects the arc from the zenith to the horizon of an observer, serving as a measure of the apparent degree of flatness of the dome of the sky.

halfbeak *Vertebrate Zoology.* a small marine and freshwater fish of the family Exocoetidae (in some schemes, of Hemirhamphidae), having a short upper jaw and a long, protruding lower jaw, a silvery, slender body, and the ability to spring above water for short distances.

half block *Computer Programming.* in some systems, a block of data transferred between main memory and the hardware buffer control unit; it consists of 128 16-byte elements.

half-breadth plan *Naval Architecture.* an engineering drawing showing an overhead view of one side of the hull of a ship; only one side need be shown as the other side is a mirror image of the side shown.

half-bridge *Electricity.* a bridge with power supplies in two arms, to replace the usual single power supply of a regular bridge.

half carry *Computer Programming.* a bit used in some computers to indicate that a carry has occurred from the low-order half of a digit to the high-order half; used to implement packed-decimal arithmetic.

half-cell *Physical Chemistry.* a single electrode along with its surrounding electrolytic solution; so called because it is one of two such electrodes forming a complete electrochemical cell, one being the anode and the other the cathode. Also, HALF-ELEMENT.

half-cell potential *Physical Chemistry.* the energy developed by each of the electrodes in an electrochemical device such as a battery; the assumed potential of a single electrode equated to the potential of the electrode when the counter electrode is the relation $H_2 = 2H^+ + 2e^-$, which is arbitrarily assigned the value of zero.

half-cock *Ordnance.* in a firearm, the position of the hammer in which it is held by a cocking notch before the full-cock position; in the half-cock position the trigger is locked and the weapon may be handled or carried in relative safety.

half column *Architecture.* an engaged column that projects from a wall by about half its diameter.

half-course *Mining Engineering.* a drift or opening driven at a 45° angle to the strike and in the plane of the seam.

half cycle *Engineering.* a time period that equals 180° of a circuit's operating frequency.

half-cycle transmission *Telecommunications.* a data transmission system that employs synchronized sources of 60-hertz power at both source and destination ends.

half-duplex *Telecommunications.* describing a method of communication in which either sending or receiving, but not both, can occur at any given time. Thus, **half-duplex circuit, half-duplex operation, half-duplex repeater.**

half-element see HALF-CELL.

half-hard *Metallurgy.* the hardness of a metallic material, intermediate between dead soft and full-hard.

half hatchet *Mechanical Devices.* a hatchet with the cutting edge as a straight, flat hammer on one edge and a broader, curved blade on the other.

half-hot switch *Electronics.* a switch that allows current to flow to an electric light or appliance, but that does not turn the device on directly.

half joist *Building Engineering.* a joist cut in two pieces along its web, thus forming a T-section; used primarily in steelwork welding. Also, CASTELLATED BEAM.

half-lap joint *Engineering.* **1.** a joint formed by halving together two pieces of wood to form a flush surface. **2.** a similar joint formed by bolting or riveting together two pieces of metal or other material.

half-life *Nuclear Physics.* **1**. the time required for half of the original atoms of a radioactive material to undergo radioactive transformation. **2**. in a living system or ecosystem, the time required for half of a radioactive substance, such as a radioactive tracer, to disintegrate by radioactive decay or to be eliminated by natural processes. *Chemistry.* the time taken for a given chemical reaction to affect half of the reactants involved.

half-loaded *Ordnance.* in an automatic weapon, a condition in which the cartridge belt or magazine is inserted and the receiver charged, but the first cartridge is not in the chamber.

half-model *Cartography.* the stereoscopic model that results when sections of two adjacent right-hand or left-hand exposures of convergent photographs overlap.

half-moon *Astronomy.* the moon in its first- or last-quarter phase, when it appears to be half-lit.

half-moon clip *Ordnance.* a revolver clip that holds the cartridges for one-half the cylinder; it is used to facilitate loading or as an adaptor for rimless ammunition.

half nut *Mechanical Devices.* a nut split lengthwise, designed to clamp onto a screw such as a lathe carriage.

half-open interval *Mathematics.* an interval on the real line of the form $(a,b] = \{x : a < x \leq b\}$ or $[a, b) = \{x : a \leq x < b\}$.

half-plane or **half plane** *Mathematics.* the portion of the real or complex plane lying on one side of some line, particularly the portion of the real or complex plane lying on one side of one of the axes. The **upper half-plane** lies above the horizontal axis; the **lower half-plane** lies below.

half-power beamwidth *Electromagnetism.* the angle across the main lobe of an antenna pattern between the two directions, which subtends the major lobe at its half-power values.

half-power frequency *Electronics.* either a high or a low frequency at which the power gain of an amplifier, network, or the like falls to 50% of its maximum or nominal response.

half-power point *Electronics.* the point on a response characteristic curve that corresponds to half the power at the maximum of the curve.

half-reaction *Physical Chemistry.* either of the two parts of an oxidation-reduction reaction in an electrochemical cell, representing oxidation only or reduction only. Also, HALF-CELL REACTION.

half-reaction time see COT 1/2.

half-rip saw *Mechanical Devices.* a carpentry handsaw used to cut both along and against the grain of wood.

half-round bit *Mechanical Devices.* a drilling bit with a cutting face slope of 4° whose cross section resembles a semicircle. Also, CYLINDER BIT, D-BIT.

half-round file *Mechanical Devices.* a file that is concave or convex on one side and flat on the other.

half-round screw *Mechanical Devices.* see BUTTONHEAD.

half-section *Cartography.* in a surveying section, any two quarter-sections that have a common boundary. Thus, **north half, east half, south half, west half.**

half-silvered *Optics.* describing a surface that is thinly coated with metallic film, so that it reflects approximately half of the incident light and transmits half.

half space *Building Engineering.* a broad step between two half flights of stairs. *Mathematics.* either of the parts into which a finite-dimensional space is divided by a hyperplane (codimension = 1).

half step see SEMITONE.

half S trap *Mechanical Devices.* a plumbing trap shaped like the upper or lower half of a letter S.

half-subtractor *Electronics.* a logic element whose inputs are two digits from a preceding stage and whose outputs are a difference digit and a borrow digit. Also, ONE-DIGIT SUBTRACTOR, TWO-INPUT SUBTRACTOR.

half tap *Electricity.* a bridge placed across conductors that does not affect their continuity.

half-tetrad analysis *Molecular Biology.* the analysis of chromosome tetrads after recombination, when only two of the four chromosomes of the original tetrad can be analyzed.

half-thickness *Physics.* the thickness of a sheet of a homogeneous substance that, when exposed to radiation, reduces (by absorption) the intensity of the transmitted radiation to one-half of the incident intensity.

half tide *Oceanography.* the stage at which the tide is halfway between high water and the following or preceding low water.

half-tide basin *Civil Engineering.* a very large lock of irregular shape that stays open during high tide to allow vessels to enter; vessels remain in the basin until the next flood tide, when they may pass through the lock's gate to the inner harbor.

half-tide level see MEAN TIDE LEVEL.

half-timbered *Building Engineering.* of or relating to a timber-framed structure in which the spaces between the timbers are filled with masonry, brickwork, plaster, or wattle.

half-time or **halftime** *Physical Chemistry.* the time required for half of a quantity of a substance to be eliminated from a system at a rate proportional to its concentration. *Nucleonics.* the period during which half the radioactive material produced by a nuclear explosion remains in the environment.

half-title *Graphic Arts.* the title of a book, set in smaller type than on the title page and standing alone on a page preceding the front matter or text or both.

halftone or **half tone** *Graphic Arts.* **1.** the product of the halftoning process, a negative or plate on which continuous tones and details are represented by a pattern of dots of varying size and shape. **2.** a print made from such a negative and plate. **3.** the gradations of gray in such an image. *Acoustics.* see SEMITONE.

halftone characteristic *Telecommunications.* in facsimile systems, the relationship between the density of the recorded copy and the density of the subject copy.

halftone screen *Graphic Arts.* in halftoning, a screen covered with a grid of very fine lines (50–300 per inch); it is placed over continuous-tone copy, which is then rephotographed to produce a halftone negative.

halftone step scale see STEP SCALE.

halftoning *Graphic Arts.* a process used to reproduce a black-and-white photograph or other artwork by rephotographing it through a screen that converts the image into a pattern of dots, which print as continuous tones of gray.

half-track or **halftrack** *Mechanical Engineering.* **1.** a drive system in which a revolving belt lays a flexible moving track beneath the rear, driving wheels on either side of a motor vehicle, while the front wheels carry tires. **2.** any vehicle having such a drive system, especially an armored military vehicle.

half-value *Radiology.* the thickness of a material that, when inserted in a narrow beam of radiation, will reduce the beam's intensity to one-half its initial value. Also, **half-value layer.**

half-wave *Electricity.* an electrical length that is equivalent to half of one wave cycle.

half-wave amplifier *Electronics.* a magnetic amplifier whose total induced voltage frequency equals the power supply frequency.

half-wave antenna *Electromagnetism.* an antenna in which the electrical length is equal to one-half of the wavelength of the signal being transmitted or received.

half-wavelength *Electromagnetism.* the electrical length corresponding to one-half of a wavelength of a transmission line or antenna element.

half-wave plate *Optics.* a plate of electro-optical material that turns the plane of polarization of a light beam.

half-wave potential *Analytical Chemistry.* the midpoint of a polarographic wave, which is independent of concentration and is characteristic for the species undergoing reaction.

half-wave rectifier *Electricity.* a circuit or device that passes either positive or negative voltages; often the output is filtered to create a constant DC output voltage from an AC input voltage. Thus, **half-wave rectification.**

half-wave transmission line *Electromagnetism.* a transmission line whose electrical length is equal to one-half of the wavelength of the signal being transmitted.

half-wave vibrator *Electricity.* a vibrator containing only one pair of contacts; it does not reverse the direction of the direct current that is interrupted by its transformer.

halfway house *Medicine.* a facility for patients, especially psychiatric patients, who do not require hospitalization but who do need a certain level of treatment and supervision before returning to general society.

half-word *Computer Technology.* a unit of memory that has a length equal to half that of a machine word.

half-word I/O buffer *Computer Technology.* a buffer in which the upper half stores the upper half of a word for input and output characters and the lower half stores other information, such as constants.

halibut *Vertebrate Zoology.* either of two large flatfish of the genus *Hippoglossus, H. hippoglossus* of the North Atlantic or *H. stenolepis* of the North Pacific; a commercially important food fish found mostly in cold, deep waters. (From a Middle English term meaning "holy fish;" so called because it was eaten on certain holy days in place of meat.)

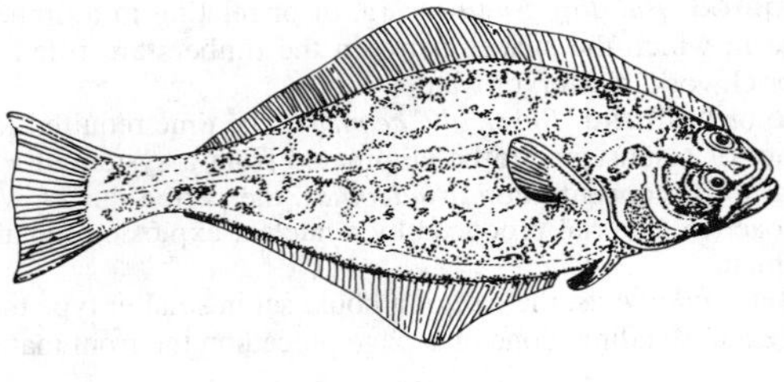

halibut

Halichondria *Invertebrate Zoology.* a genus of marine sponges in the order Halichondrida.

Halichondrida *Invertebrate Zoology.* an order of sponges with siliceous spicules in the class Demospongiae. Also, **Halichondrina.**

Halictidae *Invertebrate Zoology.* a very large family of hymenopteran insects, including sweat bees, in the superfamily Apoidea, that burrow in the ground.

halide *Chemistry.* a binary combination of a halogen with another element, such as potassium iodide, KI, or an organic compound in which halogen atoms replace one or more hydrogen atoms; for example, methyl chloride CH_3Cl.

Halimeda *Botany.* a common genus of bushy green algae in the family Udoteaceae; its calcified remains have been important to the buildup of coral reefs.

Haliotis *Invertebrate Zoology.* a genus of commercially important abalone.

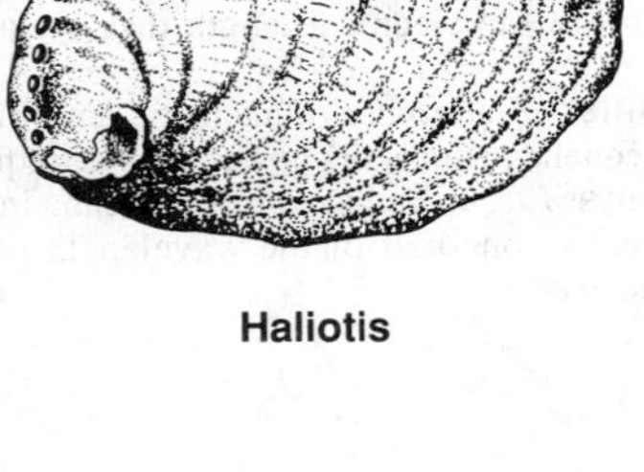

Haliotis

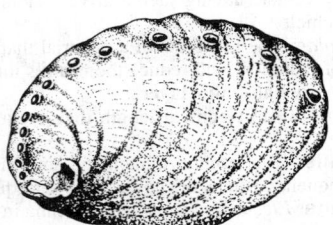

Haliphthoraceae *Mycology.* a family of fungi belonging to the order Saprolegniales that occur in marine environments and are parasitic to marine Crustacea.

Haliplidae *Invertebrate Zoology.* a family of coleopteran insects, crawling water beetles that live among aquatic vegetation and submerged debris in shallow freshwater.

Haliscomenobacter *Bacteriology.* a genus of Gram-negative bacteria that occur as nonmotile, rod-shaped cells, typically in sheath-enclosed trichomes.

halite *Mineralogy.* NaCl, a colorless to white, yellow, reddish, or blue cubic mineral having a specific gravity of 2.17 and a hardness of 2 on the Mohs scale, massive in habit or as cubic crystals, and forming deposits of considerable thickness in close association with anhydrite and gypsum, especially in Permian and Triassic rocks. Also, COMMON SALT, ROCK SALT.

halite

Halitheriinae *Paleontology.* an extinct subfamily of early Tertiary sirenians, possibly ancestral to modern dugongs.

halitosis [hal i tō´sis] *Medicine.* offensive breath; bad breath.

halitus *Physiology.* an exhalation or vapor; expired breath.

Hall, Edwin Herbert 1855–1938, American physicist; discovered the Hall effect.

Hall, Sir James 1761–1832, Scottish geologist; analyzed volcanoes and volcanic deposits.

Hall accelerator *Physics.* a type of plasma accelerator that works in accordance with the Hall effect.

Hall angle *Electromagnetism.* the ratio of the electric field arising from the Hall effect to the electric field driving the current that subsequently produces the Hall effect.

Hall antidote *Toxicology.* a solution of potassium iodide, quinine hydrochloride, and water, formerly used as an antidote for mercuric chloride poisoning.

Hall coefficient *Electromagnetism.* a constant of proportionality in the Hall effect, given by the magnitude of the current density and the magnetic induction.

Hall effect *Electromagnetism.* an electric field observed in a conductor that is transverse to both the current driven through the conductor and a magnetic field applied to the conductor externally.

Hall-effect gaussmeter *Engineering.* a gaussmeter that employs a chip of semiconductor material and the principles of the Hall effect to measure magnetic field strength.

Hall-effect isolator *Electromagnetism.* a waveguide component that causes greater losses in one direction of propagation than in the other direction, by means of the Hall effect.

Hall-effect modulator *Electronics.* a Hall-effect multiplier used to modulate an output voltage that is proportional to the product of two input voltages, signals, or currents.

Hall-effect multiplier *Electronics.* a multiplier based on the Hall effect, used in analog operations such as multiplication or root extraction.

Hall-effect switch *Electronics.* a magnetically activated switch using a Hall generator, trigger circuit, and transistor amplifier on a silicon chip.

Haller, Albrecht von [häl´ər] 1708–1777, Swiss anatomist, physiologist, and botanist; the foremost physiologist of his time; distinguished sensibility and irritability; studied circulation.

Halley, Edmond [hal´ē; hā´lē] 1656–1742, English astronomer; catalogued southern stars; accurately calculated orbits of Halley's comet and other comets, using Newton's theories.

Halley's comet *Astronomy.* the famous comet named for Edmond Halley, who discovered its approximate 76-year period, and which will return next in 2061; its nucleus measures 15 kilometers by 8 by 7.5.

Halley's comet

Hall generator *Electromagnetism.* a generator using the Hall effect to give an output voltage proportional to magnetic field strength.

Hallian *Geology.* a North American geologic stage of the Pleistocene epoch, occurring after the Wheelerian and before the Holocene.

Hall mobility *Solid-State Physics.* a quantity given by the product of the Hall coefficient and the conductivity of a conductor or semiconductor; used to quantify electron mobility of a conductor or semiconductor.

Hallopora *Paleontology.* a genus of "branching twig" bryozoans in the extinct order Trepostomata; Ordovician.

halloysite *Mineralogy.* $Al_2Si_2O_5(OH)_4$, a white, monoclinic, clay mineral of the kaolinite-serpentine group, polymorphous with dickite, kaolinite, and nacrite, and having a specific gravity of 2.0 to 2.2 and a hardness of 1 to 2.5 on the Mohs scale.

Hall plate *Electronics.* a three-dimensional configuration of any material in which the Hall effect is utilized.

Hall's theorem *Mathematics.* **1.** let G be a finite solvable group of order mn, where m and n are relatively prime. Then: (a) G has a subgroup of order m; (b) any two subgroups of order m are conjugate; (c) any subgroup of G of order k, where k divides m, is contained in a subgroup of order m. **2.** let G be a bipartite graph with bipartition v_1, v_2. Then G has a matching incident with every vertex of v_1 if and only if the v_1-deficiency def(G) satisfies def(G) ≤ 0.

hallucinate [hə loo´si nāt´] *Psychology.* to experience hallucinations.

hallucination [hə loo´si nā´shən] *Psychology.* the apparent perception of an object or action that is not actually present outside the mind.

hallucination of conception *Psychology.* the apparent perception of sounds or voices that are internal.

hallucination of perception *Psychology.* the apparent perception of sounds or voices that are external.

hallucinogen [hə loo´si nə jin] *Pharmacology.* any agent that acts on the central nervous system to produce hallucinations and other disturbances of the thought processes.

hallucinogenic [hə loo´si nə jen´ik] *Pharmacology.* producing hallucinations.

hallucinosis [hə loo´si nō´sis] *Psychology.* a condition in which a person hallucinates while fully alert, usually arising from alcohol or drug abuse, or following a physical trauma, such as surgery or childbirth.

hallux *Anatomy.* the first digit on the foot; the big toe. *Vertebrate Zoology.* the comparable digit on the foot of an animal or the claw of a bird.

hallux valgus *Medicine.* a deformity in which the tip of the big toe or its main axis is directed toward the other toes of the same foot.

Hall voltage *Electronics.* the voltage generated in a Hall plate due to the Hall effect.

Hallwachs effect *Physics.* a photoelectric effect in which a negatively charged conductor emits electrons when exposed to ultraviolet radiation.

halmeic *Geology.* describing a mineral or deep-sea sediment derived directly from sea water or formed around an organic nucleus.

halmophagous *Zoology.* feeding mainly on grass and grain stems.

halmyrolosis *Geochemistry.* in regions of little or no sedimentation, the chemical reactions occurring between sea water and sediment which has been deposited on the sea floor. Also, SUBMARINE WEATHERING.

halo *Optics.* **1.** the faintly hued ring encircling a light source, such as a star or lamp, that arises from the scattering of light by water droplets or ice crystals. **2.** a similar ring that encircles the photographic image of a bright light source. *Meteorology.* an arc or ring of light that appears around the sun or the moon, caused by the reflection or refraction of light by ice crystals suspended in the earth's atmosphere, and exhibiting whitish luminosity or prismatic coloration, with red nearest the inside and blue at the outside. *Astronomy.* **1.** see GALACTIC HALO. **2.** see COMA. *Electronics.* a glow in a cathode-ray tube that remains momentarily after the cathode-ray beam has passed. *Mineralogy.* **1.** a ring- or crescent-shaped distribution pattern about the source of a mineral or petrographic feature, resulting from the diffusion of the mineral into the surrounding ground or rock. **2.** a ring-shaped discoloration observed in a thin section of a mineral.

halo- a combining form meaning "salt," as in *halophile*.

haloalkane *Organic Chemistry.* a halogenated aliphatic hydrocarbon.

Halobacteriaceae *Bacteriology.* a family of Gram-negative, rod-shaped or spheroid, pigmented, aerobic bacteria, occurring in salt water and salted fish and requiring high concentrations of sodium chloride for growth.

Halobacterium *Microbiology.* a genus of Gram-negative, rod-shaped, red- or purple-pigmented, aerobic bacteria of the family Halobacteriaceae; one of the few bacteria without a rigid cell wall.

halo blight *Plant Pathology.* a fatal disease of legumes caused by the bacteria *Pseuomonas phaseolicola* and characterized by the appearance on plant parts of water-soaked spots encircled by yellow rings; the plant eventually withers and dies without rotting. Also, HALO SPOT.

halocarbon *Organic Chemistry.* a compound containing carbon, a halogen, and sometimes hydrogen; used as a refrigerant and propellant gas; when polymerized with hydrocarbons, halocarbons yield plastics having extreme chemical resistance and high electrical resistivity.

halocarbon plastic *Organic Chemistry.* a plastic made from halocarbon resins, characterized by extreme chemical resistance and high electrical resistivity.

halocarbon resin *Organic Chemistry.* a resin produced by the polymerization of monomers made of halogenated hydrocarbons; an example is tetrafluoroethylene.

halocline *Oceanography.* a well-defined and usually positive vertical gradient of salinity.

Halococcus *Bacteriology.* a genus of Gram-negative, sphere-shaped, nonmotile, red- or orange-pigmented bacteria of the family Halobacteriaceae, occurring in salted fish and meat and requiring a high concentration of sodium chloride for growth.

Halocypridacea *Invertebrate Zoology.* a superfamily of ostracod crustaceans, in the suborder Myodocopa, having a thin carapace.

halo effect *Psychology.* the tendency to be influenced by a single positive or outstanding personality trait in forming an overall favorable evaluation of a person. *Industrial Engineering.* the tendency to evaluate a worker on the basis of a positive performance in one particular area.

halo error *Industrial Engineering.* in worker performance rating, a rating error due to the background, attitudes, or biases of the rater.

haloform *Organic Chemistry.* CHX_3, the general name given to a trihalogenated compound formed by the reaction between acetaldehyde or methyl ketones with NaOX (X is a halogen), e.g., iodoform (HCI_3).

haloform reaction *Organic Chemistry.* the trihalogenation of acetaldehyde or methyl ketones.

halogen *Chemistry.* any of the electronegative nonmetallic elements of Group VII, which includes fluorine, chlorine, bromine, iodine, and astatine. Halogens combine directly with most metals to form salts.

halogenated hydrocarbon *Organic Chemistry.* a halogen derivative of organic hydrogen- and carbon-containing compounds.

halogenation *Organic Chemistry.* a reaction in which a halogen is introduced into a substance.

halogen mineral *Mineralogy.* any of the naturally occurring compounds that contain a halogen as the sole or principal anionic constituent.

halohydrin *Organic Chemistry.* a compound with the general formula of X–R–OH, in which X is a halide such as Cl^-.

halokinesis see SALT TECTONICS.

Halomonas *Bacteriology.* a genus of Gram-negative, aerobic bacteria that are found in saline-rich habitats as flagellated, rod-shaped cells or filaments which may contain a yellow pigment.

halomorphic soil *Geochemistry.* a locally developed soil containing neutral and/or alkali salts, usually found in relatively arid regions where the water table is near the surface.

halonate *Mycology.* describing a fungal spore that has a colored circle surrounding it like a halo.

halo of Hevelius or **halo of 90 degrees** see HEVELIAN HALO.

halo ore see FRINGE ORE.

Halopappaceae *Botany.* a family of flagellate marine algae of the order Coccosphaerales, characterized by cells enclosed within a cane-olithic coccosphere and by a group of flattened, radiating coccoliths that form spines on the coccosphere.

halophilous *Biology.* living or thriving in a saline environment. Thus, **halophile, halophily.**

halophobic *Biology.* not tolerant of a saline environment. Thus, **halophobe.**

halophyte *Botany.* any plant that lives or thrives in a saline environment, such as glassworts.

Haloragaceae *Botany.* a family of dicotyledonous aquatic, amphibious, or marsh-growing herbs in the order Haloragales, found worldwide but most common in the Southern Hemisphere and having some species often cultivated as an aquarium plant. Also, **Haloragidaceae.**

Haloragales *Botany.* an order of dicotyledonous herbs in the subclass Rosidae that are often amphibious or aquatic and have perfect or unisexual flowers generally pollinated by the wind. Also, **Haloragidales.**

halorhodopsin *Biochemistry.* a retinylidene protein that acts as the light-driven chloride ion pump of halobacteria.

Halosauridae *Vertebrate Zoology.* a small family of deep-sea fishes in the order Notacanthiformes; species are related to eels and live near the sea floor of all oceans.

halosere *Ecology.* an ecological succession that begins in a saline environment.

Halosphaeraceae *Botany.* a monogeneric family of marine green algae of the order Pyranimonadales, characterized by cells in cyst or motile phases, the cysts never reproducing themselves but always producing flagellate cells.

Halosphaeriaceae *Mycology.* a family of fungi belonging to the order Diaporthales which is composed of marine fungi that occur on wood.

halosphere *Ecology.* the marine portion of the environment; saltwater areas as a whole.

halo spot see HALO BLIGHT.

halothane *Pharmacology.* $C_2HBrClF_3$, a potent inhaled anesthetic widely used to produce and maintain general anesthesia.

halotolerent *Biology.* tolerating, though not necessarily preferring, a saline environment. Also, **haloxene.**

halotrichite *Mineralogy.* $Fe^{+2}Al_2(SO_4)_4 \cdot 22H_2O$, a colorless to yellowish water-soluble monoclinic mineral occurring in aggregates of fibrous silky crystals, and having a specific gravity of 1.895 and a hardness of 1.5 on the Mohs scale; found in pyritic ore deposits and coal veins. Also, IRON ALUM, BUTTER ROCK.

Halsted's operation *Surgery.* **1.** an operation for inguinal hernia, involving transposition of the spermatic cord above the external oblique aponeurosis. **2.** see RADICAL MASTECTOMY. (Named for William Stewart *Halsted,* 1852–1922, American surgeon.)

halt *Computer Programming.* the cessation of a sequence of machine instructions due to the execution of a **halt instruction** (program stop), an interrupt, a hang-up, or a breakpoint.

haltere *Invertebrate Zoology.* either of a pair of knobs representing the degenerate hindwings in dipteran flies, shaped like short drumsticks and acting as gyroscopic stabilizers when they vibrate during flight. Also, BALANCERS.

Halteria *Invertebrate Zoology.* a genus of suspension-feeding ciliate protozoans of the subclass Spirotricha.

halyard *Naval Architecture.* a line, rope, or tackle used for hoisting a sail, flag, yard, or gaff; it does not apply to the ropes used to hoist the heaviest sails on a square-rigged vessel, which are termed jeers.

Halysites *Paleontology.* a genus of tabulate Paleozoic corals in the extinct family Halysitidae; their tubes are circular or elliptical, and they are called "chain coral" because of their appearance in cross section.

halysitid *Paleontology.* any coral of the family Halysitidae.

Halysitidae *Paleontology.* a family of corals in the extinct order Tabulata, widespread in the Ordovician and Silurian.

ham *Food Technology.* a cut of meat from a hog's hind quarter, between the hip and hock; usually cured. *Telecommunications.* an informal term for an amateur radio operator.

HAM *Aviation.* the airport code for Hamburg, Germany.

hamada *Geology.* a nearly level, upland desert surface of bare bedrock or of bedrock covered by a thin layer of sand or pebbles that has been swept clean by wind action. *Geography.* the flat, rocky desert of the Sahara. Also, HAMMADA.

Hamal *Astronomy.* Alpha (α) Arietis, an orange-colored giant star that has an apparent brightness of 2.0 magnitude and a spectral type of K2; it lies 72 light-years away.

Hamamelidaceae *Botany.* the family of witch hazels, dicotyledonous shrubs or trees in the order Hamamelidales, found worldwide but mostly in subtropical regions, having bisexual or unisexual flowers and stellate hairs.

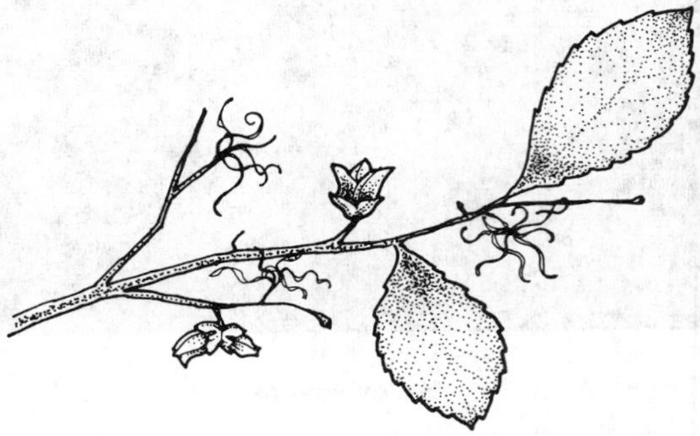

Hamamelidaceae (witch hazel)

Hamamelidae *Botany.* a subclass of ancient dicotyledonous plants in the class Magnoliopsida, having mostly small, inconspicuous flowers generally pollinated by the wind and having a poorly developed perianth or none at all.

Hamamelidales *Botany.* a small order of woody plants in the subclass Hamamelidae characterized by vessels in the wood and reduced flowers.

hamart- or **hamarto-** a combining form meaning: **1.** a defect, as in *hamartophobia.* **2.** a hamartoma, as in *hamartoblastoma.*

hamartoma *Oncology.* a benign tumorlike growth containing tissue which is normally present in that locality, but in an abnormal mixture of tissue elements.

hamartophobia *Psychology.* an irrational fear of committing a sin or making an error.

hamate *Biology.* hooked at the tip.

hamathecium *Mycology.* any of various types of fungal tissue located between the asci or spore sacs.

hambergite *Mineralogy.* $Be_2BO_3(OH)$, a grayish-white or colorless, brittle, orthorhombic mineral occurring as prismatic crystals, having a specific gravity of 2.35 to 2.37 and a hardness of 7.5 on the Mohs scale; found in syenite pegmatites.

Hamel basis *Mathematics.* a maximal linearly independent subset of a (usually infinite-dimensional) vector space X; i.e., a linearly independent set that is not properly contained in any other linearly independent set. Such a basis always exists by Zorn's lemma.

Hamilton, Alice 1869–1970, American physician; pioneer in industrial medicine and occupational health and safety.

Hamilton, Sir William Rowan 1805–1865, Irish mathematician; laid bases for quantum mechanics; formulated theory of quaternions.

Hamiltonian cycle *Mathematics.* in a graph, a cycle that includes all of the vertices of the graph. A graph that has a Hamiltonian cycle is called a **Hamiltonian graph.**

Hamiltonian dynamics *Mechanics.* the description and calculation of motion in terms of the Hamilton equations. (From Sir William Rowan *Hamilton,* 1805–1865, Scottish mathematician and astronomer.)

Hamiltonian function *Mechanics.* a function of generalized coordinates and their conjugate momenta, whose value for a conservative holonomic mechanical system is given by subtracting the Lagrangian function of the system from the summation of the products of generalized velocity and generalized momentum for all its degrees of freedom; it is equal to the total energy of the system.

Hamiltonian operator *Quantum Mechanics.* the energy operator, in quantum mechanics denoted by $\{H\}$, that gives the energy eigenstates and that is the counterpart of the Hamiltonian function in classical mechanics for the total energy of a dynamical system.

Hamiltonian path *Mathematics.* a path in a graph that includes every vertex.

Hamilton-Jacobi equation *Mathematics.* a first-order partial differential equation, not quite the most general possible, whose integrability characterizes the solubility of various related problems. Usually written in the form $\partial S/\partial t + H(\partial S^3/\partial x_i\, \partial x_i\, \partial t) = 0$, where S is to be determined and x_i represents the spatial variables. H is called the Hamiltonian function and the distinguished coordinate t often represents time.
Hamilton-Jacobi theory *Mathematics.* a systematic theory of integration of canonical differential equations. *Mechanics.* a method of analyzing mechanical systems by recasting the equations of motion into a single Hamilton-Jacobi equation.
Hamilton's equations see CANONICAL EQUATIONS.
Hamilton's equations of motion *Mechanics.* a set of partial differential equations expressing the motion of a conservative dynamical system in terms of its Hamiltonian function. Also, CANONICAL EQUATIONS OF MOTION.
Hamilton's principal function see PRINCIPAL FUNCTION.
Hamilton's principle *Mechanics.* the variational principle that of all the possible paths that a mechanical system can follow as it moves from one point to another within a given time, the actual path will be the one that will minimize (or in some cases maximize) the time integral of the Lagrangian function for the system. Also, **Hamilton's variation(al) principle.**
Hamites *Paleontology.* a genus of heteromorphic ammonoids, notable for a shell characterized by V-shaped loops; Cretaceous.
hammada see HAMADA.
hammarite *Mineralogy.* $Pb_2Cu_2Bi_4S_9$, a reddish steel-gray, opaque, metallic, orthorhombic mineral, having a specific gravity of 6.73 and a hardness of 3 to 4 on the Mohs scale; found as crystals on drusy quartz in tungsten deposits.
hammer *Mechanical Devices.* a hand tool, generally consisting of a solid metal head set transversely on a wooden handle; used for pounding nails, beating metals, and similar impact-related tasks. *Mechanical Engineering.* any of various tools or machine parts that function in a manner similar to that of a hammer. *Ordnance.* in the firing mechanism of a gun, a metal lever that pivots on its axis in order to strike the firing pin or percussion cap, thus firing the gun.
hammer drill *Mechanical Engineering.* any of various compressed-air rock drills in which the hammer strikes a free-moving piston.
hammer gun *Ordnance.* a gun with an exterior, visible hammer.
hammerhead *Mechanical Devices.* **1.** the part of a hammer that strikes the work. **2.** a part in a piano mechanism that strikes the strings. *Vertebrate Zoology.* Any of several species of sharks of the family Sphyrnidae that inhabit temperate and tropical seas; named for flattened protuberances on either side of the head, on which the eyes and nostrils are positioned. Also, **hammerhead shark.**

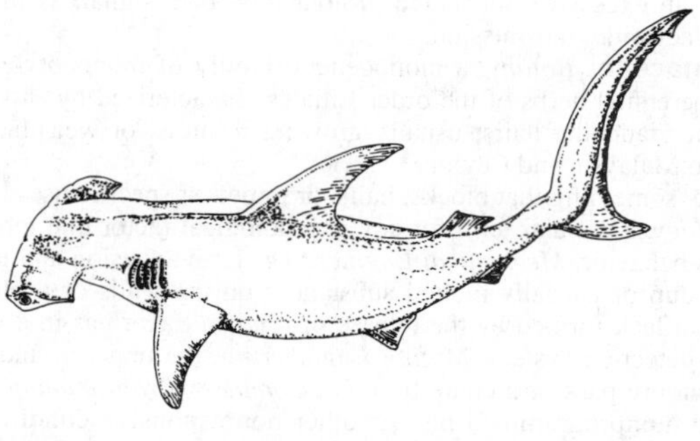

hammerhead shark

hammerhead crane *Mechanical Engineering.* a crane with a counterbalance to offset a projecting arm or jib.
hammerless gun *Ordnance.* a gun with an enclosed hammer and firing mechanism.
hammer mill *Mechanical Engineering.* any of various machines that use a series of revolving hammers to crush ore, coal, and similar materials.
hammerstone *Archaeology.* a hard, fist-sized stone that has the function of a hammer, as to flake other stone for tools, process foods, break up shells or bones, and so on.
hammer test *Metallurgy.* any of several standardized tests used to assess the impact resistance of metals and alloys.
hammertoe *Medicine.* a deformity of the toe, usually the second, in which there is permanent angular flexion.
hammer tongs *Mechanical Devices.* tongs used in hammer construction which bend at right angles to hold the workpiece.
hammer welding *Metallurgy.* forge welding, using a forging hammer.
hamming code *Telecommunications.* a data transmission code that permits error detection and correction; the extent of detection and correction capabilities depends on the degree of redundancy employed.
hamming distance *Computer Programming.* in two binary words of the same length, the number of bit positions in which the corresponding bits are different. Also, SIGNAL DISTANCE.
hammock see HUMMOCK.
Hammond, John Hays, Jr. 1888–1965, American inventor; a pioneer in remote control; devloped various telegraphic and telephonic devices.
hamper *Naval Architecture.* **1.** rigging or superstructure above the upper-deck level. Also, TOPHAMPER. **2.** a term for any necessary but cumbersome gear aboard ship.
Hamprace see MONTANA.
Hampshire *Agriculture.* **1.** a breed of swine having a long black body with a white belt. **2.** a breed of hornless dark-faced sheep raised for mutton and wool. Also, **Hampshire Down.** (From *Hampshire*, a county in England.)
Hampson unit *Radiology.* a unit of absorbed X-radiation dose, equivalent to one-quarter of the dose that will produce erythema.
ham radio *Telecommunications.* a popular term for amateur radio. See AMATEUR RADIO.
hamster *Vertebrate Zoology.* any of a family of small to medium-sized cricetid burrowing rodents having cheek pouches for food storage; native from central Europe to eastern Asia in steppes, grassland, desert borders, and other habitats; also popular as pets and used in laboratory research.

hamster

hamstring *Anatomy.* in humans, one of the three tendons that connect the hamstring muscles of the thigh to the knee. *Vertebrate Zoology.* in four-legged animals, the large tendon at the back of the hock. Also, HAMSTRING TENDON.
hamstring muscles *Anatomy.* the large muscles on the posterior surface of the thigh, including the biceps femoris, semitendinosus, and semimembranosus.
hamstring tendon see HAMSTRING.
hamulus *plural,* **hamuli.** *Vertebrate Zoology.* on a bird feather, a small filament on the underside of an anterior barbule, which assists in holding overlapping barbules together. Also, BARBICEL. *Entomology.* a hook-like structure found in insects in the order Collembola, which together with the furcula enables them to leap forward. (From a Latin word meaning "little hook.")
hancockite *Mineralogy.* $(Pb,Ca,Sr)_2(Al,Fe^{+3})_3(SiO_4)_3(OH)$, a brownish-red, translucent, brittle monoclinic mineral of the epidote group occurring in small lath-shaped crystals, having a specific gravity of 4.03 and a hardness of 6 to 7 on the Mohs scale; found with garnet, biotite, and axinite at Franklin, New Jersey.
hand *Anatomy.* the distal segment of the upper limb, including the wrist, palm, and fingers. *Robotics.* an end effector that can grip, clamp, or grasp objects. *Metrology.* a traditional measure for the height of horses based on the width of a human palm; now accepted as 4 inches.
hand-and-foot monitor *Nucleonics.* an instrument that measures the radioactivity on the hands and feet of individuals as they leave areas that might contain radioactive materials.
hand auger *Mechanical Devices.* a hand tool resembling a carpenter's bit, which bores holes in near-surface soil to retrieve samples.
hand axe *Archaeology.* a stone tool that fits in the hand and is used for chopping or hacking, usually flaked on both sides and pointed or pear-shaped; characteristic of the Lower Paleolithic. Also, **hand ax, hand-axe.**

handbarrow *Engineering.* **1.** a rectangular frame fixed with handles at both ends used by two people to transport objects; often used in the military. Also, BARROW. **2.** a wheelbarrow.
hand brake see EMERGENCY BRAKE.
hand drill *Mechanical Devices.* any small drilling device that is designed to be held and operated by hand.
hand element see MANUAL ELEMENT.
handguard *Ordnance.* a heat-insulating wood or plastic piece that fits above or around the barrel of a firearm in order to allow the shooter to hold it during firing.
handgun *Ordnance.* a firearm that is designed to be held in one hand for firing; e.g., a pistol or revolver.
handicap principle *Ecology.* the idea that elaborate, sexually selected displays and adornments act as handicaps that demonstrate the generally high fitness of the bearer.
handle *Mechanical Devices.* the part of a tool or other object that is designed to be grasped or held. *Computer Science.* in bottom-up parsing, the substring that should next be reduced as a phrase. *Mathematics.* a handle of type (q, p) is a manifold that is homeomorphic to the product $D^q \times D^p$, where D^k is the closed unit ball in a Hilbert (or Euclidean) space of dimension k. For example, a tube is a handle of type (2, 1).
hand level *Cartography.* a small leveling instrument, designed to be held in the hand, in which the spirit level is so mounted that the observer can see the bubble during the observation of an object through the telescope. Also, ABNEY LEVEL, LOCKE LEVEL.
handling *Industrial Engineering.* **1.** the process of manually or mechanically moving a manufactured object through any stage of production or distribution. **2.** the means by which this is done for any particular object.
hand loom *Mechanical Devices.* a manually driven weaving machine operated by levers and treadles.
handmade *Engineering.* made by hand, not by machine.
hand miller *Mechanical Devices.* a manually fed milling machine.
hand nut see WING NUT.
handoff or **hand-off** *Aviation.* **1.** the transfer of control or surveillance of an aircraft from one control center to another. **2.** the means by which or period in which this is accomplished.
handover line *Military Science.* a control feature, preferably following easily defined terrain features, at which responsibility for the conduct of combat operations is passed from one force to another.
handpicking *Industrial Engineering.* the process of manually selecting or removing certain items from a group. Also, **handsorting.**
hand pump *Mechanical Devices.* a small suction pump with a handle and foot valve for single-person operation.
handrail *Building Engineering.* a continuous, conveniently positioned bar that is grasped by the hand and used by people for support on stairways, balcony edges, or gallery walkways.
hand-rammed *Ordnance.* describing a cartridge that is pushed into the gun by hand.
hand-rearing *Behavior.* the raising of an animal of a nondomestic species by humans in a domestic setting. Also, **hand-raising.**
hand-reset *Electricity.* of or relating to a device, such as a relay, in which the contacts must be reset by hand.
hand rule *Electromagnetism.* any of several conventional rules for indicating certain directional relationships, particularly between magnetic and electric fields, currents, generators, and the like, by using the appropriate hand and fingers to specify the proper directions and circulations of such quantities.
handsaw *Mechanical Devices.* a sharp, serrated, manually operated, hand-held cutting instrument used on wood, stone, or metal.
hand screw *Mechanical Devices.* **1.** a clamp having two wooden jaws that are adjusted by turning two long screws. Also, **hand-screw clamp.** **2.** any screw that can be turned and tightened by hand.
handset *Telecommunications.* a hand-held telephone having the speaker and receiver set on one mount.
handset bit *Mechanical Devices.* a drill bit in which diamonds are manually placed in similarly shaped holes.
handshake *Computer Programming.* an electronic signal by which communication is acknowledged and synchronized between two components of a computer or robotic system. Thus, **handshaking.**
hand shank *Mechanical Devices.* a foundry ladle used for the controlled pouring of liquids, semiliquids, or solvents by an iron rod-support in its center.
hand template *Photogrammetry.* a template traced from the radials on a photograph onto a transparent plastic sheet, used to form the radial triangulation.
hand-template triangulation *Photogrammetry.* a graphic radial triangulation in which any form of hand template is used.
hand time *Industrial Engineering.* the portion of a work cycle during which manual elements are performed.
hand tool *Mechanical Devices.* a general term for any instrument that is used and operated by hand.
hand vise *Mechanical Devices.* a small vise used to hold objects in fixed jaws while the object is moved by turning a threaded rod.
hangar bolt *Mechanical Devices.* a headless bolt used in timber construction having a machine screw thread at one end and a tapered, lag-screw thread on the other.
hangfire *Ordnance.* **1.** a delay of ten milliseconds or less between the striking of a primer and the ignition of the cartridge in a firearm. **2.** any undesired delay in the functioning of a firing system.
hanging indentation *Graphic Arts.* a type pattern in which the first line of a paragraph is set flush left and all subsequent lines are indented, as is the style on this dictionary page. Also, **hanging indent.**
hanging numerals *Graphic Arts.* numerals typeset in Old Style.
hanging stile see PULLEY STILE.
hanging valley *Geology.* a tributary glacial valley whose mouth opens above the floor of the main valley.

hanging valley

hangover *Medicine.* an informal term for various unpleasant physical aftereffects, such as headache, nausea, fatigue, and excessive thirst, that result from the excessive use of alcohol or certain drugs. *Telecommunications.* an excessive, unwanted prolongation of a signal, as in television or facsimile transmission.
Hanguanaceae *Botany.* a monogenetic family of monocotyledonous robust perennial herbs of the order Liliales characterized by short, multicellular branching hairs; usually growing in moist or wet places and native to Malaysia and Ceylon.
hang-up something that blocks, halts, or impedes; specific uses include: *Psychology.* a popular term for any psychological factor that inhibits or restricts behavior. *Mechanical Engineering.* **1.** an adhesion and interfering buildup of partially melted substances on the walls of a blast furnace. **2.** a leak formed by the release of trapped tracer gas in a vacuum or leak-detection system. *Mining Engineering.* the underground blockage of an ore pass or a chute by rock. *Computer Programming.* an unplanned, nonprogrammed halt or other nonresponsive condition in a computer program due to a program error or a machine malfunction.
hangwire *Ordnance.* a wire that connects the fuse assembly of an aerial flare or bomb to the structure of the aircraft; it is designed to remove the safety and arm the fuse after the flare or bomb has dropped the length of the wire; it may also open a parachute or stabilizing sleeve.
Hankel, Hermann 1839–1873, German mathematician; formulated general theory of function; classified linear functions.
Hankel functions *Mathematics.* Bessel functions of the third kind.
Hankel's integral *Mathematics.* a representation of the reciprocal of the gamma function by a contour integral; given by $1/\Gamma(z) = \int_C e^t t^{-z}\, dt$, where the contour of integration C follows the lower edge of the cut from $-\infty$ to $-\delta$ ($\delta > 0$), goes around the origin in the positive (counterclockwise) direction on the circle $|t| = \delta$, and then follows the upper edge of the cut from $-\delta$ to $-\infty$. Also, **Hankel's formula.**

Hankel transform *Mathematics.* the Hankel transform of order *m* of a real function $g(r)$ is defined by $G(\rho) = \int_0^\infty g(r)J_m(\rho r)r\,dr$, where $J_m(r)$ is a Bessel function of order *m*. The inversion formula is given by $g(r) = \int_0^\infty G(\rho)J_m(\rho r)\rho\,d\rho$. Also, BESSEL TRANSFORM, BESSEL-FOURIER TRANSFORM.

hanksite *Mineralogy.* $KNa_{22}(SO_4)_9(CO_3)_2Cl$, a colorless to yellow, or gray hexagonal mineral occurring as tabular to prismatic crystals, having a specific gravity of 2.56 and a hardness of 3 to 3.5 on the Mohs scale; found with trona, halite, and borax in saline beds.

hannayite *Mineralogy.* $(NH_4)_2MG_3H_4(PO_4)_4 \cdot 8H_2O$, a yellowish, transparent, triclinic mineral occurring as slender crystals in bat guano, having a specific gravity of 2.03 and an undetermined hardness.

Hansa yellow *Organic Chemistry.* the trade name for various organic, insoluble azo pigments with good weather resistance and poor resistance to bleeding; used in nontoxic paints such as toy enamels and in emulsion paints.

Hansen, Gerhard Armauer 1841–1912, Norwegian bacteriologist; discovered Hansen's bacillus, the agent of leprosy (Hansen's disease).

Hanseniaspora *Mycology.* a genus of fungi belonging to the family Saccharomycetaceae and whose sexual or teleomorphic counterparts belong to the genus Kloeckera.

Hansen's bacillus *Bacteriology.* another name for *Mycobacterium leprae,* the rod-shaped, Gram-positive bacillus that is the causal agent of leprosy.

Hansen's disease *Medicine.* another name for leprosy. See LEPROSY.

Hansenula *Mycology.* a genus of yeast fungi belonging to the family Saccharomycetaceae, its species *H. anomala* is nonpathogenic and occurs as normal flora in the throat and digestive tract of humans.

Hantaan virus *Virology.* a virus that causes hemorrhagic fevers, believed to be transmitted through contact with infected rodents; proposed as a genus, *Hantavirus,* of the family Bunyaviridae.

H antigen *Bacteriology.* any bacterial antigen with a long filament, termed a flagellum, that it uses for locomotion; important in serologic classification of certain bacteria such as Salmonella. Also, FLAGELLAR ANTIGEN.

Hantzsch synthesis *Organic Chemistry.* a reaction in which a pyrrole compound is formed through the condensation of a β-keto ester, α-chloromethylketone, and a primary amine or ammonia.

Hanus solution *Analytical Chemistry.* a solution of iodine monobromide and glacial acetic acid, used to determine the iodine number in oils and fats.

haphephobia see HAPTEPHOBIA.

hapl- or **haplo-** a combining form meaning "single" or "simple," as in *haploid.*

haplobiont *Botany.* a plant during whose life either the sporophyte or the gametophyte generation is absent. Thus, **haplobiontic.**

haplocaulescent *Botany.* having a simple axis with reproductive organs located on the principal axis.

haploid *Genetics.* of a cell, nucleus, or organism, having a single set of unpaired chromosomes. Also, **haploidic.**

haploidiploidy *Genetics.* a genetic system in honeybees and other insects in which the males are haploid, being formed from unfertilized eggs, while the females develop from fertilized eggs and are diploid.

haploidization *Genetics.* the production of a haploid cell from a diploid by progressive chromosome loss due to nondisjunction.

haploid number *Genetics.* **1.** the number of chromosomes in a gametic cell; that is, the number of single, nonhomologous chromosomes in one set carried in each gamete. **2.** the number of chromosomes in a somatic cell of a haploid organism, such as a fungus.

Haplolepidae *Paleontology.* a family of small, primitive actinopterygian fishes in the extinct order Haplolepiformes; Upper Carboniferous.

Haplomi *Vertebrate Zoology.* an alternate name for the fish suborder Esocoidei of the order Salmoniformes, which includes the freshwater pikes and mudminnows.

haplomitosis *Cell Biology.* a rudimentary type of cell division that occurs in certain flagellates, in which nuclear chromatin granules separate into two groups.

Haplomitriaceae *Botany.* a monogeneric family of liverworts in the order Calobryales, characterized by irregular branches organized into a system of basal, prostrate, and often branched stolons; distribution is widespread.

haplont *Botany.* a plant that is haploid throughout its life, except as a zygote, as are many algae and fungi. Thus, **haplontic.**

haplophase *Biology.* the haploid stage of a life cycle.

haplopore *Paleontology.* a type of pore on cystoid echinoderms.

Haplorhini *Paleontology.* a proposed suborder of primates that would include the tarsioids and anthropoids; the remaining primates would be grouped in the suborder Strepsirhini; these two would replace the suborders Prosimii and Anthropoidea.

Haplosclerida *Invertebrate Zoology.* an order of sponges, in the class Demospongiae, with a reticulate skeleton of both siliceous spicules and spongin.

haplosis *Cell Biology.* the reduction of the chromosome number by half during meiosis.

Haplosporea *Invertebrate Zoology.* a class of sporozoan protozoans.

Haplosporida *Invertebrate Zoology.* an order of protozoans in the class Haplosporea, parasites of invertebrates and fishes.

Haplosporidium *Invertebrate Zoology.* a genus of protozoans in the order Haplosporidia; parasitic in the liver, connective tissue, or gut of various invertebrates and lower vertebrates.

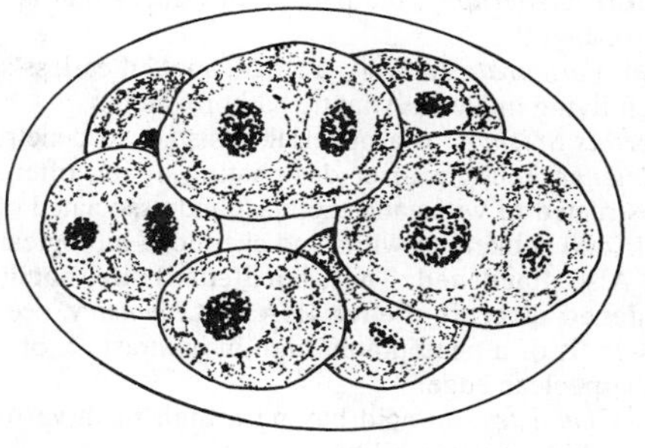

Haplosporidium

haplostele *Botany.* a type of protostele in which the core of xylem has a smooth circular outline.

Haplotaxida *Invertebrate Zoology.* in some classifications: an order of annelid worms including all families in the class Oligochaeta except Lumbriculidae and Moniligastridae.

haplotype *Genetics.* **1.** a combination of alleles of closely linked loci that are found in a single chromosome and tend to be inherited together. **2.** a combination of isoantigens that are produced by a single allele or by several closely linked genes.

Haplozoaceae *Botany.* a monotypic family of heterotrophic parasitic algae belonging to the order Blastodiniales; they commonly attach to the intestines of species of Polychaeta.

hapt- or **hapto-** a combining form meaning: **1.** touch, as in *haptephobia.* **2.** binding, as in *haptoglobin.*

hapten *Immunology.* a small molecule, having at least one of the determinant groups of an antigen, that can combine with an antibody but is not immunogenic unless it acts in conjunction with a carrier molecule.

haptephobia *Psychology.* an irrational fear of physical contact or of being touched by others.

hapteron *Botany.* a disklike holdfast; an organ that attaches the stem of various aquatic plants or marine algae to the substrate.

haptochlamydeous *Botany.* describing a plant that has the sporophylls protected by rudimentary perianth leaves.

haptocyst *Invertebrate Zoology.* any of numerous minute organelles on each tentacle of suctorians that are discharged into prey as anchors while specialized tentacles suck in contents.

haptoglobin *Hematology.* a protein that is one of the fractions of blood plasma; an acid glycoprotein that binds to free plasma hemoglobin to form a complex that cannot be filtrated by the kidneys. It is thought to be increased in infections, malignancy, and certain endocrine disorders.

haptonema *Botany.* an appendage located between the flagella of the motile cells of certain algae, used for temporarily attaching to some surface.

haptophonia *Psychology.* a delusion in which an individual hears voices or other noises emanating from a part of his body, usually as the result of a physical sensation.

haptophyta see COCCOLITHOPHORIDA.

haptor *Invertebrate Zoology.* a posterior organ of attachment in parasitic worms and flukes that has hooks or suckers or both.

haptotropism *Biology.* an orientation response to a touch or contact stimulus.

HAR *Aviation.* the airport code for Harrisburg, Pennsylvania.

harassing agent *Ordnance.* a chemical agent designed to irritate and lower the efficiency of enemy troops, usually by forcing them to wear gas masks.

harassing fire *Military Science.* fire designed to disturb enemy troops, prevent movement, and lower morale by threat of losses.

harassment *Military Science.* an action that is intended to disrupt the activities of an enemy unit, installation, or ship, rather than inflict serious casualties or damage.

harbor *Geography.* a natural or artificial shelter for ships.

harbor chart *Navigation.* a chart that is used for navigation in harbors and smaller waterways, generally having a scale larger than 1:50,000.

harbor engineering *Civil Engineering.* the application of planning and design to loading facilities for passengers and cargo of ships.

harbor line *Civil Engineering.* a boundary beyond which wharves and other harbor structures cannot extend.

harbor oscillation *Oceanography.* any nontidal vertical movement in a harbor or bay. Also, **harbor surging.**

harbor reach *Geography.* the part of an estuary that is visible from a harbor. Also, REACH.

harbor seal *Vertebrate Zoology.* a small, spotted earless seal, *Phoca vitulina,* often living in North Atlantic harbor waters.

hard *Materials Science.* of a material, resistant to penetration or wear. *Military Science.* relating to or designating a site, often underground, that is constructed to withstand the blast and associated effects of a nuclear attack, and is likely to withstand chemical, biological, or radiological attack. Also, **hardened.** Thus, **hard(ened) site, hard(ened) missile base, hardened aircraft shelter (HAS).** *Chemistry.* see HARD WATER. *Graphic Arts.* **1.** of a photograph, high in contrast. **2.** of a halftone dot, having a sharp, clean edge.

hard acid *Chemistry.* an acid having a high positive oxidation state, small size, and low polarizability.

hard anodizing *Materials Science.* a metal-coating process by anodic oxidation that requires a high current density and low-temperature electrolytes; the coating is resistant to corrosion, abrasion, and erosion.

hard automation *Robotics.* hardware that is specially designed for automatic production.

hard base *Chemistry.* an acid having low polarizability and high electronegativity.

hard beach *Civil Engineering.* a hard-surfaced portion of a beach extending into the water, used for loading and unloading landing vessels.

hardboard *Materials.* a type of fiberboard made of compressed wood chips, which is stiff and water resistant.

hardbound see HARDCOVER.

hard bronze *Metallurgy.* a copper-base alloy containing 7% tin, 3% zinc, and 2% lead.

hard carcinoma see SCIRRHOUS CARCINOMA.

hard chromium *Metallurgy.* a coating of chromium that has been placed on a material to improve its abrasion resistance.

hard coal *Mineralogy.* another term for anthracite, which is the hardest form of coal. See ANTHRACITE.

hard code *Computer Programming.* computer program statements that contain constants, absolute addresses, and other precise information that reduce the program's flexibility, limit its scope, and make it difficult to modify.

hard copy *Computer Programming.* **1.** any form of text or graphic material produced by a computer and printed on paper; a printout. **2.** the printed version of a document that also exists in computer form. (Called *hard* to contrast it with a *soft* or *software* copy; i.e., a version of the same material appearing on the computer screen or stored in memory.)

hard cosmic ray *Nucleonics.* a component of cosmic radiation that is capable of penetrating a moderate portion of a material's density.

hardcover *Graphic Arts.* of a book, encased in a binding of cloth-covered cardboard or other rigid material. Also, HARDBOUND, CASEBOUND.

hard crash *Computer Technology.* an abrupt breakdown in a computer system due to failure in a hardware or software component, allowing the users no chance for a controlled termination of activities.

hard detergent *Materials.* a term for a detergent that tends to resist biodegradation or decomposition.

hard disk *Computer Technology.* a rigid magnetic disk that provides high-speed, high-capacity, direct-access auxiliary storage.

hard-drawn wire *Metallurgy.* wire that has been drawn, but not annealed; thus, it is much harder than the starting stock.

hard drive *Computer Technology.* a disk drive that reads and writes data on hard disks; it is internal in some systems but may also be connected to an external port. Also, **hard-disk drive.**

hard edit *Computer Programming.* a process of data checking and correction that enables a computer system to reject any data that contains errors.

Harden, Sir Arthur 1865–1940, English chemist; shared the Nobel Prize for his work on fermentation of sugars.

hardenability *Metallurgy.* the measure of the propensity of a steel to harden upon quenching. It is usually expressed in terms of the depth of hardening.

hardened circuit *Electronics.* a circuit utilizing components having a high tolerance to nuclear radiation.

hardened links *Telecommunications.* transmission links that require special construction or installation to provide reliable communication in the event of a nuclear attack.

hardened steel *Metallurgy.* a steel that has been hardened by appropriate thermal treatment and quenching.

hardener *Materials Science.* any compound that reacts with a resin polymer to cure or harden it into plastic.

hardener bath *Graphic Arts.* **1.** in film developing, a chemical bath used to make the emulsion more resistant to scratches and softening. **2.** in printing, a chemical solution used to soak halftone plates in order to harden the halftone dots.

hardening the fact of becoming hard; specific uses include: *Metallurgy.* the process of increasing the hardness of a metallic part by a thermal treatment. *Histology.* the process of rendering tissue firm so that it may be more readily cut for purposes of microscopic examination. *Botany.* of plants, an increasing ability to tolerate drought or extreme temperatures brought on naturally or in horticultural practice.

hardening of the arteries see ARTERIOSCLEROSIS.

Harderian gland *Vertebrate Zoology.* an ocular gland of the anterior portion of the conjunctiva, which maintains moisture with an oily secretion in the eyes of most amphibians, reptiles, and birds in cases where a lacrimal gland is either poorly developed or absent.

hard error see PERMANENT ERROR.

hard-face *Metallurgy.* to deposit an abrasion-resistant thermal coating.

hard failure *Computer Technology.* the failure of a hardware unit.

hard fiber *Botany.* a woody leaf fiber used in making cording, twine, and some textiles.

hard-fiber elastomer *Materials Science.* an elastomer that has been vulcanized with zinc chloride.

hard freeze *Hydrology.* a freeze in which the ground surface freezes solid, heavy ice forms on small surfaces, and vegetation is destroyed.

hard glass *Materials.* **1.** glass that contains one or more boron compounds and is therefore resistant to heat and chemical actions. **2.** any glass having a high softening point.

hard ground *Graphic Arts.* an etching ground that is applied by heating the plate over a flame and spreading the ground over its surface.

hard-hammer percussion *Archaeology.* the use of a hammerstone to chip or shape a second stone. Also, **hard-hammer technique.**

hardhat *Engineering.* a protective metal or plastic hat such as those worn by construction workers or miners.

hardhead *Vertebrate Zoology.* **1.** a California freshwater fish, *Mylopharodon conocephalus.* **2.** the Atlantic croaker, *Micropogonias undulatus. Metallurgy.* a semirefined tin that contains iron.

hard hyphen *Computer Science.* in word processing, a hyphen that is part of the normal spelling of a word, as opposed to one that is inserted in a normally unhyphenated word to divide it at the end of a line.

Hardinge feeder-weigher [här′ding] *Mechanical Engineering.* a short, pivoted, belt-type conveyor that regulates the flow rate from a hopper by the weight of material per cubic foot.

Hardinge mill *Mechanical Engineering.* a widely used grinding mill that consists of a relatively flat cone at the feed end, followed by a cylindrical drum with wedge-shaped liners in the midsection, and a steep cone at the discharge end.

Hardinge thickener *Mechanical Engineering.* a device that separates liquid from a liquid-solid mixture by letting the solids sink to the bottom and the liquid pour out over the top.

hard iron *Metallurgy.* a ferrous product that retains a substantial fraction of magnetization.

hard-laid *Design Engineering.* of or relating to a rope with its strands twisted at a relatively large angle to its axis, generally at 45°.

hard landing *Aviation.* an aircraft descent so steep that it results in damage or overstressing. *Space Technology.* the impact of a spacecraft on a surface without deceleration, usually destroying the craft.

hard lead *Metallurgy.* a lead-base alloy containing either 4% or 6% antimony. Also, CHILLED SHOT.

hard lens *Optics.* a contact lens that is made of rigid plastic or silicon.

hard limit *Computer Technology.* an operating limit set by the hardware, such as memory size. *Telecommunications.* a limiting condition with negligible changes in output in the range in which the amplitude, power, or any other characteristic of the output signal is limited, where signal input value varies widely. Thus, **hard limiting.**

hard liquor *Food Technology.* a term for distilled spirits such as gin, vodka, and whiskey, as opposed to fermented beverages such as wine and beer.

hard magnetic material *Metallurgy.* a material suitable for making permanent magnets, such as an alloy based on cobalt and samarium, on iron and neodymium, or on aluminum, nickel, and cobalt.

hardness the condition of being hard; specific uses include: *Chemistry.* **1.** the quality of firmness produced by the cohesion of the particles composing a substance. **2.** the quality of water produced by soluble salts of calcium, magnesium, or other substances, which form an insoluble curd with soap and thus interfere with its cleansing power. *Materials Science.* **1.** the quality or degree of being resistant to penetration or wear. **2.** the characteristic of a solid enabling it to resist deformation by cutting, grinding, scratching, or impact by another substance. *Electromagnetism.* specifically, the quality of a material that determines the ability of X-rays to penetrate it; the hardness is frequency-dependent, so that the higher the frequency, the higher the hardness and the deeper the penetration.

hardness number *Materials Science.* a number indicating the relative hardness of a solid substance; determined by various hardness tests.

hardness test *Analytical Chemistry.* a test used to measure the amount of magnesium and calcium in water.

hard pad *Veterinary Medicine.* a disorder of dogs caused by some strains of the canine distemper virus, and characterized by encephalitis and thickening of the skin of the nose and foot pads.

hard palate *Anatomy.* the anterior part of the palate, a bony shelf consisting of the maxilla and palatine bones, which separates the oral and nasal cavities.

hard pan or **hardpan** *Agronomy.* a compacted layer of soil just below the area disturbed by plowing.

hard patch *Computer Programming.* see PATCH, def. 2.

hard radiation *Physics.* the radiation of photons with wavelengths less than 0.01 nanometer, able to penetrate deeply into most materials including metals.

hard rays *Radiology.* high-energy X rays of short wavelength and high penetrability.

hard rime see RIME.

hard rock *Geology.* **1.** igneous or metamorphic rock, as distinguished from sedimentary rock. **2.** any rock that is relatively unyielding to pressure or resistant to erosion. *Mining Engineering.* **1.** certain pre-Cretaceous sedimentary rock that is drilled relatively slowly. **2.** any rock that must be drilled or blasted for economic removal.

hard rot *Plant Pathology.* any plant disease that produces lesions with calloused surfaces; particularly, a disease of gladiolus caused by the fungus *Septoria gladioli* and characterized by hard lesions on the surfaces of its leaves and corms.

hard rubber *Materials.* rubber that has been cured (vulcanized) with a high amount of sulfur, usually 30% or more, making it a stiff and tough material.

hard sand casting *Materials Science.* a casting method that utilizes a dry sand mold containing bonding agents such as fire clay or bentonite.

hard science *Science.* a term used to describe the physical or natural sciences, such as biology or chemistry, that employ relatively precise and verifiable criteria to investigate aspects of the universe, as opposed to disciplines such psychology or anthropology, whose evaluations are less readily measurable.

hard-sectored disk *Computer Technology.* a disk that is physically divided into sectors by holes punched through the disk, in contrast to a soft-sectored disk.

hard-sphere model *Physical Chemistry.* **1.** an equation used to calculate reaction rate constants in a gas, which assumes that colliding gas molecules will retain a rigid spherical shape; this model provides fairly good approximations for some reactions involving simple molecules. **2.** a useful model of a gas at supercritical temperatures. Also, **hard-sphere (collision) theory.**

hardstand *Civil Engineering.* an open ground area with a prepared surface, pavement or gravel, used for parking vehicles, airplanes, or storage.

hard superconductor *Physics.* a superconductor, such as niobium, that is resistant to the destruction of its superconducting properties.

hard-surface *Civil Engineering.* to prepare a ground surface, e.g., by compacting, to prevent muddiness.

hardware *Mechanical Engineering.* any relatively small physical object having a certain capability or function, especially metal parts such as fasteners, locks, hinges, and nails. *Computer Technology.* the physical equipment that makes up a computer system, such as the magnetic, mechanical, electrical, or electronic devices and components, as opposed to the software programs that run or are run by the system. *Ordnance.* physical items, especially metal items, that are designed to be used in combat.

hardware compatibility *Computer Technology.* a property of computers that permits machine language routines which run on one computer to run on another with precisely the same results.

hardware description language *Computer Technology.* the notation and language that facilitates the design, documentation, simulation, and manufacturing of digital computer systems, particularly VLSI circuits.

hardware development *Design Engineering.* the design, analysis, and manufacture of hardware.

hardware diagnostic *Computer Technology.* a computer program used to check for hardware malfunctions in specific computer components.

hardware division *Computer Technology.* the mathematical operation of division implemented in logic circuits.

hardware floating point *Computer Technology.* a hardware facility that includes the machine instructions, registers, and logic circuitry for performing floating point arithmetic operations.

hardware monitor *Computer Technology.* a software or hardware system used to observe, validate, and collect data on the performance of computer hardware.

hardware multiplexing *Computer Technology.* the utilization of a single device to service multiple units by interleaving its focus between the units in such a way as to minimize the service delay perceived by the units.

hardware stage *Ordnance.* the stage in the development of a weapon in which it actually exists as a finished object, as opposed to the design stage.

hard water *Chemistry.* water that contains percentages of various minerals, such as calcium and magnesium carbonates, bicarbonates, sulfates, or chlorides, due to prolonged contact with rocky substrates and soils; such water tends to discolor, scale, and corrode materials, and has relatively little ability to produce suds from soaps or detergents.

hard-wire *Electricity.* to use wire to connect electrical devices directly, without intervening switches.

hard-wired or **hardwired** *Computer Technology.* describing a computer whose functionality is implemented in hardware rather than in software and therefore cannot be altered by programming.

hardwired numerical control *Robotics.* a numerical control system that uses built-in or hardwired instructions rather than software.

hard-wire telemetry see WIRE-LINK TELEMETRY.

hardwood *Materials.* the wood of various deciduous or broad-leaved trees, such as the birch, elm, oak, mahogany, and maple; used for furniture, cabinets, paneling, and floors. The *hardwood/softwood* distinction is based on the type of tree and does not alway reflect relative hardness; the wood of some softwood (coniferous) trees is harder than that of certain hardwoods.

hardwood bearing *Mechanical Engineering.* a film or fluid bearing made of lignum vitae or of hard maple saturated with a lubricant.

hardwood forest *Forestry.* a forest composed of hardwood trees such as oaks, maples, and hickories; characteristic of much of the eastern half of the United States.

hard X-ray *Electronics.* a highly penetrating X-ray.

hardy *Botany.* describing a plant, especially a garden plant, that is able to withstand cold winter weather in the open air.

Hardy, Godfrey 1877–1947, English mathematician; developed the Hardy-Weinberg law of distribution and the Hardy-Ramanujan formula.

Hardy plankton indicator *Engineering.* a net-type device having a metal shroud, which is used on moving ships to collect samples of plankton.

Hardy-Schulze rule *Physical Chemistry.* the principle stating that the ability of ions to cause the formation of insoluble particles (precipitates) in solution varies according to their charge.

hardystonite *Mineralogy.* $Ca_2ZnSi_2O_7$, a white, translucent, tetragonal mineral of the melilite group, massive in habit and having a specific gravity of 3.44 and a hardness of 3 to 4 on the Mohs scale; found with vesuvianite, willemite, and franklinite at Franklin, New Jersey.

Hardy-Weinberg equilibrium *Genetics.* the presence in a population of the three genotypes AA, Aa, and aa in the frequencies p^2, $2pq$, and q^2, respectively (where p and q are the frequencies of the alleles A and a); at these genotype frequencies, there is no change from one generation to the next.

Hardy-Weinberg law *Genetics.* a law stating that genotypic frequencies will reach equilibrium in a randomly mating population and will remain constant over many generations in the absence of selection or evolutionary forces.

hare *Vertebrate Zoology.* a mammal closely related to the rabbit, having long ears and hind legs, but differing from the rabbit by its generally larger size, longer hind legs, and often black ear tips; found in forests, shrub areas, grasslands, tundra, and alpine slopes in most parts of the world.

European hare

harelip *Medicine.* a congenital malformation of the upper lip resulting from the incomplete merging of tissues of the face and jaw during the embryonic processes. Thus, **harelipped.** Also, CLEFT LIP.

harelipped bat see NOCTILIONDAE.

harem *Behavior.* the mating and association of several adult females with one male. Thus, **harem male.**

Hare's hygrometer *Engineering.* an open-ended, inverted-U-shaped glass tube used to determine the relative densities of two liquids. Each end of the tube is immersed into one of the liquids, and suction is applied to the cross limb of the hygrometer; the heights to which the liquids rise above the surfaces of the liquids in their reservoirs are inversely proportional to their respective densities.

Hargreaves process *Chemical Engineering.* a process for manufacturing sodium sulfate from sulfur dioxide and sodium chloride in a countercurrent contactor.

Haring cell *Physical Chemistry.* a cell with four electrodes that measures resistance and polarization.

Harker diagram *Geology.* a diagram showing the chemical composition of an igneous rock series.

Harker-Kasper inequalities *Crystallography.* the space group-dependent inequalities among unitary structure factors that allow for the determination of the phases of certain intense reflections in the X-ray diffraction pattern of a centrosymmetric crystal.

Harker sections *Crystallography.* certain portions of the Patterson map that, depending on the space group, contain a large proportion of the readily interpretable structural information because they contain many vectors between space group-equivalent atoms. An analysis of Harker sections may be useful in structure determination.

Harkins, William Draper 1873–1951, American chemist; predicted and described the subatomic particles now known as neutrons.

Harkins' rule *Physics.* a rule stating that isotopes of an element with an odd mass number are generally less abundant than those of an element with an even mass number.

Harlechian *Geology.* a European geologic stage of the Lower Cambrian period.

harlequin [här´lə kwin] *Vertebrate Zoology.* any snake having diamond-shaped scales, including venomous snakes of the genus *Elaps.*

harlequin chromosome *Molecular Biology.* a chromosome that is derived from a cell treated with 5'-bromodeoxyuridine; when incorporated into DNA, it produces chromatin that stains lighter than normal chromatin, creating a characteristic harlequin-type pattern of alternating dark and light segments following sister chromatid exchange.

HARM *Ordnance.* a U.S. air-to-surface anti-radiation missile designed to home onto radiation sources at sufficient speed to prevent target reaction; it is powered by a single-grain reduced-smoke boost/sustain motor and delivers a fragmentation warhead at a speed over Mach 2 and a range up to 11.5 miles; officially designated **AGM-88A.** (An acronym for high-velocity anti-radiation missile.)

harman *Organic Chemistry.* $C_{12}H_{10}N_2$, bitter crystals that melt at 237–238°C; used to inhibit the growth of molds and bacteria.

harmattan *Meteorology.* a dry, dust-bearing desert breeze blowing from the northeast or east over west Africa, usually between late November and mid-March.

harmful interference *Telecommunications.* a term for any interference that significantly degrades, obstructs, or interrupts a communications service.

harmonic [här män´ik] *Acoustics.* **1.** relating to or exhibiting harmony. **2.** a tone whose frequency is an integral multiple of the fundamental frequency. *Physics.* in an oscillating or periodic system, a sinusoidal quantity having a frequency that is an integral multiple of the fundamental frequency. *Biology.* of parts, organs, and so on, working together smoothly. *Mathematics.* **1.** the eigenvalues or eigenfunctions of the Laplace (Laplace-Beltrami) operator for a given physical system; e.g., given a string of length L, the nth harmonic is a displacement function of the form $u_n(x,t) = A_n \cos(n\pi ct/L) \sin(n\pi x/L)$. The first harmonic is often called the **fundamental harmonic,** and subsequent harmonics the nth overtones or nth natural modes of vibration. **2.** a function or differential form is harmonic if it is a solution of Laplace's equation.

harmonica bug *Electronics.* a surreptitious interception technique applied to telephone lines; the target is modified so that a tuned relay will bypass the switch hook and ringing circuit when a 500-hertz tone, such as one from a harmonica, is received.

harmonic analysis *Crystallography.* the process of finding the relative amplitudes of all the significant harmonic components of a given complex waveform. *Mathematics.* the study of the eigenfunctions of the Laplace-Beltrami operator on a homogenous space, and the expression of a function on the space as a series of such eigenfunctions.

harmonic analyzer *Electronics.* an analyzer that evaluates the harmonic content of a complex wave. Also, HARMONIC WAVE ANALYZER.

harmonic antenna *Electromagnetism.* an antenna whose electrical length is equal to an integral multiple of the half-wavelength of the signal being transmitted or received.

harmonic attenuation *Electronics.* the reduction of the amplitude of undesirable harmonic components in a complex wave.

harmonic coefficient *Geodesy.* any of the coefficients of the trigonometric terms of an infinite series used in producing global models of the earth's gravitational field.

harmonic conjugates *Mathematics.* **1.** if $f(z)$ is an analytic function of a complex variable, then the functions $u(x,y) = \text{Re}\,f$ and $v(x,y) = \text{Im}\,f$ are known as harmonic conjugates. Equivalently, a pair of complex functions that satisfy the Cauchy-Riemann equations are harmonic conjugates. **2.** either of the first pair or second pair of points of a harmonic tetrad.

harmonic constituent *Oceanography.* any of the quantities in a mathematical expression of the tide-producing force, principally the variable tide-producing forces of the sun and moon. Also, HARMONIC TIDAL CONSTITUENT, PARTIAL TIDE, TIDAL COMPONENT.

harmonic content *Physics.* the combined effect of the harmonics in a periodic system after the fundamental frequency has been removed.

harmonic conversion transducer *Electronics.* a conversion transducer whose output frequency is either a multiple or submultiple of the input frequency.

harmonic detector *Electronics.* a voltmeter circuit that measures only a particular harmonic of a fundamental frequency.

harmonic distortion *Electronics.* **1.** the production of harmonics by the circuit or the device processing the signal. **2.** the malformation of the original signal resulting from the generation of harmonics. **3.** the disproportionate reproduction of the harmonic components of a signal. **4.** the ratio of the total power in the unwanted harmonics to the power in the fundamental frequency at the output of an amplifier.

harmonic division *Mathematics.* the labeling of two points on a line segment so that the points chosen are the harmonic conjugates of the endpoints.

harmonic drive *Mechanical Engineering.* a drive system designed to provide smooth motion by the use of inner and outer gear bands.

harmonic fields *Electromagnetism.* the Fourier components of any field (magnetic, electric, or the like) that is confined to a finite region of space whose characteristic dimensions are an integral multiple of half-wavelengths of the fields.

harmonic filter *Electronics.* **1.** a bandpass filter for selecting one or more harmonics of a complex input wave. **2.** a band-suppression filter that can remove one or more harmonics of a complex input wave.

harmonic folding *Geology.* any folding in which the strata remain parallel or concentric throughout the thickness of the bed, and in which there are no abrupt changes in the form of the folds with depth.

harmonic frequency *Physics.* any frequency that is a positive integral multiple of the fundamental frequency.

harmonic function *Mathematics.* a continuous real-valued function f defined on the closed unit ball B in P^n is said to be harmonic if f is differentiable at every interior point of B and if $\sum_{i=1}^{n} \partial^2 f(x)/\partial x_i^2 = 0$ at every point x in the interior of B; i.e., f satisfies (the Euclidean) Laplace's equation.

harmonic interference *Telecommunications.* in radio broadcasts, interference due to the radiation of unwanted harmonics.

harmonic law *Astronomy.* a law stating that the square of a planet's orbital period around the sun as measured in years equals the cube of its average distance from the sun as measured in astronomical units; published by Johann Kepler in 1619.

harmonic loss *Electromagnetism.* energy loss in a generator due to space harmonics of the magnetomotive force produced by armature current.

harmonic mean *Mathematics.* given a sequence $\{a_1, a_2, a_3, \ldots\}$ of positive numbers, the (nth) harmonic mean is $n(1/a_1 + \cdots + 1/a_n)$.

harmonic motion *Mechanics.* periodic motion whose path or displacement consists of one or more vibratory motions that are symmetric about an equilibrium position. Also, SIMPLE HARMONIC MOTION.

harmonic oscillator *Mechanics.* **1.** any physical system that carries out a process of simple harmonic motion; i.e., the net force or torque that acts on it is directly proportional and opposite to its displacement from equilibrium, thus restoring it to the equilibrium position. Its resulting free motion would be purely sinusoidal, assuming no other forces acting on the body. **2.** a practical example of this, such as a pendulum or a spring-mass system. Also, SIMPLE HARMONIC OSCILLATOR. *Electronics.* a crystal oscillator whose output frequency is a harmonic of the crystal frequency.

harmonic prediction *Oceanography.* a method of predicting tides and tidal currents by combining all harmonic constituents into a single tide curve.

harmonic producer *Electronics.* **1.** an oscillator that utilizes a tuning fork to establish the fundamental frequency, producing an output that is an even harmonic of this frequency. **2.** a frequency multiplier. **3.** a nonlinear circuit used in a calibrator to generate markers at integral multiples of the fundamental frequency.

harmonic progression *Mathematics.* any sequence $\{a_1, a_2, a_3, \ldots\}$ such that $\{1/a_1, 1/a_2, 1/a_3, \ldots\}$ is an arithmetic progression. Also, **harmonic sequence.**

harmonic selective ringing *Telecommunications.* selective ringing in which currents of several frequencies, tuned to one of the ringing currents, are used in order to allow the response of the selected ringer.

harmonic series *Mathematics.* any series whose terms are reciprocals of an arithmetic progression; in particular, the (divergent) infinite series $1 + 1/2 + 1/3 + 1/4 + \cdots$.

harmonic speed changer *Mechanical Engineering.* a drive mechanism designed to transmit positive motion at high ratios.

harmonic synthesizer *Mechanics.* a machine that combines components of simple harmonic motion to yield some total periodic waveform.

harmonic telephone ringer *Telecommunications.* a ringer that responds only to alternating current within a narrow frequency band.

harmonic tetrad *Mathematics.* four (collinear) points (P_1, P_2, P_3, P_4) in projective 2-space form a harmonic tetrad if their cross-ratio is equal to –1. Either of P_3 and P_4 is said to be a harmonic conjugate of the other with respect to P_1 and P_2. By theorem, the first two points are also harmonic conjugates with respect to the last two points.

harmonic tidal constituent see HARMONIC CONSTITUENT.

harmonic vibration-rotation band *Spectroscopy.* a spectral band produced by vibrational-rotational transitions in which the vibrational levels are equally spaced; overtones are multiples of the given fundamental frequency.

harmonic wave *Physics.* a sinusoidal wave.

harmonic wave analyzer see HARMONIC ANALYZER.

harmonize *Science.* to bring into harmony. *Ordnance.* **1.** to align the sights of a gun so that the curving path of the projectile will intersect the line of sight at the target. **2.** in a fighter aircraft, to adjust or align the gunsights of weapon systems in order to establish a desired pattern of fire or accurate aim at a given range.

harmony *Acoustics.* a musical sound based on the principle of the chord, that is, a combination of different tones, and the combination of sounds that these tones produce. The combination is said to be **in harmony** if the produced overtones have more than a whole step of separation, giving a sensation of consonance. *Biology.* the state or quality of working or functioning together.

harmotome *Mineralogy.* $(Ba,K)_{1-2}(Si,Al)_8O_{16} \cdot 6H_2O$, a colorless to white, yellow, or brown, translucent, monoclinic mineral of the zeolite group, occurring in cruciform twin crystals, and having a specific gravity of 2.35 to 2.5 and a hardness of 4.5 on the Mohs scale; found in basic igneous rocks.

Harnack's convergence theorems *Mathematics.* **1.** the metric space (with a special metric used for continuous functions) of functions harmonic on a given open subset of the complex plane is complete. Also, **Harnack's first convergence theorem. 2.** if $\{u_n\}$ is a monotone increasing sequence of harmonic functions defined on an open subset G of the complex plane, then either $\{u_n\}$ converges to a harmonic function or $u_n(z) \to \infty$ uniformly on compact subsets of G. Also, **Harnack's second convergence theorem.**

Harnack's inequality *Mathematics.* let u be a real-valued nonnegative function of a complex variable, harmonic on an open ball B with center a and radius R, and continuous on the closure of B. Then for all nonnegative $r < R$ and for all θ,

$$u(a)(R - r)/(R + r) \leq u\,(a + re^{i\theta}) \leq u\,(a)(R + r)/(R - r).$$

harness *Agriculture.* an apparatus consisting of straps, bands, bits, and other parts forming the working gear of a draft animal. *Engineering.* an apparatus consisting of straps and belts, used in a variety of vehicles (including flight vehicles) to secure passengers or crew members in their seats. Also, SAFETY HARNESS. *Aviation.* **1.** a system of straps used to suspend a parachutist or load from a parachute. Also, PARACHUTE HARNESS. **2.** a system of straps or similar restraints used to secure a cargo pallet or container to an aircraft floor in the absence of an inbuilt anchorage. *Electricity.* see WIRING HARNESS.

Haro galaxies *Astronomy.* blue galaxies whose spectra display bright, sharp emission lines and which are suspected of undergoing recent star-formation.

Harpacticoida *Invertebrate Zoology.* an order of primitive, aquatic, free-living copepod crustaceans.

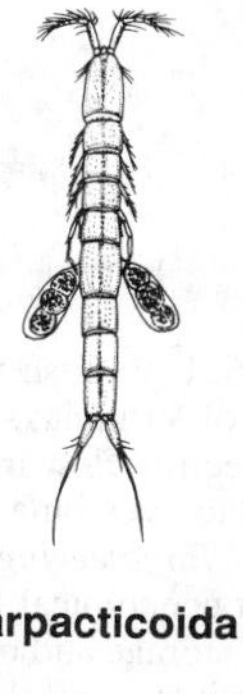

Harpacticoida

harpago *Invertebrate Zoology.* a portion of the clasper on the copulatory organ of certain male insects.

Harpellaceae *Mycology.* a family of fungi belonging to the order Harpellales occurring inside immature freshwater insects (such as blackflies and midges) that inhabit streams.

Harpellales *Mycology.* an order of fungi belonging to the class Trichomycetes which occurs inside immature freshwater insects that inhabit streams.

Harpidae *Invertebrate Zoology.* the harp shells, a family of tropical snails that frequent shallow coastal regions with sandy bottoms.

harpoon *Mechanical Devices.* a barbed spear, hand thrown or shot from a gun, used in hunting whales and other large sea creatures.

Harpoon missile

Harpoon *Ordnance.* an antiship cruise missile that can be launched from surface ships, submarines, or aircraft; it is turbojet powered, guided by active radar, and employs a 500-pound conventional warhead. Official designation, **RGM-84A** for the shipborne, **AGM-84** for the air-launched, and **UGM-84** for submarine-launched versions.

harrier *Vertebrate Zoology.* any of various slender hawks of the subfamily Circinae and famiy Acciptridae, having a small beak, long legs, and a long tail; inhabit marshes and meadows worldwide.

harrier

Harriot, Thomas 1560–1621, English mathematician and geographer; wrote statistical geography of Virginia; simplified algebraic notation.

Harris flow *Electronics.* electron flow in a cylindrical beam where a radial electric field is utilized to overcome space charge divergence.

Harris formula *Industrial Engineering.* an inventory-control formula used to determine the most economical lot size to purchase, taking into account the annual costs of storage and of order preparation.

Harris process *Metallurgy.* any of several processes to refine crude lead.

harstigite *Mineralogy.* $Ca_6(Mn^{+2},Mg)Be_4(SiO_4)_2(Si_2O_7)_2(OH)_2$, a colorless, orthorhombic mineral occurring in small prismatic crystals, and having a specific gravity of 3.05 to 3.16 and a hardness of 5.5 on the Mohs scale.

hart *Vertebrate Zoology.* a male deer, especially one that is more than five years old.

hartebeest [härt´bēst´; här´tə bēst´] *Vertebrate Zoology.* a swift, large antelope of the genera *Alcelaphus* and *Damaliscus* in the family Bovidae, inhabiting the open plains of sub-Saharan Africa; characterized by a small hump on a sloping tan back, a long bushy tail, and lyre-shaped horns that connect at their base.

Hartford loop *Mechanical Engineering.* a low-pressure steam heating arrangement in which condensed water is returned to the boiler, maintaining a steady water line.

hartite *Geology.* a white crystalline fossil resin, found in lignites.

Hartley, David 1705–1757, English physician and philosopher; the founder of associative psychology.

hartley *Telecommunications.* a unit of information based on a scale of ten.

Hartley formula *Telecommunications.* a formula that expresses an inverse relationship between time and frequency; as the time function decreases, the frequency spectrum must increase.

Hartley oscillator *Electronics.* an oscillator circuit in which the feedback is provided by an inductor voltage divider, which is part of the tuned circuit.

Hartley principle *Telecommunications.* a theory stating that the total number of information bits transmitted through a channel in a previously allotted time frame is directly proportional to the product of channel bandwidth and transmission time.

Hartline, H. Keffer 1903–1983, American physiologist; awarded Nobel Prize for studying the electrical effect of light on optic nerve cells.

Hartmann, Johannes Franz 1865–1936, German astronomer; a pioneer in spectroscopy; established the existence of interstellar matter.

Hartmann diaphragm *Analytical Chemistry.* a device used to identify compounds from their emission spectra.

Hartmann dispersion formula *Optics.* a formula relating the variation of the refractive index *n* of a material to the wavelength of light. Also, **Hartmann formula.**

Hartmannella *Invertebrate Zoology.* a genus of freeliving protozoa of the order Amoebida, found in stagnant fresh water or wet soil; some species, such as *H. hyalina,* are facultative parasites causing a primary amebic meningoencephalitis.

Hartmann flow *Physics.* the movement of an electrically conducting fluid between two parallel nonconducting walls.

Hartmann number *Physics.* a dimensionless number given to measure the effect of drag forces on the Hartmann flow of conducting fluids.

Hartmann test *Spectroscopy.* an accuracy test for spectrometers in which a fault is indicated by passing light through different parts of the entrance slit and noting any changes in the spectra produced.

Hartree, Douglas Raynar 1897–1958, English mathematician and physicist; developed methods of numerical analysis.

hartree *Atomic Physics.* a unit of energy approximately equal to 4.36×10^{-18} joules or 27.21 electron-volts; used to study atomic structure.

Hartree equation *Electronics.* an equation used to determine the lowest anode voltage at which it is theoretically possible to maintain oscillation in the different magnetron modes.

Hartree-Fock approximation *Quantum Mechanics.* an elaboration of the Hartree method, in which the many-particle wavefunction is expanded as a Slater determinant; this introduces exchange terms into the Hamiltonian.

Hartree method *Quantum Mechanics.* an iterative integral equation method of constructing an approximate solution to Schrödinger's equation for more than one charged particle, employing a product of single particle wavefunctions, each consistent with the potential deduced from the others. Also, SELF-CONSISTENT FIELD METHOD.

hartebeest

Hartree units *Atomic Physics.* the symbols universally accepted under international agreement to represent given values in atomic physics; e.g., the unit of mass is the mass of the electron, and the unit of length is the radius of the first Bohr orbit in hydrogen of infinite nuclear mass.

hartshorn *Vertebrate Zoology.* the horn of a hart (male deer), formerly used as a source of ammonium carbonate. *Inorganic Chemistry.* see AMMONIUM CARBONATE.

Harvard classification see HENRY DRAPER SYSTEM.

harvest *Agriculture.* **1.** to cut and gather crops that are ripe or mature. **2.** the time or season when mature crops are cut. **3.** the yield of a particular crop.

harvester *Agriculture.* a farm machine that is used to harvest grain. *Forestry.* a similar machine used to gather felled trees. Also, **harvesting machine.**

harvester

harvester-thresher *Agriculture.* a farm machine that performs both the harvesting and threshing processes.

harvesting *Agriculture.* the act of cutting and gathering in crops.

harvest moon *Astronomy.* a name for the full moon nearest the September equinox, which for three or four days around full rises only a little later each night. (Associated with the *harvest* season at this time.)

Harvey, William 1578–1675, English physician; discovered the physiology of the circulatory system; founder of modern embryology.

harzburgite *Petrology.* a variety of peridotite consisting primarily of olivine and orthopyroxene.

Hasche process *Chemical Engineering.* a thermal reforming method for hydrocarbon fuels in which a mixture of hydrocarbon vapor and air flows through a vessel and becomes increasingly hotter in the direction of the gas flow; partial combustion occurs, thus creating heat to crack the hydrocarbons that remain in the combustion zone.

hash *Computer Programming.* **1.** to convert a record key value into a "random" but consistent code for indexing. **2.** a code so derived. **3.** see GARBAGE. *Electronics.* **1.** an electrical noise produced within a receiver by a vibrator or a mercury-vapor rectifier. **2.** random signal interference caused by arcing or a natural environmental disturbance.

Hashimoto's struma see STRUMA LYMPHOMATOSA.

hashing *Computer Programming* **1.** the process of performing a class of operations on a set of objects, such as records or numbers, to map them into a different, usually more compact, arrangement. **2**. the process of transforming a record key into an index value for storing or retrieving the record.

hashish *Pharmacology.* $C_{21}H_{30}O_2$, a drug produced from the resin of the flowers of hemp, *Cannabis sativa;* far more potent than marijuana, it is smoked or chewed for its intoxicating effects. (From an Arabic word meaning "herb.")

hash total *Computer Programming.* the sum of values in several fields of a file, calculated for the purpose of checking the accuracy of the data in those fields.

HASP *Computer Programming.* a program used on some large multiprogramming computer systems to manage input and output processes by utilizing mass storage devices for the temporary storage of data. (An acronym for Houston Automatic Spooling Program.)

Hassel, Odd 1897–1981, Norwegian chemist; won Nobel Prize for formulating concept of conformation and applying it to chemical analysis.

hastate *Biology.* of a leaf or other structure, shaped like an arrowhead but having divergent triangular lobes at the base.

Hastelloy *Materials.* a trade name for a group of nickel-chromium-molybdenum alloys that are resistant to acids at elevated temperatures and have high strength.

hastingsite *Mineralogy.* $NaCa_2(Fe^{+2},Mg)_4Fe^{+3}(Si_6Al_2)O_{22}(OH)_2$, $Mg/(Mg+Fe^{+2})=0–0.69$, a dark-green or black, monoclinic mineral of the amphibole group, massive in habit or as prismatic crystals, having a specific gravity of 3.17 to 3.59 and a hardness of 5 to 6 on the Mohs scale; found in igneous and metamorphic rocks.

hasty *Military Science.* of or relating to military operations or activities that are conducted without extensive planning, usually while in contact or imminent contact with the enemy. Thus, **hasty breaching, hasty crossing, hasty defense, hasty minefield.**

hat *Telecommunications.* a group of symbols that have been arranged in an entirely random sequence, as if they had been put into a hat, shaken, and then drawn out one by one.

hatch *Zoology.* to emerge or cause to emerge from an egg. *Graphic Arts.* to mark with hatching. *Engineering.* **1.** a doorway or opening in a ship, aircraft, or spacecraft. Also, **hatchway. 2.** the door or lid for such an opening. Also, **hatch cover, hatchcover.**

hatch battens *Naval Architecture.* steel or wooden strips wedged into the coaming of a hatch in order to secure a tarpaulin in place of a solid hatch cover.

hatch carlings *Naval Architecture.* fore-and-aft timbers running under the coamings of hatches and serving to support the partial deck beams.

hatch coaming *Naval Architecture.* a raised flange around the edge of a hatchway onto which the hatch cover fits.

hatchet *Mechanical Devices.* a small, hand-held axe having a cutting head on one end and a hammerhead on the other.

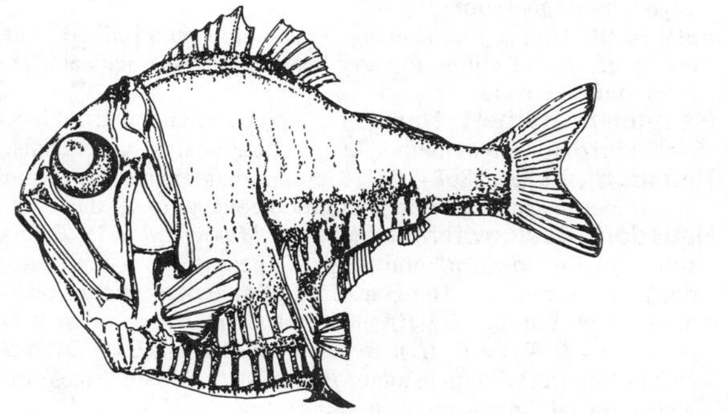

hatchetfish

hatchetfish *Vertebrate Zoology.* any member of the marine fish family Gasteropelecidae, characterized by a sharply angled head, a deep flattened body, and the ability to fly above water by flapping its large pectoral fins like wings.

hatchettite *Mineralogy.* $C_{38}H_{78}$, a soft, yellowish-white paraffin wax occurring as veinlike masses in ironstone nodules associated with coal-bearing strata or in cavities in limestone. Also, **hatchettine.**

hatching *Zoology.* the process of emerging from an egg. *Graphic Arts.* closely drawn parallel lines used to suggest shadows or textures.

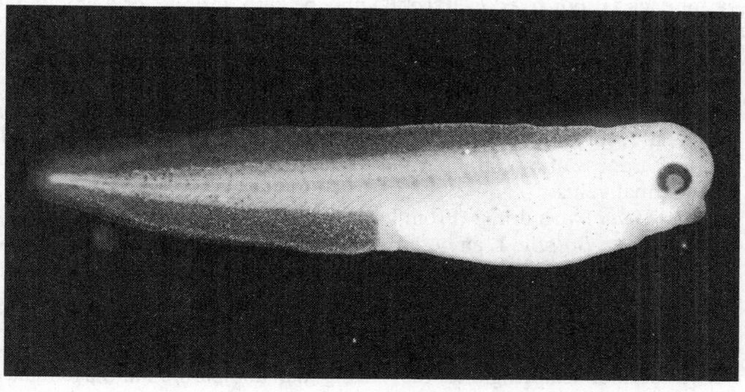

hatching stage

hatchite *Mineralogy.* $(Pb,Tl)_2AgAs_2S_5$, a lead-gray triclinic mineral occurring in minute crystals, having a specific gravity of 5.81 and an undetermined hardness; found in crystalline dolomite.

hatchling *Vertebrate Zoology.* a bird, fish, or reptile that has just emerged from an egg.

Hatch-Slack pathway *Biochemistry.* an alternative pathway to the Calvin cycle of photosynthesis, believed to be operative in a number of tropical grasses and in certain dicotyledons. Also, **Hatch-Slack-Kortschak pathway.**

HAT medium *Biotechnology.* a tissue culture medium that contains hypoxanthine, aminopterin, and thymidine and is used as a selection medium.

Hatschek process *Materials Science.* a process for manufacturing asbestos by dry-mixing cement, silica, and asbestos fibers.

hatted code *Telecommunications.* a randomized code involving an encoding section; the plain text groups are ordered alphabetically, while their code groups are ordered randomly.

haud *Meteorology.* a term used in Scotland to denote a squall.

hauerite *Mineralogy.* MnS_2, a reddish-brown or brownish-black cubic mineral of the pyrite group, occurring in octahedral or cubo-octahedral crystals, and having a specific gravity of 3.46 and a hardness of 4 on the Mohs scale; found in salt dome capping.

haughtonite *Mineralogy.* a black variety of biotite that is rich in ferrous iron.

haul *Engineering.* the quantity of material or items conveyed at one time, or caught in one draft of a net or scoop of a bucket. *Navigation.* **1.** to sail, especially in a given direction. **2.** of the wind, to shift, especially in a counterclockwise direction. Also, BACK. **3.** to move a vessel up on land, as for repairs.

haulage *Mining Engineering.* the process of underground or surface transport of cars, workers, supplies, ore, and waste. Thus, **haulage conveyor, haulage drum**

haul road *Mining Engineering.* a road built on a limited grade, usually less than 17% of climb, to carry heavily loaded trucks at a good speed. Also, **haulage road.**

Hauptman, Herbert born 1917, American chemist; with Karle, won Nobel Prize for direct method of structural analysis of crystals.

Hausdorff, Felix 1868–1942, German mathematician; formulated theory of metric spaces and neighborhood concept in set theory.

Hausdorff-Besicovitch dimension *Mathematics.* let E be a bounded subset of Euclidean R^n and $\mu^*(p, E)$ denote the p-dimensional Hausdorff measure of E . The Hausdorff-Besicovitch dimension of E is the unique real number $D_H(E)$ such that $\mu^*(p, E) = \infty$ if $p < D_H(E)$ and $\mu^*(p, E) = 0$ if $p > D_H(E)$. By theorem, $0 \le D_H(E) \le D(E) \le n$, where $D(E)$ is the fractal dimension of E. Used to compare "sizes" of sets in R^n whose fractal dimension is the same.

Hausdorff dimension *Mathematics.* let E be a subset of a metric space X with p-dimensional Hausdorff measure $\mu_p^*(E)$. Then the Hausdorff dimension of E is the supremum of all p for which $\mu_p^*(E) = \infty$.

Hausdorff maximality principle *Mathematics.* a theorem stating that every partially ordered set P contains a maximal totally ordered subset Q; i.e., Q is not a subset of any other totally ordered set; it can be shown to be equivalent to the Axiom of Choice, the well-ordering principle, Zorn's lemma, Tukey's lemma, etc. Also, **Hausdorff maximality theorem.**

Hausdorff measure *Mathematics.* let E be a bounded subset of Euclidean R^n and p a nonnegative real number, and let $\delta(E)$ denote the diameter of E. Then the p-dimensional Hausdorff (outer) measure of E is the number $\mu^*(p, E) = \sup_{\varepsilon>0} \inf\{\sum_{i=1}^{\infty} [\delta(E_i)]^p : E = \cup_{i=1}^{\infty} E_i$ and $\delta(E_i) < \varepsilon$ for each $i\}$. For fixed E, $\mu^*(p, E)$ may be zero, infinite, or a unique finite number.

Hausdorff space *Mathematics.* a topological space in which every pair of distinct points can be contained in disjoint (open) neighborhoods.

hausmannite *Mineralogy.* $Mn^{+2}Mn_2^{+3}O_4$, a brownish-black, tetragonal, mineral occurring in granular masses, having a specific gravity of 4.84 and a hardness of 5.5 on the Mohs scale; found in high-temperature hydrothermal veins.

haust. *Medicine.* a drink. (From Latin *haustus.*) Also, **Haust.**

haustorium *Botany.* **1.** an outgrowth in a parasitic plant that attaches itself to a host and penetrates the host's tissue to absorb nutrients. Also, **sucker. 2.** in nonparasitic plants, the cell of the embryo sac that absorbs food.

Hauterivian *Geology.* a European geologic stage of the Lower Cretaceous period, occurring after the Valanginian and before the Barremian.

haut mal see GRAND MAL.

Haüy, Abbé René-Just [ä wē´] 1743–1821, French mineralogist; the founder of crystallography; formulated the geometrical law of crystallization.

Haüy law *Crystallography.* a law stating that a shattered crystal will cleave so as to reveal the fundamental structure of the crystal.

haüyne *Mineralogy.* $(Na,Ca)_{4\text{-}8}Al_6Si_6(O,S)_{24}(SO_4,Cl)_{1\text{-}2}$, a blue cubic mineral of the sodalite group, usually occurring as rounded grains, having a specific gravity of 2.44 to 2.50 and a hardness of 5.5 to 6 on the Mohs scale; found in phonolites and in similar igneous rocks.

Hauzeur furnace *Metallurgy.* a furnace that is used to distilled zinc.

HAV *Aviation.* the airport code for Havana, Cuba.

HAV hepatitis A virus. Also, **HA virus.**

haven *Navigation.* a bay or harbor that offers secure refuge for a ship.

Haverhill fever *Medicine.* an acute infection caused by *Streptobacillus moniliformis,* which is commonly transmitted through the bite of a rat and is marked by acute onset, intermittent fever, rash, and inflammation of joints.

Haversian canal see OSTEONAL CANAL.

Haversian lamella *Histology.* any of the concentric layers of bone comprising an osteon.

haversine *Mathematics.* the haversine of an angle θ is the quantity $(1 - \cos\theta)/2$; i.e., half of the versine of θ. Denoted hav θ .

havgull *Meteorology.* a cold, damp sea wind that blows over Scotland and Norway in the summertime.

Havoc *Aviation.* a name used by the U.S. Navy for the A-20 and by the Royal Air Force for the P-70 (all-weather version of the A-20).

Hawaiian high see PACIFIC HIGH.

Hawaiian Islands *Geography.* an archipelago in the central Pacific containing eight major islands (Hawaii, Oahu, Maui, Kauai, Molokai, Lanai, Niihau, and Kahoolawe) and many islets.

Hawaiian region *Ecology.* a distinct zoogeographical area that includes the Hawaiian Islands; often classified within the Polynesian subregion of the Australian region rather than regarded as a separate entity.

Hawaiian-type eruption *Volcanology.* a type of volcanic eruption characterized by a copious flow of low-viscosity lava but little or no explosive activity.

hawk *Vertebrate Zoology.* any of various birds of prey in the family Accipitridae closely related to Old World vultures and eagles and characterized by a strongly hooked bill and powerful toes with hooked nails; among those found in North America are the **red-tailed hawk,** *Buteo jamaicensis,* the **red-shouldered hawk,** *B. lineatus,* the **sharp-shinned hawk,** *Accipter striatus,* and the **marsh hawk,** *Circus cyaneus. Engineering.* a small quadrangular board with a handle underneath, such as those used by masons to hold a small quantity of mortar or plaster.

red-tailed hawk

Hawk *Ordnance.* a mobile, non-nuclear, surface-to-air missile system using radar guidance that provides low to medium air-defense coverage for ground forces; officially designated **MIM-23.**

hawkfish *Vertebrate Zoology.* a small, colorful tropical marine fish of the family Cirrhitidae, known to perch on rock and coral; characterized by a small flag of skin that extends from the tip of each spine on dorsal fins and by a fringe on the anterior nostril; inhabits warm and shallow coastal areas of the Atlantic and Indo-Pacific Oceans.

Hawking, Stephen born 1942, British physicist and author; studied black holes and researched the big bang theory; known for writings on cosmology.

Hawking radiation *Astrophysics.* radiation emitted by a black hole by means of the distortion of space-time just outside its event horizon.
Haworth, Walter Norman 1883–1950, English chemist; Nobel Prize for synthesis of ascorbic acid and research in carbohydrates.
hawse *Naval Architecture.* **1.** the area near the bow where the hawsehole and hawsepipes are located. **2.** the space between the bow of a ship and a point directly above its anchor when the anchor is resting on the bottom.
hawse bolster *Naval Architecture.* in wooden ships, extra planking fitted around the hawsehole as reinforcement.
hawsehole *Naval Architecture.* a hole in the deck near the bow through which the anchor chain passes; it is the deck-side opening of the hawsepipe.
hawsepipe *Naval Architecture.* a tube through which an anchor chain passes from the deck through the side of the ship.
hawser *Naval Architecture.* a heavy line, particularly one over five inches in diameter.
Hawthorne effect *Psychology.* the principle that any group that is singled out for special study or consideration will have its performance positively affected by the knowledge that it has been so selected. *Industrial Engineering.* the principle that any major change in management style or techniques, regardless of the nature of the change, will result in at least a temporary improvement in performance. (Based on the results of the *Hawthorne* studies.)
Hawthorne studies *Industrial Engineering.* a series of experiments in which the output of the workers was observed to increase as a result of improved treatment by their managers. (Named for their site, at the Western Electric Company plant in *Hawthorne,* Illinois.)
hay *Agriculture.* a term for grass, clover, and other such plants that have been cut and dried for use as fodder.
Hayashi track *Astrophysics.* a generally downward line on the Hertzsprung-Russell diagram traced by a completely convective gas sphere as it evolves to become a star.
hay baler *Agriculture.* see BALER.
Hay bridge *Electricity.* a four-arm alternating-current bridge in which the arms adjacent to the unknown impedance are nonreactive resistors and the opposite arm is composed of a capacitor in series with a resistor; used to measure inductance.
haycock *Hydrology.* an isolated cone of ice produced by pressure or ice movement, which projects from the land or from shelf ice.
hay crimper *Agriculture.* a farm machine having two closely spaced, parallel rollers that crush freshly cut hay to ease dyeing and curing.
Hayden process *Metallurgy.* an electrochemical process to refine copper.
hay fever *Medicine.* the common term for a seasonal allergic rhinitis brought about by pollen antigens in the air and resulting in sneezing, mucous discharge from the nose, itching of the eyes, and tearing.
Hayford spheroid *Geodesy.* a reference spheroid with a semimajor axis of approximately 6,378,388 meters, a semiminor axis of approximately 6,356,909 meters, and an ellipticity of 1/297.
hayloft *Agriculture.* the upper part of a barn where hay is stored. Also, **haymow.**
hay rake or **hay-rake** *Agriculture.* a machine or hand tool used to gather hay.
Hay's test *Pathology.* a method for identifying the presence of bile salts by dropping a pinch of sublimated sulfur into urine; if the sulfur sinks, bile is present; if the sulfur floats, bile is absent. (Named for Matthew *Hay,* 1855–1932, Scottish physician.)
hazard *Industrial Engineering.* any potentially dangerous condition at an industrial site, whether a preventable condition or inherent in the nature of the work done there; industrial hazards are categorized as negligible, marginal, critical, or catastrophic, depending on the amount of personnel injury or product damage incurred.
hazard rate see FAILURE RATE.
haze *Meteorology.* fine dust, salt, mist, smoke, or other solid particles dispersed throughout a portion of the atmosphere, which are invisible to the naked eye but which diminish horizontal visibility and give the atmosphere an opalescent appearance, rendering distant objects indistinct and subdued in color.
haze droplet *Meteorology.* **1.** any small droplet of liquid that contributes to an atmospheric haze condition. **2.** an intermediate state between a condensation nucleus and a cloud or fog drop.
haze factor *Meteorology.* the ratio of the luminance of a haze condition such as mist or fog, through which an object is observed, to the luminance of the object itself.
haze horizon *Meteorology.* the top of a haze layer confined by a low-level temperature inversion, which appears to be the true horizon when viewed from above against the sky.
haze layer *Meteorology.* a layer of haze in the atmosphere that is bounded at the top by a temperature inversion and frequently extends downward to the ground.
haze line *Meteorology.* the boundary between a haze layer and the transparent, cleaner air above that layer. Also, **haze level.**
hazel sandstone *Geology.* a hard, condensed gritstone that varies to freestone, flagstone, and chert, according to the locality.
hazemeter see TRANSMISSOMETER.
Hazen-Williams formula *Fluid Mechanics.* an equation for determining the head loss in a fluid due to friction in a pipeline. Also, WILLIAMS-HAZEN FORMULA.
HB hepatitis B.
Hb hemoglobin.
H band *Histology.* the central, light-staining area of each A band in a muscle sarcomere that contains only thick filaments of myosin.
h-bar *Quantum Mechanics.* a fundamental constant equal to Planck's constant (h) divided by 2π. Symbolized $\hbar$.
H beam *Civil Engineering.* a principal horizontal, supporting structural member made of steel, shaped like an H in cross section and similar to a wide flange beam with longer flanges.
H biotin see BIOTIN.
H-bomb see HYDROGEN BOMB.
HBV hepatitis B virus. Also, **HB virus.**
HC or **H.C.** Hospital Corps.
H carrier system *Telecommunications.* a low-frequency carrier system that provides one carrier channel by means of effective four-wire transmission on a single open-wire pair.
HCF or **H.C.F.** highest common factor. Also, **hcf** or **h.c.f.**
HCG human chorionic gonadotropin.
H-chain see HEAVY CHAIN.
H-components *Mathematics.* see FRAGMENT.
HCT hematocrit.
HD heavy-duty.
hd. head; hand.
h.d. or **hd** at bedtime. (From Latin *hora decubitus.*) Also, **H.d.**
H display *Electronics.* in radar, a B display modified to include angle of elevation; a target appears as two closely spaced blips connected by a short line, the slope of which is proportional to the sine of the angle of target elevation.
HDL high-density lipoprotein.
HDN hemolytic disease of the newborn.
hDNA hybrid DNA.
hdq or **hdq.** headquarters.
HDTV high-definition television.
hdw. hardware. Also, **hdwe.**
hdwd hardwood.
HDX half duplex.
HE high explosive.
He the chemical symbol for helium.
head *Zoology.* the front or upper part of the body of an animal, containing the brain and special sense organs.
head any of various structures or features thought of as resembling the head of an animal, as by being on top, in front, or in another prominent position; specific uses include: *Anatomy.* the enlarged rounded end of a bone. *Botany.* a compact cluster of flowers growing on a single stalk, typical of composite flowers such as ragweed, daisies, and sunflowers. *Virology.* an isometric or elongated component of tailed phages that contains the DNA genome. *Engineering.* the part of a tool or weapon that is used for striking. *Naval Architecture.* a ship's toilet. *Geography.* **1.** a headland. **2.** the source of a river. *Oceanography.* the end of a rip current as it widens seaward of the breaker zone. *Fluid Mechanics.* the pressure of a liquid expressed as the height of a column of the liquid which would have that same pressure at the base of the column. Also, PRESSURE HEAD. *Electronics.* the device that reads, records, or erases data on a storage medium. *Computer Programming.* a special data item that points to the beginning of a linked list. *Graphic Arts.* **1.** the top of a page. **2.** a word or words printed at the top of a passage, usually in large boldface type, designed to summarize or call attention to the text that follows. Also, HEADING, HEADLINE. **3.** see RUNNING HEAD. *Mathematics.* the vertex toward which a directed edge is directed in a graph; the edge is directed from the tail and is denoted (u,v) with tail u and head v.
headache *Medicine.* a popular term for any pain in the head.

head crash *Computer Technology.* a failure in a disk drive that causes the head to make physical contact with the rotating disk surface, resulting in physical damage and loss of data.
head current *Navigation.* a current that is flowing in a direction opposite to the course of the vessel.
head-disk assembly *Computer Technology.* an integrated assembly consisting of the disks and the read/write heads.
header *Building Engineering.* **1.** a large beam that frames common joists, studs, or rafters and transfers their weight to parallel members. **2.** in masonry, a stone or brick laid in a wall with its short end toward the face of the wall. *Civil Engineering.* a conduit or pipe having many outlets or connected pipes that are usually parallel, serving as a central point of distribution for the contents. *Mechanical Engineering.* **1.** a machine used for managing screws, rivets, and bolts by their heads. **2.** see EXHAUST MANIFOLD. *Mining Engineering.* a machine that bores the entire section of an entry in one operation. *Electricity.* a mounting plate through which the insulated terminals or leads are brought out from a hermetically sealed piece of equipment. *Telecommunications.* the first field in a data message to be transmitted, containing information that is required for the data to be routed to its destination. *Computer Programming.* in word processing, copy that appears in the top margin on a printed page, above the main body of text.
header bond *Building Engineering.* a typical masonry bond whose face shows only headers, the center of which is placed directly over the joint of the two adjacent headers below.
header course *Building Engineering.* in masonry, a course of bricks laid as headers, i.e., with short ends toward the face of the wall.
header label see FILE LABEL.
head erosion see HEADWARD EROSION.
header record *Computer Programming.* a record that contains common or identifying information for a group of succeeding records. Also, **header table.**
header-type boiler see STRAIGHT-TUBE BOILER.
head flowmeter *Engineering.* a flowmeter in which the measurement of quantity per unit time is dependent upon change in pressure. Also, **head meter.**
head fold *Developmental Biology.* a crescentic ventral fold of the blastoderm in front of the head of the embryo.
head-foot *Invertebrate Zoology.* the combined head and foot of a gastropod mollusk such as a snail.
headframe *Mining Engineering.* **1.** the frame at the top of a shaft on which the hoisting pulley is mounted. **2.** the shaft frame, sheaves, hoisting arrangements, dumping gear, and connected works at the top of a shaft or pit. Also, **headgear.**

headframe

headful mechanism *Molecular Biology.* a method of packing DNA into a bacteriophage head that is based on the length of the packaged DNA rather than its sequence. Also, **headful packaging.**
head gap *Computer Technology.* the distance maintained between the surface of a magnetic disk and the read/write head hovering over it.
head gate *Civil Engineering.* a gate on the upstream side of a lock or conduit or at the starting point of an irrigation ditch. *Petroleum Engineering.* on oil or gas lines, the gate valve that is placed closest to the compressor or pump.
headgear *Engineering.* any of various forms of protective or functional devices worn on the head. *Medicine.* specifically, a device designed to fit over the head and serve as an external anchor and source of resistance for an orthodontic appliance.
heading *Navigation.* the direction in which a craft is pointing. It is measured clockwise through 360° from some reference direction, often true north, to the craft's longitudinal axis. *Graphic Arts.* see HEAD, def. 2. *Mining Engineering.* an active working section of an underground mine. *Metallurgy.* the process of upsetting a wire or bar to form a head, as in the manufacture of rivets.
heading angle *Navigation.* heading as measured from a reference direction clockwise or counterclockwise through 90° or 180°.
heading blasting see COYOTE BLASTING.
heading joint *Building Engineering.* **1.** the joint formed by butting the end of one timber against the side of another, parallel timber. **2.** a masonry joint formed between two stones aligned on the same course.
heading line *Navigation.* a line extending in the direction of the heading.
heading wall see FOOTWALL.
head jamb *Building Engineering.* the upper jamb of a door frame that is attached to a side member.
headland *Geography.* a high, steep promontory.
headline *Graphic Arts.* see HEAD, def. 2. *Mining Engineering.* the line, in dredging, that holds the dredge up to its digging front.
head loss *Fluid Mechanics.* a reduction in the total head (sum of velocity head and potential head) for a fluid flowing through a tube due to frictional forces within the fluid.
headman *Anthropology.* a symbolic position in an egalitarian society that has influence but carries no power to impose sanctions; often an office given to an individual who is mature, economically stable, and charismatic.
head margin *Graphic Arts.* the blank vertical space between the top of a page and the first printed element on that page.
head of a comet *Astronomy.* see COMA.
head-on collision *Mechanics.* a collision between two bodies that approach each other **head-on**, i.e., in the same linear direction at the same initial speed. Also, CENTRAL COLLISION.
head organ *Invertebrate Zoology.* an enlarged head structure in flukes that contains openings for adhesive glands.
head-per-track *Computer Technology.* describing a disk drive that has one read/write head for each track on the disk to eliminate the need for head movement.
headphone *Acoustical Engineering.* **1.** a sound-receiving device, in effect a miniature loudspeaker, used for private or enhanced listening to a tape player, radio, television, or other such sound reproduction unit. **2. headphones.** a pair of such listening devices, usually held in place by a band or attachment over the head.
head process *Developmental Biology.* an axial strand of cells in the embryo, extending forward from Hensen's node. Also, NOTOCHORDAL PLATE.
headrig *Forestry.* in a sawmill, the carriage and saw used to cut a log into slabs.
headroom *Building Engineering.* that space that lies between the head and sill of a doorway or between the ceiling and floor of an attic, allowing passage of a person standing upright.
head section *Telecommunications.* an added-on portion of a switchboard used to extend some of the trunks to the end positions, with the purpose of placing all jacks within easy grasp of the switchboard operators. Also, END SECTION.
headset *Acoustical Engineering.* **1.** a device used for phone or radio communication, consisting of a headphone or earphone for listening and a mouthpiece for speaking, with a strap or wire to hold the device to the head. **2.** any set of headphones or earphones.
head shaft *Mechanical Engineering.* a chain-driven shaft mounted at the delivery end of a chain conveyor and used to mount the sprocket that drives the drag chain.
head shield *Invertebrate Zoology.* a protective head structure in nematodes.
head sign *Transportation Engineering.* a sign above the front window of a bus or other vehicle, showing its route or destination.
headsill *Building Engineering.* a horizontal beam that lies along the head of a door or window.
head smut *Plant Pathology.* a disease of corn and sorghum caused by the fungus *Sphacelotheca reilana* and characterized by ears and tassels that are aborted, barren, or converted to a smut gall.

headspace *Ordnance.* the distance between the face of the locked bolt or breech of a gun and the surface of the chamber against which the cartridge presses prior to firing; in a weapon using a rimmed cartridge, the latter is the contact point between the rim and the chamber; in the case of a rimless cartridge, it is the contact point between the shoulder of the cartridge and the chamber.

head stepping rate *Computer Technology.* the rate at which the read/write head of a disk drive moves between tracks on a disk surface.

headstock *Mechanical Engineering.* any of a wide variety of mechanisms that support the head or endpiece of a part, such as the mechanism of a lathe that supports the revolving spindle or the mechanism of a planing machine that supports the cutter.

headstream *Hydrology.* a stream that is the source or one of the sources of a larger stream or river.

head tide *Navigation.* a head current caused by tidal action is sometimes colloquially referred to as a head tide.

head-to-head configuration *Materials Science.* the polymer-chain configuration in which monomers are bound together in alternately reversed position.

head-to-tail configuration *Materials Science.* the polymer-chain configuration characterized by bonding between regularly repeating monomers oriented in the same direction.

head-up display *Optics.* the central device in a system that projects electronically generated symbols relaying flight and attack data onto the windscreen of an aircraft; enables the pilot or air crew to receive information while keeping a clear view ahead.

headwall *Geology.* a steep slope at the head of a valley, especially the rock cliff at the back of a cirque. *Civil Engineering.* a retaining wall at the outlet of a drain or culvert serving as protection against scouring or undermining of fill or as a flow-diverting device.

headward erosion *Geology.* the lengthening of a valley or gulley, or of a stream, by erosion at the valley head. Also, HEAD EROSION, HEADWATER EROSION.

headwater *Hydrology.* **1.** the water upstream from a dam or other such structure. **2.** see HEADSTREAM. Also, **headwaters.**

headwater erosion see HEADWARD EROSION.

headway *Navigation.* movement in a forward direction. *Building Engineering.* a space of a given height underneath an arch or above a stairway that allows people to move while standing upright. *Transportation Engineering.* the distance between two consecutive vehicles traveling in the same direction along the same path; usually expressed as a time interval. *Mining Engineering.* the principal cleat in coal.

headwind *Navigation.* a wind that is blowing from the same direction as the course of a craft.

headworks *Civil Engineering.* any device or structure at the front or diversion point of a waterway to control the amount of water flowing.

heaf test see STERNEEDLE TEST.

heal *Medicine.* to restore wounded parts or to make healthy; to become well or healthy.

healing *Medicine.* a process of curing; the restoration of the normal structure and function to diseased or damaged tissue.

healing by first intention *Surgery.* the healing in which union or restoration occurs directly without the intervention of granulations.

healing by second intention *Surgery.* the healing in which union or restoration occurs after granulations have formed on the sides and base of a wound.

healing ridge *Surgery.* a firm ridge that forms onto the skin along either side of the length of a wound as it heals.

health *Medicine.* a state of physical, mental, and social well-being that permits the optimal functioning of the body and mind, including but not limited to the freedom from disease, pain, or other abnormality.

health care or **healthcare** *Medicine.* **1.** the general field concerned with the maintenance or restoration of health; medicine. **2.** any of the methods, procedures, or specialized fields within this general field. **3. health-care** or **healthcare.** of or relating to this field.

health food *Nutrition.* a food believed to be nutritious and beneficial to good health, such as whole-grain cereals, fresh fruits and vegetables, or other foods produced and prepared without chemical additives.

heathful *Medicine.* promoting good health; conducive to health.

health maintenance organization *Medicine.* any of a variety of healthcare delivery systems utilizing a group medical practice that provides an alternative to fee-for-service private medical practice; typically a prepaid, organized system providing comprehensive health care to all persons under contract, with an emphasis on preventative medicine. Also, HMO.

health physics *Nucleonics.* a field of science that is concerned with protecting individuals from the hazards associated with ionizing radiation, through such practices as the use of protective equipment and procedures in the workplace.

health psychology *Psychology.* a branch of psychology in which biomedical and behavioral techniques are combined in order to prevent disease and to improve the health practices of both individuals and society in general.

healthy *Medicine.* **1.** in a state of good health; not suffering from disease, pain, or other defect. **2.** promoting good health; healthful. A distinction is often made in which the term *healthy* is restricted to the sense of "in good health" and *healthful* is applied to things that promote good health.

heap *Computer Programming* **1**. a term for a portion of memory organized randomly, or as a stack, and used for dynamic memory allocation. **2**. in data structures, a complete binary tree in which the value stored in each node is less than (or sometimes greater than) or equal to the values of the child nodes.

heap sort *Computer Programming.* an ordering on a binary tree that imposes a heap structure on values as it orders them. Also, TREE SORT.

hearing *Physiology.* **1.** the perception of sounds. **2.** the faculty or sense by which sounds are perceived.

hearing aid *Acoustical Engineering.* **1.** any of various small, self-powered electronic devices used by persons whose hearing is impaired, typically consisting of a microphone, amplifier, receiver, and power source, and designed to fit into or just behind the user's outer ear. **2.** broadly, any device used to amplify sound for improved hearing.

hearing loss *Physiology.* partial or complete failure of the sense of hearing; deafness.

heart *Anatomy.* the muscular organ that contracts and propels blood throughout the blood vessels, consisting in mammals of two atria (right and left) and two ventricles (right and left). *Zoology.* a corresponding structure in other animals. *Botany.* the center or core of a plant, especially the solid center of a tree or the core of a vegetable. *Science.* any structure, part, or shape thought of as comparable to the human heart.

heartbeat *Physiology.* **1.** a throb or pulsation of the heart. **2.** the muscular action or pumping of the heart that occurs with autorhythmicity impressed on it by nodal tissue.

heart block *Medicine.* an impairment of normal conduction of electrical impulses controlling the heartbeat.

heart bond *Civil Engineering.* a masonry bond at a point where the headers meet in the middle of a wall with their joints covered by another header; no headers stretch across the wall.

heartburn *Medicine.* a burning sensation in the upper central area of the abdomen and lower chest, usually from irritation of the esophagus.

heart clot *Hematology.* postmortem coagulation within the heart.

heart disease *Cardiology.* any disease of the heart or coronary arterial system.

heart failure see CARDIAC FAILURE.

heart-failure cells *Pathology.* large, extravasated mononuclear macrophages containing granules of iron, present in the pulmonary alveoli, lungs, and sputum in congestive heart failure.

hearth *Building Engineering.* a stone or brick surface forming the base of a fireplace, typically extending a short distance into a room and often slightly raised above the level of the floor. *Metallurgy.* the bottom of a smelting or refining furnace. *Archaeology.* **1.** any feature or area showing evidence of human use for fire, as by the presence of ashes, charcoal, charred rock, and so on. **2.** see FIRESTARTER.

heart-hand syndrome see HOLT-ORAM SYNDROME.

hearth furnace *Metallurgy.* a furnace in which the charge rests on a hearth.

heartland *Geography.* a popular term for the central region of a continent, especially the part devoted to agriculture.

heart-lung machine *Medicine.* an apparatus consisting of a pump and an oxygenator used as a bypass to perform the functions of the heart and lungs; usually used during heart or lung surgery.

heart murmur see CARDIAC MURMUR.

heart rate *Physiology.* the action of the heart, measured in beats per minute.

heart rot or **heartrot** *Plant Pathology.* **1.** any rot that is caused by wood-decaying organisms and involves the disintegration of the supporting vascular tissue of a tree. **2.** a disease that rots the central tissue of beets and related plants; caused by the fungus *Mycosphaerella tabifica.* **3.** a rotting disease of sugarbeets caused by a boron deficiency. **4.** a fatal palm-tree disease caused by the trypanosomatid flagellate *Phytomona.*

heart valve *Anatomy.* one of several flaplike structures that separate the chambers of the heart from one another and from the large vessels that bring blood to or remove blood from the heart, including the tricuspid valve, the mitral valve, the aortic valve, and the pulmonary valve. They function to prevent a backward flow of blood.

heartwater disease *Veterinary Medicine.* an often fatal, tick-borne disease of cattle, sheep, goats, and wild ruminants, caused by the rickettsial microorganism *Cowdria ruminantium*, and characterized by high fever, anorexia, nervous symptoms, edema, hydrothorax, and accumulation of fluid in the pericardial sac. Also, COWDRIOSIS.

heart-weight rule see HESSE'S RULE.

heartwood *Botany.* the nonliving older wood in the center of a tree trunk that is harder, denser, and darker than the newer sapwood surrounding it; though no longer living, it provides strength for the tree. Also, DURAMEN.

heartworm *Invertebrate Zoology.* a parasitic, filarial nematode, *Dirofilaria immitis,* that invades the heart muscle of carnivorous mammals, especially dogs.

heat *Thermodynamics.* **1.** a form of energy that is the result of the temperature difference between the boundary of a system and its surrounding environment. This is more precisely identified as the transfer of energy and thus is often described as *heat transfer* rather than simply *heat.* **2.** a description of this form of energy; the heat transfer that is supplied to a system can be measured by the difference between its energy change and the work done: $Q = \Delta E - W$, for which Q is the heat transfer to the system, ΔE is the energy change in the system, and W is the actual work performed by the system. *Physics.* **1.** the temperature of a body, substance, or physical environment. **2.** specifically, a high temperature; the quality of being hot. *Meteorology.* a condition of relatively high temperature in the atmosphere. *Physiology.* **1.** a bodily sensation of warmth or hotness. **2.** see ESTRUS. *Engineering.* the supplying of warm air to a room, building, or other enclosed area.

heat-affected zone *Metallurgy.* in cutting or joining, the portion of the base in which properties are affected, usually deleteriously, by the heat.

heat balance *Thermodynamics.* an equilibrium condition of a thermodynamic system in which all heat transfers into and out of the system are accounted for. *Geophysics.* specifically, the balance between the heat received by the earth and its atmosphere from the sun, and the heat given off by the earth and its atmosphere.

heat barrier see THERMAL BARRIER.

heat budget *Thermodynamics.* a listing of all the sources of heat transfers for some thermodynamic system, to account for the total heat transfers into or out of the system. *Geophysics.* the amount of atmospheric heat required to raise the temperature of the water in a lake to its maximum summer temperature from its midwinter temperature.

heat capacity *Thermodynamics.* the ratio of the amount of heat transfer applied to a body to the change in temperature produced by this heat; usually expressed as the heat energy required to raise the temperature of a specific substance by 1°C at constant pressure and volume.

heat check *Metallurgy.* surface cracks caused by rapidly heating and cooling during metal working.

heat coil *Electricity.* a protective device designed to ground or open a circuit when the current rises above a specific value.

heat conduction *Thermodynamics.* the heat transfer of energy through a substance by means of the presence of a temperature difference within the substance; it is established that heat will spontaneously flow from a hotter region to a colder region.

heat conductivity see THERMAL CONDUCTIVITY.

heat content *Thermodynamics.* the relative capacity of a system to transfer heat across its boundaries to the surrounding environment; now usually expressed as the enthalpy of the system.

heat convection *Thermodynamics.* the transfer of heat energy from one place to another by means of the actual physical motion of fluid matter.

heat cramp *Medicine.* any cramp caused by dehydration and salt depletion resulting from heat exhaustion, typically occurring after strenuous physical activity in a hot environment.

heat death *Thermodynamics.* a term for the state of a system that has reached maximum entropy (i.e., it has a uniform temperature throughout and a complete lack of order). The system therefore has no further capacity to do work on its surroundings in this state.

heat development *Graphic Arts.* a process in which heat and ammonia are used to develop diazo film material.

heat dissipation see HEAT LOSS.

heat-distortion temperature *Materials Science.* the temperature at which a standard bar of plastic material deflects 0.01 inch under a specified load. Also, **heat-distortion point.**

heat engine *Thermodynamics.* a system that, with the aid of external energy sources, operates a cycle of processes to extract useful work from the system. *Mechanical Engineering.* **1.** any machine or device that functions to convert some portion of the heat transferred to it into mechanical energy. **2.** see EXTERNAL-COMBUSTION ENGINE.

heat equation *Thermodynamics.* a second-order partial differential equation that gives the temperature distribution in a region as a function of space and time, when the properties of the region, the initial temperature distribution, and the temperature at the boundaries all are known.

heat equator *Meteorology.* **1.** a line around the earth that connects all points having the highest mean annual temperature for their longitudes. Also, THERMAL EQUATOR. **2.** the parallel of latitude of 10° north, having the highest mean temperature of any latitude.

heater *Engineering.* any device or apparatus designed for heating something, as to warm the air in a room or motor vehicle, provide hot water for use in a home, and so on. *Electronics.* **1.** an element that supplies heat to an indirectly heated cathode. **2.** a resistor that converts electrical energy into heat.

heat exchange *Chemical Engineering.* a unit operation involving the heating or cooling of a single fluid, or the heating and cooling of two fluids with or without a change of state taking place.

heat exchanger *Engineering.* a device that transfers heat from one medium or system to another; for example, heat from hot fluid contained in a radiator dissipates when the metal walls of the device come in contact with cold air.

heat exhaustion *Medicine.* a condition marked by weakness, vertigo, headache, and nausea, caused by exposure to extreme heat or inability to adapt to climatic heat; often precipitated by intense physical activity.

heat filter *Optics.* a device, usually a glass filter containing a copper compound, that reduces the heat intensity in a light beam without appreciably reducing light intensity; used in lens systems to shield film from heat.

heat flow *Thermodynamics.* **1.** the transfer of energy from one substance to another as a result of a temperature difference. **2.** the amount of such transfer for a unit time.

heat-flow province *Geophysics.* a geographic area in which heat-flow characteristics are common.

heat flux *Thermodynamics.* the flow of heat across a surface of unit area in a unit amount of time; commonly expressed in units of cal/cm^2-sec or $W/(M^2\text{-}A)$.

heat gain *Engineering.* the rise in the temperature in a space that is produced by objects and activity within it, such as lights, machinery, humans, and solar radiation.

heath *Ecology.* **1.** an area of open land with low-growing evergreen shrubs and few trees, usually having acidic, poorly drained soil of low fertility; found extensively in the British Isles. **2.** any of the various low-growing evergreen shrubs common on such land.

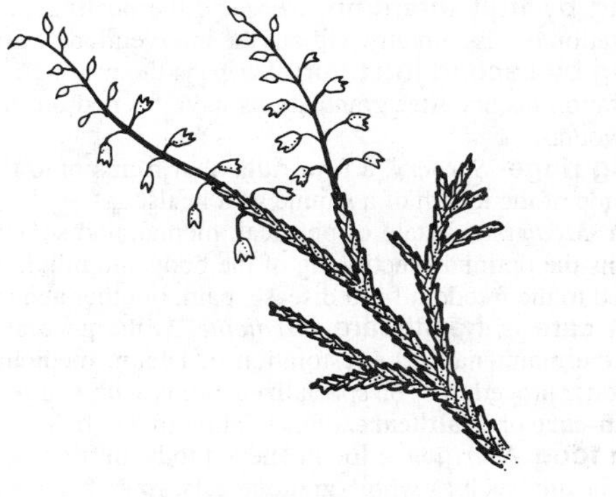

heather

heather *Botany.* any plant of the genus *Calluna,* family Ericaceae (the heath family), especially *C. vulgaris*, an evergreen heath with racemes of small purple-pink flowers; native to northern and alpine regions.

heating chamber *Engineering.* an enclosed area within a machine where materials are heated before further processing, such as the part of an injection mold where cold plastic is made molten.

heating degree-day *Meteorology.* a basis upon which the use of fuel for home heating is measured; one heating degree-day is given for each degree below 65°F of the daily mean air temperature.

heating element *Electricity.* in a cooking or heating appliance, the coiled resistor and associated ceramic insulation through which electrical energy is directed to produce heat.

heating load *Civil Engineering.* the heat per unit time that is required to maintain a specific temperature within a given enclosed space.

heating plant *Civil Engineering.* the entire system for heating an enclosed space including either a boiler, piping, and radiators or a furnace, ducts, and air outlets.

heating surface *Engineering.* a surface that absorbs heat and transfers it from one medium or system to another, such as a boiler surface having hot gas on one side and cold water on the other.

heating value see CALORIFIC VALUE.

heat island *Meteorology.* an area, such as the downtown of a major metropolitan area, in which the air temperature is generally higher than that of surrounding rural areas due to restricted air-mixing abilities; the heat that is absorbed and generated by the pavement, structures, and people in such an area is often prevented from rising and being cooled normally because of an overlying layer of haze and pollutants.

heat lamp *Electricity.* an infrared lamp that is used in situations in which heat is needed but light is not necessary, as for incubation or for keeping food warm.

heat lightning *Geophysics.* a term for the light seen from ordinary lightning that is too far away for the associated thunder to be heard.

heat loss *Physics.* the energy or power transmitted from a system in the form of heat. Also, HEAT DISSIPATION.

heat-loss flowmeter *Engineering.* an instrument that determines the quantity of fluid flow per unit time by measuring the cooling effect of the flow on an electrical sensor. Also, THERMAL-LOSS METER.

heat low see THERMAL LOW.

heat of absorption *Physical Chemistry.* the amount of heat released or absorbed when a gas is dissolved. Also, ENTHALPY OF ABSORPTION.

heat of activation *Physical Chemistry.* the increase in the heat content of a substance as it moves from a less reactive form to a more active form at constant pressure. Also, ENTHALPY OF ACTIVATION.

heat of adhesion *Physical Chemistry.* the amount of heat released or absorbed when a mixture or solution of different substances is formed.

heat of admixture see HEAT OF MIXING.

heat of adsorption *Physical Chemistry.* the amount of heat released when one substance is adsorbed on the surface of another at constant pressure. Also, ENTHALPY OF ADSORPTION.

heat of aggregation *Physical Chemistry.* the amount of heat change involved in the formation of an aggregate of matter, such as a crystal.

heat of association *Physical Chemistry.* the increase in heat content that accompanies the formation of a compound at constant pressure. Also, ENTHALPY OF ASSOCIATION.

heat of combination see HEAT OF FORMATION.

heat of combustion *Physical Chemistry.* the amount of heat released when a definite quantity of a substance is completely oxidized at constant pressure or constant volume. Also, ENTHALPY OF COMBUSTION, HEAT OF OXIDATION.

heat of compression *Physical Chemistry.* the amount of heat generated when a given volume of a gas is compressed by a specified amount.

heat of condensation *Physical Chemistry.* the amount of heat released when a definite quantity of a substance condenses at constant pressure and temperature. Also, ENTHALPY OF CONDENSATION.

heat of cooling *Physical Chemistry.* the change in heat content of a system that is cooling under constant pressure, resulting from an internal change of the system, such as an allotropic modification.

heat of crystallization *Physical Chemistry.* the amount of heat released or absorbed when a definite quantity of a substance is crystallized at constant pressure. Also, ENTHALPY OF CRYSTALLIZATION.

heat of decomposition *Physical Chemistry.* the amount of heat released or absorbed when a definite quantity of a compound decomposes into simpler molecules at constant pressure. Also, ENTHALPY OF DECOMPOSITION.

heat of dilution *Physical Chemistry.* the amount of heat released or absorbed when a definite quantity of a solvent is added to a solution of given concentration at constant pressure. Also, ENTHALPY OF DILUTION.

heat of dissociation *Physical Chemistry.* the amount of heat change involved when a molecule fragments or disintegrates. Also, ENTHALPY OF DISSOCIATION.

heat of dissolution see HEAT OF SOLUTION.

heat of emission *Electronics.* additional heat energy supplied to an electron-emitting surface to keep its temperature constant.

heat of evaporation see HEAT OF VAPORIZATION.

heat of explosion *Physical Chemistry.* the amount of heat released by a definite quantity of an explosive substance.

heat of formation *Physical Chemistry.* the amount of heat released or absorbed when a definite quantity of a substance is formed from its elements at a specific temperature and pressure. Also, HEAT OF COMBINATION.

heat of fusion *Physical Chemistry.* the amount of heat required to convert the solid form of a substance into its liquid form through melting, at constant pressure and temperature. Also, LATENT HEAT OF FUSION, ENTHALPY OF FUSION.

heat of hydration *Physical Chemistry.* the amount of heat released or absorbed by a definite quantity of a substance during a process of hydration at constant pressure and temperature. Also, ENTHALPY OF HYDRATION.

heat of hydrogenation *Physical Chemistry.* the degree of heat change involved in the reaction of hydrogen with another substance, as with unsaturated hydrocarbons. Also, ENTHALPY OF HYDROGENATION.

heat of ionization *Physical Chemistry.* the increase in heat content when a substance is totally ionized at a constant pressure. Also, ENTHALPY OF IONIZATION.

heat of linkage *Physical Chemistry.* the energy required to form bonds between atoms in a molecule, determined by the amount of energy needed to break the bonds of the molecule divided by the number of bonds in the molecule.

heat of mixing *Physical Chemistry.* the difference between the heat content of a mixture of components and the sum of the heat contents of the individual components at the same temperature and pressure. Also, HEAT OF ADMIXTURE.

heat of neutralization *Physical Chemistry.* the amount of heat released by the reaction of a strong acid with a strong base. Also, ENTHALPY OF NEUTRALIZATION.

heat of oxidation see HEAT OF COMBUSTION.

heat of reaction *Physical Chemistry.* the amount of heat that is released or absorbed when a chemical reaction takes place, with the final conditions of temperature and pressure being the same as the original conditions.

heat of recovery *Metallurgy.* the energy released in a cold-worked metal when it undergoes recovery.

heat of solidification *Physical Chemistry.* the amount of heat released by the solidifying of a liquid at its freezing point or of a gas at its sublimation point, with constant temperature and pressure. Also, ENTHALPY OF SOLIDIFICATION.

heat of solution *Physical Chemistry.* the amount of heat released or absorbed when a definite quantity of a substance dissolves in a solvent to form a solution of given concentration at constant pressure. Also, ENTHALPY OF SOLUTION, HEAT OF DISSOLUTION.

heat of sublimation *Physical Chemistry.* the amount of heat change involved when a definite quantity of a substance is converted from a solid to a gaseous state at constant pressure and temperature. Also, LATENT HEAT OF SUBLIMATION, ENTHALPY OF SUBLIMATION.

heat of transfer *Physical Chemistry.* the amount of heat released or absorbed by the transfer of a component from one ideal solution to another. Also, ENTHALPY OF TRANSFER.

heat of transition *Physical Chemistry.* the amount of heat change that occurs as a substance undergoes some change of phase at constant pressure and temperature. Also, **heat of transformation.**

heat of vaporization *Physical Chemistry.* the amount of heat change required to convert a definite quantity of a liquid substance from a liquid to a gas at constant pressure and temperature. Also, LATENT HEAT OF VAPORIZATION, ENTHALPY OF VAPORIZATION, HEAT OF EVAPORATION.

heat of wetting *Physical Chemistry.* the amount of heat change associated with the adsorption of a liquid (usually water) on another substance.

heat pipe *Mechanical Engineering.* a self-contained heat-transfer device that transports thermal energy by vaporizing a liquid inside one end near a heat source and recondensing it at the other end.

heat pump *Mechanical Engineering.* a machine that uses a refrigerant to transfer heat energy from a cold temperature source, such as the ground, air, or water, to another source, such as a building; after absorbing energy from the cold temperature source, the refrigerant is mechanically compressed, thus creating a temperature increase; the heat is then transferred to the new source using a heat exchanger.

heat radiation *Thermodynamics.* electromagnetic radiation that is in equilibrium with matter at a constant temperature.

heat rash see MILIARIA.

heat rate *Mechanical Engineering.* an expression of the conversion efficiency of thermal energy to work output.

heat reactor *Nucleonics.* a device that generates heat from nuclear fission for industrial purposes.

heat release *Thermodynamics.* the liberation of energy through a heat transfer from a system, especially as measured for a certain volume over a given time.

heat reservoir *Thermodynamics.* any body that receives or accepts a heat transfer from a heat engine and is sufficiently large to be unaffected by the transfer.

heat-resistant alloy *Metallurgy.* any of several alloys that resist stress and oxidation at elevated temperatures.

heatseal *Engineering.* to bond or weld a material to itself or another material by means of heat or heat and pressure; used commonly when joining thermoplastic materials.

heat-seeker or **heatseeker** *Ordnance.* a guided missile equipped with a heat-sensitive, infrared device for homing on heat-radiating machines or installations. Thus, **heat-seeking.**

heat shield *Space Technology.* any protective device, such as metal sheeting on a nose cone, used to protect a reentry vehicle from aerodynamic heating.

heat-shock gene *Genetics.* a gene whose activity is induced by exposure to a brief period of elevated temperture.

heat-shock protein *Biochemistry.* any protein that is synthesized by an organism in response to certain stresses, such as an abrupt rise in temperature or electromagnetic radiation; occurs in a variety of organisms including bacteria, plants, and mammals.

heat-shock puff *Genetics.* on a polytene chromosome of *Drosophila,* a localized swelling in response to a brief exposure to elevated temperature, indicating an increase in transcriptional activity.

heat sink or **heatsink** *Thermodynamics.* any body of matter, gaseous, liquid, or solid, that receives a heat transfer from its surrounding environment. *Space Technology.* **1.** on a spacecraft or reentry vehicle, any protective device or mass of metal, fuel, oil, or other material used to absorb unwanted heat. **2.** in a nuclear-powered spacecraft, any thermodynamic device, such as a radiator or condenser, used to absorb excess heat energy from the working fluid. Also, **heat dump**. *Electrical Engineering.* a metal plate used to conduct and radiate heat away from an electrical component in an electrical device.

heatsink cooling *Engineering.* a cooling of a body or system through the absorption of its thermal energy by another object, substance, or surface in close proximity to it.

heat source *Thermodynamics.* any body or system that serves to provide energy for another body or system.

heat sterilization *Engineering.* to destroy, remove, and prevent the further breeding of microorganisms by means of heat (dry or moist).

heat storage *Oceanography.* the action of the ocean as a heat reservoir, moderating the daily and seasonal temperature variations over the sea.

heat stress index *Physiology.* for an average person, the ratio of the amount of sweat evaporated under certain job conditions to the person's maximum evaporative capacity.

heatstroke *Medicine.* a severe and sometimes fatal heat-exposure syndrome marked by extremely high fever (usually 105°F or more) and extreme exhaustion; a neurological response to excessive heat.

heat thunderstorm *Meteorology.* a popular term for a convective thunderstorm occurring at the end of a hot and humid summer day.

heat time *Metallurgy.* in welding, the time during which heat is applied.

heat tinting *Metallurgy.* a metallographic technique that enhances microstructural features by oxidizing the surface of the specimen.

heat transfer *Thermodynamics.* the exchange of heat energy between a system and its surrounding environment, which results from a temperature difference and takes place by means of a process of thermal conduction, mechanical convection, or electromagnetic radiation.

heat-transfer coefficient *Thermodynamics.* the amount of heat exchanged across a unit area in a unit amount of time through a temperature difference of 1°C between the system boundary and the surrounding fluid. Also, FILM COEFFICIENT. *Materials Science.* the coefficient h, relating the rate of heat q, transfer to the contact area A, and temperature difference between two phases, $q = hA(T_1 - T_2)$, values of h being dependent upon the physical properties of the two phases, as well as their relative motions.

heat transmission see HEAT FLOW.

heat-treatable alloy *Metallurgy.* an alloy that can be hardened by heat treatment, without mechanical deformation.

heat treatment *Metallurgy.* any process that modifies the properties of a metal or alloy and follows a defined schedule of heating and cooling.

heat wave or **heatwave** *Meteorology.* a popular term for a period of two or more days of abnormally and uncomfortably hot, and usually humid, weather. *Electromagnetism.* radiation in the infrared region of the spectrum, of a frequency higher than that of radio waves.

heave *Geology.* **1.** a predominantly upward expansion or displacement of a soil surface, caused by swelling clay, overloading of an embankment, seepage pressure, or frost action. **2.** the horizontal component of separation or displacement in a fault. *Oceanography.* the vertical and horizontal movement caused in a floating body by wave action.

heaves *Veterinary Medicine.* a noninfectious chronic emphysema of the lungs of horses, which is characterized by respiratory distress due to overinflation of the alveoli and associated with moldy or dusty stable conditions.

heave to *Navigation.* to cause a vessel to maintain an approximate stationary position, especially with reference to the waves; the vessel may make some headway or leeway, but is not considered to be underway.

heavier-than-air craft *Aviation.* any flight vehicle weighing more than the air it displaces and acquiring its lift from aerodynamic forces. Also, AERODYNE.

heavier-than-air craft

heaving *Naval Architecture.* **1.** the vertical movement of a hull as a whole, as distinguished from pitching. **2.** any vertical movement of a ship in a seaway. **3.** the use of mechanical power to accomplish a shipboard task, as distinguished from handpower (or hauling). *Petroleum Engineering.* a limited or complete collapse of drillhole walls created by internal pressures.

Heaviside, Oliver 1850–1925, English physicist; important advances in electromagnetism; codiscoverer of Kennelly-Heaviside Layer.

Heaviside calculus *Mathematics.* an operational calculus applied to ordinary differential equations connected with electromagnetic problems; a forerunner to the study of distributions.

Heaviside-Lorentz system *Electromagnetism.* a system of units having the same quantities as the Gaussian system of units, with the exception that the units of charge and current are reduced by a factor of $(4\pi)^{-1/2}$ while the units of electric and magnetic field strengths are increased by a factor of $(4\pi)^{1/2}$.

Heaviside unit function *Mathematics.* the function given by $H(x) = 1$ for $x \geq 0$ and $H(x) = 0$ for $x < 0$. The derivative of the Heaviside function is the Dirac distribution.

heavy alloy *Metallurgy.* a powder metallurgy, tungsten-base alloy containing nickel, which is used for electrical contacts and radiation shielding.

heavy antiaircraft artillery *Ordnance.* antiaircraft artillery with a caliber larger than 90 mm and weighing more than 40,000 lbs in a trailed mount.

heavy antitank weapon *Ordnance.* a weapon that is capable of operating from either the ground or a vehicle, but not easily man-portable, used against armor or other material targets. Also, **heavy assault weapon.**

heavy artillery *Ordnance.* **1.** artillery with a caliber of 155 mm or larger. **2.** large howitzers and cannons in general, as opposed to smaller antiaircraft weapons. Also, **heavy field artillery.**

heavy-atom derivative *Crystallography.* the resulting crystal obtained by soaking a solution of the salt of a metal of high atomic number into a crystal of a protein. If the derivative is to be of use in structure determination, the heavy atom must be substituted in only one or two ordered positions per molecule of protein. Then the method of isomorphous replacement can be used to determine phases for the calculation of an electron density map.

heavy-atom method *Crystallography.* a method of deriving phase angles in which the phases calculated from the position of the heavy atom are used to compute the first approximate electron density map, from which further portions of the structure are recognizable as additional peaks in this map. If necessary, successive approximate electron density maps are computed to give the entire structure.

heavy bombardment *Military Science.* an intense bombardment, especially one using large aerial bombs or missiles.

heavy bomber *Aviation.* any bomber considered large and heavy for its time, such as the B-29 of World War II or the postwar B-52.

heavy-case bomb *Ordnance.* a high-explosive bomb whose container is relatively heavy compared to the weight of the bursting charge.

heavy chain *Immunology.* the heavier of the two types of polypeptide chains that occur in a normal immunoglobulin or antibody molecule. Also, H CHAIN.

heavy crude *Petroleum Engineering.* crude oil of 20° API gravity or less that contains a great proportion of viscous, high-molecular-weight hydrocarbons plus a generally high sulfur content; requires special production procedures to extract from underground formations.

heavy cruiser *Naval Architecture.* as defined by the Washington Naval Conference in the 1920s, a cruiser of up to ten thousand tons displacement, armed with up to eight-inch guns; not a current usage.

heavy drop *Military Science.* an airdrop in which trucks, artillery, or other heavy objects are dropped by parachute.

heavy-duty *Engineering.* of an object, designed to resist damage from excessive strain, harsh weather, and long wear. Also, **heavy-service.**

heavy-duty car *Mechanical Engineering.* a railway motorcar designed for heavy hauling and yard service, weighing more than 1400 pounds and powered by a 12- to 30-horsepower engine.

heavy feeder *Agronomy.* a plant that requires a large amount of fertilizer, because it consumes nutrients at a high rate.

heavy floe *Oceanography.* an ice floe that is more than 3 meters thick. Also, **heavy ice.**

heavy force fit *Design Engineering.* a fit for heavy steel parts or shrink fits in medium sections.

heavy fraction *Petroleum Engineering.* during distillation, the heavier final products produced from crude oil such as heavy gas oil, fuel oil, lubes, and asphalt. Also, END CUT.

heavy howitzer *Ordnance.* a howitzer with a caliber larger than 200 mm.

heavy hydrogen see DEUTERIUM.

heavy ion see LARGE ION.

heavy-ion linear accelerator *Nucleonics.* a machine in which an alternating voltage generates a beam of highly energized large ions; used to create transuranic elements and short-lived isotopes, and to analyze nuclear reactions, nuclear spectroscopy, and large-ion absorption.

heavy-ion source *Electronics.* a source of ionized molecules or atoms of elements that are heavier than helium.

heavy isotope *Chemistry.* a form of a chemical element that has a higher mass number than other forms of the same element due to the presence of one or more additional neutrons in the nucleus.

heavy-lift ship *Naval Architecture.* a type of ship equipped with powerful derricks and other fittings for handling massive cargoes.

heavy-liquid bubble chamber *Nucleonics.* a device in which charged particles leave a trail of tiny bubbles as they move through deuterium or an organic liquid, such as propane or Freon, at a temperature just above their respective boiling points, thus allowing researchers to analyze the particles and their interactions.

heavy machine gun *Ordnance.* a relatively heavy machine gun including .30 caliber water-cooled guns, .50 caliber machine guns, or any aircraft machine gun larger than .30 caliber.

heavy mass see GRAVITATIONAL MASS.

heavy meromyosin *Biochemistry.* the portion of a myosin molecule consisting of the globular head and part of the tail.

heavy metal *Metallurgy.* any metal or alloy of high specific gravity, especially one that has a density higher than 5 grams per cubic centimeter.

heavy-metal soap see METALLIC SOAP.

heavy-metal star *Astronomy.* see S STAR.

heavy mineral *Mineralogy.* a rock-forming mineral having a specific gravity higher than 2.9, the specific gravity of bromoform.

heavy mineral analysis *Archaeology.* a rock analysis carried out on artifacts such as potsherds to identify the materials used; the sherd is crushed and put into a viscous fluid in which the heavier minerals sink to the bottom.

heavy mineral oil *Materials Science.* mineral oil of a specified heavier density, with a specific gravity of 0.845 to 0.905.

heavy oxygen *Nuclear Physics.* an isotope of oxygen with atomic weight 18 that makes up a small portion of the oxygen-16 that is found in water, air, and rocks. Also, OXYGEN-18.

heavy-rail transit see RAIL RAPID TRANSIT.

heavy rocket *Ordnance.* a nonguided rocket with a diameter of 318 mm or larger.

heavy sea *Oceanography.* the condition of the sea when waves are running high.

heavy section car *Mechanical Engineering.* a railway motorcar that weighs approximately 1200 to 1400 pounds and is powered by an 8- to 12-horsepower engine.

heavy strand see H STRAND.

heavy tank *Ordnance.* a full-track combat tank weighing from 56 to 85 tons.

heavy-timber construction *Building Engineering.* a type of construction, used in nearly all building code applications and permits, characterized by the use of heavy timber for primary structural members. Also, MILL CONSTRUCTION.

heavy water *Inorganic Chemistry.* **1.** deuterium oxide, D_2O, water in which the two hydrogen atoms are in the form of deuterium, which has about twice the mass of normal hydrogen. There is about one part of heavy water in 6500 parts of ordinary water. Heavy water has a higher freezing and boiling point than ordinary water (3.8°C; 101.42°C) and a greater density (1.1056 at 20°C). It does not support plant and animal life. It is used in thermonuclear weapons and nuclear reactors. **2.** tritium oxide, another variety of this in which the hydrogen atoms are in the form of tritium (which is about three times the mass of normal hydrogen) rather than of deuterium.

heavy-water reactor *Nucleonics.* a nuclear reactor in which heavy water, D_2O, moderates the velocity of neutrons, thus increasing the potential for fission in unenriched uranium. Also, **heavy-water-moderated reactor.**

heavy weapon *Ordnance.* a heavy machine gun, recoilless rifle, mortar, howitzer, long-barreled cannon, or other such weapon used by the infantry.

heavyweight concrete *Materials.* a type of high-density concrete that is used for radiation shielding.

Heawood map coloring theorem *Mathematics.* a proper coloring of a map on a surface of genus $p \geq 1$ requires at most $[7 + (1 + 48p)^{1/2}]/2$ colors. The proof of this theorem, first conjectured in 1890, was completed in 1968 by Ringel and Youngs. Also, RINGEL-YOUNGS' THEOREM.

heazlewoodite *Mineralogy.* Ni_3S_2, a light bronze, opaque, metallic trigonal, mineral, massive in habit, having a specific gravity of 5.82 and a hardness of 4.0 on the Mohs scale; found in serpentine.

Hebe [hē´bē] *Astronomy.* asteroid 6, discovered in 1847 and measuring 186 kilometers in diameter; it belongs to type S. (Named for *Hebe,* the Greek goddess of youth.)

hebephrenia *Psychology.* see HEBEPHRENIC SCHIZOPHRENIA.

hebephrenic [hē bə fren´ik] *Psychology.* **1.** of or relating to hebephrenic schizophrenia. **2.** a person affected with this condition.

hebephrenic schizophrenia *Psychology.* a type of schizophrenia characterized by extreme disorders in thought, unstable or markedly silly behavior, a lack of emotion, and delusions and hallucinations; generally having its onset during adolescence. Also, HEBEPHRENIA, DISORGANIZED SCHIZOPHRENIA.

Heberden's disease *Medicine.* **1.** a degenerative joint disease of the terminal joints of the fingers, marked by inflammation, enlargement, and flexion deformities. **2.** another name for angina pectoris. (Named for William *Heberden,* 1710–1801, English physician.)

Heberden's node *Medicine.* an abnormal bony or cartilaginous nodule formed usually at the terminal finger joint, associated with degenerative osteoarthritis.

hebetic *Physiology.* of, relating to, or occurring during puberty.

hebetude [hēb´ə tood] *Psychology.* a state of emotional dullness in which the individual appears to be listless, apathetic, and completely withdrawn from his environment, commonly seen in schizophrenia. *Medicine.* apathy or dullness from any cause.

Hebrovellidae *Invertebrate Zoology.* a family of surface water bugs, hemipteran insects in the subdivision Amphibicorisae.

hecatomeric *Neurology.* describing a spinal neuron having processes that divide into two, one going to each side of the spinal cord. Also, **hecatomeral, hecateromeral, hecateromeric.**

Heckler and Koch *Ordnance.* any of several West German small arms weapons, including 7.62-mm and .223-caliber assault rifles, a 9-mm Parabellum submachine gun, a .223-caliber belt-fed light machine gun and a .50-caliber heavy machine gun.

hect- or **hecto-** a combining form meaning "one hundred."

hectare [hek´târ] *Metrology.* a basic unit of mass in the metric system, equal to 10,000 square meters or 100 ares; equivalent to 2.471 acres.

hectocotylus *Invertebrate Zoology.* a tentacle of a male cephalopod mollusk modified to fertilize eggs; in some species it breaks off when inserted under the female's mantle.

hectogram *Metrology.* a unit of mass equal to 100 grams. Also, **hectogramme.**

hectography see GELATIN DUPLICATING.

hectoliter *Metrology.* a unit of capacity equal to 100 liters. Also, **hectolitre.**

hectometer *Metrology.* a unit of length equal to 100 meters. Also, **hectometre.**

hectometric wave *Telecommunications.* a radio wave having a frequency between 3000 and 300 kilohertz, and a free-space wavelength between 1 and 10 hectometers.

hectorite *Mineralogy.* $NaO_3(Mg,Li)_3Si_4O_{10}(F,OH)_2$, a monoclinic mineral of the smectite group, having a specific gravity of 2 to 3 and a hardness of 1 to 2 on the Mohs scale.

H.E.D. *Radiology.* a unit of X-ray dosage. (An abbreviation of Haut-Einheits-Dosis, German for "unit skin dose.")

hedenbergite *Mineralogy.* $CaFe^{+2}Si_2O_6$, a green monoclinic mineral of the pyroxene group occurring in massive lamellar form or as short prismatic crystals, and having a specific gravity of 3.5 to 3.56 and a hardness of 6 on the Mohs scale; found in limestone contact zones and in iron-rich metamorphic rocks.

hedgehog *Vertebrate Zoology.* a partly insectivorous mammal of the subfamily Erinaceinae, having an elongated blunt head, and with most species having a spiny coat (as in spiny hedgehogs); found among logs, rocks, tree roots, brush piles, termite mounds, and burrows in Europe, Africa, and Asia. *Ordnance.* **1.** a portable obstacle made of crossed poles and barbed wire. **2.** an obstacle made of steel or iron bars imbedded in concrete; used to impede and damage boats, tanks, and other vehicles of a beach landing force. *Military Science.* a concentration of entrenched and fortified troops facing all directions, especially with antipersonnel, antitank, and antiaircraft capability.

hedgehog

Hedgehog *Ordnance.* an antisubmarine weapon consisting of a deck-mounted mortar launcher for 24 small contact-fused depth bombs, able to project a pattern ahead of the ship; officially identified as a 7.2-inch high-explosive projector charge.

hedge sparrow see ACCENTOR.

Hedi *Ordnance.* a U.S. missile system designed to intercept and destroy enemy intercontinental ballistic missiles in the upper atmosphere. (An acronym for high endoatmospheric defense interceptor.)

hedleyite *Mineralogy.* Bi_7Te_3, a tin-white, opaque, metallic trigonal mineral, having a specific gravity of 8.6 to 8.91 and a hardness of 2 on the Mohs scale; found with native gold and pyrrhotite in skarns and in sulfide deposits.

hedonic *Psychology.* of or relating to pleasure.

hedonic gland *Vertebrate Zoology.* in male salamanders and newts, a mucus-secreting scent gland that acts as a sexual chemoattractant affecting female olfactory nerves in courtship behavior.

hedonism [hē´də niz əm] *Psychology.* **1.** the doctrine that pleasure or happiness is the ultimate goal of human conduct. **2.** the theory that all behavior is motivated by the desire to seek pleasure and avoid pain.

hedonistic *Psychology.* relating to or characterized by hedonism.

hedonophobia *Psychology.* an irrational fear of pleasure.

hedreocraton *Geology.* the stable center of a continental craton, including both shield and platform.

hedrite *Materials Science.* a sheaflike transitional crystalline structure formed during the cooling of a polymer melt; an intermediate stage in the formation of a spherulite.

Hedvall effect I *Solid-State Physics.* a discontinuity in the temperature dependence of the chemical reaction rate at the Curie temperature.

Hedvall effect II *Solid-State Physics.* a discontinuity in the activation energy at the Curie temperature.

Hedwigiaceae *Botany.* a family of robust, dull-green mosses in the order Isobryales, characterized by its leaf cells and lack of a peristome.

hedyphane *Mineralogy.* $Pb_3Ca_2(AsO_4)_3Cl$, a yellowish-white, translucent, hexagonal mineral of the apatite group, occurring in prismatic and tabular crystals, and having a specific gravity of 5.82 and a hardness of 4.5 on the Mohs scale.

heel *Anatomy.* the hindmost part of the foot. *Engineering.* any part that resembles a shoe heel in shape, use, or location in relation to other parts. *Mechanical Engineering.* see HEEL BLOCK. *Navigation.* an inclination (about the longitudinal axis) of a vessel to one side or the other caused by wind or wave action. **Heeling** is a normal and relatively long-lasting (several minutes to several days) inclination; rolling is a momentary action and usually implies an oscillation from one side to the other; listing is an undesirable inclination caused by improper loading or by some other imbalance.

heel block *Mechanical Engineering.* a part fixed on a die shoe to minimize the deflection of a cam or punch. Also, HEEL.

heeling adjuster *Navigation.* the dip needle used to find the correct position of a heeling magnet.

heeling error *Navigation.* a variation in the compass of an iron ship that is caused by the heeling of the ship.

heeling magnet *Navigation.* a magnet that is positioned vertically in a tube and placed directly under the compass of a ship to correct heeling error.

heel plate *Civil Engineering.* a truss end plate.

heel post *Civil Engineering.* a post serving as the support for hinges of a gate or door.

Hefner-Alteneck, Friedrich Franz von 1845–1904, German engineer; invented Hefner lamp, Hefner candle, and drum armature.

Hefner candle *Optics.* a unit of luminous intensity that equals the amount of light emitted from a Hefner lamp burning amyl acetate under standard conditions or 0.9 international candle; used in the early 1900s. Also, **Hefnerkerze.**

Hefner lamp *Chemistry.* a lamp that burns amyl acetate; also burns pentyl acetate with a 4-cm flame; used as a photometric standard.

Hehner number *Analytical Chemistry.* the weight percent of water-insoluble fatty acids in oils and fats.

Heidelberg capsule *Radiology.* a pill that emits radio waves for use in measuring pH values in gastric acidity.

Heidelberg man *Paleontology.* the name given to a hominid known only from a mandible found near Heidelberg in 1907, dated to the middle Pleistocene and probably related to *Homo erectus*; possibly also related to the Ternifine remains.

heifer [hef´ər] *Agriculture.* a young cow, especially one that has not bred a calf.

height *Metrology.* the upward extent of any object or point above a given level, such as the surface of the ground or the sea. *Mathematics.* **1.** the perpendicular distance between horizontal lines or planes; an altitude. **2.** for a rational number m/n, where m and n are relatively prime, the larger of $|m|$ and $|n|$.

height above airport *Aviation.* the height of the minimum descent altitude above a given airport.

height above landing *Aviation.* the height above a helicopter landing area used for helicopter instrument approaches.

height above touchdown *Aviation.* the height of the minimum descent altitude or decision height above the elevation of the touchdown zone of the highest runway for a given airport.

height anomaly *Cartography.* any difference between the height of a point on the earth's surface above the reference spheroid and the corresponding normal height as measured along the normal plumb line.

height-change chart *Meteorology.* a chart indicating the altitude changes of a constant-pressure surface over a specified time interval; similar to a pressure-change chart.

height-change line *Meteorology.* **1.** a line indicating an equal altitude change in a constant-pressure surface over a specified previous interval of time. **2.** a line drawn on a height-change chart. Also, CONTOUR-CHANGE LINE, ISOALLOHYPSE.

height control *Electronics.* in a television, the adjustment determining the amplitude of the vertical-scanning pulses or the observed height of the picture.

height equivalent of theoretical plate *Chemical Engineering.* the height of packing in a packed fractionating column that achieves a separation equivalent to that calculated for a theoretical plate; used in sorption and distillation calculations. *Analytical Chemistry.* specifically, the efficiency in chromatography expressed in terms of the number of theoretical plates and the length of the column in which the solute is in equilibrium between the mobile and stationary phases.

height finder *Engineering.* the component in radar that finds the altitude of an aerial target. Thus, **heightfinder radar.**

height gauge *Engineering.* a gauge that comprises a micrometer or vernier scale and an upright that is graduated for direct reading; used to lay out points from a plane or base surface and to measure heights.

height of burst *Ordnance.* vertical distance from the ground or target to the point of burst.

height of eye *Navigation.* the distance above sea level of the eye of an observer making a celestial observation.

height-of-eye correction *Navigation.* a correction to a sextant altitude due to a dip of the horizon.

height-of-instrument method *Engineering.* a surveying technique used to determine heights; the surveyor notes the first sight on a point of known level, then calculates the height of any other point by subtracting the staff reading from the instrument height. Also, COLLIMATION METHOD.

height of tide *Navigation.* the difference between the tidal reference point (generally charted depth) and the actual depth of the water at any given time.

height of transfer unit *Chemical Engineering.* a parameter with dimensions of length that is related to the separation efficiency of certain countercurrent contacting devices.

height pattern *Meteorology.* the geometric characteristics of height distribution of a constant-pressure surface as shown by contour lines on a pressure chart. Also, ISOBARIC TOPOGRAPHY, PRESSURE TOPOGRAPHY.

height-position indicator *Electronics.* a radarscope that displays both the height of a target and its angular elevation slant range.

height-range indicator *Electronics.* a radarscope that displays both altitude and range measurements of a target.

heiligenschein *Optics.* the diffuse white halo that encircles the shadow of an observer when sunlight reflects off dew drops.

Heilong see AMUR.

Heimlich maneuver [hīm′lik] *Medicine.* a method of dislodging food or another object from the trachea of a choking victim, accomplished by wrapping the arms around the choking person at the belt line, then making a fist with one hand and grasping it with the other; with both hands placed against the victim's abdomen below the rib cage and slightly above the navel, the fist is forcefully pressed into the abdomen with a quick upward thrust. This maneuver may be repeated several times if necessary. (Named after the American physician Henry J. *Heimlich,* born 1920, who developed this method.)

Heine, (Heinrich) Eduard [hī′nə] 1821–1881, German mathematician; formulated the Heine-Borel Theorem of uniform continuity.

Heine-Borel property *Mathematics.* a topological space X is said to have the Heine-Borel property if every closed and bounded subset of X is compact.

Heine-Borel theorem *Mathematics.* a subset of R^n is compact if and only if it is closed and bounded.

Heine's operation see CYCLODIALYSIS.

heintzite see KALIBORITE.

Heinz bodies *Hematology.* microscopic bodies noted in red blood cells with enzyme deficiencies, identified as either cholesterinolein-based or as dead cytoplasm, resulting from oxidative injury to and precipitation of hemoglobin. (Named after Robert *Heinz,* 1865–1924, German pathologist who identified them.)

Heisenberg, Werner [hīz′ən burg] 1901–1976, German physicist; a founder of quantum and matrix mechanics; formulated the uncertainty principle.

Heisenberg algebra *Quantum Mechanics.* a matrix algebra formed by the position and momentum operators.

Heisenberg equation of motion *Quantum Mechanics.* the equation governing the time dependence of an operator in the Heisenburg representation.

Heisenberg exchange coupling *Solid-State Physics.* the exchange forces acting between electrons in neighboring atoms in a crystal that, in the Heisenberg theory, are responsible for ferromagnetism.

Heisenberg force *Nuclear Physics.* a force that exists between two nucleons that causes the particles to exchange their spin and position.

Heisenberg representation *Quantum Mechanics.* a transformation of a matrix representation of operators in which the wavefunction appears as a constant, having the effect of transformation from a stationary system of axes to a system that rotates with the wavefunction vectors. Also, **Heisenberg picture.**

Heisenberg theory of ferromagnetism *Solid-State Physics.* a theory stating that the exchange interaction between electrons in neighboring atoms depends on the relative orientations of the electron spins; when all spins are parallel the energy is minimized, and thus the spins tend to align, resulting in ferromagnetism.

Heisenberg uncertainty principle *Quantum Mechanics.* a principle stating that a component of a particle's position and its associated momentum cannot be simultanously observed with arbitrary accuracy or, in general, that measuring any quantity of a system will necessarily render some of its other quantities uncertain.

Heising modulation see CONSTANT-CURRENT MODULATION.

Heitler, Walter 1904–1981, German-Swiss chemist; with London, used wave mechanics to develop covalence theory of chemical bond.

Heitler-London covalence theory *Physical Chemistry.* the theory that two hydrogen atoms form a stable compound when their electrons spin in opposite directions, and form unstable compounds when their electrons have the same spin.

hekistotherm *Ecology.* **1.** a type of plant that can tolerate extremely cold temperatures, such as those found in the polar regions. **2.** any organism living in an extremely cold environment. Thus, **hekistothermic.**

Hektoen medium *Microbiology.* a nutrient medium used for the selection of certain enterobacteria, in which the ability to ferment sucrose, lactose, or salicin results in pink colony growth, and the strains unable to ferment any of these substrates form blue-green colonies.

Hektor *Astronomy.* asteroid 624, discovered in 1907; measuring roughly 150 by 300 kilometers, it is suspected to be two objects orbiting around a common center of mass or actually touching.

HEL *Aviation.* the airport code for Helsinki, Finland.

HeLa cells *Cell Biology.* a strain of human cancer cells that has been maintained in culture since 1951; used in the study of life processes and the cultivation of viruses. (Named for patient *Henrietta Lacks,* from whose cervical carcinoma the cells were originally obtained.)

Helaletidae *Paleontology.* a widespread family of lophiodont tapirs in the extinct superfamily Tapiroidea; Paleocene to Oligocene.

Helcionellacea *Paleontology.* a superfamily of bellerophontid gastropods of the early Cambrian; characterized by a fully coiled shell with a wider funnel shape at the mouth.

Helderbergian *Geology.* a North American geologic stage of the lowermost Devonian period, occurring after the Upper Silurian and before the Deerparkian.

Heleidae *Invertebrate Zoology.* a family of dipteran flies, biting midges in the series Nematomorpha; some are intermediate hosts of parasitic worms. Also, CERATOPOGONIDAE.

Helene 1980 S6 *Astronomy.* the temporary designation given to Helene, a moon of Saturn discovered in 1980; it has an orbital period of 2.7 days and is 35 kilometers in diameter.

Heleochloridaceae *Botany.* a family of freshwater green algae of the order Chlorococcales, characterized by spherical cells with large vacuoles and either an arrangement of four cells connected by gelatinous strands or a dendroid, gelatinous structure.

Heleomyzidae *Invertebrate Zoology.* a family of small dipteran flies, the sun flies, with saprophagous larvae. Also, **Helomyzidae.**

heli- a combining form meaning: **1.** sun, as in *helianthus.* **2.** helicopter, as in *heliport.*

heliacal [hi lī′ ə kəl] relating to or occurring near the sun. Also, **heliac.**

heliacal rising *Astronomy.* the rising of a celestial object late in morning twilight just before the rising of the sun.

helianthus *Botany*. the sunflowers; any composite plant of the genus *Helianthus*.

Heliasteridae *Invertebrate Zoology*. the sun stars, a family of sea stars that lack five-point symmetry; echinoderms in the subclass Asteroidea.

heliborne *Ordnance*. carried by helicopter.

helic- a combining form meaning "spiral," as in *helical*.

helical *Mathematics*. relating to a helix, usually a cylindrical helix.

helical antenna *Electromagnetism*. a directional broadband antenna consisting of a helical radiating element whose axis is perpendicular to a reflecting plane.

helical conformation *Materials Science*. a common polymer conformation in which atoms are regularly spaced along helixes, allowing substituents of chains to pack closely without appreciable distortion of chain bonds.

helical conveyor *Mechanical Engineering*. a conveyor for small-particle bulk materials, consisting of a horizontal shaft with affixed helical paddles that rotates within a tube filled with the materials.

helical-fin section *Chemical Engineering*. an extended surface form having a helical shape, which increases the external surface area of process-fluid tubes, thereby increasing heat-exchange efficiency.

helical-flow turbine *Mechanical Engineering*. a steam turbine in which the steam, directed tangentially inward toward buckets in the wheel rim, flows in a helical pattern.

helical gear *Mechanical Engineering*. a gearwheel in which the teeth are sections of a helix described on the wheel face as opposed to being parallel with the wheel axis; such teeth are thus set at an angle to the axis.

helical line *Electromagnetism*. a transmission line with a helical inner conductor.

helical milling *Mechanical Engineering*. milling in which the work is translated and rotated simultaneously.

helical potentiometer *Electricity*. a precision potentiometer in which the control knob must be turned several times to move the contact arm from one end of the spirally wound resistance element to the other. Also, MULTITURN POTENTIOMETER.

helical rake angle *Design Engineering*. in reamer and cutting miller applications, the angle between the axis of the reamer or cutter and a plane that is tangential to its helical cutting edge.

helical resonator *Electromagnetism*. a cavity resonator with a helical inner conductor.

helical scanning *Engineering*. a radar scanning method in which the antenna beam rotates about the vertical axis as the elevation angle moves gradually from horizontal to vertical. *Telecommunications*. **1.** in video recording, a technique in which recording heads and tape are designed to intersect at an angle in order to produce a diagonal series of tracks. **2.** a method of FAX scanning in which a helix rotates against a stationary bar to give horizontal movement of an elemental area.

helical spline broach *Mechanical Engineering*. a hole-cutting tool that produces intricate helical splines on the surface surrounding the hole.

helical spring *Mechanical Devices*. a steel, brass, or iron helical coil of uniform pitch with a uniform or gradually decreasing diameter; its expansion and compression characteristics occur axially along its length; used in locomotive axles and on watches.

helical steel support *Mining Engineering*. a continuous screw-shaped steel joist lining used for staple shafts.

helical symmetry *Virology*. a form of capsid structure in many RNA viruses, including all rod-shaped plant viruses, in which certain protein subunits form a helix.

helical thickening see SPIRAL THICKENING.

helicase *Biochemistry*. any enzyme, such as a rep protein, that unwinds a DNA double helix molecule ahead of DNA polymerase III in *E. coli*.

Helicinidae *Invertebrate Zoology*. a family of tropical terrestrial snails.

helicity *Particle Physics*. the projection of a particle's spin along its direction of motion. The helicity of a particle is described as being either left- or right-handed depending on whether its spin vector is in the direction of motion or against it.

helico- a combining form meaning "spiral," as in *helicospore*.

Helicocephalidaceae *Mycology*. a family of fungi belonging to the order Zoopagales that occur in dung, soil, and plant debris and are parasitic to nematodes.

helicoid *Invertebrate Zoology*. of certain snail shells, having the form of a flattened coil or flattened spiral.

helicoid cyme *Botany*. a coiled or spiral-shaped inflorescence with flowers on only one side of an axis.

Heliconiaceae (lobster claw)

Heliconiaceae *Botany*. a family of perennial monocotyledonous plants in the order Zingiberales, distinguished by the inverted symmetry of its flower, schizocarpic fruit, and capitate stigma.

Helicophyllaceae *Botany*. a monotypic family of medium, glaucous-green mosses of the order Orthotrichales, noted for forming extensive flat mats on rocks and tree trunks and characterized by prostrate, creeping stems with lateral branches, abundant ventral rhizoids, and narrowly bordered dimorphic leaves.

Helicoplacoidea *Paleontology*. a small class of echinoderms, known from three genera; the Helicoplacoidea are an important early group of Cambrian echinoderms, descended from the same Precambrian ancestor as the carpoids but developing independently at the base of the Cambrian, becoming widespread, then becoming extinct at the end of the lower Cambrian; ancestral to all later echinoderms except the carpoids.

helicopter *Aviation*. a wingless aircraft acquiring its lift chiefly or entirely from revolving blades driven by an engine about a near-vertical axis; a rotorcraft acquiring its primary motion from engine-driven rotors that accelerate the air downward, providing a reactive lift force, or accelerate the air at an angle to the vertical, providing lift and thrust.

helicopter assault force *Military Science*. a task force combining helicopters, supporting units, and helicopter-borne troop units for use in helicopter-borne assault operations.

helicopter drop point *Military Science*. a point where helicopters are unable to land because of terrain, but at which they can discharge cargo or troops while hovering.

helicopter support team *Military Science*. an organization equipped for employment in a landing zone, to facilitate the landing and movement of helicopter-borne troops, equipment, and supplies, and to evacuate selected casualties and prisoners of war.

helicopter team *Military Science*. the combat-equipped troops transported in one helicopter at one time.

helicopter yarding *Forestry*. a method of moving logs with a helicopter from a forest area to a wide clearing along a road; used to preserve watersheds.

Helicosphaeraceae *Botany*. a monogeneric family of flagellate marine algae belonging to the order Coccosphaerales, characterized by placoliths with an overlapping helical marginal flange.

Helicosporae *Mycology*. a type of spirally coiled spore found in certain imperfect fungi.

Helicosporida *Invertebrate Zoology*. an order of sporozoan protozoans in the class Myxosporidea; parasitic in insects.

helicotrema *Anatomy*. the connecting space in the inner ear between the scala tympani and the scala vestibuli at the tip of the cochlea.

Heligmosomidae *Invertebrate Zoology*. a family of nematode worms that are parasites of the intestinal tracts of vertebrates, especially rodents.

helimagnet *Solid-State Physics*. a material that exhibits helimagnetism.

helimagnetism *Solid-State Physics*. a property of certain metals, alloys, or salts that are composed of crystal planes possessing a magnetic moment whose alignment at low temperatures varies uniformly from atomic plane to plane and resembles that of a screw or helix.

helio- a combining form meaning "sun," as in *heliosphere*.

Heliobacterium *Bacteriology*. a genus of photosynthetic bacteria, the sole species of which possesses bacteriochlorophyll *g*.

Heliobiales *Botany*. in some systems of classification, an order encompassing most of the Alismatidae.

heliocentric *Astronomy.* measured or considered from the center of the sun. *Anthropology.* describing a group or culture which holds the belief that the sun is the center of the solar system or the universe.

heliocentric coordinate system *Astronomy.* a coordinate system based on the ecliptic plane but with its center at the sun; 0° longitude lies in the direction of the vernal equinox, and its poles are the ecliptic poles.

heliocentric Julian date *Astronomy.* a Julian calendar date referred to the sun, thereby corrected for the varying light-travel time (from the object to the earth) caused by the varying position of the earth in its orbit.

heliocentric latitude *Astronomy.* the ecliptic latitude that an object would have if viewed from the center of the sun.

heliocentric longitude *Astronomy.* the ecliptic longitude that an object would have if viewed from the center of the sun; as with ecliptic longitude, the zero point is the vernal equinox.

heliocentric orbit *Astronomy.* an orbit that has one of its foci at the sun.

heliocentric parallax *Astronomy.* the angular difference in a celestial object's position as seen from the center of the sun and the center of the earth. Also, ANNUAL PARALLAX.

heliocentric theory *Astronomy.* **1.** the theory of planetary motion proposed by Copernicus and others, postulating a motionless sun at the center of the solar system with the earth and other planets revolving around it; contrasted with the earlier geocentric theory proposed by Ptolemy and others, according to which the earth was at the center. **2.** any theory of cosmology in which the sun is central to the solar system.

Heliodinidae *Invertebrate Zoology.* a widely distributed family of small, brightly colored moths, lepidopteran insects in the suborder Heteroneura.

heliodor *Mineralogy.* a clear yellow form of beryl that is used as a gemstone.

heliogram *Telecommunications.* a communication transmitted on a heliograph.

heliograph *Engineering.* an instrument that reflects sunlight to a remote station; used especially in signaling and surveying. *Meteorology.* an instrument that records the amount and duration of sunshine, often on a strip of blueprint paper.

heliographic latitude *Astronomy.* an object's angular distance north or south of the solar equator.

heliographic longitude *Astronomy.* an object's angular distance measured westward from the standard solar meridian, which is the meridian that passed through the ascending node of the solar equator at Greenwich mean noon on January 1, 1854.

heliolite see SUNSTONE.

Heliolites *Paleontology.* a genus of massive tabulate corals in the extinct order Heliolitida; Middle Ordovician to Middle Devonian.

heliolitid *Paleontology.* **1.** of or relating to corals in the extinct order **Heliolitida. 2.** a member of this order.

heliometer *Optics.* a telescope with a movable split-lens objective, used primarily to measure the angular distance between two stars; originally used to measure the diameter of the sun.

heliopause *Astronomy.* the outer boundary of the heliosphere, where the solar (or interplanetary) magnetic environment gives way to the interstellar.

Heliopeltaceae *Botany.* a family of marine diatoms of the order Centrales, characterized by valves with radial undulations or divisions into raised and depressed sections.

heliophilous *Ecology.* living or thriving in full sunlight. Thus, **helioophile, heliolophily.** Also, **heliophilic.**

heliophobe *Ecology.* a plant that is intolerant of full sunlight and grows best in the shade. Thus, **heliophobic, heliophobous.** *Medicine.* a person who is abnormally sensitive to or has an exaggerated fear of the rays of the sun.

heliophobia *Psychology.* an irrational fear of sunlight.

heliophyll *Botany.* a plant with leaves having similar structure on both sides and arranged more or less vertically.

heliophyllite *Mineralogy.* a yellow to greenish-yellow, orthorhombic mineral with an approximate formula of $Pb_6As_2^{+3}O_7Cl_4$, occurring in massive forms and as pyramidal or tabular crystals, and having a specific gravity of 6.89 and a hardness of about 2 on the Mohs scale.

heliophyte *Ecology.* a plant that flourishes under conditions of full sunlight. Thus, **heliophytic.**

Heliornithidae *Vertebrate Zoology.* sun grebes and finfoots, a family of aquatic birds of the order Gruiformes, native to the tropics and subtropics of Central and South America, Africa, and southeast Asia.

Helios *Space Technology.* a series of two U.S.-German probes that passed within 30 million miles of the sun to study its surface, winds, magnetic field, cosmic rays, and other features.

helioscope *Optics.* a telescope with a great focal length that receives images of the sun from plane mirrors attached to it, so that the viewer's eyes are protected from the sun's glare.

helioseismology *Astrophysics.* the study of the sun's interior by analysis of the sound waves passing through it; the visible manifestation is slight distortions of the solar surface caused by the waves' vibrations.

heliosphere *Astronomy.* the region of space where solar and planetary particles and magnetic fields are dominant.

heliostat *Engineering.* an instrument that reflects the sun's rays in a continuous beam and in various directions, and which can therefore serve as a signaling station; the device is clock-driven so that it points at all times toward the sun.

heliotaxis *Biology.* a directed response of a motile organism toward (positive) or away from (negative) sunlight. Thus, **heliotactic.**

heliotherapy *Medicine.* the treatment of disease by exposing the body to sunlight.

heliotrope [hē′lē ə trōp′] *Botany.* **1.** a plant that turns or bends in response to sunlight. **2.** any of the genus *Heliotropium,* an herb or shrub of the borage family. *Mineralogy.* see BLOODSTONE. *Engineering.* a heliograph that is capable of reflecting solar rays over long distances.

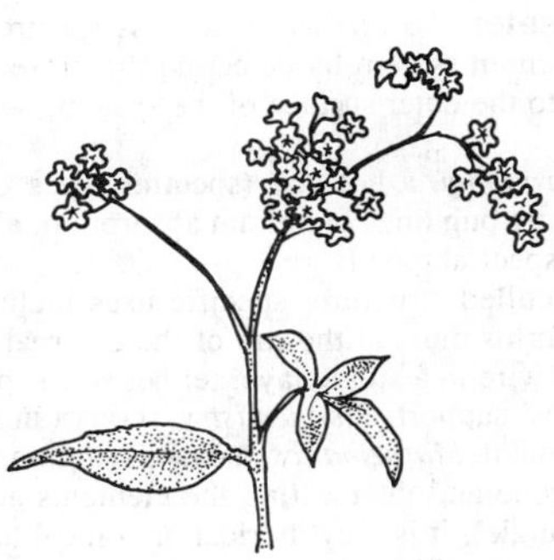

heliotrope

heliotropic [hē′lē ə trō′pik] *Botany.* turning or growing toward sunlight or other light.

heliotropic wind *Meteorology.* a wind component that adapts (by subtle shifting) to the diurnal shift of the sun's position, in consonance with the east-to-west progression of daytime surface heating.

heliotropism see PHOTOTROPISM.

Heliozoa *Invertebrate Zoology.* sun animalcules, an order of free-living aquatic protozoans in the subclass Actinopoda; spherical with radiating filopodia, they feed on other protozoans and rotifers. Also, HELIOZOIA

heliozooid *Biology.* a member of the order Heliozoa, consisting of rhizopod Protozoa distinguished by their spherical shape and stiff, radiating pseudopodia.

helipad *Civil Engineering.* an area for the launching and landing of helicopters in a heliport.

heliport *Civil Engineering.* a transportation depot designed for takeoffs and landings of helicopters.

helitron *Electronics.* an oscillator used at ultrahigh and microwave frequencies, having an output frequency that is variable over a wide range.

helium *Chemistry.* a gaseous element having the symbol He, the atomic number 2, an atomic weight of of 4.0026 (only hydrogen is lighter), and a boiling point of –268.9°C. The first element in the noble gas group, it is colorless, odorless, tasteless, and noncombustible. It occurs on earth in natural gas and makes up a small fraction of the atmosphere, and it is abundant elsewhere in the universe. Liquid helium exhibits superconductivity and near-zero viscosity. (From *Helios,* the ancient Greek god of the sun; evidence of its existence was first noted in the sun.)

helium I *Physics.* the normal component of liquid helium when cooled below the lambda point.

helium II *Physics.* the superfluid component of liquid helium, evident when the liquid is cooled below the lambda point.

helium-3 *Nuclear Physics.* an isotope of helium with mass number 3 that is found in 1.3 parts per million of the helium found in nature.

helium-4 *Nuclear Physics.* an isotope of helium with mass number 4 that makes up most of the helium found in nature.

helium burning *Nuclear Physics.* a thermonuclear process occurring in stars by which nuclei of a fully ionized helium plasma undergo fusion.
helium-cadmium laser *Optics.* a laser in which the interaction between helium and cadmium vapor produces a continuous beam of ultraviolet or blue light.
helium embrittlement *Materials Science.* a high-temperature embrittlement of stainless steel caused by the development of helium bubbles along the grain boundaries.
helium flash *Astrophysics.* the initiation of helium burning in a red giant star's core after the hydrogen there has been exhausted.
helium liquefier *Physics.* any of a number of machines capable of cooling gaseous helium to a liquid state through adiabatic expansion and external work.
helium magnetometer *Physics.* a device that can measure magnetic fields by observing the effect of the field on the lowest triplet energy level of helium atoms.
helium-3 maser *Physics.* a maser that uses gaseous helium-3 as its active medium.
helium-neon laser *Optics.* a laser in which an electrical discharge energizes helium atoms, which in turn energize neon atoms to produce a continuous red beam; the most commonly used type of laser.
helium refrigerator *Mechanical Engineering.* a refrigerator that uses liquid helium as the working fluid; provides cooling to temperatures approaching absolute zero.
helium spectrometer *Spectroscopy.* a mass spectrometer used to locate leaks in a vacuum system by detecting the presence of helium that has been applied to the outer surface of the system, where leaks are suspected.
helium star *Astronomy.* a hot star (spectral types O, B, and A) that shows abnormally strong lines of helium absorption; also, an out-of-date name for stars of spectral type B.
helix a spiral or coiled structure; specific uses include: *Anatomy.* the curved fold that forms most of the rim of the external ear. *Electricity.* a spread-out coil of wire in a single layer, either self-supporting or wound around a cylindrical support. *Biochemistry.* see DOUBLE HELIX. *Architecture.* a spiral ornament. *Mathematics.* a curve lying on the lateral surface of a cylinder or cone and intersecting the elements at a constant angle (called the helix angle); it is a cylindrical or conical helix, respectively. (From the Greek for "snail" or "coil.")
Helix *Invertebrate Zoology.* the genus of terrestrial snails that contains the edible escargot, *Helix pomatia.*

Helix (terrestrial snail)

α-helix *Biochemistry.* a secondary structure that occurs in many proteins; a right-handed helix with 3.6 amino acid residues per turn stabilized by hydrogen bonds between the imino hydrogen of each peptide bond and the carbonyl oxygen of the peptide bond four residues farther along the polypeptide chain.
helix angle *Mathematics.* a constant acute angle between a tangent to a helix and the intersecting generator of the cone or cylinder.
helix-coil transition *Molecular Biology.* a structural transformation that occurs in a protein or a nucleic acid molecule involving a change from an ordered, helical conformation to a random, disordered conformation. Also, MELTING.
helix-destabilizing proteins *Biochemistry.* proteins that bind to single-stranded regions of duplex DNA created by "breathing," causing unwinding of the helix.
helix joint *Robotics.* a joint based on a spiral screw or spline.
Helix Nebula *Astronomy.* NGC 7293, a large but faint planetary nebula in Aquarius whose name comes from its appearance in long-exposure photographs.
helix recorder *Electronics.* see LAWNMOWER, def. 2.
helix tube *Electronics.* a traveling-wave tube in which an electromagnetic wave travels along a spiral wire wound around a beam such that the wave and beam have approximately equal velocities. Also, **helical traveling-wave tube.**
hellandite *Mineralogy.* $(Ca,Y)_6(Al,Fe^{+3})Si_4B_4O_{20}(OH)_4$, a red, brown, or blackish, translucent, monoclinic mineral, having a specific gravity of 3.63 and a hardness of 5.5 on the Mohs scale; found as crystals in granite pegmatites.
Hellas Planitia *Astronomy.* an impact basin in Mars' southern hemisphere that is 1800 kilometers across, contains vast fields of sand dunes, and is more than 4 billion years old.

hellbender

hellbender *Vertebrate Zoology.* the common name for a voracious nocturnal salamander of the genus *Cryptobranchus,* living in the eastern United States under rocks in large rivers.
Hellcat *Aviation.* a popular name for the F-6F Navy fighter.
hellebore [hel´ə bôr´] *Botany.* any of the plants of the genus *Helleborus* of the buttercup family, characterized by basal leaves and clusters of flowers.
helleborein *Organic Chemistry.* $C_{37}H_{56}O_{18}$, a yellow, crystalline, poisonous solid that is obtained from the root and rhizome of certain hellebores; used as a heart stimulant.
Hellenic [hə len´ik] relating to the culture of ancient Greece.
Heller's test *Pathology.* a method of identifying the presence of albumin, blood, or dextrose in urine. For albumin: in a test tube layer cold nitric acid below the urine, and albumin will form a white coagulum between the two. For blood: add potassium hydroxide solution and heat; the earthy phosphates are precipitated and will turn red when stained by hematin if blood is present. For dextrose: add a solution of potassium hydroxide; a brownish-red precipitate will indicate sugar. (Named for Johann Florian *Heller,* 1813–1871, Austrian pathologist.)
Hellfire *Ordnance.* an advanced U.S. heliborne antitank missile powered by a reduced smoke all-boost rocket motor; it is equipped with semiactive laser homing and delivers a 20-pound hollow charge warhead at Mach 1.17; officially designated **AGM-114.**
hello-goodbye effect *Psychology.* a term for the tendency of an individual to exaggerate symptoms and anxieties at the beginning of therapy, as an indication that treatment is needed, and to minimize them at the end, as an indication that it has been effective.
Hell-Volhard-Zelinsky reaction *Organic Chemistry.* the reaction of a carboxylic acid with phosphorus or phosphorus halide, followed by hydrolysis, to form an α-halo-substituted acid.
helm *Naval Architecture.* **1.** the device that controls a vessel's rudder; it may be a tiller or a wheel. **2.** the overall steering of a ship; a certain officer is said to "have the helm."
Helmert's formula *Geophysics.* a formula for theoretical gravity based on a triaxial ellipsoid and, therefore, including a longitude term.
helmet *Engineering.* **1.** a protective headgear having a plastic or glass faceplate; used to protect a person while performing tasks such as arc welding. **2.** the hollow round headpiece on a deep-sea diving suit; attaches to the breastplate of the suit and is supplied with air through a hose. *Civil Engineering.* see PILE HELMET.
helmet-mounted display *Ordnance.* a display that presents primary information for directing firepower within the visor of a gunner's helmet. Also, VISUALLY COUPLED DISPLAY.
Helmholtz, Hermann von [helm´hōlts] 1821–1894, German physicist and physiologist; proved the law of conservation of energy; invented the ophthalmoscope; wrote on acoustics; estimated the sun's age.
Helmholtz coil *Electromagnetism.* a device used for providing a relatively uniform magnetic field, consisting of two identical circular coils on a common axis, connected in series, and separated by the radius of one of the coils.

Helmholtz double layer *Physical Chemistry.* the earliest model of the double layer at an electrode surface; it was offered in 1879 by Helmholtz, who suggested a layer of ions of one charge at a solid surface and a rigidly held layer of oppositely charged ions in the electrolyte solution.

Helmholtz energy see HELMHOLTZ FREE ENERGY.

Helmholtz equation *Physical Chemistry.* an equation stating that the electromotive force in an electrolytic cell is equivalent to the chemical reaction when a charge goes through the cell, with the addition of the product of the temperature and the derivative of the electromotive force.

Helmholtz flow *Fluid Mechanics.* a fluid flow in which there can exist free streamlines and vortex sheets.

Helmholtz free energy *Thermodynamics.* a thermodynamic property A that describes the capacity of a system to do work, which is expressed by $A = U - TS$, where U is the internal energy, T is the absolute temperature, and S is the entropy of the system. The Helmholtz free energy is also denoted by F instead of A. Also, **Helmholtz function.**

Helmholtz instability *Fluid Mechanics.* a shearing instability at the boundary between two fluids that are in relative motion.

Helmholtz-Kelvin contraction *Astrophysics.* the contraction of a star that results from the radiation of thermal energy generated solely by gravitational contraction (i.e., with no contribution from thermonuclear reactions); this process is thought to power protostars before they reach the main sequence.

Helmholtz layer see COMPACT LAYER.

Helmholtz resonator *Acoustical Engineering.* an enclosure with an opening which, as a fluid flows over a surface boundary, produces sound frequencies related to the width of the opening in the surface boundary of the object.

Helmholtz's theorem *Fluid Mechanics.* a theorem claiming that individual vortices arising in a nonviscous fluid which flows isentropically and is free of body forces always consist of the same fluid particles.

Helmholtz wave *Fluid Mechanics.* an unstable wave that propagates on the interface between two homogeneous fluids that are in relative motion.

helminth *Invertebrate Zoology.* any parasitic worm, especially those found in the intestines of vertebrates.

helminthemesis *Medicine.* the vomiting of parasitic worms.

helminthiasis *Medicine.* an infection of parasitic worms, which may be intestinal, visceral, or cutaneous.

helminthic abscess *Medicine.* a localized accumulation of pus due to the presence of parasitic worms.

Helminthocladiaceae *Botany.* a family of marine red algae of the order Nemaliales, characterized by erect gametangial thalli.

helminthogogue *Pharmacology.* any agent used to treat infestations of parasitic worms. Also, ANTIHELMINTHIC.

helminthoid *Biology.* shaped like a worm.

helminthologist *Biology.* one who studies parasitic flatworms and round worms.

helminthology *Invertebrate Zoology.* the branch of science dealing with parasitic worms.

Helminthomorpha *Invertebrate Zoology.* the largest subclass of the Diplopoda, containing seven superorders and 11 orders of millipedes.

helminthophobia *Psychology.* an irrational fear of worms or of becoming infected with worms. Also, VERMIPHOBIA.

helminthosporin *Biochemistry.* $C_{15}H_{10}O_5$, a natural product of a pathogenic fungus that has a physiological effect on plants similar to that of the gibberellins.

Helminthosporium *Mycology.* a genus of fungi belonging to the family Dematiaceae; its species are parasitic and are characterized by wormlike spores that are pluriseptate.

Helminthosporium leaf spot *Plant Pathology.* a plant disease common to grasses, caused by species of the parasitic fungi *Helminthosporium* and characterized by areas of discoloration on the foliage.

helm roof *Architecture.* a steeply pitched roof with four faces rising from gables to form a spire.

Helmstetter-Cooper model *Microbiology.* see COOPER-HELMSTETTER MODEL.

helm wind *Meteorology.* a strong and cold northeasterly wind that blows in northern England.

helo- a combining form meaning "corn" or "callus," as in *helotomy.*

Helobiae *Botany.* in some systems, a name for the order Heliobiales.

Heloderma *Vertebrate Zoology.* a single genus of reptiles of the family Helodermatidae, including the only venomous lizards known, the Gila monster and the poisonous Mexican beaded lizard.

Helodermatidae *Vertebrate Zoology.* the beaded lizards, a family of the order Squamata, including the only venomous lizards, the Gila monster and the poisonous Mexican beaded lizard; found in the southwestern United States to northwest Mexico.

Helodidae *Invertebrate Zoology.* a family of coleopteran insects with aquatic larvae, the marsh beetles, in the superfamily Dascilloidea. Also, SCIRTIDAE.

Helodontidae *Paleontology.* a family of holocephalic fishes in the suborder Helodontoidei; upper Devonian to Permian.

Helomyzidae *Invertebrate Zoology.* see HELEOMYZIDAE.

helophyte *Botany.* a marsh plant whose stems appear above the water line, while its renewal buds are in the soil or mud below the water level.

helophytia *Ecology.* differing ecological conditions caused by differences in the water level, as in a swamp or marsh.

Heloridae *Invertebrate Zoology.* a family of hymenopteran insects, solitary wasps whose larvae parasitize lacewings.

Helostamadtidae *Vertebrate Zoology.* the kissing gourami, a freshwater fish family in the suborder Anabantoidei that consists of one species, *Helostoma temminki,* found in Southeast Asia; noted for using mouth-to-mouth contact to establish territory in mate selection.

Helotiales *Mycology.* an order of fungi belonging to the class Discomycetes which includes both plant parasites and types that live from dead organic material. Also, CUP FUNGI.

Helotidae *Invertebrate Zoology.* a family of coleopteran insects, the metallic sap beetles, in the superfamily Cucujoidea, that feed on sap exuding from tree wounds.

helotism *Ecology.* a relationship between two organisms in which one "enslaves" the other for its own benefit. (From *helot,* the name of a class of slaves in ancient Greece.)

Helotrephidae *Invertebrate Zoology.* a family of minute aquatic true bugs, hemipteran insects in the subdivision Hydrocorisae; found in tropical forests.

help *Computer Programming.* in some software packages, a command that supplies the user with additional information about various features of the package. Also, **help screen.**

HELP *Robotics.* the programming language used by the Allegro robot.

helper *Behavior.* an individual other than the parents that assists in the raising of young; characteristic of many bird species.

helper factor *Immunology.* a substance secreted by the T helper cells or accessory cells that contributes to B cell activation.

helper phage *Genetics.* a virus that, upon infecting a bacterial cell, provides some vital material lacking in a defective virus, thereby allowing the defective virus to reproduce.

helper T lymphocyte *Immunology.* a white blood cell formed in the thymus that is necessary to the development of normal levels of antibody by B lymphocytes in both in vitro and in vivo situations. Also, **helper T cell.**

helper virus *Virology.* a virus that aids the development of a defective virus by enabling it to form a protein coat or by supplying or restoring the activity of a viral gene.

helplessness motive *Psychology.* a tendency to respond to failure or frustration by giving up. Also, **helplessness orientation.**

helve *Engineering.* the handle of a hammer, hatchet, axe, or the like.

Helvella *Mycology.* a genus of fungi belonging to the order Pezizales which is characterized by its saddle shape; some are toxic and can be deadly. Also, SADDLE FUNGI.

Helvellaceae *Mycology.* a family of fungi belonging to the order Pezizales, occurring primarily in forest soils and composed of some poisonous mushrooms.

Helvetian *Geology.* a European stage of the Miocene epoch, occurring after the Burdigalian and before the Tortonian.

helvite *Mineralogy.* $Mn_4^{+2}Be_3(SiO_4)_3S$, a yellow to brown, brittle, cubic mineral occurring in tetrahedral or octahedral crystals, having a specific gravity of 3.17 to 3.37 and a hardness of 6 on the Mohs scale; found in granitic pegmatites. Also, **helvine.**

HEM hostile-environment machine.

hema- or **hem-** a combining form meaning "blood," as in *hemathermal, hemanalysis.*

hemadsorption *Hematology.* the adherence of red cells to other cells, particles, or surfaces. *Virology.* the specific attachment of red blood cells to the surface of cells that have been infected with certain viruses. Also, **haemadsorption.**

hemadsorption test *Virology.* an in vitro test for detecting hemagglutinating viruses based on the adherence of red blood cells to cells of the infected tissue in the presence of hemagglutinin.

hemadsorption virus *Virology.* a former term for one of two parainfluenza viruses; **hemadsorption virus, type 1** (parainfluenza 3) causes bronchitis and pneumonia, especially in children; **hemadsorption virus, type 2** (parainfluenza 1) has been isolated from children with febrile respiratory disease.

hemafibrite see SYNADELPHITE.

hemagglutination *Immunology.* the joining together of red blood cells, caused by a specific antibody or by viruses, bacteria, or plant proteins.

hemagglutination-inhibition test *Immunology.* a test used to determine the abilitiy of a patient's serum to lessen or destroy the properties of a given virus that promotes the clumping together of the pateint's red blood cells. Also, HI TEST, HAI TEST.

hemagglutination test *Immunology.* an agglutination test in which an antibody and an antigen react with each other on the surface of red blood cells.

hemagglutinin *Immunology.* any substance, particularly an antibody, that causes red blood cells to agglutinate.

hemal *Anatomy.* **1.** of or relating to the blood or the blood vessels. **2.** ventral to the spinal axis, where the heart and great vessels are located.

hemal arch *Anatomy.* **1.** a ventral arch in the third to sixth coccygeal vertebrae that encloses the vertebral artery and vein. **2.** the skeletal elements of the thorax in humans.

hemal ring *Invertebrate Zoology.* a circulatory vessel found in certain echinoderms.

hemal sinus *Invertebrate Zoology.* a space located adjacent to the digestive tube in certain echinoderms.

hemal system *Invertebrate Zoology.* the blood vascular system in echinoderms.

hemanalysis *Hematology.* an analysis of the chemical constituents of blood.

hemangioblastoma *Oncology.* a hemangioma of the brain that is composed of blood vessel cells or angioblasts.

hemangioendothelioblastoma *Oncology.* a tumor that originates in the mesenchyme and is characterized by the formation of endothelial cells and a tendency for its cells to line blood vessels.

hemangioendothelioma *Oncology.* a tumor arising from the endothelial cells of a blood vessel.

hemangioma *Oncology.* a benign tumor that occurs commonly in infancy and childhood, composed of newly formed blood vessels and resulting from malformation of fetal angioblastic tissue.

hemangiopericytoma *Oncology.* an uncommon tumor thought to be derived from connective tissue cells, apparently arising from capillaries (pericytes).

hemapodium *Invertebrate Zoology.* the dorsal lobe of a parapodium.

hemarthrosis *Medicine.* the presence of blood in a joint, usually accompanied by pain, tenderness, and swelling.

hemat- a combining form meaning "blood," as in *hematin.*

hematemesis *Medicine.* the vomiting of blood.

hematic *Hematology.* relating to or contained in blood. Also, **hematinic.**

hematidrosis *Medicine.* an extremely rare disorder characterized by excretion of blood or blood pigment in the sweat.

hematin *Organic Chemistry.* $C_{34}H_{32}N_4O_4FeOH$, the hydroxide of ferriheme; a blue to black powder that is soluble in hot alcohol and insoluble in water; decomposes at 200°C without melting; used in biochemical research.

hematite *Mineralogy.* α-Fe_2O_3, a gray to black trigonal mineral of the hematite group, dimorphous with maghemite, usually occurring in massive, fibrous, reniform, micaceous, or granular form, having a specific gravity of 5.26 and a hardness of 5 to 6 on the Mohs scale; the most abundant iron ore, found mainly in sedimentary deposits.

hemato- a combining form meaning "blood," as in *hematocrit.*

hematoblast *Histology.* a stem cell in hematopoietic tissue.

hematocele *Medicine.* a swelling due to the effusion and collection of blood, usually into a canal or cavity of the body.

hematochrome *Biochemistry.* a red pigment that appears when green algae are subject to bright light.

hematocrit *Pathology.* **1.** the percentage in a volume of whole blood represented by red blood cells. **2.** originally, the instrument used in isolating the cells and other particulate constituents from the plasma.

hematocyst *Pathology.* a cyst that contains blood.

hematodocha *Invertebrate Zoology.* a sac found in the palpus of certain spiders that fills with hemolymph and expands during copulation.

hematogenic *Hematology.* **1.** produced by or derived from blood. Also, **hematogenous. 2.** see HEMATOPOIETIC.

hematogenous *Physiology.* **1.** produced by or derived from the blood. **2.** disseminated by the circulation or through the blood stream.

hematoid *Hematology.* resembling blood.

hematoidin *Hematology.* a substance that is apparently chemically identical with bilirubin, but which has a different site of origin, being formed locally in the tissues from hemoglobin, especially under conditions of reduced oxygen tension.

hematoidin crystals *Pathology.* blood crystals, derived from hemoglobin, with a chemical affinity to bilirubin, but originating intracellularly from within the reticuloendothelial tissue.

hematolite *Mineralogy.* $(Mn^{+2},Mg,Al)_{15}(AsO_3)(AsO_4)_2(OH)_{23}$, a brownish-red trigonal mineral occurring in rhombohedral crystals with one perfect cleavage, having a specific gravity of 3.49 and a hardness of 3.5 on the Mohs scale; found in limestone in Sweden.

hematologist [hē´mə täl´ə jist] *Medicine.* **1.** a physician specializing in the diagnosis and treatment of disorders of the blood and blood-forming tissues. **2.** a scientist who specializes in the study and treatment of the blood and the blood-forming system.

hematology [hē´mə täl´ə jē] *Medicine.* the branch of medical science that deals with the diagnosis and treatment of disorders of the blood and blood-forming tissues.

Hematology

Hematology is a protean field. Its thorough understanding demands experience of cell and molecular biology, multiple areas of normal physiology and pathophysiology, and (for clinicians) excellent medical skills. The field is usually divided into at least eight major areas. Among these are:

1) The physiology and disorders of red cells, including the hormonal and nutritional basis of erythropoiesis, syntheses of and transport across membranes, erythrocyte metabolism, globin synthesis and development, porphyrin metabolism, hemoglobin function and its mutations, and the influence of hemoglobins on bilirubin production. 2) The normal development and the disorders of phagocytes, including myelopoiesis, phagocyte trafficking, organism ingestion and killing, the pathophysiology of toxic cytokines, and phagocyte storage diseases. 3) The organization of the lymphoid system, with particular attention to the cellular basis of immunity, antibody and autoantibody function, and defects of plasma proteins. 4) Hematopoiesis, including the hematopoietic growth factors and their receptors and bone marrow transplantation. 5) The hematopoietic malignancies, including the leukemias and lymphomas (the pediatric hematologist is equally concerned with the nonhematopoietic malignancies of childhood). 6) The normal and pathophysiologic function of the coagulation system, including the fluid phase of clotting, platelet function, megakaryocytopoiesis, and the thrombotic and inflammatory disorders of the vasculature. 7) Transfusion medicine, which utilizes and develops the techniques of blood product collection, separation, and storage and their selective administration to patients. 8.) The hematologic manifestations of systemic diseases, with particular reference to AIDS and the parasitic infections.

Obviously this entire field cannot be completely mastered by a single individual. Hematologists have necessarily differentiated into sub-subspecialists, as has been the case in so many medical fields.

David G. Nathan
Robert A. Stranahan Professor of Pediatrics
Harvard Medical School
Physician-in-Chief, Children's Hospital

hematolymphangioma *Oncology.* a tumor that is composed of blood vessels and lymph vessels.

hematoma [hē′mə tō′mə] *Medicine.* a localized mass of blood, clotted or partially clotted, found outside the blood vessels in an organ, space, or tissue; caused by a break in the wall of a blood vessel.

hematometra *Medicine.* a collection of blood or menstrual fluid in the uterine cavity, resulting in distention of the uterus.

hematomyelia *Medicine.* a hemorrhage into the spinal cord, usually due to trauma. Also, **hematorrhachis.**

hematophagous *Zoology.* feeding mainly on blood.

hematophanite *Mineralogy.* $Pb_4Fe_3^{+3}O_8(OH,Cl)$, a dark red-brown opaque tetragonal mineral occurring as thin, tabular crystals, and having a specific gravity of 7.7 and a hardness of 2 to 3 on the Mohs scale; found in granular limestone.

hematopoiesis *Hematology.* the formation and development of blood cells.

hematopoietic *Hematology.* **1.** relating to or affecting the formation of blood cells. **2.** an agent that promotes the formation of blood cells.

hematopoietic tissue *Histology.* any reticular tissue that produces blood, such as the red marrow in bone.

hematopoietin *Biochemistry.* a glycoprotein that is produced by the kidneys and stimulates the formation of erythrocytes.

hematoporphyrin *Biochemistry.* $C_{34}H_{38}N_4O_6$, a porphyrin produced by adding sulfuric acid to hemoglobin *in vitro;* not found in nature. Also, **hemoporphyrin.**

hematosalpinx *Medicine.* an accumulation of blood in a uterine tube, usually associated with tubal pregnancy.

hematotropic *Hematology.* having a special affinity for or exerting a specific effect on the blood or blood cells.

hematoxylin trihydrate *Organic Chemistry.* $C_{16}H_{14}O_6 \cdot 3H_2O$, light yellow crystals that turn red in light; soluble in hot water and alcohol, melts at 100–120°C; thought to be carcinogenic; used as a stain in microscopy and as a colorant in inks.

hematoxylon see LOGWOOD.

hematuria *Medicine.* the presence of blood in the urine.

heme *Biochemistry.* $C_{34}H_{32}O_4N_4Fe$, a protoheme or iron-porphyrin complex that has a protoporphyrin nucleus, specifically one containing the oxygen-binding portion of the hemoglobin molecule.

hemelytron *Entomology.* the anterior wing of certain insects in which the basal half is hardened and the apex membranous; used especially in reference to the true bugs.

heme protein *Biochemistry.* any protein in which an iron-porphyrin functions as the active (prosthetic) group; for example, cytochromes or hemoglobin.

hemeralopia *Medicine.* day blindness; defective vision in bright light with comparatively good vision in dim light.

heme synthetase *Enzymology.* an enzyme that catalyzes the formation of any iron-porphyrin coordination complex.

hemi- a combining form meaning "half," as in *hemisphere.*

hemiacetal *Organic Chemistry.* any of a class of compounds having the =C(OH)O*R* group that results from the reaction of an aldehyde with one mole of an alcohol.

hemialgia *Neurology.* pain that affects only one side of the body.

hemianalgesia *Neurology.* insensitivity to pain on one side of the body.

hemianesthesia *Medicine.* the loss of tactile sensation on one side of the body. Also, UNILATERAL ANESTHESIA.

hemianopia *Medicine.* the loss of vision in one-half of the visual field of one or both eyes. Also, **hemianopsia.**

Hemiascomycetes *Mycology.* a class of fungi belonging to the subdivision Ascomycotina, characterized by spore sacs or asci that are not found in and protected by spore-bearing structures; it is composed mainly of yeast fungi that occur in water, soil, or decaying organic matter and live primarily off nonliving organic matter, although some are pathogenic to plants and animals or cause food spoilage.

Hemiascomycetidae *Mycology.* a former term for a subclass of fungi of the class Ascomycetes; now classified as the class Hemiascomycetes.

hemiataxia *Neurology.* a loss of coordination on one side of the body.

hemiazygos vein *Anatomy.* a vein that drains the left abdominal and thoracic wall regions; it pierces the diaphragm and empties into the azygos vein.

hemiballismus *Neurology.* sudden, severe, violent, and involuntary motor restlessness affecting only one side of the body, resulting from cerebral vascular disease. Also, **hemiballism.**

Hemibasidiomycetes see HETEROBASIDIOMYCETIDAE.

hemiblock *Cardiology.* failure in the conduction of the cardiac impulse in either the anterior (superior) division or the posterior (inferior) division of the left ventricle's conducting system.

hemicardia *Cardiology.* a congenital abnormality of the heart in which only one half of the four-chambered heart is present.

hemicellulose *Biochemistry.* $C_6H_{10}O_5$, a high-molecular-weight polysaccharide complex that functions as a structural component of plant cells; it is a polymer of D-oxylose containing side chains of other sugars.

hemicephaly *Medicine.* a congenital absence of one side of the cerebrum, caused by arrested brain development in the fetus; there may be rudimentary development of cerebellum and basal ganglia.

hemichannel see CONNEXON.

Hemichordata *Biology.* a phylum in the kingdom Animalia that contains a group of ancestral marine animals, including the classes Enteropneusta, Pterobranchia, and Graptolithina, which bear an anatomical structure similar to or homologous to the notochord of higher chordates.

Hemicidaridae *Paleontology.* a family of Mesozoic euechinoids in the extinct order Hemicidaroida.

Hemicidaris *Paleontology.* a genus of regular, pentaradially symmetrical euechinoids in the extinct order Hemicidaroida and family Hemicaridae; Lower Jurassic to Upper Cretaceous.

Hemicidaroida *Paleontology.* an extinct order of regular euechinoids in the superorder Stirodonta; Upper Triassic to Upper Cretaceous.

hemic murmur *Medicine.* a cardiac or vascular murmur that does not indicate heart or blood vessel disease; often associated with anemia.

hemicolectomy *Surgery.* the surgical removal of approximately half of the colon.

hemicrania *Neurology.* **1.** the occurrence of pain on one side of the head. **2.** see MIGRAINE. *Medicine.* partial anencephaly.

hemicryptophyte *Ecology.* a perennial plant whose buds rest on the surface of the soil hidden by scale, snow, or litter.

hemicyclic *Botany.* describing flowers that have whorled or spiral floral leaf patterns.

hemidesmosome *Cell Biology.* a specialized cell junction found on the basal surface of epithelial cells that anchors these cells to the basal lamina and morphologically resembles half a desmosome.

hemidiaphragm *Anatomy.* one lateral half of the diaphragm. *Medicine.* a condition in which the muscle of the diaphragm develops on one side only.

Hemidiscosa *Invertebrate Zoology.* an order of sponges in the subclass Amphidiscophora.

hemiepilepsy *Neurology.* epilepsy that affects only one side of the body.

Hemigastraceae *Mycology.* a family of fungi belonging to the order Protogastrales that occur on rabbit excrement.

hemiglobin see METHEMOGLOBIN.

hemihedral symmetry *Crystallography.* of or relating to a crystal that has only half of the possible symmetry elements of the crystal system to which it belongs.

hemiholohedral *Crystallography.* of or relating to a crystal that has one-half the number of planes required by the highest degree of symmetry.

hemihypertonia *Neurology.* increased muscle tone on only one side of the body, sometimes causing permanent shortening of muscle; often occurs after a stroke. Also, HEMITONIA.

hemilesion *Neurology.* a lesion on one side of the spinal cord.

Hemileucidae see SATURNIIDAE.

hemimellitic acid *Organic Chemistry.* $C_6H_3(COOH)_3$, colorless needles that are slightly soluble in water; melts at 196°C.

Hemimetabola see EXOPTERYGOTA.

hemimetabolous metamorphosis *Entomology.* an incomplete metamorphosis which involves very little change at each stage, as with exopterygote insects; the young, known as nymphs, are generally similar to the adults but lack wings.

hemimethylated DNA *Genetics.* dsDNA having only one methylated strand.

hemimorphic crystal *Crystallography.* a crystal having unlike faces at the ends of the same axis.

hemimorphite *Mineralogy.* $Zn_4Si_2O_7(OH)_2 \cdot H_2O$, a colorless, white, pale green, blue, or yellow orthorhombic mineral, having a specific gravity of 3.4 to 3.5 and a hardness of 4.5 to 5 on the Mohs scale. One of the best minerals for demonstrating polar symmetry, because the two ends of its crystals are distinctly dissimilar; found as a secondary mineral in the oxidized zones of ore deposits.

hemin *Biochemistry.* $C_{34}H_{32}O_4N_4FeCl$, a tetrapyrrolic chelate of iron in which the iron is in the ferric form.

hemiparalysis see HEMIPLEGIA.

hemiparaplegia *Neurology.* a paralysis of one side of the lower half of the body.

hemiparasite *Ecology.* a parasite that has some capacity for independent existence in the absence of a suitable host. Thus, **hemiparasitic.**

hemiparesis *Medicine.* a muscle weakness or slight paralysis affecting one side of the body.

hemiparetic *Medicine.* **1.** of or relating to hemiparesis. **2.** an individual with hemiparesis.

hemipelagic *Ecology.* of or relating to sediment that contains the remains of both marine and land organisms.

hemipelagic region see BATHYAL ZONE.

hemipelagic sediment *Geology.* a deep-sea deposit containing a small amount of terrestrial material, as well as the remains of pelagic organisms.

hemipenis *Vertebrate Zoology.* either of a pair of copulatory organs (hemipenes) found in snakes and reptiles, which during sex are turned inside out, extruded, and inserted into the female cloaca to guide sperm for internal fertilization.

Hemipeplidae see CUCUJIDAE.

Hemiphacidiaceae *Mycology.* a family of fungi belonging to the order Helotiales, characterized by colored spore cases and consisting of plant pathogens which cause such diseases as needle blight or snow blight in conifers.

hemiplegia *Medicine.* a paralysis of one side of the body.

hemiprism *Crystallography.* a prism with only two parallel faces.

Hemiprocnidae *Vertebrate Zoology.* the crested swifts, a monogeneric family of birds of the order Apodiformes, characterized by a conspicuous crest and rapid flight and found from India to New Guinea and in some South Pacific islands.

Hemiptera *Entomology.* the true bugs, a large order of insects including bedbugs, cicadas, and aphids, with mouthparts adapted for piercing and sucking and with mandibles in the form of long stylets lying in a trough-like labium.

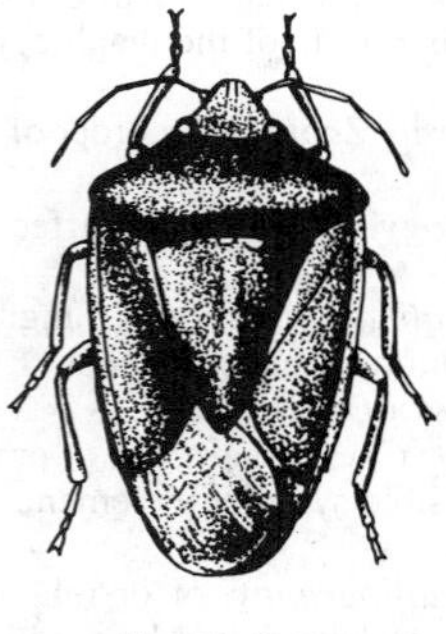

Hemiptera

hemipteran *Invertebrate Zoology.* **1.** belonging to the insect order Hemiptera. **2.** a member of this order.

Hemiramphidae *Vertebrate Zoology.* the halfbeaks, a family of usually marine fishes of the order Antheriniformes; characterized by a short triangular jaw and their ability to leap and skitter on the ocean surface; found in the Atlantic, Pacific, and Indian Oceans.

Hemiselmidaceae *Botany.* a family of marine flagellates of the order Cryptomonadales, characterized by cells with furrows along the short axis.

hemisotonic *Hematology.* having the same osmotic pressure as the blood.

hemispasm *Neurology.* a spasm that affects only one side of the body or one side of a body part.

Hemisphaeriales *Mycology.* an order of fungi formerly classified as belonging to the class Ascomycetes, now reclassified under the order Dothideales.

hemisphere *Geography.* **1.** half of the earth, traditionally divided into the Northern and Southern Hemispheres along the equator and into the Eastern and Western Hemispheres along a north-south line running 20°W and 160°E. **2.** any division of the earth into two equal parts. *Mathematics.* one of the two surfaces into which a sphere is divided by a great circle.

hemispherical candlepower *Optics.* a unit of measure used to express the luminous intensity of a hemispherical light source.

hemispherical map *Cartography.* a map that shows one-half of the earth's surface; either a polar projection bounded by the equator, or a latitudinal projection with the equator as a centerline and bounded by meridians 180° apart.

hemispheric wave number see ANGULAR WAVE NUMBER.

hemispheroid *Mathematics.* one of two halves into which a spheroid is divided by a plane of symmetry.

Hemist *Geology.* a suborder of the soil order Histosol, characterized by having an intermediate degree of partially decomposed plant fiber and a bulk density of 0.1 to 0.2, and usually requiring artificial drainage for cultivation.

hemithorax *Anatomy.* one side of the chest, or thorax.

hemitonia see HEMIHYPERTONIA.

Hemitrichia *Mycology.* a genus of fungal slime molds belonging to the class Myxomycetes; it occurs on rotting wood.

hemitropic *Crystallography.* describing a twinned crystal in which one part would be parallel to the other if it were rotated 180°.

hemizygous gene *Genetics.* a gene that is present only in a single dose or on the sex chromosome in the heterogametic sex.

hemlock *Botany.* **1.** the common name for any tree of the genus *Tsuga* in the family Pinaceae; an evergreen coniferous tree that is native to the temperate regions in the Northern Hemisphere; characterized by two bite lines beneath needlelike leaves. Also, **hemlock spruce. 2.** an herb of the species *Conium maculatum*, belonging to the carrot family Apiaceae, that is highly poisonous, having finely cut leaves and small white flowers.

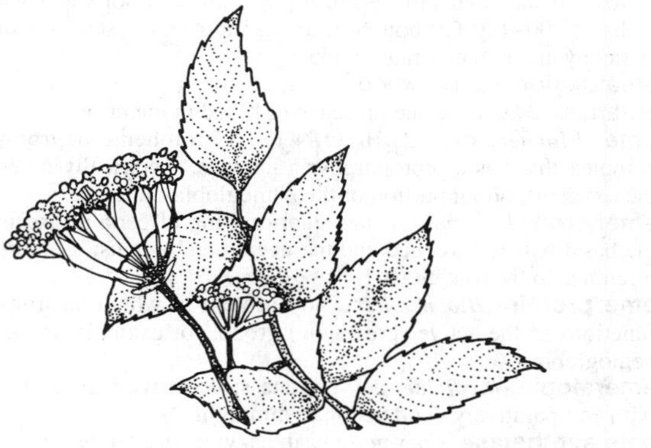

hemlock

hemming *Mechanical Engineering.* the bending of metal back upon itself to produce an edge.

hemming filter *Biotechnology.* a filter most commonly used for selective filtration, often found between two tubes that are fastened together; when centrifuged, the compound being selected passes from one tube to the other.

hemo- a combining form meaning "blood," as in *hemophilia.*

hemoaccess *Surgery.* the site of entry into a blood vessel, as one maintained for recurrent hemodialysis.

hemoagglutinating encephalomyelitis see VOMITING AND WASTING DISEASE.

hemoblastic leukemia see STEM CELL LEUKEMIA.

hemochorial placenta *Developmental Biology.* a placenta, as in humans, in which maternal blood comes in direct contact with the chorion.

hemochromatosis *Medicine.* **1.** a disorder in which excess iron is deposited in the tissues of organs such as the heart, skin, liver, and pancreas, leading to impaired functioning of these organs; it exists in both hereditary and acquired forms. **2.** any poisoning due to the ingestion of excess iron.

hemocoel *Invertebrate Zoology.* the blood-filled cavity or sinus enclosing many of the organs in insects and other arthropods; blood is pumped forward by the dorsal tubular heart, not entering the capillaries but bathing the organs directly.

hemocoelous viviparity *Entomology.* a condition in which eggs escape into the hemocoel, where the larvae develop and eventually exit through secondary body-wall openings.

hemoconcentration *Medicine.* an increase in the concentration of red blood cells in the circulating blood, usually resulting from loss of plasma from the blood stream.

hemoconia *Hematology.* small, round or dumbbell-shaped particles demonstrating Brownian movement, observed in blood platelets in a wet film of blood under darkfield microscopy. Also, MÜLLER'S DUST BODIES.

hemoconiosis *Hematology.* the presence of an abnormal amount of hemoconia in the blood.

hemocyanin *Hematology.* a blue respiratory pigment that is found in the blood plasma of many mollusks and arthropods.

hemocyte *Invertebrate Zoology.* a blood cell, especially in invertebrates.

hemocytoblast *Histology.* the undifferentiated stem cell in hematopoietic tissue that is capable of producing all other blood cells. Also, STEM CELL.

hemocytoblastic leukemia see STEM CELL LEUKEMIA.

hemocytoblastoma *Oncology.* a tumor that is composed of bone marrow-type cells.

hemocytolysis *Hematology.* the dissolution of blood cells.

hemocytoma *Oncology.* a tumor that is composed of undifferentiated blood cells.

hemocytometer *Pathology.* an instrument used to count blood cells manually, using a reservoir of uniform depth covered by a grid; the area under each square contains a known amount of diluted blood.

hemodialysis see DIALYSIS.

hemodichorial placenta *Developmental Biology.* a hemochorial placenta in which two layers of the trophoblast are interposed between fetal capillaries and maternal blood, as in rabbits.

hemodynamics *Physiology.* a branch of physiology that studies the forces related to circulation of the blood.

hemoendothelial placenta *Developmental Biology.* a placenta in which maternal blood comes in direct contact with the endothelium of the chorionic capillaries.

hemoerythrin *Hematology.* a red respiratory pigment found in the blood of earthworms.

hemoflagellate *Invertebrate Zoology.* any flagellate protozoan that parasitizes the blood of its host.

hemogenic see HEMATOGENIC.

hemoglobin [hēm´ə glō´bin] *Hematology.* the oxygen-carrying pigment of the erythrocytes, formed by the developing erythrocyte in bone marrow. It is a complex protein composed of four heme groups and four globin polypeptide chains. They are designated α (alpha), β (beta), γ (gamma), and δ (delta) in an adult, and each is composed of several hundred amino acids.

hemoglobin A *Hematology.* normal adult hemoglobin, composed of two alpha and two beta chains, $\alpha_2^A\beta_2^A$.

hemoglobin C *Pathology.* an oxygen-carrying pigment of the erythrocytes characterized by a molecular abnormality in which lysine replaces glutamic acid at position six of the beta chain, reducing the normal plasticity of the erythrocyte.

hemoglobin E *Pathology.* an oxygen-carrying pigment of the erythrocytes characterized by a molecular abnormality due to a beta chain mutation in the hemoglobin; observed mainly in southeast Asia, and linked to microcythemia, target cell formation, and mild hemolytic anemia.

hemoglobin electrophoresis *Medicine.* the technique of using an applied electric field to separate hemoglobin proteins from one another.

hemoglobinemia *Medicine.* the presence of free hemoglobin in the plasma, associated with injury to red blood cells within the blood vessels.

hemoglobin F *Pathology.* hemoglobin found in the fetus.

hemoglobin H *Pathology.* an oxygen-carrying pigment of the erythrocytes characterized by a molecular abnormality leading to precipitate migration and constituting four beta chains; possessing a high oxygen affinity, but ineffective as an oxygen conveyor.

hemoglobinopathy *Medicine.* a blood disorder caused by alteration in the structure or synthesis of hemoglobin, such as in sickle cell disease and thalassemia.

hemoglobin S *Pathology.* the most frequently seen abnormal form of hemoglobin; its beta chain contains valine with gluconic acid at position six. The heterozygous phase gives rise to sickle cell disease, while the homozygous phase is seen in sickle cell anemia.

hemoglobinuria *Medicine.* the presence of free hemoglobin in the urine.

hemogram *Pathology.* a complete written or graphic representation of the differential blood count.

hemohistioblast *Histology.* the original reticuloendothelial cell that produces hemocytoblasts.

hemokinesis *Hematology.* the flow of blood in the body.

hemokinetic *Hematology.* relating to or causing the flow of blood in the body.

hemology see HEMATOLOGY.

hemolymph *Entomology.* the bloodlike fluid moving through the hemocoelom of those invertebrates (e.g., mollusks, arthropods, and tunicates) with open circulatory systems, which combines the properties of blood and lymphlike interstitial fluid.

hemolysin *Immunology.* an antibody that is able to destroy red blood cells in the presence of a complement.

hemolysis *Physiology.* the destruction of red blood cells with the liberation of hemoglobin into the plasma.

hemolytic anemia *Medicine.* any anemia characterized by the destruction of red blood cells.

hemolytic immune body *Immunology.* an antibody that reacts with the surface antigens of red blood cells.

hemolytic jaundice *Medicine.* a yellowish discoloration of the skin and sclerae resulting from the destruction of red blood cells.

hemolytic system *Immunology.* a mixture of red blood cells covered with antibodies that are genetically identical to their surface antigens; used to measure the amount of complement that remains unbound in an antiserum dilution after the first stage of a complement fixation test.

hemomonochorial placenta *Developmental Biology.* a hemochorial placenta in which one layer of trophoblast separates the fetal capillaries from the maternal blood, as in humans and guinea pigs.

hemoparasite *Invertebrate Zoology.* a parasite that lives in the blood of an animal.

hemopathology *Medicine.* the science of the diseases of the blood.

hemopathy *Medicine.* any disorder of the blood or blood-forming tissues.

hemopericardium *Medicine.* an accumulation of blood in the pericardial sac.

hemoperitoneum *Medicine.* an effusion of blood into the peritoneal cavity.

hemopexin *Hematology.* a plasma glycoprotein whose function is the binding of free heme in plasma.

hemophilia [hēm´ə fēl´yə] *Medicine.* a genetic blood disorder in which there is a deficiency of one of the factors needed for the normal coagulation of blood, classified into two major types: **hemophilia A,** a sex-linked, recessive bleeding disorder affecting males caused by a deficiency of factor VIII, and **hemophilia B** or Christmas disease, in which there is a deficiency of clotting factor IX.

hemophiliac [hēm´ə fēl´ē yak] *Medicine.* **1.** a person affected by hemophilia. **2.** of or relating to hemophilia. Also, **hemophilic.**

hemophilic bacteria *Bacteriology.* any of various bacteria of the genera *Haemophilus, Bordetella,* and *Moraxella;* they are Gram-negative, rod-shaped, nonmotile parasites that grow especially well in culture media containing blood or that have a nutritional need for constituents of fresh blood.

hemophilioid *Medicine.* resembling classical hemophilia but of different genetic derivation, as in various hereditary or acquired hemorrhagic disorders that are not due solely to a deficiency of blood coagulation.

Hemophilus see HAEMOPHILUS.

hemophobia *Psychology.* an irrational fear of blood. Thus, **hemophobic.**

hemoporphyrin see HEMATOPORPHYRIN.

hemoptysis *Medicine.* the coughing or spitting of blood or blood-stained sputum.

hemorrhachis see HEMATOMYELIA.

hemorrhage [hēm´rij; hem´ə rij] *Medicine.* the process of bleeding; the escape of blood from the vessels, especially a large amount of blood in a short time.

hemorrhagic [hēm´ə raj´ik] *Medicine.* relating to or producing a hemorrhage.

hemorrhagic diathesis *Medicine.* a susceptibility to bleeding, caused by congenital or hereditary factors or as the result of a metabolic or nutritional abnormality, as in hemophilia, scurvy, or vitamin K deficiency.

hemorrhagic fever virus *Virology.* any of a number of viruses of the families Bunyaviridae and Flaviviridae that cause hemorrhagic diseases; generally transmitted through arthropod bites or contact with infected rodents.

hemorrhagic measles *Medicine.* a severe form of measles in which the eruption is dark in color due to bleeding into the skin.

hemorrhagic pericarditis *Medicine.* an inflammation of the pericardium, the thin, double-layered membranous sac that encloses the heart, accompanied by bleeding.

hemorrhagic pleuritis *Medicine.* an inflammation of the serous membrane covering the lungs and lining the walls of the chest cavity, accompanied by bleeding.

hemorrhagic septicemia *Veterinary Medicine.* an infectious bacterial disease of cattle and buffalo, caused by *Pasteurella multocida*; characterized by fever, loss of appetite, and profuse salivation, and associated with stress due to exhaustion, malnourishment, or lengthy transport.

hemorrhagin *Toxicology.* a poison that causes destruction of endothelial cells and blood vessels.

hemorrheology *Hematology.* the scientific study of the flow of blood and of the vessels with which the blood comes in direct contact.

hemorrhoid [hem´ə roid´] *Medicine.* **1.** a dilated, abnormally large or varicose vein in the lower rectal or anal wall. **2. hemorrhoids.** a condition marked by the presence of such veins, often leading to itching, pain, or bleeding. ((From a Greek term meaning "the flow of blood.")

hemorrhoidectomy [hem´ə roi dek´tə mē] *Surgery.* the surgical removal of hemorrhoids.

hemosiderin *Hematology.* an intracellular storage form of iron; the granules consist of an ill-defined complex of ferric hydroxides, polysaccharides, and proteins having an iron content of about 33% by weight.

hemosiderosis *Physiology.* a general increase in iron stores in tissues without tissue damage.

hemostasis *Hematology.* **1.** the stoppage of bleeding, either by the physiological properties of vasoconstriction and coagulation, or by some mechanical or surgical means. **2.** any interruption of the flow of blood through a vessel or to an anatomical area. Also, **hemostasia.**

hemostat *Medicine.* a surgical instrument or an agent used to arrest the flow of blood.

hemostatic *Medicine.* referring to a substance, procedure, or device that arrests the flow of blood.

hemostatic suture *Surgery.* a type of suture used to control the oozing of blood from raw areas.

hemostatis *Medicine.* **1.** the arrest of blood flow. **2.** the arrest or slowdown of the circulation of blood.

hemotherapy *Medicine.* the treatment of disease by the administration of blood or blood products.

hemothorax *Medicine.* an accumulation of blood in the pleural cavity.

hemotoxin *Toxicology.* any poison that attacks blood cells.

hemotrichorial placenta *Developmental Biology.* a hemochorial placenta in which three layers of trophoblast are interposed between the fetal capillaries and the maternal blood, as in rats, mice, and hamsters.

hemotroph *Hematology.* the sum total of the nutritive substances supplied to the embryo from the maternal blood during gestation. Also, **hemotrophe.**

hemotropic *Biology.* acting upon the blood or affecting blood. Also, HAEMOTROPIC.

hemotropic poison *Toxicology.* any poison that attacks the red blood cells.

hemozoin *Pathology.* a stain generated by malaria-producing parasites, obtained from the host cell's hemoglobin, and made up of polymers that allow the parasites to segregate in harmless form. Also, HAEMOZOIN.

hemp *Botany.* **1.** a tall, coarse plant, *Cannabis sativa,* native to Asia; cultivated in many parts of the world as a source of valuable fiber and for narcotic drugs such as marijuana and hashish. **2.** the coarse, tough fiber of this plant, used for making rope, fabric, and other materials. **3.** the common name for any narcotic drug, such as hashish, made from this plant.

Hemphillian *Geology.* a North American geologic stage of the Middle Miocene epoch, occurring after the Clarendonian and before the Blancan.

hen *Vertebrate Zoology.* **1.** a female bird, especially a female domesticated fowl at least one year old. **2.** the female of some aquatic animals, such as lobsters and fish.

henbane *Botany.* the common name for a plant of the species *Hyoscyamus niger*, a highly poisonous Old World herb of the nightshade family Solanaceae, characterized by sticky, hairy dentate leaves and yellowish-brown flowers. (Because its poison is especially destructive to domestic fowl.)

Hench, Philip 1896–1965, American physician; with Kendall and Reichstein, Nobel Prize for use of cortisone in treatment of arthritis.

hendecanal see UNDECANAL.

hendecane see *n*-UNDECANE.

hendecanoic acid see UNDECANOIC ACID.

Henderson-Hasselbalch equation *Biochemistry.* the equation expressed as pH = pK + log 9([base]/[acid]), which relates the pH of a solution to the pK of any dissociating species in it and to the concentrations of an acid and its conjugated base.

Henderson process *Metallurgy.* any of several extractive processes for copper based on sulfate roasting followed by leaching.

heneicosane *Organic Chemistry.* $C_{21}H_{44}$, a saturated hydrocarbon of the methane series; crystals; melts at 40°C.

henge *Archaeology.* a type of ritual monument found in the British Isles, usually consisting of a circular arrangement of stone pillars enclosed by a bank and ditch; Stonehenge is the most noted example.

Hengstebeck approximation *Chemical Engineering.* a calculation technique for estimating the distribution of non-key components in products from distillation columns.

Henicocephalidae *Invertebrate Zoology.* an archaic family of homopteran insects recognized by some authorities.

Henle, Jakob 1809–1885, German pathologist; formulated early germ theory of disease; classified diseases; wrote important texts in anatomy, physiology, and pathology.

Henle's fissures *Cardiology.* spaces within the heart's musculature filled with connective tissue.

henna *Botany.* a plant of the species *Lawsonia inermis*, an Old World tropical shrub or small tree of the family Lythraceae, having white flowers and small opposite leaves that are dried and made into a reddish-brown dye often used for hair coloring.

Henneguya *Invertebrate Zoology.* a large genus belonging to the family Myxobolidae, with over 80 species of histozoic parasites of freshwater fish.

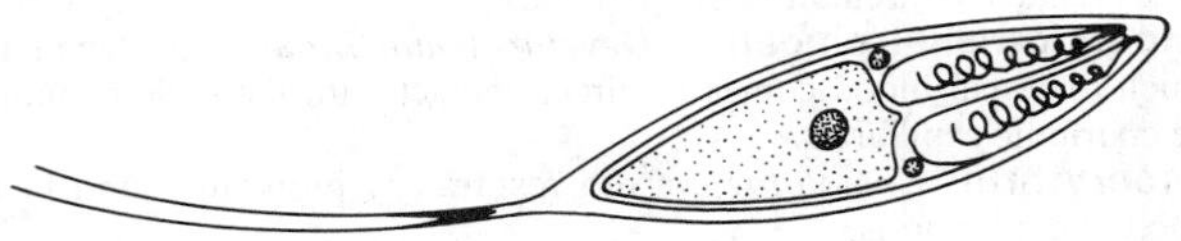

Henneguya

Hénon attractor *Chaotic Dynamics.* a type of strange attractor created by the two-dimensional mapping of repeated folding and stretching described by iteration of the Hénon map.

Hénon map *Chaotic Dynamics.* a two-dimensional iterative procedure for finding sequences of pairs of values for *x* and *y* which has characteristic stretching and folding: $x_{n+1} = 1 - ax_n 2$, $y_{n+1} = bx_n$.

Henry, Joseph 1797–1878, American physicist; improved the electromagnet; discovered electromagnetic self-induction.

Henry, William 1775–1836, English chemist; formulated and refined Henry's law on the aborption of a gas in a liquid.

henry *Electromagnetism.* an mks unit of inductance equivalent to that of an induced 1 volt in the presence of a current that is changing at a rate of 1 ampere per second. (Named for Joseph *Henry.*)

Henry Draper catalogue *Astrophysics.* a catalogue compiled at Harvard College Observatory between 1918 and 1924 that lists spectral types for about 225,000 stars.

Henry Draper system *Astrophysics.* a classification of stars by spectral type (in decreasing order of temperature: O, B, A, F, G, K, M) that was used in the Henry Draper catalogue and with modifications is still in use today.

Henry's law *Physical Chemistry.* a law stating that the amount of gas which will dissolve in a given quantity of liquid is proportional to the partial pressure of the gas above the liquid at constant temperature.

Henry the Navigator 1394–1460, Portuguese prince; at Sagres, established center for study of astronomy, geography, and navigation.

Hensen's node *Developmental Biology.* a local thickening of the blastoderm at the cephalic end of the primitive streak of the embryo of a bird or mammal. Also, PRIMITIVE KNOT.

hentriacontane *Organic Chemistry.* $C_{31}H_{64}$, crystals found in beeswax; melts at 68°C and boils at 302°C at reduced pressure.

Henyey track *Astronomy.* a nearly horizontal track on the Hertzsprung-Russell diagram that traces the path of a star in radiative equilibrium as it moves from its Hayashi track to the main sequence.

HEP *Virology.* an attenuated virus that has undergone serial passage in eggs. (An acronym for high egg passage.)

Hepadnaviridae *Virology.* a family of enveloped DNA-containing viruses that infect man, animals, and birds, causing acute and chronic hepatitis; includes hepatitis B virus.

heparin *Biochemistry.* a naturally occurring polysaccharide that has anticoagulant properties, composed of D-glucosamine-2,6-disulfate and D-glucuronic acid-2-sulfate and applied clinically for the treatment of embolism and thrombosis.

hepar lobatum *Medicine.* the condition of nodular lobules on the liver, present in syphilitic cirrhosis.

HEPAT see HIGH-EXPLOSIVE PLASTIC ANTITANK CHARGE.

hepat- a combining form meaning "liver," as in *hepatitis.*

hepatectomy *Surgery.* the surgical removal of a portion of the liver.

hepatic *Anatomy.* relating to or affecting the liver.

hepatica *Botany.* a plant of the genus *Hepatica,* growing in forest areas in most parts of the Northern Hemisphere and having dark green, leathery leaves; the two main types, designated by their leaf shapes, are the **round-lobed hepatica,** *H. americana,* and the **sharp-lobed hepatica,** *H. acutiloba.* (Going back to the Greek word for liver; suggested by the shape of the round-lobed variety.)

Hepaticae *Botany.* in some classification systems, an equivalent name for Marchantiatae.

hepatic artery *Anatomy.* the large artery that supplies blood to the liver and gallbladder.

hepatic cecum *Invertebrate Zoology.* a glandular pouch in the digestive tract, especially one of the digestive and storage glands in a sea star.

hepatic coma *Medicine.* the deep coma that may occur in a patient suffering from advanced stages of chronic or acute liver disease.

hepatic duct *Anatomy.* the duct that drains bile from the liver and joins the cystic duct to form the common bile duct.

hepatic duct system *Anatomy.* the system of ducts that drain bile from the liver, including the hepatic ducts, the gallbladder, the cystic duct, and the common bile duct.

hepatic encephalopathy *Medicine.* the impaired functions of the brain, such as memory loss, lethargy, changes in personality, and variable consciousness, that occur in patients with advanced liver disease.

hepatic lobule *Anatomy.* a small functional unit of the liver.

Hepaticopsida *Botany.* the liverworts, a class of lower green plants of the division Bryophyta, characterized by regular alternation between a short diploid sporophyte generation and a haploid gametophyte generation. Also, MARCHANTIOPSIDA.

hepatic plexus *Anatomy.* a network of autonomic nerves that accompany the hepatic artery into the liver.

hepatic portal system *Anatomy.* a system of veins carrying blood that has absorbed nutrients in capillaries of the small intestine to the liver.

hepatic vein *Anatomy.* any of several large veins that drain blood from the liver into the inferior vena cava.

hepatitis [hep´ə tī´tis] *Medicine.* an inflammation of the liver, generally involving such symptoms as jaundice, weakness, fever, loss of appetite, and gastric and abdominal disturbances.

hepatitis A *Medicine.* a self-limited viral disease of worldwide distribution that is caused by the hepatitis A virus, and is more prevalent in areas of poor hygiene and low socioeconomic standards; most cases have mild flulike symptoms or jaundice. Also, INFECTIOUS HEPATITIS.

hepatitis A virus *Virology.* an RNA virus having a single-stranded genome, probably belonging to the genus *Enterovirus* of the family Picornaviridae; it is the causative agent of hepatitis A, and is typically transmitted through the consumption of water or food contaminated with human feces. Also, INFECTIOUS HEPATITIS VIRUS, EPIDEMIC JAUNDICE VIRUS.

hepatitis B *Medicine.*a viral disease of worldwide distribution caused by the hepatitis B virus, characterized by a relatively long incubation period (averaging 90 days); infection may be severe and can result in destruction of liver cells, cirrhosis, or death. Also, SERUM HEPATITIS.

hepatitis B antigen see AUSTRALIA ANTIGEN.

hepatitis B virus *Virology.* a DNA virus of the family Hepadnaviridae, with complex, double-layered virions, a partially double-stranded genome, and three major antigens; it is the causative agent of hepatitis B, and is typically transmitted through the inoculation of contaminated serum, by unsterilized medical instruments or hypodermic needles, or by sexual contact. Also, SERUM HEPATITIS VIRUS.

hepatitis delta *Medicine.* an acute form of hepatitis of relatively recent manifestation, caused by an incomplete virus (delta virus) that links with the hepatitis B virus.

hepatitis delta virus see DELTA VIRUS.

hepatitis non-A non-B virus *Virology.* either of the two presently unclassified viruses causing non-A, non-B hepatitis; a calicivirus apparently causes the water-borne form of the disease, and a togovirus apparently causes the post-transfusion form. Also, **hepatitis C virus.**

hepatization *Pathology.* the metamorphosis of loose tissue into a liver-like, solidified mass, as occurs within the lungs in pneumonia.

hepato- a combining form meaning "liver," as in *hepatology.*

hepatocyte *Histology.* the primary functional cell of the liver.

hepatolenticular degeneration see WILSON'S DISEASE.

hepatolith *Medicine.* a gallstone in the liver.

hepatoma *Medicine.* a malignant liver tumor that usually occurs in association with cirrhosis or following infection with the hepatitis B virus.

hepatomegaly *Medicine.* an enlargement of the liver.

hepatopancreas *Invertebrate Zoology.* a glandular structure that combines the functions of the liver and pancreas.

hepatopathy *Medicine.* any disease of the liver.

hepatorenal syndrome *Medicine.* a functional renal failure occurring in patients suffering from severe liver disease.

hepatorrhagia *Medicine.* a hemorrhage from the liver.

hepatoscopy *Medicine.* an examination of the liver.

hepatosplenomegaly *Medicine.* the enlargement of the liver and spleen.

hepatotomy *Surgery.* a surgical incision of the liver.

hepatotoxic *Medicine.* destructive to the liver. Also, **hepatolytic.**

hepatotoxin *Toxicology.* any poisonous substance that destroys liver cells.

hepatoxicity *Toxicology.* the quality or property of having a poisonous or destructive effect on liver cells.

HEPES *Organic Chemistry.* $C_8H_{18}N_2O_4S$, crystals that are soluble in alcohol and water; melts at 234°C; used as a biological buffer. Also, **4-(2-hydroxyethyl)-1-piperazineethanesulfonic acid, Hepes.**

Hepialidae *Invertebrate Zoology.* a family of moths including the ghost or swift moths, lepidopteran insects in the superfamily Hepialoidea.

Hepialoidea *Invertebrate Zoology.* a superfamily of moths, lepidopteran insects in the suborder Homoneura, whose larvae are stem and root borers.

HEPP high-explosive plastic projectile.

Hepsogastridae *Invertebrate Zoology.* an archaic family of biting lice recognized by some authorities.

hept- or **hepta-** a combining form meaning "seven," as in *heptagon.*

heptachlor *Organic Chemistry.* $C_{10}H_7Cl_7$, a poisonous white solid; soluble in carbon tetrachloride; melts at 95–96°C; used as an insecticide.

heptacosane *Organic Chemistry.* $C_{27}H_{56}$, combustible crystals found in beeswax; soluble in alcohol and insoluble in water; melts at 60°C and boils at 270°C.

heptad *Science.* a group of seven items.

heptadaidecagon *Mathematics.* a polygon with seventeen sides.

heptadecane *Organic Chemistry.* $C_{17}H_{36}$, a solid combustible hydrocarbon that is soluble in alcohol and insoluble in water; melts at 23°C and boils at 303°C; used as a chemical intermediate.

***n*-heptadecanoic acid** *Organic Chemistry.* $CH_3(CH_2)_{15}COOH$, colorless crystals; melts at 61°C; soluble in alcohol and ether and insoluble in water; used in organic synthesis.

heptadecanol *Organic Chemistry.* $C_{17}H_{35}OH$, a colorless combustible liquid; slightly soluble in water; boils at 309°C; used as a chemical intermediate and in perfumes, cosmetics, and soaps.

heptagon *Mathematics.* a polygon with seven sides.

heptaldehyde *Organic Chemistry.* $C_6H_{13}CHO$, a colorless oil; slightly soluble in water and miscible with alcohol; melts at –43.3°C and boils at 153°C; used in perfumes and pharmaceuticals. Also, HEPTANAL.

heptamerous *Botany.* of a flower, having seven members per whorl. Also, 7-MEROUS.

heptane *Organic Chemistry.* $CH_3(CH_2)_5CH_3$, a volatile, irritating, colorless liquid that is soluble in alcohol and insoluble in water; boils at 98.48°C and melts at –90.7°C; toxic by inhalation; used as an anesthetic, as a solvent, and as a standard for octane ratings of gasoline.

heptanoic acid *Organic Chemistry.* $CH_3(CH_2)_5COOH$, a clear oil with an unpleasant odor; soluble in alcohol and slightly soluble in water; boils at 222°C and melts at –7°C; used as a chemical intermediate and in the production of lubricants for aircraft and brake fluids.

1-heptanol *Organic Chemistry.* $C_7H_{15}OH$, a colorless liquid that is slightly soluble in water and miscible with alcohol; melts at –34.6°C and boils at 174°C; used as a chemical intermediate, as a solvent, and in cosmetics.

2-heptanol see METHYL AMYLCARBINOL.

3-heptanol *Organic Chemistry.* $CH_3CH_2CH(OH)C_4H_9$, a toxic, combustible liquid that is slightly soluble in water; boils at 156.2°C; used as a solvent and flotation frother.

2-heptanone see METHYL AMYL KETONE.

4-heptanone *Organic Chemistry.* $(CH_3CH_2CH_2)_2CO$, a colorless liquid that is insoluble in water and miscible with alcohol; boils at 144°C and melts at –32°C; used in lacquers and as a flavoring. Also, DIPROPYL KETONE.

heptene *Organic Chemistry.* C_7H_{14}, an isomeric liquid that is soluble in alcohol; boils at 189°C (**1-heptene** boils at 93°C, **2-heptene** at 98°C, and **3-heptene** at 95°C); used as an additive in lubricants and as a surface-active agent. Also, **heptylene.**

heptode *Electronics.* an electron tube containing seven electrodes: an anode, cathode, primary control electrode, and four auxiliary elements, usually grids.

heptose *Biochemistry.* a monosaccharide that contains seven carbon atoms in a molecule and is important in carbohydrate metabolism.

heptoxide *Chemistry.* any elemental oxide that contains seven atoms of oxygen.

heptulose *Biochemistry.* any ketose produced from a seven-carbon monosaccharide.

Heraclides of Pontus c. 388–315 BC, Greek astronomer, proposed 24-hour day and solar orbits of Mercury and Venus.

herb [urb; hurb] *Botany.* **1.** a flowering, vascular seed plant that lacks a woody stem aboveground and whose aboveground parts die at the end of a season. **2.** any plant whose parts are used for medicinal purposes, the seasoning of food, or perfume.

herbaceous *Botany.* **1.** of or relating to an herb. **2.** soft and green, rather than woody.

herbage *Agriculture.* the total vegetation available for grazing animals to feed on.

herbarium *Botany.* a collection of dried and mounted or otherwise preserved plant specimens that are systematically arranged for reference.

Herbertaceae *Botany.* a family of medium to large dioecious liverworts of the order Jungermanniales, often having erect leafy axes rising perpendicularly from rhizomatous axes with scaly leaves, and occurring in New Zealand, Tasmania, and South America.

herbicide *Agriculture.* any agent, either organic or inorganic, used to destroy unwanted vegetation, especially weeds and grasses; **selective herbicides** eliminate weeds without destroying desirable crop or garden plants; **nonselective herbicides** destroy all vegetation in the given area.

herbicolous *Ecology.* **1.** living on herbs. **2.** living in a herbaceous habitat. Thus, **herbicole.**

Herbig emission stars *Astronomy.* a name for Be and Ae stars.

Herbig-Haro object *Astronomy.* small, bright areas of nebulosity found in interstellar dust clouds; they may be condensing to form stars.

herbivore *Biology.* an organism that feeds on plants, especially an animal whose diet is exclusively plants.

herbivorous *Biology.* feeding exclusively or mainly on plants.

herbivory *Biology.* the act of eating plants.

herbosa *Ecology.* vegetation consisting of grasses and herbs, without woody plants.

Herbrand base *Artificial Intelligence.* the set of all predicates whose arguments are taken from the Herbrand universe. Also, ATOM SET.

Herbrand universe *Artificial Intelligence.* in mathematical logic, a recursively defined set of terms, in which the first level is the set of all constant symbols and each subsequent level includes all functions of terms in the previous levels.

Herbst corpuscle *Vertebrate Zoology.* a pressure and vibration sense organ derived from prevertebral ganglia of the sympathetic nervous system, found in the tongues of woodpeckers and parrots, and on the palates and beaks of ducks.

Hercules *Astronomy.* a keystone-shaped constellation of the northern celestial hemisphere lying west of Lyra; it is best visible on summer evenings.

Hercules cluster *Astronomy.* **1.** a cluster of a few hundred galaxies, about half of which are ellipticals, lying at a distance of roughly 400 million light-years. **2.** the large globular cluster M 13 in Hercules.

Hercules stone see LODESTONE.

Hercules trap *Analytical Chemistry.* a liquid trap used in aquametry for collecting liquids heavier than water.

Hercules X-1 *Astronomy.* an X-ray emitting binary star and pulsar, optically identified as HZ Herculis.

Hercynian orogeny see VARISCAN OROGENY.

Hercynian geosyncline see VARISCAN GEOSYNCLINE.

hercynite *Mineralogy.* $Fe^{+2}Al_2O_4$, a black, opaque, cubic mineral in the spinel group occurring usually in finely granular form, having a specific gravity of 4.32 and a hardness of 7.5 to 8 on the Mohs scale; found in emery deposits. Also, FERROSPINEL, IRON SPINEL.

herd *Vertebrate Zoology.* **1.** in mammals, a form of social organization in which individuals of one or more kinds group together for purposes of hunting, defense, etc. **2.** to group together for puposes of hunting, defense, and so on.

herderite *Mineralogy.* $CaBe(PO_4)F$, a colorless to pale yellow or greenish white monoclinic mineral occurring in prismatic or tabular crystals, having a specific gravity of 2.95 to 3.01 and a hardness of 5 to 5.5 on the Mohs scale; found in franite pegmatites.

herd instinct *Behavior.* the tendency of certain species to congregate in large groups. Also, **herding.** *Psychology.* **1.** the desire to associate with others in a group and to take part in group activities. **2.** conformity to established beliefs or practices based on this desire.

hereditary *Genetics.* **1.** of a characteristic or trait, genetically transmitted from a parent to an offspring. **2.** of or relating to heredity. *Medicine.* describing a disease or condition that is genetically transmitted from parent to child.

The terms *hereditary, congenital,* and *familial* are similar, but in medical usage they are not exact synonyms. A *hereditary* condition is passed down by inheritance from parent to child; it thus exists at birth but is not necessarily evident at birth; for example, Wilson's disease is hereditary but its symptoms typically are not manifested before adolescence. A *congenital* condition is evident at birth but is not necessarily inherited; e.g., an infant's congenital heart disease may result from the mother's exposure to radiation during pregnancy. A *familial* condition occurs more often than is statistically normal among the members of a given family and may or may not be inherited; for example, alcoholism or schizophrenia.

hereditary *Mathematics.* a nonempty class H of sets is said to be hereditary if, whenever $T \in H$ and S is contained in T, then also $S \in H$. If E is any class of sets closed under countable unions, then the hereditary σ-ring generated by E, denoted $H(E)$, is the class of all sets that are subsets of some set in E.

hereditary disease *Medicine.* a disease that is genetically transmitted from parent to child. Similarly, **hereditary disorder.**

hereditary hemorrhagic telangiectasia *Medicine.* a genetic disorder in which there is dilation of a group of capillaries in the skin and mucous membranes that appears after puberty. Also, RENDU-OSLER-WEBER DISEASE, SUTTON-RENDU-OSLER-WEBER SYNDROME.

hereditary hypophosphatemic rickets *Medicine.* a genetically transmitted disorder in which defective bone growth or rickets is associated with a deficiency of phosphate in the blood.

hereditary mechanics *Mechanics.* the field of mechanics dealing with the dependence of current mechanical properties on deformation history, which includes mechanical hysteresis.

hereditary nephritis *Medicine.* a genetically transmitted renal disease that progresses to chronic renal failure. Also, ALPORT'S SYNDROME.

hereditary spherocytosis *Medicine.* an inherited disorder of the red blood cells in which the shape of the cells is more spherical than biconcave, predisposing to hemolytic anemia.

heredity *Genetics.* **1.** the genetic transmission of a certain characteristic or trait from a parent to an offspring. **2.** the genetic constitution of a given individual.

heredofamilial *Medicine.* describing a disease or condition occurring in several family members and thought to be hereditary. It is now more common to use the terms *hereditary* or *familial,* whichever is more appropriate. See HEREDITARY.

Hereford [hur′furd; her′ə furd] *Agriculture.* **1.** a breed of beef cattle having a red body with a white face and white markings; the **Polled Hereford** is a hornless variety of this breed. **2.** a breed of medium-sized hog with coloring similar to that of the Hereford cattle breed, developed in the United States. (From the county of *Hereford,* in England.)

Hericiaceae *Mycology.* a family of fungi of the order Agaricales that is commonly known as tooth fungi and occurs primarily on wood.

Hericium *Mycology.* a genus of fungi belonging to the family Hericiaceae.

Hering theory *Physiology.* a theory of color vision that assumes three types of cones in the retina containing three different photochemical substances, which respond to different wavelengths of light and produce different sensations of color.

heritability *Genetics.* **1.** the fraction of the total phenotypic variance that remains after exclusion of the variance due to environmental effects. **2.** more specifically, the ratio of the total genetic variance to the total phenotypic variance.

Hermann-Maugiun symbol *Crystallography.* a notation for representing the symmetry of space groups.

hermaphrodite [hur maf′rə dīt′] *Biology.* **1.** a flowering plant possessing both male and female structures on the same flower, which is the usual arrangement in most plants. **2.** an animal possessing both male and female functional reproductive organs, such as the earthworm. **3.** a unisexual animal having male and female gonads as an aberration. Thus, **hermaphroditic.**

hermaphrodite caliper *Mechanical Devices.* a drawing device having one caliper leg and one divider leg, used for layout work of drawing center lines on a shaft or parallel lines with an edge.

hermaphroditic connector *Electricity.* a connector that is the same at both mating surfaces.

hermaphroditism *Physiology.* in humans, a condition in which both ovarian and testicular tissues are present in an individual. Also, GYNANDRY.

hermatopelago see REEF CLUSTER.

hermatype *Invertebrate Zoology.* a reef-building coral.

Hermes *Astronomy.* an asteroid discovered in 1937 that passed about 780,000 kilometers from the earth, the closest approach of an asteroid ever observed; its orbit was never established well enough for it to be located again, and the object is now considered lost.

Hermesinaceae *Botany.* a monospecific family of marine flagellates of the order Ebriales that are unarmored and have a complex three-dimensional internal skeleton with four arms radiating at equal angles.

hermetic *Engineering.* made airtight, e.g., by fusion or scaling.

hermetic seal *Engineering.* a seal that is impervious to air and other fluids.

hermit crab

hermit crab *Invertebrate Zoology.* the common name for a type of marine decapod crustacean with a long, soft, coiled abdomen that lives in discarded snail shells.

Hermite, Charles 1822–1901, French mathematician; introduced Hermitian forms; the first to solve a quintic equation; proved that *e*, the base of natural logarithms, is transcendental.

Hermite polynomials *Mathematics.* the polynomials given by

$$H_n(x) = (-1)^n \exp(x^2)\, d^n\, [\exp(-x^2)]/dx^n.$$

The exponential generating function is

$$\exp(-t^2 + 2tx) = \sum_{n=0}^{\infty} H_n(x)t^n/n!.$$

$H_n(x)$ satisfies the differential equation $H_n'' - 2xH_n' + 2nH_n = 0$. The first three Hermite polynomials are $H_0(x) = 1$, $H_1(x) = 2x$, and $H_2(x) = 4x^2 - 2$. The Hermite functions $\phi_n(x) = \exp(-x^2/2)H_n(x)$ form an orthogonal set using the unweighted inner product on the interval $(-\infty, \infty)$.

Hermitian adjoint *Mathematics.* for an $n \times n$ matrix A with complex entries, the matrix A^* whose (i,j)th entry is the complex conjugate of the (j,i)th entry of A. Also, HERMITIAN CONJUGATE. If $A^* = A$, then A is called a Hermitian matrix.

Hermitian form *Mathematics.* a bilinear form f on a real or complex vector space V for which $f(\alpha, \beta)$ equals the complex conjugate of $f(\beta, \alpha)$ for all $\alpha, \beta \in V$.

Hermitian kernel *Mathematics.* a kernel $K(x, t)$ of an integral equation, transform, or operator that equals its adjoint.

Hermitian matrix *Mathematics.* a matrix that is equal to its transposed conjugate matrix; i.e., a self-adjoint matrix. The characteristic roots of a Hermitian matrix are all real, even if the matrix has complex entries.

Hermitian operator *Mathematics.* a bounded self-adjoint linear operator on a Hilbert space. *Quantum Mechanics.* any of the operators representing observable dynamical quantities whose eigenfunctions are orthogonal and that always have real eigenvalues.

Hernandiaceae *Botany.* in some classification systems, a family of tropical trees of the order Ranales, characterized by light combustible wood, alternate entire leaves, small flowers borne on panicles, and drupaceous fruits.

hernia *Medicine.* **1.** an abnormal protrusion of loop or knuckle of an organ or tissue through its containing wall. **2.** specifically, such a protrusion in the abdominal region.

hernial sac *Medicine.* a sac or protrusion of tissue that is formed by gradual pressure against a weakness in the containing wall and in which there is a herniated organ or part.

herniated disk *Medicine.* a protrusion through the surrounding fibrocartilage of an intervertebral disk, usually in the lower lumbar region.

herniated nucleus pulposus *Medicine.* the rupture or prolapse of the nucleus pulposus into the spinal canal.

herniorrhaphy *Medicine.* the surgical repair of a hernia.

herniotomy *Medicine.* an operation for the relief of hernia by cutting through the neck of the sac.

Herodotus c. 484–425 BC, Greek historian; on the basis of fossil remains, concluded that water once covered Lower Egypt.

heroin *Pharmacology.* $C_{21}H_{23}NO_5$, a narcotic analgesic drug prepared from morphine; formerly used to relieve coughing and as a numbing painkiller, it is now prohibited in the U.S. even for medicinal uses because of the danger of addiction. Also, DIACETYLMORPHINE.

heron *Vertebrate Zoology.* a predatory shorebird of the large subfamily Ardeinae, having long legs and generally having a long neck; distribution is worldwide, generally in tropics and subtropics, except for Antarctica and far northern latitudes.

herons

Hero of Alexandria c. 62 AD, Greek geometrician; applied mathematics to invention of devices using cogs, pulleys, levers, and so on.

Herophilus c. 300 BC, Greek anatomist; performed first public dissection; identified brain as seat of intellect; studied nervous system.

Hero's principle *Optics.* the observation that a ray of light, traveling from one point to another by means of reflection from a plane mirror, will always take the shortest path between the two points. (Formulated by the ancient Greek scientist *Hero* of Alexandria.)

hero worship *Psychology.* the tendency of people to seek out a leader or other authority figure to admire and adulate.

herpangina *Medicine.* an acute infectious disease caused by coxsackie virus and characterized by sudden onset of fever, loss of appetite, sore throat, difficulty in swallowing, possible abdominal pain, vomiting, and nausea.

herpes *Virology.* any virus of the family Herpesviridae. *Medicine.* any of a wide variety of human diseases associated with these viruses, including cold sores, venereal ulcerations, and encephalitis.

herpes ophthalmicus see OPHTHALMIC ZOSTER.

herpes simplex *Medicine.* any of a group of acute infections caused by herpes simplex virus type 1, characterized by blisters on the lips or nostrils and possibly on the conjunctiva or cornea, or type 2, characterized by lesions on or around the genitalia.

herpes simplex virus *Virology.* a virus of the Herpesviridae family that is latent in ganglia and can cause cold sores on the mouth or genital lesions upon activation.

Herpesviridae *Virology.* a family of enveloped DNA viruses occurring in man and cold-blooded vertebrates and invertebrates; usually transmitted through contact.

herpes zoster *Medicine.* a painful viral infection characterized by an eruption of vesicles on one side of the body along the course of a nerve; believed to be caused by activation of latent varicella-zoster virus after a previous attack of chickenpox. Also, SHINGLES.

herpetic *Medicine.* **1.** characterized by herpes or having the nature of herpes. **2.** relating to or caused by a herpesvirus.

herpetic encephalitis *Medicine.* an inflammation of the brain resulting from infection by a herpes virus.

herpetic tonsillitis *Medicine.* an inflammation of a tonsil resulting from infection by a herpes virus.

Herpetomonas *Invertebrate Zoology.* a genus of flagellate protozoans, in the family Trypanosomatidae, parasitic in the gut of insects.

Herpetomonas

Herpetosiphon *Bacteriology.* a genus of yellow- or orange-pigmented gliding bacteria of the family Cytophagaceae; occur in salt water or fresh water.

herpolhode *Mechanics.* the name given by Poinsot to the planar curve described by the tip of the angular velocity vector of a freely rotating body; it is a circle for axisymmetric bodies.

herpolhode cone see SPACE CONE.

Herpomycetaceae *Mycology.* the only family of fungi belonging to the suborder Herpomycetineae, which is composed of cockroach parasites.

Herpomycetineae *Mycology.* a suborder of fungi belonging to the order Laboulbeniales, which is characterized by its distinct mode of forming its fruiting body, known as a perithecium.

Herpotrichiellaceae *Mycology.* a family of fungi belonging to the order Chaetothyriales, which live off of decaying plants and other fungi, or on the surfaces of living plant leaves and branches.

$H^+/2e^-$ ratio *Biochemistry.* the quantity of protons that traverse the membrane while two electrons pass along the electron transport chain.

herring *Vertebrate Zoology.* **1.** a migratory, commercially important bony fish found in the Atlantic and Pacific Oceans that generally travels in extremely large and extensive schools. **2.** a common name for the order Clupeiformes, which includes herring-like fishes, tarpons, salmonids, and their relatives.

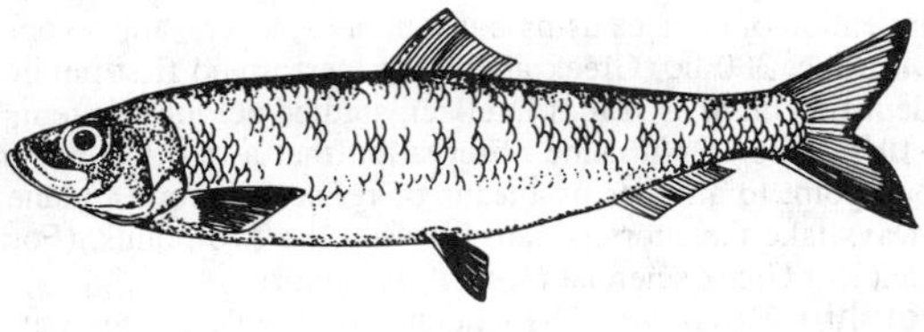

herring

Herring body *Histology.* neurosecretory granules found in cells of the neurohypophysis (pituitary glands).

herringbone bridging *Building Engineering.* see CROSS BRIDGING.

herringbone gear *Mechanical Engineering.* a gear in which the teeth slope both ways from the center line of the gear face, as if two opposite-handed helical gears were placed side by side; designed to eliminate axial thrust.

herringbone pattern *Electronics.* in a television picture, a pattern of interference that resembles the skeleton of a fish.

Herring-Nabarro creep *Materials Science.* in metals and alloys, a high-temperature creep involving stress-directed vacancy diffusion from grain boundaries under tension to those under compression.

Herschbach, Dudley born 1932, American chemist; shared Nobel Prize for new method of observing reaction dynamics.

Herschel, Caroline Lucretia 1750–1848, English astronomer, sister of William Herschel; discovered five comets and three nebulae.

William Herschel

Herschel, Sir (Fredrick) William 1738–1822, English astronomer; discovered Uranus, satellites of Uranus and Saturn, hundreds of nebulae and binary stars, and infrared radiation.

Herschel, Sir John Frederick William 1792–1871, English astronomer, mathematician, and chemist, son of William Herschel; catalogued nebulae and southern skies.

Herschel-Cassegrain telescope *Optics.* a telescope that has a primary paraboloidal mirror angled somewhat toward the axis, and an off-axis secondary hyperboloidal mirror and eyepiece.

Herschel effect *Graphic Arts.* the partial or total destruction of a photographic image by exposure to red or infrared light. When the reaction occurs in a gelatin emulsion, it is called the **latent Herschel effect;** when it is on a silver chloride negative, it is called the **visual Herschel effect.** (Named for its discoverer, English astronomer Sir John F. W. *Herschel.*)

Hershey, Alfred Day born 1908, American virologist; shared Nobel Prize for discoveries in genetic structure and replication of viruses.

Hertz, Gustav 1887–1975, German physicist; shared Nobel Prize for confirmation of Bohr model; developed isotope separation technique.

Hertz, Heinrich 1857–1894, German physicist; produced hertzian (radio) waves, confirming Maxwell's electromagnetic theory.

hertz *Metrology.* a standard unit of measurement for frequency, equivalent to one wave cycle per second.

Hertz antenna *Electromagnetism.* a half-wave antenna that is not grounded.

Hertz effect *Electronics.* the ionization of a gas produced by intense ultraviolet radiation.

Hertzian crack *Materials Science.* a localized cone-shaped crack that appears at the point of contact or impact; used to diagnose cause of fracture.

Hertzian oscillator *Electromagnetism.* a generator device that produces ultrahigh-frequency dipole radiation.

Hertz's law *Mechanics.* an expression that gives the radius of the contact area between an elastic sphere and a rigid surface against which it is pressed, in terms of the sphere's radius, the force exerted on it, and Young's modulus for the sphere's material.

Hertzsprung, Ejnar [hert´sprung] 1873–1967, Danish astronomer; discovered giant stars and the Hertzsprung gap; drew the Hertzsprung-Russell diagram.

Hertzsprung gap *Astrophysics.* a star-poor region on the Hertzsprung-Russell diagram that lies above the main sequence but below the giant branch; it is star-poor because stars evolve rapidly through the region.

Hertzsprung-Russell diagram *Astrophysics.* a two-dimensional plot of stars, in which the horizontal axis is stellar temperature (or color index) and the vertical axis is absolute magnitude (or intrinsic luminosity).

hervidero see MUD VOLCANO.

Herzberg, Gerhard born 1904, German-born Canadian physicist; awarded the Nobel Prize for determining the electronic structure of free radicals and other molecules.

Herzogiariaceae *Botany.* a monotypic family of dioecious, isophyllous, rigid, bristly liverworts of the order Jungermanniales, characterized by very thick-walled cells and native to southern South America.

Hesionidae *Invertebrate Zoology.* a family of small to moderate-sized, deep-water, polychaete annelids.

hesitation *Computer Programming.* a brief suspension of a program in order to perform part or all of another sequence of operations, as in an input or output procedure.

hesperidium *Botany.* **1.** any fruit with a leathery rind, membranous sections, and a pulpy interior. **2.** specifically, a citrus fruit such as an orange, grapefruit, or lemon.

Hesperiidae *Invertebrate Zoology.* the only family in the superfamily Hesperioidea, butterflies called skippers because of their irregular flight.

Hesperioidea *Invertebrate Zoology.* a superfamily of common butterflies, the skippers, lepidopteran insects in the suborder Heteroneura.

Hesperornis *Paleontology.* the "dawn birds," an early genus of flightless, diving seabirds in the extinct order Hesperornithiformes; about 3 feet long and widespread in both North and South America in the Upper Cretaceous.

Hesperornithidae *Paleontology.* a family of birds in the extinct order Hesperornithiformes; Cretaceous.

Hesperornithiformes *Paleontology.* an extinct order of flightless diving birds including the families Hesperornithidae, Enaliornithidae, and Baptornithidae; Cretaceous.

Hesperus *Astronomy.* the ancient Greek name for Venus when it was visible as an "evening star;" in the morning it was called Phosphorus.

Hess, Victor 1883–1964, Austrian physicist; Nobel Prize for discovery of cosmic radiation.

Hess, Walter 1881–1973, Swiss physiologist; Nobel Prize for discovering how the middle brain controls functions of internal organs.

Hess diagram *Astronomy.* a diagram that plots the relative numbers of stars in each region of the Hertzsprung-Russell diagram.

Hesse's rule *Ecology.* the principle that animals living in colder regions have a larger heart weight in proportion to total body weight, because of their greater need to maintain a body temperature above that of the environment. Also, HEART-WEIGHT RULE.

Hessian *Mathematics.* for a smooth real-valued function f on a real Hilbert space E, the quadratic form $f''(x)$. If $E = R^n$, then the Hessian f'' is represented by the matrix whose (i, j)th entry is $\partial^2 f/\partial x^i \partial x^j$.

Hessian fly *Invertebrate Zoology.* a small fly, *Phytophaga destructor,* whose larvae feed on the stems of wheat and other grains.

hessite *Mineralogy.* Ag_2Te, a metallic gray, monoclinic mineral occurring in pseudocubic crystals and in granular form, having a specific gravity of 8.3 to 8.41 and a hardness of 2 to 3 on the Mohs scale; found in hydrothermal vein deposits associated with other tellurides and gold. (Named for Germain Henri *Hess,* 1802–1850, Swiss-Russian chemist.)

Hess's law *Physical Chemistry.* a law stating that the amount of heat released or absorbed in a chemical reaction is not affected by the number of steps in the reaction; i.e., the heat change is the same whether the reaction takes place in one step or several steps. Also, CONSTANT HEAT SUMMATION; LAW OF CONSTANT HEAT SUMMATION.

hetaerolite *Mineralogy.* $ZnMn_2^{+3}O_4$, a brownish-black, opaque, tetragonal mineral occurring in massive form or as octahedral crystals, and having a specific gravity of 5.18 and a hardness of 6 on the Mohs scale.

Heteractinida *Paleontology.* an order of sponges with calcareous spicules; of the four orders in the class Calcarea, the Heteractinida is the only one that has become extinct; probably ancestral to the second group, the pharetrones; extant in the Cambrian to Carboniferous.

Heterakidae *Invertebrate Zoology.* a family of nematodes that parasitize all vertebrates except fish.

Heterakoidea *Invertebrate Zoology.* a superfamily of nematodes that parasitize all vertebrates except fish.

heterandrous *Botany.* having stamens unlike each other in length or form.

Heteraulacaceae *Botany.* a family of tropical to subtropical dinoflagellates of the order Peridiniales that are thecate and photosynthetic and are characterized by cell division in which division is coupled with fission of the parental theca.

hetero- or **heter-** a combining form meaning "other" or "different," as in *heterodyne, heteresthesia.*

heteroagglutinin *Immunology.* an antibody found in normal blood serum that is capable of agglutinating red blood cells of other species.

heteroallele *Genetics.* any of a number of different forms of a gene formed by mutations at different sites in the DNA molecule.

heteroallelic *Genetics.* of or relating to a heteroallele.

heteroantibody *Immunology.* an antibody that is capable of a cross-reaction with an antigen from another species.

heteroantigen *Immunology.* an antigen that is produced in one species and is capable of inducing an immune response in another species.

heteroauxin *Biochemistry.* $C_{10}H_9O_2N$, the standard of comparison for the activity of the larger group of plant hormones (auxins) that stimulate extension growth, cell divisions, and certain enzyme activities.

heteroazeotrope *Chemistry.* a constant boiling mixture having more than one liquid phase in equilibrium with the vapor phase at boiling point.

heterobaric *Chemistry.* having different mass numbers, as with two isotopes.

Heterobasidiomycetidae *Mycology.* a subclass of fungi in the class Phragmobasidiomycetes which includes the smut and rust fungi that are pathogenic. Also, **heterobasidiomycetes.**

heteroblastic *Developmental Biology.* originating from two or more types of tissue. *Petrology.* of or relating to a crystalloblastic texture of metamorphic rocks, in which the sizes of essential constituents are of two or more distinct orders of magnitude.

Heterocapsina *Botany.* an order of green algae in the class Xanthophyceae.

heterocarpous *Botany.* producing two different types of fruit.

Heterocera *Invertebrate Zoology.* a former classification of Lepidoptera consisting of the moths, as distinct from Rhopalocera, the butterflies; replaced by classification based on wing venation.

heterocercal *Vertebrate Zoology.* in fishes, describing a caudal fin in which the spine tilts upward, leaving most of the fin membrane below the axis of the tail and resulting in a large dorsal flange and a smaller ventral one; especially common in sharks.

Heteroceridae *Invertebrate Zoology.* a family of coleopteran insects, beetles that construct tunnels in sand or mud in shallow fresh water.

heterochain polymer *Materials Science.* a polymer in whose backbone the regular sequence of carbon atoms is interrupted by the presence of atoms of (usually) nitrogen or oxygen; characterized by high strength and toughness, and retention of mechanical properties over a wide temperature range.

Heterocheilidae *Invertebrate Zoology.* a family of nematode roundworms that parasitize marine mammals and crocodiles, in the superfamily Ascaridoidea.

heterochlamydeous *Botany.* having a perianth separated into a distinct calyx and corolla.

Heterochlorida *Invertebrate Zoology.* an order of protozoans in the class Phytomastigophora, green, plantlike, flagellates with chloroplasts.

heterochromatic *Botany.* having flowers or leaves of more than one color.

heterochromatin *Cell Biology.* one of two distinct forms of chromatin; as distinguished from euchromatin, it is relatively transcriptionally inactive, replicates late in the S phase, is highly condensed, and stains more intensely with cytological dyes.

heterochromia *Physiology.* diversity of color in a part or parts that should normally be of one color, such as the irises of the eyes.

heterochromous *Botany.* having central florets that differ in color from the marginal petals, as an aster or daisy.

heterochronic mutation *Genetics.* a mutation that affects the relative time at which developmental events occur.

heterochthonous *Science.* not indigenous to the place where it is found.

heteroclitic antibody *Immunology.* an antibody formed against one antigen that fortuitously reacts better with a second antigen than with the immunogen.

Heterocoela see SYCONOSA.

heterocoelous *Anatomy.* describing a type of vertebra found in the necks of birds, having saddle-shaped articular faces between successive vertebrae.

Heterocorallia *Paleontology.* a small, extinct group of elongate solitary corals, possibly related to the Rugosa; contains one family, the Heterophyllidae; sometimes classified as a Zoantharian order and sometimes called a subclass by itself; Upper Devonian to Upper Carboniferous.

Heterocotylea see MONOGENEA.
heterocyclic compound *Organic Chemistry.* a compound in which the ring structure is made up of more than one kind of atom, such as thiophene, C_4H_4S.
heterocyclic polymer *Materials Science.* any of a group of high-temperature-resistant polymers, including polyamides, polyphenylene oxide, and various conductive polymers; improved stability of the structure is caused by the addition of rings to form fused groups.
heterocyst *Botany.* a large, transparent, thick-walled cell that occurs at intervals along the filament in some blue-green algae.
heterocytotropic antibody *Immunology.* an antibody that is capable of attaching to cells of its own species as well as to certain cells of other species.
heterodactylous *Vertebrate Zoology.* having the first two toes of the foot curved back.
Heterodendraceae *Botany.* a monogeneric family of freshwater yellow-green algae of the order Tribonematales, characterized by uniseriate, branched filaments.
Heterodera *Invertebrate Zoology.* a genus of nematodes that parasitize the roots of many plants.

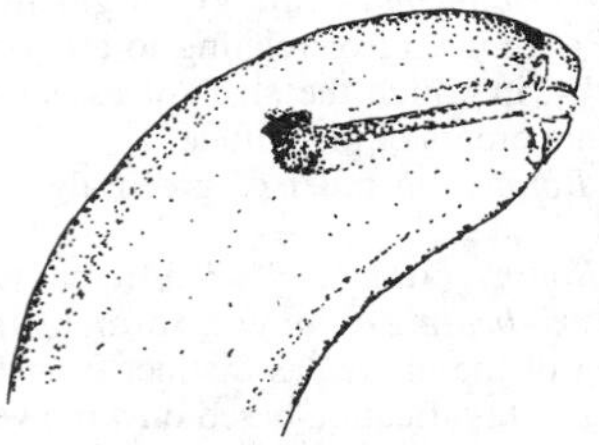

Heterodera

heterodermic graft *Surgery.* a skin graft taken from a species other than that of the recipient.
heterodesmic *Crystallography.* of or relating to a crystal structure that contains two or more types of bonding.
Heterodiniaceae *Botany.* a family of warm-temperate and tropical marine thecate dinoflagellates of the order Peridiniales, usually found at depths of 100 meters or more.
heterodont *Anatomy.* **1.** having a variety of tooth types, as do humans and most mammals. **2.** an organism having such a variety.
Heterodonta *Invertebrate Zoology.* a subclass of marine bivalve mollusks, with few hinge teeth of variable form.
Heterodontidae *Vertebrate Zoology.* the bullhead or horn sharks, a small family of primitive sharks (fewer than ten known species) having a large, rounded head and several rows of teeth. They are found in shallow waters of all seas except the Atlantic Ocean and the Mediterranean.
Heterodontoidea *Vertebrate Zoology.* a suborder of fishes of the order Heterodontiformes, including the primitive sharks of the family Heterodontidae.
heterodromous *Botany.* **1.** having asymmetric flowers of two kinds that are mirror images. **2.** having the spiral arrangement of leaves on a stem in one direction and the arrangement of stems on the branches in the opposite direction.
heteroduplex *Molecular Biology.* **1.** a double-stranded nucleic acid molecule in which some, but not all, of the bases in one strand are complementary to and paired with bases in the other strand. **2.** any duplex whose strands originate from different parent duplexes. **3.** a duplex made with one DNA strand and one RNA strand.
heteroduplex mapping *Biotechnology.* a method used to map nucleic acids when there is limited information on their nucleotide sequence; often used for double-stranded molecules of DNA.
heterodyne *Electronics.* **1.** of or relating to the combination of two different radio frequencies, one usually being a received external signal and the other a signal generated within the receiving apparatus, in order to produce electrical beats whose frequencies are equal to the sum of and the difference between the original frequencies. **2.** to produce such an effect. **3.** see BEAT-FREQUENCY OSCILLATOR.
heterodyne analyzer *Acoustical Engineering.* a frequency analyzer that mixes a known frequency band with an unknown one, thereby creating beat frequencies from which the frequency of the unknown band can be established.
heterodyne detector *Electronics.* **1.** a device that makes a radio-frequency signal audible by combining it with one produced by a local oscillator, producing an audio-frequency beat note. **2.** a linear detector and local RF oscillator combination used to detect and measure test signal frequencies. Also, SUPERHETERODYNE DETECTOR.
heterodyne frequency *Telecommunications.* periodic variations that are caused by the superposition of waves having different frequencies.
heterodyne frequency meter *Electronics.* a tunable radio-frequency meter that zerobeats an unknown frequency, or one of its harmonics, against a standard, internally generated frequency, or one of it harmonics. Also, **heterodyne wavemeter**.
heterodyne measurement *Electronics.* a measurement performed by a harmonic analyzer that utilizes a highly selective filter, at a frequency well above the highest frequency to be measured, and a heterodyne oscillator.
heterodyne oscillator *Electronics.* a signal generator that outputs the beat product of two internal oscillator outputs; the output frequency, selected by output filtering or tuning, may be the sum or difference of the oscillator frequencies.
heterodyne reception *Electronics.* radio reception, especially in telegraphy, using the beat-note process.
heterodyne repeater *Electronics.* a common type of repeater in which received signals are converted to another frequency before transmission.
heterodyne whistle *Telecommunications.* a steady, high-pitched audio tone produced when two signals of slightly different carrier frequency insert themselves into a receiver and then heterodyne to create a beat.
heteroecious *Biology.* **1.** requiring different hosts for different stages of the life cycle, as in certain parasitic rust fungi. **2.** not host-specific.
heteroerotism *Psychology.* the force that drives an individual to seek sexual gratification from the opposite sex. Thus, **heteroerotic.**
heterofermentation *Microbiology.* a fermentation process that results in more than one major end product.
heterogamete [het´ə rō gam´ēt] *Biology.* either of two gametes that can be distinguished from each other.
heterogametic sex *Genetics.* the sex that has unlike sex chromosomes and produces two kinds of gametes containing different sex chromosomes. In humans, the male is the heterogametic sex.
heterogamous [het´ə räg´ə məs] *Biology.* producing unlike gametes, which may differ in size, form, or in the chromosomes they contain. *Botany.* producing flowers of more than one type. *Genetics.* having or reproducing by unlike gametes. *Zoology.* referring to the alternation of two types of reproduction involving gametes during the life cycle of an animal.
heterogamy [het´ə räg´ə mē] *Botany.* **1.** the fact of producing flowers of two sexually different kinds in the same flower cluster. **2.** a process of sexual reproduction involving the fusion of gametes differing in size, structure, or physiology. *Genetics.* the fact of having unlike gametes.
heterogeneity [het´ə rō jə nā´ə tē] *Science.* the fact or state of being dissimilar, as in composition, source, quantity, dimensions, and so on.
heterogeneity index *Materials Science.* the breadth of the molar mass distribution or dispersivity in polymers, expressed as a ratio of weight average molecular mass to number average molar mass.
heterogeneous [het´ə rə jēn´əs; het´ə rə jē´nē əs] *Science.* differing in type, origin, constituents, or other qualities; dissimilar. *Chemistry.* composed of different substances or different phases of the same substance; e.g., a colloid. *Biology.* having a source outside the organism; having a foreign origin. *Mathematics.* having different degrees, grades, dimensions, and so on. Also, HETEROGENOUS.
heterogeneous catalysis *Chemistry.* **1.** a catalytic reaction in which the catalyst and reactants are in different phases, usually involving a solid catalyst for a gas or liquid phase reaction. **2.** a catalytic reaction that occurs at the boundary between two phases.
heterogeneous chemical reaction *Chemistry.* a reaction that takes place between two substances in different phases.
heterogeneous fluid *Fluid Mechanics.* a fluid whose density field varies as a function of the position.
heterogeneous nuclear RNA *Molecular Biology.* a population of RNA molecules, distinguished by their variable size, that are transcribed from DNA within the eukaryotic nucleus by the RNA polymerase II enzyme.
heterogeneous nucleation *Materials Science.* nucleation occurring on preexistent material such as oxide particles.
heterogeneous polymerization *Materials Science.* polymerization in which the monomer is in an emulsion or suspension phase; used to produce many polymers, including vinyl polymers and ethylene.

heterogeneous radiation *Physics.* **1.** the radiation of several different frequencies. **2.** the radiation of different types of particles. **3.** the radiation of different energies.

heterogeneous reactor *Nucleonics.* a nuclear reactor in which the fissionable material used as fuel and the substance used to moderate the velocity of the neutrons are arranged in a uniform pattern of discrete packages, so that the neutron travels through two distinct media.

heterogeneous summation *Behavior.* an increase in the strength or the likelihood of a response that results from adding or combining stimuli. Also, STIMULUS SUMMATION.

heterogeneous vaccine *Immunology.* a vaccine originating from a source other than the individual receiving it.

heterogenesis [het′ə rō jen′ə sis] *Biology.* originally, the concept that life could arise spontaneously from nonliving matter; spontaneous generation. *Genetics.* **1.** the alternation of generations. **2.** the appearance of a mutant in a population.

heterogenetic antibody see HETEROPHILE ANTIBODY.

heterogenetic antigen see HETEROPHILE ANTIGEN.

heterogenic *Surgery.* originating in a different source or species.

heterogenite *Mineralogy.* $Co^{+3}O(OH)$, a black, tetragonal, cobalt mineral, dimorphous with heterogenite-2H, occurring in mammillary masses, having a specific gravity of 4.13 to 4.72 and a hardness of 3 to 5 on the Mohs scale. Also, HETEROGENITE-3R, STAINIERITE.

heterogenite-3R see HETEROGENITE.

heterogenote *Genetics.* a partially diploid bacterium whose diploid portion contains nonidentical alleles, often resulting from transduction.

heterogenous [het′ə räj′ə nəs] see HETEROGENEOUS.

Heterogloeaceae *Botany.* a family of yellow-green algae in the order Heterogloeales, having spherical or ellipsoidal cells in a mucilaginous gelatinous mass.

Heterogloeales *Botany.* an order of freshwater, brackish, and marine yellow-green algae of the class Xanthophyceae, characterized by nonmotile cells with permanent contractile vacuoles and an eyespot.

Heterognathi *Vertebrate Zoology.* an equivalent name for the Cypriniformes, an order of soft-rayed fishes.

heterogonous relating to or characterized by heterogony.

heterogony [het′ə räg′ə nē] *Biology.* the fact of alternating one or more parthenogenetic generations with a bisexual generation. *Botany.* the condition of having both stamens and pistils that vary in length in the same flower.

heterograft see HETEROPLASTIC GRAFT.

heterohemolysin *Immunology.* an antibody that destroys red blood cells in an organism belonging to a species different from that used to obtain the antibody.

heteroimmunization *Immunology.* an immune response that is derived from antigens belonging to a species other than the species in which the immune response is activated.

heterojunction *Electronics.* a junction between two dissimilar semiconductor materials having different energy gaps between their valence and conduction bands.

heterokaryon *Mycology.* a fungal cell containing two or more nuclei that are genetically different from each other. Also, HETEROCARYON.

heterokaryosis *Mycology.* the condition in which a fungal cell possesses two or more nuclei that are genetically different from each other.

heterokont *Biology.* a biflagellated cell having flagella of unequal length.

Heterokontae *Botany.* in some classification systems, a class of yellow-green algae equivalent to Xanthophyceae.

heterolactic fermentation *Microbiology.* a bacterial fermentation in which sugars are converted into lactic acid and other products.

heterolalia see HETEROPHASIA.

heterolateral *Anatomy.* relating to or situated on the opposite side. Also, CONTRALATERAL.

heterolecithal *Cell Biology.* describing an egg with a nonuniform distribution of yolk.

heterolithic unconformity see NONCONFORMITY, def. 1.

heterologous [het′ə räl′ə gəs] *Biology.* **1.** consisting of different or noncorresponding elements, or of like elements in widely varying proportions. **2.** derived from a separate species. *Immunology.* referring to a substance (such as a serum or antigen) derived from a source or species that is genetically different from the primary substance.

heterologous antigen *Immunology.* an antigen that participates in a cross-reaction with an antibody of a species different from that of the antigen.

heterologous graft see HETEROPLASTIC GRAFT.

heterologous promoter *Biotechnology.* an antigen or antibody that does not correspond to a particular substance but that can be used as a catalyst to promote a reaction or to increase the rate of activity.

heterologous protein *Biochemistry.* any of two or more noncorresponding proteins.

heterologous stimulus *Physiology.* a stimulus that produces an effect or sensation when applied to any part of a nerve tract.

heterologous tumor *Medicine.* an abnormal mass composed of tissue unlike that in which it develops.

heterologous vaccine *Immunology.* a suspension of attenuated or killed microorganisms that is injected into an individual to induce immunity; it specifically protects against pathogens not present in the suspension.

heterolysis *Biology.* destruction by an outside agent, especially of cells by enzymes or lysins from another organism.

Heteromera *Invertebrate Zoology.* see TENEBRIONOIDEA.

heteromerous *Botany.* of a flower, having whorls that vary in number of members from other whorls.

heterometaplasia *Medicine.* a tissue transformation resulting in the development of tissue from cells that normally produce a different type of tissue.

heterometric autoregulation *Cardiology.* those intrinsic mechanisms which regulate the degree of ventricular contraction and which depend upon end-diastolic myocardial fiber length.

Heteromi *Vertebrate Zoology.* an alternate name for Notacanthiformes, an order of eel-like deep-sea fishes.

heteromixis *Mycology.* sexual reproduction in which genetically different fungal cell nuclei join.

heteromorphic *Biology.* of or relating to an organism that has different forms during different seasons, during different stages of its life cycle, or from generation to generation. *Cell Biology.* of or relating to two homologous chromosomes that differ morphologically, such as the X and Y chromosome pair. Also, **heteromorphous.**

heteromorphic bivalent *Genetics.* a chromosome pair composed of partly homologous chromosomes, such as the XY bivalent pair.

heteromorphism *Biology.* the state of being heteromorphic, or differing in form or size from the normal structure or type.

heteromorphite *Mineralogy.* $Pb_7Sb_8S_{19}$, an iron black, opaque, metallic, monoclinic mineral, massive in habit, having a specific gravity of 5.73 and a hardness of 2.5 to 3 on the Mohs scale; found in antimony mines.

heteromorphosis *Developmental Biology.* **1.** the development of one tissue from a tissue of another kind or type. **2.** the embryonic development or regeneration of a tissue or an organ at an abnormal site.

Heteromyidae *Vertebrate Zoology.* the kangaroo rats and pocket mice and related burrowing rodents, a family of the suborder Sciuromorpha found from southwestern Canada to northern South America in grasslands, wastelands, semidesert, and desert.

Heteromyinae *Vertebrate Zoology.* the spiny pocket mice, a subfamily of the family Heteromyidae that includes the genera *Liomys* and *Heteromys.*

Heteromyota *Invertebrate Zoology.* an order of marine wormlike animals belonging to the phylum Echiurida, unsegmented and bilaterally symmetrical.

Heteronema *Invertebrate Zoology.* a genus of protozoans with two flagella, widespread in ponds and brackish water and feeding on algae and bacteria.

Heteronematales *Botany.* an order of colorless, phagotrophic euglenoids with one or both flagella emergent, one of which is straight but coils or flickers at the tip.

Heteronemertea *Invertebrate Zoology.* an order of the phylum Nemertea, free-living marine ribbon worms. Also, **Heteronemertini.**

Heteroneura *Invertebrate Zoology.* a suborder of lepidopteran insects containing forms with dissimilar forewings and hindwings.

heteronomous *Biology.* subject to different modes of growth or specialization, as organs or parts.

heteropathy *Neurology.* an abnormal sensitivity to stimuli. *Medicine.* see ALLOPATHY.

Heteropediaceae *Botany.* a family of freshwater yellow-green algae of the order Tribonematales, having branched to multiseriate filaments differentiated into erect filaments and pseudoparenchymatous basal structures.

heteropelmous *Vertebrate Zoology.* having toes with bifid flexor tendons, with the branches of one going to the first and second toes and the other to the third and fourth.

heterophasia *Psychology.* the use of senseless or inappropriate words in place of the words intended. Also, HETEROPHONY.

heterophile see HETEROPHILE ANTIBODY.

heterophile agglutination test *Pathology.* an assay of serum to find heterophile antibodies produced in infectious mononucleosis; a positive result produces agglutination of sheep red cells.

heterophile antibody *Immunology.* an antibody that cross-reacts with a heterophile antigen and occurs in unrelated species of organisms, in a similar or related chemical form. Also, HETEROGENETIC ANTIBODY, HETEROPHILE.

heterophile antigen *Immunology.* an antigen that occurs (in a similar or related chemical form) in unrelated species of organisms and is thus involved in cross-reactions with antibodies. Also, HETEROGENETIC ANTIGEN.

heterophile leukocyte *Histology.* a neutrophilic leukocyte found in vertebrates other than humans.

heterophilic *Immunology.* relating to or being a heterophile antibody.

heterophony see HETEROPHASIA.

heterophyiasis *Medicine.* an infection with nematodes of the genus *Heterophyes.*

Heterophyllidae *Paleontology.* a small family of elongate solitary corals in the extinct subclass or order Heterocorallia; Upper Devonian and Lower Carboniferous.

heterophyllous *Botany.* having two or more different leaf forms on the same plant or stem.

heterophyte *Botany.* a plant that depends on other living or dead plants for its nutrients; a parasite or saprophyte. Thus, **heterophytic.**

Heteropiidae *Invertebrate Zoology.* a widely distributed family of calcareous sponges.

heteroplasia *Medicine.* the replacement of normal tissue with abnormal tissue; malposition of normal cells.

heteroplastic graft *Surgery.* a graft of tissue taken from a donor of one species and implanted in a recipient of another species. Also, HETEROGRAFT, HETEROLOGOUS GRAFT, XENOGRAFT.

heteroploidy *Genetics.* a condition or quality of a species in which the chromosome number differs from the characteristic diploid number.

heteropodal *Neurology.* of nerve cells, having various branches or processes.

heteropolar generator *Electromagnetism.* a generator whose active conductors successively pass through magnetic fields of opposite direction.

heteropoly acids *Inorganic Chemistry.* any complex of acids consisting of a metal with a specific gravity greater than 4 and phosphoric acid; e.g., phosphotungstic acid, $H_3PW_{12}O_{40}$.

heteropoly compounds *Inorganic Chemistry.* any compound made up of molybdates, tungstates, vanadates, etc., joined in polymeric chains with the anhydrides of other elements such as phosphorus.

heteropolysaccharide *Biochemistry.* a polysaccharide composed of two or more types of monosaccharides.

Heteroporidae *Invertebrate Zoology.* a family of erect, branched, colonial bryozoans in the order Cyclostomata.

Heteroptera see HEMIPTERA.

heteropycnosis *Cell Biology.* a condition in which the differing degrees of chromatin condensation that occur in chromosomes or chromosomal regions result in distinctive cytological staining properties during prophase and anaphase.

heteroscedasticity *Statistics.* nonequality of variance across populations, control levels, or other statistical groups.

heterosexual *Psychology.* **1.** sexually attracted to the opposite sex. **2.** a person who is sexually attracted to the opposite sex.

heterosexuality *Psychology.* the fact or condition of being heterosexual; sexual activity or attraction involving the opposite sex.

heterosis *Genetics.* the superiority of a heterozygote over the homozygotes with respect to one or more characteristics such as growth, survival, or fertility. Also, HYBRID VIGOR.

heterosite *Mineralogy.* $Fe^{+3}PO_4$, a rose to purple, brittle, orthorhombic mineral, massive in habit, having a specific gravity of 3.3 to 3.41 and a hardness of 4 to 4.5 on the Mohs scale; found in weathering zones of granite pegmatites.

Heterosomata *Vertebrate Zoology.* an alternate name for the fish order Pleuronectiformes, which includes the flatfishes.

Heterosoricinae *Paleontology.* a subfamily of insectivores in the suborder Soricomorpha and family Soricidae (the shrews); characterized by a short jaw and teeth similar to those of hedgehogs; Oligocene to Pliocene.

heterosphere *Meteorology.* the upper of two portions of the atmosphere as distinguished by the general homogeneity of their composition (the other portion is the homosphere); it begins at 80 to 100 kilometers above the earth, closely coinciding with the ionosphere and the thermosphere, and is characterized by variation in composition and mean molecular weight of constituent gases.

Heterospionidae *Invertebrate Zoology.* a family of poorly known marine polychaete annelids.

heterospory *Botany.* the production of two or more kinds of asexual spores, typically megaspores that grow into female gametophytes and microspores that grow into male gametophytes.

heterostemony *Botany.* the presence of more than one type of stamen in the same flower.

Heterostraci *Paleontology.* an alternative name for Pteraspida, an extinct order of jawless vertebrate fishes; Silurian and Devonian.

heterostructure *Materials Science.* a multilayer, single-crystal structure consisting of epitaxial layers of varying composition; used in semiconductor devices.

heterostyly *Botany.* a condition in which the length of the style varies in different flowers of the same species, causing the anthers of one flower to be at the same height as the stigmas of another and thus promoting cross-pollination by insects.

heterosuggestion *Psychology.* suggestion that comes from another person or persons rather than from the individual himself.

Heterotardigrada *Invertebrate Zoology.* an order of widely varied tardigrades. Also, WATER BEARS.

heterotaxis *Science.* an abnormal or nontypical arrangement, such as the transposition of parts of the body.

heterothallic *Botany.* of algae and fungi, describing the condition in which there are two mating types that act as male and female, as opposed to one self-fertile individual; having the condition of heterothallism.

heterothallism *Botany.* a seed plant having staminate and pistillate flowers borne on different individuals. *Biology.* the condition of having two or more mating types, with reproduction occurring only by individuals of different types.

heterotonia *Neurology.* the condition of having variations of tension or tone.

heterotopic *Medicine.* occurring at an abnormal place or upon the wrong part of the body. *Ecology.* occurring in a number of different habitats. Thus, **heterotopic transplantation.**

heterotopic epithelium *Medicine.* the development of epithelium at an abnormal site, as intestinal epithelium in gastric epithelium.

heterotopic pregnancy *Medicine.* a pregnancy that is not in the uterine cavity.

heterotoxic *Toxicology.* poisonous only to species that are different from the organism producing the poison.

heterotransplantation *Surgery.* the transplantation of an organ, tissue, or structure from a donor of a different species.

Heterotrichida *Invertebrate Zoology.* an order of aquatic ciliated protozoans, in the subclass Spirotrichia, often colorful, with both commensal and parasitic forms.

Heterotrichina *Invertebrate Zoology.* a suborder of protozoans in the order Heterotrichida, found in marine and salt marsh habitats.

heterotrichous *Botany.* describing certain algae that have a body divided into both prostrate and erect parts. *Zoology.* having more than one type of cilium or flagellum.

heterotroph *Biology.* an organism, such as any animal or fungus or almost any bacterium, that is unable to synthesize organic compounds from inorganic substrates for food and must consume other organisms or their organic products. Thus, **heterotrophic.**

heterotrophic effect *Biochemistry.* an allosteric interaction between different ligands, as when CTP inhibits the binding of *N*-carbamoyl phosphate with aspartate, or when ATP activates it.

heterotropia see STRABISMUS.

heterotropic enzyme *Enzymology.* an allosteric enzyme whose allosteric effector is another small molecule other than the substrate.

heterotropic ligands *Chemistry.* structurally identical ligands whose individual replacement by a different ligand creates an isomeric structure.

heteroxenous *Biology.* occupying more than one host during the life cycle.

heterozygote *Genetics.* a diploid or polyploid individual that inherits different alleles at one or more loci and therefore produces gametes of varied genetic composition. Also, HYBRID.

heterozygote advantage *Genetics.* the advantage that exists in individuals that are heterozygous at many loci over individuals that are homozygous at many loci.

HETP height equivalent of theoretical plate.

Hettangian *Geology.* a European geologic stage of the lowermost Jurassic period, after the Upper Triassic and before the Sinemurian.

Heublein method *Radiology.* a treatment that exposes the entire body to low levels of radiation for 10–20 hours each day for several days.

heulandite *Mineralogy.* $(Na,Ca)_{2-3}Al_3(Al,Si)_2Si_{13}O_{36}\cdot 12H_2O$, a colorless to white, yellow, red, or brown monoclinic mineral of the zeolite group occurring in trapezoidal crystals, having a specific gravity of 2.1 to 2.2 and a hardness of 3.5 to 4 on the Mohs scale; found in cavities in basic igneous rocks.

Heunlinger equations *Physics.* a set of equations concerned with thermoelectric and thermomagnetic effects in isothermal and adiabatic processes that relate the values of various quantities such as the Hall coefficient, electrical resistivity, thermal conductivity, and thermoelectric power.

heuristic [hyů ris´tik] *Science.* of or relating to heuristics.

heuristic function *Artificial Intelligence.* a function that estimates the value of a state, or the remaining cost to reach a goal from a state, in a search.

heuristic knowledge *Science.* knowledge of approaches that are likely to work or of properties that are likely to be true (but not guaranteed).

heuristic method *Science.* a problem-solving approach characterized by exploration and trial and error.

heuristic program *Computer Programming.* a program that provides the framework for a trial-and-error approach to problem solving, based on guesses of likely solutions.

heuristics [hyů ris´tiks] *Science.* any of various problem-solving techniques that involve the use of subjective knowledge, hunches, trial and error, rule of thumb, and other such informal but generally accurate methods. (Going back to a Greek word meaning "find" or "discover.")

heuristic search *Artificial Intelligence.* a hypothesize-and-test strategy for problem solving that uses heuristics, such as production rules, to select the best solution.

Hevea *Botany.* a genus of rubber trees of the family Euphorbiaceae; the largest source of latex used in rubber manufacturing.

Hevelian halo *Optics.* a faint, white ring appearing around the sun or moon, and having an angular radius of 90°; an optical phenomenon appearing only occasionally. Also, HALO OF HEVELIUS.

Hevelius, Johannes 1611–1687, Prussian astronomer; charted the moon; catalogued 1500 visible stars; mapped stars and constellations.

Hevelius's parhelia *Optics.* two bright reddish spots situated at points along the parhelic circle; probably due to superposition of the parhelic circle and the Hevelian halo. Also, MOCK SUNS, MOCK MOONS.

Hevesy, Georg von 1885–1966, Hungarian chemist; experimented with radioisotope tracing; shared Nobel Prize for discovery of hafnium.

hewettite *Mineralogy.* $CaV_6^{+5}O_{16}\cdot 9H_2O$, a deep-red, monoclinic mineral, occurring in aggregates of silky fibrous crystals, and having a specific gravity of 2.62 and an undetermined hardness; found in vanadium deposits in Colorado, Utah, and Peru.

Hewish, Antony born 1924, English astronomer; won the Nobel Prize for the discovery of pulsars.

Hewson, William 1739–1774, English physiologist; discovered and described the basic elements of blood coagulation and circulation.

hexa- or **hex-** a combining form meaning "six," as in *hexagon.*

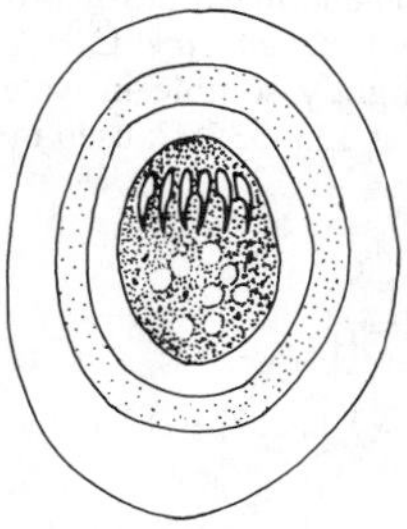

hexacanth

hexacanth *Invertebrate Zoology.* a tapeworm larva with six hooks.

hexacanth embryo see ONCOSPHERE.

hexachlorobenzene *Organic Chemistry.* C_6Cl_6, colorless crystals that are insoluble in water and slightly soluble in alcohol; melts at 231°C and boils at 323–326°C; used as a fungicide.

1,2,3,4,5,6-hexachlorocyclohexane *Organic Chemistry.* $C_6H_6Cl_6$, white to yellow powder or flakes with a musty odor; color and melting point vary with isomeric composition; toxic by inhalation and ingestion; used as an insecticide.

hexachloroethane *Organic Chemistry.* CCl_3CCl_3, toxic, colorless crystals; insoluble in water and soluble in alcohol; sublimes without melting and boils at 185°C; used as a retarding agent in fermentation, as a solvent, and in explosives. Also, PERCHLOROETHANE.

hexachlorophene *Organic Chemistry.* $(C_6HCl_3OH)_2CH_2$, a white powder that is insoluble in water and soluble in alcohol; melts at 161°C; widely used in antiseptic soaps and cosmetics.

hexachloropropylene *Organic Chemistry.* $CCl_3ClC{=}CCl_2$, a white liquid; insoluble in water and miscible with alcohol; boils at 210°C; used as a solvent, plasticizer, and hydraulic fluid.

hexacontane *Organic Chemistry.* $C_{60}H_{122}$, a saturated hydrocarbon of the methane series; combustible waxy solid; melts at 101°C.

Hexacorallia see ZOANTHARIA.

hexacosane *Organic Chemistry.* $C_{26}H_{54}$, a saturated hydrocarbon of the methane series; melts at 57°C.

hexacosanoic acid see CEROTIC ACID.

hexactin *Invertebrate Zoology.* a six-rayed spicule found in sponges.

Hexactinellida *Invertebrate Zoology.* a class of sponges in the phylum Porifer, with a skeleton composed of six-rayed siliceous spicules.

Hexactinosa *Invertebrate Zoology.* an order of sponges in the subclass Hexasterophora, with a skeleton formed of hexactins.

***n*-hexadecane** *Organic Chemistry.* $C_{16}H_{34}$, a colorless liquid; insoluble in water and soluble in alcolhol; boils at 286.5°C and melts at 20°C; used in determining ignition quality of diesel fuels and as a solvent.

hexadecanoic acid see PALMITIC ACID.

1-hexadecanol see CETYL ALCOHOL.

1-hexadecene *Organic Chemistry.* $CH_3(CH_2)_{13}CH{=}CH_2$, a colorless liquid; insoluble in water and soluble in alcohol; boils at 274°C and melts at 4°C; used as a chemical intermediate.

hexadecimal notation *Computer Programming.* the notation used to represent numbers with a radix of 16; digits from decimal 10 to 15 are represented by the letters A to F, respectively.

hexadecimal number system *Mathematics.* the representation of numbers in base 16.

hexadentate ligand *Inorganic Chemistry.* a ligand attached to a metal ion in a coordination compound at six points.

hexadiene *Organic Chemistry.* C_6H_{10}, a group of C_6-unsaturated hydrocarbons with two double bonds.

2,4-hexadienoic acid see SORBIC ACID.

hexagon *Mathematics.* a polygon figure with six sides.

hexagonal column *Meteorology.* one of many forms of ice crystals that are found in the atmosphere, characterized by a hexagonal cross section in a plane perpendicular to the length of the column.

hexagonal nut *Mechanical Devices.* a nut of six sides, commonly used to thread and bolt screws. Similarly, **hexagonal bolt.**

hexagonal platelet *Meteorology.* a small ice crystal of hexagonal tabular form, in which the distance from one side to the other is 10 times as great as the thickness perpendicular to this dimension; it forms at temperatures of –10° to –20°C by the process of sublimation.

hexagonal unit cell *Crystallography.* a unit cell in which there is a six-fold rotation axis parallel to one axis (arbitrarily chosen as *c*) and also two-fold rotation axes perpendicular to *c*. These symmetry relations dictate that the lengths of *a* and *b* are identical, the angle between *a* and *b* is 120°, and the other two angles are 90° ($a = b$, $\alpha = \beta = 90°$, $\gamma = 120°$). Rhombohedral crystals, with three-fold axes, can be referred to hexagonal unit cells.

hexagon gauge *Mechanical Devices.* a thin steel plate with an angular span of 120°, used for verifying such an angle in specific workpieces.

hexahedrite *Geology.* an iron meteorite consisting of single crystals or coarse aggregates of kamacite, whose nickel component measures less than 6%.

hexahedron *Mathematics.* a polyhedron with six faces, such as a cube.

hexahydric alcohol *Organic Chemistry.* a member of the mannitol-sorbitol-dulcitol sugar group.

hexahydrite *Mineralogy.* $MgSO_4\cdot 6H_2O$, a white or greenish-white, monoclinic mineral, occurring in columinar to fibrous form, and having a specific gravity of 1.76 and an undetermined hardness; found as a dehydration product of epsomite.

hexahydrotoluene see METHYLCYCLOHEXANE.

***n*-hexaldehyde** *Organic Chemistry.* $CH_3(CH_2)_4CHO$, a colorless liquid; immiscible with water; boils at 128.6°C; used in dyes, insecticides, and resins.

hexametapol *Organic Chemistry.* $C_6H_{18}N_3OP$, a colorless liquid that melts at 7.2°C and boils at 235°C; thought to be carcinogenic; used as a deicing additive in jet fuel and as a chemical mutagen.

hexamethonium *Pharmacology.* $C_{10}H_{24}N_2$, a ganglion-blocking agent.

hexamethonium chloride *Pharmacology.* $C_{12}H_{30}Cl_2N_2$, a dichloride salt of hexamethonium, formerly used to lower blood pressure.

hexamethylene see CYCLOHEXANE.

hexamethylenediamine *Organic Chemistry.* $NH_2(CH_2)_6NH_2$, a colorless solid that is soluble in water and slightly soluble in alcohol; melts at 42°C and boils at 205°C; used to make nylon.

Hexamita *Invertebrate Zoology.* a genus of colorless protozoans in the family Hexamitidae, with eight flagella, including free-living forms and pathogenic intestinal parasites of birds and fish.

Hexamita salmonis *Invertebrate Zoology.* an intestinal parasite of salmonid fish; a species of binucleate zooflagellates having six anterior and two trailing flagella.

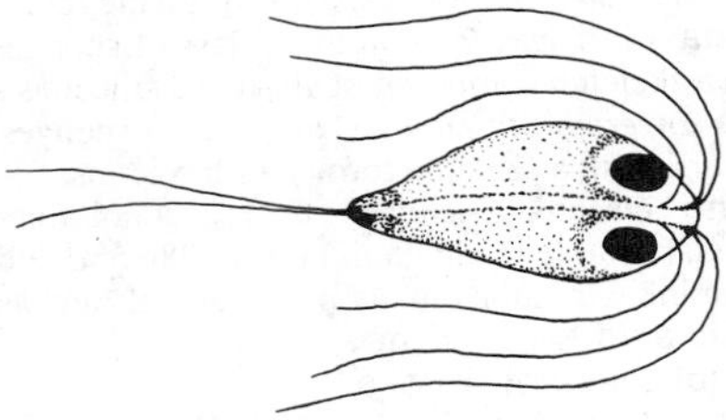

Hexamita salmonis

hexamitiasis *Veterinary Medicine.* an acute, infectious catarrhal enteritis of turkeys and other fowl that is characterized by ruffled feathers and watery diarrhea and is caused by the protozoan parasite *Hexamita meleagridis.*

hexanaphthene see CYCLOHEXANE.

Hexanchidae *Vertebrate Zoology.* the cow sharks, a family of primitive sharks in the order Hexanchiformes; characterized by their six or seven gill slits and arches and found in deep waters worldwide, though especially prevalent in the Mediterranean.

***n*-hexane** *Organic Chemistry.* C_6H_{14}, a colorless, volatile alkane that is toxic and flammable; soluble in alcohol and insoluble in water; boils at 68.74°C; used as a solvent and paint diluent, and in low-temperature thermometers.

1,6-hexanediol *Organic Chemistry.* $HO(CH_2)_6OH$, crystals that are soluble in water and alcohol; melts 42.8°C and boils at 208°C; used in gasoline refining and in nylon manufacture. Also, **hexamethylene glycol.**

2,5-hexanedione see ACETONYLACETONE.

hexanedoic acid see ADIPIC ACID.

hexanitrodiphenylamine *Organic Chemistry.* $(NO_2)_3C_6H_2NHC_6H_2(NO_2)_3$, a yellow solid that is insoluble in water and alcohol, and soluble in alkalies and acids; melts and decomposes at 238–244°C; explodes on shock or exposure to heat; used as an explosive. Also, DIPICRYLAMINE, HEXIL.

hexanoic acid *Organic Chemistry.* $CH_3(CH_2)_4COOH$, an oily, colorless or slightly yellow liquid; slightly soluble in water and soluble in alcohol; melts at –2°C and boils at 205°C; used in the manufacture of rubber chemicals, resins, artificial flavors, and pharmaceuticals. Also, CAPROIC ACID.

2-hexanone see METHYL BUTYL KETONE.

hexapetalous *Botany.* having a perianth with six petals or petal-like divisions.

hexaphenylethane *Organic Chemistry.* $(C_6H_5)_3CC(C_6H_5)_3$, the theoretical dimer of the triphenylmethyl radical, which actually has a different structure.

hexapod *Invertebrate Zoology.* **1.** an organism having six legs; a true insect. **2.** having six legs. *Robotics.* a robot with six legs that can walk over an uneven surface.

Hexapoda see INSECTA.

hexaster *Invertebrate Zoology.* a six-rayed spicule that is found in sponges.

Hexasterophora *Invertebrate Zoology.* a subclass of sponges, in the class Hexactinellida, with a skeleton of six-rayed siliceous spicules.

hexastyle *Architecture.* of a portico, having six columns.

1-hexene *Organic Chemistry.* $CH_3(CH_2)_3HC{=}CH_2$, a combustible, colorless liquid; insoluble in water and soluble in alcohol; boils at 64°C; used as a chemical intermediate and for resins, drugs, flavors, dyes, and insecticides. Also, **hexylene.**

hexil see HEXANITRODIPHENYLAMINE.

hex nut see HEXAGONAL NUT.

hexobarbital *Pharmacology.* $C_{12}H_{16}N_2O_3$, a short-acting barbiturate occurring as colorless crystals or white crystalline powder; used as an oral sedative.

hexoctahedron *Crystallography.* an isometric crystal form with 48 equal triangular faces.

hexode *Electronics.* an electron tube containing six electrodes: an anode, cathode, primary control electrode, and three auxiliary electrodes.

Hexogen see CYCLONITE.

hexokinase *Enzymology.* an enzyme of the transferase class that catalyzes the phosphorylation of hexose sugars; occurs in all tissues and exists as various isoenzymes.

hexosamine *Biochemistry.* an amino sugar of a six-carbon polyhydroxy-alcohol containing either an aldehyde or a ketone group.

hexosaminidase A *Enzymology.* an enzyme that catalyzes the separation of the *N*-acetylgalactosamine residue from certain gangliosides.

hexose *Biochemistry.* any monosaccharide that contains six carbon atoms in one molecule, such as fructose or glucose.

hexose diphosphate pathway see EMBDEN-MEYERHOF GLYCOLYSIS.

hexose monophosphate cycle *Biochemistry.* the oxidative pathway of carbohydrate metabolism in which glucose-6-phosphatase is totally degraded to form carbon dioxide.

hexose phosphate *Biochemistry.* a hexose sugar that is an intermediate in the pentose phosphate cycle; produced by the metabolism of carbohydrates in living organisms.

hexosyltransferase *Enzymology.* any of a group of enzymes of the transferase class that catalyze the transfer of a hexose group from one compound to another.

hextetrahedron *Crystallography.* a hemihedron contained under 24 similar and equal triangular faces.

hexulose *Biochemistry.* a ketose having six carbon atoms in its chain.

***n*-hexyl acetate** *Organic Chemistry.* $CH_3COOC_6H_{13}$, a combustible, colorless liquid that is insoluble in water and soluble in alcohol; boils at 169°C; used as a solvent for resins.

hexylacetylene see OCTYNE.

hexyl alcohol *Organic Chemistry.* $CH_3(CH_2)_4CH_2OH$, a colorless liquid; slightly soluble in water and miscible with alcohol; boils at 156°C; used in pharmaceuticals, perfumes, and antiseptics. Also, **hexanol.**

hexylamine *Organic Chemistry.* $CH_3(CH_2)_5NH_2$, a flammable, poisonous white liquid that is slightly soluble in water; boils at 129°C and melts at –21°C.

hexylene glycol *Organic Chemistry.* $C_6H_{14}O_2$, a colorless liquid that is soluble in water and alcohol; boils at 198°C; used in hydraulic brake fluid, in printing inks, and in textile processing. Also, 2-METHYL-2,4-PENTANEDIOL.

***n*-hexyl ether** *Organic Chemistry.* $C_6H_{13}OC_6H_{13}$, a combustible, faintly colored liquid that is slightly soluble in water; freezes at –43°C; used in solvent extraction and in cellulosic product manufacture.

hexylresorcinol *Organic Chemistry.* $C_6H_{13}C_6H_3(OH)_2$, white to yellow crystals that are slightly soluble in water and soluble in alcohol; melts at 64°C and boils at 333–335°C; used in medicine.

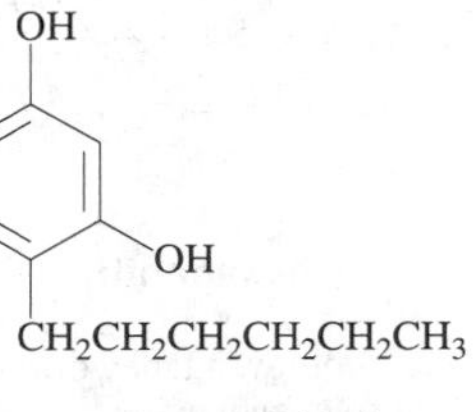

hexylresorcinol

1-hexyne *Organic Chemistry.* $C_4H_9C{\equiv}CH$, a flammable, colorless liquid that boils at 71.5°C.

Heymans, Corneille [hī´mänz] 1892–1938, Belgian physiologist; won the Nobel Prize for determining the respiratory function of the sinus aorta and the carotid sinus.

Heyrovsky, Jaroslav [hī räf´skē] 1890–1967, Czechoslovakian chemist; won the Nobel Prize for the invention of polarography.

heyrovskyite *Mineralogy.* $Pb_{10}AgBi_5S_{18}$, a lead-gray, opaque, metallic, orthorhombic mineral occurring as prismatic to acicular cyrstals, having a specific gravity of 7.17 and a hardness of 3.5 to 4 on the Mohs scale; found with galena and cosalite. Also, GOONGARRITE.

HF high-frequency; Hageman factor.

Hf the chemical symbol for hafnium.

hf. or **hf** half.

HF/DF *Ordnance.* high-frequency direction finding.

HF 1189 *Chemical Engineering.* a citrus-based solvent developed as a substitute for CFCs; used to clean circuit boards.

HF akylation *Chemical Engineering.* a refinery alkylation process in which light olefins react with isobutane in the presence of a hydrofluoric acid catalyst.

HFIR high-flux isotope reactor.

Hfr or **hfr** *Microbiology.* of or related to a bacterial donor cell that possesses a certain fertility factor, called an integrated F factor, enabling it to transfer chromosomal material to recipient bacterial cells not having this factor, and thus creating a high-frequency rate of recombination in a mixed population of donors and recipients. (An acronym for high-frequency recombination.)

hfr cell *Genetics.* a bacterial cell that has the F plasmid (the sex factor) integrated into its chromosome. Also, **Hfr cell.**

hfr strain *Genetics.* a bacterial strain that produces many more recombinants for marker genes than do its progenitors, because it consists of hfr cells. Also, **Hfr strain.**

Hg the chemical symbol for mercury. (From Latin *hydrargyrum.*)

hg or **hg.** hectogram; hectograms; hemoglobin.

HGH human growth hormone.

HGPRT *Enzymology.* hypoxanthine-guanine phosphoribosyltransferase; an enzyme that catalyzes the transfer of phosphoribosyl to hypoxanthine and guanine.

hgt. or **hgt** height.

hgwy. or **hgwy** highway.

Hha restriction enzyme *Enzymology.* a type of bacterial endonuclease that separates a DNA molecule at the following recognition site: 5'...GCGC...3' 3'...C-GCG...5'.

hhd. or **hhd** hogshead; hogsheads.

hhds. or **hhds** hogsheads.

HH objects see HERBIG-HARO OBJECT.

H-hour *Military Science.* the specific hour on D-day at which a particular operation commences.

HHS Department of Health and Human Services.

HHV higher heating value.

HI or **H.I.** humidity index.

hiatus [hī ā´təs] a break or interruption; specific uses include: *Anatomy.* an opening, hole, or cleft. *Geology.* a break or interruption in a stratigraphic sequence resulting from either a period of nondeposition or the erosion of the strata prior to the deposition of overlying beds, and indicated by the absence of normally occurring rocks or strata.

hiatus hernia *Medicine.* a hernia in which part of the stomach projects through the opening in the diaphragm through which the esophagus and the two vagus nerves pass. Also, **hiatal hernia.**

hibernaculum *Biology.* the place in which an animal hibernates or overwinters; winter quarters. *Invertebrate Zoology.* **1.** a bud of certain freshwater bryozoans that survives winter to grow into a new colony in spring. **2.** a snail's epiphragm. **3.** a protective structure occupied through winter by a hibernating insect.

hibernal *Meteorology.* of or relating to winter.

hibernate *Biology.* to pass the winter in a condition of hibernation.

hibernating spacecraft *Space Technology.* an orbiting spacecraft whose power plant is shut down, then reactivated only when signaled from ground control.

hibernation *Biology.* a dormant, sleeplike state, with a lower body temperature and slower heart and breathing rates, that is characteristic of various animals during the winter months in cold climates, such as bears, bats, certain birds, snakes, frogs, and turtles; this state tends to protect against cold weather and to reduce the need for food. (From the Latin word for winter.)

hibernation-inducing triggger *Biochemistry.* a substance present in the blood of hibernating mammals that is thought to induce the condition of hibernation at the onset of cold weather.

Hibernian orogeny see ERIAN OROGENY.

hibernoma *Oncology.* a rare benign tumor that is composed of large polyhedral cells with coarsely granular cytoplasm, usually occurring on the back or hips. (So named because it is believed to be a manifestation of a vestigial organ that is analogous to the dorsal fat pads of hibernating animals.)

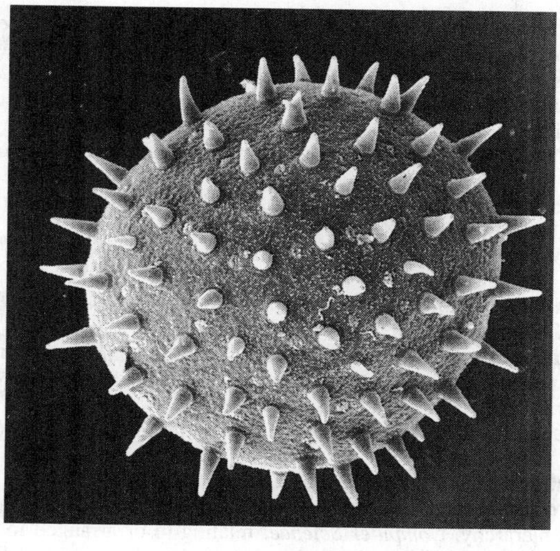

hibiscus pollen grain

hibiscus *Botany.* **1.** a woody plant of the mallow family, *Hibiscus rosasenensis,* having large, luxuriantly blooming flowers. Also, CHINA ROSE. **2.** any of various trees, shrubs, or plants of the genus *Hibiscus,* having lobate or dentate leaves and large flowers.

hickey *Graphic Arts.* a flaw on a printed page, usually a small dark spot surrounded by a blank (noninking) circle, caused by a speck of dust or dried ink on the printing plate or blanket. *Electricity.* a threaded coupling used to mount a lighting fixture to an outlet box.

hickory *Botany.* any tree of the genus *Carya,* deciduous hardwood trees of the walnut family Juglandaceae that are common to the eastern United States and are characterized by pliable wood and pinnately compound leaves; some species produce sweet edible nuts.

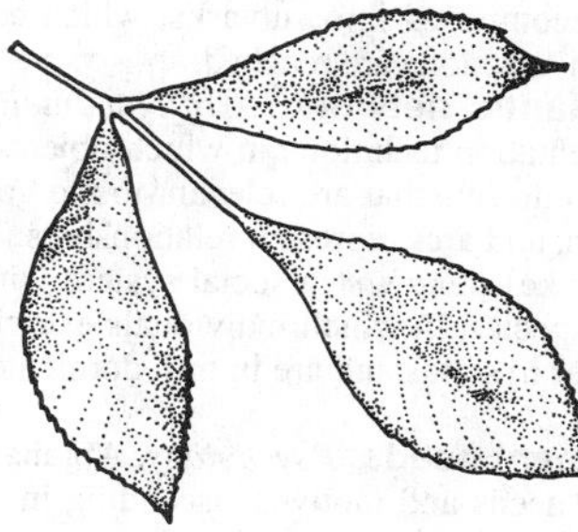

hickory

Hidalgo *Astronomy.* asteroid 944, discovered in 1920 and having a diameter of about 60 kilometers; of type D, its aphelion (9.6 AU) reaches almost to the orbit of Saturn, its orbit is highly inclined at 43° and has an eccentricity of 0.66, and (with its type D spectrum) the object is believed to be an extinct comet nucleus.

hiddenite *Mineralogy.* an emerald-green gem variety of spodumeme containing chromium.

hidden units *Artificial Intelligence.* in a neural network, a layer of artificial neurons between the layer connected to inputs and the layer connected to outputs; it may not be clear to a human observer what the individual units in this layer represent.

hidden variables *Quantum Mechanics.* the postulated unobservable parameters possessed by a system that are inaccessible to measurement and that determine the values of individual measurable quantities of the system (such as position, momentum, and spin) not specified by its wave function, and, according to standard quantum theory, not yet determined in any sense.

hidden-variables theory of the first kind *Quantum Mechanics.* a theory based on the existence of hidden variables, in which a system's hidden variables decay quickly to an equilibrium distribution, whereupon the system will behave exactly as predicted by standard quantum theory.

hidden-variables theory of the second kind *Quantum Mechanics.* a theory in which the outcome of any conceivable experiment on a system is predetermined according to its hidden variables; such a theory fundamentally contradicts standard quantum theory, and predicts certain experimental results that differ from those predicted by quantum theory.

hide *Materials.* a raw or dressed animal skin, especially that of a relatively large animal.

hidebound *Botany.* of trees, having the bark so close and tight that it impedes growth.

hidr- or **hidro-** a combining form meaning "sweat," as in *hidradenoid, hidrosis.*

hidradenitis *Medicine.* an inflammation of a sweat gland.

hidradenoma see SPIRADENOMA.

hidradenoma papilliferum *Medicine.* an uncommon tumor derived from the epithelial cells of sweat glands, usually occurring in the labia majora.

hidrosis [hī drō′sis] *Medicine.* the production and secretion of sweat. Thus, **hidrotic.**

hiemal climate *Meteorology.* a winter climate.

hierarchical [hī′ər är′ki kəl] *Behavior.* describing a social group that has a hierarchy. *Computer Science.* relating to or arranged in a ranked or graded series. Thus, **hierarchical database**, **hierarchical file**, **hierarchical storage management.**

hierarchical computer network *Computer Technology.* a form of distributed processing system having two or more levels, in which a central computer controls and distributes various functions to lower-level computers which may, in turn, further distribute work to the next lower level; the lower the level in the structure, the more narrow the responsibilities. Also, **hierarchical-distributed processing system.**

hierarchical control *Robotics.* a type of distributed control where robotic processes are arranged in a hierarchy, with those above sending control signals to those below, and the ones below sending feedback and sensor signals to those above. Also, CONTROL HIERARCHY.

hierarchical memory *Computer Technology.* a computer architecture incorporating multiple kinds of memory, with the fastest and most expensive memory such as cache memory at the top of the hierarchy and slower, more voluminous, and less expensive memory at lower levels.

hierarchical planning *Artificial Intelligence.* a form of planning in which a task is decomposed into subtasks, which are likewise decomposed, until primitive actions are reached.

hierarchical semantic network *Artificial Intelligence.* a pictorial knowledge-representation technique in which objects and their relationships are grouped into sets that are relevant to the goal; nodes represent concepts or objects, and arcs represent relationships.

hierarchy [hī′ər är′kē] *Behavior.* a social system within a group of animals of the same species, in which individuals exercise dominance over those ranking lower in status and are in turn dominated by those ranking higher.

hierarchy of (human) needs *Psychology.* Abraham Maslow's classification of human needs and motives, ascending in a pyramid from the basic level of physiological needs, such as hunger and thirst, to the highest level of self-actualization.

hieratite *Mineralogy.* K_2SiF_6, a colorless to grayish, cubic mineral having a specific gravity of 2.665 and a hardness of about 2.5 on the Mohs scale; found as stalactitic concretions in fumarolic deposits.

hieroglyph *Archaeology.* a single character or pictorial element used in hieroglyphics. *Geology.* any sedimentary structure or mark found on a bedding plane.

hieroglyphic [hī′rō glif′ik] *Archaeology.* of or relating to a system of writing using hieroglyphics.

hieroglyphics [hī′rō glif′iks] *Archaeology.* a writing system, especially the one used by the ancient Egyptians, in which meaning is expressed by standardized pictorial symbols that represent the objects and ideas being described.

hierophobia *Psychology.* an irrational fear of sacred objects and rituals. Also, HAGIOPHOBIA.

hi-fi *Acoustics.* a shorter term for high-fidelity sound reproduction or equipment.

Higgins larva *Invertebrate Zoology.* a developmental stage in the Loricifera that is close to the adult form except that the buccal cone lacks stylets and the thorax lacks spines.

Higgs bosons *Particle Physics.* hypothetical particles with zero spin that result when spontaneous symmetry breaks in the standard model.

Higgs mechanism *Particle Physics.* that part of the gauge theory of the weak force that gives mass to the W and Z bosons.

high *Meteorology.* a high-pressure, closed-circulation atmospheric weather system with a relative rotational direction opposite to that of a cyclone. Also, HIGH-PRESSURE SYSTEM, ANTICYCLONE.

high airburst *Ordnance.* the fallout-safe height of burst for a nuclear weapon that increases damage or casualties to soft targets or reduces radiation contamination at ground zero.

high-alloy steel *Metallurgy.* any of many steels that contain substantial amounts of alloying elements besides carbon, and in which manganese exceeds 1.65% and copper and silicon 0.6%.

high aloft see UPPER-LEVEL ANTICYCLONE.

high-altitude burst *Ordnance.* the explosion of a nuclear weapon at an altitude over 100,000 feet.

high-altitude method *Navigation.* a technique for establishing a circular line of position using the geographic position of the observed body as the center and the zenith distance as the radius. This is normally done when the body is near the observer's zenith.

high-altitude sickness see ALTITUDE SICKNESS.

high-altitude station *Meteorology.* a weather observing station that is established at an elevation of at least 6500 feet above sea level to monitor weather conditions that are distinct from those at sea level.

high-alumina cement *Materials.* a high-strength cement that is made with bauxite, which has a high aluminum content; used in building roads and walls. Also, ALUMINOUS CEMENT, HIGH-SPEED CEMENT.

high-angle fault *Geology.* a fault having a dip angle greater than 45°.

high-angle fire *Military Science.* fire at an angle of elevation greater than the elevation that will produce the maximum range for the specific gun and ammunition; such fire decreases in range as the angle of elevation increases, and the range table is said to be in the upper register.

high-angle gun *Ordnance.* a gun, such as an antiaircraft cannon, that is capable of firing at a high elevation.

high-angle strafing *Military Science.* strafing from an airplane in a high-angle dive, usually 45° or greater.

high anticyclone see UPPER-LEVEL ANTICYCLONE.

high band *Telecommunications.* the range of television broadcast frequencies from 174 to 216 MHz; the high band of VHF television covering channels 7–13.

high blood pressure *Medicine.* the popular term for hypertension, a condition in which the blood pressure is elevated to an abnormally high level. See HYPERTENSION.

high blower *Aviation.* a blower-type supercharger set at high rpm.

high boost see HIGH-FREQUENCY COMPENSATION.

high-burst ranging *Ordnance.* the process of adjusting gunfire by observing airbursts.

hieroglyphics

high-capacity projectile *Ordnance.* a projectile with a thin casing and a high-explosive charge, designed to be used where target penetration is unnecessary.

high-carbon chromium *Metallurgy.* chromium containing about 10% carbon.

high-carbon steel *Metallurgy.* any of various steels containing more than 0.5% carbon.

high clouds *Meteorology.* a class of clouds consisting of those that occur at altitudes higher than 20,000 feet above sea level, including the cirrus, cirrostratus, and cirrocumulus species.

high contrast *Radiology.* the degree of radiographic contrast that promotes visual differentiation of image density despite objects or object components having similar structural features. Also, SHORT-SCALE CONTRAST.

high-contrast *Graphic Arts.* having or designed to yield extreme light and dark visual areas, with few intermediate tones or values.

high core *Computer Technology.* an older term for the locations in the higher addresses of the main memory of a computer, sometimes occupied by parts of the operating system.

high-current rectifier *Electronics.* a device (solid-state, gas tube, or vacuum tube) used to convert alternating current to direct current for powering low-impedance loads.

high-current switch *Electricity.* a switch whose function is to reroute heavy current flow, usually containing a make-before-break feature in order to prevent excessive arcing.

high-cycle fatigue *Materials Science.* fatigue that occurs at stresses generally below the yield stress when a cyclic load is applied.

high-definition *Telecommunications.* describing television or FAX equipment of high fidelity, in which the reproduced image delivers an excellent approximation of the original scene or source document.

high-definition television *Telecommunications.* a television system that provides an extremely precise and sharply defined picture, due to the unusually high number of scanning lines per frame. Also, HDTV.

high-density lipoprotein *Biochemistry.* any of several lipoproteins that contain more protein than lipid and carry excess cholesterol out of the body tissues to the liver for excretion; they may also serve to stabilize the lower-density lipoproteins. Also, HDL.

high-density polyethylene *Materials.* a low-cost plastic that is almost entirely unbranched and does not absorb water; used in applications such as piping, kitchen utensils, and bottles. Also, LOW-PRESSURE POLYETHYLENE, LINEAR POLYETHYLENE.

high egg passage see HEP.

high enema *Medicine.* a rectal infusion of fluid high into the colon for cleansing or other therapeutic purposes.

high-energy bond *Physical Chemistry.* a term applied to any chemical bond that results in a decrease of the system's free energy of at least 5 kilocalories per mole.

high-energy electron diffraction *Physics.* the diffraction of electrons with energies in the range of 30 to 70 keV; used in the study of atomic and molecular structures of liquids and gases.

high-energy environment *Geology.* an aqueous sedimentary environment characterized by a high level of kinetic energy from currents, waves, or surf, and turbulent activity that prevents fine-grained sediments from settling and accumulating.

high-energy particle *Particle Physics.* a particle having mass of several hundred MeV or greater; its physical mass is given by the energy value, in MeV, divided by the square of the speed of light, in meters per second.

high-energy physics *Physics.* the area of physics that is concerned with charged particles at high energies, X rays, and gamma rays. Similarly, **high-energy astrophysics.**

high-energy-rate forging *Metallurgy.* forging performed at very high ram velocities.

high-energy scattering *Particle Physics.* a reaction that occurs when two elementary particles collide with a total energy in the range of several hundred MeV.

high-epithermal neutron range *Nucleonics.* a range of neutron energy that falls between 1000 and 100,000 electron volts.

higher fungi *Mycology.* a term for fungi belonging to the subdivisions Basidiomycotina and Ascomycotina.

higher heating value *Thermodynamics.* the maximum energy released in a combustion process when all the water in the combustion products is in the liquid state. Also, HIGH HEAT VALUE.

higher high water *Oceanography.* the higher of two high waters during a tidal day.

higher-high-water interval *Oceanography.* the amount of time between the transit (upper or lower) of the moon over the local or Greenwich meridian and the next higher high water.

higher low water *Oceanography.* the higher of two low waters during a tidal day.

higher-low-water interval *Oceanography.* the amount of time between the transit (upper or lower) of the moon over the local or Greenwich meridian and the next higher low water.

higher mode *Electromagnetism.* a waveguide mode whose frequency is higher than the lowest one.

higher-order conditioning *Behavior.* a type of classical conditioning in which the original conditioned stimulus is used as the unconditioned stimulus in a new conditioning procedure.

higher-order logic *Computer Science.* a logic that is more powerful than first-order predicate calculus, e.g., one that allows quantification over predicate symbols.

higher pair *Mechanical Engineering.* an arrangement in which mating parts of a mechanism have surface contact, rather than line or point contact.

higher plane curve *Mathematics.* any plane algebraic curve of degree greater than two.

higher than high-level language *Computer Programming.* a programming language that is designed for a specific application, such as a report generation or financial planning language.

high-expansion alloy *Metallurgy.* any of several alloys that have high coefficients of thermal expansion.

high explosive *Ordnance.* **1.** a class of explosives in which the active agent is in chemical combination and is detonated so readily that the chemical reaction is virtually instantaneous. **2. high-explosive.** of or relating to such an explosive. Thus, **high-explosive foam, high-explosive plastic.**

high explosive anti-tank *Ordnance.* a projectile exploding a shaped charge to pierce tank armor. Also, HEAT.

high-explosive plastic antitank charge *Ordnance.* an antitank projectile composed of a shaped charge and a high-explosive plastic charge; the shaped charge produces jet penetration, followed by detonation of the plastic charge.

high-explosive plastic projectile *Ordnance.* an antiarmor projectile composed of a plastic explosive in a thin case; designed to flatten against the armor just before detonation and scatter fragments at high velocity from the back of the target plate. Also, **high-explosive squash head, HESH.**

high fidelity *Acoustics.* **1.** in audio recording and reproducing applications, the sound reproduction that occurs over the full range of audible frequencies, thus approximating the sound of the original performance. **2. high-fidelity.** relating to or describing such faithful sound reproduction, or equipment that is capable of providing it. Also, HI-FI.

high-flux isotope reactor *Nucleonics.* a research reactor housed at the Oak Ridge National Laboratory in Tennessee that produces isotopes with atomic numbers higher than that of plutonium.

high foehn *Meteorology.* a condition in which the sky is clear and the air above the general surface is warm and dry, due to subsiding air in a high-pressure system above a cold surface layer. Also, FREE FOEHN.

high fog *Meteorology.* a type of fog that occurs frequently along the slopes of the California coastal mountains, when moist air overruns the peaks of the mountains and extends as low-level stratus clouds in the leeward valleys.

high frequency *Telecommunications.* an FCC designation for the range of frequencies from 3 to 30 megahertz.

high-frequency bias *Acoustics.* the introduction of a frequency several times higher than the high end of the frequency band of recorded sound, to reduce high-frequency distortion.

high-frequency carrier telegraphy *Telecommunications.* a type of carrier telegraphy in which the frequencies of the carrier currents are above the range sent through a voice-frequency telephone channel.

high-frequency compensation *Electronics.* an increase in the amplification of the high frequencies, relative to the low and middle frequencies, within a given band of frequencies. Also, HIGH BOOST.

high-frequency furnace *Engineering.* a furnace used to melt steel or other metals, using currents induced by the magnetic flux of a surrounding water-cooled coil of copper tubing.

high-frequency propagation *Telecommunications.* the propagation of high-frequency radio waves that depends solely on reflection from the ionosphere.

high-frequency recombination see HFR.

high-frequency resistance *Electricity.* the total resistance of a circuit as seen by high-frequency signals. This often varies considerably from the (DC) resistance measured by an ordinary ohmmeter. Also, AC RESISTANCE, EFFECTIVE RESISTANCE.

high-frequency titration *Analytical Chemistry.* a titration measured by the use of alternating current in the megahertz range passing between two electrodes mounted on the outside of the vessel containing the solution being analyzed.

high-frequency transformer *Electronics.* a transformer that matches impedances and transmits a frequency band in the higher ranges, usually with bandpass response.

high-frequency voltmeter *Electronics.* a voltmeter that is designed to measure high-frequency alternating currents.

high-frequency welding *Metallurgy.* resistance welding in which the heat is generated by a high-frequency electric current.

high-front shovel *Mechanical Engineering.* a power-driven shovel having a dipper stick mounted high on the boom.

high-grade *Science.* **1.** at the high end of a specified range. **2.** of superior quality.

high hat see TABLE TRIPOD.

high heat *Thermodynamics.* the amount of heat that is absorbed in the cooling medium of a calorimeter as the combustion products are cooled to the ambient temperature.

high heat value see HIGHER HEATING VALUE.

high-helix drill *Mechanical Devices.* a double-fluted drill with a helix of 35 to 40 degrees used in mining operations to bore holes in aluminum, copper, hard brass, and soft steel. Also, FAST-SPIRAL DRILL.

high-impact polystyrene *Materials.* a polystyrene rubber incorporated so as to have high impact strength, improved surface gloss, and heat resistance; used in toys, insulating liners, and containers.

high-impedance voltmeter see HIGH-RESISTANCE VOLTMETER.

high index *Meteorology.* a high value of the zonal index, indicating (in the middle latitudes) a strong westerly wind component and associated weather conditions.

high key *Graphic Arts.* of a photographic image, dominated by light tones, with few if any dark areas.

high-key lighting *Graphic Arts.* a photographic lighting technique by which shadows are reduced to a minimum, creating an image composed of light and medium tones, with no dark grays or blacks.

highland *Geography.* any elevated area, especially a mountainous region having an extensive, fairly uniform summit.

Highland *Agriculture.* a small breed of beef cattle that varies in color from red or fawn to brown or black, characterized by a thick shaggy outer coat, heavy dewlap, and wide horns. (From the *Highlands* of Scotland, where the breed originated.)

highland climate see MOUNTAIN CLIMATE.

highland glacier *Hydrology.* an ice cap or glacier system that more or less covers the highest or central portion of a mountainous area and partly reflects the irregularities of the land surface underneath. Also, **highland ice.**

Highlands *Geography.* the mountainous district of northern Scotland, between the Firth of Clyde and Nairn.

high-lead bronze *Metallurgy.* any of several copper-base, tin- and often zinc-bearing alloys, also containing from 3 to 34% lead.

high level *Electronics.* a level within the more positive of the two ranges of logic levels chosen to represent the logic states.

high-level data-link control *Computer Technology.* a standardized communications protocol that enables devices from different manufacturers to interface with each other.

high-level index *Computer Programming.* the first part of a file name that is used to specify its data category.

high-level language *Computer Programming.* a programming language that enables the user to write statements oriented toward the nature of the problem, such as equations or English-like commands, rather than in terms of machine instructions; each statement generally corresponds to multiple machine instructions.

high-level modulation *Telecommunications.* in a system, the modulation that originates at a point at which the power level is about equal to the system output.

high-level ridge see UPPER-LEVEL RIDGE.

high-level thunderstorm *Meteorology.* a thunderstorm based at an altitude of approximately 8000 feet or higher above the earth's surface that occurs dramatically over desert areas; its precipitation often evaporates before reaching the earth.

high-level trough see UPPER-LEVEL TROUGH.

high-lift truck *Mechanical Engineering.* a forklift truck with a mast, either fixed or telescoping, allowing for elevation of a load.

highlight *Electronics.* **1.** the brightest area of a reproduced image. **2.** in computer display, making a part of the image (or text) stand out by special visual effects such as color, reverse video, or distinctive frame.

highlight halftone see DROPOUT.

highlighting see SHADING.

high-line logging see SKYLINE LOGGING.

high-low bias test *Electronics.* a procedure routinely performed to test equipment over and under normal operating conditions in an attempt to detect defective units.

highly repetitive DNA *Genetics.* DNA that consists of many related or identical repeated sequences.

highly skewed propeller *Naval Architecture.* a term for a propeller whose blades are so extremely curved that the tip of one meets the base of the next.

high-mobility group protein *Biochemistry.* a group of nonhistone chromosomal proteins of high electrophoretic mobility, some of which are apparently associated with actively transcribed genes.

high-modulus fiber *Materials Science.* a polymer fiber whose macromolecules are aligned along the fiber axis, possessing a high modulus of elasticity.

high-moor bog *Ecology.* a bog that is not dependent on the water table and whose surface is covered by moss.

high-mu tube *Electronics.* a vacuum tube with an amplification factor of 30 or higher.

high-oblique photograph *Cartography.* an aerial photograph that includes the apparent horizon by intentional tilting of the axis of the camera.

high-order *Computer Programming.* of or relating to the weight or significance of a digit farthest to the left in a number. Thus, **high-order digit.**

high-pass filter *Electronics.* a filter that has a single transmission band extending from a cutoff frequency (not zero) to infinite frequency.

high-performance liquid chromatography *Analytical Chemistry.* a type of liquid chromatography in which the liquid (mobile) phase can readily be changed to affect resolution and time of analysis.

high polymer *Organic Chemistry.* a molecule of molecular weight greater than 10,000, composed of repeat units of low-molecular-weight species.

high-positive indicator *Computer Technology.* a component in some computers that is turned "on" to indicate that the result of an arithmetic operation is positive and not zero.

high-potential testing *Electricity.* a testing method in which a high-voltage source is applied to an insulating material to determine the condition of the material. Also, **high-pot.**

high-pressure center *Meteorology.* **1.** a point on a synoptic chart at which maximum pressure occurs; the center of a high-pressure system. **2.** a center of anticyclonic circulation.

high-pressure chemistry *Physical Chemistry.* the study of chemical reactions that take place at pressures of more than about 10,000 bars or atmospheres, especially for solid-state devices.

high-pressure cloud chamber *Nucleonics.* a device that keeps an ionized vapor at a high pressure in order to limit the energy range of the charged particles and make it easier to observe their movement and interactions.

high-pressure mercury-vapor lamp *Electronics.* a high-intensity discharge lamp in which light is produced by radiation discharged by mercury vapor.

high-pressure physics *Physics.* the study of high-pressure phenomena.

high-pressure process *Chemical Engineering.* a chemical process that operates at an extremely elevated pressure.

high-pressure system see HIGH.

high-pressure torch *Engineering.* a welding torch that uses a mixture of acetylene and oxygen under pressure.

high-pressure well *Petroleum Engineering.* an oil well that has a shut-in wellhead pressure of more than 2000 psia.

high Q *Electronics.* **1.** a high value for the ratio of reactance to resistance in a component or circuit. **2.** refers to a resonant circuit whose bandwidth is very small compared to the center frequency.

high-Q cavity *Electromagnetism.* a resonator cavity having very low energy losses at resonance.

high quartz *Mineralogy.* the high-temperature polymorph of quartz that is stable from 573 to 870°C. Also, BETA QUARTZ.

high-rank coal *Geology.* any coal that is more than 84% carbon, or that contains less than 4% moisture upon air-drying.

high-rank graywacke see FELDSPATHIC GRAYWACKE.

high-residual-phosphorus copper *Metallurgy.* phosphorus-deoxidized copper containing enough residual phosphorus to impair electrical conductivity.

high-resistance voltmeter *Electricity.* a voltmeter with a resistance much greater than 1000 ohms per volt, so that it requires only a minimal amount of current from the circuit in which testing is performed.

high-resolution radar *Engineering.* radar that is capable of distinguishing between and displaying two targets in close proximity.

high side *Computer Technology.* the component of a remote device that communicates with a computer.

high-side capacitance coupling *Electronics.* the process of using a capacitor to block direct current flow from the output of an oscillator or amplifier at a point of high potential.

high-speed *Computer Technology.* **1.** describing a data-communications device capable of handling more than 4800 bits per second. Thus, **high-speed printer, high-speed reader. 2.** operating at a speed that is high relative to current technology.

high-speed carry *Computer Technology.* a method for speeding up the processing of carries in parallel addition by computer logic circuits.

high-speed cement see HIGH-ALUMINA CEMENT.

high-speed excitation system *Electricity.* an excitation system that is capable of rapidly changing its voltage in direct response to variation in the excited generator field circuit.

high-speed machine *Mechanical Engineering.* **1.** broadly, any machine that operates at a high rate of speed, especially of rotational speed. **2.** specifically, a diamond drill with a minimum operating speed of 2500 revolutions per minute, compared to the maximum 1600–1800 revolutions per minute of the average diamond drill.

high-speed oscilloscope *Electronics.* an oscilloscope that can reproduce high-speed pulses accurately because of its efficient high-frequency and step-function response.

high-speed photography *Graphic Arts.* the process of taking still photographs of a moving subject using extremely short exposures, often with the aid of a strobe light.

high-speed relay *Electronics.* a relay designed to have short make or short break intervals.

high-speed steel *Metallurgy.* any of several tool steels that retain hardness at elevated temperatures, and are therefore suitable for high-speed machining.

high-strength low-alloy steel *Metallurgy.* any of several low-alloy steels that have a yield strength substantially higher than that of carbon steels; typically 275 to 760 MPa (40,000–100,000 pounds per square inch).

high-tech a shorter term for high-technology, especially in business or other nontechnical use. See HIGH TECHNOLOGY.

high technology *Science.* **1.** a general term for any form of technology that uses or requires the most advanced systems and techniques currently available, especially in such fields as telecommunications, office or factory automation, and information processing. **2. high-technology.** relating to or describing such a device or operation.

high-technology robot *Robotics.* a robot that uses powerful controllers, vision, feedback, and real-time data acquisition; a generation 2 or higher robot.

high-temperature alloy see HEAT-RESISTANT ALLOY.

high-temperature chemistry *Physical Chemistry.* the study of chemical reactions that take place at temperatures of more than 500,000 degrees Celsius.

high-temperature corrosion see HOT CORROSION.

high-temperature gas-cooled reactor *Nucleonics.* an experimental nuclear reactor in which the fuel of highly enriched uranium and thorium is cooled by pressurized helium gas.

high-temperature phenomena *Physics.* the behavior and properties of substances subjected to temperatures of about 500 K and higher.

high-temperature polymer *Materials Science.* a polymer that retains good mechanical properties for a long period of time at a high temperature, such as polyetherimides and polyphenylene oxide.

high-temperature reactor *Nucleonics.* a nuclear reactor designed to generate enough heat to power mechanical devices.

high-temperature reservoir *Thermodynamics.* a body that provides a heat transfer to a heat engine by virtue of its having a higher temperature than the heat engine, since heat transfer occurs from a hotter body to a colder one.

high-temperature superconductor *Solid-State Physics.* a substance capable of achieving superconductivity at temperatures at or above the liquid nitrogen temperature (77 K).

high-tensile bolt *Engineering.* an adjustable bolt that is controlled by the tension of a device such as a calibrated torsion wrench.

high-tensile steel *Metallurgy.* any of several steels that have a yield strength higher than a specified amount.

high-tension detonator *Engineering.* a detonator that requires a relatively high electrical charge for firing, typically 50 volts.

high-test chain *Engineering.* a heavy-duty chain made of heat-treated carbon steel.

high tide see HIGH WATER.

high vacuum *Physics.* a system that has been evacuated to a pressure below 10^{-3} torr.

high-vacuum insulation *Chemical Engineering.* a type of thermal insulation used in containers of cryogenic materials where a high vacuum is maintained between the two walls of double-walled vessels.

high-vacuum rectifier *Electronics.* a vacuum-tube rectifier in which conduction is by electrons emitted from the cathode.

high-vacuum tube *Electronics.* an electron tube that has been evacuated to such a degree that gaseous ionization cannot occur.

high velocity *Ordnance.* a relatively high muzzle velocity of a projectile, generally defined as follows: in artillery, from 3000 feet per second up to but not including 3500 feet per second; in small arms, between 3500 and 5000 fps; in a tank cannon, between 1550 and 3350 fps.

high-velocity anti-radiation missile see HARM.

high-velocity cloud *Astronomy.* a fast-moving interstellar cloud, mostly neutral hydrogen, detected in ultraviolet light.

high-velocity drop *Military Science.* an airdrop procedure in which the drop velocity is greater than 30 feet per second but lower than the free drop velocity.

high-velocity star *Astronomy.* an old star belonging to the galaxy's halo population; its orbit in the galaxy takes it far out of the galactic disk, with the result that its velocity relative to the sun is high (100 kilometers per second or so).

high-volatile bituminous coal *Geology.* a bituminous coal yielding more than 31% volatile matter.

high voltage *Electricity.* **1.** an imprecise term for voltage that is large enough to present a safety hazard, or that is considerably higher than would normally be expected in a particular application. **2. high-voltage.** relating to or having such a voltage level. Thus, **high-voltage wire, high-voltage generator,** and so on.

high-voltage direct current *Electricity.* a long-distance direct-current power transmission system that uses direct-current voltages of up to one megavolt to minimize transmission loss.

high-voltage microscope *Electronics.* an electron microscope whose accelerating voltage is in the range of 10^6 volts, as opposed to 40–100 kilovolts, resulting in increased specimen penetration, reduced specimen damage, better resolution, and more efficient dark-field operation.

high water *Oceanography.* the highest point reached by a rising tide.

high-water full and change *Geophysics.* the average interval between upper and lower lunar transit near the time of the new and full moon and the next high water.

high-water inequality *Oceanography.* the difference between the heights of two successive high waters in one lunar day.

high-water line *Oceanography.* the intersection of the shore by the plane of high water.

high-water lunitidal interval *Geophysics.* the interval between the moon's transit over the prime meridian and the following high water.

high-water mark *Oceanography.* an established reference mark on a permanent structure or object, indicating the highest observed water level at that location. *Computer Programming.* the maximum number of jobs observed in a computer's job queue over a specified period of time.

high-water platform see WAVE-CUT BENCH.

high-water quadrature see MEAN HIGH-WATER NEAPS.

high-water stand *Oceanography.* the stage of high water at which there is no change in the level of the water.

highway *Civil Engineering.* any public road, especially a route between cities, on which vehicles are permitted to travel at a higher speed than on local streets.

Hikojima serotye *Immunology.* a strain of the bacteria *Vibrio cholerae* that is characterized by having the A, B, and C antigens; a causative agent of cholera.

hilar appendix *Mycology.* in certain fungal spores, a small outgrowth that joins the spore to the sterigma or hyphal stalk.

Hilbert, David 1862–1943, German mathematician; did work in geometry, integral equations, theory of invariants, and mathematical physics.

Hilbert cube *Mathematics.* **1.** the topological space that is the countably infinite Cartesian product of the unit interval with itself; denoted by I^{ω}. **2.** the set I of all real sequences $(a_1, a_2, a_3, \ldots)$ such that $0 \le a_n \le 1/n$ for every n.

Hilbert Nullstellensatz *Mathematics.* let F be an algebraically closed extension field of a field K; let I be a proper ideal of the polynomial ring $K[x_1, \ldots, x_n]$; and let $Z(I)$ be the set of all n-tuples in F^n that are zeros of some polynomial in I. Then a polynomial f satisfies $f(a_1, \ldots, a_n) = 0$ for all $(a_1, \ldots, a_n) \in Z(I)$ if and only if $f^m \in I$ for some integer m. Also, HILBERT ZEROS THEOREM.

Hilbert-Schmidt norm *Mathematics.* let T be a Hilbert-Schmidt operator on a separable Hilbert space H, having the matrix representation (a_{ij}) with respect to some orthonormal basis of H. The Hilbert-Schmidt norm of T is $|T| = (\sum_{i,j=1}^{\infty} |a_{ij}|^2)^{1/2}$. $|T|$ can be shown to be independent of the choice of basis.

Hilbert-Schmidt operator *Mathematics.* a bounded linear operator T on a separable Hilbert space H such that $\sum_{i=1}^{\infty} \|Te_i\|^2 < \infty$, where $\{e_i\}$ is an orthonormal basis of H. If A is the (infinite) matrix representation of T, with (i,j)th entry $a_{ij} = <Te_j, e_i>$, then the linear operator T is a Hilbert-Schmidt operator if and only if $\sum_{i,j=1}^{\infty} |a_{ij}|^2 < \infty$.

Hilbert-Schmidt theorem *Mathematics.* the theorem that if T is a self-adjoint compact operator on a Hilbert space, there then exists an orthonormal basis for the space composed of eigenvectors of T.

Hilbert space *Mathematics.* a Banach space with norm $\| \ \|$ and a real-valued, symmetric inner product $< \ , \ >$ such that $<v, v> = \|v\|^2$ for all vectors v in the space; equivalently, a complete inner product space. For example: (a) R^n with ordinary dot product, (b) the space L^2, with $<f, g> = \int f(x)g(x)\,dx$. Also, COMPLETE INNER PRODUCT SPACE.

Hilbert transform *Mathematics.* a function u related to a function v by the following formulas is said to be the Hilbert transform of v (integrals are understood to be the Cauchy principal values):

$$u(y) = (1/\pi)\int_{-\infty}^{\infty} v(t)/(y - t)\,dt$$

$$v(y) = (-1/\pi)\int_{-\infty}^{\infty} u(t)/(y - t)\,dt$$

Hilbert zeros theorem see HILBERT NULLSTELLENSATZ.

Hilda group *Astronomy.* a group of 31 main-belt asteroids (named for 153 Hilda) whose relatively eccentric orbits have a 3:2 resonance with Jupiter's orbit; most are of type D or P, and many may be captured comet nuclei.

Hildebrand function *Thermodynamics.* a function of the heat of vaporization of a compound expressed in terms of the molal concentration of the vapor; approximately the same for many compounds.

Hildenbrandiaceae *Botany.* a monospecific family of marine and freshwater red algae in the order Cryptonemiales, having crustose perennial thalli and unknown gametangia.

Hildoceras *Paleontology.* a genus of ammonoids in the order Ammonitida and superfamily Hildoceraceae; characterized by an evolute, laterally compressed shell; extant in the Lower Jurassic.

hilgardite *Mineralogy.* $Ca_2B_5O_9Cl{\cdot}H_2O$, a colorless, transparent, monoclinic mineral, trimorphous with hilgardite-ITc and hilgardite-Tc, occurring as tabular crystals, having a specific gravity of 2.71 and a hardness of 5 on the Mohs scale; found in the water-insoluble residue of rock salt.

Hill, Archibald 1886–1977, English physiologist; received Nobel Prize for discovering process of heat production in the muscles.

hill *Geography.* any natural elevation of the earth's surface, usually lower than a mountain (less than 1000 feet) and less steep.

hill-climbing *Engineering.* the continuous or periodic self-adjustment of a system to achieve optimum results. *Artificial Intelligence.* a form of search in which the path of steepest ascent toward the goal is taken at each step. It is excellent if the domain is well-behaved, but can get stuck on local maxima or mesas.

hill creep *Geology.* the slow downhill movement of soil or rock waste that occurs on steep hillsides under the influence of gravity. Also, **hillside creep.**

Hilleaceae *Botany.* a monogeneric family of simple biflagellates belonging to the order Cryptomonadales, characterized by a lack of trichocysts, cells that are not compressed, and two equal to subequal flagella; they occur mostly in marine and brackish waters.

hillebrandite *Mineralogy.* $Ca_2SiO_3(OH)_2$, a white or greenish monoclinic mineral occurring in fibrous masses, having a specific gravity of 2.66 and a hardness of 5.5 on the Mohs scale; found in impure thermally metamorphosed limestones and in boiler scale.

hillock

hillock *Geology.* a general term for a small, low hill or mound.

hill plane *Cartography.* a plane that contains the positions of three ground marks representing control points.

Hill plot *Biochemistry.* a graphical representation used to determine the number of binding sites of a given type per molecule of protein.

Hill reaction *Biochemistry.* the light reaction in photosynthesis that is carried out in the presence of an artificial electron acceptor.

hill shading *Cartography.* a method of depicting a change in elevation on maps without contours by representing the shadows that would be cast by high ground if light were shining from the northwest.

Hiltner-Hall effect *Astrophysics.* a phenomenon in which light from stars becomes polarized as it travels through space to earth.

Hilt's law *Geology.* a law stating that at any point in a vertical succession in a coal field, the carbon content of the coal increases with depth.

hilum *Anatomy.* a recess or opening in an organ through which nerves, muscles, or ducts pass into the organ. Also, **hilus.** *Botany.* **1.** a scar on a seed marking the point of attachment to the seed vessel. **2.** the central point of a grain of starch, around which starch is deposited.

hilus of liver see PORTA HEPATIS.

Himalayas [him´ə lā´əz; hi mäl´yəz] *Geography.* a range of mountains in southern Asia between India and Tibet, curving 1500 miles west to east and containing many of the world's highest mountains.

Himalia [hi mäl´yə] *Astronomy.* the sixth moon of Jupiter, discovered in 1904 and measuring about 180 kilometers in diameter.

Himantandraceae *Botany.* a family of large dicotyledonous trees belonging to the order Magnoliales that produce himandrin and related alkaloids and are found only in New Guinea, the Molucca Islands, and northeast Australia.

Himanthaliaceae *Botany.* a monotypic family of clustering brown algae of the order Fucales, distinguished by a thallus reduced in size during vegetative growth but producing elongated receptacles at reproductive maturity; after its reproductive season, the plant disintegrates.

Himantopterinae *Invertebrate Zoology.* a subfamily of small, brightly colored moths; lepidopteran insects in the family Zygaenidae.

Himmelweit pipet *Biotechnology.* a Pasteur pipet that is used in the harvesting of allantoic fluid from an egg.

hind *Zoology.* situated at the back or rear; posterior. Thus, **hind leg(s), hind quarters,** and so on. *Vertebrate Zoology.* a female deer, especially a female red deer more than three years old.

HindII *Genetics.* the designation for a restriction endonuclease that cleaves DNA molecules at a GTPy/PuAC site; isolated from the bacterium *Hemophilus influenzae*. Also, **HindII restriction enzyme.**

HindIII *Genetics.* the designation for a restriction endonuclease that cleaves DNA molecules at a specific recognition site, A/AGCTT; isolated from the bacterium *Hemophilus influenzae*. Also, **HindIII restriction enzyme.**

hindbrain see RHOMBENCEPHALON.

hindered contraction *Metallurgy.* in casting, contraction upon cooling that is hindered in certain locations, thus generating stresses.

hindered polymer *Materials Science.* a polymer in which side groups prevent rotation about the carbon-carbon bond.

hindered settling *Mining Engineering.* **1.** the settling of crushed ore in a suspension of such ore and water. **2.** the settling of minerals in a thick pulp. **3.** the settling of particles through a crowded zone.

hindgut see EPIGASTER.

H indicator see H DISPLAY.

Hindley screw *Mechanical Devices.* an endless worm-shaped screw with an hourglass symmetry; used to increase the bearing area of a worm wheel, thus reducing wear. Also, HOURGLASS SCREW.

Hind's nebula *Astronomy.* a reflection nebula in Taurus that is illuminated by the irregular variable star T Tauri; officially designated **NGC 1554-5.**

Hindu-Arabic numerals see ARABIC NUMERALS.

Hindu Kush *Geography.* a mountain range in northern Pakistan and northeastern Afghanistan, west of the Pamir Knot; highest peak: Tirich Mir (25,230 feet).

hinge *Mechanical Devices.* a jointed device on which an attached part, such as a door or window, swings. *Zoology.* any similar structure occurring in nature, especially the hinge ligament of a bivalve. *Graphic Arts.* see JOINT. *Geology.* see FLEXURE.

hinged arch *Civil Engineering.* a structural arch that can rotate about its supports, about its center, or in both places.

hinged bar *Mining Engineering.* a steel bar in contact with the mine roof and perpendicular to the longwall face. Also, LINK BAR.

hinged joint *Anatomy.* a synovial joint in which a convex surface of one bone fits into a concave surface of another bone. Also, **hinge joint.**

hinge fault *Geology.* a fault in which the blocks along one wall are rotated or hinged about an axis perpendicular to the fault plane, thus causing the displacement to increase with distance from the hinge.

hinge ligament *Invertebrate Zoology.* the elastic protein membrane that holds two halves of a bivalve shell together.

hinge line *Geology.* **1.** a boundary between a stable or stationary region of the earth's crust and one undergoing changes in elevation; often formed in response to isostatic adjustment or tectonic deformation. **2.** the line along the surface of a hinge fault where the direction of displacement apparently changes. **3.** an imaginary line connecting the points of maximum curvature of folded bedding planes.

hinge moment *Aviation.* on a control surface, the tendency of an aerodynamic force to generate motion along a hinge line.

hinge plate *Invertebrate Zoology.* **1.** in bivalve mollusks, the part of a valve that supports the hinge teeth. **2.** in brachiopods, the part of the dorsal valve that has a socket.

hinge region *Immunology.* the flexible, proline-rich juncture of an immunoglobulin molecule that joins two Fab fragments and the remaining Fc fragment.

hinge tooth *Invertebrate Zoology.* in clam shells, one component of an arrangement of irregular toothlike structures that interlock to fit both halves of the clam shell together.

hinoki *Materials.* the light, soft, resilient wood of the **hinoki cypress** tree of Asia, *Chamaecyparis obtusa*; used in construction for such objects as foundations, siding, trim, and fiberboard.

Hinosan *Organic Chemistry.* $C_{14}H_{15}O_2PS_2$, a trade name for edifenphos, a clear, yellow to light brown liquid that is soluble in acetone and xylene and insoluble in water; it boils at 154°C (0.01 torr); used as a fungicide.

Hinsberg test *Analytical Chemistry.* a method of separating mixtures of primary, secondary, and tertiary amines by treatment with benzenedisulfonyl chloride; primary amines form sulfonamides soluble in water and alkali, secondary amines form insoluble sulfonamides, and there is no reaction with tertiary amines.

hinsdalite *Mineralogy.* $(Pb,Sr)Al_3(PO_4)(SO_4)(OH)_6$, a colorless or greenish trigonal mineral of the beudantite group, occurring in granular form or as pseudocubic, rhombohedral, or tabular crystals, and having a specific gravity of 3.65 and a hardness of 4.5 on the Mohs scale.

Hinshelwood, Sir Cyril 1897–1967, English chemist; shared Nobel Prize for experiments in chain reaction kinetics.

hinterland *Geology.* **1.** the terrain lying at the rear of a folded mountain chain, away from the direction of thrusting and overfolding. Also, BACKLAND. **2.** an actively moving block that forces compressed geosynclinal sediments toward the foreland.

hintzeite see KALIBORITE.

Hiodontidae *Vertebrate Zoology.* the mooneyes and goldeyes, a family of mainly nocturnal freshwater fish that resemble herring and belong to the order Osteoglossiformes; found in lakes and streams of eastern North America, where they are caught commercially.

hip *Anatomy.* the area of the trunk of the body that is lateral to and includes the hip joint. *Architecture.* the external angle formed at the junction of two sloping sides of a hip roof.

hipbone or **hip bone** *Anatomy.* the bone consisting of the ilium, ischium, and pubis that combines with the sacrum and the coccyx to form the pelvis. Also, PELVIC BONE, COXAL BONE, INNOMINATE BONE.

hip joint *Anatomy.* the ball-and-socket joint between the hipbone and the head of the femur. *Civil Engineering.* the joint of a truss top chord and an inclined head post.

Hipparchus c. 190–120 BC, Greek astronomer; discovered procession of equinoxes; compiled first star catalog; founder of trigonometry.

hipped *Anatomy.* having hips, or hips of a certain kind; used in compounds such as **broad-hipped.** *Veterinary Medicine.* having an injured hip; especially, of a horse, having a fracture at the point of the hip.

Hippias of Elis 5th century BC, Greek philosopher; invented the quadratix of Dinostratus.

Hippidea *Invertebrate Zoology.* the mole crabs or sand crabs, a superfamily of burrowing decapod crustaceans in the group Anomura.

hippo- or **hipp-** a combining form meaning "horse," as in *hippocampus.*

Hippoboscidae *Invertebrate Zoology.* the louse flies and deer flies, a family of two-winged bloodsucking insects in the suborder Cyclorrhapha, having flattened bodies, with small wings that are usually lost upon attaching to a host.

hippocampal *Anatomy.* of or relating to the hippocampus.

hippocampal sulcus *Anatomy.* a shallow cleft between the dentate gyrus and parahippocampal gyrus in the cerebrum. Also, **hippocampal fissure.**

hippocampus *Anatomy.* a complex curved structure of the cerebral cortex forming the floor of the inferior horn of the lateral ventricle. *Vertebrate Zoology.* the genus of seahorses of the family Sygnathidae, found in shallow waters worldwide amid muddy ground and algae.

Hippocastanaceae *Botany.* a family of dicotyledonous trees and shrubs distributed mostly in Eurasia and temperate North America and belonging to the order Sapindales, characterized by palmately compound opposite leaves, large showy flowers, and a large hard-coated seed; it includes the horse chestnut and buckeyes, known in cultivation.

Hippocrateaceae *Botany.* a family of tropical dicotyledonous plants in the order Celastrales, characterized by a well-developed latex system, simple, usually opposite leaves, and a well-developed nectary disk.

Hippocrates of Chios [hi päk′rə tēz] taught about 430 BC, Greek mathematician; wrote first (pre-Euclidian) *Elements of Geometry.*

Hippocrates of Kos c. 460–c. 375 BC, Greek physician; known as the father of medicine; established Hippocratic (inductive) method of diagnosis. Thus, **Hippocratism, Hippocratic.**

Hippocratic oath [hip′ə krat′ik] *Medicine.* an oath attributed to Hippocrates of Kos, embodying the duties, obligations, and ethics of the medical profession; traditionally taken by graduates of medical school during commencement ceremonies.

hippocrepiform *Biology.* horseshoe-shaped.

Hippoglossidae *Vertebrate Zoology.* in some classification systems, a family of flounders in the order Pleuronectiformes.

hip pointer *Medicine.* a contusion of the bone of the iliac crest.

Hippomorpha *Vertebrate Zoology.* a suborder of odd-toed ungulates of the order Perissodactyla; including the horses, zebras, and related forms.

hippopede *Mathematics.* the loci of points satisfying $r^2 = 4b(a–b \sin^2\theta)$ in polar coordinates. If $a \le b$, the curve has the shape of a figure eight.

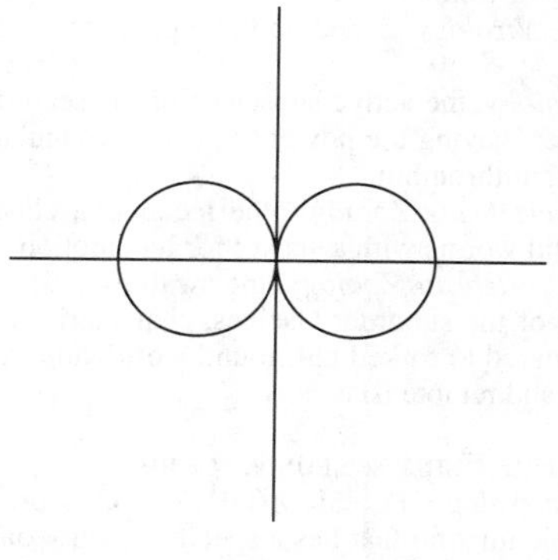

hippopede

Hippopotamidae *Vertebrate Zoology.* the hippopotamuses, a family of herbivorous even-toed ungulates of the order Artiodactyla, having a broad head, wide mouth, short bulky body, short legs, and thick skin with mucous glands for adaptation to life in water; found in sub-Saharan Africa.

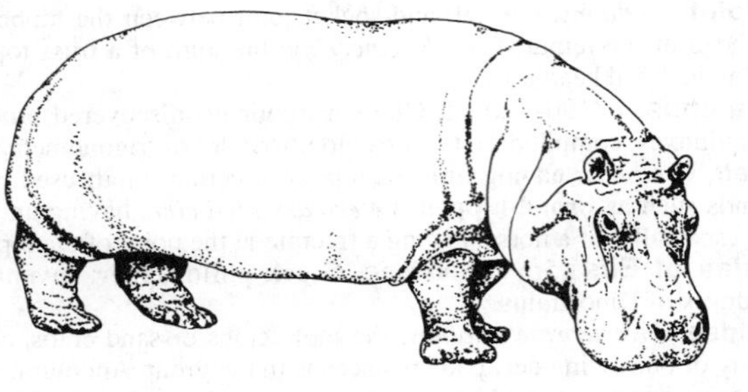

hippopotamus

hippopotamus *Vertebrate Zoology.* either one of two large, herbivorous, four-toed amphibious mammals of the family Hippopotamidae, the **river (common) hippopotamus,** *Hippopotamus amphibius,* or the smaller, less aquatic **pygmy hippopotamus**, *Choeropsis liberiensis.* (From a Greek term literally meaning "river horse.")

hippuric acid *Organic Chemistry.* $C_6H_5CONHCH_2COOH$, colorless crystals obtained from the urine of domestic animals; slightly soluble in water and alcohol; melts at 188°C; used in medicine. Also, *N*-BENZOYLGLYCIN.

Hippuridaceae *Botany.* mare's tail, a monospecific family of dicotyledonous aquatic herbs of the order Haloragidales that are perennial and rhizomatous with erect, emergent stems and are distributed chiefly in north temperate and boreal areas.

Hippurites *Paleontology.* a genus of large, aberrant, sessile bivalves in the extinct order Hippuritoida; one of the rudists, characterized by pachydont dentition and attachment to the sea floor or reef by the apex of the right valve; extant in the Cretaceous.

Hippuritoida *Invertebrate Zoology.* an order of fossil rock oysters in the class Bivalvia, subclass Heterodonta.

hip roof *Architecture.* a roof with sloping sides and sloping ends.

Hirayama family [hī´rä yä´mä] *Astronomy.* a group of asteroids that share similar orbital characteristics (distance from the sun, period, eccentricity, inclination, etc.) and which may once have been part of a single object that fragmented.

hircinol *Organic Chemistry.* $C_{15}H_{14}O_3$, crystals produced by *Loroglossum hiricinum* infected with *Rhizoctoria repens*; melts at 162.5–164°C; used as a fungicide.

HI region *Astronomy.* a cloud of neutral (un-ionized) hydrogen.

Hirschsprung's disease *Medicine.* a congenital disorder occurring in early infancy and marked by dilation of the colon, constipation, and distention of the abdomen. (Named after Harald *Hirschsprung,* 1830–1916, Danish physician.) Also, CONGENITAL MEGACOLON.

Hirst spore trap *Biotechnology.* an apparatus used to trap a sample of airborne microflora for evaluation.

hirsute [hur´soot´; hur´soot] *Biology.* **1.** covered with stiff hairs or bristles. **2.** having abundant or excessive hair.

hirsutism *Medicine.* excessive hair growth on the cheek, chin, lip, or chest, especially in women.

Hirt extraction *Virology.* a method for separating viral DNA from cellular DNA.

hirudin *Hematology.* the active substance of the secretion of the buccal glands of leeches, having the power to prevent coagulation of the blood by acting as an antithrombin.

Hirudinea *Invertebrate Zoology.* the leeches, a class of parasitic or predatory annelid worms with a sucker for locomotion and attachment.

Hirundinidae *Vertebrate Zoology.* the swallows, a family of swift-flying, agile birds of the suborder Oscines, characterized by long, pointed wings and a squared to forked tail; found worldwide except for the Arctic, Antarctica, and remote islands.

His histidine.

His' bundle or **His' band** see BUNDLE OF HIS.

hisingerite *Mineralogy.* $Fe_2^{+3}Si_2O_5(OH)_4{\cdot}2H_2O$, a brown or brownish-black monoclinic mineral that has a specific gravity of 2.67 and a hardness of 2.5 to 3 on the Mohs scale; ubiquitous as an alteration mineral in iron-rich rocks.

his operon *Genetics.* a genetic system that is responsible for the synthesis of histidine from its precursors ATP and phosphoribosyl pyrophosphate; originally described for *Salmonella typhimurium.*

Hispaniola *Geography.* a large island (about 29,400 square miles), in the West Indies between Cuba and Puerto Rico, on which the countries of Haiti and the Dominican Republic are situated.

hispid *Biology.* having a surface sparsely covered with long, rather stiff hairs or spines (usually stiff enough to penetrate the skin).

hispidulous *Biology.* somewhat or minutely hispid.

hiss *Telecommunications.* random audio-frequency noise with a continuous spectrum; in recording systems, usually detected during playback.

Hiss serum water *Microbiology.* a nutrient medium that is composed of serum, water, sugar, and a phenol red pH indicator.

hist- a combining form meaning "tissue," as in *histology.*

histamine [his´tə mən; his´tə mēn´] *Biochemistry.* $C_5H_9N_3$, a bioactive amine that is present in the secretory granules of the mast cells and their related blood basophils; formed by the decarboxylation of histidine. It is also present in ergot and other plants and may be synthesized outside the body from histidine or citric acid. Its functions include dilation of capillaries, contraction of most smooth muscle tissue, induction of increased gastric secretion (its most important use), and acceleration of heart rate.

histamine flush *Physiology.* a reddening of the skin due to vasodilation, caused by the presence of histamine.

Histeridae *Invertebrate Zoology.* a family of coleopteran insects, the hister beetles, in the order Polyphaga; larvae are predators and adults are scavengers in decaying animal material.

histidase *Enzymology.* an enzyme, produced in the liver, that catalyzes the deamination of histidine to urocanic acid.

histidine *Biochemistry.* $C_6H_9O_2N_3$, an essential amino acid that contains the imidazole ring system; common in hemoglobin, and essential for optimal growth in infants; its decarboxylation results in the formation of histamine.

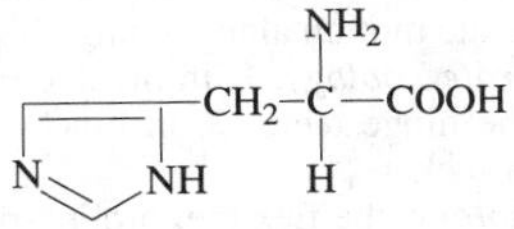

histidine

histidinemia *Medicine.* a hereditary metabolic disorder in which there is an elevation of blood histidine level, excretion of histidine in the urine, and deficiency of histidase, sometimes accompanied by speech and reading disorders and mild retardation.

histiocyte see MACROPHAGE.

histiocytoma *Medicine.* a benign skin tumor made up of phagocytic reticuloendothelial cells known as histiocytes (macrophages).

histiocytosis *Medicine.* a condition marked by the abnormal presence of histiocytes (macrophages) in the blood.

histioma see HISTOMA.

histo- a combining form meaning "tissue," as in *histocompatibility.*

histochemistry *Biochemistry.* the science that deals with the chemical changes and chemical constitution of tissues and cells.

histocompatibility *Immunology.* the ability to be accepted and remain functional; such a relationship, determined by the presence of histocompatibility antigens, exists between the genotypes of a donor and host between whom a graft generally will be accepted. Thus, **histocompatible.**

histocompatibility antigens *Immunology.* the antigens on cells that are responsible for tissue transplant rejection.

histodifferentiation *Developmental Biology.* the specialization of cells and tissues during development.

histogen *Botany.* a clearly defined area of primary tissues found in apical meristems and other areas of plants, which gives rise to a specialized tissue system in the plant body.

histogenesis *Developmental Biology.* the formation and development of tissues from undifferentiated cells of the germ layers of the embryo.

histogen theory *Botany.* a theory of development of the apical meristem, citing three histogen zones that give rise to later structures: the dermatogen, which develops into the epidermis; the periblem, which becomes the cortex; and the plerome, which gives rise to the internal tissues of the cortex.

histogram see BAR GRAPH.

histography *Histology.* a description of a body tissue.

histoincompatiblity *Immunology.* a condition in which a tissue transplant donor and recipient have differences in antigens that cause a rejection of a tissue graft. Thus, **histoincompatible.**

Histology

Histology (from the Greek *histos,* web or tissue, and *logos,* knowledge or study) is the biological science concerned with the minute or microscopic structure of cells, tissues, and organs in relation to their function. The word includes both animal and plant tissues. Garrison's history (1929) records Althanasius Kircher (1602–1680), a Jesuit priest, as the first microscopist and Marcello Malpighi (1628–1694), physician to Pope Innocent XII and professor of anatomy at Bologna, Pisa, and Messina, as the founder of histology. Leeuwenhoek and Swammerdarn are among the great and early microscopists whose observations were also of immense value in stimulating subsequent study. Leeuwenhoek alone sent 375 communications of observations to be published in the British Proceedings of the Royal Society. The first to consider cell structure was Hooke (1865), who observed the walled compartments of dead cork tissue and named them "cells." Brown (1831) described the nucleus of the cell and Schleiden and Schwann (1838,1839) were the first to propose a cell theory of plant and animal structure.

The further development of histology required the refinement of microscopes (leading to the electron microscope) and methods of cell, tissue, and organ preservation, supporting materials for delicate structures, stains and staining methods, and thin sectioning of cells and tissues by glass and diamond knives to reveal macromolecular structure. This technology has led to significant advances in our understanding of the function of cells, tissues, and organs over the past 300 years. The development of histochemical methods permitted the identification and localization of intercellular inclusions and enzymes. The isolation, study, and identification of living cell organelles, cells, other tissue components, and organ functions had their beginnings in the histologist's laboratory.

Building on a solid foundation, the scope of histology is enormous and expanding. Molecular biologic techniques will permit examination of the structural chemistry of cells, tissues, and organs. This integrative approach will ultimately lead to a definitive understanding not only of cellular, tissue, and organ structure but also of their function.

Ronald A. Bergman
Professor of Anatomy
University of Iowa

histology *Anatomy.* the specialized field of biology that studies the microscopic anatomy and function of cells and tissues. Thus, **histologist, histologic(al).**

histolysis *Pathology.* the dissolution, breakdown, or disintegration of tissue. Thus, **histolytic.**

histoma *Oncology.* any tumor that is formed from mature tissue. Also, HISTIOMA.

Histomonas *Invertebrate Zoology.* zooflagellate protozoans in the family Mastigamoebidae, exhibiting both amoeboid and flagellate phases; parasites in the liver and intestinal mucosa of poultry and other birds; *H. meleagridis* causes blackhead in the comb and wattles.

histomycosis *Medicine.* a disease of deep bodily tissues, caused by the presence of fungi.

histone *Biochemistry.* any of a group of five basic proteins characterized by a high content of lysine and arginine; found in the chromosomes of all eukaryotic cells but never in prokaryotes.

histone acetylase *Enzymology.* an enzyme that catalyzes the addition of an acetyl group (CH_3CO–) to certain amino acids of histone molecules.

histone gene *Genetics.* a gene that codes for histones, proteins of eukaryotes that bind DNA into nucleosomes, the basic subunit of chromatin.

histonelike protein *Biochemistry.* any of a class of basic proteins having a high arginine/lysine content and a low molecular weight, found in the nuclei of eukaryotic cells where they bind to DNA.

histopathology *Pathology.* the branch of science concerned with the histologic or cytologic structure of diseased tissue.

histophilous *Ecology.* living or thriving within the tissues of another organism.

histophysiology *Physiology.* a branch of physiology concerned with the relationship of function to the structure of tissues and cells.

histophyte *Ecology.* an organism that lives as a parasite within another organism.

Histoplasma *Mycology.* a genus of imperfect fungi of the class Hyphomycetes that are parasitic and are found in blood, lymph, or other tissue cells and in soil.

Histoplasma capsulatum *Mycology.* a species of fungi of the genus *Histoplasma*, the parasites that cause histoplasmosis.

histoplasmin *Pharmacology.* antigenic material that is derived from cultures of *Histoplasma capsulatum* and injected intradermally in histoplasmin testing.

histoplasmin test *Immunology.* a skin test in which an extract of the fungus *Histoplasma capsulatum* is used to diagnose histoplasmosis.

histoplasmoma *Medicine.* a rounded granulomatous density of the lung, caused by the fungus *Histoplasma capsulatum*, and often seen as a coin-shaped lesion in X rays.

histoplasmosis *Medicine.* an infection caused by inhaling or, infrequently, ingesting the fungus *Histoplasma capsulatum,* usually resulting in a primary, benign pneumonitis, **primary histoplasmosis,** but also known to progress to **progressive histoplasmosis**, a severe disseminated disease marked by fever, anemia, enlargement of the liver and spleen, and pulmonary lesions.

historical or **historic** *Anthropology.* relating to or designating human cultures within the period of historic time; based on the point at which written records and other such formal data began.

historical climate *Meteorology.* a climate of the historical period, as distinguished from geological, or prehistoric, climate.

historical data *Computer Programming.* a collection of data that is kept for possible future reference but is no longer actively maintained or updated.

historical era see HISTORICAL PERIOD.

historical geology *Geology.* the study of the history and chronological development of the earth from its origin to the present, including the evolution of its life forms.

historical linguistics *Linguistics.* the study of how language changes over time; includes techniques such as comparing written works to present forms, reconstructing protolanguage, glottochronology, and lexicostatistics. Also, DIACHRONIC LINGUISTICS.

historical materialism see ECONOMIC DETERMINISM.

historical particularism *Anthropology.* a school of thought developed in the early 1900's by Franz Boas, in which all elements of a culture are studied to determine the laws of the culture; all cultures are viewed as unique and any similarity between cultures is seen as the product of exchange.

historical period *Anthropology.* the period from the beginnings of recorded history to the present; variously defined but usually regarded as including the past 5000 to 7000 years. Also, HISTORICAL ERA.

historic chart *Cartography.* a chart used to project probable patterns of currents and winds in the future based on previously accumulated information.

historrhexis *Medicine.* the breaking up of tissues.

history *Science.* **1.** the study of the past. **2.** a record of past events. *Medicine.* see MEDICAL HISTORY.

history overlay *Cartography.* a bathymetric chart covered by a clear plastic overlay showing the sources of all the sounding data used to compile the chart.

Histosol *Geology.* an order of soils that develops in cool, poorly drained, humid areas, characterized by a composition of more than 50% undecomposed organic matter.

histotope *Immunology.* the part of the MHC molecule that is recognized by the T-cell.

histotoxic hypoxia *Toxicology.* the poisoning of a tissue by reducing the ability of its cells to utilize oxygen, as in cyanide poisoning.
histozoic *Ecology.* of or relating to an organism that lives parasitically in the tissues of another organism.
Histriobdellidae *Invertebrate Zoology.* a family of tiny polychaete worms in the group Errantia, parasites on marine crustaceans.
histrionic *Psychology.* of or relating to histrionism.
histrionism *Psychology.* a disorder in which an individual exhibits a dramatic, self-centered, and emotionally unstable personality that leads him or her to engage in attention-seeking behavior. Also, **histrionic personality (disorder).**
His-Werner disease see TRENCH FEVER.
hit *Ordnance.* the impact or point of impact on a target by a bullet, bomb, missile, or other projectile. *Computer Programming.* a comparison between two items of data, such as record keys, in which a match occurs. *Electricity.* a temporary electrical disturbance on a transmission line, such as a lightning stroke.
hitch *Geology.* a minor displacement of a vein or stratum that does not exceed the layer's thickness. *Mining Engineering.* a hole cut in a rock face to hold timber supports for underground mining.
hitchhiker's thumb see DISTAL HYPEREXTENSIBILITY.
HI test hemagglutination-inhibition test.
hit-on-the-fly printer *Computer Technology.* a printer in which the paper, the print head, or both are moving continuously.
hit probability *Ordnance.* for a target, the probability of being hit at least once out of a given number of projectiles directed at the target.
hit rate *Computer Programming* **1.** the ratio of the number of successful references to main memory to the total number of references in a hierarchical memory. **2.** the ratio of the number of records accessed in a run to the total number of records in the file; usually expressed as a percentage. Also, **hit ratio.**
hit theory *Radiology.* a theory that explains the biological effects of radiation on regions of varying sensitivity. Also, TARGET THEORY.
Hittorf, Johann Wilhelm 1824–1914, German physicist; pioneer in the study of cathode rays.
Hittorf cell *Physical Chemistry.* a special three-section electrolytic cell used in the Hittorf method; current passing through the cell causes ions to migrate between the anode compartment and the cathode compartment via a central compartment that does not experience net change. Also, **Hittorf transference cell.**
Hittorf method *Physical Chemistry.* a technique for calculating the fraction of total current carried by each ion of an electrolyte; based on the principle that if a current of known strength is passed through a cell, the change in the amount of electrolyte will indicate the number of ions gained or lost in the process of carrying current.
Hittorf number see TRANSFERENCE NUMBER.
Hittorf principle *Electronics.* the principle that discharge between electrodes in a gas at a specified pressure does not necessarily occur between the closest points of the electrodes if the distance between the points is less than the minimum required for spark potential.
HIV *Virology.* a human T-cell leukemia/lymphoma virus of the subfamily Lentivirinae that is the causative agent of the disease AIDS; it exhibits a selective affinity for helper T-cells; the in vitro infection results in cytopathic effects and cell lysis. (An abbreviation of human immunodeficiency virus.)
hives see URTICARIA.
HIVIP *Robotics.* a robot made by Hitachi for assembly work; it can match drawings with component shapes using three-dimensional visual sensors.
HIV-positive *Immunology.* having tested positive for the presence of the HIV virus in one's system.
HIV virus see HIV.
HK Hefner candle.
HKG *Aviation.* the airport code for Hong Kong, China.
hl or **hl.** hectoliter; hectoliters.
HLA *Immunology.* the major histocompatibility antigen occurring on human nucleated cells, including lymphocytes. (An abbreviation of human leukocyte antigen.)
hm or **hm.** hectometer; hectometers.
Hm manifest hypothermia.
HMG-CoA reductase *Endocrinology.* a rate-limiting enzyme in the biosynthesis of cholesterol. Also, 3-HYDROXY-3-METHYLGLUTARYL-COA REDUCTASE.
HMG proteins see HIGH-MOBILITY GROUP PROTEIN.
HMO health maintenance organization.
H Monel *Metallurgy.* a cast nickel-base alloy that contains 30% copper and 3% silicon.
HND *Aviation.* the airport code for Haneda, Tokyo, Japan.
HNL *Aviation.* the airport code for Honolulu, Hawaii.
hnRNA *Molecular Biology.* heterogeneous nuclear RNA, the primary RNA transcripts found in the nucleus of a eukaryotic cell.
Ho the chemical symbol for holmium.
H_2O the chemical formula for water.
hoar crystal *Hydrology.* any of the ice crystals that form (hoar)frost.
hoard *Archaeology.* a hidden group of artifacts, especially valuable or scarce items such as ceremonial objects or food stores.
hoarding *Behavior.* the act of storing food or other items in an animal's home or territory, for use when the natural supply is insufficient. *Psychology.* the practice of collecting a large number of similar objects of little practical value.
Hoarella *Invertebrate Zoology.* in the family Eimeriidae, a small genus of typical coccidians.
hoarfrost or **hoar frost** *Hydrology.* see FROST.
hoarseness *Medicine.* a harsh, grating quality of the voice possibly caused by an abnormal condition of the larynx such as swelling, infection, tumors, or foreign bodies.
hoary *Botany.* of leaves, having a gray or whitish color.
hob *Building Engineering.* a level projection such as a bracket in a fireplace used to warm cookware. *Mechanical Devices.* a rotary cutting tool used for cutting the teeth of worm wheels or gear wheels. *Design Engineering.* a master tap fabricated from hardened steel and used to form the shape of a plastic mold into a block of soft steel.
Hoba meteorite *Astronomy.* a 66-ton iron meteorite found in 1920 in Namibia; the world's largest known meteorite, it still lies where it fell.
hobbing *Mechanical Engineering.* the use of a hobbing machine for cutting teeth on gear blanks. *Design Engineering.* the process of forming multiple mold cavities by pressing a hob into soft steel.
hobbing machine *Mechanical Engineering.* a machine equipped with a hob having a spiral thread for cutting gear teeth; used in the production of spur, worm, and helical gears.
hobbing steel *Metallurgy.* any of several steels suitable for hobbing.
hobnail *Mechanical Devices.* a short, sharp nail with a large head; used for studding the soles of work shoes.
HOBOS *Ordnance.* a system designed to transform certain standard unpowered bombs into accurate guided weapons; it consists of a forward guidance assembly, an interconnect assembly that includes the bomb, and an aft control section that includes the autopilot. (An acronym for homing bomb system.)
hod *Mechanical Devices.* a portable trough fitted with a handle and used for carrying bricks or mortar over the shoulder.
Hodge's theorem *Mathematics.* there exists a unique harmonic form α of degree p having arbitrarily given periods on b_p independent p-cycles of a compact orientable Riemannian manifold. That is, each deRham cohomology class of p-forms on a compact orientable Riemannian manifold has a unique harmonic representative.
Hodgkin, Sir Alan Lloyd born 1914, English chemist; shared Nobel Prize for electrochemical analysis of nerve impulses.
Hodgkin, Dorothy Crowfoot born 1910, English X-ray crystallographer; Nobel Prize for determining molecular structure of vitamin B_{12}.
Hodgkin, Thomas 1798–1866, English physician; described the cancer of lymph nodes now known as Hodgkin's disease.
Hodgkin's cells see REED-STERNBERG CELLS.
Hodgkin's disease *Oncology.* a malignant lymphoma characterized by a painless, progressive enlargement of the lymph nodes, spleen, and other lymphoid tissue, often accompanied by symptoms such as anorexia, fever, anemia, and night sweats. Also, **Hodgkin's lymphoma.**
hodgkinsonite *Mineralogy.* $Mn^{+2}Zn_2(SiO_4)(OH)_2$, a bright pink to reddish-brown, brittle, monoclinic mineral occurring as pyramidal or prismatic crystals, having a specific gravity of 3.91 to 4.01 and a hardness of 4.5 to 5 on the Mohs scale; found with barite and calcite at Franklin, New Jersey.
Hodgson number *Chemical Engineering.* a procedure for predicting the metering error during pulsating gas flow, in which a surge tank is located between the pulsation source and the meter.
hodograph *Mechanics.* a plot of a curve that is traced by the tip of a vector whose magnitude and direction change in time.
hodograph method *Fluid Mechanics.* a technique used in studying two-dimensional fluid flow, whereby the Cartesian or polar components of velocity instead of the position coordinates are taken to be independent variables.

hodophobia *Psychology.* an irrational fear of traveling. (From *hodos,* a Greek word meaning "path.")

hodoscope *Nucleonics.* a network of measuring instruments, such as miniature Geiger counters or scintillation counters, that track the paths of highly charged particles in a magnetic field; used to study cosmic radiation or to study particles in accelerator experiments.

Hodotermitidae *Invertebrate Zoology.* a family of primitive harvester termites, social insects in the order Isoptera, with very long antennae.

hoe *Mechanical Devices.* a thin, broad, flat-bladed tool set on a long handle, used for loosening soil and chopping weeds. *Medicine.* a dental instrument having its cutting edge at a right angle to the blade axis; used to break down tooth structures damaged by caries.

höernesite *Mineralogy.* $Mg_3(AsO_4)_2 \cdot 8H_2O$, a white monoclinic mineral of the vivianite group, forming prismatic crystals that resemble those of gypsum, having a specific gravity of 2.57 to 2.6 and a hardness of 1 on the Mohs scale. Also, HÖRNESITE.

hoe shovel see BACKHOE.

Hofbauer cell *Histology.* a large, presumably phagocytic cell present in the chorionic villi of a placenta.

Hoffmann, Roald born 1937, American chemist; shared Nobel Prize for use of quantum model to predict course of chemical reactions.

Hoffman's nucleation theory *Materials Science.* a model explaining the kinetics of polymer crystallization; chain folding and lamellar formation are assumed to be kinetically controlled and the resulting crystal is metastable.

Hofmann amine separation *Organic Chemistry.* a technique used to separate a mixture of primary, secondary, and tertiary amines by heating with ethyl oxalate and distilling. (Named for August Wilhelm von *Hofmann,* 1818–1892, German chemist.)

Hofmann exhaustive methylation reaction *Organic Chemistry.* the thermal decomposition of a quaternary ammonium hydroxide to a terminal alkene (mostly) and water.

Hofmann mustard-oil reaction *Organic Chemistry.* the reaction of a primary amine, mercuric chloride, and carbon disulfide to produce an alkyl isothiocyanate.

Hofmann reaction *Organic Chemistry.* a degradation reaction used to produce an amine from an amide by the addition of a halogen (usually bromine) and sodium hydroxide; used in the production of nylon. Also, **Hofmann degradation.**

Hofmann rearrangement *Organic Chemistry.* the thermal rearrangement of an *N*-alkylaniline hydrohalide into aminoalkylbenzene. Also, **Hofmann-Martius rearrangement.**

Hofmann rule *Organic Chemistry.* a rule stating that when a quaternary ammonium hydroxide containing various primary alkyl radicals is decomposed, it will tend to form the least-substituted alkene.

Hofmeister, Wilhelm Friedrich Benedict 1824–1877, German botanist; pioneer in plant embryology and morphology.

Hofmeister series *Chemistry.* an arrangement of anions and cations in decreasing order of their ability to coagulate lyophilic sols.

Hofstadter, Robert born 1915, American physicist; awarded Nobel Prize for his work in electron scattering and the structure of nucleons.

hog *Agriculture.* any swine, especially a male.

hogback *Geology.* a narrow, sharp-crested, nearly symmetrical ridge formed by an outcrop of steeply inclined, resistant sedimentary strata as a result of differential erosion, and resembling the back of a hog.

hogback

högbomite *Mineralogy.* $(Mg,Fe^{+2})_2(Al,Ti)_5O_{10}$, a black, opaque, metallic, hexagonal, and trigonal mineral occurring in minute grains and rarely as tabular crystals, having a specific gravity of 3.81 and a hardness of 6.5 on the Mohs scale; found in emery deposits. Also, **hoegbomite.**

hog cholera *Veterinary Medicine.* a highly infectious viral disease of swine manifested by high fever, diarrhea, dyspnoea, nervous symptoms, conjunctivitis, and petechiae as well as ecchymosis of the skin and enlarged hemorrhagic lymph nodes. Also, SWINE FEVER.

Hogg, Helen S. born 1905, American-born Canadian astronomer; catalogued and discovered variable stars in globular clusters.

hogging *Engineering.* the process of making heavy cuts using metal machine tools. *Naval Architecture.* **1.** structural drooping at the ends of a hull; a ship that is supported by a wave amidships but not supported on the ends is subjected to the hogging stress. **2.** the process of scraping barnacles off the hull of a vessel at sea. *Aviation.* the drooping of the bow and stern on an airship.

Hogness box see TATA BOX.

hohmannite *Mineralogy.* $Fe_2^{+3}(SO_4)_2(OH)_2 \cdot 7H_2O$, a chestnut brown to burnt orange to amaranth red triclinic mineral occurring as granular aggregates, having a specific gravity of 2.2 and a hardness of 3 on the Mohs scale.

Hohmann orbit *Space Technology.* an elliptical trajectory tangent to planetary orbits and used to transfer spacecraft from one planetary orbit to another with a minimum of energy. The trajectory that intersects two orbits with the least angle generally requires the least energy. Also, **Hohmann trajectory.** (Named for Walter *Hohmann,* 1880–1945, German engineer.)

hoist *Mechanical Engineering.* **1.** any manual or powered device that is designed to raise and lower a load, generally using a block and tackle and often using a boom; for example, a crane or derrick. **2.** to move a load with such a device. Thus, **hoisting.**

hoist back-out switch *Mechanical Engineering.* a protective mechanism allowing a hoist to operate only in reverse in the case of overwind.

hoist hook *Mechanical Devices.* a hook set on a swivel and attached to a cable; used for lifting heavy loads.

hoistman *Mechanical Engineering.* a person who operates the machinery used to raise and lower heavy items such as the cargo of a ship or instruments used in a well.

hoist overspeed device *Mechanical Engineering.* a device that activates an emergency brake when a hoist operates at a speed greater than the predetermined or allowable speed. Similarly, **hoist overwind device.**

hoist slack-brake switch *Mechanical Engineering.* a device that automatically cuts off power to a hoist and sets the brake if the brake lining requires repair or the rigging must be retightened.

hoist tower *Building Engineering.* a temporary structure used to move materials during construction, sometimes made of scaffolding.

hoja blanca *Plant Pathology.* a major viral disease of rice plants, especially in Venezuela and Cuba.

Hokkaido *Geography.* the northernmost and second largest of Japan's major islands (area: about 30,200 sq. mi.).

hol- a combining form meaning "whole," as in *holistic.*

holandric *Genetics.* inherited exclusively through male descent; transmitted through genes located on the Y chromosome.

holandric genes see Y-LINKAGE.

Holarctic region *Ecology.* a zoogeographical region that includes all the Americas, and that served as a major center of evolution producing a number of unique species, such as the beaver, raccoon, grouse, and turkey. It is usually divided into the Nearctic region, which extends north from the Mexican Plateau to the Arctic Circle, and the Neotropical region, which extends south from the Mexican Plateau.

holaspis *Invertebrate Zoology.* the final larval stage of trilobite development.

Holaster *Paleontology.* an extinct genus of heart-shaped Mesozoic echinoderms in the order Holasteroida; Cretaceous.

Holasteridae *Invertebrate Zoology.* a small family of oval or heart-shaped, deep-water irregular sea urchins in the order Holasteroida.

Holasteroida *Invertebrate Zoology.* an order of irregular echinoids (sea urchins) in the subclass Euechinoidea; often oval with a thin delicate test.

Holbach, Baron d' 1723–1789, French man of letters; wrote and compiled scientific articles for Diderot and D'Alembert's *Encyclopedie.*

holcodont *Vertebrate Zoology.* an animal that has teeth attached in a long, continuous groove.

hold any action that is thought of as comparable to grasping something or keeping it in place; specific uses include: *Mechanical Engineering.* a temporary pause in a machine's motion, especially one that continues until the machine is restarted by the operator. *Telecommunications.* **1.** a status in a telephone system in which a calling party remains on the line but is not directly connected for speaking. Also, **on hold. 2.** a button or other feature that permits calls to be placed in this status. *Transportation Engineering.* **1.** an air traffic control restriction on the forward progress of an aircraft. An aircraft may be held on the ground, in one control area along its flight path, or in a landing pattern at its arrival airport. **2.** of an aircraft, to maintain such a holding pattern until given further clearance. *Naval Architecture.* the area in a vessel where the cargo is stored. *Aviation.* a similar storage compartment under or above the floor of an aircraft. *Military Science.* **1.** to maintain or retain possession of an area or position by force. **2.** in an attack, to prevent movement or redisposition of enemy forces. Similarly, **holding attack.** *Computer Programming.* to retain data for future use in one storage medium or location after copying it to another medium or location. *Electronics.* **1.** in charge-storage tubes, to maintain the equilibrium potential by means of electron bombardment. **2.** see HOLD CONTROL. *Industrial Engineering.* in micromotion studies, the action of supporting an object with one hand while the other hand does work. *Space Technology.* see HOLDING.

holdback *Mechanical Engineering.* a brake on an inclined-belt conveyor that automatically stops the loaded belt from running downward in the case of power failure. *Graphic Arts.* the process of reducing the amount of light reaching certain areas of photographic paper during the printing process; usually achieved with paper masks or by briefly placing the hand between the light source and the paper.

hold control *Electronics.* in manually adjusted television sets, a control for adjusting the frequency of the vertical or horizontal oscillator to prevent vertical or horizontal rolling of the picture. Also, HOLD.

holdenite *Mineralogy.* $(Mn^{+2},Mg)_6Zn_3(AsO_4)_2(SiO_4)(OH)_8$, a pink to red orthorhombic mineral occurring in thick tabular crystals, and having a specific gravity of 4.11 and a hardness of 4 on the Mohs scale.

Hölder, (Ludwig) Otto 1859–1937, German mathematician; devised treatment of divergent series with arithmetical summations.

Hölder continuous *Mathematics.* **1.** a function f is said to be Hölder continuous of order p at a point x_0 if there exists a constant M such that $|f(x) - f(x_0)| \leq M|x - x_0|^p$ for all x in some neighborhood of x_0. Also, **Hölder condition**. **2.** a function f is said to be **uniformly Hölder continuous** of order p in an interval (or region) if there exists a constant M such that $|f(x) - f(y)| \leq M\,|x - y|^p$ for all x and y in the interval (or region). **3.** suppose $\lim_{x\to\infty} f(x)$ exists and is denoted $f(\infty)$. f is Hölder continuous of order p at infinity if a constant M exists such that, for $|x| \geq M$, $|f(x) - f(\infty)| \leq M|x|^{-p}$.

Hölder's inequality *Mathematics.* the theorem that if p and q are real numbers greater than 1 such that $1/p + 1/q = 1$, then for real functions f and g,

$$|\int f(x)g(x)dx| \leq (\int |f(x)|^p dx)^{1/p} (\int |g(x)|^q dx)^{1/q};$$

that is, if $f \in L_p$ and if $g \in L_q$, then $fg \in L_1$ and $\|fg\| \leq \|f\|_p \cdot \|g\|_q$.

Hölder summation *Mathematics.* a method of assigning a value to a divergent series $\sum_{i=1}^{\infty} a_i$. Let $s_{0,n} = \sum_{i=1}^{n} a_i$ denote the nth partial sum; $s_{1,n} = (1/n) \sum_{i=1}^{n} s_{0,i}$; etc., with $s_{k,n} = (1/n) \sum_{i=1}^{n} s_{k-1,i}$. That is, at each stage, the new terms are defined in terms of averages of the old terms. The Hölder sum of the series $\sum_{i=1}^{\infty} a_i$ is $\lim_{n\to\infty} s_{K,n}$, where K is the least value of k for which the limit exists. If $K = 1$, the Hölder sum is the same as the Cesàro sum.

hold facility *Computer Technology.* the ability of a computer to retain the values of variables when the processing is interrupted.

holdfast *Botany.* **1.** a single-celled or multicellular suckerlike organ other than a root that attaches an alga to the substrate. **2.** a disklike structure on the tendrils of various plants that attaches the plant to a flat surface.

holding *Space Technology.* the process of halting a countdown until a defect or impediment has been cleared and the countdown can be resumed at the same point, as in "T minus 20 and holding."

holding anode *Electronics.* the auxiliary electrode in a mercury-arc rectifier.

holding beam *Electronics.* the electron beam that regenerates replacement charges for those stored on and lost from the dielectric surface in an electrostatic storage tube.

holding circuit *Electronics.* an alternate circuit that maintains sufficient current in an electromechanical relay winding to keep the relay energized after the initial current has ceased. Also, LOCKING CIRCUIT.

holding coil *Electronics.* an extra coil associated with the holding circuit in an electromechanical relay.

holding current *Electronics.* **1.** the minimum current necessary to maintain a switching device in a closed or conducting state after it is energized or triggered. **2.** the minimum current necessary to maintain ionization in a gas tube.

holding fix *Navigation.* a specified point, identified by radio aids or a landmark, that a pilot uses as a reference point in executing a hold.

holding-flush jar procedure *Microbiology.* a process for minimizing the oxygen exposure of anaerobic plates prior to incubation by temporarily placing them in an anaerobic jar and passing a stream of nitrogen or carbon dioxide through the jar.

holding furnace *Metallurgy.* a furnace in which a molten metal or alloy is held until it is time to cast it; it is not a melting furnace.

holding ground *Navigation.* the character of the bottom as it relates to the holding power of an anchor.

holding pattern *Aviation.* a circular or oval course flown by aircraft waiting to land at a given airport. Similarly, **holding procedure.**

holding point *Aviation.* a geographical reference point used as a reference point for a holding pattern.

holding time *Telecommunications.* **1.** the total amount of time that a trunk or channel is in use. **2.** the time that a telephone call occupies equipment, measured from the time a demand for service is initiated until restoration to the idle state.

hold lamp *Electricity.* an indicating lamp designed to remain lighted while a telephone connection is on hold.

hold mode *Computer Technology.* a state in which an analog computer operation is interrupted and all variables retain their values.

hold-over command *Computer Programming.* a command that is entered at the end of a record to indicate that it is to be continued to the next record.

hold queue *Computer Programming.* a queue of jobs waiting to be run in a large computer system.

holdup *Chemical Engineering.* a term for the liquid suspended in a vertical process line or vessel by upflowing vapor or gas streams.

hole *Aviation.* an air pocket, especially one that causes a flight vehicle to drop suddenly. *Solid-State Physics.* an energy level near the top of an energy band that is not occupied by an electron in a solid; this void can move through the lattice in the manner of a positively charged electron.

hole burning *Spectroscopy.* a spectroscopic technique in which two laser beams are used to observe very narrow spectral linewidths by temporarily removing ions and molecules embedded in crystalline solids from their absorption levels and observing the resulting dip in absorption. Also, **hole-burning spectroscopy.**

hole conduction *Electronics.* in a semiconductor material, the electrical conduction that occurs when electrons under the influence of an applied voltage move into holes and thereby create new holes; the apparent movement of new holes is toward the more negative terminal and is thus equivalent to a flow of positive charges toward that terminal.

hole content *Materials Science.* a term for a condition in which a semiconductor contains a relatively large number of electron holes as a result of the use of a large amount of n-type dopant.

Holectypidae *Paleontology.* a family of irregular echinoids in the order Holectypoida; Lower Jurassic to Uppermost Cretaceous.

Holectypoida *Invertebrate Zoology.* an order of primitive sea urchins in the subclass Euechinoidea, with lantern jaws and keeled teeth.

Holectypus *Paleontology.* a genus of irregular echinoids in the extinct family Holectypidae; Lower Jurassic to Upper Cretaceous.

hole gauge *Mechanical Devices.* a small, spring-action gauge that is used to measure the size of a hole with a micrometer after it is expanded and locked within the hole.

hole injection *Electronics.* the production of mobile holes in a semiconductor by applying an electric charge.

hole mobility *Electronics.* in a semiconductor, the ratio of the hole drift velocity to the intensity of the electric field that causes this motion.

hole saw *Mechanical Devices.* a circular, rotating power saw used with a pilot drill to make large holes in wood, metal, and fiber. Also, CROWN SAW.

hole-through *Mining Engineering.* the meeting of two approaching tunnel heads.

hole trap *Electronics.* an impurity that can cancel holes in a semiconductor by releasing electrons to fill them.

holiday *Engineering.* a portion of a surface that is missed during an operation such as painting or coating. *Ordnance.* a gap in a naval minefield left unintentionally during sweeping or minehunting operations.

holism *Science.* the belief that complex systems may be understood only when viewed in their entirety.

holistic *Science.* of or relating to holism or holistic medicine. Also, WHOLISTIC.

holistic medicine *Medicine.* the theories and practice of considering a patient's entire body, mind, and spirit in the treatment of disease; the application of holism to medical practice. Thus, **holistic health care.**

holistic masks *Computer Technology.* the set of characters stored in an optical character reader that represent, in theory, the exact replicas of all possible input characters.

holistic psychology see HUMANISTIC PSYCHOLOGY.

hollandite *Mineralogy.* $Ba(Mn^{+4},Mn^{+2})_8O_{16}$, a silver-gray to black, pseudotetragonal, monoclinic mineral of the cryptomelane group, usually occurring as fibrous masses, and having a specific gravity of 4.95 and a hardness of 6 on the Mohs scale; found in manganese deposits.

Hollerith, Herman 1860–1929, American mathematician; designed and built a punch-card tabulating machine for the U.S. Census.

Hollerith code *Computer Programming.* a coding scheme for representing alphanumeric characters in punched cards; a combination of one or more holes out of 12 possible positions in a card column represents one character. (Named after Herman *Hollerith.*)

Hollerith string *Computer Programming.* in the Fortran language, a character string constant. Also, **Hollerith constant.**

Holley, Robert William born 1922, American biochemist; Nobel Prize for determining structure and nucleotide sequence of nucleic acids.

Holliday model *Molecular Biology.* a theory proposed to explain the recombinational events that occur between homologous chromosomes, based on a series of chromosomal breakage and rejoining steps.

Hollinacea *Paleontology.* an extinct superfamily of dimorphic, lobate ostracods in the suborder Beyrichicopina; Early Paleozoic.

Hollinidae *Paleontology.* a family of Paleozoic lobate ostracods in the extinct superfamily Hollinacea.

hollow-casting see DRAIN CASTING.

hollow cathode *Electronics.* a discharge tube having a hollow cathode that is closed at one end; the radiation is almost entirely from the cathode glow within the cathode.

hollow-core construction *Building Engineering.* a type of panel construction in which wood faces are bonded to a framing that supports the facing at fixed intervals.

hollow-core door *Building Engineering.* a flush door made entirely of low-density, kiln-dried wood such as ponderosa pine.

hollow drill *Mechanical Devices.* a drill rod having an axial hole that allows bored materials to be removed when passed with water or air.

hollow-ended *Naval Architecture.* describing a sharp hull with concave lines near the bow and stern, producing confined "hollow" spaces at the ends of the hull; it is associated with fast, fine-lined hull designs.

hollow gravity dam *Civil Engineering.* a reinforced concrete, plain concrete, or masonry dam where the pressure of water is taken on a sloped slab or vault supported by transverse buttresses.

hollow mill *Mechanical Engineering.* a milling device with three or more revolving cutters designed to shape a cylindrical workpiece.

hollow plane *Mechanical Devices.* a plane used for bead and molding smoothing consisting of a hollowed-out cutting iron and face.

hollow point(ed) *Ordnance.* describing a bullet that has a small cavity in the nose to increase its expansion upon hitting the target. Also, OPEN POINT(ED).

hollow punch *Mechanical Devices.* a cylindrically shaped tool tapered to form a cutting edge; used to punch washers in soft materials such as leather or rubber.

hollow reamer *Engineering.* in borehole drilling, a bit used to correct the curve or deviation from the intended path.

hollow-rod drill *Mechanical Engineering.* a drill that has an axial hole through which water or air flows into the drill hole to flush out cuttings. Similarly, **hollow-rod churn drill.** Also, **hollow drill.**

hollow shafting *Mechanical Engineering.* shafting formed from hollow tubing or hollowed-out rods; designed to minimize weight, protect internal shafting, and permit internal support.

hollow tile see TILE, def. 4.

hollow-tile floor *Building Engineering.* a reinforced-concrete floor in which hollow tiles of burnt clay, brick, or concrete are laid on shuttering and then plastered; T-beams are the principal load-bearing members, reinforced with steel bars in the bottom of the span. Also, **hollow-block floor**, **hollow-brick floor**.

hollow wall see CAVITY WALL.

hollow-web girder see BOX GIRDER.

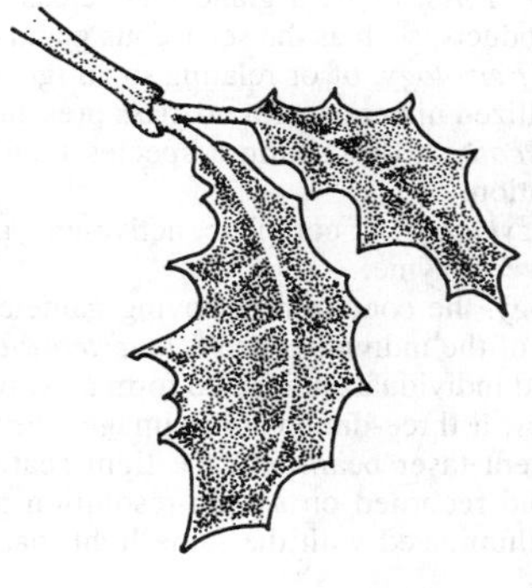

holly

holly *Botany.* any of numerous plants of the genus *Ilex;* evergreen trees or shrubs of the family Aquifoliaceae often cultivated for their ornamental foliage, with glossy spiny leaves and small, bright red berries.

Holmberg radius *Astrophysics.* the photometric radius of a galaxy used in determining the length of its major and minor axes; arbitrarily chosen to lie where its brightness falls to 26.6 magnitudes per square arcsecond.

Holme mud sampler *Engineering.* a device that collects samples of deep-sea sediment, using a rotating scoop that rests on the ocean floor.

Holmes, Arthur 1890–1965, Scottish geologist; pioneered radioactive dating; proposed convection currents as a cause of continental drift.

Holmes, Oliver Wendell 1809–1894, American physician and author; discovered the cause of and cure for puerperal fever.

Holmes scale see LIFE CHANGE SCALE.

holmium *Chemistry.* a rare-earth (lanthamide) metallic element having the symbol Ho, the atomic number 67, an atomic weight of 164.93, a melting point of about 1470°C, and a boiling point of c. 2720°C; used in spectroscopy and electrochemistry. (For *Holmia*, the Latin word for Stockholm, Sweden, where this element was identified.)

holmquistite *Mineralogy.* $Li_2(Mg,Fe^{+2})_3Al_2Si_8O_{22}(OH)_2$, $Mg/(Mg+Fe^{+2}) = 0.1–0.89$, a rare bluish-black, orthorhombic mineral of the amphibole group, dimorphous with clinoholmquistite, occurring in prismatic crystals, and having a specific gravity of 3.06 to 3.13 and a hardness of 5 to 6 on the Mohs scale; found near contacts between lithium-rich pegmatites and basic country rocks.

holo- a combining form meaning "whole," as in *hologram.*

holobenthic *Biology.* living at or near the bottom of the sea during all stages of life.

holoblast *Petrology.* a mineral that is newly grown as a result of metamorphism.

holobranch *Vertebrate Zoology.* in fishes, an "entire gill" or gill arch that has respiratory filaments on both sides of the branchial arch.

holocarpic *Botany.* **1.** of a fungus, having the entire thallus mature into a fruiting body or sporangium. **2.** lacking rhizoids and hausteria. Also, **holocarpous.**

holocellulose *Biochemistry.* the entire polysaccharide constituent of a fibrous substance, such as wood or straw.

Holocene *Geology.* **1.** the geologic epoch of the Quaternary period extending from the end of the Pleistocene to the present. **2.** referring to the rocks and deposits formed during that time. Also, RECENT.

holocentric *Genetics.* of a chromosome, containing a diffuse centromere, rather than a discrete centromere located at a definite spot on the chromosome.

Holocentridae *Vertebrate Zoology.* the squirrelfishes and soldierfishes, a family of mostly nocturnal, brightly colored fishes of the suborder Berycoidei, characterized by sharp spines in their fins and gill cover; found near coral reefs and rocky areas in tropical and warm temperate seas.

Holocephali *Vertebrate Zoology.* the chimaeras, a subclass of cartilaginous benthic fishes of the Chrondichthyes, characterized by a lack of vertebrae in the vertebral column; found worldwide, except in polar regions.

holoclastic *Petrology.* of or relating to sedimentary clastic rock.

holocoen *Ecology.* an entire environment, composed of its biocoen (living elements) and abiocoen (nonliving elements).

holocoenosis *Ecology.* the overall effect produced on a living organism by all aspects of its environment. Thus, **holocoenotic.**

holocrine gland *Physiology.* a gland that releases entire cells laden with secretory products, such as the sebaceous glands in the skin.
holocrystalline *Petrology.* of or relating to an igneous rock consisting entirely of crystallized minerals with no glass present.
holoendemic *Ecology.* describing a species that has a limited geographical distribution.
holoenzyme *Enzymology.* a complete, active enzyme consisting of an apoenzyme and a coenzyme.
hologamy *Biology.* the condition of having gametes of similar size to the somatic cells of the individual itself. *Invertebrate Zoology.* in protozoa, two full-sized individuals uniting to form a zygote.
hologram *Optics.* a three-dimensional image formed by interference between a coherent laser beam and the light scattered by the object being imaged, and recorded on a high-resolution photographic plate; viewable when illuminated with the same light that formed the image. Also, **holograph.**
holographic *Optics.* relating to or displayed by means of a holograph.
holographic interferometry *Optics.* the study of interference fringe patterns that occur when a wave that was generated previously and stored in a hologram is reconstructed and made to interfere with a comparison wave.
holographic memory *Computer Technology.* a memory technology that stores data in the form of holographic images on the surface.
holographic optical element *Optics.* a hologram designed and used to regulate transmitted light beams, rather than to display a three-dimensional image.
holography *Optics.* the techniques and practice of making holograms.
holohedral *Crystallography.* of or relating to crystals that have the full number of planes for maximum symmetry. Thus, **holohedral crystal.**
holohyaline *Petrology.* of or relating to igneous rock consisting entirely of glass.
Holometabola *Invertebrate Zoology.* a division of insect orders in the subclass Pterygota that undergo complete metamorphosis through the stages of egg, larva, pupa, and adult.
holometabolous metamorphosis *Entomology.* complete metamorphosis involving four stages: the egg, larva, pupa, and imago (adult).
holomicrography *Optics.* the technique of producing holograms through the specialized use of various kinds of microscopes.
holomictic lake *Hydrology.* a lake that undergoes a complete mixing of its waters throughout its entire depth during periods of circulation or overturn.
holomorph *Mycology.* a fungus viewed as a whole, including all of its phases (both sexual and asexual).
holomorphic *Mathematics.* analytic; the term usually refers to complex-valued functions of a complex variable. For complex mappings of higher degree, the term implies the analyticity of each dependent component with respect to the independent variables jointly. The term *holomorphic* is used in preference to *analytic* wherever the global properties of analyticity or its independence of particular coordinates are being emphasized.
holonephros see ARCHINEPHROS.
holonomic constraint *Mechanics.* in classical mechanics, a constraint given by generalized forces that are gradients of potential functions depending only on particle position; expressed as equations of constraints that impose relations between only the generalized coordinates of a mechanical system, but do not involve their differentials.
holonomic system *Mechanics.* a mechanical system in which all the constraints are holonomic, so that the system's configuration may be described completely by a set of independent generalized coordinates.
holonomy *Mathematics.* if, following some rule, tangent vectors are transported parallel to themselves along a curve on a manifold M, the rule is said to exhibit holonomy. For example, a sphere that rolls without slipping on the plane induces parallel transport along the curve on the sphere that traced the sphere's contact with the plane. Such a no-slip condition is an example of a holonomic constraint in mechanics. Each curve that starts and ends at a point x_0 of M induces a linear transformation of the tangent space at x_0. These transformations form a group, called the **holonomy group,** that contains information about the local and global properties of M.
holophotal *Optics.* **1.** having the ability to reflect or refract, without appreciable loss, almost all the rays of light emitted from a source. **2.** referring to or describing a holophote.
holophote *Optics.* a system of lenses or mirrors that gathers virtually all the light emitted from a single source and propagates it in one direction; used in lighthouse lamps.
holophyletic *Evolution.* of or relating to a direct line of descent from a single ancestral species or taxonomic group.
holophyly *Evolution.* a direct line of descent from a single ancestral species or taxonomic group.
holophyte *Botany.* **1.** a green plant that synthesizes organic compounds from carbon dioxide, water, and mineral salts by means of light absorbed by chlorophyll. **2.** a plant in which the entire, above-ground portion is dispersed.
holophytic nutrition *Botany.* the production of food by the process of photosynthesis, occurring in plants that are not parasitic, saprophytic, or phagocytic.
holoplankton *Zoology.* free-floating marine and freshwater organisms that complete their life cycle as they are carried by the surrounding water.
Holoptychidae *Paleontology.* a family of crossopterygian fishes in the extinct suborder Rhipidistia and superfamily Holoptychoidea; notable for internal nostrils and other similarities to the modern amphibian urodeles (salamanders and newts); extant in the Upper Devonian and Lower Carboniferous.
holopulping process *Chemical Engineering.* a process of making paper pulp without using sulfur compounds, in which delignification of wood fiber is obtained by alkaline oxidation of particularly thin wood chips at low pressure and temperature, followed by solubilization of the lignin portion.
holorhinal *Vertebrate Zoology.* describing a bird having a rounded anterior margin on the nasal bones.
Holospora *Bacteriology.* a genus of Gram-negative bacteria that occur as obligate parasites in species of *Paramecium.*
Holostei *Vertebrate Zoology.* a superorder of lower bony fishes of medium to large size, found in marine and freshwater habitats, having four extinct orders and two living orders including the gars and bowfins.
Holostomata *Invertebrate Zoology.* a group of families of parasitic trematode flatworms in the order Digenea, with an anterior oral sucker and a posterior ventral adhesive disk.
holostome *Invertebrate Zoology.* **1.** a member of the trematode group Holostomata. **2.** of or relating to this group.
holostratotype *Geology.* the specific sequence of strata originally designated at the establishment of a time-stratigraphic unit.
holosymmetry *Crystallography.* of a crystal structure, having the highest point group symmetry in a crystal class. Thus, **holosymmetric(al).**
holothurian *Invertebrate Zoology.* belonging to the Holothuroidea.
Holothuriidae *Invertebrate Zoology.* a family of sea cucumbers of the order Aspidochirotida, having long, slender tentacular ampullae and one gonad cluster on the left side of the dorsal mesentery.
Holothuroidea *Invertebrate Zoology.* the sea cucumbers, a class of echinoderms with a cylindrical, elastic, leathery body and oral tentacles.
Holothyridae *Invertebrate Zoology.* a family of large mites in the order Acarina.
holotonia *Neurology.* a muscular spasm that affects the entire body. Thus, **holotonic.**
holotopy *Anatomy.* the position of an organ in relation to the body as a whole.
Holotrichia *Invertebrate Zoology.* a large subclass of primitive protozoans with uniform ciliation covering the entire body.
holotype see TYPE SPECIES.
holozoic *Zoology.* obtaining nourishment by ingesting solid food particles in the manner characteristic of animals.
Holstein *Agriculture.* a common breed of dairy cattle having black-and-white markings; they are the largest milk producers of all dairy breeds. Also, **Holstein-Friesian.** (From *Holstein*, a region in Germany, and *Friesland*, a province in Holland, where the breed was developed.)
holster *Ordnance.* a carrying case designed to hold a handgun, allowing the gun to be easily removed; it is usually worn on a belt or a shoulder harness.
Holt-Oram syndrome *Medicine.* a heart disease, of autosomal dominant inheritance, marked by varying degrees of atrial or ventricular septal defect in association with skeletal deformities such as hypoplastic thumb and short forearm. It is of autosomal dominant inheritance. Also, HEART-HAND SYNDROME.
Holtz machine see TOEPLER-HOLTZ MACHINE.
Holuridae *Paleontology.* a family of chondrostean fishes in the suborder Palaeoniscoidea; Cretaceous.
Holywood, John died 1250, English mathematician; wrote astronomy and arithmetic texts; was highly influential in Europe's acceptance of Arabic numbers.

Hom *Mathematics.* the collection of morphisms between two given objects of a category. For example, if V and W are two vector spaces, then $\mathrm{Hom}(V, W)$ is the set of all vector space homomorphisms of V into W and is itself a vector space.

hom- a combining form meaning "same," as in *homaxial.*

Homacodontidae *Paleontology.* the earliest and most generalized artiodactyls, an extinct family of paleodonts belonging to the suborder Dichobunoidea; Eocene.

Homalopteridae *Vertebrate Zoology.* the hillstream loaches, a small family of freshwater fish in the suborder Cyprinoidei, having a flattened forebody and laterally espanded pelvic fins that form a suction disc on the smooth underside. They are found in fast-moving freshwater streams of Southeast Asia, where they cling to rocks and feed on algae.

Homalorhagae *Invertebrate Zoology.* a suborder of tiny marine animals in the class Kinorhyncha.

Homalozoa *Invertebrate Zoology.* a subphylum of echinoderms with bilateral rather than radial symmetry.

Homans' sign *Medicine.* a mild pain in back of the knee or calf when the foot is turned upward, symptomatic of thrombosis in the veins of the leg. (Named for John *Homans,* 1877–1954, American surgeon.)

Homaridae *Invertebrate Zoology.* the lobsters, a family of marine decapod crustaceans with a long, cylindrical armorlike carapace and pinching claws; found on rocky coasts.

Homarus *Invertebrate Zoology.* a genus of lobsters, decapod crustaceans of the family Homaridae, including the American lobster of the Atlantic coast.

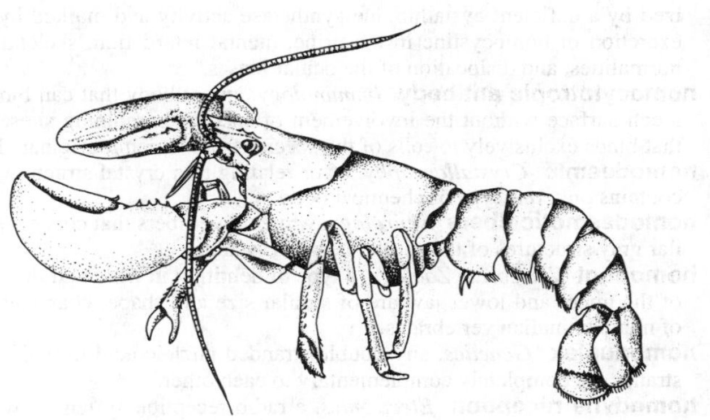

Homarus

homatropine *Organic Chemistry.* $C_{16}H_{21}NO_3$, an alkaloid, the tropine ester of mandelic acid; toxic, hygroscopic white crystals that are soluble in alcohol and slightly soluble in water; melts at 95.5°C; used medicinally, with anticholinergic effects that are similar to but weaker than those of atropine.

home *Behavior.* the particular site that an individual animal regularly returns to and occupies for resting and sleeping when it is not seeking food elsewhere. Also, HOMESITE. *Navigation.* **1.** the location of a craft's usual base of operations. **2.** to steer directly toward a navigational aid, usually by means of radio bearings. *Electronics.* any starting or reference position of an electromechanical device. *Computer Technology.* specifically, the starting position of the cursor on a visual display unit, usually in the upper left-hand corner of the screen. *Electricity.* the starting position of a stepping relay.

home address *Computer Technology.* an address written on a track of a direct-access volume, such as a disk or drum, to identify the track's address relative to the beginning of the volume.

home care *Medicine.* any medical service provided in the patient's residence. Thus, **home-care nurse.**

home computer *Computer Technology.* a personal computer adapted or intended for use in the home for nonprofessional applications.

home loop see LOCAL LOOP.

homeo- a combining form meaning "similar," as in *homeopathy.*

homeoblastic *Petrology.* of or relating to a crystalloblastic texture of metamorphic rocks, in which the sizes of essential constituents are approximately equal.

homeobox *Genetics.* a DNA sequence of about 180 base pairs, occurring near the 3' end of certain homeotic genes; it is structurally similar to certain prokaryotic and eukaryotic DNA-binding proteins.

homeogenetic induction *Cell Biology.* a phenomenon, usually occurring in plant cells, in which a differentiated cell can induce an uncommitted adjacent cell to undergo differentiation.

homeohydric *Ecology.* able to maintain approximately constant water content in the body under variable water conditions in the environment.

homeometric autoregulation *Cardiology.* the intrinsic mechanisms that regulate the degree of ventricular contraction and that are independent of the change in end-diastolic myocardial fiber length.

homeomorph *Crystallography.* a substance that has the same crystal form as another substance but not necessarily the same chemical composition.

homeomorphic *Crystallography.* relating to or characterized by homeomorphism; having the same crystal structure as another substance. *Mathematics.* two topological spaces X and Y are said to be homeomorphic if there exists a homeomorphism $f\colon X \to Y$; intuitively, Y can be derived from X by stretching, shrinking, or twisting.

homeomorphic graphs *Mathematics.* two graphs are homeomorphic if they can both be obtained from the same graph G by inserting zero or more new vertices (of degree two) into one or more edges of G. For example, any two circuit graphs are homeomorphic. Homeomorphism of graphs is an equivalence relation.

homeomorphism *Crystallography.* the fact or condition of having the same crystal form as another substance. *Mathematics.* a one-to-one continuous transformation of a topological space whose inverse is also continuous. Also, BICONTINUOUS FUNCTION, TOPOLOGICAL MAPPING.

home-on-jam *Electronics.* a radar feature that allows angular tracking of a jamming source.

homeopathy *Medicine.* a system of therapy advanced in the late eighteenth century by Dr. Samuel Hahnemann, based on the theory that "like cures like;" if the conditions produced by giving large doses of a drug to a healthy person are similar to conditions occurring as a natural consequence of disease, then that disease may be treated by the same drug in much smaller doses. Thus, **homeopathic, homeopathist.**

homeostasis *Biology.* the ability of an organism to maintain a constant internal environment, such as body temperature or fluid content, by regulating its physiological processes and by making adjustments to the external environment. Thus, **homeostatic.**

homeotherm *Biology.* a warm-blooded animal or organism that maintains a constant body temperature regardless of environmental changes. Also, HOMOTHERM.

homeothermia *Biology.* the condition of being warm-blooded.

homeothermic *Biology.* able to maintain an approximately constant body temperature in the face of fluctuating environmental temperature; warm-blooded. Also, **homeothermal.**

homeothermy *Biology.* the maintenance of constant body temperature despite changes in the environmental temperature.

homeotic mutation *Genetics.* a mutation that causes one body structure to be replaced by a different, but homologous, body structure during development.

homeotoxic *Toxicology.* poisonous to members of the same species as the organism producing the poison.

homeotoxin *Toxicology.* any poison that is toxic to members of the same species that produces the poison. Also, HOMOIOTOXIN, ISOTOXIN.

homeotype see HOMOTYPE.

home position *Robotics.* a predetermined position in a coordinate system or a fixed point on an axis.

homer *Navigation.* an airport that can locate aircraft and guide them in using radio signals.

home range *Behavior.* the general area around an animal's home throughout which it normally travels in search of food; the boundaries of this area may be established, but it is usually not defended against intrusion by others of the same species in the manner of a territory.

home record *Computer Programming.* the first record in a chain produced by the chaining method of file organization.

home signal *Civil Engineering.* a signal, commonly red, at the beginning of a block of railroad track that indicates whether the block is clear.

homesite *Behavior.* see HOME.

hometaxial-base transistor *Electronics.* a transistor in which the emitter and connector junctions are formed by a single-diffusion process in a uniformly doped silicon slice.

homichlophobia *Psychology.* an irrational fear of fog.

homilite *Mineralogy.* $Ca_2(Fe^{+2},Mg)B_2Si_2O_{10}$, a black or blackish-brown monoclinic mineral of the gadolinite group, occurring in tabular crystals, having a specific gravity of 3.36 to 3.38 and a hardness of 5 on the Mohs scale.

homilophobia *Psychology.* **1.** an irrational fear of sermons. **2.** an irrational fear that one's peers will find fault with one's appearance or actions.

homing *Behavior.* the ability of some birds or other animals to return to a specific location, such as a breeding site, often involving travel over long distances to reach the destination. Also, **homing behavior, homing instinct.** *Navigation.* the act of heading directly toward a navigational aid, usually by means of radio bearings. *Military Science.* **1.** a technique by which a missile or other vehicle directs itself, or is directed, toward a source of primary or reflected energy, or to a specified point. **2.** of or relating to a system or weapon using this technique. Thus, **homing guidance, homing mine, homing torpedo.**

homing antenna *Electromagnetism.* a directional receiving antenna used on aircraft that directs the craft to fly directly toward a target that is emitting or reflecting a signal.

homing device *Electronics.* **1.** a transmitter, receiver, or other device on a vehicle, especially an aircraft or missile, that continuously indicates the vehicle's destination or selected target. **2.** a control device, such as a remote-control television tuner, that automatically moves in the correct direction to achieve the desired result.

homing guidance *Engineering.* a matched-wavelength seeking and error-sensing flight-control system that allows a guided missile to guide itself toward an intended target.

homing relay *Electricity.* a stepping relay that returns to its predefined starting position prior to each operating cycle.

homing system *Navigation.* a navigational system that operates by homing in on a beacon, transponder, or other target.

hominid *Anthropology.* any member of the family *Hominidae*, including fossil forms and modern representatives.

Hominidae *Anthropology.* the humans, a monotypic family of the order Primates that includes the genera *Australopithecus* and *Homo;* characterized by opposable thumbs, no tail, and proportionately longer lower limbs reflecting the ability for erect bipedalism; worldwide distribution with one species, *Homo sapiens.*

homininoxious *Toxicology.* harmful to humans.

hominoid *Anthropology.* any member of the superfamily Hominoidea, including fossil forms and modern representatives.

Hominoidea *Vertebrate Zoology.* the superfamily of the Primate order that includes the fossil families, Pliopithecidae and Dryopithcidae, and the living families of apes and humans, the Hylobatidae, Pongidae, Panidae, and Hominidae.

Homo *Anthropology.* a genus of the family Hominidae that includes all modern humans and fossil species except *Australopithecus;* present grades are *H. habilis, H. erectus, H. sapiens neanderthalensis,* and *H. sapiens sapiens.*

homo- a combining form meaning "the same," as in *homogeneity. Immunology.* a prefix referring to reactants or reactions involving a single species, or derived from a similar substance. *Organic Chemistry.* a prefix indicating: **1.** a homolog differing by an increase of one CH_2 group; **2.** a homopolymer consisting of a single type of monomer.

Homobasidiomycetidae *Mycology.* in some classifications, a subclass of fungi belonging to the class Hymenomycetes and an approximate equivalent of Holobasidiomycetidae; it forms a spore-bearing structure termed a holobasidium, characterized by its lack of cross walls or septa.

homoblastic see HOMEOBLASTIC.

homobront see ISOBRONT.

homocarnosine *Biochemistry.* a dipeptide that is normally present in the brain.

homocentric *Science.* having a common center; concentric. *Optics.* describing rays that are parallel or that have the same focal point.

homocercal *Vertebrate Zoology.* of or relating to the tails of most bony fishes, in which the whole of the fin membrane is behind the spinal terminus, with the dorsal and ventral lobes being of equal or roughly equal size.

homochain polymer *Materials Science.* a polymer having only carbon atoms in the main chain; the usual condition of polymers produced by chain reactions rather than by step reactions.

homochlamydeous *Biology.* having a perianth composed of similar parts, each called a tepal.

homochromy *Zoology.* the condition of being of one color.

homochronous *Genetics.* occurring at the same age in successive generations.

homocline *Geology.* a rock unit or series of beds that exhibit a similar dip and strike.

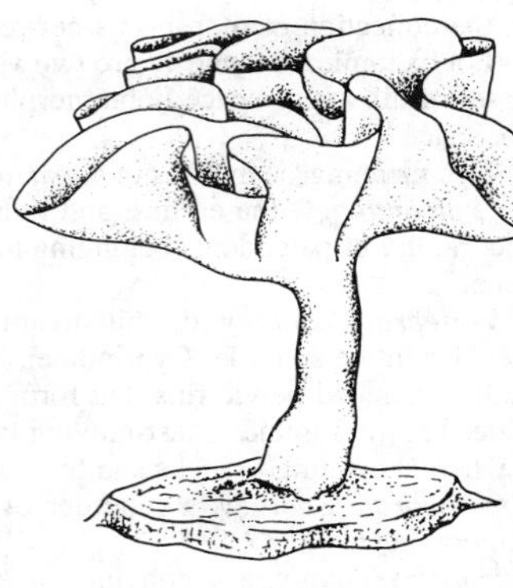

Homocoela

Homocoela *Invertebrate Zoology.* an order of sponges in the class Calcarea; the whole internal cavity is lined with flagella. Also, ASCONOSA.

homocyclic *Chemistry.* describing a closed chain or ring including only atoms of the same element.

homocyclic compound *Organic Chemistry.* a ring compound, such as benzene, that has one type of chemical atom in its structure.

homocysteine *Biochemistry.* a higher homologue of cysteine that contains an additional methylene group in a side chain.

homocystinuria *Medicine.* an inherited metabolic disorder characterized by a deficient cystathionine synthetase activity and marked by the excretion of homocystine in the urine, mental retardation, skeletal abnormalities, and dislocation of the ocular lenses.

homocytotropic antibody *Immunology.* an antibody that can bind to a cell surface without the involvement of specific combining sites, and that binds exclusively to cells of the species from which it originated.

homodesmic *Crystallography.* of or relating to a crystal structure that contains only one type of chemical bond.

homodesmotic fibers *Neurology.* white nerve fibers that connect similar gray structures of the central nervous system.

homodont *Vertebrate Zoology.* a type of dentition in which all the teeth of the upper and lower jaw are of similar size and shape; characteristic of nonmammalian vertebrates.

homoduplex *Genetics.* any double-stranded nucleic acid in which the strands are completely complementary to each other.

homodyne reception *Electronics.* a radio reception system in which the frequency of the signal from a local oscillator is adjusted to or locked into the frequency of the incoming signal to enhance its magnitude and improve reception. Also, ZERO-BEAT RECEPTION.

homoecious *Biology.* **1.** occupying the same host throughout its life cycle. **2.** host-specific.

homoeo- or **homœo-** a combining form meaning "similar," as in *homoeomerous.*

homoeomerous *Botany.* having algae that are evenly distributed throughout a lichen thallus.

homoeotype see HOMOTYPE.

Homo erectus *Anthropology.* **1.** an extinct hominid species with an upright stance, having a sloping forehead, prominent brow, large facial skeleton, and large lower jaw; examples include Java man and Peking man. **2.** any of various fossils of this species found in Europe, Asia, and Africa dating from about 1.6 million to 400,000 years ago. (From Latin for "standing man.")

homoerotism *Psychology.* the force that drives an individual to seek sexual gratification from the same sex. Thus, **homoerotic.**

homofermentation *Microbiology.* a fermentation process that results in the production of a single major end product, primarily lactic acid.

homofermentative lactobacilli *Bacteriology.* bacteria that produce only lactic acid upon fermenting carbohydrates.

homogametic sex *Genetics.* the sex in which both sex chromosomes are homologous and which therefore produce gametes that all carry only one type of sex chromosome. In human beings, the female is the homogametic sex.

homogamous *Biology.* describing a bisexual flower having simultaneous maturation of stamens and stigma.

homogamy *Biology.* positive assortative mating. *Botany.* **1.** the production of flowers of a uniform sexual type. **2.** a condition in which the anthers and stigmas of a flower mature at the same time, facilitating self-pollination.

homogeneity [hō´mō jə nē´ə tē] *Science.* the fact or condition of being homogeneous; being uniform or similar; having the same structure, properties, composition, form, and so on. Also, **homogenicity.**

homogeneity coefficient *Radiology.* the ratio of the first half-value layer to the second half-value layer of a radiation beam.

homogeneous [hō´mō jēn´yəs; hō´mō jē´nē əs] *Science.* consisting of or composed of similar elements or ingredients. *Chemistry.* being uniform in chemical structure or composition. *Physics.* having uniform properties throughout, regardless of position. *Biology.* having a similar structure because of a common origin. Also, HOMOGENOUS. *Petrology.* see HOMEOBLASTIC. *Statistics.* describing the uniformity of different populations or of their samples with respect to all or some of the data from those populations. *Mathematics.* **1.** for a vector space norm $\| \|$, the property that $\|cx\| = |c| \|x\|$ for all scalars c and all vectors x. **2.** for a vector space inner product $< , >$, the property that $<cx,y> = c <x,y>$ for all scalars c and all vectors x and y. **3.** a polynomial that is the sum of monomials, each of degree k, is said to be homogeneous of degree k. Euler's theorem is satisfied by homogeneous polynomials. **4.** more generally, let f be a function defined on an open subset D of R^n such that $f(tx) = t^k f(x)$, for every real number t and $x \in D$. Then f is said to be homogeneous of degree k. **5.** if f is a homogeneous function, then any equation which can be rewritten in the form $f(x) = 0$ is a homogeneous equation.

homogeneous atmosphere see ADIABATIC ATMOSPHERE.

homogeneous catalysis *Chemistry.* a catalytic reaction in which the catalyst and reactants are of the same phase.

homogeneous chemical reaction *Chemistry.* a reaction in which all the components are of the same phase.

homogeneous differential equation *Mathematics.* a differential equation with the property that any scalar multiple of a solution is also a solution. A **first-order homogeneous differential equation** has the form $y' = P(x, y)/Q(x, y)$, where P and Q are homogeneous functions of the same degree.

homogeneous integral equation *Mathematics.* an integral equation with the property that any scalar multiple of a solution is also a solution.

homogeneous line-broadening *Optics.* the broadening of the natural linewidth of a laser, which influences each atom or molecule and, consequently, the entire system, and usually results from a disturbance such as collision or lattice vibration.

homogeneous network *Computer Technology.* a network of similar host computers, such as those of a particular model from a single manufacturer.

homogeneous nucleation *Materials Science.* nucleation in which the material itself provides the atoms to form nuclei.

homogeneous polymerization *Materials Science.* polymerization using conventional homogeneous, free-radical or ionic-initiated addition reactions; used to produce substances very similar to natural rubber.

homogeneous radiation *Physics.* **1.** radiation that has a single frequency. **2.** radiation having a single energy.

homogeneous reactor *Nucleonics.* a nuclear reactor in which the neutrons travel through only one medium, either the fuel or a mixture of the fuel and the moderator.

homogeneous space *Mathematics.* let H be a subgroup of a (Hausdorff) topological group G. The quotient space G/H (composed of left cosets xH of H) is called the (left) homogeneous space of G with respect to H. The homogeneous space G/H is a Hausdorff space if and only if H is a closed subgroup of G.

homogeneous staining region *Molecular Biology.* a region of a genome that is composed of many repeated copies of one small nucleic acid segment.

homogeneous system *Mathematics.* a system of n linear equations in m unknowns which can be represented in matrix form as $AX = 0$, where A is the $m \times n$ matrix of coefficients, X is the m-element column vector representing the unknowns, and 0 is the n-tuple with all entries equal to zero.

homogenesis *Genetics.* reproduction that takes place by the same process in each generation. Thus, **homogenetic.**

homogenization *Science.* the fact or process of homogenizing, especially the process of homogenizing milk.

homogenize [hə mäj´ə nīz´] *Science.* to make uniform or similar; cause to be homogeneous. *Food Technology.* specifically, to reduce the size of fat globules, as in milk or cream, in order to distribute them equally throughout and produce a uniform emulsion. Thus, **homogenized milk.** *Metallurgy.* to heat an alloy for a sufficient time so that segregation is alleviated by diffusion of certain alloying elements.

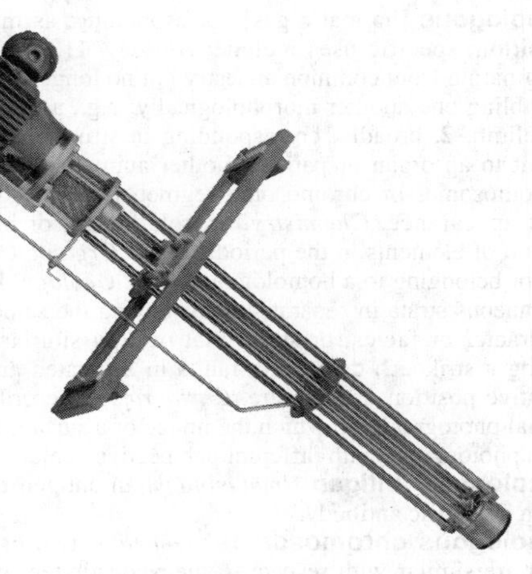

homogenizer

homogenizer *Mechanical Engineering.* a device in which substances are emulsified by being forced through an energetic shear field. Also, **homogenizing mixer.**

homogenous see HOMOGENEOUS.

homogentisase *Enzymology.* an enzyme that catalyzes the conversion of homogentisic acid to fumaryl acetoacetic acid.

homogentisic acid *Biochemistry.* $C_8H_8O_4$, an intermediate in the catabolism of phenylalanine and tyrosine.

homogonous *Botany.* relating to or characterized by homogony.

homogony *Botany.* the condition of having only one type of flower, with each flower having pistils and stamens of a uniform size.

homograft see ALLOGRAFT.

homograft rejection *Immunology.* the rejection by an individual of a tissue transplanted from a donor of the same species.

homograph [häm´ə graf] *Linguistics.* one of two or more words having the same written form but a different origin and a different meaning; some homographs are also different in sound; e.g., a *bass* fish and a *bass* singing voice, but many are not; e.g., *bear* a large animal and *bear* to carry, or a baseball *bat* and *bat* the flying mammal. (From a Greek phrase meaning "same writing.")

Homo habilis *Anthropology.* **1.** the earliest known species of *Homo,* from 2 million to 1.5 million years old, a transitional form between *Australopithecus* and *Homo,* having a cranial capacity between 440 and 680 cubic centimeters; associated with tools of the Oldowan cultural tradition. **2.** a fossil of this species found in Olduvai Gorge. (From a Latin phrase for "handy man;" coined by Louis Leakey.)

homoio- a combining form meaning "similar," as in *homoiopodal.*

homoiomerous *Botany.* describing lichens that have evenly distributed algal and fungal cells.

Homoiostelea *Paleontology.* an extinct class of biradiate homalozoan invertebrates, usually considered echinoderms but thought by some to be calcite-plated chordates with echinoderm affinities; possibly ancestral to later triradiate forms; they appeared already differentiated in the Lower Cambrian and became extinct in the Lower Devonian.

homoiotoxin see HOMEOTOXIN.

Homo kanamensis *Anthropology.* a fossil specimen identified from a jaw fragment found by Louis Leakey in western Kenya in 1932; first identified to be a separate species of *Homo,* but considered to be *Homo sapiens.* Also, KANAM MAN.

homokaryon *Mycology.* a fungal cell containing two or more nuclei that are genetically identical to each other. Also, HOMOCARYON.

homokaryosis *Mycology.* the condition in which a fungal cell contains two or more nuclei that are genetically identical to each other. Thus, **homokaryotic.**

homolateral *Anatomy.* relating to, affecting, or located on the same side. Also, IPSILATERAL.

homolecithal see OLIGOLECITHAL.

homolog see HOMOLOGUE.

homological *Science.* relating to or showing correspondence; homologous.

homological algebra *Mathematics.* the collection of rules and techniques used to extract information from a chain complex; e.g., "diagram chasing."

homologous [hə mälʹə gəs] corresponding, as in structure or relative position; specific uses include: *Biology.* **1.** of anatomical structures, originating from common ancestry but no longer functioning alike or resembling one another morphologically; e.g., a bird's wing and reptile's forelimb. **2.** broadly, corresponding in structure, position, or development to an organ or part in another animal. *Genetics.* of two or more chromosomes or chromosome segments, having the same genetic loci and appearance. *Chemistry.* of, relating to, or belonging to a single group of elements in the periodic table. *Organic Chemistry.* of, relating to, or belonging to a homologous series. *Geology.* **1.** describing contemporaneous strata in separated areas having the same general lithological character or facies, or strata that occupy similar structural positions along a strike. **2.** describing faults in separated areas having the same relative position or structure. *Cartography.* describing the condition in aerial photography in which the image of a surface feature is common to two photographs with different perspective centers.

homologous antigen *Immunology.* an antigen that elicits and reacts with a specific antibody.

homologous chromosomes *Genetics.* two or more chromosomes that are similar with respect to their constitutent genetic loci and their visible structure, and that pair during meiosis.

homologous recombination *Genetics.* the recombination of a piece of DNA into a homologous sequence. Also, GENERAL RECOMBINATION.

homologous series *Organic Chemistry.* a series of organic compounds whose structure differs regularly by some radical, usually CH_2, with each successive member having an additional radical and having a graded change in properties.

homologous serum jaundice *Medicine.* an extremely common viral disease caused by the hepatitis B virus that is endemic worldwide, characterized by fever, malaise, anorexia, and nausea, followed by clinical jaundice. The virus is shed in all body fluids by individuals with acute or chronic infections and by asymptomatic carriers; primary transmission is through blood transfusion, sharing of needles among drug users, and intimate contact. Also, TYPE B VIRAL HEPATITIS.

homologous stimulus *Physiology.* any adequate stimulus.

homologous tumor *Medicine.* a tumor whose tissue is the same type as the tissue in which the tumor develops.

homolographic map projection *Cartography.* another term for equal-area map projection; used to denote some specific projections, such as the recentered version of the Mollweide projection. Also, **homalographic map projection.**

homologue anything that is homologous; specfic uses include: *Biology.* any of a group of homologous structures. *Organic Chemistry.* any member of a homologous series. *Genetics.* either member of a pair of homologous chromosomes. Also, HOMOLOG.

homology [hə mälʹə jē] the condition of being homologous; specific uses include:. *Biology.* the fundamental similarity of structures in different organisms that no longer function in the same manner, but that result from common ancestry; e.g., the flippers of a whale and the forelegs of a land mammal. *Organic Chemistry.* the condition or relationship of members of a homologous series; the similarity of related organic compounds. *Mathematics.* a chain complex together with its associated homology groups.

homology group *Mathematics.* let C be a chain complex of R modules with differentials d_n: $C_n \rightarrow C_{n-1}$. The nth homology group of the complex is $H_n(C) = \text{Ker } d_n/\text{Im } d_{n+1}$. Also, nTH HOMOLOGY.

homology theory *Mathematics.* the study of a mathematical object (e.g., a topological space or an R-module) by means of the homology groups of its associated chain complex.

homolysis *Chemistry.* a symmetrical cleavage of a bond, such that each of the cloven molecules or atoms retains one of the bonding electrons.

homometric pair *Crystallography.* a pair of crystals with different atomic arrangements that produce identical X-ray diffraction patterns. The metal in positions in the mineral bixbyite provides an example.

homomorph *Science.* anything that is homomorphic.

homomorphic having similar form; specific uses include: *Chemistry.* of two or more molecules, having similar size and shape. No other characteristics need be shared. *Botany.* having perfect flowers that consist of only one sexual type. *Genetics.* having chromosome mates of similar size and form during synapsis of the first meiotic division.

homomorphism *Science.* the condition of being homomorphic. *Mathematics.* a transformation T from one algebraic system (group, ring, module, etc.) to another that preserves the operation(s). For example, a ring homomorphism is a mapping ϕ: $R \rightarrow S$, where R and S are rings, such that: (a) $\phi(a + b) = \phi(a) + \phi(b)$ and (b) $\phi(ab) = \phi(a)\phi(b)$ for all $a,b \in R$.

homomorphosis *Biology.* the regenerative replacement of a lost part by a similar part.

Homoneura *Invertebrate Zoology.* a suborder of moths of the insect order Lepidoptera, with nearly identical forewings and hindwings.

homonomous *Biology.* similar in function and structure; developed to a like degree (such as starfish arms).

homonym *Linguistics.* one of two or more words that have the same sound but different spellings and meanings. *Systematics.* a formal taxonomic name inadvertently given to more than one species; the more recent homonyms are considered invalid. A **primary homonym** is a name given to two or more species within the same genus; a **secondary homonym** results when a species is reclassified to a genus in which the name already exists. Thus, **homonymous.**

homonymous hemianopsia *Medicine.* a loss of vision in one-half of the visual field of both eyes, affecting the nasal half of one visual field and the temporal half of the other.

homopause *Geophysics.* an area of the atmosphere about 80 to 90 kilometers above sea level, in which the homosphere ends and the heterosphere begins.

homopetalous *Botany.* a condition in which all petals are identical.

homophobe *Psychology.* **1.** a person who has an irrational fear or hatred of homosexuals. **2.** in popular use, a person who is biased against homosexuals.

homophobia *Psychology.* **1.** an irrational fear or hatred of homosexuals. **2.** in popular use, social prejudice against homosexuals. Thus, **homophobic.**

homophone [hämʹə fōnʹ] *Linguistics.* one of two or more words that have the same sound but different meanings, and often, though not necessarily, different spellings; e.g., *so, sew, sow* or *stair, stare.*

homophonic enciphering *Computer Programming.* an enciphering method that hides statistical features of cryptographic text.

homoplastic *Evolution.* of, relating to, or characterized by homoplasy; denoting organs or parts that resemble one another in structure and function but not in origin and development; e.g., the wings of birds and insects. *Surgery.* of or relating to homoplasty or to an allograft.

homoplastic graft see ALLOGRAFT.

homoplasty *Surgery.* the operative replacement of lost parts or tissues with similar parts from another individual of the same species or from a member of another inbread strain of the same species.

homoplasy *Evolution.* a structural resemblance that is due to parallelism or convergent evolution, rather than to common ancestry.

homoploid *Genetics.* describing a cell or organism whose chromosome set exhibits the same degree of ploidy as another given cell or organism. Thus, **homoploidy.**

homopolar *Electricity.* having equal charge distribution.

homopolar bond *Physical Chemistry.* the sharing of an electron between two atoms in which there is no displacement of the negative and positive charges, so that the dipole moment equals zero.

homopolar crystal *Solid-State Physics.* a crystal in which all atomic bonds are covalent.

homopolar generator *Electronics.* a DC generator having poles of equal polarity with respect to the armature and thus requiring no commutator.

homopolymer *Organic Chemistry.* a polymer that is derived from a single repeated monomer.

homopolymer tailing *Molecular Biology.* the addition of a series of one specific deoxyribonucleotide to the end of a nucleic acid molecule.

homopolysaccharide *Biochemistry.* a polysaccharide consisting of only one type of monosaccharide.

Homoptera *Invertebrate Zoology.* an order of insects having membranous wings and piercing and sucking mouthparts, including many plant pests.

homopteran *Invertebrate Zoology.* belonging or relating to the insect order Homoptera. Also, **homopterous.**

Homo sapiens [hōʹmō sāpʹē ənz] *Anthropology.* the scientific name for human beings, the species of bipedal primates characterized by the use of tools and language. (From Latin for "wise man.")

Homo sapiens neanderthalensis see NEANDERTHAL MAN.

Homo sapiens sapiens *Anthropology.* the modern human subspecies that appeared around 100,000 years ago.

homoscedasticity *Statistics.* an equality of variance across populations, control levels, or other statistical groups.

Homoscleromorpha *Invertebrate Zoology.* a subclass believed to represent an early stage in the evolution of demosponges; contains one order and two families.

Homosclerophorida *Invertebrate Zoology.* an order of primitive sponges in the class Demospongiae, having eight-rayed siliceous spicules.

homoserine *Biochemistry.* $C_4H_9O_3N$, the homologue of serine that contains one methylene group more than serine, produced during the metabolic breakdown of cystathionine to cysteine.

homosexual *Psychology.* a person who is sexually active with or sexually attracted to individuals of the same sex; the term is sometimes restricted to males, with *lesbian* used as the corresponding female term.

homosexuality *Psychology.* the fact or condition of being homosexual; sexual activity or attraction involving people of the same sex.

homosexual panic *Psychology.* a state of anxiety produced by the fear that one is in danger of being sexually assualted by a person of the same sex or by the fear that one is perceived by others as a homosexual.

homosphere *Meteorology.* the lower of two portions of the atmosphere as distinguished by the general homogeneity of their composition (the other portion is the heterosphere); it extends from the surface to an altitude of 80 to 100 kilometers above the earth, closely coinciding with the neutrosphere, and is characterized by uniformity of composition.

homosporous *Botany.* having only one kind of asexually produced spores, as in some ferns, mosses, and algae, giving rise to a gametophyte that usually produces both male and female reproductive organs.

homospory *Botany.* the production of only one type of spore.

homothallic *Botany.* of algae or fungi, having a self-fertile thallus.

homothallism *Biology.* the condition of having only one haploid type that produces gametes that mate with each other; used most often in reference to algae, fungi, and some spores.

Homotherium *Paleontology.* one of the saber-toothed cats, an extinct genus of large carnivores in the family Felidae, distinguished by moderately long canines, very sharp cheek teeth, and disproportionately long forelimbs; widespread in the Pleistocene but there is no known association with man; probably became extinct because its prey (large herbivores) became extinct.

homotherm see HOMEOTHERM.

homothetic center see CENTER OF SIMILITUDE.

homotopy *Mathematics.* let f and g be smooth mappings of the topological spaces X into Y, and let R be the real numbers. f and g are said to be homotopic if a continuous function $H: X \times R \rightarrow Y$ exists such that $H(x, 0) = f(x)$ and $H(x, 1) = g(x)$. If for some point x_0 of X, $f(x_0) = g(x_0)$ and f and g are homotopic, then f and g are said to be homotopic with respect to x_0. H is also called a continuous deformation of f into g. Homotopy induces an equivalence relation on the class of smooth mappings between two particular spaces or manifolds.

homotopy group *Mathematics.* the nth homotopy group $H_n(X)$ on a topological space X is the group of equivalence classes (under homotopy) of mappings of the n-dimensional sphere S^n into X. Two mappings are in the same equivalence class if they are homotopic on S^n.

homotopy theory *Mathematics.* the study of the structure of a topological space by means of its homotopy groups.

homotransplant see ALLOGRAFT.

homotropic *Cell Biology.* attracting cells of a like order. Thus, **homotropism.**

homotropic effect *Biochemistry.* an allosteric interaction between identical ligands, as when ATCase binds *N*-carbamoyl phosphate to aspartate.

homotropic enzyme *Enzymology.* an allosteric enzyme whose allosteric effector is the substrate upon which it acts.

homotropous *Botany.* of a plant, having the radicle bent toward the hilum.

homotype *Systematics.* a specimen of an organism, as described by someone other than the original author, determined to be of the same species as the type specimen. Also, HOMEOTYPE, HOMOEOTYPE.

homovanillic acid *Biochemistry.* $C_9H_{10}O_4$, a product of catecholamine metabolism; elevated urinary levels of this acid are symptomatic of various medical conditions.

homozygote *Genetics.* an individual or cell that is characterized by homozygosity.

homozygous *Genetics.* of an individual or cell, having identical rather than different alleles at a given locus on both homologous chromosomes. Thus, **homozygosis, homozygosity.**

homunculus *Developmental Biology.* a miniature human form, formerly believed to be fully formed in either the sperm or ovum.

hone *Mechanical Engineering.* **1.** to sharpen a cutting tool. **2.** any of various devices and machines used for this purpose, often including one or more abrasive stones.

Honest John *Ordnance.* a nonguided nuclear-armed surface-to-surface rocket that is designed to attack ground forces; officially designated **MGR-1.**

honey *Invertebrate Zoology.* a sweet, viscous fluid consisting primarily of levulose and dextrose that is secreted by the honeybee and is stored in hives as food. *Food Technology.* this substance as used in cooking or as a sweetener.

honey bear see SLOTHBEAR.

honeybee *Invertebrate Zoology.* the common name for members of the group Apoidea, species of which are often domesticated and kept for the commercial production of honey, especially *Apis mellifera* in the hymenopteran family Apidae, a social, usually stinging insect, often with yellow and black stripes.

honeybee

honeycomb *Invertebrate Zoology.* a mass of hexagonal wax cells constructed by honeybees for brood cells and for storing honey. *Metallurgy.* a cellular metallic material resembling a natural honeycomb.

honeycomb coil *Electromagnetism.* a coil whose windings are wound in a cross-hatch pattern in order to reduce the amount of distributed capacitance.

honeycomb coral *Paleontology.* a nickname for Favosites, but applicable as a descriptive term to other hexacorals in the order Tabulata.

honeycomb lung *Medicine.* a lung marked by a spongy or honeycomb appearance as a result of the presence of many small cysts due to widespread fibrosis and cystic dilation of bronchioles.

honeycomb radiator *Mechanical Engineering.* a heat-exchange device that cools the water in an automobile by circulating the water through many small, air-cooled cells.

honeycomb structure *Geology.* a structure in rock or soil characterized by pitting or cavities forming a cell-like pattern, so that the surface resembles a honeycomb. Also, HONEYCOMB FORMATION.

honeycreeper *Vertebrate Zoology.* **1.** any of the small, colorful birds of the family Coerebidae; nectar-eating birds with deeply grooved or brush-edged tongues and a thin, down-curved bill. **2.** any similar birds in the Hawaiian family Meliphagidae. Also, **honeyeater.**

honeydew *Invertebrate Zoology.* a sweet fluid secreted by aphids and scale insects that is eaten by ants and bees.

honeydew melon *Botany.* a variety of melon of the genus *Cucumis* in the family Cucurbitaceae; a large oval fruit that has a smooth, yellowish rind, juicy, greenish flesh, and multiple seeds in a central cavity.

honeyguide *Vertebrate Zoology.* any of several species of small, tough-skinned birds of the family Indicatoridae that have the habit of leading mammals to honey in bee hives; after the larger animal departs, the honeyguide feeds on the larvae and the wax.

honeysuckle *Botany.* the common name for upright or climbing shrubs of the genus *Diervilla,* especially *D. ionicera,* having fragrant white, yellow, or red tubular flowers.

honey tube *Invertebrate Zoology.* an organ in aphids that was formerly thought to secrete honeydew.

honing *Mechanical Engineering.* **1.** the process of sharpening a cutting tool. **2.** the process of obtaining a particular finish or dimensional tolerance, as in the bore of a cyclinder, by the application of abrasive stones or silicon carbide slips held in a fitting that has both rotational and axial motion.

hood *Mechanical Devices.* **1.** a usually dome- or funnel-shaped canopy or overhang that acts as a ventilator. **2.** an enclosure or cover over a hearth forming the upper portion of a fireplace, designed to direct smoke and dust out a flue. **3.** a screen placed over a cathode-ray tube to decrease the amount of emitted light. **4.** the protecting cover to an automobile engine. **5.** a tight-fitting headpiece, often part of a protective suit such as those worn by scuba divers.

hooded anode *Radiology.* an anode of an X-ray tube encapsulated in a copper shield to diminish secondary X-ray emissions.

hoodmold *Architecture.* a molding that projects over an arched window or doorway. Also, **hoodmolding.**

hoof *plural,* **hooves** or **hoofs.** *Vertebrate Zoology.* a homologous, keratinized, epidermal nail structure that grows beneath the digits of ungulate mammals and is used in locomotion, protection, digging, and so on.

hoof-and-mouth disease see FOOT-AND-MOUTH DISEASE.

hook *Mechanical Devices.* **1.** a curved instrument made of a hard material, usually metal, used for holding, grasping, suspending, or catching. **2.** an implement used to cut grass or grain. *Surgery.* a similarly curved instrument, often having a sharp point, used to elevate, hold, or exert traction on a tissue. *Geography.* **1.** a curved piece of land, especially a recurved spit. **2.** any feature that is sharply curved or bent, especially a bend in a river. *Computer Programming.* **1.** a sequence of code in a computer program that facilitates the inclusion of additional code in order to enhance program capabilities. **2.** an entry point or other facility by means of which external programs can access a program. *Electronics.* a circuit occurrence in four-zone transistors in which hole or electron conduction occurs in opposite directions, generating voltage drops that encourage other types of conduction.

hook-and-eye hinge *Mechanical Devices.* a gate hinge consisting of an L-shaped hook secured into one member such as a gate, which is fitted into a loop secured in the other such as a gate post.

hook bolt *Mechanical Devices.* a bolt with a hook at one end and screws for a nut at the other.

Hooke, Robert 1635–1703, English experimental scientist; formulated Hooke's law; invented the compound microscope, wheel barometer, and Gregorian telescope.

Hookean deformation *Mechanics.* a deformation in which stress is directly proportional to strain.

Hookean solid *Mechanics.* an ideal elastic solid that exactly conforms to Hooke's law; that is, stress on the material is directly proportional to strain.

hooked joint *Mechanical Devices.* a joint used between the meeting edges of a door and its recessed case when an airtight or dustproof joint is needed.

Hooker, Joseph 1817–1911, English botanist, son of William Hooker; wrote the major work *Genera Plantarum* (with George Bentham) and also compiled floras of India, New Zealand, and Antarctica.

Hooker, William 1785–1865, English botanist, developed Kew Gardens; wrote major study of ferns and flora of England and Scotland.

Hooker diaphragm cell *Chemical Engineering.* a cell design used in industry for the electrolysis of brine to create chlorine and caustic soda or caustic potash, in which purified brine is introduced around the anode and passes through the diaphragm to the cathode; chlorine forms at the anode and hydrogen escapes at the cathode, leaving sodium hydroxide and residual sodium chloride in the cell liquor.

Hookeriaceae *Botany.* a mostly tropical family of complanate, shiny mosses of the order Hookeriales that grow on tree trunks, humus, and leaves, characterized by simple to branched stems, lateral sporophytes, and spreading to erect leaves.

Hookeriales *Botany.* an order of perennial mosses that are usually yellowish-green, shiny, and mat-forming and grow on soil, humus, trees, and rock.

Hooke's constant see FORCE CONSTANT.

Hooke's force *Mechanics.* a restoring force that is directly proportional to the force of displacement, so that the affected mass is returned to its equlibrium position. Also, **Hooke's law force.**

Hooke's joint *Mechanical Engineering.* a universal joint in which two forks are set at right angles and connected by a crosspiece.

Hooke's law *Mechanics.* a law stating that the stress on a solid is proportional to the strain, up to a certain (yield) point; many elastic solids closely obey this law.

hook gauge *Mechanical Devices.* a device used to measure water level, consisting of a pointed hook attached to a vernier that slides along a graduated staff; it is placed in a pan of water and pulled up just until the upward point pierces the surface of the water.

hook transistor *Electronics.* a transistor having four alternating layers of n- and p-type material; the two inner layers are thin in comparison to diffusion length. Also, **hook collector transistor.**

hookup *Electricity.* a configuration of circuits or other components for a particular purpose.

hookup wire *Electricity.* a tinned and insulated, soft-drawn copper wire that is used in low-power circuit connections.

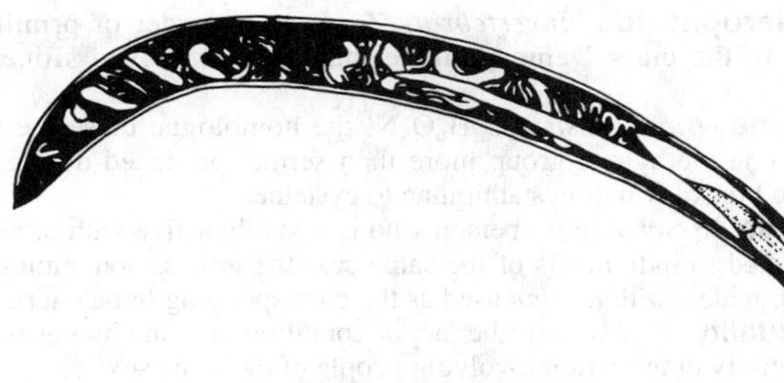

hookworm

hookworm *Invertebrate Zoology.* the common name for any of several parasitic roundworms of the family Ancylostomidae, especially *Necator americanus,* found in the intestines of mammals, including humans.

hookworm disease *Medicine.* an intestinal disease caused by a parasitic roundworm of the genus *Ancylostoma* or *Necator*; symptoms may include abdominal pain, diarrhea, colic, and nausea

hook wrench *Mechanical Devices.* a wrench with a hook at its end. Also, HAND HOOK, HAND SPANNER.

hoop *Engineering.* a circular band or ring made of a stiff material such as metal or wood, which encircles and supports upright members such as the staves of a barrel. *Civil Engineering.* an additional reinforcement in a reinforced concrete column placed around the main reinforcement.

hooped column *Civil Engineering.* a reinforced concrete column with hoops around the main reinforcement.

hoopoe *Vertebrate Zoology.* the common name for any Old World bird of the family Upupidae, usually having a fanlike crest.

hoop reinforcement *Building Engineering.* a metal rod bent into a helix, forming one continuous spiral, that is used within concrete columns to reinforce against cracking.

hoot stop *Computer Programming.* a sequence of code in a computer program that generates an audible signal to notify the user of some event, such as an error.

hop *Botany.* the common name for plants of the genus *Humulus* in the mulberry family Cannabaceae, dioecious herbaceous vines producing an inflorescence that is harvested for use in beer production. *Telecommunications.* one excursion of a radio wave from the earth to the ionosphere and back.

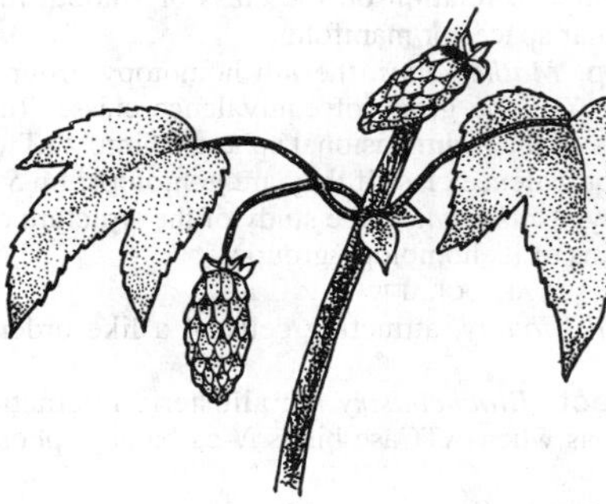

hop

HOP *Oncology.* a cancer chemotherapy regimen that includes the use of the drugs hydroxydaunomycin, Oncovin, and prednisone.

HOP high oxygen pressure.

hopeite *Mineralogy.* $Zn_3(PO_4)_2 \cdot 4H_2O$, a gray, colorless, or white orthorhombic mineral, dimorphous with parahopeite, occurring as prismatic or tabular crystals, and having a specific gravity of 3.05 to 3.06 and a hardness of 3.25 on the Mohs scale; found as a secondary mineral in zinc-bearing ore deposits.

Hopf bifurcation *Chaotic Dynamics.* a bifurcation under which a solution loses stability by oscillation about the previous solution. A branch of periodic solutions joins the branch of steady-state solutions at the bifurcation point. The eigenvalues of the linear stability analysis of the steady states near the bifurcation point include a pair of complex conjugates whose real part vanishes at the bifurcation, changing from positive to negative as the bifurcation point is crossed by scanning a parameter. For a map, a Hopf bifurcation is signaled by a pair of complex conjugate Lyapunov numbers which cross the unit circle.

Hopfner process *Metallurgy*. one of several hydrometallurgical processes for the extraction of copper, based on chlorination and electrowinning.

hop hornbeam *Botany*. the common name for trees of the genus *Ostrya* in the family Betulaceae; they are birchlike, having scaly bark and fruit that resembles hops.

Hopkins, Sir Frederick Gowland 1861–1947, English biochemist; awarded the Nobel Prize for work in discovery of the vitamin concept.

Hopkins-Cole reaction *Analytical Chemistry*. the addition of concentrated sulfuric acid to a mixture that includes a protein and glyoxylic acid; the appearance of a violet ring at the interface indicates a positive test for protein.

Hopkinson's coefficient *Electromagnetism*. a coefficient given by the average magnetic flux per turn in a coil divided by the average magnetic flux per turn of another coil that is inductively coupled to the first coil.

Hoplestigmataceae *Botany*. a little-known monogeneric family of dicotyledonous tropical African trees of the order Violales, characterized by large alternate leaves, numerous stamens, and a drupaceous fruit with a leathery exocarp.

Hoplocarida *Invertebrate Zoology*. a crustacean superorder with a single order Stomatopoda (mantis shrimps), which have a short shieldlike carapace, stalked eyes, and abdominal gills.

Hoplonemertini *Invertebrate Zoology*. an order of nemertian ribbon worms in the class Enopla, with an armed proboscis; includes both aquatic and terrestrial species.

hopper *Engineering*. a receptacle with sloping sides into which materials are temporarily loaded and then discharged through a valvelike opening in the bottom.

Hopper, Grace 1906–1992, American mathematician; pioneer in computer engineering and programming.

Grace Hopper

hopperburn *Plant Pathology*. a disease of potatoes, peanuts, and other plants, characterized by a marginal yellowing, scorching, and shriveling of foliage; caused by the toxic secretions of the leafhopper insect.

hopper car *Engineering*. a railroad freight car having the sloped sides and bottom opening of a hopper. Also, **hopper barge.**

hopper dredger *Engineering*. a dredging machine having hopper compartments for discharging the dredged material.

hopper dryer *Engineering*. in injection molding, a hopper that serves both to feed material and then to dry it, using a flow of hot air.

hoppit *Mining Engineering*. a large bucket used in shaft sinking for hoisting men, rock, materials, and tools. Also, BOWK, KIBBLE, SINKING BUCKET.

hor. or **hor** horizontal; horizon.

H^+/O ratio *Biochemistry*. the quantity of protons that traverse the membrane per oxygen atom lost when oxygen is the terminal electron acceptor.

hordeivirus group *Virology*. a group of multicomponent plant viruses characterized by rigid, rod-shaped virions and infecting primarily wheat, oats, and barley; transmitted mechanically and by seed. Also, BARLEY STRIPE MOSAIC VIRUS GROUP

hordeolum see STYE.

Hordeum *Botany*. a genus of the grass family Poaceae that contains all species of common barley.

hor. interm. at intermediate times; in the intervening hours. (From Latin *horis intermediis*). Also, **Hor. interm.**

horizon *Cartography*. the line of the apparent junction of the earth and sky, marking the limit of the area that can be surveyed from any given observation point. *Astronomy*. the apparent great circle that lies 90° away from an observer's zenith. *Navigation*. a reference point used in celestial navigation; the most common types are: apparent or visible horizon (the line where the earth and sky appear to meet); celestial or rational horizon (a great circle of the celestial sphere halfway between the observer's zenith and nadir); and sensible horizon (a great circle on the celestial sphere formed by a plane parallel to the celestial horizon and passing through the eye of the observer). *Geology*. **1.** a once horizontal, continuous plane of stratification, representing a particular level in the geologic column or the position of a stratum in the geologic time scale. **2.** a thin bed or surface within a sedimentary sequence that represents a particular portion of geologic time, and that is characterized by its distinctive fossil content or lithology. Also, GEOLOGIC HORIZON. **3.** see SOIL HORIZON. *Archaeology*. a term used to describe an artifact, art style, or other such cultural trait that has extensive geographical distribution but a limited time span. *Developmental Biology*. any of 23 specific stages of human embryonic development that occur during the 7-week period following fertilization, defined by anatomical characteristics and lasting 2 to 3 days each.

horizon bar *Navigation*. a reference line in an airplane's attitude indicator that stays parallel to, and represents, the natural horizon.

horizon distance *Astrophysics*. the farthest distance from which light would just be reaching us now if it has been traveling since the universe began; this distance steadily increases as the universe ages.

horizon effect *Artificial Intelligence*. the tendency of a search procedure, such as a game-playing program, to make innocuous moves that have the effect of postponing bad news so that it is beyond the horizon formed by the depth bound of the search.

horizon glass *Navigation*. a partly silvered piece of glass on a marine sextant through which one observes the visible horizon.

horizon marker *Archaeology*. an artifact having wide distribution at a specific time and thus useful for dating other objects found in association with it. Also, **horizon style.**

horizon mining *Mining Engineering*. a system of mine development suitable for inclined, and even faulted, coal seams, in which main stone headings are driven at predetermined levels from the winding shaft to intersect the seams to be developed.

horizon prism *Navigation*. part of a bubble sextant. *Cartography*. a prism that is inserted in the optical path of a surveying or navigational instrument so that observation of the visible horizon can be made while observing another object in order to orient the instrument accurately.

horizon problem *Astrophysics*. the problem of explaining why different regions of the universe appear so nearly identical, given that they have always been beyond each other's horizon distances and thus could not have influenced one another.

horizon sensor *Engineering*. a radiometer or other sensor that is used to align or control the axis of a spacecraft, missile, or satellite with the apparent horizon of the earth or of any other astronomical body.

horizon sweep *Cartography*. in surveying, a technique for preliminary orientation in which a sighting is taken on the most distant known point, and then clockwise angles are measured from that point to other notable objects in the landscape, to help establish the exact point of the observation.

horizon system of coordinates *Astronomy*. the coordinate system that locates celestial objects by altitude and azimuth; in addition, date, time, latitude, and longitude are also necessary to specify the object's position uniquely.

horizontal *Science*. in the plane of or parallel to the horizon or to a base line; perpendicular to the vertical; level. *Mechanical Engineering*. **1.** of or relating to a device or machine that is mounted or oriented in the horizontal plane. Thus, **horizontal lathe, horizontal-tube evaporator. 2.** of or relating to a device or machine that functions in the horizontal plane. Thus, **horizontal auger, horizontal drilling machine, horizontal crusher, horizontal pendulum.**

horizontal action mine *Ordnance.* a land mine designed to produce a destructive effect on a plane approximately parallel to the ground.

horizontal alignment *Transportation Engineering.* the precise routing of a highway in the horizontal plane, normally expressed in terms of a survey grid.

horizontal base-line method *Military Science.* the location of targets or other points by the intersection of observation lines from stations located at opposite ends of a base line.

horizontal blanking *Electronics.* in television scanning, the automatic cutoff of the electron beam during a horizontal retrace period, which prevents an extraneous line on the screen during the period. Also, HORIZONTAL RETRACE BLANKING.

horizontal blanking pulse *Electronics.* the rectangular pedestal-shaped pulse occurring between the active horizontal lines and producing horizontal blanking.

horizontal boiler *Mechanical Engineering.* a water-tube type boiler that has a bank of tubes inclining slightly toward the rear.

horizontal bombing see LEVEL BOMBING.

horizontal branch *Astrophysics.* the region on the Hertzsprung-Russell diagram of a globular cluster that lies to the left of the red giant region and has an absolute magnitude of about zero; the region contains stars that are burning helium in their cores.

horizontal centering control *Electronics.* a potentiometer used to horizontally position the image on the screen of an oscilloscope or television receiver.

horizontal circle *Engineering.* a circular, graduated plate affixed to the base of a transit or theodolite telescope in order to measure horizontal angles.

horizontal control *Cartography.* in surveying, any type of positional grid based on a network of survey stations of known positions that have been referred to a common horizontal datum; accurate reference is thereby provided for the horizontal positions of other features to be mapped.

horizontal control datum *Geodesy.* a geodetic reference point about which five quantities are known: latitude, longitude, azimuth, and semimajor and semiminor axes of the reference ellipsoid; this datum is the basis for horizontal control surveys.

horizontal control station *Cartography.* any survey station whose position has been accurately established in terms either of Cartesian coordinates or longitude and latitude. Also, **horizontal control point.**

horizontal convergence control *Electronics.* the adjustable component for varying the horizontal dynamic-convergence voltage in a color television receiver.

horizontal coordinate system *Cartography.* the set of celestial coordinates based on the celestial equator, composed of an object's altitude above or below the celestial equator and its azimuth.

horizontal coplane *Cartography.* the condition that exists when a pair of photographs is in the basal coplane and the air base is horizontal.

horizontal danger angle *Navigation.* an angle, measured horizontally between two fixed terrestrial points, that marks the position of some hazard; frequent measuring of this angle as the vessel passes through the area allows the navigator to avoid the hazard.

horizontal deflection electrode *Electronics.* a pair of electrodes that use electrostatic deflection to move the electron beam from side to side on the screen of a cathode-ray tube.

horizontal displacement see STRIKE SLIP.

horizontal distributed processing system *Computer Technology.* a distributed system in which two or more peer computers are networked together.

horizontal distribution *Computer Programming.* a method of assigning initial strings to tapes when performing the polyphase-merge sort.

horizontal drive control *Electronics.* the potentiometer used to adjust the output of the horizontal oscillator. Also, DRIVE CONTROL.

horizontal earth rate *Navigation.* a movement of a gyrocompass in which the gyro appears to rotate about its horizontal axis at the same speed as, but in the opposite direction from, the rotation of the earth. The rate of change is equal to the rate of rotation of the earth times the cosine function of the latitude.

horizontal engine *Mechanical Engineering.* an engine in which the cylinders and piston stroke are horizontal.

horizontal error *Ordnance.* the error in range, deflection, or radius that a weapon may be expected to exceed 50% of the time; in weapons making a nearly vertical approach to a target, it is expressed as circular error probable; in weapons producing elliptical dispersion patterns, it is expressed as probable error.

horizontal exposure *Archaeology.* the excavation of a site to reveal its horizontal extent, with relatively little depth.

horizontal field-strength diagram *Electromagnetism.* a diagram indicating the antenna signal intensity measured at a constant radius about the antenna in a horizontal plane.

horizontal firing *Mechanical Engineering.* the horizontal discharge of fuel and air into a boiler furnace.

horizontal flow chart *Computer Technology.* a flow chart that shows the movement of tapes, forms, and other recording media through an organization, for the entire life cycle of the media.

horizontal flyback *Electronics.* a flyback in which a television picture tube's electron beam moves from the end of one scanning line to the beginning of the next line.

horizontal fold see NONPLUNGING FOLD.

horizontal force instrument *Engineering.* an instrument that compares the intensity of the horizontal component of the earth's magnetic field to the magnetic field of a shipboard compass. Also, **horizontal vibrating needle.**

horizontal frequency see LINE FREQUENCY, def. 1.

horizontal heart *Cardiology.* the counterclockwise rotation or left deviation of the electrical axis of the heart.

horizontal hold control *Electronics.* the control for adjusting the horizontal oscillator's frequency so that the picture in a television receiver remains steady.

horizontal instruction *Computer Programming.* a wide machine-language microinstruction that is composed of independent component operations that can be executed in parallel or in a well-defined time sequence.

horizontal intensity *Geodesy.* the degree of intensity of the horizontal component of the earth's magnetic field as measured in the plane of the magnetic meridian.

horizontal linearity control *Electronics.* the control used to adjust horizontal linearity in an oscilloscope or TV receiver.

horizontal magnetometer *Engineering.* an instrument designed to detect and measure changes in the horizontal component of magnetic field intensity.

horizontal output stage *Electronics.* an output amplifier following the horizontal oscillator in a TV receiver.

horizontal output transformer *Electronics.* the output transformer in the horizontal-oscillator-output amplifier section of a TV circuit. Also, FLYBACK TRANSFORMER.

horizontal parallax *Astronomy.* the diurnal parallax of an object on the astronomical horizon. *Photogrammetry.* see ABSOLUTE STEREOSCOPIC PARALLAX.

horizontal parity check see LONGITUDINAL PARITY CHECK.

horizontal plane *Anatomy.* a transverse plane; a plane that passes through the body at right angles to the median and the frontal plane, dividing the body into upper and lower parts. *Cartography.* in surveying, a plane perpendicular to the plumb line and on which measurements are made. *Geodesy.* **1.** a plane perpendicular to the direction of gravity. **2.** a plane tangent to a reference geoid, or parallel to such a plane.

horizontal polarization *Telecommunications.* a type of radio-wave transmission in which the electric lines of force are horizontal.

horizontal-position welding *Metallurgy.* **1.** the process of making a fillet weld on the upper side of two perpendicular parts. **2.** the process of making a horizontal groove weld on a vertical surface.

horizontal pressure force *Geophysics.* the horizontal pressure gradient per unit mass.

horizontal range *Ordnance.* the horizontal distance between a gun and its target, or the point on the ground directly below the target.

horizontal resolution *Electronics.* the number of discernable picture elements or dots in a horizontal scanning line in a television or facsimile image.

horizontal retort process *Metallurgy.* a process for the smelting of zinc, based on a series of horizontal, closed refractory vessels.

horizontal retrace blanking see HORIZONTAL BLANKING.

horizontal retraction *Cartography.* in surveying, a horizontal bending of light rays because of differences in the density of air that result from temperature variations between the object and the observing instrument, leading to errors in observations.

horizontal ring drilling see HORADIAM DRILLING.

horizontal-rolled-position welding *Metallurgy.* the butt welding of two rotating pipes.

horizontal scanning *Engineering.* the rotation of a radar antenna in azimuth around the horizon. Also, SEARCHING LIGHTING.

horizontal scanning frequency *Electronics.* the frequency at which horizontal lines are scanned in TV circuits; equal to 15,750 hertz in the United States. Also, **horizontal repetition rate.**

horizontal separation see STRIKE SLIP.

horizontal separator *Petroleum Engineering.* a horizontal tank that is used to separate liquid hydrocarbons from the free oil-well gas.

horizontal silo *Agriculture.* a silo whose long axis is parallel to the ground, with openings at both ends for storing silage.

horizontal-situation indicator *Navigation.* a combination instrument that shows a pilot the actual course, as compared to the intended course, and the relation of the aircraft to the glide slope.

horizontal sweep *Electronics.* **1.** the back-and-forth sweeping of the spot on the screen of a cathode-ray tube. **2.** the circuit that produces horizontal sweep.

horizontal synchronizing pulse *Electronics.* in a video signal, the pulse that synchronizes a television's horizontal scanning-component receiver with that of the camera and triggers horizontal retracing and blanking.

horizontal transmission *Virology.* the transmission of a virus, parasite, or other pathogen from one individual or one cell to another within the same generation, as opposed to vertical transmission through the germline. Also, LATERAL TRANSMISSION.

horizontal vee *Electromagnetism.* a V-shaped antenna whose elements lie in a horizontal plane.

horm- a combining form meaning to urge or stimulate, as in *hormone,*

hormesis *Toxicology.* the stimulatory effect of small doses of a toxic substance that is inhibitory in larger doses.

Hormodin see INDOLEBUTYRIC ACID.

hormogonium *Botany.* a portion of a filament between heterocysts of certain algae that detaches as a reproductive body.

hormonal *Endocrinology.* relating to or produced by the action of a hormone or hormones.

hormone *Endocrinology.* any one of a number of biochemical substances that are produced by a certain cell or tissue and that cause a specific biological change or activity to occur in another cell or tissue located elsewhere in the body. *Biochemistry.* a synthetic substance having a similar effect. *Botany.* a plant compound that affects the growth or differentiation of plant tissue. (Going back to a Greek term meaning "to set in motion.")

Hormosiraceae *Botany.* a monotypic family of brown algae of the order Fucales, characterized by a thallus consisting of cartilaginous, dichotomously branched axes attached by a small holdfast and by inflated portions in the axes which give the thallus buoyancy or store water.

Hormotilaceae *Botany.* a family of green algae usually placed in the order Chlorococcales, characterized by spherical-type cells irregularly distributed in a mucilaginous matrix or attached to the tip of a dendroid stratified stalk, and forming filamentous colonies.

Hormuz, Strait of *Geography.* a strait connecting the Persian Gulf and the Gulf of Oman.

horn *Vertebrate Zoology.* **1.** one of the pair of hard, hollow permanent growths, usually pointed, that project upward on the head of cattle, sheep, goats, and other ungulate mammals. **2.** a similar projection on other animals, such as a rhinoceros. **3.** an antler. **4.** the bony substance of which horns are composed, or any similar hard substance such as nails, hoofs, shells, or scales. *Anatomy.* a structure resembling an animal's horn in shape. *Acoustical Engineering.* **1.** any device having or consisting of a tube that increases in cross section from one end to the other, used to direct and intensify sound waves. Also, ACOUSTIC HORN. **2.** any of the group of musical instruments of this type, such as the trombone or trumpet. *Electromagnetism.* a microwave antenna that is constructed by flaring the walls of a waveguide at the end of the guide. *Geology.* **1.** a high, sharp, steep-sided, pyramid-shaped peak left where cirques have eroded into a mountain from more than two sides around a central area. Also, GLACIAL HORN, PYRAMIDAL PEAK. **2.** the pointed end of a dune or other beach formation. *Engineering.* any of various structures or devices suggestive of an animal's horn. *Building Engineering.* a jamb that extends above a door or window frame. *Metallurgy.* a component of resistance-welding equipment. *Ordnance.* a projection from the shell of certain naval contact mines that, when broken or bent by contact, causes the mine to fire.

horn angle *Mathematics.* the configuration formed by two plane curves that are tangent at a point P and that lie on the same side of their mutual tangent line in a neighborhood of P.

horn antenna *Electromagnetism.* a microwave antenna whose radiating element is a horn that radiates directly into space.

horn arrester *Electricity.* a lightning arrester in which the spark gap has thick wire diversion horns that project upward.

hornbeam *Botany.* the common name for trees of the genus *Carpinus* in the birch family Betulaceae, distinguished by doubly serrate leaves and small angular winter buds and noted for their durable white wood, often used in woodworking. *Materials.* the hard, smooth, close-grained wood of this tree; used to make tool handles and bearings.

hornbill *Vertebrate Zoology.* a large, tropical Old World bird of the family Bucerotidae, known for its long hooked bill that often has a bony helmet on top.

hornblende *Mineralogy.* a mineral of the ferrohornblende-magnesiohornblende series, monoclinic, rock-forming minerals of the amphibole group having the general formula $Ca_2(Mg,Fe^{+2})_4Al(Si_7Al)O_{22}(OH,F)_2$, occurring as dark green to black crystals and in granular masses, found in many different types of igneous and metamorphic rocks.

hornblendite *Petrology.* an igneous rock consisting primarily of hornblende.

horn buoy *Navigation.* a floating aid to navigation that includes a horn-type sound-signaling device.

Horn, Cape *Geography.* a cape on Horn Island, Tierra del Fuego, Chile; the southernmost point in South America.

Horn clause *Artificial Intelligence.* a logical formula that is a disjunction of literals and has at most one positive literal. Viewed as a rule, a Horn clause is a rule from a conjunction of positive literals to a single positive conclusion literal.

horned dinosaur see CERATOPSIA.

horned liverwort *Botany.* the common name for members of the family Ceratophyllaceae; submersed, rootless, aquatic perennial herbs with branching stems. Also, **hornwort.**

horned owl

horned owl *Vertebrate Zoology.* an owl of the family Strigidae and the genus *Bubo,* having pointed, feathered ear tufts, a wingspan up to six feet, a sharp, hooked bill, and powerful talons; hunts medium-sized mammals and birds in forested and scrub areas worldwide.

horned toad *Vertebrate Zoology.* a small, insectivorous lizard of the genus *Phrynosoma,* characterized by a head and flattened body covered by spiny scales; found in arid areas of North America, especially in the southwestern United States. Also, **horned lizard.**

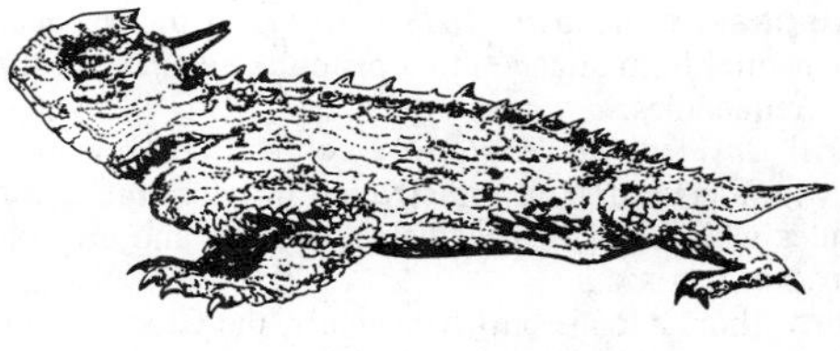

horned toad

horned-toad dinosaur *Paleontology.* a term referring to the ankylosaurs, especially Scolosaurus.

horn equation *Acoustics.* a second-order partial differential equation that gives the velocity potential in terms of distance and time for an exponential horn.

Horner's method *Mathematics.* **1.** an algorithm for approximating positive real roots of an algebraic equation, accomplished as follows: (a) Assume that the equation has a positive root (if not, substitute $-x$ for x in the equation and continue). (b) Isolate a root between nonnegative integers a and $a + 1$. (c) Substitute $x - a$ for x in the equation. Now the equation has a root between 0 and 1. (d) Isolate a root between successive multiples of 1, i.e., between $b/10$ and $(b + 1)/10$, where b is a nonnegative integer less than 10. (e) Substitute $x - b/10$ for x in the equation. The equation now has a root between 0 and 1/10. Continue this process until the desired accuracy is reached. **2.** a method of rapidly evaluating a polynomial using synthetic division.

Horner's syndrome *Medicine.* a group of symptoms including the recession of the eyeball into the orbit, drooping of the eyelid, constriction of the pupil of the eye, diminished sweating, and flushing of the face due to paralysis of the cervical sympathetic nerve fibers on the affected side. Also, **Horner's ptosis.** (Named for Johann Friedrich *Horner,* 1831–1886, Swedish ophthalmologist.)

hörnesite see HÖERNESITE.

hornet *Invertebrate Zoology.* the common name for large stinging, nest-building social wasps in the hymenopteran family Vespidae, similar in general appearance to the common wasp but up to 35 mm in length.

Hornet *Aviation.* a fighter escort plane designed to attack other fighter jets and perform bombing missions; officially designated **F/A-18.**

Horney, Karen 1885–1952, German-American psychiatrist; studied social factors in genesis of neuroses and inner conflicts.

hornfels *Petrology.* a fine-grained rock formed by contact metamorphism and composed of equidimensional fine grains, sometimes with large crystals.

hornfels

hornfels facies *Petrology.* the set of metamorphic mineral assemblages (facies) constituting rocks formed by thermal or contact metamorphism at temperatures of 250–800°C at depths of less than 10 km; it includes albite-epidote-hornfels facies, hornblende-hornfels facies, pyroxene-hornfels facies, and sanidinite facies.

hornito *Volcanology.* a small, beehive-shaped mound on the surface of a lava flow; formed from lava clot ejected through the roof of a lava tube. Also, DRIBLET CONE.

horn loudspeaker *Acoustical Engineering.* a loudspeaker consisting of an exponential horn attached to a pistonlike source for high output at low audio frequencies.

horny coral *Invertebrate Zoology.* a colorful coral, in the order Gorgonacea, with a central axial rod covered in gorgonin, a hornlike material; includes whip coral, sea feathers, sea fans, and precious red coral. Also, GORGONIAN CORAL.

Horologium [hôr´ ə lō´ jē əm] *Astronomy.* the Clock, a faint constellation of the Southern Hemisphere with no star brighter than fourth magnitude.

Horology

Newton's laws of motion and gravitation revealed that the Universe operates as a vast majestic clock whose stars and planets move with relentless regularity so that, for our ancestors, to look to the sky with the unaided eye was to know the time with ample precision. But that changed in the nineteenth century. That age brought the mass production of cheap clocks and watches, almost instantaneous global communication, factory workers punching time clocks, and time zones demanded by speeding transcontinental trains. Thus it was that the heaven's clocklike precision entered our daily lives and social structures through the doors of the industrial revolution. Time structured by religious authority and the universe's natural rhythms was replaced by the secular authority of mechanical timekeepers.

Today's best clocks are hitched to the natural rhythms of atoms—not yesterday's fragile balance wheels. Thus in a sense we have come full circle—from the natural rhythms of the stars to the natural rhythms of the atom. The move from balance wheels to atoms was underpinned by tumultuous advances in our understanding of the fundamental laws of the universe—advances embodied in the theories of relativity and quantum mechanics.

Yet to keep better time is not to know time better. Time's mystery is not lessened by ever better clocks—it is brought only more clearly into focus.

James Jesperson
Physicist
National Institute for Standards and Technology

horology *Science.* the scientific study of time measurement and the making of timepieces. Thus, **horologist.** (Derived from the Greek word for "hour.")

Horomoconis *Mycology.* a genus of fungi belonging to the family Amorphothecaceae, which includes the species *H. resinae,* commonly termed "kerosene fungus" or "creosote fungus."

horotely *Evolution.* a normal or average rate of evolutionary change for a given species within a group of related species.

horse *Vertebrate Zoology.* a mammal of the Equidae family, characterized by a slender head and a graceful body with long legs, solid hooves, and a long neck covered with a mane; domesticated throughout the world. *Geology.* **1.** a large, displaced mass of rock caught between the walls of a fault. **2.** see HORSEBACK. *Mining Engineering.* an extensive mass of low-grade ore or country rock enclosed in a mineral vein.

horseback *Geology.* **1.** a low, steep ridge of sand, gravel, or rock, such as an esker or a kame. Also, HORSE. **2.** a large body of shale, sandstone, or other foreign matter that fills a natural channel cut into a coal seam by flowing water, or any ridge or bank of foreign matter within a coal seam. Also, CUTOUT. *Mining Engineering.* a portion of the roof or floor that bulges or intrudes into the coal.

horse chestnut *Botany.* the common name for trees of the species *Aesculus hippocastanum,* an ornamental buckeye tree of the family Hippocastanaceae, having palmately compound opposite leaves, large showy flowers, and glossy brown seeds.

horse fetter see HIPPOPEDE.

horsehair *Materials.* a hair or the hair of a horse, especially from the mane or tail. *Textiles.* a sturdy, glossy fabric made from this hair.

horsehair blight *Plant Pathology.* a disease of tea and other tropical plants caused by the fungus *Marasmius equicrinis* and characterized by black, stringy bunches of fungal filaments hanging from the branches of affected plants.

Horsehead Nebula *Astronomy.* a dark cloud of dust that resembles a horse's head in profile; it is superimposed on the hydrogen emission nebula IC 434 that lies south of the star Zeta (ζ) in Orion's Belt.

horse latitudes *Meteorology.* the belts of latitude over the ocean at approximately 30–35°N and S, where winds are predominantly calm or very light and weather is hot and dry. (So named because in the days of sailing ships, horses that were being carried across the Atlantic to the West Indies would be thrown overboard if the supply of water ran short when the voyage was slowed by calm or baffling winds in these latitudes.)

horsepower *Metrology.* the unit of power in the British engineering system, equal to 550 foot-pounds or approximately 746 watts. (Originally developed by James Watt as a means of relating the work done by a steam engine to comparable work done by a horse.)

horsepox *Veterinary Medicine.* a rare contagious viral disease of equines characterized by the formation of skin eruptions on the pasterns and in the mucous membranes of the mouth; recovery within three weeks is usual.

horseradish *Botany.* the common name of the plant species *Armoracia rusticana,* a tall, coarse, white perennial herb of the order Capparales, cultivated for its pungent root that is used in condiments.

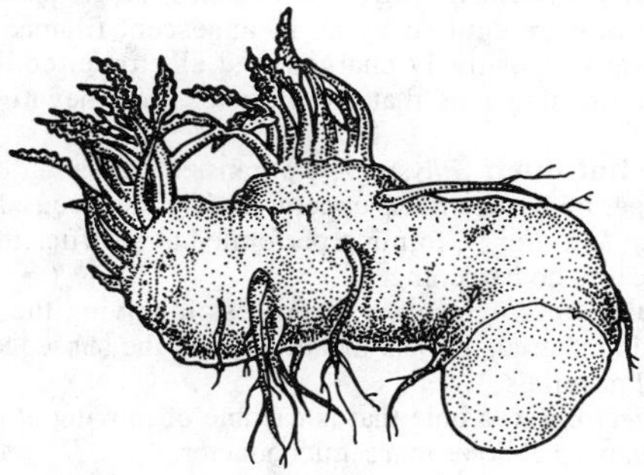

horseradish

horseradish peroxidase *Enzymology.* an enzyme that reacts with the contents of endocytic vesicles; its activity can be monitored or assayed becaused it produces substances that can be stained with osmium tetroxide and identified through electron microscopy.

horse serum *Immunology.* an antiserum derived from the blood of horses, especially tetanus antitoxin.

horseshoe bend see OXBOW.

horseshoe crab *Invertebrate Zoology.* the common name for a primitive arachnid arthropod in the family Limulida, with a dark brown, smooth, horseshoe-shaped carapace and a long, spinelike tail; found in large colonies in marine waters where it buries itself in the sandy, muddy bottom.

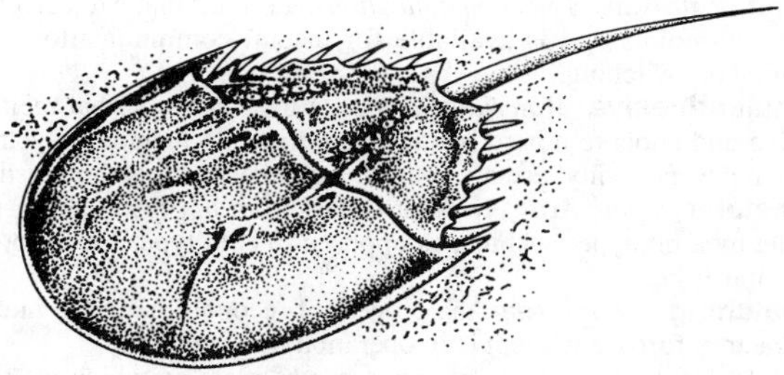

horseshoe crab

horseshoe dune see BARCHAN.

horseshoe kidney *Medicine.* a congenital condition of fusion of the two kidneys to a varying degree, usually at the lower portions.

horseshoe lake see OXBOW LAKE.

horseshoe magnet *Electromagnetism.* a permanent magnet that is U-shaped so that its poles are close together.

horsetail *Botany.* the common name for perennial, flowerless plants of the genus *Equisetum* that are related to the ferns and comprise the order Equisetales. Also, SCOURING RUSH. *Geology.* an arrangement of closely spaced minor veins and fissures that emanate from a main vein, or an ore occurring in such veins.

horsfordite *Mineralogy.* Cu_5Sb, a silver-white, probably cubic, opaque metallic mineral occurring in massive form, having a specific gravity of 8.81 and a hardness of 4 to 5 on the Mohs scale; found in Turkey.

hor. som. at bedtime. (From Latin *hora somni,* "at the hour of sleep.")

horst

horst *Geology.* **1.** an elongated block of crustal rock that has been uplifted relative to the faults along its sides. **2.** a knobby ledge of limestone underlying a thin soil.

hort. or hort horticulture; horticultural.

horticultural crop *Agriculture.* a crop that is cultivated in a nursery, orchard, or garden, rather than one that grows naturally.

horticulture *Agriculture.* **1.** the branch of agriculture that is concerned with garden crops, orchard plants, and ornamental plants. **2.** the cultivation of such plants. Thus, **horticulturist, horticultural.**

Horton number *Hydrology.* a number that expresses the relative intensity of erosion on the slopes of a drainage basin in terms of the rate of runoff and the rate of erosion. (From Robert *Horton,* 1875–1945, American engineer.)

hortonolite *Mineralogy.* $(Fe^{+2},Mg,Mn)_2SiO_4$, a magnesian manganoan variety or fayalite.

hor. un. spat. at the end of one hour. (From Latin *horae unius spatio.)* Also, **Hor. un. spat.**

hose *Mechanical Devices.* a long, flexible pipe or tube of rubber or other synthetics used to convey fluids.

hosp. or **hosp** hospital.

hospice *Medicine.* a facility that provides palliative and supportive care for terminally ill patients and their families, either directly or on a consulting basis.

hospital *Medicine.* a facility for treatment of the sick, having inpatient beds, continuous nursing services, and an organized medical staff.

hospital-acquired infection see NOSOCOMIAL INFECTION.

hospital ship *Naval Architecture.* an unarmed ship fitted as a floating hospital and marked in accordance with the Geneva conventions.

host *Biology.* **1.** the organism from which a parasite obtains its nutrition or shelter. **2.** an animal that is the recipient of a tissue graft; an organism into which an experimental graft is transplanted. *Virology.* an organism or cell culture in which a certain virus can replicate.

host computer *Computer Technology.* **1.** the central computer in a distributed computer system that provides service to remote terminals or computers. **2.** in microprogramming, the machine that interprets programs written in other machine languages. **3.** in multiprogramming, the machine responsible for allocating storage and I/O resources to virtual machines that are dependent upon it. Also, HOST PROCESSOR.

hostile-environment machine *Mechanical Engineering.* a machine that is capable of working under extreme conditions of temperature, moisture, nuclear radiation, electromagnetic radiation, or vibration. *Robotics.*specifically, a robot that can work under such conditions

hostile ice *Oceanography.* an ice canopy that does not present an outlet (such as a large ice skylight or other type of opening) for a submarine to surface.

host processor see HOST COMPUTER.

host race *Entomology.* a species of organisms on or in which another species lives for nourishment or protection.

host range *Virology.* the collection of all species of hosts that are susceptible to infection with a given virus.

host rock *Geology.* a body of rock that is older than the other rocks or mineral deposits introduced into it or formed within or adjacent to it.

hot *Physics.* having or giving off heat. *Electricity.* **1.** of a terminal or conductor, live, energized, or connected. **2.** not grounded. **3.** excited to a high energy level.

HOT *Ordnance.* a French-German antitank missile powered by a two-stage boost/sustainer rocket motor; it delivers a 13.2-pound (6 kg) hollow-charge warhead at high subsonic speed and a maximum range of 13,125 feet (4000 m). (An acronym for high-subsonic optically guided tube-launched.)

hot acid *Petroleum Engineering.* hot hydrochloric acid at 200–300°F or 93–149°C, used for acidizing an oil well when the wellbore scale is difficult to remove and the scale dissolves slowly.

hot-air engine *Mechanical Engineering.* an engine in which air, or another gas such as helium, hydrogen, or nitrogen, is used as the working fluid and is alternately heated and cooled by a furnace and regenerator.

hot-air furnace *Mechanical Engineering.* a furnace that supplies warm air to circulation ducts. Similarly, **hot-air heater.**

hot-air oven *Biotechnology.* an enclosed heating unit used for the dry-heat sterilization of glassware and other heat-resistant materials that need to be dry after treatment.

hot-air soldering *Metallurgy.* soldering in which the heat is supplied by a blast of hot air.

hot-air sterilization *Engineering.* to destroy, remove, and prevent the further breeding of microorganisms on heat-resistant materials by means of dry heat at temperatures over 160°C for a minimum of two hours.

hot atom *Nucleonics.* an atom whose kinetic energy has increased as the result of a nuclear process such as radioactive decay.

hot bed *Metallurgy.* a name for a cooling table used for hot-worked mill products.

hot belt *Meteorology.* a climatologic belt around the earth within which the annual mean temperature is higher than 20°C.

hot big bang *Astrophysics.* the standard model for the big bang theory, according to which the universe began in a singularity of extremely high temperature and density.

hot-bulb *Mechanical Engineering.* of or relating to a method of ignition in semidiesel engines, in which the fuel mixture is ignited in a separate chamber that is maintained above ignition temperature by the heat of compression.

hot carrier *Electronics.* an electron or hole carrier in a semiconductor whose energy is greater than that of majority carriers normally found in the same material.

hot-carrier diode see SCHOTTKY BARRIER DIODE.

hot cathode *Electronics.* a cathode that is directly or indirectly heated.

hot-cathode tube see THERMIONIC TUBE.

hot cell *Physics.* a heavily shielded work space used in the handling of radioactive materials for the protection of the handling technicians.

hot chamber *Metallurgy.* a type of die-casting equipment in which the metal chamber is immersed in molten metal. Thus, **hot-chamber die casting.**

Hotchkiss *Ordnance.* **1.** one of several guns invented by Benjamin Hotchkiss or produced by the Hotchkiss Company of France; the most famous were a revolving crank-operated semiautomatic 37-mm cannon introduced in 1875 and a gas-operated, air-cooled machine gun introduced in 1875. **2.** see BRANDT.

Hotchkiss drive *Mechanical Engineering.* an automobile rear suspension system in which torque is absorbed by longitudinal leaf springs.

Hotchkiss superdip *Engineering.* an extremely responsive dip needle having a magnetic needle that rotates around a horizontal axis and a nonmagnetic bar that sits at the pivot point.

hot-cold lysis *Hematology.* a term for lysis that occurs only if the material is incubated as usual and then allowed to stand overnight at room temperature.

hot-cold working *Metallurgy.* the process of working at an elevated temperature without destroying completely the deformation structure, thereby achieving some hardening.

hot configuration *Aviation.* the condition of a rocket vehicle that is ready for firing.

hot corrosion *Materials Science.* a process of electrochemical attack of a metal occurring at elevated temperatures. Also, HIGH-TEMPERATURE CORROSION.

hot cracking *Materials Science.* a fracture that occurs as material components cool and solidify at different rates; thus exerting stresses upon each other.

hot-die steel *Metallurgy.* a toll steel suitable for making forging dies.

hot dipping *Metallurgy.* the process of coating a material by dipping it into a molten metallic bath, as of tin or zinc.

hot-drawn *Engineering.* of metal wire or other materials, shaped by being heated and then pulled through an orifice; extruded.

hot editing *Robotics.* a robot debugging technique in which as many problems as possible are identified and resolved prior to initializing the robot.

hot electron *Electronics.* an electron that exceeds the thermal-equilibrium number and, for metals, has an energy greater than the Fermi level; for semiconductors, the energy is a specified amount above that of the conduction band.

hot electron triode *Electronics.* a solid-state, evaporated, thin film structure equal to a vacuum tube.

hot extrusion *Metallurgy.* extrusion performed at a sufficiently high temperature to cause no hardening.

hot-filament ionization gage *Electronics.* an ionization gauge in which electrons are emitted by an incandescent filament and, while moving toward a positively charged grid electrode, collide with gas molecules producing ions that are attracted by a negatively charged electrode.

hot flash or **hot flush** *Physiology.* a transient sensation of heat in the face and upper body sometimes experienced by menopausal women.

hot forming *Metallurgy.* forming performed at a sufficiently high temperature to cause no hardening.

hot-gas welding *Engineering.* a process involving the softening of thermoplastic materials by jets of hot air, and the subsequent fusing of the softened materials.

hot hole *Electronics.* a hole that is capable of moving at much greater velocity than normal holes in a semiconductor.

hothouse *Engineering.* **1.** a heated greenhouse used to grow and shelter tender or tropical plants, or for growing plants out of season. **2.** a drying room for newly made pottery. **3.** any room or building that is heated to high temperatures.

hot isostatic pressing *Materials Science.* a method of densifying ceramic materials, characterized by high temperatures and pressure equally applied from all directions; this increases the strength and wear resistance of ceramics. *Metallurgy.* **1.** in powder metallurgy, pressing at elevated temperatures in a pressurized gas chamber. **2.** in casting, a similar process used to decrease porosity.

hot junction *Electronics.* the heated junction of a two-junction thermocouple circuit.

hot laboratory *Nucleonics.* a research facility designed with special safety features such as remote-control equipment so that research with radioactive materials can be carried out. Also, **hot lab.**

hot lahar *Volcanology.* a flow of mud and volcanic material (lahar) occurring soon after an eruption, usually as a result of heavy rainfall.

hot line or **hotline** *Telecommunications.* a circuit that exclusively connects two points and is available for instant communication without patching or switching.

hot-melt adhesive *Materials Science.* a thermoplastic that melts upon heating and cools to a hard bond; its constituents include polyethylene, polyamides, polyvinyl acetate, waxes, and other hydrocarbon resins.

hot metal *Graphic Arts.* type that is cast from hot metal slugs, for example, by a linotype machine, as opposed to cold type. Also, **hot-metal type, hot type.**

hot patching *Engineering.* the process of reconditioning the interior of a refractory furnace while it is in operation or is still hot.

hot plate *Metallurgy.* in soldering, a heated plate on which the components to be joined rest.

hot-pressed silicon nitride *Materials.* a fully dense silicon nitride material that is highly resistant to chemical attack; used to make tube-drawing plugs and bearings for equipment lacking adequate lubrication.

hot press forge *Metallurgy.* a press in which the forging operation is performed at a sufficiently high temperature to cause no hardening.

hot pressing *Engineering.* **1.** the formation of a bond or compact by means of heat and pressure applied to any of various materials. **2.** the application of heat and pressure on cloth or paper to achieve a glossy finish. **3.** the process of forming a coherent mass by applying pressure and heat to a metal powder.

hot-pressure welding *Metallurgy.* a process of welding that is performed by the simultaneous application of heat and pressure.

hot-quenching *Metallurgy.* the process of quenching into a medium that is above ambient temperature.
hot rolling *Metallurgy.* rolling performed at a sufficiently high temperature to cause no hardening.
hot saw *Mechanical Engineering.* a power saw that is used to cut hot metal.
hot shortness *Metallurgy.* the undesirable separation of grains at their boundaries during deformation at elevated temperatures.
hot spot *Physics.* a point or small region in space having a higher temperature than its local surroundings. *Chemical Engineering.* a point or area in a reaction system in which the temperature is considerably higher than that in most of the reactor; this usually locates the reaction front for an exothermic reaction. *Forestry.* **1.** a region of a forest where fires frequently occur. **2.** the most active part of a forest fire. *Graphic Arts.* an area on an image that is too light or bright. *Volcanology.* an ancient site of volcanic activity, thought to originate from convection currents of molten mantle material along the core-mantle boundary. *Molecular Biology.* a region of DNA that is particularly susceptible to changes such as recombination or spontaneous mutation. Also, MUTATIONAL HOT SPOT.
hot spring *Hydrology.* any spring whose water temperature is above the temperature of the human body.
hot stamp *Engineering.* an indentation made in a hot piece of forged metal.
hot stamping *Graphic Arts.* a process in which a die stamps a colored foil into printed material and heat releases the color from the foil.
hot-swage *Metallurgy.* in metal working, tapering a bar at a sufficiently high temperature to cause no hardening.
hot tear *Metallurgy.* in casting, a fracture caused by hindered contraction.
hot test *Aviation.* a propulsion-system test in which the propellants are actually fired, as opposed to a cold-flow test, in which firing is simulated or omitted.
hot top *Metallurgy.* in the casting of ingots, an insulated liquid-metal reservoir placed on top of the mold to ensure adequate feeding during cooling.
hot-trim *Metallurgy.* to remove flash from a heated part.
hot type see HOT METAL.
hot well *Mechanical Engineering.* the tank or pipes in which the condensate of a steam-engine or turbine condenser is collected before it is returned to the boiler by the feed pump.
hot wind *Meteorology.* any wind that is characterized by intense heat and low relative humidity, such as a summertime desert wind.
hot wire *Electricity.* a resistive wire in an electric relay that expands when heated and contracts when cooled.
hot-wire *Mechanical Engineering.* an informal term for the process of short-circuiting the ignition of a motor vehicle in order to start the engine without a key.
hot-wire ammeter *Engineering.* an ammeter that uses the hot-wire principle to measure current density. Also, THERMAL AMMETER.
hot-wire anemometer *Engineering.* an instrument used to measure low wind velocities according to the electrical resistance of a fine platinum wire placed in the air; as the wind velocity increases, the wire is cooled and resistance decreases.
hot-wire instrument *Engineering.* any instrument or device whose operation relies on the hot-wire principle; typically, the heating effect of increasing or decreasing current running through the wire causes it to expand or contract, thus moving a pointer over a scale indicating the magnitude of the current voltage.
hot-wire microphone *Acoustical Engineering.* a microphone that uses a hot wire, which changes resistance according to the cooling or heating effect of sound waves and thereby controls the current through it.
hot-wire principle *Engineering.* the phenomenon that a current-carrying wire expands as it heats.
hot working *Metallurgy.* the process of working metal at a sufficiently high temperature to cause no hardening.
HOU *Aviation.* **1.** the airline code for Houston, Texas. **2.** the airport code for Hobby Airport, Houston.
Houben-Hoesch synthesis or **reaction** *Organic Chemistry.* the synthesis of an acylphenol from a phenol or phenolic ether by the action of an organic nitrile in the presence of hydrogen chloride and aluminum chloride.
Houdry fixed-bed catalytic cracking *Chemical Engineering.* a method for cracking petroleum distillate using a hydrosilicate of alumina in fixed-bed reactors to produce high-octane gasoline.
Houdry hydrocracking *Chemical Engineering.* a catalytic process that combines cracking and desulfurization of crude petroleum oil in the presence of hydrogen.
Hound Dog *Ordnance.* a turbojet-propelled, air-to-surface missile carried externally on the B-52; it is equipped with a nuclear warhead and can be launched for high- or low-altitude attacks against enemy targets; officially designated **AGM-28B**.
hound shark *Vertebrate Zoology.* another name for the dogfish; any of the various small sharks that often appear in schools near the shore, and whose livers are used for oil.
Hounsfield, Godfrey N. born 1919, English engineer; shared Nobel Prize for invention of computerized axial tomography (CAT scan).
hour a unit of time equal to 60 minutes or 3600 seconds.
hour angle *Astronomy.* the distance between a celestial object's hour circle and the observer's meridian, measured westward in hours, minutes, and seconds. Also, **HA.**
hour circle *Navigation.* a great circle of the celestial sphere that passes through the celestial poles. *Astronomy.* **1.** a great circle on the sky that passes through both poles and a given celestial object. **2.** the setting circle on a telescope mounting that reads out right ascension. *Cartography.* any great circle on the celestial sphere having a plane perpendicular to the plane of the celestial equator.
hourglass stomach *Medicine.* a stomach divided into two compartments by a constriction in the middle.
hourglass uterus *Medicine.* a uterus having a segment of circular muscle fibers that contract during labor; this condition causes constriction ring dystocia and, despite adequate contractions, a lack of progress in labor.
houseboat *Naval Architecture.* a boat designed primarily to be moored in one location for use as a dwelling.
house drain *Civil Engineering.* the lowest horizontal drain inside the wall of a building to receive waste water.
housefly *Invertebrate Zoology.* the common name for *Musca domestica,* a dipteran insect generally found in or near human habitations; often a disease vector.

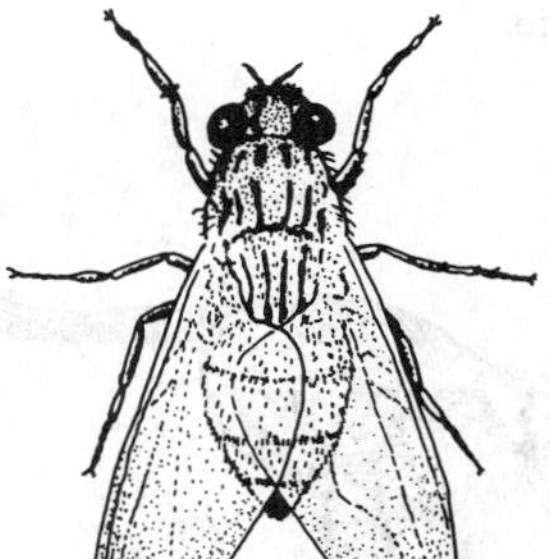

housefly

household *Archaeology.* a group of people whose members regularly eat and sleep together. Also, **household unit.**
housekeeping *Computer Programming.* the administrative or overhead operations that are secondary to the main objectives of a computer program, such as file creation, setting up of constants, and the initialization or cleaning up of storage areas. Thus, **housekeeping operation, housekeeping run.** Also, BOOKKEEPING.
housekeeping gene *Genetics.* a gene that is expressed in all cells because it provides a basic function needed in all cell types.
housemound *Archaeology.* a low earth platform used by the ancient Maya Indians as a foundation for a house.
house physician *Medicine.* **1.** the senior resident in a hospital who is in charge of service during the absence of the attending physician. **2.** a physician on call and immediately available in a hospital or other such facility.
house sewer *Civil Engineering.* the exterior horizontal extension of a house drain connecting it to the public sewer.
housing *Engineering.* a protective or supportive container or covering for a machine or instrument.
Houskeeper seal *Engineering.* a seal between copper and warm glass that allows the copper to flex as the glass shrinks during cooling.

Houssay, Bernardo 1887–1971, Argentine physiologist; Nobel Prize for discovering role of pituitary hormone in sugar metabolism.

Houston Automatic Spooling Program see HASP.

Hovercraft *Aviation.* a transportation craft that is designed to travel at a short distance above the ground or water, moving (or hovering) on a cushion of air that is held in a chamber beneath the vehicle; some of these vehicles are now over 100 tons and can travel at over 100 miles per hour.

howardite *Geology.* an achondritic stony meteorite composed largely of calcic plagioclase and orthopyroxene, and having a lower iron and calcium content than eucrite.

Howe, Samuel Gridley 1801–1876, American physician and social reformer; pioneered education for the blind and blind deaf-mute.

Howell-Jolly bodies *Pathology.* small, ovoid fragments of nuclear remains, benign and of unknown significance, appearing in erythrocytes in various anemias and leukemias after splenectomy. (Named after William H. *Howell,* 1860–1945, American physiologist, and Justin M. J. *Jolly,* 1870–1953, French histologist.)

Howe truss *Civil Engineering.* a large structural truss for spans up to eighty feet consisting of both vertical and diagonal members; may be composed of steel, timber, or both.

how explanation *Artificial Intelligence.* a facility provided in an expert system that gives the user the chain of rules it used to reach the conclusion.

howitzer *Ordnance.* a cannon having a tube length of 20 to 30 calibers that delivers projectiles with medium velocities, either by low or high trajectories; cannons with tubes exceeding 30 calibers may be considered howitzers if the high-angle fire permits range overlap between charges.

howl *Acoustical Engineering.* the buildup of a specific frequency or frequency band above an acceptable level, due to feedback.

howler *Electronics.* **1.** a device that produces an audio-frequency alarm. **2.** a sound-emitting test device. *Telecommunications.* a circuit that generates a telephone receiver off-hook tone of varying loudness.

howler monkey *Vertebrate Zoology.* any of various South American monkeys of the genus *Alouatta* in the family Cebidae, recognized by their loud resonant roars; further characterized by a long prehensile tail and a hunched posture.

howler monkey

howlite *Mineralogy.* $Ca_2B_5SiO_9(OH)_5$, a white monoclinic mineral occurring in nodular or earthy form, or as tabular crystals, having a specific gravity of 2.45 and a hardness of 3.5 on the Mohs scale; found in colemanite and borax deposits.

howl repeater *Telecommunications.* a state of telephone-repeater operation in which more energy is returned than is sent, and oscillation or singing occurs.

Howship's lacuna *Histology.* a minute depression in the surface of a bone that is being resorbed.

HP or **H.P.** high pressure; high power; horsepower; house physician.

H_p the international meteorological symbol for station elevation.

hp or **h.p.** horsepower.

HPA high-power amplifier.

HpaI *Genetics.* GTT/AAC, a restriction endonuclease derived from *Haemophilus parainfluenzae.*

h parameters *Electronics.* hybrid parameters of the four-terminal network equivalent of a transistor.

HPF or **H.P.F.** highest possible frequency; high-power field.

hpf or **h.p.f.** high-power field.

H plane *Electromagnetism.* the plane of an antenna that contains the magnetic field vector of linearly polarized radiation.

H-plane bend *Electromagnetism.* a waveguide bend in which the structure changes only in a direction perpendicular to the polarization direction.

H-plane T junction *Electromagnetism.* a T junction in a waveguide in which the structure changes only in the plane containing the magnetic field.

HPLC high-performance liquid chromatography.

HPLC FT-IR *Spectroscopy.* a combination of high-performance liquid chromatography (HPLC) analysis coupled with spectroscopic instrumentation capable of performing a Fourier transformation on the infrared spectrum (FT-IR) in a chromatographic separation.

H^+/p ratio *Biochemistry.* the quantity of protons that travel through a proton ATPase per molecule of ATP secreted at the catalytic site.

HPUra *Genetics.* an inhibitor of DNA polymerase in certain types of bacteria.

HQ or H.Q. headquarters. Also, **hq** or **h.q.**

hr. or **hr** hour; hours.

HRAF Human Relations Area Files.

H-RAS *Oncology.* a proto-oncogene containing point mutations; found in carcinoma of the colon, lung, and pancreas, and also in melanoma.

H-R diagram see HERTZSPRUNG-RUSSELL DIAGRAM.

H II region *Astronomy.* a cloud of hydrogen ionized by the ultraviolet radiation of hot stars embedded within it or lying nearby.

HRF histamine-releasing factor.

HRI height-range indicator.

HRL *Aviation.* the airport code for Harlingen, Texas.

H rod *Mechanical Devices.* a type of drill rod whose outside diameter measures 8.89 centimeters or 3.5 inches.

hrs. or **hrs** hours.

hrz. or hrz horizontal; horizon; hertz.

H.S. or **HS** house surgeon.

h.s. or **hs** at bedtime. (From Latin *hors somni.*)

H scan see H DISPLAY.

H scope *Electronics.* a radarscope that produces an H display.

HSGT high-speed ground transport.

HSL high-speed launch.

HSR heterochromatically staining region; homogeneous staining region.

HST hypersonic transport.

H strand *Genetics.* heavy strand; the strand of a dsDNA molecule having the higher buoyant density when the dsDNA is denatured and subjected to equilibrium centrifugation.

hsuan-pan *Mathematics.* a Chinese form of abacus in which each wire has two markers in the upper section and five in the lower.

H substance *Biochemistry.* a polysaccharide precursor in the biosynthesis of blood group antigens.

HSV *Aviation.* the airport code for Huntsville/Decatur, Alabama.

HT or **H.T.** high tide; high tension.

Ht total hypothermia.

ht or **ht.** height.

h.t. or **ht** at this time. (From Latin *hoc tempore.*)

HTGR high-temperature gas-cooled reactor.

HTLV-I human T-lymphotropic virus type I.

HTLV-II human T-lymphotropic virus type II.

HTLV-III human T-lymphotropic virus type III.

Huang He or **Huang Ho** see YELLOW RIVER.

huangho deposit *Geology.* a coastal-plain deposit of alluvium overlying a flood plain or other level surface beyond the normal reach of the sea, but stretching sideways into marine beds of equivalent age.

huarizo *Vertebrate Zoology.* a hybrid offspring of the female alpaca and the male llama, bred for its silky fur.

hub *Mechanical Devices.* **1.** the cylindrical center of any circular, rotating part. **2.** the part of a wheel or fan into which an axle or blades are attached. **3.** the part of a lock that causes the bolt to move. **4.** a steel punch from which a die for a coin or medal is made. **5.** a small coupling used to connect two pieces of a pipe. *Transportation Engineering.* a central airport from which a carrier's routes radiate, and through which service is funneled. A route arrangement based on a hub is sometimes called a **hub and spoke system.** *Graphic Arts.* one of a set of raised horizontal ridges on the spine of a leatherbound book. *Computer Technology.* a socket on a control panel or plugboard into which an electrical lead may be connected in order to distribute signals to multiple sites.

Hubbard tank *Medicine*. a tank designed to allow a patient to be immersed for a variety of therapeutic underwater exercises; generally used for exercising the trunk and lower limbs.

hubbing *Metallurgy*. the process of producing cavities in a die by pressing a plug into it.

Hubble, Edwin 1889–1953, American astronomer; discovered and classified galaxies outside the Milky Way; formulated Hubble's law.

Hubble classification of galaxies *Astronomy*. a classification system for galaxies, originating with Edwin Hubble and still used today with modification; it divides galaxies into spirals, ellipticals, and irregulars, with each category having several subclasses.

Hubble constant *Astrophysics*. the constant that scales the expansion velocity of the universe; measured in kilometers per second per megaparsec, its value lies between 50 and 100 and is the subject of considerable debate among cosmologists.

Hubble diagram *Astronomy*. a diagram that plots a galaxy's redshift versus its distance.

Hubble flow *Astrophysics*. the overall smooth and linear increase in the redshifts of galaxies with increasing distance.

Hubble-Sandage variable stars *Astronomy*. massive, blue, supergiant variable stars that are the most luminous in the Galaxy and are easily identified in other nearby galaxies, such as M 31 and M 33. Also, S DORADUS VARIABLES after the prototype star in the Large Magellanic Cloud.

Hubble's law *Astronomy*. a law proposed by Edwin Hubble that a galaxy's redshift is directly proportional to its distance; the constant of proportionality is the Hubble constant.

Hubble's Variable Nebula *Astronomy*. a V-shaped reflection nebula illuminated by the variable star R Monocerotis. Also, NGC 2261.

Hubble time *Astrophysics*. the age of the universe based on a given value for the Hubble constant; for a value of 50, the age is 20 billion years.

hub cannon *Ordnance*. a cannon mounted through the propeller hub of an aircraft.

hubcap *Mechanical Devices*. a metal cap placed over the end of an axle, as on an automobile or other motor vehicle.

Huber, Francois 1750–1831, Swiss naturalist; pioneer investigator of the life of honeybees.

Huber's reagent *Analytical Chemistry*. an aqueous solution of potassium ferrocyanide and ammonium molybdate, used to detect most free mineral acid; the formation of a reddish-brown precipitate or turbidity indicates a positive test.

Huble, David Hunter born 1926, American neurobiologist; with Wiesel, Nobel Prize for work on processing of visual information.

Hubl's reagent *Analytical Chemistry*. a solution of mercuric chloride and iodine in alcohol, used to determine the iodine number for oils and fats.

Hubner rhomb *Optics*. a glass rhomb that compares two illuminated surfaces as the rhomb's angles transmit two light beams, so that they emerge juxtapositioned at the rhomb's fine edge; used in photometry.

hub ring *Computer Technology*. a plastic ring fitted around the center hole of a floppy disk to prevent wear and extend the life of the disk.

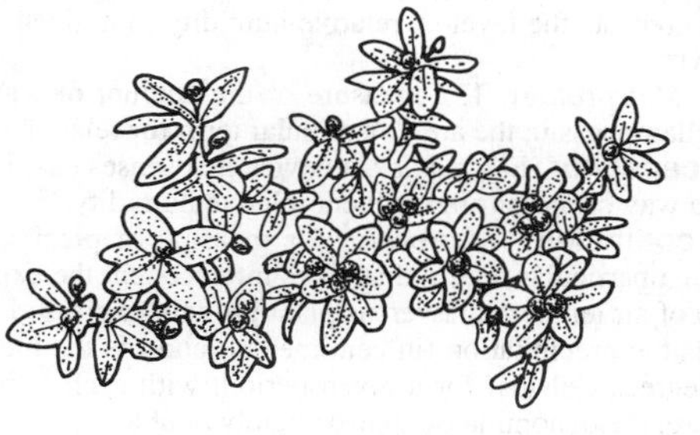

huckleberry

huckleberry *Botany*. **1.** the common name for shrubs of the genus *Gaylussacia* in the heath family Ericaceae, cultivated for their edible, dark-blue acidic berries. **2.** the fruit of such a shrub.

HUD head up display.

huddling *Behavior*. see CONTACT BEHAVIOR.

Hudson Bay *Geography*. an inland arm of the Atlantic in northeastern Canada.

Hudsonian life zone *Ecology*. a forest region, which includes Labrador and southern Alaska, that contains such vegetation as spruce and lichens, and such animals as the moose, woodland caribou, and mountain goat.

hudsonite see CORTLANDITE.

hue *Optics*. the property by which a color is perceived as resembling light of a particular wavelength in the visible spectrum, without regard to brightness or saturation; the aspect of color that is identified by words such as "red," "yellow," or "green." Colors such as white, gray, and black do not exhibit hue. *Graphic Arts*. **1.** a color within a certain range of colors, such as brownish or the various blues. **2.** a product of two colors combined, such as blue-violet.

hüebnerite *Mineralogy*. $Mn^{+2}WO_4$, a brownish-red to black monoclinic mineral occurring as aggregates of prismatic crystals, having a specific gravity of 7.12 to 7.18 and a hardness of 4 to 4.5 on the Mohs scale; found in hydrothermal veins and alluvial deposits; a principal ore of tungsten. Also, **huebnerite.**

hue control *Electronics*. in color television receivers, a control to vary the chrominance signals with respect to the burst signal. Also, PHASE CONTROL.

Huey *Aviation*. a type of helicopter extensively used by the U.S. Army in Vietnam.

Huffman code *Computer Technology*. a data-compression scheme in which each character is assigned a unique bit pattern such that the most frequently appearing characters have the fewest bits and the least frequently appearing characters have the most bits.

hugger see BACKS.

Huggins, Charles Brenton born 1901, Canadian-born American surgeon; Nobel Prize for hormone treatment of prostate cancer.

Huggins, William 1824–1910, English astronomer; made first spectroscopic observation of a nova; measured radial velocity of Sirius.

Huggins' equation *Materials Science*. an equation used to plot the relationship of viscosity data as a function of concentration; used to determine intrinsic viscosity.

Hughes formula 1189 see HF 1189.

Hughes press *Biotechnology*. a device that disrupts cells prior to downstream processing by forcing frozen cells through small openings under high pressure.

Hugoniot function *Physics*. a function specifying the possible states immediately following the passage of a shock-type wave front, and that relates pressure as a function of the specific volume of matter.

Huhner test *Pathology*. a test for male fertility; the analysis of the secretions obtained from the vaginal fornix and the endocervical canal after coitus to determine the number and condition of spermatozoa present. (Named after Max *Huhner*, 1873–1947, American urologist.)

hull *Botany*. **1.** the outer, often hard covering of a seed or fruit. **2.** the calyx of some fruits, such as the leaves on the stem of a strawberry. *Naval Architecture*. the floating shell of a vessel; it does not include the masts, rigging, and sails. *Ordnance*. the outer casing of a rocket or guided missile; the armored body of a tank. *Aviation*. a central structure of a flying boat, corresponding to the fuselage on other types of aircraft

Hull, Albert Wallace 1880–1966, American physicist; independently discovered powder method of X-ray crystallography.

Hull cell *Physical Chemistry*. an electrodisposition cell that operates within a concurrent range of known current densities.

hulsite *Mineralogy*. $(Fe^{+2},Mg)_2(Fe^{+3},Sn)BO_5$, a black monoclinic mineral occurring in small crystals or tabular masses, having a specific gravity of 4.5 to 4.62 and a hardness of about 3 on the Mohs scale; found in metamorphosed limestone.

hum *Electronics*. an interfering signal whose frequency is that of the AC power, caused by inadequate filtering of the DC power supply or by direct pickup of AC power signals.

human *Vertebrate Zoology*. **1.** a member of the species *Homo sapiens*, especially the modern subspecies *Homo sapiens sapiens*, the group of mammals characterized by an erect stance, a large, highly developed brain, and the use of tools and language. **2.** relating to or characteristic of this species. **3.** any member of the genus *Homo*.

human being *Vertebrate Zoology*. another name for a human.

human biogeography *Ecology*. the science of human population distribution.

human chorionic gonadotropin *Biochemistry*. a gonadotropic hormone that is produced by the placenta and that has biological effects similar to those of luteinizing hormone.

human cytomegalovirus group *Virology*. viruses of the subfamily Betaherpesvirinae that infect humans.

human ecology *Ecology.* the application of the principles of ecology to humans and human societies.

human embryo lung cells *Virology.* nontransformed diploid cells that are used in the production of vaccines against certain viruses, including the rabies virus.

human ethology *Behavior.* a branch of ethology that is concerned with observation of the behavior of the human species, especially for the purpose of establishing general behavioral characteristics comparable to the species characteristics of animals.

human-factors engineering *Design Engineering.* the design of machines or systems taking into account the life styles, capabilities, and limitations of human beings.

human genetics *Genetics.* the study of heredity in humans.

human geography *Geography.* the geographical study (usually comparative) of the relationships between human beings and the earth's features.

human immunodeficiency virus *Virology.* the full name for the HIV virus, the causative agent of AIDS. See HIV VIRUS.

humanistic psychology *Psychology.* an approach to psychology that emphasizes an individual's ability to direct his life and assume responsibility for his actions. Also, HOLISTIC PSYCHOLOGY.

human leuckocyte antigen *Immunlogy.* the human major histocompatibility antigen occurring on nucleated cells, including lymphocytes.

human measles immune serum *Immunology.* an antiserum derived from the blood of a person who has contracted and recovered from measles.

human natural killer cells *Immunology.* any lymphocytes that are able to lyse virally infected and tumor cells as part of the body's natural defense against pathogenic invasion and malignancy.

human paleontology see PALEOANTHROPOLOGY

human papilloma virus *Virology.* HPV, a virus that causes common warts on the feet and hands, as well as lesions of the mucous membranes of the oral, anal, and genital areas; over fifty types of HPV have been identified, and some are associated with cancerous conditions. Transmission can occur through sexual contact; interferon, laser treatment, and surgery are used to control, not cure, this virus.

human placental lactogen see CHORIONIC SOMATOMAMMOTROPIN.

human protein C *Genetics.* an anticoagulant that is produced by genetically engineered bacteria; used to inactivate coagulation cofactors 5 and 8c and mediate clot lysis by tissue plasminogen activator.

human race *Anthropology.* humans as a group; all human beings.

Human Relations Area Files *Anthropology.* an extensive cross-cultural reference system giving comprehensive information on hundreds of world cultures.

human resources *Industrial Engineering.* a general term for the various systems, procedures, and services involving the employees of an organization, including such activities as recruiting and hiring new employees, reviewing the performance of existing employees, dealing with employee grievances, complying with government laws and regulations for workers, and administering employee-benefit programs.

human-resources department *Industrial Engineering.* the division of an organization concerned with human resources; a newer term that is now generally preferred in place of the term *personnel department.*

human-resources planning *Industrial Engineering.* the development of a strategy for meeting the future staffing needs of an organization.

human rhinoviruses see COMMON COLD VIRUSES.

human T-lymphotropic virus type I *Virology.* a retrovirus of the subfamily Oncovirinae that can transform normal human peripheral-blood T cells in vitro and causes adult T-cell leukemia.

human T-lymphotropic virus type II *Virology.* a retrovirus of the subfamily Oncovirinae, in which the genome is only partly homologous with that of HTLV-I and the pathological potential is unknown.

human T-lymphotropic virus type III *Virology.* a former term for human immunodeficiency virus.

hum bar *Electronics.* a dark, horizontal bar in a television picture caused by hum interference in the video signal.

Humble gauge *Petroleum Engineering.* an apparatus used for measuring bottom-hole pressure in an oil well; through a stuffing box, a piston acts upon a helical spring in tension.

Humble relation *Petroleum Engineering.* an equation used for estimating the porosity of a formation, employing measurements and data created with a microlog or other contact-resistivity apparatus.

Humboldt, Baron Alexander von 1769–1859, German naturalist; founder of modern geography and meteorology; explored Latin America and Asia; drew the first isothermal map.

humboldtine *Mineralogy.* $Fe^{+2}C_2O_4 \cdot 2H_2O$, a yellow monoclinic mineral usually occurring as fibrous crusts, having a specific gravity of 2.28 and a hardness of 1.5 to 2 on the Mohs scale; found in brown coal and black shale. Also, OXALITE.

hum-bucking coil *Acoustical Engineering.* an additional coil wrapped around the field coil of a moving-coil loudspeaker, in opposition to the voice coil, so that hum is eliminated.

humectant *Chemistry.* a substance having hydrophilic properties and a stabilizing effect on the water content in a material, maintaining water content in a narrow range regardless of humidity fluctuations.

humeral *Anatomy.* of or relating to the humerus.

humeral circumflex artery *Anatomy.* an artery that winds around the neck of the humerus in the upper arm.

humeroradial *Anatomy.* of or relating to the humerus and the radius.

humeroscapular *Anatomy.* of or relating to the humerus and the scapula.

Hume-Rothery, William 1899–1968, English metallurgist; studied alloys; discovered Hume-Rothery rules of intermetallic compounds.

Hume-Rothery phases *Materials Science.* intermetallic phases that occur at definite valence electron-to-atom ratios in a number of different systems.

Hume-Rothery ratios *Materials Science.* a description of the fact that alloy systems exhibit different phases, designated by b-, g-, or e-phase, according to the ratio of the number of valence electrons to the number of atoms; these ratios are 1.50 for the b-phase, 1.62 for the g-phase, and 1.75 for the e-phase.

Hume-Rothery rules *Materials Science.* a set of criteria that must be satisfied in order for two elements to exhibit complete solid solubility. The rules are: (a) the crystal structure of each component must be the same. (b) the size of the atoms of each component must not differ by more than 15%. (c) the elements should not form compounds with each other; that is, there should be no appreciable difference in the electronegativities of the two elements. (d) the elements should have the same valence. Also, **Hume-Rothery (solid) solubility rules.**

humeroulnar *Anatomy.* of or relating to the humerus and the ulna.

humerus *Anatomy.* the bone of the upper arm, extending from the elbow to the shoulder.

humic *Geology.* relating to or derived from humus.

humic acid *Organic Chemistry.* any organic acid that is obtained or derived from humus.

humic coal *Geology.* a coal derived from peat through the process of humification and characterized by a high humic-acid content. Also, CHAMEOLITH, HUMULITH.

humicolous *Ecology.* describing a plant that grows in semi-arid soil. Thus, **humicole.**

humid climate *Meteorology.* a climate characterized by dense forest vegetation. Also, FOREST CLIMATE.

humidifier *Mechanical Engineering.* a device for adding moisture to the air or another gas. *Medicine.* specifically, a home or office appliance of this type, used to provide a desired level of humidity for persons having various conditions associated with hypersensitivity to dry air.

humidify *Engineering.* the process of increasing the amount of water vapor in air or other gases. Thus, **humidification.**

humidistat *Engineering.* a regulatory device that automatically measures and controls the level of relative humidity in a given space. Also, HYDROSTAT

humidity *Meteorology.* **1.** a measure of the amount of water vapor or degree of dampness in the air. **2.** a popular term for relative humidity.

humidity capacitor *Electronics.* a device that senses changes in capacitance as a way of measuring ambient relative humidity.

humidity coefficient *Meteorology.* a measure of precipitation effectiveness in a particular region that takes into account the exponential relationship of air temperature versus plant growth; expressed as $P/(1.07)^t$, where P is the precipitation (in centimeters) and t is the mean temperature (in degrees Celsius) for a given period; with each 10°C rise in air temperature, the denominator approximately doubles.

humidity element *Engineering.* the part of a hygrometer that senses the amount of water vapor in the atmosphere.

humidity index *Meteorology.* a mathematical index of the degree of water surplus over water need at a given site; used as a component of the moisture index and as a basis for the classification of dry climates.

humidity indicator *Inorganic Chemistry.* a term used for a cobalt salt, such as cobaltous chloride, $CoCl_2$, that changes color as the humidity changes. Such compounds range in color from greenish-blue when anhydrous, to pink when hydrated.

humidity mixing ratio *Meteorology.* the ratio of the mass of water vapor within a sample of moist air to the mass of the associated dry air, expressed as grams of water vapor per kilogram of air.
humidity province *Meteorology.* an area in which the precipitation effectiveness of the climate produces a particular biological consequence, particularly climatic climax forms of vegetation such as rainforest, tundra, or the like.
humid transition life zone *Ecology.* the climate and biotic communities of the northwest coniferous forest of the north-central United States.
humification *Geology.* a process in which humus or humic acids are formed from organic matter by slow oxidation.
humin *Organic Chemistry.* a pigment formed by the acid hydrolysis of a protein that contains tryptophan. *Geology.* see ULMIN.
Humiriaceae *Botany.* a mainly American family of tropical, dicotyledonous evergreen trees and shrubs in the order Linales, characterized by perfect flowers having five sepals and five petals and an intrastaminal nectary disk.
humite *Mineralogy.* $(Mg,Fe^{+2})_7(SiO_4)_3(F,OH)_2$, a white, light yellow, or dark orange to chestnut-brown, brittle, vitreous orthohombic mineral of the humite group; found as small crystals in impure marbles.
humivore *Ecology.* an organism that feeds on humus.
hummingbird *Vertebrate Zoology.* a small, brightly colored bird of the large family Trochilidae, having a bill that is long, narrow, curved, or tube-shaped, an extensile tongue, and long, rapidly beating wings capable of special maneuverability (e.g., hovering, flying backward, sideways, downward, or upward); found in North, Central, and South America wherever nectar-producing flowers bloom.

hummingbird

hummock *Geology.* **1.** a conical or rounded, usually equidimensional mound, hillock, or other small elevation. Also, HAMMOCK. **2.** a frost mound. *Hydrology.* a mound or pile of broken, fresh, or weathered floating ice that has been forced upward by pressure.
hummocked ice *Oceanography.* pressure ice that is characterized by randomly arranged hummocks, having less definite form than tented or rafted ice.
hummocky *Geology.* an uneven topographical surface, especially one characterized by hummocks.
Humod *Geology.* a suborder of the soil order *Spodosol* that has a dark to black upper layer and is characterized by a high accumulation of aluminum-rich organic matter with little iron.
hum modulation *Electronics.* the modulation of a signal by hum interference.
humodurite see TRANSLUCENT ATTRITUS.
humogelite see ULMIN.
humor *Psychology.* in ancient and medieval thought, one of the four elemental fluids of the body: blood, phlegm, black bile, and yellow bile; regarded as determining or influencing physical and mental health according to their proportion in the body. *Physiology.* any fluid occurring normally in the body
humoral *Immunology.* referring to a substance that occurs in relation to body fluids, such as circulating antibodies.
humoral antibodies *Immunology.* the proteins, produced in reaction to antigens in body fluids, that play a major factor in the processes of immune responses.
humoral immunity *Immunology.* a type of immunity that is mediated by factors in the body fluids, such as antibodies or complements.
Humox *Geology.* a suborder of the soil order *Oxisol* that develops in cool, humid climates at high altitudes and is characterized by being high in organic matter and well-drained but moist almost all year.
humpback whale *Vertebrate Zoology.* a large baleen whale of the genus *Megaptera* in the family Balaenopteridae, characterized by a humped back, ridged underside, and long knobbed fins; inhabits the oceans of the world.

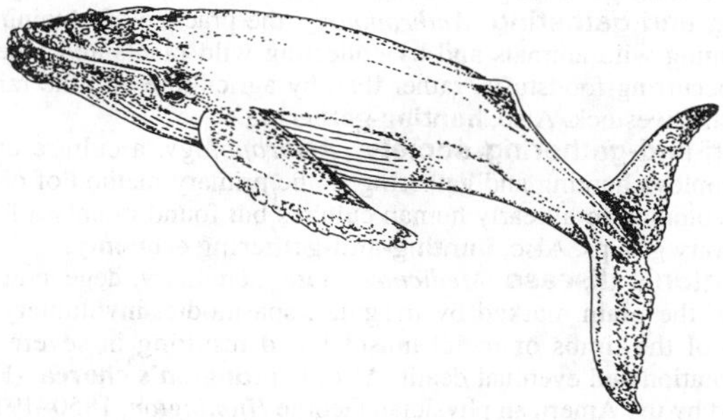

humpback whale

Humphreys series *Spectroscopy.* a set of spectral lines in the far-infrared region of the hydrogen spectrum.
Humphries equation *Thermodynamics.* an expression for the ratio of specific heats (at constant pressure to constant volume) of moist air as a function of the water-vapor pressure: $g = C_p/C_v = 1.40 - (0.1)p_w/p_a$ where p_w is the water-vapor pressure and p_a is the total atmospheric pressure.
humulith see HUMIC COAL.
Humult *Geology.* a suborder of the soil order *Ultisol* that develops in mountainous areas with high rainfall and is characterized by being well drained, having a high organic-carbon content, and a thick surface horizon.
humus *Geology.* dark, colloidal, amorphous, relatively inert organic material in soil, resulting from the decomposition of plant and animal matter.
hunchback *Medicine.* an informal and disparaging term for a person who has an abnormal hunched posture because of a deformity of the spine (kyphosis). See KYPHOSIS.
hundred-percent rectangle *Ordnance.* the rectangle in which very nearly all shots fired with the same data under the same conditions will fall; its center is the center of dispersion; its length is eight probable errors in range and its width is eight probable errors in direction. Also, RECTANGLE OF DISPERSION.
hundredweight *Metrology.* **1.** a unit of weight equal to 100 pounds. Also, SHORT HUNDREDWEIGHT. **2.** a unit of troy weight equal to 100 troy pounds. *Science.* a unit of weight equal to 100 U.S. pounds or 112 Imperial pounds.
Hund rules *Atomic Physics.* the two rules that relate the angular momentum of an atom to the number of electrons in it and their electron configuration.
Hundsdiecker reaction *Organic Chemistry.* the reaction of a silver salt, such as silver carboxylate, with bromine in carbon tetrachloride to produce an alkyl halide containing one less carbon.
hungate tube *Biotechnology.* a gas-tight, tube-shaped glass vessel with a rubber stopper that is held in place by a screw cap.
hunger *Psychology.* a desire or craving for something felt to be lacking.
hung shot *Engineering.* a delayed explosion; a shot that does not explode immediately after detonation.
Hunner's ulcer *Medicine.* a lesion of the urinary bladder occurring in chronic interstitial cystitis. (Named for Guy LeRoy *Hunner*, 1868–1957, American surgeon.)
hunt *Engineering.* to fluctuate about a midpoint due to instability; used to describe a weaving or yawing vehicle or rocket, an oscillating control surface, a wavering indicator, and the like.
Hunter, John 1728–1793, English surgeon; wrote physiologies of human teeth and uterus; developed ligature technique for aneurysms.
hunter-gatherer *Anthropology.* a member of a society that practices hunting and gathering as the means of obtaining food.
hunter's moon *Astronomy.* the full moon in October, which for three or four days around full rises only a little later each night. (Thought of as providing more light for a night hunter.)

Hunter's syndrome *Medicine.* a hereditary defect, an X-linked mucopolysaccharidosis that is caused by deficient sulfoid uronate sulfatase and characterized by dermatan sulfate and heparitin sulfate in the urine, abnormal skeletal development, retinitis pigmentosa, skin lesions, progressive deafness, and heart problems. (Named for Charles *Hunter,* English physician.)

hunting *Engineering.* the tendency of a mechanical or electrical system to oscillate about its normal position, frequency, or speed due to imperfections in the governing device or inherent instabilities in the system.

hunting and gathering *Anthropology.* the practice of obtaining food by hunting wild animals and by collecting wild plants and other naturally occurring foodstuffs, rather than by agriculture and the raising of domestic livestock. Also, **hunting-gathering.**

hunting-and-gathering society *Anthropology.* a culture or group that employs hunting and gathering as the primary method of obtaining food; common to all early human cultures but found in only a few contemporary groups. Also, **hunting-and-gathering economy.**

Huntington's disease *Medicine.* a rare, hereditary, degenerative disease of the brain marked by irregular, spasmodic, involuntary movements of the limbs or facial muscles and resulting in severe mental deterioration and eventual death. Also, **Huntington's chorea.** (First described by the American physician George *Huntington,* 1850–1916.)

huntite *Mineralogy.* $CaMg_3(CO_3)_4$, a soft white trigonal mineral occurring as chalklike masses, having a specific gravity of 2.696 and an undetermined hardness.

HU protein *Biochemistry.* a protein, commonly found in *E. coli,* that resembles the histonelike DNA binding proteins and seems to interact with DNA to produce nuclesomelike structures.

hura *Materials.* the pale, soft wood of the *Hura crepitans* tree of Central and South America; used for general carpentry and in the construction of crates, furniture, and particleboard.

hureaulite *Mineralogy.* $Mn_5^{+2}(PO_4)_2[PO_3(OH)]_2{\cdot}4H_2O$, a yellow, reddish, or gray monoclinic mineral occurring as prismatic crystals or in massive form, having a specific gravity of 3.17 to 3.19 and a hardness of 3.5 on the Mohs scale; found as a secondary mineral in granite pegmatites.

Hurler's syndrome *Medicine.* an inherited metabolic disorder in which there is an accumulation of abnormal intracellular material; characterized by skeletal deformities, mental retardation, and early death. (Named for Gertrude *Hurler,* Austrian physician.)

Huron, Lake *Geography.* the second largest of the Great Lakes (area: 23,010 square miles), between the Lower Michigan peninsula and Ontario, Canada.

Huronian *Geology.* a stratigraphic division of Proterozoic time of the Canadian Shield.

hurricane *Meteorology.* a severe tropical cyclonic storm with winds in excess of 70 miles per hour, typically originating over warm tropical seas and moving over the North Atlantic Ocean, the Caribbean Sea, the Gulf of Mexico, the eastern North Pacific Ocean, or the western South Pacific Ocean.

hurricane

hurricane beacon *Engineering.* a balloon-mounted beacon that is designed to transmit radio signals from the eye of a hurricane.

hurricane-force wind *Meteorology.* on the Beaufort wind scale, a wind having a speed of at least 64 knots (73 miles per hour); it does not necessarily occur as part of a hurricane.

hurricane lamp *Engineering.* an oil lamp or candlestick having a tall glass chimney to protect the flame from wind.

hurricane monitoring buoy *Meteorology.* a free-floating, automated weather station that is designed to be expendable and is used solely for the purpose of monitoring typhoons and hurricanes, in order to ascertain their positions and forecast their movement.

hurricane warning *Meteorology.* a local weather forecast that warns of impending hurricane-force winds; at sea, the warning signals consist of two square red flags with black centers by day, and a white lantern between two red lanterns at night.

hurricane watch *Meteorology.* a local forecast that warns of the threat of hurricane conditions, in which residents are advised to be prepared, but otherwise to continue their normal activities.

hurricane wave *Oceanography.* a sudden rise in the level of the sea along the shore, associated with a hurricane.

hurricane wind *Meteorology.* a severe wind associated with an intense tropical cyclone, as distinguished from a hurricane-force wind. Also, TYPHOON WIND.

Hurst formula *Petroleum Engineering.* a formula used in reservoir material-balance analysis by which a special relationship interrelates production data and field pressure at various times.

Hurst method *Petroleum Engineering.* a technique for calculating the bottom-hole static pressure of an oil well; a pressure buildup over a brief time span is used to present a graphic extrapolation or conclusion.

Hurter and Driffield curve see CHARACTERISTIC CURVE.

Hurthle cells *Pathology.* large, atypical eosinophilic cells occasionally seen in the thyroid gland, which seem to be more typical of parathyroid tissue.

Hurtley's test *Pathology.* a method for establishing the presence of acetoacetic acid by adding 2 ml of strong hydrochloric acid and 1 ml of fresh 1% sodium nitrate solution to any substance; the slow appearance of violet or purple color indicates acetoacetic acid.

Hurwitz polynomial *Mathematics.* a polynomial, all of whose zeros have negative real parts.

husbandry see ANIMAL HUSBANDRY.

husk *Botany.* the dry, often membranous outer covering of some seeds and fruit, as on an ear of corn.

Hutchinson, Sir Jonathan 1828–1913, British surgeon; specialized in ophthalmology, dermatology, and congenital syphilis.

Hutchinson-Gilford disease see PROGERIA.

Hutchinsonian niche see MULTIDIMENSIONAL NICHE.

hutchinsonite *Mineralogy.* $(Pb,Tl)_2As_5S_9$, a scarlet to deep cherry red, orthorhombic mineral occurring as tufts of prismatic crystals, having a specific gravity of 4.6 and a hardness of 1.5 to 2 on the Mohs scale; found in crystalline dolomite.

Hutchinson's freckle *Medicine.* a light brown patch on the skin that grows slowly and becomes dark, thick, and nodular; usually seen on the face of an elderly person; excision is recommended as it often becomes malignant. (After Sir Jonathan *Hutchinson.*)

Hutchinson's teeth *Medicine.* a deformity of the teeth of congenital syphilis marked by notched and narrowed incisal edges. (After Sir Jonathan *Hutchinson.*)

hut gene *Genetics.* a gene that is involved in histidine utilization.

Huttig equation *Thermodynamics.* an equation that relates the volume of gas adsorbed onto the surface of a nonporous solid to the volume of gas required to completely cover the surface of the solid in a unimolecular layer at constant pressure and temperature; the relation is dependent on the temperature, the heat of adsorption for the first layer, and the heat of liquefication of the adsorbate.

Hutton, James 1726–1797, Scottish geologist; founder of modern geology; studied origins of earth, valleys, and volcanoes; formulated the theory of uniformitarianism.

huttonite *Mineralogy.* $ThSiO_4$, a colorless to pale-cream monoclinic mineral of the monazite group, dimorphous with thorite, having a specific gravity of 7.1 to 7.2 and an undetermined hardness; found in beach sands as minute grains.

Huxley, Andrew Fielding born 1917, English physiologist; shared Nobel Prize for electrochemical analysis of nerve-cell discharges.

Huxley, Sir Julian 1887–1975, English biologist; studied ornithology and human development; founding director of UNESCO.

Thomas Henry Huxley

Huxley, Thomas Henry 1825–1895, English biologist; supporter of Darwin; interpreted Neanderthal fossils; coined the term *biogenesis*.

Huxley's layer *Histology*. the part of the inner epithelial root sheath of a hair follicle. (Named for T. H. *Huxley*.)

Huygenian eyepiece *Optics*. an eyepiece that consists of two plano-convex lenses separated by a space equal to half the sum of their focal lengths; this ocular is free from chromatic aberration and is commonly used in microscopes .

Huygens, Christian [hī´gənz; hoi´gənz] 1629–1695, Dutch mathematician and physicist; improved telescopic lenses; invented micrometer and pendulum clock; discovered rings of Saturn; formulated wave theory of light; discovered polarization.

Huygens' approximation *Mathematics*. a formula for approximating the length s of an arc: $s \approx (8c' - c)/3$, where c is the length of the chord on the arc and c' is the length of the chord on half the arc.

Huygens' principle *Optics*. the principle that each point of an advancing wavefront may also be considered the source of a new series of secondary waves.

Huygens' wavelet *Optics*. a secondary wave, described in Huygens' principle, that subsequently establishes the position of the wavefront.

HV or **H.V.** high voltage; high velocity.

H vector *Electromagnetism*. the magnetic field vector.

HVL half-value layer.

HW or **H.W.** high water.

Hwang Ho see YELLOW RIVER.

HWK *Anthropology*. a site at Olduvai Gorge in East Africa, the location of important finds of hominid remains. (An acronym for Henrietta Wilfrida Karongo, from the name of the wife of Louis Leakey, an early explorer of this site, and *karongo*, a Swahili word for "gully.")

hy. or hy henry; henries.

hyacinth *Botany*. any plant of the genus *Hyacinthus*; bulbous herbs of the lily family commonly cultivated as garden plants, having fragrant flowers that grow in cylindrical clusters. *Mineralogy*. see ZIRCON.

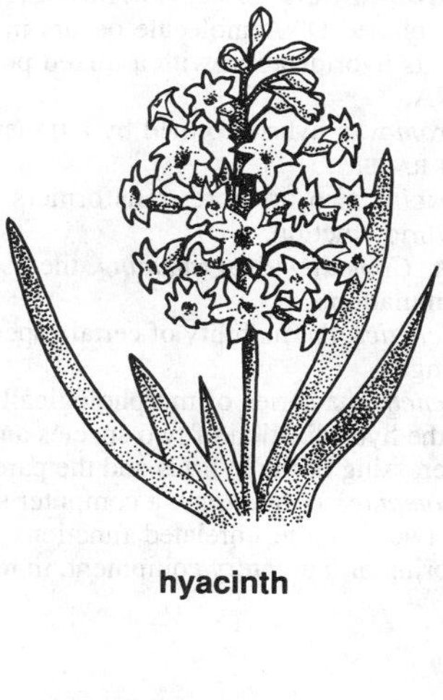

hyacinth

Hyades [hī´ ə dēz] *Astronomy*. a large, loose, open star cluster in Taurus that contains about 200 stars (about a dozen are visible to the naked eye); it lies about 150 light-years from the earth.

Hyaenidae *Vertebrate Zoology*. the hyenas and the aardwolf, a family of predatory carnivores of the suborder Feliforma, closely resembling domestic dogs, but with longer forelimbs that are more developed than their hindlimbs; native to Africa and southwest Asia. *Paleontology*. a family of carnivores in the superfamily Aeluroidea (Feloidea); represented today by the hyenas and the aardwolf; Miocene to present.

Hyaenodon *Paleontology*. a widespread genus of powerful creodonts in the extinct suborder Hyaenodontia and family Hyaenodontidae; Eocene to Miocene.

hyaenodont *Paleontology*. of or relating to creodonts of the extinct suborder Hyaenodontia.

Hyaenodontidae *Paleontology*. a family of creodonts in the extinct suborder Hyaenodontia; extant in the Early Tertiary.

hyal- a combining form meaning "glass" or "glassy," as in *hyaline*.

Hyalellidae *Invertebrate Zoology*. a family of amphipod crustaceans with no carapace, the sand hoppers, in the suborder Gammaridea.

hyaline *Biochemistry*. a horny substance found in hydatid cysts that closely resembles chitin in its properties and chemical composition. *Geology*. having a glasslike transparency.

hyaline cartilage *Histology*. a type of cartilage that appears translucent, bluish-white in the fresh condition; the cartilage cells (chondrocytes) lie in lacrinae in a matrix that contains collagen, which is masked by basophilia from the high content of proteoglycans.

hyaline cast *Pathology*. an almost-transparent urinary cast, slightly refractive, composed of homogeneous protein originating from cellular dissolution.

hyaline degeneration *Pathology*. an assortment of disintegrative processes influencing cellular change whereby the cytoplasm adopts a homogeneous, translucent eosinophilic form; seen in developed fibrous tissue, in smooth muscles of the uterus, or in the arterioles.

hyaline membrane *Histology*. **1.** the basement membrane between epithelial and connective tissues. **2.** the membrane lying between the inner fibrous sheath and outer root sheath in a hair follicle.

hyaline membrane disease *Medicine*. a frequently fatal condition of premature babies with acute respiratory distress, affecting the lining of the air spaces of the lungs, due to lung immaturity.

hyaline test *Invertebrate Zoology*. the translucent shell of some foraminiferans.

hyalinization *Pathology*. the forming of hyalin, a smooth, glassy substance resulting from some types of tissue and cell disintegration.

hyalinocrystalline *Petrology*. pertaining to a porphyritic rock texture in which the phenocrysts occur in an equal or nearly equal glassy groundmass.

hyalite *Mineralogy*. a colorless, clear or translucent variety of common opal, occurring as globular concretions and botryoidal crusts in cavities of rocks. Also, MÜLLER'S GLASS, WATER OPAL.

hyalo- a combining form meaning "glass" or "glassy," as in *hyaloplasm*.

hyalobasalt see TACHYLITE.

hyaloclastite *Geology*. a tufflike deposit formed by the flowing of basalt beneath water or ice and the resultant granulation or brecciation into small angular fragments. Also, AQUAGENE TUFF.

Hyalodictyae *Mycology*. in certain systems of taxonomy, a subdivision of the fungal spore group Dictyosporae that is characterized by transparent or hyaline spores.

Hyalodidymae *Mycology*. in certain systems of taxonomy, a subdivision of the fungal spore group Didymosporae that is characterized by transparent or hyaline spores.

Hyaloheliocosporae *Mycology*. in certain systems of taxonomy, a subdivision of the fungal spore group Helicosporae, which is characterized by transparent or hyaline spores.

hyaloid membrane *Anatomy*. a condensation of fine collagen fibers in places in the vitreous body of the eye.

hyaloophitic *Petrology*. of or relating to a texture of igneous rocks consisting primarily of a glassy groundmass; the texture is intermediate between hyalopilitic and hyalocrystalline.

hyalophane *Mineralogy*. $(K,BA)Al(Si,Al)_3O_8$, a colorless to white monoclinic mineral of the feldspar group occurring as prismatic crystals and in massive form, having a specific gravity of 2.58 to 2.88 and a hardness of 6 to 6.5 on the Mohs scale.

Hyalophragmia *Mycology*. in certain systems of taxonomy, a subdivision of the fungal spore group Phragmosporae that is characterized by pluriseptate spores.

hyaloplasm *Cell Biology.* **1.** an older term for cytosol, the portion of the cellular cytoplasm that is more fluid in nature and of a fine granular texture. **2.** see AXOPLASM. Also, **hyalotome.**

hyalopsite see OBSIDIAN.

Hyaloriaceae *Mycology.* a family of fungi belonging to the order Auriculariaceae that occurs exclusively on rotten palm wood in Brazil.

Hyaloscolecosporae *Mycology.* in certain systems of taxonomy, a subdivision of the fungal spore group Hyaloscolecosporae that is characterized by transparent or hyaline spores.

Hyaloscyphaceae *Mycology.* a family of fungi belonging to the order Helotiales that occurs on the decaying remains of higher plants.

Hyalospongidae *Paleontology.* a family of hexactinellid glass sponges, characterized by 6-rayed siliceous spicules; Cambrian.

Hyalosporae *Mycology.* in certain systems of taxonomy, a subdivision of the fungal spore group Amerosporae that is characterized by transparent or hyaline spores.

Hyalostuarosporae *Mycology.* in certain systems of taxonomy, a subdivision of the fungal spore group Stuarosporae that is characterized by transparent or hyaline spores.

hyalotekite *Mineralogy.* $(Ba,Pb,Ca,K)_6(B,Si,Al)_2(Si,Be)_{10}O_{28}(F,Cl)$, a white or gray, translucent, triclinic, pseudomonoclinic mineral occurring in crystalline masses, having a specific gravity of 3.82 and a hardness of 5 to 5.5 on the Mohs scale.

hyaluronate *Biochemistry.* an ester or salt of hyaluronic acid, an intermediate in the synthesis of the glycosaminoglycan of hyaluronic acid.

hyaluronic acid *Biochemistry.* an unbranched mucopolysaccharide of connective tissue that is composed of D-glucuronic acid and synthesized from D-glucose in the fibroblasts; serves to lubricate body tissue and block the spread of invading microorganisms.

hyaluronidase *Enzymology.* an enzyme that catalyzes the breakdown of hyaluronic acid; found in mammalian testicular and spleen tissue, in snake venom, and in certain species of *Streptococcus* and *Staphylococcus.* Also, **hyaluronate lyase.**

Hybodontoidea *Paleontology.* a suborder of Mesozoic sharks in the order Hybodontiformes; probably ancestral to the Ptychodontidae; Permian to Paleocene.

hybrid *Science.* something that is heterogeneous; a composite. *Genetics.* **1.** an offspring of two genetically different parents, or even different species. **2.** see HETEROZYGOTE. *Petrology.* **1.** an igneous rock formed by the assimilation of wall rock into a magma or by the mixing of two magmas. **2.** of or relating to a rock so formed. Thus, **hybrid rock.**

hybrid arrested translation *Molecular Biology.* an event in which the in vitro translation of a given mRNA molecule into protein is arrested due to its hybridization with a complementary strand of either RNA or DNA.

hybrid circuit *Electricity.* a circuit using two or more different types of components that perform similar functions. *Telecommunications.* an electrical or electronic circuit that allows the interconnection of two-way two-wire circuits with one-way four-wire circuits.

hybrid composite *Materials Science.* any fiber composite that contains more than one kind of fiber; generally characterized by high strength, excellent fatigue resistance, formability, and machinability.

hybrid computer *Computer Technology.* a computer that has both digital and analog capability.

hybrid DNA *Molecular Biology.* a double-stranded DNA molecule produced from single-stranded molecules of different sources by means of the formation of hydrogen bonds at their complementary nucleotide sequences.

hybrid dysgenesis *Genetics.* a syndrome of related genetic abnormalities caused by a transposable element that surfaces when certain strains of *Drosophila* mate.

hybrid engine *Space Technology.* a rocket engine that uses a liquid oxidizer to burn a solid fuel, or one that uses a combination of liquid and solid propellants. Thus, **hybrid rocket.**

hybrid enzyme *Enzymology.* a polymeric enzyme produced in heterozygous individuals that contains subunits differing in one or more amino acids, due to different genetic sequences in complementary alleles that code for different subunits of the enzyme.

hybrid integrated circuit *Electronics.* an integrated circuit composed of both integrated and microminiature discrete components.

hybrid interface *Computer Technology.* a device that connects a digital computer to an analog computer and performs the analog-digital conversion of signals.

hybrid inviability *Genetics.* the reduction of somatic vigor or survival rate in certain hybrid organisms.

hybridization the formation of a hybrid; specific uses include: *Genetics.* **1.** the pairing of individuals that belong to genetically different populations or species. **2.** the pairing of complementary RNA and DNA strands to make an RNA-DNA hybrid. **3.** the partial pairing of DNA single strands from genetically different sources. *Physical Chemistry.* the mixing of atomic orbitals of a single atom to give a new set of orbitals for that atom; this represents the blending of higher and lower energy orbitals to form orbitals of intermediate energy. *Materials Science.* the formation of a hybrid group of four electrons, giving rise to four equal tetrahedral covalent bonds.

hybridized orbital see HYBRID ORBITAL.

hybrid junction *Electricity.* a four-branch device that transfers energy from one branch to two of the remaining three branches; usually the energy is evenly divided between the two. Hybrid junctions can be made of resistors, transformers, or waveguides. Also, HYBRID-T, MAGIC-T.

hybrid magnet *Physics.* a superconducting magnet consisting of an external coil of niobium and titanium and an inner coil of niobium and tin.

hybrid manipulation language *Computer Programming.* an algebraic manipulation language that can accept and process the broadest range of mathematical expressions; includes special algorithms and notation for handling special classes of expressions, such as integrals.

hybrid merogony *Developmental Biology.* the fertilization of a section of an ovum of one species by the sperm of another species.

hybrid microcircuit *Electronics.* a microcircuit composed of diffused or thin-film elements interconnected with separate chip elements.

hybrid molecule *Biochemistry.* a nucleic acid molecule that is artificially hybridized for the purpose of studying and comparing nucleotide sequences.

hybrid network *Telecommunications.* a nonhomogeneous communication network that functions using signals of varying characteristics, e.g., analog and digital signals, two-wire and four-wire circuits.

hybridoma *Immunology.* a hybrid formed by the fusion of plasmacytoma (a primary tumor of the bone marrow) with lymphocytes, forming a distinctive antibody.

hybrid orbital *Physical Chemistry.* a combination of atomic orbitals that yields a new set of orbitals having the appropriate directional properties to account for the observed geometry of the chemical bonds. Also, HYBRIDIZED ORBITAL.

hybrid plasmid *Genetics.* a plasmid or phage that consists in part of the authentic DNA of the original genome and in part of the additional foreign sequences; produced by insertion.

hybrid problem analysis *Computer Programming.* an analysis that determines which parts of a problem are best suited for solution by a digital computer.

hybrid processing system *Computer Technology.* a distributed processing system that has both hierachical and horizontal organization and control.

hybrid programming *Computer Programming.* the use of hybrid routines to assist in a hybrid problem analysis; these routines deal with timing, function generation, simulation, and hardware diagnostics.

hybrid redundancy *Computer Technology.* a hardware redundancy scheme that is composed of fault-masking types, such as triple modular redundancy, and self-repair types, such as a standby replacement system; the system includes a bank of spare units so that if one unit fails it is automatically replaced by a spare.

hybrid relay *Electricity.* a type of relay in which solid-state elements are combined with moving contacts.

hybrid released translation *Molecular Biology.* an event in which the translation of a cloned DNA molecule occurs in a cell-free translation system due to its hybridization with a mixed population of mRNA or total cellular RNA.

hybrid ring *Electronics.* a hybrid formed by a transmission line closed in a ring. Also, RAT RACE.

hybrid set *Electricity.* two or more transformers connected to each other, forming a hybrid junction.

hybrid simulation *Computer Programming.* the use of a hybrid computer to perform simulation.

hybrid sterility *Genetics.* the inability of certain species hybrids to produce viable offspring.

hybrid swarm *Genetics.* a series of morphologically distinct individuals resulting from the hybridization of two species and subsequent interbreeding and backcrossing of the hybrids and the parents.

hybrid system *Computer Technology.* a computer system that is capable of performing two or more unrelated functions, such as word processing and monitoring of laboratory equipment, in real-time.

hybrid system checkout *Computer Technology.* the static verification of a hybrid computer that has been set up to solve a particular class of problems.
hybrid T see HYBRID JUNCTION.
hybrid tee *Electromagnetism.* a microwave hybrid waveguide junction formed by an E-H tee junction with matching internal elements so that it is reflectionless.
hybrid transformer *Electricity.* a single transformer that performs the functions of a hybrid set.
hybrid vigor see HETEROSIS.
hybrid wavefunction *Quantum Mechanics.* an approximate wavefunction constructed by adding known wavefunctions of the constituent parts of a system.
hybrid zone *Ecology.* the overlapping area of bordering populations in which interbreeding can occur.
hydantoin *Organic Chemistry.* $C_3H_4N_2O_2$, white crystals that melt at 220°C; used in pharmaceuticals, as a textile softener, and as a lubricant.
Hydatellales *Botany.* a small order of monocotyledonous flowering plants of the division Magnoliophyta, consisting of wholly or partly submerged aquatic annuals with a simplified internal anatomy.
hydathode *Botany.* a gland on the tips and margins of certain leaves through which water is lost by guttation.
hydatid *Medicine.* any structure resembling a cyst.
hydatidiform mole *Medicine.* an abnormal pregnancy in which a neoplastic mass of grapelike enlarged chorionic villi are formed in the uterus.
hydatidosis see ECHINOCOCCOSIS.
hydatogenesis *Geology.* a process in which mineral salts are crystallized or precipitated from normal aqueous solutions, as in the formation of evaporites.
Hydnaceae *Mycology.* a family of fungi belonging to the order Agaricales that is composed of some tooth fungi, wood rotting parasites, and edible mushrooms.
hydnocarpic acid *Organic Chemistry.* $C_{16}H_{28}O_2$, white crystals that melt at 60°C; formerly used to treat leprosy.
Hydnoraceae *Botany.* a family of leafless, highly modified terrestrial parasitic herbs of the order Rafflesiales, characterized by a lack of chlorophyll, a rhizomelike pilot root from which numerous haustorial roots branch out to parasitize the roots of other plants, and malodorous flowers that rise from the roots and barely reach the ground surface; native to Africa, Madagascar, and South America.
Hydopogonaceae *Botany.* a small family of slender mosses of the order Isobryales that attach to roots and twigs in aquatic habitats in northern South America; characterized by numerous short branches on elongated stems, terminal sporophytes on short lateral branches, and shiny erect leaves.
hydr- a combining form meaning "water," as in *hydrase*.
Hydra *Astronomy.* the Water-Snake, a long, relatively dim constellation of the spring sky that stretches west to east, roughly paralleling the celestial equator.
hydra *Invertebrate Zoology.* a common genus of small, sessile freshwater hydrozoans that reproduce by budding. (From the Greek myth of *Hydra,* a water serpent with many heads.)

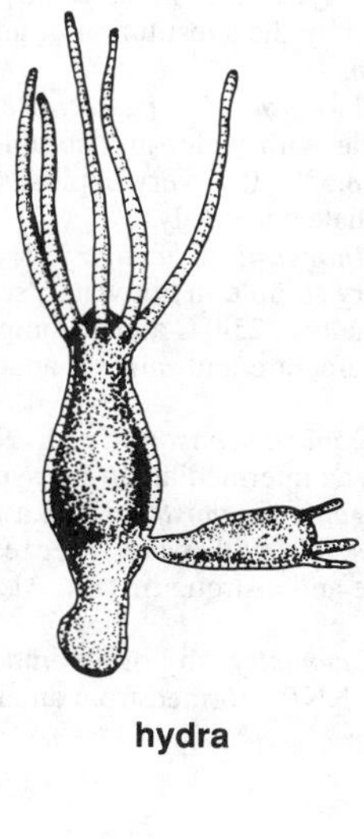

hydra

Hydrachnellidae *Invertebrate Zoology.* a family of freshwater mites, in the suborder Trombidiformes; the larvae are parasitic on aquatic insects and clams.
hydracrylic acid *Organic Chemistry.* $CH_2OH{\cdot}CH_2COOH$, a viscous liquid; soluble in water and alcohol; decomposes on heating.
Hydraenidae *Invertebrate Zoology.* a family of small coleopteran insects, mostly aquatic beetles that build protective egg cases spun from silk thread.
hydragogue *Medicine.* an agent that causes the discharge of watery fluid, as from the bowel.
hydralazine *Pharmacology.* $C_8H_8N_4$, a drug used orally or injected to lower blood pressure by dilating the blood vessels.
hydralazine hydrochloride *Pharmacology.* $C_8H_8N_4{\cdot}HCl$, an antihypertensive occurring as a white crystalline powder that is administered orally, intramuscularly, or intravenously.
Hydramoeba *Invertebrate Zoology.* naked amoebas in the family Amoebidae, lacking flagellated stages and parasitic on hydra.
hydranencephaly *Medicine.* the complete or almost complete absence of the cerebral hemispheres, with the space they normally occupy filled with cerebrospinal fluid.
Hydrangea [hi drān´jə; hī drān´jē ə] *Botany.* **1.** a genus of shrubs of the family Hydrangeaceae, widely grown as ornamental flowers. **2.** a specific plant of this genus, such as the **hydrangea bush,** *H. paniculata,* or the **dwarf hydrangea,** *H. macrophylla.*
Hydrangeaceae *Botany.* a widespread family of dicotyledonous woody plants and sometimes rhizomatous herbs of the order Rosales, often cultivated as ornamental shrubs and characterized by complex, large, many-flowered cymose inflorescences.
hydrant *Civil Engineering.* an upright fixture with a nozzle or other outlet, especially on a street to draw water for firefighting.
hydranth *Invertebrate Zoology.* in hydrozoans, the head or oral end of a polyp, with the mouth and tentacles.
hydrarch succession *Ecology.* an ecological succession that begins in an area with abundant water.
hydrargyria see MERCURY POISONING.
hydrarthrosis *Medicine.* the accumulation of fluid in a joint.
hydrase *Enzymology.* an enzyme that catalyzes either the addition or removal of water to a substrate, without hydrolyzing the substrate.
hydrastine *Organic Chemistry.* $C_{21}H_{21}NO_6$, an alkaloid isolated from golden seal, *Hydrastis canadensis*; white prisms that melt at 132°C.
hydrastinine *Organic Chemistry.* $C_{11}H_{13}O_3N$, a compound that is formed by the decomposition of hydrastine; used as a stimulant.
hydratase see HYDROLYASE.
hydrate *Chemistry.* **1.** a substance that contains water combined in the molecular form, such as $H_2SO_4{\cdot}H_2O$. **2.** a crystalline substance that contains molecules of water of crystallization, such as $Na_2SO_4{\cdot}7H_2O$. **3.** to form or become such a substance.
hydrated cellulose see HYDROCELLULOSE.
hydrated electron *Physical Chemistry.* an electron that is released in an aqueous solution by the ionization of a water molecule, so that it is surrounded by water molecules that are oriented to it. Also, AQUEOUS ELECTRON.
hydrated ion *Physical Chemistry.* an ion in an aqueous solution surrounded by water molecules that are oriented to it.
hydrated mercurous nitrate see MERCUROUS NITRATE.
hydrathalloysite see ENDELLITE.
hydration *Chemistry.* the incorporation of a water molecule into a compound, yielding a hydrate.
hydraucone *Mechanical Devices.* a spreading, conical draft tube, symmetrical with the axis of a water turbine; used in hydraulic installations.
hydraulic *Fluid Mechanics.* **1.** of or relating to liquid in motion. **2.** of or relating to the pressure created by forcing liquid through a relatively small orifice, pipe, or other channel. *Mechanical Engineering.* **1.** of or relating to a device, machine, or system that operates by hydraulic pressure. Thus, **hydraulic air compressor, hydraulic drill, hydraulic elevator, hydraulic motor. 2.** of or relating to a device, machine, or system that converts hydraulic power into mechanical power. Thus, **hydraulic actuator, hydraulic turbine.**
hydraulic accumulator *Mechanical Engineering.* **1.** an apparatus in which air or other gases function as a cushion or shock absorber in a hydraulic system. **2.** an apparatus in a hydraulic system that is used to store liquid at a constant pressure.
hydraulic amplifier *Control Systems.* a device that uses fixed or variable orifices to increase the power of a signal in a hydraulic servomechanism. Also, **hydraulic intensifier.**

hydraulic analog table *Fluid Mechanics.* an experimental or model apparatus that makes use of the hydraulic analogy wherein a shallow depth of water is made to flow steadily over a smooth horizontal surface bounded by vertical walls.

hydraulic analogy *Fluid Mechanics.* a similarity between the nature of the flow of a compressible gas and the flow of a shallow liquid; shock waves may appear under the proper conditions in calculations that include the neglect of vertical accelerations as well as certain restrictions on the specific heats of the gas.

hydraulic brake *Mechanical Engineering.* a motor vehicle brake that is applied by small pistons actuated by pressurized oil flowing through the brake line from a pedal-operated master cylinder.

hydraulic circuit *Mechanical Engineering.* a circuit or controller that operates by hydraulic control of water or fluid currents rather than by electric currents.

hydraulic classification *Mechanical Engineering.* the use of a hydraulic classifier to separate particles of different size from each other. Also, HYDRAULIC SEPARATION, HYDRAULIC SEGREGATION.

hydraulic classifier *Mechanical Engineering.* a classifying device employing gravitational force in which water rising vertically in a tube or column at a controlled rate separates lighter particles from heavier particles or smaller particles from larger ones. Also, HYDRAULIC SEPARATOR, HYDRAULIC SEGREGATOR.

hydraulic computer *Computer Technology.* a computer in which valves and fluids are used in place of electronic circuitry.

hydraulic conveyor *Mechanical Engineering.* a material handling system in which material discharged from a grinder is conveyed by a hydraulic pumping device to a disposal site.

hydraulic coupling see FLUID COUPLING.

hydraulic current *Oceanography.* a gravity flow of water through a channel, resulting from a tidally caused difference in the water levels at the two ends of the channel.

hydraulic discharge *Hydrology.* the direct release of groundwater from the zone of saturation into a body of surface water or onto land.

hydraulic dredge *Mechanical Engineering.* a dredge that operates by suction, often with a mechanical device for breaking up earth in front of the suction pipe, and with side-mounted pumps for extracting water and loose materials from within the main suction pipe.

hydraulic drive *Mechanical Engineering.* a device that uses hydrostatic or hydrodynamic means to control the ratio of motion transmittal from the drive shaft to the driven shaft.

hydraulic engineering *Civil Engineering.* the application of scientific methods and engineering principles to civil engineering problems concerned with the design, erection, and construction of sewage disposal plants, waterfronts, dams, hydroelectric powerplants, and so on.

hydraulic filling *Mining Engineering.* a process in which water is used to wash waste material into stopes to prevent failure of rock walls and subsidence.

hydraulic fracturing *Petroleum Engineering.* a production-forcing process in which high fluid pressure is applied to the face of an underground well with low permeability to force the strata apart, allowing oil to move toward the wells.

hydraulic friction *Fluid Mechanics.* the flow resistance between a fluid and its conduit, acting at the surface of contact.

hydraulic grade line *Fluid Mechanics.* the free surface of water in an open channel.

hydraulic gradient *Fluid Mechanics.* a pressure head gradient (rate of change per unit distance of flow) measured at a specific point and in a given direction, often due to frictional effects along the flow path.

hydraulic jetting *Engineering.* the use of water jets in cleaning, for example, the interiors and exteriors of heat exchangers and boilers.

hydraulic jump *Fluid Mechanics.* an abrupt change in the steady-state flow of water in a channel, accompanied by a turbulent region in which water with uniform shallow depth and high velocity suddenly enters a region of uniform high depth and low velocity.

hydraulicking *Engineering.* **1.** the use of a powerful jet of water to remove undesired deposits and materials from valuable minerals in sluice boxes; a method employed only in regions where water is plentiful. **2.** the moving of earth by means of flowing water; hydraulic excavation. *Mining Engineering.* see HYDRAULIC MINING.

hydraulic lift *Mechanical Engineering.* a lifting jack in which a small hydraulic press within its frame actuates the lifting motion; often used in automobile repair work. Also, **hydraulic jack.**

hydraulic lime *Materials.* a mixture of slaked lime and volcanic ash that hardens with water, retaining its hardness when placed under water.

hydraulic loss *Fluid Mechanics.* a loss of kinetic energy in a fluid due to the presence of internal friction.

hydraulic mining *Mining Engineering.* the process of working a placer mine by directing a stream of water against a bank of sand, gravel, or talus. Also, HYDRAULICKING.

hydraulic nozzle *Mechanical Engineering.* a device for converting fluid pressure into fluid velocity.

hydraulic packing *Engineering.* packing material that becomes self-tightening under the influence of fluid pressure.

hydraulic press *Mechanical Engineering.* a machine in which a small force applied to a small piston is transmitted and increased through hydraulic pressure to a larger piston, which in turn applies pressure to the work piece.

hydraulic pressure *Fluid Mechanics.* the pressure created by forcing liquid through a relatively small orifice, pipe, or other channel.

hydraulic profile *Hydrology.* a vertical section showing the level to which water will rise in an aquifer.

hydraulic pumping *Petroleum Engineering.* the process of using the kinetic energy of flow to drive running water up to a higher level; water flow is intermittently suspended in the supply pipeline so that a small amount of water is lifted by the velocity head of a larger amount.

hydraulic radius *Fluid Mechanics.* for a fluid flowing through a conduit, a quantity given by the cross-sectional area of the wetted portion of the conduit divided by the inside wetted perimeter of the conduit.

hydraulic ram *Mechanical Engineering.* the plunger of a hydraulic press.

hydraulic ratio *Geology.* a value that expresses the quantity of a heavy mineral in a sediment, obtained by dividing the weight of the heavy mineral by the weight of a hydraulically equivalent light mineral.

hydraulic robot *Robotics.* a robot that uses hydraulic actuators.

hydraulics *Fluid Mechanics.* the study of the dynamics of fluids, particularly water and its applications in engineering.

hydraulic scale *Mechanical Engineering.* a scale in which the weight of the load applied is converted into hydraulic pressure.

hydraulic separation or **hydraulic segregation** see HYDRAULIC CLASSIFICATION.

hydraulic separator or **hydraulic segregator** see HYDRAULIC CLASSIFIER.

hydraulic shovel *Mechanical Engineering.* a rotating shovel in which hydraulic rams and motors are utilized instead of drums and cables.

hydraulic spraying see AIRLESS SPRAYING.

hydraulic state *Anthropology.* a theory suggesting that the early development of state societies was motivated by the need for large-scale organization and cooperation in the management of irrigation projects, as in ancient Egypt.

hydraulic telemetry *Telecommunications.* a method of mechanical telemetry in which signals are transmitted by sound waves through a liquid medium, usually water.

hydraulic transport *Engineering.* the conveyance of broken or crushed materials by means of water flowing in pipes or channels.

hydrazine *Inorganic Chemistry.* **1.** H_2NNH_2, a dangerously explosive, highly toxic, carcinogenic, colorless, fuming liquid; very soluble in water and soluble in alcohol; melts at 2.0°C and boils at 113.5°C; used as a rocket fuel, reducing agent, corrosion inhibitor, and catalyst, and in water treatment. Also, **hydrazine base.** **2.** any of various compounds derived from this liquid by the substitution of an organic radical for one or more hydrogen atoms.

hydrazine hydrate *Inorganic Chemistry.* $H_2NNH_2{\cdot}H_2O$, a colorless, fuming liquid, miscible with water and soluble in alcohol; freezes at –40°C and boils at 118.5°C. It is very explosive and toxic, and is used as a chemical intermediate and catalyst.

hydrazine sulfate *Inorganic Chemistry.* $H_2NNH_2{\cdot}H_2SO_4$, a white crystalline powder; very soluble in hot water, soluble in cold water, and insoluble in alcohol; melts at 254°C and decomposes at its boiling point; used in chemical manufacture and mineral analysis, and as a fungicide and germicide.

hydrazobenzene *Organic Chemistry.* $C_{12}H_{12}N_2$, colorless crystals that melt at 132°C; used as an intermediate in the synthesis of benzidine.

hydrazoic acid *Inorganic Chemistry.* HN_3, a colorless, volatile liquid with a foul smell, miscible with water; freezes at –80°C and boils at 37°C; highly explosive and a strong irritant. Also, HYDROGEN AZIDE, HYDRONITRIC ACID.

hydrazone *Organic Chemistry.* any of a number of compounds having the general formula $R{=}NNR_2$, formed from an aldehyde or ketone by the action of alhydrazine.

hydremia *Medicine.* a condition of excessive water in the blood; an abnormal increase in the proportion of plasma volume in relation to the red blood cell volume.

hydric *Ecology.* describing organisms that live or thrive in a wet or moist environment.

hydride *Inorganic Chemistry.* a binary compound of hydrogen with another element or a complex species containing hydrogen bound to another element as in BH_4; common examples are the hydrides of boron, lithium, and sodium.

hydride descaling *Metallurgy.* the process of descaling by using a hydride in a fused salt bath.

hydrindantin *Organic Chemistry.* $C_{18}H_{10}O_6$, a compound used as a chemical reagent in the photometric determination of amino acids.

hydrindene see INDAN.

hydriodic acid *Inorganic Chemistry.* a solution of hydrogen iodide, HI, in water; a strong acid and a reducing agent; used in various chemical processes.

hydro- **1.** a combining form meaning "water," as in *hydropathy.* **2.** a combining form meaning "hydrogen," as in *hydrocracking.*

hydrobasaluminite *Mineralogy.* $Al_4(SO_4)(OH)_{10}\cdot 12\text{–}36H_2O$, a fully hydrated variety of basaluminite.

Hydrobatidae *Vertebrate Zoology.* the storm petrels, a family of small marine birds, having dark brown, black, or gray plumage on top, and white plumage underneath; found in oceans worldwide except the Arctic seas.

***o*-hydrobenzoic acid** see SALICYLIC ACID.

hydrobenzoin *Organic Chemistry.* $C_{14}H_{14}O_2$, isomeric mixture, colorless crystals; melts at 136°C; *dl*-form melts at 120°C; *meso*-form is soluble in water and melts at 139°C.

hydrobiotite *Mineralogy.* a light-green, trioctahedral mica-clay mineral in which layers of equal proportions of biotite and vermiculite are interstratified.

hydroboracite *Mineralogy.* $CaMgB_6O_8(OH)_6\cdot 3H_2O$, a white, monoclinic mineral occurring as acicular crystals or in fibrous and foliated masses, having a specific gravity of 2.15 to 2.17 and a hardness of 2 to 3 on the Mohs scale.

hydroboration *Organic Chemistry.* the addition of a compound with a B–H bond to an unsaturated hydrocarbon, producing an organoborane.

hydrobromic acid *Inorganic Chemistry.* a solution of hydrogen bromide, HBr, in water; sensitive to light and a strong acid; used in analytical chemistry and as a solvent for ore minerals.

hydrocalumite *Mineralogy.* $Ca_2Al(OH)_6[Cl_{1-x}(OH)_x]\cdot 3H_2O$, a light-green, transparent, monoclinic mineral, massive in habit, having a specific gravity of 2.15 and a hardness of 3 on the Mohs scale; found in larnite rock.

hydrocarbon *Organic Chemistry.* an organic compound that contains only carbon and hydrogen; classified, according to the arrangement of the atoms and the chemical properties of the compounds, as alicyclic, aliphatic, and aromatic; derived mostly from crude petroleum and also from coal tar and plant sources.

hydrocarbonism *Toxicology.* poisoning due to consumption of or excessive exposure to hydrocarbons. Also, **hydrocarbarism.**

hydrocast see NANSEN CAST.

hydrocaulus *Invertebrate Zoology.* in hydrozoans, the upright, often branched part of a colony.

hydrocele [hī´drō sēl´] *Medicine.* an accumulation of serous fluid in a sac-shaped cavity, especially in the sac of the tunica vaginalis testis.

hydrocellulose *Materials Science.* a gelatinous mass formed from cellulose that has been caused to react with water; used in the manufacture of paper and of various artificial fibers such as rayon and mercerized cotton.

hydrocephalic [hī´drō sə fal´ik] *Medicine.* relating to or affected by hydrocephaly. Also, **hydrocephalous.**

hydrocephalus [hī´drō sef´ə ləs] *Medicine.* an abnormal condition that may be either congenital or acquired, in which there is an increase in the volume of cerebrospinal fluid within the skull, usually accompanied by increased pressure, resulting in enlargement of the head and compression of the brain, mental deterioration, and convulsions. Also, **hydrocephaly.** (From Greek; literally "water-headed sickness.")

hydrocerussite *Mineralogy.* $Pb_3(CO_3)_2(OH)_2$, a colorless, brittle, trigonal mineral occurring as tabular crystals with hexagonal outline, and having a specific gravity of 6.8 and a hardness of 3.5 on the Mohs scale; found as secondary encrustations on native lead and galena.

Hydrocharitaceae *Botany.* a family of tropical and temperate freshwater and marine monocotyledonous plants of the order Hydrocharitales.

Hydrocharitales *Botany.* an order of aquatic plants composed of the single family Hydrocharitaceae and belonging to the subclass Alismatidae.

hydrochloric acid *Inorganic Chemistry.* a solution of hydrogen chloride, HCl, in water; a strong acid with a pungent odor, highly corrosive, toxic, and irritating; widely used as a laboratory reagent, in food processing, in the petroleum industry, as a cleanser, and for many other purposes.

hydrochlorothiazide *Pharmacology.* $C_7H_8ClN_3O_4S_2$, a diuretic used in the treatment of high blood pressure and edema.

hydrocholeresis *Medicine.* an increased output of watery bile, relatively low in specific gravity, viscosity, and content of solid matter.

hydrochoric *Biology.* dispersed by water.

hydrocinnamic acid *Organic Chemistry.* $C_6H_5CH_2CH_2COOH$, floral-smelling, white crystals; soluble in water and alcohol; melts at 46°C and boils at 280°C; used in perfumes and flavoring.

hydrocladium *Invertebrate Zoology.* in hydrozoan colonies, a small branch-bearing hydrothecae.

hydroclone *Chemical Engineering.* an apparatus for separating a solid-liquid mixture from some process source by using centrifugal force and a conical vortex.

hydrocoel *Invertebrate Zoology.* in echinoderms, the water vascular system or that part of an embryo that develops into the system.

hydrocolloid *Materials Science.* a material that combines with water to form a colloid, usually a carbohydrate polymer.

Hydrocorallina *Invertebrate Zoology.* an order of colonial hydrozoans with a coral-like skeleton but differing from most anthozoan corals in having a free-swimming medusa generation.

Hydrocorisae *Invertebrate Zoology.* a suborder of aquatic true bugs in the order Hemiptera, with short or concealed antennae.

hydrocortisone *Biochemistry.* $C_{21}H_{30}O_5$, the major glucocorticoid in humans, biosynthesized from progesterone and having strong anti-inflammatory properties.

hydrocrackate *Organic Chemistry.* a high-quality motor fuel that is produced by hydrocracking.

hydrocracker *Chemical Engineering.* a processing unit that uses high pressure to crack long hydrocarbon molecules in an atmosphere of high hydrogen content.

hydrocracking *Chemical Engineering.* a catalytic, high-pressure refinery process that involves the cracking of heavy petroleum fractions in the presence of an excess of hydrogen in which special catalysts, such as platinum on a solid base of mixed silica and alumina, are used; the process may be viewed as a combination of hydrogenation and catalytic cracking.

hydrocyanic acid *Inorganic Chemistry.* HCN, a colorless, light-sensitive liquid with an odor of bitter almonds, miscible with water and alcohol; freezes at –14°C and boils at 26°C. It is very toxic and a fire and explosion hazard. It is used in rodent and insect poisons and in the manufacture of various compounds. Also, PRUSSIC ACID, HYDROGEN CYANIDE.

hydrocyanism *Toxicology.* poisoning due to excessive exposure to hydrogen cyanide; symptoms may include cessation of breathing.

hydrocyclone *Mechanical Engineering.* see CYCLONE, def. 2.

hydrocystoma *Medicine.* an eruption of deeply seated liquid-filled sacs caused by the retention of fluid in the sweat follicles.

Hydrodamalinae *Vertebrate Zoology.* a subfamily of sea cows of the family Dugongidae, composed of the single recently extinct species of Steller's sea cow.

hydrodealkylation *Chemical Engineering.* a type of hydrogenation in petroleum refining in which heat and pressure in the presence of hydrogen are used to remove methyl or larger alkyl groups from hydrocarbon molecules; for example, used to convert low-value toluene to high-value benzene.

hydrodesulfurization *Chemical Engineering.* a catalytic procedure in which the petroleum feedstock reacts with hydrogen to reduce the content of mercaptans.

Hydrodictyaceae *Botany.* a diverse family of freshwater green algae of the order Chlorococcales, characterized by multinucleate mature cells that divide to form biflagellate zoospores which enlarge to form a mature coenobium.

Hydrodictyon *Botany.* the water net, a genus of the green algae family Hydrodictyaceae, found in freshwater lakes and slow-moving waters.

hydrodynamic equation *Fluid Mechanics.* any of a set of three partial differential equations that are used to describe the motion of a unit element of fluid subjected to different forces such as pressure gradients, gravity, and friction.

hydrodynamic lubrication *Materials Science.* a form of lubrication that is produced by bearing-surface motion.

hydrodynamic oscillator *Acoustical Engineering.* a transducer that is used for producing acoustic waves by means of modulating the flow of a fluid through an orifice with a reciprocating valve that is controlled by acoustic feedback.

hydrodynamic pressure *Fluid Mechanics.* a pressure measurement performed on a liquid in motion, equal to the difference between the pressure of the liquid and the hydrostatic pressure measured at the same point.

hydrodynamics *Fluid Mechanics.* the study of fluid motion and fluid-boundary interaction.

hydrodynamic volume *Materials Science.* the volume occupied by a molecule in solution, used to determine molecular size and intrinsic viscosity.

hydroelasticity *Fluid Mechanics.* **1.** the study of the elasticity of fluids. **2.** the elastic interaction between a body and the fluid in which it is immersed.

hydroelectric of or relating to the production of electricity by water-power. Thus, **hydroelectric generator, hydroelectric plant, hydroelectric station.**

hydroelectricity *Electricity.* electric power that is produced by hydroelectric generators.

hydroexplosion *Volcanology.* any volcanic explosion caused by a sudden generation of steam as, for example, when water comes in contact with hot lava or rock.

hydrofining *Chemical Engineering.* a catalytic-refining process to hydrogenate and desulfurize a wide range of feed stocks in fixed-bed reactors.

hydroflap *Aviation.* a hydrofoil attached at the stern of a seaplane hull and acting as a brake or rudder.

hydrofluoric acid *Inorganic Chemistry.* a colorless solution of hydrogen fluoride, HF, in water; it is a weak acid; it attacks glass and is toxic and highly corrosive. It is used in producing gasoline and aluminum and for various other purposes.

hydrofoil *Naval Architecture.* **1.** an airfoil-shaped plate fitted to the underside of a boat to provide lift at high speeds, reducing hull displacement and associated friction. **2.** a boat equipped with such a plate. *Aviation.* a similar foil or plate attached to the bottom of a seaplane.

hydrofoil

hydroforming *Chemical Engineering.* a high-temperature refining method in which naphthas contact a catalyst in the presence of hydrogen to yield high-octane aromatics.

hydroform process *Materials Science.* a method of shaping metal by forcing it over a flexible die filled with hydraulic fluid.

hydroformylation *Chemical Engineering.* the reaction created by adding hydrogen and a –CHO group to the carbon atoms across a double bond to produce oxygenated derivatives.

hydrofuge *Invertebrate Zoology.* describing the water-repellent hairs of certain aquatic insects; used for retaining a film of air.

hydrogarnet *Mineralogy.* a member of the garnet group with SiO_4 partly replaced by $(OH)_4$, and a formula of $A_3B_2(SiO_4)_{3-x}(OH)_{4x}$.

hydrogasification *Chemical Engineering.* the production of methane from pulverized coal by the reaction with hydrogen and steam at high pressures (about 1000 psig).

hydrogel *Materials Science.* a water-based gel; a gel whose liquid constituent is water.

hydrogen *Chemistry.* a nonmetallic element having the symbol H, the atomic number 1, an atomic weight of 1.00797, a melting point of about –259°C, and a boiling point of about –253°C. It is the lightest, simplest, and most abundant element in the universe, and is found in nature as a colorless, odorless, highly flammable gas in the molecular form H_2. Hydrogen also has two other isotopes, deuterium (^2H) and tritium (^3H). (From a Greek phrase meaning "to create water.")

hydrogenase *Enzymology.* an enzyme that catalyzes the oxidation of hydrogen.

hydrogenated oil *Organic Chemistry.* an unsaturated liquid vegetable oil that has had hydrogen added to it in order to make it a hydrogen-saturated solid.

hydrogenation *Chemical Engineering.* a general reaction in which hydrogen is added to the unsaturated molecules of hydrocarbons or fatty acids, normally by use of a catalyst. *Organic Chemistry.* any process in which hydrogen is combined with another substance.

hydrogen azide see HYDRAZOIC ACID.

hydrogen bacteria *Bacteriology.* aerobic bacteria that can obtain energy by oxidizing hydrogen. Also, HYDROGEN-OXIDIZING BACTERIA.

hydrogen blistering *Metallurgy.* internal defects in a steel, caused by hydrogen.

hydrogen bomb *Ordnance.* the most powerful nuclear weapon ever produced, in which the fusion of deuterium and tritium, heavy isotopes of hydrogen, releases vast amounts of energy in the form of heat; first, the explosion of an atomic bomb acts as a trigger, providing the heat and pressure needed for fusion, then a mixture of deuterium and tritium fuses in a thermonuclear reaction, rapidly releasing huge amounts of energy and causing a powerful explosion.

hydrogen bond *Physical Chemistry.* the attraction of a hydrogen atom, already covalently bonded to one electronegative atom, to a second electronegative atom of the same molecule or an adjacent molecule; these bonds are found in compounds containing such strongly electronegative atoms as fluorine, oxygen, or nitrogen.

hydrogen bonding *Physical Chemistry.* the effect of hydrogen bonds, which are the strongest intermolecular force; this is largely responsible for the unique properties of water and for the ability of molecules in living systems to maintain the appropriate structure to carry out their biochemical functions; e.g., the double helix structure of DNA.

hydrogen bromide *Inorganic Chemistry.* HBr, a toxic, colorless gas; soluble in water and alcohol; freezes at –88.5°C and boils at –67°C; a strong irritant to eyes and skin; used in organic synthesis and in various chemical processes.

hydrogen burning *Astrophysics.* the process of fusing hydrogen into helium that powers a star when it is on the main sequence.

hydrogen chloride *Inorganic Chemistry.* HCl, a colorless, fuming gas with a suffocating odor; soluble in water, alcohol, and ether; freezes at –114.8°C and boils at –84.9°C. It is a widely produced chemical used in a variety of processes, including polymerization, isomerization, alkylation, chlorination, and nitration reactions.

hydrogen cracking *Materials Science.* a cracking of welds that is caused by hydrogen precipitation.

hydrogen cyanide see HYDROCYANIC ACID.

hydrogen cyanide laser *Optics.* a gas laser in which a mixture of hydrogen cyanide moving through a sealed vessel, regulated for pressure and rate of flow, produces a beam of light at wavelengths between 311 and 337 micrometers.

hydrogen damage *Metallurgy.* any of several defects created by the presence of hydrogen in metals or alloys.

hydrogen-discharge lamp *Electronics.* a gas-discharge lamp containing hydrogen; used as an ultraviolet radiation source.

hydrogen electrode *Physical Chemistry.* a standard reference electrode, usually consisting of a platinum surface coated with platinum black that is bathed with a stream of hydrogen-gas bubbles and immersed in a solution of hydrogen ions; it has a potential of zero when the activity of all species is unity and is used to measure hydrogen-ion concentration.

hydrogen embrittlement *Materials Science.* a lowering of a material's ductility and impact strength due to dissolved interstitial hydrogen atoms; typically occurs during pickling or electron-plating processes.

hydrogen equivalent *Chemistry.* **1.** the number of replaceable hydrogen atoms in one molecule of an acid. **2.** the number of replaceable OH groups in one molecule of a base. **3.** the number of hydrogen atoms with which a given molecule can react.

hydrogen fluoride *Inorganic Chemistry.* HF, a toxic, colorless, fuming, corrosive liquid or gas; miscible with cold water and very soluble in hot water; freezes at –83.1°C and boils at 19.5°C; a strong irritant; used as a catalyst in many reactions, as an additive in liquid rocket propellants, and for various other purposes.

hydrogen half-cell *Materials Science.* a reference electrode usually consisting of a platinum wire containing adsorbed hydrogen in a standard solution of hydrogen ions.

hydrogenic ion *Atomic Physics.* an atom from which all but one of the electrons have been removed.

hydrogenic rock see AQUEOUS ROCK.

hydrogen iodide *Inorganic Chemistry.* HI, a toxic, colorless, fuming, nonflammable gas; very soluble in hot water and soluble in alcohol; freezes at –50.8°C and boils at –35°C; a strong irritant; used to make hydriodic acid.

hydrogen ion concentration see PH.

hydrogen laser *Optics.* a molecular gas laser in which hydrogen produces a beam of coherent light at a wavelength of approximately 0.6 micrometer.

hydrogen line *Spectroscopy.* a line in the spectrum of neutral hydrogen having a frequency of 1420 megahertz and used in radio astronomy to study the amount of hydrogen in the galaxy.

hydrogen loss *Metallurgy.* in powder metallurgy, the weight loss of a metal powder upon heat treatment in hydrogen.

hydrogen maser *Physics.* a maser that uses gaseous hydrogen as the active medium.

hydrogenolysis *Organic Chemistry.* a cleavage of an organic compound, with the simultaneous addition of hydrogen to each cleavage product.

Hydrogenomonas *Bacteriology.* a former genus of rod-shaped bacteria that obtain energy from the oxidation of hydrogen; the species of this genus have been reclassified into other genera, including *Aquaspirillum* and *Pseudomonas.*

hydrogenosome *Cell Biology.* a cellular organelle that contains a hydrogenase enzyme for the production of hydrogen and is found in certain protozoa and trichomonads.

hydrogenous [hī dräj´ə nis] *Chemistry.* of, relating to, or containing hydrogen.

hydrogen-oxidizing bacteria see HYDROGEN BACTERIA.

hydrogen peroxide *Inorganic Chemistry.* H_2O_2, a colorless, dense liquid; soluble in water and alcohol; freezes at –41°C and boils at 150.2°C; it is rendered unstable by impurities and is a dangerous fire and explosion hazard. It is toxic and an irritant in concentrated form, and is used widely in industry in bleaches, dyes, cleansers, antiseptics, and disinfectants.

hydrogen phosphide see PHOSPHINE.

hydrogen selenide *Inorganic Chemistry.* H_2Se, a toxic, colorless, violently reactive gas; soluble in water; freezes at –64°C and boils at –41.5°C; a dangerous fire and explosion hazard, and a strong irritant; used in making selenium compounds and semiconductors.

hydrogen sulfate see SULFURIC ACID.

hydrogen sulfide *Inorganic Chemistry.* H_2S, a toxic, colorless gas that has an offensive odor of rotten eggs; soluble in water and alcohol; freezes at –85.5°C and boils at –60.7°C; a dangerous fire and explosion hazard, and a strong irritant; used as a reagent and as a source of hydrogen and sulfur. Also, SULFURETTED HYDROGEN.

hydrogen tellurate see TELLURIC ACID.

hydrogen thyratron *Electronics.* a thyratron containing low-pressure hydrogen.

hydrogeochemical *Geochemistry.* relating to the chemical properties of ground and surface waters.

hydrogeochemistry *Geochemistry.* the study of the chemical properties of ground and surface waters and their relationship to the regional geology.

hydrogeology *Hydrology.* the scientific study of subsurface water, especially the study of the flow characteristics of underground water.

hydrograph *Hydrology.* a graph showing stage, discharge, velocity, and other characteristics of water with respect to time.

hydrographic *Hydrology.* relating to the physical features of water areas or the study of such areas.

hydrographic chart *Cartography.* a chart showing information such as water depth, bottom topography, and tides and currents; used for navigation on bodies of water.

hydrographic cruise *Oceanography.* an exploratory cruise carrying out a hydrographic survey.

hydrographic reconnaissance *Military Science.* the reconnaissance of a water area to determine depths, beach and bottom characteristics, and the location of natural and artificial obstacles.

hydrographic section *Military Science.* personnel responsible for clearing a beach of damaged boats, conducting hydrographic reconnaissance, and assisting in the removal of underwater obstacles.

hydrographic sextant *Engineering.* a small, light sextant designed so that the maximum angle that can be read is slightly greater than that on the celestial navigating sextant.

hydrographic sonar *Engineering.* a form of sonar used to form a geographic profile of the ocean floor.

hydrographic survey *Oceanography.* a survey of the physical characteristics of a body of water, especially its currents, chemical and physical properties, and bottom relief. *Navigation.* a survey of this type made specifically to determine depth, bottom topography, location of aids and impediments to navigation, and other properties in order to produce charts and maps for use in navigation, flood control, and water supply and storage.

hydrographic table *Oceanography.* a tabulation that relates seawater density to salinity, temperature, and pressure.

hydrography *Navigation.* the science that deals with the physical features of bodies of water and their adjoining shores, especially as these features affect safe navigation.

hydrohalite *Mineralogy.* $NaCl{\cdot}2H_2O$, a monoclinic mineral with a specific gravity of 1.607 and an undetermined hardness; it is formed only from the crystallization of salty water at or below a temperature of 0°C. Also, MAAKITE.

hydroherderite see HYDROXYLHERDERITE.

hydrohetaerolite *Mineralogy.* $Zn_2Mn_4^{+3}O_8{\cdot}H_2O$, a dark-brown to brownish-black tetragonal mineral occurring in massive form, having a specific gravity of 4.64 and a hardness of 5 to 6 on the Mohs scale.

hydroid *Invertebrate Zoology.* **1.** the sessile polyp form of a hydrozoan coelenterate, as opposed to the free-swimming medusa form. **2.** any member of the order Hydroida. (From the Greek myth of *Hydra,* a many-headed water serpent.)

Hydroida *Invertebrate Zoology.* an order of hydrozoan coelenterates with a sessile polyp generation that buds asexually to produce a free-swimming, sexual, medusa generation.

hydrokinetic *Fluid Mechanics.* **1.** relating to or involving fluid motion. **2.** of or relating to the science of hydrokinetics.

hydrokinetics *Fluid Mechanics.* the study of fluid motion without regard to the cause of the motion.

hydrolaccolith *Geology.* a frost mound that has a core of ice and resembles a laccolith in cross section. Also, CRYOLACCOLITH.

HYDROLANT *Navigation.* a message, issued by the U.S. Navy Hydrographic Office and broadcast by long-range radio, warning of navigational dangers in the North Atlantic. (Formed from Hydrographic Office and Atlantic Ocean.)

hydrolase *Enzymology.* an enzyme that catalyzes the hydrolysis of a substrate, including esterases, lipases, nucleotidases, peptidases, phosphatases, and proteinases.

hydrolith *Petrology.* **1.** a rock that has been chemically precipitated from solution in water, such as rock salt. **2.** a rock or deposit relatively free of organic material.

hydrolithic *Petrology.* relating to or being a hydrolith.

hydrologic or **hydrological** relating to or involving the science of hydrology.

hydrologic budget *Hydrology.* an evaluation of the balance of water in an area, such as a drainage basin or aquifer, taking into account the inflow, outflow, and storage, and the relationship among evaporation, precipitation, and runoff. Also, **hydrologic accounting, hydrologic balance, hydrologic inventory.**

hydrologic cycle *Hydrology.* the interdependent and continuous circulation of water from the ocean, to the atmosphere, to the land, and back to the ocean. Also, WATER CYCLE.

hydrologic regimen see REGIMEN.

hydrologist *Hydrology.* a scientist or technician who specializes in hydrology.

Hydrology

Hydrology literally means the science of water. It has come to denote the study of the properties, distribution, and movement of water on the land surface, in the soil and through the subsurface rocks of the earth. It may be divided into two branches, surface hydrology, which is the science of water on the ground surface, and subsurface hydrology, which is the science of water in the pores and fractures of rocks beneath the ground surface. Subsurface hydrology is also referred to as "geohydrology" or "hydrogeology," with emphasis on the flow characteristics of underground water or the geologic settings of underground water, respectively. Surface hydrology includes limnology (the study of lakes), potamology (the study of surface streams), or cryology (the study of snow and ice).

In addition to the two branches of hydrology there is also the intermediate field of hydrometeorology, which relates the fields of hydrology and meteorology. There is an increasing interest in this subject because of concerns regarding global climatic changes.

Water is essential to almost all human endeavors. Its uneven distribution on the earth both spatially and temporally influences the development of population centers. Accounts of surface water management and well construction for the extraction of ground water are found in the ancient literature, but the science of water was not put on a proper mathematical basis until the early part of this century. Up to 10–20 years ago, hydrology focused mainly on water resource evaluation and management, and water extraction and supply. However, attention is progressively turning towards water quality, because of a heightened concern for surface and ground water contamination and environmental protection. This changing emphasis from water quantity to water quality has stimulated both the basic scientific development of hydrology and the application of hydrology to practical large-scale problems. Significant advances will continue to be made in this exciting field of study, as the demand of water resources of acceptable quality increases with the world's expanding population.

Chin-Fu Tsang
Senior Scientist
Earth Sciences Division
Lawrence Berkeley Laboratory
Berkeley, California

hydrology the scientific study of the portion of the earth that is water, in liquid, frozen, or vapor form, as it moves or is distributed on the earth's surface, under the ground, or in the atmosphere.

hydrolyase *Enzymology.* a sub-subclass of enzymes of the lyase class that catalyze the removal of water from a substrate by the breaking of a carbon-oxygen bond, which leads to the formation of a double bond. Also, DEHYDRATASE, HYDRATASE.

hydrolysis [hī dräl´ə sis] *Chemistry.* **1.** a chemical reaction in which water reacts with another substance and gives decomposition or other products. **2.** a reaction of water with a salt to create an acid or base.

hydrolytic [hī´drə lit´ik] *Chemistry.* characterized by or promoting hydrolysis.

hydrolytic enzyme *Enzymology.* an enzyme that acts through hydrolysis, using water to break down a substrate.

hydrolytic process *Chemistry.* a double decomposition reaction effected by water, wherein a hydrogen atom goes to one compound and a hydroxyl group to another.

hydrolyzate *Geology.* a sediment characterized by elements, such as aluminum, potassium, and sodium, that are easily hydrolyzed and tend to accumulate in the fine-grained alteration products of primary rocks.

hydromagnesite *Mineralogy.* $Mg_5(CO_3)_4(OH)_2 \cdot 4H_2O$, a colorless or white, transparent, brittle, monoclinic mineral occurring as acicular to bladed crystals and in chalklike form, having a specific gravity of 2.25 and a hardness of 3.5 on the Mohs scale; found as an alteration product of serpentine and related rocks, sometimes in deposits with economic significance.

hydromechanics *Fluid Mechanics.* the study and application of fluids, particularly water, as a medium for transmitting forces.

hydromedusae *Invertebrate Zoology.* **1.** medusae produced as buds from certain hydroids. **2.** hydrozoa.

hydrometallurgy *Metallurgy.* the art and science of metal extraction from an aqueous solution.

hydrometamorphism *Geology.* the alteration of rock caused by the infiltration of water and the subsequent addition, removal, or exchange of materials in the absence of high temperature or pressure. Also, **hydrometasomatism.**

hydromorphic soil *Agronomy.* a type of soil formed in the presence of excess water.

hydromyelia *Neurology.* increased accumulation of spinal fluid, accompanied by dilation of the central canal of the spinal cord. Also, HYDRORACHIS.

hydronitric acid see HYDRAZOIC ACID.

HYDROPAC *Navigation.* a message, issued by the U.S. Navy Hydrographic Office and broadcast by long-range radio, warning of navigational dangers in the Pacific Ocean. (Formed from Hydrographic Office and Pacific Ocean.)

hydropathy *Medicine.* the use of water in the treatment or cure of disease.

hydropericardium *Medicine.* an abnormal accumulation of serous fluid in the pericardial cavity.

hydropharynx *Entomology.* a median mouth-part structure anterior to the labium; ducts from the salivary glands open from this structure.

Hydrophiidae *Vertebrate Zoology.* a family of sea snakes, having uniform scales and a laterally compressed tail used as a rudder for swimming; found in tropical waters of the Indian and Pacific Oceans.

hydrophile *Chemistry.* a hydrophilic substance, usually a colloid or an emulsion.

hydrophilic *Chemistry.* having an affinity for water; attracting, dissolving in, or absorbing water.

Hydrophilidae *Invertebrate Zoology.* water scavenger beetles, coleopteran insects in the superfamily Hydrophiloidea, mostly plant scavengers.

Hydrophiloidea *Invertebrate Zoology.* a superfamily of coleopteran insects, beetles in the suborder Polyphaga.

hydrophilous *Biology.* living or thriving in a wet or aquatic environment. Thus, **hydrophile, hydrophily.**

hydrophobe *Chemistry.* a substance, usually a colloid, that does not adsorb or absorb water. *Biology.* an organism that is intolerant of water or a wet environment.

hydrophobia [hī´drə fō´bē ə] *Psychology.* an irrational fear of water and other liquids. *Medicine.* another name for rabies; so called because one symptom of the disease is the inability to swallow water. See RABIES.

hydrophobic *Chemistry.* relating to or being a hydrophobe; repelling water. *Biology.* not tolerant of water or a wet environment. *Medicine.* of or relating to rabies. *Psychology.* of or relating to hydrophobia.

hydrophobic bonding *Molecular Biology.* an attraction between the hydrophobic or nonpolar portions of molecules, causing their aggregation and sequestration away from water molecules.

hydrophobic effect *Biochemistry.* the effect of nonpolar groups on lipids, tending to aggregate and exclude water from between them.

hydrophobicity *Molecular Biology.* the extent of insolubility, or resistance to wetting or hydration, of a molecule.

hydrophobophobia *Psychology.* an irrational fear of rabies.

hydrophone *Acoustical Engineering.* an electroacoustic device that converts acoustic energy to electromagnetic waves in a water medium; used to detect and register the source of underwater sounds, as in tracking the presence of a submarine. Also, **hydrophone array.**

hydrophotometer *Optics.* an instrument that measures the attenuation coefficient of collimated light passing through a column of sea water.

Hydrophyllaceae *Botany.* a family of dicotyledonous herbs and shrubs of the order Solanales, often odorous and characterized by gland-tipped hairs; they are concentrated in dry habitats of the western United States.

hydrophyllium *Invertebrate Zoology.* in certain siphonophores, a shieldlike structure covering a cormidium.

hydrophyte *Botany.* a plant that grows with its leaves partially or completely submerged in water or one that can grow only in very wet soil. Also, HYGROPHYTE.

hydropic *Medicine.* relating to or affected with dropsy.

hydroplane *Naval Architecture.* a small, light boat designed to operate at high speed on the surface of the water. *Navigation.* a horizontal rudder used in submerging or elevating a submarine. *Aviation.* **1.** a hydrofoil designed to provide lift for an aircraft while skimming the surface of the water. **2.** another name for a seaplane. *Mechanical Engineering.* of a vehicle, vehicle tire, or the like, to travel on the film of water on a wet surface with a resulting decrease in control.

hydroplanula *Invertebrate Zoology.* the larval stage of a coelenterate found between the planula and actinula stages.

hydroplasma *Cell Biology.* the liquid or watery part of the protoplasm.

hydroplasticity *Materials Science.* the occurrence of plastic behavior in a material when water is added.

hydropneumatic *Engineering.* involving or requiring the combined action of water and air or water and gas.

hydropneumatic recoil system *Ordnance.* a mechanism for absorbing the recoil energy of a gun by forcing oil through small holes and then returning the gun to battery position by means of compressed gas.

hydroponics *Botany.* a technique of growing plants by suspending them with their roots in a nutrient solution or by rooting them in an inert material such as sand and supplying them with a nutrient solution.

hydropore *Invertebrate Zoology.* in the digestive tract of echinoderms, the pore through which the left protocoel (axocoel) opens dorsally.

hydropower see HYDROELECTRIC POWER.

hydrops *Medicine.* an abnormal accumulation of serous fluid in the tissues or in a body cavity; formerly called dropsy.

hydrops fetalis *Medicine.* a gross edema in the fetal tissues of the entire body, associated with severe anemia and present in erythroblastosis fetalis.

hydropsis *Oceanography.* a subfield of oceanography that is concerned with continuous observation, collection, and reporting of oceanographic data, particularly data relevant to commercial fishing.

hydropyle *Entomology.* a specialized area on an insect eggshell through which water can pass.

hydroquinone *Organic Chemistry.* $C_6H_4(OH)_2$, combustible white crystals; soluble in water, alcohol, and ether, melts at 170°C and boils at 285°C; used in photography, medicine, paints, and motor oils.

hydroquinone

hydroquinone dimethyl ether *Organic Chemistry.* $C_6H_4(OCH_3)_2$, combustible white flakes; melts at 56°C and boils at 213°C; used in paints, dyes, cosmetics, and flavorings. Also, 1,4-DIMETHOXYBENZENE.

hydroquinone monomethyl ether *Organic Chemistry.* $CH_3OC_6H_4OH$, a combustible, waxy white solid; melts at 52.5°C and boils at 243°C; used as an antioxidant, stablizer, and chemical inhibitor.

hydrorachis see HYDROMYELIA.

hydrorachitis *Neurology.* spinal meningitis accompanied by the discharge of fluid.

hydrorhiza *Invertebrate Zoology.* in colonial hydrozoan growth, a hollow tubelike root structure that anchors the hydrozoan polyps to the substratum.

hydrosalpinx *Medicine.* a collection of serous fluid in the uterine tube, causing distention.

Hydroscaphidae *Invertebrate Zoology.* a family of aquatic coleopteran insects, the skiff beetles, in the suborder Myxophaga.

hydroscope *Optics.* an instrument that brings objects at great depths under water into view, consisting of a series of mirrors encased in a long metal tube. Thus, **hydroscopic.**

hydroseparator *Mechanical Engineering.* a mechanism for separating solids from the liquid in which they are suspended by agitating them with hydraulic pressure or a stirring impeller. Thus, **hydroseparation.**

hydrosere *Ecology.* a community that is created by plants migrating to open water, decomposing, and forming moss or peat deposits.

hydroskeleton *Invertebrate Zoology.* in many softbodied invertebrates, a water-filled body cavity controlled by muscles to serve a skeletal function. Also, HYDROSTATIC SKELETON.

hydrosol *Chemistry.* a colloidal dispersion in water; a sol whose dispersion medium is water.

hydrosoluble *Chemistry.* soluble in water.

hydrosome *Invertebrate Zoology.* the body of a compound hydrozoan. Also, **hydrosoma.**

hydrospace detection *Military Science.* the detection of underwater targets, by sonar or other means.

hydrosphere *Hydrology.* the portion of the earth that is water, including liquid water, ice, and water vapor on the surface, underground, or in the atmosphere.

hydrospinning *Metallurgy.* a shaping process in which a metal is spun using hydraulic force and formed with rollers.

hydrospire *Invertebrate Zoology.* a pouch or tube in echinoderms of the subclass Blastoidea, thought to be a part of the respiratory system.

Hydrostachyaceae *Botany.* a monogeneric family of dicotyledonous, submerged, aquatic perennial herbs of the order Callitrichales, characterized by a short, thick, tuberous stem, fibrous roots, a basal holdfast and leaves, and a central scape with a terminal spike.

Hydrostachydaceae *Botany.* in some classification systems, an equivalent name for Hydrostachyaceae.

hydrostat *Mechanical Devices.* a device designed to regulate the height of fluid in a column, reservoir, or other container.

hydrostatic *Fluid Mechanics.* of or relating to hydrostatics.

hydrostatic approximation *Meteorology.* an assumption that the atmosphere is in a state of hydrostatic equilibrium.

hydrostatic assumption *Geophysics.* the assumption, disregarding local water density and the local acceleration of gravity, that the pressure of seawater will increase by 1 atmosphere (101,325 newtons per square meter) with every 10 meters of increased depth.

hydrostatic bearing *Mechanical Engineering.* a sleeve bearing in which high-pressure oil is pumped into the area between the bearing and the shaft, creating hydrostatic pressure around the shaft, thereby raising and supporting the shaft by a film of oil.

hydrostatic equation *Physics.* an equation relating the force per unit mass to the pressure in a fluid system in hydrostatic equilibrium: $\rho F = -\text{grad}(p)$, where ρ is the local fluid density, F is the vector force per unit mass, and p is the hydrostatic pressure.

hydrostatic equilibrium *Fluid Mechanics.* the condition of a fluid system in which the center of mass of each particle is at rest relative to the fluid's container.

hydrostatic forging *Metallurgy.* forging with the pressure supplied by a fluid.

hydrostatic fuse *Ordnance.* a fuse used on depth charges or other underwater bombs; it is set off by the water pressure at a preset depth.

hydrostatic level see STATIC LEVEL

hydrostatic modulus see BULK MODULUS.

hydrostatic pressing *Engineering.* the application of liquid pressure on a container in order to compact the material inside. *Materials Science.* see ISOSTATIC PRESSING.

hydrostatic pressure *Fluid Mechanics.* the pressure, measured at a point in a fluid, due solely to the weight of the fluid in the column above the point.

hydrostatic roller conveyor *Mechanical Engineering.* a roller conveyor having liquid-weighted rolls that help control the speed of conveyance.

hydrostatics *Fluid Mechanics.* the study of liquids at rest as well as the forces and pressures associated with them.

hydrostatic skeleton see HYDROSKELETON.

hydrostatic stability see STATIC STABILITY.

hydrostatic strength *Mechanics.* the ability of a body or vessel to withstand static internal pressure.

hydrostatic stress *Mechanics.* the condition of a system in which compressive or tensile stress is equal in all directions at all points and no shear stress exists.

hydrostatic test *Engineering.* a leakage test for a drain, vessel, pipe, or other hollow object or equipment; the item is filled with a test liquid and subjected to pressure.

hydrostatic weighing *Fluid Mechanics.* a technique for determining the true density of a body, accounting for air buoyancy, in which the body is weighed in air and then weighed in a liquid of known density; the volume of the body is given by the weight difference divided by the liquid density.

hydrosulfide *Chemistry.* any compound that contains the HS– group, such as sodium acid sulfide, NaHS; formed by the action of hydrogen sulfide on certain hydroxides.

hydrotalcite *Mineralogy.* $Mg_6Al_2(CO_3)(OH)_{16}\cdot 4H_2O$, a pearly white, transparent, trigonal mineral of the hydrotalcite group, dimorphous with manasseite, occurring in foliated masses, and having a specific gravity of 2.06 to 2.09 and a hardness of 2 on the Mohs scale; found with serpentine.

hydrotheca *Invertebrate Zoology.* in some hydrozoans, a thin, translucent exoskeleton surrounding the feeding polyps.

hydrotherapy *Medicine.* the application of water, usually externally, as a treatment for disease. Thus, **hydrotherapeutic.**

hydrothermal *Geology.* of or relating to hot or heated water, to its actions, or to products related to its actions. Thus, **hydrothermally.**

hydrothermal alteration *Geology.* the alteration of rocks by the interaction of hydrothermal water with preformed solid phases.

hydrothermal crystal growth *Chemical Engineering.* the formation of simple crystals of quartz in an autoclave with an alkaline solution at elevated pressures and temperatures.

hydrothermal deposit *Geology.* a mineral deposit formed from hydrothermal solutions under widely varying conditions of temperature and pressure.

hydrothermal solution *Geology.* a hot, magmatic solution derived from the crystallization of magmas, and having a high concentration of dissolved metals.

hydrothermal stage *Volcanology.* in the crystallization of magma, a latter phase characterized by hot residual fluids highly saturated with dissolved materials.

hydrothermal synthesis *Geology.* the synthesis of minerals in the presence of heated water.

hydrothermal vent *Oceanography.* a fissure in the sea bottom through which hot aqueous solutions rise from the magma beneath the crust.

hydrothorax *Medicine.* the presence of serous fluid in one or both pleural cavities.

hydrotreating *Chemical Engineering.* an oil refinery catalytic process in which hydrogen is contacted with petroleum product streams to remove impurities.

hydrotroilite *Mineralogy.* $FeS\cdot nH_2O$, a black, finely divided colloidal material found in many muds and clays; thought to be formed by bacteria on bottoms of marine basins having reducing conditions and restricted circulation; possibly colloidal hydrous ferrous sulfide.

hydrotrope *Chemistry.* a compound that increases the aqueous solubility of various slightly soluble organic compounds. Thus, **hydrotropic.**

hydrotropism *Biology.* growth toward or in response to the influence of water.

hydrotungstite *Mineralogy.* $H_2WO_4\cdot H_2O$, a dark yellowish-green, monoclinic mineral occurring in minute tabular crystals, having a specific gravity of 4.6 and a hardness of about 2 on the Mohs scale; found as an alteration product of ferberite.

hydroureter *Medicine.* an abnormal distention of the ureter with urine or with a watery fluid, due to obstruction from any cause.

hydrous *Chemistry.* containing water.

hydrous salt *Chemistry.* a salt that contains water of crystallization.

hydroxamic acid *Organic Chemistry.* any compound that contains the –C(=O)NHOH group.

hydroxide *Chemistry.* any of various compounds that contain the OH^- ion; hydroxides of metals are usually bases, and hydroxides of nonmetals are usually acids.

hydroxide ion *Chemistry.* the OH^- anion.

hydroximic acid *Organic Chemistry.* a compound that is isomeric with a hydroxamic acid and has the general formula *R*C(=NOH)OH.

hydroxisoxazole *Organic Chemistry.* $C_4H_5NO_2$, colorless crystals that melt at 86–87°C; used as a fungicide.

hydroxy- a combining form indicating the presence of an –OH group in a compound, as in *hydroxyketone.*

hydroxy acid *Organic Chemistry.* an acid having an –OH group.

hydroxyamphetamine hydrobromide *Pharmacology.* $C_9H_{13}NO\cdot HBr$, a drug occurring as a white crystalline powder, administered orally to increase blood pressure or prepared in solution and used topically to dilate the pupil and as a nasal decongestant.

hydroxyapatite *Organic Chemistry.* **1.** $3Ca_3(PO_4)_2\cdot Ca(OH)_2$, the mineral constituent of bone, consisting of needles arranged in a rosette that are practically insoluble in water and decompose above 1100°C. **2.** an analogous compound in which Ba or Sr replaces Ca; found in phosphorite deposits and biological tissue. *Molecular Biology.* a form of this substance used in certain chromatographic procedures to separate single-stranded from double-stranded nucleic acid molecules; also used for protein separations.

***o*-hydroxybenzamide** see SALICYLAMIDE.

hydroxybenzoic acid *Organic Chemistry.* $C_7H_6O_3$, any of three crystalline hydroxy derivatives (*ortho-*, *meta-*, or *para-*) of benzoic acid.

***m*-hydroxybenzoic acid** *Organic Chemistry.* $C_6H_4(OH)COOH$, a white powder that is soluble in water and alcohol; melts at 200°C; used as an intermediate for resins, petroleum additives, and pharmaceuticals.

***o*-hydroxybenzoic acid** see SALICYLIC ACID.

***p*-hydroxybenzoic acid** *Organic Chemistry.* $C_6H_4(OH)COOH\cdot 2H_2O$, colorless crystals that are slightly soluble in water and soluble in alcohol and ether; melts at 210°C; used as an intermediate and preservative.

β-hydroxy butyric dehydrogenase *Enzymology.* an enzyme that removes hydrogen atoms from β-hydroxy butyric acid, producing acetoacetic acid.

hydroxycarbonyl compound *Organic Chemistry.* a compound having one or more carbonyl (=C=O) and hydroxy (–OH) groups.

hydroxychloroquine *Pharmacology.* $C_{18}H_{26}ClN_3O$, a drug used to prevent and treat malaria and to suppress symptoms of lupus erythematosus; it has also been used in combination with other drugs in the treatment of rheumatoid arthritis.

hydroxycinchonidine see CUPREINE.

hydroxycitronellal *Organic Chemistry.* $C_{10}H_{20}O_2$, a viscous, colorless or pale yellow liquid;slightly soluble in water and soluble in alcohol; boils at 94–96°C; used in perfumes and flavorings.

2-(hydroxydiphenyl)methane *Organic Chemistry.* $C_6H_5CH_2C_6H_4OH$, crystals; insoluble in water and soluble in organic solvents; melts at 20.2–20.9°C; used as a preservative and an antiseptic.

hydroxyl- a combining form indicating the presence of the univalent radical OH^- (i.e., the hydroxyl group) in a chemical compound, as in *hydroxylamine.*

hydroxylamine *Inorganic Chemistry.* NH_2OH, colorless crystals that decompose quickly at room temperature and explosively when heated; soluble in cold water and in acids and alcohol; melts at 33°C and boils at –56.5°C. It is used as a reducing agent and in organic synthesis. Also, OXAMMONIUM.

hydroxylamine hydrochloride *Organic Chemistry.* $NH_2OH\cdot HCl$, colorless crystals that are soluble in water, alcohol, and glycerol and melt at 152°C; used as an antioxidant, photographic developer, and a reagent in synthesis. Also, AMMONIUM HYDROCHLORIDE.

hydroxylapatite *Mineralogy.* $Ca_5(PO_4)_3(OH)$, a rare, brittle, hexagonal mineral of the apatite group, occurring as prismatic crystals or in massive form, having a specific gravity of 3 to 3.08 and a hardness of 5 on the Mohs scale; found in talc schist. Also, **hydroxyapatite.**

hydroxylase *Enzymology.* any of a class of enzymes that utilize atomic oxygen to catalyze hydroxylation reactions.

hydroxylation reaction *Organic Chemistry.* an oxidation reaction that introduces one or more hydroxyl groups into an organic compound.

hydroxyl group *Chemistry.* the univalent group –OH, occurring in many organic and inorganic compounds. Also, **hydroxyl, hydroxyl radical.**

hydroxylherderite *Mineralogy.* $CaBe(PO_4)(OH)$, a colorless, white, green, or purple monoclinic mineral, isomorphous with herderite and structurally related to the silicates of the gadolinite group; occurring as prismatic or tabular crystals, and in fibrous crusts, having a specific gravity of 2.94 and a hardness of 5 to 5.5 on the Mohs scale; found in granite pegmatites. Also, HYDROHERDERITE.

3-hydroxy-3-methylglutaryl CoA reductase *Enzymology.* an important intermediate in the biosynthesis of cholesterol and ketone bodies.

hydroxynaphthalene see NAPHTHOL.

3-hydroxy-2-naphthoic acid *Organic Chemistry.* $C_{10}H_6OHCOOH$, pale yellow rhombic leaflets that melt at 222–223°C; used in dyes and pigments and as a stabilizer.

4-hydroxy-3-nitrobenzenearsonic acid *Organic Chemistry.* $HOC_6H_3(NO_2)AsO(OH)_2$, pale yellow crystals; slightly soluble in water and soluble in alcohol; melts at 218°C; used as a reagent and in medicine.

hydroxyproline *Biochemistry.* $C_5H_9NO_3$, an amino acid that is derived from proline and occurs in gelatin and collagen.

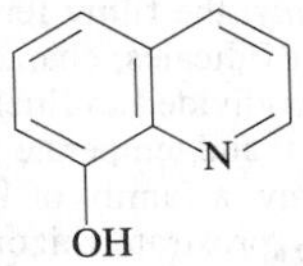

hydroxyquinoline

hydroxyquinoline *Organic Chemistry.* C_9H_6NOH, toxic white crystals; melts at 73–75°C and boils at 267°C; used as a disinfectant and in the manufacture of fungicides. Also, **8-hydroxyquinoline.**

5-hydroxytryptamine see SEROTONIN.

5-hydroxytryptophan *Biochemistry.* $C_{11}H_{12}N_2O_3$, an intermediate in the biosynthesis of the neurotransmitter serotonin; necessary for vasodilation and contraction of smooth muscle.

3-hydroxytyramine hydrobromide *Organic Chemistry.* $C_8H_{11}NO_2 \cdot HBr$, crystals; decomposes at 210–214°C; used as a source of dopamine for the synthetic production of catecholamine analogs.

hydroxyurea *Pharmacology.* $CH_4N_2O_2$, a drug that inhibits the proliferation of tumor cells by inhibiting DNA synthesis, used chiefly to treat certain types of leukemia; its main side effect is bone marrow depression.

hydrozincite *Mineralogy.* $Zn_5(CO_3)_2(OH)_6$, a white, grayish, or yellowish, monoclinic mineral occurring as masses or crusts, and having a specific gravity of 3.5 to 4.0 and a hardness of 2 to 2.5 on the Mohs scale; found as an alteration product of sphalerite. Also, ZINC BLOOM, CALAMINE, EARTHY CALAMINE.

Hydrozoa *Invertebrate Zoology.* a class of coelenterates including many small jellyfish, some corals, freshwater hydras, and marine hydroids; most are colonial, alternating sessile polyp generations with free-swimming medusae.

hydrozoan [hī′drə zō′ən] *Invertebrate Zoology.* **1.** any member of the class Hydrozoa. **2.** of or relating to this class.

hydrozoan medusae see HYDROMEDUSAE.

Hydruraceae *Botany.* a family of the order Chrysocapsales, characterized by cells in an irregularly ramified gelatinous mucilage and having a thallus with apical growth; found on rocks in cold, clear mountain rivers.

Hydrus *Astronomy.* the Lesser Water Snake, a small and inconspicuous constellation lying near the south celestial pole.

hyelophobia *Psychology.* an irrational fear of glass.

hyena *Vertebrate Zoology.* a predatory, nocturnal, carnivorous doglike mammal of the family Hyaenidae, having pointed ears and long, well-developed forelimbs and usually feeding as a scavenger; found in open country in Africa and southwest Asia.

hyena

Hyeniales *Paleontology.* an order of primitive plants formerly thought to be sphenophytes but now generally placed in the division Pteridophyta (ferns); restricted to the Devonian.

Hyeniopsida *Paleontology.* an extinct group of primitive plants in the division Pteridophyta, probably related to the Hyeniales; middle to late Paleozoic.

hyet- or **hyeto-** a combining form meaning "rain" or "precipitation."

hyetal *Meteorology.* of or relating to rain or other precipitation.

hyetal equator *Meteorology.* a line or zone that circles the earth north of the geographical equator and lies between two belts characterizing the annual time distribution for rainfall in the lower latitudes of each hemisphere; a type of meteorological equator.

hyetal region *Meteorology.* an area in which the amount and seasonal variation of rainfall are of a certain type.

hyetograph *Meteorology.* an instrument that collects and measures precipitation, recording its temporal and areal distribution.

hyetography *Meteorology.* the study of the annual variation and geographical distribution of precipitation over a given area.

hyetology *Meteorology.* a branch of meteorology that involves the study of all forms of precipitation.

Hygeia [hī jē′ə] *Astronomy.* asteroid 10, discovered in 1849 and measuring 443 km in diameter; it belongs to type C.

hygiene [hī′jēn] *Medicine.* **1.** the science of health and its preservation. Also, **hygienics. 2.** a practice or condition that is conducive to health. (Named for *Hygeia,* the ancient Greek goddess of health.)

hygienic [hī jen′ik] *Medicine.* **1.** of, relating to, or promoting health. **2.** of or relating to hygiene.

hygral strain *Materials Science.* the strain applied to a composite by an increase in the moisture concentration and consequent water absorption and swelling by the fibers.

hygric *Chemistry.* of or relating to moisture.

hygristor *Electronics.* a resistor whose resistance is dependent upon humidity; used in the measurement of relative atmospheric humidity.

hygro- or **hygr-** a combining form meaning "wet" or "moisture," as in *hygrometer, hygric.*

Hygrobiidae *Invertebrate Zoology.* a family of aquatic coleopteran insects, the screech or squeaker beetles, in the suborder Adephaga.

hygrograph *Meteorology.* a recording hygrometer.

hygrokinematics *Meteorology.* a branch of meteorology that involves the study of movement of water substance in the atmosphere.

hygrology *Meteorology.* a branch of meteorology that involves the study of the humidity in the atmosphere.

hygroma *Medicine.* a cystic swelling containing serous fluid.

hygrometer *Meteorology.* an instrument for estimating the relative humidity of air, consisting of a wet-bulb and a dry-bulb thermometer on a metal frame.

hygrometry *Meteorology.* the scientific study and calculation of the relative humidity of the atmosphere.

hygromycin *Microbiology.* an antibiotic that is effective against a wide range of microbes; produced by the bacteria *Streptomyces hygroscopicus* and *S. noboritoensis.* Also, **hygromycin A.**

hygrophobia *Psychology.* an irrational fear of dampness or moisture.

Hygrophoraceae *Mycology.* a family of fungi belonging to the order Agaricales, composed of gill mushrooms, some of which are edible.

hygrophyte see HYDROPHYTE.

hygroscopic *Chemistry.* of a substance, absorbing or attracting moisture from the air; having an affinity for moisture. *Botany.* of a plant, sensitive to moisture or responding to changes in moisture.

hygroscopic coefficient *Hydrology.* the percentage of water that will be absorbed by a completely dry mass of soil and be held in equilibrium if the soil comes in contact with a saturated atmosphere. Also, **hygroscopic capacity.**

hygroscopicity *Chemistry.* the quality or degree of absorbing moisture.

hygroscopic water *Hydrology.* moisture that adheres to soil particles and does not evaporate at ordinary temperatures. Also, **hygroscopic moisture.**

hygrothermal effect *Materials Science.* the degradation of a polymer caused by exposure to moderately high temperatures and moisture absorption; in composites, produces severe internal stresses due to expansion of the fiber inside of the matrix.

hygrothermograph *Meteorology.* an instrument that records both temperature and humidity on one graph.

Hylidae *Vertebrate Zoology.* the tree frogs, a family of small to medium-sized tree-dwelling amphibians of the suborder Neobatrachia, having large padded toes with adhesive properties for life in trees; distribution is nearly worldwide in temperate and tropical regions, with most in South America, Australia, and New Guinea.

Hylleraas coordinates *Atomic Physics.* the coordinates that constitute the various distances between two points (the sum and difference distances between two particles and the origin) in a helium atom.

hylo- a combining form meaning: **1.** wood, as in *hylophagous.* **2.** material, as in *hylotropic.*

Hylobatidae *Vertebrate Zoology.* the gibbons, a family of tailless primates of the infraorder Catarrhini, having small heads and long forearms for brachiation and aboreal life; found in southeast Asia and the Malay Archipelago.

Hylocomiaceae *Botany.* a family of large, robust, shiny mosses of the order Hypnobryales that form loose mats on humus and soil, characterized by a double costa and peristome structure and, sometimes, by frondose stems; includes some of the most common boreal forest mosses and weft-forming feather mosses of the taiga.

Hylonomus *Paleontology.* the earliest known cotylosaur, a stocky, lizardlike animal about three feet long, named "wood-dweller" because its remains were first found inside fossilized hollow logs; a genus of amniotes in the extinct suborder Captorhinomorpha and family Romeriidae, one of the first amphibians to become principally terrestrial; middle Carboniferous.

hylopathism *Medicine.* the former concept that disease is caused by changes in the composition of matter.

hylophagous *Zoology.* feeding mainly or exclusively on wood.

hylophobia *Psychology.* an irrational fear of forests.

hylotomous *Entomology.* describing those insects that are able to cut or chew through wood.

hylotropic *Physical Chemistry.* describing a substance that has the ability to change from one phase to another, as from a liquid to a gas, without any change in chemical composition.

hylotropy *Physical Chemistry.* the fact of being hylotropic; the capability of changing phase without a change in composition.

hymecromone *Organic Chemistry.* $C_{10}H_8O_3$, crystals that melt at 194–195°C; used medicinally as a choleretic and biliary antispasmotic.

hymen *Anatomy.* the thin fold of mucous membrane that partially or wholly covers over the external vaginal opening. (From *Hymen,* the Greek god of marriage.)

hymenal *Anatomy.* of or relating to the hymen.

hymenium *Mycology.* a layer of fungal tissue containing the spores, as on the gills of mushrooms.

hymeno- or **hymen-** a combining form meaning: **1.** membrane, as in *hymenology, hymenium.* **2.** hymen, as in *hymenotomy, hymenitis.*

Hymenochaetaceae *Mycology.* a family of fungi of the order Agaricales, occurring on soil, dung, wood, and plant material or in a symbiotic relationship with plants; some species are pathogenic, causing serious wood rots to trees.

Hymenogastraceae *Mycology.* a family of fungi of the order Hymenogastrales growing symbiotically with green plants.

Hymenogastrales *Mycology.* an order of fungi of the class Gasteromycetes, characterized by having a single peridium.

hymenolepiasis *Medicine.* infestation with the parasitic tapeworm of the genus *Hymenolepis,* marked by abdominal pain, bloody stools, and disorders of the nervous system.

Hymenolepis *Invertebrate Zoology.* a genus of tapeworm in the family Hymenolepididae, relatively innocuous parasites in birds, rodents, and occasionally humans.

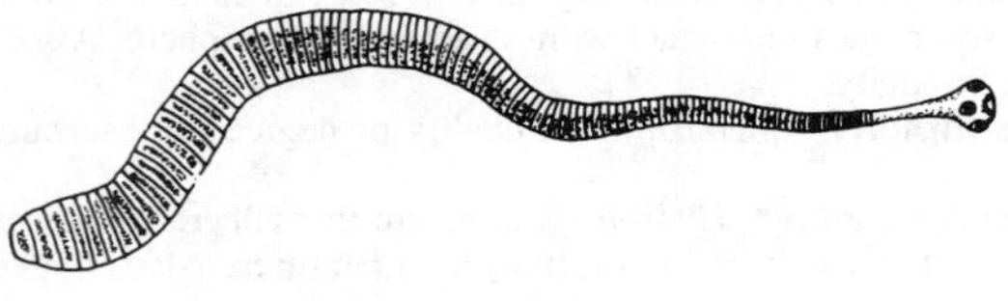

Hymenolepis

Hymenomonadaceae *Botany.* a family of freshwater and marine flagellate algae of the order Isochrysidales, distinguished by coccolith-bearing cells.

Hymenomycetes *Mycology.* a class of fungi belonging to the subdivision Basidiomycotina, characterized by the formation of basidia or spore-bearing structures located in a hymenium and supported by slender stalks.

hymenophore *Mycology.* in fungi, the part of the fruiting body that supports the hymenium or spore-bearing tissue.

Hymenophyllaceae *Botany.* the filmy ferns, a large family of mostly epiphytic ferns of the order Filicales, characterized by very thin leaves, a fine hairy rhizome, and undivided to finely dissected fronds; they are found in very humid tropical and temperate regions.

Hymenophytaceae *Botany.* a family of liverworts of the order Metzgeriales, characterized by prostrate rhizomes from which arise erect stipes that broaden into flat, horizontal fronds, by a well-defined midrib with central strands on the thallus, and by the absence of ventral scales.

Hymenopodium *Mycology.* a genus of fungi belonging to the order Moniales.

hymenopodium *Mycology.* in certain fungi, tissue located under the hymenium.

Hymenoptera *Invertebrate Zoology.* a large order of social insects including bees, ants, wasps, and sawflies, most having a clearly differentiated head, thorax, and abdomen, and four membranous wings that can be linked together by small hooks when the insects are flying.

hymenopteran *Invertebrate Zoology.* **1.** a member of the order Hymenoptera. **2.** of or relating to this order.

hymenopterism *Toxicology.* poisoning due to the sting of a bee, wasp, or hornet.

hymenopterous *Invertebrate Zoology.* of or relating to the order Hymenoptera.

Hymenostomata *Invertebrate Zoology.* a suborder of ciliated protozoans with two nuclei of unequal size.

Hymenostomatida *Invertebrate Zoology.* an order of protozoans in the subclass Holotrichia, with uniform body ciliation and a ventral buccal cavity with ciliature.

hymenotomy *Medicine.* the surgical incision, division, or dissection of the hymen.

hynobiid *Vertebrate Zoology.* **1.** a member of the family Hynobiidae, Asian land salamanders. **2.** of or relating to this family.

Hynobiidae *Vertebrate Zoology.* a family of the suborder Cryptobanchoidea that are considered the most primitive of all salamanders; characterized by complete metamorphosis, during which adults lose functional gills and teeth and become terrestrial in most genera; found throughout Asia.

hyo- or **hy-** a combining form meaning "U-shaped," as in *hyolithid, hyoid.* (From *hypsilon,* a Greek name for the letter υ.)

hyobranchium *Vertebrate Zoology.* the Y-shaped bone that supports a snake's tongue and related muscles.

Hyocephalidae *Invertebrate Zoology.* a family of hemipteran insects, true bugs in the superfamily Pentatomorpha.

hyoglossal *Anatomy.* a muscle lying at the base of the tongue that pulls the extended tongue back into the mouth.

hyoid *Anatomy.* **1.** having a U shape. **2.** pertaining to the hyoid bone.

hyoid arch *Developmental Biology.* the second branchial arch.

hyoid bone *Anatomy.* a U-shaped bone located at the base of the tongue suspended from the styloid processes of the temporal bones and located above the larynx.

hyoid tooth *Vertebrate Zoology.* any of several teeth on the tongue of some fishes.

Hyolitha *Paleontology.* a proposed class of tube-worms or early mollusks that were very common in the Cambrian; generally less than 5 cm long, the conical or pyramidal hyolithids were rare after the Devonian and became extinct at the end of the Permian.

hyomandibula *Vertebrate Zoology.* in many fishes, the upper portion of the hyoid arch, which supports the upper jaw by propping it against the braincase. Thus, **hyomandibular.**

hyomandibular cleft *Developmental Biology.* the cleft between the mandibular and hyoid arches in the developing embryo. Also, **hyoid cleft.**

hyomandibular pouch *Developmental Biology.* the first in a series of pharyngeal pouches of the vertebrate embryo, present between the mandibular and hyoid arches.

Hyopsodontidae *Paleontology.* an extinct family of small, bunodont, herbivorous mammals in the order Condylarthra; the hyopsodonts are related to the ungulates but had clawed toes; Paleocene and Eocene.

hyoscine *Pharmacology.* see SCOPOLAMINE.

hyoscyamine *Organic Chemistry.* $C_{17}H_{23}O_3N$, a white crystalline powder that melts at 108.5°C; derived from henbane, belladonna, and similar plants; the levorotatory component of racemic atropine, used medicinally as an anticholinergic alkaloid.

hyostylic *Vertebrate Zoology.* descriptive of a condition in certain fishes in which the upper jaw joint is braced only by the hyomandibular arch, and thus makes a loose attachment to the skull.

hyp- a combining form meaning: **1.** under or below, as in *hyparterial.* **2.** less than or less than normal, as in *hypalgesia.* (A variant of *hypo-,* and the opposite of *hyper-.*)

hypabyssal rock *Petrology.* an igneous rock that rose as magma from great depths but solidified as a minor intrusion, such as a dike or sill, before reaching the surface.

hypacusia *Medicine.* slightly impaired hearing. Also, **hypoacusis.**

hypalgesia *Neurology.* lowered sensitivity to pain. Also, **hypalgia.**

hypandrium *Entomology.* a plate underlying the genital region of a male insect.

hypanthium *Botany.* in perigynous flowers, an enlargement of the receptacle to which the sepals, petals, and stamens are attached.

hypantrum *Vertebrate Zoology.* the notch on the anterior part of a reptile's neural arch that functions with the hyposphere in articulation.

Hypaque *Radiology.* a trade name for diatrizoate sodium and diatrizoate meglumine contrast mediums.

Hypatia c. 370–415, Greek mathematician and philosopher; first known female mathematician; wrote on arithmetic, algebra, and conics.

hypaxial *Anatomy.* ventral to the long axis of the body.

hypaxial musculature *Anatomy.* muscles that develop ventral to the transverse processes of the vertebrae.

hypengyophobia *Psychology.* an irrational fear of taking on responsibility.

hyper *Psychology.* in popular use, a shorter form of *hyperactive*; i.e., high-strung, nervous, or agitated.

hyper- a prefix meaning: **1.** above or beyond, as in *hyperspace, hypersonic.* **2.** more than or excess, *hyperalgesia.* (The opposite of *hypo-* or *hyp-.*)

hyperacoustic zone *Geophysics.* a part of the atmosphere (altitude 100 to 160 kilometers) in which the distance between air molecules approximates a wavelength of sound, resulting in the transmission of sound at lower volume than below this zone.

hyperactive *Psychology.* relating to, characterized by, or affected by hyperactivity. Also, HYPERKINETIC.

hyperactivity *Psychology.* **1.** excessive or abnormally rapid activity; overactivity. **2.** specifically, a childhood disorder in which the individual is constantly restless and performs activities at an abnormally high rate of speed; may be caused by brain damage or by psychosis. Also, HYPERKINESIA.

hyperacusia *Medicine.* an exceptionally acute sense of hearing due to greater sensitivity of the sensory neural mechanism; auditory hyperesthesia.

hyperadrenalinemia *Medicine.* an increased amount of adrenalin in the blood.

hyperadrenalism *Medicine.* a condition resulting from overactivity of adrenal secretions or from excessive or prolonged treatment with adrenal hormones.

hyperadrenocorticism *Medicine.* the excessive secretion of adrenocortical hormones; hyperadrenalism.

hyperaldosteronism *Medicine.* a condition of muscle weakness, excessive elimination of urine, hypertension, potassium deficiency in the blood, and alkalosis produced by excessive secretion of aldosterone by the adrenal cortex.

hyperalgesia *Medicine.* a heightened or excessive sensitivity to pain. Thus, **hyperalgesic, hyperalgetic.** Also, **hyperalgia.**

hyperalimentation *Medicine.* the administration or ingestion of more than the required amount of nutrients.

hyperbaric *Medicine.* characterized by greater than normal pressure or sometimes weight; used especially to refer to gases under greater than atmospheric pressure (as in a hyperbaric chamber) or to a solution of greater specific gravity than a reference fluid (e.g., a hyperbaric medicine).

hyperbaric chamber *Medicine.* a compartment in which air pressure may be raised to greater than atmospheric pressure; used for treating gas gangrene, other anaerobic infections, or other conditions in which a high concentration of oxygen is desirable.

hyperbaric medicine *Pharmacology.* any medicinal solution having a higher specific gravity than another solution that is taken as the reference, such as a spinal anesthetic that is heavier than spinal fluid.

hyperbaric oxygenation *Medicine.* the therapeutic use of oxygen administered under greater than atmospheric pressure in a specially designed chamber, called a hyperbaric chamber.

hyperbilirubinemia *Medicine.* an excessive amount of bilirubin in the circulating blood, which may lead to jaundice, sometimes occurring in premature infants.

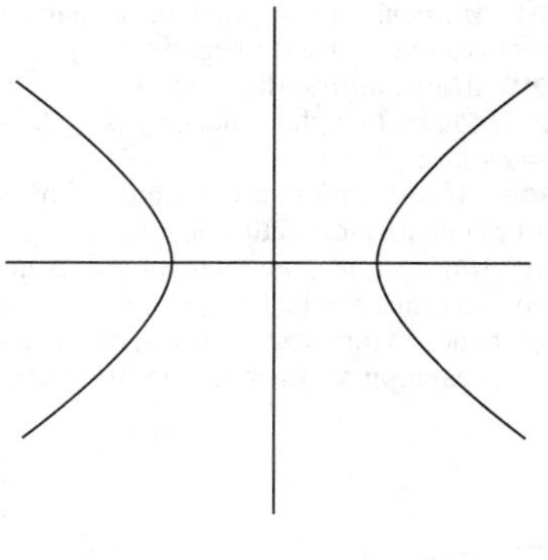

hyperbola

hyperbola [hī pur′bə lə] *Mathematics.* **1.** the set of points in the plane for which the difference of the distances from two fixed points (called the foci) is constant. **2.** equivalently, the conic section obtained by intersecting a cone with a plane such that the intersection is nonempty on both sides of the vertex. A standard form of the equation for a hyperbola in rectangular coordinates is $x^2/a^2 - y^2/b^2 = 1$. The axes of symmetry are $x = 0$ and $y = 0$. The asymptotes are $y = \pm bx/a$.

hyperbolic [hī′pur bäl′ik] *Mathematics.* of, relating to, derived from, or having the shape of a hyperbola.

hyperbolic amplitude *Telecommunications.* a signal amplitude that is measured along hyperbolic coordinates rather than Cartesian coordinates.

hyperbolic antenna *Electromagnetism.* an antenna having a reflector whose cross section is in the form of a half-hyperbola.

hyperbolic cosecant *Mathematics.* the function $f(z) = 1/\sinh z = 2/(e^z + e^{-z})$.

hyperbolic cosine *Mathematics.* the function $f(z) = (e^z + e^{-z})/2$; abbreviated cosh z; $\cosh iz = \cos z$ and $\cosh z = \cos iz$.

hyperbolic cotangent *Mathematics.* the function $f(z) = (\cosh z/\sinh z) = (e^z + e^{-z})/(e^z - e^{-z})$.

hyperbolic decline *Petroleum Engineering.* a type of decline in the gas or oil production rate; the other types of decline are harmonic and constant-percentage.

hyperbolic differential equation *Mathematics.* a second-order differential equation of the form

$$\sum_{i,j=1}^{n} A_{ij}(\partial^2 u/\partial x_i \partial x_j) + \sum_{i=1}^{n} B_i(\partial u/\partial x_i) + Cu + D = 0$$

such that: (a) A_{ij}, B_i, C, and D are all differentiable functions of $(x_1, \ldots, x_n)$, and also (b) the quadratic form $\sum_{i,j=1}^{n} A_{ij}x_i x_j$ can be reduced to the form $\sum_{i=1}^{n} F_i x_i^2$, where the F_i do not all have the same sign.

hyperbolic fix *Navigation.* a navigational fix based on hyperbolic lines of position, such as those produced by a loran system.

hyperbolic functions *Mathematics.* the hyperbolic cosecant, cosine, cotangent, secant, sine, or tangent functions.

hyperbolic geometry *Mathematics.* the geometry of 2-manifolds of constant (scalar) curvature equal to –1. Equivalently, a non-Euclidean planar geometry in which any point P not on a line l has at least two lines through it that are parallel to l. Many models exist; for example, the geometry of the upper half plane with metric $ds^2 = (dx^2 + dy^2)/y$. In this model, the geodesics are semicircles with diameters along the line $y = 0$. Quotients of this space by discrete subgroups of the modular group also have constant negative curvature. Also, LOBACHEVSKI GEOMETRY, BOLYAI GEOMETRY.

hyperbolic guidance *Ordnance.* the use of hyperbolic navigation in a missile guidance system.

hyperbolic line of position *Navigation.* a line of position obtained by measuring the difference in distance to two known, fixed points.

hyperbolic navigation *Navigation.* a radio navigation system based on the time difference between radio signals transmitted simultaneously from two or more ground stations. Also, **hyperbolic radionavigation.**

hyperbolic paraboloid *Mathematics.* the saddle-shaped surface whose points satisfy the equation in the standard form: $z = x^2/a^2 - y^2/b^2$. The sections parallel to the (y,z)-plane are congruent parabolas; those parallel to the (x,z)-plane are also congruent parabolas; and those parallel to the (x,y)-plane are hyperbolas (including a pair of intersecting lines).

hyperbolic point *Mathematics.* a point on a manifold for which the Gaussian (scalar) curvature is strictly negative.
hyperbolic secant *Mathematics.* the function $f(z) = 2/(e^z + e^{-z})$.
hyperbolic sine *Mathematics.* the function $f(z) = (e^z - e^{-z})/2$; $\sinh iz = i \sin z$ and $\sin iz = i \sinh z$.
hyperbolic space *Mathematics.* a complete, simply connected Riemannian manifold of constant curvature $K < 0$.
hyperbolic spiral *Mathematics.* the locus of points in the plane satisfying the parametric equations $x = (a \cos t)/t$ and $y = (a \sin t)/t$. The curve consists of two branches symmetric with respect to the y-axis and having the line $y = a$ as asymptote. Each of the branches spirals in toward the origin.

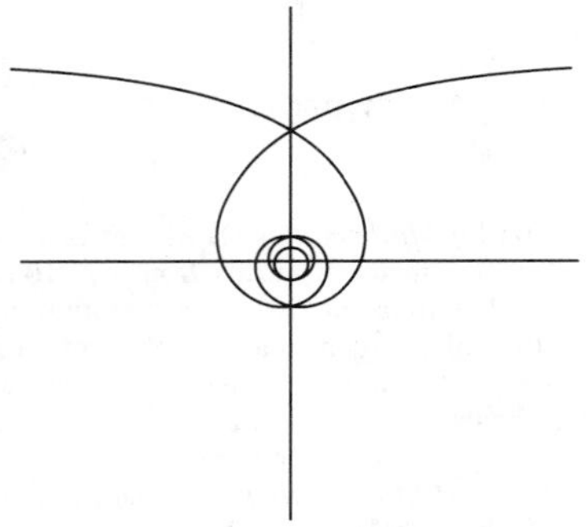

hyperbolic spiral

hyperbolic sweep generator *Electronics.* a sweep generator whose generated waveform resembles a hyperbola.
hyperbolic tangent *Mathematics.* the function $f(z) = (\sinh z)/(\cosh z) = (e^z - e^{-z})/(e^z + e^{-z})$.
hyperbolic trajectory *Space Technology.* the trajectory initiated by a spacecraft when its velocity surpasses the escape velocity of a planetary body, satellite, or star.
hyperbolic waveform *Electronics.* a waveform that is approximately a hyperbola.
hyperboloid *Mathematics.* **1.** a **hyperboloid of one sheet** is a connected tunnel-shaped surface whose points satisfy an equation of the standard form $x^2/a^2 + y^2/b^2 - z^2/c^2 = 1$. Planes parallel to the (x,y)-plane intersect the surface in ellipses; planes parallel to the (x,z)- or (y,z)-plane intersect it in hyperbolas. **2.** a **hyperboloid of two sheets** is a double-bowl-shaped surface composed of two disjoint parts, whose points satisfy an equation of the standard form $x^2/a^2 - y^2/b^2 - z^2/c^2 = 1$. Planes parallel to the (y,z)-plane intersect the surface in ellipses; planes parallel to the (x,y)- or (x,z)-plane intersect it in hyperbolas.
hyperboloid of revolution *Mathematics.* a hyperboloid whose elliptical sections are circles. It may be realized by rotating a hyperbola about one of its axes of symmetry.
hyperbranchial gland *Invertebrate Zoology.* a gland on the inner surface of the mantle in many marine snails.
hypercalcemia *Medicine.* an excess of calcium compounds in the blood.
hypercalciuria *Medicine.* an excess of calcium in the urine.
hypercanthosis see ACANTHOSIS.
hypercapnia *Medicine.* an excess of carbon dioxide in the blood.
Hypercard *Computer Programming.* a programming environment that treats information as linked structures, analogous to index cards, that can be manipulated or navigated through by the user.
hypercharge *Particle Physics.* a quantum number (Y) that is calculated by a particle's strangeness. Hypercharge is analogous to electric charge, except that hypercharge is not conserved when the particle decays.
hyperchlorhydria *Pathology.* the excessive secretion of hydrochloric acid by the stomach cells.
hypercholesteremia *Pathology.* an excess of cholesterol in the blood.
hyperchromatic *Pathology.* of or relating to hyperchromatism. *Chemistry.* staining more intensely than usual.
hyperchromatism *Pathology.* excessive pigmentation, particularly during cellular breakdown when the nucleus fills with chromatin. Also, **hyperchromia, hyperchromasia, hyperchromatosis.**
hyperchromemia *Hematology.* a high color index of the blood.
hyperchromic *Pathology.* **1.** highly or excessively pigmented, stained, or colored. **2.** of or relating to hyperchromatism (hyperchromia).
hyperchromic anemia *Pathology.* **1.** a condition of anemia that is characterized by red blood cells that are more deeply colored than usual due to their increased thickness. **2.** anemia that is related to a deficiency of vitamin B.
hyperchromic effect *Molecular Biology.* an increase in the ultraviolet absorbance exhibited by a solution of DNA or RNA when it is denatured to single strands.
hyperchromicity *Molecular Biology.* any increase in the absorption of ultraviolet light by a polynucleotide solution.
hyperchromic shift *Molecular Biology.* any change in the ultraviolet absorbance exhibited by a solution of DNA or RNA when it is denatured to single strands.
hypercoagulability *Medicine.* a condition in which the blood coagulates more readily than normal.
hypercoracoid *Vertebrate Zoology.* in teleost fishes, the upper bone at the base of the pectoral fin.
hypercube *Computer Science.* a parallel computer architecture in which many CPU's (the number of which is a power of 2 , say 2^n) are logically connected as an n-dimensional hypercube, where each processor is at a corner of the cube and is directly connected to the processors at neighboring corners. A message can be transferred from any processor to any other in a number of steps proportional to the logarithm of the number of processors.
hyperdisk *Computer Technology.* a large-capacity mass storage system that employs a disk as an overflow device.
hyperemesis *Medicine.* a condition of excessive vomiting.
hyperemesis gravedorium *Medicine.* pernicious vomiting during pregnancy.
hyperemia *Medicine.* an excess of blood in part of the body.
hyperergia *Immunology.* a condition in which a reaction to antigenic materials is more marked than normal, such as a condition of allergic hypersensitivity.
hyperesthesia *Medicine.* a greater than normal sensitivity to stimulation such as light, touch, or pain. Also, OXYPATHIA, OXYPATHY.
hypereutectic alloy *Materials Science.* an alloy experiencing the eutectic reaction and having a higher concentration of a component than the eutectic composition; in Pb-Sn solders it would contain between 61.9 and 97.56% Sn. Thus, **hypereutectoid.**
hyperextension *Medicine.* the extreme or excessive extension of a limb or part, such as the knee or elbow. Thus, **hyperextended.**
hyperfine structure *Spectroscopy.* the splitting of spectral lines into components lying very close together as a result of nuclear spin or the occurrence of isotopes.
hyperflexion *Medicine.* the overflexion of a limb or part.
hyperfocal distance *Optics.* the object distance in front of the lens, when it is focused on infinity, beyond which all objects are reasonably in focus; normally equal to 1000 times the lens aperture diameter.
hyperforming *Chemical Engineering.* a catalytic hydrogenation process to increase the naphtha octane number by removing nitrogen or sulfur compounds.
hyperfrequency waves *Electromagnetism.* microwave electromagnetic radiation having wavelengths in the range of one centimeter to one meter.
hyperfullerene structure *Chemistry.* a structure that is proposed to form along with ordinary fullerenes in a laser-vaporized carbon plume; this process could reoccur indefinitely to make a particle whose pentagons are in icosahedral alignment.
hyperfusible *Volcanology.* describing a material that is capable of lowering the melting range of magmatic fluids.
hypergammaglobulinemia *Medicine.* an excess of immunoglobulins in the blood, common in many chronic infectious diseases.
hypergene *Mineralogy.* see SUPERGENE.
hypergeometric distribution *Statistics.* the distribution of the number of individuals with a given characteristic when sampling without replacement from a finite population.
hypergeometric function *Mathematics.* any function of the form

$$\sum_{n=0}^{\infty} zn[(a_1)_{(n)} \cdots (a_v)_{(n)}]/[n!(c_1)_{(n)} \cdots (c_w)_{(n)}],$$

where $a_{(n)} = a(a+1) \cdots (a+n-1) = (a+n-1)!/(a-1)!$. Denoted by ${}_vF_w(a_1, \ldots, a_v; c_1, \ldots, c_w; x)$ or just by ${}_vF_w$. Also, GENERALIZED HYPERGEOMETRIC FUNCTION. Various special cases are significant; e.g., ${}_1F_1$ is **Kummer's function** (a solution to Kummer's equation), and ${}_2F_1$ is **Gauss' hypergeometric function** (a solution of the usual hypergeometric equation).

hypergeometric series *Mathematics.* any convergent power series such that the ratio of consecutive coefficients is a fixed rational function of *n*; in particular, the series representation of Gauss' hypergeometric function ${}_2F_1$.

hypergeusia *Neurology.* excessive sensitivity of the sense of taste. Also, **hypergeusesthesia.**

hyperglobulinemia *Medicine.* an abnormally high globulin content of the blood.

hyperglucagonemia *Medicine.* an abnormally high level of the polypeptide hormone glucagon in the blood.

hyperglycemia *Medicine.* an abnormally high concentration of sugar in the blood. Thus, **hyperglycemic.**

hyperglycinemia *Medicine.* a metabolic disorder in which glycine accumulates in body fluids.

hypergolic *Chemistry.* self-igniting; tending to ignite spontaneously. *Space Technology.* of a fuel or propulsion system, igniting spontaneously upon contact with an oxidizer.

hyperhidrosis *Medicine.* a condition of excessive sweating.

Hyperiidea *Invertebrate Zoology.* a suborder of crustaceans in the order Amphipoda, with large eyes covering most of the head, a maxilliped without a palp, and a transparent body; some are parasitic.

hyperimmune *Immunology.* referring to a state that exists following repeated injections of a single antigen which causes the production of large amounts of highly potent antibodies in the serum of an organism. Thus, **hyperimmunity.**

hyperimmune antibody *Immunology.* an antibody from a hyperimmune individual.

hyperimmune serum *Immunology.* an antiserum containing a large quantity of potent antibodies, thus providing a high degree of immunity when injected.

hyperinsulinism *Medicine.* **1.** an excessive secretion of insulin by the pancreas, causing a significant fall in the level of blood sugar. **2.** insulin shock. **3.** an excess of insulin in the blood.

Hyperion [hī pēr′ē än] *Astronomy.* the seventh moon of Saturn, discovered in 1848 and measuring 175 by 120 by 100 kilometers; it rotates in a chaotic (nonperiodic) fashion. Also, SATURN VII. (From the Greek mythical figure, the father of Helios, the sun god.)

hyperkalemia see HYPERPOTASSEMIA.

hyperkeratosis *Medicine.* an overgrowth of the cornified layer of the epidermis.

hyperkinesia or **hyperkinesis** see HYPERACTIVITY.

hyperkinetic see HYPERACTIVE.

hyperlipemia *Medicine.* an excess of lipids in the blood plasma.

hyperlogia *Psychology.* a speech disorder characterized by incessant or excessive talking.

hyperlucent *Radiology.* characterized by an increased or extreme transparency to X rays. Thus, **hyperlucency.**

Hypermastigida *Invertebrate Zoology.* an order of flagellate protozoans in the class Zoomastigophorea, commensal in the guts of roaches and termites.

hypermedia *Computer Science.* the integration in a single user interface of multiple perceptual or presentation modalities, such as a combination of text, photographs, and sound.

hypermetabolism *Medicine.* an abnormally high metabolic rate, with higher than normal heat production by the body.

hypermetamorphism *Entomology.* a type of metamorphic change found in some parasitic insects, in which the organism has two or more larval structures and habits.

hypermetamorphosis *Entomology.* in some insects, a metamorphosis with a first, tiny, active stage; a large, sluggish second stage; and if present, a legless third stage.

hypermetropia *Medicine.* farsightedness in which a refractive error causes parallel rays to focus behind the retina. Also, **hyperopia.**

hypermnesia *Psychology.* an exaggerated or abnormal ability to remember.

hypermorph *Genetics.* a mutant gene characterized by an increase in the activity it influences.

hypermotility *Medicine.* increased motility, as of the stomach or the intestines.

hypernatremia *Medicine.* an abnormally high level of the sodium ion in the blood.

hypernephroma see RENAL-CELL CARCINOMA

hypernucleus *Nuclear Physics.* a nucleus that is "strange" and has one or more hyperons in addition to nucleons; only lambda hypernuclei with one or two lambda particles have been observed.

Hyperoartii *Vertebrate Zoology.* an alternate name for the fish order Petromyzontiformes, which includes the lampreys.

hyperoid axle *Mechanical Engineering.* a gear set in some rear-axle drives that carries the pinion 1.5 or more inches below the gear's centerline.

hyperon *Particle Physics.* a particle having baryon number +1 and a nonzero strangeness quantum number; one of the first two "strange" particles observed in cosmic rays in 1947.

hyperosmia *Medicine.* an abnormally acute sense of smell.

hyperosmotic *Ecology.* having an osmotic potential (generally, a salt concentration) greater than that of the surrounding medium.

Hyperotreti *Vertebrate Zoology.* an alternate name for the fish order Myxiniformes, the hagfishes.

hyperoxemia *Medicine.* an excessive acidity of the blood.

hyperparasite *Ecology.* an organism that is parasitic on another parasite. Thus, **hyperparasitic.**

hyperparathyroidism *Medicine.* an abnormal condition produced by a hyperactivity of the parathyroids, resulting in increased resorption of calcium from the skeletal system and increased absorption of calcium by the kidneys and gastrointestinal system.

hyperpathia *Neurology.* an excessive subjective response to painful stimuli, which is often caused by lesions of the peripheral nerves or of the spinothalamic tract; a perception of pain may continue after the stimulus is removed.

hyperperistalsis *Medicine.* increased peristaltic activity of the intestines, leading at times to an increased rate of movement of food through the stomach and intestine.

hyperphagia *Pathology.* an excessive ingestion of food, as in bulimia.

hyperphosphaturia *Medicine.* an abnormal amount of phosphates in the urine.

hyperpigmentation *Medicine.* abnormally increased pigmentation of a tissue or part; especially, unusual darkening of the skin due to heredity, drugs, exposure to the sun, or adrenal insufficiency.

hyperpituitarism *Medicine.* an abnormally high production of growth hormone by the pituitary gland, possibly resulting in gigantism in children and acromegaly in adults.

hyperplane *Mathematics.* an affine subspace of a finite-dimensional vector space whose dimension is one less than that of the space. For example, a hyperplane in R^n is a set of the form

$$H = \{(x_1, \ldots, x_n) : \sum_{i=1}^{n} a_i x_i = c\},$$

where $a_1, \ldots, a_n$ and c are fixed real numbers with $a_i \neq 0$ for at least one *i*.

hyperplasia *Medicine.* an increase in the size of a tissue or organ due to an increase in the number of its constituent cells. *Plant Pathology.* a condition arising as a result of a plant disease or injury, in which the flow of water through vascular tissue becomes obstructed.

hyperploid *Genetics.* a cell or individual that contains one or more chromosomes or chromosome segments in addition to the characteristic euploid number.

hyperploidy *Genetics.* the state of being a hyperploid.

hyperpnea *Medicine.* an abnormal increase in the depth and rate of breathing.

hyperpolarization *Biochemistry.* an increase in polarity caused by a negative shift in a cell's resting potential.

hyperponesis *Neurology.* a reversible condition involving misdirected neurophysiologic reactions, characterized by excessive action potential output from the motor and premotor portions of the cerebral cortex.

hyperponetic *Neurology.* relating to or affected by hyperponesis.

hyperpotassemia *Medicine.* an excess of the potassium ion in the blood. Also, HYPERKALEMIA.

hyperproteinemia *Medicine.* an abnormally high concentration of protein in the blood.

hyperpure germanium detector *Electronics.* a variation on the lithium-drifted germanium crystal that utilizes high-purity germanium, thus allowing the device to be stored at room temperature.

hyperpycnal inflow *Hydrology.* a flowing in of water that is denser than the water into which it flows.

hyperpyrexia *Medicine.* an abnormally high fever.

hyperreflexia *Medicine.* an exaggeration of deep tendon reflexes.

hyperresonance *Medicine.* an exaggerated resonance or percussion, such as that heard in pulmonary emphysema and pneumothorax.

hypersaline *Geochemistry.* relating to geological material whose salinity exceeds that of seawater.

hypersensibility *Neurology.* excessive sensitivity to a stimulus or foreign substance.

hypersensitive *Medicine.* of or relating to hypersensitivity or hypersensibility; highly or extremely sensitive. *Psychology.* unusually sensitive to criticism or correction.

hypersensitive site *Genetics.* a short segment of DNA, commonly found in the promoter region of genes, that is very sensitive to digestion by nucleases such as DNase I.

hypersensitivity *Immunology.* a state of increased susceptibility to an antigen, such as an allergic reaction, caused by previous exposure to the antigen. *Virology.* an extreme reaction to a virus.

hypersensitization *Graphic Arts.* any of a variety of processes designed to increase the photosensitivity of a printing plate or film.

hypersensor *Electronics.* a single-component, resettable circuit breaker that functions as a majority-carrier tunneling device; used for overcurrent or overvoltage protection.

hypersexuality *Psychology.* excessive sexual activity or sexual desire.

hypersomnia *Medicine.* **1.** sleep of excessive depth or duration. **2.** an abnormal condition in which one sleeps for excessively long periods of time with intervals of normal sleep. *Psychology.* an excessive need for sleep or desire to sleep.

hypersonic *Fluid Mechanics.* relating to velocities greater than approximately five times the speed of sound in a fluid medium. *Acoustics.* relating to extremely high-frequency vibrations, greater than about 1 GHz, such as magnetorestrictive or piezoelectric phenomena.

hypersonic flight *Aviation.* flight that takes place at hypersonic speed (Mach 5 or greater).

hypersonic flow *Fluid Mechanics.* flow at very high supersonic speeds; the shock wave strength in hypersonic flow is strong enough to generate temperatures capable of melting airfoil surfaces and of causing chemical reactions and viscous interaction.

hypersonic glider *Space Technology.* an unpowered reentry vehicle designed to descend at hypersonic speeds.

hypersonic inlet *Fluid Mechanics.* a nozzle inlet that is designed to withstand and accommodate hypersonic flow.

hypersonic nozzle *Fluid Mechanics.* a nozzle that is designed to produce and endure hypersonic flow.

hypersonics *Fluid Mechanics.* the study of velocities greater than Mach 5. *Acoustics.* the use of extremely high frequencies, as in transducers that use magnetorestrictive or piezoelectric vibrations to produce sound, or sound to produce such vibrations.

hypersonic speed *Fluid Mechanics.* a speed whose Mach number is equal to or greater than 5.

hypersorption *Chemical Engineering.* a process in which activated carbon selectively adsorbs less volatile components from a gaseous mix, leaving the more volatile components unaffected.

hypersplenism *Medicine.* a pattern of reaction to excessively increased spleen activity, which may result in blood disorders such as neutropenia and thrombocytopenic purpura.

hypersteatosis *Medicine.* excessive sebaceous secretion, as in seborrhea.

hyperstereoscopy *Cartography.* a type of stereoscopic viewing in which the vertical scale is exaggerated relative to the horizontal scale along the line of sight; the details of the relief are thereby more easily visible.

hypersthene *Mineralogy.* an intermediate member of the enstatite-ferrosilite series, pyroxene group.

hypersusceptibility *Pathology.* a condition characterized by abnormally increased susceptibility to poisons, infective agents, or substances that are normally innocuous.

hypertelorism *Anatomy.* an abnormally large separation between two paired organs.

hypertely *Evolution.* a condition of evolutionary overspecialization in which the large size of an organism or body structure eventually becomes disadvantageous to the organism. *Zoology.* specifically, extreme or overdeveloped imitation of the color or markings of another animal so as to be unusable for camouflage.

hypertensin see ANGIOTENSIN.

hypertension *Medicine.* a condition of persistently high arterial blood pressure; in adults, this is usually identified as a blood pressure reading exceeding 140/90 mm Hg. The condition may be associated with other disorders (**secondary hypertension**) or may have no single identifiable cause (**essential** or **primary hypertension**). Individuals affected with hypertension are at risk for heart disease, kidney failure, and stroke.

hypertensive *Medicine.* relating to, having, or causing hypertension.

Hypertext *Computer Programming.* **1.** a particular implementation of a dynamic information system consisting of a computer-based, primary-source notebook that is continually updated by a teacher and is read and annotated by each student who uses it. **2.** an information system containing a graph structure of records; the user can move from the current record to other records that are related to it in various ways. Records may contain text, graphics, and program interfaces.

hyperthecosis *Pathology.* widespread hyperplasia, with an overabundance of lutein in the the cells of the Graafian follicles of the ovary.

hyperthermia *Physiology.* an abnormally elevated body temperature; fever. Thus, **hyperthermic.**

hyperthyroidism *Medicine.* a condition in which there is excessive secretion from the thyroid gland or excessive intake of thyroid hormone, marked by goiter, weight loss, rapid heart rate, tremor, fatigue, and increased appetite. Also, **hyperthyroidosis.**

hyperthyrotropinism *Medicine.* an excessive secretion of thyrotropic hormone by the adenohypophysis, which controls the status of the thyroid.

hypertonia *Medicine.* an excess of muscular or arterial tension. Also, **hypertonicity.**

hypertonic *Medicine.* relating to or affected by hypertonia.

hypertonic bladder *Medicine.* a condition of the bladder marked by excess muscular tension.

hypertonic contracture *Medicine.* a permanent muscle contraction due to continuous nerve stimulation in spastic paralysis.

hypertonic solution *Chemistry.* a solution having a greater osmotic pressure than another solution. *Cell Biology.* specifically, a solution that when bathing body cells causes a net flow of water out of the cells through the semipermeable membrane.

hypertoxicity *Toxicology.* the extreme toxicity of a material.

Hypertragulidae *Paleontology.* an extinct family of early ruminant artiodactyls in the infraorder Traguloidea; it flourished in the Oligocene and early Miocene in Eurasia and North America; extant from the Eocene to Miocene.

hypertrophic *Pathology.* relating to or affected by hypertrophy.

hypertrophy *Pathology.* the distention or enlargement of an organ partially due to the amplification of its component cells, but not due to a quantitative increase in cells or new growth. *Psychology.* an excessively frequent occurrence of a certain behavior pattern. Also, **hypertrophia.**

hypertrophy of the heart *Cardiology.* enlargement of the walls of the heart due to heart disease. Also, COMPENSATORY HYPERTROPHY OF THE HEART.

hypertropia *Medicine.* strabismus in which the visual axis of the eye is permanently deviated upward.

hypertropic cardiomyopathy *Cardiology.* a disease of the heart characterized by thickening of the muscular walls in the ventricular septum and left ventricle. Because there is often greater thickening in walls of the septum, the result is a narrowing of the left ventricle outflow tract.

hypertropic gastritis *Medicine.* an inflammatory condition of the stomach marked by thickening of the gastric mucous membrane and hyperplasia or enlargement of the peptic glands, as in Zollinger-Ellison syndrome.

hyperuricemia *Medicine.* an excess of uric acid in the blood.

hypervariable region *Immunology.* any of numerous areas, varying greatly in structure and composition, that occur within the V region of the heavy and light chains of immunoglobulin molecules, and are responsible for the antibody combining sites of antibody molecules. Also, **hypervariable site.**

hypervelocity *Mechanics.* extremely high velocity; a velocity many times the speed of sound.

hypervelocity armor-piercing *Ordnance.* describing an artillery projectile composed of a hard, high-density core encased in a lightweight carrier, called a sabot.

hyperventilate *Medicine.* to be affected with hyperventilation.

hyperventilation *Medicine.* **1.** a state in which there is an increased amount of air entering the pulmonary alveoli, resulting in reduction of carbon dioxide tension and leading to alkalosis. **2.** abnormally rapid, difficult, deep breathing, often caused by anxiety and producing tightness of the chest, a feeling of suffocation, dizziness, and tingling of the lips and fingers.

hypervisor *Computer Programming.* a control program that enables two or more operating systems to share a single computer system.

hypervitaminosis *Toxicology.* poisoning due to the ingestion of excessive amounts of vitamins. Also, SUPERVITAMINOSIS.

hypervitaminosis A *Toxicology.* poisoning due to the ingestion of excessive amounts of vitamin A; the symptoms may include irritability, fatigue, skin changes, hair loss, headache, and abdominal discomfort.

hypervitaminosis D *Toxicology.* poisoning due to the ingestion of excessive amounts of vitamin D; the symptoms may include headache, weakness, digestive disturbances, elevated blood pressure, and the calcification of certain tissues.

hypesthesia see HYPOESTHESIA.

hypha *plural,* **hyphae.** *Mycology.* a nonreproductive filament in fungi. Also, MYCELIAL THREAD.

hyphal body *Mycology.* in certain fungal species belonging to the order Entomophthorales, a type of hypha that forms conidiophores or spore-bearing structures.

hyphemia see OLIGEMIA.

hyphidium *Mycology.* in fungi belonging to the class Hymenomycetes, a sterile structure of hyphal origin.

Hyphochytriaceae *Mycology.* a family of fungi belonging to the order Hypochytriales, occurring in Europe and North America on freshwater algae and plant debris.

Hyphochytriales *Mycology.* an order of fungi belonging to the class Hypochytridiomycetes that are mainly aquatic, possessing flagella and producing asexual spores called zoospores.

Hyphochytridiomycetes *Mycology.* a class of fungi belonging to the subdivision Mastigomycotina, including land species and aquatic parasites of marine and freshwater algae and nonparasites that live from dead organic matter. Also, **Hyphochytriomycetes.**

hyphoid *Mycology.* describing a hyphalike fungal growth.

Hyphomicrobiaceae *Bacteriology.* a former classification for a family of budding bacteria of the order Hyphomicrobiales; the genera *Hyphomicrobium* and *Rhodomicrobium* have since been assigned to separate taxonomical groups.

Hyphomicrobiales *Bacteriology.* a former classification for an order of bacteria that reproduce by budding or by splitting in two; now classified as four separate genera.

Hyphomicrobium *Bacteriology.* a genus of budding bacteria occurring as ovoid cells in a dense clump with radiating filaments; found in soils and aquatic habitats.

Hyphomonas *Bacteriology.* a genus of Gram-negative, aerobic, budding bacteria that are made up of oval or pear-shaped cells; daughter cells possess filaments that are used for locomotion.

Hyphomycetales *Mycology.* an order of fungi belonging to the class Hyphomycetes, characterized by conidiophores that are not organized as synemmata or sporodochia.

Hyphomycetes *Mycology.* a large class of fungi belonging to the subdivision Deuteromycotina, characterized by having conidia on the substrate; its members are pathogenic to plants and animals, causing food spoilage or wood decay; some species are used to produce food, medicine, and industrial products.

hyphopodium *Mycology.* a type of hypha found in hyphal masses or mycelia of black mildew fungi.

Hyphozyma *Mycology.* a genus of fungi of the class Hyphomycetes that possess yeastlike characteristics.

hypidiomorphic *Petrology.* of or relating to an igneous rock texture in which the individual crystals are bounded partly by crystal faces and partly by surfaces resulting from growth interference with other crystals.

hypn- or **hypno-** a combining form meaning: **1.** sleep, as in *hypnagogic.* **2.** hypnosis, as in *hypnotherapy.*

Hypnaceae *Botany.* a widespread family of shiny mosses of the order Hypnobryales that form mats on soil, logs, and humus; characterized by falcate-secund, ecostate leaves, by longate-linear leaf cells, and by a hypnoid peristome.

hypnagogic imagery *Psychology.* a vivid perception that appears to someone in a hypnagogic state. Also, **hypnagogic hallucination.**

hypnagogic state *Neurology.* a sleeplike state of consciousness that immediately precedes sleep and that is sometimes accompanied by hallucinations.

hypnalgia *Neurology.* pain experienced during sleep.

Hypneaceae *Botany.* a small family of red algae of the order Gigartinales, having multibranched, uniaxial thalli and one genus regarded as parasitic; the genus *Hypnea* is used extensively as food.

Hypnineae *Botany.* in some taxonomic systems, a suborder of the moss order Hypnobryales.

hypnoanalysis *Psychology.* a treatment in which a patient is hypnotized in order to help him recall memories that are blocked when he is awake.

Hypnobryales *Botany.* an order of large, perennial, mat-forming mosses that occur commonly in forested habitats and have prostrate, creeping, and ascending stems.

Hypnodendraceae *Botany.* a family of spectacular large, glossy, dendroid mosses of the order Bryales that grow matted on humus and logs in rain forests; characterized by unbranched, erect stipes topped by branched fronds.

hypnolepsy see NARCOLEPSY.

hypnophobia *Psychology.* an irrational fear of sleep.

hypnopompic state *Neurology.* a drowsily conscious state that immediately follows sleep and is sometimes accompanied by hallucinations.

hypnosis *Psychology.* a condition in which an individual is placed in a sleeplike state, typically induced by having the subject fix his attention on a single object as a certain word or phrase is repeated to him; while in this state he may be induced to recall certain memories or reveal certain thoughts that are not forthcoming in the normal state of consciousness. *Medicine.* the induction of this state as a means of treating mental illness, anesthetizing or immobilizing a patient, and so on. *Behavior.* a motionless, trancelike state induced in an animal by the immediate threat of a nearby predator. (From *Hypnos,* the Greek god of sleep.)

hypnospore *Cell Biology.* a thick-walled resting spore.

hypnotic *Psychology.* relating to, producing, or produced by hypnosis.

hypnotic suggestion *Psychology.* the adoption of specific ideas or attitudes by an individual under the influence of hypnosis.

hypnotic trance *Psychology.* a state of hypnosis in which the subject is totally controlled by the hypnotist and is highly suggestible.

hypnotism *Psychology.* the act or practice of placing an individual in an artificial sleeplike state; hypnosis.

hypnotize *Psychology.* to place an individual in an artificial sleeplike state; affect by hypnosis.

hypnotoxin *Toxicology.* a highly potent toxin derived from the jellyfish *Physalia,* the Portuguese man-of-war; it affects the nervous system, causing motor and sensory malfunctions.

hypo *Medicine.* an informal term for a hypodermic needle or injection. *Inorganic Chemistry.* sodium thiosulfate, $Na_2S_2O_3{\cdot}5H_2O$, especially as used in photography to stabilize photographs during development.

hypo- *Biology.* a prefix meaning "under or below," as in *hypolemmal. Medicine.* a prefix meaning "less than normal," as in *hypoadrenia. Chemistry.* a combining form indicating that a compound contains an element in its lowest oxidation state, or has the lowest proportion of oxygen of any in a series of compounds. *Materials Science.* a combining form indicating that the composition of an alloy is less that the eutectic or eutectoid composition.

hypoadrenia *Medicine.* a decreased functioning of the adrenal glands. Also, **hydroadrenalism.**

hypoadrenocorticism *Medicine.* a condition in which the secretion of the adrenal cortex is abnormally diminished.

hypoalbuminemia *Medicine.* a condition in which the concentration of albumin in the blood is abnormally low.

hypoallergenic *Pharmacology.* of a cosmetic or medicinal substance, not likely to cause an allergic reaction.

hypoautomorphic see HYPOCRYSTALLINE.

hypobaric *Medicine.* **1.** characterized by less than normal pressure, usually pressure of ambient gases below 1 atmosphere. **2.** describing a solution having a specific gravity lower than that of the diluent or medium.

hypobasal *Botany.* located to the posterior of the basal wall.

hypoblast see ENDODERM.

hypobranchial musculature *Anatomy.* muscles that develop below the gills or branchial arches in the neck of a vertebrate.

hypocalcemia *Medicine.* a condition in which the level of calcium in the blood is abnormally low; this condition may cause hyperactive deep tendon reflexes, muscle and abdominal cramps, and spasm.

hypocalcification *Medicine.* an insufficient calcification of the bones or teeth.

hypocalciuria *Medicine.* a condition in which the amount of calcium in the urine is abnormally diminished.

Hypochilidae *Invertebrate Zoology.* a spider family in the order Araneida, with long, thin legs, dark spots, and two pairs of book lungs.

Hypochilomorphae *Invertebrate Zoology.* a spider suborder in the order Araneida, with two pairs of book lungs, six spinnerets, and eyes in three groups.

hypochloremia *Medicine.* a condition in which the level of chloride ions in the blood is abnormally low.

hypochlorhydria *Medicine.* a condition in which the amount of hydrochloric acid in the gastric juices of the stomach is abnormally low.

hypochlorite *Inorganic Chemistry.* **1.** a compound containing the ClO^- ion. **2.** a salt or ester of hypochlorous acid, HOCl; used in medicine.
hypochlorization *Medicine.* a decrease in the level of sodium chloride present in the diet.
hypochlorous acid *Inorganic Chemistry.* HOCl, an unstable, greenish-yellow, weak acid existing only in aqueous solution; decomposes to hydrogen chloride and oxygen. It is used as a bleach, water purifier, and disinfectant.
hypochnoid *Mycology.* describing fungi having dry, loosely intertwined hyphae whose undersides face upward.
hypocholesterolemia *Medicine.* a condition in which the level of cholesterol in the blood is abnormally low.
hypochondria [hī´pə kän´drē ə] *Psychology.* **1.** a condition, often brought on by anxiety or depression, in which an individual has an unfounded conviction that he or she is suffering from a serious illness and refuses to accept reassurances, even from physicians, that there is no basis for this concern. **2.** the tendency to be overconcerned with one's health or to exaggerate the symptoms of illness. Also, **hypochondriasis.**
hypochondriac [hī´pə kän´drē ak] *Psychology.* a person who suffers from hypochondria; someone who imagines himself to be seriously ill or is preoccupied with symptoms of illness.
hypochondriac region *Anatomy.* the upper lateral abdominal region located caudal to the left and right lower ribs on either side of the epigastric region. Also, **hypochondrium.**
hypochromic effect *Molecular Biology.* a decrease in the ultraviolet absorbance exhibited by a nucleic acid solution upon renaturation or base pairing of the nucleic acid.
hypochromicity *Physical Chemistry.* a decrease in the absorption of ultraviolet light by a polynucleotide solution.
hypochromic microcytic anemia *Medicine.* a condition in which there is a decrease in the amount of hemoglobin in the red cells and a decrease in the size of the red cells.
hypocleidium *Anatomy.* the area below the clavicles. *Vertebrate Zoology.* in birds, a skeletal structure that is a projection from the angle of the furcula, and is perhaps a rudiment of the interclavicle.
Hypocopridae *Invertebrate Zoology.* a family of beetles, coleopteran insects in the superfamily Cucujoidea.
hypocoracoid *Vertebrate Zoology.* in teleost fishes, the lower bone at the pectoral fin base.
hypocotyl *Botany.* the portion of the embryonic plant axis beneath the cotyledons, between the stem and the root.
Hypocreaceae *Mycology.* a family of fungi belonging to the order Hypocreales that is composed of plant pathogens which cause serious diseases such as canker and wilt.
Hypocreales *Mycology.* an order of fungi belonging to the class Pyrenomycetes, characterized by its bright colors; some are plant or insect parasites, while others live on dead organic matter.
hypocrystalline *Petrology.* of or relating to an igneous rock texture in which crystalline components lie in a glassy groundmass, the ratio of crystals to glass being between 7:1 and 5:3. Also, HYPOAUTOMORPHIC, SUBIDIOMORPHIC.

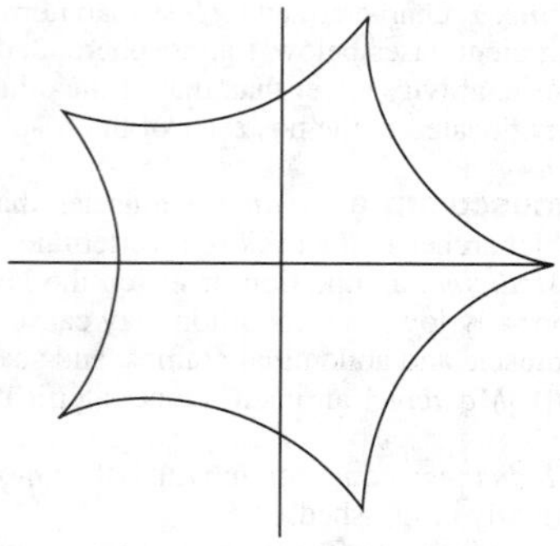

hypocycloid

hypocycloid *Mathematics.* a particular cyclic curve; the guiding curve L is a circle of radius b and a circle K of radius $a < b$ rolls along the inside of L without slipping. The hypocycloid is the path traced by a fixed point on K; for $-\infty < \phi < \infty$, the parametric representation is
$x = (b - a)\cos \phi + a \cos[(b - a)\phi/a]$, $y = (b - a)\sin \phi - a \sin[(b - a)\phi/a]$.
Hypodermatidae *Invertebrate Zoology.* the warble flies, a family of myodarian two-winged insects in the subsection Calypteratae; larvae are serious parasites in cattle and other mammals.
hypodermic *Anatomy.* relating to or located in the area beneath the skin. *Medicine.* applied or administered beneath the skin.
hypodermic needle *Medicine.* a short, hollow needle, usually part of a hypodermic syringe, that is used for injecting drugs beneath the skin and for aspiration. Thus, **hypodermic injection.**
hypodermis *Anatomy.* the area of loose connective tissue directly beneath the skin. *Botany.* a layer of cells just beneath the epidermis of plants, especially as modified to serve as a supporting or protecting layer. *Entomology.* a layer of epidermal cells that lies under and secretes the cuticle or chitin that forms the basis for an insect's exoskeleton.
hypodermoclysis *Medicine.* the subcutaneous injection of a fluid, usually a sodium chloride solution, to replace a loss of water and salt.
hypodynamia cordis *Cardiology.* the condition of diminished cardiac force.
hypoechoic *Radiology.* of or relating to tissues or structures that produce or reflect relatively few or less-dense echoes.
hypoequilibrium *Neurology.* absence of the normal tendency to experience vertigo during prolonged revolving of the body.
hypoergasia *Neurology.* abnormally decreased functional activity. Also, HYPOERGIA.
hypoergia *Neurology.* see HYPOERGASIA. *Immunology.* an abnormally low sensitivity to allergens. Also, **hypoergy.**
hypoergic *Neurology.* having less than normal energy.
hypoesthesia *Neurology.* an abnormally low sensitivity to stimuli. Also, HYPESTHESIA.
hypoeutectic alloy *Materials Science.* an alloy containing less of a component than the eutectic composition; in lead-tin solders, the tin content varies from 19 to 61.9%.
hypoeutectoid alloy *Materials Science.* an alloy containing less of a component than the eutectoid composition; in a steel, the carbon content is less than 0.8%.
hypofibrinogenemia *Medicine.* a condition in which the concentrations of fibrinogen in the blood plasma are abnormally low.
hypogammaglobulinemia *Medicine.* **1.** a condition in which the levels of the gamma fraction of serum globulin are abnormally low. **2.** broadly, any decrease in immunoglobulins.
hypogastric *Anatomy.* **1.** situated below the stomach. **2.** of or relating to the hypogastric region.
hypogastric region *Anatomy.* the pubic region; the middle portion of the inferior region of the abdomen, located below the umbilical region and between the inguinal regions. Also, **hypogastrium.**
hypogean *Botany.* a plant that lives primarily beneath the surface of the ground, or that germinates with the cotyledons remaining in the soil, such as the peanut or oak.
hypogene *Geology.* **1.** formed by ascending fluids within the earth, as ore or mineral deposits. **2.** formed beneath the earth's surface. **3.** a primary deposit.
hypogeous *Biology.* living or germinating below the soil surface.
hypogeusia *Neurology.* an abnormally weak sense of taste.
hypoglossal *Anatomy.* situated beneath the tongue.
hypoglossal nerve *Anatomy.* the twelfth cranial nerve; either of a pair of motor nerves having four major branches, which are essential for swallowing and for moving the tongue.
hypoglossal nucleus *Anatomy.* the nucleus of the origin of the hypoglossal nerve, in the midline of the medulla oblongata of the brain.
hypoglycemia [hī´pō glī sē´mē ə] *Medicine.* a condition in which the concentration of glucose in the blood is abnormally low; this can cause such symptoms as a cold sweat, piloerection, tremulousness, hypothermia, headache, irritability, confusion, hallucinations and bizarre behavior, convulsions, and coma.
hypoglycemia of piglets see BABY PIG DISEASE.
hypogonadism *Medicine.* a condition in which deficiencies in gametogenesis or a decrease in the secretion of gonadal hormones lead to inadequate gonadal function and retardation of sexual growth and development.
hypogynium *Botany.* a structure that supports the ovary in certain plants.
hypogynous *Botany.* having sepals, petals, and stamens attached beneath the pistil and free of the ovary.
hypohidrosis *Medicine.* an abnormal decrease in perspiration.
hypoid gear *Mechanical Engineering.* a gear system in which gear wheels connect nonparallel, nonintersecting shafts.

hypokinesia *Neurology.* abnormally decreased mobility, motor function, or motor activity. Also, **hypokinesis, hypomotility.**

hypokinetic *Neurology.* characterized by an abnormally low level of motor activity.

hypokinetic syndrome *Medicine.* an overall reduction in motor functions caused by brain dysfunction.

hypolemmal *Neurology.* situated beneath a sheath, as in the motor nerves under the sarcolemma or sheath of muscle fiber.

hypolimnion *Hydrology.* **1.** the relatively colder lowermost layer of water in a lake. **2.** specifically, the denser, stagnant or oxygen-poor lower layer of a thermally stratified lake, lying below the thermocline and unaffected by sunlight.

hypomagma *Geology.* a generally immobile, viscous magma that forms deep beneath a shield volcano and triggers volcanic activity.

hypomagnesemia see GRASS TETANY.

hypomania *Psychology.* a moderate mania marked by an expansive emotional state, elation, hyperirritability, and increased motor activity.

hypomenorrhea *Medicine.* an abnormal decrease in the flow or duration of menstruation.

hypomere *Developmental Biology.* **1.** the portion of the myotome that extends ventrolaterally to form body-wall muscle, innervated by the primary ventral branch of a spinal nerve. **2.** the somatic and splanchnic layers of the lateral mesoderm which give rise to the lining of the colon. *Invertebrate Zoology.* the basal part of some sponges, without flagellated cavities.

hypometabolism *Medicine.* a decreased metabolic rate.

hypomorph *Genetics.* a mutant gene that acts in the same direction as the normal allele, but with a lesser effect.

hypomorphic allele see LEAKY GENE.

Hypomyces *Mycology.* a genus of fungi belonging to the order Clavicipitales; most live from other fungi.

hyponasty *Botany.* the inward and upward bending of a plant organ, caused by the greater growth of its lower end.

hyponatremia *Medicine.* a condition in which the concentration of sodium ions in the blood is abnormally low.

hyponeural sinus *Invertebrate Zoology.* a tubular part of the coelom of many echinoderms, containing hemal vessels.

hyponeural system *Invertebrate Zoology.* part of the nervous system of echinoderms, consisting of a ring of motor nerve tissue around the mouth, and five pairs of nerves.

hyponychium *Histology.* the thickened extension of stratum corneum lying under the free edge of a fingernail.

hypoosmotic *Biology.* having an osmotic potential less than that of the surrounding medium.

hypoovarianism *Medicine.* a condition marked by a decrease in the endocrine activity of the ovaries.

hypoparathyroidism *Medicine.* a condition caused by the decrease in the secretion or absence of the secretion of the parathyroid hormones. Also, PARATHYROID INSUFFICIENCY.

hypophalangism *Medicine.* a usually congenital absence of one or more of the phalanges of a finger or toe.

hypopharynx *Anatomy.* the lowest portion of the pharynx, below the upper edge of the epiglottis and opening into the larynx and esophagus. *Entomology.* the long, tonguelike structure attached to the floor of insects' mouths, generally bearing salivary openings.

hypophosphatasia *Medicine.* a genetic metabolic condition in which the content of alkaline phosphatase in the blood and bones is abnormally low; this can lead to hypercalcemia, ethanolamine phosphatemia, and ethanolamine phosphaturia.

hypophyseal duct tumor *Medicine.* a tumor that originates from epithelial remnants of the pituitary diverticulum.

hypophysectomy *Medicine.* the removal or destruction of the hypophysis.

hypophyseoportal circulation *Neurology.* the passage of blood from hypophyseal arteries through capillaries of the hypothalamus and the neural stalk, through a second group of capillaries in the adenohypophysis. Also, **hypophysioportal circulation.**

hypophyseoportal system *Neurology.* the circulation system of the pituitary gland, or hypophysis. Also, **hypophysioportal system.**

hypophysis see PITUITARY GLAND.

hypopituitarism *Medicine.* a condition marked by a decrease in the activity of the anterior lobe of the hypophysis and a related decrease in the secretion of one or more of the anterior pituitary hormones.

hypoplankton *Biology.* planktonic organisms living close to the sea bed or lake floor.

hypoplasia *Medicine.* **1.** the decreased development of an organ or tissue, usually a result of a decrease in the number of cells. **2.** atrophy caused by the destruction of organ or tissue elements, not simply by general reduction in size. *Plant Pathology.* of or relating to any plant disease that causes an abnormal reduction in cell production and results in the underdevelopment of plant cells, organs, or tissues.

hypoplastic dwarf *Medicine.* an individual who, although normally proportioned, is of subnormal size.

hypoplastron *Vertebrate Zoology.* either one of the third set of bony plates in the plastron of most turtles.

hypoploid *Genetics.* a cell or individual that contains less than the normal diploid number of chromosomes (e.g., 45 chromosomes in humans).

hypoploidy *Genetics.* the state of being a hypoploid.

hypoponesis *Neurology.* a reversible condition involving misdirected neurophysiologic reactions, characterized by an abnormally low action potential output from the motor and premotor portions of the cerebral cortex.

hypoproliferative anemia *Medicine.* an anemia characterized by a reduction in the concentration of hemoglobin and in the amount of red blood cells as a result of an abnormally low amount of erythrocyte primordial cells.

hypoproteinemia *Medicine.* a condition in which the amount of total protein in the blood plasma is abnormally low; this condition can lead to edema and the accumulation of fluid in the serous cavities.

hypoprothrombinemia *Medicine.* a condition in which the amount of prothrombin in the blood is abnormally low. Also, FACTOR II DEFICIENCY.

Hypopterygiaceae *Botany.* a tropical and southern temperate family of glossy, often whitish-green, dendroid plants of the order Hookeriales that form loose mats on humus, logs, or tree trunk bases in lowland rainforests and are characterized by tomentose, prostrate primary stems with unbranched erect secondary stems terminated by a branched frond.

hypopus *Invertebrate Zoology.* a larval stage of some mites, in which they attach to an animal, without feeding, in order to migrate.

hypopycnal inflow *Hydrology.* a flowing in of water that is less dense than the water into which it flows. *Oceanography.* specifically, an inflow of fresh water into the denser saline water of the ocean.

hypopygium *Invertebrate Zoology.* the ninth segment of an insect, modified to bear the copulatory organs.

hypopyon *Medicine.* an accumulation of puslike fluid in the anterior chamber of the eye.

hyporeactive *Medicine.* relating to or marked by a less than normal response to stimuli.

hyporeflexia *Medicine.* a state of decreased response or strength of the reflexes.

hyposensitivity *Medicine.* **1.** abnormally low sensitivity marked by an unusually slow or weak response to a stimulus. **2.** a condition in which a reaction to a specific allergen is reduced by repeating the application and increasing the dosage of the allergen. Thus, **hyposensitive.**

hyposomnia see INSOMNIA.

hypospadias *Medicine.* an orifice caused by a defect in the wall of the urethra, causing a part of the canal to open on the undersurface of the penis or on the perineum; when present in the female, the urethra opens into the vagina.

hypostasis *Medicine.* **1.** formation of a sediment at the bottom of a liquid. **2.** congestion caused by the pooling of venous blood in a dependent part or organ. Also, **hypostatic congestion.**

hypostatic *Genetics.* of an interaction between nonallelic genes, resulting in the suppression of expression of one gene by the other.

hyposthenia *Medicine.* weakness.

hyposthenuria *Medicine.* production of urine of low specific gravity, usually caused by an inability of the kidney tubules to secrete concentrated urine.

Hypostomata *Invertebrate Zoology.* a subclass of protozoans of the class Kinetofragminophora, with tubular or somewhat flattened bodies and mouth on the ventral side, somatic cilature often reduce;. free-living and symbiotic species.

hypostome *Invertebrate Zoology.* any structure immediately below or behind the mouth, such as the oral cone of hydrozoans, or the rodlike structure in the mouth of some ticks.

hypostracum *Invertebrate Zoology.* one of three calcareous layers making up the plates (valves) of mollusk chitons.

hyposynergia *Medicine.* a decreased ability to coordinate parts of the body.

hypotarsus *Vertebrate Zoology.* in birds, a projection from the tarsometatarsal bone.

hypotelorism *Medicine.* an abnormal closeness between two body parts or organs; usually refers to abnormally close eyes.

hypotension *Medicine.* **1.** abnormally low arterial blood pressure. **2.** any kind of reduced pressure or tension.

hypotenuse [hī pät´ə nooz´] *Mathematics.* in a right triangle, the side opposite the right angle.

hypothalamic center *Anatomy.* one of several regions in the hypothalamus that control autonomic, endocrine, and somatic functions.

hypothalamic releasing factor *Biochemistry.* a hormone that is produced by the hypothalamus and stimulates the release of other hormones including the growth hormone and the luteinizing hormone.

hypothalamoneurohypophyseal tract *Anatomy.* a tract of neuronal axons that connects the hypothalamus with the neurohypophysis and stores the hormones oxytocin and vasopressin, which are produced in the hypothalamus.

hypothalamus *Anatomy.* the lower part of the diencephalon of the brain, including a portion of the ventral wall of the third ventricle and the ventricular floor; it is connected to and controls the pituitary gland.

hypothallus *Botany.* **1.** in crustose lichens, a marginal outgrowth from the thallus of hyphae. **2.** a residual film remaining after sporangia formation in slime molds of the class Myxomycetes.

hypotheca *Invertebrate Zoology.* the bottom or inner half of a diatom frustule.

hypothenar *Anatomy.* **1.** a fleshy ridge on the ulnar side of the palm that includes the muscles that control the small (fifth) finger. Also, **hypothenar eminence. 2.** of or relating to this ridge.

hypothermal *Geology.* **1.** formed at great depth, under high pressure, and at temperatures between 300 and 500°C. **2.** the environment in which such deposits are formed.

Hypothermal *Geology.* a postglacial interval characterized by little glacial expansion and moderate decreases in temperature.

hypothermal deposit *Mineralogy.* a hydrothermal mineral deposit formed at great depth (high pressure) and a temperature of 300–500°C.

hypothesis [hī päth´ə sis] *Science.* a theory or proposition that is based on certain assumptions and that can be evaluated scientifically. *Statistics.* in testing, an inference as to the value of an unknown population parameter.

hypothesis testing *Statistics.* inferential methods for assessing a hypothesis about population parameters based on a sample.

hypothesize and test *Artificial Intelligence.* a problem-solving strategy in which the best solution is found and selected from a stored or generated set of plausible solutions.

hypothyroidism *Medicine.* a decreased secretion of thyroid hormone, resulting in thyroid insufficiency.

hypotonia *Medicine.* **1.** reduced tension in any part. **2.** relaxation of the arteries. **3.** a condition marked by decreased tone in the skeletal muscles, or a decreased resistance of the muscles to passive stretching; can lead to the muscles being stretched beyond their normal limits.

hypotonic *Physiology.* describing a solution having a lower concentration of dissolved solute molecules and a lower osmotic pressure than another solution with which it is compared.

Hypotrichida *Invertebrate Zoology.* an order of protozoans in the subclass Spirotrichia; the flattened body has cilia on the ventral side and none on the dorsal side.

hypotrochoid *Mathematics.* a contracted cycloid in which the rolling circle is inside the guiding circle.

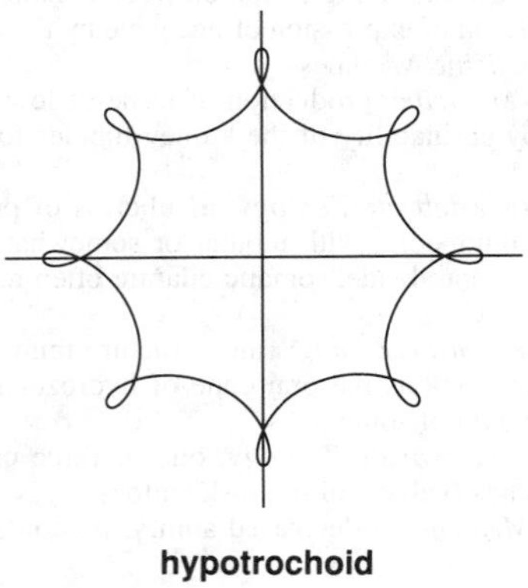

hypotrochoid

hypotype *Systematics.* a specimen used in the technical description of a species, but not considered to be a member of the type series.

hypovigility *Psychology.* an abnormal reaction to an external stimuli in which an individual either overreacts or does not react at all; most commonly seen in schizophrenia.

hypovirulence *Microbiology.* the decreased virulence of a pathogenic organism.

hypovitaminosis *Medicine.* a condition that is due to a deficiency of one or more essential vitamins. Also, AVITAMINOSIS.

hypovolemia *Medicine.* a condition in which the volume of plasma in the body is abnormally decreased. Also, OLIGEMIA, HYPHEMIA.

hypovolemic shock *Medicine.* a state of profound physical depression caused by a decrease in the volume of blood to a level that is insufficient for the maintenance of adequate cardiac output, blood pressure, and tissue perfusion; possible causes are hemorrhage or dehydration.

hypoxic encephalopathy *Medicine.* a degeneration of the brain caused by a lack of oxygen.

hypoxanthine *Biochemistry.* $C_5H_4ON_4$, a purine that occurs in transfer RNA and is formed by the deamination of adenine.

hypoxia *Medicine.* a condition characterized by abnormally low levels of oxygen in the blood and tissues. *Ecology.* a condition in which inadequate environmental oxygen is available to an organism.

hypozygal *Invertebrate Zoology.* in the endoskeleton of echinoderms of the class Crinoidea, the proximal member of a syzygial pair of brachials.

hypsi- a combining form meaning "high," as in *hypsicephalic.*

hypso- a combining form meaning "height," as in *hypsometry.*

hypsochrome *Physical Chemistry.* an atom or group whose introduction into a compound shifts the compound's absorption band toward the violet end of the spectrum.

hypsochromy *Physical Chemistry.* a process in which a compound shifts its absorption band toward the violet end of the spectrum. Also, **hypsochromatic shift.**

hypsodont *Vertebrate Zoology.* of or relating to teeth with high or deep crowns and short roots.

hypsographic *Cartography.* of or relating to the measurement of altitude and elevation. Also, HYPSOMETRIC.

hypsographic curve *Cartography.* a line on a hypsographic map or chart indicating height above or below a specified datum; for example, a contour line indicating height above mean sea level.

hypsographic map *Cartography.* a type of map that shows the topography of the earth in terms of its elevation above or below a specified datum, usually mean sea level, by means of contour lines, hachures, shading, or tinting. Also, HYSPOMETRIC MAP.

hypsographic tinting *Cartography.* a method for showing elevations on a map or chart by tinting or coloring those parts of the surface features that lie between specified elevations with graded shades of color. Also, **hypsometric tinting.**

hypsographic tint scale *Cartography.* a scale in the margin or cartouche of a map that shows which shades or colors represent which elevations. Also, **hypsometric tint scale.**

hypsography *Cartography.* the theories and techniques used during surveying and mapmaking to describe accurately the elevations of the earth's surface above and below a specific datum, usually mean sea level. Also, HYPSOMETERY.

hypsometer *Engineering.* an instrument used to measure the boiling point of a liquid in order to determine the atmospheric pressure and thus the altitude, or to calibrate a thermometer.

hypsometric see HYPSOGRAPHIC.

hypsometric formula *Geophysics.* a formula, based upon the hydrostatic equation, for determining differences in geopotential between two points or determining the pressure at one level when the pressure at another level is known.

hypsometry see HYPSOGRAPHY.

hypsophobia see ACROPHOBIA.

hypural *Vertebrate Zoology.* in fishes, one of a set of enlarged hemal arches or spines on the ventral tip of the vertebrae supporting the caudal fin.

Hyracodontidae *Paleontology.* a family of perissodactyls in the superfamily Rhinocerotoidea; small and tapirlike, the hyracodontids are ancestral to modern rhinoceroses; widespread holarctic distribution; Eocene to Miocene.

Hyracoidea *Vertebrate Zoology.* the hyraxes, conies, or dassies, an order of small, ungulate, herbivorous mammals with unique adhesive feet, a compact body about the size of a rabbit, and small rounded ears.

Hyracotherium *Paleontology.* the ancestral horse, known as "eohippus;" Hyracotherium was widespread throughout the Pangaean land mass of the Early Eocene; entire and exactly alike skeletons have been found in Europe and in North America, and fossils have also been found recently in Paleocene deposits in Mongolia; it was the earliest known perissodactyl and was the size of a small dog, about 27 cm high.

hyrax *Vertebrate Zoology.* any of several small mammals of the order Hyracoidea, having molars resembling those of a rhinoceros and incisors like a rodent.

hyrax

hyster- a combining form meaning: **1.** uterus, as in *hysterectomy.* **2.** hysteria, as in *hysterical.*

Hysterangiaceae *Mycology.* a family of fungi belonging to the order Phallales, which is composed of stinkhorns; its spores are disbursed by burrowing animals.

hysterectomy *Medicine.* the surgical removal of the uterus; one or both ovaries and oviducts may be removed at the same time.

hysteresis *Physics.* a dependence of the state of a system on its previous history, generally the retardation or lagging of an effect behind the cause of the effect. *Electromagnetism.* **1.** the inclination of a magnetic material to saturate and retain some of its magnetism after the alternating magnetic field to which it is subjected reverses polarity, causing magnetization to lag behind the magnetizing force. **2.** an analogous electrostatic action in a ferroelectric dielectric material. *Electronics.* a double-valued function, in which different values are obtained depending on whether the independent variable increases or decreases *Robotics.* a lagging condition in the compliance of a manipulator. *Nucleonics.* a condition in which the counting rate and the voltage characteristics in a counter tube are temporarily changed.

hysteresis clutch *Mechanical Engineering.* a clutch that produces torque by magnetic hysteresis.

hysteresis coefficient *Physics.* the constant of proportionality in a formula for hysteresis loss. Also, STEINMETZ COEFFICIENT.

hysteresis damping *Mechanical Engineering.* the damping of a vibration caused by mechanical hysteresis.

hysteresis error *Physics.* the maximum difference between the upscale-going and downscale-going values of the driven variable, at a common driving variable in a hysteresis loop.

hysteresis loop *Physics.* a closed curve that traces the path of a hysteretic quantity's state, especially that of a driven variable as the function of a driving variable, such as the magnetization of a ferromagnetic material as a function of an externally applied magnetic field.

hysteresis loss *Physics.* the amount of energy required to run through one complete cycle in the magnetization hysteresis loop of a ferromagnetic material, equal to the area enclosed by the loop and appearing as heat when the driving magnetic field is rapidly passed through its cycle.

hysteresis motor *Electricity.* a small synchronous motor such as a phonograph motor, for light-duty, constant-speed applications; it uses the hysteresis and eddy-current losses induced in its hardened-steel rotor to produce rotor torque.

hysteretic damping *Mechanics.* the dissipation of energy in a vibrating mechanical system or solid because of mechanical hysteresis.

hysteria *Psychology.* **1.** a disorder in which an individual develops mental or physical symptoms with no apparent organic basis, such as hallucinations, amnesia, paralysis, or uncontrolled weeping, in response to a traumatic experience. **2.** in popular use, any extreme or uncontrolled emotional reaction, as of panic, weeping, laughter, and so on.

Hysteriaceae *Mycology.* a family of fungi belonging to the order Hysteriales that is composed primarily of members living off of decaying organic matter, and some parasites of higher plants, particularly conifers.

hysteriaceous *Mycology.* **1.** relating to the fungal order Hysteriales. **2.** referring to fungi having a kind of perithium or fruiting body called a hysterothecium, which has a cleft opening.

Hysteriales *Mycology.* in some classifications, an order of spore-bearing fungi of the class Loculoascomycetes having discoid or cupulate ascocarps, which occur on decaying organic matter or are parasitic to plant material.

hysteric *Psychology.* **1.** a person affected by hysteria. **2.** of or relating to hysteria.

hysterical *Psychology.* **1.** relating to or affected by hysteria. **2.** in an unstable or uncontrolled emotional state.

hysterical anesthesia *Medicine.* a loss of tactile sensation as a result of conversion hysteria, in which the patient substitutes physical signs or symptoms for anxiety.

hysterical neurosis see HYSTERIA.

hysterical paralysis *Medicine.* a loss of movement in a part without any apparent organic neurological cause.

hysterical personality (disorder) *Psychology.* a disorder characterized by emotional outbursts, extreme self-centeredness, dramatization, and intense excitability.

hysterics *Psychology.* a popular term for an emotional outburst.

hysteriform *Psychology.* having the appearance of hysteria.

hysteritis see METRITIS.

hystero- a combining form meaning: **1.** uterus, as in *hysterogram.* **2.** hysteria, as in *hysteroepilepsy.*

hysterocarcinoma *Oncology.* cancer of the endometrium. Also, ENDOMETRIAL CARCINOMA.

hysterochroic *Mycology.* relating to fungal fruiting bodies that become discolored with age.

hysteroepilepsy *Neurology.* a severe form of hysteria characterized by episodes of epilepticlike convulsions that usually occur only in the presence of others and that stop abruptly without the confusion and lethargy which usually accompany true epilepsy.

hysteroepileptogenic *Neurology.* inducing epilepticlike hysterical episodes known as hysteroepilepsy. Also, **hysteroepileptogenous.**

hysterogenic zone *Neurology.* an area of the body that, when touched, induces an attack of hysteria. Also, **hysterogenic point.**

hysterogram *Medicine.* **1.** a radiograph of the uterus containing contrast medium. **2.** a record of the strength of uterine contractions.

hysterography *Medicine.* **1.** radiography of a uterine cavity filled with contrast medium. **2.** a procedure used to record uterine contractions.

hystero-oophorectomy *Medicine.* surgical removal of the uterus and ovaries.

hysteropexy *Surgery.* the fixation of a misplaced or abnormally movable uterus. Also, UTEROPEXY, UTEROFIXATION.

hysterorrhaphy *Surgery.* the repair of a lacerated uterus.

hysterorrhexis see METRORRHEXIS.

hysterosalpingectomy *Surgery.* the removal through surgery of the uterus and one or both uterine tubes.

hysterosalpingography *Radiology.* radiography of the uterus and uterine tubes after radiopaque material has been inserted.

hysterosalpingo-oophorectomy *Surgery.* the removal of the uterus, uterine tubes, and ovaries.

hysteroscope *Radiology.* an endoscope used to view the uterus and cervix.

hysterosoma *Invertebrate Zoology.* the posterior segment of the body in certain mite species, divided from the anterior proterosoma by a dorsal or ventral groove behind the second pair of legs.

hysterostat *Radiology.* a mechanical device containing sealed ionizing radiation material, used in intrauterine radiation therapy.

hysterotomy *Surgery.* an incision of the uterus, usually for the delivery of a fetus.

Hystrichospherida *Paleontology.* a group of unicellular organisms that may have affinities with desmids, the zygotes of freshwater Chlorophyta; they may also be cysts (aplanospores) of dinoflagellates.

Hystricidae *Vertebrate Zoology.* the Old World or terrestrial porcupines, a family of herbivorous rodents of the suborder Hystricomorpha, having a heavyset body with the upper body covered with thick elastic spines or quills; found in forests, deserts, and savannahs from southern Europe to Africa, and eastern Asia to the Philippines and Indonesia.

Hystricomorpha *Vertebrate Zoology.* the South American or porcupine rodents, a suborder that includes the porcupines, cavies, guinea pigs, capybaras, pacarana, agoutis, chinchilla, cane rats, and related families and species.

Hz or **hz** hertz. Also, **Hz.** or **hz.**

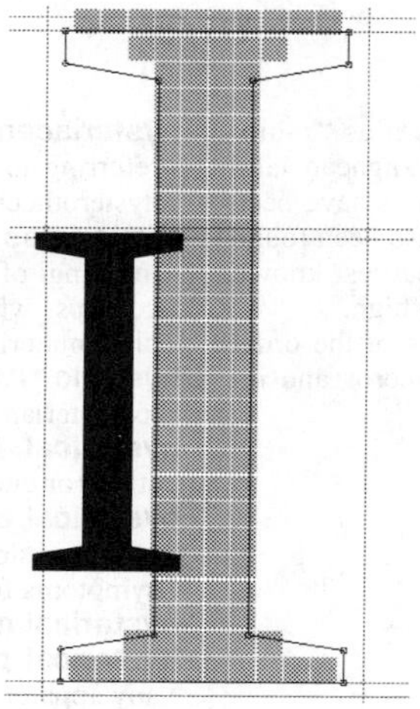

I electric current; iodine; inosine.
i intelligence; intensity.
IA international angstrom.
IAA indoleacetic acid.
Ia antigens *Immunology*. the antigens that determine an immune reaction to grafts (known as Class II histocompatibility antigens), and that occur on the surface of mouse B lymphocytes, macrophages, and accessory cells.
IAC international analysis code.
IAD *Aviation*. the airport code for Dulles International Airport, Washington, D.C.
IAEA or **I.A.E.A.** International Atomic Energy Agency.
IAF or **I.A.F.** International Astronautical Federation.
IAH *Aviation*. the airport code for Intercontinental Airport, Houston, Texas.
IAHA immune adherence hemagglutination assay.
$I^AI^BI^O$ *Genetics*. the allelic forms of the genes that determine the ABO human blood groups.
IALC instrument approach and landing chart.
IAMS or **I.A.M.S.** International Association of Microbiological Societies.
IAN indoleacetonitrile.
I & D incision and drainage.
I & O intake and output.
ianthinite *Mineralogy*. $UO_2{\cdot}5UO_3{\cdot}10H_2O$, a violet, orthorhombic mineral, having a specific gravity of 5.16 and a hardness of 2 to 3 on the Mohs scale; found as minute lathlike crystals in pitchblende.
IAP international airport.
Iapetus *Astronomy*. the eighth moon of Saturn, discovered in 1671 and measuring 1400 kilometers in diameter; its leading hemisphere is very dark (about 5% reflective), while its trailing hemisphere is bright (50% reflective).
IAS indicated airspeed; Institute of Aerospace Sciences.
iatro- a combining form meaning "medicine" or "healing."
iatrogenic *Medicine*. **1.** describing any adverse condition that is a reaction to treatment by physicians, especially to infections transmitted during therapy. **2.** originally, describing disorders caused by autosuggestion resulting from the physician's examination, manner, or discussion.
IAU or **I.A.U.** International Astronomical Union.
I.B. inclusion body.
IB *Aviation*. the airline code for Iberia.
Iballidae *Invertebrate Zoology*. a small family of brightly colored wasps, hymenopteran insects in the superfamily Cynipoidea.
I band *Histology*. a band, encompassing two adjacent muscle sarcomeres, that contains actin filaments and a Z line.
IBD infectious bursal disease virus.
I beam *Civil Engineering*. a principal horizontal supporting structural member that is constructed of steel in an I shape in cross section with short flanges.
Iberia *Geography*. a peninsula of southwestern Europe on which the countries of Spain and Portugal are located.
Ibe wind *Meteorology*. a strong wind that blows through the Dzungarian Gate in western China, characterized by a foehnlike weather condition in which the air temperature experiences a sudden rise; in the wintertime, from about –15°F to about 30°F.
ibex *Vertebrate Zoology*. **1.** the wild European goat *Capra ibex* of the family Bovidae, characterized by a short, gray-haired body and large, backward pointing, ridged horns; the few remaining goats inhabit high mountain areas, having been hunted to near extinction by the ancient Romans. **2.** the wild Asian goat *Capra aegagrus* of the family Bovidae, believed to be the ancestor of the domesticated goat.

ibex

IBF immunoglobulin-binding factor.
IBIB isobutyl isobutyrate.
ibis *Vertebrate Zoology*. **1.** any of several large, long-legged, wading birds of the family Threskiornithidae, related to the storks and herons and characterized by a long, slender, downward-curved bill; found in warm temperate and tropical regions. **2.** any of various similar birds belonging to the family Ciconiidae, particularly the wood stork.
I blood group *Immunology*. a group of red-blood-cell antigens that are identified by their reactions with the antibodies anti-I and anti-i, occurring in individuals having acquired hemolytic anemia or in the serum of persons of the rare phenotype i.
ibogaine *Organic Chemistry*. $C_{20}H_{26}N_2O$, a narcotic alkaloid obtained from the root of the *Tabernanthe iboga* plant of Zaire; crystals, soluble in alcohol and slightly soluble in water; melts at 152–153°C; used in medicine.
IBT *Biochemistry*. isatin-β-thiosemicarbazone, an antiviral agent that acts late in the replication cycle to inhibit the replication of poxviruses.
ibuprofen *Pharmacology*. $C_{13}H_{18}O_2$, a white powder; a nonsteroidal anti-inflammatory used in the treatment of rheumatoid arthritis and osteoarthritis and as an analgesic and antipyretic.
IBY International Biological Year.
IC integrated circuit; internal connection; inspiratory capacity; irritable colon.
ICAI intelligent computer-assisted instruction.
ICAO International Civil Aeronautical Organization.

Icarus [ik′ə rəs] *Astronomy.* asteroid 1566, discovered in 1949 and measuring roughly 1.7 km in diameter; it has the fastest rotation period (2 hours and 27 minutes) of any asteroid known.
ICBM see INTERCONTINENTAL BALLISTIC MISSILE.
ICD International Classification of Diseases.
ICE internal-combustion engine; Institution of Civil Engineers.
ice *Hydrology.* water in the solid state, produced by the freezing of liquid water, by the condensation of water vapor directly into crystals, or by the compaction and recrystallization of fallen snow.
ice accretion *Hydrology.* the process by which a layer of ice builds up on the exposed surfaces of solid objects.
Ice Age *Geology.* a popular name for the Pleistocene epoch, so called because of its widespread glacial ice. Also, GREAT ICE AGE.
ice age see GLACIAL EPOCH.
ice anchor *Naval Architecture.* a single-fluked anchor designed to set in ice to secure a vessel or to provide a hold for a hawser in warping it along. Also, ICE DRAG.
ice apron *Hydrology.* **1.** a thin mass of relatively immobile snow and ice that is attached to the rock cliff at the back of a cirque. **2.** ice that flows from an ice sheet over the edge of a plateau. *Civil Engineering.* a wedge-shaped structure that protects bridge piers from floating ice.
ice avalanche *Hydrology.* the sudden fall of broken glacier ice down a steep slope. Also, ICEFALL.
ice band *Hydrology.* a layer of ice within firn or snow.
ice barrier *Hydrology.* a term for various formations or obstructions of ice, such as ice front or ice shelf.
ice bay *Oceanography.* an indentation in the edge of a large ice floe or an ice shelf.
ice belt *Oceanography.* a long band of sea-ice fragments in otherwise open water or along the shore, sometimes hundreds of kilometers long and up to 100 kilometers wide.
iceberg *Oceanography.* a large floating block of freshwater ice that has broken off the edge of a glacier and been carried out to sea; about 90% of its mass lies under the water.

iceberg

iceberg storage *Computer Technology.* a system that makes extensive use of data-compression technology to permit the accommodation of a relatively large amount of data within a relatively small amount of storage space.
ice blink *Meteorology.* a term used primarily in the polar regions to denote a bright yellowish-white glare on the underside of a cloud layer; produced by the reflection of light from an ice-covered surface below.
ice boulder see GLACIAL BOULDER.
icebound *Navigation.* prevented from movement by surrounding ice.
ice boundary *Hydrology.* the boundary between fast and pack ice, or between areas having different concentrations of pack ice.
icebreaker *Naval Architecture.* a vessel fitted with a powerful bow that is specially designed to drive a navigable path through sea ice.
ice bridge *Oceanography.* river ice that is thick enough to hinder or prevent navigation.
ice buoy *Navigation.* a sturdy buoy that replaces a standard buoy during times that heavy ice is expected.
ice cake *Hydrology.* a piece of floating sea ice less than 10 meters in diameter.
ice canopy *Oceanography.* pack ice viewed from the perspective of personnel on a submarine.
ice cap or **icecap** *Hydrology.* a perennial dome-shaped or platelike mantle of ice and snow, having an area of less than 50,000 square kilometers, that covers the summit of a mountain or a flat landmass and spreads out radially under its own weight.
ice-cap climate see PERPETUAL FROST CLIMATE.
ice cascade see ICEFALL, def. 1.
ice cave *Hydrology.* any cave having a low enough temperature for ice to form and remain more or less throughout the year. Also, ICE GROTTO.
ice chart *Navigation.* a chart, such as a pilot chart, that shows the usual incidence of ice, especially in navigable waters.
ice color see AZOIC DYE.
ice concentration *Oceanography.* the ratio of ice area to water area in a given location.
ice-contact delta *Geology.* a delta formed by a stream flowing out of a glacier into a lake at the margin of the glacier ice. Also, DELTA MORAINE, MORAINAL DELTA.
ice-contact terrace see KAME TERRACE.
ice-cream plant *Agriculture.* a term for a highly palatable plant that grazing or browsing animals will seek out and consume in preference to other vegetation.
ice crust *Hydrology.* **1.** a continuous layer of ice formed on the surface of snow by the freezing of meltwater or rainwater. **2.** see ICE RIND.
ice crystal *Hydrology.* any one of a variety of crystalline shapes assumed by ice, such as hexagonal columns, hexagonal platelets, or ice needles.
ice-crystal cloud *Meteorology.* a cloud that consists entirely of ice crystals, having a diffuse and fibrous appearance.
ice-crystal haze *Meteorology.* a very fine ice fog that is composed entirely of ice crystals and may extend to altitudes as high as 20,000 feet.
ice dam *Hydrology.* blocks of floating ice that obstruct a river and may cause flooding during spring and early summer.
ice day *Meteorology.* for a given area, a 24-hour period in which the air temperature does not rise above 32°F, and ice on the surface of water does not thaw.
ice desert *Meteorology.* a polar area that is permanently covered by snow and ice and is essentially devoid of vegetation.
iced firn *Hydrology.* a mixture of ice and firn, or firn that has been saturated with meltwater and later refrozen.
ice drag see ICE ANCHOR.
ice edge *Oceanography.* the boundary between the open sea and sea ice of any kind, fast or floating.
ice erosion *Geology.* **1.** see GLACIAL EROSION. **2.** any erosion that results from water freezing in cracks of rock.
icefall or **ice fall** *Hydrology.* **1.** a chaotic breaking up of ice in a glacier. Also, ICE CASCADE. **2.** a mass of glacial ice that overhangs a precipice. **3.** see ICE AVALANCHE.
ice feathers *Hydrology.* a term for hoarfrost forming on the windward side of objects and on aircraft flying from cold to warm layers of air.
ice field or **icefield** *Hydrology.* **1.** a large sheet of land ice covering a mountainous region and consisting of many interconnected glaciers. **2.** an extensive region of unbroken pack ice. *Oceanography.* a flat sheet of sea ice more than 8 kilometers in diameter; the largest areal subdivision of sea ice.
icefish *Vertebrate Zoology.* a common name for any of a group of small, shining, nearly translucent fishes, including some species of smelt and 17 species of the family Salangidae; an Antarctic fish that lacks red blood cells
ice flow see GLACIAL FLOW.
ice flower *Hydrology.* **1.** a formation of leafy ice crystals on the surface of ice or snow that resembles a flower. Also, FROST FLOWER. **2.** a formation of ice crystals on the surface of a quiet body of water.
ice fog *Meteorology.* a fog composed of minute particles of ice that are suspended in the atmosphere, usually occurring in clear weather in an environment of extremely cold temperatures. Also, FROST FOG, RIME FOG.
ice foot *Oceanography.* sea ice that is firmly attached to the shore at the high-water line and is unaffected by tides.
ice-free or **ice free** *Hydrology.* describing a water surface that is essentially free of ice. *Navigation.* specifically, describing a harbor or other body of water that is free of ice that would impede navigation throughout the year.

ice fringe *Hydrology.* **1.** a belt of sea ice that extends a short distance from the shore. **2.** a narrow, seaward-sloping mass of coastal ice that extends less than one kilometer inland from the sea. **3.** see ICE RIBBON.

ice front *Hydrology.* the high, steep, floating vertical cliff that forms the seaward edge of a glacier.

ice gland *Hydrology.* a roughly cylindrical column of ice or iced firn within a firn field.

ice grotto see ICE CAVE.

ice gruel *Hydrology.* a slushy mass of floating ice formed by the irregular freezing together of frazil on the surface of the ocean.

ice island *Oceanography.* a large, generally level, homogeneous floating sheet of ice in Arctic regions, up to 300 square miles in area and often persisting for several decades.

ice island iceberg *Oceanography.* a conical or dome-shaped iceberg, easily mistaken for an ice-covered island.

ice jam *Hydrology.* an accumulation of blocks of river ice that obstruct part of the channel. *Oceanography.* large fragments of lake or sea ice that have thawed loose in spring and been blown against the shore.

ice keel *Oceanography.* a downward-projecting ridge on the underside of pack ice; the underwater equivalent of a pressure ridge, an ice keel may extend as much as 50 meters below sea level.

Iceland agate see OBSIDIAN.

Icelandic low *Meteorology.* a low-pressure center located between Iceland and southern Greenland, as represented on mean charts of sea-level pressure; a principal center of action in the atmospheric circulation of the Northern Hemisphere.

Icelandic type *Volcanology.* a volcanic eruption characterized by great volumes of fluid basaltic lava flowing from fissures over a broad area.

Iceland moss see CETRARIA.

Iceland spar *Mineralogy.* a very pure, transparent, and easily cleavable variety of calcite, having strong double refraction; found particularly in Iceland and used in optical instruments.

ice layer *Hydrology.* **1.** a layer of solid ice or iced firn within a mass of snow or firn. **2.** a more or less horizontal layer of ground ice.

ice load *Engineering.* in safety codes, the maximum amount of ice allowed to accumulate on an overhead wire in a power supply system (usually 1/4 to 1/2 inch in radial thickness).

ice-marginal terrace see KAME TERRACE.

ice mine *Engineering.* a waterproof mine placed in or under the ice on a lake or river; it is detonated by a pressure device on the surface to break up the ice.

ice needle *Hydrology.* an ice crystal that is characterized by an elongated hexagonal shape. Also, ICE SPICULE.

ice nucleation bacteria *Bacteriology.* bacteria that are able to initiate the transition of water to ice by functioning as nucleation sites for ice crystal formation; may contribute to frost damage in agricultural crops.

ice nucleus *Meteorology.* a particle that serves as a nucleus in the formation of ice crystals in the atmosphere.

ice pedestal *Oceanography.* a column, cone, pinnacle, or mushroom-shaped pillar of ice rising above the surface of a glacier or other ice-covered area and protected from the sun's rays by an overlying rock or mass of debris, so that the surrounding ice ablates more rapidly than the column itself.

ice pellets *Meteorology.* small, transparent or translucent, round or irregularly shaped particles of ice, having a diameter of less than 5 mm, that usually bounce and make a sound on impact with hard ground; may occur as sleet or hail. Also, SMALL HAIL.

ice period *Meteorology.* the portion of a year during which ice appears and remains over a given area.

ice pick *Mechanical Devices.* a long, thin, sharp-pointed metallic spike set in a wooden handle and used for chipping ice.

ice point *Physical Chemistry.* the true melting point of ice (or the true freezing point of water), reached when a combination of pure ice and air-saturated pure water reaches equilibrium at standard atmospheric pressure; represented by 32°F or 0°C.

ice pole *Oceanography.* the center of the Arctic pack ice, currently near 84°N and 160°W; by surface travel, the least accessible point of the pack.

ice push *Hydrology.* **1.** the lateral force exerted by expanding, shoreward-moving ice on a lake or bay. Also, ICE THRUST. **2.** the ridge of material formed by such a force. Also, LAKE RAMPART.

icequake *Hydrology.* the concussion accompanying the breakup of large ice masses.

ice ribbon *Hydrology.* a thin, white, curly deposit of ice crystals formed on plant surfaces by the freezing of moisture exuded from dead stems.

ice rind *Hydrology.* a brittle, thin, hard crust of sea ice formed on a quiet surface of water having a low salinity, either by direct freezing or from grease ice.

ice run *Hydrology.* **1.** the rapid breakup of river ice in spring or early summer. **2.** the rush of water that follows such a breakup.

ice sheet *Hydrology.* a broad, thick sheet of glacier ice that covers an extensive land area for a long period of time. Also, MANTLE ICE.

ice shelf *Oceanography.* a continuous and more or less permanent sheet of fast ice, generally level and sometimes extending hundreds of kilometers from the shore, fed by glaciers and by annual accumulation of snow; although attached to land, an ice shelf's edge is chiefly supported by the water on which it floats.

ice slush see GREASE ICE.

ice spicule see ICE NEEDLE.

ice splinters *Physical Chemistry.* minute electrically charged ice fragments that form when ice crystals are exposed to air currents; chiefly observed under laboratory conditions.

ice storm *Meteorology.* a weather condition characterized by freezing precipitation, with the attendant formation of a hazardous glaze on objects and surfaces on the ground below. Also, SILVER STORM.

ice stream *Hydrology.* a stream of ice flowing down a valley, or a high-velocity current of ice flowing within an ice sheet or ice cap.

ice thrust see ICE PUSH.

ice tongs *Mechanical Devices.* any of various-sized tongs with either claw-shaped or pointed ends, used for gripping ice cubes or blocks.

ice tongue *Hydrology.* any long, narrow extension of a glacier on land or afloat.

ice wall *Hydrology.* a cliff of ice forming the seaward edge of a glacier that is not afloat.

ice wedge see FOLIATED ICE.

ICF intracellular fluid.

ich see ICHTHYOPHTHIRIASIS.

I channel *Electronics.* the American color-television system channel used to transmit cyan-orange color information.

ichneumon [ik noo´mən] *Vertebrate Zoology.* a long-tailed mongoose, *Herpestes ichneumon,* that is found in Africa and southern Europe; believed by the ancient Egyptians to eat crocodile eggs. *Invertebrate Zoology.* any of various wasplike insects of the family Ichneumonidae whose larvae are parasitic on caterpillars and other immature insect forms. Also, **ichneumon fly**.

Ichneumonidae *Invertebrate Zoology.* a family of wasps, hymenopteran insects that are parasitic on other insects.

Ichneumonoidea *Invertebrate Zoology.* a superfamily of wasps that are parasitic on other insects; eggs are laid in or on the body of the host, which is eaten by the larvae at hatching.

ichno- a combining form meaning "track" or "footstep," as in *ichnofossil, ichnology.*

ichnofossil see TRACE FOSSIL.

ichnology see PALEOICHNOLOGY.

ichor *Geology.* a fluid believed to aid in the granitization of certain materials. *Medicine.* a thin, watery discharge, usually from a wound or an ulcer.

ichth. ichthyology.

ichthammol *Pharmacology.* a viscous, odiferous, usually reddish-brown liquid prepared from certain bituminous mineral deposits and used as a topical anti-infective for skin disease. Also, AMMONIUM ICTHYOSULFONATE, BITUMINOL.

ichthyism *Toxicology.* poisoning due to any toxin produced by a fish. Also, **ichthyotoxism**.

ichthyismus *Toxicology.* chronic poisoning due to a toxic substance produced by a fish.

ichthyo- or **ichthy-** a combining form meaning "fish" or "fishlike," as in *ichthyology, ichthyism.*

ichthyoacanthotoxin *Toxicology.* any poison produced by the spines, stings, or teeth of a fish.

Ichthyobdellidae *Invertebrate Zoology.* a family of leeches in the order Rhynchobdellae, characterized by large powerful suckers.

Ichthyodectidae *Paleontology.* a family of neopterygian teleost fishes in the extinct order Ichthyodectiformes; large predaceous fish closely related to the Osteoglossomorpha, which generally preferred a marine environment; upper Jurassic and Cretaceous.

ichthyohemotoxin *Toxicology.* a poison found in the blood of certain fish.

ichthyoid *Science.* similar to a fish, especially in structure or appearance.

ichthyology [ik´thē äl´ə jē] *Vertebrate Zoology.* the study of fish and fishlike vertebrates.
ichthyootoxin *Toxicology.* a poison found in the roe of certain fish.
ichthyophagous *Zoology.* feeding mainly or exclusively on fish.
ichthyophobia *Psychology.* an irrational fear of fish.
Ichthyophonus *Mycology.* a genus of fungi perhaps belonging to the order Entomophthorales; its species are fish parasites.
ichthyophonus infection *Veterinary Medicine.* a common fungal infection in wild and aquarium fish that is usually detected on necropsy; characteristic spherical cyst stages are observed microscopically in the smears of granulomatous lesions of liver, spleen, and muscle.
ichthyophthiriasis *Veterinary Medicine.* a common, fatal infection of fish that causes chronic inflammation of the skin and gills in the form of spotty, white lesions, due to invasion by the protozoan fish parasite *Ichthyophthirius multifilis.* Also, ICH, WHITE SPOT DISEASE.
Ichthyophthirius *Invertebrate Zoology.* a genus of ovate ciliate protozoans in the subclass Holotrichia; parasites causing the often fatal freshwater fish skin disease called ich.
Ichthyopterygia *Paleontology.* a proposed extinct subclass of diapsid marine reptiles, including ichthyosaurs and some related forms.
Ichthyornis *Paleontology.* a genus of early toothed birds in the extinct order Ichthyornithiformes; a small, ternlike bird, strong-flying and having a keeled breast-bone; upper Cretaceous.
Ichthyornithes *Paleontology.* a proposed subclass or superorder of primitive flying birds of the Cretaceous; not adopted by leading authorities.
Ichthyornithiformes *Paleontology.* an order of early flying birds in the extinct superorder Odontognathae; Cretaceous.
ichthyosarcotoxin *Toxicology.* a poison found in the muscles or flesh of certain fish.
ichthyosarcotoxism *Medicine.* poisoning that results from the ingestion of the flesh of poisonous fish, in contrast with ordinary bacterial food poisoning that might result from the ingestion of nonpoisonous fish contaminated by bacteria.
ichthyosaur *Paleontology.* any of the marine reptiles of the extinct order Ichthyosauria, ranging greatly in length (from 4 to 40 feet), and characterized by a round, tapering body, a large head, four flippers, and a vertical caudal fin.

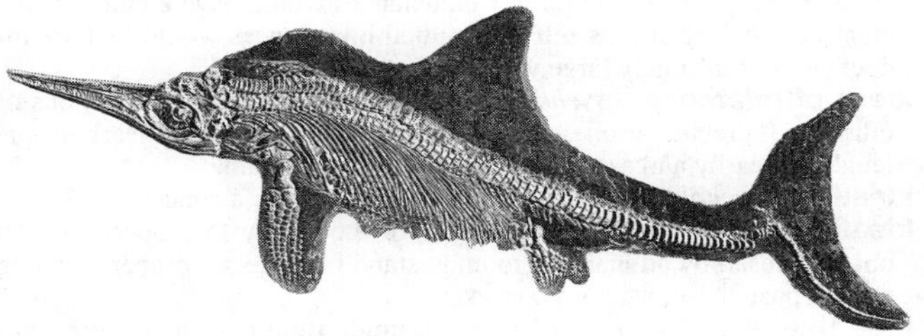

ichthyosaur

Ichthyosauria *Paleontology.* an order of highly specialized marine reptiles in the subclass Diapsida, characterized by an elongated snout and a fish-shaped body, and generally very fishlike; some authorities group them in a separate subclass, Ichthyopterygia, because of their great divergence from the basic reptilian pattern; Middle Triassic to Cretaceous.
ichthyosis *Medicine.* a keratinization disorder identified by dryness and noninflammatory fishskin-like scaling of the skin.
ichthyosis congenita *Medicine.* ichthyosis that is present at birth.
ichthyosis simplex *Medicine.* a form of ichthyosis inherited as an autosomal dominant trait; the onset in childhood is identified by the fine scales present on the trunk and extremities. Also, **ichthyosis vulgaris.**
Ichthyosporidium *Invertebrate Zoology.* a genus of microspora of the family Nosematidae, with one known species pathogenic to fish; found in North America and Europe.
Ichthyostegalia *Paleontology.* an order of primitive amphibians in the subclass Labyrinthodontia, probably the first amphibians and possibly ancestors of all later land vertebrates; Upper Devonian.
Ichthyotomidae *Invertebrate Zoology.* a family of annelid worms in the superfamily Eunicea.
ichthyotoxicology *Toxicology.* the study of natural toxins produced by fish.
ichthyotoxin *Toxicology.* any natural toxin or poison that is produced by fish. *Veterinary Medicine.* the hemolytic ingredient in eel serum.
icicle *Hydrology.* a long, slender, downward-pointing, tapering piece of ice formed by the freezing of dripping water.
icing *Hydrology.* **1.** in a permafrost region, an ice mass formed on the surface during the winter by the successive freezing of sheets of water. **2.** any accumulation or coating of ice on a solid object. *Aviation.* specifically, the freezing of atmospheric moisture on the surface of an aircraft.
icing level *Meteorology.* in aviation weather practice, the lowest altitude at which an aircraft will encounter icing conditions.
icing-rate meter *Engineering.* an instrument used to measure the rate of ice accumulation on an unheated object.
ICLARM or **I.C.L.A.R.M.** International Center for Living Aquatic Resources Management.
ICN or **I.C.N.** International Council of Nurses.
icon *Computer Programming.* a symbol representing a command or program that is displayed on a computer screen and provides easy access to the command or program when selected with a pointing device such as a mouse or a light pen.
iconic memory *Psychology.* the coding of memories by using visual imagery.
iconify *Computer Science.* to reduce a displayed window to a small icon symbol in order to free screen space. The information in the window is saved; the icon can later be selected and re-expanded into a window.
iconocenter *Electromagnetism.* the image of the reflection coefficient for an impedance-matched load plotted on an Argand diagram.
iconology the science of producing, recording, displaying, and transmitting visual images.
iconometer *Optics.* an instrument that determines the size of an object at a known distance from the instrument, or determines the distance from the instrument of an object of known size.
iconoscope *Electronics.* a television-camera tube in which an electron beam scanning a photoemissive mosaic screen generates a charge equal to the intensity of light at various points on the screen. Also, STORAGE CAMERA.
icosahedral symmetry *Virology.* a form of cubic capsid structure in certain viruses, characterized by a pattern of 5-fold, 3-fold, and 2-fold rotational symmetry. Also, 5-3-2 SYMMETRY.
icosahedron *Mathematics.* a polyhedron with twenty sides. The faces of a regular icosahedron are all congruent equilateral triangles; the regular icosahedron is one of the five Platonic solids.
Icosteidae *Vertebrate Zoology.* the ragfishes, a family in the suborder Stromateoidei that is characterized by a cartilaginous skeleton that makes the body appear limp; found in deep waters of the north Pacific Ocean.
icotype *Systematics.* a specimen collected from the same locality as the type series and/or described by the original author but not used in the final technical published description.
I.C.S. International College of Surgeons.
ICSH interstitial-cell-stimulating hormone.
ICSU or **I.C.S.U.** International Council of Scientific Unions.
ICT *Aviation.* the airport code for Wichita, Kansas.
ICT International Critical Tables.
ictal *Neurology.* of or relating to a seizure, stroke, convulsion, or other paroxysmal attack.
Ictaluridae *Vertebrate Zoology.* a family of freshwater catfishes of the order Siluriformes; a typical catfish, including the channel and bullhead catfishes, characterized by a wide head and long mouth barbels; native to North America.
icteric *Medicine.* associated with or characterized by icterus (jaundice).
Icteridae *Vertebrate Zoology.* a diverse family of passerine birds of the suborder Oscines, including blackbirds, orioles, cowbirds, and meadowlarks; found in the New World except at high latitudes.
ictero- a combining form meaning "jaundice," as in *icterohepatitis.*
icteroanemia *Veterinary Medicine.* an acute febrile disease of swine, characterized by jaundice, emaciation, anemia, and significant destruction of red blood cells.
icterogenic *Medicine.* relating to or causing jaundice.
icterohematuria *Veterinary Medicine.* an infectious disease of sheep that is characterized by jaundice and caused by the parasitic protozoan *Babesia ovis,* which destroys red blood cells.
icterus see JAUNDICE.
icterus gravis *Medicine.* jaundice marked by high fever and delirium; seen in acute yellow atrophy, it can be a rare complication of viral hepatitis and other diseases of the liver, a result of insensitivity to drugs, or the result of exposure to hepatotoxins. Also, MASSIVE HEPATIC NECROSIS OR MALIGNANT JAUNDICE.

icterus gravis neonatorum *Medicine.* extreme jaundice in the newborn; usually a form of isoimmunization associated with Rh factor or a severe and usually fatal jaundice caused by occlusion of the common bile duct, erythroblastosis fetalis, congenital syphilitic cirrhosis of the liver, or septic pylephlebitis.

icterus index *Pathology.* an index formerly used to gauge the bilirubin content in the serum by means of a color comparison with a solution of potassium dichromate.

Ictidosauria *Paleontology.* an extinct order of small, generally carnivorous cynodonts, very closely related to mammals; recent authorities have proposed the term Tritheledontia for the order; Upper Triassic and beginning of Jurassic.

ictus *Neurology.* a sudden seizure, stroke, convulsion, or other paroxysmal attack.

ICU intensive care unit; instruction control unit.

ICW interrupted continuous wave.

id *Psychology.* in Freudian theory, the part of the personality that includes the instinctive and unconscious biological impulses that seek immediate gratification, personal pleasure, or satisfaction.

ID inside diameter; inner diameter; inside dimensions; internal diameter; initial dose; intradermal.

I.D. identification; infantry division; intelligence department.

id. the same. (From Latin *idem.*)

ID3 *Artificial Intelligence.* an algorithm for learning a decision tree from a set of training examples with known classifications.

IDD insulin-dependent diabetes.

iddingsite *Mineralogy.* a mixture of silicates formed as an alteration product of olivine; found as reddish-brown patches in basic igneous rocks.

idea *Psychology.* the result of a mental process that originates internally.

ideal *Mathematics.* a left (resp. right) ideal of a ring R is a subring I of R such that, for all $i \in I$ and $r \in R$, $ri \in I$ (resp., $ir \in I$). That is, a left (resp. right) ideal of R is a subring that is closed under left (resp. right) multiplication by ring elements. An ideal that is both a left and right ideal is called a **two-sided ideal.**

ideal aerodynamics *Fluid Mechanics.* a field of study in aerodynamics wherein simplifying assumptions are made to aid in solving airflow problems. Also, **ideal fluid dynamics.**

ideal bunching *Electronics.* the conditions under which alternately accelerating and retarding electrons in a velocity-modulated tube would theoretically generate a large current peak during each cycle.

ideal copolymer *Materials Science.* a copolymer that has a completely random sequence of monomer units.

ideal cultural pattern *Anthropology.* an idealized standard or convention of culture that may differ from the actual cultural practice; reveals the values and norms of a society.

ideal dielectric *Electricity.* a dielectric having zero conductivity, in which the energy needed to establish an electric field can be recovered when the voltage is removed; a complete vacuum is the only ideal dielectric that is known. Also, PERFECT DIELECTRIC.

ideal exhaust velocity *Fluid Mechanics.* the theoretical maximum velocity of gas flow through a nozzle as the gas moves from the nozzle inlet pressure and temperature to an ambient environment.

ideal flow *Fluid Mechanics.* a two-dimensional fluid flow that is incompressible, irrotational, and inviscid.

ideal fluid *Fluid Mechanics.* a fluid that is incompressible and has no internal friction or viscosity.

ideal gas *Physical Chemistry.* a gas for which the product of the pressure and volume is proportional to the absolute temperature; realized in actual gases in extreme dilution where the forces between the molecules have a negligible contribution. Also, PERFECT GAS. *Thermodynamics.* a gas whose particles are infinitely small and have no effective interaction with each other; this is a theoretical gas that would perfectly obey the ideal gas law under all conditions.

ideal gas law *Thermodynamics.* the equation of state for a gaseous system, $pV = nRT$, in which p is the pressure of the system, V is the volume, n is the number of moles of the sample, R is the gas constant for the gas in question, and T is the absolute temperature. Many real gases will approximately obey this law at sufficiently low pressures or high temperatures.

ideal grain *Ordnance.* the granulation of a propellant that will produce the maximum projectile velocity within the pressure limits of the bore.

ideal line *Mathematics.* in projective geometry, the set of all ideal points. Also, LINE AT INFINITY.

ideal masochism see MORAL MASOCHISM.

ideal mechanical advantage *Mechanical Engineering.* the ratio of input distance to output distance; i.e., the distance that a force must travel in order to move a load for a given distance. For example, if a lever moves a distance of 4 feet to raise a load 1 foot, it has an ideal mechanical advantage of 4.

ideal molecular mass distribution *Materials Science.* a condition in which a polymer has a heterogeneity index of 2; it occurs when step-growth polymerization is complete or very near complete.

ideal network *Electronics.* a simplified circuit or network, used to simplify the analysis of a network by disregarding some of its parameters.

ideal point *Mathematics.* **1.** in projective geometry, a point at which all lines parallel to a given line are postulated to meet. **2.** a point, denoted by ∞, that is adjoined to the complex plane in the construction of the extended complex plane. Also, POINT AT INFINITY.

ideal polarized electrode *Physical Chemistry.* a hypothetical electrode at which no faradaic processes can take place regardless of the external applied potential; an electrode approaching this ideal shows large potential changes upon the passage of very small currents.

ideal rocket *Space Technology.* a theoretical rocket that would operate perfectly at a velocity equal to that of its jet gases; used to provide ideal parameters to be compared to practice. An ideal rocket would require a homogeneous and invariant propellant, observance of the perfect gas laws, a constant and steady propellant flow, no friction, no heat transfer across rocket walls, an axially directed velocity of all exhaust gases, a uniform gas velocity across every section, and chemical equilibrium in the combustion chamber and nozzle.

ideal solution *Physical Chemistry.* a solution that exhibits no attractive force between its components, that experiences no internal energy change on mixing, and that will obey Raoult's law perfectly over all temperatures and concentrations.

ideal strength *Materials Science.* the stress at which the interatomic bonds rupture.

ideal theory *Mathematics.* the study of rings, with emphasis on those properties that can be described in terms of ideals.

ideal transducer *Electricity.* a hypothetical passive transducer that transfers the maximum possible power from a specified source to a load and thus has no losses.

ideal transformer *Electricity.* a hypothetical transformer that does not store or dissipate energy; its self-inductances would have a unity coefficient of coupling and its self and mutual impedances would be pure inductances of infinitely large values.

ideas of reference *Psychology.* an irrational belief that the actions of others are directed at oneself, such as when an individual hears people laughing nearby and assumes they are laughing at him.

ideation *Psychology.* the ability to form or consider a concept.

ideational apraxia *Medicine.* a loss of the ability to properly use an object, caused by an inability to understand the object's proper meaning and purpose. Also, SENSORY APRAXIA.

idée fixe [ē´dā fēks´] *Psychology.* a single dominant idea, often unfounded or irrational, that an individual stubbornly holds to despite evidence to the contrary. Also, FIXED IDEA.

idempotent *Mathematics.* **1.** if $*$ is an operation on an object x, then x is said to be idempotent if $x * x = x$. For example, an idempotent matrix is a matrix M such that $M^2 = M$ and an idempotent group element is an element g such that $g * g = g$ (where $*$ is the group operation). In many rings, idempotents are in one-to-one correspondence with left (or right) ideals. **2.** set union and intersection satisfy idempotent laws; i.e., for any set A, $A \cap A = A$ and $A \cup A = A$. **3.** an algebraic system in which every element is idempotent is called an **idempotent system.**

identical twins *Developmental Biology.* a popular term for monozygotic twins; that is, twins developed from a single fertilized ovum that splits into equal halves early in embryonic development and therefore have the same genotype. Identical twins are of the same sex and usually resemble each other very closely both physically and mentally.

identifiability *Statistics.* of or relating to a family of probabilistic models in which no two elements generate the same probability distribution for the data.

identification *Psychology.* the process of adopting as one's own the beliefs, values, or actions of another or others. *Military Science.* **1.** the process of determining the friendly or hostile character of an unknown contact or object. Similarly, **identification maneuver. 2.** in arms control, the process of determining which nation is responsible for a detected violation of an arms control measure. *Control Systems.* a method of finding the transfer function of a system by examining its response to either an impulse or a step-function input.

identification division *Computer Programming.* one of the four main parts of a COBOL program, in which the programmer identifies the source and object programs.
identifier *Computer Programming.* a symbol used to name, identify, or specify a program or portion of data.
identifier word *Computer Programming.* a computer word that is stored in a special register and is used in a search-read sequence; each word read is compared with the identifier word stored in the register.
identity *Science.* the state of remaining the same under varying conditions. *Psychology.* a person' s concept of self; the aggregate of characteristics by which a person is recognized by him- or herself and others. *Mathematics.* **1.** see IDENTITY ELEMENT. **2.** an equation that is satisfied for all possible values of the variables. **3.** an equation relating operators, mappings, or other higher-order mathematical objects without reference to their operands.
identity crisis *Psychology.* a conflict in which an individual loses his sense of self and cannot accept or adopt the role he believes society has placed on him; typically expressed by withdrawal, rebelliousness, or negativity; often triggered by rapid social or technological changes.
identity element *Crystallography.* a symmetry element whose operation leaves unchanged anything on which it operates. *Mathematics.* a two-sided identity element for an algebraic operation $*$ is an element e of the object A (such as a group, ring, etc.) on which $*$ is defined, such that $e * a = a * e = a$ for every element a of A. If it is only true that $e * a = a$ (or $a * e = a$), then e is called a left identity (or right identity). By convention, a (two-sided) identity element for (ring, field, etc.) addition is denoted by 0, and the identity element for (ring, field, etc.) multiplication, if it exists, is denoted by 1 or by 1_A.
identity function *Mathematics.* any function f for which $f(x) = x$ for all x in the domain of definition. Denoted *Id* or *I*. Also, **identity map, identity operator.**
identity functor *Mathematics.* the functor from a category C that assigns each object of C to itself and each morphism of C to itself. Also, COVARIANT IDENTITY FUNCTOR.
identity matrix *Mathematics.* the $n \times n$ identity matrix is the matrix with 1 in every main diagonal position and 0 elsewhere.
identity stage *Psychology.* the fifth of Erik Erikson's eight stages of human development, during the adolescent years, when the person develops a characterization of himself or herself as a unique individual with a sense of purpose and direction. Also, **identity formation.**
identity unit *Computer Technology.* a logic circuit with several binary input signals and a single binary output signal such that, when all the input signals are the same (0 or 1), the output signal represents 1.
identity vs diffusion *Psychology.* the conflict or stress that can arise during the identity stage of development, if a person cannot form a sense of being a unique individual. Also, **identity vs (role) confusion.**
ideo- a combining form meaning "idea," as in *ideogram.*
ideofact *Archaeology.* an object whose function is to express or symbolize the beliefs of a people rather than to serve practical or social needs.
ideofunction *Archaeology.* the use of an object for ideological purposes; for example, the wearing of a certain special garment as part of a religious ceremony.
ideofunctional *Archaeology.* of or relating to ideofunction; used in a ceremonial way. Also, IDEOTECHNIC.
ideogram *Archaeology.* a symbol that represents a concept or object, as in the picture-writing systems of the ancient Mideast.
ideology *Anthropology.* the belief system of a society that includes values, philosophies, religions, and sciences and that may be expressed in a number of institutions such as school, church, family, and so on.
ideomotor *Physiology.* of or related to involuntary muscular action produced by mental energy. Also, IDEOKINETIC.
ideomotor apraxia *Medicine.* the inability to perform complex body movements due to an interruption between the ideation center and the limb center. Also, LIMB-KINETIC APRAXIA, TRANSCORTICAL APRAXIA.
ideophobia *Psychology.* an irrational fear of ideas.
ideotype *Systematics.* a specimen that is determined by the original author of that taxa to belong within that taxon, but not collected from the type locality. Also, IDIOTYPE.
idigbo *Materials.* the light yellow wood of the *Terminalia worensis* tree of Africa; used in the construction of furniture and joinery.
idio- a combining form meaning: **1.** personal or peculiar, as in *idiosyncrasy.* **2.** spontaneously or self-produced, as in *idiomuscular.*
idioblast *Geology.* a mineral constituent in a metamorphic rock that is formed by recrystallization and bounded by its own crystal faces.
idioblastic series see CRYSTALLOBLASTIC SERIES.
idiochromatic *Mineralogy.* of or relating to a mineral whose characteristic color is a result of its chemical composition.
idiochromatin *Cell Biology.* the chromatin of a cell that is concerned with reproductive function, as opposed to strict metabolic function.
idiocy *Psychology.* the fact or condition of being an idiot.
idiogenous see SYNGENEIC.
idioglossia *Psychology.* **1.** the development of a private language not intelligible to others, by a child or by a pair of children in close contact, such as twins. **2.** any unintelligible or invented speech. Also, **idiolalia.**
idiogram *Genetics.* a diagrammatic representation of a karyotype, based on a measurement of the chromosomes of a number of cells.
idiographic *Psychology.* relating to psychological studies or techniques that focus on a certain individual person, as opposed to people in general. Thus, **idiographic goals, idiographic emphasis.**
idiomorphic see AUTOMORPHIC.
idiomorphism *Materials Science.* a condition in which igneous rock minerals are bounded by their own crystal faces. Also, AUTOMORPHISM.
idiomuscular *Physiology.* of or related to the muscular tissue apart from neural stimulation; applied to certain muscular contractions that occur only in degenerated muscles.
idiopathic *Immunology.* referring to a disease that is of unknown cause.
idiopathic colitis *Medicine.* a primary inflammation of the colon.
idiopathic eunuchoidism *Medicine.* male hypogonadism, a primary condition marked by the presence of nonfunctioning testes.
idiopathic familial jaundice *Medicine.* a form of obstructive jaundice in which the ability to excrete conjugated bilirubin into the bile duct is decreased.
idiopathic hypercholesterolemia *Medicine.* the presence of an abnormally large amount of cholesterol in the cells and plasma of the blood, without any known cause.
idiopathic megacolon *Medicine.* the extreme enlargement and hypertrophy of the colon; often associated with constipation.
idiopathic multiple pigmented hemorrhagic sarcoma see KAPOSI'S SARCOMA.
idiopathic pulmonary hemisiderosis *Medicine.* a condition, of unknown cause, marked by hemorrhaging from pulmonary capillaries.
idiopathic thrombocytopenic purpura *Medicine.* a systemic illness marked by hemorrhages from mucous membranes, extensive ecchymoses, deficiencies in platelet count, anemia, and prostration, and often accompanied by a serum antiplatelet factor.
idiopathic ulcerative colitis *Medicine.* a primary disease marked by ulceration and bleeding of the colon and rectum, mucosal crypt abscesses, and inflammatory pseudopolyps; often leads to anemia, hypoproteinemia, and electrolyte imbalance, and is sometimes made more severe by perforation or carcinoma of the colon.
idiopathy *Medicine.* a primary disease; an illness with a spontaneous, unknown origin.
idiophase *Biotechnology.* a phase in culture production in which the biosynthetic pathways are altered and products other than the primary metabolites are formed.
idiosome *Cell Biology.* any of a number of specialized organelles or regions of a cell, such as the centrosome.
idiospasm *Neurology.* a spasm limited to a certain area of the body.
Idiospermaceae *Botany.* a monospecific family of dicotyledonous evergreen trees belonging to the order Laurales, characterized by scattered spherical cells containing volatile oils, opposite and simple leaves, large perfect flowers, and a large poisonous seed; native to Australia.
Idiostolidae *Invertebrate Zoology.* a small family of true bugs, hemipteran insects in the superfamily Pentatomorpha; found in South America and Australia.
idiosyncrasy *Medicine.* a mental, behavioral, or physical characteristic unique to an individual. *Immunology.* an individual's unique abnormal response to a drug, food, treatment, or environmental condition.
idiot *Psychology.* **1.** a term formerly used to describe a severely retarded person having a mental age of less than two years and an IQ of less than 25. **2.** in historic use, a general term used to describe a person regarded as mentally deficient; applied to various conditions of mental retardation, mental illness, autism, and the like.
idiotope *Immunology.* a determinant region of an antigen that is characteristic of one individual antigen.
idiotroph *Biotechnology.* a microorganism with a mutation in its biosynthetic pathways, making it unable to produce the valuable products for which it is being cultured.
idiot savant see SAVANT.

idiot tape *Graphic Arts.* in computerized phototypesetting, a paper tape containing keyboarded copy and some typesetting commands, but not others; usually sent by a publisher to a compositor, who adds justification and other commands.

idiotype *Immunology.* a class of antigenic determinants that are characteristic of one individual animal, or to a particular immunoglobulin molecule. *Systematics.* see IDEOTYPE.

idiotypic variation *Immunology.* a condition that exists on individual proteins, in which they posess a variety of antigens due to differences in their controlling genes.

idioventricular *Cardiology.* relating to or affecting only the ventricle and not the atria.

idioventricular rhythm *Cardiology.* the cardiac rhythm originating from one of the ventricles.

I display *Electronics.* the representation of a target on a conical-scan radar screen, in which a full circle indicates the radar antenna is pointed directly at the target; the radius of the circle is proportional to the target distance. Also, I SCAN.

IDL intermediate-density lipoprotein.

idle *Science.* not in use or operation. *Mechanical Engineering.* **1.** specifically, describing a machine, engine, or other mechanism that is running without a load. **2.** to operate in such a state.

idle current see REACTIVE CURRENT.

idle pulley see IDLER PULLEY.

idler frequency *Electronics.* a frequency, generated by merging two signals together in a parametric amplifier, that cannot be used by the circuit.

idler gear *Mechanical Engineering.* an intermediate gear placed between a driving gear and a driven gear to transmit motion between them. Also, **idle gear.**

idler pulley *Mechanical Engineering.* a movable pulley that presses against a driving belt to tighten or guide it. Also, IDLE PULLEY, GUIDE PULLEY.

idler wheel *Mechanical Engineering.* **1.** a wheel that is interposed in a gear train to reverse the direction of rotation. **2.** a wheel in a gear train that is used to change the spacing of the gear centers without affecting the ratio of the drive. **3.** a rubber roller used to transfer sound on a magnetic surface by frictional means. Also, **idle wheel.**

idle time *Computer Technology.* a time period in which a computer is switched on and available, but is not actually being used. *Industrial Engineering.* a time during which a worker is on the job but not actually working.

idling jet *Mechanical Engineering.* a part of the carburetion system in which gasoline is diffused during the idling speed of the engine or under minimal load conditions.

idling reaction *Biochemistry.* the production of ppGpp and pppGpp by ribosomes when an uncharged tRNA is present in the A site.

idling system *Mechanical Engineering.* a system for obtaining sufficient fuel and air metering with small throttle openings of an automobile carburetor in the idling position.

idocrase see VESUVIANITE.

idolatry [ī däl′ə trē] *Anthropology.* the worship or reverence of images or objects, either as representations of supernatural beings or as supernatural entities in themselves.

Idoteidae *Invertebrate Zoology.* a family of isopod crustaceans in the suborder Valvifera, characterized by a flattened body and seven pairs of legs.

idoxuridine *Pharmacology.* $C_9H_{11}IN_2O_5$, an antiviral agent applied to the eye to treat inflammation caused by a herpes virus. Also, 5-IODO-2′-DEOXYURIDINE.

id reaction *Medicine.* a secondary allergic skin reaction present in sensitized patients as a result of the spreading of allergens from a primary site of infection.

idrialite *Mineralogy.* $C_{22}H_{14}$, a green to light brown orthorhombic mineral occurring in small tabular crystals, having a specific gravity of 1.24 and a hardness of less than 2 on the Mohs scale; found mixed with cinnabar and clay in mercury ore.

IDU idoxuridine.

IE industrial engineer.

i.e. that is. (From Latin *id est.*)

IEA or **I.E.A.** International Energy Agency.

IEC International Electrotechnical Commission.

IEEE or **I.E.E.E.** Institute of Electrical & Electronics Engineers.

IEP immunoelectrophoresis.

IES or **I.E.S.** Illuminating Engineering Society.

I-E score *Psychology.* internal-external score; a measure of the extent to which a person believes or does not believe that he has effective control over the events of his life.

IEV *Aviation.* the airport code for Kiev, Ukraine.

IF *Electronics.* intermediate frequency. Thus, **IF strip, IF transformer.**

if-added method *Artificial Intelligence.* in a frame system, a procedure that is called automatically when a new value is stored into a slot; this procedure can be used as a demon to perform monitoring, display information, or propagate activity to related frames. Also, ANTECEDENT METHOD.

if and only if *Artificial Intelligence.* referring to a condition that applies in the event that, and only in the event that, some other condition applies. *Mathematics.* thus, a logic statement of the form "if p then q and if q then p" or "p if and only if q;" written symbolically as $p \Leftrightarrow q$ or p iff q. The statement $p \Leftrightarrow q$ is true if p and q are both true or both false, and is false otherwise; $\Leftrightarrow$ is also called a biconditional operator. Also, BICONDITIONAL STATEMENT.

-iformes a combining form meaning "having the form of," used especially in the names of vertebrate orders.

IFF *Military Science.* a system using electromagnetic transmissions to which equipment carried by friendly forces automatically responds, thereby distinguishing those forces from enemy forces. (An acronym for identification, friend or foe.) Thus, **IFF code.**

iff if and only if.

IFGO or **I.F.G.O.** International Federation of Gynecology and Obstetrics.

IFN interferon.

if-needed method *Artificial Intelligence.* in a frame system, a procedure that is called automatically to compute the value of a slot; for example, an if-needed method could compute a person's age from a known birth date and the current date. Also, CONSEQUENT METHOD.

IFO *Systematics.* identified flying object; that is, a moving object in the atmosphere that can be positively identified, as opposed to a UFO, one for which the identification is in question.

ifosfamide *Oncology.* $C_7H_{15}Cl_2N_2O_2P$, an antineoplastic alkylating agent.

IFR instrument flight rules.

if-removed method *Artificial Intelligence.* in a frame system, a procedure that is called automatically when a value is removed from a slot.

IFR terminal minimums *Meteorology.* in aviation weather practice, the operational weather limits, with respect to ceiling and visibility at an airport, at which an aircraft may legally approach and land under instrument flight rules.

IFR weather see INSTRUMENT WEATHER.

IF statement see CONDITIONAL STATEMENT.

if-then-else *Computer Programming.* a conditional instruction; the first part (if) establishes a condition, the second part (then) specifies an action to be executed if the condition is met, and the third part (else) specifies an alternate action to be executed if the condition is not met.

if-then statement see IMPLICATION.

IFV infantry fighting vehicle.

IG inspector general.

Ig immunoglobulin.

IgA *Immunology.* a class of immunoglobulins that occur in human fluids such as saliva and include an IgA dimer and a protein known as a secretory piece.

IgD *Immunology.* a class of immunoglobulins that occur in low concentrations in human serum and are present at the B cell surface.

IgE *Immunology.* a class of immunoglobulins that are generally elicited by an allergen and that bind to the surfaces of mast cells.

Igewsky's solution *Metallurgy.* in metallography, an etching solution for carbon steel.

IGF insulinlike growth factors.

IGF I and **II** see SOMATOMEDIN.

IgG *Immunology.* a class of immunoglobulins that represent the major class of serum immunoglobulin in humans.

igloo *Anthropology.* **1.** any of various dwellings or shelters used by the Inuit or other native peoples of Canada and Alaska. **2.** specifically, a dome-shaped winter shelter constructed of blocks of hard-packed snow. *Ordnance.* an earth-covered structure of concrete and steel used to store ammunition and explosives. (From the Inuit word for "house.")

IgM *Immunology.* a class of immunoglobulins that contain complement-fixing antibodies that are highly effective against antigens exhibiting a pattern of repeated antigenic determinants, such as bacterial cells.

ign. **1.** ignition. (From Latin *ignis.*) **2.** unknown. (From Latin *ignotus.*)

ignatia *Toxicology.* the dried ripened seeds of *Strychnos ignatii,* which contain the poisons strychnine and brucine.

igneous [ig´nē əs] *Petrology.* of or relating to a rock that was formed by solidification from molten or partly molten material; one of the three principal classifications of rocks, along with metamorphic and sedimentary. (Going back to the Latin word for "fire.")

igneous complex *Petrology.* a mass of intimately associated and approximately contemporaneous igneous rocks that differ in form or petrographic type.

igneous cycle *Volcanology.* the progress of igneous activity from a volcanic action to large and then smaller intrusions.

igneous facies *Petrology.* a part of an igneous rock body that varies in some respect, such as structure, texture, or composition, from the main rock mass.

igneous meteor *Geophysics.* any optically visible electric discharge in the atmosphere, such as lightning.

igneous petrology *Petrology.* the study of the origin, composition, and occurrence of igneous rocks.

igneous province see PETROGRAPHIC PROVINCE.

ignimbrite *Petrology.* a pyroclastic volcanic rock deposited from an ash flow; it is commonly silica-rich and may be welded. Also, FLOOD TUFF.

ignite *Chemistry.* to cause or begin combustion in a substance.

igniter *Engineering.* a blasting charge, fuse, or other device used to fire an explosive charge or to start the combustion of fuel in an engine.

igniter cord *Engineering.* a rope that moves an intense flame along at a steady rate, used to ignite safety fuses sequentially.

igniter pad *Ordnance.* a thin pad of cartridge cloth containing a black powder charge, used to facilitate ignition in separate loading ammunition.

igniter train *Ordnance.* a series of charges that transmits and amplifies the initial fire from the primer to the main charge of an explosive munition. Also, BURNING TRAIN.

igniting fuse *Ordnance.* a fuse that sets off the main charge of a munition through ignition rather than through detonation; it is used in munitions with a low explosive charge.

igniting primer *Ordnance.* in certain subcaliber gun tubes, a secondary primer that transmits the fire from the primer to the propelling charge.

ignition *Chemistry.* **1.** the point at which a substance begins combusting. **2.** the means by which a combustion process begins. *Mechanical Engineering.* see IGNITION SYSTEM.

ignition coil *Electromagnetism.* a small, open-core transformer with a high step-up turn ratio for use in an automotive ignition system.

ignition interference *Telecommunications.* a radio interference caused by the spark discharges of a vehicle's ignition system.

ignition lag *Mechanical Engineering.* the time interval between the onset of a spark and the resulting pressure rise due to combustion in an engine cylinder. Also, **ignition delay.**

ignition quality *Chemical Engineering.* the characteristic of a fuel represented by the cetane number that causes ignition when the fuel is injected into the compressed air in a diesel engine cyclinder.

ignition system *Mechanical Engineering.* the system in an internal combustion engine that produces the spark to ignite the mixture of fuel and air; it includes the battery, ignition coil, spark plugs, distributor, and associated switches and wiring. Also, IGNITION.

ignition temperature *Chemistry.* the minimum temperature required to begin and maintain combustion of a substance.

ignitor *Electronics.* **1.** an element that triggers and sustains a discharge in a switching tube. Also, PILOT ELECTRODE. **2.** an element in a mercury-pool cathode that causes conduction at a given point in an alternating-current cycle.

ignitron *Electronics.* a mercury-pool rectifier in which a high concentration of electrons, known as a cathode spot, appears on the pool surface before conduction begins.

ignitron contactor *Electronics.* a device that serves as a heavy duty switch in a resistance-welding transformer.

ignorable coordinate *Mechanics.* a term for a generalized coordinate that does not appear explicitly in the Lagrangian expression for the kinetic and potential energy of a conservative holonomic dynamical system.

ignore character *Computer Programming.* a character indicating that no action is to be performed, or an instruction requiring that normal execution of a command not occur.

IGT impaired glucose tolerance.

IGT or **I.G.T.** Institute of Gas Technology

IGU or **I.G.U.** International Geographical Union.

iguana [i gwän´ə] *Vertebrate Zoology.* a large neotropical American lizard belonging to the family Iguanidae that is insectivorous and is typically dark-colored with a serrated dorsal crest.

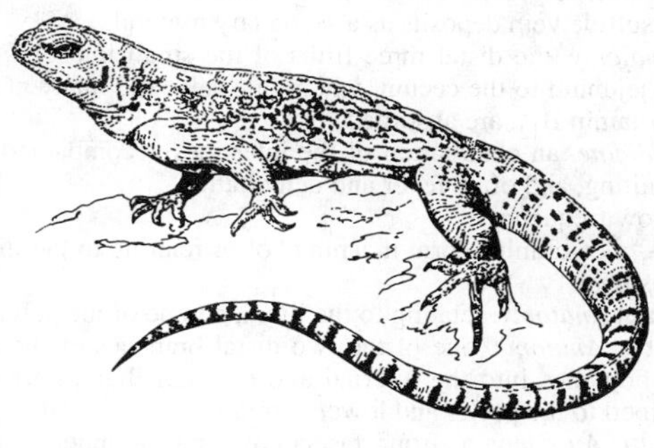

iguana

Iguanidae *Vertebrate Zoology.* a large family of arboreal lizards of the order Squamata, containing over 600 diverse, predominantly American species.

Iguanodon *Paleontology.* a genus of bipedal ornithischian dinosaurs in the suborder Ornithopoda and family Iguanodontidae; one of the largest animals of the Lower Cretaceous, ranging from 6 to 11 meters long.

IGY or **I.G.Y.** International Geophysical Year.

I.H. infectious hepatitis.

IHD ischemic heart disease.

IHD or **I.H.D.** International Hydrological Decade.

I-head cylinder *Mechanical Engineering.* an internal-combustion engine design characterized by the placement of both inlet and exhaust valves in the cylinder head.

IHP indicated horsepower.

IIL integrated injection logic.

ijolite *Petrology.* an alkalinic plutonic rock or rock group with a granitic texture, consisting of nepheline and mafic minerals, generally including sodic pyroxene, and commonly containing accessory apatite, calcite, sphene, and melanite.

Ikara *Ordnance.* an Australian long-range, antisubmarine missile powered by a dual-thrust, solid-propellant rocket motor; it is equipped with radio/radar guidance and carries a lightweight acoustical homing torpedo that is lowered into the water by parachute.

Ikeya-Seki *Astronomy.* comet 1965 VIII, a bright sun-grazing comet that was visible in daylight during October 1965, and which displayed a tail 60° long at maximum; its nucleus split into three parts at perihelion.

IL interleukin.

Il the chemical symbol for illinium.

I^2L integrated injection logic.

il- a prefix meaning: **1.** in or into, as in *illuminate.* **2.** not, as in *illegible.*

ILA International Laboratory Accreditation Cooperation.

I.L.A. International Leprosy Association.

ilang-ilang oil *Materials.* an oil derived from the blossoms of the tree *Canangium odorata*; used in perfumery. Also, YLANG-YLANG OIL.

ilarvirus *Plant Pathology.* a virus that causes the prunus necrotic ringspot virus and is transmitted through the sap or seeds of the plant.

Ilarvirus group *Virology.* a group of tripartite ssRNA-containing plant viruses with quasi-isometric particles; transmitted mechanically and through seeds and pollen over a wide host range. Also, ISOMETRIC LABILE RINGSPOT VIRUSES, TOBACCO STREAK VIRUS GROUP.

ileac *Medicine.* relating to or affecting the ileum.

ileitis *Medicine.* an inflammation of the ileum.

ileocecal *Anatomy.* relating to the ileum and the cecum.

ileocecal valve *Anatomy.* the sphincter located where the ileum of the small intestine empties into the cecum of the large intestine.

ileocecostomy see CECOILEOSTOMY.

ileocolic artery *Anatomy.* a branch of the superior mesenteric artery that supplies blood to the terminal portion of the ileum, cecum, appendix, and ascending colon.

ileocolic intussusception *Medicine.* infolding of the ileum through the ileocecal valve into the colon. Also, **ileocolic introsusception.**

ileocolitis *Medicine.* inflammation of both the ileum and colon.

ileocolostomy *Surgery.* the surgical creation of an opening between the ileum and the colon.

ileostomy *Surgery.* the surgical creation of an opening through which the ileum discharges directly to the outside of the body.

ilesite *Mineralogy.* $(Mn^{+2},Zn,Fe^{+2})SO_4 \cdot 4H_2O$, a green, transparent, monoclinic mineral of the rozenite group occurring in prismatic crystals, having a specific gravity of 2.25 to 2.26 and an undetermined hardness; found in sulfide vein deposits as a secondary mineral.

ileum *Anatomy.* the distal three-fifths of the small intestine, extending from the jejunum to the cecum, where bile salts and many nutrients, including vitamin B_{12}, are absorbed.

ileus *Medicine.* an obstruction of the intestines accompanied by severe pain, vomiting, and often fever and dehydration.

ILF infralow frequency.

ili- or **ilio-** a combining form meaning "of or relating to the ilium," as in *iliac, iliospinal.*

iliac [il′ē ak] *Anatomy.* relating to the ilium, a bone of the pelvis.

iliac artery *Anatomy.* one of the two distal branches of the abdominal aorta; it branches into the internal and external iliac arteries that distribute blood to the pelvis and lower extremity.

iliac fascia *Anatomy.* a strong fascia covering the inner surface of the iliac and psoas muscles.

iliacus *Anatomy.* an elongate muscle that stretches from the iliac fossa to the lesser trochanter of the femur; it flexes the thigh.

iliac vein *Anatomy.* the common iliac vein, formed by the union of right and left iliac veins draining blood from the lower extremity and pelvic organs; unites with its partner to form the inferior vena cava.

iliocostalis *Anatomy.* the lateral portion of the erector spinae muscles that hold the spine erect in a standing position and when walking.

iliofemoral *Anatomy.* relating to the ilium and the femur.

iliofemoral ligament *Anatomy.* a large Y-shaped ligament that connects the anterior inferior iliac spine and rim of the acetabulum to the anterior intertrochanteric line of the femur. Also, LIGAMENT OF BIGELOW.

iliolumbar ligament *Anatomy.* the thick ligament connecting the ilium to the fourth and fifth lumbar vertebrae.

iliopsoas *Anatomy.* the fused iliacus and psoas major muscles that stretch from the ilium and lumbar vertebrae to the lesser trochanter of the femur, and that flex the femur.

iliopubic *Anatomy.* relating to the ilium and the pubis.

iliotibial tract *Anatomy.* a thickened longitudinal band of fascia lata extending from the crest of the ilium downward along the lateral side of the thigh to the lateral condyle of the tibia. Also, **iliotibial band.**

ilium *Anatomy.* the flared portion of the fused os coxae of the pelvic girdle; distinct at birth, but fused with the ischium and pubis later.

Ilkovic equation *Analytical Chemistry.* a mathematical relationship used in polarography; the diffusion current is related to the diffusion coefficient, the concentration of the active substance, and the electrode parameters.

illaenids *Paleontology.* a large trilobite in the suborder Illaenina, extant in the Middle Cambrian.

Illaenus *Paleontology.* a genus of relatively large isopygous trilobites in the order Ptychopariida and suborder Illaenina; Illaenus grew as large as 5 cm; extant in the Lower Ordovician to Upper Devonian.

ill-conditioned problem *Mathematics.* a type of numerical problem in which a single, small round-off error in the data or in intermediate calculations causes significant errors in the solution.

illegal character *Computer Programming.* a character or code that is not recognized as valid by a specific program routine or a computer.

illegal operation *Computer Programming.* a process that results when a computer cannot properly perform an instruction or an undefined operation code is used.

illegitimate *Genetics.* of recombination, taking place in the absence of substantial homology.

illegitimate name *Systematics.* a term for a validly published taxonomical designation that is not in accord with the appropriate nomenclatural guidelines.

illegitimate recombination *Molecular Biology.* a recombinational event that occurs between segments of DNA that are nonhomologous or have a very limited homology.

Illiciaceae *Botany.* the star anise, a monogeneric family of dicotyledonous aromatic evergreen shrubs or small trees of the order Illicales, characterized by the presence of volatile oils; native mainly to southeastern Asia and the northern Caribbean.

illinium see FLORENTIUM.

Illinoian *Geology.* the third glacial stage of the Pleistocene epoch in North America, occurring between the Yarmouthian and Sangamon interglacial stages. Also, ILLINOISAN.

illite *Mineralogy.* any of various white to pale gray, green, or yellowish-brown monoclinic mica-clay minerals having the general formula $(K,H_3O)(Al,Mg,Fe)_2(Si,Al)_4O_{10}[(OH)_2,H_2O]$, occurring in very fine-grained masses, and having a specific gravity of 2.6 to 2.9 and a hardness of 1 to 2 on the Mohs scale; found in shales and mudstones.

illiterate *Anthropology.* not having a written language; not able to write; now regarded as a pejorative term and generally supplanted by the terms *nonliterate* or *preliterate.*

illness *Medicine.* any unhealthy condition of body or mind; sickness.

ill-posed problem *Mathematics.* a problem whose answer exhibits large or irregular variation when the initial conditions of the problem are changed only slightly.

illuminance *Optics.* luminous incidence; the amount of luminous flux striking a unit area of surface, measured in lumens per square meter and expressed by the unit of illumination known as the lux. Also, LUMINOUS FLUX DENSITY.

illuminating grenade *Ordnance.* a grenade designed to illuminate an area through incendiary action; it may also be used as an incendiary weapon.

illuminating projectile *Ordnance.* a projectile equipped with a time fuse to release a parachute flare at a preset height in order to illuminate an area. Also, STAR SHELL. Similarly, **illuminating mortar.**

illumination *Optics.* **1.** the application and distribution of light to a subject. **2.** a popular name for illuminance. *Graphic Arts.* **1.** the use of light to brighten a photographic subject, to expose photosenstive film or plates, or to create a picture of a character in a phototypesetting machine. **2.** the art of decorating letters and characters, often with images, as in an **illuminated manuscript.** *Electromagnetism.* for an antenna, the geometric distribution of power delivered to the elements or the distribution of power arriving at various portions of the reflector dish.

illumination climate *Meteorology.* **1.** the worldwide distribution of natural light from the sun and the sky as received on a given horizontal surface. **2.** the character of total illumination at a given location. Also, LIGHT CLIMATE.

illumination control *Electronics.* a device that automatically turns on an artificial light when the amount of daylight falls below a certain point.

illumination design *Design Engineering.* the use and composition of lights and lighting systems in a room, building, or area.

illumination distribution *Optics.* the orientation of light rays striking or dispersed on a surface.

illumination fire *Military Science.* fire designed to illuminate an area, usually with illuminating projectiles.

illuminometer *Optics.* a photometer that measures the intensity of light falling on a surface.

illusion *Psychology.* a misleading perception of an object or experience that leads to an incorrect conclusion about the nature of physical reality.

illusionary *Psychology.* relating to or being an illusion.

illustration *Graphic Arts.* a piece of artwork, especially a drawing or painting, printed within or adjoining a text and usually serving to clarify, decorate, or otherwise support written ideas and information.

illustration board *Graphic Arts.* a thin white cardboard, often printed with nonrepro-blue guidelines, on which artwork is positioned in preparation for photographing.

illuvial *Geology.* relating to the process of illuviation, or a soil horizon or other material added to or transported by illuviation.

illuviation *Geology.* the accumulation in a lower soil horizon of soluble or suspended minerals, clays, salts, and colloids leached down from an upper horizon.

illuvium *Geology.* the material leached by physical or chemical processes from an upper soil horizon and redeposited in a lower horizon.

ilmenite *Mineralogy.* $Fe^{+2}TiO_3$, a black, opaque, trigonal mineral of the ilmenite group occurring as tabular crystals and in massive or granular form, having a specific gravity of 4.72 and a hardness of 5 to 6 on the Mohs scale; found as a common accessory mineral in basic igneous and metamorphic rocks, and in beach sand and vein deposits; the principal ore of titanium. Also, MOHSITE, TITANIC IRON ORE.

ILS instrument landing system.

ilsemannite *Mineralogy.* a blue-black amorphous mineral with the approximate formula $Mo_3O_8 \cdot nH_2O$, occurring in earthy massive forms as an alteration product of molybdenum minerals.

ILSS interlaminar shear strength.

im- a prefix meaning: **1.** in or into, as in *immersion.* **2.** not, as in *impermeable.*

I.M. intramuscularly.

image a likeness or representation of a person or object; specific uses include: *Physics.* the profile of any object that is produced by means of focusing and reconstructing light, sound, or radiation emanating or reflecting from the object. *Psychology.* a mental view or likeness of another individual, such as one created by a child of a parent, which is formed in the unconscious and usually remains there. *Graphic Arts.* **1.** a printed reproduction of an object, such as a photograph. **2.** the printed part of a printed page. Also, **image area.** *Electricity.* an imaginary electrical counterpart of an actual object, such as an image antenna. Also, ELECTRIC IMAGE. *Acoustics.* see ACOUSTIC IMAGE. *Computer Technology.* **1.** the exact duplication of data in a different medium, such as a screen display of the contents of computer memory. **2.** in graphics, the output form of a graphics file. *Electromagnetism.* for a specified load placed on one side of a waveguide and a slotted line placed on the other side, the input reflection coefficient associated with the load. *Telecommunications.* one of two sets of sidebands that are created by modulation. *Mathematics.* let f be a mapping from the set X to the set Y. The image of X under the mapping f is the subset $\{f(x) : x \in X\}$ of Y; denoted $f(X)$. $f(x)$ is also called the image of x. If Y is a subset of $f(X)$, then the set $\{x : f(x) \in Y\}$ is called the **preimage** of Y or **inverse image** of Y. The image of X under f is also called the range of f.

image amplification *Radiology.* the increased brightness of a fluoroscopic image produced by an electron image or electron multiplier tube.

image antenna *Electromagnetism.* a fictitious antenna that exists only in a mathematical sense, thought of as located directly below the actual antenna; used to account for radiation that appears to come from the ground but is in fact reflected from the ground.

image assembly see STRIPPING.

image carrier *Graphic Arts.* a device such as a printing plate or stencil that is designed to transfer an image to a sheet of paper.

image converter *Optics.* a device that converts electromagnetic radiation into a visual replica of an image produced on a cathode-ray screen; used in the infrared, ultraviolet, and X-ray regions, as well as in the visible regions. Thus, **image converter camera**.

image converter tube see IMAGE TUBE.

image dissection *Computer Technology.* a technique used in optical character recognition, in which a mechanical or electronic transducer detects the light level in small, adjacent areas of a completely illuminated image and reads those levels sequentially into the computer, which converts the readings into digital information.

image-dissection photography *Electronics.* a technique used in high-speed photography, in which an image may be split in a variety of ways and then reassembled for viewing or for reproduction onto a master negative.

image-dissector tube *Electronics.* a type of television picture tube in which an image is swept past an aperture that dissects it section by section, instead of being scanned by an electron beam. Also, FLYING APERTURE SCANNER.

image effect *Electromagnetism.* the resultant field produced by an antenna that includes radiation reflected from the ground.

image enhancement *Computer Technology.* **1.** the use of highlighting, reverse video, shading, or blinking to emphasize a portion of a graphics image. **2.** a signal processing operation applied to image data to enhance the image, e.g., contrast enhancement.

image force *Electricity.* the electrostatic force exerted on a charge in the area of a conductor, which may be perceived as the attraction to the charge's electric image.

image frequency *Electronics.* in a heterodyne receiver, a frequency whose value is two times the local oscillator frequency minus the frequency of the desired signal. On a frequency scale, the signal and image frequencies are displaced from the local oscillator frequency by equal amounts in opposite directions.

image grammar *Computer Technology.* a system of describing shapes or images on a computer so that they fit into a coherent pattern.

image iconoscope *Electronics.* **1.** a type of iconoscope in which the charge storage function is separated from the photoemission function in order to achieve greater sensitivity. **2.** a tube that resembles a camera tube, except that the image is projected onto a photocathode and then transferred onto another material by photoelectrons.

image impedance *Electronics.* in an electric transducer, one of two impedance values associated with the input and output, respectively. When the output is terminated by the output image impedance, the impedance seen at the input is equal to the input image impedance, and vice versa.

image intensifier see LIGHT AMPLIFIER.

image interference *Telecommunications.* a failing peculiar to superheterodynes that occurs when a station broadcasting on the image frequency is received along with the desired station.

image isocon *Electronics.* a tube in a television camera that resembles a highly sensitive iconoscope, except that the image produced by a photoemitting surface is scanned by a highly charged electron beam.

image load *Electronics.* the load impedance seen by a source when an electrical transducer is inserted between it and the actual load.

image orthicon *Electronics.* a tube in a television camera that resembles a highly sensitive iconoscope except that the image produced by a photoemitting surface is scanned by a low-velocity electron beam.

image parameter design *Electronics.* a technique for designing filters that uses image impedance and image transfer functions to specify the filter operation.

image parameter filter *Electronics.* a filter whose operation is specified by its image impedances and image transfer functions.

image phase constant *Electronics.* the imaginary part of the image transfer constant in a transducer.

image plane *Optics.* a plane that lies perpendicular to the axis of an optical system in which an image is formed.

image processing *Computer Technology.* computer operations on graphic image data; pictorial information is both input and output.

image ratio *Electronics.* the relationship between the undesired signal and the desired signal received by a heterodyne receiver.

image reject mixer *Electronics.* in a heterodyne receiver, a circuit that produces a response from the desired input signal but suppresses any response from the image frequency signal without using a filter.

image response *Electronics.* the response of a heterodyne receiver to the image frequency.

image restoration *Computer Technology.* a technique of estimating an original image by assuming a known model for a blur that may appear in a picture, in order to improve the quality of the picture.

imagery intelligence *Military Science.* intelligence information derived from images produced by visual photography, infrared sensors, lasers, electro-optics, or radar sensors. Also, IMINT. Similarly, **imagery correlation, imagery exploitation, imagery interpretation.**

image space *Optics.* the region of an optical system where images are produced.

image storage array *Electronics.* a panel in a solid-state device in which an layer of light-sensitive material, such as zinc oxide, illuminates characters on a display when there is little natural light.

image table *Control Systems.* a data table for programmable controllers that contains the statuses of all inputs, registers, and coils.

image transfer constant *Electronics.* in an electric transducer, the ratio of output power to input power when operated between a source and load whose impedances are equal to the input and output image impedances, respectively. Also, TRANSFER CONSTANT.

image tube *Electronics.* a device that projects an optical image generated by electromagnetic radiation from a photosensitive surface onto a fluorescent screen. Also, ELECTRON IMAGE TUBE.

image understanding *Artificial Intelligence.* the process of building a description of an image and also of the scene it represents. Also, SCENE ANALYSIS, SHAPE RECOVERY.

imaginal disk *Entomology.* any of several masses of tissue within the pupa of a developing insect that produces adult insect organs.

imaginary axis *Mathematics.* the vertical axis in the Argand diagram of the complex plane; equivalently, all complex numbers of the form bi, where b is real.

imaginary number *Mathematics.* a number of the form $a + bi$, where a and b are real numbers, $b \neq 0$, and $i = \sqrt{-1}$. When $a = 0$, the number is said to be **pure imaginary.**

imaginary part *Mathematics.* the value of b for a complex number $a + bi$, where a and b are real numbers; a is the real part.

imaginary power see REACTIVE POWER.

imaging *Physics.* the process by which images are formed. *Radiology.* the process of producing contrast and clarity in radiological and ultrasound images.

imaging plate *Crystallography.* a surface that behaves like photographic film and can be used to store X-ray intensities as latent images in the form of color centers. The stored image is scanned by laser light. The plate is erasable and can be used many times.

imago [i mā´gō; i mä´gō] *Entomology.* the sexually mature stage in the life cycle of insects. *Psychology.* an idealized view of an important figure from one's childhood, usually a parent, that remains unaltered in later life.

imbecile [im´bə səl] *Psychology.* a term formerly used to describe a mentally retarded person with a mental age of two to seven years and an IQ between 25 and 50.
imbedding *Mathematics.* a homeomorphism that maps one topological space into a subset of another topological space. Also, **embedding.**
imbibition *Physical Chemistry.* the process by which a solid or a gel absorbs or adsorbs a liquid.
imbricate *Biology.* describing structures, such as flat pebbles, that overlap one another like shingles on a roof.
imbricate structure *Geology.* **1.** a sedimentary structure characterized by a slanting, overlapping pattern in which gravel, pebbles, or grains are all inclined in the same direction, with their long axes at an angle to the dominant bedding plane. Also, SHINGLE STRUCTURE. **2.** a tectonic structure resembling shingled tiles in which a series of more or less parallel, overlapping, minor thrust faults or reverse faults have pushed up tabular sheets or masses of rock that have the same general orientation and roughly the same amount of displacement. Also, DISTRIBUTIVE FAULTING, SCHUPPENSTRUCTURE, SHINGLE-BLOCK STRUCTURE.
imbrication *Geology.* the formation of an imbricate structure.
imbuia *Materials.* a species of durable, olive-brown wood from the Nuctandra tree of Brazil, which closely resembles walnut; used for cabinetry, flooring, and furniture.
imdtly. immediately.
IME or **I.M.E.** Institution of Mechanical Engineers. Also, **IMechE.**
Imhoff cone *Civil Engineering.* a graduated cylinder used for measuring settled solids in testing the composition of sewage.
Imhoff tank *Civil Engineering.* a tank used to treat sewage, in which digestion and settlement take place in separate compartments one below the other.
Imhotep c. 2700 BC, Egyptian physician, statesman, and architect; honored in medical tradition as the first known physician.
imidamine see ANTAZOLINE.
imidazole *Organic Chemistry.* $C_3H_4N_2$, a strongly basic heterocyclic compound; colorless crystals; melts at 89°C and boils at 257°C; used as a nonpoisonous insecticide. Also, GLYOXALINE, IMINAZOLE.

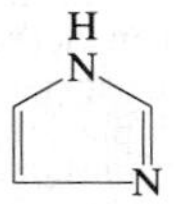

imidazole

imidazolyl *Organic Chemistry.* the radical $C_3H_3N_2$– from imidazole.
imide *Organic Chemistry.* any of a group of compounds containing the bivalent radical =NH attached only to acid radicals; the nitrogen analog of an acid anhydride.
imido- a combining form meaning "imide," as in *imidodiphenyl.*
iminazole see IMIDAZOLE.
imine *Organic Chemistry.* any of a class of compounds containing the =C=NH group.
imino- a combining form meaning "imine," as in *iminourea.*
imino acid *Organic Chemistry.* any acid having an =NH group attached to one or two carbon atoms.
imino base *Organic Chemistry.* any base containing the =C=NH group, such as guanidine.
imino compound *Organic Chemistry.* any compound having the =NH radical attached to one or two carbon atoms.
imint see IMAGERY INTELLIGENCE.
imit. imitative.
imitation *Behavior.* the act of observing the behavior of another, especially a novel or unfamiliar action, and then performing the behavior in the same manner. Also, OBSERVATIONAL LEARNING.
imitative *Biology.* **1.** of, relating to, or characterized by imitation. **2.** mimetic.
imitative deception *Electronics.* a technique in which electromagnetic radiation is used to infiltrate an enemy's radio channels in order to generate false data.
immature soil see AZONAL SOIL.
immediate access *Computer Technology.* the ability to obtain or store data directly and without delay.
immediate address *Computer Programming.* an assembly language instruction in which the address portion contains the operand itself. Also, ZERO-LEVEL ADDRESS.
immediate air support *Military Science.* air support operations that are designed to meet specific requests in the course of battle and that cannot be planned in advance. Similarly, **immediate nuclear support.**
immediate data *Computer Programming.* data that is stored immediately within, or adjacent in memory to, the instruction that uses it as an operand.
immediate hypersensitivity *Immunology.* an intense inflammatory reaction that occurs within minutes of exposure to an antigen.
immediate instruction *Computer Programming.* an instruction containing an immediate operand.
immediate memory span *Psychology.* the amount of material which, after a single presentation, can be correctly reproduced.
immediate operand *Computer Programming.* an operand whose value, rather than address, is contained within an instruction.
immediate transfusion *Medicine.* the transfer of blood from one person to another without the use of an intermediate container or anticoagulant. Also, DIRECT TRANSFUSION.
immersed *Botany.* growing under water, as a plant. *Biology.* partially or wholly sunk in the surrounding parts, as an organ.
immersion *Science.* the process of submerging or being submerged in a liquid. *Astronomy.* the disappearance of a celestial object in an eclipse or occultation. *Mathematics.* **1.** a mapping f: $X \to Y$, where X and Y are topological spaces, such that for every $x \in X$, there exists a neighborhood N of x on which f is a homeomorphism. **2.** in general, if X^q is a differentiable manifold, and f: $X^q \to Y^n$ is a differentiable mapping of rank q for every point $x \in X^q$ (so that $q \le n$), then f is called an immersion. If f is of rank n for every point $x \in X^q$, then f is called a submersion. In this case, it follows that $q \ge n$.
immersion cleaning *Metallurgy.* the process of cleaning a metallic part by immersing it in a cleansing solution.
immersion coating *Materials Science.* the process of dipping a metal or ceramic in a liquid material to coat its surface; a coating resulting from an electrochemical replacement reaction.
immersion electron microscope *Electronics.* an instrument that enlarges images of minute objects by causing them to emit low-velocity electrons, through heat, illumination, or bombardment with high-velocity electrons and then accelerating them to a high velocity in an immersion objective or a cathode lens.
immersion foot *Medicine.* a condition similar to trench foot, affecting people who have spent long periods of time standing in water.
immersion heater *Electricity.* an electric device that heats a liquid when it is directly immersed into the liquid.
immersion liquid *Optics.* any liquid used to form optical contact between a short focal length objective and a microscope slide.
immersion objective *Optics.* a microscope objective in which the space between the lens and the object being viewed is filled with an oil, whose refractive index exceeds that of air; used to reduce reflection losses and spherical aberration. Also, OIL-IMMERSION OBJECTIVE, OPTICAL IMMERSION LENS.
immersion plating *Metallurgy.* the process of applying a metallic coating to a part simply by immersing it in a solution; for instance, by electroless plating.
immersion-proof *Ordnance.* describing a piece of equipment that can be immersed for up to two hours at a depth of up to three feet in saltwater or freshwater and be ready for field use at normal effectiveness immediately after removal from the water.
immersion refractometer *Optics.* a refractometer used to determine the index of refraction of liquids; operated by immersing the prism section of the device into the liquid being measured.
immersion scanning *Engineering.* ultrasonic scanning involving the immersion of both an ultrasonic transducer and the object it will scan into a liquid suitable for coupling.
immigrant *Ecology.* an organism that moves into a community or region where it was previously not found.
immigration *Ecology.* the movement of an organism or a group of organisms into a new community or region.
immiscible *Chemistry.* not miscible; describing two liquids that do not mix, such as oil and water. Thus, **immiscibility.**
immittance *Electricity.* **1.** a general term for both impedance and admittance, as applied to transmission lines, networks, and certain measuring instruments. **2.** in linear passive networks, a response function in which one variable is a voltage and the other is a current.

immobilization *Biotechnology.* any of a variety of physical or chemical processes used to fix enzymes, bacteria, or plant or animal cell cultures; the substance or cells are permanently affixed to an appropriate support or trapped in a stable matrix.

immobilization test *Microbiology.* any test that includes the immobilization or inhibition of motility of a microorganism.

immortalization *Cell Biology.* the transformation of an in vitro cell culture line into a strain with unlimited growth potential.

immortalized cell *Cell Biology.* a cell that has been transformed, as by viral infection, and becomes capable of unlimited proliferation.

IMMS or I.M.M.S. International Material Management Society.

immun. immunity; immunization.

immune [i myoon´] *Immunology.* **1.** describing a state following contact with an antigen in which antibodies specific for that antigen occur in the body. **2.** describing an organism that resists and overcomes an infection or disease. (From Latin; literally, "free from duties; exempt.")

immune adherence *Immunology.* the adherence of antigen-antibody complexes to primary erythrocytes, macrophages, polymorphs, or blood platelets of non-primates, in the presence of a complement.

immune body see ANTIBODY.

immune complex *Immunology.* an antigen-antibody complex.

immune cytolysis *Immunology.* the rupturing of cells that have been coated with antibodies to their surface antigens, in association with complement.

immune gamma globulin *Hematology.* a concentrated preparation containing gamma globulins from a large group of human donors; used in prevention of infectious hepatitis and measles and for treatment of immunodeficient patients.

immune globulin see IMMUNOGLOBULIN

immune hemolysis *Immunology.* the rupturing of red blood cells coated with antibodies to their surface antigens, in association with complement.

immune lysin *Immunology.* an antibody or other bacterial toxin that disrupts cells in the presence of complement.

immune oposonin *Immunology.* a blood component that forms in reaction to an infection or the administration of dead cells of the infecting bacteria species, which helps bacterial phagocytosis.

immune precipitation *Immunology.* a precipitation technique used to separate a protein from a mixture with the use of a particular antibody as the separating agent.

immune protein *Immunology.* an immunoglobulin (or antibody) that forms in response to antigenic stimulation and reacts specifically with the stimulating antigen.

immune reaction *Immunology.* a reaction of the immune system to an antigen resulting in the production of antibodies or the development of cellular immunity.

immune response *Immunology.* the total response of a body to the introduction of an antigen, including antibody formation, cellular immunity, hypersensitivity, or immunological tolerance.

immune-response gene *Genetics.* any of a group of genes of the major histocompatibility complex that govern the immune response of lymphocytes to specific antigens.

immune serum see ANTISERUM.

immune surveillance *Immunology.* the process in which the immune system prevents the growth of tumor cells.

immune system *Immunology.* the cells and tissues involved in recognizing and attacking foreign substances in the body.

immunity [i myoon´ə tē] *Immunology.* a biological condition by which a body is capable of resisting or overcoming an infection or disease. *Metallurgy.* the attribute of a metal or alloy that resists corrosion.

immunization [im´yə nə zā´shən] *Immunology.* a process that increases an organism's reaction to an antigen and thereby enables the organism to resist or overcome an infection or disease.

immunization therapy *Medicine.* treatment with antiserum and antigenic substances, such as vaccines.

immuno- a combining form meaning "immune," "immunity," or "immune system," as in *immunocyte, immunodeficiency.*

immunoassay *Immunology.* a process that employs seriological methods to identify and quantify a specific biological substance, such as an antigen.

immunoblotting. *Biotechnology.* a technique that uses antibodies and protein blotting to detect a specific protein on a nitrocellulose sheet.

immunochemistry *Immunology.* a field of chemistry that specializes in the study of the chemical activity occurring in association with immunological phenomena.

immunochemotherapy *Oncology.* the treatment of cancer with a combination of immunotherapy and chemotherapy.

immunocompetence *Immunology.* the ability of the immune system to react to stimulation by an antigen and develop an effective response to it.

immunocompetent *Immunology.* having the capacity to respond to and resist invading antigens.

immunocompromised *Immunology.* having the immune response weakened by the administration of immunosuppressive drugs, by irradiation, by malnutrition, or by a disease.

immunoconglutinins *Immunology.* antibodies that are fixed to complements and formed in response to infection or the injection of antigens.

immunocyte *Immunology.* a leukocyte that is capable of mounting an immune response. Also, **immunologically competent cell.**

immunocyto-adherence *Immunology.* a process whereby antigenic particles join with specific antibody-forming cells to form clusters.

immunodeficiency [im´yə nō di fish´ən sē] *Immunology.* a condition that exists when humoral or cell-mediated immunity is weakened or lacking; a disorder characterized by deficient immune response.

immunodeficient [im´yə nō di fish´ənt] *Immunology.* lacking an adequate capacity to respond to and resist invading antigens.

immunodiffusion *Immunology.* a method used to analyze antigen and antibody mixtures by observing them as they diffuse toward each other within a supporting medium, usually a gel. Also, DIFFUSION TEST, GEL DIFFUSION.

immunodominant *Immunology.* referring to the part of an antigenic determinant that has the strongest affinity toward binding with an antibody. Thus, **immunodominance.**

immunoelectron microscopy *Immunology.* a procedure used to locate certain antigens in cells or tissues by electron microscopy.

immunoelectrophoresis *Immunology.* a technique used to separate and detect soluble antigens, whereby proteins are first separated by electrophoresis and then subjected to immunodiffusion.

immunofluorescence *Cell Biology.* a technique by which cells can be localized, labeled, and examined microscopically under ultraviolet light by the binding of a fluorescently labeled antibody.

immunogen *Immunology.* a substance capable of producing an immune response.

immunogenetics *Genetics.* a branch of genetics that utilizes both genetic and immunological analyses to study the genetic determination of antigens, antibodies, and the immune response.

immunogenic *Immunology.* of or relating to a substance that is capable of producing immunity; evoking an immune response.

immunogenicity *Immunology.* the quality that makes a substance immunogenic, or the degree to which a substance possesses this quality.

immunoglobulin *Immunology.* a class of proteins found in plasma and other body fluids that exhibits antibody activity and binds with other molecules with a high degree of specificity; divided into five classes (IgM, IgG, IgA, IgD, and IgE) on the basis of structure and biological activity.

immunoglobulin classes *Immunology.* groups of immunoglobulins that are characterized by their unique antigenic determinants occurring on their heavy chains.

immunoglobulin domain *Immunology.* a three-dimensional structure composed of one homologous region of an immunoglobulin heavy or light chain. Also, DOMAIN.

immunoglobulin gene *Genetics.* any of a class of genes that directs the production of the light and heavy chains of the immunoglobulins.

immunogold technique *Immunology.* a form of immunoelectron microscopy in which the antibodies are mixed with gold prior to use; the gold serves as an electron-dense marker.

immunogranulomatous disease *Medicine.* a disease marked by a deviation from normal immune mechanisms and believed to be associated with granulomatosis.

immunological disease [im´yə nə läj´i kəl] *Pathology.* any malady caused by the effect of antibodies, such as allergic reactions to antigens.

immunological drift see ANTIGENIC DRIFT.

immunological unresponsiveness *Immunology.* the failure of an organism to respond upon contact with an antigen.

immunologic suppression *Immunology.* an artificial lessening of antibody production that is induced through the use of drugs.

immunologic tolerance *Immunology.* the failure of an organism to produce an antibody against, or to develop cell-mediated immunity toward, a specific antigen. Also, TOLERANCE.

Immunology

Immunology is concerned with the mechanisms by which vertebrate animals defend themselves against foreign substances that may be introduced into their blood and tissues. The immune system, the protector against invaders, recognizes foreign substances (antigens) as "nonself" and responds by producing specific molecules (antibodies) that can interact with bacteria, viruses, protozoa, toxins, and so on, to neutralize their harmful effects. For survival of the species, the myriad of essential molecules normally present in the bloodstream and tissues of each organism are recognized as "self" and fail to elicit a similar response.

Antibodies are large glycoproteins found in the globulin fractions of serum and in tissue fluids. Structural studies of several immunoglobulins have revealed the location and relative dimensions of their combining sites. The key characteristic of the interaction between an antigen and antibody remains its specificity, i.e., the ability of the antibody to selectively and tightly bind its antigen even when many structurally unrelated compounds are present.

Knowledge regarding the complex processes by which antibodies are produced in vivo is extensive and, as yet, incomplete. Upon exposure to an antigen, certain cells of the lymphatic system (the B-cells whose surface immunoglobulin can bind the antigen) are stimulated to proliferate and differentiate into cells that can secrete antibodies. Antibody synthesis by B-cells involves cooperation with other cells (e.g., T-cells and macrophages). When animals are immunized, antibodies with specificities directed toward different portions of the antigen can be produced by the various clones. However, antibodies secreted by a single cell can be highly specific for a particular antigenic determinant (epitope). With hybridoma technology, it is now possible to obtain antibodies that originate from a single clone of a B-cell obtained from the spleen of an immunized animal after fusion with a mouse tumor cell. Monoclonal antibodies with specificity directed to a single epitope on the surface of infecting agents or cell types as well as to macromolecules or haptens are essentially homogeneous reagents. They can be prepared in large quantities and are finding increased use in diagnostic procedures (i.e., in various sensitive, specific, and rapid immunoassays) and in immunopurification techniques.

For therapeutic purposes, monoclonal antibodies either alone or in covalent linkage with a drug or toxin are being tested as "magic bullets." The hope is that the antibody directed to a particular antigen on a cell surface (e.g., a cancer cell) would block certain processes and selectively kill the cell. While maintaining the specificity and affinity for a particular antigen, the size of the antibody molecule is being altered to minimize undesirable side effects (e.g., allergic reactions) and to increase its ability to reach target sites. Active immunization for prophylactic purposes in the form of vaccination against infectious diseases continues with established methods. However, predictions of the essential structure as well as the ease of polypeptide synthesis permit vaccination with "protective" antigens of small molecular weight carrying the essential features but not the detrimental properties of natural antigens (e.g., the AIDS virus).

Recently, antibodies that possess catalytic activities similar to enzymes (Abzymes) have been obtained by immunizing with transition state analogs. Even more promising for future immunological research may be the production of genetically engineered antibodies in which even immunization procedures will be circumvented.

Studies on the cellular and molecular basis of immunological specificity have provided information on how it is possible for the immune system to generate specific antibodies that can recognize millions of different molecular structures from a limited gene pool. The shuffling of DNA and RNA fragments in a combinatorial process with the potential to produce 106–109 specific antibody molecules is one of the most fascinating aspects of this research. There appears to be no shortage of biosynthesizing apparatus to produce specific antibodies for molecules that are known and those that are yet to be described.

Helen Van Vunakis
Professor of Biochemistry
Brandeis University

immunologist [im′yə näl′ə jist] *Immunology.* a scientist who studies the processes and substances associated with the resistance of humans and higher animals to infections and diseases.

immunology [im′yə näl′ə jē] a branch of biology that deals with the study of the processes and substances associated with the resistance of humans and higher animals to infections and diseases.

immunomodulation *Immunology.* any process in which an immune response is altered to a desired level.

immunonephelometry *Immunology.* a method of measuring antibody or antigen particles, in which an instrument (known as a nephelometer) is used to measure the particles as they scatter through a medium.

immunopathogenesis *Immunology.* a process whereby an immune response changes the development of a disease.

immunopathology *Medicine.* **1.** the study of immune responses to disease, of immunodeficiency diseases, and of diseases that have immunological etiology or pathogenesis. **2.** the physical manifestations related to immune responses to disease or with diseases that have an immunologic etiology.

immunopotentiation *Immunology.* the artificial heightening of an immune response.

immunoradiometric assay *Immunology.* a test that determines the concentration of antigens in a specimen, using a serological method that mixes radioactive antibodies with antigens.

immunosorbent assay see ELISA.

immunosuppressant *Pharmacology.* any agent that suppresses the ability of the body's immune system to fight disease. Also, IMMUNOSUPPRESSIVE.

immunosuppression *Immunology.* the process of inhibiting a normal immune response with the use of drugs, biological agents, or chemical agents; commonly used in association with tissue transplantation or to control autoimmune diseases.

immunosuppressive *Pharmacology.* **1.** of, relating to, or inducing immunosuppression. **2.** see IMMUNOSUPPRESSANT.

immunosurveillance theory *Immunology.* a theory stating that potentially cancerous cells occur with great frequency, increasing an organism's reaction to an antigen and thereby enabling the organism to resist or overcome an infection or disease.

immunotherapy *Medicine.* treatment, both active and passive, that combines immunization with immunopotentiators and immunosuppressants, hyposensitization for allergic disorders, bone marrow transplants, and thymus implantation.

immunotoxicity *Toxicology.* toxic effects that impair the immune system.

immunotoxin *Immunology.* an antibody linked to a poisonous substance, such as bacterial exotoxins.

Imou pine see RIMU.

impact *Mechanics.* **1.** any contact, collision, or shock that transfers momentum between two bodies. **2.** to contact another body in this manner.

impact-action fuse *Ordnance.* a fuse that is actuated by the impact of a projectile or bomb against an object. Also, DIRECT ACTION FUSE, IMPACT FUSE, POINT DETONATION FUSE.

impact area *Engineering.* **1.** the area affected by any event, such as an earthquake or a nuclear emission. **2.** a section of land having clearly marked boundaries, inside which objects moving across a given range are expected to land. *Ordnance.* an area having designated boundaries within which it is intended that all ordnance will detonate on impact.

impact bar *Materials Science.* a standard object used to test the relative hardness of a plastic material, particularly its tendency to fracture upon impact.

impact breaker *Mechanical Engineering.* a device for breaking up stone in which energy captured from falling stones is combined with the action of extremely large impellers. Also, DOUBLE IMPELLER BREAKER.

impact crater *Geology.* a crater formed by the impact of an unspecified object, especially on the surface of the earth or moon.

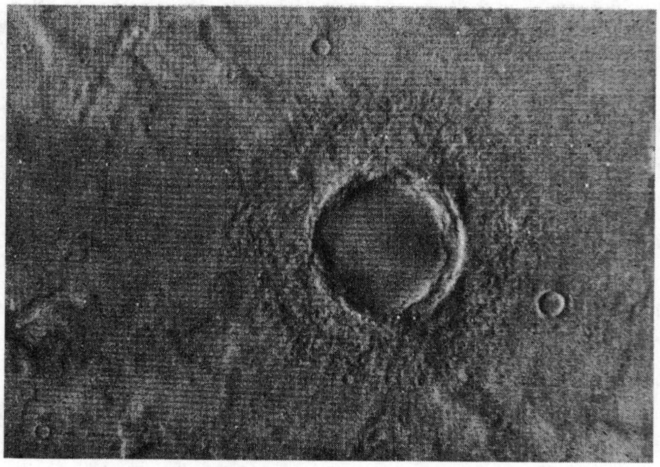

impact crater

impact crusher *Mechanical Engineering.* a machine in which soft rock is crushed by swift blows from rotating breaker bars or plates coupled with hammers.

impacted *Medicine.* firmly lodged, packed, or wedged in position.

impacted fracture *Medicine.* a fracture in which the fragmented ends of the fractured bone are wedged together.

impacted tooth *Medicine.* a tooth that is so firmly lodged or wedged against another tooth that its complete and normal eruption is unlikely; used especially to refer to such a condition for a wisdom tooth.

impact energy *Mechanical Engineering.* the amount of energy that is required to fracture hard materials such as rock. Also, IMPACT VALUE. *Materials Science.* see IMPACT STRENGTH, def. 2.

impact excitation *Electricity.* an excitation produced by a short-duration surge, such as a spark discharge.

impact extrusion *Metallurgy.* extrusion performed by impacting the stock in a confining die; heat is not necessarily applied.

impact force *Mechanics.* the force between two bodies during impact.

impact grinding *Mechanical Engineering.* a method by which various types of crushing equipment are used to grind rocks, boulders, and stones into fine particles.

impaction *Medicine.* **1.** in general, the condition of being impacted, wedged, or lodged. **2.** specifically, in obstetrics, the indentation of the fetal parts of one twin onto the surface of the other twin; this allows for the simultaneous partial engagement of both twins.

impact ionization *Electronics.* a process for increasing the number of free moving electrons in semiconductor material by bombarding it with highly energized electrons.

impactite *Geology.* a vesicular, glassy to finely crystalline material produced where a meteor has struck the earth, and consisting of fused meteoric material and slag.

impact law *Physics.* a law giving the relationship between fluid density, fluid viscosity, and particle density in the settling velocity of large particles in a fluid; the settling velocity is directly proportional to the square root of the particle diameter.

impact load *Engineering.* **1.** a sudden, powerful application of force. **2.** the effect of such force on an object. *Aviation.* the load imposed on an aircraft by impact, as in landing.

impact loss *Fluid Mechanics.* a reduction in the head of a flowing stream caused by the collision of fluid particles with each other or some obstacle in the stream's path.

impact mark see PROD MARK.

impact microphone *Acoustical Engineering.* a device used to detect small objects that strike it by receiving the vibrations caused by their impact; such microphones are used in space to record the presence of small meteoroids.

impact mill *Mechanical Engineering.* a machine for crushing rocks and minerals by means of rotating blades that project the material against steel plates.

impactor *Engineering.* a device designed to gather suspensoid samples. Also, **impactometer.** *Mechanical Engineering.* a machine that operates with striking blows.

impact parameter *Nuclear Physics.* the distance, measured perpendicularly, from the target nucleus to the first line of motion of the particle traveling toward it during a nuclear collision.

impact predictor *Space Technology.* a computer that processes data from a trajectory measuring system and continuously computes the time at which a rocket would hit the earth if its power were shut off at that moment and the remaining trajectory were ballistic in nature.

impact roll *Mechanical Engineering.* an idler roll covered by a resilient material used to protect a conveyor belt from being damaged when material is dumped upon it.

impact screen *Mechanical Engineering.* a screen designed to swing or rock forward to a particular stop point; it allows varying sizes of material to be processed.

impact strength *Materials Science.* **1.** the ability of a material to withstand shock loading. **2.** the impact stress necessary for fracturing a substance. Also, IMPACT ENERGY.

impact stress *Mechanics.* any suddenly applied stress, or the measurement of such a stress.

impact test *Materials Science.* a method of measuring the energy absorbed during fracture when a standard test specimen, usually a notched bar, is subjected to a rapid loading.

impact theory *Astronomy.* a theory widely held by planetary geologists that explains the formation of most craters on solar-system bodies by meteorite impacts.

impact transition temperature *Materials Science.* the temperature at which the behavior on impact of certain metals, polymers, and ceramics changes from brittle to ductile; it is brittle at low temperature.

impact tube see PITOT TUBE.

impact typesetting *Graphic Arts.* the process of composing lines of type by striking the characters directly onto a page using a typewriter or similar device. Also, STRIKE-ON TYPESETTING.

impact velocity *Mechanics.* the velocity between two bodies during air impact.

impact wrench *Mechanical Engineering.* a wrench operated electrically or pneumatically to provide a rapid succession of sudden torque.

impala *Vertebrate Zoology.* a swift, medium-sized African antelope, *Aepyceros melampus,* usually a yellowish-brown color; the male is characterized by ringed, lyre-shaped horns.

impala

Impatiens

Impatiens [im pā´shənz] *Botany.* **1.** the principal genus of plants of the family Balsaminaceae, including many species that are widely distributed in North America, such as the **jewelweed,** *I. capensis.* **2.** a plant of this genus. (Related to the word *impatient*; from the fact that its seeds readily burst open from a slight touch.)

IMPATT amplifier *Electronics.* an amplifier using an IMPATT diode. The operating frequency usually ranges from 5 to 100 gigahertz, and the power output reaches up to 20 watts continuous wave or 100 pulsed.

IMPATT diode *Electronics.* a device in which electron transit time and a sudden increase of current, known as avalanche breakdown, causes it to generate negative resistance; generally used as a gigahertz oscillator or amplifier. (An acronym for Impact Avalanche Transit-Time diode.)

impedance *Electricity.* the effective resistance to the flow of current at a given frequency in an alternating current circuit, due to the combined effect of resistance and reactance.

impedance bridge *Electricity.* an instrument used to measure the resistive and reactive components of an unknown impedance.

impedance coil *Electricity.* a wire coil designed to hinder the flow of alternating current in an electric circuit.

impedance compensator *Electricity.* a network that modifies the impedance of a circuit or device over a desired frequency range.

impedance coupling *Electricity.* the coupling of two signal circuits by means of an impedance or a tuned circuit, usually in audio systems.

impedance drop *Electricity.* in an AC circuit, the voltage drop due to impedance.

impedance irregularity *Electricity.* an impedance mismatch in a transmission medium, usually due to a junction between unlike sections.

impedance match *Electricity.* the matching of the impedance of a power source to the impedance of the load, by means of special transformers or networks, to produce maximum power transfer, minimum reflection, and minimum distortion.

impedance-matching network *Electricity.* a circuit made up of inductors and capacitors that aids in transferring maximum power from the source to the load.

impedance matrix *Electricity.* in a multiport network, an operator that interrelates currents and voltages at certain ports.

impedometer *Electromagnetism.* a device that can measure the impedance of a waveguide.

impeller *Mechanical Engineering.* **1.** a rotor for transmitting motion, as in a centrifugal pump, blower, turbine, agitator vessel, or fluid coupling. **2.** a rotating member of a centrifugal flow compressor or supercharger.

impeller pump *Mechanical Engineering.* a mechanical pump designed to move fluids continuously through a system.

Impennes *Vertebrate Zoology.* a superorder of birds consisting of the single order Sphenisciformes, which includes the penguins.

imperative language *Computer Programming.* a source language that is converted into the machine language instructions of an object program and is in the form of commands that direct the performance of actions.

imperative statement see EXECUTABLE STATEMENT.

imperception *Neurology.* impaired perceptive ability.

imperfect crystal *Crystallography.* a mosaic crystal in which primary extinction is negligible.

imperfect fungi *Mycology.* the common name for fungi belonging to the subdivision Deuteromycotina. Also, FUNGI IMPERFECTI.

imperfect gas see REAL GAS.

imperforate *Biology.* lacking perforations.

imperforate anus *Medicine.* the absence of an anal opening due to the presence of the anal membrane or to the absence of the anal canal. Also, PROCTATRESIA.

imperial or **Imperial** *Metrology.* of or relating to the imperial system of measurement.

imperial bushel see BUSHEL, def. 2.

imperial gallon see GALLON, def. 2.

imperial red *Inorganic Chemistry.* a pigment made from red ferric oxide, Fe_2O_3.

imperial system or **imperial measure** *Metrology.* a system of liquid and dry measure used in Great Britain and certain Commonwealth countries, in which a gallon is equivalent to 1.2 U.S. gallons and a bushel to 1.03 U.S. bushels.

impermeable *Physical Chemistry.* not permeable; not allowing liquid to pass through. Also, **impervious.**

impersonal micrometer *Astronomy.* a device used to view and follow celestial objects and reduce observational error, consisting of a vertical wire positioned in the focal plane of a transit circle and instrumentation that records the position of the wire over time.

impetigo [im´pi tī´gō] *Medicine.* a contagious skin disease, caused by staphylococci and streptococci, in which superficial flaccid vesicles erupt, usually on the face, and form thick yellowish crusts; most often present in children.

impetigo neonatorum *Medicine.* bullous impetigo of the newborn, in which lesions develop soon after birth, caused by infection with staphylococci, sometimes combined with streptococci.

impingement *Mechanics.* the fact or process of coming into physical contact with or striking an object or substance in motion.

impinger *Engineering.* a device that draws in a stream of air and impacts it on a treated glass plate in order to measure and analyze the suspensoid samples that adhere to the glass.

implant *Medicine.* **1.** to graft or insert an object or material into the body, as for purposes of therapy, replacement, diagnosis, and so on. **2.** a substance or material that is so inserted or grafted.

implantation *Medicine.* **1.** the embedding of a fertilized ovum (blastocyst) into the endometrium; this occurs six or seven days after fertilization. **2.** the inserting or grafting of material into tissues.

implanted atom *Electronics.* an atom of different material that has been placed in a semiconductor material in order to change its electrical properties.

implanted device *Electronics.* a device, such as a resistor, that has been deposited into a semiconductor material by bombardment with highly energized atoms.

implanted pacemaker *Cardiology.* an electronic device surgically implanted in the heart to regulate heartbeat; used for chronic heart block.

implementation *Computer Technology.* the various steps involved in installing a computer system. *Computer Programming.* a program that implements a design or algorithm.

implexed *Invertebrate Zoology.* indented, such as an insect's exoskeleton for muscle attachments.

implication *Mathematics.* a logic statement of the form "if p, then q" or "p implies q;" written symbolically as $p \Rightarrow q$, where p and q are statements. The statement $p \Rightarrow q$ is false when p is true and q is false, and is true otherwise; $\Rightarrow$ is also called a conditional operator. Also, IF-THEN STATEMENT, CONDITIONAL STATEMENT.

implicit function *Mathematics.* a function given by an equation of the form $F(x_1, x_2, \ldots, x_n, y) = 0$, where y is considered to be a dependent variable, is said to define y as an implicit function of the independent variables $x_1, x_2, \ldots, x_n$. If, in addition to this, an equality of the form $y = f(x_1, x_2, \ldots, x_n)$ can be derived, then y is called an explicit function of $x_1, x_2, \ldots, x_n$, and the equation $F(x_1, x_2, \ldots, x_n, y) = 0$ is said to be uniquely solvable for y.

implicit function theorem *Mathematics.* a theorem that states the conditions under which an implicit equation $F(x_1, x_2, \ldots, x_n, y) = 0$ may be solved for y as an explicit function of $x_1, x_2, \ldots, x_n$. In particular, suppose (a) X, Y, and Z are Banach spaces (such as R^n); (b) U is an open set of $X \times Y$; (c) $F\colon U \to Z$ is a continuously differentiable map such that $F(a,b) = 0$ for some point $(a,b) \in U$; and (d) the partial derivative $F_y''(a,b)$ is an isomorphism of Y onto Z. Then there exists an open set W in X with $a \in W$, an open set V contained in U with $(a, b) \in V$, and a continuously differentiable mapping $g\colon W \to Y$ such that, for all $(x,y) \in V$, $F(x,y) = 0$ if and only if $x \in W$ and $y = g(x)$. That is, if $x \in W$, the implicit equation $F(x,y) = 0$ has a continuously differentiable solution $y = g(x)$ with $(x, y) \in V$. This solution is unique on some open subset of W.

implicit parameter *Computer Science.* a parameter that is passed to a subprogram without being specified directly by the programmer, e.g., the return address.

implicit personality theory *Psychology.* the popular assumption that personality traits tend to be associated with each other in certain ways; e.g., that a person who is plain-spoken and direct is also honest.

implicit programming *Robotics.* robotic programming that is less exact in its descriptions of the tasks at hand than explicit programming.

implosion *Physics.* a violent inward collapse, as in the breaking of an evacuated glass envelope, due to external forces acting inward.

implosion weapon *Ordnance.* a nuclear weapon in which fissionable material of a less than critical mass is compressed by high-explosive charges, thus creating a supercritical mass and the ensuing nuclear explosion.

implosive therapy *Behavior.* a form of treatment for fear and avoidance responses that takes place in a comfortable, secure environment and involves confronting the anxiety-provoking response in either a real or imagined way. Also, **implosion therapy.**

import *Computer Programming.* to read a file or data created by another program. Thus, **importation.**

imposed load *Civil Engineering.* all the loads that a structure must support other than the dead loads of the structure itself.

imposition *Graphic Arts.* the process of arranging the pages of a form on a flat so that they will be in the correct order after printing and folding.

impossible process *Thermodynamics.* a term for a process that would violate the first or second law of thermodynamics, such as a process in which there is a net gain of work but no net heat transfer.

impost *Architecture.* a molding or other masonry member that carries the thrust of an arch.

impotence [im´pə təns] *Medicine.* **1.** in general, weakness or inability. **2.** specifically, the chronic inability of an adult male to achieve or maintain an erection of the penis for sexual activity. **Atonic impotence** involves disturbed neuromuscular function; **functional impotence** has a psychological basis. Also, **impotency.**

impotent [im´pə tənt] *Medicine.* affected by a condition of impotence; not able to achieve and maintain an erection.

impound *Civil Engineering.* to collect water for irrigation, flood control, or similar purposes.

impounding reservoir *Civil Engineering.* a gate-controlled storage reservoir that releases water as needed for irrigation; also used for domestic, industrial, or flood-control storage. Also, STORAGE RESERVOIR.

impregnate [im preg´nāt] *Medicine.* **1.** to make pregnant; cause to conceive. **2.** to diffuse, saturate, or permeate with another substance.

impregnated bit *Mechanical Devices.* a drill bit that is uniformly fitted with whole diamonds or diamond particles about its crown.

impregnation *Medicine.* the fact of becoming pregnant. *Engineering.* the process of soaking or treating one substance with another or forcing one substance into the porous spaces of another. *Materials Science.* specifically, the process of soaking timber with a liquid preservative to help it resist decay and water damage.

impression a mark or indentation produced by pressure; specific uses include: *Graphic Arts.* **1.** the process of printing on paper with a plate, a die, or type. **2.** a press run or printing cycle. *Geology.* **1.** the hardened form or indentation left on a soft sedimentary surface by contact with plant or animal structures, such as shells or stems. **2.** a fossil footprint, track, or burrow. Also, IMPRINT. *Metallurgy.* **1.** a cavity in a forging die used to create specially shaped parts. **2.** the part so created.

impression cylinder *Graphic Arts.* a cylinder designed to place paper in contact with an inking surface, especially the cylinder on an offset press that brings paper to the blanket. Similarly, **impression roller.**

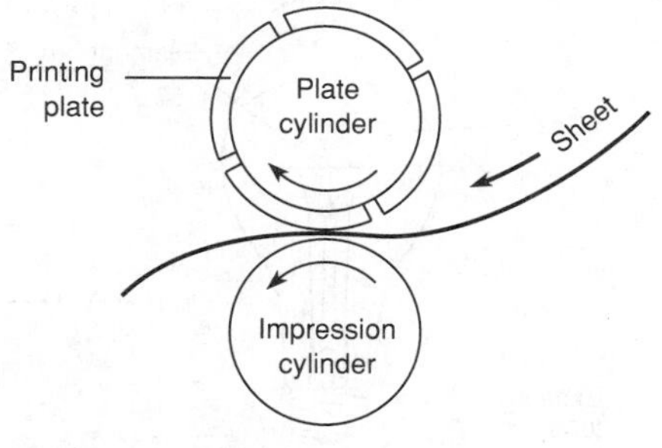

impression cylinder

imprint *Graphic Arts.* the name of a publishing company, especially as printed on the title page of a book, often along with a colophon and the place and date of publication. *Geology.* see IMPRESSION.

imprinting *Behavior.* a rapid learning process that occurs early in the lives of young animals, in which the responses are acquired quickly and are usually not reversible. *Graphic Arts.* the process of printing additional copy on a previously printed sheet, as when a local dealer's name is added to the title page of a nationally distributed brochure.

improper fraction *Mathematics.* **1.** a rational number expressed in the form a/b, whose numerator a has an absolute value greater than or equal to that of its denominator b; for example, 3/2. **2.** in general, the quotient $p(x)/q(x)$, where $p(x)$ and $q(x)$ are polynomials and the degree of $p(x)$ is greater than or equal to the degree of $q(x)$.

improper integral *Mathematics.* **1.** given a particular definition of integration (such as Lebesgue or Riemann), an integral in which the domain of integration is noncompact (for example, a half-open interval or an infinite interval). **2.** an integrand that is Lebesgue integrable but not Riemann integrable. **3.** an integral that diverges.

improper integrals of the first kind see INFINITE INTEGRAL.

improper symmetry operation *Crystallography.* any symmetry operation that converts a chiral object into its enantiomorph, such as a center of symmetry, a mirror plane, or a glide plane. Also, SYMMETRY OPERATION OF THE SECOND KIND.

improved conventional munitions *Ordnance.* munitions in which two or more antipersonnel, antimateriel, or antiarmor submunitions are delivered by an artillery warhead or projectile.

improvement cutting *Forestry.* the felling or suppression of less valuable trees in favor of more valuable tree growth, usually done in a mixed, uneven-aged forest.

improvement planting *Forestry.* any planting that is designed to improve the value of a stand.

improvement threshold *Telecommunications.* the value of the carrier-to-noise ratio below which the signal-to-noise ratio decreases faster than the carrier-to-noise ratio.

improvised *Ordnance.* describing a nonstandard weapon manufactured from available materials at or near the point of use. Thus, **improvised explosive devices, improvised grenade, improvised mine.**

impsonite *Geology.* a dull, black, infusible, asphaltic pyrobitumen with a high fixed-carbon content that is derived from the metamorphism of petroleum.

impulse *Psychology.* a stimulus that causes the mind to act. *Electricity.* a single, usually sudden flow of current or voltage in one direction for a brief time. *Physics.* a vector quantity given by the integral over time of the force acting on a body, usually in a collision in which the time interval is very brief; it is equal to the change in the momentum of the body. *Metallurgy.* in resistance welding, a pulse of heating current.

impulse disorder *Psychology.* any of various disorders characterized by immediate response to an impulse without restraint; types include addiction, pathological gambling, pyromania, and kleptomania.

impulse function see DELTA FUNCTION.

impulse generator *Electricity.* any device that produces short electrical pulses of high voltage; a common type uses inductance or capacitance to store energy and discharge it abruptly. Also, SURGE GENERATOR.

impulse modulation *Control Systems.* a signal modulation using a series of impulses that are equally spaced in time but are stronger or weaker in proportion to the amplitude of the signal as an approximation of the actual signal.

impulse noise *Electricity.* a transient electrical disturbance caused by a sudden, high-amplitude voltage pulse, such as noise caused by automobile ignitions.

impulse period see PULSE PERIOD.

impulse relay *Electricity.* **1.** a relay that completely opens and closes, even when operated with short pulses. **2.** a relay that distinguishes between strong and weak impulses, operating only on the strong ones.

impulse response *Control Systems.* the response of a system to an impulse, represented mathematically by a Dirac delta function that differs from zero for an instant but whose integral over time is unity.

impulse sealing *Engineering.* a technique for fusing plastic materials by briefly applying a pulse of intense thermal energy to the section to be sealed, then immediately cooling it.

impulse separator *Electronics.* a circuit that divides the horizontal synchronizing signals from the vertical synchronizing signals in a television receiver. Also, SYNC SEPARATOR.

impulse signaling *Telecommunications.* a method of signaling in which on-off conditions are used to represent information.

impulse solenoid *Electromagnetism.* a solenoid that operates on pulsed signals rather than continuous wave power; employed in high-speed devices such as shutters, punches, or tape drives.
impulse strength *Electricity.* the ability of insulation to withstand voltage surges lasting only microseconds.
impulse tachometer *Engineering.* a tachometer that measures the rate of pulses emitted by a shaft that generates one pulse per rotation.
impulse theory *Acoustics.* a theory stating that sound is created by a brief disturbance or pressure change in an acoustic medium, thereby creating waves of this disturbance in the medium.
impulse train *Control Systems.* an input consisting of an unending series of impulses that are equally spaced in time.
impulse transmission *Telecommunications.* a method of signaling that uses impulses of either or both polarities for transmission to indicate signal transitions.
impulse turbine *Mechanical Engineering.* a turbine that moves pressurized steam into a stationary nozzle where its potential energy is converted to kinetic energy that is directed onto blades carried by a rotor.
impulse-type telemeter *Telecommunications.* a telemeter that uses characteristics of intermittent electrical signals other than their frequency as the translating means.
impulse voltage *Electricity.* a high-voltage impulse lasting only microseconds, produced by an impulse generator and used to test the strength of insulators and power equipment against lightning and other surges.
impulsion *Psychology.* the tendency to act immediately in response to internal drives, without reflection or forethought. Also, **impulsivity, impulsiveness.**
impulsive *Psychology.* **1.** relating to or governed by impulse. **2.** of or relating to an impulse disorder.
impulsive sound equation *Acoustics.* an equation stating the relationship between the decay of short-burst sound intensity in a room and the characteristics of the room.
impure *Science.* contaminated by or mixed with a foreign substance.
impure flutter *Cardiology.* a combination of atrial flutter and another rhythm.
impurity *Science.* **1.** a foreign substance that contaminates something pure. **2.** the condition of being impure. *Solid-State Physics.* a substance that is incorporated into a semiconductor material and provides free electrons (n-type impurity) or holes (p-type impurity).
impurity band *Solid-State Physics.* an energy band that exists outside of the normal band scheme, and that arises from orbital overlap of impurities in a solid.
impurity level *Solid-State Physics.* an energy level, associated with the energy of a set of impurities in a solid, that is not part of the normal band scheme.
impurity semiconductor *Solid-State Physics.* an extrinsic semiconductor that has been doped with an impurity giving rise to impurity levels.
IMS or **I.M.S.** International Metallographic Society.
IMV intermittent mandatory ventilation.
IMViC tests *Microbiology.* a group of four biochemical tests, namely indole, methyl red, Voges-Proskauer, and citrate, used in the identification of members of the family Enterobacteriaceae.
in *Artificial Intelligence.* in a truth maintenance system, describing a proposition that is currently believed or supported.
In the chemical symbol for indium.
in- a prefix meaning: **1.** in or into, as in *incubate*. **2.** not, as in *incoherent*.
in. or **in** inch.
INA or **I.N.A.** International Neurological Association.
inactivate *Science.* to render inactive; destroy the activity of.
inactivated vaccine *Immunology.* a vaccine that has been treated by physical means (such as heating) or chemical agents, so that its component microorganisms no longer have the ability to cause disease.
inactivation *Virology.* the loss of a virus's ability to complete an infection.
inactivation center *Genetics.* an area on an X chromosome that inactivates adjacent loci in the inactivated copy of the X chromosome.
inactive current see REACTIVE CURRENT.
inactive front *Meteorology.* a weather front that is characterized by little cloudiness and an absence of precipitation. Also, PASSIVE FRONT.
inactive tartaric acid see RACEMIC ACID.
inactive volcano *Volcanology.* a volcano having no known history of eruption.
inadequate personality *Psychology.* a disorder in which an individual has little physical or emotional stamina and exhibits a marked inability to adapt to social situations.
Inadunata *Paleontology.* a class of crinoids that contains many heterogeneous forms difficult to classify; ancestral to the Flexibilia and Articulata; extant in the Lower Ordovician to Middle Triassic.
inadunate *Invertebrate Zoology.* of crinoids, having arms free of the calyx.
in-and-out bond *Civil Engineering.* a masonry bond made of vertically alternating stretchers and headers, used especially at corners.
inanimate *Science.* not alive; lacking animation.
inanition *Medicine.* a condition that is marked by weakness, exhaustion, weight loss, and decreased metabolism, as a result of a lack of food or an inability to assimilate food.
in antis *Architecture.* between antae, as columns in a portico.
inaperturate *Biology.* describing pollen grains that lack germinal pores.
inapparent infection *Virology.* a viral infection that does not give rise to cytopathic effects in cells, or to signs or symptoms in animals.
inappetence *Medicine.* lack of appetite.
inarch *Botany.* to graft a growing branch to a stock without separating the branch from its parent stock.
Inarticulata *Invertebrate Zoology.* a class of marine bivalves in the phylum Brachiopoda, characterized by shells that are held together only by muscles and soft tissues.
inarticulate *Invertebrate Zoology.* having no segments or joints.
inarticulo motis *Medicine.* at the exact point of death.
inattention see SELECTIVE INATTENTION.
inaxon *Neurology.* a nerve cell having an axon that divides into terminal filaments at a considerable distance from the cell nucleus.
inboard *Engineering.* **1.** located inside the hull or bulwarks of a boat, ship, or aircraft. **2.** closer or closest to the longitudinal axis of a ship or aircraft. **3.** generally, toward the center or inside.
inboard profile *Naval Architecture.* a side cutaway drawing of a vessel, showing the interior arrangement.
inbond *Civil Engineering.* the bricks or stones laid as the headers across a wall.
inborn *Biology.* naturally present at birth. *Behavior.* describing a behavior pattern that is instinctive and not acquired or learned.
inborn error *Genetics.* a hereditary biochemical condition that interrupts a normal metabolic process, usually resulting in a disease in the individual.
inbred *Agriculture.* of or relating to a plant or animal that is the result of inbreeding.
inbred strain *Genetics.* a stock of organisms that are essentially identical, except for sexual differences, as a result of inbreeding.
inbreeding *Genetics.* **1.** the reproduction of closely related plants or animals. **2.** the crossing of closely related plants or animals within a species.
inbreeding coefficient *Genetics.* **1.** the probability of homozygosity that results when the zygote obtains copies of the same ancestral gene. **2.** the percentage of loci at which an individual is homozygous.
inbreeding depression *Genetics.* the decrease in vigor that normally accompanies a program of extensive inbreeding.
incalescent *Physics.* growing warm; increasing in heat.
incandescence *Optics.* the glowing emission of light that a substance gives off at a high temperature.
incandescent lamp *Electricity.* a light bulb with a resistive wire filament, usually of tungsten, that can be heated until it glows white hot; the filament is enclosed in an evacuated bulb to prevent oxidation.

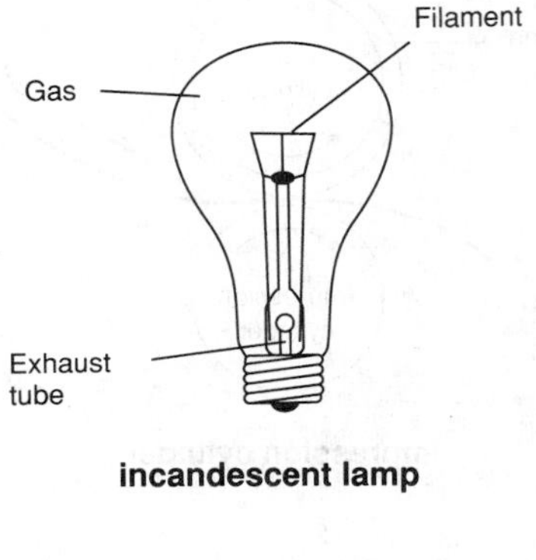

incandescent lamp

incandescent readout *Electronics.* a readout that displays data characters by energizing an appropriate combination of seven bar-shaped incandescent lamps.

incandescent tuff flow see ASH FLOW.

INCAP or I.N.C.A.P. Institute of Nutrition for Central America and Panama.

incapacitating agent *Ordnance.* an agent, usually chemical, that produces temporary physiological or psychological debilitation, thus rendering an individual incapable of effectively performing assigned duties.

incarbonization see COALIFICATION.

incarcerated hernia *Medicine.* a hernia that cannot be reduced or returned with manipulation; thus requiring surgical repair. Also, IRREDUCIBLE HERNIA.

incarnative *Surgery.* **1.** promoting the formation of granulations. **2.** an agent that promotes the formation of granulations.

incendiary [in sen´dē er ē] *Ordnance.* **1.** a chemical agent designed to cause combustion. **2.** of or relating to ammunition or weapons designed to create a burning effect at the target. Thus, **incendiary bomb, incendiary-bomb cluster, incendiary grenade, incendiary rocket.**

incentive *Behavior.* an object or event that motivates the performance of a behavior. *Industrial Engineering.* specifically, any kind of reward that stimulates a worker to perform at an above-standard level.

incentive learning *Behavior.* a type of learning in which energy is expended to achieve a goal because a reward is expected.

incentive theory *Psychology.* a theory that emphasizes the influence on behavior of external rewards and punishments, as opposed to internal drives.

inceptisol *Geology.* an order of soils characterized by one or more horizons in which mineral substances other than amorphous silica or carbonates have been altered or removed, but have not accumulated to any great degree.

incertae sedis *Systematics.* a designation for a taxon whose relationships are uncertain. (A Latin term meaning "uncertain seat.")

incest avoidance *Behavior.* any pattern of behavior or other factor that tends to prevent mating between closely related individuals and thus avoid the negative effects of inbreeding; e.g., the tendency in certain species of young males to depart from the family group.

incest taboo *Anthropology.* a social sanction that prohibits sexual relations or marriage between parent and offspring, or between siblings, present in some form in all societies; may also extend to the extended family of cousins, aunts, uncles, and grandparents, and to members of certain descent groups.

inch *Metrology.* a traditional unit of measure equal to 1/12 of a foot, equivalent to 2.54 centimeters.

inching *Electronics.* see JOGGING.

inch of mercury *Metrology.* a unit of measure for atmospheric pressure, equal to the amount of power pressure exerted by a one-inch column of mercury under standard conditions of temperature and gravity.

incidence (angle) see ANGLE OF INCIDENCE.

incidence matrix *Mathematics.* the incidence matrix of a graph G (composed of a finite number of enumerated edges and vertices) is the matrix whose (i, j)th entry is 1 if the ith vertex is incident on the jth edge, and is 0 otherwise. If G is a directed graph, nonzero entries of the incidence matrix of G are +1 and –1 if the edge is oriented away or toward the vertex, respectively. Isomorphic graphs have incidence matrices that differ only by a permutation of rows and columns.

incident *Military Science.* a brief clash or disturbance that is relatively minor and does not involve extended hostilities. *Mathematics.* **1.** a vertex and edge of a graph are said to be incident if the vertex is an endpoint of the edge. An **incidence set** of a vertex is the set of edges incident on the vertex. **2.** intersecting; e.g., a vertex of a polygon may be incident on a curve.

incidental element see IRREGULAR ELEMENT.

incidental learning *Behavior.* a type of learning that occurs without formal instruction, intent to learn, or obvious motive.

incidental memory *Psychology.* learning or remembering what occurs without intentional effort or purpose.

incidental reinforcement see ACCIDENTAL REINFORCEMENT.

incident field intensity *Electromagnetism.* the field strength of a wave as measured at the location of a receiving antenna.

incident light *Optics.* light that falls directly on a surface.

incident power *Electricity.* the product of the outgoing current and voltage from a transmitter, traveling along a transmission line to an antenna.

incident wave *Electronics.* the wave that travels toward a discontinuity in a transmission line or wave-propagating medium.

incinerate *Chemistry.* of a material, to burn or be burned until it is reduced to ashes.

incineration *Chemistry.* the process of incinerating, especially at very high temperatures.

incinerator *Engineering.* a furnace that is designed for the destruction of refuse; it may be fired by electricity, gas, oil, or solid fuel.

incipient *Medicine.* coming into existence or into an initial stage; beginning to appear.

incipient melting *Metallurgy.* the localized melting that occurs in areas with lower melting points in a metallic alloy upon heating.

incipient peneplain see STRATH.

incipient surface energy *Materials Science.* the electrostatic charge on the surface of the particles of a polymer colloid.

Incirrata *Invertebrate Zoology.* a suborder of cephalopod mollusks, eight-tentacled animals in the order Octopoda.

incised [in sīzd´] *Medicine.* having been cut with a sharp instrument, especially a surgical instrument. *Biology.* describing leaf margins formed with deep, often irregular, indentations.

incised meander *Hydrology.* an intrenched winding of a river, resulting from renewed downcutting in a period of rejuvenation.

incised wound *Surgery.* the separation of soft tissue made by a cutting instrument.

incision [in sizh´ən] *Surgery.* **1.** a cut, especially an intentional cut produced by a sharp instrument. **2.** the act of cutting.

incisional biopsy *Surgery.* the removal by cutting of a sample of a lesion tissue and sometimes adjacent healthy tissue for examination.

incisional hernia *Medicine.* a hernia that occurs through a surgical incision or scar.

incisive canal *Anatomy.* a passageway through the front of the hard palate, through which the nasopalatine nerves and branches of the greater palatine arteries pass.

incisive foramen *Anatomy.* the openings of the incisive canals into the incisive fossa.

incisive fossa *Anatomy.* a depression in the bony palate behind the central incisors that leads into the incisive foramen.

incisor [in sīz´ər] *Anatomy.* any of the four cutting teeth that lie between the canines in either jaw.

inclination *Science.* an angular deviation from the horizontal or vertical; specific uses include: *Geology.* the slope of any geological formation or surface, as measured from the upward or downward deviation from the horizontal or vertical. *Geophysics.* the angle between the horizontal and the dip of the earth's magnetic field at a given location. *Space Technology.* the degree of deviation of the orbit of an artificial satellite from an equatorial orbit. *Mathematics.* **1.** the inclination of a line of slope m in the plane is the value of arctan m; equivalently, the angle made by the line with the x-axis, measured counterclockwise from the positive direction of the x-axis. **2.** the inclination of a line with respect to a given plane in space is the angle made by the line with its orthogonal projection onto the plane; by convention, the smaller angle is used. **3.** the inclination of a plane with respect to a given plane in space is the smaller of the dihedral angles formed by the two planes.

inclination of axis *Astronomy.* the angle between a planet or moon's rotation axis and its orbit; in the case of a star, the reference plane is usually that of the celestial sphere.

inclination of planetary orbits *Astronomy.* **1.** the angle between a planet's orbit and the earth's orbit (the ecliptic). **2.** the plane of the solar system is used as the reference plane.

inclination of the wind *Meteorology.* on a synoptic weather chart, the representation of the angle between the wind direction and the isobaric contours.

incline *Science.* **1.** to lean or slant. **2.** a surface that leans or slants; a slope. Thus, **inclined.**

inclined bedding *Geology.* a bedding whose strata dip in the direction of current flow or are inclined to the principal surface of deposition.

inclined cableway *Mechanical Engineering.* a single-cable arrangement in which the slope of the track cable is steep enough to allow the carrier to descend along the cable under its own weight.

inclined drilling *Engineering.* a process of blasthole drilling that is carried out at an angle rather than vertically.

inclined extinction *Optics.* an extinction wherein the directions of vibration are sloped toward a crystal axis or the direction of cleavage.

inclined orbit *Space Technology.* an orbit, as of an artificial satellite, that is inclined with respect to the equator.

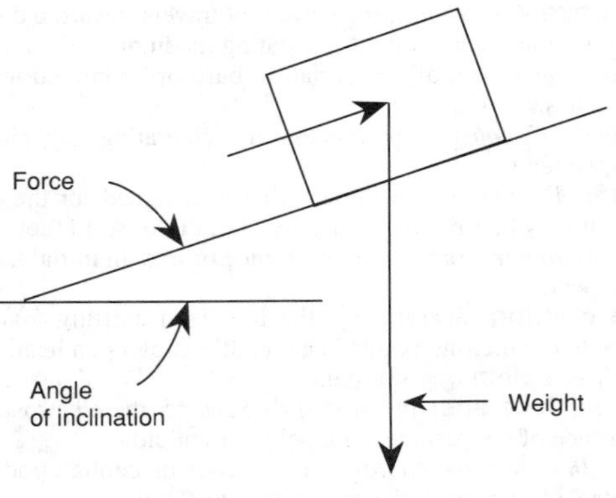

inclined plane

inclined plane *Mechanical Engineering.* a surface sloped at an angle to the horizontal (or some other reference surface), which provides a mechanical advantage for raising loads; one of the simple machines along with the lever, screw, and so on.

inclined-tube manometer *Engineering.* a manometer having an inclined leg that serves to increase the scale for more precise readings.

inclining experiment *Naval Architecture.* an experiment, performed with weights or on an inclining table, in which a ship is tipped to calculate its stability factors.

inclinometer *Engineering.* **1.** a magnetic needle that swings freely in the vertical plane; used to indicate the direction of the earth's magnetic field. **2.** an instrument that is used to determine the inclination of an embankment or a drill bore. Also, CLINOMETER. *Aviation.* an instrument that indicates the altitude of an aircraft, or the craft's angle in reference to the horizon.

inclusion the fact of including, or something that is included; specific uses include: *Crystallography.* a small, foreign body of solid, liquid, or gaseous substance contained within a crystal. *Petrology.* a fragment of older rock embedded within an igneous rock. *Metallurgy.* a particle, generally undesirable and nonmetallic, that is contained in a metallic structure but is not essential to it. *Cell Biology.* in the cytoplasm of a cell, a particulate body that represents either a stage in the multiplication of a virus or a cellular by-product, such as starch. Also, CELL INCLUSION. *Mathematics.* **1.** if X is a subset of Y, then the function f defined by $f(x) = x$ for each x in X is called the **inclusion map** (or function) of X into Y. If $Y = X$, then f is called the identity map on X. **2.** if X is a subset of Y, written $X \subseteq Y$, then X is said to be **included** in Y. If X is a proper subset of Y, written $X \subset Y$, then X is said to be **strictly included** in Y. **3.** any relation defined on a Boolean algebra that is reflexive, antisymmetric, and transitive is called an **inclusion relation.**

inclusion body *Virology.* a group or cluster of virus particles or proteins that are found in certain virus-infected cells and are visible by microscopy.

inclusion body hepatitis *Veterinary Medicine.* a viral disease of young chickens and pheasants characterized by watery blood, anemia, depression, hemorrhage, pale bone marrow, and pale combs, wattles, skin, and legs. Also, HEMORRHAGIC ANEMIA SYNDROME.

inclusion body protein *Virology.* the primary component of an inclusion body.

inclusion body rhinitis *Veterinary Medicine.* a mild respiratory disease caused by porcine cytomegalovirus infection, chiefly affecting unweaned pigs; marked by sneezing, nasal discharge, depression, and lesions in the mucous membrane of the upper respiratory tract.

inclusion complex *Chemistry.* a combination of two or more materials wherein the molecules of one are enclosed in the crystal lattice of the other.

inclusion conjunctivitis *Medicine.* an inflammation of the mucous membrane covering the anterior surface of the eyeball and lining the eyelids; caused by *Chlamydia trachomatis,* it mainly affects infants. Also, INCLUSION BLENNORRHEA, SWIMMING POOL CONJUNCTIVITIS.

inclusion cyst *Medicine.* a cyst created by the inclusion of a small piece of epithelium or mesothelium in connective tissue.

inclusion encephalitis *Medicine.* a disease, usually fatal, that seems to result from a measles virus infection and results in inflammatory reaction in both the white and gray matter of the brain. Also, SUBACUTE SCLEROSING LEUKOENCEPHALITIS, PANENCEPHALITIS.

inclusion-exclusion principle *Mathematics.* let U be a finite set; i.e., $n(U) = m$, where $n(S)$ denotes the number of elements of a finite set S. Consider t properties or attributes P_i $(i = 1, \ldots, t)$, and let A_i be the subset of U consisting of all elements that have the properties P_i. Then the number of elements of U that have none of the properties P_i is, by theorem,

$$n(\bar{A}_1 \cap \bar{A}_2 \cap \cdots \cap \bar{A}_t) = n(U) - [\sum_{i=1}^{t} n(A_i)] + [\sum_{i \neq j} n(A_i \cap A_j)] - [\sum_{i \neq j \neq k} n(A_i \cap A_j \cap A_k)] + \cdots + (-1)^t n(A_1 \cap A_2 \cap \cdots \cap A_t),$$

where each bracketed sum contains no repeated terms (set intersection is commutative). This fundamental principle of counting is called the inclusion-exclusion principle because at each stage of the computation, elements of $A_1 \cap A_2 \cap \cdots \cap A_t$ are alternately included and excluded.

inclusive fitness *Ecology.* an assessment of an organism's fitness as measured by the ability of that organism and its near relatives to reproduce successfully.

incoalation see COALIFICATION.

incoherence *Science.* the fact or condition of being disordered or unconnected. *Psychology.* the condition of being unable to think or express one's thoughts in a clear and orderly manner.

incoherent *Psychology.* relating to or affected by incoherence. *Geology.* loose or unconsolidated; applied especially to recent sediments.

incoherent interface *Metallurgy.* in a metallic structure, an interface at which there is lattice discontinuity.

incoherent light *Optics.* light waves that are not in phase, such as the light emitted from a standard light bulb or other conventional sources of light.

incoherent scattering *Physics.* a scattering of photons or particles in which the scattered entities have no apparent phase relationship and therefore do not produce a stable interference pattern.

incoherent waves *Physics.* waves that do not possess any definite phase relationship.

incommensurable *Mathematics.* **1.** a collection S of sets is said to be incommensurable if no member of S is a subset of any other member of S. **2.** two numbers a and b whose ratio is irrational are called **incommensurable numbers**. In general, if a and b are incommensurable numbers, one line segment of length a or b cannot be constructed with Euclidean methods, given the other.

incommensurate frequencies *Chaotic Dynamics.* frequencies whose ratio is an irrational number (one that cannot be written as the ratio of two integers).

incomparable elements *Mathematics.* two elements of a partially ordered set that are not related by the partial ordering. Equivalently, if the partial ordering is denoted $\leq$, then elements a and b are comparable if either $a \leq b$ or $b \leq a$; otherwise they are incomparable.

incompatibility *Science.* the unsuitability of one thing to another. *Genetics.* **1.** the restriction of fusion between gametes of certain genetic constitutions. **2.** the inability of a recipient tissue to accept a donor tissue due to genetically caused antigenic differences.

incompatible *Plant Pathology.* of or relating to an interaction between a plant host and a pathogen, in which a disease does not develop.

incompatible equations *Mathematics.* equations that do not possess a common solution. Also, INCONSISTENT EQUATIONS.

incompatible response *Psychology.* a technique of controlling aggression by inducing an emotional state, such as humor or sympathy, that is incompatible with extreme anger or aggressiveness.

incompetence *Medicine.* **1.** physical or mental inadequacy or insufficiency; the inability to perform given processes. **2.** the mental inability to distinguish right from wrong or to manage one's own affairs.

incompetent bed *Geology.* **1.** a bed that deforms plastically and flows under the stress of folding. **2.** a bed whose strata vary in thickness as a result of flowage during folding.

incompetent rock *Engineering.* in excavating underground tunnels, rock that must be braced with timbers or cement because of its softness or tendency to fragment.

incomplete abortion *Medicine.* an abortion in which part of the embryo or fetus remains in the uterus.

incomplete beta function *Mathematics.* given a beta function $\beta(m, n) = \int_0^1 x^{m-1}(1-x)^{n-1}\,dx$, where m and n are positive real numbers, the incomplete beta function is: $\beta_x(m,n) = \int_0^x t^{m-1}(1-t)^{n-1}dt$, where $0 \leq x \leq 1$.

incomplete cleavage see PARTIAL CLEAVAGE.

incomplete combustion *Chemistry.* the process of oxidizing a fuel incompletely.

incomplete dominance *Genetics.* an inability of a dominant allele to fully express itself phenotypically in a heterozygous condition, usually resulting in a phenotype that is intermediate between the dominant and recessive alleles.
incomplete gamma function *Mathematics.* given a gamma function $\Gamma(\alpha) = \int_0^\infty x^{\alpha-1}e^{-x}\,dx$, where $\alpha > 0$, an incomplete gamma function is either $\gamma(\alpha,x) = \int_0^x t^{\alpha-1}e^{-t}\,dt$ or $\Gamma(\alpha,x) = \int_x^\infty t^{\alpha-1}e^{-t}\,dt$, where $0 \le x \le \infty$.
incomplete lubrication *Mechanical Engineering.* a condition that occurs when the load on rubbing surfaces is carried by both fluid viscous film and boundary lubrication; the friction is intermediate between fluid and boundary lubrication.
incompletely linked genes *Genetics.* genes located at different loci on a chromosome that undergo recombination during crossing-over.
incomplete penetrance *Genetics.* a condition in which not all of the individuals having a particular genotype exhibit the phenotype associated with that genotype; usually attributable to environmental effects.
incomplete protein *Nutrition.* a protein that does not supply all the amino acids essential for growth and good health.
incomplete sex linkage *Genetics.* an uncommon phenomenon in which homologous segments of both the X and Y chromosomes have the same gene.
incomplete virus *Virology.* a virus that is unable to replicate fully, due to a lack of some part of its genome.
incompressibility *Mechanics.* the property of an idealized substance that is able to maintain a constant volume regardless of the extent of the pressure applied to it.
incompressibility condition *Fluid Mechanics.* a condition in which the density of a fluid is constant.
incompressible *Mechanics.* not able to be compressed; maintaining its original volume regardless of pressure.
incompressible flow *Fluid Mechanics.* the flow of a fluid that does not experience any changes in density; in practice no liquid is completely incompressible.
Inconel *Materials.* a trade name for a nickel-base alloy containing chromium, molybdenum, iron, and smaller amounts of other elements.
incongruity *Psychology.* a situation in which a person's self-image is significantly different from the ideal self he or she would like to be. Also, **incongruence.**
incongruous *Geology.* describing a drag fold in which the axis and axial surface are not parallel to those of its related main fold.
inconsistent *Artificial Intelligence.* of a logical formula, false under every interpretation.
inconsistent equations see INCOMPATIBLE EQUATIONS.
incontinence *Medicine.* the inability to control the release of urine or feces, in an individual who is past the age at which such control is normally achieved.
incontinent *Medicine.* relating to or affected by incontinence.
incoordination *Neurology.* poor muscular coordination; the inability of various muscle groups to work together harmoniously in order to perform motor activities.
incr. increase; increased.
increaser *Engineering.* **1.** a coupling piece used to connect pipes of different sizes. **2.** any object or device that increases the size or strength of another object or device.
increment *Science.* a usually small addition or increase. *Hydrology.* see RECHARGE. *Mathematics.* **1.** a small change in the value of a variable x; often denoted Δx. The **incremented variable** is $x + \Delta x$, and Δx may be either positive or negative. **2.** a change in the value of a function resulting from incrementing a variable; denoted $\Delta f = f(x + \Delta x) - f(x)$. Also, **increment of the function.**
incremental compiler *Computer Programming.* a compiler that permits each statement of the program to be treated as a separate entity, thus allowing the user to build and test a program incrementally and modify or restructure it as necessary.
incremental computer *Computer Technology.* a computer that is capable of processing changes in the actual variables.
incremental dump *Computer Programming.* a technique for protecting data, in which small amounts are written away on a storage medium at frequent intervals.
incremental frequency shift *Telecommunications.* a frequency-shift technique in which information is superimposed on other information by shifting the center frequency of an oscillator by a specific amount.
incremental hysteresis loss *Electromagnetism.* the hysteresis loss in a magnetic material that is subjected to a pulsed magnetic field, rather than a continuously varying magnetic field.
incremental induction *Electromagnetism.* the average value of the magnetic induction, considering only the upper and lower limits, at a point in a material that is simultaneously subjected to an electric and a changing magnetic field.
incremental learning *Artificial Intelligence.* the process by which a learning program can update its hypotheses as additional training instances are selected or made available by the environment.
incremental permeability *Electromagnetism.* the magnetic permeability of a substance when it is subjected to an alternating magnetizing force superimposed on a direct magnetizing force.
incremental plotter see PLOTTER.
incremental representation *Computer Programming.* the representation of a variable that is achieved by referring to a change in its value (for example, using a plus or minus before the variable) rather than its actual value.
increment borer *Forestry.* an instrument with a hollow bit that is used to extract thin radial cylinders of wood from trees having annual growth rings, in order to identify age and rate of growth.
incretin see SECRETIN.
incretion see INTERNAL SECRETION.
incross *Genetics.* to interbreed individuals of the same genotype, usually individuals that belong to the same inbred strain. Thus, **incrossing.**
incrossed *Genetics.* resulting from a process of incrossing.
inc. sed. incertae sedis.
incubate *Agriculture.* **1.** of a hen, to sit on and hatch eggs. **2.** to keep eggs in an incubator for hatching.
incubation *Vertebrate Zoology.* the act or process of maintaining eggs at a temperature and humidity conducive to development, often involving one or both parents sitting on or wrapped around the eggs. *Agriculture.* specifically, the hatching of chicken eggs naturally or by the use of an incubator. *Medicine.* **1.** the maintenance of environmental conditions for the purpose of growing or developing microbial or tissue cultures. **2.** the maintenance of enviromental conditions in order to encourage the development of an infant, usually a premature or hypoxic one. **3.** the formation of the embryo in the eggs of oviparous animals. **4.** the development, without sign or symptom, of an infection from the time it gains entry until the appearance of the first signs or symptoms. *Chemistry.* the process of maintaining a substance at a certain temperature over time to study chemical reactions. *Psychology.* see INCUBATION EFFECT.
incubation effect *Psychology.* a process of problem solving in which a solution becomes evident after an interim period during which the problem was not actively confronted.
incubation period *Developmental Biology.* the length of time necessary for the development or hatching of eggs being incubated. *Biology.* the time period necessary for any process of development. *Medicine.* the time after an infection gains entry but before the first signs or symptoms appear. Also, LATENT PERIOD.
incubator *Biotechnology.* an apparatus in which environmental conditions may be controlled, as for culturing microoganisms, for developing eggs or other living cells, or for maintaining a premature infant.

incubator

incubatory carrier *Medicine.* an individual whose body contains the organisms of a disease; such an individual does not show symptoms of the disease but is capable of transmitting the disease.

incumbent *Biology.* lying or leaning upon. *Botany.* **1.** with the anthers turned inward. **2.** having cotyledons arranged so that the back of one rests against the radicle. *Geology.* a stratum that is superimposed on or is lying above another stratum.

incurrent canal *Invertebrate Zoology.* a canal that leads water into the body of a sponge or a mollusk. Also, INHALANT CANAL.

Incurvariidae *Invertebrate Zoology.* a family of small moths, including the yucca moths, lepidopteran insects in the superfamily Incurvarioidea; found worldwide.

Incurvarioidea *Invertebrate Zoology.* a superfamily of small to minute moths in the suborder Heteroneura, characterized by wings that have microscopic spines instead of scales.

incus *Anatomy.* the middle of the three ossicles of the ear that serve with the stapes and the malleus to conduct vibrations from the tympanic membrane to the inner ear. Also, ANVIL.

IND *Aviation.* the airport code for Indianapolis, Indiana.

IND investigational new drug.

ind. independent; index; industrial.

indan *Organic Chemistry.* $C_6H_4(CH_2)_3$, a combustible, colorless liquid; insoluble in water and soluble in organic solvents; freezes at –51.4°C and boils at 177°C; derived from coal tar and used in organic synthesis. Also, HYDRINDENE.

indanthrone *Organic Chemistry.* $C_{28}H_{14}N_2O_4$, a stable blue powder; soluble in dilute alkaline solutions; decomposes at 470°C; used as a dye and pigment. Also, **indanthrene blue.**

indecomposable *Mathematics.* **1.** a nontrivial group that is not the direct product of two of its proper subgroups is called an **indecomposable group.** That is, the group of integers, simple groups, the symmetric group on n elements, and the integers modulo p^n, p a prime. **2.** a tensor that cannot be written as the tensor product of two other tensors is called an **indecomposable tensor**.

indefinite ceiling *Meteorology.* a classification applied when the ceiling value represents the vertical visibility upward into surface-based atmospheric phenomena such as fog, blowing snow, or lithometeors.

indefinite integral *Mathematics.* given a function $f(x)$, any function $F(x)$ whose derivative is $f(x)$. By the fundamental theorem of calculus, $F(x) - F(a) = \int_a^x f(t)\,dt$. Also, ANTIDERIVATIVE.

indelible ink *Materials.* any ink that cannot be removed or erased, such as India ink.

indene *Organic Chemistry.* C_9H_8, a colorless liquid; insoluble in water and miscible with most organic solvents; freezes at –2°C and boils at 182.4°C; turns yellow when left standing, and polymerizes when heated or when combined with catalysts; the resins formed from polymerization are commercial products.

indenization see INDINIATION.

indent *Science.* **1.** to create a recess or depression. **2.** a recess or depression in a surface. *Graphic Arts.* **1.** to set a line or lines of text farther in than the normal left margin, especially to do this with the first line of a paragraph. **2.** such an arrangement of lines.

indentation the fact of indenting, or something that is indented.

indentation hardness *Metallurgy.* one of several methods to test for hardness, based on pressing an indenter on a metallic specimen and assessing the impression so generated.

indentation modulus *Materials Science.* the depth of the indentation of a sphere into the surface of a gelatin; related to Young's modulus.

indented bolt *Mechanical Devices.* a toothed or notched bolt used for a secure hold in cemented grout.

indenter *Mechanical Devices.* an instrument used to depth-penetrate materials and measure hardness.

independent assortment *Genetics.* a random distribution to the gametes of genes located on different chromosomes.

independent axiom *Mathematics.* any of a collection of axioms, no one of which may be deduced from any subset of the others.

independent chuck *Mechanical Devices.* a lathe chuck in which each of its four jaws can be moved and set independently to hold irregularly shaped objects.

independent equation *Mathematics.* any of a set of equations, no one of which is satisfied by all simultaneous solutions of any subset of the others.

independent events *Statistics.* two random events A and B such that the occurrence of one does not affect the probability of the occurrence of the other; thus, $P(A \text{ and } B) = P(A)P(B)$.

independent footing *Civil Engineering.* a structural footing that supports only a concentrated load, such as a column load. Also, ISOLATED FOOTING.

independent functions *Mathematics.* a set of functions, no one of which is completely determined by the values of the others. For k functions in n independent variables ($k \leq n$), having continuous first partial derivatives, this is implied by requiring that the differentials of the k functions be linearly independent at every point of the domain under consideration. If $k > n$, the functions cannot be independent.

independent line of sighting *Ordnance.* an aiming system on a gun in which the angle of sight and elevation mechanisms work independently.

independent migration law *Analytical Chemistry.* a law stating that each ion in a conductiometric titration contributes a definite amount to the total conductance, regardless of the other ions in the electrolyte.

independent points *Mathematics.* a set $k + 1$ points $\{p_0, p_1, \ldots, p_k\}$ in a vector space of dimension $n \geq k$ is independent if the set of vectors $\{p_1 - p_0, p_2 - p_0, \ldots, p_k - p_0\}$ is linearly independent.

independent property *Thermodynamics.* a variable property of a substance, such as its temperature, that must be specified in order to establish the state of the substance. Also, **independent quality, independent variable.**

independent random variable *Mathematics.* **1.** random variables $X_1, \ldots, X_n$ on a probability space (Ω, F, P) are said to be independent if

$$P(X_1 \in B_1, \ldots, X_n \in B_n) = \prod_{k=1}^{n} P(X_k \in B_k),$$

where the sets B_k are Borel sets on the real line. In particular, for discrete random variables,

$$P(X_1 = x_1, \ldots, X_n = x_n) = \prod_{k=1}^{n} P(X_k = x_k).$$

Thus, the simultaneous probability of unrelated events is the product of the probabilities of the individual events. **2.** a countable collection of random variables $X_1, X_2, \ldots$ on a probability space (Ω, F, P) is independent if $X_1, \ldots, X_n$ are independent for every $n \geq 2$. It follows that

$$P(X_1 \in B_1, X_2 \in B_2, \ldots) = \prod_{k=1}^{\infty} P(X_k \in B_k),$$

where the sets B_k are Borel sets on the real line.

independent recoil system *Ordnance.* an artillery recoil mechanism in which the recuperator is independent of the recoil brake.

independent set *Mathematics.* **1.** a collection S of vertices and edges of a graph is said to be an independent set if no distinct members of S are adjacent. **2.** a subset of a matroid is said to be independent if it is contained in some base of the matroid. Thus bases of a matroid are maximal independent sets. **3.** a set of vectors in a vector space is said to be independent if no member of the set can be written as a linear combination of the others; also called a **linearly independent set.** Equivalently, every independent set is contained in some base of the vector space, and bases of a vector space are maximal linearly independent sets.

independent-sideband modulation *Telecommunications.* a modulation technique in which the radio-frequency carrier is reduced and two information units are transmitted, one on the upper sideband and the other on the lower.

independent-sideband receiver *Electronics.* a type of radio receiver that accepts sideband frequencies generated by modulation and has the capacity to restore the carrier signal lost through modulation. Similarly, **independent-sideband transmitter.**

independent subgoals *Artificial Intelligence.* subgoals whose solutions do not interact; for example, in theorem proving, proof of a lemma cannot hurt the proof of another lemma; but in shopping with a total cost constraint, buying an expensive item can make it impossible to buy another needed item, so these subgoals are not independent.

independent suspension *Mechanical Engineering.* an automotive suspension system in which the wheels are mounted separately on a chassis by means of springs and guide links, rather than being connected by an axle beam, allowing independent vertical movement so that a road bump affecting one of the wheels will not affect the others.

independent variable *Statistics.* the variable manipulated by the experimenter in order to study changes in the dependent variable. *Mathematics.* **1.** the argument or arguments of a function; for example, x in the equation $y = f(x)$. Also, INPUT VARIABLE. **2.** see INDEPENDENT RANDOM VARIABLE.

inderborite *Mineralogy.* $CaMg[B_3O_3(OH)_5]_2 \cdot 6H_2O$, a colorless, transparent monoclinic mineral having a specific gravity of 2.0 and a hardness of 3.5 on the Mohs scale; found as prismatic crystals with other borate minerals.

inderite *Mineralogy.* $MgB_3O_3(OH)_5 \cdot 5H_2O$, a colorless, transparent, monoclinic mineral that is dimorphous with kurnakovite, occurring in prismatic crystals and having a specific gravity of 1.78 to 1.8 and a hardness of 2.5 to 3 on the Mohs scale. It is found in borate deposits. Also, LESSERITE.

indeterminate equations *Mathematics.* a system of equations having an infinite number of solutions.

indeterminate form *Mathematics.* an expression that is undefined when evaluated at some particular value of the variables; it usually involves composite functions and an expression such as $0/0$, ∞/∞, 0^0, $\infty-\infty$, $0 \cdot \infty$, 1^∞, etc. An indeterminate form may have a well-defined value in the limit.

indeterminate truss *Civil Engineering.* a structural truss with an excess number of members or redundant members.

index *plural,* **indices.** [in′di sēz] a guide, reference, or pointer; specifc uses include: *Computer Programming.* **1.** a reference table of computer words or fields that contain record location storage addresses. **2.** a symbol or number identifying a specific element in an array. **3.** to reference an element in an array or sequence using an index, typically by forming the element address as the base address of the array plus a constant multiple of the index value. *Crystallography.* a number used to indicate the faces of a crystal or an order of diffraction. *Physics.* a value, usually dimensionless, that represents a measure of some physical effect characteristic of a substance or process, such as the index of refraction. *Engineering.* **1.** a piece of metal, wood, or the like that serves as a pointer or an indicator. **2.** to rotate (a workpiece) on a milling machine in order to repeat an operation at a different position. **3.** to provide (a workpiece) with spaces, parts, or angles using an indexing head. *Mathematics.* **1.** an element of an index set or a variable representing the elements of an index set; usually (a subset of) the natural numbers. **2.** let H be a subgroup of a group G. The index of H in G, denoted $[G : H]$, is the number of distinct right (or left) cosets of H in G. **3.** in particular, if a and m are relatively prime (notation: $(a, m) = 1$) and if g is a primitive root of m, then the integer k for which $a \equiv g^k (\text{mod } m)$ is called the index of $a(\text{mod } m)$. **4.** a winding number. **5.** the number n in the expression $a^{1/n}$. Also, INDEX OF A RADICAL. **6.** a chromatic number.

index arm *Navigation.* the movable part of a marine sextant whose adjustment allows an observer to read the altitude of an observed body from scales on the frame and the end of the arm.

index bristol *Graphic Arts.* a stiff, rugged bristol paper commonly used for index cards, postcards, menus, and the like. Also, INDEX STOCK.

index case see PROPOSITUS.

Index Catalogue *Astronomy.* a catalogue of galaxies, star clusters, and nebulae, published in two parts in 1895 and 1908 by J. L. E. Dreyer as supplements to the *New General Catalogue.*

index chart *Mechanical Engineering.* a chart indicating the arrangement of parts in a machine.

index contour (line) *Cartography.* a contour line on a map, drawn with a more heavily weighted line to distinguish it from the other contour lines on either side; this is usually done for one out of every four or five contour lines to indicate a specific multiple of the interval of change in elevation indicated by the regular contours.

index counter *Engineering.* a counter indicating the amount of recording tape that has passed on a reel, thus making it possible or easier to locate or index a section on the reel.

indexed address *Computer Programming.* an address that has been modified by an index register prior to or during the execution of an instruction.

indexed array *Computer Programming.* an array whose elements are specified by row and column positions.

indexed sequential access *Computer Programming.* a technique for rapidly accessing arbitrary records by means of index markers.

indexed sequential data *Computer Programming.* a set of files in which each record contains a key to its location, and an index computes the location of each record.

indexed sequential organization *Computer Programming.* a system of file organization in which files are arranged in a logical sequence and an index contains the key to the location of each physical file.

index ellipsoid *Optics.* the ellipsoid surface that determines the optical structure of an anisotropic substance; used in crystallography. Also, ELLIPSOID(AL) OF WAVE NORMALS, INDICATRIX.

index error *Engineering.* a constant error occurring in an instrument due to the inaccurate adjustment of the measurement mechanism, such as the vernier or index.

index finger see FOREFINGER.

index forest *Forestry.* a forest that in volume, density, and growth reaches the highest average in a given area and is then used as an index.

index fossil *Paleontology.* a morphologically distinctive fossil taxon that is abundant and widespread, but is restricted to and therefore characteristic of a limited stratigraphic range.

index hole *Computer Technology.* a hole punched in a floppy disk jacket through which the electro-optical disk drive can read the start of sector 0 on the disk.

index horizon see INDEX PLANE.

indexing *Mechanical Engineering.* a machining process in which discrete spaces, angles, or contours are provided through the use of an indexing head.

indexing head *Mechanical Engineering.* a machine tool attachment used to rotate a workpiece through any required angles so that its faces can be machined and drilled in prescribed angular relationships. Also, **index head.**

index line *Mathematics.* see ISOPLETH.

index liquid *Optics.* any of a group of chemicals with known refractive indices, used in conjunction with a microscope to determine the refractive indices of powdered minerals.

index mark *Computer Technology.* a recorded code or an actual hole or mark that identifies a starting point for each track on a disk. *Photogrammetry.* **1.** in aerial photography, a mark such as a cross or dot made on the surface being photographed; used as a reference when the photograph is viewed through monocular instruments. **2.** one of a pair of marks on the surface being photographed that combine to form a floating mark when the photograph is viewed through a stereoscope. Each index mark making up a floating mark is called a **half-mark.**

index mineral *Petrology.* a mineral whose first appearance in passing from lower to higher grades of metamorphism signals the outer limit of a metamorphic zone.

index mirror *Navigation.* a mirror that is attached to, and moves with, the index arm of a marine sextant.

index number *Statistics.* a summary measure that is relative to the same quantity or base; used for comparing prices, production, employment, and the like.

index of a radical *Mathematics.* see INDEX, def. 5.

index of aridity *Meteorology.* the mathematical expression of the precipitation effectiveness or aridity of a region, shown as $P/T + 10$, where P is the annual precipitation in centimeters and T is the annual mean temperature in °C.

index of cooperation *Telecommunications.* a mathematical formula used in rectilinear scanning and recording; the total length of the scanning/recording line is multiplied by the number of scanning/recording lines per unit length.

index of nutritional quality *Nutrition.* an expression of nutrient density in relation to caloric value.

index of precision *Statistics.* a means of evaluating estimation errors by computing the reciprocal of the standard deviation multiplied by the square root of the sample size. Also, MODULUS OF PRECISION.

index of refraction see REFRACTIVE INDEX.

index of similarity *Ecology.* the ratio of the number of entities (such as species) that is common to two comparison units (such as communities) to the total number of species found in both.

index plane *Geology.* any structural surface used as a reference plane in studying the geological structure of an area.

index plate *Mechanical Devices.* a perforated, circular plate with either graduated or circular rings of holes, used in a milling-type machine as a guide for teeth in wheel cutting. Also, DIVISION PLATE .

index point *Computer Technology.* a reference point on a disk that provides a point of origin for tracks and a measure of angular displacement around the track.

index prism *Navigation.* a prism on a bubble sextant through which a celestial body is observed.

index ratio *Electromagnetism.* the ratio of the radius of a conductor used in induction heating to its skin depth at the operating frequency.

index register *Computer Technology.* a register in a central processor, whose contents can be added to the address specified in an instruction to form the effective address.

index set *Mathematics.* in set theory, the domain I of a function x (called a family) to a set X. The range of x is called an indexed set and the value of the function x at i, called a term of the family, is denoted x_i. In a given situation, the index set I may be indicated by a parenthetical expression, such as $i \in I$.

index stock see INDEX BRISTOL.

India ink *Materials.* **1.** a solid black pigment made from lampblack mixed with glue or size. **2.** an intensely black, permanent, liquid ink made from this pigment.
indialite *Mineralogy.* $Mg_2Al_4Si_5O_{18}$, a colorless, transparent, hexagonal mineral, dimorphous with cordierite, occurring in prismatic crystals, and having an undetermined specific gravity and a hardness of 7 to 7.5 on the Mohs scale; found in sediments fused by burning coal.
indianaite *Mineralogy.* a white clay composed primarily of halloysite; found in Indiana, and used to increase translucency in porcelain.
Indiana limestone see SPERGENITE.
Indian balsam see PERU BALSAM.
Indian Ocean *Geography.* the ocean that is east of Africa, west of Australia, and south of Asia.
Indian paintbrush *Botany.* any of various semiparasitic plants of the genus *Castilleja* of the family Scrophulariaceae, several species of which grow widely in western North America, such as the **Wyoming paintbrush**, *C. linariaefolia*, or the **purple paintbrush**, *C. purpurea.*

Indian paintbrush

Indian red *Materials.* **1.** a yellowish-red soil found in the Persian Gulf; used as a pigment and for polishing objects made of silver or gold. **2.** a maroon pigment composed mainly of iron oxide; used in paints, rubber, and plastics.
Indian Runner *Agriculture.* a small breed of egg-producing ducks that are characterized by orange feet and shanks and an erect body, usually white or fawn in color. (From *India,* where the breed is believed to have originated.)
Indian spring low water *Oceanography.* a tidal datum that represents the approximate mean water level of all lower low waters at spring tides. (Originally used about 1880 in investigations of tidal variations in *Indian* waters.)
Indian subregion *Ecology.* a distinct zoogeographical region that includes most of the Indian subcontinent, extending from the Himalayas to the Ceylon subregion in the extreme south.
Indian summer *Meteorology.* a period of abnormally warm weather with sunny, hazy days and cool nights that occurs in autumn, especially in the northeastern and central United States, usually following a period of normally cool weather. Indian summer is not necessarily an annual occurrence, but in some years it may occur twice or even three times. (Named for the American *Indians,* perhaps because early colonists in the Americas associated it with regions inhabited by the Indians.)
Indian yellow *Materials.* a powdery yellow pigment composed of cobalt potassium nitrite.
India paper see BIBLE PAPER.
indican *Organic Chemistry.* $C_{14}H_{17}NO_6$, an indoxylglycoside obtained from indigo-bearing plants; yellow leaflets that are soluble in water; the anhydrous form decomposes at 178–180°C; the trihydrate melts at 57–58°C. *Biochemistry.* a substance present in animal urine that yields indigo and contains a potassium salt.
indicated airspeed *Aviation.* the direct instrument reading obtained from a differential-pressure airspeed indicator uncorrected for altitude, temperature, atmospheric density, or instrument error.
indicated air temperature *Meteorology.* the free air temperature value at a given moment, as read from an exposed (and usually shaded) thermometer. Also, OUTSIDE AIR TEMPERATURE.
indicated altitude *Aviation.* the altitude reading obtained from an altimeter, especially from a pressure altimeter adjusted to the aircraft's estimated height above mean sea level but uncorrected for instrument error or variation from standard atmospheric conditions.
indicated horsepower *Mechanical Engineering.* the horsepower delivered by an engine as shown on an indicator gauge, according to computations of the average pressure and displacement of the working fluid in the cylinders; this calculation does not take into account frictional losses.
indicated ore *Mining Engineering.* any ore for which tonnage and grade are computed partly from specific measurements and partly by projections from geologic evidence.
indicating gauge *Engineering.* **1.** any measuring device that indicates a value by means of the position of a pointer on a graduated scale. **2.** any device that gives visual evidence of change or variation in the dimensions or contour.
indicator something that points out or displays; specific uses include: *Aviation.* a device that monitors information without recording it, such as an airspeed indicator or radar indicator. *Computer Technology.* a device on a computer that signifies the existence of a particular condition. *Electronics.* a device, such as a cathode tube in a radar receiver, that displays information sent from another source. *Analytical Chemistry.* a compound that reversibly changes color depending on the pH of the solution or other chemical change.
indicator cell *Virology.* a cell that reveals the existence of a virus by responding in a specific way to its presence.
indicator electrode *Analytical Chemistry.* an electrode that measures changes in the concentration of the electrolytic solution; generally made with metal.
indicator element *Electronics.* the component within a system that displays the greatest variation in a given parameter.
indicator function see CHARACTERISTIC FUNCTION.
indicator gate *Electronics.* a signal that sensitizes or desensitizes an indicator in a cathode-ray tube, when applied to the tube's grid or cathode circuit.
indicator lamp *Electricity.* a small light bulb, usually incandescent, that is placed in a circuit to show an operator the state of the circuit.
indicator medium *Microbiology.* a nutrient culture medium containing a chemical indicator to reflect the physiological characteristics of a bacterial colony by showing changes in metabolic activity, as indicated by pH values.
indicator plant *Virology.* any species of plant that shows specific symptoms of the presence of a particular virus.
indicator species *Ecology.* a species found only in environments having certain set characteristics, and therefore indicating the nature of the environment in which it is found.
indicator tube *Electronics.* a device in which a voltage applied to its electrode varies the degree of luminances on its fluorescent screen; used as a tuning indicator in radio receivers. Also, MAGIC-EYE TUBE.
indicator unit *Electrical Engineering.* an apparatus designed to detect and indicate the presence, and sometimes the quantity, of electricity in an area or system.
indicator variable *Statistics.* a variable taking only the values 0 and 1, depending on the occurrence of a given condition; often used to include qualitative variables in linear models. Also, DUMMY VARIABLE.
indicatrix see INDEX ELLIPSOID.
indices [in´də sēz] the plural of index. See INDEX.
indicial notation *Mechanics.* the representation of vectors and tensors by indicating the row and column numbers using subscripts; that is, r_i is the *i*th component of the vector r and A_{ij} is the element in the *i*th row and *j*th column of the matrix representing the tensor A.
indicolite *Mineralogy.* an indigo-blue variety of tourmaline that is used as a gemstone. Also, **indigolite.**
indicophose *Neurology.* the subjective perception of the color indigo.
indifference point *Psychology.* the midpoint between opposites in experience or feeling, as between being happy and unhappy.
indifferent equilibrium see NEUTRAL STABILITY.
indigene [in´də jin] *Biology.* an organism that is indigenous to a given area.
indigenous [in dij´i nəs] *Science.* native to or occurring naturally in a particular area.
indigenous inclusion *Materials Science.* an oxide, sulfide, or silicate of one of the alloy components trapped in a metal during solidification.
indigenous limonite *Mineralogy.* a sulfide-derived limonite that remains fixed at the site of the parent sulfide.
indigestion *Medicine.* **1.** a lack or failure of digestion. **2.** a general term for poor digestion or abdominal discomfort, especially after eating; not a specific condition but possibly an indication of some underlying gastric or intestinal disorder.

indigo *Organic Chemistry.* $C_{16}H_{10}N_2O_2$, a dark-blue crystalline powder derived from plants or produced synthetically; slightly soluble in water and alcohol; sublimes at 300°C and decomposes at 390°C; used for dyes, inks, and paints. Also, **indigotin.** *Botany.* any of various plants, genus *Indigofera,* from which this substance is derived, native to and chiefly grown in India. *Materials.* a blue dye produced from this or similar plants or made synthetically. *Graphic Arts.* a deep blue color that is typical of this dye. Also, **indigo blue.** (From *India,* where it was first identified.)

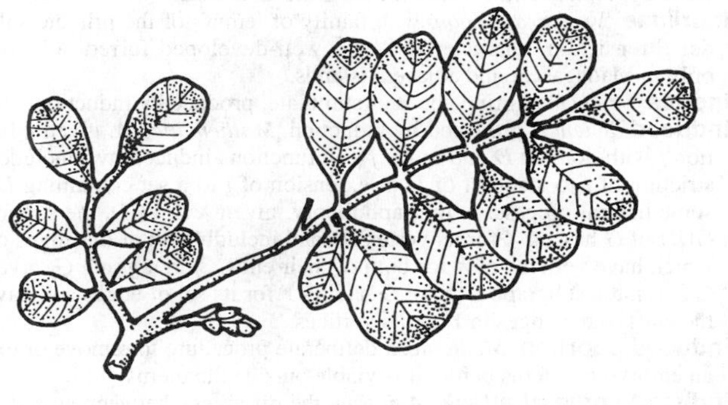

indigo

indigoid dye *Organic Chemistry.* a vat dye having indigo or thioindigo groupings; used to dye cotton and rayon.

indirect address *Computer Programming.* an address of a location that contains the storage location of the operand or another indirect address, rather than the actual operand itself. Also, SECOND-LEVEL ADDRESS.

indirect air support *Military Science.* the air support provided to land or sea forces that does not immediately assist those forces in the tactical battle.

indirect-arc furnace *Engineering.* a furnace in which the charge is heated primarily by the radiant energy coming from an arc struck between two electrodes.

indirect-band-gap semiconductor *Solid-State Physics.* a semiconductor in which the minimum in the conduction band and the maximum in the valence band occur at different locations in reciprocal space; their energy difference defines the band-gap energy.

indirect cell *Meteorology.* a closed air circulation in a vertical plane, in which the rising motion occurs at a lower potential temperature than the descending motion, with a resulting energy sink.

indirect competition *Ecology.* the use of a resource by one species that lessens its availability for use by another species.

indirect control *Computer Technology.* a relationship in which a peripheral, or off-line, unit is controlled by another unit, with the necessary help of the human operator.

indirect cycle *Nucleonics.* the portion of the nuclear generator's coolant system in which heat transferred by the coolant is transferred to a second fluid that passes the turbine where its thermal energy is converted to mechanical energy.

indirect effect *Physical Chemistry.* a term for the interaction in a solvent between solute molecules and ions formed by radiation.

indirect extrusion *Metallurgy.* extrusion performed with a die placed at the ram end of the stock. Also, BACKWARD EXTRUSION.

indirect fire *Ordnance.* fire delivered on a target that was not the point of aim for the weapons or the director.

indirect fluorescent antibody technique *Immunology.* a method used to determine the presence of either an antibody or an antibody having a developing antibody labeled with fluorochrome, which joins with an intermediate antibody or antigen, causing an increased sensitivity.

indirect heating *Engineering.* heating by convection; that is, using a primary source that transfers its heat energy to a fluid or solid, which in turn serves to heat an environment.

indirect hernia *Medicine.* a hernia that leaves the abdomen through the internal inguinal ring and passes into the inguinal canal; also called indirect inguinal hernia, external hernia, or oblique hernia.

indirect illumination *Military Science.* battlefield illumination provided by searchlights or pyrotechnic illuminants using diffusion or reflection.

indirect immunofluorescence *Immunology.* a technique in which a labeled indicator antibody reacts with an unlabeled detector antibody that has formerly reacted with an antigen.

indirect labor *Industrial Engineering.* the portion of labor input that is not directly involved in production but is required for its support, such as clerical or maintenance labor.

indirect laying *Ordnance.* the process of aiming a gun without using the target as the aiming point, either by sighting a fixed object other than the target as the aiming point or by using an aiming method other than sight, such as a gun director.

indirect lighting *Electrical Engineering.* a system of lighting in which most of the light emitted from fixtures is diffused, directed, reflected, or shielded in such a way that the majority of light reaches the eye by reflection or refraction; this results in the absence of glare and shadows.

indirectly heated cathode *Electronics.* in a thermionic vacuum tube, a cathode that is electrically insulated from the heating element. Also, EQUIPOTENTIAL CATHODE, UNIPOTENTIAL CATHODE.

indirect material *Industrial Engineering.* material put into a production process that is not used in the actual production, such as fuel, lubricant, or parts for production machinery.

indirect proof *Mathematics.* a method of proving assertions; the opposite or negation of the desired conclusion is assumed to be given and then a contradiction of some other known or given statement is deduced as a logical consequence. Also, REDUCTIO AD ABSURDUM.

indirect rays *Radiology.* the radiation formed outside the anode-ray target of a cathode-ray tube.

indirect stratification see SECONDARY STRATIFICATION.

indirect stroke *Electricity.* a stroke of lightning that induces an overvoltage in a network even though it does not strike any part of the network directly.

indirect variation see INVERSE PROPORTION.

indirect wave *Physics.* any wave that reaches a destination after experiencing a change in direction due to reflection, refraction, or diffraction.

indium *Chemistry.* a metallic element having the symbol In, the atomic number 49, an atomic weight of 114.82, a melting point of 156°C, and a boiling point of 2075°C; a soft, silver-white metal that occurs in zinc and other ores; used in automobile bearings and semiconductor devices. *Metallurgy.* this low-melting metal, extensively used for fusible alloys. (From *indigo*; its spectrum has two lines colored deep blue.)

indium antimonide *Inorganic Chemistry.* InSb, a toxic crystalline solid, melting at 535°C; used in semiconductor, infrared, and computer applications.

indium arsenide *Inorganic Chemistry.* InAs, toxic metallic crystals, insoluble in acids; melts at 943°C; used in semiconductors and lasers.

indium chloride *Inorganic Chemistry.* $InCl_3$, a toxic white deliquescent powder that is very soluble in water; melts at 586°C and sublimes at 300°C. It is derived from the action of hydrochloric acid on indium. Also, **indium trichloride.**

indium phosphide *Inorganic Chemistry.* InP, a brittle, toxic, metallic mass, very soluble in mineral acids; melts at 1070°C; used in semiconductors, lasers, and solar cells.

indium sulfate *Inorganic Chemistry.* $In_2(SO_4)_3$, a gray deliquescent powder, decomposed by heat and toxic on inhalation; soluble in cold water and very soluble in hot water.

indium telluride *Inorganic Chemistry.* In_2Te_3, black brittle crystals, melting at 667°C; used in semiconductors.

indiv. individual.

individual *Science.* existing as a distinct entity. *Biology.* a single member of a species; one organism that can exist by itself.

individual disposition see PERSONAL TRAIT.

individual distance *Behavior.* the minimum distance that an animal will tolerate between itself and another of the same species, with closer proximity resulting in aggression or withdrawal. *Anthropology.* the customary distance that members of a particular culture maintain between each other in conversation and other nonintimate social contact; this can vary from one culture to another. *Psychology.* see PERSONAL SPACE.

individual equipment *Ordnance.* clothing and equipment intended for the personal use of an individual. Similarly, **individual reserves.**

individualism *Behavior.* a strong personal attitude or action that indicates independence from group standards.

individualistic or **individualist** *Anthropology.* of or relating to a culture marked by a relatively high degree of individual freedom and commitment to personal goals, as opposed to one in which group values take precedence. Thus, **individualism.**

individual line *Telecommunications.* a line from a switching center arranged to serve a single subscriber by connecting the center to the subscriber's end telecommunication device.

individual psychology *Psychology.* a theory in which the individual is seen as the product of his own creative powers and not as victim of his past or his environment. Also, ADLERIAN PSYCHOLOGY.

individual recognition *Behavior.* the ability of an animal to identify another specific individual; usually limited to members of the same species such as parents, offspring, or territorial neighbors.

individual selection *Evolution.* a process of natural selection in which hereditary characters of individual members of a group that are selected for benefit the individuals rather than the group as a whole.

individuation *Psychology.* **1.** the process by which any of the various systems in the personality develop to the fullest capacity and become distinct from the original. **2.** a phase of development, usually between 18 and 36 months of age, during which an infant becomes increasingly independent from the mother as it gains mobility and dexterity and learns to tolerate the mother's absence.

indn. indication.

Indo-Chinese subregion *Ecology.* a distinct zoogeographical region that includes the Indochina peninsula and also much of southeastern China.

Indocin *Pharmacology.* trade name for indomethacin.

Indo-European *Linguistics.* **1.** a language that developed about 5000 years ago among a migratory people north of the Black Sea. **2.** a large language family descended from this language, including Greek, Latin, Romance languages such as French and Spanish, Germanic languages such as German, English, and Danish, Slavic languages such as Russian, Polish, and Czech, Indo-Iranian languages such as Persian and Hindi, and Celtic languages such as Irish and Scots Gaelic.

indogen *Organic Chemistry.* the radical $C_6H_4(NH)COC=$ from indigo.

indogenide *Organic Chemistry.* any compound that contains the indogen radical.

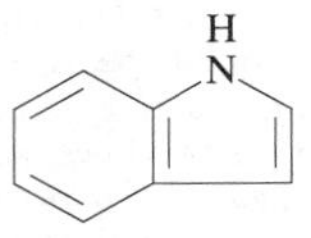

indole

indole *Organic Chemistry.* C_8H_7N, a colorless heterocyclic compound occurring in the form of white to yellowish scales that turn red on exposure to air; soluble in water, alcohol, and ether. It melts at 52°C and boils at 253–254°C. *Biochemistry.* the natural form of this substance, a carcinogen that is present in several plants including indigo; it is also obtained from coal tar and produced by the breakdown of the protein tryptophan in the intestine.

indoleacetic acid *Biochemistry.* $C_{10}H_9O_2N$, a plant hormone that stimulates cell growth, particularly in the plant tips facing away from the light.

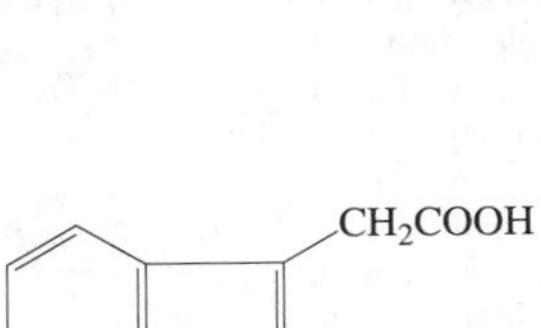

indoleacetic acid

indoleacetonitrile *Biochemistry.* $C_{10}H_8N_2$, a naturally occurring auxin, a plant growth hormone.

indolebutyric acid *Organic Chemistry.* $C_{12}H_{13}NO_2$, a white powder that is soluble in alcohol, insoluble in water, and melts at 123°C; used as a plant hormone. Also, HORMODIN.

indolent *Medicine.* **1.** painless. **2.** slow growing, inactive, or sluggish.

indole test *Microbiology.* an IMViC test used to determine whether or not a microorganism grown in peptone water or tryptone water can produce indole from tryptophan.

Indo-Malayan subregion *Ecology.* a distinct zoogeographical region that includes the Malay Peninsula and various nearby islands of the South China Sea.

indomethacin *Pharmacology.* $C_{19}H_{16}ClNO_4$, a nonsteroidal anti-inflammatory agent that is used as medication for arthritis.

Indo-Pacific faunal region *Ecology.* a geographical area that extends eastward from the east coast of Africa through the area north of Australia and south of Japan to the eastern Pacific south of Alaska.

indoxyl *Organic Chemistry.* $(C_8H_7N)OH$, yellow crystals that melt at 85°C; an oxidation product of indole that is formed by decomposition from tryptophan and excreted in the urine as indican.

Indriidae *Vertebrate Zoology.* a family of lemurs of the primate suborder Strepsirhini, characterized by a well-developed furred tail; found only in Madagascar and adjacent islands.

induce *Science.* to bring about or stimulate; produce by induction.

induced *Science.* produced by induction. *Mathematics.* **1.** given a function f with domain D and range $f(D)$, functions induced by f include restriction of f to a subset of D or extension of f to a set containing D by some natural process. **2.** in graph theory, given $X \subseteq V(G)$, the subgraph $G[X]$ of G having X for its vertex set and including all of the edges of G which have vertices of X for both of their ends. **3.** in a graph G, given $Y \subseteq E(G)$, the subgraph $G(Y)$ of G having Y for its set of edges and having the ends of the edges in Y for its vertices.

induced abortion *Medicine.* a deliberate procedure to remove or expel an embryo or a fetus before it is viable outside the uterus.

induced angle of attack *Aviation.* the difference between an airfoil's actual angle of attack and its effective angle of attack; that is, the angle between the path of the airfoil and that of the airstream passing over it.

induced anisotropy *Solid-State Physics.* the anisotropic magnetism achieved by annealing the magnetic material in a magnetic field.

induced attraction see INDUCTION FORCE.

induced charge see INDUCTIVE CHARGE.

induced current *Electromagnetism.* any current that is produced by a time-varying magnetic field.

induced dipole moment *Electricity.* the average dipole moment that is induced in an atom or molecule when a dielectric substance is brought into an electromagnetic field.

induced draft *Mechanical Engineering.* a mechanical draft produced by suction, stream jets, or fans at the point where air or gases leave a furnace or heater.

induced drag *Fluid Mechanics.* an aerodynamic resistive force that is produced as a consequence of the generation of lift; its magnitude is directly proportional to the lift.

induced electromotive force *Electromagnetism.* an electromotive force produced by a conductor moving through a magnetic field or by changing the magnetic flux that passes through the conductor's conductive loop.

induced enzyme *Enzymology.* an enzyme whose production has been or may be stimulated by another compound, often a substrate or a structurally related compound called an inducer. Also, ADAPTIVE ENZYME, INDUCIBLE ENZYME.

induced fission *Nuclear Physics.* fission incurred by the bombardment of the target nucleus by neutrons or high-energy radiation (gamma rays).

induced fit *Enzymology.* a theory used to explain enzyme activity, proposing that the substrate first induces a conformational change in the enzyme such that the catalytic and binding groups of the enzyme achieve the required active site orientations, which then permits the substrate to attach itself to the enzyme.

induced force see INDUCTION FORCE.

induced labor *Medicine.* labor that is brought on by mechanical or chemical means, as by artificially rupturing the fetal membranes or administering certain drugs that stimulate uterine contraction.

induced magnetization *Geophysics.* the magnetic field in a rock that has the same direction as, and strength proportional to, the ambient geomagnetic field.

induced mutation *Genetics.* a mutagen-initiated modification of a gene or nucleotide sequence.

induced radioactivity *Nucleonics.* a type of radioactivity that arises when a stable substance absorbs ionizing radiation. Also, ARTIFICIAL RADIOACTIVITY.

induced voltage *Electromagnetism.* any voltage that is the result of electromagnetic induction.

inducer *Enzymology.* the activating agent of an inducible system. Also, **inducer substance.**

inducer exclusion *Microbiology.* a phenomenon in microorganisms whereby a preferred metabolic substrate is able to inhibit the cellular uptake of other substrates that are less readily metabolized.

inducible enzyme see INDUCED ENZYME.

inducible promotor *Genetics.* a promotor that may be induced by some *cis-* or *trans-*acting factor.

inducible system *Genetics.* a self-regulating mechanism for allowing the transcription of a structural gene, even in the presence of a repressor substance that normally blocks transcription. An inducer substance combines with and inactivates the repressor, thereby allowing transcription to proceed.

inductance *Electromagnetism.* **1.** the proportionality constant (denoted by *L*) of a coil or other inductive component, equal to the ratio of a generated electromotive force to the rate of change of the current. **2.** the property associated with a circuit (self-inductance) or a neighboring circuit (mutual inductance) whereby an electromotive force is generated in the presence of a changing current.

inductance bridge *Electromagnetism.* a device whose design is similar to that of a Wheatstone bridge and which is used to measure the inductance of an unknown coil.

inductance standards *Electromagnetism.* a pair of identical coils wound on toroidal cores and having a highly stable and accurate inductance.

induction a process of causing or bringing about; specific uses include: *Science.* a process of reasoning in which a general conclusion is reached from specific data, especially when the conclusion does not necessarily or directly follow from this data. *Surgery.* the period from the start of anesthesia to the establishment of a depth of anesthesia that is adequate for operation. *Developmental Biology.* the influence of an organizer or evocator on the differentiation of nearby cells or on the formation of an embryo. *Virology.* the spontaneous or stimulated activation of a latent virus. *Microbiology.* see DEREPRESSION. *Molecular Biology.* see INDUCTION OF A LYSOGENIC PHAGE. *Genetics.* a change imposed upon offspring as a result of the environment's effect on the germ cells of one or both parents. *Psychology.* see MOOD INDUCTION. *Electromagnetism.* the generation of voltages, currents, electric fields, or magnetic fields by interactions among these quantities without direct contact. *Mathematics.* see MATHEMATICAL INDUCTION.

induction accelerator *Nucleonics.* a device, such as a betatron, that increases the energy of electrons by propelling them through a ring using a rapidly fluctuating magnetic field.

induction attraction see INDUCTION FORCE.

induction brazing *Metallurgy.* a process of brazing in which the heat is generated by an induced electromagnetic field.

induction burner *Engineering.* a fuel-air burner that relies on fuel supplied under pressure, which entrains the air necessary for combustion.

induction charging *Electricity.* a process in which an electrostatic charge accumulates in an inductor and is stored as a magnetic field.

induction coil *Electromagnetism.* a high-voltage step-up transformer that can change direct current into alternating current by periodic interruption of direct current through the primary winding.

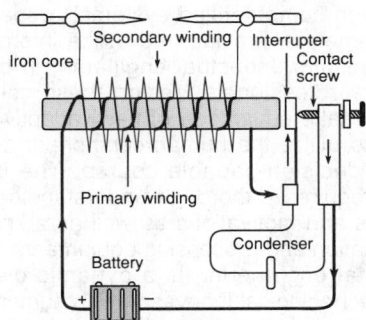

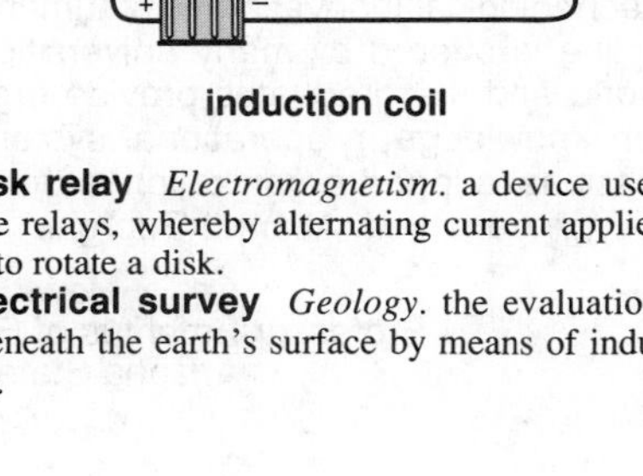

induction coil

induction disk relay *Electromagnetism.* a device used in regulating and protective relays, whereby alternating current applied to a coil produces torque to rotate a disk.

induction-electrical survey *Geology.* the evaluation of geological formations beneath the earth's surface by means of induction and electrical logging.

induction field *Electromagnetism.* the component of an electromagnetic field associated with the alternating-current component; the energy seems to be alternately carried away from and returned to the circuit with no net loss.

induction force *Physical Chemistry.* a relatively weak force of attraction that arises from the momentary interaction between a polar atom or molecule and a polarizable one that is adjacent to it. Also, INDUCED ATTRACTION, INDUCED FORCE, INDUCTION ATTRACTION, DEBYE FORCE.

induction furnace *Metallurgy.* an electric furnace in which metals generate a secondary charge as they liquefy; used for induction melting.

induction generator *Electricity.* an induction device driven at higher-than-synchronous speed by a separate external mechanical power, used to convert mechanical power to electrical power.

induction heating *Engineering.* heating by electrical induction, particularly the process of heating electrically conductive materials, such as steel, by inducing high-frequency currents within the material.

induction inclinometer see EARTH INDUCTOR.

induction loudspeaker *Acoustical Engineering.* a moving-coil loudspeaker that uses an air-suspended induction coil to move its diaphragm, thereby producing sound.

induction machine *Electricity.* an asynchronous alternating-current machine, such as an induction generator or an induction motor, composed of a magnetic circuit connected to two electric circuits; power is transferred between the two circuits by electromagnetic induction.

induction melting *Metallurgy.* a process of melting in which the heat is generated by an induced electromagnetic field. Similarly, **induction hardening.**

induction method *Geophysics.* a technique used to estimate the radioactivity in the atmosphere by means of the electrical change in a negatively charged wire exposed to the air.

induction microphone *Acoustical Engineering.* a moving-coil microphone that uses an induction coil to produce an electroacoustical signal.

induction motor *Electricity.* an alternating-current motor in which the currents in the secondary winding are created by induction; often used because of its simple construction, efficiency, and speed regulation.

induction noise *Acoustics.* sound interference in an electrical sound reproduction system, caused by electromagnetic waves, due to close-proximity electrical current flow.

induction of a lysogenic phage *Molecular Biology.* a phenomenon in which bacteriophage DNA integrated into a host cell genome leaves the chromosome and initiates a lytic cycle, usually because the host DNA has been damaged in some way.

induction period *Physical Chemistry.* the time required for a chemical reaction to advance from zero to its maximum observable rate; usually expressed as an ideal figure that is not affected by the physical limitations of the reactor system.

induction problem *Electromagnetism.* an effect of potentials and current induced in conductors of a telephone system by paralleling power lines.

induction regulator *Electricity.* a transformer with a primary winding in shunt and a secondary winding in series with a circuit; it gradually adjusts the voltage or phase relation of the circuit by changing the relative magnetic coupling of the primary and secondary windings.

induction rule shell *Artificial Intelligence.* a software package used to build expert systems that induces rules from examples of similar problems for which the outcomes are known.

induction salinometer *Engineering.* an instrument that detects the voltage of currents in seawater and is able to indicate the salinity of the water.

induction stroke see SUCTION STROKE.

induction valve *Mechanical Engineering.* **1.** a valve through which a charge is induced into the cylinder of an internal-combustion engine. **2.** see INLET VALVE.

induction variable *Computer Science.* a variable that is incremented during a loop and used to perform a similar action on multiple data. Also, LOOP INDEX, LOOP VARIABLE.

induction voltage regulator *Electromagnetism.* a transformer in which the primary winding is connected in parallel with the circuit and the secondary winding in series; the relative positions of the coils can affect the phase and the voltage in the circuit.

induction welding *Metallurgy.* welding in which the heat is generated by an induced electromagnetic field. Similarly, **induction soldering.**

inductive bias *Artificial Intelligence.* in a machine learning system, the effect of the data that are examined and the representation of learned information on the result of the learning process.

inductive charge *Electricity.* the charge that is contained on an object as a direct result of being near another charged object. Also, INDUCED CHARGE.

inductive circuit *Electricity.* a circuit that combines inductive reactance and resistance.

inductive coordination *Electromagnetism.* a combination of design, location, operation, and other parameters of electrical systems intended to minimize the amount of inductive interference.

inductive divider *Electromagnetism.* a transformer having a single input and multiple outputs, used for incorporating a desired fraction of an inductance into a circuit.

inductive interference *Telecommunications.* a condition in which the characteristics of electric supply and communications systems prevent circuits from providing satisfactory service.

inductive load *Electricity.* a load that can be represented as an inductor or a combination of an inductor and a resistor. Also, LAGGING LOAD.

inductive neutralization *Electronics.* a process by which capacitance cancels inductance in an amplifier's feedback circuit. Also, COIL NEUTRALIZATION, SHUNT NEUTRALIZATION.

inductive post *Electromagnetism.* a conductive post that is used in a waveguide oriented parallel to the electric field so as to provide additional inductive susceptance in parallel with the waveguide.

inductive reactance *Electricity.* the magnitude of the impedance of an inductor, which varies proportionately with frequency and is measured in ohms.

inductive relay *Electromagnetism.* a relay that operates on the principle of magnetic induction.

inductive spacing *Electromagnetism.* spacing of parallel transmission lines so that there is transfer of energy by mutual inductance.

inductive surge *Electromagnetism.* a surge of voltage developed across the ends of a coil due to the sudden change in current by opening or closing a switch in a circuit.

inductive tuning *Electronics.* a method for adjusting the frequency on a radio that involves moving a core in and out of a coil to change its inductance.

inductive window *Electromagnetism.* a metallic diaphragm that is inserted into a waveguide in order to introduce inductive susceptance in parallel with the waveguide.

inductometer *Electromagnetism.* a variable inductor consisting of two or more coils of known inductance connected in series; used to determine an unknown inductance in an L-C circuit by measuring the resonant frequency.

inductor *Electricity.* a coil of wire that is often wound on magnetic material.

inductor alternator *Electricity.* an early generator that produced large amounts of power at high frequencies using the principle of variable reluctance.

inductor generator *Electricity.* a generator with the field coils fixed in magnetic position relative to the armature conductors, producing electromotive force by movement of masses of magnetic material.

inductosyn *Control Systems.* a type of resolver whose output phase is proportional to the shaft angle.

inductotherm *Medicine.* an apparatus used in inductothermy.

inductothermy *Medicine.* the artificial production of fever by electric induction.

inductura *Invertebrate Zoology.* a layer of laminated shell material lining the inner lip of the aperture in gastropods.

indumentum *Biology.* any hairy covering, as on a plant or animal. *Vertebrate Zoology.* the plumage of a bird.

induration a process of hardening or state of being hard; specific uses include: *Geology.* **1.** a process in which rock material becomes hardened by heat, pressure, or cementation. **2.** the hardening of a soil horizon by chemical action, as in the formation of a hardpan or caliche. **3.** the hardened mass of rock or soil formed by this process. *Medicine.* **1.** an abnormally hard spot or place on or in the body. **2.** the quality of being hard, or the process of hardening.

Indus *Geography.* a river that rises in Tibet and flows miles through Pakistan to the Arabian Sea. *Astronomy.* the Indian, a dim constellation of the south celestial hemisphere whose brightest stars are third magnitude.

indusium *Anatomy.* **1.** any membranous covering. **2.** see AMNION. *Mycology.* in certain fungi, an outgrowth or veil located on the stalk underneath the mushroom cap or pileus.

industrial alcohol *Organic Chemistry.* any mixture of ethanol and additives for denaturing or other specific solvent purposes.

industrial climatology *Meteorology.* a field of climatology that involves the study of climate and weather on the operations of an industry, providing climatological data on which to base administrative and operational decisions that involve weather factors.

industrial diamond *Materials.* a diamond that is unsuitable as a gemstone; used as an abrasive or in cutting tools.

industrial dynamics *Industrial Engineering.* the use of systems analysis for industrial problem-solving.

industrial engineering *Engineering.* the branch of engineering that manages and improves the economical use of people and equipment through application of cost and work standards and enhancement of the working environment.

Industrial Engineering

Pioneered by early 20th century engineers such as Frederick Taylor and Frank and Lillian Gilbreth, and with the subsequent contributions of many outstanding engineers, researchers, and practitioners during the 1900s, industrial engineering has assumed an important and well recognized role within the engineering profession. That role is to design and implement technical manufacturing and business systems which will foster and enhance the economical production of quality goods and services by business organizations. The industrial engineer draws on specialized knowledge in the mathematical, physical, and social sciences, and the principles of engineering analysis and synthesis, to create work environments which support high levels of organizational effectiveness, individual and team performance, and human satisfaction.

As a systems design specialist, the engineer with industrial engineering education seeks to integrate both the technical and non-technical components of a business organization into a synergistic and meshed whole. He or she works in close cooperation with corporate management to uncover inefficiencies in operations and to create viable and cost-effective solutions. The industrial engineer employs an extensive set of design tools to accomplish the systems integration process, including the use of engineering design models, mathematical models, computer models, statistical and probabilistic models, organizational and managerial models, and human factors models.

One very important area of current concern relates to computer-integrated design and manufacturing, in which employees at all levels of the organization interface through computerized systems to specify operational parameters, provide and utilize information, and measure results. Industrial engineers, in cooperation with other professionals, design these computerized systems to take advantage of new, adaptive technologies and to utilize the human component as an intelligent and decision-capable source. The information interface requires a thorough understanding of human capabilities and motivations as well as a knowledge of communication and processing equipments.

Industrial engineering is a dynamic discipline for merging technological innovation with human concerns. The discipline is offered by many universities throughout the world and the graduates provide organizations with expert knowledge in operational systems design. The professional support organization is the Institute of Industrial Engineers, Inc., Atlanta, Georgia.

Hewitt H. Young
Professor Emeritus of Engineering
Arizona State University

industrial-frequency band *Telecommunications.* a radio-frequency band that is reserved for mobile communication by private industries other than the transportation industry.

industrial geography *Geography.* a branch of economic geography dealing with industrial activity, such as the influence of geography on plant location.

industrial humanism *Industrial Engineering.* a philosophy of management that recognizes the correlation between worker productivity and worker motivation.

industrial hygiene *Medicine.* preventative medicine that deals with the protection of the health of those involved in industry.

industrial jewel *Mineralogy.* a hard stone, such as ruby or sapphire, that is used for bearings and impulse pins in instruments and for recording needles.

industrial melanism *Ecology.* an increase in the amount of the pigment melanin found in some species inhabiting highly industrialized areas, due to selection against lighter-colored individuals in an environment darkened by pollution.

industrial meteorology *Meteorology.* a field of meteorology that involves the application of meteorological information and techniques to industrial problems.

industrial microbiology *Microbiology.* the study and large-scale production of microorganisms as related to the preparation of economically valuable and useful products, such as fermentation and pharmaceutical products, solvents, enzymes, and insecticides.

industrial microorganism *Biotechnology.* a microorganism, such as a mold, yeast, or bacterium, that is employed in industry to convert raw material into a new, commercially useful product.

industrial mineral filler *Materials.* a finely powdered mineral that is added to a material to improve certain properties.

industrial photogrammetry *Photogrammetry.* the techniques of close-range photogrammetry applied to research, production, testing, monitoring, and repair in a number of different industrial areas, such as building construction, civil engineering, mining, manufacturing, and shipbuilding.

industrial psychology *Psychology.* a branch of psychology that studies work and work environments in such areas as personnel selection, job evaluation, training, and career counseling. Thus, **industrial psychologist.**

industrial revolution *Industrial Engineering.* the vast social and economic changes that resulted from the development of steam-powered machinery and mass-production methods, beginning in the late eighteenth century in Great Britain and extending through the nineteenth century elsewhere in the world.

industrial robot *Robotics.* a programmable robot in wide use in industrial applications, having a freedom of movement similar to that of a human's waist, shoulder, elbow, wrist, and fingers.

industrial-strength *Materials.* of a commercial product, highly potent, powerful, or durable compared with other products of the same type.

industrial television *Telecommunications.* closed-circuit television used for monitoring and viewing industrial operations.

industrial waste *Engineering.* any materials that are produced by or remaining from industrial operations and then rejected.

industrial yeast *Bacteriology.* any of various yeasts, especially *Saccharomyces,* used for the fermentation process by brewers, distillers, winemakers, bakers, and chemical manufacturers.

industry *Industrial Engineering.* a particular business enterprise, or business activity in general. *Archaeology.* **1.** a large grouping of artifacts that is considered to represent or identify a particular people or culture: the Acheulian *industry.* **2.** a category of artifacts grouped by material or function: a ceramic *industry,* a cutting-tool *industry.*

industry stage *Psychology.* the fourth of Erik Erikson's eight stages of human development, during the elementary school years, when children develop competence in intellectual, social, and physical skills.

industry vs inferiority *Psychology.* the conflict that can arise during the industry stage of development, when children may doubt their competence if their efforts are unsuccessful or unrewarded.

indwelling catheter *Medicine.* a tube held in position in the urethra to drain urine from the bladder.

inedible *Nutrition.* not suitable for human consumption; poisonous or unpleasant tasting.

ineffective time *Computer Technology.* the time during which a system is not processing due to operational delays or idle time.

inelastic *Mechanics.* not having the property of elasticity; not able to return to its original shape after experiencing strain.

inelastic buckling *Mechanics.* buckling in a structural member beyond its elastic limit.

inelastic collision *Mechanics.* a collision between bodies in which some kinetic energy is lost by conversion into internal energy for one or both bodies.

inelastic cross section *Physics.* an area which measures the probability that an inelastic collision will occur between two particles.

inelasticity *Mechanics.* the property of a material, such that it is not able to return completely to its original shape after experiencing strain.

inelastic scattering *Physics.* a scattering of particles in which kinetic energy is not conserved.

inelastic stress *Mechanics.* a stress beyond the elastic range of a material.

inequality the condition of being unequal; a lack of equality; specific uses include: *Mathematics.* a mathematical statement of the form $A < B$, $A \leq B$, $A > B$, or $A \geq B$, where A and B are numerical quantities, algebraic expressions, functions, cardinal or ordinal numbers, etc. This notation is read "A is less than, less than or equal to, greater than, or greater than or equal to B," respectively. *Astronomy.* any irregularity in an object's orbital motion.

inequality of Clausius see CLAUSIUS INEQUALITY.

inequality relationships see HARKER-KASPER INEQUALITIES.

inequilateral *Biology.* asymmetrical, unequal-sided.

inermous *Biology.* unarmed, without spines or prickles.

inert *Science.* slow or unable to move or react. *Chemistry.* not reacting with other elements. *Pharmacology.* having no pharmacological action. *Ordnance.* of or relating to a munition that contains no explosive, incendiary, or chemical charge. Thus, **inert ammunition, inert bomb, inert grenade, inert mine.**

inert atmosphere *Chemical Engineering.* an atmosphere of nonreactive gas used to blanket stored reactive liquids or partly filled containers of reactive substances, and for purging process lines and vessels of reactive liquids and gases. Examples are nitrogen, carbon dioxide, helium, and argon.

inert filling *Ordnance.* a nonexplosive filling of the same weight as the appropriate explosive filling.

inert gas see NOBLE GAS.

inertia [i nur´shə] *Mechanics.* the property of a material body, due to its mass, by which it resists any change in its motion unless it is overcome by force. Thus if no outside forces are present, a stationary body will remain at rest and a moving body will continue to move in the same direction at the same speed. *Mechanical Engineering.* of a device, activated or operated by an abrupt change in velocity. Thus, **inertia starter, inertia switch.**

inertia currents *Oceanography.* **1.** currents occurring after the wind has stopped blowing over a fetch or after the water movement has left the fetch. **2.** circular currents that have a period of half a pendulum day (as recorded by a Foucault pendulum).

inertia ellipsoid *Mechanics.* an imaginary surface, stationary with respect to a rotating body, that is determined by the body's moments of inertia and that shares the same principal axes. Also, POINSOT ELLIPSOID; MOMENTAL ELLIPSOID.

inertia governor *Mechanical Engineering.* a governor that utilizes an eccentrically pivoted arm, which responds to and suppresses speed fluctuations by reason of its inertia.

inertial *Mechanics.* relating to or caused by inertia.

inertial circle *Meteorology.* a nearly circular loop in the path of an air parcel in inertial flow, under conditions of minor latitudinal displacement. Also, CIRCLE OF INERTIA. *Oceanography.* a circle described by inertial motion in a body of water, the radius being determined by the formula $R{=}C/f$, where C represents the particle velocity in a given direction and f the Coriolis parameter.

inertial confinement *Nucleonics.* the temporary confinement of a plasma by inertia that resists the thermally induced expansion process; a high-temperature core where fusions take place is maintained.

inertial flow *Fluid Mechanics.* fluid flow that is not subjected to any external forces. *Geophysics.* the flow that occurs in the absence of a pressure gradient.

inertial force *Mechanics.* **1.** see APPARENT FORCE. **2.** see FICTITIOUS FORCE.

inertial frame of reference *Mechanics.* a frame of reference that will allow Newton's laws to be valid in describing the motion of a system; the frame must either be at rest or be translating with a constant velocity. Also, NEWTONIAN OR GALILEAN FRAME OF REFERENCE, INERTIAL REFERENCE SYSTEM.

inertial guidance *Navigation.* a guidance system in which the path of a missile, aircraft, or other vehicle is adjusted in flight by internal devices independent of external information; the system measures and converts accelerations experienced into distance traveled in a certain direction.

inertial instability *Fluid Mechanics.* an instability existing between steady and turbulent flow whereby only kinetic energy is received by the disturbed region at the expense of the kinetic energy of the steady-state flow.

inertial mass *Mechanics.* the property of a body of matter that makes it resistant to a change in its motion (or lack of motion); empirically, it is equal to gravitational mass, but the fundamental reasons for this are not clear.

inertial navigation system *Navigation.* a complex of gyroscopes, accelerometers, computers, and other devices, used to measure changes in velocity and direction. These measurements, applied to a known starting point, yield a dead-reckoning position without the use of outside reference points.

inertial orbit *Astronomy.* an orbit computed on the assumption that the only forces operating are those described by Newtonian mechanics.

inertial platform *Navigation.* the device to which inertial navigation sensors are attached. It is designed to maintain the sensors in a fixed relationship to the earth.

inertial reference system see INERTIAL FRAME OF REFERENCE.

inertial space *Navigation.* referring to directions relative to the fixed stars.

inertial theory *Oceanography.* the theory concerning the effects of inertia and the Coriolis force on ocean currents.

inertia matrix *Mechanics.* an $n \times n$ matrix I corresponding to a system having n generalized coordinates; the value I_{ij} gives the independent contribution to the kinetic energy of the product of generalized velocities i and j.

inertia of energy *Physics.* the concept that any system which possesses energy will also possess inertia because of the mass-energy equivalence ($E = mc^2$).

inertia period *Oceanography.* the time required for a particle of water to complete an inertial circle.

inertia switch *Electricity.* a switch that is activated by an abrupt change in the velocity of the object to which it is mounted, thus sensing changing inertia.

inertia tensor *Mechanics.* a tensor associated with a body in a given coordinate system; the body's angular momentum is given by the product of the tensor and the angular velocity.

inertia wave *Fluid Mechanics.* a wave in a fluid wherein the only form of energy present is kinetic energy; for example, a barotropic wave or Helmholtz wave.

inertinite *Geology.* a group of high-carbon coal macerals, including micrinite, sclerotinite, fusinite, and semifusinite, that are relatively inert during carbonization.

inert primer *Engineering.* a small tube or other casing containing a detonator that does not affect the explosive charge.

inesite *Mineralogy.* $Ca_2Mn_7^{+2}Si_{10}O_{28}(OH)_2 \cdot 5H_2O$, a rose-red to flesh-red, translucent, triclinic mineral occurring in small prismatic crystals and acicular aggregates, and having a specific gravity of 3.03 and a hardness of 5.5 to 6 on the Mohs scale.

inevitable abortion *Medicine.* a state in which the cervix is dilated and vaginal bleeding is heavy and prolonged, ultimately resulting in a natural abortion.

inexact reasoning *Artificial Intelligence.* the process used by an expert system to make a decision based on partial or incomplete information. Also, APPROXIMATE REASONING.

inextensional deformation *Mechanical Engineering.* the bending of a surface in such a way that the length of any line and the curvature of any surface is unchanged at any point.

in extremis *Medicine.* at the point of death.

inf or **inf.** infantry; inferior; infinity; infuse.

inf. below; after. (From Latin *infra.*)

inface see SCARP SLOPE.

infancy *Physiology.* the early period of human life, usually measured from the end of the newborn period (the first four weeks of life) to the time of assumption of erect posture (12 to 14 months), but sometimes extended to the age of 24 months. *Geology.* the early erosional stage during which a region is first exposed to the action of surface waters, resulting in nearly level surfaces dissected by narrow stream gorges, many water-filled depressions, and an imperfect drainage system.

infant *Physiology.* a young child, especially one in the period of infancy.

infanticide *Behavior.* the killing of a young animal by its parent or by another of its own species.

infantile [in′fən tīl′] *Psychology.* **1.** occurring in or characteristic of infancy. **2.** occurring in an older individual, but more appropriate for an infant; very childish. *Medicine.* of a disease or condition, affecting infants or likely to occur in infancy.

infantile acne see NEONATAL ACNE.

infantile amnesia *Psychology.* the normal inability of an individual to remember events that happened to him or her before the approximate age of three.

infantile autism see AUTISM.

infantile behavior *Behavior.* the performing of immature behavior patterns by adult individuals; it may be a response to aggression or an element of courtship behavior.

infantile celiac disease *Medicine.* an inborn error of metabolism of infants and young children characterized by intolerance to gluten (a protein present in the grains of wheat, rye, oats, and barley), abdominal distention, the poor absorption of food, foul-smelling stools, malnutrition, failure to thrive, and growth retardation.

infantile cortical hyperstosis *Medicine.* a swelling of the soft parts of the lower jaw occurring in early infancy and accompanied by pain, irritability, and fever.

infantile diarrhea *Medicine.* an intestinal disease of infants characterized by frequent discharge of watery stools.

infantile eczema *Medicine.* a noncontagious, inflammatory skin disease occurring in infants that may be acute or chronic, characterized by varying combinations of redness, formation of papules, vesicles, pustules, scales, and crusts, and accompanied by itching and burning.

infantile genitalia *Medicine.* an abnormal condition involving the lack of development and maturation of the reproductive organs.

infantile massive spasms see JACKKNIFE SEIZURE.

infantile neuroaxonal dystrophy *Medicine.* a degenerative disease of nerve cells that occurs in infants.

infantile paralysis see POLIOMYELITIS.

infantile scurvy *Medicine.* a nutritional deficiency disease of infants resulting from poor or inadequate intake of vitamins and marked by symptoms similar to those of scurvy in adults.

infantile sexuality *Psychology.* the capacity of an infant or young child to experience and enjoy activities that are essentially sexual.

infantile spasm *Medicine.* a kind of seizure occurring in infants and young children in which there is a sudden, brief, massive contraction of body muscles.

infantilism *Medicine.* an abnormal condition in which the characteristics of childhood persist into adult life, marked by mental retardation, underdevelopment of sexual organs, and sometimes dwarfism. *Psychology.* a condition characterized by speech and voice patterns in an adult or older child that are typical of a much younger person.

infant respiratory distress syndrome *Medicine.* a disease of lung immaturity marked by difficulty in breathing, bluish discoloration of the skin, lips, and nails due to insufficient oxygen in the blood, and easily collapsible alveoli, affecting premature infants, those born of diabetic mothers, or those delivered by cesarean section.

infantry *Ordnance.* **1.** those forces that fight mainly on foot with hand-held weapons; foot soldiers. **2.** major army units equipped with rifles, machine guns, light mortars, and antiaircraft and antitank weapons. (Related to *infant*; probably because early foot soldiers were very young.)

infantry fighting vehicle *Ordnance.* an armored vehicle in which infantry can be transported and in which they can fight.

infantry gun *Ordnance.* a small artillery piece operated by infantry rather than artillery units.

infarct *Medicine.* a localized area of tissue that has suffered irreversible damage or necrosis due to obstruction in the artery supplying the tissue.

infarction *Medicine.* the development or formation of an infarct.

infauna *Zoology.* those animals living in the sediments at the bottom of bodies of water.

infect *Medicine.* to contaminate with a pathogenic organism, or to invade a body or part of the body, causing disease.

infection *Medicine.* **1.** the invasion and multiplication of pathogenic microorganisms in the body, with or without the manifestation of disease. **2.** any specific infectious disease.

infection court *Plant Pathology.* the initial area at which a pathogen contacts the surface of a plant host.

infectious *Medicine.* caused by, capable of producing, or transmitted by infection. Also, INFECTIVE.

infectious anemia of horses *Veterinary Medicine*. an infectious viral disease of equines transmitted by biting insects or contaminated surgical instruments, and characterized by fever, jaundice, anemia, and rapid weight loss.

infectious arthritis *Medicine*. an acute or chronic inflammatory disease of a joint caused by pathogenic microorganisms.

infectious bovine keratoconjunctivitis see INFECTIOUS KERATOCONJUNCTIVITIS.

infectious bronchitis *Veterinary Medicine*. an acute, highly contagious respiratory disease of chickens and other domestic birds, caused by a virus and manifested by coughing, sneezing, nasal discharge, reduced egg yield, and the production of rough or soft-shelled eggs.

infectious bursal disease *Veterinary Medicine*. a highly contagious viral disease of chickens from three to six weeks old; characterized by whitish diarrhea, dehydration, prostration, edema, and swelling of the lymphoid tissue of the cloacal bursa. Also, IBD.

infectious canine hepatitis *Veterinary Medicine*. an acute, contagious, and potentially fatal liver disease of dogs and foxes that can be clinical or subclinical; symptoms may include fever, diarrhea, anemia, leukopenia, jaundice, hepatitis, and nervousness.

infectious chlorosis *Plant Pathology*. a virus disease of plants that affects the development of chlorophyll and is characterized by a yellowing of green plant tissues.

infectious conjunctivitis *Medicine*. an infectious inflammation of the mucus membrane covering the anterior portion of the eyeball and the lining of the inner surface of the eyelids.

infectious disease *Medicine*. an abnormal condition in which there is impaired functioning of the body or any of its parts due to the presence of pathogenic microorganisms.

infectious drug resistance *Microbiology*. a condition that occurs when an antibiotic is no longer effective in repressing the growth of a microorganism, due to the presence of resistance-transfer factors transmitted to other cells during conjugation.

infectious endocarditis *Cardiology*. an inflammatory heart disease caused by the infection by pathogenic microorganisms of the lining membrane of the cardiac chambers and its valves.

infectious hematopoietic necrosis *Veterinary Medicine*. an acute viral infection of salmonid fish that is transmitted in water of 54°F (12°C) or lower; characterized by darkening of the body, pale gills, and exopthalmia.

infectious hepatitis *Medicine*. another name for hepatitis A. See HEPATITIS A.

infectious keratoconjunctivitis *Veterinary Medicine*. a term used for a group of bacterial eye diseases occurring in cattle, goats, and sheep, involving both conjunctivitis and keratitis; probably caused by *Moraxella bovis* (in cattle), *Mycoplasma agalactiae* (in goats), and *Weisseria ovis* (in sheep). Also, INFECTIOUS OPHTHALMIA, INFECTIOUS BOVINE KERATOCONJUNCTIVITIS.

infectious laryngotracheitis *Veterinary Medicine*. a highly infectious viral disease of chickens and pheasants, caused by a herpesvirus and characterized by loss of appetite, respiratory distress, tracheal exudate containing blood, and conjunctivitis.

infectious mononucleosis *Medicine*. an acute infectious herpesvirus disease caused by the Epstein-Barr virus, characterized by fever, sore throat, enlargement of the spleen and lymph nodes, and an abnormal increase in the number of monocytes in the blood.

infectious myocarditis *Cardiology*. an inflammation of the muscular tissue of the heart as a result of infection by pathogenic microorganisms.

infectious myxomatosis *Veterinary Medicine*. a highly contagious and fatal viral disease of rabbits, characterized by conjunctivitis, swelling of the head, and subcutaneous tumors. The causal myxoma virus or poxvirus is transmitted by mosquitos and fleas. Also, MYXOMATOSIS.

infectious necrotic hepatitis see BLACK DISEASE.

infectious nucleic acid *Virology*. a nucleic acid that can replicate and produce progeny virions if transmitted into a susceptible host cell.

infectious ophthalmia see INFECTIOUS KERATOCONJUNCTIVITIS.

infectious pancreatic necrosis *Veterinary Medicine*. an acute disease highly fatal in young trout, and chronic in other salmonids, in which white exudate is found in the stomach and intestines. Pancreatic acinal cells and intestinal mucosa cells show severe cytolytic necrosis.

infectious pancreatic necrosis virus *Virology*. a highly contagious virus of the genus *Birnavirus* of the family Birnaviridae, which causes acute, lethal pancreatic infection, especially in young, hatchery-raised salmonid fishes.

infectious particle *Virology*. a virus particle containing the complete viral genome, thus able to replicate successfully in a susceptible host cell.

infectious pig paralysis see TESCHEN DISEASE.

infectious porcine potyoencephalomyelitis see TESCHEN DISEASE.

infectious rhinitis *Medicine*. an infectious inflammation of the mucous membrane of the nose with excessive discharge of mucus; the common cold.

infectious unit *Virology*. the lowest number of virus particles sufficient to cause infection.

infective see INFECTIOUS.

infective endocarditis *Cardiology*. an inflammation of the lining membrane of the heart chambers due to the presence of bacteria or other microorganisms involving the valve leaflets.

infectivity *Microbiology*. the ability of an infectious or pathogenic microorganism to become established in a host. *Virology*. specifically, the ability of a virus to replicate in a susceptible host cell.

infer *Science*. to make an inference; reach a logical conclusion.

inference *Science*. a conclusion that follows logically from the available evidence but that is not the direct and incontrovertible result of that evidence. *Statistics*. a statement about a population based on a sample drawn from it.

inference engine *Artificial Intelligence*. the part of an expert or knowledge system that scans, selects, and applies rules to stored knowledge; it also determines when a solution has been found. Also, RULE INTERPRETER.

inference program *Artificial Intelligence*. a program that derives conclusions from a series of rules and stated facts or observations.

inference rule *Artificial Intelligence*. a rule that allows derivation of additional true statements from a given set of true statements.

inferential flowmeter *Engineering*. a flowmeter that determines the actual mass flow by means of another phenomenon, such as the cooling effect of flow on a heated wire or the drop in static pressure at a restriction in a pipe.

inferential liquid-level meter *Engineering*. a meter that determines the level of a liquid indirectly by means of another associated phenomenon, such as the pressure associated with certain levels.

inferior *Anatomy*. describing a structure that is positioned below or lower than another structure in the body. *Biology*. a term used in place of *posterior* when comparing structures in humans to those in other mammalian bipeds; e.g., the inferior vena cava in humans is equivalent to the posterior vena cava in quadrupeds. *Botany*. of a calyx, inserted below the ovary. *Astronomy*. situated below the horizon. *Materials*. of relatively low grade or quality.

inferior alveolar artery *Anatomy*. a branch of the maxillary artery that passes through the mandibular canal and distributes blood to teeth and muscles of the lower jaw.

inferior alveolar nerve *Anatomy*. a branch of the mandibular nerve that passes through the mandibular canal and innervates the teeth, gums, skin, and lips of the lower jaw.

inferior cerebellar peduncle *Anatomy*. a tract of nerve fibers passing lateral to the fourth ventricle, and connecting the cerebellum with the medulla oblongata and spinal cord. Also, RESTIFORM BODY.

inferior cervical ganglion *Anatomy*. a ganglion of the sympathetic trunk, lying at the level of the seventh cervical vertebra, near the base of the vertebral artery.

inferior colliculus *Anatomy*. the lower of the two pairs of rounded protrusions from the roof of the midbrain; a reflex center for auditory sensations.

inferior hypogastric plexus *Anatomy*. a network of autonomic nerves in the pelvis innervating the pelvic viscera.

inferiority *Psychology*. a mental or psychological state of feeling less adequate than others; a sense of being weak or inadequate in some fundamental way. Also, **inferiority complex**. *Anatomy*. the quality or condition of being inferior.

inferior mesenteric ganglion *Anatomy*. a mass of sympathetic neurons, located at the origin of the inferior mesenteric artery, that sends nerve fibers to the descending colon and the sigmoid colon.

inferior mirage *Optics*. a phenomenon in which the spurious image of an object appears below its actual position.

inferior nasal concha *Anatomy*. a thin bony plate with curved margins, articulating with the ethmoid, maxilla, and lacrimal and palatine bones, and forming the lower part of the lateral wall of the nasal cavity, and the mucous membrane covering the plate. Also, MAXILLOTURBINAL BONE.

inferior planet *Astronomy.* a planet that orbits closer to the sun than some other; for example, Mercury and Venus are observed from earth as inferior planets.

inferior temporal gyrus *Anatomy.* the ridge of cortical tissue that forms the inferolateral border of the temporal lobe of the brain.

inferior vena cava *Anatomy.* the large vein that delivers blood from the lower part of the body to the right atrium.

inferior vena cava syndrome *Medicine.* the obstruction of the inferior vena cava or its main tributaries by carcinoma, neoplasm, or lymphoma, causing edema and engorgement of the lumbar, renal, testicular, ovarian, suprarenal, inferior phrenic, saphenous, hemorrhoidal, and hepatic veins.

inferior vermis *Anatomy.* the portion of the vermis of the cerebellum that projects below the two cerebellar hemispheres.

inferior vestibular nucleus *Anatomy.* one of four small masses of nerve cells lying in the lateral region of the hindbrain, receiving fibers of the vestibular nerve.

infernal machine *Ordnance.* an older term for an explosive device that is disguised or concealed, usually for use in sabotage.

inferolateral *Anatomy.* situated below and to one side.

inferomedian *Anatomy.* situated in the middle of the underside.

inferoposterior *Anatomy.* situated below and behind.

inferred ore *Mining Engineering.* any ore for which quantitative estimates are based primarily on knowledge of the deposit's geological character, and for which there are few samples or measurements.

infertile *Medicine.* not fertile; exhibiting or characterized by infertility.

infertility *Medicine.* the diminished ability or inability to produce offspring. This term is contrasted with *sterility,* which describes a complete inability to produce offspring. Also, **infertilitas.**

infest *Medicine.* to live in or invade a host parasitically.

infestation *Medicine.* the act of living parasitically on the surface of a host, as with ticks, lice, or mites, or in the organs of a host, as with worms.

infilling well *Petroleum Engineering.* a well that is drilled between two or more producing wells in order to achieve greater recovery of oil from a reservoir.

infiltrate *Science.* **1.** to carry out a process of infiltration. **2.** something deposited by this process.

infiltration the process of penetration or permeation; specific uses include: *Geology.* the flow of a liquid into a solid via pores or small openings; applied specifically to the movement of water into soil or porous rock. *Hydrology.* the movement or passage of water and dissolved substances from another area into the soil. *Pathology.* **1.** the diffusion or accumulation in a tissue or cells of substances not normal or in excess of normal. **2.** the material so accumulated. *Military Science.* **1.** the movement into or through an area by small groups or individuals at extended or irregular intervals; if the area is occupied by the enemy, contact is avoided. **2.** in intelligence operations, the placing of an agent or other person in hostile territory. *Metallurgy.* in powder metallurgy, the diffusion of a liquid metal in a porous compact; for instance, diffusion of liquid copper in an iron compact.

infiltration capacity *Hydrology.* the maximum rate at which water can be absorbed by a soil at a particular point under given conditions.

infiltration gallery *Civil Engineering.* a large underground conduit of porous or perforated material constructed so that percolating water can be collected by infiltration.

infiltration vein *Geology.* a mineral vein formed in the interstices of the country rock by the action of percolating water.

infiltration water *Hydrology.* see FREE WATER.

infimum *Mathematics.* the greatest lower bound; written *inf* or *glb.*

infinite *Mathematics.* not finite; larger than any fixed finite quantity.

infinite baffle *Acoustical Engineering.* a loudspeaker enclosure that is designed to prevent sound from traveling from the front to the back of the speaker.

infinite-capacity loading *Control Systems.* a method of determining the overload protection requirements for a robotic workcenter by deliberately loading it with excessive force or weight.

infinite edge sequence *Mathematics.* a sequence of edges of an infinite graph of the form $v_0v_1, v_1v_2, \ldots$ (called a **one-way infinite edge sequence** with initial vertex v_0) or of the form $\ldots, v_{-1}v_0, v_0v_1, v_1v_2, \ldots$ (called a **two-way infinite edge sequence**).

infinite graph *Mathematics.* a graph having at least one vertex and edge set that is infinite. If the vertex and edge sets of an infinite graph are both countable sets, the graph is said to be a **countable graph;** otherwise it is an **uncountable graph.**

infinite integral *Mathematics.* **1.** an integral with an unbounded domain of integration. **2.** in particular, suppose the function f is Riemann integrable over $[a,c]$, with $I_c = \int_a^c f$ for each $c > a$. A real number I is said to be the infinite integral of f over $\{x : x \geq a\}$ if, for every $\varepsilon > 0$, there exists a real number $M(\varepsilon)$ such that if $c > M(\varepsilon)$, then $|I - I_c| < \varepsilon$. Also, IMPROPER INTEGRAL OF THE FIRST KIND.

infinite loop *Computer Programming.* a repeating sequence of instructions that never stops, usually due to a program error.

infinitely large *Mathematics.* let K be an ordered field with ordering $<$; it can be shown that K contains a subfield Q isomorphic to the field of rational numbers. If $<$ is not Archimedean, then there exists an element b of K such that $q < b$ for any $q \in Q$. Such elements b of K are said to be infinitely large.

infinitely small *Mathematics.* let K be an ordered field with ordering $<$; it can be shown that K contains a subfield Q isomorphic to the field of rational numbers. If $<$ is not Archimedean, then there exists an element a of K such that $a < q$ for any $q \in Q$. Such elements a of K are said to be infinitely small.

infinite multiplication factor *Nucleonics.* the ratio between the number of neutrons in a reactor from the current generation and the number from the previous generation when no leakage occurs.

infinite product *Mathematics.* a product with an infinite number of factors; denoted by $\Pi_{i=1}^{\infty} a_i$, where the ith factor a_i may be a real or complex numerical value, a function, etc. The value of an infinite product is defined to be $\lim_{n\to\infty} \Pi_{i=1}^{n} a_i$, provided this limit exists. Sufficient conditions for convergence are that $a_i \neq 0$ and $\Sigma_{i=1}^{\infty} \log a_i$ converges.

infinite sequence *Mathematics.* a function whose domain is the set of natural numbers or the set of counting numbers. By convention, the values of the function are written in the form a_i, where i is the function argument. If the range of the function is in some set X (such as R^n), then it is a sequence in X. Such a sequence is denoted $\{a_i\} = \{a_1, a_2, a_3, \ldots\}$.

infinite series *Mathematics.* if $\{a_i\}$ is an infinite sequence, then the infinite series (or simply the series) that is generated by $\{a_i\}$ is the sequence $S = \{s_k\}$, as defined by $s_1 = a_1$, and $s_k = s_{k-1} + a_k = \Sigma_{i=1}^{k} a_i$. If S converges, then $\lim_{k\to\infty} s_k$ is the sum of the infinite series; the a_i are the terms; and the s_k are the partial sums.

infinite set *Mathematics.* a set that has an infinite number of elements; that is, a set which is not equivalent to some natural number n; a set whose members cannot be put into a one-to-one correspondence with the set $\{1, 2, 3, \ldots, n\}$ for some natural number n.

infinitesimal generator *Mathematics.* an element α of a Lie algebra that, when multiplied by t and exponentiated, generates a one-parameter Lie group $A(t)$. α is called the infinitesimal generator of $A(t)$.

infinity *Mathematics.* **1.** a number larger than any bounded quantity; denoted ∞. **2.** the name given to the point that provides the one-point compactification of the complex plane; denoted ∞. **3.** a term used to indicate that a quantity x grows without bound; symbolized by $x \to \infty$. *Computer Programming.* a value larger than the maximum value that a given computer is capable of storing.

infinity method *Optics.* a technique in which two lines of sight are made essentially parallel by aiming them at an object located at great distance, such as at a star.

infinity transmitter *Electronics.* a device that can intercept a phone transmission from a remote location, so that a conversation can be tapped without the caller's knowing about it.

infirm *Medicine.* feeble or weak, as from disease or old age.

infirmary *Medicine.* a hospital or other institution where the sick are maintained or treated, especially one specifically for the care of members of a group or community.

infirmity *Medicine.* **1.** a state of feebleness or weakness. **2.** a disease or condition that produces such a state.

infix *Linguistics.* a morpheme that is inserted within the element to which it is attached, rather than before it (a prefix) or after it (a suffix); found in Tagalog, Samoan, and other languages.

infix notation *Computer Programming.* a type of arithmetic notation in which the operator appears between operands.

inflammation *Medicine.* a protective response of tissues affected by disease or injury, and characterized by redness, localized heat, swelling, pain, and possibly impaired function of the affected part.

inflammatory *Medicine.* relating to, causing, or characterized by inflammation.

inflammatory carcinoma *Medicine.* a rapidly spreading malignant tumor accompanied by inflammation, especially in the breast.

inflammatory tissue *Medicine.* any tissue characterized by or affected by inflammation.

inflate *Science.* to swell or distend with air or gas. Thus, **inflated, inflating, inflation, inflationary.**

inflationary universe *Astrophysics.* a theorized stage in the development of the universe about 10^{-35} second after the Big Bang when the universe expanded catastropically by a factor of about 10^{50}.

inflator *Science.* an instrument for inflating.

inflection *Linguistics.* a change in the form of a word to show a change in meaning or function; e.g., in English the addition of the **inflected form** *s* at the end of a word indicates a change from singular to plural.

inflight or **in-flight** *Aviation.* performed, occurring, or used during flight. Thus, **inflight refueling, inflight icing.**

inflight start *Aviation.* any engine ignition sequence performed during flight, such as the initial firing of a rocket stage after takeoff or the restarting of a jet engine after flameout.

inflorescence *Botany.* **1.** the process of flowering or blossoming. **2.** the structure or arrangement of flowers on a plant. Thus, **inflorescent.**

influence function *Statistics.* a function describing the effect of individual observations on a statistic.

influence fuse see PROXIMITY FUSE.

influence line *Mechanics.* a graph of a quantity such as stress or strain at a given point versus the position of a moveable load on a structure.

influence mine *Ordnance.* a mine that is actuated by the effect of a target upon some physical condition in the vicinity of the mine or radiations emanating from the mine.

influence release sinker *Ordnance.* a sinker that holds a mine at the seabed and releases it when actuated by influence from a ship.

influence sweep *Military Science.* a sweep designed to produce an influence similar to that produced by a ship and thus actuate mines.

influent *Hydrology.* **1.** referring to the process of flowing in or flowing away. **2.** see INFLUENT STREAM, def. 1.

influent flow *Hydrology.* the flow of water into the ground from a body of surface water.

influent stream *Hydrology.* **1.** a surface stream that flows into a lake, or any branch that flows into a larger stream. **2.** a stream that has developed bank storage and that loses water to the zone of saturation.

influenza [in´floo en´zə] *Medicine.* an acute, infectious, often epidemic respiratory disease in which the inhaled virus attacks epithelial cells, causing fever, headache, muscle pain, dry cough, fatigue, and physical exhaustion. Also, FLU. (Related to the word *influence.*)

influenza meningitis *Medicine.* an inflammation of the membranes of the brain and spinal cord associated with influenza infection.

influenza pneumonia *Medicine.* an inflammation of the lungs associated with influenza infection.

influenza vaccine *Immunology.* a vaccine that is used against the infectious disease known as influenza, and contains a mixture of strains of formaldehyde-inactivated influenza virus.

influenza virus *Virology.* a genus of viruses of the family Orthomyxoviridae that includes the influenza types A, B, and C, causing sporadic and epidemic influenza and pneumonia in humans, pigs, horses, and birds, with transmission by aerosols, by water, or by direct contact. Also, FLU VIRUS.

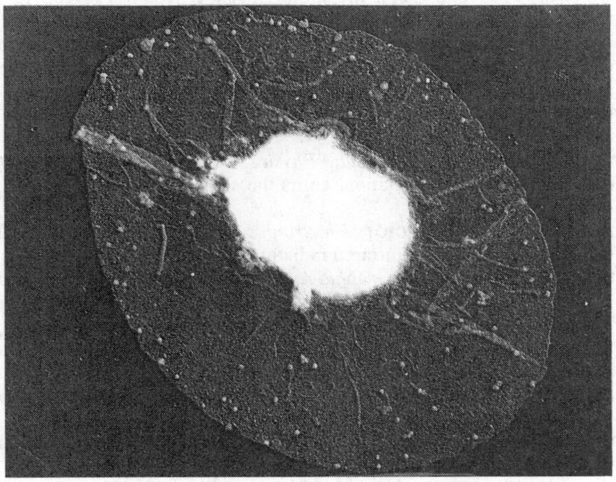

influenza virus

influx *Hydrology.* the place at which a stream flows into the sea or into another stream.

infolding *Surgery.* the enclosing of a lesion or redundant tissue within a fold of adjacent tissue; e.g., a stomach ulcer having the walls of the stomach on either side of it sutured together. *Developmental Biology.* the folding inward of a layer of tissue, as in the formation of the neural tube.

informant *Anthropology.* a person who voluntarily gives information about his or her culture to an ethnographer at a personal level.

information *Science.* a general term for any data that have been recorded, classified, organized, related, or interpreted within a certain context so that meaning is apparent. *Telecommunications.* specifically, data that are transmitted by signals via telecommunication channels.

information bit *Telecommunications.* a bit that is generated by a data source but does not function in data transmission.

information center *Telecommunications.* a facility whose purpose is to store, process, and retrieve information according to users' needs.

information channel *Telecommunications.* the transmission medium used to transmit information between spatially separated communication entities.

information content *Telecommunications.* the information that is represented by data bits in a given transmission, usually measured in hartleys.

information feedback system *Telecommunications.* an information transmission system that sends data back to the transmitter for the purpose of checking the accuracy of the transmitted data.

information flow *Computer Programming.* the movement of data into and within a system, and out to the end users.

information hiding *Computer Science.* a method of making programs modular by allowing programs to see only a set of well-defined interfaces to a data type, and not the internal implementation of the data.

information interchange *Telecommunications.* the exchange of information between two communication entities.

information management *Telecommunications.* a term applied to a body of definitions, standards, and techniques for information processing and transmission among communications systems.

information network *Computer Technology.* an electronic interconnection of dispersed libraries and information centers for the purpose of permitting greater access to the total information resources.

information processing *Computer Technology.* the computerized evaluation, analysis, and tabulation of data to generate usable information. *Psychology.* an approach to cognition that views the human brain as analogous to a computer; i.e., as a system that can accept, store, organize, and manipulate information. Thus, **information-processing system, information-processing model, information-processing theory.**

information rate *Telecommunications.* the amount of information that can be transmitted in a given time, usually expressed in bits per second.

information requirements *Computer Programming.* the actual or anticipated questions or problems presented to an information system for analysis or resolution; the end-user needs and expectations.

information-resources management *Computer Technology.* the process of managing information within an organization, including such aspects as defining, evaluating, and distributing the data for use in problem solving.

information retrieval *Computer Programming.* **1.** any technique used for storing and categorizing large blocks of data and then extracting all or selected portions of that data. **2.** specific methods used to catalog, index, and locate particular data in a database. **3.** to retrieve information from a database. Thus, **information retrieval system.**

information separator *Computer Programming.* an indicator that separates fields or items within a record.

information source *Telecommunications.* a communication entity that creates messages by making successive selections from an arrangement of symbols.

information system *Computer Technology.* a computer application consisting of a database, application programs, and manual and machine procedures for processing data. *Telecommunications.* a communications system designed to communicate information directly between users.

information system architecture *Computer Technology.* the physical structure of an information storage and retrieval system, including hardware, system and application software, procedures, and organizational interfaces between users and computer personnel.

information technology *Computer Technology.* the use of computers and telecommunications for the processing and distribution of information in digital, audio, video, and other forms.

information theory *Artificial Intelligence.* the branch of learning that is concerned with the study of information representation and transmission. *Mathematics.* a branch of mathematics that ascribes a notion of information carried by certain classes of functions, including the study of the amount of information and its efficiency of representation. Information theory forms the basis of modern signal-processing techniques.

information unit *Telecommunications.* a unit of information content, equal to a nit, bit, or hartley, depending on the base of the logarithm used.

information utility *Computer Technology.* a service that updates and provides data bank information for public use.

informed *Artificial Intelligence.* describing the quality of information provided by a heuristic; a heuristic that gives more accurate estimates is called more informed.

informed consent *Medicine.* permission to perform a specific procedure or test, given by a patient who has been properly informed of the facts and risks involved.

informosome *Cell Biology.* a ribonucleoprotein particle that is believed to pass from the nucleus into the cytoplasm, found in the egg cytoplasm of certain echinoderm and fish species.

infra- a combining form meaning "below" or "beneath," as in *infracostal, infrastructure.*

infrabranchial *Vertebrate Zoology.* of or relating to structures located below or inferior to the gills or throat region in a vertebrate animal.

Infracambrian see EOCAMBRIAN.

infracerebral gland *Invertebrate Zoology.* a glandular structure lying below the brain in annelid worms.

infraciliature *Invertebrate Zoology.* the neuromotor system of ciliates.

infraclass *Systematics.* a taxonomic rank below a subclass and above a cohort.

infraclavicle *Vertebrate Zoology.* a small bone found in the pectoral girdle of certain fish.

infraclavicular *Anatomy.* beneath a clavicle. *Vertebrate Zoology.* of or relating to an infraclavicle.

infracostal *Anatomy.* below a rib or ribs.

infraction *Medicine.* an incomplete fracture of a bone, without displacement of the fragments.

infradian *Biology.* of or relating to biological cycles or rhythms that recur less often than once per day.

infraduction *Physiology.* the downward rotation of an eye.

infradyne receiver *Electronics.* a device that raises the frequency of audio or video signals to enhance selectivity.

infraglacial see SUBGLACIAL.

infraglenoid *Anatomy.* lying below the glenoid cavity of the scapula.

infraglenoid tubercle *Anatomy.* a roughened area on the scapula, just below the glenoid cavity that anchors the origin of the triceps muscle.

infralateral tangent arcs *Meteorology.* a pair of oblique, colored arcs that are convex to the sun and tangent to the halo of 46° at points below the sun's altitude; produced by refraction in hexagonal columnar ice crystals with principal axes that are horizontal but randomly directed in the azimuth, they appear only if the elevation of the sun is less than approximately 68°.

infralow frequency *Telecommunications.* a frequency that lies in the range from 300 hertz to 3 kilohertz in the radio spectrum.

infraneritic *Oceanography.* a subdivision of the neritic zone that refers to the pelagic environment in water 40–200 meters deep.

infraorbital *Anatomy.* lying under or on the floor of the orbit.

infraorder *Systematics.* a taxonomic rank below a suborder and above a superfamily.

infrapatellar *Anatomy.* beneath the patella.

infrared [in frə red′] *Electromagnetism.* **1.** the portion of the invisible spectrum consisting of electromagnetic radiation with wavelengths in the range from 750 nanometers to 1 millimeter. **2.** relating to such rays.

infrared absorption *Electromagnetism.* the absorption of infrared radiation by a medium through which the radiation is passing; causing the excitation of vibration modes or electronic transitions.

infrared astronomy *Astronomy.* the study of astronomical objects at wavelengths from about 0.8 micrometer (8000 angstroms) to about 1000 micrometers.

infrared background radiation *Astronomy.* diffuse radiation at wavelengths of about 10 to 200 micrometers that comes from many sources, mostly dust in the solar system, around stars, and in the Milky Way, plus infrared radiation from other galaxies.

infrared beacon *Navigation.* a navigational beacon that shines in the infrared, providing greater penetration of some atmospheric conditions.

infrared binoculars *Optics.* an instrument that views and enlarges infrared images and resembles ordinary binoculars.

infrared bolometer *Electronics.* a device whose sensitivity to heat has been adjusted for infrared radiation detection, as opposed to one adjusted for microwave radiation detection.

infrared catastrophe *Quantum Mechanics.* a prediction of quantum electrodynamics that the radiation of charged particles approaches infinity at low frequencies. This problem disappeared in later refinements to the theory. Also, INFRARED PROBLEM.

infrared cirrus *Astronomy.* streaky sources of infrared radiation detected by the Infrared Astronomy Satellite at wavelengths of 60 to 100 micrometers; the cirrus is believed to be composed of clouds of dusty gas in the galaxy as well as sources in the solar system.

infrared detector *Electronics.* a device that responds to infrared radiation, commonly used to detect the presence of fire or heat in machinery, vehicles, or people.

infrared emission *Physics.* the radiation of waves having wavelengths in the range of 0.8 to 1000 micrometers.

infrared-emitting diode *Electronics.* a device in which current flowing across a junction produces radiant energy in the infrared region.

infrared excess *Astronomy.* infrared emission from a celestial object that is greater than the object would produce if it were radiating solely as a black body; the excess usually indicates the presence of dust in the object's vicinity that has been warmed as it absorbs shorter wavelength radiation from the object.

infrared film *Graphic Arts.* a specially sensitive film used in infrared photography.

infrared filter *Optics.* a filter that transmits infrared radiation while absorbing other forms of radiation; used for long-distance, medical, forensic, and unobtrusive flash photography.

infrared galaxy *Astronomy.* a dust-laden galaxy that appears bright at infrared wavelengths; many such are visible only in the infrared.

infrared heating *Engineering.* heating by means of focused, intense infrared radiation; used for baking or drying.

infrared heterodyne detector *Electronics.* a device that makes infrared signals audible by combining them with signals generated by a local oscillator.

infrared homing *Engineering.* tracking by means of the infrared radiation emitted by a target.

infrared image converter *Electronics.* a device that transforms invisible infrared images into visible ones through a system of light-sensitive components. Also, **infrared imaging device, infrared image tube.**

infrared jamming *Electronics.* the use of infrared radiation to reduce the effectiveness of heat-seeking missiles by overloading their detectors.

infrared lamp *Electricity.* a specialized incandescent lamp designed to produce radiant thermal energy in the infrared electromagnetic spectrum.

infrared laser *Physics.* a laser that emits a beam of infrared radiation.

infrared maser *Physics.* a maser that radiates or detects infrared radiation.

infrared microscope *Spectroscopy.* a microscope that uses radiation from the infrared region of the spectrum to illuminate details of objects which are opaque to visible radiation. Thus, **infrared microscopy.**

infrared nuclear magnetic resonance *Spectroscopy.* a spectroscopic technique for studying nuclei and chemical changes in which nuclear magnetic resonance is employed in a spectrometer over a range of infrared frequencies.

infrared optical material *Electromagnetism.* any material that transmits infrared radiation.

infrared phosphor *Solid-State Physics.* a type of phosphor that, after exposure to infrared radiation, emits the same spectrum as that of the dominant activator.

infrared photoconductor *Electronics.* a device that transmits electricity when exposed to infrared radiation.

infrared photography *Graphic Arts.* a photographic method requiring special film and equipment that can record the infrared spectrum, used especially for photographing in very low light.

infrared problem see INFRARED CATASTROPHE.

infrared radiation *Electromagnetism.* electromagnetic radiation lying in the infrared spectrum, with wavelengths between 750 nanometers and 1 millimeter.

infrared receiver *Electronics.* a device that intercepts infrared radiation thought to be carrying intelligence.

infrared scanner *Electronics.* a device that reacts to infrared radiation in such a way that it can be used to scan a field of view line by line.

infrared searchlight *Optics.* a device that emits infrared radiation over an area so that the scene may then be viewed through an infrared image-converter tube.
infrared soldering *Metallurgy.* soldering in which the heat is supplied by infrared radiation. Similarly, **infrared brazing.**
infrared spectrometer *Spectroscopy.* a spectrometer used to detect and measure absorption in the infrared region, especially for the quantitative analysis of organic compounds and to indicate the presence of heteroatomic elements and functional groups in molecules.
infrared spectrophotometry *Spectroscopy.* the photometric measurement of the wavelengths in the infrared region that are absorbed by molecules and that correspond to rotational and vibrational transitions.
infrared spectroscopy *Spectroscopy.* the study of the infrared radiation characteristically absorbed by molecular substances.
infrared spectrum *Electromagnetism.* the portion of the electromagnetic spectrum that lies between radio waves and visible light.
infrared star *Astronomy.* a star whose radiation is wholly or primarily at infrared wavelengths; indicates a dusty environment for the star.
infrared telescope *Optics.* a telescope that transmits and converts infrared images into enlarged, visible images.
infrared thermistor *Electronics.* a resistor that responds to heat in a predictable manner, used to measure the intensity of infrared radiation.
infrared thermometer *Engineering.* an instrument that determines the temperature of a subject by measuring the infrared energy emitted.
infrared vidicon *Electronics.* a television camera forming an electron beam that scans a charged-density pattern formed and stored on the surface of a photoconductor that is sensitive to infrared radiation.
infrared window *Geophysics.* any region from 1.0 to about 20 micrometers in the infrared part of the electromagnetic spectrum in which the earth's atmosphere is partly or largely transparent, and in which transmission of electromagnetic radiation through the atmosphere is good.
infra roentgen rays *Radiology.* the soft, poorly penetrating X-rays with a wavelength of approximately 2 angstroms, ranging between X-rays and ultraviolet rays on the electromagnetic spectrum, that are often used to treat lesions of the skin. Also, GRENZ RAYS.
infrascapular *Anatomy.* beneath the scapula.
infrasonic *Acoustics.* describing sound waves of low frequency, below the normal range of human hearing (about 20 hertz). Also, SUBSONIC.
infrasonics *Acoustics.* the study of infrasonic sound waves.
infrasound *Acoustics.* slow vibrations of about 20 hertz or less.
infraspecific *Systematics.* of or relating to taxa below the rank of species.
infraspinous *Anatomy.* beneath the spine of the scapula.
infraspinous fossa *Anatomy.* a large depression lying below the spine of the scapula.
infrasternal *Anatomy.* beneath the sternum.
infrastructure *Civil Engineering.* the basic facilities of a country or region, such as roads, bridges, water delivery, and sewage treatment. *Design Engineering.* broadly, the underlying structure of any system.
infrasubspecific *Systematics.* of or relating to taxa below the rank of subspecies.
infratemporal *Anatomy.* below the temporal fossa.
infratemporal fossa *Anatomy.* the cavity on the lateral side of the skull, bounded by the zygomatic arch laterally, the temporal bone posteriorly, the sphenoid bone medially, and the zygomatic process of the maxilla anteriorly.

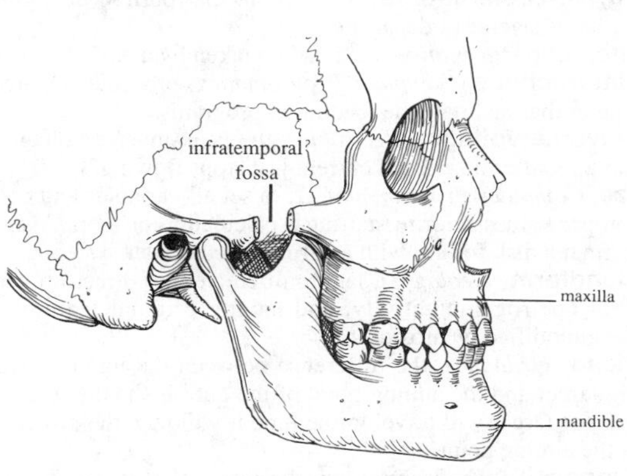

infratemporal fossa

infundibular *Biology.* funnel-shaped. Also, **infundibuliform.** *Anatomy.* of or relating to an infundibulum.
infundibular canal *Invertebrate Zoology.* a canal in cephalopods that conducts water from the mantle cavity through the funnel.
infundibulum *Anatomy.* a funnel-shaped structure, such as the stalk connecting the hypothalamus to the pituitary gland, the cavity surrounded by fimbriae at the proximal end of the oviduct, the opening at the upper end of the cochlear canal in the inner ear, and the tube that connects the frontal sinus to the middle nasal meatus. *Invertebrate Zoology.* a funnel-shaped structure, such as the siphon of a cephalopod or the flattened gastric cavity of a ctenophore.
infusion *Chemistry.* the soluble constituent that is derived by steeping a substance in water (usually hot) for a period of time. *Medicine.* **1.** the therapeutic introduction of a fluid other than blood into a vein by gravity. By comparison, an *injection* is forced in by a syringe. **2.** any medicinal preparation. *Biology.* a suspension of decaying organic or inorganic material in a liquid, used as a culture medium.
infusoria *Bacteriology.* a former term for various microorganisms, especially *Ciliophora,* that are found in brackish water and decayed organic matter.
infusorial earth *Geology.* an obsolete, technically incorrect term for diatomaceous earth.
infusoriform larva *Invertebrate Zoology.* the final larval stage in the life cycle of mesozoans.
infusorigen *Invertebrate Zoology.* a developmental stage in the life cycle of mesozoans with germ cells that produce an infusoriform larva.
Ingenhousz, Jan 1730–1799, Dutch engineer and naturalist; discovered the workings of photosynthesis and the carbon cycle.
ingesta *Biology.* the total intake of substances into the body.
ingestion *Biology.* **1.** the action of taking in food material; feeding; **2.** consumption.
Ingolfiellidea *Invertebrate Zoology.* a suborder of amphipod crustaceans found in fresh and cave waters in tropical and warm temperate regions.
ingot *Metallurgy.* a casting used as a starting stock for working or for remelting; not used in the as-cast conditions.
ingot casting *Metallurgy.* a method of mold-casting metals into approximately square cross sections, with rounded corners and tapered sides.
ingot iron *Metallurgy.* a term for commercially pure iron.
ingrain color see AZOIC DYE.
ingress *Science.* the process of entering. *Astronomy.* the instant at which a celestial object begins immersion.
ingrown *Medicine.* referring to a condition in which the free tip or edge of a hair or nail grows inwardly and is embedded in the skin.
ingrown meander *Hydrology.* a stream meander that forms when the rate of downcutting is slow enough to allow for lateral erosion, resulting in an incised meander that expands or grows continually in place.
inguen see INGUINAL REGION.
inguinal *Anatomy.* relating to or located in the inguinal region, or groin.
inguinal canal *Anatomy.* the opening from the lower abdominal wall that carries the round ligament in females and the spermatic cord in males, and through which the testes descend in a developing male fetus.
inguinal fold *Developmental Biology.* an elevation on the posterior abdominal wall of an embryo, extending caudally from the caudal pole of the gonad and containing the upper portion of the gubernaculum. Also, EPIGONAL FOLD, GUBERNACULAR FOLD.
inguinal gland *Anatomy.* any of numerous lymph nodes lying in the inguinal region, draining lymph fluid from the buttocks, external genitals, lower limb, and skin of the lower abdomen.
inguinal hernia *Medicine.* a hernia of the intestine at the inguinal or groin region, the most common type of hernia, usually repaired surgically.
inguinal ligament *Anatomy.* the tough, fibrous cord extending from the anterior superior iliac spine to the pubic tubercle.
inguinal region *Anatomy.* the region of the trunk lying between the abdomen and the thigh. Also, GROIN, INGUEN.
inhabited building distance *Engineering.* the minimum distance allowed between the site of a possible explosion and the nearest building or area of assemblage.
inhalant *Medicine.* a drug or other substance that may be taken into the body by way of the nose and trachea, or through the respiratory system.
inhalant canal see INCURRENT CANAL.
inhalant siphon *Invertebrate Zoology.* a tubular structure that brings water into the mantle of bivalve mollusks. Also, INCURRENT SIPHON.

inhalation *Physiology.* the act of drawing air or other substances into the lungs. *Pharmacology.* the act of administering a drug or combination of drugs by nasal or oral respiration.
inhalator *Medicine.* a device for enabling the body to draw in oxygen or oxygen-carbon dioxide mixtures, such as that which used in resuscitation.
inhaler *Medicine.* an apparatus for administering vapor or anesthetics by inhalation. *Mechanical Engineering.* an apparatus designed to prevent dust, smoke, or noxious gases from entering the lungs, or to enable a person with affected lungs to breathe more easily.
inhaul cable *Mechanical Engineering.* a line that pulls a bucket to dig and remove soil for cable excavating operations. Also, DIGGING LINE.
inherent ash *Mining Engineering.* the portion of the ash content of coal that is structurally part of the coal itself and cannot be removed by mechanical means.
inherent burst *Mining Engineering.* a rock burst that occurs during development.
inherent damping *Mechanical Engineering.* a method of vibration damping that utilizes the mechanical hysteresis of certain materials such as rubber and cork.
inherent density *Radiology.* the density of a processed film relative to the film base, emulsion gelatin, and other innate factors.
inherent stability *Aviation.* an aircraft's built-in stability; specifically, the aerodynamic design features that help it return to its original steady flight condition after a disturbance without corrective action.
inherent storage *Computer Technology.* a portion of the automatic data-processing system or hardware that is used for storing data and that may be controlled automatically without operator intervention.
inherent viscosity *Materials Science.* a measure of the viscosity of a polymeric dilute solution viscosity, expressed as the ratio of the natural logarithm of the relative viscosity to the concentration of the polymer in grams per 100 milliliters of solvent.
inheritance *Genetics.* the reception of a set of traits that are biologically transmitted from an individual to the offspring. *Computer Science.* the availability of procedures or data by virtue of membership in a class, e.g., in an object-oriented system. *Artificial Intelligence.* the pre-existing information received by an object or rule at start-up time or from another object or rule in the knowledge system during processing.
inheritance of acquired characteristics *Genetics.* the obsolete and discredited concept that somatic characteristics accumulated by one generation can somehow be transferred to the germ cells, and then on to succeeding generations. Also, LAMARCKISM.
inherited *Genetics.* received by inheritance; involving the transmission of hereditary characteristics. *Medicine.* of a disease or condition, genetically determined; hereditary. Thus, **inherited disorder.**
inherited attribute *Computer Science.* an attribute of a node in a parse tree that is derived from the context in which the node appears, rather than from the attributes of structural components.
inherited error *Computer Programming.* an error resulting from incorrect initial values caused by uncertain measurements, inadequate precision, or human mistakes. Also, **inherent error**.
inherited meander see ENTRENCHED MEANDER.
inhibin *Endocrinology.* a peptide hormone that is secreted by the follicular cells of the ovary or the Sertoli cells of the testis and inhibits the release of follicle-stimulating hormone from the anterior pituitary.
inhibit *Science.* to hinder, restrain, obstruct, or prohibit from doing something. Thus, **inhibited, inhibiting.**
inhibited acid *Petroleum Engineering.* an acid used in procedures of acid fracturing that has been altered chemically to lessen corrosive action on piping, though still maintaining its effectiveness in fracturing.
inhibited mud *Petroleum Engineering.* a drilling fluid that is chemically treated to prohibit the swelling of clay particles in a formation, thereby not affecting the permeability of a productive zone.
inhibit-gate *Electronics.* a device that produces an output signal only when certain signals are absent from the inputs.
inhibiting factor see RELEASE-INHIBITING FACTOR.
inhibiting input *Computer Technology.* **1.** an inhibiting signal. **2.** a terminal to which an inhibiting signal is applied.
inhibiting signal *Computer Technology.* a signal that halts the operation of a circuit in a digital computer.
inhibition *Science.* **1.** the process of inhibiting. **2.** something that inhibits. *Behavior.* a restraint on an instinctual impulse, such as eating or sex, that may interfere with or restrict specific activities. Also, **inhibitive behavior.** *Psychology.* a mental state that results in a hesitancy or blockage of action.
inhibition index *Biochemistry.* a measure of the amount of a given antimetabolite required to negate the effect of a given metabolite.
inhibitor *Chemistry.* any substance that stops or slows a chemical reaction, especially an undesirable reaction such as corrosion or decomposition; a negative catalyst. *Aviation.* specifically, a substance designed to retard the burning of a propellant; usually added as a catalyst to liquid propellants and bonded or otherwise attached to solid propellants.
inhibitor sweetening *Chemical Engineering.* a treating process for sweetening gasoline with low mercaptan content using caustic, air, and a phenylenediamine inhibitor.
inhibitory force *Behavior.* a factor that serves to eliminate or restrain the response to a stimulus.
inhibit pulse *Computer Technology.* a pulse that serves as an inhibiting signal.
inhomogeneity *Science.* the fact of being inhomogeneous; variability. *Mechanics.* specifically, the property of a solid or fluid whose physical properties vary with position, making the analysis of stress and strain in the medium much more difficult. *Astrophysics.* see ANISOTROPY.
inhomogeneous *Science.* not homogeneous; having variable properties or composition.
inhomogeneous line-broadening *Optics.* the broadening of the natural linewidth of a laser, occurring as a result of strains or imperfections in the system, or when the resonance frequencies of the atoms or molecules of the medium are not homogeneous on the same transition.
inhour *Nucleonics.* a unit quantity of a reactor's radioactivity for one hour. (A shortened form of in̲verse hour̲.)
inhour equation *Nucleonics.* an equation that compares a reactor's reactivity to parameters of the delayed-neutron emitters and the life expectance of the neutrons.
in-house *Industrial Engineering.* of or relating to work done directly by an enterprise, using its own staff and facilities, as opposed to work contracted out to others.
Iniomi *Vertebrate Zoology.* in some classifications, an order of small bony fish that includes the lantern fishes (family Myctophidae) and lizard fishes (family Synodontidae).
init. initial.
initial *Linguistics.* **1.** occurring at the beginning of a word or syllable. **2.** the first letter of a word, especially of a proper name. *Medicine.* relating to the very first stage of any condition or process. *Microbiology.* any portion of an organism that eventually develops into a distinguishing, characteristic structure.
initial aiming point *Ordnance.* a point at which a gun is aimed to establish a line of reference from which directional angles can be measured to aim at the target or other points.
initial boiling point *Chemical Engineering.* the recorded temperature when the first drop of distilled vapor liquefies and drops from the end of a condenser, according to the American Society for Testing and Materials petroleum-analysis distillation procedures.
initial condition *Meteorology.* a prescription of the state of a dynamical system at a specified time; for all subsequent times, the equations of motion and boundary conditions determine the state of the system. *Computer Programming.* **1.** see ENTRY CONDITION. **2.** the value of a variable prior to any computation or processing.
initial free space *Mechanics.* the initial volume of a gun chamber, behind the projectile, that is not filled with propellant; the spaces between and in the grains are included.
initial great-circle course *Navigation.* that portion of a great-circle course that is steered at departure.
initial heading *Navigation.* the heading taken by a craft on departure.
initial instructions *Computer Programming.* procedures stored within a computer that facilitate the loading of programs.
initial inverse voltage *Electronics.* the peak-inverse voltage that appears in a rectifier tube when current just stops flowing.
initialize *Computer Programming.* **1.** to set all variables and counters to their proper values prior to starting a calculation or loop. Also, PRESET. **2.** to format a disk for use with a particular computer system.
initial landform *Geology.* a landform created by direct orogenic, volcanic, or epeirogenic activity, and having original features that are largely unmodified by erosion.
initial lead *Ordnance.* the difference between the initial location of a moving target and the aiming point of the gun; it should be equal to the distance the target will travel in the time it will take the projectile to arrive at the aiming point.
initial mass *Space Technology.* the mass of a rocket or missile at launch.

initial mass function *Astrophysics.* the distribution of stellar masses when a galaxy, star cluster, or other stellar group forms from a cloud of gas.

initial nuclear radiation *Nucleonics.* the radiation emitted from a nuclear explosion for one minute after detonation.

initial operational capability *Ordnance.* the date at which a new system is ready to commence operations. Also, IOC.

initial point *Cartography.* the point from which any survey is begun. *Ordnance.* an easily identified point from which an aircraft begins its bomb run.

initial program load *Computer Programming.* the procedure that causes disk and tape operating systems to begin processing.

initial recoil *Ordnance.* the recoil in a small arm before the bullet leaves the muzzle.

initial reserves *Ordnance.* in an amphibious operation, the supplies unloaded immediately after the assault wave; they are used to initiate and sustain combat until higher supply installations can be established.

initial segment *Mathematics.* the set of all predecessors of an element of a partially ordered set. In particular, if X is a partially ordered set, and if $a \in X$, then the set $\{x \in X : x < a\} = s(a)$ is the **strict initial segment** determined by a, and the set $\{x \in X : x \le a\} = s(a)$ is the **weak initial segment** determined by a. Used to construct ordinal numbers, thus extending the process of counting beyond the natural numbers.

initial shot pressure *Ordnance.* the pressure in a gun barrel behind the projectile, which is required to instigate movement of the projectile against frictional forces.

initial surge voltage *Electricity.* after a widespread power failure, a spike of voltage that may occur when the power is restored, causing damage to appliances.

initial time delay *Acoustics.* a measure of the sound qualities of an area, defined as the time interval between the first signal arriving at a receiver and the first significant reflection.

initial-value problem *Fluid Mechanics.* a problem whose solution depends on the state of a system at a particular time for which the initial conditions are specified; the solution of such a problem will determine the state of the system at all times thereafter. Also, TRANSIENT PROBLEM. *Mathematics.* **1.** given an evolution equation, i.e., one whose solution evolves with time, an initial value problem is created by specifying additional requirements on the solution at time $t = 0$. **2.** an nth-order partial or ordinary differential equation in which values of the solution and its first $n - 1$ derivatives are specified at a particular value of (one of) the independent variable(s).

initial-value theorem *Mathematics.* **1.** the foundation for solving initial-value problems of ordinary differential equations. Suppose that a function $f(t)$ and its first n derivatives have Laplace transforms, and let $L\{g(t)\} = \int_0^\infty e^{-st}g(t)dt$ denote the Laplace transform of a function $g(t)$, $f^{(k)}(t)$ denote the kth derivative of $f(t)$, and $f_0^{(k)}$ denote the limit of $f^{(k)}(t)$ as t approaches 0 from the right, i.e., $f_0^{(k)}(t) = \lim_{t\to 0+} f^{(k)}(t)$. Then the initial-value theorem is

$$L\{f^{(n)}(t)\} = s^n L\{f(t)\} - \sum_{k=1}^{n} s^{n-k} f_0^{(k-1)}.$$

2. specifically, the case $n = 1$; that is, $L\{f'(t)\} = sL\{f(t)\} - f_0$, where $f_0 = f(0)$ is the initial value.

initial velocity *Physics.* the velocity of an object measured at the beginning of a process or phase of a process.

initial yaw *Ordnance.* the yaw of a projectile as it leaves the muzzle of a gun, resulting from its clearance within the bore.

initiation *Ordnance.* **1.** the action of the first element of an explosive train that, upon receiving the proper impulse, causes the detonation or combustion of the explosive. **2.** the action in a nuclear weapon that sets off a chain reaction in a fissile mass that has reached a critical state.

initiation ceremony *Anthropology.* a ceremony or rite of passage that is required to become a member of a group or enter into a new stage of life; e.g., a puberty ceremony in which certain trials or ordeals are endured to prove the worthiness of the initiate to be an adult member of the community.

initiation codon *Biochemistry.* a codon, usually AUG, that signals the first amino acid in a protein sequence.

initiation factor *Biochemistry.* any of several proteins that, along with mRNA and ribosomes, are needed to initiate protein synthesis.

initiative stage *Psychology.* the third of Erik Erikson's eight stages of human development, during the preschool years, when children develop the ability to initiate independent activities and to investigate their environment.

initiative vs guilt *Psychology.* the conflict that can arise during the initiative stage of development, if the child has a sense of guilt about self-initiated activities.

initiator something that begins a process or operation; specific uses include: *Chemistry.* any substance or molecule that begins a chemical reaction and is not one of the reactants. *Materials Science.* an end unit, such as a radical with either a free electron or ionized group, that is added to a polymer to begin the mechanism of addition polymerization. *Computer Programming.* a utility program in some computers that makes a job or part of a job ready to run. *Ordnance.* the first element of an explosive train; e.g., a detonator.

initiator codon *Molecular Biology.* a specific sequence of bases in mRNA to which a molecule of tRNA carrying a methionine residue binds in the first step of protein synthesis. Also, START CODON.

inject to carry out an injection process.

injected *Petrology.* of an intrusive igneous rock, formed when magma penetrates older rocks.

injected gas *Petroleum Engineering.* gas that has been pumped into an oil-producing formation to supply gas-drive for greater oil production.

injected hole *Mining Engineering.* a borehole into which a cement slurry or grout has been forced by high-pressure pumps and allowed to harden.

injection *Medicine.* the process of driving a substance into the skin, subcutaneous tissue, muscle, blood vessels, or body cavities, usually by means of a needle. *Mechanical Engineering.* the process of spraying fuel into the inlet manifold or cylinder of an internal-combustion engine by means of an injection pump. *Geology.* **1.** the intrusion of an ore-bearing magma into the host rock or into surrounding rocks. **2.** a sedimentary structure formed by such a process. *Electronics.* **1.** the process by which a signal is introduced into a device. **2.** the introduction of impurities into a semiconductor. *Space Technology.* **1.** the process of placing a spacecraft into a specific trajectory, or a satellite into orbit. **2.** the point after a launch when nongravitational forces become insignificant factors in the trajectory of a rocket or spacecraft. *Mathematics.* a mapping or function that is one-to-one. Also, ONE-TO-ONE FUNCTION.

injection carburetor *Mechanical Engineering.* a carburetor in which fuel delivery to the jets is maintained by pressure rather than by a float chamber. Also, PRESSURE CARBURETOR.

injection electroluminescence *Electronics.* the radiation that arises when minority charge carriers are recombined and injected into the semiconductor junction with a forward bias.

injection fluid *Petroleum Engineering.* water or gas that is injected into a reservoir formation to augment hydrocarbon production. Thus, **injected water, injected gas.**

injection folding *Petrology.* a deformative structure in a plastic layer between solidified, compacted layers, due to differential thickness changes.

injection gneiss *Petrology.* a composite rock whose banding results in whole or in part from a layer-by-layer injection of granitic magma.

injection grid *Electronics.* **1.** the electrode that controls the flow of electrons while preventing interaction between the screen and control grids in a vacuum tube. **2.** the point in a multigrid converter tube where a signal from a local oscillator is applied. **3.** in some superheterodyne receivers, the point where the oscillator signal enters the mixer stage.

injection laser *Optics.* a laser in which a forward-based semiconductor diode converts input power directly into coherent light. Also, DIODE LASER.

injection locking *Electronics.* a process that stabilizes an oscillator's frequency by injecting a weak signal whose frequency equals the oscillator's natural frequency.

injection luminescent diode *Electronics.* a device that serves as a source of visible or near-infrared light in a triggering circuit, such as a switch.

injection mold *Mechanical Engineering.* a mold that is filled with molten material by means of a heating cylinder located outside the mold.

injection molding *Mechanical Engineering.* the process of molding shapes out of softened, usually thermoplastic material that has been forced or squirted out of a heated cylinder by a plunger into a water-cooled metal mold; used, for example, in powder metallurgy.

injection pressure *Petroleum Engineering.* the pressure of the fluid forced into formations of oil for pressure or waterflood maintenance.

injection pump *Mechanical Engineering.* a unit of a diesel engine or gasoline injection system that injects measured quantities of fuel for each cylinder in turn on a compression stroke.

injection well *Petroleum Engineering.* a well created for injecting water or gas into a formation in secondary recovery, to supply extra energy for driving the remaining oil in the reservoir toward the production wells; also used in pressure-maintenance operations. *Hydrology.* see RECHARGE WELL.

injective module *Mathematics.* a module P over a ring R is said to be injective if, given any diagram of R-module homomorphisms,

$$\begin{array}{ccccc} & & g & & \\ 0 & \to & A & \to & B \\ & & f \downarrow & & \\ & & J & & \end{array}$$

with the top row exact (that is, g is a monomorphism), there then exists an R-homomorphism $h: B \to J$ so that the following diagram commutes ($hg = f$):

$$\begin{array}{ccccc} & & g & & \\ 0 & \to & A & \to & B \\ & & f \downarrow & & \downarrow h \\ & & & J & \end{array}$$

injectivity index *Petroleum Engineering.* the barrels per day of gross liquid pumped into an injection well per pounds per square inch pressure differential, this differential being the average injection pressure minus the average formation pressure.

injectivity test *Petroleum Engineering.* at various pressures, a sequence of tests of reservoir water injection rates to forecast the performance of an injection well.

injector *Mechanical Engineering.* **1.** a device that introduces feedwater into a boiler by means of a jet or stream. **2.** see FUEL INJECTOR. *Electronics.* an element that introduces a signal into a device.

injurious fighting *Behavior.* fighting that is actually intended to kill or injure the opponent, as opposed to ritualized fighting between members of the same species competing for some resource.

injury *Medicine.* any damage or wound to the body; usually applied to damage inflicted by an external force.

injury potential *Physiology.* the current flow from an uninjured nerve or muscle to an injured part.

ink *Materials.* a liquid, powder, or paste used for writing or printing on paper and other material, usually consisting of a colorant and a carrier vehicle. *Zoology.* a dark protective fluid ejected by certain cephalopods.

ink bleed *Computer Technology.* the capillary flow of ink beyond the original edges of a printed character, sometimes causing errors in an optical character recognition system. Also, **ink smudge, ink squeezeout.**

inkblot test *Psychology.* **1.** another name for the Rorschach test. See RORSCHACH TEST. **2.** any of various other psychological tests in which the subject responds to an abstract design resembling blotted ink.

ink-cap fungi *Mycology.* a species of fungi belonging to the genus *Coprinus,* which are characterized by the black fluid that their spores produce.

ink disease *Plant Pathology.* a disease of European chestnuts caused by the fungus *Phytophthora cambivora* and characterized by the appearance of black cankers and an inky fluid that seeps out of the tree trunk. Also, BLACK CANKER.

ink drum *Graphic Arts.* a solid or cored metal drum used in some inking systems to break down ink and carry it to form rollers.

ink fountain *Graphic Arts.* the reservoir on a press from which ink is distributed to the rollers.

ink-jet printer *Graphic Arts.* a nonimpact printing machine used in ink-jet printing.

ink-jet printing *Graphic Arts.* a computerized printing process in which tiny laser-activated jets, following instructions from a coded tape, spray ink in the form of characters directly onto paper.

ink knife *Graphic Arts.* a flat trowel-like tool used to mix and spread inks.

ink-receptive *Graphic Arts.* having a surface that will attract greasy inks rather than water.

ink-repellant *Graphic Arts.* having a surface that will attract water rather than greasy inks.

ink roller *Graphic Arts.* in offset lithography, one of a set of rollers designed to carry ink from an ink supply (usually in a fountain) to the printing plate, which then transfers it to the blanket.

ink sac *Invertebrate Zoology.* an organ in many cephalopods that produces and ejects an inky fluid to conceal the animal from a predator.

ink stamping *Graphic Arts.* a relief printing process in which the image is simultaneously cut below the printing surface and filled with ink; similar to letterpress, but with a stronger impression. Also, COLD STAMPING.

inland *Geography.* describing land that is away from the sea.

inland ice *Hydrology.* **1.** a continental glacier or ice sheet. **2.** the ice forming the inner part of either of these structures.

inland rules of the road *Navigation.* rules specifying the conduct of vessels in inland waters, generally inshore of the outermost buoys.

inland sea *Geography.* a sea that is entirely surrounded by land, such as the Mediterranean.

inlay *Medicine.* a dental restoration made outside of a tooth to correspond with the form of a prepared cavity and secured with some form of cement into the tooth. *Surgery.* a tissue, such as skin or bone, inserted into a tissue defect.

inlay graft *Surgery.* a mucosal or skin graft applied by spreading the graft over a stent and suturing the graft and mold into a prepared pocket. *Botany.* a graft in which the scion is inserted at a point in the stock where the bark has been removed.

inlet *Geography.* **1.** a small arm of the sea. **2.** an indentation in the shoreline of an ocean, lake, or river. **3.** a narrow opening allowing the sea to penetrate the land, often connecting with another body of water such as a lagoon. *Mechanical Engineering.* an opening through which air, fuel, or power enters a machine or system.

inlet of the pelvis *Anatomy.* the upper opening into the pelvic cavity.

inlet valve *Mechanical Engineering.* a valve through which fluid is drawn into a cylinder in a positive-displacement engine, pump, or compressor.

inlier *Geology.* an irregular area of older, exposed rock that is surrounded completely by outcrops of stratigraphically younger rocks.

in-line or **inline** *Engineering.* **1.** of a device or machine, having similar parts positioned together and in a straight line, such as cylinders arranged one behind the other in an engine. **2.** of drilling equipment, centered over a borehole.

in-line analysis *Biotechnology.* the analysis of a process using direct reading probes inserted directly into the process stream; this method eliminates errors that could be caused by the disturbance required for sample removal.

in-line assembly machine *Industrial Engineering.* an assembly machine that is a direct part of a production line, as opposed to one feeding its output into the line from the side or in a rotary production process.

in-line coding *Computer Programming.* a part of program code that is stored in the main path of a routine.

in-line engine *Mechanical Engineering.* a multicylinder engine that has a bank of cylinders aligned along a common crankcase.

in-line guns *Electronics.* three electron tubes that are aligned horizontally in a color television system, characterized by having metal plates with vertical slots over their color phosphor stripes.

in-line linkage *Mechanical Engineering.* an assembly consisting of a power steering linkage coupled with a control valve and actuator.

in-line procedure *Computer Programming.* a subroutine that is inserted into and a part of the main sequential and controlling flow of the program, as opposed to being stored elsewhere and called up as needed. Also, IN-LINE SUBROUTINE, OPEN SUBROUTINE.

in-line processing *Computer Programming.* the processing of data that has not been previously sorted or edited.

in-line subroutine see IN-LINE PROCEDURE.

in-line tuning *Electronics.* a process by which all amplifier stages are tuned to the same frequency.

in-lining see OPEN CODING.

in-motion time *Transportation Engineering.* the total time that a vehicle is moving; that is, running time minus dwell time.

I.N.N. International Nonproprietary Names.

innage *Engineering.* the quantity of goods remaining in a container after shipment. *Mechanical Engineering.* the quantity of remaining liquid in a fuel tank after operation, especially of a flight vehicle.

innate *Biology.* present at birth; inherited. *Behavior.* of or relating to a behavior that occurs naturally in all members of a species despite differences in environment or experience. *Mycology.* describing fungal structures embedded in the thallus or surface on which the fungi are growing.

innate capacity for increase see INTRINSIC RATE OF INCREASE.

innate rate of increase see INTRINSIC RATE OF INCREASE.

innate releasing mechanism *Behavior.* the sensory and neural components of an organism that combine to filter incoming stimuli and allow only the relevant stimuli to trigger behavior.

inner automorphism *Mathematics.* **1.** let g be any fixed element in a group G. The inner automorphism induced by g is the map $\alpha: G \to G$, given by $\alpha(x) = g \times g^{-1}$, where x is any element of G; that is, conjugacy by any fixed element of G is an inner automorphism. **2.** let u be a unit in a ring R with identity. The inner automorphism induced by u is the map $\alpha: R \to R$, given by $\alpha(r) = uru^{-1}$, where r is any element of R.

inner bottom *Naval Architecture.* a layer above the bottom plating of a ship's hull, protecting the interior against leaks or punctures.

inner cell mass *Developmental Biology.* the original small group of cells that segregate within the enveloping trophoblast at one pole of the hollow mammalian blastocyst and later develops into the embryo. Also, EMBRYONIC MASS.

inner core *Geology.* the central or innermost part of the earth's interior, extending from a depth of 5100 kilometers to the center of the earth at 6371 kilometers, and equivalent to the G layer.

inner-directed *Psychology.* tending to be influenced by one's own internal beliefs and values rather than those of others.

inner ear *Anatomy.* the portion of the ear that includes the semicircular canals, vestibule, and cochlea.

inner hearth see BACK HEARTH.

inner mantle see LOWER MANTLE.

inner planet *Astronomy.* a general term for the any of the terrestrial planets (Mercury, Venus, Earth, and Mars).

inner product *Mathematics.* **1.** the inner product of two vectors $u = (\alpha_1, \ldots, \alpha_n)$ and $v = (\beta_1, \ldots, \beta_n)$ in an n-dimensional real vector space is denoted $<u, v>$ or (u,v) and is a scalar equal to $\sum_{i=1}^{n} \alpha_i\beta_i$. Also, DOT PRODUCT, SCALAR PRODUCT. **2.** in general, let V be a vector space over a field F. An inner product on V is a function $f: V \times V \to F$, i.e., a scalar-valued function of two vectors satisfying the following: (a) $f(v,v) \geq 0$ with equality holding if and only if v is the zero vector; (b) $f(au,bv,w) = af(u,w) + bf(v,w)$ for vectors u, v, w and scalars a and b; and (c) if F is the field of real numbers, then $f(u,v) = f(v,u)$, and if F is the field of complex numbers, then

$$f(u,v) = \overline{f(v,u)},$$

where $\overline{f(v,u)}$ denotes the complex conjugate of $f(v,u)$. The inner product is usually denoted $<u,v>$ or (u,v).

inner product of functions *Mathematics.* an inner product on a vector space of real or complex-valued functions, usually defined by

$$<f,g> = \int f(x)\, \overline{g(x)}\, dx,$$

where $\overline{g(x)}$ denotes the complex conjugate of $g(x)$.

inner product of tensors *Mathematics.* a contraction of tensors.

inner product space *Mathematics.* a vector space on which an inner product has been defined; e.g., Hermitian space (vector space over the complex numbers) and Euclidean n-space.

inner quantum number *Atomic Physics.* a quantity J that provides an atom's entire angular momentum without its nuclear spin.

inner tube *Engineering.* an inflatable, airtight rubber tube placed inside the casing of a pneumatic tire; used to hold air under pressure.

innervation *Neurology.* the distribution of nerves in an area of the body, and the supply of nerve energy or stimulus to that area.

innidiation *Oncology.* the growth and proliferation of cells in a part of the body to which they have metastasized. Also, INDENIZATION.

innominate artery *Anatomy.* the large right branch from the arch of the aorta that divides into the right common carotid artery and right subclavian artery. Also, BRACHIOCEPHALIC ARTERY.

innominate bone see HIPBONE.

innovative behavior *Behavior.* any action that occurs spontaneously in a new situation, rather than as the result of trial-and-error learning. Also, **innovation.**

Inoceramus *Paleontology.* a genus of toothless, deeply ribbed bivalves in the subclass Pteriomorphia; Lower Jurassic to Cretaceous.

inoculant *Science.* a material introduced in an inoculation process.

inoculate *Science.* to carry out an inoculation process.

inoculation *Medicine.* the introduction of a disease agent into a healthy individual in order to produce a mild form of the disease followed by immunity. *Microbiology.* the transfer of microorganisms to a sterile medium for the purpose of initiating a culture. *Metallurgy.* the addition of a particulate material to a molted metal or alloy to modify the structure of the resulting casting.

inoculum *plural,* **inocula.** *Microbiology.* the small quantity of material containing microorganisms that is introduced into a medium for culture.

Inonotus *Mycology.* a genus of fungi belonging to the order Agaricales, characterized by the reddish-brown color of its fruiting body.

inoperable *Surgery.* unsuitable for a surgical procedure; determining factors for this status include the general physical condition of the patient, and the location of the diseased or damaged area.

Inoperculate *Mycology.* a series of fungi belonging to the order Chytridiales, characterized by the dissolution of the sporangium wall when a spore is being discharged from the sporangium.

inoperculate *Biology.* lacking an operculum.

inorder *Computer Science.* an order of visiting binary trees, in which the left subtree of a node is examined, followed by the node itself, followed by the right subtree.

inorganic *Biology.* **1.** not organic; not having the structure or undergoing processes that are characteristic of living organisms, i.e., plants and animals. **2.** not composed of living or formerly living material. *Chemistry.* **1.** not an organic substance; not a hydrocarbon or a derivative of hydrocarbon. **2.** of or relating to inorganic chemistry.

inorganic acid *Inorganic Chemistry.* any of the mineral acids, such as hydrochloric acid, sulfuric acid, carbonic acid, or phosphoric acid, as opposed to the organic or carboxyl acids.

inorganic chemistry *Chemistry.* **1.** in earlier use, the study of those elements and compounds that are not associated with living organisms. **2.** in later use, the study of all chemical substances except hydrocarbons and their derivatives. Thus in general it excludes all substances that contain carbon, although there are certain exceptions; e.g., carbon oxides, carbon sulfides, and metallic carbonates are usually included.

Inorganic Chemistry

An earlier definition of Inorganic Chemistry as the study of those chemical compounds (and elements) not associated with living organisms requires some modification as a result of recent trends and developments in the subject.

The prime focus remains a consideration of the chemistry of 103 elements, 90 of which are naturally occurring. The elements have associated metallic or non-metallic properties. Chemical combination results in the formation of a wide range of compounds, of which there are now many applications and uses. The one element that is a notable exception is carbon. However, in addition to the large number of organic compounds, there are a significant number of inorganic compounds of carbon centering around, e.g., the carbonates, cyanides, and carbides.

Chemistry has widened considerably as a result of cross-subject interests, with research more broadly based and less confined to the Inorganic/Organic/Physical subject areas. Thus there is now an extensive chemistry of organic molecules (fragments of molecules) bound to metal centers (Organometallic Chemistry). There is also an increasing appreciation of the wide-ranging and sophisticated roles trace metals play in biology, haemoglobin being a prime example (Bio-inorganic Chemistry). Significant developments have also emerged in the solid state (Materials Sciences) area, including studies on semi- and super-conductors, inorganic polymers, and so on. All these topics are generally considered as wholly or partly falling within the teaching and research of Inorganic Chemistry.

Investigations relating to molecular structure and properties, chemical bonding, reactivities, and photochemical effects, the latter including studies on solar energy, continue to be important. A wide range of physical techniques are applied in researching the subject.

A. Geoff Sykes
The University of Newcastle upon Tyne

inorganic chert *Petrology.* chert from siliceous colloids precipitated from water saturated with silica.

inorganic liquid laser *Optics.* a laser in which the active medium is an inorganic liquid, such as neodymium-selenium oxychloride.

inorganic oxidation see CHEMOSYNTHESIS.

inorganic peroxide *Inorganic Chemistry.* any inorganic compound containing two oxygen atoms joined to each other by a single bond; they easily release atomic oxygen and are therefore fire hazards.

inorganic pigment *Inorganic Chemistry.* any inorganic compound used as a coloring agent.

inorganic polymer *Inorganic Chemistry.* a polymer with the main chain made up of atoms other than carbon; possible elements include silicon, sulfur, and oxygen.

inorganic theory *Petroleum Engineering.* a theory of petroleum formation that attributes the origin of gas and oil to gaseous, hydrocarbonaceous, and volcanic emanations, rather than organic matter.

inosilicate *Mineralogy.* a group of silicates in which the silicate tetrahedrons are arranged in either single or double linear chains linked by shared oxygen atoms. Also, CHAIN SILICATE.

inosine *Biochemistry.* $C_{10}H_{12}N_4O_5$, a nucleoside formed by deamination.

inosinic acid *Biochemistry.* $C_{10}H_{13}N_4O_8P$, a phosphate ester of inosine, produced in nature by the deamination of adenylic acid.

inositol *Organic Chemistry.* $C_6H_6(OH)_6$, a sugarlike vitamin of the B complex that is found in many animal and plant tissues or produced synthetically; occurs in nine stereoisomeric forms; sweet-tasting, odorless white crystals; soluble in water and slightly soluble in alcohol; dihydrate form melts at 216°C and anhydrous form melts at 225–227°C; used in medicine.

inositol kinase *Enzymology.* an enzyme that catalyzes the phosphorylation of inositol.

inositol monophosphate *Biochemistry.* $C_6H_{13}O_9P$, *myo*-inositol 1-(dihydrogen phosphate).

inositol trisphosphate *Biochemistry.* the intracellular second messenger produced by the action of phospholipase C on membrane phosphatidylinositol phosphate in response to stimulation of cell-surface receptors by growth factors, hormones, or neurotransmitters.

Inoue-Melnick virus *Virology.* a virus isolated from patients with chronic diseases of the central nervous system; e.g., multiple sclerosis.

Inoviridae *Virology.* a family of ssDNA-containing phages with rod-shaped virions that protrude during maturation through the host cell membrane.

Inovirus *Virology.* a genus of nonlytic ssDNA-containing phages of the family Inoviridae that infect enterobacteria.

in-phase *Physics.* describing a state of coherency in which two or more periodic quantities of the same frequency have the same phase.

in-phase current *Electricity.* the component of current that is parallel to the voltage phasor. Also, RESISTIVE CURRENT.

in-phase rejection see COMMON-MODE REJECTION.

in-phase signal *Electronics.* **1.** see COMMON-MODE SIGNAL. **2.** the signal that is formed from two subcarriers of the signal used to modulate color in a television. **3.** the sideband frequencies generated by changing the characteristics of chrominance signals, whose colors range from orange to bluish-green. Also, I SIGNAL.

in-pile *Nucleonics.* referring to equipment or devices placed within a nuclear reactor.

in-pile loop *Nucleonics.* an experiment carried out inside a nuclear reactor that employs a liquid held in a closed circuit to transfer heat from the reactor core.

input to enter something into a process or system, or something that is entered; specific uses include: *Electronics.* **1.** the energy that is fed to a device. **2.** the point at which energy enters a device. **3.** to convey or transfer energy or data to a device. *Computer Programming.* **1.** data that is transferred from external storage or an input device to the computer's internal storage. **2.** to introduce data from auxiliary storage or an input device into main memory.

input admittance *Electricity.* the admittance of a device as seen from the input terminals; the reciprocal of input impedance.

input area *Computer Programming.* an area of a computer's internal storage reserved for the acceptance of input data. Also, INPUT BLOCK, INPUT SECTION.

input block *Computer Programming.* **1.** see INPUT AREA. **2.** an input buffer. **3.** a block of records treated as a unit for transferring from an external source or storage medium to internal computer memory.

input buffer register see INPUT REGISTER.

input capacitance *Electronics.* the effective capacitance between an input terminal and the reference terminal (usually ground).

input data *Computer Programming.* see INPUT, def. 1.

input device *Computer Technology.* any device used for transferring data into a computer system, such as a disk drive or keyboard. Also, **input equipment.**

input gap *Electronics.* the point at which the velocity modulation of an electron stream in a microwave tube first takes place. Also, BUNCHER GAP.

input impedance *Electricity.* **1.** the impedance of a device as seen from the output terminals **2.** the impedance at the transverse plane of a transmission line port.

input-limited *Computer Technology.* describing a program whose processing time is limited by the speed of the input unit.

input/output *Computer Technology.* **1.** the transfer of data between a computer and a peripheral device. **2.** of or relating to equipment, techniques, or media used for this process. Also, I/O.

input/output adapter see CHANNEL ADAPTER.

input/output bound see INPUT/OUTPUT LIMITED.

input/output buffer *Computer Technology.* a block of storage used to transfer data received from a peripheral unit to and from main memory.

input/output channel see DATA CHANNEL.

input/output controller *Computer Technology.* a separate logic processor that provides a high-speed interface between peripheral devices and main storage, including buffer operations, channel selection, scheduling, and interrupt handling.

input/output control system *Computer Programming.* a set of standard computer routines that initiate and control the operations and functions associated with the transfer of data between main memory and an input/output device.

input/output control unit *Computer Technology.* the logic and connectivity needed to initiate and control the flow of data to and from a specific input/output device.

input/output device *Computer Technology.* a unit that sends data to the computer for processing or that receives data from the computer, or both.

input/output handler *Computer Programming.* a program that communicates directly with an input/output device, sometimes including error detection and recovery.

input/output instruction *Computer Programming.* a program instruction that controls the transfer of data to and from input/output devices such as printers or disk drives.

input/output interrupt *Computer Technology.* an external hardware interrupt caused by an input/output device requesting a transfer of data (input) or signaling the completion of transfer of data (output).

input/output interrupt identification *Computer Programming.* the identification by the system of the device and channel that caused an input/output operation to interrupt.

input/output library *Computer Programming.* a set of standard input/output handlers and drivers, customized for various peripheral devices, which can be included in application programs.

input/output limited *Computer Technology.* describing a program whose processing time is limited by the speed of the input/output peripheral unit. Also, INPUT/OUTPUT BOUND, PERIPHERAL LIMITED.

input/output referencing *Computer Programming.* the technique of referring to input/output data symbolically by name instead of providing the exact physical location on tape or disk.

input/output register *Computer Technology.* a register that can load new data (read) or output the register's contents (write).

input/output switching *Computer Technology.* the allocation to a peripheral unit of more than one channel for communication with the central processing unit.

input/output time *Computer Science.* the amount of time spent by a program in performing input/output operations.

input/output traffic control *Computer Technology.* the controlled time-sharing of a computer's memory among peripheral devices and the central processor, allowing simultaneous input/output operations to occur.

input record *Computer Programming.* a set of related data that is organized and contained on a computer storage medium ready to be input to and processed by the computer as a unit.

input register *Computer Technology.* a special-purpose shift register that receives serially transmitted data and retains it until a complete word is accumulated, then transfers the data to a buffer, another register, or internal storage. Also, INPUT BUFFER REGISTER.

input routine *Computer Programming.* a utility routine that controls the reading of data or programs into the computer.
input section see INPUT AREA.
input signal *Robotics.* any external signal sent to the controller of a robot.
input station *Computer Technology.* a computer or intelligent terminal that is linked with a central computer system to feed manufacturing data directly to a processor, enabling real-time processing of the data.
inquartation *Metallurgy.* a step in the assaying of precious metals.
inquiline *Zoology.* an animal that lives in the home of another and shares its food.
inquiry *Computer Programming.* a request for specific data to be retrieved from either storage or a database.
inquiry and communications system *Computer Science.* a computer system that is capable of conducting on-site and long-distance data requests.
inquiry and subscriber display *Computer Technology.* a video display terminal and keyboard connected to a central computer which is part of a service enabling subscribers to gain access to specialized information or messages, such as stock prices or news reports.
inquiry display terminal *Computer Technology.* a type of video display terminal in which an inquiry placed into the computer through the keyboard is simultaneously displayed on the screen; the computer's response to the inquiry on the screen is then displayed on the screen as well.
inquiry station *Computer Technology.* a unit or device from which a request for specific data is made. Also, **inquiry unit.**
ins. inches.
insane *Psychology.* judged by legal authority to be incapable of handling one's own affairs and not legally responsible for one's actions.
insanity *Psychology.* a legal term for a condition in which an individual is judged to be insane.
inscribe *Computer Technology.* to rewrite key information onto a document, such as the check amount onto a check, so that it can be read by an optical character reader or other scanner.
inscribed *Mathematics.* **1.** a polygon (or polyhedron) is said to be inscribed in a closed configuration of curves (or surfaces) if: (a) every vertex of the polygon is incident on the configuration, and (b) the polygon and the configuration have no other points in common, so that the polygon is inside the configuration. **2.** a closed configuration of curves (or surfaces) is said to be inscribed in a polygon (or polyhedron) if: (a) every side (or face) of the polygon is tangent to the configuration, and (b) the polygon and the configuration have no other points in common, so that the configuration is inside the polygon.
inscription *Pharmacology.* the part of a prescription containing the names and amounts of the ingredients.
insect *Invertebrate Zoology.* **1.** a member of the class Insecta, such as a fly, ant, bee, wasp, mosquito, beetle, grasshopper, or flea. **2.** informally, any other arthropod resembling an insect, such as a spider or centipede. (Going back to a Greek word meaning "notched" or "cut;" with reference to an insect's segmented body.)

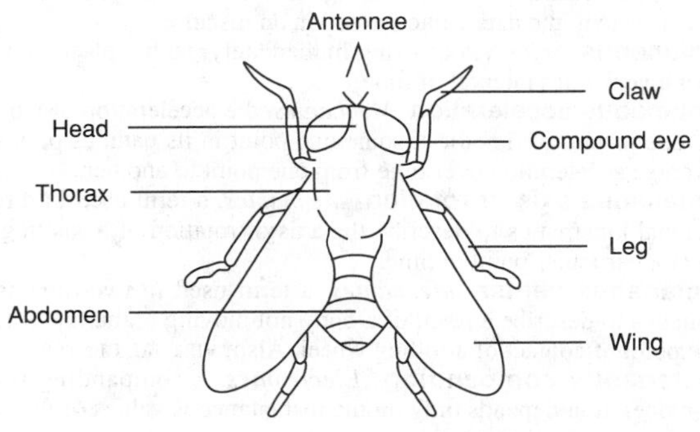

insect

Insecta *Invertebrate Zoology.* a class of air-breathing arthropods usually having a segmented body with a chitinous exoskeleton, a pair of compound eyes, a pair of segmented antennae, three pairs of mouthparts, three pairs of legs, and two pairs of wings. There are over 800,000 identified species, more than all other animal species combined.
insecticide *Materials.* any substance, either organic, inorganic, botanical, or microbial, that is used to destroy insects.
insect invader *Entomology.* a species that displaces the incumbent species through competition; for example, the fire ant from South America has displaced many native ant species in North America.
Insectivora *Vertebrate Zoology.* a very diverse order of small eutherian mammals that feed chiefly on insects, including the moles, hedgehogs, shrews, and tree shrews. The first insectivores appeared in the Lower Cretaceous.
insectivore *Biology.* an organism that eats insects, especially one that feeds mainly or exclusively on insects. *Vertebrate Zoology.* a member of the order Insectivora.
insectivorous *Biology.* feeding on insects; insect-eating. *Botany.* specifically, describing plants that capture and consume insects.
insectology *Science.* another name for entomology, the scientific study of insects.
insect pathology *Entomology.* a biological discipline embracing the general principles of pathology as applied to insects.
insect physiology *Entomology.* the study of the functional properties of insect tissues and organs.
insect picornaviruses *Virology.* small RNA-containing viruses that infect insects and resemble other members of the family Picornaviridae in size, density, and protein content.
inselberg *Geology.* a generally bare and rocky, residual, steep-sided rounded hill or small mountain in arid or semiarid regions, which rises abruptly from the extensive, flat, lowland erosion surface that surrounds it. Also, ISLAND MOUNTAIN.
insemination *Zoology.* the introduction of sperm into the genital tract of a female. *Physiology.* in humans, the introduction of seminal fluid into the vagina or cervix.
insensible *Neurology.* **1.** imperceptible to the senses. **2.** unconscious. **3.** lacking the sensory ability to perceive a stimulus.
insensitive time see DEAD TIME.
insequent stream *Hydrology.* a stream that has developed on the existing surface, but whose course is seemingly haphazard, being independent of the general form and slope of the surface.
insert to put in, or something put in; specific uses include: *Graphic Arts.* **1.** in book binding, to place one or more folded sheets within a signature. **2.** a set of pages so placed. **3.** in photography, to fit one negative inside another; a negative so attached. *Metallurgy.* **1.** in casting, a component inserted in the mold that remains intentionally attached to the resulting casting. **2.** a removable component of a die or mold.
insert bit *Mechanical Devices.* a type of drill bit in which multishaped, hard cutting points can be placed. Also, SLUG BIT.
inserted *Botany.* growing upon or attached to; united by natural growth to a larger or more important structure. *Anatomy.* attached; having an insertion.
insertion *Anatomy.* a place of attachment, as of a muscle to a bone. *Graphic Arts.* something inserted, especially a correction or alteration in existing text. *Computer Science.* placement of a new data item in its proper position in an ordered sequence, such as a list, array, or data file.
insertional inactivation *Molecular Biology.* the inactivation of a gene due to the insertion of a piece of foreign DNA into the coding sequence of the gene.
insertional mutagenesis *Genetics.* any mutation or alteration of a DNA sequence that results from the insertion of DNA.
insertion gain *Electronics.* the increase in power delivered to a load when an amplifier is inserted between the source and the load; usually measured in decibels.
insertion loss *Electronics.* the decrease in power delivered to a load when a circuit, such as a filter or attenuator, is inserted between the source and the load; usually measured in decibels.
insertion meter *Engineering.* a flowmeter that determines the mass flow of water in a stream or pipe by measuring the rotation rate of a small propeller.
insertion mutation *Molecular Biology.* a type of mutation in which one or more new bases are added between pre-existing bases on the nucleic acid chain. Also, ADDITION MUTATION.
insertion polymerization *Materials Science.* a chain propagation technique that results in the insertion of a monomer between a catalyst and the chain to which it is attached.
insertion sequence *Molecular Biology.* a mobile genetic element that can be inserted into or removed from DNA sequences by recombination; one end of its sequence is characteristically an inverted repeated sequence of the other end. Also, **insertion element.**

insertion sequence selection *Biotechnology.* a technique that distinguishes and selects foreign cells that have been inserted by gene manipulation into a general cell population.
insertion sort *Computer Science.* a method of sorting in which records are successively considered from left to right; each record is inserted into the sorted (left) portion of the file in proper order so that the left portion remains sorted. It is a good method for almost-sorted files.
insertion vector *Genetics.* a lambda bacteriophage that is genetically engineered to hold foreign DNA fragments in recombinant DNA experiments.
inset *Cartography.* **1.** a map or chart of an area that would otherwise lie beyond what is covered by a main map, set in borders and positioned within the open-water areas of the main map. **2.** a representation of a small portion of the main map at a larger scale. **3.** a representation of the main map at a smaller scale, showing its orientation to adjacent areas (called an **orientation inset**).
inshore *Geography.* close to the shore. Also, OFFSHORE.
inshore current *Oceanography.* the flow of water, as in longshore and rip currents, inside the surf zone.
inshore zone *Geology.* a beach zone of variable width that extends from the low-tide line to the farthest reach of the breaker zone.
inside caliper *Mechanical Devices.* a caliper whose straight legs end in a slightly outward orientation; used to measure inside diameters.
inside diameter *Mechanical Devices.* a line with endpoints that lie on the circumference of a spherically shaped object.
inside gauge *Design Engineering.* the inside diameter of a drill bit between its cutting points, such as of an inset diamond laid up on the inside-wall surface of a coring bit; used in oil excavation applications.
inside micrometer *Mechanical Devices.* a type of micrometer whose points are turned outward to measure the internal dimensions of an object.
inside-to-outside model *Microbiology.* a model that depicts the mechanism by which peptidoglycan is incorporated into the cell wall of growing Gram-positive bacteria such as bacilli.
inside work see INTERNAL WORK.
insight *Psychology.* **1.** the ability of an individual to understand his origin, nature, and the mechanisms that underlie his attitudes and behavior. **2.** the point at which a patient realizes his symptoms are abnormal.
insight learning *Behavior.* a type of learning in which previously acquired information is reorganized to provide a solution to a problem situation. Thus, **insight behavior.**
in situ [in si´too] in the normal or natural position; specific uses include: *Geology.* in its original location. *Archaeology.* describing an artifact recovered from the place where it was deposited by its last user. *Medicine.* **1.** in the normal location in the body. **2.** confined to the site of origin without invasion of neighboring tissue. (From a Latin term meaning "in place.")
in situ combustion *Petroleum Engineering.* a procedure for recovering oil of low API gravity and high viscosity from a field, which involves igniting the oil downhole in the formation; the heat breaks down the oil into coke and light oil, and while the coke burns, the lighter, less-viscous oil is forced ahead to the wellbores of the producing wells.
in situ composite see EUTECTIC COMPOSITE.
in situ foaming *Engineering.* the placement and combination of ingredients for foamable plastic in the place where the final product is used, as in walls where insulation is desired.
in situ hybridization *Molecular Biology.* an experimental procedure for localizing a specific gene or DNA sequence within a chromosome based on binding a complementary, radioactively labeled segment of RNA or DNA to it.
insol. insoluble.
insolation *Meteorology.* **1.** solar radiation that is received at the earth's surface. **2.** the rate at which direct solar radiation is incident upon a unit horizontal surface. *Medicine.* **1.** treatment by exposure to the sun's rays. **2.** see SUNSTROKE.
insoluble *Chemistry.* incapable of dissolving; not soluble.
insoluble anode *Chemistry.* an anode that resists oxidation during electrolysis.
insoluble residue *Geochemistry.* the mainly siliceous material that remains after a geological sample has been dissolved in hydrochloric or acetic acid.
insoluble salt *Chemistry.* a salt that is either insoluble or (more often) only slightly soluble.
insomnia [in säm´nē ə] *Medicine.* the inability to sleep; abnormal sleeplessness or frequent intermittent waking.
insomniac [in säm´nē ak] *Medicine.* a person who is unable to get adequate sleep over an extended period of time.
insomnic *Medicine.* unable to sleep; relating to or characterized by insomnia.
insonate *Radiology.* to subject to ultrasound waves.
insorption *Physiology.* the movement of a substance into the blood, especially from the gastrointestinal tract into the bloodstream.
inspect *Industrial Engineering.* to examine a machine, system, device, material, and so on for defects or abnormal conditions, or to assure that proper procedures are followed. Thus, **inspector, inspection.**
inspiration *Physiology.* the act of drawing air from the outside of the body into the lungs.
inspirator *Chemistry.* a laboratory device that automatically regulates the proportions of the constituents in a mixture of gases.
inspiratory *Physiology.* of or relating to inspiration.
inspiratory capacity *Physiology.* the amount of gas that can be inhaled from the resting end-expiratory position of the chest and expiratory muscles; equal to the sum of the tidal volume and the inspiratory reserve volume.
inspiratory reserve volume *Physiology.* the amount of gas that can be inhaled beyond the tidal volume of normal breathing.
inspirometer *Medicine.* an apparatus for measuring the force, frequency, or volume of inhalations.
inspissation *Chemistry.* the thickening of a liquid by evaporation. *Geochemistry.* specifically, the thickening of an oil deposit over time as gases and lighter fractions evaporate or oxidize, leaving behind the heavier fractions, such as asphalt.
instability the quality or condition of being unstable; specific uses include: *Physics.* the state of a body when any slight displacement decreases its potential energy. Also, UNSTABLE EQUILIBRIUM. *Electricity.* the inability of a component or circuit to be electrically stable, characterized by fluctuating current and voltage. *Control Systems.* persistent, unwanted oscillations in the output of a control system due to excessive positive feedback.
instability line *Meteorology.* a transitory nonfrontal line or band of convective activity in the atmosphere that may extend for several hundred miles; the term generally includes the developing, mature, and dissipating stages of instability, although when the mature stage consists of a line of active thunderstorms, it is known as a squall line.
instability strip *Astrophysics.* a name for a thinly populated region of the Hertzsprung-Russell diagram through which stars pass after their main-sequence lifetimes are over; it is in this region that pulsating variable stars, including Cepheids, are found.
installation *Engineering.* **1.** the fixing in position for use of any device, equipment, mechanical apparatus, or system. **2.** something so fixed. *Military Science.* a group of facilities supporting particular functions and located in the same vicinity; it may be part of a base.
installed capacity *Electricity.* the full-load continuous capacity of an electric generating unit.
instance *Artificial Intelligence.* in a frame system or object-oriented system, a frame or object representing an individual that is an example of a particular class.
instance variables *Artificial Intelligence.* in a frame system or object-oriented system, the data values stored in an instance.
instantaneous *Science.* occurring immediately; taking place in or lasting for a very brief interval of time.
instantaneous acceleration *Mechanics.* the acceleration of a body at some one instant of time or at some one point in its path, as opposed to its average acceleration over time from one point to another.
instantaneous axis of rotation *Mechanics.* a term used in three-dimensional kinematics to describe the axis of rotation of a rotating rigid body at a particular point in time.
instantaneous center *Mechanics.* a term used in two-dimensional kinematics to describe a point of a body not moving at that instant, such as the point of contact of a rolling wheel. Also, VIRTUAL CENTER.
instantaneous companding *Electronics.* a companding process whose operation depends only on the instantaneous values of the signal, independent of earlier values.
instantaneous condition *Physics.* the state of a dynamic system at a particular instant.
instantaneous cut *Engineering.* a cut in rock made by several instantaneous detonators that set off their charges simultaneously.
instantaneous description *Computer Technology.* the total configuration of a computer with an entire set of machine conditions at a given point during a computation.

instantaneous detonator *Engineering.* a detonator that instantly sets off an explosion, with no delay between the passage of electric current through the device and the explosion it causes.
instantaneous effects *Telecommunications.* the harmful impairment of transmission due to instantaneous shifts in phase or wave amplitude in a transmission line.
instantaneous field of view *Optics.* the moment in time that the field of view containing radiation is observed by an imaging system utilizing a scanning mechanism.
instantaneous frequency *Telecommunications.* the frequency or time rate of change by a phase angle whose sine is related to the amplitude of a frequency-modulated wave.
instantaneous fuse *Engineering.* a rapidly burning fuse that proceeds at a rate of several thousand feet per minute; faster than a slow fuse, but slower than a detonating fuse.
instantaneous power *Electricity.* the power at points where an electric circuit enters a region, equivalent to the rate at which energy is transmitted into the region by the circuit.
instantaneous power output *Electricity.* the rate at which power is delivered to a load at any given instant.
instantaneous readout *Telecommunications.* a characteristic of radio transmission systems in which readout by a radio transmitter is performed simultaneously with the computation of data to be transmitted.
instantaneous recording *Acoustical Engineering.* a term for direct recording of sound for which no further processing is intended.
instantaneous recovery *Mechanics.* the immediate disappearance of elastic deformation when the applied force is removed.
instantaneous sample *Telecommunications.* any of a series of instantaneous values of a wave; measurements are periodic.
instantaneous strain *Mechanics.* the immediate appearance of deformation in an elastic material when a force is applied to it.
instantaneous value *Physics.* the value of a time-varying quantity at a particular instant.
instantaneous velocity *Mechanics.* the velocity of a body at some one instant of time or at some one point in its path, as opposed to its average velocity over time from one point to another. Similarly, **instantaneous speed.**
instantiation *Computer Programming.* the act of assigning a value to a variable or creating a data structure. *Artificial Intelligence.* the process of making an instance of a pattern, rule, or frame by substituting particular facts for variables in a pattern.
instantizing see AGGLOMERATION.
instanton *Particle Physics.* a pseudoparticle representing vacuum fluctuations that exert forces on quarks; hypothesized to solve equations describing gauge fields of quantum chromodynamics.
instant-on switch *Electronics.* a device that continuously supplies all the tubes in a television set with a weak voltage, so that a picture appears almost immediately after the set has been turned on.
instant replay *Telecommunications.* the immediate rebroadcast of a recorded segment from a live television broadcast; used especially in the coverage of sporting events to show an action that was just completed, using slow motion, different camera angles, stop action, and so on.
instar *Entomology.* a developmental stage in insects, between moults or during metamorphosis.

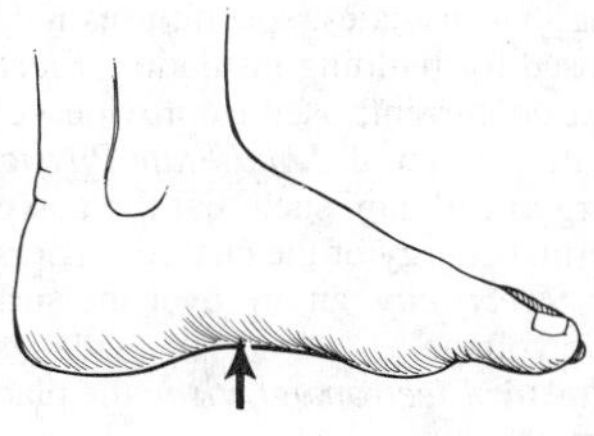

instep

instep *Anatomy.* the medial arched part of the foot.
instinct *Behavior.* an inherited pattern of behavior or tendency to action that is common to all members of a given species of the same sex and is generally based on a biological need. Also, **instinctive behavior.**
instinctive *Behavior.* relating to or done by instinct.
instinctive drift *Behavior.* the tendency for instinctive behavior to affect or interfere with learned behavior.
instinct theory *Psychology.* according to Freud, a constant psychic drive which stems from a disequilibrium, and which spurs the individual to restore equilibrium through attachment to an object that will bring fulfillment.
instinctual *Behavior.* of, relating to, or prompted by instinct. Thus, **instinctually.**
institution *Anthropology.* an organized system that promotes a fundamental human need, such as food-getting, reproduction, health, safety, or comfort; also includes secondary outgrowths of the basic needs, such as social control, education, economics, and political organization. Thus, **institutional.**
institutionalization *Psychology.* the process by which individuals adapt to the characteristic behavior patterns of the setting in which they live.
instr. instructor, instruction; instrument, instrumental.
instruction *Science.* anything that conveys knowledge or information, especially as to how to carry out an operation. *Computer Programming.* specifically, a word that specifies a computer operation; it usually includes an action code and one or more operands.
instruction address *Computer Programming.* the storage location of an instruction word.
instruction address register see PROGRAM COUNTER.
instruction area *Computer Programming.* a portion of main memory designated for storing instructions that are to be executed.
instruction code see OPERATION CODE.
instruction constant *Computer Programming.* a type of dummy instruction not intended to be executed as an instruction.
instruction control unit *Control Systems.* the hardware that handles the coded instructions within a computer control system.
instruction counter see PROGRAM COUNTER.
instruction cycle *Computer Programming.* the time required to fetch an instruction from storage, and decode and execute it.
instruction format *Computer Programming.* the layout and arrangement of the fields that make up a computer instruction.
instruction length *Computer Programming.* the number of bits that define or represent a single instruction (sometimes called a word) in the main memory of a computer; the actual length of a valid instruction varies with the type of computer, and a given computer may have multiple instruction lengths.
instruction look-ahead or **instruction pipeline** see AUXILIARY INSTRUCTION BUFFER.
instruction modification *Computer Programming.* a change in an instruction that causes the computer to perform a different operation or use a different address the next time the instruction is executed.
instruction pointer *Computer Programming.* a pointer linking one instruction to another instruction.
instruction register *Computer Technology.* a register in the central processing unit that contains an instruction after it has been read from memory. Also, PROGRAM REGISTER.
instruction set *Computer Technology.* the repertoire of operation codes for a particular computer or computer family.
instruction sheet *Industrial Engineering.* a written description of the standard method for operating a machine or performing a process, as well as the estimated time for each segment of the process, for reference by operators who are being trained in a new technique.
instruction time *Computer Programming.* the portion of an instruction cycle during which the computer is fetching and interpreting an instruction.
instruction word *Computer Programming.* a computer word, or fundamental unit of data storage, containing an instruction.
instructive hypothesis of antibody formation *Immunology.* a theory, now considered obsolete, stating that antibodies have no specificity for antigens prior to coming into contact with them. Also, **instructive theory.**
instrument *Engineering.* a measuring device designed to determine, and sometimes record, the present value of a quantity under observation. *Aviation.* **1.** specifically, a device designed to measure and usually record the condition of a flight vehicle's engines, indicators, or navigational equipment. **2.** any of certain related devices, especially when associated with navigation, such as a compass or automatic pilot. **3.** dependent upon instruments, as in **instrument flying**, often including ground support, as in an **instrument approach** or **instrument landing.** *Mechanical Devices.* a mechanical tool or device, especially one designed for precise operations. *Acoustical Engineering.* a device or apparatus used to produce musical sounds.

instrumental *Engineering.* of, relating to, or performed by or with one or more instruments.
instrumental act *Behavior.* any action that is taken to obtain a goal.
instrumental aggression *Behavior.* a type of aggressive behavior that helps an organism to obtain a goal.
instrumental analysis *Engineering.* a procedure in which an instrument monitors a component to determine whether it has completed a quantitative reaction or to determine whether there has been a change in the system's properties.
instrumental compression *Surgery.* the compression of blood vessels with instruments.
instrumental conditioning or **instrumental learning** *Psychology.* the process of teaching a specific behavior by rewarding or reinforcing occurrences of that behavior. Also, OPERANT CONDITIONING.
instrument approach procedure *Aviation.* a series of prescribed maneuvers, flown by reference to instruments, that control the movements of an aircraft from the initial approach fix until the aircraft has landed or aborted the landing.
instrumentation *Engineering.* **1.** the design and use of scientific instruments for purposes such as communication, control, computation, detection, or measurement. **2.** the ensemble of instruments contained in a given machine or system. *Aviation.* **1.** the installation or use of instruments on a flight vehicle. **2.** the assemblage of instruments on a given flight vehicle.
instrumentation amplifier *Electronics.* a device that produces a linear scaled version of an incoming voltage, noted for its high input impedance, low drift, and a high common-mode rejection over a range of frequencies.
instrument bombing *Military Science.* bombing by reference to radar, navigation aids, or other instruments rather than by visual reference to the ground.
instrument correction *Engineering.* a correction that rectifies a known error made by a testing instrument while evaluating a system.
instrumented buoy *Oceanography.* an anchored floating object to which measuring devices are attached.
instrument flight *Aviation.* flying in accordance with instrument flight rules.
instrument flight rules *Aviation.* a set of flight rules in which primary dependence is placed on instrument readings and ground-control information rather than on visual references; required for all-weather operation. Also, IFR.
instrument housing *Engineering.* a casing or covering designed to contain and protect an instrument. Also, **instrument shelter.**
instrument landing *Aviation.* an approach and landing based on instruments rather than on visual references.
instrument landing system *Aviation.* a precision system for instrument approach to an airport; it consists of a number of electronic and visual components including the localizer, glideslope, outer marker, middle marker, and approach lights.
instrument landing system localizer *Aviation.* a highly directional radio signal that indicates to an approaching aircraft whether it is on the centerline of the runway and, if it is not, in which direction the centerline lies.
instrument multiplier *Electricity.* a type of resistor used in series to extend the voltage range of an instrument.
instrument panel *Engineering.* any panel on which various instruments and controls are mounted, as in a motor vehicle. *Aviation.* specifically, a board or panel containing the switches, instruments, and other remote devices used to control an aircraft.
instrument reading time *Engineering.* the time required for an instrument to reach and indicate a final value for a quantity being measured, after a change occurs in the quantity.
instrument resistor *Electricity.* for certain measuring instruments such as the voltmeter, a multiplier resistance used to increase the range of the instrument.
instrument runway *Aviation.* a runway that is equipped with electronic or visual landing aids and for which a precision or nonprecision approach procedure has been established.
instrument shunt *Electricity.* a resistor that is connected in parallel with a device to extend the current range.
instrument system *Engineering.* **1.** the integration of one or more instruments for purposes such as communication, computation, control, detection, and measurement. **2.** the extent and arrangement of instruments within a given machine or operating system. Sensors and transducers are often included.
instrument transformer *Electricity.* a transformer that reproduces in its secondary circuit the current or voltage of its primary circuit without a change in phase relationship.
instrument weather *Meteorology.* in aviation weather practice, weather conditions of sufficiently low visibility to require utilization of instrument flight rules for aircraft operation. Also, IFR WEATHER.
insufflation *Surgery.* **1.** the act of blowing air, gas, vapor, or powder into a body cavity. **2.** any finely powdered or liquid drugs carried into the respiratory passages.
insula see ISLAND OF REIL.
insulate to cover or protect with an intervening material; specific uses include: *Materials Science.* to cover or line a space or substance with a material that prevents or retards the passage of heat, electricity, or sound. *Electricity.* to insert an intermediate nonconducting material that will provide electric isolation of two or more conductors.
insulated *Materials.* covered or protected by insulation.
insulated conductor *Electricity.* a conductor that is covered with a nonconducting material such as rubber or plastic; used in wiring to prevent short circuits, fire, and shock. Also, **insulated wire.**
insulated-return power system *Electricity.* an electric power distribution system for trains or other vehicles in which both the outgoing and return conductors are insulated from ground.
insulated-substrate monolithic circuit *Electronics.* a device designed so that all of the components on a silicon substrate are insulated from each other by a strip of silicon dioxide, rather than relying on reverse-biased positive-negative junctions to isolate components.
insulating board *Building Engineering.* a low-density, porous fiberboard characterized by low conductivity or high resistance to heat passage due to minute air spaces in the materials; used in walls and ceilings for heat and sound insulation.
insulating compound *Materials.* a liquid that is poured into cable joint boxes and allowed to solidify in order to retard the passage of heat and electricity.
insulating glass *Building Engineering.* two or three parallel panes of glass separated by a thin, dehydrated air space maintained by an airtight glass seal around the edges, thus preventing condensation and contamination within the air space.
insulating oil *Materials.* an oil having a high flash point and high dielectric strength; used as insulation in switches, circuit breakers, and oil-immersed transformers.
insulating tape *Materials Science.* tape that is impregnated or coated with an electrical or thermal insulating material such as adhesive; used to cover joints in wires or cables.
insulation the act or fact of insulating; specific uses include: *Building Engineering.* any material used to reduce or prevent the transfer of electricity, heat, cold, or sound; used primarily in walls, ceilings, and floors. *Electricity.* any nonconducting material used to provide electric isolation of two or more conductors.
insulation coordination *Electricity.* the correlation of insulating strength of equipment with expected overvoltages and with the characteristics of surge protective devices.
insulation rating *Electricity.* a measurement that rates insulating materials according to factors such as dielectric strength, tensile strength, and leakage resistance.
insulation resistance *Electricity.* a measure of the ability of an insulation material to prevent the flow of undesirable currents.
insulator something that insulates; specifc uses include: *Materials Science.* any material used for building insulation. *Electricity.* **1.** a material that conducts almost no current; used to provide voltage isolation. **2.** a body made of insulating material. *Solid-State Physics.* a material with a large forbidden energy band gap, such that the gap energy is very large compared to the thermal energy of the charge carriers.
insulator arc-over *Electricity.* an arc over the surface of an insulator induced by excessive voltage.
insulator arrangement *Electromagnetism.* the placement of insulators on a transmission mast.
insulin [in´sə lən] *Endocrinology.* a polypeptide hormone consisting of two peptide chains linked by a disulfide bridge, produced in the beta cells of the endocrine pancreas and released in response to elevated levels of blood glucose. Insulin acts to decrease blood glucose by increasing the cellular uptake and utilization of glucose; it regulates the use and storage of body fuel molecules such as glucose, amino acids, fatty acids, and ketone bodies.
insulinase *Enzymology.* an enzyme produced in the liver that destroys or inactivates insulin.

insulinlike growth factors *Biochemistry.* substances in serum that resemble insulin but that do not react with insulin antibodies; they are growth-hormone dependent and possess all the growth-promoting properties of the somatomedins. Also, NONSUPPRESSIBLE INSULINLIKE ACTIVITY.

insulin shock *Medicine.* the hypoglycemic shock resulting from an overdosage of insulin, decreased intake of food, or excessive exercise; marked by sweating, tremors, anxiety, dizziness, double vision, delirium, convulsions, and collapse.

insulin shock therapy *Medicine.* a rarely used psychiatric therapy in which large doses of insulin are used to induce convulsions, for therapeutic treatment of psychosis.

int. intelligence; intercept; interior; intermediate; internal; intersection; interval.

intaglio [in täl′yō] *Graphic Arts.* an image engraved or incised on a stone, printing plate, or other hard material. *Archaeology.* a rock art figure in the ground, produced by clearing rocks in a particular pattern that leaves a negative imprint. (From Italian *intagliare,* "to engrave.")

intaglio plate *Graphic Arts.* a printing plate designed to be directly incised with an image.

intaglio printing *Graphic Arts.* any printing method, such as gravure, in which the inked image is carried in lines incised below the surface of the plate; one of the three basic printing techniques, along with letterpress and lithography.

intake *Engineering.* **1.** an opening for air, water, fuel, or other fluid. **2.** the quantity of the fluid taken in. *Hydrology.* **1.** see RECHARGE. **2.** a term for the openings through which water passes into a well. *Nutrition.* substances or amounts of substances ingested by the body.

intake area *Hydrology.* see RECHARGE AREA.

intake chamber *Civil Engineering.* a large aqueduct that gradually narrows to an intake tunnel; designed to control desired water currents.

intake gate *Civil Engineering.* a movable barrier designed to open or close a water intake.

intake manifold *Mechanical Engineering.* in an internal-combustion engine, an assembly of pipes through which fuel flows from the carburetor or fuel injector to the intake valves.

intake stroke see SUCTION STROKE.

intake valve *Mechanical Engineering.* a valve that opens to allow air or an air/fuel mixture to enter a cylinder.

integer [in′tə jər] *Mathematics.* a positive or negative whole number or zero; an element of the set $\{\ldots, -2, -1, 0, 1, 2, \ldots\}$.

integer data type *Computer Programming.* a class of data that can assume only integer values.

integer spin *Quantum Mechanics.* a spin quantum number equal to an integer; characteristic of bosons.

integer variable *Computer Programming.* a variable of the integer data type.

integrable *Mathematics.* given a particular type of integral (e.g., Riemann, Lebesgue, Denjoy, etc.), a function f is said to be integrable on a given domain if it is sufficiently well-behaved (e.g., continuous, countably discontinuous, measurable, etc.), so that: (a) the value of the integral can be calculated over the domain, and (b) the calculated value is finite.

integral *Mathematics.* **1.** an operator that maps subsets of a (topological) space X and functions on X into the space of linear functionals on a function space of X. That is, an integral takes a subset of X (the domain of integration) and a function $f: X \to C$ (the **integrand**) and produces a number called the value of the integral of f over the domain. For example, the operator that maps a closed interval $[a, b]$ on the x-axis to the area below the graph of f on that interval is an integral. **2.** a solution to a differential equation or differential system. **3.** let s be an element of an extension ring S of a ring R. If s is a root of a monic polynomial with coefficients in R, then s is said to be integral over R. If every element of S is integral over R, then S is an **integral extension** of R.

integral action *Control Systems.* a control action in which the correcting force changes in direct proportion to the deviation.

integral calculus *Mathematics.* the study of integration, integration techniques, and applications of integrals such as area and volume problems, solutions of differential equations, etc.

integral closure *Mathematics.* the integral closure of a ring R in an extension ring S is the subring of S consisting of all elements of S that are integral over R. R is said to be **integrally closed** in S if it is equal to its integral closure in S.

integral compensation *Control Systems.* compensation in which the output is directly proportional to the input.

integral control *Control Systems.* control in which changes in the control signal are proportional to the error signal.

integral discriminator *Electronics.* a device that accepts only signals whose frequencies exceed a minimum height.

integral domain *Mathematics.* a commutative ring that has a multiplicative identity and no zero divisors.

integral dose *Nucleonics.* the entire amount of radiation absorbed by a body, generally expressed in gram-rads or gram-roentgens. Also, INTEGRATED ABSORBED DOSE, VOLUME DOSE.

integral equation *Mathematics.* **1.** an equation involving an unknown function that appears as part of an integrand. **2.** the equation $f(x) = 0$, where f is the monic polynomial (with coefficients in a ring R) that determines whether a particular element x is integral over R.

integral extension *Mathematics.* an extension ring S of a ring R such that every element of S is integral over R.

integral function *Mathematics.* **1.** see ENTIRE FUNCTION. **2.** a function whose image is a subset of the integers; i.e., that takes on only integer values.

integral-furnace boiler see WATER-TUBE BOILER.

integral hologram *Optics.* a hologram produced from a large series of photographs, each photograph rendering a single subject from a slightly different angle.

integral-mode controller *Control Systems.* a controller in which the control signal is proportional to the integral of the error signal.

integral modem *Telecommunications.* a modem built into a system or device, enabling it to communicate over a telephone line.

integral network *Control Systems.* a network that produces high gain at low input frequencies and low gain at high frequencies in order to achieve a low steady-state error rate. Also, LAGGING NETWORK, LAG NETWORK.

integral operator *Mathematics.* a (usually linear) operator whose inverse is a differential operator.

integral square error *Control Systems.* a measure of system performance based on the integral of the square of the system error over a given period of time.

integral test *Mathematics.* let f be a positive, nonincreasing continuous function defined on $t \geq 1$. Then the infinite series $\sum_{n=1}^{\infty} f(n)$ converges if and only if the following infinite integral converges:

$$\int_{-1}^{\infty} f(t)\,dt = \lim_{n\to\infty} \int_{1}^{n} f(t)\,dt.$$

integral transformation *Mathematics.* for a given function $F(x)$, a transformation of the form $f(y) = \int_a^b K(x, y)F(x)\,dx$. $K(x, y)$ is called the kernel of the transformation. Also, **integral transform.**

integral-type flange *Mechanical Devices.* a forge or casted flange on a nozzle neck, piping wall, or pressure container.

integral waterproofing *Engineering.* the waterproofing of concrete by means of adding waterproof substances to the cement as it is mixed.

integrand *Mathematics.* see ELEMENT, def. 2.

integrase *Virology.* the virus-coded enzyme that facilitates the insertion of the viral DNA into the host genome.

integrated *Science.* incorporating different elements or units that function together.

integrated absorbed dose see INTEGRAL DOSE.

integrated circuit *Electronics.* a circuit whose components are formed on a single semiconductor substrate.

integrated-circuit capacitor *Electronics.* a device added to a semiconductor during assembly that stores an electrical charge.

integrated-circuit resistor *Electronics.* a capacitor formed in an integrated circuit as part of its manufacturing process.

integrated communications system *Telecommunications.* a communications system that transmits both analog and digital signals over the same switched network.

integrated console *Telecommunications.* a console that controls the switching center apparatus of an integrated communications system.

integrated control *Agronomy.* the control or elimination of undesirable plants or insects by a combination of chemical products and natural means. Also, **integrated approach.**

integrated data processing *Computer Programming.* the organized and systematic method of acquiring and processing information that will serve several functions within an organization. Also, **integrated information processing.**

integrated drainage *Hydrology.* in an arid region, mature drainage characterized by the joining of drainage basins across intervening mountains or ridges as a result of erosion and aggradation.

integrated electronics *Electronics.* the branch of electronics concerned with integrated circuits, especially the interdependence of material, circuits, and design.

integrated fire-control system *Ordnance.* a system that combines the functions of target acquisition, tracking, data computation, and engagement control, primarily using electronic means assisted by electromechanical devices.

integrated information system *Telecommunications.* an expansion of a basic information system achieved by integrating two or more information systems.

integrated injection logic *Electronics.* a type of semiconductor device technology that uses bipolar techniques, making possible large-scale integration on a single silicon chip; in a computer, it permits higher speeds, lower power dissipation, and a greater storage capacity than in the older CMOS models. Also, MERGED-TRANSISTOR LOGIC.

integrated intensity *Crystallography.* the total intensity measured at the detector as a Bragg reflection produced by X-ray diffraction is scanned.

integrated magnitude *Astronomy.* the apparent magnitude that an extended object (such as a nebula) would have if all its light were concentrated into a starlike point.

integrated neutron flux *Nucleonics.* a quantity of radiation exposure derived from the product of the number of free neutrons per unit volume, their average speed, and the exposure time.

integrated optics *Optics.* a device that contains a system of miniature lenses, prisms, and switches on a thin-film strip, and performs operations similar to those of integrated electronic circuits; used in switching, communications, and logic components.

integrated profile see MEAN PROFILE.

integrated radiation *Optics.* the integral of the radiation flux emitted during exposure time.

integrated services digital network *Telecommunications.* a public end-to-end digital communication network that provides different types of services, such as telephone, data, electronic mail, or facsimile service.

integrated software *Computer Programming.* a group of programs that operate on the same database and exchange data with each other, often by means of splitting the screen into two or more windows.

integrated train *Mining Engineering.* a string of permanently coupled cars that travels nonstop between a particular mine and a generating plant; rotary couplers permit each one to be flipped over and dumped as the train moves slowly across a trestle.

integrated warfare *Military Science.* the conduct of military operations in which opposing forces employ nonconventional weapons in combination with conventional weapons.

integrating accelerometer *Engineering.* an accelerometer whose output signals are proportional to the velocity or to a vector in reference to time of the vehicle being monitored, rather than to the acceleration of the vehicle.

integrating amplifier *Electronics.* an amplifier that produces an output signal that is the time integral of the input signal. Also, **integrating circuit.**

integrating detector *Electronics.* a device that extracts information from a radio signal in such a way that it can be restored to its original form by an integrator.

integrating factor *Mathematics.* **1.** a factor that, when multiplied into a differential equation, results in an exact derivative (plus possibly other terms not involving derivatives). **2.** more generally, a factor that, when multiplied by a differential form, makes the differential form exact (or closed).

integrating filter *Electronics.* a circuit that suppresses signals at a given frequency; characterized by the buildup of charges and voltage on its output capacitor caused by a continuous wave of electric pulses.

integrating frequency meter *Engineering.* an instrument that sums the total number of cycles the alternating voltage has passed through in an electric power supply over a given period of time.

integrating galvanometer *Engineering.* an instrument that is similar to the d'Arsonval galvanometer, but which is able to measure changes in flux occurring over several minutes of time in addition to the integral of current over time.

integrating gyroscope *Engineering.* a gyroscope that measures and transmits the time integral of the rate of angular displacement.

integrating meter *Engineering.* any instrument that integrates a measured quantity with respect to time, such as an ampere-hour meter or watt-hour meter.

integrating network *Electronics.* a device in which output voltage is proportional to the time integral of the input voltage. Also, INTEGRATOR.

integrating-sphere photometer *Optics.* an instrument that measures the total luminous flux emitted by a light source.

integration *Engineering.* the process of combining different acts or elements into a functioning whole; coordination. *Physiology.* anabolic action or activity; assimilation. *Psychology.* the constructive assimilation of knowledge and experience into the personality. *Genetics.* the insertion, usually through recombination or additive integration, of a DNA sequence into a larger, normally genomic sequence. *Virology.* the process in which the viral DNA inserts into the host genome for replication. *Mathematics.* the process of computing an integral.

integration by parts *Mathematics.* a technique for evaluating integrals based on the Leibnitz (product) rule for derivatives. In particular, for the case of functions of one real variable, the integrand is written in the form $f(x)g'(x)$ and evaluated as

$$\int_a^b f(x)g'(x)\,dx = f(b)g(b) - f(a)g(a) - \int_a^b f'(x)g(x)\,dx.$$

integration efficiency *Genetics.* the frequency with which a specific foreign DNA segment can be incorporated into the genome of a recipient bacterium following transformation.

integration test *Computer Programming.* the highest level of developer testing, in which all hardware and software components are tested together.

integrative suppression *Genetics.* a condition in which the integration of a plasmid into a bacterial chromosome suppresses the effect of a chromosomal mutation that has inhibited the initiation of chromosome replication.

integrator *Electronics.* **1.** see INTEGRATING NETWORK. **2.** a device that mimics the mathematical process of integration in certain digital machines. *Engineering.* any system or device that integrates.

integrator gene *Genetics.* a hypothetical DNA sequence that, when stimulated, controls another group of genes.

integrity *Computer Programming.* the maintenance of accurate and consistent data and the restriction of programs to their intended purpose. *Mathematics.* a measure of graph vulnerability defined as $I(G) = \min(m(G-S) + |S|)$, where the minimum is taken over all subsets of $V(G)$, $m(G-S)$ is the order of the largest component of $G-S$, and $|S|$ is the number of vertices in S.

integrity stage *Psychology.* the final stage of Erik Erikson's eight stages of human development, during the aging years, when an individual develops a conception of his or her earlier life as a satisfactory and integrated whole.

integrity vs despair *Psychology.* the stress or despair that can arise during the integrity stage of life, if the person has a sense that earlier stages of life were unsuccessful or unfulfilling.

integrodifferential equation *Mathematics.* a functional equation involving both integrals and derivatives.

integument [in teg′yə mənt] *Anatomy.* **1.** the covering of a body part or organ. **2.** a technical name for skin. *Entomology.* the outer covering or skeleton of insects and other invertebrates; mainly composed of chitin.

integumentary *Anatomy.* relating to, composed of, or serving as skin.

integumentary musculature *Vertebrate Zoology.* in some terrestrial vertebrates, the system of voluntary muscles that are attached to the skin and allow movement of specific parts of the skin.

integumentary pattern *Anatomy.* the genetic pattern of ridges under the skin, expressed externally as fingerprints, and other prominent folds.

integumentary system *Anatomy.* a technical name for the skin.

intellect *Psychology.* the mind, thinking facility, or understanding.

intellectual *Psychology.* relating to the mind or to intelligence.

intellectualization *Psychology.* a defense mechanism by which an individual avoids the emotion or feeling evoked by a problem, by attempting to analyze it in purely intellectual terms.

intelligence *Psychology.* a general term encompassing various mental abilities, including the ability to remember and use what one has learned, in order to solve problems, adapt to new situations, and understand and manipulate one's environment. *Behavior.* as applied to the activities of animals, the capacity to show a change in behavior as the result of experience. *Computer Technology.* **1.** the ability of a device or system to process data. **2.** the ability of a device or system to solve problems more automatically or more skillfully than competitive systems. *Telecommunications.* data, information, or messages that represent some meaning to humans or machines. *Military Science.* the product resulting from the collection, analysis, and interpretation of available information concerning foreign countries or areas.

Time-exposure photograph of polar star trails. Polaris is the short bright trail just off center. The circular formation of the trails results from the rotation of Earth. D. Parker/SPL/Photo Researchers.

High-speed photograph of secondary drop formation after the impact of a falling drop containing red dye and a pool of water. The secondary drop contains a high proportion of the red dye from the original drop. SPL/Photo Researchers.

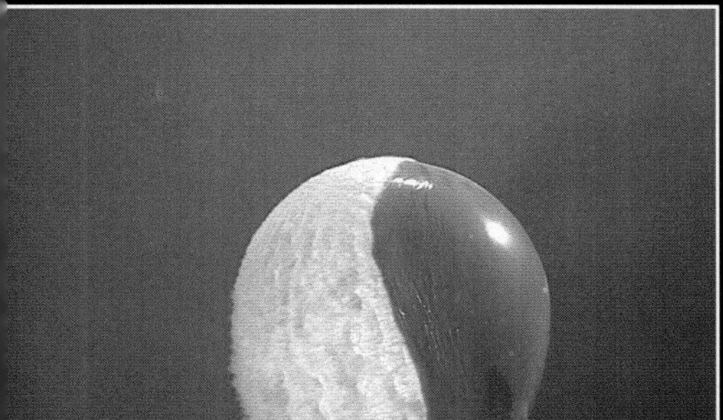

High-speed photograph of a balloon containing talcum powder after being shot with an air-gun pellet. The talcum powder temporarily retains the shape of the balloon after the surrounding material has shattered. J. Watts/SPL/Photo Researchers.

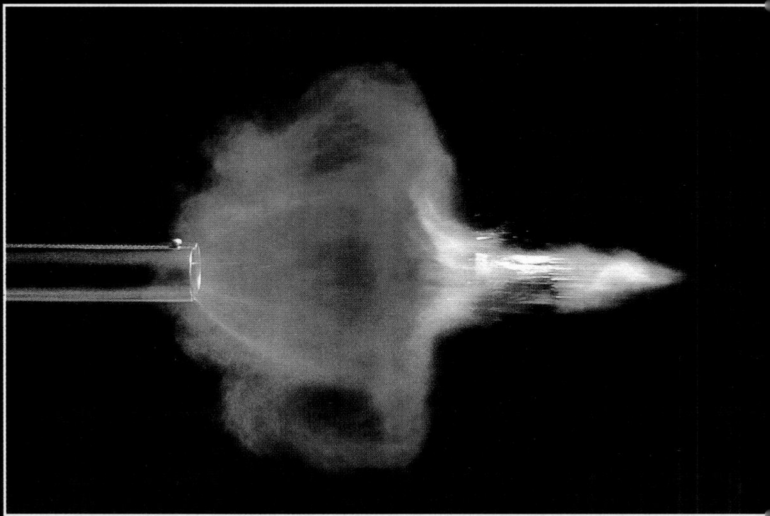

Twelve-gauge shotgun discharge approximately 4 milliseconds after detonation. Stephen Dalton/Photo Researchers.

Sedimentary rock strata in sea cliffs. Bill Evarts Photography.

Tree growth rings. San Diego Natural History Museum.

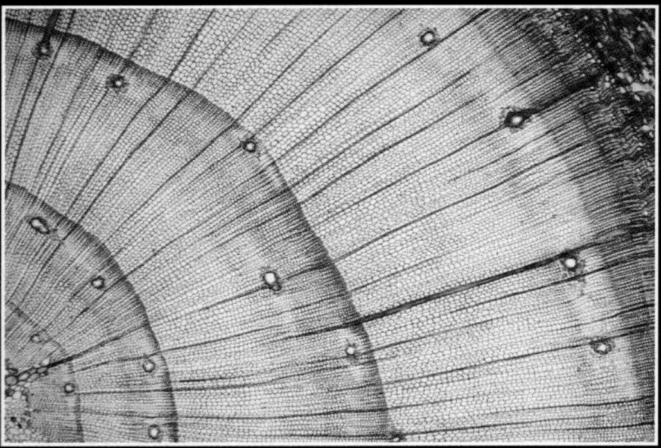

Micrograph of a cross section of the stem of a pine tree showing growth rings.
© Bruce Iverson/Science Photos.

Time-exposure photograph of a solar eclipse.
Greg Davis/The Stock Market.

intelligence quotient *Psychology.* a measure of intellectual development that is the ratio of a child's mental age to his chronological age, multiplied by 100. Also, IQ.

intelligence test *Psychology.* a method by which an individual's intellectual capacity is measured.

intelligent *Computer Science.* describing a system or device that operates with a high degree of sophistication and independence, in a manner regarded as comparable to human intelligence.

intelligent computer-assisted instruction *Artificial Intelligence.* an educational application of artificial intelligence that has representations of the subject matter; it carries on an interactive exchange with the student, and also recognizes his or her misunderstandings by analyzing mistakes.

intelligent controller *Control Systems.* any control system that can deliver a degree of humanlike response to stimuli.

intelligent machine *Computer Technology.* any computer that can solve problems by reasoning and inference. *Engineering.* any machine that can perform a function in the presence of variables.

intelligent robot *Robotics.* any robot that can be programmed to take actions or make choices based on sensory input.

intelligent terminal *Computer Technology.* a remote terminal that has processing capabilities, as opposed to a dumb terminal. Also, **intelligent work station.**

intelligibility *Telecommunications.* the percentage of speech units correctly interpreted by a listener.

INTELSAT *Telecommunications.* a satellite network under international control, used for worldwide communication. (An acronym for International Telecommunications Satellite Consortium.)

intensification *Graphic Arts.* the process of increasing the density of a photographic negative with an intensifier; usually involves bleaching and redeveloping.

intensifier something that intensifies; specific uses include: *Graphic Arts.* a chemical bath used in intensification; for example, a **chromium intensifier,** a water solution of potassium bichromate and hydrochloric acid. *Petroleum Engineering.* in the acidizing of an oil well, a solution of hydrofluoric acid that is added to hydrochloric acid in order to eliminate silica films which are insoluble in hydrochloric acid. *Linguistics.* a word or other element that gives more emphasis or force to a statement; e.g., adverbs such as *very, extremely, greatly,* or *certainly.*

intensifier electrode *Electronics.* the element in an electrostatic cathode-ray tube that allows for additional acceleration of the electron beam following deflection, so as to increase the intensity of the trace without decreasing the tube's deflection sensitivity. Also, POSTACCELERATING ELECTRODE.

intensifier image orthicon *Electronics.* a circuit that amplifies the electron stream generated by a photocathode tube before the stream strikes its target.

intensifying screen *Radiology.* a material impregnated with calcium tungstate that fluoresces during radiography; used to enhance the imaging properties of the photographic emulsion.

intensimeter *Radiology.* a device used to measure X-ray intensity, based on the variation of electric resistance of a selenium cell under influence of irradiation at different intensities.

intension *Artificial Intelligence.* an abstract description of something, as distinguished from the extension, which is the actual thing described. For example, "John wants to own the fastest car in town" is an intensional description of what John wants. Assuming that Bill actually owns the fastest car in town, it does not mean that John wants to have Bill's car (the extension).

intensionometer *Radiology.* an ionometric instrument used to measure X-ray intensity by deflection of a galvanometer needle that registers the difference in electric potential.

intensity *Chemistry.* the concentration or strength of a given solution. *Physics.* **1.** the magnitude, strength, or amount of a quantity. **2.** the power per unit area transmitted by a wave.

intensity distribution see DISTRIBUTION OF INTENSITIES.

intensity level *Physics.* a level of power per unit area expressed in decibels above an arbitrary zero level.

intensity mine circuit *Ordnance.* a circuit that is actuated when the strength of the acoustic, magnetic, or pressure field reaches a preset level relative to that experienced by the mine when no ships are in the vicinity.

intensity modulation *Electronics.* the process by which the luminance of an electron beam in a cathode-ray tube is varied in accordance with the magnitude of the signal.

intensive care unit *Medicine.* a hospital unit housing special equipment and skilled personnel for **intensive care,** the care of seriously ill patients requiring immediate and continuous attention. Also, ICU.

intensive farming *Agriculture.* a method of farming that uses large amounts of labor and resources to make each unit of land as productive as possible.

intensive forestry *Forestry.* the practice of forestry with the goal of obtaining a high level of volume and quality of output per area unit, using the finest techniques of silviculture and forest management

intensive property *Chemistry.* a property of a system, such as its pressure, temperature, or density, that is not dependent on the quantity of material present in the system. Also, **intensive quality, intensive variable.**

intention *Medicine.* a manner or process of healing. In **healing by first intention,** the union or restoration of an affected area occurs without the intervention of granulations; **healing by second intention** is the union by closure of a wound with granulations, which form from the base and both sides toward the surface of the wound.

intentionality *Behavior.* the planning or undertaking of an action in pursuit of an object or goal.

intention movement *Behavior.* the performance of the initial parts of a behavior pattern without the pattern being completed; often serves to communicate to others what the animal will do next.

intention tremor *Medicine.* a tremor of the limbs that occurs when a voluntary movement is made; present in diseases of the nervous system, such as multiple sclerosis.

inter- a prefix meaning "between" or "among," as in *interannular, interstitial.*

interaction the quality or state of mutual action; specific uses include: *Behavior.* any encounter between animals or humans in which each reacts to the presence of the other. *Pharmacology.* the action of one drug on the effectiveness or toxicity of another. *Physics.* **1.** the transfer of energy between two particles. **2.** the interchange of energy between particles and a wave motion. *Fluid Mechanics.* the nonlinear transference of fluid flow properties such as energy, momentum, and vorticity between different components of the wave spectrum. *Statistics.* dependence between the effect of an independent variable on a response and the level of a further independent variable.

interaction balance method see GOAL-COORDINATION METHOD.

interactionism *Behavior.* a theory suggesting that both personality and the specifics of a situation must be considered in order to understand behavior. Thus, **interactionist approach.**

interaction of species *Ecology.* a general term for the influence that two species have on each other.

interaction picture SEE INTERACTION REPRESENTATION.

interaction prediction method *Control Systems.* a method of coordinating subproblem solutions in plant decomposition in which a second-level controller specifies the interaction variables based on optimal conditions, and subproblems are then solved to satisfy local conditions based on the specific values of the interaction variables. Also, FEASIBLE METHOD.

interaction representation *Quantum Mechanics.* a means of representing quantum mechanics in which state vectors change in time only as a result of interactions, and the operators corresponding to physical quantities carry the remaining time dependence.

interaction space *Electronics.* the section of an electronic tube where the electrons interact with an alternating electromagnetic field.

interactive *Computer Technology.* describing a system or program that maintains an exchange with the user, alternately accepting input and then responding. Thus, **interactive graphics, interactive information system, interactive processing.**

interactive language *Computer Programming.* a programming language in which each statement is executed after the programmer types it into the keyboard; for example, BASIC or LISP.

interactive learning *Computer Science.* a process of computer-assisted instruction in which the learner's responses determine the specific nature of the instruction, particularly the level of difficulty of the material presented and the amount of repetition provided.

interactive services *Telecommunications.* **1.** telephone services that allow users to communicate directly with a central computer by pressing numbers or letters in order to carry out activities such as ordering from a catalog or making a bank deposit. **2.** communication services that are able to maintain a two-way information flow.

interactive terminal *Computer Technology.* a terminal characterized by a relatively fast response to each user request.

interactive video *Computer Technology.* a system in which the user can interact with a computer-controlled videodisk or CD/ROM display image by means of a device such as a keyboard or touch-sensitive screen. Also, **interactive videodisk system.**

interallelic complementation see INTRAGENIC COMPLEMENTATION.

interalveolar *Anatomy.* situated or occurring between alveoli.

interambulacrum *Invertebrate Zoology.* in echinoderms, the area between the rows of plates associated with the tube feet.

Interamnia *Astronomy.* asteroid 704, discovered in 1910 and measuring 338 kilometers in diameter; it belongs to type F.

interarticular *Anatomy.* **1.** located between two skeletal joints. **2.** located between two articular surfaces in a movable joint.

interatomic *Physical Chemistry.* between atoms; relating to the interaction of different atoms.

interatomic forces *Physical Chemistry.* the forces that act on an atom to cause it to attract or repel adjacent atoms; to be distinguished from intramolecular forces.

interatrial *Anatomy.* located between the two atria of the heart. Also, **interauricular.**

interatrial septal defect *Medicine.* a defect in the interatrial septum, resulting in abnormal communication between the atria.

interatrial septum *Anatomy.* the wall between the atria of the heart.

interaxial angles *Crystallography.* the angles between the unit cell axes.

interband *Molecular Biology.* the area lying between the bands in a polytene chromosome.

interbase current *Electronics.* the flow of electrons from one connection to another in the base region of a junction tetrode transistor.

interbedded *Geology.* related to a depositional sequence in which beds with different characteristics alternate with or lie between other beds.

interblock gap *Computer Technology.* the space between two blocks of records on a magnetic tape, in which no data is recorded; created when the reel of tape slows or comes to a stop between write operations.

interbrain see DIENCEPHALON.

intercalary *Science.* inserted or located between other things or parts. Also, **intercalated.**

intercalary regeneration *Cell Biology.* a process of regulated cell growth, as after a cut, when cells proliferate to fill the gap between the original cells in order to maintain the appropriate spatial relationships and positional values of the cells in the tissue.

intercalated disc *Histology.* a dark region that marks the junction between adjacent myocardial cells.

intercalated nucleus *Anatomy.* a small mass of neurons in the medulla oblongata, lying lateral to the hypoglossal nerve.

intercalating agent *Biochemistry.* a dye having a planar chromophore that can be placed between neighboring base pairs in dsDNA or ssDNA, in order to unwind the helix or lengthen it.

intercalation *Geology.* **1.** the presence of a body of rock interbedded or interlaminated with another body of different rock. **2.** a body of material so layered.

intercapillary *Anatomy.* situated among a group of capillaries.

intercapillary glomerulosclerosis *Pathology.* a degenerative diabetic condition signified by albuminuria, hypertension, retinal disease, and kidney malfunction.

intercardinal heading *Navigation.* a heading on one of the intercardinal points.

intercardinal point *Navigation.* a direction that is halfway between two adjacent cardinal points of a compass; i.e., northeast, southeast, southwest, or northwest.

intercardinal rolling error *Navigation.* a swinging of a gyrocompass caused by the craft's rolling. It is most pronounced on intercardinal headings. Also, QUADRANTAL ERROR

intercarotic *Anatomy.* between the carotid arteries. Also, **intercarotid.**

intercarpal *Anatomy.* lying between or among the carpal bones.

intercarrier channel *Telecommunications.* a carrier telegraph channel found in the frequency spectrum between carrier telephone channels.

intercarrier noise suppression *Telecommunications.* the process by which a circuit automatically suppresses the audio-frequency output of a television receiver, in order to eliminate the noise generated by an automatic volume control that is tuned between stations.

intercarrier sound system *Telecommunications.* a type of television receiver in which the video and sound signals merge to form an intermediate frequency that is equal to the difference between the frequencies of the separate signals.

intercartilaginous *Anatomy.* situated between or among cartilages.

intercavernous sinus *Anatomy.* either of two venous sinuses surrounding the infundibulum of the pituitary gland, and connecting the two cavernous sinuses.

intercellular *Histology.* of or relating to the area between cells. Thus, **intercellular space.**

intercellular cement *Histology.* a material that holds epithelial cells together.

intercellular plexus *Histology.* a network of nerve fibers surrounding the cells in a sympathetic ganglion.

intercellular substance *Histology.* any material naturally occupying the space between cells.

intercentrum *Vertebrate Zoology.* a small, bony element found wedged ventrally between successive centra of the vertebral column of reptiles and in the tail of certain mammals.

intercept *Military Science.* to carry out an interception. *Mathematics.* a point at which a given curve intersects a coordinate axis. *Engineering.* see ALTITUDE DIFFERENCE.

intercepting *Telecommunications.* **1.** the rerouting of a telephone call placed to a disconnected or nonexistent telephone number. **2.** the deliberate reception of a call or message by a station to which the call or message was not intended to be routed.

intercepting sewer *Civil Engineering.* the part of a sewer system that receives flow from transverse sewers and takes the sewage to a treatment plant or disposal unit.

interception the process of intercepting or cutting off; specific uses include: *Meteorology.* **1.** a loss of sunshine, as by interference from tall trees, buildings, or the like. **2.** the absorption of ultraviolet radiation by ozone and dust particles, depleting part of the solar spectrum. *Hydrology.* **1.** the process by which water from precipitation is collected and stored on the surface of vegetation, from where it eventually evaporates without ever reaching the ground. **2.** the water or quantity of water so collected and stored. *Military Science.* the process and act of meeting or interrupting the course of a moving vessel, aircraft, or missile. *Telecommunications.* the process of tuning in to a communication not originally intended for the listener.

intercept method *Metallurgy.* in metallography, a method used to quantitatively measure certain microstructural features.

interceptometer *Engineering.* a rain gauge that is placed under trees or foliage; its readings are compared with those of a similar instrument placed in an open space where rain falls unimpeded to the ground in order to determine how much rainfall is intercepted by vegetation.

interceptor (F-14A Tomcat)

interceptor *Aviation.* a fast, well-armed, short-range aircraft designed to identify and if necessary intercept and destroy other aircraft. *Space Technology.* a missile, especially a surface-to-air guided missile, designed to intercept and destroy other airborne objects.

intercept service *Telecommunications.* in telephony, a service that connects a subscriber directly to an intercept operator if a disconnected line is reached, provides a recorded message informing the caller that the line is no longer in service, or provides the caller with a new telephone number that replaces the one dialed.

intercept tape *Telecommunications.* a tape used for temporary storage of messages for malfunctioning trunk channels and tributary stations.

intercept trunk *Telecommunications.* a central office termination to which calls for disconnected numbers are routed for intercept service.

intercerebral *Anatomy.* between the two cerebral hemispheres.

interchange *Civil Engineering.* an area of intersection of two or more highways at different elevations, constructed so that traffic can pass from one highway to another without crossing the main stream of traffic in any of the highways.

interchangeable *Engineering.* describing parts or components manufactured to certain identical specifications, and therefore readily substituted for one another.

interchangeable lens *Optics.* a lens that can be removed from a camera so that another lens with a different characteristic, such as focal length, can be substituted.

interchange coefficient see EXCHANGE COEFFICIENT.

interchannel crosstalk *Telecommunications.* crosstalk produced between channels in a multiplex system.

interchondral *Anatomy.* between two or more cartilages.

interclavicle *Vertebrate Zoology.* a bony element lying along the midline of the body, between the clavicles in the pectoral girdle of fossil amphibians and living reptiles.

intercloud discharge see CLOUD-TO-CLOUD DISCHARGE.

intercolumniation *Architecture.* **1.** the clear space between adjacent columns. **2.** the system of spacing in a series of columns, measured in multiples of the shaft diameter.

intercom *Telecommunications.* **1.** any internal communication system for use within a building, ship, aircraft, and so on, providing a microphone for speaking and a receiver or loudspeaker for listening. **2.** specifically, a telephone system providing direct communication between telephones on the same premises. **3.** a speaking or listening device used in such a system. (A shortened form of intercommunication system.)

intercombination line *Astrophysics.* a spectral line emitted in an atomic transition between two electron levels that have different multiplicities.

intercommunicating porosity *Metallurgy.* in powder metallurgy, the porosity of a sintered compact that is continuous, making the component permeable to fluids.

intercommunicating system see INTERCOM.

intercommunication *Petroleum Engineering.* the drainage of adjacent wells resulting from the flow interconnections between reservoir areas.

intercommunication system see INTERCOM.

intercondenser *Mechanical Engineering.* a condenser that separates two stages of a multistage steam jet pump.

interconnection *Electricity.* **1.** a scheme for wiring circuit components, usually shown on a circuit diagram. **2.** the linking of two or more power generators in phase to produce more energy than a single one.

intercontinental ballistic missile *Ordnance.* a ballistic missile with a range capability in excess of 5500 km. Also, ICBM.

interconversion *Telecommunications.* the process of converting information from one format or code to another, as from ASCII to EBCDIC.

intercooler *Mechanical Engineering.* **1.** in a turbocharged engine, a heat exchanger that lowers the temperature of intake air in order to improve volumetric efficiency. **2.** any device for cooling a fluid between successive heating processes, especially for cooling a gas between successive stages of compression.

intercostal *Anatomy.* located between the ribs. *Naval Architecture.* describing a noncontinuous structural member that runs between other adjacent structures. For example, an **intercostal keelson** runs between floors, an **intercostal floor** runs between keelsons, and an **intercostal stringer** runs between frames.

intercostal muscles *Anatomy.* the sets of external and internal muscles that stretch between successive ribs, are part of the lateral thoracic wall, and elevate the ribs during breathing.

intercostal nerve *Anatomy.* a ventral branch of a thoracic nerve, passing between and innervating the intercostal muscles.

intercourse *Physiology.* sexual union; coitus.

intercrop *Agronomy.* a small-grain or sod crop, such as clover or alfalfa, grown between rows of a field crop to protect against erosion.

intercropping *Agronomy.* the growing of two or more crops at the same time in alternate rows on the same field or land. Also, INTERPLANTING.

intercrystalline *Crystallography.* situated or passing between crystals.

intercrystalline corrosion *Metallurgy.* corrosion occurring preferentially at grain boundaries.

interdemic selection *Evolution.* a process of natural selection occurring between local interbreeding populations.

interdendritic corrosion *Metallurgy.* corrosion occurring preferentially between dendrites. Also, **interdendritic attack.**

interdendritic segregation see MICROSEGREGATION.

interdendritic void see MICROVOID.

interdental *Anatomy.* situated between teeth, especially between the proximal surfaces of adjacent teeth in the same dental arch.

interdental cavity *Medicine.* a cavity occurring between adjacent teeth.

interdental fricative *Linguistics.* a sound produced by pressing the tongue against the upper teeth or between the upper and lower teeth while forcing air out, such as the [th] sound in *these* or [th] in *thin.*

interdiction *Military Science.* an action to divert, disrupt, delay, or destroy enemy military potential before it can be moved into position for use against friendly forces. Thus, **interdiction bombing, interdiction fire.**

interdiffusion *Physical Chemistry.* the mixing by molecular diffusion processes of two fluids previously separated by an impervious diaphragm.

interdigital magnetron *Electronics.* a device that generates microwaves by using an external magnetic field to channel the flow of electrons through anode segments connected at both ends of the cathode.

interdigital structure *Electronics.* a type of circuit design in which the distance between two electrodes is lengthened by an interlocking-finger structure.

interdigital transducer *Electronics.* a device that turns microwave voltage into surface acoustic waves or vice versa by applying two interlocking comb-shaped metallic patterns to a piezoelectric substrate, such as quartz or lithium niobate.

interelectrode capacitance see PLATE-GRID CAPACITANCE.

interelectrode transit time *Electronics.* the time that it takes for an electron to cross from one electrode to another.

interesterification *Organic Chemistry.* a reaction between a compound and an ester, involving the exchange of alkoxy or acyl groups; the reaction produces another ester. Also, ESTER INTERCHANGE.

interface a connection or common boundary between two spaces or things; specific uses include: *Physical Chemistry.* **1.** the area or surface that represents the boundary between two separate phases of a chemical or physical process. The three different states of matter have five possible types of interface: liquid/liquid, liquid/gas, solid/solid, solid/liquid, and solid/gas; gas/gas has no definable interface because all gases are completely soluble in one another. **2.** a situation or condition in which such a meeting of separate phases takes place. *Meteorology.* a surface that separates two fluids, across which there is a discontinuity of some fluid property in a direction perpendicular to the interface. Also, INTERNAL BOUNDARY. *Computer Technology.* the physical or logical connection between a computer and the user, a peripheral device, or a communications link. *Computer Programming.* the communication between a calling program and a subroutine. *Geology.* see SEISMIC DISCONTINUITY. *Forestry.* see BARK BAR.

interface card *Computer Technology.* a type of circuit board that connects external devices to computers.

interface connection see FEEDTHROUGH.

interface mixing *Physical Chemistry.* the process in which two immiscible or partially miscible liquids mingle at their point of first contact.

interface resistance *Thermodynamics.* **1.** the impedance of heat transfer across a boundary between two different materials, due to imperfect contact between the materials. **2.** a quantitative description of this, given by the temperature difference across the interface divided by the heat flux through it.

interfacial *Physical Chemistry.* of, relating to, or of the nature of an interface. *Crystallography.* between two faces of a crystal.

interfacial angles *Crystallography.* the angles between adjoining faces of a crystal.

interfacial energy *Physics.* a free energy appearing at the boundary of two dissimilar substances. Also, SURFACE ENERGY.

interfacial free energy *Materials Science.* the free energy of interfaces, equal to the surface tension multiplied by the surface area. Also, SURFACE ENERGY.

interfacial polarization *Optics.* the polarization of light that arises when reflection occurs at a specific angle from the surface of a dielectric.

interfacial polymerization *Materials Science.* a polymerization reaction that occurs at or near the interfacial boundary of two immiscible solutions or liquids.

interfacial pressure *Materials Science.* shrinkage stresses that occur during curing of a polymer matrix composite; the matrix shrinks more than the fibers, causing axial and lateral compression of fibers, and in extreme cases, produces fiber buckling.

interfacial shear strength *Materials Science.* the strength of the interfacial bond between the fiber or particles and the matrix in a composite.

interfacial tension *Physical Chemistry.* the surface tension at the interface of two liquids.

interfacility transfer trunk *Telecommunications.* a trunk that connects the switching centers of two or more different outlets.

interfascicular region *Botany.* the tissue region located between the vascular bundles in a stem.

interference the process of blocking, hindering, or opposing, or something that does this; specific uses include: *Physics.* the superposition of two or more waves of the same frequency emitted from coherent sources, resulting in a single wave. *Telecommunications.* in a communications system, extraneous power entering or induced in a channel that tends to interfere with reception of desired signals. *Optics.* see OPTICAL INTERFERENCE. *Genetics.* a measure of the degree to which one crossover by a chromatid affects the probability of a second crossover by that same chromatid. *Virology.* the inhibition or prevention of the replication of a virus by the presence in the host cell of another virus. *Psychology.* conflict or inhibition caused by incompatible ideas, acts, memories, or thoughts. *Ecology.* see COMPETITION.

interference analyzer *Electronics.* a device that reveals the frequency and amplitude of a spurious input pulse.

interference blanker *Electronics.* a device that permits two or more sets of radio or radar equipment to be operated simultaneously without confusion.

interference colors *Optics.* the colors formed by interference and appearing when light is reflected from a very thin film, such as soap bubbles, or when light is transmitted through a thin piece of doubly refracting crystal being viewed through a polarizing microscope.

interference competition see DIRECT COMPETITION.

interference fading *Telecommunications.* the fading of a received signal caused by different wave components traveling along slightly different paths.

interference figure *Optics.* a pattern of light and dark concentric rings displayed when birefringent crystal is viewed in convergent light between crossed Nicol prisms in a polarizing microscope.

interference filter *Electronics.* a circuit that is placed between a source of interference and a radio receiver in order to reduce or eliminate the interference. Also, **interference eliminator.** *Optics.* a filter that prevents certain light rays from being transmitted by obstructing them rather than scattering or absorbing them.

interference fit *Design Engineering.* a fit between the mating parts of a male and female assembly in which one of the parts is forced into the other to allow maximum metal overlap.

interference fringes *Optics.* a pattern of alternating light and dark bands that arises when homogeneous light beams overlap.

interference microscope *Optics.* a microscope in which a direct beam of light splits and recombines as it passes through a specimen, forming a pattern that corresponds to variations in the specimen's thickness; used to examine transparent subjects. Also, MICROINTERFEROMETER.

interference of light see OPTICAL INTERFERENCE.

interference pattern *Physics.* the resultant superposition of waves represented by a spatial distribution of the interfering quantity, such as an electric field, pressure, or particle density. *Electronics.* the pattern that interference signals form on a radar screen.

interference prediction *Electronics.* an estimate of the level of interference a given device might experience in an electromagnetic environment.

interference reduction *Electronics.* a process by which interference in an electrical circuit from outside factors, such as radio transmitters and lightning, is minimized. Also, **interference suppression.**

interference region *Telecommunications.* a location in space in which interference between wave trains takes place.

interference rejection *Electronics.* a technique that prevents unwanted signals from entering a system, generally by routing the signals to ground.

interference ripple mark see CROSS RIPPLE MARK.

interference source suppression *Electronics.* a technique that diminishes the emission of spurious signals at a radiation source.

interference spectrum *Spectroscopy.* a spectrum produced by interference of a beam of light, as through a thin film. *Electronics.* the range of frequency distribution that occurs when a device is being jammed.

interference theory *Psychology.* the concept that information may become lost or changed as a result of new information being stored.

interference time *Industrial Engineering.* the time lost when two machines are out of service at the same time and only one can be repaired at a time.

interference wave *Telecommunications.* a radio wave that transmits an interference pattern when united with the direct wave.

interferogram *Spectroscopy.* a graph showing the variation of output signal from an interferometer as the interference is varied; used in Fourier-transform infrared spectroscopy.

interferometer *Optics.* an instrument that splits a light beam into two or more beams, which then recombine to form interference; used for testing optical surfaces and for studying light.

interferometer system *Electronics.* a technique that measures the azimuth of a target by transmitting its signal to two separate antennae, then comparing the phase differences between the signals received by each antenna.

interferometry *Optics.* the design and application of interferometers, as well as the study of interference patterns produced by interferometers.

interferon [in′tər fēr′än] *Immunology.* any of a family of glycoproteins secreted by virus-infected cells, which can protect noninfected cells from replication of the virus; they also have immunoregulatory functions and can inhibit the growth of nonviral intracellular parasites.

interferon-α *Immunology.* the major interferon produced by virus-induced leukocyte cultures. *Pharmacology.* a preparation of this substance, used as an antineoplastic drug in the treatment of hairy-cell leukemia.

interfertile *Biology.* capable of interbreeding between species.

interfix *Computer Programming.* a method of describing the key words in a record without ambiguity so that very specific inquiries can be answered without false retrievals.

interflow *Hydrology.* the lateral movement of water through the upper layers of soil, above the water table, until the water reaches a stream channel.

interfluve *Geology.* an area of relatively undissected upland between adjacent streams flowing in the same general direction.

interframe compression *Telecommunications.* data compression technology designed for use in full-motion color video that does not require frame-by-frame editing.

intergalactic *Astronomy.* of or relating to space between the galaxies.

intergalactic medium *Astronomy.* a thin gas, mostly ionized hydrogen, that fills the space lying between the galaxies in a cluster; it may or may not be similar to any medium between clusters of galaxies.

interganglionic *Neurology.* located or occurring between ganglion cells.

intergelisol see PERELETOK.

intergenic crossing-over *Molecular Biology.* recombination that occurs between gene-coding sequences.

intergenic suppression *Genetics.* a type of suppressor mutation that may code for an abnormal tRNA whose anticodon reads the mutated codon either as a normal codon or as an acceptable substitute.

interglacial *Geology.* related to or formed during a period separating successive glacial epochs or two distinct glaciations, characterized by a milder climate in which the ice sheets melt or retreat.

intergradation *Ecology.* **1.** the interbreeding of two subspecies of the same species in the area where they meet. Also, **primary intergradation.** **2.** the interbreeding of two closely related species whose ranges overlap, before isolating mechanisms are fully developed. Also, **secondary intergradation.**

intergrade *Agronomy.* describing soils that contain the properties of two or more distinctive soils or soil horizons.

intergranular *Materials Science.* located or occurring between the grains of a substance, or along grain boundaries. *Neurology.* located or occurring between the granule cells of the brain.

intergranular corrosion *Materials Science.* corrosion that preferentially occurs at interfaces.

intergranular fracture *Materials Science.* a fracture in which a crack takes a path along the grain boundaries, especially when grain boundaries are weakened by segregation or inclusions.

intergranular precipitation *Materials Science.* precipitation at grain boundaries.

intergranular pressure see EFFECTIVE STRESS.

intergrowth *Mineralogy.* the interlocking of grains of two or more different minerals as a result of their simultaneous crystallization.
interhalogen *Inorganic Chemistry.* any compound composed of two different elements from the halogen group (periodic table Group VIIa); e.g., chlorine pentafluoride, ClF_5.
interionic attraction *Physical Chemistry.* the extent to which ions of the opposite charge within a solution have attraction for each other.
interior *Science.* situated inside; inward. *Anatomy.* an inner part or cavity. *Geography.* the inland part of a country or region. *Building Engineering.* **1.** the inside part of a building. **2.** of, relating to, or designed for use inside. Thus, **interior lighting, interior paint.** *Mathematics.* **1.** an **interior point** of a set *A* is a point that can be contained in an open set (or neighborhood) which is itself contained in *A*. **2.** the **interior of a set** *A* is the largest open set contained in *A*; equivalently, the set of interior points of *A*.

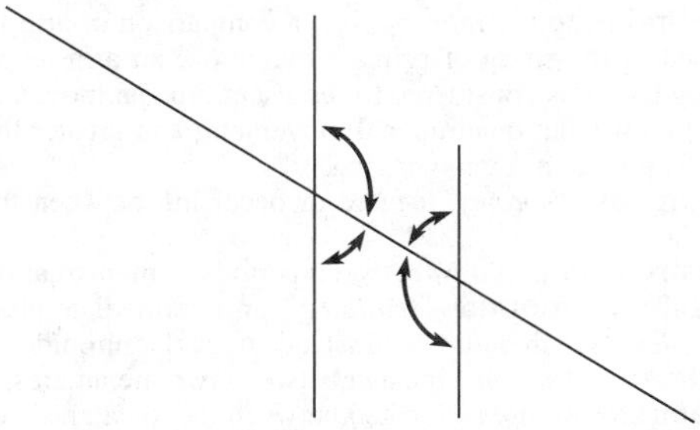

interior angles

interior angle *Mathematics.* **1.** one of four angles made by a transversal cutting two lines, and lying between the two lines. Each interior angle has the transversal for one of its sides and one of the two lines for the other side. Nonadjacent interior angles lying on opposite sides of the transversal are called **alternate interior angles. 2.** at each vertex of a polygon, the angle formed by the two adjacent sides and lying inside the polygon.
interior angle traverse *Cartography.* in surveying, a traverse that begins and ends on the same station, and measures only distances and interior angles.
interior ballistics *Mechanics.* the branch of ballistics dealing with the propulsion of a projectile within the gun barrel and the combustion of the propellant, or of the behavior of rocket propellant prior to exiting the thrust cone of a rocket. Also, INTERNAL BALLISTICS.
interior distribution *Electricity.* the pattern of electric power distribution within a building or plant.
interior drainage see CLOSED DRAINAGE.
interior label see INTERNAL LABEL.
interior orientation *Photogrammetry.* the determination of the calibrated focal length, the location of the calibrated principal point, and the calibrated lens distortion of a camera, giving the perspective of the photograph inside the camera at the instant of exposure. Also, INNER ORIENTATION.
interkinesis *Cell Biology.* a period between nuclear mitotic divisions during which no DNA replication takes place, such as the period between the first and second divisions in meiosis .
interlace *Computer Programming.* to assign successive addresses to storage locations that are physically separated on magnetic disk or memory banks in order to reduce access time.
interlaced scanning *Electronics.* a procedure for scanning a television picture in which the odd- and even-numbered lines are transmitted sequentially as two separate fields. Also, LINE INTERLACE.
interlacing arcade *Architecture.* a row of arches resting on alternate columns such that their moldings intersect and appear as if interlaced.
interlamellar spacing *Materials Science.* the distance from the center of one lamella to the center of the next lamella.
interlaminar nuclei *Anatomy.* several small masses of neurons located in the internal medullary lamina of the thalamus.
interlaminar shear strength *Materials Science.* the bond between fiber layers and the polymer matrix that contains them; determined by a short beam-bending test.
interlaminated *Science.* **1.** situated between layers. **2.** arranged in alternating layers.
interleave *Computer Programming.* to arrange sequential memory accesses in different program modules in order to reduce memory access time.
interleaved windings *Electricity.* an arrangement of transformer windings in which primary, secondary, and tertiary windings are subdivided into layers and interleaved on the same core.
interleukin *Immunology.* a generic term for a group of protein factors that affect primary cells and are derived from macrophages and T cells that have been stimulated by antigens or mitogens.
interleukin 1 *Immunology.* a protein that occurs in macrophages and accessory cells and is a major factor in the activation of T and B lymphocytes by antigens or mitogens; formerly known as leukocyte activating factor.
interleukin 2 *Immunology.* the soluble product of a stimulated T cell that initiates the activation and differentiation of other T lymphocytes regardless of antigen specificity.
interleukin 3 *Immunology.* an extracellular protein that is derived from mitogen-activated T lymphocytes or other cells and acts as a colony stimulating factor for bone marrow progenitor cells and a growth factor for mast cells.
interlobate moraine see INTERMEDIATE MORAINE.
interlock *Engineering.* a safety switch or system that prevents the operation of equipment when a hazard exists, such as the mechanism that shuts down high-voltage circuits when protective doors to the circuits are opened. *Computer Technology.* a hardware or software mechanism that coordinates processes so that they function independently without interfering with each other, but achieve proper coordination and sharing of resources. *Robotics.* **1.** a condition that prevents a machine or device from beginning a new task before the task at hand is completed. **2.** a deadlock under parallel processing.
interlocking *Medicine.* a complication of labor in twin births in which the inferior surface of the chin of one twin is hooked to that of the other, in such a way that vaginal delivery is impossible. Thus, **interlocking twins, interlocked twins.** *Transportation Engineering.* a group of railroad turnouts (switches) that are interconnected so that only one path through the system is opened at any one time, preventing conflicting train movements. Also, **interlocking plant.**
interlocking cutter *Mechanical Devices.* a milling cutter made up of two parts with alternating or overlapping teeth.
interlock relay *Electricity.* a relay having two or more armatures that are linked mechanically or interconnected electrically in such a way that the position of any one of the armatures causes or prevents the motion of another one.
interlock switch *Electricity.* an electrical switch, usually attached to a cabinet door or housing, that removes power from a circuit when the door is opened, thus providing protection against high-voltage electric shock.
interlude *Computer Programming.* the part of a program that performs preliminary computations, housekeeping, or data organization before the main program is run.
intermale aggression *Behavior.* a form of aggression that is triggered by an unfamiliar male of the same species.
intermammary *Anatomy.* between the breasts.
intermediary metabolism *Biochemistry.* the utilization of foodstuffs (nutrients) for processes of life: energy production, growth, reactions, and so on.
intermediate *Chemistry.* **1.** a substance that is formed in a middle stage of a series of chemical reactions; a "stepping stone" between a parent substance and a final product; especially important in the production of dyes, pharmaceuticals, and other organic synthetics. **2.** of or relating to such a substance or to a middle stage in a series of reactions.
intermediate annealing *Metallurgy.* annealing that precedes the final thermal treatment.
intermediate cell mass see NEPHROTOME.
intermediate code see INTERMEDIATE LANGUAGE.
intermediate contour *Cartography.* any of the contour lines drawn between index contours on a map and representing the basic interval of change in elevation. See INDEX CONTOUR.
intermediate control change *Computer Programming.* a program control change with an expected or average moderate magnitude.
intermediate control data *Computer Programming.* a data field that is of secondary significance in sorting records.
intermediate distributing frame *Electricity.* in telephone switching systems, a frame at which cross-connections between central office equipment units are made.

intermediate disturbance hypothesis *Ecology.* the hypothesis that species diversity is greatest in habitats with moderate amounts of physical disturbance, owing to the coexistence of early and late successional species.

intermediate filament *Cell Biology.* one of a diverse group of long, cytoplasmic protein filaments found in many eukaryotic cells that constitute part of the cytoskeletal structure of the cell.

intermediate fix *Aviation.* a navigational fix provided for the intermediate leg of an aircraft landing approach.

intermediate frequency *Electronics.* **1.** the frequency to which a signal is shifted before it is modulated for transmission or reception. **2.** the frequency produced by combining an incoming signal with one generated internally, chiefly found in superheterodyne circuits.

intermediate-frequency amplifier *Electronics.* in a heterodyne receiver, an amplifier that processes the intermediate-frequency signal.

intermediate-frequency interference ratio *Electronics.* in a heterodyne receiver, the relationship between the intermediate-frequency signal input at the antenna and the signal input needed for identical outputs. Also, **intermediate-frequency response ratio.**

intermediate-frequency jamming *Electronics.* a method for blocking a radar signal in which signals are transmitted to an enemy's radar receiver at frequencies in the frequency range of its intermediate-frequency amplifier.

intermediate-frequency signal *Electronics.* in a heterodyne receiver, the signal produced by mixing the incoming signal with the output of the local oscillator.

intermediate-frequency stage *Electronics.* one of the stages in the intermediate-frequency amplifier of a superheterodyne receiver.

intermediate-frequency strip *Electronics.* a unit added to a receiver that houses its intermediate-frequency stages.

intermediate-frequency transformer see TRANSFORMER.

intermediate ganglion *Anatomy.* any of several ganglia of the sympathetic nervous system lying on the rami communicantes in the cervical and lumbar regions.

intermediate glass *Materials Science.* a glass formed by the joining of intermediate oxides, such as lead oxide, with the silica chain.

intermediate horizon *Electromagnetism.* an obstacle, such as a hill or building, that lies between a radar and the radar horizon, thus screening off a portion of the radar's range.

intermediate host *Biology.* a secondary host that harbors only immature stages of a parasite, whereas a definitive (primary) host harbors the mature stage; for example, the mosquito is an intermediate host to *Plasmodium,* the agent of malaria.

intermediate infrared radiation *Electromagnetism.* infrared radiation that lies in the wavelength range of about 2.5 micrometers to 50 micrometers; this range covers most molecular vibrations.

intermediate ion *Meteorology.* an atmospheric ion having a size and mobility that are intermediate between those of a small and a large ion; they are rarely observed and not fully understood.

intermediate language *Computer Programming.* a type of program representation that lies between a machine or absolute language and a higher-order or machine-independent language; used, for example, by a compiler as an intermediate representation between parsing of the source language and generation of machine code. Also, INTERMEDIATE CODE.

intermediate lava *Volcanology.* lava containing 52–66% silica, less than acidic lava but more than basic.

intermediate layer see SIMA.

intermediate marker *Ordnance.* a natural or artificial marker that is used as a point of reference between a landmark and a mine.

intermediate moraine *Geology.* a lateral moraine that forms between adjacent glacial lobes whose margins have been pushed together. Also, INTERLOBATE MORAINE, MEDIAL MORAINE.

intermediate neutron *Nucleonics.* a neutron whose energy level ranges from 100 to 100,000 electron volts.

intermediate phase *Metallurgy.* in a phase diagram, a phase that does not extend to any pure component.

intermediate-range ballistic missile *Ordnance.* a ballistic missile having a range of 2500–5500 km.

intermediate reactor *Nucleonics.* a nuclear reactor in which fission is sustained primarily by intermediate energy neutrons.

intermediate repeater *Electronics.* a device that strengthens a fading signal anywhere along a telephone line except at the end.

intermediate result *Computer Programming.* a temporary result that is kept in working storage for use in a multistep computer operation.

intermediate-scale map *Cartography.* a map that uses a scale of between 1:200,000 and 1:500,000; used especially for military operations.

intermediate state *Physics.* a state of partial superconductivity achieved when a magnetic field is applied to a superconducting material cooled below its critical temperature. *Quantum Mechanics.* a temporary state occurring between the initial and final states in the course of a transition.

intermediate storage *Computer Programming.* a block of memory that is capable of temporarily holding calculations until they are further processed or released for output. Also, **intermediate memory storage**.

intermediate value theorem *Mathematics.* if f is a continuous real-valued function on an interval $[a, b]$ with $f(a) \leq \xi \leq f(b)$, then there is a point $x, a \leq x \leq b$ such that $f(x) = \xi$.

intermediate vector boson *Particle Physics.* a general name for the W and Z particles that carry the weak force.

intermedin *Biochemistry.* a melanocyte-stimulating hormone, named for the intermediate lobe of the pituitary gland, from which it is secreted in amphibians, reptiles, and fish.

intermembral index *Anthropology.* a comparison of arm length to leg length used in the study of primate evolution; an arm length less than 80% of leg length is considered to represent bipedal movement, from 80 to 100% to represent quadrupedal movement, and greater than 100% to represent brachiation (tree-swinging).

intermeningeal *Anatomy.* located or occurring between the meninges of the brain.

intermenstrual *Physiology.* between periods of menstruation.

intermetallic compound *Metallurgy.* an intermediate phase in which the components are metallic; for instance, nickel aluminide.

intermetameric *Anatomy.* located between two metameres.

intermetatarsal *Anatomy.* located between the metatarsals of the foot.

intermitotic *Cell Biology.* **1.** of or relating to the span of time between mitoses of a cell. **2.** referring to the interphase stage of a cell cycle.

intermittency *Chaotic Dynamics.* a form of chaos arising from a tangent bifurcation that is characterized by irregular oscillations occurring between periods of nearly periodic oscillation. The nearly periodic (laminar) intervals are of irregularly varying lengths. A "route to chaos" develops as the tangent bifurcation takes place. As the chaos becomes more fully developed, the statistical variations in irregular and laminar regions become greater. There are three types characterized by eigenvalues of the iterations of an associated map (such as for the peak heights): Type I, a real eigenvalue crosses the unit circle at +1; Type II, two complex conjugate eigenvalues cross the unit circle simultaneously; Type III, a real eigenvalue crosses the unit circle at –1.

intermittent arming device *Ordnance.* a device enabling a mine to be armed only at preset times.

intermittent claudication *Cardiology.* a condition caused by diminished blood supply to the leg muscles as a result of the narrowing of arteries of the legs; marked by pain and lameness during mobility, but absence of pain when sedentary. Also, CHARCOT'S SYNDROME.

intermittent current *Electricity.* a unidirectional current that flows and ceases to flow at intervals. *Oceanography.* an ocean current that periodically ceases to flow, usually according to a seasonal pattern.

intermittent defect *Engineering.* a flaw that is occasionally but not continually found in the operation of a device, instrument, electrical circuit, or system.

intermittent fault *Computer Science.* a fault, such as a loose connector, that occurs sometimes but not continuously.

intermittent fever *Pathology.* a fever that recurs between periods of normal temperature; characteristic of malaria.

intermittent firing *Mechanical Engineering.* a furnace-firing cycle in which fuel and air are burned in short, frequent bursts. *Ordnance.* the firing of weapons at irregular intervals.

intermittent illumination *Ordnance.* fire in which illuminating projectiles are fired at irregular intervals.

intermittent lake *Hydrology.* a lake that contains water for only part of the year rather than the entire year.

intermittent light *Navigation.* a navigational light that flashes on and off, with the periods of darkness shorter than the periods of illumination; more commonly called an occluding light.

intermittent operation *Engineering.* a condition in which the operation of a device, instrument, or system is disrupted at regular or irregular intervals.

intermittent reinforcement *Psychology.* in operant conditioning, a type of reinforcement schedule in which only certain selected occurrences of a response are reinforced. Also, PARTIAL REINFORCEMENT.

intermittent scanning *Electronics.* a process in which an antenna beam is scanned at intermittent intervals in order to make it more difficult for enemy receivers to intercept it.
intermittent schedule of reinforcement see INTERMITTENT REINFORCEMENT.
intermittent spring *Hydrology.* a spring that flows periodically, otherwise remaining dry or nonexistent. Also, **intermitting spring.**
intermittent stream *Hydrology.* **1.** a stream that flows only during certain times of the year, as when it receives water from some other source. **2.** a stream that occasionally stops flowing, as when more water is lost by evaporation or seepage than is replaced.
intermittent weld *Metallurgy.* a weld consisting of a sequence of weld beads and of unwelded spaces.
intermitter *Petroleum Engineering.* an apparatus housed in a well that can be controlled to permit wide-open flow for brief periods at various times a day and then be turned off.
intermodal *Transportation Engineering.* involving more than one mode of transportation, such as truck and rail or bus and rapid transit. Thus, **intermodal conveyance, intermodal transfer.**
intermodulation *Electronics.* a variation in the characteristics of audio signals that results when they interact while being transmitted through the same system, especially in a nonlinear transducer.
intermodulation distortion *Electronics.* **1.** the appearance of unwanted frequencies that arise when two signals interact at a nonlinear component in a system. **2.** the appearance of veiled or muffled sound quality that arises from the interaction of two unrelated tones in a recording or amplifying system.
intermodulation interference *Electronics.* the interference generated when signals of two or more frequencies are applied to a nonlinear amplifier.
intermolecular [in′tər mə lek′yə lər] *Physical Chemistry.* relating to interaction between different molecules.
intermolecular distance *Physical Chemistry.* the space between two molecules undergoing some form of interaction with each other.
intermolecular forces *Physical Chemistry.* the forces that act on an individual molecule to cause it to attract or repel other adjacent molecules; the strength of these forces in a given molecular substance at a certain temperature determines whether the substance exists as a gas, a liquid, or a solid.
intermontane *Geology.* located between or surrounded by mountains or mountainous regions. Also, **intermountain.**
intermontane glacier *Hydrology.* an alpine glacier created by the joining of several valley glaciers and lying between mountain ranges or ridges.
intermontane trough *Geology.* **1.** a subsiding region between stable or uplifting regions in an island arc formation. **2.** a basinlike area between mountain ranges or ridges that often holds an intermontane glacier.
intermural *Anatomy.* located between the walls of an organ or cavity.
intermuscular *Anatomy.* located between or among muscles.
intermuscular septum *Anatomy.* the sheets of fibrous tissue lying between muscles.
intern or **interne** *Medicine.* a recent graduate of medical school who is serving a period of residency in a hospital prior to being licensed to practice.
internal *Science.* situated or occurring on the inside. *Anatomy.* of or relating to the middle. Many anatomical structures formerly designated *internal* are now designated *medial. Psychology.* see INTERNALIZER.
internal acoustic meatus *Anatomy.* the canal that passes through the petrous portion of the temporal bone, containing the vestibulocochlear and facial nerves and the labyrinthine artery and veins.
internal arithmetic *Computer Programming.* the arithmetic operations performed by the arithmetic/logic unit of a computer.
internal balance see COADAPTATION.
internal ballistics see INTERIOR BALLISTICS.
internal boundary see INTERFACE.
internal brake see EXPANDING BRAKE.
internal broaching *Mechanical Engineering.* the process of removing material on an internal surface using a tool having teeth of increasing size in a straight line over the surface.
internal buffer *Computer Technology.* a portion of the main memory of a computer that temporarily holds data that will eventually be delivered to a processor or sent to an output device.
internal-calibration compensation *Robotics.* a method of automatically calibrating to obtain the calculations needed for repeatability.
internal capsule *Anatomy.* a thick layer of nerve fibers, separating the thalamus from the lentiform nucleus.
internal carotid artery *Anatomy.* a large artery arising as a branch of the common carotid and branching to produce the anterior and middle cerebral arteries in the cranial cavity.
internal carotid nerve *Anatomy.* a sympathetic nerve extending along the internal carotid artery from the superior cervical ganglion to the internal carotid plexus.
internal cast see STEINKERN.
internal clock *Computer Technology.* **1.** an electronic device within a computer that tells the date and time of day. **2.** a timing device used to regulate and synchronize the sequential logic systems in a computer.
internal clot *Hematology.* a blood clot formed within a blood vessel.
internal-combustion engine *Mechanical Engineering.* an engine in which the process of combustion takes place in the cylinder or cylinders, and the products of combustion serve as the thermodynamic fluid; for example, a gasoline or diesel engine.

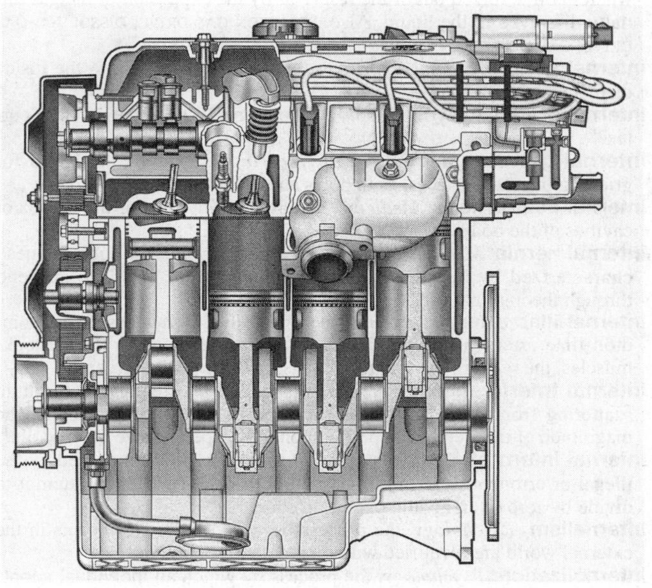

internal-combustion engine

internal conversion *Nuclear Physics.* a process in which a highly energized nucleus transmits its energy to an orbiting electron, causing the electron to be ejected from the atom.
internal cycle time see CYCLE TIME.
internal defense *Military Science.* the measures taken by a government to free and protect its society from subversion, lawlessness, and insurgency.
internal diffusion *Chemical Engineering.* the diffusion of gaseous or liquid-phase reactants into the inner pores of a catalyst. High internal diffusion gives high catalyst efficiency.
internal drainage see CLOSED DRAINAGE.
internal drift current *Oceanography.* motion caused in an underlying layer of water by shearing stresses and friction at the point where an overlying layer of different density meets the underlying layer.
internal elastic membrane *Histology.* a thin layer of elastin fibers found between the tunica media and tunica intima of small arteries.
internal energy *Thermodynamics.* a property of a thermodynamic system that includes the kinetic energies of the individual particles of the system, the interaction energies between particles, and the intrinsic energies of individual particles, but that excludes the kinetic and potential energies of the system as a whole. A change in internal energy during a given process will be equal to the difference between the heat transferred to the system and the work done by the system.
internal erosion *Geology.* any erosion that occurs within a compacting sediment as a result of the movement of water through pores.
internal failure *Materials Science.* the failure of an insulator that occurs when impurities provide donor or acceptor levels that permit electrons to be excited into the conduction band.

internal fertilization *Physiology.* **1.** broadly, the reproductive condition in animals, such as birds and mammals, when the egg is fertilized within the female's body. **2.** specifically, the union of the nuclei of the egg cell and the sperm cell, resulting in the disappearance of their nuclear membranes and the combination of their chromosomes.

internal fistula *Anatomy.* an abnormal sinus or passage between two hollow organs.

internal flue *Mechanical Engineering.* a furnace tube or fire tube that runs through the water space of a boiler.

internal force *Mechanics.* any force exerted within a system to affect another part of the same system; e.g., the force of nuclear attraction within a solid body.

internal friction *Mechanics.* a property of matter by which deformation is somewhat resisted and the conversion of applied energy to heat occurs during deformation of material. *Fluid Mechanics.* friction that is caused by the viscous forces present in a moving fluid.

internal furnace *Mechanical Engineering.* a boiler furnace containing a firebox set inside a water-cooled heating surface.

internal gas drive *Petroleum Engineering.* a primary recovery of oil wherein oil is forced out of the reservoir by the expansion of the gas initially dissolved in the liquid. Also, SOLUTION GAS DRIVE, DISSOLVED-GAS DRIVE, GAS-DEPLETION DRIVE.

internal gear *Mechanical Devices.* a gear whose teeth line the inside diameter of its rim.

internal granular layer *Histology.* the fourth layer of the cerebral cortex.

internal grinder *Mechanical Engineering.* a machine designed for grinding internal surfaces, as in holes and cylinders.

internal hemorrhage *Medicine.* hemorrhage into one of the organs or cavities of the body.

internal hernia *Medicine.* a hernia in which the pouch is subperitoneal, characterized by failure of the testicle or vaginal pouch to descend through the inguinal canal.

internal iliac artery *Anatomy.* one of two large branches of the common iliac artery in the pelvic region, carrying blood to the buttocks muscles, the rectum, the urinary bladder, and the genital organs.

internal interference *Materials Science.* an interference in the light scattering from polymer solutions caused by large polymer coils; the magnitude of interference depends upon the shape and size of the coils.

internal interrupt *Computer Programming.* an interrupt caused by the illegal or erroneous use of an instruction or data, such as an attempt to divide by zero or an invalid operation code.

internalism *Psychology.* the process by which objects or norms in the external world are identified with mental representations.

internalization *Psychology.* the process by which an individual adopts the attitudes, norms, and prejudices of another or others.

internalizer *Psychology.* an individual who generally believes that the outcome of events in his life can be affected or controlled by his own actions.

internal label *Computer Programming.* an identifying label, such as a tape label, that is contained within the data itself; usually read under program control. Also, INTERIOR LABEL.

internally disjoint *Mathematics.* in graph theory, two paths are internally disjoint if any vertex shared by both paths is an end vertex of both paths. Thus, neither path has an internal vertex on the other path.

internally fired boiler *Mechanical Engineering.* a boiler whose firebox or furnace is set inside the boiler and is surrounded by water.

internally stored program *Computer Programming.* a routine or sequence of instructions that is stored in the computer's internal memory rather than in some external storage device.

internal medicine *Medicine.* a branch of medicine that deals specifically with the diagnosis and treatment of diseases and disorders of internal structures of the body.

internal membrane *Cell Biology.* any of the numerous membrane-bound organelles and vesicles that are found within eukaryotic cells.

internal memory see INTERNAL STORAGE.

internal-mix atomizer *Mechanical Engineering.* a pneumatic atomizer in which gas and liquid are mixed within the air cap, before the gas expands through the nozzle; used in low-pressure work, as in small compressors. Also, **internal-mix spray gun.**

internal oxidation *Metallurgy.* the oxidation of a reactive element, such as aluminum, in a less reactive metal or alloy, caused by solid-state diffusion of oxygen; usually done intentionally for dispersion hardening.

internal perimysium see PERIMYSIUM.

internal phase see DISPERSE PHASE.

internal photoelectric effect *Solid-State Physics.* a phenomenon in which a photon absorbed by a semiconductor provides the energy to advance an electron from the valence band to the conduction band.

internal promoter *Genetics.* a promoter sequence that is contained within the structural coding sequence of a gene.

internal reader *Computer Technology.* a facility of the operating system that transfers jobs to the job entry subsystem.

internal reflectance see ATTENTUATED TOTAL REFLECTANCE.

internal resistance *Electricity.* the inherent resistance of a device, such as a battery or generator.

internal respiration *Physiology.* the exchange of oxygen and carbon dioxide between body cells and surrounding fluids.

internal schema *Computer Programming.* a description within a database of the database environment and of the physical storage of information in the database.

internal secretion *Physiology.* any hormone that is not discharged by a duct from the body, but is given off into the blood and lymph, and effects a response distant from the site of origin. Also, INCRETION.

internal sedimentation *Geology.* the accumulation of chemical or clastic material eroded from a relatively consolidated carbonate sediment, and the subsequent deposition of this material in secondary cavities created in the host rock by folding or internal erosion.

internal sensor *Robotics.* a sensor that is part of a robot rather than external to it.

internal sorting *Computer Programming.* the process of sorting records within a computer's main memory, without the use of external storage such as tape or disk.

internal standard *Spectroscopy.* a method for calibrating an instrument or assay, whereby a substance of known properties and concentration is added to the sample in order to produce a defined, standard spectral line, whose intensity and position can be compared with those of the spectral lines of the sample.

internal storage *Computer Programming.* the addressable data and program storage facilities that are contained within and controlled by the central processing unit. Also, PRIMARY MEMORY, MAIN MEMORY, INTERNAL MEMORY.

internal storage capacity *Computer Programming.* the amount of data, usually expressed in megabytes, that can be stored in the computer's main memory.

internal stress *Mechanics.* stress that exists in a solid when no external forces are being applied. Also, RESIDUAL STRESS, INITIAL STRESS.

internal table *Computer Programming.* a collection of organized data contained in the main memory of the computer.

internal thread *Design Engineering.* a screw thread cut inside a piece of hollow material or cylinder on its inner surface, such as that of a nut.

internal vertex *Mathematics.* in a walk $v_0, e_1, v_1, \ldots, v_{n-1}, e_n, v_n$, the vertices $v_1, v_2, \ldots, v_{n-1}$, are the internal vertices of the walk.

internal vibrator *Mechanical Engineering.* a vibrating device that is enclosed in a cylinder and drawn through wet concrete to ensure proper compaction.

internal wave *Fluid Mechanics.* a wave in a fluid whose motion is vertical and occurs beneath the surface.

internal work *Thermodynamics.* the amount of work required to separate all the particles of a thermodynamic system to infinite distances against the attractive forces that tend to hold the system together. *Industrial Engineering.* manual work that is performed by a machine operator while the machine runs automatically. Also, FILL-UP WORK, INSIDE WORK.

internal writer *Computer Technology.* a job entry subsystem facility that allows the writing of data on devices that are not directly managed by the job control program.

international air mile see AIR MILE.

international ampere *Electricity.* the standard value of the ampere as a unit of the international system of electrical units. It is the current that, when flowing through a solution of silver nitrate in water, deposits silver at a rate of 0.001118 gram per second. This is equivalent to about 0.999850 ampere.

international analysis code *Meteorology.* a code used internationally for the communication of data from synoptic chart analyses.

international angstrom *Physics.* a unit of length used in the measurement of wavelengths, defined by 1/6438.4696 of the wavelength of the red cadmium spectral line.

international annealed copper standard *Metallurgy.* a reference level of the electrical conductivity of copper, usually referred to as IACS and expressed as a percentage. Commercially pure copper ranges from 100 to 103 IACS.

International Astronomical Union *Astronomy.* the world organization for astronomers; it was founded in 1919 and holds meetings every three years.
international broadcasting *Telecommunications.* radio broadcasting for international public entertainment; frequencies used by this broadcasting service are usually between 5950 kilohertz and 21,750 kilohertz.
International Cable Code see MORSE CABLE CODE.
international call sign *Telecommunications.* a call sign assigned according to the rules of the International Telecommunication Union, which permits a receiver to identify a radio station; the nationality of the radio station is identified in the first character or the first two characters.
international candle *Optics.* a unit of luminous intensity, based on the burning of a group of carbon-filament lamps; has been replaced by candela.
international code signal see INTERNATIONAL SIGNAL CODE.
International Color System *Astronomy.* a now-obsolete system used for two-color photographic photometry of stars; it has been superseded by various photoelectric systems, chiefly among which is the UBV system.
international control frequency bands *Telecommunications.* in the United States, radio-frequency bands that are allocated to form links between stations used for international communication.
international control station *Telecommunications.* a fixed station connected with the international fixed public radio communications service.
International Date Line *Cartography.* a particular hypothetical line on the earth, coinciding basically with the 180° meridian but with deviations to avoid separating contiguous or nearby inhabitated areas; separates neighboring time zones in which the date differs by one day.
international ellipsoid of reference *Geodesy.* a reference ellipsoid with a semimajor axis of approximately 6,387,388 meters, a semiminor axis of approximately 6,356,911.9 meters, and an ellipticity of 1/297.
International Geophysical Year *Geophysics.* the period between July 1, 1957, and December 31, 1958, during which international programs of geophysical explorations of the solar and terrestrial atmospheres were conducted.
international gravity formula *Geodesy.* a formula for determining theoretical gravity on the basis of the following assumptions: that the spheroid of reference is an exact ellipsoid of revolution, with the dimensions of the international ellipsoid of reference, rotating about its minor axis once a day; that the surface of the ellipsoid is level; and that gravity at the equator is 978.049 gals.
international henry *Electromagnetism.* an international unit of electrical inductance, equivalent to 1.00049 henry.
International Ice Patrol *Oceanography.* an organization that monitors and provides warnings of dangerous icebergs.
international index numbers *Meteorology.* an internationally recognized numbering system for designating meteorological observing stations, administered by the World Meteorological Organization.
International Morse Code see MORSE CODE.
international nautical mile see NAUTICAL MILE, def. 2.
international ohm *Electricity.* the standard unit of resistance, reactance, and impedance, represented by the Greek letter Ω, derived from the standard ampere; the resistance at 0°C of a column of mercury of uniform diameter that is 106.300 centimeters long and has a mass of 14.4521 grams, equivalent to 1.00049 ohms.
International Phonetic Alphabet see IPA.
International Polar Year *Meteorology.* a designation for the years 1882 and 1932, during which many countries participated in increased observations of geophysical and meteorological phenomena in the polar regions; the concept was continued and expanded in scope as the International Geophysical Year.
International Practical Temperature Scale *Thermodynamics.* a standard temperature scale defined on the basis of certain fixed and easily reproducible points that are assigned definite temperatures; the primary fixed-point temperatures in °C include the triple point of water (0.01); the normal boiling points of water (100) and oxygen (–182.692), and the normal freezing points of gold (1064.43) and silver (961.93).
International Quiet Sun Year *Geophysics.* an international program of solar observations carried out in 1964–1965, during the minimum of the eleven-year cycle of solar activity.
international radio silence *Telecommunications.* a three-minute period of radio silence during which distress signals from ships or aircraft are listened for by marine radio stations; transmissions occur 15 and 45 minutes after each hour on the frequency of 500 kilohertz.
international rules of the road *Navigation.* rules specifying the conduct of vessels in international waters, more formally referred to as the **International Regulations for Preventing Collisions at Sea.**
international signal code *Telecommunications.* a worldwide code used for international communication, employing sequences of letters instead of phrases, words, or sentences.
international spheroid see INTERNATIONAL ELLIPSOID OF REFERENCE.
International Standards Organization *Computer Technology.* an international agency responsible for establishing standards for information exchange.
international sunspot number see WOLF NUMBER.
international synoptic code *Meteorology.* a synoptic code administered by the World Meteorological Organization in which observable meteorological elements are encoded in five-digit word lengths for transmission from one weather station to another. Also, SYNOPTIC CODE.
international system of electrical units *Metrology.* an adoption of standard unit values for the ohm, ampere, centimeter, and second, used for measuring electrical and magnetic quantities from 1893 until 1948, when it was replaced by the Giorgi system.
International System of Units *Physics.* a unit system that is used worldwide and is based on the units of the meter, second, kilogram, ampere, kelvin, candela, and mole. (From the French *Système Internationale d'Unités.*)
International Temperature Scale see INTERNATIONAL PRACTICAL TEMPERATURE SCALE.
international thread *Design Engineering.* a standardized thread size in which pitch and diameter are related, thus giving the thread a rounded root and a flat crest.
international unit *Metrology.* any unit based on the International System of Units (SI).
international volt *Electricity.* the standard unit of electric potential, derived from the international ampere. It is the voltage that will produce a current of one international ampere through a resistance of one international ohm, equivalent to 1.00034 volts.
internegative *Graphic Arts.* a black-and-white negative produced by rephotographing the image projected from a color transparency. Also, **interneg.**
internet *Computer Technology.* **1.** any network that connects other networks. **2. Internet.** a large network of this type that covers the U.S. and extends to Canada, Europe, and Asia, providing connectivity between governments, universities, and corporate networks and hosts.
internetting *Computer Technology.* the technology of connecting multiple and diverse computer subnetworks. Also, **internetworking.**
interneuron *Neurology.* any neuron located between the primary afferent neuron and the final output neuron in a neural chain.
internides *Geology.* the interior of an orogenic belt, lying farthest from the craton. Also, PRIMARY ARC.
internist *Medicine.* a specialist in internal medicine; a physician who specializes in the diagnosis and medical (as distinguished from surgical or obstetric) treatment of diseases of adults.
internodal *Anatomy.* situated or occurring between two nodes.
internode of Ranvier *Neurology.* that portion of a nerve fiber which is located between two nodes of Ranvier.
internuclear *Science.* situated between the nuclei of cells or atoms.
internuclear distance *Physical Chemistry.* the space between two nuclei in a molecule.
internuncial *Neurology.* functioning in a connector between nerve cells or nerve centers, as interneurons in a neural chain.
internuncial neuron *Neurology.* an association neuron that connects sensory and motor neurons in the spinal cord.
interocclusal *Anatomy.* situated between the occlusal surfaces of opposing teeth.
interoceptor *Physiology.* receptors, located in the viscera, that receive stimuli connected with internal body organ activities such as digestion, excretion, circulation, hunger, thirst, and sexual feelings.
interocular [in′tər äk′yə lər] *Anatomy.* between the eyes.
interocular distance *Anatomy.* the distance between the eyes, usually measured as the distance between the pupils of the eyes.
interoffice trunk *Telecommunications.* a trunk connected between two central offices.
interoperability *Military Science.* the capability of military systems, units, or forces to provide services to and accept services from other military systems, units, or forces, and to use the services exchanged to enable them to operate effectively together.
interorbital *Anatomy.* located between the eye sockets.

interparietal *Anatomy.* **1.** between the walls of a body part. **2.** between the parietal bones. **3.** between the parietal lobes of the brain.

interpass temperature *Metallurgy.* in multipass welding, the lowest temperature of a pass before the next pass is initiated.

interpenetration twin *Crystallography.* a twin pair of crystals that grow in such a way that they appear to pass through each other.

interpersonal theory *Psychology.* the theory that the most important factor in the development of an individual's personality is the nature of his or her relationships and interactions with other people. Thus, **interpersonal psychiatry, interpersonal therapy.**

interphase *Cell Biology.* **1.** the growth period of a cell that occurs between mitotic divisions, during which protein synthesis and DNA replication take place. **2.** the period between the first and second divisions of meiosis.

interphase reactor *Electricity.* a center-tapped autotransformer used to provide step-up or step-down voltage using only one winding.

interphone *Telecommunications.* an intercommunicating system using headphones and/or microphones for contact between nearby stations.

interpial *Neurology.* located between the two layers of the pia mater.

interplanar spacing *Crystallography.* the perpendicular spacings between planes through points in a crystal lattice.

interplanetary *Astronomy.* existing or occurring between planets in the solar system, as in interplanetary space. *Space Technology.* traveling or designed to travel between planets in the solar system, especially between earth and another planet or earth and the moon. Thus, **interplanetary flight, interplanetary spacecraft.**

interplanetary dust *Astronomy.* dust in the solar system; comes mostly from sublimating comets, although some is material ejected from rocky bodies by meteorite impacts.

interplanetary magnetic field *Astronomy.* the magnetic field that fills solar system space, wound into a spiral structure by the rotation of the sun; at earth's distance from the sun, it has a field strength of about 5×10^{-5} gauss.

interplanetary medium *Astronomy.* an extremely tenuous mixture of dust and ionized gas (mostly hydrogen) that pervades the solar system and carries the expanding solar magnetic field.

interplanetary probe *Space Technology.* an unstaffed instrumented spacecraft designed to land on or fly close to another planet.

interplanetary scintillation *Astronomy.* an apparent "twinkling" that distant pointlike radio sources undergo as their signals pass through regions of variable density in the interplanetary medium.

interplanetary space *Astronomy.* a term that loosely describes the region of space from the sun to the outer limits of the heliosphere, at a distance of about 100 astronomical units.

interplanetary transfer orbit *Space Technology.* a trajectory used to transfer a spacecraft from one planetary orbit to another. Also, HOHMANN TRAJECTORY.

interplanting *Forestry.* a pattern of planting in which young trees are set out among existing natural or planted forest growth.

interpluvial *Geology.* **1.** relating to an interval of geologic time during which the climate was drier than that in the pluvial periods preceding or following it. **2.** such an episode or geologic interval.

interpolar *Science.* located or traveling between two poles.

interpolation *Mathematics.* a collective term for various techniques of determining an approximate value of a function at a point in the domain between given points at which the function values are known. For example, **linear interpolation** is based on the presumption that the value being sought lies on or near a straight line joining two known values.

interpole see COMMUTATING POLE.

interposition trunk *Telecommunications.* a trunk linking two switchboard positions in order to allow communication between the two positions.

interpret *Computer Programming.* **1.** to translate data from one form into a form that can be read by another device, as from a high-level language to machine language, or from computer code to standard letters, symbols, and characters. **2.** to determine the meaning of and execute program instructions or statements by means of a program, rather than hardware.

interpretation *Artificial Intelligence.* in mathematical logic, an assignment of a universe of possible objects for terms, of particular objects for constants, of the meanings of functions from terms to terms, and of the truth values of predicates applied to terms.

interpreted code *Computer Science.* a form of program that is read and executed by an interpreter program. The interpreter reads an instruction, determines its meaning, and executes it.

interpreted knowledge *Artificial Intelligence.* knowledge that is analogous to an interpreted procedure: it can be executed, but it is also possible to examine it or reason about it.

interpreter *Computer Programming.* a program that interprets a language; this may be an interactive interpreter of a programming language, such as a LISP interpreter, or an emulation of instructions that are not present in hardware, such as a floating point interpreter.

interpretive language *Computer Programming.* a programming language that may be run on a computer having an interpreter.

interpretive programming *Computer Programming.* the process of writing programs in a language that may be precisely converted by an interpreter into actual machine-language instructions.

interproglottid gland *Invertebrate Zoology.* a small cluster of secretory cells situated along the posterior margins of segments in certain tapeworms.

interpubic *Anatomy.* between the pubic bones.

interpulmonary *Anatomy.* located between the lungs.

interpulse time *Metallurgy.* in resistance welding, the time between pulses.

interquartile range *Statistics.* the range including the central 50% of the observations in a data set.

interradial canal *Invertebrate Zoology.* one of the radially arranged digestive canals in certain jellyfish and ctenophores.

interradius *Invertebrate Zoology.* the area between two adjacent arms in echinoderms.

interrenal *Anatomy.* located between the kidneys.

interrogate *Computer Programming.* **1.** to search, count, or tally records in a file. **2.** to test the status of a terminal or system.

interrogation *Telecommunications.* the transmission of a signal that requires a reply from the receiver.

interrogator *Electronics.* a device that generates the signals needed to trigger a transponder. Also, CHALLENGER.

interrogator-responsor *Electronics.* a device that can generate the signals needed to trigger a transponder and then receive and display the reply.

interrupt *Computer Programming.* a temporary break in the normal flow of instruction processing in the central processing unit, due to a hardware signal from an external or internal source and the subsequent transfer of control to the interrupt handler.

interrupt-driven *Computer Programming.* describing a computer that uses an interrupt system to allocate the workload to the various hardware and software elements.

interrupted aging *Metallurgy.* aging performed by a sequence of two or more thermal treatments, each followed by cooling.

interrupted continuous wave *Telecommunications.* **1.** a modulation technique that uses on-and-off keying of a continuous carrier. **2.** a continuous wave that is interrupted to give several interruptions for each keyed code dot.

interrupted current *Electricity.* alternating or direct current that is interrupted intermittently by an electrical or mechanical switch.

interrupted fire *Ordnance.* fire delivered from automatic weapons in short series of bursts.

interrupted mating *Genetics.* an experimental technique used to examine the transfer of genetic information from donor to recipient during bacterial conjugation by interrupting the mating procedure at various times.

interrupted quenching *Metallurgy.* quenching performed by using two quenching media, imparting different cooling rates.

interrupted screw *Mechanical Devices.* a screw with long grooves cut at 90° angles to the thread, designed so that it is locked with a mating part by only a slight turn.

interrupter *Electricity.* an electrical or mechanical device, such as a switching diode, a switch, or a relay, that opens a circuit intermittently. *Ordnance.* **1.** in general, any barrier in a fuse that prevents transmission of the explosive effect. **2.** specifically, a safety device in a shell fuse that prevents activation until the projectile passes the muzzle of the gun.

interrupter gear *Ordnance.* in the earlier technology of a machine gun firing through the arc of an aircraft propeller, a gear that interrupted the firing mechanism while a propeller blade was in front of the muzzle.

interrupter vibrator *Electricity.* a mechanical component used to convert direct current to alternating current.

interrupt handler *Computer Programming.* a program that acknowledges an interrupt by a device and services it, e.g., by clearing the interrupt and providing data to or accepting data from the device that caused the interrupt. Also, **interrupt service routine.**

interrupting capacity *Electricity.* the highest current at a specified voltage that a device can interrupt.

interrupt mask *Computer Technology.* an internal mask register that allows a computer to enable or ignore specific interrupts by setting or clearing the corresponding mask bits. Also, MASK REGISTER.

interrupt mode see HOLD MODE.

interrupt processing *Computer Technology.* a programmed response to an interrupt, typically by acknowledging (clearing) the interrupt signal and initiating an input/output transfer.

interrupt signal *Computer Technology.* an electronic signal or flag generated when certain internal or external conditions occur, causing an interrupt.

interrupt system *Computer Programming.* a feature of some operating systems in which the source of an interrupt is immediately identified and the program is automatically altered to correct or circumvent the interrupt.

interrupt vector *Computer Programming.* a method of dispatching control to an interrupt handler, by means of a vector of interrupt handler addresses that is indexed by an interrupt number provided by the interrupting source.

interscapular *Anatomy.* located between the scapulae.

intersect *Science.* **1.** to cross or pass through, as a pair of lines. **2.** to overlap.

Intersecting Storage Rings *Nucleonics.* a container that maintains protons accelerated to a high energy level (up to 31 gigaelectron volts) by holding them in a constant magnetic field; attached to a synchrotron in Geneva, Switzerland.

intersection the place where two or more things intersect; specific uses include: *Transportation Engineering.* a crossing or juncture of two roadways; a basic source of bottlenecks, congestion, and traffic delays, and therefore a primary concern of traffic engineers. *Aviation.* a crossing or juncture of runways, taxiways, or a runway and a taxiway. *Cartography.* in surveying, the procedure determining the horizontal position of an object or point from observations from other survey stations without making any direct observations of the object itself. *Photogrammetry.* the procedure for determining the horizontal position of an object or point from intersecting lines of direction taken from photographs by graphic or mathematical means. *Mathematics.* the intersection of two sets X and Y, denoted $X \cap Y$ or sometimes XY, is the set of elements that belong to both X and Y.

intersection method *Engineering.* a method of plane-table surveying in which the plane table is set up and points are sighted at each end of a measured baseline; those mapped are located at the intersection of the rays extending from each setup. Also, TRIANGULATION METHOD.

intersection point *Civil Engineering.* the point where two straights or tangents of a railroad or highway curve would meet if they were to be completely extended.

intersection station *Cartography.* a survey station that has had its horizontal position determined by indirect observation from other survey stations, with no direct observations being made at the point of the intersection station itself.

interseeder *Agriculture.* a machine designed for interseeding.

interseeding *Agronomy.* the planting of seeds within an area of existing vegetation to establish or reestablish a desired plant species.

intersegmental *Developmental Biology.* between segments of the embryo.

intersegmental reflex *Physiology.* a reflex having sensory input at one level of the spinal cord and motor output at another level.

intersertal *Petrology.* of or relating to a texture of porphyritic igneous rock in which glassy or partly glassy groundmass fills the interstices between unoriented feldspar laths.

intersex *Physiology.* **1.** see INTERSEXUALITY. **2.** an individual who exhibits intersexuality.

intersexuality *Physiology.* the intermingling, to varying degrees, in one individual of the characteristics of each sex, including physical form, reproductive organs, and sexual behavior; may result from any of several errors in embryonic development. Also, INTERSEX.

interspace *Anatomy.* a space between two similar body parts. *Building Engineering.* any airspace between walls and structural members. *Forestry.* see BARK BAR.

interspecific *Systematics.* **1.** of or relating to members of different species. **2.** a member of a different species. *Behavior.* of or relating to interaction between individuals of different species.

interspecific aggression *Behavior.* aggression directed toward a member of a different species.

interspecific releaser *Behavior.* any stimulus, such as a behavior pattern or body feature, that is intended to communicate to members of another species rather than to the individual's own species.

interspecific territoriality *Behavior.* competition for territory between two different species with similar ecological needs.

interspersion *Ecology.* **1.** the spatial intermingling of various organisms within a given community. **2.** the level of spatial intermingling that takes place within a community.

interspinal *Anatomy.* located between two spinous processes.

interstadial *Geology.* relating to a warmer substage of a glacial stage, during which the ice retreated temporarily.

interstage transformer *Electronics.* a transfomer that couples two electronic stages together.

interstellar *Astronomy.* existing or occurring between stars. *Space Technology.* traveling or designed to travel between stars. Thus, **interstellar flight.**

interstellar communication *Telecommunications.* theoretical communication between entities in different stellar systems.

interstellar extinction *Astrophysics.* a diminution of electromagnetic radiation by dust and gas in space; can vary greatly and unpredictably from one object to the next. Also, **interstellar absorption.**

interstellar gas *Astronomy.* the gas between stars of this galaxy; most of this gas is hydrogen in various states; hot ionized clouds or HII regions with hot halos of thinner gas, cool clouds of neutral hydrogen (HI regions), and cold dense clouds of molecular hydrogen.

interstellar lines *Astrophysics.* absorption lines in the spectra of distant stars, nebulae, and galaxies that are produced when the light from them passes through elements in the interstellar medium.

interstellar medium *Astronomy.* the dust, molecular clouds, and neutral hydrogen that lie between the stars of this galaxy, generally in the plane of the Milky Way, but whose density is highly variable.

interstellar polarization *Astrophysics.* the partial polarization of starlight produced by dust grains aligned by magnetic fields in the interstellar medium.

interstellar probe *Space Technology.* an unstaffed instrumented spacecraft that is propelled beyond the solar system and is capable of gathering specific data about the interstellar environment.

interstellar scintillation *Astrophysics.* an apparent "twinkling" of the signals from distant pointlike radio sources that comes from changes in the density of the interstellar medium through which the signals have passed on their way to the earth.

interstellar space *Astronomy.* space between the stars of the Milky Way Galaxy.

intersternite *Invertebrate Zoology.* a ventral plate lying between abdominal segments of an insect.

interstice [in tur´stis; in´tər stis] a small, usually narrow space; specific uses include: *Geology.* an opening, void, or pore space in a rock or soil that is not occupied by solid material. Also, PORE. *Anatomy.* a small gap or space in a tissue or structure. *Solid-State Physics.* the unoccupied space between atoms in a crystal structure that can be occupied by an atom of suitable size.

interstitial [in´tər stish´əl] *Science.* of, relating to, or located in one or more interstices.

interstitial atom *Crystallography.* a foreign atom located at a position that is not a lattice site, and thus causing a defect in a crystal.

interstitial cell *Histology.* **1.** a specific interstitial cell (of Leydig) that lies between seminiferous tubules in the testis, contains crystalloids (of Reinke) and secretes steroid hormones. **2.** a cell present in ovarian stroma that contains lipid droplets and, in humans, secretes estrogens. **3.** a cell that is found between pinealocytes in the pineal gland and may be a type of glial cell.

interstitial cell stimulating hormone see LUTEINIZING HORMONE.

interstitial cell tumor *Medicine.* a benign tumor of the testis made up of interstitial cells, related to hypersecretion of male sex hormones.

interstitial compound *Materials Science.* an intermediate phase composed of large and small atoms in which the small atoms are located in the spaces between the large ones. Also, LAVES PHASE.

interstitialcy *Materials Science.* a point defect in which an atom occupies an interstitial site. Also, SELF-INTERSTITIAL POINT DEFECT.

interstitial diffusion *Materials Science.* the migration of small atoms through interstitial sites.

interstitial emphysema *Medicine.* the escape of air into the connective tissue of the lung, mediastinum, or subcutaneous tissue resulting from the rupture of air cells, as from an obstruction or a penetrating chest wound.

interstitial endometriosis *Medicine.* an abnormal condition in which the uterine mucosal lining invades the spaces of other pelvic cavity tissues.

interstitial-free steel *Metallurgy.* a steel that is virtually free from interstitial solute elements.

interstitial gland *Histology.* **1.** a group of Leydig cells in the testes. **2.** a group of epithelial cells in the ovaries of some mammals.

interstitial hepatitis *Medicine.* a former term for cirrhosis of the liver.

interstitial hernia *Medicine.* the abnormal protrusion of a part of an organ through the containing wall of its cavity, causing it to occupy the interspaces of a tissue and organ.

interstitial implant *Medicine.* a small radioactive tube or needle placed in the interspaces of tissues to deliver irradiation therapy.

interstitial implantation *Developmental Biology.* the complete embedding of the blastocyst within the endometrium of the uterine wall.

interstitial impurity *Solid-State Physics.* an atom of a species not normally found in a crystal, located between normal atomic sites.

interstitial inflammation *Medicine.* an inflammation that occurs mainly in supportive fibrous connective tissue or stroma of an organ.

interstitial keratitis *Medicine.* an uncommon inflammation of the cornea with formation of blood vessels primarily in the middle layer, caused by tuberculosis, leprosy, or vascular hypersensitivity, and in children as a result of congenital syphilis.

interstitial lamella *Histology.* a thin layer of bone lying between adjacent osteons.

interstitial myocarditis *Medicine.* an inflammation of the heart muscle in which there is cellular infiltration of the interstitial tissues. Also, FIEDLER'S MYOCARDITIS.

interstitial pneumonia *Medicine.* a chronic inflammation of the stroma of the lungs, often resulting in compression of the air cells.

interstitial radiotherapy *Radiology.* a treatment using radioactive implants, such as seeds or needles, introduced into the tissues. Also, **interstitial radiation therapy, interstitial radiation, interstitial therapy.**

interstitial site see VOID.

interstitial solid solution *Materials Science.* a solid solution in which small solute atoms occupy the interstices in the lattice of the solvent or parent atoms.

interstitial water *Hydrology.* subsurface water contained in the spaces of pores in rock or soil. Also, PORE WATER.

interstock *Botany.* a stock that is grafted in the area between the scion and the rootstock.

intersymbol interference *Telecommunications.* the overlapping of successively transmitted signals within a medium, such as an electromagnetic transmission line, due to its dispersion of the frequencies constituting the signal.

intersystem communications *Computer Technology.* the transfer of data between two or more processors that share peripherals and have common input/output channels or are directly linked.

intertergite *Entomology.* a dorsal plate between segments in insects.

intertidal zone *Oceanography.* the part of the shore that lies between the low and high water lines. Also, LITTORAL ZONE.

intertillage *Agronomy.* the process of cultivating between rows of growing crops, to control weeds.

intertropical convergence zone *Meteorology.* the axis, or a portion thereof, of the broad trade-wind current of the tropics, occurring as a narrow zone of low pressure somewhat north of the equator, in which air masses originating in the Northern and Southern Hemispheres converge, giving rise to cloudy, showery weather. Also, EQUATORIAL CONVERGENCE ZONE.

intertropical front *Meteorology.* a front that is said to exist within the equatorial trough, separating the air of the Northern and Southern Hemispheres. Also, EQUATORIAL FRONT, TROPICAL FRONT.

intertubercular sulcus *Anatomy.* the groove on the proximal end of the humerus, lying between the greater and lesser tubercles.

interval *Physics.* a measurement of the separation of two values, such as the distance between two positions in space or the time between two events. *Aviation.* the space between adjacent aircraft, measured from side to side. *Ordnance.* the time between one gun firing and the next gun firing, or between two rounds from one gun. *Military Science.* in a formation, the space between adjacent individuals, vehicles, or units. *Mathematics.* a set of real numbers x satisfying one of the following (where a and b are fixed real numbers): (a) $a < x < b$; denoted (a, b); called an **open interval;** (b) $a < x \leq b$ or $a \leq x < b$; denoted $(a, b]$ or $[a, b)$, respectively; called a **half-open interval;** (c) $a \leq x \leq b$; denoted $[a, b]$; called a **closed interval.**

interval arithmetic *Computer Programming.* **1.** a method by which modifications are made to significant-digit arithmetic to provide finer resolution and prevent loss of information, while retaining an indication of the degree of accuracy. **2.** the methods of reasoning about results of arithmetic operations on numbers known to lie within certain intervals; e.g., if $3 \leq x \geq 7$ and $4 \leq y \geq 10$, then $7 \leq x + y \geq 17$.

interval estimate see CONFIDENCE INTERVAL.

intervalometer *Ordnance.* an electrical device using preset data to drop a specific number of bombs at a constant interval.

interval operation *Surgery.* a surgical procedure performed between two acute attacks of a disease.

interval scale *Statistics.* a measurement reference that has equal units with an arbitrarily selected zero point, such as the Fahrenheit scale.

interval schedule of reinforcement *Psychology.* in operant conditioning, an intermittent reinforcement schedule in which responses are reinforced after a time interval from the previous reinforcement.

interval timer *Engineering.* a device operated by an electric motor, a clockwork, or an electronic or resistor-capacitor circuit that opens and closes at specific times; used for automatically controlling functions such as lighting systems.

intervascular *Anatomy.* located between blood or lymph vessels.

intervening sequence see INTRON.

interventricular foramen *Anatomy.* the small opening on each side of the third ventricle in the brain that connects to the lateral ventricles and allows passage of cerebrospinal fluid. Also, FORAMEN OF MONRO.

interventricular septal defect *Medicine.* a malformation in the interventricular septum.

interventricular septum *Anatomy.* the largely muscular, partly membranaceous wall between the ventricles of the heart.

intervertebral *Anatomy.* located between the vertebrae.

intervertebral disk *Anatomy.* a layer of fibrocartilage between adjacent spinal vertebrae.

intervillous space *Histology.* the interconecting spaces between villi in the placenta that are filled with maternal blood.

intervisibility *Ordnance.* the condition in which an uninterrupted line of sight exists.

intestinal *Anatomy.* relating to the intestine. *Medicine.* affecting or involving the intestine. Thus, **intestinal flu, intestinal angina,** and so on.

intestinal crura *Invertebrate Zoology.* the principal branches of the intestine in certain trematodes.

intestinal digestion *Physiology.* the chemical breakdown of food materials in the intestine through mixture with intestinal juices.

intestinal dyspepsia *Medicine.* any impairment of the power or function of digestion that originates in the intestines.

intestinal hormone see GASTROINTESTINAL HORMONE.

intestinal juice *Enzymology.* a fluid secreted by glands lining the intestine, containing enzymes needed for the further digestion of foods to enable them to pass into the bloodstream.

intestinal villi *Anatomy.* the tiny, fingerlike projections from the mucosal lining of the intestine that increase surface area for the uptake of the products of digestion.

intestine *Anatomy.* the portion of the digestive tract that extends from the lower opening of the stomach to the anus, composed of the small intestine and the large intestine.

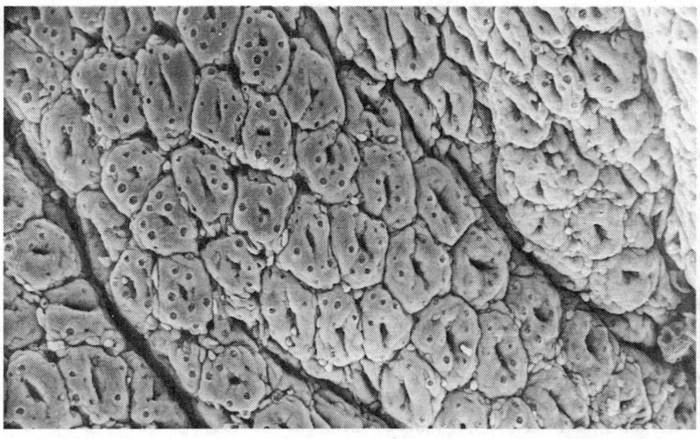

intestinal cells

intima *Histology.* the innermost layer of tissue in a blood vessel. Also, TUNICA INTIMA.

intimacy *Acoustical Engineering.* a measure of how close a performer's sound is to the audience during a live performance; it is considered adequate when all reflected sound reaches all listeners within 20 milliseconds of direct sound.

intimacy stage *Psychology.* the sixth of Erik Erikson's eight stages of human development, during early adulthood, when an individual develops the ability to establish and maintain intimate relationships with others.

intimacy vs isolation *Psychology.* the sense of isolation and self-concern that can arise during the intimacy stage of life, if an individual cannot form close and lasting relationships.

intimidation *Behavior.* aggressive behavior that causes the retreat or submission of the threatened animal.

intine *Botany.* the inner membrane of the wall of a pollen grain or spore.

intl. international.

into *Mathematics.* a function or transformation whose domain is a set X and whose range is a subset of Y is described as a function from X into Y.

intolerance *Nutrition.* a condition in which the body is unable to digest a specific substance, especially some nutrient or drug that is normally digestible by others.

intolerant *Nutrition.* of or relating to the body's inability to digest a specific substance.

intonation *Linguistics.* **1.** the variations in pitch level that characterize spoken language and that serve to emphasize or deemphasize certain elements of a sentence. English is usually identified as having four intonation levels: the normal speaking level, a level one step below normal for lesser emphasis, a level one step above for greater emphasis, and a level two steps above for special emphasis. **2.** a general pattern of pitch change that distinguishes a certain language or dialect; e.g., speakers of English typically use a rising intonation at the end to convert a statement into a question, as in "She's a doctor?" *Acoustics.* the manner of producing musical tones, especially with respect to a standard of pitch.

intoxicated *Medicine.* affected by intoxication; drunk.

intoxication *Medicine.* a condition that results from an overindulgence in alcohol; drunkenness. Typical manifestations are slurred speech, an unsteady gait, and a loss of coordination. *Toxicology.* the state of being poisoned by a drug or other toxic substance. *Psychology.* a state of being mentally or emotionally overexcited, especially being euphoric.

intoxication amaurosis *Toxicology.* a condition of impaired vision or blindness caused by systemic poisoning due to chemicals.

intra- a prefix meaning "within," as in *intraocular.*

intra-abdominal *Anatomy.* located within the abdomen.

intra-aortic balloon counterpulsation *Cardiology.* a technique that is used to support the circulation, provided by an inflatable balloon device, called an **intra-aortic balloon pump,** that is placed inside the thoracic aorta. The device is inflated during diastole and deflated during systole to alleviate excessive afterloading and improve heart performance.

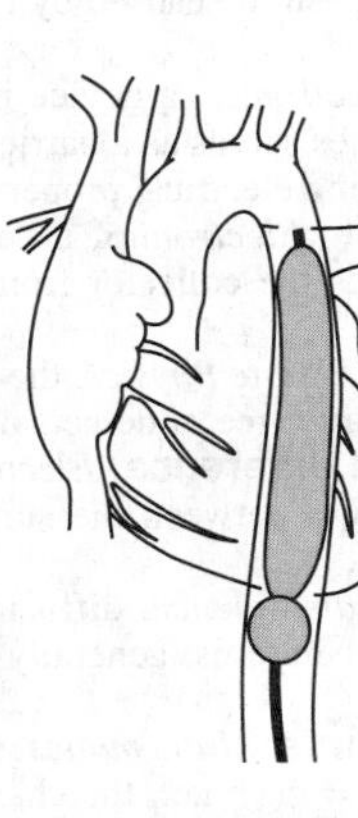

intra-aortic balloon pump

intra-arterial *Anatomy.* within an artery or arteries.

intra-atrial heart block *Medicine.* a delayed or obstructed passage of the nervous impulse through the atria of the heart, detected by widened and notched P waves in the electrocardiogram.

intrabeam viewing *Optics.* the subjection of the human eye to radiation when it is exposed to all or part of a laser beam.

intracartilaginous bone see ENDOCHONDRAL BONE.

intracavernous aneurysm *Medicine.* a localized abnormal dilation of the internal arotid artery within the avernous sinus.

intracavitary radiotherapy *Radiology.* a treatment using a radioactive element inserted into a natural cavity of the body.

intracavity absorption spectroscopy *Spectroscopy.* a spectroscopic technique in which a broadband dye laser is used to irradiate a sample whose absorption lines are then detected as dips in the emission spectrum of the laser.

intracelial *Anatomy.* located within a body cavity.

intracellular *Cell Biology.* located within a cell.

intracellular canaliculi *Cell Biology.* tiny secretory channels that are located in the cytoplasm of a cell, especially an oxyntic, or acid-secreting, cell.

intracellular digestion *Physiology.* the digestion of materials within the cell, usually through the chemical action of lysosomal enzymes.

intracellular enzyme *Enzymology.* an enzyme that functions only in the cell in which it was produced. Also, ENDOENZYME.

intracellular symbiosis *Cell Biology.* the phenomenon whereby one cell harbors another organism within it to their mutual benefit, such as *Paramecium aurelia* and its endosymbiont, the bacterium *Caenobacter taenospiralis.*

intracephalic *Anatomy.* located within the brain.

intracerebral *Anatomy.* located within the cerebrum.

intracerebral hematoma *Medicine.* the accumulation of blood within the brain tissue.

intracervical *Anatomy.* located within the canal of the cervix uteri.

intrachromosomal recombination *Genetics.* recombination resulting from a crossing over between two chromatids of the same chromosome.

intracistron complementation *Genetics.* the production of a wild-type phenotype in a cell containing two mutant sequences at different places in the same cistron in homologous genes; the remaining wild-type portions of the sequences are complementary and can still produce the wild-type polypeptide.

intraclast *Geology.* a reworked limestone fragment formed by the erosion of a penecontemporaneous sediment from within the basin and redeposited as a new sediment.

Intracoastal Waterway *Navigation.* a system of sheltered waterways, including canals, rivers, sounds, and bays, that runs parallel to the Atlantic and Gulf coasts from Manasquan Inlet in New Jersey to the Mexican border.

intracortical *Anatomy.* located within the cortex of an organ.

intracranial *Anatomy.* located within the cranial cavity.

intracranial aneurysm *Medicine.* the abnormal dilation of an artery within the skull.

intracranial angiography *Medicine.* a radiographic visualization of the blood vessels supplying the brain.

intracytoplasmic *Cell Biology.* occurring in the cytoplasm of a cell.

intrademic selection *Evolution.* a process of natural selection that occurs completely within a local interbreeding population.

intradermal *Anatomy.* **1.** located within the skin. **2.** located specifically within the dermis of the skin.

intradermal injection *Medicine.* the injection of a small amount of a drug or other agent into the outer layers of the skin using a short, very fine-gauge needle.

intrados *Architecture.* the inner curve of an arch.

intraductal *Anatomy.* located within the ducts of a gland.

intradural *Anatomy.* located within or inside the dura mater of the brain.

intraembryonic *Developmental Biology.* located or occurring inside the embryo.

intraepidermal *Anatomy.* located within the epidermis of the skin.

intraepidermal epithelioma *Medicine.* a tumor derived from epithelium within the epidermis; a form of skin cancer.

intraepithelial *Anatomy.* located within or among a group of epithelial cells.

intraesophageal *Anatomy.* located within the esophagus.

intrafascicular *Botany.* of or relating to a structure or process occurring within vascular bundles.

intraformational breccia *Petrology.* a rock formed by the cracking or breaking of partly consolidated sediment, followed by nearly contemporaneous sedimentation.

intraformational conglomerate *Geology.* a conglomerate in which the clasts are fragments eroded from newly or partially consolidated sediments and almost immediately incorporated into the accumulating sediments, so that the clasts are contemporaneous in origin with the matrix.
intraformational fold *Geology.* a minor fold restricted to a sedimentary stratum that lies between undeformed beds, and attributed to processes syngenetic with those responsible for the bed itself.
intraframe compression *Telecommunications.* data compression technology designed for use in full-motion color video requiring frame-by-frame editing.
intrafusal fiber *Histology.* a striated muscle fiber lying within a muscle spindle.
intragenic *Molecular Biology.* occurring within a gene.
intragenic complementation *Genetics.* the cooperative effect of two alleles that code for different subunits of a protein, such that the effect of a mixture of the subunits is much more efficient than either subunit alone. Also, INTERALLELIC COMPLEMENTATION.
intragenic recombination *Molecular Biology.* recombination of a DNA molecule that occurs within the structure of a gene.
intragenic suppression *Genetics.* a suppressor mutation that restores the orginial reading frame after a frameshift mutation. Also, SECOND-SITE MUTATION.
intrageosyncline see PARAGEOSYNCLINE, def. 1.
intrahepatic *Anatomy.* located within the liver.
intrajugular process *Anatomy.* a small bony process extending into the jugular foramen, and contributing to its partial division into two openings.
intraluminal *Anatomy.* within a tubular structure.
intramarginal *Biology.* within margins.
intramedullary *Anatomy.* **1.** within the bone marrow. **2.** within the medulla oblongata. **3.** within the spinal cord.
intramembranous *Histology.* within a membrane.
intramembranous ossification *Histology.* the direct formation of bone within a membrane without a preformed cartilage model.
intramodal transfer *Transportation Engineering.* a change from one line, route, or vehicle to another in the same mode of transportation.
intramolecular [in′trə mə lek′yə lər] *Physical Chemistry.* relating to the interaction of different atoms within the same molecule.
intramolecular forces *Physical Chemistry.* those forces which act to cause individual atoms to attract other adjacent atoms within the same molecule, and which are relatively strong in comparison to the forces that act between different molecules.
intramolecular lyase *Enzymology.* an enzyme that catalyzes the removal of a group from a molecule with the formation of a double bond.
intramolecular oxidoreductase *Enzymology.* an enzyme that catalyzes an intramolecular reaction through simultaneous oxidation and reduction reactions.
intramolecular transferase *Enzymology.* an enzyme that catalyzes the movement of a functional group from one location to another within the same molecule.
intramural *Anatomy.* located within the wall of an organ.
intramuscular *Anatomy.* located within the tissues of a muscle.
intramuscular injection *Medicine.* the injection of a medication or other material into a muscle, usually in the anterior thigh, deltoid, or buttocks.
intramyocardial *Anatomy.* within the myocardium (heart muscle).
intranasal *Anatomy.* within the nose.
intranatal *Medicine.* occurring during birth.
intraneural *Neurology.* within or extending into a nerve.
intraocular *Anatomy.* located within the eye.
intraocular pressure *Physiology.* the pressure of the aqueous humor that occupies the chambers of the eye.
intraoperative *Surgery.* occurring during a surgical procedure.
intraoperative cholangiography *Radiology.* the radiography of the gallbladder and bile ducts, performed during gallbladder surgery.
intraoptical light sighting system *Optics.* a target that produces and aims pulses of visible light to identify a field of view; used as a sighting device in conjunction with infrared pyrometers.
intraparietal *Anatomy.* within the parietal lobe of the brain.
intrapartum *Medicine.* occurring during labor and childbirth.
intraperitoneal *Anatomy.* **1.** within the peritoneal cavity. **2.** intramural.
intrapetiolar *Botany.* of or relating to a structure or process occurring within the petiole or between the petiole and the stem.
intrapulmonary *Anatomy.* located within a lung.
intraspecific *Ecology.* of or relating to interaction between individuals of the same species. *Systematics.* a member of the same species.
intraspecific aggression *Behavior.* aggression that is directed toward another member of the same species.
intraspinal block *Medicine.* the injection of an anesthetic drug into the spinal subarachnoid space to produce the total loss of sensation. Also, SUBARACHNOID BLOCK.
Intrasporangium *Bacteriology.* a genus of bacteria of the order Actinomycetales that carry out aerobic respiration and grow as branching substrate mycelia.
intrasternal *Anatomy.* located within the sternum.
intrastratal solution *Geochemistry.* the removal by solution of the mineral constituents of rock within a sedimentary bed. Also, DIFFERENTIAL SOLUTION.
intratelluric *Geology.* **1.** of, relating to, or constituting the period or stage of crystallization of igneous rocks prior to the eruption of the host magma. **2.** located, occurring, or formed deep within the earth.
intrathecal *Anatomy.* located within a sheath.
intrathoracic *Anatomy.* located within the thoracic cavity.
intratracheal *Anatomy.* located within the trachea.
intrauterine [in′trə yoot′ə rin] *Anatomy.* located within the uterus.
intrauterine device *Medicine.* a coil, loop, T, or triangle made of plastic or metal that is inserted within the uterus to prevent implantation of a fertilized ovum. Also, **intrauterine contraceptive device.**
intravaginal *Anatomy.* within the vagina. *Botany.* contained within a sheath, as grass branches.
intravascular *Anatomy.* situated or occurring within a vessel or vessels.
intravascular thrombolysis *Medicine.* the application of a thrombolytic agent directly to a blood vessel; primarily used to treat myocardial infarction and acute thrombotic occlusion of vessels.
intravenous [in′trə vē′nəs] *Anatomy.* located or occurring within a vein or veins.
intravenous cholangiography *Radiology.* roentgenography of the gallbladder and bile ducts following the intravenous introduction of a contrast medium that targets the liver and is excreted into the bile ducts.
intravenous injection *Medicine.* any injection into a vein.
intravenous pyelogram *Radiology.* an X-ray of the kidney and ureter following the intravenous injection of a radiopaque iodine solution that passes into the urine.
intraventricular heart block *Medicine.* the delay or obstruction of nervous conduction in the heart, occurring within the ventricles.
intravesical *Anatomy.* located within the urinary bladder.
intrazonal soil *Geology.* any of an order of soils having relatively well-developed characteristics that reflect the dominant influence of local factors, such as topography, age, or parent material, rather than characteristics attributed to climatic conditions or type of vegetation.
intrenched stream see ENTRENCHED STREAM.
intrinsic *Science.* **1.** situated entirely within the thing or substance that it acts upon, as an **intrinsic muscle. 2.** being among the essential parts or characteristics of something, as an **intrinsic trait.**
intrinsic asthma *Medicine.* asthma caused by bronchial or other infection of the respiratory tract.
intrinsic-barrier diode *Electronics.* a device in which a thin layer of silicon that is free of impurities serves as a barrier between regions in an integrated circuit with opposing electrical properties.
intrinsic-barrier transistor *Electronics.* a device in which the thin layer of silicon that separates the collector from the base is free from impurities.
intrinsic conductivity *Solid-State Physics.* the conductivity of a semiconductor or metal that is nearly free of defects or impurities.
intrinsic contact potential difference *Electricity.* the potential difference occurring in a vacuum between the surfaces of two metals in contact.
intrinsic diffusion *Materials Science.* diffusion in crystalline structures that occurs at high temperatures; generally follows a path through the principal constituents.
intrinsic equations of a curve *Mathematics.* for a given curve in R^3, the equations $\rho = f(s)$ and $\tau = g(s)$; i.e., the characterization of a curve by specifying its radius of curvature ρ and torsion τ as a function of arc length. These equations are independent of coordinate systems. Also, NATURAL EQUATIONS OF A CURVE.
intrinsic factor *Biochemistry.* a protein that is secreted by certain cells in the walls of the gastric glands of the stomach and is required for the absorption of vitamin B_{12}.

intrinsic function *Computer Science.* a simple function, such as absolute value, compiled as in-line code rather than as a subroutine call.
intrinsic grain-boundary strength *Materials Science.* the resistance to grain-boundary deformation in a metal or other solid material.
intrinsic induction *Electromagnetism.* a vector quantity given by the vector difference between the magnetic flux density at a specified point and the magnetic flux density that would exist at the same point in the absence of matter.
intrinsic layer *Electronics.* a layer of semiconductor material to which no impurities have been added.
intrinsic luminosity *Astronomy.* an object's total brightness integrated over all wavelengths.
intrinsic mobility *Solid-State Physics.* the electron mobility in a semiconductor that is free of impurities.
intrinsic motivation *Behavior.* an incentive that originates from the behavior itself rather than from an external reward or reinforcement. Thus, **intrinsically motivated behavior.**
intrinsic nerve supply *Anatomy.* the nerves that are located within or restricted to a specific organ or body part.
intrinsic parity *Particle Physics.* a quantum number that designates a particle's inherent and unchanging evenness or oddness, expressed as +1 or –1.
intrinsic photoconductivity *Solid-State Physics.* the increase in electrical conductivity associated with the intrinsic properties of a substance, usually nonmetallic, when exposed to electromagnetic radiation.
intrinsic photoemission *Solid-State Physics.* the photoemission produced in a pure crystal, free of defects and impurities.
intrinsic pressure *Physics.* a pressure that results from intermolecular attractive forces acting inward on molecules in a liquid at or near its surface.
intrinsic properties of a surface *Mathematics.* all the properties of a surface that are not dependent on the surrounding space. The description of such properties is the intrinsic geometry of the surface.
intrinsic Q see UNLOADED Q.
intrinsic rate of increase *Ecology.* a population's growth rate, derived by subtracting the instantaneous death rate from the instantaneous birth rate. Also, INNATE RATE OF INCREASE.
intrinsic reward *Psychology.* a reward that is derived from or closely tied to the behavior being rewarded.
intrinsic semiconductor *Materials Science.* an undeveloped semiconductor in which the temperature determines the conductivity. *Solid-State Physics.* a semiconductor that has no impurities to aid in the conduction of charge carriers. Also, I-TYPE SEMICONDUCTOR.
intrinsic temperature range *Solid-State Physics.* the temperature range over which the electrical properties of a semiconductor are not significantly modified by impurities and defects in the crystal.
intrinsic tracer *Nuclear Physics.* a naturally occurring isotope that is ideally suited to serve as a tracer for a given element in chemical and physical process studies.
intrinsic vacancy *Materials Science.* a point defect in a crystal; a vacant lattice site.
intrinsic variable star *Astronomy.* a star whose apparent brightness varies due to physical changes in the star, not to accidents of geometry, such as those which produce an eclipsing binary.
intrinsic viscosity *Materials Science.* the limit of the reduced and inherent viscosities as the concentration of the polymeric solute is extropolated to zero; usually used as a measure of molecular weight. Also, LIMITING VISCOSITY.
intro- a prefix meaning "inward" or "within," as in *introversion, introspection.*
introfaction *Chemistry.* a change of fluidity and wetting properties of an impregnating material caused by the introduction of an introfier.
introfier *Chemistry.* any substance that accelerates the penetrating power of a fluid or other impregnating material; a wetting accelerator.
introgression *Molecular Biology.* the incorporation of a gene from one complex into another as a result of hybridization.
introgressive hybridization *Genetics.* the integration of genes from one species into the gene pool of another.
introitus *Anatomy.* the opening or entrance into a hollow organ or cavity.
introjection *Psychology.* the act of creating a mental image of an object and then responding to the image as if it were the object.
intromission *Zoology.* the act of placing one body or body organ in another, especially the placing of the penis into the vagina for the purpose of insemination.
intromittent *Zoology.* adapted for insertion, such as the male copulatory organ.
intron *Genetics.* an apparently nonfunctional segment of DNA, ranging in size from fewer than 100 nucleotides to more than 1000, which is transcribed into nuclear RNA but is then removed from the transcript and rapidly degrades. Also, INTERVENING SEQUENCE.
introrse *Biology.* from outside inward, such as introrse dehiscence, the spontaneous opening of a ripe fruit from outside inward, or an introrse stamen, which opens toward the center of the flower.
introspection *Psychology.* the process of examining one's self and one's own actions in order to gain insight.
introversion *Medicine.* the process of turning an organ or part inward. *Psychology.* a tendency to direct one's interests and energies inward, toward one's own experiences and feelings, rather than outward toward other people and the external world.
introvert *Psychology.* **1.** a person who has an introverted personality. **2.** in popular use, a person who is withdrawn and uncommunicative. *Invertebrate Zoology.* a body part or structure that can be turned inward, back within or upon itself, such as the proboscis of a Nemertina worm or the part of a bryozoan that can be drawn within the zooecium.
introverted *Psychology.* characterized by introversion; tending to direct one's attention toward oneself and away from other people or things.
intruder *Military Science.* an individual, unit, or weapon system that presents the threat of intelligence gathering or disruptive activity in the vicinity of an operational or exercise area.
Intruder *Aviation.* a popular name for the American A-6 long-range bomber.

Intruder (A-6)

intruder operation *Military Science.* an offensive air operation over enemy territory, with the primary object of destroying enemy aircraft in the vicinity of their bases.
intrusion *Geology.* **1.** the injection or emplacement of magma under high pressure into a country rock or other preexisting rock formation. Also, INVASION. **2.** a large-scale injection and emplacement of a plastic sediment under abnormal pressure. **3.** the body of igneous or sedimentary rock formed by such processes.
intrusion breccia *Petrology.* the outer shell of a large, intrusive mass with abundant cognate and foreign inclusions.
intrusive *Petrology.* **1.** of or relating to molten material forced into older rocks or between rock layers and solidifying before reaching the surface. **2.** a rock so formed.
intrusive growth *Botany.* a type of plant tissue growth in which elongating cells grow by projection between adjacent cells or into intercellular spaces.
intrusive mountain see BATHOLITH.
intrusive sill see SILL.
intsv. intensive.
intubation *Medicine.* the insertion of a tube into a canal or into the trachea for anesthesia or to allow the passage of air.
intuitionism *Mathematics.* a philosophical view of mathematics in which finite sets are taken as given and every other object must be constructed from these sets. The **intuitionists** reject a proof of the existence of an object if no means of constructing the object in a finite number of steps is provided. In particular, a proof of existence by contradiction is disallowed. This view was vigorously espoused by the Dutch mathematician Luitzen Brouwer in the first half of the 20th century.

intumescence *Materials Science*. the swelling of a material upon heating. *Plant Pathology*. on a plant, any knobby blister or pustule formed on leaves, stems, or other areas due to a physiological disturbance. Thus, **intumescent.**

intumescent coating *Materials Science*. a material coating that swells when heated and allows moisture to escape, typically used as fireproofing.

intussusception *Medicine*. the folding inward of one part of the intestine into another, seen more commonly in children.

inulase *Enzymology*. in various fungi and higher plants, an enzyme that catalyzes the hydrolysis of inulin into the sugar fructose. Also, **inulinase.**

inulin *Biochemistry*. in numerous plant roots, a carbohydrate that yields the sugar levulose when mixed with oxygen and water; used in tests for determining renal function, and as a starch in bread for diabetics.

inunction *Medicine*. the application of an oily substance to the skin by rubbing or smearing, as with a medicinal ointment.

inundation *Hydrology*. the rise and spread of water over land that is not usually submerged.

in utero *Developmental Biology*. within the womb; not yet born.

in utero surgery *Surgery*. surgery performed on a fetus before birth, and within the uterus.

in vacuo *Physics*. occurring or located in a vacuum.

invade *Science*. to carry out an invasion.

invader one that invades; specific uses include: *Agriculture*. a weed or other undesirable plant that enters an area in which it was previously absent or only minimally present. *Ecology*. a species that moves into and colonizes a new community. Also, **invading species.**

invagination *Physiology*. the state or process of infolding or ensheathing one portion of an anatomical structure within another portion, such as occurs in the intestine during peristalsis. *Developmental Biology*. formation of a gastrula as a result of the infolding of a section of a blastula wall.

invalid *Artificial Intelligence*. of a logical formula, false under some interpretation.

invar *Materials*. an alloy composed primarily of nickel and iron; it is distinguished by a coefficient of thermal expansion which is essentially zero in the range of 160–270°C.

invariable line *Mechanics*. the line of the angular momentum vector of a rotating rigid body not acted on by external torques.

invariable plane *Mechanics*. the fixed plane which must at all times contain the tip of a freely rotating rigid body's angular velocity vector, as a result of conservation of energy and angular momentum. *Astronomy*. see INVARIANT PLANE.

invariance *Physics*. any property of physical law or quantity that is unchanged after the application of any member of a given family of transformations or operations. Also, **invariant property.**

invariance principle *Physics*. a principle stating that the laws of motion remain unchanged by certain transformations or operations, especially such a principle in general relativity theory.

invariant *Mathematics*. **1.** a subset S of a set A is said to be invariant with respect to a group G of functions acting on S if $g(x) = x$ for all $g \in G$ and for all $x \in S$. **2.** a subset S of a set A is invariant under a permutation P if, whenever $x \in S$, both $P(x) \in S$ and $P^{-1}(x) \in S$. **3.** let T be a transformation of a set S onto itself. A function f acting on S is said to be invariant with respect to T if $f(Tx) = f(x)$ for all $x \in S$.

invariant plane *Astronomy*. the plane in the solar system defined by the combined angular momentum of the planets and the sun; it lies at an angle of about 1.5° to the ecliptic. Also, INVARIABLE PLANE. *Atomic Physics*. the plane perpendicular to the total angular momentum of an atom.

invariant theory *Mathematics*. given a group G of linear transformations of a vector space V, the action of G can be extended in a unique and natural way to the dual space V' and to the mixed tensor algebra on V and V'. Invariant theory is the study of those mixed tensors that exhibit various (relative and absolute) invariance properties.

invasion *Military Science*. the process of forcefully, and usually suddenly, entering the territory of another country with a major military force. *Ecology*. a process in which a species makes a movement into an established community and colonizes that area. *Medicine*. the attack or onset of a disease. *Pathology*. the entrance of bacteria into the body or their deposition in the tissues, as distinguished from infection, or as part of the process of infection. *Oncology*. the infiltration and active destruction of surrounding tissue; a characteristic of malignant tumors. *Geology*. see INTRUSION, def. 1.

invasive *Ecology*. of a species, carrying out an invasion of another area. *Surgery*. involving penetration of the body by cutting the skin or inserting an instrument or foreign material. *Pathology*. of a microorganism, having the capacity to invade the body and circulate at will, with or without inducing disease. *Oncology*. of a malignant tumor, able to infiltrate and destroy the surrounding tissue. *Volcanology*. see AGGRESSIVE.

invasiveness *Science*. the fact of being invasive or of carrying out an invasion.

invent *Science*. to create and implement a previously unknown device or physical process. *Anthropology*. specifically, to create a new artifact. Thus, **invented, invention.**

inventory *Industrial Engineering*. the total range of goods and materials that are stocked by an organization for sale to customers, for use in a production process, or for various other activities of the organization. *Engineering*. **1.** the quantity of fissile material in a nuclear reactor. **2.** in injection molding, the amount of plastic in the heating cylinder.

inventory control *Industrial Engineering*. the management of supplies on hand of materials or products. Also, STOCK CONTROL.

inverna *Meteorology*. a southeast wind that blows near Lake Maggiore in Italy.

inverse beta decay *Nuclear Physics*. a collision of a neutrino with a neutron that produces a proton, providing evidence for the existence of the neutrino.

inverse bremsstrahlung *Atomic Physics*. the absorption of a photon by an electron in a strong electric field.

inverse Compton effect *Quantum Mechanics*. a scattering process by which energetic particles transfer energy to photons, increasing the frequency of radiation.

inverse current *Electronics*. the current that is generated when inverse voltage is applied to a circuit.

inverse curve *Mathematics*. **1.** one of two coplanar curves together with a fixed (coplanar) circle with the property that each point on one curve is the inverse point of some point on the other curve with respect to the circle. **2.** see INVERSE FUNCTION.

inverse density dependence *Evolution*. a change in the influence of an environmental factor that affects population density as that density changes, tending to enhance population growth (by decreasing deaths or increasing births) as population density increases or to retard population growth (by increasing deaths or decreasing births) as population decreases.

inverse electrode current *Electronics*. electrons that are flowing through an electrode in the direction opposite to that for which the electrode was designed.

inverse element *Mathematics*. let $*$ be an operation defined on an object A (such as a group, ring, etc.) with identity element e, and suppose $a \in A$. A **two-sided inverse element** of a is an element a^{-1} of A such that $a^{-1} * a = a * a^{-1} = e$. If it is only true that $a^{-1} * a = e$ (or $a * a^{-1} = e$), then a^{-1} is called a **left inverse** (or **right inverse**). By convention, a (two-sided) inverse element for (ring, field, etc.) addition is denoted by $-a$, and a (two-sided) inverse element for (ring, field, etc.) multiplication, if it exists, is denoted by a^{-1}.

inverse feedback see NEGATIVE FEEDBACK.

inverse function *Mathematics*. if f is a function on a space X, and if for every $y \in f(X)$ there is only one $x \in X$ such that $f(x) = y$, then f has an inverse function f^{-1} and is said to be one-to-one or injective. The inverse function of f is that function, denoted f^{-1}, such that the domain of f is the range of f^{-1}; the range of f is the domain of f^{-1}; and $f \circ f^{-1} = f^{-1} \circ f = I$, the identity function on X. That is, if $f(x) = y$, then $f^{-1}(y) = x$.

inverse function theorem *Mathematics*. a theorem that states the conditions under which the equation $f(x) = y$ may be solved for x as a function of y. Let $f: X \to Y$ be a function between two Banach spaces X and Y satisfying the following: (a) U is an open neighborhood of a in X and V is an open neighborhood in Y containing $b = f(a)$; (b) $f: U \to V$ is continuously differentiable; and (c) $f'(a)$ is an isomorphism of X onto Y. Then there exists an open neighborhood U' containing a and contained in U and an open neighborhood V' containing b and contained in V such that f^{-1} exists and is a diffeomorphism of V' onto U'. In particular, if X and Y are both R^n, then the inverse function theorem holds if the Jacobian matrix of f evaluated at a is nonsingular.

inverse image *Mathematics*. if f is a mapping on a set X and N is a subset of the range of f, then the set $\{x: f(x) \in N\} = f^{-1}(N)$ is called the inverse image of the set N. Also, PREIMAGE.

inverse Langevin function *Materials Science*. a correcting term for rubber elasticity in regions of high strain where the contour length of the polymer chain is 1/3 to 1/2 its end-to-end distance.

inverse limit *Mathematics.* let $\{B_i\}$ be a family of objects in a category B, with the property that whenever $i \le j \le k$, there are morphisms $\theta_{ij}: B_j \to B_i$ such that: (a) $\theta_{ij} \circ \theta_{jk} = \theta_{ik}$, and (b) θ_{ii} is the identity morphism. The inverse limit of the family $\{B_i\}$ is a universally attracting object in the category D whose objects are the pairs $(B,(\theta_i))$, where B is any object of B and (θ_i) is a family of morphisms $\theta_i: B \to B_i$ such that for $i \le j$, the following diagram commutes:

$$\begin{array}{ccccc} & & \theta_{ij} & & \\ \cdots \leftarrow & B_i & \leftarrow & B_j & \leftarrow \cdots \\ & \theta_i \uparrow & & \uparrow \theta_j & \\ & & B & & \end{array}$$

inverse limiter *Electronics.* a transducer whose output is constant for input that falls in a specified range and linear (or some other function) for input outside that range.

inverse matrix *Mathematics.* the inverse of an $n \times n$ matrix M (if it exists) is the matrix denoted M^{-1}, such that $MM^{-1} = M^{-1}M = I$, where I is the $n \times n$ identity matrix.

inverse network *Electricity.* a pair of two-terminal networks in which the product of their impedances is independent for a given frequency range.

inverse neutral telegraph transmission *Telecommunications.* a form of transmission that uses zero current during marking intervals and pulses of current during spacing intervals.

inverse of a number *Mathematics.* **1.** the **additive inverse** of a real or complex number α is the number $-\alpha$. **2.** the **multiplicative inverse** of a *nonzero* real or complex number α is the number $1/\alpha$; also denoted α^{-1}.

inverse operator *Mathematics.* if F is an operator on a space X, and if for every $y \in F(X)$ there is only one $x \in X$ such that $F(x) = y$, then F has an inverse operator F^{-1} and is said to be one-to-one or injective. That is, F^{-1} is the operator that is the inverse function of F.

inverse peak voltage *Electronics.* **1.** the voltage that appears across a rectifier tube or X-ray tube during the nonconducting portion of a voltage cycle. **2.** the highest instantaneous voltage that a rectifier tube or an X-ray tube can tolerate in the inverse direction without causing current to flow.

inverse piezoelectric effect *Solid-State Physics.* a change of size in a crystal brought about by application of an electric field to the crystal.

inverse points *Mathematics.* points lying on the same radial line of a circle (or sphere) such that the product of their distances to the center of the circle equals the square of the radius. Also, **inverse points with respect to the circle.**

inverse position computation *Cartography.* in surveying, the determination of the length of a line, and its forward and back azimuths, from computations based on the positions of the ends of that line.

inverse probability principle see BAYE'S RULE.

inverse problem *Control Systems.* the problem of finding the performance criteria for which a given feedback control law is optimal.

inverse proportion *Mathematics.* the relationship between two variables whose product remains constant; e.g., $k = xy$, where k is a constant. Also, INDIRECT VARIATION.

inverse ranks *Statistics.* data, usually ranked from high to low or best to poorest, that is currently ranked in the reverse order.

inverse sampling *Statistics.* sampling that is continued until a given event occurs, rather than until a given amount of data is collected.

inverse scattering theory *Physics.* a theory concerned with determining the structure of a target object by using information derived from the scattering distribution off the object.

inverse segregation *Metallurgy.* in casting, elemental segregation in which lower-melting constituents are present in larger than normal amounts at the locations of earlier freezing.

inverse spinel *Materials Science.* a spinel wherein the trivalent ions occupy tetrahedral and octahedral sites and the divalent ions occupy octahedral sites.

inverse-square law *Physics.* any relation in which a quantity is dependent on the inverse square of the distance between a source and a field point, such as an electric field or a gravitational field.

inverse Stark effect *Spectroscopy.* the splitting of absorption lines, rather than emission lines, in a strong electric field.

inverse trigonometric see ARC FUNCTION.

inverse voltage *Electronics.* voltage that is applied to the negative terminal of a rectifier when current is not flowing through it.

inverse Zeeman effect *Spectroscopy.* the splitting of absorption lines, rather than emission lines, in a static magnetic field.

inversion the fact of being reversed in position or relationship; specific uses include: *Anatomy.* a turning inward, upside down, or in an opposite direction. *Geology.* **1.** the folding back of rock strata upon themselves so that their sequence seems to be reversed. **2.** the movement of a lava flow into a ravine or valley on the flank of a volcano. *Meteorology.* **1.** an anomalous reversal in the normal temperature lapse rate, so that temperature increases rather than decreases with altitude. Also, ATMOSPHERIC INVERSION, TEMPERATURE INVERSION. **2.** an atmospheric layer in which such a reversal occurs. **3.** any departure from the usual decrease or increase of an atmospheric property with respect to altitude. *Thermodynamics.* the point at about 4°C at which the rate of change in the density of water with temperature changes sign. *Chemistry.* **1.** a reversal, such as the change of an isomeric compound to its opposite or the turning of a levo compound to a dextro. **2.** the hydrolysis of certain carbohydrates, especially sugars, in which the direction of the solution's optical rotation is reversed. **3.** the interchange of internal and external phases in an emulsion of two immiscible liquids. *Optics.* **1.** the creation of an inverted image in an optical system. **2.** the chemical conversion of an optical substance into one having no effect, or the opposite rotary effect, on the plane of polarization. *Telecommunications.* a form of speech scrambling that inverts the original frequency spectrum of the signal. *Physics.* a transformation of spatial cartesian coordinates in which all three coordinates are replaced by their negative equivalents. *Crystallography.* a process in which each point of an object is converted to an equivalent point by projecting through a common center (the **center of inversion** or **center of symmetry**) and extending an equal distance beyond this center. If the center of symmetry is taken at the origin, every point x, y, z becomes $-x$, $-y$, $-z$ by inversion. *Mathematics.* **1.** the Möbius transformation of the form $S(z) = 1/z$. **2.** in general, the process of applying an inverse function or operator; for example, Fourier and Laplace transforms and Mellin inversion formulas. **3.** the inversion of a point P with respect to a circle is the process of finding the point on the radial line through P such that the distances of the two points from the center of the circle are equal to the square of the radius. The points are inverses of each other and the center of the circle is the center of inversion. Inversion with respect to more general curves and surfaces is also possible.

inversion casting *Metallurgy.* a casting process in which an electric furnace is inverted over the mold.

inversion heterozygote *Genetics.* a heterozygote in which an inversion has changed the arrangement of the gene loci one chromosome, but not on its homologue.

inversion layer *Meteorology.* the atmospheric layer through which the reversal of an atmospheric property occurs. *Solid-State Physics.* the formation of a layer of P- or N-type semiconductor material on the surface of a semiconductor, opposite to the type of the bulk of the semiconductor, formed as the result of an applied electric field.

inversion polymorphism *Genetics.* the presence of two or more chromosome sequences, differing by inversions, in the homologous chromosomes of a population.

inversion ratio *Physics.* the ratio of the difference in populations between two nondegenerate energy states under a condition of population inversion to the population difference at equilibrium.

inversion spectrum *Spectroscopy.* spectral lines in the microwave region for certain molecules, such as ammonia, that result from transitions representing an oscillation between two molecular configurations which are mirror images of each other.

inversion symmetry *Physics.* a principle stating that the laws governing physical processes are invariant under an inversion process of transformation.

inversion temperature *Thermodynamics.* the temperature at which the Joule-Thomson coefficient changes sign for an isenthalpic expansion.

inversive complex plane see COMPLEX PLANE.

invert *Science.* to carry out a process of inversion; subject some material or system to inversion. *Chemistry.* a substance produced by inversion, such as an invert sugar. *Civil Engineering.* the base or bottom of a pipe or artificial channel.

invertase *Biochemistry.* an enzyme in yeast and the digestive juices of animals that causes the inversion of cane sugar into invert sugar.

invertebrate [in vurt′ə brət] *Invertebrate Zoology.* **1.** any animal lacking a spinal column; an animal other than a fish, amphibian, reptile, bird, or mammal. **2.** of or relating to such animals.

invertebrate pathology *Invertebrate Zoology.* the study of disease processes in invertebrates.

Invertebrate Zoology

To be called an invertebrate, an animal need have no specified shape, external covering, or interior structure. It need have only a single negative attribute—the lack of a row of small hard bones along the middle of the back. The animals that do have such a row of bones, called vertebrae, are grouped as the vertebrates and embrace the familiar fishes, amphibians, reptiles, birds, and mammals, the class to which humans belong.

The invertebrates are often spoken of as if they were a category comparable with the more conspicuous vertebrates. But vertebrates all have a common basic body plan, and they occupy only a part of a single phylum, the Chordata, sharing this major division of the animal kingdom with some invertebrates. In number of species, the vertebrates include only about 3% of the animal kingdom, whereas the invertebrates embrace more than 97% of the animals species, and they present us with so many distinctive body plans that they must be divided up into 30 or more phyla.

Only a few phyla can be mentioned in this brief account. The sponges (Porifera) take their name from their unique porous structure with which they filter microscopic food from the water that passes through their bodies. The Cnidaria (also called Coelenterata) take their name from the cnidae, or stinging cells, with which they capture their animal prey. Familiar cnidarians are sea anemones, jellyfishes, corals, sea fans, and others that have mostly delicate bodies with a central mouth encircled by tentacles armed with stinging cells. The flatworms (Platyhelminthes) are free-living little worms or parasitic flukes or tapeworms; all have flattened bodies, with a similar internal structure that distinguishes them from other soft, wormlike types that make up about half the 30 or more phyla of the animal kingdom. The cylindrical worms that compose the Nematoda swarm by the billions in soft aquatic bottoms and in soils, and are the most numerous of the wormlike types. Most familiar, however, are the Annelida (bristle worms, earthworms, and leeches) whose ringed or segmented bodies are usually divided by partitions into separate compartments containing repeated but similar organs. The Echinodermata (starfishes, sea urchins, sea cucumbers, etc.) are named for their spiny skins; they are distinguished from other spiny animals by their unique 5-sided body plan. The second largest phylum, the Mollusca, consist of more than 100,000 species of soft-bodied animals, most of them enclosed in shells (snails and clams). The giant clam is the heavyweight among invertebrates, but it sits on marine bottoms, unable to move about. At the opposite extreme are some of the cephalopod mollusks, the squids and octopuses, which have sacrificed a heavy protective shell for high mobility. The jet-propelled squids are the speediest aquatic invertebrates and giant squids measure more than 18 meters from tip to tip, exceeding in length all other invertebrates except the giant whales.

The largest phylum, by far, is the Arthropoda, with 900,000 described species of crustaceans (lobsters, shrimps, etc.), spiders, scorpions, centipedes, millipedes, miscellaneous smaller groups and predominant insects, with 800,000 described species and perhaps millions more waiting to be described. The success of arthropods must be attributed in great part to the hardened but jointed external skeleton that encloses the muscular body. The hard exoskeleton provides much more than protection and a firm surface for muscle attachment; it lends itself to the elaboration of biting jaws, piercing beaks, a variety of sense organs, and the wings of insects.

Ralph Buchsbaum
Emeritus Professor of Biology
University of Pittsburgh

invertebrate zoology *Zoology.* the branch of zoology that studies those animals lacking a spinal column.

inverted *Geology.* see OVERTURNED.

inverted amplifier *Electronics.* a vacuum tube circuit in which the output capacitance is significantly reduced by passing the incoming signals between the cathodes of two separate tubes, whose grids are grounded and act as a shield between the input and output circuits.

inverted arch *Civil Engineering.* a structural arch with a downward crown below the line of the springing; used commonly in tunnels and foundations. Also, INFLECTED ARCH.

inverted compass *Navigation.* a compass or compass repeater that is mounted facedown, typically affixed to the overhead in the master's cabin of a vessel.

inverted engine *Mechanical Engineering.* an in-line engine whose cylinders are set beneath the crankshaft; an adaptation used especially in certain aircraft to improve the pilot's line of sight.

inverted file *Computer Programming.* **1.** a directory that includes all key word or content identifiers with the corresponding set of document or data identifications and their locations. **2.** a file sorted on a different key than usual, such as a telephone book sorted by number rather than by name, to facilitate lookup using that key.

inverted image *Optics.* an image that appears to have been rotated 180° from above to below or from left to right; commonly used to mean lateral inversion. Also, REVERSED IMAGE.

inverted L antenna *Electromagnetism.* an antenna constructed of one or more horizontal wires with a vertical lead-in wire connected to one end.

inverted microscope *Optics.* a microscope in which the body, ocular, and objective are located underneath the stage, so that the specimen is placed facedown over the objective and is lit from above.

inverted Oedipus (complex) see NEGATIVE OEDIPUS.

inverted plunge *Geology.* a feature in excessively folded terranes in which the inclination of the fold axis has been carried beyond the vertical, so that the plunge is less than 90° in the direction opposite to the original attitude.

inverted repeat *Genetics.* either of two copies of a DNA sequence found in identical but inverted form, e.g., at the opposite ends of a transposon.

inverted siphon *Civil Engineering.* a pressure pipeline that extends over a depression or crosses a pass under a highway; sometimes called a sagline because of its U shape.

inverted terminal repeats *Molecular Biology.* two identical sequences of nucleotides that occur in opposite orientations at the opposite ends of some transposons.

inverted vee *Electromagnetism.* an antenna that is formed from a wire in the shape of an inverted V and either fed at the center point or fed at one end and appropriately terminated at the other.

invert-emulsion mud *Petroleum Engineering.* a drilling mud having a dispersed phase of salt or fresh water and a continuous phase of oil.

inverter *Electronics.* **1.** a device that converts a positive signal into a negative one, or vice versa. **2.** a device that converts an input pulse into an output pulse in a computer, so that when the input is high, the output is low and vice versa. *Electricity* see STATIC INVERTER.

inverter circuit see NOT CIRCUIT.

invertible *Mathematics.* having an inverse; nonsingular; regular.

inverting amplifier *Electronics.* a circuit that increases current or voltage and reverses its polarity.

inverting function *Electronics.* a type of logic circuit that inverts the input signal, so that the output signal is out of phase with the input signal.

inverting parametric device *Electronics.* a device that inverts the distribution of electrons in a laser through the time variation of three frequencies, the harmonic of the pump frequency (which initiates the inversion) and two signal frequencies.

inverting telescope *Optics.* a telescope that presents an image right side up, as opposed to other telescopic images that appear inverted.

inverting terminal *Electronics.* in an operational amplifier, the inverting input that provides a 180° phase shift at the output.

invert level *Engineering.* the lowest level of liquid in a conduit, such as a drain.

investment casting *Materials Science.* **1.** a casting method used primarily for intricate shapes, in which a wax pattern is coated with a ceramic; to form a mold after the wax is melted and drained, metal is poured into the mold. **2.** a component fabricated by investment casting.

inviable *Biology.* not viable; incapable of surviving, especially because of a genetic defect.

inviscid [in vis´əd] *Fluid Mechanics.* having no viscosity.

inviscid fluid *Fluid Mechanics.* ideally, a fluid that does not possess any viscosity and is therefore able to flow with no dissipation of energy or shearing stress.

invisible glass *Physics.* any glass that has a curved surface or a coating of material of molecular thickness to eliminate surface reflections.

invisible image *Physics.* an image in a form that cannot be seen with unaided vision.

invisible ink *Graphic Arts.* any of several types of novelty ink that become visible only when in contact with a particular environmental factor such as moisture or heat, or with an acid. Also, SYMPATHETIC INK.

in vitro [in vē´trō] *Biotechnology.* within a glass or observable in a test tube; referring to a process that takes place under artificial conditions or outside of the living organism. (From Latin; literally, "in glass.")

in vitro assay *Molecular Biology.* any experimental procedure that takes place outside of a living organism.

in vitro fertilization *Biotechnology.* a technique in which an ovum, especially a human one, is fertilized by sperm outside the body; the resulting embryo is implanted in the uterus for gestation.

in vitro replication *Biotechnology.* the replication that takes place in a synthetic medium outside of a living organism.

in vitro transcription *Biotechnology.* the transcribing of the information contained in a nucleotide sequence of DNA outside of a living organism.

in vivo [in vē´vō] *Biology.* within a living organism.

in vivo desensitization *Psychology.* behavior therapy in which the client confronts actual fear-related situations together with the therapist.

invoke *Artificial Intelligence.* to cause a procedure to be executed and applied to a given problem or data.

involucel *Botany.* a secondary involucre. Thus, **involucellate.**

involucre *Botany.* one or more whorls of small leaves or bracts at the base of an inflorescence or capitulum. Thus, **involucrate, involucral.**

involucrum *Anatomy.* a covering or sheath, such as that containing the sequestrum of a necrossed bone.

involuntary *Science.* not subject to the control or will. *Physiology.* describing a process or activity that goes on without any conscious control or direction.

involuntary muscle *Physiology.* a muscle that acts without voluntary control, such as intestinal or cardiac muscle.

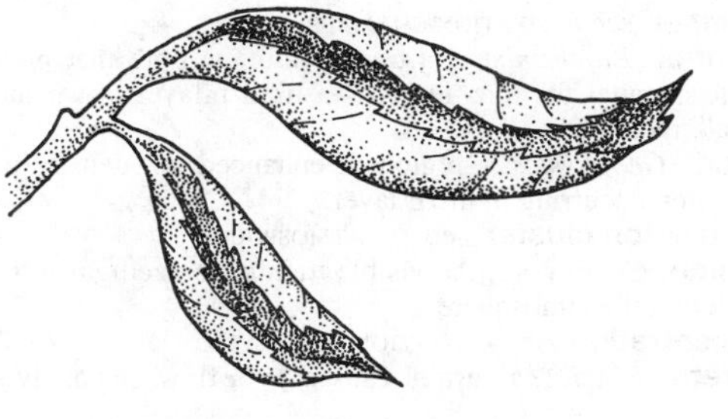

involute leaves

involute *Biology.* **1.** having margins rolled or turned in over the upper or ventral surface. **2.** rolled up tightly with hidden edges, such as leaves when first formed. *Invertebrate Zoology.* of shells, closely wound. *Mathematics.* the evolute of a curve γ_1 is a second curve γ_2 consisting of the centers of curvature of γ_1; equivalently, γ_2 is the envelope of the normals of γ_1. γ_1 is an involute of γ_2. Each normal of the involute is a tangent of the evolute, and to a given evolute there corresponds a family of involutes.

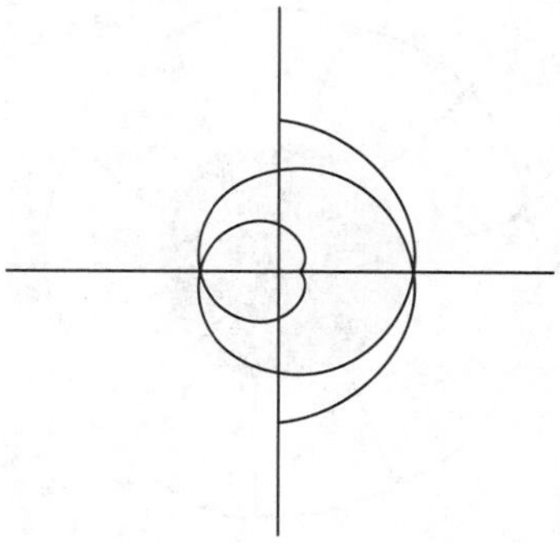

involute of a circle

involute gear tooth *Design Engineering.* a gear tooth whose flight profile is the locus of the end of a string uncoiled from a base circle.

involution *Developmental Biology.* **1.** the return of an enlarged organ, such as the postpartum uterus, to normal size. **2.** the turning inward of the edges of a part. *Cell Biology.* a step in a certain type of gastrulation where cells move to the interior of the embryo. *Medicine.* the progressive degeneration of processes and organs as a natural consequence of advancing age. *Mathematics.* a mapping in an algebra B denoted by the assignment $T \rightarrow T^*$, for all operators T on B, such that: (a) $(T + S)^* = T^* + S^*$ and $(aT)^* = aT^*$ (linearity is preserved); (b) $(ST)^* = T^*S^*$ (algebraic structure of the space is preserved); and (c) $T^{**} = T$ (involution is an isomorphism). An algebra having such an involution is called an **involutive algebra.**

involutional *Medicine.* relating to the period of approximately age 40 to 55 in women and 50 to 65 in men, associated with the onset of the progressive degeneration of organs and tissues. *Psychology.* relating to psychological reactions that occur in this period. Thus, **involutional depression, involutional melancholia, involutional psychosis.**

involution form *Cell Biology.* a body or cell that is structurally deformed or irregular; typically found under harsh, unfavorable conditions, as in old cultures.

inyoite *Mineralogy.* $Ca_2B_6O_6(OH)_{10}\cdot 8H_2O$, a colorless, monoclinic mineral having a specific gravity of 1.875 and a hardness of 2 on the Mohs scale; found as crystals and in granular form in borate deposits.

Io [ī´ō] *Astronomy.* the innermost of the four Galilean moons of Jupiter; a rocky orb with a diameter of 3640 kilometers, a density of 3.5, and several active sulfur volcanoes.

Io

Io the chemical symbol for ionium.
I/O *Computer Science*. a shorter term for input/output.
Iodamoeba *Invertebrate Zoology*. a genus of amebas that are commensal in human intestines.

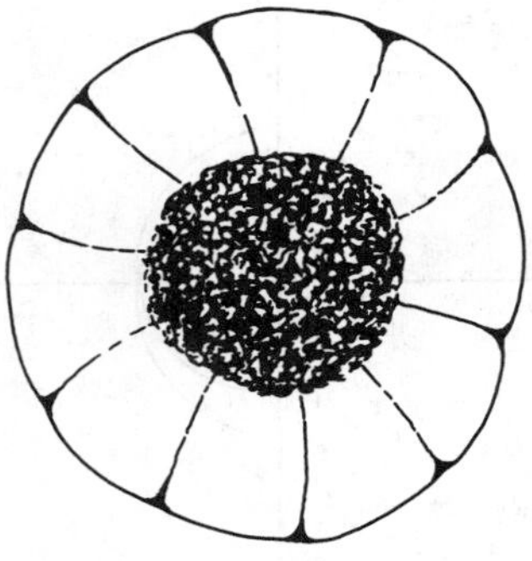

Iodamoeba

iodargyrite *Mineralogy*. AgI, a yellow hexagonal mineral occurring as aggregates of prismatic or tabular crystals and as scaly masses, having a specific gravity of 5.69 and a hardness of 1.5 on the Mohs scale; found as a secondary mineral in oxidized silver ore deposits. Also, **iodyrite.**
iodate *Inorganic Chemistry*. any compound derived from iodic acid and containing the IO_3^- ion.
iodeosin see TETRAIODOFLUORESCEIN.
iodic acid *Inorganic Chemistry*. HIO_3, colorless to pale-yellow crystals or powder; very soluble in water and decomposed by heat; toxic and an irritant; used in analytical chemistry and medicine.
iodic acid anhydride see IODINE PENTOXIDE.
iodide *Chemistry*. a compound of iodine with another element or radical, usually a binary compound with a metal.
iodide ion *Chemistry*. an iodine atom occurring in the -1 oxidation state.
iodide process *Metallurgy*. a reactive-metal-refining process based on the decomposition of metal iodides.
iodinated density-gradient medium *Molecular Biology*. a medium used for subcellular fractionation or separation of macromolecules by ultracentrifugation.
iodine [ī´ə dīn´] *Chemistry*. a nonmetallic halogen element having the symbol I, the atomic number 53, an atomic weight of 126.9045, a melting point of 113.5°C, and a boiling point of 184°C; occurs as grayish-black slates or granules; used in the production of dyes, water treatment, and medical disinfectants. *Pharmacology*. a preparation of this substance used as a topical anti-infective. (From the Greek *iodés* meaning "violet" because it sublimes to a dense violet vapor when heated.)
iodine-131 *Nuclear Physics*. a radioactive isotope of iodine with mass number 131 and a half-life of 8.04 days; used as a tracer in medical and industrial research.
iodine cyanide see CYANOGEN IODIDE.
iodine monobromide *Inorganic Chemistry*. IBr, dark gray crystals; soluble with decomposition in water; melts at 42°C and decomposes at 116°C; toxic and corrosive; used in organic synthesis.
iodine monochloride *Inorganic Chemistry*. ICl, a dark red solid or a brown oily liquid; soluble in alcohol and ether; decomposed by water and heat; used in analytical chemistry and organic synthesis.
iodine number *Analytical Chemistry*. a determination of the unsaturation of an organic compound by measuring the amount of iodine absorbed over a specific period of time. Also, **iodine value.**
iodine pentafluoride *Inorganic Chemistry*. IF_5, a toxic, colorless fuming liquid that freezes at 9.6°C and boils at 98°C; a dangerous fire risk that reacts violently with water; used as a fluorinating and incendiary agent.
iodine pentoxide *Inorganic Chemistry*. I_2O_5, a toxic, white crystalline powder; very soluble in water; decomposes at 300–350°C; used as an oxidizing agent and in organic synthesis. Also, IODIC ACID ANHYDRIDE.
iodine test *Analytical Chemistry*. a test for the presence of starch using a potassium iodide solution; a blue color indicates a positive test.
iodine trichloride *Inorganic Chemistry*. ICl_3, yellowish-brown deliquescent crystals that dissolve and decompose in water; decomposes at 77°C; toxic and corrosive to tissue; used as an antiseptic and in organic synthesis.
iodism *Toxicology*. poisoning due to the chronic ingestion of iodine.
iodoacetic acid *Organic Chemistry*. CH_2ICOOH, colorless crystals that are insoluble in water and alcohol; melts at 83°C; used in biochemical studies, for example, to inhibit enzyme activity.
iodoform *Organic Chemistry*. CHI_3, greenish-yellow crystals or powder having a penetrating odor; slightly soluble in water and alcohol; melts at 119°C and decomposes to free iodine above 200°C; used as a topical anti-infective.
iodoformism *Toxicology*. poisoning due to excessive exposure to the drug iodoform; symptoms may include skin problems.
iodometry *Analytical Chemistry*. the quantitative analysis of copper, gold, arsenic, and other elements or compounds using excess iodide ion as a reductant; the iodine freed in the associating reaction is determined by titration with potassium thiosulfate, using starch as an indicator.
iodonium *Inorganic Chemistry*. the H_2I^+ or R_2I^+ cation.
iodophor *Chemistry*. any carrier of iodine.
iodopsin *Biochemistry*. a pigment found in the cones of the retina, responsible for day or color vision. Also, VISUAL VIOLET.
iodosylbenzene *Organic Chemistry*. C_6H_5IO, a colorless, amorphous powder; soluble in hot water and alcohol; explodes at 210°C; used as an oxidizing agent. Also, **iodosobenzene.**
iodothyronine *Biochemistry*. any of several thyroid hormones, such as thyroxine and triiodothyronine, formed by the oxidative coupling of two iodotyrosines by an ether linkage in the *para* configuration, formed by the enzyme thyroid peroxidase.
iodotyrosine *Biochemistry*. a precursor of thyroxine and triiodothyronine, the thyroid hormones.
iodoxybenzene *Organic Chemistry*. $C_6H_5IO_2$, colorless needles that are soluble in water; melts at 167°C and explodes at 236–237°C; used as an oxidizing agent.
Ioffe bars *Physics*. the heavy bars used in certain controlled fusion reactors to carry the current that helps stabilize the plasma.
iojap *Genetics*. a mutant chromosomal gene that occurs in corn and causes the chloroplasts in the cells to undergo changes.
ion [ī´än´; ī´ən] *Chemistry*. an atom, radical, or molecule that has gained or lost one or more electrons and thus acquired a net negative or positive charge. In electrolysis, positive ions (cations) travel to the cathode, while negative ions (anions) travel to the anode. (Coined by Michael Faraday, from a Greek form meaning "going.")
Ion. ionic.
ion accelerator *Nucleonics*. a machine in which an electric field produced by external oscillators or amplifiers propels electrons in a straight line to produce a beam of highly charged particles.
ion-acoustic wave *Physics*. a longitudinal compression wave in the ion density of a plasma that can occur at high electron temperatures and low frequencies, and is caused by a combination of ion inertia and electron pressure.
ion atmosphere *Physical Chemistry*. a cloudlike configuration of ions that are loosely bound around an ion of the opposite charge. Also, ION CLOUD, ION CLUSTER.
ion backscattering *Solid-State Physics*. the scattering in a nearly backward direction of an ion beam incident on a film or body.
ion-beam mixing *Engineering*. the bombardment of a substance with high-energy ions so as to cause the intermixing of atoms of two different phases in the near-surface region.
ion-beam scanning *Electronics*. the process by which the mass spectrum of an ion beam is analyzed, generally by altering the electric or magnetic fields or by moving a probe in a mass spectrometer.
ion chamber see IONIZATION CHAMBER.
ion channel *Biochemistry*. a transmembrane pore that presents a hydrophilic channel for ions to cross a lipid bilayer down their electrochemical gradients.
ion cloud *Geophysics*. a region of enhanced ion density in the ionosphere, often occurring in the E layer.
ion cloud or **ion cluster** see ION ATMOSPHERE.
ion column *Geophysics*. the visible train of ionized gas left by a meteorite entering the atmosphere.
ion concentration see IONIZATION DENSITY.
ion current *Physics*. a current caused by a flow of positively charged ions.
ion-cyclotron-resonance mass spectrometer *Spectroscopy*. a mass spectrometer in which the mass distribution of orbiting ions within a magnetic field is detected by bringing ion frequencies sequentially into resonance with applied radio frequencies.
ion density *Physics*. the number of ions per unit volume.

ion detector *Analytical Chemistry.* any instrument used to determine the presence or concentration of ions.

ion emission *Physics.* the ejection of ions from the surface of a medium into the surrounding space, due to the influence of an electric field or heat.

ion engine *Space Technology.* a type of rocket engine that generates thrust from the electrostatic acceleration of ionized particles. Also, ION ROCKET.

ion exchange *Physical Chemistry.* a chemical reaction in which ions are interchanged between one substance and another, usually by means of passing a liquid through a porous, granular solid that is relatively insoluble. *Biotechnology.* the use of such a process to replace certain selected anions or cations in a solution; used in various applications to remove undesirable substances, as in water softening, or to recover desirable ones, as in the separation of valuable metals from wastes. *Geochemistry.* the replacement of inner layer and surface ions in clay minerals by ions from an adjacent solution. Also, BASE EXCHANGE, CATION EXCHANGE.

ion-exchange chromatography *Analytical Chemistry.* a separation technique using a resin (stationary phase), which may be basic or acidic, to bind ionized solutes (mobile phase) reversibly.

ion-exchange column *Biotechnology.* a column that fractionates mixtures of charged molecules and is packed with charged resins, cellulose, or other supporting compounds suitable for ion-exchange chromatography; used in laboratory or full-scale industrial processes.

ion-exchange electrolyte cell *Electricity.* a fuel cell having a chemical process involving the reversible interchange of ions at the boundary of an electrolyte solution and an ionic solid.

ion exchanger *Physical Chemistry.* the solid medium through which a liquid passes in the process of ion exchange, typically a material such as zeolite or a synthetic resin.

ion-exchange resin *Materials Science.* any of a number of (usually organic) materials that are capable of exchanging the included ions with ions in a surrounding solution; used for deionizing water or for chromatography of organic molecules.

ion exclusion *Chemistry.* a system for the separation of ionic from nonionic solutions through the use of a synthetic resin, which absorbs the nonionized solute and passes the ionic solute.

ion-exclusion chromatography *Analytical Chemistry.* a technique in which the adsorbent material is saturated with the same ions as are in the sample.

ion fractionation *Chemical Engineering.* the separation of anions or cations from an ionic solution by a membrane permeable to the desired ion, using electrodialyzers and ion-fractionation stills.

ion gauge see IONIZATION GAUGE.

ion gun see ION SOURCE.

Ionian Sea *Geography.* an arm of the Mediterranean between Italy and Greece.

ionic [ī än´ik] *Chemistry.* **1.** relating to or involving ions. **2.** occurring in the form of an ion or ions.

Ionic [ī än´ik] *Architecture.* the second major order of Greek architecture (about 450–339 BC), characterized by the spiral scrolls (volutes) on its capital.

ionic bond *Physical Chemistry.* an attractive force that draws electrons from one atom to another, thus transforming neutral atoms into electrically charged ions. Also, ELECTROVALENT BOND.

ionic bonding *Physical Chemistry.* the process by which an ionic bond occurs.

ionic channels *Molecular Biology.* membrane passages that allow certain ions to cross the membrane.

ionic charge *Physics.* the total charge carried by an ion, equivalent to the amount of charge associated with the electrons that are removed from a neutral atom or molecule in order to create the ion.

ionic conductance *Physical Chemistry.* a measure of the ability of an individual ion in an electrolyte to carry a current, assuming that no interaction between ions is taking place. Similarly, **ionic conductivity.**

ionic conduction *Solid-State Physics.* the conduction of charge through a crystal produced by the movement of ions in the crystal lattice.

ionic coupling *Biochemistry.* an intimate cytoplasmic contact between proximate cells that is mediated by gap junctions, such that electrical current injected into either cell changes the membrane potential of both.

ionic crystal *Crystallography.* a crystal held together by the electric forces between ions, as for a chemical compound that is a salt, such as sodium chloride.

ionic equilibrium *Physical Chemistry.* a condition in which the number of existing molecules dissociating into ions equals the number of new molecules recombining from ions.

ionic equivalent conductance *Physical Chemistry.* the contribution that each individual ion makes toward an electrolyte's overall ability to conduct current.

ionic focusing see GAS FOCUSING.

ionic gel *Chemistry.* a gel that contains ionic groups attached to the colloid structure, preventing the groups from diffusing out into a surrounding medium.

ionicity *Chemistry.* the ionic characteristics of a solid.

ionic lattice *Crystallography.* a lattice with symmetrically arranged ions and a good conducting power.

ionic membrane *Chemical Engineering.* a semipermeable membrane that allows the electrophoretic passage of ions when an electric field is applied.

ionic mobility *Physics.* the ratio of the average drift velocity of an ion in a solution to the magnitude of the electric field causing the drift.

ionic polarization *Materials Science.* the creation of a net dipole moment in a material when an applied electrical field displaces the cations in one direction and the anions in the opposite direction.

ionic polymerization *Materials Science.* a polymerization process carried out by means of electrically charged ions (either cations or anions) that initiate a chain polymerization reaction.

ionic radius *Physical Chemistry.* a measure used to determine the effective range of an ion in a compound, commonly based on the sum of the radii of a pair of oppositely charged ions in a crystal.

ionic ratio *Oceanography.* the ratio by weight of a major constituent of sea water to the chloride ion content; these ratios are essentially unvarying, giving rise to the principle of constant proportions.

ionic semiconductor *Solid-State Physics.* a semiconductor whose primary charge carriers are ions rather than electrons and holes.

ionic spectrum SEE SPARK SPECTRUM.

ionic strength *Physical Chemistry.* a measure of the level of electrical force in an electrolytic solution, given by the formula $I = (1/2)\sum m_i z_i^2$, in which I is the ionic strength, m is the molar concentration of the medium i, and z is the ionic charge of the medium i.

ionic strength principle *Physical Chemistry.* the concept that the amount of ionic activity in an electrolytic solution is based on the charge on the ions present rather than on their particular chemical natures.

ion implantation *Materials Science.* the acceleration of foreign ions into the surface of a material, used for surface hardening and doping semiconductors.

ion irradiation *Physics.* the bombardment of a specimen with ions.

ionium *Nuclear Physics.* a naturally occurring radioactive isotope of thorium with atomic number 90 and mass number 230.

ionization [ī´ən ə zā´shən] *Chemistry.* the process of adding an electron to, or removing an electron from, an atom or molecule so as to give a net charge; the atom is then called an ion.

ionization arc-over *Electricity.* an electric spark that is created when ionized charges build up in a medium and produce forces on the electrons.

ionization chamber *Nucleonics.* an instrument that determines the level of radiation by measuring the electrical current generated by charged particles moving through its gas-filled tube. Also, ION CHAMBER.

ionization coefficient see SPECIFIC IONIZATION.

ionization constant *Physical Chemistry.* a fixed quantity that is based on the ratio of the ions produced from a given substance to the undissociated molecules of the substance; given by $K = [C^+] [A^-] / [CA]$, where K is the ionization constant, $[C^+]$ is the cation concentration, $[A^-]$ is the anion concentration, and $[CA]$ is the concentration of the substance.

ionization cross section *Physics.* an area in which the probability that an atom or ion will undergo ionization when it collides with a particle or photon of sufficient energy is measured.

ionization degree *Physical Chemistry.* the fractional degree of ionization of acids, bases, or salts that has taken place in a solution or reaction mixture.

ionization density *Electronics.* the concentration of charged particles in a gas. Also, ION CONCENTRATION.

ionization energy *Atomic Physics.* the amount of energy required to remove an electron from a specific atom or ion to an infinite point, generally expressed in electron volts and numerically equal to the ionization potential.

ionization front *Astrophysics.* a region in space in which the interstellar gas (commonly hydrogen) changes from a mostly neutral state to a mostly ionized state, due to ultraviolet radiation from hot stars nearby.

ionization gauge *Electronics.* **1.** a device that determines the degree of vacuum in an electron tube by measuring the amount of ionization current in the tube. **2.** a device, such as a Geiger counter, that determines the amount of radiation in a medium by measuring the ionization generated by charged particles passing through a gaseous substance. Also, ION GAUGE, IONIZATION VACUUM GAUGE.

ionization potential see ION POTENTIAL.

ionization source see ION SOURCE.

ionization spectrometer see BRAGG SPECTROMETER.

ionization techniques *Spectroscopy.* a class of techniques in crystal spectroscopy wherein an ionization chamber (an X-ray detector) is appropriately placed to detect the reflected X-rays after undergoing Bragg scattering.

ionization time *Electronics.* the amount of time it takes for a gaseous substance to ionize after an ionizing property has been applied to it.

ionization vacuum gauge see IONIZATION GAUGE.

ionized atom *Chemistry.* any atom with more or fewer electrons than protons, giving a net charge.

ionized gas *Physics.* a gas composed partially or totally of ions, such as a plasma.

ionized layers *Geophysics.* the layers of enhanced ionization produced within the ionosphere by impinging cosmic radiation.

ionizing event *Physics.* any event in which ionization occurs, commonly caused when an electron or a photon with sufficient energy passes through a sample of matter.

ionizing radiation *Nucleonics.* a particle, such as a photon, that has enough energy to remove an electron from an atom or a molecule, thus producing an ion and a free electron; examples of ionizing radiation include gamma rays, X-rays, and alpha particles.

ion kinetic energy spectrometry *Spectroscopy.* a spectroscopic technique for analyzing the energy of ionic products produced when a beam of ions having high kinetic energy is passed through a field-free reaction chamber.

ion laser *Optics.* a laser in which a beam of radiation is produced from the transitions in energy levels of ions in a noble gas.

ion machining *Engineering.* the use of a beam of high-energy ions to remove foreign or undesirable materials from a surface.

ion mean life *Physical Chemistry.* a term for the time it takes for an electron to escape from an atom or molecule and attach itself to another atom or molecule.

ion microscope see FIELD-ION MICROSCOPE.

ion migration *Electricity.* the movement of the ions in an electrolyte or a semiconductor toward the electrodes when an electric current is applied to the electrodes.

ion nitriding *Materials Science.* a process of surface hardening by diffusion of nitrogen ions primarily into steel.

ionogram *Engineering.* a record produced by an ionosonde, plotting radio frequency against the round-trip time of each pulse.

ionomer *Materials Science.* a polymer that has covalent bonds in the chain and ionic bonds between the chains.

ionomer resin *Organic Chemistry.* a copolymer of ethylene and a vinyl monomer with an acid group, forming a transparent, resilient thermoplastic material used in making bottles, pipe and tubing, and electroconductive elements.

ionone *Organic Chemistry.* $C_{13}H_{20}O$, a violet-scented, light-yellow to colorless liquid; slightly soluble in water and miscible with alcohol; boils at 128°C (at reduced pressure); used in perfumery, flavoring, and organic synthesis.

ionophone *Acoustical Engineering.* a loudspeaker that creates high-frequency audio signals which are used to modulate a radio-frequency signal to a quartz tube; the modulated signal then acts directly on a sample of ionized air to produce sound.

ionophore *Biochemistry.* an antibiotic molecule that transports specific ions across cell membranes.

ionophose *Neurology.* the subjective perception of the color violet.

ionosonde *Engineering.* a pulsing radar device or system used to measure the height of ionospheric layers.

ionosphere [ī än′i sfēr] *Geophysics.* a region in the earth's atmosphere, beginning at an altitude of 70–80 kilometers and extending to an indefinite height, in which free electrons and ions produced by solar radiation are abundant and affect radio waves that propagate through it.

ionospheric *Geophysics.* relating to or taking place in the ionosphere.

ionospheric disturbance *Geophysics.* an abnormal variation of the ion density in part of the ionosphere, commonly caused by solar flares.

ionospheric D scatter meteor burst *Geophysics.* a disturbance caused by infalling meteors that allows the penetration of radio waves from the ionosphere's D layer.

ionospheric error *Telecommunications.* all systematic and random errors caused by the reception of a navigation signal after ionospheric reflection.

ionospheric propagation *Telecommunications.* the transmission of electromagnetic waves of frequencies up to 25 megahertz over great distances, by ionospheric reflection.

ionospheric recorder *Electronics.* a device that gauges frequency distribution of the reflections from different layers in the ionosphere, used in radios.

ionospheric scatter *Telecommunications.* the transmission of electromagnetic waves by means of scattering, due to variations within the ionosphere.

ionospheric storm *Geophysics.* turbulence in parts of the ionosphere, probably connected with sunspot activity, that causes dramatic changes in its reflective properties, sometimes disrupting short-wave communications.

ion pair *Nucleonics.* a pair consisting of a positive ion and a negative ion (generally an electron) with equal charges formed from an irradiated atom or molecule.

ion potential *Atomic Physics.* the minimum energy per unit charge required to remove an electron from a specific atom or ion to an infinite point, usually expressed in volts. Also, IONIZATION POTENTIAL.

ion propulsion *Space Technology.* a propulsor, usually a small thruster, used to create vehicular motion by generating a high-velocity jet of ions in an electrostatic field, then ejecting the ions behind a vehicle.

ion pump *Electronics.* a pump that ionizes gas molecules with high-energy electrons in a high-intensity magnetic field and then deposits them onto a cathode or ejects them into an auxiliary pump or ion trap.

ion retardation *Chemical Engineering.* a process based on bifunctional ion-exchange resins containing both anion and cation adsorption sites, which removes both kinds of ions from solutions.

ion rocket see ION ENGINE.

ion-scattering spectroscopy *Spectroscopy.* a surface analysis technique in which a surface is bombarded with a beam of ionized inert gas and the energy of the scattered ions is measured in order to identify surface atoms.

ion-solid interaction *Solid-State Physics.* the interaction of an energetic ion colliding with condensed matter, resulting in elastic or inelastic scattering of the ion, the release of atoms from the target, emission of photons, ionization of target atoms, or ion implantation.

ion source *Electronics.* a device that ionizes gas molecules and then focuses, accelerates, and emits them as a narrow beam. Also, ION GUN, IONIZATION SOURCE.

ion spot *Electronics.* a section of a cathode-ray tube screen that has lost much of its luminescence because of negative ion bombardment.

iontophoresis *Medicine.* the introduction of ions of soluble salts into the skin by an electric current, usually for therapeutic purposes. Also, **iontherapy.**

ion transport *Biochemistry.* the force that moves an ion across a membrane under a specific transport system energized by a concentration gradient within the ion itself, a gradient from another ion, ATP hydrolysis, or by coupling metabolism to ion transport.

ion trap *Electronics.* **1.** a system that prevents an ion spot from forming on a cathode-ray tube screen, generally by using a magnetic field to divert the beam. Also, BEAM BENDER. **2.** a type of metal electrode, generally consisting of titanium, to which ions from an ion pump are attracted.

IOP intraocular pressure.

iophenoxic acid see TRIIODOTHYROACETIC ACID.

iophobia *Psychology.* an irrational fear of poison or of being poisoned.

Iospilidae *Invertebrate Zoology.* a family of slender, semitransparent, carnivorous polychaetes widely distributed in the Atlantic and Pacific Oceans.

I/O time see INPUT/OUTPUT TIME.

Iowan *Geology.* the earliest substage of the Wisconsinan glacial stage, once considered as a separate stage between the Illinoian and Wisconsinan stages.

ioxynil *Organic Chemistry.* $C_7H_3I_2NO$, a colorless solid that melts at 205–206°C; used as an herbicide.

ioxynil octanoate *Organic Chemistry.* $C_{15}H_{17}I_2NO_2$, a waxy solid that melts at 60°C; used as an insecticide.
IP initial point; intermediate pressure.
IP or **I.P.** Institute of Petroleum.
I.P. intraperitoneally; isoelectric point.
IPA *Linguistics.* International Phonetic Alphabet, a standard alphabet intended to provide a consistent system of representing all the possible sounds of a variety of languages.
IPA or **I.P.A.** International Phonetic Association.
IPAA or **I.P.A.A.** International Psychoanalytical Association.
ipe *Materials.* the durable, heavy wood of the *Tabebuia* timber trees of Central and South America; used in the construction of railroad ties, industrial flooring, and decorative veneers.
ipecac [ip´i kak´] *Botany.* the common name for several tropical shrubs of the genus *Cephaelis,* especially the species *C. ipecacuanha. Toxicology.* a syrup prepared from the dried roots of *C. ipecacuanha* or *C. acuminata,* used to induce vomiting in cases of oral poisoning. (From a native Brazilian word meaning "vomit.")

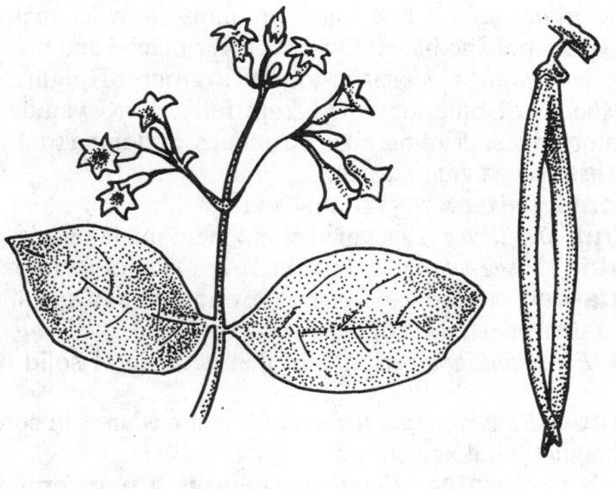

ipecac

Ipidae see SCOLYTIDAE.
IPL initial program load.
ipm or **i.p.m.** inches per minute.
Ipnopidae *Vertebrate Zoology.* a cosmopolitan family of rare benthic marine fishes of the order Myctophiformes, having a depressed snout and broad flat eyes.
IPNV or **IPN virus** infectious pancreatic necrosis virus.
ipomeamarone *Biochemistry.* a phytoalexin that is generated by sweet potatoes infected by fungi, such as block rot; it also causes liver disease in cattle.
IPPB intermittent positive pressure breathing.
IPPF or **I.P.P.F.** International Planned Parenthood Federation.
ips or **i.p.s.** inches per second.
ipsilateral see HOMOLATERAL.
IPSP inhibitory postsynaptic potential.
IPTS international practical temperature scale.
IPV inactivated poliovirus vaccine.
IQ see INTELLIGENCE QUOTIENT.
i.q. the same as. (From Latin *idem quod.*)
IR information retrieval; infrared; intelligence ratio.
Ir the chemical symbol for iridium.
ir- a prefix meaning: **1.** in or into, as in *irrigation.* **2.** not, as in *irregular.*
IRAC or **I.R.A.C.** Information Resource and Analysis Center.
IRBM intermediate range ballistic missile.
IRC inspiratory reserve capacity.
irdome *Optics.* a covering that shields an infrared detector, usually made of a material that is transparent to infrared radiation.
IRE or **I.R.E.** Institute of Radio Engineers.
I region *Immunology.* the segment of the major histocompatibility complex that contains products involved in immune reponses.
Ireland *Geography.* a large island in the North Atlantic, the westernmost of the British Isles, containing the Republic of Ireland and Northern Ireland; area: 32,597 square miles.
Irenidae *Vertebrate Zoology.* the leafbirds or fairy bluebirds, a family of brightly colored birds of the passerine suborder Oscines; found in forests of southern Asia, India, southern China, and the Philippines.
ir gene see IMMUNE-RESPONSE GENE.
IRI or **I.R.I.** Industrial Research Institute.
irid- a combining form meaning: **1.** rainbow, as in *iridescent.* **2.** the iris of the eye, as in *iridemia.*
irid. iridescent.
Iridaceae *Botany.* a family of perennial monocotyledonous herbs of the order Liliales, having fibrous roots, regular or irregular bisexual flowers, three stamens, and an inferior ovary.
iridectomy *Surgery.* the surgical removal of part of the iris.
iridescence [ēr i des´əns] *Optics.* the appearance of rainbowlike colors due to the interference of light reflected from the front and back surfaces of a thin film or thinly layered material; exhibited in nature in soap bubbles and mother of pearl.
iridescent [ēr i des´ənt] *Optics.* having the property of iridescence; displaying a shining, rainbowlike range of colors.
iridescent cloud *Meteorology.* a high cloud composed of ice crystals, typically observed up to about 30° from the sun, that exhibits brilliant patches of color, usually red and green. This phenomenon is believed to be caused by the diffraction of sunlight by supercoded droplets of water.
iridescent insect viruses *Virology.* isometric dsDNA-containing viruses that create an iridescent appearance upon the infection of insects.
iridic chloride *Inorganic Chemistry.* $IrCl_4$, a brownish-black mass that absorbs moisture from the air; soluble in water and alcohol; used in analytical chemistry, microscopy, and plating solutions.
iridium [i rid´ē əm] *Chemistry.* a metallic element having the symbol Ir, the atomic number 77, an atomic weight of 192.22, a melting point of 2443°C, and a boiling point of c. 4500°C; a silver-white corrosion-resistant metal of the platinum group, used as a catalyst and as an alloy with noble metals. *Metallurgy.* one of this group of metals used commercially to make crucibles for the manufacture of superpurity semiconductor crystals. (From the Greek word for "rainbow," because of the variety of colors it displays in solution.)
iridium chloride or **iridium tetrachloride** see IRIDIC CHLORIDE.
irido- a combining form meaning: **1.** rainbow, as in *iridocyte.* **2.** the iris of the eye, as in *iridopathy.*
iridocele *Medicine.* the protrusion of a portion of the iris through a wound or defect of the cornea.
iridocyclitis *Medicine.* an inflammation of the iris as well as the ciliary body.
iridocyclochoroiditis *Medicine.* an inflammation of the iris, the ciliary body, and the choroid.
iridocyte *Histology.* in the skin of certain animals, a specialized iridescent cell containing guanine crystals and lipophores.
iridosmine *Metallurgy.* a native iridium-base alloy containing osmium and, at times, platinum, ruthenium, rhenium, iron, or copper.
Iridoviridae *Virology.* a family of dsDNA-containing viruses that infect a wide range of insects and poikilothermic vertebrates.
Iridovirus *Virology.* a genus of viruses of the family Iridoviridae that infects insects, often creating a blue iridescence.
I.R.I.S. International Research Information Service.

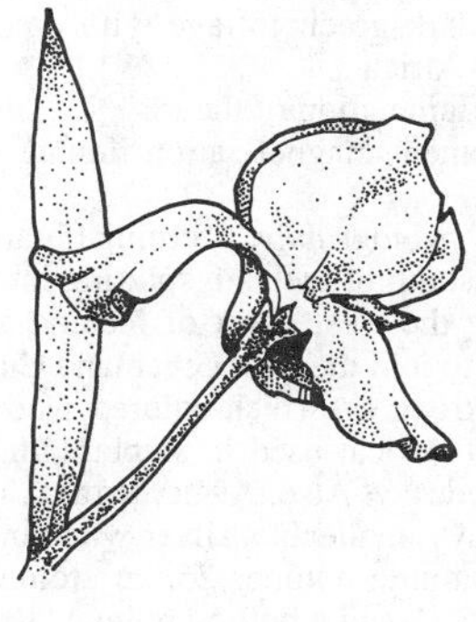

iris flower

iris *Botany.* **1.** any plant of the genus *Iris* of the family Iridaceae characterized by sword-shaped leaves and in most cases, large ornamental flowers in a variety of colors. **2.** a flower of this plant. (From Iris, the Greek goddess of the rainbow.)

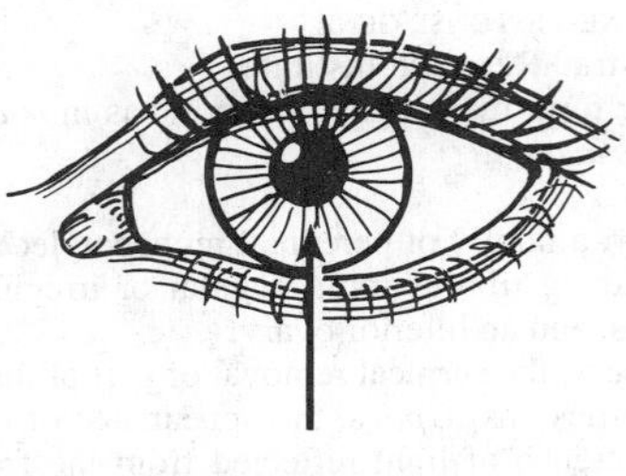

iris of the eye

iris any of various structures or devices thought to resemble the iris flower; specific uses include: *Anatomy.* the pigmented circular membrane that lies between the lens and the aqueous humor of the eye. *Electromagnetism.* a diaphragm of conductive material that partially occupies the cross section of a waveguide so as to introduce impedance. *Optics.* a device that controls the amount of light that reaches the film in a camera. Also, **iris diaphragm.**

irisation *Meteorology.* the coloration peculiar to iridescent clouds, which is also at times observed along the borders of lenticular clouds.

Irish moss *Materials.* the dried, bleached plants of the red algae *Chondrus crispus* and *Gigartina mamillosa*; used as a thickening or emulsifying agent in cooking and pharmacy.

Irish moss

Irish Sea *Geography.* an arm of the North Atlantic between Ireland and Great Britain.

Irish yew *Botany.* the yew *Taxus baccata stricta*, characterized by upright branches and dark green foliage with color variation; found in Eurasia and northern Africa.

iritis *Medicine.* an inflammation of the iris.

IRM isothermal remanent magnetization; innate releasing mechanism. Also, **irm.**

Irminger Current *Oceanography.* a terminal branch of the Gulf Stream that flows westward south of Iceland, splitting into a branch that continues northward along the west coast of Iceland and another that turns back southwestward to join the East Greenland Current.

iroko *Materials.* a strong brownish-colored wood with a coarse, open grain, found in West Africa; used in shipbuilding because of its resistance to termites and decay. Also, AFRICAN TEAK.

iron *Chemistry.* a heavy, malleable, silver-white metallic element having the symbol Fe, the atomic number 26, an atomic weight of 55.847, a melting point of 1536°C, and a boiling point c. 3000°C. *Metallurgy.* this element used as a metal; almost never used in its pure form commercially. It is the only metal that can be tempered, and it is used extensively as a base for cast steels and cast irons. *Mechanical Engineering.* any of various tools or devices made from iron or similar metals. (An Old English word; perhaps from the ancient name of a place where iron deposits were found.)

iron-55 *Nuclear Physics.* a toxic, radioactive isotope of iron with a half-life of 2.73 years.

iron-59 *Nuclear Physics.* a radioactive isotope of iron with a half-life of 44.5 days, widely used in medicine and industry.

iron acetate liquor *Materials.* a black liquor that contains 5–5.5% iron and sometimes copperas or tannin; used as a dye mordant.

Iron Age *Archaeology.* a time period, beginning about 3000 years ago, designated by the Three-Age System as following the Stone Age and the Bronze Age; defined by a general shift to the use of iron as the main material for tools.

iron alloy *Metallurgy.* a generic name for any alloy containing at least 50% iron, including steels and cast irons.

iron alum see HALOTRICHITE.

iron ammonium sulfate **1.** see FERRIC AMMONIUM SULFATE. **2.** see FERROUS AMMONIUM SULFATE.

iron arsenate see FERROUS ARSENATE.

iron bacteria *Bacteriology.* bacteria capable of transforming soluble iron compounds into insoluble ferric oxide; often the cause of a metallic flavor in water, of rust stains on laundry, and of clogged water pipes.

iron-binding protein *Biochemistry.* a serum protein, such as ferritin or transferrin, that binds iron for storage or transport.

iron black *Materials.* a fine black antimony powder that contains no iron but gives a polished-steel look to papier-mâché and plaster of Paris.

iron blue *Inorganic Chemistry.* any of a group of pigments having a range of shades of blue, obtained from ferric ferrocyanide and widely used in paints, dyes, printing inks, cosmetics, and industrial finishes.

iron bromide see FERRIC BROMIDE.

iron carbonyl see IRON PENTACARBONYL.

iron casting *Metallurgy.* a shaped component made of cast iron.

iron chloride **1.** see FERRIC CHLORIDE. **2.** see FERROUS CHLORIDE.

iron-constantan *Metallurgy.* one of several pairs of metallic materials that form a thermoelectric couple.

iron core *Electromagnetism.* a core that is made of solid or laminated iron.

iron-core coil *Electromagnetism.* a coil that has an iron core for part or all of its magnetic linkage circuit.

iron-core transformer *Electromagnetism.* a transformer that has a core constructed of iron laminations.

iron count *Chemical Engineering.* the analysis of iron compounds in a product stream that shows the extent and occurrence of corrosion.

iron-deficiency anemia *Medicine.* anemia due to inadequate intake or absorption of iron, or the excessive loss of iron; symptoms include pallor, angular stomatitis and other oral lesions, and gastrointestinal complaints.

iron dichloride see FERROUS CHLORIDE.

iron-dust core *Electromagnetism.* a core made from a fine powder of iron or other magnetic material, usually pressed into a rod shape that can be moved in and out of a coil for various degrees of inductance.

irone *Organic Chemistry.* $C_{14}H_{22}O$, a homolog of ionone that occurs naturally in the essential oil of *Iris florentina* and as a terpene from orris root; a colorless to yellowish liquid that is slightly soluble in water and boils at 85–88°C (0.06 torr); used in perfumery.

iron ferrocyanide see FERRIC FERROCYANIDE.

iron fluoride see FERRIC FLUORIDE.

iron formation *Geology.* **1.** a thin-bedded, usually finely laminated chemical sedimentary rock containing at least 15% iron of sedimentary origin. **2.** a low-grade iron ore in which segregated bands of iron minerals are intermingled with irregularly banded layers of chert or quartz.

iron foundry *Metallurgy.* a foundry producing iron casting.

iron hydroxide or **iron hydrate** see FERRIC HYDROXIDE.

ironing *Metallurgy.* the process of drawing a hollow stock between a die and a punch.

iron lung *Medicine.* an airtight respirator composed of a metal tank that encloses the body from the neck down and provides artificial respiration by contracting and expanding the walls of the chest.

iron metabolism *Biochemistry.* a group of chemical processes in which iron is used to transport oxygen (in the blood, for example) and to involve it in various biochemical activities.

iron metavanadate see FERRIC VANADATE.

iron meteorite *Astronomy.* one of the three chief classes of meteorites, whose members are typically made up of about 90% iron and 10% nickel, with other minor consituents.

iron mica see LEPIDOMELANE.

iron monoxide see FERROUS OXIDE.

iron nitrate see FERRIC NITRATE.

iron nonacarbonyl *Inorganic Chemistry.* $Fe_2(CO)_9$, yellowish metallic crystals that are insoluble in water and decompose at 80°C. Also, **iron enneacarbonyl.**

Ironoidea *Invertebrate Zoology.* a superfamily of carnivorous nematodes found in fresh water and soil.

iron ore *Geology.* a rock or deposit containing iron-bearing compounds from which metallic iron can be extracted economically.

iron oxide *Inorganic Chemistry.* **1.** see FERRIC OXIDE. **2.** see FERROUS OXIDE. **3.** any of various other iron-oxygen compounds in which iron appears in several oxidation states; colors range from reddish-brown to black; used as pigments.

iron oxide process *Chemical Engineering.* a process for removing sulfides from a gas by passing the gas through a mixture of iron oxide and wood shavings.

iron pentacarbonyl *Inorganic Chemistry.* $Fe(CO)_5$, a viscous yellow liquid that gives off carbon monoxide on exposure to air or light; insoluble in water; freezes at –21°C and boils at 102.8°C. It is toxic and a dangerous fire risk, and is used as a catalyst and in electromagnetic coils.

iron phosphate see FERRIC PHOSPHATE.

iron-porphyrin protein *Biochemistry.* a protein that contains iron and a porphyrin, such as hemoglobin.

iron pyrite see PYRITE.

iron red *Materials.* any of the red pigments derived from ferric oxide.

ironshot *Mineralogy.* describing a mineral that is streaked, speckled, or marked with spots of iron or iron ore.

iron sight *Ordnance.* any metal gunsight, as distinguished from an optical or computing gunsight.

iron sinter see PHARMACOSIDERITE.

iron soldering *Metallurgy.* the process of soldering with the heat supplied by a hot iron.

iron spar see SIDERITE.

iron spinel see HERCYNITE.

ironstone *Petrology.* **1.** a sedimentary rock rich in iron due to either chemical replacement or direct deposition as a ferruginous sediment. **2.** any iron-rich rock used for commercial smelting operations.

iron-stony meteorite see STONY-IRON METEORITE.

iron sulfate see FERROUS SULFATE.

iron sulfide see FERROUS SULFIDE.

iron-sulfur proteins *Biochemistry.* a group of proteins that contain non-heme iron and sulfur atoms and take part in a variety of electron transport processes, such as the electron transport chain and the nitrogenase system.

iron tetracarbonyl *Inorganic Chemistry.* $Fe(CO)_4$, dark green lustrous crystals that decompose at 140–150°C; insoluble in water and soluble in organic solvents.

iron trichloride see FERRIC CHLORIDE.

iron vitriol see FERROUS SULFATE.

iron whiskers *Metallurgy.* extremely thin, acicular single crystals consisting essentially of pure iron.

iron winds *Meteorology.* northeasterly winds that blow for several days at a time across Central America; most prevalent between the months of February and March.

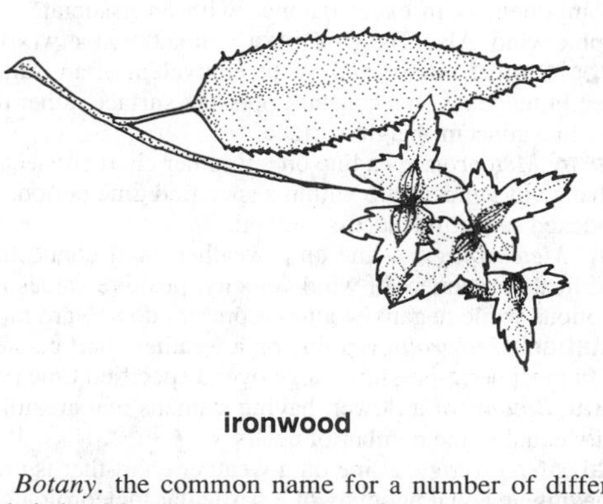

ironwood

ironwood *Botany.* the common name for a number of different trees noted for their hard heavy wood, such as those of the genus *Carpinus,* especially the species *C. caroliniana.*

irradiate *Radiology.* to expose to ionizing radiation or other radioactivity for therapeutic or diagnostic purposes.

irradiation *Radiology.* exposure to any form of radiation for therapeutic or diagnostic purposes or to increase vitamin efficiency. *Engineering.* the amount of radiant energy incident on an object. *Optics.* the illusion that causes bright objects to seem bigger than they are. *Navigation.* this effect that introduces an error when sextant readings are taken of bright objects, such as the sun.

irradiation cataract *Medicine.* a loss of transparency in the lens of the eye due to prolonged or excessive exposure to radiation.

irradiation correction *Navigation.* a correction to be applied to a sextant reading affected by irradiation; this effect is highly variable between individuals, and no set value can be given.

irradiation cystitis *Medicine.* an inflammation of the urinary bladder after exposure to radiation.

irrational equation *Mathematics.* an equation in which variables appear as part of an expression raised to a nonintegral power.

irrational number *Mathematics.* any real number that cannot be written in the form a/b, where a and b are integers and $b \neq 0$. Irrational numbers are either algebraic (roots of algebraic equations) or transcendental (all others).

Irrawaddy *Geography.* a river flowing 1250 miles through Mayanmar (Burma) into the Bay of Bengal.

irreducible element *Mathematics.* an element c of a commutative ring with identity R is an irreducible element if: (a) c is nonzero and not a unit, and (b) if $c = ab$ then either a or b is a unit. Every prime element of a ring is also irreducible and, in a unique factorization domain, every irreducible element is prime.

irreducible equation *Mathematics.* an equation formed by setting an irreducible polynomial equal to zero.

irreducible hernia see INCARCERATED HERNIA.

irreducible polynomial *Mathematics.* let $f(x)$ be a polynomial with coefficients in some ring R. $f(x)$ is said to be irreducible over R if it cannot be written in the form $f(x) = g(x)h(x)$, where $h(x)$ and $g(x)$ are also polynomials over R and neither is of degree zero. In particular, a polynomial that is irreducible over a ring R has no roots in R.

irreducible representation *Mathematics.* a representation of a group, ring, or algebra on a vector space V is irreducible if V has no subspaces that are invariant under the action of the representation.

irreducible tensor *Mathematics.* a tensor that cannot be written as the (inner) product of tensors of lower degree. Also, INDECOMPOSABLE TENSOR.

irreg. irregular.

irregular *Botany.* not symmetrical; without uniform dimensions, as a flower whose petals differ in size or shape.

irregular cleavage *Developmental Biology.* an embryonic cleavage resulting in more or less haphazard arrangements of cells.

irregular connective tissue *Histology.* connective tissue that contains collagen fibers oriented in an irregular manner.

irregular crystal *Meteorology.* a snow particle consisting of microscopic crystals that have grown together randomly; it is sometimes covered by a coating of rime. Also, AMORPHOUS SNOW.

irregular element *Industrial Engineering.* a basic work element that is of irregular frequency, but is a normal and predictable part of the work process. Also, INCIDENTAL ELEMENT.

irregular forces *Military Science.* armed individuals or groups that are not members of the regular armed forces, police, or other internal security forces.

irregular galaxy *Astronomy.* a small galaxy, composed of about 100 million to 10 billion stars, with no discernible structure. About 3% of all galaxies are irregular; most contain substantial amounts of gas and dust.

Irregularia *Invertebrate Zoology.* a group or subclass of irregular echinoid echinoderms, including sand dollars and heart urchins, that show some degree of bilateral symmetry. Also, EXOCYCLICA, EXOCYCLOIDA.

irregular movements *Statistics.* in time series analysis, the erratic, short-term fluctuations that follow no recognizable pattern and are not repetitive. Also, RESIDUAL VARIATIONS.

irregular outer edge *Ordnance.* short mine strips laid in an irregular manner in front of a minefield facing the enemy in order to disguise the nature or extent of the minefield.

irregular variable star *Astronomy.* a variable star with no detectable period in its brightness variations.

irrespirable *Toxicology.* describing gas or contaminated air that is toxic or otherwise unsuitable for breathing.

irrespirable atmosphere *Mining Engineering.* any atmosphere in a coal mine that can be entered only by workers wearing breathing apparatus.

irreversibility the fact of being irreversible; a situation in which change from one state to another is permanent and does not involve a return to the earlier state; specific uses include: *Evolution.* a condition prevailing in evolutionary change higher than the microevolutionary level, requiring that an evolutionary trend, once begun, cannot be reversed or lost. *Thermodynamics.* the condition of a thermodynamic process that cannot be reversed and that does not restore the system in question to its original state. Increased irreversibility results in an entropy increase to an isolated system.

irreversibility rule see DOLLO'S LAW.

irreversible *Science.* not reversible; not able to be reversed or counteracted. *Physical Chemistry.* **1.** relating to a process or change that cannot be reversed and that, once initiated, usually proceeds forward to completion. **2.** describing a colloidal system that cannot be restored to its original state once it has been formed in its new state, as in the formation of butter from milk.

irreversible change see IRREVERSIBLE PROCESS.

irreversible coma see BRAIN DEATH.

irreversible cycle *Thermodynamics.* a thermodynamic cycle in which there is a net heat transfer between the system in question and its surroundings, and a net amount of work done; i.e., at the end of the process there is some change of state in the system or its surroundings. This is characteristic of all actual cycles.

irreversible energy loss *Thermodynamics.* the increase in entropy, and thus reduction in energy, that characterizes an irreversible process.

irreversible process *Thermodynamics.* any thermodynamic process that cannot be reversed by simply imposing infinitesimal changes on the parameters; i.e., the system and its surrounding environment cannot be restored to their respective initial states.

irreversible reaction *Physical Chemistry.* a chemical reaction that continues forward in one direction until it has been completed, yielding a stable product that will not revert to the original reactants.

irreversible thermodynamics see NONEQUILIBRIUM THERMODYNAMICS.

IRRI or **I.R.R.I.** International Rice Research Institute.

irrigate *Agronomy.* to apply water to the soil to promote plant growth. *Medicine.* to wash out a wound or body cavity.

irrigation *Agronomy.* the artificial distribution of water to land for agricultural use. *Medicine.* the process of washing out a wound or a body cavity, such as the urinary bladder, with a stream of water or other fluid.

irrigation

irrigation canal *Civil Engineering.* an artificial course to aid in the process of irrigation.

irrigation pipe *Civil Engineering.* an artificial conduit to aid in the process of irrigation.

irrigator *Surgery.* an apparatus for performing irrigation.

irritability *Physiology.* the capability of an organism to react to a stimulus or stimuli. *Neurology.* excessive responsiveness to stimuli.

irritable bowel syndrome *Medicine.* any of the common disturbances of the bowel, such as diarrhea or constipation, that occur with abdominal pain and sometimes accompany emotional stress.

irritable colon *Medicine.* the disturbance of the colonic functions, sometimes with colicky pains and diarrhea, often accompanying emotional distress.

irritant *Science.* a biological, chemical, or physical agent, usually airborne, that stimulates a characteristic function or causes a response, particularly an inflammatory response such as the stinging of eyes.

irritation *Neurology.* **1.** the application of a stimulus to responsive tissue. **2.** the response of nerve tissue to stimuli.

irritative *Neurology.* **1.** causing irritation. **2.** dependent on or caused by irritation.

irrotational flow *Fluid Mechanics.* a fluid flow whose flow field has zero curl everywhere; i.e., flow with no circulation. Also, ACYCLIC MOTION, IRROTATIONAL MOTION.

irrotational motion see IRROTATIONAL FLOW.

irrotational strain *Geology.* a strain in which the orientation of the major axes does not change. Also, P WAVE.

irrotational vector field *Mathematics.* a vector field F in a simply connected domain D is said to be irrotational if curl F is identically zero. Equivalently, a vector field F is irrotational if there exists a function U (called the potential of the field) on D such that $F = -\text{grad } U$. Also, SINK-FREE, LAMELLAR.

Irtish or **Irtysh** *Geography.* a river rising in the Altai Mountains of northwestern China, flowing 2760 miles northwest through Siberia to join the Ob.

IRV inspiratory reserve volume.

Irvingtonian *Geology.* a geologic stage of the Lower Pleistocene epoch in southern California, occurring after the Blancan and before the Rancholabrean.

IS *Genetics.* a designation for various insertion sequences, such as IS*1*, IS*10*, and IS*50*.

I.S. intercostal space.

IS1 *Genetics.* an insertion sequence that occurs, for example, in R plasmids and in the chromosomes of *E. coli* K12; encodes two genes, *insA* and *insB*, which are required for IS*1*-mediated transposition and cointegration.

IS5 *Genetics.* an insertion sequence that occurs in *E. coli* K12 strains and has a high target specificity.

IS101 *Genetics.* a defective insertion sequence that carries no genes.

IS1000 *Genetics.* an insertion sequence that shares extensive DNA homology with Tn3.

ISA or **I.S.A.** Instrument Society of America.

Isaac Judaeus 855–955, Egyptian-Jewish physician; wrote a noted early medical text.

isabnormal *Meteorology.* a line on a weather chart drawn through points having an equal deviation from the normal value of a particular meteorological quantity.

is-a link *Artificial Intelligence.* in a frame system or semantic network, a link between an instance and its class (e.g., Fido is-a Dog) or between a class and its superclass (e.g., Dog is-a Mammal). Also, AKO, ISA.

isallobar *Meteorology.* a line on a weather chart connecting points of equal change of atmospheric pressure within a specified time period; commonly used on synoptic surface charts to depict three-hourly local pressure tendencies.

isallobaric *Meteorology.* of or relating to equal or constant pressure change, referring specifically to either the distribution of equal pressure tendency in space or the constancy of pressure tendency with time.

isallobaric wind *Meteorology.* a wind velocity in which the Coriolis force component is in exact balance with an associated accelerating geostrophic wind. Also, BRUNT-DOUGLAS ISALLOBARIC WIND.

isallohypsic wind *Meteorology.* the equivalent of an isallobaric wind, using height tendency in a constant-pressure surface rather than pressure tendency in a constant-height surface.

isallotherm *Meteorology.* a line on a weather chart connecting points of equal change of temperature within a specified time period.

ISAM indexed sequential access method.

isanabat *Meteorology.* a line on a weather chart connecting points of equal vertical component of wind velocity; positive values represent upward motions while negative values represent downward motions.

isanakatabar *Meteorology.* a line on a weather chart connecting points of equal atmospheric-pressure range over a specified time period.

isandrous *Botany.* of a flower, having stamens that are alike and are of a quantity equal to the number of petals.

isanomal *Meteorology.* a line on a weather chart that is drawn through points having an equal anomaly of a particular meteorological quantity.

isanthous *Botany.* bearing symmetric flowers.

isarithm see ISOPLETH, def. 2.

Isastraea *Paleontology.* a genus of scleractinian corals in the suborder Fungiina; extant in Jurassic and Cretaceous.

isatin *Organic Chemistry.* $C_6H_5O_2N$, yellowish-red crystals that are soluble in water, alcohol, and ether and melt at 203.5°C; used in the production of dyestuffs and pharmaceuticals.
isaurore see ISOCHASM.
ISC or **I.S.C.** International Society of Chemotherapy.
I scan see I DISPLAY.
ischemia [i skē´mē ə] *Medicine.* a condition of deficiency of oxygenation of a body part, caused by an obstruction in or the constriction of a blood vessel. Thus, **ischemic** [i skē´mik; i skem´ik].
ischemic cardiomyopathy *Cardiology.* a diminished supply of blood to the myocardium, usually resulting from coronary artery disease.
ischemic heart *Cardiology.* an inadequate supply of blood to the heart, mainly due to the narrowing of blood vessels.
ischemic necrosis *Medicine.* death of tissue that is associated with occlusion of the blood supply to the affected area.
ischemic neuropathy *Medicine.* a disorder of the nervous system associated with a reduction of blood supply to a nerve, as in atherosclerosis, diabetes, ligation of an artery, or improper use of a tourniquet.
ischemic paralysis *Medicine.* the loss of muscle function or of sensation, associated with a reduction of the blood supply to the affected area due to occlusion of the artery supplying that area.
ischial *Anatomy.* relating to the ischium.
ischialgia *Medicine.* any pain in the ischium, the inferior dorsal part of the hipbone. Also, **ischiodynia.**
ischio- a combining form meaning "os ischium" or "hip," as in *ischiopubic, ischiocapsular.*
ischiococcygeal *Anatomy.* relating to the ischium and the coccyx.
ischiofemoral *Anatomy.* relating to the ischium and the femur.
ischiopodite *Invertebrate Zoology.* the segment of a walking leg nearest the basipodite in certain crustaceans.
ischiopubic *Anatomy.* relating to the ischium and pubic region.
ischiorectal region *Anatomy.* the space between the ischium of the pelvic girdle and the rectum.
ischium [is´kē əm] *plural,* **ischia.** *Anatomy.* the posteroinferior portion of the os coxa, distinct in the fetus and fusing with the other pelvic bones during childhood.
Ischnacanthidae *Paleontology.* a family of light-armored acanthodian spiny sharks in the extinct order Ischnacanthiformes; extant in Late Silurian to Carboniferous.
Ischnacanthiformes *Paleontology.* a single-family order of acanthodian fishes of the Paleozoic.
Ischnochitonida *Invertebrate Zoology.* an order containing nine families of generally small, cosmopolitan chitons with ovate or elongated bodies; range from littoral to deep seas.
ischo- or **isch-** a combining form denoting suppression or deficiency.
ischuria *Medicine.* the suppression or retention of urine. Thus, **ischuretic.**
I scope *Electronics.* a radarscope that produces an I display.
ISCP or **I.S.C.P.** International Society of Comparative Pathology.
ISDN integrated services digital network.
Isectolophidae *Paleontology.* a family of perissodactyl tapiroids belonging to the suborder Ceratomorpha; descended, along with the Asian family Helaletidae, from the Homagalax, an early tapiroid of Asia and North America. The **isectolophids,** which are known only from the Eocene of North America, are also closely related to the ancestral horse, Hyracotherium.
isenthalpic *Thermodynamics.* not involving a change in enthalpy. Thus, **isenthalpic compression, isenthalpic expansion.**
isentrope *Thermodynamics.* a constant entropy line on any diagram.
isentropic *Thermodynamics.* relating to or having a constant value of entropy. Thus, **isentropic compression, isentropic expansion.**
isentropic chart *Meteorology.* a constant-entropy chart showing the distribution of meteorological elements on a surface of constant potential temperature, along with plotted data and analyses of pressure, wind, air temperature, and moisture.
isentropic condensation level see LIFTING CONDENSATION LEVEL.
isentropic flow *Thermodynamics.* the motion of a fluid that experiences no change in entropy in any portion as it flows.
isentropic map *Geology.* a map that depicts the constant-entropy function for different facies.
isentropic mixing *Meteorology.* any atmospheric mixing process occurring within an isentropic surface.
isentropic process *Thermodynamics.* an adiabatic and reversible thermodynamic process in which there is no change in the entropy of the system.
isentropic surface *Meteorology.* a surface in space of equal potential temperature; a constant-entropy surface.
isentropic thickness chart *Meteorology.* a chart depicting the thickness of an atmospheric layer that is bounded by two selected isentropic surfaces; the thickness of such a layer is directly proportional to the static instability within it. Also, THICK-THIN CHART.
isentropic weight chart *Meteorology.* a chart depicting the atmospheric pressure difference between two selected isentropic surfaces; the greater the pressure difference, the greater the weight of the air column that separates the two surfaces.
isethionic acid *Organic Chemistry.* $HOCH_2CH_2SO_3H$, a syrupy liquid that is very soluble in water; used, often with oleic acid, to produce detergents.
ISGE or **I.S.G.E.** International Society of Gastro-Enterology.
ISH or **I.S.H.** International Society of Hematology.
Ishigeaceae *Botany.* a monogeneric family of marine brown algae of the order Chordariales, characterized by dichotomously branched thalli with compressed to flattened axes that have a very firm consistency.
ishikawaite *Mineralogy.* a black, orthorhombic mineral with an approximate formula of $(U,Fe,Y,Ca)(Nb,Ta)O_4$, occurring in tabular crystals, and having a specific gravity of 6.2 to 6.4 and a hardness of 5 to 6 on the Mohs scale; found in pegmatites in Japan.
ISI or **I.S.I.** International Standards Institute.
isidium *Botany.* an outgrowth on the surface of the thallus of certain lichens that resembles a soredium. Thus, **isidiate.**
Ising coupling *Solid-State Physics.* a binary alloy model in which it is assumed that the two sets of constituent atoms are arranged in a regular array; the spin components of each atom along a particular axis are taken to be +1 or –1, and the interaction energy is proportional to the negative of the product of the spin components along this axis.
isinglass *Materials.* a transparent or translucent gelatinous substance derived from the swim bladders of certain fish, such as the sturgeon; the purest form of animal gelatin. It has strong adhesive properties, and is used in inks, glues, and cements. *Mineralogy.* mica, especially muscovite in thin transparent sheets.
isitizin see 1,8-DIHYDROXYANTHRAQUINONE.
island *Geography.* a piece of land that is smaller than a continent and is entirely surrounded by water. *Anatomy.* a cluster of cells; an isolated piece of tissue. Also, ISLET.
island arc *Geography.* a chain of usually volcanic islands arranged in an arc; especially common in the Pacific, most lie near the continental masses but are not a part of the continents proper.
island biogeography *Ecology.* the branch of the biological sciences that studies the distribution and abundance of species on islands.
island disease see SCRUB TYPHUS.
island fever see TSUTSUGAMUSHI DISEASE.
island flap *Surgery.* a skin flap consisting of skin and subcutaneous tissue elevated from the body but still attached by nutrient vessels and sometimes nerves.
islandic acid *Biochemistry.* a water-soluble glucan, with malonic acid residues, formed by *Penicillium islandicum.*
island mountain see INSELBERG.
island of automation *Robotics.* a stand-alone robotic system or device that works independently of any other machine or process.
island of Reil *Anatomy.* the triangular central lobe of the cerebral cortex, located in the floor of the lateral fissure. Also, the INSULA.
island structure *Materials Science.* any of various small, discrete, three-dimensional structural units that coalesce to form layers.
islet *Geography.* a small island. *Anatomy.* see ISLAND.
islet-cell carcinoma *Medicine.* a malignant cellular tumor with its origin in the pancreatic islands or islets of Langerhans.
islet-cell tumor *Medicine.* a tumor with its origin in the pancreatic islands or islets of Langerhans.
islets of Langerhans see PANCREATIC ISLETS.
ISM or **I.S.M.** International Society of Microbiologists.
ISO International Standards Organization.
iso- *Science.* a prefix meaning "equal," as in *isobaric. Chemistry.* a combining form indicating the presence of an isomeric compound, or a branched carbon chain in a molecule.
isoacceptor tRNA *Molecular Biology.* any of two or more transfer ribonucleic acids that are able to receive the same amino acid, but have different primary sequences. Also, tRNA ISOACCEPTOR.
isoacetyl thioglycolate *Organic Chemistry.* $HSCH_2COOCH_2C_7H_{15}$, a colorless liquid used as an insecticide and plasticizer, and in oils and antioxidants.

isoagglutinin *Immunology.* an antibody that is capable of clumping together the red blood cells of the same species in which it is found.
isoalkane *Organic Chemistry.* a branched-chain alkane having a methyl group bonded to its next-to-last carbon atom.
isoallele *Genetics.* an allele that is so similar to another allele that a special test is required to distinguish the two.
isoallohypse see HEIGHT-CHANGE LINE.
isoamyl acetate *Organic Chemistry.* $CH_3COOCH_2CH_2CH(CH_3)_2$, a colorless liquid having a bananalike odor; slightly soluble in water and miscible with alcohol and ether; boils at 142°C and freezes at –78.5°C; used in flavoring, perfumery, and solvents.
isoamyl alcohol *Organic Chemistry.* $(CH_3)_2CHCH_2CH_2OH$, a pungent, combustible, colorless liquid; slightly soluble in water and miscible with alcohol and ether; boils at 132°C (secondary form boils at 113°C) and freezes at –117.2°C; used as a solvent and in organic synthesis. Also, ISOBUTYL CARBINOL.
isoamylase *Enzymology.* an enzyme that catalyzes the conversion of starch into sugar.
isoamyl benzoate *Organic Chemistry.* $C_6H_5COOC_5H_{11}$, a combustible, fruity-smelling, colorless liquid; soluble in alcohol and insoluble in water; boils 262.3°C; used in perfumery and cosmetics.
isoamyl bromide *Organic Chemistry.* $(CH_3)_2CHCH_2CH_2Br$, a colorless liquid that is soluble in water and miscible with alcohol and ether; boils at 120–121°C; used in organic synthesis.
isoamyl butyrate *Organic Chemistry.* $C_5H_{11}OOCC_3H_7$, a combustible water-white liquid that is soluble in alcohol and ether and slightly soluble in water; boils at 179°C; used in flavoring extracts.
isoamyl chloride *Organic Chemistry.* $(CH_3)_2CHCH_2CH_2Cl$, a colorless liquid that is soluble in alcohol and slightly soluble in water; boils at 99.7°C; used in solvents, soil fumigants, and chemical intermediates.
isoamyl nitrite see AMYL NITRITE.
isoamyl valerate *Organic Chemistry.* $C_4H_9CO_2C_5H_{11}$, a clear, combustible liquid that smells of apples when diluted with alcohol; slightly soluble in water and soluble in alcohol and ether; boils at 203.7°C; used in fruit essences and as a flavoring agent. Also, APPLE OIL, APPLE ESSENCE.
isoantibody *Immunology.* an antibody that is capable of reacting with an antigen derived from a member of the same species.
isoantigen *Immunology.* an antigen that is capable of producing an immune response in genetically different individuals of the same species. Also, ALLOANTIGEN.
isobar *Meteorology.* a line on a map or chart connecting points of equal or constant pressure. *Fluid Mechanics.* a line or curve that represents a locus of equal pressure points in a fluid. *Atomic Physics.* any of a group of atomic species having the same mass number but different chemical or physical properties.
isobaric *Science.* being or relating to an isobar.
isobaric analog states see ANALOG STATES.
isobaric chart see CONSTANT-PRESSURE CHART.
isobaric divergence *Meteorology.* the horizontal divergence in a constant-pressure surface, expressed in a coordinate system with pressure represented as an independent variable.
isobaric map *Meteorology.* a map depicting points of equal barometric pressure in the atmosphere.
isobaric process *Thermodynamics.* a thermodynamic process that occurs at constant pressure.
isobaric spin see ISOSPIN.
isobaric surface see CONSTANT-PRESSURE SURFACE.
isobaric topography see HEIGHT PATTERN.
isobaric vorticity *Meteorology.* a representation of relative vorticity on a constant-pressure surface.
isobath *Oceanography.* a line that connects points of equal depth on a chart or diagram.
isobathytherm *Oceanography.* a map or chart line that connects points at which the temperature is the same at the same depth.
isobiochore *Ecology.* a boundary line on a map that connects regions in the world that have similar plants and animals.
isobornyl acetate *Organic Chemistry.* $C_{10}H_{17}OOCCH_3$, a colorless, combustible, pine-scented liquid; insoluble in water and soluble in mineral oil and many fixed oils; boils at 220–224°C; used as an odorant and flavoring agent.
isobornyl thiocyanoacetate *Organic Chemistry.* $C_{10}H_{17}OOCCH_2SCN$, a combustible, oily, yellow liquid with a terpenelike odor; practically insoluble in water and soluble in alcohol, benzene, and ether; boils at 95°C under reduced pressure; used as an insecticide.
isobront *Meteorology.* a line on a weather chart connecting points at which there is a simultaneous occurrence of a particular phase of thunderstorm activity. Also, HOMOBRONT.
Isobryales *Botany.* an order of mosses of the division Bryophyta, characterized by their matlike growth and by prostrate stems that are profusely to irregularly branched and frondose.
isobutane *Organic Chemistry.* $(CH_3)_2CHCH_3$, a stable, highly flammable, colorless gas; soluble in water and slightly soluble in alcohol; boils at –11.73°C and freezes at –159°C; an important constituent of natural gas; used in motor fuels, propellants, and refrigerants. Also, 2-METHYLPROPANE.
isobutene see ISOBUTYLENE.
isobutyl *Organic Chemistry.* the radical $(CH_3)_2CHCH_2^-$ from isobutane.
isobutyl acetate *Organic Chemistry.* $C_4H_9OOCCH_3$, a flammable, colorless liquid; soluble in alcohols and ether; boils at 117°C and freezes at –99°C; used as a solvent and in perfumery, flavorings, and surface protectants.
isobutyl alcohol *Organic Chemistry.* $(CH_3)_2CHCH_2OH$, a flammable, colorless liquid; soluble in water and miscible with alcohol and ether; boils at 108.1°C and freezes at –108°C; used in paints, solvents, and organic synthesis.
isobutyl carbinol see ISOAMYL ALCOHOL.
isobutylene *Organic Chemistry.* $(CH_3)_2C{=}CH_2$, a flammable, colorless volatile liquid or easily liquified gas; insoluble in water and soluble in alcohol; boils at –6.9°C and freezes at –139°C; used as a chemical intermediate, in gasolines, and in making butyl rubber. Also, ISOBUTENE, 2-METHYLPROPENE.
isobutyl isobutyrate *Organic Chemistry.* $(CH_3)_2CHCOOCH_2CH(CH_3)_2$, a colorless, combustible liquid; insoluble in water and miscible with alcohol; boils at 148.7°C and freezes at –80.7°C; used in lacquers, thinners, insect repellants, and flavorings.
isobutyric acid *Organic Chemistry.* $(CH_3)_2CHCO^-$, a toxic, colorless liquid; soluble in water and miscible with alcohol; freezes at –47°C, boils at 154.7°C, and is volatile with steam; its free acid or esters are present in many plants; used as a disinfecting and tanning agent and in solvents, flavorings, and perfumery.
isobutyryl *Organic Chemistry.* the radical $(CH_3)_2CHCO–$ from isobutyric acid.
isocaloric [ī´sō kə lôr´ik] *Nutrition.* of or relating to an equal amount of calories.
isocarb *Geochemistry.* on a coal-deposit map or diagram, a contour line connecting points of equal fixed-carbon content.
isocarboxazid *Pharmacology.* $C_{12}H_{13}N_3O_2$, a monoamine oxidase inhibitor occurring as a white or off-white crystalline powder; administered orally as an antidepressant.
isocarpic *Botany.* having a number of carpels that is equal to the number of other flower parts.
isocenter *Photogrammetry.* **1.** in aerial photography, the single point common to the actual plane of a photograph, its principal plane, and the plane of a hypothetical photograph taken from the same camera station, with the same principal distance, and assumed to be truly vertical. **2.** the point of intersection of the principal line and the isometric parallel on a photograph. **3.** the point on a photograph intersected by the bisector of the angle between the plumb line and the photograph perpendicular. *Radiology.* the point at which there is a maximum or minimum of the radiation dose.
isocenter triangulation *Cartography.* a method of aerotriangulation in which the isocenters of overlapping photographs are used as the radial centers for extending horizontal control.
isoceraunic *Meteorology.* having or representing equal frequency or intensity of thunderstorm activity. Also, **isokeraunic.**
isoceraunic line *Meteorology.* a line connecting points of equal frequency or intensity of a particular thunderstorm phenomenon, such as lightning discharges.
isocercal *Vertebrate Zoology.* of or relating to a type of fish tail in which the diminished terminal vertebrae end in the midline of the caudal fin.
isocetyl laurate *Organic Chemistry.* $C_{11}H_{23}COOC_{16}H_{33}$, an oily, combustible liquid; insoluble in water and soluble in most organic solvents; freezes at –65°C; used as a solvent and lubricant in cosmetics, textiles, and pharmaceuticals.
isochasm *Geophysics.* a line connecting points on the earth's surface at which the aurora can be seen with equal frequency. Also, ISAURORE.
isocheim *Meteorology.* a line on a weather map connecting points having the same mean winter temperature. Thus, **isocheimal.**

isochela *Invertebrate Zoology.* **1.** a chelate sponge spicule with both ends identical. **2.** a chela having two equally developed parts.

isochemical metamorphism *Petrology.* a metamorphism with minimal change in chemical composition; a theoretical process that can only be approached. Also, TREPTOMORPHISM.

isochemical series *Petrology.* a group of rocks with identical bulk-chemical composition, but with different textures or mineral contents resulting from differing degrees of metamorphism.

isochor *Physics.* a curve that graphs temperature against pressure for a constant volume of a given substance. Also, **isochore.**

isochoric *Physics.* relating to or having a constant volume. Thus, **isochoric process.**

isochromatic *Optics.* having the same color, as in the lines of interference figures formed by biaxial crystals.

isochromatic fringe pattern *Optics.* **1.** a pattern that contains lines of the same color. **2.** a figure that reveals the relative strains present in various areas of a birefringent substance; used in photoelastic stress analysis.

isochromosome *Cell Biology.* a metacentric chromosome containing two arms of equal length and identical sequence, proposed to arise during mitosis or meiosis from the transverse separation of a centromere rather than the usual longitudinal division.

isochronal test *Petroleum Engineering.* a short-time, back-pressure test for reservoirs with low permeability, which normally need extremely lengthy time periods for pressure stabilization when the wells are closed in.

isochrone *Mathematics.* the property of an inverted cycloid that a particle sliding along the curve without friction will arrive at a minimum point in the same amount of time, no matter what the starting point.

isochronic *Physics.* not varying in time; occurring at the same time or at a regular time.

isochronism *Physics.* **1.** the occurrence of isochronic events or stages of processes. **2.** the quality of being isochronous.

isochronous *Physics.* relating to processes that have a fixed period of temporal oscillation.

isochronous circuits *Electricity.* circuits that have equivalent resonant frequencies.

isochronous communications *Telecommunications.* the synchronization of a communications network from timing signals within the network.

isochronous governor *Mechanical Engineering.* a governor that maintains a constant speed, regardless of load. Also, ASTATIC GOVERNOR.

isochroous *Optics.* having the same color in every part.

Isochrysidaceae *Botany.* a family of mostly marine flagellate algae in the order Isochrysidales, characterized by rudimentary haptonema with few microtubules and a scaly cell body.

Isochrysidales *Botany.* an order of mostly marine flagellate algae of the class Prymnesiophyceae, characterized by the lack of a regular haptonema in the flagellate state and by body scales on the cells of most species.

isocitrate dehydrogenase *Enzymology.* an enzyme that catalyzes the oxidation of isocitric acid to α-ketoglutarates acid during the citric-acid cycle.

isocitric acid *Biochemistry.* a citric-acid isomer that is common in nature, especially in the stone-crop family (Crassulaceae) and in some fruits; a major component in the citric-acid cycle, which causes the oxidation of proteins, fats, and carbohydrates.

isoclasite *Mineralogy.* $Ca_2(PO_4)(OH) \cdot 2H_2O$, a colorless or white, monoclinic, dubious mineral occurring in small prismatic crystals or fibrous forms, having a specific gravity of 2.92 and a hardness of 1.5 on the Mohs scale; found with chalcedony and dolomite.

isoclinal *Science.* inclining in the same direction; of or relating to an equal direction of inclination for various items. *Geology.* of or relating to an isocline. *Cartography.* of or relating to an isoclinic line or chart. Also, **isoclinic.**

isocline *Geology.* an anticline or syncline closely folded so that the rock beds of the two limbs have the same dip.

isoclinic chart *Cartography.* a chart that has as its main feature a system of isoclinic lines, each representing a different value or degree of magnetic inclination. Also, **isoclinal chart.**

isoclinic line *Cartography.* a line on a map or chart drawn through all points on the earth's surface that have the same magnetic inclination. Also, **isoclinal line.**

isoconcentration *Chemical Engineering.* concentration values that remain constant.

isoconcentration map *Chemical Engineering.* a diagram that shows the concentration of a liquid or gas system in relation to a single system component of the system, expressed by constant-concentration contour lines.

isocracking *Chemical Engineering.* a procedure similar to hydrocracking, used to convert hydrocarbons into lower-boiling products, which operates at relatively low pressures and temperatures in the presence of a catalyst and hydrogen.

Isocrinida *Invertebrate Zoology.* an order of crinoids that are stalked throughout life.

isocyanate *Organic Chemistry.* any compound that contains the –N=C=O radical.

isocyanic acid *Organic Chemistry.* HN=C=O, a poisonous, explosive gas produced by the depolymerization of cyanic acid in a stream of CO_2; used as an intermediate in the production of urethanes and allophanates.

isocyanide *Organic Chemistry.* any compound of type R–N=C=O.

isocyanine *Organic Chemistry.* any of several dyes having two heterocyclic or quinoline rings connected by a chain of carbon atoms that have conjugated double bonds, such as cyanine blue; used as photographic sensitizers.

isocyclic *Organic Chemistry.* a ring structure containing atoms of only one element.

isodecyl chloride *Organic Chemistry.* $C_{10}H_{21}Cl$, an isomeric, colorless, combustible liquid; insoluble in water; boils at 210.6°C and freezes at –180°C; used as a solvent, extractant, and intermediate.

isodiametric *Mathematics.* having equal diameters or axes. *Cell Biology.* having similar diameters, as a cell whose length, width, and height are equal or nearly equal.

isodiapheres *Nuclear Physics.* a group of nuclides, each of which has the same ratio of neutrons to protons.

isodif *Cartography.* a line on a map or chart that connects all the points that have received equal amounts of correction, or that differ equally from a datum; used especially to map readjustment surveys from one datum to another. An isodif connecting points of equal latitude correction is an **isolat**; an isodif connecting points of equal longitude correction is an **isolong.**

isodont *Vertebrate Zoology.* **1.** possessing all teeth of the same size and shape. **2.** of a snake, having maxillary teeth of the same size.

isodose *Radiology.* a radiation dose of equal intensity to more than one body area.

isodose curve *Nucleonics.* a line on a chart or map that connects points at which an object absorbed equal doses of radiation.

isodrosotherm *Meteorology.* a line on a weather chart connecting locations having equal dew points.

isodynamic *Mechanics.* **1.** describing equal forces or forces that remain constant with time. **2.** describing an imaginary surface on which force is everywhere the same.

isodynamic line *Geophysics.* a line connecting points at which the earth's magnetic field has the same strength.

isoeffect *Radiology.* an effect centered between two reference points.

isoeffect lines *Radiology.* a graph of lines indicating those radiation doses giving equivalent biological effects.

isoelectric *Chemistry.* having a net electrical charge of zero. *Electricity.* at the same electric potential.

isoelectric focusing *Physical Chemistry.* a technique used to separate proteins in a gel or liquid medium by subjecting the medium to an electric field, so that the proteins migrate until they reach their respective isoelectric points; they then have no net charge and are relatively immobile. Also, ELECTROFOCUSING.

isoelectric point *Physical Chemistry.* **1.** a condition in which a substance has a neutral charge; the pH level above which the substance would act as a base and below which it would act as an acid. A solution of proteins or amino acids has its minimum conductivity and viscosity at the isoelectric point and thus coagulates best at this point. **2.** this pH level established as a value for a given substance; e.g., the isoelectric point of gelatin is pH 4.7.

isoelectric precipitation *Physical Chemistry.* the precipitation of a substance at the pH point where the molecule's net charge is zero, the optimal point for the coagulation of proteins.

isoelectronic *Atomic Physics.* of or relating to atoms that have the same number of electrons outside the nucleus.

isoelectronic sequence *Spectroscopy.* a series of spectra produced by different elements that have been ionized so that they have the same electronic configuration.

isoenzyme *Enzymology.* any of various structurally related forms of the same enzyme, having the same mechanism but differing from each other in chemical or immunological characteristics. Also, ISOZYME, ISOMER.

Isoetaceae *Botany.* a family of primitive, aquatic, vascular herbs that produce two types of spores: megaspores at the outer leaves and abundant microspores at the inner leaves.

Isoetales *Botany.* a monogeneric order of the class Isoetopsida, characterized by long linear leaves and a corm.

Isoetopsida *Botany.* a class of heterosporous, ligule-bearing plants of the division Lycopidiophyta.

isoeugenol *Organic Chemistry.* $C_{10}H_{12}O_2$, a pale-yellow oily liquid that has a clovelike odor; slightly soluble in water and miscible with alcohol; melts at –10°C; transform melts at 33°C and boils at 267.5°C; used in the production of perfumes and vanillin.

isofloridoside *Biochemistry.* a glycoside that serves to regulate osmotic pressure and acts as a storage unit in organisms, such as *Poterioochromonas malhamensis*.

isofluorphate see DIISOPROPYL PHOSPHOROFLUORIDATE.

isoflurane *Pharmacology.* a potent inhalational anesthetic, used for induction and maintenance of general anesthesia.

isoforming *Chemical Engineering.* a petroleum refinery process in which olefinic naphtha contacts an alumina catalyst at low pressure and high temperature to form high yields of xylene isomers with low hydrogen consumption and minimal catalyst regeneration.

isofronts-preiso code *Meteorology.* a modification of the international analysis code, in which data on sea- or surface-level isobars and fronts are encoded and transmitted.

isogal *Geophysics.* a line connecting points at which the Earth's gravitational field has the same strength.

isogamete *Biology.* one of a group of gametes which cannot be distinguished from each other.

isogamy *Biology.* **1.** the fusion of gametes of similar size, shape, and behavior. **2.** the production of gametes that are similar in size and morphologically indistinguishable, which occurs uncommonly in some algae, fungi, and protozoa.

isogeneic *Genetics.* **1.** having identical genotypes. **2.** of a tissue graft, involving a genetically identical host and donor. Also, **isogenic, isologous.**

Isogeneratae *Botany.* a class of brown algae whose members undergo an isomorphic alternation of generations.

isogloss *Linguistics.* the geographical range of a given word, pronunciation, or usage.

isogonic *Geophysics.* having equal magnetic variation. Also, **isogonal.**

isogonic chart *Cartography.* a chart characterized by a system of isogonic lines, each representing a different value or degree of magnetic declination.

isogonic line *Cartography.* a line on a map or chart drawn through all points on the earth's surface that have the same magnetic variation or declination.

isogony *Biology.* a property of hybrids that show equal expression of parental characters.

isogor *Petroleum Engineering.* a constant proportion of gas to oil.

isogor map *Petroleum Engineering.* a contour-line map of an oil reservoir depicting steadfast gas-oil ratios.

isograd *Geology.* a line on a map joining points at which metamorphism occurred under similar conditions of pressure and temperature, thus indicating metamorphic zones of equal grade.

isogradient *Meteorology.* a line on a weather chart connecting points having the same horizontal gradient of a meteorological quantity, such as temperature or pressure.

isograft *Genetics.* a tissue graft involving a genetically identical host and donor. Also, SYNGRAFT.

isogriv *Cartography.* a line on a map or chart drawn through all points on the earth's surface that have equal grivation. *Navigation.* in grid navigation, a line connecting points of equal grid variation.

isogriv chart *Cartography.* a chart characterized by a system of isogriv lines, each representing a different value or degree of grivation.

isogyre *Optics.* any of the dark, brushlike bands appearing in the interference figure of a biaxial crystal, corresponding to the directions of transmission through the crystal in which the polarization of light passing through the crystal remains unaffected.

isohaline *Oceanography.* **1.** of equal or constant salinity. **2.** a chart or map line that connects points of equal salinity.

isoheight see CONTOUR LINE.

isohel *Meteorology.* a line on a weather chart connecting points that receive equal amounts of sunshine.

isohemagglutinin *Immunology.* an antibody that reacts with an antigen present on red blood cells of different individuals of the same species.

isohemolysin *Immunology.* a substance that destroys red blood cells and is produced by an individual that has been administered red blood cells from an individual of the same species.

isohemolysis *Immunology.* a process in which red blood cells are destroyed by the interaction of isohemolysin and a specific antigen.

isoheptane see 2-METHYLHEXANE.

isohexane *Organic Chemistry.* C_6H_{14}, an isomeric, highly flammable colorless liquid; boils at 60.3°C; used to depress the freezing points of other compounds.

isohume *Geology.* a line on a map connecting points of equal moisture content in a coal bed.

isohydric *Chemistry.* of a set of solutions, having the same hydrogen ion concentration and, upon mixing, maintaining the conductivity of each solution.

isohyet *Meteorology.* a line on a weather chart connecting points that receive equal amounts of rainfall at a given time.

isohypse see CONTOUR LINE.

isohypsic chart see CONSTANT-HEIGHT CHART.

isohypsic surface see CONSTANT-HEIGHT SURFACE.

isoimmunization *Immunology.* the production of an immune response in an individual, achieved through the injection of an antigen from an individual of the same species.

isokinetic *Mechanics.* of movement or exertion, occurring at a constant speed. Thus, **isokinetic exercise.**

isolabeling *Molecular Biology.* the process of labeling both chromatids or homologous parts of daughter chromatids.

isolable *Chemistry.* of a substance, capable of being isolated. Also, **isolatable.**

Isolaimioidea *Invertebrate Zoology.* a superfamily of soil nematodes frequently found in orchards.

isolate [ī´sə lāt] to separate or set apart; specific uses include: *Chemistry.* to obtain a substance in a pure or uncombined state. *Chemical Engineering.* to separate a production unit or a part of a process by closing valves or installing line blanks to block flow so the isolated portion can be removed or repaired. *Electricity.* **1.** to set apart an electrical or electromagnetic component, circuit, or system, as from a source of electrical energy. **2.** to insulate or to shield. *Microbiology.* to obtain a pure culture of a microorganism. Thus, **isolated.**

isolate [is´ə lit] something that is set apart; specific uses include: *Microbiology.* an isolated culture or population. *Archaeology.* a single object that is found without an association to any other artifact or feature; typically lost during travel or moved by a relic hunter. *Psychology.* a person who avoids the company of others.

isolated camera *Electronics.* **1.** a type of camera that tapes a specific area of action for replay later, such as those used at sporting events for instant replay. **2.** the technique, used to provide video replay, that relies on an isolated camera.

isolated footing *Civil Engineering.* a structural footing supporting only a concentrated load, such as a column load. Also, INDEPENDENT FOOTING.

isolated location *Computer Programming.* the location of a block of storage in such a way that its contents are protected from access by a user's program or accidental corruption.

isolated point *Mathematics.* In a topological space X, an element p of a set S is an isolated point of S if there exists an open set of X that contains p but no other points of S. Also, ACNODE.

isolated set *Mathematics.* a set whose elements are all isolated points; equivalently, any set that contains none of its accumulation points. Also, DISCRETE SET.

isolated singularity *Mathematics.* A (complex) function f has an isolated singularity at a point $z = a$ if there is a positive number R such that f is analytic on $\{z : 0 < |z - a| < R\}$, i.e., on the interior of the disk of radius R, centered at a with the point a removed. Called an **isolated discontinuity** if f is a real function.

isolated system *Thermodynamics.* a system that has no interaction is not influenced in any way by interaction with its surroundings; i.e., it has rigid, impermeable, adiabatic boundaries, and no heat or work can cross those boundaries.

isolated vertex *Mathematics.* a vertex of a graph that has no edge incident on it.

isolating diode *Electronics.* a device used in an integrated circuit to pass signals coming from one direction and to block those coming from another. Also, **isolation diode.**

isolating switch *Electricity.* a switch that isolates an electric circuit from the source of power.

isolation a process or instance of isolating; specific uses include: *Medicine.* **1.** the separation of a patient from others, to avoid the spread of a contagious disease or to protect the patient from irritating environmental factors. **2.** in experimental medicine, the separation of a part for study, as by tissue culture or by interposition of inert material. *Microbiology.* the successive reproduction of microorganisms for the purpose of obtaining a pure culture. *Psychology.* **1.** a process by which the emotions attached to an unpleasant memory are stripped away; for example, when a patient, in therapy, is able to recall an attempt to harm his father without affect. **2.** the act of keeping an individual away from all external stimuli or social activities, sometimes used as a treatment for a person who is experiencing a mental disorder. *Evolution.* a condition in which interbreeding between two populations is impossible due to premating or postmating mechanisms. *Chemistry.* the process of separating a pure chemical from a natural source or mixture of chemicals, typically by precipitation or distillation. *Computer Technology.* the compartmentalization of information within a computer in such a way as to limit access to the information.

isolation amplifier *Electronics.* a circuit that increases electrical power passing through a system in order to minimize interaction between circuits that precede and follow each other. Also, BUFFER AMPLIFIER.

isolation booth *Acoustical Engineering.* a small enclosed chamber constructed of or lined with sound-absorbing material, designed to isolate a single person within from noise outside, as in recording an individual narrator or performer.

isolation experiment see EXPERIENCE DEPRIVATION.

isolation network *Electricity.* a network placed in a circuit or transmission line in order to minimize interaction between circuits on either side of the insertion point.

isolation transformer *Electricity.* a multiple-winding type transformer, with physically separated primary and secondary windings so arranged to prevent the primary circuit potential from being impressed on the secondary circuit.

isolator *Medicine.* something serving to isolate a patient from others.

isolead curve *Ordnance.* a curved line on a chart or diagram that indicates the correct lead for a projectile to hit a moving target.

isolecithal see OLIGOLECITHAL.

isoleucine *Biochemistry.* $C_6H_{13}NO_2$, one of the amino acids found in proteins, considered essential for animal growth.

isolichenin *Biochemistry.* a glucan that is present in some lichens such as Icelandic moss, *Cetraria islandica.*

isoline *Photogrammetry.* a line that represents the intersection of the plane of a vertical photograph with the plane of an oblique photograph that overlaps it.

isolith *Geology.* **1.** an imaginary line joining points of similar lithology, and excluding rocks having different characteristics, such as color, composition, or texture. **2.** an imaginary line that connects points at which the thickness of a particular class of material within a formation or other stratigraphic interval is equal. *Electronics.* an integrated circuit on a single silicon chip, characterized by components that are connected with beam leads and isolated from each other when the silicon between them is removed.

isolith map *Geology.* a map that indicates variations in aggregate thickness of a given lithologic facies as measured perpendicular to the bedding at selected points.

isoln. isolation.

isologous *Genetics.* referring to a substance that is of identical gene construction to another substance.

isolux *Optics.* a curve, line, or surface whose points all have equal light intensity. Also, **isophote.**

isomagnetic *Geophysics.* **1.** of equal magnetic force. **2.** sharing any magnetic property.

isomagnetic chart *Cartography.* any chart that represents the earth's magnetic field using a system of isogonic, isoclinic, or isodynamic lines.

isomagnetic line *Geophysics.* a line connecting points at which a magnetic property such as strength, declination, or inclination remains the same.

isomaltose *Biochemistry.* an isomer of maltose that is derived from partial hydrolysates of glycogen and amylopectin.

isomer [īs′ə mər] *Chemistry.* one of two or more substances that have the same chemical composition but differ in structural form. *Enzymology.* see ISOENZYME. *Nuclear Physics.* any of two or more nuclei having the same atomic number *A* and mass number *Z,* but different half-lives.

isomerase *Enzymology.* any of a class of enzymes that convert the substances being acted upon into their isomers.

isomeric *Science.* of or relating to an isomer; relating to or exhibiting isomerism.

isomeric transition *Nuclear Physics.* an isomer or isomers of metastable states that decay to one or more lower-lying states, emitting energy quanta in the form of gamma rays, or internal-conversion electrons from the various atomic shells.

isomerism *Chemistry.* a condition in which two or more chemical compounds possess the same number of atoms of the same elements, and thus have the same molecular formula, but differ in the arrangement of the atoms, and thus have different chemical properties. Isomerism can be described in terms of difference in structure or difference in spatial configuration. *Nuclear Physics.* the existence of excited states in two or more isomers.

isomerization *Chemistry.* a process that changes a substance into an isomer, such as butane into isobutane.

isomerous *Biology.* in flowers, having equal numbers of parts in each whorl.

isometric [īs′ə met′rik] of equal measure; specific uses include: *Physiology.* of or related to a muscle contraction that produces tension, consumes nutrients, and generates heat, but does not produce movement; for example, pushing against the wall of a building. Thus, **isometric contraction, isometric exercise.** *Thermodynamics.* relating to or having a constant volume. *Crystallography.* having three equal axes at right angles to one another; of or relating to a cubic crystal.

isometric labile ringspot viruses see ILARVIRUS GROUP.

isometric particle *Virology.* a symmetrical virus particle with identical linear proportions.

isometric process *Thermodynamics.* a thermodynamic process that occurs with constant volume.

isometric system see CUBIC CRYSTAL.

isometropia *Medicine.* a condition of equality in the refraction of the two eyes. Thus, **isometropic.**

isometry [ī säm′ə trē] the fact of being equal in dimension; specific uses include: *Biology.* body growth in which the proportionate size of body parts remains constant. Also, **isometric growth.** *Geography.* equality of elevation above sea level. *Mathematics.* a transformation that preserves distance; a diffeomorphism $f: X \to Y$ (where X and Y are metric spaces with metrics μ and λ, respectively) that leaves the metric invariant; i.e., $\mu(a,b) = \lambda(f(a), f(b))$, for all a and b in X. The isometries of X form a group (of transformations on X). When $X = Y$ (and $\mu = \lambda$), isometries other than the identity diffeomorphism may exist. These isometries also form a group. In the Euclidean plane, isometries are reflections, rotations, translations, and glide-reflections.

isomorph *Science.* **1.** something that is isomorphic to something else. **2.** two or more things that are isomorphic to each other.

isomorphic *Science.* **1.** identical or similar in shape, structure, or appearance. **2.** relating to or characterized by isomorphism.

isomorphism *Physical Chemistry.* a condition in which two or more chemical compounds have essentially the same crystalline structure; such substances are so closely similar that they can generally crystallize together to form a continuous series of solid solutions. *Mathematics.* **1.** a homomorphism between algebraic structures A and B of the same type (e.g., groups, rings, etc.) that is both one-to-one and onto. If such an isomorphism exists, it is said that A is isomorphic to B and written $A \cong B$. **2.** in more general category theory terminology, an isomorphism is a morphism $f \in \mathrm{Hom}(X,Y)$ for which there exists $g \in \mathrm{Hom}(X,Y)$ such that $f \cdot g$ is the identity morphism on the object Y, and $g \cdot f$ is the identity morphism on the object X. **3.** in graph theory, an isomorphism between graphs G and H is a bijection f from $V(G) \cup E(G)$ to $V(H) \cup E(H)$ such that $f(V(G)) = V(H)$, $f(E(G)) = E(H)$, if edge $e = uv$ in G, then $f(e) = f(u)f(v)$, and if edge $e = (u,v)$ in G, then $f(e) = (f(u), f(v))$.

isomorphous *Crystallography.* relating to or characterized by isomorphism.

isomorphous replacement method *Crystallography.* a method of deriving phases from measurements of the intensities of the Bragg reflections in the X-ray diffraction pattern from two or more isomorphous crystals.

isoneph *Meteorology.* a line on a weather chart connecting points that have the same degree of cloudiness.

isoniazid *Pharmacology.* $C_6H_7N_3O$, a drug occurring as colorless or white crystals or as a white crystalline powder, administered orally or intermuscularly to inhibit growth of the tuberculosis bacterium.

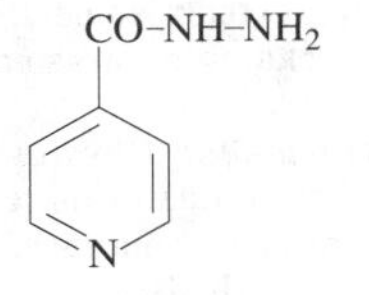

isoniazid

isooctane *Organic Chemistry.* C_8H_{18}, a flammable colorless liquid; insoluble in water and soluble in alcohol; boils at 99.3°C and freezes at –107.4°C; used as a standard fuel in gasoline performance tests.

isooctyl alcohol *Organic Chemistry.* any isomer or mixture of isomers having the formula $C_7H_{15}CH_2OH$, with eight carbon atoms forming a branched chain; a clear, combustible liquid that boils at 182–195°C; used as a chemical intermediate, emulsifier, and resin solvent. Also, **isooctanol.**

isoosmotic *Ecology.* having an osmotic potential equal to that of the surrounding medium.

isopach *Geology.* a line on a map that connects points of equal thickness within a particular stratigraphic unit or group of units. Also, **isopachous line, isopachyte.**

isopach map *Geology.* a map that depicts the varying true thickness of a rock bed or formation throughout a geographical region. Also, **isopachous map.**

isopathic principle *Psychology.* a phenomenon in which the cause of an individual's problem cures it, such as when a patient suffering from guilt over hating his brother overcomes his guilt by reliving his hate.

isopathy *Psychology.* treatment based on the isopathic principle. *Medicine.* a method of treating disease by administering products of the disease or portions of the affected organ.

isopectic *Meteorology.* a line on a weather map connecting points where ice formation begins at approximately the same time at the onset of winter.

isopentane *Organic Chemistry.* $(CH_3)_2CHCH_2CH_3$, a colorless, highly flammable liquid that is derived from petroleum; insoluble in water and soluble in ether; boils at 27.95°C and freezes at –159.6°C; used as a solvent.

isopentanoic acid *Organic Chemistry.* C_4H_9COOH, an isomeric, pungent, colorless to water-white liquid; slightly soluble in water; boils at 183.2°C and freezes at –44°C; used as an intermediate and extractant.

isoperimetric curve *Cartography.* a line on a map or chart along which there is no variation from exact scale.

isoperimetric figure *Mathematics.* a plane geometric figure whose perimeter has the same length as that of another. Also, **isoperimer.**

isoperimetric inequality *Mathematics.* suppose a plane region has boundary (or perimeter) of length l and area A; the isoperimetric inequality states that $A \le l^2/4\pi$, with equality holding if and only if the region is a circle.

isoperimetric problem *Mathematics.* in the calculus of variations, the problem maximizing (or minimizing) the integral of one function while keeping the integral of a second function constant; for example, finding the curve which encloses the greatest area within a boundary of fixed length. By the isoperimetric inequality, the solution is a circle.

isoperm *Petroleum Engineering.* **1.** a line on a reservoir map marking an area of constant permeability. **2.** an area so delineated.

isophene *Biology.* **1.** a line on a chart or map connecting points having the same frequency of occurrence of a given phenotype or variant. **2.** a line on a chart or map connecting points at which seasonal events occur at the same date.

isophorone *Organic Chemistry.* $C_9H_{14}O$, a water-white, combustible liquid; slightly soluble in water; boils at 215.2°C and freezes at –8.1°C; used in solvent mixtures.

isophotometer *Optics.* a photometer that can automatically scan all points of the image on a photographic negative, measuring and recording the optical density of those points.

isophthalic acid *Organic Chemistry.* $C_6H_4(COOH)_2$, combustible colorless crystals; slightly soluble in water and soluble in alcohol; melts at 348°C and sublimes without forming an anhydride; used in the production of polyesters, high polymers, and plasticizers.

isophyllous *Botany.* producing leaves of only one type, alike in size and shape.

isopiestic *Physics.* relating to processes that occur at equal pressures or at constant pressures.

isopiestic line *Hydrology.* a contour line indicating the potentiometric surface of an aquifer. Also, EQUIPOTENTIAL LINE.

isopleth *Meteorology.* **1.** a line on a weather chart representing constant or equal value for a given meteorological quantity with respect to either space or time. Also, **isogram. 2.** a line drawn through points on a graph at which a quantity has the same numerical value or frequency as a function of the two coordinate variables. Also, ISARITHM. *Mathematics.* the straight line that passes through corresponding values on the scales of a nomograph. Isopleths on a circular nomograph are diameters. Also, INDEX LINE.

isopluvial *Meteorology.* a line on a weather chart connecting geographical points having the same pluvial index.

isopod *Invertebrate Zoology.* any member of the order Isopoda.

Isopoda *Invertebrate Zoology.* an order of malacostracan crustaceans with a flattened body, no carapace, and seven pairs of legs; includes the sowbugs, pillbugs, and wood lice.

isopolymolybdate *Inorganic Chemistry.* a group of polymeric compounds made by acidifying a molybdate solution or by heating molybdates. Also, POLYMERIC MOLYBDATE.

isopor *Geophysics.* a line connecting regions whose magnetic properties change at an equal rate. *Cartography.* a line on a map or chart drawn through all points on the earth's surface that have equal annual magnetic change. Also, **isoporic line.**

isoporic chart *Cartography.* a chart that has as its main feature a system of isopors, each representing a different value or degree of annual magnetic change.

isopotential map *Petroleum Engineering.* a contour-line map that depicts the original or computed daily rate of oil well production in a field with numerous wells.

isoprecipitin *Immunology.* an antibody that joins with and precipitates soluble antigenic material in the cells or blood of individuals of the same species from which it originated.

isoprene *Organic Chemistry.* $CH_2=C(CH_3)CH=CH_2$, a colorless volatile liquid; soluble in alcohol and insoluble in water; boils at 34.1°C and freezes at –146°C; used in the production of polyisoprene and elastomers.

isopropanol see ISOPROPYL ALCOHOL.

isopropanolamine *Organic Chemistry.* C_3H_9NO, a combustible liquid that has a mild ammonia odor; soluble in water; melts at 1.4°C and boils at 159°C; a product of the decomposition of vitamin B_{12}; used as an emulsifying agent.

isopropenyl acetate *Organic Chemistry.* $CH_3OCOC(CH_3)=CH_2$, a flammable, water-white liquid; slightly soluble in water; boils at 97.4°C and freezes at –92.9°C; used as a solvent for paints and inks.

isopropyl *Organic Chemistry.* the radical $(CH_3)_2CH–$.

isopropyl acetate *Organic Chemistry.* $CH_3COOCH(CH_3)_2$, a colorless, aromatic liquid; soluble in water and miscible with alcohol; boils at 88.9°C and freezes at –73.4°C; used as a solvent and in perfumery.

isopropyl alcohol *Organic Chemistry.* C_3H_8O, a clear, colorless liquid; miscible with water, alcohol, ether, and chloroform; an isomer of propyl alcohol and homologue of ethyl alcohol; used as a solvent and as a basis for isopropyl rubbing alcohol. Also, ISOPROPANOL, 2-PROPANOL.

isopropylamine *Organic Chemistry.* $(CH_3)_2CHNH_2$, a colorless volatile liquid that boils at 32.4°C, freezes at –101°C, and is miscible with water, alcohol, and ether; used as a solvent and intermediate.

isopropyl-*m*-cresol see THYMOL.

isopropyl ether *Organic Chemistry.* $(CH_3)_2CHOCH(CH_3)_2$, a colorless volatile liquid that boils at 67.5°C and freezes at –60°C; used as a solvent for oils, dyes, and paint removers.

7-isopropyl-1-methyl phenanthrene see RETENE.

isopropyl rubbing alcohol *Organic Chemistry.* a preparation of 68–72% isopropyl alcohol in water; used chiefly for massaging and cleaning the skin.

isoproterenol hydrochloride *Pharmacology.* $C_{11}H_{17}NO_3{\cdot}HCl$, an adrenergic drug used to dilate the airways and stimulate the heart in treatment of bronchial asthma, shock, irregular heart rhythm, and stoppage of the heart, and of spasm of the airways during anesthesia.

Isoptera *Entomology.* a primitive order of social, ground-dwelling, herbivorous insects characterized by lack of true larval and pupal stages, two pairs of subequal wings, biting mouthparts, and the abdomen joined broadly to the thorax.

isoptic *Mathematics.* the locus of intersection of tangents to a curve C meeting at constant angle a is an isoptic of C.

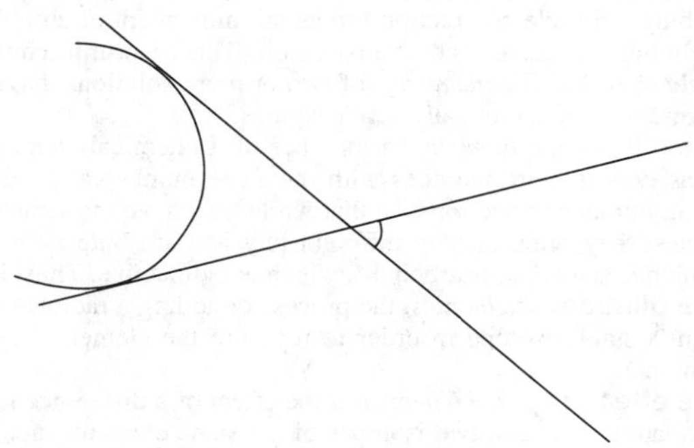

isoptic

isopulegol *Organic Chemistry.* $C_{10}H_{17}OH$, a mint-scented, combustible, water-white liquid that boils at 100°C (at 18 mm pressure); obtained from the oil of *Mentha pulegium* and used in perfumery and flavorings.

isopulse system *Telecommunications.* a pulse coding system in which the number of transmitted information pulses is delineated by special inserted pulses.

isopycnic *Meteorology.* of air, having equal or constant density, with respect to either space or time. Also, ISOSTERIC. *Cartography.* a line on a map or chart drawn through all points of equal density, especially of ocean water. Also, ISOSTERE. *Molecular Biology.* describing molecules that have the same buoyant density in ultracentrifugation.

isopycnic level *Meteorology.* a level surface in the atmosphere at an altitude of approximately 8 km, at which the air density is virtually constant in both time and space.

isopycnotic *Molecular Biology.* describing chromosomes or chromosomal regions that are not highly condensed and that stain similarly.

isopygous *Invertebrate Zoology.* possessing a pygium and a cephalon of equal sizes.

isoquinoline *Organic Chemistry.* C_9H_7ON, combustible colorless plates that melt at 24.6°C, boil at 243°C, and are soluble in most organic solvents; derived from coal tar or made synthetically, and used in the manufacture of dyes, insecticides, and pharmaceuticals, and as a chemical intermediate.

isoradial *Cartography.* in aerotriangulation, the isocenter of a photograph used as a radial for extending horizontal control.

isorhiza see GLUTINANT.

isosafrole *Organic Chemistry.* $C_{10}H_{10}O_2$, a colorless, aromatic liquid that boils at 253°C, is volatile with steam, and is soluble in alcohol, ether, and benzene; derived from safrole and used in flavorings and perfumery.

isosceles [ī säs´ə lēz] *Mathematics.* having two sides the same length. For example, an **isosceles triangle** has two sides of the same length, and the nonparallel sides of an **isosceles trapezoid** are the same length.

isoschisomer or **isochisomere** *Enzymology.* one of two or more restriction endonucleases that are isolated from different sources but that will cleave a DNA molecule at the identical recognition site. Also, **isoschizomer, isoschizomere.**

isoseismal *Geophysics.* a line connecting points at which the intensity of an earthquake was felt equally.

isoshear *Meteorology.* a line on a weather chart representing equal vertical wind shear magnitude.

isospin *Particle Physics.* a quantum number representing a fictitious spin vector, conserved in all nuclear reactions, in which charge independence of hadronic forces treats neutrons and protons as two states of one particle without electromagnetic interaction. Also, ISOBARIC SPIN, ISOTOPIC SPIN.

isospin multiplet *Particle Physics.* a collection or family of elementary particles having identical quantum numbers and closely similar masses, the members distinguished essentially only by their electric charges.

Isospondyli *Vertebrate Zoology.* an order of bony fish that includes salmon and herrings.

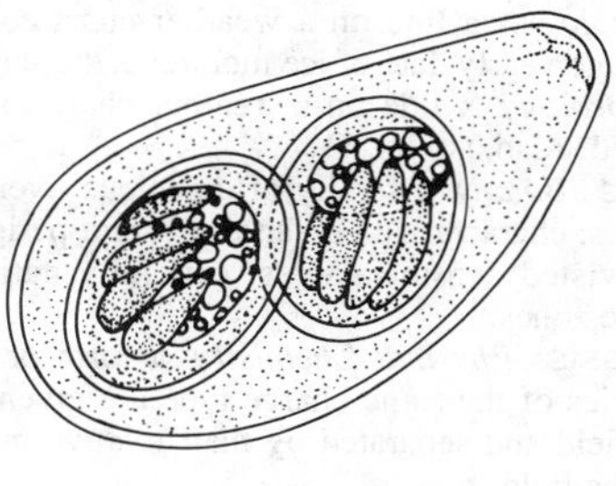

Isospora

Isospora *Invertebrate Zoology.* a genus of protozoans in the subclass Coccidia, infesting primarily fish, and rarely humans, usually without severe results.

isospore *Biology.* an asexual spore produced in only one kind.

isostasy *Geophysics.* the process by which crustal rocks float in gravitational equilibrium on the viscous mantle rocks beneath; a condition of approximate gravitational equilibrium in the earth's crust, in which the gravitational effect of continental masses extending above the surface of the geoid is counterbalanced by a lack of density in the material beneath those masses, and the gravitational effect of a lack of density in ocean waters is counterbalanced by an excess of density in the material under the oceans. Also, **isostacy.**

isostatic *Geophysics.* of, relating to, or characterized by isostasy. *Mechanics.* an imaginary line in an elastic body that is tangent to the principal directions of stress.

isostatic anomaly *Geophysics.* a condition in which elevated parts of the earth's crust are gravitationally compensated for by low-density material extending to greater-than-average depths.

isostatic compaction *Materials Science.* a ceramic-forming method in which a powdered material contained in a rubber envelope is compacted into a desired shape by pressure from an isostatic fluid.

isostatic compensation *Geology.* a process by which the earth's lithosphere adjusts to maintain a balance among units of different mass and density by compensating for an excess of mass above with a lowering of densities below, and vice versa. Also, **isostatic adjustment, isostatic correction.**

isostatic pressing *Materials Science.* a process used to shape ceramic components by encasing powder in a rubber or plastic bag and subjecting it to hydrostatic pressure. Also, HYDROSTATIC PRESSING.

isostatic surface *Mechanics.* an imaginary surface in an elastic body that is everywhere perpendicular to one of the principal directions of stress (tangent to a principal plane of stress).

isostemonous *Botany.* of a flower, having a number of stamens that is equal to the number of sepals or petals.

isostere *Cartography.* see ISOPYCNIC.

isosteric *Physics.* having equal or constant specific volume with respect to either time or space. *Chemistry.* of two or more compounds, similar in electronic arrangement. *Meteorology.* see ISOPYCNIC.

isosterism *Physical Chemistry.* a similarity in the physical properties of ions, compounds, or elements because of their identical or similar electronic arrangements.

isostrain condition *Materials Science.* a condition in loading on a composite sample in which the bonding between the layers of continuous fibers and matrix material remains intact during the stressing; the stress on the material causes uniform strain on all the composite layers.

isostrain deformation *Materials Science.* an equal deformation of both matrix and fibers of a composite, caused by loading the composite parallel to its fiber alignment.

isostress condition *Materials Science.* a condition occurring in an idealized lamellar composite structure consisting of layers of fiber and matrix in which the layers are perpendicular to the applied stress; the stress on the composite structure produces an equal stress condition on all the layers.

isostress deformation *Materials Science.* in a composite, a deformation in which the matrix and fibers experience identical stresses, caused by loading the composite perpendicular to its fiber alignment.

isostructural *Crystallography.* of or relating to compounds that crystallize with the same arrangement of atoms.

isosynthesis *Organic Chemistry.* the process of producing branched hydrocarbons by thorium oxide-catalyzed reaction of hydrogen with carbon monoxide.

isotac *Meteorology.* a line on a weather chart connecting points at which the average yearly date of ice melting is the same.

isotach *Meteorology.* a line on a weather chart connecting points of equal windspeed. Also, ISOVEL.

Isotachidaceae *Botany.* a family of dioecious liverworts of the order Jungermanniales, characterized by an erect stem perigynium and by linear, spirally twisted capsule valves; generally native to tropical and south-temperate zones.

isotachophoresis *Physical Chemistry.* a form of electrophoresis in which ion species of the same charge type are given the same velocity in an electric field and separated by their relative mobility as they migrate through the field.

isotactic polymer *Materials Science.* a vinyl polymer that has a special type of regular arrangement of its side groups, all on one side of the chain when the chain is fully extended.

isothere *Meteorology.* a line on a weather map connecting points having the same mean summer temperature. Thus, **isotheral.**

isotherm *Meteorology.* a line on a weather map connecting points having equal or constant temperature.

isothermal *Thermodynamics.* relating to or having a constant temperature. Thus, **isothermal compression, isothermal expansion.**

isothermal annealing *Metallurgy.* a thermal treatment for ferrous materials, consisting of austenitizing followed by a transformation to a soft ferrite and carbide mixture.

isothermal atmosphere *Meteorology.* an atmosphere in hydrostatic equilibrium, in which the temperature is constant with height and the barometric pressure decreases exponentially with an increase in altitude. Also, EXPONENTIAL ATMOSPHERE.

isothermal calorimeter *Thermodynamics.* a calorimeter that is maintained at a constant temperature by a heat reservoir of a large body of liquid held at its melting point or boiling point.

isothermal equilibrium *Thermodynamics.* a state in which two bodies or systems have the same temperature, so that no heat transfer takes place between them. *Meteorology.* a condition of uniform air temperature, resulting from the conduction of heat from one area to another in an atmosphere at rest, without any external influence, over a sufficient length of time. Also, CONDUCTIVE EQUILIBRIUM.

isothermal flow *Thermodynamics.* the motion of a fluid that experiences no change in temperature in any portion as it flows.

isothermal layer *Thermodynamics.* a fluid layer in which all points in the layer are at the same temperature. *Meteorology.* an isothermal atmospheric region directly above the tropopause.

isothermal magnetization *Thermodynamics.* a process in which a substance is magnetized while held at a constant temperature; the process is followed by adiabatic demagnetization so as to produce temperatures close to absolute zero.

isothermal process *Thermodynamics.* a thermodynamic process in which the temperature of the system is maintained at a constant value throughout the process.

isothermal theory *Astrophysics.* a theory postulating that galaxies grew out of small isothermal density fluctuations in the early universe.

isothermal transformation *Metallurgy.* a structural transformation that occurs at constant temperature.

isothermal treatment *Metallurgy.* any thermal treatment performed at constant temperature.

isothermobath *Oceanography.* in a diagram of a vertical plane in the ocean, a line that connects points at which the temperature is the same.

isotherm ribbon *Meteorology.* an area of crowded isotherms on an upper-level synoptic weather chart, representing an atmospheric condition characterized by an unusually high temperature gradient.

isothiocyanate *Organic Chemistry.* any compound containing the radical –NCS.

isotimic *Meteorology.* having an equal value in space at a particular time.

isotimic line *Meteorology.* a line on a given reference surface in space connecting points of equal value of some quantity; used in the analysis of synoptic charts.

isotimic surface *Meteorology.* a surface in space on which the value of a given quantity is equal at all points.

isotin β-thiosemicarbazone see THIOSEMICARBAZONE.

isotone *Nuclear Physics.* any of two or more nuclides with the same neutron number N that have a constant difference, $A - Z$, between their mass number A and their atomic number Z. *Mathematics.* an operator or function T is said to be isotone if, for u and v in the domain of T, $u \leq v$ implies $Tu \leq Tv$.

isotonia *Physiology.* a condition of equal or constant tone, tension, or activity. *Biochemistry.* the quality or condition of being isotonic. Also, **isotonicity.**

isotonic *Mechanics.* of a force or a system of forces, remaining constant with time. *Physiology.* **1.** of or relating to isotonia. **2.** specifically, describing a muscle contraction producing movement of an object, thus maintaining a relatively constant tension. Thus, **isotonic contraction, isotonic exercise.** *Biochemistry.* of two or more solutions, having equal concentrations of osmotically active solutes.

isotope *Physical Chemistry.* a member of a chemical-element family that has two or more nuclides with the same number of protons but a different number of neutrons, so that while they have the same chemical attributes, they often display different physical attributes; e.g., carbon-12, which is stable, and carbon-14, which is radioactive. Thus, **isotopic.**

isotope dilution *Nucleonics.* the process of adding a radioisotope to an element's stable isotope in order to measure the element's age, mass, and volume.

isotope effect *Physical Chemistry.* the effect of a difference in mass or mass quantity between two isotopes of the same element, such as a difference in reaction rate, vapor pressure, or equilibrium constant. The term includes effects on molecular or atomic properties; specific nuclear effects such as radioactivity are excluded.

isotope exchange reaction *Physical Chemistry.* a reaction wherein atoms that will interchange with other atoms are isotopically labeled, enabling them to be traced through the reaction.

isotope farm *Botany.* a closed greenhouse system into which isotope-containing nutrients are introduced for biochemical labeling of plants.

isotope fractionation *Nucleonics.* the process of changing the isotopic composition of an element by means of diffusion, evaporation, or chemical exchange using minute differences in the physical and chemical properties of isotopes; it may occur naturally or be induced artificially.

isotope geochemistry see RADIOGEOLOGY.

isotope geology see RADIOGEOLOGY.

isotope separation *Nucleonics.* a process in which different isotopes are separated from or added to a given element, generally by gaseous diffusion or electromagnetic radiation.

isotope shift *Spectroscopy.* a displacement or splitting of spectral lines indicating the presence of isotopes of an element.

isotopic abundance *Physical Chemistry.* the ratio of the number of atoms of a specific isotope to the total number of atoms in an element as found in nature; usually expressed as a percentage. Also, RELATIVE ABUNDANCE.

isotopic chronometer *Nucleonics.* a technique that measures the number of a given isotope and its daughter in a material to determine its age; used to date geological or archeological samples.

isotopic element *Physical Chemistry.* an element with more than one naturally occurring isotope.

isotopic enrichment *Nucleonics.* any process that alters the relative abundance of isotopes in a given element, producing a form of the element that is enriched with one isotope and depleted of another.

isotopic irradiation *Nucleonics.* the process by which a specimen, such as a tumor, is subjected to ionizing radiation, generally for therapeutic purposes.

isotopic molecule *Nucleonics.* a molecule that has a nucleus with a special isotope.

isotopic parity see G PARITY.

isotopic spin see ISOSPIN.

isotopic tracer *Physical Chemistry.* an isotopic element, usually radioactive but sometimes stable, that is incorporated into a system undergoing change to trace the course of the changes.

isotopy *Physical Chemistry.* the fact or condition of being an isotope.

isotoxin see HOMEOTOXIN.

Isotricha or **Isotrichia** *Invertebrate Zoology.* a genus of ciliate protozoans in the subclass Holotrichia that live in cattle and sheep; includes some medically significant species and some commensals.

isotropic *Science.* having physical properties that do not vary with direction. Thus, **isotropic material, isotropic fluid, isotropic radiation, isotropic turbulence, isotropic displacement.**

isotropic dielectric *Electricity.* a dielectric whose polarization always has a direction that is parallel to the applied electric field, and a magnitude that is independent of the electric field direction.

isotropic fabric *Petrology.* a rock configuration characterized by random orientation of the elements.

isotropic flux *Physics.* a flow of any quantity that has equal intensity in all directions.

isotropic noise *Electromagnetism.* random electromagnetic noise that is received from all directions at equal intensity.

isotropic universe *Astrophysics.* a universe having observed properties that appear identical in every direction.

isotropy *Science.* the quality or condition of being isotropic. Also, **isotropism.**

isotropy group *Mathematics.* let G be a group acting on a set S. For each element x of S, the subgroup $G_x = \{g \in G : gx = x\}$ is called the isotropy (sub)group of x.

isotype *Immunology.* any of a range of varying forms of antigenic determinants that occur in every normal individual in a given species. *Biology.* any of two or more distinct populations of the same or similar type. *Crystallography.* either of two crystalline compounds having analogous chemical compositions and atomic structures. Thus, **isotypic, isotypical.**

isovalent conjugation *Physical Chemistry.* a structure found in compounds with alternating double and single bonds, from which an alternative structure can be constructed, such as in benzene.

isovalent hyperconjugation *Physical Chemistry.* a structure found in two compounds with alternating double and single bonds, where both compounds have the name number of bonds, but one structure is more active chemically.

isovaleraldehyde *Organic Chemistry.* $C_5H_{10}O$, a sweet-smelling liquid aldehyde that boils at 92.5°C; present in orange, bergamot, lemon, sandalwood, citronella, peppermint, eucalyptus, and other oils.

isovaleric acid *Organic Chemistry.* $C_5H_{10}O_2$, a liquid that has an offensive odor similar to moldy cheese; boils at 176.5°C and freezes at –30°C; present in tobacco and hop oil.

isovel see ISOTACH.

isozyme see ISOENZYME.

ISR or **isr** interrupt service routine; intersecting storage rings.

Issatchenkia *Mycology.* a genus of yeast fungi belonging to the family Saccharomycetaceae which occurs in fruit and soil and reproduces by multilateral budding.

issue *Medicine.* **1.** a discharge of pus, blood, or other material. **2.** a suppurating lesion emitting such a discharge.

IST *Aviation.* the airport code for Istanbul, Turkey.

IST insulin shock therapy.

isthmospasm *Neurology.* a spasm of an isthmus of a body part, as of the fallopian tubes.

isthmus *Geography.* a narrow strip of land that connects two larger land masses and is bordered on two sides by water. *Anatomy.* **1.** a usually narrow part or passage connecting larger structures or cavities. **2.** specifically, a sharp constriction in the brain separating the mesencephalon from the rhombencephalon. *Vertebrate Zoology.* a narrow fleshy area between the sides of a fish's lower jaw. *Mathematics.* a cut edge. Also, BRIDGE.

Istiophoridae *Vertebrate Zoology.* a family of bony fish in the suborder Scombroidei, containing the billfishes, marlins, sailfishes, and spearfishes, which usually live near the surface in open seas and are swift-moving predators.

ISTM or **I.S.T.M.** International Society for Testing Materials.

ISU or **I.S.U.** International Society of Urology.

Isuridae *Vertebrate Zoology.* an equivalent name for Lamnidae, a family of mackerel sharks of the order Lamniformes.

ISV International Scientific Vocabulary.

ITA or **I.T.A.** International Tuberculosis Association.

itabirite *Geology.* a metamorphosed iron formation in which the original bands of jasper or chert have been recrystallized into visible quartz grains, and the iron occurs in thin layers of hematite, magnetite, or martite. Also, BANDED-QUARTZ HEMATITE.

itacolumite *Petrology.* a fine-grained sandstone or schistose quartzite containing mica, chlorite, and talc in interstitial, loosely interlocking grains; it is characterized by flexibility when split into thin slabs.

itaconic acid *Organic Chemistry.* $HO_2CCH_2C(=CH_2)CO_2H$, a solid acid that melts at 166–167°C; prepared by fermentation with *Aspergillus terreus.*

itai-itai disease *Toxicology.* poisoning due to the ingestion of shellfish or rice contaminated with cadmium; symptoms may include pain, anemia, and kidney and bone disorders.

italic *Graphic Arts.* a variation of a given typeface that slants to the right, *like this.*

itatartaric acid *Organic Chemistry.* $C_5H_8O_6$, a compound formed in small amounts (5.8% of the total acidity) by the itaconic-acid producing *Aspergillus niger.*

itch *Physiology.* **1.** a mildly irritating sensation on the skin that stimulates the desire to scratch; may be caused by a skin disorder, an insect bite, or a parasite, or, in the absence of external stimulation, by a mild chemical stimulator acting on nerve endings for pain. **2.** to feel such a sensation. Thus, **itching, itchy.**

-ite a combining form denoting: **1.** a mineral or rock. **2.** a part of a body or organ. **3.** a salt or ester of an acid whose name ends with *-ous.*

iter *Anatomy.* a tubular passage; a canal.

iterated integral *Mathematics.* if f is a real-valued function integrable over a subset D of R^n, then by Fubini's theorem, the integral of f over D can be calculated as an n-fold iterated integral; i.e., as successive integrals over line segments. An iterated integral over a plane region is also called a **double integral,** and an iterated integral over a region in R^3 is called a **triple integral.**

iterated kernel *Mathematics.* Let $K(x, y)$ be a given kernel of the Fredholm integral equation; i.e., $\lambda\int_a^b K(x,y)\phi(y)\,dy + f(x) = \phi(x)$, where $\phi(x)$ is to be determined. An iterated kernel K_n is a function that is defined inductively by $K_1(x, y) = K(x, y)$ and $K_n(x, y) = \int_a^b K(x, t)K_{n-1}(t, y)\,dt$. The **resolvent kernel** is $\Gamma(x, y, \lambda) = \sum_{n=1}^{\infty} \lambda^{n-1}K_n(x, y)$.

iteration *Mathematics.* see ITERATIVE METHOD. *Computer Programming.* a repetition of a sequence of events or instructions in a routine.

iterative adaptation *Evolution.* a process in which an adaptive type arises at a number of successive points in a lineage.

iterative array *Computer Programming.* a model of an abstract computing device consisting of a network of finite-state machines.

iterative deepening *Artificial Intelligence.* a search method in which a depth-first search is done with a fixed depth bound; if no solution is found, the depth bound is increased by one and the process is repeated.

iterative method *Mathematics.* a general term for numerical approximation by repeated application of a given approximation method. Frequently used iterative methods are often known by the names of their originators; e.g., the Picard-Lindelöf iteration method and Newton's method. Also, ITERATION, ITERATIVE PROCESS.

iterative process *Mathematics.* see ITERATIVE METHOD. *Computer Programming.* a method for achieving specific results by repeating a cycle of operations.

iterative routine *Computer Programming.* a looping program construct that repeats one or more computer operations.

iteron *Genetics.* any of a number of repeated, or reiterated, nucleotide sequences that occur in the area of the replication origins of certain plasmids.

Itersonilia *Mycology.* a genus of fungi of the class Hyphomycetes having mycelium that contain clamp connections; causes such plant diseases as leaf spot and canker of parsnip.

Ithomiinae *Invertebrate Zoology.* the glossy-winged butterflies, a subfamily of medium-sized lepidopteran insects with transparent patches on their wings.

ITP idiopathic thrombocytopenic purpura.

i-type semiconductor see INTRINSIC SEMICONDUCTOR.

I.U. immunizing unit; international unit.

IUCD intrauterine contraceptive device.

IUD intrauterine device.

IUPAC or **I.U.P.A.C.** International Union of Pure and Applied Chemistry.

i.v. initial velocity.

I.V. intravenous, intravenously.

Ivanov reagent *Organic Chemistry.* a reagent formed when an arylacetic acid (or its sodium salt) reacts with isopropyl magnesium halide; similar to a Grignard reagent.

IVC intravenous cholangiogram.

ivory *Materials.* **1.** a hard white dentin that forms the tusks of elephants, walruses, and other large mammals and yellows with age; used for ornaments, art objects, and piano keys. **2.** of, relating to, or resembling this material.

ivory black *Materials.* a fine black pigment that is obtained by calcining ivory.

ivory board *Materials.* a finished cardboard made by starch-pasting two or more high-quality papers together; used for printing.

IVP intravenous pyelogram.

ivs intervening sequence.

IW or **i.w.** inside width; isotopic weight.

Ixodidae *Invertebrate Zoology.* a family of hard-bodied ticks in the order Acarina, bloodsucking parasites on vertebrates.

Izod test *Metallurgy.* one of several standardized tests used to measure the impact resistance of a metallic material.

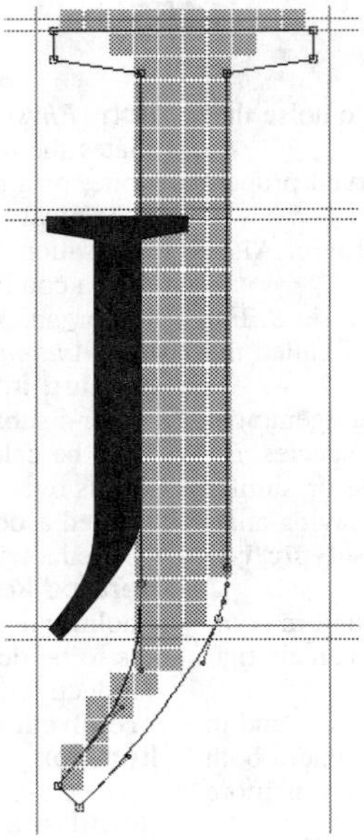

J or **j** the symbol for joule.

jaagsiekte [yäg sēk´tə] *Veterinary Medicine.* a contagious pulmonary adenomatosis of adult sheep, and sometimes pigs and goats, characterized by emaciation, loss of lung tissue, and respiratory distress. Also, **jaagzietke.**

jabiru [jab´i roo] *Vertebrate Zoology.* **1.** a large stork, *Jabiru mycteria,* of the family Ciconiidae that ranges from Mexico to Argentina, having a dark blue naked head, a red neck, a white body, and a long, heavy, and slightly upturned bill. **2.** any of various other birds of the same family.

jabiru

jaçana or **jacana** [zhä kä´nə] *Vertebrate Zoology.* any bird that is a member of the family Jacanidae, especially those of the genus *Jaçana,* such as the American jaçana, *J. spinosa.* Also, LILY TROTTER.

Jacanidae *Vertebrate Zoology.* a family of colorful tropical shorebirds of the order Charadriiformes, specialized for walking on lily pads and other vegetation and distinguished by extremely long toes.

Jacanoidea *Vertebrate Zoology.* the suborder of tropical shorebirds including the jaçanas.

jacaranda [jak´ə rän´də] *Botany.* **1.** any of various tropical American trees belonging to the genus *Jacaranda,* of the family Bignoniaceae, having clusters of bright, usually purple flowers, such as the **fern-tree jacaranda,** *J. ovalifolia.* **2.** any of various similar or related trees, including the carob wood tree.

jacinth see ZIRCON.

jack *Electronics.* a socket into which a plug can be inserted for circuit connections or switching functions. *Mechanical Engineering.* any of various mechanical, hydraulic, or pneumatic devices used to lift heavy objects through a short height. *Vertebrate Zoology.* any of various fishes of the genus *Caranx,* of the western Atlantic Ocean. (Of uncertain origin; thought to derive from the use of the man's name *Jack* as a nickname for various tools.)

jackal *Vertebrate Zoology.* any of several species of Old World wild dogs of the genus *Canis* that hunt in packs and eat carrion.

jackal

jackass *Vertebrate Zoology.* a common name for a male donkey.

jackbit *Mechanical Devices.* a drill bit of a rock drill used to blast short holes in rock cutting. Also, RIP-BIT.

jack chain *Mechanical Devices.* **1.** a chain made of light wire whose loops are set at 90° angles to each other and twisted or bent into a figure eight. **2.** a looped, toothed chain used for dragging lumber, usually from a pond to a sawmill.

jackdaw *Vertebrate Zoology.* a European bird belonging to the crow family, Corvidae, especially the common jackdaw, *Corvus monedula.*

jacket *Textiles.* a short coat worn as an outer garment. *Medicine.* a similar therapeutic or confining garment worn on the torso. *Graphic Arts.* a protective and decorative paper wrapping that is placed around the cover of a hardbound book. Also, DUSTCOVER, DUSTJACKET. *Ordnance.* **1.** the outer casing of a lead bullet, usually made of copper or steel. **2.** a steel cylinder designed to cover and reinforce the breech end of a gun or howitzer. *Mechanical Engineering.* a casing or envelope formed around the cylinders of an engine or an air compressor that allows water or coolant to flow. *Nucleonics.* see CLADDING, def. 1. (Thought to derive from the French word for "peasant," because they often wore such a garment.)

jackhammer

jackhammer *Mechanical Engineering.* a pneumatically operated impact hammer for breaking or drilling rock, pavement, or other solid materials.

Jackiellaceae *Botany.* a monogeneric family of prostrate liverworts belonging to the order Jungermanniales, characterized by brownish pigmentation, ventral intercalary branches, and rhizoids restricted to underleaf bases or swellings; found in tropical and subtropical Asia and in Australia

jack-in-the-pulpit *Botany.* a North American wild flower, *Arisaema triphyllum,* that is found widely in moist wooded areas, having a hooded spathe that extends up and over a clublike spadix with tiny greenish-yellow flowers. (So called because it gives the appearance of a preacher speaking from a pulpit.)

jackknife *Mechanical Devices.* a pocket knife with a clasp.

jackknife clam *Invertebrate Zoology.* another name for the razor clam.

jackknife position *Medicine.* an anatomic postion in which the patient is placed on the back in a semisitting position, with the shoulders raised and the thighs flexed at right angles to the abdomen.

jackknife seizure *Neurology.* a sudden, severe muscle contraction that occurs during infancy, caused by deterioration of the cerebrum and characterized by severe spasms of the neck and trunk and extension of the arms and legs. Also, INFANTILE MASSIVE SPASMS.

jack ladder *Forestry.* an endless toothed chain that is held by a V-shaped trough and that is used to transport logs from a pond or lake to a sawmill.

jackleg *Mechanical Devices.* **1.** a supportive bar that is used with a jackhammer. **2.** an informal term for a tool or device used temporarily to suit a purpose for which it is not properly intended.

jack line *Petroleum Engineering.* a steel rod or cable that joins the arms of the central pumping engine to the two or more wells which are in the pumping process.

jack-o'-lantern *Optics.* a name for various natural phenomena of light that are thought to resemble the glow of a Halloween pumpkin, such as ignis fatuus. *Mycology.* a bright orange poisonous fungus of eastern North America, *Omphalotus olearius.*

jack plane *Mechanical Devices.* a medium-sized curved plane over one foot in length; used for smoothing wooden surfaces.

jack post *Mining Engineering.* any timber that is used where a coal seam is separated by a band of rock and one bench is loaded out before the other.

jackrabbit or **jack rabbit** *Vertebrate Zoology.* a large western North American hare of the genus *Lepus* that inhabits grasslands and deserts; noted for its great speed and agility, it travels up to 35 miles per hour and can leap 20 feet; characterized by a brown-haired body, long legs, and long black-tipped ears.

jack rafter *Building Engineering.* a secondary rafter, having a length less than that of a full rafter.

jackscrew *Mechanical Engineering.* **1.** a jack that uses a screw mechanism to raise and lower heavy objects. **2.** the threaded shaft used in a jackscrew. Also, SCREW JACK.

jackshaft *Mechanical Engineering.* **1.** a short drive shaft, usually located between the clutch and the transmission. **2.** an intermediate shaft that cranks the driving wheels in locomotives having collective drive.

Jackson, John Hughlings 1835–1911, English physiologist; discovered hemiplegic epilepsy; distinguished functions of brain hemispheres.

Jackson's rule *Neurology.* a theory stating that, after a loss of neurological functions during an epileptic attack, simple neurologic functions are more quickly recovered than complex ones.

Jacksonian *Geology.* a North American geologic stage of the Middle Eocene epoch, occurring after the Claibornian and before the Vicksburgian.

Jacksonian epilepsy *Medicine.* a form of epilepsy marked by **Jacksonian seizure,** which originates in the contralateral motor cortex, resulting in spasmodic contractions spreading from one muscle group to adjacent groups, a progression known as the **Jacksonian march.**

jack staff *Naval Architecture.* a flag staff at a naval ship's bow, used to display the jack or national naval ensign when the ship is not underway.

jackstay *Naval Architecture.* **1.** a line or strap along the upper side of a sail yard, used as attachment for the head of the sail. **2.** any taut rope that supports another rope, a sail, or other weight.

jack truss *Building Engineering.* a secondary truss that supports any region not immediately beneath the peak or ridge of a hip roof.

Jacob, Francois born 1920, French biologist; with Wollman, discovered episomes; with Monod, won the Nobel Prize for proposing the concepts of messenger RNA and the operon.

Jacob's ladder *Naval Architecture.* a hanging rope ladder with wooden rungs that is used to climb aboard a ship, climb masts, and so on. (From the Biblical account of the heavenly ladder that appeared to *Jacob* in a dream.)

Jacobi, Carl [jə kō´bē] 1804–1851, German mathematician; worked on elliptic functions, calculus of variations, and functional determinants.

Jacobian *Mathematics.* the determinant of the Jacobian matrix of a function f. Also, **Jacobian determinant.**

Jacobian elliptic function *Mathematics.* a doubly periodic, meromorphic function with one simple pole and one simple zero in each period parallelogram. By scaling, every such function can be put into one of twelve normalized forms. As defined by Carl Jacobi for the real argument u and modulus m, with $0 < m < 1$, let ϕ be the real number such that $u = \int_0^{\phi} (1 - m \sin^2\theta)^{-1/2}\, d\theta$. Then the first three Jacobian elliptic functions of u with modulus m are: $\mathrm{sn}(u \mid m) = \sin\phi$, $\mathrm{cn}(u \mid m) = \cos\phi$, and $\mathrm{dn}(u \mid m) = (1 - m \sin^2\phi)^{1/2}$. The other nine Jacobian elliptic functions are written using Glaisher's notation.

Jacobian matrix *Mathematics.* let f: $R^n \rightarrow R^n$ be a function that is given by n functions f_i of n variables x_j, such that each f_i has continuous partial derivatives. Then the Jacobian matrix of f is the $n \times n$ matrix whose (i,j)th entry is $\partial f_i/\partial x_j$.

jackrabbit

Jacobi condition *Mathematics.* suppose that the Legendre condition is strictly satisfied for $y(x)$; i.e., $f_{y'y'}(x, y(x), y'(x)) > 0$ for $x \in [a, b]$. Then the integral $\int_a^b f(x, y(x), y'(x))\,dx$ is minimized only if the conjugate point x_k (a particular zero of the solution to the Jacobi differential equation) of a is greater than b; i.e., zeros of the solution of the Jacobi differential equation are outside the interval $[a, b]$; used in the calculus of variations. Also, NECESSARY CONDITION OF JACOBI.

Jacobi ellipsoid *Astrophysics.* a triaxial ellipsoidal figure maintained by a homogeneous, self-gravitating mass that is spinning rapidly and uniformly.

Jacobi identity *Mathematics.* the identity, analogous to associativity, that is satisfied by the Lie bracket:

$$[v_1, [v_2, v_3]] + [v_2, [v_3, v_1]] + [v_3, [v_1, v_2]] = 0.$$

Jacobi polynomials *Mathematics.* a family of polynomials $\{J_n^{(\alpha,\beta)}(t)\}$, which are orthogonal on the interval $[-1,1]$ with the weight function $w(t) = (1-t)^\alpha(1+t)^\beta$, given by the formula

$$J_n^{(\alpha,\beta)}(t) = {}_2F_1(\alpha + n, -n; \beta; t),$$

where ${}_2F_1$ is Gauss's hypergeometric function. The special case $\alpha = \beta = -1/2$ gives the Chebyshev polynomials.

Jacobi's differential equation *Mathematics.* the differential equation

$$d/dx(f_{y'y'}u') - u\,(f_{yy} - d/dx\,f_{yy'}) = 0$$

with boundary condition $u(x_0) = 0$, where $f(x,y(x),y'(x))$ is given. If $u(x)$ has other zeros that are greater than x_0, then the smallest of them is the conjugate point x_k required by the Jacobi condition. If $u(x)$ has no other zeros greater than x_0, then $x_k = \infty$.

jacobsite *Mineralogy.* $(Mn^{+2},Fe^{+2},Mg)(Fe^{+3},Mn^{+3})_2O_4$, a black, opaque, magnetic, cubic mineral of the spinel group, dimorphous with iwakiite, occurring in granular form or as distorted octahedra, having a specific gravity of 4.76 and a hardness of 5.5 to 6.5 on the Mohs scale.

Jacobson's organ *Vertebrate Zoology.* an olfactory sac that opens into the mouth, found in the palate of many vertebrates; it is vestigial in humans but highly developed in reptiles.

Jacobson's theorem *Mathematics.* let D be a division ring in which, for every $a \in D$, there exists a positive integer $n(a) > 1$ such that $a^{n(a)} = a$. Commutativity is not assumed. Then D is a commutative field.

Jacobsoniidae *Invertebrate Zoology.* a family of tiny beetles usually found under bark or in litter.

Jacobson radical *Mathematics.* the intersection of all the maximal ideals of a ring.

Jacquemart's reagent *Analytical Chemistry.* an aqueous solution of mercuric nitrate and nitric acid; used to test for ethanol.

jacupirangite *Petrology.* an ultramafic, plutonic holocrystalline rock of the ijolite series, mainly pyroxene with a smaller amount of nepheline.

jade *Mineralogy.* a hard, tough, compact gemstone having an unevenly distributed color ranging from dark green to greenish white, and composed of either jadeite or nephrite. Also, **jadestone.**

jadeite *Mineralogy.* $Na(Al,Fe^{+3})Si_2O_6$, a white, green, gray, or mauve monoclinic mineral of the pyroxene group occurring only in metamorphic rocks as granular or fibrous-foliated masses, having a specific gravity of 3.25 and a hardness of 6 to 7 on the Mohs scale.

jadeitite *Petrology.* a metamorphic rock composed almost entirely of jadeite with traces of feldspathoids.

jaeger [yā´gər; jā´gər] *Vertebrate Zoology.* a large, northern marine bird, genus *Stercorarius,* of the family Stercorariidae; a strong flier that harasses smaller birds to steal their prey.

jaeger

Jaeger method *Fluid Mechanics.* a method of measuring the surface tension of a liquid; a capillary tube is immersed in the liquid, and the pressure needed to force air through the tube is recorded.

jaff *Military Science.* a process of jamming using a combination of chaff and electronic techniques.

jag bolt *Mechanical Devices.* an anchor bolt with a barbed, flaring shank, designed to resist retraction. Also, RAG BOLT.

jaguar *Vertebrate Zoology.* the largest and most powerful wild cat of the Western Hemisphere, *Felis onca,* of the family Felidae, resembling a leopard in having black spots on a yellowish or buff coat, but larger and stockier; found in tropical America.

jaguar

jaguarundi *Vertebrate Zoology.* a small, dark, unspotted wild cat, *Felis yagouaroundi,* of the family Felidae, native to Central America and South America and having a short muzzle, a long neck and tail, and short legs; it resembles an otter in appearance and swimming ability. Also, OTTER CAT.

Jahn-Teller effect *Physical Chemistry.* the observation that some highly symmetric geometries of molecules are visitable.

jalap *Botany.* any of several plants of the morning glory family, especially *Exogonium purga. Materials.* the dried root of such a plant or the yellow to brown powder derived from it; used as a cathartic.

jalousie [jal´ə sē] *Building Engineering.* **1.** a window having a series of horizontal glass slats, each hinged at the top and opening out at the bottom. **2.** a window blind or shutter made of a series of horizontal slats or panels, each hinged at the top and opening out at the bottom, letting in light and air but not direct sunlight or rain.

jam *Food Technology.* a preserve, usually of whole fruit that is slightly crushed and boiled with sugar.

jam *Ordnance.* **1.** a stoppage that prevents a weapon from firing; it may be caused by overheating, faulty ammunition, improper loading or ejection, etc. **2.** to become inoperative because of such a stoppage or to cause such a stoppage. *Military Science.* to disrupt enemy use of electronic devices, equipment, or systems by transmitting electromagnetic signals or deploying confusion reflectors.

Jamaica *Geography.* an island in the West Indies, south of Cuba, entirely occupied by the nation of Jamaica (area: 11,430 square miles).

Jamaica bayberry see BAYBERRY.

jamb *Building Engineering.* the vertical member of a door or window frame.

jamb liner *Building Engineering.* a small wooden strip laid up along the edge of a window jamb to increase its width for use in thicker walls.

jamb-shaft *Building Engineering.* any of a number of small shafts applied to doors and windows along the inside arris.

jam density *Transportation Engineering.* the traffic density at which extreme congestion is encountered, so that traffic can move slowly or not at all.

James, William 1842–1910, American philosopher and psychologist; proposed neurological basis of human thought.

James concentrator *Mining Engineering.* a concentration table with a deck divided into two sections; one section contains riffles for coarse material, and the other smooth section allows settling of finer particles that will not settle on a riffled surface.

James-Lange theory (of emotion) *Psychology.* the theory that what is perceived as emotion is actually a sensation of a physiological change in the body, caused by a reaction to an external emotional stimulus. Also, **James-Lange hypothesis.** (Based on the theories of William *James* and Carl *Lange.*)

jamesonite *Mineralogy.* $Pb_4FeSb_6S_{14}$, a lead-gray to gray-black, brittle, opaque, metallic monoclinic mineral occurring in hydrothermal vein deposits as aggregates of fibrous to acicular crystals, and having a specific gravity of 5.63 and a hardness of 2.5 on the Mohs scale. Also, FEATHER ORE, GRAY ANTIMONY.

Jamieson, Thomas 1829–1913, Scottish geologist; proposed theory of postglacial uplift.

Jamin effect *Fluid Mechanics.* a substantial resistance to flow in a capillary tube due to the presence of air bubbles in the tube.

Jamin refractometer *Optics.* a refractometer that measures the refractive index of a gas by observing the fringe shift formed by two interfering light beams.

jammer *Electronics.* a device that prevents radio or radar transmissions by beaming spurious signals into an enemy's radar system.

jammer finder *Electronics.* a type of radar system that finds a target's range by focusing a narrow cone of energy, known as a pencil beam, into a jamming source. Also, BURN-THROUGH.

jamming *Electronics.* the intentional disruption of radio or radar station operation by saturating the receiver with electromagnetic radiation.

Jamoytius *Paleontology.* one of the earliest vertebrates and the oldest known anaspid, known only from near-shore marine or brackish-water deposits of the middle and upper Silurian; a genus in the order Anaspida and proposed family Jamoytiidae; related to the extinct fish Thelodus.

JAN *Aviation.* the airport code for Thompson Field, Jackson, Mississippi.

Janecke coordinates *Chemical Engineering.* a diagram that plots the mass of solvent per unit mass of solvent-free substance versus concentration on a solvent-free basis for a system in equilibrium; used for design calculations of solvent-extraction equipment.

Jansky, Karl 1905–1950, American engineer; discovered extraterrestrial radio waves, laying the foundation for radio astronomy.

jansky *Astrophysics.* the unit of flux density or "brightness" used in radio astronomy; 1 Jy = 10^{-26} watt per square meter per hertz.

Jansky noise *Astrophysics.* see COSMIC BACKGROUND RADIATION.

Jansky's classification *Hematology.* a classification of ABO blood types designated by roman numerals I to IV, corresponding with types O, A, B, and AB, respectively.

Janssen, Pierre 1824–1907, French astronomer; used spectroscopy to determine the sun's composition and to observe solar prominences.

J antenna *Electromagnetism.* a half-wave antenna that is fed at one end by a quarter-wave wire parallel to the antenna, having a configuration resembling a J.

Janthinobacterium *Bacteriology.* a genus of strictly aerobic, flagellated, rod-shaped, Gram-negative bacteria that occur in soil and water.

January thaw *Meteorology.* a period of unseasonably warm weather that is said to occur each year in late January in parts of the northeastern United States; it is associated with the occurrence of southerly winds on the back side of an anticyclone off the southeastern United States.

Janus *Astronomy.* the tenth moon of Saturn; it shares an orbit with Epimetheus and measures 220 kilometers by 160 kilometers. (Named for *Janus,* the Roman god of doors and entrances.)

Janus system *Navigation.* a docking or mooring aid for large ships that determines speed relative to the bottom in both fore-and-aft and athwartship directions. It uses four Doppler sonar speed logs aimed obliquely at the bottom in both fore-and-aft and athwartships directions. Also, **Janus configuration.**

Jaoaceae *Botany.* in some classification systems, a monogeneric family of green algae of uncertain affiliation that grow on rocks in streams in China; the pseudoparenchymatous thallus is a large, hollow sphere that is attached by simple or branched rhizoids. Also, COELODISCACEAE.

japan *Materials.* a hard, glossy black varnish with an asphalt base; used to coat wood or metal surfaces.

Japan, Sea of *Geography.* an arm of the western Pacific between Japan and northeastern Asia.

Japanese ash *Botany.* see MANCHURIAN ASH. *Materials.* see SEN.

Japanese B encephalitis *Medicine.* an epidemic encephalitis striking Japan and much of the Far East during summer months, often symptomless but sometimes having symptoms resembling those of poliomyelitis.

Japanese B encephalitis virus *Virology.* the causative agent of Japanese B encephalitis, a virus similar to agents of various types of equine encephalomyelitis but attacking a wider range of hosts.

Japanese black pine see KURAMATSU.

Japanese paper *Materials.* a decorative paper with a mottled surface that is handmade from mulberry bark; used for woodcuts, engravings, and greeting cards.

Japanese twin law *Crystallography.* a twin law specifying the relationship governing the twinning of crystals in quartz in which the two individual crystals have a composition plane of (1122).

japanning *Engineering.* the process of coating an object or part with a japan varnish.

Japan wax *Materials.* a combustible, pale yellow, waxy solid that is insoluble in water and alcohol and is obtained from the fruit of certain sumac trees, especially the species *Rhus succedanea*; it melts at 50°C and is used in candles, furniture polish, and floor wax.

Japygidae *Invertebrate Zoology.* a family of wingless carnivorous soil-dwelling insects in the order Diplura.

jar *Chemistry.* any wide-mouthed, usually glass container, such as a bell jar or Leyden jar. *Petroleum Engineering.* a tool that is composed of a series of links in the drill string to join drill cables to the drill bit; it produces a jarring impact in cable-tool drilling and originates the uneven motion on the upstroke which helps to release the bit when stuck. Also, **jar coupling**. *Electronics.* **1.** a former unit of capacitance, equivalent to 1/900 μF. **2.** in a lead-acid storage cell, the container that holds the element and electrolyte.

jargonaphasia *Neurology.* a speech defect characterized by the combination of several words into one.

Jarisch-Herxheimer reaction *Medicine.* a transient, short-term immunologic reaction commonly seen after antibiotic treatment of early and later stages of syphilis and sometimes other diseases; marked by fever, chills, headache, myalgias, and exacerbation of cutaneous lesions.

jarlite *Mineralogy.* $NaSr_3Al_3(F,OH)_{16}$, a colorless to gray, monoclinic mineral occurring as tiny, tabular crystals and in massive form, and having a specific gravity of 3.87 and a hardness of 4 to 4.5 on the Mohs scale; found in cryolite deposits.

jarosite *Mineralogy.* $KFe_3^{+3}(SO_4)_2(OH)_6$, an ocher-yellow or brown, translucent, trigonal mineral of the alunite group occurring as coatings of small, tabular crystals and in earthy form, and having a specific gravity of 2.9 to 3.26 and a hardness of 2.5 to 3.5 on the Mohs scale; found as a secondary mineral in ferruginous ores and rocks. Also, UTAHITE.

jarosite process *Metallurgy.* an electrochemical process used to refine zinc.

jarrah *Materials.* the dense, deep-red wood of the Australian tree, *Eucalyptus marginata*; used in construction and heavy framing.

jarrah dieback *Plant Pathology.* a disease of the *Eucalyptus marginata* tree caused by the fungus *Phytophthora cinnamomi,* which is characterized by the death of the plant branches from the top of the tree downward.

Jarvik-7 *Medicine.* a trade name for a four-valved artificial heart designed for implantation in the human body. (Developed by the American physician Robert *Jarvik,* born 1946.)

jasmine *Botany.* **1.** any of a numerous shrubs or vines of the genus *Jasminum,* of the olive family, whose fragrant flowers are the source of jasmine oil. **2.** any of various other plants having similar fragrant flowers.

jasmine oil *Materials.* a colorless fragrant oil that is extracted from the blossoms of various jasmine plants; used in flavoring and perfumery.

jasmone *Organic Chemistry.* $C_{11}H_{16}O$, a liquid ketone found in jasmine oil and other flower oils; it boils at 134–135°C (12 torr) and is used in making perfume.

jasper *Petrology.* a dense, generally opaque, cryptocrystalline sedimentary rock composed of quartz and containing iron-oxide impurities; it is characteristically red but may also appear yellow, green, grayish-blue, brown, or black. Also, **jasperite, jasperoid, jaspis.**

jaspillite *Petrology.* a compact, alternately banded siliceous rock that contains red jasper and at least 25% iron.

jaspoid see TACHYLITE.

JATO [jā´tō] *Aviation.* a takeoff utilizing one or more jet-producing units, such as rockets, for additional thrust. The auxiliary power system used to assist such a takeoff is called a **JATO bottle** or **JATO unit.** One rocket in such a system is called a **JATO engine.** (An acronym for jet-assisted takeoff.)

jauk *Meteorology.* a foehn in the Klagenfurt basin of Austria that may originate from a southerly direction, but develops as a northern foehn. Also, **jauch.**

jaundice [jôn´dis] *Pathology.* a syndrome characterized by a yellowish appearance of the patient due to hyperbilirubinemia and deposition of bile pigment in the skin, mucous membranes, and sclera. *Invertebrate Zoology.* see GRASSERIE. (From the French word for "yellow.")

jaundice of the newborn *Pathology.* the yellowish appearance of the skin or the sclera common in newborns due to excessive bilirubin in the blood, usually the result of immaturity of the liver.

Java *Geography*. a densely populated volcanic island in Indonesia, south of Borneo.

Java black rot *Plant Pathology*. a disease common to stored sweet potatoes that is caused by the fungus *Diplodia tubericola* and characterized by a blackening and hardening of the inside of the root.

Java cotton see KAPOK.

Java man *Anthropology*. a fossil specimen discovered in the 1890s on the island of Java; first called *Pithecanthropus erectus*, and now identified as *Homo erectus*.

javelina or **javelin** see PECCARY.

jaw *Anatomy*. either of the two tooth-bearing bones of the head, the maxilla and mandible, that form the framework of the mouth. *Robotics*. an end effector with two or more parts in opposition that can open and close to grasp and hold an object.

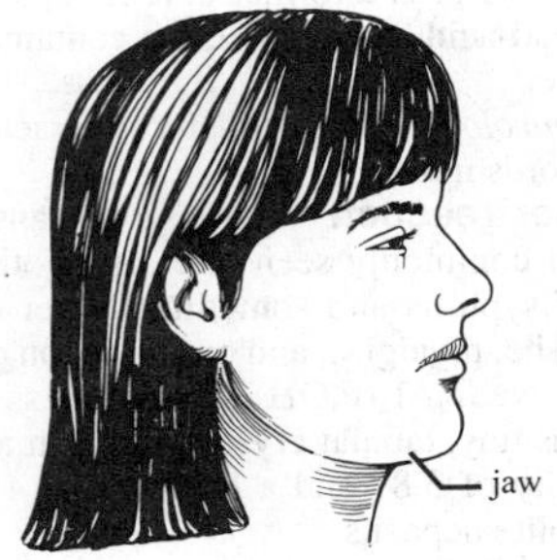

jaw

jaw clutch *Mechanical Engineering*. a type of clutch that allows for positive connection between two shafts by means of interlocking faces; the most typical kind of positive clutch; may be either spiral or square.

jaw crusher *Mechanical Engineering*. a machine that is used for crushing ore or other materials; it has a fixed vertical jaw and an inclined swinging jaw moved by a toggle joint. Also, **jawbreaker.**

jawless vertebrate *Vertebrate Zoology*. any member of the class Agnatha, such as the lamprey or hagfish; they possess a suctorial mouth that lacks biting jaws.

JAX *Aviation*. the airport code for Jacksonville International Airport.

jay *Vertebrate Zoology*. any of about 40 species of birds of the family Corvidae that are found primarily in the New World and are known for their harsh voice and boisterous and aggressive nature; characterized by a long tail and a pointed bill, often blue in color with a crested head.

jay

J bolt *Mechanical Devices*. a bolt shaped like the letter J with threads on its longer leg.

J box see JUNCTION BOX.

J-carrier system *Telecommunications*. a broadband carrier system that allows for 12 telephone channels, having a bandwidth of up to 140 kilohertz.

j chain or **J chain** *Immunology*. a short peptide chain joining individual immunoglobulin molecules in IgM pentamers and in oligomers of IgA.

JCL job-control language.

J display *Electronics*. the representation of targets on a radar scope in which the time base is depicted as a circle, target signals appear around the circumference, and distances are measured circumferentially between targets.

Jeans, James Hopwood 1877–1946, English physicist; adapted his research on rotating gases to explain stellar evolution.

Jeans length *Astrophysics*. the critical size of a disturbance in a homogeneous medium for which gravity is unable to prevent the material from detaching.

Jeans viscosity equation *Thermodynamics*. an equation that relates the viscosity of a gas to its temperature; the form of the equation is $h = kT^n$, where k is a constant, T is the absolute temperature, and n is a value characteristic of the gas.

jeep or **Jeep** *Mechanical Engineering*. a rugged, adaptable 1/4-ton vehicle capable of four-wheel drive; widely used by the U.S. military in World War II, and later adapted for civilian use. (Of uncertain origin; possibly a rendering of the initials *GP,* for general-purpose vehicle.)

jeffersonite *Mineralogy*. a variety of dark monoclinic pyroxene, usually manganoan and zincian augite or diopside.

Jeffrey diaphragm jig *Mining Engineering*. a jig with a plunger situated beneath the screen.

Jeffrey molveyor *Mining Engineering*. an arrangement consisting of a string of short conveyors on driven wheels connected together to run alongside a heading or room conveyor that allows a continuous miner to operate at all times.

Jeffrey single-roll crusher *Mining Engineering*. a simple coal crusher with toothed segments bolted to a drum that force the coal down into the crushing opening.

Jeffrey swing-hammer crusher *Mining Engineering*. a crusher encased in an iron casing with a rotating driving shaft that causes freely swinging arms to swing out and strike the coal, ore, or other minerals against the iron casing and pass through the grated bottom.

Jeffrey-Traylor vibrating feeder *Mining Engineering*. a feed chute electromagnetically vibrated in a direction oblique to its surface, in which the amplitude and frequency of vibration determine the rate of movement of rock.

Jeffrey-Traylor vibrating screen *Mining Engineering*. an electric vibrating screen operated by an oscillating armature and a stationary coil.

jejunitis *Medicine*. an inflammation of the jejunum.

jejunoileostomy *Surgery*. the formation of a new opening between the jejunum and the ileum.

jejunostomy *Surgery*. the surgical creation of a permanent opening between the jejunum and the surface of the abdominal wall.

jejunotomy *Surgery*. a surgical incision into the jejunum.

jejunum *Anatomy*. the middle portion of the small intestine, approximately eight feet in length, extending from the duodenum to the ileum.

jelly *Food Technology*. a soft, elastic food preparation, especially one of fruit juice reduced with sugar that is used as a spread for bread, filling for cakes, etc. *Science*. any substance having a similar texture or appearance to the food *jelly*; a colloidal semisolid mass.

jellyfish *Invertebrate Zoology*. the common name for the free-swimming medusa stage of hydrozoans and scyphozoans, having an umbrella-like body and marginal tentacles.

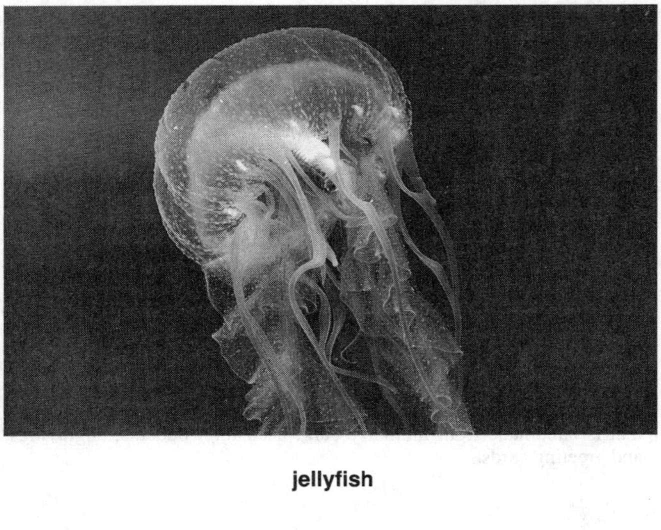

jellyfish

jelly fungus *Mycology.* any of various fungi belonging to the order Tremellales, characterized by their gelatinous fruiting bodies. Also, TREMBLING FUNGI, TREMMELLALES.

jelly lichen *Botany.* any of various lichens belonging to the genus *Nostoc* and having a gelatinous thallus and an algal component.

jelly roll motif *Crystallography.* the topology of a protein that is essentially a Greek key fold with an additional swirl and a slightly different topology of connections.

jelutong *Materials.* the resin of trees of the genus *Dyera,* used primarily in the production of rubber and chewing gum. Also, PONTIANAK GUM.

Jenner, Edward 1749–1823, English physician; developed the smallpox vaccine, thereby founding modern immunology.

Jenner, Sir William 1815–1898, English physician and pathologist; distinguished typhoid fever from typhus.

jenny *Vertebrate Zoology.* a nickname for a female donkey or female bird.

Jensen, J. Hans 1906–1973, German physicist; shared Nobel Prize for research on the shell structure of atomic nuclei.

Jensen's inequality *Mathematics.* **1.** let ϕ be a function which is convex on some region containing arbitrary points $x_1, \ldots, x_n$, and let λ_i be any nonnegative real numbers such that $\sum_{i=1}^{n} \lambda_i = 1$. Then Jensen's inequality is

$$\phi\left(\sum_{i=1}^{n} \lambda_i x_i\right) \leq \sum_{i=1}^{n} \lambda_i \phi(x_i).$$

2. in general, suppose that ϕ is a convex function on $(-\infty, \infty)$ and f is an integrable function on $[0, 1]$. Then $\phi\,[\int f(t)\,dt] \leq \int \phi\,(f(t))\,dt$. **3.** If $x_1, \ldots, x_n$, s, and t are positive numbers with $s > t$, then

$$\left(\sum_{i=1}^{n} x_i^{s}\right)^{1/s} \leq \sum_{i=1}^{n} (x_i^{t})^{1/t}.$$

That is, for $t > 0$, the sum of order t is a nonincreasing function of t.

jerboa *Vertebrate Zoology.* a small, nocturnal, social rodent of the family Dipodidae found in Old World deserts, having enlarged hindlimbs that are modified for leaping.

jerboa

jeremejevite *Mineralogy.* $Al_6B_5O_{15}(F,OH)_3$, a colorless to pale yellowish-brown, hexagonal mineral occurring in prismatic crystals, and having a specific gravity of 3.28 to 3.29 and a hardness of 7.5 on the Mohs scale. Also, EREMEYEVITE.

Jeremiassen crystallizer *Chemical Engineering.* an apparatus used for growing solid crystals in a supersaturated liquid solution and for removing them from the solution.

jerk *Physiology.* a sudden reflex or involuntary movement. *Mechanics.* a unit of measurement for the rate of acceleration, equal to 1 foot per second per second squared. *Food Technology.* to preserve food, such as beef, by marinating, cutting in strips, and drying, often in the sun.

jerkinhead *Architecture.* the hipped upper section of a roof that is itself partly hipped, with a truncated gable below. Also, CLIPPED GABLE; HIPPED GABLE.

jerk pump *Mechanical Engineering.* a fuel-injection pump having a cam-driven plunger that overruns a spill port in a cylinder wall to initiate a rapid rise in pressure, in order to atomize fuel for injection.

Jerne, Niels born 1911, Swiss immunologist; awarded Nobel Prize for formulating three fundamental theories of immunology.

jerry can *Ordnance.* a five-gallon, flat-sided can used for storing, transporting, and discharging fuel. Also, **jerrican**.

Jersey *Agriculture.* a breed of small dairy cattle, varying in color from fawn to dark gray; its milk has the highest butterfat content of any breed. (From the British island of *Jersey.*)

JES job entry subsystem.

jet *Fluid Mechanics.* a continuous stream of concentrated and well-defined incompressible or compressible fluid emitted by a nozzle orifice. *Mechanical Engineering.* any device that is powered by such a stream of fluid. *Aviation.* specifically, a jet engine or jet aircraft. (Derived from the Latin word for "throw.") *Geology.* see JET COAL.

jet aircraft *Aviation.* an aircraft, especially a fixed-wing airplane, that attains its thrust from one or more jet engines.

jet assist *Aviation.* additional thrust, beyond the power plant already in use that is given to an airplane by a jet engine.

jet-assisted takeoff see JATO.

jet bit *Mechanical Devices.* a modified bit with hydraulic jet characteristics used to accelerate a drilling operation.

jet bundle *Mathematics.* a fiber bundle over a differentiable manifold with fibers created at each point by identifying functions that have the same values for all their partial derivatives up through order n at the base point. Also, N-JET BUNDLE.

jet coal *Geology.* a hard, pure black variety of lignite exhibiting a high luster and conchoidal fracture; found as isolated masses in bituminous shale. Also, JET, CANNEL COAL.

jet compressor *Mechanical Engineering.* **1.** an axial flow compressor that uses aerodynamic fan blades evenly spaced along a rotor to draw in compressible gas along the axis of the compressor; by increasing the number of rotors and decreasing the distance between the blades and the compressor housing, the pressure of the discharge gas rises. **2.** a centrifugal-type compressor that uses aerodynamic impellers to draw in compressible gases near the axial area and discharges it over the top into an annular chamber for combustion or other use; multistage compression is achieved by using diaphragms to direct the gas down into other impeller inlets.

jet condenser *Mechanical Engineering.* a condenser that uses a spray of water to condense the exhaust steam from a steam engine, turbine, or other source.

jet drilling *Mechanical Engineering.* **1.** a drilling method that uses a chopping bit on hollow drill rods and a jet of water to wash the material fragments to the surface. **2.** a method of boring hard rock in which a hydrocarbon-oxygen jet is used to melt the rock, and the rock fragments are simultaneously flushed out with water.

jet-effect wind *Meteorology.* air whose windspeed is enhanced by channeling through an orographic configuration such as a narrow mountain pass or canyon.

jet engine *Aviation.* any of a class of reaction engines that take in outside air for use as fuel oxidizer and eject a jet of hot exhaust gases backward for thrust.

jet-flame drill *Mining Engineering.* a mining drill that uses a high-velocity flame to chip out a hole.

jet flap *Aviation.* a sheet of air or fluid ejected downward at a high speed from a thin, spanwise slot near the trailing edge of a wing in order to induce lift over the wing.

jet fuel *Materials.* any fuel used especially for jet engines; usually a kerosenelike petroleum distillate.

jet helicopter *Aviation.* a helicopter powered by one or more jet engines, which usually drive a jet rotor.

jet hole *Engineering.* a borehole drilled by a strong stream of air or fluid.

jet injector *Medicine.* a hand-held pressurized device resembling a small gun, used especially for immunizations; it injects the drug at a velocity that is sufficient to penetrate the skin. Also, **jet gun**.

jet lag or **jetlag** *Physiology.* a temporary condition characterized by fatigue and often disorientation, stemming from a disruption of normal biological rhythms following air travel across several time zones.

jet lift *Aviation.* an upward force derived directly from one or more vertical jets.

jet loop fermenter *Biotechnology.* a fermenter in which the broth is recirculated into the main fermentation vessel under pressure; the result is an improvement in gas absorption.

jet mixer *Mechanical Engineering.* a device for combining two liquids, in which one liquid is pumped through a small nozzle or jet to achieve a fine dispersion, which then impinges into the flowing stream of the second liquid.

jet molding *Engineering.* a molding process that applies most of the heat to the material to be molded while it moves through a jet or nozzle, instead of in a traditional heating cylinder.

jet nozzle *Mechanical Devices.* a type of nozzle that is uniquely designed to produce a jet of exhausted propulsion gases.

jet piercing see THERMIC BORING.

jet polishing *Materials Science.* the process of directing a jet of an electrolyte onto a small disk to cause a perforation in the center; used to thin samples for observation in a transmission electron microscope.
jet propulsion *Aviation.* the moving forward of a flight vehicle by means of duct propulsion. *Space Technology.* the moving forward of a rocket or spacecraft by means of a reaction engine; rocket propulsion.
jet pump *Mechanical Engineering.* a pump that delivers a large volume of fluid at a low lift by imparting the momentum of a high-velocity jet of steam or air to this column of fluid.
jet rotor *Aviation.* a rotor driven by some form of jet device mounted within or upon the blades, usually at the tips.
jetsam *Navigation.* pieces of cargo or other objects that are jettisoned in an emergency. Also, **jetsom.**
jet spinning *Engineering.* a method of manufacturing plastic fibers, wherein molten polymer is forced from a die lip by a direct blast or jet of hot gas; similar to melt spinning.
jet stream or **jetstream** *Meteorology.* **1.** a concentration of relatively strong winds within a narrow stream in the atmosphere, especially a quasi-horizontal concentration of strong westerly winds in the high troposphere. **2.** in upper-level charts, a similar concentration of winds having a speed of 50 knots or greater. *Aviation.* the exhaust stream of a jet or rocket.
jet thrust *Aviation.* the thrust of a fluid, as distinguished from the thrust of a propeller.
jetting *Civil Engineering.* the process of sinking piles or well points into sandy soils by means of water or air compressed jets, generally employed when regular pile driving might weaken the foundations of nearby buildings. *Materials Science.* a dynamic process that occurs during explosive welding, in which the controlled impact of the surfaces to be joined results in the breakup and removal of bond-limiting surface layers.
jetting tool *Petroleum Engineering.* equipment used downhole that forces a high-pressure, fluid stream laden with sand to clean out wellbore holes, disintegrate perforating pipe, and execute other functions.
jettison *Navigation.* to throw goods overboard in order to lighten a vessel or aircraft in distress.
jettisoned mines *Ordnance.* mines that are laid as quickly as possible in order to empty the minelayer, without regard to their condition or relative position.
jetty *Civil Engineering.* **1.** see GROIN. Also, **jetty breakwater. 2.** a small wharf or landing pier. *Architecture.* an upper part of a building that projects beyond a lower part, such as an overhanging second story. Also, JUTTY.
jet vane *Aviation.* a fixed or movable vane used directly in a jet stream to improve stability or control, especially at low speeds or under conditions where external aerodynamic controls are ineffective.
Jetway *Transportation Engineering.* a trade name for an enclosed, telescoping movable bridge used to connect an airliner with a terminal gate.
Jevons effect *Meteorology.* the effect upon the measurement of rainfall caused by the presence of the rain gauge itself, in which the gauge disturbs the flow of air that passes it, carrying away a portion of the rain that would normally be collected; a function of the wind speed and the height of the gauge from the ground. This phenomenon was described in 1861 by W. S. Jevons.
jewel *Engineering.* **1.** a bearing made of natural ruby or sapphire, or synthetic stone, such as synthetic corundum, used for gyros, precision timekeeping devices, and other instruments. **2.** a soft, metal-bearing lining used in railroad cars and the like.
Jewel Box Cluster *Astronomy.* NGC 4755, a small but populous open star cluster in the southern constellation of Crux. Also, KAPPA CRUCIS CLUSTER. (From its brightest member star.)
JFET junction field-effect transistor.
JFK *Aviation.* the airport code for John F. Kennedy International Airport, New York, New York.
J function *Geophysics.* a mathematical relation used, among others, to compare capillary pressure measurements between two or more geological formations.
J genes *Immunology.* sets of gene segments in the immunoglobulin heavy- and light-chain genes and in the genes for the chains of the T-cell receptor that are recombined during lymphocyte ontogeny and that contribute to the genes for the variable domains.
jib *Naval Architecture.* **1.** a triangular fore and aft sail set from the forestay of a vessel. **2.** the projecting arm of a crane.
jib boom *Mechanical Engineering.* the inclined or horizontal boom or strut of a crane supported by a stationary or retractable guy wire.
jib crane *Mechanical Engineering.* a crane, constructed on a vertical building column, supporting a guy-wired boom or jib that pivots 270° from side to side and supports an electric or hand-operated hoist.
jib end *Mining Engineering.* the delivery end of a conveyor system with a jib fitted to deliver the load in advance of and remote from the drive.
jig *Mechanical Engineering.* a device used to hold and guide objects, such as component parts moving through an assembly line. *Mining Engineering.* a screen that is submerged in water and vibrated in order to filter or concentrate ore. *Textiles.* in dyeing piece goods, a device that moves the cloth at full width on rollers through the dye liquor.
jig back *Mechanical Engineering.* an aerial ropeway with two suspended containers that travel in opposite directions and stop alternately at the terminals for loading and unloading.
jig borer *Mechanical Engineering.* a vertical milling device that bores holes in the surface of a work piece held by a jig.
jigger see JIGGING CONVEYOR.
jiggering *Materials Science.* the process of molding wet ceramic clay into a desired shape prior to firing, in which the clay is spread under gradual pressure into a rotating mold; used in mass production of dinnerware.
jigging *Mining Engineering.* **1.** the process of separating mineral from gangue by means of a jig. **2.** the up-and-down motion of a mass of particles in water by means of pulsion.
jigging conveyor *Mining Engineering.* an apparatus used to transport minerals by a series of steel troughs suspended from the roof of a stope, or laid on rollers on its floor, and given reciprocating motion by mechanical means. Also, JIGGER, CHUTE CONVEYOR, PAN CONVEYOR.
jig grinder *Mechanical Engineering.* a machine that is similar to a jig borer, designed to grind, rather than bore, holes in hardened-steel plates, jogs, or other parts.
jigsaw *Mechanical Devices.* an electrical hand tool that moves a thin, narrow saw blade in a vertical or horizontal reciprocating manner; used for cutting intricate curves, lines, and patterns.
jig washer *Mining Engineering.* a washer used for relatively coarse coal or minerals, in which the broken ore is pulsed vertically in water through a screen so that the heavy pieces pass through the screen into a hutch and the gangue goes over the side.
jim crow *Mechanical Devices.* **1.** a crowbar with a claw on one end. **2.** a rail bender or straightener, consisting of a strong, semielliptical dog and a heavy buttress screw. **3.** a swivel-type tool head for cutting on each stroke of a planing machine table.
jimmy *Mechanical Devices.* a short crowbar.
jimsonweed or **jimson weed** *Botany.* a poisonous annual weed, *Datura stramonium,* of the nightshade family Solanaceae, having foul-smelling leaves and white trumpet-shaped blooms that give rise to globose prickly fruits; the source of a number of medicinal alkaloids, including hyoscine, hyoscamine, and atropine. (An alteration of *Jamestown weed*; noted by early settlers in Jamestown, Virginia.) Also, **Jimson weed, jimpson weed.**

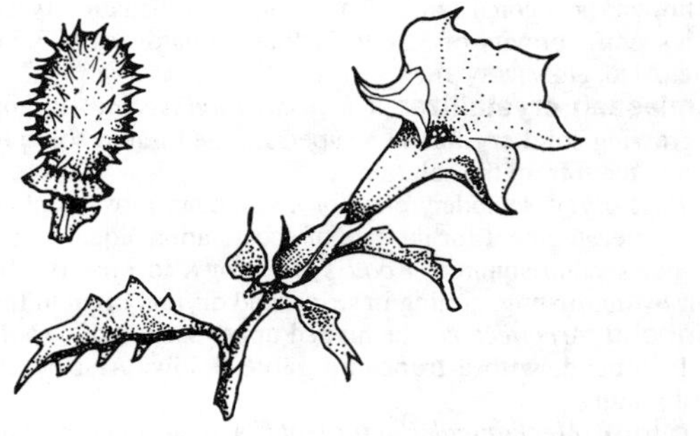

jimsonweed

J indicator see J SCOPE.
J integral *Materials Science.* a path-independent line integral employed in fracture mechanics to evaluate the strain energy release rate of a cracked body.
JIRA or **J.I.R.A.** Japan Industrial Robot Association.
jird *Vertebrate Zoology.* any of various tiny rodents of the genus *Meriones*, native to northern Africa and southwestern Asia; used as hosts in experiments regarding schistosomiasis.

jitney *Transportation Engineering.* a small bus or van that travels along a fixed route, from which it may deviate in response to passenger demand.
jitter *Telecommunications.* **1.** any instability in the timing of trains of pulsed signals. **2.** in television, a weak signal that results in sudden irregular variation, usually caused by defects in synchronization. **3.** in facsimile, blurriness or distortion in received copy.
j-j coupling *Atomic Physics.* a process for determining the wave function of a many-electron atom by first combining each particle's spin and orbital functions, and then combining the wavefunctions.
J K coryneform *Bacteriology.* any coryneform bacteria that are resistant to antibiotics.
J-K flip-flop *Electronics.* a bistable circuit whose state can be changed by various combinations of its "J" and "K" inputs.
JKT *Aviation.* the airport code for Jakarta, Indonesia.
JL *Aviation.* the airline code for Japan Airlines.
JNB *Aviation.* the airport code for Johannesburg, South Africa.
JND or **jnd** just-noticeable difference.
jnt or **jnt.** joint.
joaquinite-(Ce) *Mineralogy.* $Ba_2NaCe_2Fe^{+2}(Ti,Nb)_2Si_8O_{26}(OH,F)\cdot H_2O$, a honey-yellow to brown, monoclinic mineral of the joaquinite group, dimorphous with orthojoaquinite-(Ce); occurring in tabular to equant crystals, and having a specific gravity of 3.89 and a hardness of 5 to 5.5 on the Mohs scale.
job *Industrial Engineering.* **1.** the combined tasks, duties, and responsibilities that are considered to be a worker's regular assignment. **2.** an authorization for the manufacture of a specific quantity of an item by a specific date. Also, SHOP ORDER. *Computer Programming.* a collection of tasks representing a unit of computer work, composed of a program or several programs treated as a unit.
job analysis *Industrial Engineering.* the process of determining the requirements of a job by observing and evaluating the work performed. Also, JOB STUDY.
jobber's reamer *Mechanical Devices.* a hand or machine reamer with flutes and taper shanks.
job breakdown *Industrial Engineering.* the division of a job into its component motions or elements for analysis. Also, OPERATION BREAKDOWN.
job class *Computer Programming.* a job category that is defined according to the computer resources needed to run it; often used to control the mixture of jobs that are performed concurrently. Also, **job family, job grade.**
job classification *Industrial Engineering.* the designation of a job into one of several possible categories, commonly used to determine pay levels, skill and training requirements, etc.
job-control language see COMMAND LANGUAGE.
job-control statement *Computer Programming.* a control statement that identifies a job and describes its requirements to the operating system.
job description *Industrial Engineering.* a detailed summary of the nature and requirements of a job; used when advertising for workers, making promotions, etc.
job-entry subsystem *Computer Programming.* a system facility for spooling, queuing, and managing jobs.
job-entry system see REMOTE JOB ENTRY.
job evaluation *Industrial Engineering.* a systematic analysis of jobs, as opposed to workers, and their function and relative worth in a production process or enterprise.
job factor *Industrial Engineering.* a mental or physical requirement, skill, judgment capability, etc., for a job; used in job evaluation. Also, **job characteristic.**
job-flow control *Computer Programming.* control over the sequence of jobs that are processed on a computer, performed manually or by means of an operating system. Job-flow control ensures the efficient use of peripheral devices and the central processor.
job library *Computer Programming.* a series of user-identified partitioned data sets that serve as the primary source of load modules for a given job.
job lot *Industrial Engineering.* a large, often varied quantity of goods that are part of a single production run or sales transaction.
job-management program *Computer Programming.* system software that manages resource allocation and job submission, initiation, execution, and termination.
job mix *Computer Programming.* the set and variety of jobs being executed within a multiprogramming system at any given time.
job-oriented terminal *Computer Technology.* **1.** a data terminal designed to perform a particular application. **2.** a terminal that allows data to be transmitted directly to a computer from a data source.
job press *Graphic Arts.* a printing press designed to print on large sheets of paper.
job printer *Graphic Arts.* a printer who does a variety of work, such as letterheads and announcements, rather than just books, periodicals, and the like.
job-processing control *Computer Programming.* the part of the control program that initiates job operations, assigns input/output units, and performs functions needed to proceed from one function to another.
job schedule *Control Systems.* a control program that selects which job will be processed next.
job shop *Industrial Engineering.* **1.** intermittent production in job lots, rather than continuous production. Does not necessarily imply "make to order" production. **2.** an industrial facility in which each job done may require different machines and a different sequence of operations, rather than a unidirectional production-line type flow.
job stacking *Computer Programming.* the presentation of jobs to a system in the order in which they are to be performed.
job step *Computer Programming.* a specification for work to be performed during a job or a set of tasks. Each instruction is considered one step in a job.
job stream *Computer Programming.* a series of related jobs that are run consecutively by the operating system. Also, STREAM.
job study see JOB ANALYSIS.
job ticket *Graphic Arts.* a sheet giving the specifications and instructions for a printing job, such as the stock, trim size, and quantity; usually attached to the copy during the printing process.
Jocasta complex [jō kas′tə] *Psychology.* excessive or abnormal love of a mother for her son. (From the Greek myth in which *Jocasta* marries her son Oedipus.)
jochwinde *Meteorology.* a wind that blows through a mountain gap of the Tauern Pass in the Alps.
jodfenphos *Organic Chemistry.* $C_8H_8O_3Cl_2IPS$, crystals that are slightly soluble in water and melt at 76°C; used as an insecticide.
Joffe effect *Materials Science.* the sensitivity of the fracture of brittle solids to environmental conditions that alter the surface energy.
jog *Materials Science.* a sessile segment of a dislocation, often associated with dislocation climb or interactions between dislocations. *Graphic Arts.* to tap a stack of paper sheets in order to align their edges.
jogging *Electronics.* a technique used to generate movement in a motor by rapidly opening and closing its starter circuit. Also, INCHING.
joggle *Building Engineering.* **1.** a notch in the joining surface of a piece of building material that prevents slipping, thus forming a joint. **2.** a dowel used to join blocks of masonry.
joggled frame *Naval Architecture.* a frame with its outside edge "stepped" so that when the hull shell is applied, one edge of each plank or plate is offset from the edge of the plank or plate running parallel to it, thus creating the appearance of overlapping without actually doing so.
joggled plating *Naval Architecture.* plating set so that the edge of one row of plates is offset from the edge of the plating below it.
Johann crystal geometry *Crystallography.* the geometry of the focusing crystal of an electron-probe microanalyzer; it is less stringent than Johannson crystal geometry.
johannite *Mineralogy.* $Cu(UO_2)_2(SO_4)_2(OH)_2\cdot 8H_2O$, an emerald-green to apple-green, triclinic mineral occurring in thick tabular crystals and drusy aggregates, and having a specific gravity of 3.32 and a hardness of 2 to 2.5 on the Mohs scale; found in oxidized uraninite veins. Also, GILPINITE.
Johannsen, Wilhelm [yō hän′sən] 1857–1927, Dutch geneticist; showed that environmental adaptations are not inherited; introduced the term *gene*.
Johannsen's classification *Petrology.* a system for quantitative mineralogic classification of igneous rocks using Johannsen numbers.
johannsenite *Mineralogy.* $CaMn^{+2}Si_2O_6$, a brown, grayish, or greenish monoclinic mineral of the pyroxene group, usually occurring in massive form or as fibrous aggregates, and having a specific gravity of 3.44 to 3.55 and a hardness of 6 on the Mohs scale; found in metasomatized limestones.
Johannsen number *Petrology.* a three- or four-digit number used to define the position of an igneous rock in the Johannsen classification system; the first digit indicates the class, the second indicates the order, and the third and fourth indicate the family.

Johannson crystal geometry *Crystallography*. the geometry of a full-focusing crystal in an electron-probe microanalyzer.

Johansson block *Mechanical Devices*. a gauge block that is ground to a minimum tolerance of one hundred-thousandth of an inch. Also, JO BLOCK.

John Dory *Vertebrate Zoology*. a cosmopolitan deepwater fish of the genus *Zeus,* family Zeidae, having a deep, laterally compressed body, a long, spiny dorsal fin, and a wide mouth with telescopically protruding jaws used to capture prey.

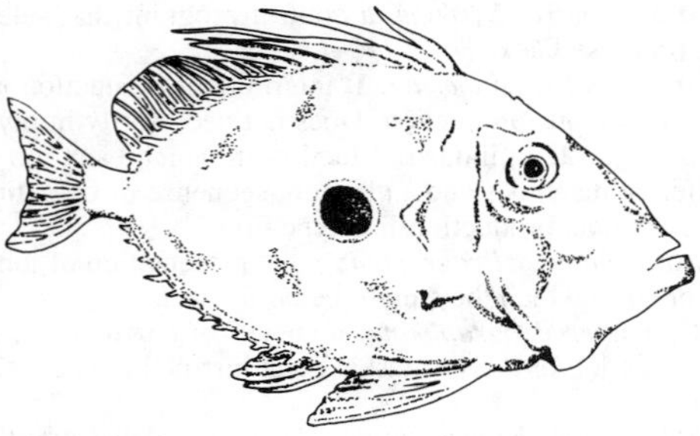

John Dory

Johne's bacillus see MYCOBACTERIUM PARATUBERCULOSIS.

Johne's disease *Veterinary Medicine*. a chronic, infectious wasting disease affecting cattle, sheep, and goats, caused by *Mycobacterium paratuberculosis johnei* and characterized by diarrhea, weight loss, and chronic enteritis affecting the ileum, cecum, and colon. (Named for Heinrich Albert *Johne,* 1839–1910, German pathologist.) Also, PARATUBERCULOSIS.

Johnson, Philip born 1906, American architect and author, theorist and practitioner of the eclectic "International Style."

Johnson, Virginia born 1925, American psychologist; with William Masters, conducted noted studies in human sexual behavior.

Johnson and Lark-Horowitz formula *Solid-State Physics*. a relation between the resistivity r of a degenerate semiconductor or metal and the number density n of the impurity atom per cubic centimeter: $r = 6270n^{-1/3}$ ohm-cm.

Johnson concentrator *Mining Engineering*. a machine consisting of a cylindrically shaped shell lined with rubber grooves parallel to the inclined axis; used to separate heavy particles from auriferous pulp.

Johnson-Mehl-Avrami analysis *Materials Science*. a collection of theories detailing the nucleation and growth kinetics of isothermal solid-state transformations.

Johnson-Mehl equation *Materials Science*. an equation to determine the volume fraction of transformed material as a function of time for a solid-state nucleation and growth transformation under isothermal constant composition conditions.

Johnson-Rahbeck effect *Physics*. an increase in frictional force between two electrodes in contact with a semiconductor, arising from the application of a potential difference between the electrodes.

Johnston's organ *Entomology*. a collection of receptors, located near the base of the antennae, that are sensitive to movement and vibration.

johnstrupite see MOSANDRITE.

join to bring, put, or come together; a place or instance of this; specific uses include: *Computer Programming*. in relational database management, to combine the information in two or more files into a third file by combining records that share a common field. *Robotics*. a function of a robot's control program that allows an activity to begin again after it has been interrupted. *Mathematics*. **1.** if A and B are subgroups (or subrings, etc.) of a group G (or ring, etc.), then the join of A and B, denoted $A \vee B$, is the minimal subgroup (subring, etc.) of G that contains both A and B. **2.** if a and b are elements of a partially ordered set L with ordering $\leq$, then the join of a and b, denoted $a \vee b$, if it exists, is $\sup\{a, b\}$; i.e., the least element of L such that $a \vee b \geq a$ and $a \vee b \geq b$. In particular, join is one of the binary operations defined on a lattice.

joiner *Building Engineering*. a carpenter, especially one who works with timber materials to form the finishings of a building.

joiner plans *Naval Architecture*. a plan indicating the layout of a ship's living areas and quarters.

joinery *Building Engineering*. **1.** a classification of various types of joints fabricated by woodworkers. **2.** the craft or trade of a joiner.

joint a connection between elements, or the place where this occurs; specific uses include: *Anatomy*. the place of union or junction between two or more bones, especially a junction that admits motion of one or more bones; an articulation. *Botany*. a part on a stem, usually thickened, from which a leaf or branch grows. Also, NODE. *Engineering*. the surface where two or more structural or mechanical shapes are joined; the juncture of two or more pieces of wood, masonry, or metal. *Robotics*. the point in a robotic arm or body where two sections are linked together and have some degree of freedom. *Graphic Arts*. either of two hinges where the sides of a casebound book are attached to the spine. Also, HINGE. *Electricity*. a temporary or permanent juncture of two wires or paths of conductive material. *Geology*. a surface fracture, break, or parting plane in a rock that occurs without displacement.

joint

joint common to or shared by two or more; specific uses include: *Military Science*. of or relating to activities, operations, organizations, etc., in which elements of more than one service of the same nation are involved. Thus, **joint amphibious operation, joint communications (network), joint deployment system, joint operations (area), joint staff, joint target list, joint task force.**

joint arm *Robotics*. a manipulator arm that has one or more joints.

joint bar *Civil Engineering*. a rigid metal rod used to connect surfaces.

joint block *Geology*. a mass of rock that is bordered by joints or that lies between adjacent joints.

joint buildup sequence *Metallurgy*. in multiple-pass welding, the sequence in which the welds are deposited.

joint capsule *Anatomy*. the saclike structure that surrounds a synovial joint, consisting of an inner synovial layer that secretes synovial fluid and an outer fibrous layer.

joint clearance *Engineering*. the distance between the united surfaces of a joint.

joint drag see KINK BAND.

jointed-arm robot *Robotics*. a robot with rotary joints along the arm that correspond to a human's shoulder, elbow, and wrist. Also, REVOLUTE-COORDINATE ROBOT.

joint efficiency *Metallurgy*. the strength of a welded joint, expressed as a percentage of the strength of the base.

jointer *Mechanical Devices*. **1.** any tool used to make joints, such as a hand tool for marking grooves in fresh cement or a plane for leveling wood surfaces before uniting them. **2.** a file for making sawteeth the same height. *Engineering*. a person who creates joints, such as a construction worker who cuts stones to fit properly.

jointer gauge *Mechanical Devices*. a gauge attachment for a bench plane or vise, used to gain the desired angle for planing.

jointer plane *Building Engineering*. a device used to true the edges of boards or for planing larger surfaces.

joint firs see EPHEDRA.

joint-ill *Veterinary Medicine*. a disease affecting neonatal swine and resulting in abscesses of the umbilicus and various organs, and arthritis in the joints of the limbs; caused by a number of different bacteria (streptococci, staphylococci, *Pasteurella, E. coli,* etc.) which may gain entrance to the animal via the navel, tonsils, and small intestine. Also, NEONATAL SEPTIC POLYARTHRITIS.

jointing *Building Engineering*. the caulking or other finishing of the exterior surface of mortar joints between bricks, stones, or timber.

jointless flooring see MAGNESITE FLOORING.

joint molecule *Molecular Biology*. a structure composed of two double-stranded DNA molecules joined by hydrogen bonding.

joint motions *Robotics*. sliding, bending, or rotary motions by a joint in a robot.

joint mouse *Pathology*. in osteoarthritis, a part of the edge of the synovial membrane that transforms into cartilage and moves freely within a joint. Also, **joint body.**

joint parameters *Robotics*. values that refer to the extensions, rotations, link lengths, offset distances, and angles between the axes of joints in a robot.

joint penetration *Metallurgy*. in welding, the minimum depth between the face of the weld and the joint.

joint pin *Mechanical Devices*. a pin that connects pieces of a knuckle joint and is held secure by a split pin.

joint plane *Geology*. the surface along which fracturing occurs or may occur to produce a joint.

joint pole *Electricity*. a pole shared by two or more kinds of utilities.

joint probability *Statistics*. the probability of the simultaneous occurrence of two or more events.

joint probability distribution *Statistics*. the probability distribution describing the simultaneous occurrence of two or more random variables.

joint ring *Mechanical Devices*. a lead ring with hollowed-out faces; used to prevent the escape of liquid or gaseous matter in pipe joints.

joint set *Geology*. a regional pattern in sedimentary or metamorphic rock, consisting of a group of parallel or nearly parallel joints.

joint space *Robotics*. the space defined by a vector of the angular and translational displacements of each joint in a robotic link.

joint system *Geology*. a system consisting of two or more intersecting joint sets of the same or different ages.

joint vein *Geology*. a small vein that is confined to one bed of rocks showing no signs of displacement.

Joinvilleaceae *Botany*. a monogeneric family of coarse, erect herbs of the order Restionales that are native to the Pacific islands; characterized by an unbranched stem that is hollow except at the nodes, alternate leaves that are well distributed along the stem, and an undifferentiated embryo.

joist *Building Engineering*. a beam or heavy piece of planking made of timber, steel, or reinforced concrete that is laid edgewise and supports floors or ceilings.

joist anchor see WALL ANCHOR.

jojoba *Botany*. a shrub of the genus *Simmondsia,* belonging to the family Buxaceae, that is native to the southwestern United States and is cultivated for its seed oil, which is used widely in the production of cosmetics and lubricants.

joking relationship *Anthropology*. a relationship between two individuals or groups in which one is allowed to tease the other with no retaliation or anger permitted; may be reciprocal or one-sided; serves to avert or alleviate potential conflicts.

Joliot-Curie, Irene [zhō´lē ō kyůr´ē] 1897–1956 (daughter of Marie and Pierre Curie), and her husband **Frédéric,** 1900–1958, French physicists; Nobel Prize for producing artificial radioactive substances.

Jolly balance *Engineering*. a spring balance that is used to determine the specific gravity of a solid by weighing it in the air and then in water or some other liquid of known density.

jolt-and-jumble tests *Ordnance*. a series of tests that are designed to simulate the shocks experienced by ammunition in transportation and handling.

jolt molding *Engineering*. a method used to shape material whereby a mold that holds a prepared batch is jerked by a mechanism to compress the substance.

Jominy end-quench test see END-QUENCH TEST.

Jonathan freckle *Plant Pathology*. a disease of stored apples characterized by small discolored dots on the fruit skin.

Jonathan spot *Plant Pathology*. a nonparasitic apple disease characterized by depressed brown spots around the stem area of the fruit.

Jones, Inigo 1573–1652, English Renaissance architect; designed Covent Garden and the Whitehall Banqueting House.

Jones-Mote reaction *Immunology*. a weak delayed hypersensitivity reaction that appears on the skin after sensitization with a small amount of soluble proteins; it appears and disappears more quickly than typical delayed reactions. Also, **Jones-Mote sensitivity.**

Jones reductor *Chemistry*. a device used to reduce solutions, consisting of a vertical tube filled with an amalgamated zinc through which the solution is passed.

Jones riffle *Mining Engineering*. an apparatus used for cutting the size of a sample by means of a hopper above a series of open-bottom pockets constructed to discharge alternately, so that each time the sample passes through, it is divided into two equal parts, continuing until the sample is reduced to the desired weight.

Jones splitter *Mining Engineering*. a device used to reduce the volume of a sample by means of a series of narrow slots or alternating chutes at the bottom of a belled, rectangular container; designed to cast material in equal quantities to opposite sides of the device.

Jones sucker rod *Petroleum Engineering*. a joining rod between a lifting or pumping device on the surface and the subsurface pump; aids in forcing oil up out of the cased hole.

Jones zone *Materials Science*. an incomplete Brillouin zone that defines quantized energy levels by both scattering intensity and geometric symmetry. Also, **Jones's zone.**

Joppeicidae *Invertebrate Zoology*. a family of very small, predatory, true bugs, hemipteran insects that live under bark or stones and in leaf litter.

joran see JURAN.

jordan [jôr´dən] *Mechanical Engineering*. a mechanized refiner that contains a rotating cone inside another cone, both having longitudinal cutters.

Jordan, Marie-Ennemond Camille [zhôr dän´] 1838–1922, French mathematician; worked on group theory; proved Cayley's Theorem; proposed Jordan curve theorem.

Jordan, Pascual [yůr´dən] 1902–1980, German physicist; a founder of quantum mechanics; developed second quantization.

Jordan algebra *Mathematics*. an algebra A over a field in which the multiplication satisfies the following rule, rather than the associative law: $(ab)a^2 = a(ba^2)$, for all $a,b \in A$.

Jordan block *Mathematics*. a square matrix with the same nonzero value λ appearing in each diagonal position, 1's on the super diagonal, and 0's elsewhere. Also, **Jordan matrix.**

Jordan canonical form *Mathematics*. a square matrix composed of basic Jordan blocks along the diagonal and 0's elsewhere. By theorem, if A is an $n \times n$ matrix over a field F and if K is the splitting field of the minimal polynomial of A over F (so that all of the roots of that polynomial lie in K), then an invertible $n \times n$ matrix C with entries in K can be found so that CAC^{-1} is in Jordan canonical form. Equivalently, two linear transformations on a vector space over a field F which have all their characteristic roots in F are similar if and only if they can be brought to the same Jordan form. Also, **Jordan form.**

Jordan curve *Mathematics*. a continuous curve (in real or complex space) without multiple points; that is, a curve having a parameterization $z = z(t)$ that is a one-to-one relation (except, possibly, for the initial and end values of t). If the initial and endpoints coincide, it is said to be a simple closed curve.

Jordan curve theorem *Mathematics*. a theorem stating that every simple closed curve separates the plane into two disjoint parts.

Jordan-Hölder theorem *Mathematics*. **1.** the theorem that any two composition series of a group are equivalent; thus, every group that has a composition series determines a unique list of simple groups. **2.** the theorem that any two normal series of a module A have refinements that are equivalent, and any two composition series of A are equivalent.

jordanite *Mineralogy*. $Pb_{14}(As,Sb)_6S_{23}$, a lead-gray, metallic monoclinic mineral occurring in tabular crystals, and having a specific gravity of 6.44 and a hardness of 3 on the Mohs scale; found in cavities in metamorphosed dolomite.

Jordan measurable *Mathematics*. a subset A of R^n is said to be Jordan-measurable if A is bounded and the set of boundary points of A can be covered by at most countably many intervals of arbitrarily small content.

Jordan's rule *Evolution*. a rule stating that the closest relations to a species are found immediately adjacent to but isolated from it by a geographical barrier.

Jordan-Wigner commutation rules *Quantum Mechanics*. the rules, applicable to fermions, which result from the replacement of commutators in the ordinary commutation rules by anticommutators.

joseite *Mineralogy*. Bi_4TeS_2, a gray-black, opaque, metallic, trigonal mineral, dimorphous with protojoseite, and having a specific gravity of 8.1 and a hardness of 2 on the Mohs scale; found as laminated masses in granular limestone.

josephinite *Mineralogy.* a nickel-iron such as awaruite, kamacite, and tetrataenite; occurring in stream gravel from Josephine County, Oregon.

Josephson, Brian David born 1940, British physicist; awarded the Nobel Prize for discovery of the Josephson effects in superconductivity.

Josephson effect *Materials Science.* the flow of superconducting electron pairs (Cooper pairs) across a thin dielectric separating two superconducting electrons, in the absence of a voltage drop. Also, **Josephson tunneling.**

JOSS *Computer Programming.* an interactive programming language, created by the Sperry-Rand Corporation in 1964 to handle quick calculations that were too complicated for a calculator. (An acronym for Johnniac Open-Shop System.)

Joule, James Prescott [jool; joul] 1818–1889, English physicist; formulated Joule's law; discovered the mechanical equivalent of heat and the first law of thermodynamics; demonstrated the Joule-Thomson effect.

joule *Metrology.* the basic unit of energy in the MKS system, equal to 10,000,000 ergs; the amount of work done by a force of one newton acting through a distance of one meter in the direction of the force. (Named for James P. *Joule.*)

Joule-Clausius velocity *Physics.* a quantity used in the description of the kinetic behavior of a gas, derived by finding the square root of the ratio of the pressure of the gas to 1/3 its density.

Joule cycle see BRAYTON CYCLE.

Joule effect *Physics.* a change in the length of a ferromagnetic substance, in a direction parallel to an externally applied magnetic field.

Joule equivalent *Thermodynamics.* a relationship between thermal energy and mechanical energy, expressed as the amount of heat transfer required to raise the temperature of one gram of water from 14.5°C to 15.5°C; equal to 4.1855 ± 0.0005 joules.

Joule heat *Materials Science.* the heat produced by current flow through an electrically resistive material.

Joule heating *Materials Science.* the evolution of heat in any material due to the flow of current through it, in which the heat produced is equal to the product of resistance and the square of the current.

Joule's law *Electricity.* the law that gives the relationship between heat dissipated when a current flows through a constant resistance circuit and the current; the heat energy per unit time equals the resistance of the circuit multiplied by the square of the current. *Thermodynamics.* the principle that the internal energy of a sample of an ideal gas depends only on its temperature, and is independent of its pressure and volume.

Joule-Thomson coefficient *Thermodynamics.* a quantitative expression of the change in temperature of a gas undergoing a Joule-Thomson expansion, given as μ in $\mu = (\partial T/\partial P)_H$, where T is temperature, P is pressure, and H is the enthalpy of the system.

Joule-Thomson effect *Thermodynamics.* the change in temperature of a sample of gas as it experiences a Joule-Thomson expansion. Also, **Joule-Kelvin effect.**

Joule-Thomson expansion *Thermodynamics.* a gas-expansion process in which an enclosed sample of gas is allowed to escape through a small aperture or porous plug into a lower pressure region; generally it is observed that the final temperature of the sample differs from the original temperature. Also, **Joule-Thomson process.**

Joule-Thomson experiment *Thermodynamics.* an experiment that consists of passing a fluid in steady flow through a porous plug in a duct under such conditions that heat transfer, kinetic energy change, and potential energy change are negligible.

Joule-Thomson inversion temperature see INVERSION TEMPERATURE.

Joule-Thomson valve *Physics.* a porous plug used for the liquefication of gases; gases under pressure expand through the plug to an area of lower temperature, thereby cooling adiabatically in accordance with the Joule-Thomson effect.

journal *Mechanical Engineering.* a term for the portion of a shaft or axle that is supported by a bearing in which it revolves.

journal bearing *Mechanical Engineering.* a cylindrical bearing that surrounds and supports a rotating shaft.

journal box *Mechanical Engineering.* a metal container that houses a journal bearing.

journal friction *Mechanical Engineering.* resistance that is caused by compression of the bearing against the rotating shaft under load.

journey *Transportation Engineering.* a passenger's total movement from origin to destination.

JOVIAL *Computer Programming.* an early computer programming language designed for command and control applications.

Jovian *Astronomy.* relating to the planet Jupiter. (From *Jove,* another name for the god Jupiter.)

Jovian moon *Astronomy.* any of the sixteen moons of Jupiter, numbered I-XVI respectively. They are Io, Europa, Ganymede, Callisto, Amalthea, Himalia, Elara, Pasiphae, Sinope, Lysithea, Carme, Ananke, Leda, Thebe, Adrastea, and Metis.

Jovian moons

Jovian planet see GAS-GIANT PLANET.

Jovian Van Allen belt *Astronomy.* any of several regions in the Jovian magnetosphere where enormous numbers of charged particles are trapped.

Joy double-ended miner *Mining Engineering.* a cutter loader for continuous mining on a longwall face, consisting of two cutting heads fixed at each end of a caterpillar-mounted chassis.

Joy extensible conveyor *Mining Engineering.* a type of belt conveyor used between a loader or continuous miner and the main transport, consisting of a head section and a tail section, each mounted on crawler tracks and independently driven.

Joy extensible steel band *Mining Engineering.* a hydraulically driven system with a steel band coiled on the drivehead that connects a continuous miner to the main transport.

Joy loader *Mining Engineering.* a machine for loading coal or ore that uses mechanical arms to gather mineral on an apron pressed into the severed material.

Joy longwall loading machine *Mining Engineering.* a modified Joy loader, 10 feet long, with a hydraulically elevated loading head fitted with mechanical gathering arms.

Joy microdyne *Mining Engineering.* a dust collector used at the return end of tunnels or hard headings to dampen and trap dust as it passes through the appliance and releases it as slurry.

Joy miner *Mining Engineering.* a continuous miner, composed of a turntable, a ripper bar, and a discharge boom conveyor; used primarily in coal headings and in extraction of coal pillars.

joystick or **joy stick** *Aviation.* an informal term for the control stick of an airplane or other vehicle. *Computer Technology.* a lever that pivots in all directions, used widely in computer games as a locator device because its action is faster than that of the directional arrow keys; it also serves as an input device in computer-aided design systems.

Joy-Sullivan hydrodrill rig *Mining Engineering.* a drill rig in which the drill is mounted on a jib or boom that can be moved and locked in any required position by hydraulic power.

Joy transloader *Mining Engineering.* a self-propelled machine with rubber tires used for loading, transporting, and dumping.

Joy walking miner *Mining Engineering.* a continuous miner with a walking mechanism instead of caterpillar tracks, making it more suitable for thin seams.

J particle *Particle Physics.* a particle made up of a charmed quark and an anticharmed quark, having a mass three times that of a proton and a lifetime 1000 times longer than expected. Also, PSI PARTICLE.

J scan see J DISPLAY.

J scope *Electronics.* a radarscope that produces a J display. Also, J INDICATOR.

Jubulaceae *Botany.* a large, cosmopolitan family of liverworts belonging to the order Jungermanniales, characterized by mostly prostrate plants that produce reddish to purplish pigments, pinnate branches, rhizoids growing from the base or underleaf middle, and incubous complicate-bilobed leaves.

judgment sampling *Statistics.* a form of nonrandom sampling in which judgment is used to decide which elements to include in the sample.

jugal *Anatomy.* **1.** of or relating to a structure or part that connects other structures or parts in a yokelike manner. **2.** of or relating to the cheek or the cheekbone.

Jugatae see HOMONEURA.

jugate *Biology.* having two leaflets originating from a common point. *Invertebrate Zoology.* having a jugum.

Juglandaceae *Botany.* a family of trees of the order Juglandales, having pinnate leaves and bearing nuts that remain closed at maturity; it includes species of walnut, pecan, and hickory trees.

Juglandales *Botany.* an order of trees that includes the family Juglandaceae and the ancestral family Rhoipteleaceae.

jugular *Anatomy.* **1.** of or relating to the throat or neck. **2.** of or relating to the jugular veins. **3.** a jugular vein.

jugular foramen *Anatomy.* an opening at the base of the skull between the temporal and occipital bones through which pass the sigmoid venus sinus and several nerves.

jugular foramen syndrome *Medicine.* a paralysis of the nerves of the tongue, pharynx, vagus, and spine, resulting from injury to a jugular foramen.

jugular process *Anatomy.* either of two processes that extend from the occipital bone toward the temporal bone.

jugular vein *Anatomy.* any of four veins, including two **external jugulars** and two **internal jugulars**, that drain blood from the head and neck.

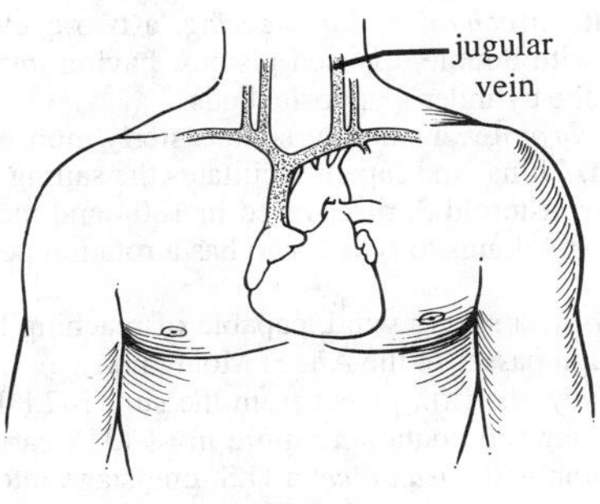

jugular vein

jugum *Botany.* **1.** a pair of opposite leaves or leaflets. **2.** a ridge on the mericarp of umbelliferous plants. *Invertebrate Zoology.* **1.** a lobe at the base of the forewing of an insect that holds fore- and hindwing together in flight. **2.** a crossbar connecting the two arms of the brachidium in certain brachiopods. (A Latin word meaning "yoke.")

Julian, Percy Lavon 1899–1975, American chemist; developed synthetic physostigmine and progesterone; researched uses of soybean.

Julian calendar *Astronomy.* the calendar established by Julius Caesar, consisting of 365 days per year (366 every fourth year) and 12 months of 30 or 31 days (28 in February, except for the fourth or leap year, when it has 29 days).

Julian date *Astronomy.* a serial number used since 1582 in astronomical calculations, equal to the number of days elapsed since January 1, 4713 BC. Also, **Julian day.**

Julianiaceae *Botany.* a family of tropical American small trees or shrubs of the order Sapindales, characterized by well-developed resin ducts, a dicotyledonous embryo, and a dry fruit with one or two hairy nuts.

Julia sets *Chaotic Dynamics.* a family of geometrical shapes, early models of fractals, generated by mapping functions of the form

$$z_{n+1} = az_n(1-z_n),$$

where z is a complex number, and then plotting the real part of z versus the imaginary part of $-z$ coloring each iterate differently.

Julida *Invertebrate Zoology.* a large order of millipedes in the subclass Helminthomorpha, having thirty to ninety trunk segments, all with poison glands. Also, **Juliformia.**

julienite *Mineralogy.* $Na_2Co(SCN)_4 \cdot 8H_2O$, a blue, tetragonal mineral that may be an artifact; occurring in crusts of needlelike crystals, and having a specific gravity of 1.65 and an undetermined hardness; found in Katanga, Zaire.

jump *Computer Programming.* to bypass the usual one-step incrementing of the program counter and pass program control to another location in the program; the address of that instruction is loaded into the program counter. Also, BRANCH.

jump discontinuity *Mathematics.* a function f is said to have a jump discontinuity at a point c if $\lim_{x \to c-} f(x)$ and $\lim_{x \to c+} f(x)$ both exist but are unequal; that is, the function has different limiting values at c when x is allowed to approach c from different directions.

jumper *Electricity.* a short piece of conductor used to make a connection between two points or terminals of a circuit, or to provide current flow around an open circuit. *Computer Technology.* **1.** a section of a circuit board designed for jumpers. **2.** to make jumper connections. *Mechanical Devices.* a device that imparts a jumping motion to a tube or drill.

jumper cable see BATTERY CABLE.

jumper tube *Mechanical Engineering.* a short tube that bypasses the fluid flow in a heater or boiler.

jump fire *Forestry.* a forest fire that starts a certain distance ahead of the main, larger fire because of burning materials carried by wind.

jump flap *Surgery.* a flap cut from the abdomen and attached to the forearm; later the flap is transferred to another location to correct a defect.

jump frequency *Materials Science.* the temperature-dependent frequency at which atoms diffusing in the solid state jump from one site to another.

jump phenomenon *Control Systems.* a phenomenon that occurs when a nonlinear system is subjected to a sinusoidal input at a constant frequency, the value of the amplitude of the forced oscillation jumps up or down as the amplitude of the input is varied through either of two fixed values, and a graph of the forced amplitude, measured against the input amplitude, follows a hysteresis loop.

jump resonance *Control Systems.* a jump discontinuity in the frequency response of a nonlinear closed-loop control system whose loop has become saturated.

jump-start *Electricity.* to start a car or truck engine that has a weak or dead battery by means of battery cables.

Juncaceae *Botany.* the rushes; a family of grasslike or rushlike plants having pithy or hollow stems, flowers with perianth, and frequently, rhizomes and hairy roots.

Juncaginaceae *Botany.* a family of monocotyledonous cyanogenic and commonly rhizomatous herbs of the order Najadales, generally growing in bogs and other very wet and often saline places in temperate and cold regions.

Juncales *Botany.* an order of the subclass Liliidae that includes the family Juncaceae.

junco *Vertebrate Zoology.* any of several North American species of small sparrows of the genus *Junco,* family Fringillidae, having dark backs, white underwings and belly, and a twittering call. (From a Spanish word for "rush;" because the bird is often found among such plants.)

junction an act or state of joining; a place where two things are joined; specific uses include: *Electricity.* **1.** a point, line, or place at which two or more semiconducting materials of opposite polarity meet, such as a P-N junction. **2.** a fitting used to join a branch waveguide to a main waveguide at an angle. **3.** see JOINT. *Electronics.* the connection between conductors or between sections of a transmission line. *Transportation Engineering.* a meeting of streets, railroads, or the like. *Cartography.* in surveying, a point at which two or more lines of levels are joined. *Anatomy.* the place of joining or meeting, as of two different organs or types of tissue. Thus, **junctional.**

junctional complex *Cell Biology.* a specialized group of intercellular junctions, found flanking the apical domain of epithelial cells and composed of tight junctions, adherent junctions, and desmosomes, that maintains the structural continuity of the epithelial sheet.

junctional nevus *Medicine.* a skin lesion containing melanocytes at the junction of the epidermis and dermis.

junction battery *Nucleonics.* a battery in which a radioactive element such as strontium-90 is used to irradiate the junction between the P-type material and the N-type material on a silicon chip.

junction bench mark *Cartography.* in surveying, a natural object or point of known elevation that is selected to be the junction of two or more lines of levels.
junction box *Electricity.* an enclosed container designed to protect wires and cables that lead into it and are there connected to form joints; also provides ease in making changes.
junction buoy *Navigation.* a buoy that marks the intersection of two channels, or the point where a channel splits into two or more legs.
junction capacitance see DEPLETION-LAYER CAPACITANCE.
junction capacitor *Electronics.* a device that utilizes capacitance in a junction between two semiconductor materials with opposing electrical properties.
junction detector *Nucleonics.* an ionization chamber in which a semiconductor generates an electric pulse that is linearly proportional to the ionizing radiation deposited at its reversed bias junction.
junction diode see JUNCTION RECTIFIER, def. 2.
junction field-effect transistor *Electronics.* a field-effect transistor in which the controlling electric field is produced by a reversed-biased P-N junction.
junction filter *Electronics.* a filter circuit that is incorporated in a junction of three or more transmission lines in order to control the transfer of signals of different frequencies between the paths that intersect at the junction.
junction isolation *Electronics.* the technique of sectioning off a component on an integrated circuit with a reversed-biased P-N junction to enhance its resistance to electron flow.
junction laser *Optics.* a laser in which the junction of a semiconductor device serves as the source of the beam of radiation.
junction loss *Telecommunications.* in telephony, the measurement of the grade of transmission that can be assigned to interaction effects that take place at trunk terminals.
junction point see BRANCH POINT.
junction pole *Electricity.* **1.** a pole at the end of a transposition section of an open wire. **2.** a pole that is shared by two adjacent transposition sections.
junction potential *Physical Chemistry.* the potential that exists at the meeting point between two solutions in an electrochemical cell, caused by the movement of some oppositely charged ions across this junction; in some types of cells it is virtually nonexistent, but in others it is significant enough to affect the accuracy of measurements of the energy developed by the electrode reactions.
junction rectifier *Electronics.* **1.** a device in which a single crystal conducting material, such as silicon, conducts current more easily in one direction than another; such devices are basic elements in injection lasers and can be used as solar cells. **2.** a device that generates a direct current by channeling the electron flow through a junction between two materials with opposing electrical properties, thus allowing it to operate faster and at higher temperatures than a conventional diode. Also, JUNCTION DIODE.
junction station *Electronics.* one of several stations that relay microwave radio signals along a line-of-sight route to the main station.
junction streamer *Geophysics.* the progressive linking together of negative charges at ever-higher altitudes in a thunderstorm prior to a lightning stroke.
junction transistor *Electronics.* a type of transistor in which the emitter and collector serve as barriers between regions with opposing electrical properties.
junction transposition *Electricity.* a transposition located at the junction pole between two transposition sections of an open wire line.
junctor *Electricity.* in crossbar systems, a circuit that lies between frames of a switching unit and terminates in a switching device on each frame.
June bug or **Junebug** *Invertebrate Zoology.* any of several large, brown beetles appearing in late spring, such as *Phyllophaga fusca.*
June solstice *Astronomy.* the moment when the sun reaches its most northerly declination on the celestial sphere; this marks the start of summer in the Northern Hemisphere and the beginning of winter in the Southern Hemisphere.
Jung, Carl Gustav [yŭng] 1875–1961, Swiss psychiatrist and psychologist; founded analytical psychology; delineated the personal and the collective unconscious.
Jungermanniaceae *Botany.* a cosmopolitan family of erect and creeping liverworts of the order Jungermanniales, characterized by pigmentation ranging from brownish to purple, lateral or ventral intercalary branches, and a weakly differentiated stem cortex.
Jungermanniales *Botany.* the leafy liverworts; an order of the division Bryophyta, distinguished by a globose antheridium on a long stalk.
Jungermanniidae *Botany.* a subclass of liverworts of the class Hepticopsida in the division Bryophyta, having little or no tissue differentiation and occurring as erect or prostrate stems with leafy appendages.
Jungian [yŭng′ē ən] *Psychology.* of or relating to Carl Jung or his theories of psychology, especially Jung's conception of the collective unconscious of humanity as a whole. Thus, **Jungian psychology, Jungian analysis, Jungian psychoanalysis.**
jungle *Ecology.* **1.** an area of dense, wild vegetation that forms in an open area of a tropical rain forest, as along a river bank or where the original tree growth has been cleared; characterized by a heavy, impenetrable undergrowth of such plants as bamboo scrub and palms. **2.** a popular term for a tropical rain forest or for any heavy, unrestrained growth of tropical vegetation. (From a Hindi word for a wild or desolate place.)
jungle yellow fever *Medicine.* a form of yellow fever, transmitted by the *Aedes leucocloenus* mosquito and various mosquitoes of the Haemagogus complex and originating in the primates of South America, Central America, and Africa.
juniper *Botany.* any plant of the genus *Juniperus,* coniferous trees and shrubs of the family Cupressaceae, especially *J. communis,* the **juniper tree.**
juniper berry *Materials.* the dried ripe fruit of the juniper tree, used for flavoring gin and formerly as a diuretic.
juniperic acid *Organic Chemistry.* $C_{16}H_{32}O_3$, a crystalline hydroxy acid found in wax from conifers such as savin and arborvitae; it melts at 95°C. Also, **juniperinic acid.**
juniper oil *Materials.* an oil derived from the wood or berries of the juniper tree; used in medicines and liquors.
juniper tar oil see CADE OIL.
junk *Naval Architecture.* a traditional Chinese sailing ship having square sails, a high stern, and usually a flat bottom.
junk DNA *Genetics.* DNA that is found within the genome of an organism and is passed from one generation to the next, but appears to have no function.
Junkers engine *Mechanical Engineering.* a two-cycle internal combustion engine with double-opposed pistons, having intake and exhaust ports located at the cylinder's opposite ends.
junk wind *Meteorology.* a south or southeasterly monsoon wind occurring in Thailand, China, and Japan; facilitates the sailing of junks.
Juno *Astronomy.* asteroid 3, discovered in 1804 and measuring 230 by 288 kilometers; it belongs to type S and has a rotation period of 7 hours, 12 minutes.
junta *Meteorology.* a strong wind, capable of reaching hurricane force, that blows through passes in the Andes Mountains.
Jupiter *Astronomy.* the fifth planet from the sun, 142,800 kilometers in equatorial diameter and containing more mass (318 earths) than all the other planets combined. *Ordnance.* a U.S. one-stage intermediate-range ballistic missile powered by a liquid-propulsion motor and equipped with all-inertial guidance, delivering a nuclear warhead at a speed of 10,000 mph and a range of 1500 miles; officially designated **PGM-19.**

Jupiter

Jupiter's comet family *Astronomy.* a comet family with about 70 members, whose aphelia lie roughly at Jupiter's distance from the sun and whose orbits are subject to frequent gravitational perturbations by Jupiter.
Jupiter's moons *Astronomy.* see the following entries:
Jupiter I see Io.
Jupiter II see EUROPA.
Jupiter III see GANYMEDE.
Jupiter IV see CALLISTO.
Jupiter V see AMALTHEA.
Jupiter VI see HIMALIA.
Jupiter VII see ELARA.
Jupiter VIII see PASIPHAE.
Jupiter IX see SINOPE.
Jupiter X see LYSITHEA.
Jupiter XI see CARME.
Jupiter XII see ANANKE.
Jupiter XIII see LEDA.
Jupiter XIV see THEBE.
Jupiter XV see ADRASTEA.
Jupiter XVI see METIS.
Jura Mountains *Geography.* a mountain range stretching between the Rhine and Rhône rivers in west central Europe.
juran *Meteorology.* a cold, snowy, and frequently turbulent wind that blows from the Jura mountains in Switzerland toward Lake Geneva, especially in the spring. Also, JORAN.
Jurassic *Geology.* the second period of the Mesozoic Era, occurring after the Triassic and before the Cretaceous Periods (from about 195 million to 135 million years ago), and the rocks formed during that time. Also, **Jura.** (Associated with the *Jura* Mountains.)

Jurassic fossils

juridical anthropology *Anthropology.* a branch of anthropology that studies laws, sanctions, and the administration of justice within a culture or in various cultures.
Jurin rule *Fluid Mechanics.* a rule stating that the height of a liquid column in a capillary tube is equal to twice the product of the surface tension and the cosine of the contact angle, divided by the product of the tube radius and the liquid weight density.
jurupaite *Mineralogy.* a magnesian variety of xonotlite.
jury rig *Engineering.* any temporary or makeshift device, rig, or piece of equipment.
jury rudder *Naval Architecture.* a temporary rudder, rigged because of damage to or loss of the permanent rudder.
Jussieu, Antoine Laurent de 1748–1836, French botanist; wrote *Genera Plantarum,* a landmark in plant classification.
Just, Ernest Everett 1883–1941, American biologist; studied fertilization in marine invertebrates.
justification *Design Engineering.* a series of economic and noneconomic factors that form the basis of using robotic technology in a manufacturing application; these factors include improved quality, efficient use of materials, performance of hazardous or undesirable tasks, advancement of manufacturing technology, and competitive market advantage. *Graphic Arts.* the process of adjusting the space within lines of type to fit a desired measure.
justified *Graphic Arts.* of type, aligned along both margins. The lines on this dictionary page are arranged in two justified columns.
justify *Graphic Arts.* to adjust word positions for printing such that the left-hand margin or right-hand margin, or both, are in alignment. *Computer Programming.* to adjust an item in a register such that the most or least significant digit is at the corresponding end of the register.
justifying space *Graphic Arts.* an extra space added between words or characters in order to create a justified line of type.
justifying typewriter *Graphic Arts.* a machine capable of producing justified type, either automatically or with some operator assistance.
just-in-time *Design Engineering.* of or relating to a material control system that uses supplies as needed, without buffering excess stock or inventory, thus eliminating the need for extra storage space. Thus, **just-in-time inventory system, just-in-time materials management.**
justly tempered scale see DIATONIC SCALE.
just-noticeable difference *Psychology.* the finest distinction that can be detected between two stimuli, such as length, loudness, or brightness. Also, DIFFERENCE THRESHOLD.
just scale *Acoustics.* a scale based on natural rather than exact intervals between tones, which results in scales having more steps, depending on the instrument (including the human voice), and having few, if any, discordant overtones in chords.
just ton *Metrology.* see TON, def. 1.
just tuning *Acoustics.* the tuning of an instrument based on natural tones and relations between these natural tones, such as fifth and seventh chords, rather than tempered tuning based on exact intervals between notes.
just-world theory *Psychology.* the concept that life is fair, that people deserve the rewards or punishments they experience because of their own actions and character.
jute *Botany.* any plant of the genus *Corchorus,* especially *C. capsularis* or *C. olitorius. Materials.* the fiber that comes from the inner bark of these shrubs, used in making burlap, gunny sacks, wrapping paper, and twine.
jute board *Materials Science.* a strong, flexible cardboard made from jute fiber; used in manufacturing shipping cartons.
Jutland *Geography.* a peninsula in northern Europe occupied by Denmark and north central Germany.
jutty see JETTY.
juvenile of or relating to youth or immaturity; a young or immature object or organism; specific uses include: *Physiology.* a young person. *Zoology.* a young animal. *Cell Biology.* a cell that is midway between the immature and mature forms. *Medicine.* denoting various diseases or conditions that primarily or exclusively affect young people. *Volcanology.* of pyroclastic material or water, derived directly from magma.
juvenile characteristics *Behavior.* any of the various physical features or behavior patterns that serve to distinguish young animals from adults of the same species.
juvenile delinquency *Psychology.* a persistent pattern in an adolescent or child of antisocial, illegal, or criminal behavior.
juvenile diabetes see JUVENILE-ONSET DIABETES.
juvenile hormone *Entomology.* neotenin; a hormone secreted by a small gland behind the brain during all insect larval stages; the insect metamorphosizes into adulthood when production of the hormone ceases.
juvenile melanoma *Medicine.* a benign, slightly pigmented skin tumor occurring in children.
juvenile-onset diabetes *Medicine.* a severe metabolic disease in which an insulin deficiency prevents the body from utilizing carbohydrates efficiently and forces it to rely on lipids and proteins; characterized by an abrupt onset during the first two decades of life, requiring mandatory insulin and dietary regulation.
juvenile rheumatoid arthritis see STILL'S DISEASE.
juvenile rift *Geology.* a stage in the breakup of a continent just prior to the beginning of actual seafloor spreading and the generation of new oceanic crust.
juvenile water *Hydrology.* water contained in ore derived from magmatic rock and thought to have come to the surface for the first time.
juvite *Petrology.* a light-colored nepheline syenite that is primarily composed of potassium feldspar with more potassium oxide than sodium oxide.
juxta- a combining form meaning "near" or "adjoining," as in *juxtaposition, juxtaspinal.*
Jynginae *Vertebrate Zoology.* the wrynecks, a subfamily of woodpeckerlike birds of the family Picidae; found in Eurasia and Africa.

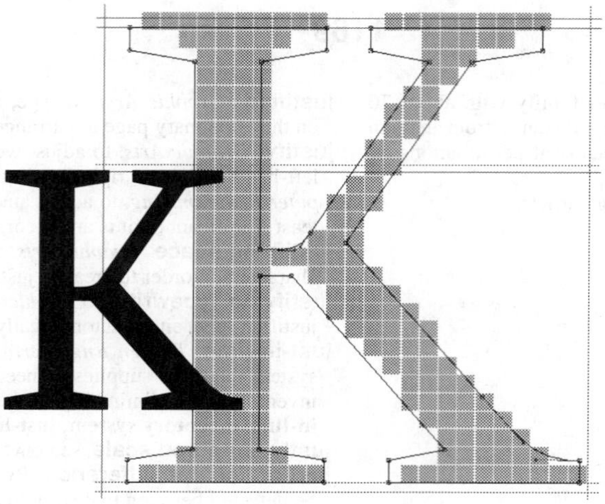

k kilo-.

k or **k.** karat; kilogram.

k a symbol for: velocity constant; mass transfer coefficient; Boltzmann constant; radius of gyration.

K Kelvin.

K a symbol for: *Chemistry.* potassium. (From New Latin *kalium.*) *Nucleonics.* the neutron multiplication constant or multiplication factor. *Computer Programming.* the number 1024 or 2^{10}, as in 20K of memory.

K a symbol for: equilibrium constant; stress-intensity factor; bulk modulus.

K. kip, kips.

K2 *Geography.* the world's second-highest mountain (or possibly the highest, according to some estimates), in the northwestern Himalayas on the China-Pakistan border. Also, MOUNT GODWIN AUSTEN.

kΩ kilohm.

kA kiloampere.

Ka. or **ka** cathode. (From Greek *kathode.*)

kaavie *Meteorology.* a term used in Scotland to denote a heavy snowfall.

kachchan *Meteorology.* a hot, dry foehnlike wind that develops in the lee side of the Sri Lanka hills during the southwest summer monsoon.

K acid *Organic Chemistry.* $C_{10}H_9O_7NS_2$, a compound that serves as an intermediate for mono- and polyazo dyestuffs.

K-A dating or **K-Ar dating** see POTASSIUM-ARGON DATING.

K-A decay *Nuclear Physics.* a type of radioactive decay in which potassium-40 captures an orbital electron and as a result changes into argon-40; used to determine the age of rock..

kaemferol *Biochemistry.* $C_{15}H_{10}O_6$, a plant pigment containing the flavone ring; occurs naturally in half of all angiosperms.

kagu

kagu *Vertebrate Zoology.* a ground-dwelling bird native to New Caledonia, characterized by rough, gray feathers and a long, pointed bill with which it probes the ground for insects; an endangered species.

Kähler manifold *Mathematics.* a complex manifold with a positive definite Hermitian metric tensor $h = h_{ij}(z)dz^i \otimes d\bar{z}^i$ whose associated fundamental two-form $\Omega = \sqrt{(-1/2)}\, h_{ij}(z)dz^i \wedge d\bar{z}^i$ is closed ($d\Omega = 0$).

Kahler's disease see MULTIPLE MYELOMA.

Kahn, Louis 1901–1974, American architect; designed the Capitol of Bangladesh and the Salk Institute (La Jolla, California).

Kahn flocculation test *Pathology.* a precipitation test formerly employed to test for syphilis, using an antigen derived from healthy cattle hearts.

kainite *Mineralogy.* $MgSO_4 \cdot KCl \cdot 3H_2O$, a colorless to gray, pink, or bluish monoclinic mineral occurring as thick tabular crystals and in irregular granular masses, having a specific gravity of 2.15 and a hardness of 2.5 to 3 on the Mohs scale; found in bedded salt deposits. It is used as a fertilizer and as a source of potassium and magnesium compounds.

kainosite-(Y) *Mineralogy.* $Ca_2(Y,Ce)_2Si_4O_{12}(CO_3) \cdot H_2O$, a yellowish-brown orthorhombic mineral occurring in prismatic crystals, having a specific gravity of 3.34 to 3.61 and a hardness of 5 to 6 on the Mohs scale; found in granite pegmatites. Also, CENOSITE.

kairomone *Ecology.* a type of pheromone that attracts other species, sometimes even natural enemies.

Kaiserstuhl disease *Toxicology.* chronic poisoning due to exposure to arsenic-containing compounds used as insecticides on grapes grown in the Kaiserstuhl wine district of Germany.

kaki *Botany.* **1.** the Japanese persimmon, *Diospyros kaki,* a tree of the family Ebenaceae. **2.** the fruit of this tree.

kakidrosis *Medicine.* a former term for bromidrosis or bromhidrosis; foul-smelling perspiration.

kakorrhaphiophobia *Psychology.* an irrational fear of failure.

kala azar *Medicine.* a chronic disease, fatal when untreated, marked by fever, fatigue, secondary infections, and enlargement of the lymph nodes, spleen, and liver; caused by the hemoflagellate *Leishmania donovani* and transmitted by a species of sandfly in parts of Asia, Africa, and South America. (A Hindi term meaning "black fever.")

Kalashnikov *Ordnance.* a Soviet assault rifle; one model, the **AK-47,** was the standard weapon of the Warsaw Pact armies.

kal Baisakhi *Meteorology.* a brief, dusty squall that heralds the arrival of the southwest monsoon over Bengal, between the months of April and June.

kale *Botany.* a variety of cabbage, *Brassica oleracea acephala,* of the order Capparales, having loose curled or wrinkled leaves that do not form a head. Also, BORECOLE.

kale

kaleidoscope [kə līd′ə skōp′] *Optics.* a tube-shaped instrument or toy containing colored bits of plastic or glass that are reflected by two plane mirrors placed at a 60° angle; when the tube is rotated and pointed toward a light source, it produces symmetrical patterns of colors that can be seen through a peephole at one end. (From a Greek phrase meaning "beautiful shape.")

kalema *Oceanography.* a very heavy winter surf on the Guinea coast of West Africa, about 10°N.

kali- *Geology.* a prefix in the name of an igneous rock, indicating an absence of plagioclase or a plagioclase content of less than 5.0%.

kaliborite *Mineralogy.* $KHMg_2B_{12}O_{16}(OH)_{10}\cdot 4H_2O$, a colorless to white, monoclinic mineral that has a specific gravity of 2.116 and a hardness of 4 to 4.5 on the Mohs scale; found as crystals and granular masses in salt deposits. Also, PATERNOITE, HEINTZITE, HINTZEITE.

kalicinite *Mineralogy.* $KHCO_3$, a soft, colorless to white or yellowish, monoclinic mineral occurring in crystalline aggregates, having a specific gravity of 2.17 and an undetermined hardness. Also, **kalicine, kalicite.**

kalinite *Mineralogy.* $KAl(SO_4)_2\cdot 11H_2O$, a colorless or white, transparent, probably monoclinic mineral occurring in fibrous form, having a hardness of about 2 to 2.5 on the Mohs scale and an undetermined specific gravity. Also, POTASH ALUM.

kaliophilite *Mineralogy.* $KAlSiO_4$, a rare hexagonal mineral, polymorphous with kalsilite, panunzite, and trikalsilite, occurring in prismatic crystals, and having a specific gravity of 2.49 to 2.67 and a hardness of 6 on the Mohs scale; found in ejected blocks of basement rocks at Mt. Somma, Italy. Also, FACELLITE, PHACELLITE.

kalium see POTASSIUM.

kallidin *Biochemistry.* either of two types of polypeptides released by blood plasma globulin.

kallidin I see BRADYKININ.

kallikrein *Enzymology.* a proteolytic enzyme that is found in tissues and plasma and catalyzes the formation of kinins from kininogens.

Kallima *Invertebrate Zoology.* a genus of butterflies whose folded wings mimic a dead leaf, with markings in various shades of brown; the Indian Leaf butterfly is a well-known example.

Kallymeniaceae *Botany.* a family of red algae in the order Cryptonemiales, having compressed or laminate branched thalli and spermatangia borne in superficial patches.

Kalman filter *Control Systems.* a linear system that minimizes the mean squared error between the desired output and the actual output when subjected to a random input. *Mathematics.* in its simplest form, a recursive technique for performing a linear least-squares fit to data. As data points are reported, the filter's estimate of the equation for the straight line is updated, as well as estimates for the error covariances relating the line's slope and intercept. At each step, the error variances for the reported data point are taken into account. Uncertainty about the validity of the straight-line assumption can also be taken into account, via "plant noise" or "model noise" error covariances that have the effect of diminishing the influence of older measurements. In its more sophisticated incarnations, the Kalman filter need not be limited to straight-line models, and can be used to estimate function values and derivatives to arbitrarily high order. A broad class of linear control theory problems are solved by the Kalman filter.

Kalotermitidae *Invertebrate Zoology.* a family of dry wood termites in the order Isoptera.

kalsilite *Mineralogy.* $KAlSiO_4$, a colorless, white or gray, hexagonal mineral, polymorphous with kaliophilite, panunzite, and trikalsilite; massive in habit, and having a specific gravity of 2.59 to 2.63 and a hardness of 6 on the Mohs scale; found in volcanic rocks in southwestern Uganda.

Kaluza theory *Physics.* a model of the nature of the universe in which the knowable four-dimensional universe is held to be the observable manifestation of an unobservable five-dimensional universe.

kamacite *Mineralogy.* (Fe,Ni), a steel-gray, alpha-nickel-iron, body-centered cubic mineral, usually containing 4–7.5% Ni, and having a hardness of 4 on the Mohs scale and a specific gravity of 7.85; found in meteorites.

Kamchatka *Geography.* a peninsula in Asia, between the Bering Sea and the Sea of Okhotsk.

Kamchatka Current *Oceanography.* a current that carries Arctic water out of the Bering Sea into the Pacific, flowing southwestward on the eastern side of the Kamchatka Peninsula down to the northern Kurile Islands, where it joins the Oyashio Current. Also, EAST KAMCHATKA CURRENT.

kame

kame *Geology.* a long, low, steep-sided mound, hummock, or irregular ridge, consisting of poorly stratified fluvioglacial drift that is primarily sand and gravel, and deposited as an alluvial fan or a delta at the terminal edge of a melting glacier.

kame delta see DELTA KAME.

kame moraine *Geology.* **1.** an end moraine containing numerous low mounds or irregular ridges of stratified sand and gravel. **2.** a group of such mounds or ridges along the front of a stagnant glacier.

Kamerlingh-Onnes, Heike 1853–1926, Dutch chemist; a pioneer in low-temperature chemistry and superconductivity; the first to liquefy helium.

kame terrace *Geology.* a steep-sided, terracelike ridge, largely stratified gravel and sand, formed by a meltwater stream flowing between a melting glacier or stagnant ice lobe and a higher valley wall or other higher ground and remaining after the ice disappears. Also, ICE-CONTACT TERRACE, ICE-MARGINAL TERRACE.

Kamptozoa see ENTOPROCTA.

kampyle of Eudoxus *Mathematics.* the graph of the equation $x^4 = a^2(x^2 + y^2)$, or in polar coordinates, $r^2\cos^4\theta = a^2$, where a is constant. Also, CURVE OF EUDOXUS.

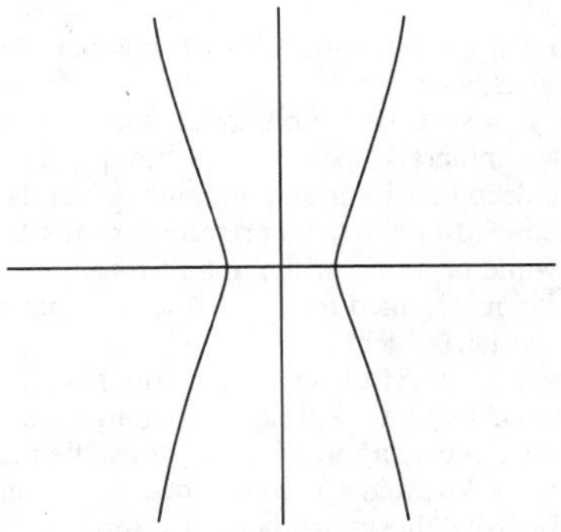

kampyle of Eudoxus

Kanagawa phenomenon *Microbiology.* a phenomenon in which strains of *Vibrio parahaemolyticus* isolated from humans display a clear hemolysis on an agar containing human erythrocytes, but not upon incubation with horse erythrocytes; most other strains do not follow this pattern.

Kanam man see HOMO KANAMENSIS.

kanamycin *Bacteriology.* a broad-spectrum antibiotic that is produced by *Streptomyces kanamyceticus* and that disrupts protein synthesis in bacteria; effective against many strains of staphylococci.

kanban system *Industrial Engineering.* a type of just-in-time inventory system that uses cards to display the status of the inventory and to control production levels. (From a Japanese word for "card.")

Kanchenjunga *Geography.* the world's third-highest mountain (28,208 feet), in the Himalayas on the Nepal-Sikkim border.

kangaroo *Vertebrate Zoology.* any of several herbivorous, marsupial mammals of the family Macropodidae having large tails, powerful hind limbs, and small forelimbs; native to Australia, New Guinea, and adjacent islands. The larger, more familiar species are known collectively as the **great kangaroos.** (An English alteration of a native Australian name for this animal.)

kangaroo

Kangaroo *Ordnance.* a Soviet strategic air-to-surface missile that is powered by a turbojet engine and is probably equipped with inertial or preprogrammed guidance, delivering a nuclear or thermonuclear warhead at a speed of Mach 1.8 and a range of 300 miles; officially designated **AS-3.**

kangaroo rat *Vertebrate Zoology.* any of various species of pouched, nocturnal, burrowing rodents of the genus *Dipodomys* in the family Heteromyidae inhabiting arid regions of North America; so named because of its pouch, upright posture, and kangaroo-like hopping movements.

Kansan *Geology.* the second glacial stage of the Pleistocene epoch in North America, occurring after the Aftonian interglacial stage and before the Yarmouthian.

Kansasii disease *Medicine.* a type of tuberculosis that is caused by a group of atypical mycobacteria and characterized by a mild pulmonary infection that may lead to severe pulmonary disease similar to tuberculosis.

k antigen *Immunology.* any antigen that forms outside a cell wall or as part of a bacterial capsule.

kaolin *Petrology.* a soft, fine, nonplastic white or nearly white rock composed of clay minerals of the kaolin group, chiefly kaolinite, and formed from the decomposition of aluminous minerals such as feldspar. Also, BOLUS ALBA. *Materials.* a ceramic clay made from this rock, which remains white or nearly white upon firing; also used as an adsorbent and in medicine. (Named for a site in China that was a noted early source of this material.)

kaolinite *Mineralogy.* $Al_2Si_2O_5(OH)_4$, a colorless or white triclinic mineral of the kaolinite-serpentine group, polymorphous with dickite, halloysite, and nacrite, occurring in compact to friable masses, and having a specific gravity of 2.62 and a hardness of 2 to 2.5 on the Mohs scale; resulting chiefly from the hydrothermal alteration or weathering of feldspars and found in situ and in sedimentary clay beds; used in the ceramic industry.

kaolinization *Geology.* the hydrothermal alteration or weathering of aluminum silicate or other clay minerals to form kaolin.

kaon see K MESON.

kaonic atom *Atomic Physics.* an atom that has a negatively charged K meson orbiting its nucleus.

Kapitsa, Pyotr 1894–1984, Russian physicist; carried out important research in magnetism; awarded the Nobel Prize for work in low-temperature physics.

Kaplan-Meier estimator *Statistics.* a distribution-free estimator of a failure or survival distribution.

Kaplan turbine *Mechanical Engineering.* a propeller-type water turbine having blades that are adjusted according to the load in order to provide for increased efficiency.

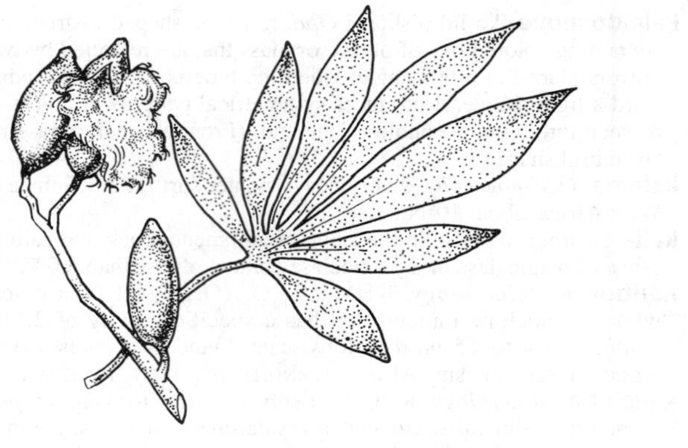

kapok tree

kapok *Botany.* a tree, *Ceiba pentandra,* of the family Bombaceae, having pods of fluffy, silky fibers that surround pea-sized seeds; native to tropical America, East Africa, and the East Indies. Also, **kapok tree.** *Materials.* the hollow, lightweight, water-resistant fibers surrounding the seeds of this tree; used to stuff mattresses, life preservers, and pillows and to make ceiling insulation and wallboard. Also, JAVA COTTON.

kapok oil *Materials.* a yellowish-green oil that is derived from seeds of the kapok tree, is soluble in alcohol and ether, and is used in foods and in making soaps.

Kaposi's sarcoma *Oncology.* a malignant neoplastic vascular proliferation characterized by soft purplish nodules that usually occur first on the toes or feet and then slowly spread over the skin, increasing in size and number. A particularly virulent form of this condition occurs in patients whose immune response has been weakened, such as AIDS patients or transplant recipients who are taking immunosuppressive drugs. (After Moritz *Kaposi* Kohn, 1837–1902, Austrian dermatologist.) Also, IDIOPATHIC MULTIPLE PIGMENTED HEMORRHAGIC SARCOMA, MULTIPLE IDIOPATHIC HEMORRHAGIC SARCOMA.

kappa the tenth letter of the Greek alphabet, written as Κ or κ.

kappa chain *Immunology.* one of the two forms of smaller polypeptide chains (known as light chains) that occur in immunoglobulins.

kappa curve *Mathematics.* the graph of the equation $x^4 + x^2y^2 = a^2y^2$, where a is constant; so named because of its resemblance to the Greek letter κ.

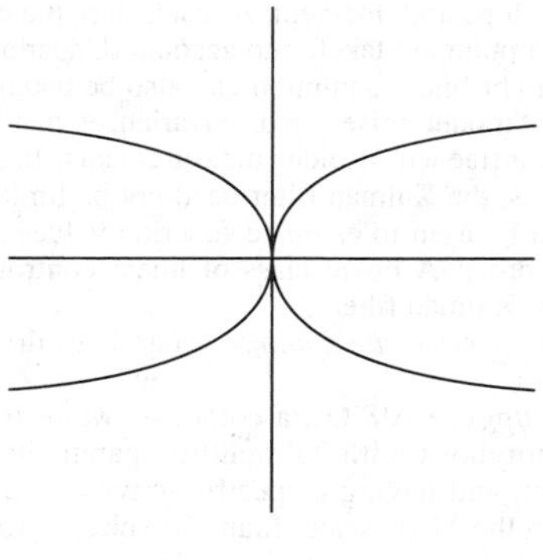

kappa curve

kappa particle *Bacteriology.* an endosymbiotic bacterium found within the cytoplasm of certain strains of *Paramecium* that produces a factor toxic to other paramecia lacking this bacteria.

Kapteyn, Jacobus 1851–1922, Dutch astronomer; analyzed the proper motions of stars; constructed a scale model of the Milky Way.

Kapteyn's Selected Areas *Astronomy.* 262 regions of the sky, each 75 arc-minutes square, in which the magnitudes, spectral types, and luminosity classes of stars have been carefully determined.

Kapteyn's star *Astronomy.* a 9th-magnitude red subdwarf star in Pictor that is one of the closest to the sun (13 light-years) and that has a high proper motion (8.8 arc-seconds per year).

kapur *Materials.* the wood of an Indonesian and Malaysian tree, *Dryobalanops camphora*; used in cabinetmaking. Also, BORNEO CAMPHOR.

karaburan *Meteorology*. a violent, dusty northeast wind that darkens the sky over central Asia during the spring and summer.

K-Ar age *Geology*. the age, in years, of a rock or mineral, based on the ratio of radioactive potassium-40 to argon-40 present in a sample. (An acronym formed from the chemical symbols *K* for potassium and *Ar* for argon.) Also, POTASSIUM-ARGON AGE.

karajol *Meteorology*. a west wind, usually of one to three days' duration, that follows rain on the Bulgarian coast.

Karakul or **karakul** [kâr′ə kəl] *Agriculture*. a breed of wool-producing sheep with fat tails and brown or black coloration whose coarse, glossy wool is used widely for fur coats. The skins of young Karakul lambs are variously known as Broadtail, Persian Lamb, and Caracul. (From *Karakul*, a village in Asia where these sheep were bred.)

karat *Metallurgy*. a value indicating the proportion of pure gold in an alloy, with 24 karats representing 100% gold.

karaya gum *Materials*. the dried exudate of an Indian tree, *Sterculia urens*; used in textile finishing and as a thickener in foods and cosmetics.

karbutilate *Organic Chemistry*. $C_{14}H_{21}N_3O_3$, an off-white solid that melts at 176–177°C and is used as a herbicide.

K-Ar dating see POTASSIUM-ARGON DATING.

karema *Meteorology*. a violent east wind that occurs on Lake Tanganyika in Africa.

Karen *Ordnance*. a Soviet naval short-range air-to-surface missile; officially designated **AS-10.**

karif *Meteorology*. a strong southwest wind that occurs on the southern shore of the Gulf of Aden during the southwest monsoon. Also, KHARIF.

Karle, Jerome born 1918, American crystallographer; with Hauptman, awarded the Nobel Prize for direct method of structural analysis of crystals.

Karl Fischer reagent *Analytical Chemistry*. a solution of iodine, pyridine, methanol, and sulfur dioxide used to determine trace quantities of water in a sample.

Karl Fischer technique *Analytical Chemistry*. the titration of a sample with a Karl Fischer reagent to determine trace amounts of water; the end point is indicated by a yellowish color change.

Kármán constant *Fluid Mechanics*. a dimensionless numerical value *k*, usually about 0.4, that is the proportionality constant in the relation $(t/r)^{1/2} = kz(\partial U/\partial z)$, where *t* is the shear stress of a liquid, *r* is the liquid density, *z* is the normal distance to a plane wall, and *U* is the mean velocity of the liquid parallel to the wall.

Kármán-Tsien relation *Fluid Mechanics*. a correction for the lift of an airfoil in transonic flow including the effects of compressibility.

Kármán vortex street *Fluid Mechanics*. a double row of line vortices arising in a fluid in the wake of a cylindrical body whose axis is perpendicular to the fluid flow, with the rows of vortices moving with a uniform velocity but with opposite rotations.

karnal bunt *Plant Pathology*. a disease of wheat caused by the fungus *Tilletia indica*.

Karnaugh map *Electronics*. a chart that displays the relationship and functions of switching circuits in a computer, commonly used to design the simplest logic circuit for a given task.

Karnian see CARNIAN.

Karp circuit *Electronics*. a device in which the microwaves travel at a speed well under the speed of light, commonly used in backward-wave oscillators.

karren *Geology*. the narrow channels or furrows, usually separated by sharp, knifelike ridges or crests, that are formed on the surface of limestone by the solvent action of rainwater. Also, LAPIES.

Karrer, Paul 1889–1971, Swiss chemist; awarded the Nobel Prize for work on flavins, carotenoids, and vitamins A and B_2.

Karrer method *Chemical Engineering*. an industrial process for the chemical synthesis of riboflavin.

karri *Materials*. the hard, red wood of the Australian tree, *Eucalyptus diversicolor*; widely used as a commercial timber.

karroo *Geology*. a dry, terraced tableland, especially of southern Africa, that supports grassy vegetation during the wet season. Also, **karoo.**

Karroo System *Geology*. a system of glaciated strata widely found in southern Africa, and formed during the Permian period.

karst *Geology*. a landscape characterized by numerous sinkholes, caves, and extensive underground drainage that is produced on limestone, gypsum, or dolomite formations by solution or dissolution. (Named for a noted area of this type near Trieste.)

karstbora *Meteorology*. a term for the bora that occurs on the Yugoslavian coast.

karst plain

karst plain *Geology*. a plain of nearly horizontal limestone strata on which karst features develop. Also, **karst plateau.**

Karumiidae *Invertebrate Zoology*. a family of termitelike beetles now included within the family Dascillidae.

karyo- or **kary-** a combining form meaning "the nucleus of a cell."

karyochrome *Neurology*. any of several varieties of nerve cells in which the nucleus stains very easily while the cytoplasm does not, and in which there are few or no Nissl bodies.

karyocyte *Cell Biology*. any nucleated cell.

karyogamy *Cell Biology*. the fusion of two nuclei, especially of gametic nuclei during sexual conjugation.

karyokinesis see MITOSIS.

karyological relict *Invertebrate Zoology*. the only existing ciliate form (order Karyorelictida) known to possess an apparently primitive dual nuclear apparatus, the macronucleus being diploid, not polyploid, and nondividing.

karyology *Biology*. the study of chromosomes, ordinarily in terms of their gross morphology and banding patterns.

karyolymph see NUCLEOPLASM.

karyolysis *Cell Biology*. the dissolution of the cell nucleus, as in necrosis, resulting in a loss of affinity for basic dyes.

karyomastigont *Invertebrate Zoology*. any member of the protozoan order Oxymonadida.

karyomegaly *Cell Biology*. the abnormal enlargement of a cell nucleus, not caused by polyploidy.

karyoplasm *Cell Biology*. the clear, apparently empty areas of the cell nucleus. Also, NUCLEAR SAP.

karyoplasmic ratio see NUCLEOPLASMIC RATIO.

karyoplast *Cell Biology*. **1.** a nucleus that has been removed from a eukaryotic cell and is surrounded by a thin layer of cytoplasm and a plasma membrane. **2.** any cell nucleus.

karyorrhexis *Cell Biology*. a degenerative change or fragmentation of the cell nucleus.

karyosome *Cell Biology*. a central body within a nucleus that may contain a mass of heterochromatin.

karyotype *Genetics*. **1.** the complete chromosome set of the nucleus of a cell. **2.** a photomicrographic display of the set of the chromosomes of an organism, typically shown in order of decreasing size.

karyotypic abnormality *Genetics*. a defect in a metaphase chromosome set.

karyotypic analysis *Genetics*. a study of all the visible traits of the chromosomes of a typical cell.

Kashmir see CASHMERE.

kasolite *Mineralogy*. $Pb(UO_2)SiO_4{\cdot}H_2O$, a yellow to brown monoclinic mineral occurring in aggregates of stout prismatic crystals, having a specific gravity of 5.83 to 6.5 and a hardness of about 4.5 on the Mohs scale; found as an alteration product of uraninite.

Kasper-Hauser *Behavior*. a term used to describe an animal that has purposely been reared under conditions of extreme isolation in order to observe the effect of this experience on its behavior. (From the noted 19th-century case of *Kasper Hauser,* a child who was thought to have been raised in isolation in a dark room.)

Kastler, Alfred 1902–1984, French physicist; awarded the Nobel Prize for discovering methods of studying Herzian resonances in atoms.

kasugamycin *Microbiology.* an atypical aminoglycoside antibiotic that is bacteriostatic, rather than bactericidal, and is also active against certain fungi. Also, **kasugamycin hydrochloride, kasumin.**

kat katal.

kata- or **kat-** a combining form meaning: **1.** down; lower; under. **2.** against. **3.** along with. **4.** very.

katafront *Meteorology.* a front at which warm air is descending relative to the frontal surface, except in the lowest layers; usually, a cold front.

katagenesis see CATAGENESIS.

katal *Enzymology.* a unit of catalytic activity, including that of enzymes; that amount of a catalyst, such as an enzyme, which brings about a reaction rate of 1 mole of substrate per second.

katallobaric *Meteorology.* indicating a decrease in atmospheric pressure. Also, **katabaric.**

Kata thermometer *Engineering.* an alcohol thermometer that measures an airflow's cooling effect, as distinguished from air temperature, by recording the time required to cool the heated thermometer a given number of degrees.

katazone see CATAZONE.

Kater, Henry 1777–1835, English physicist; invented floating collimator and Kater's pendulum; determined length of second's pendulum.

Kater's (reversible) pendulum *Mechanics.* a pendulum that can be supported from either of two movable knife edges. By comparing measurements made with the two suspension points, the acceleration of gravity can be determined without knowing the location of the center of mass of the pendulum, making it more accurate for this purpose.

Kathablepharidaceae *Botany.* a family of mostly freshwater, usually colorless flagellates of the order Cryptomonadales, characterized by two divergent, heterodynamic flagella and usually lacking trichocysts.

katharometer *Engineering.* an instrument used to analyze a gas mixture by measuring the changes in its thermal conductivity. Also, THERMAL CONDUCTIVITY CELL.

Kathlaniidae *Invertebrate Zoology.* a family of nematodes that parasitize fish, amphibians, and reptiles.

kation see CATION.

katoptric see CATOPTRIC.

katoptrite *Mineralogy.* a black, opaque, metallic monoclinic mineral occurring as tabular, elongated crystals, having a specific gravity of 4.56 and a hardness of 5.5 on the Mohs scale; found with sonolite, mostly in mines in Sweden. Also, CATOPTRITE.

katsura *Materials.* the soft, nondurable wood of a Far Eastern tree, *Cercidiphyllum japonicum*; used in making furniture, tools, and plywood.

katydid *Entomology.* a popular term for the larger and noisier nocturnal long-horned grasshoppers.

Katz, Sir Bernard born 1911, German-born English physiologist; shared Nobel Prize for research in nerve-impulse transmission.

Kauertz engine *Mechanical Engineering.* a rotary engine with vanes that serve as pistons; two of the pistons connect with a rotor and move with constant angular speed, and the other two pistons are operated simultaneously by a gear-and-crank assembly, moving with variable angular speed.

Kauffmann-White classification *Microbiology.* a nomenclature for the classification of the many serotypes of *Salmonella* bacteria.

kauri *Materials.* the brownish resin of the pine tree *Agathis australis* of New Zealand, used in varnishes and linoleum. Also, **kauri gum, kauri copal, kauri resin.**

kauri-butanol value *Analytical Chemistry.* the liquid volume of a substance required to produce turbidity when added to a solution containing a standard concentration of kauri resin dissolved in butanol.

kaus *Meteorology.* a moderate to gale-force southeasterly wind that blows in the Persian Gulf, usually between December and April, characterized by overcast skies, rain, and squalls. Also, QUAS.

kavaburd see CAVABURD.

Kawasaki disease or **syndrome** *Medicine.* a disease of unknown cause, seen in children and especially those of Japanese ancestry. Laboratory tests show a marked increase in leukocytes, mild anemia, and thrombocytosis; symptoms include fever lasting more than 5 days, a rash on the trunk, cervical lymphadenopathy, redness of the conjunctiva, changes in the oral mucosa including "strawberry tongue," and edema, redness, and peeling skin of the hands and feet. (Identified by the Japanese pediatrician T. *Kawasaki*.) Also, MUCOCUTANEOUS LYMPH NODE SYNDROME.

kay *Geology.* see KEY.

kayser *Spectroscopy.* a unit once proposed for the unit of wavenumber, equal to the reciprocal of 1 centimeter (1 cm^{-1}); formerly used in vibrational (infrared and Raman) spectroscopy.

Kayser-Fleischer ring *Pathology.* a greenish-yellow to red-gold pigmented band encircling the cornea, due to the deposition of copper; noted in Wilson's disease and other liver disorders.

Kazanian *Geology.* a European geologic stage of the Upper Permian period, occurring after the Kungurian and before the Tatarian.

kb kilobar. Also, **kbar.**

kb or **Kb** kilobase.

Kb kilobit. Also, **Kbit.**

KB kilobyte. Also, **Kbyte, kbyte.**

K band *Telecommunications.* a band of radio frequencies that extends from 10,900 to 36,000 MHz.

KB cells *Cell Biology.* a tissue culture cell line derived from abnormal epidermal cells isolated from the human nasopharynx.

kbp kilobase pair.

KBS *Anthropology.* a site at Lake Rudolf in East Africa, significant for early evidence of the genus *Homo;* dated to about 1.8 million years ago. (An acronym for Kay Behrensmeyer Site, from the name of the finder.)

kc or **kc.** kilocycle; kilocurie.

KC- *Aviation.* the U.S. military designation for tanker aircraft.

KC-135 *Aviation.* a multipurpose U.S. military tanker-transport powered by four turbojet engines and equipped for high-speed, high-altitude refueling of fighters and bombers.

kcal or **kcal.** kilocalorie.

Kcalorie *Nutrition.* the amount of heat required to raise the temperature of 1 kilogram of water through 1°C (from 14.5°C to 15.5°C); generally called "calorie" when used to express the energy value of a food.

k capture see ORBITAL-ELECTRON CAPTURE.

K-carrier system *Telecommunications.* a carrier system that allows for 12 telephone channels, having a bandwidth of up to 60 kilohertz.

K cell see KILLER CELL.

kCi kilocurie.

K-corona *Astronomy.* the part of the solar corona closest to the photosphere, extending to about two solar radii and made up of rapidly moving free electrons that emit a continuous spectrum.

kc/s kilocycles per second. Also, **kc.p.s., kc/sec.**

K damage *Military Science.* combat damage that necessitates a vehicle's destruction or causes an aircraft to fall out of control.

K-day *Military Science.* the basic date for the introduction of a convoy system on any particular convoy lane.

K display *Electronics.* the representation of targets on a radarscope in which the targets are depicted as a pair of vertical deflections of equal height when the antenna is on target; when off target, the difference in blip height indicates the direction and magnitude of the error.

kea *Vertebrate Zoology.* a stocky, olive-feathered parrot, *Nestor notabilis* of the family Psittacidae, characterized by a long-hooked upper bill; living in New Zealand forests and open country, it has an exaggerated reputation as a sheep-killer, though it does prey on sick animals or carrion.

kea

kedge *Navigation.* to move a vessel by dropping an anchor some distance away, then pulling the vessel toward the anchor by its chain.
kedge anchor *Naval Architecture.* an anchor used primarily for kedging; it may also serve as an auxiliary to the main anchor.
KEE *Artificial Intelligence.* a commercially available expert system shell that runs on LISP machines.
keel *Naval Architecture.* the main beam or girder running the length of a vessel's bottom; the "backbone" of the vessel, providing much of its longitudinal strength.
Keel *Astronomy.* see CARINA.
keel arch see OGEE ARCH.
keelboat *Naval Architecture.* a shallow freight boat formerly used on North American rivers.
keel condenser *Naval Architecture.* a steam condenser with lines placed outside the underbody near the keel.
Keeler, James 1857–1900, American astronomer; analyzed the rings of Saturn.
keel molding *Architecture.* a molding formed by two ogee curves meeting in a sharp arris shaped like a ship's keel.
keelson *Naval Architecture.* an internal longitudinal beam or girder running parallel to the keel and just above it. Also, KELSON.
Keene's cement *Materials.* the trademark name for a hard-finish white plaster that sets quickly, made by soaking plaster of Paris in a solution of alum or borax and cream of tartar.
keep *Architecture.* a great inner tower serving as the stronghold of a medieval castle. Also, DONJON.
keep-alive circuit *Electronics.* a circuit that maintains a degree of ionization in the gas of a transmit-receive tube, so that full ionization occurs more rapidly when the transmitter fires.
keeper *Electromagnetism.* an iron or steel bar that is placed across the poles of a permanent magnet so as to complete the magnetic circuit and thus prevent self-demagnetization.
Keesom, Willem 1876–1956, Dutch physicist; studied structures of liquids and compressed gases; solidified helium; discovered Keesom relationship.
Keesom relationship *Physical Chemistry.* an equation that uses the distance between interacting dipolar molecules to calculate the level of energy generated.
Keewatin *Geology.* a division of the Archeozoic rocks that make up the Canadian Shield. Also, **Kewatinian.**
Kegel karst see CONE KARST.
kehoite *Mineralogy.* $(Zn,Ca)Al_2(PO_4)_2(OH)_2 \cdot 5H_2O$, an amorphous mineral of doubtful status that occurs in white chalky masses; found at Galena, South Dakota.
Keilor skull *Paleontology.* a fossil skull from a late Pleistocene deposit of southern Australia, classified as *Homo sapiens.*
Kekulé von Stradonitz, (Friedrich) August 1829–1896, German chemist; founder of structural organic chemistry; developed benzene ring molecular model.
Kekulé structure *Organic Chemistry.* a valence-bond formula for benzene; the formula shows how the delocalization of the pi electrons in benzene results in a pattern in which the pi electrons include more than two atoms; the actual form of benzene is a composite of the two Kekulé structures. Also, **Kekulé formula.**
kelat *Materials.* the fine-textured wood of trees of the species *Eugenia* of Southeast Asia; used for building construction and in making furniture, sleepers, and bridge timbers.
Keldysh theory *Atomic Physics.* a theory that explains the relationship between energy fields and photons in multiphoton ionization.
kelectome *Surgery.* a device used in removing specimens of tissue from tumors.
K electron *Atomic Physics.* one of the electrons that is found in the shell of an atom and displays the characteristics of principal quantum number 1.
K-electron capture see ORBITAL-ELECTRON CAPTURE.
keledang *Materials.* the coarse-textured, resistant wood of trees of the species *Artocarpus* of southeast Asia; used for the construction of houses and to make pulp, interior trim, and furniture.
Kell blood group system *Immunology.* a group of red blood cell antigens that is characterized by the immune antibody known as anti-K.
kellering *Mechanical Engineering.* a process in which a contoured surface is three-dimensionally cut by tracer milling the die block or punch; a tracer guides the cutter's path around the edges of a die model.
Kellner eyepiece *Optics.* an eyepiece that has a double convex field lens and a cemented doublet eye lens.
Kellogg equation *Thermodynamics.* an equation of state relating the pressure, absolute temperature, and density of a gas: $p = RTr + S(a_n - b_nT - c_n/T^2)p^n$, where p is the pressure, R is the gas constant, T is the temperature, r is the density, and a_n, b_n, and c_n are constants.
Kelly ball *Engineering.* a metal hemisphere measuring six inches in diameter and weighing 30 pounds; used in the **Kelly ball test**, an on-site test to determine the consistency of freshly mixed concrete by measuring the ball's depth of penetration under its own weight.
keloid *Medicine.* an irregularly shaped, raised overgrowth of scar tissue at a wound site caused by the formation of excessive amounts of collagen during healing.
keloid acne *Medicine.* a chronic inflammation of the hair follicles, usually at the hairline on the back of the neck, characterized by lesions in papular or pustular form and resulting in keloidal scarring.
kelp *Botany.* the common name for any large, brown cold-water seaweed, especially of the orders Laminariales and Fucales.

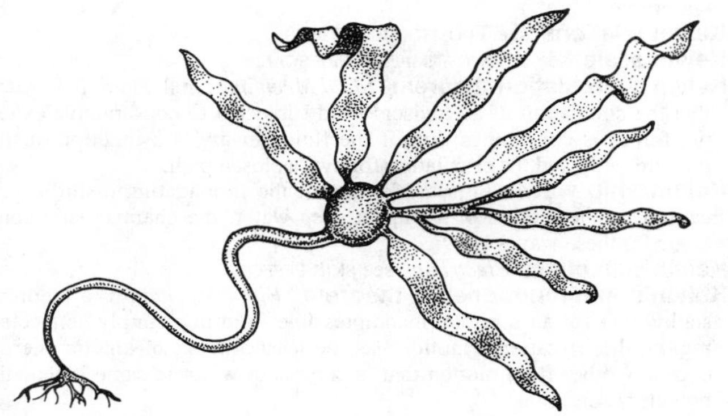

kelp

Kelsey-Sykes test *Microbiology.* a test to examine the effectiveness of a given disinfectant at each of several dilutions by incubation with a test microorganism and scoring the number of surviving cells.
kelsher *Meteorology.* a term used in England to denote a heavy rainfall.
kelson see KEELSON.
Kelt *Ordnance.* a Soviet air-to-surface stand-off missile powered by a liquid-propellant rocket engine and probably equipped with active radar or passive infrared homing, delivering a 1000-kilogram conventional warhead at a speed of Mach 1.2 and a range of 180 kilometers; officially designated **AS-5.**
Kelvin, William Thomson, Baron 1824–1907, British physicist and mathematician; a founder of thermodynamics; devised the Kelvin temperature scale.
kelvin or **Kelvin** *Thermodynamics.* **1.** the fundamental SI unit of temperature, equal to 1/273.16 of the absolute temperature of the triple point of pure water. A temperature increment of one degree kelvin (1 K) is equal to one degree Celsius (1°C). **2.** relating to or expressed by the Kelvin temperature scale.
Kelvin absolute temperature scale *Thermodynamics.* a temperature scale that is based on the absolute zero temperature (0 K) and the ice point of water (273.16 K). The scale is defined so that the ratio of the temperatures of two heat reservoirs is equal to the ratio of heat transfer from the hotter reservoir by a heat engine to the amount of heat transfer to the colder reservoir during a Carnot cycle.
Kelvin balance *Electromagnetism.* an ammeter that operates by passing the current to be measured through two coils, connected in series, one coil mounted on the arm of a balance directly above the other coil; the attractive force acting between the coils is countered by weights on the other arm of the balance.
Kelvin body *Mechanics.* an ideal body of viscoelastic material, modeled by a spring and dashpot in parallel, whose deformation is proportional to both the stress and the time it is applied.
Kelvin bridge *Electricity.* a circuit used to measure very low resistances, usually of 0.1 ohm or less. Also, KELVIN NETWORK, DOUBLE BRIDGE, THOMSON BRIDGE.
Kelvin degree see KELVIN.
Kelvin element *Materials Science.* the representation of the reaction of polymers or other materials to stress by a parallel combination of a spring and dashpot; a model for the viscoelastic behavior of a polymer.

Kelvin equation *Physical Chemistry.* a principal equation of surface chemistry, which calculates the free energy of a small particle as a function of its size and pressure. Minute droplets of a liquid have a higher vapor pressure than the bulk liquid and thus a greater rate of evaporation; larger drops decrease in free energy with their increase in size and should grow spontaneously.

Kelvin guard-ring capacitor *Electricity.* a capacitor with parallel circular plates, one plate having a guard ring separated by a small gap, used as a standard for calculating capacitance using the dimensions of the plate.

Kelvin network see KELVIN BRIDGE.

Kelvin-Planck statement *Thermodynamics.* an expression of the second law of thermodynamics, stating that it is not possible for a system to undergo a cycle of thermodynamic processes in which there is a heat transfer and an equivalent amount of work is performed, without some other change in the surrounding environment; i.e., no actual heat engine that will have an efficiency of 100% can be constructed. Also, **Kelvin statement.**

Kelvin relations see THOMSON RELATIONS.

Kelvin scale see KELVIN TEMPERATURE SCALE.

Kelvin's circulation theorem *Fluid Mechanics.* a theorem that states that the circulation of an inviscid fluid subjected to conservative external forces is constant in time if the fluid density is a function of the pressure only and the circulation follows a closed path.

Kelvin ship waves *Fluid Mechanics.* the propagating disturbances caused by the motion of a ship in deep water; the characteristic cone angle for these waves is 39°.

Kelvin skin effect *Electricity.* see SKIN EFFECT.

Kelvin's minimum energy theorem *Fluid Mechanics.* a theorem stating that for an inviscid, incompressible fluid in a simply connected region, the irrotational motion has the least amount of kinetic energy over any other fluid motion that is consistent with the same kinematic boundary condition.

Kelvin temperature scale *Thermodynamics.* a temperature scale that is accepted as an International Practical Temperature Scale and that is based on the Kelvin absolute temperature scale, within the limits of experimental accuracy. Temperature degree increments on this scale are identical to those of the Celsius temperature scale.

Kelvin time scale *Astronomy.* the time it takes for a star to contract from a cloud of gas to its current diameter, assuming that gravitational contraction is the star's only source of energy.

Kelvin wave *Oceanography.* a tidal phenomenon caused by the Coriolis effect in certain relatively confined bodies of water (the North Sea is a good example); the tidal bulge is higher to the right of the direction of advance in the Northern Hemisphere, while in the Southern Hemisphere it is higher to the left.

kelyphyte see CORONA.

kelyphytic rim *Petrology.* a secondary peripheral zone or corona consisting of concentric shells with a fibrous or radial structure of two minerals, such as pyroxene and hornblende, that sometimes forms in ultramafic igneous rocks around olivine or garnet. Also, **kelyphytic border.**

kempas *Materials.* the heavy, durable wood of the Malaysian tree, *Koompassia malaccensis.*

Kempe chain *Mathematics.* given a graph *G* on which a vertex coloring has been defined, any component of the subgraph of *G* generated by all vertices that have been assigned a particular subset of colors, together with edges joining pairs of such vertices. Kempe chains are fundamental to the Appel-Haken proof of the four-color theorem.

kempite *Mineralogy.* $Mn_2^{+2}Cl(OH)_3$, an emerald-green orthorhombic mineral occurring in small prismatic crystals, having a specific gravity of about 2.94 and a hardness of about 3.5 on the Mohs scale.

Kendall, Edward Calvin 1886–1972, American biochemist; awarded the Nobel Prize for isolating thyroxine and cortisone.

Kendall effect *Telecommunications.* in FAX transmission, a distortion caused by undesired modulation products that originate from carrier signal transmission.

Kendrew, Sir John Cowdery born 1917, English physicist; shared Nobel Prize for X-ray mapping of protein molecules.

Kennard packet *Quantum Mechanics.* a wave packet whose product of uncertainties in position and momentum is the smallest allowable by the uncertainty principle.

Kennelly, Arthur 1861–1939, American electrical engineer; independently discovered the Kennelly-Heaviside layer.

Kennelly-Heaviside layer see E LAYER.

kenophobia *Psychology.* an irrational fear of open or empty spaces. Also, TOPOPHOBIA.

Kenoran orogeny *Geology.* a Precambrian deformation of the earth's crust affecting the Archean rocks of the Canadian Shield. Also, ALGOMAN OROGENY.

kenotron *Electricity.* in telegraph circuits, a high-vacuum tube diode rectifier used in high-voltage, low-current applications.

kenozooecium *Invertebrate Zoology.* the outer, nonliving portion of a kenozooid.

kenozooid *Invertebrate Zoology.* a special type of nonfeeding member of a bryozoan colony that possesses a completely enclosed tubelike body chamber.

Kent-His bundle see BUNDLE OF HIS.

kentrolite *Mineralogy.* $Pb_2Mn_2^{+3}Si_2O_9$, a dark reddish-brown, orthorhombic mineral occurring in massive form and as tiny, brittle, prismatic crystals, having a specific gravity of 6.2 and a hardness of 5 on the Mohs scale; found with calcite and willemite at Franklin, New Jersey.

Kentrophoros *Invertebrate Zoology.* a genus of marine psammophilic forms of the family Loxodidae, having a cytostome slit on the concave body surface and the body laterally compressed, with cilia limited to the right side.

Kentucky coffee tree *Botany.* a deciduous tree, *Gymnoclades dioica,* of the family Leguminosae, whose fruits are thick woody pods containing black seeds that were formerly used in place of coffee beans.

Kenya, Mount *Geography.* a mountain (17,058 ft) in Kenya, east central Africa, near the equator.

Kenyapithecus *Anthropology.* the genus name originally given to two hominoid fossils, species *wickerii* and *africanus,* later reclassified in the genus *Ramapithecus.*

Kenyon, Dame Kathleen 1908–1978, English archaeologist; excavated Jericho and Jerusalem.

kenyte *Petrology.* a fine-grained igneous rock, essentially an olivine-bearing phonolite with phenocrysts of anorthoclase and sometimes acmite-augite and olivine; found as lava flows on Mt. Kenya and in the Antarctic.

Kepler, Johannes 1571–1630, German astronomer; formulated Kepler's three laws of planetary motion; improved telescopic lenses.

Keplerian [kep lâr′ē ən] *Astronomy.* relating to Johannes Kepler or his research and discoveries.

Keplerian motion *Astronomy.* motion in a Keplerian orbit.

Keplerian orbit *Astronomy.* an orbit involving two spherical objects and governed by gravitational forces only. Also, **Keplerian ellipse.**

Keplerian telescope *Optics.* a simple astronomical telescope having a fixed objective and a focusable eyepiece; it forms an intermediate image in the focal plane that appears inverted when viewed through the eyepiece.

Kepler's equation *Astronomy.* an equation that describes the relation between the mean anomaly and the eccentric anomaly in a planetary orbit.

Kepler's laws *Astronomy.* the three basic laws of planetary motion: **1. elliptical law.** every planet's orbit is an ellipse with the sun at one focus. **2. equal-areas law.** the radius vector from the sun sweeps across equal areas in equal time. **3. harmonic law.** the square of a planet's period measured in years equals the cube of its distance from the sun measured in astronomical units.

Kepler's supernova *Astronomy.* a bright supernova in southern Ophiuchus that was first observed in October 1604 and was visible to the naked eye for over a year. Also, **Kepler's star.**

kerabitumen see KEROGEN.

keranji *Materials.* the hard, tough wood of trees of the genus *Dialium* of southeast Asia; used for underwater construction, decorative work, flooring, and paneling.

kerat- a combining form meaning "horn," as in *keratin.*

keratectomy *Surgery.* the surgical removal of a portion of the cornea.

keratin *Biochemistry.* a fibrous scleroprotein that is the principal constituent of hair, nails, and other epidermal structures.

keratin filament *Cell Biology.* one of the several classes of intermediate filaments, composed of the structural protein keratin and found in the outer layers of the vertebrate epidermis.

keratin gene *Genetics.* any of a large family of genes that code for the keratin proteins.

keratinized tissue *Histology.* any tissue containing an abundance of keratin, such as skin, hair, or nails.

keratinocyte *Histology.* an epidermal cell that produces keratin.

keratinophilic *Biology.* attracted to or growing on hair, feathers, skin, horns, and other keratinous material; used chiefly in describing some fungi.
keratinous degeneration *Cell Biology.* the accumulation of keratin granules within any cell other than a keratinocyte.
keratitis *Medicine.* an inflammation of the cornea.
keratitis rosacea *Medicine.* an inflammation of the cornea characterized by minute infiltrates at the periphery of the cornea and the presence of small blood vessels.
kerato- a combining form meaning "horn," as in *keratogenesis.*
keratoconjunctivitis *Medicine.* an inflammation of the cornea and the conjunctiva.
keratogenesis *Medicine.* the formation of horny material.
keratohyalin *Histology.* a granular deposit in epidermal cells that is transformed into keratin.
keratoma *plural,* **keratomata** or **keratomas.** *Medicine.* a callus or callosity. *Veterinary Medicine.* a horny tumor on the inner surface of the wall of a horse's hoof.
keratomalacia *Medicine.* a degenerative condition of the cornea often resulting from severe vitamin A deficiency, marked by nightblindness and dryness of the cornea with eventual thinning and ulceration.
keratomycosis *Medicine.* a fungal infection affecting the cornea of the eye.
keratophyre *Petrology.* a fine-grained igneous dike rock that is essentially a soda trachyte, characterized by the presence of albite or albite oligoclase and chlorite, epidote, and calcite, which have modified the original character of the colored minerals.
keratoplasty *Surgery.* the surgical removal of an opaque portion of the cornea with replacement of healthy corneal tissue; corneal grafting.
keratosis *Medicine.* any skin condition marked by horny growth, such as a wart or callosity.
kerf *Engineering.* **1.** the portion of a material such as metal or wood that is removed in the course of any process such as sawing, cutting, or pressing. **2.** the space previously occupied by such removed material. *Mining Engineering.* a horizontal cut made in a seam or block of coal, often to facilitate its fall.
kerion *Medicine.* a skin disease that is characterized by nodular, circumscribed lesions covered with pustules, usually associated with tinea infections and also as a secondary complication of a fungal infection of the hair.
kermesite *Mineralogy.* Sb_2S_2O, a red, triclinic, pseudomonoclinic mineral occurring in lath-shaped crystals, and having a specific gravity of 4.68 and a hardness of 1 to 1.5 on the Mohs scale; formed from the alteration of stibnite. Also, PYROSTIBITE, PURPLE BLENDE, RED ANTIMONY.
kern *Graphic Arts.* the part of a printed character that extends beyond the body of the type and sometimes overlaps another character, especially in italic or other scriptlike type.
Kern counter see DUST COUNTER.
kernel *Botany.* **1.** the part of a seed that lies within the seed coat. **2.** a whole seed grain, as of corn. **3.** the soft, usually edible inner part of a nut or of the pit of a fruit. *Atomic Physics.* an atom that has lost its outermost or valence electrons. *Computer Programming.* a set of programs in an operating system that carry out the most elementary functions such as process and memory management, security, and input/output control. *Mathematics.* a null space.
kernel blight *Plant Pathology.* any of several barley diseases caused by fungi such as *Gibberella zeae, Helminthosporium sativum,* and *Alternaria,* and characterized by a withering, drying up, and discoloration of the grain.
kernel ice *Hydrology.* very irregular, opaque, low-density ice that forms on aircraft at subfreezing temperatures.
kernel of a homomorphism *Mathematics.* given a homomorphism ϕ from an algebraic object A to an algebraic object B (e.g., groups, rings, modules, etc.), the set $\ker\phi$ of all elements of A that are mapped to the (additive or zero) identity element of B. The kernel of a homomorphism is a subgroup (subring, submodule, etc.) of the domain A.
kernel of a linear transformation *Mathematics.* given a linear transformation T from a vector space V to a vector space W, the subspace of V consisting of all vectors that T maps to the zero vector of W; denoted $\ker T$.
kernel of an integral equation *Mathematics.* the (known) function $K(x, y)$ that appears inside the integral of a given integral equation.
kernel spot *Plant Pathology.* a disease of pecans caused by the fungus *Coniothyrium caryogenum* and characterized by dull-brown irregular dots on the kernels.
kernicterus *Pathology.* an acute neural state associated with excessive levels of bilirubin in the blood, marked by a yellow pigmentation of the basal nuclei and degenerative lesions of the lenticular nucleus and other areas of the intracranial gray matter, causing extensive degenerative alteration; a serious form of icterus neonatorum.
Kernig's sign *Medicine.* a condition present in various forms of meningitis in which the patient, lying on the back with the thigh flexed at the hip, is unable to extend the leg at the knee without extreme discomfort.
kerning *Graphic Arts.* in typesetting, the process of adjusting the space between characters to compensate for awkward kerns, as when a capital *A* or lowercase *l* follows a capital *F.*
kernite *Mineralogy.* $Na_2B_4O_6(OH)_2{\cdot}3H_2O$, a colorless, transparent, monoclinic mineral occurring as fibrous masses, having a specific gravity of 1.91 and a hardness of 2.5 to 3 on the Mohs scale; found as veins in playa deposits. Also, RASORITE.
kerogen *Geology.* a fossilized, highly insoluble, bituminous material found in sedimentary rocks, especially oil shales; it may be transformed into petroleum products by distillation. Also, KERABITUMEN, PETROLOGEN.
kerogen shale see OIL SHALE.
kerolite *Mineralogy.* a variety of talc having a randomly stacked structure. Also, CEROLITE.
kerosene *Materials.* a combustible, water-white, oily liquid with a strong odor that boils at 180–300°C; it is distilled from petroleum and is used as a fuel, as a cleaning solvent, and in insecticides. Also, **kerosine.**
kerosene fungus *Mycology.* an asexual fungus, *Hormoconis resinae,* of the family Amorphothecaceae, that often blocks fuel filters in jet engines; it is also found in soil, air, and feathers. Also, CREOSOTE FUNGUS.
Kerr cell *Optics.* a glass cell containing a transparent liquid, usually nitrobenzene, that exhibits double refraction when placed in a strong electric field; it can rotate the plane of polarization and is often used as a high-speed shutter.
Kerr constant *Optics.* a relation between the square of an electronic field and the birefringence induced by that field by the Kerr effect.
Kerr (electro-optic) effect *Optics.* **1.** a phenomenon in which a transparent substance becomes birefringent if placed in a strong electric field; the birefringence is proportional to the square of the applied field and varies according to the material used. Also, ELECTRO-OPTICAL KERR EFFECT, ELECTRO-OPTICAL BIREFRINGENCE. **2.** the elliptical polarization of plane-polarized light when reflected from the polished surface of a magnetized material. Also, MAGNETO-OPTIC (KERR) EFFECT.
kerria *Botany.* a shrub, *Kerria japonica,* of the rose family, having yellow flowers; native to Asia and widely cultivated as an ornamental.
Kerr solution *Physics.* a solution to Einstein's field equations, describing a rotating, axisymmetric black hole.
kersantite *Petrology.* a dark lamprophyre rock composed chiefly of biotite and plagioclase, with or without clinopyroxene and olivine.
kersey *Textiles.* **1.** a heavy overcoating material of wool and sometimes cotton. **2.** a coarse wool and cotton cloth used to make work clothes.
keruing *Materials.* a tree of the genus *Dipterocarpus* that produces an oil used for varnishing and for caulking boats.
kestrel *Vertebrate Zoology.* any of several small falcons, such as the European *Falco tinnunculus* or American *F. sparverius,* characterized by blue wings and head in males and red in females and noted for their ability to hover in the air against the wind.

kestrel

ket- a combining form denoting the presence of a carbonyl group, C=O.

ketal *Organic Chemistry.* formerly, referring to or describing compounds containing a carbon with two oxygen atoms and a radical attached to each oxygen; the term *acetal* is currently used.

ketch *Naval Architecture.* **1.** a two-masted rig, with the mizzenmast stepped forward of the rudderpost, as opposed to a yawl, in which the mizzenmast is abaft of the rudderpost. **2.** a vessel equipped with such a rig.

ketene *Organic Chemistry.* C_2H_2O (CH_2=C=O), a colorless, toxic gas that boils at –56°C; readily dimerizes to diketene; used as an acetylating agent.

ketene lamp *Chemical Engineering.* an electrically heated Chromel filament that hydrolyzes acetone to produce ketene.

kethoxal *Genetics.* $C_6H_{12}O_4$, a reagent that reacts with and modifies unpaired guanine residues in DNA or RNA.

ketimide *Organic Chemistry.* a compound defined as R_2=C=N*X*, where *X* is an acyl radical.

ketimine *Organic Chemistry.* a compound having –C=NH, as in a Schiff base; used as a curing agent for epoxy resins.

keto- a combining form denoting the presence of a carbonyl group, C=O.

keto acid *Organic Chemistry.* a compound that has both an acid and a ketone group.

ketoacidemia *Medicine.* the presence of keto acids in the blood.

ketoacidosis *Medicine.* acidosis accompanied by ketosis, as in diabetic acidosis.

ketoadipic acid *Biochemistry.* $C_6H_8O_5$, a product formed during the metabolism of lysine to glutaric acid.

ketogenic amino acid *Biochemistry.* any amino acid, such as leucine or phenylalanine, that gives rise to ketone bodies during its oxidation to carbon dioxide and water.

α-ketogluterate dehydrogenase *Enzymology.* a multienzyme system that catalyzes the oxidation of α-ketoglutaric acid to succinic acid as a part of the citric acid cycle.

ketohexose *Biochemistry.* any monosaccharide containing a ketone group.

ketolase *Enzymology.* an enzyme capable of splitting a carbohydrate at the carbonyl-carbon position in the molecule and which requires thiamine pyrophosphate as a coenzyme.

ketolysis *Biochemistry.* the breaking down of acetoacetic acid, acetone, or beta-hydroxybutyric acid by the removal of ketone bodies.

ketone *Organic Chemistry.* any of a class of organic compounds, such as acetone, that have the carbonyl group C=O attached to two alkyl groups; obtained by the oxidation of secondary alcohols.

ketone body *Biochemistry.* any of three compounds that arise from acetyl coenzyme A and that may accumulate in excessive amounts as a result of starvation, diabetes mellitus, or other defects in carbohydrate metabolism.

ketonemia *Medicine.* an excess of ketone bodies in the blood.

ketonuria *Medicine.* the presence of ketone bodies in the urine, as in diabetes mellitus.

ketose *Biochemistry.* **1.** a monosaccharide that has a ketone group. **2.** a ketone derivative of a polyatomic alcohol.

ketosis *Medicine.* a condition characterized by the presence of excessive quantities of ketone bodies in body tissues and fluids, as seen in diabetes mellitus and starvation.

ketosteroid *Biochemistry.* any of a group of degradation products of steroids that are excreted in urine and provide an index of androgen production in the body; the principal ketosteroids are androsterone, etiocholanolone, and estrone.

γ-ketovaleric acid see LEVULINIC ACID.

Kettering, Charles Franklin 1876–1958, American engineer; invented automobile ignition and high-compression engine.

kettle *Food Technology.* a metal container used for such purposes as boiling water (as for tea) or cooking soup. *Geology.* **1.** a steep-sided, basinlike depression in glacial drift deposits, probably formed when a slowly melting, large ice block left by a retreating glacier was subsequently buried by newer deposits, causing the overlying layers to collapse. Also, **kettle hole, kettle basin. 2.** see POTHOLE.

kettle reboiler *Chemical Engineering.* a tube-and-shell heat exchanger in which liquid is vaporized on the shell side by heat transferred from hot liquid flowing through the tubes; adequate dome space above the tube bundle allows for liquid-vapor separation.

ket vector *Quantum Mechanics.* a state vector in Hilbert space analogous to a column vector; the dual of a bra vector.

Keuper *Geology.* a European geologic stage of the Upper Triassic period in Germany, occurring after the Muschelkalk and before the Jurassic.

keV kiloelectron volt.

Kevlar *Materials.* the trade name for a stiff, strong aramid fiber developed for belting radial tires; used in manufacturing cables and webbings and to reinforce plastic composites in bulletproof vests.

Kew barometer *Meteorology.* a type of cistern barometer in which no adjustment is made for the changes of the mercury level in the cistern while changes in pressure occur; instead, a uniformly contracting scale determines the effective height of the mercury column.

Keweenawan *Geology.* the younger of two provincial series of the Precambrian period in Michigan and Wisconsin.

key *Mechanical Devices.* **1.** the common instrument that, when inserted into a lock, moves the bolt to open the lock. **2.** a device used to inhibit motion between a shaft and a hub. Also, MACHINE KEY. **3.** any device used to wind, tighten, or secure. *Building Engineering.* **1.** a piece of wood or metal wedged into a joint to stop movement. **2.** any of various means of improving bonding capacity, such as plastering forced between laths, a cotter pin driven through a protrusion, or the deliberately roughened or serrated reverse side of construction material. **3.** a backing plate affixed to a board to limit warpage. *Electricity.* a lever-type switch that opens and closes a circuit only as long as the handle is depressed. Also, SWITCHING KEY. *Electronics.* a switch that is used to transmit signals by modulating a circuit. *Computer Programming.* **1.** a field of a data record used for identifying, ordering, or retrieving the record. **2.** to input data by means of a keyboard. *Telecommunications.* a set of symbols used to encrypt or decrypt text. *Systematics.* a table or other such device used for the identification of specimens. *Graphic Arts.* the dominant tone in a photographic negative or print; in a **high-key** photograph, lighter tones prevail.

key *Geology.* a small, coral islet or cay, especially such a feature off the southern coast of Florida. Also, KAY.

key access *Computer Programming.* a way of using a key to locate a particular record in a file or in memory.

keyaki *Materials.* the coarse-textured hardwood of the Far Eastern tree *Zelkova serrata;* used in the construction of temples and shrines, furniture, ships, and veneers.

key auto-key cipher *Telecommunications.* a type of stream cipher in which the cryptographic bit stream generated by a given time is determined by the cryptographic bit stream generated at an earlier time.

key bed *Geology.* **1.** an easily identifiable, well-defined stratum or body of strata used to facilitate correlation in field mapping or other geological work because of its distinctive characteristics. **2.** a bed whose upper or lower surface is used as a datum in the drawing of structure-contour maps. Also, INDEX BED, MARKER BED.

keyboard *Mechanical Engineering.* a set of keys or levers designed to be depressed and used as a control or input device in operating a machine such as a typewriter, computer, or piano. *Computer Technology.* to enter or manipulate data by means of such a device.

keyboard entry *Computer Programming.* a data item that is manually entered into computer memory from a keyboard.

keyboarding *Computer Technology.* the process of using a computer keyboard to enter or manipulate data.

keyboard inquiry *Computer Programming.* an interrogation of the status of a program, disk storage, or other aspects of the computer system via a terminal keyboard.

keyboardless typesetter *Computer Technology.* an automated typesetting machine, now considered obsolete, that receives its input from paper tape that must be punched on a separate keyboard entry machine.

keyboard lockout *Computer Technology.* a property of a keyboard that prevents it from transmitting characters while the circuit is busy with other transmissions.

keyboard lockup *Computer Programming.* a condition in which keystrokes typed on a keyboard are ignored by the computer.

keyboard printer *Computer Technology.* a simple input/output device combining a keyboard and a printer that echo-prints the keyed-in data; also used to display computer output.

keyboard send/receive *Electronics.* a device combining a teleprinter with a transmitter and receiver to send and receive information.

key bounce *Computer Technology.* the repetitions of a character transmitted by one keystroke due to mechanical bouncing of the key lever.

key cabinet *Electronics.* a system that informs a telephone subscriber which lines are busy and allows different lines to be connected to various telephone stations.

key change *Computer Programming.* the occurrence of a record whose key differs from that of its immediate predecessor during the processing of records in key sequence.
key click *Telecommunications.* a transient signal, resembling either a chirp or a click, that is sometimes produced with the opening or closing of a radiotelegraph key.
key compression *Computer Programming.* a data compression technique applied to record keys.
key day see CONTROL DAY.
key-disk machine *Computer Technology.* a data entry machine in which data entered via a keyboard is recorded directly on a magnetic disk.
key-driven calculator *Computer Technology.* an electromechanical desk calculator, now considered obsolete, in which the digit keys, arranged in columns, drive the accumulator dials for displaying results.
keyed clamp *Electronics.* a device that uses a control signal to hold current or voltage to a fixed level.
keyed mortise and tenon *Building Engineering.* in woodworking, a joint that is pierced to receive a tapered key to draw the joint up tightly via an extended tenon; primarily used in nonglue applications for heavy furniture.
keyed sequential access method *Computer Programming.* a technique for storing and retrieving records in a file on a direct-access device either sequentially, by accessing the records in key order, or directly, by using the key value for a particular record.
keyer *Electricity.* a device used to change the output of a transmitter or to shift its frequency in order to transmit a specific type of intelligence. *Electronics.* **1.** a circuit that modulates by keying. **2.** a radar modulator.
keyer adapter *Electronics.* a device that, when triggered by a modulated signal, generates a direct current output signal at an amplitude that varies with the modulation; used in radio facsimile transmission.
Keyes equation *Thermodynamics.* an equation of state of a gas used to correct the molecular volume term in the van der Waals equation.
Keyes process *Chemical Engineering.* a distillation process that adds benzene to a constant-boiling 95% alcohol-water solution to obtain absolute (100%) alcohol.
key field *Computer Programming.* the field in a data record that contains the unique identifier.
keyhole *Mechanical Engineering.* a hole or slot into which a key is inserted. *Metallurgy.* a hole-and-slot notch, resembling a keyhole. *Mechanical Devices.* see KEYHOLE SAW. *Ordnance.* **1.** describing a shot in which the bullet rotates off its long axis in flight, usually due to insufficient spin. Thus, **keyhole shot, keyholing. 2.** the elongated hole in the target made by a keyholing bullet.
keyhole saw *Mechanical Devices.* a long, slender, fine-toothed saw with a 6-inch to 16-inch blade, used for cutting tight internal curves.
keyhole specimen *Metallurgy.* a specimen containing a keyhole; usually used in impact strength testing.
key informant see INFORMANT.
keying *Building Engineering.* the establishment of a mechanical bond in a construction joint. *Electricity.* **1.** the breaking up of signals into intervals of varying duration. **2.** the modulation of a signal carrier by varying its amplitude or frequency. *Graphic Arts.* **1.** the process of preparing a manuscript for typesetting by marking with symbols its various sections, headlines, and so on. **2.** in page makeup, the process of preparing a dummy showing the correct placement and position of artwork. *Computer Technology.* the process of using a keyboard to enter or manipulate data.
keying frequency *Telecommunications.* in facsimile, the greatest number of occurrences per second of a black-line signal during the scanning operation.
keying interval *Telecommunications.* one of the set of intervals in a periodically keyed transmission system.
keying wave see MARKING WAVE.
key job *Industrial Engineering.* **1.** an essential job in a given production process. **2.** a characteristic job used in establishing job factor comparisons.
keyless ringing *Telecommunications.* ringing on a manual switchboard that takes place automatically when a call setup connection is established.
keyline *Graphic Arts.* an outline drawn on artwork to indicate either its colors (in some simple forms of color separation) or the position of an element such as type or halftoning.
keylock switch *Electricity.* a switch that can only be operated by using a key.
key operator *Artificial Intelligence.* an operator that can be determined a priori to be necessary in a search or planning task.
keypad *Computer Programming.* a supplementary cluster of keys, usually numerical, integrated on one side of a keyboard or on a small separate keyboard.
key plate *Graphic Arts.* the most important color plate in a color separation, usually black.
key point *Military Science.* a concentrated site or installation, the destruction or capture of which would seriously affect the war effort or the success of operations. Similarly, **key area, key terrain.**
keypunch *Computer Technology.* a keyboard-activated device that punches holes in punch cards.
key seat *Mechanical Engineering.* **1.** a recessed groove or space in either a shaft or hub fabricated so as to receive a key. **2.** the opening in a keylock. **3.** the hollowed area that provides a driving surface for a key at its end. Also, KEYWAY. *Petroleum Engineering.* a vertical groove along the side of the oil well hole resulting from improper handling of the pipe.
keyseater *Mechanical Engineering.* a machine that creates a keyway or groove in a mechanical part where a key is designed to fit.
keyshelf *Telecommunications.* on a manual telephone switchboard, a shelf on which keys for switchboard operation are mounted.
keystone *Architecture.* the central voussoir of a semicircular arch.
keystone distortion *Telecommunications.* the distortion of a projected or televised image so that it appears wider at the top than at the bottom, or vice versa; caused by the angled orientation of the electron beam in the television camera tube relative to the principal axis of the tube. Also, **keystoning.**
keyswitch *Computer Technology.* a switch that is turned on and off by keyboard entry.
key telephone system *Telecommunications.* a small telephone system that connects two or more lines to an equal or larger number of special telephones or stations which are equipped with special pushbuttons, or keys. Calls can be answered at any of the stations under key control. From each station, idle lines can be selected for calls outgoing from the key telephone system to the public switched network.
key telephone unit *Telecommunications.* a modular unit providing basic or add-on functionality in a key telephone system.
key-to-disk system *Computer Technology.* a data entry technique that permits simultaneous data entry from multiple keyboards into different sections on the disk.
key-to-tape system *Computer Technology.* a data entry machine that uses buffers to merge records entered simultaneously from multiple keyboards onto a magnetic tape.
key transformation *Computer Programming.* a function that maps a set of keys into a set of integers that can be used to index the corresponding records, as in a hashing technique.
key value *Computer Programming.* the actual characters or digits comprising a key.
key verify *Computer Programming.* to check the validity of data punched on a card by retyping the data sequence on a verifier machine.
keyway *Mechanical Engineering.* see KEY SEAT. *Building Engineering.* an interlocking groove or channel that supplies reinforcement in a wood or cement joint.
keyword or **key word** *Computer Programming.* **1.** one of the informative words in a title or document that may describe the content of that document. **2.** a part of a command operand that consists of a particular character string, e.g., REWIND. **3.** see RESERVED WORD.
keyword-in-context index *Computer Programming.* a computer-generated listing of document titles arranged alphabetically by the keywords in the titles; usually the keyword is aligned in a column in the middle of the line with left and right contexts and an identifying reference number.
keyword-out-of-context index *Computer Programming.* a computer-generated listing of document titles arranged alphabetically by the keywords in which the titles appear in full under each keyword.
keyword parameter *Computer Programming.* a parameter consisting of a keyword followed by one or more values identified by the keyword.
keyword search *Computer Programming.* a search for records or articles that contain or are described by specified keywords or combinations of keywords.
K factor *Nucleonics.* a unit of quantity for the level of gamma ray energy produced by a specific type of emitter, measured in roentgens per hour at a distance of one centimeter from the source, with a decay rate of one per millicurie. *Cartography.* see BASE-HEIGHT RATIO.

kG kilogauss.
kg or **kg.** kilogram; keg.
kg-cal kilogram-calorie.
kgf kilogram-force.
kg-m kilogram-meter.
kgps or **KGPS** kilograms per second. Also, **kg/sec.**
K gun *Ordnance.* a K-shaped gun used by the U.S. Navy to fire depth charges.
kg-wt kilogram-weight.
Khaki Campbell *Agriculture.* a small breed of egg-producing duck developed in England that is khaki and bronze in color; the breed has set egg-production records. (From Adele *Campbell,* a 19th-century British duck breeder.)
khamsin *Meteorology.* a hot, dry, dusty desert wind that blows over Egypt and the Red Sea from a southerly or southeasterly direction ahead of low-pressure systems moving eastward over North Africa or the southeastern Mediterranean Sea.
kharif see KARIF.
khellin *Pharmacology.* $C_{14}H_{12}O_5$, a drug prepared from the fruit of a Mediterranean plant, *Ammi visnaga,* that relaxes smooth muscle; formerly used to treat angina and asthma and to lower blood pressure.
KHI *Aviation.* the airport code for Karachi, Pakistan.
Khorana, H. Gobind born 1922, Indian-born American biochemist; shared Nobel Prize for research in nucleic acids and the genetic code.
Khyber Pass *Geography.* a mountain pass connecting Pakistan and Afghanistan.
kHz kilohertz.
kibble see HOPPIT.
kick *Ordnance.* **1.** the backward movement of a gun caused by the force of the propellant gases; it may be used as a synonym for recoil or as a description of the effect of recoil upon the shooter. **2.** to move in such a manner. *Petroleum Engineering.* downhole pressure greater than that exerted by the weight of the drilling mud, which can cause loss of circulation; if the gas pressure is not curtailed by increasing the mud weight, a kick can force out the column of drilling mud, resulting in a blowout.
kickback *Mechanical Engineering.* the sudden backward movement of an internal-combustion engine as it is being started, or of a piece of material as it is being fed into a machine.
Kickback *Ordnance.* a Soviet air-launched cruise missile with a single 350-Kt nuclear warhead; officially designated **AS-16.**
kickdown *Mechanical Engineering.* **1.** the action of changing to a lower gear in an automotive vehicle. **2.** the equipment that shifts an automotive vehicle into a lower gear.
Kickellales *Mycology.* an order of nonparasitic fungi occurring in dung and soil; consists of a single family, **Kickellaceae.**
kicker *Building Engineering.* a concrete plinth that extends about 2 inches above a concrete floor, thus forming the base of a concrete wall or column.
kick over *Mechanical Engineering.* to begin to fire, as in an internal-combustion engine.
kickpipe *Building Engineering.* a short pipe that protects an electrical cable, especially at the point where it becomes exposed from a floor.
kickplate *Building Engineering.* a metal plate fastened to the bottom of a door, wall, cabinet, or stair riser to prevent marring of the finish from shoe marks.
kick press see PENDULUM PRESS.
Kick's law *Engineering.* a law stating that the amount of energy needed to crush a given quantity of solid material to a particular fraction of its original size is the same, whatever the original size of the feed substance.
kick starter *Mechanical Engineering.* a starter that functions by a downward kick of a pedal, as on most motorcycles.
kick wheel *Engineering.* a potter's wheel or a wheel on any other apparatus that is operated by using a foot pedal.
Kickxellales *Mycology.* an order of fungi belonging to the class Zygomycetes; some parasitize other fungi, while others live from dead organic matter in soil and dung.
Kidd blood group (system) *Genetics.* a hereditary human blood group that is encoded by the Jk gene located on chromosome 2. *Immunology.* the red blood cell antigens of this group, characterized by a reaction to the antibody anti-Jk(a) and the serum anti-Jk(b). (From the name of the person in which it was first identified.)
Kidder, Alfred Vincent 1885–1963, American archaeologist; improved field techniques; studied the southwestern U.S. and lowland Maya area.
kidney *Anatomy.* either of two organs in the lumbar region that filter water and wastes from the blood, excrete the end-products of body metabolism in the form of urine, and regulate the concentrations of hydrogen, sodium, potassium, phosphate, and other ions in the extracellular fluid. *Zoology.* a corresponding organ in vertebrates, or an analogous organ in invertebrates.

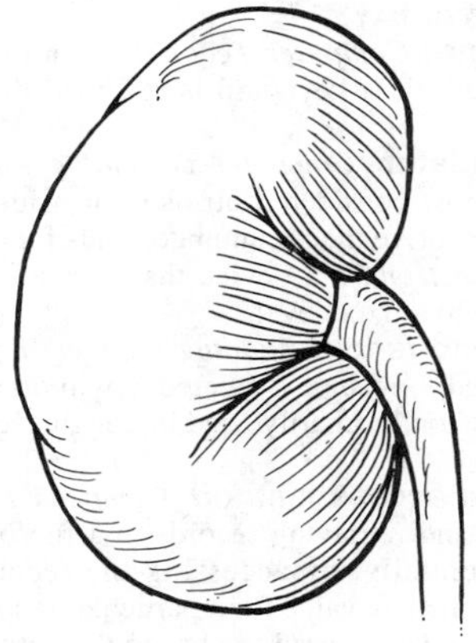

kidney

kidney bean *Botany.* a common bean, *Phaseolus vulgaris,* several varieties of which are cultivated for their purple, kidney-shaped edible seeds.
kidney joint *Electromagnetism.* a type of waveguide connector used in radar systems that allows for movement of the waveguide.
kidney ore *Mineralogy.* a form of the mineral hematite, Fe_2O_3, found in compact, kidney-shaped masses, concretions, or nodules.
kidney stone *Medicine.* a renal calculus.
kidney worm disease *Veterinary Medicine.* a parasitic disease of pigs caused by *Stephanurus dentatus*, the intermediate host of which is the earthworm; infestation involves the liver, kidney and surrounding tissue, and lungs, and results in emaciation and stiffness. Also, SWINE KIDNEY WORM INFECTION.
Kienbock unit see X-RAY UNIT.
kieselguhr see DIATOMACEOUS EARTH.
kieserite *Mineralogy.* $MgSO_4 \cdot H_2O$ a white, monoclinic mineral of the kieserite group, occurring in granular masses, and having a specific gravity of 2.57 and a hardness of 3.5 on the Mohs scale; found in some salt deposits.
kievitone *Biochemistry.* a phytoalexin released by the French bean, *Phaseolus vulgaris,* when attacked by certain fungi such as *Rhizoctonia solani.*
Kikuchi lines *Crystallography.* the parallel lines observed in the scattering of electrons from a solid crystal.
Kiliani reaction *Organic Chemistry.* the reaction that synthesizes glucose from arabinose, or, in general, higher monosaccharides from lower ones.
Kilimanjaro *Geography.* the highest mountain of Africa (19,340 feet), located in northeastern Tanzania.
kill *Metallurgy.* to treat molten steel with a strong deoxidizer such as silicon or aluminum. *Petroleum Engineering.* **1.** to prevent a well blowout during drilling operations. **2.** to halt oil production in order to recondition a well. *Electricity.* to cut off a circuit, or render it dead. *Archaeology.* see KILL SITE.
killed *Computer Science.* of a subexpression, having any previously computed value invalidated by redefinition of a component of the subexpression.
killed steel *Metallurgy.* a steel that has been thoroughly deoxidized.
killed vaccine *Immunology.* a suspension of inactivated microorganisms, administered as antigens to induce immunity.
killer cell *Immunology.* an effector cell that mediates antibody-dependent cell-mediated cytoxicity, such as a cytotoxic T cell, neutrophil, leucocyte, or macrophage. Also, K CELL.
killer circuit *Electronics.* **1.** a device that shuts down some aspect of a system's operation, as by disabling the audio in a television receiver. **2.** a device that produces blanking pulses that temporarily interrupt a signal or an electron beam, as in a radar system.

killer factor *Biochemistry.* any of various protein toxins secreted by certain strains of the yeast-forming fungus *Saccharomyces cerevisiae* that are encoded by a dsRNA element and a K_1 toxin before adhering to a vulnerable cell wall and passing into its cytoplasmic membrane, where the toxin inhibits membrane activity.
killer paramecia *Invertebrate Zoology.* strains of *Paramecium aurelia* containing the toxin paramecin, which kills other types of paramecia.
killer plasmid *Mycology.* any of certain toxic chemicals secreted by some yeast fungi that can kill other fungi.
killer pulse *Electronics.* a signal generated by a killer circuit, used to temporarily disable a circuit by interrupting its signal or electron beam.
killer whale *Vertebrate Zoology.* any of several aggressive whale species of the genus *Orcinus* that are black with white patches and attack seals, porpoises, and other small whales; found in all oceans.

killer whale

killer yeast *Mycology.* a yeast fungus belonging to the species *Saccharomyces cerevisiae* that secretes a toxic protein capable of killing other yeasts.
killfish **1.** see CYPRINODONTIDAE. **2.** see TOPMINNOW.
Killing form *Mathematics.* In the theory of Lie algebras, the symmetric bilinear form κ given by $\kappa(X,Y) = \mathrm{tr}(\mathrm{ad}\ X\ \mathrm{ad}\ Y)$. By theorem, if the Killing form is nondegenerate, the Lie algebra is semisimple.
Killing tensor *Mathematics.* **1.** a symmetric covariant tensor field on a manifold M whose symmetrized covariant derivative vanishes on an open subset of M. **2.** a symmetric covariant tensor field on a manifold M that is constant along geodesics on M. **3.** a symmetric contravariant tensor field on M obtained by raising indices on a tensor field that satisfies definition 1 or 2. **4.** a symmetric contravariant tensor field on M whose induced function on the cotangent bundle $T*M$ is in involution with the (Hamiltonian) function induced by the contravariant form of the Riemannian metric tensor.
Killing vector *Mathematics.* a vector field on a manifold M corresponding to the infinitesimal generator of an isometry of M; a Killing tensor of rank 1.
killing zone *Military Science.* an area in which a commander plans to force enemy concentration in order to destroy the enemy with conventional weapons or the tactical employment of nuclear weapons.
kill probability *Military Science.* a measure of the probability of destroying a target.
kill site *Archaeology.* a location that is identified as having been used for butchering, as evidenced by animal bones in association with tools.
kiln *Engineering.* a heated enclosure, often a refractory-lined cylinder, that is used for drying, burning, baking, or firing materials such as ore, cement, bricks, or ceramics.
kilo- a prefix meaning one thousand or 10^3, as in *kilometer.*
kiloampere *Metrology.* a metric unit of electrical current equal to one thousand amperes.
kilobar *Metrology.* a metric unit of pressure equal to 1000 bars, or 14,500 pounds per square inch.
kilobase *Genetics.* in nucleic acids, a unit of length equal to 1000 bases.
kilobit *Computer Technology.* a unit of storage equal to 1024 (2^{10}) bits.
kilocalorie *Metrology.* a unit of energy equivalent to 1000 small calories. Also, KILOGRAM-CALORIE.
kilocurie *Metrology.* a unit of radioactivity equal to 1000 curies.
kilocycle *Metrology.* a unit of 1000 cycles, usually per second, applied to the frequency of electromagnetic waves.
kilodyne *Metrology.* a unit of force equal to 1000 dynes.
kiloelectron volt *Metrology.* a unit of energy equivalent to 1000 electron volts.
kilogauss *Metrology.* a unit of magnetic induction equal to 1000 gauss.
kilogram *Metrology.* a basic unit of mass, equal to 1000 grams or the mass of 1 liter of water at a temperature of 4°C; equivalent to 2.205 pounds. Also, **kilogramme.**
kilogram-calorie see KILOCALORIE.
kilogram-equivalent weight *Chemical Engineering.* a mass unit 1000 times the gram-equivalent weight.
kilogram-force or **kilogram force** *Metrology.* a unit of force equal to the weight of a mass of 1 kilogram measured at sea level with standard gravity. Also, **kilogram-weight.**
kilogram-meter see METER-KILOGRAM.
kilohertz *Metrology.* a unit of frequency equivalent to 1000 hertz or 1000 cycles per second.
kilohm *Metrology.* a metric unit of electrical resistance equal to one thousand ohms.
kilojoule *Metrology.* a unit of energy equivalent to 1000 joules.
kiloliter *Metrology.* a unit of capacity equal to 1000 liters. Also, **kilolitre.**
kilomega- see GIGA-.
kilomegacycle see GIGACYLE.
kilomegahertz see GIGAHERTZ.
kilometer [ki läm´ə tər; kil´ə mē´tər] *Metrology.* a basic unit of length equal to 1000 meters; equivalent to 0.62 mile. Also, KILOMETRE.
kilometric wave *Telecommunications.* a low-frequency band, with a range of 30–300 kilohertz.
kiloparsec *Astronomy.* one thousand parsecs, or 3262 light-years.
kiloton *Metrology.* a unit of explosive power equivalent to the yield of 1000 metric tons of trinitrotoluene (TNT).
kiloton weapon *Ordnance.* a nuclear weapon having a yield of 1 to 999 kilotons.
kilovar *Metrology.* a unit of reactive power equal to one thousand volt-amperes reactive.
kilovolt *Metrology.* a unit of potential difference in volts equal to one thousand volts.
kilovolt-ampere *Metrology.* a metric unit of power that gives true power in a direct-current circuit and the apparent power in alternating-current equal to one thousand volt-amperes.
kilovoltmeter *Electricity.* a voltmeter whose scale is calibrated to measure potential difference in thousands of volts.
kilowatt *Metrology.* a unit of power equivalent to 1000 watts.
kilowatt-hour *Metrology.* a unit of energy or work equal to one thousand watt-hours.
Kimberley reefs *Geology.* extensive gold-bearing reefs overlying the Main and Bird reefs in southern Africa. Also, BATTERY REEFS.
kimberlite *Petrology.* a highly serpentinized variety of mica peridotite with phenocrysts of olivine and phlogopite, subordinate melilite, and small amounts of pyroxene, apatite, perovskite, and opaque oxides.
Kimmeridgian *Geology.* a European geologic stage of the Middle Upper Triassic period, occurring after the Oxfordian and before the Portlandian.
kimzeyite *Mineralogy.* $Ca_3(Zr,Ti)_2(Si,Al,Fe^{+3})_3O_{12}$, a dark-brown cubic mineral of the garnet group, occurring in dodecahedral crystals, having a specific gravity of 4.00 and a hardness of about 7 on the Mohs scale; found in carbonatite and shoshonitic basalt.
KIN *Aviation.* the airport code for Kingston, Jamaica.
kinanesthesia *Neurology.* the inability to sense movement. Thus, **kinanesthesic, kinanesthetic.**
kinase *Enzymology.* an enzyme in the body that transfers a phosphate group from a nucleoside triphosphate to another molecule.
Kinderhookian *Geology.* a North American provincial series of the lowermost Mississippian, occurring after the Chautauquan of the Devonian period and before the Osagian.
K indicator see K SCOPE.
kine- or **kin-** a combining form meaning "movement," as in *kinematics, kinanesthesia.*
kinematic *Mechanics.* of or relating to kinematics; describing or predicting the course of motion. Also, **kinematical.**
kinematical diffraction *Crystallography.* a diffraction theory in which it is assumed that the incident beam is not affected, apart from the results of simple diffraction, by its passage through the crystal. This type of diffraction is assumed in most crystal structure determinations by X-ray diffraction. Also, **kinematical theory.**

kinematic boundary condition *Fluid Mechanics.* a condition stating that the normal component of fluid velocity at a solid boundary must vanish at the boundary, and if the boundary is taken as a fluid surface, then the velocity at the surface is given by the vector difference between the velocities across the interface.

kinematic constraint see CONSTRAINT.

kinematic errors *Robotics.* inaccuracies that occur when parts are handled or robots move in relation to processing of parts so that the dynamics of all aspects of motions are considered except for force and mass.

kinematic fluidity *Fluid Mechanics.* a quantity given by the reciprocal of the kinematic viscosity.

kinematic pair see PAIR.

kinematics *Mechanics.* a branch of mechanics that is concerned with the study of the geometry of motion, without regard for the effects of mass or force on the system.

kinematic similarity *Fluid Mechanics.* a relationship of similarity between two fluid flow systems, in which corresponding velocities and velocity gradients appear in corresponding locations in the same ratios.

kinematic viscosity *Fluid Mechanics.* the absolute viscosity of a fluid divided by its density.

Kineosporia *Bacteriology.* a genus of bacteria of the order Actinomycetales that occur as substrate mycelium, with the sporangia each containing a single zoospore and developing on the hyphae.

kineplasty *Surgery.* the amputation of an extremity wherein the muscles and tendons in the stump are arranged to allow them to be used in conjunction with the moving parts of a prosthetic apparatus.

kinescope see PICTURE TUBE.

kinescope recording *Telecommunications.* a former method of television recording, in which images on the surface of a television monitor were filmed for later broadcasts.

kinesiatrics *Medicine.* the use of systematic movements as a form of therapy. Also, **kinesitherapy, kinesiotherapy.**

kinesics *Behavior.* the study of body language.

kinesio- or **kinesi-** a combining form meaning "movement."

kinesiology *Physics.* the study of the mechanics of human movement.

kinesiotherapy *Medicine.* the treatment of disease by movement or exercise.

kinesis *Physiology.* movement or action in response to a stimulus. *Behavior.* the movement of an animal away from unfavorable conditions or into favorable ones within its environment.

kinesthesia *Physiology.* **1.** the perception of body position, muscle movement, weight, or equilibrium. **2.** specifically, such perception by proprioceptors located in the skeletal muscles, tendons, and joints. Also, **kinesthesis.**

kinesthetic apraxia see MOTOR APRAXIA.

kinet- a combining form meaning "movement," as in *kinetism.*

kinetic *Mechanics.* involving or causing motion; having to do with motion and its effects.

kinetic chain length *Materials Science.* the number of times that given monomer molecules add on to the initiator radicals during their lifetime; used to calculate the degree of polymerization during free radical addition polymerization.

kinetic complexity *Genetics.* the complexity of a specific length of DNA, as indicated by its reassociation kinetics in a renaturation experiment.

kinetic energy *Mechanics.* the energy inherent in the motion of a body; in classical mechanics, it is proportional to the square of its linear and angular velocities.

kinetic energy ammunition *Ordnance.* ammunition that is designed to inflict damage through the effect of its kinetic energy upon impact, as by shattering or piercing.

kinetic equilibrium see DYNAMIC EQUILIBRIUM.

kinetic friction *Mechanics.* the force of friction that exists between two surfaces that are sliding against each other. Also, SLIDING FRICTION.

kinetic lead *Ordnance.* in gunnery, the adjustment for the relative motion of the target in computing the lead angle.

kinetic momentum *Mechanics.* the classical momentum of a body, neglecting electromagnetic momentum.

kinetic potential see LAGRANGIAN.

kinetic pressure *Fluid Mechanics.* the kinetic energy per unit volume in a fluid, equal to $(1/2)rv^2$, where r is the fluid density and v is the fluid velocity.

kinetic reaction *Mechanics.* the reactive force that an accelerated body exerts against the accelerating object; it is equal and opposite to this force, as stipulated by Newton's third law.

kinetics *Mechanics.* a branch of mechanics that is concerned with the relationship between motion and force and mass.

kinetic stress *Physics.* the pressure in a fluid system due to the presence of a velocity distribution of molecules within that system.

kinetic theory *Physics.* a theory explaining the behavior of physical systems as a result of the combined actions of the molecules constituting the system. Also, MOLECULAR THEORY.

kinetin *Biochemistry.* $C_{10}H_9ON_5$, the highly potent plant growth factor cytokinin 6-furfurylaminopurine, found in tobacco callus tissue.

kinetism *Physiology.* the ability to perform or initiate muscular action.

kineto- a combining form meaning "movement," as in *kinetoplasm.*

kinetochore *Cell Biology.* a platelike structure that develops on the surface of the centromere and attaches the chromosomes to the mitotic spindle; it is essential for chromosomal movement during mitosis.

kinetodesma *Invertebrate Zoology.* in protozoans, one of the fine strands, just below and parallel to the surface, connecting the cilia.

Kinetofragminophora *Invertebrate Zoology.* a class of protozoans having isolated kineties in the oral region and bearing cilia but not compound ciliary organelles; includes the human parasite *Balantidium coli.* Also, **Kinetofragminophorea.**

kinetoplasm *Cell Biology.* the highly contractile portion of a cell's cytoplasm; generally refers to the chromatophilic substances of nerve tissue.

kinetoplast *Cell Biology.* a complex interlocked network of DNA circles located within mitochondria and often associated with the base of a flagellum in certain organisms such as trypanosomes.

Kinetoplastida *Invertebrate Zoology.* an order of small, flagellate protozoans, in the class Zoomastigophorea; some are pathogenic parasites of humans and other vertebrates.

kinety *Invertebrate Zoology.* in the infraciliature of ciliate protozoans, a longitudinal unit consisting of cilia, basal bodies, and the connecting kinetodesma. Also, **kinety system.**

kingbird *Vertebrate Zoology.* any of several small, aggressive birds of the genus *Tyrannus* in the family Tyrannidae, known to attack larger birds such as hawks or crows; characterized by dark backs with a white underbelly and a white undertail and a red or orange crown; inhabits open country of the Americas.

king closer *Building Engineering.* a three-quarter-length brick used to finish a masonry course at its end. Also, **king closure.**

kingdom *Systematics.* the highest of the principal or obligatory ranks in the taxonomic hierarchy, above phylum or division.

Kingella *Bacteriology.* a genus of aerobic bacteria of the family Neisseriaceae, found in the human upper respiratory tract as rod-shaped cells, occurring in pairs or chains.

kingfish *Vertebrate Zoology.* any of various large marine food fish, including the white croaker, *Genyonemeus lineatus,* of California; a game fish, *Seriola grandis,* of Australia and New Zealand; and several fish of the drum family, especially of the genus *Menticirrhus,* of the U.S. Atlantic coast.

Kingfish *Ordnance.* a Soviet air-to-surface strategic cruise missile believed to be powered by a solid-propellant rocket engine and equipped with inertial guidance and terminal homing, delivering a 350-kiloton nuclear warhead or a 1000-kilogram conventional warhead at a speed of Mach 3 and a range of 200 miles; officially designated **AS-6.**

kingfisher *Vertebrate Zoology.* any of over 200 species of brightly colored, crested birds of the family Alcedinidae that capture fish with their long, stout beaks and are distributed worldwide.

kingfisher

kingpin *Mechanical Engineering.* **1.** a pin that connects the stub axle to the fixed axle beam of an automobile and is tilted to the vertical to allow for caster action. **2.** any of various vertical hinges and swivel pins, usually supported at both top and bottom ends.

king post *Building Engineering.* the principal vertical member of a king post truss. *Naval Architecture.* **1.** a deck support pillar in a ship's hold, located along the center line. **2.** a mastlike pole on a cargo ship, serving to support a cargo boom.

king post truss *Building Engineering.* a wooden roof truss that lies vertically between two abutted principal rafters and across the base of a triangular roof truss or tie beam.

king rod *Building Engineering.* a vertical steel rod that supports ceiling loads and prevents sagging of the tie beam.

king salmon see CHINOOK SALMON.

king's blue see COBALT BLUE.

king snake or **kingsnake** *Vertebrate Zoology.* any smooth-scaled snake with bright bands of the genus *Lampropeltis* in the family Colubridae; native to the southern and central United States and noted for constricting its prey, which includes rattlesnakes.

Kingston valve *Naval Architecture.* **1.** a valve connecting the ballast tank of an older type submarine to the open sea, and used in surfacing and submerging operations. **2.** a valve on the bottom of a ship that closes in response to water pressure.

kinic acid see QUINIC ACID.

kinin *Biochemistry.* any of a family of small peptide autacoids that are potent vasodilators and mediators of the inflammatory response, e.g., bradykinin and kallidin.

kininogen *Biochemistry.* either of two plasma α_2-globular proteins that are substrates of kallikrein and precursors of kinins.

kinin system *Biochemistry.* a humoral amplification system for inflammation, involving the activation of substrate protein molecules.

kink *Engineering.* a tight loop in a wire or wire rope that permanently deforms and damages the wire. *Medicine.* a popular term for muscle stiffness or soreness, as in the neck or back.

kink band *Geology.* a deformation band that occurs in crystals or foliated rocks in which the orientation of the lattice or foliation has been altered or deflected as a result of slippage or gliding. Also, JOINT DRAG.

kink instability *Physics.* an instability occurring in a plasma, in which the ionized gas along with its magnetic field becomes unstable and forms a hitch or kink that becomes progressively larger. Also, SAUSAGE INSTABILITY.

kino *Pharmacology.* the dried juice of the *Pterocarpus marsupium* tree of south Asia, used as an astringent. Also, **kino gum.**

kinocilium *Cell Biology.* a single cilium that is often present behind the stereocilia of a mammalian auditory hair cell.

kinoplasm *Cell Biology.* a cellular component that is active in the formation of filamentous structures such as cilia and spindle fibers.

kinoplastic *Cell Biology.* of or relating to kinoplasm.

Kinorhyncha *Invertebrate Zoology.* a phylum or class of free-living, segmented, marine invertebrates that live in subtidal mud and feed principally on diatoms.

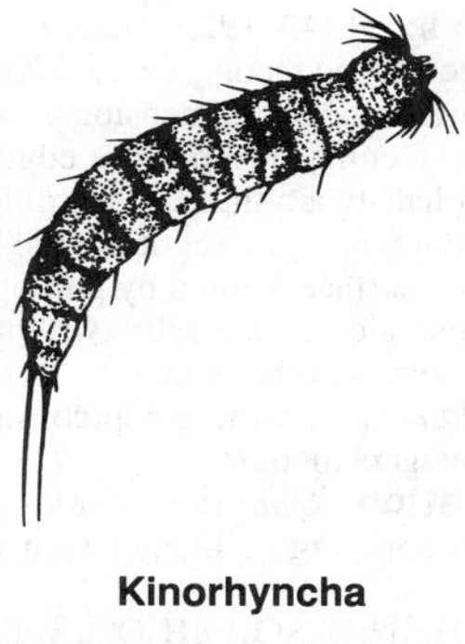

Kinorhyncha

Kinosternidae *Vertebrate Zoology.* the musk turtles, an aquatic family of the infraorder Cryptodira, characterized by a domed oval shell and found throughout North and South America.

kin recognition *Behavior.* the ability of an animal to distinguish between its relatives and nonrelated members of its own species.

kin selection see GROUP SELECTION.

Kinsey, Alfred 1894–1956, American biologist; conducted noted early studies of human sexual behavior.

kinship system *Anthropology.* the institution involving descent, its organization, and the way in which various kin are affiliated with each other.

kinzigite *Petrology.* a coarse-grained metamorphic rock consisting primarily of garnet and biotite, with lesser amounts of quartz, feldspar, mica, cordierite, muscovite, and sillimanite.

kip *Metrology.* one kilopound (one thousand pounds).

kipper *Food Technology.* a herring or other fish that has been filleted and then cured by salting, drying, and smoking.

Kipper *Ordnance.* a Soviet airborne anti-ship cruise missile powered by a turbojet engine and probably equipped with autopilot midcourse guidance and active radar terminal homing, delivering a 1000-kilogram conventional warhead at a speed of Mach 1.2 and a range of 120 km; officially designated **AS-2.**

KIPS Knowledge Information Processing System.

kipuka *Volcanology.* a tract of open land surrounded by lava flow. (A Hawaiian word meaning "opening.")

kipuka see STEPTOE.

Kirchhoff, Gustav 1824–1887, German physicist; with Bunsen, discovered spectral analysis; explained Fraunhofer lines; formulated Kirchhoff's laws of electrical circuits.

Kirchhoff assumption *Materials Science.* the theory that in-plane displacements are linear functions of the thickness of a laminated fibrous composite; used to analyze strain-displacement relationships in laminates.

Kirchhoff's equations *Thermodynamics.* a set of partial differential equations which state that the partial derivative of the enthalpy change of a system, with respect to the temperature during a reaction at constant pressure or volume, is equal to the change in the specific heat at constant pressure or volume.

Kirchhoff's formula *Thermodynamics.* an equation that is valid over specified temperature ranges, relating the vapor pressure p to the temperature T, of a substance: $\log p = a - (b/T) - c \log T$, where a, b, and c are constants.

Kirchhoff's law *Thermodynamics.* a law stating that in thermodynamic equilibrium, and depending on the conditions imposed, the emission and absorption by a body are related. For one direction and one wavelength, the emissivity and absorptivity of the body are equal. *Electricity.* either of two basic principles of electrical circuits that are based on this law. **Kirchhoff's first law** (**Kirchhoff's current law**) states that at any given time, the sum of the instantaneous values of current flowing into a given point or node equals the sum of the instantaneous values flowing out of that point. **Kirchhoff's second law** (**Kirchhoff's voltage law**) states that at any given time, the sum of all voltage rises in a closed loop equals the sum of all voltage drops in a circuit.

Kirkbyacea *Paleontology.* an extinct superfamily of Paleozoic ostracods; Silurian to Permian.

Kirkbyidae *Paleontology.* a large family of straight-hinged ostracods in the extinct superfamily Kirkbyacea; Silurian to Permian.

Kirkendall effect *Metallurgy.* a phenomenon in which a marker placed at the interface of a diffusion couple moves toward the alloy region.

Kirkidium *Paleontology.* one of the later genera of articulate brachiopods in the extinct order Pentamerida; large and nonstrophic, Kirkidium is found in Silurian deposits of Europe and North America.

Kirkwood, Daniel 1814–1895, American astronomer; discovered Kirkwood asteroid gaps.

Kirkwood-Brinkely's theory *Mechanics.* a theory for predicting the results of a blast at high altitude, based on the measured ground effects.

Kirkwood gaps *Astronomy.* orbits in the asteroid belt that contain few or no asteroids because objects there are perturbed out of the orbit by gravitational resonances with Jupiter.

kirovite *Mineralogy.* $(Fe^{+2},Mg)SO_4 \cdot 7H_2O$, a variety of melanterite containing magnesium.

Kirsten propeller *Naval Architecture.* a propeller having a vertical axis, with blades so geared that each blade makes half a revolution around its own axis per full revolution of the whole propeller.

Kiruna method *Mining Engineering.* a borehole-inclination survey method in which the electrolytic deposition of copper from a solution is used to make a mark on the inside of a metal container.

kish *Metallurgy.* free carbon that forms in a molten hypereutectic cast iron upon cooling.

kiss impression *Graphic Arts.* in offset lithography, a very light impression, made with little pressure on the paper.

kissing gourami *Vertebrate Zoology.* a Southeast Asian gourami, *Helostoma temminicki,* known for pressing its protrusile fleshy lips against those of another.

kiss-roll coating *Engineering.* a method for coating a substrate web in which the coating roll conveys a metered film of coating material; part of the film transfers to the web, while a portion remains on the roll.

Kistiakowsky-Fishtine equation *Physical Chemistry.* an equation that relates the latent heat of vaporization for a pure compound to its molal volume; used when vapor pressure or other critical data are not known.

Kitasato, Shibasaburo 1852–1931, Japanese bacteriologist; worked on diphtheria and tetanus antitoxins; isolated bubonic plague germ.

Kitasatoa *Bacteriology.* a genus of soil bacteria of the family Actinoplanaceae that produce club-shaped sporangia, each containing a chain of motile spores; similar to *Streptomyces* in phage sensitivity and wall type.

Kitchen *Ordnance.* a Soviet air-to-surface strategic cruise missile powered by a liquid propellant rocket engine and probably equipped with an inertial guidance system, delivering a 350-kiloton nuclear warhead or 1000-kilogram conventional warhead at a speed of Mach 3.5 and a range of 200 miles; officially designated **AS-4.**

kitchen midden *Archaeology.* a deposit in which the remains of cooking equipment, food, and food storage are evident, formed by the accumulation of refuse near a dwelling or campsite.

kite *Vertebrate Zoology.* any of several hawk species of the family Accipitridae that inhabit warm areas worldwide; characterized by a slender build with a small head, long narrow wings, and a deeply forked tail; appears buoyant and graceful in flight. *Ordnance.* in naval mine warfare, a device that, when towed, submerges and planes at a preset level without sideways displacement.

kite balloon *Aviation.* an elongated captive balloon having an empennage consisting of gas-filled lobes, which incline the balloon's longitudinal axis to the wind and thereby increase lift.

kite observation *Meteorology.* an atmospheric sounding taken by instruments carried aloft attached to a kite or kite balloon.

kit fox *Vertebrate Zoology.* the smallest North American fox, *Vulpes velox* or *V. macrotisis* of the family Canidae, inhabiting the plains of North America; characterized by yellow to gray fur, a black tipped tail, large ears, sensitive hearing, and nocturnal hunting. (Probably a shortened form of *kitten fox,* because of its small size.)

kit fox

kitol *Organic Chemistry.* $C_{40}H_{60}O_2$, a compound that is found in whale-liver oil and other liver oils, which yields vitamin A when heated to 200°C.

kitting *Design Engineering.* the predisposing of all required parts onto a pallet for delivery to an assembly area.

kittiwake *Vertebrate Zoology.* either of two gray gulls of the genus *Rissa,* family Laridae; the **black-legged kittiwake,** *R. tridactyla,* of North Atlantic coastal regions, spends the most time of any gull over the ocean, coming inland only to nest on narrow cliff ledges; the **red-legged kittiwake,** *R. brevirostris,* has a red bill and lives on the Bering Sea.

Kittyhawk *Aviation.* a popular name for the P-40 pursuit (fighter) aircraft.

kiva *Archaeology.* a specialized room or chamber of the ancient southwestern U.S., used for ceremonies or other community activities; completely or partly underground and containing benches and fire pits.

Kiviat graph *Computer Technology.* a circular chart used to evaluate computer performance that has various axes representing different states, with the distance from the center of the circle representing percentage of time in the state; a point on the circumference means 100 percent.

kiwi [kē´wē´] *Vertebrate Zoology.* a small, flightless, nocturnal, burrowing bird of the family Apterygidae, inhabiting New Zealand forests and having grayish-brown hairlike feathers, a long bill, and vestigial wings.

kiwi nuclear reactor *Nucleonics.* a 1950s-to-1960s reactor design that tested nuclear reactors used in rocket engines.

kJ kilojoule

KJ or **k.j.** knee jerk.

Kjeldahl flask [kel´däl´] *Analytical Chemistry.* a special type of long-necked flask used in the Kjeldahl method.

Kjeldahl method *Analytical Chemistry.* a test for the determination of nitrogen in organic compounds, using a series of steps to convert nitrogen to ammonia, which is distilled into a standard acid and titrant.

KK damage *Military Science.* combat damage that causes an aircraft to disintegrate immediately upon infliction.

KL *Aviation.* the airline code for KLM Royal Dutch Airlines.

kl or **kl.** kiloliter.

klapperstein see RATTLESNAKE ORE.

Klaproth, Martin 1743–1817, German chemist; discovered uranium, zirconium, and titanium; analyzed many elements and compounds.

klaxon *Acoustical Engineering.* **1.** an electronic horn generating a loud, repetitious sound that varies in frequency; used for police or ambulance sirens and other such warning signals. **2.** a similar hand-operated horn formerly used on automobiles. (From a former trademark.)

klebelsbergite *Mineralogy.* $Sb_4^{+3}O_4(OH)_2(SO_4)$, a yellow, transparent, orthorhombic mineral occurring as tufts of acicular crystals with stibnite, and having a specific gravity of 4.62 and a hardness of 3.7 on the Mohs scale.

Klebs, Edwin [klebz´; klāps] 1834–1913, German pathologist; discovered Klebs-Löffler (diphtheria) bacillus.

Klebsiella *Bacteriology.* a genus of Gram-negative, facultatively anaerobic, nonmotile, rod-shaped bacteria of the family Enterobacteriaceae; found in the intestinal or respiratory tract of humans and animals; a common cause of urinary, pulmonary, and wound infections.

Klebsiella pneumoniae *Bacteriology.* a species of *Klebsiella* that is a cause of pneumonia in alcoholics, diabetics, and other adults with altered defenses; some strains capable of nitrogen fixation.

Klebs-Löffler bacillus *Bacteriology.* the specific etiologic agent of diphtheria, *Corynebacterium diphtheriae,* which also causes various skin infections.

Klebsormidiaceae *Botany.* a family of unbranched green algae of the order Klebsormidiales, characterized by short filaments made up of cylindrical cells and rounded at one or both ends.

Klebsormidiales *Botany.* an order of unbranched filamentous green algae of the class Chlorophyceae, distinguished by motile cells with two flagella, flagellar basal bodies associated with a strip of microtubules, and the persistence of an interzonal spindle during cytokinesis.

Klebsormidium *Botany.* a genus of green algae of the family Klebsormidiaceae, characterized by chloroplasts containing a pyrenoid traversed by thylakoids and also by asexual reproduction.

Klein, (Christian) Felix 1849–1925, German mathematician; developed non-Euclidean geometric theory; devised Klein bottle.

Klein bottle *Mathematics.* a two-dimensional, nonorientable topological space obtained by identifying opposite edges of a rectangle, after giving the rectangle a half-twist; that is, regarding opposite sides of the twisted rectangle as being the same set of points. It may be visualized as a continuous, one-sided surface formed by pulling the small end of a tapering tube through one side of the tube (without making a hole) and then flaring it so as to join the other end.

Klein four group *Mathematics.* the group consisting of the direct sum of two copies of the integers modulo 2.

Klein-Gordon equation *Quantum Mechanics.* a version of the Schrödinger equation applying to integer spin particles that includes special relativity.

kleinite *Mineralogy.* $Hg_2N(Cl,SO_4)\cdot nH_2O$, a yellow to orange, hexagonal mineral occurring in short prismatic crystals, having a specific gravity of about 8 and a hardness of 3.5 to 4 on the Mohs scale; found in mercury deposits.

Kleinmann-Low Nebula *Astronomy.* an extended infrared source with a temperature of about 70 kelvins in the molecular cloud that lies directly behind the Orion Nebula.

Klein-Nishina formula *Quantum Mechanics.* the expression for the amount of energy per electron per unit time scattered by the Compton process.

Klein paradox *Quantum Mechanics.* the conclusion, following Dirac's electron theory, that an electron can penetrate a potential barrier of over twice its rest energy by changing to a negative energy state; this event occurs with significant probability only through barriers no deeper than the Compton wavelength.

Klein-Rydberg method *Physical Chemistry.* a technique for calculating the potential energy as a function of distance between nuclei of a diatomic molecule by determining the molecule's vibration and rotations.

Kleinschmidt monolayer technique *Genetics.* a laboratory procedure in which DNA molecules are spread as a thin, positively charged protein film on the surface of water, then transferred to precoated copper grids for study under the electron microscope. Also, **Kleinschmidt spreading technique.**

Klein's hypothesis *Astronomy.* a theory stating that the visible universe is just part of a larger "metagalaxy" structure.

Klein's reagent *Chemistry.* a saturated solution of cadmium borotungstate, used in the separation of minerals by specific gravity.

klendusity *Botany.* the ability of a plant to avoid infection because of some type of protective structure that prevents or hinders inoculation by a microorganism or virus.

Klenow fragment *Molecular Biology.* the larger of two portions of the DNA polymerase I enzyme that remain after proteolytic cleavage, containing the DNA polymerase activity and a 3'–5' exonuclease activity but lacking the 5'–3' exonuclease activity which is present in the complete enzyme.

klepto- a combining form meaning "theft," as in *kleptomania.*

kleptobiosis *Behavior.* the scavenging of food by one species from the nests or supplies of another species.

kleptogamy *Behavior.* a behavior pattern in which males from outside a group or lacking in status within the group attempt to mate with females from the group, often by means of some deceptive behavior.

kleptomania [klep′tə mā′nē ə] *Psychology.* an irrational impulse to steal; usually the items stolen have symbolic significance to the individual but relatively little monetary value. (Going back to the Greek word for "thief.")

kleptomaniac [klep′tə mā′nē ak] *Psychology.* a person who is affected by kleptomania.

kleptoparasite *Behavior.* an animal that obtains its food mainly by scavenging the food supplies of another species, especially one with which it habitually lives in close proximity. Thus, **kleptoparasitism.**

kleptophobia *Psychology.* an irrational fear of being robbed.

kliachite see CLIACHITE.

Kligler's iron agar *Microbiology.* a solid nutrient medium that contains iron and is used for distinguishing between members of the bacterial family Enterobacteriaceae.

Klinefelter syndrome *Genetics.* a human syndrome marked by diminuitive testes and mental retardation; caused by the presence of one extra X chromosome (XXY) in the male karyotype. Also, XXY TRISOMY.

Kline flocculation test *Pathology.* a microscopic precipitation test for identification of antibodies found after infection with syphilis.

K lines *Crystallography.* characteristic X-ray frequencies of radiation from atoms as a result of excitation of electrons in the K-shell. Such X-rays are designated **K**α, **K**β, and **K**γ, and are all doublets.

Klinkenberg correction *Petroleum Engineering.* a conversion of model laboratory measurements of air permeability, made on core material from a formation into equivalent liquid-permeability values.

klinokinesis *Entomology.* avoidance behavior seen in some insects that normally fly straight but turn rapidly when there is any unfavorable stimulus, including a subtle change in temperature or humidity.

klinotaxis *Biology.* the response of a motile organism that moves its head from side to side symmetrically in moving toward the stimulus, comparing the intensity of the stimulus on each side of its body. *Entomology.* any deliberate movement of an insect in relation to a stimulus, such as movement directly toward or away from light.

klint *Geology.* **1.** a previously buried, resistant coral reef or bioherm that stands in relief after the softer surrounding rocks have eroded and exposed it. **2.** a nearly vertical coastal cliff or abrasion precipice in Denmark and Sweden, several meters high and more than 100 meters long.

klintite *Geology.* the dense, resistant dolomite that constitutes a klint and provides its strong, durable framework.

klipfish see CLINIDAE.

klippe *Geology.* an isolated rock mass or part of a nappe that was separated from the surrounding rock or main nappe by faulting and subsequent erosion.

Klitzing, Klaus von born 1943, German physicist; Nobel Prize for discovery of quantum Hall effect.

KL Nebula see Kleinmann-Low Nebula.

klockmannite *Mineralogy.* CuSe, a bluish-black to slate-gray, hexagonal mineral occurring in granular aggregates, having a specific gravity of 5.99 and a hardness of 2 to 3 on the Mohs scale; found in uranium ores.

Kloeckera *Mycology.* a genus of fungi belonging to the class Hyphomycetes that includes species which cause the fermentation of glucose sugar in fruit.

Kloedenellacea *Paleontology.* a superfamily of dimorphic ostracods in the extinct suborder Kloedenellocopina; extant in the Early Paleozoic, especially Silurian.

Kloedenellocopina *Paleontology.* a suborder of ostracods belonging to the extinct order Palaeocopida; they differ from their relatives the Beyrichicopina in having one valve much larger than the other; especially abundant in the Early Paleozoic.

KL-ONE *Artificial Intelligence.* a framelike representation language that allows a new concept to be classified within a hierarchy where it best fits.

kloof wind *Meteorology.* a cold, southwesterly wind that blows near Simons Bay, South Africa.

K/L ratio *Nuclear Physics.* the ratio of the number of electrons captured from the *K* shell, nearest the nucleus, to the number captured from the *L* shell, the second nearest the nucleus, during inverse beta decay.

kludge *Computer Technology.* **1.** a computer whose design and performance characteristics are questionable. **2.** of or relating to a set of mismatched components that have been assembled into a system. **3.** a program or device that has been put together hastily or in a crude or slipshod manner.

Klug, Aaron born 1926, British biochemist; awarded the Nobel Prize for research in nucleic acid–protein complexes.

Kluyvera *Bacteriology.* a genus of Gram-negative bacteria of the family Enterobacteriaceae that are found in soil, sewage, and food and may be opportunist pathogens in humans.

kluyver effect *Biochemistry.* a phenomenon in which particular yeasts are only able to ingest a given disaccharide aerobically; mechanics of the effect are not known.

Kluyveromyces *Mycology.* a genus of yeast fungi of the family Saccharomycetaceae found in numerous environments such as foods, beverages, plants, soil insects, and sea water.

Klydonograph *Electricity.* an instrument designed to make a photographic record of electrical power surges, such as those generated by lightning striking an electrical power line.

klystron *Electricity.* an evacuated electron beam tube that is used for ultrahigh-frequency applications such as generation and amplification of microwave and electromagnetic energy; it incorporates an electron gun, one or more cavities, and an apparatus for modulation of the electron gun beam.

klystron generator *Electricity.* a klystron tube used as a generator with its second cavity used to feed waves to a waveguide.

klystron repeater *Electricity.* a klystron tube used to amplify incoming signals from one circuit and feed them into a following circuit.

km or **km.** kilometer.

kMc kilomegacycle.

K meson *Particle Physics.* a meson having zero spin, a nonzero strangeness quantum number, a mass of approximately 494 MeV, and a mean lifetime of 1.24×10^{-8}. It is the lightest hadron that contains a strange quark. Also, KAON.

kmole kilomole.

K Monel *Metallurgy.* a heat-treatable nickel-base alloy nominally containing 29.5 copper, 2.7 aluminum, and 0.5 titanium; UNS N05500.

KMPS or **kmps** kilometers per second. Also, **km.p.s.** or **km/sec.**

kn knots.

knallgas reaction *Biochemistry.* a process in which oxygen removes electrons from hydrogen gas and forms part of a system that generates energy for growth in certain bacteria.

knap *Geology.* **1.** the top or summit of a hill. **2.** any small hill or rise of ground. *Mining Engineering.* **1.** to break up rock. **2.** to chip out gangue manually in order to improve the grade of ore. A person who does this is a **knapper.** *Archaeology.* to make stone tools from flint or other stone by chipping or breaking off pieces.

knapsack problem *Mathematics.* the problem of partitioning a positive integer into a sum of positive integers selected without repetition from a given finite set; equivalently, given a collection of objects whose sizes are known, the problem of determining whether any subset of the objects fills a given knapsack exactly.

knebelite *Mineralogy.* $(Fe^{+2},Mg,Mn)_2SiO_4$, a manganoan variety of fayalite.

knee *Anatomy.* **1.** the articulation between the thigh and the leg, consisting of three condyloid joints, twelve ligaments, three bursae, and the patella. **2.** any structure that is bent like a knee. *Vertebrate Zoology.* a similar joint or region in a quadruped or bird. *Mechanical Engineering.* an adjustable component of a knee-and-column milling machine that supports the table and saddle and moves up and down on the column. *Metallurgy.* a supporting structure used in resistant welding.

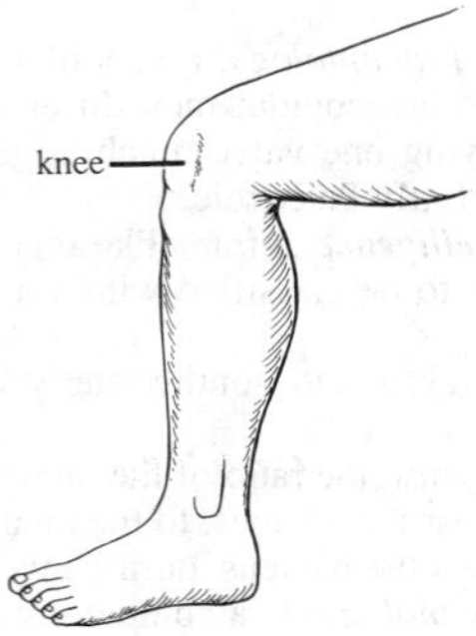

knee

knee brace *Building Engineering.* a diagonal member or stiffener used to brace the angle between two joined members, as between a joist and a rafter. Also, **knee.**

kneecap see PATELLA.

knee-chest position see GENUPECTORAL POSITION.

knee frequency see BREAK FREQUENCY.

kneeler *Building Engineering.* a stone used to support inclined masonry where it changes direction, such as at the corner of an arch.

knee rafter *Building Engineering.* a rafter using a brace to maintain the angle between a principal rafter and a tie beam.

knee tool *Mechanical Engineering.* a tool holder that is shaped like a knee; used especially for simultaneous turning and internal cutting operations.

knee wall *Building Engineering.* a partition that forms a side wall to support roof rafters beneath a pitched roof.

Kneriidae *Vertebrate Zoology.* a small family of cylindrical freshwater fish of the suborder Chanoidei, found in running water in tropical and southeastern Africa.

knick or **knickpoint** *Geology.* any break or interruption indicating a point of abrupt change in slope, particularly in the longitudinal profile of a stream, river, or its valley, that marks the intersection of the established longitudinal profile with the more recent profile resulting from glacial erosion, rejuvenation, or the exposure of resistant strata. Also, BREAK, REJUVENATION HEAD, ROCK STEP.

knife *Mechanical Devices.* a sharp, thin-bladed steel instrument whose edge can be either straight or serrated; used for cutting or slicing.

knife bayonet *Ordnance.* a bayonet with a handle and short, wide blade, thus able to be used as a knife.

knife coating *Engineering.* a method of coating a continuous-web substrate wherein the thickness of the coating is regulated by the distance between a movable knife or bar and the substrate.

knife-edge *Design Engineering.* a hardened-steel sharp edge, similar to that of a knife, that permits fine balance or acts as a fulcrum for a lever arm when used in a measuring instrument.

knife-edge bearing *Mechanical Engineering.* a rigid steel wedge that serves as a balance beam or level arm fulcrum in an equipment assembly in order to reduce friction; often used on weighing and testing machines.

knife-edge cam follower *Mechanical Engineering.* a cam follower with a knifelike edge or point for cam profiling determinations.

knife-edge refraction *Electromagnetism.* a lessening of the atmospheric signal attenuation when a radio wave is diffracted by an obstacle having sharply defined edges such as a building.

knife file *Mechanical Devices.* a thin, tapered triangular file resembling a knife in cross section.

knifefish *Vertebrate Zoology.* any of about 50 species of fishes in the families Apteronotidae, Gymnotidae (carnivorous species), and Rhamphichthyidae (herbivorous) found in fresh waters of Central and South America; typically blade-shaped, it is propelled by a long anal fin for great mobility in all directions, and navigates by emitting low voltage.

knife key *Mechanical Devices.* a key with prongs for tightening the joints of a compass, equipped with a knife blade at one end and a file about the midpoint.

knife-line attach *Metallurgy.* the intergranular corrosion of a welded joint.

knife switch *Electricity.* a switch in which contact is made by sliding a blade between two contacts to close the circuit.

knik wind *Meteorology.* a strong southeasterly wind that blows in the Matanuska Valley of Alaska, usually in the wintertime.

knismogenic *Neurology.* causing a tickling sensation.

knit lines *Materials Science.* an injection molding defect caused by incorrect temperature or improper mold design; it usually imparts a laminar or folded appearance to the finished product.

KNM-ER *Anthropology.* the identification for either of two important hominid fossils from Lake Rudolf in East Africa: **KNM-ER 1470,** a transitional *Homo habilis* specimen dated to 1.9 million years ago; and **KNM-ER 3733,** a *Homo erectus* dated to 1.5 million years ago, one of the most complete *H. erectus* ever found. (From their location at Kenya National Museum, East Rudolf.)

knob *Design Engineering.* a rounded handle or protuberance mounted on a shaft to facilitate manual manipulation rotation to open, close, tighten, or loosen. *Geology.* **1.** a rounded hill or small mountain, especially a steep, isolated hill. **2.** a projection arising from the top of a hill or mountain. **3.** a mass of resistant rock projecting from the side of a hill or mountain. *Anatomy.* any bulbous mass or protuberance.

knob-and-tube wiring *Electricity.* a former wiring method that used open insulated wire on solid insulators.

knock *Mechanical Engineering.* a metallic sound in an internal combustion engine, resulting either from faulty fuel combustion or from a mechanical malfunction, usually worn bearings.

knocked-on atom *Solid-State Physics.* an atom in a solid that is struck by an energetic particle passing through the solid and consequently recoils, which may displace the atom from its lattice position.

knock intensity *Petroleum Engineering.* the degree of knock or detonation recorded while testing gasoline combustion.

knock-knee *Medicine.* a deformity in which the knees are abnormally close together and the space between the ankles is increased.

knockmeter *Mechanical Engineering.* an instrument used to test fuels for knock intensity.

knock-off *Mechanical Engineering.* **1.** the automatic discontinuation of a machine's operation. **2.** a device that automatically discontinues a machine's operation.

knockout *Mechanical Engineering.* **1.** a perforated piece in a plastic or metal form that can be pushed out when a hole is desired. **2.** a mechanism that forces a piece from a mold.

knockout drops *Organic Chemistry.* a popular name for chloral hydrate, $CCl_3CH(OH)_2$, because of its powerful sedative effect. See CHLORAL HYDRATE.

knockout vessel *Chemical Engineering.* a vessel or trap used for removing fluid droplets from flowing gases.

knock rating *Petroleum Engineering.* the rating of gasolines by their tendency to knock; usually a fixed number on a set scale.

Knoevenagel reaction *Organic Chemistry.* the ammonia or amine-catalyzed reaction of an aldehyde with a compound that has a hydrogen alpha to two activating groups.

knoll *Geology.* **1.** a low, rounded hill or mound, or the rounded top of a larger hill. **2.** a sea mound that rises less than 1000 meters above the seafloor. **3.** see REEF KNOLL.

Knoop hardness [noop] *Metallurgy.* a measure of microhardness by the Knoop indentation test.

Knoop indentation test *Metallurgy.* a standardized test to measure microhardness, often of individual constituents of a metallic structure.

Knoop indenter *Metallurgy.* the indenter used for the Knoop indentation test.

knopite *Mineralogy.* a cerium-bearing variety of perovskite.

Knorr synthesis [nôr] *Organic Chemistry.* the formation of pyrrole derivatives from α-amino ketones and carbonyl compounds having active α-methylene groups.

knot *Engineering.* an intertwining of the ends or parts of one or more ropes, threads, or the like so that they cannot be easily separated. *Forestry.* a hard mass of wood at the place where a branch is connected to a tree trunk. *Anatomy.* a knoblike swelling or protuberance, such as a node. *Organic Chemistry.* a chiral structure having rings of 50 or more members in a knotlike arrangement. *Mathematics.* **1.** a curve in space formed by looping and intertwining a piece of string and then joining the ends together. **2.** a closed curve in space that is topologically equivalent to a circle. **3.** more generally, any imbedding of an n-sphere into an $(n + 2)$-sphere. *Computer Technology.* see DEADLOCK. *Metrology.* a unit of speed that is equivalent to one nautical mile per hour (1.852 kilometers per hour); widely used in measuring ship or aircraft speed.

knot grass see BIRD GRASS.

knot theory *Mathematics.* the mathematical study of knots, including their classification and methods for determining whether one knot may be continuously deformed into another.

knowbot *Artificial Intelligence.* a knowledge-based agent that attempts to accomplish a goal for a user, such as to find desired materials in a remote library. (An acronym for knowledge robot.)

knowledge *Artificial Intelligence.* **1.** a general term for the aggregation of facts, principles, and other information that is characteristic of human intelligence. **2.** a level of capability or sophistication in a machine system that is regarded as analogous to human knowledge.

knowledge acquisition *Artificial Intelligence.* the process of collecting knowledge for an expert system from a human domain expert and formalizing it.

knowledge-acquisition bottleneck *Artificial Intelligence.* the concept that knowledge acquisition is the major impediment and most time-consuming part of the development of an expert system.

knowledge base *Artificial Intelligence.* the collection of knowledge of an expert system, including facts, rules, heuristics, and procedures.

knowledge-based system *Artificial Intelligence.* a computer system that stores and uses a very large amount of information about an application and serves as an intelligent assistant, rather than as an expert.

knowledge-based management system *Artificial Intelligence.* one of the three segments of an expert system that automatically organizes, controls, propagates, and updates information in the knowledge store.

knowledge coding *Artificial Intelligence.* the process of producing computer-readable code from knowledge representations such as rules, frames, semantic networks, and decision tables.

knowledge engineer *Artificial Intelligence.* a person who develops an expert system by gathering knowledge of the domain from human experts and other sources, organizing the knowledge, choosing appropriate software tools, codifying the knowledge in machine-processable form, and verifying the quality of the result.

knowledge engineering *Artificial Intelligence.* the process of translating knowledge gained from a human expert into a representation that can be coded and integrated into a computer system.

knowledge information processing system *Artificial Intelligence.* a proposed fifth-generation computer concept that would have symbolic inference capabilities, large knowledge bases, high processing speed, and an extremely friendly user interface.

knowledge refining *Artificial Intelligence.* the capability of an expert system to examine its own performance and, by learning, improve it for future reference.

knowledge representation *Artificial Intelligence.* a technique for structuring human knowledge about a subject or specialty so that it can be coded for input into a computer; examples are rules, frames, semantic networks, propositional logic, and predicate calculus.

knowledge source *Artificial Intelligence.* **1.** any kind of information about a problem that can be used to help solve it. **2.** in a blackboard system, a procedure or set of rules triggered by the addition of data to the blackboard, which examines that data, adding its own conclusions to the blackboard or taking other action.

known traffic *Transportation Engineering.* a flight that is known to air traffic controllers at a given control center by direct communication between the center and the flight.

knuckle *Anatomy.* the dorsal aspect of a phalangeal joint, especially of the metacarpophalangeal joints of the flexed fingers. *Design Engineering.* a protruding angle at the intersection of two surfaces or members, as in a roof or ship hull. *Mining Engineering.* the top of a grade or hill on a track over which mine cars are hauled.

knuckle joint *Anatomy.* any joint that forms a knuckle. *Mechanical Engineering.* a joint between parts that allows movement in only one plane. Thus, **knuckle-joint hinge, knuckle-joint press.**

knuckle pin *Mechanical Devices.* the wood or metal rod in a knuckle joint that joins the eye to the knuckles. Also, WRIST PIN.

knuckle-walking *Anthropology.* a method of locomotion used by certain African apes in which weight is partially supported by specially adapted pads on the back of the fingers.

Knudsen, Martin [noot´sən; knoo´sən] 1871–1949, Danish physicist and oceanographer; studied sea water; showed relationship of chlorinity, salinity, and density.

Knudsen cell *Physical Chemistry.* a device that is used to measure the amount of vapor escaping from a system in which a liquid is in equilibrium with its vapor; used to determine vapor pressure.

Knudsen cosine law *Physics.* a law relating the number of molecules in a gas in equilibrium striking or leaving a differential element of area dS on the wall of a solid angle dw to the angle of incidence with the normal to the surface: $N = [(dSnc)\cos q dw]/4p$, where n is the particle density and c is the average speed of the particles.

Knudsen gauge *Engineering.* a device used to measure extremely low pressures by determining the force of a gas on a cold plate that is next to an electrically heated plate.

Knudsen-Langmuir equation *Chemical Engineering.* an equation giving the quantitative rate of molecular distillation related directly to the vapor saturation pressure and the square root of the molecular weight, and inversely to the square root of the solution temperature.

Knudsen number *Fluid Mechanics.* a dimensionless quantity that is used in studying low-density gas flow, given by the mean free path of molecules divided by a characteristic length.

Knudsen's equation *Physics.* an equation giving the volume (at unit pressure) of a gas that flows through a tube of diameter d and length L at pressures low enough to be considered free molecular flow, expressed as $q = (2p)^{1/2} pd^3/6L\,(\rho)^{1/2}$, where p is the pressure difference across the tube, and ρ is the density of the gas at unit pressure.

Knudsen's tables *Oceanography.* a set of hydrographic tables that aid in computing the results of seawater chlorinity titrations and hydrometer readings and in converting the results to values of salinity, density, and sigma-T.

Knudsen vacuum gauge *Engineering.* an instrument used to determine negative or vacuum gas pressures; a rotatable vane moves from the pressure of heated molecules in proportion to the concentration of molecules in the system.

knurl *Engineering.* to roughen a surface with small ridges or knobs, either for decoration or in order to provide a strong grip, as on the head of a "milled" screw.

knurling tool *Mechanical Devices.* a lathe tool for knurling a surface.

koala *Vertebrate Zoology.* an arboreal Australian marsupial, *Phascolarctus cinereus,* that feeds exclusively on the leaves and shoots of certain eucalyptus trees and has gray fur, large hairy ears, and sharp claws. (From a native Australian name for this animal.)

koala

kobellite *Mineralogy.* $Pb_{22}Cu_4(Bi,Sb)_{30}S_{69}$, a blackish-gray, opaque, orthorhombic, metallic mineral occurring in massive form and as prismatic crystals, and having a specific gravity of 6.48 and a hardness of 2.5 to 3 on the Mohs scale; found in granite pegmatite veins.

Koch, Robert [kōk] 1843–1910, German physician and bacteriologist; awarded the Nobel prize in 1905 for his work on tuberculosis; formulated Koch's postulates.

Robert Koch

Kochab *Astronomy.* β-Ursae Minoris, a second-magnitude orange star in the bowl of the Little Dipper.

Koch curve *Mathematics.* **1.** a fractal curve, having fractal dimension equal to log 4/log 3, constructed from an equilateral triangle P_0 by the following recursive procedure: At the *n*th step, divide every side of the polygon P_{n-1} into three equal segments of length a_n. To each of the center segments, attach an equilateral triangle having sides of length a_n so that one side of the triangle is the corresponding center segment. The boundary of the region thus obtained is the polygon P_n. The Koch curve is the limit, as *n* goes to infinity, of the polygons P_n. Also, **triadic Koch curve, snowflake curve.** The area enclosed by a Koch curve is a **Koch island. 2.** a **generalized Koch curve** is obtained from a regular polygon P_0, called the initial polygon or initiator. A (not necessarily convex) polygonal curve or generator having *N* edges of equal length $r < 1$ is chosen. At the *n*th step of the construction, each side of the polygon P_{n-1} is replaced by the generator, reduced and displaced so its endpoints coincide with those of the side it replaces; the polygon thus obtained is P_n. The generalized Koch curve is the limit, as *n* goes to infinity, of the polygons P_n; it has the fractal dimension log *N*/log(1/*r*). The region bounded by the generalized Koch curve is a **generalized Koch island.** (Named for the mathematician Helge von *Koch.*)

Kocher, (Emil) Theodor [kōk′ər] 1841–1917, Swiss surgeon; awarded the Nobel Prize for his work in the physiology, pathology, and surgery of the thyroid; also designed many surgical instruments and performed the first operation to removed an enlarged thyroid gland.

Kocher's forceps *Surgery.* a type of surgical forceps having sharp points at the tips and transverse serrations along the full edge for holding tissues during operation or for compressing bleeding tissue.

Koch's postulates *Pathology.* four criteria that were established by Robert Koch to identify the causative agent of a particular disease: (a) the microorganism or other pathogen must be present in all cases of the disease; (b) the pathogen can be isolated from the diseased host and grown in pure culture; (c) the pathogens from the pure culture must cause the disease when inoculated into a healthy susceptible laboratory animal; (d) the pathogen must be isolated from the new host and shown to be the same as the original organism.

Kodak *Optics.* the trade name for a widely used series of portable cameras introduced by George Eastman in 1888 and subsequently manufactured by the Eastman Kodak Company.

Kodel *Textiles.* the trade name for a brand of polyester fiber that is used in cotton-polyester blended fabrics.

kodiak bear

kodiak bear *Vertebrate Zoology.* a variety of grizzly bear, *Ursus arctos middendorffi*, the largest of all living land carnivores at 10 feet tall and up to 1700 pounds. (Found on the *Kodiak* Islands of Alaska.)

Koebe function *Mathematics.* a member of the one-parameter family of univalent functions given by $f(z) = z/(1 - e^{i\alpha}z)^2$; in particular, the function $f(z) = z/(1 - z)^2 = z + 2z^2 + 3z^3 + 4z^4 + \cdots$, which attains the maximum coefficient values allowed by deBranges' theorem.

koechlinite *Mineralogy.* Bi_2MoO_6, a greenish-yellow orthorhombic mineral occurring in lathlike twinned crystals, having a specific gravity of 8.26 and an undetermined hardness; found with quartz and other bismuth minerals.

koembang *Meteorology.* a dry, foehnlike southeasterly or southerly wind of Java; a result of the passage of the east monsoon through gaps in mountain ranges and its descent on the leeward side.

koenenite *Mineralogy.* $Na_4Mg_9Al_4Cl_{12}(OH)_{22}$, a colorless to dark-red, trigonal, very soft mineral occurring in indistinct crystals, having a specific gravity of 1.98 and a hardness of about 1.5 on the Mohs scale; found in potash mines.

Kofoidiniaceae *Botany.* a family of large, marine dinoflagellates of the order Noctilucales that are nonthecate, motile, and phagotrophic and have an adult cell with a laterally flattened expanded hyposome.

Kohlberg, Lawrence 1927–1987, American educator and social psychologist; developed an important theory of moral development.

Lawrence Kohlberg

Köhler, Georges born 1946, German immunologist; shared the Nobel Prize for development of monoclonal antibodies.

Köhler, Wolfgang 1887–1967, German-born American psychologist; studied perceptual organization and learning among apes.

Köhler illumination *Optics.* a method for illuminating a microscope specimen that does not involve the use of lenses for light adjustment, often employing a pair of iris diaphragms that are adjusted separately for brightness and contrast.

kohlrabi *Botany.* a variety of cabbage, *Brassica oleraceae gongylodes*, having a greatly enlarged, turnip-shaped, edible stem. Also, STEM CABBAGE, TURNIP CABBAGE.

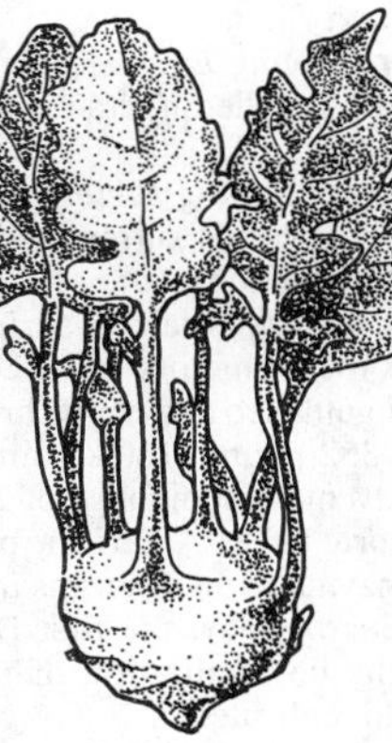

kohlrabi

Kohlrausch method *Physical Chemistry.* a system that measures the resistance in an electrolytic solution to determine the ability of the solution to conduct electricity. (From Friedrich *Kohlrausch,* 1840–1910, German physicist.)

Kohlrausch's law *Physical Chemistry.* the statement that the ability of an electrolytic solution to conduct electricity is the sum of the conductivities of its ionic components.

kohm kilohm.

Kohoutek [kə hō´tek] *Astronomy.* comet 1973 XII, discovered near Jupiter's orbit and predicted to be a splendid naked-eye sight at perihelion; although it did not brighten as much as predicted, the comet was intensively studied and provided much information about comets in general. (Named for Lubos *Kohoutek,* born 1935, Czech astronomer.)

koi *Vertebrate Zoology.* any of various common carp that are bred for large size and bright colors.

koilo- a combining form meaning "hollow."

koilonychia *Medicine.* a condition in which the fingernails are malformed, spoon-shaped or concave, sometimes symptomatic of iron-deficiency anemia.

koilorrhachic *Medicine.* a deformity of the spine in which the lumbar curvature is concave anteriorly.

kojibiose *Biochemistry.* a disaccharide that is formed from the enzymatic degradation of some plant carbohydrates, and occurs in the teichoic acids of certain streptococci.

kojic acid *Organic Chemistry.* $C_6H_6O_4$, a crystalline substance that is produced by the action of certain molds on starches and sugars; melts at 152–154°C; used as an antifungal and antimicrobial agent.

koktaite *Mineralogy.* $(NH_4)_2Ca(SO_4)_2 \cdot H_2O$, a colorless or white monoclinic mineral occurring in commonly twinned, fibrous or acicular crystals, having a specific gravity of 2.09 and an undetermined hardness; found as pseudomorphs after gypsum in the tailings of a lignite mine in Czechoslovakia.

Kolbe, (Adolf Wilhelm) Hermann 1818–1884, German chemist; pioneer in integration of organic and inorganic chemistry.

kolbeckite *Mineralogy.* $ScPO_4 \cdot 2H_2O$, a blue to gray, transparent monoclinic mineral occurring in short prismatic, brittle crystals, and having a specific gravity of 2.35 to 2.44 and a hardness of 3.5 to 5 on the Mohs scale; found in altered variscite nodules.

Kolbe hydrocarbon synthesis *Organic Chemistry.* the aqueous electrolysis of salts of carboxylic acids that yields carbon dioxide and hydrocarbons by a free radical path. Also, **Kolbe electrolysis, Kolbe electrolytic synthesis.**

Kolbe-Schmidt synthesis *Organic Chemistry.* the reaction of sodium phenoxide with CO_2 to yield sodium salicylate, which produces salicylic acid when acidified. Also, **Kolbe-Schmidt reaction.**

Kölliker, Rudolph Albrecht von 1817–1905, Swiss physiologist; fundamental contributions to histology; improved microscopic techniques.

Kollsman number *Aviation.* the number at which a pressure altimeter is preset, corresponding to the barometric pressure at a given point on the earth's surface. Also, ALTIMETER SETTING.

Kollsman window *Aviation.* a small window on the front of a pressure altimeter in which the Kollsman number is indicated.

Kolmer test *Pathology.* a modification of the Wassermann test; a quantitative assay for syphilis using an antigen and a fixation rather than flocculation method.

Kolmogorov, Andrei Nikolaevich [kōl´mə gôr´äf] 1903–1987, Russian mathematician; known primarily for his work in probability theory.

Kolmogorov-Arnold-Moser theorem *Physics.* a theorem stating that oscillatory motions in conservative dynamical systems persist through the addition of small perturbations to the system.

Kolmogorov consistency conditions *Mathematics.* let (Ω_F, F_F, P_F) be a probability space, where Ω_F is a (finite) Cartesian product of copies of the real line, indexed over some finite set F of real numbers. A family $\{P_{F\alpha}\}$ of such probability measures is said to satisfy the Kolmogorov consistency conditions if, given any two finite sets F_1 and F_2 of real numbers with F_1 contained in F_2, P_{F2} coincides with P_{F1} when restricted to the sets of Ω_{F1} that are independent of coordinates in $F_2 - F$.

Kolmogorov forward equation see FOKKER-PLANK EQUATION.

Kolmogorov inequalities *Mathematics.* suppose $\{X_k\}$ is a countable sequence of independent random variables in a probability space (Ω, F, P), having finite variances σ_k and such that the X_k are uniformly bounded by a constant c. Let $S_k = \sum_{i=1}^{k} X_i$. Then, for every $\varepsilon > 0$ and integer n, the following inequalities hold:

$$1/\varepsilon^2 \sum_{k=1}^{n} (\sigma_k)^2 \geq P(\max_{k \leq n} |S_k + ES_k| \geq \varepsilon) \geq 1 - (\varepsilon + 2c)^2/[\sum_{k=1}^{n} (\sigma_k)^2],$$

where ES_k denotes the expected value of S_k.

Kolmogorov inertial subrange *Fluid Mechanics.* the wavenumber range of the turbulent energy spectrum where viscous dissipation balances cascade of energy from large scale eddies.

Kolmogorov-Smirnov test *Statistics.* a distribution-free test to determine whether an empirical distribution function represents a sample from a specified continuous distribution.

Komodo dragon *Vertebrate Zoology.* the largest living lizard (up to 10 feet long and 250 pounds), a carnivorous lizard of the family Varanidae. (Native to the island of *Komodo* in southeastern Asia.)

Komodo dragon

kona *Meteorology.* a strong southwesterly wind that blows over Hawaii and usually brings rain, occurring approximately five times a year on the southwest slopes, in the lee of the prevailing northeast trade winds. *Geography.* the leeward side of an island, or the side away from the trade winds.

kona cyclone *Meteorology.* a strong, slow-moving cyclone system that develops in subtropical latitudes during the winter months. Also, **kona storm.**

Kondo alloy *Metallurgy.* an alloy that exhibits the Kondo effect.

Kondo effect *Metallurgy.* the increase in electrical resistivity of some magnetic materials as the temperature is decreased.

Kondo temperature *Metallurgy.* the temperature below which the Kondo effect is observed.

kongsbergite *Mineralogy.* (Ag,Hg), a cubic mercurian variety of silver.

Konig erecting prism *Optics.* a roofed prism used to erect the image in some binoculars.

Königsberg bridge problem *Mathematics.* the famous problem of determining whether it is possible to traverse seven bridges in Königsberg, Germany, without using the same bridge twice; i.e., whether there exists an Eulerian trail for that configuration of bridges and streets. Euler proved that no such route exists; his name is given to the terms Eulerian circuit and Eulerian trail because of his work on this problem.

König's minimax theorem *Mathematics.* if G is a bipartite graph, then the maximum number of edges in a matching of G equals the minimum size of a set S of vertices of G such that every edge of G is incident with at least one member of S.

konimeter *Engineering.* an instrument used to measure the amount and concentration of dust in the air. (From Greek *kóni,* "dust.")

koninckite *Mineralogy.* a yellow, transparent, tetragonal mineral with an approximate formula of $Fe^{+3}PO_4 \cdot 3H_2O$, occurring as small spherical aggregates, having a specific gravity of 2.40 and a hardness of 3.5 on the Mohs scale.

koniscope *Engineering.* an instrument used to reveal the presence of dust particles in the air.

Konowaloff rule *Physical Chemistry.* the observation that vapor pressure rises when there is a greater proportion of vapor to liquid in a mixture in a closed system.

Koobi Fora *Archaeology.* the site of important hominid finds on the northeastern shore of Lake Rudolf, in northern Kenya.

kookaburra *Vertebrate Zoology.* the largest of the kingfishers, belonging to the Alcedinidae family; a stout, gray-brown bird with a long, broad bill; sometimes called a "laughing jackass" for its wild, raucous call; inhabits Australian forests, savannas, and open country.

kookaburra

Koonungidae *Invertebrate Zoology.* a family of small, blind crustaceans in the order Anaspidacea.

Kopeloff modification *Microbiology.* a procedure for the staining of anaerobic bacterial cells, using a modification of the Gram stain method, in which safranin is the counterstain.

kopfring *Ordnance.* a metal ring attached to the nose of a bomb in order to reduce its penetration of land or water.

kophemia *Neurology.* a type of aphasia characterized by the inability to recognize words spoken in a language with which the subject is familiar. Also, LOGOKOPHOSIS, WORD DEAFNESS.

Koplik's spots *Medicine.* small red spots with a bluish white center that are visible on the lingual and buccal mucosa at the end of the premonitory stage of measles, a day or two before the rash appears. Also, **Koplik's sign.**

kopophobia *Psychology.* an irrational fear of becoming exhausted or fatigued.

Köppen, Wladimir 1846–1940, Russian-born German climatologist; developed climatic classification system; pioneer in paleoclimatology.

Köppen-Supan line *Meteorology.* an isotherm on a synoptic weather chart connecting geographical locations whose average temperature is 50°F (10°C) for the warmest month of the year.

koppite see PYROCHLORE.

Kopp's asthma see LARYNGISMUS STRIDULUS.

Kopp's law or **rule** *Physical Chemistry.* the statement that the heat capacity of a solid compound, at standard temperature and pressure, is approximately equal to the sum of the heat capacities of the elements in this compound; there are exceptions for certain of the light elements. (From Hermann *Kopp,* 1817–1892, German physical chemist.)

Korea *Geography.* a long (600 miles), mountainous peninsula in eastern Asia, south of northeastern China.

Korfmann arch saver *Mining Engineering.* a machine for withdrawing steel arches using a controlled hydraulic system instead of drawing them by hand or by winches.

Korfmann power loader *Mining Engineering.* a cutter-loader device equipped with four drilling heads surrounded by one cutter chain that can cart and load in both directions.

Kormoran *Ordnance.* a German air-to-surface anti-shipping missile powered by twin boost rocket motors and a central sustainer; equipped with midcourse inertial guidance and active or passive terminal homing, it delivers an advanced 352-pound (160 kg) armor-piercing warhead at a speed of Mach 0.95 and a maximum range of 23 miles (37 km).

Kornberg, Arthur born 1918, American biochemist; shared Nobel Prize for discovering enzymatic mechanisms in the synthesis of DNA.

Kornberg enzyme *Enzymology.* the first DNA-directed polymerase enzyme discovered; the determination of its functional mechanics showed how DNA could replicate.

kornelite *Mineralogy.* $Fe_2^{+3}(SO_4)_3 \cdot 7H_2O$, a pale pink to violet, water-soluble monoclinic mineral occurring in lathlike or acicular crystals and fibrous masses, having a specific gravity of 2.3 and an undetermined hardness; found as a secondary mineral with coquimbite and other sulfates.

kornerupine *Mineralogy.* $Mg_4(Al,Fe^{+3})_6(Si,B)_4O_{21}(OH)$, a white to green, orthorhombic mineral occurring in fibrous aggregates and prismatic crystals, having a specific gravity of 3.27 and a hardness of 6 to 7 on the Mohs scale; found with quartz, tourmaline, zircon, and garnet.

Korsakoff's syndrome *Psychology.* a disorder in which an individual cannot remember newly acquired information; most commonly found in chronic alcoholics or victims of serious head injuries. Also, **Korsakov's psychosis, Korsakov's syndrome.** (Named for Sergei *Korsakoff* (Korsakov), 1874–1920, Russian physician.)

Korshun method *Analytical Chemistry.* a test used to determine the amount of hydrogen and carbon in organic compounds; the sample is cracked in a shortage of oxygen and then oxidized in an excess of oxygen.

Korteweg-de Vries equation *Fluid Mechanics.* the nonlinear equation describing the motion of an incompressible, inviscid gravity wave of finite amplitude; its solutions are solitary waves.

Korteweg-de Vries soliton *Physics.* a single-oscillation soliton.

Kort nozzle *Naval Architecture.* a protective cylindrical fairing forming a duct around a propeller; used for high-thrust, low-speed applications.

Koserella *Bacteriology.* a genus of Gram-negative bacteria of the family Enterobacteriaceae that metabolize cellibiose and are found in clinical specimens.

Koser's citrate medium *Microbiology.* a bacteriological nutrient medium that is used for the citrate test in bacteria.

Kosmoceras *Paleontology.* a genus of ammonoids in the order Ammonitida and superfamily Stephanocerataceae; found worldwide in Middle Jurassic deposits.

kosmochlor *Mineralogy.* $NaCr^{+3}Si_2O_6$, an emerald-green, monoclinic mineral of the pyroxene group, having a specific gravity of 3.60 and a hardness of 6 on the Mohs scale; found as polycrystalline aggregates in meteorites. Also, UREYITE.

kossava *Meteorology.* a cold wind that descends from the east or southeast in the area of the Danube "Iron Gate" through the Carpathians, characterized by squall-like weather conditions.

Kossel, Albrecht 1853–1927, German biochemist; awarded the Nobel Prize for research in cell chemistry.

Kossel, Walther 1888–1956, German chemist; developed theory of chemical bonds.

Kossel lines *Spectroscopy.* the curves on photographic film that result from the intersection of a cone of X-rays and planes of a single crystal.

Kossel-Sommerfield law *Spectroscopy.* a law stating that for an isoelectronic sequence, the arc spectra resemble each other, especially in their multiplet structure.

Koster's stain *Microbiology.* a bacteriological stain that detects *Brucella* species in mammalian tissues, using a safranin stain and a carbol methylene blue counterstain.

Koszul connection *Mathematics.* a connection on a differentiable manifold given in terms of a covariant differentiation operator ∇ with the following properties:

$$\nabla_{X+Y} Z = \nabla_X Z + \nabla_Y Z$$
$$\nabla_X(Y+Z) = \nabla_X Y + \nabla_X Z$$
$$\nabla_{fX} Y = f \cdot \nabla_X Y$$
$$\nabla_X(fY) = f \cdot \nabla_X Y + X(f) \cdot Y.$$

Use of the Koszul connection (read "del") permits many of the manipulations in the Ricci calculus to be represented in a coordinate-free manner.

kotoite *Mineralogy.* $Mg_3B_2O_6$, a colorless, transparent, orthorhombic borate mineral, isostructural with jimboite, and having a specific gravity of 3.04 and a hardness of 6.5 on the Mohs scale; found as granular masses in a granite-dolomite contact zone.

köttigite *Mineralogy.* $Zn_3(AsO_4)_2 \cdot 8H_2O$, a carmine, monoclinic mineral of the vivianite group, occurring as prismatic crystals and fibrous crusts, and having a specific gravity of 3.33 and a hardness of 2 to 3 on the Mohs scale. Also, KOETTIGITE.

Kourbatoff's reagents *Metallurgy.* in metallography, a reagent for carbon microanalysis.

Kovalevsky, Sofya 1850–1891, Russian mathematician; known for her comments on the rotation of a solid body about a fixed point.

Kovar *Materials.* the trade name for a type of borsilicate glass that contains iron, nickel, cobalt, and manganese; it is heat-resistant and is used for making gastight windows, machine sealings, and metal attachments to glass.

Kovat's retention indexes *Analytical Chemistry.* a scale of numerical values for alkanes, used to identify a compound in gas chromatography.

Kozeny-Carmen equation *Fluid Mechanics.* an equation that applies to laminar flow of fluids through a particulate medium.

KP or **K.P.** kitchen police.

kpc kiloparsec. Also, **kparsec.**

kph kilometers per hour.

kr kilorad.

Kr the chemical symbol for krypton.

Kraemer equation *Materials Science.* the mathematical definition for the intrinsic viscosity of a series of polymers of different molecular weights in a given solvent.

Krafft-Ebing, Richard von 1840–1902, German psychiatrist; studied sexual psychoses.

kraft paper *Graphic Arts.* a sturdy tan paper made from unbleached sulfate pulp, often used in combination with other fibers to make container board, wrapping and bag papers, and the like. Also, **kraft.** (A German word meaning "strength.")

kraft process or **kraft pulping** see SULFATE PULPING.

Krakatao *Geography.* a small volcanic island between Sumatra and Java; an active volcano, site of an enormous eruption in 1883. Also, **Krakatau, Krakatoa.**

Krakatao winds *Meteorology.* an easterly moving wind layer over the tropic zone, typically at altitudes between 18 and 24 kilometers, having a depth of at least 6 kilometers.

Krameriaceae *Botany.* a monogenetic family of dicotyledonous, tanniferous, hemiparasitic shrubs and perennial herbs of the order Polygalales, native to dry regions of North and South America.

Kramers, Hendrik 1894–1952, Dutch physicist; developed Kramers-Kronig equations; worked in X-rays and kinetic theory of gases.

Kramers-Konig relation *Optics.* an integral expression that relates the refractive index and extinction coefficient of a material to its reflectance spectrum.

Kramer's theorem *Solid-State Physics.* a theorem stating that in the presence of an external electrostatic field, the states of a system having an odd number of electrons are at least twofold degenerate.

K-RAS *Oncology.* a proto-oncogene that contains point mutations and is found in acute myelogenous and lymphoblastic leukemia, carcinoma of the thyroid, and melanoma.

Krasovski ellipsoid of 1938 *Geodesy.* a reference spheriod having a semimajor axis of approximately 6,378,245 meters and an ellipticity of 1/298.3.

K ratio *Space Technology.* in a jet engine or rocket engine, the ratio of the propellant surface area to the nozzle throat area.

K ration *Ordnance.* a lightweight, packaged ration that was used during World War II under emergency field conditions.

kraurosis *Medicine.* a dry, shriveled condition of a part, especially the vulva, occurring most often in older women.

Krause rolling mill *Metallurgy.* a type of rolling mill that achieves high reductions in each pass.

Krause's corpuscles *Anatomy.* small, encapsulated sensory nerve endings located in the skin, mucous membranes, conjunctiva, and heart. (Named after W. J. F. *Krause,* 1833–1910, German anatomist.) Also, CORPUSCULA BULBOIDEA, END BULBS OF KRAUSE.

Krause-Wolfe graft see WOLFE'S GRAFT.

krausite *Mineralogy.* $KFe^{+3}(SO_4)_2 \cdot H_2O$, a pale lemon-yellow, transparent, monoclinic mineral occurring in short prismatic crystals, and having a specific gravity of 2.84 and a hardness of 2.5 on the Mohs scale; found with other sulfates in borax mines.

krebiozen *Pharmacology.* a substance isolated from the blood of horses previously injected with *Actinomyces bovis* and believed by some to be an effective treatment for cancer; it is not used clinically in the United States.

Krebs, Sir Hans Adolf 1900–1981, German-born British biochemist; awarded the Nobel Prize for studies in intermediate metabolism including Krebs cycle.

Krebs cycle *Biochemistry.* the cyclic mechanism by which complete oxidation of acetyl-coenzyme A is effected. Also, TRICARBOXYLIC ACID CYCLE, CITRIC ACID CYCLE.

Krebs-Henseleit cycle *Biochemistry.* a series of reactions that produce urea, providing the major route for removal of ammonia produced in the metabolism of amino acids in the liver and kidneys.

Krebspest *Invertebrate Zoology.* a fatal fungal disease of crayfish.

Krein-Milman theorem *Mathematics.* the theorem that in a locally convex topological vector space *X*, any compact convex subset *K* is identical with the intersection of all convex sets containing its extreme points; i.e., *K* is the closed convex hull of its extreme points.

kremersite *Mineralogy.* $(NH_4,K)_2Fe^{+3}Cl_5 \cdot H_2O$, a red, orthorhombic mineral occurring in psuedo-octahedral crystals, having a specific gravity of 2.00 to 2.17 and an undetermined hardness; found in fumaroles on Mt. Etna and Mt. Vesuvius, Italy.

Kremser formula *Chemical Engineering.* an equation used to calculate the number of equilibrium stages in an absorber, stripper, or distillation column, which assumes that the equilibrium concentrations of the two contacting phases, such as liquid and vapor, are linearly proportional to each other.

krennerite *Mineralogy.* $AuTe_2$, a silver-white to yellowish-white orthorhombic mineral occurring in short prismatic crystals, having a specific gravity of 8.63 and a hardness of 2 to 3 on the Mohs scale. Also, WHITE TELLURIUM.

kreotoxin *Toxicology.* any basic poison produced by microorganisms in meat.

kreotoxism see MEAT POISONING.

kribergite *Mineralogy.* $Al_5(PO_4)_3(SO_4)(OH)_4 \cdot 4H_2O$, a white, triclinic, chalklike mineral having a specific gravity of 1.92 and an undetermined hardness; found in filling crevices in cupriferous pyrite.

krill *Invertebrate Zoology.* a general term for the various species of shrimplike crustaceans that form the major part of the diet of the baleen whale.

Kroeber, Alfred 1876–1960, American anthropologist; a founder of modern anthropology, particularly in the theory of culture.

Krogh, August 1874–1949, Danish physiologist; Nobel Prize for discovering regulation of capillaries' vasomotor mechanism.

kröhnkite *Mineralogy.* $Na_2Cu^{+2}(SO_4)_2 \cdot 2H_2O$, an azure-blue, monoclinic mineral occurring in granular masses and in short prismatic crystals, and having a specific gravity of 2.9 and a hardness of 2.5 to 3 on the Mohs scale; found as a secondary mineral in copper ore deposits. Also, **kroehnkite.**

Kroll ladle *Metallurgy.* a refractory-lined container used to transfer molten metal.

Kroll process *Metallurgy.* a commercial process to extract titanium from titanium tetrachloride by reacting it with magnesium.

Kromdraai *Anthropology.* a well-known site for species of australopithecines near Johannesburg, South Africa.

Kronecker, Leopold 1823–1891, German mathematician; explored relations between algebra, number theory, and elliptical functions.

Kronecker delta *Mathematics.* the function symbolized as δ_{ij}, whose value is 1 if $i = j$ and 0 if $i \neq j$; a special case of the permutation symbol. Also, **Kronecker symbol.**

Kronecker product *Mathematics.* Given a $p \times q$ matrix A and an $m \times n$ matrix B, their Kronecker product $A \otimes B$ is a $pm \times qn$ matrix whose entries are obtained in blocks as follows: each entry a_{ij} of A is multiplied by the matrix B and the blocks are arranged in the same relative positions as the a_{ij}; i.e., $A \otimes B$ consists of blocks of the form $a_{ij}B$, for $i = 1, 2, \ldots, m$ and $j = 1, 2, \ldots, n$.

Kronig-Penney model *Solid-State Physics.* an idealized one-dimensional model of a crystal for which the wave equation can be solved exactly; the potential energy in the model is represented by an infinite sequence of periodic square wells.

Kron's method of tearing see DIAKOPTICS.

Krukenberg's tumor *Oncology.* a carcinoma of the ovary, usually metastatic from cancer of the gastrointestinal tract.

Krull intersection theorem *Mathematics.* Let I be an ideal of a commutative ring R with identity, and let A be a Noetherian R-module. If an R-module B can be written in the form $B = \cap_{n=1}^{\infty} I^nA$, then $IB = A$.

Krull-Schmidt theorem *Mathematics.* Let G be a group satisfying both the ascending and descending chain conditions on normal subgroups. Suppose G may be written in two distinct ways as the direct product of indecomposable normal subgroups: $G = G_1 \times \cdots \times G_s = H_1 \times \cdots \times H_t$. Then the following results hold: (a) $s = t$; (b) the G_i and H_i can be reindexed so that $G_i \cong H_i$ for every i ($\cong$ denotes isomorphism); and (c) for every $r \leq t$ and after the reindexing of (b), $G = G_1 \times \cdots \times G_r \times H_{r+1} \times \cdots \times H_t$.

krummholz *Ecology.* the small, stunted tree growth that is characteristic of alpine forests at the margins of the timberline.

Krupp ball mill *Mining Engineering.* an ore pulverizer in which the grinding action is generated by chilled iron or steel balls of various sizes moving against each other, and a die ring.

Kruse's bacillus *Bacteriology.* a former name for the bacterial species *Shigella sonnei*, an intestinal pathogen of primates.

Kruskal's algorithm an algorithm for solving the minimal connector problem on a connected graph G with n vertices and weighted edges. See GREEDY ALGORITHM.

Kruskal-Wallis test *Statistics.* in nonparametric statistics, a significance test that is an extension of Wilcoxon's rank-sum test; used to compare three or more unmatched random samples of measurements.

kryokonite see CRYOCONITE.

krypton *Chemistry.* a gaseous element that has the symbol Kr, the atomic number 36, an atomic weight of 83.80, a melting point of –156.6°C, and a boiling point of –152.9°C; a colorless and odorless noble gas used in lasers and fluorescent lamps. The earth's atmosphere is about one-millionth part krypton. (From a Greek word meaning "hidden" or "secret," because it is present in air in such minimal amounts.)

krypton lamp *Electricity.* a very powerful arc lamp that uses krypton gas for light emission.

KSAM keyed sequential access method.

K scan see K DISPLAY.

K scope *Electronics.* a radarscope that produces a K display. Also, K INDICATOR.

k selection *Evolution.* the selection of traits that stress competitive capability and slow population growth, appropriate for populations at or near the carrying capacity of the habitat.

K shell *Atomic Physics.* the lowest energy level, occupied by orbits of electrons closest to the nucleus, having electrons with the principal quantum number 1.

K-14 sight *Ordnance.* a gyroscopic computing gunsight with a mechanical range-control system.

K-18 sight *Ordnance.* a gyroscopic computing gunsight with an electrical range-control system.

K-space see RECIPROCAL SPACE.

k-space see WAVE-VECTOR SPACE.

KSR keyboard send/receive.

K star *Astronomy.* a cool, orange-red star whose strongest spectral lines are neutral calcium and other metals and whose surface temperature falls in the range 5000 to 3600 kelvins.

kt or **kt.** karat; kiloton; knot.

K-T extinction *Paleontology.* a term for a period about 65 million years ago, at the end of the Cretaceous era, when dinosaurs and numerous other species became extinct.

K theory *Mathematics.* a theory that associates an Abelian group $K_n(R)$ with each ring R and nonnegative integer n. For $n = 0$, the group $K_0(R)$ is called the Grothendieck group of R; its elements are generalizations to R-modules of the notion of dimension of a vector space, and the group characterizes how these quantities add when direct sums of modules are taken. $K_1(R)$ is called the Whitehead group of R; its elements are generalized determinants of R-module homomorphisms. The version of K-theory that associates groups with rings is **algebraic *K* theory.**

Kubelka-Munk model *Optics.* a theory of reflectance proposing that portions of the light passing through a homogeneous sample are scattered and absorbed, thus causing the light to attenuate in both directions.

kudzu *Botany.* a hairy twining vine of the species *Pueraria lobata* or *Pueria thunbergiana* in the family Leguminosae, having thick woody roots and clusters of purple flowers in erect racemes from leaf axils; used for cattle forage and erosion control, but present as a serious weed in the southeastern United States. Also, **kudzu vine.**

Kuehneosauridae *Paleontology.* a family of gliding lizards known only from deposits in England and New Jersey; now classified in the squamatan infraorder Eolacertilia; Triassic and Jurassic.

Kuhliidae *Vertebrate Zoology.* the aholeholes, a family of marine and freshwater fish of the order Perciformes, having a compressed oblong body, a large protrusible mouth, and scaly fin sheaths; found in the tropical Indo-West Pacific region.

Kuhn, Bernard Friedrich 1762–1825, Swiss geologist; studied moraines; first to propose former existence of more extensive glaciation.

Kuhn, Richard 1900–1967, Austrian chemist; Nobel Prize (declined) for research in vitamins and carotinoids; isolated vitamins B_2 and B_6.

Kuiper, Gerard 1905–1973, Dutch-born American astronomer; discovered moons of Neptune and Uranus; directed Ranger space project.

kukersite *Geology.* an organic-rich sediment characterized by a high concentration of the remains of the alga *Gloexapsamorpha prisca*; found in the Ordovician rocks of Estonia.

Kula ring *Anthropology.* a complex trading system among the Trobriand Islanders identified by Bronislaw Malinowski, involving the ceremonial exchange of necklaces and armbands between trading partners.

kulterkreis *Anthropology.* a school of thought of the early 1900s maintaining that the history of a cultural complex can be traced back to a central area of influence which diffused outward into many places; it assumes that there are few independent inventions and that all of culture derives from a few sources. (A German term meaning "culture circle.")

Kummer, Ernst 1810–1893, German mathematician; worked in ideal numbers; extended theory of Gaussian complex numbers.

Kummer's function see CONFLUENT HYPERGEOMETRIC FUNCTION.

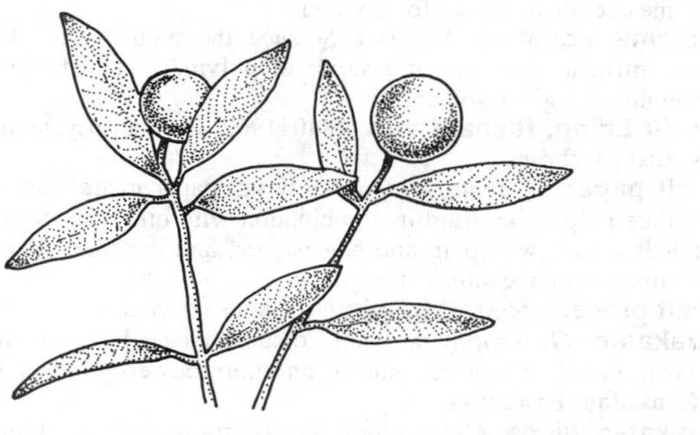

kumquat

kumquat [kum´kwät´] *Botany.* **1.** any of several evergreen shrubs and trees of the genus *Fortunella* in the family Rutaceae that bear fragrant white flowers and orange citruslike fruits the size of large olives. **2.** the fruit of this tree.

Kundt, August Adolph 1839–1894, German physicist; best known for experiments with sound and light waves; devised the Kundt tube.

Kundt effect see FARADAY EFFECT.

Kundt rule *Spectroscopy.* a rule stating that for a solution, an increase in the refractive index produces a displacement in the absorption bands toward the red end of the spectrum.

Kundt's constant *Optics.* a measure of the strength at which light, traveling through matter, rotates in the direction of an applied magnetic field (Faraday effect), equivalent to the ratio of the Faraday effect to the product of path length and the magnetization of the material.

Kundt tube *Acoustics.* a glass tube partially filled with dust; wave patterns can be formed in the dust by various methods of sound transmission into the tube, as by a loudspeaker at one end.
Kungurian *Geology.* a European geologic stage of the Lower or Middle Permian period, occurring after the Artinskian and before the Kazanian.
kunzite *Mineralogy.* a clear, lilac-colored gem variety of spodumene.

kunzite crystal

Kupffer cells *Immunology.* fixed phagocytes that occur on the lining of the liver sinusoids.
kuramatsu *Materials.* the heavy, resistant wood of the Far Eastern tree *Pinus thunbergii*; used for the construction of houses and furniture, and for carvings. Also, JAPANESE BLACK PINE.
Kuratowski's lemma *Mathematics.* an alternative form of Zorn's lemma stating that each linearly ordered subset of a partially ordered *X* set is contained in a maximal linearly ordered subset of *X*; equivalent to the axiom of choice, the well-ordering principle, the Hausdorff maximality principle, Tukey's lemma, etc.
Kuratowski's theorem *Mathematics.* the theorem that a graph is planar if and only if it contains no subgraph homeomorphic to K_5 (the complete graph on five vertices) or $K_{3,3}$ (the bipartite graph having three vertices in each vertex set). A corollary states that a graph is planar if and only if it contains no subgraph contractible to K_5 or $K_{3,3}$. K_5 and $K_{3,3}$ are sometimes called **Kuratowski graphs.**
Kurchatov, Igor 1903–1960, Soviet nuclear physicist; developed first Soviet atomic bomb.
kurchatovium see ELEMENT 104.
Kurdjumov-Sachs orientation relation *Materials Science.* the mutual parallelism between close-packed planes and the directions of two crystal structures aligned with each other.
Kurie plot *Nuclear Physics.* a straight-line graph used to analyze beta decay by plotting the square root of the Coulomb-corrected beta-particle intensity versus the beta-particle energy on a linear scale; useful for determining the upper energy limit by extrapolation. Also, FERMI PLOT, FERMI-KURIE PLOT.
kurnakovite *Mineralogy.* $MgB_3O_3(OH)_5 \cdot 5H_2O$, a white or colorless, transparent, triclinic mineral, dimorphous with inderite, occurring in large blocky crystals and granular aggregates, and having a specific gravity of 1.83 and a hardness of 3 on the Mohs scale; found in borate deposits.
Kuroshio Countercurrent *Oceanography.* a current that flows southwestward, counter to and on the south side of the Kuroshio Current, about 70 km south of Japan between 155 E and 160 E.
Kuroshio Current *Oceanography.* a fast-flowing (2–4 knots), narrow, and relatively deep current that flows northeastward from Taiwan to about 150 E, along the coast of Japan. Also, JAPAN CURRENT.
Kuroshio Extension *Oceanography.* an ill-defined eastward-flowing current that carries warm water from the Kuroshio Current into the North Pacific.
Kuroshio system *Oceanography.* a circular chain of ocean currents in the Northwest Pacific, including part of the North Equatorial Current, the Kuroshio Current, the Kuroshio Extension, and the North Pacific Current.
Kürschák, József 1864–1933, Hungarian mathematician; extended concept of absolute value to abstract fields.
Kurthia *Bacteriology.* a genus of Gram-positive, obligately aerobic bacteria, commonly found in manure and stagnant water and as a contaminant of meat; long chains of rods in the exponential growth phase, but generally develop coccoid forms or short rods in the stationary phase.
Kurtidae *Vertebrate Zoology.* a family of deep-bodied fishes composing the suborder Kurtoidei, distinguished by a modified supraoccipital crest and expanded ribs partially encasing the gas bladder.
Kurtoidei *Vertebrate Zoology.* the humpheads or forehead brooders, a monofamilial suborder of deep-bodied fishes found in India, China, Australia, and New Guinea.
kurtosis *Statistics.* the degree of peakedness of a frequency curve.
kuru *Pathology.* a progressive, fatal viral malady of the nervous system, found among the native people of New Guinea, possibly associated with cannibalism; on autopsy, the brain shows spongelike encephalopathies.
Kusch, Polykarp born 1911, German-born American physicist; Nobel Prize for precisely determining magnetic moment of the electron.
Kusnezovia *Bacteriology.* a genus of manganese-depositing budding bacteria found in mud, but not yet obtained in pure culture.
kutnohorite *Mineralogy.* $Ca(Mn^{+2},Mg,Fe^{+2})(CO_3)_2$, a white or pale pink, translucent, trigonal mineral of the dolomite group, occurring in irregular masses. Also, **kutnahorite.**
Kutorginida *Paleontology.* an early order of brachiopods of the Lower Cambrian that has been included in both the Articulata and the Inarticulata; characterized by biconvex calcareous shells and the absence of teeth or sockets; the group may eventually be considered a distinct class.
Kutta-Joukowski airfoil *Fluid Mechanics.* an airfoil designed to maintain streamline flow past the trailing edge of the structure.
Kutta-Joukowski theorum *Fluid Mechanics.* the lift, L, of an airfoil is given by $L = pTVl$, where p is the density, T is the circulation, V is the velocity, and l is the length of the airfoil.
kV or **kV.** or **kv** kilovolt.
kVA or **kva** kilovolt-ampere.
kVp kilovolts peak.
kW or **kW.** or **kw** kilowatt.
kwashiorkor [kwä´shē ôr´kôr] *Medicine.* a nutritional deficiency disease of children, occurring mainly in the tropics and subtropics; caused by inadequate intake of proteins and characterized by anemia, edema, pot belly, impaired growth, skin changes, fatty changes in the liver cells, and weakness. (From a native word used in Ghana.)
kWh or **kwh** kilowatt-hour. Also, **kW-hr.**
KWIC index see KEYWORD-IN-CONTEXT INDEX.
KWOC index see KEYWORD-OUT-OF-CONTEXT INDEX.
kyanite *Mineralogy.* Al_2SiO_5, a blue, white, or light green, triclinic mineral occurring as long-bladed crystals, having a specific gravity of 3.53 to 3.67 and a hardness of 4 to 7.5 on the Mohs scale; found in metamorphic rocks. Also, DISTHENE, CYANITE, SAPPARE.
kyanization *Chemical Engineering.* a method of preserving wood and preventing decay by saturating it with a mercuric chloride solution.
kymatism see MYOKYMIA.
kymatology *Oceanography.* the study of waves and wave motion.
kymograph *Medicine.* an instrument used to graphically record variations or undulations, as of the heart and blood vessels, consisting of a paper-covered revolving drum upon which a stylet or other writing point inscribes the curve. *Industrial Engineering.* a similar device for measuring very short work time intervals. Thus, **kymogram, kymography.**
kynurenic acid *Biochemistry.* $C_{10}H_7O_3N$, an intermediate in tryptophan metabolism in which tryptophan is dehydroxylated to a form that can be excreted.
kyphos *Medicine.* a convex prominence of the thoracic spine.
kyphoscoliosis *Medicine.* a rotary lateral curvature of the spine that deforms and disables, leading progressively from impaired lung to impaired heart functions.
kyphosis *Medicine.* an abnormal curvature of the spine with backward convexit.
kyrohydratic point *Oceanography.* the eutectic temperature of a salt; the temperature of crystallization of a salt in brine enclosed by frozen seawater.
kyrtorrhachic *Medicine.* having a curvature of the lumbar spine with backward concavity.
kytoon *Aviation.* a captive balloon that is similar to a kite balloon but smaller and with a thin fabric empennage rather than gas-filled lobes; often used to support an antenna. (A term coined from *kite* and *balloon.)*
Kyushu *Geography.* the southernmost main island of Japan.

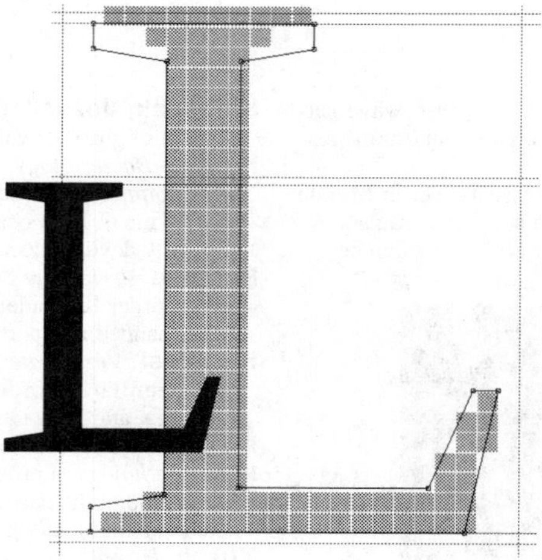

L Lagrangian; luminosity.
L *Electricity.* the symbol for inductance.
L. lactobacillus; lethal; coefficient of induction.
l. pound (from Latin *libra*); liquid; liter; locus; lambert.
La lanthanum.
labbé *Meteorology.* a moderate to strong southwest wind of southeastern France that occurs usually during March, characterized by moist, mild, and cloudy to rainy weather conditions. Also, **labé, labech.**
labdanum oil *Materials.* a yellow oil, soluble in alcohol and ether, that is derived from the resin of various rockroses of the genus *Cistus*; used in perfumes and fumigating substances.
label *Computer Programming.* **1.** a symbolic name or identifier used in a program to reference a particular instruction, an entry point to a sequence of instructions, a data area, or a format specification. **2.** see LABEL RECORD. **3.** to create a label. *Architecture.* a squared hoodmold.
label constant see LOCATION CONSTANT.
label data type *Computer Programming.* a class of data used to mark locations in a program.
labeled graph *Mathematics.* a finite graph whose vertices are numbered; a graph G of n vertices together with a one-to-one mapping ϕ, called a **labeling,** from the vertex set of G onto the set $\{1, 2, 3, \ldots, n\}$. Two labeled graphs (G_1, ϕ_1) and (G_2, ϕ_2) are isomorphic if there is an isomorphism between them that preserves the labeling of the vertices.
labellum *Botany.* **1.** the enlarged third petal in the corolla of an orchid; it is twisted by torsion of the ovary from the upper to the lower position. **2.** two fused staminodes in flowers of the family Zingiberaceae. *Invertebrate Zoology.* **1.** the elongate section of the upper lip in certain beetles and true bugs. **2.** in flies, either of a pair of sensitive, fleshy lobes formed by the expanded end of the lower lip.
label record *Computer Technology.* a record added to a tape reel, disk pack, or other mass storage media to provide information about its contents, such as record length, block size, or volume number.
labetadol *Pharmacology.* $C_{19}H_{24}N_2O_3$, an α- and β-adrenergic blocking agent that is used in treating hypertension.
labia *Anatomy.* fleshy edges; usually designating the labia majora and labia minora.
labial gland *Anatomy.* one of many small mucous or serous glands located in the submucosa of the lips. *Invertebrate Zoology.* a salivary gland draining at the base of the lower lip in certain insects.
labial hooks *Entomology.* spines located on the labial pads of dragonfly nymphs, used to catch tadpoles or other prey.
labial palp *Invertebrate Zoology.* **1.** one of a pair of thin, fleshy appendages on either side of the mouth of some bivalve mollusks. **2.** a jointed appendage attached to the lower lip of certain insects.
labial papilla *Invertebrate Zoology.* a sensory, hairlike projection near the mouth of nematodes.
labia majora *Anatomy.* a pair of elongated, fleshy folds running downward and backward from the mons pubis in the female.
labia minora *Anatomy.* a pair of small folds of skin located on either side between the labia majora and the opening of the vagina.
Labiatae *Botany.* the mint family; a family of aromatic herbs and shrubs of the order Laminales, including species of mint, thyme, and sage, that secrete a pungent oil; characterized by square stems, a four-lobed ovary that becomes four one-seeded fruits, and a corolla divided into two lips. Also, LAMIACEAE.
labiate *Anatomy.* possessing flat, liplike parts. *Botany.* **1.** liplike or having lips. **2.** having the petals of the corolla form two lips, as in flowers of the Lamiaceae.
labile *Psychology.* given to emotional instability. *Chemistry.* of a compound, capable of changing state when subjected to heat or radiation.
labile oscillator *Electronics.* a device that generates an alternating current whose frequency is subject to signals sent by remote control.
lability *Psychology.* a condition in which an individual's emotions are extremely fluid and volatile, usually the result of an organic brain disorder.
labio- or **labi-** a combining form meaning "lip," especially lips of the mouth, as in *labiodental, labialism.*
labiochorea *Neurology.* a condition characterized by jerky or grimacing movements of the lips, accompanied by stammering or stuttering.
labiodental fricative *Linguistics.* the sound produced when pressing the lower lip against the teeth and expelling air (in a hiss) such as that made when pronouncing the letter *f* or *v*.
labioplasty see CHEILOPLASTY.
labite *Mineralogy.* $MgSi_3O_6(OH)_2{\cdot}H_2O$, a hydrous basic silicate of magnesium, possibly chrysotile.
labium *plural,* **labia.** *Anatomy.* a lip or lip-shaped part, especially any of the folds of skin bordering the vulva. *Invertebrate Zoology.* **1.** the lower lip in insects, often modified for various purposes. **2.** the inner margin of the opening of a gastropod shell. *Botany.* the lower lip of a bilabiate corolla.
labium majus *Anatomy.* one of the two external folds of skin forming the lateral border of the vulva.
labium minus *Anatomy.* one of the two folds of skin lying between the vaginal orifice and the labia majora; smaller than the folds of the labia majora.
labor *Medicine.* the process by which the fetus and the placenta are expelled from the uterus to the outside of the birth canal.
laboratory *Science.* a place in which scientific experimentation and observation are performed under controlled conditions.
laboratory coordinate system *Mechanics.* a stationary frame of reference with respect to laboratory apparatus; often taken to be an inertial frame.
Laborde map projection *Cartography.* a conformal projection similar to the transverse Mercator projection but projecting a spheroid rather than a sphere onto a plane; best suited to regions of considerable elongation from the meridian.
labor specialization *Industrial Engineering.* the division of a task into separate operations in order to increase output.
Laboulbeniales *Mycology.* an order of fungi belonging to the subdivision Ascomycotina, consisting of insect parasites that generally do not harm their hosts.
Laboulbeniineae *Mycology.* a suborder of fungi belonging to the order Laboulbeniales and characterized by cell walls that grow around the fertile cell of the fruiting body, which eventually produces asci.
Labrador *Geography.* a peninsula in eastern Canada between the Atlantic Ocean and Hudson Bay.
Labrador Current *Oceanography.* a cold, weakly saline current that flows southward from the Davis Strait along the coasts of Labrador and Newfoundland, turning east and joining the North Atlantic Current southwest of Newfoundland.

labradorite *Mineralogy.* a gray, blue, green, or brown, brittle triclinic plagioclase feldspar containing 50–70% of the anorthite molecule, occurring in basic igneous rocks, and having a specific gravity of 2.69 to 2.27 and a hardness of 6. to 6.5 on the Mohs scale; the feldspar sometimes exhibits variously colored luster as the natural material cools in the course of formation. Also, PLAGIOCLASE.

Labrador Sea *Geography.* an arm of the North Atlantic Ocean between Labrador and Greenland.

Labridae *Vertebrate Zoology.* the wrasses, a cosmopolitan family of carnivorous bony fish of the suborder Labroidei, having some species that act as cleaner fish on tropical reefs.

Labrinthomorpha *Invertebrate Zoology.* a phylum of disease-bearing protozoa.

Labroidei *Vertebrate Zoology.* a suborder of tropical coastal marine fishes of the order Perciformes, including the wrasses, odacids, and parrotfishes.

labrum *plural,* **labra** *Biology.* a lip or a liplike part. *Invertebrate Zoology.* **1.** the upper lip of arthropods. **2.** the outer edge of a gastropod shell. *Anatomy.* a ring of cartilage around the edge of the joint surface of a bone.

laburnum *Botany.* any of various small, ornamental trees of the genus *Laburnum,* characterized by long clusters of pendulous yellow flowers. *Materials.* the wood of this tree.

labyrinth [lab´ə rinth] *Anatomy.* **1.** a system of interconnecting canals, channels, or cavities. **2.** the internal ear, consisting of both bony and membranous portions.

labyrinth fish *Vertebrate Zoology.* any of various freshwater fishes of the order Labyrinthi, having a labyrinthine structure above each gill chamber that allows them to breathe air when out of the water; found in southeastern Asia and Africa.

labyrinthine reflexes *Physiology.* involuntary actions that depend on stimuli from the labyrinth of the inner ear, including the righting reflex and the placing and hopping reflexes.

labyrinthitis *Medicine.* inflammation of the labyrinth of the inner ear.

Labyrinthodontia *Paleontology.* a subclass of early amphibians descended from the rhipidistian fishes and possibly ancestral to all later amphibians; characterized by a labyrinthine infolding of the dentine, conspicuous fangs, and well-developed limbs; extant in the Devonian to Lower Jurassic.

Labyrinthulea *Invertebrate Zoology.* a class of marine protozoans.

Labyrinthulida *Invertebrate Zoology.* an order of marine protozoans, some of which are parasitic.

Labyrinthulomycetes *Mycology.* in some classifications, a class of fungi belonging to the division Myxomycota, consisting of certain aquatic organisms of uncertain taxonomic affiliation and including both algae parasites and organisms that live off dead organic matter.

lac *Materials.* a resinous substance that is excreted and deposited by certain scale insects on the twigs of various Indian jungle trees, especially the soapberry and acacia of India and Burma; used in making shellac, abrasives, sealing wax, and a type of bright red dye (**lac dye**).

Lacaille, Nicolas de [la kā´] 1713–1762, French astronomer; measured lunar and solar parallaxes; catalogued 10,000 southern stars.

Lacaille's constellations *Astronomy.* fourteen Southern Hemisphere constellations: Antlia Pneumatica, the Air Pump; Apparatus Sculptoris, the Sculptor's Studio (Sculptor); Caelum, the Chisel; Circinus, the Drawing Compasses; Equuleus Pictoris, the Painter's Easel (Pictor); Fornax Chemica, the Chemists' Furnace (Fornax); Horologium Oscillatorium, the Pendulum Clock (Horoscopium); Microscopium, the Microscope; Mons Mensa, the Table Mountain (Mensa); Norma et Regula, the Square and Level (Norma); Octans Hadleianus, the Octant of Hadley (Octans); Pyxis Nautica, the Nautical Compass (Pyxis); Reticulum Rhomboidalis, the Net (Reticulum); and Telescopium, the Telescope. (Named for Nicolas de *Lacaille.*)

LaCasitan *Geology.* a North American Gulf Coast stage of the Upper Jurassic period, occurring after the Zuloagan and before the Durangoan of the Cretaceous period.

laccal *Biochemistry.* $C_{17}H_{31}C_6H_3(OH)_2$, a compound found in the sap of lacquer trees.

laccate *Biology.* having a varnished appearance.

Lacciferinae *Invertebrate Zoology.* the lac insects, a subfamily of homopteran insects that secrete a substance from which shellac is made.

laccolith *Geology.* a lens-shaped, concordant body of igneous rock that intrudes into a layer of sedimentary rocks, producing a structure with a distinctly domed roof and a generally flat floor. Also, CISTERN ROCK, LACCOLITE.

lacerate [las´ə rāt´] *Medicine.* to tear tissues; to wound.

lacerated *Botany.* of a leaf, having variously cut edges, as if torn into sections.

laceration *Medicine.* a wound resulting from the tearing of tissues.

Lacerta *Astronomy.* the Lizard, a small and faint constellation lying between Cygnus and Andromeda in the northern celestial hemisphere.

Lacertid see BL LACERTAE OBJECT.

Lacertidae *Vertebrate Zoology.* the Old World runners, a diverse family of small- to medium-sized lizards of the infraorder Scincomorpha; found in Eurasia and Africa.

Lachnocladiaceae *Mycology.* a family of fungi belonging to the order Agaricales, having a dimitic hyphal system and occurring on soil, dung, wood, and plant material or in a symbiotic relationship with plants; some species are parasitic to vascular plants or other fungi.

Lachnospira *Bacteriology.* a genus of Gram-negative, anaerobic, curved rod-shaped bacteria of the family Bacteroidaceae; found in cattle rumen.

lachrymal canaliculitis *Medicine.* an inflammation of the lacrimal duct.

L acid *Organic Chemistry.* $C_{10}H_6(OH)SO_3H$, a white solid that is soluble in water; used as an azo dye intermediate.

laciniate *Botany.* of a leaf, having jagged edges; cut into narrow, often pointed, irregular lobes.

lac insect *Invertebrate Zoology.* an insect, *Laccifer lacca,* of Southeast Asia that secretes the sticky substance known as lac, used in making shellac.

Lacistemataceae *Botany.* a family of dicotyledonous, tanniferous shrubs and small trees of the order Violales, characterized by distichous leaves and small flowers borne in catkinlike spikes or racemes; native to tropical American regions.

lac operon *Genetics.* a gene system involved in the metabolism of lactose in *E. coli*; consists of an operator gene and three structural genes, ***lacZ, lacY,*** and ***lacA,*** which encode the enzymes used in the breakdown of lactose. Also, **lactose operon.**

lacquer [lak´ər] *Materials.* **1.** a substance made of resins, cellulose esters, shellac, or similar material in a solvent such as ethyl alcohol; it forms a glossy finish and is used as a protective or decorative coating on metals, woods, plastics, and other products. **2.** a similar natural coating derived from certain Japanese and Chinese trees of the cashew family, applied especially to wood surfaces in order to obtain a polished finish. **3.** to apply such a coating.

lacrim- or **lacrimo-** a prefix meaning "tear," as in *lacrimation.*

lacrimal [lak´rə məl] *Anatomy.* of or related to tears.

lacrimal apparatus *Anatomy.* the anatomical structures that produce and conduct tears, including the lacrimal gland, lacrimal canal, lacrimal sac, and nasolacrimal duct.

lacrimal bone *Anatomy.* one of two paired bones located in the medial wall of the orbits.

lacrimal canal *Anatomy.* a canal that is formed by the lacrimal sulcus of the maxilla, the lacrimal bone, and inferior nasal concha, and that contains the nasolacrimal duct. Also, NASAL CANAL.

lacrimal canaliculus *Anatomy.* one of two small ducts that conduct tears from the punctum of the eye into the nasolacrimal duct.

lacrimal fluid *Medicine.* the watery solution secreted by the lacrimal glands; tears.

lacrimal gland *Anatomy.* one of the the tear-secreting glands, located in the upper lateral corner of the orbit.

lacrimal punctum *Anatomy.* a small opening in the inner corner of each eye that drains off tears.

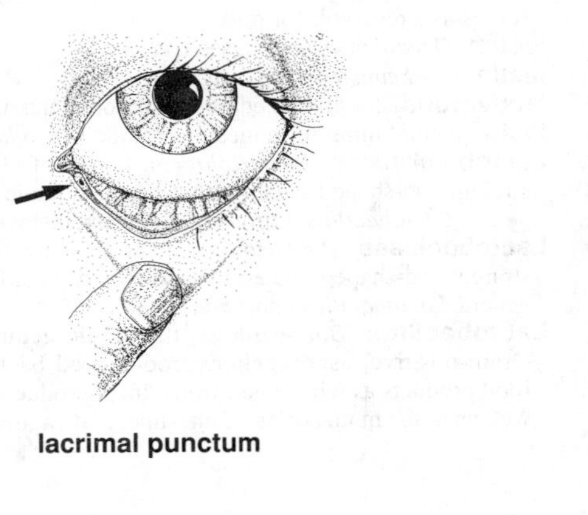

lacrimal punctum

lacrimal sac *Anatomy*. the upper portion of the nasolacrimal duct, into which the lacrimal canaliculi empty.

lacrimation *Physiology*. the secretion of tears by the lacrimal gland.

lacrimator *Toxicology*. any substance that irritates the eyes, such as tear gas.

lacrimoid *Biology*. resembling a teardrop.

lacroixite *Mineralogy*. $NaAl(PO_4)F$, a pale yellowish monoclinic mineral occurring as crystals, having a specific gravity of 3.29 and a hardness of 4.5 on the Mohs scale; it is found with morikite, apatite, and tourmaline.

La Crosse virus *Virology*. a virus of the genus *Bunyavirus* that can cause human encephalitis.

lact- a combining form meaning "milk," as in *lactation*.

lactalbumin *Biochemistry*. a protein that is soluble in water and coagulable by heat; occurs in milk and reassembles serum albumin.

lactam *Organic Chemistry*. a cyclic amide that is produced by removing one molecule of water from an amino acid. Also, LACTIM.

β-lactamase *Enzymology*. any bacterial enzyme that destroys the activity of certain β-lactam antibiotics by hydrolyzing the β-lactam bond in the molecule.

Lactarius *Mycology*. a genus of fungi of the family Russulaceae, sometimes known as "milk caps," characterized by the liquid or latex exuded when cut; some are edible.

lactase *Nutrition*. an enzyme for digesting lactose into glucose and galactose; it can be irretrievably lost in adults, producing lactose intolerance.

lactase deficiency syndrome *Medicine*. an intolerance for milk or milk products due to a deficiency of lactase; marked by gastrointestinal symptoms such as bloating, flatulence, and diarrhea.

lactate *Physiology*. to secrete milk. *Organic Chemistry*. a salt or ester of lactic acid, containing the radical Me·CHOH·COO=; used as a food emulsifier or flour conditioner.

lactate dehydrogenase *Enzymology*. a glycolytic enzyme that is widely distributed in all body tissues and catalyzes the reversible reaction of pyruvate to lactate in the presence of NADH. Also, **lactic dehydrogenase.**

lactation *Physiology*. the secretion of milk, especially in the nourishment of an infant.

lacteal *Anatomy*. of or relating to milk. *Physiology*. any of numerous small lymphatic vessels that absorb the products of lipid digestion in the small intestine.

lactescent *Physiology*. becoming milky, or forming a milky fluid.

lacti- a combining form meaning "milk," as in *lactiferous*.

lactic acid *Biochemistry*. $C_3H_6O_3$, the hydroxy acid that is formed from pyruvic acid when glycolysis proceeds under anaerobic conditions.

lactic acid bacteria *Bacteriology*. any of a number of Gram-positive bacteria, such as *Lactobacillus* species, that are capable of lactic acid fermentation of sugar substrates used commercially in dairy products.

lactic acid fermentation *Biochemistry*. the aerobic or anaerobic fermentation of carbohydrates with lactic acid as a major end product.

lactide *Organic Chemistry*. any of a group of intramolecular cyclic esters formed by self-esterification from two or more molecules of a hydroxy acid.

lactifer *Botany*. any of various plant cells or cell series that contain latex and produce a milky, latex-based liquid.

lactiferous *Medicine*. producing milk.

lactiferous duct *Anatomy*. a tubular channel that carries milk from the mammary gland. *Botany*. a channel in a plant that transports latex produced by the plant.

lactiferous sinus *Anatomy*. a dilated portion of a lactiferous duct that serves as a reservoir for milk.

lactim *Organic Chemistry*. see LACTAM.

lactin *Biochemistry*. see LACTOSE.

lactivorous *Zoology*. feeding mainly or exclusively on milk.

lacto- a combining form meaning "milk," as in *lactoprotein*.

Lactobacillaceae *Bacteriology*. a family of Gram-positive, curved or straight rod-shaped bacteria that produce lactic acid; includes a single genus, *Lactobacillus*. Formerly, **Lactobacteriaceae.**

Lactobacilleae *Bacteriology*. a former term for a tribe of curved or straight rod-shaped bacteria of the family Lactobacillaceae; now of the genera *Lactobacillus* and *Eubacterium*.

Lactobacillus *Bacteriology*. the single genus of Lactobacillaceae; Gram-positive, asporogenous, rod-shaped bacteria occurring in such food products as wine, beer, fruits, meat products, and dairy products, as well as in the human mouth, intestine, and vagina.

Lactococcus *Bacteriology*. a proposed genus to include Gram-positive bacteria that occur as facultatively anaerobic, nonmotile, coccoid cells in certain dairy products.

lactoferrin *Biochemistry*. an iron-binding protein that is commonly found in bile, milk, saliva, and tears, functioning as an iron transport protein.

lactoflavin see RIBOFLAVIN.

lactogen *Endocrinology*. any agent that stimulates lactation, such as prolactin or placental lactogen.

lactogenesis *Medicine*. the origin or formation of milk.

lactogenic *Medicine*. stimulating the production of milk.

lactogenic hormone see PROLACTIN.

lactoglobulin *Biochemistry*. a globulin occurring in milk; insoluble in water but soluble in saline solutions.

lactometer *Food Engineering*. a hydrometer used to measure the specific gravity of batches of milk.

lactonase *Enzymology*. an enzyme that hydrolyzes lactones to hydroxy acids.

lactone *Organic Chemistry*. any of a group of cyclic compounds derived by intramolecular loss of water from a hydroxyl and a carboxyl group of a hydroxy acid, leading to the formation of a cyclic ester.

lactonitrile *Organic Chemistry*. C_3H_5ON, a combustible, toxic, colorless to straw-colored liquid; soluble in water and alcohol and insoluble in petroleum ether and carbon disulfide; boils at 182–184°C and freezes at about 40°C; used as a solvent and a chemical intermediate.

lactoperoxidase *Biochemistry*. an iron porphyrin (heme) protein, commonly found in milk, saliva, and tears, that functions as an inhibitor of bacterial growth.

***p*-lactophenetide** see LACTOPHENIN.

lactophenin *Pharmacology*. $C_{11}H_{15}NO_3$, a bitter crystalline powder derived from phenetidin and lactic acid; formerly used to treat pain and fever. Also, *p*-LACTOPHENETIDE.

lactophenol cotton blue *Mycology*. a chemical composition used to stain fungi prior to microscopic examination.

lactopoiesis see GALACTOPOIESIS.

Lactoridaceae *Botany*. a monospecific and morphologically isolated family of dicotyledonous shrubs usually placed in the order Magnoliales; characterized as polygamodioecious and native to the Juan Fernandez Islands off the coast of Chile.

lactose *Biochemistry*. $C_{12}H_{22}O_{11}$, a disaccharide of galactose and glucose that is present in milk; used as a laxative and a diuretic, and in infant feeding formulas. Also, LACTIN, MILK SUGAR.

lactose intolerance *Medicine*. an inability to assimilate lactose due to the loss of the lactase enzyme from the intestines; common in many adult Asians, blacks, and people of Mediterranean origin, and less frequently seen in other Caucasians; marked by stomach pain, gas, and diarrhea after ingesting milk products. Also, MILK INTOLERANCE.

lactosuria *Medicine*. the presence of lactose in the urine, occurring during pregnancy or lactation and in premature babies or infants.

lactotoxin *Toxicology*. a toxin formed in milk.

lacuna [lə koo´nə] *plural,* **lacunae** or **lacunas.** a gap or missing part; specific uses include: *Anatomy*. any of the minute cavities in bones. *Botany*. an airspace in the hollow center of certain plant stems or root tissues. *Invertebrate Zoology*. **1.** a space in the tissue of many invertebrates that serves in place of circulatory vessels. **2.** in Acanthocephala, any of numerous spaces and chambers in the hypodermis.

lacunar system *Invertebrate Zoology*. in echinoderms, a series of intercommunicating spaces and channels through which watery blood flows.

lacunosus *Meteorology*. a variety of cloud having a honeycomb or netlike appearance due to the occurrence of characteristically rounded spaces, often with fringed edges, between cloud elements.

lacustrine *Hydrology*. of, related to, produced by, or inhabiting lakes. Thus, **lacustrine fauna, lacustrine terrace, lacustrine sand.**

lacustrine sediments *Geology*. mineral or organic matter that has been deposited in lakes.

lacustrine soil *Geology*. soil that has been formed by deposits in a lake that has become extinct; characterized by a uniformity of texture and variable chemical composition.

Lacydoniidae *Invertebrate Zoology*. a small family of marine polychaete worms in the group Errantia.

LAD language acquisition device.

ladar *Optics*. a system in which a light beam tracks the speed, altitude, direction, and range of a missile; used instead of a microwave radar beam. (An acronym for laser detection and ranging.)

ladder *Mechanical Devices.* a device commonly used for climbing, composed of two parallel sides joined by rungs or crosspieces that serve as steps.

ladder attenuator *Electronics.* an attenuator constructed as a ladder network.

ladder diagram *Control Systems.* a controller-logic diagram showing the flow of power through a network of relay contacts arranged in horizontal rows (rungs) between two vertical columns (rails) that represent the power supply.

ladder fire *Ordnance.* in gunnery, a method of establishing range by firing a series of shots at different elevations.

ladder network *Electronics.* a network composed of an alternation of serial and shunt branches. Also, SERIES-SHUNT NETWORK.

ladder polymer *Materials Science.* a heat-resistant polymer that has a characteristic ladderlike molecular structure of adjacent ring groups, such as pyrolyzed polyacrylonitrile.

ladder tape *Mechanical Devices.* a tape support for a venetian blind, consisting of intervals of woven plastic strips with narrow cross strips serving as a ledge for the blind slats.

ladder track *Transportation Engineering.* a track that extends at an angle from a main track, providing access to a series of parallel tracks, as at one end of a rail yard.

laddic *Electronics.* a device that structurally resembles a ladder and performs logic functions in a computer by utilizing the flux change that occurs when the ladder's rungs are magnetized by the opposite polarity.

Ladenberg f value see F VALUE.

Ladinian *Geology.* a European geologic stage of the Middle Triassic period, occurring after the Anisian and before the Carnian.

ladle *Mechanical Devices.* **1.** a large, deep spoon with a long handle, used for transporting or pouring liquids. **2.** a wrought-iron vessel lined with a refractory material, used for conveying molten metal.

ladr or **LADR** see LINEAR ACCELERATOR-DRIVEN REACTOR.

ladybug *Invertebrate Zoology.* a popular name given to various small, brightly colored beetles of the family Coccinellidae, having short legs and black, yellow, or reddish markings; known predators of plant pests, including aphids. Also, **ladybird.**

ladyfish see BONEFISH.

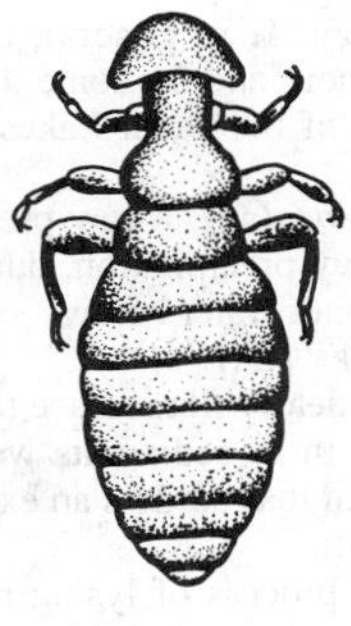

Laemobothridae

Laemobothridae *Invertebrate Zoology.* a family of chewing lice in the order Mallophaga, parasitic on aquatic birds.

Laënnec, René 1781–1826, French physician; invented the stethoscope.

Laënnec's cirrhosis see PORTAL CIRRHOSIS.

Laetoli site *Anthropology.* a hominid fossil bed in Tanzania in which australopithecine fossils and a unique trail of footprints were preserved in volcanic ash, dating to 3.6 million years ago; discovered on a 1978 expedition led by British anthropologist Mary Leakey.

Laetrile [lā´ə tril´] *Pharmacology.* $C_{14}H_{15}NO_7$, the trademark name for a substance derived from amygdalin that is alleged to cure cancer by releasing cyanide into the body; not used clinically in the United States.

lag an interval or lapse of time; specific uses include: *Physics.* the difference in time or phase between two events or waves. *Electronics.* **1.** a phenomenon in which an electrical-charge image persists on a camera tube for a short interval, generally equivalent to a few frames of film or tape. **2.** the period of time between the transmission and reception of a signal. *Transportation Engineering.* for a particular traffic flow, the delay between the beginning of a traffic light and the appearance of a green light.

Lag. flask (From Latin *lagena.*)

lagan *Navigation.* a heavy object that is thrown overboard to lighten a vessel's load, but buoyed so that its location is noted for later recovery.

lag angle *Physics.* a negative phase angle arising from a delay in angular placement of one of two sinusoidally varying quantities having the same frequency.

lag bolt see COACH SCREW.

lag deposit see LAG GRAVEL.

lagena *Vertebrate Zoology.* a small, blind, tubular projection from the underside of the sacculus in the inner ear of vertebrates.

Lagenidiaceae *Mycology.* a family of fungi belonging to the order Lagenidales; most are parasitic of algae, with some living off nematodes; found in freshwater and land environments.

Lagenidiales *Mycology.* an order of fungi belonging to the class Oomycetes and composed of aquatic organisms that live off nonliving organic matter or parasitize algae or lower animals.

lag fault *Geology.* an overthrust composed of rocks that are thrust differentially, leaving the upper part of the geologic section behind. Also, TECTONIC GAP.

lagging *Materials Science.* a heat-insulating plaster that is made of asbestos and magnesium, used on piping and process equipment. *Civil Engineering.* sheathing of wood, wire mesh, or corrugated metal that is placed in an excavation area to prevent cave-ins.

lagging coil *Electricity.* a small coil designed to compensate for the lagging current in the voltage coil of an AC watt-hour meter.

lagging current *Electricity.* AC current that lags the applied voltage, occurring when voltage is applied to an inductive load.

lagging load see INDUCTIVE LOAD.

lagging network see INTEGRAL NETWORK.

lagging strand *Molecular Biology.* the strand of DNA that is replicated discontinuously with the addition of short segments (Okazaki fragments), which are then linked together.

lag gravel *Geology.* **1.** an accumulation of coarse-grained material that is dragged along a river bottom or left behind when finer particles have been winnowed or carried away by water currents. Also, LAG DEPOSIT. **2.** a residual accumulation of coarse, hard rock and mineral debris that remains on a surface after winds have blown away the finer material.

lag network see INTEGRAL NETWORK.

Lagomorpha *Vertebrate Zoology.* the rabbits, hares, and pikas, an order of terrestrial mammals distinguished from rodents by teeth structure and motion, a fused tibia and fibula, and a spiral valve in the cecum; found in almost all parts of the world.

lagoon *Geography.* **1.** a small body of coastal water, usually brackish, separated from the sea by a low, narrow strip of land. **2.** the expanse of water inside a coral reef or atoll.

Lagoon Nebula *Astronomy.* a large emission nebula in Sagittarius, also known as **M 8**.

Lagoon Nebula

lag phase *Microbiology.* the phase following the inoculation of overgrown bacteria into a new culture medium; characterized by a lack of cell multiplication while the bacteria become adjusted to the new environment. Also, **lag, lag period.**

Lagrange, Joseph 1736–1813, French mathematician; formulated the theory of recurring series and calculus of variations; wrote *Mécanique Analytique*; worked in celestial mechanics.

Lagrange bracket *Mechanics.* the sum over a system's generalized coordinates of the Jacobian of the coordinate and its conjugate momentum with respect to a given pair of variables.

Lagrange equations *Mechanics.* the set of differential equations describing a conservative dynamical system, $d/dt(\{\partial\}\{L\}/\{\partial\}\{\dot{q}\}_i) = \{\partial\}\{L\}/\{\partial\}q_i$, which hold for each generalized coordinate q_i, where $\{L\}$ is the system Lagrangian and $\{\dot{q}\}_i$ is the time derivative of q_i; these are one of the most basic equation sets of classical mechanics. Also, **Lagrange's equations of motion.**

Lagrange-Hamilton theory *Mechanics.* the application of Lagrangian mechanics to continuous or large statistical systems, using a continuous analogue of the standard Lagrangian.

Lagrange polynomials *Mathematics.* a sequence of polynomials $L_i(x)$ appearing in a general technique for writing down a polynomial $P(x)$, of degree at most n, passing through a set of $n + 1$ given points; i.e., $P(x_i) = y_i$ for $i = 0, 1, 2, \ldots, n$. In particular, $P(x) = \Sigma_{i=0}^{n} y_i L_i(x)$, and

$$L_i(x) = \prod_{k=0}^{n} (x - x_k)/(x_i - x_k).$$

The $L_i(x)$ are polynomials of degree n, with $L_i(x_i) = 1$ and $L_i(x_j) = 0$ for $i \neq j$. Lagrange polynomials are useful in interpolation problems.

Lagrange's theorem *Mathematics.* If H is a subgroup of a group G, then $|G| = |G{:}H||H|$, where $|G{:}H|$ is the index of H in G and $|G|$ and $|H|$ are the orders of G and H, respectively. That is, the order of a subgroup always divides the order of the group. It is not assumed that G or H has finite order. In particular, if G is a finite group, then the order $|a|$ of any element a of G divides the order of G.

Lagrange stream function *Fluid Mechanics.* a scalar function of position whose constant values identify streamlines; the function is applicable for incompressible, steady, two-dimensional flow.

Lagrangian *Mechanics.* **1.** for a discrete mechanical system, the quantity equal to the system's kinetic energy minus its potential energy, regarded as a function of the generalized coordinates, generalized velocities, and time. **2.** for a continuous system, the integral of the Lagrangian density over the volume of the system. Also, KINETIC POTENTIAL.

Lagrangian current measurement *Oceanography.* the direct observation of a current's speed or direction by means of a recording device, usually a parachute drogue, that follows the movement of a water mass through the ocean.

Lagrangian density *Mechanics.* for a continuous mechanical system or field, the differential expression for the difference between the kinetic and potential energies per unit volume or length; the volume integral of this quantity is the Lagrangian function as it appears in Lagrange-Hamilton theory.

Lagrangian function see LAGRANGIAN.

Lagrangian method *Fluid Mechanics.* one method of defining the equation of fluid motion in the presence of deformable bodies, whereby fluid volume elements are considered to flow and there is no flow across a volume element boundary.

Lagrangian multiplier *Mathematics.* the name given to a constant λ that appears in the process for obtaining extrema of functions of several variables by application of the following theorem. Let f and g be continuously differentiable real-valued functions on a region containing a point c in R^n. Suppose that there exists a neighborhood of c such that $f(x) \geq f(c)$ or $f(x) \leq f(c)$ for all points x in this neighborhood that also satisfy the constraint $g(x) = 0$. If the derivative of g at c is nonzero, i.e., $g'(c) \neq 0$, then there exists a real number λ such that $f'(c) = \lambda g'(c)$. This technique is known as the method of **Lagrangian multipliers** or **Lagrange's method.**

Lagrangian points *Astronomy.* five points in the orbital plane of two large bodies at which any small (essentially massless) object can remain in equilibrium relative to the two large bodies.

Lagriidae or **Lagriinae** *Invertebrate Zoology.* a subfamily of beetles, coleopteran insects in the superfamily Tenebrionoidea; found in leaf litter.

lag time a general term for the delay between an action and a desired effect. *Electricity.* **1.** a time displacement between two waves of the same frequency, expressed in degrees. **2.** in control circuits, a delay between the trigger pulse and the operation of the circuit.

Laguerre, Edmond 1834–1886, French mathematician; worked with continued fractions; formulated Laguerre's differential equations.

Laguerre polynomials *Mathematics.* **1.** a family of $L_n(t)$ orthogonal polynomials that are solutions of Laguerre's differential equations for positive integral values of the parameter. $L_n(t) = (-1)^n e^t/n!d^n(e^{-t}t^n)/dt^n$. **2.** more generally, a normalized special case of the hypergeometric function ${}_1F_1$; given by

$$L_n^{\alpha}(z) = \Gamma(\alpha + n+1)/n!\Gamma(\alpha+1)\,{}_1F_1{-n}\,;\,\alpha+1;\,z).$$

Laguerre's differential equation *Mathematics.* the ordinary differential equation $xy'' + (1 - x)y' + ay = 0$, where the parameter a is constant.

Lagynoidea *Invertebrate Zoology.* a superfamily of foramaniferan protozoans, both marine and freshwater. Also, **Lagynacea.**

LAH lithium aluminum hydride.

lahar *Geology.* **1.** a mass flow of mud mixed with hot volcanic debris and water down the sides of a volcano. **2.** a deposit of such a mixture.

LAI leaf area index.

laid paper *Materials Science.* a type of paper that is watermarked with a pattern of closely spaced parallel lines crossed by more widely spaced perpendicular lines; used for writing and printing.

laissez-faire group [lā´sā fâr´] *Psychology.* a term for a therapy group in which the therapist does not take the initiative and exercises minimal control.

laitance *Materials.* **1.** a milky scum that exudes from grout or mortar when tiles are pressed into place. **2.** a similar scum that forms on new cement or concrete, usually due to excess water.

lake *Hydrology.* an extensive body of standing fresh or salt water that occupies a depression in the earth's surface and is more or less surrounded by land. *Materials Science.* any of numerous organic dyes that are insoluble in water, produced by the interaction of an oil-soluble organic dye, a precipitant, and an absorptive inorganic substrate; used in inks, decorative metal coatings, rubber, plastics, and food coloring.

lake ball *Ecology.* a spherical mass of filamentous material composed of living or dead vegetation, sand, and fine-grained minerals, produced on a lake bottom by the action of waves. Also, BURR BALL, HAIR BALL.

lake breeze *Meteorology.* a wind that blows from the surface of a large lake to the shoreline, usually during the afternoon, as a result of the difference in surface temperatures between the land and the water.

lake copper *Metallurgy.* copper found in nature or extracted from ore from the Lake Superior region.

lake effect *Meteorology.* **1.** in general, the effect of a lake on the weather adjacent to its shore and for some distance downward. **2.** in the United States, the effect of the Great Lakes on the weather of the surrounding region.

lake-effect storm *Meteorology.* a low-pressure system that develops over a lake, causing heavy precipitation, due to the interaction of cool, unstable air with the warmer water below.

lake peat see SEDIMENTARY PEAT.

lake plain *Geology.* **1.** a nearly flat surface that marks the floor of a former lake that was filled in as sediments were deposited by inflowing streams. **2.** such a lake bed that borders an existing lake.

lake rampart see ICE PUSH.

laking *Cell Biology.* the process of lysing red blood cells, resulting in the release of hemoglobin.

laky blood *Hematology.* a term for blood containing at least some lysed erythrocytes.

Lalande cell *Electricity.* a type of wet cell that contains a zinc anode and a cupric-oxide cathode with an electrolyte of sodium hydroxide in an aqueous solution.

laliophobia *Psychology.* an irrational fear of speaking in front of others.

lalo- a combining form meaning "speech," as in *lalognosis.*

lalognosis *Neurology.* the ability to understand speech.

lalopathy *Medicine.* any form of speech defect.

laloplegia *Medicine.* a paralysis of the muscles that control speech, other than the tongue.

LAMA *Robotics.* a robot language, developed at MIT, that generates task programs using linguistic descriptions rather than interaction with program operators or trainers.

Lamarck, Jean Baptiste 1744–1829, French naturalist; founder of invertebrate paleontology; formulated Lamarckism, the Lamarckian theory of evolution.

Lamarckism *Evolution.* a theory of evolution stating that changes in an organism's environment cause variations in the use and disuse of various organs, resulting in changes in their size and function, and that these variations can be inherited and passed on to offspring.

Lamaze method [lə mäz´] *Medicine.* a widely used psychophysical technique for preparing a prospective mother for ease of childbirth, through such methods as instruction in the physiology of pregnancy and delivery, exercises to increase strength and control of muscles involved in labor, and training in techniques of breathing and relaxation. (Originated by Fernand *Lamaze,* 1890–1957, French physician.)

lamb *Vertebrate Zoology.* a young sheep, especially one less than one year old and lacking permanent teeth. *Agriculture.* the meat of such an animal.

Lambda (λ) Scorpii *Astronomy.* Shaula, a B2-type star of 1.6 magnitude that lies about 270 light-years away and represents the stinger in the Scorpion's tail.

Lamb, Willis Eugene, Jr. born 1913, American physicist; awarded the Nobel Prize for discovering the hyperfine structure of the hydrogen spectrum.

lambda the eleventh letter in the Greek alphabet, Λ or λ.

lambda calculus *Mathematics.* a mathematical formalization of the rules of substitution; used for modeling the process of substituting values for bound variables.

lambda chains *Immunology.* one of the two forms of smaller polypeptide chains (known as light chains) that occur in immunoglobulins.

lambda expression *Mathematics.* an expression used to define a function in the lambda calculus; e.g., $f(x) = x + 1$ is defined by $\lambda x(x + 1)$.

lambda hyperon *Particle Physics.* the lightest strange baryon particle, with a mass of 1115 MeV, and one of the slowest to decay, with a lifetime of 2.6×10^{-10} seconds.

lambda insertion vector *Molecular Biology.* a lambda vector that is to be used as a lysogen.

lambda particle *Physics.* any of a family of neutral baryons with charm +1 or strangeness –1, and isotopic spin.

lambda phage *Virology.* a temperate bacterial virus that infects *Escherichia coli* and is able to integrate into the host DNA or to initiate a lytic cycle, copying its own genetic information, lysing the host cell, and releasing new viruses; used as a model vector in gene cloning.

lambda point *Physics.* the transition temperature, approximately 2.19 K, below which helium I changes into helium II. *Thermodynamics.* the temperature at which the specific heat of a substance has a sharp peak; this is observed in many second-order transitions.

lamb dysentery *Veterinary Medicine.* infection and inflammation of the large and small intestines in newborn lambs that is caused by *Clostridium perfringens,* characterized by diarrhea and associated with the overdistension of the stomach with milk.

Lambeosaurus *Paleontology.* a genus of medium-sized duck-billed ornithopods in the family Hadrosauridae; the most advanced hadrosaurs, characterized by a large cephalic crest whose function is still unclear; extant in the Upper Cretaceous.

Lambert, Johannn Heinrich [läm´bərt] 1728–1777, German mathematician and astronomer; worked with noneuclidean geometry, conics, and the theory of π.

lambert [lam´bərt] *Optics.* a unit of brightness that equals the amount of flux radiated into a hemisphere by 1 sq cm of a diffusing surface having a luminance of 1 stilb and equal to π times the luminance in candles per sq cm. (Named for J. H. *Lambert.*)

Lambert bearing *Navigation.* a bearing on a Lambert conformal chart; it will be very close to a great circle bearing.

Lambert conformal projection *Cartography.* a conical type of conformal map projection in which all the geographic meridians are shown as straight lines that meet in a common point beyond the map's boarders, and the geographic parallels are shown as a series of arcs of circles having this common point as their center; this projection may have one or two standard parallels along which exact scale is maintained.

Lambert course *Navigation.* a course laid out on a Lambert conformal chart; it will approximate a great circle course.

Lambert's law *Optics.* a set of laws relating to the angle of illumination. The first, **Lambert's cosine law of intensity,** states that when a source is viewed obliquely, its intensity decreases as the cosine of the angle between the viewing direction and the direction perpendicular to the surface. The second, **Lambert's cosine law for illumination,** states that a parallel beam which obliquely illuminates a plane covers a larger area than it does in a perpendicular position, causing illuminance to decrease as the cosine of the angle. Also, BOUGUER'S LAW, BOUGUER-LAMBERT LAW.

Lambert surface *Thermodynamics.* a surface that diffuses radiation perfectly; the directional energy leaving a surface is equal in all directions.

Lambornella *Invertebrate Zoology.* a genus of small ciliates of the family Tetrahymenidae that are parasitic upon larvae of treehole-breeding mosquitoes.

lamB protein *Biochemistry.* a protein in the outer membrane of *E. coli* that is the receptor for bacteriophage lambda and is active in the uptake of certain solutes, such as maltose, maltodextrins, and other sugars and amino acids.

Lamb shift *Atomic Physics.* a small difference in the energy levels in the hydrogen atom which would otherwise be the same; according to the Dirac theory, explained by the principle of quantum electrodynamics.

Lamb-shift source *Nucleonics.* a device that generates a beam of polarized ions from atoms with only one electron.

lambswool *Vertebrate Zoology.* the hairy coat of a sheep less than one year old.

Lamb wave *Electromagnetism.* an electromagnetic wave that propagates on the surface of a solid whose thickness is comparable to the wavelength of the wave.

Lamé, Gabriel [la mā´] 1795–1870, French mathematician and civil engineer; introduced curvilinear coordinates.

Lamé constants *Mechanics.* two physical constants that linearly relate stress to strain in an isotropic linear elastic solid.

Lamé equation *Mathematics.* a second-order differential equation with five singularities; developed by Gabriel Lamé in an attempt to classify the ordinary differential equations that occur after separation of variables of the partial differential equations of mathematical physics. Most, but not all, of the separated equations occur as degenerate cases of the Lamé equation.

lamella *plural,* **lamellae** *Anatomy.* a thin sheet or plate of tissue, as found in compact or lamellar bone. *Mycology.* the gill of a mushroom. *Materials Science.* a thin plate crystal that frequently forms during certain three-phase reactions, such as the eutectic or eutectoid reaction. *Building Engineering.* any of a series of wood, metal, or reinforced-concrete members joined in a crisscross pattern, forming a vault.

lamellar bone *Histology.* a major type of bone, consisting of thin layers of mineralized bone deposits.

lamellar crystal *Crystallography.* a polycrystalline substance in which the crystals are arranged in platelike fashion, as in mica.

lamellar gill see PHYLLOBRANCHIATE GILL.

lamellar settlers *Biotechnology.* a tank having inserted plates or tubes that shorten the distance a particle has to settle, thereby increasing the settling efficiency.

lamellar structure *Materials Science.* the platelike crystals that form under certain conditions, as when an alloy solidifies; a characteristic microstructure of certain alloys and also certain polymers.

lamellate *Mycology.* of or relating to a growth that possesses lamellae or mushroom gills.

lamelli- a combining form meaning "thin plate," as in *lamelliform.*

Lamellibranchiata see BIVALVIA.

lamellipodium *Cell Biology.* a thin, flattened extension of a migrating cell that projects outward from the leading edge of the cell and may or may not adhere to the substratum.

Lamellisabellidae *Invertebrate Zoology.* a family of beard worms, mollusks in the phylum Pogonophora composed of threadlike, tentacled animals living in tubes.

Lamiaceae *Botany.* a family of aromatic dicotyledonous herbs and shrubs of the order Laminales, having simple leaves and secreting a pungent oil. Also, LABIATAE.

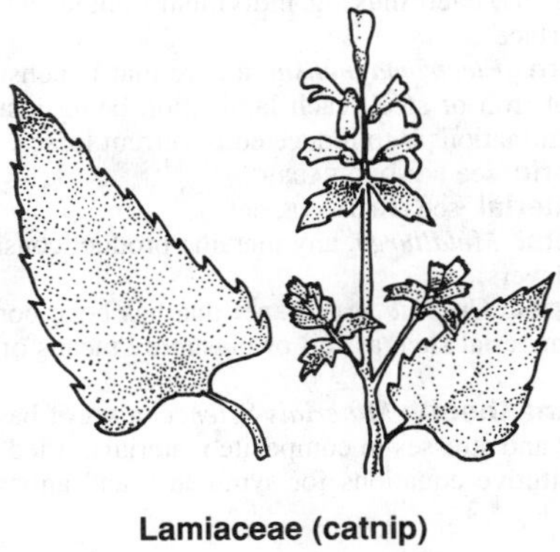

Lamiaceae (catnip)

Lamiales *Botany.* an order of dicotyledonous plants of the subclass Asteridae, having opposite leaves and four ovules in the ovaries.

lamina *Anatomy.* a thin membrane or flat sheet of tissue. *Botany.* **1.** the flattened portion of a leaf or blade. **2.** any thin, flat plant organ, such as a petal. *Geology.* the thinnest clearly recognizable layer of a sediment or sedimentary rock, differing from other layers in color, composition, or particle size; usually 1 millimeter or less in thickness. Also, LAMINATION. *Materials Science.* the unit building block of laminated materials, thin sheets of material that are stacked on one another, cured, and bonded into a composite.

lamina cribrosa *Anatomy.* the cribriform plate of the ethmoid bone.

Laminales *Botany.* an order of dicotyledonous plants of the subclass Asteridae with opposite leaves and four ovules in the ovary.

laminal placentation *Botany.* the positioning of the ovules on the inner surface of the ovary. Also, **laminate placentation.**

laminar *Science.* consisting of or arranged in thin layers or plates. *Aviation.* see LAMINAR FLOW.

laminar boundary layer *Fluid Mechanics.* the layer of fluid in contact with the surface of a body in which it is immersed; fluid velocity is relatively low in this layer.

laminar flow *Fluid Mechanics.* the streamline flow of a viscous, incompressible fluid in which fluid particles travel along well-defined separate lines. *Meteorology.* a nonturbulent flow. Also, SHEET FLOW.

laminar-flow airfoil *Aviation.* a low-drag airfoil designed to maintain laminar flow over a high percentage of the chord about itself; often relatively thin, especially along the leading edge, with most of its bulk near the center of the chord.

laminar flow control (system) *Aviation.* a method for reducing drag by maintaining laminar boundary layers, using a compressor to suck in a small amount of external boundary layer flow through the perforated or slotted airfoil skin.

Laminariaceae *Botany.* a kelp family of the order Laminariales, characterized by a discoid or branching hapterous holdfast and usually extensive sori that cover most of the blade surfaces; distributed mostly in the North Pacific.

Laminariales *Botany.* an order of brown algae, commonly known as kelp, of the class Phyophyceae.

laminarin *Biochemistry.* the chief carbohydrate food reserve of the brown algae.

Laminariophyceae *Botany.* a class of algae of the division Phaeophyta.

laminar sublayer *Fluid Mechanics.* a fluid layer of streamline flow that exists beneath a turbulent boundary layer.

laminar wing. *Aviation.* a wing designed with laminar-flow airfoils.

laminate [lam´i nāt´] *Materials Science.* **1.** a material formed by bonding together two or more sheets of reinforcing fibers, usually with heat or pressure; plywood, fiberglass, and reinforced plastics produced by this technique are widely used in industry. Also, LAMINATED COMPOSITE, LAMINATED MATERIAL. **2.** to form or create such a material; to carry out the process of lamination.

laminated *Materials Science.* **1.** describing a material that is composed of thin layers that have been pressed tightly together, often using resins or heat. Thus, **laminated plastic, laminated glass, laminated wood,** and so on. **2.** covered with a thin sheet of transparent plastic or similar protective material.

laminated clot *Hematology.* a blood clot formed by successive deposits, giving it a layered appearance. Also, STRATIFIED CLOT.

laminated composite see LAMINATE, def. 1.

laminated contact *Electricity.* a switch contact composed of two or more laminations, each making individual contact with the opposite conducting surface.

laminated core *Electromagnetism.* a core that is constructed of laminated plates of iron or steel, each lamination being insulated; the purpose for the laminations is to reduce eddy current losses.

laminated fabric see BONDED FABRIC.

laminated material see LAMINATE, def. 1.

laminated metal *Metallurgy.* any metallic product consisting of two or more bonded layers.

laminated paper *Graphic Arts.* paper that has been bonded to another paper or to a material such as foil or plastic by means of heat, pressure, or adhesive.

laminated plane theory *Materials Science.* a set of basic assumptions about bonding and stresses in composite materials; it led to the development of constitutive equations for symmetric and antisymmetric laminates.

laminated safety glass *Materials Science.* a type of glass made by alternating layers of glass and plastic material, so that if the outside layer of glass breaks it will be held together by the attached layer of plastic; used in automobile windshields.

laminated spring *Mechanical Devices.* a spring made of superimposed strips, plates, or leaves that are firmly anchored at one end, forming a uniform beam or cantilever.

laminated wood *Materials.* alternating layers of wood that are bonded together with their grains running at right angles to each other on successive layers, in order to give added strength.

lamina terminalis *Anatomy.* a thin sheet of tissue that forms the anterior wall of the third ventricle.

lamination *Science.* the act or process of arranging in a succession of thin layers. *Geology.* see LAMINA. *Materials Science.* the process of forming a material by pressing or bonding together several thin layers. *Medicine.* an arrangement in the form of plates or layers.

laminectomy *Surgery.* the surgical removal of a vertebral lamina, such as the removal of the posterior arch.

laminin *Biochemistry.* a glycoprotein in the extracellular matrix that allows epithelial cells to adhere to connective tissues.

lamio see SENGKUANG.

Lami's theorem *Mechanics.* a theorem stating that if three forces acting on a particle are in equilibrium, each is proportional to the sine of the angle between the other two.

Lamnidae *Vertebrate Zoology.* the mackerel sharks, a family of fast swimming, dangerous sharks of the order Lamniformes, including the great white shark and the mako shark, characterized by a conical-pointed head, a long snout, large pectoral fins, and a general mackerel or tuna shape; found in cold temperate to tropical seas worldwide.

lamp *Engineering.* any appliance that yields light or heat, such as an electric lamp. *Graphic Arts.* specifically, any artificial light source used in photography.

lampadite *Mining Engineering.* a hydrated oxide of manganese and copper.

lamp bank *Electricity.* an arrangement of incandescent lamps, mounted either in parallel or in series, that is used as a resistance load in electrical tests.

lampblack *Materials.* a black material made by burning low-grade oils or other carbon-rich substances in a chamber with limited air and collecting the resulting soot; widely used as a black pigment in various products such as inks, paints, and crayons, and for other industrial purposes.

lampbrush chromosome *Genetics.* a transcriptionally active, long-lived chromosome that is found in the primary oocytes of vertebrates, characterized by decondensed loops of DNA that resemble an antique lampbrush.

lamp cord *Electricity.* a standard type of cable having two insulated wires, used to transfer 115-volt power to simple household appliances.

lampholder *Electricity.* **1.** a standard support or post with internal and external attachments for wiring the luminaire and bracket. **2.** a device that electromechanically accepts one or more crystals or fuses in a way that simplifies insertion or removal.

lamphouse *Engineering.* **1.** a small, light housing behind the head of a film projector that contains a carbon arc lamp, a concave reflector behind the arc, and cooling devices. **2.** a box that is pierced with a hole and that contains an electric lamp and concave mirror; used as a concentrated light source in a photographic enlarger, microscope, or other equipment.

lamping *Mining Engineering.* in prospecting, the use of a portable ultraviolet lamp to detect fluorescent minerals.

lampman *Mining Engineering.* a worker responsible for maintaining, cleaning, and servicing miner's lamps.

lamprey *Vertebrate Zoology.* a jawless, eel-like vertebrate of the genus *Petromyzon* that uses its suctorial mouth to parasitize fish; distributed in temperate and subarctic regions in fresh and salt water. Also, **lamprey eel, lamper (eel).**

lamprey

Lampridiformes *Vertebrate Zoology.* an order of bony fish composed of 35 species characterized by a unique type of protrusible upper jaw. Also, LAMPRIFORMES.

Lamprocystis *Bacteriology.* a genus of aquatic, photosynthetic, anaerobic, motile, spheroid bacteria of the family Chromatiaceae, requiring sulfide and vitamin B12 for maintenance and growth.

Lampropedia *Bacteriology.* a genus of Gram-negative, aerobic, spheroid bacteria of undecided classification, occurring in environments containing organic matter, such as stagnant, polluted waters.

lamprophyllite *Mineralogy.* $Na_2(Sr,Ba)_2Ti_3(SiO_4)_4(OH,F)_2$, a translucent, brown monoclinic mineral occurring in elongated, tabular crystals, having a specific gravity of 3.44 to 3.53 and a hardness of 2.5 on the Mohs scale; found in nepheline syenites.

lamprophyre *Petrology.* a dark, igneous dike rock or rock group, normally rich in alkalis, characterized by a porphyritic texture, and having an abundance of mafic minerals in both the phenocrysts and fine-grained groundmass.

Lampyridae *Invertebrate Zoology.* the fireflies and glowworms, a family of predacious beetles; bioluminescent coleopteran insects in the superfamily Cantharoidea.

lamziekte *see* BOTULISM.

LAN *Computer Science.* local area network.

LAN *Aviation.* the airport code for Lansing, Michigan.

Lanarkian *Geology.* a European geologic stage of the lower Upper Carboniferous period, occurring after the Lancastrian and before the Yorkian.

lanarkite *Mineralogy.* $Pb_2(SO_4)O$, a white, greenish, or gray monoclinic sulfate of lead, having a specific gravity of 6.92 and a hardness of 2 to 2.5 on the Mohs scale; found with anglesite and leadhillite (into which it easily alters), as at Leadhills, Lanarkshire, Scotland.

lanate *Biology.* covered with fine, wool-like filaments.

lanatoside *Pharmacology.* a stable and easily absorbed glycoside that is used as a cardiotonic where digitalis is recommended.

Lancashire boiler [lang´kə shir] *Mechanical Engineering.* a type of steam boiler, cylindrical in form, with two furnace tubes that are set lengthwise and have internal grates at the front; hot gases travel from the tubes to the front along a bottom flue and then return to the chimney along side flues. Also, FIRE TUBE BOILER.

Lancastrian *Geology.* a European geologic stage of the lower Upper Carboniferous period, occurring after the Viséan and before the Lanarkian.

Lance *Ordnance.* a mobile, storable, surface-to-surface guided missile with nuclear and conventional capability, designed to support the Army corps with long-range fire; officially designated **MGM-52C.**

lance *Medicine.* to cut into or open, as an abscess or boil.

lance door *Mechanical Engineering.* an opening in a boiler furnace that provides for the insertion of a hand lance.

Lancefield group *Microbiology.* one of a number of streptococci categories determined using a serological technique to identify a specific carbohydrate antigen found in the streptococcus cell wall. (Named for American bacteriologist Rebecca C. *Lancefield,* 1895–1981.)

lancelet *Zoology.* a small fishlike creature of the subphylum Cephalochordata, characterized by a long, slender, compressed body pointed at each end.

lanceolate *Biology.* lance-shaped; much longer than broad; widened above the base and tapering toward the apex, such as an elongated, simple leaf.

Lanceolidae *Invertebrate Zoology.* a family of marine amphipod crustaceans with inflated bodies, in the suborder Hyperiidea.

lancet *Medicine.* a small, pointed surgical knife with a double-edged blade. *Architecture.* a sharply pointed window characteristic of the Early English style of Gothic architecture. Also, **lancet window.**

lancet arch see ACUTE ARCH.

lancetfish *Vertebrate Zoology.* a large, marine fish of the family Alepisauridae, characterized by daggerlike teeth.

Lanchester balancer *Mechanical Engineering.* an instrument having two interlocking gears with eccentric masses that are moved by a crankshaft; used to balance four-cylinder engines.

land *Geography.* the solid part of the earth's surface, usually including the polar icecaps. *Aviation.* **1.** to alight on the earth or another surface. **2.** to deliver cargo, passengers, or troops to the ground, either directly or by parachute. *Design Engineering.* **1.** the top portion of the tooth of a cutting tool, behind the cutting edge. **2.** the flat top of a gear tooth or the flat bottom between two teeth of a cutting edge wheel. **3.** the space between flutes or grooves in drills, taps, reamers, or other cutting tools. *Electronics.* the conductive portion of a printed circuit board to which components are attached. Also, PAD, TERMINAL AREA. *Acoustical Engineering.* the surface space wedged between grooves on a recording disk. *Engineering.* **1.** in equipment for molding plastics, the horizontal bearing surface of a flash mold that allows excess melt to escape; the bearing surface at the top of the screw flight in a screw extruder; the surface of an extrusion die, parallel to the melt flow. **2.** the surface between consecutive grooves of a phonograph record, or a diffraction grating. *Ordnance.* one of the raised ridges between the grooves of a rifled bore.

land accretion *Civil Engineering.* the process of land reclamation using such methods as draining extensively, sowing reeds or other plants to encourage natural silt deposits, and filling, dumping, or dredging mud or other deposits.

land and sea breeze *Meteorology.* the daily cycle of local winds on seacoasts, caused by differences in the surface temperature between the land and sea; the land breeze blows from the land to the sea and the sea breeze blows from the sea to the land.

Landau, Lev Davidovich [lan´dou] 1908–1968, Soviet physicist; awarded the Nobel Prize for formulating the theory of superfluidity.

Landau damping *Physics.* the damping of a longitudinal wave in a plasma caused by a transfer of energy from the wave to particles whose velocity equals the phase velocity of the wave.

Landau fluctuations *Nucleonics.* differences in the energies lost by various particles, arising from variations in the number of collisions and variations in the amount of energy lost with each collision.

Landau levels *Solid-State Physics.* the quantized energy levels that exist for electrons in a conductor subjected to a magnetic field.

Landau symbols *Mathematics.* two symbols, o and O, that indicate order of magnitude. Suppose $X = (x_n)$ and $Y = (y_n)$ are two nonzero sequences in R^p and R^q, respectively (constant sequences are allowed), and consider three cases: (a) $\lim_{n\to\infty}(x_n/y_n) = 1$. Then the sequences X and Y are said to be of the same order of magnitude, or equivalent; denoted $X \sim Y$ or $(x_n) \sim (y_n)$. (b) $\lim_{n\to\infty}(x_n/y_n) = 0$. Then X is said to be of lower magnitude than Y; denoted $X = o(Y)$ or $x_n = o(y_n)$. (c) If there is a positive constant K such that $|x_n| \le K\,|y_n|$ for all sufficiently large n, then X is said to be dominated by Y; denoted $X = O(Y)$ or $x_n = O(y_n)$. Note that $X = O(Y)$ if either $X \sim Y$ or $X = o(Y)$.

landblink *Meteorology.* a yellowish glow that is sometimes observed over snow-covered land in the polar regions.

land breeze *Meteorology.* a light wind that blows from the land to the sea, usually at night, alternating with a sea breeze that blows in the opposite direction during the day; it occurs when the temperature of the sea surface is warmer than that of the adjacent land.

land bridge *Geography.* a strip of land connecting two landmasses, especially one that is intermittently exposed and submerged.

Land camera see POLAROID CAMERA.

land-capacity classfication *Agronomy.* the grouping of soil types into classes and subclasses according to their capacity for use and the treatments needed for sustained use.

land-control operations *Military Science.* the employment of ground forces, supported by naval and air forces as appropriate, to achieve military objectives in vital land areas; such objectives include destruction of opposing ground forces, securing key terrain, protection of lines of communication, and establishing local military superiority.

land crab *Invertebrate Zoology.* any crab that lives mostly on land, especially those of the family Gecarcinidae.

land diameter *Ordnance.* the diameter of a rifled bore as measured between two opposing lands.

land drainage *Civil Engineering.* a method used by water authorities to reclaim areas of low-lying marsh for practical use.

Landé Γ permanence rule [län dā´] *Atomic Physics.* a rule stating that there is no dependence of the strength of an external magnetic field on the sum of the energy shifts produced in a system of particles whose spin-orbit interaction is calculated over a series of states having identical spin and angular momentum quantum numbers, but different total angular momentum.

Landé g factor *Nuclear Physics.* a dimensionless factor used in calculating the measure of the ratio of the angular momentum of a subatomic particle to its magnetic moment in the presence of spin-orbit coupling. Also, *G*-FACTOR.

Landé interval rule *Atomic Physics.* a rule stating that when the interaction between an electron's spin and orbit is weak, the energy levels of each are split, with each having a different angular momentum, so that the interval between successive energy levels is proportional to the larger of combined angular momentum values.

Landenian *Geology.* a European geologic stage of the Upper Paleocene epoch, occurring after the Montian and before the Ypresian of the Eocene epoch.

landesite *Mineralogy.* $(Mn^{+2},Mg)_9Fe_3^{+3}(PO_4)_8(OH)_3 \cdot 9H_2O$, a brown orthorhombic mineral occurring in octahedral crystals, having a specific gravity of 3.03 and a hardness of 3 to 3.5 on the Mohs scale; found as an alteration product of reddingite.

landfall *Navigation.* **1.** the first sighting of land by an approaching vessel. **2.** the land sighted.

landfill *Civil Engineering.* **1.** a low area of land that is built up by deposits of layers of solid refuse covered by soil. **2.** the solid refuse itself. **3.** to create usable land by this means.

landform *Geography.* any surface feature of the earth, including large-scale features such as plains and mountains and minor features such as hills and valleys.

landform map see PHYSIOGRAPHIC DIAGRAM.

land hemisphere *Geography.* the half of the globe having its center in northwestern France and containing most of the earth's dry land.

Landholt fringe *Optics.* a fringe whose main axes are at right angles to each other and that appears across a darkened field when an intensely bright light is observed through two Nicol prisms.

landing *Mining Engineering.* the top or bottom of a slope, shaft, or inclined plane. *Civil Engineering.* a platform connecting two flights of stairs. *Military Science.* of or related to the operations of an amphibious force, including troops landed from ships, boats, amphibious vehicles, or aircraft. Thus, **landing area, landing attack, landing beach, landing craft, landing diagram, landing force, landing group, landing schedule, landing ship, landing site.** *Navigation.* a place for the loading and unloading of passengers and cargo. *Aviation.* **1.** to carry out the process of bringing a flight vehicle down to the earth or another surface. **2.** an instance of this process. **3.** of or relating to this process; used to form a number of compounds such as **landing approach, landing check, landing weight.**

landing angle. *Aviation.* the angle between the fuselage reference line and the horizontal at the moment of touchdown.

landing area *Aviation.* the part of an airport, flight deck, or other surface or facility used primarily for the takeoff and landing of aircraft.

landing beam *Aviation.* a directional radio beam used to mark the glide path in an instrument landing system.

landing brief *Transportation Engineering.* the planned course of action for an aircraft's approach and landing.

landing craft *Naval Architecture.* a small vessel used in amphibious assault; the landing craft is capable of beaching itself, allowing men or vehicles to move directly ashore down a bow ramp.

landing deck *Aviation.* the flight deck of an aircraft carrier. *Aviation.* a circular path followed or described by an aircraft before entering a landing pattern; used especially for landing on an aircraft carrier.

landing direction indicator *Navigation.* a device on the ground that indicates which direction landings and takeoffs should take.

landing flap *Aviation.* a flap designed to increase lift during landing; typically located behind the rear beam or spar of the wing.

landing flare *Navigation.* the adoption by an aircraft of a nose-up attitude immediately before landing, so that the main landing gear touches down before the nose gear.

landing gear *Aviation.* the collection of devices including wheels, skis, and floats that enable a flight vehicle to land and move about on land, water, or other surfaces. *Space Technology.* on certain rockets, a reentry apparatus consisting of a parachute and release mechanism.

landing light *Aviation.* a floodlight on an aircraft used to enhance visibility during night landings; located on the leading edge of the wing, below the nose of the fuselage.

landing pattern *Aviation.* the flight path followed by an aircraft between approach and touchdown.

landing ship *Naval Architecture.* a ship capable of beaching itself during amphibious operations to offload men, vehicles, or equipment.

landing stage *Civil Engineering.* a platform, often floating, on which materials are disembarked.

landing strip *Aviation.* a designated portion of a landing area used for landing and takeoff of aircraft in a specific direction; may include more than one runway.

landing tee see WIND TEE.

landline *Electricity.* a communication cable that connects two ground locations, as opposed to a submarine cable.

landlocked *Geography.* **1.** of a body of water, entirely or nearly surrounded by land. **2.** of a land region, having no access to a waterway.

landmark *Geography.* any permanent artificial or natural object or monument that marks a land boundary. *Navigation.* a conspicuous object on land that can be used as an unofficial navigational aid.

land mile see MILE, def. 1.

land mine *Ordnance.* see MINE, def. 1.

land mobile service *Telecommunications.* a mobile service between base stations and mobile stations or between two land mobile stations.

land mobile station *Telecommunications.* a transportable station in the land-mobile service.

land navigation *Navigation.* the art and science of finding one's way across land or ice.

land pebble see LAND-PEBBLE PHOSPHATE.

land-pebble phosphate *Geology.* a small, spherical fragment of sedimentary rock composed mainly of calcium phosphate that occurs below the surface in clay, gravel, or compacted beds; mined extensively as a source of phosphate fertilizer. Also, LAND PEBBLE, LAND ROCK, MATRIX ROCK.

Landrace *Agriculture.* a breed of hog developed in Denmark, having white hair, a long body, and drooping ears.

land reclamation *Civil Engineering.* the accretion of marshy land or sea bed by means of extensive drainage, plant sowing to encourage natural silt deposits, and filling, dumping, or dredging of mud or other deposits.

land rock see LAND-PEBBLE PHOSPHATE.

Landry-Guillain-Barre syndrome *Medicine.* a neurological syndrome associated with viral and Mycoplasma infections, marked by paresthesias of the limbs, weakness or a paralysis in which the muscles manifest loss of tone and diminished tendon reflexes, and increased protein in the cerebrospinal fluid without an increase in cell count.

landscape *Geography.* **1.** the general characteristics or aspect of an area. **2.** an area having certain characteristics.

landscape agate *Mineralogy.* a white or gray chalcedony with inclusions of irregular arrangements of manganese oxide that have a slight resemblance to a landscape.

landscape architecture *Civil Engineering.* the planning and designing of gardens and outdoor areas.

landscape ecology see AREOGEOGRAPHY.

landside *Agriculture.* the part of a plow bottom that guides the plow and holds back the furrow slice.

land sky *Meteorology.* a term used primarily in the polar regions to denote the relatively dark appearance of the underside of a cloud layer that occurs over land not covered with snow.

landslide *Geology.* **1.** any perceptible downward mass movement of soil or rock under the influence of gravity. **2.** the mass itself.

landslide track *Geology.* the exposed path in rock or earth produced by a landslide.

Landsteiner, Karl 1868–1943, American pathologist born in Austria; awarded the Nobel Prize for the discovery of blood types; codiscoverer of Rh factor.

land surveyor *Civil Engineering.* a person professionally engaged in the measurement of land and buildings for topographical purposes, substantiating existing boundaries or establishing new ones.

land tie *Civil Engineering.* a tie rod holding a sheet pile or other retaining wall to a buried deadman or stay pile.

land transportation frequency bands *Telecommunications.* a cluster of radio-frequency bands between 25 and 30,000 megahertz; used for land transportation radio services by railroads, limousine services, and trucking companies.

land-use classes *Civil Engineering.* the various categories and types of conditions, uses, and factors that are present in a particular region.

lane *Civil Engineering.* **1.** a narrow country road usually hedged on either side. **2.** a narrow strip on a freeway for single line traffic.

lane marker *Ordnance.* in land-mine warfare, a sign used to mark a lane through a minefield; entrance and exit markers may be referenced to a landmark or an intermediate marker.

Lane's law *Astronomy.* a law stating that if a star is assumed to behave like a perfect gas, its radius is inversely proportional to its temperature.

langbanite *Mineralogy.* $(Mn^{+2},Ca)_4(Mn^{+3},Fe^{+3})_9Sb^{+5}Si_2O_{24}$, an iron-black trigonal mineral occurring in prismatic crystals, having a specific gravity of 4.6 to 4.8 and a hardness of 6.5 on the Mohs scale.

langbeinite *Mineralogy.* $K_2Mg_2(SO_4)_3$, a colorless, yellowish, reddish, or greenish cubic mineral with a vitreous luster, having a specific gravity of 2.83 and a hardness of 3.5 to 4 on the Mohs scale; found in salt deposits and often used in fertilizer manufacture and as a source of potassium compounds.

Langerhans, Paul [läng′ər häns′] 1847–1888, German anatomist; discovered islets of Langerhans.

Langerhans cell *Histology.* **1.** a cell covering the chorionic villi in the placenta that forms the syncytial trophoblast that in turn synthesizes hormones. **2.** a star-shaped cell of mammalian epidermis that appears to be part of the mononuclear phagocyte system.

Lange's nerve *Invertebrate Zoology.* one of the paired nerve cords found in the wall of the radial peripheral canal in asteroids.

Langevin, Paul [länzh′van] 1872–1946, French physicist; made contributions in magnetism, relativity, kinetic theory of gases, and development of sonar.

Langevin function *Electromagnetism.* a mathematical function, $L(x) = \coth x - 1/x$, which appears in the classical expression for the paramagnetic susceptibility of a collection of magnetic dipoles.

Langevin ion see LARGE ION.

Langevin ion-mobility theories *Electronics.* two separate theories that explain the actions of ions in gases; in the first, atoms and ions are believed to interact through a hard-sphere collision; the second argues that in addition to the repulsion that occurs when atoms and ions come into close contact, there is also an attraction between them that arises from the polarization of the atoms in the ion's field.

Langevin ion-recombination theory *Electronics.* a theory that predicts the rate at which negative and positive ions will recombine in an ionized gas, based on the assumption that there is mutual attraction between ions with opposing polarities and that their velocities are determined by their mobility.

Langevin radiation pressure *Acoustics.* the pressures and tensile forces that occur on the boundary surfaces of a fluid volume due to the nonlinearity of the sound field equation.

Langevin theory of diamagnetism *Electromagnetism.* a theory stating that diamagnetism results from currents caused by the Larmor precession of electrons inside atoms.

Langevin theory of paramagnetism *Electromagnetism.* a classical theory that considers a substance as a collection of noninteracting, permanent magnetic dipoles having a Boltzmann distribution of alignments in an externally applied field.

Langhian *Geology.* a European geologic stage of the Middle Miocene epoch, occurring after the Burdigalian and before the Seravillian.

langite *Mineralogy.* $Cu_4^{+2}(SO_4)(OH)_6 \cdot 2H_2O$, a greenish-blue monoclinic mineral, dimorphous with wroewolfeite, occurring in small, twinned crystals and having a specific gravity of 3.28 to 3.31 and a hardness of 2.5 to 3 on the Mohs scale; found in the oxidized zone of copper deposits.

Langley, Samuel 1834–1906, American physicist and inventor; built first successful heavier-than-air flying machine; invented the bolometer.

langley *Physics.* a unit of energy flux equivalent to 1 calorie per square centimeter; commonly used in radiation studies.

Langley trap *Materials Science.* a phenomenon in which physical crosslinks or permanent loops and entanglements exist in a polymer network; a form of trapped entanglement.

Langmuir, Irving 1881–1957, American chemist; awarded the Nobel Prize for his work in surface chemistry; invented gas-filled tungsten lamp and atomic-hydrogen welding.

Langmuir adsorption isotherm *Physical Chemistry.* an equation that gives the amount of substance adsorbed on a surface as a function of the pressure exerted; it assumes that the molecules are adsorbed as a single layer, that there is no interaction between adsorbed molecules, and that temperature is constant; used chiefly for the adsorption of gaseous materials on a solid surface. Also, **Langmuir (isotherm) equation.**

Langmuir adsorption model *Physical Chemistry.* a formula that describes the pressure dependence of gas adsorption on a solid surface; expressed as $C_i = afn/(1 + af)$, where C_i is the concentration of adsorbed gas molecules, a is an equilibrium constant, f is the gas fugacity, and n is the concentration of potential adsorption sites on the surface.

Langmuir-Blodgett film *Physical Chemistry.* a monomolecular film produced by dipping a solid in water having a compressed surface layer of molecules that are amphiphilic; i.e., they have polar ends that readily interact with water and nonpolar ends that resist water; thus their alignment is highly regularized.

Langmuir dark space *Electronics.* the area around the negatively charged probe that is inserted into the positive column of a glow discharge tube, from which little light emanates.

Langmuir effect *Solid-State Physics.* an effect in which ions are produced from atoms having a low ionization potential when these are brought into contact with a hot metal having a high work function.

Langmuir plasma frequency *Physics.* the natural frequency of oscillation for electrons in a plasma.

Langmuir probe *Physics.* an instrument that measures electron density and temperature in a plasma.

Langmuir wave *Physics.* a longitudinal, electrostatic wave that propagates in a plasma due to variations in the electron density of the plasma.

language *Linguistics.* **1.** the organized system of speech sounds that humans use to communicate with one another, or the written representation of these speech sounds. **2.** any of various other forms of expression that are regarded as analogous to this system, such as sign language, body language, the sounds made by animals, and so on. *Computer Programming.* a defined set of symbols and a set of rules and conventions for combining them into sentences that may be used to describe and process information.

language converter *Computer Technology.* a device that physically translates data stored on one type of media (such as punched cards) into another (such as magnetic tape).

language family *Linguistics.* a group of languages that show close historical connections.

language subset *Computer Programming.* a subset of elements of a programming language that can be used on its own; usually designed for small computers that are unable to handle the full language.

language theory *Mathematics.* a mathematical modeling of the grammar of a language, usually in terms of automata theory. Applications of this include machine translation and the construction of higher-level programming languages.

language translator *Computer Programming.* a computer program that accepts programs written in one language as input and produces equivalent programs in another language as output.

languet *Invertebrate Zoology.* a tonguelike process or part, such as that dangling down the pharynx of a tunicate.

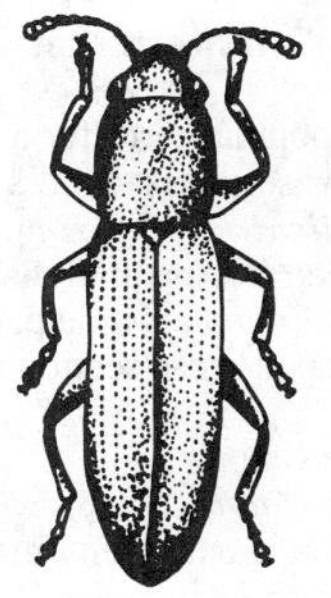

Languriidae

Languriidae *Invertebrate Zoology.* the lizard beetles, a large family of plant-eating coleopteran insects in the superfamily Cucujoidea.

Laniatores *Invertebrate Zoology.* a suborder of arachnids in the order Phalangida, found worldwide.

Laniidae *Vertebrate Zoology.* the shrikes, a family of small to medium-sized birds of the passerine suborder Oscines, characterized by a large head and the use of the bill to kill prey, although the prey may also be impaled on a thorn or nail; found worldwide except in South America, Australia, and oceanic islands.

Lankesterella *Invertebrate Zoology.* a genus of protozoans in the class Sporozoa that are parasites of frog blood and are transmitted by leeches.

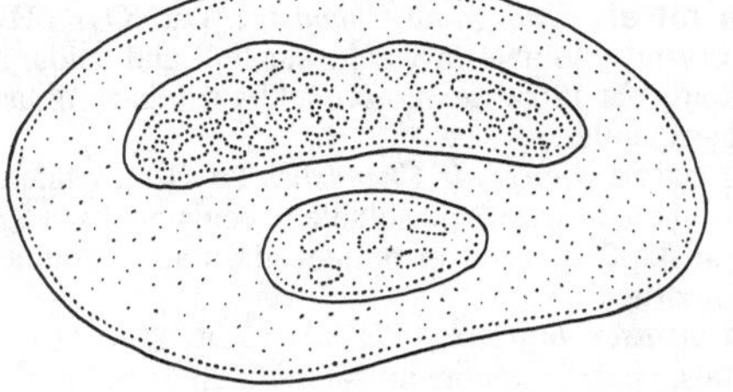

Lankesterella

lanolin or **lanoline** *Materials.* a yellowish fatty substance derived from sheep wool; it does not react readily with acids and alkalis, and emulsifies easily in water; widely used in cosmetics, ointments, soaps, and waterproof coatings. Also, WOOL FAT, WOOL WAX.

lanosterol *Biochemistry.* $C_{30}H_{50}O$, a triterpenic sterol found in wool fat and yeast; a parent compound that is converted in several steps to cholesterol.

lansan *Meteorology.* a strong, southeasterly trade wind that blows over New Hebrides and the East Indies.

lansfordite *Mineralogy.* $MgCO_3{\cdot}5H_2O$, a colorless or white, monoclinic mineral occurring in small prismatic crystals, having a specific gravity of 1.69 to 1.7 and a hardness of 2.5 on the Mohs scale; found as a recent product in some coal mines.

L antenna *Electromagnetism.* an antenna consisting of a long horizontal wire that is fed at one end by a shorter vertical length of wire.

lantern *Engineering.* a small portable lamp in which light is provided by electric batteries or by a fuel such as kerosene, especially one that has a transparent case to enclose and protect the light. *Architecture.* a rooftop structure (such as a church tower) with open or windowed walls.

lantern fish *Vertebrate Zoology.* a member of the deep-sea fish family Myctophidae, noted for the numerous luminous organs on the head and body.

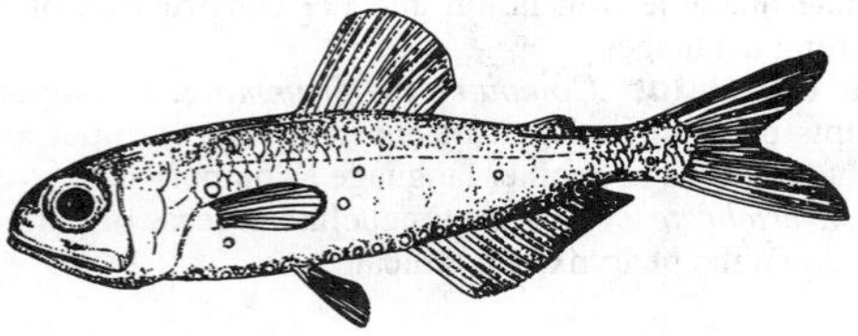

lantern fish

lantern jaw *Anatomy.* a popular term for a prominent, protruding lower jaw. *Invertebrate Zoology.* see ARISTOTLE'S LANTERN.

lantern pinion *Mechanical Engineering.* a type of gear pinion with cylindrical bars set between two parallel disks.

lantern ring *Mechanical Engineering.* a packing ring or sleeve for a rotating shaft with an H-shaped cross section that feeds pumped lubricants to bearing surfaces.

lanthana see LANTHANUM OXIDE.

lanthanide contraction *Atomic Physics.* a phenomenon in which the radii of atoms in the rare-earth elements are observed to decrease slightly as their atomic numbers increase.

lanthanide series *Chemistry.* the rare-earth elements, having atomic numbers 57 through 71; shiny metals that are corroded by both acids and water and burn in air and oxygen.

lanthanite-(LA) *Mineralogy.* $(La,Ce)_2(CO_3)_3{\cdot}8H_2O$, a colorless, white, pink, or yellow orthorhombic mineral occurring in lathlike crystals or in earthy form, having a specific gravity of 2.69 to 2.74 and a hardness of 2.5 to 3 on the Mohs scale; found in oxidized zinc ores.

lanthanum [lan′thə nəm] *Chemistry.* a metallic element having the symbol La, the atomic number 57, an atomic weight of 138.9055, a melting point of 920°C, and a boiling point of 3454°C; a silver-white malleable metal used in lanthanum salts, electronic devices, and rocket propellants. (From a Greek term meaning "to lurk" or "to lie hidden"; in reference to its being discovered in combination with other minerals after having long been undetected.)

lanthanum chloride *Inorganic Chemistry.* $LaCl_3{\cdot}7H_2O$, white hygroscopic crystals; soluble in water, alcohol, and acids; decomposes at 91°C; used to prepare lanthanum.

lanthanum nitrate *Inorganic Chemistry.* $La(NO_3)_3{\cdot}6H_2O$, white hygroscopic crystals; soluble in water, alcohol, and acids; melts at 40°C and decomposes at 126°C; explosive and a fire hazard; used as antiseptic and in gas mantles.

lanthanum oxide *Inorganic Chemistry.* La_2O_3, a white or pale brown powder; insoluble in water and soluble in acids; melts at 2307°C and decomposes at 4200°C; used in optical glass and ceramics. Also, **lanthanum trioxide.**

lanthanum sulfate *Inorganic Chemistry.* $La_2(SO_4)_3{\cdot}9H_2O$, colorless to white crystals; slightly soluble in water and hydrochloric acid and insoluble in alcohol; decomposed by white heat; used in the atomic weight determination of lanthanum.

Lanthonotidae *Vertebrate Zoology.* a family of reptiles of the infraorder Anguinomorpha containing one species, the earless monitor lizard of Sarawak, Borneo.

LANTIRN *Ordnance.* a night navigation and target-locating system that consists of a pod that is attached to aircraft for attack targets and presents an FLIR image of the ground, radar-measured altitude, and laser range finding and target designation. (An acronym for low-altitude navigation and targeting infrared system for night.)

lanugo *Anatomy.* the covering of fine, soft hair on the body of the fetus. Also, DOWN.

lanyard *Naval Architecture.* **1.** a line running between the deadeyes at the foot of a ship's shroud; the lanyard is tightened to set up the shroud. **2.** any strong, relatively short cord.

lap *Engineering.* any part or member that overlaps another to provide reinforcement. *Metallurgy.* in metal hot-working, an undesirable defect caused by the folding of a small portion of the material without welding. *Materials Science.* a rotating wheel or disk having a fine abrasive material on its surface, used for polishing gems, metals, glass, and other materials. *Military Science.* in naval mine warfare, the section or strip of an area assigned to a single minesweeper or formation of minesweepers for a run through the area.

LAP leukocyte alkaline phosphatase; lyophilized anterior pituitary (tissue).

lapachol *Biochemistry.* $C_{15}H_{14}O_3$, a derivative of naphthoquinone that naturally occurs in members of the genus *Bignonaceae,* a hardwood timber native to Argentina and Paraguay. Also, **lapachoic acid.**

laparo- or **lapar-** a combining form meaning "abdominal wall," as in *laparoscopy, laparectomy.* (From a Greek word meaning "flank.")

laparocolostomy *Surgery.* the creation of a permanent opening into the colon by an incision in the wall of the abdomen.

laparocolotomy *Surgery.* an incision of the colon through the abdominal wall to remove a foreign object or benign growth. Also, COLOTOMY.

laparocystectomy *Surgery.* the removal of a cyst by an abdominal incision.

laparocystotomy *Surgery.* **1.** the removal of the contents of a cyst through an incision in the abdominal wall. **2.** see SUPRAPUBIC CYSTOTOMY.

laparogastrostomy *Surgery.* the creation of a permanent gastric bistula through the abdominal wall.

laparogastrotomy *Surgery.* an incision into the stomach through the abdominal wall.

laparohepatotomy *Surgery.* an incision of the liver through the abdominal wall.

laparohysterectomy *Surgery.* the removal of the uterus through an incision in the abdominal wall.

laparorrhaphy *Surgery.* **1.** a suture or repair of the abdominal wall. **2.** see CELIORRHAPHY.

laparoscope *Surgery.* an instrument comparable to an endoscope that is used to examine the peritoneal cavity. Also, PERITONEOSCOPE.

laparoscopy *Medicine.* a visual examination of the contents of the abdominal cavity by use of an endoscope. Also, CELIOSCOPY.

laparotome *Surgery.* a knife used in laparotomy.

laparotomy *Surgery.* a surgical incision through a part of the abdominal wall. Also, CELIOTOMY.

lap course *Military Science.* in naval mine warfare, the true course to be cleared during a run along a lap.

lap dissolve *Telecommunications.* a transition from one scene to another in a motion-picture film or videotape in which one scene fades down and out while the next scene simultaneously fades up and in.

lapel microphone *Acoustical Engineering.* a small microphone positioned close to a person's mouth, usually attached to a piece of clothing such as a jacket lapel; used in television interviews, theater performances, and other such situations in which a larger, more visible microphone is undesirable.

lapidary *Science.* **1.** the art of cutting, polishing, and engraving precious stones. **2.** of or relating to this art. **3.** a person who is skilled in this art. (From the Latin word for "stone.")

lapidicolous *Ecology.* describing an organism that lives under or among stones. Thus, **lapidicole.**

lapies see KARREN.

lapilli *Volcanology.* small fragments of pyroclastic material, ranging from 2 to 64 mm in diameter, that are blown out from an erupting volcano.

lapilli tuff *Geology.* a hardened deposit composed of fragments of pyroclastic material in a matrix of fine-grained tuff.

lapinized vaccine *Immunology.* a vaccine that is produced by passing live microorganisms through a rabbit until they lose their virulence for their original host. (From *lapin,* the French word for rabbit.)

lapinized virus *Virology.* a virus adapted to growth in rabbits.

lapis lazuli *Petrology.* a blue, especially azure-blue, semitranslucent to opaque crystalline rock consisting primarily of lazurite and calcite, with lesser amounts of hauyne, sodalite, pyrite inclusions, and other minerals; used as a semiprecious stone and thought to be the original sapphire. Also, LAZULI.

lap joint *Engineering.* a joint in which two parts or members overlap and are fastened together by plugs, bolts, rivets, or welding.

Laplace, Pierre Simon de [lä pläs′] 1749–1827, French astronomer and mathematician; applied Newton's theories to celestial mechanics; formulated nebular hypothesis on origin of solar system.

Pierre Simon de Laplace

Laplace azimuth *Cartography.* in surveying, a geodetic azimuth determined by use of the Laplace equation from a previously observed astronomic azimuth.

Laplace-Beltrami operator *Mathematics.* a canonical second-order partial differential operator constructed from the metric tensor of a given (pseudo-) Riemannian manifold. The spectrum of the Laplace-Beltrami operator contains information about the geometry of the manifold. In addition to scalar-valued functions, the Laplace-Beltrami operator acts on covariant and contravariant tensor fields; in particular, on differential forms. The operator is given by $\Delta = d\delta + \delta d$, where d is the covariant differential and δ is the divergence operator. In coordinates, $\Delta f = g^{ij} f_{,i,j}$, where the commas indicate covariant differentiation and the summation convention applies.

Laplace condition *Geodesy.* a condition of divergence between an astronomical azimuth and a geodetic azimuth resulting from the deflection of the vertical.

Laplace control *Geodesy.* the control and correction of divergence between the astronomical azimuth and the geodetic azimuth by application of the Laplace equation.

Laplace determinism *Mechanics.* see CAUSALITY.

Laplace irrotational motion *Fluid Mechanics.* the motion of a fluid that has zero circulation and is incompressible and nonviscous.

Laplace operator *Mathematics.* the Laplace-Beltrami operator on n-dimensional Euclidean space; usually written

$$\Delta = \nabla^2 = \sum_{i=1}^{n} \partial^2/\partial x_i^2,$$

where ∇ is the del operator. Also, LAPLACIAN.

Laplace's equation *Acoustics.* a relationship for the speed of sound in a gas, as given by $c = (\gamma p/\rho)$, where γ is the ratio of specific heats of the gas, p is the gas pressure, and ρ is the density. *Mathematics.* the elliptic partial differential equation for a scalar-valued function on Euclidean space that sets the sum of all the nonmixed second partial derivatives equal to zero.

Laplace's expansion *Mathematics.* a method expanding the determinant D of an $n \times n$ matrix as a sum of products of determinants of minors of the matrix and their complementary minors; useful for large values of n. For example, suppose the second and third columns of a 4×4 matrix are selected as the (fixed) basis for a Laplace expansion. Starting with the first and second rows, there are six 2×2 submatrices of the two selected columns. Each of these submatrices has a unique complementary matrix, namely the matrix formed from the elements of columns one and four (i.e., the columns not selected) and the rows not included in the submatrix. Form the determinants of these two matrices and multiply them together. There will be six such products in all. Attach a sign to each of these products according to the following rule: any upper left square submatrix is assigned a positive sign, and each row or column transposition from that reverses the sign. In this example, arriving at columns two and three from one and two requires two transpositions, so the sign is positive for the product of the submatrix from rows one and two (columns two and three) and its complementary matrix. For the submatrix from rows one and three, the sign is negative, etc. The determinant of the 4×4 matrix is the signed sum of the six products of the 2×2 matrices.

Laplace station *Cartography.* a survey station at which the Laplace azimuth is determined.

Laplace transform *Mathematics.* an integral transformation that is related to the Fourier transform and that has similar properties. Let $f(t)$ be a complex-valued function of a nonnegative real variable t that is integrable over $(0,\infty)$ and such that $|f(t)| \leq Ke^{ct}$ for some constants K and c (i.e., f is exponentially bounded). Then the Laplace transform of $f(t)$ is defined to be

$$L\,\{f(x)\}(s) = \int_0^\infty e^{-st} f(t)dt.$$

The Laplace transform is especially useful in solving ordinary differential equations because it converts derivatives with respect to t into algebraic expressions in s.

Laplacian see LAPLACE OPERATOR.

Laplacian speed of sound *Fluid Mechanics.* the phase speed of sound in a fluid medium, in which it is assumed that compressions and rarefactions in the fluid occur with no gain or loss of heat.

Laporte selection rule *Atomic Physics.* a rule stating that the transition of an electric dipole can take place only between states of opposite parity.

lappet *Invertebrate Zoology.* **1.** any of several small flaps on the edge of a jellyfish. **2.** an earlike process on the head of planarian worms. *Zoology.* a wattle, as seen on domestic fowl.

lapping *Metallurgy.* in metal finishing, abrading a surface by rubbing it with a tool containing a fine abrasive. *Electronics.* **1.** a process by which quartz crystal plates are passed over a flat plate covered with a liquid abrasive to raise them to their final frequency. **2.** a process by which products, such as semiconductor blanks, are ground and polished so that they are an exact thickness or smoothness.

lapse line *Meteorology.* a graphical representation on a chart indicating the relationship between temperature and altitude in the atmosphere.

lapse rate *Meteorology.* the rate of decrease of an atmospheric variable, usually temperature, with increasing altitude.

lap shear test *Materials Science.* a test of the shear strength of a joint, in which the bond is tested to failure.

lap siding *Building Engineering.* overlapping beveled boards used in construction as clapboard siding or roof and wall shingles.

lapstrake *Naval Architecture.* planking laid so that each plank's lower edge overlaps the upper edge of the plank below.

lapsus calami *Psychology.* a slip of the pen; a seemingly inadvertent error in writing that reveals some unconscious wish or thought.

lapsus linguae [lap′sis ling′gwī; lap′sis ling′gwə] *Psychology.* a slip of the tongue; a seemingly inadvertent error in speech that reveals some unconscious wish or thought.

laptop computer *Computer Technology.* a small, portable personal computer with an internal power supply, usually weighing less than ten pounds.

lap track *Military Science.* in naval mine warfare, the center line of a lap; ideally, it is the track to be followed by the sweep or detecting gear.

lap weld *Metallurgy.* a weld made between two overlapping parts.

lap width *Military Science.* in naval mine warfare, the swept path of the ship or formation divided by the percentage coverage being swept.

lap winding *Electricity.* in a commutator AC machine, a winding that completes all turns under a given pair of poles before moving to the next pair.

lapwing *Vertebrate Zoology.* an Old World wading bird of the family Charadriidae, characterized by a markedly slow beat of the wings, dark, glossy plumage, and distinctive white markings.

Laramide orogeny *Geology.* a Late Cretaceous and Early Tertiary series of crustal deformations in western North America that affected the eastern Rocky Mountains. Also, **Laramian orogeny, Laramide orogeny.**

Laray viscometer *Graphic Arts.* an apparatus used to determine the viscosity of ink.

larboard [lär´bərd; lär´bôrd´] *Navigation* a former expression designating the left side of a vessel, now usually called *port*, which is preferred to avoid confusion with "starboard."

larch *Botany.* the common name for any deciduous, coniferous tree of the genus *Larix*, characterized by short, linear needles arranged spirally on terminal shoots.

larch canker *Plant Pathology.* a very destructive disease of larch trees and sometimes pine and fir trees, caused by the fungus *Dasyscypha wilkommi* and characterized by bowl-like cankers on the tree limbs.

lard *Materials.* a preparation of the purified internal fat of a hog; a soft, white mass with a faint odor and a bland taste; soluble in ether and insoluble in water; melts at 36–42°C; used in cooking and in ointments and perfumes.

larderellite *Mineralogy.* $(NH_4)B_5O_6(OH)_4$, a white monoclinic mineral occurring as a crystalline powder, having a specific gravity of 1.905 and an undetermined hardness; found with amonioborite in boric acid lagoons in Tuscany, Italy.

lardite see AGALMATOLITE.

Lardizabalaceae *Botany.* a family of dicotyledonous twining woody vines and sometimes arborescent shrubs of the order Ranunculales, characterized by alternate compound leaves, small trimerous flowers borne in racemes, and a berry or fleshy follicle fruit; native to Southeast Asia and Chile.

lard oil *Materials.* a colorless or yellowish liquid with a distinctive odor that is derived by cold-pressing lard; composed of olein with a small amount of the glycerides of fatty acids; used as a lubricating agent and in soaps.

large deviation theory *Statistics.* the study of the long-run decay rate of the probabilities of certain rare events.

large intestine *Anatomy.* the broader and shorter portion of the intestine, including the cecum, the ascending, transverse, descending, and sigmoid colons, and the rectum.

large ion *Meteorology.* an ion characterized by a large mass and low mobility, resulting from the attachment of an Aitken nucleus to a small ion. Also, HEAVY ION, SLOW ION, LANGEVIN ION.

Large Magellanic Cloud *Astronomy.* an irregular galaxy with a mass of about 10^{10} suns that orbits the Milky Way Galaxy at a distance of 160,000 light-years.

large nuclei *Oceanography.* particles with a radius greater than 10 cm of concentrated seawater or crystalline salt in the air above the water.

large number hypothesis *Physics.* a hypothesis stating that the approximate equality of two numbers on the order of 10^{40} has a physical basis: the ratio of the electrostatic to the gravitational force between a proton and an electron, and the ratio of the age of the universe to the time required for light to cross an elementary-particle diameter.

large polaron *Solid-State Physics.* a defect produced by the interaction of an electron in a semiconductor or crystalline insulator with ions constituting the lattice in its immediate neighborhood. The electron and the lattice deformation around it move as one unit in response to the imposition of an external electric field.

large scale *Meteorology.* a term for a scale, such as the scale of the long-wave pattern of the higher troposphere, in which the curvature of the earth is taken into account.

large-scale convection *Meteorology.* a pattern of organized vertical motion in the atmosphere, often associated with major weather systems such as hurricanes or migratory cyclones.

large-scale integration *Electronics.* the construction of over 100 different devices mounted onto a single substrate, in order to carry out an extensive range of operations.

large-scale map *Cartography.* any map or chart with a scale of 1:75,000 or larger.

large-systems control theory *Control Systems.* a branch of control-systems theory that deals with special problems inherent in the design of control algorithms for complex systems.

Largidae *Invertebrate Zoology.* a family of brightly colored, plant-feeding true bugs, hemipteran insects in the superfamily Pentatomorpha.

Laridae *Vertebrate Zoology.* the gulls and terns, a worldwide family of long-winged water birds of the order Charadriiformes.

larixine see MALTOL.

larixinic acid see MALTOL.

lark *Vertebrate Zoology.* any of numerous, usually small songbirds of the family Alaudidae, characterized by a crested head and long, pointed wings; the skylark is the best known species; found worldwide.

horned lark

Larmor, Sir Joseph 1857–1942, British mathematical physicist; developed the Larmor formula and Larmor procession.

Larmor formula *Electromagnetism.* a formula that gives the rate at which energy is radiated by a nonrelativistic, accelerated charged particle: $F(a) = 2q^2a^2/3c^3$, where q is the charge in electrostatic units, a is the acceleration, and c is the speed of light.

Larmor frequency *Electromagnetism.* the angular frequency of Larmor precession, given by $-qB/2mc$, where q is the charge in electrostatic units, B is the magnetic induction, m is the mass of the particle, and c is the speed of light.

Larmor orbit *Electromagnetism.* the motion of a charged particle in a uniform magnetic field that causes the particle to travel in a circular path in a plane perpendicular to the magnetic field.

Larmor precession *Electromagnetism.* a rotation superimposed on the motion of a system of particles having equal charge and mass when in a magnetic field.

Larmor radius *Electromagnetism.* the radius of the Larmor orbit of a charged particle.

Larmor's theorem *Electromagnetism.* for a system of charged particles in a central field of force, if the particles have the same charge-to-mass ratio, their motion in a uniform magnetic field will be the same (to the first order of B) as that in the absence of a magnetic field except for the angular (Larmor) precession.

larnite *Mineralogy.* β-Ca_2SiO_4, a gray monoclinic mineral occurring as crystals and granular masses, having a specific gravity of 3.28 and a hardness of about 6 on the Mohs scale; found in contact zones of metamorphosed limestone. Also, BELITE.

larr linear accelerator-regenerator reactor.

larsenite *Mineralogy.* $PbZnSiO_4$, a colorless or white orthorhombic mineral occurring in thin prismatic crystals, having a specific gravity of 5.9 and a hardness of 3 on the Mohs scale; found in lead-zinc deposits.

Larson-Miller parameter *Materials Science.* one of several parameters used to relate stress, temperature, and rupture time in creep.

larva [lär´və] *plural,* **larvae** [lär´vē] *Invertebrate Zoology.* an independent, often free-living, developmental stage that undergoes changes in form and size to mature into the adult; especially common in insects and aquatic organisms. (From a Latin word meaning "ghost" or "mask.")

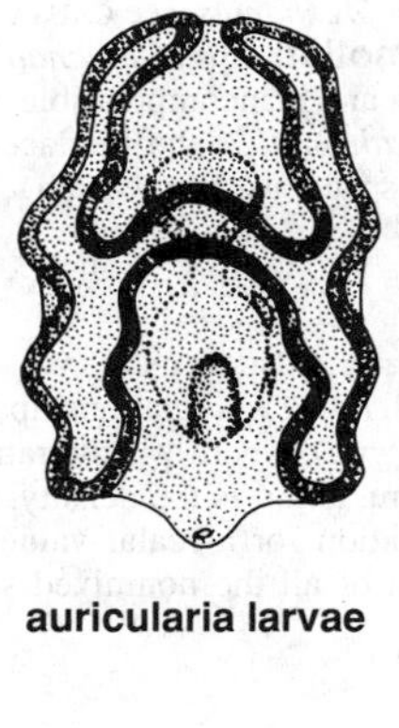

auricularia larvae

Larvacea *Invertebrate Zoology.* a class of planktonic tunicates that retain a tail, a dorsal nerve cord, and a notochord throughout life.
larvae the plural of larva. See LARVA.
Larvaevoridae see TACHINIDAE.
larval *Invertebrate Zoology.* relating to or in the form of larvae.
larva migrans *Invertebrate Zoology.* a parasitic larva of a fly that produces a creeping eruption in the dermis of its host. *Medicine.* see CREEPING ERUPTION.
larvarium *Entomology.* a shelter woven by insect larvae, consisting of leaves, twigs, pine needles, and dirt dampened with saliva; or less commonly, a tube or hammock of spun silk fiber.
larvate *Medicine.* describing a disease or a disease symptom that is concealed or hidden.
larvicide *Materials.* any chemical agent that is used to destroy larvae, especially harmful insect larvae, such as copper acetoarsenite.
larvikite or **laurvikite** *Petrology.* an alkali syenite with abundant phenocrysts of two feldspars, often intimately intergrown and composing 90% of the rock; it features a fine blue color and is used as an ornamental building material. Also, BLUE GRANITE.
larviporous *Invertebrate Zoology.* of some insects and mollusks, producing larvae rather than eggs.
larvivorous *Invertebrate Zoology.* feeding exclusively or primarily on larvae.
laryng- a combining form meaning "larynx," as in *laryngitis.*
laryngalgia *Medicine.* any pain in the larynx.
laryngeal [lə rinj′ē əl] *Anatomy.* of or related to the larynx. *Medicine.* affecting or involving the larynx. Thus, **laryngeal cancer.**
laryngeal pouch *Vertebrate Zoology.* a lateral, saclike expansion of the larynx cavity that is greatly developed in certain monkeys.
laryngectomy [lâr′in jek′tə mē] *Medicine.* the surgical removal of the larynx, as in the treatment of laryngeal cancer.
larynges the plural of larynx. See LARYNX.
laryngismus or **laryngismus stridulus** *Medicine.* a spasmodic contraction of the larynx, lasting a few seconds, followed by a crowing noise upon inspiration; a disease of children. Also, CROWING CONVULSIONS, KOPP'S ASTHMA, PSEUDO CROUP.
laryngitic *Medicine.* of or relating to laryngitis.
laryngitis [lâr′in jī′tis] *Medicine.* an inflammation of the larynx, commonly accompanied by swelling of the vocal cords and hoarseness or loss of voice. It may be caused by a cold, by the use of tobacco or alcohol, or by the breathing of irritating fumes.
laryngo- a combining form meaning "larynx," as in *laryngospasm.*
laryngography *Radiology.* a radiographic study of the larynx after introduction of a contrast medium.
laryngology *Medicine.* the branch of medical science that deals with the study of the larynx, its diseases and treatments.
laryngopathy *Medicine.* any disease of the larynx.
laryngopharyngeal *Anatomy.* of or related to the voice box and the throat.
laryngopharynx *Anatomy.* the lower part of the pharynx that lies above the epiglottis and opens into the larynx and esophagus.
laryngophone *Acoustical Engineering.* a throat-contact microphone designed to pick up the voice directly, so as to minimize extraneous noise.
laryngoplasty *Surgery.* plastic surgery of the larynx.
laryngoplegia *Medicine.* a paralysis of the larynx.
laryngoscope *Medicine.* a tubular instrument with illuminating and telescopic devices for direct examination of the larynx.
laryngoscopy *Medicine.* an examination of the larynx by means of a laryngoscope.
laryngospasm *Medicine.* a spasmodic narrowing or closure of the glottis; laryngismus stridulus.
laryngostasis see CROUP.
laryngostomy *Surgery.* the surgical creation of an artificial opening into the larynx.
laryngotracheal *Anatomy.* of or relating to the larynx and the trachea.
laryngotracheal groove *Developmental Biology.* a depression in the canal at the ventral wall of the embryonic pharynx that develops into the respiratory tract.
laryngotracheitis *Medicine.* an inflammation of the larynx and the trachea.
laryngotracheobronchitis *Medicine.* an acute inflammation of the upper respiratory tract involving the larynx, trachea, and bronchi, occurring mostly in young children and marked by high fever, toxemia, a hoarse, barking cough, and difficulty in breathing.

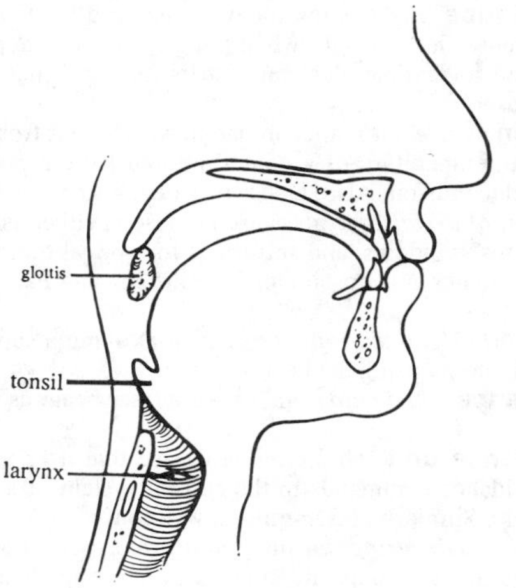

larynx

larynx [lâr′inks] *plural,* **larynges** [lâr′in jēz′] *Anatomy.* the expanded upper part of the trachea that contains the vocal cords and produces vocal sounds.
LAS *Aviation.* the airport code for Las Vegas, Nevada.
LASCR light-activated silicon controlled rectifier.
LASCS light-activated silicon controlled switch.
laser [lā′zər] *Physics.* a device that emits a high-intensity, narrow-spectral-width, highly directional or near-zero-divergence beam of light by stimulating electronic, ionic, or molecular transitions to higher energy levels and allowing them to fall to lower energy levels. Lasers are capable of producing intense light and heat when focused at close range, and they have many applications in industry, medicine, telecommunications, scientific research, and military operations. (An acronym for light amplification by stimulated emission of radiation.)
laser altimeter *Navigation.* an altimeter that determines altitude above ground level by reflecting a laser beam off the ground.
laser amplifier *Electronics.* an amplifier whose operation is based on light amplification by stimulated radiation.
laser anemometer *Engineering.* an anemometer that projects two perpendicular laser beams and calculates wind velocity on the basis of consequential changes in the velocity of one or both beams.
laser beam *Optics.* the bright stream of light created by a laser; widely used in surgery, communications, weapons systems, printing, and various industrial processes.
laser-beam printer *Graphic Arts.* a high-speed phototypesetting machine having a laser printer as its output device.
laser camera *Optics* a camera in which a laser beam is divided into two separate beams; one scans the ground and the other reflects images back from the ground; used in airplanes for night photography.
laser ceilometer *Engineering.* an instrument that measures the time it takes for a light pulse from a ground laser to travel straight to a cloud ceiling and reflect back to a receiving photomultiplier; the time is converted into a cathode-ray display which gives cloud-base height.
laser cooling *Atomic Physics.* a method to slow atoms in an atomic beam to very low velocities.
laser designator *Ordnance.* a device that emits a laser beam to mark a specific place or object; often used in conjunction with a laser seeker on a homing missile as part of a laser guidance system.
laser disk see OPTICAL DISK.
Laser Doppler anemometer *Biotechnology.* an optical device that measures fluid movement based on the scattering of laser light by small particles in the fluid.
laser Doppler velocimeter *Optics.* an instrument that determines the velocity of an electromagnetic wave by measuring the frequency of the wave produced by the Doppler effect.
laser drill *Optics.* a device in which a beam of light from a ruby laser burns tiny holes, no more than 0.00025 cm, into hard materials such as tungsten and gemstone.
laser earthquake alarm *Engineering.* a proposed earthquake early warning system that incorporates two lasers with beams at right angles placed over a known geologic fault, providing a constant surveillance of the distance across the fault.

laser flash tube *Electronics.* a device that is filled with xenon to provide an intense flash of light when a high voltage is applied to its electrodes; used to reverse electron distribution and increase the energy level in a laser.
laser fusion *Nucleonics.* a technique in which light from a laser is used to raise the temperature of a deuterium and tritium pellet to 10^9°C in order to induce fusion. Also, LASER-INDUCED FUSION.
laser glass *Materials.* the glass used in lasers, either as a passive material for lenses, windows, and substrates for optical thin film coatings or as an active material for oscillators, amplifiers, and Faraday rotators and isolators.
laser glazing *Metallurgy.* the process of alloying a superficial layer of a metallic material using the heat of a laser.
laser guidance *Navigation.* the use of a laser beam as a guide to a target.
laser guidance unit *Ordnance.* a device that uses a laser seeker to provide guidance commands to the control system of a missile, projectile, or bomb. Similarly, **laser-guided weapon.**
laser gyro *Engineering.* an integrating rate gyroscope in which two laser beams move in opposite courses over a ring-shaped path created by three or more mirrors, thus measuring rotation without the aid of a spinning mass.
laser hardening *Materials Science.* a process using a laser for heat treating material surfaces; used to produce a hard martensite layer that increases the wear resistance of steel parts.
laser heterodyne spectroscopy *Spectroscopy.* a spectroscopic technique used to obtain high resolution in astronomical and atmospheric observations by mixing a sample signal with a laser signal to produce a signal in the radio-frequency range.
laser-holography storage *Computer Technology.* a variety of laser memory in which data are stored in a holographic substrate by a laser beam; the data are read by a low-energy laser beam.
laser illuminator *Ordnance.* a device for enhancing the illumination in a zone of action by irradiating it with a laser beam.
laser-induced breakdown *Materials Science.* damage that causes any permanent change in the optical or mechanical properties of optical components used in laser devices, including such changes as localized melting, fractures, coating delamination, and increases in optical scattering and absorption.
laser-induced fusion see LASER FUSION.
laser-induced nuclear polarization *Nuclear Physics.* an optical pumping process that uses a polarized laser light to force the spin vectors of an ensemble of nuclei to point in one direction.
laser interferometer *Optics.* an interferometer that utilizes a laser's monochromaticity or its long coherence length to study light.
laser intrusion detector *Engineering.* a photoelectric intrusion detector in which a laser produces an extremely thin and almost invisible beam of light to encircle the guarded area.
laser jamming *Military Science.* an electronic countermeasure employing a continuous-wave laser to jam an enemy laser receiver, thus preventing it from interfering with laser rangefinders, radars, and tracking equipment.
laser machining *Materials Science.* the use of a laser beam as a source of directed energy to cut and shape material by localized heating; applied in a wide range of hole-drilling, through-cutting, marking, scribing, micromachining, and three-dimensional shaping operations.
laser memory *Computer Technology.* a computer memory in which a laser beam is used to record and subsequently read data; enables the storage of large quantities of data in a very small volume.
laser photocoagulator *Medicine.* a device that produces a controlled and intense beam of light; used in surgical procedures for the coagulation of tissue, as in ophthalmology.
laser printer *Graphic Arts.* a high-speed printing device that projects laser beams onto an image area, yielding an electrostatic charge that attracts a metallic pigmented powder; usually used as an output device for a word processor.
laser radiation detector *Electronics.* a device that senses the coherent light in a laser beam.
laser rangefinder *Optics.* an instrument that measures the distance of far-off objects by calculating the time required for a flash of laser light to travel from a source to the object, and back to the source.
laser ranging *Ordnance.* a method of calculating the distance to a target by means of a laser rangefinder.
laser recorder *Telecommunications.* an instrument that reproduces an image by use of a modulated laser beam.
laserscope *Engineering.* **1.** a surgical instrument in which a projecting laser beam is used to open channels or to destroy diseased tissue. **2.** a high-power pulsed laser used with scanning and imaging instruments to discern objects and project three-dimensional images of them on a viewing screen.
laser seeker *Ordnance.* a device using a direction-sensitive receiver that detects the energy reflected from a laser-designated target and defines the direction of the target relative to the receiver.
laser seismometer *Engineering.* a laser interferometer system that senses the earth's seismic strains by measuring changes in the distance between two granite piers placed at opposite ends of an empty pipe; a laser beam, such as helium-neon, makes a round trip through the pipe.
laser-solid interaction *Solid-State Physics.* the interaction between a concentrated coherent beam of light and a solid, usually resulting in thermal effects.
laser spectroscopy *Spectroscopy.* the use of a laser as an intense, monochromatic source of light; commonly used in Raman (vibrational) spectroscopy.
laser spectrum *Physics.* a spectrum indicating all wavelengths of coherent radiation which various types of lasers are capable of producing, ranging from infrared to ultraviolet wavelengths.
laser storage *Computer Technology.* an auxiliary storage device that uses a laser beam to record data onto a metallic surface.
laser target designating system *Ordnance.* a system, consisting of a laser designator and display and control components, used to direct laser energy to a target in order to mark the target for a laser seeker or other seeker.
laser terrain profile recorder *Cartography.* an electronic surveying instrument that emits a continuous-wave laser beam to measure the distance between an airplane and the ground; used in aerial surveys to produce a printed profile of the terrain below the plane.
laser threshold *Electronics.* the least amount of pumping energy needed to trigger a laser beam.
laser tracker *Ordnance.* a device that locks on to the reflected energy from a laser-designated target and defines the direction of the target relative to itself.
laser tracking *Engineering.* a tracking technique in which echoed coherent light gives the range and direction of a target.
laser transit *Engineering.* a transit in which a laser is placed above the sighting telescope to focus a thin visible beam onto a little target at the survey site.
laser-triggered switch *Electricity.* a high-voltage, high-power switch in which a spark gap is triggered into conduction by a laser beam.
laser velocimeter *Optics.* an instrument that measures velocity by aiming a steady stream of coherent light generated by a laser at objects traveling at right angles to it. Also, DIFFRACTION VELOCIMETER, OPTICAL DIFFRACTION VELOCIMETER.
laser welding *Materials Science.* a fusion-welding process in which a laser beam is focused through a lens onto the area to be welded and provides the heat to melt the components.
lasing *Optics.* the amplification of light by the stimulated emission of radiation.
Lasiocampidae *Invertebrate Zoology.* the lappet moths and tent caterpillars; a family of lepidopteran insects in the suborder Heteroneura.
Lasky's law *Materials Science.* a rule for determining the depletion of mineral resources stating that as the cumulative tonnage of mineralized rock grows at a constant geometric rate, the average grade of the accumulated tonnage decreases at an arithmetical rate.
Lass process *Materials Science.* a process for forming and drying molded fiber products, in which large split-screened felting dies are dipped in a fiber slurry and the slurry is drawn in to accrete the part. The dies are withdrawn and heated air is pulled through the aperture to dry the molding, and the dies are separated to obtain the part.
last-in, first-out *Industrial Engineering.* an inventory management method in which the items most recently received are the first to be used; used where deterioration of stored inventory is not a concern. *Computer Programming.* see FILO.
last-in, last-out *Computer Programming.* see FIFO.
last-mask read-only memory *Computer Technology.* a read-only memory in which the final mask used in the fabrication process writes the data pattern that will be read when the memory is accessed.
last quarter moon *Astronomy.* the lunar phase that occurs when the moon has completed three-fourths of its monthly orbit of the earth, about a week after the full moon.
lat. latitude.

latch *Mechanical Devices.* any of various door-closing devices that fit into a cavity, notch, or hook in the doorframe. *Electronics.* a bistable circuit whose state is changed by applying a signal to its input, remaining unchanged even after the input is removed.
latch bolt *Mechanical Devices.* a spring-action bolt with a beveled head.
latch-in relay *Electricity.* a relay that keeps its contacts in the last position assumed without requiring coil energization. Also, LOCKING RELAY.
latch-up *Electronics.* **1.** the failure of the collector voltage in a circuit to return to the supply voltage when a transistor is switched from saturation to cutoff. **2.** the failure of a stable circuit mode to return to a previous mode after a stimulus, such as radiation, is removed. **3.** the characteristic of certain amplifiers to remain in their saturation mode after the maximum voltage has been exceeded. **4.** the switching that occurs when an improper voltage is applied in an electrical circuit.
late *Science.* belonging to the more advanced or recent phase of some development or process, such as a geological period or a cultural era.
late blight *Plant Pathology.* any of a variety of fungal plant diseases in which symptoms, such as decay and leaf spot, do not occur until late in the growing season.
late gene *Virology.* a gene in a viral nucleic acid that is expressed late in the virus replication cycle.
latency the fact of being latent; specific uses include: *Medicine.* the status of a disease or condition when there is a seeming lack of effect or activity. *Physiology.* the time interval between the occurrence of a stimulus and the occurrence of a response to that stimulus. *Psychology.* see LATENCY PERIOD. *Computer Technology.* **1.** the time delay between the instant a request is made for an item of data and the instant the actual transfer of data starts. **2.** the time required for a disk or drum to rotate to the position at which the desired sector is under the read/write heads.
latency period *Medicine.* the period of a disease or condition after exposure has occurred but before there has been any evident repsonse. *Psychology.* in psychoanalytic theory, the fourth stage of psychosexual development, from infancy to adolescence, when sexual urges are repressed or dormant. Also, **latency stage, latency phase**. *Robotics.* the time elapsed between the giving of a command and the action taking place as commanded.
latent *Science.* present but not visible or active; dormant. (From a Latin word meaning "hidden.")
latent bud *Botany.* a bud, especially an axillary bud, that remains dormant until conditions are favorable to its development.
latent content *Psychology.* the hidden meaning in dreams or in the content, given by the person in psychoanalysis.
latent heat *Thermodynamics.* the amount of thermal energy that is absorbed or released by a unit amount (usually one mole) of a substance in the process of a phase change under conditions of constant pressure and temperature.
latent heat of fusion see HEAT OF FUSION.
latent heat of sublimation see HEAT OF SUBLIMATION.
latent heat of vaporization see HEAT OF VAPORIZATION.
latent homosexuality *Psychology.* a condition in which an individual's homosexual tendencies remain dormant or unacknowledged.
latent image *Graphic Arts.* an invisible image produced on a photosensitive emulsion by exposure; made visible by development.
latent infection *Virology.* any viral infection that does not produce evident symptoms but may manifest itself at a later time.
latent instability *Meteorology.* a type of conditional instability that occurs if a rising parcel of air reaches a critical atmospheric level at which the environment is warmer than the parcel.
latent learning *Behavior.* previously acquired information or behavior that is not revealed until the organism is later placed in a new situation.
latent load *Mechanical Engineering.* the extra load on an air conditioner required to cool moisture that condenses and must be removed from the conditioned space.
latent magma *Volcanology.* a highly viscous, nearly solid magma that is found beneath the earth's crust; densely packed under high pressure during the latter stages of volcanic activity, it may flow again when pressure decreases.
latent period *Medicine.* see INCUBATION PERIOD, def. 2.
latent structure analysis *Statistics.* a study of data based on the assumption that the observed characters are indicators of some underlying structure.
latent travel demand *Transportation Engineering.* the projected number of trips that would be generated if travel were more convenient, less expensive, or otherwise improved.
latent virus see OCCULT VIRUS.
latent-virus infection *Medicine.* an infection resulting from a virus that leaves its DNA within a host cell, resulting in molecules capable of subsequent production of viral proteins which destroy the cell.
laterad *Anatomy.* toward the side.
lateral *Science.* of, relating to, situated on, or directed toward a side. *Anatomy.* **1.** denoting a position that is relatively farther from the median plane or midline of the body or of a structure. **2.** of or related to a side. *Engineering.* a pipe branching out from the primary part of the system in a gas distribution or transmission network. *Mining Engineering.* **1.** a hard heading in horizontal mining that branches off a horizon along the strike of the seams. **2.** a horizontal mine working.
lateral area *Mathematics.* in traditional solid geometry, the area of a surface, excluding any bases.
lateral axis *Naval Architecture.* the horizontal line through a vessel's center of gravity. *Aviation.* the side-to-side axis of a flight vehicle, about which it rotates when pitching.
lateral body *Virology.* a part of the structure of vertebrate poxvirus particles whose function is unknown.
lateral bud *Botany.* a bud that develops on the side of a stem.

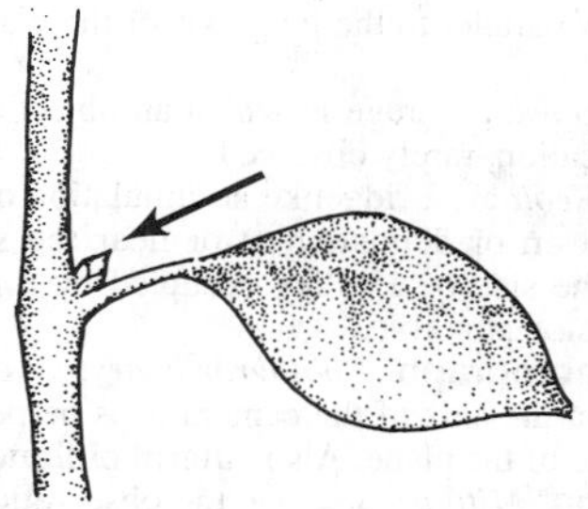

lateral bud

lateral compliance *Engineering.* the feature of a stylus founded on the requisite force to move it from side to side in following the grooves of a phonograph record.
lateral control *Aviation.* **1.** control over the lateral movement or attitude of a flight vehicle. **2.** also, **lateral controls.** the control surfaces, such as ailerons, used to maintain lateral control.
lateral decubitus *Surgery.* the position of an individual lying on the side. *Radiology.* a radiologic examination performed while the patient lies on one side, with the X-ray beam directed horizontally and the film directed vertically.
lateral deflection angle *Ordnance.* in gunnery, the angular difference between the azimuth of the target upon firing and the azimuth of the point at which the gun must be aimed in order to hit the target. Similarly, **lateral deflection scale, lateral deflection setting.**
lateral deviation *Ordnance.* the horizontal distance between the point of impact or burst and the gun-target line.
lateral erosion *Geology.* the wearing away of the banks of a meandering stream in a horizontal direction as the stream swings from side to side and undercuts its banks.
lateral extensometer *Engineering.* an instrument that determines the stresses on a photoelastic plate by measuring the change in the plate's thickness, resulting from stress at different points.
lateral extrusion *Metallurgy.* extrusion in which the product exits sidewise.
lateral face *Mathematics.* in traditional solid geometry, any face of a prismatoid that is not a base.
lateral fault *Geology.* a fault along which strike separation has taken place. Also, STRIKE-SEPARATION FAULT.
lateral hermaphroditism *Medicine.* the presence of ovarian tissue on one side and testicular tissue on the other side in the same individual.
lateralia *Invertebrate Zoology.* paired sensory bristles on the lateral sides of the head of jaw worms of the phylum Gnathostomulida.
laterality *Behavior.* the tendency to use one hand or side in preference to the other. *Psychology.* the dominance of the right brain or left brain. *Neurology.* the preferential use of one side of the body, as in relying upon only the right or only the left hand, foot, ear, or eye.
lateral jump *Ordnance.* in gunnery, the horizontal angle between the plane of departure of the shell and the plane of fire.

lateral lemniscus *Anatomy.* a tract of neural fibers extending through the lateral portions of the medulla oblongata and pons.

lateral ligature *Surgery.* any substance such as a wire or thread, applied to a blood vessel to control but not stop the distal blood flow.

lateral line *Vertebrate Zoology.* a faint line that is visible on the lateral side of the bodies of cyclostomes, fish, and certain aquatic amphibians, which marks the lateral-line organs. *Invertebrate Zoology.* a line of cell bodies of circular muscle that extends the length of the bodies of certain annelids.

lateral-line organ *Vertebrate Zoology.* one of a series of small sense organs along the lateral sides of the body of cyclostomes, fish, and certain aquatic amphibians that is sensitive to water movements and the pressure of currents against the body.

lateral-line system *Vertebrate Zoology.* the system of sense organs and nerves along the lateral side of the bodies of cyclostomes, fish, and certain aquatic amphibians that apprises the animal of water movements and the pressure of currents against the body.

lateral magnification *Optics.* the ratio of image size to the size of the imaged object, as measured in a plane perpendicular to the optical axis and read as a negative when the image is upside down. Also, MAGNIFICATION.

lateral meristem *Botany.* dividing tissue, such as the cambium or cork cambium, that runs parallel to the long axis of the plant part in which it occurs.

lateral mirage *Optics.* a mirage in which an object appears off to one side of its exact location; rarely observed.

lateral moraine *Geology.* a ridgelike accumulation of fallen rock fragments transported on or deposited at or near the side margins of a glacier, or along the side of a valley occupied by a glacier. Also, SIDE MORAINE, VALLEY-SIDE MORAINE.

lateral-oblique photograph *Photogrammetry.* an oblique aerial photograph taken when the axis of the camera is as perpendicular as possible to the flight line of the plane. Also, **lateral oblique**.

lateral observation *Military Science.* the observation of gunfire from a point significantly to the right or left of the line of fire.

lateral parity check *Computer Technology.* an odd or even parity check across the width of a magnetic tape or other storage medium.

lateral planation *Geology.* the leveling of the land adjacent to a stream or a flood plain as a result of the lateral erosion of a stream against its banks.

lateral pointing correction *Ordnance.* the component of the lateral deflection angle that is not due to target movement; it may be due to wind, drift, or lateral adjustment correction.

lateral quadrupole *Electromagnetism.* an electric or magnetic quadrupole that produces a field of two equal and opposite dipoles lying next to each other and separated by a small distance that is perpendicular to the dipole directions.

lateral recording *Engineering.* a form of disk recording in which the cutting stylus removes a thread from the surface of a blank disk; the groove modulation is parallel to the surface of the recording medium, which causes the cutting stylus to move from side to side while recording.

lateral root *Botany.* a root branch originating from a main root.

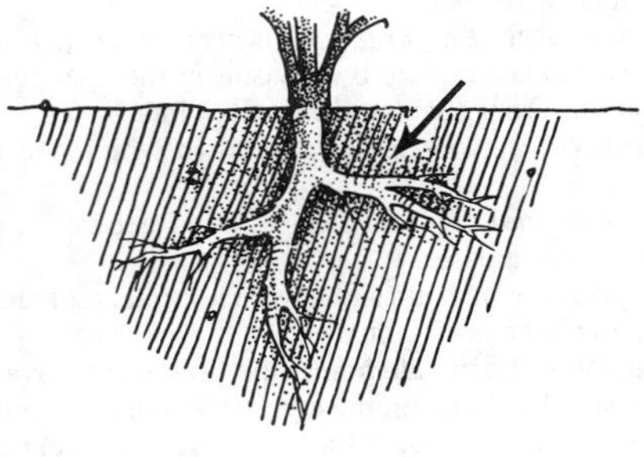

lateral root

lateral secretion *Geology.* a theory that explains the formation of certain ore deposits as a result of the leaching of adjacent wall rock.

lateral separation *Navigation.* the spacing of aircraft at the same altitude by requiring that the aircraft fly on different routes or in different locations.

lateral shear interferometer *Optics.* an interferometer in which a portion, usually half, of the power contained in a wavefront is divided from that wavefront and displaced laterally by a small amount, then recombined with the original beam to create an interference pattern.

lateral spread *Ordnance.* a technique used to place the mean point of impact of two or more units 100 meters apart on a line perpendicular to the gun-target line.

lateral storage see BANK STORAGE.

lateral support *Civil Engineering.* a horizontal buttress to a structure that provides greater resistance to lateral forces to prevent buckling.

lateral surface see CYLINDRICAL SURFACE.

lateral system *Navigation.* a buoyage system in which the characteristics of the buoys indicate the direction of danger relative to the course that should be followed.

lateral transmission see HORIZONTAL TRANSMISSION.

lateral transverse aberration *Optics.* the chromatic or spherical aberration of an image in a plane perpendicular to the optical axis.

lateral ventricle *Anatomy.* either of two paired horseshoe-shaped cavities in the cerebral hemispheres that contain cerebrospinal fluid.

laterite *Geology.* a reddish soil residue produced by weathering under temperate to tropical conditions and composed primarily of hydrated iron or aluminum oxides.

lateritic soil *Geology.* **1.** any soil that contains or is developed on laterite. **2.** any reddish soil developed in temperate to tropical regions as a result of weathering. Also, LATOSOL.

laterization *Geology.* the complex weathering process by which a soil or rock is converted into laterite through the removal of silica and bases and the increase of aluminum or iron oxides. Also, **lateritization**.

laterlog *Engineering.* a technique for measuring downhole resistivity, in which electric current is made to flow radially through the formation in a sheet of predesigned thickness; used for evaluating the resistivity in hard-rock reservoirs as a way to measure subterranean structural characteristics.

latero- a combining form meaning "side."

late-type star *Astrophysics.* a star of the spectral class K or M; the term dates from a time when astronomers thought that cool red stars like these were at a later stage of evolution than the hot blue stars of spectral types O and B (called *early-type*).

latewood *Forestry.* a term for the part of a wood growth ring formed late in the growing season (i.e., in summer). Latewood contains smaller, thicker-walled cells. Also, SUMMERWOOD.

latex *Materials.* **1.** a milky, generally white substance that is excreted by certain plants, such as the milkweed, poppy, dandelion, and guayule, and especially by rubber trees; used in making rubber, gutta-percha, balata, and similar elastic materials. Natural latex is highly unstable and is preserved by the addition of a small percentage of ammonia. **2.** a similar synthetic material, a colloidal suspension of fine particles of plastic or rubberlike material in water; used in paints, adhesives, and rubber products. **3.** relating to or containing this substance. Thus, **latex cement.**

LATEX *Computer Science.* a text-formatting program, consisting of macros for the TEX program.

latex agglutination test *Immunology.* a test used to determine the degree to which antibodies will aggregate or clump together, using polystyrene latex particles coated with antigen.

latex paint *Materials.* any of various paints containing water as the solvent and originally containing latex as the binder, but now usually made with other materials such as polyvinyl acetate or acrylic resins. Latex paints are widely used as interior household paints because of their quick-drying properties, lack of odor and flammability, and ease of application and cleaning.

latex vessel *Botany.* one of a group of elongated cells set end to end to form a duct in latex-producing plants. Also, **laticiferous vessel.**

lath *Civil Engineering.* a thin wooden strip of timber used to provide strength for plaster, trellis, or slats.

lathe *Mechanical Devices.* a machine that produces rotary work; the workpiece is rotated while cutting tools cut into its side, end, or interior.

lathe bed *Mechanical Devices.* the lathe portion that supports the headstock, tailstock, and carriage, consisting of a smooth-faced, rigid box girder with leg supports. Also, **lathe carrier**.

lathing hammer *Mechanical Devices.* a small face hammer with a long blade and a grooved, crosshatched head; used for trimming and nailing laths. Also, **lathing hatchet**.

Lathridiidae *Invertebrate Zoology.* a family of tiny coleopteran insects in the superfamily Cucujoidea, consisting of scavenger beetles that eat spores, molds, and stored grains.

lathyrism *Toxicology.* poisoning due to ingestion of peas of the genus *Lathyrus*. The toxic ingredient, β-aminopropionitrile, inhibits the enzyme lysyl oxidase. Symptoms may include spastic paraplegia, pain, hyperesthesia, and paresthesia.

lathyrogenic *Toxicology.* describing any poison capable of producing the symptoms of lathyrism.

laticifer *Botany.* a cell or a linked group of cells containing latex, found in various plant tissues.

laticiferous *Botany.* having or producing latex.

Latimeria *Vertebrate Zoology.* the genus name for the coelacanth of the family Latimeriidae and the order Crossopterygii.

Latimeriidae *Vertebrate Zoology.* a monospecific family of fish of the order Coelacanthiformes, containing the coelacanth or tassel-finned fish, a lobefin marine fish of South Africa and the Comoro Islands.

Latin rectangle *Mathematics.* an $r \times n$ arrangement (i.e., r rows and n columns, $r \leq n$) of the numbers 1, 2, 3, . . . , n in such a way that no number appears twice in the same row or in the same column. If $r = n$, the array is called a **Latin square**. It is always possible to append $n - r$ rows to any given $r \times n$ Latin rectangle to form an $n \times n$ Latin square.

latissimus dorsi *Anatomy.* the widest muscle of the body, extending from the spine to the humerus and adducting and rotating the arm.

latitude *Navigation.* the angular distance north or south of the equator, measured from 0° at the equator to 90° at the poles and labeled North or South. *Cartography.* **1.** the distance north or south of the equator on a sphere or spheroid, measured as either an angle or a linear distance. **2.** the angle measured at the center of a sphere between the plane of that sphere's equator and a line to a point on the sphere's surface. **3.** in surveying, the north-south component of a traverse course.

latitude effect *Geophysics.* a variation of a quantity with latitude, especially an increase in cosmic-ray intensity with increasing magnetic latitude.

latitude factor *Navigation.* the change in latitude along a celestial line of position for a given change in longitude.

latitude line *Navigation.* a line of position extending approximately east and west, normally the result of observing a body on or near the celestial meridian.

latitude method *Navigation.* a technique for determining latitude by a meridian altitude or an observation of Polaris.

latitude variation *Geophysics.* a periodic change in the latitude of a given location that is caused by wandering of the poles.

latitudinal *Science.* **1.** of or relating to breadth or latitude. **2.** running crosswise.

latitudinal band *Cartography.* any latitudinal strip, designated by accepted units of linear or angular measurement, that circumscribes the earth.

latosol see LATERITIC SOIL.

latrappite *Mineralogy.* $(Ca,Na)(Nb,Ti,Fe)O_3$, a black, opaque orthorhombic mineral of the perovskite group occurring in small cubic crystals, having a specific gravity of 4.4 and a hardness of 5.5 on the Mohs scale; found in carbonatite deposits.

Latrobe, Benjamin 1764–1820, English-born American architect; participated in the design and construction of the U.S. Capitol.

latrodectism *Toxicology.* poisoning by venom from a spider of the genus *Lathrodectus*, especially the black widow spider; symptoms may include muscular spasms, respiratory paralysis, and death.

LATS LONG-ACTING THYROID STIMULATOR.

latten *Metallurgy.* **1.** a brasslike alloy used for ornamental applications. **2.** tin plate.

lattice [lat′is] *Civil Engineering.* an open structure of intersecting diagonal laths, usually of wood or light metal. *Navigation.* a pattern formed by two or more sets of intersecting lines, such as overlapping loran lines. *Nucleonics.* in a nuclear reactor, an arrangement of components forming a regular geometrical pattern. *Materials Science.* a regular three-dimensional arrangement of points. *Crystallography.* see CRYSTAL LATTICE. *Mathematics.* a partially ordered set A in which each pair of elements has both a greatest lower bound and a least upper bound. If, in addition, every nonempty subset of A has both a greatest lower bound and a least upper bound, then A is said to be a **complete lattice.**

lattice beam see LATTICE TRUSS.

lattice constant *Crystallography.* one of the characteristic dimensions of the unit cell of crystal. Also, LATTICE PARAMETER.

lattice defect *Materials Science.* an imperfection in the regular three-dimensional arrangement of points in a crystal. Defects are generally vacant or interstitial sites; they scatter electrons, thus reducing the electrical conductivity of metals.

lattice drainage pattern see RECTANGULAR DRAINAGE PATTERN.

lattice dynamics *Solid-State Physics.* the study of vibrations in a crystal lattice due to thermal effects arising from temperature changes. Also, CRYSTAL DYNAMICS.

lattice energy *Solid-State Physics.* the energy associated with the construction of a crystal lattice relative to the energy of all constituent atoms separated by infinite distances.

lattice filter *Electronics.* a four-terminal filter in which each of the two input terminals is connected by a branch to each of the two output terminals.

lattice-gauge theory see GAUGE THEORY.

lattice girder *Civil Engineering.* a girder that has an intersecting diagonal framework connecting parallel flanges, rather than a solid plate web.

lattice hypothesis *Immunology.* a concept stating that in a precipitin test, antigens and antibodies having a valency of at least two may combine in varying portions to form giant insoluble aggregates (known as lattices).

lattice image *Materials Science.* an image produced from a single set of atomic planes by combining a transmitted beam and a diffracted beam in a transmission electron microscope.

lattice network *Electricity.* a network consisting of four branches connected in series to form a mesh, in which two nonadjacent junction points serve as input terminals and the remaining two serve as output terminals.

lattice parameter see LATTICE CONSTANT.

lattice points *Materials Science.* the points that constitute the three-dimensional array of a crystal. One or more atoms are associated with each lattice point, and the surroundings of each lattice point are identical.

lattice polarization *Solid-State Physics.* the electric polarization in a crystal lattice resulting from the displacement of ions from their equilibrium positions.

lattice reactor *Nucleonics.* a nuclear reactor in which the fissionable material used as fuel and the substance used to moderate the velocity of the neutrons are arranged in a uniform pattern of long rods, so that the neutron travels through two distinct media, rather than one.

lattice scattering *Solid-State Physics.* the electron scattering that occurs within a crystal due to vibrations in the lattice structure.

lattice translations *Materials Science.* in the unit cell of a material, the integral multiples of lattice constants along directions parallel to crystallographic axes.

lattice truss *Civil Engineering.* a diagonally intersecting support framework. Also, LATTICE BEAM.

lattice vibration *Solid-State Physics.* a regular vibration of atoms in a crystal lattice about their equilibrium positions.

lattice wave *Solid-State Physics.* a disturbance that propagates through a crystal lattice in which the displacement of one atom occurs in concert with those of its neighboring atoms.

lattice winding *Electricity.* a winding composed of three lattice coils, used for electric machines.

Lattorfian see TONGRIAN.

latus rectum *Mathematics.* **1.** a chord of a conic that passes through a focus point perpendicular to the axis of symmetry. **2.** the length of such a chord.

laubmannite *Mineralogy.* $Fe_3^{+2}Fe_6^{+3}(PO_4)_4(OH)_{12}$, a green or brown orthorhombic mineral occurring as fibrous botryoidal aggregates, having a specific gravity of 3.33 and a hardness of 3.5 to 4 on the Mohs scale.

laudanidine *Organic Chemistry.* $C_{20}H_{25}O_4N$, an opium alkaloid consisting of white crystals that are insoluble in water, soluble in alcohol, and slightly soluble in ether; melts at 182–185°C; used as an analgesic. Also, TRITOPINE.

laudanine *Organic Chemistry.* $C_{20}H_{25}O_4N$, an opium alkaloid consisting of orthorhombic crystals that melt at 166°C; a strong tetanic poison.

laudanosine *Organic Chemistry.* $C_{21}H_{27}O_4N$, one of the opium alkaloids consisting of toxic yellow crystals; insoluble in water; the levorotatory form melts at 89°C. Also, **laudanine methyl ether.**

Laue, Max von [lou′ə] 1879–1960, German physicist; awarded the Nobel Prize for his work in X-ray diffraction; contributed to the theory of relativity.

Laue camera *Crystallography.* an apparatus used to record the diffraction pattern of white (polychromatic) radiation produced by a stationary crystal.

Laue condition *Crystallography.* the satisfaction of the three Laue equations.

Laue equations *Crystallography.* equations showing that the path length differences of waves diffracted by atoms separated by one lattice translation must be an integral number of wavelengths for diffraction (i.e., reinforcement) to occur; further, that this condition must be true simultaneously in all three dimensions:

$$PD_1 = \lambda h_1 \lambda,\ PD_2 = h_2 \lambda,\ PD_3 = h_3 \lambda,$$

where λ is the wavelength of the radiation, *PD* represents the path difference, and *h* is the order of diffraction.

Laue method *Crystallography.* a method that demonstrates the diffraction symmetry of a crystal, whereby a diffraction photograph is produced by sending a beam of X-rays, with a wide range of wavelengths ("white" X-rays), along a principal axis of a stationary crystal.

laughing gas *Inorganic Chemistry.* a popular name for nitrous oxide, N_2O, especially when used in anesthesia. See NITROUS OXIDE. (So called because it has the effect of inducing spontaneous laughter or giggling if administered in less than an anesthetizing amount.)

Laugiidae *Paleontology.* a family of early Triassic sarcopterygian fishes in the suborder Coelacanthiformes, becoming extinct in the Jurassic; closely related to the family that includes the living coelacanth genus *Latimeria,* whose discovery in 1939 contradicted the belief that the lineage had died out in the Cretaceous.

Laumomier's ganglion see CAROTID GANGLION.

laumontite *Mineralogy.* $CaAl_2Si_4O_{12}{\cdot}4H_2O$, a white to pinkish or brownish monoclinic mineral of the zeolite group, having a specific gravity of 2.20 to 2.41 and a hardness of 3 to 4 on the Mohs scale; found in cavities of basic igneous rocks and in veins in schists and slates.

launch *Navigation.* to float a newly constructed or newly repaired ship. *Space Technology.* **1.** to send forth a rocket vehicle under its own power. **2.** an instance or result of this process. **3.** of or related to this process; used to form compounds such as **launch control, launch crew, launch point. 4.** to give an added boost to a space probe for flight into space just prior to separation from its launch vehicle. *Civil Engineering.* in the construction of a bridge, to slowly maneuver a prebuilt structure to meet a piece already in place and joining it to that part.

launch complex *Space Technology.* the site, ground facilities, equipment, and sometimes ground crew used to launch a rocket vehicle.

launcher *Ordnance.* **1.** a structural device designed to support and hold a missile in position for firing. **2.** a device attached to a rifle for firing a grenade. *Space Technology.* any of a variety of devices used to support and often direct a rocket vehicle or missle that is being launched. Thus, **launch time.**

launching *Navigation.* the process in which a projectile or vehicle is sent forth into the air or the water, especially a missile or rocket. Thus, **launching platform, launching tube, launching site.** *Electromagnetism.* the process of transferring energy from a coaxial cable to a waveguide.

launching angle *Space Technology.* the acute angle between the horizontal and the longitudinal axis of a rocket or missile at launch.

launching base *Space Technology.* an air base containing one or more launch sites.

launching cradle *Civil Engineering.* a supporting framework that supports a ship's hull during launching.

launching pad *Space Technology.* a temporary or permanent platform constructed for the takeoff and landing of aircraft or space vehicles. Also, **launch pad.**

launching ways *Civil Engineering.* a sloping track or guideway on which a ship is constructed, and down which it slides into the water when launched.

launch site *Space Technology.* **1.** the delimited area from which a launch takes place. **2.** the components of a launching base, such as the platform, ramp, and rack. **3.** broadly, a launching base.

launch vehicle *Space Technology.* a rocket or other vehicle used to carry a probe, satellite, or payload to a desired point in space or in the atmosphere. Also, BOOSTER.

launch window *Space Technology.* a postulated interval of time and space during which a missile or space vehicle must be launched in order to carry out its specific mission; usually based on atmospheric conditions, relative positions of the launch site and target, and other transitory factors.

launder *Engineering.* a trough or channel that is inclined along its length and used for conveying a liquid, such as water used in mining or building construction. *Metallurgy.* a refractory-lines channel used for conveying molten metal; as from a furnace to a ladle or mold.

launder screen *Mining Engineering.* a screen used for sizing and dewatering small sizes of anthracite.

launder separation process *Mining Engineering.* a process in which water is used to separate heavy gravity product from the lighter products that flow above it.

Lauraceae *Botany.* a family of aromatic trees and shrubs of the order Magnoliales, usually having evergreen, alternate leaves, and small, green or yellow bisexual or unisexual flowers.

Laurales *Botany.* an order of flowering, mostly woody plants of the class Magnoliopsida.

Laurasia *Geology.* the hypothetical supercontinent or landmass that fragmented millions of years ago to form the present-day continents of the Northern Hemisphere, including North America, Greenland, and Eurasia, excluding India.

Lauratonematidae *Invertebrate Zoology.* a family of free-living marine nematodes in the superfamily Enoploidea; some females have a cloaca.

laurdalite *Petrology.* a nepheline syenite with rhombic anorthoclase and traces of biotite, pyroxene, and amphibole.

laurel *Botany.* **1.** the common name for various trees or shrubs of the genus *Laurus,* especially the **true laurel,** *L. nobilis,* a small evergreen tree that is native to Europe, having dark green, glossy leaves. **2.** any of various other woody plants with laurel-like foliage.

laurel-leaf blade *Archaeology.* a distinctive stone blade of the Upper Paleolithic period in Europe, characterized by a long, very thin form and delicate workmanship.

Laurell rocket test *Immunology.* a method used to detect and quantify soluble antigens, whereby the antigens are observed in a charged medium (usually agar) containing antibodies, and then immunodiffused with the antibodies; the antigens create rocketlike patterns of precipitation. Also, ROCKET IMMUNOELECTROPHORESIS.

laurence *Optics.* a light that wavers above a torrid surface on a calm, clear day as a result of the unevenness in the refraction of light.

Laurence-Moon-Biedl syndrome *Medicine.* a recessive hereditary disorder occurring more frequently in males and resulting in degenerative disease characterized by obesity, hypogonadism, mental retardation, polydactyly, and visual disturbances. Also, **Laurence-Biedl syndrome.**

Laurent half-shade plate *Optics.* a device in which a quartz plate covering half of a plane-polarized beam is used to determines its rotation.

Laurentian Plateau see CANADIAN SHIELD.

Laurentians *Geography.* a range of low mountains in eastern Canada, between Hudson Bay and the St. Lawrence River. Also, **Laurentian Mountains.**

Laurentide ice sheet *Hydrology.* a continental ice sheet that at its maximum completely covered North America east of the Rockies and south from the Arctic, including the upper half of the eastern and midwestern United States.

Laurent series *Mathematics.* if a function $f(z)$ of a complex variable is analytic in a punctured disk or annular region centered at a point a, then $f(z)$ may be represented within this region by the Laurent series

$$f(z) = \sum_{n=-\infty}^{\infty} c_n (z-a)^n,$$

$$\text{where } c_n = {}^1\!/\!_{2\pi i} \int_C f(z)\,(z-a)^{-(n+1)}dz$$

and C is any closed path encircling a counterclockwise within the region. Sometimes called the **Laurent expansion** of f. The c_{-1} coefficient is known as the *residue* of f. If, in addition, f is analytic on the entire disk, then the Laurent and Taylor series for f coincide.

Laurer's canal *Invertebrate Zoology.* a canal in certain flukes that connects the oviduct to the ventral surface of the body.

lauric acid *Organic Chemistry.* $C_{12}H_{24}O_2$, a combustible fatty acid occurring as the glyceride in many vegetable fats, consisting of colorless needles; insoluble in water and soluble in ether; melts at 44–47°C and boils at 225°C; used in alkyd resins, wetting agents, detergents, cosmetics, insecticides, and food additives. Also, DODECANOIC ACID.

lauric aldehyde see LAURYL ALDEHYDE.

laurionite *Mineralogy.* PbCl(OH), a colorless orthorhombic mineral, dimorphous with paralaurionite, occurring in tabular crystals, and having a specific gravity of 6.24 and a hardness of 3 to 3.5 on the Mohs scale; found in lead slags exposed to sea water.

laurite *Mineralogy.* RuS_2, a black, opaque, cubic mineral of the pyrite group occurring as small cubes, octahedra, or grains, having a specific gravity of 6 to 6.99 and a hardness of 7 to 7.5 on the Mohs scale; found in platinum concentrates.

Lauritsen electroscope *Electricity.* an electroscope that uses a metalized quartz fiber as the sensitive element.

lauryl alcohol *Organic Chemistry.* $C_{12}H_{26}O$, a combustible, colorless solid alcohol with a floral odor, derived from fatty acids in coconut oil; insoluble in water and soluble in alcohol; melts at 24°C and boils at 259°C; used in synthetic detergents, lube additives, pharmaceuticals, rubber, textiles, and perfumes. Also, DODECYL ALCOHOL.

lauryl aldehyde *Organic Chemistry.* $CH_3(CH_2)_{10}CHO$, a combustible, colorless solid or liquid with a strong fatty floral odor; insoluble in water and soluble in alcohol; melts at 44°C and boils at 142–143°C (22 torr); used in perfumes and as a flavoring agent. Also, DODECANAL, LAURIC ALDEHYDE.

lauryl mercaptan *Organic Chemistry.* $C_{12}H_{26}S$, a combustible water-white or pale yellow liquid with a mild odor; insoluble in water and soluble in methanol, ether, and acetone; boils at 165–169°C (39 torr); used to make elastomers, fungicides, insecticides, and plastics. Also, DODECYL MERCAPTAN, THIODODECYL ALCOHOL.

lausenite *Mineralogy.* $Fe_2^{+3}(SO_4)_3 \cdot 6H_2O$, a white monoclinic mineral of undetermined hardness and specific gravity, occurring in lumpy aggregates of fibers resulting from a fire in the United Verde Mine in Jerome, Arizona.

Lauson engine *Engineering.* a single-cylinder engine in which screening tests were performed before the L-series of lube-oil tests were devised.

lautarite *Mineralogy.* $Ca(IO_3)_2$, a rare, colorless or yellow, transparent monoclinic mineral having a specific gravity of 4.5 to 4.52 and a hardness of 3.5 to 4 on the Mohs scale; found as crystals in nitrate deposits in the Atacama Desert, Chile.

lautite *Mineralogy.* CuAsS, a black or steel-gray, opaque, metallic, orthorhombic mineral occurring in tabular or prismatic crystals and in massive or fibrous form, having a specific gravity of 4.9 and a hardness of 3 to 3.5 on the Mohs scale; found in mesothermal vein deposits.

LAV lymphadenopathy-associated virus.

lava *Volcanology.* **1.** a molten mass of rock material that is extruded by a volcano or through a fissure in the earth. **2.** the rock that is formed by the cooling and solidifying of such molten materia. (From a Neopolitan dialectal term; going back to a Latin word meaning "sliding" or "falling.")

lava

lava blisters *Volcanology.* small, hollow, steep-sided swellings produced on the surface of some lava flows by gas bubbles that push up the lava's viscous surface.

lava column *Volcanology.* a shaft of solidified or fluid lava found in a volcanic vent. Also, MAGMA COLUMN.

lava cone *Volcanology.* a cone-shaped structure formed by the accumulation of lava flowing from a vent or fissure.

lava dome see SHIELD VOLCANO.

lava field *Volcanology.* a wide expanse of lava that often forms along the base or on the sides of a volcano.

lava flow *Volcanology.* **1.** a stream or sheet of molten lava that flows laterally over the surface of the earth from a volcanic vent or fissure. **2.** the mass of solidified rock produced by the congealing of such a flow.

lava fountain *Volcanology.* a plume of incandescent molten lava sprayed vertically into the air from a volcanic vent or fissure.

lava fountain

lavage *Medicine.* the washing out of a cavity or a hollow organ, such as the stomach or lower bowel.

lava lake *Volcanology.* a lake of molten, partially solidified, or solidified lava occurring within a volcanic crater or other depression.

lavalier microphone *Acoustical Engineering.* a small broadcasting microphone that is worn on a cord around the neck. Also, **lavaliere microphone.**

Laval nozzle *Aviation.* a converging-diverging nozzle used in certain steam turbines and rockets. Also, **de Laval nozzle.** (Named for its inventor, the Swedish engineer Carl Gustaf Patrik de *Laval*.)

lava plateau *Geology.* a broad, elevated tableland or flat-topped highland that is underlain by a thick succession of lava flows, such as the Columbia Plateau of the northwestern United States.

lava tube *Geology.* a long, hollow opening beneath the surface of a solidified lava flow, formed by the withdrawal of the molten portion after the crust has solidified. Also, **lava tunnel.**

lavedulan *Mineralogy.* $NaCaCu_5^{+2}(AsO_4)_4Cl \cdot 5H_2O$, a lavender to turquoise-blue, orthorhomic mineral occurring as aggregates of fine flakes, having a specific gravity of 3.54 and hardness of 2.5 to 3 on the Mohs scale; found at the Blanca Mine, Freirina, Chile. Also, FREIRINITE.

lavender *Botany.* any of various Old World plants and shrubs belonging to the genus *Lavandula* of the family Labiatae, especially *L. officinalis,* having long, narrow pale green leaves and fragrant flowers that are a distinctive shade of purple. *Materials.* the dried flowers of this plant used as a scent or preservative. (From the Latin word for "wash;" from its early use in perfuming water for bathing.)

lavender oil *Materials Science.* an essential oil used in perfumery, derived from several species of lavender.

lavendulan *Mineralogy.* $NaCaCu_5^{+2}(AsO_4)_4Cl \cdot 5H_2O$, a translucent, lavender-blue orthorhombic mineral occurring as crusts and aggregates of fine flakes, having a specific gravity of 3.54 and a hardness of 2.5 to 3 on the Mohs scale; found with erythrite, olivenite, and conichalcite.

lavenite *Mineralogy.* $(Na,Ca)_2(Mn^{+2},Fe^{+2})(Zr,Ti)Si_2O_7(O,OH,F)_2$, a colorless, yellow, or brown, translucent monoclinic mineral occurring in prismatic crystals and massive form, having a specific gravity of 3.51 to 3.547 and a hardness of 6 on the Mohs scale; found in alkali rocks and their pegmatites.

Laveran, Charles 1845–1922, French surgeon; awarded the Nobel Prize for discovering the parasites that cause malaria and sleeping sickness.

Laves phase see INTERSTITIAL COMPOUND.

laveur *Surgery.* an instrument used to perform lavage or irrigation.

Antoine Lavoisier

Lavoisier, Antoine [lə vwä zyä´] 1743–1794, French chemist; the founder of modern chemistry; proved the law of conservation of energy; formulated oxygen theory of combustion; devised first chemical equation and a system of nomenclature.

law *Science.* a scientific principle that is invariable under certain stated conditions; for example, Boyle's law holds that the product of the pressure of a gas times the volume of the gas will remain constant if tempersature remains constant.

Lawless, Theodore K. 1892–1971, American physician; known for his research in dermatology.

lawn *Agriculture.* an open stretch of mowed grass, as near a house or in a park. *Engineering.* a fine gauze sieve that is used, for example, to filter clay. *Molecular Biology.* an expanse of bacterial colonies that completely covers the surface of a petri plate. (From a Celtic word meaning "heath.")

lawn *Textiles.* a sheer linen or cotton fabric. (Named after *Laon,* France, once a linen-making center.)

lawnmower or **lawn mower** *Mechanical Devices.* a motor-driven or hand-operated machine used to cut grass. *Electronics.* a radar-preamplifier noise limiter that reduces "grass" (a pattern on the display screen of small deflections from the base line, having the appearance of a cross section of lawn).

lawn plate *Microbiology.* a petri plate in which a confluent population of a given microorganism completely covers the surface of the plate.

law of conservation of energy see CONSERVATION OF ENERGY.

law of conservation of mass see CONSERVATION OF MASS.

law of constant angles *Crystallography.* a law covering the interfacial angles in all crystals of a given sort stating that angles between corresponding faces have a constant value.

law of constant heat summation see HESS'S LAW.

law of corresponding state *Chemistry.* a law stating that when two substances have ratios of pressure, temperature, or volume to their respective critical properties, and two of these ratios are equal, the third ratio must equal the other two.

law of cosines *Mathematics.* for a triangle with angles α, β, and γ and sides *a, b,* and *c* opposite the angles, respectively, $c^2 = a^2 + b^2 - 2ab \cos \gamma$. This reduces to Pythagoras' theorem when $\gamma = 90°$, so that $\cos \gamma = 0$. Also, COSINE THEOREM.

law of effect *Behavior.* a theory stating that reward strengthens a stimulus-response connection, while punishment weakens it.

law of electric charges *Electricity.* a law stating that like charges repel, while unlike charges attract.

law of electrostatic attraction see COULOMB'S LAW.

law of equal areas see EQUAL-AREAS LAW.

law of errors see CENTRAL-LIMIT THEOREM

law of exercise *Behavior.* a theory stating that repetition of an act promotes learning and makes subsequent performance of that act easier, other conditions being equal.

law of exponents *Mathematics.* the rules for combining expressions with exponents. Specifically, $a^pa^q = a^{p+q}$, $a^p/a^q = a^{p-q}$, $(a^p)^q = a^{pq}$, $(ab)^p = a^pb^p$, and $(a/b)^p = a^p/b^p$, which hold for positive numbers *a* and *b*, and positive integers *p* and *q*. The rules can be extended to all real *a* and real $b \neq 0$ and rational *p* and *q,* provided *a* and *p* are not negative when *p* or *q* is a rational fraction with an even denominator. They can be extended to complex numbers *a, b, p,* and *q* or to any mathematical expressions as long as division by zero is avoided and certain other restrictions are followed.

law of flotation *Fluid Mechanics.* a law stating that an object floating in a fluid displaces its own weight of fluid.

law of gravitation or **gravity** see NEWTON'S LAW OF GRAVITATION.

law of interfacial angle *Crystallography.* a law that in all crystals of a given sort, angles between corresponding faces have a constant value; applies to one particular form of a polymorphous crystalline material.

law of large numbers *Statistics.* any of several theorems regarding the convergence of the sample average to the mean of the underlying population; for example, Bernoulli's theorem.

law of magnetism *Electromagnetism.* a law stating that common magnetic poles repel each other, whereas unlike magnetic poles attract each other.

law of mass action see MASS ACTION LAW.

law of minimum *Biology.* a law stating that plants have minimum requirements of certain mineral salts, and if a soil does not supply this minimum, the plants requiring those particular salts cannot grow, regardless of the abundance of other nutrients.

law of rational indices see RATIONAL INDICES.

law of signs *Mathematics.* the product or quotient of two nonzero numbers with like signs is positive, while the product or quotient of two nonzero numbers with differing signs is negative.

law of sines *Mathematics.* for a triangle with angles α, β, and γ and sides *a, b* and *c* opposite the angles respectively, $a/\sin \alpha = b/\sin \beta = c/\sin \gamma = 2R$, where *R* is the radius of the circumscribed circle. Also, SINE THEOREM.

law of storms *Meteorology.* a statement about the general behavior of storms, describing the rotation of winds about the center of a cyclone and the way in which the disturbance tends to travel over the earth's surface; used in navigation.

law of superposition *Geology.* the law of geologic chronology stating that in any undisturbed sequence of sedimentary strata, the oldest stratum will be at the base, and the youngest stratum will be at the top.

law of tangents *Mathematics.* for a triangle with angles α, β, and γ and sides *a, b,* and *c* opposite the angles, respectively,

$$(a - b)/(a + b) = \tan[(\alpha - \beta)/2]/\tan[(\alpha + \beta)/2] = \tan[(\alpha - \beta)/2]/\cot(\gamma/2).$$

Also, TANGENT THEOREM.

law of thermodynamics see FIRST (SECOND, THIRD) LAW OF THERMODYNAMICS.

Lawrence, Ernest Orlando 1901–1958, American physicist; awarded the Nobel Prize for inventing and developing the cyclotron.

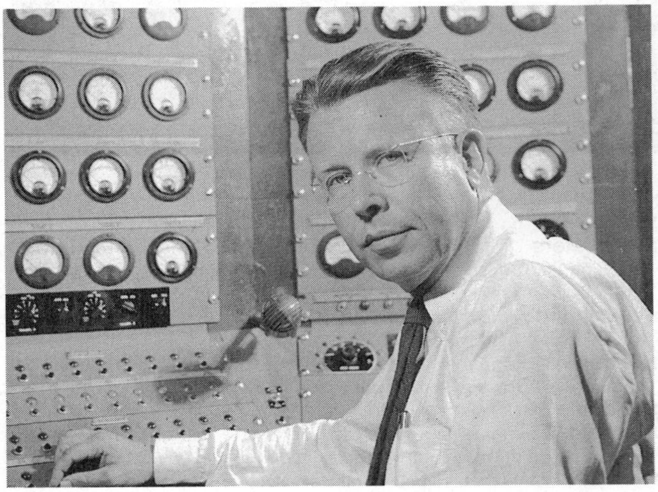

Ernest O. Lawrence

lawrencite *Mineralogy.* $(Fe^{+2},Ni)Cl_2$, a soft brown or green trigonal mineral having a specific gravity of 3.162 and an undetermined hardness; found as filling fissures in iron meteorites.

lawrencium *Chemistry.* a short-lived radioactive element having the symbol Lr, the atomic number 103, and an atomic weight of 257; made by bombarding californium with boron ions.(From the *Lawrence* Radiation Laboratory, Berkeley, California, founded by Ernest O. Lawrence.)

laws of motion see NEWTON'S LAWS OF MOTION.

Lawson criterion *Physics.* a relation between the minimum confinement time and particle density needed for a nuclear fusion reaction to take place.

lawsonite *Mineralogy.* $CaAl_2Si_2O_7(OH)_2{\cdot}3H_2O$, a colorless or grayish-blue orthorhombic mineral, dimorphous with partheite, having a specific gravity of 3.05 to 3.12 and a hardness of 6 on the Mohs scale; found as prismatic crystals in low-grade, regionally metamorphosed schists, especially glaucophane schists.

LAX *Aviation.* the airport code for Los Angeles International Airport, Los Angeles, California.

laxative *Pharmacology.* **1.** any mild agent that aids emptying of the bowels without pain or violent action; a cathartic or purgative. **2.** of, relating to, or having the effect of such an agent.

lay *Design Engineering.* the dominant direction, length, or degree of twist of the strands in a rope or wire cable. *Ordnance.* **1.** to direct or adjust the aim of a weapon. **2.** to set a weapon for a given range, given direction, or both. **3.** to drop one or more bombs or mines onto the surface from an aircraft. **4.** to spread a smoke screen from an aircraft.

Layard, Sir Austen Henry 1817–1894, British diplomat, author, and archaeologist; excavated Nineveh and other Mesopotamian sites.

lay-down bombing *Military Science.* a very low-level bombing technique using delay fuses or similar devices to allow an attacking aircraft to escape the effects of its bomb.

layer *Ecology.* one of the four horizontal zones of vegetation; i.e., tree layer, shrub or bush layer, field or herb layer, and ground or moss layer. Thus, **layering.** *Geology.* any tabular body of rock, ice, sediment, or soil lying more or less parallel to the surface on or against which it was formed, and distinctly limited above and below. *Geophysics.* **1.** any of the concentric zones inside the earth that are defined by abrupt changes in seismic properties. **2.** any of the distinct regions in the ionosphere. **3.** any of the flat geological structures. *Metallurgy.* a weld layer parallel to the base. *Computer Programming.* **1.** a collection of related functions that compose one level of a hierarchy of functions; each layer, except for the most primitive, specifies its own functions and assumes that lower-level functions are provided. **2.** a subset of data that has a logical internal relationship.

layer architecture *Computer Technology.* a methodology used in designing computer hardware or software such that system components can be implemented in a modular fashion to allow for changes in one layer without affecting the others. Also, **layered architecture.**

layer capacitance see CATHODE INTERFACE CAPACITANCE.

layer depth *Oceanography.* the thickness of the mixed layer; the distance from the water's surface to the top of the main thermocline.

layer depth effect *Geophysics.* a weakening of a sound beam or seismic pulse as it spreads while passing from a positive gradient layer to an underlying negative layer.

layered compound *Materials Science.* describing any compound in which molecular layers alternate between two or more different types of atoms, as between metal atoms and organic molecules or between metal atoms and semiconductors; important in superconductivity and photoelectric systems.

layer impedance see CATHODE INTERFACE IMPEDANCE.

layering *Ecology.* the distribution of plants in a given community, such as mosses, shrubs, or trees in a bog area. *Agriculture.* a method for propagating plants, in which portions of their stems or branches are covered with soil so that they will take root while still attached to the parent plant. Also, **layerage.**

layering granulation *Materials Science.* in ceramics processing, a system for producing granules of a required uniform size in which a finely dispersed powder is added to a liquid in a rotating pan or drum and subjected to random rolling motion. As the granules begin to agglomerate, they roll to the upper layer of the mix and are discharged with a well-defined size.

layer line *Crystallography.* the series of straight lines formed by the arrangement of diffraction spots on a cylindrical film surrounding a crystal rotated or oscillated about a principal axis that are perpendicular to the axis of rotation.

layer of no motion *Oceanography.* a layer of water assumed to be at rest at some depth.

layer structure *Crystallography.* a crystal structure in which flat molecules or groups of atoms or ions are arranged in parallel, flat layers or planes.

layer winding *Electricity.* a form of winding that is used in most transformers, in which adjacent turns are laid side by side in layers along the length of the coil form, with a thin sheet of insulation between each layer.

lay off *Engineering.* to draw at full size; used especially in ship and aircraft design.

layout *Architecture.* a plan showing the arrangement of objects and spaces in a structure. *Industrial Engineering.* the physical arrangement of workers and machinery in a production system. Also, WORK FLOW LAYOUT. *Graphic Arts.* **1.** a sketch or plan showing the arrangement of elements on a page or of pages in a book, often including instructions for the stripper or printer. **2.** the process of creating such a plan.

layout character *Computer Programming.* any character used to control the positioning of data on a printer, terminal, or other display device.

layover *Transportation Engineering.* a rest period for a crew between the end of one scheduled flight and departure on another flight, or in the course of an extended flight.

lay ratio *Electricity.* the ratio of the axial length of one complete turn of the helix formed by the core of the cable or wire of a stranded conductor to the mean diameter of the cable.

layup *Engineering.* **1.** the process of assembling veneers for pressing into plywood. **2.** the process of forming a bonded material by applying alternate layers of material and a binder.

lazaret *Naval Architecture.* **1.** a storage compartment below decks aft. **2.** a quarantine hospital or detention cell. Also, **lazarette.**

lazuli see LAPIS LAZULI.

lazulite *Mineralogy.* $MgAl_2(PO_4)_2(OH)_2$, a deep violet- or azure-blue to bluish green monoclinic mineral of the lazulite group, having a hardness of 5.5 to 6 on the Mohs scale and a specific gravity of 3.08 to 3.10; found as crystals and granular masses in aluminous high-grade metamorphic rocks and granite pegmatites. Also, BERKEYITE, BLUE SPAR, FALSE LAPIS.

lazurite *Mineralogy.* $(Na,Ca)_{7-8}(Al,Si)_{12}(O,S)_{24}[(SO_4),Cl_2,(OH)_2]$, a blue or violet-blue cubic mineral of the sodalite group, massive in habit, having a specific gravity of 2.38 to 2.45 and a hardness of 5 to 5.5 on the Mohs scale; found in contact metamorphosed limestones, it is the chief mineral constituent of the semiprecious stone lapis lazuli.

lazy H antenna *Electromagnetism.* a directional antenna that is composed of two vertically stacked collinear elements, giving rise to both vertical and horizontal directivity.

lb. or **lb** pound. (From Latin *libra.*)

LBA linear bounded automaton.

L band *Telecommunications.* a band of radio frequencies with a range of 390 to 1550 megahertz.

LBB *Aviation.* the airport code for Lubbock, Texas.

lbf pound-force.

LBG *Aviation.* the airport code for Le Bourget, Paris.

lbs. or **lbs** pounds.

lb.s.t. static thrust in pounds.

LBW low birth weight.

LC landing craft.

l.c. or lc lowercase; lowercase letters.

L capture *Nuclear Physics.* an interaction that occurs during beta decay in which a nucleus captures an electron from the *L* shell of α atomic electrons.

LCAT lecithin-cholesterol acyltransferase.

LCD liquid crystal display.

L.C.D. or **l.c.d.** lowest common denominator; least common denominator.

L cells *Cell Biology.* a cultured sarcoma cell line that is isolated from mouse connective tissue and produces sarcomas when it is injected into healthy mice.

L.C.F. or **l.c.f.** lowest common factor.

LC filter **1.** a filter composed of inductors and capacitors. **2.** a filter composed of an L network of one inductor and one capacitor.

L chain light chain.

LCI landing craft, infantry.

L.C.M. or **l.c.m.** lowest common multiple; least common multiple.

LCN load classification number.

L/C ratio *Electricity.* a ratio of inductance to capacitance, used in selecting suitable component values for tuning a circuit.
LD lethal dose.
ld. load.
LD_{50} *Toxicology.* lethal dose 50; the amount of a given toxic substance that will elicit a lethal response in 50% of the test organisms. Also, MEDIAN LETHAL DOSE.
LDA *Navigation.* an aircraft navigational aid that is used for nonprecision instrument approaches; not part of a complete Instrument Landing System, and not aligned with the runway. (From localizer-type directional aid.)
L.D.A. left dorsoanterior.
L ders lines *Metallurgy.* deformation bands visible on the surface of a specimen.
LDH lactate dehydrogenase; lactic dehydrogenase.
L display *Electronics.* the representation of a target on a radarscope in which the target is depicted as two horizontal blips, with one veering to the right and the other to the left of a central vertical time base; when the radar antenna is on target, both blips are of equal amplitude, with distance shown by the position of the signal along the scope's baseline. Also, L SCAN.
LDL low-density lipoprotein.
L-dopa *Pharmacology.* $C_9H_{11}NO_4$, the levorotary isomer of dopa, occurring naturally as a whitish crystalline powder and administered orally to relieve symptoms of parkinsonism. Also, LEVODOPA.
L.D.P. left dorsoposterior.
LDPE low-density polyethylene.
L/D ratio length to diameter ratio; lift-drag ratio.
leachate *Chemical Engineering.* the solution or soluble material that results from a leaching process.
leaching *Chemical Engineering.* the process of separating a soluble substance from a solid by washing or by the percolation of water or other liquid through the substance, as in coffee-making. *Geochemistry.* specifically, the natural or artificial removal of soluble substances from rock, ore, or layers of soil by the action of percolating substances. Also, LIXIVIATION.
lead [led] *Chemistry.* a metallic element having the symbol Pb, the atomic number 82, an atomic weight of 207.2, a melting point of 327.4°C, and a boiling point of 1755°C; a soft, heavy solid occurring naturally in galena. *Metallurgy.* this substance in the form of a heavy, corrosion-resistant metal, principally used for storage batteries, cable sheathing, bearings, ammunition, and low-melting alloys such as solders. Lead presents a health hazard, and its use is restricted or prohibited in certain applications. *Engineering.* a small body of this metal that is attached to a sounding line and used as a plummet to determine the depth of water. *Graphic Arts.* **1.** in hot-metal typesetting, the metal strips of various thicknesses used to create space between lines of type. **2.** in phototypesetting, the space created by computer between lines of type. (From Old English; probably going back to a Celtic word for this substance.)
lead [lēd] *Physics.* **1.** an advance in time. **2.** an advance in angular displacement giving rise to a positive phase angle. *Oceanography.* a navigable channel through pack ice. *Ordnance.* **1.** to aim a projectile-type weapon ahead of a moving target, making all necessary adjustments. **2.** the distance ahead of a moving target at which a weapon must be aimed in order to hit the target. **3.** in an explosive train, a column of high explosive designed to transmit the charge from the detonator to a booster charge. *Design Engineering.* the distance a screw advances into a nut when given a single complete turn. *Electricity.* **1.** a connecting conductor between a winding and its termination. **2.** the angle, expressed in degrees, by which one alternating quantity, usually current or voltage, precedes another in time (as opposed to lag); in a perfect capacitor, the current leads the applied voltage by 90°. *Geology.* **1.** a small, narrow, uniformly trending passage in a cave. **2.** see LEDGE, def. 1. **3.** see LODE.
lead-206 *Nuclear Physics.* the stable isotope of lead formed as the end product of the uranium decay series
lead-207 *Nuclear Physics.* the stable isotope of lead formed as the end product of the actinium decay series.
lead-208 *Nuclear Physics.* the stable isotope of lead formed as the end product of the thorium decay series.
lead acetate *Organic Chemistry.* $Pb(C_2H_3O_2)_2 \cdot 3H_2O$, poisonous white crystals or flakes having a sweet taste; soluble in water and slightly soluble in alcohol; decomposes at 280°C and loses water at 75°C; used in the dyeing of textiles, in medicines, and in the manufacture of varnishes and pigments.
lead-acid battery *Electricity.* a rechargeable storage battery based on positive and negative lead plates submerged in sulfuric acid.
lead angle *Design Engineering.* the angle made by the tangent to a helix with the plane normal to the axis of the helix. *Physics.* a positive phase angle arising from an advance in angular placement of one of two sinusoidally varying quantities having the same frequency. *Metallurgy.* in cutting tools, the helix angle of the flutes. *Ordnance.* in aiming at a moving target, the angle between the gun-target line and the line from the gun to the aiming point.
lead antimonate *Inorganic Chemistry.* $Pb_3(SbO_4)_2$, a noncombustible, orange-yellow powder that is insoluble in water; used as a stain and pigment. Also, NAPLES YELLOW, ANTIMONY YELLOW.
lead arsenate *Inorganic Chemistry.* $Pb_3(AsO_4)_2$, white crystals that are very slightly soluble in water and soluble in nitric acid; decomposes at 1000°C; highly toxic and a carcinogen; used in insecticides and herbicides. Also, LEAD ORTHOARSENATE.
lead azide *Inorganic Chemistry.* $Pb(N_3)_2$, colorless, highly sensitive needlelike crystals; highly explosive and usually handled under water; used as a detonator for explosives.
lead-base babbitt *Metallurgy.* any of several bearing alloys containing 75–85% lead.
lead borate *Inorganic Chemistry.* $Pb(BO_2)_2 \cdot H_2O$, a toxic, noncombustible, white powder that is insoluble in water; loses its water at 160°C; used in paints, lead glass, and ceramic coatings.
lead bromide *Inorganic Chemistry.* $PbBr_2$, a white powder, toxic on inhalation; slightly soluble in hot water and insoluble in alcohol; melts at 373°C and boils at 916°C.
lead bronze *Metallurgy.* one of a few copper-base alloys used only for shaped castings, containing 20–40% lead.
lead bullet *Ordnance.* a small-arms bullet made of lead or a high percentage of lead.
lead carbonate *Inorganic Chemistry.* **1.** $PbCO_3$, toxic white crystals, decomposing in hot water and soluble in acids; decomposes at 315°C. **2.** **basic lead carbonate.** $2PbCO_3 \cdot Pb(OH)_2$, a toxic white powder that is insoluble in water and soluble in acids; decomposes at 400°C; used as a pigment. Also, LEAD SUBCARBONATE, LEAD FLAKE.
lead-chamber process *Chemical Engineering.* the first large-scale process for producing sulfuric acid, in which a mixture of sulfur dioxide, air, steam, or water, and an oxide of nitrogen is contacted in lead-lined chambers, so that a chemical reaction of the ingredients with steam produces sulfuric acid.
lead chart *Ordnance.* a chart indicating the lead necessary to hit a moving target, based on variables such as the target's range, speed, and direction of travel.
lead chloride *Inorganic Chemistry.* $PbCl_2$, toxic white crystals; slightly soluble in hot water and insoluble in cold water and alcohol; melts at 501°C and boils at 950°C; used in making lead salts and as an analytical reagent.
lead chromate *Inorganic Chemistry.* $PbCrO_4$, yellow crystals; soluble in strong acids and insoluble in water; melts at 844°C; toxic and a carcinogen; used in pigments and organic analysis. Also, CHROME YELLOW.
lead citrate *Biotechnology.* a heavy metal salt used as a dye or stain to prepare specimens for transmission electron microscopy.
lead compensation *Control Systems.* the feedback compensation that is used to stabilize or improve a system's transient response.
lead computer *Ordnance.* a device, usually placed in the mount or sights of an airplane machine gun, that automatically calculates the lead necessary to hit a moving target, based on the tracking of the target by the gun.
lead curve *Civil Engineering.* on a railroad turnout, the curve of the diverging line's rail between the actual switch and the frog.
lead cyanide *Inorganic Chemistry.* $Pb(CN)_2$, a toxic white or yellowish powder; slightly soluble in water; decomposes in acids; used in metallurgy.
lead dioxide *Inorganic Chemistry.* PbO_2, toxic brown crystals; insoluble in water; decomposes at 290°C; a dangerous fire risk in contact with organic substances; used as an oxidizing agent, in storage batteries, and in the textile industry. Also, PLUMBIC ACID, LEAD PEROXIDE, BROWN LEAD OXIDE.
leade *Ordnance.* the part of a gun barrel that is just ahead of the chamber; it is slightly enlarged and tapered to accept the bullet. Also, LEDE, THROAT.
leaded *Materials.* containing lead, as gasoline or glass.
leaded alloy *Metallurgy.* any alloy that is not in solid solution and contains lead as a separate constituent; for example, free-cutting brass.

leaded gasoline *Materials.* gasoline to which tetraethyl lead has been added to increase the octane number; its use has now been curtailed or prohibited in many areas because it is a major source of air pollution. New vehicles manufactured for sale in the United States are now required to use unleaded gasoline.

lead encephalopathy *Medicine.* a disease of the brain resulting from chronic lead poisoning, characterized by epileptic convulsions, delirium, hallucinations, and other cerebral disturbances.

lead equivalent *Radiology.* an equivalent thickness of pure lead that would shield radiation to the same degree as a material under consideration.

leader *Behavior.* an animal that goes ahead of its group or otherwise directs the group's movements. *Engineering.* an unrecorded strip at the beginning or end of a reel of tape or film that permits handling without damaging the recorded material. *Computer Programming.* a record that precedes a group of detail records and provides information about the group that is not present in the records themselves. *Geophysics.* the high-ion-density channel in the atmosphere through which the first part of a lightning stroke propagates.

leader label see HEADER.

leader peptide *Molecular Biology.* an amino acid sequence found at the amino-terminal end of a secretory protein that guides the protein to the endoplasmic reticulum membrane, where it is usually enzymatically removed. Also, **leader sequence peptide.**

leaders *Graphic Arts.* a row of dots or dashes used to lead a reader's eye from one side of a page to the other, as in a table of contents.

leader sequence *Genetics.* an untranslated segment of mRNA found at the 5' end and containing regulatory information. Also, SIGNAL SEQUENCE.

leader stroke *Geophysics.* the first stroke in a lightning discharge, consisting of a weakly luminous stepped leader that proceeds from a cloud to the ground and creates an ionized channel for the much more powerful return stroke.

lead flake see LEAD CARBONATE, def. 2.

lead fluoride *Inorganic Chemistry.* PbF_2, colorless crystals; very slightly soluble in water; melts at 855°C and boils at 1290°C; toxic and a strong irritant; used in optics and for various other industrial purposes.

lead fluorosilicate *Inorganic Chemistry.* $PbSiF_6 \cdot 2H_2O$, colorless crystals that are very soluble in hot water and are decomposed by heat; toxic and a strong irritant; used in lead refining. Also, LEAD SILICOFLUORIDE.

lead foil *Metallurgy.* a very thin sheet or strip of commercially pure lead.

lead formate *Organic Chemistry.* $Pb(CHO_2)_2$, brownish-white needles that are soluble in water and decompose at 190°C; used as a reagent in analytical determinations.

lead-free gasoline see UNLEADED GASOLINE.

lead glass *Materials.* glass that contains lead oxide and that has a high refractive index and optical dispersion; used in the manufacture of optical glass, in high-quality crystal glassware, and for radiation shielding.

lead halide *Inorganic Chemistry.* any binary compound of lead and one of the halogen elements (periodic table Group VIIa), such as bromine or iodine.

leadhillite *Mineralogy.* $Pb_4(SO_4)(CO_3)_2(OH)_2$, a yellowish-, greenish-, or grayish-white monoclinic mineral, trimorphous with macphersonite and susannite, having a specific gravity of 6.55 and a hardness of 2.5 to 3 on the Mohs scale; found as crystals and granular masses in oxidized lead ore at Leadhills, Scotland.

lead hydroxide *Inorganic Chemistry.* $Pb(OH)_2$, a noncombustible, white amorphous powder that is slightly soluble in hot water; decomposes at 145°C; absorbs carbon dioxide from the air. Also, **lead hydrate.**

lead-I-lead junction *Solid-State Physics.* a Josephson junction that is constructed of two pieces of lead and an insulating layer of lead oxide between them.

lead-in *Electricity.* a single-wire feed line in a shortwave receiving antenna. Also, DOWN-LEAD.

leading *Graphic Arts.* **1.** originally, the space created by strips of lead or other metal between lines of type. **2.** a similar space created by computer on photocomposed type. *Industrial Engineering.* a management function that motivates employees to achieve goals.

leading character elimination *Computer Programming.* a data compression technique for collections of character strings in alphabetic order; each string is stored as an integer representing the number of leading characters in common with those of the previous string, followed by the unique suffix.

leading coefficient *Mathematics.* the (nonzero) coefficient of the highest-order term of a polynomial in one variable.

leading current *Electricity.* an alternating current that leads the applied voltage, occurring when voltage is applied to a capacitive load.

leading edge *Design Engineering.* the edge of the surfaces or inset cutting points on a bit that are oriented in the same direction as the bit's rotation. *Physics.* the rising portion of a pulse.

leading-edge pulse *Electrical Engineering.* the first major transition from the pulse baseline to occur after a given time.

leading-edge slat *Aviation.* an auxiliary airfoil attached to the leading edge of a wing; used to increase lift at large angles of attack.

leading end *Computer Technology.* the end of a tape, document, or other medium that is read first.

leading fire *Ordnance.* fire delivered ahead of a moving object to compensate for its travel.

leading light(s) *Navigation.* a light or lights that indicate a path to be followed.

leading limb *Astronomy.* a former term for the western limb of a celestial object.

leading load *Electricity.* **1.** a load that is mostly capacitative, so that its current leads the voltage that is applied to the load. **2.** see CAPACITIVE LOAD.

leading pad *Computer Programming.* the group of characters, such as zeros, used to fill the unused space at the left end of a data field.

lead-in groove *Design Engineering.* the initial, blank, spiral groove adjacent to the outer surface of a disk recording whose pitch is greater than that of the recorded grooves; it acts as a lead-in to move the stylus to the first recorded groove. Also, LEAD-IN SPIRAL.

leading stone see LODESTONE.

leading strand *Molecular Biology.* the strand of DNA that grows continuously during DNA replication.

leading truck *Mechanical Engineering.* the pivoting front wheel assembly of a locomotive.

leading zeros *Mathematics.* all zeros (if any) to the left of the first nonzero entry in the decimal representation of a real number; in particular, for a real number of absolute value less than one, the zeros (if any) immediately to the right of the decimal point.

lead-in insulator *Electricity.* a tubular insulator inserted through a wall, through which lead-in wire is brought into a building. Also, **lead-in tube.**

lead-in spiral see LEAD-IN GROOVE.

lead iodide *Inorganic Chemistry.* PbI_2, toxic golden-yellow crystals; soluble in hot water; melts at 402°C and boils at 954°C; used in bronzing, printing, photography, and cloud seeding.

lead joint *Engineering.* a pipe joint created by caulking with molten lead or lead wool.

lead-lag network *Control Systems.* a compensating network in which the phase of sinusoidal response lags behind a sinusoidal input at low frequencies and leads it at high frequencies.

lead line see SOUNDING LINE.

lead lining *Engineering.* sheeting made of lead that is used to line the inside surfaces of equipment and vessels to help prevent corrosion.

lead molybdate *Inorganic Chemistry.* $PbMoO_4$, a toxic yellow powder; insoluble in water and alcohol and soluble in nitric acid; melts at 1060–1070°C; used in analytical chemistry.

lead monoxide *Inorganic Chemistry.* a compound appearing in two forms (litharge, massicot) with the same formula, PbO, but different physical properties.

lead network see DERIVATIVE NETWORK.

lead nitrate *Inorganic Chemistry.* $Pb(NO_3)_2$, colorless to white crystals that are soluble in water and alcohol; decomposes at 470°C; toxic, a strong oxidant, and a fire hazard in contact with organic substances; used in matches, explosives, photography, tanning, and engraving, and as a mordant and oxidizer.

lead oleate *Organic Chemistry.* $Pb[CH_3(CH_2)_7HC{=}CH(CH_2)_7COO]_2$, a toxic white powder; insoluble in water and soluble in alcohol, ether, and turpentine; used in varnishes, lacquers, paint driers, and high-pressure lubricants.

lead orthoarsenate see LEAD ARSENATE.

lead-out groove *Design Engineering.* a blank spiral groove at the innermost point of a disk recording that prevents the stylus from locking out to the center point or spindle. Also, THROW-OUT SPIRAL.

lead-over groove *Design Engineering.* a groove cut between recordings on a disk to allow the stylus to move from one selection to the next. Also, CROSSOVER SPIRAL.

lead patenting *Materials Science.* a steel-rod or wire production process in which an austenitized rod is cooled at a controlled rate in molten lead (about 550°C), usually as part of a multistrand continuous operation to develop a microstructure giving high strength and good ductility.

lead peroxide see LEAD DIOXIDE.

lead phosphate *Inorganic Chemistry.* $Pb_3(PO_4)_2$, a toxic white powder; insoluble in water and soluble in acids; melts at 1014°C; used as a plastics stabilizer. Also, **(normal) lead orthophosphate.**

lead pigment *Chemistry.* any of various lead compounds, such as lead carbonate, chromates, silicates, thiosilicates, and oxides, used to give color to another substance.

lead plant *Botany.* the shrub *Amorpha canescens* of the legume family, having a gray cast to the leaves and twigs; found in North America.

lead poisoning *Toxicology.* an acute or chronic intoxication with lead or lead salts caused by prolonged ingestion or absorption of lead-containing materials and resulting in gastrointestinal and mental disturbances, anemia, constipation, abdominal pain, convulsions, and coma. Also, PLUMBISM, SATURNISM, COLICOPLEGIA.

lead polyneuropathy *Medicine.* a disease of the peripheral nerves, affecting mainly the wrist and hand, that results from chronic lead poisoning and is marked by weakness, abnormal sensations of the skin, and pain.

lead pursuit course *Ordnance.* an interceptor vector designed to follow a course of flight at a predetermined point ahead of a target.

lead rail *Civil Engineering.* on a railroad turnout, the section of a diverging line's rail that lies between the rails of the main track.

lead resinate *Organic Chemistry.* $Pb(C_{20}H_{29}O_2)_2$, combustible, toxic brown lumps or yellow-white powder or paste; insoluble in most solvents; used in paint and varnish driers and as a textile waterproofing agent.

lead screw *Mechanical Engineering.* a horizontal screw that changes rotation to longitudinal motion, for moving the tool carriage of a lathe when cutting threads or guiding the cutter of a disk recorder across the surface of an ungrooved disk.

lead shield *Radiology.* a lead barrier that protects an X-ray technician or a patient from radiation.

lead silicate *Inorganic Chemistry.* $PbSiO_3$, noncombustible, toxic white crystals; insoluble in water; melts at 766°C; used in ceramics and fireproofing fabrics. Also, **lead metasilicate.**

lead silicofluoride see LEAD FLUOROSILICATE.

lead-soap lubricant *Materials Science.* any of various lubricants made of lead salts saponified with fats; an extreme-pressure lubricant that is hard at low temperatures, becoming somewhat fluid when heated by friction.

lead sodium thiosulfate *Inorganic Chemistry.* $PbS_2O_3 \cdot 2Na_2S_2O_3$, toxic, small, heavy, white crystals that are insoluble in water; used in matches. Also, **lead sodium hyposulfite.**

lead solder *Metallurgy.* any of several solders containing at least 50% lead.

lead stearate *Organic Chemistry.* $Pb(C_{18}H_{35}O_2)_2$, a combustible, toxic white powder; insoluble in water and soluble in alcohol; melts at 100–115°C; used as a lacquer and varnish drier and in high-pressure lubricants.

lead subcarbonate see LEAD CARBONATE, def. 2.

lead sulfate *Inorganic Chemistry.* $PbSO_4$, rhombic white crystals that are a strong irritant to tissue; slightly soluble in hot water; melts at 1170°C; used in storage batteries and paint pigments.

lead sulfide *Inorganic Chemistry.* PbS, toxic metallic crystals or black powder; insoluble in water; melts at 1114°C; used in ceramics, infrared detectors, semiconductors, and as a source of lead. Also, PLUMBOUS SULFIDE.

lead sulfide cell *Electronics.* a device that converts energy from light in order to measure the intensity of infrared radiation.

lead susceptibility *Chemical Engineering.* an increase in the octane number of gasoline due to lead additives, such as tetraethyl lead.

lead tack *Civil Engineering.* **1.** a lead strip used to adhere a lead pipe to a means of support. **2.** a lead strip folded over the edge of metal flashing.

lead telluride *Inorganic Chemistry.* PbTe, a toxic crystalline solid; insoluble in water and most acids; melts at 917°C; single crystals are used as a photoconductor and semiconductor in thermocouples.

lead tetraacetate *Organic Chemistry.* $Pb(CH_3COO)_4$, combustible, colorless to slightly pink crystals that are soluble in benzene and chloroform; melts at 175°C; used as an oxidizing agent in organic synthesis and as a laboratory reagent.

lead tetroxide *Inorganic Chemistry.* Pb_3O_4, a bright red powder that is insoluble in water and alcohol; decomposes at 500°C; used in storage batteries, paints, glazes, and varnishes, and for various other purposes. Also, **red lead (oxide).**

lead thiocyanate *Inorganic Chemistry.* $Pb(SCN)_2$, toxic white to pale yellow crystals that are slightly soluble in cold water; decomposes in hot water and the solid decomposes at 190°C; used in cartridge powder, matches, and dyeing. Also, **lead sulfocyanate.**

lead time *Industrial Engineering.* **1.** the time required between the beginning of a projector process and its completion. **2.** specifically, the elapsed time between successive events in a production process.

lead titanate *Inorganic Chemistry.* $PbTiO_3$, a toxic pale-yellow solid, insoluble in water; prepared by the interaction of lead oxide or carbonate and titanium dioxide at high temperatures; used as a paint pigment.

lead tungstate *Inorganic Chemistry.* $PbWO_4$, a toxic white powder that is insoluble in water; used as a pigment. Also, **lead wolframate.**

lead vanadate *Inorganic Chemistry.* $Pb(VO_3)_2$, a toxic yellow powder that is insoluble in water; used as a pigment and to make vanadium compounds. Also, LEAD METAVANADATE.

lead wire *Engineering.* a heavy wire that links the cap wires of an explosive device to the firing switch.

leadwort *Botany.* any shrub of the genus *Plumbago,* characterized by spikes of blue, white, or red flowers.

lead zirconate titanate *Materials.* a material that forms piezoelectric crystals; used in transducers and computer memory units.

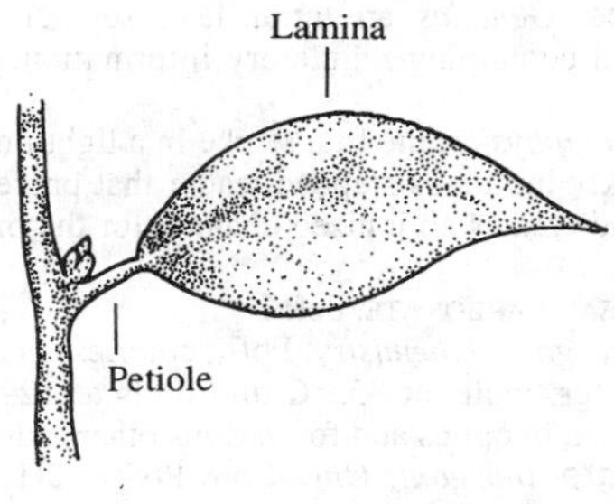

leaf

leaf *Botany.* an outgrowth of a plant stem, consisting of the lamina, petiole, and leaf base, that grows from a node; a leaf usually contains chlorophyll and is the primary site of photosynthesis. *Computer Programming.* the node, or vertex, at the end of a path in a tree data structure. *Building Engineering.* **1.** a sliding hinged or detachable flat part of a door or partition that is a separately movable piece. **2.** one of two halves of a double door window. *Graphic Arts.* **1.** a part of a book having a page printed on either side. **2.** also, **leafing.** a thin layer of gold or other metal, often stamped into another material as on a book cover.

leaf and dart *Architecture.* a design consisting of a closely set, alternating series of leaf-shaped and pointed forms; used especially in decorative moldings.

leaf and dart

leaf beetle *Invertebrate Zoology.* any of various brightly colored beetles of the family Chrysomelidae that feed on the leaves of plants and whose larvae infest the roots, leaves, and stem.

leafbird *Invertebrate Zoology.* any of various passerine birds of the genus *Chloropsis,* which are related to the bulbuls and often kept as pets; found in Asia.

leaf blight *Plant Pathology.* any of a variety of plant diseases that cause browning, defoliation, and eventually death of the plant, without rotting.

leaf blotch *Plant Pathology.* **1.** any of various plant diseases that produces spotted or streaked areas on the foliage. **2.** the areas affected by such a disease.

leaf bud *Botany.* an embryonic structure, usually on a stem, that produces only leaves. Also, **leaf buttress.**

leaf butterfly *Invertebrate Zoology.* any of several butterflies of the genus *Kallima,* characterized by wings that resemble dead leaves; found in southern Asia, Australia, and the East Indies.

leaf cast *Plant Pathology.* any of several diseases of conifers characterized by a shedding of needles.

leaf coral *Invertebrate Zoology.* a red coral of the species *Bossea orbigniana,* characterized by calcified, jointed stems; it is found as seaweed along the Pacific coast of the U.S.

leaf curl *Plant Pathology.* any fungus or viral plant disease characterized by foliage that becomes crinkled or curled.

leaf cushion *Botany.* the permanent base of some leaves, especially those of conifers and of certain extinct plants, that remains after the leaf has fallen.

leaf drop *Plant Pathology.* a diseased condition of plants in which the plant leaves fall prematurely.

leaf fiber *Botany.* any of the various fibers that are derived from plant leaves, especially for the manufacture of cordage or hemp.

leaf fish *Vertebrate Zoology.* any of some ten species of freshwater fish of the family Nandidae, characterized by a small, long, flat body with numerous dorsal and anal spines, and a large protrusible mouth; a ravenous predator that simulates a floating leaf as it awaits prey; found in tropical South America.

leaf frog see CENTROLENIDAE.

leaf gap *Botany.* a point at which the leaf trace diverges into the leaf; a region or parenchyma tissue in the vascular cylinder of the stem above the point of departure of the leaf trace or traces.

leafhopper *Invertebrate Zoology.* a common name for members of the insect family Cicadellidae, crop pests that suck plant juices.

leaflet *Botany.* **1.** a small or immature leaf. **2.** a single part of a compound leaf.

leaflet projectile *Ordnance.* a projectile used to carry paper leaflets; usually light-cased and equipped with a fuse that opens it before impact with the ground, thus distributing the leaflets. Also, **leaflet bomb.**

leaf lettuce *Botany.* a lettuce with loosely clustered leaves, often curled with red-tinged ends; widely used in salads and as a garnish.

leaf miner *Invertebrate Zoology.* an insect larva that burrows into leaves and eats the leaf tissue.

leaf mold *Geology.* a layer composed primarily of partially decayed vegetable matter that accumulates on the surface of a soil.

leaf mosaic *Botany.* the natural arrangement of leaves on a tree, bush, or vine to maximize light exposure and minimize shading.

leaf mottle *Plant Pathology.* a fungus disease affecting a wide range of plants, characterized by semitransparent spots on the leaves.

leaf-nosed *Vertebrate Zoology.* in vertebrates, having a prominent triangular scale or flap over the tip of the nose.

leaf-nosed bat *Vertebrate Zoology.* any of several bats from the families Phyllostomatidae, Rhinolophidae, and Hipposideridae, characterized by a leaflike scale over the tip of the nose; found in both the Old and New Worlds.

leaf primordium *Botany.* an immature leaf arising from the meristematic tissue.

leaf roll *Plant Pathology.* any of several viral plant diseases that cause an upward or inward curling of plant leaves and stunting and necrosis of the phloem.

leaf rot *Plant Pathology.* any plant disease characterized by a breakdown of leaf tissue.

leaf rust *Plant Pathology.* any plant disease, especially common to cereal grains, that is caused by rust fungi and characterized by reddish-brown lesions on the affected leaf blades and sheaths.

leaf scald *Plant Pathology.* a disease of sugarcane caused by *Bacterium albilineans,* which results in the infection of vascular tissues, streaking of plant parts, and subsequent withering of the plant.

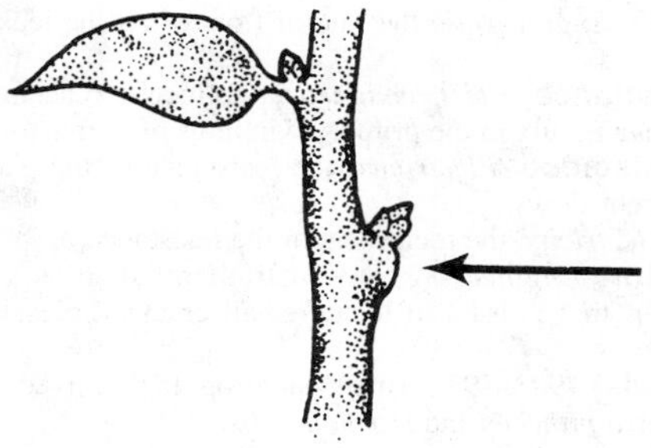

leaf scar

leaf scar *Botany.* a mark left on a stem at the point where a leaf has fallen off.

leaf scorch *Plant Pathology.* a fungal leaf disease giving the leaf a burnt appearance.

leaf sight *Ordnance.* a hinged sight that may be folded into a horizontal position when not in use and raised into a vertical position for sighting; usually used as the rear sight of small arms weapons.

leaf spot *Plant Pathology.* any plant disease characterized by well-defined, diseased, discolored spots on the leaves; caused by fungi, bacteria, viruses, or environmental factors.

leaf spring *Mechanical Devices.* a spring composed of thin, laminated plates independently and securely anchored at one end, forming a cantilevered beam. Also, LAMINATED SPRING, FLAT SPRING.

leaf stripe *Plant Pathology.* any of various plant diseases characterized by discolored linear markings on the foliage.

leaf trace *Botany.* a vascular bundle leading from the stele to the leaf and connecting the leaf with the vascular system of its stem.

leaf warbler *Vertebrate Zoology.* any of various small, green or brown warblers of the genus *Phylloscopus* that feed on insects on the leaves of trees.

league *Metrology.* **1.** a historic measure of distance not now in technical use, varying in length but typically three miles; widely used in preindustrial Europe. **2.** see SQUARE LEAGUE. (From the name of an ancient European measure of distance.)

leak *Engineering.* an unwanted and slow escape or entrance of something, such as the movement of water through a hole in a pipe, the escape of electricity above or through an insulating material, or the loss of neutrons through diffusion from the core of a nuclear reactor. *Plant Pathology.* a fungus disease of various fruits and vegetables, characterized by a watery breakdown of the plant tissue.

leakage *Engineering.* any act or process of leaking; an unwanted and slow escape or entrance of particles or material. *Physical Chemistry.* specifically, a condition in an ion-exchange process in which some of the unwanted ions are not removed by the solid, but remain in the liquid phase.

leakage current *Electronics.* **1.** the unwanted stray current that flows across the surface of an insulator or an insulating material. **2.** the current flowing between electrodes in a tube that does not flow across the interelectrode space. **3.** current before a voltage is applied. **4.** current that is traveling between energized parts of a circuit and ground. *Electricity.* an undesirable, small direct current flowing through the dielectric in a capacitor.

leakage factor *Electromagnetism.* a quantity given by the total amount of magnetic flux in a transformer or electric rotating machine divided by the amount of useful flux passing through the secondary winding or armature.

leakage flux *Electromagnetism.* magnetic field lines that are not useful because they extend around or beyond the useful region of interest. *Nucleonics.* the number of neutrons that escape permanently from the surface of a reactor core within a specified time.

leakage halo *Geochemistry.* a pattern of anomalous distributions of one or more elements dispersed outward during formation of an ore body or other such concentration, usually by solutions moving below the earth's surface.

leakage indicator *Electricity.* a test instrument used to measure the amount of current leakage from an electric system or circuit.

leakage inductance *Electromagnetism.* the self-inductance that is caused by leakage flux in a transformer.

leakage radiation *Electromagnetism.* any form of electromagnetic radiation other than the intended radiating system.

leakage rate *Engineering.* the rate of flow of all the leaks from a vessel.
leakage reactance *Electromagnetism.* inductive reactance that results from the leakage flux in the primary windings of a transformer.
leakage resistance *Electricity.* the resistance of the path over which leakage current flows.
leakance *Electricity.* the reciprocal of the resistance of insulation.
leak detector *Engineering.* any instrument, such as a helium mass spectrometer, that is used to locate small cracks, holes, or ruptures in vessel walls.
Leakey, Louis 1903–1972, British anthropologist, lived in Kenya; discovered *Kenya pithecus* and *Homo habilis.*

Louis and Mary Leakey

Leakey, Mary Douglas born 1913, wife of Louis Leakey, British anthropologist living in Kenya; discovered *Proconsul africanus* and *Zinjanthropus boisei.*
Leakey, Richard born 1944, son of Louis and Mary Leakey, British anthropologist living in Kenya; discovered *Homo erectus.*
leaking mode *Geophysics.* an imperfectly confined seismic wave that loses energy at the upper and lower boundaries of the layer on which it is propagating. Also, LEAKY WAVE.
leak-off rate *Petroleum Engineering.* the rate at which the fluid used in hydraulic fracturing permeates the surrounding reservoir formation.
leaky *Electricity.* of or relating to a capacitor with leakage current that is far above normal.
leaky gene *Genetics.* an allele that functions imperfectly as compared with a wild-type allele; prevalent among induced gene mutations. Also, HYPOMORPHIC ALLELE.
leaky mutant gene *Genetics.* a mutant gene that fails to change completely the action of a wild-type gene and shows some residual wild-type activity.
leaky wave see LEAKING MODE.
leaky-wave antenna *Electromagnetism.* a wideband, directional microwave antenna whose beam direction varies with the frequency.
lean *Materials Science.* describing a mixture that has a relatively small percentage of a desired or active component, as compared with a normal or rich mixture.
lean fuel mixture *Materials Science.* a fuel mixture composed of a high amount of air in proportion to its fuel content. Also, **lean mixture.**
leaning-wheel grader *Civil Engineering.* a grader that is fitted with skewed wheels, designed to cut and spread soil.
lean-to *Architecture.* a small, appurtenant structure with a single-pitched roof. *Building Engineering.* a single-sloped roof whose peak is supported by the wall of a higher adjacent structure.
leapfrog system *Mining Engineering.* a system using self-advancing supports on a longwall face in which alternate supports are advanced on each web of coal removed.
leapfrog test *Computer Programming.* a check of computer memory by a program that performs a series of arithmetic and logic operations on a block of memory, copies itself to another, noncontiguous block, and repeats the test until the entire memory has been tested.
leap second *Astronomy.* a second sometimes inserted at the end of a calendar year to keep atomic clock time in close agreement with the rotation of the earth, which is slowing gradually.
leap year *Astronomy.* **1.** in the Gregorian calendar, a year lasting 366 days rather than 365, with February 29th (**leap day**) added as the extra day; this occurs in years whose last two digits are evenly divisible by 4; e.g., 1996. **2.** a year with an extra day in any other calendar.
learboard *Building Engineering.* a board upon which a lead gutter is mounted. Also, LAVERBOARD.
learned [lurnd] *Behavior.* acquired through experience or conditioning; not innate.
learned behavior *Behavior.* any change in behavior that results from experience.
learned helplessness *Psychology.* a condition in which an individual becomes passive and depressed in response to an adverse or uncontrollable event, because he believes that his actions are ineffective. *Behavior.* a conditioned response in which an animal that has repeatedly been prevented from avoiding a negative stimulus will not seek to avoid it later when avoidance is possible.
learning *Behavior.* a relatively permanent change in behavior that results from practice or experience rather than from maturation or development of the organism.
learning by analogy *Artificial Intelligence.* a form of machine learning that attempts to understand a new domain more rapidly by making analogies with a domain that is already understood.
learning control *Control Systems.* a type of automatic control in which the control parameters and algorithms are modified by what actually happens in the system.
learning curve *Psychology.* a graph of educational performance with the horizontal axis showing the number of trials and the vertical axis showing the actual change in a subject's performance. *Industrial Engineering.* a pattern in the productivity of workers exposed to a new technology or process, in which they show a lowered rate at first and then disinct improvement as familiarity increases.
learning disability *Psychology.* **1.** an abnormal condition that often affects children of normal or above-normal intelligence, characterized by difficulty in understanding or using spoken or written language; thought to be related to slow development of perceptual motor skills. **2.** in popular use, any inability to learn new material at the average or expected rate.
learning disposition *Psychology.* the predisposition determining the range of things that an animal can learn and the rate at which it can learn them. Similarly, **learning capacity.**
learning machine *Computer Technology.* a computer that is capable of altering its behavior based on its analysis of its past experience.
learning set *Behavior.* **1.** an approach to problem situations in which the subject assumes that a specific method can be discovered to solve the problem. **2.** a factor that determines which kinds of responses will be made in a particular kind of problem situation.
learning theory *Behavior.* a group of concepts regarding the learning process, including theories of association, conditioning, and cognition.
leased facility *Telecommunications.* a communication facility that is reserved for use by a single leasing customer.
leased line see DEDICATED LINE.
least-action principle see MAUPERTUIS' PRINCIPLE.
least commitment strategy *Artificial Intelligence.* a search strategy in which a commitment to a particular choice is deferred as long as possible, so that constraints on possible values can be posted and propagated. When a variable is fully constrained, a choice is made, thus significantly reducing search and backup for some problems.
least common denominator *Mathematics.* given a finite set of rational numbers, the least common multiple of their denominators.
least common multiple *Mathematics.* the smallest positive integer that is divisible by each integer in the given set. The least common multiple of the elements $a_1, \ldots, a_n$ in a unique factorization domain R is an element c of R such that: (a) $a_1, \ldots, a_n$ are all divisors of c, and (b) if $a_1, \ldots, a_n$ are also divisors of another element x, then c is also a divisor of x. This is denoted $[a_1, \ldots, a_n]$. It can be shown that $[a_1, a_2] = a_1a_2/(a_1, a_2)$, where (a_1a_2) is the greatest common divisor of a_1 and a_2. Also, L.C.M.
least-constraint principle see GAUSS' PRINCIPLE.
least element see MINIMUM.
least-energy principle *Mechanics.* a principle stating that the stable equilibrium of a system occurs in the configuration having the least potential energy. Also, PRINCIPLE OF LEAST ENERGY.

least frequently used *Computer Programming.* an algorithm used by operating systems for memory management, in which a block of data that requires memory replaces the memory block that has been accessed least often.

least recently used *Computer Programming.* an algorithm used for memory management in which a block of data that requires memory replaces the memory block that has not been accessed for the longest time.

least significant *Computer Programming.* referring to the bit, character, or digit in the rightmost position of a word or number, the position of least importance or weight. Thus, **least significant bit, least significant character.**

least-squares method *Statistics.* a method of model parameter estimation that selects as a best estimate one that minimizes the sum of the squares of deviations of the data points from the values predicted based on the model. *Crystallography.* the use of this method in crystal structure analyses, in which atomic coordinates and other parameters may be fitted to the observed intensities; ideally, there should be at least 10 measurements for each parameter to be determined. In a similar way, the least-squares criterion can be applied to the computation of a plane through a group of atoms.

least upper bound *Mathematics.* a least upper bound (if it exists) for a subset B of a partially ordered set A (with ordering $\leq$) is an element s of A such that: (a) s is an upper bound of B, and (b) $s \leq c$, where c is any other upper bound of B.

least-work theory *Mechanics.* a theory stating that when forces are applied to an elastic system, the deflections of its parts are distributed so as to minimize the total work done by the forces.

leather *Materials.* **1.** a dressed animal skin, cured by the action of tannins or other processes to make it soft and flexible; widely used since ancient times for shoes, belts, gloves, outer clothing, and many other items. Hides from cattle are the most widely used source; others include sheep, pigs, sharks, and reptiles. **2.** relating to, containing, or made from this material.

leather rot *Plant Pathology.* a disease of strawberries caused by the fungus *Phytophthora cactorum* and characterized by a hardening and decay of the berries.

leathery cure *Materials.* an overcured state of vulcanized rubber in which the mixture is stiffer, harder, and less extendable than at optimal cure.

leathery state *Materials.* in the heating of polymers, a state in the glass transition temperature range in which the mechanical behavior of the polymer becomes sluggish; the polymer can be extensively deformed but will slowly return to its original state when the stress is removed.

leaving group *Organic Chemistry.* any group that can be displaced from a carbon atom.

lebek *Botany.* **1.** the tropical tree *Albizzia lebbeck* of the legume family, having pinnate leaves and greenish-yellow flowers; found in Asia and Australia. **2.** the durable wood of this tree used in building construction.

lebensspur see TRACE FOSSIL.

Lebesgue, Henri-Léon [lə beg′] 1875–1941, French mathematician; noted for the Lebesgue integral.

Lebesgue convergence see FATOU-LEBESGUE THEOREM.

Lebesgue integral *Mathematics.* for an integrand that is a real-valued function of a real variable, an integral defined by partitioning the range of the integrand function and summing up the corresponding areas below the graph of the function. It differs from the Riemann integral, which proceeds by partitioning the domain of the integrand. The Lebesgue integral exists for measurable integrands, whereas the Riemann integral exists only for integrands that are continuous except on a set of measure zero.

Lebesgue measure *Mathematics.* the Lebesgue measure on Euclidean R^n is the measure that arises by taking as the measure of a parallelepiped the usual definition of volume (the n-fold product of the lengths of the edges). Lebesgue measure is defined from Borel measure by a process called extension. The sigma algebra of Lebesgue measurable sets is larger than the sigma algebra of Borel sets.

Lebesgue number *Mathematics.* let X be a compact metric space with an open cover. The Lebesgue number of the cover is the positive real number λ such that any subset of X of diameter less than λ is completely contained in a member of the cover.

Lebesgue-Stieltjes integral *Mathematics.* an integral with the form $\int_a^b f(x)dg(x)$, where g is a nondecreasing left-continuous real-valued function that coincides with the Riemann-Stieltjes integral for continuous integrands f, but which also exists for measurable integrands f.

Leboyer method [lə boi yä′] *Medicine.* a theory of delivery focusing on the elimination of stress and birth trauma to the infant. It emphasizes a gentle, controlled delivery in a quiet, dimly lit room; careful and gentle handling of the infant, with no unnecessary intervention in the delivery or artificial stimulation of breathing; and immediate bonding between the mother and the child. (Formulated by the French obstetrician Frederick *Leboyer*.)

Lecanicaphaloidea *Invertebrate Zoology.* an order of tapeworms in the subclass Cestoda that parasitize rays and sharks.

Lecanidiales *Mycology.* in some classifications, an order of fungi of the subdivision Ascomycotina living nonparasitically on leaves or bark.

lecanora *Botany.* the type genus of the family Lecanoraceae, crustaceous lichens having an apothecia in which the disk is surrounded by a pale margin; sometimes used to make dyes or for food.

Lecanoraceae *Botany.* a family of lichens of the order Lecanorales, having a thin thallus that is crustaceous or squamulose and mostly superficial apothecia with a distinct rim.

Lecanorales *Botany.* an order of lichens of the class Ascolichenes having discoid, marginate apothecia and a thalline layer affording a protecting amphithecium.

Le Châtelier, Henri [lə shat′əl yā′] 1850–1936, French chemist; worked in high-temperature thermometry; formulated Le Châtelier's principle.

lechatelierite *Mineralogy.* SiO_2, a natural fused silica or silica glass formed by melting quartz sand at high temperatures generated by lightning strikes and meteoric impacts; found as fulgurites, irregular tubes of sand fused by lightning, and at Meteor Crater, Arizona. (Named for Henri *Le Châtelier*.)

Le Châtelier's principle *Physics.* a principle stating that if a system in equilibrium is disturbed by some external influence, the system will react in such a way as to alleviate the disturbance.

lecher wires [lek′ər] *Electromagnetism.* two long, insulated, parallel wires separated by a short distance that are made to resonate by sliding a shorting bar over the wires, forming a microwave electromagnetic transmission line that may be used as a tuned circuit, as an impedance matching device, or to measure wavelengths. (Named for the Austrian physicist E. *Lecher*.)

Lecideaceae *Botany.* a family of crustaceous lichens of the order Lecanorales having no thalloid rim surrounding the apothecia, which can be soft or carbonaceous.

lecithin *Biochemistry.* a phosphoglyceride that is the major component of cell membranes, consisting of esters of glycerol with two molecules of long-chain aliphatic acids and one of phosphoric acid.

lecithinase *Enzymology.* an enzyme that hydrolyzes lecithin at different positions.

lecithinase A *Enzymology.* phospholipase A1 and phospholipase A2, enzymes that remove fatty acids from lecithin.

lecithinase C *Enzymology.* phospholipase C, an enzyme that hydrolyzes lecithin to produce a diglyceride and a phospho-compound.

lecithinase D *Enzymology.* phospholipase D, an enzyme that hydrolyzes lecithin to produce phosphatidic acid and an organic compound.

Lecithoepitheliata *Invertebrate Zoology.* in some classifications, an order of neoophoran platyhelminth turbellarians having both marine and freshwater species.

Leclanche cell see DRY CELL.

lecontite *Mineralogy.* $(NH_4,K)Na(SO_4)\cdot 2H_2O$, a colorless, orthorhombic mineral having a specific gravity of 1.745 and a hardness of 2 to 2.5 on the Mohs scale; found as prismatic crystals and granular masses in bat guano.

Le Courbusier [lə kôr′byoo zyāy′] the professional name of Charles Edmond Jeanneret, 1887–1965, Swiss architect.

lectin *Biochemistry.* any of a large group of hemagglutinating proteins found principally in plant seeds. Certain lectins cause agglutination of erythrocytes of certain blood groups; others stimulate the proliferation of lymphocytes.

lectotype *Systematics.* a specimen selected from the type series by a subsequent author to be the holotype because the original author did not designate one.

Lecythidaecae *Botany.* a family of tropical trees of the order Lecythidales, characterized by simple, alternate leaves without stipules and solitary, bisexual flowers in spikes and racemes.

Lecythidales *Botany.* an order of tropical dicotyledonous trees of the subclass Dilleniidae, characterized by entire leaves, separate petals, valvate sepals, centrifugal stamens, and a syncarpous, inferior ovary.

LED light-emitting diode.

Leda *Astronomy*. Jupiter's thirteenth moon, discovered in 1974 and measuring about 15 km in diameter.

lede see LEADE.

ledeburite *Metallurgy*. the eutectic phase of the iron-carbon system, consisting of austenite and iron carbide.

Lederberg, Joshua born 1925, American geneticist; shared the Nobel Prize for the discovery of transduction.

Lederman, Leon M. born 1922, American physicist; discovered the upsilon particle; shared Nobel Prize for work with subatomic particles.

ledge *Geology*. **1.** a narrow, elongated, usually horizontal projection of rock formed on a rock wall or cliff face. Also, LEAD, LEDGE ROCK. **2.** an underwater ridge of rock, or a platform of resistant rocks along a coast. *Mining Engineering*. **1.** a projecting outcrop or vein, commonly of quartz, that is supposed to be mineralized. **2.** a mass of rock that constitutes a valuable mineral deposit. *Engineering*. **1.** an elevated molding or edge. **2.** a narrow shelf that protrudes from the side of a vertical building or structure. **3.** see LEDGER. *Building Engineering*. a horizontal timber across the back of a batten door or on a framed or braced door.

ledged-and-braced door *Building Engineering*. a door that is similar in appearance to a batten door, having diagonal framing and braces across the rear between battens.

ledged door see BATTEN DOOR.

ledger *Engineering*. a horizontal member, such as a piece of lumber, that helps to support a scaffolding. Also, LEDGE. *Building Engineering*. see LEDGER PLATE.

ledger balance *Telecommunications*. a facility linked with message-switching equipment to check the number of addresses received against those that are sent.

ledger beam *Building Engineering*. a reinforced-concrete beam with projected ledges to receive the ends of joists or similar members.

ledger board *Building Engineering*. an attachment applied to studding to carry joists.

ledge rock *Geology*. see LEDGE, def. 1.

ledger plate *Building Engineering*. a wooden strip laid horizontally across the tops of studding to add joist support. Also, LEDGER.

ledger strip *Building Engineering*. a wooden piece attached to a beam face to support the ends of joists.

Ledian see AUVERSIAN.

Ledoux bell meter *Engineering*. a manometer used to evaluate the difference in pressure between two points generated by one of various kinds of flow measurement instruments, such as a pitot tube.

Leduc current *Electricity*. a pulsed direct current that has a duty cycle of 1:10; used in electrotherapy.

lee *Meteorology*. a side or part that is sheltered from the wind. *Navigation*. the side of a vessel away from the wind, or the area on this side.

Lee, Tsung Dao born 1926, American physicist, born in China; shared the Nobel Prize with Chen Ning Yang for showing that the principle of conservation of parity does not hold true in all cases.

Tsung Dao Lee and Chen Ning Yang

Lee, Yuan T. born 1936, American chemist; shared the Nobel Prize for his work in reaction dynamics.

Leeaceae *Botany*. a monogeneric family of Old World tropical trees and shrubs of the order Rhamnales, characterized by simple to alternate leaves and hypogynous flowers with segments united basally into a tube.

leeboard *Naval Architecture*. either of two broad, flat planes attached to the hull of a sailing ship at the midpoint, so that the plane on the lee side can be lowered into the water to prevent the ship from moving downwind unintentionally.

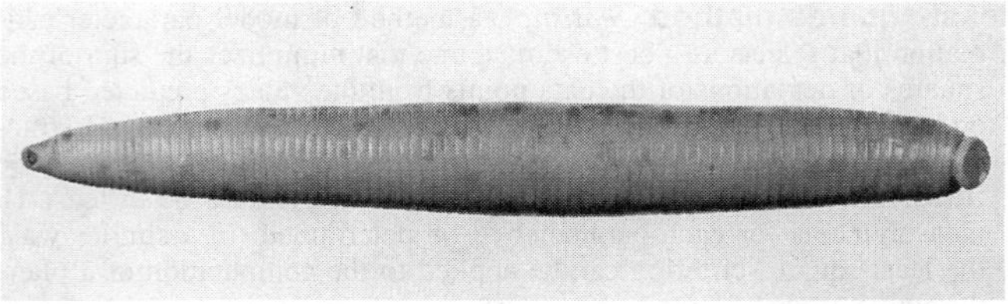

leech

leech *Invertebrate Zoology*. the common name for various predatory and bloodsucking parasites belonging to the annelid class Hirudinea. *Medicine*. **1.** also, **medicinal leech.** a particular worm of this type, *Hirudo medicinalis*, that was used in the former practice of leeching. **2.** a former name for a physician. (From an Old English word originally meaning "to pull.")

leech *Naval Architecture*. either of the side edges of a square sail. (From a Middle English word probably originally meaning "to tie.")

leechee see LITCHI.

leeching *Medicine*. the use of medicinal leeches to draw blood from patients; an extensive practice in earlier times that is now largely discontinued, though a few specialized uses of this technique still exist.

LEED low-energy electron diffraction.

lee dune *Geology*. a dune that is built up on the leeward side of an obstacle.

lee eddies *Fluid Mechanics*. the small, turbulent eddies or currents that are produced in the immediate vicinity behind an obstacle in the path of a flowing fluid.

Lee-Enfield *Ordnance*. one of several bolt-action .303 caliber carbines and repeating rifles widely used by the British Army, beginning in 1869 (carbine) and 1895 (rifle).

leek *Botany*. the common name for the species *Allium porrum* or *A. ampeloprasum*, a bulbous plant of the amaryllis family that resembles and is related to the onion, but has a milder flavor, and whose stems and leaves are often used as a flavoring in soups and stews.

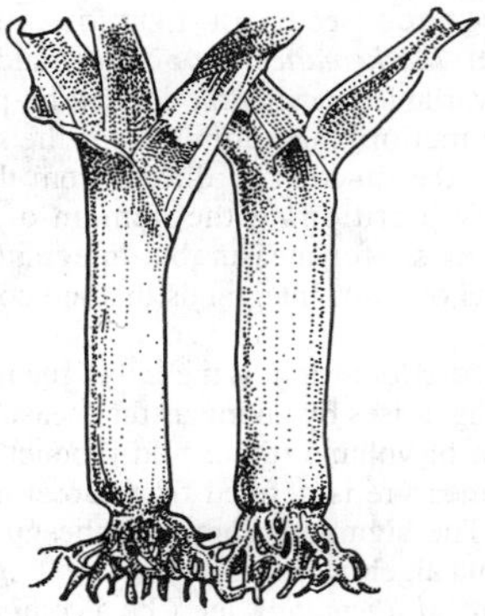

leek

lee shore *Meteorology*. a shore toward which the wind is blowing. *Navigation*. specifically, a shore toward which the wind is blowing and moving a ship.

lee tide see LEEWARD TIDAL CURRENT.

lee trough see DYNAMIC TROUGH.

Anton van Leeuwenhoek

Leeuwenhoek, Anton van [lā´vən hook´] 1632–1723, Dutch microscopist; designed noted early microscopes; observed cells, tissues, and microorganisms.

leeward *Meteorology.* away from the wind; the opposite of windward.

leeward tidal current *Oceanography.* a tidal current setting in the direction in which the wind is blowing. Also, LEE TIDE.

lee wave *Fluid Mechanics.* a wave disturbance that remains stationary with respect to a fixed boundary in the path of a flowing fluid.

leeway *Navigation.* the distance a craft makes downwind, especially if that is not the direction it is heading; measured in either distance or number of degrees off the desired course. *Engineering.* freedom of movement or operation; extra space, time, materials, and so on.

LE factor *Pathology.* a substance in a particle of the circulating blood serum that can provoke the formation of a lupus erythematosus cell.

LEFM linear elastic fracture mechanics.

left associative operator *Computer Science.* an operator in an arithmetic expression such that if there are two adjacent occurrences of the operator, the left one should be done first.

left-brain *Psychology.* of or relating to the concrete thought processes associated with the left hemisphere of the brain, such as calculation, organization, and logic.

left-continuous function *Mathematics.* a real function f is said to be left-continuous at a point c if $\lim_{x \to c-} f(x) = f(c)$; that is, $f(x)$ approaches $f(c)$ from below, or from the left, or for $x < c$. A right-continuous function is similarly defined.

left-hand *Design Engineering.* of or relating to those tools used in production or manufacturing work, such as drills, cutters, and screw threads, that are designed to rotate clockwise as they cut to the left. Thus, **left-hand screw, left-hand tool.**

left-hand derivative *Mathematics.* a real function f is said to have a left-hand derivative at a point c if $\lim_{x \to c-} |f(x)-f(c)|/|x-c|$ exists and is finite; i.e., the limit exists as x approaches c from below, or for $x < c$. A right-hand derivative is similarly defined.

left-handed *Neurology.* having a tendency to use the left hand more frequently and with greater dexterity than the right. Thus, **left-handedness.** *Design Engineering.* of or relating to screw threads that allow coupling only by turning in a counterclockwise direction.

left-handed coordinate system *Mathematics.* **1.** a system visualized in Euclidean space by extending the left thumb and the forefinger at a right angle and the left middle finger at right angles to both of them. Then the positive x, y, and z axes are taken in the directions of the thumb, forefinger, and middle finger, respectively. It is so named because a screw with a left-handed thread placed along the z axis will advance in the positive z direction when it is turned so that the positive x axis rotates directly into the positive y axis. **2.** more generally, on a Riemannian n-manifold, where the coordinate functions have gradient vectors, the coordinate system is left-handed if the determinant of the gradients (in the given order) is negative.

left-hand polarization *Electromagnetism.* for a circularly or elliptically polarized electromagnetic wave, the polarization that rotates in a counterclockwise direction when observed along a direction of propagation.

left-hand rule *Electromagnetism.* **1.** a rule stating that for a conductor carrying a current, if the thumb of a person's left hand points along the direction of electron flow (opposite to the conventional current direction), the fingers will curl along the direction of the magnetic field lines. **2.** a rule stating that if the left middle finger, index finger, and thumb are set mutually perpendicular, then they will describe the relative directions of the current, magnetic field direction, and the force on a current-carrying conductor in a magnetic field, respectively.

left-hand screw *Mechanical Devices.* a screw that is inserted into a threaded bolt or hole in a counterclockwise direction.

left-hand taper *Electricity.* a taper in which there is a greater resistance in the counterclockwise half of the operating range of a variable resistor than in the clockwise half, looking from the shaft end.

left heart bypass *Cardiology.* a shunt that diverts the flow of blood from the pulmonary veins directly to the aorta, avoiding the left atrium and the left ventricle; used in cardiac surgery.

left-invariant *Mathematics.* a geometrical object (i.e., an object with a transformation law), such as a vector field, defined on an open set of a group, is said to be left-invariant if the transformation law corresponding to translation by left multiplication gives a new value for the object equal to the value it already has at the left-translated point. Accordingly, a left-invariant object whose value is given at the identity element of the group is defined on (an open neighborhood of the identity element of) the entire group.

left-justified *Graphic Arts.* describing spacing within a line of type that is arranged or adjusted so that the first character is in a desired position, usually flush left.

left-justify see JUSTIFY.

left-laid *Design Engineering.* in a wire or fiber optic cable, the lay of strands that are twisted to the left in bunches, as distinguished from individual wires or fibers twisted to the right.

left-lateral fault *Geology.* a strike-slip fault in which the relative displacement appears to be offset to the left when viewed across the fault plane. Also, LEFT-SLIP FAULT, SINISTRAL FAULT.

left parasternal impulse *Cardiology.* the cardiac impulses felt on the chest at the left sternal border.

left recursion *Computer Science.* in top-down parsing, an infinite recursion caused by a grammar rule whose right-hand side begins with the nonterminal symbol on the left-hand side.

left rudder *Navigation.* a command to turn a craft to the left by putting the rudder over to port.

left-slip fault see LEFT-LATERAL FAULT.

left value *Computer Programming.* the memory location of a symbolic variable in a program.

leg *Anatomy.* **1.** the lower limb of the body. **2.** specifically, the portion of the lower limb between the knee and ankle. *Zoology.* an animal limb used for walking or support.

leg *Engineering.* any object or part that resembles a human or animal leg, either structurally or functionally, such as an upright member or branch of a joint or derrick. *Transportation Engineering.* **1.** a single airborne portion of a flight, from takeoff to landing. **2.** an approximately straight line portion of an aircraft's landing approach to an airport. *Computer Programming.* a sequence of instructions between one branch point and the next in a computer routine. *Geophysics.* a single cycle of a periodic motion in a wave train of a seismogram.

legal-move generator *Artificial Intelligence.* a program that generates the set of legal moves or actions that can be taken from a given position, as in a board game.

legend *Anthropology.* a popular narrative, typically dealing with a hero or great event, passed down by tradition and believed to have historical accuracy, though without verification. *Graphic Arts.* an explanatory caption. *Cartography.* a key to the symbols on a map, chart, or table.

legendary *Anthropology.* expressed in or according to a legend.

Legendre, Adrien [lə zhän´dər] 1752–1833, French mathematician; worked in number theory and elliptic functions; the first to state the theory of least squares.

Legendre condition *Mathematics.* a test corresponding to the second derivative test of single-variable calculus; used in the calculus of variations. Suppose a function $y(x)$ satisfies the Euler equation, so that it extremizes the integral $\int_a^b f(x, y(x), y'(x))dx$. Suppose that, for $x \in [a, b]$,

$$f_{y'y'}(x, y(x), y'(x)) \geq 0 \text{ (or } \leq 0).$$

Then the extremal must be a minimum (or maximum) on $[a, b]$. Also, OPTIMIZATION CONDITION OF LEGENDRE.

Legendre equation *Mathematics.* **1.** a differential equation having the form

$$(1 - x^2)y'' - 2xy' + n(n + 1)y = 0,$$

where the parameter n is any nonnegative real number and y is a function of x. Also, **Legendre's differential equation. 2.** a differential equation of the form

$$(1 - x^2)y'' - 2xy' + [n(n + 1) - m^2/(1 - x^2)]y = 0,$$

where m and n are nonnegative integers. Also, ASSOCIATED LEGENDRE EQUATION.

Legendre function *Mathematics.* **1.** any solution of the Legendre equation; when n is an integer, the solutions are called *Legendre polynomials.* **2.** solutions of the associated Legendre equation are usually called **associated Legendre functions** and are denoted $P^m_n(x)$. Associated Legendre functions are related to the Legendre polynomials by the equation

$$P^m_n(x) = (1 - x^2)^{m/2}\, d^m/dx^m[P_n(x)].$$

Legendre polynomials *Mathematics.* a family of mutually orthogonal solutions of the Legendre equation in the case where n is a nonnegative integer. They are the only linearly independent solutions of the Legendre equation that are bounded at $x = \pm 1$. Usually denoted by $P_n(x)$, they satisfy a Rodriguez formula: $P_n(x) = d^n/dx^n\, [(x^2 - 1)^n]$ and the recursion formula:

$$(n + 1)P_{n+1}(x) = (2n + 1)x\, P_n(x) - nP_{n-1}(x).$$

$P_n(x)$ is a polynomial of degree n having exact n real simple zeros in the open interval (–1,1) and in which the leading coefficient is $(2n)!/2^n(n!)^2$. The ordinary generating function of the Legendre polynomials is given by

$$(1 - 2xz + z^2)^{-1/2} = \sum_{n=0}^{\infty} P_n(x)z^n.$$

Legendre's theorem *Mathematics.* a theorem relating spherical and plane trigonometry. The area of a spherical triangle with small sides and hence small spherical excess is nearly equal to that of a plane triangle having the same sides. Further, each angle of the plane triangle is less than the corresponding angle of the spherical triangle by 1/3 the spherical excess.

Legendre symbol *Mathematics.* in elementary number theory, the symbol $(x \mid p)$, which equals 1 if an integer x is a quadratic residue of an odd prime p, and equals –1 if x is a quadratic nonresidue.

Legendre transformation *Mathematics.* a transformation of a convex function $f(x)$, constructed as follows: for each real number p, the function $F(p, x) = px - f(x)$ has a maximum with respect to x at some unique point $x(p)$. Then $g(p) = F(p, x(p))$ is the Legendre transformation of $f(x)$.

leghaemoglobin *Botany.* a hemoglobinlike protein in the nitrogen-fixing nodules of legumes that functions to maintain a low concentration of oxygen. Also, **leghemoglobin.**

Leghorn *Agriculture.* a breed of chicken that lays white eggs and is characterized by a yellow beak and skin, a single comb, and nonfeathered shanks. (From the name Livorno, a city in Italy, altered in English to *Leghorn.*)

Legionellaceae *Bacteriology.* a taxonomic family of Gram-negative, pathogenic bacteria that occur as aerobic, flagellated rods or filaments, use amino acids as both carbon and energy sources, and occur in aquatic habitats, including domestic water systems.

legionellosis *Pathology.* the medical name for Legionnaire's disease. See LEGIONNAIRE'S DISEASE.

Legionnaire's disease or **legionnaire's disease** *Pathology.* a severe contagious disease that is caused by the bacterium *Legionella pneumonophila,* characterized by influenzalike symptoms followed by high fever, chills, headache, pleurisy, pneumonia, and sometimes death. (First specifically identified among people attending a convention of the American *Legion* at Philadelphia in 1976.) Also, LEGIONELLOSIS.

legrandite *Mineralogy.* $Zn_2(AsO_4)(OH)\cdot H_2O$, a yellow to nearly colorless monoclinic mineral occurring in aggregates of long, prismatic crystals, having a specific gravity of 3.98 and a hardness of 4.5 to 5 on the Mohs scale; found in limestone.

legume [leg´yoom´] *Botany.* **1.** any plant of the family Leguminosae. **2.** a dry, dehiscent fruit of the Leguminosae, such as a bean, pea, or lentil, that develops from the splitting of a single carpel into two valves, each bearing seeds alternately along the ventral margin.

legumin see AVENIN.

Leguminosae *Botany.* a family of dicotyledonous herbs, trees, and vines of the order Rosales, characterized by pinnately compound leaves, fruits that are pods, and root nodules that contain nitrogen-fixing bacteria of the genus *Rhizobium.* Also, FABACEAE.

leguminous *Botany.* relating to or being a plant that bears legumes.

leg wire *Engineering.* either of two wires forming part of an electric blasting cap. Also, **leg.**

lehiite *Mineralogy.* a mixture of apatite and crandallite.

Lehn, Jean-Marie born 1939, French chemist; awarded Nobel Prize for analysis of crown ethers and synthesis of cryptands.

lehr *Materials Science.* a long, annealing furnace in which residual stresses are removed from glass made in the float-glass process. The lehr directly follows the float bath, and the glass proceeds through it on rollers.

Leibnitz, Gottfried Wilhelm [lib´nits] 1646–1716, German philosopher and mathematician; developed the theory, methods, and notation of differential and integral calculus.

Leibnitz's formula *Mathematics.* a formula used to compute the nth derivative of the product of two functions. Let D^k denote the differential operator d^k/dx^k. Then

$$D^n(fg) = \sum_{k=0}^{n} \binom{n}{k} D^{n-k}(f)D^k(g).$$

For $n = 1$, this is also known as **Leibnitz's product rule.**

Leibnitz's rule *Mathematics.* a rule for differentiation under the integral sign: Let $\phi(\alpha) = \int_u^v f(x, \alpha)dx$, where $f(x, \alpha)$ and $\partial f/\partial\alpha$ are continuous in both x and α for $u \le x \le v$ and $a \le \alpha \le b$, and where u and v may depend on the parameter α. Then Leibnitz's rule for differentiation is:

$$d\phi/d\alpha = \int_u^v \partial f/\partial\alpha\, dx + f(v, \alpha)\, \partial v/\partial\alpha - f(v, \alpha)\, \partial v/\partial\alpha.$$

(If u and v are constant, then the last two terms are zero.)

Leibnitz's series *Mathematics.* an infinite series expression for the value of $\pi/4$:

$$\pi/4 = \sum_{k=0}^{\infty} (-1)^k/(2k + 1) = 1 - 1/3 + 1/5 - 1/7 + \cdots .$$

Leibnitz's test *Mathematics.* The alternating series $\sum_{k=1}^{\infty} (-1)^k a_k$ converges if the a_k are positive real numbers which monotonically approach zero. Also, ALTERNATING SERIES TEST.

Leicester [les´tər] *Agriculture.* the smallest breed of long-wool sheep, having no horns and a white face; used widely for breeding. Also, ENGLISH LEICESTER, LEICESTER LONGWOOL. (From *Leicester,* a county in England.)

Leidenfrost point *Thermodynamics.* the lowest temperature that a heated body may have when submerged in boiling water while still maintaining a vapor layer over its entire surface; heat conduction from the body to the water is a minimum at this point.

Leidenfrost's phenomenon *Thermodynamics.* the formation of a vapor film between a liquid and a surface upon which the liquid is dropped, if the surface is at a temperature greater than the critical point of the liquid; the result is that the surface is insulated from the liquid and is not made wet by it.

leifite *Mineralogy.* $Na_2(Si,Al,Be)_7(O,OH,F)_{14}$, a transparent, colorless or white trigonal mineral occurring in acicular, striated crystals, having a specific gravity of 2.57 and a hardness of 6 on the Mohs scale; found in alkali pegmatites.

Leifson's flagella stain *Microbiology.* a bacteriological stain that coats the flagella on a bacterial cell, rendering them visible by light microscopy.

Leigh light *Ordnance.* a British airborne searchlight for illumination of U-boats detected by radar, enabling surprise depth-charge attack.

leightonite *Mineralogy.* $K_2Ca_2Cu^{+2}(SO_4)_4\cdot 2H_2O$, a rare, pale-blue, triclinic, pseudo-orthorhombic mineral occurring in lathlike crystals, having a specific gravity of 2.95 and a hardness of 3 on the Mohs scale; found at Chuquicamata, Chile.

Leighton tube *Biotechnology.* a test tube having a flat, rectangular part near the closed end, at which a single layer of cells can form.

Leiodidae *Invertebrate Zoology.* a family of coleopteran insects in the superfamily Staphylinoidea; round carrion beetles found in decaying vegetable matter.

Leiognathidae *Vertebrate Zoology.* the ponyfishes, a family of mostly marine schooling fishes of the order Perciformes, characterized by a naked bony head, large eyes, and a small protrusible mouth; found in the Indo-West Pacific.

leiomyofibroma *Oncology.* a benign tumor composed of smooth muscle and fibrous tissue.
leiomyoma *Oncology.* a benign tumor composed of smooth muscle tissue, frequently seen in the uterus.
leiomyosarcoma *Oncology.* a malignant neoplasm containing large spindle cells of nonstriated smooth muscle cells.
leiosporous *Mycology.* of or relating to a fungal growth that possesses smooth pores.
Leishman-Donovan bodies *Pathology.* spherical bodies found in the reticuloendothelial cells, particularly those of the spleen and liver, of patients diagnosed with kala-azar; essentially the nonflagellate intracellular form of *Leishmania donovani,* the parasite primarily responsible for this disease. (Named for Sir William *Leishman,* 1865–1926, English army surgeon, and Charles *Donovan,* 1863–1951, Irish physician.)
Leishmania *Invertebrate Zoology.* a genus of tiny, flagellated protozoans that are disease-causing parasites of humans and other animals.

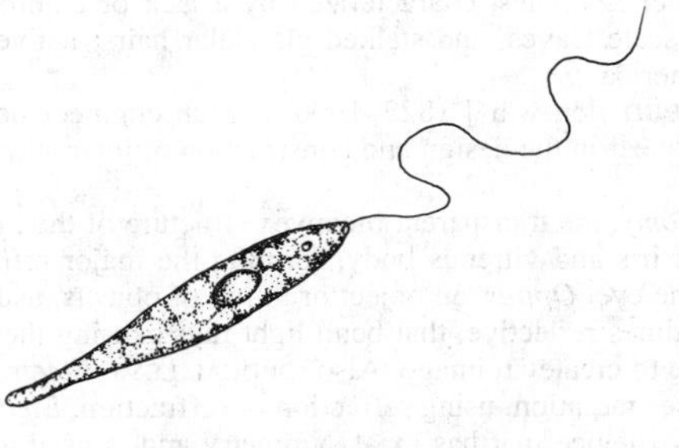

Leishmania

Leishmania donovani *Invertebrate Zoology.* the parasitic species of flagellated protozoan that causes kala-azar.
Leishmania infantum *Invertebrate Zoology.* the parasitic species of flagellated protozoan that causes infantile leishmaniasis.
leishmaniasis *Pathology.* an infection caused by protozoa of the genus *Leishmania* and transmitted by the bite of a sandfly. Each of the three main forms, visceral, cutaneous, and mucocutaneous, has distinctive manifestations and distribution.
leishmaniasis recidivans *Pathology.* a recurring infection caused by *Leishmania* and resembling skin tuberculosis. The skin ulcers partially heal, then spread at the edge of the scar.
Leitneriaceae *Botany.* a monospecific family of light- and soft-wooded dicotyledonous shrubs comprising the order Leitneriales, characterized by deciduous leaves and confined to the coastal plain of the southeastern United States.
Leitneriales *Botany.* an order of flowering, woody plants of the subclass Hamamelidae, characterized by alternate, entire leaves, and flowers in axillary catkins developing into drupes.
Lejeuneaceae *Botany.* a large and complex family of prostrate, brownish liverworts of the order Jungermanniales, characterized by exclusively terminal branches, no stolons or flagella, complicate-bilobed leaves, and usually large and conspicuous underleaves; most common in tropical and subtropical regions.
lek *Behavior.* a communal area in which males assemble during the mating season to attract females; males control small territories in the lek that are used for courtship and copulation. *Psychology.* an individual's need for social association with others, through friendship, cooperative relationships, groups and organizations, and the like.
Lelapiidae *Invertebrate Zoology.* a family of calcareous sponges in the order Sycettida, found in shallow warm seas.
L electron *Atomic Physics.* any of the electrons that are found in the L shell of an atom and display the characteristics of the principal quantum number 2.
Leloir, Luis Federico born 1906, Argentine biochemist born in France; awarded the Nobel Prize for the discovery and analysis of sugar nucleotides.
LEM see LUNAR MODULE.
Lemaitre, Georges-Henri 1894–1966, Belgian astrophysicist; formulated Lemaitre model; research in relativity theory and cosmic rays.
Lemaitre model *Astrophysics.* a mathematical model for the universe in which the rate of expansion decreases steadily after its birth in a Big Bang explosion.
Lemaneaceae *Botany.* a monogeneric family of freshwater red algae belonging to the order Nemaliales, characterized by a thallus having a prostrate filamentous basal portion and an upright, cylindrical pseudoparenchymatous portion where gametangia are formed.
Lembophyllaceae *Botany.* a family of shiny, rigid, robust mosses of the order Hypnobryales that form loose mats on tree trunks, humus, and rocks; characterized by creeping prostrate stems that are regularly branched and sometimes have secondary stems or branches that form dendroid or frondose mats.
Leminorella *Bacteriology.* a genus of Gram-negative bacteria of the family Enterobacteriaceae that metabolize arabinose and xylose and are found in human feces and urine.
lemma *Botany.* the lower of a pair of bracts enclosing the florets of grasses, frequently bearing an awn. *Mathematics.* an initial result to be proven for use in establishing a theorem.
LEMMA *Robotics.* a remote teleoperator designed to perform assembly, maintenance, and repair activities in space.
lemming *Vertebrate Zoology.* a small, herbivorous, circumpolar rodent of the genus *Lemmus* or *Dicrostonyx,* having small ears and a short tail. In conditions of food shortage or overpopulation, they participate in huge migrations that are popularly believed to result in mass drownings in the sea, though this is now known to be atypical.

lemming

Lemnaceae *Botany.* a monocotyledonous family of aquatic plants of the order Arales having a reduced green, leaflike or globose thallus and minute, unisexual flowers.
lemniscate (of Bernoulli) *Mathematics.* a fourth-order curve with equation in Cartesian coordinates: $(x^2 + y^2)^2 - 2a^2(x^2 - y^2) = 0$, where $a > 0$. The curve resembles a figure-eight on its side with double point at the origin and is a special case of a Cassinian oval.

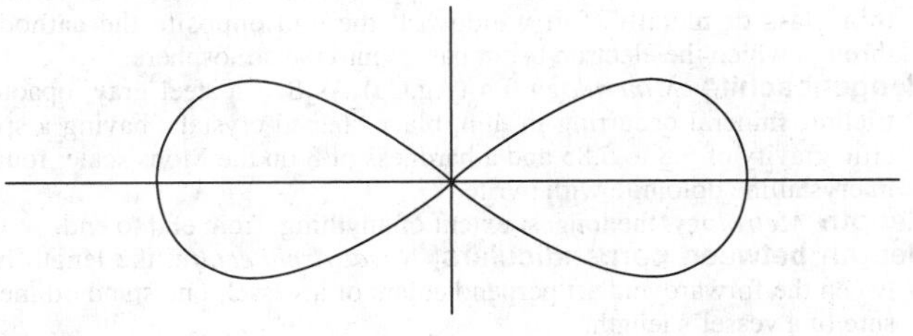

lemniscate

lemniscus *Anatomy.* any of several bundles of sensory nerve fibers that pass from sensory nuclei to the thalamus.
lemon *Botany.* **1.** the common name for the species *Citrus limon,* a small evergreen tree of the order Sapindales that produces an acid citrus fruit. **2.** the small, bright yellow, edible fruit of this tree, widely used for drinks and flavorings.
lemon grass *Botany.* any of various lemon-scented grasses of the genus *Cymbopogon* that yield lemon-grass oil; found in tropical regions.
lemon-grass oil *Materials.* a dark yellow to brown essential oil with a heavy lemon odor, obtained from the distilling of certain types of lemon grass, especially *Cymbopogon citratus*; used in flavorings and in perfumes.
lemon oil *Materials.* **1.** a fragrant yellow essential oil expressed from the peel and rind of lemons, widely used in flavorings, perfumes, and furniture polishes. **2.** a similar product that is manufactured synthetically.

lemur

lemur *Vertebrate Zoology.* a nocturnal, arboreal, monkeylike vertebrate of the primate family Lemuridae, having a long, furry tail and large eyes; once widespread but now largely confined to Madagascar.

Lemuridae *Vertebrate Zoology.* the lemurs, a family of monkeylike primates of the suborder Lemuroidea; inhabiting the forests of Madagascar and the Comoro Islands.

Lemuriformes *Paleontology.* an infraorder of primates that includes the families Lemuridae and Lorisidae, the lemurs of Madagascar and bushbabies of Africa.

Lemuroidea *Vertebrate Zoology.* a suborder of primates containing the lemurs, tarsiers, and lorises.

Lena *Geography.* a river in eastern Siberia rising in the Baikal Mountains and flowing about 2700 miles north and northeast to the Arctic Ocean.

Lenard, Phillipp 1862–1947, German physicist; pioneer in photoelectricity; awarded Nobel Prize for work with cathode rays.

Lenard rays *Electronics.* the stream of electrons that passes through a thin glass or metallic foil window in a vacuum tube.

Lenard tube *Electronics.* an early type of electron-beam tube, with a thin glass or metallic foil window at the end opposite the cathode, through which the electron beam passes into the atmosphere.

lengenbachite *Mineralogy.* $Pb_6(Ag,Cu)_2As_4S_{13}$, a steel gray, opaque, triclinic mineral occurring in thin, blade-shaped crystals, having a specific gravity of 5.8 to 5.85 and a hardness of 3 on the Mohs scale; found in crystalline dolomite with pyrite.

length *Metrology.* the longest extent of anything, from end to end.

length between perpendiculars *Naval Architecture.* the length between the forward and aft perpendiculars of a vessel; one standard measure of a vessel's length.

lengthened dipole *Electronics.* an antenna element that has lumped inductance to compensate for end loss.

lengthening joint *Engineering.* a joint between two pieces of material that are in alignment.

lengthening reaction *Physiology.* the elongation of an extensor muscle that permits the flexion of a limb.

length of a curve *Mathematics.* the length of the graph of $y = f(x)$ from a to b is given by $\int_a^b [1 + (f'(x)^2]^{1/2}\,dx$, if it exists. More generally, if a curve k in Euclidean R^n has the parametric representation

$$x_k = \phi_k(t),\ \ a \le t \le b,\ \ k = 1, 2, \ldots, n,$$

where the $\phi_k(t)$ are continuously differentiable functions, then the length of k between a and b is given by

$$\int_a^b [\sum_{k=1}^{n} \phi_{k'}^2(t)]^{1/2} dt.$$

length of bore *Ordnance.* the length of a weapon's bore measured from the rear face of the tube or barrel to the muzzle; expressed in inches or calibers.

length of lay *Design Engineering.* the distance measured along a line parallel to the axis of a rope in which the strand completes one turn around the axis of the rope, or the wire completes one turn around the axis of the strand.

length of record *Meteorology.* the period of time during which observations have been maintained without interruption at a meteorological station; used as a frame of reference for climatic data at that site.

length of thaw *Meteorology.* the interval of time during which the air temperature over a given area has remained above 32°F, thus inducing the melting of ice and frozen water bodies.

length on waterline *Naval Architecture.* the length of a vessel on its load waterline.

Lennard-Jones potential (function) *Physical Chemistry.* an approximate equation that predicts the potential force between two molecules on the basis of the distance between their centers; it is a difference of two terms, each involving reciprocal distances, one representing repulsion and the other attraction.

Lennoaceae *Botany.* a family of dicotyledonous, fleshy-root parasites of the order Lamiales, characterized by a lack of chlorophyll, spirally arranged scale leaves, and stalked glandular hairs; native to North and South America.

Lenoir, Jean [len wär′] 1822–1900, French engineer born in Luxembourg; pioneer in the design and construction of internal-combustion engines.

lens *Anatomy.* the transparent biconvex structure of the eye that lies between the iris and vitreous body, forming the major refracting mechanism of the eye. *Optics.* an object or group of objects, usually refractive but sometimes reflective, that bend light rays causing them to converge or diverge to create an image. Also, OPTICAL LENS. *Electronics.* a device that focuses radiation, using refraction or diffraction. *Electromagnetism.* a magnetic device that has axial symmetry and is used to cause beams of charged particles to converge. *Geology.* any geologic deposit or body bounded by at least one curved, converging surface, so that it is thick in the middle and thins out toward the edges, giving it a lenslike shape. *Telecommunications.* a dielectric or metallic structure that refracts radio waves to produce a desired radiation pattern; commonly used in microwave radio-system transmitters and repeaters. (A special use of the word *lentil*; from the similarity in shape.)

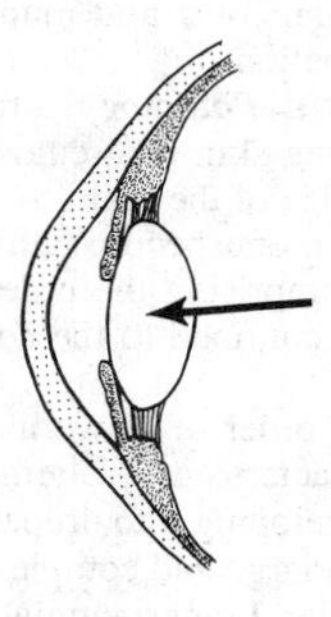

lens

lens antenna *Electromagnetism.* a directional microwave antenna that uses a dielectric lens in front of a dipole or horn radiator to focus the radiated energy into a concentrated beam.

lens coating *Optics.* the thin layer of material that is evaporated onto a glass lens surface to reduce the amount of light the lens will reflect, thereby increasing its transmittance and reducing possible lens flare.

lens element *Optics.* a single lens within a multi-lens system.

lens equation *Optics.* the relationship of the position of an image to the position of the object and focal length of the lens. Also, THIN LENS EQUATION, GAUSSIAN LENS FORMULA.

lens placode *Developmental Biology.* a thickened area of ectoderm that lies over the optic vesicle in the early embryo and from which the lens develops.

lens-shaped *Microbiology.* describing a microbial colony or structure that resembles a lens in shape.

lens shim *Optics.* the fragment of material that positions and focuses a lens.

Lentibulariaceae *Botany.* a family of cosmopolitan insectivorous herbs of the order Scrophulariales that are aquatic or grow in very wet places, either rooted in the substrate or rootless and free floating, with submerged photosynthetic organs; includes the bladderwort species.

lentic *Ecology.* of or relating to freshwater habitats characterized by calm waters, such as lakes, ponds, swamps, or bogs. Also, **lenitic.**

lenticel *Botany.* a spongy area in the stems of woody plants that allows for the exchange of gases between the plant and the atmosphere.

lenticle *Geology.* a rock stratum or rock bed that thins at the edges, thus having a somewhat lens-shaped appearance; characterisitic of most undeformed strata.

lenticular *Optics.* **1.** an object shaped like a double convex lens. **2.** relating to or describing lenses.

lenticular galaxy *Astronomy.* an S0-type galaxy containing old stars and little or no gas.

lenticularis *Meteorology.* a species of clouds having a characteristic lens- or almond-shaped appearance, with elements sharply outlined.

lentiginose *Anatomy.* freckled; covered with numerous small round to oval brown spots.

lentil *Botany.* **1.** the species *Lens culinaris* or *L. esculenta*, a plant of the Leguminosae, having flattened biconvex seeds. **2.** the edible seeds of this plant, used as a food and as animal fodder since ancient times. *Geology.* **1.** a minor stratigraphic unit of limited geologic extent that thins out in all directions. Also, TONGUE. **2.** any lens-shaped rock or ore body.

lentil

Lentinula *Mycology.* a genus of fungi belonging to the order Agaricales; the species *L. edodes* is an edible mushroom known as the shiitake, which is cultivated in Japan and China.

Lentinus *Mycology.* a genus of fungi belonging to the order Agaricales that occurs on wood.

Lentivirinae *Virology.* a subfamily of the nononcogenic viruses of the family Retroviridae, one member of which causes AIDS.

lentogenic strain *Virology.* a mild or avirulent virus strain.

Lenz's law *Electromagnetism.* a law stating that an induced electromotive force in a conductor is always polarized in a direction so as to oppose the change that causes the induced electromotive force.

Leo *Astronomy.* the Lion, a conspicuous northern constellation that dates from antiquity and lies in the zodiac between Cancer and Virgo; its brightest star is 1st magnitude Regulus.

Leo I galaxy *Astronomy.* a dwarf, elliptical galaxy 5900 light-years in diameter with an absolute magnitude of about –11 that lies roughly 720,000 light-years in the direction of the constellation of Leo.

Leo II galaxy *Astronomy.* a dwarf, elliptical galaxy 4200 light-years in diameter with an absolute magnitude of about –9.5 that lies roughly 720,000 light-years in the direction of the constellation of Leo.

Leo Minor *Astronomy.* the Lesser Lion, a faint constellation of the northern hemisphere that contains no star brighter than 4th magnitude and lies between Leo and Ursa Major.

Leonardian *Geology.* a North American provincial series of the Lower Permian period, after the Wolfcampian and before the Guadalupian.

Leonardo da Vinci 1452–1519, Italian artist and scientist; investigated many areas of knowledge, including aeronautics, anatomy, physiology, botany, marine biology, astronomy, meteorology, and paleontology.

Leonids *Astronomy.* an annual meteor shower with a radiant in Leo that reaches peak activity around November 17; derived from Comet Tempel-Tuttle, the Leonids have about 10 meteors per hour in most years, but at 33-year intervals (1999 is next) can produce "storms" with thousands of meteors an hour.

leonite *Mineralogy.* $K_2Mg(SO_4)_2 \cdot 4H_2O$, a colorless or yellowish monoclinic mineral occurring in tabular and irregular grains, having a specific gravity of 2.20 and a hardness of 2.5 to 3 on the Mohs scale; found in oceanic salt deposits.

leopard

leopard *Vertebrate Zoology.* **1.** a large species of carnivore, *Panthera pardus,* of the cat family, characterized by a tawny or buff coat with black spots arranged in rosettes; an endangered species found in forests in Asia and Africa. **2.** any of various related cats that resemble this animal.

Leotiaceae *Mycology.* a family of fungi belonging to the order Helotiales that occurs on decaying plant material.

Leotichidae *Invertebrate Zoology.* a small family of true bugs in the superfamily Leptopodoidea; hemipteran insects found on bat guano in caves in Asia.

LEP *Virology.* low egg passage, a term referring to the number of times a given strain of viruses has undergone serial transition in eggs.

Lepadomorpha *Invertebrate Zoology.* a suborder of goose barnacles, crustaceans in the order Thoracica.

leper *Medicine.* a former term describing a person who is afflicted with leprosy.

Leperdita *Paleontology.* a genus of thick-shelled ostracods in the extinct order Leperditicopida; relatively large, often longer than 8 mm; Ordovician to Carboniferous.

Leperditicopida *Paleontology.* an order of large, thick-shelled ostracods, ranging from 4 to 8 mm long; Ordovician to Carboniferous.

Leperditillacea *Paleontology.* a superfamily of middle Paleozoic ostracods in the extinct order Leperditicopida.

Lepiceridae *Invertebrate Zoology.* a family of coleopteran insects, Horn's beetles, found along streams in Central America. Also, CYATHOCERIDAE.

Lepicoleaceae *Botany.* a monogeneric family of large, light green to gray-green branched liverworts of the order Jungermanniales; common in temperate areas worldwide.

lepidine *Organic Chemistry.* $C_{10}H_9N$, a colorless oil with a quinoline-like odor that is derived from cinchona bark; slightly soluble in water and soluble in alcohol, ether, and benzene; boils at 266°C; used in organic synthesis.

lepido- or **lepid-** a combining form meaning "scale" or "flake," as in *lepidophyllous.*

lepidoblastic *Petrology.* of or relating to a foliated or schistose metamorphic rock whose texture is characteristically flaky or scaly due to the parallel orientation during recrystallization of minerals such as mica and chlorite.

Lepidocentroida *Invertebrate Zoology.* a grouping of echinoids now included in several different subfamilies.

lepidocrocite *Mineralogy.* γ-$Fe^{+3}O(OH)$, a ruby to reddish brown orthorhombic mineral, polymorphous with akaganeite, feroxyhyte, and goethite, having a specific gravity of 3.85 to 4.09 and a hardness of 5 on the Mohs scale; found with limonite in iron ore. Also, GLOCKERITE.

Lepidodendrales *Botany.* an order of Devonian and Carboniferous woody fossil plants of the class Lycopodineae, characterized by dichotomous branching in both the crown and the rootlike supports.

Lepidolaenaceae *Botany.* a family of medium to large brown, red, or purplish liverworts of the order Jungermanniales, distinguished by endosporic spore germination and incubous leaves; restricted to the southern hemisphere.

lepidolite *Mineralogy.* $K(Li,Al)_3(Si,Al)_4O_{10}(F,OH)_2$, a pink to purple monoclinic mineral of the mica group occurring as tabular crystals and scaly aggregates, having a specific gravity of 2.8 to 3.3 and a hardness of 2.5 to 4 on the Mohs scale; found almost exclusively in granite pegmatites. Also, LITHIUM MICA, LITHIONITE, LITHIA MICA.

lepidomelane *Mineralogy.* a black to dark brown variety of biotite, rich in ferric iron; found commonly in iron-rich feldspathic igneous rocks. Also, IRON MICA.

Lepidophylloides *Paleontology.* a form genus of the leaves of the lycopsid tree Lepidodendron; Carboniferous.

lepidophyllous *Botany.* having leaves that are scaly.

Lepidopleurida *Invertebrate Zoology.* an order of primitive mollusks in the group Polyplacophora, having several pairs of gills under the posterior mantle.

Lepidoptera *Invertebrate Zoology.* a large order of scaly-winged insects including the butterflies, skippers, and moths, often brightly colored and having a coiled sucking proboscis.

lepidopteran *Invertebrate Zoology.* **1.** of or relating to the insect order Lepidoptera. **2.** a member of this order. (From a Greek term meaning "scaled wings.")

lepidopterist *Invertebrate Zoology.* **1.** a person who studies or is an expert on insects of the order Lepidoptera. **2.** a person who collects specimens of such insects, especially butterflies.

lepidopterology *Invertebrate Zoology.* the study of insects of the order Lepidoptera. Thus, **lepidopterological.**

lepidopterous *Invertebrate Zoology.* of or relating to the insect order Lepidoptera.

Lepidosaphinae *Invertebrate Zoology.* a family of bugs, the scale insects, in the order Homoptera, including harmful plant pests.

Lepidosauria *Vertebrate Zoology.* the subclass of scaly reptiles containing the tuatara and all lizards, snakes, and amphisbaenians; characterized by the lack of an antorbital opening and by a skull with two temporal openings on each side with reduced bony arcades. *Paleontology.* any of various extinct forms of this group, dating back to the Late Paleozoic.

Lepidosirenidae *Vertebrate Zoology.* a monospecific family of South American lungfishes belonging to the order Lepidosireniformes; characterized by an elongated body and branched vascular filaments in the ventral fins of the males.

Lepidosireniformes *Vertebrate Zoology.* the lungfishes, an order of bony fishes characterized by a cylindrical body and a paired swim bladder that acts as a functional lung.

Lepidostrobus *Paleontology.* a form group of cones (strobili) of the Lepidodendrales; the cones were 1 to 3 inches across and up to a foot in length and were borne in the crown of distal branches at the top of the plant; Carboniferous.

lepidote *Botany.* covered with scalelike hairs.

Lepidotrichidae *Invertebrate Zoology.* a family of silverfish, primitive wingless insects in the order Thysanura.

Lepidoziaceae *Botany.* a very large liverwort family of the order Jungermanniales; characterized by small to robust, leafy, prostrate to erect plants with reddish-brown pigments, pinnate to plumose branches, and sometimes having leafy branches arising from a prostrate rhizome system; most common in south temperate and tropical regions.

Lepiosteiformes *Vertebrate Zoology.* a former term for the fish order Semionotiformes, containing the gars.

Lepiota *Mycology.* a genus of fungi belonging to the order Agaricales; the species *L. procera*, known as the parasol mushroom, is edible.

Lepiotaceae *Mycology.* a family of fungi belonging to the order Agaricales, which includes the edible parasol mushrooms and lives on decaying organic matter.

Lepismatidae *Invertebrate Zoology.* a family of silverfish, primitive wingless insects, in the order Thysanura, having small or no eyes; of worldwide distribution.

Lepisostei *Vertebrate Zoology.* a former term for the order Semionotiformes.

Lepisosteidae *Vertebrate Zoology.* the gars, a family of freshwater fish of the order Semionotiformes, characterized by an elongate head and body and a vascularized swim bladder; found in North and Central America and Cuba.

lepisphere *Petrology.* a microspherical aggregate of silica crystals with radial orientation and scaly terminations on the outer surface.

Lepista *Mycology.* a genus of fungi belonging to the order Agaricales, including some edible species; occurs in temperate regions.

Leporidae *Vertebrate Zoology.* a cosmopolitan family of herbivorous mammals in the order Lagomorpha, including the rabbits and hares; characterized by long, muscular hindlimbs, movement by quadrupedal hopping, and long ears.

Leporipoxvirus *Virology.* a genus of viruses in the subfamily Chordopoxvirinae that infect hares, rabbits, and squirrels, causing tumors.

Lepospondyli *Paleontology.* a subclass of small Paleozoic amphibians known only from European and North American deposits; it is composed of three orders, each of which may have independently evolved from early labyrinthodonts: Aistopoda, Nectridea, and Microsauria; distinguished from the labyrinthodonts by the absence of infolded dentine in the teeth.

lepospondylous *Vertebrate Zoology.* having vertebrae whose centra are spool-shaped and formed as single structures by direct deposition of bone around the notochord.

leproma *Medicine.* the characteristic cutaneous nodular lesion of leprosy, the chronic infectious disease caused by the bacillus *Mycobacterium leprae.*

lepromatous leprosy *Medicine.* the malignant form of leprosy characterized by the presence of large numbers of mycobacterium in the diffuse cutaneous lesions, a negative lepromin reaction, and deformities of the extremities in advanced cases.

lepromin test *Immunology.* a skin test used to diagnose leprosy, in which the test antigen is a suspension of *Mycobacterium leprae* that is made from the skin lesions of a lepromatous patient.

leprophobia *Psychology.* an irrational fear of leprosy.

leprosarium *Medicine.* a hospital for the care and treatment of persons afflicted with leprosy.

leprose *Botany.* scaly in form or appearance.

leprosy [lep′rə sē] *Medicine.* a chronic infectious disease known since ancient times, resulting in lesions of the skin, mucous membranes, peripheral nervous system, and bones, caused by the bacillus *Mycobacterium leprae.* Also, HANSEN'S DISEASE.

Leptaena *Paleontology.* a genus of articulate brachiopods in the extinct order Strophomenida and superfamily Plectambonitacea; Ordovician to Devonian.

Leptaleinae *Invertebrate Zoology.* a subfamily of largely arboreal ants in the family Formicidae, found in tropical and subtropical forests.

Leptictidae *Paleontology.* a family of small, archaic eutherian mammals in the extinct order Leptictida; the leptictids were a short-lived group that left no known descendants; Paleocene to Middle Oligocene.

Leptinidae *Invertebrate Zoology.* a family of beetles, coleopteran insects in the superfamily Staphylinoidea.

leptite *Petrology.* an obsolete term for a fine-grained, quartzofeldspathic metamorphic rock with little or no foliation; in modern usage, such a rock might be described as metamorphosed rhyolite or felsite.

lepto- or **lept-** a combining form meaning "thin" or "slight," as in *leptotene.*

leptoaporangium see LEPTOSPORANGIUM.

Leptocardii see CEPHALOCHORDATA.

leptocephalous larva *Vertebrate Zoology.* the small, translucent, small-headed, leaf-shaped larva of freshwater eels.

leptocercal *Vertebrate Zoology.* of or relating to a long, slender, evenly tapering tail on a fish.

Leptochoeridae *Paleontology.* a family of artiodactyl mammals in the suborder Suina and extinct superfamily Entelodontoidea; formerly classified as palaeodont dichobunoids; Oligocene.

Leptodactylidae *Vertebrate Zoology.* a large, diverse family of neotropical frogs of the suborder Neobatrachia, many with direct development without a tadpole stage; native to the American tropics and Australia.

leptodactylous *Vertebrate Zoology.* possessing long, narrow, tapering fingers or toes, as certain birds.

Leptodiridae *Invertebrate Zoology.* a family of carrion beetles, coleopteran insects in the superfamily Staphylinoidea.

Leptodiscaceae *Botany.* a family of large marine dinoflagellates of the order Noctilucales, characterized by phagotropism by means of a cytosome and reproduction by binary fission.

leptogeosyncline *Geology.* a deep oceanic trough filled with only minor sedimentary accumulations and associated with volcanism.

Leptograptid fauna *Paleontology.* a fossil assemblage of the Ordovician based on the planktic graptoloid Leptograptus, which became extinct at the end of the Ordovician.

leptokurtic distribution *Statistics.* a type of distribution curve that is more heavily concentrated about the mean than the normal distribution, and thus has a large kurtosis.

Leptolegniellaceae *Mycology.* a family of fungi belonging to the order Saprolegniales that lives off keratin or is a parasite to animals, diatoms, or desmids.

Leptolepidae *Paleontology.* a family of neopterygian fishes in the order Leptolepiformes; probably ancestral to the advanced teleosts, the modern bony fish; Jurassic and Cretaceous.

Leptolepiformes *Paleontology.* an extinct order of neopterygian fishes in the division Teleosti; Jurassic and Cretaceous.

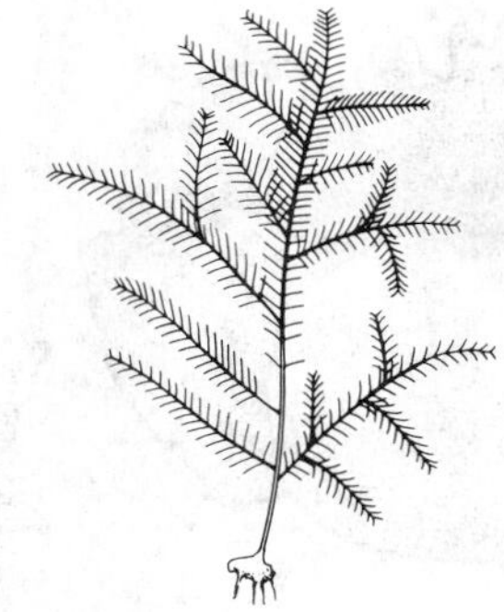

Leptomedusae

Leptomedusae *Invertebrate Zoology.* a suborder of colonial hydrozoans with hydrotheca in the order Hydroida. Also, CALYPTOBLASTEA.

leptomeninges *Anatomy.* the pia mater and arachnoid meninges considered together.

leptomeningitis *Neurology.* an inflammation of the pia and arachnoid membranes of the brain or spinal cord.

Leptomitaceae *Mycology.* a family of fungi belonging to the order Leptomitales consisting of nonpolar water molds that occur in waterlogged soil, bogs, and streams.

Leptomitales *Mycology.* an order of fungi belonging to the class Oomycetes that occurs in aquatic environments and is primarily nonparasitic, living on dead organic matter.

Leptomonas *Invertebrate Zoology.* a genus of flagellate protozoans in the family Trypanosomatidae, having one flagellum; parasitic in the digestive tract of insects.

leptomorphic see XENOMORPHIC.

lepton *Particle Physics.* **1.** a particle that does not experience the strong nuclear force; specifically electrons, muons, and tau leptons and their respective neutrinos and antiparticles. **2.** originally, any light particle.

lepton conservation *Particle Physics.* a principle stating that the net number of leptons before and after an interaction must remain the same.

leptonema see LEPTOTENE.

lepton era *Astrophysics.* a period that began 10^{-4} second after the Big Bang and ended when the universe was about 10 seconds old, during which time the universe had a temperature of about 10^{12} kelvins and consisted mainly of leptons and photons.

leptonic decay *Particle Physics.* a process in which a hadron decays solely into leptons.

leptophos *Organic Chemistry.* $C_{13}H_{10}BrCl_2O_2PS$, a white solid that is slightly soluble in water and melts at 70°C; used as an insecticide.

leptophyll *Ecology.* a classification of leaf size that includes leaves having a surface area of less than 25 mm.

Leptopodidae *Invertebrate Zoology.* a family of predatory, spiny, true bugs in the superfamily Leptopodoidea.

Leptopodoidea *Invertebrate Zoology.* a superfamily of true bugs, hemipteran insects in the subdivision Geocorisae.

leptoquark *Particle Physics.* a hypothetical elementary particle having a mass of approximately 100–300 GeV that would decay into a lepton and a quark.

leptoquark boson *Particle Physics.* a charged-vector boson having a mass of approximately 1015 GeV that is responsible for proton decay by transforming a quark to a lepton; postulated in relation to grand unified theories.

Leptosomatidae *Vertebrate Zoology.* the cuckoo rollers, a monospecific family of large arboreal birds of the order Coraciiformes, named for their tumbling, acrobatic, courtship flight; found in Madagascar.

Leptosphaeria *Mycology.* a genus of fungi belonging to the order Pleosporales, characterized by narrow ascospores and including both nonparasites and parasites that cause such plant diseases as cane blight and glume blotch.

Leptospira *Bacteriology.* a genus of coil-shaped, motile, aerobic bacteria of the family Leptospiraceae; they occur as free-living organisms in fresh water, salt water, and soil, and as parasites, often pathogenic, in animals and humans.

Leptospiraceae *Bacteriology.* a family of bacteria, of the order Spirochaetales, that occur as aerobic, helical cells utilizing long-chain fatty acids or fatty alcohols as metabolic substrates; consists of a single genus, *Leptospira.*

leptospiral jaundice see WEIL'S DISEASE.

leptospirosis *Medicine.* any of various illnesses caused by infection with bacteria of the genus *Leptospira*, found in the urine of a wide variety of wild and domestic animals and transmissible to humans through contact with the urine or contaminated soil or water; clinical syndromes vary from mild carrier state to fatal disease.

leptosporangium *Botany.* a sporangium that is derived from a single parent cell in a meristem, as in true ferns.

Leptostomaceae *Botany.* a monogeneric family of mosses of the order Bryales that forms dense cushions on tree trunks, branches, and rocks; found in the Southern Hemisphere, particularly in New Zealand; characterized by stems united by dense red-green tomentum, terminal sporophytes, and loosely erect branches that twist around the stems when dry.

Leptostraca *Invertebrate Zoology.* a group of primitive malacostracan crustaceans, small shrimplike animals with a bivalve carapace.

Leptostromataceae *Mycology.* a family of fungi belonging to the order Sphaeropsidales, having black fruiting bodies called pycnidia; some are pathogens of fruit trees.

leptotene *Cell Biology.* the first of the five stages of meiotic prophase in a cell, during which the chromosomes appear as thin, discrete threads. Also, LEPTONEMA.

Leptothrix *Bacteriology.* a genus of Gram-negative, motile, rod-shaped, sheathed bacteria covered with manganese or iron oxides and occurring in sludge and polluted or fresh water.

Leptotrichia *Bacteriology.* a genus of Gram-negative, rod-shaped, nonmotile, anaerobic bacteria of the family Bacteriodaceae; found in the human mouth.

Leptotyphlopidae *Vertebrate Zoology.* a family of blind, slender snakes of the superfamily Typhlopoidea, found in Africa, southwestern Asia, tropical Central and South America, and the southwestern United States.

Lepus *Vertebrate Zoology.* the genus containing the hares and jackrabbits. *Astronomy.* the Hare, an inconspicuous southern constellation of the winter sky lying directly south of Orion.

Lepyrodontaceae *Botany.* a monogeneric family of yellowish, glossy mosses of the order Isobryales that form dense mats on tree bark, moist earth, and rocks in Australasia and southern South America; characterized by inconspicuous creeping stems with many erect stout branches having abundant tomentum below.

Lerch's theorem *Mathematics.* **1.** suppose $\int_0^c x^n f(x)dx = 0$ for all $n = 1, 2, 3, \ldots$. If $f(x)$ is continuous on the interval $[0, c]$, then $f(x) = 0$ for all x in $[0, c]$. **2.** If f(x) and (x) are continuous functions and if the Laplace transforms $Lf(x)$ and $Lg(x)$ are equal for all $x > 0$, then $f(x) = g(x)$ for all $x > 0$.

Lernaeidae *Invertebrate Zoology.* a family of copepod crustaceans in the suborder Caligoida, external parasites on fish.

Lernaeopodidae *Invertebrate Zoology.* a family of crustaceans in the superfamily Lernaeopodoida, parasitic on the gills of fish.

Lernaeopodoida *Invertebrate Zoology.* a superfamily of fish maggots, crustaceans parasitic on the gills of fish.

l.e.s. local excitatory state.

lesbian *Psychology.* a woman who is sexually active with or sexually attracted to members of her own sex. (From the Greek island of *Lesbos,* the home of the poet Sappho and her followers.)

lesbianism *Psychology.* the fact or condition of being a lesbian; sexual attraction of women for other women.

Lesch-Nyhan syndrome *Medicine.* a rare inherited disorder in male children transmitted as an X-linked recessive, characterized by an abnormally high level of uric acid in the blood, mental retardation, and compulsive self-mutilating behavior.

lesion *Biology.* a part of a tissue, often epithelial, marked by a diseased change in structure; in many cases, it causes a small wound or a small fissure in the tissue.

Leskeaceae *Botany.* a family of dull mosses of the order Isobryales that form mats on tree trunks and bases and also on rocks; characterized by creeping, often tangled stems, irregular branching, lateral sporophytes, and a diplolepideous peristome.

Leskeineae *Botany.* a suborder of mosses of the order Hypnobryales, having squarrose leaves and mostly undifferentiated alar cells; found in creeping to scattered mats.

Lespedeza *Botany.* a genus of plants of the Leguminosae characterized by alternate, trifoliolate leaves, small purple flowers, and fruit that is a pod.

lesser ebb *Oceanography.* the weaker of two ebb currents of a tidal day.

lesser flood *Oceanography.* the weaker of two flood currents of a tidal day.

lesserite see INDERITE.

lesser omentum *Anatomy.* that part of the peritoneal omentum that extends from the liver to the lesser curvature of the stomach.

lesser panda see PANDA, def. 2.

Lessoniaceae *Botany.* a kelp family of the order Laminariales, characterized by a blade that splits longitudinally, resulting in multiple blades borne in various arrangements; it contains the species *Macrocystis,* a primary source of alginate, that forms extensive beds from Alaska to Baja California.

leste *Meteorology.* a sirocco wind that blows in an easterly or southeasterly direction from the Atlantic coast of Morocco to Madeira and the Canary Islands, characterized by hot, dry, and dusty conditions.

Lestidae *Invertebrate Zoology.* a family of damselflies, insects in the order Odonata.

Lestoniidae *Invertebrate Zoology.* a family of plant-eating true bugs, hemipteran insects in the superfamily Pentatomorpha, found only in Australia.

letdown *Aviation.* **1.** the gradual gliding descent of a flight vehicle preparatory to its approach and landing. **2.** also, **letdown procedure.** a procedure for reducing altitude characterized by the orderly adoption of successively lower flight levels. Gear and flaps usually are put down when airspeed permits.

letdown reflex *Medicine.* the release of milk from the alveoli of the breast into the ducts. Also, MILK EJECTION, MILK LETDOWN.

lethal *Pathology.* causing death; able to cause death.

lethal area *Ordnance.* a figure that indicates the expected casualty producing effect of an explosion detonated under specific conditions; expressed in terms of area.

lethal dose *Toxicology.* the amount of a substance that is sufficient to cause death. Also, FATAL DOSE.

lethal dose 50 see LD_{50}.

lethal gene *Genetics.* a mutant gene that causes the early death of an organism.

lethality *Pathology.* the capability of a disease, drug, or other agent to cause death.

lethal radius *Ordnance.* the distance from point of impact or ground zero at which a missile, bomb, or similar weapon can be expected to destroy targets or cause lethal casualties.

lethargy *Neurology.* abnormal drowsiness or stupor. *Nucleonics.* a unit of quantity for the energy lost by a neutron; equal to the natural logarithm of the ratio between its initial energy and its energy at any point at which it begins to lose energy. (From the Greek myth of *Lethe,* a river in Hades that caused forgetfulness in those who drank from it.)

letovicite *Mineralogy.* $(NH_4)_3H(SO_4)_2$, a colorless or white, transparent, triclinic mineral occurring in small pseudohexagonal crystals, having a specific gravity of 1.83 and an undetermined hardness; formed during the burning of coal mine waste dumps.

letter *Linguistics.* a character used in an alphabet that normally represents one or more sounds of a spoken language.

letter code *Computer Technology.* a Baudot code function that directs machines to shift to lower case; used to cancel errors, causing the receiving terminal to print nothing.

Letterer-Siwe disease *Medicine.* an acute, progressive, and fatal disease of infancy and childhood in which proliferating histiocytes invade the spleen, liver, and bone marrow, with enlargement of the spleen, liver, and lymph nodes, accompanied by pupuric rash and anemia.

letterpress *Graphic Arts.* any printing process in which the inked type or other printing element is raised above the surface of the plate; one of the three basic printing techniques, along with lithography and intaglio. Also, RELIEF PRINTING.

letter-quality *Computer Technoogy.* describing printed copy of better quality, comparable to that obtained on a typewriter.

letter-quality printer *Computer Technology.* a printer that produces good-quality printed copy.

letterset *Graphic Arts.* a hybrid printing process using a letterpress plate to impress the image on an offset blanket, which then transfers it to the paper. (From *letter*press and off*set.*)

letterspacing *Graphic Arts.* in typesetting, the addition of extra horizontal spaces between characters (l i k e t h e s e w o r d s) to make a word or line fit a desired measure or to achieve a certain design effect.

letters shift *Telecommunications.* **1.** the movement of a teletypewriter carriage that allows the printing of alphabetic characters in a linear fashion. **2.** the key or control that initiates this action.

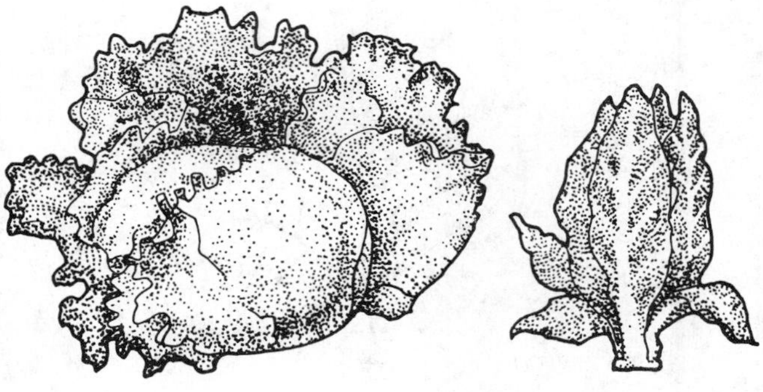

lettuce

lettuce *Botany.* the common name for any of a number of species of the genus *Lactuca* of the order Asterales, especially *L. sativa,* occurring in many varieties, whose leaves are widely grown as a food crop.

Leu leucine.

leuc- a combining form meaning "white," as in *leucite.*

Leucaltidae *Invertebrate Zoology.* a family of calcareous tubular or branching sponges in the order Leucettida.

Leucascidae *Invertebrate Zoology.* a family of massive calcareous sponges in the order Leucettida.

Leucettida *Invertebrate Zoology.* an order of calcareous sponges in the subclass Calcinea, that may be encrusting, massive, or lobose in shape.

leucine *Biochemistry.* $C_6H_{13}O_2N$, an aliphatic branched amino acid that is essential for healthy growth in infants and for nitrogen equilibrium in adults; obtained from the hydrolysis of protein.

leucine amino peptidase *Enzymology.* an enzyme that hydrolyzes proteins from the N-terminal end.

Leucippus c. 475 BC, Greek philosopher; laid the foundation for the atomic theory of matter developed by his student, Democritus.

leucite [loo´sīt] *Mineralogy.* $KAlSi_2O_6$, a white or gray, vitreous, tetragonal mineral, having a specific gravity of 2.47 to 2.50 and a hardness of 5.5 to 6 on the Mohs scale; found in potassium-rich basic lavas as disseminated grains and crystals. Also, AMPHIGENE, GRENATITE, VESUVIAN, WHITE GARNET.

leucite phonolite *Petrology.* a volcanic rock consisting of leucite, orthoclase, and mafic minerals.

leucitite *Petrology.* a fine-grained basalt, porphyritic extrusive or hypabyssal igneous rock consisting primarily of leucite and pyroxene and featuring little or no feldspar.

leuco- a combining form meaning "white," as in *leucophore.*

Leucobryaceae *Botany.* a family of tropical, subtropical, and temperate forest mosses of the order Dicranales that are whitish and form dense cushions on tree trunks and bases, logs, and soil; characterized by an unusual leaf anatomy in which large, empty hyaline cells surround a network of chlorophyllose cells.

leucocidin *Immunology.* an antivirulence factor produced as an extracellular bacterial product that kills white blood cells and may have the ability to destroy other cells.

leucocratic *Petrology.* describing igneous rocks of a light color that generally contain less than 30% dark minerals.

Leucocytozoon *Invertebrate Zoology.* a genus of protozoans in the order Eucoccidiida, sporozoans parasitic in birds, transmitted by blood-sucking insects.

Leucodontaceae *Botany.* a temperate and subtropical family of robust, dull mosses of the order Isobryales that form loose mats on trees and rocks; characterized by creeping stems with many stiffly ascending, curved, or pendant branches.

Leucodontineae *Botany.* a moss family of the order Isobryales having foliated branches and bearing catkins.

leucon *Invertebrate Zoology.* a sponge characterized by a complex branching canal system and by choanocytes that are restricted to small internal chambers.

Leuconostoc *Bacteriology.* a genus of Gram-positive, sphere-shaped, nonpathogenic bacteria of the family Streptococcaceae; found in dairy products and in fermenting vegetables and beverages.

Leucopaxillus *Mycology.* in some classifications, a genus of fungi belonging to the family Tricolomataceae that occurs in temperate and subtropical climates.

leucopenia see LEUKOPENIA.

leucophanite *Mineralogy.* $(Na,Ca)_2BeSi_2(O,OH,F)_7$, a glassy greenish to yellowish triclinic, pseudo-orthorhombic mineral occurring in short prismatic or tabular crystals, having a specific gravity of 2.96 to 2.98 and a hardness of 4 on the Mohs scale; found in nepheline-syenite pegmatites. Also, **leucophane.**

leucophore *Histology.* a pigment cell that appears white.

leucophosphite *Mineralogy.* $KFe_2^{+3}(PO_4)_2(OH)\cdot 2H_2O$, a brown or green monoclinic mineral occurring in short prismatic crystals and lamellar masses, having a specific gravity of 2.948 and a hardness of 3.5 on the Mohs scale; found with rockbridgeite and frondelite in pegmatites.

leucoplast *Botany.* a colorless plastid that is found in the cells of underground roots, stems, and storage organs, around which starch forms.

Leucosiidae *Invertebrate Zoology.* the purse crabs, a family of malacostracan crustaceans.

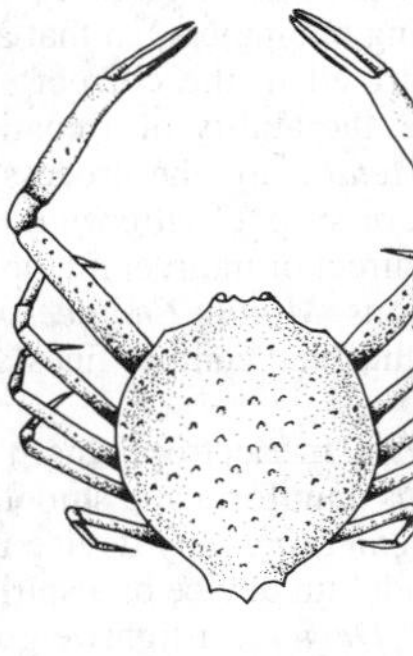

Leucosiidae

leucosin *Biochemistry.* a protein that is coagulable by heat, found in various cereal grains such as wheat and barley.

Leucosoleniida *Invertebrate Zoology.* an order of calcareous sponges in the subclass Calcaronea, of wide distribution.

leucosphenite *Mineralogy.* $BaNa_4Ti_2B_2Si_{10}O_{30}$, a white to grayish-blue monoclinic mineral occurring in wedge-shaped or tabular crystals, having a specific gravity of 3.05 to 3.089 and a hardness of 6.5 on the Mohs scale.

Leucospidae *Invertebrate Zoology.* a family of parasitic wasps, hymenopteran insects in the superfamily Chalcidoidea.

leucosporous *Mycology.* describing a fungal growth that possesses white spores.

Leucothoidae *Invertebrate Zoology.* a family of marine amphipod crustaceans found mainly in or on sponges and tunicates as parasites or commensals.

Leucothrix *Bacteriology.* a genus of gliding bacteria of the family Leucotrichaceae, occurring in aquatic habitats, where they grow nonparasitically on marine algae.

Leucotrichaceae *Bacteriology.* a family of gliding bacteria of the order Cytophagales; characterized by cylindrical or ovoid cells with long filaments; occurring in water that contains decomposing algae.

leucovorin see FOLINIC ACID.

leucovorin calcium *Pharmacology.* $C_{20}H_{21}CaN_7O_7\cdot 5H_2O$, the calcium salt of leucovorin (folinic acid), used to treat deficiencies of folic acid, such as megaloblastic anemia, and diseases caused by medications that inhibit the action of folic acid in the body. Also, CALCIUM LEUCOVORIN, CALCIUM FOLINATE.

leucoxene *Mineralogy.* any of various opaque, fine-grained, chalky-white alteration products of ilmenite occurring in some igneous rocks.

leu-enkephalin see ENKEPHALIN.

leuk- a combining form meaning "white," as in *leukemia.*

leukapheresis *Medicine.* the removal of white cells from a donor's blood with the return to the donor of the remainder of the extracted blood.

leukemia [loo kē´mē ə] *Medicine.* a progressive, malignant disease of the blood-forming organs, characterized by uncontrolled proliferation of immature and abnormal white blood cells in the bone marrow, the blood, the spleen, and the liver. Leukemia is classified clinically on the basis of the character of the disease (*acute* or *chronic*), the type of cell involved (*myeloid, lympoid,* or *monocytic*), and the increase or nonincrease in the number of abnormal cells in the blood (*leukemic* or *aleukemic*). *Veterinary Medicine.* see LEUKOSIS.

leukemoid *Medicine.* closely resembling the blood changes of leukemia but caused by other conditions.

leukencephalitis *Neurology.* an inflammation of the white matter of the brain.

leuko- a combining form meaning "white," as in *leukocyte.*

leukocidin *Biochemistry.* a substance produced by pathogenic bacteria and is toxic to some leukocytes, killing the cells with or without lysis.

leukocyte [loo´kə sīt´] *Histology.* a white or colorless cell of the blood, having a nucleus and either granular or nongranular cytoplasm; leukocytes function as bacterial or viral phagocytes, as detoxifiers of toxic proteins, and in the development of immunities. Also, WHITE BLOOD CELL.

leukocytic cream see BUFFY COAT.

leukocytic crystals see CHARCOT-LEYDEN CRYSTALS.

leukocytology *Hematology.* the branch of hematology dealing with the morphology and pathology of leukocytes.

leukocytolysin *Hematology.* an agent that leads to the destruction or disruption of leukocytes. Also, LEUKOLYSIN.

leukocytolysis *Hematology.* the breaking down or destruction of leukocytes.Also, LEUKOLYSIS.

leukocytolytic *Hematology.* **1.** pertaining to, characterized by, or causing leukocytolysis. **2.** an agent that causes leukocytolysis. Also, LEUKOLYTIC.

leukocytopoiesis *Physiology.* the formation of leukocytes or white blood cells.

leukocytosis *Hematology.* an abnormally large number of leukocytes in the blood, as during hemorrhage, infection, inflammation, or fever.

leukocytotropic *Hematology.* having a special affinity for leukocytes.

leukoderma *Medicine.* an acquired localized lack of pigment in the skin, produced by a specific substance or by dermatosis. Also, **leukodermia.**

leukodystrophy *Neurology.* a group of diseases of the central nervous system, characterized by lesions of the white matter of the brain, especially the cerebrum; a congenital defect in the building up and maintenance of myelin.

leukoencephalitis *Medicine.* an inflammation of the cerebral white matter.

leukoencephalopathy *Neurology.* any of a group of diseases that affect the white matter of the brain, occurring especially in infants and children.

leukokinetics *Hematology.* the study of the production, circulation, and destruction of leukocytes.

leukolysin see LEUKOCYTOLYSIN.

leukolysis see LEUKOCYTOLYSIS.

leukolytic see LEUKOCYTOLYTIC.

leukoma *Medicine.* a dense, opaque, white spot on the cornea.

leukomyelitis *Neurology.* an inflammation of the white matter of the spinal cord.

leukomyoma see LIPOMYOMA.

leukonychia *Medicine.* a benign, congenital condition marked by the existence of white spots or patches under the nails.

leukopedesis *Hematology.* the outward passage of leukocytes through intact vessel walls.

leukopenia *Hematology.* an abnormal reduction in the number of leukocytes in the blood; specifically, an instance in which the count is 5000 per cubic millimeter or less. Also, LEUCOPENIA.

leukoplakia *Medicine.* **1.** a white patch on a mucous membrane that will not rub off. **2.** a mucous membrane disease characterized by abnormal thickening and whitening of the epithelium, affecting the cheeks, tongue, gums, penis, or vulva and caused by chronic inflammation from mechanical and chemical irritation; called **smoker's patch** when the oral cavity is involved.

leukorrhea *Medicine.* a white or yellowish discharge from the vagina containing mucus and pus cells.

leukosarcoma *Medicine.* a type of malignant lymphoma in which large numbers of abnormal, immature white blood cells from lymphoid tissue are present in the circulating blood. *Oncology.* the development of leukemia in a person having a lymphocytic type of malignant lymphoma.

leukosarcomatosis *Oncology.* a condition characterized by the development of multiple sarcomas that are composed of leukemic cells.

leukosis *Medicine.* an abnormal proliferation of leukocyte-forming tissues. *Veterinary Medicine.* a general name for malignant diseases of hemo- and lymphopoietic cells in animals; the synonym of leukemia in humans; e.g., **avian leukosis, bovine leukosis, canine leukosis,** etc.

leukotriene *Endocrinology.* any member of the family of lipoxygenase metabolites of arachidonic acid; an autacoid hormone that may act as a mediator of the allergic response or as a chemotactic factor. (From *leuko*cytes + *triene,* indicating three double bonds.)

leukovirus *Virology.* a former term for certain retroviruses of the family Oncovirinae.

leuneburgite see LÜNEBURGITE.

leuprolide *Oncology.* a gonadotropin-releasing hormone analogue that is used as an antineoplastic agent in prostate cancer.

Levalloisian [lev´ə lōz´ē ən] *Archaeology.* **1.** of or relating to the Levallois technique. **2.** describing tools made by this method.

Levallois technique [lə val wä´] *Archaeology.* a distinctive method of stone tool-making in which flakes are removed by percussion from a preshaped core, with little other modification; characteristic of Acheulian cultures. (From *Levallois,* a suburb of Paris where early evidence of this was found.)

levan *Biochemistry.* $C_6H_{10}O_5$, a polysaccharide that upon hydrolysis yields levulose; obtained from various grasses.

levante *Meteorology.* a Spanish term for an east to northeast wind that blows along the coast and inland from southern France and the Straits of Gibraltar; it is usually accompanied by a low-pressure system over the western Mediterranean Sea, causing mild, humid, rainy weather conditions.

levantera *Meteorology.* a persistent easterly wind that originates in the Adriatic zone and is usually accompanied by cloudy weather.

Levantine Basin *Oceanography.* a large basin of the Eastern Mediterranean between Asia Minor, Egypt, Crete, and Cyprus.

levator *Physiology.* any muscle that raises a part of the body. *Medicine.* a surgical instrument used for raising the depressed portion in a fracture of the skull.

leveche *Meteorology.* a Spanish term for the sirocco; a hot, dusty, southerly wind ahead of a low-pressure system, moving from the Sahara Desert over the southeast coast of Spain.

levee [lev´ē] *Hydrology.* **1.** a naturally occurring, broad, low ridge of sand and coarse silt deposited along the banks of a stream or river during flooding. Also, NATURAL LEVEE. **2.** a ridge of sediment that borders one or both sides of a deep-sea channel or fan valley. *Agronomy.* a low ridge or bank of earth used to control the flow of water during irrigation.

level *Engineering.* **1.** parallel to the ground or to some other horizontal surface; having a flat or even surface. **2.** the horizontal position of something; e.g., the height of water in a reservoir. *Metrology.* an expression of the value or magnitude of something in relation to a graded scale; e.g., the intensity of a sound or the amount of pollutants in the air. *Civil Engineering.* **1.** a horizontal plane or surface, especially an imaginary line established as perpendicular to a plumb line. **2.** any of various instruments used to establish the horizontal or to measure horizontal distances. *Mechanical Devices.* specifically, a device shaped from a glass tube filled with spirit, on which a bubble of air and spirit vapor rises to the top of the tube when it is laid flat; used to find a level surface or plane; used in surveying, building construction, and so on. Also, SPIRIT LEVEL. *Mining Engineering.* any horizontal passage or drift through which a shaft is sunk to work a mine. *Electronics.* **1.** the intensity of an electrical signal. **2.** the range of a quantity in relation to a given reference point, expressed as an absolute value in ohms, volts, or the like, or as the ratio to a reference value. **3.** the charge value that can be held in a particular storage element of a charge storage tube and separated from other charge values. **4.** a voltage that remains stable over an extended time. *Acoustics.* the relative intensity of a sound. *Acoustical Engineering.* specifically, in sound recording, the relative intensity of a person's voice. *Electricity.* a bank of contacts, as on a stepping relay. *Telecommunications.* **1.** in an information coding system, the number of bits in each character. **2.** in a modulation scheme, the total number of discrete signal elements that are transmitted. **3.** an amplitude of a signal, usually stated in comparison to some reference of amplitude or power such as decibels above or below one milliwatt. *Computer Programming.* **1.** the position or the degree of subordination in a hierarchical organization. Thus, **first-level storage. 2.** the status of a data item in the COBOL language that indicates whether it includes additional items.

level above threshold *Physiology.* **1.** the level of a stimulus above the threshold of perception for an individual observer. **2.** the increase in the level of a stimulus that is necessary before it is noticeably different to an individual observer.

level bombing *Military Science.* the process of dropping bombs from an aircraft in level flight as opposed to nonlevel methods such as dive bombing, glide bombing, or loft bombing.

level compensator *Electronics.* **1.** a device that minimizes the effects of amplitude variations in an incoming signal. **2.** in a telegraph circuit, a device that prevents oscillations in the amplitude of an input signal from affecting the output of the receiving system.

level converter *Electronics.* a circuit that changes the output levels of a logic device to different input levels that are required by the logic device that follows it.

leveler *Engineering.* any equipment designed for smoothing and leveling land, such as a drag or scraper.

level error *Astronomy.* in a meridian circle or other transit telescope, the angular deviation of the axis from the horizontal.

level fold see NONPLUNGING FOLD.

level indicator *Engineering.* any device that shows the variable level of a fluid. *Acoustical Engineering.* any of various devices used to show voltage levels during sound recording; this may be a neon lamp, volume-unit meter, or cathode-ray tuning indicator.

leveling *Engineering.* the process of aligning a device, such as a gun mount, launcher, or sighting equipment, so that all vertical and horizontal angles will be determined in the correct vertical and horizontal planes. *Materials Science.* the ability of a coating to create a smooth, even film on a surface. *Metallurgy.* the process of improving the flatness of rolled plate, sheet, or strip. *Cartography.* in surveying, the act or process of measuring, by direct or indirect methods, vertical distances in order to determine elevations. *Mining Engineering.* the process of measuring rises and falls, heights, and contour lines. *Industrial Engineering.* see PERFORMANCE RATING.

leveling action *Metallurgy.* in electroplating, the performance feature of the electrolyte that yields a uniform and smooth plate.

leveling instrument *Engineering.* any device used to fix a horizontal line of sight, such as a pendulum device or a spirit level.

leveling rod *Mechanical Devices.* a lightweight pole or rod having a flat face that is graduated in clearly marked linear units starting with zero at the bottom; used in surveying operations as a means of reading gradations of height as viewed from a theodolite a distance away while it is held vertically. Also, LEVEL ROD.

leveling rule *Mechanical Devices.* the evenly spaced, marked gradations on a leveling rod. Also, LEVEL RULE.

leveling screw *Engineering.* a screw that adjusts an instrument or piece of equipment and makes it level.

level line *Cartography.* in surveying, a set of measured differences of elevation, given in the order in which they were measured, and the set of geodetic points to which the measurements refer. *Geodesy.* a line tangent to an equipotential surface of the earth's gravity field.

level measurement *Engineering.* the distance above or below a reference level of the surface of a body of liquid or other substance. *Electronics.* the strength of an electronic signal in decibels or nepers.

level of burst *Ordnance.* the altitude at which a projectile bursts.

level-off position *Aviation.* the point at which an aircraft is pulled out of an ascent or descent and into level flight.

level of free convection *Meteorology.* the atmospheric level at which an air parcel lifted in a dry adiabatic environment becomes saturated and is further lifted adiabatically, becoming warmer than its surroundings in a conditionally unstable atmosphere.

level of nondivergence *Meteorology.* a region of zero horizontal velocity divergence in the atmosphere; it separates the major regions of convergence and divergence in a cyclonic-scale weather system and is usually near the 500-mb level.

level of operation *Transportation Engineering.* an economic model of an air carrier's operation, used to estimate return on fleet investment, competitive effects, and related factors.

level of service *Transportation Engineering.* the frequency and capacity of service provided by a carrier on a given route.

level of significance *Statistics.* in testing, the long-run probability of committing a Type I error in a hypothetical sequence of experiments like the one performed.

level of supply *Aviation.* the amount of fuel, equipment, and other supplies currently available or directed to be held in anticipation of future needs; the basis of planning and control of aviation supply operations.

level point see POINT OF FALL.

level rod see LEVELING ROD.

level rule see LEVELING RULE.

level set *Computer Programming.* a software package revision in which the executable programs have been replaced by improved versions.

level shifting *Electronics.* the process of changing the logic level at the point of connection between two different semiconductor logic systems.

levels of consciousness *Neurology.* the clinically differentiable degrees of awareness and alertness, including alert wakefulness, lethargy, obtundation, stupor, and coma.

levels-of-processing approach *Psychology.* an approach to memory stating that information is processed at different levels depending on how it is organized, associated with other memories, and connected to an emotional experience.

level surface *Engineering.* a surface that is perpendicular to the plumb line at each point. *Geophysics.* see EQUIPOTENTIAL SURFACE.

level valve *Mechanical Engineering.* a valve that is regulated with the use of a level; it moves through a maximum arc of 180°.

level width *Quantum Mechanics.* the uncertainty in energy (full width at half maximum) of an unstable state due to the finite lifetime of the state.

Levenstein process *Chemical Engineering.* a process used to produce mustard gas from ethene and sulfur chloride.

lever *Engineering.* one of the six simple machines, a device consisting of a rigid bar that pivots about a stationary or fulcrum point; used to transmit and enhance power or motion, as in prying, raising, or dislodging an object. A **first-class lever** has the fulcrum between the load and the effort; a **second-class lever** has the load between the fulcrum and the effort; a **third-class lever** has the effort between the fulcrum and the load. A **compound lever** combines the effort of two or more simple levers. *Mechanical Devices.* any of various machines that employ such a device. (Going back to a Latin word meaning "to lift.")

lever-action *Ordnance.* of a rifle, reloading by means of a hand-operated lever arm connected to the trigger housing.

leverage *Mechanics.* the mechanical advantage of a lever; equal to the length of the arm to which force is applied divided by the length supporting the load.

leveret *Vertebrate Zoology.* a young hare less than one year old.

Leverette function *Fluid Mechanics.* a dimensionless quantity used in the study of two-liquid flow in a porous medium; given by $(x/e)^{1/2}(p/s)$, where x is the permeability, e is the porosity, p is the capillary pressure, and s is the surface tension acting between the two liquids.

lever law see LEVER RULE.

Leverrier, Urbain 1811–1877, French astronomer; developed the formula for calculating changes in the earth's orbit; predicted the existence of Neptune.

lever rule *Materials Science.* a technique for determining the amount of each phase in a two-phase region on a phase diagram. A horizontal line is drawn across the two-phase region, the compositions at the two ends of the line arc are designated as A and B, and the original composition is C. The relative amount of the phase whose composition is B is given by $(C - A)/(B - A)$, similarly the amount of the phase whose composition is A is given by $(C - A)/(B - A)$. Also, LEVER LAW.

lever shears *Mechanical Devices.* shears that operate on the lever principle, such that input force from the handles relates to the output force at the cutting edges.

lever switch *Electricity.* a switch that features a lever-shaped operating handle.

Levich equation *Physical Chemistry.* an equation that describes the effect of several variables (rotation rate, concentration, kinetic viscosity) on the current at a rotating-disk electrode.

Levi-Civita, Tullio 1873–1941, Italian mathematician and physicist; formulated the concept of parallel displacement.

Levi-Civita connection *Mathematics.* on a (pseudo-) Riemannian manifold, the unique symmetric (i.e., torsion-free) connection that is compatible with the metric tensor; that is, the covariant derivative of the metric tensor is identically zero. The connection coefficients are the Christoffel symbols of the second kind.

Levi-Civita symbol *Mathematics.* another term for a permutation symbol.

levigate *Botany.* having a smooth, polished surface devoid of hair or down. *Chemistry.* to reduce a substance to a powder by wet-grinding and fractional sedimentation.

Levi-Montalcini, Rita born 1909, Italian-American physiologist; with Stanley Cohen, shared the Nobel Prize for her research on nerve growth factor.

Levinthal's medium *Microbiology.* a medium containing V factor and X factor; used to isolate species of Haemophilus.

levirate *Anthropology.* the practice in some societies that requires a man to marry the widow of his brother.

Lévi-Strauss, Claude born 1908, French anthropologist; the founder of structural anthropology.

levitated vehicle *Mechanical Engineering.* any vehicle that moves at a vertical distance above an electrically conducting path, usually made possible by magnetic action.

levitation *Physics.* the raising of an object without physically touching it, usually by means of electric or magnetic fields. *Mining Engineering.* a froth-flotation process in which particles are raised to the surface of a pulp by activating them so that they cling to the air-water interface of a rising or coursing air bubble. *Surgery.* a support system for severe burn victims that uses sterile, warm, humidified air at a pressure sufficient to support the patient above the surface of the bed.

levitation melting *Metallurgy.* melting a metal or alloy that is not in contact with a crucible or mold, but is suspended in an atmosphere by the action of an electromagnetic field. Also, **levitation heating.**

Levitsky's theorem *Mathematics.* a theorem stating that every left (or right) nil ideal in a left Noetherian ring is nilpotent.

Leviviridae *Virology.* the family of bacteriophages containing the sole genus *Levivirus.*

Levivirus *Virology.* the sole genus of viruses of the family Leviviridae, which have been identified from enterobacteria.

levo- a combining form meaning "left," as in *levorotation.*

levocardia *Medicine.* a condition in which there is a reverse position of the other viscera (situs inversus), but in which the heart is normally situated on the left.

levodopa see L-DOPA.

levophobia *Psychology.* an irrational fear of objects on an individual's left.

levorotation *Optics.* the process of turning a plane of polarized light counterclockwise, or to the left.

levorotatory [lev´ō rōt´ə tôr´ē] *Science.* turning or rotating to the left.

levorotatory activity *Materials Science.* a property exhibited by some materials that rotate the plane of polarization of linearly polarized light. Levorotatory activity is exhibited as clockwise rotation of the light as viewed by an observer facing the approaching light beam (as compared with dextrorotatory activity, which is counterclockwise). Also, **levorotatory optical activity.**

levorotatory form *Physical Chemistry.* the isomer of a compound that rotates the plane of vibration of plane-polarized light counterclockwise, or to the left, to be distinguished from the dextro form which involves clockwise rotation.

levorotatory isomer *Biochemistry.* an isomer that turns the plane of polarization of light to the left.

levotropic cleavage *Developmental Biology.* spiral cleavage in which the third division twists spirally in a counterclockwise direction.

levulinic acid *Organic Chemistry.* $C_5H_8O_3$, combustible, colorless crystals; soluble in water and alcohol and insoluble in aliphatic hydrocarbons; boils at 245–246°C and melts at 33–35°C; used as a chemical intermediate. Also, γ-KETOVALERIC ACID, ACETYLPROPIONIC ACID.

levulose see FRUCTOSE.

levulose tolerance test *Pathology.* a method of ascertaining the ability of the liver, in normal hepatic function, to assimilate and retain large levels of fruit sugar or fructose.

levyne *Mineralogy.* $(Ca,Na_2,K_2)Al_2Si_4O_{12}\cdot 6H_2O$, a white or light-colored trigonal mineral of the zeolite group, having a specific gravity of 2.09 to 2.16 and a hardness of 4 to 4.5 on the Mohs scale; found as tabular crystals in cavities in basalt. Also, **levynite.**

Lewandowsky-Lutz disease see EPIDERMODYSPLASIA VERRUCIFORMIS.

Lewis, Gilbert Newton 1875–1946, American chemist; developed the electron theory of valence; established a definitive means of identifying acids and bases.

lewis *Mechanical Devices.* an iron dovetail tenon that fits into a cut mortise; used for lifting heavy stones.

Lewis acid *Chemistry.* any chemical substance that can accept a pair of electrons. Also, ELECTROPHILE.

Lewis base *Chemistry.* any chemical substance that can donate a pair of electrons. Also, NUCLEOPHILE.

Lewis blood group *Genetics.* an inherited human blood group that is due to an antigen coded for by the *Le* gene, which is located on chromosome 19. The antigens responsible for this blood group are found both on red blood cells and in body fluids.

lewis bolt *Mechanical Devices.* a foundation bolt with an enlarged head and a tapering, jogged tail inserted into masonry and held with lead.

Lewis gun *Ordnance.* a .303 caliber, light machine gun that was used in World War I; it was air-cooled and gas-operated with a top-mounted, horizontal drum magazine; the first machine gun to be fired from an aircraft. Also, **Lewis machine-gun.**

lewisite *Mineralogy.* $(Ca,Fe^{+2},Na)_2(Sb,Ti)_2O_7$, a dark brown cubic mineral of the stibiconite group, having a specific gravity of 4.95 and an undetermined hardness.

Lewis number *Physics.* a dimensionless value commonly used in the study of combined heat and mass transfer, given by the diffusivity divided by the diffusion coefficient.

Lewis theory *Chemistry.* the theory that any chemical substance that can accept a pair of electrons is an acid, and any chemical substance that can donate a pair is a base.

lewistonite see CARBONATE-FLUORAPATITE.

LEX *Computer Science.* a software tool for constructing a lexical analyzer from a regular grammar and actions associated with the grammar productions.

LEX *Aviation.* the airport code for Lexington, Kentucky.

Lexell's Comet *Astronomy.* a comet that was discovered in 1770 by Charles Messier, but named for Anders Johann Lexell, the astronomer who calculated its orbit; several gravitational encounters with Jupiter have changed the comet's orbit so that it no longer comes close enough to the earth to be seen.

lexeme *Linguistics.* a single word or other meaningful form in a language. *Computer Science.* a basic symbol in a computer language; e.g., a variable name would be a lexeme for a grammar of a programming language.

lexical *Linguistics.* **1.** of or relating to the vocabulary of a given language. **2.** of or relating to the isolated meaning of individual words, as opposed to their use in sentences. *Artificial Intelligence.* in natural language processing, of or relating to information contained in the lexicon. *Computer Science.* of or relating to information that can be determined by static examination of a program, i.e., without running it.

lexical ambiguity *Artificial Intelligence.* in natural language, ambiguity caused by multiple definitions or parts of speech for a single word; e.g., *right* can mean "the opposite of left" or "without errors; correct;" *view* can be either a noun or a verb.

lexical analysis *Computer Science.* parsing and conversion to the internal form of the simplest elements of a language, ususally specified by a regular grammar, such as variable names, numbers, etc.

lexical analyzer *Computer Science.* a program that performs lexical analysis and outputs the internal form of lexemes.

lexical memory *Psychology.* a memory for certain words one may recognize, without memory of their meanings.

lexical scoping *Computer Science.* a convention in a block-structured programming language that a variable can only be referenced within the block in which it is defined and in blocks contained within that block; thus, the scope of a variable is completely determined at compile time.

lexicographer [leks´i käg´rə fər] *Linguistics.* a person who writes, edits, or compiles dictionaries. (From a Greek term meaning literally "a writer of words.")

lexicographic [leks´i kə graf´ik] *Linguistics.* **1.** of or relating to the making of dictionaries. **2.** involving or based on information contained in a dictionary.

lexicographic order *Mathematics.* an order induced on sequences whose elements come from an ordered set S (with order $<$). Two sequences $(a_1, a_2, a_3, \ldots)$ and $(b_1, b_2, b_3, \ldots)$ are said to be written in lexicographic order if either $a_i = b_i$ for every i, or if $a_N < b_N$, where N is the least value of i for which $a_i \neq b_i$. The latter includes the case in which $(a_1, a_2, a_3, \ldots)$ has fewer elements than $(b_1, b_2, b_3, \ldots)$; for example, S is the alphabet with the usual order. (So named because it represents the same order in which words appear in a dictionary.)

lexicography [leks´i käg´rə fē] *Linguistics.* **1.** the practice or procedure of writing, editing, or compiling dictionaries. **2.** the principles and methods involved in dictionary making.

lexicon *Linguistics.* the compilation of all the words and other meaningful forms of a language, along with their definitions, usually in the form of a dictionary or glossary. *Artificial Intelligence.* in a natural language processing system, a table of words with their parts of speech, root words, relations to other words or phrases, semantic features, definitions, and so on, in machine-processable form.

lexicostatistics *Linguistics.* a technique used to determine an approximate time at which two languages split by comparing a list of about 100 basic terms to determine how many cognates appear; a rate has been determined based on the study of several written languages, according to which about 66% will be cognates after about 1000 years of separation.

Leyden jar [lī´dən] *Electricity.* an early capacitor of high voltage and low capacitance that has a glass jar as the dielectric. (From the city of *Leyden,* where it was developed.)

Leyden's crystals see CHARCOT-LEYDEN CRYSTALS.

Leydig, Franz von [lī´dig] 1821–1908, German anatomist; the founder of comparative histology; discovered Leydig cells and the Leydig gland.

Leydig cells *Histology.* any of the interstitial cells of the testis. *Endocrinology.* interstitial cells of the testes that produce and secrete testosterone in response to the secretion of luteinizing hormone by the anterior pituitary gland.

LF line feed.

L.F. or **l.f.** low frequency.

Lf limit flocculation.

L.F.A. left frontoanterior.

L.F.D. least fatal dose.

L-form *Microbiology.* a variant phase of certain bacteria consisting of a spheroid or ellipsoid body without a rigid cell wall, formed either spont a normal cell. Also, L-PHASE VARIANT.

L.F.P. left frontoposterior.

L.F.T. left frontotransverse.

LFU least frequently used.

LGA *Aviation.* the airport code for La Guardia, New York, New York.

LGW *Aviation.* the airport code for Gatwick, London.

LH luteinizing hormone. *Aviation.* the airline code for Lufthansa German Airlines.

LHD Doctor of Humanities; Doctor of Humane Letters.

L-head engine *Mechanical Engineering.* an internal combustion engine that has intake and exhaust valves on one side of the engine block, with valves that are operated by a single camshaft.

lherzolite *Petrology.* a variety of peridotite containing abundant olivine with orthopyroxene and clinopyroxene.

L'Hôpital, Guillaume de [lō´pē täl´] 1661–1704, French mathematician; published the first textbook of differential calculus, containing L'Hôpital's rule (based on the work of Bernouilli).

L'Hôpital's rule *Mathematics.* if $f(x)$ and $g(x)$ are differentiable in an interval (a, b) except possibly at a point c if f and g both approach 0 as $x \to c$, and if $\lim_{x\to c} f(x) = \lim_{x\to c} g(x) = 0$, and if $g'(x) \neq 0$ for $x \neq c$, then

$$\lim_{x\to c} f(x)/g(x) = \lim_{x\to c} f'(x)/g'(x),$$

whenever the limit on the right exists. The process may be repeated until the limit on the right can be evaluated. If $\lim_{x\to c} f(x) = \pm\infty$ and $\lim_{x\to c} g(x) = \pm\infty$, the rule is also valid. This is useful in evaluating indeterminate forms.

L'Hôspital [lō´pē täl´] another spelling of L'Hôpital.

LHR *Aviation.* the airport code for Heathrow, London.

LH-RF luteinizing hormone-releasing factor.

LH-RH luteinizing hormone-releasing hormone.

l'Huilier's formula [lə wē yäz´] *Mathematics.* given a spherical triangle with sides a, b, and c, corresponding angles α, β, and γ, spherical excess $\varepsilon = \alpha + \beta + \gamma - \pi$, and $s = (a + b + c)/2$, the equation $(\tan (s/2) \tan [(s - a)/2] \tan [(s - b)/2] \tan [(s - c)/2])^{1/2} = \tan (\varepsilon/4)$; used in solving spherical triangles. Also, **l'Huilier's equation**.

LHV lower heating value.

Li lithium.

li. link.

LIA leukemia-associated inhibitory activity.

liana *Botany.* the common name for any of various woody climbing plants commonly found in tropical rain forests. Also, **liano, liane**.

Liapunov exponent see LYAPONOV EXPONENT.

Liapunov function see LYAPONOV FUNCTION.

Liapunov numbers see LYAPONOV NUMBERS.

Liapunov stable see LYAPONOV STABLE.

Liapunov stability criterion see LYAPONOV STABILITY CRITERION.

Lias see LIASSIC.

Liassic *Geology.* a Middle European geologic series of the Lower Jurassic period. Also, LIAS.

lib. or **Lib.** a pound. (From Latin *libra.*)

Libby, Willard 1908–1980, American chemist; awarded the Nobel Prize for the discovery of carbon-14 and for radio-carbon dating.

Libby effect *Geochemistry.* the increase in atmospheric carbon-14 that has occurred since about 1950 as a result of the detonation of thermonuclear devices.

libeccio *Meteorology.* an Italian term for a southwest wind, especially a turbulent wind that blows year round in northern Corsica, characterized by storminess in the wintertime.

Libellulidae *Invertebrate Zoology.* the largest family of dragonflies, insects in the suborder Anisoptera.

liber *Botany.* the phloem or bast of a plant.

Liberator *Aviation.* a popular name for the B-24 bomber.

libethenite *Mineralogy.* $Cu_2^{+2}(PO_4)(OH)$, a dark-green orthorhombic mineral having a specific gravity of 3.97 and a hardness of 4 on the Mohs scale; found as equant to acicular crystals in the oxidation zone of copper-bearing ore deposits.

libidinal [li bid´ə nəl] *Psychology.* of or relating to the libido.

libido [li bē´dō] *Psychology.* Freud's term for the psychic energy that is fueled by sexual drive and instinctual desires.

Libman-Sacks endocarditis *Medicine.* the presence of nonbacterial wartlike lesions on the heart valves and endocardium, associated with lupus erythematosus. Also, ATYPICAL VERRUCOUS ENDOCARDITIS.

LiBr lithium bromide.

Libra [lē´brə] *Astronomy.* the Balance, a dim southern constellation lying in the zodiac between Virgo and Scorpius.

librarian *Computer Programming.* a utility program that maintains and provides access to all programs and routines that make up the operating system.

library any of various compilations or assemblages thought of as comparable to the collection of books in a library; specific uses include: *Computer Programming.* **1.** a collection of routines residing in the mass storage of a computer system, available for reference by a user's program. Thus, **library routine. 2.** any collection of related files. *Computer Technology.* a repository for demountable recorded media, such as disk packs.

libration *Physics.* any rotational oscillation, such as the spinning motion of a molecule, that does not have sufficient energy to make complete rotations.

libration in latitude *Astronomy.* the libration caused by the tilt of the moon's rotation axis relative to its orbital plane.

libration in longitude *Astronomy.* the libration caused by the difference between the moon's varying orbital velocity and its constant rotation rate.

libriform *Botany.* resembling or having the form of liber or bast; woody, elongated, and thick-walled, as certain cells.

Libytheidae *Invertebrate Zoology.* snout butterflies, a family of lepidopteran insects in the suborder Heteroneura.

LIC low-intensity conflict.

lice *Invertebrate Zoology.* the plural form of *louse*; the common name given to various biting or sucking ectoparasites composing the orders Anoplura and Mallophaga.

Liceaceae *Mycology.* a family of fungi belonging to the order Liceales that is composed of plasmodial slime molds.

Liceales *Mycology.* an order of fungi belonging to the subclass Myxogastromycetidae that is composed of plasmodial slime molds.

licensed practical nurse *Medicine.* a person who is licensed to perform certain services to the sick under the supervision of a registered nurse. Also, **licensed vocational nurse.**

lichee see LITCHI.

lichen [lī´kən] *Botany.* the common name for any member of the group Lichenes, occurring as fungal and algal cells in symbiotic union and growing in various forms on rocks or trees.

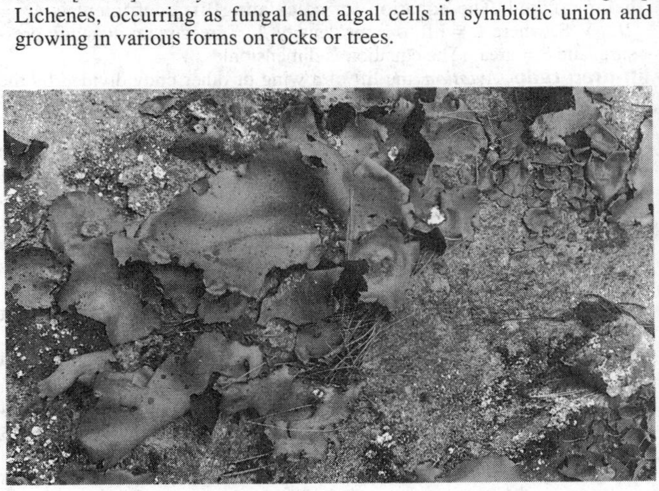

lichens

Lichenes *Botany.* a division of complex organisms consisting of fungi and algae living in such close symbiotic union as to be classified as a separate, thallophytic plant group.

Lichenes Imperfecti *Botany.* a class of lichens, some of whose members have no known method of sexual reproduction.

lichenification *Medicine.* a thickening and hardening of the skin due to prolonged scratching or irritation.

lichenin *Biochemistry.* a polysaccharide in the walls of lichen fungi that is hydrolysed to glucose by the enzyme lichenase.

lichenology *Botany.* a branch of biology devoted to the study of lichens.

lichenometry *Geology.* a method for determining the age of geomorphic features by measuring the diameter of lichens growing on the exposed surfaces of rocks.

lichenophagous *Zoology.* feeding mainly or exclusively on lichens.

Lichnophorina *Invertebrate Zoology.* a suborder of ciliate protozoans, in the suborder Heterotrichida, having an hourglass body shape.

Lichtenberg figure *Electricity.* a figure on a photographic plate whose appearance and dimensions are indicative of the polarity and crest value of a voltage surge.

licorice [lik´ə ris; lik´ə rish] *Botany.* **1.** the common name for the species *Glycyrrhiza glabra,* a perennial herb of the family Leguminosae. **2.** the dried roots of this plant. *Materials.* a flavoring derived from these roots, widely used to flavor candy, soft drinks, medicine, cigarettes, and so on.

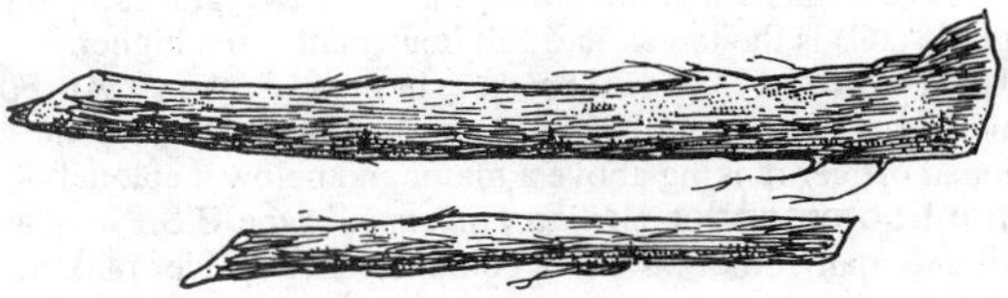

licorice

lidar *Optics.* an instrument that tracks atmospheric activities, such as cloud formations, by aiming a powerful infrared pulse at the object and measuring the reflection and scattering; also used to track weather balloons and rocket trails. (An acronym for light detection and rangefinding; on the model of *radar.*)

Liddell-Hart, B(asil) H(enry) 1895–1970, British author and military strategist; formulated modern techniques of mechanized warfare.

lidocaine *Organic Chemistry.* $C_{14}H_{22}ON_2$, a white or slightly yellow crystalline powder; insoluble in water and soluble in alcohol, ether, or chloroform; melts at 68–69°C and boils at 180–182°C (4 torr); used as a local nerve-blocking agent. Also, **lignocaine.**

lie *Medicine.* a term for the position of the long axis of the fetus in relation to that of the mother.

Lie, Marius Sophus [lē] 1842–1899, Norwegian mathematician; made major contributions to transformation groups and various other areas.

Lie algebra *Mathematics.* a vector space endowed with a Lie bracket.

Lieberkuhn's organelle *Cell Biology.* an organelle found associated with the buccal cavity beneath the pellicle of some hymenostomes.

Liebig, Justus von [lē´bik] 1803–1873, German chemist; a founder of organic and agricultural chemistry; revised chemical notation; discovered chloroform (independently) and chloral.

liebigite *Mineralogy.* $Ca_2(UO_2)(CO_3)_3 \cdot 11H_2O$, an apple-green orthorhombic mineral having a specific gravity of 2.41 and a hardness of 2.5 to 3 on the Mohs scale; found as secondary granular or scaly coatings in uranium mines. Also, URANOTHALLITE.

Liebman effect *Optics.* an effect that makes it more difficult for an individual to discern contrasting forms when they have identical luminance and divergent chromaticities than when they have divergent luminaries, but identical chromaticity.

Lie bracket *Mathematics.* an operation, denoted [,], that is bilinear, anticommutative, distributive with respect to addition, and satisfies the Jacobi identity. Also, COMMUTATOR.

Lie derivative *Mathematics.* in the Lie algebra of vector fields on a manifold, the Lie bracket $[v, w]$ is called the Lie derivative of w in the direction of v; denoted by $L_v w$. The Lie derivative obeys Leibnitz's rule with respect to tensor multiplication, commutes with tensor contraction, and $L_v f = v(f)$ for all functions f. By means of these rules, the Lie derivative is considered to be a linear differential operator on tensor fields of all types on the manifold. It has several geometric interpretations.

lie detector *Mechanical Devices.* a device that is used to investigate whether or not a person is telling the truth, as in a criminal inquiry, by detecting changes in body functions such as pulse, blood pressure, and perspiration, which can be stimulated by the emotional stress of lying in response to a question. The scientific accuracy of lie-detector tests is debatable and the results of such tests are generally not admissible as evidence in court.

Lie group *Mathematics.* a group G that is also a differentiable manifold such that the mapping ϕ given by $\phi(x, y) = xy^{-1}$ for all $x,y \in G$ is differentiable; i.e., the differentiable structure is compatible with the group operation.

Lienard-Wiechert potentials *Electromagnetism.* the advanced and retarded scalar and vector potentials produced by a point charge in motion, expressed in terms of the advanced and retarded position and velocity of the charge.

lierne *Architecture.* in Gothic vaulting, a tertiary rib, usually decorative rather than structural.

Liesegang banding *Geology.* nested, colored bands formed in a fluid-saturated rock by rhythmic precipitation. Also, **Liesegang rings.**

Lieskeela *Bacteriology.* a genus of rod-shaped, nonmotile, sheathed bacteria covered with manganese or iron oxide; found in mud and water.

lieutenant *Military Science.* **1.** in the U.S. Army, Air Force, and Marine Corps, a commissioned officer ranking above a noncommissioned officer or warrant officer and below a captain; there are two grades: second lieutenant is the lower, and first lieutenant is the higher. **2.** in the U.S. Navy and Coast Guard, a commissioned officer ranking above an ensign and below a lieutenant commander; there are two grades: lieutenant junior grade (JG) is the lower, and full lieutenant is the higher.

lieutenant colonel *Military Science.* in the U.S. Army, Air Force, and Marine Corps, and in the British Army and many other armies, a commissioned officer ranking above a major and below a colonel.

lieutenant commander *Military Science.* in the U.S. Navy and Coast Guard, and many other navies, a commissioned officer ranking above a lieutenant and below a commander.

lieutenant general *Military Science.* in the U.S. Army, Air Force, and Marine Corps, and many other armies, a commissioned officer ranking above a major general and below a general; the insignia is three stars.

LIF left iliac fossa; leukocyte inhibitory factor.

life *Biology.* the fact of being alive; the condition that distinguishes organisms such as humans, animals, and plants from inorganic matter and from dead organisms. Organisms that have life generally share powers and functions such as the following: a specific and identifiable structure or organization; the ability to move from one location to another or to carry on internal movement; the capacity for metabolism, reproduction, and growth; the ability to detect the conditions of the surrounding environment and respond to them; and the ability to adapt to long-term changes in this environment. *Medicine.* a medical determination that this condition exists in a human being; the definition is subject to many legal, ethical, and philosophical considerations, particularly in establishing the point at which life begins and the point at which it ends.

lifeboat *Naval Architecture.* **1.** a boat equipped with emergency supplies and carried aboard ship for the crew and passengers in case it becomes necessary to abandon ship. **2.** an emergency boat kept ashore and launched by rescue crews during shipwrecks. **3.** a seaworthy boat carried by a ship and kept ready for use in emergency rescue operations.

life-change scale *Psychology.* a method of quantifying stress developed by Thomas Holmes and others, assigning relative values to various significant events that may occur in a person's life, such as divorce, marriage, pregnancy, retirement, loss or change of job, a significant change in income, and so on. Also, **life-events scale.**

life-change unit *Psychology.* the relative weight assigned to an event in an individual's personal life in measuring its stressful effect; e.g., death of one's spouse is given a maximum value of 100 and vacation is given a value of 13.

life cycle *Biology.* **1.** in any organism, the entire sequence of events in its growth and development from the time of zygotic formation until the time of gamete formation. **2.** the progressive succession of changes through which an organism goes from the start of its development as a fertilized egg until its death as an adult.

life expectancy *Biology.* the average period of time that an organism is expected to survive at any given point in its life cycle.

life flight *Medicine.* the emergency transportation of a patient, particularly an accident victim, to a medical facility by helicopter or airplane.

life form *Ecology.* the characteristics of an organism, such as its structure, appearance, habits, and history.

life history *Evolution.* the significant stages in the life cycle of an organism, with special emphasis placed on the strategies that affect its ability to survive and reproduce.

life instinct see EROS.

life preserver *Engineering.* a flotation device employed to prevent a person from drowning by giving the upper body buoyancy in the water.

life raft *Naval Architecture.* an inflatable raft carried aboard a ship for escape in case the crew and passengers are forced to abandon ship.

life span *Biology.* the time from birth to death for a certain individual. *Statistics.* the average or projected length of life for the individuals within a certain group, based on statistics on the actual length of life of such individuals. *Engineering.* see LIFETIME.

life-support system *Engineering.* a controlled environment that provides the biological conditions necessary to maintain vital functions.

life table *Ecology.* a tabular expression of the complete mortality and reproductive statistics of a given population.

life test *Engineering.* a test used to estimate the expected length of use or serviceability of a product or device, by simulating on an accelerated basis the conditions it will encounter.

lifetime *Biology.* the length of life of an individual; life span. *Engineering.* the period of useful service of a machine or device.

life zone *Ecology.* one of the various divisions of the earth's surface into distinct regions that have relatively uniform conditions of climate, soil, and plant and animal life.

LIFO see FILO.

lift to raise something to a higher level, or a device that does this; specific uses include: *Fluid Mechanics.* the total amount of force that acts on a body perpendicular to the undisturbed airflow as the body moves through a fluid medium. Also, AERODYNAMIC LIFT. *Aviation.* the upward force exerted on a balloon or airship by the lighter-than-air gas within it. *Space Technology.* see LIFT-OFF. *Mechanical Engineering.* **1.** an endless chain or belt device with buckets, scoops, arms, or trays used to raise powders, granules, or solid objects to a higher level. **2.** an enclosed platform or car that transports people or materials up or down in a building shaft. Also, ELEVATOR. *Graphic Arts.* a stack of paper, especially the quantity of paper normally placed on a press or copier at one time. *Mathematics.* let B be a bundle over a manifold M. The lift (or **lifting**) of any object in M, such as a vector field, curve, diffeomorphism, and so on, is a corresponding object in B with the property that the projection of the object in B back to M is the original object. For example, let X be a manifold with cotangent bundle $T * X$. The lift of a diffeomorphism f: $X \to X$ is the mapping $T * f$ from $T * X$ into itself given by

$$(T * f)(x, p_x) = (f^{-1}(x)(f * p)_f^{-1}(x)).$$

lift-and-lower *Robotics.* a single axis of motion for a pick-and-place robot.

lift bridge *Civil Engineering.* an apparatus that opens by vertical ascension.

lift coefficient *Aviation.* a coefficient of the lift of a given wing or other body, represented by the symbol C_L and calculated by dividing the lift by the freestream dynamic pressure and by the representative area being studied. The equation for calculating lift coefficient is $C_L = 2L/\rho V^2 S$, where L = lift, ρ = air density, V = freestream dynamic pressure, and S = area. (The equation is dimensionless.)

lift-drag ratio *Aviation.* the lift of a wing or other body divided by the drag, or the lift coefficient divided by the drag coefficient; the primary measure of the efficiency of an aircraft. Also, L/D, L/D RATIO.

lifter case *Mining Engineering.* a sleeve or tubular part fastened to the lower end of the inner tube of a core barrel. Also, INNER-TUBE EXTENSION, CORE-CATCHER CASE.

lifter flight *Design Engineering.* in rotary dryer applications, the lifting and showering of solid particles through the gas-drying stream when they are passed through a cylinder that has spaced plates or projections attached to its inside surfaces.

lift fan *Aviation.* a turbo-fan that is designed to provide lift or lift thrust for an aircraft.

lifting block *Mechanical Engineering.* a combination of pulleys and ropes used for lifting heavy weights with minimal effort.

lifting condensation level *Meteorology.* the atmospheric level at which a moist air parcel lifted in a dry adiabatic environment becomes saturated. Also, ISENTROPIC CONDENSATION LEVEL.

lifting magnet *Electromagnetism.* an electromagnet that typically has a strong magnetic induction and lifts or moves objects by being placed in direct contact with the object before it is lifted.

lifting reentry *Space Technology.* the return of a space vehicle into the atmosphere using aerodynamic lift rather than free-fall in order to slow the vehicle's descent and thereby increase control of its temperature and landing site.

lift-off *Space Technology.* in a vertical ascent, the action of a rocket vehicle as it separates from the launch pad. (In an ascent other than vertical, such an action is called a takeoff.) Also, LIFT.

lift pump *Mechanical Engineering.* a pump in which a liquid is lifted (by a vacuum, for example), as distinguished from being forced upward.

lift-slab construction *Civil Engineering.* a method of forming reinforced concrete floor and roof slabs at ground level and installing them after the process by means of lifting jacks.

lift truck *Mechanical Engineering.* a dolly or truck used for lifting or moving, especially of palletized loads.

lift valve *Mechanical Engineering.* a valve that moves in a direction perpendicular to the plane of the valve seat.

lig. ligament; ligamentum.

ligament *Histology.* a type of white fibrous connective tissue that surrounds and holds bones together at joints. *Engineering.* the solid material piece of a tube sheet between adjacent holes. (Going back to a Latin word meaning "to tie.")

ligament of Bigelow see ILIOFEMORAL LIGAMENT.

ligamentous *Histology.* relating to or in the form of a ligament.

ligamentum *plural,* **ligamenta.** *Histology.* the anatomical term for a ligament.

ligand *Chemistry.* a functional group, atom, or molecule that is attached to the central atom of a coordination compound, particularly a nonmetallic substance that combines with another substance in solution to form a coordination compound. Also, COMPLEXING AGENT.

ligand membrane *Chemistry.* a solvent that is immiscible both in water and in another reagent and that acts as both an ionic extractant and a complexing agent.

ligase *Enzymology.* one of the six main classes of enzymes that serve to catalyze the bond formation between two compounds using ATP or other energy donors. Also, SYNTHETASE.

ligation *Surgery.* the tying off of a blood vessel or duct with a suture or wire ligature. *Molecular Biology.* the formation of a phosphodiester bond joining two nucleotides.

ligature *Surgery.* any substance such as catgut, cotton, silk, or wire used to tie a vessel or strangulate a part. *Graphic Arts.* a type character consisting of two or more conjoined letters, such as *æ* or *ffi*.

ligg. ligaments or ligamenta.

light something that makes things visible; specific uses include: *Optics.* **1.** a type of electromagnetic radiation to which the organs of sight react, ranging from about 400 to 700 nm in wavelength and propagated at 186,282 miles per second. **2.** a similar radiation that does not affect the organs of sight, including X-rays, ultraviolet, visible, infrared, radio waves, and radar sources. *Architecture.* a window or windowpane.

light absorption *Optics.* the value that expresses the ability of an object to take in light without reflection or transmission.

light-activated silicon controlled rectifier *Electronics.* a device that uses light rather than current to control its operations, used to convert alternating current into direct current. Also, PHOTO-SCR, PHOTOTHYRISTOR.

light adaptation *Physiology.* the adjustment in sensitivity of the dark-adapted eye to a high-light level, such as that occurring when a person emerges into daylight after being in a darkened movie theater.

light air *Meteorology.* a term for a wind of 1–3 miles per hour.

light amplifier *Electronics.* a circuit that responds to a light image by producing a brighter version of the image. Also, IMAGE INTENSIFIER.

light antiaircraft artillery *Ordnance.* conventional antiaircraft artillery, usually under 90 mm and weighing 20,000 lbs or less in a trailed mount with on-carriage fire control.

Light Armored Vehicle *Ordnance.* an eight-wheeled armored reconnaissance car used by U.S. military forces.

light artillery *Ordnance.* **1.** artillery with a caliber of less than 155 mm. **2.** smaller artillery in general, such as light antiaircraft weapons. Also, **light field artillery.**

light beacon *Navigation.* a navigational beacon that emits a visible light signal.

light-beam galvanometer see D'ARSONVAL GALVANOMETER.

light-beam oscillograph *Electromagnetism.* an apparatus used to detect low-frequency oscillation; constructed of a moving coil galvanometer with a small mirror mounted on it that can reflect a beam of focused light onto a moving sheet of photographic film.

light-beam pickup *Acoustical Engineering.* a pickup for a phonograph wherein the signal is transmitted to the transducer by means of a beam of light.

light blasting *Engineering.* the use of explosives to break up surface rock or boulders.

light bomber *Aviation.* any bomber aircraft that is considered to be relatively light in weight, especially one whose gross weight is less than 100,000 pounds, such as the B-26 or B-45.

light breeze *Meteorology.* a term for a wind of 4–7 miles per hour.

light carrier injection *Electronics.* **1.** a process by which the intensity of a light beam is varied so as to bring it into harmony with a modulating signal. **2.** a process by which the intensity of light is varied, generally at audio-frequency range, in motion picture sound systems. Also, LIGHT MODULATION.

light-case bomb *Ordnance.* a general-purpose bomb with a light metal casing that has a high charge-weight ratio; designed to cause damage by the force of its explosion.

light chain *Immunology.* the smaller of the two polypeptide chains that occur in immunoglobulins. Also, L CHAIN.

light chopper *Electronics.* a device that interrupts a light beam directed at a phototube in order to facilitate amplification of the tube's output.

light climate see ILLUMINATION CLIMATE.

light cone *Physics.* the set of all points in space-time that are reached by signals traveling at the speed of light from a specified point.

light crude *Petroleum Engineering.* a crude oil of light color that flows freely at atmospheric temperatures, having predominant low-viscosity, low-molecular-weight hydrocarbons.

light cruiser *Naval Architecture.* a cruiser of under 6000 tons displacement, mounting no larger than 6-inch guns, as defined by naval treaties between the World Wars. The distinction between light and heavy cruisers is no longer applied to modern ships.

light curve *Astronomy.* the changes in brightness of a variable star plotted over time.

light distribution photometer *Optics.* a device that gauges the luminosity of a light source from a number of directions.

light-drawn *Metallurgy.* describing the attribute of a material that has been cold reduced less than a hard-drawn product.

lighted buoy *Navigation.* a floating navigational aid that includes a light.

light-emitting diode *Electronics.* a semiconductor diode that converts electrical energy into light when an alternating current is applied. Also, SOLID-STATE LAMP.

light end *Petroleum Engineering.* any of the more volatile, lower-boiling products of petroleum refining, such as butane and propane.

lightening *Medicine.* the feeling of reduced distention of the abdomen during the latter stages of pregnancy (two or three weeks before labor) as the uterus descends into the pelvic cavity.

lighter *Mechanical Devices.* any of various devices used to light something, especially such a device used to light cigarettes. Thus, **lighter fluid.**

lighter *Naval Architecture.* an open, unpowered, bargelike vessel used to transfer cargo between the shore and a ship, or between two ships.

lighterage *Industrial Engineering.* the use of lighters in loading and unloading ships and transporting goods for short distances.

lighter-than-air *Aviation.* of or relating to lighter-than-air craft or to gases such as helium that are used to support them.

lighter-than-air craft *Aviation.* an aircraft, such as a dirigible, that weighs less than the air it displaces, so that its lift is due to aerostatic buoyancy.

light exposure *Optics.* a measure of the total light reaching a surface. Also, EXPOSURE.

lightface *Graphic Arts.* a style of type that has thinner, lighter characters than the regular weight of the same typeface.

light freeze *Meteorology.* a condition characterized by a brief drop in surface air temperature to below 32°F, adversely affecting tender plants.

light-gating cathode-ray tube *Electronics.* a cathode-ray tube in which the electron beam alters the transmission or reflection properties of the screen by exposing it to an external light source.

light gun *Electronics.* a device that generates electricity from light in order to assist and direct specific computer functions.

light-harvesting complex *Biochemistry.* a set of photosynthetic pigment molecules that absorb light and channel the energy to the photosynthetic reaction center, where photosynthesis occurs.

light homing *Ordnance.* the process of guiding a missile by using light emanating from the target.

light horse *Ordnance.* a historic term for the section of the cavalry carrying light arms and equipment.

lighthouse *Navigation.* a tower or other structure on land that flashes a bright light to warn ships of dangerous areas, guide them in certain routes, and so on.

light howitzer *Ordnance.* a howitzer with a bore diameter greater than 30 mm but not exceeding 125 mm.

light hydrogen *Nuclear Physics.* the ordinary light hydrogen isotope, consisting of one proton and one electron. Also, PROTIUM.

lighting branch circuit *Electricity.* a circuit aboard ships that supplies energy to lighting outlets and to various portable appliances of fewer than 600 watts.

lighting-off torch *Engineering.* an iron rod wrapped in an oil-drenched cloth, used to ignite a fuel-oil burner.

light-inspection car *Mechanical Engineering.* a railway motorcar that weighs between 400 and 600 pounds and has a capacity of 650 to 800 pounds.

light ion see SMALL ION.

light list *Navigation.* a publication that catalogs marine navigational lights and sound signals and related information.

light list number *Navigation.* the number assigned to a given marine navigational light or sound-emitting device in the light list.

light machine gun *Ordnance.* **1.** in U.S. military usage, any machine gun of .30 caliber or less, excluding fully submachine guns, automatic rifles, and machine pistols. **2.** in general usage, any machine gun or fully automatic rifle weighing approximately 20 to 30 pounds with its mount.

light meromyosin *Biochemistry.* the terminal tail fragment of a myosin molecule.

light metal *Metallurgy.* a metal, such as aluminum beryllium, or titanium, having density below about 5 grams per cubic centimeter.

light meter *Engineering.* a device that detects and measures incident white light or illumination.

light microsecond *Electromagnetism.* the distance traversed in free space by light in one-millionth of a second.

light mineral *Mineralogy.* **1.** a rock-forming mineral with a specific gravity less than 2.85. **2.** any light-colored mineral.

light mineral oil *Materials.* a mineral oil of a specified lighter density, with a specific gravity of 0.818 to 0.880.

light modulation see LIGHT CARRIER INJECTION.

light modulator *Electronics.* a device that reproduces an optical sound track on motion picture film by combining source light with an optical system, and then varying the resulting light beam.

light-negative *Electronics.* describing a material whose ability to conduct electricity decreases when the material is exposed to light.

Lightning *Aviation.* a popular name for the P-38 pursuit (fighter) aircraft.

lightning *Geophysics.* a luminous, high-current, electric discharge with a long path that flows between a cloud and the ground, between two clouds, or between two parts of the same cloud.

lightning

lightning arrester *Electricity.* a protective device that allows voltages higher than a certain value to be discharged to ground.

lightning bug see FIREFLY.

lightning channel *Geophysics.* the superheated, irregular, ionized pathway in the atmosphere created by the passage of a lightning bolt.

lightning conductor *Electricity.* **1.** any pathway that lightning can follow to ground, such as the trunk of a tree, a power line, or a communication antenna. **2.** see LIGHTNING ROD.

lightning discharge *Geophysics.* the neutralization of a large electric potential in the atmosphere by means of a lightning stroke.

lightning flash *Geophysics.* the observable light or luminosity that accompanies a lightning discharge.

lightning generator *Electricity.* a generator that produces surge voltages similar to lightning, for testing insulators and other high-voltage components.

lightning protection *Electricity.* any means used to protect personnel and equipment from the hazardous effects of lightning damage.

lightning rod *Electricity.* a grounded metal electrode placed on the roof of a building to protect the structure from lightning.

lightning stone see FULGURITE.

lightning surge *Electricity.* a transient electric disturbance in an electric circuit produced by lightning.

lightning switch *Electricity.* a manually operated switch that is employed during electrical storms to connect a radio antenna to ground rather than to the radio receiver. Also, GROUND SWITCH.

light oil *Materials.* **1.** a fractional distillate from coal tar, boiling at a range of 110–210°C; a source of benzene, solvent naphthas, toluene, and phenol. **2.** any of various oils from other sources with the same boiling range.

light-operated switch *Electronics.* a device that is activated by pulses or beams of light, such as a light-activated silicon-controlled switch.

light pen *Electronics.* a pencil-like device that senses light pulses on a cathode-ray tube screen. *Computer Technology.* such a device that is connected to a computer display unit and used by an operator to change or modify an image displayed by touching the pen to the screen.

light pillar see SUN PILLAR.

light pipe *Optics.* a cylindrical filament that is formed from a transparent material with a high index of refraction and that is used to conduct light through flexible pathways.

light pollution *Optics.* a term for the presence of unwanted or harmful light, such as objectionably bright neon signs or streetlights. *Astronomy.* specifically, the presence of artificial light in the night sky to such an extent that it impairs the observation or photography of faint astronomical objects.

light-positive *Electronics.* describing a material whose ability to conduct electricity increases when the material is exposed to light.

light projector *Optics.* a device that generates beams of light over long distances.

light-proof *Optics.* not allowing light to penetrate; impervious to light.

light-rail rapid transit *Transportation Engineering.* a light-rail transit system operating on its own exclusive right-of-way.

light-rail transit *Transportation Engineering.* a system using self-propelled electric rail cars, usually on conventional railway lines.

light rate *Astrophysics.* the factor (2.512, or the fifth root of 100) by which one magnitude differs in brightness from another.

light ray *Optics.* a line in space matching the direction of the flow of radiant energy.

light-red silver ore see PROUSTITE.

light reflex *Physiology.* an involuntary change in the constriction or dilation of the pupil that occurs in response to changes in light intensity.

light relay see PHOTOELECTRIC RELAY.

light repair see PHOTOREACTIVATION.

light ruby silver see PROUSTITE.

light scattering *Optics.* a change in the spatial distribution of a beam of light when it interacts with a surface; a process that causes no change in wavelength.

light section car *Mechanical Engineering.* a railway motorcar that weighs between 750 and 900 pounds and that is propelled by a 4- to 6-horsepower engine.

light sector *Navigation.* a horizontal arc in which a navigational light is visible or has a distinctive color.

light-sensitive *Electronics.* describing a material that exhibits photoelectric effects when exposed to light. *Medicine.* having eyesight that is unusually or abnormally sensitive to daylight or ordinary room light.

light-sensitive detector see PHOTODETECTOR.

lightship *Navigation.* a distinctive vessel that is moored at a charted position and displays a major navigational light.

light source *Optics.* a term for a source of visible forms of electromagnetic radiation such as the sun, a laser, a light bulb, or a light-emitting diode. *Graphic Arts.* any artificial or natural source of light used in photography, such as sunlight, a flash bulb, or a reflector.

light stability *Computer Technology.* the resistance of an image in optical character recognition to changes in its appearance after exposure to radiant energy.

light station *Navigation.* the complex of buildings associated with a lighthouse.

light-time *Astronomy.* the distance between two objects measured by the time it would take light to travel between them.

light transmission *Optics.* the process by which light passes through a material without being absorbed or reflected.

light valve *Electronics.* a device that varies its light transmission in relation to the application of an external electrical quantity, such as voltage, current, or an electron beam.

light water *Materials Science.* **1.** ordinary water, H_2O, as distinguished from heavy water (deuterium oxide or tritium oxide). **2.** a fire-fighting agent consisting of a water solution of perfluorocarbon compounds mixed with a water-soluble thickener of the polyoxyethylene type. *Nucleonics.* water in which only the two hydrogen isotopes have a proton.

light waterline *Naval Architecture.* the waterline at which a vessel floats when empty of cargo.

light-water reactor *Nucleonics.* a nuclear reactor that uses ordinary water to remove heat from the reactor core; used to generate electricity .

light watt *Optics.* a unit of light intensity in which the wavelength determines the ratio of luminescence.

lightweight concrete *Materials.* a concrete of low weight and good insulating capability, containing lightweight aggregates such as vermiculite or perlite. Also, **lightweight aggregate.**

light-year *Astronomy.* the distance that light would travel in a vacuum in one year, 9.46 trillion kilometers or 5.8 trillion miles; used in measuring astronomical distances.

Ligiidae *Invertebrate Zoology.* a family of primitive terrestrial isopod crustacea in the suborder Oniscoidea.

lignans *Organic Chemistry.* plant products of low molecular weight, representing the dimer intermediate between monomeric propylphenol and lignin; formed mainly by oxidative linking of *p*-hydroxyphenylpropene units, which may be bridged with oxygen.

ligneous *Biology.* formed of or resembling wood; woody.

lignicolous *Botany.* growing on or in wood, as some fungi.

lignify *Botany.* of plant cell walls, to harden, thicken, and become woody due to the deposition of lignin.

lignin *Cell Biology.* a complex polymer that forms an extensive network within the cell walls of certain plants and that confers strength and rigidity to the cell wall; one of the chief substances found in wood. *Materials Science.* a brown to transparent crystalline form of this substance, derived from paper-pulp sulfate liquor; used as a corrosion inhibitor, adhesive, coating, and fertilizer, and as a binder of compressed wood. (From the Latin word for wood.)

ligninase *Enzymology.* any one of several enzymes that break down lignin, the major component of wood.

lignin net *Cell Biology.* the structure that lignin forms in a cell.

lignin plastic *Organic Chemistry.* a plastic made from lignin resins; used as a binder or extender.

lignin sulfonate *Organic Chemistry.* a metallic sulfonate salt occurring as a tan to dark brown powder, derived from the lignin of sulfite, softwood pulp-mill liquids; it is nonhygroscopic and is insoluble in all organic substances; it decomposes above 200°C; used as a dispersing agent, extender, and ore flotation agent. Also, **lignosulfonate.**

lignite *Geology.* a brownish-black coal at the intermediate stage of coalification between peat and bituminous coal.

lignite A see BLACK LIGNITE.

lignite B see BROWN LIGNITE.

lignite wax see MONTAN WAX.

lignitous coal *Mineralogy.* any coal containing 75–84% carbon.

lignocellulose *Biochemistry.* a heterohexosan that is formed in the cell walls of woody plants; composed of cellulose and lignin.

lignosa *Botany.* woody plant matter.

lignum vitae or **lignumvitae** *Botany.* the common name for members of the genus *Guaiacum,* hardwood trees that produce exceptionally dense, tough wood and a resin known as gum guaiac.

ligulate *Botany.* **1.** shaped like a strap. **2.** bearing ligules.

ligule *Botany.* **1.** a strap-shaped corolla found especially in the Compositae. **2.** an appendage on the upper sheath of certain grass leaves.

likelihood function *Statistics.* the probability of observing a given sample regarded as a function of the unknown parameter in a population. *Mathematics.* a probability density function, except that its integral over the sample space need not be normalized to equal 1.

likelihood ratio *Statistics.* the ratio between the likelihood at the maximum value under the null hypothesis to the likelihood at its maximum value under the alternative hypothesis. The null hypothesis is rejected for small values of the statistic.

lilac *Botany.* any of various trees or shrubs belonging to the genus *Syringa* of the family Oleaceae, especially *S. vulgaris,* the **common lilac,** widely grown as an ornamental tree or shrub for its large, showy purple or white flowers.

Liliaceae *Botany.* an extensive family of monocotyledonous herbs belonging to the order Liliales, having a six-membered perianth, six stamens, and narrow leaves with parallel veins. It includes such common ornamental flowers as the lilies, hyacinths, and tulips; wildflowers such as the asphodel, trillium, and dogtooth violet; and agricultural plants such as asparagus, aloe, and the onion and related plants.

Liliales *Botany.* an order of monocotyledonous plants belonging to the subclass Liliidae, characterized by actinomorphic flowers and syncarpous ovaries.

Liliidae *Botany.* a subclass of the Liliopsidae, characterized by cyclic flowers with six-membered perianths.

Liliopsida *Botany.* in some classification systems, another name for the Monocotyledoneae, whose members have a single cotyledon and parallel-veined leaves.

lillianite *Mineralogy.* $Pb_3Bi_2S_6$, a steel-gray, opaque, metallic orthorhombic mineral occurring in long platy crystals, having a specific gravity of 7.02 to 7.16 and a hardness of 2 to 3 on the Mohs scale; found in tungsten deposits.

Lilliputian hallucination see MICROPSIA.

Lilly controller *Mechanical Engineering.* a device on steam or electric winding engines that protects equipment and personnel against hazards such as motor overspeed, overwind, and other conditions.

LILO see FIFO.

lily *Botany.* the common name for various members of the genus *Lilium,* perennial bulbous herbs of the family Liliaceae, which grow widely as garden flowers or wildflowers throughout the world. Among the many familiar species or subspecies are the **tiger lily,** *L. tigrinum*; the **leopard lily**, *L. pardilinum*; the **Madonna lily,** *L. candidum*; and the **Easter lily,** *L. longiflorum eximium.*

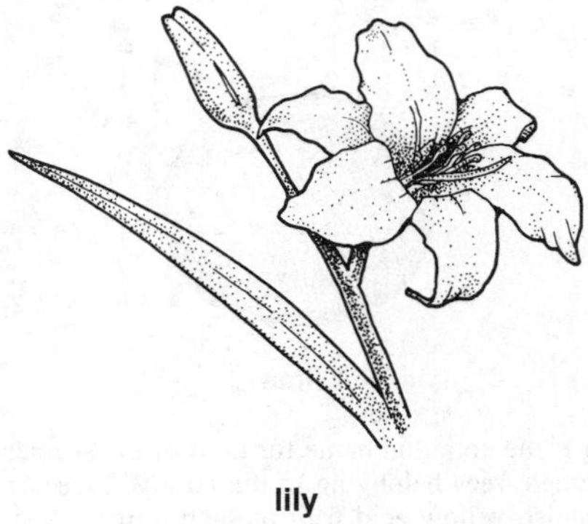

lily

lily family *Botany.* the common name for the family Liliaceae.

lily of the valley *Botany.* a common garden flower, *Convallaria majalis,* of the family Liliaceae; noted for its clusters of small, bell-shaped white flowers.

lily trotter see JACANA.

LIM *Aviation.* the airport code for Lima, Peru.

Limacidae *Invertebrate Zoology.* slugs, a family of shell-less gastropod mollusks including many plant pests.

Limacodidae see EUCLEIDAE.

limaçon [lim´ə sōn´] *Mathematics.* a conchoid of a circle of radius $a/2$ and center at $(a/2, 0)$, with the origin as pole; equation in polar coordinates: $r = a \cos \theta + b$. When $a < b$, the origin is an isolated point of the limaçon. When $a > b$, the origin is a double point. When $a = b$, the origin is a cusp. Also, **limaçon of Pascal.**

limax ameba *Invertebrate Zoology.* any of various bacteria-eating, generally small, soil-dwelling amebae that resemble a slug in appearance and movement.

limb *Anatomy.* **1.** an arm or leg. **2.** a segment of or extension from a structure. *Botany.* a primary branch or bough of a tree. *Design Engineering.* the graduated margin of an arc or circle in an instrument used for measuring angles. *Geology.* one of the two sides of a fold. Also, FLANK. *Astronomy.* the apparent edge of the sun, a planet, or a moon.

limb brightening *Astronomy.* a brightening of the solar limb at X-ray, extreme ultraviolet, and radio wavelengths, arising because the observer's line of sight intercepts more of the very hot inner corona and less of the cooler photosphere as it moves away from the center of the sun's disk.

limb bud *Developmental Biology.* an ectodermally covered mesenchymal outgrowth on the embryonic flank that develops into either the forelimb or hindlimb.

limb darkening *Astronomy.* a darkening of the solar limb at optical wavelengths, arising because the observer's line of sight passes through shallower, cooler, and more absorbing layers of the photosphere as it moves away from the center of the sun's disk; a similar effect occurs with Jupiter.

limber *Naval Architecture.* one of two drainage channels running fore-and-aft on either side of the keelson; water collected in the bottom of the hull runs through the limbers toward the bilge pump. *Ordnance.* **1.** a two-wheeled vehicle, originally pulled by horses, used to support the trail section of a gun carriage in transit. **2.** to attach a gun to a limber.

limber board *Naval Architecture.* a short board covering a length of the limbers on either side of the keelson. The limber boards can be removed in order to clean out the limbers.

limber hole *Naval Architecture.* one of a series of longitudinal drain holes through a ship's lower frames on either side of the keelson, allowing water to flow toward the bilge pump.

limberneck *Veterinary Medicine.* a colloquial name for botulism in birds, intoxication by the toxin of *Clostridium botulinum*; typical symptoms are loss of muscle strength in the neck, wing, and leg muscles.

limbic system *Anatomy.* a term for the portions of the central nervous system, including the hippocampus, amygdaloid nucleus, and portions of the midbrain, that are associated with certain autonomic functions and various emotions including pleasure, fear, and happiness.

LIMBO antisubmarine mortar *Ordnance.* a British ship-based, medium-range, triple-barreled mortar system; it includes a predictor that processes the ship's sonar data to produce a three-dimensional firing pattern of depth charges ahead of the target submarine.

limburgite *Petrology.* a dark, porphyritic igneous rock with abundant phenocrysts of olivine and pyroxene in a glassy alkali-rich groundmass.

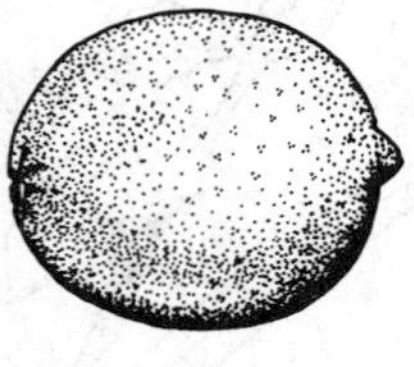

lime

lime *Botany.* **1.** the common name for trees of the species *Citrus aurantifolia,* evergreen trees belonging to the family Rutaceae. **2.** the edible, green or greenish-yellow acid fruit of such a tree, often used to flavor beverages and foods.

lime *Materials.* **1.** the chemical substance calcium oxide, CaO, a solid in the form of white to grayish pebbles, produced by heating crushed limestone to a temperature of about 900°C. It is one of the most commonly produced materials, and is widely used in building, agriculture, and metal refining, and in water and waste treatment. **2.** produced from or containing lime. Thus, **lime glass, lime mortar, lime putty,** and so on.

limed rosin see CALCIUM RESINATE.

lime kiln *Chemical Engineering.* a rotary or vertical, cylindrical furnace in which limestone is heated and contacted countercurrently with heat to convert it to lime.

limelight *Engineering.* an intense light source derived by applying an oxyhydrogen flame to a cylinder of lime; formerly used widely in lighthouse beacons and theater spotlights.

limen *plural,* **limina** *Science.* a threshold or boundary.

limen insulae *Neurology.* the point of junction on the inferior surface of the cerebral hemisphere between the cortex of the insula and the cortex of the frontal lobe.

lime oil *Materials Science.* a colorless to greenish-yellow oil distilled from the juice or crushed fruit of the lime tree; used in extracts, flavorings, perfumes, and soaps.

lime-silicate see CALC-SILICATE.

limestone *Mineralogy.* a sedimentary rock composed primarily of calcium carbonate, formed from the skeletons of marine microorganisms and coral; used in building and in the manufacture of lime.

limestone

liminal [lim′i nəl] *Neurology.* barely perceptible to the senses; at the threshold of perception to a sensory stimulus.

limit *Mathematics.* an accumulation point. In particular, if p is an accumulation point of a sequence of points $x_1, x_2, x_3, \ldots$, then we write $\lim_{n\to\infty} x_n = p$, or sometimes simply $\lim = p$. If $\liminf x_n = \limsup x_n = c$, then $\lim_{n\to\infty} x_n$ exists and is equal to c. Also, **limit of a sequence.**

limit cycle *Ecology.* the normal, periodic fluctuations in the population size of a species functioning as predators and prey. *Chaotic Dynamics.* a representation of a simple closed curve; a periodic solution in the phase space of a dynamical system.

limited-degree-of-freedom robot *Robotics.* a robot that is not capable of orienting and positioning its end effector with more than 6 degrees of freedom.

limited distance modem *Computer Technology.* see LINE DRIVER.

limited resource *Ecology.* a needed resource that is relatively scarce as compared to the demand for it.

limited-sequence robot see FIXED-SEQUENCE ROBOT.

limited space-charge accumulation diode *Electronics.* a device that prevents microwave frequencies from building up to any appreciable degree during their cycle, thus limiting transit time while increasing maximum oscillation frequency.

limited war *Military Science.* armed conflict involving the overt engagement of the military forces of two or more nations that is of lesser scope than general war, yet more protracted than incidents.

limiter *Electronics.* a transducer, such as a shunt-polarized diode between resistors, in which the output above a critical value of the input does not vary; used particularly to limit the amplitude of a radio, telephone, or recording device.

limit inferior *Mathematics.* **1.** if $X = (x_n)$ is a sequence of real numbers that is bounded below, then the limit inferior of $X = (x_n)$, if it exists, is the supremum of those real numbers u with the property that there are only a finite number of natural numbers n such that $x_n < u$. This is denoted by $\liminf X$, $\liminf x_n$, or $\underline{\lim}\, x_n$. **2.** the limit inferior of a function f at a point c is

$$\lim_{\varepsilon\to 0} (\text{glb}\{f(x)\colon |x - c| < \varepsilon \text{ and } x \neq c\}).$$

This is denoted by $\liminf_{x\to c} f(x)$ or $\underline{\lim}_{x\to c} f(x)$.

limiting amino acids *Nutrition.* the essential amino acids that are present in less than the minimum amounts relevant to total nitrogen required for growth and maintenance of good health.

limiting current *Electronics.* in a semiconductor, the very slight current that occurs in a P-N junction when an external potential is applied so that the negative connection is on the P side and the positive connection is on the N side (reverse bias). It occurs because a very small number of electron holes are present on the N side among the predominating number of electrons, while a small number of electrons are driven from the P side. *Materials Science.* the maximum current that can flow in an electrochemical reaction such as corrosion or electrodeposition; limitation is due to the availability of ions in the electrode.

limiting current density *Metallurgy.* in electroplating, the maximum current density above which polarization or other undesirable phenomena occur.

limiting deep drawing ratio *Materials Science.* the maximum value for the ratio of the blank-to-punch diameter that will produce a successful component in metal fabrication in which a flat sheet is shaped into a cylindrical or boxlike shape using a punch.

limiting factor *Ecology.* any environmental factor that prevents an organism or population from reaching its full potential of distribution or activity.

limiting magnitude *Astronomy.* **1.** the faintest stellar magnitude that is detected by eye or with a given instrument. **2.** the faintest magnitude for which a catalogue of celestial objects is complete.

limiting viscosity (number) *Physical Chemistry.* the ratio of a solution's viscosity to the concentration of its solute, extrapolated to zero concentration of the solute. Also, INTRINSIC VISCOSITY.

limit number *Mathematics.* a transfinite ordinal number that does not have an immediate predecessor.

limit of a function *Mathematics.* **1.** a function $f(x)$ is said to have a limit L at a point c if for every number $\varepsilon > 0$, there exists a number $\delta > 0$ such that $|x - c| < \delta$ implies that $|f(x) - L| < \varepsilon$, where x is in the domain of f and $x \neq c$; denoted $\lim_{x\to c} f(x) = L$. **2.** if one considers $\lim_{x\to c} f(x)$ only for $x < c$, then this value is said to be the **limit on the left** or **left-hand limit** of f at c. This is denoted by $\lim_{x\to c-} f(x)$, $\lim_{x\uparrow c} f(x)$, or $f(c^-)$. **3.** similarly, if one considers $\lim_{x\to c} f(x)$ only for $x > c$, then this value is said to be the **limit on the right** or **right-hand limit** of f at c. This is denoted by $\lim_{x\to c+} f(x)$, $\lim_{x\downarrow c} f(x)$, or $f(c^+)$.

limit of fire *Ordnance.* **1.** the boundary marking the area on which gunfire can be delivered. **2.** the angular limits for firing safely at aerial targets.

limit of proportionality see PROPORTIONAL LIMIT.

limit point *Mathematics.* an accumulation point. Also, CLUSTER POINT.

limits of integration *Mathematics.* the end points of the interval over which integration is taking place; i.e., the values a and b in the notation $\int_a^b f$.

limit superior *Mathematics.* **1.** if $X = (x_n)$ is a sequence of real numbers that is bounded above, then the limit superior of $X = (x_n)$, if it exists, is the infimum of those real numbers v with the property that there are only a finite number of natural numbers n such that $x_n > v$. This is denoted by lim sup X, lim sup x_n, or $\overline{\lim}\, x_n$. **2.** the limit superior of a function f at a point c is

$$\lim_{\varepsilon\to 0} (\text{lub}\{f(x) : |x - c| < \varepsilon \text{ and } x \neq c\}).$$

This is denoted by $\limsup_{x\to c} f(x)$ or $\overline{\lim}_{x\to c} f(x)$.

limivorous *Invertebrate Zoology.* of or relating to animals, particularly worms or bivalves, that ingest mud to feed on the organic detritus it contains.

Limnanthaceae *Botany.* a family of dicotyledonous, annual, succulent herbs of the order Geraniales, noted for producing mustard oils; occurring in moist places in North America.

Limnocharitaceae *Botany.* a family of monocotyledonous perennial aquatic herbs of the order Alismatales that are free-floating or rooted in the substrate and emergent, have a well-developed latex system, and are native to tropical and subtropical regions.

limnology *Biotechnology.* the study of freshwater ecosystems, including the chemical, physical, and biological aspects.

Limnoriidae *Invertebrate Zoology.* a family of isopod crustaceans, the gribbles in the suborder Flabellifera, that burrow into and destroy underwater timber.

limonene *Organic Chemistry.* $C_{10}H_{16}$, a colorless, optically active liquid monoterpene that has a lemonlike odor and is derived from the oils of certain citrus fruits, caraway, peppermint, and spearmint; boils at 176°C; used for flavoring, in perfume, and as a solvent or wetting agent. Also, CITRENE, DIPENTENE.

limoniform *Botany.* having the shape of a lemon.

limonite *Mineralogy.* a group of brown or yellowish-brown, naturally occurring hydrous iron oxides (mostly goethite), commonly formed as alteration products of iron-bearing minerals and the chief constituent of bog iron ore. Also, BROWN HEMATITE, BROWN IRON ORE.

Limousin *Agriculture.* a large breed of beef cattle originating in France.

limpet *Invertebrate Zoology.* the common name for marine gastropod mollusks with ridged conical and tentlike shells.

limpet mine *Ordnance.* a mine designed to cling to the underwater hull of a ship for subsequent detonation.

Limulacea *Invertebrate Zoology.* a group of horseshoe crabs, primitive arachnids in the family Limulida.

Limulida *Invertebrate Zoology.* a family of primitive arachnids including horseshoe crabs.

limulin *Biochemistry.* a lectin in the horseshoe crab that binds to specific sites in lipopolysaccharides and is able to agglutinate a variety of Gram-negative bacteria such as *E. coli*.

Limulodidae see PTILIIDAE.

Limulus *Invertebrate Zoology.* a genus of primitive arachnid arthropods, including the king crab, the largest horseshoe crab known, which is up to two feet across. Also, XIPHOSURA.

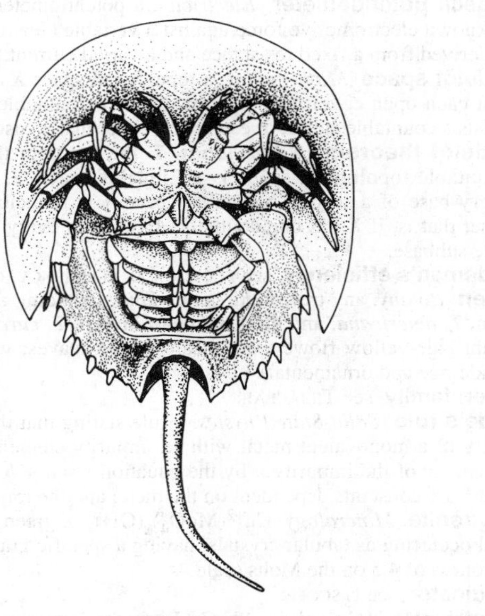

Limulus

limulus amoebocyte lysate test *Microbiology.* an assay to detect the presence or quantity of bacterial lipopolysaccharide, based on the ability of this molecule to cause gelation of a blood cell, or amoebocyte lysate of the horseshoe crab, *Limulus polyphemus*.

lin. linear; lineal.

linac *Nucleonics.* a machine that generates a beam of highly energized particles by propelling them in a straight line with energy from an electromagnetic field; such machines can accelerate a particle's energy to 30 MeV. Also, LINEAR ACCELERATOR.

Linaceae *Botany.* a family of dicotyledonous herbs and shrubs of the order Linales, having bisexual flowers and found primarily in cymes and capsular fruit.

Linales *Botany.* an order of herbs and woody plants of the class Magnoliophyta, characterized by alternate or, less frequently, opposite leaves, having hypogynous, usually perfect flowers with axile or apical-axile ovaries.

linalool *Organic Chemistry.* $C_{10}H_{18}O$, a combustible, colorless, pleasant-smelling terpene alcohol found in bergamont and rosewood; soluble in water and ether; boils at 195–199°C; used for flavoring and in perfumes.

linalyl acetate *Organic Chemistry.* $C_{12}H_{20}O_2$, the liquid acetic acid ester of linalool; a combustible, clear, colorless oil that is insoluble in water and soluble in alcohol and ether; boils at 220°C; used in perfumes and as a flavoring agent.

Lincoln *Agriculture.* the largest breed of sheep, having long, coarse wool and a white face. Also, **Lincoln Longwool.** (From *Lincoln*, a county in England.)

Lincoln index *Ecology.* a method of estimating the population size of a certain group of animals by capturing and marking individuals, releasing them into their environment, and then recapturing another sample later. The overall population is calculated from the number of original captives that are recaptured in the second sample.

lincomycin *Microbiology.* an antibiotic produced by the bacterium *Streptomyces lincolnesis*; it inhibits protein synthesis primarily in Gram-positive bacteria.

Lind, James 1716–1812, Scottish surgeon; discovered that fresh vegetables and fruit (especially citrus fruits) can cure scurvy.

lindackerite *Mineralogy.* $H_2Cu_5^{+2}(AsO_4)_4 \cdot 8\text{–}9H_2O$, a light- or apple-green, monoclinic mineral occurring as rosettes of small lathlike crystals, having a specific gravity of 3.20 to 3.27 and a hardness of 2 to 3 on the Mohs scale; found as a secondary mineral with erythrite.

Lindbergh pump *Surgery.* a perfusion device used to keep whole organs alive outside the body. (Named for the American aviatior Charles *Lindbergh,* 1902–1974.) Also, CARREL-LINDBERGH PUMP.

Lindblad resonance *Astrophysics.* the resonance that occurs when a star orbiting in a spiral galaxy repeatedly encounters a galactic spiral density wave with a frequency that is an even multiple of the star's epicyclic orbit around the center of the galaxy.

Lindeck potentiometer *Electricity.* a potentiometer that balances an unknown electromotive force against a variable electromotive force that is derived from a fixed resistance and a varied current.

Lindelöf space *Mathematics.* a topological space X with the property that each open cover of X is reducible to a countable cover; i.e., there exists a countable subset of each open cover that is also a cover.

Lindelöf theorem *Mathematics.* **1.** the theorem that every second countable topological space is a Lindelöf space. **2.** the theorem that every base of a second countable space X is reducible to a countable base; that is, if X has one countable base, then every base has a countable subbase.

Lindeman's efficiency see ECOLOGICAL EFFICIENCY, def. 1.

linden *Botany.* any tree of the genus *Tilia,* including the **American linden,** *T. americana,* and the **European linden,** *T. europaea,* having fragrant pale-yellow flowers and heart-shaped leaves; widely grown as a shade tree and ornamental.

linden family see TILIACEAE.

Linde's rule *Solid-State Physics.* a rule stating that the electrical resistivity of a monovalent metal with an impurity element is related to the valence v of the impurity n by the equation $r = a + b\,(n - 1)^2$, where a and b are constants dependent on the metal and the impurity.

lindgrenite *Mineralogy.* $Cu_3^{+2}(MoO_4)_2(OH)_2$, a green, monoclinic mineral occurring as tabular crystals, having a specific gravity of 4.26 and a hardness of 4.5 on the Mohs scale.

L indicator see L SCOPE.

lindströmite *Mineralogy.* $Pb_3Cu_3Bi_7S_{15}$, a lead-gray to tin-white, orthorhombic mineral occurring in striated prismatic crystals, and having a specific gravity of 7.01 and a hardness of 3 to 3.5 on the Mohs scale; found with quartz and bismuth at Gladhammer, Sweden.

line *Science.* a continuous extent of length without thickness or depth, either straight or curved. *Mathematics.* **1.** the mathematical description of this, a set of points $(x_1, \ldots, x_n)$ in Euclidean space R^n that satisfy n equations of the form $x_i = c_i\,t + d_i$; i.e., each coordinate of a point on a line is a linear function of the same single parameter. The line is said to be parameterized or in parametric form. Also, STRAIGHT LINE. **2.** in any geometric system of dimension n, the set of points lying in the intersection of $n - 1$ hyperplanes described by a consistent system of $n - 1$ linear equations whose coefficient matrix has rank $n - 1$. **3.** an axiomatic object in a geometric formulation, satisfying the requirements that any two lines determine a point (including the ideal point) and any two points determine a line. *Transportation Engineering.* a track or route that connects two terminals, with stops or stations in between. *Telecommunications.* **1.** a device, such as a cable, that transmits signals or data from one point to another. **2.** a telephone connection. *Electronics.* **1.** any device that conducts electrical energy. **2.** a device that joins two electrical components. *Electricity.* a unit of electric flux equivalent to one Maxwell. *Military Science.* **1.** a defensive position, front, or series of fortifications. **2.** an arrangement of troops, battleships, etc. *Naval Architecture.* see LINES. *Genetics.* a group of identical, homozygous, pure-breeding diploid or polyploid organisms, distinguished from others of the same species by some unique phenotype and genotype. *Cell Biology.* a collection of cells produced by continuously growing a certain cell culture in vitro; it usually contains a number of individual clones. *Botany.* a unit of plant measurement equal to 2.117 mm.

linea alba *Anatomy.* the white line of fibrous connective tissue extending from the xiphoid process to the pubis.

lineage *Anthropology.* a group of people who are descended from a single common ancestor and are able to trace that kinship through known forebears. *Genetics.* **1.** a series of cells that are related by an uninterrupted sequence of cell divisions. **2.** a linear evolutionary sequence from a particular older species through the intermediate to a more recent species.

lineal *Anthropology.* directly related in a single line, such as a grandfather, father, and son. Thus, **lineal kin, lineal relative, lineal descent.**

lineament [lin´ē ə mənt] *Geology.* an extensive straight or gently curved topographic feature that appears to reflect the crustal structure of a region. *Astronomy.* any straight-appearing line on the solid surface of a moon or planet, whether positive (such as a range of mountains) or negative (a valley).

line-and-staff organization *Industrial Engineering.* a table of organization that distinguishes between those functions involved in direct production (line functions) and those that support the organization in an advisory capacity, as in marketing or research, without contributing directly to production (staff functions). This is based on the military distinction between line (combat) officers and staff officers.

line and trunk group *Telecommunications.* a group that is made up of four-wire circuits, intertoll trunk groups, and automatic branch exchange trunks.

linea nigra *Physiology.* a name given to the linea alba when it becomes pigmented during pregnancy.

linear *Science.* **1.** of, relating to, or in the form of a line. **2.** involving mesaurement only in terms of length. *Graphic Arts.* expressed or depicted mainly in straight lines. *Control Systems.* describing a condition in which output varies in direct proportion to the input. *Robotics.* describing movement in a straight line by a robot. *Artificial Intelligence.* describing a form of reasoning that proceeds directly from known information to an obvious conclusion, without insight or intuition. *Geology.* see LINEAMENT.

linear absorption coefficient *Crystallography.* the intensity, I, of a beam after passing through a thickness T of an absorbing crystal is given by the equation $I = I_0 \exp(-\mu t)$, where I_0 is the intensity of the incident beam and μ the total linear absorption coefficient for the primary beam (with units of cm^{-1}); μ is a function of wavelength and atomic number. The above equation may be rewritten in terms of a mass absorption coefficient (μ/ρ) (in units of cm^2/g), with ρ the density.

linear accelerator *Nucleonics.* an apparatus that generates a beam of highly energized particles by propelling them in a straight line with energy from an electromagnetic field; such machines can accelerate a particle's energy to 30 MeV.

linear accelerator breeder *Nucleonics.* a device that uses a linear accelerator to produce neutrons to fuel a nuclear reactor.

linear accelerator-driven reactor *Nucleonics.* a nuclear reactor that combines a linear accelerator and a subcritical reactor target of natural, depleted, or slightly enriched natural uranium or thorium to produce a net amount of energy.

linear accelerator-regenerator reactor *Nucleonics.* a device that uses neutrons energized in a linear accelerator to increase the radioactivity of a nuclear fuel before it is placed in a reactor.

linear actuator *Mechanics.* any device that produces linear motion, such as a hydraulic or pneumatic cylinder, solenoid, or linear motor.

linear algebra *Mathematics.* the study of homomorphisms of (finitely generated) free modules; especially homomorphisms of finite dimensional vector spaces, including the characterization of matrices representing the same homomorphism relative to different bases.

linear array *Electromagnetism.* an array of antenna dipoles arranged end-to-end along a common straight line.

linear birefringence *Optics.* the difference in the refractive index of a material according to the orientation of linear polarized light passing through it.

linear bounded automaton *Computer Programming.* a one-tape Turing machine whose tape length is bounded by a linear function of the length of the input.

linear chain polymer *Materials Science.* a polymer whose molecular chains have no branches or cross links.

linear circuit see LINEAR NETWORK.

linear cleavage *Geology.* the tendency of metamorphic rocks to break into long, horizontal fragments.

linear coefficient of thermal expansion see COEFFICIENT OF THERMAL EXPANSION.

linear collision cascade *Solid-State Physics.* an ejection of target atoms produced by bombarding the target atoms with a projectile; the energy released into the system serves to eject the target atoms.
linear combination *Mathematics.* in a vector space or module, an expression of the form $a_1x_1 + \cdots + a_mx_m$, where the a_i are scalars, and the x_i represent vectors or unknown quantities. An equation that has the form $a_1x_1 + \cdots + a_mx_m = b$ is called a **linear equation**, and an inequality of the form $a_1x_1 + \cdots + a_mx_m \le b$ (or $\ge b$) is called a **linear inequality.** Also, **linear (algebraic) expression.**
linear comparator *Electronics.* a device that generates an output voltage or current when it receives a steady stream of electrical signals at a given frequency. Also, CONTINUOUS COMPARATOR.
linear conductor antenna *Electromagnetism.* an antenna whose elements all lie along a straight line.
linear control *Telecommunications.* in television, a control for adjusting the variation of scanning speed during the trace interval in order to minimize distortion.
linear control system *Control Systems.* a linear system in which the inputs change to meet certain criteria over a period of time.
linear definition see LINE CONVERSION.
linear density *Materials Science.* in the microstructure of a material, the number of atoms that have their centers located on a given length of a line in a certain direction. *Physics.* a quantity that is given by the amount of a substance per unit length, commonly used in one-dimensional studies.
linear detection *Electronics.* a technique for extracting intelligence from a transmitted voltage so that its value remains unchanged.
linear differential equation *Mathematics.* an ordinary or partial differential equation of the form $Lu = f$, where L is a **linear differential operator,** that is, an operator having the property that $L(au + \mathrm{b}v) = \mathrm{a}Lu + \mathrm{b}Lv$ for all functions u, v and scalars a, b. The space of solutions of a linear differential equation is the space of solutions of the homogeneous equation $Lu = 0$ plus any particular solution of the equation $Lu = f$.
linear displacement *Robotics.* a displacement that is proportional to the amount of torque.
linear elastic fracture mechanics *Materials Science.* the study of polymer fracture conditions or the conditions for crack propagation to occur in glassy polymers when Hooke's law is followed; it focuses mainly on fracture toughness and total work of fracture.
linear elasticity *Materials Science.* a behavior of a material when the strain is proportional to the applied stress.
linear electrical constants (of a uniform line) *Electricity.* the series resistance, series inductance, shunt conductance, and shunt capacitance for a unit length of line. Also, **linear electrical parameters.** *Thermodynamics.* see COEFFICIENT OF THERMAL EXPANSION.
linear electrical parameters see TRANSMISSION-LINE PARAMETERS.
linear expansion *Physics.* the one-dimensional expansion (increase in length) of a body. *Thermodynamics.* see COEFFICIENT OF LINEAR EXPANSION.
linear feedback control *Control Systems.* feedback control in a linear control system.
linear flow structure see PLATY FLOW STRUCTURE.
linear function see LINEAR TRANSFORMATION.
linear functional *Mathematics.* an element of the dual of a vector space V over a field F is called a linear functional on V into F; equivalently, a linear transformation from a vector space V to its field of scalars F.
linear hull *Mathematics.* given an arbitrary set S of elements of a vector space V, the intersection of all subspaces of V which contain S. Also, **linear span.**
linear integrated circuit *Electronics.* a type of integrated circuit in which the output level varies in direct proportion to the input level, generally assumed to be an analog device.
linear interpolation *Control Systems.* a control function that automatically defines the continuum of points in a straight line based on only two commanded coordinate positions. *Mathematics.* a technique for approximating the value of a function between two known values based on the assumption that the three values lie on the same straight line.
linearity the fact of being linear; specific uses include: *Physics.* a condition in which the change in value of one quantity is directly proportional to that of another quantity. *Radiology.* the ability to maintain a constant exposure for a given number of milliampere seconds. *Mathematics.* the properties of a linear transformation f on a vector space that $f(a + c) = f(a) + f(c)$ and $f(ra) = rf(a)$, where a and c are vectors and r is any scalar.
linearity control *Electronics.* a potentiometer used in television receivers to correct image linearity.
linearization *Control Systems.* **1.** the modification of a system for easier analysis so that its outputs are approximately linear functions of its inputs. **2.** the process of using the values of a linear system to approximate a nonlinear system when the deviations from linearity are small.
linear-logarithmic intermediate-frequency amplifier *Electronics.* a lin-log amplifier used as an intermediate-frequency amplifier in a radar receiver, to prevent it from overloading by high input signals, thus making it impervious to jamming.
linearly dependent *Mathematics.* a subset X of a given algebraic object A (such as an Abelian group, module, vector space, etc.) is linearly dependent if $n_1x_1 + \cdots + n_kx_k = 0$ is true when $n_i \neq 0$ for at least one i, where $x_1, \ldots, x_k$ are finitely many distinct elements of X, and where the n_i are integers indicating repeated addition if A is a group, ring elements if A is a module over a ring, scalars if A is a vector space, etc. Equivalently, X is a linearly dependent set if at least one element of X can be expressed as a linear combination of one or more of the other elements of X.
linearly disjoint *Mathematics.* let C be an algebraically closed field with subfields K, E, and F such that K is properly contained in $E \ll F$. E and F are said to be linearly disjoint over K if every subset of E that is linearly independent over K is also linearly independent over F (or equivalently, every subset of F which is linearly independent over K is also linearly independent over E).
linearly graded junction *Electronics.* a type of transitional region in a semiconductor in which the concentration of impurities varies evenly instead of changing abruptly.
linearly independent *Mathematics.* a subset X of a given algebraic object A (such as an Abelian group, module, vector space, etc.) is linearly independent if $n_1x_1 + \cdots + n_kx_k = 0$ always implies that $n_i = 0$ for all i, where $x_1, \ldots, x_k$ are distinct elements of X, and where the n_i are integers indicating repeated addition if A is a group, ring elements if A is a module over a ring, scalars if A is a vector space, etc.
linearly ordered set *Mathematics.* a partially ordered set A with partial ordering $\le$, such that either $a \le b$ or $b \le a$ for all elements a and b of A. Also, CHAIN.
linear manifold *Mathematics.* any subspace of a vector space.
linear mapping see LINEAR TRANSFORMATION.
linear meter *Engineering.* an analog or digital meter that displays a measured value directly proportional to the quantity under test.
linear model see ADDITIVE MODEL.
linear modulation *Telecommunications.* modulation in which the amplitude of the modulation envelope is directly proportional to the amplitude of the intelligence signal at all modulation frequencies.
linear molecule *Physical Chemistry.* a type of molecule, such as carbon dioxide, in which the bonds between the atoms are at a 180° angle.
linear motor *Electricity.* a type of alternating-current induction motor in which the motion between the stator and rotor is linear rather than rotary. *Engineering.* an electric motor that has in effect been split so that the stator is laid out flat. Thus, the stator can become the track of a magnetically levitated motor or vehicle.
linear network *Electricity.* a network in which the magnitude of the electrical elements is constant with respect to the current. Also, LINEAR CIRCUIT.
linear operator see LINEAR TRANSFORMATION.
linear order *Mathematics.* a partial ordering $\le$ on a set A, such that either $a \le b$ or $b \le a$ for all elements a and b of A. Also, TOTAL ORDER, SIMPLE ORDER.
linear oscillator see HARMONIC OSCILLATOR.
linear oxidation *Materials Science.* an oxidation that occurs linearly with the time when the scale on the surface of a material cracks allowing the corrosive medium direct access to the metal.
linear parallax see ABSOLUTE STEREOSCOPIC PARALLAX.
linear parallel texture *Petrology.* a rock texture that is characterized by the orientation of the constituents parallel to a line rather than a plane.
linear perspective *Graphic Arts.* a technique of representing three-dimensional objects and space on a two-dimensional surface, using lines that radiate from one or more vanishing points; based on the optical illusion that parallel lines seem to converge as they recede into the distance away from the observer.
linear-phase *Electronics.* describing a filter or other device in which the change in the phase of a sinusoidal signal is proportional to the frequency.

linear planning *Artificial Intelligence.* planning in a domain in which subgoals are independent; the subgoals can therefore be accomplished in any order.

linear polarization *Optics.* the amplitude of an electric field that vibrates along a line perpendicular to the direction of the propagation of a ray of light. Also, PLANE POLARIZATION.

linear polymer *Organic Chemistry.* a polymer molecule having few branches or bridges between chains.

linear power amplifier *Acoustical Engineering.* in audio circuits, an amplifier that delivers the signal to the speakers at a level proportional to the level of the input signal.

linear programming *Mathematics.* the technique of maximizing or minimizing a linear function subject to a set of linear inequalities called **linear constraints.**

linear programming model *Transportation Engineering.* a mathematical model for maximizing or minimizing a system of linear equations with linear inequalities, employed to determine the optimum scheduling of aircraft and crews.

linear-quadratic-Gaussian problem *Control Systems.* an optimal-state regulator problem in which Gaussian noise is present in both the state and measurement equations, and the expected value of the quadratic performance index must be minimized.

linear rectifier *Electronics.* a device that converts alternating current to direct current without affecting the strength of the signal.

linear regression *Statistics.* a model describing the variation of a certain response variable as a linear combination of the effects of a set of predictors and a random error term. Also, MULTIPLE REGRESSION.

linear regulator problem *Control Systems.* an optimal control problem that describes the system to be controlled as a set of linear differential equations and the performance index to be minimized as the integral of a quadratic function of the system state and its control functions. Also, OPTIMAL REGULATOR PROBLEM, REGULATOR PROBLEM.

linear repeater *Electronics.* a device that produces an output signal whose instantaneous value is strictly proportional to the input signal.

linear resolution *Artificial Intelligence.* a form of resolution in which one of the two clauses resolved is the result of the previous resolution step.

linear restoring force *Mechanics.* a force that acts to bring a system into equilibrium; a linear function of the system's displacement from equilibrium.

linear scanning *Engineering.* the movement of a trace on a cathode-ray tube display that moves with a constant linear or angular velocity, as in television or radar.

linear space *Mathematics.* another term for a vector space; that is, a module over a field.

linear stability analysis *Chaotic Dynamics.* a procedure of approximating the dynamics of a system in the vicinity of a known solution in order to find the stability of the solution with respect to infinitesimal perturbations. The variables are written as a sum of the known solution and a small perturbation, the nonlinearities are linearized in terms of the perturbations, and the eigenvalues and eigenvectors of the resulting linear problem are found.

linear Stark effect *Atomic Physics.* a phenomenon in which the spectrum of radiation that is absorbed or emitted by a hydrogenlike atom splits into equidistant spacings when the specimen is placed in an electric field.

linear strain *Mechanics.* the elongation per unit length resulting from the application of a load. Also, LONGITUDINAL STRAIN.

linear sweep *Electronics.* the movement of a cathode-ray beam across a television receiver at a uniform velocity from one side to another before snapping back to its original position.

linear-sweep delay circuit *Electronics.* a circuit in which the relationship between a periodic alternating current and a calibrated direct-current voltage determines how long the input signal is held.

linear-sweep generator *Electronics.* a device that produces a linear sweep.

linear system *Control Systems.* a term for a system in which the outputs are components of a vector that is equal to the value of a linear operator applied to a vector whose components are the system's inputs. *Mathematics.* a mathematical description of this system, a set of n linear equations in m unknowns x_i over a given field K: $a_{i1}x_1 + \cdots + a_{im}x_m = b_i$, where $i = 1, 2, \ldots, n$. Such a system has a solution if and only if the matrix equation $AX = B$ has a solution X, where A is the $n \times m$ matrix with entries a_{ij}, and B is the $n \times 1$ column vector with entries b_i. Also, SYSTEM OF LINEAR EQUATIONS.

linear system analysis *Control Systems.* an analysis based on a model in which the system inputs, as applied at the input terminals, and the system outputs, as measured or observed at the output terminals, are mapped linearly.

line art *Graphic Arts.* any black-and-white illustration that can be reproduced without screening; illustrated line copy.

linear taper *Electricity.* a characteristic of certain potentiometers, whose resistance varies in linear proportion to the rotation; used in electronic circuits for adjusting and aligning the circuit.

linear time base *Electronics.* the variable that causes the electron beam in a cathode-ray tube to travel at a constant rate of speed along a horizontal time scale.

linear transducer *Electronics.* **1.** a transducer whose output and input are related by a linear function, such as a linear algebraic differential or an integral equation. **2.** a transducer in which the output is proportional to the input at any frequency.

linear transformation *Mathematics.* let A and B be vector spaces over a field K. A function $f: A \rightarrow B$ is called a linear transformation if, for all $a,c \in A$ and $r \in K$, $f(a + c) = f(a) + f(c)$ and $f(ra) = rf(a)$. Linear transformations are used extensively in the study of vector spaces (modules over a field, especially the field of real or complex numbers). Also, LINEAR MAPPING, LINEAR FUNCTION, LINEAR OPERATOR.

linear unit *Electronics.* a device in an analog computer, in which the alteration of one of the input signals causes a proportional alteration in the output signal and is not affected by the values of the other signals.

linear variable-differential transformer see DIFFERENTIAL TRANSFORMER.

lineate *Science.* marked with lines, especially with vertical or lengthwise parallel lines. Also, **lineated.**

lineation *Science.* **1.** the process of dividing something into lines. **2.** a pattern or arrangement of lines. **3.** an outline or delineation consisting of lines. *Geology.* any linear structure in or on a rock, such as ripple marks or flow lines.

line balance *Electricity.* **1.** a device used to couple a balanced system or device to an unbalanced one. **2.** the degree of electrical similarity between conductors on various portions of a line.

line-balance converter see BALUN.

line balancing *Design Engineering.* the process of adjusting material flows to multiple production facilities, to optimize overall productivity.

line blow *Meteorology.* a strong wind on the equatorward side of a high-pressure system as a result of slight windshift direction during the blow.

line brattice *Mining Engineering.* a partition placed in an opening to divide it into intake and return airways.

linebreeding or **line breeding** *Genetics.* the mating of related animals with their offspring, in order to achieve or maintain certain desired characteristics. Thus, **linebred.**

line broadening *Spectroscopy.* any widening of spectral lines indicating an increase in the range of wavelengths emitted or absorbed by a sample. *Astrophysics.* any process that produces such a widening, such as turbulence or Doppler effects.

line capacity *Transportation Engineering.* the maximum number of vehicles or passengers that can be carried over a section of line during a given time period.

linecasting *Graphic Arts.* the process of casting an entire line of hot-metal type at once rather than individual characters, using a **linecasting machine** such as a Linotype.

line characteristic distortion *Telecommunications.* teletypewriter transmission distortion that is caused by the effects of changing current transitions in wire circuits on the lengths of received signal impulses.

line circuit *Electricity.* in telephone equipment, an interface circuit between a line and a switching system.

line clinometer *Engineering.* a device that measures slope angles of drill rods when fastened between rods.

line conditioning *Telecommunications.* the reduction of the differences among amplitude and/or phase delays over specific frequency bands; usually accomplished using loading coals.

line conductor *Electricity.* a wire or cable that carries current along a route and is supported by poles or other structures.

line-controlled blocking oscillator *Electronics.* a circuit that generates large amounts of power by drawing on the alternating current produced by an oscillator with an open transmission line to its regenerative circuit.

line conversion *Graphic Arts.* a photograph of continuous-tone copy made with line film and yielding a line image. Also, LINEAR DEFINITION.

line copy *Graphic Arts.* copy that contains only black and white, with no intermediate gray tones, and can thus be reproduced without a halftone screen; continuous tones may be simulated in line copy through the use of various-sized dots.

line cord *Electricity.* a cord having a two- or three-pronged plug at one end, used to connect a receiver or appliance to a power source. Also, POWER CORD.

line-cord resistor *Electricity.* an asbestos-enclosed wire-wound resistor that is integrated into a line cord along with the two regular wires.

line cut *Graphic Arts.* a reproduction that is made from line copy, especially a photoengraving that contains only lines, with no gradations of tone.

line defect *Materials Science.* a dislocation in a crystalline material. If the dislocation results from slip, the defect is the line on the slip plane between the slipped and not-yet-slipped regions.

line density *Transportation Engineering.* a transportation line's intensity of use, often expressed as a ratio of the line's length to the population it serves.

line discipline *Computer Programming.* the set of rules that specify precisely how data are to be transmitted between locations in a communications network. Also, LINK PROTOCOL.

line displacement *Astrophysics.* the change in a spectral line's wavelength due to the Doppler effect; approaching sources have spectral lines displaced toward the blue end of the spectrum, receding ones toward the red.

line dot matrix *Computer Technology.* a line printer that employs a dot matrix printing mechanism. Also, PARALLEL DOT CHARACTER PRINTER.

line drilling *Mining Engineering.* a method of drilling and broaching for the primary cut in which deep, closely spaced holes are drilled in a straight line by means of a reciprocating drill, and the webs between the holes are removed with a drill or a flat broaching tool.

line driver *Computer Technology.* a device designed to amplify signals for transmission over relatively short distances, typically less than five miles. Also, LIMITED-DISTANCE MODEM. *Electronics.* a circuit that transfers data between logic circuits and a two-wire transmission line.

line drop *Electricity.* the lowering of the voltage supplied to a load, sometimes produced by a dropping resistor.

line-drop compensator *Electricity.* a device that allows an increase in the voltage-regulating relay to increase output voltage by an amount that compensates for the impedance drop in the circuit between the regulator and a specified place on the circuit.

line-drop signal *Telecommunications.* on a manual switchboard, a signal that is linked to the state of a subscriber line.

line editor *Computer Technology.* a primitive text editor in which text is displayed and manipulated on a line-by-line basis; editing operations are limited and specified for lines identified by number.

line element *Mathematics.* let X be a manifold; a line element at a point x of X is any one-dimensional vector subspace of the tangent space T_x at x. A line element may be viewed as the point x together with a specified direction.

line engraving *Graphic Arts.* a line cut, or the process of printing with line cuts.

line equalizer *Electricity.* an equalizer containing inductance and/or capacitance; inserted in a transmission line in order to alter the line frequency response.

line facility *Telecommunications.* a transmission line with distributed amplification that aids in offsetting attenuation.

line fault *Electricity.* an open or short circuit in a transmission line that causes partial or total loss of the power or signal at the output end.

line feed *Computer Technology.* **1.** the advancement of paper through a printer one line at a time by means of a special signal sent by the computer. **2.** the line feed character in the ASCII character set.

line fill *Telecommunications.* the ratio of the number of connected main telephone stations on a given line to the main station capacity of the line.

line filter *Electricity.* **1.** a device placed in an AC power cord to eliminate transient unwanted spikes that can cause damage to electronic equipment. **2.** a filter placed between a power line and a unit of electrical equipment to prevent the transmission of noise signals.

line filter balance *Telecommunications.* a network that maintains phantom group balances when one side of the group has a carrier system.

line finder *Telecommunications.* in a step-by-step rotary or crossbar switching system, a mechanism that automatically locates an unused telegraph or telephone circuit.

line-finder shelf *Telecommunications.* the type of equipment needed for connecting any of the associated calling telephones to a selector or connectors that will receive the dial pulses from the calling phone. There are normally 20 line finders.

line-finder switch *Telecommunications.* in a telephone system, a switch for grasping selector apparatus that provides a dial tone.

line flux *Electromagnetism.* a local inductive field of a telephone or power line.

line-formula method *Organic Chemistry.* a system of string notation for chemical formulas, in which elements, functional groups, and ring systems are represented in linear form.

line frequency *Electronics.* the number of times per second that a scanning spot horizontally traverses a fixed vertical line in a television picture tube. Also, HORIZONTAL FREQUENCY. *Electricity.* the frequency of a power-line voltage.

line functions *Industrial Engineering.* those portions of an industrial operation which are directly involved in production, as opposed to support functions.

line gale see EQUINOCTIAL STORM.

line gauge see TYPE GAUGE.

line haul *Transportation Engineering.* an operating system in which vehicles serve all or selected stops along a fixed route.

line hydrophone *Acoustical Engineering.* a directional hydrophone (directional underwater transducer) that is one of a line of such devices.

Lineidae *Invertebrate Zoology.* a family of mostly marine, widely distributed nematodes.

line impedance *Electromagnetism.* the impedance of a transmission line as measured across its terminals.

line integral *Mathematics.* if, for the condition $a \leq t \leq b$, $g(t)$ is a rectifiable path in R^n and f is a function defined and continuous on $g(t)$, then the line integral of f along $g(t)$ is the Lebesgue-Stieltjes integral given by $\int_a^b f(g(t))dg(t)$. This line integral is also denoted by $\int_g$ or $\int_g f(x)dx$.

line interlace see INTERLACED SCANNING.

line item *Computer Programming.* a data item that is considered to be on the same level as a set of related data items in a hierarchy for a data processing application. *Industrial Engineering.* a type of expenditure that appears explicitly on a separate line of a budget with its own title.

line lengthener *Electromagnetism.* a device that effectively changes the electrical length of a transmission line or waveguide with no change to the other electromagnetic or physical properties.

line level *Telecommunications.* the signal level of a particular position on a transmission line, usually measured in decibels.

line link *Telecommunications.* in a crossbar switching system, a frame that holds multiple telephone lines. Also, **line-link frame.**

line location *Electricity.* the placement of two or more power or communication lines along the same route in such a way as to avoid unnecessary crossings or conflicts.

line loop *Telecommunications.* in a telephone circuit, a portion of a user's telephone set and the wires that provide an electrical circuit to a central office.

line-loop resistance *Electricity.* the metallic resistance of the line wires that extend from a telephone to the central office.

line loss *Electricity.* the total power loss on a transmission or distribution line.

line lubricator see LINE OILER.

lineman *Engineering.* **1.** a person who maintains and repairs power lines. **2.** a person who carries the tape in a surveying party.

line microphone *Acoustical Engineering.* a directional microphone that consists of a tube with holes through which sound enters.

line misregistration *Computer Technology.* in optical character recognition, the improper appearance of a line of characters or digits gauged with respect to a real or imaginary horizontal baseline.

line mixer see FLOW MIXER.

linen *Textiles.* a fabric woven from flax; used since ancient times for clothing, bedding, tablecloths, and so on.

line noise *Telecommunications.* noise that originates in a transmission line, caused by inductive interference from the power line or poor connections within the line.

line number *Computer Programming.* a number associated with a particular line of a program, specifying its position in a sequence.

line of action *Mechanical Engineering.* the locus of contact points as gear teeth profiles are engaged on a wheel. *Mechanics.* see LINE OF APPLICATION.

line of aim *Ordnance.* the line from the eye of the shooter or operator through the sight of a gun toward the aiming point.

line of application *Mechanics.* the line containing the vector representing a force, specifying the direction of the force but not its magnitude.
line of apsides *Astronomy.* the major axis of an elliptical orbit.
line of battle *Military Science.* a line formed by troops or ships for delivering or receiving an attack.
line of bore *Ordnance.* an imaginary straight line forming the continuation of the axis of the bore of a gun.
line of centers *Mechanics.* the straight line through the centers of mass of two bodies at the instant of impact. Also, LINE OF IMPACT.
line of code *Computer Programming.* a single statement or line in a program.
line of collimation *Optics* the imaginary line that crosses the optical center of the objective's glass and the cross-hair intersection of the diaphragm in a surveying telescope.
line of constant scale *Photogrammetry.* any line on a photograph that is parallel to the photograph's true horizon or isometric parallel. Also, **line of equal scale.**
line of cusps *Astronomy.* on the disk of the moon or a planet, a line connecting the two cusps.
line of defense *Military Science.* **1.** a barrier that can be used as a defense against attack; e.g., a crevice, a river, or a wall. **2.** the forces available for a nation's defense; the standing army is usually designated as the first line of defense, and the various reserve forces are second, third, and so on.
line of departure *Military Science.* **1.** in land warfare, a line used to coordinate the departure of attack or scouting units. Also, START LINE. **2.** in amphibious warfare, an offshore line that is marked to assist assault craft in landing on designated beaches at scheduled times. *Ordnance.* the imaginary line along which a projectile is traveling at the instant it clears the muzzle of a gun, or, in the case of a missile or bomb, at the moment of launching.
line of electrostatic flux *Electricity.* a unit of electric flux that is equivalent to the electric flux associated with a charge of one statcoulomb. Also, **line of electrostatic induction.**
line of elevation *Ordnance.* the continuation of the axis of the bore of a gun that is ready to fire.
line of fall *Mechanics.* the line tangent to a trajectory at the altitude of firing.
line of fire *Ordnance.* a straight, horizontal line extending from the muzzle of a gun in the direction of a projectile fired from the gun.
line of flight *Mechanics.* the direction of motion followed by a body in flight.
line of flow *Fluid Mechanics.* the particular path followed by an element of a moving fluid.
line of force *Physics.* a field line in a force field, in which a tangent to the line will indicate the direction of the force.
line of impact *Mechanics.* **1.** see LINE OF CENTERS. **2.** the instantaneous line of flight of a projectile at impact that determines the angle of impact. *Ordnance.* a line tangent to the trajectory of a projectile at the point of impact or burst. Also, LINE OF ARRIVAL.
line of inversion *Astronomy.* in a pair of sunspots, a line that separates regions of opposite magnetic polarity.
line of levels *Cartography.* in surveying, a continuous series of measured differences in elevations between survey stations, given in the order in which they were measured.
line of nodes *Astronomy.* a line between the ascending and descending nodes of an orbit, marking the intersection of the orbital plane and some reference plane, usually the ecliptic.
line of observation *Ordnance.* the line from a position finder to a target at the moment of a recorded observation.
line of position *Navigation.* a line, determined by observation or measurement, on which a craft may be assumed to be located. *Ordnance.* **1.** a straight line from the gun or position-finding instrument to the position of the target. **2.** in air intercept, a reference line that originates at a target and extends outward at a predetermined angle.
line of sight *Ordnance.* a straight line from the axis of the bore of a gun to the intended target. *Electromagnetism.* in radar applications, the direct, unobstructed straight path between the radar and the target. *Cartography.* a straight line between any two points on the earth's surface visible to each other; a line in the direction of a great circle, but not following the curvature of the earth. *Optics.* any such line from a sighting or surveying instrument to the object being sighted.
line-of-sight velocity *Astronomy.* the component of an object's total space velocity that is directly toward or away from the observer.
line of soundings *Navigation.* a series of soundings made by a vessel underway; normally used to help determine position by comparison with charted depths.
line of strike *Geology.* see STRIKE.
line of thrust *Mechanics.* a line tangent to the forces in an arch or support structure.
line of tunnel *Engineering.* the width of a tunnel as measured between the sides of the tunnel.
line oiler *Mechanical Engineering.* an apparatus that injects a small amount of lubricating oil into the line used to conduct air or steam to an air- or steam-activated machine. Also, LINE LUBRICATOR.
Lineolaceae *Bacteriology.* in some classification systems, a family of bacteria that includes certain members of the order Caryophanales.
lineolate *Biology.* marked with fine lines.
line pack *Engineering.* the quantity of fluid residing in a pipeline network at any given time.
line pad *Electronics.* an object that is placed between the program amplifier and the transmission line in a radio broadcast system to ensure a consistent supply of electrical power.
line pair *Spectroscopy.* a particular spectral line and the internal standard line to which it is compared; used in spectrographic analysis to establish the concentration of a substance.
line pipe *Mechanical Devices.* a special type of recessed pipe with a larger-than-normal tapered-thread coupling that is used in high-test plumbing applications.
line planting *Forestry.* **1.** a pattern of planting in which young trees are set out in parallel rows, usually at regular intervals, on cleared land. **2.** a method of planting in which a line of trees is set out in narrow lanes interspersed by undergrowth at regular intervals; used often in wet tropical forest areas.
line printer *Computer Technology.* a device that prints an entire line of characters as one unit.
line profile *Astrophysics.* a graph of spectral line intensity versus wavelength.
line pulsing *Electronics.* a method of transmitting information through a communications line, in which energy is built up in an artificial line over an extended period before it is discharged into the transmitter in short bursts for brief intervals.
liner *Engineering.* a sleeve that is placed within another object in order to protect a delicate or sensitive interior surface. *Naval Architecture.* **1.** a large passenger ship in regular open-ocean service. **2.** a large sailing warship, or ship of the line, so called in the nineteenth century. **3.** a strip placed under a plate to make it flush with a plate that it overlaps. *Mining Engineering.* **1.** a foot piece for uprights in timber sets. **2.** a timber support. *Metallurgy.* the partial melting of a heterogeneous alloy.
line radiation *Electromagnetism.* electromagnetic radiation from a power line caused mainly by corona pulses that give rise to radio interference.
liner bushing *Mechanical Devices.* a permanently installed bushing found in a jig, coupled with reusable wear bushings. Also, MASTER BUSHING.
line reflection *Telecommunications.* the reflection of a signal at the terminus of a transmission line.
line regulation *Electricity.* the maximum amount that an output voltage or a current of a power supply will change with a specified change in the line voltage.
line relay *Electricity.* a relay that is activated by signals on a line.
lines *Naval Architecture.* **1.** the ropes or cables forming the rigging of a vessel. **2.** the hull form of a vessel, as in "swift lines;" so called from the lines of a ship's plans, showing its hull form.
line scanner *Engineering.* an imaging system used in remote-sensing platforms to successively scan and record radiation from a swath of terrain in a direction perpendicular to the flight-path direction.
line segment *Mathematics.* a connected, bounded subset of a line.
line-sequential color television *Telecommunications.* a color television system in which an entire line is one color with colors changing from line to line; the sequence is red, blue, and then green.
line shafting *Mechanical Engineering.* assembled overhead shafting that is used to transmit and distribute power from a concentrated center to individual machines, using pulley wheels and belts.
line side *Electricity.* any of a series of terminal connections to an external source.
line skew *Computer Technology.* in optical character recognition, a form of line misregistration in which the string of characters to be read appears in a uniformly slanted condition with respect to the base line.

line source *Optics.* an idealized source of light that can be treated one-dimensionally, such as a very narrow, illuminated slit.

line-space lever *Mechanical Engineering.* a lever on a typewriter that is pressed to rotate the carriage to the next line on the page.

line spectrum *Spectroscopy.* **1.** the spectrum characteristic of an atom or ion, consisting of a series of discrete spectral lines that correspond to changes in energy levels of electrons. **2.** a radiation spectrum in which the studied quantity takes on discrete values.

line-spectrum sound *Acoustics.* sound that consists primarily of a fundamental frequency and harmonics of this frequency, and which therefore can be represented as a series of related lines when frequency is displayed on the horizontal axis versus sound pressure level or some other quantitative measurement on the vertical axis.

line speed *Telecommunications.* the maximum rate at which signals can be transmitted over a channel, usually expressed in bauds or bits per second.

line-speed method *Ordnance.* a method of calculating the future position of a moving target by multiplying the ground speed of the target by the flight time of the projectile, taking into consideration the flight direction of the target.

line squall *Meteorology.* a squall occurring along a squall line.

line-stabilized oscillator *Electronics.* a circuit that generates alternating current of a frequency related to the power frequency.

line storm see EQUINOCTIAL STORM.

line strength *Atomic Physics.* a measure of the intensity in a given spectrum line generated by atoms or ions in a gas.

line stretcher *Electromagnetism.* a section of waveguide that can be shortened or lengthened to provide impedance matching. *Mechanical Devices.* a section of coaxial line used in telecommunications with a varied length to provide tuning.

line switching *Electronics.* the process of connecting or disconnecting the line voltage from electronic equipment. Also, CIRCUIT SWITCHING.

line synchronizing pulse see HORIZONTAL SYNCHRONIZING PULSE.

line transducer *Electronics.* a device in which the motion of a foil diaphragm produces electrical signals along a cable.

line transformer *Electricity.* a transformer used to interconnect the transmission line and terminal devices to achieve isolation, line balance, or impedance matching.

line trap *Electricity.* any band-rejection filter used in a transmission line to notch out signals at specified frequencies.

line tuning *Electricity.* the process of adjusting the resonant frequency of a tuned circuit by varying the capacitance and leaving the inductance constant, or vice versa.

line turnaround *Telecommunications.* in half-duplex communication, the process of changing the direction of modem and communication channel transmissions to the opposite direction.

line-use ratio *Telecommunications.* in facsimile broadcasting, the ratio of the available line to the total length of the scanning line.

line voltage *Electricity.* **1.** the voltage provided by a power line at the outlet or point of use; for the United States, usually 110–125 volts. **2.** the voltage between the lines of a supplying power system.

line-voltage regulator *Electricity.* a device that counteracts variations in the power-line voltage, delivering a constant voltage to the connected load.

line vortex *Fluid Mechanics.* a fluid flow about a line in approximate circular paths; the velocity is an inverse proportional function of the distance to the line, thus giving rise to an infinite concentration of vorticity on the line.

Lineweaver-Burke plot *Biochemistry.* a plot of $1/v$ against $1/S$ for an enzyme-catalyzed reaction, where v is the initial rate and S is the substrate concentration.

linewidth *Atomic Physics.* the spread in wavelength of a spectral line, measured as a half-width.

lingu- a combining form meaning: **1.** tongue, as in *lingualis.* **2.** language, as in *linguistics.*

lingua *Anatomy.* see TONGUE.

lingua franca *Linguistics.* a language that is adopted by agreement between two or more distinct groups for trade, diplomacy, or social convention.

lingual *Anatomy.* **1.** of, relating to, or near the tongue. **2.** of a tooth surface, directed toward the tongue. *Linguistics.* **1.** of or relating to language or languages. **2.** articulated with the tongue, especially the tip of the tongue, as *d* or *n.*

lingual artery *Anatomy.* a branch of the external carotid artery that supplies blood to the tongue.

lingual gland *Anatomy.* any of several small glands located on the underside of the tongue.

lingual nerve *Anatomy.* a branch of the mandibular nerve that innervates the floor of the mouth and the tongue.

lingual tonsil *Anatomy.* a mass of lymphoid tissue located at the posterior margin of the tongue.

Linguatulida see PENTASTOMIDA.

Linguatuloidea *Invertebrate Zoology.* the tongue worms, a suborder of eyeless, parasitic, bloodsucking arthropods with no respiratory or circulatory system; found in the respiratory system of many vertebrates.

linguist *Linguistics.* a person who studies or is an expert in linguistics.

linguistic *Linguistics.* of or relating to language or linguistics.

linguistic model *Computer Programming.* a pattern recognition technique in which a pattern is specified by combining a set of subpatterns in accordance with a specific grammar or syntax.

linguistic relativity *Psychology.* a theory stating that language not only reflects the environment but also shapes one's perception of it.

linguistics the scientific or systematic study of language.

Linguistics

Linguistics, the science of language, is made up of a number of closely linked subdisciplines. Most important among these are syntax and semantics, which are concerned with the principles for concatenating words into phrases and for assigning them appropriate meanings; morphology, which deals with the formation and inflection of words, and phonology, whose subject is the sounds of speech. Linguistics studies phenomena in these and other domains as they manifest themselves in particular languages—e.g., Chinese, Walpiri, English, etc.—with special attention to those aspects that are common to all human language (universal grammar).

Although linguistics has a history that goes back 2500 years to the Sanskrit grammarians of ancient India, it was recognized only recently that linguistics is a proper subfield of cognitive psychology (Chomsky). A crucial fact here is that speakers readily understand and produce infinitely many utterances that they have never encountered before. They can do this because they have access to a certain body of knowledge; and knowledge is the subject matter of cognitive psychology. A speaker's knowledge of a language includes knowledge of words and of principles that determine how words combine into phrases and what these combinations mean. That words and phrases are not given to us directly, but are part of our linguistic knowledge, is shown by the fact that we apprehend and produce words and phrases only in a language we have mastered. Words, phrases, and their meanings are thus perceptual illusions that affect only individuals who have command of the relevant language.

While the words of a language are learned, not all aspects of linguistic knowledge can be learned. There are no experiences common to young children everywhere that would teach them that the noises grownups around them make are composed of words and phrases. The knowledge that young children manifest of the existence of words, phrases, etc., must therefore be available to them as part of their genetic endowment. It is by virtue of this endowment that children—unlike young chimpanzees—are able to acquire a language.

Morris Halle
Professor of Linguistics
Massachusetts Institute of Technology

linguistic stock *Linguistics.* a parent language and any other dialects or separate languages that developed from it.

lingula *Anatomy.* **1.** any tongue-shaped structure or process. **2.** the anterior tongue-shaped portion of the cerebellum.

Lingulacea *Invertebrate Zoology.* a superfamily of burrowing brachiopods, in the order Lingulida, with an elongated biconvex shell.

lingulate *Biology.* tongue-shaped.

Lingulella *Paleontology.* an extinct genus of inarticulate burrowing brachiopods in the superfamily Lingulacea; extant in the Lower Cambrian and Ordovician.

Lingulida *Invertebrate Zoology.* an order of inarticulate bivalve brachiopods living in vertical burrows on the sea floor.

linguo- a combining form meaning "tongue," as in *linguodental.*

linguodental *Anatomy.* of or relating to the tongue and teeth. *Linguistics.* articulated by the tongue and teeth, as a hard *th.*

linguoid current ripple see LINGUOID RIPPLE MARK.

linguoid ripple mark *Geology.* an aqueous ripple mark with tongue-shaped projections pointing into the current and formed by water moving along the bottom of shallow streams. Also, CUSPATE RIPPLE MARK, LINGUOID CURRENT RIPPLE.

liniment *Medicine.* any liquid or semiliquid preparation to be used on the skin, as to ease pain or irritation by producing a counterirritation.

lining *Design Engineering.* a layer of material set on the inner side of something, as reinforcement or decoration. *Engineering.* the process of marking a surface with lines. *Graphic Arts.* a method of designing type so that all the characters of a given point size will align, regardless of font.

lining bar *Mechanical Devices.* a crowbar with a wedge, pinch, or point at its working end.

lining fitch *Mechanical Devices.* a hog's-hair brush with an angled form that is used with a wooden straightedge to paint lines.

lining sight *Mining Engineering.* an instrument that consists essentially of a plate with a slot in the middle, and the means of suspending it vertically; used with a plumbline for directing the courses of underground drifts or headings.

lining wheel *Mechanical Devices.* a small, metal cylinder with assorted, serrated wheels to make various-sized lines.

link something that connects one part or thing with another; specific uses include: *Design Engineering.* **1.** any of the rings in a chain. **2.** a connecting part in the moving portion of a machine, usually with a pivot at its ends. *Telecommunications.* **1.** a communication system that connects two points. **2.** a channel or circuit that is connected in tandem with other channels or circuits. *Transportation Engineering.* the part of a transportation network that connects two routes, lines, systems, or modes. *Cartography.* in surveying, any combination of lines or parts of lines of levels that, when considered as a whole, give a continuous act of leveling directly from one junction bench mark to another without passing through or over any other junction bench marks. Also, LINK OF LEVELS. *Civil Engineering.* **1.** a unit of length equal to 7.92 inches, or 1/100 of a surveyor's chain. **2.** an electrical insulator. Also, ISOLATOR. *Computer Programming.* see LINKAGE.

linkage the fact of linking or connecting different elements; specific uses include: *Genetics.* the association during inheritance of two or more nonallelic genes. *Mechanical Engineering.* an assembly of four or more rods for transmitting motion, usually in the same plane or in parallel planes. *Computer Programming.* a routine that interfaces two separately coded routines and through which control information is passed. Also, LINK.

linkage disequilibrium see DISEQUILIBRIUM.

linkage editor *Computer Programming.* a system service program used to convert and link object code modules into a single module that is in a form suitable for loading and execution. It does this by assigning actual storage locations to modules, relocating addresses within modules to reflect their actual addresses, and resolving references between modules. Also, **link editor.**

linkage group *Genetics.* a group of genes that are located on the same chromosome.

linkage map *Genetics.* a diagram showing the relative positions of the loci of known genes on the chromosomes of a species.

link chute adaptor *Ordnance.* a device that allows a link ejection chute to be used with a machine gun; it also moves ejected links into the chute.

link circuit *Electromagnetism.* a closed loop circuit that has two coils of wire of a few turns each; used to couple two circuits that are separated by a considerable distance.

link control message *Telecommunications.* a communication sent over a link of a network to prepare the link to handle transmissions.

link coupling *Electromagnetism.* the coupling of two circuits by means of a link circuit.

linked *Science.* relating to or displaying linkage.

linked ammunition *Ordnance.* individual cartridges connected by interlocking metal links, thus forming an ammunition belt for use in certain automatic weapons. Similarly, **link belt.**

link ejection chute *Ordnance.* a device attached to a machine gun through which links from linked ammunition are ejected toward a desired point after being separated from the cartridges.

link encryption *Telecommunications.* in a given communications network, the end-to-end encryption within a single point-to-point link.

linker *Molecular Biology.* a segment of oligodeoxyribonucleotides that contains the specific sequence of a given restriction endonuclease cleavage site and can be joined to another fragment of DNA to facilitate cloning.

linker-delinker *Ordnance.* a machine that assembles or disassembles a linked ammunition belt.

Linke scale *Meteorology.* a device that is used to appraise the perceptible blueness of the sky, consisting of a set of eight cards displaying various shades of blue and numbered evenly from 2 to 16; the observer compares the shade of the sky to these cards, using odd numbers if the sky color is judged to lie between any of the given shades. Also, BLUE-SKY SCALE.

link field *Computer Programming.* **1.** the first word of a message buffer, used to locate the next buffer on the message queue. **2.** a field in a linked list record that points to the next record.

link group *Telecommunications.* a cluster of links that use the same multiple terminal equipment.

linking loader *Computer Programming.* a service program that combines the functions of a program loader with those of a linkage editor.

linking number *Molecular Biology.* the number of times that one strand of a double helix structure crosses the other strand within a given length of DNA.

link list *Computer Programming.* a data structure formed as a chain of records, each of which contains a pointer to the next record or a nil value indicating the end of the list.

link-loading machine *Ordnance.* a device that loads individual cartridges into interlocking metal links.

link mechanism see BAR LINKAGE.

link rods *Mechanical Devices.* the auxiliary or articulated rods that connect a part of a master rod of a radial engine.

links *Mathematics.* see CHORD.

link stretch *Ordnance.* in belted ammunition, the change in separation between individual cartridges caused by stretching as the belt is fed into the gun.

Link trainer *Aviation.* the trade name for various mechanical training machines that simulate flying conditions on the ground; widely used in the U.S. for pilot training in World War II and after. (Developed by Edwin *Link,* born 1904, American engineer.)

link V belt *Design Engineering.* a V-belt of rubberized fabric links attached by metal fasteners.

lin-log amplifier *Electronics.* a circuit that regulates its level of amplification by responding to a low-amplitude signal in a linear manner and a high-amplitude signal in a logarithmic manner.

Linnaean [lə nē´ən] *Systematics.* relating to Carolus Linnaeus or his system of biological classification.

Linnaean classification *Systematics.* a system of classification of organisms, devised by Linnaeus and employing his technique of binomial nomenclature, based on morphological similarities arranged in a hierarchy from most general to most specific.

linnaeite or **linneite** *Mineralogy.* $Co^{+2}Co_2^{+3}S_4$, a steel-gray, cubic mineral of the linnaeite group, occurring in granular masses and octahedral crystals, and having a specific gravity of 4.5 to 5.0 and a hardness of 4.5 to 5.5 on the Mohs scale; found with other sulfide minerals in hydrothermal veins. Also, COBALT PYRITES.

Linnaeus, Carolus [lə nē´əs] a Latinized version of the name of Carl von Linné, 1707–1778, Swedish botanist; developed the modern scientific method of classification, including binomial nomenclature.

Linnik interference microscope *Optics.* a type of microscope in which a semi-reflective mirror divides the light into two beams, with one focused onto the specimen's surface and the other focused onto a comparison surface, and then recombines the two beams; used to examine the surface structure of reflecting specimens.

linoleic acid *Biochemistry.* $C_8H_{14}O_2S_2$, a fatty acid containing 18 carbon atoms and two double bonds. *Nutrition.* a polyunsaturated fatty acid that is essential for growth, healthy skin, and general well being; though widely distributed in plants and in animals, it cannot be synthesized by mammals and therefore is an essential dietary constituent. Also, **linolic acid.**

linolenic acid *Biochemistry.* $C_{18}H_{30}O_2$, a fatty acid that is common in glycerophosphatides and in plant fats, and is an essential dietary constituent for mammals; contains 18 carbon atoms and three double bonds.

linolenyl alcohol *Organic Chemistry.* $C_{18}H_{34}O$, a combustible, colorless solid that melts at –5°C to –2°C and boils at 148–150°C (1 torr); used in paints.

linoleum *Materials.* a hard floor covering composed of burlap or canvas coated with a mixture of linseed oil, resinous gums, and powdered cork, with pigments usually added to provide color and design.

linoleum block *Mechanical Devices.* a thick, resilient, cork piece of linoleum mounted to a wood block and engraved with a design pattern.

linoleum knife *Mechanical Devices.* a type of short-bladed knife fixed into a wooden handle, having a curved, pointed end and used mainly for cutting linoleum.

Linotype *Graphic Arts.* **1.** the trade name of a linecasting machine that uses a keyboard to assemble type matrices. **2. linotype.** to set type on a Linotype or similar machine.

Linotype

linseed *Materials.* the seed of the flax plant, *Linum usitatissimum,* the source of linseed oil.

linseed cake *Materials.* a cake or mass that is the residue when the oil is extracted from linseed; used as feed for farm animals. Also, **linseed oil meal.**

linseed oil *Materials.* a yellowish substance obtained from linseed, a drying oil that polymerizes on exposure to air; used in linoleum, in paints and varnishes, and for various other applications.

lint *Textiles.* **1.** the longer fibers that are separated from cottonseed during the first stage of ginning. **2.** in popular use, a ball or clump of tiny fibers or yarn, such as may accumulate on the surface of clothing in a washing machine or clothes dryer. *Surgery.* an absorbent, soft, loosely woven material used for surgical dressings; a specially finished fabric woven in sheets. Also, PATIENT LINT, SHEET LINT.

lintel *Architecture.* a horizontal structural member (of wood, stone, or steel) that spans an opening and supports the wall above it. Also, DOORHEAD.

linter *Textiles.* **1.** a machine for removing lint from cloth. **2.** a machine used to strip off fuzz linters from cottonseed after ginning.

linters *Textiles.* the short, fuzzy fibers that cling to cottonseed after it has first been ginned and that are removed by a second ginning; used in making paper, cushions, plastics, and various other products.

lintin *Surgery.* a loose fabric of absorbent cotton that is used in dressing wounds.

LIOCS logical input/output control system.

lion

lion *Vertebrate Zoology.* a large, carnivorous, mostly noctural cat of the species *Panthera leo,* native to Africa and southwestern Asia, and having a tawny body, a tufted tail, and a shaggy darker-colored mane in the male.

lioness *Vertebrate Zoology.* a female lion.

lionet *Vertebrate Zoology.* a small or young lion.

Liopteridae *Invertebrate Zoology.* a small family of parasitic wasps, hymenopteran insects in the superfamily Cynipoidea.

Liouville, Joseph [lyoo´vēl´] 1809–1882, French mathematician; the founder of Liouville's *Journal*; the first to construct a proven set of transcendental numbers.

Liouville equation *Physics.* an equation stating that the density of a system of fluid particles in phase space will remain constant.

Liouville-Neumann series *Mathematics.* a series solution $\phi(x)$ to the Fredholm integral equation $\lambda \int_a^b K(x,y)\,\phi(y)dy + f(x) = \phi(x)$, constructed from iterated kernels; that is,

$$\phi(x) = \sum_{n=1}^{\infty} \lambda^n \int_a^b K_n(x,y) f(y)dy,$$

where the kernel $K(x,y)$ is real, continuous, and not identically zero for $x, y \in [a,b]$. Also, NEUMANN SERIES.

Liouville's theorem *Mathematics.* a theorem stating that a bounded harmonic function on R^2 is constant.

lip *Anatomy.* **1.** either of two muscular folds forming the anterior boundaries of the mouth. **2.** any liplike fold bordering a cavity, groove, or depression. **3.** either of the two folds (labia majora) of the female external genital organs, or either of the two folds (labia minora) at the inner surfaces of the labia majora. *Design Engineering.* the projecting rim of a hollow container; a short spout. *Mechanical Devices.* the cutting edge of a fluted drill formed by its intersection and the lip clearance angle, extending from the chisel edge at the web to its circumference.

lip- a combining form meaning "fat," as in *lipid.*

Lipalian *Geology.* a former identification of an interval of geologic time between the Precambrian and the Cambrian, represented in the strata by a widespread unconformity.

Liparidae see LIMANTRIIDAE.

liparite see RHYOLITE.

lipasarcoma *Oncology.* a rare malignant tumor typically occurring in elderly individuals, usually found in retroperitoneal and mediastinal fat deposits.

lipase *Enzymology.* an enzyme in the digestive organs that breaks down fat by catalyzing the hydrolysis of ester linkages in triglycerides.

lipectomy *Surgery.* the surgical removal of subcutaneous adipose tissue. Also, ADIPECTOMY.

lipemia *Medicine.* a condition in which an excess of fat or lipid is present in the blood. Also, LIPOIDEMIA.

Liphistiidae *Invertebrate Zoology.* a family of spiders, arachnids in the order Liphistiomorphae, whose abdomen shows evidence of segmentation.

Liphistiomorphae *Invertebrate Zoology.* a suborder of spiders, arachnids in the order Araneida, whose abdomen shows evidence of segmentation.

lipid *Biochemistry.* any of a group of fats and fatlike substances including fatty acids, neutral fats, waxes, and steroids; all contain aliphatic hydrocarbons, are water insoluble, and are easily stored in the body, serving as a source of fuel. Also, LIPIN, LIPOID.

lipid bilayer *Histology.* the basic structure of a biological membrane, composed of two layers of phospholipids.

lipid droplet *Cell Biology.* an organic substance that is stored in fat cells and is characterized by its solubility in nonpolar solvents.

lipid histiocytosis *Medicine.* a generalized multiplication of histiocytes containing lipids such as cholesterol, phospholipid, or kerasin.

lipid metabolism *Biochemistry.* the oxidation of lipids in the body to produce metabolic energy.

lipid nephrosis *Medicine.* a chronic disease of the kidneys marked by protein in the urine, abnormally low content of albumin in the blood, accumulation of fluid in the tissue spaces, and an excess of cholesterol in the blood.

lipidosis *Medicine.* any of various disorders characterized by abnormal concentrations of lipids in the tissues. Also, LIPOIDOSIS.

lipid pneumonia *Medicine.* a pulmonary disease resulting from the inhalation of oily or fatty substances, especially liquid petrolatum, or from accumulated endogenous lipid material in the lungs.

lipid proteinosis *Medicine.* a hereditary disorder in which there are extra cellular deposits of a phospholipid-protein complex in various parts of the body, including the skin and air passages.

lipid storage disease *Medicine.* a childhood lipidosis characterized by the accumulation of large, lipid-containing cells throughout the macrophage system and the nervous system.

lipin see LIPID.

Lipmann, Fritz Albert 1899–1986, American biochemist, born in Germany; studied phosphates; awarded the Nobel Prize for the discovery of coenzyme A.

lipo- a combining form meaning "fat," as in *lipoprotein.*

lipoblast *Cell Biology.* a specialized connective tissue that develops into a fat cell. Thus, **lipoblastic.**

lipoblastoma *Oncology.* a benign fatty tumor composed of a mixture of embryonal lipoblastic cells and mature fat cells; the tumor cells are arranged in lobules and occur most often in children.

lipochondroma *Oncology.* a benign tumor that is composed of mature cartilage and fatty tissue. Also, BENIGN MESENCHYMOMA.

lipochrome *Biochemistry.* any fat-soluble hydrocarbon pigment, such as lutein, chromophane, carotin, or xanthophyll.

lipocortin *Endocrinology.* an intracellular protein that is believed to be the endogenous inhibitor of phospholipase A_2, thus inhibiting the release of arachidonic acid and the subsequent formation of its active metabolites.

lipodystrophy *Medicine.* **1.** any disturbance of the metabolism of fat. **2.** any of various conditions due to defective metabolism of fat, resulting in the absence of subcutaneous fat.

lipofibroma *Oncology.* a benign tumor that is composed of fatty and fibrous tissue.

lipofuscin *Biochemistry.* a fatty hydrocarbon pigment that accumulates in certain animal tissue upon aging.

lipogenesis *Physiology.* the formation or depostion of body fat; the transformation of carbohydrates or protein into body fat.

lipogranuloma *Medicine.* a nodule of lipoid material; a foreign body inflammation of adipose tissue containing granulation tissue and oil cysts.

lipoic acid *Biochemistry.* $C_8H_{14}O_2S_2$, a coenzyme involved in acyl group and hydrogen transfer reactions; a component of pyruvate dehydrogenase.

lipoid *Biochemistry.* **1.** fatlike; resembling fat. Also, **lipoidal. 2.** see LIPID.

lipoidemia see LIPEMIA.

lipoidosis see LIPIDOSIS.

lipolysis *Biochemistry.* the decomposition of fat into fatty acids and glycerol, which are then frequently released into plasma.

lipolytic hormone *Endocrinology.* any of the hormones that induce the release of fatty acids from tissue lipids, such as the catecholamines, glucagon, and growth hormone.

lipoma *Oncology.* a benign tumor that is usually composed of mature fat cells, although some lipomas are composed wholly or partly of fetal fat cells.

lipomatoid *Oncology.* resembling a lipoma.

lipomatosis *Medicine.* a condition characterized by abnormal localized, or tumor-like, accumulations of fat in the tissues.

lipomelanotic reticulosis *Medicine.* a disease in which there is enlargement of the lymph nodes with proliferation of histiocytes and macrophages containing fat or melanin; often secondary to an extensive dermatitis.

Lipomycetoideae *Bacteriology.* a subfamily of oxidative yeasts of the family Saccharomycetaceae that reproduce by budding.

lipomyohemangioma *Oncology.* a benign tumor that is composed of vascular, adipose, and muscle tissue.

lipomyoma *Oncology.* a benign tumor that is composed of muscle and fatty tissues. Also, LEUKOMYOMA.

lipopenia *Medicine.* a deficiency of fats in the body.

lipopexia *Medicine.* the accumulation of fat in the tissues.

lipophilic *Chemistry.* having an affinity for fats and oils.

lipophore *Histology.* a pigment cell containing any of various fat-soluble pigments.

lipopolysaccharide *Biochemistry.* a water-soluble compound or molecule in which polysaccharides and lipids are linked; an important component of the outer membranae of Gram-negative bacteria.

lipoprotein [lip´ə prō´tēn] *Biochemistry.* a cholesterol carrier whose density is determined by the ratio of lipid to protein. **High-density lipoproteins** (HDL), containing more protein than lipid, carry cholesterol from tissues back to the liver, reducing the risk of heart disease; **low-density lipoproteins** (LDL) carry cholesterol from the liver to tissues, increasing the risk of heart disease.

liposarcoma *Oncology.* a rare malignant tumor occurring in elderly individuals, usually found in retroperitoneal and mediastinal fat deposits. Also, **lipomyxoma.**

liposoluble *Chemistry.* soluble in fats.

liposome *Cell Biology.* a synthetic membrane vesicle made from phospholipids and used for in vitro study of membrane-defined events such as transport, or for the delivery of substances to a cell.

liposome-swelling assay *Biochemistry.* a technique for establishing the rate at which ions or molecules diffuse through porin channels before being incorporated into liposomes.

Lipostraca *Paleontology.* an order of nonmarine branchiopods in the subclass Sarostraca; distinguished by its terminally bifurcated caudal segment; known only from the Rhynie chert of the Middle Devonian.

liposuction *Surgery.* the removal of subcutaneous fat using a suction-pump device. Also, SUCTION LIPECTOMY.

lipotrophic *Biochemistry.* of, relating to, characterized by, or causing lipotrophy.

lipotrophin *Biochemistry.* a hormone that stimulates lipolysis.

lipotrophy *Biochemistry.* an increase of bodily fat.

lipotropic *Biochemistry.* **1.** acting on fat metabolism by hastening the removal or decreasing the deposit of fat in the liver. **2.** a biologically active agent that has this effect.

β-lipotropin *Endocrinology.* a polypeptide prohormone product of the pro-opiomelanocortin gene that is found in cells of the adenohypophysis and is the precursor molecule for the endogenous opioid β-endorphin. Also, **lipotropin, lipotropic hormone.**

lipotropism *Biochemistry.* the fact or condition of being lipotropic. Also, **lipotropy.**

Lipotyphla *Vertebrate Zoology.* the suborder of insectivores containing the hedgehogs, moles, solenodons, and shrews.

lipovitellin *Biochemistry.* a lipoprotein of molecular weight 400,000 that is found in the yolks of eggs.

lipoxidase see LIPOXYGENASE.

lipoxygenase *Enzymology.* an enzyme of the oxidoreductase class that catalyzes the addition of oxygen to the double bonds of unsaturated fatty acids, forming hydroperoxides. Also, LIPOXIDASE.

lipoxysm *Toxicology.* poisoning due to ingestion of fatty acids.

lipper *Oceanography.* **1.** a slight ruffling of the ocean surface. **2.** a fine spray from small waves.

Lippershey, Hans 1587–1619, Dutch optician; credited with the invention of the first practical telescope and compound microscope.

Lippich prism *Optics.* a type of Nicol prism that masks half the field of view when mounted onto a polarmeter, thus allowing an individual to determine the character of polarized light leaving the instrument.

Lippmann, Gabriel 1845–1921, French physicist; awarded the Nobel Prize for inventing a method of taking color photographs of the light spectrum.

Lippmann effect *Physics.* a change in the surface tension on an interface between two immiscible liquid conductors when a voltage is applied across the interface.

Lippmann electrometer see CAPILLARY ELECTROMETER.

Lippmann fringes *Optics.* fringes that appear in electromagnetic radiation, produced by light which is reflected off a fine-grained photographic emulsion with a mercury-coated backing; formerly used in color photography.

LIPS *Artificial Intelligence.* a measure of the processing speed of computers used in artificial intelligence applications. (An acronym of logical inferences per second.)

Lipschitz continuous *Mathematics.* **1.** a function $f: X \to Y$ between Banach spaces (such as R^n) with norms $\| \ \|$ is said to be Lipschitz continuous at $x_1 \in X$ if there is a constant M such that $\|f(x) - f(x_1)\| \leq M \|x - x_1\|$ for all x in some neighborhood of x_1. Lipschitz continuity is a special case of Hölder continuity. Other common usages are to say that f is **Lipschitzian** or that it satisfies a **Lipschitz condition. 2.** f is said to be **uniformly Lipschitz continuous** in a subset U of X if there is a constant M such that $\|f(x) - f(y)\| \leq M \|x - y\|$ for all x, y in U.

Lipschütz body *Pathology.* a cell or centrocyte found in the lesions of herpes simplex, containing granules of varying sizes or multiples. (Named for Benjamin *Lipschütz,* 1878–1931, Austrian dermatologist.)

Lipscomb, William Nunn, Jr. born 1919, American chemist; awarded the Nobel Prize for the study of the structure and bonding mechanisms of boranes.

lip-sync or **lipsync** [lip′singk′] *Telecommunications.* **1.** in motion pictures or television, to synchronize speech or singing that is recorded separately from the picture, so that the lip movements of the actor or singer appear to coincide naturally with the sound. **2.** the process by which this is done. Also, **lip-synch, lipsynch.** (A shortened form of lip synchronization.)

liptinite see EXINITE.

liq. liquid; liquor.

Liq. *Pharmacology.* solution.

liq. pt. liquid pint.

liquefacient *Physical Chemistry.* promoting a process of liquefaction.

liquefaction *Physical Chemistry.* a change from the gaseous phase to the liquid phase.

liquefied *Chemical Engineering.* converted to the liquid phase.

liquefied hydrocarbon gas see LIQUEFIED PETROLEUM GAS.

liquefied natural gas *Petroleum Engineering.* natural gas in the liquid state.

liquefied petroleum gas *Petroleum Engineering.* a liquefied or compressed gas consisting of flammable hydrocarbons such as propane or butane, obtained as a by-product from petroleum refining or from natural gas; widely used as a heating and cooking fuel, as in mobile homes or in areas not connected to gas pipelines. Also, LIQUEFIED HYDROCARBON GAS, COMPRESSED PETROLEUM GAS.

liquefier *Chemical Engineering.* any device or system used to liquefy gases by means of compression, expansion, or heat exchange.

liquefy *Chemical Engineering.* to convert a substance to the liquid phase.

liquid *Physics.* one of the three fundamental states of matter, along with solids and gases. Unlike a solid (and like a gas) a liquid has constituent particles that are free to move past each other rather than being fixed in a given shape or position. Unlike a gas, it is relatively difficult to compress, and it lacks the capability to expand without limit to fill the space available. Most liquids will assume the shape of a container in which they are confined and will seek the lowest level available. (From a Latin word meaning "to be fluid; flow.")

liquid air *Chemical Engineering.* air that has been liquefied by compression or cooling.

liquid bright gold see LIQUID GOLD.

liquid-bubble tracer *Fluid Mechanics.* a method of studying fluid flow by observing the motion of immiscible droplets of the same density as that of the fluid under study.

liquid carburizing *Metallurgy.* carburizing in a liquid environment.

liquid chromatography *Analytical Chemistry.* a separation technique in which the moving phase is a liquid and the stationary phase is a solid or liquid on a solid support; components of a mixture in solution are selectively adsorbed.

liquid column gauge see U-TUBE MANOMETER.

liquid compass *Engineering.* a magnetic compass card that is suspended in a liquid in order to reduce the weight of the card on the point and to limit card oscillations.

liquid-cooled *Mechanical Engineering.* describing a device that is cooled by a liquid, especially an internal-combustion engine cooled by water, glycol, or oil that circulates through internal passages to a heat exchanger or liquid source. Thus, **liquid-cooled engine.**

liquid cooling *Mechanical Engineering.* the use of a circulating liquid to cool equipment and hermetically sealed components.

liquid crystal *Crystallography.* a substance that displays optical properties characteristic of crystals but other properties characteristic of liquids; commonly seen as an intermediate stage in crystalline forms that do not melt directly into a liquid, and in living organisms, such as the refracting portion of muscle fiber and axons of nerve cells. *Materials Science.* any of various industrial subtances having this form.

liquid crystal display *Electronics.* a screen in which a liquid crystal material sandwiched between two sheets of glass exhibits computer data when a voltage appears across leads that are connected to character-forming segments etched onto its inner glass surface.

liquid-dielectric capacitor *Electricity.* a capacitor with the plate assembly mounted in a tank filled with a liquid dielectric, capable of high capacity and low losses.

liquid-dominated hydrothermal reservoir *Geology.* a geothermal system, such as a geyser or hot spring, that primarily produces superheated water.

liquid-drop model *Nuclear Physics.* an idealized model in which the properties of a nucleus are portrayed as an incompressible and charged drop of liquid that is stable against small deformations by virtue of stabilizing surface tension but disrupted by strong nuclear forces.

liquid extraction see SOLVENT EXTRACTION.

liquid-filled porosity *Geology.* in porous rock or sand formations, a condition under which pore spaces are filled with water, liquid petroleum, pressure-liquefied butane or propane, or tar.

liquid filter *Chemical Engineering.* a porous material that has a degree of rigidity and is used as a barrier to strain solids from a liquid stream.

liquid flow *Fluid Mechanics.* the flow of a substance when it is in a liquid phase.

liquid flow region *Materials Science.* the state or region above the rubber flow region for linear amorphous polymers, in which the temperature is high enough for the polymer to flow readily with a behavior similar to that of a conventional liquid.

liquid fluorine *Physics.* a fluorine gas that has been liquefied by a process of intense pressurization and cooling; used as a cryogenic propellent.

liquid fuel see LIQUID PROPELLANT.

liquid fuse unit *Electricity.* a fuse in which the element is in liquid.

liquid-gas chemical reaction see LIQUID-VAPOR CHEMICAL REACTION.

liquid glass see SODIUM SILICATE.

liquid gold *Materials.* **1.** a suspension of gold particles in an oil, applied as a gilding layer to ceramics. Also, LIQUID BRIGHT GOLD. **2.** an informal term for any liquid believed to be highly valuable, such as oil.

liquid helium *Physics.* helium gas that has been cooled to a temperature of at least 4.2 K, becoming liquefied; if cooled to below 2.19 K, it exhibits superconducting properties.

liquid holdup *Fluid Mechanics.* a condition wherein gas and liquid flow simultaneously through a vertical pipe and the gas has a greater fluid velocity than the liquid; gas slippage may occur in this condition.

liquid hydrocarbon *Organic Chemistry.* a liquid compound containing only carbon and hydrogen that has been prepared from a gas by pressure or temperature, especially from butane, propane, ethane, or methane; often used as fuel.

liquid hydrogen *Physics.* hydrogen gas that has been cooled to a temperature of at least 20.46 K, becoming liquified; used as a rocket fuel.

liquid-hydrogen bubble chamber *Nucleonics.* a device in which charged particles moving through liquid hydrogen kept slightly above the boiling point leave a train of bubbles that reveal interactions between protons and charged particles.

liquid-in-glass thermometer *Engineering.* a thermometer in which the thermally sensitive element is a liquid, usually mercury or alcohol, enveloped in glass.

liquid-in-metal thermometer *Engineering.* a thermometer in which the thermally sensitive element is a liquid enveloped in metal.

liquid-jet cutting *Materials Science.* a nontraditional process for machining wood or other material, using a jet of liquid under high pressure; eliminates crushing or deformation of the material, as well as the generation of dust.

liquid junction *Physical Chemistry.* the interface between two electrolytes that differ in composition or concentration of one or more ionic species. Thus, **liquid junction electromotive force (emf).**

liquid laser *Optics.* a type of laser that combines rare earth ions, dissolved molecules, and an organic dye solution to generate a coherent beam of light.

liquid lattice model *Materials Science.* a model showing how solvent and polymer molecules fit into the sites available on a three-dimensional lattice.
liquid level control *Engineering.* the ability to maintain a constant distance between the surface of a liquid and a reference point.
liquid limit *Geology.* the minimum amount of water that, when mixed with a given sediment, will cause the sediment to flow plastically.
liquid-liquid chemical reaction *Chemistry.* a chemical reaction that involves two or more liquid reactants. Also, **liquid-liquid reaction.**
liquid-liquid extraction *Chemical Engineering.* a process in which one or more components are removed from a liquid mixture by intimate contact with a second liquid that is nearly insoluble in the first liquid and dissolves the impurities.
liquid measure *Metrology.* any system that is for the measurement of liquids, especially the traditional system of gallons, quarts, pints, and so on.
liquid-metal attack *Materials Science.* a form of corrosion in which a liquid metal attacks a solid at high-energy locations, such as grain boundaries. The liquid metal lowers the surface energy, thus making it easier for cracks to form.
liquid-metal corrosion *Materials Science.* any degradation of materials by liquid metals caused by chemical reactions combined with physical phenomena, including liquid-metal attack, intergranular attack, liquid-metal embrittlement, corrosion by contaminants, and erosion-corrosion.
liquid-metal embrittlement *Metallurgy.* the loss of ductility of a metal or alloy in a liquid-metal environment.
liquid-metal fuel cell *Electricity.* a voltaic cell that is used to convert chemical energy into electrical energy, using molten potassium and bismuth as reactants and a molten salt electrolyte.
liquid-metal nitriding *Metallurgy.* a case-hardening process in which a molten mixture of sodium and potassium salts is used as a source of nitrogen; gives the steel a thin, wear-resistant case and improves the fatigue properties of parts not subject to high wear or crushing loads.
liquid-metal nuclear fuel *Nucleonics.* a nuclear fuel composed of a solution of uranium or plutonium and a molten metal, such as bismuth.
liquid methane *Physics.* methane gas that has been cooled to a temperature of at least –161°C, becoming liquefied.
liquid nitrogen *Physics.* nitrogen gas that has been cooled to a temperature of 77.4 K or below, thus becoming liquefied.
liquid oxygen *Physics.* oxygen gas that has been cooled to a temperature of at least 160.2 K or below, thus becoming liquefied; used in rocket propellants and other high-performance fuels.
liquid paraffin see MINERAL OIL.
liquid penetrant test *Engineering.* the use of a liquid, especially a fluorescent liquid, to test for cracks in a component. It often is used with black light.
liquid petrolatum see MINERAL OIL.
liquid-phase epitaxy *Solid-State Physics.* the bulk growth of crystals, typically in semiconductor manufacturing, whereby the crystal is grown from a rich solution of the semiconductor onto a substrate in layers, each of which is formed by supersaturation or cooling.
liquid-phase hydrogenation *Chemical Engineering.* a method for hydrotreating an olefinic hydrocarbon with hydrogen in which liquid feed saturated with hydrogen flows through a high-pressure reactor, generally packed with a catalyst, at moderately high temperatures.
liquid-phase joining *Materials Science.* any technique for joining materials in which a liquid phase is involved, including the joining of metals and ceramics, the sintered metal powder process, the active metal brazing process, and the ceramic frit process.
liquid-phase sintering *Materials Science.* the progressive fusion of a ceramic material during the firing process; as it proceeds the proportion of glassy bond increases and the porosity of the fired product decreases.
liquid pint see PINT.
liquid poison *Nucleonics.* a liquid that absorbs neutrons and can be injected to a reactor's cooling system to shut the reactor down.
liquid propellant *Space Technology.* a rocket fuel in liquid form. Also, LIQUID FUEL.
liquid-propellant rocket engine *Space Technology.* a rocket engine using a liquid form of propellant; generally larger and more complex than a solid-propellant engine or motor.
liquid quart see QUART.
liquid rheostat *Electronics.* a device in which electrical resistance is varied by raising or lowering metal plates in a conductive liquid, usually water. Also, WATER RHEOSTAT.
liquid scintillation detector *Nucleonics.* a scintillation device in which a liquid, such a *p*-terphenyl dissolved in toluene, emits visible light when bombarded with particles or irradiated with ultraviolet light or X-rays.
liquid seal *Chemical Engineering.* a level of water or other liquid above the end of a discharge pipe that prevents the escape of gases, noxious vapors, and the like.
liquid-sealed meter *Chemical Engineering.* a positive-displacement gas meter in which the gas flows through one of four rotating chambers that are partially filled with water. Also, DRUM METER.
liquid semiconductor *Electronics.* a material in a solid or liquid state that generates variable resistance when an electric charge is applied.
liquid-solid chemical reaction *Chemistry.* a chemical reaction that involves at least one solid reactant and at least one liquid reactant. Also, **liquid-solid reaction.**
liquid-solid equilibrium see SOLID-LIQUID EQUILIBRIUM.
liquid-sorbent dehumidifier *Mechanical Engineering.* a device that dehumidifies air or a gas by means of a recirculating liquid, which absorbs the water vapor and is made to release it outside.
liquid sound *Linguistics.* the type of sound made when the tip of the tongue touches the alveolar ridge, as in the sound of [l] or [r].
liquid sulfur dioxide-benzene process *Chemical Engineering.* a process in which sulfur dioxide and benzene are used to dewax lubricating oils.
liquidus *Physical Chemistry.* in a two-component mixture, the temperature at which the first solid begins to form.
liquidus line *Thermodynamics.* a curve indicating the temperature at which fusion is completed in a two-component mixture at various concentrations of the mixture as the temperature is raised.
liquid-vapor chemical reaction *Chemistry.* a chemical reaction that involves at least one liquid reactant and at least one vapor reactant. Also, LIQUID-GAS REACTION.
liquid-vapor equilibrium *Physical Chemistry.* the equilibrium relationship between the liquid and vapor phases of a partially vaporized compound or multicomponent mixture at a specified temperature and pressure in a closed system.
liquid-water content see FREE-WATER CONTENT.
liquiform *Materials Science.* resembling a liquid.
liquor *Chemical Engineering.* **1.** any aqueous solution of one or more chemical compounds. **2.** the liquid overflow from an extraction process. *Pharmacology.* any solution of a medication in water. *Anatomy.* a general term for any of certain bodily fluids, such as the cerebrospinal fluid. *Food Technology.* a popular term for a distilled alcoholic beverage such as gin or whiskey, or for alcoholic beverages in general. (Related to the Latin word for "liquid.")
liquor finish *Metallurgy.* a bright finishing typical of wet-drawn wire.
lirella *Botany.* a linear apothecium having a ridge in the middle, found in some lichens.
Liriopeidae *Invertebrate Zoology.* a family of dipteran insects, the phantom craneflies, in the suborder Orthorrhapha, with black and white banded legs.
liroconite *Mineralogy.* $Cu_2^{+2}Al(AsO_4)(OH)_4 \cdot 4H_2O$, a light blue to yellowish green, monoclinic mineral occurring in glassy wedge-shaped crystals, and having a specific gravity of 2.93 to 3.01 and a hardness of 2 to 2.5 on the Mohs scale; found in the oxidized zones of copper deposits.
LIS *Aviation.* the airport code for Lisbon, Portugal.
liskeardite *Mineralogy.* $(Al,Fe^{+3})_3(AsO_4)(OH)_6 \cdot 5H_2O$, a soft, white, possibly orthorhombic mineral having a specific gravity of 3.01 and an undetermined hardness; found coating quartz and pyrite at Liskeard, Cornwall, England.
lisle [līl] *Textiles.* **1.** a fine cotton thread that is high- and hard-twisted and at least two-ply. **2.** knit goods made of this thread, often hosiery or gloves. (Named for *Lisle,* France, where it was first made.)
lisp *Linguistics.* **1.** a speech pattern characterized by an imperfect pronunciation of sibilant sounds such as [s] and [z], typically by substituting [th] or [th] for these sounds. **2.** to pronounce words with such a pattern.Thus, **lisping.** Also, PARASIGMATISM.
Lisp or **LISP** *Artificial Intelligence.* an artificial intelligence programming language developed for symbolic data manipulation. (An acronym for list processing.)
Lisp interpreter *Artificial Intelligence.* **1.** a procedure that can evaluate a Lisp expression at run time to determine its value: the eval function. **2.** the user interface of Lisp, which reads an expression input by the user, evaluates it, and prints the result. Also, READ-EVAL-PRINT LOOP.

Lissajous figures [le´sə joo´] *Physics.* the plane curves formed by the composition of two sinusoidal waveforms in perpendicular directions. (Named for Jules Antoine *Lissajous,* 1822–1880, French physicist.)

Lissamphibia *Vertebrate Zoology.* the subclass containing the recent, living amphibians, including frogs, toads, salamanders, newts, and caecilians; characterized by the presence of relatively few cranial bones and by pedicellat teeth.

lissencephaly see AGYRIA.

Lissocarpaceae *Botany.* a monogeneric family of dicotyledonous small trees of the order Ebenales, native to tropical South America and characterized by alternate simple leaves and an indehiscent, somewhat fleshy fruit.

list a series of items written together or otherwise grouped in some manner; specific uses include: *Computer Programming.* **1.** an ordered set of items of data. **2.** in the FORTRAN language, a set of data items to be read or written. **3.** a linked list. **4.** to print a program, file, or input data. *Artificial Intelligence.* in Lisp, a linked-list structure, composed of cons cells, whose external representation is the elements of the list enclosed in parentheses.

list *Engineering.* to tilt toward the horizontal; lean to one side.

listening station *Military Science.* a radio or radar station that gathers information about the enemy.

Joseph Lister

Lister, Joseph 1827–1912, English surgeon; founder of antiseptic surgery; developed absorbable catgut ligatures and the drainage tube.

Lister, Martin 1638–1712, English naturalist; studied fossil shells; first to propose systematic geologic mapping.

Listeria *Bacteriology.* a genus of Gram-positive, spheroid, motile, aerobic bacteria of undecided classification found in human and animal feces; *L. monocytogenes* causes such diseases as meningoencephalitis and meningitis.

listeriosis *Medicine.* the infection caused by the bacillus *Listeria monocytogenes,* characterized by symptoms resembling infectious mononucleosis or acute meningitis. *Veterinary Medicine.* a rare disease induced by bacterium of the genus *Listeria,* causing encephalitis and abortion in cattle and sheep; known also as **circling disease** in sheep.

listing *Computer Programming.* **1.** a printed copy of a program or file. **2.** a printed copy of a program produced by a compiler or assembler, often annotated with information about the compiled program. *Agronomy.* see BEDDING.

list processing *Computer Programming.* the manipulation of lists through the modification of pointers or cursors to the affected and adjacent items, rather than physically moving the items.

list structure *Computer Programming.* a data structure in which each element contains a pointer or cursor to the successor element, and possibly another to the predecessor element, in order to facilitate list processing.

LIT *Aviation.* the airport code for Little Rock, Arkansas.

litchi [lē´chē] *Botany.* **1.** any of various trees of the genus *Litchi* in the soapberry family Sapindaceae, especially *L. chinensis,* an evergreen tree native to China that is grown in warm climates for its edible fruit. **2.** the fruit of this tree, consisting of a single seed surrounded by a sweet chewy pulp, enclosed in a brittle, brown papery shell. Also, LEECHEE, LICHEE, LYCHEE.

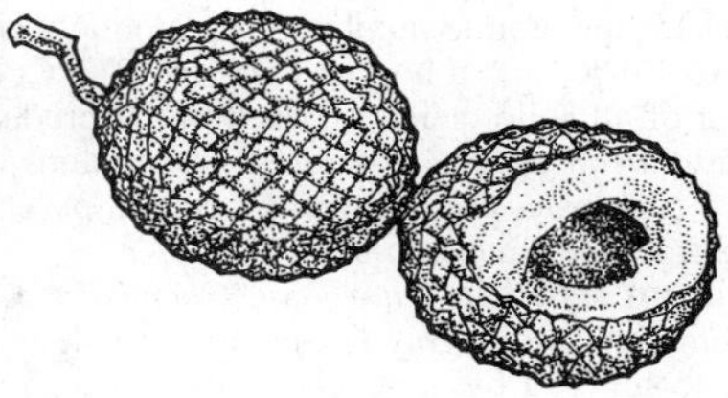

litchi

litchi nut *Food Technology.* dried litchi fruit, a popular Asian dessert.

liter [lē´tər] *Metrology.* the basic unit of capacity in the metric system, equal to the volume of 1 kilogram of distilled water at a temperature of 4°C; equivalent to 1.06 liquid quarts or 0.9 dry quart. Also, LITRE.

literal *Artificial Intelligence.* in a logical formula, a predicate or the negation of a predicate.

literal notation *Mathematics.* the use of letters to denote numerical quantities. Thus, **literal constant, literal expression, literal equation.**

literal operand *Computer Programming.* an instruction operand that specifies the value as a constant rather than as a variable name or memory address.

liter-atmosphere *Physics.* a unit of energy equal to the product of a volume of 1 liter and a pressure of 1 atmosphere.

lith- a combining form meaning "stone" or "rock," as in *lithic.*

litharenite *Petrology.* sandstone having a greater content of rock fragments than feldspar, typically 25% fine-grained rock fragments, less than 10% feldspar, and less than 75% quartz, quartzite, and chert.

litharge [li thärj´; lith´ərj] *Inorganic Chemistry.* PbO, lead monoxide in the form of toxic yellow crystals that are insoluble in water and soluble in acid and alkalis; melts at 886°C; used in storage batteries; in glass, ceramics, and pottery; and in paints, varnishes, and enamels. Also, YELLOW LEAD OXIDE.

lithia see LITHIUM OXIDE.

lithia mica see LEPIDOLITE.

lithiasis *Medicine.* a condition characterized by the formation of calculi, or stones, in the body.

lithic *Science.* of or relating to stone. *Petrology.* having an abundance of rock fragments; applied to a sedimentary or pyroclastic rock. *Archaeology.* **1.** describing a stone artifact. **2.** also, **Lithic.** describing the first developmental period in New World chronology, preceding the Archaic period and characterized by the use of flaked stone tools and hunting and gathering activity. *Medicine.* of or relating to calculi. *Chemistry.* of or relating to lithium. (Going back to the Greek word for "stone.")

-lithic *Archaeology.* a combining form used to form the names of cultural phases: Neolithic, Mesolithic.

Lithic Age see STONE AGE.

lithic culture *Anthropology.* a culture in which tools and weapons are manufactured primarily from stone.

lithic graywacke *Petrology.* a low-grade graywacke with an abundance of unstable materials; specifically, a sandstone with less than 75% quartz and chert, 15–75% detrital clay matrix, and rock fragments in greater abundance than feldspar grains.

lithics *Archaeology.* the process or industry of making stone tools.

lithic tuff *Geology.* a tuff containing fragments of previously formed nonpyroclastic rocks.

lithification *Geology.* **1.** a processes in which loose, unconsolidated sediment is converted into coherent, solid rock. Also, **lithifaction. 2.** a gradual process in which coal changes into bituminous shale or other rock.

lithionite see LEPIDOLITE.

lithiophilite *Mineralogy.* $LiMn^{+2}PO_4$, a salmon pink or brown, orthorhombic mineral, isomorphous with triphylite, occurring in massive form or as large anhedral or subhedral crystals, having a specific gravity of 3.34 and a hardness of 4 to 5 on the Mohs scale; found in granite pegmatites.

lithiophorite *Mineralogy.* $(Al,Li)Mn^{+4}O_2(OH)_2$, a dark-blue to black, opaque, hexagonal or monoclinic, pseudohexagonal mineral occurring in botryoidal masses and as fine scales, having a specific gravity of 3.14 to 3.4 and a hardness of 3 on the Mohs scale.

Lithistida *Paleontology.* an order of demosponges characterized by a skeleton that remains intact after the death of the animals because of the interlocking pattern of the siliceous spicules; Cambrian to the present.

lithium [lith´ē əm] *Chemistry.* a very soft, silvery metallic element having the symbol Li, the atomic number 3, an atomic weight of 6.941, a melting point of 179°C, and a boiling point of 1317°C; an alkali metal, and the lightest of all solid elements; used in the production of tritium and alloys. *Pharmacology.* any of various preparations of lithium salts, such as lithium carbonate, used in treating certain psychological conditions. (Formed from the Greek word for "stone.")

lithium aluminum hydride *Inorganic Chemistry.* $LiAlH_4$, a white powder that sometimes turns gray on standing; stable in dry air at room temperature, decomposed by atmospheric moisture and by heat above 125°C; violently reactive and a dangerous fire risk that may ignite spontaneously; it has many industrial uses, especially as a reducing agent.

lithium amide *Inorganic Chemistry.* $LiNH_2$, a white crystalline solid that has the odor of ammonia and that decomposes in water to give off ammonia; melts at 380–400°C; a dangerous fire hazard; used in the synthesis of organic compounds, including antihistamines and other pharmaceuticals.

lithium battery see LITHIUM CELL.

lithium borohydride *Inorganic Chemistry.* $LiBH_4$, a white crystalline powder that is soluble in water; decomposes at 275°C; a dangerous fire and explosion hazard; used as a source of hydrogen and as a reducing agent.

lithium bromide *Inorganic Chemistry.* LiBr, white crystals or a white to pinkish granular powder; very soluble in water, alcohol, and ether; melts at 550°C and decomposes at 1265°C; used in pharmaceuticals, as a drying agent, and for other industrial purposes.

lithium carbonate *Inorganic Chemistry.* Li_2CO_3, a white powder that is slightly soluble in water and insoluble in alcohol; melts at 723°C and decomposes at 1310°C. The aqueous solution is a strong irritant. It has many industrial uses, especially in nucleonics and the paint industry.

lithium cell *Chemistry.* an electrolytic cell used in the production of metallic lithium. *Electricity.* a cell having an anode made of lithium and an organic electrolyte with no water; commonly used for low-power applications such as cameras and calculators.

lithium chloride *Inorganic Chemistry.* LiCl, white deliquescent crystals; very soluble in water, alcohol, and ether; very readily absorbs moisture from the air; melts at 605°C and boils at 126°C; used in air conditioning and welding, for salt baths, and in the production of soft drinks and mineral waters.

lithium citrate $Li_3C_6H_5O_7 \cdot 4H_2O$, a white powder that is soluble in water and slightly soluble in alcohol; loses its water at 105°C; used in beverages and pharmaceuticals, and as a clay defocculant.

lithium-drifted germanium crystal *Electronics.* a device that detects gamma-radiation and high-energy electrons by processing lithium ions through a germanium crystal.

lithium-drifted silicon detector *Nucleonics.* an instrument that measures radiation doses by applying a reverse bias to a P-N junction fabricated with an N-type lithium region and a P-type silicon region.

lithium fluoride *Inorganic Chemistry.* LiF, a fine white powder that is slightly soluble in water and insoluble in alcohol; melts at 845°C and boils at 1676°C; a strong irritant; used in ceramics, as a component of fuel for molten salt reactors, and for various other purposes.

lithium fluoride dosimetry *Nucleonics.* a method for calculating radiation doses by measuring the thermoluminescence generated by an irradiated lithium fluoride when heated.

lithium glass *Radiology.* a glass containing lithium that produces photons of low voltage used in grenz ray tubes.

lithium halide *Inorganic Chemistry.* any compound consisting of lithium and an element from the halogen group (periodic table Group VIIa), such as fluorine or chlorine.

lithium hydride *Inorganic Chemistry.* LiH, a white translucent crystalline mass that decomposes in cold water; melts at 680°C; a dangerous fire hazard, igniting spontaneously in moist air; used in organic synthesis and as a reducing agent.

lithium hydroxide *Inorganic Chemistry.* LiOH, colorless to white crystals that absorb water and carbon dioxide from the air; soluble in water and slightly soluble in alcohol; melts at 450°C and decomposes at 924°C; used in storage batteries and lubricating greases.

lithium iodide *Inorganic Chemistry.* $LiI \cdot 3H_2O$, white crystals that absorb water from the air but lose water on heating; soluble in water and alcohol; used in air conditioning, as a catalyst, and in nucleonics. Also, **lithium iodide trihydrate.**

lithium mica see LEPIDOLITE.

lithium molybdate *Inorganic Chemistry.* Li_2MoO_4, white hygroscopic crystals, very soluble in water; melts at 705°C; used as a catalyst in petroleum cracking and in steel coating.

lithium nitrate *Inorganic Chemistry.* $LiNO_3$, a colorless powder that is soluble in water and alcohol; melts at 264°C and decomposes at 600°C; a strong oxidizing agent at the elevated temperature and a dangerous explosive hazard; used in pyrotechnics, heat-exchange media, and ceramics.

lithium oxide *Inorganic Chemistry.* Li_2O, a colorless to white powder that absorbs water and carbon dioxide from air; soluble in water; melts above 1700°C; used in ceramics and to manufacture lithium salts. Also, LITHIA.

lithium perchlorate *Inorganic Chemistry.* $LiClO_4$, colorless crystals that absorb water from the air and decompose on heating; soluble in water and alcohol; melts at 236°C and decomposes at 430°C; a dangerous fire and explosive hazard in contact with organic substances; used as a solid rocket fuel.

lithium star *Astrophysics.* any giant star of spectral type G, K, or M whose spectrum shows abnormally high amounts of lithium.

lithium stearate *Organic Chemistry.* $LiC_{18}H_{35}O_2$, white crystals that are insoluble in water and melt at 220°C; used in cosmetics, waxes, greases, and as a lubricant in powder metallurgy.

lithium sulfate *Inorganic Chemistry.* $Li_2SO_4 \cdot H_2O$, colorless crystals that are very soluble in water; used in ceramics and pharmaceuticals.

lithium-sulfur battery *Electricity.* a storage battery whose cells have sulfur dioxide for the cathode and lithium foil for the anode; suggested for use in storing reserve electric power for meeting peak demands at power plants.

lithium tetraborate *Inorganic Chemistry.* $Li_2B_4O_7 \cdot 5H_2O$, a white crystalline powder that loses its water at 200°C; very soluble in water and insoluble in alcohol; used in ceramics, spectroscopy, metal refining, and degassing.

lithium titanate *Inorganic Chemistry.* Li_2TiO_3, a white powder that is insoluble in water; used in enamels and glazes.

litho *Graphic Arts.* **1.** lithography. **2.** lithograph. **3.** lithographic.

litho- a combining form meaning "stone" or "rock," as in *lithophile.*

Lithobiomorpha *Invertebrate Zoology.* an order of centipedes having fifteen pairs of legs and long antennae.

lithocholic acid *Biochemistry.* $C_{24}H_{40}O_3$, a monohydroxylated steroid carboxylic acid that is formed by intestinal bacteria and can be mildly toxic to the liver.

lithoclase *Geology.* any rock fracture that occurs naturally.

lithocyst *Botany.* a plant cell containing a cystolith.

lithocyte *Invertebrate Zoology.* a specialized sensory cell containing a sand grain, used in perceiving an organism's orientation.

Lithodemiaceae *Botany.* a family of marine planktonic diatoms of the order Centrales, usually growing in chains and having valves with two slitted openings.

Lithodidae *Invertebrate Zoology.* a family of decapod crustaceans including the king crabs, in the superfamily Paguridae.

lithodomous *Zoology.* living in or burrowing in rock.

lithofacies *Geology.* **1.** a subdivision of a designated stratigraphic unit that is distinguished from adjacent and nearby subdivisions by its lithological characteristics. **2.** the collective lithological characteristics of any sedimentary rock, especially those indicating the environment of deposition.

lithofacies map *Geology.* a stratigraphic map showing the total areal variation in lithological characteristics of a given stratigraphic unit.

lithogenesis *Pathology.* the fabrication of calculi. *Petrology.* the origin and formation of rocks, especially sedimentary rocks.

lithogeochemical survey *Geochemistry.* a geochemical survey in which the chemical composition of rocks, and sometimes soils, is studied, usually to explore for mineral deposits.

lithograph *Graphic Arts.* a printed artwork produced by lithography.

lithographic limestone *Geology.* an extremely fine-grained, dense, compact, pale yellow or grayish limestone that exhibits conchoidal fracture. Also, **lithographic stone.**

lithographic plate *Graphic Arts.* a printing plate, formerly made of stone but now usually metal or plastic, that has been treated to accept ink in certain areas and repel ink in others.

lithographic texture *Geology.* a calcareous sedimentary rock texture characterized by small, uniformly sized particles and by an extremely smooth appearance.

lithography [li thäg′rə fē] *Graphic Arts.* any of several printing processes, such as offset lithography, in which the image areas of a plate are treated (photographically or directly by hand, as with a lithographic crayon) to accept greasy inks and repel water, while the nonimage areas accept water and repel ink. Thus, **lithographic, lithographer.**

lithoid *Science.* resembling stone or a stone.

lithologic map *Geology.* a geologic map that gives descriptions of the rock types of a designated area.

lithology *Geology.* the study and description of the general, gross physical characteristics of a rock, especially sedimentary clastics, including color, grain size, and composition. Thus, **lithologic, lithological.**

litholysis *Medicine.* the dissolution of a calculus, e.g., a gallstone or kidney stone.

lithometeor *Meteorology.* any dry particulate matter that is suspended in the atmosphere, such as dust, haze, smoke, or sand.

lithomorphic *Geology.* of a soil, having characteristics derived from its parent material.

lithopedion *Medicine.* a retained fetus that has undergone calcification.

lithophagous *Zoology.* swallowing stone or gravel, often as an aid to digestion.

lithophile *Geochemistry.* an element or elements having a strong tendency to combine chemically with oxygen, thus tending to concentrate in the silicate phase of meteorites or in the silicate crust of the earth. Also, OXYPHILE. *Ecology.* an organism that lives or thrives on rocks or stones, as some algae and mosses. Thus, **lithophilous, lithophilic, lithophily.**

lithophysa *plural,* **lithophysae.** *Geology.* a large, hollow, spherical structure, generally consisting of concentric or radial shells, that occurs in glassy basaltic rocks. Also, STONE BUBBLE.

lithophysae in obsidian

lithophyte *Botany.* any plant that grows on rocks or stones. *Invertebrate Zoology.* any polyp having a hard or stony structure, such as a coral. Thus, **lithophytic.**

lithopone *Materials Science.* a white pigment consisting of zinc sulfide, zinc oxide, and barium sulfate; formerly widely used as a pigment for paints, inks, paper, and leather.

lithosere *Ecology.* an ecological succession beginning on an exposed rock.

lithosiderite see STONY-IRON METEORITE.

lithosol *Geology.* a group of azonal soils, consisting of recently weathered rock fragments that develop on steep slopes, and characterized by shallow depth to bedrock. Also, SKELETAL SOIL.

lithospar *Mineralogy.* a naturally occurring mixture of spodumene and feldspar.

lithosphere *Geology.* the solid, outer layer of the earth, consisting of the crust and upper mantle. Also, GEOSPHERE.

lithostatic pressure see GROUND PRESSURE.

Lithostrotion *Paleontology.* a genus of colonial, generally cerioid rugose corals belonging to the extinct superfamily Zaphrenticae and family Lithostrotionidae; extant in the Carboniferous.

lithostyle *Invertebrate Zoology.* a small sense organ found in certain medusoid organisms; used to perceive orientation.

lithotomy *Surgery.* the surgical removal of a calculus, especially from the bladder.

lithotomy position *Surgery.* a surgical position in which the patient lies on the back with the hips and knees flexed, and the thighs abducted and externally rotated.

lithotope *Geology.* **1.** a surface or area of uniform sedimentation. **2.** a rock, sediment, or body of sediments exhibiting a relatively uniform or persistent depositional environment.

lithotripsy *Surgery.* the crushing of a bladder stone using a lithotripter. Thus, **lithotriptic.**

lithotripter *Surgery.* an instrument designed to crush bladder stones, administering a high-energy shock wave that disintegrates the stone, which is then passed in the urine. Also, **lithotriptor.**

lithotrite *Surgery.* any of various instruments used for crushing a urinary calculus. Thus, **lithotrity.**

lithotroph *Biology.* an organism that synthesizes organic nutrients from carbon dioxide, nitrogen, and other inorganic materials; an autotroph.

lithotype *Geology.* a visually observable band in humic coal that is analyzed on the basis of its physical characteristics rather than its botanical origin.

lithuria *Medicine.* an excess of uric acid or its salts in the urine.

litmus *Chemistry.* **1.** a pigment that is derived from certain lichens (such as *Roccella tinctoria)* and that is widely used as a test of acidity or alkalinity; it turns blue in alkaline solutions and red in acid solutions. **2.** of or relating to this substance. (From an older phrase meaning literally "colored moss.")

litmus milk *Bacteriology.* an indicator medium to detect the bacterial production of acid or alkali, consisting of skim milk and a litmus color indicator solution.

litmus paper *Chemistry.* a strip of paper impregnated with litmus material, used as a chemical indicator.

litmus test *Chemistry.* the testing of a solution by immersing a strip of litmus paper in it to determine its level of acidity or alkalinity. *Science.* in popular use, any single test that definitively establishes the nature or quality of something.

Litopterna *Paleontology.* an order of South American hoofed, herbivorous mammals; widespread from the Paleocene to the Pliocene, with one family, the Macraucheniidae, surviving until the Middle Quaternary; probably descended from a primitive didolodont.

litorin *Endocrinology.* a bioactive peptide that has been isolated from frog skin and has homology with the bombesin family of peptides.

lit-par-lit *Geology.* describing a layered or foliated rock structure in which the layers have been intruded by many thin, roughly parallel sheets or tongues of igneous material.

litre see LITER.

litter *Biology.* multiple offspring produced at a single birth by a multiparous animal. *Forestry.* a spongy layer of twigs, leaves, bark, and organic debris covering the floor of a forest. Also, **litter fall.** *Agriculture.* **1.** to give birth to a litter. **2.** material such as straw or sawdust that is used as bedding for livestock. *Surgery.* a stretcher or portable bed used for the transportation of the sick or wounded.

Little Bear see URSA MINOR.

little brother *Meteorology.* a subsidiary tropical cyclone that sometimes follows a more severe disturbance.

little cherry disease *Plant Pathology.* a virus disease of sweet cherries that causes small, angular fruits which fail to mature.

Little Dog see CANIS MINOR.

Little Dipper *Astonomy.* a group of seven bright stars in the constellation Ursa Minor whose outline resembles a water dipper.

Little Fox see VULPECULA.

Little Horse see EQUULEUS.

Little Ice Age *Geology.* an interval of about 5500 years marked by extended climatic deterioration and the general expansion of mountain glaciers, especially in the Alps, Scandinavia, Iceland, and Alaska.

Little John *Ordnance.* a surface-to-surface nuclear artillery rocket system using a 318-mm rocket; officially designated **MGR-3A.**

little leaf *Plant Pathology.* any of various plant diseases that cause chlorotic, malformed, and underdeveloped leaves.

Little Lion see LEO MINOR.

little peach disease *Plant Pathology.* a fatal virus disease of peaches, characterized by yellow leaves and dwarfed fruit that is delayed in ripening.

Little Ruler see REGULUS.

Little's disease *Medicine.* a congenital spastic stiffness of the limbs, a form of cerebral spastic paralysis associated with disorders such as fetal anoxia, illness of the mother during pregnancy, and birth trauma; chacterized by muscular weakness, walking difficulties, and usually convulsions and mental deficiency. Also, SPASTIC DIPLEGIA. (Named for William John *Little*, 1810–1894, English physician.)

littoral [lit´ə rəl] *Ecology.* **1.** of or relating to the shore of a lake, sea, or other body of water. **2.** see LITTORAL ZONE.

littoral current *Oceanography.* a longshore current produced by waves striking the shore at an acute angle; it moves parallel to the shore, usually within the breaker zone.

littoral drift *Geology.* the materials, including gravel, sand, and shell fragments, transported along the shore by the action of waves and currents moving parallel to the coastline. Also, (LONG)SHORE DRIFT.

littoral sediments *Geology.* accumulations of littoral drift deposited along the shore.

littoral transport *Geology.* the process by which littoral drift is moved along the shore parallel to the coastline.

littoral zone *Ecology.* **1.** the area of shallow fresh water in which light penetrates to the bottom and nurtures rooted plants. **2.** the shoreline where a normal cycle of exposure and submersion by the tides occurs.

Littorina *Paleontology.* a still-extant genus of gastropods in the superfamily Littorinacea; they first appeared in the Eocene.

Littorinacea *Paleontology.* a superfamily of gastropods in the order Mesogastropoda; extant from the Eocene to the present.

Littorinidae *Invertebrate Zoology.* a family of marine gastropod mollusks with globular spiral shells, the periwinkles, in the order Pectinibranchia.

Littrow grating spectrograph *Spectroscopy.* a spectrograph in which a lens and a plane grating are used in conjunction to collimate and focus incident light.

Littrow mounting *Spectroscopy.* **1.** an arrangement of prism and mirror in a spectrometer by which monochromation is achieved, either by rotating the prism and mirror simultaneously or by rotating only the mirror. **2.** the arrangement of the components in a Littrow grating spectrograph. The refracted beam is reflected back through the prism.

Littrow prism *Optics.* a type of prism that can be used as either a telescope or a collimeter when combined with a lens.

lituate *Botany.* having a part that is forked or whose ends turn slightly outward.

Lituolacea *Invertebrate Zoology.* a superfamily of foraminiferan protozoans in the suborder Testularia.

lituus *Mathematics.* the plane curve whose equation in polar coordinates is $r^2q = a$, where a is constant. (A Latin name for a crook-shaped staff whose shape this curve resembles.)

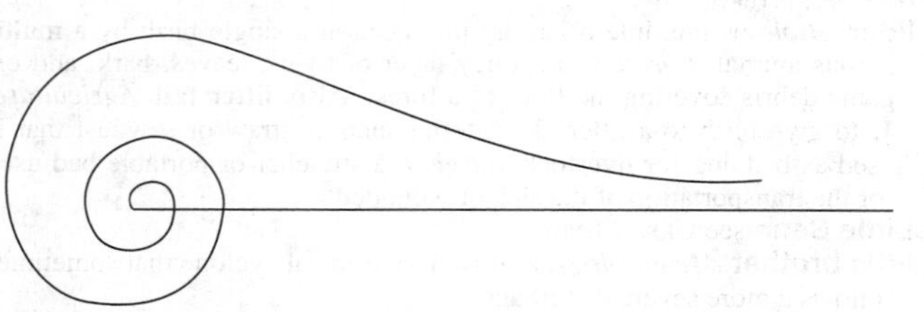

lituus

litz wire *Electricity.* a conductor made up of fine-stranded wire with each strand individually insulated; the strands are woven together in such a way that each strand successively takes all possible positions in the cross section of the conductor, minimizing skin effect and radio-frequency resistance.

live *Biology.* of an organism, having life; alive. *Science.* active or energized in some way that is comparable to organic life. *Electricity.* electrically connected to a voltage source. *Telecommunications.* describing broadcasts that are sent directly at the time of production instead of being recorded for future broadcast. *Acoustics.* describing an area that allows or produces significant sound reflection, such as a concert hall.

live ammunition *Ordnance.* ammunition containing an explosive or chemical charge as opposed to drill or dummy ammunition. Similarly, **live round.**

live attenuated virus *Virology.* a virus with low virulence used as a vaccine, such as the Sabin poliovirus vaccine.

live axle *Mechanical Engineering.* **1.** an axle that turns in bearings and has wheels rigidly fixed to it. **2.** an axle that is mounted on suspension springs together with its wheels, as opposed to independent suspension.

livebearers *Vertebrate Zoology.* any fish of the family Poeciliidae, which bear live young rather than laying eggs.

live chassis *Electronics.* in a television or radio receiver, a frame to which one side of the alternating-current line is connected.

live data *Computer Programming.* **1.** actual data that are employed in the final test of a computer program, rather than test data. **2.** data whose values will be used later in the program's execution.

live end *Acoustical Engineering.* the portion of a sound-recording area where a significant amount of sound reflection is allowed and controlled, as by reflectors that induce echo delay, the dissipation of sound by reflective diffusion, and exponential decay of sound.

live-end dead-end *Acoustical Engineering.* a sound control room that is selectively reflective or nonreflective to sound, with a dead (nonreverberant) end near the sound source being recorded, and a live end in which reverberations are allowed and controlled.

live load *Mechanics.* a structural load that moves or fluctuates in strength. Also, MOVING LOAD.

live oak *Botany.* an evergreen oak tree, *Quercus virginiana,* having shiny, oblong leaves and a short, broad trunk; native to and widely growing in the southern United States.

live oak with Spanish moss

liver *Anatomy.* the largest organ in the body, located in the abdominal cavity below the diaphragm and consisting of four lobes and up to 100,000 lobules; performs many vital metabolic processes, including the storage and filtration of blood, the secretion of bile, the excretion of bilirubin, the synthesis and degradation of amino acids, and the conversion of sugars into glycogen. *Materials Science.* a thick, rubbery mass that forms in paint, ink, or the like from a chemical reaction between a pigment and the vehicle or from the partial polymerization of the vehicle, as a result of which the material becomes unusable.

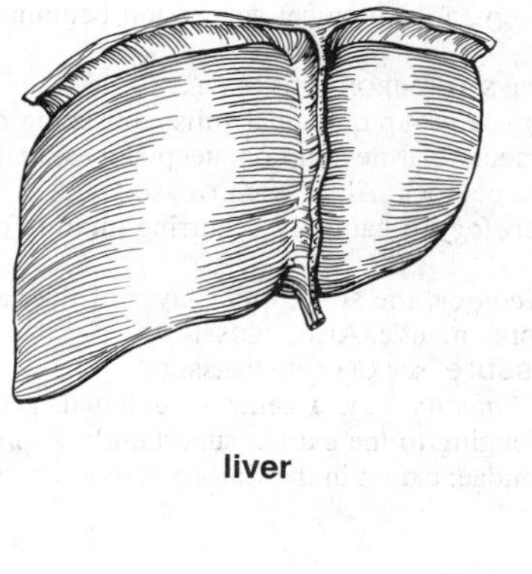

liver

liver failure *Medicine.* the inability of the liver to function effectively, evidenced by severe jaundice, impaired mental functioning, and coma.
liver fluke *Invertebrate Zoology.* a parasitic trematode flatworm that invades the liver of vertebrates.
livering *Materials Science.* the formation of a liver mass.
live-roller conveyor *Mechanical Engineering.* a conveying machine consisting of a series of rollers over which an object moves as power is transmitted to the rollers via a chain or belt.
live room *Acoustics.* a room that produces considerable reverberation of sound due to its configuration and the surface materials used there.
liver phosphorylase *Enzymology.* a liver enzyme that catalyzes the conversion of glycogen into glucose 1-phosphate.
liver rot *Veterinary Medicine.* a disease of cattle and sheep, caused by infection from the liver fluke and characterized by weight loss, sluggishness, and damage to parts of the liver.
liverwort *Botany.* any member of the class Hepaticae, mosslike plants that grow and aid decay on damp ground, rocks, or tree trunks.

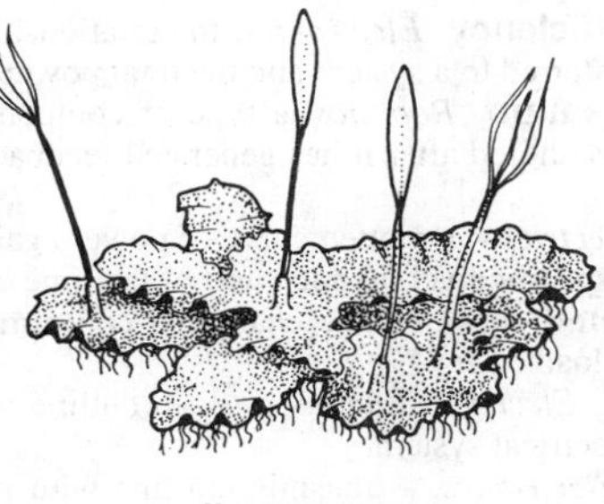

liverwort

liverwurst *Food Technology.* a soft, spreadable sausage made with liver, usually from pork. Also, **liver sausage.**
live spring see PERENNIAL SPRING.
live steam *Mechanical Engineering.* steam that is delivered at full operating pressure, directly from a boiler.
livestock *Agriculture.* a collective term for domestic animals that are produced or kept mostly for farm, market, or ranch purposes, including cattle, hogs, sheep, horses, and goats. Also, STOCK.
live stream see PERENNIAL STREAM.
live system see PRODUCTION SYSTEM.
live vaccine *Immunology.* an attenuated living bacterium or virus administered to induce an immune response, containing a pathogen that either causes zero or acceptable reactions, gives cross protection, or is the actual pathogen against which protection is wanted.
liveware *Computer Technology.* a term for the people involved in operating computers, as if they were another component of the systems, along with the hardware and software.
livid *Pathology.* discolored, especially black and blue, as from the effects of contusion or congestion. Thus, **lividity.**
living archaeology *Archaeology.* the study of the living patterns of existing hunter-gatherer and farming societies whose traditions extend from ancient predecessors, such as the Australian aborigines.
living floor *Archaeology.* any surface that indicates use as a house or camp area, as evidenced by signs of cooking, sleeping, or working at household tasks.
living fossil *Biology.* a species that has persisted to the present time with little or no change over a long period of geological time.
Livingstone sphere *Engineering.* a spherical clay atmometer, used to simulate evaporation from plant growth.
livingstonite *Mineralogy.* $HgSb_4S_8$, a lead-gray, metallic, monoclinic mineral occurring as tiny, elongated needles and in fibrous masses, having a specific gravity of 4.8 to 5 and a hardness of 2 on the Mohs scale; found in low-temperature hydrothermal veins with stibnite and gypsum.
livor mortis *Pathology.* **1.** the discoloration of skin at the point of death as a result of the discontinuance in circulation. **2.** the inaction and coagulation of the blood due to gravity.
livre *Metrology.* a French unit of mass, equal to 0.5 kilogram.
lixiviate *Chemical Engineering.* to separate a soluble substance from a solid by washing or by the percolation of water or other liquid through the substance. *Geochemistry.* see LEACHING. Thus, **lixiviation.**
lixivium *Chemical Engineering.* any alkaline substance obtained by leaching ashes or other powdered substances.

lizard

lizard *Vertebrate Zoology.* any of numerous species of reptiles of the suborder Sauria in the order Squamata, characterized by a fused lower jaw, external ears, eyes with movable lids, and usually four legs and a tapering tail; commonly found in deserts and other dry regions.
lizard-hipped dinosaur *Paleontology.* a general term referring to the dinosaurian order Saurischia.
Ljungstrauom heater [yŭng´sträm´] *Mechanical Engineering.* a continuous, regenerative heat-transfer air heater with slow-moving rotors to confine hot or cold gases to opposite sides of the rotor housing.
Ljungstrauom steam turbine *Mechanical Engineering.* a radial outward-flow turbine defined by two opposed rotation rotors.
Lk *Molecular Biology.* the linking number in DNA; refers to the number of times that one strand wraps around the other in any dscccDNA.
LK virus *Virology.* a type of equine herpesvirus.
llama [lä´mə] *Vertebrate Zoology.* **1.** a domesticated, herbivorous member of the family Camelidae found in the Andes, resembling a camel without a hump and having long, shaggy fur used as wool. **2.** a general term referring to the llama, guanaco, alpaca, and vicuna.

llama

Llandeilian *Geology.* a European geologic stage of the Middle Ordovician period, occurring after the Llanvirnian and before the Caradocian.
Llandoverian *Geology.* a European geologic stage of the Lower Silurian period, occurring after the Ashgillian of the Ordovician period and before the Tarannon. Also, VALENTIAN.
Llanvirnian *Geology.* a European geologic stage of the Middle Ordovician period, occurring after the Arenigian and before the Llandeilian.
llebetjado *Meteorology.* a hot squally wind that descends from the Pyrenees and lasts for a few hours, blowing over northeastern Spain.
LLL or **L.L.L** left lower lobe.
LLL circuit low-level-logic circuit.
Lloyd Morgan's canon *Behavior.* the principle that behavior should be interpreted as the outcome of the least complex function that is applicable to the situation, rather than one that is higher on the cognitive scale, such as insight or reason. (Proposed by Conway *Lloyd Morgan*, 1852–1936, an early comparative psychologist.)

Lloyd's mirror interference *Optics.* the pattern produced by the interference of two beams of light emanating from the same source, one which reflects from a mirror to an observation screen, the other which travels directly to the screen.

LLQ or **L.L.Q.** left lower quadrant.

LLTV low light television.

lm lumen.

LM see LUNAR MODULE.

L.M. light minimum; linguomesial.

L/M *Nuclear Physics.* the ratio of the number of internal conversion electrons emitted from the *L* shell to the number emitted from the *M* shell during the deexcitation of the nucleus.

L.M.A. left mentoanterior.

L meson *Particle Physics.* a K meson resonance with approximate mass 1781 MeV.

LMF lymphocyte mitogenic factor.

LMFBR liquid metal cooled fast breeder reactor.

LMG light machine gun.

lm-hr lumen-hour.

L.M.P. left mentoposterior.

LMP last menstrual period.

LMR liquid metal reactor.

lm-sec lumen-second.

LMT local mean time.

L.M.T. left mentotransverse.

lm/w lumen-per-watt.

L-MYC *Oncology.* a proto-oncogene that is amplified in carcinoma of the lung.

ln natural logarithm *(logarithm natural.)*

Ln lanthanide.

L network *Electronics.* a four-terminal circuit composed of two branches, one branch connected between an input terminal and an output terminal, and the other branch connected either between the two input terminals or between the two output terminals. (The circuit diagram resembles an inverted letter L).

LNG liquefied natural gas. Also, **LN gas.**

LNG ship *Naval Architecture.* a ship designed to transport liquefied natural gas.

LNPF lymph node permeability factor.

LO local oscillator.

L.O.A. left occipitoanterior.

loach *Vertebrate Zoology.* any of various European or Asian nocturnal bottom fish of the family Cobitidae, having a wormlike to flattened body.

load a thing that is put in place or carried; to put such a thing on or in something; specific uses include: *Mechanics.* the imposed force of pressure, weight, and so on that is supported by a structure or body. *Engineering.* **1.** to put materials into place for use or transport, such as fuel in a fuel tank, explosives in a bore hole, or cargo into a vehicle. **2.** the sum of goods so placed in a given operation. *Ordnance.* the charge or explosive that is placed in a weapon for firing. *Industrial Engineering.* the amount of work scheduled for a manufacturing facility, expressed in work hours or units of production. *Electrical Engineering.* the amount of electric power that is used by a machine or circuit as it performs its function. *Electricity.* any item that consumes electricity, such as a household appliance, street lighting, or an industrial electrical motor. *Computer Programming.* to insert a diskette into a disk drive or mount a magnetic tape on a tape drive. *Telecommunications.* the device that accepts the useful signal output from a signal source, such as an amplifier or oscillator.

load-and-carry equipment *Mechanical Engineering.* earthmoving equipment that is used to load and transport materials.

load-and-go *Computer Programming.* an operational method with no stops between the load and execution phases of a program run, sometimes including an assembly or compile step.

load average *Computer Programming.* an average of the utilization or backlog of a resource over time, e.g., the average number of jobs waiting for CPU service.

load balancing *Computer Programming.* the process of assigning work to computers in a network or to processors in a parallel computer in order to maintain an even distribution of the load on the system.

load-bearing *Mechanical Engineering.* describing a component of a structure that bears some or all of its weight and applied loads.

load-break switch *Electricity.* a cutout that provides a means for interrupting load currents.

load-carrying capacity *Mechanical Engineering.* **1.** the maximum load at which a mechanism can perform satisfactorily. **2.** the maximum pressure that a lubricating oil can withstand.

load cast *Geology.* a bulbous swelling or projection of coarse clastic material that extends downward from the base of a stratum into a finer-grained, softer, underlying material, such as mud or clay. Also, **load casting.**

load-casted ripple see RIPPLE LOAD CAST.

load cell *Electricity.* a test instrument that measures pressure by applying the unmeasured pressure to a piezoelectric crystal and determining the voltage across the crystal.

load characteristic *Electrical Engineering.* the reaction between the instantaneous values of the voltage and the current, between a pair of terminals in a device. Also, DYNAMIC CHARACTERISTIC, OPERATING CHARACTERISTIC.

load chart *Industrial Engineering.* a control chart used for work assignment and scheduling.

load circuit *Electronics.* **1.** a circuit that constitutes the power-consuming section of a system. **2.** a circuit that helps transfer power to another system.

load circuit efficiency *Electronics.* the relationship between the useful power transferred to a system and the total power transferred.

load compensation *Robotics.* a type of compensation in which an output signal is altered after it has generated feedback. Also, LOAD STABILIZATION.

load curve *Electricity.* a plotted curve of power against time that shows the value of a specific load for each unit of the time covered.

load deflection *Mechanics.* the change in conformation of a body resulting from a load on it.

load diagram *Electrical Engineering.* an outline or plan showing the layout of an electrical system.

loaded line *Electronics.* a transmission line with inductors and/or capacitors added at regular intervals to improve its transmission characteristics over a wider frequency band.

loaded Q *Electricity.* the value of *Q* of a loaded line when coupled under working conditions. Also, WORKING *Q*.

loaded wheel *Engineering.* a grinding wheel whose pores are clogged with the material being ground.

loader something or someone that loads; specific uses include: *Mechanical Engineering.* a mechanical shovel, vehicle, or other machine that conveys bulk materials such as coal, ore, or rock. *Ordnance.* a mechanical device that loads cartridges into a gun. *Computer Programming.* a system program that copies other programs into main memory from auxiliary storage devices.

load factor *Electricity.* the ratio of the average load over a time period to the peak load for that period. *Mechanical Engineering.* the ratio of the average operating load of an engine or other machine to its maximum rated load, during a specified time period. *Transportation Engineering.* the average number of vehicles on a given road or passengers using a given line for a certain time period.

load impedance *Electronics.* the impedance of the circuit connected to a source or to the output of an electrical transducer.

loading the act of someone or something that loads; specific uses include: *Engineering.* **1.** the process of putting materials in place for use or for transportation to another site. **2.** the buildup of a material, for example on a cutting tool. *Ordnance.* the act of placing a charge or explosive in a weapon for firing. *Electricity.* the insertion of reactance (a loading coil) in a circuit to improve its transmission characteristics in a given frequency band. *Acoustical Engineering.* the process of placing material at the front or back of a loudspeaker in order to change its acoustic characteristics. *Chemical Engineering.* a condition that occurs in distillation or absorption processes, in which the vapor-liquid capacity of packed towers or bubble-plate columns is limited because the tower becomes loaded with water due to a high vapor flow rate. *Nucleonics.* the act of placing fuel in a nuclear reactor. *Metallurgy.* **1.** In powder metallurgy, filling of a die cavity with powder. **2.** in cutting, the process of accumulating undesirable material on the tool. *Computer Programming.* the act of placing a diskette into a disk drive or mounting a magnetic tape on a tape drive.

loading algorithms *Design Engineering.* a mathematical expression used to determine routing of material to processes.

loading angle *Ordnance.* the angle of elevation required to correctly load ammunition into a particular weapon.

loading coil *Electromagnetism.* a coil that is inserted into a circuit so as to lessen the effect of line capacitance and reduce distortion.

loading density *Engineering*. the quantity of explosives in a drill hole, usually expressed in pounds per linear foot. *Ordnance*. **1.** in artillery projectiles, bombs, warheads, and similar weapons, the quantity of explosive per unit volume, usually expressed as grams per cubic centimeter. **2.** in small arms ammunition, the weight of powder in a cartridge case divided by the maximum weight of powder the case could hold if filled to the base of a seated bullet, expressed as a percent.

loading disk *Electromagnetism*. a conductive disk that is mounted on top of a vertical antenna so as to increase the antenna's electrical length.

loading gate *Ordnance*. a hinged cover on the magazine or breach of a weapon; it is closed except for loading or unloading.

loading head *Mechanical Engineering*. the part of a loader that gathers the bulk elemental materials.

loading pan *Mining Engineering*. a box or scoop into which broken rock is shoveled.

loading routine see INPUT ROUTINE.

loading station *Mechanical Engineering*. a device that receives material and places it upon a conveyor.

loading tray *Engineering*. a tray with a sliding bottom used to fill multicavity molds simultaneously with the plastic charge. *Ordnance*. **1.** a trough-shaped carrier used to facilitate the loading of heavy projectiles into the breech of a gun. **2.** a hollow slide designed to guide ammunition into the breech of some automatic weapons.

loading weight *Engineering*. a powder's weight when placed in a container.

load isolator *Electromagnetism*. a coaxial or waveguide device that effectively passes electromagnetic energy in one direction (toward the load) but attenuates it in the other direction.

load leveling *Transportation Engineering*. the process of distributing weight equally in various parts of a freight vehicle. *Electricity*. a means of reducing the large fluctuations that occur in electricity demand, as by storing surplus electricity for use during periods of high demand.

load line *Electronics*. **1.** a type of connection in which inductors or capacitors are added at given points to change the characteristics of a transmission line. **2.** a diagram in which a straight line is drawn through a series of tube or transistor characteristic curves demonstrating how an output signal current will be altered by an input signal voltage when a given load resistance is used. *Naval Architecture*. any of a set of water lines marked on the side of a ship to indicate the maximum depth of loading under a given condition. Load lines differ according to salt or fresh water, time of year, and other safety-related factors.

load loss *Electricity*. a power loss in a transformer that is incident to the carrying of a specific load.

load metamorphism see STATIC METAMORPHISM.

load module *Computer Programming*. the output from the linkage editor; a program that is in a format suitable for loading into main memory for execution.

load oil *Petroleum Engineering*. oil that is reintroduced into a well for such purposes as hydraulic fracturing, shooting, or swabbing; used to stimulate production.

load point *Computer Technology*. the starting point of the recording area on a volume of magnetic tape.

load power *Electricity*. the total power supplied by a controller to a load.

load regulation *Electricity*. the maximum amount that an output voltage will change with changes made in load conditions, given as either a percentage of the voltage or an absolute value.

load shedding *Electricity*. the process of deliberately removing loads from a power source when abnormal conditions threaten the integrity of the system.

load stabilization see LOAD COMPENSATION.

loadstone see LODESTONE.

load stress *Mechanics*. stress on the surface in contact with a load; a pressure that is used to predict the wear expected.

load time *Computer Programming*. **1.** the amount of time that is required for link editing or loading a program into memory for execution. **2. load-time.** referring to something that happens during link editing or loading, such as **load-time error.**

load transfer *Materials Science*. the transfer of the stress caused by a load applied to a fibrous composite to the fibers contained in the matrix.

load water *Petroleum Engineering*. water injected into an oil well to prime it following acidizing operations.

load waterline *Naval Architecture*. the waterline at which a vessel floats in its full load condition. This varies according to several safety-related factors.

load water plane *Naval Architecture*. the water plane of a vessel carrying its full cargo.

loafing parlor *Agriculture*. an area in a dairy barn or yard in which cows can roam freely after they have been milked or fed within a confined space.

Loa loa infection see LOIASIS.

loam *Geology*. a rich soil, composed of a mixture of sand, silt, and clay in more or less equal proportions, and usually containing humus.

loaming *Geochemistry*. the process of testing samples of surface material for traces of a desired metal in order to locate near-surface ore bodies.

Loasaceae *Botany*. a family of dicotyledonous New World herbs and sometimes shrubs or small trees of the order Violales, characterized by coarse, sometimes stinging hairs, usually lobed leaves, and a capsular fruit with abundant oily endosperm.

LOB line of balance.

Lobachevsky or **Lobachevski, Nikolai** 1792–1856, Russian mathematician; independently developed the first systematic non-Euclidean geometry.

Lobachevsky geometry see HYPERBOLIC GEOMETRY.

lobar *Physiology*. of, relating to, or affecting a lobe, especially of the lungs.

Lobaria *Botany*. the lungworts, a genus of foliaceous lichens of the family Stictaceae.

lobar pneumonia *Pathology*. an acute infectious disease, usually caused by pneumococcal bacteria, that is characterized by fever, pleuritic pains, chills, cough, blood-stained sputum, and difficult or labored breathing, accompanied by inflammation and consolidation of one or more lobes of the lungs.

lobar sclerosis *Medicine*. the hardening of tissues of a portion or all of a lobe of the brain, due to proliferation of connective tissue generally caused by chronic inflammation.

Lobata *Invertebrate Zoology*. an order of comb jellies with a helmet-shaped body, in the class Tentaculata.

lobate *Biology*. divided into or having lobes.

lobate rill mark *Geology*. a flute cast that is formed by the action of water currents.

lobe *Biology*. **1.** a usually rounded division or segment, such as a part of a leaf, petal, or calyx cut less than halfway to the center. **2.** any blunt prominence arising from a surface. *Anatomy*. a relatively well-defined portion of an organ, especially of the brain, lungs, and glands; lobes are demarcated by their shape and by fissures, sulci, and connective tissue. *Design Engineering*. a rounded projection, as on a cam wheel or other noncircular geared wheel. *Electromagnetism*. a region that contains a significant amount of radiation emitted by an antenna. *Hydrology*. **1.** see GLACIAL LOBE. **2.** a long, rounded recess of a lake.

lobectomy *Surgery*. the surgical removal of a lobe of an organ or gland.

lobed impeller meter *Engineering*. a displacement meter that divides a fluid stream into discrete units by rotating, meshing impellers pushed by interlocking gears.

lobefin fish *Vertebrate Zoology*. any member of the mostly extinct subclass Sarcopterygii, including lungfish, bichirs, and the coelacanth, whose paired fins have thick fleshy bases suggesting limbs.

lobefin fish

lobe-half-power width *Electromagnetism*. an angular measurement that is subtended by the half-power values of the major lobe of a radiating antenna, measured in the plane that contains the major lobe.

lobelia *Botany*. any of various herbaceous plants of the genus *Lobelia*, family Lobeliaceae, having long clusters of red, white, blue, or yellow flowers, such as the **great blue lobelia,** *L. siphilitica,* the **edging lobelia,** *L. erinus,* or the **cardinal flower,** *L. cardinalis*. *Pharmacology*. the dried leaves and tops of the Indian tobacco, *L. inflata,* whose properties are similar to those of nictotine. (Named for Matthias da *Lobel*, 1538–1616, Flemish botanist and physician.)

Lobeliaceae *Botany.* a family of usually herbaceous plants characterized by simple alternate leaves and irregular two-lipped flowers.
lobeline *Pharmacology.* $C_{22}H_{27}NO_2$, a drug prepared from the herb Indian tobacco, *Lobelia inflata*; used in certain antismoking preparations and formerly used to restore normal breathing in treatment of asthma, shock, and whooping cough.
lobe penetration *Electromagnetism.* the amount of penetration of a radar system that is otherwise not limited by the system's characteristics such as the sweeping angle, or angle of elevation.
lobe switching see BEAM-LOBE SWITCHING.
lobing *Electromagnetism.* the effect of constructive and destructive interference of antenna radiation due to reflections from the ground, typically resulting in the formation of energy lobes in a vertical plane.
loblolly pine *Botany.* an evergreen conifer, *Pinus taeda,* of the family Pinaceae, characterized by its dark gray bark and stout, often twisted needles; native to the southeastern United States.

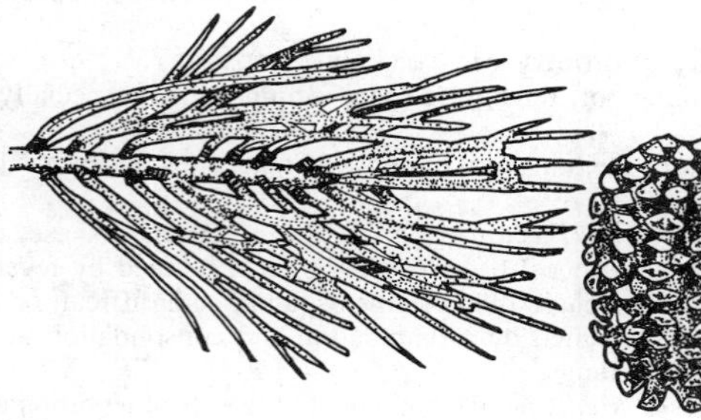

loblolly pine

Loboa *Mycology.* a genus of fungi perhaps belonging to the order Entomophthorales; the species *L. loboi* causes the skin disease lobomycosis.
lobomycosis *Medicine.* a chronic localized skin mycosis of fibrous nodules or keloids caused by the fungus *Loboa loboi*.
lobopodia *Invertebrate Zoology.* thick, wide, tonguelike or fingerlike pseudopodia with a rounded end.
Lobosa *Invertebrate Zoology.* a subclass of protozoans with thick, wide pseudopodia, including the Amoebas.
Lobotidae *Vertebrate Zoology.* the tripletails, a family of perchlike marine and freshwater fish of the order Perciformes, having rounded dorsal and anal fins that together with the caudal fin give the appearance of a three-lobed fin; found in most warm seas.
lobotomy [lə bät′ə mē] *Surgery.* **1.** in psychosurgery, the surgical incision of all the fibers of a lobe of the brain. **2.** any surgical incision into a lobe.
Lobry de Bruyn-Ekenstein transformation *Organic Chemistry.* the isomerization of aldose sugars in a dilute alkaline medium; results in an epimeric pair and 2-ketohexose.
lobster *Invertebrate Zoology.* any of various mostly rock-dwelling decapod crustaceans in the family Homaridae, especially those of the genera *Homarus* or *Panuliris*; commercially important as food.
lobule *Biology.* **1.** a small lobe. **2.** a primary division of a lobe. Thus, **lobular.**
LOC loss of consciousness.
LOCA loss-of-coolant accident.
local *Science.* restricted to or relating to one area or part only; not general. *Medicine.* affecting or located in just one specific area of the body. *Transportation Engineering.* describing rail or subway service that makes all or most station stops. *Mathematics.* a property (such as connectedness, flatness, etc.) is said to be local or to hold locally on a topological space X if every point x of X has a neighborhood which has the property under consideration. Sometimes it is required that the neighborhood be homeomorphic to a space that has the given property.
local accuracy *Robotics.* accuracy relative to a specific point within the working envelope of a robot.
local action *Electricity.* an effect caused by impurities in an electrode that can result in potential differences between different parts of the electrode.
local anesthetic *Pharmacology.* any drug that prevents a feeling of pain in the part of the body to which it is administered.
local angular momentum *Meteorology.* angular momentum about an arbitrarily located vertical axis that is fixed with respect to the earth.
local apparent noon *Navigation.* the moment when the sun is on the upper branch of the local meridian.
local apparent time *Navigation.* time as determined by the (real) sun using the local meridian as the reference point.
local area network *Computer Science.* a network covering a relatively limited geographic area, such as one or more buildings at one location. Also, LAN.
local arm see ORION ARM.
local attraction *Navigation.* a local anomaly in the magnetic field, usually caused by a large mass of iron or iron ore.
local base level see TEMPORARY BASE LEVEL.
local battery *Electricity.* a battery that supplies power for telegraphic station recording instruments.
local-battery telephone set *Electronics.* a type of telephone set that draws its transmitter current directly from a battery or other current supply circuit, its signaling current coming from either a local hand generator or a centralized power source.
local buckling *Mechanics.* **1.** buckling that is limited to parts of the cross section of a member. **2.** a failure of one component of a structure. Also, CRIPPLING.
local cable *Telecommunications.* cable that is manufactured for the termination of circuits at locations where wiring is actually run inside each unit.
local causality see DETERMINISTIC.
local cell *Electricity.* a galvanic cell resulting from differences in potential between neighboring areas on the surface of a metal immersed in an electrolyte.
local change *Oceanography.* the time rate of change of a scalar quantity (such as salinity, temperature, or pressure) at a given place.
local circuit *Telecommunications.* a circuit that provides access to a main or auxiliary circuit from any station.
Local Cluster see LOCAL GROUP.
local coefficient of heat transfer see LOCAL HEAT TRANSFER COEFFICIENT.
local control *Telecommunications.* a type of radio transmission control in which the control functions are performed directly at the transmitter location.
local controller see FIRST-LEVEL CONTROLLER.
local coordinate system *Mathematics.* let (h,U) be a chart on a topological space (or manifold) T. The **local coordinates** of a point $x \in U$ are the components (in R^n) of $h(x)$, and the chart itself is called a local coordinate system. For example, suppose that a space X is locally Euclidean of dimension d and ϕ is the required homeomorphism (called a **coordinate map**) of some connected open subset U of X onto an open subset of R^d. If $r_1, \ldots, r_d$ are coordinate functions on R^d, then the set of coordinate functions $x_i = r_i \cdot \phi$ together with the open set U are the local coordinate system of X. If x is a point of U such that $\phi(x) = 0$, then the induced local coordinate system is said to be centered on x.
local datum *Cartography.* the point of reference of the geodetic control that is used exclusively in a small area; usually identified by a proper name.
local derivative *Fluid Mechanics.* the partial derivative of a function with respect to time, $\partial f/\partial t$, measured at a fixed point in a fluid; here, f is a thermodynamic property of the fluid.
local device *Computer Technology.* equipment that is interfaced directly with a computer in a single location.
local distortion *Mathematics.* for a function f analytic at a point z_0, the value of $|f'(z_0)|$.
local enhancement *Behavior.* a process by which an animal is able to locate some object or place more readily if it observes others reacting to this same object or place.
local extra observation *Meteorology.* an aviation weather observation taken at specified intervals, typically every 15 minutes, including sky condition, visibility, ceiling, atmospheric phenomena, and any related remarks; taken when local meteorological conditions are below aircraft operational weather limits.
local extremum *Mathematics.* a local maximum or local minimum. Also, RELATIVE EXTREMUM.
local first selector *Telecommunications.* in telephony, a switching stage of a step-by-step switching system that sends a dial tone back to a caller.
local flap *Surgery.* a surgical flap formed from tissue neighboring the defect to be corrected.
local forecast *Meteorology.* a weather forecast highlighting conditions over a city, airport, or other limited area.

Local Group *Astronomy.* a group of about four dozen galaxies (mostly dwarf ellipticals) to which the Milky Way belongs; other members are the Large and Small Magellanic Clouds and the Great Galaxy in Andromeda (M 31).

local heat transfer coefficient *Thermodynamics.* the heat transfer coefficient applied to an infinitesimal area at a particular point on a surface; the value is given by the infinitesimal amount of heat transfer dQ (transferred by fluid flow), divided by the product of the infinitesimal area dA and the temperature difference between the fluid and the surface. Also, LOCAL COEFFICIENT OF HEAT TRANSFER.

local horizon *Astronomy.* **1.** the apparent or visible horizon. **2.** the lower boundary of the observed sky or the upper outline of terrestrial objects, including nearby obstructions or irregularities.

local hour angle *Astronomy.* the hour angle of the vernal equinox for an observer at a given place and time. *Navigation.* angular distance west of the local meridian, measured through 360°.

local immunity *Immunology.* a resistance to infection or disease that develops in a particular area of the body.

local infection *Virology.* an infection in an isolated area involving only a few cells of the host.

local intelligence *Computer Technology.* the storage capability and processing power built into a terminal so that it need not be connected to a computer in order to perform certain tasks.

local invariance *Physics.* a property of a physical law that remains unchanged under a set of symmetry transformations, even when the transformations are chosen independently at every point of space and time.

locality *Science.* the specific place or point at which something is located. *Physics.* the specification of a point or body in space or time by use of coordinates.

locality of reference *Computer Programming.* the tendency of a program to reference a subset of its pages in a given time interval; used as a basis for memory management in paged memory computer systems.

localization *Medicine.* **1.** the determination of the site or place of any process or lesion. **2.** the restriction of a condition, such as an infection, to a circumscribed or limited area. *Computer Programming.* the process of enforcing a physical structure upon a set of elements so that a given element has a higher probability of being in a particular part of the structure than in others. *Mathematics.* let $S = R - P$ be the set of all elements of a commutative ring R with identity, which are not in some prime ideal P of R. Then the ring of quotients $S^{-1}R$ (i.e., all elements of the form $s^{-1}r$, where $s \in S$ and $r \in R$) is called the localization of R at P. Thus, **localized.**

localize *Science.* to carry out a process of localization; make something local.

localized state *Quantum Mechanics.* an electron state whose uncertainty in space is less than the extent of the material in which the electron resides.

localized vector *Mechanics.* a vector which is not free, having either or both of its line and point of application specified.

localizer *Radiology.* **1.** an instrument that is used to detect and fix the location of a metallic foreign body in the eyeball by roentgenography. **2.** see COLLIMATOR. *Navigation.* a directional radio signal that indicates the proper approach path to a runway. When an approaching aircraft wanders off the centerline of the runway, it displays a "fly left" or "fly right" indication.

local lesion *Virology.* the death of, or changes in, the cells around the original entry point of the infecting virus.

local level *Navigation.* the level or horizontal plane at any given geographical location.

local line *Telecommunications.* a telephone line that has the local central office as its termination point.

local loop *Telecommunications.* a telephone pair that ends at the local central office. Also, HOME LOOP.

local lunar time *Astronomy.* the hour angle of the Moon, increased by 12 hours.

locally arcwise connected *Mathematics.* a topological space X is said to be locally arcwise connected if for each point $x \in X$ and each neighborhood V_x of x, there exists an arcwise connected neighborhood U_x of x that is contained in V_x. If X is locally arcwise connected, then it is locally connected; the converse is not necessarily true. X can be locally arcwise connected without being arcwise connected, and vice versa.

locally compact *Mathematics.* a topological space X is said to be locally compact if every point of X has neighborhood whose closure is compact. Locally compact does not imply compact; for example, Euclidean spaces are locally compact but not compact.

locally connected *Mathematics.* a topological space X is said to be locally connected if any neighborhood of any point $x \in X$ contains a connected neighborhood of x. If X is locally connected, it is not necessarily locally arcwise connected, and X may be connected without being locally connected. X is **locally simply connected** if every point $x \in X$ has at least one simply connected neighborhood.

locally convex *Mathematics.* a topological vector space is said to be locally convex if it has a topological base consisting of convex sets. Locally convex does not imply convex.

locally Euclidean *Mathematics.* a Hausdorff space X is said to be locally Euclidean of dimension d if each point of X has an open neighborhood that is homeomorphic to an open subset of Euclidean space R^d.

locally finite *Mathematics.* a covering U of a topological space X is said to be locally finite if for every point $x \in X$ there exists a neighborhood N_x of x which has a (nonempty) intersection with only a finite number of members of U.

locally one to one *Mathematics.* a function f on a space X is said to be locally one to one if every point $x \in X$ has a neighborhood on which f is one to one.

local Mach number *Aviation.* the Mach number of a specific section of a moving body, as distinguished from the remote Mach number of the whole body.

local magnetic anomaly *Geophysics.* a region where the magnetic field of the earth differs from the average expected for that location due to some local disturbance. Also, **local magnetic disturbance.**

local maximum *Mathematics.* a real-valued function f on an open subset U of a topological space X is said to have a local maximum at $a \in U$ if there exists a neighborhood N_a of a, contained in U, such that $f(x) \leq f(a)$ for all $x \in N_a$. f is said to have a **strict local maximum** at a if there exists a neighborhood N_a of a, contained in U, such that $f(x) < f(a)$ for all $x \in N_a$, $x \neq a$. Also, RELATIVE MAXIMUM.

local mean time *Navigation.* mean time that uses the local meridian as its reference point.

local meridian *Astronomy.* the great circle that passes through the center of the earth, the celestial poles, and an observer. *Navigation.* the meridian that passes through the observer or some other particular place on the earth's surface, serving as the reference for local time.

local minimum *Mathematics.* a real-valued function f on an open subset U of a topological space X is said to have a local minimum at $a \in U$ if there exists a neighborhood N_a of a, contained in U, such that $f(x) \geq f(a)$ for all $x \in N_a$. f is said to have a **strict local minimum** at a if there exists a neighborhood N_a of a, contained in U, such that $f(x) > f(a)$ for all $x \in N_a$, $x \neq a$. Also, RELATIVE MINIMUM.

local networking *Robotics.* the system of communication that links together the components of a single robot.

local oscillator *Electronics.* **1.** a circuit that generates alternating current for the equipment directly connected to it, as opposed to one connected to a distant transmitter. **2.** a device whose output of alternating current is combined with the incoming radio signal before the intelligence it is carrying is removed; used in superhetrodyne receivers. Also, BEAT OSCILLATOR.

local oscillator radiation *Electronics.* the radiation emitted by the fundamental and harmonic frequencies of a signal produced in a local oscillator.

local peat *Geology.* a type of peat that forms within a groundwater environment. Also, AZONAL PEAT, BASIN PEAT.

local preheating *Metallurgy.* in metal fabrication or joining, the heating of selected portions of the part.

local register see GENERAL REGISTER.

local relief *Geology.* a measure of the vertical difference in elevation between the highest and lowest points of a land surface in a designated area. Also, RELATIVE RELIEF.

local ring *Mathematics.* a commutative ring R with identity satisfying one of the following equivalent conditions: (a) R has a unique maximal ideal. (b) All nonunit elements of R are contained in some proper ideal M of R. (c) The nonunits of R form an ideal. For example, if p is prime and $n \geq 1$, then the integers modulo p^n form a local ring, with the unique maximal ideal being the subring generated by p.

local second selector *Telecommunications.* in a step-by-step switching system, a switching stage that connects a local first selector to a connector switch.

local service *Transportation Engineering.* a transport service that stops at every station or stop on a route.

local sidereal noon *Astronomy.* the moment when 0 hours right ascension crosses a given meridian.

local sidereal time *Astronomy.* the right ascension of the point in the sky that is crossing the observer's meridian at any given moment. *Navigation.* the location of the vernal equinox west of the local meridian, expressed in time through 24 hours.

local solidification time *Materials Science.* the time required for a particular location in a casting to solidify.

local solution *Mathematics.* a function which satisfies a system of functional equations only in a neighborhood of some fixed point.

local standard of rest *Astrophysics.* a frame of reference centered on the sun in which the space velocities of all stars within about 150 light-years average out to zero.

local storage *Computer Technology.* the small set of storage registers that provide high-speed addressable program or data storage.

local storm *Meteorology.* any weather phenomenon of mesometeorological scale, such as a thunderstorm, squall, or tornado.

local structural discontinuity *Mechanics.* the effect of a void, inclusion, dislocation, etc. on stress flow.

Local Supercluster *Astrophysics.* an immense collection of clusters of galaxies, including the Local Group and the Virgo-Coma cluster, that extends for perhaps 100 million light-years.

local thermodynamic equilibrium *Astrophysics.* an assumption that the matter at a given point in a star is in balance with the radiation flowing through the point.

local traffic *Transportation Engineering.* vehicle traffic that is involved in short trips within a local area, as opposed to traffic making longer trips along arterials. Traffic flow is frequently improved by segregating local traffic from through traffic. *Aviation.* aircraft that operate entirely within sight of one airport's control tower, as in touch-and-go landing practice.

local transformation *Mathematics.* given a vector field V on a manifold, a local transformation of a neighborhood U maps each point of U to a point that lies a fixed value along the integral curves of V.

local trunk *Telecommunications.* a trunk between private exchange and local switchboards or switching systems, or a trunk between long-distance and local switchboards or switching systems.

local variable *Computer Programming.* a variable that can be referenced in only the program block in which it is defined.

local vertical *Geodesy.* at any point on the earth's surface, the direction of the acceleration of gravity, which, because of the deflection of the vertical, may or may not conform to a straight line perpendicular to the reference surface at the same point. *Navigation.* the vertical line, from the center of the earth to the zenith, at any given geographical location.

local war *Military Science.* see LIMITED WAR.

local wind *Meteorology.* any wind differing from winds that are appropriate to the general pressure distribution in an area, usually developing as a result of local thermal or topographic effects on an air mass.

locant *Organic Chemistry.* the number or letter immediately preceding a chemical symbol, as in 2H, indicating the position of an atom or group in a molecule.

locate mode *Computer Programming.* a method of providing access to data for input/output control routines by transferring the data addresses rather than the data themselves.

locating *Mechanical Engineering.* a term for the process of bringing together the appropriate contact points between a workpiece and a tooling device.

locating back *Photogrammetry.* a flat surface in an aerial camera against which the film is held, either by means of a vacuum or air pressure, so that the emulsion lies exactly in the focal plane. A locating back that uses a vacuum is a vacuum back, and one that uses air pressure is a pressure back.

locating hole *Mechanical Engineering.* a hole used for accurately positioning a part in relation to a cutting tool or to other parts.

locating surface *Design Engineering.* an area on a part used for setting alignment with other parts. *Mechanical Engineering.* specifically, a surface used by automated manufacturing and assembly machines for indexing or positioning parts.

location *Computer Programming.* a term for any addressable place in memory where an instruction or data item is stored.

location constancy *Psychology.* the autonomic modification by the perceptual system of objects and their distance, depending on the individual's location or the change in the position of the object.

location constant *Computer Programming.* an explicit value that identifies an instruction in a computer program; used to reference the instruction from other parts of the program, such as in a branch instruction. Also, LABEL CONSTANT.

location counter *Computer Programming.* any counter maintained by a compiler or assembler to indicate the current location at which instructions or data are being placed.

location dimension *Design Engineering.* a measurement that specifies the distance of one feature of an object with respect to another.

location fit *Design Engineering.* the assemblying of parts by mating one piece with another into a precise position between the two.

location notice *Mining Engineering.* a written sign visibly displayed on a claim, indicating the locator's name and describing the claim's extent and boundaries.

location parameter *Statistics.* a parameter whose variation causes a translation in a probability distribution.

location plan *Mining Engineering.* a scale map of a projected mine development that indicates proposed shafts, works in relation to existing surface features, etc.

location work *Mining Engineering.* any labor required by law in order to establish ownership.

locator a person or thing that locates; specific uses include: *Engineering.* any instrument or process by which the location of an object is determined, such as a radar system that locates airborne aircraft. *Mining Engineering.* a person who locates a mining claim.

Loc. dol. to the painful spot. (From Latin *loco dolenti.*)

locellus *Botany.* a compartment in the ovary of some legumes or in a pollen sac.

lochia *Medicine.* the vaginal discharge of mucus, blood, and tissue debris during the first week or two following childbirth.

loci the plural of *locus.*

lociation *Ecology.* a ranked category used in the classification of vegetation.

lock a mechanical device for fastening and securing something, or the process of operating such a device; specific uses include: *Mechanical Devices.* a door, lid, or cabinet fastening device with a cylinder and a movable bolt accessed by turning a key in the cylinder to secure or open the assembly to which it is a part. *Ordnance.* **1.** the position of the safety mechanism that prevents a loaded weapon from being fired. **2.** to set the safety mechanism in such a position. **3.** see GUNLOCK. *Computer Science.* **1.** to set a lock variable to prevent other processes from accessing shared data during a critical region. **2.** see LOCK VARIABLE. *Electronics.* to latch onto and automatically track a target, using a radar beam. *Metallurgy.* in forging, the condition in which the line on which the flash is formed is not in a single plane. *Hydrology.* a section of a waterway at which the level of the water can be raised or lowered.

locks

lockalloy *Metallurgy.* the trade name of a noncommercial alloy for aerospace application, based on beryllium.

lock bolt *Engineering.* **1.** the bolt portion of a lock. **2.** a bolt that is fixed with a locking collar rather than a nut. **3.** a bolt used to adjust or lock machine parts.

lock chamber *Civil Engineering.* the enclosed section of a canal or river where the level can be altered to account for the change in altitude and through which ships or barges may pass.

locked-coil rope *Mechanical Devices.* a steady, durable rope with concentric wires layered around a central core. Also, LOCKED-WIRE ROPE.
locked groove *Design Engineering.* a blank, continuous finishing groove on the surface of a disk recording, placed at the end of the modulated grooves to prevent further movement of the needle or stylus beyond it. Also, concentric groove.
locked oscillator *Electronics.* a circuit that converts direct current into alternating current at a frequency synchronized with the frequency of an external signal.
locked oscillator detector *Electronics.* a frequency-modulation detector to which a local oscillator is locked; the phase difference between the two is proportional to the frequency deviation, and the output voltage is proportional to the phase difference.
locked-rotor current *Electricity.* the steady-state current taken from a line with the rotor locked and the rated voltage or frequency applied to the motor.
Locke level see HAND LEVEL.
lock frame *Ordnance.* a mechanical component used in some firearms to assist in unlocking the bolt from the barrel, thus cushioning the shock of recoil.
lock front *Mechanical Devices.* the plate on a door lock through which the bolt or bolts project.
lock gate *Civil Engineering.* the ingress or egress through which ships or barges must pass and which acts as a barrier for the water in the upper or lower sections of the lock chamber.
lock-in *Electronics.* **1.** a state in which two oscillating systems are synchronized, as when a self-excited oscillator is locked-in with a standard-frequency generator. **2.** to shift and automatically retain one or both frequencies of two interconnected oscillating systems, so that both frequencies are synchronized.
lock-in amplifier *Electronics.* a circuit that detects and measures faint electromagnetic radiation at radio or optical frequencies through some form of automatic synchronization with an external reference signal.
locking any process in which something is locked; specific uses include: *Electronics.* to control the frequency of an oscillator by applying a signal whose frequency is consistent.
locking circuit see HOLDING CIRCUIT.
locking fastener *Mechanical Devices.* a device which prevents loosening of threaded fasteners such as a seating lock, quick-release nut, or spring stop.
locking-jaw pliers *Mechanical Devices.* a type of pliers whose jaws connect at a sliding rivet that can be locked in a fixed position. Also, **locking pliers.**

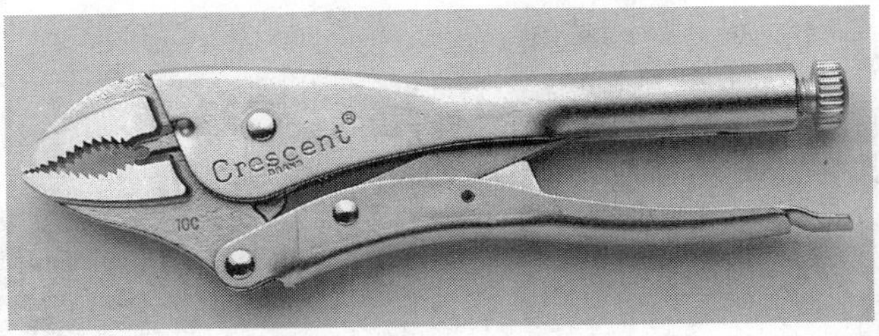

locking-jaw pliers

locking lugs *Ordnance.* metal projections from the bolt of a small arms weapon that are designed to lock the bolt to the barrel and seal the breech before firing.
locking plate *Mechanical Devices.* a narrow, geared wheel with notches about its rim attached to a striking train used to engage another mechanism by rotation of a watch or other timing instrument.
locking relay see LATCH-IN RELAY.
locking ring *Mechanical Devices.* a screw collar which connects the tube end and jacket of small caliber guns.
lockjaw see TRISMUS.
lock joint *Mechanical Devices.* any type of joint, such as a dovetail, in which particular parts are interlocked with or without the use of a fastener or bond.
lock mortise chisel *Mechanical Devices.* a swan-necked chisel with a long, squared blade and an upturned cutting edge used for cutting deep blind recesses, such as a door mortise lock, and for removing waste from deep mortises.
locknut *Mechanical Devices.* **1.** a self-locking nut. **2.** a nut screwed against another nut or a washer to prevent either from loosening due to vibration. Also, JAM NUT. **3.** a nut used with an electrical or piping conduit about its end at a junction box or fitting to secure it.
lock-on *Electronics.* **1.** the procedure by which a radar system tracks a target in one or more coordinates, such as range, bearing, and elevation. **2.** the precise moment at which a radar system begins to follow a target automatically. *Ordnance.* a term signifying that a tracking or target-seeking system is following its target automatically.
lockout *Computer Programming.* **1.** a technique used to prevent simultaneous access to critical data by two separate programs in a multiprogramming environment. **2.** the prevention of access to a memory block by a processing unit while input/output operations are taking place in the block. **3.** the denying of access of a terminal to a computer until log-in information has been verified or the preventing of simultaneous input and output operations. **4.** the temporary suppression of an interrupt. *Telecommunications.* **1.** in a telephone circuit under the control of two voice-operated instruments, a condition in which excessive local circuit noise or uninterrupted speech from either one or both of the calling parties prevents one or both of the parties from getting through. **2.** in mobile communications, the arrangement of control circuits in such a way that only one receiver can feed the system at a time, to avoid distortion.
lockout circuit *Electronics.* a type of circuit that activates an electric circuit by responding to only one of a number of input circuits at any given time.
lock rail *Building Engineering.* an intermediate horizontal member of a door between the vertical stiles at the height of the lock.
lockset *Mechanical Engineering.* **1.** an assembly consisting of knobs, plates, and locking mechanisms that forms a complete locking system. **2.** a template used to cut a hole so that a lock system will fit it exactly.
lock stile *Building Engineering.* in a door frame, the stile upon which a lock is fastened.
lock up *Graphic Arts.* the process of locking metal type elements into a chase. *Mechanical Engineering.* a sudden halt in the rotation of a wheel.
lock-up relay *Electricity.* a relay that locks in the energized position by means of permanent magnet bias, or by auxiliary contact that keeps the coil energized until the circuit is interrupted.
lock variable *Computer Science.* a variable that is set to indicate that some process is using shared data and that no other process may use it.
lock washer *Mechanical Devices.* a solid, split-ring, or tooth-lock washer designed to maintain a firm grip on a nut.
Lockyer, Joseph Norman 1836–1920, English astronomer; studied solar chemistry; found helium on the sun and named this new element.
loco disease *Veterinary Medicine.* the poisoning of livestock by the ingestion of locoweed, giving rise to disordered vision, delirium, convulsions, and often death.
locomotion *Science.* **1.** the ability to move from place to place. **2.** the process or means of moving from place to place.
locomotive *Mechanical Engineering.* **1.** a self-propelled vehicle, powered by fuel, compressed air, or electricity, used for moving loads on railroad tracks. Also, ENGINE. **2.** of or relating to such a vehicle. *Science.* of or relating to locomotion.

locomotive

locomotive boiler *Mechanical Engineering.* an internally fixed horizontal fire-tube boiler integrated with a furnace, used on steam locomotives.

locomotive crane *Mechanical Engineering.* a steam-powered crane mounted on a flatbed railroad car or on a special chassis with flange wheels.

locomotive gradient *Mining Engineering.* the incline set by law for a locomotive haulage; while the maximum limit is 1 in 15, for practical purposes it is 1 in 25.

locomotive haulage *Mining Engineering.* any coal ore, workers, and materials that are transported by locomotive-hauled mine cars.

locomotor *Physiology.* **1.** of or relating to locomotion. **2.** relating to or affecting the locomotive apparatus of the body.

locomotor ataxia see TABES DORSALIS.

locomotor system *Zoology.* physical parts and processes that enable an organism to move independently from one place to another.

loco weed or **locoweed** *Botany.* any selenium-containing legume of the genera *Astragalus* and *Oxytropis,* which poison domestic animals.

loctal base see LOKTAL BASE.

loculate *Biology.* divided into compartments, chambers, or cavities, such as an ovary, anther, or fruit.

locule *Botany.* a small chamber, cavity, or compartment in a plant in which specialized structures may grow, such as an ovary. Also, **loculus.**

loculicidal *Botany.* of or relating to a type of dehiscence in which the locule is bisected.

Loculoanoteromycetidae *Mycology.* a subclass of fungi belonging to the class Loculoascomycetes which are characterized by ascocarps that are determinate in size and generally round or oval; it is composed of plant parasites or members which live off of nonliving organic matter.

Loculoascomycetes *Mycology.* a class of fungi belonging to the subdivision Ascomycotina which are characterized by an ascocarp that is formed by cell division or by the interweaving of the hyphae of vegetative mycelium.

Loculoedaphomycetidae *Mycology.* a subclass of fungi belonging to the class Loculoascomycetes which are characterized by ascocarps that are determinate in size and of various shapes.

Loculoparenchemycetidae *Mycology.* a subclass of fungi belonging to the class Loculoascomycetes which are characterized by ascocarps that are determinate in growth, form a layer, and are generally asymmetric.

Loculoplectascomycetidae *Mycology.* a subclass of fungi belonging to the class Loculoascomycetes which primarily occur on decaying organic matter in tropical regions, although some members are parasitic to plants.

locus *Science.* the particular site of something; a location. *Genetics.* the site on a chromosome where a particular gene is normally located. *Mathematics.* the set of points or values satisfying a given equation or set of conditions.

locus of control *Psychology.* an individual's sense of the extent to which the events in his or her life are under his or her internal control, rather than the result of external causes.

locust *Invertebrate Zoology.* any of various grasshoppers of the family Locustidae, noted since ancient times for traveling in huge swarms and causing extensive destruction of agricultural crops. *Botany.* any of various North American trees of the genus *Robinia,* especially the **black locust,** *R. pseudodacacia,* which has pinnate leaves and clusters of white flowers. *Materials.* the wood of such a tree, widely used in construction and in industry.

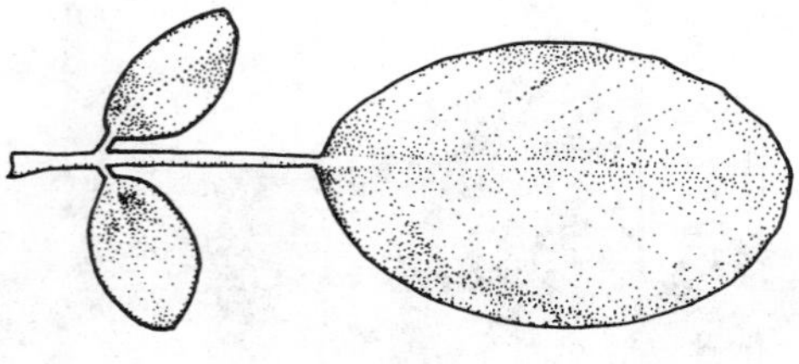

locust

Locustidae *Invertebrate Zoology.* a family of migratory, short-horned grasshoppers, serious pests that swarm in vast numbers and strip large areas of vegetation.

Lodderomyces *Mycology.* a genus of yeast fungi belonging to the family Saccharomycetaceae; only one species, *L. elongisporus,* is found in soil and fruit juice.

lode *Geology.* **1.** in general, any mineral deposit occurring within consolidated rock. **2.** specifically, a body of ore that has commercial value; often used in the names of specific sites; e.g., the Comstock Lode.

lode claim *Mining Engineering.* the legal acquisition of a particular vein or lode, and of the adjoining surface.

lodestar *Astronomy.* any star that serves as a guide to navigation, especially the North Star (Polaris). (From an earlier use of *lode* meaning "to guide.")

lodestone *Mineralogy.* a naturally occurring magnetic iron oxide, (Fe_3O_4, the mineral magnetite), exhibiting polarity. Also, HERCULES STONE, LEADING STONE.

lodgepole pine *Botany.* a tall pine tree of western North America, *Pinus contorta,* having short needles and twisted, usually closed cones. (From its use as timber.)

lodging *Plant Pathology.* an abnormal condition of cereal plants, in which they collapse before being harvested due to lack of light, nitrogen or moisture excess, frost injury, parasite attack, or breaking due to weather (such as wind).

lodicule *Botany.* a minute scale at the base of the ovaries of certain grasses.

lodos *Meteorology.* a southerly wind that occurs on the Black Sea coast of Bulgaria.

lodranite *Geology.* a stony, iron-bearing meteorite that is composed of bronzite and olivine, and enclosed in a fine network of nickel-iron.

lod score *Genetics.* in an experiment, a quantity indicating the statistical significance of a departure of the actual results from the expected results; a logarithmic value of the odds of the departure being due only to chance.

Loeb, Jacques 1859–1924, American physiologist; pioneer in chemical physiology and zoology; developed artificial fertilization.

loellingite see LÖLLINGITE.

loess [les; lō´əs] *Geology.* an extremely fertile, fine-grained loam composed of quartz, feldspar, horneblende, mica, and clay; deposited by the wind during the Pleistocene Age. It originates in arid regions from glacial outwash. It is normally yellowish-brown and has a widely varied calcium-carbonate content. (Going back to a German word meaning "loose.")

loess

loess kindchen *Geology.* a concretion of calcium carbonate occurring in loess whose shape somewhat resembles a child's head.

loeweite see LÖWEITE.

Loewi, Otto [lē´vē; lō´ē] 1873–1961, German pharmacologist; awarded the Nobel Prize for his discoveries regarding the chemical transmission of nerve impulses.

Löffler, Friedrich [lef´lər] 1852–1915, German bacteriologist; discovered the causes of diphtheria (with Klebs) and of foot-and-mouth disease (with P. Frosch).

Löffler's serum *Microbiology.* a solid nutrient medium typically used to make slope cultures of bacterial species such as *Coryneform diphtheriae,* and composed of blood serum, nutrient broth, and glucose.

Löffler's syndrome *Medicine.* a pulmonary disorder of relatively short duration, characterized by transient infiltrates of the lungs, low fever, and an increased number of eosinophils in the peripheral blood.

loft bombing *Military Science.* a method of bombing in which the delivery plane approaches the target at a very low altitude, makes a definite pullup, and releases the bomb at a predetermined point during the pullup by tossing it onto the target.

log *Forestry.* **1.** a relatively large section of a felled tree before it has been cut into lumber. **2.** to cut trees for lumber. *Navigation.* **1.** originally, a piece of wood attached to a measured line and thrown overboard, used to calculate the distance traveled by a ship. **2.** thus, any of various similar instruments for measuring speed or distance traveled through the water. **3.** a record of the pertinent details of conducting a craft on a particular trip. *Computer Technology.* a record of a sequence of operational activities on a computer, such as tape mounts, terminal log-ins, service requests, and job processing time. *Telecommunications.* a daily written record of radio and television broadcast data, as required by the Federal Communications Commission. *Mining Engineering.* a record of the characteristics of rock and other materials penetrated in drilling.

log or **log.** logarithm.

log- a combining form meaning "word" or "speech."

-log a combining form used to denote kinds of discourse or records, as in *analog, catalog.*

logagnosia *Neurology.* any of various word defects caused by disorder or disease of the central nervous system, such as aphasia or alogia.

logamnesia *Neurology.* receptive aphasia, characterized by the inability to recognize spoken or written words.

Logan, Sir William Edmond 1798–1875, Canadian geologist; studied coal deposits; first director of the Canadian Geological Survey.

loganberry *Botany.* **1.** a trailing blackberry plant, *Rubus ursinus-loganobaccus,* of the family Rosaceae. **2.** the large reddish-purple fruit of this plant; used in jams, jellies, juices, and wine. (Named for the American horticulturist James *Logan,* 1841–1928.)

Loganiaceae *Botany.* a family of dicotyledonous herbs, shrubs, and trees characterized by usually opposite, simple leaves and regular bisexual flowers in panicles or cymes, whose members lack a latex system.

Logan slabbing machine *Mining Engineering.* a machine with three cutting chains, two horizontal and one mounted vertically, that shear off the coal at the back of the cut, and a short conveyor that moves the coal onto the face conveyor.

logaphasia *Neurology.* a type of motor aphasia characterized by the inability to express one's thoughts in spoken words.

logarithm [lô´gə rith´əm; läg´ə rith´əm] *Mathematics.* **1.** the real number x satisfying the equation $b^x = a$, where the base b is a real number greater than zero and not equal to one; the inverse function of $f(x) = b^x$. Notation: $x = \log_b a$; read as: "log of a (to the base b)." If $b = 10$, the logarithm is called a **common logarithm** or **Briggs' logarithm.** If $b = e$, the logarithm is called a **natural logarithm, Napierian logarithm,** or **hyperbolic logarithm.** $\mathrm{Log}_e a$ is often written $\ln a$; read as: "natural log of a." **2.** if $z = e^w$, where w and z are complex, the natural logarithm of z is the inverse of the exponential function; $w = \ln z = \ln r + i(\theta + 2\pi k)$, where $z = re^{i(\theta+2\pi k)}$. In this case, $\ln z$ has infinitely many branches; the principal branch is usually defined as $\mathrm{Ln}\, z = \ln r + i\theta$. **3.** In general, any real function f, not identically zero, that satisfies the functional equation $f(x) + f(y) = f(xy)$ for all real x and y is called a **logarithmic function.**

logarithmic *Mathematics.* relating to or expressed as a logarithm.

logarithmically convex function *Mathematics.* a positive, real-valued function $f(s)$ such that $\log f(x)$ is a convex function. A logarithmically convex function is convex, but not conversely.

logarithmic amplifier *Electronics.* a circuit that produces an output signal whose amplitude increases at a rate proportional to the logarithm of the input signal's amplitude. Also, **logamp.**

logarithmic coordinate paper *Mathematics.* graph paper printed with two perpendicular coordinate axes marked in a logarithmic scale, and with the sets of mutually perpendicular lines parallel to the coordinate axes spaced according to the logarithmic scale.

logarithmic coordinates *Mathematics.* a rectilinear coordinate system in two coordinate axes marked with a logarithmic scale; the coordinates of a point are antilogarithms of the standard Cartesian coordinates.

logarithmic curve *Mathematics.* the graph in the plane of $y = \log_b x$. The curve lies entirely in the right half-plane and passes through the point (1,0). If $b > 1$, the curve is asymptotic to the negative y-axis and is monotone increasing. If $0 < b < 1$, the curve is asymptotic to the positive y-axis and is monotone decreasing.

logarithmic decrement *Physics.* the natural logarithm of the ratio of one amplitude of an underdamped oscillation to that of the next amplitude of the same polarity.

logarithmic differentiation *Mathematics.* let f be a differentiable function which is positive in some domain D. The logarithmic derivative of f is $d(\ln f(x))/dx = f'(x)/f(x)$. Logarithmic differentiation is used as a computation technique when it is difficult to compute $f'(x)$ directly.

logarithmic diode *Electronics.* a device that has a precise logarithmic relationship between current and voltage.

logarithmic equation *Mathematics.* **1.** a particular type of transcendental equation in which: (a) the variable occurs only in the arguments of logarithms, and (b) logarithms appearing in the equation all have the same argument $a(x)$, an expression in the variable. The substitution $y = \log_b a(x)$ leads to an algebraic equation in y. **2.** in general, any equation involving a logarithmic function of a variable.

logarithmic fast time constant *Electronics.* a constant false alarm rate scheme that uses a logarithmic intermediate-frequency amplifier followed by a fast-time constant circuit.

logarithmic growth *Microbiology.* see EXPONENTIAL GROWTH.

logarithmic multiplier *Electronics.* a device that produces an output that is proportional to the product of two inputs by entering each variable to a logarithmic function generator and then adding the outputs together and entering them into an exponential function generator.

logarithmic oxidation *Materials Science.* in the buildup of oxide scale on the surface of a material, a logarithmic relation that applies in cases of thick scale containing voids, or when a limiting thickness is approached; expressed as: $w = k \log(ct + A)$, where w = weight of scale, t = time of exposure, and k, c, and A are constants.

logarithmic potential *Physics.* any potential function which is proportional to the logarithm of some coordinate.

logarithmic profile of velocity *Fluid Mechanics.* the relation $v = K \ln(z/z_0)$, where v is the mean fluid velocity parallel to a plane boundary, z is the perpendicular distance to the boundary, z_0 is the roughness length, and K is a quantity dependent upon the shearing stress, the Karman constant, and the pressure.

logarithmic scale *Mathematics.* a labeling of a line so that each point labeled x is a distance proportional to $\log_b x$ from some fixed reference point, where x varies and b is a fixed base value.

logarithmic series *Mathematics.* the Maclaurin series expansion $\ln(1+x) = x - x^2/2 + x^3/3 - x^4/4 + \cdots$, where $-1 < x \leq 1$.

logarithmic spiral *Mathematics.* the graph of the equation in polar coordinates: $r = ae^{k\theta}$, where k and a are fixed, $a > 0$, and $-\infty < \theta < \infty$. The logarithmic spiral begins at the origin and spirals outward in a counterclockwise direction, intersecting rays from the origin at a constant angle $\alpha = \cot^{-1} k$. When $k = 0$, i.e., $\alpha = \pi/2$, the curve degenerates to a circle. Also, EQUIANGULAR SPIRAL.

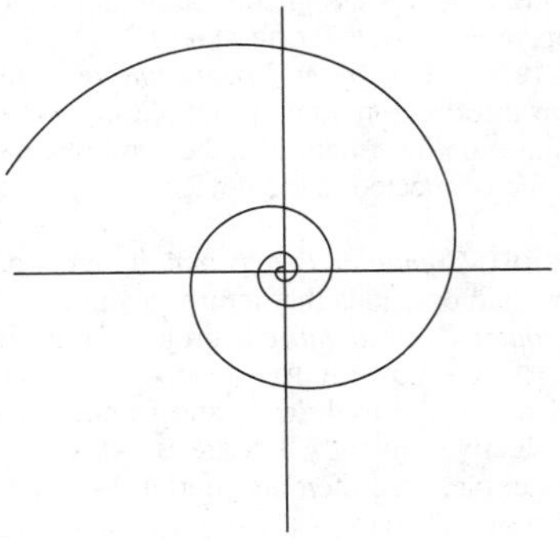

logarithmic spiral

logarithmic strain see TRUE STRAIN.

logarithmic transformation *Statistics.* the conversion of data to a logarithmic scale.

logarithmic velocity profile *Meteorology.* the variation with altitude of the mean windspeed in the surface boundary layer.

logbook *Navigation.* a book for recording details of a trip made by a ship or aircraft.

log chip *Navigation.* the chip of wood that is used as an indicator at the end of a log line.

log-dec logarithmic decrement.

logger *Forestry.* **1.** a person who works in logging. **2.** a tractor that is used in logging. **3.** a machine used to haul logs. *Engineering.* an automatic recorder that logs various information at specified intervals. *Electronics.* any of various electronic devices used to obtain an output proportional to the logarithm of the inputs.

loggia [lō´zhə; lō´jē ə] *Architecture.* **1.** a porch or gallery set behind an open arcade or colonnade and attached to a building. **2.** a balcony. (From an Italian word meaning "lodge.")

logging *Forestry.* **1.** the process of felling and the extraction of timber. **2.** the practice of cutting down trees and moving the logs. *Navigation.* the process of maintaining a log.

logic *Science.* the use of correct reasoning, or the principles involved in correct reasoning. *Mathematics.* the branch of mathematics that formulates and studies principles of reasoning; includes axiomatic theories, predicate calculus, statement calculus, Boolean algebra, and symbolic logic. *Computer Programming.* the algorithm or decision procedures used by a program. *Computer Technology.* a collection of circuits that perform logic functions, with the significant signal levels restricted to a number of discrete levels (usually two), rather than changing over a continuous range.

logical *Science.* **1.** of or relating to the various forms of logic. **2.** making proper use of logic; employing correct reasoning.

logical addition *Mathematics.* (binary) addition in a Boolean algebra.

logical comparison *Computer Programming.* a logic operation to determine whether two character or bit strings are equal; returns 1 (true) if both operands are the same, and 0 (false) if they are different.

logical connectives *Mathematics.* the terms or mathematical symbols that are used to connect or to modify mathematical statements: *not*, $\sim$; *and*, $\wedge$; *or*, $\vee$; *if . . . then* (or *implies*), $\Rightarrow$; and *if and only if*, $\Leftrightarrow$. Also, SENTENTIAL CONNECTIVES.

logical consequence *Artificial Intelligence.* a logical formula Q is a logical consequence of a formula P if and only if in any interpretation in which P is true, Q is also true. Written $P \rightarrow Q$ ("P implies Q").

logical construction *Computer Programming.* a simple logical property that is used to determine the data type of a sequence of data; for example, the first bit could distinguish numeric data from non-numeric data.

logical contouring *Cartography.* the process of interpolating data from field notes to produce accurate contour lines on a map without the need for running a level line for every contour, the contours are spaced proportionally between established spot elevations.

logical data independence see DATA INDEPENDENCE.

logical data type *Computer Programming.* a scalar data type whose values are either *true* or *false*. Also, BOOLEAN DATA TYPE.

logical decision *Computer Programming.* the selection of an appropriate path in a program based on the result of logic operations.

logical device *Computer Programming.* referring to a number or name, internal to a program, that is assigned to an input/output device to facilitate device independence. Also, LOGICAL UNIT.

logical device table *Computer Programming.* a table used to keep track of pertinent information about input/output operations on a logical device, containing such information as the symbolic name of the device, the name of the file connected to the device, and the addresses of associated buffers.

logical expression *Computer Programming.* an expression composed of logic operators and operands that returns a value of *true* or *false*.

logical file *Computer Programming.* a group of logical records.

logical flow chart see FLOW CHART.

logical form *Science.* the traditional and standard form that an argument takes in deductive logic: "All A are B. All C are A. Therefore, all C are B." For example, "All men are mortal. Socrates is a man. Therefore, Socrates is mortal." Also, LOGICAL STRUCTURE.

logical inferences per second see LIPS.

logical input/output control system *Computer Programming.* a comprehensive set of routines provided to handle file creation, deletion, modification, and retrieval.

logical instruction *Computer Programming.* a machine instruction that performs a logic operation or a combination of such operations, such as "and," "or," and "not."

logical multiplication *Mathematics.* (binary) multiplication in a Boolean algebra.

logical page *Computer Programming.* a fixed-length block of data or program code, as viewed by the software accessing it.

logical record *Computer Programming.* a group of logically related data items, independent of the physical manner of storage.

logical security *Computer Technology.* the internal computer system schemes used to protect against unauthorized access to data and other computer resources.

logical shift *Computer Technology.* a shift operation in which bits that are shifted out are lost and zero bits are shifted in.

logical structure see LOGICAL FORM.

logical sum *Computer Programming.* a process of addition in which the sum is 1 when either one or both input variables are 1, and the sum is 0 when the input variables are both 0.

logical sum see INCLUSIVE.

logical symbol *Computer Programming.* a graphical symbol used to represent any one of the logic operators.

logical unit see LOGICAL DEVICE.

logical value see BOOLEAN VALUE.

logic board *Computer Technology.* a circuit board in which integrated logic components are surface mounted.

logic board

logic bomb *Computer Science.* unauthorized coding placed within a computer program and designed to be triggered, causing damage or printing a message, upon the occurrence of a certain event, such as the date being Friday the 13th.

logic card *Electronics.* a frame upon which a number of electronic devices, such as transistors, capacitors, and magnetic cores, are mounted in a configuration designed to carry out a specific logic function.

logic circuit *Computer Technology.* an electronic circuit that implements a logical operation.

logic decoder see DECODER, def. 1.

logic design *Computer Technology.* the process of specifying an interconnection of logic elements in computer hardware in order to implement a desired function.

logic diagram *Computer Technology.* a presentation in graphics of a logic design, showing the elements and their interconnection.

logic element *Computer Technology.* a basic building block of a logic circuit that performs a simple logic function on its input, such as an "and" gate.

logic error *Computer Programming.* a programming error that is due to faulty reasoning by the programmer and that produces incorrect or unexpected results.

logic function see BOOLEAN FUNCTION.

logic gate *Electronics.* see GATE, def. 1.

logician [lə jish´ən] *Science.* a person who studies logic or who is skilled in the use of logical reasoning.

logicism [lə jish´ən] *Mathematics.* the principle or doctrine that mathematics can be reduced to logic, in the sense that mathematical concepts can be defined in terms of logical concepts, and mathematical theorems can be derived through logical deduction.

logic level *Electronics.* one of two voltages whose values have been arbitrarily selected to represent the binary numbers 1 and 0 in a given computer system.

logic operation *Computer Programming.* **1.** an operation in which one or more inputs, each having one of two values, are evaluated in accordance with a rule to produce an output, having only one of two values and indicating the result of the evaluation. **2.** any computer operation that is performed in accordance with a logic instruction, usually on entire computer words. Also, **logical operation.**

logic operator *Computer Programming.* any operator that performs a Boolean operation on single-bit operands, such as "and," "or," "not" operators. Usually the operation is performed simultaneously on corresponding bit positions of whole machine word operands.

logic programming *Artificial Intelligence.* a form of programming, as in PROLOG, in which programs take the form of logical statements, usually as Horn clauses, and the solution to a desired problem is found by a backchaining search using the database of facts and the logical axioms. *Robotics.* in robotic programming, the use of conditional jumps or wait instructions by a manual programming unit from predetermined data files.

logic-seeking printer *Computer Technology.* a line printer that decreases its printing time by examining each line and skipping over sequences of blank spaces.

logic swing *Electronics.* the difference between the voltage designated as 1 and the voltage designated as 0 in a logic circuit.

logic symbol *Computer Programming.* any symbol that represents a logic operation or the symbol form of a relational operator.

Logic Theorist *Artificial Intelligence.* one of the earliest artificial intelligence programs, by Newell and Simon, which proved theorems in set theory from Newton's *Principia Mathematica.*

logic word *Computer Technology.* a computer word that represents a collection of bits, each of which is treated as a Boolean value.

logistic curve *Statistics.* an S-shaped increasing curve often used to represent population growth.

logistic equation *Ecology.* an equation for population growth through time, starting from a minimum number of individuals to a maximum number in accordance with the carrying capacity of the region, described by an elongated S-shaped curve.

logistic map *Chaotic Dynamics.* an iterative map on the interval $0 \leq x < 1$ specified by $x_{n+1} = \lambda x_n(1-x_n)$, which is a paradigmatic iterative map containing period doubling and chaos for different values of the parameter l. Period doubling occurs for $3.0 < \lambda < \lambda_\infty = 3.569944\ldots$, and chaos with interspersed periodic windows appears for $\lambda > \lambda_\infty$.

logistic regression *Statistics.* a method used to study the probability of an event of interest as a function of a set of independent variables.

logistics *Military Science.* the process of planning and executing the movement and maintenance of military forces, including the design, development, acquisition, movement, and maintenance of equipment and supplies, the movement and evacuation of personnel, the construction, maintenance, and operation of facilities, and the acquisition or furnishing of services. Similarly, **logistic assessment, logistic assistance, logistic implications test, logistics sourcing, logistic support** (medical).

log line *Navigation.* the line on which a measuring log is suspended behind a ship.

log loader *Forestry.* a diesel tractor equipped with hydraulic grabbers; used for moving logs.

log loader

log-mean temperature difference *Thermodynamics.* a quantity given by $T_{LM} = (T_2 - T_1)/\ln(T_2/T_1)$, where T_1 and T_2 are two absolute temperatures; it is used in heat transfer calculations involving the cooling or heating of one fluid by another without mixing the fluids.

log-normal distribution *Statistics.* a distribution of a random variable X such that the natural logarithm of X is normally distributed.

logo [lō´gō] *Graphic Arts.* **1.** a symbol used to identify something such as a product or organization. **2.** a group of characters set in a distinctive type associated with a particular corporation or trademark, such as the script used to spell *Coca-Cola.* **3.** see LOGOTYPE.

LOGO *Computer Programming.* a high-level interactive programming language designed for educational and graphics applications.

logo- a combining form meaning "word" or "speech," as in logotype, logorrhea.

log-off *Computer Technology.* the process that a computer user employs to terminate an interactive session and release assigned resources. Also, **logout.**

logokophosis see KOPHEMIA.

logomania *Medicine.* an uncontrollable talkativeness to the degree that a manic state is manifested.

log-on *Computer Technology.* the process of gaining access to a computer and its resources by supplying required information, such as a password, with a specific sequence of commands. Also, **log-in.**

logopathy *Neurology.* any speech disorder caused by a defect in the central nervous system.

logoplegia *Neurology.* paralysis of the speech organs, characterized by an inability to speak despite being able to understand and remember the words that are intended to be spoken.

logorrhea *Medicine.* a pattern of incoherent, uninterrupted speech, seen in manic episodes of bipolar disorder. *Psychology.* excessive or compulsive talkativeness.

logotype *Systematics.* **1.** a type determined from a written description when there is neither a specimen nor an illustration of a specimen. **2.** see GENOLECTOTYPE. *Graphic Arts.* **1.** a type element containing two or more letters commonly used together, especially short words such as "and" or "the." Also, LOGO. **2.** see LOGO, def. 2.

log-periodic antenna *Electromagnetism.* an antenna array whose elements have lengths and relative spacing that increase logarithmically so as to produce a broadband spectrum.

log-phase growth *Biotechnology.* a stage of microbial growth where all foodstuffs and other requirements are present in excess and cell growth is limited only by the intrinsic limits of the organism.

log rule *Forestry.* a table giving the estimated amount of lumber in board feet that can be sawed from logs of different lengths and diameters.

log run *Forestry.* the run of the log or the total marketable lumber that can be cut from the saw; with softwoods, it includes the lumber of all grades; with hardwoods, it excludes certain lower grades.

log volume *Forestry.* the cubic volume of a log inside the bark.

logwood *Botany.* a tree, *Haematoxylon campechianum*, of the legume family that is found in the West Indies and Central America. *Materials.* the heavy brownish-red heartwood of this tree, which yields a black dye and a stain once used in microscopy. Also, HEMATOXYLON.

loiasis *Medicine.* a chronic filarial disease that is caused by the worm *Loa loa*, transmitted by the bite of an infected African deer fly, and characterized by transient cutaneous swelling, itching, and pain due to the action of migrating adult worms. Also, CALABAR SWELLING, LOA LOA INFECTION.

loktal base *Electronics.* the base in a type of vacuum tube whose pins fit snugly into the eight holes in a corresponding socket, so that the two lock firmly together. Also, LOCTAL BASE.

loliferous *Botany.* producing or growing leaves.

löllingite *Mineralogy.* $FeAs_2$, a silver-white to steel-gray, orthorhombic mineral of the löllingite group, occurring in large, prismatic crystals and in massive form, having a specific gravity of 7.43 and a hardness of 5 to 5.5 on the Mohs scale; found in mesothermal vein deposits and pegmatites. Also, LOELLINGITE, LEUCOPYRITE.

lollipop *Molecular Biology.* a term for a stem and loop structure arising from the binding of complementary nucleic acid bases within a single strand.

lolly ice *Oceanography.* seawater frazil, ice crystals kept in suspension by water turbulence.

lomasome *Cell Biology.* a membrane-bound vesicle whose function has not been clearly determined, occurring between the plasma membrane and the cell wall of certain plant cells.

lombarde *Meteorology.* an easterly wind that blows along the French-Italian border; originating in the High Alps, it occurs as a violent wind that forms snowdrifts in mountain valleys and as a very dry and gentle breeze in the plains. (Named for the Italian region of *Lombardy.*)

loment *Botany.* a dry, one-celled fruit that divides transversely into many segments; a legume fruit that is conspicuously constricted between the seeds. Also, **lomentum.**

Lomonosov Ridge *Geography.* a submarine ridge that nearly bisects the Eurasian (Arctic) Basin. From the Lincoln Sea, north of Greenland, it stretches to the Laptev Sea, north of central Siberia.

LON *Aviation.* the airline code for the city of London, England; London's principal airports are Heathrow (LHR), Gatwick (LGW), and Luton (LTN).

Lonchaeidae *Invertebrate Zoology.* a family of tiny flies, the lance-flies, dipteran insects in the subsection Acalyptratae.

London, Fritz 1900–1954, American physicist, born in Germany; contributed to the theories of covalent bonding and superconductivity.

London equations *Solid-State Physics.* the equations derived from London superconductivity theory that relate the time derivative and the curl of the electric current vector to the electric and magnetic fields.

London force(s) *Physical Chemistry.* **1.** a general term for the universal forces of attraction that occur between fluctuating dipoles in atoms and molecules that are very close together. **2.** see DISPERSION FORCE. Also, **London effect.** (Identified by Fritz *London* in 1930.)

London hammer *Mechanical Devices.* a light type of cross-pane hammer. Also, EXETER HAMMER.

Londonian see YPRESIAN.

London penetration depth *Solid-State Physics.* the effective depth to which a magnetic field may penetrate into a superconductive material that excludes those fields from the bulk.

London screwdriver *Mechanical Devices.* a tapered, flat-bladed screwdriver used for large screw threads.

London superconductivity theory *Solid-State Physics.* a theory of superconductivity that assumes a two-fluid electron model, whereby the superfluid (superconductive) component of electrons behaves as a superconductor and the remainder behave as a collection of electrons in the normal state.

London superfluidity theory *Physics.* a theory that treats helium-4 as a Bose-Einstein gas and that attributes its superconducting properties to a fraction of its particles existing in a ground state.

lone wolf *Behavior.* a term for any animal of a social species that lives in isolation.

long *Materials Science.* **1.** of a mixture, having a relatively high percentage of a desired or active component. **2.** of clay, very plastic.

Long, Crawford 1815–1878, American physician; the first to use ether as an anesthetic.

long. or **lon.** longitude.

long-acting thyroid stimulator *Endocrinology.* an IgG immunoglobulin that is found in the serum of patients with Graves' disease and believed to be an autoantibody to the TSH receptor of the thyroid gland; it causes prolonged and unregulated stimulation of the TSH receptor, resulting in overproduction of thyroid hormones and causing hyperthyroidism.

long-baseline system *Telecommunications.* a system in which the distance separating ground stations is almost the same as the distance to the target that is being tracked.

longboat *Naval Architecture.* the longest boat carried by a sailing ship.

long bone *Anatomy.* one of the elongate bones of the limbs, consisting of a shaft, or *diaphysis,* and two extremities, or *epiphyses;* the long bone establishes the height or length of the limb.

long-chain branching *Materials Science.* a condition in which side chains on the main polymer chain are relatively long, causing a decrease in the melt viscosity; found in polyethylene and silicone polymers, among others.

long column *Civil Engineering.* a pillar that, by its slender design, is known to buckle under the stress of overloading, and therefore must be compensated for in construction.

long-day plant *Botany.* a plant that requires a long photoperiod in which to blossom.

long-distance loop *Telecommunications.* a line that links a subscriber's station directly to a long-distance switchboard or switching system.

long-distance navigation *Navigation.* **1.** position-finding employing only aids that are usable at a long range. **2.** any navigational technique usable on a long trip, such as a long-range chart.

long-distance xerography *Telecommunications.* in facsimile, a system that uses a cathode-ray scanner at the transmitting terminal, while at the receiving terminal a lens projects a cathode-ray image onto a selenium-coated drum of a xerographic copying machine.

long division see DIVISION ALGORITHM.

lon gene *Genetics.* a gene that encodes an ATP-dependent protease in *E. coli.*

longeron *Aviation.* any heavy longitudinal structural member of a fuselage, nacelle, or similar body, usually running continuously across a number of points of support. (A lighter member serving a similar purpose is called a *stringer.*)

longevity *Biology.* the condition or quality of being long-lived.

longevity in vitro *Virology.* a measurement of infectivity of plant viruses.

long-flame burner *Engineering.* a burner that has a long flame due to the air and fuel not readily mixing.

long-focus photographic astrometry *Astronomy.* the measurement of stellar positions with telescopes whose long focal lengths provide a large photographic plate scale and enhanced accuracy.

long gun *Ordnance.* a term applied to long-barrelled firearms such as rifles or shotguns to distinguish them from short arms such as pistols and revolvers.

long-haul carrier system *Telecommunications.* in telephony, an intercity transmission system of multiplexed communication channels.

long-haul radio *Telecommunications.* in radio, a microwave system that is able to transmit telephone and telegraph signals over distances in excess of 6500 kilometers (4000 miles).

long hole *Mining Engineering.* a term for an underground borehole or blasthole exceeding 10 feet in depth.

long-hole drill *Mining Engineering.* a rotary or percussive-powered drill for drilling long holes.

long-hole jetting *Mining Engineering.* a system to remove coal primarily by drilling a hole down the pitch of the vein, replacing the drilling head with a jet cutting head, and then retracting the drill column with the jets in operation.

Longhorn or **longhorn** *Agriculture.* a type of cattle having very long, curving horns; widely raised for beef in the U.S. in the 19th century, now used chiefly for breeding. Also, TEXAS LONGHORN.

long hundredweight *Metrology.* a unit of weight used in Great Britain, equal to 112 pounds or 0.05 long ton.

Longhurst-Hardy plankton sampler *Engineering.* a metal-shrouded net that is used to capture plankton for study and analysis.

longi- a combining form meaning "long," as in *longilineal.*

Longidorinae *Invertebrate Zoology.* a subfamily of nematodes in the superfamily Dorylaimoidea, including several important plant parasites.

long ink *Graphic Arts.* viscous printing ink, so called because it stretches into long strings when compressed and released.

Longipennes *Vertebrate Zoology.* an equivalent name for the bird order Charadriformes.

longitude [län´ji tood´] *Geography.* a distance either east or west on the earth's surface, measured and expressed as the angle between a given meridian and a reference meridian (usually the prime meridian of Greenwich, England); sometimes measured as a linear distance. *Astronomy.* see CELESTIAL LONGITUDE. *Engineering.* in surveying, the east-west component of a traverse course.

longitude factor *Navigation.* the change in longitude along a celestial line of position for a given change in latitude.

longitude line *Navigation.* a line of position that extends in a generally north-south direction, normally the result of observing a body on or near the prime vertical.

longitude method *Navigation.* a technique for determining longitude by observing a body near the prime vertical.

longitude of the ascending node *Astronomy.* the angular distance measured counterclockwise along the ecliptic from the vernal equinox to the point in an object's orbit where the object, moving northward, intersects the ecliptic.

longitudinal [län´ji tood´ə nəl] *Science.* **1.** of or relating to longitude or length. **2.** running lengthwise.

longitudinal aberration *Optics.* the chromatic or spherical aberration of an image near the focal plane which varies with its position along the optical axis.

longitudinal acceleration *Mechanics.* the acceleration or component of acceleration in the longitudinal direction of an object or projectile.

longitudinal axis *Aviation.* a line running nose-to-tail along a flight vehicle's center of gravity.

longitudinal baffle *Chemical Engineering.* a sheet of metal, in fixed-tube-sheet construction with multipass shells, attached to the tube bundles and sealed against the shell to provide support and to guide the fluid flow on the shell side.

longitudinal bulkhead *Naval Architecture.* a bulkhead running fore and aft, approximately parallel with the sides of the ship.

longitudinal center of gravity *Naval Architecture.* the point along a ship's length at which its weight balances forward and aft, and therefore the center around which it pitches.

longitudinal circuit *Electricity.* a circuit made up of one or more communication conductors in parallel that returns through ground.

longitudinal coastline see CONCORDANT COASTLINE.

longitudinal coefficient *Naval Architecture.* the submerged volume of a hull, divided by the product of the area of its midship section and its length on the waterline; also called the prismatic coefficient.

longitudinal control *Aviation.* **1.** control over the longitudinal attitude (pitch) or movement of a flight vehicle. **2.** a control surface or device, such as a control column or side stick, designed to control pitching movement.

longitudinal current *Electricity.* a current that flows in the same direction in the two wires of a parallel pair using the earth for a return path.

longitudinal data *Statistics.* data collected on constant experimental units over a period of time. Also, PANEL DATA.

longitudinal direction *Metallurgy.* in metal working, the principal direction in which the material is deformed.

longitudinal drum boiler *Mechanical Engineering.* a boiler having a horizontal drum axis that is parallel to its tubes and is located in the same plane.

longitudinal dune *Geology.* a narrow, elongated sand dune that lies parallel to the direction of the prevailing wind.

longitudinal fault *Geology.* a fault whose strike is parallel to the structural trend of the surrounding region.

longitudinal flow reactor *Chemical Engineering.* a theoretical reactor system, usually tubular, operating at high flow velocity in which there is little back mixing of reactants and products along the reactor length.

longitudinal frame *Naval Architecture.* a primary structural support running fore and aft.

longitudinal framed ship *Naval Architecture.* a ship with a hull structure based primarily on longitudinal framing.

longitudinal joint *Building Engineering.* a joint that secures two pieces of timber along their length.

longitudinal magnetostrictance *Electromagnetism.* a change in the length of a ferromagnetic substance that occurs parallel to an applied magnetic field.

longitudinal mass *Physics.* the quotient of force acting on a relativistic particle parallel to the particle's velocity divided by the acceleration resulting from the force.

longitudinal mode *Optics.* a term for the modes of a laser separated in frequency by the speed of light in the cavity divided by two times the cavity length.

longitudinal-mode delay line *Electronics.* a device in which a signal's output is delayed by passing it through a bar of magnetic material, such as nickel, that expands and constricts in direct proportion to an applied magnetic field.

longitudinal parity *Telecommunications.* parity that is associated with bits recorded on one track of a data block to indicate whether there is an odd or even number of bits in the block.

longitudinal parity check *Telecommunications.* **1.** in a given message, the count for even or odd parity of all the bits; a precaution against transmission error. **2.** a parity check of one track of a block of data words reloaded on multiple tracks of a magnetic tape or disk. Also, HORIZONTAL PARITY CHECK.

longitudinal profile see STREAM PROFILE.

longitudinal quadrupole *Electromagnetism.* an electric or magnetic quadrupole that produces two equal and opposite dipoles separated by a small distance.

longitudinal redundancy check *Computer Programming.* a method of checking errors in a computer system, in which the bits in each track or row of a record will total a given even (or odd) number, in the absence of error.

longitudinal resistance seam welding *Metallurgy.* resistance seam welding parallel to the throat depth of the welding machine.

longitudinal separation *Navigation.* spacing of aircraft operating at the same altitude by a minimum distance expressed in terms of miles or time.

longitudinal sequence *Metallurgy.* in welding, the weld sequence in the longitudinal direction.

longitudinal stability *Aviation.* the ability of an aircraft to return to its plane of symmetry following a change in motion, such as pitching or a variation in longitudinal and normal velocity.

longitudinal strain see LINEAR STRAIN.

longitudinal stream see SUBSEQUENT STREAM.

longitudinal vibration *Mechanics.* cyclic motion entirely along the long dimension of an object. Also, COMPRESSION WAVES.

longitudinal wave *Physics.* a wave in which the disturbance mechanism acts in the direction of the wave propagation.

Long Lance torpedo *Ordnance.* a Japanese oxygen-driven torpedo used during World War II, delivering a 1100-pound warhead at a speed of 36 knots and a maximum range of 43,500 yards; officially designated torpedo **Type 93.**

longleaf pine *Botany.* a tall pine tree of the southeastern United States, *Pinus palustris*; having bright green, densely bunched needles that can be up to 18 inches long. *Materials.* the hard, heavy wood of this tree, widely used for timber. Also, **Georgia pine.**

longleaf pine

long-life items *Ordnance.* a term for ordnance items that have an average service life of at least five years.

long-line azimuth survey *Cartography.* a type of survey that makes use of ground photography and airborne strobe lights to measure azimuths not visible between ground stations. Also, LOLA SURVEY.

long-line current *Electricity.* positive electricity flowing through earth from an anodic to a cathodic area that returns along an underground metallic structure.

long-line effect *Electronics.* a phenomenon that takes place when an oscillator coupled to a transmission line oscillates at two frequencies, causing the current to jump from one to the other.

long-lines engineering *Telecommunications.* engineering with the ultimate goal of modernizing, developing, and/or expanding long-haul, point-to-point communications outlets that employ microwave, radio, fiber optics, or wire circuits.

long nipple see LONG SCREW.

long-nose pliers *Mechanical Devices.* a type of pliers having long, tapered jaws.

long-period comet *Astronomy.* a comet whose orbital period is greater than 200 years.

long-period interspersion *Molecular Biology.* a sequence of DNA distinguished by segments of moderately repetitive DNA alternating with sequences of nonrepetitive DNA.

long-period variable star *Astronomy.* a pulsating variable star with an M-, R-, or N-type spectrum, a range in brightness of about six magnitudes, and a period lasting 200 to 600 days. Also, MIRA STAR, MIRA VARIABLE.

long-period wave *Oceanography.* a wave whose period is greater than 5 minutes, usually caused by a seismic disturbance or a storm.

long-persistence screen *Electronics.* a type of fluorescent screen that holds an image for several seconds after the electron-beam that produced it has passed.

long-pillar work *Mining Engineering.* a mining technique in which large pillars of coal are left as the face is advanced, and the remainder of the pillars are removed together.

long play *Mechanical Engineering.* a videotape recording speed that is slower than standard speed to allow for twice the usual recording time; for example, four hours instead of two hours for a T-120 tape.

long-playing record *Acoustical Engineering.* a phonograph record made to be played at 33-1/3 rpm. Also, LP.

long primer *Ordnance.* a primer encased in a long container; used to ignite a propelling charge.

long-range forecast see EXTENDED FORECAST.

long-range materiel requirements *Ordnance.* ordnance requirements projected as necessary for operations or other concepts five to ten years in the future.

long-range order *Crystallography.* the periodicity in a crystal lattice whose period extends over many unit cells. *Solid-State Physics.* any regular repetitive arrangement of the atoms in a solid that extends over a significant distance.

long-run frequency *Statistics.* the number of occurrences of an event in a large number of trials.

long-scale contrast *Radiology.* a broad range of image densities on a radiograph, with very little change in tones. Also, LOW CONTRAST.

long screw *Mechanical Devices.* a six-inch nipple with a longer-than-standard thread. Also, LONG NIPPLE.

longshore bar *Geology.* a low, elongated ridge of sand built mainly by wave action that lies generally parallel to the shoreline and is submerged at high tide. Also, BALL, BARRIER BAR, OFFSHORE BAR.

longshore drift see LITTORAL DRIFT.

longshore trough *Geology.* a wide, elongated, shallow depression of the sea floor lying parallel to the shore.

long span *Engineering.* an open wire span whose length is greater than 250 ft (76 meters).

long-span steel framing *Civil Engineering.* a framing system used for slabs, beams, girders, or bridges where there is a greater distance between supports than can be spanned with rolled beams; cantilevered suspension spans can be used in this system.

long string *Petroleum Engineering.* the last string of casing in a well, set just above or through the producing zone; it is the longest casing in a well, running from the pay zone to the surface, and the smallest in diameter. Also, OIL STRING, PRODUCTION STRING.

long-tail pair *Electronics.* a circuit with two tubes, having a common cathode resistor.

long terminal repeat *Molecular Biology.* an identical sequence of several hundred nucleotides found in the same orientation at both ends of a eukaryotic transposon.

long-term memory or **long-term store** *Behavior.* the part of the human memory system that stores most of the information that has been learned along with rules for processing it, characterized by a slow rate of decay and a large capacity. Also, SECONDARY MEMORY.

long-term repeatability *Robotics.* the ability to move a robotic body over the same path repeatedly for a long period of time.

long tom *Mining Engineering.* **1.** a trough, longer than a rocker, for washing auriferous earth. **2.** any portable sluice used to make rough concentrates during treatment of alluvial sands or gravels.

Long Tom *Ordnance.* **1.** originally, a 155-mm self-propelled gun, designated M2, that was used by the U.S. military in World War II; it weighed 15 tons, had a barrel length of 23 feet, and accurately fired a 95-lb. projectile at a range of over 14 miles. **2.** a similar gun, designated M-53, that is currently used by the U.S. Army and Marine Corps; it is full-tracked and served by a crew of six. **3.** sometimes applied to any long pivot gun or any large gun of long range.

long ton *Metrology.* a unit of weight used in Great Britain, equal to 2240 pounds or 1.016 metric tons.

long-tube vertical evaporator *Chemical Engineering.* a common type of natural-circulation evaporator consisting of a vertical shell-and-tube heat exchanger in which hot fluid on the shell side causes the cold fluid on the tube side to rise and evaporate.

long-waisted *Naval Architecture.* of or related to a ship in which there is a long open deck between the poop deck and the forecastle.

longwall *Mining Engineering.* a mining method in which the faces are advanced from the shaft toward the boundary, and the roof is allowed to cave in behind the miners as work progresses. Also, **longwall system.**

longwall coal cutter *Mining Engineering.* a compact machine driven by electricity or compressed air, and with its jib at right angles to its body, for cutting into the coal on relatively long faces.

longwall peak stoping *Mining Engineering.* an underland procedure for stoping below 5000 feet.

longwall retreating *Mining Engineering.* a longwall working system in which there are all three roadways in the solid coal seam, the waste areas are left behind, developing headings are driven close to the boundary or limit, and the coal is taken out by the longwall retreating toward the shaft.

longwall stope see FLAT-BACK STOPE.

long wave *Meteorology.* a wave of significant amplitude lying in the major zone of westerlies and extending for thousands of kilometers. Also, MAJOR WAVE, PLANETARY WAVE. *Hydrology.* see SHALLOW-WATER WAVE. *Telecommunications.* a wave that has a length greater than 1000 meters; corresponds to frequencies below 300 kilohertz.

long-wavelength infrared radiation *Electromagnetism.* electromagnetic radiation having wavelengths greater than 8 micrometers.

long-wave radio *Telecommunications.* radio that can transmit and/or receive frequencies below 550 kHz.

long-wire antenna *Electromagnetism.* a directional antenna whose element length is several wavelengths longer than its operating wavelength.

Lonsdaleia *Paleontology.* a genus of colonial, cerioid rugose corals in the extinct suborder Columnariina and family Axophyllidae, characterized by their lonsdaleoid dissepiments; Carboniferous.

lonsdaleite *Mineralogy.* carbon in the form of a pale brown to yellow, hexagonal mineral, polymorphous with diamond, graphite, and chaoite, having a specific gravity of 3.3 to 3.51 and an undetermined hardness; found as pseudocubic crystals in meteorites.

loo *Meteorology.* a hot wind that blows from the west in India.

loofah *Botany.* any of several tropical vines of the genus *Luffa*; produces a large elongated fruit whose fibrous interior is used as a sponge.

look *Ordnance.* in mine warfare, a period during which a mine circuit is capable of receiving an influence.

lookahead see INSTRUCTION LOOKAHEAD.

look and feel *Computer Technology.* the user-discerned characteristics of a computer system that set it apart from others, often a copyrightable design.

look angle *Engineering.* the angle at which a radar, optical instruction, or space radiation detector functions most effectively. *Space Technology.* the elevation and azimuth that a satellite is predicted to reach at a certain time.

look-back time *Astrophysics.* the elapsed time since light that is now arriving at the earth left its celestial source; this is 8 minutes for the sun and over 2 million years for the Great Galaxy in Andromeda.

look box *Chemical Engineering.* a strong glass tube or panel coupled to a pipe or process vessel, to show the level, stability, color, or quality of a process stream. Also, GAUGE GLASS, SIGHT GLASS, LIQUID LEVEL GAUGE.

looking-glass self *Psychology.* the concept that an individual's self-image is dependent on the reflected opinions of others.

lookout *Building Engineering.* a horizontal wood framing member that overhangs a gable in a roof, extending along the principal rafters.

lookout station *Engineering.* any structure that is equipped with facilities for remote observation, as along a shore or in forested areas.

lookout tower *Engineering.* a tower from which habitual watches are kept over fire conditions and marine operations; there is usually a small building on the top.

look-through *Electronics.* **1.** a technique in which a signal is interrupted at irregular intervals in order to monitor the "victim" signal during a jamming operation. **2.** a technique that permits viewing or monitoring the "victim" signal during breaks in a jamming operation.

look-up *Computer Programming.* a process in which a table is searched until a value corresponding to a specific value is found.

look-up table *Computer Programming.* a collection of data stored in table form for ease of reference.

loom *Meteorology.* a glow of light that is observed below the horizon under abnormally high refraction conditions in the troposphere when air density decreases more rapidly than height in the standard atmosphere. *Navigation.* the glow in the sky of a light or lights located below the horizon. *Textiles.* a manually operated or power-driven apparatus for weaving yarn or thread into cloth.

looming *Meteorology.* **1.** a mirage effect that is produced by abnormally high refraction in the lower atmosphere, permitting the observation of objects that are usually below the horizon; it occurs when air density decreases more rapidly with height than in the normal atmosphere. **2.** a phenomenon in which objects that are observed through fog or mist appear to be enlarged.

Loomis-Word diagram *Spectroscopy.* a graph in which differences between observed wavenumbers and extrapolated wavenumbers belonging to one branch of overlapping rotational bands are plotted with respect to arbitrary running numbers for that branch; used to assign lines to overlapping rotational bands.

loon *Vertebrate Zoology.* a diving, aquatic, fish-eating, monogamous bird of the order Gaviiformes; found in the northern North Hemisphere and characterized by legs placed far back, giving it a clumsy walk. (Possibly from an earlier word imitative of the sound of its call; influenced by the similar word *loon* "a crazy or feeble-minded person.")

loon

Loon *Ordnance.* an early American cruise missile, based on the German VI; adapted for launch from submarines.

loop a curved shape formed by a thread, wire, or the like that folds or doubles upon itself; specific uses include: *Physics.* a closed curve in graphical form, such as a hysteresis loop or a momentum-space representation of a harmonic oscillator. *Engineering.* **1.** a tape whose ends have been spliced together, resulting in continuous play. **2.** a closed circuit of pipe containing materials to be tested under varying conditions. *Electricity.* **1.** a complete electrical circuit. **2.** a set of branches that form a closed current path, where the omission of any branch breaks up the closed path. Also, MESH. *Electromagnetism.* a coupling loop that extends into a waveguide to deliver or receive electromagnetic energy. *Computer Programming.* **1.** a sequence of instructions in a program that can be repeated as long as a specified condition is met. **2.** an infinite sequence of this type. *Aviation.* **1.** a flight maneuver in which an airplane rotates nose-up through a complete circle in a vertical plane, maintaining a horizontal lateral axis; a normal loop regains level flight when it completes a maneuver at the original heading and at a slightly higher altitude. **2.** the path followed or described in such a maneuver. *Molecular Biology.* a single-stranded segment of DNA, the ends of which are paired together by complementary base sequences, causing the unpaired region to loop out. *Biotechnology.* a device used for the manipulation of small quantities of solid or liquid material during inoculation in bacteriology or mycology. *Mathematics.* **1.** let X be an arcwise connected topological space and $x_0 \in X$ a fixed point. A loop based at x_0 is a continuous function f: $[0,1] \rightarrow X$ such that $f(0) = f(1) = x_0$. **2.** a loop on a graph G is an edge that is incident on only one vertex, that is, whose endpoints are identical.

loop antenna *Electromagnetism.* a directional antenna in which the radiating element is bent into one or more complete turns of a conductor and commonly has a tuning capacitor connected across its ends.

loopback switch *Electronics.* a switch that, when closed, tests a telephone line by reflecting received signals from the end of the line to the sender.

loop body *Computer Programming.* a set of statements to be performed repeatedly within the range of a loop.

loop checking *Telecommunications.* the transmission of signals from a central office in order to test local loop integrity.

loop circuit *Telecommunications.* a communications circuit that is shared by more than two parties; in the operation of a teletypewriter, every machine prints all the data entered on the loop.

loop control *Control Systems.* a photoelectric control system used to regulate the position for a loop of material being transferred from one line to another. Also, PHOTOELECTRIC LOOP CONTROL.

loop coupling *Electromagnetism.* a method of extracting or delivering electromagnetic energy to a waveguide by means of inserting a coupling loop into the guide.

loop dialing *Telecommunications.* a return-path dialing method that sends pulses out over one side of an interconnecting line and returns them over the other side.

looped pipeline *Petroleum Engineering.* a pipeline accompanied by a second, parallel pipeline whose purpose is to increase the throughput capacity of the system; both lines serve the same source and destination.

looper *Invertebrate Zoology.* a descriptive term for a caterpillar that has no center legs and moves by anchoring the front of the body and moving the rear portion up to the anchored front in a looping motion. Also, MEASURING WORM, INCHWORM.

loop exit *Computer Programming.* **1.** a statement that terminates execution of a loop. **2.** the act of exiting a loop.

loop fermenter *Biotechnology.* a bioreactor in which the culture is continuously recycled through two chambers to increase both the rate of oxygen transfer and the production volume.

loop filter *Electronics.* a circuit that passes the audio or video components of a frequency-modulated signal, but blocks the components of the signal that transmitted it.

loop flow see PARALLEL FLOW.

loop gain *Control Systems.* in a feedback control system, the ratio of the magnitude of a primary feedback signal to the magnitude of the actuating signal.

loop head *Computer Programming.* the initial instruction of a loop, which may evaluate the condition for terminating the loop and may also increment variables used to control the looping process.

loop index *Computer Science.* a variable that is incremented during a loop and used to perform a similar action on successive data. Also, LOOP VARIABLE, INDUCTION VARIABLE.

looping *Engineering.* the process of laying a duplicate pipeline along the whole or a portion of the original pipeline in order to increase capacity. *Entomology.* the movement made by numerous hairless moth caterpillars such as inchworms, in which they bring their front and hind legs together due to a lack of prolegs in the middle.

looping mill *Metallurgy.* a bar-rolling mill in which the bar exiting from one stand enters a second stand and reverses the rolling direction.

loop invariant *Computer Programming.* a proposition, involving variables of a program, that is true for every execution of a loop; used in program verification.

loop lake see OXBOW LAKE.

loop-mile *Electricity.* a length of wire in a mile of two-wire line.

loop model *Biochemistry.* a model expressing the phenomenon that electrons circle through the interior of an energy-transducing membrane, beginning and ending at the same point on the surface of the membrane, while protons move through the membrane, beginning at one side and ending at the other.

Loop Nebula see TARANTULA NEBULA, CYGNUS LOOP.

loop of Henle *Anatomy.* the long U-shaped portion of the nephron extending between the proximal tubule and distal convoluted tubules.

loop prominence *Astronomy.* arches and coils of high-temperature gas that form above the sun's limb after great flares have erupted, and in which gas drains down to the solar surface.

loop pulsing *Telecommunications.* the regular, momentary suspension of the direct-current path at the sending end of a transmission line. Also, DIAL PULSING.

loop rating *Hydrology.* a graph of river stage with respect to discharge, in which some values for discharge are higher at certain stages when the river is rising than when it is falling, thus describing a loop with each rise and fall of the river.

loop seal *Chemical Engineering.* an immersed tube used to control the rate of liquid drawoff from process or storage vessels, while preventing the escape of vapor.

loopstick antenna see FERRITE-ROD ANTENNA.

loop structure *Computer Programming.* the representation in a structured flowchart of the repetitive execution of a set of instructions and the logic for exiting the loop.

loop test *Electricity.* a test used to locate a fault in the insulation of a conductor when the conductor can be made a part of a loop.

loop transfer function *Control Systems.* in a feedback control system, the ratio of the Laplace transform of a primary feedback signal to the Laplace transform of the actuating signal. Also, **loop ratio.**

loop transmission *Telecommunications.* **1.** in a given communications system, a series connection of terminal circuits. **2.** in telephony, signals, circuits, and systems between a customer's telephone and the central office.

loop transmittance *Control Systems.* **1.** transmittance between the source and sink created by splitting a specified node in a signal flow graph. **2.** a node that has been inserted in a specified branch in such a way that the transmittance of the branch is unchanged.

loop traverse *Cartography.* in surveying, a closed traverse that begins and ends at the same station.

loop tunnel *Civil Engineering.* a tunnel that has been folded back on itself in order to change grade.

loop-type pit bottom *Mining Engineering.* an arrangement at the bottom of a pit in which loaded cars are fed to the cage on one side only; the empties are returned by a loop roadway to their original place.

loopway *Mining Engineering.* a double-track loop on which mine cars may pass in a main single-track haulage plane.

loose butt hinge *Building Engineering.* a hinge fabricated so that one leaf can be lifted from the other, allowing the door to be easily removed.

loose connective tissue *Histology.* an irregular connective tissue containing collagenous fibers and an abundance of liquid-to-semiliquid matrix.

loose coupling *Electricity.* coupling of a degree significantly less than critical coupling, in which there is very little transfer of energy. Also, WEAK COUPLING.

loose-detail mold *Engineering.* a plastics mold in which some parts are removed with the molded piece.

loose fit *Design Engineering.* a fit of machine parts that allows enough clearance for joined members to move freely.

loose ground *Mining Engineering.* any broken, fragmented, or loosely cemented bedrock material requiring support because of its tendency to slough from sidewalls into a borehole. Also, BROKEN GROUND.

loose-joint hinge *Mechanical Devices.* a door hinge with two knuckles, one having a vertical pin in its center that fits into a corresponding hole in the other. Also, **loose-joint butt.**

loose kernel smut *Plant Pathology.* a disease of a plant kernel caused by smut fungi, and characterized by a gall containing spores that burst.

loose list *Computer Programming.* a list containing some calls that are empty. Also, THIN LIST.

loosely coupled *Computer Science.* describing a parallel computer architecture consisting of multiple CPUs that communicate with each other via a network or I/O channels, but are otherwise relatively independent. Thus, **loosely coupled computer.**

loose-pin hinge *Mechanical Devices.* a knuckle hinge with a removable pin. Also, **loose-pin butt.**

loose-powder hot pressing *Materials Science.* in the powder processing of metals or ceramics, the densification of loosely packed powders by the simultaneous application of heat and pressure.

loose pulley *Mechanical Engineering.* a pulley that turns freely on a shaft, used with a fast pulley in belt-driven machinery; the driving belt is shifted from one pulley to the other to cause the machine to start or stop.

loose rate *Industrial Engineering.* an allowed time that is greater than necessary for a normal operator to achieve standard performance, thereby permitting less than optimal effort.

Looser-Milkman's syndrome see MILKMAN'S SYNDROME.

loose round *Ordnance.* a defective cartridge with a bullet loose in the cartridge case.

loose running see FREE RANGING.

loose-snow avalanche *Hydrology.* a type of avalanche in snow that lacks cohesion, starting at one localized point and widening as it slides downhill.

loosestrife *Botany.* any of various plants belonging to the family Primulaceae, especially those of the genus *Lysimachia,* generally having loose clusters of yellow, white, or rose flowers, such as the **golden** or **garden loosestrife,** *L. vulgaris,* the **tufted loosestrife,** *L. thyrsiflora,* or the **whorled loosestrife,** *L. quadrifolia. Medicine.* the leaves or flowers of such a plant, formerly regarded as having tranquilizing properties. (Said to be named for King *Lysimachos* of Thrace, who according to legend used a sprig of this plant to soothe a raging bull; the English word is an altered version of this to reflect the idea of *loosing strife.*)

looting *Archaeology.* the unscientific and often illegal act of plundering archaeological sites for profit, as by the sale of precious stones, art objects, and other curios.

LOP line of position.

L.O.P. left occipitoposterior.

Lopadorrhynchidae *Invertebrate Zoology.* a small family of aquatic polychaete annelid worms in the group Errantia.

loparite-(Ce) *Mineralogy.* $(Ce,Na,Ca)_2(Ti,Nb)_2O_6$, a brown to black, pseudocubic, possibly orthorhombic mineral of the perovskite group occurring in pseudocubic twinned crystals, having a specific gravity of 4.77 and a hardness of 5.5 on the Mohs scale; found in alkalic massifs of the Kola Peninsula, USSR.

lopezite *Mineralogy.* $K_2Cr_2^{+6}O_7$, an orange-red, transparent, water-soluble, triclinic mineral occurring in small spherical aggregates, having a specific gravity of 2.69 and a hardness of 2.5 on the Mohs scale; found in cavities in nitrate rock.

Lophiaceae *Mycology.* a family of fungi belonging to the order Melanommatales which live off decaying woody plants, primarily in temperate regions.

Lophialetidae *Paleontology.* a short-lived Asian family of perissodactyl ceratomorph mammals in the superfamily Tapiroidea; similar to the deperetellids, lophiodonts, and isectolophids; extant in the Middle to Upper Eocene.

Lophiidae *Vertebrate Zoology.* a family of marine food fishes composing the suborder Lophioidei in the order Lophiiformes; characterized by a depressed and very broad head and body, a wide large mouth, and naked skin; found in the Atlantic, Indian, and western Pacific Oceans, where they live in muddy bottoms of the continental slope.

Lophiiformes *Vertebrate Zoology.* an order of deep-sea fish containing the anglerfishes; characterized by dorsal fins modified into line and bait devices for capturing prey.

lophine *Organic Chemistry.* $C_{21}H_{16}O_2$, colorless needles that are insoluble in water and melt at 275°C; a strong base that forms metallic derivatives and is used as a fluorescent neutralization indicator.

Lophiodontidae *Paleontology.* a small, short-lived family of perissodactyl ceratomorphs belonging to the superfamily Tapiroidea; the **lophiodonts** were the European cousins of the Asian deperetellids and lophialetids and of the North American isectolophids; extant in the Lower to Upper Eocene.

Lophiostomataceae *Mycology.* a family of fungi belonging to the order Melanommatales, characterized by brightly colored ascospores and living off the decaying remains of herbaceous or woody plants and old fungi.

lopho- or **loph-** a combining form meaning "ridge" or "tuft," as in *lophodont, lophophore.*

lophocercous *Vertebrate Zoology.* having a ridgelike caudal fin lacking fin rays.

Lophocoleaceae *Botany.* a cosmopolitan family of minute to robust liverworts of the order Jungermanniales, characterized by reddish or brownish plants that are prostrate to erect and have irregular, variable branches, three-lobed wide-mouthed perianths, and large bifid underleaves.

lophocyte *Invertebrate Zoology.* a special cell of uncertain function found in certain sponges.

Lophodiniaceae *Botany.* a family of freshwater biflagellates of the order Peridiniales that are thecate and photosynthetic and are characterized by numerous plastids, with an eyespot and a thin theca composed of hexagons.

lophodont *Vertebrate Zoology.* the type of dentition found in elephants and horses, in which the enamel and dentine are intricately folded to form transverse ridges across the tooth.

Lophogastrida *Invertebrate Zoology.* a suborder of marine crustaceans in the order Mysidacea, with seven pairs of gills.

lophophore *Invertebrate Zoology.* a feeding organ with ciliated tentacles, found in certain aquatic groups including Phoronida, Bryozoa, and Brachiopoda.

Lophosoriaceae *Botany.* a monotypic family of primitive trunkless tree ferns of the order Filicales that grow at high elevations in tropical America; characterized by short-creeping rhizomes with long golden hairs and fronds that grow up to four meters long and that are divided four times.

Lophotidae *Vertebrate Zoology.* the crestfishes, a monogeneric family of rare marine fish of the order Lampriformes, characterized by a long ribbonlike body, small deciduous scales, and an ink sac.

lophotrichous *Cell Biology.* describing flagella that are arranged in a tuft at the pole of a cell.

Lophoziaceae *Botany.* a family of primitive liverworts of the order Jungermanniales, characterized by brownish to reddish pigmentation, lateral intercalary branches, strongly differentiated stems, and scattered rhizoids often in a dense ventral mat; they develop best in colder areas of the Northern Hemisphere.

lopolith *Geology.* a large, concordant igneous intrusion having a lens- or saucer-shaped cross section as a result of the sagging of the underlying country rock.

lopping shears *Mechanical Devices.* long-handled pruning shears with one straight and one curved cutting edge; used for cutting branches from shrubs and trees.

loquat *Botany.* a small evergreen tree, *Eriobotrya japonica,* having yellow, plumlike fruit with a tart taste; native to China and Japan, but also grown in North America.

loran or **Loran** [lôr´an] *Navigation.* a widely used radio navigation system in which hyperbolic lines of position are determined by measuring the difference of time in receiving signals from two fixed, synchronized transmitters. (An acronym for long-range navigation.)

loran A *Navigation.* a former (now discontinued) loran system that operated in the frequency band 1750–1950 kHz.

loran B *Navigation.* an experimental loran system.

loran C *Navigation.* the current loran system. It operates in the frequency band 100–110 kHz.

loran chain *Navigation.* a group of loran transmitting stations consisting of a master and two or more slaves. They are located so that signals from the master and at least two slave stations can be received throughout the coverage area.

loran chart *Navigation.* a navigational chart that has been overprinted with loran position-fixing lines.

loran D *Navigation.* a readily transportable, low-power version of loran C designed for military tactical use.

lorandite *Mineralogy.* $TlAsS_2$, a vivid red, monoclinic mineral occurring in tabular or prismatic crystals, having a specific gravity of 5.53 to 5.54 and a hardness of 2 to 2.5 on the Mohs scale; found in arsenic-antimony deposits.

loran fix *Navigation.* a relatively accurate position that has been determined by the use of loran.

loran lattice tables *Navigation.* a set of tables that provide coordinates that enable a navigator to construct straight line representations of Loran's hyperbolic lines of position.

loran rate *Navigation.* the frequency channel and repetition that are used to identify a given pair of loran stations.

loranskite-(Y) *Mineralogy.* an inadequately described black mineral with an approximate formula of $(Y,Ce,Ca)ZrTaO_6$, a specific gravity of 3.8 to 4.8, and a hardness of 5 to 6 on the Mohs scale.

Loranthaceae *Botany.* a family of dicotyledonous plants of the order Santalales, characterized by regular, bisexual and unisexual flowers, broad, generally opposite, entire leaves, and ovules embedded in a central placenta.

lorate *Biology.* strap-shaped, such as in certain leaves.

lordosis *Medicine.* an abnormal forward curvature of the lumbar spine. Also, SWAY-BACK, SADDLE BACK, HOLLOW BACK.

Lorentz, Hendrik Antoon 1853–1928, Dutch physicist; developed the Lorentz transformations and Fitzgerald-Lorentz contraction; shared the Nobel Prize for discovery of the Zeeman effect.

Lorentz-Boltzmann equation *Physics.* an approximation to the Boltzmann transport equation for states that are near equilibrium.

Lorentz electron *Electromagnetism.* an electron that is treated as a damped harmonic oscillator; the model is used to account for the frequency dependence of the real and imaginary parts of the index of refraction for a specified substance.

Lorentz equation *Electromagnetism.* an equation of motion for a charged particle in which the Lorentz force is equated to the rate of change of the momentum of the particle.

Lorentz factor *Crystallography.* a factor used in analyses of X-ray diffraction intensity data from a crystal. It takes into account the time that it takes for a given Bragg reflection (represented as a reciprocal lattice point with finite size) to pass through the surface of the sphere of reflection. The value depends on the scattering angle and on the geometry of the measurement of the reflection.

Lorentz-FitzGerald contraction *Astrophysics.* an effect of special relativity in which all moving objects contract in the dimension parallel to their motion.

Lorentz force *Electromagnetism.* the force acting on a charged particle in the presence of a magnetic field: $F_L = q(E + v \times B)$.

Lorentz force density *Electromagnetism.* for a system of charged particles, the force per unit volume that is acting on a charge density or on a current.

Lorentz frame *Physics.* any one of the inertial coordinate systems, with three space coordinates and one time coordinate, in which each frame is in uniform motion with respect to all the other frames, and the interval between any two events is the same in all frames; used to describe the motions of relativistic objects.

Lorentz gas *Electronics.* a model in which ions in a completely ionized gas are presumed to be stationary; therefore, any interactions between electrons are ignored.

Lorentz gauge *Electromagnetism.* any gauge whereby the divergence of the magnetic vector potential and the quantity given by partial derivative of the scalar potential divided by the speed of light (in Gaussian units) identically sums to zero.

Lorentz group *Mathematics.* the group of all linear transformations of the vector space R^4 that leave the quadratic form $-x_0^2 + x_1^2 + x_2^2 + x_3^2$ invariant. The Lorentz group is isomorphic to $O(1, 3, R)$, a real form of the complex orthogonal group $O(4)$.

Lorentzian *Spectroscopy.* of or relating to a distribution of intensities as a function of frequency centered about a specific frequency for a pulse emission of finite duration. Lorentzian curves differ from Gaussian curves.

Lorentz invariance *Physics.* a property or relationship that is not affected by the Lorentz transformation and remains the same in any Lorentz frame.

Lorentz line-splitting theory *Atomic Physics.* a theory that attempts to explain the normal Zeeman effect in which the spectral lines of a light source are split into three lines, one corresponding to zero-field frequency and the other two corresponding to the zero-field frequency shifted up and down by the Larmor frequency.

Lorentz-Lorenz equation *Optics.* the equation defining molecular refraction as $R = (n^2 - 1/n^2 + 2)(M/\rho)$, where R is the molecular refraction, n is the refractive index, M is the molecular mass, and ρ is the density.

Lorentz number *Physics.* the ratio of the velocity of a fluid to the velocity of light. *Solid-State Physics.* a quantity according to the Wiedermann-Franz law, given by thermal conductivity in a metal divided by the product of the electrical conductivity and the absolute temperature of the metal: $L = K/Ts$.

Lorentz theory of light sources *Atomic Physics.* a theory stating that light is emitted from atoms through electron vibrations; the model assumes that the electrons are damped harmonic oscillators.

Lorentz transformation *Physics.* a relation between measurable quantities that refer to two inertial frames of reference in relative motion.

Lorentz unit *Spectroscopy.* a wavenumber unit used to measure the difference between a ground-state spectral line and its Zeeman components.

Lorenz, Konrad 1903–1989, Austrian zoologist; awarded the Nobel Prize for his pioneering work in ethology; studied the process of behavioral imprinting.

Lorenz attractor *Chaotic Dynamics.* the strange attractor that appears in the phase space for the chaotic solution of the Lorenz equations for parameter values $\sigma = 10$, $b = 8/3$, and $r = 28$, though similar attractors also appear for relatively large values of r so long as $\sigma > 1 + b$ and $b > 0.5$.

Lorenz curve *Statistics.* a graphical method that is used to study the concentration of a transferrable character, such as income, in a given population.

Lorenz equations *Chaotic Dynamics.* a simplified model devised by E. Lorenz as part of a study of the weather through convective hydrodynamics in the presence of horizontal and vertical temperature variations. The three equations are: $dX/dt = -\sigma X + \sigma Y$, $dY/dt = rX \cdot Y - XZ$, and $dZ/dt = XY - bZ$, and they appear for a wide range of the parameters. A notable characteristic is an inversion symmetry; i.e., if $X(t)$, $Y(t)$, $Z(t)$ is a solution, then $(-X(t), -Y(t), -Z(t))$ is also a solution.

Lorenz map *Chaotic Dynamics.* a cusp-shaped mapping created by E. Lorenz by plotting pairs of successive peaks of the variable z for the Lorenz attractor. The near tangency of the map to $x_{n+1} = x_n$ creates the characteristic (Type I) intermittency of the chaotic Lorenz attractor.

lorettoite *Mineralogy.* $Pb_7O_6Cl_2$, a honey-yellow to reddish-yellow, tetragonal mineral of uncertain status, possibly an artifact, having a specific gravity of 7.39 and a hardness of 2.5 to 3 on the Mohs scale; found with other lead minerals at Loretto, Tennessee.

lorica *Invertebrate Zoology.* the hard external covering that functions as an exoskeleton in protozoans and rotifers.

Loricariidae *Vertebrate Zoology.* a family of more than 400 species of South American armored catfishes belonging to the order Siluriformes, characterized by an iris structure that occludes the pupil in light and contracts it in darkness, and by antennalike growths on the heads of some males.

Loricata see POLYPLACOPHORA.

loricate *Invertebrate Zoology.* covered with a lorica. Also, **loricated.**

Loricifera *Invertebrate Zoology.* a phylum of minute marine aschelminths that adhere tightly to gravel substrates; the cosmopolitan *Nanaloricus mysticus* is typical.

lorikeet see LORY.

loris *Vertebrate Zoology.* either of two slow-moving, nocturnal lemurs belonging to the family Lorisidae; the **slender loris**, *Loris tardigradus,* is found in Sri Lanka and India, and the larger **slow loris**, *Nycticebus coucang,* an endangered species, is found in various areas of Southeast Asia.

loris

Lorisidae *Vertebrate Zoology.* the loris, potto, and angwantibo, a family of primates of the suborder Strepsirhini, native to Asia, Africa, India, and Ceylon.

lorry *Mechanical Engineering.* **1.** in British English, an informal term for a truck, especially a large truck. **2.** any of various small carts or trucks running on rails, as in a mine or factory. (From an older English word meaning "pull" or "drag.")

lory *Vertebrate Zoology.* a small, brightly plumed parrot of the Loriidae family, having a brush-tipped tongue for collecting nectar and pollen; inhabits forested tropical areas of Australia, New Guinea, and the Pacific Islands. Also, LORIKEET.

lory

Loschmidt number *Physics.* the number of molecules contained in one cubic centimeter of an ideal gas at standard atmospheric pressure and a temperature of 0°C, approximately equal to 2.687×10^{19}. (Calculated by Joseph *Loschmidt,* 1821–1895, Austrian chemist.)

loseyite *Mineralogy.* $(Mn^{+2},Zn)_7(CO_3)_2(OH)_{10}$, a bluish-white or brownish monoclinic mineral occurring in aggregates of lathlike crystals, having a specific gravity of 3.27 and a hardness of about 3 on the Mohs scale.

losing stream see INFLUENT STREAM.

loss a failure, shortfall, or detriment; specific uses include: *Engineering.* power dissipated in a system without performing its desired function. *Telecommunications.* a reduction in the level of signal strength during its conveyance from one point to another in a network, usually expressed in decibels. *Statistics.* **1.** in decision theory, a function describing the penalty for taking a certain action as a function of the unknown quantities. **2.** in estimation, a measure of the distance between the estimator and the correct population value.

loss angle *Electromagnetism.* a measure of power loss in an inductive or capacitive component when the current and voltage are out of phase by 90°.

loss compliance *Materials Science.* a mechanical property of viscoelastic materials, in which the out-of-phase ratio of strain to stress is related to energy dissipated, defined by the equation $g = S_0(J' \sin wt - J'' \cos wt)$, where J' and J'' are the compliances for viscoelastic behavior, g is the strain, and S_0 is the sinosoidal amplitude.

loss cone *Physics.* a cone confined by magnetic mirrors in the velocity space of particles in a plasma, from which particles with velocity are not trapped by the mirrors and are lost from the system.

loss current *Electromagnetism.* in an inductor, the component of current that is in phase with the voltage and consequently is responsible for the resistive power losses of the coil.

losser *Electricity.* a dielectric material that dissipates energy, placed in a circuit to prevent oscillation. Thus, **losser circuit.**

loss evaluation *Electricity.* a measure of the decrease in signal power from point to point in a transmission.

loss factor *Electricity.* **1.** the ratio of average power loss to peak-load power loss during a given period of time. **2.** the power factor of a material multiplied by its dielectric constant, determining the amount of heat generated in the material.

loss-in-weight feeder *Mechanical Engineering.* a device used to control the rate of output of a granulated or powdered solid from a feed hopper; as the weight of the hopper contents decreases, the device causes the discharge chute to open more widely, compensating for the associated loss of flow.

lossless junction *Electromagnetism.* a perfectly reflecting waveguide junction.

lossless material *Physics.* any substance that ideally does not dissipate any energy (acoustic or electromagnetic) as heat.

loss modulation see ABSORPTION MODULATION.

loss modulus *Materials Science.* an imaginary part of the complex modulus that describes the dissipation of elastic energy into heat when a material is deformed sinosoidally; it is the product of the storage modulus and the tangent of the loss angle.

loss mutant *Genetics.* an alternate term for deletion mutation.

loss-of-coolant accident *Nucleonics.* a decrease or cessation in the quantity or flow rate of coolant to or from a reactor core, resulting in a fuel heat-up followed by a rupture of the fuel cladding and release of fission products; if no emergency core cooling system is present, the exposed fuel rods would melt from fission-product heating, accompanied by possible meltdown and the release of hydrogen from chemical reactions between molten fuel and the steam-water mixture.

loss of head *Fluid Mechanics.* the reduction in the available energy of a flowing fluid, measured between two points and due to such causes as friction, obstacles, and bends.

loss-of-head gauge *Engineering.* a gauge located on a rapid sand filter that monitors loss of head during filtration.

loss rate the rate at which something is lost or depleted; specific uses include: *Ecology.* a process of population depletion due to death and emigration. *Archaeology.* the pattern of loss for a cultural object; large objects are less subject to loss than small ones, and valuable items are more likely to be searched for than items considered to have little value.

lossy *Electricity.* causing a significant loss or dissipation of energy.

lossy attenuator *Electromagnetism.* a waveguide attenuator that is constructed from dissipative material, deliberately introducing a transmission loss.

lossy line *Electricity.* a transmission line having intentionally high attenuation.

lossy material *Physics.* a substance that tends to dissipate more acoustic or electromagnetic energy than is considered normal for the class of substance.

lost circulation *Petroleum Engineering.* a serious condition in which drilling mud that has been pumped through a drill pipe into a well does not return to the collar of the drill hole, having instead been lost in crevices or porous formations.

lost hole *Petroleum Engineering.* a well that has suffered a serious incident, such as a blowout, and can no longer be worked.

lost motion *Mechanical Engineering.* **1.** any mechanical motion during which no useful work is accomplished. **2.** the delay between the movement of a driver and that of a follower.

lost stream *Hydrology.* **1.** a dried-up stream in an arid region. **2.** see SINKING STREAM.

lost time *Acoustical Engineering.* a term for recording time that could be used more effectively with variable recording speed than with a constant velocity recorder.

lost-wax process *Metallurgy.* an investment casting process in which the mold is built up around a pattern made of wax; used to produce extremely precise molds. The technique has been used since ancient times in sculpture, as in the Classical Greek era. (So called because the mold is subjected to heat to melt away the wax before the casting is made.)

lost work *Thermodynamics.* an effect associated with every irreversible process, representing the difference between the heat transfer to the system and the amount of actual work done. In a reversible process, the lost work is zero.

lot *Civil Engineering.* an area, parcel, or allotment of land that is determined by a survey or topographically noted and recorded. *Industrial Engineering.* a standard production quantity of a good or product.

L.O.T. left occipitotransverse.

lot-by-lot inspection *Industrial Engineering.* the inspection of random samples that are taken from successive production lots produced by a single continuous process.

lotic *Ecology.* of or relating to freshwater habitats characterized by swift-moving currents, such as rivers and streams.

lot line *Civil Engineering.* the borderline or boundary of an assessed parcel of land.

lot number *Industrial Engineering.* a number assigned to all members of a given production lot.

lot plot method *Industrial Engineering.* a quality-control sampling procedure based on a frequency plot of defects found in a random sample of fifty items from each lot.

lot size *Industrial Engineering.* the number of workpieces or products ordered from a manufacturer. Also, ORDER QUANTITY.

lot splitting *Industrial Engineering.* the process of expediting production by breaking a critical order into smaller parts, moving some parts ahead quickly, and allowing the rest of the order to proceeed at a normal rate.

lot tolerance percent defective *Industrial Engineering.* in quality control, the highest acceptable percentage of defectives in a lot.

lotus *Botany.* **1.** a plant described in Greek legends as producing a dreamy, sleeplike state in those who ate its fruit. **2.** a popular name for several different kinds of plants identified with this, especially various waterlilies of the family Nymphaeaceae, such as the **Egyptian lotus,** *Nymphaea lotus*, the **American** or **yellow lotus,** *Nelumbo lutea*, or the **Hindu** or **sacred lotus,** *N. nucifera*, the national flower of India. **3.** any of various plants of the genus *Lotus* of the family Leguminosae, having yellow, red, purple, or white flowers. *Architecture.* a decorative motif derived from lotus plants or blossoms, as in Egyptian columns.

loudness *Acoustics.* the relative extent to which a sound can be heard; the degree of auditory sensation produced by a sound. *Acoustical Engineering.* see LOUDNESS CONTROL.

loudness analyzer *Electronics.* an instrument that creates a display on a cathode-ray screen depicting the amplitude of airborne sounds at various bands of the audio spectrum.

loudness control *Acoustical Engineering.* a device in a sound reproduction system that compensates for the human ear's loss of sensitivity to low and high frequencies at low volume levels. Similarly, **loudness switch.**

loudness level *Acoustics.* the relative intensity of a sound source, measured in decibel units and expressed in terms of an equally loud reference tone heard at the same place.

loudness unit *Acoustics.* a sound intensity unit relating a given sound to a known reference level expressed in units of phons by the expression $10 \log_{10}(I_1/I_2)$, where I_2 represents the reference level of 0.0002 dyne/cm^2.

loudspeaker *Acoustical Engineering.* any of various electronic devices that convert electromagnetic waves into acoustic waves, in order to provide audible sound for a radio, tape player, phonograph, television, public-address system, or the like.

Lou Gehrig's disease *Medicine.* a popular name for amyotrophic lateral sclerosis; the disease is identified with its most famous victim, the baseball star Lou Gehrig (1903–1941). See AMYOTROPHIC LATERAL SCLEROSIS.

loughlinite *Mineralogy.* $Na_2Mg_3Si_6O_{16}\cdot 8H_2O$, a pearly white, orthorhombic mineral having a specific gravity of 2.165 and an undetermined hardness; found as fibrous veinlets in dolomtic oil shale.

louping ill *Veterinary Medicine.* ovine encephalomyelitis; a tick-borne viral disease of predominantly sheep and cattle, resulting in muscle spasms, lack of coordination, nervous symptoms, and paralysis.

louse *plural,* **lice.** *Invertebrate Zoology.* **1.** the common name for any of various parasitic wingless insects of the orders Anoplura or Mallophaga. **2.** any of various arthropods that suck the blood of animals or the juices of plants. **3.** any of various nonparasitic insects such as the wood louse or book louse.

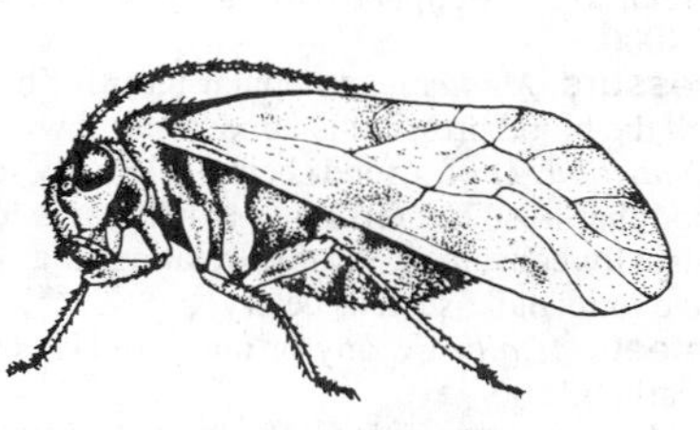

louse (book louse)

louver *Engineering.* an assembly of angled slats placed over an opening or window to allow ventilation and yet protection from the elements. *Acoustical Engineering.* an arrangement of slats or barriers in a loudspeaker, serving to protect the unit and often to enhance its acoustic characteristics.

lovage *Botany.* a European plant of the parsley family, *Levisticum officinale*, having coarsely toothed compound leaves; it produces a yellow-brown essential oil known as **lovage oil** or **levisticum oil,** which is used in flavoring and perfumes.

LOVAL *Virology.* a term for baculovirus particles that have been incorporated in occlusion bodies during the infection of insect larvae and are then released in alkali. (An acronym for larval occluded virus alkali-liberated.)

lovebird *Vertebrate Zoology.* a popular name for any of various small tropical parrots, especially certain species of the genus *Agapornis*, that are noted for an affectionate attachment to their mates and are often kept as pets; native to Africa.

Lovell, Sir Bernard born 1913, English radioastronomer; at Jodrell Bank, he built the first fully steerable radiotelescope.

Love wave *Geophysics.* a surface seismic wave that has a horizontal motion at right angles to the direction of propagation.

Lovibond tintometer *Optics.* an instrument that determines the intensity of colors in liquids or solids by measuring the reflected or transmitted light and matching it against three basic colors.

lovozerite *Mineralogy.* $Na_2Ca(Zr,Ti)Si_6(O,OH)_{18}$, a red, brown, or black, opaque, possibly trigonal mineral of the lovozerite group, massive to granular in habit, having a specific gravity of 2.3 to 2.7 and a hardness of 5 on the Mohs scale; found in nepheline-syenite.

low *Meteorology.* an area of lower barometric pressure where air is rising, thus resulting in inclement weather. Also, LOW-PRESSURE SYSTEM, DEPRESSION, CYCLONE.

low airburst *Ordnance.* the fallout-safe height of burst for a nuclear weapon that maximizes damage to or casualties on a surface target.

low-alloy steel *Metallurgy.* a steel containing alloy elements in moderate amounts, often below 1%, in addition to carbon and manganese.

low aloft see UPPER-LEVEL CYCLONE.

low-altitude alert *Transportation Engineering.* a controller's warning given to an aircraft crew when the aircraft approaches dangerously close to the ground.

low-altitude bombing *Military Science.* horizontal bombing with a release height between 900 and 8000 feet. Also, LOW-LEVEL BOMBING.

low-angle bombing *Military Science.* the process of dropping bombs from an aircraft that is diving at a slight angle, as opposed to the greater dive angles used in glide bombing or dive bombing.

low-angle fault *Geology.* a fault having a dip of less than 45°.

low-angle fire *Ordnance.* fire delivered at an angle of elevation below the elevation corresponding to the maximum range of the particular gun and ammunition.

low-angle grain boundary *Metallurgy.* in the microstructure of a polycrystalline material, a grain boundary between grains having approximately the same orientation.

low-angle loft bombing *Military Science.* a method of loft bombing in which the release angle is less than 35° above the horizontal.

low-angle thrust see OVERTHRUST.

low band *Telecommunications.* **1.** in television, the band that includes channels 2 to 6, ranging from 54 megahertz to 88 megahertz. **2.** in videotape recording, a technique that uses the 5.5 to 6.5 megahertz band. **3.** describing the apparatus for that type of recording.

low-birth-weight *Medicine.* a term used to describe an infant whose weight at birth is less than 2500 grams (6.25 pounds); such infants are so identified because they are at greater risk for certain conditions.

low blower *Aviation.* a blower-type supercharger set at low rpm.

low brass *Metallurgy.* a copper-base alloy normally containing 20% zinc; UNS C24000.

low blood pressure *Medicine.* a popular term for hypotension, a condition in which the blood pressure is abnormally low.

low-calorie *Nutrition.* **1.** of a food, being relatively low in calories for the type of food that it is. **2.** referring to a diet that is low in calories relative to that usually consumed. **3.** of or relating to a food that is low in calories relative to its bulk, such as celery.

low-carbon steel *Metallurgy.* any of many steels containing less than about 0.25% carbon.

low clouds *Meteorology.* a class of clouds consisting of those that occur at altitudes lower than 6000 feet above sea level, including the stratus and stratocumulus species.

low contrast *Radiology.* a condition in which there is small definition between tones on an X-ray. Also, LONG-SCALE CONTRAST.

low-core *Computer Technology.* a former term that refers to low memory, the locations with the lower addresses in a computer's main memory; usually restricted to access only by the operating system.

low-definition television *Telecommunications.* a television system that uses fewer than 200 scanning lines per complete frame.

low-density lipoprotein *Biochemistry.* a lipoprotein containing more lipids than protein and carrying cholesterol from the liver to various tissues.

low-dimensional solid *Materials Science.* a solid with a high degree of anisotropy, caused by a molecular or crystal structure consisting of chains, so that although the chains are assembled in a three-dimensional lattice, interactions along the chains are much stronger than those between the chains, resulting in reduced dimensionality.

low egg passage see LEP.

löweite *Mineralogy.* $Na_{12}Mg_7(SO_4)_{13}\cdot 15H_2O$, a colorless to pale-yellow, water-soluble, trigonal mineral occurring in massive granular form, and having a specific gravity of 2.36 to 2.38 and a hardness of 2.5 to 3 on the Mohs scale; found in oceanic salt deposits. Also, LOEWEITE.

Lowell, Percival 1855–1916, American astronomer; established the Lowell Observatory; predicted the existence of Pluto.

low-energy electron diffraction *Solid-State Physics.* electron diffraction from single crystal surfaces; the approximate energy range of the electrons is 5 to 500 electron volts.

low-energy environment *Geology.* an aqueous sedimentary environment characterized by a low level of kinetic energy, standing water, or a general lack of wave or current action that permits the settling and accumulation of fine-grained sediments.

low-energy np and pp scattering *Nuclear Physics.* a type of collision between a neutron and a proton, or between two protons, each with energy less than 10 MeV, in which the total kinetic energies of the two systems remain the same; usually occurring in the nucleus of a hydrogen atom.

low-energy physics *Physics.* the study of processes that occur at the nuclear level, involving energies on the order of a few MeV or less.

Lowenhertz thread *Design Engineering.* a non-U.S. standard screw thread whose angle between flanks is measured on an axial plane of 53.3 degrees, 8 minutes, with a height at approximately three-fourths the pitch, and the width of flat-set top and bottom at one-eighth times the pitch.

Lowenstein-Jensen medium *Microbiology.* a solid bacteriological culture medium used for the isolation of species of *Mycobacterium* from clinical specimens, including tubercle bacilli.

Lower or **lower** *Geology.* **1.** denoting an early division of a period or system, as in *Lower* Cambrian. **2.** referring to rocks or strata that normally occur below those of later formations of the same subdivision of rocks.

lower atmosphere *Meteorology.* the part of the atmosphere in which most weather phenomena occur; consists of the lower troposphere and the lower stratosphere.

Lower Austral life zone *Ecology.* the southern portion of the Australian zoogeographical region, defined as a separate entity.

lower bound *Mathematics.* a number less than or equal to every number in a given set. A lower bound (if it exists) for a nonempty subset B of a partially ordered set A (with ordering $\leq$) is an element d of A such that $d \leq b$ for every $b \in B$. Lower bounds are not necessarily unique; however, if $c \leq d$ for every other lower bound c of B, then d is a **greatest lower bound** of B .

lower branch *Astrophysics.* the half of an observer's meridian circle that passes through the nadir. *Navigation.* that portion of a terrestrial or celestial meridian that passes through the antipode or nadir of a given location.

Lower Cambrian *Geology.* the first subdivision of the Cambrian period, terminating about 540 million years ago.

lower chord *Civil Engineering.* the bottom longitudinal members of a truss.

Lower Cretaceous *Geology.* the first epoch of the Cretaceous period, occurring from about 140 to 120 million years ago.

lower critical solution temperature *Materials Science.* the minimum point at which phase separation will invariably occur in a polymer solution.

lower curtate *Computer Technology.* the adjacent lower rows on the bottom portion of a punched card, usually the punch positions for the digits 1 through 9.

Lower Devonian *Geology.* the first epoch of the Devonian period, occurring from about 400 to 385 million years ago.

lower fungi *Mycology.* fungi belonging to the division Myxomycota, containing the subdivisions Mastigomycotina and Zygomycotina.

lower half-power frequency *Electronics.* the frequency on an amplifier response curve that is less than the frequency for peak response.

lower heating value *Thermodynamics.* the maximum energy released in a combustion process when all the water in the combustion products is in the vapor state.

lower high water *Oceanography.* the lower of the two high waters of a tidal day.

lower integral *Mathematics.* let f be a real-valued function that is bounded on an interval $[a,b]$, and let $a = x_0 < x_1 < x_2 < \cdots x_{n-1} < x_n = b$ be a partition of the interval into n parts. The **lower Riemann integral** of $f(x)$ on $[a,b]$ is $\sup\{\sum_{k=1}^{n} y_k (x_k - x_{k-1})\}$, where y_k is the minimum value of f on the interval $[x_{k-1}, x_k]$, and where the supremum is taken over all posssible (finite) partitions of $[a,b]$.

Lower Jurassic *Geology.* the first epoch of the Jurassic period, occurring from about 185 to 170 milllion years ago.

lower larynx see SYRINX.

lower limb *Astronomy.* the limb of a celestial object (usually the sun or moon) that is closer to an observer's horizon.

lower low water *Oceanography.* the lower of the two low waters of a tidal day.

lower mantle *Geology.* the part of the earth's mantle that extends about 1000 kilometers below the surface, and is equivalent to the D layer. Also, INNER MANTLE, MESOSPHERE.

Lower Mississippian *Geology.* the first epoch of the Mississippian period, commencing about 350 million years ago.

lower motor neuron *Anatomy.* the motor neuron in a pyramidal pathway whose cell bodies are located in the nuclei of cranial nerves and in the anterior columns of gray matter of the spinal cord.

lower motor neuron disease *Medicine.* a disease resulting from injury to lower motor neurons and marked by paralysis of muscles, diminished or absent reflexes, sclerosis, and progressive muscle atrophy.

lower nephron nephrosis *Medicine.* the degeneration of the kidneys, usually due to severe injury or shock, primarily affecting the distal parts of the renal tubules.

Lower Ordovician *Geology.* the first epoch of the Ordovician period, occurring from about 490 to 460 million years ago.

lower pair *Mechanical Engineering.* a link in a mechanism in which the mating parts have surface contact.

Lower Paleolithic *Archaeology.* the earliest part of the Paleolithic period, extending from about 2.5 million years ago to about 100,000 years ago; characterized by the first use of crude stone tools, the practice of hunting and gathering, and the development of social units, settlements, and structures.

Lower Pennsylvanian *Geology.* the first epoch of the Pennsylvanian period, commencing about 310 million years ago.

Lower Permian *Geology.* the first epoch of the Permian period, occurring from about 275 to 260 million years ago.

lower pitch limit *Acoustics.* the lowest audible pitch at a given intensity.
lower plate see FOOTWALL.
lower punch *Metallurgy.* in metalworking, the lower portion of a die.
lower semicontinuous function *Mathematics.* **1.** an extended real-valued set function μ defined on a class C of sets is continuous from below at a set S in C, if for every increasing sequence $\{S_n\}$ of sets in C for which $\lim_{n\to\infty} S_n = S$, we have $\lim_{n\to\infty} \mu(S_n) = \mu(S)$. It can be shown that if such a set function μ is finite, nonnegative, and additive on a ring R, and μ is continuous from below at every subset S of R, then μ is a measure on R. **2.** a real-valued function f defined on a domain D in R^n is said to be lower semicontinuous at a point c of D if $f(c) \leq \lim\inf_{x\to c} f(x)$. Then f is said to be lower semicontinuous on D if it is lower semicontinuous at every point of D.
lower sideband *Telecommunications.* in carrier transmission, the difference between the instantaneous values of the carrier frequency and the modulating frequency, or the band of frequencies that is lower than the carrier frequency.
lower-sideband up-converter *Electronics.* a circuit that amplifies low-frequency radio signals, produced by amplitude modulation (AM), by infusing them with high-frequency alternating current.
Lower Silurian *Geology.* the first epoch of the Silurian period, commencing about 420 million years ago.
Lower Sonoran life zone *Ecology.* the biogeographical region of deserts and savannahs of the southwestern part of the United States and northern Mexico.
lower transit *Astronomy.* the moment when a circumpolar object crosses the observer's meridian beneath the pole. Also, LOWER CULMINATION.
lower-triangular matrix *Mathematics.* a square matrix whose (i,j)th entry is zero if $j > i$. If, in addition, the diagonal entries $(i = j)$ are zero, the matrix is said to be **strictly lower-triangular.**
Lower Triassic *Geology.* the first epoch of the Triassic period, occurring from about 230 to 215 million years ago.
lower variation *Mathematics.* Let μ be a signed measure on a measure space (X, A, μ), and let the sets A and B form the Hahn decomposition of X with respect to μ. The lower variation of μ, denoted μ^-, is the set function defined for every measurable set E by $\mu^-(E) = \mu(E \cap B)$. The **total variation** of μ is the sum of the upper and lower variations of μ.
lower yield point *Metallurgy.* in a steel subjected to tension, the lowest stress measured after the beginning of plastic deformation.
lowest required radiating power *Telecommunications.* the smallest power output of an antenna that is able to maintain a predetermined type of broadcast service.
lowest safe waterline *Mechanical Engineering.* the water level in a boiler tank or drum that is the lowest point for safe operation.
lowest useful high frequency *Telecommunications.* the lowest frequency effective at a predetermined time for ionospheric propagation of radio waves between any two points. It is determined by conditions such as antenna gain, noise conditions, and transmitter power.
low-expansion alloy *Metallurgy.* any of several alloys that have a low coefficient of thermal expansion.
low explosive *Materials Science.* a term for an explosive that under normal conditions deflagrates (burns rapidly); e.g., gunpowder or fireworks.
low frequency *Telecommunications.* **1.** a band of frequencies that range from 30 kHz to 300 kHz. **2. low-frequency.** of or relating to such a frequency.
low-frequency antenna *Electromagnetism.* an antenna designed to transmit or receive radiation at frequencies of less than 300 kilohertz.
low-frequency current *Electricity.* alternating current at the lower end of the frequency scale for radio waves, between 30 and 300 kilohertz.
low-frequency cutoff *Electronics.* the cutoff frequency at the lower end of the passband of a filter or amplifier.
low-frequency gain *Electronics.* an increase in an amplifier's output voltage that occurs at frequencies below the maximum value in gain.
low-frequency impedance corrector *Electricity.* a network that is connected to a basic network so that the combination will simulate at low frequencies the sending-end impedance of a line.
low-frequency induction furnace *Engineering.* an induction furnace where commercial current flow is induced in the charge to be heated.
low-frequency loran see LORAN C.
low-frequency padder *Electronics.* a device that adjusts the calibration of a superheterodyne receiver at the low-frequency end of the tuning range.
low-frequency propagation *Electromagnetism.* the propagation of radio waves at frequencies between 30 and 300 kilohertz.
low-frequency spectrum *Spectroscopy.* for atoms and molecules, spectral lines or bands in the microwave region.
low-frequency transconductance *Electronics.* the ratio between the change in the anode current of a vacuum tube and the change in the control-grid voltage that produced it, at frequencies low enough for the two to be considered in phase.
low-frequency tube *Electronics.* a vacuum tube in which the transit time of the electrons is much shorter than the period (reciprocal of frequency) of the signals applied to it.
low-grade *Science.* **1.** at the low end of a specified range. **2.** of inferior quality. *Mining Engineering.* describing any ore that has a relatively low content of minerals. Also, see COARSE, LEAN.
low heat value see LOWER HEATING VALUE.
low-helix drill *Mechanical Devices.* a double-fluted twist drill with a lower-than-average helix. Also, SLOW-SPIRAL DRILL.
low-hydrogen electrode *Metallurgy.* in arc welding, a composite electrode that provides a low-hydrogen atmosphere during operation.
Lowiaceae *Botany.* a monogeneric family of monocotyledonous, perennial, tanniferous herbs of the order Zingiberales, characterized by sympodial rhizomes, vessels only in the roots, and distichous leaves; native to China, the Malay Peninsula, and some Pacific islands.
Lowie, Robert H. 1883–1957, Austrian-born American anthropologist; a founder of ethnology; studied North American Indians.
low-impedance measurement *Electronics.* a technique used to indirectly measure the opposition to alternate current flow when it is too weak to measure directly.
low-impedance switching tube *Electronics.* a device that is able to transmit a signal with little loss of intelligence, because while it has high static impedance (well over 10,000 ohms), its negative dynamic impedance is at or below zero; commonly used as a relay.
low index *Meteorology.* a low value of the zonal index, indicating (in the middle latitudes) a weak westerly wind component and associated weather conditions.
low-intensity atomizer *Mechanical Engineering.* an electrostatic atomizer that employs the principle of Rayleigh instability, in which the presence of a charge in the surface counteracts surface tension.
low-intensity conflict *Military Science.* a limited politico-military struggle to achieve political, economic, or psychological objectives; it is usually of extended duration, confined to a specific geographic area, and constrained in weaponry, tactics, and level of conflict.
low-key *Graphic Arts.* having predominantly dark tones, with few highlights.
lowland *Geography.* land that is lower or flatter than adjacent land, usually less than 600 feet above sea level.
low level *Electronics.* the lower of the standard levels of a logic circuit.
low-level *Electronics.* **1.** having an amplitude below that which is available in a comparable system. **2.** of or relating to that which is primitive, such as a low-level computer language.
low-level condenser *Mechanical Engineering.* a condenser that has a direct-contact water-cooled steam flow and uses a pump to remove liquid from a vacuum space.
low-level counting *Nucleonics.* the process of measuring tiny amounts of radioactivity that are produced by long-lived radioactive isotopes, such as carbon-14, or by nuclear explosions.
low-level language *Computer Programming.* a machine-dependent programming language, such as an assembly language, consisting of mnemonics directly corresponding to the machine-language instructions of the computer.
low-level logic circuit *Electronics.* a device that performs problem-solving functions in a computer at a low current or voltage level.
low-level modulation *Electronics.* the process of varying the amplitude, frequency, or phase of a signal in a section of a system where the power level is lower than that at the system's output.
low-lift truck *Mechanical Engineering.* a lift truck that is manually or electronically operated to raise or lower a load enough to make it mobile.
lowmoor bog *Geology.* a bog lying at, or slightly above or below, the water table.
lowmoor peat *Geology.* a type of peat that contains little or no sphagnum and occurs on low-lying moors or swamps.
low-noise amplifier *Electronics.* a circuit that generates few spurious signals when the desired signal is missing or weak; e.g., a field-effect transistor operating in an audio preamplifier.

low-order *Computer Programming.* of or relating to the relatively less significant digits, positioned farthest to the right in a number. Thus, **low-order digit.**
low-order burst *Ordnance.* a term for the incomplete detonation of an explosive.
low-pass *Electricity.* describing the bandpass of a filter that attenuates the output to a given cutoff frequency.
low-pass/bandpass transformation see FREQUENCY TRANSFORMATION.
low-pass filter *Electricity.* a filter that attenuates signals above a given cutoff frequency, but passes signals below that frequency, for the most part, unaffected.
low population zone *Engineering.* an area of low population density, usually required around nuclear facilities and testing areas.
low-power cartridge *Ordnance.* a cartridge producing a muzzle velocity of less than 1850 feet per second.
low-pressure area *Mechanical Engineering.* the place in a bearing where the area or space for the lubricant is greatest and the pressure is the least.
low-pressure center *Meteorology.* **1.** a point on a synoptic chart at which minimum pressure occurs; the center of a low-pressure system. **2.** a center of cyclonic circulation.
low-pressure fluid flow *Fluid Mechanics.* the flow of a fluid, especially a gas, at any pressure below atmospheric pressure.
low-pressure permanent molding *Materials Science.* a process used for cast iron, steel, and aluminum alloys, in which liquid metal enters the mold cavity through a refractory-lined tube or stalk from the bottom of the mold by means of low-pressure air.
low-pressure polyethylene see HIGH-DENSITY POLYETHYLENE.
low-pressure system *Meteorology.* see LOW.
low-pressure torch *Engineering.* a torch in which the amount of acetylene drawn into the flame is regulated by the velocity of an oxygen jet in the mixing chamber.
low-pressure well *Petroleum Engineering.* an oil or gas well in which shut-in pressures in the wellhead remain below 2000 pounds per square inch absolute (1.38×10^7 newtons per square meter).
low-Q filter *Electronics.* a filter with a low ratio between the passband width and the center frequency.
low quartz *Mineralogy.* SiO_2, the trigonal polymorph of quartz stable below 573°C, having less symmetry than quartz polymorphs stable at higher temperatures; a common constitutent of most rock types. Also, ALPHA QUARTZ.
low-rank graywacke *Petrology.* graywacke that contains little or no feldspar.
low-rank metamorphism *Geology.* a metamorphic grade that occurs under low to moderate temperature and pressure conditions.
low-reflection film *Optics.* a type of film that is placed over a glass surface so that it will transmit a greater portion of incident light than it reflects.
low-relief *Graphic Arts.* of a letterpress printing plate, having a printing image that is raised only slightly higher than the plate surface.
low-residual-phosphorus copper *Metallurgy.* a phosphorus-deoxidized copper that has a sufficiently low residual phosphorus to cause minimal or no impairment to the electrical conductivity.
low-resolution *Computer Technology.* describing output devices such as CRTs or printers that produce images which are not sharply defined.
Lowry-Brönsted theory see BRÖNSTED-LOWRY THEORY.
Lowry process *Engineering.* a procedure for wood preservation that starts at atmospheric pressure and when performed in a vacuum introduces preservative into the wood.
low-shaft furnace *Metallurgy.* one of several types of shaft furnaces.
low side *Computer Technology.* a component of a distributed system that communicates with remote terminals, rather than with the host computer.
low-speed wind tunnel *Engineering.* a wind tunnel with a maximum speed of 200 miles (480 kilometers) per hour.
low-technology robot *Robotics.* a simple robot having only two or three degrees of freedom, fixed or adjustable stops, and a limited sequence of movements.
low-temperature carbonization *Chemical Engineering.* a process for carbonizing coal or other fuel of high volatile content in an air-free oven using moderate temperatures to produce coke or char, tar, and gas.
low-temperature creep *Materials Science.* the gradual deformation of a crystalline material, such as a metal or ceramic, under an applied stress at temperatures that are low relative to its melting point.
low-temperature flexibility test *Materials Science.* a test of the brittleness of a polymer such as an adhesive, which measures the temperature at which it first exhibits brittle characteristics as it is cooled. The most common form is the mandrel bend test, in which a strip of the polymer is bent around a rod of selected diameter at a set rate at different temperatures to determine where fracture begins.
low-temperature hygrometry *Engineering.* the study of water vapor at low temperatures.
low-temperature physics *Physics.* the study of the properties of matter and energy at extremely low temperatures, usually below 90 K, the boiling point of oxygen.
low-temperature production *Physics.* the various means and techniques used to produce extreme low temperatures, 90 K and lower.
low-temperature reservoir *Thermodynamics.* a body that receives a heat transfer from a heat engine by virtue of its having a lower temperature than the heat engine.
low-temperature separation *Chemical Engineering.* the separation of components of mixtures at low temperatures by means of enhanced differences in boiling points, freezing points, or adsorptive capacities; normally requires refrigeration cycles to reach operating temperatures.
low-temperature steel see CRYOGENIC STEEL.
low-temperature thermometry *Physics.* the study of the various techniques and devices used to measure extremely low temperatures.
low tide *Oceanography.* the lowest limit reached by a falling tide.
low-tide terrace *Geology.* a relatively flat, horizontal area of a beach lying near the low-water line.
low transverse abdominal incision *Surgery.* a horizontal, curved incision into one side of the lower abdomen.
low velocity *Ordnance.* a muzzle velocity of less than 2500 feet per second.
low-velocity *Ordnance.* denoting firearms or artillery with muzzle velocities of less than 2500 feet per second.
low-velocity drop *Military Science.* a drop procedure in which the drop velocity does not exceed 30 feet per second.
low-velocity layer *Geophysics.* a layer in the earth in which the wave velocity is lower than that of the layers above or below it.
low-volatile bituminous coal *Geology.* a bituminous coal yielding less than 15–22% volatile matter.
low voltage *Electricity.* **1.** voltage that is too low to cause electric shock. **2. low-voltage.** of or relating to such voltage.
low-voltage contact material *Materials Science.* a material used to transmit an electric current as junctions between two conductors; used in circuits with up to a few volts and in the milliampere range of currents, generally for control or information transmission.
low-voltage electrical insulator *Materials Science.* a ceramic insulator used in electrical systems at low voltage, such as electrical porcelains and ceramics of steatite, forsterite, zircon, and cordierite.
low-voltage relay *Telecommunications.* a relay that responds to a drop in voltage in an active telephone line.
low water see LOW TIDE.
low-water datum *Geodesy.* the approximate plane of mean low water, used as the standard datum plane of reference for adjacent areas.
low-water fuel cutoff *Mechanical Engineering.* a float device that causes the shutdown of a fuel supply and burner system when the low-level line in the boiler tank is violated.
low-water inequality *Oceanography.* the difference in height between two successive low waters in one tidal day.
low-water lunitidal interval *Geophysics.* the elapse of time between the transit of the moon for an observer and the next low tide in a specific location.
LOX or **lox** liquid oxygen.
Loxodes *Invertebrate Zoology.* a genus of ciliate protozoans in the subclass Holotrichia, with an oral opening at the base of the proboscis.

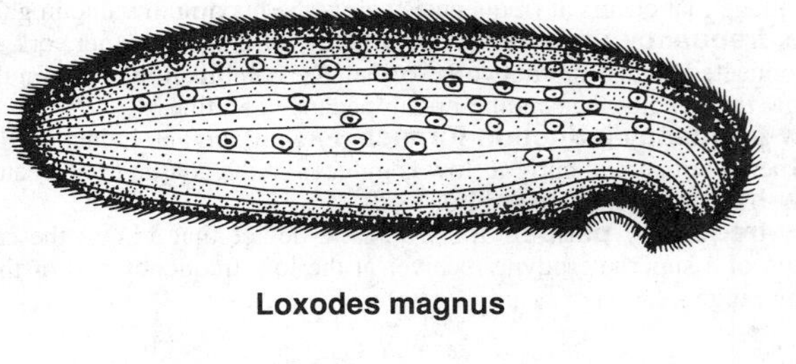

Loxodes magnus

loxodont *Vertebrate Zoology.* having molar teeth in which shallow depressions are present between the ridges. Also, **loxodontic.**

loxodromic spiral *Mathematics.* any curve on a given surface of revolution that intersects the meridians of the curve at some constant angle a, where $a \neq 90°$. Also, **loxodrome.**

loxolophodont *Vertebrate Zoology.* having molar teeth in which crests connect three of the tubercles, and the posterior, inner tubercle is reduced or absent.

Loxonematacea *Paleontology.* a superfamily of gastropods in the order Mesogastropoda, with shells that are generally multiwhorled, slender, and high-spired; extant from the Ordovician to Middle Jurassic.

loxoscelism *Toxicology.* poisoning by venom from spiders of the genus *Loxosceles*, such as the brown recluse spider, characterized by skin lesions, fever, and severe gastrointestinal disturbances.

Loxsomataceae *Botany.* a small family of rare, medium-sized ferns of the order Filicales, characterized by creeping hairy rhizomes, leathery dissected fronds, and sori attached to leaf margins; found in New Zealand and northern South America.

lozenge *Medicine.* a medicated tablet or disk. *Surgery.* a triangular mark showing tissue to be excised in plastic surgery. *Architecture.* a diamond-shaped decorative figure or motif.

lozenge file *Mechanical Devices.* a die-forming file consisting of four sides, each with cutting teeth.

LP *Acoustical Engineering.* a long-playing record; a record that is made at 33-1/3 rpm.

LP or **L.P.** low pressure. Also, **l.p.**

L pad *Acoustical Engineering.* a volume control designed to present the same impedance throughout its entire range by using an L-network in which both elements are adjusted simultaneously.

LPF process *Mining Engineering.* the act of recovering metals from tailings by a process of leaching, precipitation, and flotation, in that order.

LPG liquefied petroleum gas. Also, **LP gas.**

L-phase variant see L-FORM.

LPL *Aviation.* the airport code for Liverpool, England.

LPM or **lpm** lines per minute.

LPN licensed practical nurse.

lpp gene *Genetics.* a gene that encodes for the Braun lipoprotein in *E. coli.*

LPP group *Microbiology.* a nontaxonomic classification of filamentous cyanobacteria, consisting of two groups, A and B, into which genera are categorized based on the morphology of their cells and trichomes.

LP record *Acoustical Engineering.* see LP.

l-process *Nuclear Physics.* a process by which certain light nuclides are synthesized during the decomposition of heavier nuclides.

LPS or **lps** *Biochemistry.* lipopolysaccharide, a major constituent of the outer membrane in the cells of Gram-negative bacteria.

l^p space *Mathematics.* for $p > 0$, a sequence of complex numbers c_n is said to belong to the space l^p if $\sum_{n=1}^{\infty} |c_n|^p < \infty$; also denoted $\boldsymbol{l_p}$.

L^p space *Mathematics.* for $p > 0$, $L^p(X)$ is the space of measurable functions f defined almost everywhere on a measure space X such that $|f|^p$ is integrable. For $p \geq 1$, the norm on $L^p(\mathbf{X})$ is given by $\|f\|_p = (\int_X |f|^p)^{1/p}$. With this norm, $L^p(X)$ is a Banach space; also described by saying that f is an L^p function or that f is in L^p.

lpW or **l.p.w.** lumen per watt; lumens per watt.

LQG problem see linear-quadratic-Gaussian problem.

Lr the chemical symbol for lawrencium.

LRAM long-range attack missile.

LRBM long-range ballistic missile.

LRF luteinizing hormone-releasing factor.

LRRP lowest required radiating power.

LRT light-rail transit.

LRU least recently used.

LS left septum; lumbosacral.

L.S. licentiate in surgery.

L.S.A. left sacroanterior; Licentiate of Society of Apothecaries.

LSA diode limited space-charge accumulation diode.

L.Sc.A. left scapuloanterior.

L scan see L DISPLAY.

L scope *Electronics.* a radarscope that produces an L display. Also, L INDICATOR.

LS coupling see RUSSELL-SAUNDERS COUPLING.

LSD *Organic Chemistry.* the popular term for the hallucinogenic chemical lysergic acid diethylamide. Also, **LSD-25.** See LYSERGIC ACID DIETHYLAMIDE.

L shell *Atomic Physics.* the electron orbit of the atom, second closest to the nucleus, in which electrons with principal quantum number 2 occur.

LSI large-scale integration.

LSM landing ship, medium.

L.S.P. left sacroposterior.

LST landing ship, tank.

L.S.T. left sacrotransverse.

L statistic *Statistics.* a statistic that is obtained as a linear combination of order statistics.

L strand *Genetics.* light strand; the strand of a dsDNA molecule having the lower buoyant density when the dsDNA is denatured and subjected to CsCl density gradient centrifugation.

LT *Aviation.* the airline code for LTU International Airways; lymphotoxin.

L.T. or **Lt** low-tension. Also, **lt.**

Lt. lieutenant.

LTA lighter than air.

Lt. Com. or **Lt. Comdr.** lieutenant commander. Also, **LTC.**

Lt. Col. lieutenant colonel.

LTE local thermodynamic equilibrium.

L-1 test *Engineering.* a test that is used to evaluate the detergency of heavy-duty lubricating oils; performed in a single-cylinder Caterpillar diesel engine.

L-2 test *Engineering.* a test used to determine the oiliness of an engine lubricant; performed in a single-cylinder Caterpillar diesel engine.

L-3 test *Engineering.* a test used to measure the stability of crankcase oil at high temperatures and under harsh operating conditions; executed in a four-cylinder Caterpillar engine.

L-4 test *Engineering.* a test used to evaluate crankcase oil oxidation stability, engine deposits, and bearing corrosion; performed in a six-cylinder spark-ignition Chevrolet engine.

L-5 test *Engineering.* a test used to judge corrosiveness, ring sticking, detergency, and oxidation stability properties of lubricating oils; performed in a General Motors diesel engine.

LTF lymphocyte transforming factor.

Lt. Gen. lieutenant general. Also, **LTG.**

LTH or **LtH** luteotropic hormone.

Lt. Inf. light infantry.

LTM or **ltm** long-term memory.

LTN *Aviation.* the airport code for Luton, London.

LTPD see LOT TOLERANCE PERCENT DEFECTIVE.

LTS launch telemetry station; launch tracking system.

lt-yr light-year; light-years.

Lu the chemical symbol for lutetium.

l.u.b. least upper bound.

lubb *Cardiology.* a syllable representing the first cardiac sound as heard by auscultation; marks closure of the ventrical and tricuspid valves.

lubb-dupp *Cardiology.* syllables representing the combined first and second heart sounds of a normal heartbeat.

lubber's line *Navigation.* a reference line on a direction-reading instrument, such as a compass, that indicates the heading.

lubber's line error *Navigation.* an error due to incorrect alignment of the lubber's line on a compass, pelorus, or radar display.

lube *Materials Science.* short for lubricate, lubricating, and so on.

lube oil see LUBRICATING OIL.

lubimin *Biochemistry.* a phytoalexin generated by the potato plant.

lubricant *Materials.* **1.** any substance, such as grease, oil, silicone, or graphite, that reduces friction between two interacting surfaces. **2.** of or relating to such a substance. Thus, **lubricant additive.**

lubricate *Materials Science.* to apply lubricant to reduce friction between two interacting surfaces.

lubricated gasoline *Materials.* gasoline to which a lubricant has been added.

lubricating *Materials Science.* serving to lubricate; acting as a lubricant. Used to form many compound terms; for example, **lubricating film, lubricating grease, lubricating solid, lubricating gas,** and so on.

lubricatlng oil *Materials Science.* any oil used as a lubricant, especially one refined from crude petroleum and used on metallic surfaces.

lubrication *Materials Science.* the fact or process of lubricating two interacting solid surfaces.

lubricator *Engineering.* anything that lubricates.

lubricity *Materials Science.* the relative ability of a substance to act as a lubricant.

lubricous *Materials Science.* describing a substance that has a relatively high ability to act as a lubricant.

Lucanidae *Invertebrate Zoology.* the stag beetles, a large family of coleopteran insects in the superfamily Scarabaeoidea.

Lucibacterium *Bacteriology.* a former term for a genus of motile bacteria now assigned to the species *Vibrio harveyi.*

luciferin *Biochemistry.* a pigment found in animals capable of bioluminescence; when undergoing oxidation, it is capable of producing heatless light.

luciferous *Science.* bringing or providing light.

Luciocephalidae *Vertebrate Zoology.* a family of freshwater perciform fish of the order Luciocephaloidei, containing one species, the pikehead of the Malay Archipelago.

Lucke virus *Virology.* a proposed genus of the family Herpesviridae, containing viruses of mammals that show a large degree of serological cross-reactivity and some genetic homology with the type species human herpesvirus.

Luckiesh-Moss visibility meter *Engineering.* a meter with two variable density filters that reads the reduction in visibility of a barely discernible object.

Lucretius c. 99–55 BC, Roman philosopher and didactic poet; wrote *De rerum natura,* a six-volume scientific inquiry that anticipated modern atomic theory.

Lucy *Anthropology.* the name for a skeleton of *Australopithecus afarensis* found by Donald Johanson and Tom Grey at the Hadar site in Ethiopia in 1974; one of the first with both cranial and postcranial remains and thought to be one of the earliest hominid fossils. (Named by association with a popular song of the period, the Beatles' "Lucy in the Sky with Diamonds.")

Luddite *Industrial Engineering.* **1.** a member of any of various groups of British workers who rioted and attempted to destroy textile machinery in the early period of the Industrial Revolution. **2.** a person who is regarded as hostile to or ignorant of technological and scientific innovation. (Said to be from Ned *Ludd,* a workman who was an originator of the movement.) Thus, **Luddism.**

LU decomposition *Mathematics.* a decomposition of a matrix A as a product of a lower-triangular matrix L times an upper-triangular matrix U. The decomposition is usually made unique by choosing the diagonal elements of L to be equal to 1. The decomposition $A = LU$ is important in numerical methods for solving systems of linear equations; it is a generalization of the Cholesky decomposition.

Ludian *Geology.* a European geologic stage of the uppermost Eocene epoch, occurring after the Bartonian and before the Tongrian of the Oligocene epoch.

ludlamite *Mineralogy.* $(Fe^{+2},Mg,Mn^{+2})_3(PO_4)_2{\cdot}4H_2O$, a green, monoclinic mineral occurring as tabular crystals and in massive form, having a specific gravity of 3.15 to 3.16 and a hardness of 3.5 on the Mohs scale; found in the oxidized zone of ore deposits and in granite pegmatites.

Ludlovian *Geology.* a European geologic stage of the Upper Silurian period, occurring after the Wenlockian and before the Gedinnian of the Devonian period.

ludwigite *Mineralogy.* $Mg_2Fe^{+3}BO_5$, a blackish-green, orthorhombic mineral of the Ludwigite group, occurring in fibrous masses or granular form, and having a specific gravity of 3.86 and a hardness of 5 on the Mohs scale; found in contact metamorphic deposits.

Ludwig's angina *Medicine.* a painful inflammation and pus formation on the floor of the mouth, often due to tooth infection. (Named for Wilhelm Friedrich von *Ludwig,* 1790–1865, German surgeon.)

Luenberger observer *Control Systems.* a compensator that is driven by the inputs as well as the measurable outputs of a control system.

lueneburgite see LÜNEBURGITE.

luer fitting *Biochemistry.* a type of joint that forms an airtight connection between two tubular parts of an apparatus.

lueshite *Mineralogy.* $NaNbO_3$, a black, monoclinic, pseudocubic mineral of the perovskite group, dimorphous with natroniobite, having a specific gravity of 4.44 and a hardness of 5.5 on the Mohs scale; found as pseudocubic crystals at the contact of a cancrinite syenite and pyrochlore-bearing carbonatite.

lug *Mechanical Devices.* an irregularly casted projection shaped for its intended function on a metal part; used as a handle, a supporting piece, or a connecting piece.

luganot *Meteorology.* a strong south-to-southeast wind that blows off Lake Gorda, Italy.

lug bolt *Mechanical Devices.* **1.** a bolt that has a long, flat extension or hook instead of a head. **2.** a bolt used for fastening a lug. Also, STRAP BOLT.

Luger (pistol) *Ordnance.* the U.S. name for a 9-mm toggle-action automatic pistol that was the official German Army sidearm from 1908 to 1938.

Luggin probe *Physical Chemistry.* a device containing an electrolyte and a reference electrode with a small orifice placed close to the working electrode that serves to minimize ohmic losses in potential measurements.

lug nut *Mechanical Devices.* a large nut that fits on a heavy bolt, used especially to attach a wheel to a motor vehicle.

Lugol solution *Chemistry.* an aqueous solution of potassium iodide and iodine, used medicinally.

lug wrench *Mechanical Devices.* a wrench used for loosening or tightening lug nuts.

LUHF least useful high frequency.

Luidiidae *Invertebrate Zoology.* a family of sea stars, echinoderms in the suborder Paxillosina, having suckerless, pointed tube feet.

Luisian *Geology.* a North American geologic stage of the Miocene epoch, occurring after the Relizian and before the Mohnian.

Lukasiewicz notation see POLISH NOTATION, def. 1.

LUL or **L.U.L.** left upper lobe.

Lulu *Ordnance.* a nuclear depth bomb, deliverable by fixed-wing aircraft, helicopter, or ASROC rocket.

lumb- a combining form meaning "loins" or "lower back," as in *lumbago.*

lumbago *Medicine.* a pain, especially a chronic pain, in the middle and lower back, usually caused by muscle strain, arthritis, vascular insufficiency, or a herniated disk.

lumbar *Anatomy.* **1.** of or relating to the lower back. **2.** a lumbar vertebra, nerve, artery, and so on.

lumbar artery *Anatomy.* any of several paired arteries that branch from the abdominal aorta and carry blood to the abdominal wall.

lumbarization *Medicine.* a condition in which the first segment of the sacrum is not fused with the second, resulting in one additional articulated vertebra, with the sacrum consisting of only four segments instead of five.

lumbar nerve *Anatomy.* any of five pairs of nerves that emerge from the spinal cord between the lumbar vertebrae and contribute to the formation of the lumbosacral plexus.

lumbar plexus *Anatomy.* **1.** a network of nerves formed by the ventral branches of the second to fifth lumbar nerves in the psoas major muscle; the branches of the first lumbar nerve may also be included. **2.** a lymphatic network in the lumbar region.

lumbar puncture see SPINAL PUNCTURE.

lumbar vertebrae *Anatomy.* five massive vertebrae that form the portion of the vertebral column between the thorax and the sacrum.

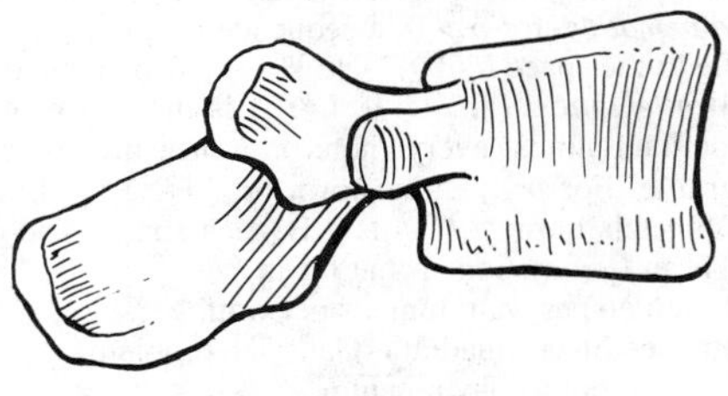

lumbar vertebrae

lumber *Materials.* **1.** a collective term for wood that has been sawed into appropriate sizes for building and other uses. **2.** to cut such wood and prepare it for use or sale. Thus, **lumbering.**

lumbo- a combining form meaning "lower back," as in *lumbosacral.*

lumbodorsal fascia see THORACOLUMBAR FASCIA.

lumbosacral plexus *Anatomy.* a network of interconnected spinal nerves from the lumbar and sacral regions of the spine.

Lumbricidae *Invertebrate Zoology.* a family of earthworms, annelids in the order Oligochaeta; hermaphrodites that reproduce by mutual cross-fertilization.

Lumbriclymeninae *Invertebrate Zoology.* a subfamily of tube-dwelling marine annelids.

lumbricoid *Invertebrate Zoology.* relating to or resembling an earthworm. Also, **lumbrical.**

Lumbriculidae *Invertebrate Zoology.* a family of aquatic annelid worms in the order Oligochaeta.

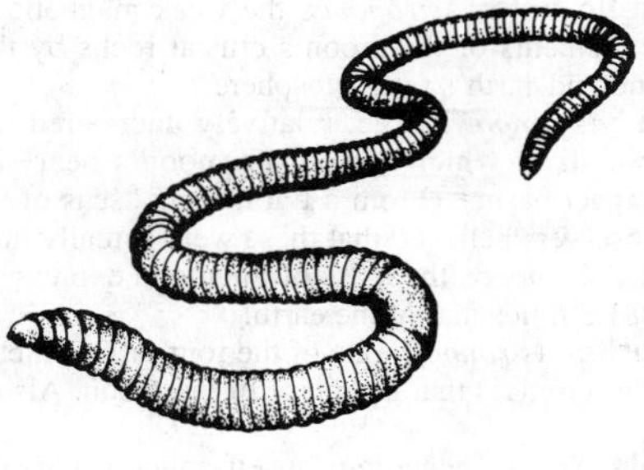

Lumbricus

Lumbricus *Invertebrate Zoology.* a genus of common earthworms.
Lumbrineridae *Invertebrate Zoology.* a family of marine polychaetes that superficially resemble earthworms, in the superfamily Eunicea.
lumen *Anatomy.* a general term referring to the space within a tubular or hollow organ such as an artery, a vein, or the intestine. *Optics.* a unit of luminous flux equal to the quantity of light emitted per unit solid angle by a point source having the intensity of one candle.
lumen hour *Optics.* a unit of luminous energy that equals the quantity of light emitted by a one-lumen lamp operating for one hour.
lumen per watt *Optics.* a measure of the efficiency of a light source.
Lumière, Auguste [loo myâr´] 1862–1954, and his brother **Louis-Jean,** 1864–1948, French chemists and inventors; pioneers in color photography and motion-picture technology; the first to show a theatrical film.
luminaire *Electricity.* a complete lighting unit consisting of a lamp with the parts designed to distribute the light, to position and protect the lamp, and to connect the unit to a power supply.
Luminal *Organic Chemistry.* $C_{12}H_{12}O_3N_2$, a trade name for phenobarbital; a crystal that is present in three forms: a stable form, whose melting point is 174°C; and two unstable forms, whose melting points are 156–157°C and 166–167°C; all three forms produce salts with alkali metals and with complex metallic derivatives; this drug, administered orally, has a hypnotic and sedative effect. Also, ETHYLPHENYLBARBITURIC ACID, GARDENAL.
luminance *Optics.* **1.** a measure of the brightness of a surface, equal to the luminous flux leaving, arriving at, or passing through a unit area of surface per unit solid angle. Also, BRIGHTNESS. **2.** broadly, the quality or state of emitting light.
luminance channel *Telecommunications.* in a color television system, any path that carries the luminance signal.
luminance factor *Optics.* the luminance of a surface expressed as a ratio of the luminance of a perfectly reflecting surface.
luminance primary *Telecommunications.* in a color television system, the one of three transmission primaries whose amount determines the luminance of a color.
luminance signal *Telecommunications.* in a color television system, the signal that controls the brightness of the picture.
luminescence *Physics.* the radiation of light from a body produced by means other than heat; luminescence can be subdivided into fluorescence and phosphorescence.
luminescence spectroscopy *Spectroscopy.* the spectroscopic study of materials that emit light by means that are not temperature dependent.
luminescent *Optics.* exhibiting luminescence; giving off light.
luminescent center *Solid-State Physics.* a defect in a transparent crystal lattice structure that emits radiation or is observed by its luminescent properties.
luminescent screen *Electronics.* an element in a cathode-ray tube or television picture tube that glows when struck by an electron beam and retains its glow for a long time afterward.
luminol *Organic Chemistry.* $C_8H_7N_3O_2$, white to yellow crystals that are insoluble in water and soluble in alcohol; melts at 329–332°C; used in analytical chemistry and as a chemiluminescent agent.
luminophor *Physics.* a luminescent material that emits radiation by absorbing and converting a portion of the incident energy. Also, FLUORPHOR, FLUOR, PHOSPHOR.
luminosity *Optics.* the fact or quality of giving off light. *Astrophysics.* the absolute brightness of a celestial object.
luminosity classes *Astrophysics.* eight distinct regions on the Hertzsprung-Russell diagram into which stars naturally group: bright supergiants, supergiants, bright giants, normal giants, subgiants, main-sequence (or dwarf) stars, subdwarfs, and white dwarfs.
luminosity factor *Optics.* the ratio of the luminous flux emitted by a light source at a given wavelength to the total energy emitted.
luminosity function *Astrophysics.* the number of stars or galaxies with a given intrinsic brightness in a given volume of space. *Optics.* a measure of the reactions of the eye to monochromatic light at different wavelengths. Also, LUMINOSITY CURVE, SPECTRAL LUMINOUS EFFICIENCY, VISIBILITY FUNCTION.
luminosity-volume test *Astrophysics.* a cosmological test that indicates whether evolution has occurred within a given sample of galaxies.
luminous *Optics.* radiating or reflecting light; shining.
luminous cloud see SHEET LIGHTNING.
luminous coefficient *Optics.* one of several factors that multiply the tristimulus values of a color to equal a total that represents the color's subjective luminance as perceived by the eye.
luminous efficiency *Optics.* the ratio of the luminous flux of an energy source to the total radiant flux; for electric lamps, usually expressed in lumens per watt.
luminous emittance *Optics.* the emittance of visible radiant energy; generally used in relation to self-luminous sources.
luminous energy *Optics.* the total amount of radiant energy generated by visible electromagnetic waves, measured in lumen-hours or lumen-seconds.
luminous flux *Optics.* the rate of light emission within a unit solid angle of one steradian by a point source having a uniform luminous intensity of one candela; measured in lumens.
luminous intensity *Optics.* the density of luminous flux per unit solid angle when measured in a given direction in relation to the emitting source. Also, **light intensity.**
luminous mass *Astrophysics.* the part of a galaxy's total mass that is luminous; now thought in most cases to be only about 10% of the total that is actually present.
luminous meteor *Meteorology.* any of numerous atmospheric phenomena other than lightning that appear as luminous patterns in the sky, including halos, coronas, rainbows, aurorae, and their variations.
luminous nebula see EMISSION NEBULA.
luminous paint *Materials.* a paint that has the ability to radiate visible light when exposed to ultraviolet light due to the presence of phosphors. Thus, **luminous pigment.**
luminous quantities *Optics.* those physical quantities such as luminous efficacy, luminous energy, luminous flux, and luminous emittance that can be detected by the human eye and can be adjusted to account for various responses at various wavelengths of light.
luminous range *Navigation.* the maximum distance at which a light can be seen, based on the intensity of the light and existing atmospheric conditions.
luminous range diagram *Navigation.* a graph used to convert the nominal range of a light to its luminous range.
luminous sensitivity *Electronics.* of a phototube, the ratio between the current that is generated by the phototube and the amount of light striking it.
Lummer, Otto Richard [lùm´ər] 1860–1925, German physicist; conducted important investigations in optics and thermal radiation; did research that led to the quantum theory.
Lummer-Brodhun cube *Optics.* a device in which two prisms are used to simultaneously observe both sides of a white diffuse plaster screen lit by light sources whose intensities are being compared. Also, **Lummer-Brodhun sight box.**
Lummer-Gehrcke plate *Optics.* a type of interferometer that produces multiple reflections of electromagnetic waves passing through a thick plate of glass or quartz with parallel surfaces.
lump *Systematics.* the concept of lowering the rank of one or more taxa and including them in some superior taxon.
lump coal *Mining Engineering.* any bituminous coal lumps remaining after an initial screening, often designated by the size of the mesh used for screening.
lumpectomy *Medicine.* the surgical removal of a hard mass of tissue or tumor, without removal of large amounts of surrounding tissue, as from the breast.
lumped constant *Electricity.* the representation of a distributed reactance or resistance in a single element.
lumped-constant network *Electricity.* a set of lumped-constant components interconnected for a special purpose.
lumped discontinuity *Electromagnetism.* in waveguide analysis, the equivalent total capacitance, inductance, and resistance (represented as discrete components) for a discontinuity in a guide.

lumped element *Electromagnetism.* an equivalent capacitance, inductance, or resistance of an electrical circuit.

lumped impedance *Electromagnetism.* the impedance of a transmission line represented and concentrated at a single point, as opposed to the impedance associated with stray or distributed fields.

lumping *Systematics.* the practice of distinguishing taxa only to a certain level of resolution, ignoring minor variation, which makes possible the placement of the taxa within a established classification system.

lumpy skin disease *Veterinary Medicine.* an acute viral disease of cattle in eastern Africa that is characterized by fever, inflammation of the lymph glands, and the appearance of lumps in the skin and mucous membranes; caused by the Neethling pox virus.

lumpy wool *Veterinary Medicine.* an infectious exudative dermatitis of sheep, caused by the bacterium *Dermatophyilus congolensis*; occurring in the tropics or in other areas where there are prolonged wet periods; characterized by the formation of hard, yellowish scabs that are bound to the wool and prevent shearing; recovery often occurs during dry periods. Also, DERMATOPHILOSIS, DERMATOPHILUS INFECTION.

Luna *Astronomy.* **1.** the Roman goddess of the moon. **2.** another name for the moon.

lunabase *Astronomy.* a former term for the material making up the lunar maria.

lunacy *Medicine.* a popular term for insanity or mental illness, especially in older use. (Going back to the Latin word for "moon;" from the former belief that a person's mind could be affected by the phases of the moon.)

Luna program *Space Technology.* a series of lunar probes launched by the Soviet Union between 1959 and 1976. An individual vehicle in this series was called a **lunik.**

lunar *Astronomy.* **1.** of or relating to the moon; used in compounds to mean "to the moon" (as in **lunar flight, lunar probe, lunar spacecraft**) or "around the moon" (as in **lunar orbit, lunar satellite**). **2.** a former term for the now-superseded method of deriving longitude in navigation.

lunar appulse *Astronomy.* the close approach of the moon to a planet or a bright star.

lunar atmosphere *Astronomy.* an extremely thin gas (less than a billionth of the earth's surface atmospheric pressure) that is composed mainly of neon, helium, molecular hydrogen, and argon.

lunar atmospheric tide *Meteorology.* an atmospheric tide that is caused by the gravitational pull of the moon; it is extremely small in amplitude and is detectable only by careful statistical analysis of a long record.

lunar caustic *Chemistry.* $AgNO_3$, silver nitrate that is used to cauterize tissues.

lunar crater *Astronomy.* a much-encompassing term for any roughly circular or bowl-shaped depression in the moon's surface; nearly all are formed by meteorite impact, though many have been modified by subsequent landslides and volcanic or tectonic activity.

lunar crust *Astronomy.* the outermost layer of the moon, about 45 kilometers thick on the near side and 75 kilometers thick on the far side; it is composed primarily of anorthositic gabbro rocks and has an average density of 3.0 grams per cubic centimeter.

lunar day *Astronomy.* the time between one sunrise and the next for an observer at one location on the moon; it is equal in length to the synodic month: 29.5306 days.

lunar distance *Navigation.* the observed angle between the moon and another heavenly body.

lunar eclipse *Astronomy.* an eclipse occurring when all or part of the moon passes through the earth's shadow at full moon; three kinds of eclipses are possible: total (when the moon goes entirely into the umbral shadow), partial (when only part of the moon is in the umbral shadow), and penumbral (when the moon misses the umbral shadow, but passes through the penumbra).

lunar ephemeris *Astronomy.* a tabulation of lunar positions or physical phenomena for specific dates and times.

lunar excursion module see LUNAR MODULE.

lunar grid *Astronomy.* a supposed network of geological lineations on the moon that if it existed would imply global tectonics, but which is more probably an artifact of sunlight falling obliquely across radial grooves from the Imbrium Basin impact.

lunar inequality *Astronomy.* see INEQUALITY.

lunarite *Astronomy.* an older term for material making up the lunar highlands, the light-colored regions on the moon.

lunar libration see LIBRATION IN LATITUDE, LIBRATION IN LONGITUDE.

lunar magnetic field *Astronomy.* the weak magnetic field induced in the metallic elements of the moon's crustal rocks by the interaction of the solar wind and earth's magnetosphere.

lunar maria *Astronomy.* large, relatively uncratered expanses of dark basalt, nearly all of which are on the moon's near side and most of which fill impact basins. (From a Latin term, "seas of the moon;" early telescopic observers believed that these were actually lunar oceans.)

lunar mass *Astronomy.* the mass of the moon expressed as 7.35×10^{25} grams, or 1/81.3 times that of the earth.

lunar meteorite *Astronomy.* one of the four or five meteorites found on earth (all in Antarctica) that is a piece of the moon. Also, **lunar meteoroid.**

lunar module *Space Technology.* a self-contained auxiliary vehicle designed to disengage from a larger spacecraft and carry astronauts to and from the moon. Also, LUNAR EXCURSION MODULE.

lunar month *Astronomy.* the interval between one new moon and the next, approximately 29 1/2 days; the time period of one complete revolution of the moon around the earth.

lunar noon *Astronomy.* the moment when the sun crosses the merdian for an observer on the moon.

lunar nutation *Astronomy.* see NUTATION.

lunar parallax *Astronomy.* the difference, averaging 3422.5 arc-seconds, in the moon's apparent celestial position as seen from the center of the earth and when it lies on the horizon for an observer at the surface.

lunar polarization *Astronomy.* the partial polarization of sunlight reflected from the lunar surface, with the bright highlands regions showing a stronger effect than the dark maria, and both reaching a maximum at the quarter phases.

lunar rainbow see MOONBOW.

lunar rover *Space Technology.* a battery-powered vehicle used by U.S. astronauts to explore the moon's surface. Also, **lunar roving vehicle.**

lunar rover

lunar satellite *Space Technology.* a satellite designed to make one or more orbits around the moon.

lunar tide *Oceanography.* the harmonic tidal constituent due solely to the tide-producing forces of the moon.

lunar year *Astronomy.* a year composed of twelve lunations or synodic months (354.3672 days).

lunate *Anatomy.* moon- or crescent-shaped.

lunate bar *Geology.* a submerged or partly submerged crescent-shaped sand bar, often found near the entrance to a harbor or at the mouth of a stream.

lunation *Astronomy.* the time between one new moon and the next, 29.5306 days. Also, SYNODIC MONTH.

Lundegardh-type respiration [lun´də gärd´] *Biochemistry.* respiration in which electrons are coupled to the uptake of anions and travel from the inside to the outside of the cytoplasmic membrane.

Lundegardh vaporizer *Analytical Chemistry.* a device used to vaporize a sample solution for analysis by emission flame photometry. Thus, **Lundegardh vaporizer macroanalysis.**

lune *Mathematics.* one of four areas on the surface of a sphere into which two (intersecting) great circles divide the sphere. Also, SPHERICAL LUNE.

Luneberg lens *Electromagnetism.* a dielectric sphere that has a radially varying index of refraction so that uhf radiation can be focused to a point on the surface of the sphere if the incident radiation is a plane wave; conversely, plane waves may be generated by a point feed diametrically opposite on the sphere's surface.

lüneburgite *Mineralogy.* $Mg_3B_2(PO_4)_2(OH)_6 \cdot 5H_2O$, a white to brownish-white monoclinic mineral occurring in masses of minute pseudohexagonal tablets, and having a specific gravity of 2.05 to 2.07 and a hardness of about 2 on the Mohs scale; found in salt basin deposits. Also, LEUNEBURGITE.

lunette *Architecture.* a semicircular area or opening, especially at the base of a vault. *Geology.* a broad, low, crescentic mound or ridge of fine silt and clay produced by dust-laden winds. *Ordnance.* a towing ring on a gun carriage, trailer, or similar vehicle, used to attach it to another vehicle for towing.

lung *Anatomy.* one of the two spongy organs of respiration that occupy the lateral cavities of the chest; the right lung is composed of three lobes, the left lung of two lobes; each lung consists of an external serous coat, subserous areolar tissue, and lung parenchyma made up of lobules bound together by connective tissue.

lung bud *Developmental Biology.* an outgrowth from the foregut that gives rise to the trachea, bronchi, and all the branchings that make up the tracheobronchial tree.

lungfish *Vertebrate Zoology.* any of several species of freshwater fishes that possess a lung attached to the gut and breath air, as well as having gills.

lungfish

lung-governed breathing apparatus *Engineering.* any underwater breathing apparatus in which the diver's need for oxygen determines the rate at which it is supplied.

lungworm *Invertebrate Zoology.* the common name for any of various parasitic nematodes that infest the lungs of vertebrates, principally domestic animals.

lunisolar precession *Astronomy.* a westward motion of the equinox of about 50 arc-seconds per year caused by the gravitational pull of the moon (accounting for about 2/3 the effect) and sun (about 1/3) on the equatorial bulge of the earth.

lunisolar tides *Oceanography.* harmonic tidal constituents derived partly from the lunar and solar tides and partly from the lunisolar synodic fortnightly constituent.

lunitidal interval *Oceanography.* the period of time between the transit (upper or lower) of the moon over the local or Greenwich meridian and a subsequent specified phase of the tide; the lag time between the passage of the moon and the arrival of a tide.

lunk *Telecommunications.* a line that terminates in an automatic dial exchange and whose function it is to serve as an access line for subscribers; in automatic dial exchange equipment, it functions as a trunk.

Lunulariaceae *Botany.* a monotypic family of large, typical liverworts of the order Marchantiales, characterized by dichotomous branching, simple elevated pores in the epidermis, a single layer of air chambers in the thallus, and thick parenchymatous ventral tissue; a common greenhouse species.

lupine *Botany.* the common name for members of the genus *Lupinus,* leguminous herbs having alternate, palmate leaves, an upright stem, and pea-shaped flowers in racemes.

lupinidine see SPARTEINE.

lupinosis *Veterinary Medicine.* the poisoning of various domestic animals by ingestion of plants in the genus *Lupinus*, characterized by joint contractures, scoliosis, or kyphosis.

lupus see LUPUS ERYTHEMATOSUS.

Lupus *Astronomy.* the Wolf, a moderately bright constellation of the southern celestial hemisphere that lies next to the Milky Way and adjoins Centaurus.

lupus erythematosus *Medicine.* a chronic inflammatory disease of uncertain origin, affecting many systems of the body, characterized by a rash on the face and other areas exposed to sunlight, involving the vascular and connective tissues of many organs, and accompanied by serologic abnormalities.

Lupus Loop *Astronomy.* an old supernova remnant about 1500 light-years away in Lupus that is detectable at radio and X-ray wavelengths.

lupus vulgaris *Medicine.* a rare, cutaneous form of tuberculosis of the skin, with nodular lesions on the face, especially the nose and ears.

LUQ or **L.U.Q.** left upper quadrant.

Luria, Salvador born 1912, Italian-born American biologist; with Delbruck, shared the Nobel Prize for research on bacterial viral infections.

Luria and Delbruck experiment *Genetics.* a test of changes in population structure to distinguish a direct modification by an environmental agent from a spontaneous mutation that is selected by an environmental agent; used to study the role of the environment in the origin of new strains.

Luschka's bursa see PHARYNGEAL BURSA.

lusec *Physics.* a unit of pumping power commonly used in describing the evacuation capability of a vacuum pump, equivalent to the power associated with a leak rate of 1 liter per second at a pressure of 1 millitorr.

Lusin's theorem *Mathematics.* suppose that E is a Borel set of finite measure in a locally compact space, and f is a Borel measurable function on E. Then, for every $\varepsilon > 0$, there exists a compact set C contained in E such that: (a) $E - C$ has measure less than ε, and (b) f is continuous on C.

Lusitanian *Geology.* a European geologic stage of the Upper Jurassic period, occurring after the Oxfordian and before the Kimmeridgian.

luster *Optics.* **1.** the appearance of a surface that relies on reflected light. **2.** the ability of mineral surfaces to reflect light.

luster mottling *Geology.* the shimmering appearance of a broken surface of certain sandstones caused by the reflection of light from the cleavage faces of large calcite crystals containing small amounts of other minerals.

lutaceous *Geology.* of, related to, or having the fine-grained texture of a sedimentary rock formed from mud.

lute *Materials.* a soft, earthy packing mixture used for closing or sealing apertures, joints, or porous surfaces in order to make them resistant to liquids or gases. Also, LUTES, LUTING.

lutecite *Mineralogy.* a type of fibrous chalcedony that exhibits characteristic optical anomalies. Also, **lutecin.**

luteic acid *Biochemistry.* a water-soluble glucan that is formed by *Penicillium luteum* and contains malonic acid residues.

lutein *Biochemistry.* **1.** any lipochrome. **2.** specifically, a yellow, crystalline carotenoid occurring in the leaves and petals of various plants, in the egg yolk of animals, and in their corpus luteum; also present in algae, especially in the classes Chlorophyceae and Phaeophyceae; used for the biochemical analysis of the carotenoids. Also, **luteine.**

luteinization *Physiology.* a process in which an ovary follicle that has just produced an ovum transforms into a corpus luteum, which secretes progesterone in preparation for and during pregnancy.

luteinizing hormone *Endocrinology.* a gonadotropic glycoprotein hormone produced by the gonadotrophs of the anterior pituitary gland; stimulates ovulation and progesterone secretion by the corpus luteum in the female and stimulates the production of testosterone by the Leydig cells in the male. Also, INTERSTITIAL CELL STIMULATING HORMONE.

luteinizing hormone-releasing hormone see GONADOTROPIN-RELEASING HORMONE.

luteoma *Medicine.* **1.** a granulosa-theca cell tumor whose cells resemble those of the corpus luteum. **2.** an ovarian tumor of cell origin in which luteinization has occurred.

luteose *Biochemistry.* a type of β-linked glucan generated by *Penicillium luteum.*

Luteovirus group *Virology.* a group of plant viruses that can affect various monocotyledonous or dicotyledonous plants and that can cause significant loss of crop plants.

lutes see LUTE.

Lutetian *Geology.* a European geologic stage of the Middle Eocene epoch, occurring after the Cuisian and before the Auversian.

lutetium *Chemistry.* a soft rare-earth (lanthanide) metallic element having the symbol Lu, the atomic number 71, an atomic weight of 174.967, a melting point of 1660°C, and a boiling point of 3360°C; used in nuclear technology. Also, **lutecium.** (From *Lutetia,* an ancient name for the city of Paris; it was discovered there by Georges Urbain in 1907.)

Lutheran blood group *Genetics.* an inherited human blood group that is produced by an antigen coded for by the *Lu* gene, which is located on chromosome 19.

luting see LUTE.

lutite *Geology.* any fine-grained, consolidated sedimentary rock derived from or composed of mud and its associated materials.

Lutjanidae *Vertebrate Zoology.* the snappers, a family of marine fish of the suborder Percoidei, characterized by a long, compressed body without anterior scales, nostrils on the side of the head, and a protrusible upper jaw.

lux *Optics.* a unit of illumination equal to one lumen per square meter.

Luxemburg effect *Telecommunications.* a nonlinear effect that takes place in the ionosphere; the modulation on a strong carrier wave is transferred to another carrier as it passes through the same area.

luxury gene *Genetics.* a gene that codes for a specialized function and is synthesized, usually in large amounts, in particular cell types.

LVDT linear variable-differential transformer.

L.V.H. left ventricular hypertrophy.

LVN licensed vocational nurse.

LVT landing vehicle, tracked.

L wave *Geophysics.* a surface earthquake wave, without respect to type.

Lwoff, André [lwôf] born 1902, French microbiologist; studied protozoan morphogenesis; Nobel Prize for analyzing lysogenic bacteria.

LWR light water reactor.

lx lux (From Latin *lux, lucis.*)

LY *Aviation.* the airline code for El Al Israel Airlines.

Lyapunov exponent [lē ap´ə näf] *Chaotic Dynamics.* the eigenvalues of the linearized dynamics of a nonlinear problem in the vicinity of a particular solution. For stable steady-state solutions, all of these eigenvalues are negative. For periodic solutions, the largest exponent is equal to zero. These eigenvalues signal stability or instability if all eigenvalues have negative real parts or if at least one eigenvalue has a positive real part, respectively. The largest Lyapunov exponent is a measure of the divergence or convergence of adjacent trajectories of (or near) solution. $\lambda > 0$ indicates divergence and chaos; $\lambda < 0$ indicates decay of transients leading to periodic or quasiperiodic attractors.

Lyapunov function *Mathematics.* a function of a (vector) quantity and its derivatives that is positive-definite for all times t, but has a negative-definite derivative with respect to t. Such a function, if it exists, can demonstrate the stability of a mapping of the (vector) quantity.

Lyapunov numbers *Chaotic Dynamics.* the eigenvalues q for a solution of an iterative function (mapping) that satisfy $f_{n+1} = qf_n$. Constant solutions have $q = 1$, and periodic solutions (solutions that are periodic after N iterates are called N-periodic) have $q = 1$ for f_{n+N} of qf.

Lyapunov stable *Mathematics.* a fixed point x_0 of a mapping F is Lyapunov stable if, for each $\varepsilon > 0$, there exists a $\delta > 0$ such that $|x - x_0| < \delta$ implies that $|F^n(x) - F^n(x_0)| < \varepsilon$ for all $n > 0$. Here the notation $F^n(x)$ means $F(F(F(\dots F(x))))$; i.e., F is applied n times.

Lyapunov stability criterion *Control Systems.* a method of using the sign-definitive properties of a Lyapunov function to determine the stability of a nonlinear system.

lyase *Enzymology.* one of the six primary classes of enzymes, composed of those that catalyze the cleavage of C–C, C–O, C–N or other bonds without a hydrolysis or oxidation reaction; two molecules are formed, one or both of which contain a double bond. Also, SYNTHASE.

Lycaenidae *Invertebrate Zoology.* a family of butterflies, including gossamers, blues, coppers, and hairstreaks, lepidopteran insects in the superfamily Papilionoidea.

Lycaeninae *Invertebrate Zoology.* a subfamily of small papilionid butterflies in the family Lycaenidae.

lychee see LITCHI.

lychnisc *Invertebrate Zoology.* a type of sponge spicule with six points.

Lychniscosa *Invertebrate Zoology.* an order of sponges, in the subclass Hexacterophora, with a skeletal lattice of joined lychnisc spicules.

Lycogala *Mycology.* a genus of fungi belonging to the class Myxomycetes which is composed of slime molds that include the species *L. epidendrum,* found on rotting logs and forming fruiting bodies resembling puffballs.

lycopene *Biochemistry.* a red carotenoid pigment found in tomatoes; the parent compound of the carotenoids.

Lycoperdaceae *Mycology.* a family of fungi belonging to the order Lycoperdales which is composed of puffballs and occurs worldwide.

Lycoperdales *Mycology.* an order of fungi belonging to the class Gasteromycetes, found in soil and in rotting wood; includes the puffballs and earthstars.

lycoperdon nut *Mycology.* the fruiting body of fungi belonging to the order Elaphomycetales, which occurs underground.

lycoperdonosis *Medicine.* a respiratory disease caused by the inhalation of spores of the puffballs *Lycoperdon pyriforme* and *Lycoperdon bovista.*

lycopersicin see TOMATINE.

lycophore larva *Invertebrate Zoology.* a ciliated tapeworm larva with ten hooks and large frontal glands. Also, DECACANTH LARVA.

lycopod *Botany.* any creeping or erect, mosslike plant of the genus *Lycopodium,* such as the ground pine.

Lycopodiales *Botany.* the club mosses, an order of the Lycopodiopsida occurring as herbs with branching stems and numerous small leaves without ligules.

Lycopodiophyta *Botany.* a division of the subkingdom Embryobionta characterized by independent sporophytes with branching roots and stems and small leaves (microphylls) with a single vascular bundle in a spiral arrangement on stems. Also, **Lycopsida.**

Lycopodiopsida *Botany.* a subdivision of the Lycopodiophyta, many of whose members are now extinct, characterized by long, slender spirally arranged leaves that bear sporangia, and roots borne along the stem; the club mosses. Also, LYCOPODINEAE, PLEUROMEIALES.

Lycoriidae *Invertebrate Zoology.* a family of small dipteran dark-winged flies, insects in the suborder Orthorrhapha.

Lycosidae *Invertebrate Zoology.* a family of arachnids in the suborder Dipneumonomorphae, wolf spiders and hunting spiders that actively pursue prey.

Lycoteuthidae *Invertebrate Zoology.* a family of small, open ocean squid that feed on fish, crustaceans, and other squid.

Lyctidae *Invertebrate Zoology.* powder post beetles, a family of wood-boring coleopteran insects including serious pests in structures and furniture.

Lyddane-Sachs-Teller relation *Solid-State Physics.* a relation stating that the ratio for an ionic crystal of the static dielectric constant to the dielectric constant in the high-frequency limit is equal to the squared ratio of the frequency of longitudinal optical phonons with zero wave vectors to the frequency of transverse optical phonons having large wave vectors.

lydian stone see BASANITE.

lydite see BASANITE.

lye *Inorganic Chemistry.* a strong solution of sodium hydroxide, NaOH, or potassium hydroxide, KOH, in water; used widely in industry, as in drain and oven cleaners. Also, CAUSTIC SODA, CAUSTIC POTASH.

Lyell, Sir Charles 1797–1875, British geologist; wrote the landmark work *Principles of Geology*; supported the theories of James Hutton and Charles Darwin.

Sir Charles Lyell

Lygaeidae *Invertebrate Zoology.* a family of true bugs, often bright colored hemipteran insects in the superfamily Lygaeoidea, including the chinch bug, a harmful cereal-eating pest.

Lygaeoidea *Invertebrate Zoology.* a superfamily of true bugs, hemipteran insects in the group Pentatomorpha.

Lyginopteridaceae *Botany.* an extinct family of the order Lyginopteridales that includes pteridosperms with one or two vascular traces at the base of the petiole.

Lyginopteridales *Botany.* an order of fossil plants in the class Pteridospermae.

Lyginopteridatae *Botany.* in some classification systems, an alternate name for Pteridospermae.

lying-over theorem *Mathematics.* let S be an integral extension ring of a ring R and let P be a prime ideal of R. Then there exists a prime ideal Q in S such that $Q \cap R = P$. Q is then said to **lie over** P in S.

lying panel *Building Engineering.* a door panel with greater width than height.

Ly loci *Genetics.* a cluster of genes that code for antigens found on different varieties of T and B lymphocytes.

Lyman-alpha radiation *Spectroscopy.* radiation associated with the line in the Lyman series having a wavelength of approximately 1215 angstroms.

Lyman band *Spectroscopy.* a relatively wide spectral band in the ultraviolet region of the hydrogen spectrum, extending from 125 to 161 nanometers.

Lyman continuum *Spectroscopy.* in the ultraviolet region of the hydrogen spectrum, a continuous region of wavelengths (or frequencies) that begins at the Lyman limit and represents transitions of a hydrogen atom from the ground state to the ionized state.

Lyman ghost *Spectroscopy.* in a spectroscopic instrument, a false spectral line in the ultraviolet region of the hydrogen spectrum resulting from a combination of periodicities produced by the rulings on a diffraction grating.

Lyman limit *Spectroscopy.* the limiting wavelength near which the distance between consecutive lines of the Lyman series decreases and approaches a continuum.

Lyman series *Spectroscopy.* a set of spectral lines in the far ultraviolet region of the hydrogen spectrum that result from electron transitions from outer orbits to the first orbit.

Lymantriidae *Invertebrate Zoology.* tussock moths, a family of lepidopteran insects in the superfamily Noctuoidea, including the gypsy moth and other harmful defoliators of trees.

Lyme disease *Medicine.* an acute, recurrent inflammatory disorder that is caused by the spirochete *Borrelia burgdorferi,* transmitted by ticks commonly found in heavily wooded areas in summer, and is marked by the appearance of a rash; if untreated by antibiotics, it can lead to arthritis, heart abnormalities, and disorders of the nervous system. (After the town of Old *Lyme,* Connecticut, where it was first reported.)

Lymexylidae *Invertebrate Zoology.* a family of beetles associated with ambrosia fungi.

Lymexylonidae *Invertebrate Zoology.* ship timber beetles, a family of woodboring coleopteran insects in the superfamily Lymexylonoidea.

Lymexylonoidea *Invertebrate Zoology.* a superfamily of woodboring beetles, coleopteran insects in the suborder Polyphaga.

lymph [limf] *Histology.* the clear extracellular fluid that is collected by the lymphatic vessels and filtered by the lymph nodes.

lymph- a combining form denoting a relationship to lymph, lymphoidal tissue, lymphatics, or lymphocytes, as in *lymphedema.*

lymphadenectomy *Surgery.* the surgical removal of one or more lymph nodes, usually in association with the surgical removal or destruction of a cancer.

lymphadenitis *Medicine.* an inflammation of the lymph nodes.

lymphadenoid goiter see STRUMA LYMPHOMATOSA.

lymphadenoma *Oncology.* a tumor of lymphoid tissue.

lymphadenomatosis *Medicine.* a condition marked by the presence of numerous enlarged lymph nodes, as in lymphosarcoma and Hodgkin's disease; malignant lymphoma.

lymphadenopathy *Medicine.* any disease affecting the lymph nodes.

lymphadenopathy-associated virus *Virology.* another term for the AIDS virus. Also, LAV.

lymphadenosis *Medicine.* the proliferation of lymphoid tissues.

lymphagogue *Pharmacology.* any agent promoting the production of lymph, a clear fluid that circulates in the body.

lymphangiectasis *Medicine.* the dilation of the lymphatic vessels. Also, TELANGIECTASIA LYMPHATICA.

lymphangiectomy *Medicine.* the surgical removal of a lymphatic vessel.

lymphangioadenography *Radiology.* radiographic study of the lymphatic channels and lymph nodes following their opacification by the injection of an oily radiopaque material into a lymphatic vessel. Also, **lymphography.**

lymphangioendothelioma *Medicine.* a tumor composed of small masses of endothelial cells and collections of tubular structures thought to be derived from lymphatic vessels.

lymphangiofibroma *Medicine.* a benign tumor whose characteristic tissue is made up of lymphangiomatous and fibromatous parts.

lymphangiogram *Radiology.* an X-ray record visualizing the lymphatic vessels. Also, LYMPHOGRAM.

lymphangiography *Radiology.* roentgenography of the lymphatic vessels after injection of a radiopaque medium.

lymphangioma *Medicine.* a mass of lymphatic vessels that are often greatly dilated.

lymphangiomyomatosis *Medicine.* a disorder occurring in women of child-bearing age in which there is progressive proliferation of smooth muscle in the interspaces of tissues of the lungs, lymph nodes, and thoracic duct.

lymphangiosarcoma *Oncology.* a malignant tumor derived from vascular tissue resembling lymphatics.

lymphangitis *Medicine.* an inflammation of the lymphatic vessels.

lymphatic [lim fat′ik] relating to lymph or the lymphatic system.

lymphatic system *Anatomy.* the system of lymphatic vessels, lymph nodes, and masses of lymphatic tissue, such as the spleen, that collects lymphatic fluid from the tissues, filters it, and returns it to the venous system; also transports fats and proteins to the blood system and restores to the circulation more than half of the fluid that filters out of the capillaries during normal exchange across capillary walls.

lymphatic tissue *Histology.* tissue consisting of lymphocytes enclosed in a network of reticular and collagenous fibers. Also, LYMPHOID TISSUE.

lymphatic vessel *Anatomy.* one of the systems of vessels through which lymph flows.

lymphatology *Medicine.* the study of the lymph and lymphatic system.

lymphatolysis *Medicine.* the destruction of lymphatic tissue.

lymphedema *Medicine.* a swelling of the extremities with the accumulation of lymph, due to obstruction of the lymphatic vessels or lymph nodes.

lymph heart *Vertebrate Zoology.* one of several thin, pulsating lymph sacs found in fishes, amphibians, reptiles, and birds.

lymph node *Anatomy.* one of the many rounded to oval masses of lymphatic tissue distributed along the lymphatic vessels that filter lymph. Also, **lymph gland.**

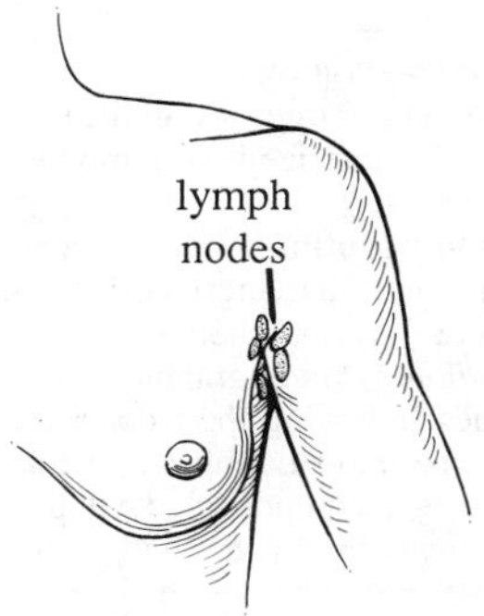

lymph nodes

lympho- a combining form indicating a relationship to lymph, lymphoidal tissue, lymphatics, or lymphocytes, as in *lymphography.*

lymphoblast *Histology.* a large cell of lymphoid tissue that gives rise to a lymphocyte.

lymphoblastic reticulosarcoma *Oncology.* a form of malignant lymphoma in which the chief cell type is a cancerous reticulum cell.

lymphoblastoma *Oncology.* a form of malignant lymphoma whose characteristic tissue is composed of lymphoblasts.

lymphoblastosis *Medicine.* a condition in which an excessive number of lymphoblasts are present in the peripheral blood. Also, ACUTE LYMPHOCYTIC LEUKEMIA.

lymphocystic disease *Veterinary Medicine*. a chronic viral infection of marine or freshwater species, resulting in the development of usually external benign growths and white nodules on the cornea.

lymphocyte [lim′fə sīt′] *Histology*. an agranular leukocyte that is concentrated in lymphoid tissue and is active in immunological responses in the body, including the production of antibodies.

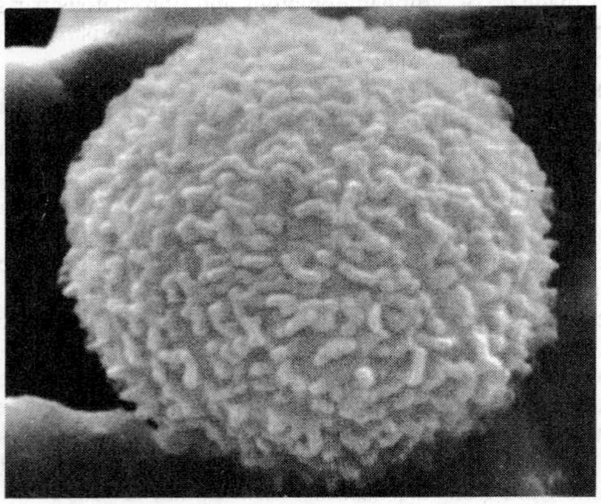

lymphocyte

lymphocytic horiomeningitis *Medicine*. a viral disease of mice, sometimes transmitted to man, in which there is lymphocytic infiltration of the meninges resulting in a type of cerebral meningitis.

lymphocytic leukemia *Medicine*. **1.** an acute childhood form of leukemia in which lymphoblasts occur in blood-forming tissues, and abnormal cells are present in bone marrow. **2.** a chronic form of leukemia, occurring in the elderly, in which mature lymphocytes are present in the blood.

lymphocytic lymphoma *Oncology*. a malignant tumor of the lymphatic tissue in which the predominant cell type belongs to the lymphocytic series.

lymphocytoma *Medicine*. a mass of mature lymphocytes, grossly resembling a tumor.

lymphocytopenia *Medicine*. a reduction in the number of lymphocytes in the circulating blood.

lymphocytosis *Medicine*. an excess of normal lymphocytes in the blood.

lymphogram see LYMPHANGIOGRAM.

lymphogranuloma *Oncology*. a condition, such as sarcoidosis or Hodgkin's disease, characterized by lymphadenopathy and multiple granulomas in lymph nodes.

lymphogranuloma venereum *Medicine*. a venereal infection usually caused by chlamydia and characterized by genital ulceration and regional inflammation of the lymph nodes.

lymphography *Radiology*. radiographic study of the lymphatic channels and lymph nodes following their opacification by injection of an oily radiopaque material into a lymphatic vessel.

lymphoid cell *Histology*. a mononucleocyte that resembles a leukocyte.

lymphoid follicle *Immunology*. a small area of the cortex of a lymphoid organ, composed primarily of small lymphocytes and sometimes having a reactive structure known as a germinal center, which forms in response to antigenic stimulation.

lymphoid leukosis *Veterinary Medicine*. an egg-transmitted form of lymphocytic leukemia found in chickens, caused by an RNA virus; symptoms in infected birds include depression, a decrease in egg laying, and formation of tumors in the bursa, liver, and spleen.

lymphoid tissue see LYMPHATIC TISSUE.

lymphokine *Immunology*. an extracellular molecule, generally produced by T-lymphocytes, that is associated with the reactions of cell-mediated immunity.

lymphoma *Oncology*. a general term for any of the various tumors, usually malignant, of the lymphoid tissues.

lymphomatosis *Oncology*. the widespread involvement of the viscera by malignant lymphoma.

lymphomyxoma *Medicine*. a soft, benign neoplasm that contains adenoid in loose, areolar connective tissue.

lymphon *Immunology*. the entire immune system of a body, including all lymphoid tissue.

lymphopoiesis *Physiology*. the formation of lymphocytes.

lymphoproliferative *Oncology*. a reference to the proliferation of lymphoid cells.

lymphosarcoma *Oncology*. a malignant tumor of the lymph nodes, with proliferation of lymphoblasts and lymphocytes, resulting in enlargement and infiltration of various tissues. *Veterinary Medicine*. see BOVINE LEUKOSIS.

lymphosarcomatosis *Oncology*. a condition marked by numerous, widespread masses or lesions of lymphosarcoma.

lymphostasis *Medicine*. the stoppage of lymph flow.

lymphotoxin *Immunology*. a lymphokine that kills nonlymphocytic cells.

lymph sinus *Anatomy*. one of the channels within a lymph node.

lymph system see LYMPHATIC SYSTEM.

Lynen, Feodor 1911–1979, German chemist; awarded the Nobel Prize for his research on the formation and mechanisms of cholesterol and fatty acids.

Lyngbya *Bacteriology*. a genus of cyanobacteria that are found in freshwater and marine environments, one species of which produces several toxins.

lynx *Vertebrate Zoology*. **1.** a small carnivorous European and Asian cat of the family Felidae that is related to the bobcat of North America; characterized by relatively long legs, a stubby tail, a spotted coat, and often tufted ears. **2.** a North American cat distinguished from the bobcat by its larger size, larger claws, longer ear tufts, and a black tail tip. Also, CANADA LYNX.

lynx

Lynx *Astronomy*. a faint constellation of the northern sky lying between Ursa Major and Auriga whose brightest stars are 4th magnitude.

lyocratic *Physical Chemistry*. describing a colloid that owes its stability to the affinity of its particles for the liquid in which they are dispersed.

Lyomeri *Vertebrate Zoology*. an older term for the deep-sea, eel-like gulpers currently included in the order Anguilliformes.

Lyon hypothesis *Genetics*. a hypothesis suggesting that random inactivation of all but one of the X chromosomes in somatic cells of the female is the mechanism that achieves dosage compensation of X chromosomes in the homogametic sex of mammals; the process so described is known as the **Lyon phenomenon.**

lyonization *Genetics*. the process by which all X chromosomes of the cells in excess of one are inactivated on a random basis.

lyophilic *Chemistry*. of a material, readily forming colloidal suspensions in a liquid.

lyophilization *Chemical Engineering.* the process of dehydrating a frozen material for storage in which the water content is converted to vapor in a vacuum by sublimation and then desorption, leaving the material as a porous solid. Also, FREEZE DRYING.

lyophobic *Chemistry.* of a material, occurring in a colloidal state but tending to repel liquids.

Lyopomi *Vertebrate Zoology.* an alternate name for the deep-sea eel order Notacanthiformes.

Lyot, Bernard-Ferdinand [lyō] 1897–1952, French astronomer; invented the solar coronagraph.

Lyot Division *Astronomy.* a former term for the Maxwell Gap in Saturn's rings.

lyotropic *Physical Chemistry.* of or relating to a series of ions, salts, or radicals placed in descending order relative to the magnitude of their effect on a given solid.

lyotropic liquid crystal *Physical Chemistry.* a liquid crystal substance produced from two or more components when one of the components is polarized.

Lyra *Astronomy.* the Lyre, a small but conspicuous constellation of the northern celestial hemisphere that contains the 1st-magnitude star Vega.

lyrebird *Vertebrate Zoology.* either of two species of Australian mountain forest-dwelling birds of the family Menuridae, having a long sweeping lyre-shaped tail; known for the elaborate courtship dance of the male, who displays his beautiful tail over himself and sings rhythmically; ground-dwelling insectivores, they nest in large domelike grass structures.

Lyrids *Astronomy.* an annual meteor shower that peaks in activity around April 21st, when about 10 to 15 meteors an hour appear to come from the Lyra-Hercules region; another, sparser Lyrid shower occurs around June 15th.

Lys lysine.

lys- a combining form meaning "lysis" or "decomposition," as in *lysate.*

Lysaretidae *Invertebrate Zoology.* a family of carnivorous polychaete worms in the superorder Eunicea, found in tropical and subtropical waters.

lysate *Biochemistry.* a population of phage particles that are released from host cells through the lytic cycle.

lyse *Cell Biology.* to cause cell destruction by lysins or to undergo lysis.

Lysenko, Trofim Denisovich [lī seng′kō] 1898–1976, Russian biologist; proposed a controversial theory of acquired characteristics that was adopted unsuccessfully by Soviet agronomists.

Lysenkoism *Genetics.* a concept proposed by T. D. Lysenko, now largely discredited, arguing that acquired characteristics can be inherited and that alteration of a species' environment can alter its character.

lysergic acid *Organic Chemistry.* $C_{16}H_{16}N_2O_2$, an amphoteric crystalline compound derived from ergot alkaloids that is slightly soluble in water and in most organic solvents; it melts (and decomposes) at 240°C; used in medical research and pharmaceutical manufacturing.

lysergic acid diethylamide *Organic Chemistry.* $C_{15}H_{15}N_2CON(C_2H_5)_2$, a colorless, tasteless, odorless derivative of lysergic acid, commonly abbreviated as LSD; a hallucinogen and serotonin antagonist that is habit-forming and may have mutagenic effects. Also, **lysergide.**

lysi- a combining form meaning "lysis" or "decomposition."

Lysianassidae *Invertebrate Zoology.* a large worldwide family of amphipod crustaceans.

lysigenous *Biology.* referring to intercellular spaces that are formed by the destruction or dissolution of cells. Thus, **lysigenous canal.**

lysimeter *Engineering.* an instrument that measures the amount of water-soluble matter in soil.

lysin *Immunology.* an antibody or bacterial toxin that kills or ruptures cells.

lysine *Biochemistry.* $C_6H_{14}N_2O_2$, an amino acid that is essential for optimal growth in children and for the maintenance of nitrogen equilibrium in adults.

Lysiosquillidae *Invertebrate Zoology.* a family of malacostracan crustaceans.

lysis *Cell Biology.* the process of breaking open a cell by disruption of the plasma membrane. *Medicine.* **1.** the disintegration or destruction of cells by a specific lysin. **2.** the gradual abatement of the symptoms of a disease. *Surgery.* the freeing of an organ from adhesions.

lysis from without *Cell Biology.* the lysis of a host cell when large numbers of virions adsorb to its surface.

lysis inhibition *Cell Biology.* a phenomenon in which phage lysis of an infected host bacterial cell is delayed following superinfection of the cell with a second bacteriophage of the same strain.

Lysithea [lis′ə thē′ə] *Astronomy.* the tenth moon of Jupiter, discovered in 1938 and measuring about 40 kilometers in diameter.

lyso- a combining form meaning "lysis" or "decomposition."

Lysobacter *Bacteriology.* a genus of unicellular bacteria that are capable of gliding motility.

lysogen *Virology.* a bacterial or other prokaryotic cell carrying a phage in a state of lysogeny.

lysogenic bacterium *Bacteriology.* a bacterial cell that harbors the genome of a bacteriophage which persists in the host cell, integrated into the chromosomal DNA, and which retains the capacity to lyse the host cell.

lysogenic conversion *Microbiology.* an alteration in one or more phenotypic characteristics of a bacterial cell due to bacteriophage infection, resulting from the expression of viral genes or the inactivation or activation of one or more host genes.

lysogenic pathway *Biotechnology.* the developmental sequence of a phage that results not in lysis but in the phage becoming attached to a specific site on the chromosome of the host cell, thus forming a lysogen.

lysogenic virus *Virology.* a special type of virus bacterial cell interaction maintained by a complex cellular regulatory mechanism.

lysogenization *Microbiology.* a process by which the genome of a bacteriophage is established in a host bacterial cell, either integrated into the host genome or as an extrachromosomal plasmid that is replicated along with the host genome, and in which it retains its capacity to lyse the cell.

lysogeny *Microbiology.* the phenomenon in which a temperate bacteriophage infects a bacterium, integrating its DNA with that of the host cell and replicating along with the host cell.

lysosome *Cell Biology.* an organelle containing hydrolytic enzymes that degrade macromolecules and other materials taken up by a cell during endocytosis.

lysozyme *Enzymology.* a protein that functions as an enzyme, by dissolving bacteria found in fluids such as saliva and tears.

lyssacine *Invertebrate Zoology.* an early developmental stage of the skeletal network of hexactinellid sponges.

Lyssacinosida *Invertebrate Zoology.* an order of sponges whose skeletons consist of six-rayed spicules.

lyssophobia *Psychology.* an irrational fear of the dementia associated with rabies.

Lystrosaurus *Paleontology.* a genus of herbivorous, semi-aquatic dicynodont therapsids in the extinct infraorder Pristerodontia; about the size of a horse, *Lystrosaurus* inhabited Antarctica, southern Africa, and India in the Lower Triassic, thus supplying evidence for the time of separation of the several parts of Gondwanaland.

lysyl oxidase *Enzymology.* an enzyme of the oxidoreductase class that catalyzes the oxidation of lysine residues, especially in collagen and elastin; important to the health of connective tissues and arteries. Also, PROTEIN LYSINE 6-OXIDASE.

Lythraceae *Botany.* a widespread family of tropical dicotyledonous herbs, shrubs, and trees of the order Myrtales that often produce alkaloids and usually have internal phloem; species include the crape myrtle tree and the henna plant.

lytic cycle *Virology.* the lysis of a host cell by a virulent phage releasing progeny phages that infect other host cells and repeat the cycle.

lytic infection *Microbiology.* the invasion of a host cell by a bacteriophage, leading to growth of the bacteriophage and lysis of the cell.

lytic pathway *Biotechnology.* the sequence of reactions, including DNA replication and the assembly of progeny phages, that result in the completion of a lytic cycle.

lytic phage *Virology.* a phage that produces a lytic infection in its host and cannot induce a lysogenic state.

lytic response *Cell Biology.* a process in which a bacterial cell ruptures following infection by a bacteriophage, causing the release of a newly replicated set of virus particles into the environment.

lyticum *Bacteriology.* a genus of Gram-negative, rod-shaped bacteria that occur as endosymbionts in certain strains of *Paramecium aurelia.*

lytic virus *Virology.* a virus that produces infection and subsequent lysis of susceptible cells.

lytoceratid *Paleontology.* of or relating to the ammonoids of the order Lytoceratida, which arose from the phylloceratids at the end of the Triassic, thrived in the Jurassic, and persisted into the Late Cretaceous; the genus *Lytoceras* was as large as 1.5 meters in diameter; generally more evolute (loosely coiled) than the phylloceratids.

D-lyxitol see ARABITOL.

LZ landing zone.

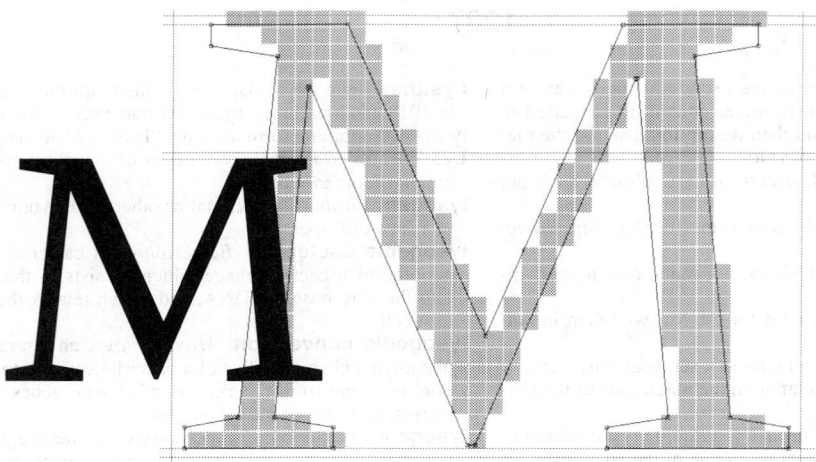

m mass; medium; mega-.
m or **m.** meter.
m *Electricity.* a symbol for magnetic pole strength.
μ the Greek letter mu (capital: M).
M male, masculine; Mach.
M *Computer Technology.* a symbol for megabyte.
M-1 *Ordance.* **1.** a gas-operated, semiautomatic .30-caliber rifle, the standard U.S. Army rifle of World War II and the Korean War. **2.** the standard U.S. Army battle tank.
M-1A1 tank *Ordance.* a later version of the M-1 tank with a 120-mm gun; the **M-1A2** version has heavier armor.
M-14 or **M14** *Ordnance.* a fully automatic, gas-operated .30-caliber rifle; a standard U.S. military weapon.
M-16 or **M16** *Ordnance.* a lightweight automatic rifle that is a standard U.S. military weapon, firing either .30- or .223-caliber ammunition.
M-40 *Ordnance.* a 7.62-mm Marine sniper rifle.
mA or **ma** *Electricity.* the symbol for milliampere.
μA *Electricity.* the symbol for microampere.
Ma the symbol for masurium, now called technetium.
MA or **M.A.** master of arts; meter angle; military academy.
MA *Psychology.* mental age.
MAA methanearsonic acid.
maakite see HYDROHALITE.
maar *Volcanology.* a broad, flat-floored volcanic crater, generally formed by a single explosive eruption and frequently filled with water.
MAC military airlift command; membrane attack complex; multiplexed analogue components; multiaccess computing.
M.A.C. maximum allowable concentration.
macadam [mə kad′əm] *Civil Engineering.* road-making material utilizing uniformly sized stones rollered into layers and finished with asphalt. (Developed by the Scottish engineer John *McAdam,* 1756–1836.)
macadamia [mak′ə dā′mē ə] *Botany.* any tree of the genus *Macadamia* of the family Proteaceae, especially the species *M. ternifolia,* a tropical evergreen tree with whorled leaves and long clusters of pink flowers, grown commercially as a food crop. (Named for John *Macadam,* an Australian chemist.)
macadamia nut *Botany.* the edible seed of the fruit of the macadamia tree.

macadamia nut

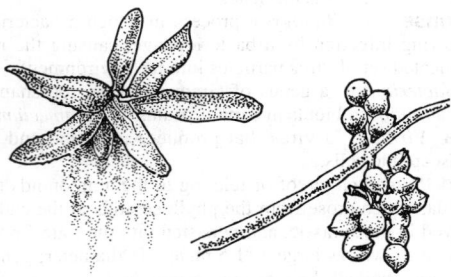

macaque [mə kak′] *Vertebrate Zoology.* any member of 12 short-tailed monkey species of the genus *Macaca* in the family Cercopithcidae, characterized by doglike faces, long limbs, and dark fur; some of moderate size (up to 28 inches and 18 pounds), both arboreal and ground-dwelling, and diurnal; found in Asia, Gibraltar, and North Africa and including the rhesus monkey.

macaque

Macassar oil *Materials.* **1.** an oil originally derived from materials from the region around Macassar, a seaport of Indonesia; once widely used as a hairdressing. **2.** any of various similar hairdressing preparations.
macaw *Vertebrate Zoology.* any of numerous tropical, fruit-eating birds of the order Psittaciformes, including some of the largest and showiest parrots; native to Central and South America.
macawood *Materials.* the economically valuable wood of tropical American trees of the genus *Platymiscium.* Also, QUIRA.
Macbeth illuminometer *Optics.* a type of photometer that measures light intensity by matching it against a comparison lamp after it has been balanced by a Lummer-Brodhum sight box, and corrects for color differences by passing the light through calibrated optical filters.
MACC *Oncology.* an acronym for a chemotherapy regimen for cancer treatment that includes the use of the drugs methotrexate, Adriamycin, cyclophosphamide, and CCNU (lomustine).
MacCullagh's formula *Physics.* a formula expressing the gravitational or electrostatic potential due to a distribution of mass or electric charge.
Mace *Materials Science.* the trade name for a liquid tear gas mixture sprayed from a pressurized container to temporarily disable a person; used to resist criminal assault, subdue rioters, and so on. Also, **Chemical Mace.** *Ordnance.* a surface-to-surface missile that can deliver a conventional or atomic warhead at a range of over several hundred miles; a newer version of the MGM-1C Matador, with an improved guidance system, longer-range, low-level attack capability, and higher-yield warhead; officially designated **MGM-13.**

macedonite *Mineralogy.* $PbTiO_3$, a black, tetragonal mineral usually occurring as irregular grains, having a specific gravity of 7.82 and a hardness of 5.5 on the Mohs scale; found in amazonite quartz syenite veins. *Petrology.* **1.** a fine-grained basaltic rock consisting of orthoclase, sodic plagioclase, biotite, olivine, and rare pyriboles in a green vitreous or chloritic base. **2.** any trachyte bearing olivine and biotite.

maceral *Geology.* any of the microscopic, organic constituents of coal, as distinguished from the macroscopic constituents seen in hand specimens.

maceration *Medicine.* the softening and discoloration of tissues of the fetus characterizing its degeneration and disintegration after death in the uterus.

macgovernite see MCGOVERNITE.

Mach, Ernst [mäk] 1838–1916, Austrian physicist; developed a now-standard method for measuring high speeds of objects moving through air and other gases.

Mach or **mach** *Fluid Mechanics.* an expression of the speed at which an object is moving through a fluid medium according to the Mach number, which relates the speed of the object to the speed of sound. An airplane traveling at Mach 1 is moving at the speed of sound, Mach 2 is twice the speed of sound, and so on. (From Ernst *Mach.*)

mach. machine; machining; machinist.

Machaeridea *Invertebrate Zoology.* a former classification of echinoderms, now considered ancestral to modern echinoderms or as aberrant barnacles.

Mach angle *Fluid Mechanics.* the half-angle associated with the Mach cone.

Mach cone *Fluid Mechanics.* a conical surface that coincides with the superposition of all the shock wave phase fronts generated by a body traveling through a fluid medium at supersonic speeds.

Mach disks see DIAMONDS.

machete *Mechanical Devices.* a large, heavy knife with a broad blade approximately 2 to 3 feet in length, used to clear paths through vegetation. Also, **machette.**

Mach front see MACH STEM.

Machiavellian [mak′ē ə vel′ē ən] *Psychology.* of or relating to behavior in which a person purposely manipulates or misleads others in order to achieve his or her own personal goals. Thus, **Machiavellianism.** (From the famous description of such actions by the Italian political theorist Niccolo *Machiavelli,* 1469–1527.)

Machilidae *Invertebrate Zoology.* a family of bristletails, primitive wingless insects in the order Thysanura.

machinability *Materials.* the quality or degree of being machinable.

machinability index *Materials.* a standardized measure of machinability, usually relative to the machinability of a reference material.

machinable *Materials.* able to be shaped or cut with machine tools. Also, **machineable.**

machine *Mechanics.* **1.** any device that transmits or modifies energy. **2.** see SIMPLE MACHINE. *Mechanical Engineering.* **1.** an assembly of interrelated parts, each with a definite motion and separate function; used to perform a specific task. **2.** relating to, produced by, or designed for use with one or more machines; mechanical. **3.** to make or finish an object with a machine or machine tools. *Computer Technology.* a real or conceptualized computer.

machine address see ABSOLUTE ADDRESS.

machine attention time *Industrial Engineering.* the time during which an operator must be watching, but not actively running, a machine. Also, SERVICE TIME.

machine available time *Computer Technology.* the amount of time during which a computer is up and ready for productive processing, not including maintenance time.

machine balancing *Design Engineering.* the process of providing balance to all rotating parts in a machine.

machine bolt *Mechanical Devices.* a heavyweight, blunt-ended bolt with standardized threads and a square, flat or hexagonal head, commonly used in various mechanical assemblies.

machine check *Computer Programming.* an automatic or programmed equipment check.

machine code *Computer Programming.* absolute machine-language instructions.

machine control *Design Engineering.* the use of various computer controls to initiate actions of machine tools.

machine cut *Mining Engineering.* a slot or groove, either vertical or horizontal, made by a coal cutter in a coal seam in preparation for shot firing.

machine cycle *Computer Programming.* **1.** a computer cycle such as fetch, execute, or interrupt. **2.** the time during which a computer can perform a particular sequence of operations such as a loop.

machine cycle time *Industrial Engineering.* the time required for a machine to perform one complete cycle.

machine-dependent *Computer Technology.* describing a program or a language that works on only one type of computer and requires features unique to that specific computer. Also, HARDWARE-DEPENDENT.

machine design *Design Engineering.* the use of engineering principles applicable to machinery to advance its use, function, and operating characteristics.

machine diagnostics *Mechanical Engineering.* information about the operating state of a machine which can be used to analyze the functional behavior of the machine.

machine drill *Mechanical Devices.* any mechanically driven diamond, rotary, or percussive drill.

machine element *Design Engineering.* any of the machined parts or components designed for various machine devices, including gears, bearings, fasteners, screws, pipes, springs, rivets, locks, and bolts.

machine error *Computer Technology.* an error during processing caused by a machine malfunction, as opposed to an error caused by the software or the operator.

machine forging *Metallurgy.* forging performed in vertical or horizontal mechanical equipment.

machine gun *Ordnance.* **1.** an automatic weapon firing small-arms ammunition of .60 caliber or less; as opposed to an automatic rifle, it is capable of longer sustained fire than an automatic rifle and is usually fired from a mount rather than hand-held. **2.** to attack with machine gun fire.

machine-gun microphone see LINE MICROPHONE.

machine-hour *Industrial Engineering.* a measure of machine availability or production effort, based upon the use of a given machine for one hour, two machines for 30 minutes, or an equivalent combination.

machine idle time *Industrial Engineering.* the portion of a work cycle during which a machine is not in operation, for reasons such as conflicting demands for service or a wait for completion of manual work.

machine-independent *Computer Technology.* **1.** referring to a programming language or program created in terms of the problem without regard for the devices used to process them. **2.** referring to the ability to execute a program successfully on a variety of computers.

machine instruction *Computer Programming.* an instruction, written in machine code, that a computer can recognize and execute directly.

machine interruption *Computer Technology.* a suspension in program processing caused by an exception condition detected by the computer hardware.

machine key *Mechanical Devices.* see KEY.

machine language *Computer Technology.* the set of machine instructions that a particular model of computer can execute. Also, **machine language code.**

machine learning *Computer Programming.* the ability of a machine to improve its behavior based on previous trials and past performance.

machine load *Industrial Engineering.* **1.** the scheduled work load for a machine in a given time period. **2.** the percentage of maximum load at which a machine is actually used.

machine logic *Computer Technology.* the internal design of a computer, using the principles of truth tables and the relationships among the logic elements.

machine oil *Materials.* any of various oils used to lubricate machine parts.

machine operator *Computer Technology.* a person who loads programs and data into a computer or controls the console during program runs; a computer operator.

machine-oriented programming system *Computer Technology.* a system that uses a language oriented to a specific computer's internal language.

machine pacing *Industrial Engineering.* the control of the work rate by means of machinery rather than workers.

machine performance *Design Engineering.* a measure of the operation of a machine relative to specifications.

machine pistol *Ordnance.* a pistol capable of fully automatic fire.

machine rating *Mechanical Engineering.* the amount of power that can be drawn or delivered by a machine without its overheating.

machine ringing *Telecommunications.* telephone ringing that is initiated either by an operator or by mechanical means, and that continues automatically until the receiver answers the call or until the call itself is abandoned.

machinery *Mechanical Engineering.* a group of parts or machines that are arranged to perform a particular operating or machining function.

machinery noise *Acoustics.* background noise of operating machinery that is inherent in a particular sound-recording location.

machine screw *Mechanical Devices.* any of various blunt-ended screws having a slotted or recessed head and standardized threads, used for joining two or more pieces of metal with or without a nut.

machine script *Computer Technology.* the representation of data in machine-readable form.

machine-sensible *Computer Programming.* capable of being sensed or translated by the machine into bit patterns for computer storage; in a form that is readable by the computer. Thus, **machine-sensible data, machine-sensible information, machine-sensible program.** Also, **machine-readable, machine-recognizable.**

machine shop *Engineering.* a workshop in which machines are used to cut, form, or shape metals and other substances.

machine shot capacity *Engineering.* in injection molding, the greatest amount of thermoplastic resin a single stoke of the injection ram can displace.

machine-spoiled time *Computer Technology.* any computer time that is wasted as a result of a computer malfunction during production runs.

machine steel *Metallurgy.* unalloyed medium-carbon steel.

machine taper *Mechanical Engineering.* a taper that connects a tool, arbor, or center to its mating part and keeps them accurately aligned.

machine tool *Mechanical Engineering.* any stationary power-driven machine designed primarily for shaping and sizing metal parts.

machine-tool accuracy *Design Engineering.* the geometric behavior of a machine or machine tool relative to specification.

machine-tool control *Computer Programming.* the direct control of machine tools by computer. *Control Systems.* see NUMERICAL CONTROL.

machine translation see MECHANICAL TRANSLATION.

machine utilization *Engineering.* the time when a machine is actually operating within a given period, often expressed as a percentage of the total elapsed time.

machine wall *Mining Engineering.* the face being cut into by a coal-cutting machine.

machine welding *Metallurgy.* welding performed by machine, but not completely automated.

machine word *Computer Science.* a unit of information in a machine-organized computer.

machining *Mechanical Engineering.* the process of using a machine to remove material from a workpiece to achieve a desired size and shape.

machining center *Mechanical Engineering.* equipment that removes metal under computer control, usually using several axes and a variety of tools and operations.

machining stress *Mechanical Engineering.* the residual stress that is directly caused by machining.

machinist's file *Mechanical Devices.* a cross-cut file used for quick removal of rough metal.

machinist's hammer see BALL-PEEN HAMMER.

machinist's vise *Mechanical Devices.* a metal-gripping vise having thick, cast-iron jaws lined with serrated steel and either a fixed or swivel base. Also, ENGINEER'S VISE, FITTER'S VISE.

Mach line *Fluid Mechanics.* **1.** a line of a shock wave that lies on the surface of the Mach cone and connects all the pressure disturbances emanating from the source having a common phase. **2.** see MACH WAVE.

Machlovirus group *Virology.* a group of plant viruses that have a narrow host range and are transmitted by leafhoppers; the type species is maize chlorotic dwarf virus.

Machmeter *Engineering.* an instrument that measures the ratio between the speed of a body or the flow of a fluid and the speed of sound in the same medium. Also, **Mach indicator.**

Mach number *Fluid Mechanics.* a dimensionless quantity given by the ratio of the velocity of a source in a fluid medium (e.g., an airplane in flight) to the velocity of sound in the same fluid.

machozooid *Invertebrate Zoology.* in some hydroid colonies, a defensive polyp with stinging cells and little or no gastrovascular cavity; also, **machopolyp.**

Mach reflection *Fluid Mechanics.* the reflection of a shock wave from a rigid boundary, resulting in the reduction of the reflected shock wave strength and the reflection angle to the smaller of the two theoretically possible values for each of these quantities.

Mach's principle *Physics.* a principle stating that the inertial properties of any piece of matter can be determined only by reference to all the other matter in the universe.

Mach stem *Fluid Mechanics.* the formation of a shock wave from the superposition of direct and reflected shock waves produced by spherical shock reflecting from a planar surface. Also, **Mach front.**

Mach wave *Fluid Mechanics.* the shock wave produced by a body traveling at supersonic speeds.

Mach-Zehnder interferometer *Optics.* a type of interferometer that splits a beam of light into two less-bright, but usually equal, beams that travel along differing paths and are later recombined to form an interference pattern. Also, **Mach refractometer.**

M acid *Organic Chemistry.* $C_{10}H_5NH_2(OH)SO_3H$, gray needles that are slightly soluble in cold water and soluble in hot water and alcohol; used in making dyes.

MAC INH or **MAC inh** membrane attack complex inhibitor.

mackayite *Mineralogy.* $Fe^{+3}Te_2O_5(OH)$, a green, tetragonal mineral occurring in pyramidal to prismatic crystals, having a specific gravity of 4.86 and a hardness of 4.5 on the Mohs scale; found in the oxidized portion of pyrite and native tellurium veins at Goldfield, Nevada.

Mackenzie *Geography.* a river in northwestern Canada that flows from the Great Slave Lake about 1100 miles north and west to the Beaufort Sea.

mackerel *Vertebrate Zoology.* an important food fish of the marine family Scombridae, being streamlined, swift swimmers related to tunas, having a green body with blue bars and a silvery underside, and found in all tropical and temperate seas.

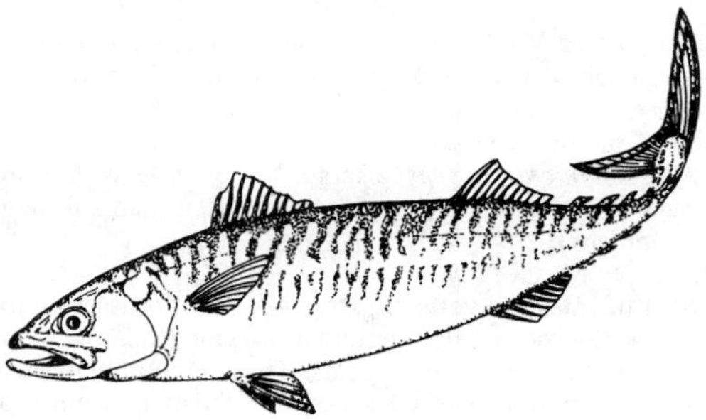

mackerel

mackerel shark *Vertebrate Zoology.* fast-swimming sharks of the chondrichthyan family Lamnidae, characterized by a conical-pointed head, a large mouth, and relatively few but still well-developed teeth; found from cold temperate to tropical seas.

mackerel sky *Meteorology.* an arrangement of cirrocumulus or small-element altocumulus clouds in a regular pattern in the sky, resembling the pattern on a mackerel's back.

mackinawite *Mineralogy.* $(Fe,Ni)_9S_8$, a bronze-colored, opaque tetragonal mineral occurring as crystals and irregular grains, having a specific gravity of 4.3 and a hardness of about 2.5 on the Mohs scale; found as a product of hydrothermal activity in mineral deposits and serpentinized rocks.

Macky effect *Meteorology.* a pronounced reduction in the dielectric strength of air in the presence of water drops.

Maclaurin, Colin 1698–1746, Scottish mathematician; developed Newton's theorems; introduced a method of generating conics.

Maclaurin expansion *Mathematics.* a special case of the Taylor series expansion of a function $f(x)$; expansion is done about the point $x_0 = 0$.

Maclaurin spheroid *Astrophysics.* the stable shape that a rotating mass has when it is homogenous and self-gravitating.

Maclaurin's theorem *Mathematics.* let $f(x)$ be infinitely differentiable in an interval $(-a,a)$, and let $f^{(n)}$ denote the nth derivative of f. Suppose that $\lim_{n\to\infty}[x^n f^{(n)}(tx)/n!] = 0$, where $-a < x < a$ and $0 < t < 1$. Then $f(x)$ may be represented by the **Maclaurin series,**

$$f(x) = \sum_{k=0}^{\infty} x^k f^{(k)}(0)/k!\ .$$

macle *Mineralogy.* a twinned crystal of diamond, composed of two (or occasionally three) flat crystals; used for certain industrial applications such as wiredrawing, and sometimes as gems. *Crystallography.* any twin crystal.

Macleod, John J. R. [mə kloud′] 1876–1935, Scottish physiologist; shared the Nobel Prize (with F. G. Banting) for the discovery of insulin.

Macleod equation *Fluid Mechanics.* a relationship expressing the fact that the fourth root of the surface tension of a liquid is proportional to the difference between the liquid density and its vapor density at saturation conditions; the proportionality constant depends upon the values of Sugden's parachor and the molecular weight.

MacMichael degree *Fluid Mechanics.* a unit used in viscosity measurement in a rotational (Couette) viscometer; the size of the unit depends on the stiffness of the inner cylinder suspension.

Macoma *Paleontology.* a genus of deep-burrowing clams in the order Veneroida; Pleistocene to the present.

MACOP-B *Oncology.* an acronym for a chemotherapy regimen for cancer treatment that includes methotrexate, Adriamycin, cyclophosphamide, Oncovin, predenisone, and bleomycin.

Macquer's salt see POTASSIUM ARSENATE.

macr- a combining form meaning: **1.** large, as in *macrurous.* **2.** excessively or abnormally large, as in *macrencephaly.*

macrandrous *Botany.* of a species, having male plants of normal size, as distinguished from nannandrous species, in which the male plants are considerably smaller than the females.

Macraucheniidae *Paleontology.* a family of camel-like hoofed mammals in the order Litopterna, having a nasal opening on top of the head that indicates adaptation to living in water; found only in South America, the Macraucheniidae date back to the Paleocene and were the last surviving litopterns, persisting into the Middle Quaternary.

macrencephaly *Neurology.* a congenital anomaly marked by an abnormally large brain. Also, MACROENCEPHALY.

Macristiidae *Vertebrate Zoology.* the family of oceanic fishes sometimes assigned to the order Ctenothrissiformes.

macro *Computer Programming.* an instruction written as part of a source language that will expand into multiple source language instructions when compiled. Also, MACROINSTRUCTION.

macro- a combining form meaning: **1.** large, as in *macroscopic.* **2.** excessively or abnormally large, as in *macrocephalic.*

macroanalytical balance *Engineering.* a very accurate balance scale, typically capable of weighing as much as 200 grams to the nearest 0.1 milligram.

macroassembler *Computer Programming.* a program that translates assembly language, including user defined and created macros, into machine language. Also, **macroassembler, macroassembly program.**

macrobead *Biotechnology.* a large round particle often used as the stationary phase in affinity chromatography, when large molecules or whole cells are being separated.

macrobiotic *Science.* having a long life span; long-lived. *Nutrition.* describing a dietary program advocating the restrictive use of certain foods, such as one rich in whole grain cereals; maintained by proponents as a diet promoting good health, but regarded by many nutritionists as deficient in essential proteins.

macroblepharia *Medicine.* an abnormally large eyelid.

macrobrachia *Medicine.* an abnormal size or length of the arm.

macrocephalus *Medicine.* an individual with an abnormally large head.

macrocephaly *Medicine.* an abnormally large head.

macrocheiria *Medicine.* abnormally large hands.

macroclastic *Petrology.* rock composed of fragments visible to the naked eye.

macroclimate *Meteorology.* the general climate over a broad area of the earth's surface, as distinguished from mesoclimate or microclimate.

macrocode *Computer Programming.* a coding language containing macros.

macroconidium *Mycology.* a large sexual spore or conidium. Also, MACROSPORE.

macroconsumer *Ecology.* any large organism that ingests smaller organisms.

macrocosm *Science.* a very large or very general frame of reference.

macrocrania *Medicine.* an exceptionally large cranium in relation to the face, a characteristic of hydrocephalus.

macrocrystalline *Petrology.* **1.** of or relating to the texture of holocrystalline rocks in which crystals are visible to the naked eye. **2.** of or relating to the texture of a recrystallized sedimentary rock with grains or crystals of a diameter greater than 0.75 millimeter.

macrocycle *Organic Chemistry.* an organic macromolecule containing a cyclic element of usually more than 15 atoms. Thus, **macrocyclic.**

Macrocypracea *Invertebrate Zoology.* a superfamily of marine ostracod crustaceans, in the suborder Podocarpa, with variously modified thoracic legs.

macrocyst *Medicine.* **1.** a massive cyst. **2.** in the study of molds, a huge spore container or the gamete of a slime mold.

macrocyte *Hematology.* an abnormally enlarged erythrocyte.

macrocytic anemia *Hematology.* a decrease in red blood cells with the remaining cells being larger than normal.

macrocytosis *Hematology.* an accelerated production of macrocytes.

macrodactyly *Medicine.* abnormally large fingers and toes.

Macrodasyida *Invertebrate Zoology.* an order of tiny, marine littoral, hermaphroditic worms in the phylum Gastrotricha, having anterior, lateral, and posterior adhesive suckers. Also, **Macrodasyoidea.**

macrodefinition *Computer Programming.* a specification or statement that defines a macro.

Macrodontia *Invertebrate Zoology.* a genus of South American longhorn beetles having saw-toothed mandibles; native to tropical forests, they are among the world's largest insects.

macrodontia *Medicine.* a developmental disorder characterized by increase in the size of the teeth.

macroencephaly see MACRENCEPHALY.

macroenzyme *Enzymology.* any very large enzyme.

macroesthesia *Neurology.* a perception of all objects as larger than their actual size.

macroetching *Metallurgy.* in metallography, etching to reveal structural features visible at no or low magnification.

macroevolution *Evolution.* the process of evolutionary change as manifest over the course of geological time in biological events above the level of species. Also, TRANSPECIFIC EVOLUTION.

macroexpansion *Computer Programming.* a sequence of statements produced from a macrogeneration operation.

macrofacies *Geology.* see FACIES TRACT.

macrofauna *Zoology.* animals that are large enough to be seen with the naked eye. *Ecology.* **1.** widely distributed fauna. **2.** the fauna of a macrohabitat.

macrofiber *Microbiology.* a helically twisted chain of bacterial cells, formed under certain culture conditions by *Bacillus subtillis.*

macroflora *Botany.* plants that are large enough to be seen with the naked eye. *Ecology.* **1.** widely distributed flora. **2.** the flora of a macrohabitat.

macro flow chart see FLOW CHART.

macrofollicular adenoma *Medicine.* a disease of the thyroid gland, or other gland, characterized by large spaces which contain colloid. Also, COLLOID ADENOMA.

macrofossil *Paleontology.* a fossil that is large enough to be studied directly, without the aid of a microscope.

macrogamete *Biology.* the larger gamete, always female, produced by a heterogamous organism.

macro generation *Computer Programming.* the production of a sequence of machine instructions by a macroassembler.

macrogenerator see MACROPROCESSOR.

macrogenesis see SALTATION.

macrogeography *Geography.* large-scale, usually highly theoretical geographical study.

macroglia *Histology.* the portion of the neuroglia that includes two types of astrocytes and oligodendrocytes.

macroglobulin *Biochemistry.* a plasma globulin with molecular weight over 400 kDa.

macroglobulinemia *Medicine.* a condition characterized by an abnormal increase of macroglobulins in the blood.

macroglossia *Medicine.* an abnormally large tongue.

macrograph *Graphic Arts.* a usually photographic reproduction whose size is equal to, nearly equal to, or up to ten times as large as the object reproduced. *Metallurgy.* a recorded image of a metallic structure obtained at no or low magnification; usually below 10×.

macrogyria *Neurology.* an abnormal coarseness or breadth of the brain substance, resulting in expansion in the size of the gyri. Also, PACHYGYRIA.

macrohabitat *Ecology.* a habitat that contains a variety of environments and ecological niches capable of supporting a wide range of plants and animals.

macroinstruction *Computer Programming.* **1.** see MACRO. **2.** a machine-language instruction containing several microinstructions.

Macrolepidoptera *Invertebrate Zoology.* a former classification of large moths and butterflies, now used for descriptive purposes rather than for taxonomy.

macro library *Computer Programming.* a collection of commonly used macros available to multiple users.

macrolide *Organic Chemistry.* any member of a class of natural compounds produced by various species of *Streptomyces* and containing highly oxygenated large lactone rings, including methmycin, erythromycin, and carbomycin; used as antibiotics. Thus, **macrolide antibiotic.**

macrolinguistics *Linguistics.* the branch of linguistics concerned with language in its broadest sense, including associated cultural and behavioral features.

macrolymphocyte *Histology.* an enlarged lymphocyte.

macromastia *Medicine.* exceptionally large breasts, usually unrelated to hormonal stimuli.

macromechanics see COMPOSITE MACROMECHANICS.

macromelia *Medicine.* an abnormal increase in the size of one or more limbs.

macromere *Developmental Biology.* one of the large blastomeres formed by the unequal cleavage of a fertilized egg.

macrometeorology *Meteorology.* the study of large-scale weather features and characteristics in the atmosphere, such as general circulation and weather types.

macromolecular hypothesis *Materials Science.* the postulate or theory that organic colloids of high viscosity are composed of long chains; an important concept in the development of modern polymer science.

macromolecule *Organic Chemistry.* an enormous molecule, often containing hundreds or thousands of atoms. Thus, **macromolecular.**

Macromonas *Bacteriology.* a genus of large, Gram-negative, motile, cylindrical bacteria of uncertain classification that oxidize sulfur compounds and contain sulfur granules; occur in fresh waters having low concentrations of oxygen.

macromyeloblast *Histology.* a large myeloblast.

macronucleus *Invertebrate Zoology.* the larger of two nuclei found in many protozoans. Also, MEGANUCLEUS, VEGETATIVE NUCLEUS.

macronutrient *Biochemistry.* a chemical element required in relatively large amounts for the normal physiologic processes of the body, including such elements as carbon, hydrogen, oxygen, nitrogen, potassium, calcium. Also, **macroelement.**

macro parameter *Computer Programming.* the symbolic or literal operands of a macrostatement; the macro parameters are substituted into specific locations in the macro definition to create a complete open subroutine.

macrophage *Immunology.* a large cell that ingests material by an engulfing process (known as phagocytosis) and is active in immune responses. Also, MONONUCLEAR PHAGOCYTE.

macrophage activating factor *Immunology.* an extracellular molecule, generally produced by T-lymphocytes, that enhances the ability of macrophages to destroy microbes and tumor cells.

macrophagy *Zoology.* the process or characteristic of feeding on pieces of food that are large in relation to body size.

macrophotograph *Graphic Arts.* a photographic macrograph. Also, PHOTOMACROGRAPH.

macrophreate *Invertebrate Zoology.* any member of the echinoderm order Comatulida, free-living feather stars.

macrophyllous *Botany.* of a plant, having large or long leaves.

macrophyte *Botany.* a large plant, especially a large aquatic plant. Thus, **macrophytic.**

macropinocytosis *Cell Biology.* a process by which extracellular fluids or solutes enter a cell within membrane vesicles that have pinched off the plasma membrane and are large enough to be distinguished under the light microscope.

macroplankton *Biology.* plankton of relatively large body size, 20–200 mm in diameter.

macropodia *Medicine.* abnormally large feet.

Macropodidae *Vertebrate Zoology.* a family of herbivorous marsupial mammals in the order Diprotodontia that includes the kangaroos and wallabies of Australia.

macropodous *Botany.* **1.** of a plant, having a long stem. **2.** of a plant embryo, having a long hypocotyl.

macropore *Materials Science.* in porous materials, a pore of relatively large size. *Geology.* a minute opening in soil that is too large to hold water by capillarity. Also, MEGAPORE.

macroporous resin *Materials Science.* any resin characterized by relatively large intermolecular spaces. Also, MACRORETICULAR RESIN.

macro processor or **macroprocessor** *Computer Programming.* a program that replaces each macro with the sequence of machine instructions it represents; often part of the assembler system. Also, MACROGENERATOR.

macroprogramming *Computer Programming.* the process of generating machine-language code using macros.

macroprolactinoma *Oncology.* a large pituitary adenoma that secretes high levels of prolactin.

macroprosopia *Medicine.* an abnormally large face in relation to the size of the cranium.

macropsia *Medicine.* an illusion in which objects appear larger than their actual size.

macropterous *Zoology.* having large, well-formed wings or, less frequently, fins.

macrorelief *Geography.* major variations in elevation or topography.

macroreticular resin see MACROPOROUS RESIN.

macrorheology *Mechanics.* rheology in which the microscopic properties of the fluid or solid are averaged or ignored.

Macroscelidea *Vertebrate Zoology.* in some classifications, a distinct order of insectivorous mammals containing the family Macroscelididae.

Macroscelididae *Vertebrate Zoology.* the elephant shrews, a family of small, insectivorous African mammals.

macroscopic [mak′rə skäp′ik] *Science.* **1.** visible without special magnification; able to be seen by the naked eye. Also, MEGASCOPIC. **2.** considered in terms of large elements or units of measurement; considered from a large or general perspective.

macroscopic anisotropy *Mining Engineering.* in electrical downhole logging, the phenomenon wherein current flows more readily along sedimentary strata beds rather than perpendicular to them.

macroscopic property *Thermodynamics.* an attribute or observable property of a thermodynamic system that can be characterized by a single quantity which is applicable to the system as a whole, or which is a function of position within the system and is continuous so long as it does not vary rapidly over microscopic distances, except at the boundaries of the system. Examples of macroscopic properties are temperature, pressure, volume, density, viscosity, and concentration. Also, THERMODYNAMIC PROPERTY. *Nucleonics.* in a nuclear reactor, any factor that can be treated separately from other factors.

macroscopic state *Physics.* any state of a system as described in terms of its macroscopic properties. Also, **macrostate.**

macroscopic theory *Physics.* any theory associated with the behavior of a body or system formulated in terms of measurable properties, but not taking into account atomic processes within the system.

macrosegregation *Materials Science.* a variation in the composition in a material over large distances, generally due to nonequilibrium solidification.

macrosonic *Acoustics.* having to do with sound that has an intensity greater than 1 watt per square centimeter, which is used in underwater transmissions and which can be very dangerous to human ears and other living tissue. Thus, **macrosound.**

macrosoporangium *Botany.* an ovule in which macrospores are produced. Also, MEGASPORANGIUM.

macrospore *Botany.* the larger of the two types of spores produced by heterosporous plants; the spore carrying the female gamete. Also, MEGASPORE. *Mycology.* see MACROCONIDIUM.

Macrostomida *Invertebrate Zoology.* an order of free-living aquatic flatworms in the class Turbellaria.

Macrostomidae *Invertebrate Zoology.* a family of free-living aquatic flatworms, in the order Macrostomida, found on surfaces of plants, rocks, or sediment.

macrostructure *Metallurgy.* the gross structure of a metal, on a scale that is observable with little or no magnification.

macrostylous *Botany.* of a plant, having long styles and short stamens.

macro system *Computer Programming.* a programming system having the symbolic capabilities of an assembly system and the capability to develop many-for-one- or macroinstructions.

macrotectonics see MEGATECTONICS.

Macrotermitinae *Invertebrate Zoology.* a subfamily of termites in the family Termitidae.

macrotherm or **macrothermophyte** see MEGATHERM.

macrotome *Engineering.* a device used in making large anatomical sections of a specimen.

macrotrichia *Entomology.* any of various types of single or branched larger hollow bristles, usually arising from a pit or socket in an insect's epidermis and cuticle.

Macrouridae *Vertebrate Zoology.* the grenadiers or rattail fishes, a cosmopolitan family of deep-sea bottom fishes of the order Gadiformes; characterized by a long, tapering tail with a caudal fin, which makes them appear tailless.

Macroveliidae *Invertebrate Zoology.* the predatory true bugs, a family of hemipteran insects in the subdivision Amphibicorisae, found in damp areas or on mosses.

Macrura *Invertebrate Zoology.* a group of decapod crustaceans with well-developed tail fans, in the suborder Reptantia, including the true lobsters and spiny lobsters.

macrurous *Zoology.* **1.** having a long tail; used especially for crustaceans. **2.** of or relating to Macrura.

macula *plural,* **maculae.** *Anatomy.* **1.** any spot or patch that is different in color than the surrounding tissue. **2.** see MACULA RETINAE. *Medicine.* see MACULE, def. 1. *Astronomy.* a dark spot on the solid surface of a planet or satellite.

macula densa *Medicine.* a small area of the hepatic tubule where the cell nuclei are larger and in closer proximity than neighboring cells.

macula lutea *Medicine.* a yellow spot on the surface of the retina, located slightly below and beside the optic disk; it includes the fovea, the retinal area producing sharply focused vision and containing cone photoreceptors that provide color vision.

macular *Anatomy.* of or relating to the macula retinae. *Medicine.* relating to or characterized by the presence of macules.

macula retinae *Anatomy.* an irregular, yellowish depression on the retina, lateral to and slightly below the optic disk; the site of absorption of short wavelengths of light. Also, MACULA.

maculate *Biology.* spotted or blotched.

maculation *Biology.* the pattern of markings, especially spots, on a plant or an animal.

macule *Medicine.* **1.** a discolored spot on the skin not raised above the surface of the skin, as a freckle. Also, MACULA. **2.** a moderately dense scar of the cornea that is visible without magnification.

maculopapular *Medicine.* both macular and papular, as an eruption consisting of both macules and papules.

maculose *Geology.* of or related to certain contact-metamorphic rocks characterized by a spotted or knotted appearance.

mad *Medicine.* **1.** a popular term, especially in former use, for someone who is mentally ill or mentally disturbed. **2.** of a dog or other animal, suffering from rabies; rabid.

MAD mutual assured destruction; magnetic anomaly detection.

MAD *Aviation.* the airport code for Madrid, Spain.

Madagascar *Geography.* a large island (226,658 square miles) off the southeastern coast of Africa.

madarosis *Medicine.* a loss of the eyebrows or eyelashes.

madder *Botany.* **1.** a Eurasian plant, *Rubia tinctorum,* having tough, prickly leaves and clusters of greenish-yellow flowers. **2.** the root of this plant, the source of many natural dyes. *Materials.* a dye or coloring obtained from madder root.

madder family *Botany.* a popular name for the large plant family Rubiaceae, which includes the madder and gardenia, as well as coffee and quinine. See RUBIACEAE.

madder lake *Materials.* a purple-red pigment, originally obtained from madder root, used in paints and inks.

Madeira [mə dēr′ə; mə dâr′ə] *Geography.* **1.** a group of volcanic islands in the Atlantic Ocean, about 350 miles off the northwestern coast of Morocco. **2.** the chief island of this group. **3.** a river in western Brazil, formed by the confluence of streams at the Bolivia-Brazil border and flowing over 2000 miles northeast to the Amazon. *Food Technology.* also, **madeira.** a fortified wine similar to sherry, made chiefly on the islands of Madeira.

Madelung constant *Solid-State Physics.* a dimensionless constant determined solely by the geometry of an ionic array that determines the Coulomb energy of an ionic crystal; defined by the relation $a = -f/(e/R)$, where f is the Coulomb energy per ion, e is the electric charge, and R is the nearest-neighbor spacing.

Madelung indole synthesis *Organic Chemistry.* a method for producing indoles (2-benzopyrroles) from a base-catalyzed thermal cyclization of *N*-acyl-*O*-toluidide.

MADGE microwave aircraft digital guidance equipment.

madistor *Electronics.* a cryogenic semiconductor device that acts as a switch by steering or controlling injection plasma by means of transverse magnetic fields.

madras [mad′rəs] *Textiles.* a light cotton fabric produced in various patterns, especially a distinctive pattern of multicolored plaids and stripes, often used for shirts, dresses, jackets, neckties, and so on. (From Madras [mə dräs′], a city in India associated with the manufacture of this cloth.)

MADRE magnetic drum receiving equipment.

Madreporaria see SCLERACTINIA.

madreporite *Invertebrate Zoology.* a sieve plate that allows water to pass into and out of the water vascular system of echinoderms.

Madsen impedance meter *Acoustical Engineering.* a device that employs a Wheatstone bridge circuit to measure acoustic impedance in human hearing.

MADT microalloy-diffused transistor.

maduromycetes *Mycology.* any of certain fungi belonging to the order Actinomycetales that are characterized by having the sugar madurose.

madurose *Biochemistry.* a sugar present in certain spore-forming organisms, such as *Actinomadura, Excellospora,* or *Microtetraspora.*

maelstrom [māl′strəm] *Oceanography.* **1.** a strong, turbulent, and often destructive current caused by the meeting between a high sea and a tidal current setting in an opposing direction; sometimes characterized by an eddying, whirlpool-like flow. **2. Maelstrom.** originally, a current of this type that occurs frequently near the island of Moskoe off the west coast of Norway; noted as a danger to ships.

Maestrichtian *Geology.* a European geologic stage of the Upper Cretaceous period, occurring after the Campanian and before the Danian of the Tertiary period.

maestro [mī′strō] *Meteorology.* a northwesterly wind that brings clear weather to the Adriatic region, especially in the summer. (From the Italian word for "master.")

MAF macrophage activating factor.

Maffei galaxies *Astronomy.* two galaxies (a large elliptical and a medium-size spiral) discovered at infrared wavelengths near the galactic equator in Cassiopeia and which are heavily obscured by Milky Way dust and gas. Also, **Maffei system.**

mafic *Petrology.* of or relating to an igneous rock that is composed primarily of dark-colored ferromagnesian minerals. (An acronym formed from Mg, magnesium, and Fe, iron.)

mafic front see BASIC FRONT.

mafic mineral *Mineralogy.* any of the dark-colored ferromagnesian rock-forming silicates; more generally, synonymous with any dark mineral.

mag. magnesium; magneitc; magnetism; magneto; magnitude.

Mag. large (From Latin *magnus.*)

MAG maximum available gain.

magamp see MAGNETIC AMPLIFIER.

magazine any of various objects or devices thought of as resembling the pages of a magazine; specific uses include: *Ordnance.* **1.** a compartment or area for storing powder, ammunition, or explosives. **2.** the part of a gun that holds ammunition prior to chambering; it may be removable or fixed. **3.** of or relating to a magazine or a weapon that uses a magazine. Thus, **magazine-fed, magazine gun, magazine pocket, magazine weapon.** *Computer Technology.* a document input hopper that presents documents to a feed mechanism for reading.

magazine filler *Ordnance.* an attachment on a rifle magazine that holds and positions the cartridge clip.

magazine safety *Ordnance.* on an automatic pistol, a device that prevents firing unless the magazine is in place.

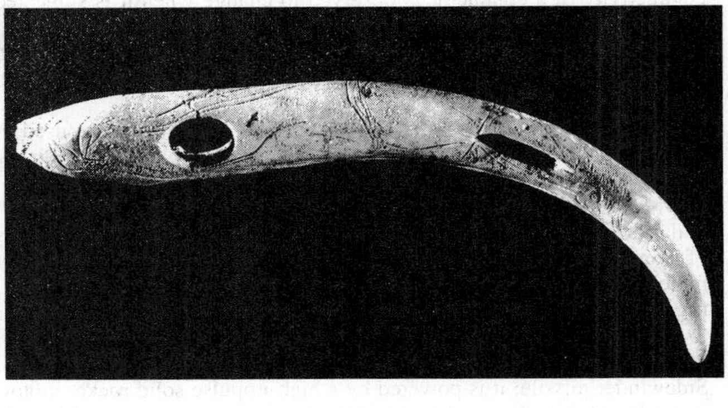

Magdalenian artifact

Magdalenian [mag′də lē′nē ən] *Archaeology.* the final major European culture of the Upper Paleolithic period, from about 15,000 to about 10,000 years ago; characterized by composite or specialized tools, tailored clothing, and especially notable geometric and representational cave art. (From La *Madeleine* or *Magdalene,* a characteristic site in southwest France.)

Ferdinand Magellan

Magellan, Ferdinand 1480?–1521, Portuguese navigator; commanded the first expedition that circumnavigated the globe.

Magellan, Strait of *Geography.* a channel between the South American mainland and Tierra del Fuego.

Magellanic Clouds [maj´i lan´ik] *Astronomy.* two relatively small irregular galaxies that orbit the Milky Way Galaxy at distances of roughly 160,000 light-years (the **Large Magellanic Cloud**) and 190,000 light-years (the **Small Magellanic Cloud**).

Magellanic stream *Astronomy.* an elongated cloud of neutral hydrogen that reaches from the Magellanic Clouds almost to the Milky Way Galaxy and that is thought to have been pulled out by tidal action between the Milky Way and the Clouds.

Magellanic system *Astronomy.* the cloud of neutral hydrogen that envelops both Magellanic Clouds.

Magelonidae *Invertebrate Zoology.* a family of polychaete worms in the group Sedentaria, widely distributed in shallow water.

magenstrasse *Anatomy.* a name sometimes given to the grooved path that food takes along the lesser curvature of the stomach and into the gastroduodenal junction. (A German term meaning "stomach street.")

magenta [mə jen´tə] *Graphic Arts.* a red-blue color, one of the three subtractive primary colors (along with yellow and cyan) used in color separation.

magenta printer *Graphic Arts.* a positive print made from a green color-separation negative.

magentic stepping motor see STEPPER MOTOR.

Maggi-Righi-Leduc effect *Physics.* an effect in which the thermal conductivity of a conductor is observed to change when it is subjected to a magnetic field.

maggot *Entomology.* the common name for the legless, soft-bodied larva of insects in the order Diptera, especially parasites and those found in decaying matter.

maggot therapy *Medicine.* a treatment in which maggots are used to remove dead tissue from suppurative infections; the allantoin in their secretions promotes healing.

maghemite *Mineralogy.* γ-Fe_2O_3, a highly magnetic, brown, cubic mineral, massive in habit, having a specific gravity of 4.90 and a hardness of 5 on the Mohs scale; dimorphous with hematite; found as a secondary mineral in Gossans.

magic *Anthropology.* the performance of rituals to manipulate others or to affect the laws of nature.

Magic *Ordnance.* a French air-to-air missile that is similar to the U.S. Sidewinder missile; it is powered by a high-impulse solid rocket motor, is equipped with infrared homing, and delivers a conventional 27.6-pound (12.5 kg) warhead at a speed of Mach 3 and a range of 0.2 to 6.2 miles (0.32–10 km); officially designated **R550.**

magic-angle method *Materials Science.* in stereochemical configurational analysis, a technique in which the oriented sample is spun around an axis to reduce line broadening due to anisotropic conditions.

magic-eye tube or **magic eye** see INDICATOR TUBE.

Magicicada septendecim see SEVENTEEN-YEAR LOCUST.

magic numbers *Nuclear Physics.* a set of integers 2, 8, 20, 28, 50, 82, and 126 that represent the number of particles in an extra-stable atomic nucleus that has only closed or completed shells of protons and neutrons.

magic spot *Biochemistry.* either of a pair of unusual regulatory nucleotides (ppGpp and pppGpp) occurring in microorganisms and believed to function as alarmones. (So named because they were originally discovered as spots on an autoradiogram.)

magic square *Mathematics.* a square array of integers, usually positive, having the property that the sum of each row, column, and diagonal is the same constant number, called the **magic constant.**

magister of sulfur *Chemistry.* a sulfur obtained by reacting acids with hyposulfites or polysulfides.

maglev *Transportation Engineering.* a rail system in which vehicles are suspended above or below a guide rail by means of magnetic levitation. (An acronym for magnetic levitation.)

magma *Geology.* the naturally occurring, mobile mass of molten rock material generated within or beneath the earth's surface, and from which igneous rocks are derived. *Chemistry.* a suspension of finely divided material in a small amount of water. Thus, **magmatic.**

magma chamber *Geology.* a subsurface reservoir filled with magma. Also, **magma reservoir.**

magma column see LAVA COLUMN.

magma geothermal system *Geology.* a geothermal system in which heat from a large reservoir of magma is transported to the earth's surface.

magma province see PETROGRAPHIC PROVINCE.

magmatic differentiation *Petrology.* **1.** the process by which several types of igneous rocks are derived from a common magma. **2.** the process by which ores solidify from a magma. Also, **magmatic segregation.**

magmatic digestion see ASSIMILATION.

magmatic emanation *Volcanology.* any mixture of gases and liquids given off by magma.

magmatic rock *Petrology.* a rock that has solidified from magma.

magmatic stoping *Geology.* a process whereby magma gradually works its way upward by detaching blocks of country rock and engulfing them.

magmatic water *Hydrology.* water contained in or derived from magma or molten rock.

magmatism *Petrology.* **1.** the development, movement, and solidification of magma to form igneous rock. **2.** a theory stating that granite, to a large extent, has crystallized from magma rather than formed through granitization.

magmosphere see PYROSPHERE.

magn [mag´ən] *Electromagnetism.* a unit of absolute permeability equivalent to one henry per meter.

magn- a combining form meaning "large" or "great," as in *magnafacies.*

magnafacies *Geology.* a primary, continuous, homogeneous belt of deposits, marked by similar lithological and paleontological characteristics, and extending diagonally through several time-stratigraphic units or time planes.

Magnel, Gustave 1889–1955, Belgian civil engineer; a pioneer in the development of prestressed concrete.

magnesia *Inorganic Chemistry.* another term for magnesium oxide, MgO. *Materials Science.* any material containing magnesia. Thus, **magnesia brick, magnesia cement.**

magnesia magma see MILK OF MAGNESIA.

magnesia mixture *Analytical Chemistry.* an aqueous solution of ammonia, magnesium sulfate, and ammonium chloride; used to detect phosphorus.

magnesian calcite *Mineralogy.* $(Ca,Mg)CO_3$, a variety of calcite, $CaCO_3$, in which magnesium ions are substituted randomly in solid solution for calcium within the calcite structure. Also, MAGNESIUM CALCITE.

magnesian limestone *Petrology.* limestone containing a significant amount of magnesium; specifically, limestone composed of at least 90% calcite, less than 10% dolomite, an equivalent of 1.1–2.1% magnesium oxide, and an equivalent of 2.3–4.4% magnesium carbonate.

magnesian marble *Petrology.* a metamorphosed magnesian limestone with magnesium silicates, generally containing up to 15% dolomite.

magnesiochromite *Mineralogy.* $MgCr_2O_4$, a black, opaque, metallic, cubic mineral of the spinel group, occurring in granular masses (or rarely as octahedral crystals), having a specific gravity of 4.2 and a hardness of 5.5 on the Mohs scale; found in olivine-rich igneous rocks.

magnesiocopiapite *Mineralogy.* $MgFe_4^{+3}(SO_4)_6(OH)_2 \cdot 20H_2O$, a yellow, pearly, triclinic mineral of the copiapite group, occurring as scaly to granular aggregates, having a specific gravity of 2.08 to 2.17 and a hardness of 2.5 to 3 on the Mohs scale; found as a secondary mineral formed by the oxidation of sulfides.

magnesioferrite *Mineralogy.* $MgFe_2^{+3}O_4$, a black, cubic, strongly magnetic mineral of the spinel group, occurring in granular masses or as twinned octahedral crystals, having a specific gravity of 4.44 to 4.6 and a hardness of 5.5 to 6.5 on the Mohs scale; found in crystalline limestone.

magnesite *Mineralogy.* $MgCO_3$, a colorless, white, or brown trigonal mineral of the calcite group, occurring in compact to granular masses as an alteration product of magnesium-rich rocks, having a specific gravity of 3.0 to 3.1 and a hardness of 3.75 to 4.25 on the Mohs scale; used principally in basic refractory bricks for high-temperature furnaces.

magnesite flooring *Building Engineering.* a hard, continuous screed that is applied to floor surfaces through a composition of cement, magnesium compounds, sawdust, and sand. Also, JOINTLESS FLOORING.

magnesium [mag nēz´ē əm] *Chemistry.* a silvery alkaline-earth metallic element having the symbol Mg, the atomic number 12, an atomic weight of 24.305, a melting point of 650°C, and a boiling point of 1107°C. The lightest of all structural metals, magnesium is used in alloys for structural parts, in pyrotechnics, and in batteries. (Going back to a phrase meaning "stone of Magnesia;" from the name of a place in ancient Greece.)

magnesium acetate *Organic Chemistry.* $Mg(C_2H_3O_2)_2 \cdot 4H_2O$, a colorless crystalline solid; soluble in water and in dilute alcohol; melts at 80°C; used as a dye fixative, deodorant, disinfectant, and antiseptic.

magnesium amide *Inorganic Chemistry.* $Mg(NH_2)_2$, a gray powder that is decomposed by water or heat; yields ammonia on reaction with water; used as a catalyst for polymerization.

magnesium anode *Electricity.* a bar of magnesium that is buried underground and connected to underground cable to prevent corrosion of the cable due to electrolysis.

magnesium arsenate *Inorganic Chemistry.* $Mg_3(AsO_4)_2 \cdot xH_2O$, a toxic white powder; insoluble in water; used as an insecticide. Also, ARSENIC ACID, MAGNESIUM SALT.

magnesium benzoate *Organic Chemistry.* $Mg(C_7H_5O_2)_2 \cdot 3H_2O$, a white powdery solid; soluble in alcohol and in hot water; melts at 200°C.

magnesium biphosphate see MONOBASIC MAGNESIUM PHOSPHATE.

magnesium bomb *Ordnance.* an incendiary bomb in which the burning agent is magnesium. Similarly, **magnesium flare.**

magnesium borate *Inorganic Chemistry.* $3MgO \cdot B_2O_3$ **(magnesium orthoborate)** or $Mg(BO_2)_2 \cdot 8H_2O$ **(magnesium metaborate),** colorless transparent crystals or white powder; slightly soluble in water and soluble in alcohol; used as an antiseptic, preservative, and fungicide.

magnesium bromate *Inorganic Chemistry.* $Mg(BrO_3)_2 \cdot 6H_2O$, white crystals that lose water at 200°C and decompose at the boiling point; soluble in water and insoluble in alcohol; a dangerous fire risk in contact with organic substances; used as an analytical and oxidizing agent.

magnesium bromide *Inorganic Chemistry.* $MgBr_2 \cdot 6H_2O$, colorless, bitter-tasting crystals that absorb water from the air; soluble in water and slightly soluble in alcohol; melts at 172.4°C; used in organic synthesis and medicine.

magnesium calcite see MAGNESIAN CALCITE.

magnesium carbonate *Inorganic Chemistry.* $MgCO_3$, a light, bulky white powder; very slightly soluble in water; loses CO_2 even on gentle heating; used extensively in the manufacture of food products, pharmaceuticals, cosmetics, and magnesium salts.

magnesium cell *Electricity.* an electric cell that uses magnesium or one of its alloys for the negative electrode.

magnesium chlorate *Inorganic Chemistry.* $Mg(ClO_3)_2 \cdot 6H_2O$, a white, bitter-tasting powder; absorbs water from the air; soluble in water and slightly soluble in alcohol; melts at 35°C and decomposes at 120°C; a dangerous fire hazard in contact with organic substances and a strong oxidizing agent; used as a defoliant and desiccant.

magnesium chloride *Inorganic Chemistry.* **1.** $MgCl_2$, white toxic crystals that absorb water from the air; soluble in water and alcohol; melts at 714°C and boils at 1412°C. **2.** $MgCl_2 \cdot 6H_2O$, the hexahydrate form, colorless toxic crystals, melting at 116-118°C and decomposing at the boiling point. Magnesium chloride is used for a variety of industrial purposes, as in pharmaceuticals and disinfectants, in fire extinguishers and fireproofing material, in ceramics, refrigeration, textiles, and paper manufacture, and as a source of metallic magnesium.

magnesium-copper-sulfide rectifier *Electronics.* a metallic disk rectifier consisting of magnesium in contact with copper sulfide. Also, COPPER SULFIDE RECTIFIER.

magnesium dioxide see MAGNESIUM PEROXIDE.

magnesium fluoride *Inorganic Chemistry.* MgF_2, white fluorescent crystals; insoluble in water and alcohol; melts at 1261°C and boils at 2239°C; a strong irritant; used in ceramics, glass, and optics. Also, **magnesium flux.**

magnesium fluorosilicate *Inorganic Chemistry.* $MgSiF_6 \cdot 6H_2O$, a white crystalline powder that loses water to the air; soluble in water; decomposes at 120°C; a strong irritant; it has various industrial uses, as in ceramics, glassmaking, and waterproofing. Also, MAGNESIUM SILICOFLUORIDE.

magnesium formate *Organic Chemistry.* $Mg(CHO_2)_2 \cdot 2H_2O$, combustible, colorless crystals; soluble in water and insoluble in alcohol and ether; melts at 100°C; used in analytical chemisty.

magnesium front see BASIC FRONT.

magnesium gluconate *Organic Chemistry.* $Mg(C_6H_{11}O_7)_2 \cdot 2H_2O$, a combustible, odorless, tasteless, white powdery solid; soluble in water; used in pharmaceutical manufacturing.

magnesium halide *Inorganic Chemistry.* a compound consisting of magnesium bonded to one or more halogen atoms.

magnesium hydride *Inorganic Chemistry.* MgH_2, a white crystalline substance that ignites spontaneously in air and decomposes violently in water; also decomposes in a vacuum at 280°C.

magnesium hydroxide *Inorganic Chemistry.* $Mg(OH)_2$, a white powder that is nearly insoluble in water and alcohol; loses its water at 350°C; used in medicine as an antacid and laxative, in food processing, and as an intermediate in manufacturing magnesium metal. Also, **magnesium hydrate.**

magnesium hyposulfite see MAGNESIUM THIOSULFATE.

magnesium iodide *Inorganic Chemistry.* $MgI_2 \cdot 8H_2O$, a white, crystalline, deliquescent powder that is discolored by air and light; soluble in water, alcohol, and ether; decomposes at 41°C; used in medicine.

magnesium-iron mica see BIOTITE.

magnesium lactate *Organic Chemistry.* $Mg(C_3H_5O_3)_2 \cdot 3H_2O$, a white crystalline powder that is soluble in water and slightly soluble in alcohol and ether; used in medicine.

magnesium lime see LIMESTONE.

magnesium-manganese dioxide cell *Electricity.* an electric cell that uses a magnesium-manganese based electrolyte where there are two oxygen atoms per molecule of the oxide substance.

magnesium mesosilicate see MAGNESIUM TRISILICATE.

magnesium methoxide *Organic Chemistry.* $Mg(OCH_3)_2$, colorless crystals that decompose on warming; used as a reagent to remove small amounts of water from alcohols. Also, **magnesium methylate.**

magnesium nitrate *Inorganic Chemistry.* $Mg(NO_3)_2 \cdot 6H_2O$, white deliquescent crystals; soluble in water and alcohol; melts at 89°C and decomposes at 330°C; a strong oxidant and a dangerous fire and explosion hazard in contact with organic substances; used in pyrotechnics and in concentrating nitric acid.

magnesium oleate *Organic Chemistry.* $Mg(C_{18}H_{33}O_2)_2$, a combustible yellow powdery solid; insoluble in water and soluble in alcohol and ether; used in varnish dryers and dry cleaning solvents.

magnesium orthophosphate see MAGNESIUM PHOSPHATE.

magnesium oxide *Inorganic Chemistry.* MgO, a colorless to white powder existing in two forms, light and fluffy **(light magnesium oxide)** or dense (**heavy magnesium oxide**); slightly soluble in water and soluble in acids; melts at 2852°C and boils at 3600°C; noncombustible; toxic when inhaled. Magnesium oxide has a wide range of industrial purposes, as in refractories, in fertilizers and food additives, in semiconductors and electrical insulation, and in antacids and laxatives.

magnesium perchlorate *Inorganic Chemistry.* $Mg(ClO_4)_2 \cdot 6H_2O$, white deliquescent crystals; very soluble in water; decomposes at 185–190°C; a dangerous fire and explosion hazard in contact with organic materials; used as a drying and oxidizing agent.

magnesium peroxide *Inorganic Chemistry.* MgO_2, a white powder that is insoluble in water and soluble in dilute acids; decomposes on heating; a strong oxidant and dangerous fire hazard; reacts with atmospheric moisture and acidic substances; used as a bleaching and oxidizing agent and in medicine. Also, MAGNESIUM DIOXIDE.

magnesium phosphate *Inorganic Chemistry.* a general term applied to several compounds; see the specific entries MONOBASIC MAGNESIUM PHOSPHATE, DIBASIC MAGNESIUM PHOSPHATE, and TRIBASIC MAGNESIUM PHOSPHATE.

magnesium salicylate *Organic Chemistry.* $Mg(C_7H_5O_3)_2 \cdot 4H_2O$, a colorless, efflorescent, crystalline powder; soluble in water and alcohol; used as an anti-infective medicine.

magnesium salt see MAGNESIUM ARSENATE.

magnesium silicate *Inorganic Chemistry.* $3MgSiO_3 \cdot 5H_2O$, a fine white powder; insoluble in water and alcohol; used in ceramics, glass, paints, and varnishes, as a rubber filler, and for many other purposes.

magnesium silicofluoride see MAGNESIUM FLUOROSILICATE.

magnesium-silver chloride cell *Electricity.* a reserve primary cell that uses an electrolyte containing magnesium and silver chloride; becomes activated when water is added.

magnesium stearate *Organic Chemistry.* $Mg(C_{18}H_{35}O_2)_2$, an inflammable, tasteless, odorless, soft white powdery solid that is insoluble in water and alcohol; melts at 88.5°C; used as a tablet lubricant in pharmaceutical preparations, a dietary supplement, a drier in paints and varnishes, and an emulsifier in cosmetics.

magnesium sulfate *Inorganic Chemistry.* **1.** $MgSO_4$, colorless crystals with a bitter, salty taste, very soluble in water; decomposes at 1124°C. **2.** $MgSO_4 \cdot 7H_2O$, the heptahydrate form; loses $6H_2O$ at 150°C and $7H_2O$ at 200°C. Also, EPSOM SALTS. Magnesium sulfate has a wide variety of industrial uses, as in fireproofing, textiles, ceramics, and fertilizers, and in medicine as a cathartic, anticonvulsant, and anti-inflammatory agent.

magnesium sulfite *Inorganic Chemistry.* $MgSO_3 \cdot 6H_2O$, a white crystalline powder; slightly soluble in water and insoluble in alcohol; loses its water at 200°C and decomposes at the boiling point; used in manufacturing paper pulp.

magnesium thiosulfate *Inorganic Chemistry.* $MgS_2O_3 \cdot 6H_2O$, colorless to white crystals; very soluble in water and insoluble in alcohol; loses $3H_2O$ at 170°C and decomposes at the boiling point; used in medicine. Also, MAGNESIUM HYPOSULFITE.

magnesium trisilicate *Inorganic Chemistry.* $Mg_2Si_3O_8$, a fine white powder; insoluble in water and alcohol; used as an industrial odor absorbent, decolorizing agent, antioxidant, and antacid. Also, MAGNESIUM MESOSILICATE.

magnesium tungstate *Inorganic Chemistry.* $MgWO_4$, white crystals that are insoluble in water and alcohol and soluble in acids; used in luminescent paint and X-ray screens. Also, MAGNESIUM WOLFRAMATE.

magnesium wolframate see MAGNESIUM TUNGSTATE.

magneson *Organic Chemistry.* $C_{12}H_9N_3O_4$, a brownish-red solid that is soluble in sodium hydroxide; used in identifying magnesium and molybdenum.

Magnesyn *Electricity.* the trade name for a two-pole, permanently magnetized, freely rotating rotor within a three-phase, two-pole, delta-connected stator device; used to provide the current direction of rotation within a repeater unit.

magnet *Electromagnetism.* a substance composed of ferromagnetic or ferrimagnetic material having domains that are aligned in such a way as to produce a net magnetic field outside the substance or to experience a torque when placed in an external magnetic field.

magnet grate *Mining Engineering.* a series of magnetic bars situated so as to attract and thereby remove tramp iron from pulverized or granulated dry solids that pass over them; used as protection for crushing or grinding equipment.

magnetherm process *Materials Science.* a process for the production of magnesium, in which the metal ore and carbon are placed in a furnace consisting of a vertical iron shell lined with refractory. Heat for the reaction is produced internally by passing an electric current through the fused charge residue.

magnetic *Electromagnetism.* **1.** or or relating to magnets or magnetism. **2.** having the properties of a magnet. **3.** having the ability to become magnetized.

magnetically hard alloy *Metallurgy.* an alloy that can be magnetized permanently; examples include Alnico, cobalt-samarium, or iron-neodymium.

magnetically soft alloy *Metallurgy.* an alloy that can be magnetized by an electromagnetic field, but returns to its original state upon removal of the field; e.g., iron-silicon or iron-nickel.

magnetic amplifier *Electronics.* an iron-core device with a variable impedance that can be used to control power delivered to a load. The impedance can be varied with a relatively small amount of current in the signal winding. Also, MAGAMP.

magnetic amplitude *Navigation.* the angle, measured north or south, between magnetic east or west and some other point on the celestial sphere.

magnetic analysis inspection *Metallurgy.* a nondestructive quality control method based on changes in magnetic flux.

magnetic anisotropy *Electromagnetism.* a directional dependence of magnetic properties exhibited by certain materials.

magnetic annealing *Metallurgy.* an annealing process that is carried out while the material is subjected to a magnetic field.

magnetic annual change *Geophysics.* the change in the geomagnetic field in one year.

magnetic annual variation *Geophysics.* the annual change in the azimuth of magnetic north or south for a given location.

magnetic anomaly see LOCAL MAGNETIC ANOMALY.

magnetic anomaly detector *Electromagnetism.* a magnetometer that is mounted on an aircraft; used to detect the magnetic field of submerged submarines.

magnetic azimuth *Navigation.* from any given point of observation, the angle between the vertical plane through the observed object and the vertical plane indicated by a compass needle, influenced by no transient or artificial magnetic disturbances, measured clockwise through 360° from magnetic north.

magnetic bay *Geophysics.* a brief magnetic variation, typically lasting about one hour.

magnetic bearing *Navigation.* the horizontal direction of one point to another, measured clockwise from magnetic north through 360°.

magnetic bias *Electromagnetism.* a steady magnetic field used in a magnetic circuit.

magnetic blowout *Electromagnetism.* a permanent magnet that is placed near the contacts of a switch so that the arcing length is increased and the arc is thus more readily extinguished.

magnetic bottle *Physics.* a magnetic field that is configured to confine a plasma in a controlled-fusion experiment.

magnetic brake *Mechanical Engineering.* a friction brake that is activated by an electromagnet.

magnetic-bridge sensor *Robotics.* a noncontact sensor capable of identifying metals or materials with metallic properties.

magnetic bubble *Solid-State Physics.* see BUBBLE.

magnetic bubble electromagnetic pulse *Electromagnetism.* a slow pulse of electromagnetic radiation that is produced by the expanding conducting plasma shell of bomb material from a nuclear explosion.

magnetic card *Computer Technology.* **1.** a storage device consisting of one or more cards with magnetized surfaces on which data is recorded, so that it is machine-readable. **2.** in word processing, a paper or plastic card that serves as a medium for data encoding and reading.

magnetic card file *Computer Technology.* a direct-access storage device, held in one or more magazines, containing magnetic cards upon which data is stored. Once addressed, the cards are transported from the magazine past a read/write head.

magnetic cell *Electronics.* a basic unit of magnetic storage, capable of containing a single bit.

magnetic character *Computer Technology.* a character printed on a document using ink that can be read by automatic scanners.

magnetic character reader *Computer Technology.* a data entry device that reads and interprets characters printed with magnetic ink.

magnetic chart *Cartography.* a special-purpose navigational map that uses isogonic lines to show the distribution of the earth's magnetic field; facilitates conversion between magnetic and true bearings. Also, ISOMAGNETIC CHART.

magnetic chuck *Mechanical Engineering.* a chuck on whose surface alternate elements at rest are polarized by electromagnets.

magnetic circuit *Electromagnetism.* a system producing magnetic flux lines that form closed paths, commonly by means of ferromagnetic materials.

magnetic clutch see MAGNETIC FLUID CLUTCH.

magnetic compass *Navigation.* a compass that works on the principle that a free-moving magnet will align itself with the earth's magnetic lines of force.

magnetic confinement *Physics.* the use of magnetic fields to contain a plasma within a region of space.

magnetic constant *Metrology.* in the centimeter-gram-second system of units, the unit of absolute permeability of free space, equal to $4\pi \times 10^{-7}$ henries per meter.

magnetic core *Electronics.* a variously shaped piece of magnetic material, often of iron oxide or ferrite, that is placed in a fixed spatial arrangement with current-carrying conductors and used as a logic element or for storing data. *Electromagnetism.* specifically, the iron core within a coil, transformer, or electromagnet.

magnetic core storage *Computer Technology.* an early storage technology that used selective polarization of magnetic cores to store data; magnetic cores were arranged in matrices to form the memory.

magnetic coupling *Electromagnetism.* the magnetic induction arising between two or more circuits or systems.

magnetic course *Navigation.* a course that takes magnetic north as its reference point.

magnetic Curie temperature *Solid-State Physics.* for a given ferromagnetic material, the temperature above which ferromagnetism is destroyed and the material becomes paramagnetic.

magnetic cutter *Acoustical Engineering.* an electromagnetic device that is driven by magnetic force to cut grooves in a lacquer-based master record.

magnetic daily variation see MAGNETIC DIURNAL VARIATION.

magnetic damping *Electromagnetism.* mechanical damping by means of the motion of a conductor through a magnetic field.

magnetic declination *Geophysics.* the difference between magnetic north and true north.

magnetic deflection *Electronics.* in a cathode-ray tube, the deflection of the electron beam by the action of an electromagnetic field produced by a coil around the neck of the tube.

magnetic delay line *Electronics.* a computer delay line whose operation is based on the velocity of propagation of magnetic waves.

magnetic deviation *Navigation.* see DEVIATION.

magnetic diffusivity *Electromagnetism.* a quantity given by the partial derivative of the magnetic field strength with respect to time divided by the Laplacian of the magnetic field strength.

magnetic dip *Navigation.* the angle between the horizontal and the magnetic lines of force at a given place on the surface of the earth. This effect causes the needle of a compass to depress below the horizontal.

magnetic dipole *Electromagnetism.* a system that has a north magnetic pole and a south magnetic pole, separated by a small distance, giving rise to a magnetic field in the vicinity of the poles.

magnetic dipole antenna *Electromagnetism.* a loop antenna that radiates electromagnetic radiation in response to current circulating through the loop.

magnetic dipole density see MAGNETIZATION, def. 2.

magnetic dipole moment *Electromagnetism.* in a magnetic dipole system, a vector whose cross product with the magnetic induction is equivalent to the torque exerted on the dipole. *Physics.* see MAGNETIC MOMENT.

magnetic direction *Cartography.* the direction of a line or course expressed as angular distance from magnetic north.

magnetic disorder scattering *Metallurgy.* a property of electrical conductivity in metals and alloys, representing the contribution to electrical resistivity made by the random orientations of magnetic moments above the Curie point (T_C) of ferromagnetics or the Néel temperature (T_N) of antiferromagnetics.

magnetic displacement see MAGNETIC INDUCTION.

magnetic diurnal variation *Geophysics.* the daily changes in the earth's magnetic field. Also, MAGNETIC DAILY VARIATION.

magnetic domain *Materials Science.* in the microstructure of materials with magnetic properties, the regions where atoms are aligned to produce a magnetic field in one direction. When no magnetic flux is present outside the specimen, the magnetic domains aligned in one direction are balanced by domains of opposite alignment. *Solid-State Physics.* see FERROMAGNETIC DOMAIN.

magnetic double refraction *Optics.* linear birefringence that is induced in a vapor by a magnetic field perpendicular to a light beam. Also, VOIGHT EFFECT.

magnetic-drum receiving equipment *Electronics.* a type of radar that detects targets beyond the horizon by means of ionospheric reflection and very low power.

magnetic earphone *Acoustical Engineering.* an earphone whose operation is based on the principle of magnetic induction, whereby sound is created by a moving diaphragm.

magnetic element *Engineering.* the portion of an instrument or device that produces or is acted upon by magnetism. *Geophysics.* any of seven numerical characteristics that together define a magnetic field: declination, inclination, total intensity, vertical intensity, horizontal intensity, north component, and east component.

magnetic equator *Geophysics.* an imaginary line circling the earth, following the path where the dip of the geomagnetic field is zero. Also, ACLINIC LINE, DIP EQUATOR.

magnetic ferrite see FERRITE.

magnetic-ferroelectric or **magneticferroelectric** *Solid-State Physics.* being both ferromagnetic and ferroelectric; having both spontaneous magnetic and electric polarization in a crystal.

magnetic field *Electromagnetism.* a vector field occupying physical space wherein magnetic forces may be detected; typically in the presence of a permanent magnet, current-carrying conductor, or an electromagnetic wave. Also, MAGNETIC FORCE.

magnetic field aging *Materials Science.* a process for producing anisotropic magnets from chromium-cobalt-iron alloys; a magnetic field is applied during the initial aging treatment below the spinodal temperature, causing the decomposed particles to elongate and align in the direction of the field.

magnetic field bay *Geophysics.* a small magnetic disturbance of short duration, usually an hour, so called because of the V- or bay-shaped mark it makes on a magnetic record.

magnetic field sensor *Engineering.* an instrument that employs a reed switch and a magnet to determine whether a magnetic field is present.

magnetic field strength *Electromagnetism.* a vector field used to describe magnetic phenomena, having the property that the curl of the field is equal to the free current density vector in the meter-kilogram-second system of units. Also, **magnetic field intensity.**

magnetic film see MAGNETIC THIN FILM.

magnetic filter *Chemical Engineering.* an industrial device consisting of steel grids that are magnetized by electric current or a permanent magnet; used to separate iron or other magnetic particles from liquids.

magnetic filtration *Materials Science.* the process of separating ferromagnetic or paramagnetic particles from a mass stream of nonmagnetic materials, which may involve either conventional magnets or superconductors.

magnetic flaw detector *Electromagnetism.* a device that sprays magnetic particles onto the surface of a magnetized body; flaws in the body's magnetic field will appear as black marks where the ink is concentrated.

magnetic fluid *Materials Science.* a fluid composed of solid magnetic particles of subdomain size colloidally dispersed in a liquid carrier; used in inertial dampers, fluid-cooled loudspeakers, and ink-jet printing.

magnetic fluid clutch *Mechanical Engineering.* a friction clutch in which connection between the drive and the driven member is engaged by electromagnetic force. Also, MAGNETIC CLUTCH.

magnetic flux *Electromagnetism.* **1.** the lines of magnetic force arising from and found in the vicinity of a magnetized body. **2.** the integral over a specified area of the perpendicular component of a magnetic field.

magnetic flux density see MAGNETIC INDUCTION.

magnetic focusing *Electromagnetism.* the use of magnetic fields to focus a beam of charged particles.

magnetic force see MAGNETIC FIELD.

magnetic-force welding *Metallurgy.* welding in which pressure is created by a magnetic field.

magnetic force parameter *Physics.* a dimensionless quantity that is commonly used in magnetofluid dynamics, given by the product of the square of the magnetic field strength, the square of the magnetic permeability, the electrical conductivity, and a characteristic length, divided by the product of the fluid velocity and the mass density.

magnetic forming *Metallurgy.* a forming process using stresses generated by a magnetic field.

magnetic gap *Electromagnetism.* a small nonmagnetic section in an otherwise magnetic circuit, such as an air gap.

magnetic grenade *Ordnance.* a small explosive charge equipped with a time fuse and encased in a container to which magnets have been attached; when thrown at or placed on tanks or other armored targets, the magnets hold the charge in position until it is detonated by the fuse. Also, **magnetic charge.**

magnetic gripper *Robotics.* an end effector that uses magnets to handle metallic components.

magnetic hardness comparator *Engineering.* a device that compares the magnetic properties of a test metal and a metal of known hardness while in an induction coil.

magnetic head *Computer Technology.* an electromagnetic device that reads, records, or erases data on a magnetic disk.

magnetic heading *Navigation.* a heading that takes magnetic north as its reference point.

magnetic hysteresis *Electromagnetism.* hysteretic effects displayed by ferromagnetic substances when placed in an alternating magnetic field: the magnetization lags behind the changes in the external field.

magnetic induction *Electromagnetism.* a quantity whereby the force on a charged particle passing through a magnetic field with velocity v is given by $F = qv \times B$. Also, MAGNETIC DISPLACEMENT, MAGNETIC FLUX DENSITY.

magnetic induction gyroscope *Engineering.* a gyroscope in which water doped with salts exhibiting nuclear paramagnetism is acted on by alternating-current and direct-current magnetic fields.

magnetic ink *Materials Science.* a specially prepared ink whose magnetic qualities will activate an electronic character-reading device, as in check processing and other banking procedures.

magnetic inspection paste *Metallurgy.* a paste used to prepare magnetic analysis inspection oil. Similarly, **magnetic inspection powder.**

magnetic iron ore see MAGNETITE.

magnetic latitude *Geophysics.* the angular distance that a location lies north or south of the magnetic equator.

magnetic leakage *Electromagnetism.* magnetic field lines that pass through regions of space where they are inaccessible and are thus not useful in an application.

magnetic lens *Electromagnetism.* an apparatus that produces a magnetic field with axial symmetry that causes beams of charged particles to converge.

magnetic lines of force *Electromagnetism.* geometric lines in space that indicate the vector directions of the magnetic induction field; the concentration of the lines (lines per unit area) gives a relative degree of the field strength. Also, **magnetic lines of flux.**

magnetic local anomaly *Geophysics.* a magnetic anomaly that is visible in a profile of a geologic structure.

magnetic loudspeaker *Acoustical Engineering.* any loudspeaker that operates by the vibration of a diaphragm in response to a varying magnetic field, such as a moving-coil loudspeaker. Similarly, **magnetic earphone.**

magnetic Mach number *Physics.* a dimensionless quantity given by the fluid velocity of a plasma divided by the velocity of an Alfvén wave in the fluid.

magnetic material *Electromagnetism.* any material exhibiting ferromagnetism.

magnetic memory see MAGNETIC STORAGE.

magnetic meridian *Geophysics.* at any location, the line that points toward magnetic north.

magnetic microphone *Acoustical Engineering.* a microphone in which sound waves strike a diaphragm, causing an armature to vary the reluctance of a magnetic circuit and thus produce an electroacoustic signal.

magnetic mine *Ordnance.* a mine that responds to the magnetic field of the target.

magnetic mine detector *Ordnance.* an electrical device for locating metallic mines.

magnetic minehunting *Military Science.* the process of using magnetic detectors to determine the presence of mines or minelike objects.

magnetic mirror *Physics.* a magnetic field configuration used to confine a plasma by reflecting the charged particles into the central portion of a controlled fusion experiment.

magnetic modulator *Electronics.* a modulator having a magnetic circuit as the modulating element.

magnetic moment *Physics.* a property associated with a magnetic domain or magnetized body, equal to the maximum torque experienced by the body divided by the magnetic induction acting on the body, generally measured in units of magnetons. Also, DIPOLE MOMENT, MAGNETIC DIPOLE MOMENT. *Nuclear Physics.* see BOHR MAGNETON.

magnetic monopole *Electromagnetism.* a hypothetical particle that would carry a magnetic charge much like an electric monopole carries an electric charge.

magnetic multipole *Electromagnetism.* a configuration of magnetic dipoles, from a series of common configurations, either static or oscillating, that produce a characteristic electromagnetic field.

magnetic multipole field *Electromagnetism.* the characteristic electromagnetic field generated by a magnetic multipole.

magnetic needle *Electromagnetism.* a freely suspended, narrow permanent magnet that tends to align itself in a direction parallel to an existing magnetic field.

magnetic north *Geophysics.* the direction indicated by the north end of a magnetic compass needle.

magnetic number *Physics.* a dimensionless quantity used in the study of magnetofluid dynamics, given by the square root of the magnetic force parameter.

magnetic observatory *Geophysics.* a station equipped to make measurements of changes in the strength and direction of the Earth's magnetic field.

magnetic octupole moment *Electromagnetism.* a quantity given by the integral over volume of the product of the divergence of the magnetization, the third power of the distance from the origin, and a spherical harmonic Y_3^m; used to characterize a magnetization distribution.

magnetic oscillograph *Electromagnetism.* an instrument that can record a particular component of the earth's magnetic field.

magnetic Oseen number *Physics.* a dimensionless quantity used in magnetofluid dynamics, equal to $0.5(1 - N_{AL}^2)R_M$, where N_{AL} is the Alfvén number and R_M is the magnetic number.

magnetic-particle test *Metallurgy.* a nondestructive quality control method for ferromagnetic materials, based on fine particles applied to a magnetized part.

magnetic pendulum *Electromagnetism.* a permanent magnet that is suspended and free to pivot in a horizontal plane and will oscillate in that plane if it is disturbed and released while in a magnetic field having a horizontal component.

magnetic permeability *Materials Science.* the ratio between the magnetic field and the inductance or magnetization; expressed as $m = B/H$, where m = magnetic permeability, B = flux density, and H = the strength of the magnetic field.

magnetic polarization *Materials Science.* polarization caused by the asymmetrical nature of certain molecules or crystal structures, as in water, organic polymers, and certain ceramic crystal structures.

magnetic pole *Geodesy.* either of two points on the earth's surface at which the magnetic dip is exactly 90° where the meridians join; i.e., where the magnetic field is vertical.

magnetic pole strength *Electromagnetism.* **1.** a quantity given by the force due to magnetic induction divided by the magnitude of the induction (magnetic field strength). **2.** a quantity given by the magnetic moment divided by the distance between the poles of a magnetized body.

magnetic potential *Electromagnetism.* the amount of work required to move a magnetic pole of specified strength from a reference point (commonly at an infinite distance) to a point in question. Also, MAGNETIC SCALAR POTENTIAL.

magnetic potentiometer *Engineering.* an instrument used to explore the distribution of magnetic potential in a field, consisting of a flexible solenoid that is wound on a nonmagnetic base; used in conjunction with a ballistic galvanometer.

magnetic power *Electromagnetism.* a measure of the relative strength of a given magnetic field.

magnetic pressure *Physics.* the pressure associated with a magnetic field, equal to the square of the magnetic induction divided by twice the magnetic permeability.

magnetic pressure transducer *Engineering.* a pressure transducer that converts pressure changes into magnetic reluctance or inductance through the movement of a portion of a magnetic circuit by a pressure-sensing element such as a bourdon tube, bellows, or a diaphragm.

magnetic prime vertical *Geophysics.* the great circle passing through the magnetic east and west points on the horizon at a given location.

magnetic printing *Electronics.* the transfer of recorded information from a section of a magnetic recording medium to another section of the same or a different medium when these sections are brought into close proximity. Also, MAGNETIC TRANSFER, CROSSTALK.

magnetic probe *Electromagnetism.* a small coil that is inserted in a magnetic field to measure changes in field strength.

magnetic profile see MAGNETIC SURVEY.

magnetic prospecting *Mining Engineering.* a form of geophysical prospecting in which variations in the earth's magnetic field are used to find deposits of magnetic materials such as iron, nickel, or titanium; performed as a ground or airborne survey.

magnetic pulley *Metallurgy.* a magnetic device that is used to separate metals from dry products such as sand or refuse.

magnetic pumping *Electromagnetism.* a technique for heating a plasma to very high ion temperatures by applying an oscillating electromagnetic field.

magnetic quadrupole lens *Electromagnetism.* a magnetic lens that generates a field from a magnetic quadrupole formed from a circularly symmetrical arrangement of two dipoles.

magnetic quantum number *Atomic Physics.* a quantum number that corresponds to the component of an angular momentum in a specified direction, usually that of a magnetic field; it measures the angular momentum component in units of Planck's constant divided by 2π.

magnetic recorder *Acoustical Engineering.* any device used to record sound on magnetic tape.

magnetic recording *Acoustical Engineering.* **1.** the process of recording sound or signals on magnetized areas of a medium such as tape or wire. **2.** the product of such a process.

magnetic refrigerator *Physics.* a device used to maintain temperatures below 1 kelvin; its operation depends on the demagnetization of a paramagnetic salt through isentropic thermodynamic processes.

magnetic relaxation *Physics.* a process in which a magnetic system approaches equilibrium through time.

magnetic reproducer *Electronics.* a device that converts recorded magnetic signals into electric signals. Thus, **magnetic reproduction.**

magnetic resonance *Physics.* a frequency matching phenomenon in which the spin systems of certain atoms will readily absorb energy from an externally applied, alternating magnetic field, when the frequency of the magnetic field is the same as a frequency characteristic of the spin system.

magnetic resonance imaging *Electromagnetism.* an imaging technology in which the object to be viewed is exposed to a radio-frequency pulse while in the presence of a varying magnetic field; used to produce cross-sectional images, as for medical diagnostics and research.

magnetic reversal *Geophysics.* a change in geomagnetic field polarity.

magnetic Reynolds number *Electromagnetism.* a dimensionless quantity used in hydromagnetics, given by the product of the fluid velocity of a plasma and a characteristic length divided by the magnetic diffusivity.

magnetic rigidity *Physics.* a property inherent in a conducting fluid in which there exist restoring forces that oppose axial displacements in the presence of a magnetic field. *Electromagnetism.* a quantity that expresses the momentum of a charged particle moving perpendicularly to a magnetic field, given by the product of the magnetic induction and the radius of curvature of the particle's motion.

magnetic rotation see FARADAY EFFECT.

magnetics *Electromagnetism.* the science of magentism; the branch of physics that is concerned with the study of magnetic phenomena.

magnetic saturation *Electromagnetism.* a condition in which a body is magnetized and exhibits no further increase of magnetization upon further increase in an external magnetic field.

magnetic scalar potential see MAGNETIC POTENTIAL.

magnetic scanning *Spectroscopy.* in a mass spectrometer, the sorting of ions for analysis by varying the strength of the magnetic field while keeping the electrostatic field constant.

magnetic scattering *Physics.* a technique used to scatter neutrons in which neutron magnetic moments interact with the magnetic moments of the atoms or particles in the scattering medium.

magnetic secular change *Geophysics.* a change in a magnetic field that varies slowly over many years.

magnetic separation *Materials Science.* a technique that uses magnetic solids and an external magnetic field to separate materials or compounds. *Metallurgy.* specifically, a process for separating mineral ores in order to concentrate paramagnetic minerals such as magnetite, hematite, and ilmenite, from diamagnetic materials such as quartz and feldspar. Thus, **magnetic separator.**

magnetic shell *Electromagnetism.* two layers of magnetic charge of opposite sign, separated by an infinitesimal distance.

magnetic shunt *Electromagnetism.* a piece of ferromagnetic material that is inserted into a gap in a magnetic circuit to a certain depth in order to divert a portion of the magnetic field away from the circuit.

magnetic soundtrack *Acoustical Engineering.* any soundtrack that is recorded onto a magnetic tape, as in an audio cassette recording or the audio portion of a motion picture.

magnetic spark chamber *Nucleonics.* an instrument that measures the momentum and trajectories of charged particles by tracking the sparks they form while moving through an inert gas under the influence of a magnetic field.

magnetic spectrograph *Nucleonics.* a magnetic spectrometer that produces a permanent record showing the distribution of intensity versus the distribution of momentum in a beam of charged particles.

magnetic spectrometer *Nucleonics.* a device that measures the momentum of charged particles or the distribution of their intensity versus their momentum when they are placed in a magnetic field strong enough to make their paths proportional to their momentum.

magnetic star *Astronomy.* a star, usually of spectral type A, that has a strong (but often variable) magnetic field.

magnetic station *Cartography.* a ground station from which a series of magnetic observations have been made; usually indicated by a monument on which the magnetic value of the station has been indicated in addition to the latitude, longitude, and elevation.

magnetic steel *Metallurgy.* a class of magnetically hard steels that retain their magnetism when removed from a magnetic field.

magnetic storage *Computer Technology.* any storage technology that makes use of the magnetic properties of materials.

magnetic storage plate *Computer Technology.* magnetic storage in the form of a ferrite plate with a grid of holes through which read-in and read-out wires are threaded.

magnetic storm *Geophysics.* a disturbance in the geomagnetic field that usually lasts several days and is caused by the effects of charged particles emitted by solar flares striking the earth's magnetosphere.

magnetic strain energy *Solid-State Physics.* the potential energy of a magnetic domain, considering mechanical stress and a magnetic field, associated with the magnetostriction of the domain.

magnetic stress tensor *Physics.* a tensor whose divergence gives part of the magnetic field force on a unit volume of conducting fluid.

magnetic superconductor *Solid-State Physics.* a material that exhibits ferromagnetic properties above a certain temperature (the ferromagnetic Curie point) and superconductive properties below that temperature.

magnetic survey *Geophysics.* a series of measurements of the magnetic field at several locations, typically local or regional. Also, MAGNETIC PROFILE.

magnetic susceptibility *Electromagnetism.* the ratio of the magnetization of a substance to the applied magnetic field strength; in an isotropic medium, the susceptibility is a scalar quantity; otherwise it would be a tensor. *Physical Chemistry.* a measure of the extent to which a magnetic field influences a compound, per gram of the compound. Also, SPECIFIC SUSCEPTIBILITY.

magnetic tape *Electronics.* a magnetic recording medium on which data can be stored for later retrieval.

magnetic tape core *Electronics.* a toroidal core formed by winding thin magnetic tape in a tight, continuous spiral.

magnetic tape file operation *Computer Programming.* an operation in which magnetic tape is used for the storage of data for sequential file updating; also used as an interim storage medium in off-line conversion input to magnetic tape and in operations such as sorting, in which working tapes are used.

magnetic tape group *Computer Technology.* a group of magnetic tape decks built into a single cabinet. Each deck can operate independently, or the decks may be arranged to share one or more interface channels to communicate with a central processor.

magnetic tape librarian *Computer Technology.* a person who stores magnetic tape files on a computer and formats the installation program on the library tape.

magnetic tape parity *Computer Technology.* an automatic verification method used during the reading or writing of magnetic tape data in which a parity bit is produced and added to each character.

magnetic tape reader *Electronics.* an electronic device that converts data stored on a magnetic tape into a series of electrical impulses.

magnetic tape unit *Computer Technology.* an input/output device composed of a tape drive, read/write heads, and related controls. Also, **magnetic tape station.**

magnetic temporal variation *Geophysics.* any change in the earth's magnetic field that is a function of time.

magnetic thermometer *Solid-State Physics.* a paramagnetic salt whose paramagnetic susceptibility is measured at low temperatures (below about 1 K) and whose temperature is then derived by the Curie law: the product of the susceptibility and the temperature is a constant; the temperature can be calculated from the value of the constant and the measured susceptibility.

magnetic thin film *Solid-State Physics.* a flat sheet or cylindrical surface of ferromagnetic material of less than 5 micrometers in thickness and having uniaxial magnetic isotropy; commonly used for computer memory storage. Also, MAGNETIC FILM.

magnetic track *Navigation.* the direction of a vessel's track relative to magnetic north.

magnetic transducer *Electromagnetism.* a device consisting of a magnetic circuit that has a variable-reluctance path and a coil that surrounds this path; used to transform mechanical into electrical energy.

magnetic transfer see MAGNETIC PRINTING.

magnetic variation *Geophysics.* any change in the geomagnetic field.

magnetic viscosity *Physics.* the property of a conducting fluid responsible for the damping of motion of charged particles perpendicular to an applied magnetic field.
magnetic wave *Solid-State Physics.* the spreading of magnetization in a solid from a small region where an abrupt change in the magnetic field took place.
magnetic-wave device *Electromagnetism.* a device whose operation depends on magnetoelastic or magnetostatic wave propagation through a dielectric material; used in microwave delay lines.
magnetic well *Physics.* a magnetic field configuration designed to confine a plasma in a controlled fusion experiment so that the plasma cannot escape in any direction.
magnetic wind direction *Meteorology.* the direction from which the wind blows with respect to magnetic north.
magnetic wire *Materials Science.* wire made of magnetic material and used for magnetic recording.
magnetic X-ray scattering *Electromagnetism.* a process whereby the electric and magnetic fields of incident X-rays interact with electronic magnetic moments, producing magnetic radiation.
magnetism *Physics.* the fact of being magnetic; any of the processes and behaviors associated with magnetic fields.
magnetite *Mineralogy.* $Fe^{+2}Fe_2^{+3}O_4$, a strongly magnetic (sometimes with polarity), opaque, black, cubic mineral of the spinel group, occurring in granular to massive form and as octahedral crystals, and having a specific gravity of 5.175 and a hardness of 5.5 to 6.5 on the Mohs scale; widespread as an accessory mineral in igneous and metamorphic rocks. Also, MAGNETIC IRON ORE.
magnetization *Electromagnetism.* **1.** the process by which a magnetic material is magnetized. **2.** an expression of the extent to which a magnetic material is magnetized, given by the magnetic moment per unit volume. Also, MAGNETIC DIPOLE DENSITY, MAGNETIZATION INTENSITY.
magnetization intensity see MAGNETIZATION, def. 2.
magnetizing current *Electricity.* the current that flows through the field windings of a generator or motor, establishing the magnetic field in the core. Also, EXCITING CURRENT.
magnetizing force see MAGNETIC FIELD STRENGTH.
magneto [mag nēt´ō] *Electromagnetism.* an electric generator containing one or more permanent magnets, which provide the magnetic flux.
magneto- [mag net´ō] a combining form meaning "magnetic," as in *magnetoacoustics.*
magnetoacoustics *Physics.* a branch of physics that deals with the effects of magnetic fields on acoustic phenomena.
magnetoaerodynamics *Physics.* the study and application of properties of highly ionized air and gases and subsequent interactions with bodies passing through them.
magneto anemometer *Engineering.* a cup anemometer whose shaft is coupled to a magnet; the resultant frequency and amplitude voltage generated are proportional to the wind speed.
magnetobremstrahlung see SYNCHROTRON EMISSION.
magnetocaloric effect *Thermodynamics.* the reversible heating and cooling of a ferromagnetic substance when the magnetization is changed. Also, THERMOMAGNETIC EFFECT.
magnetocardiograph *Medicine.* an instrument that generates electrical signals proportional to magnetic pulses emanating from electrical activity in the heart.
magnetochemistry *Physical Chemistry.* the branch of chemistry that studies the relationship between the magnetic properties of a substance and its atomic or molecular structure.
magnetoelastic coupling *Solid-State Physics.* the mutual coupling between the magnetization of a solid and the stresses and the strains associated with magnetostriction.
magnetoelasticity *Solid-State Physics.* the dependence of the magnetization of a ferromagnetic material on the elastic strain imposed on the material.
magnetoelectric effect *Solid-State Physics.* an effect in which an electric field is produced in a substance (such as chromic oxide) when the substance is subjected to a static magnetic field.
magnetoelectricity *Solid-State Physics.* the generation of an electric field in certain substances in the presence of a static magnetic field. *Electromagnetism.* specifically, currents and voltages that are generated by means of magnetic phenomena, as in the secondary windings of a transformer.
magnetoencephalograph *Medicine.* an instrument that records magnetic signals proportional to electroencephalographic waves emanating from electrical activity in the brain. Thus, **magnetoencephalography.**
magnetofluid *Physical Chemistry.* any liquid that becomes less fluid under the influence of a magnetic field.
magnetofluiddynamics or **magnetofluid dynamics** see MAGNETOHYDRODYNAMICS.
magnetogasdynamics or **magnetogas dynamics** see MAGNETOHYDRODYNAMICS.
magnetograph *Electromagnetism.* a device that detects and records the direction of a magnetic field by means of three independent variometers, which read the three perpendicular directional components.
magnetohydrodynamic arcjet *Space Technology.* an electromagnetic propulsion system in which a plasma is heated in an electric arc and then adiabatically expanded through a nozzle and accelerated again by a crossed electric and magnetic field.
magnetohydrodynamic electromagnetic pulse *Electromagnetism.* a highly ionized shock wave that propagates through a conductive fluid in the presence of a magnetic field, such that large currents can develop on the surface of the shock front.
magnetohydrodynamic generator *Electricity.* a device that produces electrical energy from an electrically conducting gas flowing through a transverse magnetic field.
magnetohydrodynamic instability *Physics.* an instability that causes a plasma to expand as it enters a region of weaker magnetic field.
magnetohydrodynamics *Physics.* the branch of physics concerned with the study of the motion of electrically conducting fluid (such as plasmas or liquid metals) and their interactions with magnetic fields. Also, MAGNETOFLUIDDYNAMICS, MAGNETOGASDYNAMICS.
magnetohydrodynamic stability *Physics.* a plasma condition in which fluctuations in density, pressure, or other properties tend to die out rather than increase in time.
magnetohydrodynamic turbulence *Physics.* the turbulent movements of a plasma in a magnetic field wherein velocities and pressures change in an irregular manner.
magnetohydrodynamic wave *Physics.* a plasma oscillation in the presence of a strong magnetic field whose frequency is much less than the ion cyclotron frequency characteristic of the magnetic field.
magnetoignition system *Electromagnetism.* an ignition system in which the voltage required to generate the current in the primary winding of the ignition coil is produced by a set of permanent magnets.
magnetoionic duct *Geophysics.* a channel along geomagnetic field lines that allows radio waves to propagate.
magnetoionic theory *Geophysics.* a theory that explains and predicts the effects of geomagnetic field and atmospheric ionization on radio waves.
magnetoionic wave component *Geophysics.* either of the two elliptically polarized waves that occur when a linearly polarized wave strikes the ionosphere and is divided in two by the magnetic field.
magnetomechanical factor *Physics.* a quantity given by the magnetic dipole moment of a particle divided by the product of the angular momentum and the quantity $e/2mc$, where e is the electron charge in electrostatic units, m is the electron mass, and c is the speed of light. Also, **magnetomechanical ratio.**
magnetomechanics *Physics.* the study of the interaction between the magnetization of a substance and the mechanical properties of the substance.
magnetometer *Engineering.* an instrument used to measure the magnitude and sometimes the direction of a magnetic field.
magnetomotive force *Electromagnetism.* the work required to carry a magnetic pole of unit strength once around a magnetic circuit.
magneton *Physics.* a unit of magnetic moment commonly used in the study of magnetic processes at the atomic or molecular level.
magneton number *Physics.* in a paramagnetic or ferromagnetic substance, the ratio of the magnetic moment per atom, molecule, or ion to the Bohr magneton.
magnetooptic *Optics.* relating to the effect of magnetism on the propagation of light. Also, **magnetooptical.**
magnetooptic Faraday effect see FARADAY EFFECT.
magnetooptic Kerr effect see KERR EFFECT.
magnetooptic material *Optics.* any material that displays the magnetooptic Kerr effect, the magnetooptic Faraday effect, or double refraction when influenced by a magnetic field.
magnetooptic modulator *Optics.* a device that modulates the intensity of a light beam by passing it through a suitable crystal, such as a crystal of yttrium-iron garnet, which is subjected to an oscillating magnetic field for optical rotation. Also, **magnetooptic shutter.**
magnetooptics *Optics.* the study of magnetooptic phenomena.

magnetopause *Astrophysics.* the outer edge of a planet's magnetosphere, where it borders the magnetosheath.
magnetoplasmadynamic *Electromagnetism.* describing currents that are generated by directing a beam of ionized gas through a magnetic field. Thus, **magnetoplasmadynamics.**
magnetoplumbite *Mineralogy.* $Pb(Fe^{+3},Mn^{+3})_{12}O_{19}$, a black, opaque, strongly magnetic, hexagonal mineral of the magnetoplumbite group occurring in steep pyramidal crystals, and having a specific gravity of 5.52 and a hardness of 6 on the Mohs scale.
magnetoresistance *Materials Science.* an electrical resistance of certain materials, particularly those that have high carrier mobility, that can change in accordance with a change in the magnetization of the body.
magnetoresistivity *Electromagnetism.* the phenomenon of change in the resistivity associated with the change in the magnetization of the body.
magnetoresistor *Electronics.* a semiconductor whose resistance is proportional to the strength of the applied magnetic field.
magnetosheath *Astrophysics.* the region of a planet's magnetic domain that lies between the magnetopause and a bow shock due to the solar wind; the fields within it are disordered and turbulent.
magnetosphere *Astrophysics.* a region of space surrounding a planet in which charged particles are controlled by the planet's magnetic field. Thus, **magnetospheric.**
magnetospheric plasma *Geophysics.* a weak plasma that pervades a planet's magnetosphere.
magnetospheric ring current *Geophysics.* a current flowing in the ring of charged particles that surrounds the earth.
magnetostatic *Electromagnetism.* having to do with magnetic properties that do not depend on the motion of conductors or changing magnetic fields. Thus, **magnetostatic field.**
magnetostatic mode *Solid-State Physics.* a spin wave in a ferromagnetic material in the limit of long wavelengths.
magnetostatics *Electromagnetism.* a branch of physics dealing with magnetic phenomena that are constant in time.
magnetostatic shielding *Electromagnetism.* the shielding of certain electromagnetic components by means of enclosures of high-permeability materials so as to prevent magnetodynamic interaction with other components.
magnetostriction *Electromagnetism.* the change in size of a magnetic substance when subjected to an external magnetic field.
magnetostriction transducer *Acoustical Engineering.* a transducer device consisting of a rod-shaped magnetostrictive material wrapped in a coil that causes the rod to longitudinally change size when alternating current is driven through the coil; the device changes an alternating current to sound energy at the same frequency and forms the sound energy into a beam.
magnetostrictive *Acoustical Engineering.* describing an acoustic device whose operation is based on the energy produced by the change in the dimensions of a magnetostrictive material, such as an electrically charged metal rod. Thus, **magnetostrictive loudspeaker, magnetostrictive microphone.**
magnetostrictive delay line *Electromagnetism.* a delay line component that is constructed of magnetostrictive material whose delay is the amount of time required for a shock wave to travel through the length of the material at the speed of sound.
magnetostrictive filter *Electronics.* a filter network that uses magnetostrictive elements to form filters whose impedance characteristic is the inverse of that of a crystal.
magnetostrictive oscillator *Electronics.* an oscillator whose frequency is controlled by a magnetostrictive element.
magnetostrictive resonator *Solid-State Physics.* a rod of ferromagnetic material that is made to sustain longitudinal elastic vibrations by a high-frequency current in a helical coil wrapped about the rod, through the joint operation of the Joule effect and of the Villari effect.
magnetostrictor *Electromagnetism.* a device that converts electrical signals to mechanical vibrations by means of magnetostriction.
magneto switchboard exchange *Telecommunications.* a manual telephone exchange arranged so that calling and clearing by subscribers and operators are accomplished by magnetoelectric generators.
magnetotail *Geophysics.* the long streamer of plasma that is drawn away from a planet's magnetosphere by the solar wind.
magnetotaxis *Biology.* the movement of an organism toward or away from a magnetic field.
magnet power *Electromagnetism.* the electric power supplied to the coils of an electromagnet.
magnetron *Electronics.* an electric tube that produces alternating-current power output through the interaction of electrons with the magnetic field of a circuit element in crossed steady electric and magnetic fields.
magnetron oscillator *Electronics.* an oscillator that contains an electron tube in which electrons are accelerated to provide a high-energy electron stream.
magnetron pulling *Electronics.* a frequency shift in a magnetron due to factors that vary the standing waves or the standing-wave ratio on the radio-frequency lines.
magnetron pushing *Electronics.* a frequency shift in a magnetron due to the faulty operation of the modulator.
magnetron vacuum gauge *Electronics.* a vacuum gauge consisting of a magnetron that is operating beyond cutoff in the vacuum being measured.
magni- a combining form meaning "large" or "great," as in *magnicellular, magnification.*
magnicellular *Cell Biology.* describing any tissue composed of large cells, such as certain masses of gray matter in the central nervous system. Also, MAGNOCELLULAR.
magnification the fact of magnifying or being magnified; specific uses include: *Optics.* the difference between the size of an image created by a lens and the object being imaged. *Psychology.* maladaptive thinking in which the individual tends to exaggerate greatly the negative effect or importance of events occurring in relation to him- or herself.
magnifier see SIMPLE MICROSCOPE.
magnify *Science.* to make greater in size or apparently greater. *Optics.* to cause to appear larger by the use of lenses or mirrors.
magnifying glass *Optics.* a device that creates an image larger than the object being viewed.
magnifying power *Optics.* the difference between the size of an image as seen by the eye through an instrument and the size of the object as seen by the unaided eye.
magnistor *Electronics.* a device that uses magnetic field effects on injection plasmas in semiconductors.
magnitude the size or dimensions of something; specific uses include: *Astronomy.* a logarithmically based unit used in measuring the optical brightness of celestial objects and defined so that 1 magnitude equals the fifth root of 100, or approximately 2.512; for historical reasons, numerically lower magnitudes (which can extend into negative values) are brighter than numerically larger magnitudes. *Geophysics.* a measure of the amount of energy released by an earthquake, as on the Richter scale. *Mathematics.* see ABSOLUTE VALUE.
magnitude at opposition *Astronomy.* the magnitude that a solar system object has when it is 180° away from the sun in the sky and seen with its disk in "full" phase; it often shows an abrupt peak in brightness due to the opposition effect.
magnitude method *Ordnance.* in gunnery, a system of adjusting range based on the magnitude and direction of the deviation.
magnitude ratio *Astronomy.* the step in brightness, equal to the fifth root of 100 (2.512 approximately), that represents a 1-magnitude difference.
magnocellular see MAGNICELLULAR.
magnoferrite see MAGNESIOFERRITE.
magnolia *Botany.* any plant of the genus *Magnolia,* trees and shrubs characterized by large, ornamental flowers and alternate, entire evergreen or deciduous leaves; found mainly in the temperate Northern Hemisphere. (Named for French botanist Pierre *Magnol,* 1638–1715.)

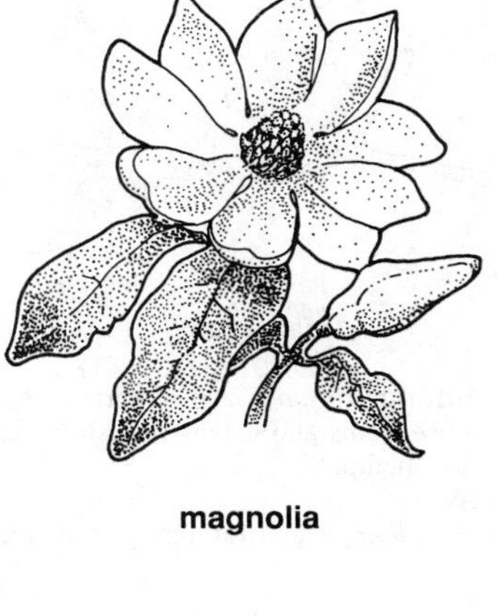

magnolia

Magnoliaceae *Botany.* a family of dicotyledonous shrubs and trees of the order Magnoliales having alternate, simple, pinnately veined leaves and large, ornamental flowers that are usually solitary and bisexual. Also, **magnolia family.**
Magnoliidae *Botany.* a subclass of dicotyledonous flowering plants of the class Magnoliopsida, characterized by an apocarpous gynaecium, a well-developed perianth, and spirally arranged floral parts.
Magnoliophyta *Botany.* the angiosperms or flowering plants; a broad group of vascular seed plants, having roots, stems, leaves, a well-developed vascular system, and flowers with ovules enclosed in carpels.
Magnoliopsida *Botany.* the dicotyledons, a class of flowering plants of the division Magnoliophyta having a pair of embryonic leaves. Also, **Magnoliatae.**
magnon *Solid-State Physics.* an important elementary excitation in which the direction of magnetization in a ferromagnetic material, or that of a sublattice moment in an antiferromagnetic material, is spatially nonuniform and propagates as a wave. Also, SPIN WAVE.
magnophorite see RICHTERITE.
magnum *Anatomy.* **1.** large or great. **2.** a former term for the os capitatum, the largest bone of the carpals.
Magnus, Heinrich G. 1802–1872, German physicist; demonstrated that oxygen content is higher in arterial blood than in venous blood.
Magnus effect *Fluid Mechanics.* a phenomenon wherein a force acts on a rotating cylinder whose axis is perpendicular to the flow of fluid in which it is immersed; the force is perpendicular to both the flow direction and the cylinder axis. Also, **Magnus force.**
Magnus moment *Fluid Mechanics.* the torque that acts on a rotating cylinder in a fluid due to the Magnus effect.

magpie

magpie *Vertebrate Zoology.* a member of any of several cosmopolitan genera of long-tailed birds of the family Corvidae, especially of the genus *Pica*; they are intelligent, gregarious, noisy, and omnivorous, and usually have black and white plumage.
maguey *Botany.* any of several agave plants of Mexico, the source of a stiff white fiber and of the alcoholic drinks tequila and pulque.
mahogany *Botany.* any of various trees of the Meliaceae family of tropical America and Africa, especially *Swietenia mahagoni* and *S. macrophylla,* grown commercially for their hard, reddish-brown wood. *Materials.* the hard, strong wood of these trees, which has an attractive appearance and can be highly polished; it is regarded as one of the finest woods and is widely used for furniture, cabinetry, and boats.

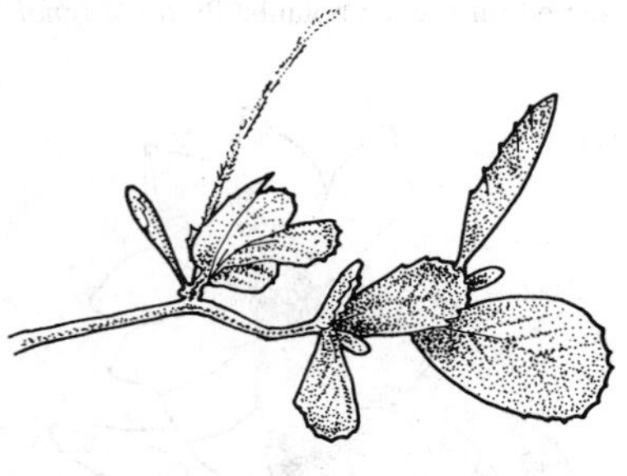

mahogany

mahuang or **ma huang** *Botany.* any of various Asian shrubs of the genus *Ephedra,* whose stems and leaves furnish ephedrine. (A Chinese term meaning "yellow hemp.")
maidenhead see HYMEN.
maieusiophobia *Psychology.* an irrational fear of childbirth.
mail *Telecommunications.* any of various applications of communications technology that are regarded as similar to communication by means of letters sent through the postal system or through an interoffice mail system; for example, electronic mail or voice mail.
mailbox *Computer Programming.* **1.** a file, associated with an individual user, that contains electronic mail messages that have been received but not yet read. **2.** an area of memory dedicated to data addressed to specific peripheral units or other processors.
mailer *Computer Programming.* a program that allows the user to read and send electronic mail.
maim *Medicine.* to disable by a wound or to dismember by means of violence.
Maimonides [mī män′ə dēz′] 1135–1204, Jewish scholar and philosopher, born in Spain; in *Guide of the Perplexed,* he tried to reconcile Judaism with Aristotle's doctrines.
main *Engineering.* the principal feeder for water, gas, electricity, or air entering or leaving a system. *Naval Architecture.* of or relating to a main mast.
main attack *Military Science.* the principal attack or effort into which the commander throws the full weight of the offensive power at his disposal; usually directed against the chief objective of the campaign or battle.
main band *Molecular Biology.* in density gradient separation, a peak that contains the majority of the DNA and excludes the satellite DNA.
main bang *Electronics.* a transmitted pulse in a radar system.
main battle tank *Ordnance.* an armored fighting vehicle weighing at least 16.5 metric tons unladen, and armed with a 360° traverse gun of at least 75-mm caliber.
main body *Naval Architecture.* a term for the hull of a ship.
main control unit *Control Systems.* the control unit that directs other control units in a complex system.
main crosscut *Mining Engineering.* the crosscut within a mining field that traverses the entire field and penetrates all deposits.
main deck *Naval Architecture.* the principal deck that forms the structural top of the vessel's hull; in vessels of three or more complete decks, it usually lies below the uppermost complete deck.
main diagonal *Mathematics.* Let M be an $m \times n$ matrix with elements a_{ij}. The elements a_{ii}, for $i = 1, 2, 3, \ldots, \min(m, n)$, are said to form the main diagonal of M. Also, PRINCIPAL DIAGONAL.
main distributing frame *Electronics.* a distributing frame that carries on one side (vertically) all outside lines and on the other (horizontally) all connections of the outside lines to the central equipment. Also, **main frame type B.**
Maindroniidae *Invertebrate Zoology.* the silverfish, a family of primitive wingless insects in the order Thysanura.
main entry *Mining Engineering.* the principal entry or entries into a coalbed from which cross entries, room entries, or rooms are turned.
main exciter *Electricity.* an exciter that supplies some or all of the necessary power for exciting the principal electric machine in a system.
main firing *Engineering.* the use of current supplied by a transformer fed by a main power supply to fire a round of ammunition.
mainframe or **main frame** *Computer Technology.* **1.** a large, expensive computer with high data-handling or computational capacity. **2.** originally, the combination of a central processor and primary memory of a computer system with storage racks and control panels.
main group element *Chemistry.* any element in the *s* or *p* blocks of the periodic table. Also, REPRESENTATIVE ELEMENT.
main haulage *Mining Engineering.* the segment of a haulage system in which the material is transported from the secondary haulage system to the shaft or mine opening.
main joint see MASTER JOINT.
mainland *Geography.* the major part of a continent or other landmass, as distinguished from an offshore island and sometimes from a projecting piece of land such as a cape or peninsula.
main line *Transportation Engineering.* a major artery of a transportation system.
main line of resistance *Military Science.* a line at the forward edge of the battle position that is used to coordinate the fire of all units and supporting weapons, including air and naval gunfire.
mainmast or **main mast** *Naval Architecture.* the largest of a ship's masts; in two-masted vessels, it is the first from the bow; in vessels with three or more masts, it is the second from the bow.
main memory see INTERNAL STORAGE.
main path *Computer Programming.* the primary sequence of operations performed by a computer in the execution of a routine, dictated by the logic of the program and the data.

main program *Computer Programming.* the part of a program entered first and from which other routines and procedures are called and controlled.

main return *Mining Engineering.* the principal return airway of a mine, through which the total quantity of air flows.

mainsail *Naval Architecture.* the lowest and largest sail on the main mast of a ship.

main-scheme station *Cartography.* an established survey station that serves as the point from which a survey is extended. Also, PRIMARY STATION, PRINCIPAL STATION.

main sequence *Astronomy.* the location on the Hertzsprung-Russell diagram where most stars lie; it runs diagonally from the lower right (stars of low temperature and luminosity) to the upper left (high temperature and luminosity).

main-sequence fitting *Astronomy.* the process of determining the distance to a star cluster by overlaying its main sequence on the standard zero-age main sequence and noting the difference between the cluster's apparent magnitude and the zero-age main sequence's absolute magnitude.

main shaft *Mechanical Engineering.* the principal shaft of an engine or motor system; the line of shafting that receives power directly from the engine or motor and conducts power to other parts of the system.

main stage *Space Technology.* **1.** in a multistage rocket vehicle, the stage housing the propulsion unit or the stage producing the greatest amount of thrust. **2.** in a single-stage rocket vehicle, the main propulsion unit, as distinguished from veniers, roll-control, and other motors. **3.** in a nonballistic missile such as the Atlas, the sustainer engine after separation of the boost motor(s). **4.** in a large ballistic missile, the time in which first-stage propulsion is generating 90% or more of maximum thrust.

main station *Telecommunications.* a telephone station that has a distinct call number and is connected to a central office.

main storage see INTERNAL STORAGE.

main stream *Hydrology.* the largest or principal stream of a given area or drainage system. Also, MASTER STREAM, TRUNK STREAM.

mainstream *Psychology.* to carry on a process of mainstreaming.

mainstreaming *Psychology.* the practice of placing children who have physical, mental, or emotional problems in conventional schools and classrooms rather than educating them separately.

main supply route *Military Science.* the route or routes within an area of operations upon which the bulk of traffic flows in support of military operations.

main sweep *Electronics.* the longest-range scale available on certain fire-control radar.

maintainability *Industrial Engineering.* **1.** the degree of ease of keeping a given system in working order. **2.** the degree to which a device can be restored to a specified standard within a given time, or the time required to restore it completely.

maintainability engineering *Industrial Engineering.* the branch of engineering that deals with the effects of product failures and seeks to minimize such effects.

maintenance the act or fact of maintaining something; specific uses include: *Industrial Engineering.* the routine process of keeping a machine or facility in proper operating condition; this can be preventive or correctional. *Military Science.* **1.** all supply and repair action taken to keep a force in condition to carry out its mission. **2.** the routine work required to keep a facility in continuous operation.

maintenance coating *Materials Science.* a general term for any coating applied for maintenance purposes, including corrosion-erosion protection, heat reflection, and fire endurance.

maintenance energy *Biochemistry.* the energy utilized by a cell for any objective other than growth, e.g., to maintain osmotic pressure.

maintenance kit *Industrial Engineering.* a collection of items used to inspect and repair or replace faulty components in equipment.

maintenance of materiel *Military Science.* all action taken to retain materiel in a serviceable condition or to restore it to serviceability.

maintenance ration *Biology.* the minimum level of consumption necessary to support physiological functions, allowing nothing for growth or reproduction.

maintenance rehearsal *Behavior.* the technique of repeating material without thinking consciously about it, in order to make it part of long-term memory. Also, TYPE I REHEARSAL.

maintenance respiration *Biology.* the minimum level of respiration necessary to support physiological functions, allowing nothing for growth or reproduction.

maintenance routine *Computer Programming.* a program designed for the routine preventative maintenance normally performed by a service engineer.

maintenance time *Computer Technology.* the time used for hardware maintenance, including preventative and corrective maintenance. Also, ENGINEERING TIME.

maintenance vehicle *Industrial Engineering.* a vehicle used to transport parts, equipment, and personnel for maintenance purposes.

main thermocline *Oceanography.* the deepest thermocline, essentially unaffected by seasonal variations in surface temperature.

mais complex *Bacteriology.* three species of mycobacteria, *Mycobacterium avium, M. intracellulare,* and *M. scrofulaceum,* that are considered as a group for clinical diagnostic purposes.

maize [māz] *Botany.* any plant variety of the species *Zea mays,* especially the colored ornamental variety known as Indian corn, occurring as large ears on a tall cereal grass.

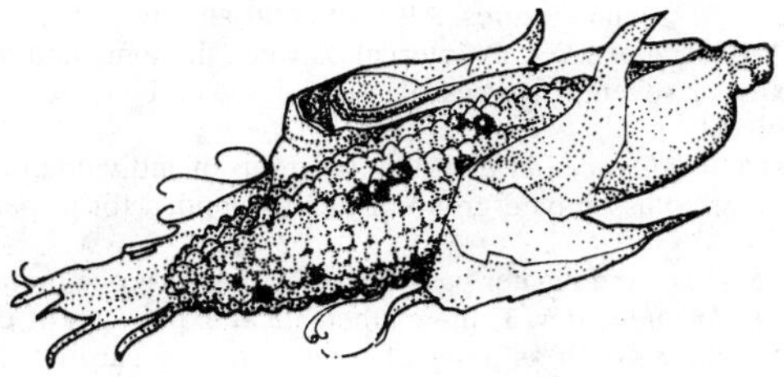

maize

Maj major.

Majac mill *Mechanical Engineering.* a dry-grinding mill having a chamber fitted with two opposing horizontal jets, into which material is continuously fed by a screw conveyor.

Maj Gen major general.

Majidae *Invertebrate Zoology.* the spider crabs, a family of slow-moving decapod crustaceans in the section Brachyura that camouflage themselves by attaching seaweed or sessile animals, such as sea anemones, to their shells.

major *Military Science.* in the United States Army, Air Force, or Marine Corps, and most other armies, a commissioned officer ranking above a captain and below a lieutenant colonel. *Acoustics.* relating to a major chord, scale, interval, etc.

Majorana force *Nuclear Physics.* an attractive force between protons and neutrons that is sufficient to account for the existence of stable nuclei; used to explain certain phenomena that take place when elements interact, such as why some nucleons change their positions, but not their spins.

Majorana neutrino *Particle Physics.* a massless particle that is self-charge conjugate and obeys the Dirac equation with mass equal to zero.

major arc *Mathematics.* the longer of the two arcs into which a circle is cut by a chord.

major assembly *Mechanical Engineering.* the complete assembly of a component to be installed in a system or major piece of equipment.

major axis *Mathematics.* the longer of the two axes of symmetry of an ellipse.

major chord *Acoustics.* a chord having a major third interval between the root and the next note above it.

major combination *Ordnance.* a major mechanical item, such as a tank or airplane, that is completely equipped and ready for independent operation, with all necessary equipment, armament, and spare parts.

major control data *Computer Programming.* the most significant data field by which records are sorted.

major cycle *Computer Technology.* **1.** the maximum access time of a recurring serial-storage element. **2.** in a dynamic store, a complete revolution of the storage medium.

major datum see PREFERRED DATUM.

major defect *Industrial Engineering.* a mid-level classification of seriousness of product defects; a defect that severely limits the ability of the product to serve its intended use.

major diameter *Design Engineering.* the largest diameter of a screw thread, measured at the crest for a male (external) thread and at the root for a female (internal) thread.

major diatonic scale see MAJOR SCALE.

major element *Geology.* any of the eight elements that each make up 1% or more of the earth's crust: oxygen, silicon, aluminum, iron, calcium, sodium, potassium, and magnesium. *Petrology.* any element making up 5% or more of a given rock.
major fold *Geology.* a dominant fold or large-scale area usually associated with a system of smaller folds.
major gene *Genetics.* a gene whose phenotypic effects are much greater than those of other modifying genes.
major general *Military Science.* in the United States Army, Air Force, or Marine Corps, and most other armies, a commissioned officer ranking above a brigadier general and below a lieutenant general; the insignia is two stars.
major groove *Molecular Biology.* one of the two types of grooves found in a right-handed double helix, running between the sugar phosphate backbones of each strand.
major histocompatibility complex *Immunology.* the set of genes that code for the most important histocompatibility and related markers.
major hysteria *Psychology.* hysterical seizures in patients previously diagnosed with actual seizures. Also, HYSTEROEPILEPSY.
major interval *Acoustics.* an interval between the tonic and the second, third, sixth, or seventh degrees of a major scale. Thus, for example, a **major third.**
major item *Ordnance.* an end item, a group of individually classified end items, or an assembled group of items intended for a specific tactical purpose.
majority *Science.* the greater part or number; at least one more than half of the total. *Mathematics.* **1.** the mathematical expression of this; a subset M of a set S (such as a set of voters) whose cardinal number is greater than the cardinal number of its complement $S - M$. **2.** a logic operator on a set of statements that has the value true if more than half the statements are true, and the value false otherwise.
majority carrier *Electronics.* in a semiconductor, the carrier whose concentration is higher than that of the other carrier; holes are the majority carrier in P-type semiconductors, and electrons in N-type semiconductors.
majority emitter *Electronics.* an electrode from which majority carriers flow into the interelectrode region of a transistor.
majority gate *Computer Technology.* a logic element that executes a majority operation; i.e., it outputs "true" if more than half the inputs are true, and "false" if not.
major joint see MASTER JOINT.
major key *Acoustics.* a key whose harmony is based on the major scale. Also, **major mode.**
major light *Navigation.* a highly reliable, high-intensity navigational light in a fixed location.
major lobe *Electromagnetism.* for a directional antenna, the lobe of radiation that contains the direction of maximum amplitude.
major mineral *Nutrition.* any mineral required by adult humans in amounts greater than 100 mg per day, such as calcium, phosphorus, magnesium, and sodium.
major-motion axis *Robotics.* the widest arc along which a manipulator can move.
major node *Electricity.* a point of contact between a large number of conductors in an electrical circuit.
major nuclear power *Military Science.* any nation that possesses a nuclear striking force capable of posing a serious threat to every other nation.
major operation see MAJOR SURGERY.
major planet *Astronomy.* a planet; that is, any of the nine large heavenly bodies of our solar system: Mercury, Venus, Earth, Mars, Jupiter, Saturn, Uranus, Neptune, or Pluto.
major premise *Science.* the first statement in a formal logical argument. It contains the term (the major term) that is the predicate in the conclusion of the argument. In the argument "All dogs have four legs; Nellie and Daisy are dogs; therefore, Nellie and Daisy have four legs," the statement "All dogs have four legs" is the major premise, and "have four legs" is the major term.
major repair *Mechanical Engineering.* a term for repair work on items that require special tools, a major overhaul, or a significant replacement of parts.
major scale *Acoustics.* a diatonic musical scale in which the third and seventh pitches are half steps above their preceding pitches while all others are whole step intervals; thus a major scale has two whole steps, then a half step, then three whole steps, then a half step. Also, MAJOR DIATONIC SCALE.
major sort key see PRIMARY KEY.
major surgery *Surgery.* a surgical procedure generally involving greater risk and a longer recovery time than a minor procedure and that typically requires a general anesthetic and/or respiratory assistance. Also, MAJOR OPERATION.
major term *Science.* the term that is the predicate in the conclusion of a formal logical argument.
major tranquilizer *Pharmacology.* any antipsychotic agent.
major trough *Meteorology.* a long-wave trough found in the large-scale pressure pattern of the upper troposphere.
major wave see LONG WAVE.
major weapon system *Military Science.* a weapon system or subsystem that is determined by the Department of Defense as being vital to the national interest.
Makapansgat *Anthropology.* a South African cave site north of Pretoria; the location of a group of australopithecine fossils about 3 million years old.
Makarov pistol *Ordnance.* a Soviet double-action blowback-operated 9-mm automatic pistol with an 8-round magazine and a muzzle velocity of 1070 feet per second.
makata see SEPETIR.
make-and-break circuit *Electricity.* a circuit, as in a doorbell, whose contacts are alternately opened and closed as needed for electrical impulse transfers.
make-before-break *Electricity.* a switch or relay in which an open contact must close before a closed one opens, thus making one circuit before breaking another.
make contact *Electricity.* a stationary contact on a relay that is normally open; its circuit closes when the relay is energized.
make fast *Navigation.* **1.** to fasten or secure. **2. makefast.** anything to which a vessel is tied, such as a buoy or bollard.
make off *Navigation.* to stand off a coast.
makeready *Graphic Arts.* the process of preparing a press for printing, including adjustments to the printing elements or paper supply once a press run begins.
make-safe *Ordnance.* the process of carrying out one or more actions necessary to prevent or interrupt the complete function of a weapons system; disarming.
make the land *Navigation.* the arrival of a craft in sight of land after an open-ocean pasage.
makeup *Graphic Arts.* **1.** the design of copy on a page. **2.** the process of arranging the type and illustrations to be printed on a page.
makeup time *Computer Programming.* the part of available time used for reruns following operating errors.
makeup water *Chemical Engineering.* the water added to a recycle process to replace losses due to evaporation or leakage.
making current *Electricity.* the maximum value reached by the current during the first cycle after a switch or circuit breaker is closed.
making way *Navigation.* a term for the condition of a vessel moving under control.
Makinoaceae *Botany.* a monotypic family of liverworts of the order Metzgeriales, characterized by thallose plants without secondary pigments or central strands and with dichotomous branching, ventral scales, and brownish-purplish rhizoids on the median-ventral surface.
makore *Materials.* the dark reddish-brown wood of the *Minusops leckeli* tree of Ghana, Nigeria, and the Ivory Coast. Also, CHERRY MAHOGANY.
Maksutov system *Optics.* a telescope in which the aspheric correction plate of the Schmidt telescope is replaced by a meniscus lens to overcorrect for spherical aberration, so as to offset the undercorrected spherical aberration of the mirror. Also, BOUWERS-MAKSUTOV SYSTEM, MENISCUS-SCHMIDT TELESCOPE.
mal- a combining form meaning "bad," "ill," or "abnormal," as in *malfunction, malady, malformation.*
Malabar Coast *Geography.* the southwestern coast of India.
malabsorption *Medicine.* the impaired absorption of nutrients in the gastrointestinal tract.
malac- or **malaco-** a combining form meaning: **1.** soft, often abnormally soft, as in *malacoma.* **2.** softening, as in *malacia.*
Malachiidae see MELYRIDAE.
malachite *Mineralogy.* $Cu_2^{+2}(CO_3)(OH)_2$, a bright-green monoclinic mineral, dimorphous with georgeite, having a specific gravity of 4.05 and a hardness of 3.5 to 4 on the Mohs scale; found typically in the oxidation zone of copper deposits as fibrous botryoidal masses and acicular crystals; used as an ornamental and gem mineral.

malachite green *Organic Chemistry.* $C_{23}H_{25}ClN_2$, toxic green crystals; soluble in water, ethyl, methyl, and amyl alcohol; prepared by condensing benzaldehyde with dimethylaniline; used to dye textiles and stain bacteria, as a plant fungicide, and as an antiseptic.

malacia *Medicine.* **1.** the softening or deterioration of any body tissue due to disease. (Often used with combining forms to denote specific conditions, such as *osteomalacia.)* **2.** an abnormal craving for highly spiced foods. (From a Greek word meaning "softness" or "weakness.") Thus, **malacic, malacoid.**

Malacobothridia *Invertebrate Zoology.* a subclass of trematode worms with flexible suckers; parasites of both vertebrates and invertebrates.

Malacocotylea see DIGENEA.

malacology *Invertebrate Zoology.* the scientific study of mollusks.

malacoma *Pathology.* any morbidly soft part or spot.

malacoplakia *Pathology.* the development of soft yellowish patches on the mucous membranes of hollow organs, such as the urinary bladder. Also, **malakoplakia.**

malacoplakia cell *Pathology.* a rare type of soft cytoplasmic lesion found on the mucous membrane of internal organs such as the urinary bladder; occurs more often in women.

Malacopoda *Invertebrate Zoology.* a former classification of brachiopods, now contained in the phylum Onychophora.

Malacostraca *Invertebrate Zoology.* a subclass of crustaceans including shrimp, crabs, lobsters, hermit crabs, sow bugs, and others.

malacostracan *Invertebrate Zoology.* **1.** belonging to the subclass Malacostraca. **2.** any member of this subclass.

maladie des jambes *Pathology.* a disease of Louisiana rice growers, probably a strain of beriberi.

maladie du sommeil see SLEEPING SICKNESS.

maladjustment *Psychology.* the failure to cope with or adjust to the everyday demands of life. Thus, **maladjusted.**

malady *Medicine.* a popular term for an illness or disease, especially a chronic illness. Also, **maladie.**

Malafon *Ordnance.* a French ship-based antisubmarine surface-to-surface or surface-to-underwater missile that is ramp-launched by two jettisonable solid-propellant boosters and controlled by radio guidance during its unpowered flight; it delivers a 1157-pound acoustic homing torpedo up to a range of 11 miles.

Malagasy subregion *Ecology.* a distinct zoogeographical region that includes the island of Madagascar (Malagasy) and the surrounding ocean areas.

malaise [mə lāz´; mə lās´] *Medicine.* a nonspecific feeling of debility or discomfort. (From a French term literally meaning "bad ease.")

Malaprade reaction *Organic Chemistry.* a reaction that utilizes periodic acid (HIO_4) to split the bond between two hydroxylated carbon atoms.

Malapteruridae *Vertebrate Zoology.* a family of bony fish of the order Siluriformes, containing two species of electric catfishes found in west Africa and the Nile River; characterized by thick, fleshy bodies that produce strong electrical currents.

malar *Anatomy.* of or related to the cheek or cheekbone.

malaria [mə lâr´ ē ə] *Pathology.* an infectious tropical disease caused by protozoa of the genus *Plasmodium,* transmitted to humans by the bite of the *Anopheles* mosquito. It is characterized by prostration, high fevers, shaking chills, sweating, and anemia, and may follow a chronic or relapsing course after the initial illness. (From an Italian phrase literally meaning "bad air;" in earlier times it was believed that the disease was contracted by breathing foul air from a swamp or marsh.)

malariacidal *Medicine.* destructive to malarial plasmodia.

malarial [mə lâr´ ē əl] *Pathology.* relating to, caused by, or affected by malaria.

malaria pigment *Pathology.* a very dark brown hematin, found mainly in the spleen and liver, created as a by-product of the presence of the malarial parasite.

malariology *Pathology.* the study of malaria. Thus, **malariologist.**

malar stripe *Vertebrate Zoology.* in birds, an area extending back and down from the corner of the mouth.

Malassezia *Mycology.* a genus of yeast fungi belonging to the class Hyphomycetes; found as normal skin flora on man and other animals. Also, PITYROSPORUM.

malassimilation *Medicine.* the incomplete assimiliation of nutrients, usually caused by the failure of the gastrointestinal tract to absorb or digest one or more nutrients due to various causes.

malate *Organic Chemistry.* a salt of malic acid that contains the radical $-OCO{\cdot}CH_2{\cdot}CHOH{\cdot}COO-$.

malathion [ma´lə thī´ än] *Organic Chemistry.* $C_{10}H_{19}O_6PS_2$, a combustible yellow-brown liquid that is slightly soluble in water and is miscible with most polar organic solvents, boiling at 156–157°C (0.7 torr). *Agriculture.* a preparation of this liquid that is widely used as an insecticide because of its low toxicity to mammals; especially effective against the Mediterranean fruit fly.

Malawi, Lake see NYASA, LAKE.

Malayan filariasis *Pathology.* an infection with the filarial worm *Brugia malayi,* transmitted to humans by a mosquito bite. The symptoms range from asymptomatic adenitis to periodic attacks of fever and lymphangitis, to elephantiasis. The condition is common in Malaysia, Borneo, and Indonesia.

Malay Archipelago *Geography.* a large group of islands between Southeast Asia and Australia, including Sumatra, Borneo, Java, Celebes, and many smaller islands; usually the Philippines and sometimes New Guinea are considered part of it.

Malayo-Polynesian family *Linguistics.* the major language group of the Pacific Ocean that includes Polynesian, Indonesian, and Melanesian.

Malay Peninsula *Geography.* a peninsula of Southeast Asia, divided between Malaysia and Thailand.

malchite *Petrology.* a fine-grained, dioritic lamprophyre featuring small phenocrysts of hornblende, labradorite, and sometimes biotite, occurring in a groundmass of the same minerals along with andesine and some quartz.

Malcidae *Invertebrate Zoology.* a family of true bugs, hemipteran insects in the superfamily Pentatomorpha.

Malcodermata see CANTHAROIDEA.

Malcopterygii see CLUPEIFORMES.

Maldanidae *Invertebrate Zoology.* the bamboo worms, a family of mud-ingesting polychaete worms in the group Sedentaria.

Maldaninae *Invertebrate Zoology.* a subfamily of mud-ingesting polychaete worms in the family Maldanidae.

mal de mer [mal´də mâr´] *Medicine.* a French term for seasickness. See SEASICKNESS.

maldonite *Mineralogy.* Au_2Bi, a pinkish or silver-white, cubic metallic mineral occurring in granular form and as octahedral crystals, having a specific gravity of 15.46 and a hardness of 1.5 to 2 on the Mohs scale; found in gold mines.

male *Biology.* **1.** of, relating to, or being the sex that fertilizes the ovum of the female and begets offspring. **2.** an organism of the sex or sexual phase that normally produces spermatozoa and fertilizes female eggs. *Botany.* referring to a plant or structure that performs the function of fertilization in the reproductive process. *Engineering.* describing the one of two parts shaped to fit into the other part (the female), a corresponding hollow part. Thus, **male fitting.**

maleate *Organic Chemistry.* a salt or ester of maleic acid.

male care *Behavior.* a term used to describe actions of parental care performed by males.

male climacteric *Physiology.* the male life period during which physiological changes occur that end his further ability to reproduce; similar to female menopause.

male connector *Electricity.* an electrical connector with one or more pins or prongs that extend into a female connector.

male heterogamety *Genetics.* the phenomenon whereby the male individuals of certain species possess a pair of nonhomologous sex chromosomes (X and Y in humans), and hence produce two types of gametes, one carrying each of the two chromosomes.

male homogamety *Genetics.* the phenomenon whereby the male individuals of certain species possess homologous sex chromosomes, and hence produce only one type of gemete with reference to sex determination.

maleic acid *Organic Chemistry.* $HO_2CCH{=}CHCO_2H$, toxic, colorless crystals, soluble in water, alcohol, and acetone; melts at 130–131°C; partly converts to fumaric acid at room temperatures that are just above its melting point; an important polyfunctional compound. It is made commercially in large quantities from the catalytic oxidation of benzene; used in the synthesis of organic compounds, in fabric dyeing, and as a preservative for oils and fats. Also, *cis*-BUTENEDIOIC ACID.

maleic anhydride *Organic Chemistry.* $C_4H_2O_3$, colorless needles that are soluble in water, alcohol, and acetone; melts at 53°C and boils at 202°C; used to form polyester resins, in the manufacture of tartaric and fumaric acid, and in pesticides.

maleic hydrazide *Organic Chemistry.* $C_4H_4N_2O_2$, a toxic crystalline solid; soluble in hot water and slightly soluble in hot alcohol; melts with decomposition at 297°C; used as an herbicide.

malenclave *Hydrology.* a body of contaminated groundwater that is surrounded by uncontaminated water.
male pseudohermaphroditism see ANDROGYNY.
Malesherbiaceae *Botany.* a monogeneric family of dicotyledonous herbs or half-shrubs of the order Violales, often having glandular, malodorous hairs; native to dry habitats in the Andes.
male sterility *Physiology.* the inability of the male reproductive organs to fertilize the ovum as a result of a failure to produce living spermatozoa, an abnormality in spermatozoa production, or another cause. *Botany.* the condition or characteristic of a male plant producing no pollen or infertile pollen. Thus, **male-sterile.**
malezal swamp *Ecology.* a swamp resulting from the drainage of water over a wide plain with a slight slope.
malformation *Science.* any result of atypical development, especially a structural defect in a living body.
malfunction *Science.* **1.** to fail to function; operate improperly or abnormally. **2.** an abnormality or defect in the operation of a system or mechanical device.
malfunction routine see DEBUGGING ROUTINE.
malic *Botany.* relating to or obtained from apples. *Chemistry.* relating to or derived from malic acid.
malic acid *Biochemistry.* HOOC–CH_2–CHOH–COOH, an intermediate metabolite (between fumarate and oxaloacetate) in the Krebs cycle.
malic dehydrogenase *Enzymology.* an enzyme in the citric acid cycle that converts malic acid into oxaloacetate.
malic enzyme *Enzymology.* an enzyme that catalyzes the breakdown of malic acid into pyruvate acid and carbon dioxide.
malignancy *Pathology.* **1.** the condition or quality of being malignant. **2.** a failure to respond to treatment. *Oncology.* a malignant tumor.
malignant [mə lig´nənt] *Pathology.* tending to become progressively worse and cause death. Thus, **malignant disease.** *Oncology.* describing a tumor that is anaplastic, invasive, and metastatic; that is, it has primitive cellular growth characterized by a lack of differentiation, it moves into and destroys surrounding tissue, and it spreads to other parts of the body. (Going back to a Latin term meaning "to act harmfully.")
malignant adenoma *Oncology.* a well-defined, cancerous, glandlike tumor.
malignant catarrhal fever *Veterinary Medicine.* a fatal viral disease of cattle, buffalo, and deer, manifested by high fever, a white appearance to the cornea, and catarrhal mucopurulent inflammation of upper respiratory and alimental epithelia with discharge from the eyes and nose.
malignant edema *Veterinary Medicine.* an acute, generally fatal wound infection that progresses to toxemia of domestic and sometimes wild animals, caused by bacteria of the genus *Clostridium* and characterized by anorexia, high fever, and soft swellings and exudate near the site of the injury. Also, GAS GANGRENE.
malignant embolus *Oncology.* a group of cells that have broken off from a cancerous tumor and traveled through the blood stream to another site. Also, CANCER EMBOLUS.
malignant endocarditis see ULCERATIVE ENDOCARDITIS.
malignant glaucoma *Pathology.* a severe eye disease that progresses rapidly and worsens despite surgical removal of the iris.
malignant hypertension *Pathology.* an abnormal condition marked by extremely elevated blood pressure, resulting in hemorrhages and kidney failure.
malignant melanoma *Oncology.* a malignant neoplasm that develops from cells that form melanin. Also, NEVOCARCINOMA.
malignant pustule *Pathology.* a type of anthrax resulting from contamination of skin wounds with *Bacillus anthracis*; dead tissue develops at the site, surrounded by swelling and intense redness. Also, CUTANEOUS ANTHRAX.
malignant somatitis see CALF DIPHTHERIA.
malignin *Oncology.* a protein segment found in the serum of patients with malignant glial tumors.
malingerer *Medicine.* a person who consciously feigns an illness or injury, or exaggerates the symptoms of an actual ilness or injury, with the intent of misleading others.
Malinowski, Bronislaw 1884–1942, Polish-born British anthropologist; founded social anthropology; best known for his research in New Guinea and the Trobriand Islands.
malladrite *Mineralogy.* Na_2SiF_6, a pale-rose, trigonal mineral occurring as crusts of minute crystals, and having a specific gravity of 2.71 to 2.76 and an undetermined hardness; found incrusting lava on Mt. Vesuvius, Italy.
mallard *Vertebrate Zoology.* a wild duck, *Anas platyrhynchos,* of the family Anatidae, characterized by bright orange feet and an irridescent green head with a white neck ring and a brown breast; found throughout the Northern Hemisphere and considered to be the ancestor of most domestic ducks.

mallard

mallardite *Mineralogy.* $Mn^{+2}SO_4 \cdot 7H_2O$, a pale-rose, water-soluble, monoclinic mineral of the melanterite group, having a specific gravity of 1.85 and a hardness of about 2 on the Mohs scale; found in crusts as an oxidation product in mines.
malleability *Materials Science.* the quality or degree of being malleable.
malleable [mal´ə bəl] *Materials Science.* capable of being extended or shaped by hammering or by other external pressure. *Metallurgy.* specifically, describing a metal or alloy that can be substantially deformed in compression without fracture. Thus, **malleable brass.**
malleable cast iron *Metallurgy.* a ternary alloy of carbon and silicon with iron that has high tensile strength, is ductile, and is very good for machining.
malleable iron *Metallurgy.* a white cast iron subjected to a thermal treatment that eliminates iron carbide from its structure, thereby improving malleability.
malleable pig iron *Metallurgy.* pig iron suitable for the production of malleable iron.
malleablize *Metallurgy.* to anneal white cast iron so that it becomes partially or totally malleable.
malleate *Metallurgy.* to beat or shape with a hammer.
malleate trophus *Invertebrate Zoology.* a type of crushing or chewing organ in rotifers.
malleation *Neurology.* rapid, repetitive twitching of clenched hands with hammerlike beating against the thighs.
mallee *Botany.* any of various dwarf Australian eucalyptus trees, such as *Eucalyptus dumosa,* sometimes forming large tracts of scrubwood.
mallein *Virology.* a preparation of concentrated cultures or extracts of the glanders bacillus, *Pseudomonas mallei*; used in the **mallein skin test** for chronic glanders disease.
malleolus *Anatomy.* a rounded projection from a bone; for example, the **malleolus lateralis,** which protrudes from the fibula on the outside of the ankle.
malleoramate trophus *Invertebrate Zoology.* a grinding and chewing organ in rotifers.
mallet *Mechanical Devices.* a hammer having a barrel-shaped head made of wood, rubber, or other relatively soft material; used for driving another tool or striking a surface without marring it.
Mallet, Robert 1810–1881, Irish engineer; originated the scientific study of earthquakes.
malleus *Anatomy.* a bone of the middle ear, attached to the tympanic membrane and having a club-shaped head that connects with the incus.
Mallodendraceae *Botany.* a monogeneric family of freshwater or brackish yellow-green algae in the order Heterogloeales, characterized by naked cells in a thick gelatinous stipe.
Mallomonas *Botany.* a genus of green algae of the family Synuraceae, characterized by complex bristles borne by scales on the cell envelope and by a reduced flagellus.

Mallophaga *Invertebrate Zoology.* the biting lice, an order of small, flattened, wingless insects that are ectoparasites of birds and mammals.
Mallory bodies *Pathology.* **1.** sizeable deposits of eosinophilic material noted in the impaired hepatic cellular structure of cirrhotic livers, generally as a result of alcoholism. **2.** bodies in the lymph spaces and epidermal cells in scarlet fever. (Named for Frank Burr *Mallory,* 1862–1941, American pathologist.)
Mallory bonding *Design Engineering.* a bonding process in which silicon chips are polished and hermetically sealed to polished glass plates through the use of heat and electricity across the assembly.
Mallory-Weiss syndrome *Medicine.* the vomiting of both red and dark blood due to torn mucous membranes at the top of the stomach, near the esophageal outlet. The tear may result from previous vomiting, or may have other causes, and surgical repair may be required.
mallow *Botany.* any of various plants of the genus *Malva,* family Malvaceae, including the **musk mallow** and other garden plants.
mallow family see MALVACEAE.
malm *Materials.* an artificial mixture of clay and chalk used to make bricks.
Malm *Geology.* a Middle European geologic series of the Upper Jurassic period, occurring after the Dogger and before the Cretaceous period.
malnourished *Medicine.* suffering from malnutrition.
malnutrition [mal′noo tri′shən] *Medicine.* any of various disorders of nutrition, caused either by an unbalanced or insufficient diet (**primary malnutrition**) or by defective assimilation of foods (**secondary malnutrition**).
malocclusion [mal′ə kloo′shən] *Medicine.* the failure of the teeth to close normally to produce an effective bite.
malonic acid *Organic Chemistry.* $HO_2CCH_2CO_2H$, white crystals that are soluble in water, alcohol, and ether; melts at 132–134°C with decomposition to ocetric acid and CO_2; used in manufacturing pharmaceuticals.
malonic acid synthesis *Organic Chemistry.* a synthetic procedure for producing carboxylic acids from malonic ester (ethyl malonate).
malonyl *Organic Chemistry.* the organic group $-OC{\cdot}CH_2{\cdot}CO-$, from malonic acid.
malonyl thiourea see THIOBARBITURIC ACID.
malonyl urea see BARBITURIC ACID.
Malpighamoeba *Invertebrate Zoology.* a genus of naked amebae of the family Entamoebidae that are obligate parasites in the intestine and Malpighian tubules of insects.
Malpighiaceae *Botany.* a tropical family of dicotyledonous trees and shrubs of the order Polygalales, typically having opposite simple leaves, indehiscent fruit, and five-petalled flowers.
Malpighi, Marcello [mal pē′gē] 1628–1694, Italian anatomist; wrote noted early anatomical descriptions; discovered capillary blood vessels.
Malpighian corpuscle [mal pig′ē ən] see GLOMERULUS, def. 2.
Malpighian layer *Histology.* the stratum germinativum layer of the epidermis.
Malpighian tubule *Entomology.* any of numerous blind excretory or secretory tubes that open into the posterior portion of the intestine in most insects and certain other arthropods.
malposition *Medicine.* an abnormal position, especially of the teeth.
malpractice *Medicine.* a legal term for improper practice; the negligence resulting from a lack of professional knowledge, experience, or skill that can reasonably be expected of others in the profession, or from a failure to exercise reasonable care or judgment, causing injury or harm to the patient.
malpresentation *Medicine.* a faulty or abnormal presentation of the fetus.
malt *Food Technology.* **1.** germinated barley or other grain, used in brewing and distilling. **2.** any alcoholic beverage that is fermented from malt, such as beer or malt liquor. **3.** to produce malt by soaking grain and allowing it to germinate.
Malta fever see BRUCELLOSIS.
malt agar *Microbiology.* a solid nutrient medium containing malt extract.
maltase *Enzymology.* an enzyme that hydrolyzes maltose to glucose.
malted *Food Technology.* made, treated, or fortified with malt.
Malter effect *Solid-State Physics.* a phenomenon in which a conductor that is coated with a surface film of a nonconducting substance has a large coefficient of secondary emission.
malt extract *Microbiology.* a preparation derived from malt that is used as a component in nutrient medium for culturing a number of microorganisms.
Malthus, Thomas [mal′thəs] 1766–1834, English economist; he originated the idea that population when unchecked tends to increase at a much faster rate than food supplies.
Malthusian [mal thoo′zhən; mal thoo′zē ən] relating to or based on the theories of Thomas Malthus.
Malthusianism *Ecology.* the theory that population increases much more rapidly than the food supply unless held in check by epidemics, wars, famine, or similar phenomena.
Malthusian parameter *Ecology.* the rate of population increase considering death rate, birth rate, and age distribution.
malt liquor *Food Technology.* a beer having a high alcohol content, usually 6–8% by volume.
maltol *Organic Chemistry.* $C_6H_6O_3$, a platelike or needlelike solid of the pyrone group that is derived from larch bark and has a caramel odor; soluble in water; melts at 161–162°C; used in bakery products and as a microchemical reagent. Also, LARIXINE, LARIXINIC ACID.
maltose *Biochemistry.* $C_{12}H_{22}O_{11}$, a sugar made up of two linked glucose molecules that results from the hydrolysis of starch and glycogen. Also, **malt sugar.**
maltose phosphorylase *Enzymology.* an enzyme that catalyzes the phosphorylation and splitting of maltose to produce glucose and glucose 1-phosphate.
maltosuria *Medicine.* the appearance of maltose in the urine.
maltotriose *Biochemistry.* a sugar that consists of three glucose molecules, resulting from the action of amylase on starch or glycogen.
malunion *Medicine.* the healing of fractured bones in a faulty position.
Malus, Étienne L. [mə loos′] 1775–1812, French engineer and physicist; studied optics; discovered light polarization by reflection.
Malus' cosine-squared law *Optics.* a law stating that the intensity of light transmitted by a pair of polarizers is proportional to the angle between their transmission axes.
Malus' law of rays *Optics.* the theory that after any number of reflections or refractions, the normal congruence of a set of rays, being a group of rays for which can be found a common surface that is orthogonal to each and every ray in the group, is preserved. Also, THEORY OF MALUS AND DUPIN.
Malvaceae *Botany.* the mallow family, dicotyledonous herbs, shrubs, and trees of the order Malvales having alternate simple leaves and mostly regular bisexual flowers with a valvate calyx; includes the mallow, hibiscus, hollyhock, cotton plant, okra, and rose of Sharon.

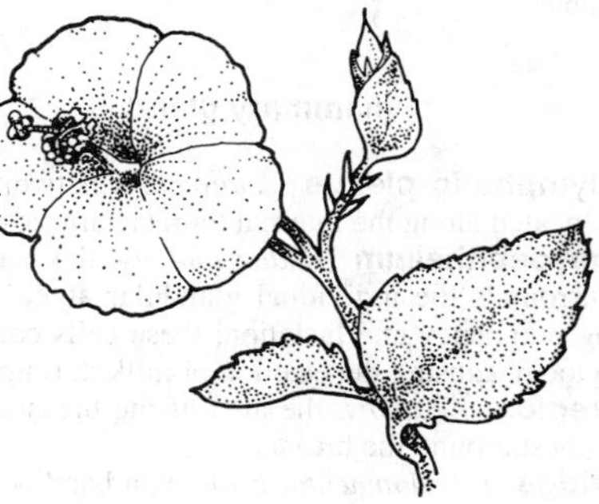

Malvaceae (hibiscus)

malvaceous *Botany.* of or relating to the family Malvaceae (mallow family).
Malvales *Botany.* an order of dicotyledonous plants of the subclass Dilleniidae having flowers with a valvate calyx and a syncarpous pistil.
mamelon *Anatomy.* any nipplelike elevation or protrusion, as on the umbilicus. *Volcanology.* a small, rounded mound of slowly extruded viscous, siliceous lava that overlies a volcanic vent. (A French word meaning "nipple.")
mamilla see MAMMILLA.
mamillary see MAMMILLARY.
mamm- a combining form meaning "breast."
mamma *plural,* **mammae.** *Anatomy.* the breast; the modified cutaneous glandular structure containing, in the female, the mammary gland.
mammal *Vertebrate Zoology.* any member of the higher vertebrate class Mammalia, including humans; characterized by live birth, body hair, and mammary glands in the female that secrete milk for feeding the young. (From the Latin word for "breast.")

Mammalia *Vertebrate Zoology.* a large class of vertebrates possessing the following general characteristics: a dentary bone that articulates with the squamosal; a mandible composed of one bone; three middle ossicles; a muscular diaphragm separating the abdominal and thoracic cavities; enucleate red blood cells; epidermal hair; and young nourished with milk from the female's mammary glands.

mammalian [mə măl′yən] *Vertebrate Zoology.* **1.** an organism of the class Mammalia; a mammal. **2.** of or relating to this class.

mammaplasty see MAMMOPLASTY.

mammary [mam′ə rē] *Anatomy.* of or relating to the mamma or mammary gland.

mammary artery see THORACIC ARTERY.

mammary fold see MAMMARY RIDGE.

mammary gland *Anatomy.* any of two or more organs of milk secretion overlying the pectoris major muscle on either side of the chest in mature female mammals, and present in rudimentary form in males and children; in the human female, each gland is composed of 15 to 25 lobes arranged radially around the nipple and separated by connective and adipose tissue, with each lobe having an excretory duct opening on the nipple. Also, **mamma.**

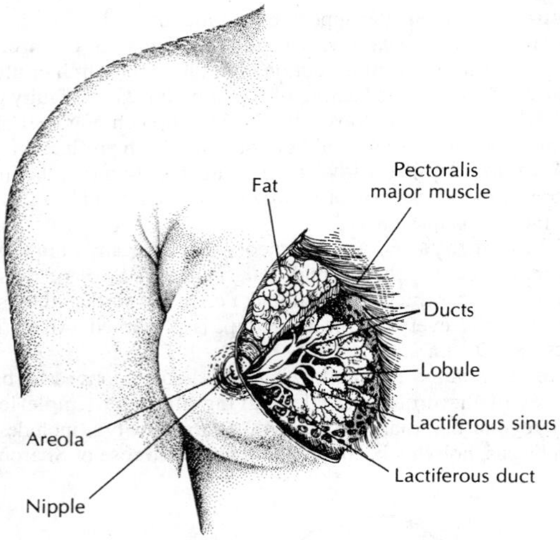

mammary gland

mammary lymphatic plexus *Anatomy.* a network of lymph nodes and vessels located along the internal thoracic arteries.

mammary myoepithelium *Endocrinology.* the network of epithelial cells that surrounds the individual glandular alveoli of the mammary gland during pregnancy and lactation; these cells contract when stimulated by oxytocin, causing the ejection of milk during suckling.

mammary region *Anatomy.* the area of the breast, or the area on the front of the chest around the breasts.

mammary ridge *Developmental Biology.* a bandlike thickening of ectoderm in the embryo that extends on either side from just below the axilla to the inguinal region; the mammary glands develop from primordia in the thoracic part of the ridge, with the balance of the ridge disappearing. Also, MAMMARY FOLD, MILK LINE.

mammilla *Anatomy.* **1.** the nipple. **2.** any nipplelike structure.

Mammillaria *Botany.* a genus of cacti, including the pincushion cactus.

mammillary *Anatomy.* **1.** of or relating to the mammilla. **2.** resembling a nipple. Also, MAMILLARY. *Mineralogy.* describing minerals that occur in rounded masses formed by groups of radiating individual crystals, and that resemble breasts or portions of spheres.

mammillary body *Anatomy.* a small nipple-shaped cell group in the hypothalamus.

mammillary line *Anatomy.* an imaginary line running vertically through the center of the nipple, defining a body axis.

mammillary process *Anatomy.* a small bony protuberance on the dorsal margin of each of the lumbar vertebrae and usually on the twelfth thoracic vertebra.

mammillitis *Veterinary Medicine.* a disease of dairy cows, caused by a herpesvirus and resulting in ulcers on the teats and udder.

mammo- a combining form meaning "breast."

mammogram [mam′ə gram′] *Radiology.* an X-ray photograph of the breast.

mammography [mə mäg′rə fē] *Radiology.* the process of making an X-ray photograph of the breast, especially in order to investigate for the presence of abnormal or malignant growth.

mammoplasia *Physiology.* the development of breast tissue. Also, MASTOPLASIA.

mammoplasty *Surgery.* the repair, reduction, or augmentation of the breast. Also, MAMMAPLASTY.

mammoth *Paleontology.* a general term for extinct elephants of the Pliocene and Pleistocene, in the suborder Mammutidea; different from modern elephants in having longer tusks and more pointed skulls; some species, such as *Mammuthus armeniacus* (14 feet at the shoulder), were larger than modern elephants, and some were smaller.

mammoth

Mammutidae *Paleontology.* the mammoths, a family of proboscideans in the extinct suborder Mammutoidea; Miocene to Holocene, becoming extinct only in the most recent glacial period. Also, **Mammutoidae.**

Mammutinae *Paleontology.* a subfamily of proboscideans in the extinct family Mammutidae; Pliocene and Pleistocene.

man *Anthropology.* **1.** a human; a member of the species *Homo sapiens.* **2.** a male member of this species. The term *human* is now generally preferred to *man* in contexts referring to the species in general rather than to the male sex in particular.

MAN *Aviation.* the airport code for Manchester, England.

Man a handful. (From Latin *manipulus.*)

man. manual.

mana *Anthropology.* a supernatural force said to inhabit people or objects and imbue them with extraordinary strength or good fortune; a concept originating in Polynesia.

management *Industrial Engineering.* the attainment of organizational goals by means of planning, organization, and efficient use of resources, especially human resources.

management by objectives *Industrial Engineering.* a management technique based on setting measurable objectives for employees, departments, or projects, and then using the objectives to direct and evaluate performance.

management information system *Telecommunications.* a communication system in which data are processed for use in decision-making by individuals in management.

management science *Industrial Engineering.* the application of quantitative techniques to the solution of managerial problems.

manakin *Vertebrate Zoology.* any of numerous brightly colored birds of the family Pipridae, inhabiting forests from Mexico to South America; known for intricate courtship dances performed by groups of males.

manandonite *Mineralogy.* $Li_2Al_4(Si_2AlB)O_{10}(OH)_8$, a colorless or white, transparent orthorhombic mineral of the kaolinite-serpentine group, having a specific gravity of 2.78 to 2.89 and a hardness of about 2.5 on the Mohs scale; found as lamellar aggregates in pegmatites.

manasseite *Mineralogy.* $Mg_6Al_2(CO_3)(OH)_{16}{\cdot}4H_2O$, a white, pale blue, or gray, pearly, transparent hexagonal mineral of the manasseite group, dimorphous with hydrotalcite, and having a specific gravity of 2.05 and a hardness of 2 on the Mohs scale; found as foliated masses with serpentine and in iron ore skarns.

manatee *Vertebrate Zoology.* any of several aquatic mammals of the family Trichechidae that feed on water plants and are characterized by two front flippers and a flat, oval tail; all species are endangered; found in the warm, shallow waters of tropical America and Africa. It is commonly called a **sea cow.** (Probably from a native American word, but influenced by *mano,* the Spanish word for hand, because of its large flippers.)

Manchurian ash *Materials.* the heavy, moderately hard wood of the *Fraxinus mandshurica* tree of Asia, which is used for firewood and for the construction of such objects as boats, furniture, and musical instruments.

Manchurian subregion *Ecology.* a distinct zoogeographical region that includes Manchuria and other areas south of Siberia extending to the Pacific Ocean, including the Korean Peninsula.

mandala *Psychology.* a concentric symbol originally derived from Indian Tantric art; used by C. G. Jung as a symbol of the self and its striving for wholeness.

mandarin *Botany.* **1.** a small Chinese tree of the species *Citrus reticulata,* bearing a variety of citrus fruit that is characterized by a loose, yellow-to-orange rind. **2.** the fruit of this tree. Also, **mandarin orange. 3.** any plant of the lily genus *Disporum* or *Streptopus,* having drooping flowers and red berries.

mandarin oil *Materials.* a fragrant yellow essential oil expressed from the peel and rind of the mandarin orange; used as a flavoring and in medicine.

mandatory layer *Meteorology.* an atmospheric layer sandwiched between any two mandatory levels.

mandatory level *Meteorology.* any of a number of constant-pressure levels in the atmosphere that are defined by analyzing data from upper-air observations. Also, **mandatory surface.**

Mandelbrot, Benoit born 1924, Polish-born American mathematician; his work popularized fractal geometry.

Mandelbrot set *Mathematics.* the set of points in the iterated function system such that z does not tend to infinity where $z \rightarrow z^2 + c$ for an initial $z = 0$ and c is any complex number.

mandelic acid *Organic Chemistry.* $C_6H_5CH(OH)COOH$, large, toxic, white crystals with a faint odor that are slightly soluble in water and alcohol, and soluble in ether; melts at 117–119°C; used in organic chemistry and as a urinary antiseptic. Also, AMYGDALIC ACID.

mandelonitrile *Organic Chemistry.* $C_6H_5CH(OH)CN$, an oily yellow liquid; insoluble in water and soluble in alcohol, chloroform, and ether; boils and decomposes at 170°C.

Mandelstam plane *Particle Physics.* a technique for plotting energy as a function of the before and after scattering angle of a two-particle collision.

Mandelstam representation *Particle Physics.* an expression giving the complex representation of a two-particle collision in which the variables are the center-of-mass energy and the scattering angle; such an expression is derived from the crossing principle.

mandible [man′də bəl] *Anatomy.* the U-shaped bone that forms the lower jaw. *Vertebrate Zoology.* **1.** the lower part of the bill in birds. **2.** also, **mandibles.** the upper and lower parts of the bill. *Invertebrate Zoology.* in arthropods, one of the first mouthparts, used to pierce and suck food.

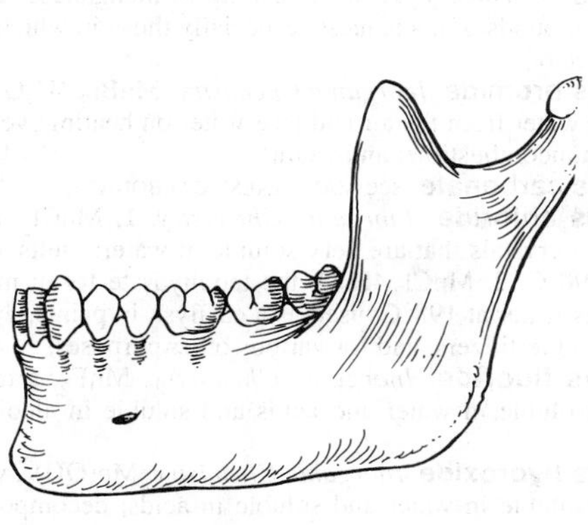

mandible

mandibular arch *Developmental Biology.* the first arch in the branchial arch series; in primate vertebrates it forms the jaws. Also, **mandibular process.**

mandibular cartilage *Developmental Biology.* a cartilage bar in the mandibular arch that forms a temporary supporting structure in the embryonic mandible; its proximal end gives rise to the cartilaginous primordium of the malleus. Also, MECKEL'S CARTILAGE.

mandibular fossa *Anatomy.* a deep depression in the temporal bone at the root of the zygoma, in which the end of the mandible or jawbone rests.

mandibular nerve *Anatomy.* a division of the trigeminal nerve that passes through the foramen ovale.

Mandibulata *Invertebrate Zoology.* a subphylum of arthropods, with mandibles and one pair of antennae, including the Crustacea, Insecta, and Myriapoda. Also, UNIRAMIA.

mandrel *Mechanical Engineering.* a shaft or bar that is inserted into a piece of work to support or hold it during machining. *Metallurgy.* any of several tools used in metal-working processes such as bending, piercing, and machining.

mandrel drawing *Metallurgy.* a method for bench-drawing tubing to a specific size, in which a mandrel is placed into the tube to be drawn, and both tube and mandrel are drawn through the die. The outside diameter of the tube is determined by the die, and the inside diameter by the mandrel.

mandrel hanger *Petroleum Engineering.* an apparatus used to prevent blowouts by securing a liquid- or gas-tight seal between the tubing and the tubing head in an oil well.

mandrel press see ARBOR PRESS.

mandrill *Vertebrate Zoology.* a large, fierce, tailless monkey of the genus *Mandrillus* in the family Cercopithcidae, characterized in the male by vivid red and blue coloring in the nose and rump; found in western Africa. The species *M. sphinx* is endangered.

mandrill

mandrin *Surgery.* a guide or stylet for soft catheters.

maneb *Organic Chemistry.* $Mn[SSCN(CH_2)_2NHCSS]$, a brown powder that is partially soluble in water and soluble in chloroform; it decomposes when heated, and is used as a fungicide for foliage.

maneuver *Military Science.* **1.** a movement to place ships, troops, materiel, or fire in a better location with respect to the enemy. **2.** a tactical exercise carried out in imitation of war; it may be executed on the ground, in the air, at sea, or on a map. **3.** the operation of a ship, aircraft, or vehicle to change directions or perform other desired movements. Also, EVASIVE MANEUVER.

maneuverability *Navigation.* the ability of a vessel to take evasive action.

maneuvering area *Transportation Engineering.* the zone around a vehicle into which the driver can quickly maneuver in order to avoid accident or delay.

maneuvering board *Navigation.* a plotting sheet, consisting of a number of concentric circles, used for plotting relative movement. One's own craft is considered to be located at the center, and other craft are plotted, over time, relative to it.

maneuvering craft *Navigation.* under the Rules of the Road at Sea, the vessel that normally has the responsibility to avoid collision with another vessel (called the **privileged vessel**). In any encounter, the maneuvering craft is determined by relative position and vessel type.

manganate *Inorganic Chemistry.* any salt with manganese in the anion or, more specifically, any salt of manganic acid, H_2MnO_4.

manganese [mang′gə nēz′] *Chemistry.* a hard, brittle, grayish metallic element that has the symbol Mn, the atomic number 25, an atomic weight of 54.938, a melting point of 1245°C, and a boiling point of 1962°C. *Metallurgy.* this element as a metal, which is never used commercially in its pure state; used extensively as an alloying element for ferrous and nonferrous alloys. (An alternate form of a Modern Latin word for *magnesia.*)

manganese acetate *Organic Chemistry.* $Mn(C_2H_3O_2)_2{\cdot}4H_2O$, combustible pink crystals; soluble in water and alcohol; melts at 80°C; used in dyeing textiles, as a drier in paint and varnish, and in fertilizers.

manganese-aluminum *Metallurgy.* a master alloy used for the production of commercially important aluminum-base alloys.

manganese black see MANGANESE DIOXIDE.

manganese borate *Inorganic Chemistry.* MnB_4O_7, a reddish-white powder that is insoluble in water; used as a varnish and oil drier.

manganese-boron *Metallurgy.* a master alloy used in the manufacture of bronze products.

manganese brass *Metallurgy.* a copper-base alloy containing nominally 28.8% zinc and 1.2% manganese.

manganese bromide see MANGANOUS BROMIDE.

manganese bronze *Metallurgy.* any of several copper-zinc alloys containing 0.5 to 3.5% manganese.

manganese carbonate *Inorganic Chemistry.* $MnCO_3$, white to rose crystals that slowly turn light brown in air; insoluble in water and decomposed by heat; used as a pigment, varnish drier, and food additive.

manganese chloride see MANGANOUS CHLORIDE.

manganese citrate *Organic Chemistry.* $Mn_3(C_6H_5O_7)_2$, a combustible white powder that is soluble in water when sodium citrate is present; melts at 80°C; used as a feed additive and food supplement.

manganese difluoride see MANGANOUS FLUORIDE.

manganese dioxide *Inorganic Chemistry.* MnO_2, black crystals or powder; insoluble in water and decomposed by heat; an oxidant that may ignite organic substances; used in batteries, fireworks, and matches, as a catalyst and reagent, and for various other purposes. Also, MANGANESE PEROXIDE, MANGANESE BLACK.

manganese fluoride **1.** see MANGANIC FLUORIDE. **2.** see MANGANOUS FLUORIDE.

manganese gluconate *Organic Chemistry.* $Mn(C_6H_{11}O_7)_2{\cdot}2HOH$, a pink powder or granules that are soluble in water and insoluble in alcohol; used in producing vitamin tablets and medicines, and as a dietary supplement and feed additive.

manganese halide *Inorganic Chemistry.* any compound consisting of manganese and an element from the halogen group (periodic table Group VIIa), such as fluorine or chlorine.

manganese heptoxide *Inorganic Chemistry.* Mn_2O_7, a highly explosive dark red oil that decomposes at 55°C and explodes at 95°C; it is formed by the reaction of permanganates with concentrated sulfuric acid.

manganese hydroxide or **manganese hydrate** **1.** see MANGANIC HYDROXIDE. **2.** see MANGANOUS HYDROXIDE.

manganese hypophosphite *Inorganic Chemistry.* $Mn(H_2PO_2)_2{\cdot}H_2O$, pink crystals or powder; soluble in water and insoluble in alcohol; loses its water above 150°C; a dangerous fire and explosive hazard when heated or with strong oxidizing agents; used as a food additive and dietary supplement.

manganese iodide see MANGANOUS IODIDE.

manganese lactate *Organic Chemistry.* $Mn(C_3H_5O_3)_2{\cdot}3H_2O$, a pale red monoclinic solid that is soluble in water and alcohol; used in medicines.

manganese linoleate *Organic Chemistry.* $Mn(C_{18}H_{31}O_2)_2$, a combustible, dark brown solid that is soluble in linseed oil; used in making pharmaceuticals and as a varnish and paint drier.

manganese monoxide see MANGANOUS OXIDE.

manganese naphthenate *Organic Chemistry.* a combustible brown resinous solid that is soluble in mineral spirits and hardens when exposed to air; melts at 130–140°C; used as a paint and varnish drier.

manganese nodule *Geology.* a small, irregular, brown to black concretionary mass composed primarily of manganese salts and manganese-oxide minerals that occurs in abundance on the deep-sea floor. Also, PELAGITE.

manganese oleate *Organic Chemistry.* $Mn(C_{18}H_{33}O_2)_2$, a flammable, brown granular solid that is insoluble in water and soluble in ether and oleic acid; used as a paint and varnish drier.

manganese oxalate *Organic Chemistry.* $MnC_2O_4{\cdot}2H_2O$, a white crystalline powder that is very slightly soluble in water and soluble in dilute acids; used as a paint and varnish drier.

manganese oxide **1.** see MANGANIC OXIDE. **2.** see MANGANOUS OXIDE.

manganese peroxide see MANGANESE DIOXIDE.

manganese poisoning *Toxicology.* poisoning due to the inhalation of manganese dust; symptoms may include mental disorders, tremor, and inflammation of the respiratory system. Also, MANGANISM.

manganese resinate *Organic Chemistry.* $Mn(C_{20}H_{29}O_2)_2$, a flammable brownish solid or flesh-colored powder that is insoluble in water and soluble in hot linseed oil; used as a varnish and oil drier.

manganese sesquioxide see MANGANIC OXIDE.

manganese silicate see MANGANOUS SILICATE.

manganese-silicon *Metallurgy.* a master alloy used for the production of certain low-alloy steels.

manganese star *Astronomy.* a star of spectral type B5 to F5, showing a spectrum with an abnormally large abundance of manganese relative to iron.

manganese sulfate see MANGANOUS SULFATE.

manganese sulfide see MANGANOUS SULFIDE.

manganese sulfite see MANGANOUS SULFITE.

manganese trifluoride see MANGANIC FLUORIDE.

manganic *Chemistry.* **1.** of or relating to manganese. **2.** describing various compounds of manganese, especially those in which the element has a valence of 3.

manganic fluoride *Inorganic Chemistry.* MnF_3, a toxic, red crystalline solid, decomposed by water or by heat; it attacks glass when hot; used as a fluorinating agent. Also, MANGANESE TRIFLUORIDE, MANGANESE FLUORIDE.

manganic hydroxide *Inorganic Chemistry.* $Mn(OH)_3$, a brown powder that easily loses water to form MnO(OH); insoluble in water and decomposed by heat; used as a pigment. Also, HYDRATED MANGANIC OXIDE.

manganic oxide *Inorganic Chemistry.* Mn_2O_3, a very hard black or brown powder; insoluble in water and decomposed by heat; flammable and toxic; found in nature as manganite.

manganism see MANGANESE POISONING.

manganite *Mineralogy.* $Mn^{+3}O(OH)$, a steel-gray to iron-black, opaque monoclinic mineral, trimorphous with feitknechtite and groutite, occurring in prismatic crystals and massive form, and having a specific gravity of 4.33 and a hardness of 4 on the Mohs scale; found worldwide in deposits formed by meteoric waters, as a hydrothermal vein mineral, and in sedimentary deposits; a minor ore of manganese. Also, GRAY MANGANESE ORE.

manganolangbeinite *Mineralogy.* $K_2Mn_2^{+2}(SO_4)_3$, a pale rose-red cubic mineral occurring in small tetrahedral crystals, having a specific gravity of 3.02 and an undetermined hardness; found at Mt. Vesuvius, Italy.

manganosite *Mineralogy.* $Mn^{+2}O$, an emerald-green cubic mineral of the periclase group occurring in granular masses and octahedral crystals that blacken upon exposure, having a specific gravity of 5.36 and a hardness of 5.5 on the Mohs scale; found at Franklin, New Jersey, and in dolomite in Sweden.

manganous *Chemistry.* **1.** of or relating to manganese. **2.** describing various compounds of manganese, especially those in which the element has a valence of 2.

manganous bromide *Inorganic Chemistry.* $MnBr_2{\cdot}4H_2O$, red crystals that absorb water from the air and lose water on heating; very soluble in water and noncombustible; an irritant.

manganous carbonate see MANGANESE CARBONATE.

manganous chloride *Inorganic Chemistry.* **1.** $MnCl_2$, rose-colored deliquescent crystals that are very soluble in water; melts at 650°C and boils at 1190°C. **2.** $MnCl_2{\cdot}4H_2O$, the tetrahydrate form; melts at 58°C and loses its water at 198°C; used as a catalyst, in paints, dyes, pharmaceuticals, and fertilizers, and for various other purposes.

manganous fluoride *Inorganic Chemistry.* MnF_2, a toxic reddish powder; insoluble in water and acids and soluble in alcohol; melts at 856°C.

manganous hydroxide *Inorganic Chemistry.* $Mn(OH)_2$, white to pink crystals; insoluble in water and soluble in acids; decomposed by heat; found in nature as pyrochroite.

manganous iodide *Inorganic Chemistry.* **1.** MnI_2, a pink deliquescent crystalline mass; soluble in water; turns brown in air and light; melts under vacuum at 638°C. **2.** $MnI_2{\cdot}4H_2O$, the tetrahydrate form, rose-colored deliquescent crystals; soluble in water and decomposed by heat.

manganous oxide *Inorganic Chemistry.* MnO, a green powder; insoluble in water and soluble in acids; converts to Mn_3O_4 on heating in air; used as a catalyst, in ceramics, paints, and colored glass.
manganous silicate *Inorganic Chemistry.* $MnSiO_3$, red crystals or a yellowish-red powder; insoluble in water; melts at 1323°C; used as a glass and ceramic glaze colorant. It is found in nature as rhodonite.
manganous sulfate *Inorganic Chemistry.* $MnSO_4 \cdot 4H_2O$, rose-red translucent crystals that lose water to the air; soluble in water and insoluble in alcohol; used as a fertilizer and feed additive, in medicine, and for various other purposes.
manganous sulfide *Inorganic Chemistry.* MnS, green crystals; nearly insoluble in water and decomposed by heat; used in steel manufacture.
manganous sulfite *Inorganic Chemistry.* $MnSO_3$, a grayish-black or brownish-red powder, insoluble in water. Also, **manganous sulfite normal.**
mange [mānj] *Veterinary Medicine.* any of several contagious skin diseases caused by burrowing parasitic mites, resulting in intense itching, inflammation of the skin, and often hair loss; commonly affects domestic animals and sometimes also affects humans. (Going back to a French word meaning "to eat.")
manger [mān´jər] *Agriculture.* **1.** a box in a stable or barn from which farm animals eat. **2.** a building housing farm animals. *Astronomy.* **Manger** or **the Manger.** an open cluster in Cancer with about 80 stars, faintly visible to the naked eye on clear dark nights. Also, PRAESEPE, BEEHIVE CLUSTER, M 44.
Mangin mirror *Optics.* a mirror formed by coating the back surface of a concave meniscus len; used in search lights and aircraft gunsights to reduce the spherical aberration of a mirror telescope.
mangle *Metallurgy.* the process of squeezing metal plates between rollers.
mangle gearing *Mechanical Engineering.* gearing that produces reciprocating motion by a means of a pinion that rotates in one direction and drives a rack with teeth at the ends and on both sides.
mango *Botany.* **1.** an evergreen tree of the species *Mangifera indica* in the family Anacardiaceae, native to Asia and now widely grown elsewhere in warm climates, including the southeastern U.S. **2.** the edible, sweet, oblong fruit of this tree.

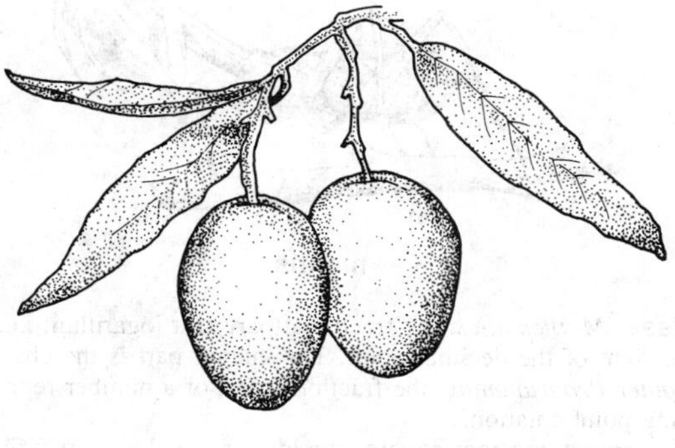

mango

mangrove *Botany.* any tree of the genus *Rhizophora,* or other genera of the family Rhisophoraceae, whose species are mostly low trees noted for their dense, interlacing above-ground roots; found in marshes or tidal shores.
mangrove swamp *Ecology.* a swamp at the fringe of tropical saltwater systems characterized by an abundance of low to tall trees, especially mangrove trees, that have dense root systems ranging well beyond their trunks.
Manhattan Project *Engineering.* the unofficial name of the secret project to design and produce the first atomic bomb (1942–1945).
manhole *Engineering.* a vertical hole with a removable cover through which a person may enter a sewer or boiler for maintenance.
manhole coaming *Naval Architecture.* a raised flange or low bulwark around a manhole in a ship's deck plating.
man-hour *Industrial Engineering.* a measure of production effort based on one person working for one hour, two people working for 30 minutes, or an equivalent combination.
mania *Psychology.* a condition in which an individual exhibits intense euphoria, agitation, and hyperactivity, and often becomes consumed with grandiose ideas or schemes. (From the Greek word for "madness.")
maniac *Psychology.* **1.** a person affected with mania or exhibiting manic behavior. **2.** in popular use, a violently insane or irrational person.
manic [man´ik] *Psychology.* of, relating to, or affected by mania.
manic-depressive *Psychology.* **1.** relating to or affected by manic-depressive psychosis. **2.** a person affected in this way.
manic-depressive psychosis or **disorder** *Psychology.* a major affective disorder in which an individual alternates between states of deep depression and extreme elation. Also, BIPOLAR AFFECTIVE DISORDER. *Genetics.* this disorder which appears to be hereditary, specifically associated with a dominant allele on the short arm of the X chromosome.
Manidae *Vertebrate Zoology.* the pangolins or scaly anteaters, a family of mammals composing the order Pholidota, characterized by a lack of teeth, a highly extensible tongue, and a diet that consists mostly of termites; found in Africa and Asia.
manifest *Transportation Engineering.* a full record of the passengers or cargo aboard an aircraft, including the point of origin, destination, and flight information.
manifest content *Psychology.* the elements of a dream that are consciously recalled by the dreamer.
manifest covariance *Physics.* a property of an expression written in terms of four-vectors, tensors, Lorentz invariant numbers, and operators, making it obvious that the special theory of relativity is applicable.
manifest deviation see STRABISMUS.
manifold *Engineering.* a pipe or chamber that has multiple openings to allow passage of a fluid. *Mathematics.* a Hausdorff topological space in which every point has a neighborhood that is homeomorphic to R^n. Sometimes the requirement that the space be Hausdorff is omitted, although this allows the dimensionality of the manifold to vary from one region to another. Also, TOPOLOGICAL MANIFOLD.
manifolding *Engineering.* the process of gathering several fluid inputs to one chamber (intake manifold) or of splitting a single fluid into several streams (distribution manifold).
manifold of states *Atomic Physics.* a set of states that can be considered to represent an operator or a Lie group of operators.
manifold pressure *Mechanical Engineering.* the pressure in the intake manifold of an internal-combustion engine.
manikin *Medicine.* a model of the body, usually with movable or removable members and parts, used to illustrate anatomy for teaching purposes.
Manila hemp *Materials.* a strong, fibrous material made from the leafstalks of the abaca, *Musa textilis*; used in making ropes and fabrics. Also, **manilla.**
Manila paper *Materials.* a strong, durable, light-brown or buff-colored paper that is used for envelopes, wrapping paper, and similar purposes; it was originally made of Manila hemp but is now made of various other fibers.
Manila resin *Materials.* a type of resin extracted from various trees of the Phillipines and East Indies; used in inks, paints, and varnishes.
manioc see CASSAVA.
manipulated variable *Computer Programming.* in control systems, a quantity or condition that is changed by the computer, based on system feedback, to cause a shift in the value of some regulated condition.
manipulation *Medicine.* the skillful or dextrous use of the hands in therapeutic or diagnostic procedures, such as in physical therapy, changing the position of a fetus, or palpation. *Behavior.* any actions by an animal that somehow change or affect its environment; e.g., burrowing, nest building, or cracking open nuts and shells.
manipulator *Engineering.* any of a variety of devices used for the remote handling of hazardous materials.
manipulator-oriented language *Robotics.* a computer language specifically developed to control the end effector of a robot.
manipulator-type robot *Robotics.* a robot with a manipulative function, such as pick-and-place or loading and unloading; usually having a universal manipulator with five degrees of freedom.
manjak or **manjack** *Materials.* a type of asphaltite found especially in Barbados; used in varnishes and for insulating cables.
mankato stone *Petrology.* limestone with more than 49% calcium carbonate, 4–5% alumina, and some silica.
man-machine system *Engineering.* a system in which a worker and a machine must work together in order for the system to function properly.
man-minute *Industrial Engineering.* a measure of production effort based on one person working for one minute, two people working for 30 seconds, or an equivalent combination.

manna *Materials.* a dried, sweetish substance exuded from the flowering ash tree, *Fraxinus ornus,* and other European ash trees; formerly used as a laxative. (From the term used in the Bible to describe the white food miraculously supplied to the starving Israelites in the wilderness.)

manna lichen *Botany.* any of several lichens of the genus *Lecanora,* particularly *L. esculenta,* that are eaten by humans and other animals; found in the African and Arabian deserts.

mannan *Biochemistry.* a complex polysaccharide composed of mannose residues and often found in plant cell walls.

manned *Engineering.* occupied, driven, or operated by one or more persons. *Space Technology.* specifically, describing a space vehicle that is occupied by one or more persons who perform some useful function, often including control of the vehicle.

Mannesmann mill *Metallurgy.* a mill used in the Mannesmann process, consisting of two rolls with slightly inclined axes.

Mannesmann process *Metallurgy.* a process in which a billet is pierced using a Mannesmann mill as a preliminary operation in the manufacture of seamless tubing.

Mannich reaction *Organic Chemistry.* a general reaction between an amine, an aldehyde (or ketone), and a highly nucleophilic carbon atom, in which the nucleophilic carbon is added to an immonium ion intermediate.

mannikin

mannikin *Vertebrate Zoology.* any of several small weaverbirds of the family Estrildidae that inhabit open country in Africa, Asia, and Australia.

Manning equation *Fluid Mechanics.* an empirical formula for predicting the Chezy coefficient; used in estimating the velocity under turbulent flow conditions in open channels.

mannitol *Organic Chemistry.* $C_6H_{14}O_6$, an odorless, combustible, white rhomboidic, needlelike, or prismatic solid with a slightly sweet taste; soluble in water and slightly soluble in alcohol; melts at 165–167°C; used to thicken or stabilize food products, in medicines and pharmaceutical products, and as an analytical reagent for boron.

mannitol hexanitrate *Organic Chemistry.* $C_6H_8(ONO_2)_6$, a combustible, highly explosive, colorless, needlelike solid; insoluble in water and soluble in alcohol and ether; melts at 112–113°C and explodes at 120°C; used in explosive caps, and also medicinally (when mixed with at least ten parts of carbohydrate). *Medicine.* this substance used as a vasodilator, as in cases of cardiac or urinary insufficiency.

mannopine *Biochemistry.* a type of Ri plasmid present in hairy root disease.

mannose *Biochemistry.* a monosaccharide stereoisomeric with glucose found in glycolypids and glycoproteins.

Mann-Whitney U test *Statistics.* in nonparametric analysis, a two-sample test to determine if the samples were drawn from populations with equal location parameters. Also, RANK-SUM TEST.

mano *Archaeology.* a loaf-shaped handstone that is rubbed against a flat stone or other surface to grind seeds and other substances into a powder.

manocryometer *Thermodynamics.* an apparatus that measures the effect of pressure on the melting point of a substance.

manometer *Engineering.* **1.** a double-legged or U-shaped tube used to measure differential pressure between two fluids. **2.** an instrument that measures the elastic pressures of gases and vapors.

manometry *Engineering.* the art of measuring gas and vapor pressures using manometers.

manostat *Engineering.* a device used to maintain a constant pressure in an enclosure or chamber.

man-portable *Ordnance.* a term used to describe equipment designed to be carried by one person over a long distance without seriously impairing the performance of normal duties; such equipment usually weighs 31 pounds (14 kg) or less.

Man. pr early in the morning. (From Latin *mane primo.*)

man-process chart *Industrial Engineering.* a chart showing the work steps or activities of a worker, including such factors as work time, equipment used, and distance moved.

mansard roof see GAMBREL ROOF.

M-A-N scavenging system *Mechanical Engineering.* a system in which waste oil and gases are removed from the cylinder of an internal combustion engine; the gases circulate in a loop with the exhaust ports located above the intake ports on the same side of the cylinder.

mansfieldite *Mineralogy.* $AlAsO_4 \cdot 2H_2O$, a white to pale-gray, transparent, orthorhombic mineral of the variscite group occurring as porous, cellular masses of spherulitic fibers, having a specific gravity of 3.02 to 3.03 and a hardness of 3.5 to 4 on the Mohs scale; found with scorodite.

Manson, Sir Patrick 1844–1922, Scottish parasitologist; the first to show that mosquitoes carried the malarial parasite.

mansonelliasis *Medicine.* a usually symptomless infection by the tropical parasite *Mansonella ozzardi,* which inhabits the human mesentery.

mansonia *Materials.* the fine-textured, strong wood of the *Mansonia altissima* tree of Africa, used as a substitute for walnut in the construction of furniture and flooring.

mantel *Building Engineering.* a projecting shelf or slab of wood, stone, or other material above a fireplace. Also, **mantelpiece.**

Mantidae *Invertebrate Zoology.* the praying mantises, a diverse family of predatory orthopteran insects, having large eyes and a pair of large grasping legs; widely distributed on all continents. Also, MANTODEA.

M antigen *Immunology.* **1.** a specific cell-surface antigen of group A beta hemolytic streptococci. Also, **M protein. 2.** a bacterial antigen of a mucoid microorganism.

mantis *Invertebrate Zoology.* the common name for various insects composing the order Mantidae, characterized by protective coloration and an elongated prothorax with the front legs modified for seizing prey; the front legs are typically held erect as if in a position of prayer. Also, **praying mantis, mantid.**

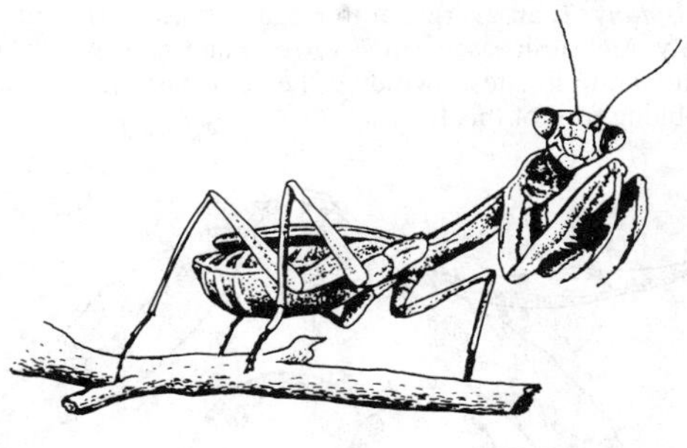

mantis

mantissa *Mathematics.* the fractional part of a logarithm; i.e., the part to the right of the decimal point. The integer part is the characteristic. *Computer Programming.* the fractional part of a number represented in floating-point notation.

mantle something that covers, shields, or envelops; specific uses include: *Anatomy.* an enveloping covering, such as the brain mantle or pallium. Also, PALLIUM. *Vertebrate Zoology.* **1.** a layer of neural tissue that develops just external to the ependymal layer of the neural tube in embryonic development. **2.** the back, wings, and scapulars of a bird. *Invertebrate Zoology.* a layer of tissue that covers the body and secretes and lines the shell of many invertebrates. *Engineering.* **1.** a loose hood designed as a protective covering. **2.** a gauzelike hood or pouch made of refractory material that radiates light when heated over a flame to incandescence; used in camping and military lanterns. *Metallurgy.* in a blast furnace, a heavy steel ring that supports the furnace shell and brickwork. *Mining Engineering.* see COVER. *Geology.* the part of the earth's interior lying below the crust and above the core, extending to a depth of 3480 kilometers below the surface.

mantle cavity *Invertebrate Zoology.* the space between the mantle and body tissues in brachiopods and mollusks, usually freely open to water or air and sometimes containing respiratory organs.

mantled gneiss dome *Geology.* in metamorphic terrains, a dome having a core of remobilized gneiss enclosed by a concordant sheath of regionally metamorphosed, overlying rock.

mantle ice see ICE SHEET.

mantle isthmus *Invertebrate Zoology.* the narrow passage that leads into the mantle cavity.

Chaos

Butterfly departing cocoon.
Myron J. Dorf/The Stock Market.

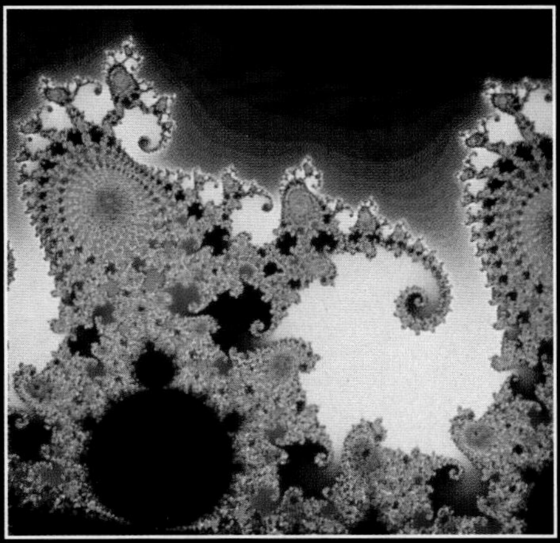

Computer graphics image created by fractal geometry. Courtesy of International Business Machines Corporation.

Computer graphics image of scenery created by fractal geometry.
© Bruce Iverson/Science Photos.

Computer graphics image entitled *Chaos Canyon* created by fractal geometry.
Gregory Sams/SPL/Photo Researchers.

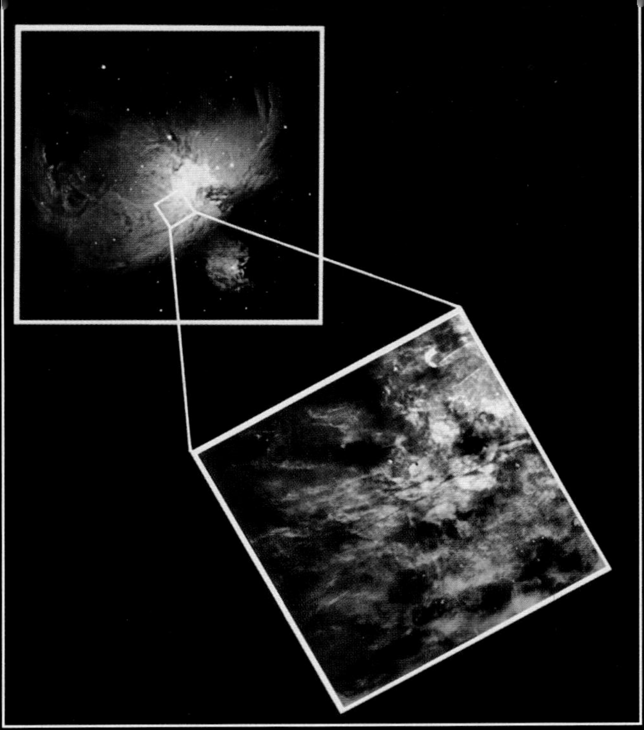

Region of the "Great Nebula" in Orion, which is where new stars are formed out of interstellar gas. Image was made with NASA's Hubble Space Telescope. Courtesy of NASA.

Computer-generated view of Hurricane Allen. The three-dimensional image is created by computer calculations of data gathered by two satellite images of the hurricane. Science Source/Photo Researchers

mantle lobe *Invertebrate Zoology.* one of the expanded dorsal or ventral flaps on the mantle of bivalve mollusks.
mantle rock see REGOLITH.
mantlet *Ordnance.* **1.** originally, a movable shelter used for protection by soldiers attacking a fortress. **2.** in modern usage, a shield, armor, or similar protection at the front of a gun or attached to the front of a tank.
manto *Geology.* a bedded, flat-lying, horizontal sedimentary or igneous ore body.
Mantodea see MANTIDAE.
Mantoux test *Immunology.* an intradermal test for tuberculosis; a hypersensitive reaction indicates a previous or current infection.
man-transportable *Ordnance.* a term used to describe equipment that is designed to be transported on wheeled, tracked, or air vehicles, but that may be handled by one or more individuals over distances of 100 to 500 meters; the upper weight limit of such equipment is approximately 65 pounds per individual.
manual *Science.* of, relating to, or performed by the hand or the hands. *Engineering.* specifically, performed or operated by hand rather than by machine. (From Latin *manualis,* derived from *manus,* the hand.)
manual control *Agronomy.* a term for the removal or control of unwanted plants by hand rather than by the use of herbicides.
manual-control unit *Computer Programming.* a small, hand-held machine that is used to program instructions for robot movements. Also, PROGRAMMING UNIT.
manual element *Industrial Engineering.* in work-motion studies, a distinct, measurable segment of work that is performed by hand or by means of hand tools, rather than by machine.
manual exchange *Telecommunications.* any exchange in which calls are controlled by an operator.
manual input *Computer Programming.* the manual entry of data.
manually controlled work see EFFORT-CONTROLLED CYCLE.
manual operation *Computer Programming.* the processing of data in a system through nonautomatic techniques.
manual ringing *Telecommunications.* a method of continuous ringing of a telephone; the ringing signal continues only as long as the ringing key is operated and held in the ringing position.
manual spinning *Metallurgy.* in metal forming, a process that employs a rotating mandrel.
manual telephone set *Electronics.* a telephone set that does not contain a dial; when the receiver is lifted, the operator is signalled.
manual tracking *Engineering.* a form of target tracking in which required power is supplied manually through the tracking handwheels.
manual word generator *Computer Technology.* a device that allows the direct entry of a word into memory following a manual operation.
manubrium *Anatomy.* **1.** a handlelike structure. **2.** that part of the sternum or of the malleus which resembles a handle. *Invertebrate Zoology.* **1.** the portion of the body bearing the oral-anal opening in hydrozoan polyps. **2.** the elongated hanging portion of a medusa that bears the oral-anal opening. *Botany.* a cylindrical cell in certain algae that protrudes inward from the center of each of the eight shields that make up the globular antheridium.
manufacturing cell *Design Engineering.* the collective plant machinery and equipment used to process related parts.
manufacturing lead time see LEAD TIME.
manufacturing progress function *Industrial Engineering.* the improvement of production efficiency that occurs with the passage of time.
manufacturing resource planning *Industrial Engineering.* a material-requirements planning system that encompasses an entire organization.
manuport *Archaeology.* a stone or other such object found at a site where it could not have occurred naturally, indicating that it must have been transported by a human.
manure *Agriculture.* **1.** animal waste that is spread on or mixed with the soil as a fertilizer, especially the excrement of cattle, chickens, horses, or sheep. **2.** any other natural waste material that is used in this way.
manure salts *Inorganic Chemistry.* a mixture of salts containing 20–30% potash and a high proportion of chloride; used in fertilizers.
manus *Anatomy.* a technical term for the hand. See HAND. *Invertebrate Zoology.* the prehensile appendage or claw of an arthropod.
manuscript *Graphic Arts.* **1.** a piece of writing that is handwritten, rather than printed or typed. **2.** the original text of a work that is submitted by an author to a publisher. **3.** any version of a text other than the final printed version.
manus valga *Medicine.* a deformed hand that is turned inward and adducted at the wrist. Also, CLUBHAND.
Manx syndrome *Genetics.* an inherited condition of taillessness in domestic cats, determined by a dominant autosomal gene and lethal during embryonic development. (Originally *Manx* cats, from the Isle of Man.)
many-body force *Physics.* the force acting on a single body due to the presence of two or more other bodies.
many-body problem *Mechanics.* the problem of predicting the motions of three or more mutually interacting bodies, which has yet to be solved analytically. Also, N-BODY PROBLEM, THREE-BODY PROBLEM.
many-body theory *Physics.* a theory for a system of several bodies or particles in which physical quantities associated with the system are derived without knowing the specific behavior of each individual body.
MAO monoamine oxidase.
map *Cartography.* **1.** an accurate representation on a plane surface, usually at an established scale and relative to a coordinate reference system, of selected features of the earth's surface (or of another planetary body). **2.** the accurate representation on a plane surface, usually at an established scale and relative to a coordinate reference system, of all or part of the celestial sphere. **3.** to engage in surveying activities with the intent of making a map, or to prepare a map for publication. *Genetics.* **1.** the ordering of genes in a eukaryotic chromosome. **2.** the determination of the arrangement of DNA sequences in a gene or cluster of genes. *Computer Programming.* **1.** to change information from one form to another. **2.** to establish a correspondence between the elements of one set and those of another. **3.** to apply a transformation to each element of a sequence. *Mathematics.* **1.** a function. **2.** in graph theory, a connected plane graph containing no isthmus. The four-color theorem was originally formulated in terms of maps.
map accuracy (standards) *Cartography.* standards established in the United States to ensure horizontal and vertical accuracy in printed maps; for maps of scales larger than 1:20,000, at least 90% of the horizontal features must be placed within 1/30 inch (0.85 mm) of their true geographic locations as indicated by the map projection, and at least 90% of the elevations indicated by contours must be accurate within one-half of the basic contour interval.
map-controlled *Cartography.* using maps, rather than geodetic or photographic information, to position cartographic features.
map distance *Genetics.* the distance between two genes on a chromosome; usually measured in Morgan units or map units.
map exercise *Military Science.* an exercise in which a series of military situations is stated and solved on a map.
maple *Botany.* the common name for any tree of the genus *Acer* of the order Sapindales; a broad-leafed deciduous tree grown commercially for its sap, which is used as a food flavoring and sweetener. *Materials Science.* the wood of any of these trees, especially that of *Acer saccharinum,* the **sugar maple** or **hard maple,** which has been widely used in North America since colonial times for firewood, furniture, flooring, and many other purposes.

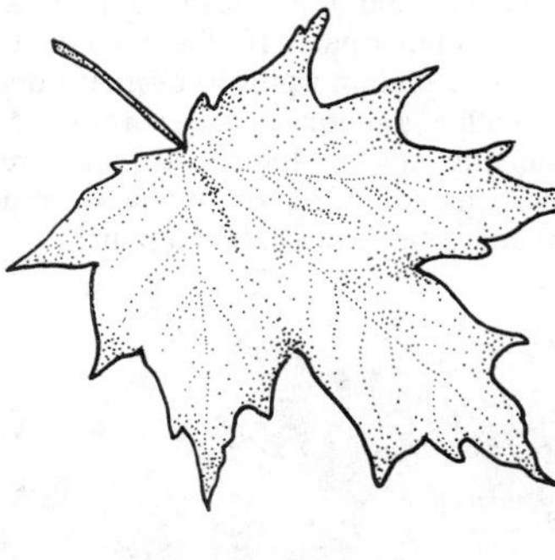

maple

maple syrup urine disease *Medicine.* a congenital failure to metabolize amino acids, resulting in mental and physical retardation, sporadic acidosis, lack of appetite, and a maple syrup odor in the urine.
map parallel see AXIS OF HOMOLOGY.
mapping *Cartography.* **1.** the art or science of cartography. **2.** the physical process of making maps. *Mathematics.* **1.** a function or multiple-valued relation. The terms "map" or "mapping" are used most often in an algebraic context; "function" is used most often in an analytic context. **2.** in topology, a continuous function.
mapping function *Artificial Intelligence.* in Lisp, a function that applies a specified function to each element of a list.

mapping unit *Agronomy.* a soil or combination of soils represented on a map to show soil types, phases, associations, and complexes.

map plotting *Meteorology.* **1.** the transcription of weather information onto maps or diagrams. **2.** specifically, the process of decoding synoptic reports and entering the data in station-model form on synoptic charts. Also, **map spotting.**

map projection *Cartography.* any system of lines representing parallels and meridians on the spherical surface of the earth or a section of the surface of the earth, drawn on a plane surface, and determined by means of geometrical construction or mathematical computation. *Navigation.* such a representation of the earth used in navigation.

map unit *Genetics.* a measure of the distance between two linked genes on a chromosome, equal to one centimorgan, which is defined as the distance such that crossing-over between the two genes during meiosis would result in 1% of the products being recombinants.

maquis [mä kē′] *Ecology.* a type of vegetation that is composed of low shrubs with hard, drought-resistant, needlelike leaves; typically found in Mediterranean climates.

MAR microanalytical reagent; memory address register.

Marafivirus group *Virology.* a group of plant viruses with isometric particles that infect a narrow host range through persistent transmission by leafhoppers.

marage *Metallurgy.* to heat treat certain specialty steels in order to harden them by the precipitation of intermetallic compounds.

maraging steel *Materials Science.* a highly alloyed steel, hardened both by martensite and by age hardening.

Marantaceae *Botany.* a family of monocotyledonous plants of the order Zingiberales, having rhizomes or tubers, simple, entire leaves that are pinnately veined, and bisexual flowers with one functional stamen and pollen sac.

marantic clot *Hematology.* a blood clot formed because of enfeebled circulation and general malnutrition.

marantic endocarditis *Cardiology.* an inflammation of the tissue lining the heart, sometimes including the heart valves, often associated with fatal disease. Also, NONBACTERIAL THROMBOTIC ENDOCARDITIS.

marasmic *Medicine.* of or referring to marasmus.

marasmus *Medicine.* protein-calorie malnutrition occurring chiefly in the first year of life; characterized by growth retardation and progressive wasting of fat and muscle, but with continued appetite and mental alertness.

Marattiaceae *Botany.* the single family constituting the order Marattiales; terrestial ferns characterized by compound, massive sporangia and fleshy, circinate fronds.

Marattiales *Botany.* an order of large ferns having eusporangiate sporangia on the dorsal side of their circinate leaves and showing no secondary thickening.

Marattiopsida *Botany.* one of three subdivisions of the fern division Filicophyta, containing the single order Marattiales; a terrestrial plant having massive sporangia, and often, fleshy and divided fronds.

Marauder *Aviation.* a popular name for the B-26 bomber.

marble *Petrology.* **1.** a metamorphic rock that consists of calcite or dolomite, typically with a granoblastic saccharoidal texture; widely used in buildings, monuments, and sculptures. **2.** commercially, any crystallized carbonate rock, including true marble and certain limestones, that is sufficiently hard and coherent to accept a polish.

marble

marble bones *Medicine.* a common name for the disease osteopetrosis. Also, **marble bone disease.**

marble state see STATUS MARMORATUS.

marbling *Engineering.* the process of creating the appearance of marble in a surface or medium.

Marburg or **Marburg-Ebola virus** *Virology.* a severe and often fatal virus of the family Filoviridae; causes fever, diarrhea, vomiting, and mental confusion in humans, and fatal infection in monkeys and guinea pigs. (From the city of *Marburg* in Germany and the *Ebola* River in Africa, where noted outbreaks of the disease occurred.)

marcasite *Mineralogy.* FeS_2, a white to brass-yellow orthorhombic mineral of the marcasite group, dimorphous with pyrite, occurring in massive, granular, or globular form and as tabular crystals, having a specific gravity of 4.89 to 4.92 and a hardness of 6 to 6.5 on the Mohs scale; it is a low-temperature mineral found in sediments and vein deposits.

marcescent *Botany.* of plant parts, withering without dropping off.

Marcgraviaceae *Botany.* a family of mostly climbing dicotyledonous shrubs and vines of the order Theales, having alternate, entire, leathery leaves and a terminal inflorescence with colored bracts.

Marchantiaceae *Botany.* a widespread family of medium to large typical liverworts of the order Marchantiales, characterized by repeatedly dichotomous branching, usually compound barrel-shaped pores in the epidermis, and unique stalked male receptacles and cup-shapped gemmae receptacles.

Marchantiales *Botany.* an order of liverworts of the class Marchantiopsida, having a thalloid gametophyte and sexual organs borne on stalks.

Marchantiidae *Botany.* a subclass of liverworts of the class Marchantiopsida, characterized by a regular alternation of generations, a flask-shaped archegonia made up of a venter and a neck, and a ribbon-shaped or rosette-shaped thallus.

Marchantiopsida *Botany.* a class of green plants commonly known as liverworts, characterized by rhizoids on the lower surface and a supine thallus. Also, HEPATICOPSIDA.

Marconi, Guglielmo [mär kō′nē] 1874–1937, Italian physicist; shared the Nobel Prize with K. F. Braun for developing wireless telegraphy.

Guglielmo Marconi

Marconi antenna *Electromagnetism.* a quarter-wave antenna that is grounded at its input end.

Marcq St.-Hilaire method *Navigation.* a means of determining position by celestial observation that compares the observed altitude of a body with an altitude derived by computation from an assumed position. Also, ST.-HILAIRE METHOD, ALTITUDE INTERCEPT METHOD.

Mardsen chart *Cartography.* a chart used to show meteorological information, especially regarding areas over the oceans.

mare [mār] *Vertebrate Zoology.* a female horse, especially when mature.

mare [mä′rā] *plural,* **maria** [mä′rē ə] *Astronomy.* one of the large dark expanses of basalt on the moon and Mars, many of which fill impact basins; for example, **Mare Nubium** (Sea of Clouds), **Mare Serenitatis** (Sea of Serenity), and **Mare Tranquillitatis** (Sea of Tranquility). (From the Latin word for "sea;" so named because Galileo believed they were seas when he first viewed them through a telescope.)

marekanite *Geology.* a small, rounded to subangular body of obsidian that occurs in masses of perlite.

Marek's disease *Veterinary Medicine.* a lymphomatosis that occurs in chickens due to the presence of the herpes virus, mainly affecting the peripheral nerves and visceral organs. (Identified by József *Marek,* 1868–1952, Hungarian veterinarian.) Also, NEURAL LYMPHOMATOSIS.

Marfan's syndrome *Genetics.* a hereditary human disorder that is caused by an autosomal dominant allele; an individual having this allele develops elongated bones, especially in the fingers and toes, often with ocular and cardiovascular abnormalities and other deformities. (Identified by Bernard-Jean Antonin *Marfan,* 1858–1942, French pediatrician.) Also, ARACHNODACTYLY.

margarine *Food Technology.* a food product meant to resemble and replace butter, made from vegetable oils, sometimes with animal fats.

margarite *Geology.* a string of spherical globulites, commonly occurring in glassy igneous rocks. *Mineralogy.* $CaAl_2(Al_2Si_2)O_{10}(OH)_2$, a yellow, pink, or pale green, brittle, monoclinic mineral of the mica group occurring in platy to scaly masses and rarely in tabular crystals, having a specific gravity of 3.0 to 3.1 and a hardness of 3.5 to 4.5 on the Mohs scale; found in metamorphic emery deposits and mica schists.

Margaritiferidae *Invertebrate Zoology.* a family of aquatic bivalve gastropod mollusks, in which some species are a source of freshwater pearls; found in the Northern Hemisphere.

Margarodidae *Invertebrate Zoology.* a family of homopteran insects, the giant scales, coccids, and ground pearls; the females secrete a pearly covering of waxy scales.

margarosanite *Mineralogy.* $Pb(Ca,Mn^{+2})_2Si_3O_9$, a colorless, transparent triclinic mineral having a specific gravity of 4.33 and a hardness of 2.5 to 3 on the Mohs scale; found in lamellar or columnar masses with franklinite at Franklin, New Jersey.

margin *Science.* **1.** the outer limit or edge of something, and sometimes the adjoining surface or area as well. **2.** a spare amount that is allowed for contingencies in an estimate or calculation (**margin of error**) or in an operation or design (**margin of safety**). *Graphic Arts.* any of the blank spaces set up as a border around the image area on a printed page. *Geography.* the boundary area around the edges of a body of water.

marginal *Science.* **1.** relating to or on the outer limit or edge of something. **2.** close to the lower or minimum limit.

marginal blight *Plant Pathology.* a disease of lettuce caused by the bacterium *Pseudomonas marginal,* characterized by the discoloration and rotting of leaf margins.

marginal checking *Electronics.* a preventive maintenance procedure that is used to detect circuit weaknesses and malfunctions by varying a circuitry's operating conditions. Also, **marginal testing.**

marginal chlorosis *Plant Pathology.* a virus disease that is common in peanut plants, characterized by yellow or pale leaf margins.

marginal dimensionality *Physics.* the largest number of spatial dimensions necessary to describe the nonlinear effects of the behavior of a substance near a critical point.

marginal escarpment *Geology.* the seaward slope of a marginal plateau, having a gradient around 1:10.

marginal fault see BOUNDARY FAULT.

marginal fissure *Geology.* a magma-filled rock fracture along the edge of an igneous intrusion.

marginal geosyncline see PARAGEOSYNCLINE, def. 2.

marginal placentation *Botany.* the development of ovules near the margins of carpels, as in the Leguminosae.

marginal plateau *Geology.* a relatively level, horizontal shelf whose topography is similar to, but deeper than, that of the continental shelf it borders.

marginal probability see UNCONDITIONAL PROBABILITY.

marginal relay *Electricity.* **1.** a relay with a small current difference between the on and off voltages. **2.** a relay that operates in response to prescribed changes in the value of the coil current or voltage.

marginal sea *Geography.* **1.** the ocean waters adjacent to a landmass, especially the territorial waters of a state. **2.** an ocean arm that is adjacent to and nearly enclosed by a landmass.

marginal sinus *Anatomy.* a set of discontinuous venous spaces around the edge of the placenta. Also, **marginal lakes.** *Developmental Biology.* bow-shaped lymph sinuses that separate the capsule from the cortical parenchyma of a lymph node, and from which lymph flows into the cortical sinuses.

marginal stability *Chaotic Dynamics.* a property of attractors that can be perturbed infinitesimally in some directions in phase space to other attractors.

marginal thrust *Geology.* one of a series of thrust faults along the edge of an igneous intrusion that crosses the intrusion and its wall rock. Also, **marginal upthrust.**

marginal trench *Geology.* see TRENCH, def. 3.

marginal ulcer *Medicine.* an ulcer that sometimes forms after gastric surgery in which two ends of intestine have been rejoined in a procedure called anastomosis. Also, ANASTOMOTIC ULCER.

marginal value theorem *Ecology.* a proposed decision-rule (derived from theoretical exploration) for a predator foraging from patches of prey which it depletes, stating that the predator should leave all patches at the same rate of prey extraction.

marginate *Botany.* having a distinct border.

margin line *Naval Architecture.* the deepest line at which a vessel may safely float, defined as 3 inches below the side of the bulwark deck.

margin of safety *Design Engineering.* a design criterion, usually the ratio between the load that would cause the failure of a component or assembly and the load that is imposed upon it during normal use.

margin trowel *Mechanical Devices.* a trowel with a flat, rectangular blade; used to apply and smooth plaster or other materials, especially in confined spaces.

marialite *Mineralogy.* $3NaAlSi_3O_8{\cdot}NaCl$, a colorless to white, gray, pink, or brown tetragonal mineral forming one end-member of the scapolite group, isomorphous with meionite, occurring in coarse, prismatic crystals and granular masses, and having a specific gravity of 2.5 to 2.62 and a hardness of 5.5 to 6 on the Mohs scale; found in regionally metamorphosed rocks and contact zones.

mariculture *Agriculture.* the use of the sea for farming or agricultural purposes, such as the growing of shrimp on a commercial basis. Also, OCEAN FARMING.

Marie's ataxia *Medicine.* any of several inherited diseases that cause lack of motor coordination and unsteadiness; they manifest late in life and progress slowly. (For Pierre *Marie,* 1853–1940, French physician.)

Marie's disease *Medicine.* a chronic condition producing enlargements of the head, facial features, hands, feet, and thorax.

marigold *Botany.* **1.** any of several composite plants, especially of the genus *Tagets,* characterized by golden flowers and strong-scented foliage, and yielding an oil that repels root parasites. **2.** any of several plants of the genus *Calendula,* the pot marigold.

marigold

marigram *Oceanography.* a graphic representation of the rise and fall of the tide in the form of a curve, the height represented by ordinates and the time represented by abscissas.

marigraph *Engineering.* an instrument that automatically records the rise and fall of the tide.

marijuana [mâr′ə wä′nə] *Botany.* the common name for herbaceous plants of the species *Cannabis sativa* in the family Cannabaceae, whose dried leaves and flowers, when smoked, produce a narcotic effect. Also, **marihuana.**

marina *Civil Engineering.* a waterway basin with moorings for small crafts and boats.

marine *Oceanography.* of or relating to the sea. *Military Science.* a member of a corps trained and equipped for service on land and sea, especially for amphibious operation.

marine abrasion *Geology.* the erosion of the seafloor by the movement of wave-agitated sediment. Also, WAVE EROSION.

marine arch see SEA ARCH.

marine biocycle *Ecology.* a division of the biosphere that includes all marine organisms.

marine biology *Biology.* a branch of biology that deals with the living organisms that inhabit the sea. Thus, **marine biologist.**

marine boiler *Naval Architecture.* a steam boiler specifically suited to marine engine use.

marine bridge see SEA ARCH.

marine cave see SEA CAVE.

marine climate *Meteorology.* a regional climate that is influenced primarily by the sea, as distinguished from a continental climate; it occurs where the prevailing winds blow onshore and is characterized by small daily and annual temperature ranges. Also, OCEANIC CLIMATE.

marine-cut platform *Geology.* a uniformly, gently sloping surface produced by water erosion or other marine processes. Also, **marine-cut terrace.**

marine cycle see SHORELINE CYCLE.

marine distiller *Naval Architecture.* a steam distiller designed to provide feed water for a steam engine.

marine engine *Naval Architecture.* an engine designed for use in propelling a vessel.

marine engineering *Engineering.* the branch of engineering concerned with the production of propelling machinery and auxiliary equipment for use on ships.

marine environment *Ecology.* the area that takes up about three-fourths of the earth's surface and includes its oceans and seas, all of which have a system of tides and waves, a high saline content, and a variety of life forms

marine forecast *Meteorology.* a weather forecast of particular interest to operators of maritime transportation or to coastal area residents, highlighting weather elements such as wind, visibility, storm warnings (when applicable), and general atmospheric weather condition.

marine geology see GEOLOGICAL OCEANOGRAPHY.

marine glue *Materials.* an adhesive used in ship carpentry, usually consisting of rubber, shellac, and pitch.

marine iguana *Vertebrate Zoology.* a large lizard, *Amblyrhynchus cristatus*, of the family Iguanidae, having a diet of seaweed and chacterized by a powerful tail used in swimming, salt-removing nasal glands, and a slowed heart rate for diving; found on the shores of the Galapagos Islands.

marine iguana

marine ivy *Botany.* the vine *Cissus incisa* of the grape family, having three leaflets or three-lobed leaves and black fruit; native to the southern U.S. and grown as a houseplant.

marine littoral faunal region *Ecology.* a division of the earth's surface that includes all marine animals.

marine marsh *Ecology.* a flat piece of land near the sea, usually covered with water at high tide.

marine meteorology *Meteorology.* the study of oceanic areas, including island and coastal regions, to serve the needs of air and surface navigation over the oceans.

marine microbiology *Microbiology.* the branch of microbiology dealing with sea microorganisms.

marine plain see PLAIN OF MARINE EROSION.

marine propeller *Naval Architecture.* a propeller that produces the thrust to drive a vessel through the water by giving momentum to a column of water that it displaces. Also, SCREW.

marine railway *Civil Engineering.* a track that slopes down into the water, along which a supportive cradle can move. A vessel is positioned above the cradle and hauled up the track out of the water to permit work on the underwater hull.

marine rainbow *Optics.* an arc of colors, formed by the reflection and refraction of sun rays, that appears in ocean spray. Also, SEA RAINBOW.

Mariner program *Space Technology.* a series of solar-powered probes of Venus, Mars, and Mercury that were launched by NASA between 1962 and 1973; provided the first close-up television pictures of another planet (Venus).

marine salina see SALINA.

Marinesco-Sjogren-Garland syndrome *Medicine.* a hereditary disorder of the brain and spinal cord, causing incoordination, nystagmus, and other signs of motor dysfunction.

Marinesian see BARTONIAN.

marine snow *Oceanography.* downward-drifting particles of living and dead organisms, including some inorganic material; sometimes suspended and concentrated at density boundaries such as thermoclines.

marine stack *Geology.* see STACK, def. 1.

marine swamp see SWAMP.

marine terminal *Civil Engineering.* the actual facility in a port that serves for loading and unloading ships.

marine terrace *Geology.* **1.** a narrow coastal strip covered by sand, silt, or fine gravel that slopes gently seaward. **2.** a wave-cut platform that is exposed by the lowering of the sea level or by uplift along the coast. **3.** a relatively flat, narrow surface dropping off to a steep embankment, formed along a seacoast by the merging of a wave-built terrace and a wave-cut platform. Also, SEA TERRACE, SHORE TERRACE.

marine traffic *Navigation.* all vessels in an area, especially a harbor, its approaches, or other restricted waters.

marine transgression see TRANSGRESSION.

marine weather observation *Meteorology.* a weather observation taken from a ship at sea in accordance with procedures established by the World Meteorological Organization.

marine zonation *Oceanography.* a system that divides the ocean into two levels and many subdivisions: the Benthic division, which includes the sea bottom, and the Pelagic division, which includes the open water.

Mariotte, Edmé 1620–1684, French physicist; studied the properties of fluids and gases, the barometer, color, and plant physiology.

marita *Invertebrate Zoology.* an adult trematode fluke.

maritime air *Meteorology.* an air mass over an extensive body of water, characterized by high moisture content in its lower levels.

maritime frequency bands *Telecommunications.* radio frequencies that are used for communication between ships, or between ships and coastal stations.

maritime law *Navigation.* the laws and regulations that apply at sea or afloat, including those bearing on the responsibilities of the masters and navigators of a vessel.

maritime mobile service *Telecommunications.* radio service that allows communication between ships, or between ships and coastal stations.

maritime polar air *Meteorology.* any polar air that becomes unstable, possessing a higher moisture content after passing over warmer water.

maritime position *Navigation.* the geographical location of a craft at sea.

maritime tropical air *Meteorology.* a very warm and humid air mass, typically occurring over tropical and subtropical oceans.

marjoram *Botany.* the common name for any of several plants of the genus *Origanum* of the family Labiatae, especially the species *O. majorana,* whose ground dried leaves are used as a food seasoning.

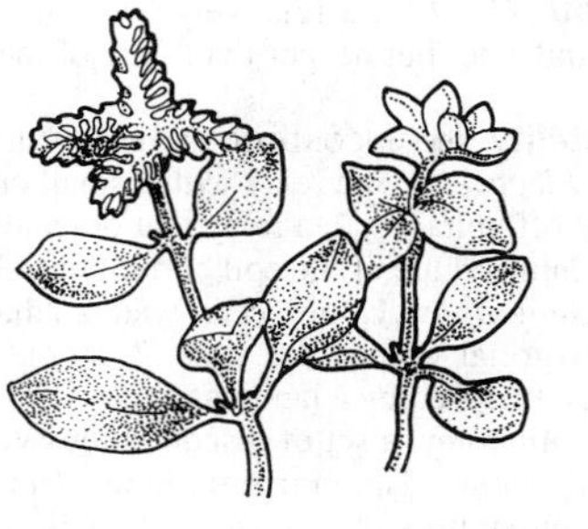

marjoram

marjoram oil *Materials.* a yellowish essential oil extracted from the sweet marjoram plant, *Majorana hortensis,* used in toilet soaps and flavorings and in salad oils.

mark a sign or symbol used to identify or designate something; specific uses include: *Computer Programming.* a symbol or physical signal delimiting the beginning or end of a unit of data, such as a field, word, data item, file, record, or block. *Geology.* see MOLD, def. 1. *Telecommunications.* **1.** in a telegraph receiver, the signal that closes the circuit or prints a character on a teletypewriter. **2.** any closed-circuit condition. **3.** a small piece of tape on the floor of a studio that indicates the positioning of performers, props, or scenery. *Ordnance.* a designation followed by a serial number that is used to identify models of military equipment; sometimes abbreviated by the letter M or Mk; e.g., U.S. Rifle Caliber .30, M1. *Military Science.* **1.** in artillery and naval gunfire support, a call for fire on a specified location to indicate targets or orient the observer; it may also refer to a report made by the observer. **2.** in close air support and air interdiction, a term used by the air control agency to indicate the point of weapon release.

Markarian galaxies *Astronomy.* galaxies having spectra with strong continuum radiation in the ultraviolet. (Discovered by the astronomer B.N. *Markarian.*)

mark detection *Computer Technology.* a class of character-recognition systems that obtain information in the form of marks placed in specific boxes or windows on paper or cards, such as the marks on computer-scored examinations.

marker a sign or device used to identify or designate something; specific uses include: *Military Science.* a visual or electronic aid used to mark a designated point; a variety of markers are used in mine warfare. *Genetics.* **1.** an allele that has a well-known phenotypic expression and locus; used as a reference point in mapping a new genetic mutant. **2.** a specific molecule of known size that is used to calibrate and compare movements of other molecules on electrophoretic gels. *Telecommunications.* a control circuit in a crossbar switching system that selects and electrically designates links connecting the stages of switches to carry out connections between lines, trunks, and service circuits.

marker beacon *Navigation.* an electronic aid to navigation that transmits a vertical fan- or bone-shaped pattern; used primarily in an Instrument Landing System.

marker bed *Geology.* **1.** a geologic formation having distinctive, easily recognizable characteristics that enable it to serve as a reference or datum, or to be tracked over long distances. Also, **marker, marker horizon. 2.** see KEY BED.

marker buoy *Navigation.* a buoy marking a specific location, such as a hazard, as distinguished from a buoy marking the edge of a channel.

marker pulse *Telecommunications.* **1.** in a teletypewriter, the signal interval that allows for the operation of the teletypewriter selector unit. **2.** a synchronizing pulse used in time-division communications systems.

marker rescue *Molecular Biology.* the restoration of normal gene function as a result of the replacement of a defective gene with a normal one by recombination.

market exchange *Anthropology.* the economic system in which the price or amount in an exchange is based on supply and demand involving unrelated parties. Also, COMMERCIAL EXCHANGE.

marketing *Industrial Engineering.* any or all of the activities involved in the transfer of finished goods from the producer to the ultimate buyer or consumer, including advertising, promotion, selling, and delivery.

marketing character type *Psychology.* a term for an individual who values or defines himself primarily in terms of his acceptance by others.

market research *Industrial Engineering.* the process of gathering and anlayzing information for the purpose of marketing a specific product, especially a new or proposed product.

mark-hold *Telecommunications.* in telegraphy, the transmission of a steady mark to show that a channel is open.

Mark-Houwink equation *Physical Chemistry.* the relationship between the intrinsic viscosity of a solution and the molecular weight of its linear polymers.

marking and spacing intervals *Telecommunications.* in transmission circuits, the intervals that correspond to the respective closed and open positions.

marking behavior *Behavior.* the use by animals of chemical secretions, songs, or visual displays in order to advertise a territory or social partner.

marking current *Electricity.* the magnitude and polarity of line current during the time that a receiving mechanism is operating.

marking-end distortion *Telecommunications.* the end distortion that serves to elongate a marking pulse.

marking fire *Military Science.* fire placed on a target for the purpose of identification.

marking gauge *Mechanical Devices.* a measuring device, consisting of a wood or steel beam with a sliding stock, that guides a steel pin a set distance from the edge of a workpiece; used to mark a line distance inside of and parallel to an edge.

marking pulse *Electricity.* the signal interval during which a teletypwriter selector unit is operating.

marking wave *Electricity.* the emission that occurs while active portions of telegraphic-code characters are being transmitted. Also, KEYING WAVE.

Markov, Andrei Andreyevich [mär´käf] 1856–1922, Russian mathematician; best known for his work on probability theory.

Markov chain *Mathematics.* a Markov process having a countable state space; that is, the ranges of all the random variables involved contain at most a countable number of values.

Markovian [mär kō´vē ən] relating to Andrei Markov or his theories.

Markovian process see MARKOV PROCESS.

Markov inequality *Statistics.* the statement that if X is a nonnegative random variable, then $\Pr(X \geq t) \leq E(X)/t$ for all $t > 0$.

Markovnikoff's rule *Organic Chemistry.* a rule stating that when an acid is added to the carbon-carbon double bond of an alkene, the hydrogen of the acid attaches itself to the carbon that already holds the greater number of hydrogens.

Markov process *Mathematics.* a stochastic process $\{X(t),\ t \geq 0\}$ with state space B, such that the probability that $X(t + \tau)$ occurs (i.e., has a value in B for $\tau \geq 0$), given that $X(s)$ has occurred for all $s \leq t$, is equal to the probability that $X(t + \tau)$ occurs given only that $X(s)$ has occurred. That is, $P(X(t + \tau) \in B \mid X(s),\ s \leq t) = P(X(t + \tau) \in B \mid X(t))$. Intuitively, in a Markov process, the probability of any future event is dependent only on the probability of the present event, and not on past events. If B is a discrete set (having finitely or at most countably many elements), the process is called a Markov chain. *Engineering.* the use of this process (i.e., the principle that the probability of an event is dependent only on the outcome of the last previous event) in various practical applications, as in analyzing the order or chaos of a system, creating traffic-generation models, and so on.

mark reading *Computer Technology.* a method of data input in which photoelectric means are used to detect marks made on preformatted documents.

mark sensing *Computer Technology.* a method of data input in which marks, usually made with a soft graphite pencil, are electronically read.

mark-space multiplier *Electronics.* a multiplier in which one input variable is used to control the mark-to-space ratio of a respective rectangular wave; its amplitude is made proportional to the other input variable, and the output is proportional to the average value of the signal. Also, TIME-DIVISION MULTIPLIER.

mark-to-space ratio *Electronics.* the ratio of the duration of a square wave's positive-amplitude part (represented by a mark) to its negative-amplitude part (represented by a space).

mark-to-space transition *Telecommunications.* the process of switching from a marking impulse to a spacing impulse.

marl *Geology.* a calcerous clay, or a mixture of clay and particles of calcite and dolomite, usually derived from shell fragments.

marlin *Vertebrate Zoology.* any of several species of oceanic sportfishes of the family Istiophoridae, having a long, spearlike snout and a streamlined body; found in warm to tropical waters worldwide.

marline *Naval Architecture.* a light, two-stranded tarred cord, laid-up left-handed and used in various shipboard work.

marlinespike *Naval Architecture.* a pointed tool for separating strands of rope; used in splicing lines. Also, **marlinspike, marlingspike.**

marlstone *Petrology.* **1.** an indurated rock similar to marl, with clay materials and calcium carbonate; considered to be an earthy or impure argillaceous limestone. Also, **marlite. 2.** a hard, ferruginous rock from England's Middle Lias.

marly *Geology.* of, related to, or resembling marl.

Marmara, Sea of *Geography.* a sea separating European and Asian Turkey; connected with the Aegean Sea via the Dardanelles and with the Black Sea via the Strait of Bosporus.

marmatite *Mineralogy.* a dark-brown to black ferroan variety of sphalerite.

Marmen effect *Materials Science.* the shape-memory of some alloys, in which the process of regaining the original shape is associated with the transformation of the deformed martensitic phase to the higher-temperature parent phase.

Marmor *Geology.* a North American geologic stage of the Middle Ordovician period, occurring after the Whiterock and before the Ashby.

marmoset

marmoset *Vertebrate Zoology.* an insectivorous, arboreal primate of the family Callithricidae, characterized by claws instead of nails on most digits; found in Central and South America.

marmot *Vertebrate Zoology.* a thick-bodied, burrowing rodent of the genus *Marmota* in the family Sciuridae, having coarse fur, a short tail, and tiny ears.

marquenching see MARTEMPERING.

marriage theorem *Mathematics.* let G be a bipartite graph with bipartition v_1, v_2. Then G has a perfect matching (marriage) if and only if $|v_1| = |v_2|$ and the v_1-deficiency def(G) satisfies def(G)0, where $|x|$ denotes the cardinality of the set X. Also, FROBENIUS' THEOREM.

married failure *Ordnance.* a moored naval mine lying on the seabed and still connected to its sinker due to failure of the release mechanism.

marrite *Mineralogy.* $PbAgAsS_3$, a metallic, gray monoclinic mineral occurring as equant to tabular crystals, having a specific gravity of 6.23 and a hardness of 3 on the Mohs scale; found in dolomite at Valais, Switzerland.

marrow see BONE MARROW.

marrubium *Botany.* a plant, *Marrubium vulgari,* of the family Labiatae.

Mars *Astronomy.* the fourth planet from the sun, having a diameter of 6787 kilometers, a density of 3.94, two moons, Phobos and Deimos, and a temperature that ranges from –191 to –24°F; it orbits the sun every 687 days at an average distance of 228 million kilometers. (Named for *Mars,* the Roman god of war, because of its reddish appearance; the color red was associated with war.)

Mars

Marsden chart *Meteorology.* a system for showing the distribution of meteorological data on a chart, using a Mercator map projection in which global latitudes between 80° north and 70° south are divided into squares of 10° latitude by 10° longitude; each square is numbered to indicate position, and can be subdivided into quarter-squares or into 100 one-degree subsquares numbered from 00 to 99, in order to provide a precise geographical position (to the nearest whole degree).

marsh *Ecology.* a tract of treeless wetland that supports a dense variety of vegetation, principally grasses.

marshalling *Military Science.* **1.** in an amphibious or airborne operation, the process by which participating units assemble, establish camps near embarkation points, complete readiness for combat, or prepare for loading. **2.** the process of assembling, holding, and organizing supplies or equipment, especially transportation vehicles, for onward movement.

Marshall line see ANDESITE LINE.

marsh gas *Geochemistry.* a colorless, odorless, tasteless gas (mainly methane, CH_4) that is produced by decaying vegetation and other organic materials in swamps and marshes.

marshite *Mineralogy.* CuI, a colorless to pale yellow to reddish-brown cubic mineral occurring as tetrahedral crystals, having a specific gravity of 5.68 and a hardness of 2.5 on the Mohs scale; found with cuprite, cerrussite, and iron at Broken Hill, Australia.

Marsh test *Analytical Chemistry.* a test used to determine the presence of arsenic in a sample by mixing the sample with zinc and adding a dilute acid; the gas produced is passed through a heated gas tube, and a black deposit forms if the test is positive. Also, **Marsh-Berzelius test.**

Marsileales *Botany.* a monofamilial order of ferns of the Pteropsida group; shallow-water plants having erect leaves on long stalks, creeping rhizomes, and sporangia enclosed in sporocarps.

Marsipobranchii *Vertebrate Zoology.* a former term for the class Cephalaspidomorphi, the living jawless fish.

Mars pigments *Inorganic Chemistry.* a group of pigments made by adding a suspension of calcium hydroxide, $Ca(OH)_2$, in water to ferrous sulfate, $FeSO_4 \cdot 7H_2O$, and heating the precipitate; depending on the temperature, the color may be yellow, orange, brown, red, or violet.

Mars program *Space Technology.* a series of Soviet Martian probes. Mars 6 (1973) was the first probe to enter the Martian atmosphere.

marstochron *Industrial Engineering.* a time-study device consisting of a motor-driven paper tape on which an observer places marks at the endpoints of motions; elapsed time is determined by measuring the distances between the marks.

Marsupalia *Vertebrate Zoology.* a superorder of mammals that includes all marsupials in the infraclass Metatheria.

marsupial [mär soo´pē əl] *Vertebrate Zoology.* **1.** any mammal belonging to the infraclass Metatheria, including kangaroos, opossums, and related animals that generally lack a placenta and have a pouch on the female abdomen in which to carry and nourish the young. **2.** of or relating to such mammals. (Derived from a Greek word meaning "bag" or "pouch.")

marsupial anteater see BANDED ANTEATER.

marsupial frog *Vertebrate Zoology.* a South American tree frog of the genus *Nototrema* in the family Hylidae, whose eggs are incubated in a large dorsal brood pouch in the female and develop into tadpoles or directly into young froglets.

marsupialization *Surgery.* the creation of a pouch, especially a procedure used to treat a large cyst when excision is not feasible; the opening is enlarged with an obturator that is gradually withdrawn as the cyst cavity grows smaller.

marsupial mole *Vertebrate Zoology.* a blind, burrowing, Australian mole of the family Notoryctidae that eats insects, earthworms, and plant tubers and is characterized by a two-chambered marsupium that opens posteriorly.

Marsupicarnivora *Vertebrate Zoology.* the order of mammals containing the carnivorous and omnivorous marsupials, including opposums, marsupial mice, Tasmanian devils, banded anteaters, and marsupial moles.

Marsupites *Paleontology.* an extinct genus of stemless crinoids belonging to the family Marsupitidae; distinguished by a large, globular calyx; widespread in the Upper Cretaceous.

marsupium [mär soo´pē əm] *Vertebrate Zoology.* **1.** an external pouch in marsupial mammals, containing the mammary gland and teats and formed by a fold of skin that receives and protects the newborn during late embryonic and postnatal development. **2.** an external pouch in certain vertebrates (such as fish and frogs) that houses fertilized eggs and developing young.

Marte *Ordnance.* an Italian heliborne radar-guided antiship missile.

Martel *Ordnance.* a French-British air-to-surface missile powered by a solid-propellant, boost-sustainer motor; it delivers a 331-pound (150-kg) conventional warhead at a speed approaching Mach 1 and a range of 18.6 to 37.2 miles, depending upon launch altitude.

Martello tower *Ordnance.* a circular fortification used for coastal defense. (From a famous fortification of this type in Corsica.)

martempering *Metallurgy.* the process of hardening an austenitized ferrous alloy by controlled quenching to a temperature above the M_s.

marten *Vertebrate Zoology.* a semiarboreal, weasel-like carnivore of the genus *Martes* in the family Mustelidae, having fine, soft, heavy fur that is often used commercially.

marten

martensite *Metallurgy.* a microstructural constituent produced by a diffusionless phase transformation; extensively induced in steel for hardening, it also occurs in nonferrous alloys such as some copper base.

martensite range *Metallurgy.* the temperature range between the point at which matersite begins to form and the point at which formation is complete.

martensitic steel *Metallurgy.* any steel that contains martensite.

martensitic structure *Metallurgy.* a metallic structure consisting of martensite.

martensitic transformation *Metallurgy.* the diffusionless transformation of a metallic structure to martensite.

Martian [mär′shən] *Astronomy.* **1.** relating to, located on, or originating from the planet Mars. **2.** a supposed being from the planet Mars.

Martian maria *Astronomy.* regions of dark, windblown dust on Mars. (So named because they were thought to resemble the lunar maria when viewed from the earth.)

Martian probe *Space Technology.* any unmanned spacecraft designed to explore and record conditions on or near Mars, as in NASA's Mariner and Viking programs and the Soviet Mars program.

Martin, Archer J. P. born 1910, British biochemist; shared the Nobel prize for developing the partition chromatography process.

martin *Vertebrate Zoology.* **1.** a small European swallow of the family Hirundinidae, characterized by bluish-black plumage with a white rump and belly and a forked tail. **2.** any of various related birds of North America, such as the widely distributed **purple martin,** *Progne subis.* (Associated with St. *Martin* of Tours, the patron saint of France.)

martingale *Statistics.* a sequence of random variables having finite expectation and the property that $E\{X_{n+1}/X_1, \ldots, X_n\} = X_n$.

martite *Mineralogy.* hematite, α-Fe_2O_3, occurring in iron-black octahedral and dodecahedral crystals, pseudomorphous after magnetite.

martonite *Materials.* a powerful type of tear gas composed of four parts bromoacetone and one part chloroacetone.

Marvin sunshine recorder *Meteorology.* a device that records the amount of sunshine during a given period based on the differential warming of air inside two bulbs, one of which is clear and the other painted black (thus absorbing more sunshine).

Marx circuit *Electricity.* an electric circuit of an impulse-type, high-voltage direct-current generator in which capacitors are used and charged in parallel through a high-resistance network. Also, **Marx generator.**

Marx effect *Solid-State Physics.* a regressive effect in which the energy of photons released in the photoelectric effect is reduced when additional light of lower frequency is incident on the illuminated surface.

Maryland *Agriculture.* a medium-sized breed of hog with black and white coloration, crossbred in the United States from the Landrace and Berkshire breeds.

mAs milliampere-second.

mascagnite *Mineralogy.* $(NH_4)_2SO_4$, a colorless to yellowish-gray orthorhombic mineral occurring as crusts and stalactite forms and in rare, prismatic crystals, having a specific gravity of 1.77 and a hardness of 2 to 2.5 on the Mohs scale; a sublimation product of volcanic activity.

maschaloncus *Oncology.* a tumor of the axilla.

Mascheroni, Lorenzo 1750–1800, Italian mathematician; studied geometrical constructions in which only the compass is used.

Mascheroni's constant see EULER'S CONSTANT.

Maschke's theorem *Mathematics.* Let $K(G)$ be the group algebra of a finite group G over a field K. (a) If K has characteristic 0, then $K(G)$ is semisimple. (b) If K has prime characteristic p, then $K(G)$ is semisimple if and only if p does not divide the order of G. That is, the group algebra of a finite group is semisimple if the characteristic of the field does not divide the order of the group.

mascon *Geology.* a large, high-density concentration of mass below the lunar surface.

masculine pelvis *Anatomy.* a female pelvis with a wedge-shaped inlet and narrowing of the anterior segment, resembling the male pelvis.

masculine protest *Psychology.* Alfred Adler's term for actions taken as a result of a distorted view that the male is the superior sex, in which a woman becomes aggressive and domineering, while a man exaggerates his maleness out of a desire to avoid femininity.

masculinize *Physiology.* to cause a female or a pubertal male to develop the secondary sexual characteristics of a mature male, as by using injections of androgens.

maser [māz′ər] *Electronics.* a device that can produce coherent electromagnetic radiation in the microwave frequency band; used in atomic clocks and to amplify weak radio signals. (An acronym for microwave amplification by stimulated emission of radiation.) Also, **MASER.**

maser amplifier *Electronics.* **1.** a low-noise oscillator that operates by induced emission. **2.** a maser that increases another maser's output.

maser source *Astrophysics.* a source of radio and microwave radiation that displays the typical characteristics, such as narrow and highly polarized spectral lines and a high brightness temperature, of radiation from a maser.

mask something that serves to cover or conceal; speciifc uses include: *Graphic Arts.* **1.** in photographing copy, for example in platemaking, an opaque material used to cover or screen a certain area. **2.** in color separation, a special filter used to correct a color imbalance. *Mechanical Devices.* a translucent or opaque screen used to cover part of the surface of a photograph or TV picture tube to conceal particular edges or modify the size or shape of an image. *Metallurgy.* a protective device used for selected coating. *Computer Technology.* a pattern of characters or bits used to control the retention or elimination of another pattern; for example, bits corresponding to a 1-bit in the mask are retained or considered. Also, EDIT MASK. *Electronics.* **1.** the viewing screen on an oscilloscope. **2.** to override one signal with a stronger one.

masked *Chemistry.* describing an element or radical combined in a compound so that its usual properties are hidden or subdued; it is prevented by a preliminary reaction from giving its usual reaction.

masked mRNA *Genetics.* quantities of mRNA that are present in a cell but are inactivated and are therefore neither digested nor transcribed.

masked search *Computer Technology.* a search in which only the bits enabled by a mask are considered for matching.

mask face *Medicine.* the expressionless, immobile facial aspect often seen in diseases of the nervous system, such as Parkinson's disease.

masking a process of covering or concealing; specific uses include: *Engineering.* the process of covering part of an object or opening, often to prevent or limit it from being covered by a substance such as paint. *Graphic Arts.* **1.** the process of covering or screening all or part of an illustration or other copy. **2.** the process of correcting color imbalance through the use of masks. *Acoustics.* a phenomenon in which the threshold level of audibility for a sound is raised by the presence of another sound in close proximity. *Military Science.* in electronic warfare, the use of additional transmitters to hide the source or purpose of a particular electromagnetic radiation. *Computer Programming.* the process of extracting desired information from a group of characters or bits while suppressing the undesired information.

masking agent *Chemistry.* a substance that hides or subdues the properties of another substance.
mask matching *Computer Programming.* a method used in character recognition that attempts to match or correlate a specimen character to each set of masks representing known characters.
mask out *Computer Science.* **1.** to remove unwanted data by performing an AND operation with a mask. **2.** to turn off an interrupt by clearing its corresponding bit in the interrupt mask register.
mask register *Computer Technology.* **1.** a register used to contain a mask to be used, e.g., in searching for data. **2.** see INTERRUPT MASK.
mask word *Computer Programming.* a word that modifies the identifier word and the input word used in a search comparison in a logical AND operation.
Maslow, Abraham 1908–1970, American psychologist; pioneer in humanist psychology; formulated the concept of self-actualization.
Maslow's hierarchy of (human) needs see HIERARCHY OF NEEDS.
masochism [mas´ə kiz´əm] *Psychology.* **1.** a sexual deviation in which a person derives gratification from suffering pain, abuse, and humiliation. **2.** any tendency to direct pain or humiliation against oneself or to seek or welcome suffering. (From Leopold von Sacher-*Masoch,* a 19th-century Austrian novelist who portrayed this condition.)
masochist [mas´ə kist] *Psychology.* a person affected by masochism; someone who derives sexual pleasure from his or her own pain or suffering. Thus, **masochistic.**
mason *Civil Engineering.* a person whose work or profession is building with brick, stone, tile, or the like.
masonry *Civil Engineering.* a general term for stone or stonework of any type, usually cast or formed, including ceramic brick, tile, concrete, glass, mud, adobe, and the like.
masonry-bonded hollow floor *Building Engineering.* a type of hollow wall in which the face brick is separated from the backing by an air space bridged by headers that bond the two surfaces together.
masonry cement *Materials Science.* any of several special types of cement that are more workable and plastic than Portland cement and are used especially in masonry; e.g., Portland cement mixed with hydrated lime, crushed limestone, or granulated slag.
masonry dam *Civil Engineering.* a dam constructed of brick or stone masonry laid and fitted with mortar.
masonry drill *Mechanical Devices.* a drill with a hard, tungsten-carbide tip brazed with brass or copper; used to drill holes in brick, stone, concrete, and ceramic tile.
masonry nail *Mechanical Devices.* a round, square or fluted nail used especially for fastening window frames to masonry joints and concrete.
masonry saw *Mechanical Devices.* a saw with carbide-tipped teeth for sawing through brick, stone, and cement.
mason's hammer *Mechanical Devices.* a hammer used in masonry work, consisting of one hammering face and one chiseling face for cutting bricks and stone. Also, BRICKLAYER'S HAMMER.
mason's putty *Mechanical Devices.* a mixture of lime putty and Portland cement, used for making fine joints.
Mason's theorem *Control Systems.* a formula that defines the overall transmittance of a signal flow graph in terms of the transmittances of the paths shown on the graph.
Mas. pil. pill mass (From Latin *massa pilularum.*)
masquerading *Computer Programming.* the use of another person's account, e.g., to allow access to data or programs or to send a message under the other person's name.
mass *Mechanics.* a fundamental property of an object, which makes it resist acceleration, and which determines its gravitational attraction. This property can be generally regarded as equivalent to the amount of matter in the object. *Mass* and *weight* are not synonymous; mass is not affected by the forces acting on an object, but weight is a relative property that can change according to the gravitational force exerted on the object. *Metrology.* a measurement of this property in an actual object, as compared to a reference sample used in the SI system (a cylinder of platinum-iridium alloy). *Military Science.* **1.** the concentration of combat power at a given location. **2.** a military formation with units spaced at less than the normal distances and intervals.
mass absorption coefficient *Physics.* a quantity given by the linear absorption coefficient divided by the density of the medium through which radiation is passing. This is denoted by μ_m in the following relationship: $I = I_0 \exp(-\mu x) = I_0 \exp(-\mu_m \rho x)$ where I is the intensity of the transmitted X-ray beam, I_0 is the intensity of the incoming beam, μ is the linear absorption coefficient, μ_m is the mass absorption coefficient, ρ is the density, and x is the thickness of the material.
mass action law *Physical Chemistry.* a rule stating that in a chemical system at a constant temperature, the rate of chemical reaction is directly proportional to the concentration of the reacting substances. Also, GULDBERG AND WAAGE LAW; LAW OF MASS ACTION.
massage *Medicine.* a systematic therapeutic friction, stroking, and kneading of the body. *Computer Programming.* to manipulate input data to produce a desired format.
mass-analyzed ion kinetic spectrometry *Spectroscopy.* a form of ion kinetic spectrometry consisting of the analysis of both the mass and energy of ionic products.
Massariaceae *Mycology.* a family of fungi belonging to the order Melanommatales which live off decaying woody plants, bark, or dead roots; found primarily in temperate regions.
Massarinaceae *Mycology.* a family of fungi belonging to the order Pleosporales which live off decaying plant material; found primarily in northern temperate regions.
massasauga *Vertebrate Zoology.* the common name for *Sistrurus catenatus*, a small rattlesnake of the family Viperidae that feeds on mice and amphibians; found in moist areas from New York state to Texas and into Mexico.
mass attraction vertical *Geophysics.* the vertical direction defined solely by the force of gravity at a location, apart from any secondary forces due to the moving earth.
mass budget or **mass balance** *Engineering.* see BALANCE.
mass casualties *Military Science.* a high number of casualties incurred in a relatively short period of time, so as to overwhelm the available medical and logistic support capabilities.
mass communication *Telecommunications.* any form of communication that reaches a large section of the population, such as TV or radio.
mass defect *Nuclear Physics.* the difference between the mass of an atom and the sum of the masses of its individual components; used to calculate the binding energy in a nucleus.
mass discrepancy *Astronomy.* the apparent difference between the mass of a galaxy as measured by summing all its visible components and the mass as measured by the galaxy's motions; the latter is ten times greater than the former.
mass-distance *Transportation Engineering.* the bulk (mass) hauled by a vehicle times the distance traveled.
mass divergence *Fluid Mechanics.* a measure of the mass flux out of a unit volume in a fluid medium, given by the divergence of the quantity rv, where r is the fluid density and v is the velocity vector.
massed fire see CONCENTRATED FIRE.
mass-energy conservation *Physics.* a fundamental principle stating that energy cannot be created or destroyed, and that mass carries an inherent energy equal to the mass times the square of the speed of light in a vacuum.
mass-energy equivalence *Nuclear Physics.* the principle of interconversion of mass and energy given in the equation $E = mc^2$, where E is the energy in ergs, m is the mass in grams, and c is the velocity of light in centimeters per second, stating that the total mass-energy before a reaction is equal to the total mass-energy after the reaction, or that mass can be converted into energy and vice versa.
mass-energy relation *Nuclear Physics.* an equation that comes from the theory of relativity, stating that $E = mc^2$, where E is the energy associated with the mass m and c is the speed of light in a vacuum.
mass erosion *Geology.* all processes whereby soil and rock materials fall and are carried together downslope by the direct application of gravitational body stresses. Also, GRAVITY EROSION.
masseter *Anatomy.* the short, thick masticatory muscle that lifts the mandible, closing the jaw.
mass extinction see FAUNAL EXTINCTION.
Massey formula *Atomic Physics.* a relation that gives the probability for a secondary electron to be emitted from an excited atom at the surface of a metal.
mass flow *Fluid Mechanics.* the mass of a fluid substance that passes a specified unit area in a unit amount of time.
mass-flow bin *Engineering.* a steep-walled bin designed so that the withdrawal of any solid causes the flow of the entire solid in the bin with no stagnant regions.
mass flowmeter *Engineering.* an instrument that measures the mass of fluid flowing through a pipe per unit of time.
mass flow sensor *Biotechnology.* a sensor used to measure fluid flows; one highly effective type uses the scatter of ultrasonic radiation by the fluid to provide a sensor that has no contact with the fluid being measured.

mass formula *Nuclear Physics.* an equation that gives the atomic mass of a nuclide as a function of its mass number and the atomic mass unit; defined to be 1/12 the mass of the carbon-12 nucleus.

mass-haul curve *Civil Engineering.* in roadway construction, a graph used to calculate the amount of excavation in a cutting that is available for removal and use as fill.

mass heaving *Geology.* the general expansion of the ground as a result of freezing.

mass hysteria *Psychology.* a condition in which a large group of people exhibit excitement or anxiety, irrational behavior, or unexplained symptoms of illness.

massicot *Inorganic Chemistry.* lead monoxide, PbO, in the form of a yellow powder, formed by the oxidation of a bath of metallic lead in such a way that the oxide is not melted. *Mineralogy.* this substance in the form of a yellow orthorhombic mineral dimorphous with litharge, having a specific gravity of 9.56 to 9.64 and a hardness of 2 on the Mohs scale; found in earthy masses as an oxidation product of primary lead minerals.

Massieu function *Thermodynamics.* a function given by the equation $J = -A/T = -U/T + S$, where A is the Helmholtz free energy, T is the absolute temperature, and S is the entropy.

massif *Geology.* a large block of rock or other structural feature within an orogenic belt that is usually older and more rigid than the surrounding rocks.

massive *Mineralogy.* **1.** of or relating to the crystal habit or appearance of a mineral. **2.** specifically, describing a mineral that appears to be physically isotropic and lacking in structure, i.e., that occurs as compact material without form or distinguishing features. *Petrology.* **1.** describing a metamorphic rock with constituents that do not feature parallel orientation and are not arranged in layers. **2.** describing an igneous rock with a homogeneous texture over a wide area without layering, foliation, cleavage, or similar features. *Geology.* of or relating to a mineral deposit having a large concentration of ore in one area. *Paleontology.* in reference to corals and archaeocyathids, the relatively compact, solid corallum formed by some colonial groups of animals, as distinguished from a branching or chain-shaped form; in vertebrates, often refers to thick bone formation, as in the cephalic boss of elephants and pachycephalosaurs.

massively parallel *Computer Science.* describing a supercomputer system that makes use of a number of different processors to create a large pool of available memory, so that thousands of separate operations can be carried out simultaneously. Thus, **massively parallel computer, massive parallelism.**

massive retaliation *Military Science.* a military counterattack strategy involving the use of nuclear weapons.

mass law of sound insulation *Engineering.* a law stating that the sound-insulating quality of a partition is proportional to its weight; a doubling of weight will reduce transmitted sound by about five decibels.

massless *Physics.* describing an elementary, such as a photon, that has zero rest mass.

mass loss *Astrophysics.* the outflow of particles and gas from a star, occurring at varying rates throughout its lifetime.

mass-luminosity relation *Astrophysics.* the semiempirical equation that specifies how bright a star of a given mass will be if it lies on the main sequence of the Hertzsprung-Russell diagram.

mass-market paperback *Graphic Arts.* a paperbound book printed in a small, standard-sized format and designed to be sold in bookstores and in a variety of other retail outlets, such as supermarkets, discount stores, and newsstands; formerly called a **pocketbook**.

mass-memory unit *Computer Technology.* a peripheral device used to store large amounts of data, such as magnetic tape or disk units.

mass movement *Geology.* any movement of a section of the land surface as a complete unit, especially under the influence of gravity.

mass number *Nuclear Physics.* the total number, A, of protons and neutrons in the nucleus of an atom; written as a superscript before the elemental symbol; e.g., ^{238}U. Also, NUCLEAR NUMBER.

mass operator *Quantum Mechanics.* an operator added to the Lagrangian of a quantized field, representing the contribution to a particle's mass due to self-interaction.

mass polymerization *Materials Science.* a poly(vinyl chloride) production process in which only monomer and initiator are present. The monomer undergoes about 12% conversion in one reactor, and then is transferred to a second for completion. The grains lack the pericellular membrane of suspension resins, but generally mass and suspension resins can be used in the same ways.

mass production *Industrial Engineering.* a type of technology in which identical products are manufactured in very large quantities.

mass ratio *Space Technology.* **1.** the ratio of the mass of a rocket's propellant charge to the total mass of the rocket when charged with the propellant. Also, PAYLOAD-MASS RATIO. **2.** less commonly, the ratio of a rocket's takeoff mass to its remaining mass at burnout. (Using the first method, a 4-ton rocket with a 1-ton payload and 2 tons of propellant would have a mass ratio of 2:7; using the second method, it would have a mass ratio of 7:5.)

mass reflex *Medicine.* in spinal cord injury, a reflex exhibited by the entire area controlled by the portion of the spinal cord that has been injured; results include flexor muscle spasms, incontinence, priapism, hypertension, and profuse sweating.

mass renormalization *Quantum Mechanics.* the process of including a mass operator in a quantized field theory in order to cancel infinities.

mass resistivity *Electricity.* the product of the volume resistivity and the density of a material at a specified temperature.

mass shift *Nuclear Physics.* the degree of displacement that arises from the difference in mass of individual isotopes.

mass spectrograph *Spectroscopy.* a mass spectroscope in which the ions fall on a photographic plate, thus showing the particle distribution.

mass spectrometer *Spectroscopy.* an instrument for producing ions in a gas and analyzing them according to their charge/mass ratio, thus providing a mass spectrum of the material.

mass spectrometry *Analytical Chemistry.* a method of identifying a material by vaporizing it with an electric discharge and exposing the gas to a beam of electrons; the ions produced are then separated, as by deflection through a magnetic field, according to their mass-to-charge ratios for identification by an ion detector.

mass spectroscope *Spectroscopy.* an instrument used to measure the masses of atoms by sending a beam of ions through electric and magnetic fields, where they are deflected according to mass.

mass spectrum *Particle Physics.* a plot of the mass-to-charge ratio of elementary particles, sorted by their isotopic mass.

mass stopping power *Nucleonics.* the decrease in kinetic energy of an ionizing particle traversing matter per unit of surface density.

mass storage see AUXILIARY STORAGE.

mass-to-charge ratio *Analytical Chemistry.* the ratio, m/z, of the mass of an ion being analyzed to its ionic charge; used in mass spectroscopy.

mass transfer *Astrophysics.* the process by which one evolved member of a binary star system passes gaseous material to its companion star.

mass-transfer rate *Physics.* the amount of matter that is passing across a specified area or region per unit time.

mass transport *Fluid Mechanics.* the motion of a given amount of fluid or an amount of material carried by a fluid or transported by eddy or molecular diffusion from one point to another.

mass velocity *Fluid Mechanics.* a quantity given by the mass flow rate of a fluid in a pipe divided by the cross-sectional area of the pipe.

mass wasting *Geology.* a general term for the process whereby loosened soil and rock material are transported downslope under the direct influence of gravity.

mast *Naval Architecture.* **1.** a pole rising from the hull of a vessel used to support a sail or sails. **2.** a pole or tower rising from the hull of a vessel, supporting halyards, electronic equipment, observation positions, and so forth. **3.** a disciplinary or other formal administrative assembly aboard a naval ship. *Engineering.* **1.** a long, thin upright pole held in place by guy lines. **2.** a pole that functions as an antenna or antenna support. **3.** a single pole used as a drill derrick.

mast *Botany.* the fruit of forest trees such as oak and beech used as fodder for hogs and other animals.

mast- a combining form meaning "breast," as in *mastalgia*.

Mastacembeloidea *Vertebrate Zoology.* a suborder of perciform, eel-like, freshwater fish found in tropical South Africa and southern Asia.

mastadenoma *Medicine.* a benign tumor of the breast.

Mastadenovirus *Virology.* a genus of viruses of the family Adenoviridae that infect mammals and cause a wide range of diseases and infections in humans.

mastalgia *Medicine.* any breast pain.

mastatrophy *Medicine.* the atrophy of the mammary gland.

mastax *Invertebrate Zoology.* the muscular pharynx in rotifers, bearing the masticatory trophi.

mast cell *Histology.* a basophilic cell found in connective tissue, containing numerous cytoplasmic granules; it degranulates during anaphylactic shock and releases histamine and other substances.

mastectomy [mas tek´tə mē] *Surgery.* the surgical removal of a breast. Also, **mammectomy.**

master *Acoustical Engineering.* a high-quality initial or original copy of a tape, record, or disk recording that serves as the template for subsequent copies of this recording. Also, MASTER RECORD. *Engineering.* **1.** a device that regulates subordinate devices. **2.** the key member of a system, such as the master cylinder in an automotive braking system. *Navigation.* the captain of a ship.

master alloy *Metallurgy.* a generic name for any of several alloys that are melted into a base to increase the contents of certain alloying elements.

master antenna television system *Telecommunications.* an antenna configuration that services a group of television receivers.

master arm *Engineering.* a component that manipulates the motions of another component, called the slave arm, which copies every action of the master arm except for occasional variations in scale or force.

master batch *Materials Science.* a prepared mixture having a high percentage of an additive that is essential to a product being manufactured and is added in uniform small amounts to the product during the compounding operation; used in various manufacturing operations, such as adding color pigments to plastics, rubber, or paints.

masterbatching *Materials Science.* the process of adding a master batch to a product during compounding.

master clock *Computer Technology.* a control device that produces timing signals, often called clock pulses, to synchronize the operation of a computer. Also, MASTER TIMER.

master compass *Navigation.* the part of a remote-indicating compass system that determines direction and transmits it to the repeaters.

master control *Telecommunications.* the control center of any broadcasting facility. *Computer Programming.* the part of the operating system responsible for controlling computer-hardware devices.

master control interrupt *Computer Technology.* a signal generated by an input/output device, the processor, or the operator that indicates a request for a memory access to read or write more data or program segments.

master copy *Graphic Arts.* **1.** the most up-to-date or most correct version of a manuscript or document, considered the official text from which other copies are made. **2.** see MASTER PROOF.

master cylinder *Mechanical Engineering.* **1.** the hydraulic pump in an automotive braking system that is put into motion by the brake pedal, and then supplies hydraulic fluid to the brakes for each wheel. **2.** the container that holds the fluid to operate the mechanism of hydraulic brakes or the clutch.

master data *Computer Programming.* a set of data that is seldom altered and supplies basic data for processing operations.

master drawing *Graphic Arts.* the drawing, often an original, against which all subsequent copies or printed reproductions are judged and proofed.

master equation *Atomic Physics.* an equation that establishes the rate of change of a population at a specified energy level in terms of other energy levels and transition probabilities.

master file *Computer Programming.* **1.** a file containing relatively permanent information that is periodically updated, thus serving as a main reference file of information. **2.** the most recently updated file to be used as the basis for the next update run.

master gain control *Electronics.* **1.** the control of gain over an entire amplifying system rather than varying the gain of individual inputs. **2.** a control on an amplifier that can adjust the gain of two or more channels simultaneously.

master gauge *Mechanical Devices.* **1.** a fine measuring device used as a gauge standard with fixed hole locations to which other similar parts are measured for precision or part positions. **2.** a wrought pipe joint made from an enlarged pipe end to receive a male or spigot end of an adjacent pipe length.

master gene *Genetics.* a gene that controls other genes, particularly in a cell of specialized function.

master glass negative see CALIBRATION PLATE.

master group *Telecommunications.* in a telephone system, 600 voice channels that are multiplexed together and treated as a single unit.

master gyro compass *Navigation.* a master compass in a gyro system.

master instruction tape see MASTER PROGRAM FILE.

master joint *Geology.* a large and persistent plane of division that passes with regularity and parallelism through a number of beds. Also, MAJOR JOINT, MAIN JOINT.

master layout *Design Engineering.* a master template used as a permanent standard for other templates.

master mechanic *Engineering.* a thoroughly skilled supervisory mechanic who is in charge of the maintenance and installation of equipment.

master mode see SUPERVISOR MODE.

master multivibrator *Electronics.* a master oscillator that uses a multivibrator unit.

master oscillator *Electronics.* an oscillator used to establish the carrier frequency of the output of a transmitter or an amplifier.

master oscillator-power amplifier *Electronics.* a transmitter or signal generator using an oscillator followed by one or more stages of radio-frequency amplification.

master phonograph record *Acoustical Engineering.* a metal negative counterpart produced by electroforming from a disk recording cut from the master tape recording; used as the original template for the subsequent production of record copy templates. Also, MOTHERS.

master plan-position indicator *Electronics.* a radar-system plan-position indicator that controls remote indicators or repeaters.

master production schedule *Industrial Engineering.* a statement of the number of items to be produced and the time at which their production is to be completed.

master program file *Computer Programming.* a magnetic tape containing all the programs needed for a particular production run or related runs, usually kept aside as a backup to the working program file. Also, MASTER INSTRUCTION TAPE.

master proof *Graphic Arts.* a page proof that is the official record of corrections and alterations to be made by the compositor or printer.

master record *Computer Programming.* a record in a master file that may be updated by a change record during the next file update. *Acoustical Engineering.* see MASTER.

master scheduler see OPERATING SYSTEM.

master-slave hypothesis *Genetics.* a hypothesis to explain the large differences in DNA content among varius eukaryotes in which it is proposed that each gene may be represented 100 to 100,000 times in a cluster; one copy is the master gene and the tandem duplicates are slave genes that can mutate independently, but fold back upon the master, match their base sequences, and are corrected if incongruities are present.

master-slave manipulator *Engineering.* any of various devices that mimic the motions and sensitivity of the human hand; used by an operator to handle remote or hazardous objects.

master-slave system *Computer Technology.* a system or computer configuration in which one computer (the master) controls all input/output and scheduling functions and transmits tasks to a slave computer.

master station *Navigation.* the main radio-transmitting station in a system of two or more synchronized stations used for position finding.

master stream see MAIN STREAM.

master switch *Electricity.* a switch that controls the operation of devices such as contactors and relays.

master synchronization pulse *Telecommunications.* in telemetry, a pulse that differs from other telemetering pulses in its amplitude and duration; used to indicate the end of a sequence of pulses.

master system tape *Computer Technology.* a tape containing operating-system programs, usually kept aside as a backup to the working-system tape or disk.

master tape *Computer Technology.* a tape containing data that is not to be overwritten, such as a tape containing a master file.

master terminal *Computer Technology.* a terminal within a network that can communicate with all the other terminals.

master timer see MASTER CLOCK.

mastery motive *Psychology.* a tendency to respond to frustration or failure with renewed effort. Also, **mastery orientation.**

masthead *Graphic Arts.* a block of type on the editorial page of a newspaper or the contents page of a periodical that lists the publication's name and address, usually with other information such as the editorial staff, publisher, date of publication, and so on.

masthead bombing *Military Science.* the process of dropping bombs from a very low altitude, especially as used against ships.

mastic *Botany.* a tree or shrub of Mediterranean regions, *Pistacia lentiscus. Materials.* **1.** a yellowish, aromatic resin obtained from this tree, used as a chewing gum base and in making varnishes and adhesives; formerly used as a dental cement and pharmaceutical coating. **2.** any of various pasty adhesives or sealing preparations that are similar to natural mastic.

mastic asphalt *Materials.* a mixture of asphalt with stone chips, sand, or other material; used in paving and roofing.
masticate *Physiology.* to chew food in preparation for swallowing and digestion. *Chemical Engineering.* to make rubber softer and more plastic by the chemical addition of oxygen from the surrounding air.
mastication *Physiology.* the process of chewing food. *Materials Science.* a process by which a substance, such as crude natural rubber, is permanently softened by severe mechanical working; usually carried out on roll mills or in internal mixers.
masticatory *Pharmacology.* any remedy that is to be chewed but not swallowed.
mastic oil *Materials.* a colorless essential oil that is used in medicine.
Mastigamoebidae *Invertebrate Zoology.* a family of amoeboid protozoans with one or two flagella.
Mastigomycotina *Mycology.* a subdivision of fungi belonging to the division Eumycota, characterized by spores having flagella.
mastigonemes *Invertebrate Zoology.* the delicate branching outgrowths along the flagellum of certain flagellates.
Mastigophora *Invertebrate Zoology.* a group of protozoans characterized by flagella used chiefly for locomotion; includes free-living and parasitic, marine and freshwater forms.
Mastigophoraceae *Botany.* a monogeneric family of oceanic, large, rigid, brownish and reddish liverworts of the order Jungermanniales, distinguished by a rudimentary perianth and exclusively terminal branching.
mastitis *Medicine.* an inflammation of the mammary glands.
mast-mounted sight *Ordnance.* a sensor system mounted on a scout helicopter that finds and designates targets day or night in all kinds of weather while the helicopter hovers undetected behind cover.

mast-mounted sight

masto- *Anatomy.* a combining form meaning "breast," as in *mastopathy.*
mastocarcinoma *Medicine.* a malignant tumor of the breast.
mastocytoma *Veterinary Medicine.* a tumor consisting of mast cells that involves the skin, subcutaneous tissue, and sometimes the muscle; common in dogs, but also affects cats, oxen, mice, and humans.
mastocytosis *Medicine.* **1.** a group of rare diseases characterized by the invasion of mast cells into tissue. **2.** an elevation in the number of mast cells in the bone marrow.
mastodon *Paleontology.* an extinct family of proboscidean mammal, shorter and stockier than elephants and characterized by long jaws, tusks in both jaws, and low-crowned teeth; extant in the Miocene to Holocene, becoming extinct around 7000 years ago.
Mastodontidae *Paleontology.* the name formerly given to the mastodons, now generally grouped in the family Gomphotheriidae.
mastodynia *Medicine.* any mammary pain. Also, MASTALGIA.
mastoid *Anatomy.* **1.** resembling a breast. **2.** of or related to the mastoid process of the temporal bone. **3.** see MASTOID PROCESS.
mastoid antrum *Anatomy.* a space in the mastoid part of the temporal bone, connecting with the mastoid cells posteriorly and connecting anteriorly with the middle ear.
mastoid canaliculus *Anatomy.* a small channel that carries the auricular branch of the vagus laterally through the mastoid process.
mastoid cell *Anatomy.* any of several small, interconnecting openings in the mastoid process of the temporal bone.
mastoid foramen *Anatomy.* an opening at the dorsal end of the mastoid process of the temporal bone through which a vein and an artery pass.
mastoid fossa *Anatomy.* a small, triangle-shaped depression between the external acoustic meatus and the posterior area of the zygomatic process of the temporal bone. Also, SUPRAMEATAL TRIANGLE.
mastoiditis *Medicine.* an infection in the mastoid process of the temporal bone, spreading from an acute middle ear infection and resulting in bone inflammation in severe cases.
mastoid process *Anatomy.* a cone-shaped extremity of the temporal bone, extending forward and downward just posterior to the external acoustic opening. Also, MASTOID.
mastopathy *Medicine.* any disease of the mammary gland.
mastopexy *Medicine.* the surgery performed to correct a pendulous breast.
mastoplasia *Medicine.* the production of breast tissue. Also, MAMMOPLASIA.
mastoplasty *Medicine.* the surgical repair of the breast. Also, MAMMOPLASTY.
mastoptosis *Medicine.* a sagging of the breast.
mastorrhagia *Medicine.* a hemorrhage from the mammary gland.
mastoscirrhus *Medicine.* a hardening of the breast tissue, commonly in breast cancer.
Mastotermitidae *Invertebrate Zoology.* a family of termites in the order Isoptera, with a single species found in tropical Australia.
mastotomy *Medicine.* a surgical incision of the breast.
mast step *Naval Architecture.* the supporting foundation of a mast.
Masurca *Ordnance.* a French ship-based, surface-to-air missile powered by tandem two-stage, solid-propellant motors and equipped with beam-riding guidance (Model 2) or semiactive radar homing (Model 3); it delivers a 105-pound (48 kg) conventional warhead at up to 31 miles (50 km).
mat *Civil Engineering.* the surface of steel or concrete positioned below a post. *Graphic Arts.* a type matrix. *Mining Engineering.* an accumulation of debris, such as broken mine timbers, rock, and earth, that is a byproduct of the caving system of mining.
Matador *Ordnance.* see MACE.
Matanuska wind *Meteorology.* a strong and gusty northeasterly wind that blows during the winter months near Palmer, Alaska.
match to compare two or more items to check for identity or similarity; specific uses include: *Computer Programming.* **1.** a data processing operation in which lists of items are matched against one another on the basis of some key. **2.** a comparison of keys or records that are identical. *Immunology.* to select compatible donors and recipients for transplantation or transfusion. (From an earlier word meaning "mate.")
match *Materials.* **1.** a short piece of wood, paper, or plastic whose tip is covered with a material that ignites through friction. **2.** any of various other igniting devices, such as a charge of gunpowder wrapped in paper. (From an earlier word for the wick of a candle.)
matched-die molding *Materials Science.* the process of forming shaped articles of reinforced plastics by pressing preforms between matching male and female sections.
matched edges *Engineering.* in machining equipment, die face edges that are machined at right angles in order to provide die alignment.
matched filter *Electronics.* a filter designed to present a matched load to the source connected at its input and to provide an output impedance so that the load connected to its output will be matched as well.
matched impedance *Electricity.* a condition in which the impedance of two interconnected networks, components, or transmission lines is the same when measured from the interconnection nodes; e.g., the output impedance of an audio amplifier and the impedance of an audio speaker. Matched impedance allows maximum power transfer from one network to another and prevents transmission-line reflections.
matched load *Electronics.* a load whose impedance value results in the maximum absorption of energy from the signal source.
matched-metal molding *Materials Science.* the use of two close-fitting metal molds mounted in a hydraulic press to form reinforced-plastic articles.

matched pairs *Statistics.* units in control and experimental groups in which each unit is paired with one that closely matches it in terms of potential extraneous influences; one member of each pair is assigned to the experimental group, and the other member to the control group. Also, **matched-pairs sample.**

matched pulse intercepting *Telecommunications.* a method used to intercept calls on party lines in a terminal-per-line office; a ground pulse is matched in time with the intercepted station's particular ringing frequency.

matched termination *Electricity.* a termination that does not cause wave reflection from a waveguide or transmission line; its impedance is matched to the characteristic impedance of the line.

matched terrace see PAIRED TERRACE.

matched transmission line *Electricity.* a transmission line in which the connections between system components are matched to avoid signal reflection within the line.

matching a process of comparing two or more items to check for identity or similarity; specific uses include: *Computer Programming.* a technique that is used to verify coding, in which actual specified codes are machine-compared against a group of master codes. *Electricity.* the process of connecting networks so that their impedances are matched. *Navigation.* the superimposition of signal traces on a cathode-ray tube as part of position finding with the now discontinued loran-A. *Mathematics.* a set M of edges in a graph G such that no two edges of M are incident with the same vertex of G. If in addition, the edges in M are incident with all of the vertices of G, then M is called a 1-factor or **perfect matching** of G.

matching diaphragm *Electromagnetism.* a diaphragm of thin metal with a slit that is inserted into a waveguide transverse to the axis of the guide for impedance matching purposes.

matching impedance *Electricity.* the impedance value required to attain matched impedance.

matching loss see MISMATCH LOSS.

matching number *Mathematics.* the number of edges in a maximum size matching in a graph G.

matching problem *Artificial Intelligence.* the problem of determining which parts of a program's knowledge are relevant to a given task or real-world situation.

matching section *Electromagnetism.* a quarter wave or half-wavelength section of a transmission line inserted between a transmission line and a load for matching purposes.

matching stub *Electromagnetism.* a device placed on a transmission line connected to an antenna to add inductive or capacitive reactance for the purpose of matching impedance.

match plate *Metallurgy.* in casting, a plate on which patterns are placed to make molds.

match processing *Computer Programming.* the comparing of two or more strings or substrings.

mate *Biology.* **1.** to pair individuals of different sexes for breeding. **2.** to copulate. **3.** either member of a breeding pair.

mate guarding see GUARDING.

mate killer *Invertebrate Zoology.* an animal, such as the black widow spider or certain mantises, that kills its mate (usually the male) during or shortly after mating.

material balance *Chemical Engineering.* an equation denoting the sum of all substances entering a process and all those leaving it within a given time period.

material culture *Anthropology.* the physical setting and tangible objects of a culture, such as dress, housing, tools, foods, and refuse.

material-handling robot *Robotics.* a robot designed to grasp, transport, and position materials during the manufacturing process.

materialize *Computer Science.* **1.** to store in memory as a discrete data value. **2.** to make a copy of otherwise transient data, such as a value in a register.

material particle see POINT PARTICLE.

material-processing robot *Robotics.* a robot that alters the material it works on by cutting, forming, heat-treating, or transforming it in some way during the manufacturing process.

material requirements planning *Industrial Engineering.* a production and inventory management technique designed to minimize inventory costs by ordering quantities based on production schedules, keeping just enough inventory on hand to meet immediate production needs.

materials handling *Industrial Engineering.* the processes through which raw materials and manufactured goods are transported, positioned, and stored for industrial and commercial operations.

Materials Science and Engineering

Materials Science and Engineering (M.S. & E.) is the branch of technology dealing with the production and optimization of the material goods that provide the foundation of our technological age. Metals and alloys, ceramics and glasses, cements and concrete, polymers and plastics, composites, semiconductors, superconductors, and graphite and diamond are all within the purview of the materials scientist or materials engineer.

For all these diverse materials, the common theme involves the structure-property-performance paradigm; to understand the *properties* of materials, either alone or in combination so that their *performance* in some specific structural or electrical function is optimized, their *structure* must be understood and controlled. Structure includes the basic arrangement of atoms in crystalline or amorphous materials, the spatial distribution of defects and phases (the microstructure), and the size and shape of the final component.

M.S. & E. is extraordinarily multidisciplinary and interdisciplinary; metallurgists, ceramists, polymer scientists, condensed matter physicists, chemists, crystallographers, and engineers are all practitioners.

The term M.S. & E. first began to be used in the 1950s when metallurgists, chemists, and solid-state physicists were required to collaborate in developing and producing the new semiconductor materials silicon and germanium; ceramics (inorganic non-metallics) and polymers were added shortly after to the classes of materials for which the structure-property-performance paradigm was important. The field is now recognized as one of the enabling technologies for continued progress in the next decade and century.

A. H. Heuer
Kyocera Professor of Ceramics
Case Western Reserve University

materials science the scientific study of the structure, properties, and performance of metals, ceramics, polymers, and semiconductors. Also, **materials science and engineering.**

materials selection *Materials Science.* the determination of the appropriate material for a particular purpose, done at two levels: selection among various categories of materials and selection within a category for the optimum material.

materials testing reactor *Nucleonics.* a nuclear reactor that is designed primarily to evaluate the reaction of materials subjected to ionizing radiation.

material well *Chemical Engineering.* in the process of the transfer molding of plastics, a heated chamber that contains the plastic to be forced into the mold plus additional material to maintain pressure during the cure time.

materia medica *Pharmacology.* the branch of medicine that deals with drugs, their sources, preparation, and uses; pharmacology.

materiel or **matériel** [mə tēr′ē el′] *Military Science.* **1.** the items necessary to equip, operate, maintain, and support military activities without distinction as to administrative or combat applications; includes vehicles, weapons, spare parts, and support equipment, but excludes real property, installations, and utilities. **2.** of or relating to materiel. Thus, **materiel inventory (objective), materiel pipeline, materiel readiness, materiel release confirmation, materiel release order, materiel requirements,** and so on.

maternal *Biology.* of or relating to a female parent.

maternal effect *Genetics.* the short-term, often embryonic effect of a mother's phenotype or genotype on the phenotype of an immediate offspring.

maternal immunity *Immunology.* a resistance to infection or disease of a humoral type that is acquired by newborn animals from their mothers.
maternal impressions *Psychology.* the experiences, feelings, and thoughts of a woman during pregnancy, which some theorize may have a direct influence on the fetus.
maternal imprinting *Behavior.* the attachment of a female to her young resulting from their close proximity shortly after birth.
maternal inheritance *Genetics.* the genetic effects that are transmitted through the maternal line, usually through cytoplasmic factors such as the mitochondrion and chlorophast.
maternal mortality rate *Medicine.* the death rate of mothers per 1000 live or still births due to problems of pregnancy, delivery, or postpartum recovery.
maternal placenta *Developmental Biology.* the maternally contributed part of the placenta, derived from the decidua basalis. Also, PARS UTERINA PLACENTAE.
maternity *Biology.* **1.** the fact or condition of motherhood. **2.** the condition of being pregnant.
mat foundation *Civil Engineering.* an interwoven network of steel used within a concrete slab to reinforce it.
math. mathematical, mathematician.
math coprocessor *Computer Technology.* an auxiliary processor that speeds up numerical operations by converting the operands to floating point representation and executing floating point arithmetic. Also, FLOATING POINT PROCESSOR.
Mathematica *Artificial Intelligence.* a widely used program for symbolic algebraic manipulation by computer.
mathematical relating to or based on the use of mathematics.
mathematical biology *Biology.* a discipline that encompasses all applications of mathematics, computer technology, and quantitative theorizing to biological systems, and the underlying processes within the systems.
mathematical check *Computer Programming.* a programmed check that uses mathematical relations; e.g., verifying that $A \times B = B \times A$.
mathematical climate *Meteorology.* an early statement of the earth's general climatic pattern, using the annual cycle of the sun's inclination as its basis; includes three latitudinal zones (now known as Temperate, Frigid, and Torrid), bounded by the Arctic and Antarctic circles and the Tropics of Cancer and Capricorn.
mathematical ecology *Ecology.* the application of mathematical theories and techniques to the study of ecology.
mathematical geography *Geography.* a branch of geography concerned with the dimensions, figure, and movements of the earth.
mathematical geology *Geology.* a branch of geology in which the probability distributions of values of random variables are studied in order to obtain information about geological processes.
mathematical induction *Mathematics.* a technique of deductive reasoning based on the **principle of mathematical induction**, as given in the following theorem: a subset S of the set N of natural numbers that contains 0 is actually N if either of the following holds: (a) $n \in S$ implies that $n + 1 \in S$ for all $n \in N$, or (b) $m \in S$ for all $0 \leq m \leq n$ implies that $n + 1 \in S$ for all $n \in N$. Proofs by mathematical induction proceed as follows: Let $A(n)$ be a mathematical statement depending on the natural number n. Suppose that one can show: (a) $A(0)$ is true (called the **basis of induction**), and (b) when it is assumed that $A(n)$ is true, it follows that $A(n+1)$ is also true (called the **inductive step**). Then by the principle of mathematical induction, $A(n)$ is true for all natural numbers n. Also, INDUCTION, COMPLETE INDUCTION.
mathematical logic see LOGIC.
mathematical model *Mathematics.* a mathematical representation of a condition, process, concept, etc. Variables are defined to represent inputs, outputs, and intrinsic states; equations or inequalities are used to describe interaction of the variables and constraints on the problem.
mathematical physics *Physics.* a discipline in physics in which physical systems are described in terms of mathematical systems, as in statistical thermodynamics or probability theory.
mathematical probability *Mathematics.* the idealization of the proportion of times that a certain result will occur in repeated trials of an experiment. Also, A PRIORI PROBABILITY.
mathematical subroutine *Computer Programming.* a subroutine written to perform a mathematical function, such as sine, cosine, or square root, usually provided in the program library or in a math coprocessor.
mathematical table *Mathematics.* a listing of the values of one or more functions at a sequence of argument values.

Mathematics

Originally mathematics consisted of two parts: arithmetic (the art of counting) and geometry (the measuring of the earth). Since hasty conclusions about numbers and pictures can lead to dangerous errors, mathematics has had to become more and more precise, and by now it is sometimes identified with logic (the science of reasoning).

Mathematics studies concepts whose names are in everybody's vocabulary, but whose precise definitions can be quite elusive. Examples: the theory of groups explains what "symmetry" means, lattice theory is concerned with the concept of "arrangement," and set theory clarifies the meanings of "finite" and "infinite."

The three major parts of mathematics are algebra (a symbolic outgrowth of arithmetic), geometry (the study of shape), and analysis (the abstract version of calculus). This old-fashioned three-way classification is not yet dead. It can be made to include topology under geometry (as in the study of the properties of a plane that is not rigid, but that can be bent and stretched), but it does not include the study of the foundations of mathematics, such as mathematical logic and set theory.

Applied mathematics reasons about abstract concepts with the intention of making predictions about concrete special instances. Pure mathematics tries to start from as few assumptions as possible and often comes to surprising conclusions by elegant proofs. Applied mathematics is usually pursued as a practical subject, and pure mathematics as an aesthetic one.

Logic is the floor on which the structure of mathematics is built, but, at the same time, logic is studied by mathematical methods, and, as such, might be considered a part of applied mathematics. Similar close relatives of mathematics, which are parts of the subject at the same time as they are applications of it, are statistics (the applied theory of probability) and computer science (the applied theory of combinatorics).

Paul Halmos
Professor of Mathematics
Santa Clara University

mathematics the scientific study of quantities, including their relationships, operations, and measurement, expressed by numbers and symbols.
Mathieu, Émile L. [ma tyoo´] 1835–1890, French mathematician; studied the vibrations of an elliptic membrane and introduced Mathieu functions.
Mathieu equation *Mathematics.* a differential equation having the form $y'' + (\alpha + \beta \cos 2x)y = 0$, where y is a function of x, and α and β are constants. The general solution is

$$y(x) = Ae^{\mu x}\phi(x) + Be^{-\mu x}\phi(-x),$$

where $\phi(x)$ has period 2π. The Mathieu equation occurs in problems of wave motion with elliptical boundaries, such as the vibrations of an elliptical drum head.
Mathieu functions *Mathematics.* periodic solutions to the Mathieu equation; denoted $ce_n(x)$ and $se_n(x)$. They occur when $\mu = 0$ in the general solution. For example, when $\beta = 0$ in the Mathieu equation, these functions take the form

$$ce_{2n}(x) = \sum_{k=0}^{\infty} A_k \cos 2kx, \quad ce_{2n+1}(x) = \sum_{k=0}^{\infty} B_k \cos 2(k+1)x,$$

$$se_{2n}(x) = \sum_{k=0}^{\infty} C_k \sin 2kx, \text{ and } se_{2n+1}(x) = \sum_{k=0}^{\infty} D_k \sin 2(k+1)x.$$

matico *Botany.* the pepper species *Piper angustifolium,* a tropical dicot with a pungent aroma, grown commercially for its oil-bearing leaves, which are used as a stimulant.

matildite *Mineralogy.* $AgBiS_2$, a gray to iron-black, opaque, metallic, hexagonal mineral occurring in massive or granular form, having a specific gravity of 6.9 and a hardness of 2.5 on the Mohs scale; found in medium- to high-temperature vein deposits and in pegmatites.

matinal *Meteorology.* a morning wind; an east wind. (From the French word for "morning.")

mating *Biology.* the pairing or coupling of individuals for reproduction.

mating system *Ecology.* the breeding pattern of a population, including such factors as the number of simultaneous mates, the length of the mating bond, and the degree of inbreeding.

mating type *Genetics.* a subpopulation of an organism that is able to mate only with another subpopulation of the same species which is genetically distinct; two such strains that can interact sexually are called **complementary mating types.** A few species, such as some fungi, possess more than two distinct mating types.

mating-type gene *Genetics.* a gene sequence that determines recognition and selection between mating types.

mating-type switch *Genetics.* a change of mating type in a lower organism, caused by the excision of a certain mating-type gene sequence from its genome.

matlockite *Mineralogy.* PbFCl, a colorless, yellow, or green transparent, brittle, tetragonal mineral occurring as aggregates of tabular crystals, having a specific gravity of 7.12 and a hardness of 2.5 to 3 on the Mohs scale; found with other secondary lead minerals.

Matoniaceae *Botany.* a family of medium-sized terrestrial ferns of the order Filicales, having fanlike fronds, a forked leaf stalk, netted veins, and a hairy rhizome.

Matra R-530 *Ordnance.* a French air-to-air, rocket-propelled missile guided by semiactive radar.

matriarchal [mā´trē är´kəl] *Anthropology.* relating to or being a matriarchal society. Also, **matriarchic.**

matriarchal society *Anthropology.* a society in which the woman's family heads the clan or lineage; rules of matrilineal descent and matrilocal residence are usually followed. Also, **matriarchy**.

Matricaria *Botany.* a genus of often aromatic herbs of the family Compositae having alternate, pinnate leaves and white or yellow flowers.

matric forces *Geology.* pressures exerted on soil water as a result of the attraction between solid surfaces and water, the attraction of water molecules for each other, and the polar nature of water.

matrifocal family *Anthropology.* a family that consists of a mother and her children.

matrilineal *Genetics.* descended through the female line.

matrilineal descent *Anthropology.* the descent system in which kinship is traced from the mother's ancestors, children belong to the female's lineage, and members believe that they are descended from a common female ancestor.

matrilocal residence *Anthropology.* a practice in which a married couple resides with the woman's family.

matrix *plural,* **matrices** [mā´trə sēz´] or **matrixes**. something that represents the point or location from which another thing originates or develops; specific uses include: *Computer Programming.* **1.** an array of symbols arranged in rows and columns. **2.** an array of circuit elements arranged and designed to perform a specific function, such as the conversion from one number system to another. **3.** a spreadsheet made up of row and column elements. *Electronics.* **1.** a rectangular array of intersections of input-output leads, with elements connected at some of the intersections. **2.** the section of a color television transmitter that converts red, green, and blue camera signals into camera difference signals and vice versa in the receiver. *Mathematics.* **1.** a rectangular array of numbers or scalars; for example,

$$\begin{matrix} 3 & 2 & 0 & -1 \\ 1 & 4 & 6 & 2 \\ -3 & 1 & 3 & 0 \end{matrix}$$

2. thus, an $m \times n$ matrix over a ring R is a rectangular array of the form

$$\begin{matrix} a_{11} & a_{12} & a_{13} & \cdots & a_{1n} \\ a_{21} & a_{22} & a_{23} & \cdots & a_{2n} \\ \vdots & & & & \vdots \\ a_{m1} & a_{m2} & a_{m3} & \cdots & a_{mn} \end{matrix}$$

where the matrix entries are elements of a given ring R, and where there are m (horizontal) rows and n (vertical) columns. If $n = m$, then the matrix is said to be an $n \times n$ **square matrix.** A square matrix in which $a_{ij} = a_{ij}$ is said to be symmetric. If $m = 1$, the matrix is called a **row matrix** (or row vector); if $n = 1$, the matrix is called a **column matrix** (or column vector). An **arbitrary matrix** is usually noted by a capital letter (e.g., M) or by (a_{ij}), indicating that the (i,j)th entry (row i, column j) of the matrix is the element a_{ij} of R. Two $m \times n$ matrices (a_{ij}) and (b_{ij}) are equal if and only if $a_{ij} = b_{ij}$ for all i and j. The $n \times n$ matrix with 1's on the main diagonal and 0's elsewhere is called the $n \times n$ **identity matrix**; the $n \times n$ matrix with all entries 0 is then called the $n \times n$ **zero matrix.** *Engineering.* a recessed mold into which an item is formed or cast. *Geology.* broadly, the naturally occurring material in which a fossil, pebble, crystal, or other element is embedded. *Petrology.* see GROUNDMASS, def. 2. *Histology.* **1.** the intercellular material of a tissue. **2.** the portion of the epithelium that gives rise to a fingernail or toenail. *Archaeology.* the soil or material in which an excavation is conducted, or in which archaeological objects or sites are held. *Graphic Arts.* **1.** in linecasting or stereotyping, a mold used to cast type. Linecaster matrices are made of brass; stereotype matrices are made of fiber and adhesive. **2.** in phototypesetting, a master negative from which the characters of a font are projected.

matrix algebra *Mathematics.* **1.** a set of rules for performing arithmetic operations on matrices: (a) If $A = (a_{ij})$ and $B = (b_{ij})$ are two $m \times n$ matrices, then their **matrix sum** $A + B$ is defined to be the $m \times n$ matrix (c_{ij}), where $c_{ij} = a_{ij} + b_{ij}$. (b) If $A = (a_{ij})$ is an $m \times n$ matrix and $B = (b_{ij})$ is an $n \times p$ matrix, then the **matrix product** AB is defined to be the $m \times p$ matrix (c_{ij}), where

$$c_{ij} = \sum_{k=1}^{n} a_{ik}\, b_{kj}\,.$$

Matrix multiplication is associative and distributive over matrix addition, but is not, in general, commutative. (c) If $A = (a_{ij})$ is an $m \times n$ matrix over a ring R, then scalar multiplication by $r \in R$ is defined as $rA = (ra_{ij})$ and $Ar = (a_{ij}\, r)$. **2.** an algebra whose elements are matrices.

matrix algebra tableau *Mathematics.* the current matrix at some point in an iterative stage.

matrix-array camera *Electronics.* a solid-state television camera with a rectangular array of pixels or light-sensitive elements.

matrix calculus *Mathematics.* when the elements of a matrix are differentiable functions of a common set of variables, differentiation and integration of these matrices can be defined. The rules for performing these processes on sums and various kinds of products of matrices constitute matrix calculus.

matrix cracking stress *Materials Science.* the point at which the matrix in a composite fails or fractures under an applied load.

matrix effect *Analytical Chemistry.* the effect the matrix (the material used to hold the compound) has on an analysis, e.g., on the spectral lines of a metallic oxide during emission spectroscopy.

matrix element *Quantum Mechanics.* an element from the matrix representing an operator over a specified function space; the matrix element O_{ij} is given by the product of the conjugate of the ith basis wavefunction, the operator, and the jth basis wavefunction, from left to right.

matrix fibers see BICONSTITUENT FIBERS.

matrix game *Mathematics.* a finite, two-person zero-sum game. Matrix games may be characterized by the pay-off matrix $A = (a_{ij})$. The row player R and column player C independently select a row i and column j of the pay-off matrix. Then the corresponding matrix element a_{ij} determines the amount that R obtains from C; if $a_{ij} < 0$ then R must pay $-a_{ij}$ to C. Various theorems exist for determining optimal strategies.

matrix isolation *Spectroscopy.* a technique in which reactive samples are kept in a very cold, inert environment while undergoing spectroscopic examination.

matrix mechanics *Quantum Mechanics.* a quantum theory, developed in the Heisenberg representation, in which operators are represented by matrices.

matrix of a linear transformation *Mathematics.* the representation of a linear transformation in a particular basis. Suppose that U and V are finite dimensional vector spaces over the same scalar field F with bases B_U and B_V, respectively. If T is a linear transformation from U to V, then there is a matrix A, dependent on B_U and B_V, such that $T(u) = Au$ for all u in U. That is, application of the linear transformation T to elements of U is the same as multiplication by the matrix A. T is said to be represented by the matrix A (with respect to the bases B_U and B_V).

matrix of a quadratic form *Mathematics.* the representation of a quadratic form in a particular basis. That is, the matrix of the quadratic form

$$f = \sum_{i=1}^{n} \sum_{j=1}^{n} a_{ij}x_i x_j$$

is the $n \times n$ (symmetric) matrix $A = (a_{ij})$. A quadratic form is often written in the form $f = XAX^T$, where X^T denotes the transpose of the row matrix $X = (x_1, \ldots, x_n)$.

matrix of a system of linear equations *Mathematics.* the $n \times m$ matrix A with the entries a_{ij}, where the a_{ij} are the coefficients of a system of n linear equations in m unknowns

$$x_i\colon a_{i1}x_1 + \cdots + a_{im}x_m = b_i \text{ (with } i = 1, 2, \ldots, n).$$

Such a system has a solution if and only if the matrix equation $AX = B$ has a solution X, where B is the $m \times 1$ column vector with entries b_i.

matrix of eigenvalues see EIGENMATRIX.

matrix porosity *Geology.* the proportion of the total volume of the matrix or finer-grained section of a rock that is occupied by pore spaces.

matrix printing *Computer Technology.* a high-speed printing process that prints different patterns of dots by a precise selection of wire ends from a matrix of wire ends.

matrix protein *Microbiology.* any of several major proteins found in the outer membrane of Gram-negative bacterial cells.

matrix rock see LAND-PEBBLE PHOSPHATE.

matrix sound system *Acoustical Engineering.* a quadraphonic sound system capable of mixing four input channels into two stereo channels for recording or broadcasting, and then restoring the stereo signal to its original quadraphonic format for playback.

matrix spectrophotometry *Spectroscopy.* the photometric measurement of the energy given off by a sample as it is irradiated in sequence at more than one wavelength.

matrix theory *Mathematics.* the study of matrices of linear transformations and bilinear and quadratic forms, especially similarities, canonical forms, and transformations.

matrix velocity *Geophysics.* the speed of sound in a given rock formation.

matrocliny *Genetics.* inheritance in which the offspring resemble the mother more than the father.

matroid *Mathematics.* a pair (E,B), where E is a nonempty finite set and B is a nonempty collection of subsets of E (called bases) such that: (a) no base is properly contained in another base, and (b) if B_1 and B_2 are bases and if e is any element of B_1, then there is an element f of B_2 with the property that $(B_1 - \{e\}) \cup \{f\}$ is also a base. It then follows as a consequence of property (b) above that any two bases of a matroid contain the same number of elements; this is the rank of the matroid. Also, CYCLE SYSTEM.

matte [mat] *Materials Science.* a term applied to paint, paper, photographs, or other surfaces that have a dull, relatively rough finish, especially as opposed to such materials having a smooth, glossy finish. *Metallurgy.* in the smelting of certain copper, lead, nickel, and other nonferrous ores, an intermediate product rich in sulfur.

matte-coated paper see PIGMENTED PAPER.

matte dip *Metallurgy.* in metal finishing, an etchant used to create a dull surface.

matter *Physics.* an aggregate of material particles possessing inertia and capable of occupying space.

matter era *Astronomy.* a period beginning about 10,000 years after the Big Bang and continuing to today, in which the universe is dominated by matter.

matte smelting *Metallurgy.* the process of producing a matte by smelting an ore or ore concentrate.

Matteuci effect *Physics.* an effect by which a voltage difference is developed across the poles of a permanent magnet when it is twisted in a magnetic field.

Matthias's rules *Solid-State Physics.* a set of empirical rules relating superconductive transition temperatures of metals and alloys to their positions in the periodic table and composition percentages.

Matthiessen's rule *Solid-State Physics.* a rule claiming that the resistivity of a metallic crystal is composed of two additive parts: the resistivity due to crystal imperfections and the resistivity due to thermal agitation of the metal ions.

mattock *Mechanical Devices.* an instrument shaped like a pickax, but having one broad end; used to loosen soil in digging.

maturase *Biochemistry.* a protein specified by the second intron in the box gene of yeast mitochondria, thought to be involved in splicing the RNA transcript of the gene.

maturation *Biology.* **1.** the process of coming to full development. **2.** specifically, the final series of changes in the growth and formation of germ cells.

maturation division *Molecular Biology.* nuclear division in which a desired adjustment is made in the chromosomal complement of a nucleus.

maturation of behavior *Behavior.* behavioral development that occurs not as a result of practice or experience, but rather in response to the development of the central nervous system of an organism.

mature *Biology.* **1.** fully grown and fully developed sexually. **2.** ripe. *Geology.* of or related to a structure or region that has undergone maximum development and accentuation of form.

matureland *Geology.* a topography that has reached the mature stage in the erosion cycle.

mature soil see ZONAL SOIL.

maturity the fact or state of being mature; final development; specific uses include: *Geology.* **1.** in the topographic development of a landscape or area, the second stage in the cycle of erosion, characterized by maximum topographic differentiation. Also, TOPOGRAPHIC MATURITY. **2.** in the development of a shore or coast, the stage that starts when a profile of equilibrium is reached. **3.** the extent to which a clastic sediment has been differentiated and evolved from its parent rock. *Hydrology.* in the development of a stream, the stage at which the stream reaches its maximum efficiency, characterized by the attainment of a profile of equilibrium and a velocity sufficient to carry the load delivered to it.

maturity index *Geology.* the measure of how far a clastic sediment has progressed toward attaining compositional stability.

maturity-onset diabetes *Medicine.* non-insulin-dependent diabetes; typically, onset occurs after 40 years of age.

MATV master antenna television system.

maucherite *Mineralogy.* $Ni_{11}As_8$, a reddish, silver-gray tetragonal mineral occurring in massive and granular form, having a specific gravity of 7.95 to 8.0 and a hardness of 5 on the Mohs scale; found in hydrothermal veins with other nickel arsenides and sulfides.

Mauchly, John W. 1907–1980, American engineer and physicist; co-inventor of the first electronic computer (ENIAC) and later models.

maul *Mechanical Devices.* a heavy, wooden-headed hammer, used especially for driving wedges or piles. Also, MALL. *Archaeology.* a similar prehistoric implement; a blunt, double-sided stone mallet or clubhead that has a groove in which a handle may be attached.

Mauler *Ordnance.* a mobile, lightweight, surface-to-air missile designed for defense against low-flying aircraft in forward areas; officially designated **XMIM-46A**.

Maunder minimum *Astronomy.* a period of time lasting from approximately 1645 to 1710 when the sun was unusually free of sunspots, and related solar activity (such as flares and terrestrial aurorae) was abnormally low.

Maunoir's hydrocele *Medicine.* a hydrocele forming in the cervical area.

Maupertuis, Pierre de [mō pər twē′] 1698–1759, French mathematician and astronomer; studied the properties of curves; formulated the principle of least action (Maupertuis's principle).

Maupertuis's principle *Mechanics.* a principle stating that for a system whose total mechanical energy is conserved, the path to be taken for the system from one configuration to another is the one whose action has the least value relative to all other possible paths to and from the same configurations. Also, PRINCIPLE OF LEAST ACTION, LEAST-ACTION PRINCIPLE.

Maurer's dots *Microbiology.* aberrations, such as dots or streaks, that are found under appropriate staining conditions in erythrocytes infected with *Plasmodium falciparum.* Also, **Maurer's clefts.** (Named for Georg *Maurer,* born 1909, German-born Sumatran physician.)

Mauriac syndrome *Medicine.* a syndrome characterized by dwarfism, enlarged liver, obesity, and delayed sexual maturity, accompanied by diabetes mellitus.

Mauritius hemp [mō rē′shəsl] *Botany.* a tropical agavelike plant of the species *Furcraea foetida* that is often grown commercially for its tough hemplike fibers. Also, CABUYA.

Mauro, Alexander 1921–1989, American biophysicist; a pioneer in the development of the radio-frequency cardiac pacemaker.

Maury, Matthew F. 1806–1873, American naval officer and oceanographer; studied ocean winds and currents.

Mauser [mouz´ər] *Ordnance.* **1.** any of various German firearms manufactured by the Mauser Company and widely used during World Wars I and II, including pistols, revolvers, carbines, and rifles. **2.** a German recoil-operated, air-cooled 20-mm aircraft cannon that was manufactured by Mauser in large quantities during World War II; officially designated **MG-151**. An antiaircraft version was designated **Flak 38**.

Maverick *Ordnance.* an air-to-surface missile with launch and leave capability, having television, laser, or IR guidance; designed for use against small, hard targets such as tanks, armored vehicles, and field fortifications; officially designated **AGM-65.**

max- or **maxi-** an informal combining form indicating great size or scope, especially in comparison with others of the same type, as in *maxflow, maxiquake.*

max. maximum.

Maxam-Gilbert DNA sequencing *Molecular Biology.* a widely used method to determine the exact nucleotide sequences in a specific DNA molecule that involves ordering a collection of radioactively labeled DNA fragments by size.

max-flow min-cut theorem *Mathematics.* in a transport network, the value of any maximal flow is equal to the capacity of any minimal cut.

maxi cell or **maxicell** *Biotechnology.* a cell of *Escherichia coli* or other bacteria that has been irradiated with ultraviolet light; used for studies of plasmid transcription.

maxilla [mak sil´ə] *plural,* **maxillae** [mak sil´ē] *Anatomy.* the upper jawbone that houses the upper teeth and helps to form the orbit, the nasal cavity, and the palate. *Invertebrate Zoology.* one of a pair of mouthparts behind the mandibles in most arthropods.

maxillary [mak´sə lâr´ē] *Anatomy.* of or relating to a jaw, jawbone, or maxilla. Thus, **maxillary artery, maxillary nerve.**

maxillary artery *Anatomy.* the artery that carries blood to the jaws, teeth, chewing muscles, ears, sinuses, nose, and palate.

maxillary gland *Invertebrate Zoology.* an excretory organ of some crustaceans, located in the head and emptying near the base of the second maxillae.

maxillary hiatus *Anatomy.* a rough opening on the medial surface of the maxillary sinus, mostly filled with parts of several bones.

maxillary nerve *Anatomy.* a branch of the trigeminal nerve going through the foramen rotundum; it innervates the skin of the face and scalp, the mucous membrane of maxillary sinuses and nasal cavity, and the teeth.

maxillary palp *Invertebrate Zoology.* a sensory process on the head of an arthropod or polychaete.

maxillary process *Developmental Biology.* the dorsal process formed by bifurcation of the first branchial arch in the embryo, which joins with the ipsilateral median nasal process in the formation of the upper jaw, not including the premaxilla. Also, PROCESSUS MAXILLARIS.

maxillary sinus *Anatomy.* one of a pair of large cavities in the maxillary bone that connect with the nasal cavity.

maxilliped *Invertebrate Zoology.* in crustaceans, one of the paired appendages on the first one to three thoracic segments.

maxillofacial [mak´sil ō fā´shəl] *Anatomy.* of, relating to, or affecting the jaws and the face. Thus, **maxillofacial surgery.**

maxillofacial polymer *Materials Science.* a polymer that is used to replace portions of the flexible tissue of the face, such as the nose, ear, cheek, and eye orbit. The four main types now in commercial use are silicones (whose use is under debate), polyurethanes, polyvinyl chlorides, and chlorinated polyethylenes.

maxilloturbinal bone see INFERIOR NASAL CONCHA.

Maxim [maks´əm] *Ordnance.* **1.** a recoil-operated, water-cooled machine gun first produced in 1884; used by many countries in the early 20th century. **2.** one of several other Maxim-type machine guns used by Germany during and after World War I. (Developed by Hiram *Maxim,* 1840–1916, American-born British inventor.)

maximal *Mathematics.* let A be a partially ordered set with ordering $\leq$. An element a of A is **maximal in A** if for all $c \in A$, $a \leq c$ implies $a = c$. That is, a is maximal in A if $c \leq a$, for every c in A which is comparable to a. For example, the ideals of a ring R form a partially ordered set with containment as the partial ordering. A proper ideal M of R is said to be a **maximal ideal** of R if whenever I is an ideal of R such that $M_c I_c R$, then either $M = I$ or $I = R$. Maximal subgroups, covers, etc., are similarly defined. A given set may have many maximal elements, or it may have none at all.

maximal breathing capacity *Physiology.* the volume of air breathed when a subject breathes as deeply as possible for 15 seconds, as measured by a spirometer without a carbon dioxide absorber.

maximal planar graph *Mathematics.* a planar graph having the property that no additional edges can be added so that the resulting graph is still planar.

maximal principle see MAXIMUM-MODULUS THEOREM.

maximal with respect to property P *Mathematics.* given a property P of sets and a set S whose subsets may or may not satisfy property P, a subset A is maximal with respect to property P if A satisfies property P and A is no proper subset of another subset of S that satisfies property P.

maximax criterion *Mathematics.* in decision theory, any criterion for adopting a strategy that will maximize the maximum outcome (such as profit).

maxim criterion *Mathematics.* in decision theory, any criterion for adopting a strategy that will maximize a minimum outcome (such as profit).

maximization *Mathematics.* the process of determining a (relative or absolute) maximum of a real-valued function.

maximum *Mathematics.* **1.** maximal element. **2.** a real-valued function f on a topological space X has attained a **relative maximum** (or strict relative maximum) value at a point $a \in U$ if there exists a neighborhood U of a such that $f(x) \leq f(a)$ (resp. $f(x) < f(a)$) for all points x in U. **3.** If $f(x) \leq f(a)$ (resp. $f(x) < f(a)$) for all points x in X, then f is said to have an **absolute maximum** (or strict absolute maximum) at a.

maximum allowable working pressure *Mechanical Engineering.* the maximum gage pressure, at a specified temperature, in a pressure vessel, by which the set pressure for a relief valve is determined.

maximum-and-minimum thermometer *Engineering.* a thermometer designed to measure both the maximum and minimum temperatures within a given time interval.

maximum angle of inclination *Mechanical Engineering.* the maximum angle at which a conveyor may be inclined and still be able to deliver a certain quantity of bulk material within a given amount of time.

maximum available gain *Electronics.* the amplification from a circuit whose input and output impedances are appropriately matched to source and load.

maximum average power output *Electronics.* in television, the highest radio-frequency output power that can occur, averaged over the longest repetitive modulation cycle.

maximum belt slope *Mechanical Engineering.* a slope value beyond which the material on the belt will tend to roll downhill; in the mining of coal and ore, belt conveyers can function up to about 18°.

maximum belt tension *Mechanical Engineering.* the value that expresses a combination of the starting and operating tensions of a conveyor.

maximum condition *Mathematics.* a module A is said to satisfy the maximum condition on submodules if every nonempty set of submodules of A contains a maximal element. By theorem, A satisfies the ascending chain condition on submodules if and only if A satisfies the maximum condition on submodules.

maximum continuous load *Mechanical Engineering.* the total load that a boiler can maintain for a designated length of time.

maximum credible accident *Nucleonics.* the worst type of nuclear reactor accident that can be expected from a foreseeable combination of factors, such as equipment malfunctions, operator errors, and natural disasters.

maximum demand *Electricity.* the greatest average value of the power or current consumed by an electrical installation over a period of time.

maximum depression *Ordnance.* the greatest angle below horizontal at which an artillery gun can still deliver effective fire.

maximum ebb *Oceanography.* the highest speed of an ebb tidal current.

maximum effective range *Ordnance.* the maximum distance at which a weapon may be expected to be accurate while achieving the desired result.

maximum elevation *Ordnance.* the greatest vertical angle at which a gun can be laid.

maximum flood *Oceanography.* the highest speed of a flood tidal current.

maximum gradability *Mechanical Engineering.* the steepest slope a vehicle can negotiate in a low gear.

maximum keying frequency *Electronics.* in a facsimile system, the frequency in hertz that is equal to the spot speed divided by twice the horizontal scanning spot dimension.

maximum likelihood *Statistics.* a principle of estimation based on choosing the value of the unknown parameter under which the probability of obtaining an observed sample is highest.

maximum-likelihood criterion *Statistics.* in decision making, the process of considering only the event most likely to occur while ignoring all other events, and choosing the action that will produce the optimal result associated with this most likely event.

maximum-likelihood estimator *Statistics.* an estimator obtained by the method of maximum likelihood.

maximum modulating frequency *Electronics.* in a facsimile system, the highest picture frequency required for transmission without lowering the quality of the reproduction.

maximum-modulus theorem *Mathematics.* an analytic function defined on a closed, bounded, simply connected domain D assumes its maximum absolute value (the modulus) on the boundary of D. Also, MAXIMAL PRINCIPLE.

maximum operating frequency *Computer Technology.* the highest clock rate or repetition during which modules perform reliably in continuous operations, under worst-case conditions, with no special trigger-pulse requirements.

maximum ordinate *Mechanics.* the peak altitude attained by a projectile, measured from its firing altitude.

maximum payload capacity *Transportation Engineering.* the heaviest load that an aircraft can safely carry or that its load space can accommodate.

maximum performance *Industrial Engineering.* the level of performance that leads to maximum production.

maximum permissible concentration *Medicine.* the greatest level of exposure to a harmful substance for eight hours daily that can be sustained throughout a working career without harmful effects or departure from a normal healthy state; this is used in some countries to define health standards for the work environment. Also, **maximum allowable concentration.**

maximum permissible dose *Medicine.* the greatest amount of radiation exposure that a person or specific body part can receive without adverse affects, according to the latest guidelines. *Military Science.* specifically, the amount of radiation that a military commander prescribes as the highest cumulative dose of nuclear radiation allowable for members of his command.

maximum power transfer theorem *Electricity.* a theorem stating that maximum power transfer from source to load is achieved when the impedances of the source and load are conjugates; this means that the resistive components must be equal and the reactive components must be equal in magnitude, but one must be inductive and the other capacitive.

maximum producible oil index *Petroleum Engineering.* a figure representing the maximum amount of oil that can be produced by water drive per bulk formation volume.

maximum production life *Mechanical Engineering.* the span of time during which a cutting tool operates at maximum efficiency.

maximum range *Ordnance.* the greatest distance a weapon can fire without consideration of dispersion.

maximum retention time *Electronics.* the maximum amount of time a charge-storage tube can hold information.

maximum signal level *Electronics.* in an amplitude-modulated facsimile system, the level corresponding to the higher signal amplitude, that of either copy black or copy white.

maximum strain criterion *Materials Science.* a failure criterion used to predict the strength of a fiber-reinforced composite under different conditions; it describes the maximum allowable tensile strains on fibers before failure.

maximum stress theory *Materials Science.* a theory stating that failure occurs in unidirectional fiber-reinforced composites because of stress inequality, or that yielding occurs in a material when any of the stresses in multistress loading is equal to or greater than the yield strength of the material determined in axial loading.

maximum subsidence *Geology.* the greatest depth to which the earth's crust sinks in a basin.

maximum sustainable yield *Ecology.* the maximum crop or yield that can be harvested repeatedly from a population without diminishing the stock.

maximum thermometer *Engineering.* a thermometer designed to register the highest temperature attained during a given time interval.

maximum thrust *Ordnance.* in missile testing, the highest recorded thrust on the thrust-time trace.

maximum unambiguous range *Electromagnetism.* the range beyond which the echo from a pulsed radar signal returns after generation of the next pulse.

maximum undistorted power output *Electronics.* the maximum power that is delivered by a tube or transistor with distortion not exceeding a certain percentage. Also, **maximum useful output.**

maximum usable frequency *Telecommunications.* **1.** in radio transmission by ionospheric reflection, the upper limit of the frequencies transmitted by reflection from regular ionized layers. **2.** in the ionosphere, the highest frequency at which signals are reflected back to earth.

maximum-wind and shear chart *Meteorology.* a synoptic weather chart in which the altitudes of the maximum wind speed, the maximum wind velocity (and sometimes direction), and the wind velocity at mandatory levels both above and below the level of maximum wind are plotted.

maximum-wind level *Meteorology.* the altitude at which the maximum wind speed occurs, determined by a winds-aloft observation.

maximum-wind topography *Meteorology.* the topography of the surface of maximum wind speed .

maximum with respect to property P *Mathematics.* given a property P of sets and a finite set S whose subsets may or may not satisfy property P, a subset A is maximum with respect to property P if A satisfies property P and no subset of S with more elements than A will also satisfy property P.

maximum work criterion see TSAI-HILL CRITERION.

maximum working area *Industrial Engineering.* the maximum space within the reach of a worker at a work station.

maximum zonal westerlies *Meteorology.* the average west-to-east component of wind over a continuous 20° belt of latitude in which this average is a maximum; in winter, it is usually found at approximately 40°–60° north latitude.

James Clerk Maxwell

Maxwell, James Clerk 1831–1879, Scottish physicist; best known for his work on electricity and magnetism.

maxwell or **Maxwell** *Electromagnetism.* a unit of magnetic flux in the centimeter-gram-second electromagnetic system of units, equivalent to the flux that produces one abvolt in a one-turn circuit when the flux is reduced to zero at a uniform rate in one second.

Maxwell body *Mechanics.* an ideal body of viscoelastic material modeled by a spring and dashpot in series whose deformation is proportional to both the stress and the time it is applied.

Maxwell-Boltzmann distribution *Physics.* a function that gives the likelihood of an atom or molecule of a gas in thermal equilibrium having a value of a specified variable (such as velocity or energy) within a certain infinitesimal range.

Maxwell-Boltzmann statistics *Physics.* the statistical treatment of classical identical particles, as distinguished from Fermi-Dirac or Bose-Einstein statistics.

Maxwell bridge *Electricity.* a four-arm, alternating-current bridge used to determine the inductance of a coil in terms of resistance and capacitance, or of a capacitance in terms of resistance and inductance.

Maxwell distribution *Physics.* a distribution function of velocities of particles in a system in which the particles are assumed not to interact and obey classical mechanics.

Maxwell effect *Optics.* a type of birefringence that arises in a viscous liquid with anisotropic molecules, when the components within the velocity gradient are perpendicular to the velocity of the liquid.

Maxwell element *Materials Science.* a representation of the response of a polymer to stress by a spring and dashpot placed in series; a model for the viscoelastic behavior of a polymer.

Maxwell equal area rule *Thermodynamics.* a rule stating that, on a pressure-volume graph at temperatures for which a portion of the constant-temperature curve has a positive slope bounded by negative slopes, a horizontal line drawn at the equilibrium vapor pressure separates two regions of equal area: the portion above the horizontal line bounded by the isotherm and the portion below the horizontal line and the isotherm.

Maxwell field equations *Electromagnetism.* four partial differential equations that are used to describe the behavior of electromagnetic fields.

Maxwell formula *Optics.* a formula stating that the refractive index, n, of a material is equal to the square root of the product of the material's dielectric permittivity and its magnetic permeability.

Maxwellian *Physics.* relating to James Clerk Maxwell or to theories and principles based on his research.

Maxwellian distribution law *Physics.* a relation that gives the average number N of atoms or molecules in a gas at thermal equilibrium possessing a speed within an infinitesimal range of c to $c + dc$:

$$N = 4\pi(m/2\pi kT)^{3/2}c^2\exp(-mc^2/2kT),$$

where m is the particle mass, k is the Boltzmann constant, and T is the absolute temperature.

Maxwellian equilibrium *Physics.* the condition of a system of particles (or a gas) in which the velocity distribution resembles a Maxwell distribution that is specified by the temperature of the environment with which the system is in contact.

Maxwellian gas *Physics.* a gas whose particles interact through a force that is inversely proportional to the fifth power of the distance between them.

Maxwell liquid *Fluid Mechanics.* a liquid that is subject to deformation; its rate of deformation is given by a term that is proportional to the shearing stress plus another term that is proportional to the rate of change of the shearing stress. Also, **Maxwell body.**

Maxwell primaries *Optics.* the primary colors, cyan, green, and magenta, that appear in a colorimetry system devised by James Clerk Maxwell.

Maxwell relation *Electromagnetism.* a relation stating that the dielectric constant of a substance is equal to the square of the index of refraction.

Maxwell's coefficient of diffusion *Fluid Mechanics.* a quantity that determines the difference of mean velocities of the particles in the gases in a system in which two gases are allowed to mix, and the contribution of the concentration gradient to this velocity difference.

Maxwell's demon see DEMON OF MAXWELL.

Maxwell's electromagnetic theory *Electromagnetism.* a theory that mathematically predicts how electromagnetic waves propagate through media where microscopic (atomic) scale processes may be neglected.

Maxwell's law *Electromagnetism.* a law stating that a circuit with movable portions will tend to move into an arrangement such that there is maximum magnetic flux linkage throughout the circuit.

Maxwell's (reciprocal) theorem *Mechanics.* a theorem stating that if a load on a linear elastic structure is moved to a new point, the deflection at the old point will then equal the deflection that the new point had before the load was moved.

Maxwell's stress functions *Mechanics.* three functions of position that fully determine the stress in a body.

Maxwell's stress tensor *Electromagnetism.* a second rank tensor whereby the product with a unit vector that is normal to the surface of a body will give the force per unit area acting across the body's surface at the point in question due to an electromagnetic field.

Maxwell's theory of light *Optics.* a theory stating that light consists of electromagnetic waves which propagate at a speed equal to the dielectric permittivity and magnetic permeability of the medium in which they are propagated.

Maxwell's thermodynamic relations *Thermodynamics.* a class of mathematical identities relating partial identities of thermodynamic properties; the most familiar are the four mathematical identities that relate the pressure, volume, entropy, and thermodynamic temperature for a system in equilibrium.

Maxwell triangle *Optics.* a graph that displays color-matching chromaticity values. Also, X,Y CHROMATICITY DIAGRAM.

Maxwell-turn *Electromagnetism.* a unit of flux linkage in the cgs electromagnetic system of units, equivalent to the flux linkage of a single-turn loop of wire through which one maxwell of magnetic flux passes.

Maxwell-Wagner mechanism *Electricity.* two parallel metal alloy plates composing a two-layer capacitor, in which one layer constitutes diminishing conductivity while the other layer constitutes finite conductivity and diminishing electric capacity.

Mayacaceae *Botany.* a monogeneric family of freshwater aquatic herbs of the order Commelinales that are free-floating or submerged and rooted to the substrate, having numerous spirally arranged leaves; native to tropical and warm-temperate America and tropical western Africa.

Mayall's object *Astronomy.* a peculiar galaxy in Ursa Major that has an elongated shape and is surrounded by a ring.

mayaro fever *Medicine.* undifferentiated type fever caused by the genus *Alphavirus;* associated with epidemics in South America.

Maybach, Wilhelm 1846–1929, German engineer; constructed (with G. Daimler) the first Mercedes automobile.

Mayer, Julius R. von 1814–1878, German physicist; independently discovered the law of the conservation of energy.

Mayer, Maria Goeppert 1906–1972, German-born American physicist; shared the Nobel Prize (with J. H. D. Jenson) for work on nuclear shell structure.

mayer *Thermodynamics.* a unit of heat capacity, equivalent to that of a substance whose temperature is increased by 1°C when one joule is absorbed by the substance.

Mayer condensation theory *Physics.* a theory concerned with critical state condensation, according to which a chemically saturated system of molecules is assumed to be composed of clusters of molecules rather than of individual molecules.

mayfly

mayfly *Invertebrate Zoology.* a common name for a slender, fragile flying insect in the order Ephemeroptera, having triangular membranous forewings and smaller hindwings and undergoing an interim moult unique among insects; most of the lifespan is spent as an aquatic nymph.

Mayo, Charles Horace 1865–1939, and his brother **William James Mayo**, 1861–1939; American physicians and surgeons who established the Mayo Clinic and the Mayo Foundation.

Mayomyzon *Paleontology.* a genus of ostracoderms in the extinct subclass Cephalaspidomorpha and order Petromyzontiformes, known from the Pennsylvanian (Upper Carboniferous) assemblage of Mazon Creek, Illinois; Mayomyzon is similar to the modern lamprey but has a primitive branchial skeleton.

Mayorella *Invertebrate Zoology.* a genus of flattened, naked amebae of the family Hyalodiscidae that feed on bacteria, algae and small protozoans; found in freshwater, marine, and leaf-litter situations.

Mayr, Ernst born 1904, American zoologist; developed the synthetic theory in evolution.

maytansine *Oncology.* $C_{34}H_{46}ClN_3O_{10}$, an antineoplastic that is extracted from species of Maytenus, a genus of tropical American trees and shrubs.
maz-, mazo- a combining form denoting a relationship to the breast, as *mazoplasia.*
mazaedium *Botany.* the powdery mass of spores produced by certain lichens.
maze *Psychology.* a patterned device frequently used in intelligence and learning experiments, consisting of pathways that must be followed from an entrance to an exit and blind alleys which, if taken, lead away from the desired goal.
mazopathy *Medicine.* a diseased condition of the breast.
mazoplasia *Medicine.* the deterioration of the mammary glandules with increased cell production in surrounding tissue.
Mazza process *Materials Science.* a process for the manufacture of pipes, using fiber-reinforced cement. Typically, the fiber used is asbestos, but cellulose derived from wood pulp or newsprint may also be used; the binder is Portland cement.
Mazzoni's corpuscle *Anatomy.* a sensory nerve ending that produces tactile sensation and resembles Krause's corpuscles.
mb millibar; millibarn.
m.b. mix well. (From Latin *misce bene.*)
MB megabyte.
MB or **M.B.** Bachelor of Medicine. (Latin *medicinae baccalaureus.*)
mbd million barrels per day.
MBO management by objectives.
MBR memory buffer register.
MBT mercaptobenzothiazole.
mc millicurie.
mc or **MC** metric carat.
mC millicoulomb.
μC *Electricity.* the symbol for microcoulomb.
Mc megacycle.
M.C. master of surgery. (From Latin *magister chirurgiae.*)
MCA middle cerebral artery; 3-methylcholanthrene.
McAdam, John L. 1756–1836, Scottish engineer; originated the macadamized road surface, which is named for him.
McArdle's disease *Medicine.* a hereditary human disease caused by a mutation of a gene located on chromosome 11, resulting in a deficiency in the production of the enzyme skeletal muscle glycogen phosphorylase. Also, **McArdle's syndrome.** (Named for Brian *McArdle,* 20th-century British neurologist.)
MCAT Medical College Admissions Test.
McBurney's incision *Surgery.* an incision that is made parallel to the fibers of the external oblique muscle of the abdomen, and one-third of the distance between the anterior superior iliac spine and the umbilicus; the underlying muscles are then split but not cut. (Named for Charles *McBurney,* 1845–1913, American surgeon.)
McBurney's point *Anatomy.* a small area on the abdomen at which extreme tenderness is felt in appendicitis; located between the navel and the right iliac bone, about two inches above the iliac bone. (Named for Charles *McBurney.*)
McCartney bottle *Biotechnology.* a small, narrow-necked, cylindrical, screw-cap glass vial or bottle.
McClintock, Barbara 1902–1991, American geneticist; awarded the Nobel Prize for the discovery of mobile genetic elements.

Barbara McClintock

McClung Toabe egg-yolk agar *Microbiology.* a solid bacteriological medium used to identify various anaerobic microorganisms, especially *Clostridia,* by detecting the presence of certain enzymatic activities.
McCollum, Elmer 1879–1967, American biochemist; the first to name vitamins with a letter system.

Cyrus Hall McCormick

McCormick, Cyrus Hall 1809–1884, American inventor and industrialist; invented the first successful reaping machine in 1831.
McCoy, Elijah 1843–1929, American inventor and engineer; invented the automatic lubricating machine.
mcd millicurie-destroyed.
McDowell, Ephraim 1771–1830, American surgeon; performed the first recorded ovariotomy.
M center *Solid-State Physics.* a color center in an ionic crystal composed of two adjacent F centers; an electron trapped at two adjacent negative ion vacancies.
mcf thousand cubic feet.
MCF macrophage chemotactic factor.
McFayden-Stevens reduction *Organic Chemistry.* an important reaction applicable only to aromatic aldehydes, in which *N*-aroylsulfonylhydrazides are converted into aryl aldehydes.
mcfd thousands of cubic feet per day.
mcg. microgram.
mcgovernite *Mineralogy.* $(Mn^{+2},Mg,Zn)_{22}(As^{+3}O_3)\ (As^{+5}O_4)_3(SiO_4)_3(OH)_{20}$, a brown, translucent, trigonal mineral occurring in massive and granular forms, having a specific gravity of 3.72 and an undetermined hardness; found with zincite, franklinite, and calcite. Also, **macgovernite.**
MCH mean corpuscular hemoglobin.
MCHC mean corpuscular hemoglobin concentration.
M.Ch.E. master of chemical engineering.
MCI *Aviation.* the airport code for Kansas City International Airport, Kansas City, Missouri.
MCi megacurie.
mCi millicurie.
mCi-hr. millicurie hour.
McIntosh and Fildes' anaerobic jar *Biotechnology.* a cylindrical, gas-tight jar with a sidearm containing an indicator such as methylene blue solution; used to detect anaerobiosis.
McKellar hypothesis *Archaeology.* a principle stating that very small items, when no longer useful, will be discarded at their original location rather than disposed of elsewhere; thus such an item found at a specific site can be presumed to have been actually used at that site. (Formulated by Judith *McKellar,* American archaeologist.)
McKinley, Mount *Geography.* the highest mountain in North America (20,320 feet), in south central Alaska.
McLafferty rearrangement *Organic Chemistry.* the molecular rearrangement involving a transient six-membered ring that occurs in mass spectrometry when hydrogen is transferred during the production of odd-electron fragment ions.

McLeod gauge *Fluid Mechanics.* an instrument that uses vacuum pressure to measure the height of a capillary column of mercury. (Named after Herbert *McLeod,* 1841–1932, British chemist.)

McMath solar telescope *Optics.* the telescope at Kitt Peak, Arizona, in which an 80-inch rotating mirror at the top reflects the sun's rays into a long fixed tube.

MCMI Millon clinical multiaxial inventory.

McMillan, Edwin M. born 1907, American physicist; awarded the Nobel Prize for the discovery of plutonium and neptunium.

McNally tube *Electronics.* a single-cavity velocity-modulated microwave tube designed to produce low-power UHF oscillation.

M-component hypergammaglobulinemia *Medicine.* a type of hypergammaglobulinema marked by an increased amount of monoclonal immunoglobulin in the blood.

M contour *Control Systems.* a line connecting those points on a Nyquist diagram that have the same magnitude as the primary feedback ratio.

MCPA or **MCP** (4-chloro-2-methylphenoxy)acetic acid.

MCPB 4-butanic acid.

mc.p.s. megacycles per second.

McQuaid-Ehn test *Metallurgy.* in metallography, a test to reveal the austenitic grain size of a ferrous alloy.

MCT mean circulation time.

MCV mean corpuscular volume.

MCW modulated continuous wave.

MD muscular dystrophy; medical department.

MD or **M.D.** doctor of medicine. (From Latin *medicinae doctor.*)

Md the chemical symbol for mendelevium.

M.D.A. motor discriminative acuity; right mentoanterior. (From Latin *mento-dextra anterior.*)

M damage *Military Science.* mobilization damage; damage that immobilizes a combat vehicle.

M-day *Military Science.* mobilization day; a designation of the day on which mobilization for a mission or operation commences or is due to commence.

M-derived filter *Electronics.* a filter derived from a constant-*k* filter, in which some elements are multiplied by a factor *M*, which lies between 0 and 1; having more elements, this filter provides sharper cutoff and more uniform attenuation in the pass region.

M-design bit *Mechanical Devices.* a long-shank, box-threaded core bit used in similarly shaped core barrels.

M-design core barrel *Design Engineering.* in drilling applications, a double-tube core barrel with a 2.5° taper-core lifter carried inside a short tubular sleeve.

MDF main distribution frame.

M display *Electronics.* a radar display in which an adjustable pedestal is moved along the baseline to allow accurate range measurement; a type of A display. Also, M SCAN.

mdn. median.

M.D.P. right mentoposterior. (From Latin *mento-dextra posterior.*)

MDR minimum daily requirement.

MDS minimum discernible signal.

M.D.T. right mentotransverse. (From Latin *mento-dextra transversa.*)

MDW *Aviation.* the airport code for Midway Airport, Chicago, Illinois.

ME or **M.E** mechanical engineer; medical examiner.

Me the chemical symbol for methyl.

Me *Acoustics.* an electroacoustic modifier which, for a speaker amplification system, describes the directivity of sound by the equation $Me = 10 \exp(M_S + M_L)$, with M_S representing the difference in decibels between a microphone's sensitivity toward the sound source and toward the loudspeaker, and M_L representing the difference in decibels between the sound intensity of the loudspeaker toward the listener and the microphone.

meacon *Electromagnetism.* to transmit false signals by means of a meaconing system.

meaconing *Electromagnetism.* a communications system that receives radio navigational information and then instantly transmits confusing signals of the same carrier frequency so as to elude enemy navigators. (An acronym for misleading beacon.)

Mead, Margaret 1901–1978, American anthropologist; researched the effects of culture on personality development.

meadow *Agriculture.* an expanse of grassland used for pasture or as a hayfield. *Ecology.* an expanse of open land that is mainly covered with grasses, especially such an area at a high elevation.

meadow grass *Botany.* a grass of the genus *Poa,* especially the Kentucky bluegrass *P. pratensis.*

meadowlark

meadowlark *Vertebrate Zoology.* a sharp-billed, short-tailed North American songbird of the genus *Sturnella* in the family Icteridae; characterized by a plump body and often having a yellow breast with brown speckled plumage above; noted for its clear, tuneful song.

meadow mouse *Vertebrate Zoology.* any of many short-tailed rodents of the genus *Microtus* and allied genera; found in fields and meadows in temperate areas of the Northern Hemisphere. Also, **meadow vole.**

meadowsweet *Botany.* any plant of the genus *Spiraea* of the rose family, especially *S. latifolia,* characterized by white or pink flowers, or any plant of the closely related genus *Filpendula.*

meager set *Mathematics.* a set of the first category; i.e., a set that can be represented as the countable union of nowhere dense (or countable) sets.

mealy see FARINACEOUS.

mealybug *Invertebrate Zoology.* the common name for any scaly insect of the family Pseudococcidae that produces a powdery substance covering the dorsal surface; harmful plant pests.

Mealy machine *Computer Programming.* a sequential machine in which the output is determined by both the present state and the input.

mean *Statistics.* the sum of all data values divided by their number. *Mathematics.* **1.** given a set of numerical quantities, a representative value, such as arithmetic average, geometric mean, harmonic mean, or expected value. **2.** the arithmetic average (arithmetic mean).

mean anomaly *Astronomy.* the mean orbital motion of a body multiplied by the time that has elapsed since it last passed periapsis.

mean-average boiling point *Chemical Engineering.* the mean of the molal average boiling point and the cubic volumetric average boiling point; used to predict physical properties of petroleum mixtures.

mean calorie *Thermodynamics.* a unit of energy, equivalent to 1/100 of the heat required to raise the temperature of one gram of water from 0°C to 100°C.

mean camber *Aviation.* the mean of an airfoil's upper and lower cambers; may be graphically represented by the **mean camber line,** a line along the cross section of the airfoil marking the midpoint between its upper and lower surfaces.

mean carrier frequency *Electronics.* the average carrier frequency of a transmitter; corresponds to the resting frequency in a frequency-modulated system.

mean chart *Meteorology.* any synoptic weather chart in which isopleths of the mean value of a given meteorological element are drawn. Also, MEAN MAP.

mean chord *Aviation.* a chord equal to the sum of an airfoil's chord lengths divided by the number of chord lengths or, equivalently, to the airfoil's area divided by its span. This is also known as the **mean geometric chord,** or sometimes, **mean aerodynamic chord (MAC)** for positive stability; for example, the center of gravity is always placed ahead of the 1/4 chord of the wing MAC. Tail surfaces bring the effective airplane MAC off the wing MAC.

mean current *Oceanography.* the average flow rate and direction of an ocean current, as measured over a considerable period of time at a specified point.

mean curvature *Mathematics.* at a given point on a surface, the arithmetic average of the principal curvatures at that point.

mean daily motion *Astronomy.* the angle an object would traverse in a day if its orbital motion were perfectly uniform.

mean density of matter *Astrophysics.* the mass of the universe divided by its volume.

mean depth *Hydrology.* the average depth of water in a stream, calculated by dividing the cross-sectional area of the stream by its width at the surface.

meander *Hydrology.* **1.** one of a series of bends, twists, curves, or other such shifts in the course of a stream or river. **2. meanders.** a series of these shifts. **3.** to move in such a pattern. (From the ancient name of a river in Asia Minor that was noted for its twisting course.)

meander

meander belt *Geology.* a zone along a valley floor that confines a meandering stream as it shifts its channel.

meander core *Geology.* **1.** a hill that is completely or almost completely encircled by a stream meander. Also, ROCK ISLAND. **2.** an isolated hill that is the remnant of a meander spur. Also, CUTOFF SPUR.

meander corner *Cartography.* in surveying, a corner station established on the meander line of a body of water at the point where a quarter-section line intersects the meander line. Also, SPECIAL MEANDER CORNER.

meandering stream *Hydrology.* a stream that has a pattern of consecutive meanders. Also, SNAKING STREAM.

meander line *Cartography.* in surveying, a traverse of the bank or shore of a permanent natural body of water at the level of mean or ordinary high water.

meander niche *Geology.* a crescent-shaped opening formed on a cave wall by stream erosion. Also, WALL NICHE.

meander plain *Geology.* a level or gently sloping plain formed as a result of lateral accretion or by the meandering process.

meander scar *Geology.* **1.** a crescent-shaped indentation on the face of a bluff or valley wall formed by the lateral expansion of a stream meander undercutting the bluff. Also, **meander scarp. 2.** an area that was once inhabited by a meandering stream.

meander spur *Geology.* an undercut, high land structure that projects out into the concave part of, and is enclosed by, a stream meander.

mean difference *Statistics.* the average distance of observations from some measure of central tendency; a measure of variation. Also, **mean deviation.**

mean diurnal high-water inequality *Oceanography.* the difference in height between the mean of all high waters and the mean of the higher high waters, as calculated for a given place over a period of 19 years.

mean diurnal low-water inequality *Oceanography.* the difference in height between the mean of all low waters and the mean of the lower low waters, as calculated for a given place over a period of 19 years.

mean draft *Naval Architecture.* the average of a vessel's draft forward and aft.

mean effective pressure *Mechanical Engineering.* the equivalent average pressure exerted by the piston throughout the stroke in an air compressor; used in the evaluation of positive-displacement machinery. Also, MEAN PRESSURE.

mean equator and equinox *Astronomy.* a reference system for celestial position indicating that it has been corrected for precession only.

mean free path *Physics.* the average distance traveled by a particle between collisions with another particle in the system; more specifically: *Acoustics.* the average distance between reflections of sound waves in an enclosure, given by the equation $d = 4V/S$, with V equal to the volume and S equal to the area. *Materials Science.* the average distance between collisions of electrons in the lattice of a material, an indicator of conductivity in metals. A long mean free path permits high mobilities and high conductivities.

mean height of burst *Ordnance.* the average of the heights of burst of a group of shots fired with the same firing data.

mean higher high water *Oceanography.* the average height of all the daily higher high waters at a given place, usually over a period of 19 years.

mean high water *Oceanography.* the average height of all high waters at a given place, usually over a period of 19 years.

mean high-water lunitidal interval *Oceanography.* the average period of time between the transit (upper or lower) of the moon over the local or Greenwich meridian and the next high water.

mean high-water neaps *Oceanography.* the average height of all high waters at quadrature (neap tides) at a given place, usually over a period of 19 years.

mean high-water springs *Oceanography.* the average height of all high waters at syzygy (spring tides) at a given place, usually over a period of 19 years.

mean latitude see MIDDLE LATITUDE.

mean life *Physics.* the average amount of time an unstable radioisotope exists before it decays, equal to the reciprocal of the decay constant.

mean lower low water *Oceanography.* the average height of all the daily lower low waters at a given place, usually over a period of 19 years.

mean low water *Oceanography.* the average height of all low waters at a given place, usually over a period of 19 years.

mean low-water lunitidal interval *Oceanography.* the average period of time between the transit (upper or lower) of the moon over the local or Greenwich meridian and the next low water.

mean low-water neaps *Oceanography.* the average height of all low waters at quadrature (neap tides) at a given place, usually over a period of 19 years.

mean low-water springs *Oceanography.* the average height of all low waters at syzygy (spring tides) at a given place, usually over a period of 19 years.

mean map see MEAN CHART.

mean motion *Astronomy.* the unvarying angular velocity that an object moving in an elliptical orbit of a given size would need to make one revolution.

mean neap rise *Oceanography.* the height of mean high-water neaps above the tidal datum.

mean normal stress *Mechanics.* the algebraic mean of the three principal stresses in a solid.

mean orbital elements *Astronomy.* the elements of a reference orbit chosen to approximate a real orbit that is perturbed by the presence of another object.

mean parallax *Astronomy.* the distance, derived by means of statistical studies of brightnesses and motions, for a group of stars whose individual distances are unmeasurable.

mean place *Astronomy.* an object's celestial position as determined for a given mean equator and equinox viewed from the barycenter of the solar system.

mean point of impact *Ordnance.* the point whose coordinates are the arithmetic means of the coordinates of the separate points of impact or burst of a finite number of projectiles fired or released at the same aiming point under a given set of conditions. Also, CENTER OF BURST.

mean point of impact error *Ordnance.* the distance between the target and the mean point of impact.

mean power (of a radio transmitter) *Electronics.* in normal operation, the power supplied to the antenna transmission line by a transmitter, averaged over a time sufficiently longer than the period of the lowest frequency encountered in the modulation.

mean pressure see MEAN EFFECTIVE PRESSURE.

mean profile *Astrophysics.* the shape of a pulsar's pulse as determined by averaging several pulses.

mean proportional *Mathematics.* the geometric mean; i.e., for two nonnegative real numbers a and b, the value $(ab)^{1/2}$; so named because $a{:}(ab)^{1/2} = (ab)^{1/2}{:}b$.

mean range *Oceanography.* the difference in height between mean high water and mean low water.

mean rise interval *Oceanography.* the average period of time between the transit (upper or lower) of the moon over the local or Greenwich meridian and the middle of the period of flood tide.

mean rise of tide *Oceanography.* the height of mean high water above the tidal datum.

mean river level *Hydrology.* at any point on a river, the average height of the surface for all stages of the tide over a given period, usually 19 years.

means *Mathematics.* in a proportional relationship $a/b = c/d$, the quantities b and c are the means; a and d are called the extremes.

mean sea level see SEA LEVEL.

means-ends analysis *Psychology.* a problem-solving technique in which the solver moves toward the solution by developing subgoals, each of which assists in moving toward the end. *Artificial Intelligence.* a machine reasoning technique that compares a starting point to a goal and attempts to reduce the difference between them by the application of relevant operators.

mean sidereal time *Astronomy.* the hour angle of the mean equinox for a given observer.

mean solar day *Astronomy.* the period during which the earth makes one rotation on its axis in relation to the mean sun; a period of 24 hours. Thus, **mean solar minute, mean solar second.**

mean solar time *Astronomy.* time measured according to the hour angle of the mean sun. Also, **mean time.**

mean specific heat *Thermodynamics.* for a specified temperature range, the average value of the specific heat of a substance over the range.

mean speed *Transportation Engineering.* the average speed of a number of vehicles, defined as the sum of their individual speeds divided by the number of vehicles.

mean sphere depth *Geodesy.* the depth to which water would uniformly cover the surface of the earth, if the earth's solid surfaces were smoothed and were parallel to the surface of the geoid.

mean spherical intensity *Optics.* the luminosity of a light source averaged from all directions.

mean spring rise *Oceanography.* the height of mean high-water springs above the tidal datum. Also, SPRING RISE.

mean square deviation see VARIANCE.

mean stress *Mechanics.* the algebraic mean of the local greatest and smallest values assumed by a variable stress.

mean sun *Astronomy.* a theoretical sun used to provide standard measurements of time; it moves along the celestial equator at a uniform rate equal to the average motion of the actual sun.

mean temperature *Meteorology.* the average air temperature over a given time period such as a day, month, or year, as recorded by a thermometer that is properly exposed to the flow of free air around it.

mean temperature difference *Chemical Engineering.* the overall temperature difference between the hot and cold fluids in a heat exchanger.

Meantes *Vertebrate Zoology.* a former name for eel-like salamanders in the southeastern United States, currently classified in the suborder Sirenoidea.

mean tide level *Oceanography.* the reference plane halfway between high water and mean low water. Also, HALF-TIDE LEVEL.

mean time between failures *Computer Technology.* a measure of system reliability; an estimate of the ratio of system-operating time to the number of failures that occur during that time.

mean time to failure *Engineering.* the average amount of time expected to elapse prior to the first failure of a piece of equipment.

mean time to repair *Engineering.* the average amount of time expected to elapse between repairs to a piece of equipment.

mean time to restore *Transportation Engineering.* the average time needed to return to normal service after a failure; determined by dividing the total maintenance time by the number of failures.

mean trajectory *Mechanics.* the ballistic path of a projectile to the point of impact, neglecting wobbles.

mean value *Mathematics.* let $f(x)$ be integrable on $[a,b]$. The mean value of f on (a,b) is $\int_a^b f(x)/(b-a)dx$.

mean value theorem of differential calculus *Mathematics.* **1.** suppose a real function f is continuous in $[a,b]$ and is differentiable in (a,b). Then there is a point $x_0 \in (a,b)$ such that

$$f(x_0) = [f(b) - f(a)]/(b - a).$$

2. in general, suppose that the real functions f and g are continuous in $[a,b]$ and differentiable in (a,b), and $g'(x) \neq 0$ in (a,b). Then there exists a point $x_0 \in (a,b)$ such that

$$f'(x_0)/g'(x_0) = [f(b) - f(a)]/[g(b) - g(a)].$$

For $g(x) = x$, this is the ordinary mean value theorem. Also, GENERALIZED MEAN VALUE THEOREM.

mean value theorem of integral calculus *Mathematics.* **1.** suppose a real function f is integrable over $[a,b]$ and real numbers m and M exist so that $m \leq f(x) \leq M$ in $[a,b]$. Then there exists a number μ with $m \leq \mu \leq M$ such that

$$\int_a^b f(x)dx = \mu(b - a).$$

2. in general, suppose the real functions f and g are integrable over $[a,b]$, $g(x)$ is either strictly nonnegative or nonpositive in $[a,b]$, and real numbers m and M exist such that $m \leq f(x) \leq M$ in $[a,b]$. Then there exists a real number μ with $m \leq \mu \leq M$ such that

$$\int_a^b f(x)g(x)dx = \mu\int_a^b g(x)dx.$$

For $g(x) = x$, this is the ordinary mean value theorem. Also, GENERALIZED FIRST MEAN VALUE THEOREM OF INTEGRAL CALCULUS. **3.** suppose the real function $f(x)$ is monotonic and bounded and the real function $g(x)$ is integrable over $[a,b]$. Then there is at least one real number ξ with $a \leq \xi \leq b$ such that

$$\int_a^b f(x)g(x)dx = f(a)\int_a^\xi g(x)dx + f(b)\int_\xi^b g(x)dx.$$

Also, SECOND MEAN VALUE THEOREM OF INTEGRAL CALCULUS.

mean velocity *Physics.* **1.** the average velocity of particles in a system of particles. **2.** the average velocity of a body over a specified time interval, given by the time integral of the velocity divided by the total amount of time.

measles *Medicine.* **1.** a highly contagious infectious disease caused by a paramyxovirus, characterized by a runny nose, enlarged neck glands, light sensitivity, muscle aches, general discomfort, and fever in the early stages. A rash of red papules follows, appearing first on the face and neck and then spreading to the trunk and limbs. It is common among children, but also occurs in nonimmune adults. Also, RUBEOLA. **2.** an infection with the larvae of tapeworms that occurs commonly in domestic animals.

measles immune globulin *Immunology.* a sterile solution of globulins prepared from immune serum globulin and administered to induce passive immunity against measles.

measles-mumps-rubella vaccine see MMR.

measles vaccine *Immunology.* a suspension of the measles virus that has been cultivated in a laboratory or made inactive; used to induce active immunity against measles.

measles virus *Virology.* a virus of the genus *Morbillivirus,* family Paramyxoviridae, that causes measles in humans. Also, RUBEOLA VIRUS.

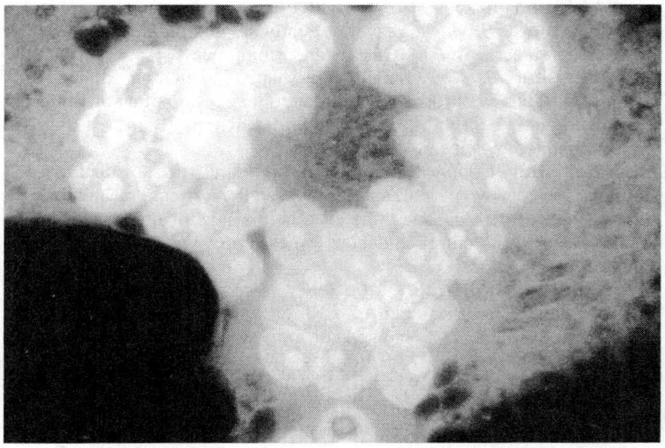

measles virus

measurable function *Mathematics.* **1.** let f be a function from a measure space (X,A,m) to a measure space (Y,B,μ). If the inverse image under f of any B-measurable set in Y is an A-measurable set in X, then f is said to be an **A-measurable function**. **2.** suppose (Y,B,μ) is the real number with the usual definition of distance. Then the (extended) real-valued function f defined on a measure space (X,A,m) is measurable if, for every real number a, $\{x \in X: f(x) \geq a\}$ is an A-measurable set.

measurable set *Mathematics.* the measurable sets of a measure space (X,A,m) are the members A of the sigma field A. Also, **A-measurable set**. The measure of such a set A is denoted $m(A)$. Any set that is not a member of A is said to be nonmeasurable with respect to m.

measurable space *Mathematics.* a set together with a sigma algebra of its subsets; the addition of a measure makes it a measure space.

measure *Science.* **1.** to use standard units to determine the dimensions of something. **2.** a quantity obtained by such a process. *Mathematics.* a (positive) measure m on a sigma field A over a topological space X is a countably additive (positive) set function from A into the extended (nonnegative) real numbers. The measure m is said to be sigma finite (usually written σ-finite) if X is expressible as a countable union of sets $A_i \in A$, such that $m(A_i) < \infty$; m is said to be finite if $m(X) < \infty$.

measured angle *Cartography.* an angle as read directly from surveying instruments, uncorrected for any distorting conditions.

measured mile *Civil Engineering.* a distance of a mile that has been measured and marked. *Navigation.* a length of one nautical mile that has been accurately measured and is marked in some manner, usually by ranges ashore.

measure of central tendency *Statistics.* a descriptive statistic that conveys information about the central location of a distribution, such as a mean, mode, or median of the distribution.

measure of variation *Statistics.* a descriptive statistic that conveys information about the spread of a distribution, such as the range, variance, mean deviation, or standard deviation of the distribution.

measure ring *Mathematics.* **1.** a Boolean s-ring S endowed with a positive measure μ; denoted (S,μ). **2.** In particular, let S_μ be the Boolean s-ring derived from (S,μ) by classifying all sets differing by a set of measure zero to be equal. If (X,S,μ) is a measure space, then (S_μ,μ) is a measure ring. (S_μ,μ) is known as the measure ring associated with X or simply the measure ring of X.

measure space *Mathematics.* a (measurable) space on which a measure is defined; in particular, a (usually topological) space X together with a sigma field A of X, endowed with a measure m; denoted (X,A,m). Sometimes no distinction is made between measure space and measurable space.

measure zero *Mathematics.* **1.** a subset A of a measure space (X,A,m) is said to be of measure zero if A is measurable and if $m(A) = 0$. **2.** In particular, a subset A of (Euclidean) R^n has measure zero if A can be contained in a countable collection of n-dimensional rectangles of arbitrarily small total volume.

meatotomy *Medicine.* the dilation of the urinary meatus by incision.

meat poisoning *Toxicology.* poisoning due to the ingestion of toxins produced by microorganisms in meat; symptoms may include severe gastrointestinal disturbances. Also, KREOTOXISM.

meatus *Anatomy.* an opening, especially of a canal or passageway in the body.

mech. mechanical; mechanism; mechanics.

mechanic *Mechanical Engineering.* a person who repairs and maintains machines, motors, or the like, particularly vehicle engines.

mechanical *Engineering.* relating to, derived from, using, or produced by mechanical devices, machinery, or tools. *Physics.* relating to, determined by, or affected by physical forces. *Graphic Arts.* in photolithography, a sheet of stiff paper on which the copy for a given plate has been mounted.

mechanical acne *Medicine.* aggravation of existing acne lesions by mechanical factors such as rubbing, pinching, or pulling.

mechanical advantage *Mechanical Engineering.* for a machine that transmits mechanical energy, the ratio of the output force to the input force; i.e., the work produced by the machine, divided by the force applied to it. Actual machines often provide a mechanical advantage that is greater than unity; however, the greater the mechanical advantage, the greater the distance that the input force must move in relation to the output force.

mechanical alloying *Metallurgy.* a process for producing composite metal powders with controlled, extremely fine microstructures, involving balanced repetitive welding, fracturing, and rewelding of a mixture of powder particles in a dry, highly energetic ball charge.

mechanical antidote *Toxicology.* an antidote that prevents or retards the absorption of a poison.

mechanical area *Building Engineering.* those areas within a building that include equipment rooms and passages through which air circulates and where piping, communication, hoisting, conveying, and electrical services are installed.

mechanical balance *Analytical Chemistry.* a balance in which the sample weight is determined by comparison with a calibrated weight.

mechanical birefringence *Optics.* the phenomenon by which normally isotropic materials are made birefringent by the application of mechanical stress. Also, STRESS BIREFRINGENCE, PHOTOELASTICITY.

mechanical classification *Mechanical Engineering.* a sorting method in which mixtures of particles of different sizes are separated into fractions by water.

mechanical classifier *Mechanical Engineering.* a machine used to identify or classify mixtures of particles of different sizes.

mechanical comparator *Mechanical Engineering.* a comparator that utilizes a rack, pinion, and pointer or a parallelogram arrangement to amplify movement.

mechanical damping *Acoustical Engineering.* mechanical resistance that acts to decrease the amplitude of vibrational motion and thus decrease the amount of sound produced by the vibrating object.

mechanical design *Engineering.* the design of machines, mechanical devices, and the like using principles and techniques of both design engineering and mechanical engineering.

mechanical device *Engineering.* a tool, apparatus, or machine; often but not always distinguished from an electrical device.

Mechanical Devices

The use of mechanical devices can be traced back as far as the making of fire by striking rocks together to create sparks. Later, as trade came into practice, a balance was invented for the purpose of measuring weight. After the steam engine became a reality in the late 18th century, steam-powered locomotives shortened travel time on land and steamships made a shorter voyage possible on water. Mechanical devices such as generators followed. Motors replaced horses and manpower. Speed of production increased manyfold within a short time. These events were part of what is now known as the Industrial Revolution.

Two world wars accelerated the invention of mechanical devices. Modern aircraft engines made it possible to drop bombs from the air. Mechanical devices were now being used to enhance destructive power rather than to raise creative efficiency. The launch of Sputnik, the landing on the moon, and outer space exploration all required ingenious mechanical devices, ranging from huge rockets to tiny precision-control mechanisms, to accomplish a successful mission.

In the environmental control area, mechanical devices such as pumps are used to drain or irrigate the land, fans are used to ventilate mines, and refrigerators are used to store food.

During recent years, a new bioengineering field has opened up for better health care. Mechanical devices such as heart-lung machines, artificial limbs, and kidney machines permit vital functions to be maintained in seriously injured or diseased patients .

Depending upon the motivation, mechanical devices can be used either for destructive purposes or for improving the living standard of billions of people on this earth.

Charles E. S. Ueng
Professor of Civil Engineering
Georgia Institute of Technology

mechanical dewaxing *Petroleum Engineering.* a method of separating solid wax particles from oil by forcing cooled oil through wax presses.
mechanical dialer see AUTOMATIC DIALER.
mechanical differential analyzer *Computer Technology.* an analog computer that uses interconnected mechanical surfaces to solve differential equations.
mechanical draft *Mechanical Engineering.* a draft that depends upon the use of mechanical devices such as fans.
mechanical drawing *Graphic Arts.* a drawing made with the aid of an instrument such as a compass or square, as opposed to a freehand drawing.
mechanical dysmenorrhea *Medicine.* painful menstruation due to obstruction of the menstrual flow by clots or malposition of the uterus.
mechanical efficiency *Mechanical Engineering.* the ratio of the brake or useful power to the indicated power developed in the cylinder of an engine.
mechanical engineering *Engineering.* the branch of engineering concerned with the conception, research, design, operation, and maintenance of machines.

Mechanical Engineering

The practice of mechanical engineering long predates its name. It goes back to the design and manufacture of machines and an understanding of their function and motion in the early years of military engineering and transportation by wheeled vehicles. Conversion of appreciable energy for use in manufacture came much later with water wheels. Limited in location, they gradually were supplanted in the 1800's by the steam engine. The large stationary engines of the second half of the 1800's and those that powered locomotives and steamships permitted rapid industrial development and captured the imagination of the public. They led to the identification of the broad field of endeavor called mechanical engineering and the establishment of mechanical engineering societies.

Steam and then gas turbines were developed to generate still greater power both locally and at a distance for transmission to points of use. Transportation needs led to the internal combustion engine, the jet engine, and the rocket engine. Environmental control, including refrigeration, heating, and air conditioning, added to the list along with complete systems such as steam power plants for generating electrical power. A wide variety of useful machines and mechanical devices were designed, developed, and improved.

Prompted by failures of boilers due to avoidable causes, experience gained in practice was recorded and made available to all through codes and standards and in publications addressed to the engineer in practice. Similarly, the mechanical engineering science base was developed and presented in an ever-increasing number of journals devoted to basic studies. Fundamental engineering sciences such as applied mechanics, bioengineering, tribology, thermodynamics and heat transfer have mechanical engineering journals of their own, as do the applied areas. Engineering science bases are being developed for integrated manufacturing systems and other aspects of mechanical engineering that do not yet have a satisfactory underpinning.

Daniel C. Drucker
Professor of Mechanical Engineering
University of Florida

mechanical equation of state *Metallurgy.* any of several equations that relate various parameters of plastic deformation, such as stress, strain, and temperature.
mechanical equivalent of heat *Thermodynamics.* a unit quantity of thermal energy in terms of mechanical work units, based on the fact that heat transfer and work are interconvertible in fixed ratio; now recognized as the specific heat capacity of water, which is 4.186 kJ/kg.
mechanical equivalent of light *Optics.* the relationship between the power of a monochromatic light, with a wavelength of approximately 555 nanometers (the most sensitive point of a phototopic vision), and the light's luminous flux, measured in lumens.
mechanical erosion see CORRASION.
mechanical filter *Electronics.* an electromechanical filter designed to have sharp cutoff frequency discrimination. Also, MECHANICAL WAVE FILTER. *Petroleum Engineering.* a filter used to remove suspended floc and undissolved solids from treated waterflood water; it consists of a steel shell packed with granules of materials such as sand, gravel, anthracite coal, or graphitic ore.
mechanical galvanization *Materials Science.* a cold-welding process for zinc coating of steel parts, which results in a thin, uniform zinc coating of controlled thickness.
mechanical gripper *Robotics.* a gripper with movable, fingerlike components.
mechanical hygrometer *Mechanical Engineering.* a hygrometer that mechanically links a pointer to an organic material, which expands and contracts with changes in moisture in the surrounding gas.
mechanical hysteresis *Mechanics.* **1.** the effect that a system's previous history has upon the current relationship between its configuration and its stored energy. **2.** see ELASTIC HYSTERESIS.
mechanical impedance *Mechanics.* the ratio of the complex excitation of a system to its complex response at a given frequency.
mechanical inoculation *Virology.* the process of inoculating a plant by rubbing a viral extract on the leaves so that the virus can enter through the small wounds in the leaf cuticle and epidermal cell walls.
mechanical instability see ABSOLUTE INSTABILITY.
mechanical lift dock *Civil Engineering.* a drydock in which a vessel is positioned on a cradle and lifted vertically out of the water.
mechanical linkage *Mechanical Engineering.* a set of rigid bodies joined together at pivots by means of pins or other small devices.
mechanical mass *Quantum Mechanics.* the portion of a particle's measured mass that supposedly is due to its own intrinsic properties.
mechanical metallurgy *Metallurgy.* the study of the effects of mechanical forces upon metals.
mechanical modulator *Electronics.* a device that varies some characteristic of a carrier wave by physically moving or changing a circuit element in order to transmit information.
mechanical mucking *Mining Engineering.* the use of machines to remove dirt or stone from a tunnel or mine.
mechanical mule *Ordnance.* a half-ton infantry light weapons carrier used by the U.S. military; it is powered by a 17-hp engine, with a top speed of approximately 25 mph and a range of around 50 miles.
mechanical oil valve *Petroleum Engineering.* a float-operated valve that controls the flow of liquid out of gas-oil separator tank systems in an oil lease.
mechanical pencil *Mechanical Devices.* a fine lead pencil whose lead is accessed by turning or pressing a screw.
mechanical plating *Metallurgy.* a process of applying a metallic coating by embedding a fine powder on a surface.
mechanical press *Mechanical Engineering.* a press whose slide is operated by mechanical means.
mechanical property *Mechanics.* any property of a material that governs its response to external forces.
mechanical pulping *Materials Science.* the recovery of cellulose fibers in wood; the finely ground wood is made into splint, newsprint, manila papers, and tissues.
mechanical pump *Mechanical Devices.* a pump in which fluid is conveyed by contact with a moving rotor or piston in a pumping machine.
mechanical reactance *Mechanics.* the imaginary part of mechanical impedance.
mechanical rectifier *Electricity.* a device, such as a synchronous vibrator, that converts AC current into DC by mechanical action.
mechanical refrigeration *Mechanical Engineering.* the utilization of a refrigerant for cycles of refrigerating thermodynamics and a mechanical compressor wherein heat is removed.
mechanical resistance see RESISTANCE.

mechanical rotational impedance see ROTATIONAL IMPEDENCE.
mechanical rotational reactance see ROTATIONAL REACTANCE.
mechanical rotational resistance see ROTATIONAL RESISTANCE.
mechanical scale *Mechanical Devices.* a scale that utilizes various levers with precise fulcrums, thus allowing heavy items to be weighed using counterweights or counterpoises.
mechanical seal *Mechanical Engineering.* a mechanical assembly in which a leakproof seal is formed between flat, rotating surfaces to prevent leakage from high-pressure pumps.
mechanical sediment see CLASTIC SEDIMENT.
mechanical separation *Mechanical Engineering.* a separation procedure in which solid particles or liquid drops are removed from a gas or liquid, or are separated into fractions by size, or both, by means of gravity separation, centrifugal action, and filtration.
mechanical setting *Mechanical Engineering.* the production of beaded material by setting diamonds in a bead mold and placing a cast or powder metal to embed them and form a bead crown.
mechanical shovel *Mechanical Engineering.* a loader limited to level or slightly graded drivages; the shovel is swung over the machine, and the load is delivered into containers or vehicles such as mine cars.
mechanical splice *Engineering.* a splice between two wire ropes made by pressing metal sleeves over the junction.
mechanical stage *Engineering.* a microscope stage fitted with a mechanical device that is used to position the slide.
mechanical stepping motor *Electricity.* a motor that spins in precise, predetermined increments or steps.
mechanical sweep *Military Science.* in naval mine warfare, any sweep used with the intention of physically contacting the mine or its appendages.
mechanical system *Mechanical Engineering.* a device that converts on-site electrical energy into motion energy; applications include electrical, hydraulic, and pneumatic. *Robotics.* any robotic assembly.
mechanical telemetry *Telecommunications.* telemetry that is effected by short mechanical hook-ups, or by propagation of pressure and/or acoustic waves through water.
mechanical tilt *Electronics.* **1.** the vertical tilt of the mechanical axis of an antenna. **2.** the angle of tilt displayed on the tilt indicator dial.
mechanical torque converter *Mechanical Engineering.* a torque converter that transmits power only with incidental losses.
mechanical translation *Computer Technology.* the translation of human languages performed by computers or other devices. Also, MACHINE TRANSLATION.
mechanical turbulence *Meteorology.* any irregular movement of air in the lower atmosphere that is caused by a physical obstruction, such as a skyscraper.
mechanical twin *Crystallography.* a crystalline twin formed by mechanical deformation.
mechanical unit *Metrology.* any unit defined in terms of length, time, mass, or other physical qualities.
mechanical vibration *Mechanics.* **1.** vibration, often complex, occurring in machines and structures. **2.** the oscillatory motion about an equilibrium position of a particle or body.
mechanical wave filter *Electronics.* see MECHANICAL FILTER.
mechanical weathering *Geology.* the breakdown and fragmentation of rock as a result of natural physical forces without chemical changes. Also, DISINTEGRATION, PHYSICAL WEATHERING.
mechanical working *Materials Science.* the process of subjecting a material to some type of plastic deformation, such as rolling or forging.
mechanical yielding prop *Mining Engineering.* a steel prop used to control yield by the action of friction between two sliding surfaces or telescopic tubes. Also, FRICTION YIELDING PROP.
mechanics *Physics.* a branch of physics that focuses on motion and on the reaction of physical systems to internal and external forces. (From a Greek word meaning "machine.")

Mechanics

Classical mechanics is the oldest of all the physical sciences. It grew out of man's curiosity about the physical laws governing the motion of terrestrial and celestial bodies.

Early developments were based on extensive observation and measurements that were made by such ingenious pioneers as Galileo Galilei, Nicolaus Copernicus and Johannes Kepler. Their contributions formed an indispensable part of the foundation on which Sir Isaac Newton proclaimed in 1687 his three laws of motion and the law of universal gravitation. These discoveries have been acclaimed as the greatest of all times.

Numerous concentrated studies have ensued, some with the ablest minds expounding on the philosophical bases of the profound significances of space, time, mass, force, momentum, energy, inertial frame of reference, etc. They have fortified advances of the subject, refined the criteria of validity of Newtonian mechanics, and led to the development of the theory of relativity in 1915 by Albert Einstein for motions involving velocities that may approach the speed of light and for the quest to combine the gravity and electrodynamic forces into a unified theory.

With this growth has come the new era of modern astrophysics and cosmology in which the concepts of singularities (black holes) and nonlinear phenomena (solitons) have emerged. In another direction concerning the behavior of atoms and subatomic particles, classical mechanics provided the conceptual analogies needed for the development of quantum mechanics by Max Planck.

In the history of science, physical conceptions often arose from observations before the required mathematical tools were available for calculation of solutions. For mechanics, Newton was inspired to "invent" calculus in order to solve the differential equations that appeared in his laws of motion; likewise was Gottfried Leibniz. The elegant formulation introduced by Joseph-Louis Lagrange and the canonical formulation by Sir William Hamilton based on the principle of least action stimulated the development of calculus of variation. Subsequent extensions of particle mechanics, led by Leonhard Euler and Augustin Louis Cauchy, to studies of fluid mechanics and solid mechanics have promoted developments of the theory of partial differential equations.

Recent advances in nonlinear mechanics have been spectacular. The development of soliton theory and the mathematics of chaos have opened new colorful chapters of modern applied mathematics and mechanics. Physically, it is remarkable that quite similar nonlinear phenomena have been observed in common to various systems in different disciplines. Solitons, for example, have been observed to occur not only in shallow water as was first discovered by John Scott Russell in 1834, but also in the ocean along the internal layer of pycnocline, in ionized gases, nonlinear optics and acoustics, super-conducting materials, in the theory of fundamental particles, in astrophysical media, and even in biochemically active macromolecules.

Progress in nonlinear mechanics has been vigorous and it undoubtedly will furnish a unified approach towards the science of nonlinear systems, whatever their origin.

Theodore Yaotsu Wu
Professor of Engineering Science
California Institute of Technology

mechanism *Mechanical Engineering.* **1.** the part of a machine containing two or more pieces that are arranged for the motion of one piece to force the motion of the others. **2.** any assembly of moving parts that perform one functional motion, usually part of a large machine. **3.** the structure or arrangement of parts in a machine. **4.** broadly, any machine or mechanical part. *Psychology.* see DEFENSE MECHANISM.

mechanization *Mechanical Engineering.* the use of power machines to replace manual labor.

mechanize *Mechanical Engineering.* to use machinery, especially as opposed to human or animal operation. *Industrial Engineering.* to equip a facility with machinery, especially when this replaces manual labor or simple machines. *Military Science.* to equip a military force with armed and armored vehicles that can be used for fighting, as opposed to **motorize**, which means to equip with transportation vehicles.

mechanized *Industrial Engineering.* **1.** equipped or performed with machinery. **2.** heavily dependent upon machinery; automated.

mechanized gun *Ordnance.* a gun mounted on and fired from a motor vehicle.

mechano- a combining form denoting a machine or mechanical process.

mechanocaloric effect *Physics.* an effect that is observed in helium II, whereby a temperature gradient is always accompanied by a pressure gradient, and vice versa.

mechanochemical effect *Physical Chemistry.* a change in the size of certain polymers, such as photoelectrolytic gels, brought on by changes in their chemical environment.

mechanochemistry *Physical Chemistry.* the branch of physical chemistry that studies how polymers convert mechanical energy into chemical energy.

mechanocyte see FIBROBLAST.

mechanomotive force *Mechanics.* the root mean square value of a cyclic force.

mechanophotochemistry *Physical Chemistry.* the branch of chemistry that studies the changes occurring in certain polymers when they are exposed to light.

mechanoreceptor *Physiology.* a receptor cell that is sensitive to mechanical stimulation, such as touch, pressure, or tension.

mechanotherapy *Medicine.* the use of mechanical devices and apparatus in the treatment of disease or symptoms, especially as an aid in performing physical therapy.

mechlorethamine *Oncology.* a nitrogen mustard compound used as an antineoplastic to treat Hodgkin's disease, mycosis fungoides, and other lymphomas.

Meckel, Johann F. 1781–1833, German anatomist; wrote extensive early descriptions of birth defects.

Meckel's cartilage see MANDIBULAR CARTILAGE.

Meckel's diverticulum *Developmental Biology.* a blind sac or pouch, the remnants of the omphalomesenteric duct of the embryo, which is on the ileum a short distance above the cecum and may be attached to the umbilicus.

meconin *Organic Chemistry.* $C_{10}H_{10}O_4$, a colorless to white, needlelike solid; soluble in hot water, alcohol, and ether; melts at 102°C; the neutral component of opium.

meconium *Developmental Biology.* the first intestinal discharges of the newborn infant, greenish in color and consisting of epithelial cells, mucus, and bile.

meconium ileus *Medicine.* an intestinal blockage in newborns caused by thickened meconium; seen frequently in cystic fibrosis.

Mecoptera *Invertebrate Zoology.* the scorpion flies, an order of predaceous four-winged insects that live in damp areas, having the head elongated into a beak with chewing mouthparts.

mecystasis *Physiology.* a process during which a muscle fiber increases in length, resists stretch, contracts, and relaxes, and manifests the same tension as before elongation.

med. medicine; medium.

MED or **M.E.D.** minimal effective dose; minimal erythema dose.

medallion *Architecture.* an ornamental tablet or panel resembling a large coin or medal, usually oval or circular in shape and decorated with figures in relief. Thus, **medallion molding.**

Medawar, Sir Peter B. born 1915, British zoologist; shared the Nobel Prize with M. Burnet for work on transplanting human organs.

Medea complex *Psychology.* the unconscious wish of a mother for the death of her children as a means for revenge against her husband. (From the Greek myth of *Medea,* who killed her own children after her husband deserted her for another woman.)

Medeolariaceae *Mycology.* the single family of fungi of the order Medeolariales, containing species that are exclusively parasitic to the *Medeola virginiana* of North America.

Medeolariales *Mycology.* an order of fungi of the class Discomycetes, characterized by asymmetric, ribbed ascospores.

medfly or **Medfly** see MEDITERRANEAN FRUIT FLY.

media a plural of *medium*; specific uses include: *Telecommunications.* the various means of communication that reach a wide audience, such as newspapers and television. This sense of *media* is sometimes regarded as a collective singular, but in formal usage it should be construed as plural. *Medicine.* middle, especially in Latin terms such as *otitis media. Anatomy.* see TUNICA MEDIA. *Histology.* the tunica muscularis in the wall of a blood vessel. *Entomolology.* a longitudinal vein in the midportion of the insect wing.

media- or **medi-** a combining form meaning "middle."

media conversion *Computer Technology.* the changing of data from one storage medium to another, as from magnetic disk to optical disk.

media conversion buffer *Computer Technology.* a temporary storage area used when converting data between storage media that have different reading and recording speeds.

mediad *Anatomy.* in the direction of a median plane or line.

medial *Anatomy.* toward the middle, or close to the midline of the body.

medial arteriosclerosis *Medicine.* a disease of the medium-sized outlying arteries, in which the medial tissues become fibrous and calcific. Also, **medial calcinosis.**

medial lemniscus *Anatomy.* a tract of nerve fibers passing through the middle of the medulla oblongata. Also, REIL'S RIBBON.

medial moraine *Geology.* **1.** a long, narrow moraine, usually formed by the joining together of adjacent lateral moraines, that is transported in or upon the middle of a glacier and parallel to its sides. **2.** a moraine that developed as a result of the glacial abrasion of a rocky outcrop near the middle of a glacier. **3.** the ridge of material left down the center of a glacial valley after the glacier has disappeared. Also, **median moraine.**

medial necrosis *Medicine.* the death of cells in the tunica media of an artery.

median a point, position, or value in the middle; specific uses include: *Statistics.* the value of the middle item when data are arranged in order of size; a measure of central tendency. *Anatomy.* situated in the median plane or the midline of a body or structure; medial. *Mathematics.* **1.** a line that joins a vertex of a triangle to the midpoint of the opposite side. **2.** given a finite set of values $a_1 \leq \cdots \leq a_n$, the value $\frac{1}{2}(a_n/2 + a_{(n+1)}/2)$, if n is even, or the value $a_{(n+1)}/2$, if n is odd; i.e., the middle element of the sequence.

median effective time *Virology.* the time that is required to produce a response in 50% of the individuals or cell cultures to which a viral dose is administered. Also, **ET_{50}.**

median effective dose *Pharmacology.* the amount of a drug that is required to produce a response in 50% of subjects to whom it is given. *Virology.* the amount of virus required to produce a response in 50% of the individuals or cell cultures to which it is administered. Also, **ED_{50}.**

median infective dose *Virology.* the amount of virus that is required to infect 50% of the individuals to whom it is administered. Also, **ID_{50}.**

median lethal concentration *Virology.* the concentration of a virus that is required to kill 50% of the subjects to whom it is administered. Also, **LC_{50}.**

median lethal dose *Virology.* the dose of a virus that is fatal to 50% of the subjects to whom it is administered. *Pharmacology.* the quantity of a substance that will kill 50% of the population exposed to it. *Military Science.* the amount of nuclear radiation or of a chemical agent that would be fatal to 50% of the exposed personnel in a given period of time. Similarly, **median incapacitating dose**.

median lethal time *Virology.* the period of time required for 50% of the subjects to whom a viral dose is administered to die. *Microbiology.* the amount of time necessary for half of a large microorganism population to die after the application of a measured harmful substance or agent, such as radiation or a drug.

median mass *Geology.* a structural unit or block between two orogenic belts of similar age that is less deformed than the bordering structures thrust away from it. Also, BETWIXT MOUNTAINS, ZWISCHENBIRGE.

median maxillary cyst *Medicine.* a sac enclosed in the connective tissue of the maxilla, close to the incisive canal.

median nasal process *Developmental Biology.* either of the two limbs of the horseshoe-shaped elevation bounding a nasal pit in the embryo; these limbs, together with the ipsilateral maxillary process, form half of the upper jaw.

median particle diameter *Geology*. in a rock or sediment, the diameter of a particle that is larger than the diameters of 50% of the particles and smaller than the diameters of the other 50%.

median plane *Anatomy*. the imaginary longitudinal plane passing through the middle of the body from front to back, dividing it into right and left halves.

median rift valley see RIFT VALLEY, def. 2.

median strip *Transportation Engineering*. a central reservation dividing the two carriageways of a roadway, serving as a buffer zone between traffic flowing in opposite directions.

median survival time *Virology*. the duration that 50% of the subjects can survive following the administration of an injurious agent, such as a virus, a drug, or radiation.

median tissue culture infective dose *Virology*. the dose of a virus that will infect 50% of susceptible tissue culture cells. Also, **$TCID_{50}$**.

mediastinitis *Medicine*. an inflamed condition in the mediastinum.

mediastinoscope *Medicine*. an endoscope used to examine the mediastinum.

mediastinoscopy *Medicine*. examination of the mediastinum using an endoscope inserted through an anterior midline incision just above the thoracic inlet.

mediastinum *Anatomy*. **1.** the group of organs and tissue between the lungs, from the sternum in front to the vertebral column behind, and from the thoracic inlet above to the diaphragm below; includes the heart, its large vessels, and pericardium; the trachea and bronchi; the esophagus; the thymus; several lymph nodes; the thoracic duct; and other structures and tissues. **2.** any middle divider or septum.

medical *Medicine*. **1.** relating to or involving the science or practice of medicine. **2.** relating to medicine, as opposed to surgery. **3.** therapeutic or curative; medicinal. **4.** relating to health in general.

medical anthropology *Anthropology*. an approach to medicine in which the traditional belief system of a patient's culture is incorporated as a valid aspect of treatment; e.g., a shaman may be consulted in conjunction with a conventional physician.

medical bacteriology *Medicine*. the study of microorganisms as they affect the health of humans.

medical climatology *Medicine*. the study of climate as it affects human health and activities, and the treatment of disease.

medical entomology *Medicine*. the discipline that examines the insects that cause disease in humans.

medical ethics *Medicine*. the moral standards that guide the practice of medicine.

medical examiner *Medicine*. **1.** a selected medical professional with a forensic pathology background who researches certain types of deaths for city, county, or state officials. **2.** a doctor who performs medical examinations for a public or private employer.

medical flat *Biotechnology*. a flat-sided bottle with a screw-on cap; available in a wide variety of sizes and used for storing specimens.

medical geography *Medicine*. the scientific study of the effects of locale and climate upon human health.

medical history *Medicine*. **1.** the part of a patient's background that is relevant to diagnosis and treatment of the present condition, including hereditary factors, occupational history, social environment, and past medical experiences. **2.** a written transcript of such information. Also, **medical records.**

medical imaging *Medicine*. any of a variety of processes used to produce images of body tissue and structures for diagnostic purposes.

medical microbiology *Medicine*. the analysis of disease-producing microorganisms and their effects on humans.

medical model *Psychology*. **1.** the theory that psychological disorders and behavioral abnormalities have physiological causes. **2.** the use of medical terminology to define elements of the practice of psychiatry or psychology; e.g., describing an abnormal psychological condition as a "disease" or an individual undergoing treatment as a "patient."

medical mycology *Medicine*. the investigation of fungi invading the human body and their relation to disease.

medical parasitology *Medicine*. the study of the relationship between parasites and human health.

medical protozoology *Medicine*. an area of study that examines the role of protozoa in human disease.

medical radiography *Medicine*. the use of X-ray film images to diagnose disease and study its effects on internal structures.

medical technician *Medicine*. a worker in a medical laboratory who has received formal training in laboratory techniques in a two-year associate degree program at a community college, a vocational technical school, or a private school.

medical technologist *Medicine*. a worker in a medical laboratory who is skilled in the performance of clinical laboratory procedures used in the diagnosis of disease and evaluation of patient progress. In general, the technologist has completed four years of specialized education in medical technology.

medicarpin *Biochemistry*. a compound that functions as a isofluridoside and as a phytoalexin in a number of flora, including the broad bean.

medication *Medicine*. **1.** a material used as medicine. **2.** a therapy employing medicine. **3.** the process of absorbing a medicine.

medicinal *Medicine*. **1.** possessing healing capacities. **2.** of or relating to medicine; medical.

medicinal oil see MINERAL OIL.

medicine *Medicine*. **1.** the discipline dedicated to preventing and treating disease, and to restoring and maintaining health. **2.** specifically, the treatment of disease or injury without the use of surgery. **3.** any material used as disease or pain therapy. (From a Latin word meaning "to heal.")

Medicine

Medicine is the discipline that is concerned with the preservation of health and the diagnosis and treatment of human disabilities. Individuals who are licensed to practice medicine or surgery are called medical doctors (M.D.s), or simply doctors who treat illness in patients. Other health care professionals include doctors of osteopathy and licensed nurse practitioners.

Throughout human history, some individuals have undertaken responsibility for the cure of illnesses, often by invoking rituals or administration of potions of dubious efficacy. Modern medicine awaited the scientific renaissance in Europe with study of human anatomy by Vesalius (1514–1564), physiology by Harvey (1578–1657), and later microbiology by Pasteur (1822–1895) and pathology by Virchow (1821–1902). Acquisition of new knowledge continues to accelerate as the profound insights from other disciplines such as genetics, biochemistry, and molecular biology provide new tools for diagnosis, and new drugs and devices for treatment.

Although the trends are for a longer life span, each new generation encounters new sets of disorders. Fewer deaths occur in infancy, many infectious diseases are controlled by immunization, and smallpox has been eradicated. Microorganisms change in virulence and prevalence. The plague of the middle ages is past, as is the pandemic of influenza in 1918, but AIDS (acquired immune deficiency disease) is out of control in the early 1990s. As the population ages, degenerative diseases increase in frequency. Changing technology advances the number of conditions that can be cured, but it can also contribute to disease through perturbation of the environment, such as exposure to water and air pollutants.

Challenges to modern medicine include more effective prevention or cure of cancer, understanding normal and abnormal mental processes (including addictions), and preventing premature births. Since much illness is aggravated by poverty, attention to the standard of living of all individuals is a responsibility of physicians as well as other members of society.

Mary Ellen Avery, M.D.
Thomas Morgan Rotch Professor of Pediatrics
Harvard Medical School

medicine man *Anthropology.* in certain tribal societies, especially among the North American Indians, a priestly figure believed to have supernatural powers; a shaman.

medico- a combining form meaning "medical" or "medicinal."

medicochirurgic *Surgery.* of or relating to medicine and surgery.

medieval *Architecture.* of or relating to the European Middle Ages (usually regarded as the period from the 5th to the 15th centuries); in architecture, refers to the Byzantine and pre-Romanesque periods but especially to the Romanesque and Gothic.

medina quartzite *Mineralogy.* a form of quartz containing 98% orthorhombic silica; used as a source for silica refractories.

mediolateral *Anatomy.* relating to or located in the middle and on one side.

medionecrosis *Medicine.* necrosis of the tunica media of an artery, commonly the aorta, often leading to its rupture.

Mediterranean *Geography.* relating to or characteristic of the Mediterranean Sea, the Mediterranean subregion, or a Mediterranean climate.

mediterranean see MESOGEOSYNCLINE.

Mediterranean anemia *Medicine.* the β strain of thalassemia.

Mediterranean Belt see ALPIDES.

Mediterranean climate *Meteorology.* a local climate characterized by long, sunny, hot, dry summers and rainy winters; found on the coasts and islands of the Mediterranean Sea and also in other parts of the world, such as the Southern California coast. This climate is especially favorable for growing fruit. Also, ETESIAN CLIMATE.

Mediterranean faunal region *Ecology.* a division of the earth's surface that includes all sea-dwelling creatures found off the coast of northern France down to the equator.

Mediterranean fever see BRUCELLOSIS.

Mediterranean fruit fly *Invertebrate Zoology.* a small, black and white dipteran insect, *Ceratitis capitata,* that severely damages citrus and other fruit crops by implanting eggs that hatch within the fruit; a major agricultural pest, particularly in California. Also, MEDFLY.

Mediterranean scrub see CHAPPARAL.

Mediterranean Sea *Geography.* the great inland sea stretching 2400 miles west to east between southern Europe, North Africa, and southwestern Asia.

Mediterranean subregion *Ecology.* a distinct zoogeographical region that includes the areas of southern Europe, Asia Minor, and Africa north of the equator, and also the Mediterranean Sea and the adjacent offshore region of the Atlantic Ocean.

medium *plural,* **media** or **mediums.** the location in which something takes place or the means by which it takes place; specific uses include: *Chemistry.* **1.** the carrier in which a chemical reaction occurs. **2.** a porous barrier that lets liquids pass while retaining most of the solids. *Bacteriology.* a substance used in the culture of bacteria. *Histology.* a preparation used in treating tissue samples. *Physics.* a physical environment in which phenomena occur; examples are a fluid or solid. Also, SEPTUM. *Electricity.* any substance that transmits impulses. *Computer Technology.* the material or basis on which data and instructions are recorded, such as magnetic tape, floppy disk, or paper. *Telecommunications.* a means of communication, especially of mass communication. *Graphic Arts.* **1.** the material with which an artist works. **2.** a size of printing paper, typically $18^1/_2 \times 23^1/_2$ inches. *Military Science.* in air intercept, a height between 2000 and 25,000 feet.

medium-altitude bombing *Military Science.* horizontal bombing with the height of release between 8000 and 15,000 feet.

medium-angle bombing *Military Science.* a type of bombing with the weapon release at an angle of 35–75° above the horizontal.

medium antiaircraft artillery *Ordnance.* conventional antiaircraft artillery having a caliber of 90 mm or larger but weighing no more than 40,000 pounds in a trailed mount, excluding on-carriage fire control.

medium artillery *Ordnance.* artillery of more than 105 mm but less than 155 mm caliber.

medium-atomic demolition munition *Ordnance.* a low-yield, team-portable, atomic demolition munition that can be detonated by a remote control or a timer device.

medium-carbon steel *Metallurgy.* any of several steels containing about 0.15–0.30% carbon.

medium-delay fuse *Ordnance.* a fuse delaying the explosion of a charge for a period of time between that of a short-delay fuse and that of a long-delay fuse; the delay is normally four to fifteen seconds, but may range up to two minutes.

medium frequency *Telecommunications.* an FCC-designated band of frequencies between 300 kilohertz and 3 megahertz.

medium-frequency propagation *Telecommunications.* propagation at broadcast frequencies in which skip is not of primary importance.

medium-frequency tube *Electronics.* an electron tube that operates at medium frequencies, where the transit time between electrodes is less than the voltage oscillation period.

medium howitzer *Ordnance.* a howitzer with a bore diameter greater than 125 mm but not exceeding 200 mm.

medium-range ballistic missile *Ordnance.* a ballistic missile having a range of approximately 1000–1500 km.

medium-range forecast *Meteorology.* a weather forecast for an approximate period of two to five days, or sometimes as much as one week in advance.

medium-scale integration *Electronics.* **1.** the placement of several circuits (usually less than 100) on a single semiconductor chip. **2.** the functioning of integrated circuits as simple, independent logic systems.

medium-scale map *Cartography.* any map or chart having a scale larger than 1:600,000 but smaller than 1:75,000.

medius *Anatomy.* of or related to the middle; especially of or related to an anatomical structure located between two other structures.

MEDLARS Medical Literature Analysis and Retrieval System.

medulla *plural,* **medullae.** *Anatomy.* **1.** the inner portion of a structure. **2.** see MEDULLA OBLONGATA. *Botany.* **1.** see PITH. **2.** the center of the lamina or stipe of brown algae. **3.** the central layer of a lichen thallus.

medulla oblongata *Anatomy.* a truncated cone of nerve tissue rising from the spinal cord and forming the lower portion of the brain stem; it contains many ascending and descending nerve tracts and regulates vital functions such as respiration, circulation, and certain special senses.

medullary *Anatomy.* relating to the marrow or to any medulla.

medullary canal see NEURAL CANAL.

medullary carcinoma *Medicine.* **1.** a cancerous tumor with little structure and unorganized composition. **2.** a malignant growth with an appearance similar to that of marrow or brain tissue.

medullary cord *Anatomy.* a strand of thick lymphoid tissue between the cavities in the middle of a lymph node. *Developmental Biology.* a strand of primordial cells in the medulla of the embryonic gonads that connects with a mesonephric tubule, and from which the rete ovarii or the rete testis develop. Also, RETE CORD, SEX CORD.

medullary fold see NEURAL FOLD.

medullary groove see NEURAL GROOVE.

medullary plate see NEURAL PLATE.

medullary ray *Botany.* an extension of pith occurring between vascular bundles in the primary tissue of the stem. Also, PRIMARY RAY.

medullary sheath *Botany.* a ring of small, thick-walled cells surrounding the pith in certain plants.

medullary tube see NEURAL TUBE.

medullectomy *Surgery.* the surgical excision of the medulla of an organ.

medulloepithelioma *Oncology.* **1.** a growth in the eye, resembling a retinoblastoma, that can be either benign or malignant, and is seen in children. **2.** an extremely rare cancerous growth in the central nervous system, made up of tissue resembling undeveloped nerve epithelium.

Medullosaceae *Paleontology.* a family of seed ferns of the division Pteridospermophyta, having large, spirally arranged petioles with many vascular bundles; Carboniferous and Permian.

medusa *Invertebrate Zoology.* **1.** a jellyfishlike, free-swimming sexual stage in the life cycle of certain coelenterates, characterized by an umbrella fringed with tentacles and a central, dangling tube bearing the combined mouth and anus. **2.** broadly, a jellyfish or hydra. (From the Greek gorgon *Medusa,* whose snaky locks were said to resemble jellyfish tentacles.)

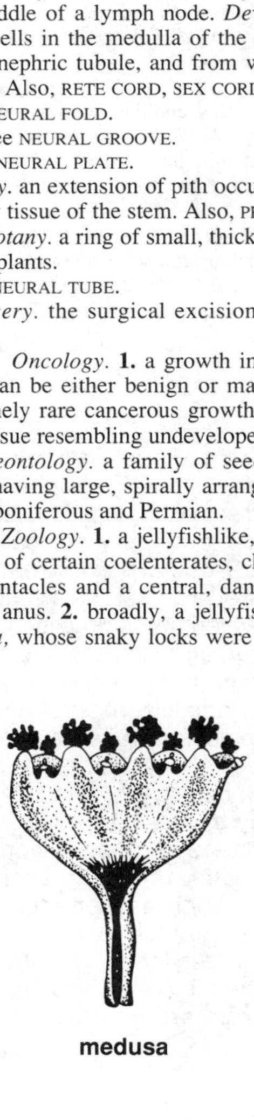

medusa

Medusagynaceae *Botany.* a monospecific family of shrubs and trees in the order Theales, having phloem stratified into hard and soft layers and winged seeds; native to the Seychelles.

Medusandraceae *Botany.* a monospecific family of dicotyledonous trees in the order Santalales, having secretory canals containing a yellow liquid and thick hairs on lower leaf surfaces; native to African rain forests.

medusoid *Invertebrate Zoology.* resembling or characteristic of a medusa or jellyfish.

meerschaum see SEPIOLITE.

Meerwein arylation *Organic Chemistry.* a method for producing aromatic compounds by the reaction of an aryldiazonium salt in the presence of Cu^{2+} salts with vinyl derivatives.

Meerwein-Ponndorf reduction *Organic Chemistry.* a reaction that utilizes the equilibrium that exists between alcohols and carbonyl compounds in the presence of aluminum alkoxides to reduce unsaturated aldehydes and ketones to primary and secondary alcohols, respectively.

Meesiaceae *Botany.* a family of matted mosses of the order Bryales that grow on soil, peat, or rotting wood, especially in the Northern Hemisphere; characterized by erect simple or forked stems with terminal sporophytes, abruptly curved capsules, very long setae, and some peristome features.

meeting rail *Building Engineering.* the rail in a double-hung window, each sash of which meets the rail of the other when the window is closed.

meeting stile *Building Engineering.* the stile of a door furthest from its hinges.

meg megohm.

MEG magnetoencephalograph.

mega- or **meg-** a combining form meaning: **1.** 1 million, as in *megabyte.* **2.** very large, as in *megagamete.* (A variation of *megalo-.)*

megabar *Metrology.* a unit of pressure equal to 1 million bars.

megabit *Computer Technology.* a unit of measure consisting of 2^{20}, or approximately 1 million, binary bits.

megabyte *Computer Technology.* a unit of measure consisting of 2^{20}, or approximately 1 million, bytes.

megacanthopore *Invertebrate Zoology.* a large spine or tube that projects from the mature region of a bryozoan colony.

megacephaly or **megacephalia** see MEGALOCEPHALY.

Megachilidae *Invertebrate Zoology.* a cosmopolitan family of large, long-tongued, mostly solitary bees.

Megachiroptera *Vertebrate Zoology.* the Old World fruit bats, a suborder containing the family Pteropodidae, characterized by orientation by vision rather than sonar, and a diet of sweet juicy fruit.

Megachytriaceae *Mycology.* a family of fungi of the order Chytridiales, occurring on land and in the water and living primarily off decaying vegetable debris.

megacin *Microbiology.* a bacteriocin that is produced by the bacterium *Bacillus megaterium.*

megacolon *Medicine.* a partially or wholly enlarged colon, sometimes hereditary, with constipation or obstructing fecal material.

megacryst *Petrology.* any grain or crystal in an igneous or metamorphic rock that is larger than the surrounding groundmass.

megacurie *Nucleonics.* a unit of radioactivity equal to 1 million curies.

megacycle see MEGAHERTZ.

megacyclothem *Geology.* a cycle or combination of related, repeated sequences of beds deposited during a single sedimentary cycle.

Megadermatidae *Vertebrate Zoology.* a family of carnivorous bats of the suborder Microchiroptera; known as "false vampire bats" because of their dental structure, even though they do not drink blood; found in Asia, Australia, and Africa.

megadyne *Metrology.* a unit of force equal to 1 million dynes.

megaelectron volt *Physics.* a unit of energy equivalent to 10^6 electron volts.

megaelectron volt-curie *Nucleonics.* a unit of radioactive power equivalent to the power of one curie generating an average power of 10^6 electron volts per disintegration.

megafauna *Ecology.* the largest arbitrary size categorization of animals in a community.

megaflop *Computer Technology.* 1 million floating point operations per second.

megagametophyte *Botany.* in seed plants, the female gametophyte, produced from the megaspore.

megagauss *Electromagnetism.* 10^6 gauss, a unit of magnetic field strength.

megagauss physics *Physics.* a branch of physics concerned with the production and application of magnetic fields on the order of 1 million gauss.

megagram *Metrology.* a unit of mass equal to 1 million grams; a metric ton. Also, **megagramme.**

megahertz *Physics.* a unit of frequency equal to 10^6 hertz or 10^6 cycles per second. Also, MEGACYCLE.

megajoule *Metrology.* a unit of energy equal to 10^6 joules.

megakaryocyte *Histology.* a giant cell of bone marrow that produces blood platelets.

megakaryophthisis *Pathology.* an atrophy of the megakaryocytes.

megalith *Archaeology.* a structure or arrangement of large stone slabs, especially those found in northern and western Europe from the Mesolithic period, such as Stonehenge.

megalithic *Archaeology.* of or relating to megaliths: *megalithic* tombs or monuments, *megalithic* architecture.

megalo- or **megal-** a combining form meaning: **1.** enlarged or abnormally large, as in *megalocardia.* **2.** very large, as in *megalopolis.*

megaloblast *Pathology.* a sizable, nucleated, developing cell; the forerunner of deviations in the red blood cell sequence, found almost solely in cases of pernicious anemia.

megaloblastic anemia *Pathology.* a blood condition in which large immature abnormal red blood cells are found in the bone marrow.

megaloblast of Sabin see PRONORMOBLAST.

megalocardia *Medicine.* the fact or condition of having an abnormally large heart. Also, CARDIOMEGALY.

megalocephaly or **megalocephalia** *Medicine.* an abnormally large head. Also, MEGACEPHALY, MEGACEPHALIA.

Megalodontoidea *Invertebrate Zoology.* the sawflies, a superfamily of hymenopteran insects in the suborder Symphyta.

megalogastria *Medicine.* enlargement or abnormally large size of the stomach; gastric hypertrophy.

megalomania [meg´ə lō mā´nē ə] *Psychology.* an irrational, exaggerated sense of one's own abilities and importance. Thus, **megalomaniac.**

Megalomycteroidei *Vertebrate Zoology.* the large-nose fishes, a suborder of the order Lampriformes and comprising the family Megalomycteridae; found in the depths of the Atlantic and Pacific Oceans.

megalopia see MACROPSIA.

megalopolis [meg´ə läp´ə ləs] *Geography.* a large region of continuous urbanization; a giant conurbation. The best-known megalopolis in the U.S. is the urban belt stretching from Boston to Washington; a comparable belt in Japan extends from Tokyo to Osaka. Thus, **megalopolitan.**

megalops *Invertebrate Zoology.* a large-eyed larval stage in the development of certain crabs, similar in appearance to the adult form.

Megaloptera *Invertebrate Zoology.* in some classifications, a suborder of insects of the order Neuroptera including the Dobsonflies, fishflies, and alderflies.

Megalopygidae *Invertebrate Zoology.* a family of medium-sized to large moths of the suborder Heteroneura, having broad wings and a thick wooly body.

megalosphere *Invertebrate Zoology.* the large-chambered shell of a sexual individual of certain foraminiferan protozoans.

megaloureter *Medicine.* an inherited enlargement of a ureter without apparent cause.

-megaly *Medicine.* a combining form denoting an unusually or abnormally enlarged condition of an organ or body part.

Megamerinidae *Invertebrate Zoology.* a family of dipteran flies in the subsection Acalyptratae.

megaparsec *Astronomy.* a unit used in cosmology to measure distances, equal to 1 million parsecs, or 3,262,600 light-years.

megaphone *Acoustics.* an open-ended conical horn that is used to amplify the human voice.

megaphyll *Botany.* a large leaf having a branching vein system, believed to have evolved from early plants having only branches.

megaphyllous *Botany.* having large leaves, with traces accompanied by leaf gaps.

megaplankton *Biology.* plankton of the largest classification of body size, greater than 200 mm in diameter.

megapode *Vertebrate Zoology.* any member of the family Megapodiidae. Also, BRUSH TURKEY, SCRUB FOWL.

Megapodiidae *Vertebrate Zoology.* the brush turkeys, a family of pheasantlike birds of the order Galliformes, found in Australia and Polynesia and noted for incubating their eggs in large mounds of decaying vegetation.

megapore *Geology.* see MACROPORE.

megaprosopous *Medicine.* having a large face.

megarad *Nucleonics.* a unit of absorbed radiation equal to 1,000,000 rads.

megarectum *Medicine.* a grossly distended rectum.

megaripple *Geology.* a large sedimentary structure composed of sand, whose shape resembles that of a water wave.

megasclere *Invertebrate Zoology.* in some sponges, a large supporting spicule.

Megasphaera *Bacteriology.* a genus of Gram-negative, anaerobic, spheroid bacteria of the family Veillonellaceae; found in the human intestines and in the rumen of sheep and cattle.

megasporangium see MACROSPORANGIUM.

megaspore *Botany.* see MACROSPORE.

megasporocyte *Botany.* the diploid cell that, after undergoing meiosis, produces four haploid megaspores.

megasporophyll *Botany.* a leaf or leaf structure on which macrosporangia develop.

megastructure *Architecture.* a very large high-rise building or complex, often used for a variety of purposes such as apartments with offices, stores, and cultural or recreational facilities.

megatectonics *Geology.* the study of the structural and deformational relationships, movements, origin, and evolution of the very large features of the earth. Also, MACROTECTONICS.

Megatheriidae *Paleontology.* a large family of xenarthran mammals in the extinct superfamily Megalonychoidea, common in the Miocene of South America; Oligocene to Pleistocene.

megatherm *Ecology.* **1.** a plant that does not tolerate low temperatures and that flourishes in a tropical environment. **2.** any organism living in a hot, moist environment. Thus, **megathermic, megathormophyte.** Also, MACROTHERM.

megathermal period see CLIMATIC OPTIMUM.

Megathymiinae *Invertebrate Zoology.* a subfamily of butterflies, lepidopteran insects in the family Hesperiidae, including the giant skippers.

megaton [meg´ə tun´] *Physics.* a unit of explosive power or energy equivalent to the yield of 10^6 metric tons of trinitrotoluene (TNT); used in describing the output of a nuclear weapon. Thus, **megatonnage.**

megaton weapon *Ordnance.* a nuclear weapon with a yield measured in millions of tons of trinitrotoluene (TNT) explosive equivalents. Also, **megaton bomb.**

megatron see DISK-SEAL TUBE.

magavitamin *Medicine.* a dose of vitamins vastly exceeding the amount required for nutritional balance. Thus, **megavitamin therapy.**

megavolt *Electricity.* a unit of electricity that is equivalent to 1 million volts.

megawatt electric *Nucleonics.* a unit of quantity for the electric power generated by a nuclear reactor, as distinguished from its thermal power.

megawatt thermal *Nucleonics.* a unit of quantity for the thermal power generated by a nuclear reactor, as distinguished from the electric power generated.

megawatt year *Electricity.* a unit of electricity that is equivalent to the energy produced by one megawatt of power in one year.

megestrol acetate *Oncology.* $C_{25}H_{32}O_4$, a synthetic progestin that occurs as a white, crystalline powder, used as an antineoplastic to treat recurrent or metastatic endometrial carcinoma and breast cancer.

megohm *Electricity.* a unit of electricity that is equivalent to 1 million ohms.

megohmmeter *Electricity.* an instrument used to measure high-voltage resistance of electrical materials, calibrated in megohms, gigohms, and teraohms.

megrim see MIGRAINE.

Mehli's gland *Invertebrate Zoology.* any of the large unicellular glands surrounding the ootype chamber in flatworms.

Meibomian cyst see CHALAZION.

Meibomianitis *Medicine.* an infection of the tarsal, or Meibomian, glands of the upper eyelid.

Meig's syndrome *Medicine.* a fibrous tumor of the ovary, combined with a buildup of fluid in the abdomen and lungs.

Meijer transform [mī´ər] *Mathematics.* an integral transform of a function $f(x)$; given by

$$F(y) = \int_0^\infty (xy)^{1/2} K_n\,(xy) f(x)\,dx,$$

where K_n is a modified Bessel function.

meiocyte *Cell Biology.* a cell that is undergoing or about to enter meiosis, such as a primary oocyte.

meiofauna *Zoology.* a classification of animals that are intermediate in size between those that can easily be seen with the naked eye (macrofauna) and those that are microscopic (microfauna).

meioflora *Botany.* a classification of plants that are intermediate in size between those that can easily be seen with the naked eye (macroflora) and those that are microscopic (microflora).

meiogenic *Genetics.* promoting or causing meiosis.

meiosis [mī ō´sis] *Genetics.* the process of nuclear division in cells during which the chromosome complement is reduced by half, an essential step in the formation of gametes in animals and spores in plants; often contrasted with mitosis. Also, MIOSIS.

meiotic [mī ät´ik] *Genetics.* relating to, characteristic of, or characterized by meiosis.

meiotic drive *Genetics.* any of several mechanisms that alter the normal events of meiosis so that a heterozygous individual does not produce equal numbers of gametes, each containing the alternative alleles of a gene.

Meisenheimer complexes *Organic Chemistry.* anionic intermediates that are formed during certain bimolecular nucleophilic substitution reactions.

Meissner, Fritz Walther [mīz´nər] 1882–1974, German physicist; described the Meissner effect.

Meissner effect *Solid-State Physics.* a phenomenon in which a superconductor becomes perfectly diamagnetic as magnetic field lines are pushed out of the specimen after being cooled through the transition point.

Meissner oscillator *Electronics.* an oscillator that uses an independent tank circuit inductively coupled to the input and output circuits of an oscillator in order to obtain the proper feedback and frequency.

Meissner's corpuscle *Anatomy.* any of numerous medium-sized encapsulated nerve endings in the dermal papillae of the skin that transmit minute touch sensations; notably prevalent on the palms and soles. (Named for Georg *Meissner,* 1829–1905, German physiologist.)

Meissner's plexus see SUBMUCOSAL PLEXUS.

Meitner, Lise 1878–1968, German physicist; discovered (with Otto Hahn) radioactive protactinium.

Mekong *Geography.* a river in Southeast Asia that rises in the Tibetan highlands and flows about 2500 miles southeast through Indochina to the South China Sea.

mel *Acoustics.* a unit of pitch; a 1000-hertz tone at 40 decibels has a pitch of 1000 mels.

MEL *Aviation.* the airport code for Melbourne, Australia.

melamine *Organic Chemistry.* $C_3H_6N_4$, a toxic, nonflammable, colorless to white monoclinic prismatic solid; slightly soluble in water, very slightly soluble in ethanol, and insoluble in organic solvents; slowly sublimes when heated and melts at 347°C with decomposition; used to make melamine resins and to give wet strength to paper.

melamine resin *Materials Science.* any of various amino resins made from melamine and formaldehyde, used as an adhesive and in various coatings and insulators.

Melamphaidae *Vertebrate Zoology.* the big-scale fishes, a family of the order Beryciformes, found in deep seas of all major oceans and characterized by large cycloid scales.

Melampsoraceae *Mycology.* the rust fungi, a family of parasitic fungi of the order Uredenales occurring primarily on fir trees; sometimes pathogenic to cottonwoods and balsam poplars.

melan- a combining form meaning "black," as in *melanin.*

melancholia [mel´ən käl´yə] *Psychology.* a severe state of depression characterized by brooding, low self-esteem, and feelings of dejection and foreboding. Also, MELANCHOLY.

melancholic [mel´ən käl´ik] *Psychology.* **1.** one of the four basic personality types defined in ancient and medieval times, described as being characteristically sad; said to result from the predominance of black bile over the other bodily humors. **2.** of or relating to this personality type. **3.** a person of this personality type. Also, **melancholic type. 4.** of or relating to melancholia.

melancholy [mel´ən käl´ē] *Psychology.* **1.** a sad, depressed, or gloomy state of mind or outlook. **2.** affected with, exhibiting, or characterized by melancholy. **3.** see MELANCHOLIA.

Melanconiaceae *Mycology.* the sole family of fungi contained in the order Melanconiales, including numerous pathogens of plants.

Melanconiales *Mycology.* an order of fungi of the class Coelomycetes, having cnidophores that are densely bunched on acervulus.

Melandryidae *Invertebrate Zoology.* a family of beetles in the superfamily Tenebrionoidea, associated with fungi and rotting wood.

Melanesia *Geography.* the islands of the western Pacific lying south of the equator from New Guinea eastward through Fiji.

mélange [mā länj′; mā länzh′] *Geology.* a mappable mixture of rock material, generally consisting of exotic and native fragments of various sizes, origins, and geologic ages embedded in a fine-grained, sheared matrix of more tractable material. Also, BLOCK CLAY.

melanic *Biology.* describing an individual that is darker than normal for a species or subpopulation, because of melanism. *Geology.* see MELANOCRATIC.

melaniline see DIPHENYLGUANIDINE.

melanin [mel′ə nin] *Biochemistry.* any of the polymers, formed by the oxidation of the amino acid tyrosine, that make the natural pigmentation of hair, skin, and eyes in vertebrates.

melanism *Biology.* a darkness of color that is due to an abnormal development of melanin. *Genetics.* a hereditary tendency to produce increased melanin pigmentation; common in birds and mammals.

melano- a combining form meaning "black," as in *melanophore.*

melanoblast *Histology.* **1.** an undifferentiated cell that develops into a melanocyte or melanophore. **2.** an immature pigment cell. **3.** a cell that actively produces melanin.

melanoblastoma *Oncology.* a cancerous tumor containing melanin.

melanocarcinoma *Oncology.* a carcinoma containing melanin. Also, MALIGNANT MELANOMA.

melanocerite *Mineralogy.* a black or brown hexagonal mineral having an approximate formula of $(Ce,Ca)_5(Si,B)_3O_{12}(OH,F) \cdot nH_2O$, occurring as rhombohedral crystals, and having a specific gravity of 4.13 to 4.29 and a hardness of 6 on the Mohs scale; found in alkali pegmatites.

melanocratic *Geology.* describing dark-colored rocks, especially igneous rocks composed of at least 60% dark minerals. Also, CHROMOCRATIC, MELANIC.

melanocyte *Histology.* any of the dendritic clear cells of the epidermis that synthesize melanin.

melanocyte-stimulating hormone *Endocrinology.* a polypeptide hormone that is produced in the anterior or intermediate lobe of the pituitary gland; formed by cleavage of the prohormone product of the proopiomelanocortin gene, it induces the formation of the pigment melanin in the melanocytes of the skin and may also have a role in the processes of learning and memory. Also, α-MELANOTROPIN, MELANOPHORE-DILATING PRINCIPLE.

melanocyte-stimulating hormone-inhibiting factor *Endocrinology.* a tripeptide cleavage product of hypothalamic oxytocin (carboxy terminus) that inhibits the release of melanocyte-stimulating hormone; it is found in lower vertebrates and probably also occurs in mammals. Also, **melanocyte-stimulating hormone release-inhibiting factor.**

melanocyte-stimulating hormone-releasing factor *Endocrinology.* a peptide cleavage product of hypothalamic oxytocin (amino terminus) that is found in lower vertebrates and stimulates the release of melanocyte-stimulating hormone. Also, MELANOLIBERIN.

melanocytoma *Oncology.* **1.** any neoplasm or hamartoma composed of melanocytes. **2.** specifically, a benign growth on the optic disk, observed primarily in black persons.

melanocytosis *Pathology.* a condition in which large numbers of phagocytes that have absorbed melanin accumulate in the tissues.

melanoderma *Pathology.* an abnormal increase in the amount of melanin in the skin.

Melanogastraceae *Mycology.* a family of fungi beloning to the order Melanogastrales, characterized by an underground fruiting body and by living symbiotically with plants.

Melanogastrales *Mycology.* an order of fungi belonging to the class Gasteromycetes, characterized by a stationary, rounded, dark fruiting body and a lack of capillitium and clamp connections.

melanogen *Biochemistry.* a compound that forms as a colorless intermediate in the metabolic pathway of tyrosine to melanin, and that occasionally converts to melanin without being catalyzed by enzymes.

melanogenesis *Biochemistry.* the production of melanin.

melanoglossia *Medicine.* a darkened or black-appearing tongue.

melanoid *Medicine.* **1.** having a dark appearance. **2.** a melaninlike pigment.

melanoliberin see MELANOCYTE-STIMULATING HORMONE-RELEASING FACTOR.

melanoma [mel′ə nō′mə] *Oncology.* **1.** a tumor, usually of the skin, that is composed of melanocytes. **2.** a malignant melanoma, or melanocarcinoma.

melanomatosis *Medicine.* the production of melanomas in multiple body areas.

Melanommataceae *Mycology.* a family of fungi belonging to the order Melanommatales; found living off of decaying woody plants and tree bark, primarily in temperate regions.

Melanommatales *Mycology.* an order of fungi of the class Loculoascomycetes, characterized by an ascocarp peridium composed of layers of thick-walled, small, compact cells and having symmetric ascospores.

melanophage *Histology.* a phygocytic cell that engulfs melanin. Thus, **melanophagous.**

melanophlogite *Mineralogy.* SiO_2-containing organic compounds, a colorless or yellow to brown, vitreous, cubic or tetragonal, pseudocubic mineral occurring as crusts of rounded aggregates, having a specific gravity of 2.01 to 2.05 and a hardness of 6.5 to 7 on the Mohs scale; found on sulfur crystals from Sicily.

melanophore *Histology.* a chromatophore that contains melanin.

melanophore-dilating principle see MELANOCYTE-STIMULATING HORMONE.

melanophore hormone see MELANOCYTE-STIMULATING HORMONE.

melanosarcoma *Oncology.* a malignancy of the connective tissue that contains melanin.

melanose *Plant Pathology.* **1.** a disease of grapevines caused by the fungus *Septoria ampelina,* which attacks the leaves and causes them to drop off. **2.** a disease of citrus trees and fruits caused by the fungus *Diaporthe citri,* characterized by crusty, dark, gummy, raised markings on leaves, fruits, and twigs.

melanosis *Pathology.* **1.** any abnormal presence or development of black or dark pigment. **2.** a disorder involving this.

melanosis coli *Pathology.* the presence of abnormal blackish-brown hematins on the colon, caused by a gathering of indeterminate pigment-laden macrophages within the lamina propria.

melanosis iridis *Pathology.* a deviant coloration of the iris due to infiltration of melanoblasts.

melanosome *Cell Biology.* an intracellular organelle that is found in melanocytes, containing aggregates of tyrosinase and the pigment melanin.

Melanosporaceae *Mycology.* the single family of fungi of the order Melanosporales; occurs in soil, usually along with other fungi that provide it with nutrients.

Melanosporales *Mycology.* an order of fungi of the class Pyrenomycetes, characterized by one-celled brown ascospores having a germ pore and by cell walls that are soft and semitransparent.

melanostibiite *Mineralogy.* $Mn(Sb^{+5},Fe^{+3})O_3$, a black, trigonal mineral occurring in porous aggregates of small crystals, and having a specific gravity of 5.63 and a hardness of about 4 on the Mohs scale; found with calcite and tephroite in dolomitic rocks. Also, **melanostibian.**

Melanotaenium *Mycology.* a genus of smut fungi of the order Ustilaginales, which parasitizes plants.

melanotekite *Mineralogy.* $Pb_2Fe_2^{+3}Si_2O_9$, a black, opaque, metallic, orthorhombic mineral with two cleavages, occurring as prismatic crystals and in massive form, having a specific gravity of 5.7 to 6.19 and a hardness of 6.5 on the Mohs scale.

melanotic cancer *Oncology.* a cancerous melanoma.

melanotic freckle *Oncology.* a black lesion on the cheek that contains aberrant melanocytes and may develop into a melanoma.

α-melanotropin see MELANOCYTE-STIMULATING HORMONE.

melanovanadite *Mineralogy.* $Ca(V_2^{+5}V_2^{+4})O_{10}{\cdot}5H_2O$, a rare, black, opaque, triclinic mineral having a specific gravity of 2.55 to 2.9 and a hardness of 2.5 on the Mohs scale; found as aggregates of prismatic crystals in black shale near Cerro De Pasco, Peru.

melanterite *Mineralogy.* $Fe^{+2}SO_4{\cdot}7H_2O$, a green to blue, water-soluble, monoclinic mineral of the melanterite group, usually occurring in fibrous to earthy aggreates, having a specific gravity of 1.9 and a hardness of 2 on the Mohs scale; found as a product of pyrite and marcasite decomposition.

melanuria *Pathology.* the passage of black pigment in the urine, common in persons with widespread malignant melanoma; the urine appears normal when passed but becomes dark upon standing.

melaphyre *Petrology.* an altered variety of basalt, especially basalt of Carboniferous or Permian age.

Melasidae see EUCNEMIDAE.

Melasmia *Mycology.* a genus of fungi of the class Coelomycetes, the sexual counterpart of which constitutes the genus *Rhytisma.*

Melastomataceae *Botany.* a tropical family of herbs, shrubs, and trees belonging to the order Myrtales, characterized by simple, opposite or whorled leaves and by regular, usually bisexual flowers with an inferior ovary.

melatonin *Endocrinology.* a biogenic amine secreted by the pineal gland that appears to suppress the secretion of luteinizing hormone and growth hormone by the anterior pituitary gland and to inhibit numerous endocrine functions including gonad development. *Biochemistry.* $C_{13}H_{16}N_2O_2$, a preparation of this hormone, used experimentally to decrease the pigmentation of the skin in amphibians.

melawis see RAMIN.

Meleagridae *Vertebrate Zoology.* the turkeys, a family of large birds of the order Galliformes inhabiting the forests of the eastern United States and Mexico in a wild form and also widely raised as a domesticated fowl.

M electron *Atomic Physics.* an electron that is found in the M shell of an atom and displays the characteristics of the principal quantum number 3.

melena *Pathology.* a condition marked by the excretion of dark stools containing blood that has been altered by gastric fluids. Also, TARRY STOOLS.

meletin see QUERCETIN.

melezitose *Biochemistry.* $C_{18}H_{32}O_{16}$, a trisaccharide from manna, from the sap of poplars and conifers, that on hydrolysis yields glucose and turanose.

meli- a combining form meaning "sweet" or denoting a relationship or resemblance to honey.

Meliaceae *Botany.* a family of tropical trees and shrubs of the order Sapindales, characterized by alternate, usually pinnate leaves without stipules and by syncarpous, usually bisexual flowers borne in panicles, cymes, spikes, or clusters.

Melianthaceae *Botany.* a family of tanniferous dicotyledonous African shrubs and trees of the order Sapindales, characterized by alternate and pinnately compound leaves, hypogenous flowers borne in racemes, and an extrastaminal nectary disk.

melilite *Mineralogy.* any of a group of variously colored tetragonal silicates of the akermanits-gehlenite isomorphous series, found in calcium-rich basic eruptive rocks and contact metamorphosed limestones.

melilitite *Petrology.* an extrusive or hypabyssal rock that consists of melilite and pyroxene or other mafic minerals, with small amounts of feldspathoids; sometimes contains plagioclase, but rarely olivine.

Melinae *Vertebrate Zoology.* the badgers, a subfamily of mammals of the family Mustelidae of North America and Eurasia.

melinite *Engineering.* a high explosive that contains picric acid and is similar to lyddite.

Melinninae *Invertebrate Zoology.* a subfamily of sessile, tube-dwelling, polychaete worms of the family Ampharetidae, having a conspicuous dorsal membrane.

melioidosis *Veterinary Medicine.* a highly fatal bacterial disease affecting primarily wild rodents in the tropics, caused by *Pseudomonas pseudomallei* and resembling glanders; characterized by supperating or caseous lesions in lymph nodes and viscera.

Meliolaceae *Mycology.* the sooty molds, a family of fungi of the order Meliolales.

Meliolales *Mycology.* an order of fungi of the class Pyrenomycetes, containing various black molds and mildews that form dark spots on plant leaves and stems without harming them.

meliphanite *Mineralogy.* $(Ca,Na)_2Be(Si,Al)_2(O,OH,F)_7$, a yellow, red, or colorless, tetragonal mineral occurring in thin tabular or short pyramidal crystals, and having a specific gravity of 3.0 and a hardness of 5 to 5.5 on the Mohs scale; found in nepheline-syenite pegmatites. Also, **meliphane.**

melissic acid *Organic Chemistry.* $CH_3(CH_2)_{28}CO_2H$, a combustible, crystalline solid that is found in beeswax; insoluble in water and soluble in hot alcohol and benzene; it melts at 94°C; used in biomedical research.

melissophobia see APIPHOBIA.

melitose or **melitriose** see RAFFINOSE.

Melittidae *Invertebrate Zoology.* a family of solitary bees of the superfamily Apoidea that nest in the soil or in cavities in wood.

melituria *Medicine.* the appearance of sugar in the urine.

Mellin transform *Mathematics.* an integral transform of a function $f(t)$; given by

$$f(z) = \int_0^\infty t^{z-1} f(t)dt.$$

The inversion is given by

$$f(t) = 1/2\pi i \int_{-i\infty}^{i\infty} t^{-z} f(z)dz.$$

mellitate *Organic Chemistry.* any salt or ester of mellitic acid.

mellite *Mineralogy.* $Al_2[C_6(COO)_6]\cdot 18H_2O$, a honey-yellow to reddish tetragonal mineral of resinous to vitreous luster, occurring in rare prismatic or pyramidal crystals, having a specific gravity of 1.64 and a hardness of 2 to 2.5 on the Mohs scale; usually found as nodules in brown coal.

mellitic acid *Organic Chemistry.* $C_6(CO_2H)_6$, a colorless needlelike solid that is soluble in water and alcohol; melts at 287°C.

melo- a combining form meaning: **1.** limb, as in *melodidymus.* **2.** cheek, as in *meloplasty.*

melodeon *Electronics.* in countermeasures reception, a broad-band panoramic receiver that presents all electromagnetic radiation as vertical pips on a frequency-calibrated radar screen.

Meloidae *Invertebrate Zoology.* the blister beetles, a cosmopolitan family of beetles in the superfamily Meloidea whose body fluids cause blistering in hosts.

Meloidea *Invertebrate Zoology.* a superfamily of beetles in the suborder Polyphaga.

melon *Botany.* **1.** the common name for various trailing plants belonging to the gourd family, *Cucurbitaceae,* that produce large, juicy fruits having a thick skin and many seeds. **2.** any fruit that is grown from such a plant, such as a cantaloupe or watermelon.

melon- a combining form meaning "cheek," as in *meloncus.*

meloncus *Oncology.* a tumor of the cheek.

melonite *Mineralogy.* $NiTe_2$, a reddish-white, metallic, trigonal mineral occurring as hexagonal lamellae and indistinct foliated particles, having a specific gravity of 7.72 and a hardness of 1 to 2 on the Mohs scale; found with other secondary lead minerals.

melos- a combining form meaning "limb," as in *melosalgia.*

melosalgia *Medicine.* pain in the limbs, especially the lower limbs.

Melosiraceae *Botany.* a family of epiphytic, epilithic, planktonic, and benthic diatoms of the order Centrales, characterized by a well-developed pervalvar axis, a marginal ring of labiate processes without external projections, and cell growth in mucilaginous or spiny chains.

melphalan *Oncology.* a nitrogen mustard compound used as an antineoplastic agent to treat multiple myeloma as well as some carcinomas of the breast and ovaries.

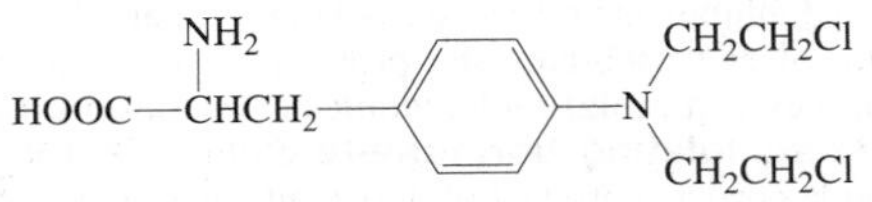

melphalan

melt *Chemistry.* to change a substance from solid to liquid form by means of heat or by the application of heat.

melt *Vertebrate Zoology.* another name for the spleen, especially of a large farm animal.

meltback transistor *Electronics.* a junction transistor in which the junction is formed by melting a doped semiconductor, which then recrystallizes.

melt-blown extrusion *Materials Science.* a process for conveying synthetic fibers directly from extrusion to nonwoven fabric forms.

meltdown *Nucleonics.* the melting of a major portion of the core of a nuclear reactor, caused by inadequate cooling of the fuel elements and often leading to a radiation leak.

melter *Engineering.* any device or chamber used to melt materials, especially metals.

melt extraction *Metallurgy.* any of several processes for the manufacture of rapidly solidified alloys.

melt extractor *Engineering.* in injection molding, a device that separates molten feed from partially molten pellets.

melt flow *Materials Science.* the irreversible bulk deformation of polymeric material under conditions of higher temperature and pressure, in which there is slippage of molecular chains past one another.

melt fracture *Materials Science.* a problem encountered when extruding plastic above its critical shear rate; irregularities range from rippled surface to annular fractures.

melt index *Materials Science.* the number of grams of thermoplastic resin heated to 190°C that can be forced through a 0.0825-inch (2.0955-millimeter) opening by a 2160-gram force in 10 minutes.

melting *Chemistry.* the process of changing from a solid to a liquid by heating. *Genetics.* **1.** the denaturation of double-stranded DNA, resulting in single-stranded DNA, caused by an increase in temperature. **2.** see HELIX-COIL TRANSITION.

melting furnace *Engineering.* a glassmaker's furnace that melts the glass for frit production.

melting level *Meteorology.* the altitude at which ice crystals and snowflake particles melt as they descend through the atmosphere.

melting loss *Metallurgy.* the loss of certain elements during the melting process.

melting point *Thermodynamics.* the temperature at which a solid begins to liquefy. The melting point of most substances is slightly dependent on the pressure; pure metals will melt at a constant temperature, whereas alloys tend to melt over a range.

melting profile *Molecular Biology.* a graph plotting the thermal denaturation of a nucleic acid in solution as a function of temperature.

melting range *Metallurgy.* the temperature range between the beginning and the end of the melting of an alloy.

melting rate *Metallurgy.* in arc welding and consumable arc remelting, the rate at which the material is melted. Also, **melt-off rate.**

melting ratio *Metallurgy.* in melting, the ratio between the weight of the metallic charge and the weight of the fuel.

melting temperature *Molecular Biology.* the temperature at which the process of thermal denaturation of a nucleic acid solution has reached the halfway point.

melt instability *Mechanical Engineering.* surface irregularities occurring in the flow of molten plastic through a die.

melt loading *Ordnance.* a process by which solid explosive is melted for loading into bombs and projectiles. Also, CAST LOADING.

melton *Textiles.* a heavy cloth, usually made of wool, that is tightly woven and finished with a smooth face; used to make overcoats, hunting jackets, and the like. Also, **melton cloth.** (Named for a town in England where it was produced.)

melt polymerization *Materials Science.* a commercial production process for polycarbonates, in which bisphenol A or one of its derivatives is reacted with a carbonic acid ester such as diphenyl carbonate in the presence of a catalyst. The phenol by-product is distilled off under reduced pressure during the reaction, leaving only the polymeric product containing the residual catalyst.

melt spinning *Materials Science.* in polymer fiber production, a process in which molten polymer is extracted through a spinneret after melting in an extruder or heating on a metal grid. The molten filaments pass into a vertical chimney, where they are cooled by a cross-current of air and attenuated to the desired diameter.

melt stability test *Materials Science.* a test of the stability of hot-melt pressure-sensitive adhesives; a specimen is held molten at the selected temperature in a suitable container and monitored for changes in viscosity and color. Similarly, **melt viscosity test.**

melt strength *Materials Science.* the mechanical properties of a molten plastic material.

meltwater or **melt water** *Hydrology.* water formed from the melting of snow or ice, especially glacier ice.

Melusinidae *Invertebrate Zoology.* a family of dipteran flies in the suborder Orthorrhapha.

Melyridae *Invertebrate Zoology.* a family of predatory flower beetles, coleopteran insects in the superfamily Cleroidea.

Melzer's reagent *Mycology.* a stain used to study the structures of fungal spores and spore sacs.

MEM *Aviation.* the airport code for Memphis, Tennessee.

MEM macrophage electrophoretic mobility.

member an individual or part that belongs to a larger group of similar entities; specific uses include: *Taxonomy.* any individual that belongs to a given taxon. *Anatomy.* **1.** any part of the body that is distinct from the rest in function or position. **2.** specifically, a limb. *Botany.* any distinct structural part of a plant. *Engineering.* a structural part of a whole, such as a beam or a wall in a building. *Geology.* a minor rock-stratigraphic unit that constitutes a part or subdivision of a geologic formation. *Mathematics.* **1.** either side of an equation. **2.** one of three primitive or undefined notions in the Gödel-Bernays formulation of axiomatic set theory, along with class and equality. Intuitively, a class is a collection A of objects or elements for which it is possible to determine whether or not x is a member (or element) of A. The expression $x \in A$ is used to represent "x is a member (or element) of A."

Membracidae *Invertebrate Zoology.* the treehoppers, a family of homopteran insects that feed on sap.

membrane a thin covering; specific uses include: *Histology.* a thin layer of tissue that surrounds or covers a body part, lines a cavity, separates cavities, or joins adjacent structures. *Cell Biology.* the thin covering of a cell or cell part. *Mechanics.* a thin shell having negligible bending rigidity. *Physical Chemistry.* any barrier separating two fluids that allows transport between the fluids only by sorption and diffusion. *Building Engineering.* a weather-resistant roof covering that consists of several layers of felt and bitumen embedded in asphalt; used primarily in built-up roofing that requires a moderate degree of flexibility.

membrane distillation *Chemical Engineering.* a separation technique that uses a temperature difference across a membrane to distill the required fluid.

membrane filtration *Chemical Engineering.* the use of membranes with different pore sizes to act as separation filters through which various materials can either pass or be retained.

membrane keyboard *Computer Technology.* a keyboard made up of two sheets of plastic containing conductive ink; when pressed, the conductive back of the membrane moves a few thousandths of an inch, completing the circuit and transmitting the keyed character.

membranelle *Invertebrate Zoology.* **1.** in some ciliate protozoans, an undulating membrane formed by a row of fused cilia and used for locomotion and feeding. **2.** the band of cilia in a tornaria larva.

membrane osmometry *Materials Science.* a widely used technique for determining the molecular weight of polymers by measuring the osmotic pressure of polymer solutions, using a semipermeable membrane that separates the polymer solution and the solvent.

membrane potential *Cell Biology.* a difference in electrical potential that is obseved between the two sides of a cell membrane.

membrane separation *Chemical Engineering.* the use of a membrane to remove components from a mixture, especially for mixtures of compounds with similar chemical or physical properties.

membrane stress *Mechanics.* the total components of stress in the surface of a membrane, plate, or shell.

membrane transport *Molecular Biology.* the movement of molecules across a biological membrane.

membrane waterproofing *Civil Engineering.* the application of impervious material to a foundation to act as a barrier against moisture.

membranous relating to, resembling, or characteristic of a membrane.

membranous labyrinth *Anatomy.* a series of interconnecting, fluid-containing canals composed of thin tissue and located within the bony labyrinth of the inner ear.

membranous pregnancy *Medicine.* a pregnancy developing within the uterus but outside the chorionic sac.

membranous urethra *Anatomy.* the portion of the male urethra that is distal to the prostate gland and proximal to the corpus spongeosum.

membron *Molecular Biology.* a functioning and regulatable translating complex of a polysome and a particular area on the surface of a biological membrane.

MEMC methoxyethylmercury chloride.

meme *Anthropology.* any single feature of a culture that can be transmitted from one generation to another; analagous to a gene in genetic transmission.

memex *Computer Technology.* a name for an early device envisioned by Vannevar Bush to store private files that could be consulted, and from which related items from several sources could be extracted; numerous related items would be joined in one or more information trails.

memomotion study *Industrial Engineering.* a time-motion study in which a slow-speed motion picture camera produces a time-lapse record of activities or processes for the purpose of analyzing the elements of the activity and improving the operation.

memorization *Behavior.* the transfer of information from short-term to long-term memory.

memory *Psychology.* **1.** a complex mental function, including the phases of memorizing, retention, recall, and recognition, that enables an individual to reproduce what he has learned or experienced. **2.** the totality of remembered experience and learning. *Computer Technology.* the various facilities of a computer that store data, including internal, external, and archival forms. *Immunology.* an altered immunological response to a specific antigen resulting from a prior exposure to that antigen, by which large amounts of antibodies are quickly produced.

memory address register see ADDRESS REGISTER.

memory alloy *Materials Science.* a term for certain alloys that possess a unique property after undergoing a sophisticated thermomechanical treatment; if any such metal is deformed into a new shape, it will return to its original shape upon heating.

memory bank *Computer Technology.* a random access memory segment containing a number of contiguous storage locations used in a computer in conjunction with other segments containing a similar number of locations. *Psychology.* see MEMORY, def. 2.

memory buffer register *Computer Technology.* a special-purpose register that acts as a buffer store through which all data transmitted between main memory and external devices must pass.

memory cache see CACHE.

memory card *Computer Technology.* an integrated circuit used as part of computer main memory.

memory cell *Immunology.* any of the cells of the immune system that do not respond to an initial encounter with a foreign antigen but, upon a second exposure to the same antigen, recognize it and initiate a secondary immune response.

memory contention *Computer Technology.* a situation that occurs when two or more processors or channels attempt to access the same memory unit simultaneously.

memory core see MAGNETIC CORE.

memory dump *Computer Programming.* **1.** a printout of the contents of memory or of a particular part of memory. **2.** a transfer of memory from one location to another.

memory dump routine *Computer Programming.* a utility program that prints the contents of memory, often as part of a debugging program.

memory effect *Materials Science.* a phenomenon that occurs as a thermal property of polymers. When a metastable glass is heated to a temperature near T_g, the volume increases, sometimes overshooting the equilibrium value. The lower the initial excess entropy, the greater the overshoot in volume.

memory element *Computer Technology.* a device that stores one item of information, implemented electronically or with the assistance of the magnetic, optical, or acoustic properties of a storage medium.

memory expansion card *Computer Technology.* an integrated circuit board that can be added to a computer to increase the internal storage capacity.

memory hierarchy *Computer Technology.* an organization of memory into multiple levels with respect to access speed, capacity, and cost.

memory management *Computer Technology.* the control and allocation of memory resources for efficient operation.

memory map *Computer Programming.* a schematic representation of the use of a memory, often produced as a by-product during the compilation of a program. Also, **memory map list.**

memory mapping *Computer Technology.* a process of managing input/output devices in which the interface registers of the devices are assigned memory addresses.

memory model *Artificial Intelligence.* a model of the organization and function of memory in human cognition, such as a representation of concepts, episodes, and events.

memory overlay see OVERLAY.

memory port *Computer Technology.* the path through which data are transferred between main memory and the central processing unit or channel.

memory power *Computer Technology.* a characteristic of memory that refers to the relative access speed of a particular memory design.

memory protection *Computer Technology.* see STORAGE PROTECTION, def. 2.

memory reference instruction *Computer Programming.* a machine instruction in which one or more operand addresses refer to a location in memory.

memory register *Computer Technology.* a circuit or part of a device that stores information represented by several bits. Also, DISTRIBUTOR.

memory trace *Physiology.* see ENGRAM.

memotron *Electronics.* a storage tube capable of displaying successive transients until intentionally erased.

men- a combining form meaning "month," as in *menacme.*

MEN multiple endocrine neoplasia.

MEN-1 *Oncology.* a tumor-suppressor gene that is found to be altered in tumors of the parathyroid, pancreas, pituitary, and adrenal cortex.

menacme *Physiology.* **1.** the height of menstrual activity. **2.** the period in a woman's lifetime marked by menstrual activity.

Menaechmus 4th century BC, Greek mathematician; discovered and described the conic sections.

menarche [mə när′kə] *Physiology.* the first menstruation, marking the beginning or establishment of the menstrual function. Thus, **menarchal, menarcheal.**

Gregor Mendel

Mendel, Gregor Johann 1822–1884, Austrian monk and botanist whose experiments represented the beginning of the science of genetics; formulated Mendel's laws of heredity.

Mendel, Lafayette B. 1872–1935, American biochemist; studied digestion and nutrition; independently discovered vitamin A.

Mendeleev (Mendeléeff), Dmitri Ivanovich [men′də lā′yef] 1834–1907, Russian chemist; formulated the periodic law of chemical elements; devised the periodic table.

mendelevium [men′də lē′vē əm] *Chemistry.* a synthetic radioactive element having the symbol Md, the atomic number 101, and an atomic weight of 256; produced by bombarding einsteinium with alpha particles in a cyclotron, mendelevium has a half-life of 1.5 hours, decaying by spontaneous fission. (Named for Dmitri *Mendeleev.*)

Mendelian [men dēl′yen; men dēl′ē ən] *Genetics.* **1.** of or relating to Gregor Mendel or to his research and experimentation. **2.** according to the laws of heredity proposed by Mendel.

Mendelian genetics *Genetics.* the body of knowledge concerning the inheritance of chromosomal genes following the patterns discovered by Gregor Mendel.

Mendelian population *Genetics.* a group of individuals of the same species who have no barrier to interbreeding and who share a common gene pool.

Mendelian ratio *Genetics.* a segregation ratio of phenotypes that results from a hybrid cross in accordance with the predictions made using the Mendelian laws of inheritance.

Mendelism *Genetics.* the inheritance of chromosomal genes according to Mendel's laws of heredity.

Mendel's laws *Genetics.* the two fundamental laws governing heredity that were proposed by Gregor Mendel: **1. the law of segregation,** which states that factors governing a character are separated at gametogenesis. It is now understood that meiosis separates the two alleles of any one gene into separate gametes, so that each gamete has one of each pair of alleles. **2. the law of independent assortment,** which states that factors (now called genes at different loci on different chromosomes) governing two separate genetic characters assort independently of one another. It is now understood that this is due to the behavior of different pairs of homologous chromosomes separating independently into different gametes during gametogenesis. Also, **Mendelian laws.**

mendip *Geology.* **1.** a buried hill that is uncovered as an inlier. **2.** a coastal plain hill that originated as an offshore island.

mendipite *Mineralogy.* $Pb_3Cl_2O_2$, a colorless or white, orthorhombic mineral occurring as fibrous masses, having a specific gravity of 7.24 and a hardness of 2.5 on the Mohs scale.

Mendonciaceae *Botany.* a family of tropical twining and usually tanniferous dicotyledonous herbs of the order Scrophulariales, characterized by jointed twigs and anomalous secondary growth; combined with Acanthaceae in some classifications.

Mendosicutes *Bacteriology.* a division of bacteria of the kingdom Prokaryotae containing a single class, the Archaebacteria; characterized by having cell walls lacking in muramic acid.

mendozite *Mineralogy.* $NaAl(SO_4)_2 \cdot 11H_2O$, a colorless or white, monoclinic mineral occurring as fibrous crusts and masses, having a specific gravity of 1.73 to 1.765 and a hardness of about 3 on the Mohs scale.

meneghinite *Mineralogy.* $Pb_{13}CuSb_7S_{24}$, a blackish to lead-gray, opaque, metallic, orthorhombic mineral occurring in slender prismatic crystals and in massive form, having a specific gravity of 6.36 and a hardness of 2.5 on the Mohs scale; found in contact metasomatic deposits and hydrothermal veins.

Menelaus' theorem *Mathematics.* if A, B, and C are the vertices of an arbitrary triangle, and if D, E, and F are the points on the sides BC, CA, and AB, respectively, or on the extensions of those sides, but not at vertices of the triangle, then the points D, E, and F, called **Menelaus points,** are collinear if and only if $(BD)(CE)(AF) = -(DC)(EA)(FB)$, where (BD), and so on, denotes the directed length of the segment BD. The trigonometric form of Menelaus' theorem states that the three Menelaus points are collinear if and only if $(\sin BAD)(\sin CBE)(\sin ACF) = -(\sin DAC)(\sin EBA)(\sin FCB)$, where $(\sin BAD)$, etc., is the directed measure of angle BAD.

Menger's theorem *Mathematics.* the maximum number of edge-disjoint paths (paths having no edges in common) connecting two distinct vertices v and w of a connected graph G is equal to the minimum number k of edges in a vw-disconnecting set.

menhaden *Vertebrate Zoology.* any marine fish of the genus *Brevoortia,* especially *B. tyrranus,* resembling a compact shad; common along the eastern coast of the U.S. and commercially important as a source of fertilizer and oil.

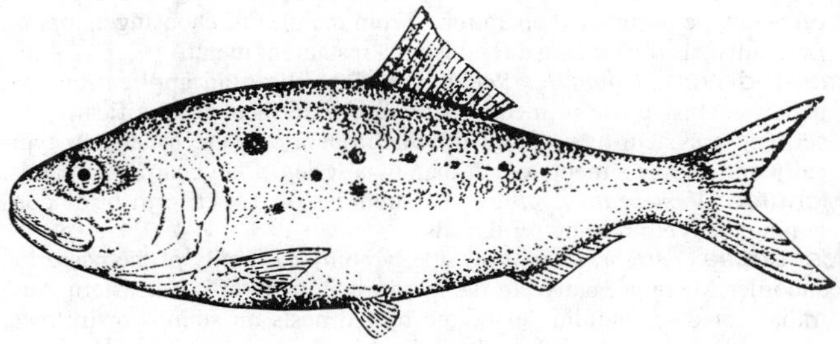

menhaden

menhaden oil *Materials.* a yellowish-brown or reddish-brown drying oil derived by cooking or pressing the body of the menhaden fish; used for making soap, linoleum, paint, and varnish.

menhir *Archaeology.* a large, upright stone, especially one that serves as a monument or marker; widely distributed in prehistoric Europe.

Meniere's disease [mān yârz′] *Medicine.* an accumulation of fluid in the endolymphatic space of the inner ear, resulting in dizziness, nausea, vomiting, ringing in the ear, and deafness. Also, **Meniere's syndrome**. (Named after the French physician Prosper *Meniere,* 1799–1862.)

mening- a combining form denoting a relationship to a membrane, especially the meninges, as in *meningitis.*

meninges [men′in jēz′] *singular,* **meninx.** *Anatomy.* the three membranes that cover the brain and the spinal cord; from the outside they are the dura mater, the pia mater, and the arachnoid. (From a Greek word meaning "membrane.")

meningioma *Oncology.* a hard, slow-growing, usually vascular tumor of the meninges; usually observed in adults over 30 years old. Also, **meningoma, meningeoma.**

meningiomatosis *Oncology.* a condition characterized by the formation of numerous meningiomas.

meningism or **meningismus** *Neurology.* a condition in which there are symptoms of meningitis, including headache and neck stiffness, without actual infection of the meninges; usually occurs in children with such febrile infections as pneumonia and tonsillitis. Also, DUPRÉ'S DISEASE OR SYNDROME.

meningitis [men′in jī′tis] *plural,* **meningitides.** *Pathology.* inflammation of the mengines. When the condition affects the dura mater, the disease is termed **pachymeningitis;** when the arachnoid and pia mater are involved, it is called **leptomeningitis** or **meningitis proper.**

meningitophobia *Psychology.* an irrational fear of meningitis or brain disease.

meningo- a combining form denoting a relationship to a membrane, especially the meninges, as in *meningococcus.*

meningoblastoma *Oncology.* a malignant melanoma of the meninges.

meningocele *Pathology.* an extrusion of brain or spinal cord membranes through an abnormal opening in the skull or spinal column.

meningocephalitis or **meningocerebritis** see MENINGOENCEPHALITIS.

meningococcal meningitis *Pathology.* a severe infectious disease caused by the meningococcus bacteria and characterized by nasal congestion, headache, neck muscle rigidity, light sensitivity, vomiting, convulsions, skin eruptions, and acute skin sensitivity. Also, CEREBROSPINAL MENINGITIS.

meningococcemia *Medicine.* the presence of meningococci in the bloodstream.

meningococcus *plural,* **meningococci.** *Bacteriology.* an individual microorganism of the bacterial species *Neisseria meningitidis,* a major cause of meningitis.

meningocyte *Neurology.* a large phagocytic interstitial cell of the meninges.

meningoencephalitis *Neurology.* an inflammation of the brain and meninges. Also, MENINGOCEPHALITIS, MENINGOCEREBRITIS.

meningoencephalocele *Pathology.* an extrusion of the membranes and spinal cord through an unnatural opening in the skull.

meningoencephalomyelitis *Pathology.* an infection of the membranes and tissues of the brain and spinal cord.

meningoencephalopathy *Pathology.* any illness involving the brain and its surrounding membranes. Also, ENCEPHALOMENINGOPATHY.

meningogenic *Neurology.* arising in the meninges.

meningomyelitis *Neurology.* an inflammation of the spinal cord and its membranes.

meningomyelocele *Pathology.* an extrusion of a portion of membrane and spinal cord tissue through an unnatural opening in the spinal cord. Also, MYELOMENINGOCELE.

meningopathy *Neurology.* any disease of the meninges, the three membranes that envelop the brain and spinal cord.

meningothelium *Histology.* the arachnoid epithelium that surrounds and covers the brain.

meningovascular *Physiology.* relating to the veins and arteries within the meninges.

meningovascular syphilis *Pathology.* an unusual complication seen in the second or third stage of syphilis, involving chronic inflammation of the meningeal blood vessels.

meninx *Anatomy.* the singular form of *meninges.*

meninx primitiva *Vertebrate Zoology.* the single sheath of tissue surrounding the nerve cord in fishes.

meniscectomy *Surgery.* the surgical removal of the meniscus of a joint, usually the knee.

meniscitis *Medicine.* an infection or irritation of the crescent-shaped cartilage of certain joints.

meniscotheriid *Paleontology.* any member of the family Meniscotheriidae.

Meniscotheriidae *Paleontology.* an extinct family of rabbit-sized protoungulate mammals in the order Condylarthra; distinguished by true selenodont teeth, which later became characteristic of the artiodactyls; widespread in Europe and North America in the Paleocene and Early Eocene.

meniscus lens *Optics.* an element in a single lens in which the curvature centers of both of its surfaces lie on the same side of the lens so that both surfaces are concave or both are convex.

meniscus-Schmidt telescope see MAKSUTOV SYSTEM.

Menispermacae *Botany.* a family of woody dicotyledonous herbs, vines, and shrubs of the order Ranunculales having alternate simple leaves and small, regular, unisexual flowers.

Menninger, Charles F. 1862–1953, American psychiatrist; with his sons Karl and William, founded the Menninger Clinic.

Menninger, Karl A. 1893–1990, American psychiatrist; introduced improved psychiatric care for mental patients.

Menninger, William Claire 1899–1966, American psychiatrist; with his father and brother, established the Menninger Foundation.

meno- a combining form denoting a relationship to the menses.

menometrorrhagia *Medicine.* an erratic or profuse bleeding during and between menstrual episodes.

menopause *Physiology.* **1.** the termination of the menstrual cycle in the human female, typically occurring between the ages of 45 and 55. **2.** the portion of life during which this occurs.

menoplania *Medicine.* menstrual flow occurring outside the normal cycle.

Menoponidae *Invertebrate Zoology.* a family of biting and chewing lice that parasitize birds, including domestic fowl, and consume blood and feathers.
menorrhagia [men´ə rā´jē ə] *Physiology.* excessive uterine bleeding occurring at the regular intervals of menstruation, with the period of flow being of greater than usual duration.
menorrhalgia [men´ə ral´jə] *Medicine.* difficult or painful menstruation; dysmenorrhea.
menorrhea *Physiology.* the normal discharge of the menses.
menostasis *Medicine.* absence of the menstrual period; amenorrhea.
menostaxis *Medicine.* an unusually protracted menstrual period.
Menotyphla *Vertebrate Zoology.* in some classifications, a suborder of the order Insectivora, including the elephant shrew and tree shrew.
Mensa *Astronomy.* the Table Mountain, a small and faint constellation near the southern celestial pole.
menses *Physiology.* the flow of blood, mucus, and tissues that occurs during menstruation.
menstrual [men´stroo əl; mens´trəl] *Physiology.* of or relating to menstruation or the menstrual cycle.
menstrual age *Developmental Biology.* the age of a human embryo or fetus, measured from the time elapsed since the onset of the mother's last normal menstruation.
menstrual cycle *Physiology.* the female reproductive cycle, characterized by a regular, recurring change in the uterine lining that ends with menstrual bleeding; most commonly 26 to 29 days in duration, with a mean of 28 days.
menstrual period *Physiology.* the period of menstruation, usually lasting 4 to 5 days.
menstruate [men´stroo āt´] *Physiology.* to discharge blood, mucus, and tissues from the uterus at monthly intervals.
menstruation [men´stroo ā´shən] *Physiology.* the periodic discharge of blood, mucus, and tissues from the uterus associated with the cyclical phenomenon of ovulation. (Going back to the Latin word for "month;" because it occurs about once a month.)
menstruum *Materials Science.* an older term for a solvent. (From the traditional belief that menstrual blood had exceptional properties as a solvent.)
mensuration *Mathematics.* the measurement of such geometric quantities as length, area, volume, etc.
mental *Psychology.* of or relating to the contents, function, or structure of the mind. (From Latin *mens,* "mind.")
mental *Anatomy.* relating to the chin. (From Latin *mentum,* "chin.")
mental aberration *Psychology.* any mental disorder.
mental age *Psychology.* a level of intellectual development expressed as the numerical age at which the statistically average child reaches that level; one of the elements used to determine IQ.
mental deficiency see MENTAL RETARDATION.
mental disorder *Psychology.* any specific form of mental illness.
mental health *Psychology.* a state of mind characterized by psychological and behavioral well-being and an ability to cope with the emotional demands of everyday life.
mental hygiene *Psychology.* the practice of maintaining mental health, or the methods and techniques used to do this.
mental illness *Psychology.* a psychological or behavioral disorder that significantly impairs the emotional and social functioning of the individual. Thus, **mentally ill.**
mentality *Psychology.* **1.** mental power or capacity. **2.** see MENTAL SET.
mentally retarded *Psychology.* suffering from mental retardation; in clinical use this term is now usually replaced by a more precise description of the individual's condition.
mental process *Psychology.* any of the various active processes of the mind, such as the registering or recalling of information.
mental retardation *Psychology.* a condition in which an individual exhibits below-average intellectual abilities and immature behavior, characterized by an IQ below 70; the four categories are **mild mental retardation** (IQ 55–69), **moderate** (IQ 40–54), **severe** (IQ 25–39) and **profound** (IQ under 25). Also, MENTAL DEFICIENCY.
mental set *Psychology.* the tendency to apply preconceptions or familiar procedures to new problems rather than seeking an original solution. Also, MENTALITY.
mental telepathy see TELEPATHY.
mentation *Psychology.* an older term for mental processes.
Mentha *Botany.* the mints, a genus of labiate plants.
Menthaceae *Botany.* in some classifications, an equivalent name for the Labiatae.
menthane *Organic Chemistry.* $C_{10}H_{20}$, a colorless liquid that is insoluble in water; boils at 169°C; a saturated hydrocarbon that is the parent substance of many terpenes.
menthene *Organic Chemistry.* $C_{10}H_{18}$, a colorless liquid that is found in certain essential oils and prepared from menthol; insoluble in water; boils at 167°C.
menthol [men´thôl; men´thäl] *Organic Chemistry.* $C_{10}H_{20}O$, a combustible, white crystalline solid derived from peppermint oil; slightly soluble in water and soluble in alcohol; melts at 42°C; used for its flavor and fragrance in perfumes, cough drops, liqueurs, cigarettes, chest rubs, and chewing gum. Also, PEPPERMINT CAMPHOR.
menthone *Organic Chemistry.* $C_{10}H_{18}O$, a combustible, colorless, oily liquid having a slight peppermint odor; slightly soluble in water and soluble in organic solvents; boils at 207°C; used for flavoring.
menthyl *Organic Chemistry.* the group $C_{10}H_{19}$, derived from menthane.
mento- a combining form meaning "chin," as in *mentoplasty.* (From Latin *mentum,* "chin.")
mentoplasty *Surgery.* plastic surgery of the chin.
Mentrier's disease *Medicine.* an inflammation of the stomach with enlargement of the gastric rugae and changes in the gastric mucosa.
mentum *Anatomy.* the Latin word for chin. *Botany.* a chinlike protruberance in certain orchids, formed by the sepals and the base of the column. *Entomology.* **1.** the median, basal portion of the labium in insects. **2.** an outgrowth between the mouth and foot in certain gastropods.
menu *Computer Science.* **1.** a displayed list of options from which a user can choose. **2.** specifically, a display of commands, such as Enter, Copy, Insert, Move, Delete, Print, that can be selected by the user to carry out the designated operation. (From the idea of choosing an option from a list, as if choosing a food from a restaurant menu.)
menu-driven *Computer Science.* **1.** describing an application that makes extensive use of menus as the user interface. **2.** specifically, describing a system or program in which operations are usually or typically initiated by a menu, rather than by a series of keyboard commands.
Menurae *Vertebrate Zoology.* the suborder of passeriform birds containing the lyrebirds and scrubbirds.
Menuridae *Vertebrate Zoology.* the lyrebirds, a family of the passerine suborder Menurae, native to damp forested areas of southeastern Australia; noted for building elaborate roofed nests on stumps or in trees, and for the showy, lyre-shaped tail display of mating males.
Menyanthaceae *Botany.* a family of cosmopolitan aquatic or semiaquatic dicotyledonous herbs of the order Solanales, characterized by a well-developed system of intercellular canals in the stem with scattered vascular bundles.
mepacrine hydrochloride see QUINACRINE HYDROCHLORIDE.
méplat *Ordnance.* the flat nose formed on a projectile by truncation of the ogive or point fuse.
meprobamate *Pharmacology.* $C_9H_{18}N_2O_4$, an oral sedative occurring as a white powder; used to relieve anxiety and tension, in the treatment of musculoskeletal disorders, and formerly as an anticonvulsant in petit mal epilepsy.
mEq or **meq** milliequivalent.
mer *Materials Science.* the molecule from which a polymer is produced after the double covalent bond has been broken
MER methanol extraction residue.
mer. meridian.
mer- a combining form meaning: **1.** part or partial, as in *meropia.* **2.** thigh, as in *meralgia.*
-mer *Organic Chemistry.* a combining form designating: **1.** a member of a class of compounds. **2.** the unit molecule from which a polymer is produced.
Merak *Astronomy.* β-Ursae Majoris, the pointer star in the Bowl of the Big Dipper that lies farther from the pole.
meralgia *Neurology.* pain in the thigh.
meralluride *Pharmacology.* $C_{16}H_{22}H_9N_6O_7$, a diuretic drug derived from mercury; formerly administered in its derivative form, **meralluride sodium,** to promote the production of urine.
Meramecian *Geology.* a North American provincial series of the Upper Mississippian period, occurring after the Osagian and before the Chesterian.
meranti *Materials.* a soft, pinkish wood from trees of the *Hopea* and *Shorea* species, used as a substitute for mahogany in cabinetmaking and design work.
meraspis *Invertebrate Zoology.* the intermediate growth stage in trilobites, usually taken as corresponding to the formation of thoracic segments and the pygidium.

merawan *Materials.* the yellowish to brown, durable wood of trees of the *Hopea* genus, which are indigenous to Malaya; used for furniture and interior decorating.

merbromin *Organic Chemistry.* $C_{20}H_8Br_2HgNa_2O_6$, a toxic, odorless, green crystalline compound that is soluble in water and insoluble in alcohol; used as a local antiseptic and germicide solution.

Mercalli scale *Geophysics.* a 12-step system used for measuring the intensity of earthquake waves; in its North American adaptation, it is called the **modified Mercalli scale.** (Named after Giuseppe *Mercalli,* 1850–1914, Italian seismologist.)

mercallite *Mineralogy.* $KHSO_4$, a colorless or sky-blue, orthorhombic mineral occurring as stalactites formed of minute tabular crystals, and having a specific gravity of 2.32 and an undetermined hardness; found in fumaroles in the crater of Mt. Vesuvius, Italy.

mercaptal *Organic Chemistry.* a group consisting of $=C(SR)_2$.

mercaptide *Organic Chemistry.* an organic compound containing the sulfhydryl (–SH) group.

mercapto- or **mercapt-** *Chemistry.* a combining form indicating the presence of a mercapto (–SH) group; equivalent to *sulfhydryl-* or *thio-*.

mercaptoacetic acid see THIOGLYCOLIC ACID.

2-mercaptobenzoic acid see THIOSALICYLIC ACID.

mercaptobenzothiazole *Organic Chemistry.* C_7H_5NS, a combustible yellowish powder that is insoluble in water and soluble in alcohol; melts at 164–175°C; used as a corrosion inhibitor and pressure additive in oils and greases, as a fungicide, and as a vulcanization accelerator for rubber.

mercapto compound see SULFHYDRYL COMPOUND.

mercaptoethanol *Organic Chemistry.* $HSCH_2CH_2OH$, a toxic, combustible, clear to white liquid with an unpleasant odor; miscible in water and in most organic solvents; boils at 157°C; used in the manufacture of dyes, pharmaceuticals, rubber chemicals, insecticides, and the like.

mercapto group the univalent group –SH. Also, **mercapto radical.**

mercaptol *Organic Chemistry.* a compound formed when sulfhydryl groups are added to a ketone, producing a compound with the formula $R_2C(SR)_2$.

2-mercaptopropionic acid see THIOLACTIC ACID.

6-mercaptopurine *Oncology.* an antineoplastic antimetabolite used to inhibit the synthesis of DNA in treating acute lymphocytic leukemia as well as acute and chronic granulocytic leukemias. Also, **mercaptopurine.**

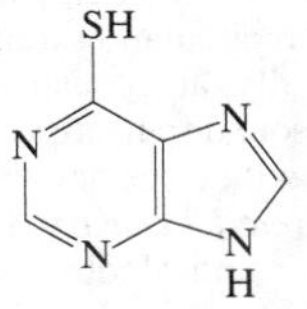

6-mercaptopurine

mercaptosuccinic acid see THIOMALIC ACID.

Mercator, Gerardus (Gerhard Kremer) 1512–1594, Flemish geographer and cartographer; formulated the Mercator projection.

Mercator chart *Cartography.* any chart based on Mercator projection; primarily used for marine navigation. Similarly, **Mercator map.**

Mercator line see RHUMB LINE.

Mercator projection *Cartography.* a map projection in which the surface of a sphere, such as the earth, is developed on a cylinder that is tangent along the equator. The equator is represented by a straight line true to scale, the geographic meridians are represented by a set of parallel straight lines that are spaced according to their distance apart at the equator and perpendicular to the equator line, and the geographic parallels are represented by a second set of straight lines perpendicular to those representing the meridians and therefore parallel to the equator. The scale increases with the distance from the equator, and conformality with this scale is achieved by mathematical means; i.e., both the meridians and the parallels are expanded at the same ratio with increase in latitude. Also, **Mercator's projection, Mercator map projection.**

Mercator sailing *Navigation.* determination of elapsed or required course and distance by computation that yields a result similar to that obtained by plotting on a Mercator chart.

Mercer, John 1791–1866, English chemist and calico printer; developed the mercerization process.

Mercer engine *Mechanical Engineering.* a revolving-block engine composed of two pistons and four rollers (two per piston) in a single cylinder.

mercerize *Textiles.* to treat cotton yarn or fabric under tension with caustic soda (aqueous sodium hydroxide), which swells the material and thereby greatly increases its strength, luster, and dye affinity. Thus, **mercerization, mercerizing, mercerized.**

merchant marine *Naval Architecture.* **1.** the merchant ships of a nation. **2.** the officers and crews of such vessels.

merchant mill *Metallurgy.* a former designation of a mill consisting of several stands, each containing three rolls.

merchant ship *Naval Architecture.* a ship engaged in trade or other commercial carrying activity.

mercurial [mur kyůr′ē əl] *Chemistry.* relating to or containing mercury. *Medicine.* caused by mercury. Thus, **mercurial tremor.** *Pharmacology.* a pharmaceutical preparation of mercury.

mercurial horn ore see CALOMEL.

mercurialism see MERCURY POISONING.

mercurial nephrosis *Medicine.* a kidney disease caused by a chronic use of mercurial diuretics or by exposure to mercury over a long period.

mercuric [mur kyůr′ik] *Chemistry.* **1.** relating to or containing mercury. **2.** describing various compounds of mercury, especially those in which the element has a valence of two.

mercuric acetate *Organic Chemistry.* $Hg(C_2H_3O_2)_2$, a toxic, light-sensitive white powder; soluble in water, alcohol, and acetic acid; used as an oxidizing agent for amines and in the production of pharmaceutical products.

mercuric arsenate *Inorganic Chemistry.* $HgHAsO_4$, a yellow powder that is insoluble in water; toxic and strongly irritant; used as an additive in waterproof and antifouling paints.

mercuric barium iodide *Inorganic Chemistry.* $HgI_2 \cdot BaI_2 \cdot 5H_2O$, yellow to reddish crystals that absorb water from the air; soluble in water and alcohol; toxic and a strong irritant; used in Rohrbach's solution and in testing for alkaloids.

mercuric benzoate *Organic Chemistry.* $Hg(C_7H_5O_2)_2 \cdot H_2O$, a toxic, light-sensitive, white crystalline or powdery solid; slightly soluble in water and alcohol; melts at 165°C; used in medicines.

mercuric bromide *Inorganic Chemistry.* $HgBr_2$, white, light-sensitive crystals; slightly soluble in water and soluble in alcohol; melts at 236°C and boils at 322°C; toxic and strongly irritant; used in medicine.

mercuric chloride *Inorganic Chemistry.* $HgCl_2$, white crystals that are soluble in water and alcohol; melts at 276°C and boils at 302°C; highly toxic and corrosive; used in the manufacture of mercury compounds, in organic synthesis, as a reagent and catalyst, as a fungicide, insecticide, and wood preservative, and for many other purposes.

mercuric cyanate see MERCURY FULMINATE.

mercuric cyanide *Inorganic Chemistry.* $Hg(CN)_2$, toxic, colorless, transparent crystals that darken in light and decompose on heating; soluble in water and alcohol; used as an antiseptic, in photography, and in making cyanogen.

mercuric fluoride *Inorganic Chemistry.* HgF_2, transparent crystals that are moderately soluble in water and alcohol; decomposes at 645°C; highly toxic and a strong irritant; used in organic synthesis.

mercuric iodide *Inorganic Chemistry.* HgI_2, red crystals that are very slightly soluble in water, turning yellow on heating to 127°C; highly toxic and a strong irritant; used in analytical reagents and formerly used in medicine. Also, RED MERCURIC IODIDE.

mercuric lactate *Organic Chemistry.* $Hg(C_3H_5O_3)_2$, a toxic white powder that is soluble in water and decomposes when heated; used as a fungicide.

mercuric nitrate *Inorganic Chemistry.* $Hg(NO_3)_2$, colorless crystals or a white powder; absorbs water from the atmosphere and decomposes on heating; soluble in water and insoluble in alcohol; highly toxic and a dangerous fire hazard in contact with organic substances.

mercuric oleate *Organic Chemistry.* $Hg(C_{18}H_{33}O_2)_2$, a toxic yellow-to-red liquid or solid; insoluble in water and slightly soluble in alcohol and ether; used as an antiseptic.

mercuric oxide *Inorganic Chemistry.* HgO, occurring in two forms: **1. red mercuric oxide.** a bright red powder; insoluble in water and alcohol and decomposed by heat; highly toxic and a fire hazard in contact with organic substances; used in paints, pharmaceuticals, perfumes, cosmetics, and for many other purposes. **2. yellow mercuric oxide.** an orange-yellow powder; slightly soluble in water; darkens in air and decomposes when heated; highly toxic and a combustible; used as an antiseptic and to make mercury compounds.

mercuric phosphate *Inorganic Chemistry.* $Hg_3(PO_4)_2$, a white or yellowish, highly toxic powder; insoluble in water and alcohol.

mercuric salicylate *Organic Chemistry.* $(OH_2C_6H_4CO_2)_2Hg$, a tasteless, odorless, toxic white powder tinged with yellow or pink; insoluble in water and alcohol; formerly used in the treatment of syphilis.

mercuric stearate *Organic Chemistry.* $(C_{17}H_{35}CO_2)_2Hg$, a toxic yellow powder; insoluble in water and alcohol and soluble in fatty oils; used as a germicide.

mercuric sulfate *Inorganic Chemistry.* $HgSO_4$, a highly toxic, white crystalline powder; decomposed by water or by heat; soluble in acids and insoluble in alcohol; used in various chemical processes.

mercuric sulfide *Inorganic Chemistry.* HgS, occurring in two forms: **1. black mercuric sulfide.** a highly toxic black powder; insoluble in water and alcohol and decomposed by heat; used as a pigment. **2. red mercuric sulfide.** a fine, bright red powder; insoluble in water and alcohol; highly toxic; used as a pigment.

mercuric thiocyanate *Inorganic Chemistry.* $Hg(SCN)_2$, a white, highly toxic powder; insoluble in water; decomposes on heating; used in photography and fireworks. Also, **mercuric sulfocyanate, mercuric sulfocyanide.**

Mercurochrome *Pharmacology.* $C_{20}H_8Br_2HgNa_2O_6$, the trade name for preparations of merbromin, a topical antibacterial solution.

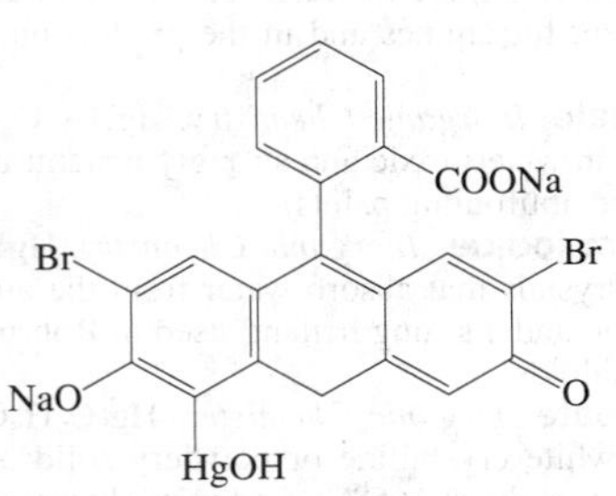

Mercurochrome (merbromin)

mercurous [mur kyůr´əs] *Chemistry.* **1.** relating to or containing mercury. **2.** describing various compounds of mercury, especially those in which the element has a valence of one.

mercurous acetate *Organic Chemistry.* $Hg_2(C_2H_3O_2)_2$, highly toxic, colorless plates or scales; decomposed by boiling water and by light into mercury and mercuric acetate; slightly soluble in water and insoluble in alcohol and ether; used as an antibacterial medicine.

mercurous bromide *Inorganic Chemistry.* Hg_2Br_2, highly toxic white powder or colorless crystals; slightly soluble in water and insoluble in alcohol; sublimes at 345°C; darkened by light; turns yellow on heating.

mercurous chlorate *Inorganic Chemistry.* $Hg_2(ClO_3)_2$, highly toxic white crystals that decompose on heating; explodes on contact with flammable substances.

mercurous chloride *Inorganic Chemistry.* Hg_2Cl_2, white rhombic crystals or powder; insoluble in water and alcohol; sublimes at 400°C; stable in air but darkens on exposure to light; used as a fungicide and pesticide, and once widely used as a cathartic.

mercurous chromate *Inorganic Chemistry.* Hg_2CrO_4, a red powder that is slightly soluble in water; decomposes on heating; highly toxic and a fire hazard in contact with organic substances; used to color ceramics green.

mercurous iodide *Inorganic Chemistry.* Hg_2I_2, a toxic, bright-yellow, water-insoluble powder that turns greenish and decomposes in light; on heating it changes through dark yellow and orange to red, then sublimes; formerly used in medicine as an antibacterial agent. Also, YELLOW MERCUROUS IODIDE.

mercurous nitrate *Inorganic Chemistry.* $HgNO_3{\cdot}2H_2O$, colorless, efflorescent, light-sensitive crystals; decomposes in water and melts at 70°C; highly toxic and explosive if shocked or heated; used as an analytical agent. Also, HYDRATED MERCUROUS NITRATE.

mercurous oxide *Inorganic Chemistry.* Hg_2O, a highly toxic black powder that is soluble in acids and insoluble in water; it decomposes when heated to 100°C.

mercurous sulfate *Inorganic Chemistry.* Hg_2SO_4, a highly toxic, white to yellow crystalline powder; slightly soluble in water; decomposes on heating; used in batteries.

Mercury *Astronomy.* the closest planet to the sun; it is 4878 kilometers in diameter, has a density of 5.43, and orbits the sun once every 88 days at a distance of 57.9 million kilometers. (Named for *Mercury,* the Greek messenger of the gods, who was a symbol of speed, because it has the fastest orbit of the planets.)

mercury *Chemistry.* a silvery-white metallic element having the symbol Hg, the atomic number 80, an atomic weight of 200.59, a melting point of –38.87°C, and a boiling point of 356.58°C; an extremely heavy liquid (the only common metal to exist in this state at ordinary temperatures) with extremely high surface tension; used in thermometers, amalgams, mercury vapor lamps, and as a neutron absorber. (Named by medieval alchemists for the Greek god *Mercury,* because of its unusual property of flowing easily and quickly.)

mercury acetate **1.** see MERCURIC ACETATE. **2.** see MERCUROUS ACETATE.

mercury arc *Electronics.* an electric discharge through ionized mercury vapor, giving off a bright bluish-green glow that contains strong ultraviolet radiation.

mercury-arc rectifier *Electronics.* a gas tube rectifier in which mercury vapor is the conducting medium. The cathode may be either a hot cathode or a pool-type cathode. Also, MERCURY-VAPOR RECTIFIER.

mercury arsenate see MERCURIC ARSENATE.

mercury ballistic *Navigation.* a system of reservoirs and connecting tubes containing mercury that is used to make a gyrocompass point to the north.

mercury barium iodide see MERCURIC BARIUM IODIDE.

mercury barometer *Engineering.* an instrument used to measure atmospheric pressure by noting the height mercury rises into an open-ended glass tube inverted into a dish of mercury. Also, TORRICELLIAN BAROMETER.

mercury benzoate see MERCURIC BENZOATE.

mercury bichloride see MERCURIC CHLORIDE.

mercury bromide **1.** see MERCURIC BROMIDE. **2.** see MERCUROUS BROMIDE.

mercury cell *Electricity.* a tiny dry cell containing a mercuric-oxide cathode and a zinc anode in a potassium-hydroxide electrolyte; used in miniature electronic products such as cameras and hearing aids.

mercury chlorate see MERCUROUS CHLORATE.

mercury chloride **1.** see MERCURIC CHLORIDE. **2.** see MERCUROUS CHLORIDE.

mercury chromate see MERCUROUS CHROMATE.

mercury cyanide see MERCURIC CYANIDE.

mercury delay line *Electronics.* a delay line in which mercury is used as the storage medium for a circulating train of waves or pulses. *Acoustical Engineering.* specifically, an acoustic delay line in which mercury is used as the medium for sound transmission.

mercury fluoride see MERCURIC FLUORIDE.

mercury fulminate *Organic Chemistry.* $Hg(CNO)_2$, a gray crystalline solid; soluble in hot water and alcohol, and slightly soluble in cold water; explodes at its melting point when dry; used to initiate explosions in caps and detonators. Also, MERCURIC CYANATE

mercury iodide **1.** see MERCURIC IODIDE. **2.** see MERCUROUS IODIDE.

mercury-manganese star *Astrophysics.* a star of spectral type B8 or B9 having abnormal abundances of elements such as manganese, mercury, strontium, gallium, and yttrium.

mercury manometer *Engineering.* a manometer that utilizes mercury as the instrument fluid to regulate or register changes in pressure or flow.

mercury monochloride see MERCUROUS CHLORIDE.

mercury naphthenate *Organic Chemistry.* a toxic yellowish liquid that is soluble in mineral oils; used in making gasoline antiknock compounds and as an antimildew agent in paints.

mercury nitrate or **mercury pernitrate** see MERCURIC NITRATE.

mercury oleate see MERCURIC OLEATE.

mercury oxide **1.** see MERCURIC OXIDE. **2.** see MERCUROUS OXIDE.

mercury phosphate see MERCURIC PHOSPHATE.

mercury poisoning *Toxicology.* poisoning due to ingestion of or excessive exposure to mercury or its salts; symptoms may include nervous ills (such as tremors and incoordination), a metallic taste in the mouth, sore gums, vomiting, and severe damage to the gastrointestinal tract. Also, MERCURIALISM, HYDRARGYRIA.

mercury-pool cathode *Electronics.* a pool-type cathode in which the pool is liquid mercury; electrons are emitted from an arc spot on the pool.

mercury-pool electrode see CALOMEL ELECTRODE.

mercury porosimetry *Materials Science.* a technique used to measure the porosity of a substance, in which mercury is forced into the pores of the sample at levels of steadily increasing pressure.

Mercury program *Space Technology.* the first series of United States manned space flights (1961–1963), developed to test the effects of rocket launch, weightlessness, and reentry; each of its six suborbital and orbital flights carried a single crew member.

mercury pycnometer *Materials Science.* a device for measuring the volume or density of irregularly shaped materials. Because mercury is a nonwetting liquid, mercury pycnometers are useful for determining the volume of ceramic materials.

mercury sulfate or **mercury persulfate** see MERCURIC SULFATE.

mercury sulfide see MERCURIC SULFIDE.

mercury switch *Electricity.* an electric switch consisting of a large globule of mercury in a level tube, having electrodes attached in such a way that tilting the tube causes the mercury to move and thereby make or break the circuit.

mercury thermometer *Engineering.* a thermometer in which the thermally sensitive liquid is mercury.

mercury-vapor lamp *Electronics.* a lamp that contains mercury vapor, producing a luminous blue-green light when the mercury is ionized by the flow of electric current.

mercury-vapor rectifier see MERCURY-ARC RECTIFIER.

mercury-vapor tube *Electronics.* a gas tube containing mercury vapor as the active gas; when the vapor is ionized, a luminous glow is produced. Also, **mercury tube.**

mercury-wetted reed switch *Electronics.* a reed switch containing a pool of mercury, in which the reeds are wetted by the mercury through capillary action.

-mere a combining form denoting a segment or part, as in *blastomere, cytomere.*

merganser *Vertebrate Zoology.* any of various fish-eating, diving ducks belonging to the family Anatidae, having a crested head and a slender, hooked bill with serrated margins; widely distributed North American species are the **common (American) merganser**, *Mergus merganser*; the **red-breasted merganser**, *M. serrator*; and the **hooded merganser**, *Lophodytes cucullatus.* Also, GOOSANDER.

merganser

merge *Computer Programming.* **1.** a process that combines two or more ordered files into one similarly ordered file. **2.** to carry out this process. Also, COLLATE, MESH. *Transportation Engineering.* **1.** to carry out a merging process. **2.** a place where merging occurs, such as when a car enters the traffic on a major highway from an on-ramp.

merged-transistor logic see INTEGRATED INJECTION LOGIC.

merge search *Computer Programming.* a sequential table-lookup technique that requires both the file and the table that is to be searched to be ordered in the same sequence.

merge sort *Computer Programming.* a sort routine that divides items in a list into a sublist, sorts the items in each sublist, and then merges the resulting sorted sublists into a single, sorted list.

merging *Transportation Engineering.* the process by which a car approaching from the side joins a traffic stream or by which two traffic streams are joined into one. Smoothness in merging is an important factor in reducing bottlenecks.

merging control system *Transportation Engineering.* a system that restricts the flow of merging traffic in order to minimize congestion.

Merian's formula *Oceanography.* a formula for the period of a seiche: $T=2L/ngh$, where L is the length of the basin (measured in the direction of wave motion), n is the number of nodes, g is the acceleration due to gravity, and h is the depth of the water.

mericarp *Botany.* a portion of a schizocarp that contains one seed; it results from the splitting of the mature schizocarp.

mericlinical chimera *Biology.* a chimera in which one type of cell does not completely surround the other type. Also, CHIMERA.

meridian *Cartography.* **1.** a great circle through the geographical poles of the earth, functioning as a north-south reference line and from which azimuths and differences in longitudes are determined. **2.** a plane perpendicular to the geoid or sphereoid and defining such a line. *Astronomy.* a great circle on the celestial sphere that passes through both celestial poles, the north and south points on an observer's horizon, and the observer's zenith.

meridian altitude *Astronomy.* the altitude a celestial object has when it crosses an observer's meridian.

meridian angle *Astronomy.* the angle measured eastward along the celestial equator from an observer's meridian to that of an object.

meridian distance *Cartography.* the perpendicular distance, in a horizontal plane, between a given point on the earth's surface and a meridian of reference (usually the observer's local meridian).

meridian line *Cartography.* the line representing the intersection of the plane of the celestial meridian with the plane of the horizon, with an astronomical azimuth of 0° or 180°; used to determine horizontal directions.

meridian observation *Astronomy.* the observation of a star's declination (and, occasionally, brightness) when it crosses an observer's meridian.

meridian passage *Astronomy.* the moment when a celestial object crosses, or transits, an observer's meridian.

meridian photometer *Astronomy.* a telescopic instrument designed to measure a star's brightness when it transits the meridian.

meridian sailing *Navigation.* the process of sailing directly north or south along a given meridian.

meridian telescope *Optics.* a telescope designed to view objects that appear in the plane of the meridian; such telescopes include the transit telescope and the zenith telescope.

meridian transit see TRANSIT CIRCLE.

meridional *Science.* of or related to a movement or direction between the poles of an object, usually north-south.

meridional cell *Geophysics.* a convection cell whose main motion is in a north-south direction.

meridional circulation *Meteorology.* the atmospheric circulation found in a vertical plane along a meridian, having only north or south components.

meridional difference *Cartography.* on a Mercator map projection, the difference between the distances from the equator of any two parallels, measured in minutes of longitude. The distances are subtracted if the parallels are on the same side of the equator, or added if they are on different sides.

meridional flow *Meteorology.* an atmospheric air flow pattern characterized by a highly pronounced meridional (north-south) component accompanied by a weaker than normal zonal (east-west) component.

meridional front *Meteorology.* a cold front that separates successive migratory subtropical high-pressure systems in the South Pacific, occurring as a high arc with longitudinal meridians as chords.

meridional index *Meteorology.* a measure of air movement along meridians, averaged around a given latitude circle.

meridional orthographic map projection *Cartography.* a map projection in which the plane of the map is parallel to the plane of a selected meridian. The outer meridian is therefore a full circle, the central meridian is a straight line, and the other meridians are arcs of ellipses, while the geographic parallels are straight lines.

meridional part *Cartography.* on a Mercator projection, the length of the arc of a meridian between the equator and a given parallel, expressed in minutes of longitude at the equator.

meridional parts *Navigation.* the amounts by which the length of a degree of latitude must be increased to be proportional to the increase in the length of a degree of longitude on a Mercator chart, expressed in units of minutes of the equator.

meridional plane *Cartography.* any plane that contains the polar axis of the earth.

meridional ray *Optics.* a ray that appears in an optical system in the plane as the optical axis.

meridional wind *Meteorology.* the wind component along a local meridian, considered separately from the zonal wind.

Meridosternata *Invertebrate Zoology.* a suborder of echinoderms including certain deep-water sea urchins.

merino or **Merino** *Agriculture.* a breed of fine-wool sheep, *Ovis aries,* having a white, wool-covered face and a pink muzzle and legs; developed in Spain by the Moors, and bred for the high quality and quantity of their wool. The **American Merino** and **Delaine Merino** are varieties of this breed. *Textiles.* **1.** the wool of this sheep. **2.** fabric or yarn made from this wool.

merino

merinthophobia *Psychology.* an irrational fear of being bound or tied up.

merisis *Biology.* growth in size resulting from cell division.

merismite *Petrology.* a variety of chorismite characterized by irregular penetration of the diverse units.

merismopedia *Bacteriology.* a genus of blue-green algae characterized by growth of the cells in a rectangular sheet conformation.

meristem *Botany.* a region of plant tissue consisting of undifferentiated cells that are capable of developing into specialized plant tissue.

meristic *Biology.* **1.** pertaining to a change in number or in geometric relation of parts of an organism, as in **meristic variation. 2.** of or relating to being divided into segments or parts.

meristoderm *Botany.* in certain brown algae, the outer portion of the stipe, in which activity resembling that of the meristem takes place.

meristogenous development *Mycology.* the development of fungal structures from one hypha.

merit *Electronics.* a term for a device's performance rating for a particular aspect of its operation, governing which applications are suitable for that device, as in signal-to-noise merit or gain-bandwidth merit.

Merkel's corpuscles *Anatomy.* small sensory nerve endings in the submucosa of the tongue and mouth, consisting of two flattened epithelial cells enclosed in a sheath and separated by a disklike extension of the sensory nerve cells.

merlon *Architecture.* the solid, upstanding part of a parapet, between embrasure openings.

mermaid's wine glass *Botany.* a green algae, *Acetabularia crenulata,* found in warm seas and distinguished by a goblet-shaped cap on a slender stalk.

Mermithidae *Invertebrate Zoology.* a family of long, thin nematodes with immature forms that parasitize insects, snails, and slugs.

Mermithoidea *Invertebrate Zoology.* a superfamily of long, thin nematodes, with free living adults and immature forms that are parasitic on invertebrates.

mero- a combining form meaning: **1.** part or partial, as in *merogony.* **2.** thigh, as in *meralgia.*

meroblastic *Developmental Biology.* undergoing cleavage in which only part of the ovum participates; partially dividing.

merocrine *Physiology.* of or related to glands that release fluid cellular products through the cell membrane without the loss of cytoplasm; characteristic of digestive glands.

merodiastolic *Cardiology.* referring to a part of the diastole or partially diastolic.

merogony *Developmental Biology.* the development of a portion of an ovum that lacks a female nucleus.

merohedral *Crystallography.* a crystal with symmetrical suppression of half, three-quarters, or seven-eighths of the faces found on the holohedral form. Also, MEROSYMMETRICAL.

meromictic lake *Hydrology.* a stratified lake that does not undergo complete mixing of its water during periods of circulation, especially a lake in which the noncirculating bottom layer does not mix with the circulating upper layer.

meromorphic function *Mathematics.* a function of a complex variable that is analytic everywhere in its domain of definition except at a finite number of poles.

meromyarian *Invertebrate Zoology.* any nematode having only two to five longitudinal muscle cells in each quadrant of the body.

meromyosin *Biochemistry.* either of two fragments of the myosin molecule isolated by treatment with proteolytic enzymes: heavy **H-meromyosin** or light **L-meromyosin.** H-meromyosin contains the subfragment responsible for the ATPase activity of myosin; L-meromyosin makes up the major part of the rodlike backbone of the molecule.

meromyxis *Genetics.* a phenomenon in which only a portion of the genome of an organism is transferred, rather than an entire haploid complement.

Meropidae *Vertebrate Zoology.* the bee eaters, a family of colorful birds of the order Coraciiformes that are noted for aerial insect-catching; found in tropical and warm temperate regions from Europe to Australia.

meroplankton *Biology.* temporary members of the planktonic community composed of floating developmental stages, such as eggs and larvae, that are present for only a portion of the life cycle of the organism.

merosporangium *Mycology.* a threadlike sac or sporangium that contains spores forming a chain.

Merostomata *Invertebrate Zoology.* a class of marine arthropods, including the horseshoe crabs and king crabs, having six pairs of abdominal appendages bearing exposed gills.

merosymmetrical see MEROHEDRAL.

-merous a combining form meaning having parts of the kind or number specified in the initial element; e.g., *pentamerous,* "having five parts."

Merozoa see CESTODA.

merozoite *Invertebrate Zoology.* an asexually produced stage in the life cycle of some parasitic sporozoan protozoans, the individuals of which disperse and infect additional cells within the host.

merozygote *Bacteriology.* a partially diploid bacterium, resulting from the genetic transfer of a portion of a donor bacterial cell chromosome to a recipient cell that is already in possession of a complete genome. Thus, **merozygotic.**

Merrifield, R. Bruce born 1921, American biochemist; awarded the Nobel Prize for developing an automated method for making peptides.

Merrifield resin *Organic Chemistry.* a chloromethylated polystyrene resin used as a solid support upon which peptides are synthesized.

merrihueite *Mineralogy.* $(K,Na)_2(Fe^{+2},Mg)_5Si_{12}O_{30}$, a greenish-blue, hexagonal mineral of the osumilite group, having a specific gravity of 2.64 to 2.87 and an undetermined hardness; found as small grains only in meteorites.

merrillite *Mineralogy.* the mineral whitlockite as found in minute quantities in meteorites.

mersawa *Materials.* the pale, coarse, even-textured wood of the trees of the Asian genus *Anisoptera*; used for light construction, flooring, furniture, and plywood.

Mersenne's law *Mechanics.* a law stating that the fundamental harmonic frequency of a tense string, vibrating with small amplitudes, varies as the square root of tension divided by the length and the square root of the mass per unit length.

Mersey yellow coal see TASMANITE.

Mertensian mimicry *Behavior.* a form of mimicry in which the model species is somewhat offensive, while the mimic is actually lethal.

Merthiolate [mur thī´ə lāt´] *Pharmacology.* $C_9H_9HgNaO_2S$, a trade name for preparations of thimerosal, an antimicrobial agent used as a topical anti-infective.

Merton grating *Optics.* a diffraction grating produced by slicing a helical thread on a cylinder and smoothing over any errors by slicing a second thread down further on the same cylinder.

merwinite *Mineralogy*. $Ca_3Mg(SiO_4)_2$, a colorless to pale-green, monoclinic mineral massive to granular in habit, and having a specific gravity of 3.15 and a hardness of 6 on the Mohs scale; found in contact zones with larnite and spurrite.
Merychippus *Paleontology*. a genus of North American horses arising from *Parahippus* in the early Miocene and persisting until the early Pliocene, when they gave way to *Nannippus*.
Merycoidodontidae *Paleontology*. a family of artiodactyl mammals in the extinct superfamily Merycoidodontoidea; Eocene to Pliocene; may have formed large herds in the Oligocene and Miocene.
Merycoidodontoidea *Paleontology*. an extinct superfamily of artiodactyls in the suborder Tylopoda; Eocene to Pliocene.
mes- a combining form meaning "middle" or "intermediate."
mesa [mā´sə] *Geography*. **1.** a flat-topped hill or tableland, larger than a butte, rising steeply from a surrounding plain. Also, PLATEAU MESA. **2.** a level terrace set among mountains and bounded by higher land on one or more sides. Also, BENCH MESA. *Electronics*. a flat, raised area on an electronic device. (From the Spanish word for "table.")
Mesacanthus *Paleontology*. a genus of acanthodian fishes in the extinct order Acanthodiformes, characterized by large scales and an unusual series of spines; lower to middle Devonian.
mesaconic acid *Organic Chemistry*. $HO_2CC(CH_3){=}CHCO_2H$, a needlelike or prismatic solid; soluble in water; melts at 202°C; an unsaturated, dibasic acid. Also, METHYLFUMARIC ACID.
mesa device *Electronics*. any device that is produced by diffusing the entire semiconductor wafer and then etching all but selected areas to produce a raised area (mesa). Thus, **mesa construction.**
mesa diode *Electronics*. a diode produced by diffusing the entire wafer and then etching all areas except the island or mesa at each junction site.
mesarch *Botany*. a growth pattern occurring primarily in ferns, in which the primary xylem (protoxylem) is surrounded by the younger, subsequent xylem (metaxylem).
mesarch succession *Ecology*. an ecological succession that originates in a habitat with a moderate amount of water.
mesarteritis *Medicine*. an infection of the middle lining of an artery.
mesa transistor *Electronics*. a diffused planar transistor in which the silicon area has been etched in steps so that the base and emitter regions appear as plateaus above the collector.
mesaxon *Neurology*. the point at which a pair of parallel and overlapping membranes completely enclose the axon of a neuron.
mescal [mə skal´] *Botany*. a common name for *Lophophora williamsii*, a cactus species of the southwestern U.S. and northern Mexico that is the source of the hallucinogenic substance mescaline. Also, PEYOTE.
mescal button *Botany*. the dried top of the cactus species *Lophophora williamsii*, used as a hallucinogen.
mescaline [mes´kə lēn´] *Organic Chemistry*. $(CH_3O)_3C_6H_2CH_2CH_2NH_2$, a highly toxic crystalline solid; soluble in water, alcohol, benzene, and chloroform and insoluble in ether; melts at 35–36°C and boils at 180°C (12 torr); a hallucinogenic alkaloid produced by mescal cactus *Lophopora williamsii*, used in biochemical and medical research.
mescalism *Toxicology*. poisoning due to ingestion of mescaline.
mesectoblast see ECTOMESODERM.
mesectoderm *Developmental Biology*. **1.** a region of embryonic tissue that contributes to both ectoderm and mesoderm. **2.** that part of the mesenchyme derived from ectoderm, such as cephalic mesenchyme derived from placodes in a vertebrate embryo.
Meselson-Radding model *Genetics*. a model for explaining how chromosomes can recombine by tristing and isomerization of the strands, without the occurrence of small breaks or nicks in the strands as proposed in an earlier model by R. Holliday.
Meselson-Stahl experiment *Molecular Biology*. an experiment described in 1958 that confirmed the semiconservative mechanism by which DNA replicates.
mesencephalic flexure see CRANIAL FLEXURE.
mesencephalon [mes´en sef´ə län] *Anatomy*. the part of the brain located below the thalamus and above the pons, developed from the middle of the three primary cerebral vesicles of the embryonic neural tube; includes the superior and inferior colliculi and cerebral peduncles. Also, MIDBRAIN.
mesenchymal cell [mes´en kī´məl] *Developmental Biology*. in mesenchymal tissue, an undifferentiated cell that can develop into any of various specialized connective tissue cell types.
mesenchymal epithelium *Developmental Biology*. a layer of squamous epithelium lining spaces around the brain, in the inner ear, and the chambers of the eyeball.
mesenchymal hyalin *Pathology*. a translucent albuminoid lubricant resulting from degeneration, occurring in the fabric of the connective tissue, blood and lymphatic vessels.
mesenchyme [mes´en kīm´] *Developmental Biology*. embryonic connective tissue that is made up of loosely associated, often stellate cells, and a dispersed extracellular matrix. Also, **mesenchymal tissue.**
mesenchymoma *Oncology*. a tumor containing a combination of mesenchymal tissues.
mesendoderm *Developmental Biology*. tissue that is modified within the embryo to become endoderm and mesoderm.
mesenteric [mes´en târ´ik] *Anatomy*. of or related to the mesentery.
mesenteric artery *Anatomy*. one of two blood vessels that supply the intestinal organs; either the **superior mesenteric artery** or the **inferior mesenteric artery.**
mesenteric lymphadenitis *Medicine*. inflamed lymph nodes in the mesentery, causing symptoms mimicking those of acute appendicitis.
mesenteron *Anatomy*. the midgut. *Entomology*. the middle part of the alimentary canal of insects, into which the digestive glands open.
mesentery [mes´en târ´ē] *Anatomy*. **1.** any membranous fold attaching various organs to the body wall. **2.** specifically, the peritoneal fold attaching the small intestine to the dorsal abdominal wall. *Invertebrate Zoology*. a vertical membrane that partitions the body cavity of sea anemones and similar animals.
mesentoderm *Developmental Biology*. the inner layer of gastrula that has not differentiated into mesoderm and entoderm.
Meseta *Geography*. the high plateau that covers most of Spain.
mesethmoid *Anatomy*. a median segment of the ethmoid bone.
MESFET metal semiconductor field-effect transistor.
mesh to interweave or interlock; an interwoven or interlocked structure; specific uses include: *Textiles*. any knit or woven fabric having an open texture. *Building Engineering*. a material made of small, interlocking metal links, often used as backing support in building construction and repair. *Design Engineering*. **1.** a screen size or the number of openings between screen material, usually expressed in terms of the number of squares per linear inch. Also, MESH SIZE. **2.** the size of particles that pass through a screen or sieve. *Mechanical Engineering*. **1.** the working contact of the teeth of gears or of a gear and rack. **2.** to make such a contact. *Computer Programming*. **1.** an arrangement of interconnecting lines forming a grid or more complex shape, often drawn automatically. **2.** see MERGE. *Electricity*. see LOOP, def. 2.
MeSH *Medicine*. a reference guide published by the National Library of Medicine for use with MEDLARS, a medical database. (An acronym for Medical Subject Headings.)
mesh analysis *Electricity*. a method of analyzing current flow in a circuit, in which the current for each mesh is indicated, as opposed to the current of each branch.
mesh connection see DELTA CONNECTION.
mesh current *Electricity*. the current analyzed and defined by mesh circuit analysis that is assumed to exist over all cross sections of a loop in an electric network.
mesh impedance *Electricity*. the effective resistance of a circuit loop.
mesh size *Design Engineering*. see MESH, def. 1.
mesh weld *Metallurgy*. seam weld that is only slightly thicker than the sheet.
mesic *Ecology*. **1.** of or relating to organisms that require moderate amounts of moisture. **2.** describing a habitat with moderate moisture.
mesic atom *Particle Physics*. a short-lived atom in which a negative muon has displaced a normal electron.
mesic molecule *Particle Physics*. a molecule in which a negative muon has displaced a normal electron.
mesityl oxide *Organic Chemistry*. $(CH_3)_2{=}CHCOCH_3$, a toxic, flammable, colorless, oily liquid with a honeylike odor; soluble in water, alcohol, and ether; boils at 130–131°C; used as a solvent and insect repellent, and in the manufacture of paints and lacquers.
mesmerism [mez´mə riz´əm] *Psychology*. an older term for hypnosis. (From F. A. *Mesmer*, 1734–1815, an early practitioner of hypnosis.)
Mesnieraceae *Mycology*. a family of tropical fungi of the order Pleosporales, occurring immersed in living or decaying plant leaf tissue.
meso- *Science*. a combining form meaning "middle" or "intermediate," as in *mesocarp, mesophyte*.
meso- *Chemistry*. a combining form designating an optically inactive isomeric compound.
Mesoamerica *Archaeology*. a cultural area roughly between central Mexico and Costa Rica, the location of several notable pre-Columbian civilizations.

mesoappendix *Anatomy.* the piece of peritoneal tissue that connects the appendix to the ileum.

mesobilirubin *Biochemistry.* a compound formed by the reduction of bilirubin.

mesobilirubinogen *Biochemistry.* a reduced form of bilirubin, formed in the intestine, which on oxidation forms stercobilin.

mesoblast *Developmental Biology.* the mesoderm, especially in the early stages of embryos. Thus, **mesoblastic.**

mesocardium *Anatomy.* a double layer of membrane in the developing embryo that attaches the heart to the body wall and the foregut.

mesocarp *Botany.* the middle layer of the pericarp of a ripened ovary or fruit, such as the fleshy part of a peach.

mesocephalic *Anthropology.* in anthropometry, having a skull width that is between 75 and 80 percent of the length. Also, **mesocranic.** *Anatomy.* of or relating to the mesencephalon.

mesocercaria *Entomology.* a developmental stage in certain parasitic trematodes that occurs in the second intermediate host.

mesoclimate *Meteorology.* **1.** the climate over a small portion of the earth's surface, where it may differ from the general climate of the area where it occurs. **2.** see MESOTHERMAL CLIMATE.

mesoclimatology *Meteorology.* the study of mesoclimates.

mesocolon *Anatomy.* the peritoneal fold that fixes the colon to the dorsal abdominal wall.

meso compound *Organic Chemistry.* a diastereomer that is not optically active because it has certain elements of symmetry which make it nondissymmetric.

mesocratic *Petrology.* of or relating to igneous rocks that are composed of a fairly equal proportion of light and dark minerals; intermediate in color between leucocratic and melanocratic.

mesocrystalline *Petrology.* of or relating to crystalline rocks having crystals of a diameter intermediate between those of microcrystalline and macrocrystalline rocks, ranging between 0.20 and 0.75 millimeters.

mesocyclone *Meteorology.* a region of low pressure circulating within a convective storm.

mesocytoma *Oncology.* a tumor of the connective tissue.

mesoderm *Developmental Biology.* the middle of the three primary germ layers of the embryo, which gives rise to the connective tissues, body musculature, blood, cardiovascular and lymphatic systems, most of the urogenital system, and the lining of the pericardial, pleural, and peritoneal cavities.

mesodermal tumor *Medicine.* a growth in the muscle cells of the uterus, consisting of more than one type of tissue and seen in older women.

mesodontic *Anatomy.* having medium-sized teeth.

mesogaster *Anatomy.* a part of the rudimentary mesentery in the embryo that envelops that portion of the enteric canal that will later become the stomach. Also, **mesogastrium.**

Mesogastropoda see PECTINIBRANCHIA.

mesogeosyncline *Geology.* a geosyncline that lies between two continents, receiving clastics from both. Also, MEDITERRANEAN.

mesoglea *Invertebrate Zoology.* the noncellular, gelatinous layer found between the ectoderm and endoderm in certain sponges and coelenterates. Also, **mesogloea.**

mesoglia *Neurology.* non-neural cells that originate in the mesoderm of the embryo and play a structural and hygienic role within the central nervous system; classified as oligodendroglia or microglia.

mesoglioma *Oncology.* a slow-growing tumor composed predominantly of oligodendroglial cells.

Mesohippus *Paleontology.* a genus of early horses slightly larger than *Hyracotherium* but still only the size of a sheep; characteristic of the Oligocene in North America; one line migrated to Eurasia in the Early Miocene but became extinct at the beginning of the Pliocene.

mesohyl *Invertebrate Zoology.* a gelatinous mesenchymal layer of tissue in sponges, consisting of a protein matrix containing skeletal material and ameboid cells.

mesoinositol or **meso-inositol** see MYOINOSITOL.

mesoionic compound *Organic Chemistry.* an ionic compound whose molecules are superimposable on their mirror images even though they contain chiral centers.

mesolamella *Invertebrate Zoology.* the noncellular, gelatinous membrane found between the epidermis and gastrodermis in hydrozoans.

mesolite *Mineralogy.* $Na_2Ca_2Al_6Si_9O_{30}\cdot 8H_2O$, a white or colorless, monoclinic mineral of the zeolite group, occurring as tufts of acicular crystals, and having a specific gravity of 2.26 and a hardness of 5 on the Mohs scale; found in amygdaloidal basalts and similar rocks.

Mesolithic *Archaeology.* **1.** a time period in human history from about 10,000 BC to 8000 BC, following the Upper Paleolithic and preceding the Neolithic; characterized by a broad use of food resources, and localized populations. **2.** of or relating to this period. Also, **mesolithic.**

mesomere *Vertebrate Zoology.* the middle region of the mesoderm, lying between the hypomere and the epimere. *Developmental Biology.* in a cleaving embryo, a blastomere of the intermediate size between micromeres and macromeres.

mesomerism *Chemistry.* a term used to describe the chemical structure of compounds that cannot be accurately represented by a single formula, implying that the true structure lies between certain approximations. Also, RESONANCE. *Materials Science.* the representation of the electronic structure of a molecular system in terms of contributing structures.

mesometeorology *Meteorology.* the study of fairly large-scale atmospheric phenomena.

mesometrium *Anatomy.* the lower portion of the broad ligament of the uterus, located below the mesovarium and composed of layers of peritoneum that enclose the uterus and attach it to the lateral wall of the pelvis.

mesomorph *Psychology.* a person of the mesomorphic body type.

mesomorphic *Psychology.* describing a body type classification for a person having a strong, muscular physique; relating to mesomorphy. *Physical Chemistry.* see MESOMORPHOUS.

mesomorphic phase see MESOMORPHISM.

mesomorphism *Physical Chemistry.* a state of matter that is intermediate between a true crystalline solid and a true liquid; this is the arrangement that characterizes liquid crystals. Also, MESOMORPHIC PHASE, MESOPHASE.

mesomorphous *Physical Chemistry.* relating to or characterized by mesomorphism.

mesomorphy *Psychology.* the extent to which a person conforms to the mesomorphic body type, said to be associated with an active, forceful, aggressive personality.

meson [mes´än; mē´sän; mā´sän] *Particle Physics.* a generic name for a group of strongly interacting particles having baryon number zero and ranging in mass from 140 MeV to near 10 GeV; all are unstable and decay to the lowest mass states that are accessible.

mesonasal *Anatomy.* located in the middle of the nose.

meson capture *Particle Physics.* a process in which a meson is trapped by the nucleus of an atom, around which the meson circulates in a tightly bound orbit until it decays.

mesonephric duct *Developmental Biology.* a duct in the embryo draining the mesonephric tubules; this becomes the ductus deferens in the male, and vestigial in the female. Also, WOLFFIAN DUCT.

mesonephric ridge *Developmental Biology.* a ridge that comprises the entire urogenital ridge in early human embryos; later in development a more medial genital ridge, the potential gonad, arises from it. Also, **mesonephric fold.**

mesonephric tubule *Developmental Biology.* any of a number of tubules making up the mesonephros.

mesonephroma *Oncology.* any tumor of mesonephric origin.

mesonephroma ovarii *Medicine.* a comparatively uncommon tumor of the ovary, arising from mesonephric tissues that may have developed erroneously in the ovary during the embryonic stage.

mesonephros *Developmental Biology.* the middle of three renal organs appearing in the evolution of vertebrates. Also, WOLFFIAN BODY.

mesonic atom see MESIC ATOM.

mesonic molecule see MESIC MOLECULE.

mesonic X-ray *Particle Physics.* an X-ray emitted from a muonic atom or molecule due to a change in the energy state of the muon.

meson resonance *Particle Physics.* a meson that decays under the influence of strong interaction with a half-life on the order of 10^{-23}, so that it cannot be detected and its formation can only be inferred from indirect measurements.

meson spectrum *Particle Physics.* the entire range of meson particles, based on their mass, ranging from the light mesons, made of up, down, or strange quarks, to the heaviest known mesons, which include B mesons and Y mesons.

Mesonychidae *Paleontology.* a family of protoungulate mammals in the order Condylarthra, recently proposed as evidence that modern ungulates and carnivores are descended from a common ancestor; the mesonychids, which ranged from 3 to 12 feet in length, may also have been ancestral to whales; Paleocene to Oligocene.

mesopause *Meteorology.* the upper boundary of the mesosphere corresponding to the level of minimum temperature at 70–80 km.

mesopeak *Meteorology*. the temperature maximum at about 50 km in the mesosphere.

mesopelagic zone *Oceanography*. the pelagic environment between 200 and 1000 meters deep.

mesophase see MESOMORPHISM.

Mesophellaceae *Mycology*. a family of fungi of the order Lycoperdales found only in South America.

mesophile *Biology*. a cell that grows at intermediate temperatures.

mesophilic *Biology*. describing organisms that live or thrive in moderate environmental conditions, such as moderate moisture or temperature, neutral soil conditions, and so. Thus, **mesophily, mesophilous.**

mesophyll *Botany*. the parenchymal, usually photosynthetic tissue found between the upper and lower epidermal layers of a leaf.

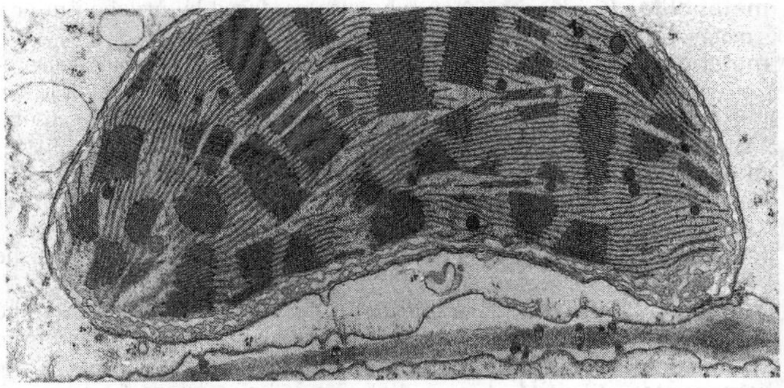

mesophyll

mesophyte *Botany*. a plant that flourishes in an environment with moderate amounts of moisture. Thus, **mesophytic.**

mesopic vision *Physiology*. the ability to see at low levels of light.

mesoplankton *Biology*. a classification of plankton of intermediate body size, between macroplanktons and microplanktons.

mesopore *Materials Science*. in porous materials, a pore with an intermediate size, ranging from 2 to 50 nanometers, as compared with micropores (less than 2 nm) and macropores (greater than 50 nm).

mesopterygium *Vertebrate Zoology*. the middle cartilaginous rod supporting the base of each pectoral fin in sharks, skates, and rays.

Mesoptychiaceae *Botany*. a monotypic liverwort family of the order Jungermanniales, characterized by robust, often erect plants with numerous scattered rhizoids and having strongly succubous leaf insertion with the upper half horizontal; restricted to the Northern Hemisphere.

mesorchium *Vertebrate Zoology*. the membranous extension of the peritoneum that supports the testes in vertebrates.

mesosalpinx *Anatomy*. the upper part of the broad ligament of the uterus, located above the mesovarium and enfolding the uterine tubes.

Mesosauria *Paleontology*. an order of anapsid reptiles that are significant as examples of a return to a marine environment by a group of amniotes, which generally continued to evolve toward a more terrestrial way of life; **mesosaurs** appear only in Permian strata in southern Africa and South America, and may have inhabited a single saltwater basin formed by juxtaposition of the two continents before they drifted apart; their taxonomic position is still unclear.

mesosere *Ecology*. a community of plants that appears temporarily in an environment when there is a moderate amount of moisture in the soil.

mesosiderite *Geology*. a stony, iron-bearing meteorite consisting of angular fragments cemented together by a nickel-iron matrix. Also, GRAHAMITE.

mesosoma *Invertebrate Zoology*. the middle portion of the body in an invertebrate, especially the anterior portion of the abdomen in certain arthropods.

mesosome *Microbiology*. a portion of the cell membrane of certain bacteria that is believed to take part in numerous cellular activities such as respiration, septum formation, and possibly DNA replication. Also, PLASMALEMMOSOME.

mesosphere *Meteorology*. **1.** the atmospheric shell that extends from the top of the stratosphere to the mesopause. **2.** the atmospheric shell between the top of the ionosphere and the bottom of the exosphere. *Geology*. see LOWER MANTLE.

mesostasis *Geology*. the most recently formed glassy or aphanitic interstitial material of an igneous rock.

mesosternum *Anatomy*. the main part or body of the sternum.

Mesostigmata *Invertebrate Zoology*. a large suborder of mites, arachnids in the order Acarina; most species are parasitic on various insects and other animals.

Mesosuchia *Paleontology*. an extinct suborder of reptiles belonging to the order Crocodilia, including several advanced types of protosuchians; widespread and abundant in the Mesozoic.

Mesotaeniaceae *Botany*. the saccoderm desmids, a family of mostly single-celled freshwater algae of the order Conjugales.

Mesotardigrada *Invertebrate Zoology*. an order of tardigrades containing a single species found in hot springs in Japan.

mesotheca *Invertebrate Zoology*. the middle lamina of bifoliate bryozoan colonies.

mesothelioma *Oncology*. a rare tumor, developing from the cells lining the pleura and peritoneum.

mesothelium *Anatomy*. the lining of the closed body cavities such as the pericardium, pleura, and peritoneum; a single layer of cells derived from the mesoderm and made up of large flattened epithelial cells.

mesotherm *Ecology*. a type of plant that does not tolerate extreme temperatures and that flourishes in an environment with a moderate climate.

mesothermal *Mineralogy*. of or relating to a hydrothermal mineral deposit formed at considerable depth (intermediate pressure) and in the intermediate temperature range of 200–300°C.

mesothermal climate *Meteorology*. a temperate climate with moderate temperatures.

mesothermic *Ecology*. living or thriving in a temperate climate. Thus, **mesothermophyte.** Also, **mesothermophilous.**

mesothorax *Invertebrate Zoology*. the middle of the three segments comprising the thorax in insects.

mesotil *Geology*. a partially plastic or somewhat brittle material derived from the chemical weathering of glacial debris beneath a partially drained area.

mesotrophic *Ecology*. of a lake or other body of fresh water, having an intermediate amount of plant nutrients and therefore moderately productive.

mesovarium *Anatomy*. the part of the broad ligament of the uterus, located between the mesometrium and mesosalpinx, that is drawn out to enclose and hold the ovary in place.

Mesoveliidae *Invertebrate Zoology*. a family of small, predatory true bugs that live on the surface of water in ponds or other wet places.

Mesozoa *Invertebrate Zoology*. a small phylum of ciliated worms, endoparasitic in marine invertebrates.

mesozoan *Invertbrate Zoology*. belonging to the phylum Mesozoa.

Mesozoic [mes´ə zō´ik] *Geology*. **1.** the geologic era extending from the end of the Paleozoic era to the beginning of the Cenozoic era, dating from approximately 225 to 65 million years ago; includes the Triassic, Jurassic, and Cretaceous periods. **2.** the rocks formed during this era.

mesozone *Petrology*. the middle depth zone of metamorphism, characterized by temperatures between 300 and 500°C, high hydrostatic pressure, and high shear stress; it includes rock formations such as mica schists, garnetiferous and staurolite schists, amphibolite, and various types of marbles, quartzites, and gneisses.

mesozooid *Invertebrate Zoology*. a type of bryozoan heterozooid that produces slender tubes.

mesquite [mes´kēt; mə skēt´] *Botany*. any of various members of the genus *Prosopis*, such as the **common mesquite,** *P. juliflora,* or the **honey mesquite,** *P. glandulosa,* pod-bearing trees or shrubs that form extensive thickets and are used for livestock feed or as cooking fuel.

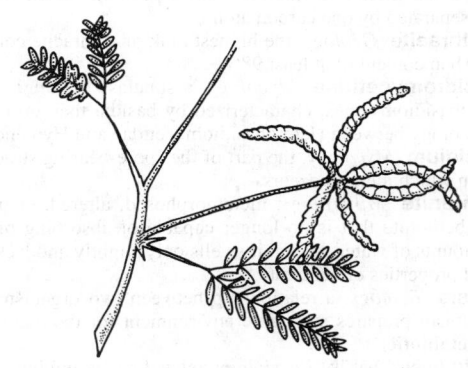

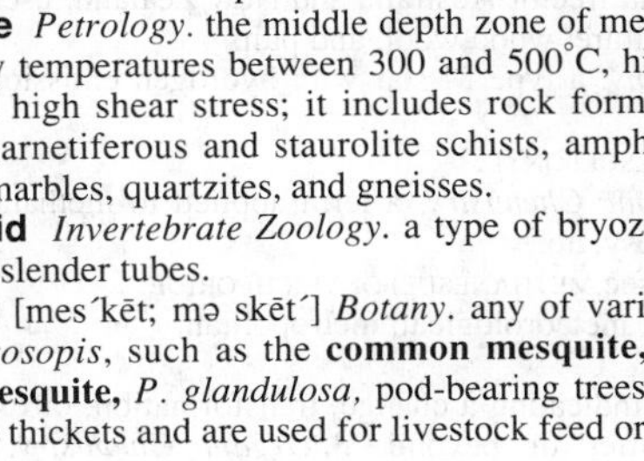

mesquite

message *Computer Programming.* **1.** a transmitted series of words or symbols that conveys information. **2.** a request to an object in an object-oriented programming system. *Telecommunications.* **1.** an ordered selection of words and symbols used for communicating information. **2.** an arbitrary amount of information that possesses a defined or implied start and finish. **3.** the original modulating wave in a communication system.

message blocking *Telecommunications.* **1.** the process of linking several messages into a single transmission or record in order to reduce the delays caused by frequent changes in the direction of the transmissions. **2.** a grouping and transmitting of a plurality of message elements, designed to increase the efficiency of equipment usage.

message half-life see MESSENGER RNA HALF-LIFE.

message interpolation *Telecommunications.* the placement of data between syllables or during pauses in speech on a busy voice channel, without breaking down the voice connection or affecting the voice transmission.

message queuing *Computer Programming.* the process of controlling the way in which messages are handled through temporary storage until the messages are processed or routed to a destination.

message trailer *Telecommunications.* the final portion of a data communication, indicating the end of the message and often including control information such as a check character.

messenger *Science.* anything that carries information. *Oceanography.* a small, cylindrical metal weight that is attached to an oceanographic wire and lowered into the sea in order to activate various oceanographic devices. *Telecommunications.* a steel cable that supports multipair aerial communications cables between poles.

messenger RNA *Molecular Biology.* any single-stranded RNA molecule that encodes the information necessary to synthesize a given protein.

messenger RNA decay *Molecular Biology.* a normal cellular process involving the degradation of messenger RNA (mRNA), which is normally unstable.

messenger RNA half-life *Molecular Biology.* the length of time required for a given amount of mRNA to be reduced or degraded by half, the time varies from 5 minutes in prokaryotes to as long as 24 hours in eukaryotes.

Messier, Charles [mes´ē ā; mes´yā] 1730–1817, French astronomer; discovered 21 comets; published catalogue of nebulae.

Messier catalogue *Astronomy.* a list of 109 star clusters, galaxies, and nebulae compiled by the French comet-hunter Charles Messier.

Messier number *Astronomy.* the identifying number that each object in the Messier catalogue bears, such as M 1 (the Crab Nebula).

messmate *Materials.* the moderately durable, pale brown wood of the *Eucalyptus obliqua* tree of Australia and New Zealand; used for general construction, furniture, wood wool, and pulp.

Me star *Astronomy.* a type M star with hydrogen emission lines in its spectrum.

mesyl see METHYLSULFONYL.

mesylates *Organic Chemistry.* a term applied to methane sulfonates, such as benzyl mesylate.

mesyl chloride see METHANESULFONYL CHLORIDE.

met. meteorology, meteorological; metropolitan.

Met methionine.

meta- **1.** a prefix indicating a change, transformation, or exchange. **2.** a prefix meaning "after" or "beyond." **3.** *Organic Chemistry.* a prefix designating a substitution or addition to two positions in the benzene ring that are separated by one carbon atom.

meta-anthracite *Geology.* the highest rank of anthracite coal, having a fixed carbon content of at least 98%.

Metabasidiomycetidae *Mycology.* a subclass of fungi of the class Phragmobasidiomycetes, characterized by basidia that are intermediate in morphology between Heterobasidiomycetidae and Hymenomycetes.

metabasidium *Mycology.* the part of the spore-bearing structure or basidium in which meiosis occurs.

metabentonite *Mineralogy.* metamorphosed, altered, or partially indurated bentonite that is no longer capable of absorbing or adsorbing large amounts of water, so that it swells only slightly and lacks the usual colloidal properties of bentonite.

metabiosis *Ecology.* a relationship between two organisms in which one organism prepares a suitable environment for the other to inhabit. Thus, **metabiotic.**

metabolic [met´ə bäl´ik] *Physiology.* related to, resembling, or affected by metabolism.

metabolic detoxification *Toxicology.* a reduction in the toxic properties of a substance by chemical changes induced in the body, producing a compound that is less poisonous.

metabolic disorder *Medicine.* any disruption of normal metabolic processes due to the congenital absence of an enzyme. Also, **metabolic disease.**

metabolic inhibition test *Microbiology.* any test that examines the ability of a substance, such as a toxin, to inhibit normal metabolic activity in a cell population.

metabolic mutant *Genetics.* a mutant in which a gene coding for an enzyme involved in metabolism has been damaged.

metabolism [mə tab´ə liz´əm] *Physiology.* the sum of all the chemical and physical processes within a living organism, including anabolism and catabolism. (Going back to a Greek word meaning "changing.")

metabolite *Biochemistry.* any substance produced by metabolism or a metabolic process.

metabolizable energy *Nutrition.* the energy used in biological processes; the potential energy stored as proteins, fats, and carbohydrates in cells that is converted into usable form for the maintenance, activity, and growth of living cells.

metabolize [mə tab´ə līz´] *Physiology.* to change by metabolism.

Metacapnodiaceae *Mycology.* a family of fungi of the order Chaetothyriales, occurring on the leaves and branches of higher plants.

metacarpal *Anatomy.* **1.** of or related to the metacarpus. **2.** any of the bones of the hand between the wrist and the fingers.

metacarpus *Anatomy.* the portion of the hand located between the wrist and the fingers, containing five cylindrical bones that extend from the carpus to the phalanges.

metacenter *Fluid Mechanics.* the point of intersection of a line connecting the center of gravity and the equilibrium center of buoyancy and a vertical line that passes through the center of buoyancy for a floating body, when the body is slightly displaced from its equilibrium position.

metacentric *Genetics.* describing a chromosome that has a centrally located centromere.

metacentric diagram *Naval Architecture.* a diagram showing changes in a vessel's metacentric height in relation to all possible changes in its draft.

metacentric height *Naval Architecture.* the vertical distance between a ship's center of gravity and its metacenter; a great metacentric height indicates a high margin of stability.

metacercaria *Invertebrate Zoology.* the encysted, infectious larval stage of a parasitic trematode.

metacestode *Invertebrate Zoology.* the encysted larval stage of a tapeworm that occurs in an intermediate host.

metacetaldehyde see METALDEHYDE.

metacharacter or **meta-character** *Computer Science.* a kind of control character indicated by holding down a special key while typing a character; it may be used for an action that is more powerful than the corresponding control character.

metachromatic *Physical Chemistry.* of a substance, appearing to be differently colored depending on the wavelength of light in which it is seen. *Materials Science.* **1.** of certain dyestuffs, having different properties or coloring effects depending on the material to which they are applied. **2.** of certain groups of materials, assuming different colors when dyed with the same dyestuff.

metachromatic granule *Cell Biology.* a cytoplasmic body, thought to be a storage vacuole, that exhibits dense staining characteristics upon treatment with basic aniline dyes.

metachromatic leukodystrophy *Medicine.* an inherited disease characterized by failure of metabolic processes, resulting in the collection of metachromatic lipids in the white matter areas of the central and peripheral nervous system, and a great increase of sulfatides in the nervous tissues and in urine, causing dementia and paralysis.

metachromatism *Chemistry.* the fact or condition of being metachromatic. Also, **metachromasia.**

metachronal rhythm *Cell Biology.* a pattern of coordinated ciliary movement in ciliated animals or cell surfaces that facilitates the movement of fluids past the ciliated surface.

metachrosis *Vertebrate Zoology.* the ability of some fishes and reptiles to change color at will by expanding special pigment cells.

metacinnabar *Mineralogy.* HgS, a black, opaque, metallic, cubic mineral of the sphalerite group, trimorphous with cinnabar and hypercinnabar, occurring in massive habit or as rare, twinned tetrahedral crystals, and having a specific gravity of 7.65 and a hardness of 3 on the Mohs scale; an ore of mercury found in near-surface mercury deposits.

metaclass *Computer Science.* in an object-oriented system, a class that describes the structure of classes.
metacneme *Entomology.* a secondary mesentery in certain anthozoans.
metacompiler *Computer Technology.* a compiler used to compile syntax-oriented compilers.
metacone *Vertebrate Zoology.* the lateral, posterior cusp of an upper molar tooth in a mammal.
metaconid *Vertebrate Zoology.* the medial, anterior cusp of a lower molar tooth in a mammal.
Metacopina *Paleontology.* a suborder of ostracods in the order Podocopida, having features that are transitional between the other two podocopid suborders, Podocopina and Platycopina; early Paleozoic to early Mesozoic, except that the modern genus *Saipanetta* may be a living descendant.
metacortandralone see PREDNISOLONE.
metacryst *Petrology.* any large crystal formed by recrystallization in metamorphic rock, such as garnet or staurolite formed in mica schist. Also, **metacrystal.**
metacyclic form *Entomology.* any of the forms of a parasitic trypanosome protozoan that are produced in an intermediate host and infect the final host.
META-DENDRAL *Artificial Intelligence.* an expert system used as a learning element for DENDRAL; an automatic rule-formation tool.
meta-directing group *Organic Chemistry.* any group, such as the nitro and carboxyl groups, that directs an incoming electrophile into the *meta* position during electrophilic substitution on the benzene ring.
metadyne *Electronics.* a rotating direct-current motor and generator used as a power amplifier on servo systems. Also, AMPLIDYNE.
metafemale see SUPERFEMALE.
metaformaldehyde see SYM-TRIOXANE.
metagalaxy *Astrophysics.* a synonyn for the universe.
metagenesis *Biology.* the phenomenon in which one generation of certain plants and animals reproduces asexually, followed by a sexually reproducing generation; alternation of sexual and asexual generations.
metagenetic *Biology.* of or relating to metagenesis. Also, **metagenic.**
metahalloysite *Mineralogy.* another name for the less hydrous form of halloysite.
metaharmosis see METHARMOSIS.
metahewettite *Mineralogy.* $CaV_6^{+5}O_{16}\cdot 3H_2O$, a deep-red monoclinic mineral occurring as nodular aggregates of lathlike, elongated crystals, and having a specific gravity of 2.94 and an undetermined hardness; found as an impregnation in sandstone in Colorado and Utah.
metahohmannite *Mineralogy.* $Fe_2^{+3}(SO_4)_2(OH)_2\cdot 3H_2O$, a mineral occurring as an orange powder resulting from the partial dehydration of hohmannite; found at Chuquicamata, Chile.
metaigneous *Petrology.* describing metamorphic rock that is formed from igneous rock.
metaknowledge or **meta-knowledge** *Artificial Intelligence.* knowledge about knowledge; e.g., knowledge about the kinds of things that are known, about the form of knowledge, or about how a system should apply stored knowledge.
metal *Chemistry.* any of a class of elements that generally are solid at ordinary temperatures, have a grayish color and a shiny surface, and will conduct heat and electricity well. In an electrolytic solution in its pure state, a metal will form positive ions. Metals constitute about three-fourths of the known elements and can form alloys with each other and with nonmetals. Common metals include copper, gold, silver, tin, iron, lead, aluminum, and magnesium. Also, METALLIC ELEMENT. *Materials Science.* an alloy or mixture that is composed of metals. *Astrophysics.* in the spectrum of a star, any element that is heavier than helium.
metal. or **metall.** metallurgy; metallurgical.
metal alkyl *Organic Chemistry.* any compound consisting of an alkyl radical and a metal.
metalanguage *Computer Programming.* a formal language used to describe the syntax of another language.
metal arc welding *Metallurgy.* any of several welding processes based on an electric arc struck between a metallic electrode and the part.
metal-cased bullet *Ordnance.* a lead bullet encased in a jacket of harder metal, except for its base; the metals used include copper, copper-nickel, and steel. Also, **metal-jacketed bullet.**
metal chelate polymer *Materials Science.* one of the three main classes of inorganic polymers, in which metal chelates from part of the polymer chain; the **metal chelates** include metal acetonylacetonates, polyphthalocyanines, and bicyclopentadienyl iron. Also, COORDINATION POLYMER.
metal-cluster compound *Chemistry.* a compound in which two or more metal atoms are within bonding distance of each other, and each is in turn bonded to at least two other metal atoms.
metal coating *Metallurgy.* the process of coating any surface with a metal or alloy.
metaldehyde *Organic Chemistry.* $(CH_3CHO)_n$, a polymer of acetaldehyde (in which *n* represents 4–6); a flammable, white prismatic solid; insoluble in water and slightly soluble in alcohol and ether; decomposes at temperatures above 80°C and melts at 246°C in a sealed tube; used for fuel and to destroy pests such as snails and slugs. Also, METACETALDEHYDE.
metal detector *Electronics.* an electronic device used to locate concealed metal objects, usually by radiating a high-frequency electromagnetic field and detecting the change that is produced in that field by the object. It is used for such purposes as screening people entering a building, aircraft, or other area to detect the presence of concealed metal weapons such as handguns or knives, or searching an area for the presence of buried coins or other valuable objects.
metal distribution ratio *Metallurgy.* in electroplating, the ratio between the maximum and the minimum thickness of the deposit.
metal-enhanced star formation *Astrophysics.* a theory postulating that stars form preferentially in regions of molecular clouds where there are greater concentrations of heavy elements ("metals").
metal-film resistor *Electricity.* a fixed resistor that employs a metal or metal-oxide film on an insulating substrate such as mica, ceramic, or quartz, possessing very high temperature stability; used for higher precision resistance values.
metal forming *Metallurgy.* the process of changing the shape of a metallic stock without removing material.
metal fouling *Ordnance.* metal residue that is deposited in the bore of a gun by the jacket or rotating bands of a projectile.
metal-fume fever *Toxicology.* poisoning that is caused by the inhalation of fumes of metallic salts or dust; the symptoms may include sudden onset of thirst and a metallic taste in the mouth, followed by fever, muscle aches, chills, headache, perspiration, and an elevated white blood cell count. Also, BRASS CHILL, BRAZIER'S CHILL, BRASS FOUNDER'S AGUE, ZINC CHILLS.
metal halide lamp *Electronics.* a high-intensity discharge lamp in which light is produced by adding metal halide salts to a discharge tube containing a high-pressure arc in mercury vapor.
metaliding *Metallurgy.* the deposition of a metal or alloy from a fused salt electrolyte.
metal inert-gas welding *Metallurgy.* metal arc welding that is performed in a protective inert atmosphere.
metal-insulator semiconductor *Solid-State Physics.* a semiconductor that has a thin insulating layer, usually less than one micrometer thick, between the substrate and the metal contacts.
metal-insulator transition *Solid-State Physics.* a transformation process whereby a material is converted from the metallic state to a nonmetallic state by alterations in pressure or temperature, or by alloying or doping with additives.
metal lath *Engineering.* a meshlike lath that is made of metal and used as a base for plastering.
metal leaf *Metallurgy.* an extremely thin metallic foil, formed by beating.
metallic *Science.* of, relating to, or consisting of metal.
metallic luster *Optics.* a brilliant luster that is characteristic of metals in a compact state; often used to describe a property of metals.
metallic bond *Physical Chemistry.* a force that holds atoms together in a metal, formed by the attraction between positively charged metal ions and mobile free electrons surrounding them; this accounts for many of the characteristic properties of metals and alloys. Also, **metallic bonding.**
metallic circuit *Electricity.* a complete circuit in which the earth is not used as ground, as in a two-wire telephone line.
metallic element see METAL.
metallic glass *Metallurgy.* a metal that has been rapidly cooled from the liquid state so that it cannot crystallize. Metallic glasses have excellent magnetic properties, can be produced in the form of thin tapes that are stacked together to produce larger cores, and combine high strength and good ductility. Also, GLASSY ALLOY.
metallic insulator *Electromagnetism.* a section of transmission line, an odd number of quarter-wavelengths long, that is short-circuited and isolates frequencies having the same quarter-wavelength.
metallic melanism see ARGYRIA.

metallic noise *Electronics.* an unwanted current that appears at a given point in the metallic circuit of a telephone system.

metallic nuclear fuel *Nucleonics.* an isotope of a metallic element or alloy that is capable of fission; used in nuclear reactors as a power source.

metallic rectifier *Electronics.* a rectifier consisting of one or more metal disks under pressure contact with semiconductor coatings such as selenium, oxides, or sulfates. Also, DRY-DISK RECTIFIER.

metallic soap *Organic Chemistry.* any soap that contains a metal heavier than sodium, such as aluminum, lead, or zinc; it is insoluble in water, and is used in lubricating greases and as a paint drier and gel thickener. Also, HEAVY-METAL SOAP.

metalliferous *Mineralogy.* of or relating to mineral deposits that yield metals.

metal line *Materials Science.* a horizontal line on refractories. Also, FLUX LINE.

metallization *Materials Science.* the application of active metal coatings that are electrochemically anodic to the substrate, forming a protective barrier against corrosion. A common form of this is galvanizing.

metallize *Engineering.* to cover or impregnate a material with metal.

metallized capacitor *Electricity.* a fixed capacitor whose plates consist of a metal film, such as silver, deposited directly onto the dielectric.

metallized dye *Materials Science.* a soluble dye containing any of various metals; applied to wool in an acid bath to salt out the dye onto the fibers.

metallized-paper capacitor *Electricity.* a capacitor whose plates consist of a metal film deposited over a paper material.

metallized resistor *Electricity.* a resistor constructed by depositing a thin, uniform layer of high-resistance metal on an insulating form.

metallized wood *Materials Science.* wood that has been impregnated with molten metal to increase its strength and provide conductivity.

metallizing *Metallurgy.* the process of producing a metallic, conducting coating on a nonmetallic part, as in the production of integrated circuits.

metallocarboxypeptidase *Enzymology.* an enzyme that catalyzes the hydrolysis of a polypeptide molecule at the carboxyl end of the molecule, which contains one or more metal atoms, such as carboxypeptidase A.

metallocene *Organic Chemistry.* any of a class of usually crystalline compounds that are cyclopentadienyl derivatives of a transition metal or metal halide; the metallocene of iron (ferrocene) melts at 173°C and is used as a catalyst or reducing agent, a UV absorber, or an antiknock agent.

metallocycle *Organic Chemistry.* a cyclic compound containing a metal atom, often one of the transition metals.

metalloenzyme *Enzymology.* an enzyme that contains a bound metal ion as an integral part of its structure.

metallogenic province *Geology.* a region in which a particular assemblage of mineral deposits or one or more types of mineralization occurs. Also, **metallogenetic province, metallographic province.**

Metallogenium *Bacteriology.* a genus of spheroid, budding iron bacteria of undecided classification found in plankton and bottom deposits in freshwater ponds and lakes.

metallograph *Optics.* an instrument used for optical metallography, having a camera for the analysis and photography of a metal or alloy.

metallographic test *Metallurgy.* any of several tests based on metallography.

metallography *Metallurgy.* the science or the study of metals and their structures using microscopes or X-ray diffraction.

metalloid *Chemistry.* **1.** an element having both metallic and nonmetallic properties, such as boron. **2.** a nonmetallic element that can combine with a metal to produce an alloy.

metallo-organic compound see ORGANO-METALLIC COMPOUND.

metallophobia *Psychology.* an irrational fear of metals.

metalloporphyrin *Biochemistry.* a porphyrin that chelates with a variety of metals and serves as a component in a number of important pigments, such as the chlorophylls, cytochromes, and heme.

metalloprotease *Enzymology.* a protein-digesting enzyme that is complexed to one or more metal atoms. Also, **metalloproteinase.**

metalloprotein *Biochemistry.* a protein with a metal ion bound to it.

metallothionein *Biochemistry.* a cadmium-binding protein involved in cell detoxification mechanisms for various heavy metals; present in mammalian tissues including those of the liver, brain, and lungs.

metallurgical [met´ə lur´ji kəl] **1.** of or relating to metals and alloys. **2.** of or relating to the science or practice of metallurgy.

metallurgical coke *Metallurgy.* a low-sulfur coke, suitable for the smelting of iron ores.

metallurgical dust *Metallurgy.* any finely particulate material derived from a metallurgical operation.

metallurgical engineering *Engineering.* the branch of engineering dealing with metals and alloys, including methods of extraction and use. Thus, **metallurgical engineer.**

metallurgical fume *Metallurgy.* a metallurgical dust condensed from the vapor phase.

metallurgical-grade silicon *Materials.* silicon that is refined but is not as pure as that used in semiconductors; mostly used for adding silicon as an alloying element.

metallurgy [met´ə lur´jē] the scientific study of metals and the technology of metal processing.

Metallurgy

Metallurgy, as defined by the first use of metals (copper), began in eastern Turkey (Cayonu Tepesi) at the end of the eighth millennium (BC). Sometime during the next two millennia, metallurgy, as defined by the chemical reduction of metals from minerals by the action of heat and charcoal, can be considered to have been introduced, since there was available a relatively inexhaustible supply of the ingredients (minerals in the form of exploitable ores and charcoal) from which metals could be extracted. Two forms of metallurgy next evolved: the craftsman responsible for smelting the minerals to produce the metals (the smelter) and the craftsman responsible for fashioning an artifact (the smith).

The smelter perfected the efficiency of his operation and, most likely with feedback from the smith, further discovered that additions of other ingredients (at first, other minerals followed by additions of other metals, i.e., alloying) allowed him to produce metals that were harder and tougher. The smith developed various methods to produce artifacts, using forging and various casting techniques.

The evolution of these two metallurgies continued for perhaps as much as five millennia until a new technology was discovered (again in the Middle East) and developed at the end of the second millennium (BC): the beginning of the Iron Age. At this time, two new metallurgy crafts evolved: the craftsman responsible for the reduction of iron ores and the blacksmith. The production of iron objects required different technologies for both smelting and artifact manufacture.

Few changes occurred until the late nineteenth century (AD) when developments in chemistry permitted a better understanding of what was going on inside metals. Chemistry was the predominant basis for further developments within metallurgy until discoveries in physics (X-rays, among others) allowed another approach to understanding the behavior of metals. Thereafter, chemistry and the emerging new physics began to change metallurgy by consolidating the various segments into what is now termed the science and engineering of materials (basically, an understanding of the relation between the structure of solids and their properties).

Robert Maddin
University Emeritus Professor of Metallurgy
University of Pennsylvania

metalmark *Invertebrate Zoology.* any of various small, usually tropical butterflies of the family Riodinidae, so named because of the metallic-looking flecks on their wings.

metal matrix composite *Materials Science.* a composite material for which the matrix is a metal, containing reinforcing fibers or particles, generally of tungsten, boron, graphite, or silicon carbide.

metal mining *Mining Engineering.* the industry involved in supplying the various metals and associated products.

metal-nitride oxide semiconductor *Solid-State Physics.* a semiconductor that contains two insulating layers: the first layer (on the silicon substrate) is usually a thin coating of silicon dioxide, and the second layer is a nitride coating such as silicon nitride.

metal-organic chemical vapor deposition *Solid-State Physics.* a process whereby a surface layer is produced by the deposition on a substrate of a volatile organo-metallic compound that is transported to the surface through the gas phase at elevated temperatures.

metal oxide resistor *Electricity.* a metal-film resistor whose resistance element is composed of a metal oxide film deposited onto an insulating substrate.

metal oxide semiconductor *Solid-State Physics.* a type of metal-insulator semiconductor structure whose insulating layer is an oxide of the substrate material.

metal oxide semiconductor field-effect transistor *Electronics.* a field-effect transistor in which the electric field is due to a charge on a capacitor formed by a metal electrode and an oxide layer separating it from the body of the semiconductor. Also, **metal oxide semiconductor transistor.**

metal powder *Metallurgy.* a particulate metal or alloy, such as a pigment or a component of conductive plastic composites; used principally for powder metallurgy fabrication.

metal-rich star *Astrophysics.* a star whose spectrum shows abnormally great abundances of metals.

metal-slitting saw *Mechanical Engineering.* a milling saw having teeth on its side and around its circumference; used for deep slotting.

metal spinning *Mechanical Engineering.* see SPINNING.

metal spraying *Engineering.* a technique of applying a coat of molten metal onto a surface using a device that atomizes the metal in a flame of hydrogen and oxygen.

metal trim *Building Engineering.* any of various finishings made from pressed metal sheeting that are fastened into position around a door or window.

metal vapor laser *Optics.* an ion laser in which a vaporized metal (such as cadmium, calcium, copper, or lead) is combined with a buffer gas (such as helium) to produce a beam of coherent light.

metalworking *Metallurgy.* the process of working or heating metals to produce desired shapes or objects.

metamale see SUPERMALE.

metamathematics *Mathematics.* the study of formal mathematical proofs through the axioms of sentential logic. This includes the study of the formal properties of formal languages, including deducibility, independence, and consistency. Also, **metalogic.**

metamer *Organic Chemistry.* any of two or more isomers of a compound.

metamere *Invertebrate Zoology.* any of the linearly arranged units or segments into which the body of a metameric organism, such as an annelid or arthropod, is divided, each segment often containing the same organs in the same arrangement as the other segments. *Developmental Biology.* a mesoblastic segment in a developing embryo.

metameric *Science.* of or relating to metamerism. *Invertebrate Zoology.* of or relating to an animal, such as an earthworm, whose body is made up of segments in a linear arrangement. Also, **metameral.**

metamerism *Invertebrate Zoology.* the property of having similarly formed and functioning body segments in a linear arrangement, or the formation of such segments. *Chemistry.* isomerism that occurs when different groups are attached to the same atom.

metamict *Mineralogy.* describing a mineral containing radioactive elements, whose original crystal structure has been disrupted by internal alpha particle bombardment while the external morphology remains intact.

metamict transformation *Mineralogy.* the effects resulting from natural radiation damage that causes a mineral to undergo crystalline-to-amorphous transition.

metamorphic [met´ə môr´fik] *Biology.* of or relating to a process of metamorphosis. *Science.* of or relating to any significant change in form. Also, METAMORPHOUS.

metamorphic differentiation *Petrology.* any process by which minerals in an initially uniform parent rock segregate during metamorphism into lenses and bands.

metamorphic facies *Petrology.* the set of metamorphic mineral assemblages (facies) composed of rocks crystallized during metamorphism under chemical equilibrium with respect to restricted ranges of physical conditions, such as temperature and hydrostatic pressure. Also, DENSOFACIES, MINERAL FACIES.

metamorphic facies series *Petrology.* the set of metamorphic facies that characterize a given area; such sets are represented by curves on a pressure-temperature diagram to indicate the range of various types of metamorphism and metamorphic facies.

metamorphic grade *Geology.* the degree of metamorphism as measured by comparing the amount of original parent rock to that of metamorphic rock. Also, RANK.

metamorphic overprint see OVERPRINT.

metamorphic rock *Petrology.* any rock altered from a pre-existing solid rock by extreme changes in temperature, pressure, shearing stress, or chemical environment.

metamorphic rock reservoir *Geology.* an oil reservoir formed when secondary porosity results from fracturing and weathering.

metamorphic zone *Geology.* see AUREOLE.

metamorphism *Petrology.* any process imposing mineralogical, chemical, or structural changes upon solid rock in response to extremely different conditions and forces than those under which the rock originated.

metamorphosis [met´ə môr´fə sis] *Biology.* **1.** a phase in the life cycle of many animals, during which there is a rapid transformation from the preadult to the adult form, such as the transformation from larva to adult in insects. **2.** any significant change of shape or structure. *Medicine.* any of various normal or pathological processes in which alteration or degeneration of tissue occurs, such as fatty metamorphosis.

metamorphous see METAMORPHIC.

metamyelocyte *Histology.* an immature granular leukocyte that contains many cytoplasmic granules and a bean-shaped nucleus.

metanauplius *Entomology.* a larval stage of certain decapod crustaceans.

metaneeds *Psychology.* greater needs; Abraham Maslow's term for higher human values such as self-fulfillment and artistic or creative achievement.

metanephridium *Entomology.* an excretory structure consisting of a ciliated tubular body that opens into the coelom.

metanephrine *Biochemistry.* a metabolite of epinephrine that is excreted in the urine and found in certain tissues.

metanephros *Developmental Biology.* the third and most caudally located of the three renal organs appearing in the evolution of the vertebrates; in mammals it develops into the adult kidney.

metanilic acid *Organic Chemistry.* $C_6H_4(NH_2)SO_3H$, a colorless, needlelike solid; soluble in water, alcohol, and ether; used in the production of sulfa drugs and dyes.

metaphase [met´ə fāz´] *Cell Biology.* one of the sequential stages of both cell mitosis and meiosis, characterized by the presence of chromosomes at the equatorial region of the cell between the spindle poles.

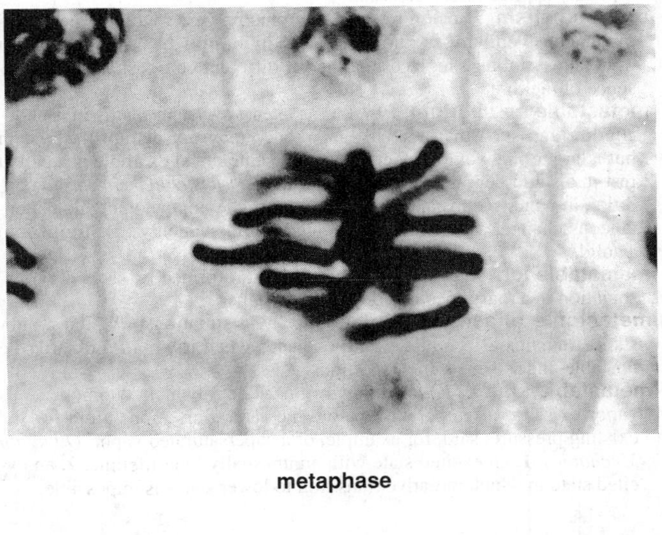

metaphase

metaphloem *Botany.* a primary phloem formed from the protophloem, consisting of sieve tubes, parenchyma, and various fibers.
metaphysics a systematic investigation of the ultimate nature of being or reality; originally considered an activity of science but later regarded as unverifiable and separated into philosophy. Thus, **metaphysical, metaphysician.**
Metaphyta *Botany.* in some classifications, a kingdom set up to include mosses, ferns, and other multicellular plants.
metaphyte *Botany.* any multicellular plant.
metaplasia *Pathology.* any abnormal alteration in the tissue of adult cells that transforms them into another form.
metaplasm *Cell Biology.* a nonliving by-product material resulting from cellular activity; an ergastic material, such as starch or fat.
metapodialia *Anatomy.* any of the bones of the metacarpus or metatarsus.
metapodium *Invertebrate Zoology.* the posterior portion of the foot of a mollusk.
metapodosoma *Invertebrate Zoology.* in mites and ticks, the portion of the body that bears the two posterior pairs of legs.
metaprotein *Biochemistry.* a denatured protein that is insoluble in water and soluble in weak acids or alkalis.
metapterygium *Vertebrate Zoology.* the cartilaginous element supporting the base of the pectoral and pelvic girdle in sharks, skates, and rays.
metaquartzite *Petrology.* a quartzite whose formation is due to metamorphic recrystallization.
metarheology *Mechanics.* an approach to rheology combining the assumptions of macrorheology with the application of some concepts of microrheology.
Metarhizium *Mycology.* a genus of fungi of the class Hyphomycetes that are pathogenic to many insects and are thus used as a form of biological control against insects.
metaripple *Geology.* a large, asymmetrical ripple produced when conditions change distinctly during the course of its formation.
metarossite *Mineralogy.* $CaV_2^{+5}O_6 \cdot 2H_2O$, a water-soluble, light-yellow triclinic mineral occurring as flaky masses and veinlets, having a specific gravity of 2.45 and an undetermined hardness; found as a dehydration product of rossite in carnotite-bearing sandstones.
metarule or **meta-rule** *Artificial Intelligence.* a rule that governs how other rules are to be used or altered.
metascope *Electronics.* an infrared receiver capable of converting invisible infrared rays into visible communication signals; used with an infrared source to read maps in darkness.
metascutellum *Entomology.* the upper portion of the third thoracic segment in insects.
metasediment *Geology.* a sediment or sedimentary rock whose characteristics indicate that it has undergone metamorphism.
metasicula *Paleontology.* the rear part of the chitinous exoskeleton of graptolites, from which polyps were budded off.
metasideronatrite *Mineralogy.* $Na_2Fe^{+3}(SO_4)_2(OH) \cdot H_2O$, a yellow, orthorhombic mineral occurring as aggregates of prismatic crystals, having a specific gravity of 2.68 and a hardness of 2.5 on the Mohs scale; found with other sulfates at Chuquicamata, Chile.
metasomatic *Petrology.* of or relating to the process of metasomatism; typically applied to ore deposits.
metasomatism *Petrology.* a metamorphic process involving nearly simultaneous capillary solution and deposition, by which one mineral or mineral assemblage replaces another of different composition in the absence of melting.
metastable equilibrium *Physics.* a state of a system that is in pseudo-equilibrium, such that a small disturbance may not disrupt the system but a larger one would render it unstable; a practical example is a ball at rest in a slight depression at the top of a hill. *Physical Chemistry.* a condition in which a substance appears to be stable but can undergo a spontaneous change, as when a supercooled liquid suddenly transforms into a solid.
metastable ion *Analytical Chemistry.* an ion formed by secondary dissociation in a mass spectroscopy analyzer tube.
metastable phase *Physical Chemistry.* a situation in which a substance remains as a liquid, solid, or vapor when it would ordinarily be unstable under these same conditions.
metastable state *Thermodynamics.* the state in which a homogeneous vapor exists at a temperature below the saturation temperature for the existing pressure; said, for example, of a supersaturated vapor. *Quantum Mechanics.* **1.** an excited state with an unusually long lifetime. **2.** an excited state in which radiative transition to lower states is impossible.
metastasis [mə tas′tə sis] *Pathology.* the spreading of a disease from one organ or part to another that is not directly connected with the first organ or part; this may be due to the transfer of microorganisms or to the transfer of cells. *Oncology.* **1.** specifically, the movement of tumor growth from an original primary location in the body to a secondary location via the lymph system or blood circulation. **2.** any growth or tumor that develops in this way. *Physics.* a fundamental transition of an electron to a different orbital, or a nucleon to a different bound state.
metastasize [mə tas′tə sīz′] *Pathology.* to spread from a primary site to a distant part of the body by metastasis. Thus, **metastasized.**
metasternum *Entomology.* the ventral plate of the third thoracic segment in insects.
metastoma *Entomology.* the lower lip of arthropods.
Metastrongylidae *Invertebrate Zoology.* the lungworms, a family of parasitic nematodes of the superfamily Metastrongyloidea; several genera are particularly dangerous to humans and domestic animals.
Metastrongylus *Invertebrate Zoology.* a genus of nematodes of the family Metastrongylidae. *M. elongatus,* a species found in the lungs of hogs, is a host of the swine influenza virus.
metatalc see PROTOENSTATITE.
metatarsal *Anatomy.* **1.** of or relating to the metatarsus. **2.** any of the bones of the metatarsus.
metatarsalgia *Medicine.* pain and tenderness in the metatarsal area of the foot.
metatarsus *Anatomy.* the area of the foot between the instep and the toes, containing five long bones that extend from the tarsus to the phalanges.

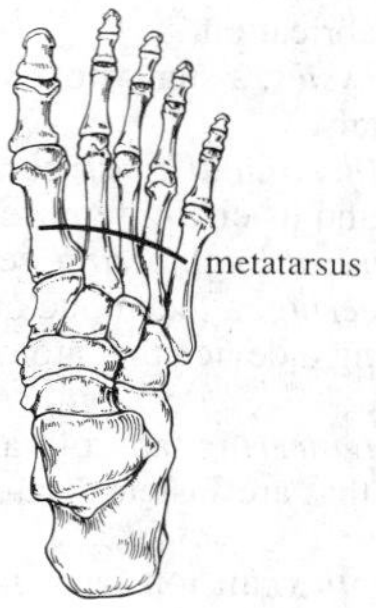

metatarsus

metate *Archaeology.* a flat grinding stone used as a base against which another stone is rubbed to process seeds, nuts, and other substances.
Metatheria *Paleontology.* an infraclass of therian mammals that includes all marsupials, extinct and living; the earliest known marsupials are from the Cretaceous of North America; they had migrated into South America by the end of the Cretaceous, and into Europe and North Africa in the Eocene. *Vertebrate Zoology.* in some classifications, an infraclass of living mammals containing the marsupials; coextensive with Marsupialia.
metathesis *Chemistry.* a reaction in which double decomposition occurs, shown by the general equation $AB + CD = AD + BC$.
metathetic *Chemistry.* of, relating to, characteristic of, or produced by metathesis. Also, **metathetical.**
metathorax *Entomology.* the third and hind segment of an insect's thorax.
metatitanic acid see TITANIC ACID.
metatorbernite *Mineralogy.* $Cu^{+2}(UO_2)_2(PO_4)_2 \cdot 8H_2O$, a green, strongly radioactive, tetragonal mineral of the meta-autunite group, occurring in sheaflike aggregates of tabular crystals, and having a specific gravity of 3.7 to 3.8 and a hardness of 2.5 on the Mohs scale; it is among the most common supergene uranium minerals.
Metatremellales *Mycology.* the single order of fungi of the subclass Metabasidiomycetidae; found in tropical to temperate regions on decaying organic matter.
metatroch *Invertebrate Zoology.* a ring of cilia behind the mouth in trochophore larvae.
metatroph *Biology.* a metatrophic organism.
metatrophia *Medicine.* **1.** atrophy from malnutrition. Also, METATROPHY. **2.** a change in diet.
metatrophic *Biology.* utilizing organic matter for food.

metatrophy *Biology.* **1.** the state of being metatrophic. **2.** metatrophic nutrition. *Medicine.* see METATROPHIA, def. 1.

metatypical *Oncology.* describing a tumor that is composed of elements of the tissue on which it develops, but having those elements arranged in an atypical manner. Also, **metatypic.**

metavariable *Computer Programming.* one of the language elements used in formal descriptions of computer languages, analogous to the parts of speech used to describe natural languages.

metavariscite *Mineralogy.* $AlPO_4 \cdot 2H_2O$, a pale-green monoclinic mineral, dimorphous with variscite, occurring as small, tabular crystals and in massive form, and having a specific gravity of 2.54 and a hardness of about 3.5 on the Mohs scale; found with variscite.

metavauxite *Mineralogy.* $Fe^{+2}Al_2(PO_4)_2(OH)_2 \cdot 8H_2O$, a colorless or white, monoclinic mineral, dimorphous with paravauxite, having a specific gravity of 2.35 and a hardness of 3 on the Mohs scale; found as aggregates of acicular crystals encrusting quartz crystals in tin mines in Bolivia.

metavoltine *Mineralogy.* $K_2Na_6Fe^{+2}Fe_6^{+3}(SO_4)_{12}O_2 \cdot 18H_2O$, a yellowish-brown, or orange- to greenish-brown, hexagonal mineral occurring as scaly aggregates of tabular crystals, having a specific gravity of 2.40 to 2.51 and a hardness of 2.5 on the Mohs scale; found with other secondary sulfate minerals.

Metaxyaceae *Botany.* a monotypic family of trunkless tree ferns of the order Filicales, characterized by a short-creeping rhizome with golden hairs, pinnate fronds up to two meters long, and round sori on lower leaf surfaces; native to low elevations in tropical America.

metaxylem *Botany.* a primary xylem formed from the protoxylem, having wider vessels and tracheids.

metazeunerite *Mineralogy.* $Cu(UO_2)_2(AsO_4)_2 \cdot 8H_2O$, a green, tetragonal mineral of the meta-autunite group, occurring in platy crystals and crystal aggregates, and having a specific gravity of 3.64 and a hardness of 2 to 2.5 on the Mohs scale; found as a secondary mineral with torbernite and uranospinite.

Metazoa *Zoology.* a division of the animal kingdom whose members include all the multicelled animals.

metazoan *Zoology.* **1.** any member of the division Metazoa. **2.** of or relating to this division.

metazoea *Invertebrate Zoology.* a larval stage in certain marine decapod crustaceans, which then metamorphoses into megalops.

metencephalon *Anatomy.* the portion of the brain consisting of the pons and the cerebellum, lying between the medulla oblongata and the mesencephalon; formed by the specialization of the rhombencephalon in the developing embryo. Also, EPENCEPHALON.

met-enkephalin see ENKEPHALIN.

meteor *Astronomy.* **1.** the brief, bright streak of light that results when a piece of silicate or metallic rock enters the atmosphere at high speed from space and burns up. **2.** the object that causes this light.

meteor bumper *Space Technology.* any device, such as a thin shield, used to protect a probe or spacecraft from meteors.

meteor crater *Geology.* a crater formed by the impact of a meteorite.

meteor crater

Meteoriacae *Botany.* a moss family of the order Isobryales, growing in loose mats and characterized by a hairy calyptra, creeping or prostrate stems, well-developed leaves, extensive branches, and lateral sporophytes.

meteoric *Astronomy.* relating to or involving meteors.

meteoric ionization *Astronomy.* the heat-generated ionization of air molecules that occurs when a meteor enters the atmosphere.

meteoric scatter *Telecommunications.* a type of scatter propagation in which trails of meteors aid in scattering radio waves downward to the earth.

meteoric stone see STONY METEORITE.

meteoric water *Hydrology.* water recently derived from the atmosphere.

meteorism *Medicine.* the distention of the abdomen due to gas in the intestines or peritoneal cavity. Also, TYMPANITES.

meteorite

meteorite *Astronomy.* a solid mass of mineral or rock matter that has fallen to the earth's surface from outer space without being completely vaporized in the atmosphere.

meteorogram *Meteorology.* a graphical representation of meteorological variables over time.

meteorograph *Meteorology.* an instrument that simultaneously measures and records various meteorological conditions such as temperature and barometric pressure.

meteoroid *Astronomy.* **1.** a solid, relatively small mass of mineral or rock traveling through space. **2.** specifically, such a body that has not entered the atmosphere of the earth or another planet.

meteorolite see STONY METEORITE.

meteorological *Meteorology.* of or relating to meteorology, or to weather and other atmospheric phenomena.

meteorological balloon *Meteorology.* a sturdy balloon that is used to transport into the upper atmosphere radiosondes and other instruments that send measurements from various altitudes to a recording station on the ground.

meteorological chart *Meteorology.* a graphical representation of weather phenomena at a given instant.

meteorological check point *Ordnance.* an arbitrary point used in gunnery as a basis for determining meteorological corrections that are then transferred to the target or targets.

meteorological correction *Ordnance.* adjustment to the firing data of a gun to compensate for the effect of weather and other atmospheric phenomena on the flight of the projectile.

meteorological data *Meteorology.* the reported values of weather characteristics such as wind, temperature, and air density.

meteorological datum plane *Ordnance.* the plane on which the meteorological station used in making artillery adjustments is located.

meteorological equator *Meteorology.* **1.** the parallel of latitude 5°N. **2.** the axis of the barotropic current characterizing the low troposphere in equatorial regions. Also, EQUATORIAL TROUGH.

meteorological frequency bands *Telecommunications.* a cluster of radio and microwave frequency bands that are used solely for the purpose of weather forecasting.

meteorological instrumentation *Meteorology.* **1.** any instrument or device used to obtain weather information. **2.** the ensemble of such devices in a particular weather station or laboratory.

meteorological minima *Meteorology.* the minimum values of meteorological elements prescribed for certain flight operations.

meteorological optics *Optics.* the branch of physics or meteorology that attempts to characterize or explain optical phenomena observed in the atmosphere. Also, ATMOSPHERIC OPTICS.

meteorological range *Meteorology.* a consistent measure of the visual range of a target that eliminates effects of luminance, which vary from observer to observer. Also, STANDARD VISIBILITY.

meteorological rocket *Space Technology.* a rocket vehicle designed for upper-air observation in the portion of the lower atmosphere that is inaccessible to balloons, particularly between 100,000 and 250,000 feet. Also, ROCKETSONDE.

meteorological satellite *Space Technology.* an artificial satellite instrumented to collect data on the atmosphere; used to prepare and transmit actual data and forecasts.

meteorological solenoid *Meteorology.* a hypothetical tube in space formed by the intersection of a set of surfaces of constant pressure and a set of surfaces of a constant specific volume of air. Also, SOLENOID.

meteorological tide *Oceanography.* a change in water level caused by meteorological factors, such as barometric pressure and wind.

meteorologist [met´ē ə räl´ə jist] *Meteorology.* **1.** a scientist who studies and reports on the conditions of the earth's atmosphere. **2.** in popular use, any weather forecaster.

meteorology [met´ē ə räl´ə jē] **1.** the scientific study of the earth's atmosphere, especially as this relates to weather and climate. **2.** the weather conditions of a particular place. (From a Greek term meaning literally "the study of things in the air.")

Meteorology

Meteorology is the scientific study of the physical and chemical phenomena of the atmosphere, especially as they relate to weather and climate. Prior to the last third of the 20th century the term was used primarily in reference to the study of the winds, temperature, clouds, and other elements associated with weather. The term is now commonly used to include all scientific study of the role of the atmosphere in the global climate system, although some prefer the term *atmospheric science.*

Prior to World War II the application of the science of meteorology to the practice of weather prediction was very limited. The development of routine upper-air observations by balloons and aircraft, together with the development of high-speed computers, provided the basis for a revolution in the science of weather prediction in the postwar period. Numerical models now provide useful large-scale weather predictions up to a week in advance for much of the globe.

Beginning in the 1970s the focus of meteorology began to shift from the study of weather to study of the global climate and the impact of human activities on the global atmosphere. Meteorologists are now concerned with problems such as acid rain caused by sulfur emissions from power plants, depletion of the ozone layer by fluorocarbon emissions, and global warming owing to increasing concentrations of carbon dioxide and other radiatively active trace gases.

These problems, and other aspects of global change, require that the atmosphere, the oceans, and the biosphere be treated as interacting parts of the climate system. Meteorologists can no longer concern themselves solely with the atmosphere.

James R. Holton
Professor of Atmospheric Sciences
University of Washington

meteor shower *Astronomy.* a visual phenomenon that is observed when a group of meteors encounter the earth's atmosphere and their luminous paths appear to diverge from a single point in the sky.

meteor storm *Astronomy.* an abnormally rich meteor shower.

meteor stream *Astronomy.* the orbital path of the meteors making up a shower, which is usually that of a parent comet.

meter *Metrology.* a basic unit of length in the metric system, roughly equivalent to 39.37 inches. It is currently defined as the distance that light will travel under vacuum in a period of 1/299,792,458 of a second. From 1960 to 1983, it was defined as 1,650,763.73 wavelengths of the orange-red light from the isotope krypton 86, measured under vacuum conditions. Originally, it was supposed to represent one ten-millionth of the distance from the North Pole to the equator along a given meridian. Also, METRE. *Engineering.* **1.** any instrument that is designed to measure the value of a quantity under observation. **2.** to observe and usually measure an object, event, or process. (Going back to a Greek word meaning "to measure.")

meterage *Metrology.* the practice or process of measuring.

meter-ampere or **meterampere** *Telecommunications.* the measure of the strength of a radio transmitter, equal to the height of the antenna times the maximum antenna current.

meter-atmosphere *Physics.* the depth that the earth's atmosphere would have, if it were replaced by an atmosphere of a specified gas having a uniform temperature of 0°C and a uniform pressure of 760 mm Hg (standard atmospheric pressure).

meter bar *Metrology.* **1.** a bar exactly one meter long, used as a standard for measurement. **2.** specifically, such a bar composed of platinum and iridium, kept at the International Bureau of Weights and Measures near Paris; formerly accepted as the official standard for the length of a meter. *Engineering.* a metal bar with inlet and outlet fittings at either end; used to mount a gas meter.

meter density *Engineering.* in an energy distribution system, the number of meters per unit length or unit section.

metered flow *Design Engineering.* a flow system that is controlled to allow a specific quantity to flow in a specified time or in harmony with an event.

meter factor *Engineering.* a factor used in conjunction with a meter in order to correct for ambient conditions such as temperature change.

metering pin *Engineering.* a valve plunger that measures and controls the rate of flow of a gas or liquid, as in the carburetor of an internal-combustion engine. Also, **metering rod.**

metering pump *Engineering.* a plunger-type pump that measures the flow of liquid that passes through it in order to control the liquid.

metering screw *Mechanical Engineering.* a conveyor section or extrusion-type screw fitter used to feed pulverized or doughy material at a constant rate.

metering valve *Mechanical Engineering.* a valve that momentarily delays application of the front disk brakes of a vehicle until the rear drum brakes are activated.

meter-in operation *Robotics.* an operation that uses a flow control valve between the pump and the load.

meter-kilogram-second *Metrology.* of or relating to a system of measurement in which the meter is the basic unit of length, the kilogram is the basic unit of mass, and the second is the basic unit of time. A **meter-kilogram-second-ampere** system also includes the ampere as the basic unit of electric current. Also, MKS.

meter-out operation *Robotics.* an operation in which a control valve is used to restrict return flow from a cylinder.

meter relay *Electricity.* a sensitive coil relay containing a contact-bearing pointer that moves either toward or away from a fixed contact attached to a meter scale.

meter run *Engineering.* **1.** a straight section of pipe through which fluid can flow without obstruction. **2.** the length of such a section, usually a specified length before and after a meter.

meter sensitivity *Engineering.* the accuracy with which a meter can measure the smallest quantity of change such as in voltage or resistance.

meter sizing factor *Fluid Mechanics.* a dimensionless quantity given by the product of the flow coefficient and the square of the orifice bore diameter, divided by the square of the inside diameter of a pipe with a circular cross section through which fluid is forced to flow; used in calculations involving fluid flow rates for pipes in which flowmeters measure drops in pressure.

meter-ton-second system *Metrology.* a variation of the meter-kilogram-second system in which the metric ton rather than the kilogram is used as the basic unit of mass.

metestrus *Physiology.* the short period following estrus during which the phenomena of estrus subside.

methacrolein *Organic Chemistry.* CH_2=C(CH_3)CHO, a flammable liquid that is soluble in water; boils at 68°C; used to make resins and copolymers.

methacrylate ester *Organic Chemistry.* CH_2=C(CH_3)COOR, an ester that readily undergoes vinyl polymerization and forms a raw material for methyl methacrylate products.

methacrylic acid *Organic Chemistry.* CH_2=C(CH_3)COOH, a toxic, combustible, colorless liquid; soluble in water, alcohol, and ether; melts at 15–16°C; used in making resins and polymers.

methacrylonitrile *Organic Chemistry.* CH_2=C(CH_3)CN, a flammable, toxic, colorless liquid; slightly soluble in water and soluble in acetone; boils at 90.3°C and melts at –38.8°C; used to make coatings and plastics.

methadone *Pharmacology.* $C_{21}H_{27}NO{\cdot}HCl$, an addictive narcotic with actions similar to those of morphine and heroin; used as a painkiller and to suppress withdrawal symptoms in the treatment of heroin addiction, because its own withdrawal symptoms are regarded as comparatively mild. Also, **methadone hydrochloride.**

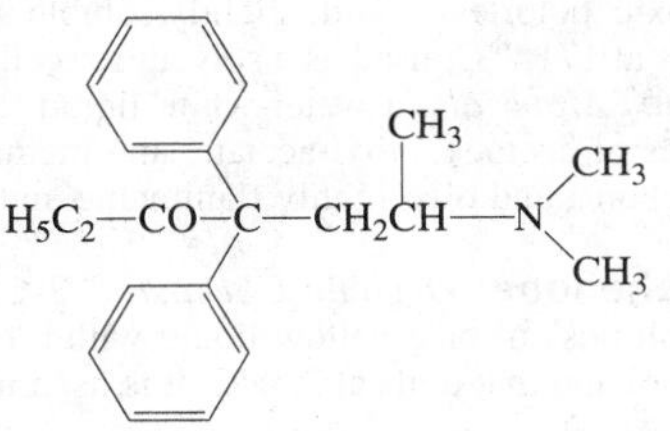

methadone

methadyl acetate *Pharmacology.* $C_{23}H_{31}NO_2$, a narcotic analgesic.

methallyl alcohol *Organic Chemistry.* CH_2=C(CH_3)CH_2OH, a flammable, colorless liquid that is soluble in water and alcohol; boils at 115°C; used as a chemical intermediate. Also, METHYLALLYL ALCOHOL.

methamphetamine *Pharmacology.* $C_{10}H_{15}N$, white crystals or a white powder; a central nervous system stimulant used in medicine as an appetite suppressant and in the treatment of narcolepsy and hyperkinesis; it is also widely abused as an illegal drug for its stimulant properties, especially in its hydrochloride form, $C_{10}H_{15}N{\cdot}HCl$.

methane *Organic Chemistry.* CH_4, a flammable, explosive, colorless, odorless, tasteless gas that is slightly soluble in water and soluble in alcohol and ether; boils at –161.6°C and freezes at –182.5°C. It is formed in marshes and swamps from decaying organic matter, and it is a major explosion hazard in mines. Methane is a major constituent (up to 97%) of natural gas, and it is used as a source of petrochemicals and as a fuel. Also, METHYL HYDRIDE.

methanearsonic acid *Organic Chemistry.* $CH_3AsO(OH)_2$, a white, leaflike solid; very soluble in water; melts at 160°C; used as an herbicide in cotton crops.

methanedisulfonic acid see METHIONIC ACID.

methane drainage see FIREDAMP DRAINAGE.

methane indicator *Mining Engineering.* a portable instrument used to quantify the methane content of mine air at a particular location. Similarly, **methane monitor.**

methane monooxygenase *Enzymology.* an enzyme that is present in methanotrophs and catalyzes the oxidation of various substances, such as methane to methanol and propene to 1,2-epoxypropane.

methane-oxidizing bacteria *Bacteriology.* any bacteria that obtain energy by oxidizing methane.

methanesulfonic acid *Organic Chemistry.* CH_3SO_3H, a liquid that is soluble in water, alcohol, and ether; freezes at 17–20°C and boils at 167°C (10 torr); used as a catalyst in chemical reactions.

methanesulfonyl chloride *Organic Chemistry.* CH_3SO_2Cl, a pale yellow liquid that is insoluble in water and soluble in most organic solvents; boils at 164°C and freezes at –32°C; used as a reagent in preparing mesylate derivatives. Also, MESYL CHLORIDE.

methanethiol *Organic Chemistry.* CH_3SH, a flammable, water-white liquid or colorless gas with a strong, unpleasant odor; slightly soluble in water and soluble in alcohol and ether; boils at 5.96°C and freezes at –121°C; used in organic synthesis and as a catalyst, as well as in jet fuel additives and fungicides. Also, METHYL MERCAPTAN.

Methanobacillus omelianskii *Bacteriology.* two bacterial species originally thought to be a single species, a *Methanobacterium* that forms methane from carbon dioxide and hydrogen and a second, unidentified bacterial organism.

Methanobacteriaceae *Bacteriology.* a family of rod-shaped or spheroid, anaerobic, methane-producing bacteria that are found in the rumen of sheep and cattle and in mud, sewage, and sludge.

Methanobacteriales *Bacteriology.* an order of methane-forming bacteria made up of those species whose cell walls contain pseudomurein.

Methanobacterium *Bacteriology.* a genus of methanogenic, strictly anaerobic bacteria that occur as rod-shaped cells and form methane from carbon dioxide and hydrogen.

Methanobrevibacter *Bacteriology.* a genus of mesophilic bacteria of the family Methanobacteriaceae, coccoid or rod-shaped cells that utilize either carbon dioxide and hydrogen or formate to form methane.

methanochondrion *Cell Biology.* a membrane-bound organelle that is sometimes observed in certain methane-producing bacteria and is believed to play a role in methanogenesis.

Methanococcus *Bacteriology.* a genus of Gram-negative bacteria composed of strictly anaerobic, coccoid cells, forming methane from carbon dioxide and hydrogen or from formate, and occurring in salt-marsh or marine habitats.

methanogen *Biochemistry.* a microorganism that produces methane gas.

methanogenesis *Biochemistry.* the production of methane under anaerobic conditions. *Biotechnology.* the final step in the artificial production of methane, during which hydrogen and bicarbonate are converted to methane.

Methanogenium *Bacteriology.* a genus of Gram-negative bacteria that grow as peritrichously flagellated, coccoid cells with a distinct cell wall.

methanoic acid see FORMIC ACID.

methanol see METHYL ALCOHOL.

methanolate see METHOXIDE.

methanol dehydrogenase *Enzymology.* an enzyme that oxidizes methanol to formaldehyde.

Methanolobus *Bacteriology.* a genus of methane-producing bacteria that occur as motile cocci, each possessing a single flagellum and a distinct cell wall.

Methanomicrobiaceae *Bacteriology.* a family of bacteria of the order Methanomicrobiales, consisting of genera that are able to derive energy from the formation of methane in anaerobic environments.

Methanomicrobiales *Bacteriology.* an order of bacteria, containing three families of bacteria capable of producing methane.

Methanomicrobium *Bacteriology.* a genus of Gram-negative bacteria of the family Methanomicrobiaceae whose cells are typically rod-shaped; found in the rumen of some animals and in marine sediments.

Methanoplanaceae *Bacteriology.* a family of methane-producing bacteria containing the single genus *Methanoplanus.*

Methanoplanus *Bacteriology.* a genus of bacteria of the family Methanoplanaceae, resembling species of the genus *Methanogenium,* and surrounded by a distinctive cell wall.

Methanosarcina *Bacteriology.* a genus of strictly anaerobic bacteria of the family Methanosarcinaceae that occur as coccoid to irregularly shaped cells and are able to form methane from a number of substrates.

Methanosarcinaceae *Bacteriology.* a family of bacteria of the order Methanomicrobiales that can utilize several substrates for methane formation, including acetate, methanol, and carbon dioxide and hydrogen.

Methanospirillum *Bacteriology.* a genus of methane-producing, strictly anaerobic bacteria that occur as flagellated rods or filaments in sewage sludge.

Methanothermaceae *Bacteriology.* a family of Gram-positive bacteria of the order Methanobacteriales, containing the genus *Methanothermus.*

Methanothermus *Bacteriology.* a genus of nonmotile, rod-shaped, Gram-positive, thermophilic bacteria of the family Methanothermaceae that produce methane from the metabolism of carbon dioxide and hydrogen.

Methanothrix *Bacteriology.* a genus of Gram-negative bacteria that occur as rod-shaped cells or sheath-enclosed filaments, and derive energy from the formation of methane from acetate; found in sewage sludge.

methanotroph *Bacteriology.* any bacterial organism that is able to utilize methane as the sole source of both carbon and energy.

methanotrophy *Biochemistry.* a process that takes place under aerobic and microaerobic conditions in certain types of bacteria, in which methane is the only source of energy.

metharmosis *Geology.* any changes that occur in a sediment after it has been buried, but before weathering begins. Also, METAHARMOSIS.
Methedrine *Pharmacology.* the trade name for a brand of methamphetamine.
methemoglobin *Hematology.* a compound formed from hemoglobin by oxidation of the ferrous to the ferric state with essentially ionic bonds. A small amount of methemoglobin is present in the blood normally, but injury or toxic agents can convert a larger proportion of hemoglobin into methemoglobin, which does not function reversibly as an oxygen carrier. Also, **methaemoglobin.**
methemoglobinemia *Medicine.* the appearance of methemoglobulin in the bloodstream.
methemoglobinuria *Medicine.* the excretion of methemoglobulin in the urine.
methene see METHYLENE.
methidathion *Organic Chemistry.* $C_6H_{11}O_4N_2PS_3$, a toxic, colorless crystalline solid; insoluble in water and soluble in most organic solvents; melts at 39°C; used as an insecticide and miticide on commercial crops.
methide *Organic Chemistry.* a methyl compound of a metal.
methine see METHYLIDINE.
methionic acid *Organic Chemistry.* $CH_2(SO_3H)_2$, a colorless hygroscopic crystalline solid that is soluble in water and alcohol and melts at 96–100°C; it is used in organic synthesis. Also, METHANEDISULFONIC ACID.
methionine *Biochemistry.* $C_5H_{11}NO_2S$, a naturally occurring amino acid; slightly soluble in water; essential in the mammalian diet, furnishing both methyl groups and sulfur needed for normal metabolism.
methionine adenosyltransferase *Enzymology.* an enzyme that catalyzes the conversion of methionine into the sulfonium salt *S*-adenosylmethionine. Also, **methionine activating enzyme.**
methionine-enkephalin *Biochemistry.* a biomolecule that functions as an opiate in the human brain.
method the means, procedure, or technique used to carry out some process; specific uses include: *Artificial Intelligence.* a procedure for performing a specific function; especially, a procedure that implements the response to a message in an object-oriented or frame system.
method of loci *Psychology.* a technique for remembering items of information by visualizing each item in a specific imagined location.
method of mixtures *Thermodynamics.* a technique in which a sample of a solid substance is placed in a calorimeter and as the solid melts, the liquid temperature decreases; used for determining the heat of fusion of a substance of known specific heat.
method of moments *Statistics.* an estimation principle based on selecting the population parameters that match population moments with sample moments.
methods analysis *Industrial Engineering.* the study of the constituent parts of an operation in order to make the system more efficient.
methods design *Industrial Engineering.* the design of a work process to produce more efficient performance.
methods engineering *Industrial Engineering.* the branch of engineering that analyzes and designs production methods in order to improve their efficiency.
methods study *Industrial Engineering.* an analysis of production methods, undertaken to identify weaknesses or increase efficiency.
methods-time management *Industrial Engineering.* the analysis of the basic motions in a manual operation and the assignment of predetermined time standards based on the nature of the motion and the setting in which it is performed, including such motions as reach, move, turn, grasp, position, disengage, and release; used to improve performance.
methone see 5,5-DIMETHYL-1,3-CYCLOHEXANEDIONE.
methotrexate *Pharmacology.* $C_{20}H_{22}N_8O_5$, a drug that inhibits the proliferation of malignant cells, used in the treatment of leukemia and other cancers; also used to treat psoriasis and as an immunosuppressant. Also, AMETHOPTERIN.
methoxide *Organic Chemistry.* a class of compounds consisting of a metal and the methoxy radical. Also, METHANOLATE.
methoxy- *Organic Chemistry.* a prefix that indicates the CH_3O– group.
methoxybenzene see ANISOLE.
***p*-methoxybenzoic acid** see ANISIC ACID.
methoxycarbonyl group *Organic Chemistry.* the CH_3COO– group.
methoxychlor *Organic Chemistry.* $Cl_3CCH(C_6H_4OCH_3)_2$, a toxic, white crystalline solid; insoluble in water and soluble in alcohol; melts at 89°C; an insecticide that is related to DDT, but is not as toxic to mammals.
2-methoxyethanol *Organic Chemistry.* $CH_3OCH_2CH_2OH$, a toxic liquid used as a solvent for resins, dyes, and cellulose acetate, and as a dye for leathers.
methoxyethylmercury chloride *Organic Chemistry.* $CH_3OCH_2CH_2HgCl$, white crystals that melt at 65°C; used as a fungicide.
methoxyl *Organic Chemistry.* the $–CH_3O$ group.
6-methoxymellein *Biochemistry.* a phytoalexin secreted by the carrot plant when exposed to such substances as cupric or mercuric acid.
methoxymethyl chloride *Organic Chemistry.* CH_3OCH_2Cl, a liquid that is soluble in alcohol, ether, acetone, and chloroform; boils at 59°C; a highly reactive substance which undergoes solvolysis. Also, CHLOROMETHOXYMETHANE, METHYL CHLOROMETHYL ETHER.
methyl *Organic Chemistry.* the akyl group CH_3–, which is derived from methane.
methyl abietate *Organic Chemistry.* $C_{21}H_{32}O_2$, a combustible, colorless-to-yellow liquid that is soluble in alcohol and boils at 365°C; used as a solvent and plasticizer.
methyl acetate *Organic Chemistry.* CH_3COOCH_3, a flammable, highly explosive, colorless liquid ester with a fragrant odor that is soluble in water and boils at 54.05°C; used in paint removal compounds and as a lacquer solvent.
methyl acetoacetate *Organic Chemistry.* $CH_3COCH_2CO_2CH_3$, a combustible, toxic, colorless liquid; slightly soluble in water and soluble in alcohol; boils at 171.7°C; used as a solvent for cellulose esters.
methyl acetone *Materials.* a water-white liquid that is composed of various mixtures of acetone, ethyl acetate, and methanol; miscible with water, hydrocarbons, and oils; highly flammable and toxic; it is used as a solvent.
methyl acetophenone *Organic Chemistry.* 2-$CH_3C_6H_4COCH_3$, a combustible, colorless or pale yellow liquid with a fragrant odor that is soluble in alcohol and that boils at 214°C; it is used in perfumes and flavorings.
methyl acrylate *Organic Chemistry.* CH_2=$CHCO_2CH_3$, a toxic, volatile, colorless liquid; slightly soluble in water; boils at 80.5°C; used to prepare acrylic polymers and as a chemical intermediate.
methylal *Organic Chemistry.* $CH_3OCH_2OCH_3$, a flammable, volatile, colorless liquid; soluble in water and miscible in alcohol and ether; boils at 42.3°C; used as a solvent and in the manufacture of perfumes, adhesives, and protective coatings.
methyl alcohol *Organic Chemistry.* CH_3OH, a toxic, flammable, colorless liquid that is miscible with water, alcohol, and ether; it boils at 64.5°C; used as an antifreeze, as a denaturant for ethanol, and in the manufacture of formaldehyde and other chemical products. Also, METHANOL, CARBINOL, WOOD ALCOHOL.
methylallyl alcohol see METHALLYL ALCOHOL.
methyl allyl chloride *Organic Chemistry.* CH_2=$C(CH_3)CH_2Cl$, a flammable, volatile, toxic, colorless to straw-colored liquid with a strong penetrating odor; boils at 73°C; used in the production of insecticides, plastics, and pharmaceuticals.
methylamine *Organic Chemistry.* CH_3NH_2, a flammable, colorless gas; soluble in water, alcohol, and ether; boils at –6.79°C and freezes at –92.5°C; used as a chemical intermediate.
***N*-methyl-*p*-aminophenol** *Organic Chemistry.* 2-$CH_3NHC_6H_4OH$, a platelike solid; soluble in alcohol and benzene; melts at 96°C; used as a photographic developer.
methylamyl acetate *Organic Chemistry.* $CH_3COOCH(CH_3)CH_2CH(CH_3)_2$, a toxic, colorless liquid with a mild odor; insoluble in water and soluble in alcohol; boils at 146.3°C; used as a nitrocellulose lacquer solvent.
methylamyl alcohol *Organic Chemistry.* $(CH_3)_2CHCH_2CH(CH_3)OH$, a colorless liquid; slightly soluble in water and miscible with most common organic solvents; boils at 131.8°C; used as a solvent for dyes, oils, and resins, and in organic synthesis.
methyl-*n*-amyl carbinol *Organic Chemistry.* $CH_3(CH_2)_4CHOHCH_3$, a combustible, colorless liquid with a mild odor; miscible with most common organic liquids; boils at 160.4°C; used as a solvent for synthetic resins and as an ore-flotation frothing agent. Also, 2-HEPTANOL.
methyl-*n*-amyl ketone *Organic Chemistry.* $CH_3CH_2CH_2CH_2CH_2COCH_3$, a combustible water-white liquid; insoluble in water and miscible with organic solvents; boils at 150.6°C and freezes at –35°C; used as a solvent, flavoring, and perfume ingredient. Also, 2-HEPTANONE.
***N*-methylaniline** *Organic Chemistry.* $C_6H_5NHCH_3$, an oily, colorless to reddish brown liquid that boils at 196°C and is soluble in alcohol and ether; used as an acid acceptor and solvent.
methyl anion *Organic Chemistry.* the group CH_3^-. Also, CARBANION.

α-methylanisalacetone *Organic Chemistry.* $CH_3OC_6H_4CH{=}CHCOCH_2CH_3$, a combustible, white to yellowish solid with a strong odor that is soluble in alcohol and melts at 60°C; used as a food flavoring.

methyl anthranilate *Organic Chemistry.* $H_2NC_6H_4CO_2CH_3$, a combustible, crystalline solid or pale yellow liquid with bluish fluorescence and a grapelike odor; slightly soluble in water and soluble in alcohol; boils at 256°C and melts at 23.8°C; used in flavorings and perfumes.

2-methyl anthraquinone see TECTOQUINONE.

methyl arachidate *Organic Chemistry.* $CH_3(CH_2)_{18}CO_2CH_3$, a waxlike solid; insoluble in water and soluble in alcohol and ether; melts at 45.8°C and boils at 215–216°C (10 torr); it is used in medical research and as a reference standard for gas chromatography. Also, METHYL EICOSANOATE.

methylarsinic sulfide *Organic Chemistry.* CH_3AsS, a colorless solid that is insoluble in water and melts at 110°C; used as a fungicide. Also, RHIZOCTOL.

methylate *Organic Chemistry.* to subject to a process of methylation.

methylated cap *Molecular Biology.* a terminal 7-methylguanosine residue that is found at the 5' end of eukaryotic mRNA molecules and is required for efficient protein synthesis.

methylation *Organic Chemistry.* **1.** the process of replacing an atom (usually a hydrogen atom) with a methyl group. **2.** the process of denaturation; i.e., the addition of methanol to alcohol, making it unfit for consumption by humans.

2-methylaziridine *Organic Chemistry.* see PROPYLENEIMINE.

methyl behenate *Organic Chemistry.* $CH_3(CH_2)_{20}CO_2CH_3$, a combustible waxlike solid; insoluble in water and soluble in alcohol and ether; melts at 53.2°C and boils at 215.5°C (10 torr); used as chemical intermediate and as a reference standard for gas chromotography. Also, METHYL DOCOSANOATE.

methylbenzene see TOLUENE.

methylbenzethonium chloride *Organic Chemistry.* $C_{27}H_{44}O_2Cl{\cdot}H_2O$, colorless, odorless crystals with a bitter taste; soluble in alcohol; melts at 161–163°C; used as a topical anti-infective medication.

methyl benzoate *Organic Chemistry.* $C_6H_5CO_2CH_3$, a toxic, combustible, fragrant, colorless oily liquid that is insoluble in water and soluble in alcohol and ether; boils at 198.6°C and freezes at –12.3°C; used as a solvent for cellulose esters and resins, and as a flavoring.

methyl *ortho*-benzoylbenzoate *Organic Chemistry.* $C_6H_5COC_6H_4CO_2CH_3$, a platelike solid that is soluble in alcohol and ether; melts at 52°C and boils at 352°C.

α-methylbenzyl acetate *Organic Chemistry.* $C_6H_5CH(CH_3)OOCCH_3$, a combustible, colorless liquid with a floral odor; insoluble in water and soluble in alcohol, glycerol, and mineral oil; boils at 109°C (18 torr); used as a food flavoring and in making perfumes.

α-methylbenzyl alcohol *Organic Chemistry.* $C_6H_5CH(CH_3)OH$, a combustible, colorless liquid; slightly soluble in water and soluble in alcohol, glycerol, and mineral oil; boils at 204°C and melts at 20.7°C; used as a laboratory reagent and in perfumes and flavorings.

α-methylbenzylamine *Organic Chemistry.* $C_6H_5CH(CH_3)NH_2$, a colorless liquid with a mild ammonia odor; slightly soluble in water and soluble in most organic solvents; boils at 189°C and freezes at –65°C; used in organic synthesis and as an emulsifying agent.

α-methylbenzyl ether *Organic Chemistry.* $C_6H_5CH(CH_3)OCH(CH_3)C_6H_5$, a combustible, straw-yellow liquid with a faint odor; very slightly soluble in water and soluble in most organic solvents; boils at 286.3°C and freezes at –30°C; used as a solvent and as a softener for synthetic rubbers.

methyl blue *Organic Chemistry.* $C_{37}H_{27}N_3O_3S{\cdot}2NaSO_3$, a dark blue powder or dye that is used as an antiseptic and as a biological and bacterial stain.

methyl bromide *Organic Chemistry.* CH_3Br, a volatile, toxic, colorless gas or liquid with a burning taste and an odor that is similar to chloroform; insoluble in water; boils at 3.46°C, freezes at –94°C, and burns in oxygen; it is used as a fumigant and an extraction solvent for vegetable oils.

2-methyl-1-butanol *Organic Chemistry.* $CH_3CH_2CH(CH_3)CH_2OH$, a combustible, colorless liquid; slightly soluble in water and miscible with alcohol and ether; boils at 128°C and freezes below –70°C; used as a solvent and additive for oils and paints.

3-methyl-2-butanone see METHYL ISOBUTYL KETONE.

2-methyl-1-butene *Organic Chemistry.* $CH_3CH_2CCH_3{=}CH_2$, a colorless, highly flammable liquid with an unpleasant odor; insoluble in water and soluble in alcohol; boils at 31.11°C and freezes at –137.52°C; used in organic synthesis and pesticides.

methyl butyl ketone *Organic Chemistry.* $CH_3COC_4H_9$, a flammable, colorless liquid that is soluble in alcohol and ether and boils at 127.2°C; used as a solvent. Also, 2-HEXANONE, PROPYLACETONE.

methylbutynol *Organic Chemistry.* $HC{\equiv}CCOH(CH_3)_2$, a colorless liquid; soluble in water, alcohol, ether, and benzene; boils at 104°C; used as a solvent, as a stabilizer for chlorinated organic compounds, and in organic synthesis.

α-methylbutyraldehyde *Organic Chemistry.* $CH_3CH_2CH(CH_3)CHO$, a combustible liquid that is insoluble in water and soluble in alcohol and ether; boils at 92–93°C; used in flavorings.

methyl butyrate *Organic Chemistry.* $CH_3CH_2CH_2COOCH_3$, a colorless, flammable liquid; slightly soluble in water and soluble in alcohol; boils at 102°C and freezes at –92°C; used in solvent mixtures for nitrocellulose and as a flavoring.

methyl caprate *Organic Chemistry.* $CH_3(CH_2)_8COOCH_3$, a combustible, colorless liquid that is insoluble in water and soluble in alcohol and ether; it boils at 224°C and freezes at –13.3°C; it is used as an intermediate for detergents, emulsifiers, and wetting agents. Also, METHYL DECANOATE.

methyl caproate *Organic Chemistry.* $CH_3(CH_2)_4CO_2CH_3$, a combustible, colorless liquid; insoluble in water and soluble in alcohol and ether; boils at 151.2°C and freezes at –71°C; used as a chemical intermediate for detergents, emulsifiers, and wetting agents. Also, METHYL HEXANOATE.

methyl caprylate *Organic Chemistry.* $CH_3(CH_2)_6COOCH_3$, a combustible, colorless liquid; insoluble in water and soluble in alcohol and ether; boils at 192°C and freezes at –37.3°C; used as a chemical intermediate for detergents, emulsifiers, and wetting agents. Also, METHYL OCTANOATE.

methyl carbonate *Organic Chemistry.* $CO(OCH_3)_2$, a flammable, toxic, colorless liquid with a mild, pleasant odor; it is insoluble in water and soluble in most organic solvents; it melts at 0°C and boils at 90.6°C; it is used in organic synthesis and as a solvent. Also, DIMETHYL CARBONATE.

methylcatechol see GUAIACOL.

methyl cation *Organic Chemistry.* the group CH_3^+, a primary carbocation.

methylcellulose *Organic Chemistry.* a combustible, grayish-white powder that is derived from cellulose polymers which have been methylated; insoluble in warm water, alcohol, and ether; used as a thickener, binder, adhesive, and sizing agent; also used for various purposes in medicine.

methyl chloride *Organic Chemistry.* CH_3Cl, a flammable, narcotic, colorless compressed gas or liquid with a faintly sweet odor; slightly soluble in water and soluble in alcohol; boils at –23.7°C and freezes at –97.6°C; used as a refrigerant, catalyst carrier, and methylating agent. Also, CHLOROMETHANE.

methyl chloroacetate *Organic Chemistry.* $ClCH_2CO_2CH_3$, a toxic, combustible, colorless liquid with a sweet odor; insoluble in water and miscible with alcohol and ether; boils at 131°C and freezes at –32.7°C; used as a solvent and chemical intermediate.

methyl chloroform see 1,1,1-TRICHLOROETHANE.

methyl chloroformate *Organic Chemistry.* $ClCO_2CH_3$, a flammable, colorless liquid that is decomposed by water and methanol; soluble in ether; boils at 71.4°C; used in military poisons (lachrymators) and in insecticides.

methyl chloromethyl ether see METHOXYMETHYL CHLORIDE.

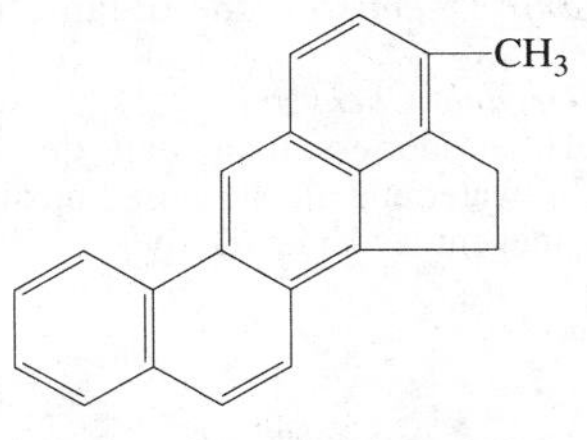

3-methylcholanthrene

3-methylcholanthrene *Biohemistry.* a highly carcinogenic polycyclic aromatic hydrocarbon synthesized by pyrolitic degradation of a cholic acid, deoxycholic acid, or cholesterol; widely used in laboratory studies of chemical carcinogenesis.

methyl cinnamate *Organic Chemistry.* $C_6H_5CH{=}CHCO_2CH_3$, a combustible, white crystalline solid with a strawberrylike odor; insoluble in water and soluble in alcohol and ether; boils at 259.6°C and melts at 34°C; used in perfumes and as a flavoring.

methyl *para*-cresol *Organic Chemistry.* 3,4-$(CH_3)_2C_6H_3OH$, a needlelike solid that is soluble in alcohol and ether and melts at 66°C.

methyl cyanide see ACETONITRILE.

methyl cyanoacetate *Organic Chemistry.* $NCCH_2CO_2CH_3$, a combustible, colorless liquid; soluble in water, alcohol, and ether; boils at 203°C and freezes at –22.5°C; used in organic synthesis and in the production of pharmaceuticals and dyes.

methylcyclohexane *Organic Chemistry.* $CH_3C_6H_{11}$, a flammable, colorless liquid that boils at 100.8°C and freezes at 126.9°C; used as a solvent for cellulose esters. Also, HEXAHYDROTOLUENE.

methylcyclohexanol *Organic Chemistry.* $CH_3C_6H_{10}OH$, a toxic, combustible, colorless mixture of isomers with a menthol-like odor; boils at 163–175°C; used as a solvent for cellulose esters.

methylcyclohexanone *Organic Chemistry.* $CH_3C_5H_9CO$, a toxic, combustible, white to pale-yellow liquid with an odor similar to that of acetone; soluble in alcohol and ether; boils at 160–170°C; used as a solvent and in lacquers.

methylcyclopentadiene dimer *Organic Chemistry.* $C_{12}H_{16}$, a combustible, colorless liquid; insoluble in water and very soluble in alcohol, ether, and benzene; boils at 200°C; used in making dyes, pharmaceuticals, plasticizers, and fuels.

methylcyclopentane *Organic Chemistry.* $CH_3C_5H_9$, a colorless, flammable liquid; immiscible with water and soluble in alcohol and ether; boils at 72°C and freezes at –142.5°C; used in organic synthesis and as an extractive solvent.

methylcyclopropane *Organic Chemistry.* $CH_3C_3H_5$, a gas that is soluble in alcohol and ether; boils at 4°C.

methyl decanoate see METHYL CAPRATE.

methyl 3,4-dichlorocarbanilate see SWEP.

methyl *N*-(3,4-dichlorophenyl)carbamate *Organic Chemistry.* $C_8H_7Cl_2NO_2$, a toxic, white crystalline solid; insoluble in water and soluble in acetone; melts at 113°C; used as an herbicide. Also, SWEP.

***N*-methyldiethanolamine** *Organic Chemistry.* $CH_3N(C_2H_4OH)_2$, a combustible, colorless liquid; miscible with water and benzene; boils at 247.2°C and freezes at –21°C; used in organic synthesis and as an absorbent for acid gases.

methyldihydromorphinone hydrochloride see METOPON.

2-methyl-1,3-dioxolane *Organic Chemistry.* $C_4H_8O_2$, a combustible, water-white liquid; soluble in water; boils at 81°C; used as a solvent and extractant for oils, fats, and waxes.

methyl docosanoate see METHYL BEHENATE.

methyldopa *Pharmacology.* $C_{10}H_{13}NO_4{\cdot}1.5H_2O$, an orally effective antihypertensive occurring as a white to yellowish white fine powder.

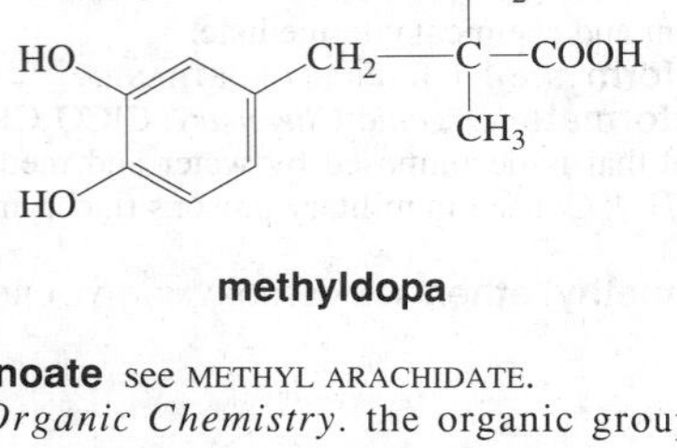

methyldopa

methyl eicosanoate see METHYL ARACHIDATE.

methylene *Organic Chemistry.* the organic group $-CH_2^-$. Also, METHENE.

methylene blue *Organic Chemistry.* $C_{16}H_{18}N_3SCl{\cdot}3HOH$ or $(C_{16}H_{18}N_3SCl)_2{\cdot}ZnCl_2{\cdot}HOH$, an odorless, toxic, dark green crystalline or powdery solid; soluble in water and alcohol; used medicinally and as a bacteriological stain, a reagent, and a textile dye.

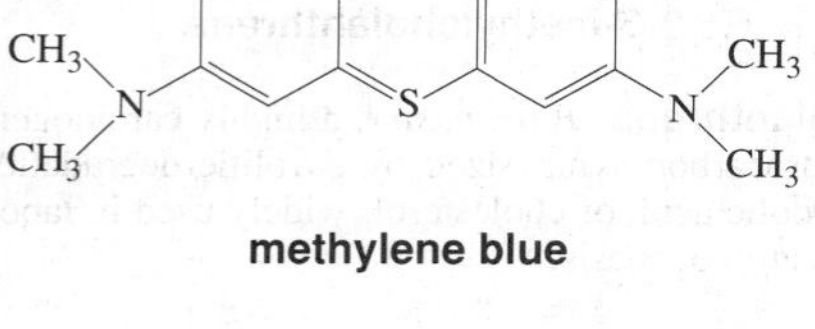

methylene blue

methylene bromide *Organic Chemistry.* CH_2Br_2, a clear, colorless liquid; slightly soluble in water and miscible with alcohol and ether; boils at 97°C and solidifies at –52°C; used in organic synthesis and as a solvent.

methylene chloride *Organic Chemistry.* CH_2Cl_2, a nonflammable, toxic, carcinogenic, narcotic, colorless, volatile liquid; slightly soluble in water and soluble in alcohol and ether; boils at 40.1°C and freezes at –97°C; used in paint removers, solvent degreasing, and plastic processing. Also, DICHLOROMETHANE.

methylene iodide *Organic Chemistry.* CH_2I_2, a yellow liquid; insoluble in water and soluble in alcohol and ether; it boils at 180°C (with decomposition); used in separating mineral mixtures and as an X-ray contrast medium. Also, DIIODOMETHANE.

methyl ester *Organic Chemistry.* any ester having the general formula $RCOOCH_3$, in which the hydrogen of a carboxylic acid has been replaced by a methyl group; usually, a group of fatty esters that are derived from vegetable oils, tallow, or similar substances, in which alkyl groups range from C_8 to C_{18}; used as lubricants.

methylethylcellulose *Organic Chemistry.* $(C_9H_{17}O_5)_n$, an alkylated form of cellulose occurring as a combustible, odorless, white to cream-colored powdery or fibrous solid; used as an emulsifier and stabilizer in the production of textiles, photographic films, and various plastics.

methyl ethyl ketone *Organic Chemistry.* $CH_3COCH_2CH_3$, a flammable, colorless liquid with an odor similar to that of acetone; soluble in water, alcohol, and ether and miscible with oils; boils at 79.6°C and freezes at –86.4°C; used as a solvent and as a reagent in organic synthesis. Also, BUTANONE.

methyl formate *Organic Chemistry.* HCO_2CH_3, a flammable, colorless liquid; soluble in water, alcohol, and ether; boils at 31.8°C and freezes at –99.8°C; used in manufacturing cellulose acetate, in organic sythesis, and as a fumigant and larvicide.

methylfumaric acid see MESACONIC ACID.

2-methylfuran *Organic Chemistry.* $CH_3(C_4H_3O)$, a flammable, colorless liquid with an odor similar to that of ether; insoluble in water and soluble in alcohol and in ether; it boils at 63.2–65.6°C and freezes at –88.68°C; used as a chemical intermediate.

methyl 2-furoate *Organic Chemistry.* $C_4H_3OCO_2CH_3$, a combustible, colorless liquid that turns to yellow when exposed to light; insoluble in water and soluble in alcohol and ether, boils at 181.3°C; used as a solvent and in organic synthesis.

α-methylglucoside *Organic Chemistry.* $C_7H_{14}O_6$, a combustible, odorless, white crystalline solid; soluble in water, slightly soluble in alcohol, and insoluble in ether; melts at 168°C and boils at 200°C; used as a plasticizer for resins and in oil varnishes and polyurethane foam.

***N*-methylglycine** see SARCOSINE.

methylglyoxal bypass *Biochemistry.* a series of reactions that can occur outside the Meyerhof-Parnas pathway, lying between triosephosphates and pyruvate.

methyl green *Organic Chemistry.* $C_{27}H_{35}BrClN_3$, a green powder that is soluble in water; used as a biological stain and in dyeing and printing textiles.

methylheptane *Organic Chemistry.* C_8H_{18}, a flammable, colorless liquid that occurs in two isomers (2-methyl- and 4-methylheptane); insoluble in water and soluble in alcohol and ether; boils at 117–122.2°C; used as a chemical intermediate.

methylheptenone *Organic Chemistry.* $(CH_3)_2C{=}CH(CH_2)_2COCH_3$, a combustible, colorless liquid; insoluble in water and miscible with alcohol and ether; boils at 173–174°C and freezes at –67.1°C; used in organic synthesis, perfumes, and flavoring.

2-methylhexane *Organic Chemistry.* $(CH_3)_2CH(CH_2)_3CH_3$, a flammable, colorless liquid; insoluble in water and soluble in alcohol; boils at 90°C and freezes at –118.5°C; used in organic synthesis. Also, ISOHEPTANE.

methyl hexanoate see METHYL CAPROATE.

methyl hexyl ketone *Organic Chemistry.* $CH_3COC_6H_{13}$, a colorless liquid with a camphorlike taste; insoluble in water and soluble in alcohol and ether; boils at 173.5°C and melts at –16°C; used in perfumes and as a solvent. Also, 2-OCTANONE.

methyl hydride see METHANE.

methyl 12-hydroxystearate *Organic Chemistry.* $C_{19}H_{38}O_3$, a combustible, white waxy solid; insoluble in water and slightly soluble in organic solvents; melts at 48°C; used in making inks, adhesives, and cosmetics.

methylidine *Organic Chemistry.* the HC= group. Also, METHINE.

3-methylindole see SKATOLE.

methyl iodide *Organic Chemistry.* CH_3I, a toxic, narcotic, flammable, colorless liquid that turns brown when exposed to light; slightly soluble in water and miscible with alcohol and ether; boils at 42°C and freezes at –66.1°C; used in organic synthesis and microscopy.

2-methylisoborneol *Organic Chemistry.* $C_{11}H_{20}O$, a compound that melts at 170°C; it is metabolized by the bacteria *Streptomyces* and *Actinomadura,* causing an objectionable odor in water supplies.

methyl isobutyl ketone *Organic Chemistry.* $(CH_3)_2CHCH_2COCH_3$, a flammable, colorless liquid; slightly soluble in water and miscible with most organic solvents; boils at 115.8°C and freezes at –85°C; used as a solvent for paints, varnishes, and lacquers, as a denaturant for alcohol, and in the extraction of uranium from fission products. Also, 3-METHYL-2-BUTANONE.

methyl isothiocyanate *Organic Chemistry.* $CH_3N{=}C{=}S$, toxic, colorless crystals; partially soluble in water and soluble in alcohol and ether; boils at 120°C and melts at 35°C; used as an insecticide and chemical weapon. Also, **methyl mustard oil.**

methyl lactate *Organic Chemistry.* $CH_3CH(OH)CO_2CH_3$, a combustible, colorless liquid; soluble in water, alcohol, and ether; boils at 144.8°C and freezes at –66°C; it is used as a solvent for cellulose compounds, lacquers, and stains.

methyl laurate *Organic Chemistry.* $CH_3(CH_2)_{10}CO_2CH_3$, a combustible, water-white liquid derived from coconut oil; insoluble in water; boils at 262°C; used as a chemical intermediate for detergents, emulsifiers, and stabilizers, and as a flavoring.

methyl linoleate *Organic Chemistry.* $CH_3(CH_2)_4CH{=}CHCH_2CH{=}CH(CH_2)_7CO_2CH_3$, a combustible, colorless oily liquid; insoluble in water and soluble in alcohol and ether; boils at 192°C (4 torr) and freezes at –35°C; used as a chemical intermediate, a reference standard in gas chromatography, and in biochemical research.

Methyl Mercaptan *Organic Chemistry.* the trade name for methanethiol.

methylmercury cyanide *Organic Chemistry.* CH_3HgCN, a toxic, crystalline solid; soluble in water and organic solvents; melts at 95°C; used as a fungicide and seed disinfectant. Also, **methylmercury nitrile.**

methyl methacrylate *Organic Chemistry.* $CH_2{=}C(CH_3)CO_2CH_3$, a flammable, colorless, volatile liquid; slightly soluble in water and soluble in most organic solvents; boils at 101°C and freezes at –42°C; it forms vinyl polymers that are widely marketed under such trade names as Plexiglas and Lucite.

methyl methylthiomethyl sulfoxide *Organic Chemistry.* $CH_3SOCH_2SCH_3$, a liquid that boils at 92°C; used as a reagent in the synthesis of ketones and aldehydes or their derivatives, such as thioenol ether or thioketal.

methylmorphine see CODEINE.

methyl myristate *Organic Chemistry.* $C_{13}H_{27}CO_2CH_3$, a combustible, colorless liquid; insoluble in water; boils at 155–157°C (7 torr) and melts at 17.8°C; used as a chemical intermediate and as a reference standard for gas chromatography.

1-methylnaphthalene *Organic Chemistry.* $CH_3C_{10}H_7$, a combustible, colorless liquid derived from coal tar; insoluble in water and soluble in alcohol and ether; boils at 240–243°C and freezes at –22°C; used in organic synthesis.

methyl nitrate *Organic Chemistry.* CH_3ONO_2, a colorless, narcotic liquid that is slightly soluble in water and soluble in alcohol and ether; boils (explodes) at 66°C; used as a rocket propellant.

methyl nonanoate *Organic Chemistry.* $CH_3(CH_2)_7CO_2CH_3$, a combustible, colorless liquid with a fruity odor; insoluble in water and soluble in alcohol and ether; boils at 213.5°C and freezes at –35°C; used in perfumes and flavorings, and as a chemical intermediate and reference standard for gas chromatography.

methyl nonyl ketone *Organic Chemistry.* $CH_3COC_9H_{19}$, a combustible, oily liquid that is soluble in alcohol and boils at 225°C; used in food flavorings and perfumes. Also, 2-UNDECANONE.

Methylobacterium *Bacteriology.* a genus of rod-shaped bacteria characterized by the use of methane or single carbon compounds as the sole source of both energy and carbon.

Methylococcaceae *Bacteriology.* a family of Gram-negative, aerobic, rod-shaped or spheroid bacteria that utilize one-carbon organic compounds, such as methane or methanol, as a carbon source.

Methylococcus *Bacteriology.* a genus of aerobic bacteria, containing a complex system of stacked membranes; distinguished by the ability to metabolize either single carbon compounds or compounds containing methyl groups that are not directly linked to one another.

methyl octanoate see METHYL CAPRYLATE.

methyl oleate *Organic Chemistry.* $CH_3(CH_2)_7CO_2CH_3$, a combustible, clear to amber liquid; insoluble in water and soluble in alcohol; boils at 218.5°C (20 torr) and freezes at –19.9°C; used as a chemical intermediate, a plasticizer and softener, and a chromatographic reference standard.

methylol reaction *Materials Science.* a reaction that is used to produce amino resins, in which additional compounds (known as **methylol derivatives**) are formed and subsequently condense.

methylol riboflavin *Organic Chemistry.* an orange to yellow powder that is soluble in water and insoluble in alcohol and ether; used as a dietary supplement and in medicines.

methylol urea *Organic Chemistry.* $H_2NCONHCH_2OH$, a colorless prismatic solid; soluble in water and methanol and insoluble in ether; melts at 111°C; used to treat textiles and wood and in manufacturing resins and adhesives.

Methylomonadaceae *Bacteriology.* a former term for Methylococcaceae.

Methylomonas *Bacteriology.* a genus of Gram-negative bacteria that are obligately methylotrophic, occur in the form of motile, flagellated rod-shaped cells, and contain a complex intracellular system of stacked membranes.

Methylophaga *Bacteriology.* a proposed bacterial genus of methylotrophic, Gram-negative organisms that grow as strictly aerobic, rod-shaped cells in marine habitats.

Methylophilus methylotrophus *Bacteriology.* a species of methylotrophic bacteria that is cultured on a large scale for use as a source of dietary protein for animals.

methyl orange *Organic Chemistry.* $(CH_3)_2NC_6H_4N{=}NC_6H_4SO_3Na$, an orange-yellow powder that is soluble in hot water and insoluble in alcohol; used as a pH indicator: red in acid and yellow in basic.

Methylosinus trichosporium *Bacteriology.* a species of obligately methylotrophic bacteria occurring as rod-shaped cells having polar flagella, and forming exospores.

methylotroph *Bacteriology.* any organism that is able to metabolize single-carbon compounds as the sole source of both carbon and energy. Thus, **methylotrophic.**

methylotrophy *Biochemistry.* the use of methanol as an energy source in yeasts. *Bacteriology.* the fact of being a methylotroph.

methyl oxide see DIMETHYL ETHER.

methyl palmitate *Organic Chemistry.* $CH_3(CH_2)_{14}CO_2CH_3$, a combustible, colorless liquid; insoluble in water and soluble in alcohol and ether; boils at 163–164°C (4 torr) and melts at 29.5°C; used as a chemical intermediate and in medical research.

3-methylpentane *Organic Chemistry.* $CH_3CH_2CH(CH_3)CH_2CH_3$, a flammable, colorless liquid; insoluble in water, soluble in alcohol, and slightly soluble in ether; boils at 64°C; used as a solvent and chemical intermediate.

2-methyl-2,4-pentanediol see HEXYLENE GLYCOL.

2-methylpentanoic acid *Organic Chemistry.* $CH_3CH_2CH_2CH(CH_3)COOH$, a combustible water-white liquid; soluble in water; boils at 196.4°C and freezes at –85°C; used in making plasticizers, metallic salts, and vinyl stabilizers.

methylpentene polymer *Organic Chemistry.* a thermoplastic polymer of 4-methyl-1-pentene.

methyl pentose *Organic Chemistry.* $C_6H_{12}O_5$, describing sugars that contain 6 carbons but only 5 hydroxy groups, such as rhamnose (**6-deoxy-L-mannose**) or fucose (**6-deoxy-L-galactose**).

methylphenol see *m*-CRESOL.

methyl phenylacetate *Organic Chemistry.* $C_6H_5CH_2CO_2CH_3$, a combustible, colorless liquid with a honeylike odor; insoluble in water and soluble in alcohol; boils at 218°C; it is used to flavor tobacco and in perfumery.

methyl phenyl ether see ANISOLE.

methyl phenyl ketone *Organic Chemistry.* $C_6H_5COCH_3$, a colorless prismatic solid; soluble in alcohol and ether; melts at 20°C; a potent lachrymator that is used extensively as a tear gas by law enforcement agencies. Also, ACETOPHENONE.

methylphosphonic acid *Organic Chemistry.* $CH_3PO(OH)_2$, a white hygroscopic solid that is soluble in water, alcohol, and ether; melts at 103–104°C; used in organic synthesis.

2-methylpropene see ISOBUTYLENE.

methyl propionate *Organic Chemistry.* $CH_3CH_2CO_2CH_3$, a flammable, colorless liquid that is slightly soluble in water and soluble in most organic solvents; boils at 78–79°C; it is used as a solvent and flavoring.

methyl propyl carbinol *Organic Chemistry.* $CH_3CH_2CH_2CHOHCH_3$, a combustible, colorless liquid that is soluble in water and miscible with alcohol and ether; boils at 119.3°C and freezes at –75°C; used as a solvent in paints and lacquers and as a pharmaceutical intermediate. Also, 2-PENTANOL.

methyl propyl ketone see PENTANONE.

***N*-methyl-2-pyrrolidone** *Organic Chemistry.* C_5H_9NO, a liquid that is soluble in water, ether, and acetone and boils at 203°C; used as a solvent for petroleum and resins and in PVC spinning.

methyl red *Organic Chemistry.* $(CH_3)_2NC_6H_4N{=}NC_6H_4CO_2H$, a dark red powder or violet prismatic solid that is insoluble in water and soluble in alcohol and ether; melts at 180°C; formerly used as an acid-base indicator.

methyl red test *Microbiology.* an IMViC test in which bacteria is inoculated into a buffered glucose-peptone broth containing methyl red; in a positive reaction, the broth remains red after incubation due to the presence of acid metabolic products; used to differentiate among Enterobacteriaceae.

5-methylresorcinol see ORCIN.

methyl ricinoleate *Organic Chemistry.* $C_6H_{13}CH(OH)CH_2CH{=}CH(CH_2)_7CO_2CH_3$, a combustible, colorless liquid derivative of oleic acid; insoluble in water and soluble in alcohol and ether; boils at 225–227°C (15 torr) and melts at –4.5°C; used as a lubricant, wetting agent, and plasticizer.

methyl rubber *Organic Chemistry.* a former term for commercially produced synthetic rubber (isoprene) based on butadiene.

methyl salicylate *Organic Chemistry.* $C_6H_4OHCO_2CH_3$, a toxic, combustible, colorless (when pure) liquid with an odor of wintergreen; slightly soluble in water and soluble in alcohol and ether; boils at 222.2°C and freezes at –8.3°C; used in perfumes, as a UV-absorber in suntan lotions, as an antipyretic and antiseptic, and as a flavoring agent. Also, WINTERGREEN OIL.

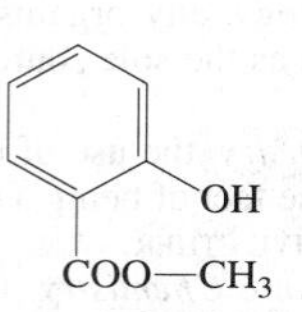

methyl salicylate

3-methylsalicylic acid *Organic Chemistry.* $C_8H_8O_3$, a white crystalline solid that is soluble in alcohol and ether and melts at 165–166°C; used as a chemical intermediate.

methyl silicone *Organic Chemistry.* $[(CH_3)_2SiO]_n$, a class of silicone compounds that occur as volatile, colorless oils with boiling points ranging from 118°C to 134°C; used as transformer liquid and brake fluid. Also, DIMETHYL SILICONE.

methyl stearate *Organic Chemistry.* $C_{17}H_{35}CO_2CH_3$, a combustible semisolid; insoluble in water and soluble in alcohol and ether; crystals melt at 37.8°C and liquid boils at 181–182°C (4 torr); used as a chemical intermediate.

3-methylstyrene *Organic Chemistry.* $H_2C{=}CHC_6H_4CH_3$, a combustible, colorless liquid; very slightly soluble in water and soluble in ether and methanol; boils at 170–171°C and freezes at –76.8°C; used as a solvent and chemical intermediate. Also, VINYLTOLUENE.

α-methylstyrene *Organic Chemistry.* $C_6H_5C(CH_3){=}CH_2$, a combustible, colorless liquid; soluble in ether; boils at 165.38°C and freezes at –23.21°C; used to make polystyrene resins.

methyl sulfate see DIMETHYL SULFATE.

methyl sulfide *Organic Chemistry.* $(CH_3)_2S$, a flammable, colorless volatile liquid with an unpleasant odor; insoluble in water and soluble in alcohol and ether; boils at –98.3°C and freezes at –83°C; used as a solvent and as an odorant for natural gas. Also, DIMETHYL SULFIDE.

methylsulfonyl *Organic Chemistry.* the organic group $CH_3SO_2^-$. Also, MESYL.

4-methyl-5-thiazoleethanol *Organic Chemistry.* C_6H_9NOS, a thick, oily liquid that is soluble in water, alcohol, and ether; used in making vitamin B_1 and as a sedative.

methylthiouracil *Pharmacology.* $C_5H_6N_2OS$, a thyroid suppressant occurring as a white crystalline powder; used in treatment of hyperthyroidism, especially to prepare patients for surgery or to maintain patients who are poor surgical risks.

methyl transferase *Enzymology.* one of a class of enzymes that catalyze the conversion of *S*-adenosyl methionine to *S*-adenosyl homocysteine, by transferring a methyl group to a suitable acceptor molecule.

α-methyl-*para*-tyrosine *Organic Chemistry.* $(4\text{-}HOC_6H_4)CH_2CH(NH_2)CO_2CH_3$, a solid that is soluble in alcohol and methanol and melts at 136°C.

methyl violet *Organic Chemistry.* $(CH_3)_2NC_6H_4C[C_6H_4N(CH_3)_2]_2Cl$, green crystals that are soluble in water and chloroform and partially soluble in alcohol and glycerol; used as a topical antibacterial agent, pH indicator, biological stain, and dye. Also, GENTIAN VIOLET.

methysergide *Pharmacology.* $C_{21}H_{27}N_3O_2$, an ergot alkaloid derivative used in the treatment of migraine headaches.

Metis [mē´tis] *Astronomy.* the sixteenth moon of Jupiter, discovered in 1979–1980 and measuring 40 km in diameter.

metoclopramide *Pharmacology.* $C_{14}H_{22}ClN_3O_2$, a white crystalline substance used as an aniemetic and in the treatment of upper gastrointestinal tract problems.

metolazone *Pharmacology.* $C_{16}H_{18}ClN_3O_3S$, a thiazide diuretic used in the treatment of hypertension and as a diuretic.

Metonic cycle [mə tän´ik] *Astronomy.* a time interval lasting 235 lunations (about 19 years) after which lunar phases recur on the same days of the year.

metope *Architecture.* any of the square spaces between triglyphs in a Doric Frieze; may be plain or decorated.

metopic *Anatomy.* of or relating to the forehead.

metopon or **metopon hydrochloride** *Pharmacology.* $C_{18}H_{21}NO_3{\cdot}HCl$, a narcotic painkiller derived from morphine. Also, METHYLDIHYDROMORPHINONE HYDROCHLORIDE. *Physiology.* an archaic term for the anterior portion of the frontal lobe of the brain.

metoprine *Oncology.* $C_{11}H_{10}Cl_2N_4$, an antineoplastic agent.

metraterm *Invertebrate Zoology.* the distal portion of the uterus in trematodes.

metre *Metrology.* the British spelling of meter. See METER.

metria *Medicine.* any inflammation of the uterus during the period following childbirth.

metric *Metrology.* **1.** of or relating to measurement. **2.** of or relating to the meter or to the metric system of measurement. *Mathematics.* a distance function. Specifically, given a set X, a nonnegative real-valued function d defined on $X \times X$ such that for all $x, y \in X$: (a) $d(x,y) = 0$ if and only if $x = y$; (b) $d(x,y) = d(y,x)$ (symmetry); and (c) $d(x,z) \leq d(x,y) + d(y,z)$ (triangle inequality). The quantity $d(x,y)$ is called the distance between x and y and d is also called a distance function. The topology on X generated by unions of open spheres of infinite radius is said to be induced by the metric d. An **infinitesimal metric** defines the distance between arbitrarily close points. A **global metric** defines the distance between two points in terms of the coordinate functions of the points. Roughly speaking, a global metric is obtained from an infinitesimal metric by integration; an infinitesimal metric is obtained from a global metric by differentiation.

metrical *Metrology.* of or relating to the meter or the metric system.

metrication *Metrology.* **1.** in general, the use of metric measurements. **2.** specifically, the process of converting from the traditional English system of measurement to the metric system, as in a nation that has retained the traditional system, such as the United States.

metric camera *Photogrammetry.* a camera in which the interior orientation is known, stable, and reproducible.

metric centner *Metrology.* a unit of weight equal to 100 kilograms.

metric horsepower *Physics.* a unit of power equivalent to 753.49875 watts.

metric hundredweight *Metrology.* a unit of weight equal to 50 kilograms.

metricize *Science.* to express in or convert into the metric system of measurement. Also, **metricate.**

metric mile *Metrology.* a metric distance of 1500 meters, comparable to a mile (1609 meters). Also, **metrical mile.**

metric on X see DISTANCE FUNCTION.

metric ounce see MOUNCE.

metrics *Metrology.* the metric system, or the use of the metric system.

-metrics a combining form meaning "the science of measuring," as in *biometrics, econometrics.*

metric space *Mathematics.* a space X endowed with a metric.

metric system *Metrology.* a standard system of measurement using decimal units, in which the meter is the basic unit of length, the gram is the basic unit of mass, and the liter is the basic unit of volume or capacity; used in science and as the measuring system of most nations.

metric tensor *Mathematics.* a continuous, 2-covariant tensor field g defined on a smooth manifold X, such that: (a) g is symmetric, and (b) for each $x \in X$, the bilinear form g_x is nondegenerate. Such a manifold and metric tensor are together called a Riemannian manifold or a Riemannian structure. This is usually expressed by the equation

$$ds^2 = \Sigma_{i,j}\, g_{ij}(x)dx^i dx^j,$$

where the $g_{ij}(x)$ represent the components of g_x in a coordinate basis. Also, FUNDAMENTAL TENSOR.

metric thread gearing *Design Engineering.* a system of interchangeable gears that provide feeds for cutting metric and module threads.

metric ton *Metrology.* a unit of mass that is equal to 1000 kilograms, equivalent to 2204.62 pounds. Also, MEGAGRAM, TONNE.

Metridiidae *Invertebrate Zoology.* **1.** a family of sea anemones found along shallow coasts of the North Pacific and North Atlantic. **2.** a family of marine copepod crustaceans.

metrifonate *Pharmacology.* $C_4H_8Cl_3O_4P$, an organophosphorus compound used as an insecticide.

metritis *Medicine.* any of several varieties of inflammation of the uterus, named according to the part of the organ affected. Also, UTERITIS, HYSTERITIS.

metrizable space *Mathematics.* a topological vector space is said to be metrizable if its topology can be induced by a metric d that is translation invariant, that is,

$$d(x,y) = d(x + z, y + z).$$

metro- *Metrology.* a combining form meaning "measurement," as in *metrology. Medicine.* a combining form meaning "uterus," as in *metrocarcinoma, metrofibroma.*

metrocarcinoma *Oncology.* a carcinoma of the endometrium.

metrocytosis *Medicine.* a condition characterized by the formation of uterine cysts.

metrofibroma *Oncology.* a leiomyoma of the uterus.

metrogenous *Medicine.* coming from the uterus.

metrology **1.** the science of measurement or of weights and measures. **2.** a specific sytem of measurement, such as the metric system.

metronidazole *Organic Chemistry.* $C_6H_9N_3O_3$, cream-colored crystals; soluble in water and slightly soluble in ethanol; melts at 161°C; used as an antiprotozoal and antibacterial.

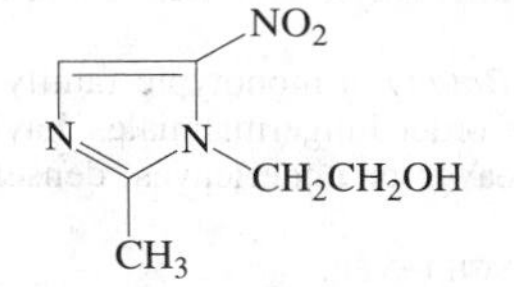

metronidazole

metroperitoneal *Medicine.* relating to the uterus and peritoneum or involved with the uterine and peritoneal cavities.

metropolis *Geography.* **1.** the largest or most important city in a country or region. **2.** any large or important city.

metropolitan area *Geography.* an area composed of a large city together with its surrounding suburbs and outlying communities.

metrorrhagia *Medicine.* a condition of erratic uterine bleeding, particularly between menstrual periods.

metrorrhea *Medicine.* a free or abnormal discharge from the uterus.

metrorrhexis *Medicine.* a rupture of the uterus. Also, HYSTERORRHEXIS.

metrosalpingitis *Medicine.* an inflammation of the uterus and either or both Fallopian tubes.

metrostaxis *Medicine.* a slight but persistent bleeding from the uterus.

Metschnikowia *Mycology.* a genus of fungi belonging to the family Spermophthoraceae that parasitizes and is pathogenic to plants and animals; found in terrestrial and aquatic environments.

Metschnikowiaceae *Mycology.* a family of fungi belonging to the order Endomycetales that parasitizes and is pathogenic to animals and plants. Also, NEMATOSPORACEAE, SPERMOPHTHORACEAE.

M. et sig. mix and write a label. (From Latin *misce et signa.*)

metula *Mycology.* an outer branch of a conidiophore from which extensions, known as phialides, radiate.

meturedepa *Oncology.* $C_{11}H_{22}N_3O_3P$, an antineoplastic agent.

Metrology

Metrology is the field of knowledge concerned with measurement. While all scientific disciplines impact on human life in one way or another, the interaction with weights and measures runs like a continuous line through everybody's life. At the beginning of a human life is the scale in the maternity ward and, on the other end, the tape measure in the undertaker's pocket.

Measurements were on the minds of the first hunters on the Serengeti Plain when they tried to determine their distance from a herd of wildebeests. Measurements were on the minds of the first farmers when they traded grain for cattle. And measurements are on the minds of kings and presidents when they impose a tax on the pint of beer that you drink and on the gallon of gasoline that you pump.

While the concept of mass, length, and volume is abstract, the effective communication of an extent of mass, length, and volume requires the establishment of a set of arbitrary units that everybody would agree with. Human nature being as it is, this has never been a trivial task. First attempts to do so were pushed by the early merchants who carried weights and measures with them. However, as soon as the first ruler got the idea of charging taxes on merchandise sold in and transported through his territory, the setting up of standard units became a matter for state authorities.

Even the king's authority could not, however, preclude confusion. In an English statute dating from about 1300, a gallon was specified as holding eight pounds of wheat. When the importation of wine from France became a flourishing business, this measure became an Excise wine gallon which is the prototype of a standard U.S. gallon. In England, however, the Excise gallon was abolished in 1824 in favor of an Exchequer ale gallon holding ten pounds of water, which was later adopted as the standard Imperial gallon.

In the modern world, metrology has lost none of its significance. In today's markets, the capability of the manufacturing industry to successfully compete depends on the ability to integrate manufacturing technology with on-line measurement of critical process parameters. Production metrology is the key component of the Total Quality Management concept. Some technologies, such as microelectronics, would not have reached their present level of sophistication without advances in production metrology.

Metrology transcends the planet that we live on. The in-space metrology has evolved from the early efforts to successfully navigate the seas of the world into a complex scientific discipline. For deep space navigation, the unit of time can now be determined with an accuracy better than one second in 300,000 years.

Metrology has come a long way from the building of Stonehenge to the landing on the moon. Its development will not stop there.

Jaromir J. Ulbrecht
Director, Technical Programs
National Institute of Standards and Technology

Metzgeriaceae *Botany.* a cosmopolitan family of liverworts of the order Metzgeriales, characterized by thallose, prostrate plants with primarily dichotomous branching from the ventral side of the sharply defined midrib and by unistratose wings.

Metzgeriales *Botany.* a liverwort order of the subclass Hepaticopsida having a long, flat gametophyte and archegonia that develop behind a growing apex.

Metzgeriopsaceae *Botany.* a monotypic family of epiphyllous liverworts belonging to the order Jungermanniales, having a vegetative shoot as a thallus and no leaves or underleaves, densely pinnate branching, and sparse rhizoids.

Meuse fever see TRENCH FEVER.

Meusnier de la Place, J. [men yā′] 1754–1793, French mathematician; formulated the theorem of curvature at a point on a surface; designed the dirigible balloon.

Meusnier's theorem *Mathematics.* suppose a given plane intersects (cuts) a surface at a point P, making an angle $\gamma \neq 0$ with the normal to the surface at P. The curvature k at P of the oblique section thus formed is $k = (\cos \gamma)k_n$, where k_n is the curvature of the normal section.

MeV or **Mev** or **mev** megaelectron volt.

mevalonate *Biochemistry.* the anionic form of mevalonate acid.

mevalonic acid *Biochemistry.* $C_6H_{12}O_4$, a six-carbon organic acid that serves as an intermediate in cholesterol biosynthesis.

mew gull *Vertebrate Zoology.* the small gull *Larus canus,* found in northwestern North America and Eurasia. Also, **mew.**

MEX *Aviation.* the airport code for Mexico City.

mexacarbate *Organic Chemistry.* $C_{12}H_{18}N_2O_2$, a toxic, colorless to brownish crystalline solid; insoluble in water and soluble in alcohol and acetone; melts at 85°C; used as an insecticide.

Mexican bean bettle *Invertebrate Biology.* the ladybird beetle *Epilachna varivestis* that feeds on the leaves of the bean plant; introduced to the U.S. from Mexico.

Mexican fireplant *Botany.* the plant *Euphorbia heterophylla* of the spurge family, characterized by red or mottled red and white bracts; found from the central U.S. to central South America.

Mexican freetailed bat *Vertebrate Biology.* any of several bats of the genus *Tadarida* that eat insects and inhabit limestone caves; found in Mexico and the southwestern U.S.

Mexican fruit fly *Invertebrate Biology.* the bright-colored fly *Anastrepha ludens* whose larvae are a serious pest of citrus fruits and mangoes; found in Mexico, Central America, and Texas.

Mexican subregion *Ecology.* a distinct zoogeographical region that includes southern Mexico and the adjacent countries of central America.

Mexico, Gulf of *Geography.* a wide arm of the North Atlantic nearly enclosed by the U.S., Mexico, and Cuba.

Meyer, Julius Lother 1830–1895, German chemist; demonstrated the relation between atomic weights and properties of the elements.

Meyer atomic volume curve *Atomic Physics.* a graph that plots the atomic volumes of the elements against their atomic numbers.

Meyerhof, Otto 1884–1951, German biochemist; awarded the Nobel Prize for his work on lactic acid production in muscles.

meyerhofferite *Mineralogy.* $Ca_2B_6O_6(OH)_{10}{\cdot}2H_2O$, a colorless to white triclinic mineral occurring as prismatic crystals and in fibrous form, having a specific gravity of 2.12 and a hardness of 2 on the Mohs scale; found as thick tabular crystals in colemanite deposits.

Meyliidae *Invertebrate Zoology.* a family of free-living marine nematodes in the superfamily Desmoscolecoidea.

Meziridae *Invertebrate Zoology.* a family of true bugs, hemipteran insects in the superfamily Aradoidea.

mezzotint *Graphic Arts.* an engraving process in which an entire copper or steel plate is first roughened, then burnished in certain areas to produce light and shade.

Mf *Metallurgy.* the temperature at which a martensitic transformation is completed upon cooling.

MF or **mf** medium frequency.

mF millifarad; millifarads.

μF *Electricity.* the symbol for microfarad.

Mflop *Computer Technology.* a million floating point operations per second.

M. ft. let a mixture be made. (From Latin *mistura fiat.*)

Mg the chemical symbol for magnesium.

Mg or **Mg.** megagram.

mg or **mg.** or **mgm** milligram.

μg microgram.

mgal milligal.

M 31 galaxy *Astronomy.* another name for the Andromeda galaxy.

M 82 galaxy *Astronomy.* an unusual galaxy in Ursa Major believed to be undergoing widespread star formation.

mgd million gallons per day.

mgh milligram-hour.

MgO magnesium oxide.

mh or **mh.** millihenry.

MHA-TP microhemagglutination assay.

MHC major histocompatibility complex.

MHD magnetohydrodynamic.

M.H.D. minimum hemolytic dose.

mho *Electricity.* see siemens. (From *ohm* spelled backward.)

M.H.W. or **m.h.w.** mean high water.

MHz megahertz.

mHz millahertz.

MI mitotic index, the fraction of a population of cells in mitosis at any given time.

MI myocardial infarction; mitral insufficiency.

mi or **mi.** mile; mileage; mill.

MIA *Aviation.* the airport code for Miami International.

Miacidae *Paleontology.* an extinct family of arboreal, carnivorous mammals in the suborder Fissipedia and superfamily Miacoidea; descended from insectivores and probably directly ancestral to the modern carnivores, the **miacids** had bigger brains than the creodonts; Paleocene and Eocene.

Miacoidea *Paleontology.* an extinct superfamily of carnivores in the suborder Fissipedia; Holarctic distribution; extant in the Paleocene and Eocene.

miagite see CORSITE.

miargyrite *Mineralogy.* $AgSbS_2$, an iron-black to steel-gray, monoclinic mineral with cherry-red streaks, having a specific gravity of 5.25 and a hardness of 2.5 on the Mohs scale; found as thick tabular crystals in low-temperature hydrothermal vein deposits.

miarolithite *Petrology.* a variety of chorismite with miarolitic cavities or remnants of such cavities.

miarolitic *Petrology.* of, relating to, or characterized by small, irregularly shaped cavities in igneous rocks, normally with small granites.

mica *Mineralogy.* a group name for various monoclinic, pseudohexagonal phyllosilicates that have the general formula $XY_{2\text{-}3}Z_4O_{10}(OH,F)_2$ (or $XY_3Si_4O_{12}$), where X=Ba,Ca,Cs(H_3O)K,Na,(NH_4); Y= Al,Cr^{+3}, Fe^{+2}, Fe^{+3},Li,Mg,Mn^{+2},Mn^{+3},V^{+3},Zn; Z=Al,Be,Fe^{+3},Si; characterized by low hardness and very perfect basal cleavage; commonly found as flakes or scales in igneous and metamorphic rocks. (From a Latin word meaning "crumb" or "grain.")

mica book *Mineralogy.* a crystal of generally large and irregular shape, in which the cleavage plates have the appearance of the pages of a book. Also, BOOK.

mica capacitor *Electricity.* a fixed capacitor that uses thin sheets of mica as a dielectric between adjacent plates; used in stable and high-Q circuits.

micaceous *Mineralogy.* relating to, consisting of, or containing mica.

micaceous arkose *Petrology.* a sandstone composed of 25–90% feldspars and feldspathic crystalline rock fragments, 10–50% micas and micaceous metamorphic rock fragments, and up to 65% chert, quartz, and metamorphic quartzite.

mica schist *Petrology.* a schist consisting principally of mica and quartz, with a characteristic foliation due to the parallel arrangement of the mica flakes.

micelle *Molecular Biology.* a spherical arrangement formed by a group of lipid molecules in an aqueous environment.

Michael addition *Organic Chemistry.* a synthetic procedure in which carbanions are added to α,β-unsaturated carbonyl compounds. Also, **Michael reaction.**

Michaelis, Leonor 1875–1945, German-born American chemist; best known for his work on enzyme-catalyzed reactions.

Michaelis constant *Biochemistry.* a constant in Michaelis kinetics that represents the substrate concentration at which the velocity of an enzyme-catalyzed reaction is half the maximal velocity.

Michaelis-Menten equation *Enzymology.* the rate equation $v = V[S]/K_m + [S]$, where v is the initial velocity of the reaction, V is the maximum velocity, [S] is the concentration of the substrate, and K_m is the Michaelis constant.

Michelinia *Paleontology.* a genus of colonial corals in the extinct order Tabulata and suborder Favositina; tabulae incomplete; extant in the Lower Devonian to Upper Permian.

Michel parameter *Particle Physics.* a number, equal to 3/4 in any two-component neutrino theory before radiative corrections are taken into account, that represents the muon-decay-momentum spectrum in an equation.

Michelson, Albert A. 1852–1931, American physicist; awarded the Nobel Prize for designing the interferometer to measure the speed of light.

Albert A. Michelson

Michelson interferometer *Optics.* an interferometer in which a white-light source, a beam splitter, a compensator plate with a thickness equal to that of the beam splitter, and a pair of mirrors split and recombine beams to form an interference pattern, especially in the Michelson and Morley experiment.

Michelson-Morley experiment *Optics.* the experiment, first performed satisfactorily in 1887, that attempted to measure the speed, relative to the earth, of the "ether wind" through which light was then believed to propagate; the failure to find evidence of such a wind was important to the development of the special theory of relativity.

Michelson stellar interferometer *Optics.* an instrument that employs a system of mirrors to direct two parallel beams of light from an astronomical object into a telescope, in order to measure the distance at which interference fringes are visible; used to determine the angular separation of the components of double stars.

Michigan, Lake *Geography.* the third largest of the Great Lakes (area: 22,400 square miles), between the states of Michigan and Wisconsin.

Michler's ketone see TETRAMETHYLDIAMINOBENZOPHENONE.

miconazole nitrate *Organic Chemistry.* $C_{18}H_{15}Cl_4N_2O^+NO_3^-$ or $C_{18}H_{15}Cl_4N_3O_4$, white crystals that are very slightly soluble in water and melt at 182°C; used as a topical antifungal.

MICR magnetic-ink character recognition.

Micractiniaceae *Botany.* a family of freshwater, planktonic green algae of the order Chlorococcales, characterized by solitary or gregarious spherical cells with long bristles and a cup-shaped, pyrenoid-containing chloroplast.

Micraster *Paleontology.* an extinct genus of atelostomate echinoids in the superorder Microstomata and order Spatangoida; characterized by a heart shape and compact apical system; *Micraster* is one of the most thoroughly studied Mesozoic echinoids; Upper Cretaceous to Paleocene.

Micrasterias *Botany.* a genus of unicellular desmids belonging to the family Desmidiaceae, distinguished by an intricate lobing in the semicell.

micrencephaly *Medicine.* an abnormally small brain.

micrinite *Petrology.* an opaque, granular, cool particle with medium hardness and no plant-cell structure.

micrite *Petrology.* **1.** a semiopaque crystalline limestone matrix of chemically precipitated carbonate mud, featuring crystals less than 4 micrometers in diameter. **2.** a limestone composed chiefly of microcrystalline calcite with less than 1% allochems.

micro- a combining form meaning "small," "very small," or "minute." *Mathematics.* a prefix indicating 10^{-6}.

microabscess *Medicine.* a very small, localized pus accumulation.

microadenoma *Oncology.* a small adenoma of the anterior pituitary gland.

microaerobic *Microbiology.* **1.** of or related to an environment containing oxygen at a concentration lower than that present in the atmosphere. **2.** see MICROAEROPHILIC.

microaerophilic *Microbiology.* of or related to an organism requiring oxygen for growth but at a concentration lower than that present in the atmosphere. Also, MICROAEROBIC.

microaggregate *Hematology.* a microscopic collection of particles, such as platelets, leukocytes, and fibrin, that occurs in stored blood.

microalloy-diffused transistor *Electronics.* a microalloy transistor whose base region is diffused into the wafer prior to the emitter- and collector-electrode production by alloying.

microalloyed steel *Metallurgy.* a strong, wrought ferrous material, used especially to build merchant-ship hulls.

microalloy transistor *Electronics.* a transistor whose tiny emitter and collector electrodes are formed by alloying a thin film of impurity to opposite facing collector and emitter pits.

microammeter *Electricity.* a sensitive ammeter with a scale calibrated for measuring current values in microamperes.

microampere *Metrology.* a unit of current equal to one-millionth of an ampere.

microanalysis *Analytical Chemistry.* an analysis involving procedures that employ very small quantities of materials.

microanatomy *Anatomy.* the study of microscopic structures of the tissues; histology.

microaneurysm *Medicine.* a microscopic dilation of an artery, as may be observed in hemorrhagic diseases such as purpura.

microangstrom *Metrology.* a unit of length equal to one-millionth of an angstrom.

Microascaceae *Mycology.* the sole family of fungi belonging to the order Microascales, occurring on soil, dung, or plant and animal debris.

Microascales *Mycology.* an order of fungi belonging to the subdivision Ascomycotina; generally nonparasitic and found in dung and soil.

microautoradiograph *Optics.* an image of a small radioactive subject produced by placing the subject close to or in contact with a radiation-sensitive emulsion.

Microbacterium *Bacteriology.* a genus of Gram-positive, aerobic bacteria that occur as heat-resistant, rod-shaped cells and are frequently found in dairy products.

microbalance *Engineering.* a small balance sensitive to one microgram that can weigh loads up to 0.1 gram.

microbar *Metrology.* a unit of pressure equal to one-millionth of a bar.

microbarm *Geophysics.* any very small quasiperiodic change in atmospheric pressure.

microbarogram *Meteorology.* a record of minute fluctuations in atmospheric pressure.

microbarograph *Meteorology.* an aneroid barograph that records very small atmospheric pressure variations.

microbe *Microbiology.* **1.** broadly, any microorganism. **2.** specifically, any of those microorganisms, especially bacteria, protozoa, and fungi, that are capable of causing disease in humans and animals.

microbial corrosion *Materials Science.* any corrosion of metals that is accelerated or induced by microorganisms, especially bacteria and fungi.

microbial floc *Biotechnology.* the particles formed by the natural aggregation of bacterial, yeast, or microalgal cells; used in fermentation processes to make clarification of medium and separation and recycling of cells easier.

microbial genetics *Genetics.* the genetic study of microorganisms.

microbial insecticide *Microbiology.* a substance used selectively against insects and containing microorganisms that are pathogenic to those particular insects.

microbial leaching *Biotechnology.* the use of bacteria of the genus *Thiobacillus* to improve the extraction of valuable metals from low-grade ores.

microbicide *Toxicology.* a preparation for killing microbes.

microbiocoenosis *Microbiology.* a mixed collection of microorganisms that exist together in a natural community, characteristically undergoing short periods of stability.

microbiological assay *Nutrition.* the examination or analysis of a biological compound, such as a vitamin or an amino acid, that is based on measuring the growth of microorganisms for which the compound is an essential growth factor.

Microbiology

The biological sciences are subdivided by the kind of organism being studied (bacteriology, botany, mycology, zoology, virology), the scientific approach (ecology, genetics, immunology, molecular biology, physiology, systematics), or the applications (agriculture, industry, medicine). Microbiology is an "artificial" classification based on the microscopic size of the life forms, which include bacteria, microfungi, protozoa, unicellular algae, and viruses. However, even more than by the size of the subject matter, the study of this diverse group is distinguished by the methods used to study them, techniques first developed by Louis Pasteur and Robert Koch in the nineteenth century.

In lay language, microbes are often called "germs" or "bugs." This negative connotation reflects the fact that the best known microorganisms cause disease, putrefaction, and decay. Less well known are the so-called "beneficial" microorganisms involved in food production, recycling of nutrients in the ecosystem, and the production of economically useful drugs and other metabolites.

The tools and techniques of microbiology form the foundation for much of modern biology and medicine. Contemporary biologists classify their disciplines with a bewildering array of hybrid rubrics such as fungal genetics, protozoan systematics, clinical microbiology, industrial mycology, and bacterial ecology (to name only a few), which reflect the diversity of specialties which rely on microbiological expertise. Nowadays, many scientists who work with microbial systems do not call themselves microbiologists at all but choose a name like biotechnologist, biochemist, molecular biologist, infectious disease specialist, public health worker, or chemical engineer.

Basic research by microbiologists has led to enormous advances in immunology and genetic engineering. Microbes can be genetically altered and used as hosts for the production of a wide range of hormones and proteins of medicinal and industrial value. Diagnostic medicine has been revolutionized. Cloning techniques developed in bacteria are used for applications as broad as the dissection of the human genome and the analysis of DNA from long dead species.

The widespread use of either microbial systems or microbial techniques means that the exact circumscription of microbiology as a science is difficult. Simultaneously, this wide scope ensures that the study of the smallest living things is one of the greatest of the biological sciences.

Joan W. Bennett
Professor of Cell and Molecular Biology
Tulane University
Past President, American Society for Microbiology

microbiology *Biology.* the scientific study of organisms too small to be seen by the naked eye, including protozoans, algae, fungi, bacteria, viruses, and rickettsiae.

microbiophobia *Psychology.* an irrational fear of microbes or germs.

microbiota *Microbiology.* the microscopic organisms found in a given region.

Microbispora *Bacteriology.* a genus of Gram-positive bacteria of the order Actinomycetales that are found in the soil as slowly growing filaments, forming branched aerial and substrate mycelia.

microbistatic *Microbiology.* of or related to a substance or agent that inhibits at least certain types of microbial growth and reproduction.

microbody *Cell Biology.* a membrane-bound, cytoplasmic organelle that is found in eukaryotic cells and contains a number of oxidative enzymes, such as a peroxisome.

microbreccia *Geology.* a poorly sorted sandstone consisting of relatively large, angular sand particles set in a matrix of fine silt or clay.

microbrenner *Surgery.* a needle-pointed electric instrument used for cautery.

microburst *Meteorology.* a dangerous vertical gust with a core of about 1.5 miles in diameter, in which downward velocities of 20 miles a second can occur.

microcalorie *Metrology.* a unit of heat equal to one-millionth of a calorie.

microcanonical ensemble *Physics.* a conceptual collection of identical systems of particles, each having an energy that lies within a specified infinitesimal range.

microcapsule *Chemical Engineering.* a capsule ranging in size from 5 to 200 micrometers with walls made from thin coatings, such as proteins, polymers, and waxes, and enclosing materials such as solvents, plasticizers, acids, colorants, and medicines; characterized by a phase separation between the core and the wall.

microcardia *Medicine.* an abnormally small heart.

microcarrier *Biotechnology.* a small bead of silica, glass, or dextran; used as a growth site for the culture of anchorage-dependent cell lines.

microcarrier cell culture *Cell Biology.* a method for culturing large numbers of cells that grow attached to the surface of small plastic beads or microcarriers and are maintained in suspension by gentle agitation.

microcentrum see CENTROSOME.

microcephalus *Medicine.* a person whose head is abnormally small in proportion to the body.

microcephaly *Medicine.* an abnormally small head.

microceratous *Invertebrate Zoology.* possessing short antennae.

microcercous cercaria *Invertebrate Zoology.* a trematode larval stage with a short broad tail.

microchannel plate *Electronics.* an array of small, aligned, cylinder-shaped electron multipliers used for image intensification, as in telescopes or night-viewing binoculars.

microchemistry *Biochemistry.* the study of chemical reactions that deal with minute quantities of substances, using apparatus of small size.

microchip *Electronics.* see CHIP.

microchip shield *Materials.* a special type of gem diamond that shields microchips against large amounts of undesired and potentially destructive heat; conducts heat up to 50% more efficiently than natural diamond and is less susceptible to laser damage.

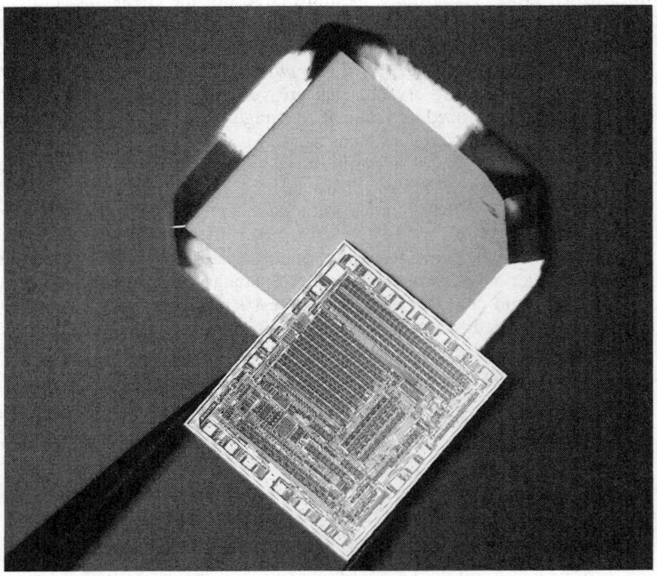

microchip shield

Microchiroptera *Vertebrate Zoology*. a suborder of small, usually insectivorous bats, ranging through all but the coldest faunal regions worldwide.

microchronometer *Horology*. a spring-driven clock capable of measuring time intervals as short as 0.03 second; used to time micromotion studies.

microcircuit *Electronics*. a minute electrical circuit on silicon; an integrated circuit.

microcircuitry *Electronics*. the formation of minute electrical circuits on silicon chips; integrated circuitry.

microcirculation *Physiology*. the flow of blood in the entire system of finer vessels, measuring 100 micrometers or less in diameter.

microcirculatory system *Anatomy*. that part of the circulatory system that can be seen only with a microscope, such as the capillaries, arterioles, and venules.

microclimate *Meteorology*. the climate of a small area, extending from the surface of the earth to a height at which the effects of the character of the underlying surface can no longer be distinguished from the general local climate.

microclimatology *Meteorology*. the study of a microclimate.

microcline *Mineralogy*. $KAlSi_3O_8$, a clear, white, pale-yellow, green, or red triclinic mineral of the feldspar group, dimorphous with orthoclase, occurring as short prismatic crystals and in massive form, and having a specific gravity of 2.55 to 2.63 and a hardness of 6 to 6.5 on the Mohs scale; found in plutonic rocks and crystalline schists.

microcneme *Invertebrate Zoology*. minute septa found in certain sea anemones.

Micrococcaceae *Bacteriology*. a family of Gram-positive, spheroid, nonmotile, aerobic or facultatively anaerobic bacteria, characterized by division in more than one plane during reproduction and by subsequent formation of clusters or packets.

micrococcal nuclease *Enzymology*. an endonuclease that hydrolyzes RNA or DNA to produce 3'-phosphorus and 3'-phosphooligonucleotides.

Micrococcus *Bacteriology*. a genus of Gram-positive bacteria of the family Micrococcaceae that grow as facultatively anaerobic cocci in soil and water and reproduce by binary fission.

microcode *Computer Programming*. **1.** a set of control instructions written to extend the capability of hard-wired machine microoperations. **2.** see COMPUTER CODE, def 2.

microcode store see CONTROL MEMORY.

Microcoleus *Bacteriology*. a genus of cyanobacteria that grow as filamentous cells in salt marshes, in mudflats, or on damp soil.

microcolony *Microbiology*. a small colony of bacterial cells.

microcomplement fixation *Microbiology*. a method of comparing the amino acid sequence of a given protein with that of the same protein isolated from a different source by comparing the complement-fixing ability of the two proteins; useful in the taxonomic classification of microorganisms.

microcomputer *Computer Technology*. a complete computer system containing CPU logic, memory for storing data and programs, input/output interfaces, and timing circuits to synchronize the flow of data; at first, all contained on a single board; the smallest and least costly class of computers.

microcomputer development system *Computer Technology*. a computer system designed to aid design, development, and test of microcomputer-based applications; the system includes debugging aids, peripheral interface test capabilities, text editors, hardware emulators, and software engineering tools.

microconstituent *Materials Science*. a microscopically small phase in an alloy.

microcontroller *Electronics*. a microcomputer or microprocessor used to control a process in data handling, communications, manufacturing, or other fields.

microcopy *Graphic Arts*. a microscopic copy of printed material, such as microfilm or microfiche.

microcoquina *Petrology*. a detrital limestone consisting entirely or primarily of sand-size cemented particles of shell detritus.

Microcoryphia *Invertebrate Zoology*. the bristletails, a primitive order of jumping insects with a flattened body and reduced mouthparts.

microcosm *Science*. a very small or very specific frame of reference, especially one that has the same characteristics as a larger entity of which it is a part. *Ecology*. a very small and localized ecosystem, such as a single pond. (From a Greek term meaning literally "small world.")

microcosmic salt see SODIUM AMMONIUM PHOSPHATE.

Microcotyloidea *Invertebrate Zoology*. a family of trematode flatworms parasitic on the gills of marine fishes.

microcoulomb *Electricity*. one millionth of a coulomb.

microcrack toughening *Materials Science*. a mechanism for increasing the toughness of ceramics by introducing small cracks.

microcrystal *Crystallography*. any solid substance whose crystalline components are visible only through a microscope.

microcrystalline *Crystallography*. of or relating to a microcrystal.

Microcyclus *Mycology*. a genus of fungi belonging to the order Dothideales.

Microcyprini *Vertebrate Zoology*. an archaic name for the marine fish order Cypriniformes.

microcyst *Medicine*. **1.** an extremely small cyst. **2.** in bacteriology, a type of resting cell derived from the vegetative cells of certain mycobacterales. **3.** an inactive cell found in certain slime molds.

Microcystis *Bacteriology*. a genus of phototrophic cyanobacteria occurring as spherical cells that contain gas vacuoles and grow as mucilaginous colonies in aquatic environments.

microcyte *Medicine*. an unusually small red blood cell of less than five micrometers in diameter.

microcythemia *Medicine*. the presence of abnormally small red blood cells.

microcytic anemia *Medicine*. a decrease in red blood cells, with those present being unusually small.

microcytosis *Medicine*. the presence of abnormally small red blood cells in the blood.

microdactyly *Medicine*. abnormally small fingers and toes.

microdensitometer *Spectroscopy*. an instrument used to measure the photographic density of spectral lines on a negative that are too faint to be seen with the naked eye.

microdermatome *Surgery*. a surgical device used to cut very thin sections of skin.

microdiagnostic program *Computer Technology*. a set of instructions for driving a microprocessor and detecting faults.

microdissection *Biology*. the technique of dissecting under a microscope, using fine mechanically manipulated instruments.

Microdomatacea *Paleontology*. an extinct superfamily of gastropods in the order Aspidobranchia; their shells were relatively high-spired and trochiform; Ordovician to Permian.

microduplex alloy *Materials Science*. an alloy consisting of a mixture of microconsituents that are distributed uniformly, so that boundaries are predominantly interphase.

microelectrode *Analytical Chemistry*. a minute electrode or one that is used in microelectrolysis. *Biotechnology*. specifically, a glass electrode with a metal tip whose diameter is a fraction of a millimeter; used to study the properties of cells.

microelectrolysis *Physical Chemistry*. the electrolysis of very small amounts of material.

microelectronic *Electronics*. of or relating to microelectronics.

microelectronics *Electronics*. the special methods and techniques used in producing miniature circuits. Also, **microminiaturization.**

microelectrophoresis *Analytical Chemistry*. an analytical technique that uses a microscope to observe and measure the velocity of ions or other charged particles immersed in a solution moving toward oppositely charged electrodes .

microelement *Electronics*. a small component, such as a capacitor or resistor, or a combination of components that is mounted on a wafer and used in microelectronic circuits.

microencapsulation *Chemical Engineering*. the coating of a material in capsules in the micrometer size range and made of glass, silica, or various high polymers or proteins, such as gelatin or albumin.

microendemic *Ecology*. describing a species that has a very limited geographical distribution.

microenvironment *Ecology*. the environmental conditions that exist in a small, localized habitat.

microevolution *Evolution*. **1.** the process of evolutionary change usually viewed over a short period of time, such as changes in gene frequencies, chromosome structure, or number. **2.** the evolutionary change resulting from gene mutation.

microfacies *Petrology*. the composition or characteristic features of a sedimentary rock seen only in a thin section under a microscope.

microfarad *Electricity*. one millionth of a farad.

microfauna *Zoology*. animals that are too small to be seen with the naked eye. *Ecology*. **1.** fauna limited to a localized habitat. **2.** the fauna of a microhabitat.

microfeature *Cartography.* any feature, natural or manmade, that can be identified on an aerial photograph but is too small to be included on a map of a given scale.

microfibril *Molecular Biology.* **1.** a very small fiber or filament that is often a part of a compound fiber. **2.** a basic structural unit of the walls of plant cells, consisting of cellulose, chitin, mannan, or xylan.

microfiche [mīk′rə fēsh] *Graphic Arts.* a card or flat sheet of microfilm bearing images, usually negative, of printed matter; used in libraries and office filing systems.

microfilament *Cell Biology.* an intracellular fiber found in eukaryotic cells, composed principally of the protein actin and concerned with cell structure and movement.

microfilaria *Invertebrate Zoology.* a minute larval stage of filarial nematodes.

microfilm *Graphic Arts.* a film bearing images, usually negative, of printed materials that have been greatly reduced in size.

microfiltration *Biotechnology.* a method of sterile filtration, clarification, or cell harvesting that removes particles in the 0.1–10.0 micrometer range.

microfissure *Metallurgy.* any microscopic crack.

microflake *Archaeology.* a tiny scar on the surface of a stone tool; this may indicate use of a specific type, such as for cutting or scraping.

microflora *Botany.* plants that are too small to be seen with the naked eye. *Ecology.* **1.** flora limited to a localized habitat. **2.** the flora of a microhabitat.

microfluid *Fluid Mechanics.* a fluid whose deformation depends in part on the local effects caused by particles in the fluid.

microfluoroscope *Engineering.* a fluoroscope that optically enlarges the very fine-grained fluorescent screen.

microforge *Engineering.* an optical-mechanical device that utilizes a micromanipulator to position needles or pipets in the field of a low-power microscope.

microform *Graphic Arts.* **1.** an arrangement of reduced printed material on microfilm or microfiche. **2.** a device that arranges reduced printed material for transfer to microfilm or microfiche.

microfossil *Paleontology.* a fossil whose typical form is microscopic in size; some of the more important groups are the Foraminiferida, Radiolaria, and Chitinozoa.

microgamete *Cell Biology.* a small, usually flagellated, motile male gamete.

microgametocyte *Biology.* a cell that gives rise to microgametes by division, such as *Plasmodium,* the causative agent of malaria.

microgametophyte *Botany.* the male gametophyte that develops from a microspore in plants with two types of spores.

microgamont *Cell Biology.* a cell that divides to produce microgametes.

microgamy *Biology.* the process of sexual reproduction by fusion of the microgametes with macrogametes.

microgastria *Medicine.* an abnormally small stomach.

microgauss *Metrology.* a unit of magnetic induction equal to one-millionth of a gauss.

microgenesis *Medicine.* the abnormally small development of a part.

microgenitalism *Medicine.* abnormally small external sex organs.

microgeography *Geography.* small-scale empirical geographic studies.

Microglaenaceae *Mycology.* a family of fungi belonging to the order Melanommatales that live as lichens on rocks, mosses, or old vegetation in regions of cold to cool temperatures.

microglia *Neurology.* small nonneural interstitial cells of various forms that are migratory and act as phagocytes to remove waste products of nerve tissue.

microgliocyte *Neurology.* the early cell that develops into a microglia cell, probably arising from the mesoderm.

microglioma *Oncology.* a tumor composed of microglial cells.

microgliomatosis *Oncology.* a condition that is characterized by the formation of tumors containing microglia. Also, RETICULUM CELL SARCOMA OF THE BRAIN.

microglobulin *Immunology.* a protein that is precipitated by half-saturated neutral ammonium sulfate (known as a globulin) and has a relatively low molecular weight (40,000 or below).

microglossia *Medicine.* an abnormally small tongue.

micrognathia *Medicine.* abnormally small jaws, referring to either the mandible or the maxilla or both.

microgram *Metrology.* a unit of mass equal to one-millionth of a gram. Also, **microgramme.**

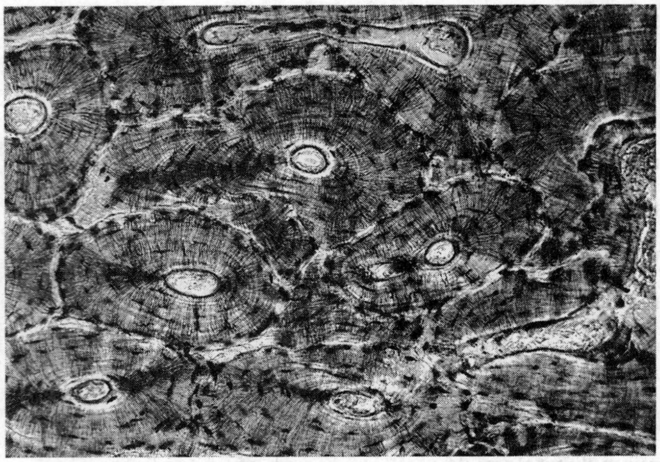

micrograph of a bone

micrograph *Optics.* a photograph or other such image of a minute object, organism, or specimen as seen through a microscope. *Graphic Arts.* a device that can execute extremely small writings or engravings.

micrographic *Science.* resembling small written characters. *Graphics.* of or relating to micrographics.

micrographics *Graphic Arts.* the technique of photographing material in minute form to produce microfilm and microfiche.

microgravity *Astronomy.* the extremely weak gravity that occurs on board a spacecraft in orbit around the earth.

microgyria *Medicine.* the abnormally small convolutions of the brain present in polymicrogyria.

microgyrus *Neurology.* an abnormally small convolution of the brain.

microhabitat *Ecology.* a small, specialized habitat that contains a limited range of plants and animals, often found in an isolated area.

microhardness *Metallurgy.* local hardness measured with a small and lightly loaded indenter, e.g., the hardness of a single constituent of a microstructure.

microhenry *Metrology.* a unit of electrical inductance equal to one-millionth of a henry.

microhm *Electricity.* a unit of current equivalent to one-millionth of an ohm.

Microhylidae *Vertebrate Zoology.* a family of frogs of the order Anura, ranging from squat, burrowing, small-headed, ant- and termite-eating toads to arboreal tree dwellers; noted for the free-swimming larvae with terminal mouths or for the direct development from terrestrial eggs to froglets; found worldwide.

microhysteresis *Solid-State Physics.* a hysteresis effect wherein the walls of the magnetic domains in a crystal lag behind a changing applied magnetic field or an elastic stress, due to the presence of defects and dislocations in the lattice structure.

microimage *Computer Technology.* an image stored on microfilm that must be magnified for visual analysis; the contents of one frame of microfilm.

microindentation hardness testing *Materials Science.* hardness tests using very small indenting forces; the resulting indentation is of microscopic size.

microinjection *Cell Biology.* a technique for the introduction of a variety of macromolecules into a cell via an ultrafine needle.

microinstruction *Computer Programming.* a low-level control mechanism stored in a control word in a CPU that causes a single data-register state transition; examples include "add," "shift," and "store." In a microcoded machine, one machine instruction is executed by a sequence of microinstructions.

microinterferometer see INTERFERENCE MICROSCOPE.

microinvasion *Medicine.* the infiltration of microscopic cancer cells into surrounding tissue from a malignant tumor.

microjustification *Computer Technology.* the capability of some word processing systems to vary the space between words or letters by small increments.

microlayer *Science.* any extremely thin layer. *Oceanography.* specifically, the thin layer just beneath the surface of the water, within which there is interaction between the water and the air.

Microlepidoptera *Invertebrate Zoology.* a former classification of small butterflies and moths, now used for descriptive purposes.

microlite *Mineralogy.* $(Ca,Na)_2Ta_2O_6(O,OH,F)$, a pale-yellow, reddish, brown, or green, weakly to moderately radioactive cubic mineral of the pyrochlore group, isomorphous with pyrochlore, and having a specific gravity of 4.3 to 6.4 and a hardness of 5 to 5.5 on the Mohs scale; found as octahedral crystals or irregular masses in pegmatites. The crystals are revealed with polarized light.

microliter *Metrology.* a unit of capacity equal to one-millionth of a liter. Also, **microlitre.**

microlith *Archaeology.* a thin blade or blade fragment with sharp cutting edges, usually geometric in shape and set into a wooden handle or shaft. *Mineralogy.* see MICROLITE.

microlithiasis *Medicine.* the generation of very small stones in an organ.

microlithiasis alveolaris pulmonum *Medicine.* a lung condition in which tiny concretions are formed in the alveoli, appearing like sand upon X-ray examination.

microlithology *Petrology.* the study of rock characteristics by means of a microscope.

microlitic *Petrology.* of or relating to a porphyritic igneous rock texture, characterized by a glassy groundmass consisting of an aggregate of differently oriented or parallel microlites.

microlock *Electronics.* **1.** a satellite telemetry system that employs phase-lock techniques in the receiving equipment to achieve high sensitivity. **2.** a lock upon a minitrack radio transmitter by a tracking station. **3.** the system by which this lock is effected.

microlog *Petroleum Engineering.* a drill-hole electric log utilizing electrodes mounted at short distances from one another in the face of a rubber-padded microresistivity sonde; used to determine the permeability of a selected formation.

Micromalthidae *Invertebrate Zoology.* the telephone pole beetles, a family of North American coleopteran insects in the superfamily Cantharoidea; found in rotting wood.

micromania *Psychology.* a delusion that one's body or some part of it has diminished in size; generally a symptom of an organic depression.

micromanipulation *Biology.* the techniques and practice of microdissection, microvivisection, microisolation, and microinjection.

micromanipulator *Engineering.* a device used to manipulate cells being viewed by a microscope.

micromere *Developmental Biology.* one of the small blastomeres that is formed as a result of the unequal cleavage of a fertilized ovum.

micrometastasis *Oncology.* the spread of cancer cells from the primary tumor to distant sites where they form microscopic secondary tumors.

micrometeorite *Astronomy.* a meteorite with a diameter of less than 0.1 millimeter.

micrometeorite penetration *Space Technology.* damage inflicted on the outer skin of a space vehicle by minute particles moving through space at high speeds.

micrometeoroid *Astronomy.* a meteoroid with a diameter of less than 0.1 millimeter.

micrometeorology *Meteorology.* the study of the smallest-scale physical and dynamic occurrences within the atmosphere.

micrometer [mī kräm´ə tər] *Engineering.* **1.** an instrument used to measure small distances or angles in a telescope or microscope. **2.** a caliper used to make precise measurements.

micrometer [mīk´rō mē´tər] *Metrology.* a unit of length equal to one-millionth of a meter; formerly called a micron.

micrometer drum sextant *Navigation.* a marine sextant that uses a micrometer drum attached to an endless tangent screw on the end of the index arm to produce a precise and easily determined reading.

micromethod *Microbiology.* any procedure using extremely small quantities of material.

micromicrocurie see PICOCURIE.

Micromonospora *Bacteriology.* a genus of Gram-positive bacteria of the order Actinomycetales that occur as strictly aerobic, filamentous cells in soil and water and reproduce by spore formation.

micromotion study *Industrial Engineering.* the process of recording events with a motion-picture camera at normal or faster than normal speed, often using a microchronometer to record elapsed time; used for analyzing short, rapid movements that occur too quickly to be observed normally.

micron [mīk´rän] *Metrology.* a former term for a unit of length equal to one-millionth of a meter; now replaced by micrometer.

micronekton *Zoology.* a classification of marine creatures, such as pelagic crustaceans, that are intermediate between free-swimming creatures (nekton) and feebler swimming creatures (plankton).

microneme *Invertebrate Zoology.* any of the numerous, elongate bodies found in the apical end of parasitic protozoans, which are thought to secrete material that helps the infective body of the protozoan penetrate a host cell.

Micronesia *Geography.* the small islands of the western Pacific lying north of the equator, including the Mariana Islands, the Caroline Islands, and the Marshall Islands.

micronize *Materials Science.* to reduce a material to a very fine powder, with the resulting particles being no larger than microscopic size. Thus, **micronization.**

micronucleus *Invertebrate Zoology.* the smaller, reproductive nucleus found in multinucleate protozoans.

micronutrient *Biochemistry.* any essential dietary nutrient, including vitamins and trace minerals, required by the body only in small quantities. *Agronomy.* any of various chemicals necessary in extremely small amounts for plant growth, such as manganese, iron, and zinc.

microoperation *Computer Programming.* an elementary computer operation executed on data stored in registers, usually performed during one clock pulse.

microorganism *Microbiology.* a general term for any microscopic organism, including algae, bacteria, fungi, protozoa, and viruses.

micropaleontology *Paleontology.* the study of microscopic fossils or of certain features of larger fossils that lend themselves to microscopic study.

micropegmatite *Petrology.* a micrographic holocrystalline rock or portion of a rock in which quartz and feldspar penetrate each other.

Micropeltidaceae *Mycology.* a family of fungi belonging to the order Pleosporales that live on decaying organic material or as lichens on the surfaces of plants.

microperf *Computer Technology.* a type of continuous form printer paper whose page separations and sprocket-hole strips along the edge are finely perforated so that, when separated and trimmed, each page looks like a single sheet of typing paper.

microperoxidase *Biochemistry.* the part of a cytochrome molecule that retains its heme group and has peroxidase activity.

microperthite *Mineralogy.* a variety of alkali feldspar consisting of microscopic intergrowths of albite exsolution lamellae in orthoclase or microcline; common in gneisses.

Micropezidae *Invertebrate Zoology.* a family of dipteran flies, the stilt-legged flies in the subsection Acalyptratae.

microphage *Histology.* any small phagocyte, especially a neutrophil.

microphagy *Biology.* feeding on minute organisms or particles of food that are small in relation to their body size.

microphakia *Medicine.* an abnormally small crystalline lens of the eye.

microphobia *Psychology.* an irrational fear of small objects.

microphone *Acoustical Engineering.* an electronic device that serves to convert airborne sound waves into electromagnetic waves for the purpose of amplifying, recording, or broadcasting the sound.

microphone transducer *Acoustical Engineering.* the element within a microphone that converts some form of mechanical energy, such as the vibrations of a diaphragm, into equivalent electromagnetic energy.

microphonics *Electronics.* the noise that occurs when elements in an electron tube, component, or system vibrate.

microphotograph *Graphic Arts.* a photographic micrograph. Also, PHOTOMICROGRAPH.

microphotometer *Engineering.* a photometer capable of measuring light emitted, transmitted, or reflected by minute objects.

microphthalmus *Medicine.* an abnormally small eye that may have other defects such as retinal scarring, lens opacity, or choroid defects.

microphyllous *Botany.* having small or reduced leaves.

microphyric *Petrology.* of or relating to a porphyritic rock texture characterized by phenocrysts with a maximum dimension of 0.2 millimeter. Also, **microporphyritic.**

Microphysidae *Invertebrate Zoology.* a family of minute predatory true bugs, hemipteran insects in the subfamily Cimicimorpha.

microphyte *Ecology.* **1.** a very small or microscopic plant. **2.** a plant that has been dwarfed by poor environmental conditions, such as spruce trees on the tundra. Thus, **microphytic.**

micropinocytosis *Cell Biology.* a process by which extracellular fluids or solutes are taken up into the cytoplasm of a cell via tiny membrane vesicles, which are derived from the plasma membrane and are visible only under the electron microscope.

micropinosome *Cell Biology.* a tiny vesicle containing extracellular fluid that has budded off the plasma membrane into the cytoplasm of a cell during micropinocytosis.

micropipet *Biotechnology.* a fine-pointed pipet used for microinjections or to measure small volumes of liquids (0.5 ml or less) with a high degree of accuracy.

microplasmodesmata *Microbiology.* tiny channels or pores found in certain filamentous procaryotes, such as the hyphae of certain actinomycetes.

micropoikilitic *Petrology.* of or relating to an igneous rock with a poikilitic texture which is visible only under a microscope.

Micropolyspora *Bacteriology.* a genus of Gram-positive bacteria of the order Actinomycetales, forming aerial or substrate mycelia and occurring in soil or compost.

micropore *Materials Science.* a pore of the smallest classifiable size, below 2 nanometers. *Geology.* a pore that is small enough to hold water against the pull of gravity in addition to inhibiting water flow.

microporosity *Materials Science.* the porosity occurring on a microscopic scale.

microporous barrier *Chemical Engineering.* a membrane, used in membrane-separation processes, having pore sizes in the 20 to 100 angstrom range.

microprobe *Spectroscopy.* a device used to irradiate a sample by focusing a beam of electrons on a very small area in order to spectroscopically analyze the resulting emitted X-rays. *Surgery.* specifically, a minute probe, such as one used in microsurgery for exploration.

microprobe spectrometry *Spectroscopy.* a technique for spectroscopic analysis of a sample using a microprobe.

microprocessing unit *Electronics.* a housing that contains a microprocessor chip along with its memory, input/output interface, buffer, clock, and driver circuits.

microprocessor *Computer Technology.* the central processing unit of a microcomputer, usually manufactured on a single semiconductor chip containing anywhere from a few thousand to several million or more logical elements.

microprogram *Computer Programming.* a set of microinstructions implemented in special circuits that directly control the sequencing of computer circuits; one microprogram corresponds to each type of basic operation.

microprogrammable instruction *Computer Programming.* an instruction that requires no access to main memory so that all operations, such as shift, skip, or input/output commands, can be written in a single microinstruction.

microprogrammed control unit *Computer Technology.* a control unit whose microoperation steps are stored in microcode memory.

microprogramming *Computer Programming.* the process of designing and implementing the control functions of a computer in microcode for the interpretation and execution of machine instructions.

microprolactinoma *Oncology.* a small pituitary adenoma that secretes prolactin.

micropsia *Medicine.* a delusional condition in which the patient perceives all objects as being much smaller than they really are. Also, LILLIPUTIAN HALLUCINATION.

Micropterigidae *Invertebrate Zoology.* a family of tiny diurnal, pollen-eating moths, lepidopteran insects in the superfamily Micropterygoidea, having no proboscis.

micropulsation *Geophysics.* a small geomagnetic variation with a frequency of less than one second to several seconds.

micropycnometer *Engineering.* a pycnometer whose capacity ranges from 0.25 milliliter to 1.6 milliliters and whose weighing precision is within 1/10,000.

Micropygidae *Invertebrate Zoology.* a rare family of echinoids, echinoderms in the order Diadematoida; found in the Indo-West Pacific.

micropyle *Botany.* a small opening in the integument of an ovule through which the pollen tube usually enters; it is carried over in the seed as a point or scar found between the hilum and the point of radicle. *Invertebrate Zoology.* **1.** a tiny opening in the protective covering of an insect's egg that allows a sperm to enter. **2.** the area of a sponge gemmule not covered by a membrane, through which the gemmule germinates.

microradiogram *Physics.* an image produced by a certain type of X-ray microscope that maps several levels of a crystal specimen into a single focal plane.

microradiograph *Optics.* an image that is produced on an emulsion by X-rays and shows a detailed image of a very small object or area.

microradiography *Analytical Chemistry.* the use of monochromatic radiation to produce a photographic image of an object whose detail is too fine to be viewed by the unaided eye; this image is then examined with a microscope or enlarged through projection; the technique is used to study the surfaces of solids.

microradiometer *Electronics.* a sensitive detector of heat and infrared radiation in which a thermopile is supported on and connected directly to the moving coil of a galvanometer.

microrefractometry *Optics.* the measurement of refraction through the use of a microscope.

microrelief *Geography.* minor variations in elevation or topography.

microresistivity survey *Petroleum Engineering.* any downhole resistivity survey of an oil-bearing formation.

microrheology *Mechanics.* rheology in which the heterogeneous properties of the medium on the microscale are considered.

Microsauria *Paleontology.* a large order of amphibian reptiles in the extinct subclass Lepospondyli; some microsaurs were primarily aquatic and some terrestrial; some genera are similar to captorhinomorph cotylosaurs, but probably only superficially, as a result of adaptive convergence.

microsclere *Invertebrate Zoology.* a minute spicule in some sponges.

microscope *Optics.* any of a variety of instruments that are designed to visually enlarge objects that are too small be seen by the naked eye, or too small to be seen in detail; the major types include optical, acoustic, electron, and X-ray microscopes.

microscope stage *Optics.* the part of a microscope upon which the subject is placed for viewing.

microscopic *Optics.* of or relating to an object that is too small to be visible without the use of a microscope.

microscopical diagnosis *Pathology.* the microscopic analysis of a specimen taken from a patient.

microscopic anisotropy *Petroleum Engineering.* in electrical downhole logging, a condition under which electric current flows most readily along water-filled interstices, commonly parallel to sedimentary bed strata.

microscopic reversibility *Physics.* a principle stating that for a system in equilibrium, all molecular processes occur at the same rate as their reverse processes. Also, REVERSIBILITY PRINCIPLE.

microscopic state *Physics.* the state of a system as specified in detail (limited by the Heisenberg uncertainty principle) wherein the properties of each constituent are known. Also, **microstate.**

microscopic stress *Metallurgy.* a residual stress that fluctuates on a microscopic scale.

microscopic theory *Physics.* a theory surrounding processes of atoms and molecules occurring over distances on the order of 10^{-10} meters or less.

microscopist *Science.* a person who is skilled in the use of a microscope.

Microscopium *Astronomy.* the Microscope, a small and dim constellation of the southern celestial hemisphere that lies south of Capricornus.

microscopy [mī kräs´kə pē; mīk´rə skōp´ē] *Optics.* the technology involving the use and applications of microscopes.

microsecond *Metrology.* a unit of time equal to one-millionth of a second.

microsegregation *Materials Science.* the presence of concentration differences in a material over extremely short distances, generally due to nonequilibrium solidification.

microseism *Geophysics.* a small seismic wave unrelated to earthquakes and having a period of less than 10 seconds.

microseismic instrument *Mining Engineering.* a device used to monitor roof strata and supports for subaudible vibrations that precede rock failure.

microseptum *Invertebrate Zoology.* an incomplete or imperfect mesentery in anthozoans.

microsere *Ecology.* an ecological succession that occurs within a small, localized habitat and that often fails to reach a climax.

Microsoft Disk Operating System see MS-DOS.

microsomal enzyme *Enzymology.* one of the enzymes associated with the microsome fraction of ultracentrifuged homogenized cells, consisting of ribosomes and pieces of rough endoplasmic reticulum.

microsome *Cell Biology.* a small vesicle derived from membrane fragments of the endoplasmic reticulum and obtained by cell homogenization.

microspectrograph *Spectroscopy.* a microspectroscope in which the resulting spectrum is recorded on photographic film.

microspectroscope *Spectroscopy.* a compound microscope to which a spectroscope has been attached in order to study spectral phenomena related to microscopic objects.

microspherulitic *Petrology.* of or relating to an igneous rock with a spherulitic texture seen only under a microscope.

microspike *Cell Biology.* one of many thin protrusions that form continually on the surface of a migrating cell; thought to function as environmental sensing devices, they can adhere to the substratum or retract into the cell.

Microspora *Invertebrate Zoology.* a phylum of sporozoan protozoans that parasitize other animals.

Microsporaceae *Botany.* a monogeneric family of free-floating, freshwater green algae of the order Ulotrichineae.

Microsporales *Botany.* an order of green algae composed of the monogeneric family Microsporaceae and belonging to the class Chlorophyceae; they are characterized by an unbranched thallus with cell walls that are composed of H-shaped overlapping pieces, the abundant storage of starch, and the positioning of the nucleus in a cytoplasmic bridge across the middle of the central vacuole.

microsporangium *Botany.* the structure that produces microspores in heterosporous plants.

microspore *Botany.* the smaller of the two types of spores produced in heterosporous plants that develops into the male gametophyte.

Microsporida *Invertebrate Zoology.* a phylum containing intracellular protozoan parasites commonly known as microsporidia.

Microsporidae see SPHAERIIDAE.

Microsporidia *Invertebrate Zoology.* an order of protozoans in the class Cnidosporidia, parasites within the cells of fish and arthropods, including silkworms and bees.

microsporidiosis *Veterinary Medicine.* ringworm caused by a species of *Microsporum.*

microsporocyte *Botany.* the diploid cell that, after meiosis, produces four microspores. Also, MOTHER CELL.

microsporophyll *Botany.* a small leaf or leaflike structure bearing microsporangia.

microstress *Materials Science.* stress in the microscopic structure of a material.

microstrip *Electromagnetism.* a thin, striplike transmission line used for transmitting microwave frequencies; typically mounted on a flat dielectric substrate that is mounted on a ground plane.

microstructure *Science.* the microscopic structure of a material or an organism.

microstylolite *Petrology.* a stylolite characterized by surface relief of less than 1 millimeter, indicating differential solution between two mineral grains.

microsurgery *Surgery.* any of various surgical procedures using small specialized instruments and performed under a microscope or other magnifying device.

microswitch *Robotics.* a touch sensor switch that detects the presence or absence of an object.

Microsyopidae *Paleontology.* a proposed family of primates in the extinct suborder Plesiadapiformes and superfamily Paramomyoidea; the genus *Microsyops* was formerly classified as a dermopteran, with Mixodectes and others; extant in the Paleocene and Eocene.

microtektite *Geology.* a microscopic tektite that is found in polar ice and ocean sediments.

Microtetraspora *Bacteriology.* a genus of actinomycete bacteria that form substrate mycelia, occur in soil, and reproduce by spore formation.

Microthamniaceae *Botany.* a monogeneric family of green algae of the order Ctenocladales, characterized by branched filaments, linear cells with a nucleus, and a long parietal laminate chloroplast; usually found attached to other algae, mosses, or aquatic angiosperms.

Microthamnion *Botany.* the genus comprising the green algae family Microthamniaceae, characterized by branched filaments, linear cells, and a thallus attached to the substrate by a bulbous basal cell.

Microtheciellaceae *Botany.* a monogeneric family of small, epiphytic mosses of the order Orthotrichales that form small mats, have prostrate stems with short, lateral branches, and a weak, single costa; found in Thailand and Laos.

microtherm *Ecology.* an organism that lives or thrives in a cold climate. Also, **microthermophyte.**

microthermal climate *Meteorology.* a cold climate with a temperature efficiency index of 32–63.

microthermic *Ecology.* living or thriving in a cold climate. Also, **microthermophilous.**

Microthrix parvicella *Bacteriology.* a species of bacteria that has been isolated as filamentous cells growing in activated sludge.

microthrowing *Physical Chemistry.* a process in which an electroplating solution deposits minute amounts of metal into a tiny crevice that is a few millimeters in dimension.

Microthyriaceae *Mycology.* a family of fungi belonging to the order Melanommatales that live on the surfaces of decaying plant material; found primarily in tropical regions.

microtia *Medicine.* severe hypoplasia or anaplasia of the pinna of the ear, with a blind or absent external auditory meatus.

Microtinae *Vertebrate Zoology.* a subfamily of Holarctic rodents including the lemmings, voles, and muskrats.

microtiter plate *Biotechnology.* a plastic plate that consists of 8 rows of 12 flat-bottom wells, each with a volume of 400 microliters; used for bioassays.

microtome *Engineering.* an instrument used to cut thin sections of specimens for microscopic examination.

microtomy *Biology.* the technique of cutting thin sections of specimens with a microtome for examination with the microscope.

microton *Nucleonics.* a machine that increases the energies of electrons to several megaelectron volts by propelling them in a circular motion with successive applications of radio-frequency voltages.

microtototherm *Ecology.* **1.** a plant that can tolerate relatively low temperatures. **2.** any organism living in a cold environment. Thus, **microtothermic.**

microtrabeculae *Cell Biology.* a network of protein filaments found in the cytoplasm of a cell, forming a cytoskeletal lattice that has been proposed to be loosely associated with most of the cytoplasmic proteins.

Microtragulus *Paleontology.* a genus of didelphoid marsupials in the extinct family Argyrolagidae; Microtragulus, which resembled a small kangaroo, is known only from the Upper Miocene and Lower Pleistocene of South America.

microtrichia *Invertebrate Zoology.* small hairs on the wings of insects.

microtubule *Cell Biology.* a type of filamentous protein polymer found in the cytoplasm of eukaryotic cells, composed of the protein tubulin, and involved in cell movement and cell structure.

microtubule-organizing center *Cell Biology.* a specialized region of a cell that nucleates the assembly of microtubules, often containing a pair of centrioles surrounded by pericentriolar material.

microvasculature *Anatomy.* the system of minute blood vessels that perfuse the body's tissues.

microvillus *Cell Biology.* a fingerlike projection found on the surface of certain animal cells, especially epithelial cells; they increase the cell surface area in large numbers and thus enhance the rate of absorption.

Microviridae *Virology.* a large family of icosahedral ssDNA-containing bacteriophages containing a single genus, *Microvirus.*

Microvirus *Virology.* ssDNA phages with isometric unenveloped particles and knoblike spikes; the sole genus of the family Microviridae.

microvitrain *Geology.* thin vitrainlike bands, ranging in thickness from around 0.05 mm to 2 mm, that occur in clarain.

microvoid *Materials Science.* a microscopic void that forms in a metallic material when a high stress causes separation of the metal at grain boundaries or at interfaces between the metal and inclusions. Also, INTERDENDRITIC VOID.

microvolt *Electricity.* one-millionth of a volt.

microvoltmeter *Electronics.* a device used to measure voltage in increments of one-millionth of a volt.

microvolts per meter *Electromagnetism.* a conventional method of expressing the intensity of a radio-transmitter signal, given by the intensity in microvolts measured at a specified point divided by the effective height of the antenna in meters.

microwatt *Metrology.* a unit of power equal to one millionth of a watt.

microwave *Electromagnetism.* electromagnetic radiation having a free-space wavelength between 0.3 and 30 centimeters, corresponding to frequencies of 1–100 gigahertz. *Engineering.* see MICROWAVE OVEN.

microwave acoustics *Acoustics.* the study and use of acoustic resonators that vibrate at microwave frequencies, such as a wavemeter that uses a closed resonant cavity rather than a lumped-circuit resonator to eliminate losses due to radiation and to minimize resistive losses.

microwave antenna *Electromagnetism.* an antenna that is designed to operate in the microwave region of the spectrum, typically consisting of an open-ended waveguide directed at a parabolic reflector.

microwave attenuator *Electromagnetism.* a waveguide device that is used to reduce the intensity of transmitted microwaves by means of absorbing a portion of the microwave energy in a lossy material.

microwave bridge *Electromagnetism.* a waveguide circuit for microwaves similar in function to an electrical bridge in determining impedances.
microwave circuit *Electromagnetism.* a circuit that is designed to operate at microwave frequencies; typically involves waveguides, attenuators, various junctions, and the like.
microwave communication *Telecommunications.* message transmission that employs highly directional microwave beams.
microwave detector *Electronics.* an electronic device that can locate radio waves in the frequency range of 1000 megacycles and upward; it reacts to the presence of microwaves by changing its resistance or some other parameter.
microwave device *Electronics.* any electronic device that can transmit or receive microwaves.
microwave early warning *Engineering.* a high-power long-range radar that gives high resolution and handles high traffic; used for early warning of missiles.
microwave filter *Electromagnetism.* a resonant cavity that is designed to pass certain frequencies and reject or absorb others; used in a transmission line or waveguide and operated at microwave frequencies.
microwave frequency *Physics.* a frequency lying within the range of approximately 10^9 to 10^{11} hertz or equivalently having a wavelength (in free space) on the order of 30 cm to 1 mm.
microwave heating *Electromagnetism.* the use of microwaves to heat an object, particularly food as in a microwave oven, by irradiating the object; heating is achieved by microwave energy absorption.
microwave impedance measurement *Engineering.* the parameters derived from the impedance at lower frequencies and the ratios of electric or magnetic field amplitudes of microwave transmissions.
microwave integrated circuit *Electronics.* an integrated circuit designed for use in microwave frequencies. Also, MIC.
microwave landing system *Navigation.* a precision electronic airport approach and landing system that operates in the microwave spectrum, consisting of an azimuth station, elevation station, and precision distance measuring equipment.
microwave maser *Physics.* a maser whose output is coherent electromagnetic radiation in the microwave region.
microwave network *Telecommunications.* a network of microwave repeaters that relay messages among one another over considerable distances by means of highly directional microwave beams.
microwave noise standard *Engineering.* an electrical noise generator that can produce measurable noise intensities at microwave frequencies; used to calibrate other noise sources.
microwave optics *Electromagnetism.* the study and application of principles common to both microwave and optical phenomena.
microwave oven *Engineering.* an electrically operated oven utilizing high-frequency electromagnetic waves to vibrate molecules of food, thus generating the heat required to warm or cook the food in a shorter period of time than conventional electric or gas ovens.
microwave position-fixing *Navigation.* the determination of a craft's position by bearings to microwave transmitter stations.
microwave pumping *Electromagnetism.* in a maser, the irradiation of a stimulation medium by microwave energy of an appropriate frequency so as to invert the population (to cause the number of excited atoms in the medium to exceed the number of unexcited atoms).
microwave reflectometer *Electromagnetism.* a device used to measure the forward and reflected power from a discontinuity in a waveguide; consists of two coupling detectors mounted on either side of the waveguide.
microwave repeater *Telecommunications.* a repeater station or tower in a microwave relay system. The repeaters, usually spaced 20 to 35 miles apart, pick up, amplify, and pass on the signals sent over a microwave network. Also, **microwave relay.**
microwave solid-state device *Electronics.* a semiconductor device used to generate or amplify electromagnetic energy at microwave frequencies.
microwave spectrometer *Spectroscopy.* a device used to measure and record the intensity of microwave radiation that is absorbed or emitted by a sample as a result of rotational transitions.
microwave spectroscope *Spectroscopy.* a device used to study the intensity of microwave radiation that is either absorbed or emitted by a sample.
microwave spectrum *Electromagnetism.* the range of the electromagnetic spectrum, from about 1 to 100 gigahertz, found between radio waves and infrared waves.
microwave thermography *Medicine.* the measurement of temperature by detection of microwave radiation emitted by heated tissue.
microwave transmission line *Electromagnetism.* a transmission line that is capable of transmitting microwave frequencies; commonly in the form of a microstrip.
microwave tube *Electronics.* an electron tube, such as a klystron or magnetron, that generates wavelengths in the 30 centimeters to 0.3 centimeter range.
microwave wavemeter *Electromagnetism.* a device, usually a resonant cavity of variable dimensions, that when tuned to resonance can measure the free-space wavelength of a microwave signal.
microworld *Artificial Intelligence.* a small, restricted, artificial model of the world used for the initial development or testing of artificial intelligence techniques or programs.
microzoon *plural,* **microzoa.** *Zoology.* a microscopic animal, especially a protozoan.
microzoospermia *Medicine.* a condition in which the live sperm found in ejaculated semen are smaller than normal.
micrurgy *Science.* the use of microscopic instruments in a magnified field.
mictic *Biology.* **1.** requiring or produced by sexual reproduction. **2.** of or relating to eggs that without fertilization develop into males and with fertilization develop into amictic females, as occurs in rotifers.
micturate *Physiology.* to pass or release urine; urinate.
micturation *Physiology.* the act of passing urine; urination.
M.I.D. minimum infective dose.
mid- a combining form meaning "center" or "between," as in *midaxilla, midcarpal.*
mid. middle.
Midas *Space Technology.* a system used to track the trajectories of two objects at the same time, usually employing two Cotar antenna systems and two receivers at each station.
MIDAS or **Midas** missile defense alarm system.
mid-Atlantic ridge *Geology.* the midoceanic ridge extending in a generally north-south direction through the Atlantic Ocean.
midaxillary line *Anatomy.* a line that runs vertically downward from a point situated midway between the dorsal and ventral boundaries of the axilla.
midbody *Cell Biology.* a dense structure at the center of the thin intercellular bridge connecting two daughter cells during the final phases of cytokinesis. The midbody contains densely packed microtubules embedded in a dense matrix material.
midbrain see MESENCEPHALON.
midclavicular line *Anatomy.* a line that runs vertically downward from the midpoint of the clavicle.
midcourse *Space Technology.* the part of a space vehicle's course between its launching site and objective, especially within the transfer trajectory between its initial earth orbit and its first objective in space.
midcourse correction *Space Technology.* an alteration in the speed or direction of a space vehicle made after launch and before reentry.
midcourse guidance *Space Technology.* the guidance of a rocket between the end of the launching phase and some discretionary point at which terminal guidance begins.
midden *Archaeology.* a large refuse heap that is associated with intense human occupation, containing such materials as discarded artifacts, food remains, charcoal, bones, and so on.
Middle Ages the period in European history roughly from the year 450 or 500 to about the year 1450 or 1500, regarded as an intermediate period between the earlier Classical era and the modern world in terms of scientific and cultural development.
Middle America see MESOAMERICA.
middle body *Naval Architecture.* the portion of a ship's hull close to the midship section.
Middlebrook 7H-9 broth *Microbiology.* a bacteriological culture medium used for the growth of species of Mycobacterium; contains either Tween 80 or glycerol.
Middle Cambrian *Geology.* the geologic epoch occurring between the Upper and Lower Cambrian, beginning approximately 540 million years ago.
middle clouds *Meteorology.* a class of clouds that occur at altitudes between 6000 and 20,000 feet above sea level, including the altostratus, altocumulus, and nimbostratus species.
middle core *Computer Technology.* the locations in the internal storage of a computer that contain medium addresses; this area is usually used for application-program workspace.

Middle Cretaceous *Geology.* the geologic epoch occurring between the Upper and Lower Cretaceous, beginning approximately 120 million years ago.
Middle Devonian *Geology.* the geologic epoch occurring between the Upper and Lower Devonian, beginning approximately 385 million years ago.
middle ear *Anatomy.* the portion of the ear between the tympanic membrane and the cochlea, containing the tympanic cavity and its bones, the malleus, incus, and stapes.
middle ground *Navigation.* a shoal with a navigable channel on both sides.
middle ground buoy *Navigation.* any buoy that marks off a middle ground.
Middle Jurassic *Geology.* the geologic epoch occurring between the Upper and Lower Jurassic, beginning approximately 170 million years ago.
middle lamella *Cell Biology.* an intercellular layer between adjacent plant cells.
middle latitude *Geography.* the latitude of a point that is midway between two parallels on the same side of the equator. Also, MEAN LATITUDE.
middle-latitude sailing or **mid-latitude sailing** *Navigation.* the determination of elapsed or required course and distance by computation using difference of longitude for the mid-latitude of the course.
middle lobe syndrome *Medicine.* a collapse of the right middle lobe of the lung, with prolonged infection. Also, BROCK'S SYNDROME.
middle marker *Navigation.* a marker beacon along the glide slope of an Instrument Landing System; normally located at or near the point of the decision height.
Middle Mississippian *Geology.* the geologic epoch occurring between the Upper and Lower Mississippian.
middle oil *Organic Chemistry.* the fraction of oil distilled from coal tar at 200–250°C, yielding primarily naphthalene. Also, CARBOLIC OIL.
Middle Ordovician *Geology.* the geologic epoch occurring between the Upper and Lower Ordovician, beginning approximately 460 million years ago.
Middle Paleolithic *Archaeology.* the intermediate part of the Paleolithic period, from about 100,000 years ago to about 35,000 years ago; characterized by the development of a variety of stone tools and the first symbolic use of artifacts and sites.
Middle Pennsylvanian *Geology.* the geologic epoch between the Upper and Lower Pennsylvanian.
Middle Permian *Geology.* the geologic epoch occurring between the Upper and Lower Permian, beginning approximately 260 million years ago.
middle plate see NEPHROTOME.
middle rail *Building Engineering.* a rail between the top and bottom rails in a door or panel, or just above the bottom rail when a top rail is not used.
Middle Silurian *Geology.* the geologic epoch occurring between the Upper and Lower Silurian.
middle-third rule *Civil Engineering.* a rule that allows the computation of the effect of all forces, including weight, on a center portion of a wall or arch, equal to 1/3 of the total thickness, to determine an absence of tensile stress; used in the design and load calculations of unreinforced masonry, brick, or concrete walls.
middletone or **middle tone** *Graphic Arts.* any neutral tone, intermediate between a shadow and a highlight.
Middle Triassic *Geology.* the geologic epoch occurring between the the Upper and Lower Triassic, beginning approximately 215 million years ago.
middle-ultraviolet lamp *Electronics.* a mercury-vapor lamp, such as a sunlamp, that produces radiation in the wavelength band of 2800–3200 angstrom units.
middling *Mining Engineering.* an ore product that is intermediate between a concentrate and a tailing.
mid-extreme tide *Oceanography.* a reference level halfway between extreme high water and extreme low water at a particular location.
midfan *Geology.* the area of an alluvial fan between the head and the outer, lower margins.
mid-frequency gain *Electronics.* the gain of an amplifier at the frequency corresponding to the maximum of its frequency response curve.
midge *Invertebrate Zoology.* the common name for the tiny dipteran flies of the family Chironomidae, including many that bite and some carriers of parasites.
midget *Medicine.* a person who is completely normal in development but of extremely small stature. Also, NORMAL DWARF.
midget impinger *Mining Engineering.* a dust-sampling device that is identical in principle and design to a standard impinger but smaller, requiring only a 12-inch head of water for operation.
midgut *Developmental Biology.* the portion of the embryonic digestive tube into which the yolk sac opens; located between the foregut and the hindgut.
MIDI *Computer Programming.* software that enables synthesizers and samplers to communicate; also provides recording and editing functions, voice libraries, transcription features that print music in standard notation, and voice editors to help create voice sounds. (An acronym for Musical Instrument Digital Interface.)
MIDI data *Computer Programming.* the digitized representations of musical performance parameters such as tone, duration, pitch, and loudness.
mid-latitude *Cartography.* the single parallel on a Mercator map projection that conforms to the scale indicated on the map.
mid-latitude jet *Meteorology.* a westerly jet-stream wind found in the mid-latitudes of the Northern Hemisphere; usually the predominant factor influencing the weather.
mid-latitudes *Geography.* the zones of marked but moderate seasonal change on either side of the equator; variously measured between 35° and 55° or between 23-1/2° and 66-1/2°; the latter pair is coterminous with the temperate zones.
mid-life crisis *Psychology.* a popular term for a period of emotional stress in middle age, typically involving a heightened awareness of one's own mortality and a reassessment or redirection of goals and lifestyle.
midline *Anatomy.* an imaginary line that divides the body into right and left halves, passing approximately through the sagittal suture of the skull.
midline incision *Surgery.* a vertical incision in the midline of the anterior abdominal wall.
midnight sun *Astronomy.* a term for the sun when it is visible for 24 hours in the arctic and antarctic regions during their summers.
mid-ocean canyon see DEEP-SEA CHANNEL.
mid-oceanic ridge *Geology.* a continuous, seismic mountain range extending along the floor of the world's major oceans; characterized by a central rift valley and generally rugged topography; believed to be the source of new crustal material.
mid-ocean rift see RIFT VALLEY, def. 2.
midpoint *Mathematics.* the point on a line segment that divides the segment into two parts of equal length.
mid-range materiel requirements *Military Science.* materiel required as soon as possible to meet current operational or administrative needs or to meet concepts to be implemented within the next five years.
midrib *Botany.* the main or central vein of a leaf.
midsagittal *Anatomy.* **1.** relating to or located along the midsagittal line of the body. **2.** relating to or located in the midsagittal section of the brain.

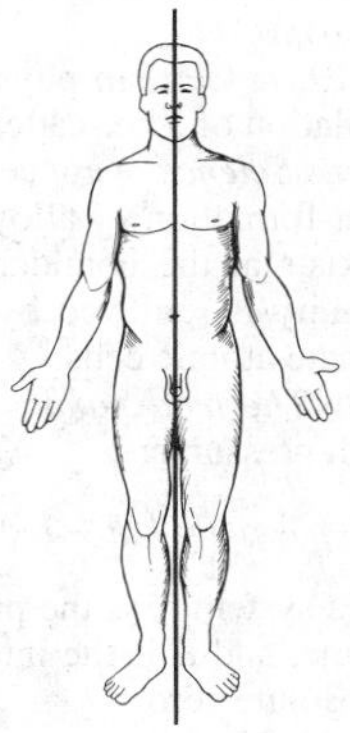

midsagittal line

midsagittal line *Anatomy.* the vertical plane through the midline that divides the body into right and left halves for purposes of description or classification. Also, **midsagittal plane.**

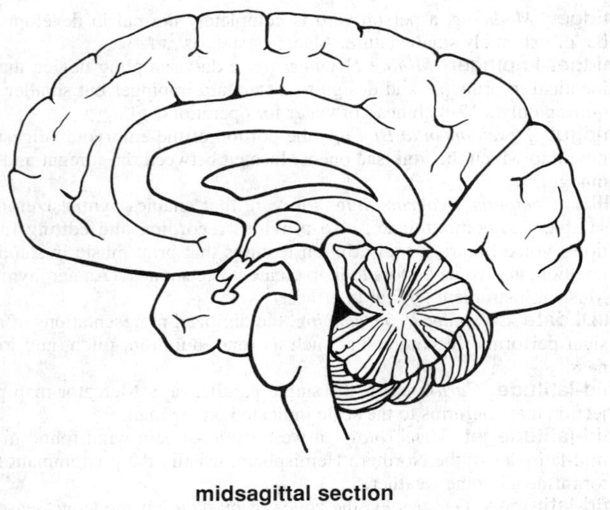

midsagittal section

midsagittal section *Anatomy.* a view or description of a part of the body, such as the brain, in which it is divided in half along the midsagittal line (the midline).

midship see MIDSHIPS.

midshipman *Military Science.* a naval cadet or junior naval officer. *Vertebrate Zoology.* a toad fish of the genus *Porichthys* of the family Batrachoidiformes that has numerous small luminous organs on its underside and produces a humming noise with its air bladder. (From the fact that these small organs were thought to resemble the row of buttons on the coat of a naval officer.) Also, SINGINGFISH.

midships *Navigation.* relating to or located in the middle part of a ship or aircraft; amidships. Also, MIDSHIP.

midship section *Naval Architecture.* **1.** the cross section of a ship's hull at the midship point. **2.** the widest part of a ship, including the midship frame and all adjacent frames of equal width.

mid-square generator *Computer Programming.* a computer procedure that generates random numbers by squaring a number (for example, an 8-digit decimal number squares to a 16-digit number) and then keeps the middle 8 digits to generate the next number.

midwater *Oceanography.* the ocean beneath the surface layers but above the benthic zone.

Midwayan *Geology.* a North American Gulf Coast stage of the Paleocene epoch, occurring after the Navarroan of the Cretaceous period and before the Sabinian.

midwife *Medicine.* a person who specializes in the practice of assisting in childbirth.

midwifery *Medicine.* the practice of assisting in the process of childbirth.

midwife toad *Vertebrate Zoology.* a small toad of the genus *Alytes* in the family Discoglossidae known for breeding behavior in which the male wraps egg strings around its hindlegs and cares for them until they hatch. Also, OBSTETRICAL TOAD.

Mie, Gustav [mē] 1868–1957, German physicist; developed the accurate electromagnetic calculation of Mie scattering.

Miedema model *Materials Science.* a concept that provides an empirical description of the heat formation of alloys: a discontinuity in electron density originally exists at the boundary between two dissimilar metals, and this discontinuity is removed by expanding or contracting the volume of the appropriate atomic cells.

Mie-Grüneisen equation *Thermodynamics.* an equation of the state of matter that is useful at high pressures:

$$V(p - p_o) = k(E - E_0),$$

where V is the volume of a system, p is the pressure, k is a quantity that depends only on the volume, and E is the internal energy; the subscript 0 refers to conditions at absolute zero.

miersite *Mineralogy.* (Cu,Ag)I, a canary-yellow cubic mineral occurring as tetrahedral or cubo-octahedral crystals, having a specific gravity of 5.64 and a hardness of 2.5 on the Mohs scale; found at Broken Hill, Australia.

Mie scattering *Optics.* the mathematical explanation for the scattering of light by homogeneous particles of any size.

Mies van der Rohe, Ludwig [mēz´van də rō´ə] 1886–1969, German-American architect; director of the Bauhaus School, designed the Seagram Building (New York City).

MIF melanocyte-stimulating hormone-inhibiting factor.

MIG *Aviation.* a designation for various Soviet jet fighter aircraft of the Korean War and after. Also, **Mig, MiG.** (An acronym from the names of its original designers, Artem Mikoyan and Mikhail Gurevich.)

MIG metal inert gas welding.

migma *Geology.* a mixture of solid and molten rock materials that is mobile or potentially mobile.

migmatite *Petrology.* a composite rock of crystalline texture, consisting of thin, alternating layers or lenses of metamorphic rock and granitic rock.

migmatization *Petrology.* the process by which migmatite is derived metamorphically.

Mignon delusion [min yōn´; mēn yōn´] *Psychology.* a false belief that one is secretly the child of famous or distinguished parents rather than of one's actual parents. (From the opera *Mignon*, by French composer Ambroise Thomas, 1811–1896.)

migraine [mī´grān] *Neurology.* a neural condition characterized by a severe recurrent vascular headache, usually on one side of the head and often accompanied by nausea, vomiting, and photophobia; sometimes preceded by sensory disturbances. Allergic reactions, excess carbohydrates or iodine in the diet, alcohol, bright lights, or loud noises may trigger attacks, which occur more frequently in women than in men. Also, HEMICRANIA, MEGRIM.

migrainoid *Neurology.* resembling a migraine headache.

migrant *Zoology.* a living organism that moves from one region to another.

migrate *Science.* to carry out a process of migration. *Zoology.* specifically, of an animal, to take part in a regular group movement from one location to another.

migration the act or process of moving from one region or place to another; specific uses include: *Zoology.* **1.** in general, any movement of an animal from one location to another. **2.** specifically, a predictable, recurring group movement that is characteristic of the members of a given species, and that occurs regularly in response to seasonal changes in temperature, precipitation, and so on. The process usually involves a round-trip movement between two areas, to seek a more suitable breeding place, a greater food and water supply, or other more favorable environmental conditions. *Anthropology.* a comparable seasonal movement from one site to another by human groups, as to avoid extreme weather conditions in winter or summer. *Geology.* the movement of a sand dune caused by a continuous transfer of sand from its windward to leeward side. Also, SHIFTING. *Hydrology.* the gradual shifting downstream of a meander system, including the enlargement of the curves and the widening of the course. *Petroleum Engineering.* the movement of liquid or gaseous hydrocarbons from their source through permeable formations or porous rock into reservoir rocks. *Physical Chemistry.* the movement of charged particles through an electrolytic medium under the influence of an electric field. *Solid-State Physics.* the movement of defects in a crystal due to thermal, mechanical, or electrical influences. *Computer Programming.* **1.** the movement of data between on-line and off-line storage areas depending on frequency of use and accessibility requirements. **2.** the shifting of procedures and functions from purely software to firmware, a combination of hardware and software. **3.** the off-loading of certain functions such as communications handling and database management from the host computer to smaller, special-purpose machines.

migration current *Physical Chemistry.* a current that is produced when positively charged ions (cations) migrate to the surface of an anode.

migration inhibition factor *Immunology.* a lymphokine that inhibits the migration of macrophages, possibly by increasing their adhesiveness. Also, **migration inhibitory factor**.

migration length *Nucleonics.* a distance that equals the square root of a neutron's migration area.

migratory *Zoology.* **1.** of or relating to animals that carry on a process of migration, such as salmon, bats, whales, caribou, and many species of birds. **2.** of or relating to behavior that occurs in connection with such movement. *Meteorology.* of or relating to a pressure system within the westerlies and moving in a general west-to-east direction.

migratory aptitude *Organic Chemistry.* the relative speeds of migration of two groups in a reaction as measured by the difference in the amount of product obtained.

migratory dune see WANDERING DUNE.

migratory restlessness *Behavior*. various incessant movements, such as hopping or flapping, characteristically made by birds in captivity as a manifestation of their readiness to migrate.

Mikulicz's disease [mik´yoo lich´ēz] *Medicine*. **1.** originally, a chronic swelling of the lacrimal and salivary glands without pain or complication. **2.** the presence of similar lacrimal and salivary swelling in association with diseases such as leukemia, lupus erythematosus, tuberculosis, lymphoma, sarcoidosis, and Sjögren's syndrome; in such contexts it is referred to as **Mikulicz's syndrome.** (Named for Johann von Mikulicz-Radecki, 1850–1905, Polish surgeon.)

mil *Metrology*. a unit of length equal to 0.001 inch, equivalent to 0.0254 millimeter. *Mathematics*. the arc subtended by a chord of length 1 on a circle of radius 1000 (in any units).

mil. military.

Milan *Ordnance*. a French-German wire-guided antitank missile with a range of 6650 feet.

milarite *Mineralogy*. $K_2Ca_4Al_2Be_4Si_{24}O_{60}{\cdot}H_2O$, a colorless to greenish, brittle, vitreous hexagonal mineral of the osumilite group occurring as prismatic crystals, having a specific gravity of 2.46 to 2.61 and a hardness of 5.5 to 6 on the Mohs scale; found in pegmatites.

mild abrasive *Materials Science*. an abrasive material, such as chalk, that has a hardness of 1 to 2 on the Mohs scale.

mildew *Plant Pathology*. a white growth on plant surfaces produced by parasitic fungi. *Mycology*. a common term for the parasitic fungi, generally of the families Erysiphaceae and Peronosporaceae, that produce this whitish substance. *Materials*. a similar growth produced by these fungi on products made from plant or animal tissue that are exposed to moisture, such as books, clothing, or leather goods.

mildewcide *Toxicology*. any of various products used to retard or prevent the growth of mildew; widely used compounds include benzoic acid, formaldehyde, cresols, phenols, and organic derivatives or salts of copper, zinc, and mercury.

mild mental retardation *Psychology*. a level of mental retardation characterized by intellectual abilities roughly equal to those of a child of eight and an IQ ranging from 55 to 69; those affected generally are capable of performing simple tasks and of developing reasonable social and communicative skills.

mild steel *Metallurgy*. any of several steels containing up to 0.25% carbon.

mile *Metrology*. **1.** a traditional unit of length that is equal to 5280 feet; equivalent to 1.609 kilometers. Also, STATUTE MILE. **2.** see NAUTICAL MILE. (From the Latin word for *thousand*; originally based on a reckoning of the distance that a Roman soldier could march in 1000 double paces, which was estimated to be 5000 feet.)

mileage *Metrology*. a given distance expressed in miles, or the miles traveled in a given time. *Mechanical Engineering*. the number, or the average number, of miles that a motor vehicle can travel on a certain quantity of fuel.

mileage check *Transportation Engineering*. a measured distance of exactly one mile along a highway, indicated by signposts, so that motorists can check this distance against their actual speedometer and odometer readings to determine the accuracy of the readings.

mileage marker *Transportation Engineering*. see MILEPOST.

mileage number *Navigation*. numbers that indicate the mileage from a designated reference point displayed by some navigation markers on major American rivers such as the Mississippi.

milepost *Transportation Engineering*. **1.** one of a series of posts or markers set along a highway or other thoroughfare to indicate distance in miles. **2.** a single post or sign indicating the mileage to a specific place. *Civil Engineering*. a marker used in surveying to record the distance of a mile.

Miles and Misra's method *Microbiology*. a method used to determine the number of viable bacterial cells in a suspension, by incubating a drop of each of several dilutions of the suspension on agar plates and subsequently counting the growing colonies.

miles of relative movement *Navigation*. the miles of motion by one craft relative to those of another.

miles on course *Navigation*. the distance in miles traveled, or expected to be traveled, on a given course.

mil formula *Ordnance*. a mathematical relationship that is used in gunnery to determine the approximate distance between two points; expressed as $n = W/R$, where n is the angular measurement in mils between two points, W is the lateral distance in yards between the points, and R is the mean distance in thousands of yards to the points; the U.S. military defines a mil as 1/6400 of a circle.

miliaria *Medicine*. a syndrome caused by the obstruction of the sweat ducts and the resulting escape of sweat into adjacent tissues, forming a rash. Also, HEAT RASH, PRICKLY HEAT.

miliary *Biology*. **1.** characterized by lesions the size of millet seeds. **2.** made up of small and numerous projections.

miliary tuberculosis *Pathology*. tuberculosis in which the blood spreads the bacilli from one point of infection, producing small tubercles in other parts of the body.

Milichiidae *Invertebrate Zoology*. a family of dipteran flies in the subsection Acalyptratae, associated with manure.

milieu interieur *Physiology*. the blood and lymph that bathe the cells of the body. (From a French term meaning "interior world," coined by the 19th-century French physiologist Claude Bernard.)

milieu therapy *Psychology*. a method of treating mental disorders in which the patient's immediate environment is altered in such a way as to provide new stimulation and promote recovery.

Miliolacea *Invertebrate Zoology*. a superfamily of marine foraminiferans in the suborder Miliolina.

Miliolidae *Invertebrate Zoology*. a family of marine, benthic foraminiferans in the superfamily Miliolacea.

Miliolina *Invertebrate Zoology*. a suborder of marine and brackish water foraminiferans with a calcite test that is an important constituent of limestone.

militarily significant fallout *Military Science*. radioactive contamination capable of reducing the effectiveness of combat personnel.

military *Military Science*. of or relating to organizations, operations, procedures, or materiel related to the military forces of a nation. Used to form many compound nouns, such as **military censorship, military convoy, military currency, military education, military geography, military governor, military intelligence, military intervention, military occupation, military objective, military operation, military police, military requirement, military research, military service, military specification, military strategy, military target, military traffic, military training**, and so on.

military aircraft *Aviation*. any aircraft designed and used for military purposes, as distinguished from private and commercial aircraft.

military capability *Military Science*. the ability to achieve a specified wartime objective, including four major components: force structure, modernization, readiness, and sustainability.

military characteristics *Ordnance*. the characteristics of equipment that enable it to perform desired military functions, including physical and operational characteristics but not technical characteristics.

military damage assessment *Military Science*. an assessment of the effects of an attack on a nation's military forces to determine remaining military capability and aid in planning for recovery and rebuilding.

military deception *Military Science*. actions executed to mislead foreign decisionmakers, causing them to derive and accept desired perceptions of military capabilities, intentions, or operations.

military department *Military Science*. any of the departments within the Department of Defense; i.e., the Department of the Army, Department of the Navy, or Department of the Air Force.

military designed vehicle *Ordnance*. a vehicle having military characteristics resulting from military research and development; designed primarily for combat or tactical use in the field. Similarly, **military-type vehicle.**

military engineering *Engineering*. the branch of engineering dealing with the design and construction of both offensive and defensive military works.

military geology *Engineering*. the application of earth science technology to military concerns such as terrain analysis, road and airfield construction, and water supply.

military grid *Cartography*. two sets of parallel lines that intersect at right angles and form squares, which are then superimposed on maps or charts used in military activities; used to identify ground locations and to measure directions and distances to other locations.

military motor vehicle *Ordnance*. a wheeled or tracked vehicle, either self-propelled or towed, that is designed to transport military cargo or personnel.

military necessity *Military Science*. the principle that a participant in a military conflict has the right to use any measures that are required to successfully conclude a military operation, as long as such measures are not forbidden by the laws of war.

military satellite *Space Technology*. any artificial satellite used by the armed services for surveillance, communication, navigation, geodesy, or other purposes.

Military Science

Modern military science has its origins in the search for rationality in the waging of war among military intellectuals of the eighteenth century. Among the first exponents of the scientific school was Henry Humphrey Evans Lloyd, a Briton who in 1766 wrote that, properly conceived, the conduct of war "is founded on certain and fixed principles, which are by their nature invariable; the application of them can only be varied, but they are themselves constant."

Further force was given this approach by Antoine, Baron de Jomini, an influential interpreter of the conduct of the Napoleonic Wars in the early nineteenth century, and whose influential ideas were contained in his *Summary of the Art of War* (1836), widely adopted as a model for textbooks in the plethora of military academies that sprang up in the Western world after 1815. Though Carl von Clausewitz's *On War* (1832) suggested that war was more of a historical than a scientific phenomenon (i.e., one concerned more by change than by constants), *On War's* precepts were widely interpreted in the scientific spirit by later generations of soldiers. Since the nineteenth century, the approach of military science has governed in Western military thought.

In general, and in order to make the principles apply at all times, and in all conditions and circumstances, the foundation of military science has held to simplified precepts. The "principles of war" vary among different military organizations, but most are similar to those of the United States Army: the Objective, the Offensive, Mass, Maneuver, Surprise, Security, Economy of Force, Unity of Command, and Simplicity. Such principles are supposed to lie in the background of military history much as the principles of politics lie in the background of political history. In practice, military science alters with social-political, technological, and organizational changes, though in theory its foundations never change.

The positive influence of military science has been the pursuit of the rational in the waging of war; its negative influence has been to foster resistance to change when new developments seemingly contradict established doctrines.

Larry H. Addington
Professor of History
The Citadel

military science **1.** the strategy and tactics involved in warfare and in military operations. **2.** the study or analysis of such activities.
military technology *Engineering.* the technology required to develop and use ordnance and to plan and carry out military operations.
military training route *Transportation Engineering.* a route along which airspace is set aside for military training flights.
military transport vehicle *Ordnance.* a nonfighting vehicle designed to transport personnel and materiel or to tow other vehicles in close support of combat and tactical operations.
military vessel *Naval Architecture.* a vessel designed for combat or other wartime use.
militia *Ordnance.* a part-time military (or paramilitary) organization available for mobilization in an emergency.
milium see WHITEHEAD.
milk *Physiology.* **1.** the white or yellowish fluid secreted by the mammary glands of female mammals for the purpose of feeding their young. **2.** any substance resembling this fluid. *Chemistry.* a suspension or emulsion, such as milk of magnesia.
milk-alkali syndrome *Medicine.* an excess of calcium in the blood and urine due to the prolonged ingestion of milk and other alkalis, associated with severe kidney problems. Also, BURNETT'S HYPERCALCEMIA.
milk-drinker's syndrome see MILK-ALKALI SYNDROME.
milk ejection see LETDOWN REFLEX.
milkers' nodule *Medicine.* paravaccinia or pseudocowpox, a papilloma of the fingers or palm acquired from lesions on the udder of a cow infected with poxvirus.
milk fever *Medicine.* **1.** an elevation of body temperature at the onset of lactation following delivery; once considered normal, it is now recognized that temperatures in excess of 99° usually indicate infection. **2.** a fever commonly associated with the use of unpasteurized milk. **3.** a paralytic illness affecting cows preceding delivery of calves.
milkfish *Vertebrate Zoology.* a herringlike fish, *Chanos chanos,* of the family Chanidae; found in the warm ocean waters of southeastern Asia.
milk glass *Materials Science.* an opaque white glass having a milky appearance.
milking *Agriculture.* the process of extracting milk from a cow or other domestic animal. *Surgery.* the pressing out of the contents of a tubular part, such as the urethra, by running the fingers along it.
milk intolerance see LACTOSE INTOLERANCE.
milk leg *Pathology.* a painful swelling of the leg that occurs shortly after childbirth, caused by thrombosis of the large veins.
milk letdown see LETDOWN REFLEX.
milk line see MAMMARY RIDGE.
Milkman's syndrome *Medicine.* a widespread malabsorption disease of the bones, in which transparent striping is evident in the long and flat bones. (Named for the American radiologist Louis Arthur *Milkman,* 1895–1951.) Also, LOOSER-MILKMAN'S SYNDROME.
milk of magnesia *Pharmacology.* $Mg(OH)_2{\cdot}H_2O$, a suspension of 7–8.5% magnesium hydroxide in water; used as an antacid and laxative.
milk ring test *Microbiology.* a test for the detection of brucellosis infection in cattle by assaying the milk for the presence of antibodies to the microorganism *Brucella abortus.*
milk sickness *Toxicology.* acute poisoning due to the ingestion of milk or flesh from animals suffering from trembles; symptoms may include weakness, vomiting, and death.
milksnake *Vertebrate Zoology.* a species of harmless kingsnake, *Lampropeltis triangulum,* characterized by black, brown, and white bands or saddles; found in the Americas. (So called because of the former belief that it could extract milk from cows, since it is often found in barns, where it hunts rodents.)

milksnake

milk sugar see LACTOSE.
milk tooth *Anatomy.* a popular term for one of the temporary (deciduous) teeth later replaced by the permanent teeth.
milk vetch *Botany.* the plant *Astragalus glycyphyllos* of the legume family that is believed to increase the secretion of milk in goats; found in Europe.
milkweed *Botany.* a popular name for various plants that produce a milky juice or latex, especially members of the genus *Asclepias,* many species of which are widely found in North America, such as the **common milkweed,** *A. syriaca,* the **swamp milkweed,** *A. incarnata,* and the **butterfly weed,** *A. tuberosa.*
milkweed beetle *Invertebrate Zoology.* any of many small, elongated, black-spotted, red beetles of the genus *Tetraopes* that inhabit the milkweed; found in North America.
milkweed bug *Invertebrate Zoology.* any of many red and black bugs, such as *Oncopeltus fasciatus,* that feed on the juice of milkweed.
milkwood *Botany.* any of various trees that secrete a milky juice, such as the Jamaican *Pseudomedia spuria.*

milkwort *Botany.* any of various plants of the genus *Polygalla,* widely distributed worldwide, such as the **field milkwort,** *P. sanguinea,* or the **orange milkwort,** *P. lutea.* (From the early belief that cows eating such plants would have an increased flow of milk.)

milky disease *Invertebrate Zoology.* a bacterial disease of Japanese beetle larvae caused by *Bacillus popilliae* and *B. lentimorbus.*

milky quartz *Mineralogy.* a nearly opaque variety of crystalline quartz with a somewhat greasy luster and a milk-white color, resulting from numerous minute fluid inclusions. Also, GREASY QUARTZ.

Milky Way *Astronomy.* as observed from earth, a broad, silvery band of diffuse light that circles the entire sky and is made up of countless stars too faint to be seen individually.

Milky Way Galaxy *Astronomy.* the galaxy we inhabit, which is thought to be a type Sb or Sc spiral containing about 100 billion stars; it is about 130,000 light-years in diameter and the sun lies about two-thirds of the way from the center, on the inner edge of one spiral arm.

mill *Industrial Engineering.* **1.** a machine or device for grinding grain into flour and other cereal products. **2.** a factory in which paper, steel, or textiles are manufactured. *Mechanical Devices.* any machine or device for grinding or crushing solid substances, such as pepppercorns or coffee beans. *Mining Engineering.* a place where ore is crushed or concentrated, or the machinery to perform these functions.

millboard *Materials Science.* a strong, stiff paper board used in book binding and box making.

mill construction see HEAVY-TIMBER CONSTRUCTION.

milled describing an object or material made by milling, such as *milled* soap.

Millepora *Invertebrate Zoology.* a genus of coralline hydrozoans of the order Milleporina, characterized by a smooth calcareous surface with many perforations; they are also called **fire corals** because of their powerful sting and resemblance to coral.

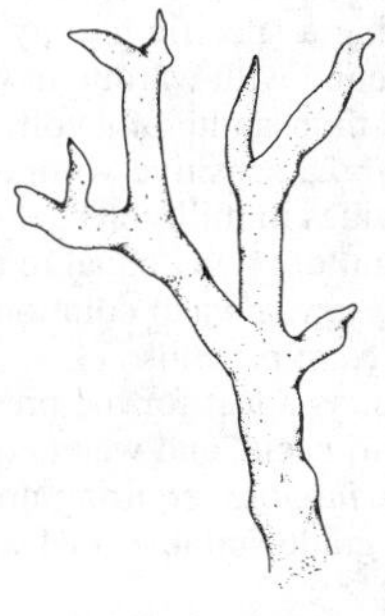

Millepora

miller see MILLING MACHINE.

Miller, William Hallowes 1801–1880, British mineralogist and crystallographer; best known for his work on crystallography.

Miller-Abbott tube *Medicine.* a twin-channeled tube used to examine the small intestine, containing a balloon at the end which can be inflated to clear obstructions.

millerbird *Vertebrate Zoology,* a rare, small, gray-brown warbler, *Acrocephalus familiaris,* characterized by a thin bill and found only on the Hawaiian islet of Nihoa.

Miller-Bravais indices *Crystallography.* in the hexagonal lattice (c unique) there are three axes perpendicular to c inclined at 120° to one another. Therefore three indices, hki, rather than the usual two Miller indices hk, are used in the hexagonal case, where $i = (-h + k)$.

Miller bridge *Electronics.* a bridge circuit that measures the amplification factors of vacuum tubes.

Miller code *Electronics.* an internal electrical-pulse coding system for binary representation.

Miller effect *Electronics.* in a vacuum tube, the increase in the effective grid-cathode capacitance due to the charge induced electrostatically on the grid by the anode through the grid-anode capacitance.

Miller generator see BOOTSTRAP INTEGRATOR.

Miller indices *Crystallography.* the plane with Miller indices h, k, and l makes intercepts a/h, b/k, and c/l with the unit-cell axes a, b, and c. The law of rational indices states that the indices of the faces of a crystal are usually small integers, seldom greater than three.

Miller integrator *Electronics.* a capacitor integrator that uses capacitance produced by the Miller effect.

millerite *Mineralogy.* NiS, a brass- to bronze-yellow, opaque, metallic, trigonal mineral occurring as tufts and masses of interwoven capillary crystals, having a specific gravity of 5.41 to 5.5 and a hardness of 3 to 3.5 on the Mohs scale; found as a low-temperature mineral in limestones, dolomites, and carbonate ore veins.

Miller law see LAW OF RATIONAL INDICES.

Miller process *Metallurgy.* in the refining of gold, a process in which chlorine is bubbled through the molten, impure gold. The base metals form volatile chlorides or go into the slag.

millet *Botany.* **1.** the popular name for several cereal grasses that are grown as a food crop for their edible seeds, especially the species *Setaria italica.* **2.** the grain of these grasses.

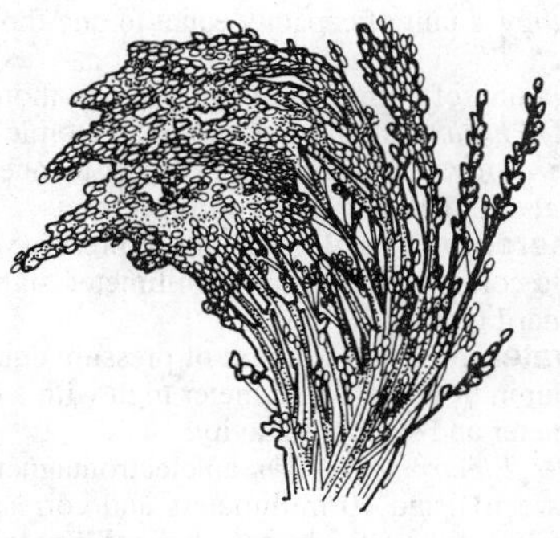

millet

mill file *Mechanical Devices.* a single-cut flat file about 10–12 inches long, with one or two rounded edges; used for lathe work and the fine-finishing and sharpening of tools; includes mill and circular saws, knives, axes, and shears.

mill finish *Metallurgy.* the finish of a wrought product as it leaves the mill without further processing.

milli- *Mathematics.* a prefix indicating "one thousand" or "one-thousandth" (10^{-3}).

milliammeter *Electricity.* an ammeter with a scale calibrated for measuring current values in milliamperes.

milliampere *Electricity.* one-thousandth of an ampere.

milliampere-second *Nucleonics.* a unit of quantity for the radiation dose received from an X-ray.

milliangstrom *Metrology.* a unit of length equal to one-thousandth of an angstrom.

millibar *Metrology.* a unit of pressure equal to one-thousandth of a bar, or 1000 dynes per square centimeter.

millibarn *Nuclear Physics.* a unit of measure equal to one-thousandth of a barn.

millicurie *Metrology.* a unit of radioactivity equal to one-thousandth of a curie.

millicurie-destroyed *Nucleonics.* a unit of quantity for a radiation dose that equals the time it takes a sample to lose one millicurie of radioactivity; in radon-222, for example, the period is 133 milligram-hours.

millicurie-of-intensity-hour see SIEVERT UNIT.

millicycle *Physics.* a unit of frequency equal to one-thousandth of a hertz; a millihertz.

millidarcy *Physics.* a unit of fluid permeability equal to one-thousandth of a darcy.

millidegree *Metrology.* a unit of measure equal to one-thousandth of a degree.

milliequivalent *Chemistry.* one-thousandth of the equivalent weight of an element or compound.

millier *Metrology.* 1000 kilograms; a metric ton.

millifarad *Electricity.* a unit of capacitance equal to one-thousandth of a farad.

milligal *Metrology.* a unit of gravitational acceleration equal to one-thousandth of a gal.

milligauss *Electromagnetism.* a unit of magnetic field-flux density equal to one-thousandth of a gauss.

milligram *Metrology.* a unit of mass equal to one-thousandth of a gram. Also, **milligramme.**

milligram-hour *Nucleonics.* a unit of quantity for a radiation dose equal to the radiation emitted by a source with an equivalent radium content of one milligram per hour.
millihenry *Electromagnetism.* a unit of inductance equal to one-thousandth of a henry.
millihertz *Physics.* a unit of frequency equal to one-thousandth of a hertz or one thousand cycles per second.
milli-k *Nucleonics.* a unit of quantity for the reactivity of a reactor.
Millikan, Robert A. 1868–1953, American physicist; studied cosmic rays; awarded the Nobel Prize for measuring the electrical charge carried by the electron.
Millikan meter *Electronics.* an integrating ionization chamber; used for cosmic ray measurements.
Millikan oil-drop experiment *Atomic Physics.* an early experimental method for determining the charge of an electron in which the terminal velocities of oil droplets, ionized by an X-ray beam, are measured.
milliliter *Metrology.* a unit of capacity equal to one-thousandth of a liter. Also, **millilitre.**
millilux *Optics.* a unit of illumination equal to one-thousandth of a lux.
milli-mass-unit *Physics.* one-thousandth of an atomic mass unit.
millimeter *Metrology.* a unit of length equal to one-thousandth of a meter; equivalent to 0.039 inch. Also, **millimetre.**
millimeter of mercury *Metrology.* a unit of pressure equal to the pressure exerted by a column of mercury 1 millimeter high at a temperature of 0°C and standard gravity.
millimeter of water *Metrology.* a unit of pressure equal to the pressure exerted by a column of water 1 millimeter high with a density of 1 gram per cubic centimeter and standard gravity.
millimeter wave *Electromagnetism.* an electromagnetic wave having a wavelength between 1 and 10 millimeters and corresponding to a frequency range of 300 and 30 gigahertz. Also, **millimetric wave.**
millimicron see NANOMETER.
millimole *Chemistry.* one-thousandth of a mole.
milling *Mechanical Engineering.* **1.** the act or process of producing plane or shaped surfaces by the use of a milling machine. **2.** the mechanical treating of solid materials, such as grain, to produce a powder. *Mining Engineering.* the process of removing valueless material and harmful constituents from ore.
milling cutter *Mechanical Devices.* a rotary steel cutter used in milling machines for shaping and dressing metal surfaces.
milling machine *Mechanical Engineering.* **1.** a machine tool that shapes workpieces by passing them through the periphery of a milling cutter. Also, MILLER. **2.** a mobile grinding apparatus used to trim the edge of a paved surface and remove excess material.

pavement milling machine

milling planer *Mechanical Engineering.* a planer that uses a rotary cutter rather than single point tools.
milling stone *Archaeology.* a stone slab or basin that is used to process seeds, nuts, and other such foods by rubbing or pounding them against this object with another stone. Also, GRINDING STONE.
milling system see CHUTE SYSTEM.
milling test *Materials Science.* a test of thermal degradation in polymers, which involves passing heated samples continuously between rolls or rotors held at close clearance.

Millington-Sette equation *Acoustics.* an equation in which the reverberation time of sound in a chamber is calculated by the formula

$$t = 60V/(1.086c - \sum S_i \ln(1 - \alpha),$$

where V is the room volume, S_i is the room surface area, c is the sound velocity, and α is the theoretical decay rate. Also, **Millington reverberation formula.**
milliohm *Metrology.* a unit of electrical resistance equal to one-thousandth of an ohm.
million *Mathematics.* the number 1,000,000.
million floating-point operations per second *Computer Technology.* a measure of the computing power of large computers and array processors. Also, MFLOP, MEGAFLOP.
million instructions per second *Computer Technology.* a measure of processing speed, the average number of machine language instructions that a computer executes in one second. Also, MIPS.
millipede *Invertebrate Zoology.* the common name for members of the arthropod class Diplopoda, characterized by a cylindrical body composed of 20 to more than 100 segments, each having two pairs of legs. Also, **millepede.**
Millipore filter *Biotechnology.* a filter with holes ranging from 0.005 to 8 micrometers; often used when presterilizing equipment.
millirad *Nucleonics.* a unit of quantity for a radiation dose that equals one-thousandth of a rad.
millirem *Metrology.* a unit of absorbed radiation equal to one-thousandth of a rem.
milliroentgen *Nucleonics.* a unit for a dose of electromagnetic radiation, as from a gamma ray, equal to one-thousandth of a roentgen.
millisecond *Computer Technology.* one-thousandth of a second.
millisecond delay cap *Engineering.* a delay cap whose interval between passing of current and explosion ranges between 0.02 and 0.4 second. Also, SHORT-DELAY DETONATOR.
millisite *Mineralogy.* $(Na,K)CaAl_6(PO_4)_4(OH)_9 \cdot 3H_2O$, a white, light gray, or greenish tetragonal mineral occurring as fibrous chalcedonic crusts or spherules, having a specific gravity of 2.83 and a hardness of 5.5 on the Mohs scale; found with wardite in variscite nodules.
millivolt *Electricity.* one-thousandth of a volt.
millivoltmeter *Electricity.* a sensitive voltmeter with a scale calibrated for measuring voltage values in millivolts.
milliwatt *Metrology.* a unit of power equal to one-thousandth of a watt.
Millon's reagent *Chemistry.* a water solution of mercuric nitrate that is used to detect the presence of proteins.
Millon's test *Biochemistry.* a test for the presence of proteins that uses a mixture of mercury, nitric acid, and water.
mill ore *Mining Engineering.* ore requiring preliminary treatment before a marketable grade or a grade suitable for further treatment is obtained. Also, SECOND-CLASS ORE.
millrace *Civil Engineering.* a direct stream of water that, by sheer force, drives a mill wheel.
mill run *Mining Engineering.* **1.** a test of the quality of a given quantity of ore by actual milling of a sample. **2.** the mineral obtained as a result of such a test.
Mills, Robert 1781–1855, American architect and engineer; designed the Treasury Building and the Washington Monument.
mill scale *Metallurgy.* a scale of ferric oxide formed on steel or iron during hot fabrication.
Mills cross *Electromagnetism.* a high-resolution radio-telescope antenna having two collinear or phased arrays with an intersecting lobe.
Mills grenade or **bomb** *Ordnance.* a British high-explosive, 5-pound grenade, widely used during World War I. (Invented by Sir William *Mills,* 1856–1932.)
millstone see BUHRSTONE.
millwork *Materials Science.* a collective term for materials prepared by milling, especially ready-made wooden materials such as doors and cabinets.
millwright *Engineering.* a person who designs, erects, and maintains the machinery at a mill.
Milne-Edwards, Henri 1800–1885, French zoologist; investigated the physiology of mollusks, crustaceans, and anthozoans.
milo *Botany.* a grain sorchum characterized by white, yellow, or pinkish seeds; grown primarily in Africa, Asia, and the U.S. Also, **milo maize.**
Milstein, Cesar born 1927, Argentine-British immunologist; awarded the Nobel Prize for discovering a method for producing monoclonal antibodies.

miltonia *Botany.* any of several tropical orchids of the genus *Miltonia,* characterized by sprays of showy, flat, variously colored flowers; found in the Americas. (Named for Charles W.W. Fitzwilliam, Viscount *Milton,* 1786–1857, English horticulturist.)

Mimas *Astronomy.* the first moon of Saturn, discovered in 1789 and believed to be made largely of water-ice; it is 392 km in diameter and has a 130-km crater, Herschel, centered on its leading face.

MIMD [mim´dē] *Computer Science.* a kind of parallel computer architecture, such as one involving multiple loosely coupled CPUs, in which different instructions are executed on different data simultaneously. (An acronym of Multiple Instruction, Multiple Data.)

mimeograph *Graphic Arts.* **1.** a reprographic device that uses an impressed stencil stretched over an inked drum to print copies. **2.** a printed piece produced on such a device.

mimetic *Crystallography.* of or relating to the inclination of a crystal to copy itself, as the inclination to twin. *Petrology.* of or relating to a tectonite characterized by a deformation fabric that reflects and is influenced by a preexisting anisotropic structure. *Zoology.* describing a species that attacks another species.

mimetic crystallization *Petrology.* during metamorphism, any process of recrystallization or neomineralization that reproduces preexisting structures; for example, bedding or schistosity. Also, FACSIMILE CRYSTALLIZATION.

mimetic labor *Medicine.* pains resembling labor pains but unaccompanied by effacement and dilation of the cervix; false pains.

mimetic twin *Crystallography.* a twinned crystal that mimics the symmetry of a higher system.

mimetite *Mineralogy.* $Pb_5(AsO_4)_3Cl$, a yellow to orange or white pseudohexagonal, monoclinic mineral of the apatite group, having a specific gravity of 7.24 to 7.28 and a hardness of 3.5 to 4 on the Mohs scale; found as prismatic crystals and in globular form in the oxidized zones of lead-bearing ore deposits.

mimic *Behavior.* an organism or species that takes on the appearance of another species or object in order to deceive a predator or prey organism.

mimicry *Behavior.* a resemblance of one organism to another or to a nonliving object that serves to deceive a predator or prey organism.

Mimidae *Vertebrate Zoology.* the mockingbirds and thrashers, a family of birds of the passerine suborder Oscines, found in forests, brushlands, and deserts of North and South America.

mimosa *Botany.* **1.** any of several plants, shrubs, or trees of the genus *Mimosa* of the legume family, characterized by small flowers in a globular head or cylindrical spike and sensitive leaves; found in tropical or warm regions. **2.** any of various related plants, especially those of the genus *Acacia* or *Albizzia.*

mimosine *Organic Chemistry.* $C_8H_{10}N_2O_4$, a crystalline solid derived from the tropical shrub *Leucaena glauca* that melts (decomposes) at 228–229°C; used as a depilatory agent.

Mimosoidae *Botany.* a subfamily of herbs, shrubs, and vines of the family Leguminosae, characterized by alternate, pinnate leaves and regular flowers, usually having numerous stamens.

min. minute; minimum; mineralogy; minim.

minable *Mining Engineering.* capable of being mined, especially profitably mined. Also, **mineable.**

Minamata disease *Toxicology.* a severe neurologic disorder of inhabitants of Minamata, Japan, following consumption of seafood heavily contaminated with alkyl mercury compounds; symptoms may include nervous disorders, permanent nerve damage, and death.

minaret *Architecture.* in a mosque, a tall slender tower with balconies, used to call the faithful to prayer.

minasragrite *Mineralogy.* $V^{+4}O(SO_4)\cdot 5H_2O$, a soft blue monoclinic mineral occurring as efflorescences and granular aggregates, having a specific gravity of 2.03 and an undetermined hardness; found with patronite.

mind *Psychology.* the total of all the mental processes, both conscious and unconscious, within an individual.

Mindel *Geology.* **1.** the second glacial stage of the Pleistocene epoch in the Alps, occurring after the Günz-Mindel interglacial stage. **2.** European geologic stage of the Pleistocene epoch, occurring after the Günz and before the Riss.

Mindel-Riss *Geology.* the second interglacial stage of the Pleistocene epoch in the Alps, occurring after the Mindel and before the Riss.

mind-mapping *Artificial Intelligence.* a knowledge organization tool used to evoke ideas from one or more users by placing a topic in the center of a blank space and branching out with related ideas.

mine *Mining Engineering.* an opening or excavation in the earth used to extract minerals, coal, or metallic ore. *Military Science.* a tunnel dug under or into an enemy position, to gain entry or to emplace explosives. *Ordnance.* **1.** an explosive or other material, normally encased, that is laid on the ground or buried underground; designed to destroy or damage vehicles, boats, or aircraft, or to kill, wound, or incapacitate personnel; it may be detonated by the action of its victim, by the passage of time, or by controlled means. Also, LAND MINE. **2.** a similar device, encased in a watertight container, that is laid on or below the surface of the water; designed to damage or sink ships, or to deter them from entering an area. Also, NAVAL MINE. **3.** to place such a device. **4.** of or relating to such a device. Thus, **mine clearance, mine defense, mine disposal, mine spotting, mine warfare, mine weapons,** and so on.

mine characteristic *Mining Engineering.* **1.** the plotted curve of the relation between pressure, p, and volume, Q, in the ventilation of a mine of resistance R, expressed as $p = RQ^2$. **2.** the point of intersection of this curve with the pressure characteristic of the fan, indicating the optimum pressure and volume for ventilating the mine with the fan.

mine detector *Ordnance.* an electromagnetic device that locates buried or concealed land mines.

mined volume *Mining Engineering.* in mine subsidence, the product of the mine area and the mean thickness of the bed or the extracted portion of the bed.

minefield *Ordnance.* an area of land or water in which a number of mines have been placed.

minefield density *Military Science.* in land-mine warfare, the average number of mines per meter of minefield front, or the average number of mines per square meter of minefield.

minefield gap *Ordnance.* a part of a minefield in which no mines have been laid, in order to allow a friendly force to pass safely. Similarly, **minefield lane.**

minelayer *Ordnance.* a vessel equipped for the laying of marine mines.

mineral *Geology.* a naturally occurring inorganic substance of inorganic or possibly organic origin that has a definite chemical composition and an orderly internal structure, crystal form, and characteristic chemical and physical properties.

mineral biological function *Nutrition.* the state of dynamic equilibrium in the body maintained by physiological processes between the intake and excretion of any mineral or any mineral constituent such as iron, calcium, or sodium.

mineral caoutchouc see ELATERITE.

mineral charcoal see FUSAIN.

mineral deposit *Geology.* an accumulation, concentration, or mass of naturally occurring mineral material, generally having economic or potential economic value.

mineral economics *Mining Engineering.* the study and application of the management, control, and finance processes involved in mineral discovery, development, exploitation, and marketing

mineral facies see METAMORPHIC FACIES.

mineral fuel *Materials.* a general term for fuels extracted from the earth, such as coal, oil, or peat.

mineral green see CUPRIC CARBONATE.

mineralization *Geology.* the process by which the organic components of an organism are replaced by inorganic materials.

mineralize *Geology.* to convert to, or impregnate with, mineral matter.

mineralizer *Geology.* a magmatic gas or fluid that aids in the crystallization and/or concentration of various ore minerals.

mineral land *Mining Engineering.* land that is more valuable for mining than for agriculture or other uses.

mineralocorticoid *Biochemistry.* a hormone, secreted by the adrenal cortex such as aldosterone, that maintains normal blood volume, promotes sodium and water retention, and increases urinary excretion of potassium and hydrogen ions.

mineralogenetic epoch *Geology.* an interval of geologic time during which mineral deposits were formed.

mineralogenetic province *Geology.* a geographic region having conditions favorable for the concentration of useful mineral deposits.

mineralogical of or relating to mineralogy.

mineralogical phase rule *Mineralogy.* a rule stating that the maximum number of mutually stable phases (minerals) possible in a natural mineralogical system (rock) at arbitrary temperature and pressure is equal to the number of components. Also, GOLDSCHMIDT'S PHASE RULE.

mineralogist *Mineralogy.* a person who examines, analyzes, and classifies minerals, specifically their mode of origin, occurrence, composition, properties, and possible uses.

Mineralogy

The solids, predominantly crystalline and inorganic, formed by natural processes, are called minerals. About 3000 species are now known and approximately 50 new species are recognized each year in materials found on earth, its moon, and in meteorites. Mineralogy is the scientific study of the physical and chemical properties of minerals: their external morphology, internal structure, and classification; their distribution, occurrence, and associations; the processes governing their growth and change under varying conditions; their synthesis and experimental determination of their stability in the laboratory; and their industrial and commercial uses.

The importance of mineralogy to mankind lies in the economic utilization of minerals. Ore deposits are natural concentrations of specific minerals in restricted regions from which useful metallic elements or nonmetallic materials may be extracted.

The first book exclusively devoted to minerals, *De Lapidibus,* was written by Theophrastus about 300 B.C. and only 16 or so "minerals" were listed. The title "Father of Mineralogy," however, is usually assigned to Georgius Agricola, who summarized the state of knowledge of mineralogy, geology, mining and metallurgy in two important works published in 1546 and 1556. Major advances in mineralogy occurred with the discovery of the systematic arrangement of atoms in minerals as observed with X-ray diffraction by Max von Laue in 1912 and the nondestructive determination of their chemical constituents with the Electron Microscope X-ray Analyzer invented by R. Castaing in 1949.

Rare minerals now recognized at the surface as having originated at depths in the earth, at least 450 kilometers, have unique properties. Most of the mineralogy of the interior of the earth, although unknown, is, therefore, of special interest to modern man. The entire range of conditions to the center of the earth, however, can be reproduced in the laboratory; hence, the exciting prospect of deducing the entire mineralogy of the earth is now at hand.

Hatten S. Yoder, Jr.
Director Emeritus, Geophysical Laboratory
Carnegie Institution of Washington

mineralogy the science that studies naturally occurring inorganic substances, or minerals; it sometimes also includes organic substances, such as those occurring naturally in fossil fuel deposits.

mineraloid *Mineralogy.* a naturally occurring, usually inorganic, mineral-like material that is amorphous and lacks a highly ordered atomic arrangement and characteristic external form; for example, opal or volcanic glass (the naturally occurring liquids water and mercury are also usually included).

mineral oil *Materials.* a colorless liquid petroleum derivative with little discernible odor or taste, widely used as a lubricant and laxative and for various other purposes. Also, LIQUID PETROLATUM, LIQUID PARAFFIN, MEDICINAL OIL, WHITE MINERAL OIL.

mineral processing *Mining Engineering.* procedures that are used to raise the mineral concentration of ore or other mineral-bearing products by means of wet or dry crushing and grinding.

mineral resources *Geology.* the valuable mineral deposits of a region, including those presently recoverable, those recoverable in the future, and those not yet discovered.

mineral soil *Agronomy.* soil composed mainly of mineral matter with less than 20% organic matter, normally having a surface organic layer less than 30 cm thick.

mineral spirits *Chemistry.* a volatile distillation product of petroleum used as a thinner for paints and varnishes.

mineral spring *Hydrology.* a spring containing enough dissolved mineral salts to give its water a distinguishable taste.

mineral suite *Mineralogy.* **1.** a collection of related minerals from one deposit. **2.** a collection of minerals taken as representative of a particular site. **3.** a collection of mineral specimens of one species taken from several locations.

mineral tanker *Naval Architecture.* a vessel that is designed to carry crude petroleum or dry bulk products in liquid suspension. Also, SLURRY TANKER.

mineral water *Hydrology.* water that contains a high quantity of dissolved mineral salts or gases, especially such water that is considered healthful to drink.

mineral wax see OZOCERITE.

mineral wool *Materials.* fine fibers of rock, glass, or a combination of these materials, made into blocks or boards and used as heat or sound insulators.

miner's horn *Mining Engineering.* a metal spoon or horn for collecting the ore particles in gold washing.

miner's inch *Mining Engineering.* an imprecise and usually locally defined quantity for the flow of water through an opening, most generally taken to amount to 2274 cubic feet, or 64.39 cubic meters, in 24 hours.

miner's self-rescuer *Mining Engineering.* a pocket-sized gas mask allowing respiration in a mine atmosphere.

mine run *Mining Engineering.* the total unscreened output of a mine. Also, RUN-OF-MINE.

mine sterilizer *Ordnance.* a device that makes an armed mine safe after a preset number of days.

mine strip *Ordnance.* in land-mine warfare, two parallel mine rows laid simultaneously six meters or six paces apart.

minesweeper or **mine sweeper** *Naval Architecture.* a naval vessel designed to detect and destroy underwater explosive mines; often built of wood to minimize their magnetic effects. *Ordnance.* a heavy piece of equipment rolled in front of a tank to explode land mines.

minesweeping *Ordnance.* the technique of searching for or clearing mines using mechanical or explosion equipment that physically removes or destroys the mine or produces the influence fields necessary to activate it.

mine thrower *Ordnance.* the tracks on the deck of a minelayer that allow mines to be dropped over the stern.

minette *Petrology.* a syenitic lamprophyre consisting of biotite phenocrysts in a groundmass of orthoclase and biotite.

minewatching *Military Science.* a countermeasure employed in naval mine warfare; it includes the detection, recording, and tracking of potential minelayers, as well as the detection, location, and identification of mines during the actual minelaying.

mini- a combining form denoting something smaller than usual for something in its class, as in *minicomputer.*

miniature electron tube *Electronics.* a small electron tube whose electrode leads project through the glass bottom in positions corresponding to those of the pins in a 7- or 9-pin base.

miniaturization *Electronics.* the process of compacting the physical size of a system, package, or component without decreasing its performance or efficiency.

minicartridge *Computer Technology.* a small, encased reel-to-reel tape cartridge.

minicell *Microbiology.* a minute cell formed by the abnormal, unequal cell division of a parent cell and that lacks chromosomal DNA, although it may contain plasmid DNA.

mini cells *Biotechnology.* a small piece of protoplasm derived from a cell that contains less than the haploid number of chromosomes.

minichromosome *Virology.* a variation of DNA viruses of the host cell in which the viral DNA combines with histones to form particles similar in structure to the chromatin of the host cell.

minicomputer *Computer Technology.* a class of small and inexpensive but fast computers with higher performance and a more powerful instruction set than microcomputers, having a wide selection of programming languages.

minidisk *Computer Technology.* a term applied to a $5\frac{1}{4}$-inch flexible diskette in order to differentiate it from the 8-inch flexible diskette.

mini-f plasmid *Genetics.* any small, self-replicating plasmid that is constructed from a fragment of the f plasmid.

minigene *Genetics.* a chromosomal gene that directs the production of the variable regions of the heavy and light chains of immunoglobulins.

minim *Metrology.* **1.** the smallest unit of apothecaries' liquid measure in the U.S., equal to 0.00167 dram and equivalent to 0.0616 milliliter. Also, DROP. **2.** a similar measure used in Great Britain, equivalent to 0.059 milliliter. Also, IMPERIAL MINIM.

minimal brain dysfunction syndrome see ATTENTION-DEFICIT DISORDER.

minimal connector problem *Mathematics.* the problem of forming a tree on *n* vertices using the minimum number of edges. The solution is given by Kruskal's algorithm. Also, GREEDY ALGORITHM.

minimal equation *Mathematics.* the equation obtained by setting the minimal polynomial of a square matrix equal to zero.

minimal lethal dose *Toxicology.* the smallest amount of a toxic substance that will kill a specified number of a specified species and size of test animal within a certain time.

minimal medium *Microbiology.* a basic culture medium that is not supplemented with certain growth factors, generally permitting the growth of only prototrophic microorganisms.

minimal polynomial *Mathematics.* **1.** the minimal polynomial of an $n \times n$ square matrix A is the unique monic polynomial of minimum degree that annihilates A; i.e, that equals the zero matrix when evaluated at A. The minimal polynomial of A divides the characteristic polynomial of A, and every eigenvalue of A (a root of the characteristic equation) is also a root of A. **2.** suppose K is a finite extension of a field F, and that the element a of K is algebraic over F; i.e., a satisfies some polynomial with coefficients in F. The minimal polynomial for a over F is the (unique) monic polynomial f such that $f(a) = 0$ with coefficients in F of least positive degree. By theorem, the degree of this polynomial equals the degree of $F(a)$ over F, where $F(a)$ is the intersection of all subfields of K that contain both F and a.

minimal principle see MINIMUM PRINCIPLE.

minimal realization *Control Systems.* in linear system theory, the set of differential equations of the smallest dimension having an input/output transfer function matrix equal to a given matrix function.

minimal surface *Mathematics.* **1.** a surface whose mean curvature at each point is zero. **2.** given a bounded curve, any surface of least area with boundary equal to the given curve.

minimax *Artificial Intelligence.* **1.** an algorithm for determining the backed-up value of a node in a game tree from the values of child nodes one move below: if it is the player's move, the maximum of the values below is taken; if the opponent's move, the minimum. **2.** to perform a search in a game tree, using the minimax algorithm to determine the best move.

mini-maxi regret *Control Systems.* in decision theory, a criterion for choosing a strategy based on the smallest maximum difference between the payoff for that strategy and the best hindsight choice.

minimization *Psychology.* a form of negative thinking in which the individual tends to underestimate greatly the significance of positive events occurring in relation to himself or herself. *Mathematics.* the process of determining a (relative or absolute) minimum of a real-valued function.

minimization principle *Physics.* a principle involved with the progression of one state of a physical system to another state; requires that the progression take the path of least resistance and that the potential energy of the system tend toward a minimum value.

minimize *Telecommunications.* a designation of a condition in which normal message and telephone traffic is reduced in order to put through emergency messages.

mini-Mu *Virology.* a defective phage Mu with a shortened genome but having intact ends; used as a cloning vector.

minimum *Science.* the least amount or quantity possible or assignable. *Mathematics.* **1.** a smallest element. Let A be a partially ordered set with ordering $\leq$. An element a of A is a minimum of A if for all $c \in A$, $a \leq c$. Also, LEAST ELEMENT. **2.** A real-valued function f on a topological space X attains a **relative minimum** (or **strict relative minimum**) value at a point $a \in U$ if there exists a neighborhood U of a such that $f(x) \geq f(a)$ (or $f(x) > f(a)$) for all points x in U. **3.** If $f(x) \geq f(a)$ (or $f(x) > f(a)$) for all points x in X, then f is said to have an **absolute minimum** (or **strict absolute minimum**) at a.

minimum-access coding *Computer Programming.* a code-optimization technique that minimizes computer time used in transferring data to and from storage in those computers whose access time depends on memory location; no longer required with modern random access memories. Also, **minimal-latency coding.**

minimum-access programming *Computer Programming.* a technique for minimizing the waiting time for fetching instructions or data from a serial access storage device; no longer needed with modern random access memories.

minimum-altitude bombing *Military Science.* horizontal or glide bombing with the height of release under 900 feet; it includes masthead bombing.

minimum attack altitude *Military Science.* the lowest altitude that permits the safe conduct of an air attack and minimizes effective enemy counteraction.

minimum bend radius *Metallurgy.* the minimum radius around which a metallic product can be bent without fracture.

minimum condition *Mathematics.* a module A is said to satisfy the minimum condition on submodules if every nonempty set of submodules of A contains a minimal element. By theorem, A satisfies the descending chain condition on submodules if and only if A satisfies the minimum condition on submodules.

minimum configuration *Computer Technology.* **1.** a computer system consisting of only the essential components and smallest memory required for complete processing, input, and output. **2.** the smallest configuration a manufacturer will sell.

minimum crossing altitude *Navigation.* the lowest altitude at which an aircraft may cross certain fixes while flying in the direction of a higher minimum en route instrument altitude.

minimum daily requirements *Nutrition.* the former U.S. standards for daily intake of calories and of various nutrients, now superseded by the **recommended dietary allowance.**

minimum descent altitude *Navigation.* the lowest authorized altitude, in feet above mean sea level, from which a circle-to-land maneuver or final approach may be made, using a standard instrument approach where there is no electronic glide slope.

minimum deviation *Optics.* the ray that crosses a prism with the least angular deviation so that it may be used to determine the refractive index of the prism.

minimum discernible signal *Electronics.* the input signal with the smallest amplitude that can be detected by a receiver and produced as a discernible output signal.

minimum-distance code *Telecommunications.* **1.** a binary code in which the signal distance is always above or equal to a predetermined minimum value. **2.** in a redundant coding scheme, the minimum difference in adjacent code values that will retain the desired extent of error detection or error correction capabilities.

minimum ebb *Oceanography.* the lowest speed of a continuously flowing ebb current.

minimum elevation *Ordnance.* the lowest elevation of a weapon at which a projectile fired from the weapon will safely clear all obstacles between the weapon and the target.

minimum en route instrument altitude *Navigation.* the lowest published altitude for flying between radio fixes; the lowest altitude that will ensure adequate radio reception throughout the airway segment and clear all obstacles.

minimum firing current *Electricity.* in electric blasting caps, the minimum value below which firing cannot occur.

minimum flood *Oceanography.* the lowest speed of a continuously flowing flood current.

minimum flying speed *Aviation.* the lowest steady speed at which a given aircraft can maintain altitude under given conditions or the lowest steady speed at which a given aircraft is stable and controllable.

minimum function see VECTOR SUPERPOSITION MAP.

minimum gliding angle *Aviation.* the shallowest angle at which an aircraft can maintain a glide under given conditions without added engine thrust.

minimum holding altitude *Navigation.* the lowest prescribed altitude for a holding pattern.

minimum IFR altitudes *Navigation.* the minimum prescribed altitudes for IFR operation as published on aeronautical charts and in the Federal Aviation Regulations.

minimum inhibitory concentration *Biotechnology.* the lowest possible concentration of an antibiotic in a dilution series that shows no growth of the test organism after incubation for a set time and temperature.

minimum ionizing speed *Atomic Physics.* the minimum velocity required for a specified charged particle to ionize an atom or molecule in a gas.

minimum lethal dose *Toxicology.* the smallest amount of a toxic substance that will be fatal under specified conditions. Also, TOXIC UNIT, TOXIN UNIT.

minimum-loss attenuator *Electronics.* an attenuator matching two unequal resistive impedances with minimum power loss to the signal. Also, **minimum-loss pad.**

minimum-loss matching *Electronics.* the process of using a resistive circuit as a transducer to achieve matching to a source and load impedance at its input and output terminals, respectively, causing the minimum possible power loss. The transducer is an L network (q.v.) of two resistors.

minimum metal condition *Design Engineering.* the maximum amount of material that can be shaved off a machine part.

minimum normal burst altitude *Ordnance.* the altitude above terrain below which air-defense nuclear warheads are not normally detonated.

minimum obstruction clearance altitude *Navigation.* the lowest published altitudes for flying between radio fixes on VOR airways or off-airway route segments that provide obstacle clearance and adequate signal coverage within 25 statute miles of a VOR.

minimum-phase system *Control Systems.* a linear system in which the poles and zeros of the transfer function all have negative or zero real parts.

minimum principle *Physics.* any of various principles formulated to account for the observation that certain important quantities tend to be minimized when a physical process takes place; e.g., Maupertuis' principle of least action or Gauss' principle of least constraint.

minimum range *Ordnance.* **1.** the minimum range setting of a gun at which a projectile will clear an obstacle or friendly troops between the gun and the target. **2.** the minimum distance to which a gun can fire from a given position.

minimum reflux ratio *Chemical Engineering.* the ratio of the reflux flow rate to the vapor flow rate that will result in the number of theoretical plates approaching infinity for a given distillation.

minimum residual radioactivity weapon *Ordnance.* a nuclear weapon designed to reduce unwanted effects from fallout, rainout, and burst-site radioactivity.

minimum resolvable temperature difference *Thermodynamics.* the smallest distinguishable change in temperature that can be measured by any device.

minimums *Navigation.* a designation of the least favorable weather conditions established for certain operations, such as VFR flight or IFR landing. Also, **minima.**

minimum safe distance *Military Science.* the sum of the radius of safety and the buffer distance in a nuclear explosion.

minimum signal level *Electronics.* in a facsimile system, the level corresponding to the lowest signal amplitude of either copy white or copy black.

minimum thermometer *Engineering.* a thermometer designed to record the lowest temperature attained during a set interval of time.

minimum tillage *Agronomy.* a method of plowing that breaks up the soil without turning it over completely. Also, **minimum till.**

minimum turning circle *Transportation Engineering.* the diameter of the circle that a vehicle forms when making its tightest turn.

minimum vectoring altitude *Navigation.* the lowest altitude at which an IFR aircraft will be given heading guidance by a radar controller except as otherwise authorized for radar approaches, departures, and missed approaches.

minimum warning time (nuclear) *Military Science.* the sum of system reaction time and personnel reaction time in the event of a nucelar attack.

minimum wetting rate *Chemical Engineering.* the minimum liquid flow rate necessary to maintain a wetted vertical surface.

mining *Mining Engineering.* the technique and business of mineral discovery, extraction, and exploitation.

mining effect *Ordnance.* the earth movement and resultant destruction or damage caused by an underground explosion.

mining engineering *Engineering.* the branch of engineering dealing with the discovery, development, exploitation, cleaning, sizing, and dressing of ores and minerals.

Mining Engineering

Mining Engineering is the integrated application of multiple scientific and engineering disciplines to the extraction of natural materials from the earth's crust. Mining engineering provides the engineering support needed to operate mines. Major subdisciplines include mine feasibility evaluation, mine planning and design, and engineering of production and of auxiliary operations.

A prime objective of modern mining engineering is to efficiently move vast quantities of rock, soil, frequently water, and, for underground mines, air. Rock handling starts with rock excavation, most commonly by drilling and blasting. Large-scale breaking and fragmenting of rock requires a clear understanding of the mechanics of inducing rock failure. Conversely, the mechanics of rock behavior are equally important in maintaining the stability of the rock surrounding excavations, be it slopes of open pit mines, or roof, floor, and walls around underground openings. Rock mechanics provides the scientific basis for engineering rock excavation and structures in rock.

The requirement to move large volumes of soil and broken rock places a premium on optimizing the haulage system, starting with the initial selection of the system. Computer simulations, operations research, linear, nonlinear, and dynamic programming, and queuing theory are applied to design haulage systems. The continuous computer monitoring of haulage systems is now becoming widespread.

In many mines, especially deep ones, more water is handled than rock. The design and operation of mine drainage systems involves hydrology, hydraulics, and power engineering.

Deep hot underground mines move even more air, by weight, than either rock or water. Mine ventilation engineers apply fluid mechanics, thermodynamics, and air conditioning theories and practices to provide adequate environmental conditions throughout the constantly changing network of airways that constitute an underground mine.

For surface mining in particular, although by no means exclusively, environmental engineering has become part of the daily practice of mining engineering. Environmental considerations impact mining engineering from early feasibility evaluations, through mine planning, design, and operations, until permanent mine closure and abandonment. Environmental and reclamation requirements continue to become ever more stringent and restrictive. Making mining compatible with preservation of the environment poses a major challenge. Responding to this challenge involves disciplines as diverse as archaeology, biology, law and regulatory requirements, wildlife management, soil science, meteorology, plant sciences, and toxicology.

The environmental aspects of mining engineering may well be the ultimate confirmation of the mining engineer as integrator of multiple disciplines; more and more, at the peak of a career, the mining engineer is likely to be manager of a large multidisciplinary team of professionals — an upgrade rather befitting the jack-of-all-trades mining engineer of earlier times.

Jaak Daemen
Professor of Mining Engineering
Mackay School of Mines
University of Nevada, Reno

mining geology *Mining Engineering.* the study of the structures, modes of formation, and occurrence of mineral deposits as well as the geologic considerations of mine planning.

miniscule *Graphic Arts.* a lowercase letter.

minitrack *Electronics.* a ground-based system for tracking satellites using interferometers. Also, **minitrack radio.**

minium *Mineralogy.* $Pb_2^{+2}Pb^{+4}O_4$, a red or brownish-red tetragonal mineral occurring in powdery form, having a specific gravity of 8.9 to 9.2 and a hardness of 2.5 on the Mohs scale; found as an oxidized product of galena and other lead minerals.

minivan *Mechanical Engineering.* a type of passenger vehicle, larger than a station wagon but smaller than a conventional van, typically with seats for six to eight people.

minivet *Vertebrate Zoology.* any of several bright-colored cuckoo shrikes of the genus *Pericrocotus* in the family Campephagidae; found in the forests of tropical Asia.

minivirus *Virology.* a small isometric virus particle found in insect virus infections.

mink *Vertebrate Zoology.* **1.** a semiaquatic, carnivorous, weasel-like member of the family Mustelidae, having partially webbed feet and a soft thick coat highly prized by furriers; found in the Northern Hemisphere. **2.** the fur of this animal.

mink

Minkowski, Hermann 1864–1909, Russian-born German mathematician; introduced the concept of the four-dimensional nature of space-time.

Minkowski electrodynamics *Electromagnetism.* an electromagnetic theory that accounts for the electric and magnetic polarization in matter; the theory is compatible with the theory of special relativity.

Minkowski metric *Physics.* a metric tensor that allows one to consider Minkowski space-time using all real coordinates, denoted by x_m (m = 1, 2, 3, 4), where $x_1 = x$, $x_2 = y$, $x_3 = z$, and $x_4 = ct$ (instead of $x_4 = ict$); the elements of the tensor are g^{ij}, where $g^{11} = g^{22} = g^{33} = 1$, and $g^{44} = -1$.

Minkowski's inequality *Mathematics.* **1.** the triangle inequality for the l_p norms; that is,

$$[\sum_{i=1}^{n} (x_i + y_i)^p]^{1/p} \le (\sum_{i=1}^{n} x_i^p)^{1/p} + (\sum_{i=1}^{n} y_i^p)^{1/p},$$

where $x_i, y_i \ge 0$ and $p \ge 1$. Equality holds if and only if the vectors $[x_i]$ and $[y_i]$ are dependent. Minkowski's inequality is a consequence of Hölder's inequality. **2.** the triangle inequality for the L^p norms; that is,

$$\{\int |f(x) + g(x)|^p dx\}^{1/p} \le \{\int |f(x)|^p dx\}^{1/p} + \{\int |g(x)|dx\}^{1/p}.$$

3. if both A and B are positive definite $n \times n$ matrices, then $[\det(A + B)]^{1/n} \ge (\det B)^{1/n} + (\det B)^{1/n}$. Equality holds if and only if $B = cA$ for some $c \ge 0$. Also, **Minkowski's inequality for determinants.**

Minkowski space *Mathematics.* a four-dimensional flat manifold over the real numbers with the (infinitesimal) metric given by $ds^2 = -dt^2 + dx^2 + dy^2 + dz^2$.

Minkowski space-time *Physics.* a completely flat four-dimensional space used in the theory of special relativity; three dimensions represent spatial dimensions and the fourth coordinate is devoted to time.

Minnesota *Agriculture.* any of several breeds of hog that were developed in the United States for meat production; **Minnesota Number 1** is red in color and was crossbred from the Tamworth and the Danish Landrace breeds; **Minnesota Number 2** is white with black spots and was crossbred from the Large White and the Poland China breeds.

Minnesota Multiphasic Personality Inventory *Psychology.* a widely used personality inventory intended to uncover a subject's unconscious attitudes; the subject indicates agreement or disagreement with various statements and the results are scored in such a way as to assess personality. (Developed at the University of Minnesota.)

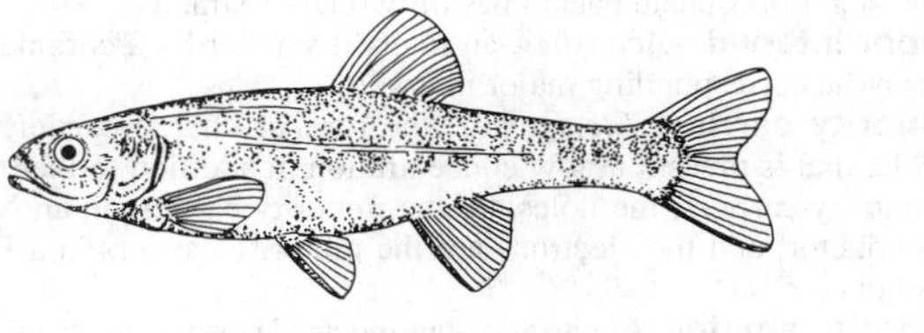

minnow

minnow *Vertebrate Zoology.* **1.** any of numerous types of small, freshwater fish belonging to the family Cyprinidae, having worldwide distribution, with many species in North America; includes the carps, goldfishes, and daces. **2.** a general name for any of various unrelated small fishes.

Minoan *Archeology.* of or relating to the ancient civilization of the island of Crete that dates from about 3000 to 1100 BC.

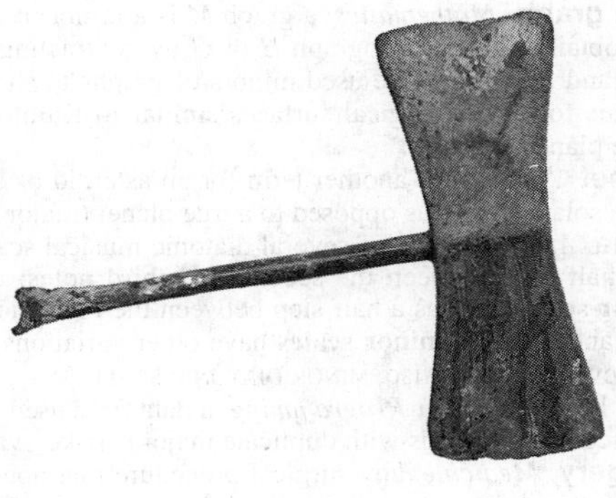

Minoan artifact

minor lesser in size or importance, especially the lesser of two similar items or groups; specific uses include: *Acoustics.* relating to a minor chord, scale, interval, and so on. *Mathematics.* the determinant of a square submatrix of a matrix A; that is, the determinant of a square matrix obtained from A by deleting one or more rows and/or columns of A. If A is a square matrix, then a **principal minor** of A is one whose main diagonal is a subset of the main diagonal of A. A signed minor, such as those appearing in the Laplace expansion, is called a cofactor of A.

minor arc *Mathematics.* the shorter of the two arcs into which a circle is cut by a chord.

minor axis *Mathematics.* the shorter of the two axes of symmetry of an ellipse.

minor bend *Electromagnetism.* a bend in a waveguide of rectangular cross section in which the waveguide axis remains in a plane that is parallel to the shorter side of the guide.

minor chord *Acoustics.* a chord having a minor third interval between the root and the next note above it.

minor control change *Computer Programming.* a change in the value of the least significant level of control or data field.

minor control data *Computer Programming.* the least significant data field by which records are sorted.

minor cycle *Computer Technology.* a single sequence of events or the time required to execute the sequence; the time required to transfer a complete machine word, including the word separator, in a serial-access storage system. Also, WORD TIME.

minor defect *Industrial Engineering.* the least serious classification of product defects; a defect that has little or no effect on the ability of the product to serve its intended use.

minor diameter *Design Engineering.* the diameter of a cylinder as measured from the root of a male (external) thread to the crest of a female (internal) thread.

minor diatonic scale see MINOR SCALE.

minor element *Geology.* any chemical element that is found in small amounts in rocks of the earth's crust.
minor focal epilepsy a form of epilepsy in which the attack is of a brief and minor nature, often consisting of only the aura, with no loss of consciousness. Also, PARAEPILEPSY.
minor groove *Molecular Biology.* the smaller of the two types of spiral grooves found on a right-handed DNA double helix; located between the sugar phosphate backbones of two DNA strands.
minor interval *Acoustics.* an interval smaller by a chromatic half step than the corresponding major interval.
minority carrier *Electronics.* the type of charge carrier, electron or hole, that is present at low concentration in addition to the preponderant majority carriers; the holes are the minority carriers in an N-type semiconductor, and the electrons are the minority carriers in a P-type semiconductor.
minority emitter *Electronics.* an electrode from which minority carriers flow into the interelectrode region of a transistor.
minor key *Acoustics.* a key whose harmony is based on the minor scale. Also, **minor mode.**
minor light *Navigation.* an automatic, unmanned navigational light that is mounted on a fixed structure and burns at low to moderate intensity.
minor lobe *Electromagnetism.* any radiation lobe of an antenna pattern other than the major lobe.
minor loop *Control Systems.* in a feedback control system, a small continuous network that contains both forward elements and feedback elements.
minor of a graph *Mathematics.* a graph M is a minor of a graph G if M can be obtained from a subgraph H of G by contracting edges of H. Robertson and Seymour have used minors of graphs to show that there are theorems for all topological surfaces similar to Kuratowski's theorem for the plane.
minor planet *Astronomy.* another term for an asteroid or other smaller body in the solar system, as opposed to a true planet (major planet).
minor scale *Acoustics.* any of several diatonic musical scales in which there is a half step between the second and third notes; a **natural** or **pure minor scale** also has a half step between the fifth and sixth notes; **harmonic** and **melodic minor scales** have other variations between the sixth and seventh notes. Also, MINOR DIATONIC SCALE.
minor sort key *Computer Programming.* a data field used during a sort routine to sequence records with duplicate major sort-key values.
minor surgery *Medicine.* any surgical procedure that does not require respiratory assistance or general anesthesia.
minor switch *Electricity.* the use of a single, moving step switch for party-line telecommunication applications.
minor trough *Meteorology.* a rapidly moving small pressure trough associated with a migratory cyclonic disturbance in the lower troposphere.
Minot, George R. [mī´nät] 1885–1950, American physician; awarded the Nobel Prize for his research on liver treatment for anemia.
minoxidil *Pharmacology.* a peripheral vasodilator used in the treatment of severe hypertension; also applied topically to promote hair growth in some types of baldness.
mint *Botany.* **1.** any of various aromatic, labiate plants of the genus *Mentha,* having worldwide distribution and characterized by opposite leaves and small, whorled flowers; widespread species of North America include the peppermint, *M. piperita,* the spearmint, *M. spicata,* and the **field mint,** *M. arvenis.* **2.** any of various similar plants of other genera of the family Labiatae.
mint family *Botany.* the popular name for the family Labiatae. See LABIATAE.
minuano *Meteorology.* a cold southwest wind of the coastal region of southern Brazil that occurs between June and September.
minuend *Mathematics.* in a subtraction problem $a - b = c$, the number a from which the subtrahend b is subtracted.
minus *Mathematics.* **1.** an indication of subtraction; that is, $a - b$ is read "a minus b." **2.** an additive inverse; that is, usually $-a$ is read "minus a" rather than "the additive inverse of a." **3.** if G is a graph and S is a set of edges of G, then $G - S$ is the graph formed by erasing the edges of S from G. No vertices are erased in this operation. **4.** if G is a graph and S is a set of vertices of G, then $G - S$ is the graph formed by erasing from G the vertices in S from G and erasing all edges incident with at least one vertex in S.
minus angle see ANGLE OF DEPRESSION.
minus-cement porosity *Geology.* the proportion of the total volume of a sedimentary material that would be occupied by pore spaces if the material did not contain chemical cement.
minus sieve *Metallurgy.* in powder metallurgy, the fraction of a metallic powder that passes through a sieve of specified size.
minus strands *Molecular Biology.* **1.** a term designating DNA or RNA coding strands. **2.** single-stranded viral genomes that are complementary to the viral mRNA.
minute *Metrology.* the sixtieth part of an hour; 60 seconds. *Mathematics.* a unit of angular measurement equal to 1/60 of a degree.
Minuteman *Ordnance.* a three-stage, solid-propellant, intercontinental ballistic missile equipped with a nuclear warhead and directed by an all-inertial guidance and control system; the Minuteman III can deliver weapons against three separate targets; officially designated **LGM-30.**
minute mutation *Genetics.* in *Drosophila,* a dominant mutation that is lethal when homozygous; affected heterozygotes develop slowly into small, sterile or less-fertile adults whose wing veins exhibit an abnormal pattern.
minyulite *Mineralogy.* $KAl_2(PO_4)_2(OH,F)\cdot 4H_2O$, a colorless to white orthorhombic mineral having a specific gravity of 2.46 and a hardness of 3.5 on the Mohs scale; found as acicular aggregates in glauconitic phosphate deposits.
Miocene [mī´ə sēn] *Geology.* **1.** the fourth of the five geologic epochs of the Tertiary period, extending from the end of the Oligocene epoch to the beginning of the Pliocene epoch. **2.** the system of rocks formed during this epoch.

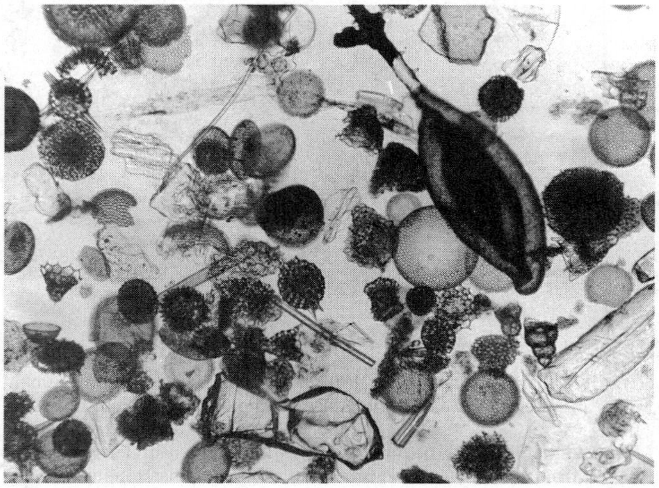

Miocene fossils

Miocidaris *Paleontology.* a genus of echinoids belonging to the subclass Cidaroidea and the family Miocidaridae; the only echinoid known to have survived the great extinction at the end of the Permian, and probably the ancestor of all post-Paleozoic echinoids; its descendants thrived in the Mesozoic and today are probably as numerous as at any time in the past; *Miocidaris* itself became extinct in the Triassic.
miogeosycline *Geology.* the portion of an orthogeosyncline located adjacent to the craton, in which volcanism is not associated with sedimentation. Also, **miomagmatic zone.**
miosis *Physiology.* the contraction of the pupil of the eye. *Medicine.* **1.** a condition in which an excessive contraction of the pupil takes place, resulting in abnormally small pupils. **2.** the stage of a disease during which the intensity of the symptoms diminishes. *Genetics.* see MEIOSIS.
miotic *Physiology.* **1.** of or relating to miosis, the contraction of the pupil of the eye. **2.** any agent that causes such a contraction. *Pharmacology.* specifically, a drug that is used to bring about this contraction, as in the treatment of glaucoma.
MIPS see MILLION INSTRUCTIONS PER SECOND.
Mira *Astronomy.* Omicron (o) Ceti, a reddish M-type star that varies in brightness from about 3rd to 10th magnitude over a period of about 330 days. (From a Latin word meaning "the Wonderful.")
mirabilite *Mineralogy.* $Na_2SO_4\cdot 10H_2O$, a white or colorless monoclinic mineral occurring as acicular crystals and as a fibrous efflorescence, having a specific gravity of 1.46 to 1.49 and a hardness of 1.5 to 2 on the Mohs scale; found typically in deposits from saline lakes, playas, and hot springs. Also, GLAUBER'S SALT.

miracidium *Invertebrate Zoology.* a free-swimming ciliated larval stage of a trematode that hatches, sometimes inside a snail, from an egg that has been deposited in water.
mirage [mə razh´] *Optics.* an illusion that arises when rays of light are bent or reflected by a layer of hot air at various densities; typically seen at sea, in the desert, or at other hot surfaces, often as a pool of water or a mirror in which a faroff object appears inverted. (From a French word meaning "to look at in wonder;" related to the word for *miracle.*)
Mirage *Aviation.* any of a series of French supersonic delta-wing combat aircraft.
mirage effect *Telecommunications.* the reception of radio waves at distances that exceed normal range; caused by abnormal refraction as a result of certain meteorological conditions.
Miranda *Astronomy.* the fifth moon of Uranus, discovered in 1948 and measuring 484 km in diameter; it has a heavily cratered surface, a bulk compositon that is mostly ice, and a spectacular vertical cliff 9 km high.
Mira variable *Astrophysics.* see LONG-PERIOD VARIABLE STAR.
mire *Geology.* **1.** wet, spongy earth. **2.** a small area of marshy, swampy, or boggy ground.
mired [mī´rəd] *Thermodynamics.* a measure of color temperature, equal to the reciprocal of a color temperature of 10^6 kelvins. (An acronym for micro-reciprocal degree.)
Miridae *Invertebrate Zoology.* a worldwide family of plant-eating true bugs in the order Hemiptera.
Miripinnatoidei *Vertebrate Zoology.* a suborder of scaleless marine fish in the order Lampriformes, found in open seas; includes the hairyfish and tapetails or ribbonbearers. Also, **Miripinnati.**
mirror *Optics.* **1.** a surface that creates a reflected image. **2.** specifically, a smooth, highly polished surface thinly coated with metal or other material that is capable of reflecting light without appreciable diffusion; usually plane, spherical (convex or concave), or paraboloidal.
mirror coating *Optics.* a very thin layer of metal, such as silver, that is evaporated onto a polished surface, such as glass, to create a reflective surface, or two layers of dielectric materials evaporated onto a glass surface in a vacuum to form a reflector of a narrow band of wavelengths.
mirror galvanometer *Electricity.* a galvanometer with a small mirror attached to a moving element so that a beam of light can act as a pointer.
mirror image *Optics.* an inverted image in handedness, from left to right or right to left, such as the image formed by a mirror.
mirror-image programming *Robotics.* a method of robotic programming used in metalworking in which the instructions for the *x* and *y* axes are reversed in order to create a mirror image of the workpiece.
mirror interference *Optics.* an interference between two beams of light when one or both are reflected off a mirror at a slight angle.
mirror interferometer *Engineering.* an interferometer in which the sea surface acts as a mirror to reflect radio waves toward an antenna, thus interfering with waves arriving from the source.
mirror nephoscope see CLOUD MIRROR.
mirror nuclei *Nuclear Physics.* a pair of nuclei that differ only in the number of protons and neutrons, so that each would be transformed into the other simply by changing its protons into neutrons or vice versa.
mirror optics *Optics.* an optic system based wholly or in part on mirrored elements.
mirror plane *Crystallography.* a symmetry element for which the corresponding symmetry operation resembles reflection in a mirror, coincident with the plane. It converts a chiral object or structure into its enantiomorph.
mirror scale *Engineering.* a scale having a mirror through which the user looks to read the pointer. It eliminates parallax.
mirror transit circle *Engineering.* a plane mirror that is mounted on a horizontal east-west axis, attached to a large circle, and calibrated to determine the mirror's position such that light from a star can be reflected into fixed horizontal telescopes pointed due north and south.
mirror yeasts *Mycology.* certain yeasts that form spores called ballistospores, characterized by a mirror image of their growth that they create on the lid of petri dishes. Also, SHADOW YEASTS.
mirror zone *Materials Science.* a smooth zone visible on the fracture surface of glass, near the origin of the fracture.
MIRV a vehicle, such as a nuclear warhead, carried by a delivery system that can directly place one or more such vehicles toward each of several separate targets. (An acronym for multiple independently targetable reentry vehicle.) Also, **M.I.R.V.**
MIS management information system; metal-insulator semiconductor.
mis- a combining form meaning "hatred of," as in *misanthropy.*
misanthrope [mis´ən thrōp´] *Psychology.* a person who has a general hatred of other people. Also, **misanthropist** [mis an´thrə pist].
misanthropy [mis an´thrə pē] *Psychology.* a general hatred of other people. Thus, **misanthropic** [mis´ən thräp´ik].
miscarriage *Medicine.* the expulsion of a fetus before it is viable, usually between the third and seventh months of pregnancy; spontaneous abortion.
misch metal *Metallurgy.* a natural mixture of rare earth elements, used in steelmaking to remove oxygen and sulfur.
Mischococcaceae *Botany.* a monogeneric family of freshwater, yellow-green algae in the order Mischococcales, having colonies borne in mucilaginous stipes.
Mischococcales *Botany.* the largest order of yellow-green algae in the class Xanthophyceae, characterized by cells with a distinct wall and an absence of contractile vacuoles and eyespots.
Mischococcus *Botany.* the genus of freshwater flagellate organisms composing the family Mischococcaceae, characterized by cells united in colonies borne on mucilaginous stipes.
miscibile [mis´ə bəl] *Chemistry.* describing two or more liquids that are able to mix with or dissolve into each other in various proportions; the opposite of immiscible.
miscibility [mis´ə bil´i tē] *Chemistry.* the quality of being miscible.
miscibility gap *Physical Chemistry.* the region between complete miscibility and immiscibility on an equilibrium phase diagram, where this partial miscibility region is a function of temperature and other conditions; common with liquids, such as oil and water, copper, and lead but also observed in alloys.
miscible-phase displacement *Petroleum Engineering.* the introduction of an oil-miscible driving fluid, such as gas or liquefied petroleum gas, into a well to increase oil recovery.
miscible-slug process *Petroleum Engineering.* the introduction of liquefied petroleum gas into a well to increase oil recovery.
misclosure *Cartography.* **1.** in surveying, a condition that exists when the results of a survey fail to agree, in part or in whole, with the results of an earlier survey, an arbitrarily assigned starting value, or the results of a theoretical calculation. **2.** the amount of the difference between values determined by a survey and values established earlier by surveying or other means.
MISD architecture *Computer Technology.* a parallel processor architecture in which several independent instruction sequences act on the same data stream. (An acronym for Multiple-Instruction, Single-Data stream.)
misenite *Mineralogy.* a grayish-white monoclinic mineral with an approximate formula of $K_2SO_4 \cdot 6KHSO_4$, occurring as lathlike or acicular crystals, and having a specific gravity of 2.32 and an undetermined hardness; found as an efflorescence in fumaroles on Cape Miseno, Italy.
miser *Petroleum Engineering.* a tubular bit used in well boring that is fitted with a valve at the bottom and a screw for forcing earth upward. Also, MIZER.
misfeed *Ordnance.* the incorrect feeding of ammunition, as in a magazine-fed or belt-fed automatic gun.
misfire *Chemistry.* **1.** to fail to ignite properly. **2.** an instance of this. *Electronics.* a failure to establish an arc between the main anode and the cathode in a mercury-pool-cathode tube at the intended conducting time.
misfit stream *Hydrology.* a stream whose size is not in proportion to the area in which it is found, especially a stream that appears too small to have eroded the valley in which it flows.
mismatch a failure to match properly; specific uses include: *Electricity.* a condition in which the impedances of two interconnected networks are not matched. *Molecular Biology.* the occurrence, during DNA replication, of a nucleotide on one strand that is not complementary to the corresponding nucleotide on the other strand.
mismatch factor see REFLECTION FACTOR.
mismatching *Electricity.* see MISMATCH.
mismatch loss *Electronics.* the loss in signal-to-noise output relative to that of a matched filter, caused by a filter that is not matched in response with the transmitted signal. Also, MATCHING LOSS.
mismatch repair *Molecular Biology.* a DNA error-correcting system that can detect and replace mismatched base pairs from a nucleic acid molecule.
miso- a combining form meaning "hatred of," as in *misogyny.*
Misodendraceae *Botany.* a monogeneric family of dicotyledonous dioecious shrublets of the order Santalales that are hemiparasitic on branches of the *Nothofagus* species; native to temperate, forested regions of South America.

misogamist [mi säg´ə mist] *Psychology.* a person who has a general hatred of marriage.
misogamy [mi säg´ə mē] *Psychology.* a general hatred of marriage.
misogynist [mi säj´ə nist] *Psychology.* a person who has a general hatred of women.
misogyny [mi säj´ə nē] *Psychology.* a general hatred of women.
misologia or **misology** *Psychology.* a hatred or fear of speaking or arguing.
misoneism *Psychology.* the avoidance of change and dislike of new subjects, events, and people.
misopedia *Psychology.* a hatred of children, especially one's own children.
misregistration *Computer Technology.* in character recognition, the improper appearance of a character, line, or document with respect to a horizontal base reference line.
misrun *Metallurgy.* in casting, a defective product caused by premature local solidification.
missanda *Materials.* the hard, coarse wood of the *Erythophleum gineense* and *E. ivorense* trees of Africa, used for heavy construction.
missed abortion *Medicine.* the death of a fetus without expulsion from the uterus, in which fetal heart tones or other indications of life have been absent for at least eight weeks.
missed approach *Navigation.* a prescribed maneuver that is carried out by a pilot when an instrument approach is aborted.
missed approach point *Navigation.* a specified point in an instrument approach at which a missed approach must be executed if the required visual reference cannot be seen.
missed round *Engineering.* a blasting round in which some or all of the explosive fails to detonate.
mis-sense codon *Genetics.* a codon that normally specifies one amino acid but is altered so that it specifies a different amino acid. Also, **mis-sense mutant.**
mis-sense mutation *Genetics.* a mutation in which a codon undergoes a nucleotide change such that it codes for a different amino acid, often resulting in the production of a nonfunctional protein.
missile [mis´əl; mis´īl] *Ordnance.* **1.** in general, any object that is designed to be thrown, dropped, or shot at a target, such as a grenade, bomb, or artillery shell. **2.** specifically, a ballistic missile or guided missile. **3.** of or relating to a missile. (Going back to a Latin word meaning "to throw.")
missile checkout *Ordnance.* the process of checking all components of a missile to ensure that they are capable of functioning properly.
missile control system *Ordnance.* a system that maintains altitude stability and corrects deflections of a missile or similar vehicle.
missile guidance system *Ordnance.* a system that correlates flight information with target data in order to determine the desired flight path of a missile and communicate the necessary commands to the missile flight control system. Also, **missile guidance control.**
missile impact predictor group *Military Science.* a military unit that provides facilities for determining the point of missile impact. Similarly, **missile position tracking group.**
missile intercept zone *Military Science.* a subdivision of the destruction area in which surface-to-air missiles have the primary responsibility for destruction of enemy aircraft.
missile launcher *Ordnance.* the equipment from which a guided missile is propelled.

missile launcher

missile monitor *Ordnance.* a mobile, electronic, fire-distribution system used by Army air defense units; it employs digital data to communicate within the system and with other missile monitor systems, thus providing the air defense commander with a means of monitoring and correcting missile defense actions.
missile plume *Electromagnetism.* a region that follows a missile in which electromagnetic disturbances are prevalent and thus make the missile more easily detected.
missile site radar *Engineering.* a phased array radar located at a missile launch site to aid interceptor missiles to their targets.
missing error *Computer Programming.* an error that occurs when a program calls a subroutine that is not available in the library.
missing mass problem *Astrophysics.* a problem raised because the mass of a galaxy as measured by its gravity and motions is ten times greater than its mass as measured by its luminous matter.
missing mass spectrometer *Particle Physics.* a device that measures the mass spectrum of a particle system by making measurements of the recoil of protons.
mission *Military Science.* **1.** the task of a military command, together with the purpose of the task. **2.** one or more aircraft ordered to accomplish a particular task.
Mississippi *Geography.* the chief river of the United States, flowing about 2350 miles from northern Minnesota to the Gulf of Mexico.
Mississippian *Geology.* **1.** a North American geologic period of the Paleozoic era, roughly equivalent to the Lower Carboniferous period in Europe, extending from the end of the Devonian period to the beginning of the Pennsylvanian period. **2.** the system of rocks formed during this period.
Missnay-Schardin effect *Ordnance.* an effect in which a metal plate accelerates from the face of an exploding charge in a manner that allows the plate to function as a solid missile projectile.
Missouri *Geography.* the longest river in the United States, flowing about 2700 miles (including ultimate tributaries) from the Rockies in southwestern Montana to join the Mississippi near St. Louis, Missouri.
Missourian *Geology.* a North American provincial series of the lower Upper Pennsylvanian, occurring after the Desmoinesian and before the Virgilian.
miss rate *Computer Science.* in a cache memory system, the frequency of accesses to memory locations that are not present in the cache.
mist *Meteorology.* an aggregate of microscopic and hygroscopic water droplets suspended in the atmosphere, producing a thin, grayish veil that reduces visibility.
mistake *Computer Programming.* a term for a human action in computer activity that causes incorrect results, such as a programming error, incorrect logic or mathematics, or the use of improper operations.
mistbow see FOGBOW.
mist droplet *Meteorology.* a particle of mist, whose size is between those of a haze droplet and a fog droplet.
mist extractor *Engineering.* a device that removes liquid mist from a gas stream.
mist fracture *Materials Science.* a small radial ridge on the surface of a smooth fracture, indicating a deviation from the original plane of the fracture that can be caused by intersection with an inclusion or a shift in the direction of the applied stress.

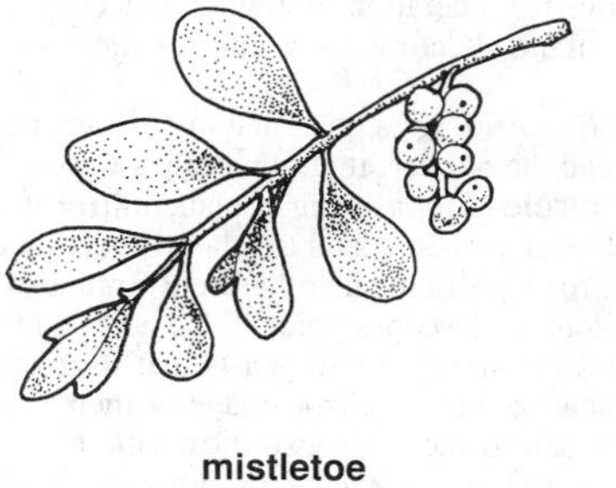

mistletoe

mistletoe *Botany.* **1.** the popular name for the species *Viscum album,* a semiparasitic green shrub of the family Viscaceae, having thick, alternate leaves, small, regular flowers, and white glutinous berries. Also, **European mistletoe. 2.** any of various similar plants of other genera of Viscacea, such as the **American mistletoe,** *Phoradendron flavescens.* (The "toe" ending is thought to be a corruption of an earlier word for "twig.")

mistral [mis′trəl; mi sträl′] *Meteorology.* a cold, dry, strong, squally north wind that blows down the Rhone Valley and out into the Mediterrenean Sea, resulting from a combination of the basic circulation, fall wind, and jet-effect wind. (From a local French name; literally, "master wind.")

mistranslation *Molecular Biology.* a term for the incorrect insertion of an amino acid residue into a growing polypeptide chain during protein synthesis.

MIT monoiodotyrosine.

Mit. send. (From Latin *mitte.*)

Mitchell *Aviation.* a popular name for the B-25 bomber.

Mitchell, Maria 1818–1889, American astronomer; studied the satellites of planets; established the orbit of a newly discovered comet.

Mitchell, Peter born 1920, British chemist; awarded the Nobel Prize for his research on plant and animal cellular energy transfer.

mite *Invertebrate Zoology.* the common name for most members of the arthropod order Acarina, including many species that are significant parasites on livestock and poultry or that are destructive to crops and vegetation.

miter *Design Engineering.* **1.** an oblique surface formed on the surface of a piece of wood or other material, to be joined with or placed against a similar surface on another piece. **2.** to create such a surface in a material. **3.** to join two pieces in this way. Also, MITRE.

miter bend *Design Engineering.* a bend in a length of pipe made by mitering, or angle-cutting, and then joining the pipe ends.

miter block *Building Engineering.* a block of wood recessed along one edge with saw-cuts above the recess inclined at a 45° angle to its face; used to guide a saw when cutting moldings for a miter joint.

miter box *Engineering.* a troughlike device containing vertical slots at various angles that hold the saw steady when cutting moldings to make mitered joints.

miter clamp *Mechanical Devices.* a clamp that holds a miter joint during gluing operations.

miter gate *Civil Engineering.* a canal lock or drydock gate consisting of paired leaves that swing out from either side to join in the middle of the channel.

miter gauge *Mechanical Devices.* a saw-setting tool for angled cuts in wood or metal.

miter gear *Mechanical Devices.* a bevel gear that is used at a right angle with a second, interchangeable gear to drive shafts at right angles to each other.

miter joint *Design Engineering.* a perpendicular joint in which the mating ends are beveled.

miter saw see DOVETAIL SAW.

miter square *Mechanical Devices.* a measuring device that consists of a steel blade and a hardwood stock used to mark both 45° halves of a miter joint or to check the accuracy of an angle when the miter joint has been cut.

miter valve *Mechanical Devices.* a 45° seated valve relative to its axial plane.

mithramycin *Oncology.* an antibiotic derived from *Streptomyces plicatus* bacteria, used as an antineoplastic agent in the treatment of various carcinomas; also used to treat hypercalcemia secondary to cancer.

mithridatism *Toxicology.* the acquisition of a tolerance for a particular poison by ingesting small but steadily increasing doses of it. (From *Mithridates* IV, king of Pontus, who is said to have taken small doses of poisons in order to become immunized against them.)

miticide *Toxicology.* an agent used for killing mites; typically a sulfur preparation.

mito- a combining form meaning "thread," as in *mitochrondria.*

mitocarcin *Oncology.* an antibiotic derived from species of *Streptomyces* bacteria, used chiefly as an antineoplastic agent.

mitochondria [mīt′i kän′drē ə] *singular,* **mitochondrion.** *Cell Biology.* self-replicating organelles, bounded by two membranes, that are found in the cytoplasm of all eukaryotic cells and produce cellular energy in the form of ATP via the oxidative phosphorylation reactions.

mitochondrial *Cell Biology.* of or relating to mitochondria.

mitochondrial DNA *Genetics.* the DNA of mitochondria, typically a circular molecule of single-stranded DNA consisting of about 50 gene sequences; the DNA of *Drosophilia melanogaster* and several other mitochondria have been completely sequenced.

mitochondrial Eve *Anthropology.* according to the single-origin hypothesis, a single female ancestor who lived about 200,000 years ago in Africa, from whom every living human is descended; based on mitochondrial evidence passed on through the female line.

mitochondrial RNA *Molecular Biology.* RNA molecules, ribosomal, messenger, or transfer, encoded by mitochondrial DNA, that are transcribed within the mitochondria and remain there.

mitocromin *Oncology.* an antibiotic derived from *Streptomyces viridochromogenes* bacteria, used chiefly as an antineoplastic agent.

mitogen *Genetics.* any of a class of substances that stimulate mitosis and cell transformation, especially lymphocyte transformation.

mitogenesis *Cell Biology.* the induction of cell division or mitosis.

mitogenetic *Cell Biology.* causing or inducing cellular mitosis.

mitogenic *Cell Biology.* of or relating to cellular mitosis.

mitomalcin *Oncology.* an antibiotic derived from *Streptomyces malayensis* bacteria, used chiefly as an antineoplastic agent.

mitomycin *Microbiology.* any of a group of three antibiotics (mytomycin A, B, and C) produced by *Streptomyces caespitosus,* that are used against certain cancers; due to its high toxicity, it is used only when other treatments have failed.

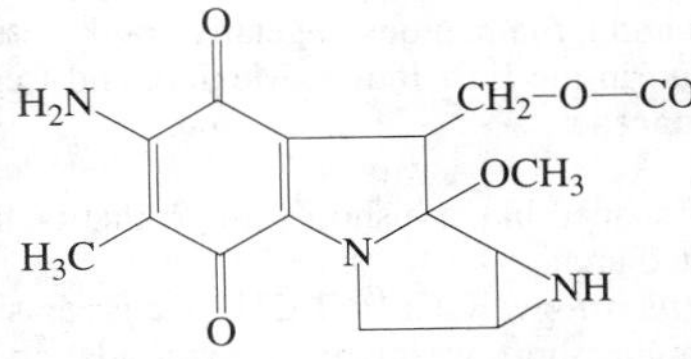

mitomycin C

mitomycin C *Virology.* an antitumor antibiotic that induces lysogenic phage development by inhibiting cell DNA synthesis but not viral DNA synthesis.

mitoplast *Cell Biology.* a cell mitochondrion from which the outer mitochondrial membrane has been removed, but which remains surrounded by the inner membrane.

mitosis [mī tō′sis] *Cell Biology.* a process of nuclear division in eukaryotic cells that apportions a diploid set of all chromosomes to each of two daughter cells during cell division and that is distinct from the process of meiosis. Also, KARYOKINESIS.

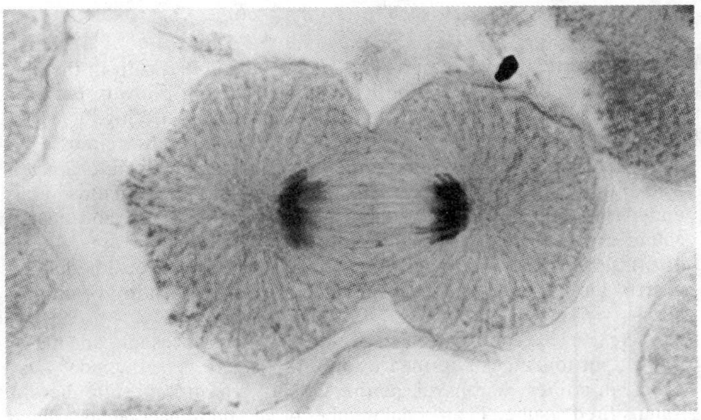

mitosis

mitosper *Oncology.* an antineoplastic agent derived from *Aspergillus glaucus* molds.

mitosporangium *Mycology.* a fungal sac or sporangium in which mitosis occurs.

mitotane *Oncology.* $C_{14}H_{10}C_{14}$, an antineoplastic substance used for the palliative treatment of inoperable adrenocortical carcinoma.

mitotic [mī tät′ik] *Cell Biology.* of or relating to the process of mitosis.

mitotic index *Cell Biology.* the fraction of cells in a cell population that are undergoing mitosis at a given time.

mitotic inhibitor *Cell Biology.* a substance, such as colchicine, that prevents cellular mitosis, usually by inhibiting the formation or operation of the spindle. Also, **mitotic poison.**

mitotic recombination *Molecular Biology.* the recombination, or crossing-over, of homologous chromosomes during mitosis. Also, **mitotic segregation.**

mitotic spindle *Genetics.* a collection of microtubules that develops in many eukaryotic cells during mitosis; the tubules are attached to the centromeres of sister chromatids and are responsible for separating the chromatids in an organized manner.

mitral [mī´trəl] *Anatomy.* **1.** shaped like a bishop's miter. **2.** of or relating to the mitral valve of the heart.

mitral commissurotomy *Medicine.* the surgical division of mitral valve leaflets that have grown together, impeding the opening.

mitral stenosis see MITRAL VALVE STENOSIS.

mitral valve *Anatomy.* one of the four valves of the heart, situated between the left atrium and the left ventricle.

mitral valve atresia *Cardiology.* the failure of ventrical valve development with the absence of an opening in the valve or the absence of valvular structure.

mitral valve stenosis *Cardiology.* a diminished opening between the left atrium and ventricle of the heart.

Mitrastemnataceae *Botany.* a monogeneric family of dicotyledonous root parasites belonging to the order Rafflesiales, characterized by a lack of chlorophyll and a filamentous vegetative body resembling a fungal mycelium; found in the East Indies, Mexico, and Central America. Also, **Mitrastemonaceae.**

mitre see MITER.

mitriform *Biology.* shaped like a bishop's miter; that is, having roughly the shape of a pointed arch.

mitscherlichite *Mineralogy.* $K_2Cu^{+2}Cl_4{\cdot}2H_2O$, a greenish-blue tetragonal mineral occurring as short, prismatic, or pyramidal crystals, having a specific gravity of 2.42 and a hardness of 2.5 on the Mohs scale; found as stalagmites on the floor of the crater of Mt. Vesuvius, Italy.

Mitscherlich's law of isomorphism *Chemistry.* an empirical law stating that substances having similar chemical compositions and properties produce similar if not identical crystalline forms.

Mitsubishi process *Materials Science.* a copper-smelting process in which concentrate and an oxygen-enriched blast are blown through vertical lances into a circular furnace; the smelting furnace and a top-lanced converting furnace are joined by refractory troughs through which matte and slag flow continuously.

Mittag-Leffler's theorem *Mathematics.* suppose that $f(z)$ is a meromorphic function having a countable number of poles $z = z_n$ and corresponding residues b_n. Then f may be written in the form

$$f(z) = f(0) + \sum_n b_n((z - a_n)^{-1} + (a_n)^{-1}).$$

mittelschmerz [mit´əl shmurts´] *Medicine.* abdominal pain in the area of the ovary during ovulation, typically occurring midway between menstrual periods. (From German; literally, "middle suffering.")

Mitteniaceae *Botany.* a monotypic family of small, delicate mosses of the order Bryales that form loose tufts on tree bases and soil; characterized by a persistent protonema that forms chains of vesiculose cells under low light conditions; found in forests of Australia and nearby countries and in southern South America.

Mitteniales *Botany.* an order of mosses of the class Bryopsida that appear to glow due to cells that reflect light from a backing of chloroplasts.

mix *Science.* to combine or blend two or more substances into one mass.

mixed put together or formed by mixing; specifc uses include: *Ordnance.* in artillery and naval gunfire support, a term indicating that the rounds fired resulted in an equal number of air and impact bursts; **mixed air** indicates that the majority of the bursts were air bursts; **mixed graze** indicates that the majority were impact bursts.

mixed acid see NITRATING ACID.

mixed anhydride *Organic Chemistry.* an anhydride derived from two different acids.

mixed aniline point *Physical Chemistry.* the lowest temperature at which a solution will form from a mixture of aniline, heptane, and another hydrocarbon.

mixed arthritis *Medicine.* a painful inflammation of joints due to more than one type of arthritis.

mixed-base notation *Mathematics.* a positional system, such as the biquinary system, for representing numerical values using more than one base or radix. Each position may represent a different number base. A number represented in this way is called a **mixed-base number.**

mixed bud *Botany.* a bud that contains both rudimentary leaves and flowers.

mixed cloud *Meteorology.* a cloud that is composed of both ice crystals and water droplets.

mixed-code expression *Computer Programming.* an expression containing operands of different data types, permitted by some language processors and causing a syntax error condition in others.

mixed congruential generator *Computer Programming.* a type of random-number generator that computes a random number as the residue of a linear transformation of the previous number in which b, the coefficient of the modolus m, is not equal to zero (as with multiplicative congruential generators).

mixed crystal *Crystallography.* a crystal made up of two or more isomorphous compounds.

mixed current *Oceanography.* a current associated with a marked difference in speed and duration between the two successive ebb or flood currents of a tidal day.

mixed cycle *Mechanical Engineering.* an internal combustion engine cycle that combines Otto-cycle and diesel-cycle combustion in high-speed compression-ignition engines.

mixed deafness *Medicine.* a diminished ability to hear, caused by more than one type of ear disease or malfunction.

mixed decimal *Mathematics.* any decimal number of absolute value greater than 1; such a number is written as the sum of its integer and decimal parts.

mixed distribution *Statistics.* a distribution obtained as a compound (or weighted average) of conditional distributions.

mixed-entry decision table *Computer Programming.* a decision table whose stubs include both limited-entry conditions (yes, no, or don't care) and extended-entry conditions, for which values are entered in the condition portion of the decision rules directly.

mixed farm *Agriculture.* a farm that produces a variety of crops or livestock. Thus, **mixed farming.**

mixed flow *Chemical Engineering.* the flow of a mixture consisting of two or more phases. Also, **mixed-phase flow.**

mixed-flow impeller *Mechanical Engineering.* an impeller used for a pump, compressor, or agitator that combines radial and axial flow principles.

mixed forest *Forestry.* a forest, crop, or stand composed of two or more species of trees; generally defined as a forest with at least 20% tree growth that is not the single most common tree.

mixed gland *Physiology.* a gland that has both endocrine and exocrine parts, such as the pancreas.

mixed graph *Mathematics.* a graph that contains both directed and undirected edges.

mixed highs *Telecommunications.* a method for reproducing very fine picture detail in color television by transmitting high-frequency components as part of luminance signals for achromatic reproduction.

mixed indicator *Analytical Chemistry.* a mixture of two indicators that allows for a more accurate end-point determination in an acid-base titration.

mixed infection *Medicine.* the infection of an organ or tissue by multiple microorganisms.

mixed laterality *Behavior.* the preferred use of the right hand or side for some activities and the left for others. *Psychology.* a situation in which neither the right nor left brain is dominant.

mixed layer *Oceanography.* a surface layer of isothermal water that often exists above the thermocline; the water in this layer is mixed by wave action or thermohaline convection.

mixed-layer mineral *Mineralogy.* a mica-clay mineral composed of alternating layers of two or more different mica or clay minerals, arranged in a regular or random interstratified structure.

mixed lymphocyte reaction *Immunology.* a technique used to determine the degree of incompatibility between the histocompatibility antigens of two donors, whereby lymphocytes are transformed to blast cells in mixed cultures of blood leukocytes from each donor. Also, **mixed leukocyte reaction.**

mixed minefield *Ordnance.* a minefield containing both antitank and antipersonnel mines.

mixed nerve *Physiology.* a nerve that has both sensory and motor nerve fibers.

mixed nucleus *Meteorology.* a condensation nucleus that contains both soluble hygroscopic matter and insoluble but wettable matter.

mixed number *Mathematics.* any nonintegral rational number of absolute value greater than 1 that is denoted by the juxtaposition of its integer and fractional parts rather than as an improper fraction; for example, $5\frac{1}{2}$.

mixed ore *Geology.* any ore composed of both oxidized and unoxidized minerals.

mixed potential *Physical Chemistry.* the amount of energy a material can generate when two or more electrochemical reactions occur simultaneously.

mixed salvo *Ordnance.* a series of shots, some falling short of the target and some falling beyond it in uneven distribution.

mixed schizophrenia *Psychology.* a form of schizophrenia having symptoms from two or more of the four basic categories of schizophrenia, so that a classification in one of these categories cannot be made.

mixed state *Quantum Mechanics.* a system that is not a single state and may be any of a number of states, each with some probability; such a system is usually presumed to be in some particular state, but the state is unknown for experimental reasons or because the system is only described statistically.

mixed tide *Oceanography.* a tide that exhibits both diurnal and semidiurnal characteristics; characterized by large inequalities in the heights (and sometimes the durations) of either the two high tides or the two low tides of a tidal day; common in many areas of the Pacific Ocean.

mixed vaccine *Immunology.* a vaccine that contains antigens from several different species of pathogens and thus provides protection against several diseases simultaneously.

mixed water *Volcanology.* a combination of volcanic and atmospheric waters.

mixer something that mixes; specific uses include: *Mechanical Devices.* an electical appliance used to mix foods. *Electronics.* a circuit or device that combines two or more input signals or frequencies to produce a single output.

mixer-settler *Chemical Engineering.* a liquid-liquid extraction device that mixes phases and then allows the liquids to settle and separate.

mixer tube *Electronics.* an electron tube that performs the frequency-conversion function of a converter in a superheterodyne receiver when it is supplied with voltage or power from an external oscillator.

mixing a process in which something is mixed; specific uses include: *Chemical Engineering.* the mechanical agitation of the ingredients of a mixture in order to blend, cool, heat, react, or coat them. *Electronics.* the process of combining two or more signals, such as the outputs of multiple microphones, in order to produce special effects.

mixing extruder *Materials Science.* a device for mixing polymers as well as incorporating additives such as color concentrates, fillers, reinforcing fibers, stabilizers, lubricants, antioxidants, foaming agents, and crosslinking agents. Convection is the dominant mixing mechanism in mixing extruders.

mixing length *Fluid Mechanics.* the distance that a turbulent wake eddy travels before mixing with the ambient fluid. Also, PRANDTL MIXING LENGTH.

mixing ratio *Meteorology.* the dimensionless ratio of a mass of water vapor to a unit of mass of dry air, in a system of moist air.

mixing transformation *Mathematics.* a function on a measure space (to itself) such that the image of any measurable set is uniformly distributed throughout the space.

mixite *Mineralogy.* $BiCu_6^{+2}(AsO_4)_3(OH)_6{\cdot}3H_2O$, a whitish or emerald-to blue-green hexagonal mineral of the mixite group occurring as aggregates of acicular crystals and in fibrous masses, having a specific gravity of 3.79 and a hardness of 3 to 4 on the Mohs scale.

Mixodectidae *Paleontology.* a proposed family of dermopteran mammals in the extinct superfamily Plagiomenoidea; known only from the Paleocene of North America.

mixolimnion *Hydrology.* in a meromictic lake, the low-density, freely circulating upper layer.

mixoploid *Cell Biology.* describing an organism or a population of cells having a variable number of chromosomes in each cell.

mixotrophic *Biology.* obtaining nutrition by combining autotrophic and heterotrophic mechanisms, such as an insectivorous plant.

mixt. mixture.

mixtite see DIAMICTITE.

mixture *Science.* the mass that results from the thorough blending of two or more substances. *Pharmacology.* any combination of different drugs, or of a drug with other ingredients, that do not react chemically with one another.

mixture ratio *Space Technology.* in the propellant of a rocket or space vehicle, the ratio of fuel to oxidizer (or oxidizer to fuel) expressed in terms of mass flow rate.

Mizar *Astronomy.* Zeta (ζ) Ursae Majoris, a 2nd-magnitude star at the bend in the Big Dipper's handle; it makes a close visual pair with 4th-magnitude Alcor.

mizer see MISER.

mizzenmast *Naval Architecture.* **1.** the third mast from the bow in a vessel of three or more masts. **2.** the small rear mast of some two-masted vessels, such as a ketch or yawl.

mizzonite *Mineralogy.* an intermediate member of the marialite-meionite isomorphous series, scapolite group. Also, DIPYRE, DIPYRITE.

mk. mark.

MKC *Aviation.* the airport code for Downtown Kansas City Airport, Kansas City, Missouri.

MKE *Aviation.* the airport code for Milwaukee, Wisconsin.

MKS or **M.K.S.** meter-kilogram-second. Also, **mks** or **m.k.s.**

MKSA or **mksa** meter-kilogram-second-ampere.

MKSA system *Metrology.* a system of measurement in which the meter is the basic unit of length, the kilogram is the basic unit of mass, the second is the basic unit of time, and the ampere is the basic unit of electric current.

MKS system *Metrology.* a system of measurement in which the meter is the basic unit of length, the kilogram is the basic unit of mass, and the second is the basic unit of time.

MK system *Astrophysics.* a system of spectral classification developed by W.W. Morgan and P.C. Keenan that assigns stars to one of six luminosity classes.

ml or **ml.** milliliter.

mL millilambert.

MLA left mentoanterior. (From Latin *mento-laeva anterior.*)

MLA or **M.L.A.** Medical Library Association; Modern Language Association.

m-lange *Geology.* a mappable mixture of rock material that generally consists of an assortment of exotic and native fragments of various sizes, origins, and geologic ages embedded in a fine-grained, sheared matrix of more tractable material. Also, BLOCK CLAY.

MLB *Aviation.* the airport code for Melbourne, Florida.

MLC mixed lymphocyte culture.

MLD or **M.L.D.** median lethal dose; minimum lethal dose.

MLF multilateral force.

M.L.P. left mentotransverse. (From Latin *mento-laeva transversa.*)

MLW mean low water.

mm or **mm.** millimeter; millimeters.

MM necessary changes being made. (From Latin *mutatis mutandis.*)

mM millimolar.

M.M. mucous membranes.

MMA monomethyl aniline.

M meter *Meteorology.* any instrument that is used to measure moisture in the atmosphere.

mmf or **m.m.f.** millimillifarad; magnetomotive force.

mm Hg millimeter of mercury.

mmho millimho.

M mode *Acoustics.* a method of ultrasonic medical tomography in which sweeping motions at modulated frequencies are used to form detailed images of organs such as the heart.

MMPI Minnesota Multiphasic Personality Inventory.

MMR *Immunology.* an active immunizing agent consisting of a combination of measles, mumps, and rubella live vaccine; used in the immunization of children.

M.M.Sc. Master of Medical Science.

MN magnetic north.

Mn the chemical symbol for manganese.

M'Naghten rule or **M'Naghten test** [mik nô´tən] *Psychology.* the traditional principle that a person is considered to be legally insane if that person does not understand the nature and consequences of his or her actions. (From a famous 19th-century British case in which Daniel *M'Naghten* was acquitted of murder for this reason.)

M-N blood groups *Genetics.* a system of human blood groups caused by the presence or absence on erythrocytes of two antigens, designated M and N, that are coded for by two codominant alleles.

mnemonic [nə män´ik] *Psychology.* **1.** a particular device or technique used to improve one's memory or one's ability to memorize specific material. **2.** relating to memory or the improvement of memory. Thus, **mnemonic device.**

mnemonic code *Computer Programming.* an easy-to-recall assembly language code that uses a name for an operation, such as "ADD" for addition.

mnemonics [nə män´iks] *Psychology.* the use of various devices, strategies, and other artificial aids to improve memory.

Mnesarchaeidae *Invertebrate Zoology.* a superfamily of New Zealand moths, lepidopteran insects in the suborder Homoneura.

Mniaceae *Botany.* a family of loosely aggregated mosses of the order Bryales that grow on humus, logs, rock crevices, and open tundra; characterized by large oblong to pointed leaves, a single strong costa ending near the apex, and elongate setae which are sometimes clustered; found in Northern Hemisphere forests.

MNL *Aviation.* the airport code for Manila, Philippines.

MNNG *Organic Chemistry.* $CH_3N(NO)C(=NH)NHNO_2$, yellow crystals that melt at 118°C with decomposition; used experimentally as a source of dizomethane and as a mutagen and carcinogen. Also, ***N*-methyl-*N'*-nitro-*N*-nitrosoguanidine.**

MNOS metal-nitride oxide semiconductor.

MO master oscillator.

Mo the chemical symbol for molybdenum.

MOA monoamine oxidase.

moa *Vertebrate Zoology.* any of the flightless birds of the family Dinornithidae, extinct since the end of the 18th century; related to the kiwi but resembling the ostrich; found in New Zealand.

moat *Geology.* **1.** a continuous or discontinuous ringlike depression around the base of many seamounts. Also, SEA MOAT. **2.** a valleylike depression between the rim of a volcanic cone and the lava dome, along the inner side. *Hydrology.* **1.** a deep glacial trench or a channel at the margin of a dwindling glacier that resembles a moat. **2.** see OXBOW LAKE.

MOB *Aviation.* the airport code for Mobile/Pascagoula, Alabama.

mobbing *Behavior.* a form of harassing behavior that is directed toward a predator, including surrounding the predator in a group and emitting loud cries; found in many bird species.

mobile [mō´bəl] *Science.* able to move or to be moved about. *Graphic Arts.* describing a printing ink that tends to flow freely.

mobile artillery *Ordnance.* artillery designed to be easily transportable and readily usable, usually firing with wheels and other traveling equipment still attached. Similarly, **mobile mount.**

mobile belt *Geology.* a long, relatively narrow region of tectonic activity within the earth's crust.

mobile crane *Mechanical Engineering.* a cable-controlled crane that is moved on crawlers or rubber-tired carriers to which it is mounted.

mobile genetic element *Molecular Biology.* any fragment of nucleic acid that is able to move into and out of other nucleic-acid sequences; a transposon.

mobile hoist *Mechanical Engineering.* a hoist that is mounted on a pair of pneumatic-tire road wheels in order to move it from one site to another.

mobile loader *Mechanical Engineering.* a machine with a self-propelling power system that is used to load coal, minerals, dirt, or other materials.

mobile mine *Ordnance.* in naval mine warfare, a mine that is propelled to its laying position by a torpedo or similar device; it then sinks and operates like a mine.

mobile phone see CELLULAR PHONE.

mobile radio *Telecommunications.* radio service between a stationary location and at least one mobile radio station, or between mobile stations.

mobile relay station *Telecommunications.* a type of base station in which the base receiver automatically activates the base-station transmitter, which then retransmits all signals that are sent to the base receiver, extending the range of mobile units; two frequencies are required for operation.

mobile robot *Robotics.* a robot attached to a movable platform that takes it to its assigned work area.

mobile station *Telecommunications.* a radio station that is used while in motion or during stops at unspecified locations.

mobile systems equipment *Computer Technology.* the computers that are found on planes, ships, or vans.

Mobilina *Invertebrate Zoology.* a suborder of motile ciliate protozoans, including commensals and parasites of mollusks and fish.

mobility the fact of being mobile; specific uses include: *Engineering.* **1.** the degree to which a material is able to flow. **2.** the degree to which an analytical balance is able to react to minute changes in load. *Physics.* in an electrical conductor, the ratio of the drift speed and the applied electric field. *Solid-State Physics.* see DRIFT MOBILITY.

mobility ratio *Petroleum Engineering.* the ratio of driving-fluid mobility to driven-fluid mobility in an oil or gas reservoir.

mobility tensor *Physics.* a second-rank tensor that, when multiplied by the electric field vector in a plasma, reveals the average velocity of the charged particles.

mobility threshold *Engineering.* the minimum load that an instrument, such as an analytical balance, is able to register.

mobilization *Military Science.* **1.** in general, the act of assembling and organizing national resources to support national objectives in time of war or other emergency. **2.** specifically, the process by which all or part of a nation's military forces are brought to a state of readiness for war or other national emergency. **3.** of or relating to mobilization. Thus, **mobilization exercise.** *Geology.* any process whereby solid rock becomes sufficiently soft and plastic to permit flowage or geochemical migration of the mobile components.

Mobiluncus *Bacteriology.* a genus of Gram-negative bacteria that grow anaerobically in the human vagina as curved, rod-shaped cells possessing several subpolar flagella.

Möbius, August F. [mā´bē əs; mō´bē əs] 1790–1868, German mathematician; introduced homogeneous coordinates, projective transformations, and Möbius strip. Also, **Moebius.**

Möbius band *Mathematics.* a nonorientable surface that may be visualized by giving a rectangular strip a half-twist and gluing the two ends together. Also, **Möbius strip.**

Möbius cell *Materials Science.* an electrolytic device for refining silver concentrate that has been processed to yield doré metal. In the Möbius cell the anodes are made of doré metal, the electrolyte is a mixture of silver and copper nitrates, and the cathodes are rolled silver, stainless steel, or graphite. The silver deposited is at least 99.9% pure.

Möbius function *Mathematics.* the function $\mu(n)$, defined on the positive integers by: (a) $\mu(n) = 1$ if $n = 1$, (b) $\mu(n) = (-1)^t$ if n is a product of t distinct primes, and (c) $\mu(n) = 0$ if p^2 divides n for some prime p.

Möbius resistor *Electricity.* a nonreactive resistor in which an assembly consisting of aluminum strips placed on opposite sides of a length of a dielectric ribbon is twisted one half-turn and its ends joined, with leads soldered on opposite surfaces of the resulting loop.

Möbius transformation *Mathematics.* a conformal mapping of the complex plane of the form $f(z) = ax + b/cx + d$, where a, b, c, and d are complex numbers satisfying $ad - bc \neq 0$. The special case of $a = d = 0$ and $b = c = 1$ is called the **Möbius inversion**, or simply inversion. Such transformations form a group. Also, LINEAR FRACTIONAL TRANSFORMATION, BILINEAR TRANSFORMATION, HOMOGRAPHIC TRANSFORMATION.

Mobulidae *Vertebrate Zoology.* a marine fish family of the order Myliobatiformes, containing the manta rays and devil fish; found in all tropical and temperate oceans.

MOCA *Oncology.* an acronym for a chemotherapy regimen for cancer treatment that includes the use of the drugs methotrexate, Oncovin, cyclophosphamide, and Adriamycin.

Mochokidae *Vertebrate Zoology.* the upside-down catfishes, a family of scaleless, freshwater African fish of the order Siluriformes, having several barbels, forked tails, and long fins; noted for swimming on their backs for long periods while eating algae on the undersides of leaves.

mock fog *Meteorology.* a simulation of a fog caused by anomalous atmospheric refraction.

mock infection *Virology.* the inoculation of an organism with a placebo or solution that does not contain infectious virus particles; used as a control in virus infection experiments.

mocking *Behavior.* the imitating by one species of sounds made by another species. Also, VOCAL MIMICRY.

mockingbird *Vertebrate Zoology.* a common American songbird, *Mimus polyglottos*, of the family Mimidae; characterized by a long tail, a slender body, a sharp bill, and brown or blue plumage with white underpatches; known for its ability to imitate other birds.

mockingbird

mock orange *Botany.* any of various shrubs of the genus *Philadelphus* of the saxifrage family, especially *P. coronarious*; characterized by fragrant white flowers; cultivated widely.
mock silver *Metallurgy.* **1.** an aluminum-base alloy containing copper and tin, or copper and silver. **2.** a brass containing 45° copper.
mock sun see PARHELION.
mock-up *Engineering.* a scale model, often full-size, of a structure, apparatus, or vehicle; used for study, training, or testing and to determine if the apparatus can be manufactured easily and economically.
modal action pattern *Behavior.* the smallest recognizable unit of behavior that can be identified as a recurrent pattern in a given species.
modal analysis *Electrical Engineering.* a method of decomposing a system of equations into independent or orthogonal variables; frequently used to decompose the spatial variations of electromagnetic waves into components, each of which satisfies the boundary conditions, or to analyze the behavior of a control system.
modal class *Statistics.* the class interval that contains the most observations.
modal length *Virology.* the standard length of rod-shaped virus particles; can be used with plant viruses to distinguish between groups of viruses.
modal logic *Computer Science.* a logic that allows modeling of such things as knowledge and belief ("John believes that Mary knows his phone number"), possibility ("It might rain tonight"), and necessity ("Water flows downhill").
modal personality *Anthropology.* a premise of culture and personality studies in which the personality traits for a group are defined according to the highest statistical incidence of these traits in individuals of the group.
modal time *Industrial Engineering.* the elapsed time value that occurs most often in the study of an element or an operation.
mode a manner, method, or state of acting or doing; specific uses include: *Physics.* a state of a vibrational system oscillating at one of the possible resonant frequencies. *Electromagnetism.* a form of propagation that describes the manner in which an electromagnetic wave travels; characterized by indicating the field pattern in a plane perpendicular to the propagation direction. *Telecommunications.* in a communication, the nature of the information, such as literal language or digital data. *Computer Technology.* a method of operation, such as binary mode or alphanumeric mode. *Computer Programming.* the form or data type of a variable or expression. *Statistics.* the most frequently occurring value in a distribution, or, for grouped data, the class interval into which the largest number of cases fall; a measure of central tendency. *Petrology.* the actual mineral composition of an unaltered igneous rock, expressed in terms of percentages of total weight or volume. *Archaeology.* the most specific category of classification for artifacts, representing items within the same type and variety that share further common attributes.
mode bit *Computer Programming.* a bit that can be set to control the mode of operation of the central processing unit or a peripheral device.
mode eddy *Oceanography.* a more or less circular, slow-moving body of water, up to 400 kilometers across and 3 kilometers deep, especially common in the Western Atlantic and the Western Pacific; in the Northern Hemisphere, the weak internal circulation in the upper part of the eddy moves clockwise, while in the lower part of the eddy it moves counterclockwise.
mode filter *Electromagnetism.* a waveguide filter that can separate different transmission modes of the same frequency.
mode jump *Electronics.* a sudden change in oscillation frequency and power level caused by a change in the mode of magnetron operation from one pulse to the next.
model *Science.* a pattern, plan, replica, or description designed to show the structure or workings of an object, system, or concept. *Behavior.* the species or structure from which a mimic takes its deceptive appearance or the sound a mimic uses for its imitative vocalization. *Computer Programming.* **1.** a mathematical representation of a process, system, or device. **2.** in mathematical logic, an interpretation that satisfies a formula. *Artificial Intelligence.* an interpretation under which a given logical formula is true. *Statistics.* the probabilistic structure underlying a random experiment, often summarized by a set of parameters. *Photogrammetry.* **1.** the three-dimensonal image produced when light shown through a stereoscopic pair of photographs is combined and focused on a viewing screen. **2.** the three-dimensional image produced by the use of a holograph.
model atmosphere *Meteorology.* any theoretical representation of the atmosphere.
model base *Cartography.* the line, or the length of the line, at the scale of a stereoscopic model, joining the respective centers as reproduced by the stereoscopic instrument.
model basin *Engineering.* a tank of water used for hydrodynamic tests on scale models of ships, seaplanes, and the like.
model coordinates *Photogrammetry.* the coordinates of any point in a model that define that point's position relative to the air base from which the photographs were taken or to the axes used by the instrument that produces the model.
model datum *Photogrammetry.* the level of the model representing the sea-level datum for the portion of the earth's surface that the model represents.
model-following problem *Control Systems.* the process of finding a control method that will make the response of a given system as close as possible to that of a model system, given the same input.
modeling *Behavior.* the process of learning and performing a new behavior based upon observation of others and imitation of their actions. *Psychology.* a form of behavior therapy in which the subject observes and then imitates appropriate behavior by another person.
modeling methodology *Design Engineering.* the reasoning and procedure associated with the production of a machine, process, or product.
mode-locked laser *Optics.* a laser in which numerous modes of oscillation, with tightly spaced wavelengths, are timed to produce a pulse of light that lasts less than a microsecond.
model reduction *Control Systems.* the process of reducing the size of the model used by an active control system in order to increase the speed with which the system can compute control commands.
model reference system *Control Systems.* **1.** an ideal system with optimal responses. **2.** a computer simulation in which a model system and an actual system are subjected to the same stimuli and responses are compared; then the parameters of the actual system are adjusted to minimize the differences between the outputs of the two systems.
model scale *Cartography.* the ratio of the distance between two points as measured on a model to the corresponding distance between the real points as measured on the earth's surface.
model symbol *Computer Programming.* any of various standard geometric shapes used in graphical models, such as program flow charts and diagrams.
modem *Electronics.* a device connected between a digital circuit and an analog communication system, such as a telephone, to transmit and convert the received analog signal to digital data. (An acronym for <u>mo</u>dulator/<u>dem</u>odulator.)
mode number *Electronics.* **1.** the number of cycles during which an electrode traveling at an average speed remains in the drift space of a reflex klystron. **2.** the number of radians of phase in a magnetron's microwave field divided by 2π, moving once about the anode.
mode of vibration *Mechanics.* a vibration of a coupled system, which occurs in such a way that every part of the system vibrates with the same frequency.
moderate breeze *Meteorology.* a wind whose speed measures 13–18 mph on the Beaufort wind scale.
moderate gale *Meteorology.* a wind whose speed measures 32–38 mph on the Beaufort wind scale.
moderate mental retardation *Psychology.* a level of mental retardation characterized by an IQ ranging from 40 to 54; those affected have limited learning abilities but can learn to take care of themselves and perform simple tasks.
moderator *Nucleonics.* a material that slows the velocity of neutrons so that they may be absorbed by nuclei; ordinary water, heavy water, and graphite are the most commonly used.
modern algebra *Mathematics.* the study of algebraic systems such as groups, rings, fields, modules, vector spaces, etc. Also, ABSTRACT ALGEBRA.
modern control *Robotics.* a type of robotic control in which the dynamics of a process and the constraints on measuring it are used to create an optimal control system.
mode shift *Electronics.* a change in the mode of magnetron operation during a pulse.
mode skip *Electronics.* the failure of a magnetron to fire on each successive pulse.
mode switch *Computer Technology.* a switch that changes the mode of operation of a device; for example, a control on a mechanical desk calculator that sets its normal response to user input. *Electronics.* a microwave control device that is used to change the microwave power transmission mode in a waveguide.

mode transducer *Electronics.* a device that is designed to transform an electromagnetic wave from one mode of propagation to another. Also, **mode transformer.**

modifer gene *Genetics.* a gene that changes or augments the phenotypic expression of a nonallelic gene.

modifiability *Mechanical Devices.* the ability of a machine or device to be modified after it has been produced. *Robotics.* specifically, the ability to change or alter a robot's program.

modification *Engineering.* a temporary or permanent change made to an object or process in order to improve design or performance, or to correct a defect. *Behavior.* see BEHAVIOR MODIFICATION.

modification enzymes *Enzymology.* bacterial enzymes (methylases) that attach methyl groups at special sites on the bacterial DNA, thereby protecting it from digestion by indigenous endonucleases that digest foreign DNA injected into the cell by a bacteriophage.

modification kit *Engineering.* a group of tools or components used to modify the performance or design of an object or process.

modified Bessel functions *Mathematics.* Bessel functions of imaginary argument. A commonly used form is given by

$$K_n(x) = i^{-n} J_n(ix),$$

where J_n is the Bessel function of order n and x is positive and real.

modified constant-voltage charge *Electricity.* a method of charging a storage battery in which voltage of the charging current remains constant over a fixed resistance inserted into the battery circuit; a high-voltage characteristic occurs at the battery terminals as the charge increases.

modified gunmetal *Metallurgy.* a form of gunmetal that contains 2.5% lead.

modified Hankel functions *Mathematics.* Hankel functions of imaginary argument. A commonly used form is given by

$$K_n(x) = \pi\, i/2(i)^n H_n^{(1)}(ix),$$

where $H_n^{(1)}$ is the first Hankel function of order n and x is positive and real; sometimes included in the definition of modified Bessel functions. Also, MACDONALD FUNCTIONS.

modified index of refraction *Meteorology.* the sum of the refractive index of the atmosphere at a given height and the ratio of the height to the radius of the earth.

modified Lambert conformal chart *Cartography.* a chart based on the modified Lambert conformal map projection. Also, NEY'S CHART.

modified Lambert conformal map projection *Cartography.* the Lambert conformal map projection as modified for use in polar maps and charts, with two standard parallels, and with all parallels forming complete concentric circles. Also, NEY'S MAP PROJECTION.

modified Lewis acid *Physical Chemistry.* an acid that accepts halide ions.

modified polyconic map projection *Cartography.* the regular polyconic projection modified so that the scale is exact along two standard meridians, equidistant from the central meridian, rather than along only one. Also, RECTANGULAR POLYCONIC MAP PROJECTION.

modified radical mastectomy *Medicine.* the surgical removal of the entire breast including nipple, areola, overlying skin, and axillary lymph nodes.

modifier something that changes or alters; specific uses include: *Computer Programming.* a quantity that is used to modify the address of an operand in an instruction; for example, the cycle index in a loop sequence. *Materials Science.* an oxide that, when added to a glass, disrupts the glassy network, eventually causing crystallization.

modifier gene *Genetics.* a gene that changes or augments the phenotypic expression of a nonallelic gene.

modify to change the form or effect of something; specific uses include: *Computer Programming.* **1.** to alter an instruction in a way that changes its interpretation and execution, either temporarily or permanently. **2.** to alter a program according to a given parameter.

modillion *Architecture.* an embellished bracket used in series to support the corona under a cornice, especially of the Corinthian order.

moding *Electronics.* a fault in magnetron oscillation in which it oscillates in an undesired mode or modes.

modioliform *Paleontology.* having the shape that is distinctive of the genus *Modiolus.*

Modiolus *Paleontology.* a still-extant genus of pteriomorphic bivalves in the subclass Isofilibranchia and order Mytiloida; *Modiolus* is generally a burrower but in the past may have also sometimes attached itself byssally to seaweed. It has a distinctive shape, which is also found in a few other genera.

modiolus *Anatomy.* the central post around which the spiral bones of the cochlea are formed.

Mod. praesc. *Medicine.* in the way directed. (From Latin *modo praescripto.*)

Modula *Robotics.* a modular programming language similar to Pascal that can be used to program a robot.

Modula-2 *Computer Programming.* a high-level programming language developed for systems programming that supports modularity and contains constructs for parallel programming; derived from Pascal.

modular *Science.* composed of standardized units or sections. *Robotics.* relating to or designating a robotic system composed of units with interchangeable parts that can be combined in various ways to perform a wide variety of tasks.

modular arithmetic *Mathematics.* in elementary number theory, addition and multiplication of residue classes. Intuitively, this is accomplished by forming the sum or product in the usual way, dividing the result by a fixed positive integer m (the **modulus**), and using the remainder as the answer to the computation. Identical results are obtained by using representatives of the congruence classes (mod m). In this context, congruence classes are also referred to as residue classes. The integers (mod m) form a commutative ring with unit element. If m is a prime, then the integers (mod m) form a field. Also, ARITHMETIC MODULO M, ARITHMETIC MOD M.

modular circuit *Electronics.* a circuit designed and built as separate modules that are connected to perform the function.

modular compilation *Computer Programming.* the translation of portions of a program into machine language, followed by link editing, as opposed to translation of the complete program.

modular design *Robotics.* a system design that uses standardized software and hardware products.

modular form *Mathematics.* **1.** an analytic function $F(z_1,z_2)$ of two complex variables defined for $\text{Im}(z_1/z_2) > 0$ such that: (a)

$$F(\lambda z_1,\lambda z_1) = \lambda^r F(z_1,z_2),$$

where λ is any constant (i.e., homogeneity of degree r); and (b)

$$F(dz_1 + cz_2, bz_1 + az_2) = F(z_1,z_2),$$

where a, b, c, and d are any constants such that $ad - bc = 1$. **2.** alternatively, a meromorphic function $F(\tau)$ defined for $\text{Im}\,\tau > 0$ is called a modular form of degree r if it satisfies

$$F((a\tau + b)/(c\tau + d)) = (c\tau + d)^{-r} F(\tau),$$

where a, b, c, and d are any constants such that $ad - bc = 1$. A modular form of degree 0 is a **modular function** or **automorphic function.**

modularity *Computer Programming.* the property of a program that is composed of a collection of well-defined, logically self-contained segments or subroutines.

modular programming *Computer Programming.* the design of a program as a set of logically discrete and self-contained parts or modules; used to break up large programs into smaller tasks, thus improving programming accuracy and production. Also, STRUCTURED PROGRAMMING.

modular robot *Robotics.* a nonservo robot that has seven axes; used for parts transfer, machine loading and unloading, and pressworking.

modular structure *Electronics.* **1.** an assembly using integral multiples of a given length to form electronics components. **2.** an assembly created from modules.

modulate *Electronics.* to change in some characteristic way the amplitude, frequency, or phase of a wave or the velocity of the electrons in an electron beam.

modulated amplifier *Electronics.* the amplifier stage in a transmitter in which the modulating signal is introduced and modulates the carrier.

modulated carrier *Telecommunications.* a radio-frequency carrier in which the frequency or amplitude has been varied by, and in accordance with, the conveyed intelligence.

modulated continuous wave *Telecommunications.* a wave in which the carrier is modulated by a steady audio-frequency tone.

modulated Raman scattering *Spectroscopy.* a spectroscopic technique in which wavelength variation and externally applied perturbations are used to lower symmetry in order to analyze the light scattered by certain crystals.

modulated stage *Electronics.* the radio-frequency stage to which the modulator is coupled, in which the carrier wave is modulated in accordance with the system of modulation and the characteristics of the modulating wave.

modulated structure *Crystallography.* a crystal in which true three-dimensionall lattice periodicity is lost but which can be described as an (imaginary) basic structure with space group symmetry combined with periodic deformation (modulation). The diffraction pattern gives information on each of these two components.
modulating codon see MODULATION CODON.
modulating electrode *Electronics.* in a cathode-ray tube, an electrode to which a potential is applied to control the magnitude of a beam current.
modulating signal *Telecommunications.* a wave that varies some characteristic of the carrier, such as the frequency, phase, or amplitude. Also, **modulating wave.**
modulation *Mechanical Engineering.* the process of regulating the air to fuel ratio in a burner for varying the load conditions on a boiler. *Telecommunications.* a process in which a characteristic of one wave is varied in accordance with the characteristic of another wave. *Molecular Biology.* a process of more frequent translation of particular sequences of messenger RNA.
modulation code *Telecommunications.* a code that causes variation in a given signal in accordance with a specific scheme; usually implemented in order to alter a carrier wave to transmit data.
modulation codon *Molecular Biology.* a codon that functions as a regulatory agent, coding for a rare tRNA and causing an interruption or a change in the rate of protein translation. Also, MODULATING CODON.
modulation crest *Telecommunications.* the maximum amplitude of an amplitude-modulated wave.
modulation-doped structure *Solid-State Physics.* a semiconductor heterostructure in which conduction electrons are spatially separated from their parent donor or acceptor impurity atoms.
modulation envelope *Telecommunications.* a curve drawn through the peaks of a graph that shows how the waveform of a modulated carrier depicts the waveform of the intelligence carried by the signal.
modulation factor *Telecommunications.* a mathematical formula for determining the percentage of modulation of an amplitude-modulated wave. The ratio of half the difference between the maximum and minimum amplitudes to the average amplitude is multiplied by 100.
modulation index *Telecommunications.* in frequency modulation with a sinusoidal modulating wave, the ratio of the frequency deviation to the frequency of the modulating wave. Also, RATIO DEVIATION.
modulation meter *Engineering.* a device that is used to measure the modulation of a modulated wave train, with readings expressed as a percentage.
modulation rise *Electronics.* an increase of the modulation percentage due to nonlinearity of a tuned amplifier; usually, the last intermediate-frequency stage of a receiver.
modulation spectroscopy *Spectroscopy.* the measurement and analysis of changes in transmittance or reflectance spectra induced by an externally applied perturbation, such as a magnetic field or a change in pressure.
modulation transformer *Acoustical Engineering.* an audio-frequency transformer used for matching impedances and transmitting such frequencies to the output stage of an audio amplifier.
modulation with a fixed carrier *Telecommunications.* phase modulation with a pilot carrier.
modulator *Electronics.* a circuit or device that varies some characteristic of a carrier signal in accordance with the waveform of a modulating signal. *Biochemistry.* see EFFECTOR.
modulator crystal *Optics.* a crystal that modulates light by electro-optic or magnetic effects, most commonly a Pockels cell.
modulator-demodulator see MODEM.
modulator glow tube *Electronics.* a cold-cathode recorder tube used for facsimile and sound-on-film recording to provide a modulated high-intensity point-of-light source.
module [mäj´əl; mäj´ool] a distinct unit or component; specific uses include: *Engineering.* any of various standards or units of measurement used in building, design, and civil engineering. *Building Engineering.* a structural unit that is designed to be joined with others. *Space Technology.* one of the individual, self-contained units of a spacecraft, having a distinct function. *Electronics.* **1.** an assembly of self-contained, interconnected components that constitutes an identifiable electronic device, instrument, or piece of equipment. **2.** a complete subassembly of such a unit. *Computer Programming.* a discrete, logical component of a program. *Computer Technology.* **1.** an interchangeable plug-in item containing electronic components that complete, enhance, or expand processing capability or memory capacity. **2.** a single memory bank. *Mathematics.* Let R be a ring. An (left) R-module is an additive abelian group A together with a (left) multiplication (sometimes called scalar multiplication) by members of R such that, for all $r, s \in R$ and all $a, b \in A$, (a) $r(a + b) = ra + rb$, (b) $(r + s)a = ra + sa$, and (c) $r(sa) = (rs)a$. If R has an identity element 1_R and if $1_R a = a$ for all $a \in A$, then A is said to be a unitary R-module. Right modules are similarly defined for scalar multiplication on the right. If R is a division ring (skew field), the module is called a vector space.
modulo *Mathematics.* a general term for the formation of congruence classes (a particular type of equivalence class) for a given modulus within an algebraic object. For example, if H is a subgroup of a group G, then two elements a and b of G are said to be congruent modulo H if $ab^{-1} \in H$; denoted $a \equiv b \pmod{H}$. In particular, if G is the integers under addition and the subgroup H consists of all multiples of a fixed integer m, then the resulting congruence classes are referred to as "the integers (mod m)" and the rules of modular arithmetic are used for computations. Two integers are said to be equivalent mod m if their difference is equal to a multiple of m.
modulo-N check *Computer Programming.* a procedure used to verify computational accuracy by repeating the calculations in modulo-N arithmetic and comparing remainders. Also, RESIDUE CHECK.
modulo-two adder *Computer Technology.* an electronic circuit that functions as an exclusive-or gate, returning a 1 if the inputs are different and a 0 if they are the same.
modulus [mäj´ə ləs] *Mathematics.* **1.** the number in a modular arithmetic. **2.** the subgroup used to form congruence classes within an algebraic object. Also, **modulus of the congruence. 3.** given a complex number $a + bi$, the quantity $(a^2 + i^2)^{1/2}$. Also called the absolute value or absolute magnitude of the complex number. The modulus of a function of a complex variable is similarly defined. **4.** the norm of a vector in Euclidean n-space. **5.** the value k in the elliptic integral of the first kind

$$u = F(k,\phi) = \int_0^{\phi}(1 - k^2\sin^2\theta)^{-1/2}d\theta.$$

Denoted by k = mod$u$. *Mechanics.* see MODULUS OF ELASCTICITY.
modulus of compression see BULK MODULUS.
modulus of continuity *Mathematics.* a former way of expressing the relationship between ε and δ in the definition of (uniform) continuity. The modulus of continuity of a given function f on the interval $[a,b]$ is

$$g(\delta) = \sup\{|f(x_1) - f(x_2)| : |x_1 - x_2| < \delta \text{ and } a \le x_1, x_2 \le b\}.$$

modulus of deformation *Mechanics.* the modulus of elasticity of materials not having elastic proportionality.
modulus of distance see DISTANCE MODULUS.
modulus of elasticity *Materials Science.* the stress per unit elastic strain, expressed as a ratio between the stress placed on a material and the strain, or dimensional response to stress. The most commonly encountered modulus of elasticity is Young's modulus: $E = F(s,e) = F(F/A,\Delta L/L)$. The modulus of elasticity is a measure of the stiffness of a material. Also, ELASTIC MODULUS.
modulus of elasticity in shear see SHEAR MODULUS.
modulus of precision see INDEX OF PRECISION.
modulus of resilience *Mechanics.* the strain energy per unit volume that an elastic material will store in uniaxial stress at the material's elastic limit.
modulus of rigidity *Mechanics.* see SHEAR MODULUS.
modulus of rupture in bending *Mechanics.* the fictitious maximum stress in a bending member at failure, computed as though the member retained linear elasticity.
modulus of rupture in torsion *Mechanics.* the fictitious maximum shear stress in a twisting member at failure, computed as though the member retained linear elasticity.
modulus of the complex number see ABSOLUTE VALUE.
modulus of torsion see TORSIONAL MODULUS.
modulus of volume elasticity see BULK MODULUS.
modus ponens *Artificial Intelligence.* a rule of logical inference: if P is true and $P \rightarrow Q$, conclude Q. (From Latin; literally, "a method of putting in place.")
MODY maturity-onset diabetes of youth.
Moellerella *Bacteriology.* a genus of Gram-negative bacteria of the family Enterobacteriaceae that are found in human feces and may be associated with diarrhea.
Moeritheriidae *Paleontology.* a family of subungulate proboscidean mammals in the extinct suborder Moeritherioidea; the first of the elephants, tapir-sized, with canines that foreshadow the development of tusks in later genera; Eocene and Oligocene.

mofette *Volcanology.* a small vent from which carbon dioxide along with some nitrogen and oxygen is emitted in an area of late-stage volcanic activity.
mohavite octahedral borax see TINCALCONITE.
Mohawkian *Geology.* a North American geologic stage of the Middle Ordovician period, occurring after the Chazyan and before the Cincinnatian.
Mohnian *Geology.* a North American geologic stage of the Miocene period, occurring after the Luisian and before the Delmontian.
Moho *Geophysics.* a shorter term for the Mohorovičić discontinuity.
Mohole *Geology.* a hole bored in the earth's crust into the region beneath the Mohorovičić discontinuity, for geological research. (An acronym for Mohorovičić discontinuity hole.)
Mohole project *Geology.* a U.S. project proposed to obtain samples of the earth's crust and mantle by drilling off the coast of Hawaii to the Mohorovičić discontinuity, at a total depth of approximately 9755 meters; Phase I succeeded in drilling to a depth of several hundred meters off the coast of California in 1961; Phase II, involving the study of deep-ocean drilling, was abandoned in 1966. Also, **Mohole drilling.**
Mohorovičić, Andrija [mō´hō rō´və chich] 1857–1936, Serbian seismologist; discovered the discontinuity separating the earth's mantle from its outer crust.
Mohorovičić discontinuity *Geophysics.* the boundary in the earth between the crust and the mantle that is marked by an abrupt change in the propagation velocity of seismic waves; approximately 10 kilometers below the ocean basins and 30 kilometers below the continents.
Mohr, Carl Friedrich 1806–1879, German chemist; developed titration techniques.
Mohr cubic centimeter *Chemical Engineering.* a unit of volume used in saccharimetry that equals 1.00238 cubic centimeters. (Named for Carl Friedrich *Mohr.*)
Mohr liter *Chemistry.* a unit of capacity equal to 1000 Mohr cubic centimeters.
Mohr's circle for stress *Mechanics.* a graphical representation of the components of stress in a solid; each point gives the normal and shear components of stress on a plane in the solid.
Mohr's salt see FERROUS AMMONIUM SULFATE.
Mohr titration *Analytical Chemistry.* a test formerly used to determine the concentration of chlorides in a solution by titrating the sample with silver nitrate.
Mohs, Friedrich [mōz] 1773–1839, German mineralogist; originated the Mohs scale of hardness.
Mohs scale *Mineralogy.* a standard of ten minerals by which mineral hardness may be rated; from the softest to the hardest on a scale of 1 to 10: talc (1), gypsum (2), calcite (3), fluorite (4), apatite (5), orthoclase (6), quartz (7), topaz (8), corundum (9), and diamond (10). Also, **Mohs hardness scale.**
moiety [moi´i tē] *Anthropology.* a unilineal descent group that is one of two such divisions in a society; the members of a group believe themselves to be descended from a common ancient ancestor. *Chemistry.* a specific section of a molecule, usually complex, that has a characteristic chemical effect or property.
moiré [mwä rā´; môr´ā] *Graphic Arts.* an undesirable blotch or pattern of lines or dots that may appear on a printed piece when a halftone screen is superimposed over a lined original or another halftone. *Telecommunications.* Also, **moiré pattern. 1.** an unwanted optical effect occurring when one set of closely spaced lines is imperfectly imposed over another. **2.** in television, disturbance brought about by interference beats of similar frequencies.
moiré effect *Materials Science.* the pattern formed by combining two sets of regular patterns. *Optics.* an effect created by crossing two fine screens laid one on top of the other.
moiré fringe *Materials Science.* the appearance of dark fringes created when two ruled patterns are superimposed on each other; used to measure the displacement of one pattern with respect to the other. *Optics.* the interferogramlike fringes produced by the moiré effect.
moiré interferometry *Engineering.* a visual method of measuring the deformations of a surface by superimposing a reference grid that conforms with the surface and a diffraction grid that does not.
Moissan, Henri 1852–1907, French chemist; awarded the Nobel Prize for isolating fluorine and developing the electric arc furnace.
moissanite *Mineralogy.* α-SiC, a dark-green to black hexagonal mineral, having a specific gravity of 3.1–3.22 and a hardness of 9.5 on the Mohs scale; found as small hexagonal plates in meteoric iron. Also, ARTIFICIAL CARBORUNDUM.
moist air *Meteorology.* **1.** air with a high relative humidity. **2.** air consisting of a mixture of dry air and any amount of water vapor.
moist climate *Meteorology.* a climate in which the seasonal water surplus counteracts the seasonal water deficiency, resulting in a moisture index greater than zero.
moisture *Physical Chemistry.* water or other liquid that is dispersed through a gas or a solid, or condensed on a solid, as a vapor or as very fine droplets. *Meteorology.* **1.** specifically, the water vapor content of the atmosphere, or the total water substance present in a given volume of air. **2.** the quantity of precipitation or of precipitation effectiveness that is present. *Thermodynamics.* a thermodynamic property that represents the fraction of mass present in the liquid phase; the moisture, M, is numerically equal to one minus the quality X; $M = (1 - X)$.
moisture content *Engineering.* the water content of a substance, determined using prescribed methods under specified conditions and stated as a percentage of the wet or dry weight.
moisture factor *Meteorology.* one of the simplest measures of precipitation effectiveness, given as P/T, where P is precipitation and T is mean temperature.
moisture inversion *Meteorology.* the layer in which there is an increase with height of the moisture content of the air, or the altitude at which the increase begins.
moisture loss *Mechanical Engineering.* the resulting differential in heat content between the moisture content of air in a boiler and the ambient environment outside of it.
moisture-retentive *Agronomy.* describing soil that retains a high percentage of water.
Moivre, Abraham de 1667–1754, French mathematician; known for his use of calculus to study probability.
mojarra *Vertebrate Zoology.* a small, silvery perchlike fish of the family Gerreidae, characterized by a protrusible mouth used to probe the bottom for food; found predominantly in tropical marine waters.
Moko disease *Plant Pathology.* a disease of bananas caused by the bacterium *Pseudomonas solanacearum,* characterized by drooping and shriveling of the plant, following loss of turgidity.
mol see MOLE.
MOL manned orbiting laboratory.
mol. molecule; molecular.
mola *Vertebrate Zoology.* any of many thin, silvery fish of the family Molidae found in tropical and temperate seas.
molal *Physical Chemistry.* **1.** of a solution, containing one mole of solute per one kilogram of solvent. **2.** see MOLAR.
molal average boiling point *Physical Chemistry.* the theoretical boiling point of a mixture, arrived at by averaging the boiling points of the individual mole fractions in a mixture.
molal heat capacity see MOLAR HEAT CAPACITY.
molality *Physical Chemistry.* a description of the concentration of a solution, expressed by the number of moles of solute per kilogram of solvent.
molal quantity *Chemistry.* the number of moles, expressed by weight in units such as gram-mole, that is equal to the molecular weight.
molal solution *Physical Chemistry.* a solution that contains one mole of solute for each kilogram of solvent.
molal specific heat see MOLAR SPECIFIC HEAT.
molal volume see MOLAR VOLUME.
molar *Anatomy.* a large, posterior tooth that is used for grinding food and supporting the jaw; there are usually 12 molars in the adult human mouth.
molar *Chemistry.* **1.** of a solution, containing enough solvent so that one mole of a solute will dissolve to make one liter of the solution. **2.** describing a physical quantity of some substance in terms of one mole of the substance; i.e., in terms of the relationship between the amount of substance and its formula weight. Also, MOLECULAR.
molar absorptivity *Analytical Chemistry.* absorptivity defined in terms of concentration expressed as moles per liter.
molar activity see TURNOVER NUMBER.
molar attraction constant *Materials Science.* a constant based on a series of values for the groups found on polymers; used to determine the solubility parameters of polymers and solvents.
molar conductivity *Physical Chemistry.* the ratio of conductivity in an electrolytic solution to the fraction of moles per unit volume in the solution. Also, MOLECULAR CONDUCTIVITY.
molar dispersion *Optics.* the variation in the molar refraction of a compound at two separate light-beam wavelengths.
molar gas constant see GAS CONSTANT.

molar heat capacity *Physical Chemistry*. the quantity of heat needed to raise the temperature of one mole of a substance by one degree. Also, MOLECULAR HEAT CAPACITY, MOLAL HEAT CAPACITY.

molarity *Physical Chemistry*. a description of the concentration of a solution, expressed by the number of moles of solute per liter of solution.

molar latent heat *Physical Chemistry*. the amount of latent heat per mole of a substance.

molar mass *Physical Chemistry*. the mass in grams of one mole of a given substance; the practical unit of measurement is g mol^{-1}.

molar refraction *Optics*. a formula that shows the refractive index of a compound when modified by the molecular weight and density of the compound. Also, LORENTZ-LORENZ MOLAR REFRACTION.

molar solution *Physical Chemistry*. a solution that contains one mole of solute for each liter of the solution.

molar specific heat *Physical Chemistry*. the ratio between the quantity of heat needed to raise the temperature of one mole of a given substance by one degree and the quantity of heat needed to raise the temperature of one mole of a reference substance, such as water, by one degree at a given temperature. Also, MOLAL SPECIFIC HEAT; MOLECULAR SPECIFIC HEAT.

molar surface *Physical Chemistry*. the area occupied by a sphere of one mole of a substance.

molar volume *Physical Chemistry*. the volume that one mole of a given substance occupies in the form of a solid, liquid, or gas at standard temperature and standard atmospheric pressure. Also, MOLAL VOLUME; MOLECULAR VOLUME; MOLE VOLUME.

Molasse *Geology*. **1.** a sedimentary sequence deposited during the Miocene period in southern Germany, subsequent to the rising of the Alps; composed primarily of soft green sandstone associated with marl and conglomerates. **2. molasse.** any similar sedimentary sequence.

molc. molar concentration.

mold *Engineering*. a hollow form or pattern into which molten metal or plastic is poured to form a desired shape when the material cools. *Acoustical Engineering*. the metal negative counterpart used to press record albums in the disk recording production. *Building Engineering*. a thin board or zinc sheet cut to a particular profile and used to run cornices. *Geology*. **1.** an original mark or depression made by a shell or other body on a sedimentary surface. **2.** a soft, friable humus-rich soil suitable for plant growth. *Graphic Arts*. a negative impression of a type character or line into which molten metal is poured to cast positive type. *Paleontology*. the sediment or volcanic material that surrounds a dead organism (an **external mold**) or that fills any void inside the organism when its soft parts decay (an **internal mold**); corresponding artificial molds may be produced by various techniques. *Mycology*. a common term for fungi that have a cottony or wooly appearance.

moldability *Materials Science*. the relative ability of a material to be molded into shape.

moldavite *Mineralogy*. a translucent, olive to pale green tektite from western Czechoslovakia, whose surface has been etched by solution. Also, MOLDAUITE, PSEUDOCHRYSOLITE.

moldboard *Agriculture*. the part of a plow bottom just behind the share that turns the furrow slice. Also, **moldboard bottom.**

moldboard plow *Agriculture*. a common type of plow that uses a moldboard to turn the furrow slice.

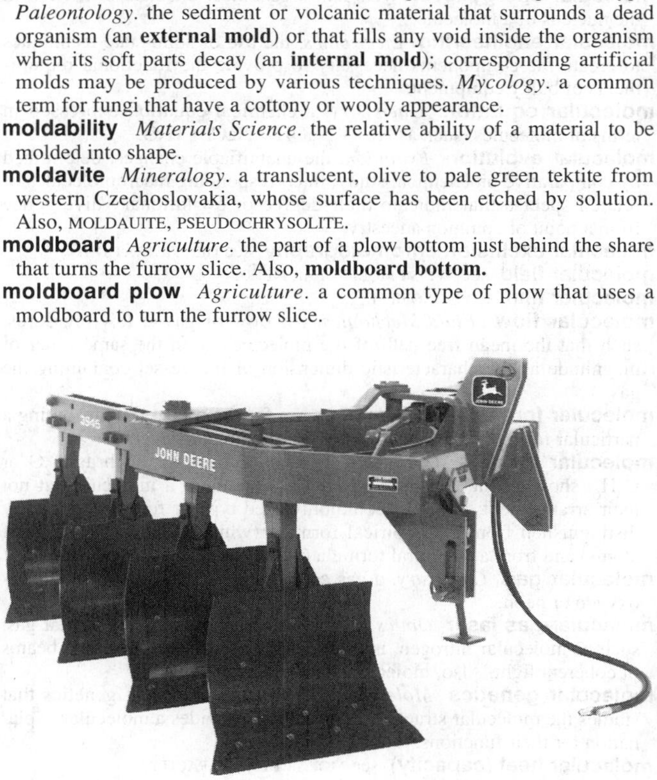

moldboard plow

mold constant *Materials Science*. a constant, denoted by the letter B in Chvorinov's rule stating that $t_s = B(F(V,A))^2$ where t_s is the total solidification time, V is the volume of the casting, and A is the surface area of the casting in contact with the mold.

molded breadth *Naval Architecture*. the maximum width of a hull as measured across its frames, excluding the thickness of the shell plates or planking.

molded brick *Mechanical Devices*. a brick fabricated from a mold, as distinguished from a wire-cut brick; used for ornamental work.

molded capacitor *Electricity*. a capacitor that is encased in a protective plastic-insulating material to keep out dust and moisture.

molded depth *Naval Architecture*. the depth of a hull as measured from the upper deck beam to the keel.

mold efficiency *Engineering*. in the molding of objects, the amount or percentage of total turn-around time used for formation, cooling, and ejection from the molds.

mold inclusion *Metallurgy*. in the casting of metals, an inclusion defect that may appear as a result of a reaction between the molten metal and the mold material or the mechanical spalling of the mold.

molding *Architecture*. a continuous decorative band of curved or rectangular profile that is applied to or carved into a surface such as a wall or doorjamb. *Medicine*. adaptation of the fetal head to the size and shape of the birth canal.

molding cycle *Engineering*. **1.** all the operations involved in forming a molded object. **2.** the total amount of time required to complete a molding operation.

molding press *Metallurgy*. in powder metallurgy, a press used to make a compact.

molding pressure *Engineering*. the amount of pressure required to force molten material into a mold.

molding shrinkage *Engineering*. the difference in size between a mold and the mold cavity when measured at room temperature.

mold loft *Engineering*. a room or building in which the components of the hull of a ship or aircraft are laid out and assembled.

mold shrinkage *Materials Science*. in a molding process, the initial shrinkage that occurs while the part is forming in the mold.

mold steel *Metallurgy*. any steel suitable for making molds.

mole

mole *Vertebrate Zoology*. any small insectivorous mammal of the family Talpidae, living underground and having velvety fur, a long, naked muzzle, very small eyes, and strong forefeet for burrowing. *Mechanical Engineering*. an egg-shaped mechanical device that is pulled behind a subsoil plow and used in tunnel excavation. (From the Middle English name for this animal.)

mole *Medicine*. **1.** a congenital discolored spot elevated above the skin; a nevocytic nevus. **2.** broadly, any skin blemish. **3.** a fleshy mass or tumor formed in the uterus by the degeneration or abortive development of an ovum. (From an Old English word.)

mole *Chemistry*. the fundamental SI unit used to measure the amount of a substance; it is equal to the amount of a substance containing the same number of given elementary entities (such as atoms, molecules, ions, or electrons) as there are atoms of carbon in 12 grams of carbon-12. One mole of atoms contains Avogadro's number of atoms (6.02×10^{23}). It is most often used to express molecular weight. (From *molecule*.)

mole *Civil Engineering*. **1.** a massive breakwater or berthing facility, usually built of stone and extending from the shore to deep water. **2.** the area enclosed or protected by such a structure. (From Latin for *dam*.)

molecular *Chemistry*. **1.** relating to, consisting of, or caused by molecules. **2.** see MOLAR.

molecular adhesion *Physical Chemistry*. the force that causes solids or liquids to bind to each other; generally applied to adhesion between different materials.

molecular amplitude *Analytical Chemistry.* the difference between molecular rotations at the the extreme values caused by the longer and the shorter wavelengths.

molecular association *Physical Chemistry.* the binding together of two or more different molecules in the same system.

molecular asymmetry *Physical Chemistry.* see ASYMMETRY.

molecular attraction *Physical Chemistry.* any force that draws molecules toward one another.

molecular beam *Physics.* a narrow beam of neutral molecules, typically found in a vacuum.

molecular-beam apparatus *Physics.* a device in which magnetic or electric fields or other influences are imposed on a molecular beam in a vacuum in order to analyze the effects on the beam.

molecular-beam epitaxy *Solid-State Physics.* a technique in crystal growth in which a molecular beam is directed at a crystalline substrate in a vacuum, resulting in the formation of crystals whose orientation is related to the orientation of the substrate.

molecular binding *Solid-State Physics.* the forces that are responsible for holding together atoms or molecules in a solid.

molecular biology *Biology.* the branch of biology that attempts to interpret biological events in terms of the physicochemical properties of molecules in a cell.

Molecular Biology

Molecular Biology emerged as a distinct discipline in 1953, upon the discovery of the DNA double helix (the giant molecule in which hereditary information is inscribed). It combined the research agendas of two separate groups of biologists, one concerned with the three-dimensional structure of living molecules and the other with the storage, transmission, and expression of hereditary information.

By the late 1960s, molecular biologists had worked out the detailed structures of such proteins as enzymes and contractile and skeletal elements of the cell, as well as of such aggregates of proteins and nucleic acids as viruses, chromosomes (the carriers of the hereditary information), and ribosomes (the engines of protein synthesis). Moreover, they had identified the gene (the element of heredity, as a stretch of DNA double helix in which the structure of a protein molecule is inscribed via a genetic code; they had deciphered the genetic code (universal for all creatures), and they had elucidated the mechanism by which cells transcribe the genetic messages and translate them into the encoded protein molecules. Finally, they had uncovered the process by which hereditary information is replicated faithfully during cell reproduction (except for rare copy errors, or gene mutations).

During the 1970s and 1980s, the results of molecular-biological research came to be exploited in all branches of pure and applied biology. The invention of the powerful experimental techniques of recombinant DNA and gene cloning allowed the physical isolation, direct chemical modification, and reintroduction into living cells of virtually any gene. The methods for the rapid determination of the sequence of the molecular building blocks of protein and DNA molecules provided insights into their function and their evolutionary relations. These developments gave rise to a novel industry, styled "Biotechnology," and transformed Molecular Biology into the fundamental life science.

Gunther S. Stent
Professor of Molecular Biology
University of California, Berkeley

molecular circuit *Electronics.* a circuit whose components are indistinguishable from one another.

molecular clock hypothesis *Evolution.* a hypothesis stating that spontaneous molecular mutations occur with such regularity that a direct relationship can be determined between the degree of molecular divergence of two related phylogenetic branches and the amount of time that has passed since the divergence of those branches from the common ancestor.

molecular cloud *Astronomy.* a large interstellar cloud with temperatures low enough to permit atoms to form molecules; these are the Galaxy's main star-forming regions.

molecular conductivity see MOLAR CONDUCTIVITY.

molecular crystal *Crystallography.* a crystal whose lattice structure is formed from an arrangement of molecules.

molecular diamagnetism *Physical Chemistry.* the response of a compound to a magnetic field in which the response opposes the field.

molecular diameter *Physical Chemistry.* a measure of a molecule, assumed to be spherical in shape, calculated by multiplying its numerical value by a factor determined for the specific compound or element in which it appears.

molecular diffusion *Fluid Mechanics.* mass transport in a fluid by random molecular motion.

molecular dipole *Physical Chemistry.* a molecule that has two equal and opposite electric charges or poles separated by a short distance; this condition may be either intrinsic or generated by an external electronic field.

molecular distillation *Chemistry.* a distillation process performed at high vacuum and low temperature, with the condensation and evaporation surfaces within the mean molecular path distance, so as to cause the least damage to molecular composition.

molecular dynamics *Physical Chemistry.* the study of the action of forces on molecular bodies in motion or at rest; includes the study of time-dependent and temperature-dependent properties of microstructure systems, often using computer simulations.

molecular effusion *Fluid Mechanics.* the flow of free-molecule mass transport through pores and orifices.

molecular electronics *Electronics.* the branch of electronics that deals with the processing of a segment of semiconductor material so that it can perform the function of a complete electronic circuit.

molecular energy level *Quantum Mechanics.* the state of motion in the nucleus of a molecule.

molecular engineering *Electronics.* the use of solid-state techniques to create the components necessary to provide the functional requirements of bulky equipment.

molecular equation *Chemistry.* a chemical equation expressed in terms of molecules, such as $HCl + NaOH = NaCl + H_2O$.

molecular evolution *Evolution.* the quantifiable changes, determined by comparative biochemical study, in corresponding macromolecules of related species that indicate their degree of evolutionary divergence from a point of common ancestry.

molecular exclusion chromatography see GEL FILTRATION.

molecular field see WEISS MOLECULAR FIELD.

molecular film see MONOFILM.

molecular flow *Fluid Mechanics.* the flow of gas at low pressures, such that the mean free path of the molecules is on the same order of magnitude as the characteristic dimension of the vessel containing the gas.

molecular forensics *Oncology.* a term for the process of attributing a particular form of cancer to a particular carcinogenic agent.

molecular formula *Chemistry.* a chemical formula, such as H_2O or C_6H_6, showing the number and kinds of atoms in a molecule but not their arrangement; the most commonly used type of formula, generally distinguished from an empirical formula (which only shows kinds of atoms) and from a structural formula (which shows arrangement).

molecular gas *Chemistry.* a gas containing only one species, such as oxygen or neon.

molecular-gas laser *Optics.* a type of laser in which a molecular gas, such as molecular nitrogen, is used instead of atoms to generate beams of coherent light. Also, **molecular laser.**

molecular genetics *Molecular Biology.* the branch of genetics that studies the molecular structure of genes and provides a molecular explanation for their functions.

molecular heat (capacity) see MOLAR HEAT CAPACITY.

molecular heat diffusion *Thermodynamics.* heat transfer by conduction attributed to the motion of molecules.

molecular hybridization *Molecular Biology.* the controlled formation of double-stranded nucleic acids in vitro by base pairing between complementary single strands of nucleic acids, DNA-DNA, RNA-RNA, or DNA-RNA.

molecularity *Physical Chemistry.* the number of reacting molecules involved at separate points in a chemical reaction, regardless of the order or the mechanism by which the reaction takes place.

molecular-line astronomy *Astrophysics.* the study of celestial objects (mainly clouds of gas and dust) by means of radio emission from their molecules.

molecular luminescence *Spectroscopy.* luminescent light emitted from a molecular medium, the spectra of which are studied by spectroscopic means.

molecularly doped polymer *Materials Science.* any of various polymers consisting of solid solutions of molecules doped within an otherwise electrically inactive polymer matrix; used in electrophotography.

molecular magnet *Physical Chemistry.* a measure of the polarity in a molecule that has two equal and opposite electric charges or magnetic poles, either intrinsic or generated by an external electronic field.

molecular mass *Chemistry.* **1.** the sum of the atomic masses of the constituent atoms of a molecule. **2.** see RELATIVE MOLECULAR MASS.

molecular modeling *Chemistry.* a technique of simulating chemical and molecular structures or processes by computer.

molecular optics *Optics.* the science or the study of the propagation of light and other optical phenomena, such as refraction or scattering, through molecules in gases, liquids, and solids.

molecular orbital *Physical Chemistry.* the region of high probability that is occupied by an individual electron as it travels with a wavelike motion in the three-dimensional space around one of two or more associated nuclei; an indicator of the electron's energy level.

molecular orbital theory *Physical Chemistry.* an explanation of bond formation as the result of a region of electron density brought about by the overlap of atomic orbitals of similar energy in adjoining molecules.

molecular paramagnetism *Physical Chemistry.* the property that draws molecules, such as oxygen, into a position parallel to the lines of force in a magnetic field.

molecular pathology *Pathology.* the branch of pathology concerned with biochemical and biophysical cellular dynamics underlying certain diseases.

molecular physics *Physics.* a branch of physics devoted to the study of the behavior and interaction of molecules.

molecular polarizability *Materials Science.* the relative ease with which a material experiences molecular polarization. *Physical Chemistry.* the percentage of dipole moment in a molecule caused by an external magnetic field.

molecular polarization *Materials Science.* polarization caused by the asymmetrical nature of certain molecules or crystal structures.

molecular pump *Mechanical Engineering.* a pump that operates at very low vacuum; used to carry away molecules of spent gas through friction between the molecules in a rapidly rotating disk or drum.

molecular relaxation *Physical Chemistry.* a reduction in the energy level of a molecule from a high state to a lower or ground state.

molecular replacement method *Crystallography.* the use of noncrystallographic symmetry, and the constraints on the phases it causes, to solve a protein crystal structure.

molecular rotation *Optics.* the product of the exact angular rotation of polarized light times an optically active compound's molecular weight.

molecular sieve *Chemistry.* any substance used to filter out molecules of a certain size. *Analytical Chemistry.* specifically, a ceramic material or other substance with controlled lattice spacing, used to separate volatile mixtures by trapping larger molecules.

molecular sieve chromatography see GEL FILTRATION.

molecular specific heat see MOLAR SPECIFIC HEAT.

molecular spectroscopy *Spectroscopy.* the measurement and analysis of the electromagnetic radiation absorbed or emitted by molecular species.

molecular spectrum *Spectroscopy.* any spectrum produced by transitions within molecules as radiant energy is either emitted or absorbed.

molecular stopping power *Nucleonics.* the energy that an ionizing particle loses per molecule as it transverses a compound.

molecular structure *Physical Chemistry.* the assumed conformation of a molecule, as established by the forces that operate within it; depicted or calculated by various methods such as quantum mechanics or spectrum analysis.

molecular theory see KINETIC THEORY.

molecular theory of yielding *Materials Science.* a postulate of yield behavior relating to the plastic deformation of glassy polymers.

molecular vibration *Physical Chemistry.* the specific frequencies that describe the motions of the atoms within a given molecule.

molecular volume *Chemistry.* the volume occupied by 1 mole of a substance in gaseous form at standard temperature and pressure; it equals molecular weight divided by density.

molecular weight *Chemistry.* **1.** the sum of the atomic weights of all atoms in a molecule. **2.** see RELATIVE MOLECULAR MASS.

molecular-weight distribution *Organic Chemistry.* the frequency with which the various molecular-weight chains occur in a homologous polymeric system.

molecule *Chemistry.* the smallest unit of matter of a substance that retains all the physical and chemical properties of that substance, consisting of a single atom or a group of atoms bonded together; e.g., Ne, H_2, H_2O. (Formed from a Latin phrase meaning literally "a tiny mass.")

mole fraction *Chemistry.* in a system of mixed components, the ratio of the amount of a given component to the system, as expressed in moles.

mole percent *Chemistry.* any percentage expressed in terms of moles rather than by weight.

mole ratio *Physical Chemistry.* the relative amounts of reacting substances, expressed in moles, that are required to produce a given amount of product in a chemical reaction.

mole volume see MOLAR VOLUME.

Molgula *Paleontology.* a genus of ascidians comprising several dozen species still known worldwide; they are generally spherical and soft-bodied, having a long siphon and an almost transparent tunic; the fossil record is very incomplete.

Molidae *Vertebrate Zoology.* the molas or ocean sunfishes, a tropical and subtropical fish family of the order Tetraodontiformes, characterized by teeth fused into a single beak on each jaw; found in the Atlantic, Indian, and Pacific Oceans.

molimen *Medicine.* an extensive and laborious effort to carry out a normal function.

molinate *Organic Chemistry.* $C_9H_{17}NOS$, a yellowish liquid that is slightly soluble in water; used as an herbicide.

Molisch test *Biotechnology.* a quantitative test that indicates the presence of carbohydrates in a sample by the formation of a violet ring at the junction of the test chemicals.

Møller scattering *Quantum Mechanics.* the scattering of electrons by electrons in a magnetized material.

Mollicutes *Bacteriology.* a class of bacteria of the division Tenericutes, composed of cells bounded by a triple-layered membrane; different from other bacteria in that they lack a true cell wall; generally parasitic in plants or animals and sometimes pathogenic.

Mollier diagram *Thermodynamics.* a diagram showing enthalpy as a function of entropy.

Mollisol or **mollisol** *Geology.* one of the 10 major soil classifications, an order of soils characterized by a thick, dark, humus-rich, base-saturated surface layer generally overlying a lighter-colored or browner B horizon; developing in subhumid to semiarid climates, the Mollisols include some of the most important agricultural soils.

moll thermopile *Engineering.* a type of thermophile that strengthens and extends the life span of an instrument.

Molluginaceae *Botany.* a family of dicotyledonous herbs and shrubs of the order Caryophyllales, often having unusual secondary thickening, giving alternating rings of phloem and xylem or concentric vascular bundle rings; native to warm regions, especially in Africa.

Mollusca *Invertebrate Zoology.* a phylum of bilaterally symmetrical invertebrates, including snails, clams, octopuses, squids, and others; soft-bodied animals, most with a calcium carbonate shell secreted by the mantle, a muscular foot for locomotion, and gills.

molluscacide *Toxicology.* any agent that kills snails or other mollusks. Also, **molluscicide.**

molluscum contagiosum *Virology.* a common, benign viral infection of the skin by a poxvirus, characterized by the formation of lesions that change from flesh-colored or gray to white. The disease is found all over the world, more commonly in children, and is transmitted from person to person by direct or indirect contact.

mollusk or **mollusc** *Invertebrate Zoology.* any member of the phylum Mollusca.

Molossidae *Vertebrate Zoology.* the free-tailed bats, a family of the suborder Microchiroptera characterized by a long, stout tail, no nose leaf, and a broad muzzle; most species root in colonies in caves; distribution is worldwide.

Molotov cocktail [mäl′i täf; mō′lə täf] *Ordnance.* a crude incendiary grenade consisting of a glass container filled with flammable liquid and a wick for ignition. (Named for the Soviet leader V. M. *Molotov,* 1890–1986, because of its early use by Bolshevik revolutionaries.)

molozonide *Organic Chemistry.* a class of ozone-derived compounds containing the –C–O–O–O–C– ring.

Molpadida *Invertebrate Zoology.* an order of sea cucumbers, holothurian echinoderms in the subclass Apodacea, with a stout body tapering to a tail.

Molpadidae *Invertebrate Zoology.* a family of sea cucumbers, holothurian echinoderms in the order Molpadida.

molt *Zoology.* **1.** to shed the hair, outer skin, feathers, or horns before replacement of parts by new growth. **2.** an act, instance, or process of molting.

molten-salt reactor *Nucleonics.* a nuclear reactor that is fueled by dissolving fluoride salts in a coolant of molten salts.

molting hormone see ECDYSONE.

moltinism *Genetics.* a process involving polymorphic strains of insects of the same species that undergo different numbers of larval molts.

mol. wt. molecular weight.

molybdate *Inorganic Chemistry.* any salt derived from a molybdic acid, H_2MoO_4; used in testing, especially for the detection of heavy metal ions.

molybdate orange *Materials.* a fine orange powder that is a solid solution of lead molybdate, lead chromate, and lead sulfate; used as a pigment in paints, inks, and plastics.

molybdenite *Mineralogy.* MoS_2, a metallic, lead-gray hexagonal mineral with a perfect cleavage, trimorphous with jordisite and molybdenite-3R, having a specific gravity of 4.62 to 5.06 and a hardness of 1 to 1.5 on the Mohs scale; found as scales and foliated masses in high-temperature veins, porphyry, and contact metamorphic deposits, and in granites and pegmatites; the principal ore of molybdenum.

molybdenosis *Toxicology.* poisoning, especially of cattle, caused by ingestion of excessive amounts of molybdenum; symptoms include weakness and diarrhea.

molybdenum [mə lib′də nəm] *Chemistry.* a heavy metallic element having the symbol Mo, the atomic number 42, an atomic weight of 95.94, a melting point of 2617°C, and a boiling point of 5560°C; highly conductive and resistant to heat; used in high-temperature alloys, resistors, and audio magnets. (From an earlier word for *lead*; because of its resemblance to lead.)

molybdenum cast iron *Metallurgy.* cast iron to which small amounts of a molybdenum compound have been added to increase tensile strength and durability.

molybdenum dioxide *Inorganic Chemistry.* MoO_2, a toxic, lead-gray powder; insoluble in water and most solvents; used to make pigments.

molybdenum disilicide *Inorganic Chemistry.* $MoSi_2$, a toxic, dark-gray, crystalline powder having properties of both ceramics and metals; resistant to corrosion and stress; used in electrical resistors and protective coatings.

molybdenum disulfide *Inorganic Chemistry.* MoS_2, a toxic black crystalline powder, insoluble in water; melts at 1185°C; used as a lubricant and catalyst. Also, **molybdenum sulfide, molybdic sulfide.**

molybdenum pentachloride *Inorganic Chemistry.* $MoCl_5$, a green-black solid that absorbs water from the air and reacts with both water and air; melts at 194°C and boils at 268°C; used in chlorination, as a fire retardant, and for various other purposes.

molybdenum sesquioxide *Inorganic Chemistry.* Mo_2O_3, known in the hydrated form, $Mo(OH)_3$, a gray-black powder, insoluble in water; used as a catalyst, metal protectant, and feed additive.

molybdenum steel *Metallurgy.* any of several steels to which a molybdenum compound has been added to increase tensile strength and durability.

molybdenum trioxide *Inorganic Chemistry.* MoO_3, a toxic powder that turns yellow on heating, then sublimes; slightly soluble in water, it reacts with acids and bases to form polymeric compounds. Also, **molybdenum anhydride.**

molybdic acid *Inorganic Chemistry.* any of various acids derived from molybdenum, especially H_2MoO_4, white or yellowish crystals. Commercial molybdic acid is a solution of ammonium molybdate or molybdenum trioxide.

molybdite *Mineralogy.* MoO_3, a green to yellow, orthorhombic mineral having a specific gravity of 4.72 and an undetermined hardness; found as aggregates of flat needles or plates associated with molybdenite. Also, **molybdine, molybdic ocher.**

molybdoenzyme *Enzymology.* any enzyme that contains the metal molybdenum.

molybdophosphoric acid see PHOSPHOMOLYBDIC ACID.

molybdophyllite *Mineralogy.* $Pb_2Mg_2Si_2O_7(OH)_2$, a colorless to pale-green trigonal mineral occurring as foliated aggregates resembling mica, having a specific gravity of 4.72 and a hardness of 3 to 4 on the Mohs scale.

molybdous *Chemistry.* **1.** of or relating to molybdenum. **2.** describing various compounds of molybdenum, especially those in which the element has a valence of 4.

molysite *Mineralogy.* $Fe^{+3}Cl_3$, a soft, yellow or brownish-red, deliquescent, hexagonal mineral, having a specific gravity of 2.9 and an undetermined hardness; found as sublimation coatings in areas of volcanic activity.

molysmophobia see MYSOPHOBIA.

moment *Science.* a brief but inexact period of time, or a specific point in time. *Mechanics.* any of various measurements or expressions of the tendency of a system to rotate about a given point; e.g., the moment of force, moment of inertia, or static moment. *Aviation.* **1.** the tendency of aerodynamic forces to induce an aircraft to rotate about its center of gravity. **2.** a measure of this tendency, equal to the product of the craft's force (or mass or velocity) and the perpendicular distance between the point or axis of rotation and the line of action of the force. Also, AERODYNAMIC MOMENT. *Robotics.* the degree of movement of a robotic part around a point or axis. *Statistics.* the expectation of a product formed by raising a random variable to a specified power; the *r*th moment of a random variable X is the expectation of $E(X^r)$.

momental ellipsoid see INERTIA ELLIPSOID.

moment arm *Mechanics.* for a body rotating about a given point due to the application of force, the perpendicular distance from this point to the line of application of the force.

moment coefficient *Aviation.* a dimensionless coefficient similar to the coefficients for lift, drag, and thrust but also including a characteristic length in addition to the area; the span is used for the coefficient of rolling or yawing moment, and the chord for the coefficient of pitching moment.

moment diagram *Mechanics.* see BENDING MOMENT DIAGRAM. *Aviation.* a plot of the coefficient of aerodynamic moment versus coefficient of lift or angle of incidence.

moment of force *Mechanics.* a description of torque, i.e., the tendency of a force to cause an object to rotate about a given point, expressed as $T = F \times r$, where T is the torque, F is the force vector, and r is the position vector from the point of origin to the point of application of the force.

moment of inertia *Mechanics.* the tendency of a body rotating about a fixed axis to resist a change in this rotating motion, expressed as the integral over the body's volume of its density, multiplied by the square of the distance to the axis. Also, AXIAL MOMENT OF INERTIA, ROTATIONAL INERTIA.

moment of momentum see ANGULAR MOMENTUM, def. 1.

moment sensor *Robotics.* an instrument designed to measure the force implemented at a remote point in a robotic system.

momentum *plural,* **momenta.** *Science.* the general effect of an ongoing motion or process. *Mechanics.* a vector quantity that is conserved in collisions between particles and in closed systems; in classical mechanics it is equal to the mass times the velocity of a body, or the vector sum of this product over all the components of a system.

momentum coefficient see BOUSSINESQ COEFFICIENT.

momentum conservation see CONSERVATION OF MOMENTUM.

momentum density *Physics.* a quantity given by the amount of momentum per unit volume in a system of particles.

momentum transport hypothesis *Fluid Mechanics.* a hypothesis in which it is assumed that in transport due to turbulent eddy diffusion, the conservation of momentum still applies.

momentum wave function *Quantum Mechanics.* the Fourier transform of a spatial wave function, representing the probability distribution of a system in momentum space.

Momertz-Lentz system *Mining Engineering.* an arrangement of two winding engines alongside the top of a shaft, with the shaft collar forming a common foundation. The ropes are nearly vertical with reduced oscillation.

momilactone *Biochemistry.* a phytoalexin produced by rice plants.

Momotidae *Vertebrate Zoology.* the motmots, a family of colorful birds of the order Coraciiformes, found in neotropical forests and characterized by a long serrated bill and weak flight.

mon- a combining form meaning "one" or "single." *Chemistry.* specifically, a combining form indicating: **1.** the presence of a single atom, molecule, radical, etc. **2.** uniformity or similarity of composition, size, structure, properties, etc.

monacid see MONOACID.

monactine *Invertebrate Zoology.* a single-rayed sponge spicule.

monad *Chemistry.* a univalent radical or element. *Biology.* any simple, single-celled organism such as a single-celled protozoan or coccus. *Genetics.* in meiosis, one member of a tetrad. (A Greek word meaning "unity.") Thus, **monadic, monadical, monadal.**

monadelphous *Botany.* having stamens that are united into a bundle by the joining of their filaments.

monadic boolean operation *Computer Programming.* a boolean operation with a single operand, such as NOT.

monadic operation see UNARY OPERATION.

Monadidae *Invertebrate Zoology.* a family of flagellated protozoans in the order Kinetoplastida.

monadnock *Geology.* a residual, isolated hill of resistant rock that projects conspicuously above the surface of a peneplain; often a remnant of rock that has withstood erosion. Also, TORSO MOUNTAIN.

monadnock

monalbite *Mineralogy.* a monoclinic high-temperature form of albite, $NaAlSi_3O_8$, stable in the approximate temperature range 1000–1150°C and forming an isomorphous series with sanidine.

monandrous *Anthropology.* of a woman, having one husband at a time. *Zoology.* of an animal, having one mate at a time. *Botany.* of a plant, having only one stamen.

monandry *Anthropology.* the practice or condition of being monandrous.

monanthous *Botany.* bearing only a single flower on each stalk.

monarch butterfly *Invertebrate Zoology.* a large butterfly, *Danaus plexippus,* characterized by its deep orange color and distinctive black-and-white markings.

monarch butterfly

Monasaceae *Mycology.* a family of fungi of the order Ascosphaerales occurring in starchy materials, root nodules, and soil; used in Asian food-processing operations.

monatomic see MONOATOMIC.

monaural sound *Acoustical Engineering.* sound that is reproduced on a single channel and thus heard without the sense of separation and depth provided by stereophonic sound.

monaxon *Neurology.* a neuron that has only one axon.

monazite *Mineralogy.* $(Ce,La,Nd,Th)PO_4$, a yellow, red-brown, yellow-brown, green, or white monoclinic mineral of the monazite group, occurring as tabular crystals, having a specific gravity of 4.6 to 5.4 and a hardness of 5 to 5.5 on the Mohs scale; found as a detrital mineral and in veins, metamorphic rocks, and pegmatites; a principal ore of the rare earths and thorium.

monazite

monchiquite *Petrology.* a mafic lamprophyre consisting of phenocrysts of olivine, pyroxene, and usually biotite, augite, or amphibole within a groundmass of glass or analcime.

Mönckeberg's arteriosclerosis see MEDIAL ARTERIOSCLEROSIS.

Mond process *Metallurgy.* a process for nickel extraction and refining, based on the formation and thermal decomposition of nickel carbonyl.

Monel *Metallurgy.* the trade name for a white copper-nickel alloy that is resistant to acid and corrosion. Also, **monel metal.**

Monera *Biology.* in some systems, a kingdom containing all prokaryotic, unicellular organisms, such as bacteria and blue-green algae.

monesthetic *Physiology.* relating to or affecting only one sense or sensation.

monestrous *Zoology.* describing an animal that has a single estrus cycle extending throughout the breeding season.

monetite *Mineralogy.* $CaHPO_4$, a pale-yellowish to white, triclinic mineral having a specific gravity of 2.93 and a hardness of 3.5 on the Mohs scale; found as aggregates of minute crystals with other phosphates on several Caribbean islands.

money machine *Electronics.* an informal name for an automated teller machine. See ATM.

Monge, Carlos [mäng´gä´] 1884–1970, Peruvian physician and pathologist; studied high-altitude medicine in the Andes Mountains; diagnosed chronic mountain sickness.

Monge, Gaspard [mōnzh] 1746–1818, French mathematician; applied calculus to study geometry; helped establish the metric system.

Monge cone *Mathematics.* a first-order partial differential equation imposes at each point of the domain a relation among the partial derivatives of, and hence the normal to, any solution of the partial differential equation. The allowable normals are all perpendicular to a cone-shaped surface with vertex at the selected point, so the solution must be tangent to the cone at that point. This cone is called the Monge cone. In the quasilinear case, the Monge cone degenerates to a line. Also, CHARACTERISTIC CONE.

Monge's disease see MOUNTAIN SICKNESS.

mongolian spot *Medicine.* a congenital dark bluish or mulberry-colored round or oval spot seen on the sacral region of some newborns, most often among Asians and dark-skinned races, and disappearing before 5 years of age.

mongolism see DOWN SYNDROME.

Mongoloid or **mongoloid** *Anthropology.* **1.** a term used to describe various groups of people residing or originating in the Far East and having physical characteristics that generally include dark eyes with epicanthic eyefolds, straight black hair, and medium-colored skin tones; applied to such groups as the Mongols, Chinese, Japanese, Koreans, and Vietnamese, and to some extent to the Eskimos and American Indians. **2.** of or relating to such people. *Medicine.* a former term for an individual with Down syndrome.

mongoose *Vertebrate Zoology.* any ferretlike mammal of the genus *Herpestes* in the family Viverridae, especially *H. edwardsi*; noted for eating rats and snakes; found in Asia and Africa and introduced in New Zealand, Hawaii, and the West Indies.

mongoose

mongrel *Vertebrate Zoology.* a dog of mixed or unkown breed. *Biology.* any plant or animal derived from the crossing of breeds or varieties.

Monhysterida *Invertebrate Zoology.* an order of widely distributed nematodes of the subclass Chromadoria; found in the sea, in fresh water, and in soil.

Monhysteroidea *Invertebrate Zoology.* a superfamily of free-living nematodes in the order Monhysterida.

monic polynomial *Mathematics.* a polynomial whose leading coefficient is +1.

Monilia *Mycology.* **1.** the former name for a genus of fungi now called *Candida.* **2.** a genus of imperfect fungi of the family Moniliaceae.

Moniliaceae *Mycology.* a family of colorless or light-colored imperfect fungi of the order Moniliales, including both nonparasites and parasites of plants and animals. Also, MUCEDINACEAE.

monilial [mə nil′ē əl] *Mycology.* relating to fungi of the order Moniliales. *Medicine.* caused by fungi of the genus *Monilia (Candida).*

Moniliales *Mycology.* in some classifications, an order of imperfect fungi that composes the subdivision Basidiomycotina and the class Hyphomycetes, including several skin parasites.

monilial vaginitis *Medicine.* an inflammation of the vagina caused by infection with yeastlike fungi of various species, chiefly *Candida albicans.*

moniliasis see CANDIDIASIS.

moniliform *Biology.* constructed with contractions and expansions at regular alternating intervals, giving the appearance of a string of beads.

Moniligastrida *Invertebrate Zoology.* a suborder of annelid worms containing one widespread family of sometimes giant earthworms, mostly found in India and Burma.

Monilinia *Mycology.* a genus of cup fungi of the order Heliotales; the species *M. fructicola*, *M. laxa*, and *M. fructigena* cause brown rot in stone fruits.

Moniliophthora *Mycology.* a genus of fungi belonging to the order Agaricales; the species *M. roreri* infects the pods of cacao.

Monimiaceae *Botany.* a loosely knit family of dicotyledonous woody plants of the order Laurales that often accumulate aluminum and are characterized by opposite, simple leaves and flowers borne in cymes; native to tropical and subtropical regions, especially of the Southern Hemisphere.

monimolimnion *Hydrology.* the high-density, stable, stagnant or noncirculating lower layer of a meromictic lake.

monimolite *Mineralogy.* a yellow, gray-green, or dark-brown, cubic mineral with an approximate formula of $(Pb,Ca)_3Sb_2O_8$, occurring as octahedral crystals, having a specific gravity of 5.94 to 7.29 and a hardness of 4.5 to 6 on the Mohs scale; found with tephroite in calcite veins.

monitor to check or evaluate something on a constant or regular basis; an instrument designed for such a check; specific uses include: *Engineering.* **1.** to measure a condition in a system by means of meters or instruments. **2.** any instrument that periodically measures or regulates any condition in a system that must be maintained within prescribed limits. *Electronics.* **1.** to check by means of a receiver the operation of a telephone, radio, television, or similar transmitter in order to ascertain certain characteristics, such as the quality of transmission or fidelity to a prescribed frequency band. **2.** the device used for monitoring, or the person who watches the device. *Computer Technology.* **1.** a hardware device that measures electrical events such as pulses or voltage levels in a digital computer. **2.** a video display unit. *Computer Programming.* **1.** to oversee a program during execution. **2.** a control program within the operating system that manages the allocation of system resources to active programs. **3.** a program that measures software performance. *Surgery.* **1.** to keep close and constant watch of a condition or function, such as the breathing of a patient during surgery. **2.** a device used to record or examine data, such as one that records the physiologic signs of a patient.

monitor board *Telecommunications.* in telephony, a console from which a supervisory operator can intercept calls that are being handled by other operators.

monitor control dump *Computer Programming.* a memory printout that is produced regularly after a program has been executed.

monitor display *Computeer Technology.* the capability of some processors to temporarily halt the central processing unit and display information from main storage and internal registers.

monitoring amplifier *Electronics.* a power amplifier used to evaluate and supervise a program.

monitoring key *Electronics.* a device that allows an operator to listen on a telephone circuit without impairing the transmission.

monitoring system *Design Engineering.* a system that collects the output of sensors and human impact to construct a history of events of a process or machine.

monitor lizard *Vertebrate Zoology.* any of various large carnivorous lizards of the family Varanidae, thought to give warnings in the presence of crocodiles.

monitor operating system *Computer Programming.* the part of an operating system that monitors and controls hardware efficiency. Also, **monitor program.**

monitor printer *Telecommunications.* in a communications center, a teleprinter that checks incoming signals.

monkey *Vertebrate Zoology.* **1.** any member of the order Primates other than humans, lemurs, apes, gibbons, orangutans, and chimpanzees. **2.** specifically, any of the longer-tailed primates such as the langurs and macaques.

monkey

monkey chatter see SIDE-BAND INTERFERENCE.

monkey-faced owl see BARN OWL.

monkeypox *Medicine.* a mild, epidemic skin disease occurring in captive monkeys; may be transmitted to humans, in whom it causes a condition clinically similar to smallpox.

monkey wrench *Mechanical Devices.* an adjustable screw wrench having one fixed jaw and one adjustable jaw positioned at right angles to a handle.

monkfish *Vertebrate Zoology.* any of several bottom-dwelling sharks of the genus *Squatina,* family Squatinidae, characterized by large, lateral, raylike pectoral and ventral fins; found in tropical and warm temperate oceans.

monk seal *Vertebrate Zoology.* a member of any of three endangered species of small, dark brown seals belonging to the genus *Monachus: M. monachus* of the Mediterranean, *M. tropicalis* of the Caribbean, or *M. schauinslandi* of the Hawaiian Islands.

monkshood *Botany.* a plant, *Aconitum napellus,* of the family Ranunculaceae, whose flowers have a large, hood-shaped sepal.

mono- a combining form meaning "one" or "single." *Chemistry.* specifically, a combining form indicating: **1.** the presence of a single atom, molecule, radical, etc. **2.** uniformity or similarity of composition, size, structure, properties, etc.

monoacetate *Organic Chemistry.* a compound that contains one acetate group, such as an ester or a salt.

monoacetin see ACETIN.

monoacid *Chemistry.* an acid containing only a single replaceable hydrogen atom. Thus, **monoacidic.** Also, MONACID.

monoallelic *Genetics.* of a polyploid individual, having identical alleles at any given locus.

monoamine *Organic Chemistry.* any amine containing only one amino group, e.g., serotonin, dopamine, or norepinephrine.

monoamine oxidase *Enzymology.* the flavin-containing amine oxidase, important in the catabolism of epinephrine and tyramine.

monoamine oxidase inhibitor *Pharmacology.* any of a group of antidepressive drugs that block the oxidative deamination of monoamines.

monoaminergic *Neurology.* of or relating to neurons that secrete monoamine neurotransmitters.

monoatomic *Chemistry.* of a molecule, consisting of a single atom. Also, MONATOMIC.

monoatomic gas *Chemistry.* any gas composed of single atoms.

monobasic *Chemistry.* of an acid, having one replaceable hydrogen atom per molecule.

monobasic calcium phosphate *Inorganic Chemistry.* $Ca(H_2PO_4)_2 \cdot H_2O$, colorless triclinic crystals; soluble in water and acids; loses its water at its melting point of 109°C and decomposes at 203°C; used in baking powders, fertilizers, and mineral supplements, and as a plastic stabilizer. Also, CALCIUM BIPHOSPHATE, PRIMARY CALCIUM PHOSPHATE.

monobasic magnesium phosphate *Inorganic Chemistry.* $Mg(H_2PO_4)_2 \cdot 2H_2O$, a white, crystalline, hygroscopic powder that is soluble in water and acids and insoluble in alcohol; decomposes on heating; used in plastics and fireproofing.

monobasic sodium phosphate *Inorganic Chemistry.* NaH_2PO_4, a white crystalline powder, or the hydrated form $NaH_2PO_4 \cdot H_2O$, white crystals; both very soluble in water; used in dyeing, electroplating, acid cleansers, baking powders, and for many other purposes.

Monobathrida *Paleontology.* an order of camerate crinoids; generally pelmatozoan; Lower Ordovician to Upper Permian.

monoblast *Histology.* an immature monocyte, usually found in the bone marrow and spleen.

monoblastic leukemia see ACUTE MONOCYTIC LEUKEMIA.

monoblastoma *Oncology.* a tumor containing monoblasts and monocytes.

Monoblepharidaceae *Mycology.* a family of microscopic fungi of the order Monoblepharidales living in soil and water and on organic material such as fruit and twigs.

Monoblepharidales *Mycology.* an order of fungi of the class Chytridiomycetes; found in aquatic habitats as nonparasites that live on dead organic matter.

monobloc projectile *Ordnance.* an armor-piercing projectile consisting of a single piece of heat-treated steel; it may have a false ogive.

monobrid circuit *Electronics.* an integrated circuit in which multiple monolithic integrated circuit chips are interconnected or interwired with thin film components, to form a larger single-package circuit.

monocable *Mechanical Engineering.* an aerial ropeway in which a single endless rope is used to both support and move loads.

monocarpic *Botany.* describing a plant that fruits once and then dies.

monocarpous *Botany.* describing a plant having a single carpel or ovary.

monocentric *Genetics.* of a chromosome, having only one centromere.

Monocentridae *Vertebrate Zoology.* the pinecone fishes, a monogeneric family of Indo-Pacific fish of the order Beryciformes, characterized by spiny or ctenoid scales, large eyes, and luminous organs under the lower jaw.

Monoceros *Astronomy.* the Unicorn, a large but faint constellation straddling the celestial equator just east of Orion.

Monoceros Loop *Astronomy.* a supernova remnant about 50,000 years old that lies in the galactic plane northeast of the Rosette Nebula in Monoceros.

Monoceros R2 molecular cloud *Astronomy.* a large, active star-forming region in Monoceros.

monocharge electret *Electronics.* a foil electret that has electrical charges of the same sign on both surfaces.

monochasium *Botany.* a cyme with a main axis producing a single lateral branch.

monochlamydous *Botany.* describing plants whose perianth has only a single whorl.

monochromat *Medicine.* a person affected by monochromatism.

monochromatic [män′ō krə mat′ik] *Optics.* **1.** relating to or describing one color or hue. **2.** relating to or describing electromagnetic radiation with one wavelength or with a narrow range of wavelengths. *Medicine.* relating to or affected by monochromatism. *Microbiology.* referring to staining with only one dye at a time.

monochromatic filter see BIREFRINGENT FILTER.

monochromatic interference *Optics.* the interference of two or more beams of light of the same wavelength.

monochromatic neutron beam *Nucleonics.* a beam of neutrons in which the energy values are extremely limited.

monochromatic radiation *Electromagnetism.* electromagnetic radiation of an extremely narrow band.

monochromatic temperature scale *Thermodynamics.* a temperature scale that is defined according to the amount of power radiated by a blackbody at a particular wavelength.

monochromatism *Medicine.* complete color blindness, in which all colors of the spectrum appear as neutral grays with varying shades of light and dark. Also, MONOCHROMIA.

monochromator *Optics.* a source of monochromatic light characterized by thin entrance and exit slits that control the spectral width of emitted light, a prism, and additional lenses and mirrors to control the path of the light.

monochrome [män′i krōm′] *Optics.* relating to or describing different shades of the same color, such as gradations of black.

monochrome channel *Electronics.* in a color television system, the path for carrying the monochrome signal.

monochrome signal *Electronics.* a signal wave in television transmission that controls the luminance values in the picture, but not the chromaticity values.

monochrome television see BLACK-AND-WHITE TELEVISION.

monochromophilic *Microbiology.* stainable by only one kind of stain.

monocistronic *Molecular Biology.* describing a mRNA molecule that encodes the information necessary for synthesis of a single polypeptide chain.

Monocleaceae *Botany.* a monogeneric family of liverworts characterized by plants with dichotomous branching, a thallus without differentiation of midrib, epidermal pores or photosynthetic air chambers, and a large gametophyte.

Monocleales *Botany.* the single, monogeneric order of liverworts of the order Marchantiidae, characterized by a gametophyte that is one of the largest among the liverworts.

monoclimax *Ecology.* a type of climax community in which there is only one controlling factor, especially a climatic factor.

monoclimax theory *Ecology.* the theory that over time and within a climatic region, there will be a single regional climax community.

monoclinal *Geology.* of, relating to, or being a monocline.

monocline *Geology.* a portion of a stratigraphic sequence that is not part of an anticline or syncline, but dips from the horizontal in one direction.

monoclinic system *Crystallography.* a unit cell in which there is a two-fold rotation axis parallel to one cell axis (usually chosen as *b*); as a result there are no restrictions on the axial ratios, but the angles made by *b* with *a,* and *b* with *c,* must be 90° ($\alpha = \gamma = 90°$). Also, **monoclinic unit cell.**

monoclinous *Botany.* describing a flower that is a hermaphrodite, having both stamens and pistils. Thus, **monoclinism.**

monoclonal *Biology.* referring to a substance that is derived from a single clone or family of cells; derived from a single cell.

monoclonal antibody *Biotechnology.* a homogeneous antibody that is produced by a clone of antibody-forming cells and that binds with a single antigenic determinant.

monocolpate pollen *Botany.* single-furrowed pollen, characteristic of monocotyledons and gymnosperms.
monocoque [män´ə käk´] *Aviation.* **1.** a type of design or structure in which all or most of the stresses are carried by the skin, as in an airplane fuselage or rocket body. **2.** designed or built according to the structural principles of monocoque. Thus, **monocoque boom, monocoque design.** (From a French phrase meaning "single shell.")
monocot *Botany.* a monocotyledonous plant.
monocotyledon [män ə kät´ə lēd´ən] *Botany.* any of the great variety of plants that bear only only one seed leaf (cotyledon), usually also characterized by leaves with parallel venation and flower parts in groups of three and generally lacking cambium tissue; one of the two major divisions of the angiosperms.
Monocotyledoneae *Botany.* a class of plants having a single cotyledon and parallel-veined leaves; in some classification systems, an equivalent name for the Liliopsida.
monocotyledonous [män ə kät´ə lēd´ə nəs] *Botany.* **1.** having only one cotyledon. **2.** of or relating to the Monocotyledoneae.
monocrepid *Invertebrate Zoology.* a branched, knobby, siliceous deposit on a sponge monaxon spicule.
monocular [mə näk´yə lər] *Optics.* relating to, using, or designed for one eye. Thus, **monocular microscope.**
monocular vision *Medicine.* vision in one eye only.
monoculture *Agronomy.* the practice of using a given piece of land for growing the same crop each year rather than rotating different crops.
Monocyathea *Paleontology.* a class of regular archaeocyathids, usually solitary but sometimes colonial; Lower and Middle Cambrian.
monocyclic *Botany.* of flower parts, arranged in a single whorl. *Chemistry.* describing a molecular structure that contains only one ring.
monocyesis *Medicine.* pregnancy with a single fetus.
Monocystis *Invertebrate Zoology.* a genus of parasitic gregarine protozoa of the order Eugregarinida, found in the coelom and seminal vessels of the earthworm and related worms.

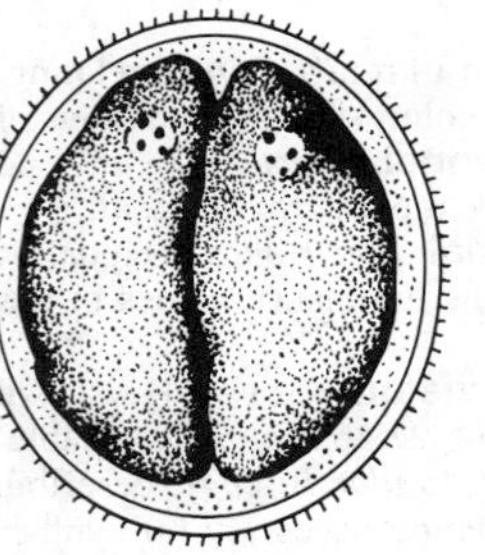

Monocystis

monocyte *Hematology.* a large, agranular leukocyte with an ovoid or kidney-shaped nucleus; formed in the bone marrow, monocytes migrate into connective tissue and become macrophages.
monocytic angina see INFECTIOUS MONONUCLEOSIS.
monocytic leukemia *Medicine.* leukemia in which the predominating leukocytes are identified as monocytes.
monocytoma *Oncology.* a tumor composed of mononuclear leukocytes.
monocytopenia *Medicine.* an abnormal decrease in the number of monocytes in the blood.
monocytosis *Medicine.* an increased proportion of monocytes in the blood.
Monod, Jacques 1910–1976, French biologist; awarded the Nobel Prize for introducing (with F. Jacob) the concepts of messenger RNA and the operon.
monodactylous *Zoology.* having a single digit or claw per extremity.
Monodellidae *Invertebrate Zoology.* a family of malacostracan crustaceans found in cool cave waters.
monodelphic *Vertebrate Zoology.* having a single, undivided uterus.
monodermoma *Oncology.* a tumor that has developed from the tissues of a single germinal layer.
monodisperse colloidal system *Chemistry.* a colloidal system having particles of the same size, interaction, and shape.
monodispersity *Organic Chemistry.* a substance that occurs as suspended particles of nearly uniform size.
monodomous *Entomology.* living as a community in a single nest.
Monodontidae *Vertebrate Zoology.* a family of marine mammals of the order Odonotoceta that includes the white whale (or beluga) and the narwhal; native to cold northern seas.
monodromy theorem *Mathematics.* let D be a simply connected domain and let $f_0(z)$ be analytic in some disk D_0 with center at a point z_0 of D. If the function element (f_0, D_0) can be continued analytically along every piecewise linear, continuous curve (or polygonal line) in D, then this continuation leads to a single-valued analytic function in D.
monoecious [mə nē´shəs] *Botany.* having both pistil-bearing and stamen-bearing flowers in a single plant. *Zoology.* having both male and female sex organs; hermaphroditic.
monoecy *Biology.* the condition or characteristic of being monoecious.
Monoedidae see COLYDIIDAE.
monoenergetic gamma rays *Physics.* a beam of gamma rays whose photon energy is found to lie within a very narrow band.
monoester *Organic Chemistry.* an ester that contains only one ester group.
monoethanoalamine see ETHANOALAMINE.
monofier *Electronics.* a complete master oscillator-power amplifier system housed in a single evacuated tube envelope.
monofilm *Physical Chemistry.* a type of film that has a thickness of only one molecule, as may be formed by a surface layer of alcohol or long-chain fatty acids on water; used as a water evaporation retardant. Also, MOLECULAR FILM; MONOLAYER; MONOMOLECULAR FILM.
monofuel see MONOPROPELLANT.
monogamous [mə näg´ə məs] *Anthropology.* relating to or following the practice of monogamy. Also, **monogamic.**
monogamy [mə näg´ə mē] *Anthropology.* a marital status in which only one spouse is permitted at a time. *Behavior.* a mating pattern in which males and females are paired one to one for at least one reproductive season; usually associated with parental care by both parents. *Biology.* asexual reproduction.
monogastric *Vertebrate Zoology.* possessing a stomach with a single compartment, as do humans.
Monogenea *Invertebrate Zoology.* a subclass of trematodes that are primarily ectoparasites of fishes and amphibians. Also, **Monogenoidea.**
monogenesis *Evolution.* the hypothetical evolution of all life from a single pair or a single cell. *Anthroplogy.* the hypothetical evolution of all humans from a single set of ancestors, such as Adam and Eve. *Biology.* the production of only male or only female offspring.
monogenetic *Biology.* relating to or characterized by monogenesis. *Invertebrate Zoology.* of certain trematodes (e.g., those of the subclass Monogenea), having only one generation in the life cycle, without an intermediate asexual phase.
monogenic *Genetics.* of a characteristic, controlled by one pair of genes. *Biology.* bearing only males or only females.
monogeosyncline *Geology.* a long, narrow, deeply subsided single geosyncline along a continental margin that receives its sediments from the borderland on its oceanic side.
monogerm *Botany.* having a single seed that develops into a single plant.
monoglyceride *Organic Chemistry.* **1.** a compound consisting of one molecule of fatty acid esterified to a glycerol group. **2.** a glyceride molecule possessing free hydroxyl groups that act as water-solubilizing groups, making the molecule an effective emulsifying agent; used in cosmetics and lubricants.
monograph *Science.* a scholarly treatise on a single subject. *Biology.* a scholarly description of a single species.
Monograptus *Paleontology.* a genus of uniserial, scandent graptolites in the order Graptoloidea, distinguished by their single stipes; the monograptids represent the culmination of the graptoloids' evolutionary tendency toward fewer stipes and scandent growth; extant in the Lower Silurian to the uppermost Lower Devonian.
monogynous *Botany.* of a flower, bearing only one carpel or pistil.
monohybrid *Genetics.* in a genetic study, an individual that is heterozygous for a single pair of alleles at one locus.
monoid *Mathematics.* a nonempty set G together with a binary operation $*$ defined on all of G, such that: (a) G is closed under $*$; (b) $*$ is associative; and (c) an element e of G exists that is a two-sided identity element, i.e., $a * e = e * a = a$ for every $a \in G$. If G satisfies only the first requirement, it is a groupoid; if G satisfies the first two requirements, it is a semigroup. G is a group if, in addition, G satisfies (d) for every $a \in G$, there exists an element $a^{-1} \in G$ that is a two-sided inverse element of a, i.e., $a * a^{-1} = a^{-1} * a = e$.
monoideism *Psychology.* an obsession with a single idea or thought and the inability to think of anything else.

monoiodotyrosine *Endocrinology.* a metabolic precursor of the thyroid hormones triiodothyronine (T_3) and thyroxine (T_4).
monokaryotic *Cell Biology.* containing one nucleus per cell.
monokine *Virology.* any of the soluble mediators of immune responses that are not antibodies or complement components, and that are produced by mononuclear phagocytes.
monolayer see MONOFILM.
monolayer culture *Biotechnology.* a single layer of cells growing on the surface of the support matrix or a culture medium.
monolinuron *Organic Chemistry.* $C_9H_{11}N_2O_2Cl$, a colorless solid, slightly soluble in water; melts at 75°C; used as an herbicide on crops and ornamental plants. Also, MSMA.
monolith *Architecture.* **1.** a column, obelisk, statue, or the like that is made from a single block of stone. **2.** a large block of stone, especially when used in building or sculpture. *Civil Engineering.* a sizable solid block without joints, as in reinforced-concrete constructions.
monolithic *Science.* of, relating to, or made from a single uniform piece of material.
monolithic ceramic capacitor *Electronics.* a capacitor made up of thin dielectric layers interleaved with staggered metal-film electrodes; it is compressed to form a solid monolithic block.
monolithic filter *Telecommunications.* a filter, mounted or electroplated on a single quartz or ceramic substrate, that separates telephone communications sent over the transmission line at the same time.
monolithic integrated circuit *Electronics.* a circuit formed in a single piece of the substrate material, in which physically separate circuit components are electrically interconnected to form the final circuit.
monomania *Psychology.* **1.** a psychological disorder characterized by an abnormal preoccupation with one idea or subject. **2.** loosely, any inordinate preoccupation with a single idea, subject, goal, or the like.
Monomastigales *Botany.* an order of marine and freshwater green flagellate algae of the class Prasinophyceae, composed of the single family Monomastigaceae and characterized by one or two flagella attached laterally or at one pole of the cell and an organic scale covering on the cells and sometimes on the flagella.
monomer *Chemistry.* a relatively simple compound, usually containing carbon, that is able to combine in long chains with other like or unlike molecules to produce very large polymers.
monomeric *Chemistry.* relating to, characteristic of, or resembling a monomer.
monomerous *Botany.* of or relating to a flower with a single member in each whorl. Also, 1-MEROUS.
monomial *Systematics.* of a name, consisting of a single word or term. *Mathematics.* a polynomial having exactly one term; that is,

$$a x_1^{k_1} x_2^{k_2} \ldots x_n^{k_n},$$

where a is an (nonzero) element of the ring of coefficients and the x_1, $x_2, \ldots, x_n$ are indeterminates. The **degree of the monomial** is the sum $k_1 + k_2 + \cdots + k_n$.
monomineralic *Petrology.* consisting wholly or chiefly of one mineral.
Monommidae *Invertebrate Zoology.* a family of beetles, coleopteran insects in the superfamily Tenebrionoidea, that feed on decaying plant material.
monomode fiber *Materials Science.* an optical fiber that offers only one path, or mode, for light propagation. These fibers give the best performance in terms of the simultaneous achievement of high information rate and low loss.
monomolecular film or **monomolecular layer** see MONOFILM.
monomorphic *Biology.* **1.** having or exhibiting only a single form. **2.** of the same or similar structure (as another organism or part).
monomorphic locus *Genetics.* a locus that is occupied by a single allele in a population, or one in which the most common allele has a frequency greater than 95%.
Mononchoidea *Invertebrate Zoology.* a superfamily of predatory nematodes found in soil and fresh water.
mononeuritis *Neurology.* an inflammation of a single nerve.
mononeuropathy *Neurology.* a disease affecting a single nerve.
mononuclear *Cell Biology.* having a single nucleus.
mononuclear phagocyte see MACROPHAGE.
mononucleosis [mä′nō noo′klē ō′sis] *Medicine.* **1.** the presence of an abnormally large number of mononuclear leukocytes in the blood, as in glandular fever. **2.** see INFECTIOUS MONONUCLEOSIS.
monooxygenase *Biochemistry.* an enzyme of the oxidoreductase class that catalyzes the incorporation of an oxygen atom into a substrate molecule and reduces the other atom into water.
monoparesis *Neurology.* the slight or incomplete paralysis of a limb or part of a limb.
Monopectinate *Invertebrate Zoology.* pectinate (comblike) along one side only.
monopetalous *Botany.* **1.** having a corolla of united petals, as in the morning glory. **2.** having a single petal in the corolla.
monophagous *Zoology.* feeding only on one kind of food. Also, MONOTROPHIC.
monophasia *Neurology.* a form of aphasia characterized by the ability to speak only one word or phrase, which is repeated constantly.
monophenol monooxygenase see TYROSINASE.
Monophisthocotylea *Invertebrate Zoology.* an order of trematode parasites in the subclass Monogenea.
Monophlebinae *Invertebrate Zoology.* a subfamily of homopteran insects in the superfamily Coccoidea.
monophonic sound see MONAURAL SOUND.
monophyletic *Systematics.* of or relating to a clade. *Evolution.* of or relating to any group sharing a single ancestral form.
monophyletic theory *Hematology.* the theory that all forms of blood cells, both red and white, have their origin in one and the same form of primordial blood cell (hemocytoblast), with the several types of cells arising by a process of differentiation. Also, UNITARIAN THEORY.
monophyllous *Botany.* consisting of or having a single leaf.
monophyodont *Vertebrate Zoology.* having a single set of teeth that are not replaced at a later stage of growth.
monopinch *Electronics.* the application of monopulse techniques in which the error signal is used to discriminate against jamming signals.
Monopisthocotylea *Invertebrate Zoology.* a suborder of trematodes with a hooked posterior adhesive disk; ectoparasites of fish, amphibians, and crustaceans.
Monoplacophora *Invertebrate Zoology.* a small class of primitive marine mollusks with a single dorsal shell and signs of segmentation.
monoplane *Aviation.* an airplane having only one main wing or lifting surface, usually divided into two parts by the fuselage; the most common general type of airplane.

monoplane

monoplegia *Medicine.* the paralysis of a single limb or a single muscle group.
monoploid *Genetics.* **1.** a somatic cell or individual having one set of chromosomes. **2.** the basic chromosome number in a polyploid mutant.
monopodial *Botany.* having stems branching from a single main axis.
monopodium *Botany.* the single primary axis from which all lateral branches develop.
monopole see MAGNETIC MONOPOLE.
monopole antenna *Electromagnetism.* an antenna that acts as half of a dipole antenna, with the other half formed by the electrical image in the ground plane; usually in the form of a vertical tube or a helical element, on which the current distribution forms a standing wave.
monopropellant *Materials Science.* a propellant that combines fuel and oxidizer in a single substance, especially a liquid fuel for rockets. *Space Technology.* using such a propellant. Thus, **monopropellant rocket, monopropellant engine.**
monopulse radar or **monopulse tracking** *Engineering.* radar capable of obtaining directional information with great accuracy; four overlapping pencil beams (two for azimuth, two for elevation) and special circuitry are arranged so that when the target is at center, voltage dissipates. Also, SIMULTANEOUS LOBING, STATIC SPLIT TRACKING.

Monopylina *Invertebrate Zoology.* a suborder of radiolarian protozoans in the order Oculosida.

monorail *Transportation Engineering.* a railroad system that uses a single overhead rail or a single large rail over which trains are held in place by side wheels. *Civil Engineering.* **1.** a deep rail track placed on sleepers over the ground to accommodate concrete. **2.** an overhead rail used for material hauling in a plant or at a construction site.

Monoraphidineae *Botany.* a monofamilial suborder of marine and freshwater diatoms of the order Pennales, distinguished by a fully developed raphe on one of two valves.

monorchid *Anatomy.* having only one testicle evident, the other being either missing or undescended. Thus, **monorchidism.**

Monorhina *Vertebrate Zoology.* archaic term used to designate lampreys, hagfishes, and certain extinct jawless fishes.

monorhinic *Anatomy.* having only one nasal cavity. Also, **monorhinal.**

monosaccharide *Biochemistry.* a simple sugar, i.e., one that cannot be split into smaller units by the action of dilute acids, consisting of a colorless crystalline substance having a sweet taste and the general formula CH_2O.

monoscope *Electronics.* **1.** a pick-up tube in which an electron beam scans a target, parts of which have different secondary emission characteristics. **2.** a cathode-ray tube that produces a stationary picture used to test television equipment. Also, **monotron.**

monosepalous *Botany.* **1.** having united sepals. **2.** having a single sepal.

Monosigales *Botany.* in some classifications, an alternate name for the Choanoflagellida.

monosiphonous *Biology.* having a single central tube, as in the thallus of certain filamentous algae or the hydrocaulus of some hydrozoans.

monosodium acid methanearsonate *Organic Chemistry.* CH_4AsNaO_3, a white crystalline solid; soluble in water; melts at 132–139°C; used as an herbicide.

monosodium glutamate *Organic Chemistry.* $NaCO_2(CH_2)_2CHNH_2CO_2H$, a white crystalline powder that is soluble in water and alcohol; decomposes at its melting point and is used as a flavor enhancer. Also, MSG, SODIUM GLUTAMATE.

monosodium phosphate or **monosodium hydrogen phosphate** see MONOBASIC SODIUM PHOSPHATE.

Monosoleniaceae *Botany.* a monotypic family of medium-sized liverworts of the order Marchantiales, that are native to India, China, Taiwan, and Japan; characterized by dichotomous branching and a thallus with a distinct midrib, wing, and lobate margin, but lacking epidermal pores.

monosomy *Genetics.* a condition in a diploid cell or organism in which only one chromosome of one pair is missing. Thus, **monosomic.**

monospasm *Neurology.* a spasm of a single muscle, muscle group, or limb.

monospermous *Botany.* producing or having a single seed.

monospore *Botany.* an undivided or simple spore, formed by certain red algae for the purpose of asexual reproduction.

monostable *Electronics.* of a circuit or system, having only one stable state and a quasi-stable state into which it can be driven for a limited period of time. Thus, **monostable circuit.**

monostable multivibrator *Electronics.* a multivibrator that has one stable and one unstable state.

monostat *Engineering.* a type of manometer used to regulate pressures within an enclosure.

monostichous *Botany.* arranged vertically on one side of a stem, as are some flowers.

monostome *Entomology.* a trematode cercaria larva having only one mouth or sucker.

Monostroma *Botany.* a genus of green algae of the family Monostromataceae, characterized by an enlarged zygote, erect clustered fronds, and no rhizoids; found in brackish water.

Monostromataceae *Botany.* a family of green algae of the order Ulvales, characterized by clusters of erect fronds; contains the genus *Monostroma* and is common in marine and brackish waters.

monostromatic *Biology.* having the cells in a single layer.

Monostylifera *Invertebrate Zoology.* a suborder of marine ribbon worms, in the phylum Rhyncocoela, having a single stylet on the proboscis.

monosynaptic *Neurology.* of a neuronal pathway, having only one synapse.

monotectic *Materials Science.* describing a three-phase reaction in which one liquid transforms into a solid and a second liquid on cooling.

monoterpene *Organic Chemistry.* a class of terpenes containing two isoprene units; an odorous component of essential or volatile plant oils, its molecular structure contains a single saturated or unsaturated carbon ring.

monotheism *Anthropology.* a religious system in which there is only one god, and other supernatural beings, if any, are creations of and subordinate to that supreme being.

monotheistic *Anthropology.* relating to or believing in monotheism.

Monotomidae see RHIZOPHAGIDAE.

monotone *Acoustics.* a single tone with no variation in pitch. *Mathematics.* **1.** a sequence $\{x_n\}$ of real numbers that satisfies exactly one of the following: (a) $x_i \leq x_{i+1}$ for all i; or (b) $x_i \geq x_{i+1}$ for all i. In the former case, the sequence is said to be **monotone nondecreasing** or **monotone increasing;** in the latter case, the sequence is said to be **monotone nonincreasing** or **monotone decreasing.** **2.** a real-valued function f that satisfies exactly one of the following: (a) f never decreases; i.e., if $x \leq y$, then $f(x) \leq f(y)$. In this case, f is said to be monotone nondecreasing (or monotone increasing). (b) f never increases; i.e., if $x \leq y$, then $f(x) \geq f(y)$. In this case, f is said to be monotone nonincreasing (or monotone decreasing). Also, MONOTONIC.

monotone convergence theorem *Mathematics.* **1.** let $\{x_n\}$ be a monotone increasing sequence of real numbers. Then the sequence converges if and only if it is bounded, in which case $\lim(x_n) = \sup\{x_n\}$. **2.** if an increasing sequence of positive integrable functions f_n converges almost everywhere to a function f and if $\lim_{n\to\infty} \int f_n$ exists, then f is integrable and $\lim_{n\to\infty} \int f_n = \int f$.

monotone decreasing function see DECREASING FUNCTION.

monotone family *Mathematics.* a monotone family is a family F of sets satisfying the following conditions: (a) if $E_n \in F$ and E_n is contained in E_{n+1} for $n = 1, 2, 3, \ldots$, then $\cup_{n=1}^{\infty} E_n \in F$. (b) if $F_n \in F$ and F_n contains F_{n+1} for $n = 1, 2, 3, \ldots$, then $\cap_{n=1}^{\infty} F_n \in F$.

monotonic [män´ə tän´ik] *Acoustics.* relating to or occurring in a monotone. *Science.* having the quality of a monotone; not changing. *Mathematics.* see MONOTONE.

monotonicity *Electronics.* in an analog-to-digital converter, a condition in which the output increases as the input voltage increases.

monotonicity theorem *Mathematics.* suppose A and B are $n \times n$ Hermitian matrices, that B is positive semidefinite, and that the eigenvalues $l_k(A)$ of A and the eigenvalues $l_k(A + B)$ of $A + B$ have been arranged in increasing order. Then $l_k(A) \leq l_k(A + B)$ for all $k = 1, 2, \ldots, n$.

monotonic reaction *Metallurgy.* in a binary system, an isothermal transformation in which, upon cooling, a liquid decomposes into a solid and another liquid of different composition.

monotonic reasoning *Artificial Intelligence.* a system of reasoning in which values or facts that have been determined cannot be altered or removed during that session.

Monotremata *Vertebrate Zoology.* the order of egg-laying mammals constituting the subclass Prototheria, and containing the duckbill platypus and the echidna (spiny anteater); the most primitive order of living mammals, known only from Australia, New Guinea, and a few adjacent islands.

monotreme *Vertebrate Zoology.* any animal of the Monotremata.

monotrichous *Microbiology.* of or relating to a bacterium with a single polar flagellum per cell.

monotron *Electronics.* a cathode-ray tube that produces a stationary picture; used to test television equipment.

Monotropaceae *Botany.* a family of dicotyledonous, perennial herbs of the order Ericales, characterized by scaly, creeping rootstocks, alternate opposite leaves, and anthers opening by lengthwise slits.

monotrophic see MONOPHAGOUS.

monotropic *Crystallography.* existing in only one stable crystalline form, all other forms being unstable under all conditions. *Immunology.* affecting only one particular kind of bacterium, virus, or tissue.

monotropy *Crystallography.* the state of being monotropic; the property possessed by substances that are stable in only one form.

monotropy coefficient *Fluid Mechanics.* a coefficient used in an equation for developing a fluid velocity profile; the quantity is related to the ratio of velocity coefficients A_y/A_x.

monotype *Systematics.* a taxon that contains only one subordinate taxon, such as a family containing only a single genus, or a genus containing only a single species. *Graphic Arts.* a single print made from a glass or metal plate on which a picture is painted or drawn.

Monotype *Graphic Arts.* a typesetting machine that uses a keyboard to encode a paper tape, which in turn directs a device to cast individual type characters.

monotype metal *Metallurgy.* a lead-base alloy containing 16% antimony and 8% tin; used to cast Monotype characters.

monotypic *Systematics.* of or relating to a monotype; having only one representative.

monounsaturated fatty acid *Nutrition.* a fatty acid that contains one double bond in the alkyl chain.

monovalent *Chemistry.* see UNIVALENT.

monoxide *Chemistry.* a binary compound containing one atom of oxygen per molecule, such as carbon monoxide.

monozygotic twins *Biology.* twins that develop from a single fertilized ovum that split into two at some stage of development of the embryo; they are genetically identical and therefore always of the same sex.

mons [mänz] *plural,* **montes** [män´tēz] *Anatomy.* an elevation or eminence, such as the mons pubis. *Astronomy.* the name given to a mountain or any large positive relief feature on a planet or moon.

monsoon *Meteorology.* a seasonal wind caused primarily by a greater annual variation of temperature over a large land area than over neighboring ocean surfaces, causing an excess of pressure over the continents in the winter and a deficit in the summer; monsoons are strongest in Asia, but also occur on the coasts of tropical regions.

monsoon climate *Meteorology.* a region with a long winter-spring dry season that includes a cold season followed by a short hot season right before the rains, a summer and early autumn rainy season that is variable but usually very wet, and a secondary maximum of temperature right after the rainy season; India is the best-known example.

Monsoon Current *Oceanography.* a seasonally variable current in the northwest Indian Ocean, in summer flowing generally eastward as the northern segment of the North Indian gyre, from the Arabian Sea to an area southeast of Sri Lanka.

monsoon fog *Meteorology.* an advection fog produced by a monsoon that moves warm moist air over a colder surface.

monsoon low *Meteorology.* a seasonal low-pressure system over a continent in the summer and over the adjacent sea in the winter.

mons pubis *Anatomy.* the mound of fatty tissue over the symphysis pubis in the female. Also, **mons veneris.**

monstatic radar *Engineering.* a radar system in which a transmitter and receiver are located at the same site and share an antenna.

monster *Medicine.* a fetus or infant with such pronounced developmental anomalies as to be grotesque and usually nonviable.

Monstrilloida *Invertebrate Zoology.* an aberrant order of tiny copepod crustaceans whose adults lack mouthparts and a gut.

Monsur medium *Microbiology.* a culture medium containing tellurite and bile salts, used for isolation of *Vibrio cholerae.*

Montadale *Agriculture.* a breed of mutton sheep crossbred in the United States from the Cheviot and Columbia breeds.

montage [män täj´] *Graphic Arts.* **1.** a photograph comprising several pieces of copy, sometimes including type; a composite photograph. **2.** in film or television, an image or series of images made up of several shots that are juxtaposed or superimposed.

Montana or **Montana Number** *Agriculture.* a breed of hog that was crossbred in the United States from the Hampshire and the Danish Landrace breeds. Also, HAMPRACE.

montane *Ecology.* a section of a mountainous region below the timberline having cool, moist temperatures and dominated by evergreen trees.

montanite *Mineralogy.* $Bi_2Te^{+6}O_6 \cdot 2H_2O$, a rare, yellowish to greenish or white, possibly monoclinic mineral occurring in soft earthy to compact form, having a specific gravity of 3.79 and an undetermined hardness; found at Highland, Montana.

montan wax *Materials.* a dark-brown wax extracted from lignite, used in polishes and paints and for various other purposes. Also, LIGNITE WAX.

Mont Blanc *Geography.* the highest peak in the Alps (15,771 ft), on the French-Italian border.

montebrasite *Mineralogy.* $LiAl(PO_4)(OH,F)$, a white to grayish-white, triclinic mineral of the amblygonite group, occurring as equant to short prismatic crystals or in cleavable masses, having a specific gravity of 2.98 to 3.03 and a hardness of 5.5 to 6 on the Mohs scale; found in granite pegmatites.

Monte Carlo method *Statistics.* a method of approximating the solution of a mathematical problem by investigating the properties of a random process or system, often using computer-generated random numbers. Similarly, **Monte Carlo experiment.** (Named for the famous gambling casino at *Monte Carlo,* in an allusion to the randomness of such a method.)

Montelius, Oscar 1843–1921, Swedish archaelogist; constructed a chronology for prehistoric Europe; known for his work on the Bronze Age.

Montenegro test *Medicine.* a test for leishmaniasis involving the intradermal injection of *Leishmania* promastigote antigens; a positive reaction (i.e., the development of a palpable nodule within 48 to 72 hours) indicates delayed hypersensitivity, but not necessarily immunity, to *Leishmania* organisms.

Monterey Jack *Food Technology.* a mild, moist, usually pale-yellow cheddar cheese. Also, JACK CHEESE.

Monterey pine *Forestry.* a California pine, *Pinus radiata,* commonly used in reforestation because of its adaptation to fire: its cones open to germinate only in the heat of a forest fire.

montes the plural of *mons.*

Montgolfier, Joseph [mänt gälf´yā] 1740–1810, and his brother **Jacques,** 1745–1799, French inventors; developed the first practical hot-air balloon.

montgomeryite *Mineralogy.* $Ca_4MgAl_4(PO_4)_6(OH)_4 \cdot 12H_2O$, a colorless to deep-green, monoclinic mineral of the mongomeryite group, occurring as lathlike crystals, having a specific gravity of 2.46 and a hardness of 4.0 on the Mohs scale; found in variscite nodules and pegmatites.

Montgomery's tubercles *Anatomy.* a condition of swollen sebaceous glands in the mammary areolae during pregnancy. (Named for William Fetherstone *Montgomery,* 1797–1859, Irish obstetrician.)

Montian *Geology.* a European geologic stage of the Paleocene epoch, occurring after the Danian and before the Thanetian.

Monticellidae *Invertebrate Zoology.* a family of tapeworms that parasitize catfish.

monticellite *Mineralogy.* $CaMgSiO_4$, a colorless or gray, orthorhombic mineral, occurring in massive to granular habit or as prismatic crystals, having a specific gravity of 3.08 to 3.27 and a hardness of 5.5 on the Mohs scale; found in limestone and dolomite as a metamorphic or contact metasomatic mineral.

monticulus *Anatomy.* **1.** a slight elevation in a surface. **2.** an elevated portion of the superior vermis projecting from the cerebellum.

montiniculous *Ecology.* living or thriving in mountainous habitats. Thus, **monticole.**

Montlivaltia *Paleontology.* an extinct genus of scleractinian corals in the suborder Faviina; solitary, cup-shaped or horn-shaped, and large (up to about 10 centimeters across); worldwide distribution; extant in the Lower Jurassic to Cretaceous.

montmorillonite *Mineralogy.* $(Na,Ca)_{0.3}(Al,Mg)_2Si_4O_{10}(OH)_2 \cdot nH_2O$, a white, yellowish, or greenish monoclinic mineral of the smectite group, very fine-grained and claylike in habit, having a specific gravity of 2 to 3 and a hardness of 1 to 2 on the Mohs scale; occurs as the major component of bentonite clay deposits and also in soils, sedimentary and metamorphic rocks, and some mineral deposits.

montroydite *Mineralogy.* HgO, a red to reddish-brown, vitreous, orthorhombic mineral having a specific gravity of 11.2 to 11.3 and a hardness of 2.5 on the Mohs scale; found as aggregates of small prismatic crystals with other mercury minerals.

monument *Engineering.* in surveying, a stone or structure set to mark the corner or line of a boundary.

monuron *Organic Chemistry.* $ClC_6H_4NHCON(CH_3)_2$, an odorless, white crystalline solid; very slightly soluble in water and partially soluble in alcohol; melts at 175°C; used as a wide-spectrum herbicide.

monzonite *Petrology.* a granular plutonic rock, or rock group, intermediate in composition between syenite and diorite, and consisting of roughly equal amounts of orthoclase and plagioclase, with smaller amounts of mafic minerals such as amphibole, pyroxene, or biotite, and trace amounts of quartz.

mood *Psychology.* a distinctive overall feeling or mental state such as anger, anxiety, or happiness. *Behavior.* **1.** a basic classification of behavior, such as feeding, grooming, or fleeing. **2.** see MOTIVATION. (Going back to an Old English word meaning "mind" or "spirit.")

mood induction *Behavior.* a behavior pattern in which an action being performed by one or some members of a group influences others to change from another behavior to this action.

Moody formula *Mechanical Engineering.* a mathematical formula of the relationship of the efficiency of a failed turbine to that of a model turbine.

Moody friction factor *Fluid Mechanics.* an improvement on the relationship between the Reynolds number, fluid flow, and the friction factor wherein surface roughness is taken into consideration.

moon

moon *Astronomy.* **1.** the natural satellite of the earth; it has a diameter of 3476 kilometers, a mass 0.0123 that of the earth, and orbits at an average distance of 384,500 kilometers from earth. Also, **Moon. 2.** this body during a particular month or phase; for example, the full moon, new moon, and so on. **3.** any similar body orbiting another of the major planets. **4.** a lunar month, or any period of one month.

moonbeam *Optics.* a ray of moonlight.

moon-blind *Veterinary Medicine.* affected by moon blindness.

moon blindness *Veterinary Medicine.* a disease of horses, characterized by chronic or recurring inflammation of the eye leading to opacity and eventual blindness.

moonbow *Optics.* an arc that arises from the refraction and reflection of moonlight on rain drops or mist; its colors are usually very difficult to detect. Also, LUNAR RAINBOW.

mooneye *Veterinary Medicine.* the eye of a horse affected by moon blindness. Thus, **moon-eyed.** *Vertebrate Zoology.* any of various large-eyed, herringlike fishes of the family Hiodontidae, especially *Hirodon tergisus.*

Mooney-Rivlin equation *Materials Science.* an equation derived from statistical information on random chain motion and the restraining power of cross-links in elastomers.

Mooney unit *Chemical Engineering.* a unit of plasticity for unvulcanized rubber.

moon illusion *Optics.* an optical illusion in which the moon seems larger when it appears on the horizon than when it appears higher in the sky.

moonlight *Optics.* the light of the moon; i.e., sunlight that is reflected by the moon, especially toward the earth.

moon pillar *Meteorology.* a halo formed by a projected vertical element of light through the moon by a halo.

moon pool *Petroleum Engineering.* on a drilling ship, the opening through which drilling occurs, usually located amidship.

moonquake *Astronomy.* a seismic disturbance on the moon; most such occurrences are small, have deep foci, and occur when the tidal effects of the earth and sun are strongest: at times of perigee, apogee, new moon, and full moon.

moonrise *Astronomy.* the rising of the moon above the horizon, or the time at which this occurs.

moonscape *Astronomy.* the general appearance of the surface of the moon.

moon shot *Space Technology.* a launch aimed at landing a rocket or spacecraft on the moon.

moonstone *Mineralogy.* a semitransparent to translucent variety of micro- or cryptoperthitic alkali feldspar, having a bluish- or milky-white pearly iridescence, and used as a gemstone.

moor *Engineering.* to fix into place a vessel or flight vehicle, using cables, buoys, anchors, or chains.

moor *Ecology.* a tract of wasteland that is permanently wet and covered with grasses, low shrubs, and peat deposits; found extensively in the British Isles. Also, **moorland.**

moor coal *Geology.* a friable lignite or brown coal characterized by good cleavage.

Moore, Stanford 1913–1982, American biochemist; won the Nobel Prize for work with W. H. Stein on the structure of a human enzyme.

Moore code *Telecommunications.* a teleprinter code that consists of seven binary digits for each letter.

moored mine *Ordnance.* a contact or influence-operated mine that is held below the surface of the water by a mooring attached to a sinker or anchor on the bottom.

moored mine

mooreite *Mineralogy.* $(Mg,Zn,Mn^{+2})_{15}(SO_4)_2(OH)_{26} \cdot 8H_2O$, a colorless, glassy, monoclinic mineral having a specific gravity of 2.47 and a hardness of 3 on the Mohs scale; found as aggregates of small tabular crystals with pyrochroite at Sterling Hill, New Jersey.

Moore machine *Computer Technology.* a sequential machine in which the output is dependent on the present state of the machine rather than on the input.

Moore-Smith convergence *Mathematics.* a concept of convergence defined on a net. In particular, if X is a net indexed by a directed set D, it is said that X converges to $x \in X$ if for any neighborhood of x, there exists an $\alpha_0 \in D$ such that x_α is in the neighborhood whenever $\alpha \geq \alpha_0$.

mooring *Navigation.* **1.** the process of securing or tying down a vessel or aircraft. **2.** a place where this is done. **3. moorings.** the equipment, such as tackle, used to moor a vessel or aircraft. *Naval Architecture.* a permanent position, usually in a harbor, designed to hold a vessel securely in position; it generally includes a mooring buoy that is attached to the chain of a permanent anchor.

mooring anchor *Naval Architecture.* **1.** a heavy permanent anchor used to anchor a buoy, lightship, or other nonmoving vessel. **2.** a similar anchor used to hold a vessel at a mooring. **3.** a second anchor used to hold a vessel's position.

mooring buoy *Naval Architecture.* a buoy that is secured to bottom with permanent anchors and used only for mooring ships, not for navigational purposes as other buoys are.

mooring line *Naval Architecture.* a line used to secure a vessel alongside a pier or other vessel. Standard mooring lines are named by position and function, e.g., bow check, stern spring, and so on.

mooring mast *Aviation.* a pole or mast designed to anchor a lighter-than-air craft such as a dirigible.

mooring screw *Naval Architecture.* an anchor that resembles a sharp screw with broad flanges; used to moor boats, especially in soft-bottomed lakes, rivers, and so on. Also, SCREW ANCHOR.

mooring winch *Naval Architecture.* a powered winch used to haul in and let out mooring lines.

moose *Vertebrate Zoology.* a large ruminant herbivore, *Alces american,* of the deer family (Cervidae), characterized by an ungainly form with humped shoulders, long legs, a large head with an overhanging snout, and broadly palmated antlers; found in forests of the Northern Hemisphere.

moose

MOPA master oscillator-power amplifier.

MOPP *Oncology.* an acronym for a chemotherapy regimen for cancer treatment that includes the use of the drugs mechlorethamine, Oncovin, procarbazine, and prednisone.

mor *Geology.* an accumulation of organic matter on the surface of soil that develops under cool, moist conditions. Also, ECTOHUMUS, RAW HUMUS.

Moraceae *Botany.* a family of dicotyledonous trees, shrubs, vines, and herbs having alternate, simple leaves, unisexual flowers, and usually containing latex.

morainal delta see ICE-CONTACT DELTA.

morainal lake *Hydrology.* a glacial lake that occupies a depression produced by the irregular deposition of drift in a moraine of a continental glacier.

moraine

moraine *Geology.* any mound, ridge, or other distinct accumulation of unsorted glacial drift, incorporated in or transported on a glacier, or deposited by direct glacial action.

moraine bar *Geology.* a terminal moraine that rises out of deep water a distance from the shore, thus serving as a bar.

moraine kame *Geology.* one of a group of kames having the same properties, topography, constitution, and position as a terminal moraine.

moraine plateau *Geology.* a more-or-less flat region within a hummocky moraine, generally at the same elevation as or slightly higher than the summits of surrounding knobs.

moral idiocy *Psychology.* Eugen Bleuler's term for an inability to feel sympathy or compassion for other people, while still being capable of other emotions.

moral masochism *Psychology.* a type of masochism in which unconscious sexual desires and the reactivation of the Oedipus complex create a need for punishment which is characterized by self-destructive behavior or the provocation of punishment from authority figures. Also, IDEAL MASOCHISM.

moravite *Mineralogy.* $Fe_2(N,Fe)_4Si_7O_{20}(OH)_4$, a black chlorite mineral.

Moraxella *Bacteriology.* a genus of Gram-negative bacteria that grow as aerobic, rod-shaped, parasitic cells on the mucous membranes of animals.

moray or **moray eel** *Vertebrate Zoology.* any of various predatory eels of the family Muraedinae; characterized by brightly colored skin, a thick body, sharp teeth, and a good sense of smell; found in shallow reefs and rocks of warm and tropical oceans.

morbidity *Medicine.* **1.** the state of being diseased. **2.** the rate of disease or proportion of diseased persons in a community.

morbific *Medicine.* causing disease.

morbilliform *Biology.* resembling an eruption of the measles.

Morbillivirus *Virology.* a group of viruses of the family *Paramyxoviridae,* having hemagglutinin activity that causes acute diseases including measles in humans and infections in cattle, dogs, and goats.

MORC Medical Officers Reserve Corps.

morcellation *Surgery.* the division of a solid tissue or mass into small fragments to facilitate its removal.

Morchella *Mycology.* a genus of fungi belonging to the order Pezizales, including the edible species *M. esculenta,* commonly known as the morel; found in the leaf litter of forests.

mordant *Chemistry.* a chemical used to fix color onto a textile fiber; such a substance, often metallic, combines with a dye to make an insoluble compound that stains a fiber or tissue better than the dye alone.

mordant dye *Materials.* a textile dye that requires the use of a mordant to effectively bind it to the textile fiber.

Mordellidae *Invertebrate Zoology.* a family of beetles that feed on flowers and rotting wood; found worldwide.

mordenite *Mineralogy.* $(Ca,Na_2K_2)Al_2Si_{10}O_{24}\cdot7H_2O$, a colorless to white, yellowish, or pinkish orthorhombic mineral of the zeolite group occurring as minute crystals or fibrous aggregates, having a specific gravity of 2.12 to 2.15 and a hardness of 3 to 5 on the Mohs scale; found in cavities in igneous rocks.

Mor. dict. in the manner directed. (From Latin *more dicto.*) Also, **more dict.**

Morea *Geography.* a former name of the Peloponnesus.

morel *Mycology.* an edible mushroom of the genus *Morchella,* especially *M. esculenta,* having a spongelike top on its stem.

morel

morenosite *Mineralogy.* $NiSO_4 \cdot 7H_2O$, a green, water-soluble, orthorhombic mineral occurring as efflorescent crusts, having a specific gravity of 1.95 to 1.97 and a hardness of 2 to 2.5 on the Mohs scale; derived by oxidation of nickel-bearing sulfides.

Morera's stress functions *Mechanics.* three functions of position that fully determine the stress in a body.

Morera's theorem *Mathematics.* Let $f(z)$ be a single-valued and continuous function of a complex variable, defined in a domain D. Suppose further that the line integral $\oint_a^b f(z)dz$ is independent of path for any smooth curve γ in D. Then $f(z)$ is analytic in D.

mores *Ecology.* a term for groups of organisms that prefer the same environment and share the same reproductive season.

Moreton wave *Astrophysics.* a shock wave on the sun's photosphere that is produced by a large solar flare and travels at roughly 1000 km per second.

Morgagni, Giovanni B. 1682–1771, Italian anatomist; originated pathologic anatomy; discovered heart diseases.

Morgan, Lewis Henry 1818–1881, American anthropologist; studied kinship systems; originated the theory of social evolution.

Morgan, Thomas Hunt 1866–1945, American geneticist and zoologist; awarded the Nobel Prize for the development of chromosome theory of heredity.

Thomas Hunt Morgan

Morganella *Bacteriology.* a genus of Gram-negative bacteria of the tribe Proteeae that may be found in the human intestine; characterized by the ability to deaminate phenylalanine.

Morgan equation *Thermodynamics.* a modified version of the Ramsay-Shields equation that relates the surface tension of a liquid, the molecular weight, and the specific volume to the temperature; the modification is that the molar surface energy is quadratically, rather than linearly, dependent on the temperature.

morganite see VOROBIEVITE.

Morgan's classification of galaxies *Astronomy.* a classification scheme invented by W. W. Morgan that uses the integrated spectrum of the stars of a galaxy together with its shape (real and apparent) and its degree of central concentration.

Morgan unit *Genetics.* a unit of map distance on a chromosome, used to express the relative distances between genes based on the frequency of crossover between the genes; one Morgan unit equals a crossover value of 100%, while a centimorgan equals a crossover value of 1%.

morgue *Medicine.* a mortuary; a place for holding dead bodies before identification and disposal.

moribund *Biology.* **1.** in a dying or deathlike state. **2.** in a state of suspended life functions; dormant.

Moridae *Vertebrate Zoology.* an alternate name for the family Eretmophoridae, codlike fishes in the order Gadiformes.

morin *Organic Chemistry.* $C_{15}H_{10}O_7 \cdot 2H_2O$, combustible, colorless to yellow needles; soluble in boiling water and alcohol; melts (decomposes) at 285°C; used as a dye and a luminescence indicator.

Morinaceae *Botany.* in some classification systems, a segregate family from the Dipsacaceae, a family of perennial herbs in the order Dipsacales that occur in the Balkans to Himalaya and China.

Morinae *Vertebrate Zoology.* the subfamily of the Eretmophoridae containing the deep-sea codfish.

Moringaceae *Botany.* a monogeneric family of xerophytic deciduous trees of the order Capparales, characterized by a stout trunk, gummy bark, unicellular hairs, and the production of mustard oils; found in Africa and across the Middle East to India.

Moringuidae *Vertebrate Zoology.* the spaghetti eels, a small family of the order Anguilliformes, characterized by a very long, scaleless body with reduced dorsal and anal fins; found in the Indo-Pacific and western Atlantic oceans.

morinite *Mineralogy.* $NaCa_2Al_2(PO_4)_2(F,OH)_5 \cdot 2H_2O$, a pink to white or colorless monoclinic mineral having a specific gravity of 2.96 and a hardness of 4 to 4.5 on the Mohs scale; found as prismatic crystals and crystalline masses in pegmatites.

Morisette expansion reamer *Mining Engineering.* a reaming device with three tapered lugs or cutters.

mormon cricket *Invertebrate Zoology.* a flightless, cricketlike orthopteran insect, *Anabrus simplex,* that is a harmful plant pest in the arid western United States. (So called because it is found in areas where the Mormons settled.)

Mormyridae *Vertebrate Zoology.* the elephant fishes, a family of freshwater African fishes of the order Ormyriformes, many of which have elongated snouts.

Mormyriformes *Vertebrate Zoology.* an order of freshwater African fishes, containing the families Mormyridae and Gymnarchidae; sometimes includes the suborder Notopteroidei.

morning glory *Botany.* any of various twining plants, especially of the genera *Ipomoea* and *Convovulus*, characterized by cordate leaves and funnel-shaped flowers of various colors, which often open only in the morning.

morning sickness *Medicine.* nausea and vomiting affecting pregnant women during the first few months of pregnancy, especially in the morning.

morning star *Astronomy.* a heavenly body that is especially visible just before sunrise; usually the planet Venus, but on occasion the term may also refer to Jupiter or Mercury.

moron *Psychology.* a term formerly used to describe a mildly retarded person with an IQ of 50 to 70.

moropus *Paleontology.* a genus of clawed perissodactyls of the family Chalicotheriidae of the American Miocene; attained the size of the modern horse.

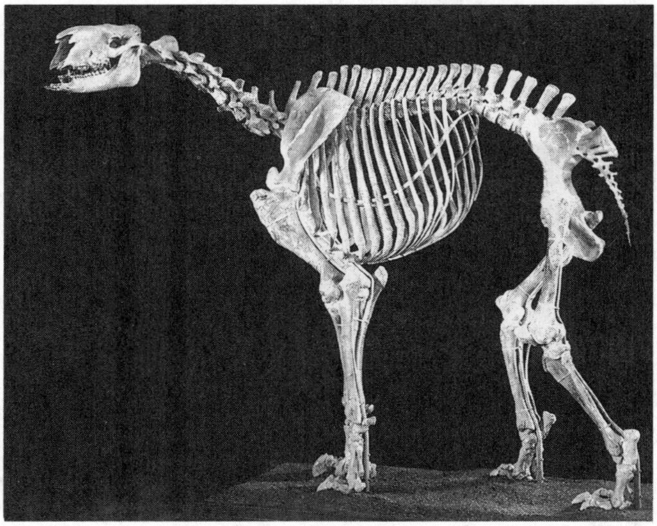

moropus skeleton

Moro reflex *Physiology.* a characteristic movement similar to an embrace that an infant makes when startled by a loud noise or loss of support; the startle reflex. (Identified by Ernst *Moro,* 1874–1951, German pediatrician.)

morph [môrf] *Biology.* the form, shape, or structure of an individual or a species. *Genetics.* any of two or more phenotypic varieties in a polymorphic population. *Mycology.* any stage in a fungus, including the sexual and asexual stages.
morph- or **morpho-** a prefix meaning "form."
morphactins *Organic Chemistry.* a class of substances that affect plant growth and development; derived from fluorine 9-carboxylic acids. (An acronym for morphologically active substance.)
morphallaxis *Physiology.* the renewal of lost tissue or body part by reorganization of the remaining parts of the body.
morpheme *Lingusitics.* the smallest unit of meaning in a language; e.g., the word *unfriendly* is made up of three morphemes: *un-, friend,* and *-ly.*
Morphidae *Invertebrate Zoology.* a family of large lepidopteran insects with metallic blue wings, including the Morpho butterflies.
morphine [môr´fēn] *Pharmacology.* $C_{17}H_{19}NO_3$, the most important and most addictive narcotic alkaloid of opium, obtained by extraction and occurring as a bitter, white crystalline powder or as white acicular crystals; used chiefly in medicine as an analgesic and sedative.

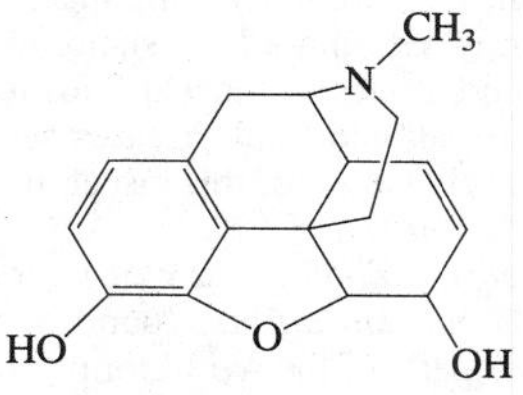

morphine

morphinism *Medicine.* a morbid condition caused by an excessive dose or habitual use of morphine.
morphism *Mathematics.* a member of a class Hom(X,Y) of mappings between two objects X and Y of a category.
morphogene *Genetics.* a gene that directly or indirectly controls the development of the form and structure of an organism, such as a gene coding for the production of a polypeptide hormone.
morphogenesis *Developmental Biology.* the emergence of shape in cells, tissues, or the entire embryo.
morphogenetic movements *Cell Biology.* the movement by cells during animal or plant development, such as the invagination of cells during embryonic gastrulation.
morphogenetic region *Geology.* a region in which the climatic conditions influence predominant geomorphic processes, imparting to the landscape distinctive regional characteristics that contrast with those of regions influenced by different climatic conditions.
morpholine *Organic Chemistry.* C_4H_9NO, a toxic, flammable, colorless liquid; soluble in water and organic solvents; boils at 128.9°C and freezes at –4.9°C; used as a solvent, an additive to waxes and polishes, a corrosion inhibitor, and a chemical intermediate.

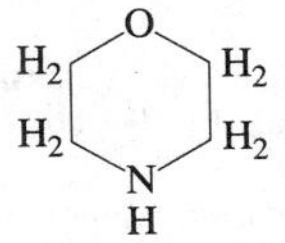

morpholine

morphological analysis *Artificial Intelligence.* in natural language processing, the process of removing affixes and reducing words to root words and affixes; e.g., *running* becomes *run + ing.*
morphological astronomy *Astrophysics.* an approach to the study of the universe that seeks to inventory all possible kinds of celestial objects, both known and as-yet unknown.
morphological stability *Materials Science.* the property of a material relating to the dynamics of the spontaneous shape of the interface that separates phases during a phase transformation.
morphological subunit *Virology.* the structural subunit of a virus particle that can be seen under the electron microscope.
morphology *Linguistics.* the study of the forms that language can take, as by the addition of a suffix to a root word. *Biology.* **1.** the study of the form and structure of organisms, especially their external form. **2.** the form and structure of a single organism considered as a whole.
morphosan *Organic Chemistry.* $C_{17}H_{19}NO_3$, a solid morphine derivative that does not produce the unpleasant aftereffects of morphine; used medicinally.
morphosis *Biology.* the sequence of development or change in an organism or in any of its parts.
morphospecies *Systematics.* the designation of a species based completely on morphological characteristics.
morphotropism *Crystallography.* the similarities of two crystals of different substances. Thus, **morphotropic.**
morphotype *Biology.* an individual of one form of a polymorphic species.
morph ratio cline *Ecology.* a gradual change in the frequency of appearance of certain basic body shapes and structures that accompanies changes in geographical location and ecological conditions.
Morquio's syndrome *Medicine.* a form of mucopolysaccharidosis causing excretion of keratin sulfate in the urine; clinical features include progressive deafness, dwarfism (growth stops at age 6), a waddling gait, mild corneal clouding, and broad-mouthed facies with spacing between the teeth. (Named for Luis *Morquio,* 1867–1935, Uruguayan physician.)
Morrowan *Geology.* a North American provincial series of the lowermost Pennsylvanian period, occurring after the Chesterian of the Mississippian and before the Atokan.
Morse, Samuel F. B. 1791–1872, American inventor; developed the first telegraph; invented the Morse code for sending messages.
Morse cable code *Telecommunications.* a code in which negative and positive current impulses of the same length represent dots and dashes, and a space is represented by the absence of current. Also, INTERNATIONAL CABLE CODE.
Morse code *Telecommunications.* either of two telegraphic systems of clicks and spaces, short and long sounds, or flashes of light that are used to represent the letters of the alphabet, numbers, or phrases. Also, INTERNATIONAL MORSE CODE, CONTINENTAL CODE.
Morse equation *Physical Chemistry.* an equation that shows the potential energy in a diatomic molecule at a specific electronic state.
Morse potential *Physical Chemistry.* the amount of potential energy in a diatomic molecule, based on the distance between the nuclei at a specific electronic state.
Morse theory *Mathematics.* the study of homotopy properties of a manifold X by analysis of the second derivative at critical points of a smooth function f on X. Applications include the study of global properties of classical and quantum physical systems, the computation of Feynman path integrals when there are conjugate points between the endpoints, and the occurrence of singularities in the elastic frequency distributions of crystals.
Mor. sol. in the usual way. (From Latin *more solito.*)
mortality rate *Medicine.* the frequency of death; the proportion of deaths in a specified number of the population.
mortar *Materials.* **1.** in general, any material used to bind bricks or stones together. **2.** specifically, a mixture of cement, lime, or gypsum plaster with sand and water that is used for this purpose or for general plastering. **3.** to apply such a mixture. *Science.* a strong, bowl-shaped vessel in which a substance is crushed to a powder using a pestle. *Ordnance.* a muzzle-loading, indirect fire weapon with a tube length of 10 to 20 calibers; it usually has a shorter range than a howitzer and employs a higher angle of fire.
mortar structure *Petrology.* a crystalline rock structure resulting from dynamic metamorphism, characterized by a mica-free aggregate of fine grains of quartz and feldspar that fills the interstices between or forms the borders of the larger, rounded grains, thus resembling stones set in mortar. Also, CATACLASTIC STRUCTURE, PORPHYROCLASTIC STRUCTURE.
Mortillet, Gabriel de 1821–1898, French archaelogist; studied the French Paleolithic; invented the system of epochs of cultural history.
mortise [môrt´is] *Engineering.* a recess cut into a piece of timber to receive a tenon and form a mortise-and-tenon joint. *Graphic Arts.* an opening cut in a printing plate or negative to insert revised copy.
mortise and tenon *Mechanical Devices.* an interlocking joint, such as a dovetail joint, in which a tenon is housed in or secured to a mortise. Also, **mortise joint.**
mortise bolt *Mechanical Devices.* a long, heavy bolt housed in a door mortise and flushed with its edge.
mortise chisel *Mechanical Devices.* a chisel with a thick, stiff blade and wide side edges, used with a mallet to shape and cut mortises. Also, HEADING CHISEL, FRAMING CHISEL.
mortise lock *Mechanical Devices.* a lock fitted into a mortise, usually in the edge of a door.

mortise pin *Mechanical Devices.* a joint-securing or locking device consisting of a tapered wood pin used between both members of a mortised joint or through an extended tenon.

mortise wheel *Mechanical Devices.* a cast-iron or steel wheel with mortises or slots to receive wooden cogs rather than iron teeth.

mortising machine *Mechanical Engineering.* a machine having an auger and a chisel that is used to produce a rectangular mortise in wood.

mortlake see OXBOW LAKE.

Morton, William T. G. 1819–1868, American dentist; tested the use of ether as an anesthetic.

morula *Developmental Biology.* the mass of rounded blastomeres resulting from the early cleavage divisions of the zygote.

Moruloidea see MESOZOA.

morvan *Geology.* a region characterized by an uplift of hard rock bordered by a sloping land of older rock and representing the intersection of two peneplains.

MOS metal oxide semiconductor.

mosaic [mō zā´ik] *Science.* a surface design made up of small pieces. *Architecture.* a decorative surface formed by inlaying small pieces of variously colored material such as stone, tile, or marble. *Cartography.* a group of overlapping aerial photographs that have been trimmed and assembled to match and to form a continuous photographic representation of a portion of the earth's surface. *Electronics.* an iconoscope screen composed of millions of tiny, silver, light-sensitive globules on a sheet of ruby mica. *Petrology.* **1.** a rock texture in which individual grains or fragments form a regular pattern with relatively straight or slightly curved boundaries or contacts. **2.** of or relating to such a texture. Also, CYCLOPEAN. *Virology.* a form of leaf variegation consisting of a pattern of patchy color distribution; symptomatic of some plant virus infections. *Biology.* an organism or part made up of tissues or somatic cells of genetically different types, caused by gene or chromosome mutations.

mosaic cleavage see DETERMINATE CLEAVAGE.

mosaic development *Developmental Biology.* a condition in fertilized eggs of certain species, such as the sea urchin, in which the cells of early stages have developed cytoplasm which determines the parts that are to develop.

mosaic disease *Plant Pathology.* a virus disease affecting a wide range of plants and characterized by a yellow and green mottling of leaves and sometimes curling, dwarfing, or narrowing of the leaves.

mosaic egg *Cell Biology.* an egg with an asymmetrical chemical composition that results in a mosaic pattern of embryonic development, in which different cells or groups of cells follow separate pathways of differentiation.

mosaic evolution *Evolution.* the evolutionary change of different adaptive components of the phenotype of an organism at different times or at different rates in an evolutionary sequence.

mosaicism *Genetics.* a phenomenon in which a mutation in early development results in an organism composed of two or more distinct cell lines of different chromosomal or genetic composition.

mosaic spread *Crystallography.* the divergence of an X-ray beam scattered by a crystal; believed to be caused by the irregularity of orientation of small blocks of unit cells in the crystal that may have varying sizes, but are very small on a macroscopic scale. The misalignment of these blocks of unit cells is small, on the order of 0.2° to 0.5° for most crystals. This imperfection is convenient because diffracted intensities from crystals with such a mosaic structure are more intense than those from "perfect" crystal.

mosaic structure *Crystallography.* a crystal structure in which each minute domain is bounded by its discontinuity lattice at the interface with other domains of like composition.

mosandrite *Mineralogy.* $(Na,Ca,Ce)_3Ti(SiO_4)_2F$, an alteration product of rinkite; a very rare, reddish-brown, yellowish-brown, or dull-green, weakly radioactive, monoclinic mineral having a specific gravity of 2.93 to 3.5 and a hardness of 5 on the Mohs scale. Also, JOHNSTRUPITE.

mosasaur *Paleontology.* of or relating to the large marine lizards of the family Mosasauridae; probably related to the varanids, the mosasaurs ate fish and mollusks and retained almost no trace of their terrestrial ancestry; upper Cretaceous.

moschellandsbergite *Mineralogy.* Ag_2Hg_3, a silver-white, metallic, cubic mineral occurring as dodecahedral crystals and in massive and granular form, having a specific gravity of 13.6 to 13.71 and a hardness of 3.5 on the Mohs scale.

Moschidae *Vertebrate Zoology.* the musk deer, in some classification systems, a family in the suborder Tylopoda; placed by others as a subfamily in Cervidae and called Moschinae.

Moseley, Henry G. J. 1887–1915, British physicist; formulated Moseley's law for the frequency of X-ray spectra.

Moseley's law *Spectroscopy.* a mathematical relationship used to show that all elements can be arranged according to the frequencies of their X-ray spectra into one continuous series that corresponds with the order of their atomic numbers.

mosesite *Mineralogy.* $Hg_2N(Cl,SO_4,MoO_4,CO_3)\cdot H_2O$, a yellow to green cubic mineral occurring as octahedral crystals, having a specific gravity of 7.72 and a hardness of 3.5 on the Mohs scale; found with native mercury and cinnabar.

MOSFET metal oxide semiconductor field-effect transistor.

mOsm milliosmol.

MOS memory *Computer Technology.* a semiconductor memory technology used in integrated memory circuits. (Acronym for metal oxide semiconductor.)

mos oncogene *Oncology.* an oncogene that produces sarcomas, found in the Moloney sarcoma virus.

mosque [mäsk] *Architecture.* an Islamic place of worship.

mosquito [mə skēt´ō] *Invertebrate Zoology.* any of various two-winged insects of the dipteran fly subfamily Culicinae; characterized by long legs, thin branched antennae, and a long suctorial proboscis with piercing stylets, used by the females of most species to suck blood; includes many carriers of protozoan and viral diseases such as malaria and yellow fever. (From the Latin name for this insect; literally, "the little fly.")

mosquito boat see PT BOAT.

mosquito fish *Vertebrate Zoology.* any of various fishes that feed on mosquite larvae, as *Gambusia affinis;* found in the southeastern U.S. and introduced to other parts of the world for mosquito control.

moss *Botany.* **1.** the common name for tiny plants of the class Musci, characterized by small, flowerless leafy stems with the sexual organs located at the ends; they reproduce by spores and grow in clusters on tree trunks, rocks, and moist ground. **2.** a growth of such plants.

moss agate *Mineralogy.* a white or nearly translucent variety of agate containing manganese or iron oxides that produce dark, mosslike dendritic inclusions.

Mössbauer, Rudolf L. born 1929, German physicist; awarded the Nobel Prize for his work on gamma rays and the discovery of recoilless resonance absorption.

Mössbauer effect *Nuclear Physics.* the emission of a gamma ray from a nucleus bound in a solid such that the recoil energy is imparted not to the emitting atom but to the whole lattice, with the result that the relatively large recoiling mass has a recoil velocity of practically zero. Also, RECOILLESS EMISSION.

Mössbauer spectroscopy *Spectroscopy.* the measurement and analysis in crystalline solids of small shifts in nuclear energy levels that correspond to gamma-ray emission and resonant reabsorption frequencies. Energy is added to the system through the Doppler effect; used in structural and qualitative analysis of at least 42 elements; mostly for study of iron and tin compounds.

Mössbauer spectrum *Spectroscopy.* a graph of the relative number of gamma rays per second that pass a gamma-ray-absorbing sample, plotted with respect to the relative velocity of the gamma-ray-emitting source and the sample.

Moss' classification *Hematology.* a classification of ABO blood types designated by roman numerals I to IV, corresponding to types AB, A, B, and O, respectively.

moss forest see TEMPERATE RAIN FOREST.

mossite *Mineralogy.* $Fe(Nb,Ta)_2O_6$, a tapiolate mineral consisting of an oxide of iron and tantalum; isomorphous with tapiolite.

moss pink *Botany.* the phlox *Phlox subulata,* characterized by showy pink to purple flowers; found in the eastern U.S.

moss rose *Botany.* the rose *Rosa centrifolia muscosa,* characterized by a mosslike growth on the calyx and stem.

mossy zinc *Metallurgy.* zinc granules produced by quenching molten zinc in water.

MOST metal oxide semiconductor transistor.

most general unifier *Artificial Intelligence.* a unifier of a set of literals such that any other unifier can be derived from it by substituting terms for its variables.

most probable position *Navigation.* the location of a craft judged to be most accurate in the absence of a fix.

most significant character *Computer Programming.* the character of a word, number, or signal that has the greatest value or significance; usually the left-most character. Similarly, **most significant bit, most significant digit.**

Motacillidae *Vertebrate Zoology.* the pipits and wagtails, a family of birds of the passerine suborder Oscines; found in grasslands, fields, marshes, and on shores worldwide, with most in the Old World.

moth *Invertebrate Zoology.* the common name for a nocturnal insect belonging to the lepidopteran suborder Heteroneura; it differs from the butterfly in having a stout body, smaller, less colorful wings, and antennae that are usually hairy, branched, or featherlike.

mothball *Military Science.* **1.** to prepare and protect military equipment in order to store it for a long period of time. **2.** of or relating to equipment that has been stored in such a manner. Thus, **mothball fleet.**

mother *Acoustical Engineering.* a term for the mold of a recording made from the master record and used as a template for subsequent record copies.

motherboard *Computer Technology.* a printed circuit board with an interconnecting assembly to which other circuit cards, boards, and peripheral control units are connected. Also, BACKPLANE.

mother cell see MICROSPOROCYTE.

Motherellaceae *Mycology.* a family of fungi of the order Mucorales, having some species that are distinguished by a garlic odor and zonate growth pattern; found in soil, plant debris, dung, and stream insects.

mother liquor *Crystallography.* the solution from which crystals are obtained.

mother lode *Geology.* **1.** the principal lode or vein passing through a district or country. **2.** the original deposit of ore from which a placer is derived.

mother-of-pearl *Materials.* a hard, iridescent substance forming the inner layer of certain mollusk shells; used for making buttons, beads, and the like.

mother-of-pearl clouds see NACREOUS CLOUDS.

motherwort *Botany.* the plant *Leonorus cardiaca* of the mint family, characterized by cut leaves with a whorl of lavender flowers in the axils; native to Europe and introduced as a weed into the U.S. (It was once thought to be useful in treating diseases of the uterus.)

motif [mō tēf′] *Science.* a recurring and characteristic form or theme. *Materials Science.* **1.** the atom group associated with each lattice point in a crystal. **2.** in a polymer microstructure, the smallest group of atoms in the chain that, under the operation of a rotation-translation operator, will generate all other atoms in the chain.

motile *Biology.* capable of spontaneous movement, such as spermatozoa or certain bacteria. *Psychology.* describing an individual whose sensory images focus on his own movements.

motile colony *Bacteriology.* any colony of bacterial cells that is able to move across the surface of a culture plate, such as colonies of *Bacillus circulans.*

motilin *Endocrinology.* a peptide hormone that is secreted by the enterochromaffin cells of the duodenum and regulates intestinal motility during intervals of fasting.

motility *Physiology.* the capacity for spontaneous movement.

motion *Science.* **1.** the act of moving; the passage of a body from one place to another. **2.** a bodily movement. **3.** the ability to move. *Mechanics.* a change in the position of a physical system over time.

motional *Mechanics.* relating to or involving motion.

motional electromotive force *Electromagnetism.* any electromotive force that is generated in a circuit by the motion of the circuit through a magnetic field.

motional impedance *Electronics.* in a transducer, such as a telephone receiver or relay, the part of the input electrical impedance that is due to the motion of the mechanism; the difference between the input electrical impedance when the system is oscillating and the same impedance when the system is stopped or blocked.

motional induction *Electromagnetism.* an induced electromotive force caused by the movement of a circuit transversely through the flux of a magnetic field.

motion analysis see MOTION STUDY.

motion cycle *Industrial Engineering.* one complete sequence of motions involved in a repetitive work process.

motion dynamics *Computer Programming.* a computer graphic function in which objects may be moved and rotated in 3 dimensions with respect to a stationary observer, or the objects may remain stationary and the observer may move around them and zoom in for more detail.

motion economy *Industrial Engineering.* the minimization of human motions involved in an industrial process in order to simplify and improve the effectiveness of the process.

motion parallax *Robotics.* the apparent change in an object's position when viewed as a sequence of pictures.

motion picture *Graphic Arts.* a series of still photographs placed in succession on a strip of film so as to give the illusion of movement when the film is run through a projector.

motion-picture projector *Engineering.* a mechanical device used to project motion pictures on a screen at the same speed they were filmed; also used to play the accompanying sound tracks.

motion register *Computer Technology.* the mechanism that controls the movement of tape on a tape drive.

motion sensing *Robotics.* the ability of a robot to sense the speed and direction of a moving object.

motion sequence programming *Design Engineering.* the sequence of motions by which a robot or machine is guided from a controller.

motion sickness *Medicine.* nausea, vomiting, vertigo, and general malaise caused by rhythmic or irregular movement, such as seasickness, airsickness, or car sickness.

motion specification *Design Engineering.* the description of motions generated to guide a machine.

motion study *Industrial Engineering.* the analysis of the human movements involved in an industrial process, for the purpose of eliminating unnecessary movements and arranging those that remain in the most effective sequence. Also, MOTION ANALYSIS.

motivated forgetting *Psychology.* the act of forgetting something that is embarrassing or painful.

motivation *Behavior.* any energy or force that brings about behavior in an organism.

motive *Behavior.* any condition in an organism that initiates or directs its behavior to achieve a certain goal. *Psychology.* **1.** an individual's stated or apparent reason for acting. **2.** the individual's true conscious or unconscious reason for acting.

motoneuron see MOTOR NEURON.

motor *Electricity.* a device that converts electrical energy into mechanical energy using forces exerted by magnetic fields on current-carrying conductors. Also, ELECTRIC MOTOR. *Mechanical Devices.* in general, any relatively small engine, such as the internal-combustion engine in an automobile. *Physiology.* of or relating to structures, such as muscles or nerves, that are involved in or cause movement; sometimes refers to nerves that innervate glands.

motor alexia *Medicine.* a condition in which the patient understands what he sees written or printed but cannot read it aloud.

motor aphasia *Medicine.* an impairment of the ability to speak and write, due to a lesion of the cortical center.

motor apraxia *Medicine.* a loss of the ability to make proper use of an object, although its proper nature is known. Also, KINESTHETIC APRAXIA.

motor area *Anatomy.* the part of the cerebral cortex that, upon stimulation by electrodes, causes the contraction of voluntary muscles. Also, **motor cortex.**

motor ataxia *Medicine.* an inability to control or coordinate movements of the muscles.

motor board see TAPE TRANSPORT.

motorboat *Naval Architecture.* any small motor-driven vessel, especially a boat fitted with an outboard motor.

motorboating *Electronics.* a repetitive popping or puffing sound in faulty equipment that resembles a motorboat sound, caused by excessive positive feedback.

motor branch circuit *Electricity.* a branch circuit that supplies energy to one or more motors and their associated controllers.

motor cell *Physiology.* an efferent neuron, especially one of the cells of the spinal cord that has its axon continued into a motor nerve fiber. *Botany.* see BULLIFORM CELL.

motor-converter *Electricity.* an electromechanical device that combines an induction motor connected to an AC supply with a synchronous converter that is connected to a DC circuit; used for converting alternating current to pulsating direct current.

motorcycle *Mechanical Engineering.* a two-wheeled motor vehicle that is larger and heavier than a bicycle; often used on highways and for long-distance transportation.

motor effect *Electromagnetism.* the phenomenon of a repulsive force acting between two conductors carrying current in opposite directions.

motor element *Acoustical Engineering.* the portion of an acoustic transducer, such as a loudspeaker or microphone, that vibrates to produce sound.

motor end plate *Neurology.* the large formation at the end of the axon of a motor neuron that contacts the muscle cell and provides for synaptic transmission of the motor-command nerve impulse to the muscle. Also, MYOCEPTOR.

motor-generator set *Electricity.* an electromechanical device in which a low-voltage motor is coupled to a high-voltage generator; used to convert power-source voltages to other desired voltages or frequencies.

motor grader see AUTOPATROL.

motorized unit *Military Science.* a unit equipped with motor transportation that moves all the personnel, weapons, and equipment at the same time.

motor meter *Engineering.* an integrating meter that registers the number of revolutions made by a motor whose speed depends on the quantity of power flowing through the circuit to which it is connected; thus the measurement reflects the energy consumed by the circuit.

motor nerve *Physiology.* any peripheral efferent nerve that stimulates muscle contraction.

motor neuron *Physiology.* a neuron possessing a motor function; an efferent neuron conveying motor impulses in a pyramidal pathway, whose cell bodies are located in the nuclei of cranial nerves and in the anterior columns of gray matter of the spinal cord. Also, MOTONEURON.

motor noise *Acoustics.* the background noise of an electric motor that is present at a particular sound-recording location.

motor parameters *Robotics.* values that describe the actions of motors, such as back emf constant, load torque, peak torque, stall torque, motor inertia, and speed-torque curve.

motor point *Neurology.* **1.** the point at which a motor nerve is connected to the muscle that it innervates. **2.** a point on the skin over a muscle at which an electric stimulus will cause the muscle to contract.

motor ship *Naval Architecture.* any ship powered by an internal combustion motor (rather than a steam engine).

motor speech area *Anatomy.* the region of the cerebral cortex concerned with speech production.

motor system *Physiology.* a system composed of the motor area of the cortex in the frontal lobes, the large pyramidal cells from which motor impulses arise, the long axons of these cells that pass through lower centers and terminate in the spinal cord, and the final motor neurons that carry impulses to the muscles involved in an action.

motor torpedo boat see PT BOAT.

motor truck *Mechanical Engineering.* an automotive vehicle that is used for freight transport.

motor unit *Anatomy.* one motor neuron and the muscle fibers with which it synapses.

motor vehicle *Mechanical Engineering.* any automotive vehicle with rubber tires that does not run on rails.

Mott, Sir Nevill F. born 1905, British physicist; developed the theory of photographic process; awarded the Nobel Prize for his work on semiconductor devices.

Mottelson, Ben Roy born 1926, American-born Danish physicist; awarded the Nobel Prize for his work on the structure of the atomic nucleus.

Mott insulator *Materials Science.* certain solids that function as insulators based on Mott's argument that for every solid, a critical bandwidth value exists, below which even the outer electrons do not spread into bands but remain localized around a particular core.

mottle *Virology.* a diffuse form of the mosaic symptom in which two or more colors develop in somewhat ill-defined areas with indistinct boundaries between them.

mottled *Geology.* of a soil, sediment, or sedimentary rock that is irregularly marked with spots or patches of different colors.

mottled iron *Metallurgy.* cast iron containing both gray and white areas.

mottle-leaf *Plant Pathology.* **1.** see MOSAIC DISEASE. **2.** a plant disease caused by a deficiency of zinc that results in chlorosis of the leaves and retardation of plant growth.

mottling *Medicine.* a condition marked by spots or patches of color.

mottramite *Mineralogy.* $PbCu^{+2}(VO_4)(OH)$, a green to brown orthorhombic mineral of the descloizite group occurring as pyramidal or prismatic crystals and in granular and fibrous forms, having a specific gravity of about 5.9 and a hardness of 3 to 3.5 on the Mohs scale; found as a secondary mineral in the oxidized zones or in ore deposits.

Mott scattering *Quantum Mechanics.* the scattering of identical particles due to coulomb interactions; this scattering differs from Rutherford scattering in the indistinguishability of the particles.

MOU memorandum of understanding.

Mougeotia *Botany.* a genus of green algae of the family Zygnemataceae, characterized by cells with a single axile laminate chloroplast and isogamous conjugation; occurring in quiet bodies of water.

moulin [moo lan′] *Hydrology.* a nearly vertical, cylindrical hole or shaft that has been worn in the ice of a glacier by the swirling motion of meltwater as it pours down from the surface. (From the French word for *mill,* because the falling water makes a similar rushing sound.)

moulin pothole see GIANT'S KETTLE.

Moulton, Forest R. 1872–1952, American astronomer; with T. Chamberlin, proposed the planetesimal theory of the origin of the solar system.

mounce *Metrology.* a unit of mass equal to 25 grams. Also, METRIC OUNCE.

mound *Geology.* **1.** any isolated, low, rounded natural hill or knoll, usually of earth. Also, TUFT. **2.** a structure built by fossil colonial organisms. *Archaeology.* **1.** a gradual accumulation of debris upon which a continuously occupied settlement is built, or which is the by-product or remains of some activity. **2.** a constructed earthwork or fortification, especially one with a geometric or animal form.

mount *Engineering.* **1.** the support on which an apparatus or instrument is designed to rest. **2.** to attach something onto such a support. *Ordnance.* specifically, **1.** a carriage or stand upon which a weapon is placed. **2.** to place a weapon on such a carriage or stand, or to put a weapon into position for use. *Military Science.* to assemble and prepare a military force for an operation. Similarly, **mounting, mounting area.**

mountain *Geography.* a natural elevation of the earth's surface, rising steeply from the surrounding level to a summit; usually defined as having an elevation at least 2000 feet above its surroundings.

mountain and valley winds *Meteorology.* a system of diurnal winds, prevalent in calm, fair weather conditions, that blow uphill and upvalley during the day and downvalley during the night.

mountain artillery *Ordnance.* light artillery designed to be used and transported in mountainous terrain.

mountain ash *Botany.* any of many small trees of the genus *Sorbus* of the rose family; characterized by clusters of small, white flowers and bright-red berries.

mountain avens *Botany.* an evergreen plant of the genus *Dryas* of the rose family, characterized by showy, single, white or yellow flowers; found in alpine regions of the Northern Hemisphere.

mountain beaver see SEWELLEL.

mountain blue *Inorganic Chemistry.* the mineral azurite, when ground and used as pigment. Also, COPPER BLUE.

mountain breeze *Meteorology.* the night-time component of mountain and valley winds that blows downvalley at night. Also, MOUNTAIN WIND.

mountain building see OROGENY.

mountain climate *Meteorology.* the climate of relatively high elevations, typically characterized by decreased pressure and temperature, reduced oxygen availability, and increased isolation. Also, HIGHLAND CLIMATE.

mountain cork *Mineralogy.* a white or gray variety of asbestos, usually tremolite, that resembles cork in weight and texture, is composed of thick, interwoven fibers, and floats on water. Also, **mountain leather.**

mountain cranberry *Botany.* the low-growing shrub *Vaccinium vitis-idaea* of the heath family, having tart, red edible berries; found in northern regions. Also, LINGONBERRY.

mountain daisy *Botany.* the sandwort *Arenaria groenlandica* of the pink family, characterized by numerous small leaves and small, white flowers; found in rocky soil in Greenland and North America. Also, MOUNTAIN SANDWORT.

mountain effect *Electromagnetism.* an effect whereby radio waves are reflected by mountainous terrain and thus may be received with error by radar direction finding methods.

mountain-gap wind *Meteorology.* a localized wind moving through a mountain gap.

mountain goat *Vertebrate Zoology.* a ruminant mammal, *Oreamnos americanus,* of the family Bovidae, more closely related to the antelope than the sheep (unlike the true goats); characterized by short, backcurved horns, a thick, white shaggy coat, and hooves specially adapted to grip rocky surfaces; found in the mountains of northwestern North America. Also, ROCKY MOUNTAIN GOAT.

mountain holly fern *Botany.* the evergreen fern *Polystichum lonchitis,* characterized by stiff, leathery fronds; found in North America, Europe, and Asia.

mountain howitzer see MOUNTAIN ARTILLERY.

mountain laurel *Botany.* **1.** the evergreen plant *Kalmia latifolia,* characterized by glossy, dark, oblong leaves with pointed ends, pink or white flowers, and poisonous berries; found in eastern North America. **2.** the flower of this plant; state flower of Connecticut and Pennsylvania.

mountain lion *Vertebrate Zoology.* the large wildcat *Felis concolor,* characterized by a gray or tawny coat with a white throat and belly, a slender body, long legs, and a round, rather small head; it is distinguished by a wild, haunting cry. Once found widely in forests throughout North America, it is now chiefly found only in protected areas in the West. Also, COUGAR, PANTHER.

mountain lion

mountain mahogany see OBSIDIAN.

mountain pediment *Geology.* a plain formed by a combination of erosion and deposition at the base of a desert mountain range, surrounding the mountain such that, from a distance, it appears to be a broad, triangular mass above which the mountain projects.

mountain range see RANGE.

mountain sandwort see MOUNTAIN DAISY.

mountain sickness *Medicine.* nausea and difficult breathing caused by being at high altitudes. Also, MONGE'S DISEASE.

mountain system *Geography.* a group of mountain ranges united by proximity and often orogenesis or morphology.

mountain wave *Meteorology.* a wavelike effect, characterized by severe updrafts and downdrafts, that results when rapidly moving air impinges upon a steep-rising mountain slope.

mountain wind see MOUNTAIN BREEZE.

mountain wood *Geology.* a compact, gray-to-brown type of asbestos that resembles dry wood in appearance. Also, ROCK WOOD. *Mineralogy.* a fibrous clay mineral such as sepiolite or palygorskite.

mountant *Microbiology.* any substance in which a specimen is placed for microscopic examination.

mourning dove *Vertebrate Zoology.* a wild pigeon, *Zendidura macroura carolinensis,* of the family Columbidae, noted for its sorrowful cooing; native to the woodlands and open country of North America.

mouse *plural,* **mice.** *Vertebrate Zoology.* **1.** the smaller rodents of the genus *Mus* and related genera in the family Muridae, such as the **house mouse,** *Mus musculus,* as well as genera of the family Zapodidae. **2.** a general term used to denote any small rodent. *Computer Technology.* a device, thought of as somewhat resembling a mouse in appearance and movement, that allows the user to control cursor movement on a video display screen by rolling the device over a flat surface. It is also used to select commands, designate text blocks, and for other functions.

mousebird *Vertebrate Zoology.* any of several small, long-tailed African birds of the genus *Colius,* constituting the family Coliidae.

mouse button *Computer Technology.* a switch on the back of a mouse device that transmits commands to the computer.

mouse click *Computer Technology.* the action of pressing a mouse button to make a selection.

mouse deer *Vertebrate Zoology.* a very small tropical artiodactyl of the family Tragulidae, characterized by a lack of horns or antlers. Also, CHEVROTAIN.

mouse hole *Petroleum Engineering.* a hole drilled into the side of a borehole to hold a joint of drill pipe; the joint will later be connected to the drill string and replaced in the mouse hole with the next joint of drill pipe.

mouse trap *Engineering.* a device that consists of an inward-facing valve attached to an open bottom; used to retrieve inaccessible objects.

Mousterian *Archaeology.* **1.** a Middle Paleolithic culture that is defined by the development of a wide variety of specialized tools, such as spear points. **2.** a term for the Middle Paleolithic period in Europe. **3.** of or relating to this culture or period. (From Le *Moustier,* a site in France where characteristic flint scraping tools were found.)

Mousterian industry *Archaeology.* a progressive stage in the manufacture of stone tools; characterized by the use of flaking techniques to produce cutting and scraping tools with sharp edges or points.

mouth *Anatomy.* the opening through which a human or animal takes in food; the upper opening of the digestive tract, containing the tongue and teeth. *Zoology.* any similar food cavity in other organisms.

mouth any opening that resembles or functions like the human mouth, providing an entrance to or exit from a larger thing; specific uses include: *Geography.* **1.** the outfall of a river or stream into another body of water. **2.** the entrance to a valley or an underground cavity. *Acoustical Engineering.* the open end of a horn, such as a trombone, from which sound emerges. *Mining Engineering.* the location at which a mine shaft or tunnel emerges at the surface.

mouthbreeder *Vertebrate Zoology.* any of the fishes of the genera *Tilapia* and *Haplochromis* that hatch and tend their young in the mouth.

mouth-to-mouth resuscitation *Medicine.* a procedure used as part of cardiopulmonary resuscitation in which the victim's mouth is covered by the mouth of the person administering aid, rhythmically forcing air into the victim's lungs.

movability *Robotics.* the ability of a robot with sensors to transfer parts and carry out assembly or other actions requiring complex motions.

movable-active tooling *Mechanical Engineering.* any tool or equipment in a robotic system that operates and moves using its own power.

movable bridge *Civil Engineering.* any bridge that can be moved to allow the unhindered passage of boats or other crafts; often constructed in steel plate and strengthened by an interstice of welded joists.

movable contact *Electricity.* a relay contact that is moved directly by an actuating system to engage or disengage one or more stationary contacts. Also, ARMATURE CONTACT.

movable-head disk drive *Computer Technology.* a storage unit consisting of magnetic disks with circular recording tracks and a set of read/write heads for each disk surface; the heads move in and out to access data on any of the circular tracks. Also, MOVING-HEAD DISK DRIVE.

movable-passive tooling *Mechanical Engineering.* any equipment that moves but requires no power to operate, such as workpieces, templates, and robotic systems.

movable platen *Engineering.* the part at the back of an injection-molding machine to which the mold is attached.

move generator *Artificial Intelligence.* in a search, a procedure that enumerates the possible moves or operator applications from a given state, e.g., all the legal moves from a given chess position.

movement *Medicine.* **1.** an act of moving; motion. **2.** an act of defecation.

movement picture see DEFORMATION PLAN.

movement plan see DEFORMATION PLAN.

move mode *Computer Programming.* a method of providing access to data for input/output control routines by transferring the data into and out of memory areas designated by the operating program, as distinguished from locate mode.

move operation *Computer Programming.* the process of transferring data from one location to another.

moving area *Telecommunications.* part of a television picture frame in which the intensity has undergone a change since the previous frame.

moving average *Statistics.* the weighted average of the k most recent observations of data constituting a time series.

moving bed *Chemical Engineering.* the granulated solids in a process vessel used for mass transfer, employing a catalyst or other solid that moves with the process stream and is then separated for reuse.

moving-bed catalytic cracking *Chemical Engineering.* a petroleum-refinery process used to crack crude oil in the presence of a catalyst that is in motion between the reactor and a regenerator.

moving-boundary electrophoresis *Physical Chemistry.* a form of electrophoresis in which the movement of the solvent is unrestricted, so that all the particles of a given material move at the same rate, forming a definite boundary.

moving cluster *Astronomy.* a nearby, open star cluster whose distance can be determined by a careful study of the individual proper and radial motions of the stars.

moving-cluster parallax *Astronomy.* a distance worked out for a moving cluster.

moving coil *Electricity.* a current-carrying coil in a magnetic field.

moving-coil galvanometer *Engineering.* a galvanometer having a moving coil that is suspended or pivoted in a magnetic field; small currents of electricity that flow through the coil are indicated and measured by the resulting motion of the coil.

moving-coil instrument *Electricity.* a measuring instrument that derives its action from the force on a moving coil suspended in a magnetic field; a pointer is usually attached to the coil to serve as an indicator. Also, D'ARSONVAL METER.

moving-coil loudspeaker *Acoustical Engineering.* a widely used type of loudspeaker, consisting of a cone-shaped diaphragm attached to a coil of wire (voice coil) in a magnetic field produced by a permanent magnet; when a current is passed through the voice coil, the force generated moves the coil and diaphragm, thus producing sound. Also, ELECTRODYNAMIC LOUDSPEAKER, MOVING-CONDUCTOR LOUDSPEAKER.

moving-coil meter *Electricity.* the use of a pivoted coil in a meter as the moving element.

moving-coil microphone *Acoustical Engineering.* a microphone in which a diaphragm and a wire coil are attached to each other and suspended in a magnetic field; sound waves striking the diaphragm cause the coil to move across the field, thus producing an electric current in the coil; typically used in CB radios.

moving-coil voltmeter *Engineering.* a voltmeter that is constructed like a moving-coil galvanometer and used to measure direct current.

moving-conductor loudspeaker see MOVING-COIL LOUDSPEAKER.

moving constraint *Mechanics.* a type of constraint whose position changes with time.

moving-crystal technique *Materials Science.* any of various methods for producing single crystals and shaping them from molten material.

moving frame *Mathematics.* in differential geometry, a family of basis tangent vector fields defined smoothly on an open set of a manifold. Also, REPERE MOBILE, ENNUPLE. Usually accompanied by a dual basis of 1-forms. On a (pseudo-) Riemannian manifold, the basis vectors may be orthogonal, constituting an **orthogonal moving frame.**

moving grid *Radiology.* a grid that is moved continuously throughout the making of a radiograph to eliminate the grid lines occurring with the use of a stationary grid.

moving-head disk drive see MOVABLE-HEAD DISK DRIVE.

moving-iron meter *Engineering.* an instrument that depends on the ability of a moving coil to attract a piece of soft-iron connected to a pointer, in order to measure current.

moving-iron voltmeter *Engineering.* a voltmeter that depends on the attraction of two soft-iron vanes (one fixed in place and one attached to a pointer) to a coil carrying electric current.

moving load see LIVE LOAD, def. 2.

moving-magnet voltmeter *Engineering.* a voltmeter relying on the alignment of a permanent magnet with the magnetic field produced by a moving coil and a second permanent magnet.

moving-mold continuous casting *Metallurgy.* a metal-casting process in which molten metal is continuously poured inside formed molds that are constantly in motion, solidifying as it moves along the molds.

moving sidewalk *Civil Engineering.* a mechanically aided walkway that conveys a pedestrian forward at a slow speed.

moving-target indicator *Electronics.* a Doppler-radar device that displays moving targets and suppresses the display of stationary targets.

MOW *Aviation.* the airport code for Moscow.

mower *Mechanical Devices.* **1.** a farm machine with a cutting bar or rotary blades, used to cut forage for hay. **2.** see LAWNMOWER.

mower-conditioner *Agriculture.* a mower equipped with intermeshing cleated, conditioning rolls that aerate the soil for better conditioning and faster drying.

mower-conditioner

moxa *Surgery.* a tuft of soft, combustible material, such as cotton or wool, burned on the skin as a cautery and counterirritant; widely used in Asia.

Moxaverine *Pharmacology.* the trade name of the drug eupavarin; used to prevent or relieve spasm.

moxibustion *Surgery.* the practice of igniting a moxa on the skin to produce cautery and counterirritation. Also, BYSSOCAUSIS.

Mozambique Channel [mō´zam bēk´] *Geography.* a 950-mile-long channel of the Indian Ocean, between southeast Africa and Madagascar.

Mozambique Current *Oceanography.* a current that is part of the western boundary current system of the Indian Ocean, flowing southward between Madagascar and the African mainland and along the coast to around 30°S.

mp, mp., or **m.p.** melting point.

MP or **M.P.** military police.

m.p. *Pharmacology.* in the manner prescribed. (From Latin *modo prescripto.*)

mpc megaparsec.

MPD or **M.P.D.** maximum permissible dose; magneto-plasmadynamic.

mpg miles per gallon.

mph or **m.p.h.** miles per hour.

M.P.H. Master of Public Health.

M phase *Genetics.* the mitosis period of the cell cycle; it is the shortest period and results in one cell producing two identical daughter nuclei.

mphps miles per hour per second.

MPM, mpm, or **m.p.m.** meters per minute.

MPO myeloperoxidase.

M protein see M ANTIGEN.

MPS mononuclear phagocyte system.

MPU microprocessing unit.

MPX multiplex.

MQ metol-quinal.

MQ register *Computer Technology.* a multiplier-quotient register in the arithmetic-logic unit that holds the multiplier prior to multiplication, the least significant digits of the product after multiplication, and the quotient during division.

MR map reference.

mR or **mr** milliroentgen.

mrad millirad.

MRBM medium-range ballistic missile.

MRC Medical Reserve Corps.

M.R.D. minimum reacting dose.

M region *Astrophysics.* the designation for a region on the sun that emits streams of charged particles for a period of at least 27 days (one solar rotation), thus creating periodic magnetic disturbances on earth.

mrem millirem.

MRF melanocyte-stimulating hormone releasing factor.

MRI magnetic resonance imaging.

mRNA messenger RNA.

mRNA decay *Molecular Biology.* a normal cellular process by which species of messenger RNA are degraded.

MRP see MATERIAL REQUIREMENTS PLANNING.

MRP II see MANUFACTURING RESOURCE PLANNING.

MRV multiple reentry vehicle, a vehicle, such as a nuclear warhead, carried by a delivery system that directs more than one such vehicle towards one target. Also, **M.R.V.**

MS multiple sclerosis.

M.S. Master of Science; Master of Surgery.

Ms *Metallurgy.* the temperature at which the martensitic transformation begins upon cooling.

mS millisiemens.

ms millisecond.

ms. manuscript.

m/s meters per second; meters per second per second.

ms2 *Virology.* an ssRNA-containing, unenveloped virion of type species Levivirus.

MSAT Minnesota Scholastic Aptitude Test.

MSBS *Ordnance.* French submarine-launched, medium-range ballistic missiles; **M20** is a two-stage rocket carrying a single thermonuclear warhead to a range of 3000 km; **M4** has three stages and delivers six nuclear warheads in a range of 4000 km or greater.

MSc. Master of Science.

M scan see M DISPLAY.

M scope *Electronics.* a radarscope that produces an M display.

MSD most significant digit.

MS-DOS [em´es´däs´] *Computer Technology.* an operating system developed by the Microsoft Corporation that provides software services for single-user microprocessors; it has become the standard operating system for IBM and IBM-compatible microcomputers. (An acronym for Microsoft Disk Operating System.)
MSG monosodium glutamate.
M.Sgt. master sergeant.
MSH melanocyte-stimulating hormone.
M shell *Atomic Physics.* the electron orbit of an atom, third closest to the nucleus, in which electrons whose principal quantum number is 3 occur.
MSH-IF melanocyte-stimulating hormone-inhibiting factor.
MSI medium-scale integration.
M.S.L. or **m.s.l.** mean sea level; midsternal line.
MSMA monosodium acid methanearsonate.
MSN master of science in nursing. *Aviation.* the airport code for Madison, Wisconsin.
MSP see MONOBASIC SODIUM PHOSPHATE.
MS star *Astronomy.* an M star with a spectrum rich in zirconium oxide (ZrO) absorption bands.
M star *Astronomy.* a reddish star with a spectrum dominated by bands of molecular absorption, principally titanium oxide (TiO), and having a surface temperature of 3500 kelvin or less.
MSTS Military Sea Transportation Service.
MSW Master of Social Work; Master of Social Welfare; magnetic surface wave.
MSY *Aviation.* the airport code for New Orleans International, New Orleans, Louisiana.
M synchronization *Graphic Arts.* the synchronization of a camera flash and shutter in which the shutter delays opening in order to allow the flash to burn to its brightest.
MT megaton; megatons; machine translation.
M.T. metric ton; medical technologist; membrana tympani.
mT maritime tropical air.
mtambara *Materials.* the plain, nondurable wood of a tree of Africa, *Cephalosphaera usambarensis,* used for furniture and plywood.
MTBF mean time between failures.
mtDNA mitochondrial DNA.
MTF modulation transfer function.
MTI moving target indicator.
MTM methods-time measurement.
MTOC microtubule-organizing center.
MTS *Acoustical Engineering.* multitrack stereo, a feature in a television receiver or videocassette recorder that allows for the reception and output of separate right and left audio channels rather than a single monaural source.
MTU methylthiouracil.
MTY *Aviation.* the airport code for Monterrey, Mexico.
M-type backward-wave oscillator *Electronics.* an ordinary or noncrossed field backward-wave oscillator that is insensitive to load variations, in which electrons interact with a radio-frequency wave that is traveling opposite to the electron beam.
mu *Molecular Biology.* a unit for mapping distances between gene loci or chromosomes.
Mu *Virology.* a temperate bacteriophage that contains linear double-stranded DNA and infects certain enterobacteria.
mU milliunit.
M.u. Mache unit.
MUC *Aviation.* the airport code for Munich, Germany.
Mucedinaceae see MONILIACEAE.
mucic acid *Organic Chemistry.* $HO_2C(CHOH)_4CO_2H$, a combustible, white crystalline powder; soluble in water and insoluble in alcohol; melts (decomposes) at 210°C; used as a chemical intermediate, a substitute for tartaric acid, and a metal ion sequestrant. Also, GALACTARIC ACID.
mucigen *Biochemistry.* any mucus-producing agent.
mucilage [myoo´si lij] *Materials.* **1.** any of various water-soluble, viscous adhesives derived from gum, glue, or similar sources. **2.** any of various gelatinous secretions from certain plants, especially from seaweed.
mucin *Biochemistry.* any mucoprotein that raises the viscosity of the medium surrounding the cells that secrete it.
mucinosis *Medicine.* the accumulations of mucins in the skin, due to abnormal metabolic changes, the degeneration of neoplasms, or localized cysts.
muck *Geology.* a dark, finely divided, well-decomposed organic material with a high percentage of mineral matter that accumulates on the surface in certain poorly drained areas, such as permafrost regions. *Civil Engineering.* excavated earth, scrap rock, or rubble often used as landfill material.
muco- *Zoology.* of or relating to mucous. Also, **muc-, muci-.**
mucocele *Medicine.* **1.** an enlarged cavity containing mucus. **2.** a polyp containing mucus.
mucocutaneous *Anatomy.* of, relating to, or affecting the mucous membranes and the skin; used especially in reference to the areas at which mucous membranes and skin meet at the oral, nasal, vaginal, and anal openings.
mucoid *Biochemistry.* **1.** of, relating to, or resembling mucus. **2.** a shimmering quality in the appearance of a bacterial colony, frequently caused by excessive amounts of water in the cell envelope.
mucolytic *Biochemistry.* destroying or dissolving mucin, or an agent that so acts.
mucopeptide *Biochemistry.* a cross-linked, polysaccharide-peptide complex of indefinite size found in the inner walls of all bacteria.
mucopolysaccharide see GLYCOSAMINOGLYCAN.
mucopolysaccharidosis *Medicine.* any illness caused by the incomplete breakdown of various mucopolysaccharide sulfates in the body, in which the compounds are passed in urine and deposited in body tissues.
mucoprotein *Biochemistry.* a glycoprotein, present in all connective and supporting tissues, that forms a highly viscous solution and often functions as a lubricant.
mucopurulent *Medicine.* made up of both mucus and pus.
Mucor *Mycology.* a genus of fungi belonging to the class Zygomycetes that occurs on various types of organic matter and soil and includes species which are parasitic to stored grains.
Mucoraceae *Mycology.* a family of fungi belonging to the order Mucorales that occurs in soil, dung, plant debris, or stored grains and lives off nonliving organic matter, although some species are parasites.
Mucorales *Mycology.* an order of fungi belonging to the class Zygomycetes that is composed of some plant and animal pathogens and occurs in soil, dung, and decaying organic matter.
mucoran *Biochemistry.* a glucan, containing fucose, mannose, and glucuronic acid, that binds to polymers containing insoluble glucosamine in cell walls to maintain its form.
mucormycosis *Medicine.* a fungal disease caused by organisms of the genus *Mucorales,* affecting persons who have been weakened by chronic diseases such as diabetes; it affects the respiratory tract initially, then spreads by spores to other organs.
mucosa *Histology.* an epithelial membrane that produces mucus at its free border.
mucosal disease *Pathology.* a malady afflicting cattle caused by the virus of bovine diarrhea; characterized by fever, diarrhea, ulcerations, and a drop in milk yield.
mucosanguineous *Medicine.* containing blood and mucus.
mucous [myoo´kəs] *Physiology.* **1.** relating to or resembling mucus. **2.** secreting mucus.
mucous cell *Physiology.* a glandular cell that secretes mucus.
mucous colitis *Medicine.* the most common form of colitis; also, SPASTIC COLITIS, IRRITABLE COLON, FUNCTIONAL BOWEL DISTRESS.
mucous connective tissue *Histology.* a type of embryonic loose connective tissue having few cells, found in the early embryo and in the umbilical cord.
mucous degeneration *Medicine.* a decomposition of epithelial tissue with the accumulation of mucus.
mucous epithelium *Developmental Biology.* an epithelium containing cells that secrete mucus, as in the respiratory tract and gastric mucosa.
mucous gland *Physiology.* a gland that secretes mucus, as the salivary gland.
mucous membrane *Histology.* the mucus-secreting epithelial lining of cavities and canals that opens to the exterior of the body.
mucro *Biology.* an abrupt, sharp terminal tip of process, such as of a leaf.
mucron *Cell Biology.* a specialized region at the anterior aspect of certain parasitic cells that attaches the parasite to its host.
mucronate *Biology.* terminated abruptly by a sharp terminal tip or process.
mucus [myoo´kəs] *Physiology.* a thick, slimy fluid of mucous membranes and glands, composed of mucin, water, desquamated cells, leukocytes, and various inorganic salts; secreted by glands lining the nasal, esophageal, and other body cavities.

mud *Geology.* **1.** a slimy, sticky, or slippery mixture of finely divided particles of clay or silt together with water and possibly other materials, whose consistency ranges from semifluid to soft and plastic. **2.** any very wet, soft soil or mass of earth.

mud auger *Mechanical Devices.* a diamond-point bit with the wings of its tip forming a spiral. Also, CLAY BIT, MUD BIT.

mud ball *Geology.* a rounded mass of mud or mudstone occurring in a sedimentary rock due to weathering and the breakup of clay deposits.

mud bit see MUD AUGER.

mud blasting *Engineering.* the use of mud in attaching an explosive charge to rock to reduce fragmentation.

mud breccia see DESICCATION BRECCIA.

mudcap *Engineering.* the mud used to cover an explosive charge and attach it to a boulder, transmitting part of the energy of the explosion to the boulder.

mud cone *Volcanology.* a small cone of sulfurous mud built on the surface of a mudflow or around the opening of a mud volcano or mud geyser. Also, PUFF CONE.

mud crack or **mudcrack** *Geology.* an irregular fracture formed by shrinkage of clay, silt, or mud, as water evaporates from the surface under the influence of atmospheric conditions. Also, DESICCATION CRACK, SHRINKAGE CRACK, SUN CRACK.

mud crack

mud-crack polygon *Geology.* a nonsorted, polygonal ground pattern outlined by mud cracks. Also, MUD POLYGON.

mud eel *Vertebrate Zoology.* the salamander *Siren lacertina* of the family Trachystomata, characterized by external gills, tiny front legs, and no hind legs; found in shallow waters in the southeastern U.S. Also, GREATER SIREN.

mudfish *Vertebrate Zoology.* any of various fishes, such as the bowfin or mummichog, that live in muddy waters.

mud flat or **mudflat** *Geology.* a relatively level, sandy, or silty strip along a shore or around an island. Also, FLAT.

mudflow or **mud flow** *Geology.* a slow to rapid flowing mass of predominantly water-saturated, fine-grained earth material.

mudhen or **mud hen** see COOT.

Mudlac system *Genetics.* a system for causing gene or operon fusions using a defective form of bacteriophage Mu.

mudlark *Vertebrate Zoology.* a black-and-white Australian lark, *Grallina cyanoleuca,* of the family Grallinidae, named for its practice of making mud nests. Also, **mud-nest builders.**

mudlump or **mud lump** *Geology.* a broad, low mound or swelling of silt or thick, plastic clay that is thrust upward from below by the loading of newly formed deltaic deposits, forming a small, short-lived island.

mudminnow *Vertebrate Zoology.* any of various small, carnivorous fishes of the genera *Umbra* and *Novumbra;* found in muddy streams and pools.

mud pit see SLUSH PIT.

mud polygon *Geology.* **1.** a nonsorted, polygonal ground pattern in which the centers lack vegetation, but the outlining fissures contain peat and plants. **2.** see MUD-CRACK POLYGON.

mud pot *Volcanology.* a hot geyser that discharges boiling mud. Also, **mud geyser.**

mud puppy *Vertebrate Zoology.* a large aquatic salamander of the genus *Necturus* in the family Proteidae, having external gills and characterized by gray to rusty-colored skin with dark spots; native to North America and Europe.

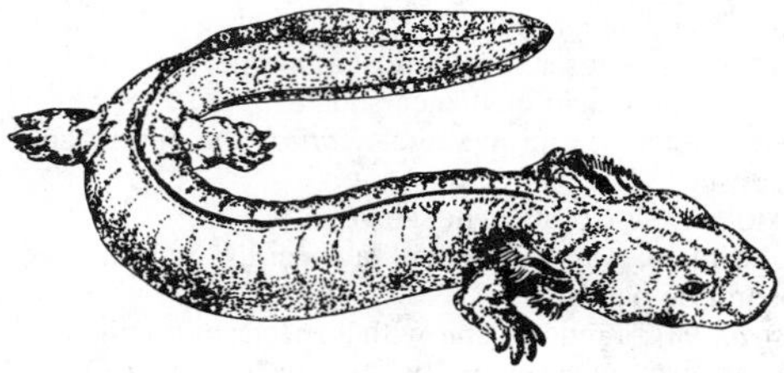

mud puppy

mud-removal acid *Petroleum Engineering.* a solution of hydrochloric and hydrofluoric acids, demulsifiers, inhibitors, and surfactants; used to clean a drill-hole face by dissolving drilling-mud clays.

mudsill *Civil Engineering.* the lowest sill in a construction; so called because it is usually embedded in the soil.

mudskipper *Vertebrate Zoology.* any of various gobies of the genera *Periophthalmus* and *Boleophthalmus;* distinguished by their tendency to stay out of the water on mud flats for extended periods of time and springing about when disturbed. Also, **mudspringer.**

mudslide *Geology.* a slow-moving mudflow that moves primarily by sliding over a discrete boundary shear surface.

mud snake *Vertebrate Zoology.* the snake *Farancia abacura,* characterized by iridescent black and red coloring and the use of its stiff tail tip to position prey for swallowing; found in the southern U.S.

mud still *Engineering.* a distillation apparatus used to separate oil, water, and other substances found in mud.

mudstone *Geology.* an indurated clayey mud or argillaceous sedimentary rock that has the composition and texture of shale, but is much less fissile.

mud volcano *Volcanology.* a usually conical accumulation of mud and rock fragments ejected by volcanic gas pressure. Also, HERVIDERO, MACALUBA.

MUF or **muf** maximum usable frequency.

mu factor *Electronics.* a measure of the relative effect of the change in voltage of two electrodes upon the current in the circuit of any specified electrode.

muffle furnace *Metallurgy.* a small electric or gas furnace that heats the outside of a chamber containing the charge; used for annealing, hardening, or enameling metals.

muffler *Acoustical Engineering.* a device used to reduce noise emanating from a machine, normally attached to a ventilation discharge pipe. *Mechanical Engineering.* specifically, such a device in a motor vehicle.

mugearite *Petrology.* a dark, fine-grained, extrusive, or hypabyssal basaltic igneous rock consisting of oligoclase, alkali feldspar, and olivine, with accessory apatite and opaque oxides.

mugger *Vertebrate Zoology.* a large freshwater crocodile, *Crocodylus palustris,* of the family Crocodylidae, native to southeast Asia. Also, MARSH CROCODILE.

muggy *Meteorology.* an informal term used to describe warm and very humid weather.

mugho pine *Botany.* a flat, shrubby pine, *Pinus mugu mugo,* that is cultivated as an ornamental; native to Europe. Also, **mugo pine.**

Mugilidae *Vertebrate Zoology.* the mullets, a family of marine fish composing the suborder Mugiloidei; characterized by a large, blunt-snouted head and heavy scales; a valuable food fish found in all tropical and temperate seas.

Mugiloidei *Vertebrate Zoology.* the suborder of perciform fish containing only the mullet family Mugilidae.

mugwort *Botany.* any of several weedy plants of the genus *Artemisia,* especially *A. vulgaris,* characterized by small, greenish flower heads and aromatic leaves.

muhimbi *Materials.* the fine-textured, heavy wood of the *Cynometra alexandri* tree of Africa; used for heavy construction and flooring strips.
muhuhu *Materials.* the aromatic, heavy, durable wood of the *Brachylaena hutchinsii* tree of East Africa; used for decorative carvings and flooring strips.
mulberry *Botany.* the common name for members of the genus *Morus,* characterized by alternate, simple leaves, unisexual flowers found in catkins, and a purple, usually edible berry.

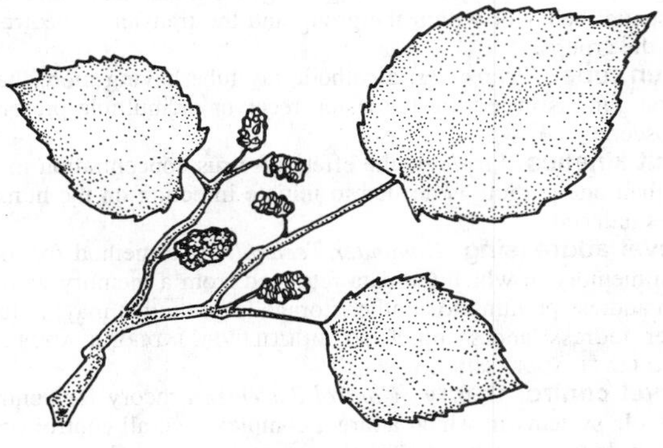

mulberry

mulch *Agriculture.* a protective covering that is spread on the ground around plants to inhibit evaporation and weed growth, control soil temperature, and enrich the soil. It may be organic material such as leaves, peat, or wood chips, or inorganic material such as plastic sheeting.
mulch tillage or **mulch planting** *Agronomy.* a system of planting in which crop residues are purposely left on or just below the soil surface to form a layer of mulch.
mule *Vertebrate Zoology.* the sterile hybrid animal resulting from a mating of a jackass (male donkey) and a mare (female horse).
mule deer *Vertebrate Zoology.* a deer, *Odocoileus hemionus,* of the family Cervidae, characterized by a stocky build, large ears, and antlers that fork differently from those of Old World deer; found in forests, woods, and deserts of western North America. (Its ears are thought to resemble those of a mule.)

mule deer

mull *Engineering.* to grind or pulverize. *Geology.* an alkaline, granular, forest humus that is incorporated with underlying mineral matter. *Geography.* a promontory or headland, especially in Scotland.
muller *Engineering.* a flat-bottom pestle that is used to crush rock or grind pigments.
Müller, Hermann Joseph [myoo′lər] 1890–1967, American geneticist; awarded the Nobel Prize for his discovery that X-rays can produce mutations.
Müller, Johannes Peter 1801–1858, German physiologist; discovered fetal pronephric ducts in the frog.
Müller, Karl Alex born 1927, Swiss physicist; awarded the Nobel Prize for his research on high-temperature superconductivity.
Müller, Paul H. 1899–1965, Swiss chemist; awarded the Nobel Prize for his research on insecticidal qualities of DDT.
Müller-Hinton medium *Microbiology.* a microbial culture medium containing beef infusion, casamino acids, and starch.
müllerian duct see PARAMESOEPHRIC DUCT.
müllerian duct cyst *Medicine.* a benign cyst that can form anywhere in the area from the cervix to the uterus and is made up of mesonephric tissues.
müllerian duct inhibitory factor see ANTIMÜLLERIAN FACTOR.
Müllerian mimicry *Behavior.* a form of mimicry in which the warning coloration of a certain species imitates that of another unpalatable or dangerous species.
müllerianoma *Oncology.* a tumor that originates in the müllerian duct.
müllerian regression factor see ANTIMÜLLERIAN FACTOR OR HORMONE.
Müller matrices *Optics.* the matrices that are used as a mathematical framework for calculating the change in the polarization state of light caused by optical elements such as polarizers.
Müller's dust bodies see HEMOCONIA.
Müller's glass see HYALITE.
Müller's larva *Invertebrate Zoology.* a free-swimming, ciliated larva of certain polyclad flatworms.
mullet *Vertebrate Zoology.* **1.** a silvery-blue fish of the Mugilidae family, especially *Mugil cephalus,* the **common** or **striped mullet**; characterized by a stout body, blunt head, small mouth, and wholesome, flavorsome flesh; found close to the shore in nearly all temperate and tropical waters. Also, **gray mullet. 2.** a small, brightly colored fish of the Mullidae family, characterized by a small mouth and weak teeth; found in warm seas. Also, **red mullet.**
Mullidae *Vertebrate Zoology.* the goatfishes, a family of marine coastal fishes of the order Perciformes, having elongated bodies with a curved dorsal profile and a flat belly; found in tropical and warm-temperate regions.
Mulliken, Robert S. 1896–1986, American chemist; awarded the Nobel Prize for his theory of molecular orbitals.
mulling *Engineering.* any of various processes used to mix clay and sand, often with water.
mullion *Geology.* in folded sedimentary and metamorphic rocks, a linear structure consisting of columns of rock that appear to intersect.
mullite *Mineralogy.* $Al_6Si_2O_{13}$, a rare, colorless to pink orthorhombic mineral occurring as prismatic crystals, having a specific gravity of 3.03 to 3.16 and a hardness of 6 to 7 on the Mohs scale; found in aluminous inclusions in volcanics and in slags; resistant to heat and corrosion, synthetic mullite is used in refractories and in the glass industry.
mull technique *Spectroscopy.* an infrared spectroscopic technique for scanning solids in which a solid sample is ground to a powder and mixed with a liquid, such as paraffin or mineral oil, that is specifically chosen not to absorb infrared wavelengths in the region being studied.
multi- a prefix meaning more than one; usually refers to many.
multiaccess computer *Computer Technology.* a computer system in which data and processing capabilities are simultaneously accessible from two or more terminals. Also, MULTIPLE-ACCESS COMPUTER.
multiaccess computing *Computer Technology.* the ability of a host computer to be accessed through more than one terminal.
multiaddress see MULTIPLE-ADDRESS.
multianode tube *Electronics.* an electron tube having at least two main anodes and a single cathode. Also, **multianode tank.**
multiaspect *Computer Programming.* of or relating to a search operation using a combination of aspects or characteristics.
multibus *Computer Technology.* of or relating to a system architecture with a main, high-speed central processing unit bus and a separate medium-speed input/output bus.
multicellular *Biology.* describing tissues, organs, or organisms consisting of many cells.
multicellular horn *Electromagnetism.* a horn antenna array in which several horns are placed on a common surface and fed by a waveguide; their spacing is one wavelength apart.
multicentered arch see BASKET-HANDLE ARCH.
multichain polymer *Materials Science.* a step-reaction polymer produced by the polymerization of a bifunctional monomer with a small amount of a different monomer; the result is a linear polymer with multiple chains but no network structure.

multichamber centrifuge *Biotechnology.* a modified centrifuge consisting of a series of concentric tubular sections that form a continuously enlarging passage for liquid flow; commonly used to clarify fruit juices and beer.

multichannel communication *Telecommunications.* communication that takes place along two or more channels over the same path.

multichannel field-effect transistor *Electronics.* a field-effect transistor in which voltages are applied to the gate to control the current flow channel space; multiple channels permit the handling of higher currents without frequency response reduction.

multichannel loading *Telecommunications.* a condition of a multichannel communications system in which all channels are active.

multichannel telephone *Telecommunications.* a telephone system in which two or more communications channels are carried over only one cable.

multichip microcircuit *Electronics.* a microcircuit composed of interconnected active and passive chip-type components.

multicipital *Biology.* having many heads or branches arising from one point.

multicollector electron tube *Electronics.* an electron tube in which electrons travel to multiple electrodes.

multicomponent analysis *Spectroscopy.* the simultaneous treatment and interpretation of emission or absorption spectra of a sample containing several constituents; the individual spectra of the components must be identified and separated.

multicomponent polymerization *Materials Science.* the copolymerization of three or more monomers, occurring with many propagation and termination reactions.

multicomponent virus *Virology.* a virus in which the infectious genome is shared by two or more particles. Also, **multipartite virus.**

multicomputer array *Design Engineering.* a group of interconnected computers working in accordance with any number of software packages.

multicomputer system *Computer Technology.* a system composed of two or more interconnected computers, either in a master-slave relationship or in a peer relationship.

multicopy inhibition *Genetics.* the inhibition of transposition of a Tn*10* element in the presence of a multicopy plasmid carrying IS10-R.

multicopy plasmid *Genetics.* any of numerous copies of a certain plasmid that are found in a cell.

multicoupler *Electronics.* a device that connects and matches the impedances of several receivers to one antenna.

multicycle *Geology.* of or relating to a landscape or landform produced by or during more than one cycle of erosion and bearing traces of former conditions.

multideck *Engineering.* of an apparatus, having two or more vertical levels, surfaces, or compartments.

multideck clarifier *Engineering.* an apparatus used to extract pollutants from waste water.

multidimensional derivative *Mathematics.* let f be a function of n variables into an m-dimensional space (so that f has m component functions). If f satisfies the usual differentiability conditions, then the generalized multidimensional derivative of f is the matrix whose (i,j)th entry is $\partial f_i/\partial x_j$. Equivalently, the derivative in a coordinate basis.

multidimensional niche *Ecology.* the range of conditions, resource levels, and densities of other species within which an organism, a population, or a species can survive, grow, and reproduce.

multidrop line *Telecommunications.* **1.** a communications system configuration that uses only one line to serve multiple terminals. **2.** a communication line with several subsidiary controllers that share time on a line under the control of a central site. Also, MULTIPOINT LINE.

multielectrode tube *Electronics.* an electron tube containing more than three electrodes associated with a single electron stream.

multielement array *Electromagnetism.* a directive antenna array having several radiating elements.

multielement parasitic array *Electromagnetism.* an antenna having several parasitic radiating elements arranged so as to produce a highly directive beam.

multienzyme complex *Enzymology.* a complex aggregation of different enzymes found in cells that together catalyze a coordinated sequence of reactions.

multifid *Biology.* divided into many lobes.

multiflora rose *Botany.* the climbing rose *Rosa multiflora,* characterized by hooked prickles and dense clusters of fragrant flowers; found in Japan and Korea.

multifunction processing *Computer Technology.* an approach to faster processing in a sequential computer in which certain processing and control hardware is replicated to allow simultaneous execution of several operations on a single processor.

multigene family *Genetics.* any group of genes that have evolved from a single ancestral gene, characterized by the existence of multiple copies, close linkage, sequence homology, and related phenotypes.

multiglandular *Anatomy.* derived from or related to several glands. Also, PLURIGLANDULAR.

multigroup model *Nucleonics.* a model for neutron behavior that basically groups neutrons by their energy ranges, but also takes into account their spatial behavior within the group and the transfer of neutrons between the groups.

multigun tube *Electronics.* a cathode-ray tube having more than one electron gun, as in a color television receiver or multiple presentation oscilloscope.

multi-hit kinetics *Virology.* the effect of virus concentration in which more than one particle is needed to initiate infection on the number of plaques induced.

multilevel addressing *Computer Technology.* a method for locating data in memory in which the data retrieved from a memory word is itself an address pointing to another storage location that may, in turn, be another address, and so on. Also, MULTILEVEL INDIRECT ADDRESSING, SECOND-LEVEL ADDRESSING.

multilevel control theory *Control Systems.* a theory for controlling large-scale systems in which a large, complex, overall control problem is broken down into a number of simpler and more easily managed subproblems that still satisfy the overall system objectives and constraints.

multilevel indirect addressing see MULTILEVEL ADDRESSING.

multilevel transmission *Telecommunications.* the transmission of digital information in which at least three levels of voltage are recognized as meaningful.

multiline appearance *Telecommunications.* the ability of a telephone to receive or originate additional calls at the same time that it is engaged in a primary voice call, or to bring an additional party or parties to the primary call.

multilinear algebra *Mathematics.* the study of functions of several variables that are linear with respect to each variable.

multilinear cultural evolution *Anthropology.* a theory proposing that individual cultures in various settings have parallel patterns of development, and that cross-cultural comparisons of various culture types will reveal the underlying laws of culture.

multilinear form *Mathematics.* a form that is linear with respect, in turn, to each subset of its arguments.

multilinear function *Mathematics.* a function that is linear in each of its arguments individually.

multiline cultivar *Plant Pathology.* a mixture of isolines that differ by a single gene in their reaction to an infecting agent.

multilinked list *Computer Programming.* a file structure in which each record contains two or more pointers to other records in the list.

multilist organization *Computer Programming.* a chained file organization in which each chain is divided into enclosed segments, permitting faster searching.

multilocus theory *Genetics.* a theory that explains the existence of alleles of correlated function at separate loci.

multi-male group *Behavior.* a form of social organization in which several mature males live within the same group along with females and young.

multimer *Molecular Biology.* a protein molecule composed of more than one polypeptide chain.

multimeter see VOLT-OHM-MILLIAMMETER.

multimotion actuator *Robotics.* a large actuator with nonrotating cylinders that can move a manipulator in controlled linear and rotary directions.

multinomial *Mathematics.* the sum of two or more monomials.

multinomial coefficient *Mathematics.* the quantity $n!/(k_1!k_2!k_r!)$, where $\sum_{i=1}^{r} k_i = n$; denoted $(k_1!k_2!n! \ldots k_r!)$. The case $r = 2$ is the usual definition of binomial coefficient.

multinomial distribution *Mathematics.* the probability distribution of a random variable taking on a finite number of possible values. In particular, suppose n successive trials have r mutually exclusive outcomes with respective probabilities $p_1, p_2, \ldots, p_r$. Then the probability that there are k_i occurrences of the ith outcome for each i (without respect to order) is given by $f(k_1, k_2, \ldots, k_r) = (k_1!k_2!n! \ldots k_r!)p_1^{k_1}p_2^{k_2} \ldots p_r^{k_r}$, where $\sum_{i=1}^{r} k = n$. The case $r = 2$ is the usual binomial distribution.

multinomial theorem *Mathematics.* let $(p_1, p_2, \ldots, p_r)$ be any finite sequence of nonzero real numbers. Then for every positive integer n, $(p_1 + p_2 + \cdots + p_r)^n = \sum(k_1!k_2!n! \ldots k_r!)p_1^{k_1}p_2^{k_2} \ldots p_r^{k_r}$, where the sum is taken over all possible r-tuples $(k_1, k_2, \ldots, k_r)$ such that $\sum_{i=1}^{r} k_i = n$. (If 0^0 is defined to equal 1, the condition that p_i be nonzero can be dropped.) The case $r = 2$ is the usual binomial theorem.

multinomial trials *Statistics.* a group of independent and identically distributed trials, each having a finite number of possible outcomes. The Bernoulli trials are a special case arising when there are only two possible outcomes.

multinuclear *Science.* having more than one nucleus. Also, **multinucleate, multinucleated.**

multipactor *Electronics.* a high-power, high-speed microwave switching device in which a radio-frequency electric field drives a thin electron cloud between two parallel plane surfaces in a vacuum.

multiparous *Botany.* of or relating to a cyme with many axes.

multipartite genome *Virology.* a viral genome that is shared by more than one nucleic acid molecule.

multipass printing *Computer Technology.* a function of some dot matrix printers in which the characters are printed more than once with slight offset adjustments of the paper between each printing, thus achieving higher print quality.

multipass sort *Computer Programming.* a routine used to sort files that are larger than available main memory by using a secondary memory device such as a disk or tape.

multipath assignment *Transportation Engineering.* the practice of dividing traffic assignments among competing routes serving the same origin-destination pair.

multipath cancellation *Telecommunications.* the complete cancellation of signals due to the relative amplitude and phase differences of components coming in over separate paths.

multipath transmission *Electromagnetism.* an effect whereby two or more paths are taken by radio signals received by an antenna; the signals usually have different amplitudes and phases that result in a distorted signal.

multiphonon emission *Solid-State Physics.* a nonradiative recombination process wherein the energy of an electron-hole recombination is dissipated in the form of lattice vibrations.

multiphoton absorption *Atomic Physics.* a process in which an atom increases its energy by absorbing two or more photons.

multiphoton ionization *Atomic Physics.* a process in which an atom becomes ionized when it absorbs two or more photons.

multiple *Science.* having several or many parts, elements, features, and so on. *Mathematics.* a quantity a (such as a real or complex number, vector, group element, etc.) is said to be a multiple of another quantity b if $a = nb$ for some (positive) integer n. If a and b are vectors or complex numbers, the product is understood to be scalar multiplication. If a and b are elements of an additive group, the product is understood to be n-fold addition (b added to itself n times). *Electricity.* **1.** a group of terminals arranged in parallel to create circuitry that is accessible at a number of points at which connections can be made. **2.** to create such an arrangement.

multiple abstract variance analysis *Psychology.* a research design for estimating the relative effects of heredity and environment, as in personality characteristics.

multiple-access computer see MULTIACCESS COMPUTER.

multiple-access network *Computer Technology.* a flexible computer network that allows every station to access the network at any time; priority schemes are established in case of contention for network use.

multiple accumulating registers *Computer Technology.* two or more accumulating registers contained in an arithmetic-logic unit so that several arithmetic results may be maintained simultaneously.

multiple activity process chart see GANG CHART.

multiple-address *Computer Programming.* of or relating to an instruction format having more than one address field. Also, MULTIADDRESS.

multiple-address code *Computer Programming.* program instructions that have two or more address fields.

multiple-address computer *Computer Technology.* a computer that can interpret and execute machine instructions with two or more address fields.

multiple-address instruction *Computer Programming.* an instruction including an operation code and more than one address.

multiple alleles *Genetics.* a group of variations of a genetic-sequence coding for a product at the same locus on a chromosome, such as the three alleles that determine the ABO human blood groups.

multiple appearance *Electricity.* in a telephone switchboard, a single-line circuit that appears before a minimum of two operators for a unique jack arrangement.

multiple-arch dam *Civil Engineering.* a lightweight construction barrier that stabilizes weak foundations, consisting of parallel recurring arches whose axes ascend approximately 45° to form a reinforcement.

multiple-beam antenna *Electromagnetism.* an antenna array that is designed to radiate several directive beams in different directions.

multiple-beam interference *Optics.* the interference between beams of light from a common source, in which each beam is a multiply reflected fraction of the incident light and refers to the vector sum of all the beams transmitted.

multiple-beam interferometer *Optics.* an interferometer that reflects a beam a number of times between two parallel plane surfaces to greatly reduce the linewidth of interference fringes and increase the precision of measurement.

multiple-beam spectrometer *Spectroscopy.* an instrument used in spectroscopy in which several beams of different wavelengths are generated.

multiple birth *Medicine.* the birth of two or more offspring produced in the same gestation period.

multiple birth (quintuplets)

multiple bond *Physical Chemistry.* any bond in which there is more than one shared electron pair, such as a double bond. Also, **multiple bonding.**

multiple call transmission *Telecommunications.* the process of routing calls to two or more stations, performed by an operator who switches the message to the multiple-address processing unit on a routing line segregator.

multiple channel queuing *Transportation Engineering.* a queuing system in which vehicles arrive on several independent paths, each forming a separate queue.

multiple connector *Engineering.* a flowchart symbol showing that numerous operations or flow lines intersect or are distributed from that point in a system.

multiple-contact sensing surface *Robotics.* a large number of single-contact sensors concentrated on a single surface.

multiple covalent bond see MULTIPLE BOND.

multiple cropping *Agronomy.* the practice of growing two or more crops consecutively on the same field within the same year.

multiple cross glide *Materials Science.* a method in which a crystal deforms plastically under strain through regenerative multiplication; common for most alkali halide-type crystals.

multiple-current hypothesis *Oceanography.* a hypothesis proposing that the Gulf Stream is a system of many quasipermanent currents and meanders, as distinguished from the traditional view that it is a shifting and highly variable single current.

multiple edges *Mathematics.* in a graph, edges that join the same pair of vertices and, if directed in the same direction, are multiple edges.

multiple-effect evaporation *Chemical Engineering.* a system of evaporation in which hot vapor produced in one vessel is condensed in the calandria of another evaporator, in which the pressure has been artificially lowered to produce further evaporation.

multiple epidermis *Botany.* an outer dermal tissue composed of several cell layers.

multiple factors *Genetics.* a series of two or more pairs of genes that are responsible for the development of complex, quantitative characters, such as size and yield.

multiple fault see STEP FAULT.

multiple fracture *Materials Science.* a phenomenon in which a two-component body or portion of a body contains a series of essentially parallel cracks throughout its length.

multiple fruit *Botany.* a fruit that develops from several flowers or from a cluster, rather than from a single flower.

multiple gene inheritance *Genetics.* the inheritance of a certain characteristic that entails many nonallelic genes.

multiple gun *Ordnance.* a group of guns emplaced, mounted, or adjusted together and fired as a single unit.

multiple hereditary exostoses *Medicine.* an inherited disease exhibiting benign bony growths at the ends of the long bones.

multiple idiopathic hemorrhagic sarcoma see KAPOSI'S SARCOMA.

multiple-impulse welding *Metallurgy.* welding in which more than one electric current pulse is applied during each cycle.

multiple independently targetable reentry vehicle see MIRV.

multiple inheritance *Artificial Intelligence.* the inheritance of properties from more than one parent class in a frame system or object-oriented system.

multiple integral *Mathematics.* an integral over a subset of dimension at least 2 of an n-dimensional space.

multiple intelligence(s) theory *Psychology.* a theory holding that intelligence is composed of a number of measurable abilities and capacities.

multiple-key access *Computer Programming.* a search technique that involves one major search field and one or more minor search fields.

multiple launch rocket system *Ordnance.* a mobile, unguided 227-mm artillery rocket system fired in rippled salvoes, having a range over 30 km.

multiple-length see MULTIPRECISION.

multiple-length arithmetic see MULTIPLE PRECISION ARITHMETIC.

multiple-length number *Computer Programming.* a number used in high-precision arithmetic that is two or more computer words in length; two or more times the normal significant digits.

multiple midstop *Robotics.* a peripheral device that is part of a robotic system in which several positions are accessed by a swinging and "pick-and-place" method.

multiple modulation *Telecommunications.* a series of modulation processes in which a modulated wave from one process becomes the modulating wave for the next. Also, COMPOUND MODULATION.

multiple module access *Computer Technology.* a control unit that establishes priorities among multiple computers requesting memory access.

multiple myeloma *Medicine.* widely dispersed cancerous growths of plasma cells in bone marrow and bone tissue, appearing on an X-ray as pushed-out areas and resulting in bone fractures and pain, with eventual spread to other tissues.

multiple neurofibromatosis *Genetics.* a hereditary human disorder, inherited as a simple autosomal dominant. Affected individuals exhibit disfiguring skin defects caused by benign tumors that grow at the ends of peripheral nerves, and a reduced fertility of approximately 50% of unaffected people.

multiple personality *Psychology.* a rare dissociative disorder in which an individual develops two or more separate and distinct personalities; each personality is independent of the other(s) and dominant at a different time, and the original personality often has no knowledge of the others or memory of their experiences.

multiple point *Mathematics.* a point through which a curve passes more than once. A point on a curve that is not a multiple point is a **simple point**.

multiple-precision see MULTIPRECISION.

multiple precision arithmetic *Computer Programming.* an arithmetic operation in which two or more computer words are used to represent the operands and results.

multiple pregnancy *Medicine.* a pregnancy that results in the birth of more than one child, either from the fertilization of one ovum, which later divides, or of more than one ovum.

multiple programming see MULTIPROGRAMMING.

multiple reentry vehicle see MRV.

multiple reflection *Geophysics.* a seismic wave that has undergone several reflections inside the earth.

multiple reflection echo *Electromagnetism.* a radar signal that is reflected off an object after it has been reflected from a target, resulting in false bearing and range.

multiple regression LINEAR REGRESSION.

multiple resonance *Electricity.* a circuit with resonances at more than one frequency.

multiple root *Mathematics* a root of multiplicity greater than 1. For example, c is a multiple root of a polynomial $p(x)$ if $p(x)$ can be written in the form $p(x) = (x - c)^k q(x)$, where $q(x)$ is a polynomial and k is an integer greater than 1.

multiple scattering *Physics.* a scattering process in which a particle suffers several scattering collisions.

multiple sclerosis *Medicine.* a chronic illness caused by lesions in the white matter of the central nervous system that degenerate the myelin sheath, marked by lack of muscle coordination, muscle weakness, speech problems, paresthesia, and visual impairments.

multiple series *Engineering.* the connecting of two or more series of circuits in parallel. Also, PARALLEL SERIES.

multiple-stage rocket see MULTISTAGE ROCKET.

multiple star system *Astronomy.* a gravitationally bound group of two or more stars that orbit a common center of mass.

multiplet *Spectroscopy.* a spectral feature that actually consists of a group of closely spaced but related single lines; often seen in nuclear magnetic resonance.

multiple target generator *Electronics.* in radar countermeasures, a device that produces multiple false targets on enemy radar.

multiplet intensity rules *Spectroscopy.* for a spin-orbit multiplet, a set of rules stating that the sum of the intensities of the spectral lines having a common initial energy level or a common final level is related to the total angular momentum of the initial or final level, respectively.

multiple tropopause *Geophysics.* an effect in which the boundary of the troposphere and stratosphere appears as several overlapping discontinuities instead of as a single one.

multiple-tube method *Microbiology.* a procedure used to estimate the number of viable microorganisms in a given liquid suspension.

multiple-tuned antenna *Electromagnetism.* a low-frequency antenna having one horizontal section and several tuned vertical sections.

multiple twin quad *Electricity.* a quad cable in which four conductors are arranged in two twisted pairs, which are wound together to form a single cable.

multiple-unit steerable antenna see MUSA.

multiple-valued *Mathematics.* describing a relation between sets that is allowed to have more than one range value assigned to a given domain element. Functions may be considered to be multiple-valued in certain circumstances; for example, a function of a complex variable may have more than one branch.

multiple valued logic *Computer Science.* a logic that allows truth values other than true and false, e.g., unknown.

multiplex *Telecommunications.* describing a communications system able to send several messages (signals) over one transmission path at the same time. Thus, **multiplex operation, multiplex transmission,** and so on. *Cartography.* a device that projects aerial photographs onto a surface in three dimensions; used to produce topographic maps.

multiplexer *Electronics.* a device that allows the transmission of two or more signals on a single line or in a single frequency channel.

multiplex holography *Optics.* a technique in which a tungsten-filament light bulb illuminates a cylindrical hologram as it rotates or as an observer moves around it.

multiplexing *Computer Technology.* the process of interweaving two or more lower-speed data streams into a single high-speed channel for simultaneous transmission.

multiplex mode *Computer Technology.* a method of communicating with multiple users by scanning incoming transmissions, organizing them by speed, and utilizing differences between the computer's operating speed and the transmission speed.

multiplex stereo *Acoustical Engineering.* FM radio broadcasting that permits the reception of both right and left channels of a stereophonic recording from a single broadcast signal.

multiplicand *Mathematics.* in a multiplication problem $ab = c$, the number a that is to be multiplied by the multiplier b.

multiplication *Mathematics.* **1.** an operation that, for integers, consists of the addition of a quantity a (the product) to itself as many times as there are units in another quantity b (the multiplier) to obtain a third quantity c (the product); i.e., an operation of repeated addition. This is symbolized by such forms as $a \times b = c$, $ab = c$, or $a \cdot b = c$. **2.** any of various other operations that are analogous to the multiplication of integers; e.g., scalar multiplication, convolution. *Science.* any process in which something increases greatly in quantity. *Nucleonics.* the factor by which a nuclear reactor increases the strength of its neutron source. *Electronics.* a current flow increase through a semiconductor due to an increase in carrier activity.

multiplication factor *Nucleonics.* **1.** the ratio between the number of neutrons present in a reactor in a given generation to those present in the preceding generation. **2.** the ratio between the average number of neutrons created by fission in one lifetime to the total number of neutrons absorbed or leaked during the same lifetime.

multiplication principle *Statistics.* if an experiment can have n_1 distinct outcomes of the first trial, n_2 outcomes of the second trial, and so on, the number of different sequences of outcomes of k trials is the product of the number of outcomes of each trial: $(n_1)(n_2) \cdots (n_k)$.

multiplication rule *Statistics.* a rule for deriving the probability of the union of two events A and B from the conditional and the marginal probability: $P(A \text{ and } B) = P(A)P(B/A)$.

multiplicative congruential generator *Computer Programming.* a type of random-number generator that computes a random number as the residue of a linear transformation of the previous number such that b, the coefficient of the modulus m, is equal to zero.

multiplicative function *Mathematics.* any function f for which $f(xy) = f(x)f(y)$ whenever f is defined on x, y and xy.

multiplicative model *Statistics.* a model in which the effect of a set of independent variables on a dependent variable is described as the product of individual effects.

multiplicative set *Mathematics.* a nonempty subset of a ring that is closed under multiplication.

multiplicity *Science.* the fact of having great variety or of being multiple in nature. *Physics.* the value $(2S + 1)$ that represents the total number of ways to couple (vectorially) the orbital angular momentum L to the spin angular momentum S in an atom or molecule. *Mathematics.* **1.** suppose c is a root of a polynomial $f(x)$ and that m is the greatest nonnegative integer ($m \leq \deg f$) such that $f(x) = (x - c)^m g(x)$. Then the root c is said to be of multiplicity m. Multiplicity is usually defined with respect to the ring of polynomial coefficients. **2.** the dimension of the eigenspace of a square matrix A corresponding to a particular eigenvalue λ is called the **geometric multiplicity** of λ; that is, the maximum number of linearly independent eigenvectors associated with λ. The multiplicity of λ as a zero of the characteristic polynomial of A is called the **algebraic multiplicity** of the eigenvalue λ. The geometric multiplicity of λ is always less than or equal to the algebraic multiplicity of λ. When the term multiplicity is used without qualification in connection with an eigenvalue, it is usually taken to mean algebraic multiplicity.

multiplicity of infection *Virology.* the ratio of the number of infectious virions to the number of susceptible cells in a culture.

multiplicity reactivation *Virology.* an interaction or complementation between two related inactivated viruses in which the sites of inactivation must be in different parts of the genomes of the two viruses.

multiplier *Mathematics.* in a multiplication problem $ab = c$, the number b by which the multiplicand a is to be multiplied. *Electricity.* a resistor used in series with a voltmeter to increase voltage range. Also, **multiplier resistor.** *Electronics.* **1.** a device with two or more inputs and an output that represents the product of the input quantities. **2.** a device that increases the current of an electron beam by secondary emission.

multiplier-quotient register see MQ REGISTER.

multipling *Telecommunications.* **1.** the usage of multidrop lines to make allowances for changes in telephone service patterns. **2.** the connections of lines on trunks to two or more switchboards to permit service by more than one attendant or operator.

multiploid virus *Virology.* a virus populated by similar particles that vary in number of genomes.

multiply *Mathematics.* to carry out any form of multiplication. *Science.* to increase the number or quantity of something, especially a great or rapid increase. *Zoology.* to breed. *Botany.* to propagate.

multiply connected *Mathematics.* connected but not simply connected. Thus, **multiply connected region.**

multiply defined symbol *Computer Programming.* an error message returned by a compiler or assembler indicating that the same label has been used for more than one instruction or location.

multipoint line see MULTIDROP LINE.

multipoint linkage analysis *Genetics.* a technique of using crossover frequencies among several genes to map the loci on a chromosome.

multipolar *Electromagnetism.* having more than one pair of magnetic poles. *Anatomy.* having more than two poles, processes, or dendrites. Thus, **multipolarity.**

multipole *Electromagnetism.* any of a series of configurations of electric or magnetic poles, either static or oscillating.

multipole radiation *Physics.* a radiation field expressed in terms of electric and magnetic multipole fields, each of which is characterized by the indices l and m and by its parity.

multipole transition *Physics.* any transition between energy states caused by the absorption or emission of energy from a photon of multipole radiation.

multipolymer system *Materials Science.* the mixing of polymers to form an immiscible system with strong interfacial adhesion between phases, as in a multilayer film.

multiport memory see SHARED MEMORY.

multiprecision *Computer Programming.* the use of two or more computer words (or three or more words) to represent a single number for greater precision. Thus, **multiprecision arithmetic, multiprecision number.** Also, MULTIPLE-PRECISION, MULTIPLE-LENGTH.

multiprocessing *Computer Technology.* the simultaneous processing of parts of a program by two or more processors.

multiprocessor *Computer Technology.* a computer system consisting of two or more processing units that are closely coupled; for example, a parallel processor. Also, **multiprocessing system.**

multiprocessor control *Control Systems.* a method of control in which a computer or network executes programs concurrently on more than one central processing unit.

multiprogramming *Computer Programming.* the simultaneous or time-sequenced execution of two or more programs on a single processor by overlapping and interleaving instructions and data or by rapid switching between programs. Also, MULTIPLE PROGRAMMING, SOFTWARE MULTIPLEXING.

multiprogramming executive control *Computer Programming.* the monitoring functions of a multiprogramming system that maintain system integrity, control system resources, and schedule jobs. Also, EXECUTIVE-SYSTEM CONCURRENCY.

multipropellant *Space Technology.* a type of liquid rocket propellant composed of at least two substances, usually a fuel and an oxidizer, which are fed separately into the combustion chamber of the rocket.

multipurpose movement *Behavior.* a term for any behavior pattern that has more than one function.

multipurpose projectile *Ordnance.* a projectile designed to deliver a variety of payloads; the loads, such as smoke or leaflets, are contained in canisters that are loaded into the projectile through a removable base plug.

multiring structure *Astronomy.* the remnant of a large impact basin which has concentric rings of mountain peaks.

multirole aerial target *Ordnance.* a programmable or remote-controlled airborne target that is designed to simulate enemy aircraft as well as missiles.

multirole aerial target

multirole programmable device *Robotics.* a device with a programmable memory that can store data for positioning and sequencing the movements of a robot.

multiscrew extrusion *Materials Science.* a method of plastic extrusion in which a polymer is propelled through two screws turning side by side in opposite directions. Also, TWIN-SCREW EXTRUSION.

multisensitive *Immunology.* **1.** sensitive to more than one antigen. **2.** allergic to more than one allergen. Thus, **multisensitivity.**

multiseriate *Biology.* being arranged in several rows.

multiset *Computer Science.* see BAG.

multisite suppressor see SUPERSUPPRESSOR.

multistage *Engineering.* having a series or sequences of stages.

multistage rocket *Space Technology.* a space vehicle possessing two or more rocket units, with each successive unit firing after the stage behind it has reached burnout (at which point the latter is usually jettisoned). Also, MULTIPLE-STAGE ROCKET.

multistatic radar *Engineering.* a radar system having several antenna lobes, each engaged in succession, thus allowing the tracking of a target without moving the antennas.

multisynaptic see POLYSYNAPTIC.

multisystem coupling *Computer Technology.* an arrangement of two or more computer systems directly linked electronically and functioning as a single computer.

multisystem network *Computer Technology.* an arrangement of two or more computer systems linked by a network and functioning as separate host computers.

multitasking see MULTIPROGRAMMING.

multithreading *Computer Technology.* a code using two or more processes or processors, which may be of different types and may be active simultaneously.

multitilocular *Biology.* having many small chambers or vesicles.

multitrack recording *Acoustical Engineering.* a sound recording technique that employs multiple microphones, with one or more microphones placed near each instrument or group of instruments.

multitrack recording system *Acoustical Engineering.* a system having two or more recording paths for one medium, allowing, for example, the recording of a melody at one time and the addition of harmony or accompaniment at a later time. Also, **multitracking.**

multitrack stereo see MTS.

Multituberculata *Paleontology.* an extinct order of protoeutherian mammals characterized by large incisors and longitudinal rows of tooth cusps; possibly the earliest herbivorous mammals, but a side-branch not closely related to any other mammals; Jurassic to Eocene.

multituberculate *Paleontology.* **1.** relating to or belonging to the order Multituberculata. **2.** describing the type of tooth characteristic of the Multituberculata, having two or three longitudinal rows of simple conical cusps. *Anatomy.* having many tubercules.

multiuser system *Computer Technology.* a computer system that can be accessed by users at multiple terminals.

multivariate analysis *Statistics.* any of several methodologies used in the analysis of data taken simultaneously on many variables.

multivariate normal distribution *Statistics.* a generalization of the normal distribution to the case of *n* random variables. The parameters are a vector of marginal means and a covariance matrix. All marginal distributions are also multivariate normal.

multivibrator *Electronics.* a relaxation oscillator consisting of two stages coupled in such a way that the input of each one is derived from the output of the other.

multivoltine *Biology.* producing more than one brood in a single season.

multivolume file *Computer Technology.* a large file stored on two or more physical storage units such as disks or magnetic tape reels.

mummification *Medicine.* the shriveling and drying of body tissue, as observed in dry gangrene or in a dead fetus.

mummy *Plant Pathology.* a shriveled, dry fruit that results from the effects of a fungus disease.

mumps *Medicine.* a contagious viral disease causing enlargement of the parotid and other salivary glands, passed by direct contact or airborne droplets and sometimes resulting in the spread of the disease to the testes or ovaries, or in mumps encephalitis.

mumps orchitis *Medicine.* an inflammation of the testes caused by infection with the mumps virus.

mumps virus *Virology.* the *Paromyxivirus,* a common virus of the family Paramyxoviridae that infects humans, particularly children, causing fever, orchitis, and parotitis.

Münchausen syndrome [mün´chou zən] *Psychology.* a disorder characterized by repeated simulation of physical symptoms as a means to gain medical attention or hospital admission. (Named for the German adventurer Baron von *Münchausen,* 1720–1797, because of his exaggerated accounts of his own exploits.) Also, **Munchausen syndrome.**

mundic see PYRITE.

muninga *Materials.* the golden to reddish-brown wood of the *Pterocarpus angolensis* tree of Africa; used for furniture.

munitions *Ordnance.* a general term for materiel used in war, especially weapons and ammunition.

Munkiellaceae *Mycology.* a family of fungi of the order Pleosporales living within decaying plant material or as parasites below the leaf cuticle in plants.

Munro's point *Surgery.* a site in the abdominal wall that is usually not near major organs and is thus suitable for puncture; located midway between the umbilicus and the left anterior superior iliac spine. (Named for John *Munro,* 1858–1910, American surgeon.)

Munsell chroma see CHROMA.

Munsell color system *Optics.* a color designation system that specifies the relative degrees of the three variables of color: hue, value, and chroma.

Munsell hue *Optics.* the portion of the Munsell color system that identifies a color without considering its lightness or saturation.

Munsell value *Optics.* the level at which an object's apparent reflectance equals the perceptual steps occurring during normal observation conditions.

muntin *Building Engineering.* **1.** a piece of vertical framing that separates door panels. **2.** see SASH BAR.

muon *Particle Physics.* an unstable second-generation lepton, having a mass 207 times that of an electron and a mean lifetime of about 2.2×10^{-6}.

muonic atom *Particle Physics.* an atom in which an electron is replaced by a negatively charged muon that orbits close to or within the nucleus.

muonium *Particle Physics.* an exotic atom that is a light, unstable isotope of hydrogen with a mass 0.11 times that of a hydrogen atom and a mean lifetime of 2.2 microseconds; formed when a positively charged muon and an electron are bound by their mutual electrical attraction.

muon lepton number *Particle Physics.* an additive quantum number that appears to govern the interactions of leptons and muons among themselves and with all other known particles.

mu particle *Bacteriology.* a bacterial endosymbiont of strains of *Paramecium aurelia* that confers the killer trait to the paramecium.

mu phage *Virology.* a temperate DNA-containing tailed phage that appears able to integrate at any site on the bacterial hosts and replicates at a rapid rate.

MUPO maximum undistorted power output.

Muraedinae *Vertebrate Zoology.* the moray eels, a family of predatory and scavenging nocturnal marine eels of the order Anguilliformes, characterized by strong jaws and fanglike teeth; found worldwide in tropical and temperate seas.

mural thrombus *Pathology.* a blood clot on a diseased area of the heart wall.

muramic acid *Biochemistry.* $C_9H_{17}NO_7$, a compound in bacterial cell walls, consisting of lactic acid and glucosamine joined by an ether linkage.

muramidase *Enzymology.* an enzyme that lyses cell walls.

Murchison, Sir Roderick I. 1792–1871, British geologist; the first to differentiate the Silurian and Devonian systems.

Murchisoniacea *Paleontology.* a long-lived superfamily of prosobranch gastropods belonging to the order Archaeogastropoda; the typical **murchisoniid** shell is high-spired and multiwhorled; extant in the Lower Ordovician (possibly Upper Cambrian) to Upper Triassic.

Murdock, George P. 1897–1985, American anthropologist; pioneer in ethnographic works and cross-cultural studies of world culture.

murein *Biochemistry.* the stiff layer within a bacterial cell in which the peptidoglycan chains are covalently crosslinked.

mureinoplast *Microbiology.* any Gram-negative bacterial cell from which the outer membrane has been removed while the peptidoglycan cell wall remains intact.

muriatic acid *Inorganic Chemistry.* an older name for hydrochloric acid, HCl. See HYDROCHLORIC ACID.

Muricacea *Invertebrate Zoology.* a superfamily of gastropod mollusks in the order Prosobranchia.

muricate *Zoology.* covered with sharp points.

Muricidae *Invertebrate Zoology.* a family of gastropod mollusks in the order Neogastropoda, including predaceous rock snails.

Muridae *Vertebrate Zoology.* the Old World rats and mice, the largest recent family of mammals, belonging to the suborder Myomorpha and found in every habitat except tundra.

muriform *Biology.* **1.** resembling the arrangement of courses in a brick wall, especially having both horizontal and vertical septa. **2.** resembling a rat or mouse.

Murinae *Vertebrate Zoology.* a large subfamily of the family Muridae, containing approximately 98 genera and 457 species of Old World mice and rats.

murine *Vertebrate Zoology.* of or relating to mice or rats. *Immunology.* affecting mice or rates.

murine plague *Veterinary Medicine.* an infection transmitted to the rat by fleas and caused by *Pasteurella pestis*; transmissible from rat to rat and from rats to humans.

murine typhus *Medicine.* a contagious disease spread by the fleas of rodents, similar to epidemic typhus but less severe.

murmur *Medicine.* an abbreviated sound heard at intervals during auscultation, arising in the heart or blood vessels.

muropeptides *Biochemistry.* the products forming when an enzyme cleaves glycosidic bonds in peptidoglycan digestion.

Murphy, William Parry 1892–1987, American physician; awarded the Nobel Prize for research on the use of liver in the treatment of anemia.

Murray Valley encephalitis *Medicine.* a viral form of brain inflammation that caused an epidemic in the Murray Valley area of Australia in 1950 and 1951; it has since been reported occasionally in Australia and New Guinea.

murre [mùr] *Vertebrate Zoology.* either of two black and white northern seabirds of the genus *Uria* in the family Alcidae, having a slender head and a long bill.

murre

murrelet *Vertebrate Zoology.* any of various small, black and white diving seabirds of the family Alcidae inhabiting the islands and coasts of the North Pacific. Also, SEA SPARROW.

MUSA *Electromagnetism.* a receiving antenna whose directivity may be steered electrically by varying the phase contributions of the individual units. (An abbreviation for multiple-unit steerable antenna.)

Musaceae *Botany.* a family of large, tropical monocotyledonous plants of the order Zingeberales having rhizomes or corms, large rolled, pinnate leaves, and unisexual flowers with five functioning stamens, and bearing a fleshy, usually edible fruit.

Musca *Invertebrate Zoology.* the common housefly, a genus of flies of the family Muscidae that are notorious for spreading human intestinal infections such as bacillary or amebic dysentery by contaminating human food. *Astronomy.* the Fly, a small southern constellation lying due south of the Southern Cross; formerly named Apis, the Bee.

muscarine *Organic Chemistry.* $C_8H_{19}NO_3$, a poisonous alkaloid that is chemically related to choline, occurring in certain mushrooms, such as *Amanita muscaria,* or as a ptomaine of decaying fish.

muscarinic *Organic Chemistry.* of or relating to muscarine. *Toxicology.* relating to, affected by, or symptomatic of muscarinism.

muscarinism *Toxicology.* poisoning due to the ingestion of muscarine; symptoms may include gastrointestinal disturbances, liver and kidney damage, or death.

Muschelkalk *Geology.* a European geologic stage of the Middle Triassic period, occurring after the Bunter and before the Keuper.

Muscicapidae *Vertebrate Zoology.* the Old World flycatchers, a diverse family of birds of the passerine suborder Oscines, characterized by an aerial style of taking prey, though they also glean in trees and catch on the ground.

Muscidae *Invertebrate Zoology.* a family of dipteran insects, in the subsection Calyptratae, including houseflies and stable flies.

muscle *Anatomy.* an organ composed of muscle tissue, which by contraction produces the movements of the body; the heart muscle and all muscles that are capable of voluntary contraction are termed striated (or striped) muscle; all involuntary muscles except the heart are termed smooth muscle.

muscle clot *Hematology.* a clot formed by the coagulation of muscle plasma.

muscle fiber *Histology.* the functional unit or cell of muscle tissue.

muscle plasma *Hematology.* a liquid that is expressible from muscle tissue; it clots spontaneously.

muscle tissue *Histology.* a basic type of tissue, composed of cells that are rich in contractile fibers.

muscone *Organic Chemistry.* $CH_3C_{15}H_{27}O$, an odor-bearing constituent extracted from the scent glands of the male musk deer that is slightly soluble in water and soluble in alcohol and ether, boiling at 130°C (0.5 torr); used as a perfume ingredient.

muscovite *Mineralogy.* $KAl_2(Si_3Al)O_{10}(OH,F)_2$, a colorless, gray, green, red, or light-brown pseudohexagonal, monoclinic mineral of the mica group, occurring as lamellar masses and disseminated scales, having a specific gravity of 2.7 to 3.1 and a hardness of 2 to 2.5 on the Mohs scale; found in granites and pegmatites. Also, COMMON MICA, WHITE MICA.

Muscovy duck *Agriculture.* a large breed of meat-producing duck, *Cairina moschata,* having a long, low, white body and orange feet; originating in South America, it is now widely domesticated. Also, MUSK DUCK.

muscular dystrophy *Pathology.* a genetic degenerative disorder characterized by weakness and atrophy of muscle without involvement of the nervous system. Also, MYODYSTROPHY.

muscularis externa *Histology.* the concentric layers of smooth muscle that compose much of the wall of the digestive tract.

muscularis mucosae *Histology.* a thin layer of smooth muscle fibers lying between the mucosal epithelium and connective tissue layer of certain organs such as the gut.

muscular system *Anatomy.* the collection of muscles that control movement of the body.

musculocutaneous *Anatomy.* of or related to both muscle and skin.

musculocutaneous nerve *Neurology.* lateral branches of the brachial plexus, located on either side and innervating the upper limbs.

musculoskeletal system *Anatomy.* the bones and muscles of the body, providing both support and body movement.

mushmelon see MUSKMELON.

mushroom *Mycology.* any of various fungi of the order Agaricalles, especially an edible one characterized by its fleshy, umbrella-shaped fruiting body; poisonous species are properly called *toadstools.*

mushroom

mushroom anchor *Naval Architecture.* an anchor having a rounded, cupped bottom and resembling an inverted mushroom in overall appearance; used for heavy or semipermanent moorings.
mushroom cloud *Nucleonics.* the distinctive mushroom-shaped cloud formed in the atmosphere as a result of a nuclear explosion.
mushroom poisoning *Toxicology.* poisoning due to the ingestion of toxic mushrooms (toadstools); symptoms may include gastrointestinal disturbances, liver and kidney damage, central nervous system damage, and death. Also, MYCETISM, MYCETISMUS.
musical acoustics *Acoustics.* the study of the relationship between music and the principles of acoustics, including such objective aspects of music as the properties of musical instruments, the formal definition of musical sounds, and the manner in which such sounds are detected by the ear.
Musical Instrument Digital Interface see MIDI.
musicogenic *Neurology.* of or relating to a reaction triggered by musical sounds.
music synthesizer *Computer Technology.* a device that produces computer-generated musical sounds and vocal tones.
music wire *Metallurgy.* high-carbon steel wire, having a very high tensile strength; used for the production of mechanical springs.
Musidoridae *Invertebrate Zoology.* a family of dipteran flies in the series Brachycera.
musk *Biochemistry.* any of various substances having a strong, penetrating odor, such as that produced by the musk glands of the male musk deer or similar animals. *Botany.* any of various plants having a musky odor, such as the monkey flower.
musk ambrette *Organic Chemistry.* $C_{12}H_{16}N_2O_5$, pale yellow crystals; insoluble in water, slightly soluble in alcohol, and soluble in ether; melts at 84–86°C; used in perfumes.
musk deer *Vertebrate Zoology.* a hornless Asian deer, *Moschus moschifers,* of the family Cervidae, characterized by short tusklike canines and an abdominal gland that produces musk during the breeding season.
musk duck see MUSCOVY DUCK.
muskeg *Ecology.* **1.** a bog characterized by scattered and stunted evergreens. **2.** broadly, any area of wetland vegetation.
musk gland *Vertebrate Zoology.* any gland secreting musk, especially the odor-bearing gland of a male musk deer. Also, **musk bag.**
Muskhelishvili's method *Mechanics.* a technique of determining the stresses and strains in an elastic body loaded in plane strain or plane stress, by the use of complex analysis and conformal mapping involving singular integral equations.
musk ketone *Organic Chemistry.* $C_{14}H_{18}N_2O_5$, yellow crystals; insoluble in water and slightly soluble in alcohol; melts at 134.5–136.5°C; used in perfumes.
muskmelon *Botany.* **1.** a plant, *Cucumis melo,* of the gourd family. **2.** the edible fruit of this plant, having a round or oblong shape and juicy aromatic flesh that is orange, white, or green. Also, MUSHMELON.
musk ox or **muskox** *Vertebrate Zoology.* a large, heavy, herbivorous wild ox, *Ovibos moschatus,* of the family Bovidae; characterized by its long, shaggy coat and downwardly curved horns; found in Greenland and the Arctic regions of North America.

musk ox

muskrat *Vertebrate Zoology.* a medium-sized semiaquatic rodent of the family Cricetidae, characterized by a long scaly compressed tail, glossy dark fur, and webbed feet; found in the United States and Canada.
musk turtle *Vertebrate Zoology.* any of several aquatic North American turtles, belonging to the genus *Sternotherus* of the family Kinosternidae, that emit a musky secretion for protection.
musk xylene *Organic Chemistry.* $C_{12}H_{15}N_3O_6$, white to yellow needles; insoluble in water, slightly soluble in alcohol, and soluble in ether; melts at 110°C; used as synthetic musk in perfumes. Also, **musk xylol.**
Musophagidae *Vertebrate Zoology.* the touracos or plantain eaters, a family of colorful birds of the order Cuculiformes; found in heavy forests of Africa.
Muspiceoidea *Invertebrate Zoology.* a superfamily of nematodes that are parasitic on fish and other vertebrates.
mussel *Invertebrate Zoology.* any bivalve mollusk, especially an edible marine bivalve of the family Mytilidae living attached to wharf pilings, sea walls, and rocks in coastal seas, or a freshwater bivalve of the family Unionidae whose shell provides mother-of-pearl.
mussitation *Physiology.* a movement of the lips without utterance of articulate sounds.
mustang *Vertebrate Zoology.* a small, sturdy feral horse of the U.S. plains and western deserts, originally descended from Spanish stock.
Mustang *Aviation.* a popular name for the F-51 escort fighter.
mustard *Botany.* the common name for any of several species of the genus *Brassica,* notably **black mustard** (*B. nigra*) and **white mustard** (*B. alba*), the principal sources of mustard seed, and **leaf mustard** (*B. juncea*), whose leaves are used as a food. *Food Technology.* an aromatic paste or powder made from the crushed seeds of the mustard plant; used as a condiment or seasoning, and also used medicinally as a plaster or poultice.

mustard

mustard family see CRUCIFERAE.
mustard gas *Organic Chemistry.* $(ClCH_2{\cdot}CH_2)_2S$, a yellow prismatic solid; insoluble in water and soluble in alcohol and ether; melts at 13–14°C and boils at 215–217°C. *Toxicology.* a preparation of this substance used as a chemical warfare agent, having numerous debilitating effects; e.g., it penetrates the skin and forms hydrochloric acid, causing severe blistering. Also, YELLOW CROSS.
mustard oil see ALLYL ISOTHIOCYANATE.
Mustilidae *Vertebrate Zoology.* the weasels, a family of carnivorous mammals in the suborder Caniformia having a long, slender body and usually well-developed anal scent glands; includes weasels, martens, mink, skunks, badgers, and otters. *Paleontology.* the ancestors of this family; generally small, but one Miocene genus, *Megalictis,* grew to the size of a black bear; arose from the Miacidae in the Eocene.
mutable *Genetics.* capable of or tending toward mutation. Thus, **mutability.**
mutable gene *Genetics.* a gene that has a higher than usual rate of mutation; usually detected by somatid mutations producing mosaics.
mutable site *Genetics.* a chromosome locus where a mutation can be detected.
mutagen [myoot´ə jən] *Genetics.* any chemical or physical agent that increases a mutation rate above the frequency of the spontaneous rate.
mutagenesis *Genetics.* the introduction into a gene of an alteration that changes the structure or function of the gene product.
mutagenic [myoot´ə jen´ik] *Genetics.* inducing genetic mutation.
mutagenicity *Genetics.* the capability of inducing genetic mutation.

mutan *Biochemistry.* a glucan present in dental plaque, and formed by some types of *Streptococcus mutans.*

mutant [myoot´ənt] *Genetics.* **1.** an individual carrying a specific mutation of a gene that is expressed in the organism's phenotype. **2.** produced by mutation.

mutarotation *Chemistry.* a change over time of the optical rotation of light in a freshly prepared solution of an optically active substance such as a reducing sugar.

mutase *Enzymology.* any enzyme that catalyzes an isomerization in which a group is transferred from one part of a molecule to another.

mutation [myoo tā´shən] *Genetics.* **1.** the process by which a gene or some other DNA sequence undergoes a change in structure. **2.** a gene or other DNA sequence that has undergone a structural change. **3.** an individual that has undergone a mutational change and expresses this change in the phenotype.

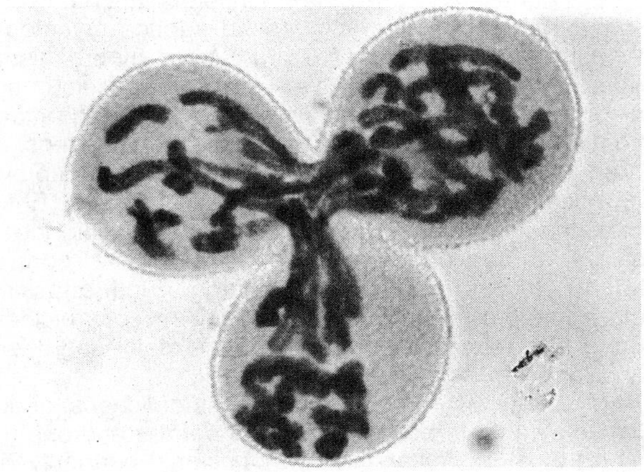

mutation

mutational *Genetics.* of or relating to mutation.

mutational hot spot see HOT SPOT.

mutational load *Genetics.* the cumulative harmful effects in a population of recurring mutations that accumulate over time.

mutation rate *Genetics.* the number of mutation events per gene or organism per unit of time.

mutator gene *Genetics.* a mutant gene whose activity increases the spontaneous mutation rate of one or more other genes.

mutator phage *Virology.* any phage that, upon infection, prompts a rate increase of mutation in the host cell.

mute *Medicine.* **1.** physically unable to speak. **2.** a person who is unable to speak. *Telecommunications.* a control on a television set that allows the viewer, when desired, to turn off the sound temporarily, as during a commercial.

muthmannite *Mineralogy.* (Ag,Au)Te, a brass-yellow, opaque, metallic mineral occurring as tabular crystals, having a hardness of 2.5 on the Mohs scale and an undetermined specific gravity.

Mutillidae *Invertebrate Zoology.* a family of ants, hymenopteran insects in the superfamily Scolioidea, including the velvet ants.

muting circuit *Electronics.* **1.** an electronic circuit that mutes a receiver's output when radio-frequency carriers at or below a specified intensity are reaching the first detector. **2.** a circuit that makes a receiver insensitive during the time that its associated transmitter is on.

muting switch *Electricity.* a switch or relay used with an automatic tuner, silencing the receiver during transmission periods or at other times when reception is not desired.

mutism *Medicine.* an inability or refusal to speak.

muton *Molecular Biology.* the smallest unit of DNA that is capable of mutation.

mutton *Agriculture.* the meat of adult sheep, especially those that are at least one year old.

mutual exclusion *Computer Programming.* in a multiprogramming system, a means of protecting shared data from disorderly changes by not allowing two processes to access shared data at the same time.

mutual-exclusion rule *Physical Chemistry.* the principle that if there is a center of symmetry in a molecule, then transition is permitted in either its Raman effect or infrared emission, but not in both.

mutual inductance *Electromagnetism.* the inductance that is shared between two neighboring circuits due to their shape, relative proximity, and orientation, equal to the ratio of the electromotive force induced in one circuit to the rate of change of current in the other circuit.

mutualism *Ecology.* **1.** a mutually beneficial relationship between two species, especially an obligate mutually beneficial relationship without which neither can survive; for example, the yucca moth feeds only on yucca fruits, and the yucca plant must be pollinated by the yucca moth. **2.** any relationship between organisms of different species.

mutuality of phases *Chemistry.* a rule stating that if two phases in a reaction are in equilibrium with a third phase at a certain temperature, then they are in equilibrium with each other at that temperature.

mutually exclusive events *Statistics.* a set of events having the property that the occurrence of one precludes the occurrence of the others.

muzzle *Ordnance.* the opening at the discharge end of the tube or barrel of a gun or similar weapon. *Vertebrate Zoology.* the snout or projecting nose and jaws of an animal. *Mechanical Devices.* an assembly of straps or bars placed over an animal's snout, usually to prevent it from biting.

muzzle blast *Ordnance.* the hot air and gases that surge from the muzzle of a gun as the projectile exits. Similarly, **muzzle flash.**

muzzle brake *Ordnance.* a device attached to the muzzle of a weapon that utilizes escaping gas to reduce recoil.

muzzle compensator *Ordnance.* a device attached to the muzzle of a weapon that utilizes escaping gas to control muzzle movement.

muzzle energy *Ordnance.* a projectile's kinetic energy at the instant it leaves the muzzle of a gun, usually expressed in foot pounds.

muzzle velocity *Ordnance.* the velocity of a projectile in relationship to the muzzle at the instant the projectile leaves the weapon.

mV millivolt.

μV microvolt.

Mv a chemical symbol for mendelevium.

MV megavolt.

MV mean variation; motor vessel.

M.V. veterinary physician. (From Latin *medicus veterinarius.*)

MVA or **Mva** megavolt-ampere.

MVBR multivibrator.

MVL51 phage *Virology.* a proposed species of the genus *Plectrovirus* with bullet-shaped virions in which the host survives infection.

MVL3 phage group *Virology.* a proposed member group of L3 *Acholeplasmavirus* marked by isometric virions with a short tail and a linear dsDNA genome.

mW or **mw** milliwatt.

Mw megawatt.

MW(E) megawatt electric. Also, **MWE, MWe.**

MW(Th) megawatt thermal.

MWYr megawatt year.

Mx maxwell.

M-X *Ordnance.* a U.S. intercontinental ballistic missile designed to replace the Minuteman; it has a range of 13,000 km and can deliver ten 300-kiloton multiple independently targetable reentry vehicles. Also, **MX Missile.**

my. mayer.

my- a combining form meaning "muscle."

My myopia.

myalgia *Medicine.* pain in the muscles.

myasthenia [mī´əs thē´nē ə] *Medicine.* **1.** a muscular debility characterized by excessive muscle fatigue and weakness. **2.** any muscle weakness or lack of muscle function.

myasthenia gravis *Medicine.* the inability of certain nerves and muscles to function together due to the existence of antibodies to the chemical receptors that conduct nerve impulses to the muscles. A specific muscle group may be affected, or the disease may be generalized. Commonly the muscles of the head and neck are affected, and transient respiratory problems are seen.

myasthenia reaction *Medicine.* a gradual muscle reaction failure when repeated electrical stimulus is applied.

myasthenic crisis [mī´əs then´ik] *Medicine.* the abrupt onset of breathing difficulties in myasthenia gravis patients, necessitating respiratory support; the crisis is usually temporary and fever may be present.

myatonia *Medicine.* a loss or absence of muscle tone.

myautonomy *Medicine.* a condition in which muscular contraction aroused by stimulation is so delayed that it appears to occur independently.

myb *Genetics.* an oncogene originally identified as the transforming determinant in avian myeloblastosis virus.

myc *Genetics.* a gene originally discovered in an oncovirus of chickens, homologous to a gene located on the long arm of human chromosome 8.

myc- a combining form meaning "fungus."

myc. mycology.

Mycelia Sterilia see AGONOMYCETALES.

mycelium *Biology.* a matted mass of fungal filaments (hyphae) that forms the vegetative body of a fungus.

Mycetaeidae see ENDOMYCHIDAE.

mycetangium *Entomology.* a specialized region in certain insects in which symbiotic fungi are carried and transmitted through reproductive tissue from one generation of insects to the next generation.

mycetism or **mycetismus** see MUSHROOM POISONING.

myceto- or **mycet-** a combining form meaning "fungus."

mycetocyte *Entomology.* in certain insects that live on specialized diets, a type of highly specialized cell containing symbiotic microorganisms that provide an internal source of nondietary nutrients or digestive assistance.

mycetome *Entomology.* a specialized structure consisting of mycetocytes.

Mycetophagidae *Invertebrate Zoology.* a family of beetles, coleopteran insects in the superfamily Cucujoidea, that feed primarily on mushrooms.

Mycetozoa *Biology.* the slime molds, an intermediate organism sometimes regarded as a plant or fungus, with a life cycle alternating between plantlike and animal-like forms; the botanical name is Myxomycetes.

Mychodeaceae *Botany.* a monogeneric family of red algae of the order Gigartinales, having flattened, multibranched, uniaxial thalli with a pinnate or radial arrangement of the lateral axes and with spermatangia formed superficially in compact bunches; native to southern Australia.

mychorrhizae see MYCORRHIZA.

myco- a combining form meaning "fungus."

Mycobacteriaceae *Bacteriology.* a family of Gram-positive, rod-shaped, aerobic bacteria of the order Actinomycetales; occur in dairy products, soil, and water, and are parasites to lower animals and humans; its single genus, **Mycobacterium,** contains species causing such diseases as leprosy and tuberculosis.

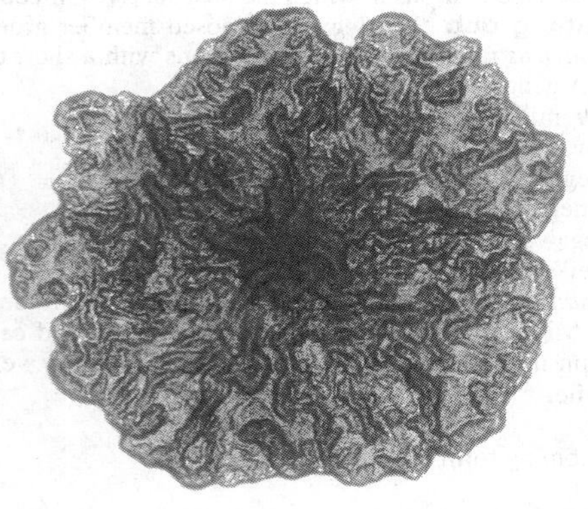

Mycobacterium

mycobacterial disease *Medicine.* any illness caused by organisms of the genus *Microbacterium,* such as tuberculosis and leprosy.

mycobacteriophage *Virology.* a temperate and virulent bacteriophage with a hexagonal head and noncontractile tail that can infect one or more *Mycobacterium* species.

mycobactin *Biochemistry.* a siderophore, formed by most mycobacterium, that chelates iron and facilitates iron transport into the cell.

mycobiont *Botany.* the fungal element in a lichen.

mycodermatitis *Medicine.* an inflammation of the skin caused by a fungus.

mycol. mycology.

mycolic acids *Biochemistry.* the high-molecular weight, alpha-branched, ß-hydroxy fatty acids that are components of the cell membranes of all *Mycobacteria.* The basic structure is R^2CH(OH)CHR^1 COOH, where R^1 is a C_{20} to C_{24} linear alkane, and R^2 is a structure consisting of 30 to 60 carbon atoms. These carbons are in an assortment of cyclopropane rings, carbon double-bonds, methyl branches, or carbon-oxygen functions.

mycologist *Mycology.* a scientist who specializes in the study of fungi.

Mycology

The term derives from *mykes,* the Greek word for "fungus" (itself a Latin word meaning "mushroom"). Fungi include the common mildews, molds, mushrooms, and yeasts. Most mycologists consider themselves botanists, and the Latin names given to fungi and to fungal groups are governed by the International Code of Botanical Nomenclature. The term microbiology is now often used as a substitute for bacteriology, though some mycologists (particularly those who work with yeasts and molds) often call themselves microbiologists.

Out of mycology grew the science of plant pathology (phytopathology), originally called applied mycology, though many plant diseases are caused by organisms other than fungi. Medical and veterinary mycology are also important fields of study, since many fungi attack humans and other animals as parasites, pathogens, or predators. A large pharmaceutical industry employs mycologists to search fungi for sources of drugs. Entomopathogenic fungi (fungi on insects) are being intensively studied by mycologists as biological control agents, as are mycopathogenic fungi (fungi attacking other fungi), and fungi used as living mycoherbicides. Fungi are even more important as the major degraders and rotters of organic matter.

Mycology also embraces the study of lichens, an intimate symbiosis between a fungal partner and one (or more) algal or prokaryotic cyanobacterial partners; the Latin name applied to the lichen duality applies to the fungus as well as to the lichen, the algal or cyanobacterial components bearing their own names. Mycetozoa (slime molds, myxomycetes) are often studied by mycologists, but they are protists, not true fungi. Some, particularly aquatic, true fungi are of such simple structure that they, too, may be classified as protists in some classifications.

Richard P. Korf
Professor of Mycology
Cornell University

mycology [mī käl´ə jē] *Botany.* the study of fungi.

mycomycin *Microbiology.* an antibiotic used against *Mycobacterium tuberculosis,* the causative agent of tuberculosis.

MYC oncogene *Oncology.* a proto-oncogene that produces sarcomas and carcinomas, found in the avian myelocytomatosis virus; it is translocated in Burkitt's lymphoma and amplified in carcinoma of the lung, breast, and cervix.

mycoparasite *Mycology.* any fungus that parasitizes other fungi.

mycophagous *Biology.* of an organism, feeding mainly or exclusively on fungi.

mycophenolic acid *Organic Chemistry.* $C_{17}H_{20}O_6$, needles that are insoluble in cold water, soluble in alcohol, and slightly soluble in ether; melts at 141°C; an antibiotic substance produced by the species of *Penicillium.*

Mycoplana *Bacteriology.* a genus of Gram-negative bacteria that occur as branching filaments in the soil and reproduce by fragmenting into flagellated, rod-shaped cells.

mycoplasmal pneumonia see EATON AGENT PNEUMONIA.

Mycoplasmataceae *Bacteriology.* a family of bacteria of the order Mycoplasmatales, possessing no cell walls and needing sterols for maintenance and growth.

Mycoplasmatales *Bacteriology.* an order of Gram-negative, usually nonmotile bacteria of the class Mollicutes that are bounded by a triple-layered membrane but lack a cell wall.

mycoplasmavirus *Virology.* any of the bacteriophages from a variety of families that infect members of the Mycoplasmatales.

mycoplasmosis *Pathology.* an infection from a pleuropneumonia-type organism, such as those of the family Mycoplasmataciae.
mycoprotein *Biotechnology.* a single-cell protein or food product consisting of fungal mycelium; commercially produced for human consumption.
mycorrhiza *Ecology.* the symbiotic relationship between certain nonpathogenic or weakly pathogenic fungi and the living cells of roots of certain higher plants. Also, MYCHORRHIZAE.
mycosamine *Organic Chemistry.* $C_6H_{13}NO_4$, a nitrogen-containing amino sugar of the antifungal antibiotics nystatin and amphotericin.
mycoside c *Biochemistry.* a complex glycolipid that contains acid and a polysaccharide moiety, and is found in mycobacterial cell walls.
mycosis *Medicine.* any illness caused by a fungus.
mycosis fungoides *Pathology.* a particularly malignant reticulosis occurring in the skin and progressing in chronic stage to the lymph nodes; characterized by viscerally invasive, large ulcerated tumors that are extremely painful and usually fatal.
Mycosphaerella *Mycology.* a genus of fungi of the order Dothideales that causes such plant diseases as clover spot and strawberry leaf spot.
Mycota see EUMYCOTA.
mycotoxicosis *Toxicology.* poisoning due to the presence of bacterial or fungal toxin, as with ergotism.
mycotoxin *Toxicology.* any poison produced by a fungus.
mycotrophic *Botany.* of or relating to a plant that is living in symbiosis with a fungus.
mycovirus *Virology.* an isometric dsRNA virus that infects and replicates in cells of fungi.
mycovore *Zoology.* feeding mainly or exclusively on fungi.
Myctophidae *Vertebrate Zoology.* the lanternfishes, a family of arctic to antarctic fish characterized by luminous organs present on the head and sides of the body.
Myctophiformes *Vertebrate Zoology.* a large order of marine fishes, including lantern fishes, lizard fishes, pearleyes, and daggertooths.
Myctophoidei *Vertebrate Zoology.* in some classification systems, a suborder within the large fish order Myctophiformes.
Mydaidae *Invertebrate Zoology.* a family of large dipteran insects, the mydas flies, in the series Brachycera.
mydatoxine *Toxicology.* a deadly ptomaine poison found in decaying flesh.
mydriasis *Medicine.* an anomalous dilation of the pupil.
mydriatic *Pharmacology.* any drug that dilates the pupil of the eye. *Physiology.* of an action or agent, dilating the pupil of the eye.
myectomy *Surgery.* the surgical excision of a tumor made up of a muscle or muscle group.
myel- or **myelo-** *Anatomy.* a combining form denoting relationship to the bone marrow, to the spinal cord, or to myelin.
myelalgia *Neurology.* pain in the spinal cord.
myelencephalitis *Neurology.* inflammation of both the brain and spinal cord. Also, MYELOENCEPHALITIS.
myelencephalon *Developmental Biology.* the posterior of the two brain vesicles that are formed by the specialization of the rhombencephalon in the developing embryo.
myelin *Neurology.* the substance of the cell membrane of Schwann's cells that coils to form the myelin sheath. Also, WHITE SUBSTANCE OF SCHWANN. *Histology.* any of various lipid substances found in normal or pathologic tissue, differing from fats in being doubly refractive.
myelinated *Neurology.* covered with a myelin sheath; said of a nerve fiber.
myelin basic protein *Neurology.* a basic protein that makes up about 30% of myelin proteins; the level is elevated in acute cerebral infarction and acute exacerbation of multiple sclerosis.
myelinoclasis *Neurology.* destruction of the myelin of a nerve fiber, often seen in encephalitis.
myelinolysin *Neurology.* a hypothetical substance, allegedly present in the serum of patients with multiple sclerosis, that is capable of destroying myelin.
myelinolysis *Neurology.* deterioration of the myelin sheath of a nerve cell.
myelinopathy *Neurology.* any disease that causes the degeneration of myelin.
myelin sheath *Neurology.* a layer of fatty material myelin that surrounds certain nerve fibers; it has a high proportion of lipid to protein, and serves as an electrical insulator.
myelitis [mī´ə lī´tis; mī´lī´tis] *Pathology.* **1.** an inflammation of the spinal cord. **2.** an infection of the bone marrow.
myeloblast *Developmental Biology.* an immature cell that occurs in the bone marrow, but not usually in the peripheral blood; as the most primitive precursor in the granulocytic series, the myeloblast has fine, evenly distributed chromatin and a nongranular basophilic cytoplasm. Also, GRANULOBLAST.
myeloblastemia *Medicine.* the presence of myeloblasts in the blood stream.
myeloblastic *Developmental Biology.* of or relating to myeloblasts.
myeloblastoma *Oncology.* **1.** a growth composed of myeloblasts. **2.** a type of leukemia originating in the bone marrow.
myeloblastosis *Oncology.* an overabundance of immature bone marrow cells in the blood stream. Also, **myeloblastic leukemia.**
myelocele *Anatomy.* the middle channel of the spinal cord. *Medicine.* a bulging of the spinal cord seen in spina bifida.
myelocyte *Histology.* a motile cell in bone marrow that develops into granular leukocytes.
myelocytoma *Oncology.* leukemia in which myelocytes are generated from lymphoid and myeloid material. Also, CHRONIC MYELOGENOUS LEUKEMIA.
myelocytosis *Medicine.* a profusion of bone marrow cells in the blood.
myelodysplasia *Medicine.* a malformation of the spinal cord.
myeloencephalitis see MYELENCEPHALITIS.
myelofibrosis *Pathology.* a replacement of bone marrow by fibrous tissue, occurring either in association with a myeloproliferative disorder or secondary to another unrelated condition. Also, **myelosclerosis.**
myelogram *Pathology.* **1.** a radiographic depiction of the spinal cord. **2.** a visual display of the numerical computation of cells noted in a stained preparation of bone marrow. Thus, **myelography**.
myeloid *Anatomy.* **1.** of, related to, derived from, or resembling bone marrow. **2.** of or related to the spinal cord. **3.** resembling myelocytes, but not derived from bone marrow.
myeloid metaplasia *Oncology.* the presence of myeloid tissue in extramedullary sites; specifically, a condition that is characterized by anemia, an enlarged spleen, and the presence in the blood of nucleated erythrocytes and immature granulocytes, as well as extramedullary blood cell formation in the liver and the spleen.
myeloid tissue *Histology.* red bone marrow consisting primarily of a latticework of fibers that contain fat cells and both immature and mature blood cells.
myelolipoma *Oncology.* a rare benign tumor of the adrenal gland composed of adipose tissue, lymphocytes, and primitive myeloid cells.
myeloma *Oncology.* a tumor composed of malignant plasma cells.
myelomalacia *Medicine.* an unnatural loss of firmness in the spinal cord.
myelomenia *Neurology.* menstrual hemorrhage into the spinal cord.
myelomeningitis *Medicine.* an inflamed condition of the spinal cord and its membranes.
myelomeningocele *Medicine.* an opening in the bony spine with part of the spinal cord and membranes protruding.
myelomonocyte *Histology.* **1.** a monocyte developing in bone marrow. **2.** a developing blood cell that is intermediate between a monocyte and a granulocyte.
myelopathy *Neurology.* any disease or dysfunction of the spinal cord.
myeloperoxidase *Biochemistry.* a green heme (iron porphyrin orchlorin) peroxidase that is found in the lysosomal granules of myeloid cells, responsible for generating potent bacteriocidal activity. Also, VERDOPEROXIDASE.
myelophthisic anemia *Hematology.* anemia in which the blood-generating tissues are displaced. Also, **myelopathic anemia.**
myelophthisis *Neurology.* a shriveling of the spinal cord. *Hematology.* a reduction of the cell-forming functions of the bone marrow.
myeloplegia *Neurology.* spinal paralysis, caused by a lesion of the spinal cord.
myelopoiesis *Physiology.* the formation of bone marrow or the cells arising from it.
myelosarcoma *Oncology.* a sarcoma composed of myeloid tissue or bone marrow cells.
myelosclerosis *Medicine.* a hardening of the spinal cord.
myelosis *Oncology.* **1.** the abnormal proliferation of bone marrow tissue that produces the blood changes of myelocytic leukemia. **2.** the formation of a spinal cord tumor.
myelosuppression *Medicine.* the inhibition of bone marrow activity. Thus, **myelosuppressive.**
myenteric plexus *Histology.* the part of the enteric plexus located within the muscular lining of the esophagus, stomach, and intestines.

myenteron *Histology.* the layer of smooth muscles in the intestinal wall. Thus, **myenteric.**

Mygalomorphae *Invertebrate Zoology.* a suborder of spiders including the American tarantulas, trapdoor spiders, and purse-web spiders.

myiasis *Medicine.* the presence of fly maggots on the body.

Mylabridae see BRUCHIDAE.

Myliobatidae *Vertebrate Zoology.* the bat rays and stingrays, a mostly tropical family of cartilaginous fishes having a large, flattened body.

mylonite *Petrology.* a hard, compact, chertlike rock without cleavage, having a streaked or banded structure produced by extreme granulation and deformation during fault overthrusting or dynamic metamorphism.

mylonitic structure *Petrology.* the flowlike structure characteristic of mylonites, formed by intense microbrecciation and shearing.

mylonitization *Geology.* the deformation of rock produced by intense microbrecciation, resulting from the application of mechanical forces in a definite direction without any appreciable chemical alteration of granulated materials. Also, **mylonization.**

Mymaridae *Invertebrate Zoology.* the fairy flies, a family of minute, delicate wasps, hymenopteran insects parasitic on insect eggs.

myo- a combining form meaning "muscle."

myoblast *Developmental Biology.* an embryonic cell that becomes a muscle.

myoblastoma *Oncology.* a benign tumor composed of soft tissue cells.

myobradia *Neurology.* a slow, sluggish reaction of muscle to electric stimuli.

myocardial *Cardiology.* of or related to the myocardium.

myocardial infarction *Cardiology.* a cessation of blood supply to an area of the myocardium, usually caused by a clot in the coronary arteries. Also, **myocardial infarct.**

myocardial ischemia *Cardiology.* a deficient blood supply to the myocardium owing to obstruction or constriction of the coronary arteries.

myocardiopathy *Cardiology.* any noninflammatory disease of the myocardium.

myocarditis *Cardiology.* any inflammation of the myocardium.

myocardium *Cardiology.* the middle and thickest layer of the heart wall, composed of cardiac muscle.

myocardosis *Cardiology.* a broad category of diseases of the myocardium that are not inflammatory, but may be caused by hypertension, coronary sclerosis, and hyperthyroidism.

myocele *Medicine.* hernia of the muscle.

myocellulitis *Medicine.* inflammation of cellular tissue and muscle.

myoceptor see MOTOR END-PLATE.

myoclonia *Neurology.* any disorder characterized by myoclonus.

myoclonic epilepsy *Neurology.* a slowly progressive form of epilepsy beginning in childhood, characterized by myoclonus, impaired voluntary motion, delayed mental development, and sometimes mental deterioration.

myoclonus *Neurology.* a brief, shocklike contraction of a muscle or group of muscles. Thus, **myoclonic.**

myocoel *Developmental Biology.* the cavity found within a myotome or somite.

myocomma *Histology.* **1.** a muscle segment or myotome, specifically in fish. **2.** the septum between two adjacent myotomes.

myocyte *Histology.* **1.** any contractile cell. **2.** a muscle cell.

myocytoma *Oncology.* a tumor composed of muscle cells.

Myodaria *Invertebrate Zoology.* in some classifications, a group of dipteran flies characterized by antennae consisting of three segments. Thus, **myodarian.**

Myodocopa *Invertebrate Zoology.* an order or suborder of ostracod crustaceans, marine planktonic bivalves.

myodynia *Medicine.* pain in a muscle.

myodystrophy see MUSCULAR DYSTROPHY.

myoelastic fiber *Histology.* an elastic fiber found with smooth muscle fibers in the respiratory passages.

myoelectric potential *Physiology.* the electrical potential created by muscle movement.

myoepithelial cell *Cell Biology.* any of various contractile epithelial cells that facilitate the secretion of sweat, saliva, or milk from their respective glands.

myofascitis *Medicine.* an inflamed state of the fascia surrounding and supporting muscle tissue, especially the fascia that connects muscle to bone.

myofibril *Cell Biology.* a fiber, composed of a bundle of myofilaments, that is found in striated muscle cells and is responsible for muscle contraction.

myofibroma *Oncology.* a tumor composed of fibrous tissue and muscle cells.

myofibrosis *Medicine.* replacement of muscle tissue by fibrous tissue.

myofilament *Cell Biology.* a thin, threadlike component of a myofibril, composed principally of the proteins actin and myosin.

myofrisk *Invertebrate Zoology.* a contractile structure surrounding the spines of certain radiolarians.

myoglobin *Biochemistry.* an oxygen-transporting and storing pigment in muscle cells; it contains a heme (iron porphyrin) where oxygen binds. Also, **myohemoglobin.**

myoglobin gene *Genetics.* the gene that codes for myoglobin, an oxygen-binding protein found in vertebrate muscle.

myoglobulinura *Medicine.* the excretion of myoglobulin in the urine, seen after exercise or trauma.

myograph *Medicine.* an instrument used to record muscle contraction. Thus, **myogram, myography.**

myohematin *Biochemistry.* the cytochrome of muscle tissue, an iron-containing catalyst of tissue oxidation.

myoinositol *Biochemistry.* a sugarlike vitamin of the B complex, found in plant and animal tissues, that promotes growth of several species of bacteria and may have lipotropic activity. Also, MESOINOSITOL.

myokemia see MYOKYMIA.

myokinase see ADENYLATE KINASE.

myokymia *Neurology.* a condition characterized by brief, persistent quivering of muscles. Also, MYOKEMIA, KYMATISM.

myolipoma *Oncology.* a tumor containing muscle and fat.

myology *Medicine.* the scientific study of muscles.

myolysis *Medicine.* any disintegration or destruction of muscle tissue.

myoma *Oncology.* a benign fibroid tumor of the uterus.

myomalacia *Medicine.* the loss of firmness in muscles.

myomatosis *Oncology.* a condition characterized by the formation of numerous myomas.

myomectomy *Surgery.* **1.** the surgical excision of a myoma. **2.** see MYECTOMY.

myomere *Developmental Biology.* see MYOTOME.

myometritis *Medicine.* an inflammation of the uterine muscle.

myometrium *Histology.* the smooth muscle forming the wall of the uterus.

Myomorpha *Vertebrate Zoology.* the largest suborder of rodents, containing true mice, rats, voles, and jumping mice; native to virtually every habitat.

myomotomy *Surgery.* a surgical incision into a myoma.

myoneme *Invertebrate Zoology.* a contractile fiber in some ciliate protozoans.

myoparalysis *Medicine.* paralysis of a muscle.

myoparesis *Medicine.* weakness or mild paralysis of a muscle.

myopathic facies *Medicine.* a facial appearance resulting from lack of muscle tone; the eyelids droop and the lips project abnormally.

myopathy *Medicine.* any disease affecting a muscle. Thus, **myopathic.** Also, **myopathia.**

myopericarditis *Medicine.* myocarditis combined with pericarditis.

myophagia *Pathology.* a condition in which diseased muscle cells, or portions of such, are ingested and destroyed by a phagocyte.

myophagism *Pathology.* atrophy of muscular tissue.

Myophorella *Paleontology.* a genus of shallow-burrowing bivalves in the superfamily Trigoniacea; trigonally ovoid with rows of prominent tubercles; extant in the Jurassic and Lower Cretaceous.

myopia *Medicine.* faulty vision caused by an error of refraction in which rays of light entering the eye are brought into focus in front of the retina, usually as a result of the eyeball being too long from front to back. Thus, **myopic.** Also, NEARSIGHTEDNESS.

myoplasty *Surgery.* the surgical repair of muscle tissue.

Myoporaceae *Botany.* a family of dicotyledonous small trees and shrubs of the order Scrophulariales, characterized by the production of iridoid compounds and scattered secretory cavities that make the leaves appear dotted; native to Australia, the West Indies, and South America.

Myopsida *Invertebrate Zoology.* a suborder of cephalopod mollusks, shallow water squid with well developed eyes and ten tentacles, of great commercial and ecological importance.

myopsychopathy *Medicine.* any lack of proper muscular function associated with a psychological disease.

myorrhaphy *Surgery.* a suture of divided muscle.

myorrhythmia *Medicine.* a slow, periodic involuntary contraction of relaxed muscles, typical of patients with encephalitis-induced parkinsonism.

myosarcoma *Oncology.* a cancerous growth arising from muscle-producing cells.
myosclerosis *Medicine.* hardening of the muscle.
myoseism *Neurology.* an irregular, spasmodic muscle contraction.
myoseptum see MYOCOMMA, def. 2.
myosin *Biochemistry.* a globin that is the most abundant protein in muscle, occurring primarily in the A band, and forming the main constituent of the thick filaments of muscle fibers.
myositis *Medicine.* an inflamed condition of muscles, especially of voluntary muscles.
myositis ossificans *Medicine.* an inflammation process that involves the calcification of muscle tissue.
myospasm *Medicine.* the cramping of a muscle.
myosteoma *Oncology.* an osteogenic sarcoma of soft tissues.
myosynovitis *Medicine.* an inflamed state involving muscles and synovial membranes.
myotasis *Physiology.* the stretching of a muscle. Thus, **myotatic.**
myotis *Vertebrate Zoology.* a bat, *Myotis lucifugus,* of the family Vespertilionidae, characterized by a shiny brown furry body; the most common North American bat, inhabiting forests and cities in most parts of the world.
myotome *Developmental Biology.* the muscle foundation or part of the muscle in an embryo that develops into striated skeletal muscles. Also, MYOMERE. *Anatomy.* any group of muscles innervated by a single segment of the spinal cord. *Surgery.* an instrument used to cut or dissect muscle or muscular tissue.
myotonia *Physiology.* a tonic muscle spasm or transient stiffness after a muscle contraction.
myotonic *Physiology.* relating to tonic muscle movements, as distinguished from myokinetic responses.
myotonic dystrophy *Pathology.* a congenital disease involving tonic muscle spasms, wasting of muscles, cataracts, and cardiac deformities.
Myoviridae *Virology.* a family of linear dsDNA-containing bacteriophages with a long, complex, contractile tail, consisting of a central tube surrounded by a coaxial contractile sheath, and a head that is isometric or elongated. At present the family contains a single genus, representing the T-Even Phage Group.
myrcene *Organic Chemistry.* $C_{10}H_{16}$, a combustible, yellow, oily liquid; insoluble in water and soluble in alcohol and ether; boils at 167°C; used in perfumery and flavoring. Also, OIL OF BAY LEAVES.
myria- a combining form meaning "10,000" or "very many."
myriametric waves *Electromagnetism.* electromagnetic radiation having wavelengths in the range between 10 and 100 kilometers, corresponding to frequencies of 30 to 3 kilohertz.
Myriangiaceae *Mycology.* a family of fungi of the order Myriangiales, living off nonliving organic matter such as leaves, twigs, or fruit.
Myriangiales *Mycology.* an order of fungi of the class Loculoascomycetes, composed primarily of plant parasites.
Myriapoda *Invertebrate Zoology.* an informal designation for arthropods (**myriapods**) having more than three pairs of walking legs.
Myricaceae *Botany.* a family of trees and shrubs of the order Myricales having alternate simple, or pinnate leaves, unisexual flowers in axillary spikes or catkins, and fruit which is a small nut.
Myricales *Botany.* a monofamilial order of dicotyledonous plants of the subclass Hamamelidae.
myricetin *Organic Chemistry.* $C_{15}H_{10}O_8$, yellow crystals; soluble in alcohol; melts at 357°C; used to inhibit the enzyme adenosinetriphosphatase. Also, CANNABISCETIN.
myricin *Organic Chemistry.* $C_{30}H_{61}C_{16}H_{31}O_2$, a substance that is extracted from beeswax by crystallization.
Myrientomata see PROTURA.
myringitis *Medicine.* an inflamed eardrum.
myringoplasty *Surgery.* repair of the eardrum.
Myriniaceae *Botany.* a family of small, delicate mosses of the order Hypnobryales that form glossy mats on tree trunks and rocks; characterized by creeping, regularly branched stems, leaves with a slender costa on the upper portion, and long, smooth setae.
Myrionemataceae *Botany.* a family of minute discoid brown algae of the order Chordariales, being epiphytic on larger algae and sea grasses and characterized by thalli composed of a prostrate layer of tight filaments and erect, simple filaments of limited length.
Myriotrichiaceae *Botany.* a monogeneric family of small, tufted brown algae of the order Dictyosiphonales, characterized by creeping filaments that bear erect filaments; occurring in the North Atlantic, the Mediterranean, and the Adriatic Sea.
Myriotrochidae *Invertebrate Zoology.* a family of sea cucumbers having digitate tentacles and wheel-shaped spicules.
Myristicaceae *Botany.* a widespread family of tropical trees and shrubs of the order Magnoliales, characterized by their aromatic volatile oils; various species produce nutmeg, mace, and timber.
myristic acid *Organic Chemistry.* $C_{13}H_{27}CO_2H$, a white, oily crystalline solid; soluble in alcohol, chloroform, acetone, and benzene; melts at 54.4°C, and boils at 326.2°C; used in soaps, cosmetics, and food additives.
myristic alcohol *Organic Chemistry.* $C_{13}H_{27}CH_2OH$, a leaflike solid; melts at 39°C and boils at 289°C; soluble in alcohol, ether, acetone, chloroform, and benzene. Also, 1-TETRADECANOL.
myristica oil see NUTMEG.
myristylation *Molecular Biology.* the process of adding myristic acid to a protein so that the protein can attach to a membrane.
myristyl mercaptan see 1-TETRADECANETHIOL.
Myrmecophagidae *Vertebrate Zoology.* the anteaters, a family of the mammalian order Edentata, characterized by a long tubular snout, no teeth, a very long sticky tongue, and a diet of ants and termites; native to the new world tropics.
myrmecophagous *Zoology.* feeding mainly or exclusively on ants.
myrmecophile *Ecology.* an organism that lives in an ant's nest. Thus, **myrmecophilous.**
myrmecophyte *Ecology.* a plant that has a symbiotic relationship with ants, often with a means to house them. Thus, **myrmecosymbiosis.**
myrmekite *Petrology.* an intergrowth of vermicular quartz and plagioclase feldspar, generally replacing potassium feldspar; during the later stages of consolidation, it forms in igneous rocks.
myrmekitic *Petrology.* of, relating to, or characteristic of myrmekite or its texture, which features an intergrowth of vermicular quartz and plagioclase feldspar.
Myrmeleontidae *Invertebrate Zoology.* a family of insects in the order Neuroptera, with larvae called ant lions that dig a pit and wait for prey to fall in.
Myrmicinae *Invertebrate Zoology.* a subfamily of ants, insects in the family Formicidae.
Myrothamnaceae *Botany.* a monogenetic family of glabrous, aromatic, dioecious shrubs of the order Hamamelidales, with small leaves that become fragile and black and fold up like a fan in dry weather, then expand and become green again after rain; native to Africa.
myrrh [mur] *Materials.* a fragrant gum resin obtained from certain small trees of the genus *Commiphora,* found in Africa and Arabia; used since ancient times in making perfumes and incense.
Myrsinaceae *Botany.* a family of evergreen dicotyledons of the order Myrtales, having mostly alternate leathery leaves, small mostly bisexual flowers, and a superior ovary.

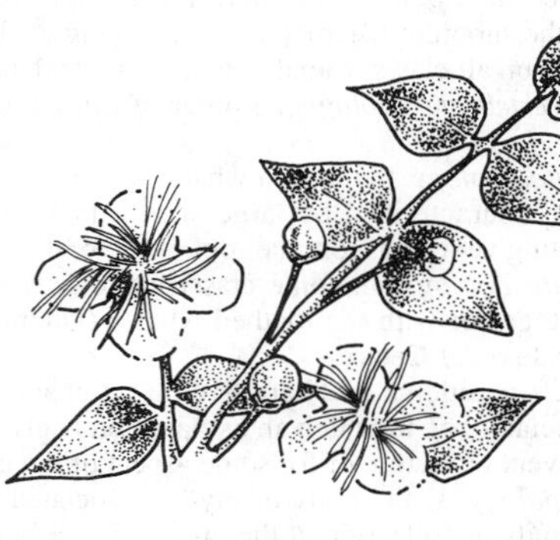

Myrtaceae

Myrtaceae *Botany.* the myrtle family, dicotyledonous trees and shrubs of the order Myrtales, characterized by opposite, simple leaves; regular, bisexual flowers in umbels, cymes, panicles, or racemes; an inferior ovary; and anthers that usually open by slits. The family includes the eucalyptus, myrtles of the genus *Myrtus,* guava, clove, and allspice.
myrtaceous *Botany.* relating to, belonging to, or resembling the family Myrtaceae.
Myrtales *Botany.* an order of dicotyledons of the subclass Rosidae having opposite, entire, simple leaves; flowers with a compound pistil; and bicollateral vascular bundles.

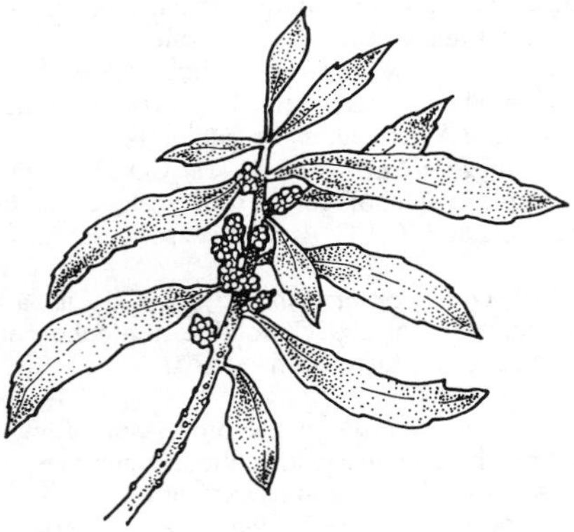

myrtle

myrtle *Botany.* **1.** any plant of the genus *Myrtus,* of the family Myrtaceae, especially *M. communis,* a southern European shrub having evergreen leaves, fragrant white flowers, and aromatic berries. **2.** any of various unrelated plants, such as the California laurel or the periwinkle.

myrtle oil *Materials Science.* a fragrant oil derived from the flowers and leaves of the European myrtle, *Myrtus communis*; used in perfumes.

Mysidacea *Invertebrate Zoology.* the opossum shrimps, an order of carnivorous shrimplike crustaceans having biramous thoracic appendages.

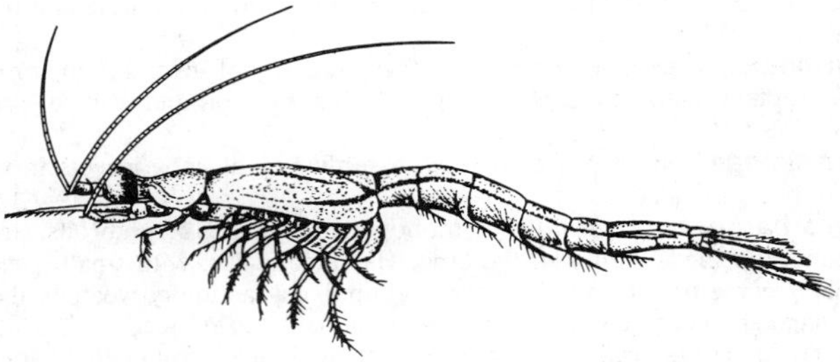

Mysidacea

mysis *Invertebrate Zoology.* a larval stage of certain crustaceans, characterized by biramous thoracic appendages. Also, SCHIZOPOD LARVA.

mysophobia *Psychology.* an irrational fear of contamination; usually accompanied by compulsive washing. Also, MOLYSMOPHOBIA.

Mystacinidae *Vertebrate Zoology.* the short-tailed bats, a family of insectivorous bats of the suborder Microchiroptera, having well-developed ear pinnae and talons on all claws; found only in New Zealand.

Mystacocarida *Invertebrate Zoology.* an order of primitive, wormlike crustaceans.

Mysticeta *Vertebrate Zoology.* the baleen whales, an order of mammals found in all oceans, characterized by large size, a lack of teeth, and baleen plates projecting ventrally from the roof of the oral cavity.

Mysticeti *Vertebrate Zoology.* in some classifications, a suborder of baleen whales that together with the toothed whales (Odontoceta) constitute the single whale order Cetacea.

myth *Anthropology.* a traditional oral narrative that is characteristic of a certain culture, especially one in which an explanation is given for some significant natural event or feature or for some aspect of the culture.

mythology *Anthropology.* **1.** the body of myths associated with a certain culture. **2.** the study or collection of the myths of a certain culture.

Mytilacea *Invertebrate Zoology.* a suborder of mussels, bivalve crustaceans in the order Filibranchia.

Mytilidae *Invertebrate Zoology.* a family of mussels, bivalve crustaceans in the order Anisomyaria.

myx- or **myxo-** *Zoology.* a combining form meaning "mucus" or "slime."

myxadenitis *Medicine.* a condition of inflamed mucous glands.

myxadenoma *Oncology.* an epithelial tumor that is structurally similar to a mucous gland. Also, MUCINOUS ADENOMA.

myxedema *Medicine.* the bodily changes resulting from a dysfunction of the thyroid gland.

Myxicolinae *Invertebrate Zoology.* the feather-duster worms, a subfamily of sessile polychaetes in the family Sabellidae.

Myxidium *Invertebrate Zoology.* a genus of protozoans parasitic in reptiles, fish, and amphibians.

Myxinidae *Vertebrate Zoology.* the scavenger hagfishes, the single famiy of fish in the order Myxiniformes.

Myxiniformes *Vertebrate Zoology.* an order of eel-like, scaleless, jawless fish of the class Agnatha; found worldwide in temperate oceans.

myxobacter *Bacteriology.* any member of the order Myxobacterales.

Myxobacterales *Bacteriology.* an order of Gram-negative, aerobic gliding bacteria occurring in animal dung, decomposing plant material, and soil as cylindrical rods with rounded or tapered ends.

Myxobolus *Invertebrate Zoology.* a genus of protozoans parasitic in freshwater fish and amphibians, causing boil disease in fish.

myxochondrofibrosarcoma *Oncology.* a cancerous growth made up of myxomatous, fibrous, chondromatous, and sarcomatous tissues.

myxochondroma *Oncology.* a chondroma having connective tissue that is similar to primitive mesenchymal tissue.

Myxochrysidaceae *Botany.* a monospecific family of plasmids of the order Chrysamoebidales, characterized by a complicated life history in which reproduction occurs when the protoplasm transforms into numerous small cysts.

Myxococcaceae *Bacteriology.* a family of gliding bacteria of the order Myxobacterales, occurring in soil and producing a distinctive type of resting cell called **myxospore.**

myxoenchondroma *Oncology.* a chondroma having areas of myxomatous degeneration.

myxoendothelioma *Oncology.* an angioendothelioma that has undergone myxomatous degeneration.

myxoflagellate *Invertebrate Zoology.* a flagellate stage in the life cycle of slime molds.

Myxogastromycetidae *Mycology.* a subclass of fungi of the class Myxomycetes, consisting of plasmodial slime molds.

myxoglioma *Oncology.* a glioma that has undergone myxomatous degeneration.

myxolipoma *Oncology.* a lipoma having areas of myxomatous degeneration.

myxoma *Oncology.* a soft, jellylike tumor that is composed of connective and mucoid tissue and that may grow to over 30 cm in diameter. Also, **myxoblastoma.**

Myxomycetes *Biology.* the slime molds, a class of microorganisms on the borderline of the plant and animal kingdoms; found in a variety of environments including dung, rotting wood, tree bark, soil, and humus. *Mycology.* in mycological classifications, a class of fungi of the division Myxomycota.

Myxomycophyta *Botany.* an order of simple microorganisms exhibiting characteristics of both plants and animals, which are sometimes classified as fungi.

Myxomycota *Mycology.* a division of fungi including slime molds and organisms similar to slime molds which have primitive amebalike colonial stages somewhere in their life cycles.

myxomyoma *Oncology.* a myoma that has undergone myxomatous degeneration.

Myxophaga *Invertebrate Zoology.* a suborder of beetles, insects in the order Coleoptera.

Myxophyceae *Botany.* in botanical classifications, the slime mold family of simple microorganisms of the order Myxomycophyta.

Myxosarcina *Bacteriology.* a genus of cyanobacteria that reproduce by both binary and multiple fission, exhibit a gliding motility, and have the ability to carry out photosynthesis accompanied by oxygen formation.

myxosarcoma *Oncology.* a partially deteriorated tumor composed of sarcoid and mucoid tissues.

Myxosporida *Invertebrate Zoology.* an order of protozoans in the subclass Cnidosporidia, having bivalve spores and living as endoparasites in aquatic vertebrates.

Myxosporidea *Invertebrate Zoology.* a class of endoparasitic protozoans in the subphylum Cnidospora.

Myxozoa *Invertebrate Zoology.* a phylum of parasitic protozoans that form infective cysts.

Myzopodidae *Vertebrate Zoology.* the sucker-footed bat, a monotypic family of the suborder Microchiroptera, having sessile suction disks on one finger and on the sole of the foot; noted for roosting in the curled leaves of bananas and similar plants.

myzorhynchus *Invertebrate Zoology.* an apical sucker on the head of certain tapeworms.

Myzostomaria *Invertebrate Zoology.* a class of aberrant annelid worms that are probably related to the polychaetes; parasites of echinoderms.

MZ monozygotic.

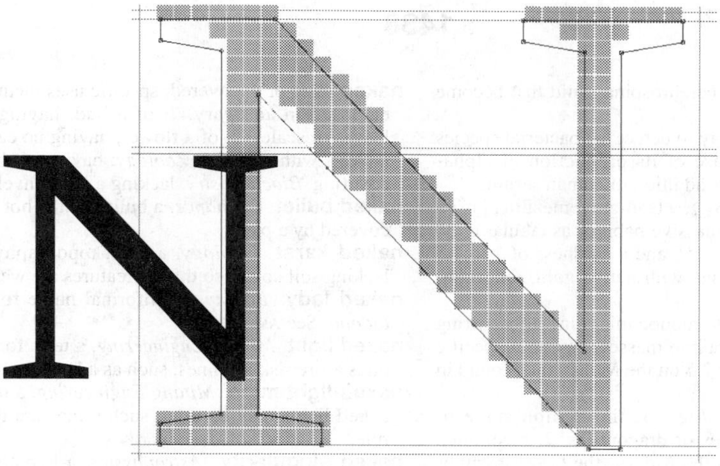

N north; newton; neutron number.

N *Chemistry*. the symbol for nitrogen. *Molecular Biology*. a symbol for the haploid chromosome number.

n *Computer Programming*. **1.** in sorting, the number of records to be processed; file size. **2.** more generally, the number of items to be processed or the size of a problem.

n or **n.** nano-; negative; neutron; north, northern; number.

n- *Organic Chemistry*. a prefix indicating a normal (unbranched) alkyl group.

NA not applicable; not available.

NA or **N.A.** neutral axis; numerical aperture.

N.A. North America.

Na *Chemistry*. the symbol for sodium. (From Latin *natrium*.)

NAA National Aeronautic Association.

Nabidae *Invertebrate Zoology*. the damselbugs, a family of predaceous insects in the superfamily Cimicimorpha that suck the blood of soft-bodied insects, including plant pests.

Nabothian cyst *Medicine*. a cyst that develops when the glands of the cervix uteri become blocked and distended with retained secretory material. Also, **Nabothian follicle, Nabothian gland.** (Named for Martin *Naboth*, 1675–1721, German anatomist.)

Naccariaceae *Botany*. a family of marine red algae of the order Nemaliales, characterized by gametangial thalli that differ markedly from the tetrasporangial thalli.

nacelle [nə sel′] *Aviation*. a streamlined enclosure designed to house and protect something such as the crew, engine, or landing gear. An airship gondola is a nacelle, as is an airplane cockpit.

nacre [nā′kər] *Invertebrate Zoology*. another name for mother-of-pearl, the hard, pearly internal layer of certain marine shells such as the pearl oyster.

nacreous *Optics*. describing minerals or other substances having an iridescent luster resembling that of mother-of-pearl. Also, PEARLY.

nacreous clouds *Meteorology*. rarely seen clouds of unknown composition that occur at altitudes between approximately 20,000 and 35,000 feet in the northern latitudes; they often resemble pale cirrus during the day, and at sunset exhibit all the colors of the spectrum, increasing in brilliance as the sky darkens. Also, MOTHER-OF-PEARL CLOUDS.

nacreous layer see MOTHER-OF-PEARL.

nacrite *Mineralogy*. $Al_2Si_2O_5(OH)_4$, a colorless or white monoclinic mineral of the kaolinite-serpentine group, having a specific gravity of 2.6 and a hardness of 2 to 2.5 on the Mohs scale; polymorphous with kaolinite, dickite, and halloysite; found in compact to friable masses as a hydrothermal mineral in ore deposits.

NAD *Enzymology*. nicotinamide adenine dinucleotide.

NAD⁺ *Enzymology*. the oxidized form of nicotinamide adenine dinucleotide.

n-address instruction *Computer Programming*. an instruction that contains *n* number of address parts.

Na-Dene *Linguistics*. a Native American language family that includes Navajo, Apache, and many other languages of the Pacific Northwest; in some schemes, one of three major American language families, the other two being *Amerind* and *Eskimo-Aleut*.

NADH *Enzymology*. the reduced (hydrogen ion-carrying) form of nicotinamide adenine dinucleotide.

nadide see NICOTINAMIDE ADENINE DINUCLEOTIDE.

nadir [nā′dēr′; nā′dər] *Astronomy*. the point on the celestial sphere, blocked from view by the earth, that lies directly below the position of the observer, opposite the zenith. *Science*. the lowest point or position.

nadir point see PHOTOGRAPH NADIR.

nadir radial *Cartography*. a line or direction with its origin at a nadir point on a photograph.

nadorite *Mineralogy*. $PbSb^{+3}O_2Cl$, a brown to yellow orthorhombic mineral occurring as thin tabular or prismatic crystals, having a specific gravity of 7.02 and a hardness of 3.5 to 4 on the Mohs scale.

NADP *Enzymology*. nicotinamide adenine dinucleotide phosphate.

NADP⁺ *Enzymology*. the nonreduced form of nicotinamide adenine dinucleotide phosphate.

NADPH *Enzymology*. the reduced form of the coenzyme nicotinamide adenine dinucleotide phosphate, required for the synthesis of fatty acids via the phosphogluconate pathway.

NADPH hydrogenase *Enzymology*. a flavoprotein, originally thought to be active in a respiratory pathway, that catalyzes the oxidation of NADPH. Also, OLD YELLOW ENZYME, WARBURG'S OLD YELLOW ENZYME.

Nadsonia *Mycology*. a genus of yeasts of the family Saccharomycetaceae, occurring on fruits and slime fluxes and characterized by bipolar budding and the formation of asci.

Naegeli-type leukemia *Oncology*. monocytic leukemia in which many of the cells are similar to myeloblasts.

Naeglaria *Invertebrate Zoology*. a genus of amoeboflagellates of the order Schizopyrenida, including species that are invasive pathogens, such as *N. fowleri*, an agent of meningoencephalitis.

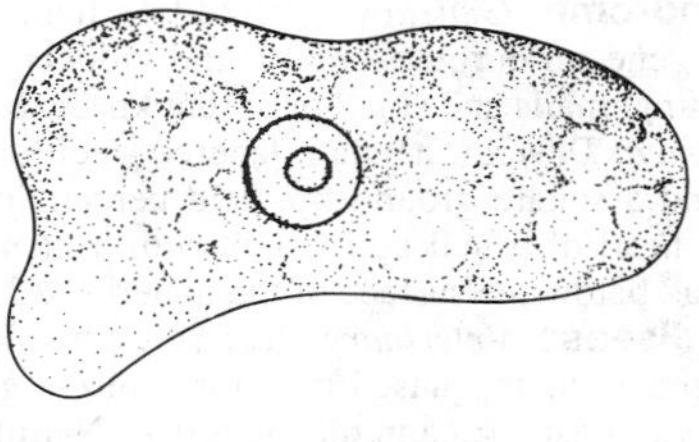

Naeglaria

Naetrocymbaceae *Mycology*. a family of fungi belonging to the order Chaetothyriales, occurring inconspicuously on the surfaces of plant leaves and branches in tropical to temperate regions.

nafoxidine hydrochloride *Pharmacology*. $C_{29}H_{31}NO_2{\cdot}HCl$, an agent that inhibits the action of estrogen, used in the treatment of breast cancer.

nagana *Veterinary Medicine*. any of several trypanosomiases that infect horses and other animals in Africa through the tsetse fly, causing a variety of ailments, including damage to vital organs. (Originally a Zulu word.)

nagarose see SUBTILISIN.

nagatelite *Mineralogy*. a phosphatian variety of allanite.

Nageotte bracelets *Neurology*. transverse bands covered with circular spines, located on myelinated axons of the nodes of Ranvier. (Named for Jean *Nageotte*, 1866–1948, French histologist.)

Nageotte cells *Pathology.* cells of the cerebrospinal fluid that become greatly increased in number in disease.

Nagler's reaction *Bacteriology.* an assay to detect the bacterial species *Clostridium perfringens*, which, because of its production of alpha-toxin, produces a dense turbidity upon its addition to human serum.

nagyagite *Mineralogy.* $Pb_5Au(Te,Sb)_4S_{5-8}$, a lead-gray, metallic, possibly orthorhombic mineral, occurring in massive habit or as tabular crystals, having a specific gravity of 7.41 to 7.55 and a hardness of 1 to 1.5 on the Mohs scale; found in vein deposits with native gold, tellurides, and sulfides.

nahcolite *Mineralogy.* $NaHCO_3$, a white monoclinic mineral occurring as small, prismatic crystals and in crystalline masses, having a specific gravity of 2.21 to 2.24 and a hardness of 2.5 on the Mohs scale; found in oil shale and evaporite beds.

naiad [nī´ad; nā´ad] *Invertebrate Zoology.* **1.** the nymph stage of hemimetabolous insects such as mayflies or dragonflies. **2.** a common name for any freshwater mussel. (From the *Naïads,* the Greek nymphs of rivers and springs.)

Naiadites *Paleontology.* a genus of nonmarine bivalves in the order Pterioida; very widespread in the Carboniferous, a distribution thought to be explainable only if it attached itself byssally to drifting vegetation, but no fossil has been found still thus attached; index fossil, extant in the Upper Carboniferous.

naif [nä ēf´] *Mineralogy.* describing a gemstone having a natural luster when uncut. Also, **naife.**

nail *Mechanical Devices.* a thin, tapered piece of metal of various thickness and length, having a flattened head at one end and a point at the other; designed to be pounded into a material with a hammer or other blunt tool, usually to fasten objects together. *Anatomy.* a flattened area of horny cutaneous cells on the dorsal side of the ends of the fingers and toes. *Surgery.* a pin or fine rod used to fix fractured bones together.

nail claw *Mechanical Devices.* a two-pronged lever opposite the face of a hammer or as a separate tool; used to grip and pry nailheads.

nailer *Building Engineering.* a strip or block of wood that is used as a backing under objects or surfaces through which nails are driven.

nail extension *Surgery.* a method for extending a distal fragment of a fracture by inserting into it a nail or pin, and putting weight on it.

nailhead *Mechanical Devices.* the flat, circular uppermost part of a nail, designed to be struck by a hammer or mallet so the nail point is driven into the material.

nailhead molding *Architecture.* a molding decorated with low, four-sided pyramids that resemble antique nailheads.

nailhead spot *Plant Pathology.* a tomato rot disease caused by the fungus *Alternaria tomato* and characterized by small, dark, depressed spots on the fruit that resemble the heads of nails.

nailhead striation *Geology.* a glacial striation having a definite blunt head or point of origin, and generally narrowing or tapering in the direction of ice movement. Also, **nailhead scratch.**

nail-patella syndrome *Genetics.* a hereditary human disease caused by a dominant gene on chromosome 9. Individuals with this disease have deformed fingernails and small or absent kneecaps.

nail set *Mechanical Devices.* a small, tapered steel tool consisting of a knurled shaft and a square ground tip, held perpendicular to a surface and against the head of a nail and struck with a hammer or mallet to drive the nailhead below the surface of the material. Also, **nail punch.**

Nairobi sheep disease *Veterinary Medicine.* a highly fatal tick-borne disease of sheep and goats, caused by a virus of the genus *Nairovirus;* occurring mainly in East Africa in the vicinity of Nairobi, Kenya.

Nairovirus *Virology.* a genus of viruses of the family Bunyaviridae that infect a wide range of vertebrates via ticks.

naive psychology *Psychology.* popular or folk beliefs about the causes of behavior.

Najadaceae *Botany.* a family of aquatic, monocotyledonous plants of the order Najadales having narrow, opposite leaves and unisexual, solitary flowers in axils.

Najadales *Botany.* an order of aquatic and semiaquatic, monocotyledonous flowering plants of the subclass Alismatidae.

NAK *Computer Programming.* negative acknowledgment, a character in the ASCII set.

Nakayama's lemma *Mathematics.* if J is an ideal in a commutative ring R with identity 1_R , then the following conditions are equivalent: (a) J is contained in every maximal ideal of R; (b) $1_R - j$ is a unit for every $j \in J$; (c) if A is a finitely generated R-module with $JA = A$, then $A = 0$; (d) if B is a submodule of a finitely generated R-module A such that $A = JA + B$, then $A = B$.

naked bare or uncovered; specific uses include: *Botany.* **1.** of a seed, not enclosed in an ovary. **2.** of a bud, having no covering of protective leaves or scales. **3.** of a flower, having no calyx or perianth. **4.** of a stalk or leaf, without hairs. *Zoology.* having no hair, shell, feathers, or other covering. *Biochemistry.* lacking a lipid envelope.

naked bullet *Ordnance.* a bullet that is not encased in a metal jacket or covered by a patch.

naked karst *Geology.* a karst topography that develops in a region lacking soil cover, so that its features are well exposed.

naked lady *Botany.* an informal name for the flower *Amaryllis belladona.* See AMARYLLIS.

naked light *Mining Engineering.* a term for any open flame that constitutes a fire risk in mines, such as a lamp, match, or burning cigarette.

naked-light mine *Mining Engineering.* a nongassy coal mine in which naked lights may be used; such mines are the exception rather than the rule.

naked singularity *Astrophysics.* a hypothetical singularity (in space-time) that has no event horizon associated with it.

naked virus *Virology.* any unenveloped virus, such as a picornavirus.

nakhlite *Geology.* an achondritic, stony, meteorite composed entirely of a mixture of diopside and olivine crystals.

naled *Organic Chemistry.* $(CH_3O)_2P(O)OCH(Br)CBr(Cl)_2$, in pure form, a white solid that melts at 26°C; in technical form, a moderately volatile liquid that boils at 110°C (0.5 torr); insoluble in water and very soluble in aromatic solvents. *Agriculture.* a commercial preparation of this substance, used to control insects and mites on a variety of crops and farm animals.

nalidixic acid *Pharmacology.* $C_{12}H_{12}N_2O_3$, a synthetic antibiotic adminstered orally in the treatment of urinary infections caused by certain Gram-negative bacteria.

Namanereinae *Invertebrate Zoology.* a subfamily of mostly freshwater annelids in the family Nereidae.

name table *Computer Programming.* a table that is created and maintained by a compiler or assembler, and contains symbol names, their machine addresses, and other information. Also, SYMBOL TABLE.

NAMH or **N.A.M.H.** National Association for Mental Health.

Namurian *Geology.* a European geologic stage of the Carboniferous period, occurring after the Viséan and before the Westphalian.

Nancy *Electronics.* a visual-blinker communications system for signaling between ships at night, using invisible infrared rays.

Nancy receiver see INFRARED RECEIVER.

NAND *Mathematics.* a logic operator that operates on a finite set of statements; the value of the operator is true if at least one of the statements is false, and is false otherwise (i.e., if all the statements are true). Also, SHEFFER STROKE. *Computer Technology.* the binary logic operator that represents NOT AND; it is true if at least one operand is false, and false only if all operands are true.

NAND circuit *Electronics.* a logic circuit in which the output is at the low level only when all inputs are at the high level, and is high otherwise; the inverse of an AND circuit. Also, **NAND gate.**

Nandidae *Vertebrate Zoology.* the leaffishes, a family of predatory freshwater fish of the order Perciformes, having deep bodies and large heads with large mouths; noted for their habit of floating near the surface, looking like a drifting leaf.

nanism *Medicine.* a condition of dwarfism or extreme undersize, from whatever cause.

nannandrous *Biology.* describing certain species, especially species of algae, in which the male of the species is markedly smaller than the female.

Nannizzia *Mycology.* a genus of perfect (sexual) fungi termed *Microsporum* in their asexual phase and belonging to the order Gymnascales; found in soil and composed primarily of skin parasites.

nannoplankton *Biology.* minute plankton; the smallest plankton, including algae, bacteria, and protozoans. Also, **nanoplankton.**

nano- a combining form meaning: **1.** very small, minute, as in *nanozooid.* **2.** one billionth (10^{-9}), as in *nanosecond.* Also, **nanno-.** (From the Greek word for "dwarf.")

Nanobryaceae *Botany.* a monotypic family consisting of a minute African moss of the order Dicranales that has few leaves and a perennial protonema, giving rise to annual stems.

nanocephalous *Medicine.* having a head that is abnormally small. Also, MICROCEPHALIC.

nanogram *Metrology.* a unit of capacity equal to one-billionth of a gram. Also, **nanogramme.**

nanoid *Medicine.* resembling a dwarf; dwarfed or dwarfish.

nanoinstructions *Computer Technology.* in microprogramming control organization, the decoded microinstructions stored in lower-level control memory.

nanoliter *Metrology.* a unit of volume that is equal to one-billionth of a liter. Also, **nanolitre.**

nanometer *Metrology.* a unit of length that is equal to one-billionth of a meter. Also, **nanometre.**

nanophanerophyte *Botany.* a classification of plants that have renewal buds less than 2 meters above the ground.

nanosecond *Metrology.* a unit of time that is equal to one-billionth of a second.

nanostorage *Computer Technology.* the lower level of microprogram control memory, in which the nanoinstructions are stored. Also, **nanomemory.**

nanozooid *Invertebrate Zoology.* **1.** a dwarf zooid. **2.** a bryozoan heterozooid that has only one tentacle.

Nansen, Fridtjof, 1861–1930, Norwegian polar explorer, author, and oceanographer.

Nansen bottle *Oceanography.* an oceanographic sampling device having valves at both ends that are designed to close when the bottle is inverted and thus trap a water sample; made of a metal alloy that reacts little with seawater. (Named for Fridtjof *Nansen.*)

Nansen cast *Oceanography.* a set of Nansen-bottle water samples and temperature readings made by using several bottles on the same cable and arranging each one so that, when it is tripped by a messenger, it sends another messenger on to the next bottle.

nap *Textiles.* the fuzzy finish produced by napping. *Botany.* a similar texture on the surface of a plant or leaf.

NAP net aboveground production.

napalm [nā´päm´] *Materials.* **1.** a substance, consisting of aluminum soaps or sodium palmitate, that is added to gasoline or oil in order to form a highly incendiary jellylike material. **2.** the material thus formed, used in bombs and flamethrowers. *Ordnance.* to attack with a napalm bomb.

napalm bomb *Ordnance.* an incendiary bomb containing napalm, a thickened gasoline; used primarily as an antipersonnel weapon.

nape *Anatomy.* the back of the neck.

napex *Anatomy.* the region of the scalp just below the occipital prominence.

naphtha [naf´thə; nap´thə] *Organic Chemistry.* **1.** a general term applied to various refined, partially refined, or unrefined products of petroleum, not less that 10% of which distill below 175°C and not less than 95% of which distill below 240°C; freezes at –73°C and boils at 30–60°C; flammable and a dangerous fire risk; used as a gasoline ingredient, as a source of synthetic natural gas, as a paint and varnish thinner, in dry cleaning fluids, and for various other purposes. **2.** another term for petroleum itself. **3.** either of two mixtures obtained by the fractional distillation of coal tar: **heavy naphtha,** a deep amber to dark red liquid that boils at 160–220°C; used in the manufacture of synthetic resins, as a solvent, in soaps, rubber cements, and cleansers, and for a variety of other purposes; or **solvent naphtha,** a yellowish or colorless to white liquid that boils at 160°C; used mainly as a solvent.

naphthacene *Organic Chemistry.* another name for tetracene, $C_{18}H_{12}$. See TETRACENE.

naphthalene *Organic Chemistry.* $C_{10}H_8$, a white, volatile, monoclinic or platelike solid having a strong coal-tar odor; insoluble in water and soluble in alcohol, ether, and benzene; melts at 80.55°C and boils at 218°C; consisting of two benzene rings, it serves as the starting point for many organic syntheses. Also, TAR CAMPHOR.

naphthaleneacetamide *Organic Chemistry.* $C_{12}H_{11}NO$, a colorless solid that melts at 183°C; used to stimulate growth of root cuttings and to control the growth of apples and pears.

α-naphthaleneacetic acid *Organic Chemistry.* $C_{10}H_7CH_2COOH$, odorless white crystals; slightly soluble in water and alcohol and soluble in acetone, ether, and chloroform; melts at 132–135°C; used commercially to promote the growth of roots in plant cuttings and to minimize premature loss of fruit.

α-naphthalenesulfonic acid *Organic Chemistry.* $C_{10}H_7SO_3H{\cdot}2H_2O$, a combustible, crystalline solid that is soluble in water and alcohol; melts at 90°C; used in the manufacture of chemicals and disinfectant soaps.

naphthalize *Organic Chemistry.* to mix or saturate with naphtha. Thus, **naphthalization.**

naphthanthracene see BENZANTHRACENE.

naphthene see CYCLOPARAFFIN.

naphthenic acid *Organic Chemistry.* any of the carboxylic acids obtained from alkali washes of the gas-oil fractions of petroleum; used chiefly in the production of metallic naphthenates for paint driers and cellulose preservatives.

naphthionic acid *Organic Chemistry.* $C_{10}H_6(NH_2)SO_3H$, a white crystalline or powdered solid that is soluble in water and slightly soluble in methanol and pyrimidine; used to manufacture dyes.

α-naphthol *Organic Chemistry.* $C_{10}H_7OH$, a combustible, colorless or yellow solid that occurs in prismatic or powdered form; insoluble in water and soluble in alcohol; melts at 96°C and boils at 278°C; used in dyes, perfumes, and organic synthesis. Also, 1-HYDROXYNAPHTHALENE.

β-naphthol *Organic Chemistry.* $C_{10}H_7OH$, a combustible white solid that occurs in lustrous, bulky leaflets or powdered form; almost insoluble in water and soluble in alcohol, oils, and alkaline solutions; melts at 121.6°C and boils at 285°C; used in dyes, pigments, insecticides, and pharmaceuticals. Also, 2-HYDROXYNAPHTHALENE.

naphtholism *Toxicology.* poisoning due to the ingestion of naphthol; the symptoms may include anemia, jaundice, convulsions, and coma. Also, **naphthol poisoning.**

β-naphtholsulfonic acid see SCHAEFFER ACID.

1-naphthol-4-sulfonic acid see NEVILE-WINTHER ACID.

1,2-naphthoquinone *Organic Chemistry.* $C_{10}H_6O_2$, orange-red crystals that are soluble in alcohol, ether, and benzene; melts and decomposes at 115–120°C; used as a chemical reagent and intermediate.

1,4-naphthoquinone *Organic Chemistry.* $C_{10}H_6O_2$, a combustible yellow powder; slightly soluble in water and soluble in ethanol, ether, benzene, and acetic acid; melts at 123–126°C; used to regulate polymerization for rubber and polyester resins, as a component in the synthesis of dyes and pharmaceuticals, and as a fungicide and algicide.

naphthoresorcinol see 1,3-DIHYDROXYNAPHTHALENE.

β-naphthoxyacetic acid *Organic Chemistry.* $C_{10}H_7OCH_2COOH$, a crystalline solid; soluble in water, alcohol, and acetic acid; melts at 156°C; used to regulate plant growth.

naphthyl *Organic Chemistry.* the radical $C_{10}H_7$.

α-naphthylamine *Organic Chemistry.* $C_{10}H_9N$, combustible white crystals that turn red upon exposure to air; slightly soluble in water and soluble in alcohol; melts at 50°C and boils at 301°C; used in the manufacture of dyes and agricultural chemicals.

β-naphthylamine *Organic Chemistry.* $C_{10}H_9N$, a highly toxic, white to reddish crystalline solid; soluble in water; used mainly in the production of azo dyes.

β-naphthyl methyl ether *Organic Chemistry.* $C_{10}H_7OCH_3$, combustible white, crystalline scales; insoluble in water and soluble in alcohol; melts at 72°C and boils at 274°C; used in perfumery.

***N*-1-naphthylphthalamic acid** *Organic Chemistry.* $C_{10}H_7NHCO C_6H_4COOH$, combustible noncorrosive crystals; almost insoluble in water and slightly soluble in acetone, benzene, and ethanol; melts at 185°C; unstable in solutions above pH 9.5 or at temperatures above 180°C; used as an herbicide.

α-naphthylthiourea *Organic Chemistry.* $C_{10}H_7NHCSNH_2$, a toxic, odorless gray powder, insoluble in water and slightly soluble in most organic solvents; melts at 198°C; used as a rodenticide. Also, ANTU.

Napier, John 1550–1617, Scottish mathematician; invented logarithms; pioneer in the use of decimal notation.

Napier diagram *Navigation.* a device for plotting the deviation of a magnetic compass graphically; used to create a deviation table, or directly to convert between magnetic and compass bearings or courses.

Napier's analogies *Mathematics.* formulas relating the sides a, b, and c and the angles α, β, and γ of a spherical triangle:

$$\tan(c/2)\cos((\alpha-\beta)/2) = \tan((a+b)/2)\cos((\alpha+\beta)/2);$$

$$\tan(c/2)\sin((\alpha-\beta)/2) = \tan((a-b)/2)\sin((\alpha+\beta)/2);$$

$$\cot(\gamma/2)\cos((a-b)/2) = \tan((\alpha+\beta)/2)\cos((a+b)/2);$$

$$\cot(\gamma/2)\sin((a-b)/2) = \tan((\alpha-\beta)/2)\sin((a+b)/2).$$

Napier's rules *Mathematics.* formulas relating the sides a and b and hypotenuse c to the angles α and β of a right spherical triangle. They may be summarized as follows: the cosine of any part of a right spherical triangle is equal to the product of the sines of the nonadjacent parts, and also is equal to the product of the cotangents of the adjacent parts; the right angle is not counted when determining adjacency, and the arms of the right angle are replaced by their complements (or cofunctions taken) when writing the formulas. Also, **Napier's rules of circular parts.**

napiform *Botany.* shaped like a turnip; applied especially to roots.
Naples yellow see LEAD ANTIMONATE.
napoleanite see CORSITE.
Napoleonville *Geology.* a North American Gulf Coast stage of the Miocene epoch, occurring after the Anahauc and before the Duck Lakean.
nappe *Geology.* a large, sheetlike rock unit that has been transported some distance forward over other rocks, either by thrust faulting, recumbent folding, or both. *Mathematics.* either of the two halves of a cone which meet at the vertex.
napping *Textiles.* a fabric treatment in which a machine containing small wires brushes the surface of the fabric, raising the loose ends of the fibers to produce a distinctive fuzzy finish, the nap. This process changes the appearance and the hand of the fabric, and also increases its insulating ability.
narbonnais [när´bə nā´] *Meteorology.* a north wind in the Roussillon region of southern France that is associated with an influx of arctic air and that may be very stormy, with heavy falls of rain or snow. Also, **narbonés.** (Named for the city of *Narbonne.*)
narceine *Organic Chemistry.* $C_{23}H_{27}NO_8 \cdot 3H_2O$, a white, odorless, bitter-tasting alkaloid that is slightly soluble in water and soluble in alcohol; melts at 176°C; used in medicine.
narcissism [när sis´iz əm] *Psychology.* **1.** a extreme form of self-love, in which an individual has an unduly high regard for his or her own deeds and physical attributes. **2.** an early stage of development at which an infant cannot yet distinguish an object, such as the mother's breast, from his or her own body. (Based on the Greek myth of *Narcissus,* a youth who fell in love with his own reflection in a pool of water.)
narcissistic personality (disorder) *Psychology.* a disorder characterized by an inflated sense of self-importance, fantasies of success, wealth, power, and ideal love, an unrealistic sense of actual accomplishments, and immature emotional reactions.
narcissus [när sis´is] *Botany.* **1.** any of a variety of plants of the genus *Narcissus* of the family Amaryllidaceae, which are native to the Old World and now widely cultivated in North America for their showy white or yellow flowers, including the daffodil, the jonquil, the **poet's narcissus,** *N. poeticus,* or the **paper white narcissus,** *N. tazetta.* **2.** the flower of any of these plants. (From the Greek myth of *Narcissus,* a youth who fell in love with his own reflection in a pool of water. As he could not bear to leave the pool, he wasted away and after his death was transformed into this flower.)

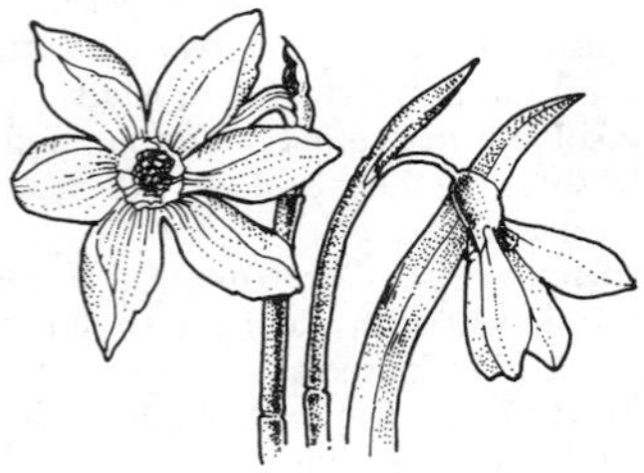

narcissus

narco- or **narc-** a combining form denoting a relationship to sleep, stupor, a stuporous state, or narcosis.
narcoanalysis *Psychology.* a type of analysis that uses sleep-inducing drugs to help patients discuss traumatic experiences. Also, **narcotherapy.**
narcolepsy [när´kə lep´sē] *Medicine.* a disorder characterized by recurrent, uncontrollable, brief episodes of sleep, often associated with hypnagogic hallucinations, cataplexy, and sleep paralysis; of unknown cause but possibly hereditary. The condition is presently incurable but can be effectively treated with amphetamines and other stimulant drugs. Also, HYPNOLEPSY, PAROXYSMAL SLEEP.
narcoleptic [när´kə lep´tik] *Medicine.* **1.** a person who suffers from narcolepsy. **2.** relating to or affected by narcolepsy.
Narcomedusae *Invertebrate Zoology.* a suborder of hydrozoans in the order Trachylina, having a lobed umbrella margin.
narcose *Neurology.* in a state of stupor.
narcosis *Neurology.* a state of reversible depression of the function of the central nervous system, marked by profound stupor, unconsciousness, or insensibility; typically produced by narcotic drugs.
narcosis therapy *Medicine.* the inducement of prolonged sleep, usually about 20 hours per day for a period of two weeks, as a treatment for anxiety, exhaustion, and extreme agitation. Also, SLEEP THERAPY.
narcosynthesis *Psychology.* a psychotherapeutic treatment in which light anesthesia is used.
narcotic *Pharmacology.* **1.** any of a group of substances, including the drugs opium, heroin, morphine, and codeine, that have a strong tendency to produce depressant effects on the central nervous system, such as dulling of the senses, insensitivy to pain, and sleepiness, and that when taken in large doses can produce euphoria, stupor, coma, or even death. Narcotics are prescribed clinically to relieve pain or induce sleep; they are noted for producing physical and psychological dependence and their use is regulated in the U.S. and other countries. **2.** of or relating to a substance that produces insensibility or stupor.
nares *singular,* **naris.** *Anatomy.* the openings of the nasal cavity.
nargusta *Materials.* the medium-textured, high-density wood of the *Terminalia amazonia* tree of Central and South America, used for flooring, furniture, boats, and plywood.
naringin *Organic Chemistry.* $C_{27}H_{32}O_{14}$, a crystalline bitter-tasting bioflavonoid whose dihydrate melts at 171°C; soluble in acetone, alcohol, or a warm solution of acetic acid and water; extracted from flowers and the rind of grapefruit and other fruit; used in beverages.
Narizian *Geology.* a North American geologic stage of the Upper Eocene epoch, occurring after the Ulatisian and before the Fresnian.
narrative method *Psychology.* a therapeutic procedure in which the subject is asked to describe his thoughts or experiences in his own way without specific suggestions or interruptions.
narrow-angle glaucoma *Medicine.* a condition of primary glaucoma in which an increase in intraocular pressure occurs when the flow of the aqueous humor from the eye is prevented by the contact of the iris with the trabecular drainage meshwork and peripheral cornea; can cause excavation and degeneration of the optic disk; damage to nerve fiber bundles can lead to defects in the field of vision. Also, ANGLE-CLOSURE GLAUCOMA.
narrow band *Acoustical Engineering.* a frequency band whose width is greater than 1% of its center frequency and less than one-third of an octave. *Telecommunications.* **1.** a circuit that can support digital transmission systems of no more than 2400 bits per second. **2.** a frequency span of less than 300 hertz. Thus, **narrow-band.**
narrow-band amplifier *Acoustical Engineering.* a device that increases a signal's strength within a narrow range of frequencies, less than one-third of an octave.
narrow-band frequency modulation *Telecommunications.* a frequency-modulated broadcasting system with a maximum permissible deviation of no more than 15 kilohertz, used for two-way voice communication.
narrow-bandpass filter *Electronics.* a device that admits frequencies from a bandwidth considerably smaller than the band's average frequency.
narrow-band path *Telecommunications.* a communications path with a maximum bandwidth of 20 kilohertz.
narrow-band pyrometer *Engineering.* a pyrometer having a color filter that allows only a finite band of wavelengths to pass through to a photoelectric detector. Also, SPECTRAL PYROMETER.
narrow-beam antenna *Electromagnetism.* a highly directional antenna that can produce a conelike beam subtending only a small angle.
narrow-gap spark chamber *Nucleonics.* a device used to produce a spark between two parallel plates that are 6–10 mm apart.
narrow gauge *Civil Engineering.* any railway gauge that is narrower than the standard gauge of 1.435 meters. Thus, **narrow-gauge railway.**
narrows *Geography.* a narrowing but usually navigable section of a river, strait, or sound.
narrow-sector recorder *Electronics.* a device that locates and records the source of electromagnetic disturbances in the atmosphere by focusing its rotating antenna on one small area at a time.
narrow-spectrum antibiotic *Microbiology.* an antibiotic that acts only against a narrow range of microorganisms.
narsarsukite *Mineralogy.* $Na_2(Ti,Fe^{+3})Si_4(O,F)_{11}$, a colorless to yellow or brown tetragonal mineral occurring as prismatic or tabular crystals, having a specific gravity of 2.78 and a hardness of 6 to 7 on the Mohs scale; found with aegirine and quartz in the Sweetgrass Hills, Montana.
NARTC *Bacteriology.* nalidixic acid-resistant, thermophilic strains of *Campylobacter* bacteria.
narthex *Architecture.* a large arcaded porch or vestibule at the western entrance of a church.

narwhal [när′wəl] *Vertebrate Zoology.* an arctic whale, *Monodon monoceros,* of the family Monodontidae, closely related to the white whale; characterized by a single pair of teeth, with the left one in males developing into a straight, screw-shaped tusk 6–8 feet long. Also, **narwhale, narwal.**

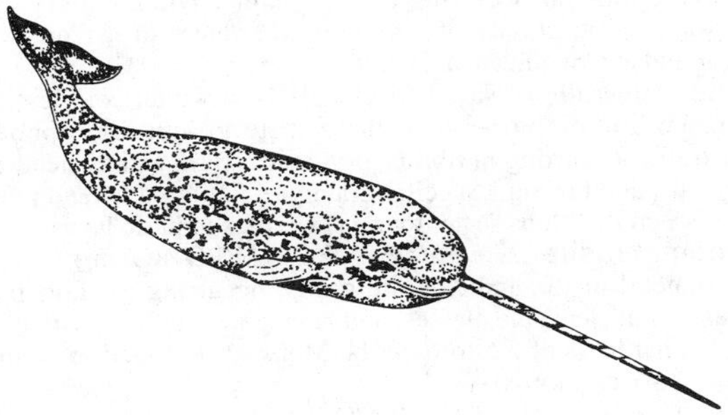

narwhal

n-ary code *Telecommunications.* a code that uses *n* distinguishable types of code elements.
n-ary composition *Mathematics.* given a set *X*, a function from X^n to *X*; i.e., a correspondence between the set of *n*-tuples with entries in *X* and the elements of *X*.
n-ary pulse-code modulation *Telecommunications.* a pulse-code modulation in which the code for each element is made up of any one of *n* distinguishable types of code elements.
n-ary tree *Mathematics.* a directed tree in which the out-degree of each vertex is at most *n*, where *n* is some fixed positive integer; equivalently, a rooted tree in which each vertex has at most *n* successors.
NAS *Aviation.* the airport code for Nassau, Bahamas.
NAS or **N.A.S.** National Academy of Sciences.
nas- a combining form meaning "nose," as in *nasal.*
NASA or **N.A.S.A.** National Aeronautics and Space Administration.
nasal *Anatomy.* of or relating to the nose.
nasal bone *Anatomy.* either of two elongated rectangular bones that meet to form the bridge of the nose.
nasal breadth *Anthropology.* in anthropometry, a measurement of the horizontal distance from one nostril wing to the other. Also, **nasal width.**
nasal canal see LACRIMAL CANAL.
nasal cavity *Anatomy.* either of two cavities, one on each side of the nasal septum.
nasal crest *Anatomy.* a ridge on the midline between the nasal cavities, created by the junction of the two maxillae and the two palatine bones.
nasal height *Anthropology.* in anthropometry, a measurement of the vertical distance from the nasion to the subnasale. Also, **nasal length.**
nasal index *Anthropology.* an anthropometric formula in which the nasal breadth is first multiplied by 100 and then divided by the nasal height.
nasal process of the frontal bone *Anatomy.* a small, uneven tip of bone that extends downward from the middle of the frontal bone to meet with the nasal bones and the frontal processes of the maxillae.
nasal process of the maxilla *Anatomy.* a bony process in the embryo formed by the splitting of the first branchial arch.
nasal septum *Anatomy.* the structure that divides the nasal cavities, formed of bone and cartilage and covered with a mucous membrane.
nasal sound *Linguistics.* a sound made when the uvula is open and air flows through the nose, such as in the sound of [n], [ng], or [m].
nascent *Chemistry.* of an element, in its nascent state. Thus, **nascence, nascency.** *Biology.* just born or just coming into existence.
nascent state *Chemistry.* the usually abnormally active state of an element in atomic form immediately following its release from a compound, as when atomic oxygen is released from hydrogen peroxide. Also, **nascent condition.**
n'aschi *Meteorology.* a northeasterly wind occurring in winter on the Iranian coast of the Persian Gulf as a result of the outflow of cold air from central Asia. Also, **nashi.**
nasion *Anthropology.* an anthropometric landmark located at the junction of the intranasal and nasofrontal sutures, corresponding roughly to the depression at the root of the nose just below the level of the eyebrows.
nasion-basion line *Anthropology.* in anthropometry, the junction of the nasion and the anterior margin of the foramen magnum in the median plane (basion).
nasion-menton distance *Anthropology.* in anthropometry, the measurement between the nasion and the tip of the chin (menton).
nasitis *Medicine.* an inflammation of the nose.
Nasmyth, James 1808–1890, Scottish engineer; invented the steam hammer, steam pile driver, and planing machine.
Nasmyth's membrane *Anatomy.* a two-layered cuticle that covers newly erupted teeth and is abraded by mastication.
naso- a combining form meaning "nose," as in *nasopharynx.*
nasofrontal *Anatomy.* relating to the nasal and frontal bones.
nasogastric *Anatomy.* of or relating to the nose and the stomach.
nasogastric tube *Surgery.* a tube that is inserted through the nose into the stomach; used to aspirate the stomach contents.
nasolacrimal *Anatomy.* of or relating to the nose and lacrimal apparatus.
nasolacrimal canal *Anatomy.* the bony passage containing the nasolacrimal duct.
nasolacrimal duct *Anatomy.* the duct that transports liquid from the lacrimal sac of the eye to the nasal cavity.
nasolacrimal groove *Developmental Biology.* an ingrowth parallel and medial to the nasomaxillary groove of the embryo that becomes the site of the nasolacrimal duct.
nasonite *Mineralogy.* $Pb_6Ca_4Si_6O_{21}Cl_2$, a white, hexagonal mineral occurring as granular masses, having a specific gravity of 5.4 to 5.55 and a hardness of 4 on the Mohs scale; found at Franklin, New Jersey.
nasopalatine duct *Developmental Biology.* a duct located between the oral and nasal openings of an embryo.
nasopharyngeal *Anatomy.* of or relating to the nasopharynx.
nasopharyngitis *Medicine.* an inflammation of the nasopharynx.
nasopharynx *Anatomy.* the part of the pharynx located above the level of the soft palate.
nasoscope see RHINOSCOPE.
nasotracheal *Anatomy.* of or relating to the nose and trachea.
nasotracheal intubation *Surgery.* the insertion of a tube through the nose into the trachea to serve as an airway.
Nassellaria *Paleontology.* an order of radiolarians having solid opaline skeletons, often in a striking circular, latticework pattern; very important stratigraphically where calcareous microfossils have disintegrated; extant from the Cretaceous to the present.
nasselline *Paleontology.* relating or belong to the order Nassellaria.
Nassi-Shneiderman diagram *Computer Programming.* in structural program design, a graphic tool used to specify program logic.
Nassula *Invertebrate Zoology.* a genus of brightly colored aquatic ciliate protozoans.

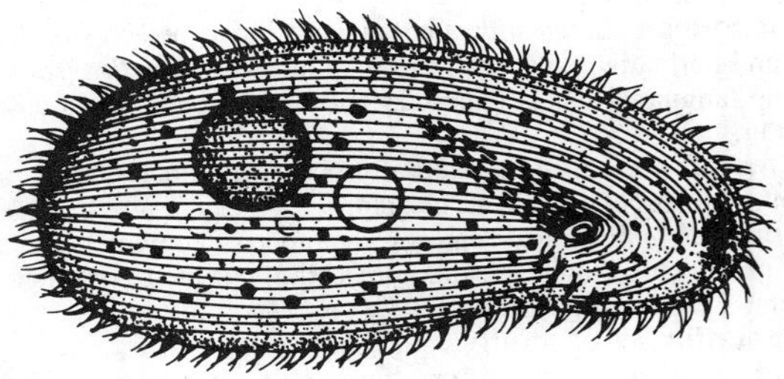

Nassula

nastic *Botany.* relating to a response of a plant part (e.g., growth or a loss of turgidity) to external stimuli that is independent of the direction of origin of such stimuli. Thus, **nastic movement, nastic growth.**
nasturtium [nə stur′shəm] *Botany.* the popular name for a group of perennial plants of the genus *Tropaeolum* of the family Tropaeolaceae, widely grown as a garden plant in North America for its showy orange, red, or yellow flowers.
nasute or **nasuti** *Entomology.* in certain termite species, an individual with an elongated head and noselike process from which a fluid that is used to defend the colony or as a cement in nest construction exudes. Also, SOLDIER TERMITE.
Nasutitermitinae *Invertebrate Zoology.* a subfamily of termites in the family Termitidae, having a noselike projection on the head.
nat. natural, naturalist; native; national.

natal [nāt´əl] *Physiology.* relating to or occurring at birth.

Natalidae *Vertebrate Zoology.* a monogeneric family of funnel-eared bats of the suborder Microchiroptera, characterized by large funnel-shaped ear pinnae, a glandular structure called the **natalid organ** on the adult male's forehead, and long, soft fur; found from Northern Mexico to Brazil.

natant *Biology.* swimming or floating, as the leaf of an aquatic plant.

Natantia *Invertebrate Zoology.* the shrimps and prawns, a suborder of free-swimming decapod crustaceans having long antennae, a laterally compressed body, and slender legs.

Nathans, Daniel born 1928, American biologist; awarded the Nobel Prize for research on molecular genetics.

Nathansohn's theory *Oceanography.* the theory that nutrient salts move cyclically in the ocean, being absorbed by plants in the euphotic surface layer, accumulating on the sea floor as dead plants and animal bodies sink to the bottom, and returning to the surface layer through diffusion and vertical circulation of the water; current research indicates that this theory may apply to shallow-water deposition, but that most organic materials are recycled continuously in the upper layers and that only small amounts of organic materials reach the ocean floor below the upper bathyal depths (1000–2000 meters).

Naticacea *Invertebrate Zoology.* a superfamily of marine gastropod mollusks in the order Prosibranchia.

Naticidae *Invertebrate Zoology.* a family of carnivorous marine gastropod mollusks, including moon shell snails and natica, in the order Pectinibranchia, with globose shells and a long retractile proboscis used for drilling other shells; eggs are deposited in sandy ribbons called sand collars.

national security *Military Science.* a condition of relative national safety that may be provided by a military or defense advantage over another nation, a favorable foreign relations position, or a defense position capable of resisting hostile action.

native *Anthropology.* relating to or describing a culture that is indigenous to a particular locale, especially a preliterate culture. *Biology.* an original or indigenous organism. *Geology.* an element when it is found uncombined with other elements in nature, such as copper, gold, or sulfur.

native asphalt *Geology.* naturally occurring exudations or seepages of asphalt in a liquid or semiliquid state. Also, NATURAL ASPHALT.

native coal see COKEITE.

native conformation *Biochemistry.* the natural three-dimensional shape of the polypeptide chain of a native protein, noted for its stability and limited variety.

native DNA *Molecular Biology.* any double-stranded nondenatured DNA extracted from a known biological source, with no additional probe or marker material.

native element *Geochemistry.* any natural element found uncombined in a nongaseous state.

native language *Linguistics.* the language that a person first learns to speak and normally converses in as a child. *Computer Programming.* a machine language that is characteristic of and peculiar to a specific computer.

native metal *Metallurgy.* a naturally occurring, uncombined metallic element, such as copper, silver, or gold.

native mode *Computer Technology.* the mode of computer operation in which the machine language program being executed is written in the computer's native language.

native paraffin see OZOCERITE.

nativism *Psychology.* the theory that man's most basic or important ideas are inborn rather than the product of experience. *Behavior.* the theory that certain basic behavior patterns are inborn or are unique to a particular species, rather than acquired through experience.

nativist **1.** a person who follows or supports nativism. **2.** of or relating to nativism.

natremia *Medicine.* the presence of excessive amounts of sodium in the blood.

natric horizon *Geology.* a soil horizon with the same properties as an argillic horizon, but exhibiting a blocky, columnar, or prismatic structure.

natrium *Chemistry.* the Latin name for sodium, from which the chemical symbol Na is derived.

natriuresis *Physiology.* the excretion of an elevated amount of sodium in the urine.

natriuretic *Pharmacology.* **1.** of, relating to, or causing natriuresis. **2.** any agent that causes natriuresis.

natroalunite *Mineralogy.* $NaAl_3(SO_4)_2(OH)_6$, a white or grayish to reddish, vitreous, trigonal mineral of the alunite group, occurring in massive form or as rhombohedral crystals, and having a specific gravity of 2.6 to 2.9 and a hardness of 3.5 to 4 on the Mohs scale; formed by solfataric processes on volcanic rock.

natrochalcite *Mineralogy.* $NaCu_2^{+2}(SO_4)_2(OH)\cdot H_2O$, an emerald-green monoclinic mineral occurring as transparent pyramidal crystals, and having a specific gravity of 3.49 and a hardness of 4.5 on the Mohs scale; found at Chuquicamata, Chile.

natrolite *Mineralogy.* $Na_2Al_2Si_3O_{10}\cdot 2H_2O$, a white, colorless, or reddish orthorhombic mineral of the zeolite group, dimorphous with tetranatrolite, occurring in fibrous or massive form and as slender acicular crystals, and having a specific gravity of 2.20 to 2.26 and a hardness of 5 to 5.5 on the Mohs scale; found in cavities in basic lavas.

natromontebrasite *Mineralogy.* $(Na,Li)Al(PO_4)(OH,F)$, a white triclinic mineral of the amblygonite group, occurring as short prismatic crystals and in cleavable masses, and having a specific gravity of 3.04 to 3.1 and a hardness of 5.5 to 6 on the Mohs scale; found in granite pegmatites. Also, FREMONTITE.

natron *Mineralogy.* $Na_2CO_3\cdot 10H_2O$, a colorless to white, yellow, or gray, water-soluble, monoclinic mineral occurring as efflorescent crusts, and having a specific gravity of 1.44 and a hardness of 1 to 1.5 on the Mohs scale; found most commonly in solution in soda lakes.

natron lake see SODA LAKE.

Natronobacterium *Bacteriology.* a genus of archaebacteria that are halophilic and alkalophilic.

Natronococcus *Bacteriology.* a genus of archaebacteria that thrive in salt-rich environments.

natrophilite *Mineralogy.* $NaMn^{+2}PO_4$, a rare, deep-yellow, orthorhombic mineral occurring in granular masses or as rare prismatic crystals, and having a specific gravity of 3.41 and a hardness of 4.5 to 5 on the Mohs scale; found in granite pegmatite.

Natta, Giulio 1903–1979, Italian chemist; awarded the Nobel Prize for work on stereospecific polymerization.

natural abundance *Nucleonics.* the ratio of abundance of each naturally occurring isotope of an element.

natural adhesive *Materials Science.* a general term for adhesive substances obtained from nature, such as dextrin, gum arabic, gum acacia, gelatin, and casein.

natural aging *Materials Science.* the process of aging a precipitation-hardenable alloy at room temperature after solution annealing and quenching.

natural antenna frequency *Electromagnetism.* the lowest resonant frequency that a given antenna can carry without being capacitively or inductively loaded.

natural antibody *Immunology.* an antibody that is produced in the serum of an individual without known or deliberate antigenic stimulus.

natural arch *Geology.* **1.** a natural bridge created by erosion. **2.** a landform similar to a natural bridge but that does not span a ravine or valley; formed by other than erosive agencies. **3.** see SEA ARCH.

natural arch

natural asphalt see NATIVE ASPHALT.

natural binary coded decimal system *Computer Programming.* a common four-bit coding system that uses the first 10 binary numbers to represent the decimal digits 0 through 9.

natural bitumen *Geology.* a naturally occurring mineral pitch, tar, or asphalt.

natural boundary *Mathematics.* if a function f has a property such as continuity or differentiability on an open set Ω, the natural boundary of Ω is the set of boundary points of Ω at which f fails to exhibit the given property.

natural bridge *Geology.* **1.** an archlike rock formation that spans a ravine or valley, or an opening where a stream abandoned a meander and broke through the meander neck. **2.** in an underground cave, a remnant of a tunnel roof that has collapsed. **3.** see SEA ARCH.

natural childbirth *Medicine.* a general term for labor and delivery that are accomplished with minimal or no medical intervention in the form of drugs, anesthesia, surgery, and so on.

natural circulation reactor *Nucleonics.* a nuclear reactor that is cooled by the free circulation of a coolant, typically water; the circulation is produced by differences in density due to temperature gradients within the coolant.

natural classification *Systematics.* a classification based on some conception of underlying natural order.

natural coke *Geology.* a coal that has been carbonized by natural combustion or by contact with an igneous intrusion. Also, BLACK COAL, CARBONITE, CINDER COAL.

natural control see BIOLOGICAL CONTROL.

natural convection *Thermodynamics.* heat transfer by fluid motion resulting from density gradients in the fluid. Also, FREE CONVECTION.

natural coordinates *Fluid Mechanics.* a system of curvilinear coordinates that are mutually perpendicular; used in dynamic fluid problems, particularly in Lagrangian hydrodynamics, in which one axis is tangent to the instantaneous velocity vector, another axis lies to the left in the horizontal plane normal to the tangent axis, and the third axis lies in the z-direction.

natural deduction *Artificial Intelligence.* a nonresolution theorem-proving system that proceeds in a goal-directed manner, breaking goals into subgoals and maintaining a distinction between conclusions and their antecedent hypotheses.

natural detail *Cartography.* items on a map that represent features of the earth's surface formed exclusively by the forces of nature, such as rivers or mountains, as opposed to artificial features, such as highways or buildings.

natural draft *Fluid Mechanics.* the flow of gases through a vertical duct, such as a chimney, without the aid of mechanical devices but instead depending on thermal gradients in the gases.

natural-draft cooler see ATMOSPHERIC COOLER.

natural-draft cooling tower *Mechanical Engineering.* a cooling tower in which the natural convection of upwardly moving air comes in contact with the water vapor to be cooled.

natural draw ratio *Materials Science.* in polymer fibers, the ratio of the cross-sectional areas on either side of the neck, formed when the fiber is cold drawn by stretching.

natural equation of a curve see INTRINSIC EQUATIONS OF A CURVE.

natural error *Cartography.* in surveying, any error that occurs as a result of variation in an environmental force such as wind, temperature, humidity, gravity, magnetic declination, or refraction.

natural experiment *Statistics.* an experiment whose design was not controlled by the experimenter, but rather happened as the result of some natural process.

natural foods *Nutrition.* a general term referring to various foods that are thought of as being produced as they would be in nature, without human intervention; as by not being processed, not containing artificial ingredients, not being grown with chemical fertilizers, and so on.

natural frequency *Electronics.* **1.** the lowest resonant frequency produced by a component without assistance from inductance or capacitance. **2.** the frequency at which a component responds to an abrupt change in signal or voltage. **3.** the frequency at which a device produces it highest rate of energy. **4.** the frequency that causes a device to vibrate at its highest amplitude.

natural fuel reactor see NATURAL URANIUM REACTOR.

natural function generator see ANALYTICAL FUNCTION GENERATOR.

natural gas *Materials.* a colorless, highly flammable substance that is a mixture of gaseous hydrocarbons, occurring naturally beneath the surface of the earth, often with or near petroleum deposits. It typically consists of 80–85% methane and 10% ethane, with the balance being propane, butane, and nitrogen. It is widely used as a fuel in the U.S. and throughout the industrialized world.

natural gasoline *Materials.* gasoline that is recovered from natural gas.

natural glass *Geology.* a naturally occurring amorphous, vitreous, inorganic substance that has solidified from magma too quickly to crystallize.

natural harbor *Geography.* a coastal inlet where the adjacent land configuration provides protection from weather and the sea.

natural history *Science.* **1.** a collective term used, especially formerly, for those sciences that involve the study of the world of nature, such as botany, zoology, and geology. **2.** the wildlife and natural features of a given region.

natural immunity *Immunology.* a nonspecific resistance to infection or disease that results from the genetic makeup of the host.

natural interference *Telecommunications.* electromagnetic interference that originates from terrestrial phenomena or from natural disturbances outside the earth's atmosphere.

naturalist *Science.* **1.** a person who is an expert on the world of nature, such as a botanist or zoologist. **2.** a person who studies or observes the world of nature as a hobby or pastime.

naturalized *Ecology.* describing an alien species that has become successfully established.

natural killer cells *Immunology.* lymphocytes present in nonimmunized normal individuals that are cytotoxic for a variety of cells.

natural language *Computer Programming.* **1.** a human language that has developed naturally over an extended period of time, such as English or Chinese, as contrasted with formal computer language. **2.** a computer language whose form is a limited subset of a human language.

natural-language generation *Artificial Intelligence.* a production of natural-language statements by a computer.

natural-language interaction *Computer Programming.* the use of a natural language user interface with a computer.

natural-language processing *Artificial Intelligence.* processing of speech or machine-readable text in a natural language; includes machine translation, natural-language understanding, content extraction, summarization, and natural-language generation.

natural-language text processing *Computer Programming.* the computerized recognition and interpretation of natural-language text; often used in library retrieval systems.

natural-language translation *Artificial Intelligence.* a conversion of ordinary natural-language statements into a set of computer instructions or into another natural language, e.g., translation from German to English by computer.

natural-language understanding *Artificial Intelligence.* an understanding of statements in a natural language such as English by producing a computer representation of their meaning.

natural law function generator see ANALYTICAL FUNCTION GENERATOR.

natural linewidth *Spectroscopy.* the width of a spectral line resulting from the finite lifetime of the excited state.

natural load *Hydrology.* a term for the amount of sediment transported by a stable stream.

natural logarithm *Mathematics.* the inverse of the complex exponential function $w = e^z$; denoted ln z or log z. $\ln x = \int_1^x (1/t)dt$ for positive real numbers. Also, NAPIERIAN LOGARITHM.

natural mapping *Mathematics.* let $\sim$ be an equivalence relation on a set X. A mapping ϕ that maps each element x of X to its equivalence class as induced by $\sim$ is called a natural or canonical mapping. A natural mapping may be a homomorphism, isomorphism, and so on, depending on the context.

natural number *Mathematics.* **1.** any nonnegative integer; an element of the set $\{1, 2, 3, 4, \ldots\}$. Sometimes 0 is included. **2.** in a set theoretic context, a natural number is an element of the minimal successor set ω.

natural period *Physics.* for a vibrational system, the period of free oscillation of the system.

natural pressure cycle *Metallurgy.* in metal forming, a schedule according to which pressure is built up proportionally to the buildup of stresses.

natural radio-frequency interference *Geophysics.* a disturbance in radio communications by radio signals that come from natural sources on earth or outside the atmosphere.

natural remanent magnetization *Geophysics.* all the remanent magnetization that a rock possesses in its original resting place.

natural resource *Ecology.* any material that exists in nature independently of human industry and that is somehow utilized by humans, such as water, petroleum, minerals, forests, fish, and wild game.

natural rubber *Materials.* the raw gum material from the rubber tree.

natural science *Science.* a science that is concerned with objects or processes that occur in nature and that can be objectively observed and measured, such as physics, chemistry, biology, or geology.

natural-season see AIR-SEASON.

natural selection *Evolution.* a theory of evolutionary change asserting that (a) every organism naturally displays slight variations from others of its kind; (b) all organisms produce more offspring than can possibly survive; (c) a struggle for limited natural resources consequently ensues, as a result of which the individuals that survive and reproduce will be those whose natural variations adapt them better to their specific environment; and (d) the favorable variations will accumulate in subsequent generations, resulting in divergence and eventual speciation. *Genetics.* this process as manifested at the genetic level, according to which some genotypes under the influence of a defined set of environmental parameters are more successful in leaving progeny than others.

natural steel *Metallurgy.* any steel made directly from an ore or cast iron.

natural taxon *Systematics.* a taxon recognized in a natural classification.

natural transformation *Mathematics.* let C and D be categories with covariant (or contravariant) functors S and T: $C \to D$. A natural transformation α: $S \to T$ (or β: $S \to T$) maps the functors S to the functors T as follows: to each object C of C is assigned a morphism α_C: $S(C) \to T(C)$ of D (or a morphism β_C: $S(C) \to T(C)$ of D) in such a way that for every morphism f: $C \to C'$ of C, the following diagrams commute:

(covariant case)

$$\begin{array}{ccc} & \alpha_C & \\ S(C) & \rightarrow & S(T) \\ \downarrow S(f) & & \downarrow T(f) \\ S(C') & \rightarrow & S(T') \\ & \alpha_{C'} & \end{array}$$

(contravariant case)

$$\begin{array}{ccc} & \beta_C & \\ S(C) & \rightarrow & S(T) \\ \uparrow S(f) & & \uparrow T(f) \\ S(C') & \rightarrow & S(T') \\ & \beta_{C'} & \end{array}$$

If α_C (or β_C) is an equivalence for every C in C, then α (or β) is a **natural isomorphism** (or **natural equivalence**) of the covariant (or contravariant) functors S and T. Natural transformations frequently appear in disguised form in specific categories; e.g., a statement may be made that a certain ring homomorphism is natural, with no mention of functors. This can usually be interpreted to mean that there are two functors and a natural transformation between them, though many useful results can be proved without the use of functors.

natural transmission *Telecommunications.* a signal sent when there is no traffic flow.

natural tunnel *Geology.* a nearly horizontal cave that is open at both ends. Also, TUNNEL CAVE.

natural uranium reactor *Nucleonics.* a reactor that uses unenriched natural uranium containing both uranium-238 and uranium-239 isotopes. Also, NATURAL FUEL REACTOR.

natural ventilation *Mining Engineering.* the weak and inconsistent ventilation produced naturally within a mine by the difference in air density between upcast and downcast shafts.

natural ventilation pressure *Mining Engineering.* the pressure difference across the shaft-bottom doors that is caused by the difference in air density between the upcast and downcast shafts.

natural wavelength *Telecommunications.* a wavelength that corresponds to the natural frequency of a tuned circuit. *Electromagnetism.* the wavelength that corresponds to a given natural antenna frequency.

natural well *Geology.* a sinkhole or other natural opening that extends below the water table and from which groundwater can be withdrawn.

natural width of energy level *Physics.* an energy spread given by the difference between the energies associated with emission, absorption, or the scattering cross section of an excited atom or molecule when measured at one-half of the maximum value of the intensity distribution.

nature *Science.* **1.** a general term for all aspects of the physical world other than humans, such as animal and plant life, features of the earth, and so on. **2.** also, **Nature.** an abstract entity regarded as regulating or epitomizing the general activities of plants and animals, especially higher animals; used in expressions such as *Nature* is cruel or the balance of *nature*.

nature-nurture issue or **controversy** *Behavior.* a dispute over the relative importance of heredity (nature) versus environment (nurture) in the development of behavior.

nature-nurture ratio *Psychology.* a statistic that indicates the relative contributions of heredity and environment to personality traits.

Naucoridae *Invertebrate Zoology.* a widely distributed family of aquatic predaceous hemipteran insects, the water creepers, having a broad, flat, oval body and greatly enlarged front femora.

naujaite *Petrology.* a coarse, hypidiomorphic-granular variety of nepheline syenite with a poikilitic texture, rich in sodalite, and containing microcline plus lesser amounts of albite, analcime, acmite, and sodium amphibole.

Naumanniella *Bacteriology.* a genus of bacteria of the family Pseudomonadaceae, composed of rod-shaped cells and found in iron-containing water or soil.

naumannite *Mineralogy.* Ag_2Se, an iron-black, metallic, pseudocubic, orthorhombic mineral occurring as cube-shaped crystals and granular masses, and having a specific gravity of 7.4 to 7.79 and a hardness of 2.5 on the Mohs scale; found in hydrothermal veins with other selenides, sulfides, and gold.

naupliar eye *Invertebrate Zoology.* the single median eye characteristic of the nauplius larva, composed of three or four small ocelli containing a few photoreceptors.

nauplius *Invertebrate Zoology.* a free-swimming larval stage of most marine crustaceans, with a single eye, two pairs of antennae, and one pair of mandibles.

nausea *Medicine.* an unpleasant sensation in the stomach or abdomen, with an accompanying urge to vomit; not a specific condition but an effect of motion sickness, food poisoning, emotional stress, and so on.

nautical **1.** of or relating to ships, especialy ocean-going ships. **2.** of or relating to navigation.

nautical almanac *Navigation.* a book of tables containing data needed for celestial navigation, primarily for marine use.

nautical astronomy see NAVIGATIONAL ASTRONOMY.

nautical chain *Metrology.* a unit of length equal to 15 feet.

nautical chart *Navigation.* a representation of a portion of the earth's surface intended for maritime use.

nautical mile *Metrology.* **1.** a traditional measure of distance used in navigation, equal to one minute of a great circle of the earth; established as 6080 feet. Also, GEOGRAPHICAL MILE, SEA MILE. **2.** a standardized measure of this distance, the **international nautical mile**, equal to 6076 feet; equivalent to 1852 meters.

nautical twilight *Astronomy.* the time between sunset or sunrise and the moment when the center of the sun is 12° below the horizon.

Nautilidae *Invertebrate Zoology.* a family of cephalopods, mollusks in the subclass Nautiloidea; it contains only the genus *Nautilus.*

Nautiloidea *Invertebrate Zoology.* a subclass of cephalopod mollusks with only one extant genus, *Nautilus*, which has a many-chambered external shell and two pairs of gills.

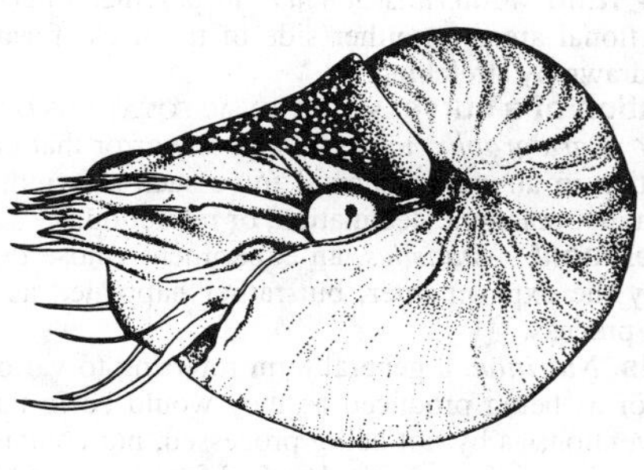

Nautilus

Nautilus *Invertebrate Zoology.* **1.** a genus of cephalopod mollusks of the South Pacific and Indian Oceans with a spiral chambered shell. Also, **chambered nautilus, pearly nautilus. 2.** a genus of cephalopod mollusks in the order Octopoda, similar to octopuses but distinctive in that the female lives in a thin, white, spiral shell, coming and going freely. Also, **paper nautilus.** (From a Greek word meaning "sailor.")

Nautococcaceae *Botany.* a family of green algae of the order Tetrasporales, characterized by solitary cells or irregular colonies with no gelatinous matrix and contractile vacuoles in the vegetative cells; species are found floating on water, among aquatic plants, and in soil.

nav. naval; navigable; navigation.

Navaglobe *Navigation.* an early (1945) hyperbolic radio aid operating in the 90 to 110 kilohertz band; it used three vertical antennas that broadcast in alternating pairs to produce three overlapping figure-eight patterns. Comparing the amplitudes of the signals yielded the bearing of the transmitting station. Cross bearings were needed to fix a position.

Navaho [näv´ə hō´] *Ordnance.* a rocket-boosted, ramjet-powered, supersonic, strategic surface-to-surface cruise missile with a cruising altitude of 100,000 feet and an operating range of approximately 4000 miles; officially designated **SM-64.**

Navaho or **Navajo sandstone** *Geology.* a fossil dune formation of the Jurassic period that is found in the Colorado Plateau of the United States.

naval **1.** of or relating to warships or ships in general. **2.** of or relating to a navy.

naval architect *Naval Architecture.* a designer of ships and other vessels, and their equipment and fittings. The term is general, and has no specifically military connotation.

naval architecture *Engineering.* a branch of engineering concerned with the design and production of marine vessels and structures, such as ships, barges, and submarines.

Naval Architecture

Naval Architecture is the branch of engineering that governs the design and construction of ships and other marine vehicles. For centuries ship design and construction was very much an intuitive process. The experienced eye and good judgment of the mast shipwright contributed much more to a successful design than the conscious application of specific engineering principles. In simpler times, when the world moved at a slower pace, there was no need to encumber the process with such things as hydrodynamics, structural analysis, fluid mechanics, and materials science; aesthetics and form were the yardsticks by which good ship design was measured.

In fact, an early treatise on the subject, Chapman's *Architectura Navalis Mercatoria,* consists almost exclusively of pictorial representations of various ships of the day and is virtually devoid of technical content. The term, "Naval Architecture," which may well be linked to the title of Chapman's work, has been in existence for centuries but only in the past one hundred and fifty or so years has it matured to its present level of scientific sophistication.

The Industrial Revolution provided the impetus for this development, presaging as it did a period of great technological progress. During this period, the use of first iron and then steel allowed for the building of bigger and bigger ships, and the introduction of steam propulsion led to the concept of reliable, regularly scheduled ocean service. Generally higher levels of military and commercial activity combined with a rapid increase in the use of fossil fuels and other natural resources were able to precipitate the development of many different types of vessels. Of necessity, and in response to the demands of a technologically more complex world, Naval Architecture shed its quasi-artistic image and became a bona fide scientific discipline that utilizes a variety of powerful analytical tools to ensure the strength, performance, reliability, and safety of today's ships and marine vehicles.

Donald V. Walter
Chief Naval Architect
National Steel and Shipbuilding Company

naval gunfire liaison team *Military Science.* personnel and equipment required to coordinate and advise ground or landing forces on naval gunfire employment.

naval gunfire operations center *Military Science.* the agency established in a ship to control the execution of plans for the employment of naval gunfire, process requests for naval gunfire support, and assign ships to forward observers.

naval gunfire spotting team *Military Science.* the unit of a shore fire control party that designates targets, controls commencement, cessation, rate, and types of fire, and spots fire on the targets.

naval meteorology *Meteorology.* a subscience of meteorology that studies atmospheric phenomena over the oceans, the influence of the ocean surface on those phenomena, and the effect of such phenomena on both shallow and deep sea water.

naval ship *Naval Architecture.* a ship specifically designed for and used in naval service; includes auxiliaries as well as combatants.

naval stores *Ordnance.* any articles or commodities used by a naval ship or station, including equipment, supplies, fuel, and ammunition. *Materials.* a term for various resinous substances obtained from conifers (especially pines), such as turpentine, rosin, and pitch, that were originally used for maintaining wooden ships.

navar *Navigation.* a navigational radar.

NAVAREAS *Navigation.* a system in which the oceans of the world are divided into navigation areas, in which appropriate countries take responsibility for broadcasting warnings of navigational dangers.

Navarroan *Geology.* a North American Gulf Coast stage of the Upper Cretaceous period, occurring after the Tayloran and before the Midwayan of the Paleocene epoch.

Navascreen *Navigation.* an air traffic control system that uses a screen to display information from radar and other sources.

nave *Architecture.* the main interior part of a church, especially the long central aisle of a basilica.

NAVEAM *Navigation.* a warning of dangers to navigation in the waters of the eastern Atlantic Ocean and the Mediterranean Sea.

navel *Anatomy.* a popular name for the umbilicus, the depression in the center of the abdomen indicating the point at which the umbilical cord was attached.

navel-ill *Veterinary Medicine.* a condition characterized by infected unhealed navels in chicks, poults, and other young fowl, caused by apportunitic bacteria, coliforms, staphylococci, and others. Also, OMPHALITIS.

Navicula *Botany.* a large genus of freshwater and marine pennate diatoms in the family Naviculaceae.

Naviculaceae *Botany.* a diverse family of freshwater and marine diatoms of the order Pennales, characterized by symmetrical valves with a raphe and central nodule but no fibulae; most species are solitary and benthic or live unattached among larger plants.

navicular *Anatomy.* **1.** having a shape resembling a boat. **2.** the central bone of the ankle, which articulates with the talus and the three cuneiform bones.

navicular cells *Pathology.* boat-shaped protoplasmic masses that contain glycogen and are conspicuous in the exfoliate of the cervix uteri.

Navier, Claude 1785–1836, French physicist; pioneer in mathematical analysis of engineering; originated the Navier-Stokes equation.

Navier's equation *Mechanics.* an equation of equilibrium or motion for a linear elastic solid subjected to body force, expressed as a vector partial differential equation for the displacement vector.

Navier-Stokes equation *Fluid Mechanics.* the nonlinear partial differential equation of motion applicable to an incompressible, viscous fluid and fundamental to all aspects of fluid dynamics.

navigable *Navigation.* **1.** of a body of water, deep and wide enough to be navigated by ships, especially passenger or cargo ships. **2.** of a craft, able to be steered or directed.

navigable airspace *Navigation.* the airspace that is at or above the minimum safe flight altitude, including safe altitudes for takeoffs and landing approaches.

navigable semicircle *Navigation.* a reference to the half of a cyclonic storm area in which the rotary and progressive motions of the storm tend to counteract each other; the winds are in such a direction as to blow a vessel away from a storm track; it is on the left side of the storm track in the Northern Hemisphere, and on the right side in the Southern Hemisphere.

navigate *Navigation.* to direct the movement of a ship, airplane, or other craft from one place to another.

navigating officer *Navigation.* the officer aboard a vessel with the principal immediate responsibility for navigation.

Navigation

Navigation is the science of guiding man's movements from place to place. The advancement of civilization has been both the cause and the result of this movement. In the 16th century navigation became the *sine qua non* of exploration and progress.

"Navigation" is sometimes also used to embrace the ships, canals, and ports that facilitate movement. Here we will deal only with the methods used in the guidance of these.

Man's innovative capabilities early developed an ever-growing list of tools for navigating: a global coordinate system for specifying location (latitude, longitude, charts); a measuring system, borrowed from astronomy, for specifying time; a system for defining direction (azimuth); and an earth-related measuring system for distance (the nautical mile).

Concurrently the instruments to utilize these coordinate systems developed; the cross staff, the magnetic compass, and the knotted line log were early ones and their skillful use was the art of the navigator. Accuracies, initially only crude, improved with the passage of time and with technological advances that resulted from the demands of navigation.

An interesting example of this demand was the need for determining accurate longitude and the consequent development, in the 18th century, of the marine chronometer. This timepiece is simply a model of the rotation of the earth, and is essential to using the longitude coordinates, which revolve with the earth.

In modern times electronics has invaded every aspect of the field. Early radio provided a means of getting accurate time almost anywhere in the world. Later, quartz watches practically replaced the need for radio time. Electronic time-of-arrival pulse systems greatly reduced the need for celestial navigation, but these in turn were replaced by data supplied from man-made satellites. Development continues apace, but the exigencies of travel still demand the skill or art of the navigator.

Thomas D. Davies
Rear Admiral, United States Navy, retired
President, The Navigation Foundation

navigation *Science.* any or all of the various processes used in determining position and directing movement from one place to another, especially the movement of a craft in water or air. *Behavior.* the process by which migrating species travel over great distances and arrive at a specific site, as when birds make a round trip between certain winter and summer homes, or when salmon return to a certain breeding place. This ability is known by ethologists to involve a sensing of various environmental factors, such as solar position, but otherwise is not clearly understood. *Design Engineering.* the process of guiding one's way through a network or system. *Computer Programming.* movement through a database, record by record, using pointers as guides to locate the desired record; many database management systems provide automatic navigation.

navigational relating to or involved in navigation.

navigational aid *Navigation.* any device, such as a buoy, radio beacon, instrument, chart, or the like, that is used to assist in position-fixing or conduct of a craft.

navigational almanac *Navigation.* a book of tables containing data needed for celestial navigation.

navigational astronomy *Navigation.* the part of general astronomy that is directly applicable to position-finding, such as coordinates of certain celestial bodies, time, and the apparent motion of celestial bodies.

navigational instrument *Aviation.* an instrument or device that indicates or is used to determine a flight vehicle's position or direction of flight, such as a compass, sextant, directional gyro, driftmeter, air-position indicator, and radio or radar devices.

navigational planets *Navigation.* a term for the planets of Venus, Mars, Jupiter, and Saturn; so-called because of their historic use in navigation.

navigational plot *Navigation.* marks that are drawn by a navigator on a chart or plotting sheet in order to mark position and keep track of progress.

navigational triangle *Navigation.* a spherical triangle on the celestial or terrestrial sphere that is used to solve problems in position-finding or great-circle courses.

navigation dam *Civil Engineering.* a dam that raises a water level sufficiently to provide enough depth for vessels to navigate a channel.

navigation dome see ASTRODOME.

navigation lights *Navigation.* lights displayed on the wingtips of a plane in flight, or on each side of a vessel underway. The starboard (right) light is green; the port (left) light is red.

navigation receiver *Electronics.* a device that receives and contrasts radio signals from transmitters at known locations in order to establish a ship's position.

navigator *Navigation.* **1.** the person who directs the movement of a craft from one place to another. **2.** the system of gyroscopes and accelerometers that constitutes the mechanical portion of an inertial navigation system.

navite *Petrology.* a porphyritic olivine-diabase containing phenocrysts of olivine, augite, and basic plagioclase (labradorite) in a holocrystalline, diabasic groundmass.

NAVSPASUR *Navigation.* U.S. naval space surveillance system, including radars and communications for the detection and tracking of objects in space.

NAVSTAR *Navigation.* a system of 24 navigational satellites used for electronic position-finding.

Nazarov cyclization *Organic Chemistry.* a cyclization reaction accomplished by adding acids to an unsaturated acyclic compound.

NB mark well. (From Latin *nota bene.*)

Nb niobium.

NBC or **N,B,C** *Ordnance.* nuclear, biological, and chemical (weapons or missile warheads).

N-body problem see MANY-BODY PROBLEM.

NBS or **N.B.S.** National Bureau of Standards.

N-channel *Electronics.* a path for conduction of electrons in a P-type semiconductor.

NCI or **N.C.I.** National Cancer Institute.

n-connected graph *Mathematics.* a graph whose connectivity is greater than or equal to some fixed positive integer n; i.e., a graph that remains connected even if any $n-1$ edges are removed, but for which there exists a disconnecting set of n edges.

NCRP or **N.C.R.P.** National Committee on Radiation Protection and Measurements.

N curve *Electronics.* a line that plots voltage against current in a negative-resistance device, the slope of which resembles the letter N.

Nd neodymium.

NDA or **N.D.A.** National Dental Association.

NDGA nordihydroguaiaretic acid.

n-dimensional space *Mathematics.* **1.** any vector space whose basis has n elements, where n is a positive integer. **2.** a manifold of dimension n. The Euclidean plane is a 2-space, for example.

N display *Electronics.* the representation of targets on a radar screen in which the amplitude of the blips indicates direction, while their movement along the baseline, in conjunction with an adjustable pedestal signal, indicates distance; a type of A display. Also, N SCAN.

Ne the chemical symbol for neon.

ne- or **neo-** a prefix meaning: new; recent; young.

NEA negative electron affinity.

NEACP *Ordnance.* national emergency airborne command post; a U.S. aircraft from which national leaders can exercise command in time of emergency.

neallotype *Systematics.* a specimen of the opposite sex designated as a replacement for the original type specimen, in the absence of that original specimen.

Neandertal *Paleontology.* another spelling of Neanderthal. See NEANDERTHAL MAN.

Neanderthal [nē an´dər thôl´] *Anthropology.* **1.** of or relating to Neanderthal man. **2.** a member of this subspecies.

Neanderthal man [nē an′dər thôl′] *Anthropology.* any of a group of hominids who lived in Europe, Africa, and the Near East between about 125,000 years ago and about 35,000 years ago; now often considered a separate species, *Homo neanderthalensis*, but formally classified as a subspecies, *Homo sapiens neanderthalensis*. They were distinguished from anatomically modern humans, *Homo sapiens,* by a prominent brow ridge, a receding chin, midfacial prognathism, and a cranial capacity (1450 cc) slightly larger than the modern average. (Named for the valley in northwestern Germany where the first Neanderthal skeleton was found in 1856.)

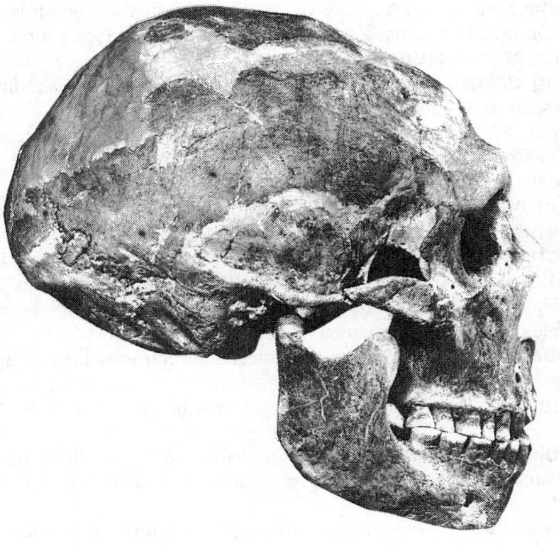

Neanderthal skull

neap *Oceanography.* relating to or designating neap tide, the tide that has the least range between high and low water.

neap range *Oceanography.* the mean semidiurnal range of the tide during a period of neap tides, generally 10–30% less than the mean tidal range; measured by taking the mean difference between neap high and low water.

neap rise *Oceanography.* the height of neap high water above the tidal datum.

neap tidal currents *Oceanography.* tidal currents that move more slowly than usual because of the decreased range of neap tides.

neap tide *Oceanography.* a twice-monthly tide of minimal range that occurs when the earth, sun, and moon are at right angles to each other, decreasing the total tidal force exerted on the earth.

nearctic *Ecology.* of or relating to the Nearctic region or its fauna.

Nearctic region *Ecology.* a zoogeographic region of the earth consisting of the area of the Americas that extends north from the Mexican Plateau to the Arctic Circle, and including the Canadian, Californian, Rocky Mountain, and Allegheny subregions.

near-end crosstalk *Telecommunications.* **1.** in disturbed channels, interference that is propagated in a direction opposite to that of the propagation of the current in the channel causing the interference. **2.** in carrier telephone repeater stations, interference that occurs when output signals of one carrier overlap onto the same end of the other repeater.

nearest-neighbor analysis *Molecular Biology.* an examination of the base sequences in a given nucleic acid by determining the frequency with which a given base occurs adjacent to a second given base.

near field *Acoustics.* **1.** a space in which the sound from a nearby source does not behave predictably because particle velocity may not be in the same direction as wave motion. **2.** a region in the sound field that is closer than a wavelength away from the source.

near-field diffraction see FRESNEL DIFFRACTION.

near-field scanning optical microscope *Optics.* a microscope that measures the intensity of light passing through an optical fiber as it systematically moves over a specimen, thus allowing a magnified image of the surface structure of the specimen to be produced on a cathode-ray screen.

near-infrared radiation *Electromagnetism.* electromagnetic radiation with wavelengths in the range of about 0.75 to 2.5 micrometers.

near-infrared reflectance spectroscopy *Spectroscopy.* spectroscopic techniques involving infrared radiation from the near infrared band (approximately 0.75 to 3.0 micrometers) wherein reflection from a surface is acceptable, as opposed to the necessary use of a diffraction grating for wavelengths of the normal infrared region.

near-infrared spectrophotometry *Analytical Chemistry.* the analysis of a chemical substance by measuring its absorption or attenuation of electromagnetic radiation at monochromatic wavelengths from 0.78 to 3.0 micrometers.

near letter quality *Computer Technology.* an output of a dot-matrix or other printer that is enhanced by multiple printing to approach the quality of a laser or daisy-wheel printer.

near miss *Ordnance.* the impact or burst of an explosive missile that misses the target but is near enough to cause effective damage; usually applied to aerial bombs. *Navigation.* an instance in which two aircraft or other craft narrowly avoid colliding with each other. *Artificial Intelligence.* in human-guided training of a machine learning system, a carefully chosen example that is only slightly different from a previous example, making it easy for the program to determine the difference between the new example and its existing model and to determine the significance of this difference.

near net shape *Metallurgy.* the attribute of a shaped part, whether cast, forged, or sintered from powder, that requires little machining in its finishing operation.

near point *Physiology.* the nearest point at which the eye can clearly see an object.

nearshore *Oceanography.* an indefinite area between the shoreline and the offshore zone beyond the breakers.

nearshore circulation *Oceanography.* circulation of water caused by nearshore currents, coastal currents, and the upwelling associated with some currents.

nearshore current system *Oceanography.* collectively, the currents operating shoreward of the coastal current system.

nearside *Astronomy.* the hemisphere of the moon that faces earth.

nearsightedness *Medicine.* the popular name for myopia, a condition in which vision is clear only at a short distance. Thus, **near-sighted.**

near stars *Astronomy.* the stars nearest the sun within some specified limit.

nearthrosis *Medicine.* a new joint, either a false joint that arises from a fracture, or a new joint that is the result of a total joint replacement operation.

near-ultraviolet radiation *Electromagnetism.* ultraviolet radiation having wavelengths in the range of about 300 to 400 nanometers.

near wilt *Plant Pathology.* a disease of peas caused by the fungus *Fusarium oxysporum pisi* and characterized by wilted plants and red discoloration; it affects scattered plants and develops more slowly than true wilt.

neat cement *Materials.* cement that is mixed with water only, without the addition of sand.

neatline *Cartography.* any of the lines on a map, frequently concurrent with parallels and meridians, that form the border of the area to be depicted. Also, SHEET LINE.

neat plaster *Materials.* plaster containing no admixture other than hair or fiber.

neat's-foot oil *Materials.* a pale yellow oil derived by boiling the feet and shinbones of cattle; used primarily in treating and preserving leather. (From an older use of the word *neat* to mean a cow or ox.)

Nebaliacea *Invertebrate Zoology.* an order of primitive, shrimplike, marine malacostracan crustaceans, with a bivalve shell and an articulated rostrum on the head.

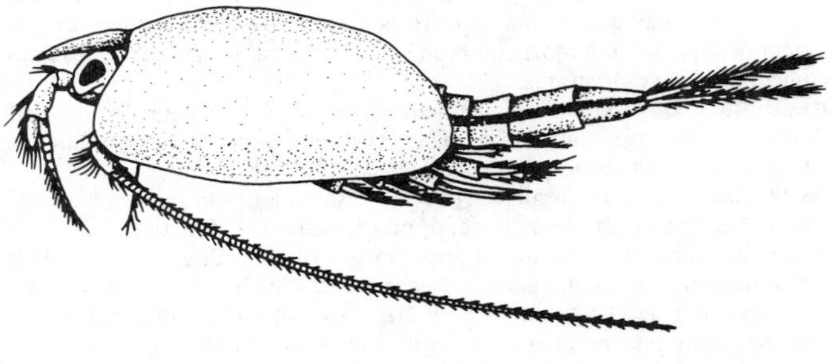

Nebaliacea

Nebelwerfer *Ordnance.* a family of German surface-to-surface unguided rockets used in World War II.

nebenkern *Cell Biology.* a region in the middle section of a spermatozoon that contains an array of mitochondria.

Nebraskan *Geology.* the first glacial stage of the Pleistocene epoch in North America, occurring before the Aftonian interglacial stage.

Nebraskan drift *Geology.* rock material transported during the Nebraskan glacial stage, now buried below the Kansan drift in Iowa.

nebula [neb′yə lə] *plural,* **nebulae** or **nebulas.** *Astronomy.* **1.** an immense cloudlike mass of dust particles and gases in interstellar space; nebulae may be seen in silhouette or illuminated by the light of nearby stars. **2.** in earlier use, a galaxy or star cluster outside the Milky Way galaxy. *Medicine.* a slight corneal opacity that can be seen only by oblique illumination and that seldom interferes with vision. (From a Greek word meaning "cloud" or "mist.")

nebula

nebular [neb′yə lər] *Astronomy.* of or relating to a nebula or nebulae. Also, **nebulous.**

nebular hypothesis *Astrophysics.* **1.** a model for the formation of the solar system in which the sun and planets condense from a cloud (or nebula) of gas and dust. **2.** specifically, such a theory formulated by the French astronomer Pierre Simon Laplace, according to which the cooling of a nebula left behind rings of matter that formed the planets and a central core that became the sun.

nebular lines *Astrophysics.* spectral lines from forbidden transitions.

nebular redshift *Astronomy.* an older term for the redshift exhibited by nearly all galaxies.

nebular transitions *Astronomy.* see FORBIDDEN TRANSITIONS.

nebulite *Petrology.* a migmatite with a textural fabric characterized by indistinct, streaky, lenticular inhomogeneities or inclusions.

nebulitic *Petrology.* **1.** characterized by textural elements with indistinct boundaries. **2.** of or relating to nebulite.

nebulosity *Astronomy.* **1.** cloudlike matter in space; a nebula. **2.** the fact of being a nebula.

nebulosus *Meteorology.* a cloud species found primarily in the genera cirrostratus and stratus that appears as a nebulous veil showing no distinct details; it is the most common species of stratus, and produces halo phenomena in cirrostratus.

necessary bandwidth *Telecommunications.* for a specified class of emission, the minimum value of the occupied bandwidth for transmitting information in the system being used.

neck *Anatomy.* **1.** the region between the shoulders and the head. **2.** any narrowed part of an organ, bone, or other anatomical structure.

neck any structure, feature, or area thought of as analogous to the neck of a human or animal; specific uses include: *Geography.* a narrow part between two larger parts, such as a narrow isthmus or channel. *Engineering.* any narrow area or section, such as the thinnest portion of an extruded filament or the thin, lower portion of a balloon envelope. *Oceanography.* the rip current proper, the long and narrow middle section of a rip current's length, where the current moves fastest. *Geology.* see PIPE, def. 1. *Volcanology.* an erosional remnant of the vertical intrusion joining the magma chamber and the vent of a volcano. Also, CHIMNEY. *Virology.* on a tailed phage, the area linking the head with the tail.

neck-down *Metallurgy.* **1.** a localized reduction of area occurring in tensile deformation. **2.** in casting, a mold component that is inserted in the riser.

Neckeraceae *Botany.* a family of mosses in the order Isobryales, characterized by rounded, undulate leaves, usually apiculate at the apex.

necking *Materials Science.* a localized plastic deformation resulting in a decrease in the cross-sectional area of a small position of a specimen when it is subjected to uniaxial tensile stress; fracture generally occurs in the necked region because of the increase in the true stress. *Metallurgy.* in metal working, a partial reduction in the size of a tube.

necking architecture see GORGERIN.

necking down *Materials Science.* the process of locally reducing the cross section of a metallic product by tensile deformation.

neck rot *Plant Pathology.* a disease of onions caused by species of the fungus *Botrytis* and characterized by a decay of the foliage directly above the bulb.

necr- or **necro-** *Medicine.* a combining form associated with death or with a dead body, cells, or tissue.

necrectomy *Surgery.* the surgical removal of necrotic or dead tissue.

Necridium *Microbiology.* a special cell in certain filamentous bacteria, that, by its death and lysis, facilitates the fragmentation of the trichome and release of the hormogonia.

necrobacillosis *Pathology.* a bacterial infection occurring in cattle, horses, hogs, and rabbits, causing tissue death of various kinds, diphtheria with abscesses, and gangrenous dermatitis. Also, SCHMORL'S BACILLUS.

necrobiosis *Medicine.* a skin disease marked by swelling, basophilia, and distortion of collagen bundles, usually associated with diabetes mellitus.

Necrolestidae *Paleontology.* a single-genus family of South American marsupials of the lower Miocene, *incertae sedis.*

necrophagous *Zoology.* feeding mainly or exclusively on carrion.

necrophilia *Psychology.* a desire to have sex with a dead body.

necrophilism *Psychology.* a desire to be near dead bodies.

necrophilous (character) type *Psychology.* an individual who displays abnormal fascination with the dead or an obsessive excitement toward corpses.

necrophobia *Psychology.* an irrational fear of dead bodies.

necrophoric behavior *Behavior.* a behavior pattern in which an animal carries around the corpse of a dead animal of the same species.

necropsy see AUTOPSY.

necrosis *Medicine.* the localized death of cells, tissue, or an organ in response to disease or injury.

necrotic *Medicine.* referring to or associated with necrosis.

necrotic enteritis *Veterinary Medicine.* an enterotoxemic disease in chickens caused by *Clostridium perfringens* types A and C, characterized by sudden onset with diarrhea, explosive mortality, and confluent mucosal necrosis of the small intestine.

necrotic ring spot *Plant Pathology.* a virus disease of cherry trees in which small, dark, water-soaked rings appear on the foliage and often drop out, leaving holes.

necrotic somatitis see CALF DIPHTHERIA.

necrotize *Medicine.* to cause necrosis.

necrotizing enterocolitis *Medicine.* the severe ulceration and necrosis of the ileum and colon in premature infants in the neonatal period, which can be caused by perinatal intestinal ischemia and bacterial invasion.

necrotoxin *Toxicology.* any poison that causes cell necrosis.

necrotroph *Biology.* an organism that feeds on dead tissues or cells.

necrotrophic fungi *Mycology.* fungi that derive their nutrients from dead organic matter.

Necrovirus group *Virology.* a group of plant viruses that have a wide host range and are transmitted mechanically and by Chytrid fungi; the type species is tobacco necrosis virus.

necrozoospermia *Medicine.* a condition in which the spermatozoa are dead or motionless. Also, **necrospermia.**

nect- or **necto-** a prefix meaning swimming.

nectar *Botany.* a sugar-containing fluid secreted by the nectaries of most flowers and by certain leaves to attract birds or insects for pollination.

nectarine *Botany*. the common name for a variety of fruit of the *Prunus persica*, having smooth skin and grown as a food crop.

Nectariniidae *Vertebrate Zoology*. the sunbirds or spiderhunters, a family of small colorful birds of the Passerine suborder Oscines, characterized by a long tubular tongue used to obtain nectar from flowers; found in Europe, Africa, southern Asia, New Guinea, and Australia.

nectary *Botany*. the structures in flowers that secrete nectar.

nectin *Biochemistry*. a protein that gives rise to the "stalk" portion of the mitochondrial proton.

nectocalyx see NECTOPHORE.

nectochaeta *Invertebrate Zoology*. the free-swimming larva of some polychaete flatworms, with rings of cilia and three pairs of parapodia.

Nectonematoidea *Invertebrate Zoology*. an order of marine hairworms in the class Nematomorpha, with rows of swimming bristles; adults are parasites of crabs.

nectophore *Invertebrate Zoology*. the free-swimming bell-shaped medusa of a siphonophore colony. Also, NECTOCALYX.

nectosome *Invertebrate Zoology*. the part of a siphonophore colony that produces nectophores.

Nectria *Mycology*. a genus of fungi belonging to the order Hypocreales which includes species that are pathogenic to plants, causing such diseases as apple canker, beech bark disease, and coral spot.

Nectridea *Paleontology*. an order of aquatic, newtlike amphibians in the extinct subclass Lepospondyli; the nectrideans were widespread in the Carboniferous and Permian.

Nectriodaceae *Mycology*. a family of fungi belonging to the order Sphaeropsidales that is characterized by brightly colored spore-bearing structures and includes numerous pathogens of plants and insects. Also, ZYTHIACEAE.

NED no evidence of disease.

need *Behavior*. a deficiency that stimulates an individual to seek out what is basic to his or her physical or emotional survival.

Needham, John 1713–1781, British biologist; proponent of theory of spontaneous generation.

Needham's sac *Invertebrate Zoology*. a large sac of the male genital system in cephalopods that stores spermatophores and opens into the mantle cavity.

needle *Mechanical Devices*. a sharp, slender metal implement with a hole or eye at the wide end to hold thread; used for sewing or knitting fabric. *Engineering*. **1.** any of various narrow, pointed objects or devices similar to a sewing needle. **2.** the pointer on a meter, dial, or other indicator. *Acoustical Engineering*. a stylus located on the tonearm of a disk phonograph player that transmits vibrations from the record grooves to a transducer for conversion to audio signals. *Botany*. the slender, pointed leaf of the Coniferae. *Geology*. a pointed, elevated, detached mass of rock formed by erosion. *Hydrology*. a long, thin snow crystal that is at least five times longer than it is wide.

needle bearing *Mechanical Devices*. a type of bearing with small rollers 1/4 inch in diameter that are contained in a flanged, antifriction cup, thus allowing individual rollers to bear directly on the shaft.

needle biopsy see ASPIRATION BIOPSY.

needle blow *Engineering*. a blow-molding process in which air is injected into the molten material in a mold by means of a hollow needle.

needle coal *Mineralogy*. a lignite consisting of a needlelike, fibrous mass of palm stems formed in vascular bundles.

needle culture see STAB CULTURE.

needle dam *Civil Engineering*. a fixed frame dam built up by heavy timbers, or needles, placed side by side vertically.

needle file *Mechanical Devices*. a small, delicate hand file with a long, knurled tang as a handle used for precision metalwork and for fine-finishing and sharpening tools. Also, **needle hand file.**

needlefish *Vertebrate Zoology*. any of about 60 species of very elongated fish of the family Belonidae; characterized by a silvery body with long slender jaws and sharp teeth; found primarily in marine tropical or temperate waters.

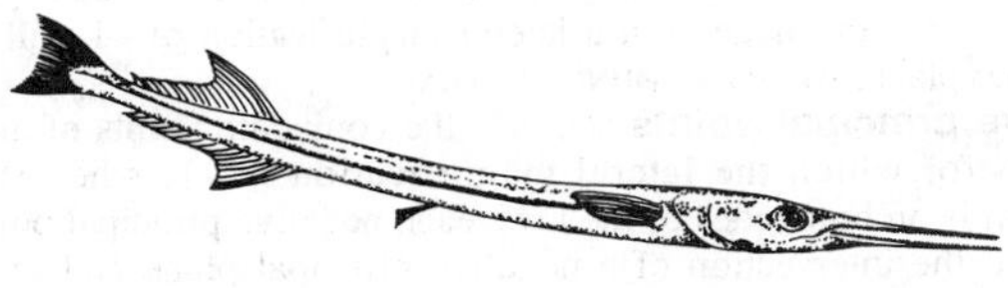

needlefish

needle gap *Electronics*. the gap of air between two needle-pointed electrodes across which a charge of electricity leaps when a high voltage is applied to one of the needles.

needle gun *Ordnance*. any of various obsolete firearms that employed an ignition system in which a needle pierced the base of the cartridge in order to strike the primer and set off the charge.

needle ice see FRAZIL ICE.

needlenose(d) pliers *Mechanical Devices*. a snipe-nosed pliers with thin, tapering jaws used for placing small washers or nuts onto fittings and for assembling delicate wiring.

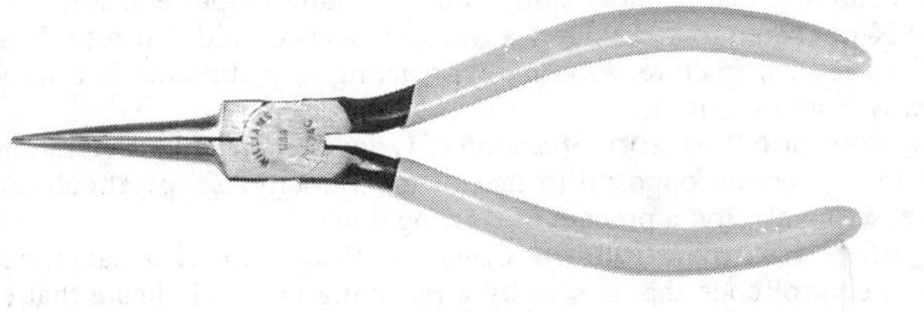

needlenose pliers

needle nozzle *Mechanical Engineering*. a hydraulic turbine nozzle with an element that converts the pressure in the pipe from the reservoir to the turbine into a smooth jet of variable diameter and discharge at a constant velocity.

needle ore *Mineralogy*. **1.** iron ore, usually present in small amounts, that has a very high metallic luster and is separable into slender filaments resembling needles. **2.** the mineral aikinite.

needle stone see HAIRSTONE.

needle valve *Mechanical Devices*. a valve with a long, tapering point that passes through a circular hole in a plate or pipe end and controls gaseous flow through a jet by the insertion or removal of a needlelike pin.

needle weir *Civil Engineering*. a fixed frame weir built up by heavy timbers, or needles, placed side by side vertically allowing the water level to be adjusted by the addition or removal of the timbers.

needling *Mining Engineering*. the process of cutting holes, notches, or ledges into the surface of a coal or rock wall such that it can receive timber supports.

need to know *Military Science*. a security criterion that requires the custodians of classified information to establish, prior to disclosure, that the intended recipient must have access to the information in order to carry out his official duties.

Néel, Louis E. F. [nā el′] born 1904, French physicist; awarded the Nobel Prize for his research on magnetic properties that applied to computer memories.

Néel's theory *Solid-State Physics*. a theory concerning antiferromagnetic materials in which a crystal lattice is composed of two or more interpenetrating sublattices and each atom in one sublattice is only affected by the fields of neighboring atoms in the other sublattice.

Néel temperature *Solid-State Physics*. the temperature at which a certain metal, alloy, or salt experiences a phase transition from the antiferromagnetic to the paramagnetic state. Also, **Néel point.**

Néel wall *Solid-State Physics*. a boundary that separates two magnetic domains in a magnetic thin film, whereby the magnetization vectors are parallel to the film surface while passing through the boundary.

neencephalon see NEOENCEPHALON.

Nef reaction *Organic Chemistry*. the formation of aldehydes and ketones from primary and secondary nitro compounds, respectively, by converting the compounds to salts and then treating the salts with sulfuric acid.

Nefud *Geography*. an extensive desert area (50,000 square miles) of northern Saudi Arabia.

negate *Science*. to cause to be negative; bring into or be in a negative condition.

negation *Mathematics*. a mathematical statement modified by the word "not." In particular, if P represents a logical statement, then the negation of P, denoted $\sim P$, is the statement that is true if and only if P is false. *Computer Programming*. a formation of the arithmetic negative or logical negative (NOT) of an operand.

negation as failure *Artificial Intelligence*. in a logic programming language such as Prolog, accepting a negative conclusion as true if its positive version cannot be proved, e.g., concluding that a man does not have a wife if no wife can be found.

negative being in a condition or direction opposite to that which is termed to be positive; specific uses include: *Mathematics.* designating a real quantity that is less than zero or that is to be subtracted. *Electricity.* having the type of electric charge in which there is an excess of electrons in relation to protons, as in a body of resin that has been rubbed with silk. *Chemistry.* describing an element or compound that tends to gain electrons. *Physics.* describing the part of an electric cell from which the current flows into the cell. *Graphic Arts.* a photographic image on which the normal tonal values are reversed, so that areas which are lighter in actuality appear darker in the image, and vice versa. *Medicine.* describing a test, investigation, or the like in which the sign or condition that is under study is not found to be present; for example, an X-ray taken to ascertain if a fracture has occurred that reveals no indication of a fracture. *Physiology.* reacting to a stimulus in a direction away from the stimulus.

negative acceleration *Mechanics.* **1.** any instance of acceleration that is in a direction opposite to that of the velocity. **2.** specifically, a decrease in velocity; a process of slowing down.

negative acknowledgment *Computer Programming.* a data transmission control code that is sent by a receiving unit to indicate that errors were detected in the frame just received.

negative adaptation see SENSORY ADAPTATION.

negative afterimage *Physiology.* a visual effect persisting after viewing an object and subsequently gazing at a neutral background; the light and dark areas of the original image are reversed in the afterimage.

negative angle *Mathematics.* an angle subtended by rotating a ray in the clockwise direction instead of the counterclockwise direction.

negative area see NEGATIVE ELEMENT.

negative binomial distribution *Statistics.* in Bernoulli trials, the distribution of the number X of trials necessary to observe a specified number k of successes. If $k - 1$, it is the geometric distribution.

negative booster *Electricity.* a booster that is used to reduce the electric potential difference between two points on a grounded return.

negative catalysis *Chemistry.* a reduction in the rate of a reaction due to the presence of a catalyst.

negative catalyst *Chemistry.* a catalyst that reduces the rate of a reaction.

negative charge *Electricity.* an electric charge about a material body, characterized by a field having more electrons than protons. Also, NEGATIVE ELECTRICITY.

negative complementation *Genetics.* a complementation of alleles that allows a mutant subunit to inactivate the wild-type subunit with which it is associated in a multimeric protein.

negative conditioning see NEGATIVE REINFORCEMENT.

negative control *Molecular Biology.* a type of gene regulation in which the binding of a specific repressor protein to a site on a DNA molecule inhibits initiation of transcription of the adjacent gene.

negative crystal *Crystallography.* a uniaxial crystal whose extraordinary wave travels faster than its ordinary wave.

negative dihedral *Aviation.* a downward inclination of an aircraft wing or other support surface. Also, ANHEDRAL.

negative electricity see NEGATIVE CHARGE.

negative electrode see CATHODE.

negative-electron affinity material *Electronics.* a material whose surface is treated with a substance such as cesium to reduce the surface barrier, raise the conduction band above vacuum level, and provide a more conducive environment for negatively charged electrons.

negative element *Geology.* a large structural feature or part of the earth's crust characterized, over long intervals of geologic time, by a tendency to sink in relation to adjacent positive elements; such relative shifting may be accomplished by downward movement, by erosion, or by upward movement that is significantly less powerful or frequent than that of the surrounding elements. Also, NEGATIVE AREA.

negative feedback *Science.* a kind of feedback that creates equilibrium between input and output in a system or process. *Control Systems.* feedback that is 180° out of phase with the input signal in order to decrease amplification and stabilize it in relation to time and frequency, resulting in a reduction in noise and distortion. Also, INVERSE FEEDBACK, REVERSE FEEDBACK, STABILIZED FEEDBACK. *Ecology.* the responses in a homeostatic system that inhibit or modify inputs into that system which promote instability. *Industrial Engineering.* a type of management model in which an actual condition is compared with a goal, and if there is a sufficient difference between the two, management acts to lessen the difference. *Psychology.* in popular use, a response to one's behavior that is hostile or critical.

negative g *Mechanics.* **1.** an acceleration that is opposite to that of gravity. **2.** see DECELERATION.

negative glow *Electronics.* a light that appears between the cathode and the Faraday dark space in a glow-discharge tube.

negative-grid thyratron *Electronics.* a tube in which current flows when the negative electric potential of the cathode drops below a given value with respect to the grid.

negative impedance *Electronics.* an impedance whose real part is a negative resistance.

negative-impedance repeater *Electronics.* a device that uses the characteristics of negative impedance to amplify voice-frequency signals weakened by impedance encountered along a telephone line.

negative incentive *Behavior.* an object or condition away from which behavior is directed.

negative interference *Genetics.* the tendency for a certain chromosomal recombination and crossover to increase the likelihood of another similar event in its vicinity.

negative ion *Chemistry.* an atom or group of atoms that have gained one or more electrons and thus have a negative charge; an anion. *Physics.* any negatively charged particle.

negative-ion vacancy *Crystallography.* an empty position in a crystal lattice that would normally be filled by a negatively charged ion.

negative landform *Geology.* a relatively low-lying topographic form, such as a valley, basin, or plain.

negative lens see DIVERGING LENS.

negative meniscus lens see DIVERGING MENISCUS LENS.

negative modulation *Electronics.* **1.** the transmission of a television signal in which an increase in brightness causes a corresponding reduction of power. **2.** the transmission of a facsimile in which an increase in brightness causes a reduction in frequency. Also, NEGATIVE TRANSMISSION.

negative movement *Geology.* **1.** a downward movement of the earth's crust relative to an adjacent part. **2.** a relative lowering of the sea level with respect to the land.

negative nodal points *Optics.* two points situated on the axis of an optical system in such a way that an incident ray passing through one point forms an emergent ray passing through the other point, creating an angle with the axis having the same magnitude as but opposite sign of the angle formed by the original ray.

negative number *Mathematics.* a number that is less than zero or less than some other specified reference point.

negative Oedipus (complex) *Psychology.* a reversal of the Oedipus complex in which the erotic attraction is to the same-sex parent and the hostility is directed toward the opposite-sex parent.

negative pedal *Mathematics.* **1.** let O be a point not on a given curve γ. The negative pedal of γ with respect to O is the envelope of lines perpendicular to OP at each point P of γ. **2.** any curve obtained from a given curve by iteration of the above process.

negative phase *Immunology.* the quantitative drop in antibodies experienced immediately after the administration of a second or later dose of an antigen to a primed animal.

negative phase sequence *Electricity.* a phase sequence opposite to the normal phase order in a polyphase system.

negative picture phase *Electronics.* a condition in which an increase in brightness causes a television picture signal's voltage to veer in a negative direction.

negative pion *Particle Physics.* a negatively charged pion particle.

negative plate *Electricity.* the plate structure connected to the negative terminal of a storage battery.

negative potential *Electricity.* electrostatic potential that is less than that of the earth.

negative practice *Behavior.* a therapy technique for eliminating small motor habits such as tics by performing the behaviors repeatedly.

negative pressure *Physics.* a gauge pressure that is below atmospheric pressure; the negative value of the number of mm Hg below 760 mm Hg.

negative principal planes *Optics.* two planes situated perpendicularly to the axis of an optical system in such a way that if an object is located in one, an image with a lateral magnification of -1 will form in the other plane. Also, ANTIPRINCIPAL PLANES.

negative principal points *Optics.* the conjugate points of an optical system for which the lateral magnification is -1; when the same medium is on both sides of the lens, each negative principal point is situated at the intersection of a negative principal plane and the optical axis. Also, ANTIPRINCIPAL POINTS.

negative rain *Meteorology.* rain that exhibits a negative electrical charge.

negative rake *Mechanical Engineering.* the orientation of a cutting tool so that the angle formed by the leading face of the tool and the surface behind the cutting edge is greater than 90°.

negative reciprocity *Anthropology.* any attempt to take advantage of another in a trade, in which something is expected for nothing or for a lesser amount than its true worth.

negative reinforcement *Behavior.* the presentation of a punishment or withholding of a reward following the performance of an undesired behavior in order to decrease the frequency of the behavior. Thus, **negative reinforcer.**

negative resistance *Electronics.* a characteristic of a two-terminal device where an increase in voltage causes a decrease in current.

negative-resistance device *Electronics.* a device that exhibits negative resistance, at least for a certain range of current and voltage.

negative-resistance oscillator *Electronics.* a device in which a parallel-tuned resonant circuit and a vacuum tube combine to provide the negative resistance required for continuous generations of alternate current.

negative-resistance repeater *Electronics.* a device that boosts a fading signal along a telephone line by employing the property of negative resistance.

negative sanction *Anthropology.* a form of punishment or restraint imposed by a society for violation of social norms or standards; may be a formal sanction such as imprisonment or an informal one such as ridicule or isolation.

negative shoreline see SHORELINE OF EMERGENCE.

negative staining *Optics.* a method of preparing material for electronic microscopy to demonstrate the three-dimensional and surface features of cells, bacteria, viruses, and other small objects by staining the ground rather than the objects.

negative-strand virus *Virology.* any virus that contains negative-sense RNA genome, such as an arenavirus or bunyavirus; major groups are the myxo-paramyxo-rhabdovirus groups.

negative supercoiling see DNA TWISTING.

negative temperature *Thermodynamics.* a nonequilibrium state of a thermodynamic system in which there are more particles having higher energies than those having lower energies, as exemplified by population inversion in a laser or maser.

negative temperature coefficient *Physics.* a condition that is indicated by a decrease in some physical quantity as the temperature is increased.

negative temperature gradient *Acoustics.* a temperature that falls below a positive gradient in a region where temperature decreases as depth increases, causing the occurrence of a shadow zone, sound channel, or similar phenomenon because of the bending of sound away from higher velocities and toward lower velocities, as described in Snell's law.

negative terminal *Electricity.* the terminal of a battery or other voltage source with an abundance of electrons.

negative test *Computer Programming.* a type of exception test, consisting of invalid data or procedures, that is expected to fail.

negative transfer *Behavior.* a process by which an individual inhibits his ability to learn a new skill or task by applying techniques learned in a similar but unrelated task or skill; e.g, when techniques used to hit a baseball are applied to golf.

negative transmission see NEGATIVE MODULATION.

negative zero *Computer Science.* in a one's-complement number representation, a representation of the number zero by all one-bits.

negativism *Psychology.* an attitude or behavior in which an individual does the opposite of what is expected, such as talking after he has been asked to be quiet. Also, COMMAND NEGATIVISM, CONTRASUGGESTIBILITY, CONTRARIETY. *Medicine.* resistance or complete opposition to advice or commands; for example, in catatonic schizophrenia, the individual lowers the arms if asked to raise them.

negator see NOT GATE.

negatoscope *Radiology.* a device used to display radiographic negatives.

negatron *Electronics.* **1.** a four-electron vacuum tube that has the property of negative resistance. **2.** see DYNATRON.

Negri bodies [nā′grē] *Pathology.* spherical inclusion bodies noted in cytoplasm and used as a determinant in the diagnosis of the disease rabies when observed in the nerve cells upon death. (Named for Adelchi *Negri,* 1876–1912, Italian physician.)

Negroid or **negroid** *Anthropology.* **1.** a term used to describe groups of people residing or originating in sub-Saharan Africa and having physical characteristics that include light brown to black skin, dark eyes, and coarse, tightly coiled hair. **2.** of or or relating to such people.

Neididae *Invertebrate Zoology.* the stilt bugs, a family of slender hemipteran insects with long, threadlike antennae and slender stiltlike legs.

neighborhood *Mathematics.* **1.** a small region surrounding a point. **2.** let p be a point in a topological space X. A neighborhood of p is any open set of X that contains p. **3.** more generally, a neighborhood of p is any subset of X that contains an open set containing p. **4.** an **ε-neighborhood** of p is any neighborhood of p that can be contained in an open ball of radius less than or equal to ε.

neighborhood stability *Ecology.* the capacity of an ecologic or taxonomic unit to withstand minor disturbances with little or no effect.

Neill's parabola *Mathematics.* the graph in the plane of the equation $ax^3 - y^2 = 0$. The curve lies in the right half-plane, has a cusp at the origin, and has two branches symmetric about the positive x-axis. Also, SEMICUBICAL PARABOLA.

Neisseria *Bacteriology.* a genus of Gram-negative bacteria of the family Neisseriaceae, that occur as aerobic, coccoid cells in pairs, are mammalian parasites, and include the agents of gonorrhea and meningitis.

Neisseriaceae *Bacteriology.* a family of Gram-negative, nonmotile, aerobic, spheroid or rod-shaped bacteria that are parasitic and pathogenic in humans and other warm-blooded animals.

nektobenthos *Ecology.* those organisms that swim actively just above the bottom of the sea, a lake, or another body of water.

nekton *Ecology.* a collective term for all organisms that swim actively in open water, independent of water currents. Thus, **nektonic.**

NEL no effect level

N electron *Atomic Physics.* one of the electrons that is found in the N shell of an atom and displays the characteristics of the principal quantum number 4.

NELIAC *Computer Programming.* a programming language that was based on Algol 58 and used on many second-generation computers. (An acronym for the Navy Electronics Laboratory International Algol Compiler.)

Nelson diaphragm cell *Chemical Engineering.* an electrolytic diaphragm cell with a graphite anode used to produce caustic soda and very pure chlorine.

nelsonite *Petrology.* a hypabyssal rock, or rock group, consisting primarily of ilmenite and apatite, with or without rutile.

Nelumbonaceae *Botany.* a family of aquatic plants of the order Nymphaeales that have roots, perfect flowers, and triaperturate pollen.

Nemagraptus *Paleontology.* a genus of early graptoloids in the suborder Didymograptina and family Nemagraptidae; most species are characterized by multiramous, pendent to reclined stipes; Ordovician.

Nemaliaceae *Botany.* a family of marine red algae of the order Nemaliales, characterized by erect gametangial thalli with filamentous axes composed of elongate cells occurring intertwined or as a bundle; the gametangial phase of some species is used as a food source in the Orient and Polynesia.

Nemaliales *Botany.* an extremely diverse order belonging to the red algae division Rhodophyta; classified in some older systems as Nemalionales.

nemalite *Mineralogy.* a fibrous variety of brucite containing ferrous oxide.

nemata *singular,* **nema.** *Invertebrate Zoology.* **1.** threadlike tubular projections or filaments or one who bears such a filament. **2. Nemata.** see NEMATODA.

Nemataceae *Botany.* a family of mosses of the order Hookeriales.

nematath *Geology.* a nonorogenic submarine ridge across an Atlantic-type ocean basin, composed of undeformed continental crust that has been stretched across a gap in the oceanic or continental crust.

Nemathelminthes *Invertebrate Zoology.* in older classifications, a phylum which included the Nematoda, the Nematomorpha, and the Acanthocephala; roughly equivalent to Aschelminthes.

nematic *Physical Chemistry.* **1.** a type of liquid crystal having an irregular threadlike structure that exhibits little viscosity. **2.** relating to or describing such a liquid-crystal phase. Thus, **nematic phase.**

nematic state *Materials Science.* a state of matter in which the anisometric molecules are regularly arranged in one direction and randomly arranged in the remaining directions.

nematoblastic see FIBROBLASTIC.

nematocalyx see NEMATOPHORE.

Nematocera *Invertebrate Zoology.* a series of dipteran insects in the suborder Orthorrapha, having slender bodies with long antennae; includes mosquitos, fungus gnats, and crane flies.

nematocerous *Invertebrate Zoology.* belonging to or resembling members of the dipteran suborder Nematocera.

nematocide *Toxicology.* an agent that is destructive to nematodes, especially those that are harmful to agricultural plants.

nematocyst *Invertebrate Zoology.* a stinging cell in coelenterates, consisting of a capsule and coiled threadlike tube that can be everted to penetrate and inject toxin into a victim; some turbellarians and gastropods ingest nematocysts and use them in their own bodies for defense.

Nematoda *Invertebrate Zoology.* the roundworms and threadworms, a phylum of smooth-skinned unsegmented worms with a long cylindrical body shape tapered at the ends; includes free-living and parasitic forms, both aquatic and terrestrial.

nematode [nēm´ə tōd´] *Invertebrate Zoology.* **1.** any member of the phylum Nematoda, such as the hookworm, pinworm, or trichina. **2.** of or relating to this phylum.

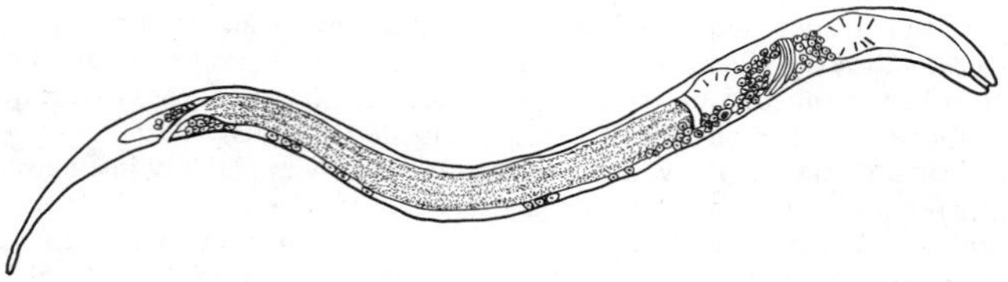

nematode

nematodesma *Cell Biology.* any of several microtubular bundles found in the cytopharynx of ciliated animals, which help to maintain the structure of the cytopharyngeal walls.

Nematodonteae *Botany.* mosses of the subclass Eubrya, with a single peristome having only faint transverse barring or no barring at all on its peristome teeth.

nematogen *Invertebrate Zoology.* the parasitic, sexual, adult phase of mesozoans.

nematogenic *Physical Chemistry.* describing a solid material that when heated forms a nematic phase. Thus, **nematogenic solid.**

Nematognathi *Vertebrate Zoology.* an equivalent name for the catfish order Siluriformes.

nematoid *Invertebrate Zoology.* resembling a nematode roundworm in form or habits.

Nematoidea see NEMATODA.

nematology *Invertebrate Zoology.* the scientific study of nematodes. Thus, **nematological.**

nematomorph *Invertebrate Zoology.* a member of the phylum Nematomorpha, the horsehair worms or hairworms.

Nematomorpha *Invertebrate Zoology.* small phylum of pseudocoelomate wormlike animals that are free-living as adults and parasitic on arthropods when juveniles.

nematophagus fungus *Mycology.* a large, varied group of fungi that obtain nutrients from nematodes and are found in decaying organic matter and soils.

Nematophora see CHORDEUMIDAE.

nematophore *Invertebrate Zoology.* in corals, an organ that carries nematocysts or stinging cells.

Nematophytales *Paleontology.* an order of large branching plants that are possibly intermediate between brown algae and vascular plants; extant in the Silurian and Devonian.

nematosphere *Invertebrate Zoology.* the enlarged end of a tentacle in some sea anemones.

Nematospora *Mycology.* a genus of fungi belonging to the family Spermophthoraceae, characterized by needle-shaped ascospores; composed primarily of plant parasites and pathogens.

nematozooid *Invertebrate Zoology.* a defensive zooid in a hydroid or siphohonophore, bearing many stinging cells.

nem. con. no one contradicting. (From Latin *nemine contradicente.*)

nem. dis. no one dissenting. (From Latin *nemine dissentiente.*)

nemere *Meteorology.* a stormy, cold autumn wind in Hungary.

Nemertea see RHYNCOCOELA.

nemertean *Invertebrate Zoology.* belonging to the phylum Rhynchocoela.

Nemertina see RHYNCOCOELA.

Nemesis *Astronomy.* a hypothesized dark companion star of the sun that has an elongated, distant orbit and whose gravitational perturbations periodically send great numbers of comets through the inner solar system, some colliding with the earth.

Nemestrinidae *Invertebrate Zoology.* a family of dipteran insects in the suborder Orthorrhapha, the hairy flies; the larvae are parasitic on other insects.

Nemichthyidae *Vertebrate Zoology.* the family of snipe eels, having elongated, slender bodies and a long beaklike snout; found in mid-water (to 2000 meters) in the Pacific, Atlantic, and Indian Oceans.

Nemognathinae *Invertebrate Zoology.* a subfamily of beetles, coleopteran insects in the family Meloidea, with greatly enlarged maxillae that form a tube.

neo- *Organic Chemistry.* a prefix indicating a hydrocarbon in which at least one carbon atom is connected directly to four other carbon atoms. *Chemistry.* a prefix formerly used to designate a newly discovered isomer of a previously well-known compound.

Neoanthropinae *Paleontology.* an anthropological term for a subdivision of the family Hominidae that includes sapient forms more recent than the Neanderthals.

neoantigen *Oncology.* an antigen that is acquired after a cell has been infected by a cancer-causing virus.

neoantimosan see STIBOPHEN.

neoautochthon *Geology.* a stable basement or autochthon formed where a nappe has ceased moving and has become defunct.

neobehaviorism *Psychology.* a modified view of behaviorism that regards actions as being affected by internal psychological states as well as by external stimuli. Thus, **neobehaviorist.**

neoblast *Invertebrate Zoology.* one of the large undifferentiated cells of an annelid that migrate to a wound site and participate in repair or regeneration.

Neocathartidae see BATHORNITHIDAE.

neocerebellum *Anatomy.* the part of the cerebellum that has evolved more recenctly, as opposed to the paleocerebellum; in humans, it refers to the large lateral portions of the cerebellum that receive input primarily from the pons and cerebral cortex.

Neocomian *Geology.* a European stage of the Lower Cretaceous period, after the Portlandian of the Jurassic period and before the Aptian.

neocortex *Anatomy.* the six-layered portion of the cerebral cortex, showing stratification and organization characteristic of the most highly evolved type of cerebral tissue. Also, NEOPALLIUM.

neocryst *Crystallography.* in an evaporite, an individual crystal of a secondary mineral.

neo-Darwinism *Evolution.* a modern theory of evolution that combines Darwin's mechanism of natural selection with later discoveries in Mendelian and population genetics.

neodymia see NEODYMIUM OXIDE.

neodymium *Chemistry.* a metallic element having the symbol Nd, the atomic number 60, an atomic weight of 144.24, a melting point of 1024°C, and a boiling point of c. 3030°C; a soft, malleable, yellow rare-earth element of the lanthanide series that is highly flammable and easily tarnished, and has high electrical resistivity; used in electronics, alloys, and astronomical lenses and lasers. (From Latin; literally, "new didymium;" based on the name of another substance formerly thought to be an element.)

neodymium chloride *Inorganic Chemistry.* **1.** $NdCl_3$, violet crystals, soluble in water and alcohol and insoluble in ether; melts at 784°C and boils at 1600°C. **2.** $NdCl_3 \cdot 6H_2O$, the hexahydrate form, red crystals that are soluble in water and alcohol; melts at 124°C and loses water at 160°C.

neodymium glass *Materials.* a colored glass containing neodymium; used for light amplifiers, astronomical lenses, and solid-state lasers.

neodymium glass laser *Optics.* a laser that uses solid-state components along with neodymium-doped glass to produce a beam of light almost identical to the YAG laser, except that it cannot emit a continuous beam.

neodymium-iron-boron magnet *Metallurgy.* a hard magnetic alloy of the named elements; the strongest commercially available magnet.

neodymium oxide *Inorganic Chemistry.* Nd_2O_3, a light-blue powder, soluble in acids and insoluble in water; melts at about 1900°C; absorbs water and carbon dioxide from the air; used in glassmaking, ceramics, and color TV tubes, and as a catalyst. Also, NEODYMIA.

neoencephalon *Anatomy.* the part of the brain that has evolved more recently, including the cerebral cortex and its dependencies. Also, NEENCEPHALON.

neoevolutionism *Anthropology.* an approach to the study of culture maintaining that the development of technology and the resulting efficient use of energy are the primary factors in increasing the complexity of a culture.

neo-Freudian *Psychology.* **1.** a later psychoanalytical theorist, such as Alfred Adler or Karen Horney, who accepts the basic principles of the Freudian system but whose own theory differs in some significant respect, generally in placing greater emphasis on social factors and less on instinctual drives. **2.** of or relating to such revisions of Freud's theory. Thus, **neo-Freudianism.**

Neogastropoda *Invertebrate Zoology.* an order of marine gastropod mollusks with a single comblike gill and a gill-like sense organ; many are carnivorous, some poisonous.

Neogea *Ecology.* another name for the Neotropical zoogeographical region.

Neogene *Geology.* an interval of geologic time that incorporates the Miocene and Pliocene epochs; equivalent to the Upper Tertiary period.

neogenesis *Geology.* the formation of new minerals by any process, such as by diagenesis or metamorphism. Also, **neoformation.**

neoglaciation *Geology.* the readvance and renewed growth of mountain glaciers during the Little Ice Age.

Neognathae *Vertebrate Zoology.* a superorder of birds including nearly all living birds except the ostrich, cassowary, emu, rhea, moa, and kiwi; characterized as flying birds with a bony keeled sternum.

Neogregarinida *Invertebrate Zoology.* an order of sporozoan protozoans in the subclass Gregarinia; parasites of insects.

neohexane *Organic Chemistry.* C_6H_{14}, a highly flammable, colorless, volatile liquid that boils at 49.7°C; derived from the thermal or catalytic union of ethylene and isobutane recovered from refinery gases; used in high-octane fuels. Also, 2,2-DIMETHYLBUTANE.

Neolampadoida *Invertebrate Zoology.* an order of rare, very small sea urchins found in deep tropic and temperate seas.

Neolithic *Archaeology.* **1.** a time period late in the Stone Age, after about 8000 BC; characterized by the use of ground and polished tools, the manufacture of pottery, and the practice of farming. **2.** of, or relating to, this period. Also, **neolithic.**

neolocal residence *Anthropology.* the practice in which a married couple sets up a household independent of either spouse's family.

neologism *Linguistics.* a newly invented word or phrase, or a new meaning given to an already existing word.

neomagma *Geology.* magma formed by the partial or complete fusion of preexisting rocks.

neomin see NEOMYCIN.

neomineralization *Geochemistry.* a type of recrystallization in which the mineral constituents of a rock are transformed into entirely new minerals.

neomorph *Genetics.* a mutant allele that causes a qualitatively different effect than that produced by the wild-type allele.

neomycin *Microbiology.* a broad-spectrum antibacterial antibiotic that is produced by the bacterium *Streptomyces fradiae.* Also, NEOMIN, NYACYNE.

neon *Chemistry.* an inert element having the symbol Ne, the atomic number 10, an atomic weight of 20.179, a melting point of –249°C, and a boiling point of –246°C; a colorless, odorless, tasteless noble gas that ionizes in electric discharge tubes; used in fluorescent lighting, electronics, and lasers. *Electronics.* see NEON LAMP. (From the Greek word for "new.")

neonatal *Medicine.* newborn; referring to the time immediately after birth and through the first 28 days of life.

neonatal impetigo see IMPETIGO NEONATORUM.

neonatal line *Anatomy.* a line in the enamel and dentin of a deciduous tooth representing the position of the enamel and dentin at birth.

neonatal mortality rate *Medicine.* the statistical ratio of the number of deaths that occur in the first 28 days of life divided by the number of live births that occur in the same population during the same time period.

neonatal myasthenia *Medicine.* a temporary myasthenia that afflicts children of myasthenic women; the condition lasts from a week to a month and is marked by difficulty in swallowing and sucking.

neonatal septic polyarthritis *Veterinary Medicine.* see JOINT ILL.

neonate *Medicine.* a newborn.

neonatologist *Medicine.* a physician who specializes in care of the newborn.

neonatology *Medicine.* the medical specialty dealing with disorders of the newborn.

Neonemataceae *Botany.* a family of freshwater yellow-green algae of the order Tribonematales, being uniseriate or multiseriate filaments with exterior mucilage layers.

neon-helium laser *Optics.* a laser using a mixture of neon and helium gases to produce a continuous-wave visible red beam.

neon lamp *Electronics.* a tube that generates a bright red glow (or, if treated with mercury, bright blue) when the neon gas inside it is ionized by an electric current; commonly used in outdoor signs and as an indicator light. Also, **neon bulb, neon tube.**

neon oscillator *Electronics.* a device that generates alternating current through a charging process controlled by the interaction of a capacitor and ionized neon gas. Also, **neon-bulb oscillator.**

neoophoran level *Invertebrate Zoology.* a level of organization in turbellarian flatworms in which the female reproductive system contains yolk glands and the eggs are accompanied by special yolk cells.

neopallium see NEOCORTEX.

neopalynology *Botany.* the study of extant pollen, pollen spores, and the components of microsporangia.

neopentane *Organic Chemistry.* C_5H_{12}, a highly flammable colorless gas or volatile liquid; insoluble in water and soluble in alcohol; boils at 9.5°C; used in manufacturing butyl rubber.

neophilia *Behavior.* the tendency of an animal to approach an unfamiliar object or situation.

neophobia *Psychology.* an irrational fear of new situations, places, or things. Also, CAINOPHOBIA. *Behavior.* the tendency of an animal to avoid or retreat from an unfamiliar object or situation.

neoplasia *Oncology.* the process of tumor formation; the proliferation of cells under conditions that would not elicit similar growth in normal cells.

neoplasm *Biology.* any new and abnormal growth. *Oncology.* specifically, growth that shows cellular proliferation, that grows at a more rapid rate than normal, that continues to grow after the instigating factor is no longer present, that shows a lack of structural organization and coordination with the normal tissue, and that usually creates a mass of tissue, which may be either benign or malignant.

neoplastic *Biology.* relating to or tending to produce neoplasms.

neoplastigenic *Oncology.* having a tendency to form neoplasms.

neoprene *Organic Chemistry.* $(CH_2ClC{=}CHCH_2)_n$, a synthetic elastomer available in solid, latex, or flexible foam form; the first chemically produced substitute with properties similar to natural rubber. *Materials Science.* this material in the form of a synthetic rubber produced by the polymerization of chloroprene and characterized by resistance to oils, heat, light, and oxidation; used for items such as shoes, gaskets, paints, putties, and conveyer belts. Also, POLYCHLOROPRENE.

Neopseustidae *Invertebrate Zoology.* a family of lepidopteran insects, tiny moths in the superfamily Eriocranioidea.

Neoptera *Invertebrate Zoology.* a major division of the subclass Pterygota composed of all insects that fold their wings over the abdomen when at rest; includes all winged insects except Odonata, Ephemeroptera, and Plecoptera.

Neopterygii *Vertebrate Zoology.* another name for the Actinopterygii, a subclass of ray-finned fishes of the class Osteichtyes.

Neorhabdocoela *Invertebrate Zoology.* an order of flatworms in the Rhabdocoela, including aquatic and terrestrial forms, some free-swimming, some parasitic.

Neorickettsia *Bacteriology.* a genus of Gram-negative bacteria, occurring as coccoid, nonmotile cells that grow intracellularly as canine parasites.

Neornithes *Vertebrate Zoology.* a subclass of birds of the class Aves, including all living birds as well as birds known from the Cretaceous Period.

neosome *Geology.* in a composite rock or mineral deposit, a geometric element that appears to be younger than the main rock mass.

neossoptile *Vertebrate Zoology.* in young birds, a down feather which forms the natal plumage.

neostigmine *Pharmacology.* $(C_{12}H_{19}N_2O_2)^+$, a drug used to treat the symptoms of myasthenia gravis, to prevent and treat malfunction of the digestive or urinary tract after surgery, and to reverse the effects of certain muscle relaxants used in surgery. Available as **neostigmine bromide** and **neostigmine methylsulfate,** both occurring as white crystalline powders.

neostomy *Surgery.* the surgical introduction of an artificial opening into an organ or between two organs.

neostratotype *Geology.* a stratotype established after the originally designated stratotype has been destroyed or is otherwise unusable.

Neo-Synephrine *Pharmacology.* $HClHOC_6H_4CHOHCH_2NHCH_3$, the trade name for preparations of phenylephrine hydrochloride used as a nasal spray.

neotectonic map *Geology.* a map depicting the post-Miocene structural history of the earth's crust as well as the associated geologic structures.

neotectonics *Geology.* the study of the most recent structures and structural history of the earth's crust, from the end of the Miocene epoch to the present.

neotenic *Invertebrate Zoology.* of termites, being the newly developed queen or king of a colony. *Zoology.* **1.** the retention of some larval characteristics in an otherwise adult organism. **2.** the attainment of sexual maturity in a larval stage. Also, **neoteinic.**

neotenin *Entomology.* a hormone secreted by the corpus allatum that controls growth in the insect during its larval stages. Also, JUVENILE HORMONE.

neoteny *Zoology.* a condition common to certain salamanders in which larvae attain full sexual maturity before metamorphosizing completely, or when larval characteristics linger indefinitely beyond the larval stage.

Neotremata *Invertebrate Zoology.* an order of lamp shells, brachiopod bivalves in the class Inarticulata.

Neotricholeaceae *Botany.* a family of dioecious liverworts of the order Jungermanniales, characterized by incubous leaves that form inflated water sacs and by endosporic spore germination; occur mainly in China and Japan.

neotropical *Ecology.* of or relating to the Neotropical region or its fauna.

Neotropical region *Ecology.* a zoogeographic region of the earth consisting of the area of the Americas that extends south from the Mexican Plateau, and includes the Mexican, Antillian, Brazilian, and Chilean subregions.

neotype *Systematics.* a specimen that is designated as a replacement for the original type specimen, in the absence of that original specimen. Also, PROXYTYPE.

neounitarian theory of hematopoiesis *Histology.* the theory that all blood cells, red and white, arise from a common cell, the hemocytoblast. Also, MONOPHYLETIC THEORY.

neovolcanic *Petrology.* of or relating to extrusive igneous rocks formed during or after the Tertiary period.

NEP net ecosystem production.

Nepenthaceae *Botany.* a family of dicotyledonous carnivorous plants of the order Sarracieniales, having alternate leaves, either simple, sessile, or petioled, whose elongated midrib ends in a cylindrical, liquid-containing pitcher.

Nepenthales *Botany.* an order of insectivorous perennial herbs and shrubs of the subclass Dilleniidae, sometimes epiphytic or climbing.

neper *Physics.* a dimensionless unit used to compare two values of some quantity, the number of nepers being equal to the natural logarithm of the ratio of the two values. (Named for the Scottish physicist John *Neper* or *Napier*.)

nephanalysis *Meteorology.* an analysis of the types and amount of clouds and precipitation on a synoptic chart in which cloud systems are identified as entities in relation to the pressure pattern and fronts.

nephcurve *Meteorology.* a line bounding a significant portion of a cloud system in nephanalysis.

nepheline *Mineralogy.* $(Na,K)AlSiO_4$, a colorless, white, yellow, or red hexagonal mineral having a specific gravity of 2.55 to 2.67 and a hardness of 5.5 to 6 on the Mohs scale; found in alkalic igneous rocks as prismatic crystals and in massive form; sometimes used in place of feldspar in the glass industry. Also, **nephelite.**

nepheline basalt see OLIVINE NEPHELINITE.

nepheline monzonite *Petrology.* a hypabyssal or plutonic igneous rock with roughly equal amounts of plagioclase and orthoclase, and containing small amounts of nepheline.

nepheline phonolite *Petrology.* the volcanic or hypabyssal equivalent of nepheline syenite.

nepheline syenite *Petrology.* a coarse-grained, intrusive igneous rock consisting primarily of alkali feldspar, nepheline, and mafic minerals.

nephelinite *Petrology.* a dark, fine-grained or porphyritic igneous rock consisting primarily of nepheline and pyroxene; resembles basalt in general appearance.

nepheloid zone *Oceanography.* a cloudy zone that sometimes occurs where there is a strong boundary current at the junction of the sea floor and the continental rise, the turbid water holding in suspension the finest clay and organic particles.

nephelometer *Optics.* any of various instruments that measure the intensity of light scattered by particles suspended in liquid, in order to determine the amount of material present in the liquid.

nephelometry *Optics.* a method of determining the amount of material suspended as small particles in liquid by measuring the light reflected and scattered by the suspensoids or by measuring the decrease in the intensity of a transmitted light beam.

nepheloscope *Engineering.* a laboratory instrument used to condense or expand moist air in order to approximate the formation of clouds.

nephology *Meteorology.* the study of clouds.

nepholometer *Geophysics.* an instrument that measures the amount of cloudiness of a medium; used to study scattering properties.

nephometer *Geophysics.* a convex mirror having one central and five radial parts; used to estimate cloud coverage at any given time.

nephoscope *Geophysics.* an optical instrument used to observe the speed and direction of clouds.

nephr- or **nephro-** a prefix meaning "kidney."

nephradenoma *Oncology.* an adenoma of the kidney.

nephrectomy *Medicine.* the surgical removal of a kidney.

nephridioblast *Invertebrate Zoology.* the ectodermal cell that gives rise to a nephridium in some flatworms.

nephridioduct *Invertebrate Zoology.* the duct of a nephridium that connects the nephrostome and the nephridiopore and often serves as both excretory and genital outlet.

nephridiopore *Invertebrate Zoology.* the external opening of a nephridium.

nephridium *Invertebrate Zoology.* the excretory organ of many invertebrates, a tubule opening to the exterior through a nephridiopore.

nephrite *Mineralogy.* a tough, compact, greenish or bluish variety of actinolite that supplies much of the less valuable type of jade.

nephritic *Medicine.* referring to or associated with nephritis.

nephritis *Medicine.* an inflammation of the kidney.

nephroabdominal *Anatomy.* of or related to the kidney and the abdominal wall.

nephrocalcinosis *Pathology.* a condition marked by the presence of random deposits of calcium phosphate in the kidney causing renal deficiencies.

nephrocele [nef′rə sēl′] *Medicine.* a hernia of the kidney.

nephrocyte *Invertebrate Zoology.* a specialized migratory cell in insects that filters nitrogenous wastes from body fluids and excretes them from the body.

nephrogenic *Developmental Biology.* forming kidney tissue.

nephrogenic cord *Developmental Biology.* a longitudinal cord in the embryo, formed from fused or never separated nephrotome plates, that forms the mesonephric and the metanephric tubules.

nephrogenic tissue *Developmental Biology.* the tissue of the nephrotomes that furnishes the material from which the three kidneys arise in vertebrate embryos.

nephroid *Mathematics.* the graph in the plane of the curve with parametric equations $x = a(3 \cos \phi - \cos 3 \phi)$ and $y = a(3 \sin \phi - \sin 3 \phi)$, where a is constant. The nephroid is an epicycloid of two cusps.

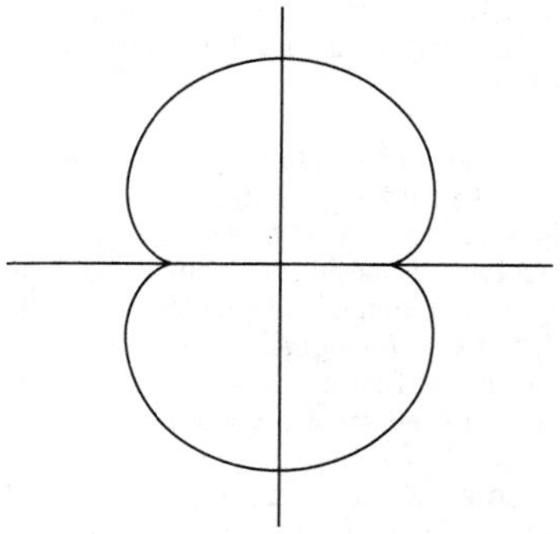

nephroid

nephrolith [nef′rə lith] *Medicine.* a kidney stone.

nephrolithiasis *Pathology.* the occurrence of calculi deposits in the kidney.

nephrolithotomy *Medicine.* an incision into the kidney in order to remove a kidney stone.

nephrology [nə fräl´i jē] *Medicine.* the branch of medicine dealing with the kidney, including its anatomy, physiology, and pathology. Thus, **nephrologist.**

nephrolysin *Biochemistry.* any agent that impairs kidney function. Also, NEPHROTOXIN.

nephrolysis *Medicine.* **1.** the process of relieving the kidney of inflammatory adhesions while retaining the capsule. **2.** the destruction of renal cells.

nephroma *Medicine.* a tumor that develops from renal tissue.

nephromalacia *Medicine.* a softening of the kidney.

nephromegaly *Medicine.* a severe hypertrophy of the kidney.

nephromixium *Invertebrate Zoology.* a combined coelomic funnel and nephridium, functioning as both an excretory organ and a genital duct.

nephron [nef´rän] *Anatomy.* the functional unit of the kidney; it consists of the renal glomerulus, Bowman's capsule, the proximal convoluted tubule, the loop of Henle, and the distal convoluted tubule. (From the Greek word for the kidney.)

nephropathy *Medicine.* **1.** see NEPHROSIS. **2.** any disease of the kidney. Also, RENOPATHY.

nephropexy *Surgery.* the surgical stabilization of a floating or mobile kidney.

Nephropsidae *Invertebrate Zoology.* see HOMARIDAE.

Nephropsidea *Invertebrate Zoology.* the lobsters and crayfish, a superfamily of decapod crustaceans with a cylindrical carapace and a pair of large claws.

nephroptosis *Medicine.* a prolapse of the kidney.

nephropyosis *Medicine.* a suppuration of the kidney.

nephrorrhaphy *Medicine.* nephropexy accomplished by suturing the kidney.

nephrosclerosis *Medicine.* a hardening of the kidney caused by the overgrowth and contraction of the interstitial connective tissue.

nephroscope *Medicine.* an instrument inserted into an incision in the renal pelvis for viewing the inside of the kidney.

nephroscopy *Medicine.* a visualization of the kidney using a nephroscope.

nephrosis [nə frō´sis] *Medicine.* any renal disease marked by degeneration, particularly one featuring deteriorative lesions of the renal tubules and marked by edema (noninflammatory), albuminuria, and decreased serum albumin.

nephrosonography *Radiology.* a visualization of the kidney by scanning with ultrasonic waves.

Nephrostome *Invertebrate Zoology.* the ciliated opening of a metanephridium into the coelom.

nephrostomy [nə frôs´tə mē] *Surgery.* the creation of a fistula into the kidney pelvis.

nephrotic [nə frät´ik] *Medicine.* associated with, caused by, or resembling nephrosis.

nephrotic edema *Medicine.* the edema that develops during nephrosis and in the middle stages of diffuse nephritis.

nephrotic syndrome *Medicine.* a condition marked by edema, albuminuria, decreased plasma albumin, doubly refractile bodies in the urine, and an increase in the blood cholesterol level.

nephrotome *Developmental Biology.* one of the segmented divisions of the mesoderm that connects the somite with the lateral plates of unsegmented mesoderm and becomes the source of much of the urogenital system. Also, INTERMEDIATE CELL MASS, MIDDLE PLATE.

nephrotomogram *Radiology.* a cross-sectional X-ray study of the kidney obtained by nephrotomography.

nephrotomography *Radiology.* a visualization of the kidney by tomography after an intravenous introduction of a contrast medium.

nephrotomy *Medicine.* a surgical incision into the kidney.

nephrotoxic *Medicine.* toxic or destructive to the renal cells.

nephrotoxicity *Toxicology.* the property of being poisonous to the kidneys.

nephrotoxin *Toxicology.* any poison that is toxic to the kidneys.

nephsystem see CLOUD SYSTEM.

Nephytyidae *Invertebrate Zoology.* a family of rapid-crawling annelids, in the Errantia, with well-developed sense organs and an eversible pharynx.

Nepidae *Invertebrate Zoology.* a family of sticklike hemipteran insects, the water scorpions, in the superfamily Nepoidea, with a beak capable of piercing skin and a long abdominal tube that allows breathing while crawling in shallow water.

Nepoidea *Invertebrate Zoology.* a superfamily of hemipteran insects, true bugs in the subdivision Hydrocorisae.

nepotism *Behavior.* a type of behavior in which an animal benefits its close relatives.

Nepovirus group *Virology.* a group of multicomponent ssRNA-containing plant viruses that have a wide host range and are transmitted naturally by seed and via soil nematodes; type member tobacco ringspot virus.

Nepticulidae *Invertebrate Zoology.* a family of minute moths, the only family in the superfamily Nepticuloidea, with spines on the wings; larvae are leaf miners on many deciduous trees.

Nepticuloidea *Invertebrate Zoology.* a superfamily of lepidopteran insects with one family, Nepticulidae.

Neptune *Astronomy.* the eighth planet of the solar system; it has a diameter of 48,600 kilometers, a density of 1.76, and a family of 8 moons and several rings; it orbits the sun once every 165 years at an average distance of 4.497 billion kilometers. (Named for the Roman god of the sea.)

neptunian dike *Geology.* a sedimentary dike that has infilled an undersea fissure or hollow.

neptunic rock *Geology.* **1.** any rock formed in the sea. **2.** a general term for all sedimentary rocks.

neptunism *Geology.* the former theory that all rocks of the earth's crust are composed of material deposited from or crystallized out of water. Also, **neptunianism, neptunian theory.**

neptunite *Mineralogy.* $KNa_2Li(Fe^{+2},Mn^{+2})_2Ti_2Si_8O_{24}$, a reddish-black, monoclinic mineral occurring as prismatic crystals, and having a specific gravity of 3.19 to 3.23 and a hardness of 5 to 6 on the Mohs scale; found with benitoite in San Benito County, California.

neptunium *Chemistry.* a radioactive element having the symbol Np, the atomic number 93, an atomic weight of 237.048, and a melting point of 640°C; a silvery-white metal found naturally in traces, with weighable amounts produced as a by-product in plutonium production; isotope 237 of neptunium is used in neutron detection instruments. (Named for the planet *Neptune,* which is the next planet after Uranus as this element is next after uranium in the periodic table.)

neptunium decay series *Chemistry.* a synthetic radioactive decay series beginning with neptunium-237 and ending with bismuth-209.

Nereid [nēr´ē id] *Astronomy.* the second moon of Neptune, discovered in 1949 and measuring 340 kilometers in diameter. (Named after the *Nereids,* the 50 daughters of *Nereus,* a Greek sea god.)

Nereidae *Invertebrate Zoology.* a family of large, predaceous clam or rag worms, polychaetes with a long, multisegmented body, four eyes, and well-developed parapodia; some types are burrowing, some are free-swimming.

Nerillidae *Invertebrate Zoology.* a family of primitive worms in the class Archiannelida, with well-developed parapodia and setae.

Neritacea *Invertebrate Zoology.* a superfamily of primitive snails, gastropod mollusks in the order Aspidobranchia, with a single gill.

neritic *Oceanography.* of or relating to the pelagic area of the shallow marine environment near shore or over a continental shelf, from low-tide mark down to a depth of about 200 meters, characterized by the abundant biotic activity fostered by photosynthesis and the mineral-rich environment.

Neritidae *Invertebrate Zoology.* a family of primitive snails, freshwater and terrestrial, in the order Archeogastropoda, with an operculum and a turbinate shell.

Nernst, Walther H. 1864–1941, German physicist and chemist; received Nobel Prize for research on heat changes in chemical reactions.

Nernst approximation formula *Thermodynamics.* for a gas reaction, a formula that specifies the equilibrium constant, based on certain assumptions and approximations of the Nernst heat theorem.

Nernst bridge *Electricity.* a four-arm bridge that utilizes capacitors instead of resistors to measure capacitance at high-frequency levels.

Nernst diffusion layer *Physical Chemistry.* an approximation used in treatments of electrodes in stirred solutions, in which a very thin layer of solution adjacent to the electrode is assumed to be stationary and penetrated only by diffusion.

Nernst effect see ETTINGSHAUSEN-NERNST EFFECT.

Nernst equation *Physical Chemistry.* an equation demonstrating that the electromotive force developed in an electrochemical cell is determined by the activities of the reacting species, the reaction temperature, and the standard free-energy change of the overall reaction.

Nernst heat theorem *Thermodynamics.* a theorem stating that for a homogeneous thermodynamic system, as the temperature approaches absolute zero, the rate of change of the entropy approaches zero and the entropy itself converges to a minimum value.

Nernst lamp *Electricity.* an electric lamp whose active element is a wire or rod that is formed of magnesium oxide mixed with oxides of rare metals such as zirconium oxide and heated to a brilliant white incandescence by the application of a current through it.

Nernst law (of distribution) *Physical Chemistry.* see DISTRIBUTION LAW.

Nernst layer *Biotechnology.* a thin layer of unstirred solvent that surrounds a particle in a stirred solution. Also, **Nernst-Planck layer.**

Nernst-Lindemann calorimeter *Engineering.* a calorimeter having a platinum heating coil and a heating reservoir of high thermal conductivity; used to determine specific heat at low temperatures.

Nernst statement *Thermodynamics.* the first statement of the third law of thermodynamics, as credited to Walther Nernst; i.e., the statement that for constant-temperature reversible processes, the change in the entropy of a homogeneous system approaches zero as the temperature approaches absolute zero. Also variously called **Nernst-Planck relation, Nernst-Simon statement,** and so on according to its attribution.

Nernst theory *Physical Chemistry.* the principle that in an electrolytic cell an equilibrium will develop between the tendency of the electrode to dissolve and form ions in the solution and the tendency of the ions in the solution to build up on the electrode.

Nernst-Thomson rule *Physical Chemistry.* the rule that little interaction occurs between anions and cations in a solvent with low electrical conductibility, and that the reverse is true in a solvent with high electrical conductibility.

Nernst zero of potential *Physical Chemistry.* the potential energy that is equivalent to the reversible equilibrium in an electrolytic system between charged hydrogen ions and hydrogen gas at a pressure of one standard atmosphere.

nerol *Organic Chemistry.* $C_{10}H_{18}O$, a combustible colorless monoterpene; insoluble in water and soluble in alcohol; boils at 103–105°C (9 torr); derived from isomerization of geraniol and used as a component of orange blossom oil. Also, 3,7-DIMETHYL-2,6-OCTADIENE-1-OL.

nerolidol *Organic Chemistry.* $C_{15}H_{26}O$, a combustible straw-colored sesquiterpene alcohol that boils at 114°C (1 torr) and is very soluble in alcohol; derived naturally from Peru balsam, orange flower, and ylang ylang or manufactured synthetically; used in perfumes and flavoring. Also, PERUVIOL.

nerve *Anatomy.* a macroscopic cordlike structure containing a collection of nerve fibers that transmit impulses between the central nervous system and other parts of the body; the nerve fibers are surrounded by a connective tissue sheath (epineurium) that encloses bundles (funiculi or fasciculi) of nerve fibers, each bundle being surrounded by its own sheath of connective tissue (perineurium).

nerve agent *Ordnance.* a potentially lethal chemical agent that interferes with the transmission of nerve impulses.

nerve block *Physiology.* the temporary prevention of sensory nerve impulses from reaching the center of consciousness, either by chemical injection or electrical means, as a method of anesthesia.

nerve cell *Cell Biology.* a cell of the nervous system; a neuron.

nerve cement see NEUROGLIA.

nerve control *Biology.* the process of regulating life functions in multicellular organisms by transmitting messages through a system of neurons.

nerve cord *Invertebrate Zoology.* one of the ventral paired cords of nervous tissue characteristic of many invertebrates.

nerve deafness *Medicine.* the deafness caused by a lesion of the auditory nerve or of the central neural pathways.

nerve ending *Anatomy.* the point at which a nerve fiber terminates either in a sensory receptor or synapses with other neurons or a motor effector.

nerve fiber *Cell Biology.* an axon or dendrite that branches out from the cell body of a neuron and conducts nerve impulses either toward or away from the cell body; it may be covered by a sheath or neurilemma.

nerve gas *Toxicology.* any of a group of toxic organophosphorus compounds used in chemical warfare.

nerve graft *Surgery.* the insertion of human or animal nerve tissue between the exposed ends of a severed nerve to form a bridge between the two ends so that the nerve may regenerate across it.

nerve growth factor *Endocrinology.* a multimeric protein, the beta subunit of which is required for the proper development and maintenance of the sensory neurons of the dorsal root ganglion and of the postganglionic sympathetic neurons.

nerve impulse *Physiology.* the electrochemical stimulus that moves along a nerve fiber.

nerve net *Invertebrate Zoology.* in coelenterates, an irregular network of nerve cells without any central control; impulses pass in any or all directions resulting in a generalized response.

nerve suture see NEURORRHAPHY.

nerve tracing *Medicine.* a chiropractic technique that is claimed to be a way of exploring peripheral nerves by noting the patient's response to pressure.

Nervi, Pier Luigi 1891–1979, Italian architectural engineer; developed ferro-cement and built vast stadiums and exhibition halls.

nervous *Biology.* **1.** of or relating to the nerves. **2.** originating in, caused by, or affected by the nerves. *Medicine.* describing a condition affecting or involving the nerves. *Psychology.* in popular use, uneasy, apprehensive, or agitated, as if from heightened activity of the nervous system.

nervous breakdown *Psychology.* a popular term, not now in technical use, for the sudden appearance of neurotic or psychotic symptoms of such severity that they impair the individual's ability to carry on the functions of daily life.

nervous irritability *Neurology.* **1.** the ability of a nerve to transmit electric impulses. **2.** the property of nerve tissue that permits excitement by a stimulus.

nervous system *Anatomy.* the intricate network of structures controlling the actions and reactions of the body and permitting response to conditions of the surrounding environment; it involves the activity of billions of specialized nerve cells (neurons), about half of which are in the brain. In humans and other vertebrates this system has three main parts: the central nervous system, which consists of the brain and the spinal cord; the peripheral nervous system, which consists of pairs of nerves that originate in the brain and spinal cord; and the autonomic nervous system (sometimes considered part of the peripheral system), which regulates involuntary processes such as breathing and digestion. *Invertebrate Zoology.* a similar but less complex system in invertebrate organisms.

nervous tissue *Histology.* the tissue of the nervous system, consisting of neurons, their supporting cells (neuroglia), and connective tissue with a rich vascular supply.

nervure *Invertebrate Zoology.* a vein in an insect's wing, a chitinous structure that stiffens the wing. Also, **nervule.**

nesidioblastoma *Oncology.* an islet-cell tumor of the pancreas.

Nesiotinidae *Invertebrate Zoology.* a family of biting lice in the order Mallophaga; parasitic on penguins.

nesistor *Electronics.* a type of transistor that is characterized by an inverse relationship between current and voltage, and by the influence of an electric field on current flow.

Nesmejanow reaction *Organic Chemistry.* the production of an arylmercuric chloride from an aryldiazonium chloride by treatment with mercuric chloride and copper powder suspended in acetone.

Nesophontidae *Paleontology.* an extinct family of small, shrewlike, insectivorous mammals that lived in the West Indies in the Pleistocene; some authorities place them with the zalambdodonts and others with the Erinaceomorpha (Lipotyphla).

nesosilicate *Mineralogy.* one of a group of silicate minerals having a characteristic structure in which individual SiO_4-tetrahedra are linked only by ionic bonds from interstitial cations. Also, ORTHOSILICATE.

nesquehonite *Mineralogy.* $Mg(HCO_3)(OH)\cdot 2H_2O$, a colorless to white, monoclinic mineral occurring in radiating groups of prismatic crystals, and having a specific gravity of 1.85 and a hardness of 2.5 on the Mohs scale; a recent product formed under atmospheric temperature-pressure conditions.

ness see PROMONTORY, def. 1.

Nessler's reagent *Analytical Chemistry.* a solution of mercuric iodide in potassium iodide used to detect ammonia; a brown color indicates a positive reaction.

Nessler tube *Analytical Chemistry.* a flat-bottomed, high-quality, calibrated quartz tube filled with sample solutions for use in spectrophotometry.

nest *Vertebrate Zoology.* **1.** a receptacle or spot prepared by a bird or reptile to hold its eggs and in which to hatch and rear its young. **2.** to build a nest or to settle in a nest in order to lay eggs. *Geology.* a small, isolated mass or concentration of some relatively conspicuous element of a geologic feature within another formation. *Computer Programming.* **1.** to insert a command or group of commands within (that is, anywhere between the beginning and end of) another logical group of commands. **2.** broadly, to insert any structure (such as data, or a loop or subroutine) within another similar structure.

nested *Volcanology.* **1.** of volcanic cones, craters, or calderas that occur one within another. **2.** of two or more intersecting calderas that were formed at different times or by different explosions.

nested loop *Computer Programming.* a loop that is entirely embedded within another loop; the inner loop executes a prescribed number of times for each iteration of the outer loop.

nested sets *Mathematics.* let (X_α) be a family of sets indexed over a totally ordered set A (e.g., the real numbers, positive integers, and the like). The sets X_α are then said to be nested if, for all α and β in A, $\alpha > \beta$ implies that $X_\alpha \supseteq X_\beta$. Equivalently, a family of sets is said to be nested if, given any two sets in the family, one is necessarily contained in the other.

nested subroutine *Computer Programming.* a subroutine that is called only from within another subroutine.

nesting *Computer Programming.* the inclusion of a group of instructions within another group, such as a DO loop or a set of IF statements.

Nestor *Astronomy.* asteroid 659, discovered in 1908 and measuring 115 kilometers in diameter; it is a Trojan asteroid and belongs to type C. (Named for a legendary Greek leader noted for his wisdom.)

nest planting *Forestry.* a pattern of planting in which a number of tree seedlings are put close together in a prepared hole.

nest relief *Behavior.* a behavior pattern among birds in which the female and male alternate in the care and feeding of their young in the nest.

net *Materials.* a sturdy, open-mesh fabric formed by knotting or weaving cord; used for various purposes, such as for catching fish. *Telecommunications.* a group of intercommunicating stations or facilities; a network. *Geology.* a pattern of horizontal ground whose mesh is between a circle and a polygon. *Petrology.* a stereographic projection from a sphere, forming a coordinate system of meridians and parallels used to study the orientation and distribution of planes and points related to structural petrology. *Mathematics.* an indexing of a set (such as elements of a space X) by elements from a directed set rather than a totally ordered set; used to generalize the notion of convergence. Also, MOORE-SMITH SET.

net *Metrology.* describing or being a total amount, especially a total that makes allowance for certain reductions or subtractions from a larger overall amount.

net aboveground production *Ecology.* the total accumulation of organic matter by the aboveground parts of plants, as measured over a given period of time. Similarly, **net aerial production.**

net assimilation rate *Ecology.* the net amount of dry matter acquired by a given plant over a certain period of time.

net balance *Hydrology.* the change in the mass of a glacier from the time of minimum mass in one year to the time of minimum mass in the next year. Also, **net budget.**

net blotch *Plant Pathology.* a disease of barley caused by the fungus *Helminthosporium teres* and characterized by irregular spotting on the leaves.

net dietary protein energy ratio *Nutrition.* the protein content of a diet or food expressed mathematically as protein energy multiplied by net protein utilization divided by total energy.

net ecosystem production *Ecology.* the total amount of energy accumulated from organic matter by an ecosystem, minus the energy expended in respiration.

net energy *Nutrition.* metabolizable energy measured by combustion of food in a bomb calorimeter. The net energy may be more than the energy produced when the food is metabolized in the human body.

net floor area *Building Engineering.* a building's gross floor area minus the circulation area, i.e., where people walk, and mechanical area, i.e., where equipment is located.

net flow area *Design Engineering.* the net area attributed to liquid or material flow after a burst occurs from a ruptured element, such as a pipe.

net-load capacity *Robotics.* the net amount of material, by weight or mass, that a robot can handle without being overloaded.

net loss *Telecommunications.* **1.** the total loss of a transmission circuit. **2.** the ratio of the power of the input of a transmission circuit to the power of the output, usually expressed in decibels.

net power flow *Electromagnetism.* the resultant electromagnetic power flowing in a waveguide as determined by the difference between the powers propagating in opposing directions.

net primary production *Ecology.* the total amount of energy acquired by green plants during photosynthesis, minus the energy they lose through respiration.

net production *Ecology.* the total amount of energy or nutrients taken in by an organism minus the energy expended in respiration.

net production efficiency *Ecology.* the percentage of energy acquired by an organism that is channeled into growth and reproduction.

net production rate *Ecology.* the amount of energy assimilated by a system, minus the amount of energy that is lost through respiration, decomposition, predation, and so on.

net protein utilization *Nutrition.* a ratio used to measure the biological value of protein in terms of the amount of dietary protein retained in the body under specific experimental conditions.

net reproductive rate *Ecology.* the total amount of young that a newly born female can expect to bear during a lifetime.

net secondary production *Ecology.* the total amount of energy accumulated by a primary consumer (plant eater), minus the energy it uses in respiration.

net shape *Metallurgy.* the attribute of a shaped part, whether cast, forged, or sintered from powder, that requires virtually no machining in its finishing operation

net-shape processing *Mechanical Engineering.* any of a variety of manufacturing processes that produce components with finished functional surfaces (such as the airfoil portion of forged blades or the teeth of forged gears), so that additional metal removal is not required.

net slip *Geology.* the distance between two points on either side of a fault that were adjacent to each other before movement occurred, as measured on or parallel to the fault surface.

net sweep *Ordnance.* in naval mine warfare, a sweep by two ships using a netlike device that is designed to collect drifting mines or scoop them up from the sea bottom.

net thrust *Aviation.* the gross thrust of a jet engine minus the drag created by the momentum of the incoming air.

nettle *Botany.* **1.** the common name for various plants of the genus *Urtica* that bear stinging hairs, such as the **stinging** or **great nettle**, *U. dioica*. The top leaves and shoots of young nettle plants are sometimes eaten as food. **2.** any of various other plants with stinging hairs, such as the **horse nettle,** *Solanum carolinense*.

nettle

nettle family *Botany.* another name for the family Urticaceae, including nettles of the genus *Urtica* and various other plants.

net ton *Metrology.* see TON, def. 1.

net tonnage *Naval Architecture.* a vessel's internal space usable for carrying cargo or passengers; equal to gross tonnage minus spaces allocated to engines, crew quarters, and so forth.

net-veined *Biology.* having a network of veins, such as a leaf or an insect wing.

network a netlike combination or pattern in which different elements are joined; specific uses include: *Telecommunications.* **1.** a number of radio or television stations that are connected or affiliated so that they can broadcast the same program at the same time. **2.** a system of telephone lines, switches, and signal repeaters that connect all users so that communication can take place. *Computer Technology.* a loosely coupled group of functional units, such as computers. The computers, called nodes of the network, exchange messages over communication links. *Electricity.* **1.** the interconnection of electric elements such as resistors, coils, and capacitors. **2.** of or relating to such a system. Thus, **network admittance, network analysis, network impedance, network synthesis, network theory.** *Mathematics.* a weighted digraph; i.e., a directed graph, each of whose arcs has been assigned a quantity (called a weight or capacity) such as a real number, vector, or tensor. The theory of networks has applications in management and engineering sciences.

network analyzer *Computer Technology.* a computer with appropriate sensors that is used to investigate signal conditions during network switching or when a fault occurs, characteristics of network traffic, and other characteristics.

network architecture *Telecommunications.* the logical organization of hardware and software elements that make up switching nodes, transmission links, terminal nodes, and operations and administrative systems, along with their interfaces and communications protocols that together define a telecommunications network.

network constant *Electricity.* a value of the resistance, inductance, mutual inductance, or capacitance in a network.

network covalent substance *Physical Chemistry.* a substance that is a three-dimensional array of covalently bonded atoms; such materials tend to have greater hardness and higher melting points; e.g., diamonds, quartz.

network data structure *Computer Programming.* a relationship between records in a database in which a child record can have more than one parent record. Also, PLEX STRUCTURE.

network filter *Electricity.* the use of elements such as coils, resistors, and capacitors in such a way that a relatively small signal attenuation occurs at a certain frequency while great attenuation applies to other frequencies.

network former *Materials Science.* any ionic material that will form a network with other ionic materials to form the coordination structure required for a glass.

networking *Computer Technology.* **1.** a method of linking computer systems by means of communication lines. **2.** the designing of distributed processing system connectivity. *Industrial Engineering.* in popular use, the practice of sharing information, advice, and services among various individuals.

network layer *Computer Technology.* layer three of the seven-layer Open System Interconnection (OSI) model for network architectures; it provides functions of internal network operations, including addressing and routing.

network master relay *Electricity.* a relay that is used to close an alternating-current, low-voltage network protector.

network modeling *Design Engineering.* the process of developing a procedure to simulate the behavior of a network.

network phasing relay *Electricity.* two relays, one of which is a master relay, that are used to limit the closure of a network protector to a predetermined arrangement between the voltage and the network voltage.

network polymer *Materials Science.* any polymer that becomes interconnected with sufficient internal bonds to form a large three-dimensional network.

network relay *Electricity.* a voltage or power type relay designed to protect and control a low-voltage AC network.

network transfer admittance *Electricity.* a complex quantity representing the ratio of current that flows from the output terminals of a network to the voltage across the input terminals.

Neuberg's fermentations *Biochemistry.* three types of alcohol fermentations catalyzed by yeast: neutral to slightly acid pH, with sodium bisulfite, and at an alkaline pH.

Neufeld reaction see QUELLUNG REACTION.

Neumann, Carl Gottfried [noiˊmən] 1832–1925, German mathematician; orginated logarithmic potentials.

Neumann approximation method *Mathematics.* an iterative method for solving the Fredholm integral equation of the second kind

$$\phi(x) = f(x) + \lambda\int_a^b K(x,y)\phi(y)dy.$$

The solution, called a *Neumann series,* is of the form

$$\phi(x) = f(x) + \sum_{n=1}^{\infty} \lambda^n \int_a^b K^{(n)}(x,y)f(y)dy,$$

where $K^{(n)}$ is the $(n-1)$th iterated kernel and $K^{(1)} = K(x,y)$.

Neumann boundary condition *Mathematics.* in a Neumann problem, the condition that $\partial u/\partial n = h$ on the boundary ∂D of some bounded domain D in R^n on which the given function h is defined, where $\partial/\partial n$ is the normal derivative with respect to ∂D.

Neumann function *Mathematics.* any solution to a Neumann problem.

Neumann-Kopp rule *Thermodynamics.* a rule claiming that the heat capacity of one mole of a substance is approximately equal to the sum of the heat capacities of the gram-atom amounts of the elements times the number of atoms in the molecule of the substance.

Neumann problem *Mathematics.* a boundary-value problem of the second kind for the Laplace equation $\Delta u = F(u_1, \ldots, u_n)$; i.e., it is required to find a function u subject to the condition that $\partial u/\partial n = h$ on the boundary ∂D of some bounded domain D in R^n on which the given function h is defined, where $\partial/\partial n$ is the normal derivative with respect to ∂D. Neumann problems occur in potential theory applications.

Neumann's formula *Electromagnetism.* a formula that gives the mutual inductance M_{12} between two neighboring closed circuits C_1 and C_2, given by $(m/4\pi)$ times the double integral over the current path elements ds_1 and ds_2 of the quantity $1/r$, where r is the distance between the current elements, and M is the permeability of the intervening medium.

Neumann's principle *Crystallography.* a principle stating that the physical properties of a crystal are never of lower symmetry than the symmetry of the external form of the crystal.

neur- or **neuro-** a prefix meaning "nerve."

Neuradaceae *Botany.* a family of prostrate, densely hairy annual herbs of the order Rosales, characterized by mucilage ducts in the pith and native to deserts in Africa to the Middle East and India.

neuragmia *Neurology.* the tearing or rupturing of a nerve trunk.

neural *Anatomy.* of or relating to a nerve or the nervous system.

neural arc *Physiology.* a pathway for nerve impulses including two or more neurons, one an afferent neuron and another an efferent neuron.

neural canal *Developmental Biology.* the canal formed by the arches of successive vertebrae that encloses the spinal cord and meninges. Also, CANALIS VERTEBRALIS, MEDULLARY CANAL, SPINAL CANAL.

neural computing *Computer Science.* a type of computing dealing with brainlike machines in which the neurons are modeled by electronic circuits or software simulations of mathematical models.

neural crest *Developmental Biology.* embryonic cells derived from the margins of the neural plate that give rise to pigment cells, cranial and spinal ganglia, and many other structures.

neural ectoderm *Developmental Biology.* the region of the ectoderm that develops into the neural tube. Also, NEURECTODERM.

neural fold *Developmental Biology.* the raised edge of a neural plate, as the plate rolls to form the neural tube. Also, MEDULLARY FOLD.

neuralgia [noo ralˊjə] *Neurology.* a paroxysmal pain that extends along one or more nerves; varieties are distinguished according to either the site affected, such as brachial or facial neuralgia, or the cause, as anemic or diabetic neuralgia. Also, NEURODYNIA.

neuralgic [noo ralˊjik] *Neurology.* of or relating to pain in an area that is served by a nerve.

neural groove *Developmental Biology.* the groove produced by the invagination of the neural plate of the embryo during the formation of the neural tube. Also, MEDULLARY GROOVE.

neural lymphomatosis *Veterinary Medicine.* **1.** a widespread contagious disease of domestic poultry belonging to the avian leukosis complex, caused by a herpes virus, and characterized by depression, paralysis, and enlargement of the peripheral nerves by lymphoid cell infiltration. **2.** a former name for Marek's disease.

neural network *Artificial Intelligence.* a computational network, often for pattern recognition, composed of mathematically defined elements that are thought to approximate the working of biological neurons; often composed of a layer that receives and organizes inputs, a hidden layer, and an output layer in which individual neurons identify particular patterns. Networks can be trained by back-propagation.

neural plate *Developmental Biology.* a thickened plate of ectoderm in the embryo that develops into the neural tube. Also, MEDULLARY PLATE.

neural spine *Anatomy.* the medial dorsal spinous process of a vertebra, to which the muscles of the back attach.

neural tube *Developmental Biology.* the epithelial tube developed from the neural plate and forming the central nervous system of the vertebrate embryo. Also, CEREBROMEDULLARY TUBE, MEDULLARY TUBE.

neuraminic acid *Biochemistry.* $C_9H_{17}NO_8$, a 9-carbon amino sugar; acid; its acylated derivatives are sialic acids which occur in glycoproteins, glycolipids, and polysaccharides.

neuraminidase *Virology.* a glycoprotein enzyme that interacts with linkages between terminal sialic acids and sugar residues on the surface of paramyxoviruses and orthomyxoviruses.

neurapraxia *Medicine.* a failure of conduction in a nerve in the absence of structural changes, due to blunt injury, compression, or ischemia; return of function normally ensues.

neurasthenic neurosis *Psychology.* a condition in which an individual feels constantly fatigued or suffers mild physical discomforts caused by some internal conflict, such as unresolved anger or a sense of failure. Also, **neurasthenia**.

neuratrophy *Neurology.* atrophy of muscle or other tissue due to disease, dysfunction, or impaired nutrition of a nerve. Also, **neuratrophia.**

neurectomy *Medicine.* the surgical removal of a segment of a nerve.

neurenteric canal *Developmental Biology.* the canal interconnecting the neural tube and the archenteron of the embryo of some vertebrates.

neurilemma *Histology.* a very thin layer covering the axon or myelin sheath of peripheral nerve fibers.

neurilemmoma *Medicine.* a benign neoplasm in which the primary elements are structurally identical to Schwann cells; neoplastic cells proliferate within the endoneurium, and the perineurium forms the capsule. Also, **neurinoma.**

neurine *Biochemistry.* $C_5H_{13}NO$, a poisonous ptomaine, vinyl trimethyl ammonium hydroxide, found in decaying fish, in fungi, and in the brain and other normal tissues.

neuristor *Electronics.* a device that simulates the actions of a human nerve cell by generating signals with uniform velocity and without attenuation.

neuritic *Neurology.* affected by neuritis.

neuritis *Medicine.* an inflammation of a nerve, characterized by neuralgia, hyperesthesia, anesthesia or parasthesia, paralysis, muscular atrophy in the area fed by the affected nerve, and by abolition of the reflexes.

neuroanatomy *Anatomy.* the division of anatomy involving the study of the nervous system.

neuroarthopathy *Medicine.* any disease of joint structures associated with disease of the central or peripheral nervous system.

neuroastrocytoma *Medicine.* a glioma primarily made up of astrocytes, similar to an astrocytoma; while it is usually found in the floor of the third ventricle and temporal lobes, it can be found anywhere in the central nervous system.

neurobehavioral *Neurology.* relating to the status of the nervous system during observation of behavior.

neurobiology *Neurology.* the biology of the nervous system, including its structure, function, biochemistry, and pathology. Thus, **neurobiologist.**

neurobiotaxis *Evolution.* a theory stating that, over the course of phylogenetic evolution, nerve tissue and ganglia tend to migrate toward the regions of an organism's anatomy that receive the greatest stimulation.

neuroblast *Developmental Biology.* an embryonic cell that produces neurons.

neuroblastoma *Oncology.* a sarcoma that is composed of primitive neural tissue and occurs most often in children under ten years of age, developing in either the autonomic nervous system or the adrenal medulla; widespread metastases are common.

neurobrucellosis *Medicine.* brucellosis with neurologic involvement, characterized by symptoms of meningitis, encephalitis, or neuritis.

neurocele *Anatomy.* another term for the central cavity of the cerebrospinal axis.

neurochemistry *Neurology.* the branch of neurology that is concerned with the chemistry of the nervous system.

neurochorioretinitis *Medicine.* inflammation of the choroid coat of the eye, the retina, and the optic nerve.

neurochoroiditis *Medicine.* an inflammation of the choroid coat of the eye and the optic nerve.

neurocirculatory *Anatomy.* of or related to the nervous and circulatory systems.

neurocirculatory asthenia *Medicine.* a psychosomatic stress disorder of nervous and circulatory abnormalities marked by fatigue, palpitation, dyspnea, rapid pulse, precordial pain, and anxiety; observed in soldiers on active duty. Also, DA COSTA'S SYNDROME, SOLDIER'S HEART.

neurocranium *Anatomy.* the part of the cranium that envelops the brain.

neurocutaneous *Anatomy.* of or related to the nerves of the skin.

neurocyte see NEURON.

neurocytology *Neurology.* the branch of neurology that is concerned with the cellular components of the nervous system.

neurocytolysin *Toxicology.* a poison found in the venom of certain snakes that is toxic to nerve cells.

neurocytoma *Oncology.* a brain tumor that is composed of undifferentiated nerve cells. Also, **neuroepithelioma.**

neurodermatitis *Medicine.* a nonspecific pruritic skin disorder presumed to result from prolonged vigorous scratching, rubbing, or pinching, sometimes forming polymorphic lesions. Also, **neurodermatosis.**

neurodiagnosis *Neurology.* the diagnosis of diseases and disorders of the nervous system.

Neurodontiformes *Paleontology.* a suborder of conodonts in the order Conodontophorida, characterized by teeth composed of layers of fibers; widespread throughout the Paleozoic.

neurodynia see NEURALGIA.

neuroelectricity *Physiology.* the electrical activity produced by nervous system cells.

neuroendocrine *Biology.* relating to both the nervous and endocrine systems, structurally and functionally.

neuroendocrinology *Biology.* the study of the structural and functional interrelationships between the nervous and endocrine systems.

neuroepidermal *Biology.* pertaining to both nerves and epidermis, structurally and functionally.

neuroepithelioma *Medicine.* a brain tumor consisting of undifferentiated cells of nervous origin. Also, **neurocytoma.**

neurofibril *Cell Biology.* one of many fine fibers present in the axon and dendrites of a neuron.

neurofibroma *Oncology.* a benign, moderately solid, nonencapsulated tumor that originates from an abnormal proliferation of Schwann cells and that includes nerve fibers. Also, SCHWANNOMA, FIBRONEUROMA.

neurofibromatosis *Medicine.* a familial condition that is marked by small, individual, pigmented skin lesions, sometimes known as cafe-au-lait spots or pigmented nevi, which become numerous, slow-growing, often subcutaneous neurofibromas, and which are present along peripheral nerves. These are sometimes related to acoustic neurinomas or other intracranial neoplasms. Also, VON RECKLINGHAUSEN'S DISEASE, NEUROMATOSIS, MULTIPLE NEUROFIBROMAS.

neurofibrosarcoma *Oncology.* a malignant, moderately solid, nonencapsulated tumor that originates from a proliferation of Schwann cells and that includes nerve fibers; a malignant form of schwannoma.

neurofilament *Cell Biology.* one of several types of intermediate filaments that occur in nerve cells and act to strengthen the nerve axon.

neurogenesis *Developmental Biology.* the development of nervous tissue.

neurogenic *Neurology.* **1.** originating in the nervous system. **2.** forming nervous tissue or stimulating nervous energy. *Medicine.* describing a condition developing in, or caused by, the nervous system or nerve impulses. Also, **neurogenetic.**

neurogenic bladder *Medicine.* the abnormal functioning of the bladder due to damaged nerves.

neurogenic shock *Medicine.* shock produced when the nervous system causes vasodilation.

neurogenic ulcer see NEUROTROPHIC ULCER.

neuroglia *Neurology.* the supporting cells of the central nervous system, consisting of a fine web of tissue made up of the macroglia and the microglia. Also, NERVE CEMENT.

neuroglial cell see GLIAL CELL.

neuroglioma *Oncology.* a tumor that is composed of neuroglial cells. Also, **neurogliocytoma.**

neurography *Neurology.* a written study or description of the nervous system.

neurohemal organ *Zoology.* a bulbous structure consisting of clusters of nerve endings that work in conjunction with blood vessels to store and release neural hormones into the blood.

neurohormonal *Neurology.* of or relating to hormones that originate in the nervous system.

neurohormone *Endocrinology.* a substance that is known to exist and exert hormonal effects outside the central nervous system; it is also found within the neurons of the central nervous system, where it may function as a neurotransmitter rather than a hormone.

neurohumor *Neurology.* a chemical substance produced in a neuron and capable of activating or modifying the function of a nearby neuron, muscle, or gland.

neurohypophysis *Anatomy.* the posterior lobe of the pituitary gland; the storage and release site for vasopressin, oxytocin, and neurophysins. Also, PARS NERVOSA.

neuroleptanalgesia *Medicine.* an intense analgesic and amnesic condition caused by a combination of neuroleptic drugs and the inhalation of a weak anesthetic.

neuroleptic *Pharmacology.* of or relating to the effects on cognition and behavior of antipsychotic drugs that reduce confusion and agitation in psychotic patients.

neurologia see NEUROLOGY.

neurologic *Neurology.* of or relating to neurology.

neurologist [nůr äl´ə jist] *Medicine.* a specialist in neurology or in the treatment of disorders of the nervous system.

Neurology

In current usage, the term "neurology" refers to a specialty field in medicine concerned with the diagnosis and treatment of nervous system disorders. However, in its broadest sense, neurology means the study of the nervous system in both health and disease, and this implies a much broader purview, one that is best embraced by the collective term, "the neurosciences," and which includes neurobiology, neurophysiology, neuroanatomy, neuropathology, neuropsychology, neuropsychiatry, and so forth.

The nervous system is composed of two major subdivisions, the central nervous system (spinal cord, brain, and retina) and the peripheral nervous system (nerve fiber networks through which the central system receives messages from and sends messages to functional components of the body).

Modern society tends to worship computers for their complexity. However, computers do not have any intelligence that was not created for them by the human intellect. The physical universe is marvelously complex, but it cannot study and analyze either its own self or its biological inhabitants. The remarkable cognitive properties of the human brain permit it to study and decipher almost anything in its physical or biological surroundings, including the mental operations, both normal and abnormal, of other human brains.

In addition, a given human brain can introspectively examine and analyze its own mental operations and productions. Thus there is nothing known to science that is more complicated than the normal human nervous system, except perhaps the abnormal (diseased) human nervous system; achieving a better understanding of both is the quintessential challenge in neurology.

In summary, neurology is a discipline in which the human nervous system studiously examines both its own and other nervous systems, in both health and disease, toward the ultimately unattainable goal of completely deciphering the incomprehensibly complex organizational plan conferred upon these systems by another even more complicated and mysterious entity we call God.

John W. Olney
Professor of Psychiatry and Neuropathology
Washington University School of Medicine

neurology [nůr äl′ə jē] *Medicine.* the study of the structure, function, and pathology of the nervous system and the treatment of its disorders. Also, NEUROLOGIA.

neurolysis [nůr äl′ə sis] *Neurology.* **1.** the release of a nerve sheath from compression by cutting it lengthwise. **2.** the destruction or dissolution of nerve tissue. **3.** the relief of tension upon a nerve obtained by stretching. *Surgery.* the operative breaking up of perineural adhesions.

neurolytic *Neurology.* causing destruction of nerve substance or nerve structures.

neuroma *Oncology.* a tumor or new growth that is composed primarily of nerve cells or nerve fibers; a tumor that grows from a nerve.

neuromast *Vertebrate Zoology.* in fish and amphibians, one of a cluster of microscopic sensory organs that determine changes in pressure and current; they are freely distributed on the skin, but usually follow nerve paths and are most common on the lateral line.

neuromatosis see NEUROFIBROMATOSIS.

neuromere *Developmental Biology.* any of the series of transitory segmental elevations in the wall of the neural tube of the developing embryo.

neuromittor *Neurology.* a terminal structure at the peripheral end of a neuron that transfers a stimulus to the receptor terminal of an adjoining neuron.

neuromuscular *Biology.* relating to both nerves and muscles.

neuromuscular junction *Anatomy.* the area where synapse occurs between a motor nerve fiber and the striated muscle fiber that it innervates; the expansion of the terminal branch of the axon forms the motor end plate and the opposed region of the surface of the muscle fiber forms the postsynaptic region.

neuromuscular spindle *Neurology.* a sensory end organ in skeletal muscle in which afferent and a few efferent nerve fibers terminate.

neuromyasthenia *Medicine.* muscular weakness, usually associated with emotional stress.

neuromyelitis *Medicine.* an inflammation of nerves and of the spinal cord. Also, MYELONEURITIS.

neuron *Cell Biology.* one of the two principal types of nerve cells in the nervous system, usually consisting of a cell body, an axon, several dendrites, and specialized axon terminals, that is able to receive and conduct electrical impulses. Also, NEUROCYTE, NERVE CELL.

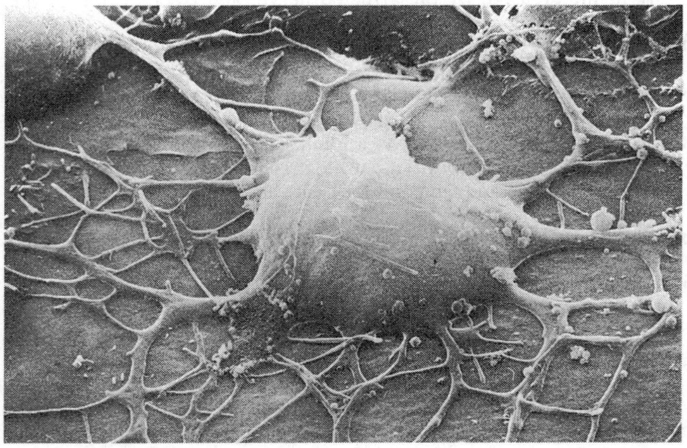

neurons

neuron doctrine *Biology.* the concept that the neuron is the basic structural and functional unit of the nervous system, and that it acts upon other neurons through the synapse.

neuronitis *Medicine.* a formerly used term describing a condition of unknown origin that is marked by the breakdown of nerve fibers, is sometimes associated with inflammatory cell reaction, and is found in the more proximal part of the peripheral nervous system.

neuropathy *Medicine.* any abnormal change or functional disturbance of the peripheral nervous system.

neuropharmacology *Medicine.* the branch of pharmacology dealing with the effects of drugs upon the nervous system.

neurophysiology *Physiology.* the discipline involving the study of the makeup and function of the nervous system; physiology of the nerves.

neuropil *Histology.* the dense network of interwoven axons and dendrites of neurons and neuroglial cells found in the central nervous system and parts of the peripheral nervous system.

neuroplasm *Neurology.* the unstructured cytoplasm of the fibrils of nerve cells.

neuroplasmic *Neurology.* of or relating to neuroplasm.

neuroplasty *Surgery.* plastic surgery of a nerve.

neuropodium *Cell Biology.* one of the terminal branches of an axon, from which electrical impulses are passed on to other cells.

neuropore *Developmental Biology.* a temporary opening at the anterior or posterior end of the neural tube of the early embryo.

neuropsychology *Psychology.* the study of the relationship between the nervous system and behavior.

neuropsychopathy *Medicine.* a diseased condition of the nerves and mind.

Neuroptera *Invertebrate Zoology.* an order of insects that includes the lacewings, ant lions, and dobsonflies, having long antennae, biting mouthparts, and two pairs of membraneous, many-veined wings; some larvae feed on insect pests.

neuroradiology *Medicine.* radiology of the nervous system.

neurorrhaphy *Surgery.* the suturing of a divided or separated nerve. Also, NERVE SUTURE.
neurorrhexis *Medicine.* the rupture of a nerve.
neurosarcoma *Oncology.* a malignant tumor that is composed of nerve tissue, connective tissue, and vascular tissue.
neuroscience *Neurology.* any of the branches of science that deal with the anatomy, physiology, biochemistry, or molecular biology of the nervous system, especially as related to behavior and learning.
neurosclerosis *Medicine.* the hardening of a nerve or nerve center.
neurosecretion *Physiology.* **1.** the mechanism by which nerve cells synthesize and release hormones, vasopressin, neurotransmitters, and other substances into the bloodstream. **2.** the product of this mechanism; a substance released through neurosecretion.
neurosis [noo rō´sis] *plural,* **neuroses.** *Psychology.* a general term applied to a variety of relatively mild disorders or conditions, typically characterized by anxiety states and phobias but not involving an inaccurate sense of reality or affecting the entire personality.
neuroskeleton see ENDOSKELETON.
neurosome *Neurology.* **1.** the cytoplasmic mass or body of a nerve cell. **2.** one of the minute granules in the ground substance of the cytoplasm of a neuron.
neurospasm *Neurology.* the nervous twitching of a muscle, caused by dysfunction of a nerve.
Neurospora *Mycology.* a genus of fungi belonging to the subdivision Ascomycotina that is found on decaying or burnt vegetation; widely used in physiological and genetic research.
Neurospora crassa *Mycology.* a species of fungi belonging to the genus *Neurospora* that causes pink bread mold; widely used in biochemical genetic research and development.
neurostatus *Neurology.* the state or condition of neural symptoms as described in a case history.
neurosurgeon *Medicine.* a physician who specializes in surgery of the nervous system.
neurosurgery *Medicine.* surgery of the nervous system.
neurosyphilis *Medicine.* the nervous system symptoms of syphilis, including tabes dorsalis, general paresis, and meningovascular syphilis.
neurotagma *Neurology.* the linear arrangement of the components of a nerve cell.
neurotensin *Endocrinology.* a peptide hormone of the gastrointestinal tract and the hypothalamus having diverse biological effects on the vasculature, digestive system, and anterior pituitary.
neuroterminal *Neurology.* an end organ of a peripheral nerve.
neurotic [noo rät´ik] *Psychology.* **1.** relating to or affected by neurosis. **2.** a person who exhibits symptoms of neurosis.
neurotic disorder see NEUROSIS.
neuroticism [noo rät´i siz´əm] *Psychology.* **1.** a tendency to become anxious in relatively nonstressful situations, measured in some types of psychological testing to determine if an individual is likely to develop a neurosis. **2.** the fact of being neurotic.
neurotic personality *Psychology.* an individual whose traits could develop into a neurosis if he or she is placed in a stressful situation, such as a fearful individual developing agoraphobia after losing a job.
neurotome *Surgery.* a needlelike knife used for dissecting nerves. *Developmental Biology.* see NEUROMERE.
neurotoxicity *Toxicology.* the property of being toxic to nerve tissue.
neurotoxin *Neurology.* any toxin that has the ability to damage or destroy nerve tissues.
neurotransmitter *Neurology.* any of a group of chemical substances that transmit nerve impulses across a synapse.
neurotrauma *Neurology.* any mechanical injury of a nerve.
neurotripsy *Surgery.* the surgical crushing of a nerve.
neurotrophic ulcer *Medicine.* an ulcer due to a nervous disorder or to emotional factors. Also, NEUROGENIC ULCER.
neurotropic *Biology.* having an affinity for nerve tissue.
neurotropic virus *Virology.* any virus that has a predilection for and causes infection in nerve tissues, such as the rabies virus.
neurotubule *Neurology.* any of the long, narrow tubules found in neurons that, along with neurofilaments, form neurofibrils, and assist in the growth and transport of intercellular materials.
neurovaricosis *Medicine.* a varicose condition of nerve fibers.
neurovascular *Biology.* pertaining structurally and functionally to both the nervous and vascular structures.
neurulation *Developmental Biology.* the formation of the neural plate in the early embryo, followed by its closure with the development of the neural tube.
neuston *Biology.* minute organisms, such as mosquito larvae, that float or swim on the surface of water.
neuter *Biology.* having no sexual organs; asexual. *Zoology.* having imperfectly developed sexual organs in the adult, as the worker bee. *Agriculture.* **1.** to sterilize an animal by castrating or spaying. **2.** an animal that has been sterilized.
neutral describing an inactive or indefinite state, especially an intermediate state in a frame of reference that has two active or definite states; specific uses include: *Military Science.* describing a nation or group that takes neither side in a war or other conflict between two opposing forces. *Chemistry.* describing a substance that is neither acid nor alkaline. *Electricity.* **1.** the condition of having no net electric charge, when an equal number of electrons and protons are present. **2.** relating to or in such a condition; not charged. *Biology.* having no sexual organs; neuter. *Mechanical Engineering.* the position or condition of a gear system or other arrangement of moving parts in which the parts are not engaged.
neutral atmosphere *Physical Chemistry.* an environment that neither oxidizes nor reduces the materials contained within it.
neutral atom *Atomic Physics.* an atom in which there is the same number of electrons and protons, so that there is no net electric charge.
neutral axis *Mechanics.* the line in a cross section of a beam subjected to bending, at which the strain is zero.
neutral beam *Physics.* a narrow beam of uncharged particles, usually in a vacuum.
neutral conductor *Electricity.* in a polyphase system, a conductor that does not carry current, unless the system becomes unbalanced.
neutral current interaction *Particle Physics.* a weak interaction in which no electric charge is exchanged between the colliding particles.
neutral-density filter *Optics.* a filter that diminishes the intensity of transmitted light without significantly changing its color; used to decrease the amount of light entering a camera if the use of a smaller stop is not possible. Also, GRAY FILTER, NEUTRAL FILTER.
neutral equilibrium *Mechanics.* the condition of a system in equilibrium, such that any net force or torque that the system experiences will move the system away from its original equilibrium state to a new equilibrium state.
neutral estuary *Geography.* an estuary where freshwater inflow balances outflow and evaporation, thus maintaining a nearly constant level of salinity.
neutral fats *Biochemistry.* fats made up entirely of triglyceride esters, and having no free fatty acids.
neutral fiber *Mechanics.* the line of zero strain in a longitudinal section.
neutral filter see NEUTRAL-DENSITY FILTER.
neutral flame *Chemistry.* a gas flame that is neither oxidizing nor reducing.
neutral granulation *Chemistry.* a form of propellant granulation in which the grain's surface area stays constant throughout the burning process.
neutral ground *Military Science.* an intermediate area that is not held by either of two opposing forces or that is protected from military action by agreement of the forces. *Electricity.* an intentional ground applied to the neutral conductor or neutral point of a circuit, transformer, motor, generator, or other system.
neutralism *Science.* the fact of being neutral. *Ecology.* an interaction between two species in which neither one is affected by the presence or activities of the other.
neutralization the act or process of making neutral; specific uses include: *Chemistry.* the process of neutralizing a solution. *Virology.* the process of inactivating or preventing the expression of an infectious virus through the action of a specific antibody. *Ordnance.* in mine warfare, an external procedure that renders a mine incapable of firing upon passage of a target, although it may remain dangerous to handle.
neutralization equivalent *Chemistry.* the equivalent weight of an acid or base.
neutralization fire *Military Science.* fire that is delivered to render the target ineffective or unusable.
neutralization number *Analytical Chemistry.* the number of milligrams of potassium hydroxide needed to neutralize the acid in 1 gram of oil; a measure of the acidity of an oil sample.
neutralization of lenses *Optics.* the procedure of finding the power of a lens by combining it with trial lenses; combination with a lens of equal but opposite power produces a resultant having no power.
neutralization test *Immunology.* a test measuring an antibody's ability to prevent or inhibit the biological effects of a toxin or microorganism.

neutralize to make neutral; specific uses include: *Chemistry.* to make a solution neutral by adding a base to an acid solution or an acid to a basic (alkaline) solution. *Electricity.* the process of removing positive feedback in radio-frequency amplifiers. *Electronics.* to use a component or an auxiliary signal to counteract conditions that may cause a circuit to oscillate.

neutralized radio-frequency stage *Electronics.* a circuit added to a unit's feedback to prevent oscillation and allow the unit to operate as an amplifier.

neutralizer something that makes a thing neutral; specific uses include: *Chemistry.* any agent that neutralizes a solution.

neutralizing antibody *Immunology.* an antibody that inhibits or prevents the biological effects of a soluble antigen or living microorganism.

neutralizing capacitor *Electronics.* a device that recycles a signal in a radio receiver or transmitter from the plate circuit to the grid circuit, so that continuous oscillation does not occur.

neutralizing tool *Electricity.* a nonconducting, screwdriverlike tool that is used to adjust small, variable capacitors and coils in order to neutralize an amplifier. Also, TUNING WIND.

neutralizing transformer *Electromagnetism.* transformer producing counterelectromotive forces that cancel out undesired longitudinal voltage and allows normal operation during inductive disturbances.

neutralizing voltage *Electronics.* the voltage that is passed from the grid to the anode and vice versa to eliminate continuous oscillation.

neutral molecule *Physical Chemistry.* **1.** any molecule that has no net electric charge. **2.** a system of two ions, one an anion and the other a cation, in an electrolytic medium.

neutral mutation *Genetics.* **1.** a mutation that has no effect on the protein coded for by that gene. **2.** a mutation that gives no selective advantage or disadvantage to the organism in which it occurs.

neutral object see SUBSTITUTE OBJECT.

neutral operation *Telecommunications.* a system of teletype operation in which marking signals are formed by current pulses of one polarity, either positive or negative, and spacing signals are formed by the reduction of the current to exactly or approximately zero.

neutral particle *Particle Physics.* any particle that has neither a positive nor negative electric charge.

neutral point *Electricity.* a point with a potential equal to the point of junction of a group of equal nonreactive resistances connected to the main terminals of the system. *Physics.* a point in a vector field that, by symmetry considerations, has no directional properties because all vectors are canceled by an opposing vector of equal magnitude at the point in question. *Fluid Mechanics.* the point in a streamline field where a line of divergence intersects a line of convergence. Also, HYPERBOLIC POINT. *Metallurgy.* in a rolling process, the point at which the speed of the work equals the peripheral speed of the roll. *Optics.* any of several points in the atmosphere where the degree of polarization of diffuse radiation equals zero.

neutral red *Organic Chemistry.* $(CH_3)_2NC_6H_3N_2C_6H_2CH_3NH_2{\cdot}HCl$, a green powder that produces a red color when dissolved in water or alcohol; used as a biologic stain and an acid-base indicator. *Virology.* a photoreactive dye that stains living cells unless they are infected with a virus, in which case a plaque becomes visible.

neutral region *Astrophysics.* a region on the sun's photosphere (usually between two sunspots of opposite magnetic polarity) where the longitudinal magnetic field strength reaches zero.

neutral relay *Electricity.* a relay with a motion characteristic of its armature and independent of the current flow direction through its coil. Also, NONPOLARIZED RELAY.

neutral shoreline *Geology.* a shoreline whose essential features are not influenced by changes in the absolute or relative level of land and water.

neutral stability *Meteorology.* the state of an unsaturated or saturated column of air in the atmosphere when its environmental lapse rate of temperature equals the dry-adiabatic lapse rate or the saturation-adiabatic lapse rate, respectively, so that a parcel of air displaced vertically will experience no buoyant acceleration. Also, INDIFFERENT EQUILIBRIUM. *Control Systems.* a condition in which the motion of a system neither increases nor decreases.

neutral stress *Hydrology.* in a saturated soil, that portion of the total normal stress resulting from the presence of interstitial water. Also, PORE PRESSURE, PORE-WATER PRESSURE.

neutral surface *Mechanics.* in a linear elastic beam subjected to bending, the surface at which the longitudinal or bending stress equals zero.

neutral temperature *Electronics.* the temperature in a thermocoupler when its hot junction reaches its highest point, while its cold junction remains at 0°C.

neutral transmission *Telecommunications.* in telegraph systems, a method in which a mark is delineated by current on the line and a space delineated by the absence of current. Also, UNIPOLAR TRANSMISSION.

neutral wave *Physics.* any wave whose amplitude remains constant in time and space; the amplitude is neither damped nor amplified.

neutrino *Particle Physics.* a stable fundamental particle of the lepton family of particles, with a spin of 1/2; it exists in four and possibly six states, can carry momentum and energy, has no charge, has a very small or zero rest mass, and has a very weak interaction with matter. The particle accounts for the continuous energy distribution of β-particles, and serves to conserve the angular momentum of the β-decay process.

neutron *Physics.* an elementary particle having zero charge, a rest mass of 1.00894 amu (about the same as that of a proton), and a spin quantum number of 1/2. Neutrons are found to be naturally present in all nuclei with a mass number greater than 1 (i.e., all known nuclei except the lightest isotope of hydrogen). They are used to produce fission and other nuclear reactions that release atomic power.

neutron absorber *Nucleonics.* a material that effectively absorbs neutrons while passing through without allowing reemission.

neutron activation analysis see ACTIVATION ANALYSIS.

neutron albedo *Nucleonics.* the probability that a neutron will return through a surface after entering a region through the surface.

neutron-antineutron oscillations *Particle Physics.* the theoretical oscillations that occur between the state of a neutron and the state of an antineutron.

neutron binding energy see BINDING ENERGY.

neutron bomb *Ordnance.* a nuclear bomb designed to release radiation consisting mainly of neutrons, intended to cause extensive personnel loss but relatively little structural or vehicular damage and minimal radioactive fallout.

neutron bottle *Nucleonics.* an evacuated vessel used to store ultracold neutrons sometimes equipped with multipolar magnetic configurations to extend the storage times.

neutron capture *Nuclear Physics.* a process in which a nucleus absorbs a neutron and as a result emits radiation or undergoes fission.

neutron-capture cross section *Nuclear Physics.* an area, measured in barns, in which a beam of bombarding neutrons is most likely to be captured by the sample nuclei.

neutron chopper *Nucleonics.* a mechanical device that interrupts the output beam of neutrons to provide pulses of neutrons.

neutron counter *Nucleonics.* a device that counts the number of neutrons passing through a region of interest.

neutron cycle *Nucleonics.* the history of neutron processes in a reactor beginning with fission processes and ending when all neutrons have been absorbed.

neutron detector *Nucleonics.* a device that is capable of detecting neutrons passing through a medium by observing the recoil of charged particles or release of gamma radiation induced by neutrons.

neutron diffraction *Physics.* the interference of neutrons that have scattered off of atoms or molecules, the neutron wavelength (given by the de Broglie relation) being roughly 10^{-10} m.

neutron diffractometer *Physics.* a device for achieving neutron diffraction; used to determine crystal structures by measuring the diffraction angles.

neutron drip *Astrophysics.* an extra abundance of free electrons in a highly compressed degenerate star.

neutron economy *Nucleonics.* the efficiency of use of neutrons that are lost to leakage or useless absorption.

neutron embrittlement *Metallurgy.* one of several radiation damage phenomena, consisting of ductility impairment caused by exposure to a strong flux of neutrons.

neutron excess *Nuclear Physics.* the amount by which neutrons outnumber protons in a nucleus.

neutron flux density *Nucleonics.* the number of neutrons passing through a given unit area in a given amount of time.

neutron hardening *Nucleonics.* an effect in which the average energy of diffusing neutrons through a medium is increased because slower neutrons are preferentially absorbed, thus decreasing the absorption cross section.

neutron howitzer *Nucleonics.* a device that produces a collimated stream of neutrons; it consists of a neutron source encased in a paraffin block with a small-bore hole from which the beam exits.

neutron inelastic scattering reactions *Nucleonics.* the inelastic emission of neutrons having low energies from a specimen that is bombarded by high-energy neutrons.

neutron irradiation *Nucleonics.* the bombardment of neutrons onto a substance.

neutron logging *Engineering.* the analysis of the water content of soil and rocks in a borehole by means of neutron-gamma bombardment and the measurement of the resulting radiation. Also, **neutron well logging, neutron gamma well logging.**

neutron magnetic moment see NUCLEAR MAGNETIC MOMENT.

neutron number *Nuclear Physics.* the total number of neutrons, denoted by *N,* in the nucleus of an atom.

neutron optics *Physics.* the study and application of neutron diffraction.

neutron radiography *Nucleonics.* a radiography process in which conventional X-ray film is placed next to a surface that radiates appropriately detectable radiation when neutrons bombard the surface.

neutron reflection *Physics.* the direct (specular) reflection of neutrons, as when the Bragg law for the de Broglie wavelength is applicable.

neutron scattering *Nuclear Physics.* the changes in direction (deflection) and energy (attenuation) of a neutron as it collides with a target nucleus; in elastic scattering, there is no change of internal energy of the colliding particles; in inelastic scattering, the neutron may interact with the target nucleus, as in capture or fission, but the sum of the kinetic and internal energies is unaltered.

neutron shield *Engineering.* a radiation shield used to guard personnel from neutron irradiation; constructed of very light hydrogenous materials capable of absorbing the neutrons.

neutron spectrometer *Nucleonics.* an instrument that measures the energy of a source of neutrons.

neutron spectrometry *Nuclear Physics.* a technique used to analyze a nucleus that involves bombarding it with neutrons and then observing the resonances in the reactions or the range of particles or rays it emits.

neutron spectrum *Nuclear Physics.* a graph that shows the distribution of neutrons at various energy levels, such as the number of neutrons ejected during a nuclear reaction.

neutron star see PULSAR.

neutron therapy *Medicine.* a form of radiation therapy, no longer in use, in which boron was injected into a tumor that was first irradiated with a neutron beam. Also, **neutron-capture therapy.**

neutron transport theory *Nucleonics.* a theory based on the Boltzmann transport equation for the diffusion of neutrons through a substance.

neutron turbine *Nucleonics.* a device used to slow neutrons that consists of a paddle wheel with curved blades that move in the same direction as the neutrons but with a slower velocity.

neutron velocity selector *Nucleonics.* a device that isolates neutrons of a particular energy range and makes them available for detection purposes.

neutropenia *Medicine.* the presence of abnormally small numbers of neutrophils in the blood. Also, **neutrophilic leukopenia.**

neutrophil *Histology.* a large leukocyte, containing a lobed nucleus and abundant cytoplasmic granules, that stains with neutral dyes.

neutrophilia *Biology.* an affinity for neutral dyes, such as that demonstrated by white blood cells containing granules that do not stain with either acid or basic dyes.

neutrophilic leukemia see GRANULOCYTIC LEUKEMIA.

neutrosphere *Meteorology.* the atmospheric shell from the earth's surface upward in which the atmospheric constituents are, for the most part, electrically neutral.

nevada *Meteorology.* a cold wind descending from a mountain glacier or snowfield.

Nevadan orogeny *Geology.* a Jurassic to Early Cretaceous crustal deformation and accompanying metamorphism and plutonism that affected the western part of the North American Cordillera.

névé see FIRN FIELD.

Nevile-Winther acid *Organic Chemistry.* $C_{10}H_8O_4S$, transparent plates that are very soluble in water and melt and decompose at 170°C; used as a dye intermediate. Also, 1-NAPHTHOL-4-SULFONIC ACID.

nevocarcinoma see MALIGNANT MELANOMA.

Nevskia *Bacteriology.* a genus of aquatic bacteria of the family Caulobacteraceae that grow as stalked, rod-shaped cells.

nevus *Medicine.* any congenital pigmented lesion of the skin; a birthmark; usually benign, a nevus may become cancerous.

nevus arachnoideus *Medicine.* a nevus that appears as a central, elevated red dot the size of a pinhead, from which small blood vessels radiate, giving it the appearance of a spider. Also NEVUS ARANEOSUS, SPIDER ANGIOMA, SPIDER NEVUS.

nevus flammeus *Medicine.* a congenital benign tumor made up of blood vessels, appearing in the form of a flat, maplike, red or purplish patch, frequently covering large areas of the face or upper trunk. Also, PORT-WINE STAIN.

nevus sebaceus *Medicine.* a congenital nevus that occurs as a solitary lesion on the scalp or face, marked by progressive changes with age; in children it appears as a flat patch that is yellowish to pale brown, following puberty the lesion becomes thickened and rough, and later in life it may develop other nodules with a tendency to become cancerous. Also, **nevus sebaceus of Jadassohn.**

nevyanskite *Metallurgy.* an iridium-base natural alloy containing substantial amounts of osmium

newberyite *Mineralogy.* $MgHPO_4 \cdot 3H_2O$, a colorless orthorhombic mineral occurring as crystals in bat guano, having a specific gravity of 2.10 to 2.12 and a hardness of 3 to 3.5 on the Mohs scale. Also, NEWBERITE.

newborn *Medicine.* **1.** recently born. **2.** a recently born infant.

Newcastle disease *Veterinary Medicine.* a deadly infectious viral disease affecting fowl and other birds, somewhat resembling bronchitis and resulting in discharge from the nostrils, pneumonia, a drop in egg production, diarrhea, tremors, and paralysis. (Associated with the city of *Newcastle,* England.) Also, AVIAN PNEUMOENCEPHALITIS.

Newcastle disease virus *Virology.* an infectious poultry disease of the family Paramyxoviridae that causes respiratory ailments in poultry and occasionally in humans. Also, AVIAN PNEUMOENCEPHALITIS VIRUS.

Newcomb, Simon 1835–1909, American astronomer and mathematician; studied six planets' orbits; determined the velocity of light.

newel or **newel-post** *Architecture.* **1.** the central shaft of a circular staircase. **2.** the principal post at the head or foot of a set of stairs.

newel post *Civil Engineering.* a cylindrical pillar terminating a wing wall of a bridge or viaduct.

New General Catalogue *Astronomy.* a list of more than 7500 star clusters, nebulae, and galaxies compiled in 1888 by J. L. E. Dreyer.

new global tectonics *Geology.* a comprehensive theory that relates the concepts of continental drift, sea-floor spreading, transform faults, and underthrusting of the crust to the movement of lithospheric plates delineated by the major seismic belts of the earth.

New Guinea *Geography.* the world's second-largest island (area: 307,000 square miles), in the southwestern Pacific just north of Australia.

New Hampshire *Agriculture.* a breed of chicken that lays light brown eggs and is characterized by chestnut-red plumage and yellow skin and shanks, developed from the Rhode Island Red.

new inflationary cosmology *Astrophysics.* a model for the evolution of the early universe that includes a period of "inflation," an extremely brief episode during which the universe expanded by a factor thought to be as large as 10^{50}.

Newlands, John Alexander 1837–1898, English chemist; contributed to the development of the true periodic table by Mendeleev.

Newlands' law of octaves *Chemistry.* a law stating that when chemical elements are arranged in order of increasing atomic weight, the first element is similar to the eighth, the second similar to the ninth, the third similar to the tenth, and so on. (Named for John *Newlands.*)

new math(ematics) *Mathematics.* a method of teaching mathematics courses, especially in U.S. schools in the 1960s and 1970s, that in comparison with traditional mathematics instruction places greater emphasis on mastering mathematical concepts, such as set theory, and less on practicing computation skills.

new moon *Astronomy.* the moment when the center of the moon passes through a plane perpendicular to the ecliptic on which lie the centers of the earth and the sun. Also, **New Moon.**

New Red Sandstone *Geology.* the red sandstone facies of the Permian and Triassic Systems of the British Isles.

new snow *Meteorology.* a snowfall that has recently occurred and overlies the snow accumulation from a previous snowfall.

Newson's boring method *Mining Engineering.* a method of boring small shafts of up to 5-1/2 feet in diameter, similar to chilled-shot drilling but performed on a larger scale.

newsprint *Graphic Arts.* the cheapest and most common paper stock, made mostly of ground wood and used chiefly for printing newspapers.

New Stone Age see NEOLITHIC.

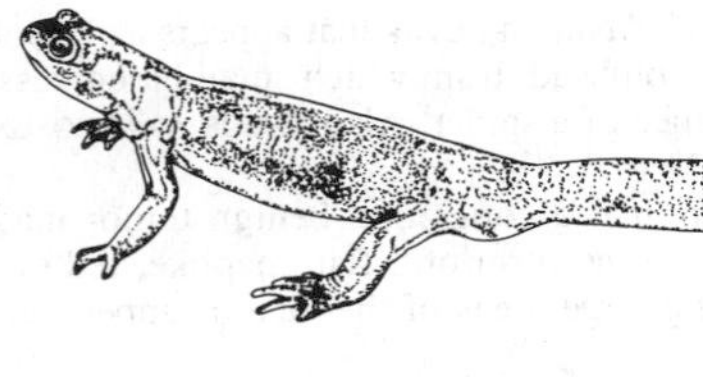

newt

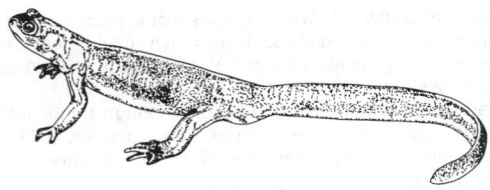

newt *Vertebrate Zoology.* any one of a family of semiaquatic amphibians of the genera *Triturus* and *Notophthalmus* in the family Salamandridae, characterized by an aquatic larval stage, stout bodies with well-developed limbs, with adults having lungs and lacking gills; found in temperate regions in or near water in North America, Europe, North Africa, and Asia. The brightly colored **eastern** or **red newt,** *N. viridescens,* is widely found in North America. (From an earlier English word *ewt*; the phrase *an ewt* was taken to be *a newt*.)

Newton, Sir Isaac 1642–1727, English mathematician, physicist, and astronomer; discovered the basic laws of motion (Newton's laws) and the nature of light and color; originated the theory of universal gravitation; invented calculus.

Isaac Newton

newton *Metrology.* the basic unit of force in the meter-kilogram-second system, equal to the amount of force needed to produce an acceleration of one meter per second per second in a mass of one kilogram. (Named for Sir Isaac *Newton*.)

Newton-Cotes formula *Mathematics.* an approximation formula giving the value of the definite integral $\int_a^b f(x)dx$, where $f(x)$ is known at the $(n + 1)$ points $\{a + jh : j = 0, 1, 2, \ldots, n$ and $h = (b - a)/n\}$. The formula is given by

$$\int_a^b f(x)\,dx = (nh/P_n) \sum_{j=0}^{n} f(a + jh)\,p_{jn} + R_n(f),$$

where p_{jn} is a weight assigned to $f(a + jh)$, $P_n = \sum_{j=0}^{n} p_{jn}$, and $R_n(f)$ denotes the integration error. The values of n and the weights p_{jn} can be varied to make $R_n(f)$ as small as desired.

Newtonian *Mechanics.* **1.** relating to Sir Isaac Newton or his theories. **2.** relating to Newtonian mechanics or to systems that conform to it.

Newtonian attraction *Mechanics.* the mutual attraction of any two bodies, defined and governed by Newton's law of gravitation.

Newtonian-Cassegrain telescope *Optics.* a telescope that uses both a Cassegrain focus and a Newtonian focus interchangeably to reflect light.

Newtonian flow *Fluid Mechanics.* a fluid flow in which stress is proportional to the deformation rate.

Newtonian fluid *Fluid Mechanics.* a fluid with viscosity whose stress at any point is proportional to the rate of strain with respect to time measured at the same point; the proportionality constant is the coefficient of viscosity.

Newtonian focus *Optics.* the type of focus that is obtained in a telescope when a prism or plane mirror positioned diagonally in the telescope tube diverts converging light rays to the side of the tube near its open end.

Newtonian frame of reference see INERTIAL FRAME OF REFERENCE.

Newtonian friction law *Fluid Mechanics.* a law stating that for Newtonian fluids, the shear stress is proportional to the rate of shearing.

Newtonian mechanics *Mechanics.* the study of force and motion on the basis of the theories of Sir Isaac Newton, including especially the assumptions that time is an absolute quantity, and that distance can be absolutely measured rather than being dependent on the time and point of observation.

Newtonian potential *Physics.* any potential function that gives rise to a force which follows an inverse-square law: a potential that is proportional to the inverse of the distance between the source and the field position.

Newtonian relativity *Mechanics.* the concept that only the relative position of two bodies affects their behavior; i.e., there is no intrinsic difference between different orientations to a point or between different points in space.

Newtonian speed of sound *Acoustics.* as described in Newton's second law of motion, the speed of sound, used as a frame of reference from which the particle velocity can be derived.

Newtonian telescope *Optics.* a telescope in which a secondary plane mirror or prism is used to divert light rays from the primary mirror to the side of the tube, where the rays then enter the eyepiece or a camera.

Newtonian velocity *Mechanics.* the velocity of an object within an inertial reference frame.

Newtonian viscosity *Fluid Mechanics.* the viscosity of a Newtonian fluid, the coefficient of which relates the shear stress to the shearing rate.

Newton-Raphson formula *Mathematics.* suppose the equation $f(x) = 0$ has an approximate root at $x = c$. Then the formula $x = c - f(c)/f'(c)$ gives a better approximation. Newton's method is based on repeated applications of the Newton-Raphson formula.

Newton's equations of motion *Mechanics.* Newton's second law of motion expressed as mathematical equations for the problem to be solved.

Newton's first law *Mechanics.* the law that a body at rest will continue at rest, and a body in motion will continue in uniform motion along a straight line, unless the body is acted upon by some outside force.

Newton's identities *Mathematics.* suppose $r_1, r_2, \ldots, r_n$ are the n (not necessarily distinct) roots of the polynomial equation

$$x^n + a_1x^{n-1} + a_2x^{n-2} + \cdots + a_n = 0,$$

$$\text{and that } s_k = r_1^k + r_2^k + \cdots + r_n^k$$

(the kth moment of the roots). Then:

$$s_k + a_1s_{k-1} + a_2s_{k-2} + \cdots + a_{k-1}s_1 + ka_k = 0 \text{ for } k = 1, 2, \ldots, n,$$

$$\text{and } s_k + a_1s_{k-1} + a_2s_{k-2} + \cdots + a_{n-1}s_{k-n} = 0 \text{ for } k > n.$$

Newton's law of cooling *Thermodynamics.* a law stating that the rate of heat transfer from a body is proportional to the difference in temperature between it and its surroundings; not true law in the same respects as the first and second laws of thermodynamics.

Newton's law of gravitation *Mechanics.* the law that any two bodies having mass will act upon each other with equal and opposite attracting forces along the line that joins them. The magnitude of this force is directly proportional to the product of the bodies' masses, and inversely proportional to the square of their distance of separation.

Newton's law of resistance *Fluid Mechanics.* a law stating that the resistive force that acts on a body traveling through a fluid is proportional to the velocity of the body, provided that the velocity is reasonably moderate; at high velocities, the resistance becomes nonlinear.

Newton's laws of motion *Mechanics.* the three laws proposed by Newton governing mechanics; the physical realm where these laws are accurate is now described as Newtonian mechanics.

Newton's method *Mathematics.* an iterative technique for approximating the roots of a (nonlinear) equation of the form $f(x) = 0$. The iteration proceeds as follows: let x_0 be an initial estimate for the value of a root of $f(x)$. Then for each $n > 0$, let $x_{n+1} = x_n - f(x_n)/f'(x_n)$, until the desired accuracy is attained. The convergence of Newton's method can depend strongly on the accuracy of the initial estimate. Also, **Newton-Raphson method.**

Newton's rings *Optics.* the bright and dark rings that form around the point of contact of a slightly convex lens and a flat sheet of glass, caused by interference in the air between the surfaces; observable using monochromatic light.

Newton's rule *Mathematics.* a rule for finding the range of the real roots of a polynomial with real coefficients. Suppose

$$\sum_{k=0}^{n} a_k x^k = 0,$$

with a_k real and $a_0 > 0$, is a polynomial of degree n. Any real number b with the property that

$$f(x) > 0, f'(x) > 0, \ldots, f^{(n-1)}(x) > 0 \text{ for all } x > b$$

is an upper bound for all the real roots of $f(x)$. A real number a is a lower bound for the real roots of $f(x)$ if $-a$ is an upper bound for the roots of $f(-x)$.

Newton's second law *Mechanics.* the law that a body acted upon by a net force will accelerate, and that it will do so in the direction of the force. The acceleration is proportional to the force, and inversely proportional to the mass of the body. This can be expressed by the equation $F = ma$, in which m is the mass of the body, a is its acceleration, and F is the net external force acting on the body.

Newton's square-root method *Mathematics.* an application of Newton's method to the problem of finding the square root of a positive real number a. In this case the formula takes the form $x_{n+1} = 1/2\,(x_n + a/x_n)$. This method allows for manual computation of a square root to the desired number of decimal places; the technique resembles long division.

Newton's theory of lift *Fluid Mechanics.* the theory that forces acting on an airfoil are the result of the impact of fluid particles on the body.

Newton's theory of light see CORPUSCULAR THEORY OF LIGHT.

Newton's third law *Mechanics.* the law that every action will produce an equal and opposite reaction; thus, when two bodies interact, the force of the first body upon the second is equal and opposite to the force of the second body upon the first.

New World *Geography.* the Americas, as opposed to Europe, Africa, and Asia.

New World monkey see PLATYRRHINE MONKEY.

New World porcupine see PORCUPINE, def 2.

new yellow enzyme *Enzymology.* a flavoprotein, found in the cytoplasm of the kidney, brain, and liver, that catalyzes the reaction D-amino acid + H_2O + O_2 = 2-keto acid anion + NH_3 + H_2O_2. Also, **D-amino acid oxidase.**

New Zealand subregion *Ecology.* a distinct zoogeographical region that includes New Zealand and the surrounding ocean areas.

nexin *Biochemistry.* a substance serving as a connecting link between the outer pairs of microtubules in cilia and flagella.

next-event file *Computer Programming.* in discrete-event simulation, a file that maintains a list of all events to be processed and serves as a clock by updating simulated time.

next instruction register see PROGRAM COUNTER.

nexus *Telecommunications.* in a given communications system, a connection or interconnection.

Ney-Allen Nebula *Astronomy.* an extended source of infrared emission that lies in the Orion Nebula near the Trapezium stars and probably comes from dust around newborn stars.

Neyman, Jerzy 1894–1981, Russian-born British statistician; proposed a theory of sample surveys.

Neyman-Pearson lemma *Statistics.* in testing a simple hypothesis against a simple alternative, the likelihood ratio test achieves the maximum probability of a correct decision under the alternative hypothesis (power) for a given probability of a correct decision under the null hypothesis.

Ney's chart see MODIFIED LAMBERT CONFORMAL CHART.

Ney's map projection see MODIFIED LAMBERT CONFORMAL MAP PROJECTION.

NF1 *Oncology.* a tumor-suppressor gene that is found to be altered in neurofibromatosis type 1.

***N*-formylmethionine** *Biochemistry.* the amino acid residue that catalyzes protein synthesis in bacteria.

ng nanogram.

N galaxy *Astronomy.* a galaxy with an extremely bright (and in some cases slightly variable) nucleus surrounded by a faint envelope.

NGF new growth factor; nerve growth factor.

NHP nominal horsepower.

NH_2 terminal see AMINO TERMINAL.

Ni nickel.

Nia *Mycology.* a genus of fungi belonging to the family Torrendiaceae; the only species, *N. vibrissa*, is found in marine environments on submerged wood.

niacin [nī´ə sən] *Biochemistry.* $C_6H_5NO_2$, an acid of the vitamin B complex that is required for the synthesis of nicotinamide, and that acts to reduce plasma cholesterol. Niacin is used for the treatment of pellagra. Also, NICOTINIC ACID, VITAMIN B_7.

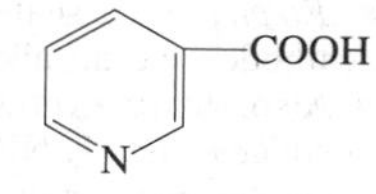

niacin

niacinamide see NICOTINAMIDE.

Niagaran *Geology.* North American provincial series of the Middle Silurian period, occurring after the Alexandrian and before the Cayugan.

niangon *Materials.* a reddish-brown, herringbone-patterned African wood of the *Tarrieta utilis* tree; used in furniture making and interior decoration.

nibble *Computer Programming.* half a byte, usually four adjacent bits in a computer word. Also, NYBBLE.

nibbling *Mechanical Engineering.* a sheet-metal cutting process in which a rapidly reciprocating punch slots sheets along a desired path.

nicad battery see NICKEL-CADMIUM BATTERY.

niccolite see NICKELINE.

niche *Architecture.* a hollowed recess in a wall, usually for displaying an urn or sculpture. *Ecology.* the unique position occupied by a particular species, conceived both in terms of the actual physical area that it inhabits and the function that it performs within the community.

niche glacier *Hydrology.* a small glacier that occupies a funnel-shaped or irregular hollow in the slope of a mountain.

Nichol's chart *Control Systems.* a chart that plots the curves of the frequency control ratio where the magnitude M or argument is constant when the ordinate is the logarithm of the magnitude of the open-loop transfer function and the abscissa is the open-loop phase angle.

Nicholson, Seth Barnes 1891–1963, American astronomer; studied sunspots; discovered four satellites of Jupiter.

Nicholson's hydrometer *Engineering.* a hydrometer designed to determine the relative density of a solid.

Nichols radiometer *Engineering.* an instrument that measures the amount of pressure exerted by a light beam.

Nichols' treponema *Bacteriology.* a virulent bacterial strain of *Treponema pallidum*, which are Gram-negative microorganisms that grow in vitro on rabbit epithelial cells.

Nichrome *Materials.* a trade name for a nickel-chromium alloy mainly containing chromium iron; used in the wiring of furnace elements because of its high electrical resistance and stability at high temperatures.

nick *Molecular Biology.* a break or cut in a single strand of a double-stranded nucleic acid due to the breaking of a phosphodiester bond between adjacent nucleotides, as distinguished from a gap, in which one or more of the nucleotides are missing from the strand.

nickase *Genetics.* an enzyme that produces unwinding in a DNA molecule by initiating single-stranded breaks.

nickel *Chemistry.* a metallic element having the symbol Ni, the atomic number 28, an atomic weight of 58.70, a melting point of 1455°C, and a boiling point of 2900°C; a malleable, silver-white transition metal having excellent resistivity to corrosion and tarnish. *Metallurgy.* this element in the form of a metal extensively used for electroplating, as a base for specialty alloys, and as an alloying element in many steels and in some nonferrous products. (Coined by the Swedish minerologist A. F. Cronstedt, 1722–1765, from a word meaning "copper demon" used for this metal by miners, since it appeared to contain copper but did not.)

nickel-63 *Nuclear Physics.* a radioactive isotope of nickel with a half-life of 100 years, produced synthetically in a nuclear reactor.

nickel acetate *Organic Chemistry.* $Ni(C_2H_3O_2)_2 \cdot 4H_2O$, toxic, carcinogenic, green monoclinic crystals are soluble in water and alcohol and decompose at 250C; used as a mordant in dyeing textiles.

nickel acetylacetonate *Organic Chemistry.* $(C_5H_7O_2)_2Ni$, green crystals that are soluble in water and alcohol and insoluble in ether; melts with decomposition at 240°C; an important catalyst in alkylation and condensation reactions.

nickel-aluminum bronze *Metallurgy.* any of several copper-base alloys containing 2–15% aluminum and up to 6% nickel.

nickel ammonium sulfate *Inorganic Chemistry.* $NiSO_4(NH_4)_2SO_4 \cdot 6H_2O$, toxic dark-green crystals that are decomposed by heat; soluble in water and insoluble in alcohol; used in electroplating. Also, AMMONIUM NICKEL SULFATE, DOUBLE NICKEL SALTS.

nickel-antimony glance see ULLMANNITE

nickel arsenate *Inorganic Chemistry.* $Ni_3(AsO_4)_2 \cdot 8H_2O$, a very toxic yellow-green powder, soluble in acids and insoluble in water; used as a catalyst in soap-making. Also, **nickel orthoarsenate.**

nickel-cadmium battery *Electricity.* a sealed storage battery with a nickel anode, a cadmium cathode, and an alkaline electrolyte; widely used in cordless appliances. Also, NICAD BATTERY.

nickel carbonate *Inorganic Chemistry.* **1.** $NiCO_3$, light-green rhombic crystals that are insoluble in water; decomposes on heating. **2.** $2NiCO_3 \cdot 3Ni(OH)_2 \cdot 4H_2O$, **nickel carbonate basic,** light-green crystals or brown powder, insoluble in cold water and decomposes in hot water; used in electroplating, in ceramic colors and glazes, and as a catalyst.

nickel carbonyl *Inorganic Chemistry.* $Ni(CO)_4$, a colorless, volatile liquid or needles, insoluble in water and soluble in alcohol and many other organic solvents; melts at –25°C and boils at 43°C. It is a dangerous fire hazard that explodes on heating, and it is toxic and carcinogenic; used in the production of nickel and nickel coatings. Also, **nickel tetracarbonyl.**

nickel chloride *Inorganic Chemistry.* $NiCl_2$, yellow deliquescent scales, soluble in alcohol; melts at 1001°C and sublimes at 973°C; used in electroplating and chemical reagents.

nickel-chromium steel *Metallurgy.* any of several low-alloy steels containing nickel and chromium.

nickel cyanide *Inorganic Chemistry.* $Ni(CN)_2 \cdot 4H_2O$, a highly toxic light-green powder, insoluble in water and acids; loses its water at 200°C and then decomposes; used in metallurgy and electroplating.

nickel delay line *Electronics.* a device that retards vibrations in an audio signal by passing it through an element made of nickel.

nickel formate *Organic Chemistry.* $(CHO_2)_2Ni \cdot 2H_2O$, green crystals that are partially soluble in water and insoluble in alcohol; lose water at 130–140°C and decompose at 180–200°C; used in the production of nickel catalysts.

nickeline *Mineralogy.* NiAs, a pale copper-red, metallic, hexagonal mineral of the nickeline group, having a specific gravity of 7.78 and a hardness of 5 to 5.5 on the Mohs scale; found as reniform masses or disseminated in high-temperature veins and in norites and peridotites; a major ore of nickel. Also, NICCOLITE.

nickel iodide *Inorganic Chemistry.* NiI_2, a hygroscopic black crystalline powder, or $NiI_2 \cdot 6H_2O$, blue-green crystals; soluble in water and alcohol; sublimes at 797°C; toxic on ingestion.

nickel-molybdenum steel *Metallurgy.* any of several low-alloy or full-alloy steels containing nickel and molybdenum.

nickel nitrate *Inorganic Chemistry.* $Ni(NO_3)_2 \cdot 6H_2O$, green, monoclinic, deliquescent crystals; soluble in water and alcohol; melts at 56.7°C and boils at 136.7°C; a dangerous fire hazard and strong oxidant; used in nickel plating and the preparation of catalysts and brown ceramic coloring.

nickelocene *Organic Chemistry.* $(C_5H_5)_2Ni$, toxic, carcinogenic, dark-green crystals that melt at 171–173°C; insoluble in water and soluble in most organic solvents; decomposes in alcohol, acetone, ether, and air; used as a catalyst and complexing agent.

nickel oxide *Inorganic Chemistry.* NiO, dark-green cubes that are obtained by heating nickel above 400°C in the presence of oxygen; insoluble in water and acids; melts at 1984°C; used to make nickel salts and fuel cell electrodes, and in porcelain painting. Also, NICKEL PROTOXIDE, GREEN NICKEL OXIDE.

nickel phosphate *Inorganic Chemistry.* $Ni_3(PO_4)_2 \cdot 8H_2O$, a green powder, insoluble in water and soluble in acid; decomposes on heating; used in electroplating. Also, **nickelous phosphate, trinickelous orthophosphate.**

nickel plating *Metallurgy.* the process of coating a material with virtually pure nickel.

nickel protoxide see NICKEL OXIDE.

nickel silver see GERMAN SILVER.

nickel steel *Metallurgy.* any low-alloy or full-alloy steel containing nickel.

nickel-vanadium steel *Metallurgy.* any of several low-alloy steels that contain nickel and vanadium.

nicking *Molecular Biology.* the process of breaking the phosphodiester bond on one of two strands of a native DNA molecule by either physical or chemical damage.

nick-translation technique *Molecular Biology.* an in vitro technique used to synthesize a strand of DNA, usually radioactively labeled, by moving a given nick along a DNA strand in the direction of synthesis, using enzymes.

Nicol, William 1768–1851, Scottish physicist; invented the Nicol prism for investigating polarized light.

Nicolas of Cusa 1401–1464, German scholar and theologian; wrote that the earth is in motion and not the center of the universe.

Nicoletiidae *Invertebrate Zoology.* a family of wingless insects in the order Thysanura.

Nicolle, Charles J. H. 1866–1936, French physician; awarded the Nobel Prize for the discovery that lice transmit typhus.

Nicol prism *Optics.* a prism in which a birefringent calcite crystal, cut diagonally in half and cemented with balsam, produces plane polarized light when a beam of unpolarized light passes through it.

Nicomachinae *Invertebrate Zoology.* a subfamily of mud-ingesting sessile annelids in the family Maldanidae.

Nicomedes 3rd century BC, Greek mathematician; discovered the conchoid curve.

nicotinamide *Biochemistry.* $C_6H_6ON_2$, a B-complex vitamin that is an amide derivative of nicotinic acid and lacks the vasodilator action of niacin; occurs naturally in the body and is used in the treatment of pellagra. Also, NIACINAMIDE.

nicotinamide adenine dinucleotide *Enzymology.* a coenzyme that is found widely in nature and is involved in numerous enzymatic reactions in which it serves as an electron carrier by being alternately oxidized (NAD^+) and reduced (NADH); required in the glycolytic and other metabolic pathways. Also, DIPHOSPHOPYRIDINE NUCLEOTIDE.

nicotinamide adenine dinucleotide phosphate *Enzymology.* a coenzyme, found in all living cells and composed of NAD with an extra phosphate group attached, that acts as a hydrogen acceptor, being reduced to NADPH, in which form it is an important source of reducing power in the cell.

nicotine *Organic Chemistry.* $C_{10}H_{14}N_2$, a thick, combustible, hygroscopic water-white oil that turns brown on exposure to air; soluble in water and alcohol, and boils and decomposes at 247°C; a toxic addictive alkaloid derived from tobacco and used as an insecticide. It is one of the major contributors to the harmful effects of smoking. (Named for Jean *Nicot*, 1530–1600, French ambassador to Portugal, who is regarded as the person who introduced tobacco to France.)

nicotinic acid see NIACIN.

nicotinism *Toxicology.* poisoning due to ingestion or inhalation of excessive amounts of nicotine; symptoms may include stimulation then depression of the nervous system, followed by death due to respiratory paralysis. Also, TABAGISM.

nictitating membrane *Vertebrate Zoology.* a third eyelid behind the upper and lower eyelids of certain reptiles and amphibians; consists of thin folds of skin which move backward or outward, and which moisten and protect the eye.

nidamental gland *Zoology.* a gland that secretes gelatinous materials which envelop certain kinds of eggs.

NIDD non-insulin-dependent diabetes.

nidicolous *Zoology.* **1.** of or relating to birds that need to live in the nest for a considerable period after hatching in order to mature. **2.** of or relating to birds that share the nests of birds of other species.

nidulant *Botany.* **1.** lying only partially embedded in a nestlike receptacle or completely free, as do sporangia. **2.** lying free in pulp, as do some seeds.

Nidulariaceae *Mycology.* a family of fungi belonging to the order Tulostomatales, commonly known as Bird's Nest Fungi; occurs on soil, dead wood, or dung, primarily in tropical regions.

Nidulariales *Mycology.* an obsolete term formerly used for an order of fungi belonging to the class Gasteromycetes; it has been reclassified within the order Tulostomatales.

nidus *Medicine.* **1.** the nucleus of a nerve. **2.** the point of origin or focus of a morbid process.

Niemann-Pick disease *Medicine.* an inherited disorder of lipid metabolism that is marked by an accumulation of phospholipid (sphingomyelin) in histiocytes in the liver, spleen, lymph nodes, and bone marrow; involvement with the brain may occur at a late stage, with red macular spots that are less common than those in Tay-Sachs disease. (Named for Albert *Niemann*, 1880–1921, and Ludwig *Pick*, 1868–1944, German physicians.) Also, SPHINGOMYELIN LIPIDOSIS.

Niepce, Joseph N. 1765–1833, French physicist; made permanent heliotypes and world's first photograph.

Nier, Alfred O. C. born 1911, American physicist; developed the mass spectrograph for nuclear research; analyzed isotopes.

nieve penitente *Hydrology.* a jagged pinnacle or spike of snow or firn produced by differential melting and evaporation as a result of exposure to intense solar radiation. Also, PENITENT, SNOW PENITENTE, SUN SPIKE.

nif gene *Genetics.* a gene that codes for nitrogen-fixing enzymes; found in bacteria that live in root nodules of legumes.

Niger *Geography.* a river in western Africa that flows westward about 2600 miles from Guinea to the Gulf of Guinea.

nigeran *Biochemistry.* a glucan present in the hyphal walls of *Aspergillus* and *Penicillium* and soluble in hot water.

niggliite *Mineralogy.* PtSn, a metallic, silver-white, hexagonal mineral of the nickeline group, having a specific gravity of 13.44 and a hardness of 3 to 5.5 on the Mohs scale; found as minute grains with other sulfides in late hydrothermal veining.

night-ape see BUSHBABY.

night blindness *Medicine.* a decreased ability to see in reduced illumination. Also, NYCTALOPIA, NOCTURNAL AMBLYOPIA.

nightcrawler *Invertebrate Zoology.* a popular term for any large earthworm active on the soil surface at night.

nightglow *Astronomy.* a faint luminosity observed in the night sky, caused by charged particles from the sun striking atoms and molecules in the earth's atmosphere.

nighthawk *Vertebrate Zoology.* any of several North American and European birds of the family Caprimulgidae, having a hawklike appearance, a rounded head, a short bill, large eyes, and a wide mouth.

nightingale *Vertebrate Zoology.* any of several species of Eurasian thrushes of the family Muscicapidae, known for the males' exquisite and usually nocturnal songs; having brown plumage, a white underbelly, and a short reddish tail. (From an earlier English word meaning literally "the night singer.")

nightingale

Nightingale *Aviation.* a popular name for the C-9A military transport version of the DC-9-30.

nightjar *Vertebrate Zoology.* a common European nocturnal bird of the family Caprimulgidae. Also, GOATSUCKER.

nightlatch *Building Engineering.* a door lock that operates from the interior by turning a knob and from the outside by turning a key.

night lizard *Vertebrate Zoology.* a small, terrestrial lizard of the family Xantusiidae, native to North and Central America.

nightmare *Psychology.* a dream that causes feelings of fright, anxiety, or helplessness and that usually abruptly awakens the sleeper.

nightmark *Navigation.* a distinctive object that can be seen at night and can thus be used as an aid to navigation.

night-sky camera *Optics.* a camera that photographs the trails of stars appearing due to the diurnal rotation of the earth and any breaks in the trails due to cloud cover; used to determine whether cloud breaks have affected individual astronomical photographs taken at specific moments in time.

night sweat *Medicine.* abnormal perspiration that occurs during sleep; a common symptom of tuberculosis.

night vision *Physiology.* the ability to see in dim light or to perceive dimly lit objects.

night-vision binoculars *Optics.* binoculars in which a television camera collects images and then projects them on television picture tubes built into the device.

night-vision telescope *Optics.* a telescope that employs television-type sensors, photomultiplier tubes to amplify existing visual light, or infrared recording to greatly increase its effectiveness at night.

night visual range *Optics.* the maximum distance at which a point source of specific luminance can be seen by an observer at night under certain atmospheric conditions. Also, PENETRATION RANGE.

night wind *Meteorology.* a dry squall that occurs at night in southwest Africa and the Congo; it is also loosely applied to other diurnal local winds.

nigre [nī´gər] *Chemical Engineering.* a dark-colored, highly hydrated and strongly alkaline layer of undesirable soap formed between the neat soap layer and the caustic solution during soap manufacturing.

nigrescent *Biology.* having a blackish color.

nigrosine *Materials.* any of a group of deep-blue or black dyes used in making ink and shoe polish, and in dyeing leather, textiles, wood, and furs.

NIH or **N.I.H.** National Institutes of Health.

nihilism [nī´ə liz´əm] *Psychology.* **1.** the persistent denial of the existence of the self, of particular things, or of everything, seen as a delusion in some forms of schizophrenia. **2.** the pessimistic belief that treatment procedures or prescribed medications are useless.

nihilist [nī´ə list] *Psychology.* a person who believes in or is affected by nihilism.

Nike [nī´kē] *Ordnance.* a series of U.S. Army, rocket-boosted, surface-to-air guided missiles designed to search, intercept, and destroy enemy aircraft or missiles. (Named for the ancient Greek goddess of victory.)

Nike-Ajax *Ordnance.* the first operational U.S. supersonic antiaircraft guided missile; it had a top speed of Mach 2, a range of 25 miles, and a ceiling of 60,000 feet. It had a conventional HE warhead, and was guided by radar command; officially designated **MIM-3.**

Nike-Hercules *Ordnance.* a U.S. Army, air defense artillery surface-to-air guided missile system that provides nuclear or conventional defense against manned bombers and air-breathing missiles; it has a top speed of Mach 3, a range of over 75 miles, and a ceiling of 150,000 feet; officially designated **MIM-14.**

Nike-Zeus *Ordnance.* a U.S. Army, surface-to-air guided missile system designed to intercept enemy intercontinental ballistic missiles; it carried a nuclear warhead and had a top speed of Mach 4. Neither Zeus nor successor Nike X was ever deployed.

niklesite *Petrology.* a pyroxenite containing the pyroxenes diopside, enstatite, and diallage, with small amounts of chromite, ilmenite, magnetite, and olivine.

Nikolsky sign *Medicine.* a condition in which there is ready separation of the outer layer of the skin from the basal layer in response to a sliding or rubbing pressure on the area involved. It may occur in pemphigus and in conditions such as certain hereditary blistering skin diseases, scalded skin syndrome, adult toxic epidermal necrolysis, and thermal burns. (Named for P. V. *Nikolsky,* 1858–1940, Russian dermatologist.)

nil *Science.* having no value; nothing. *Mathematics.* A left, right, or two-sided ideal I of a ring R is said to be nil if every element of I is nilpotent. Every nil right or left ideal of R is contained in the radical of R. Every nilpotent ideal is nil, but the converse is not true. *Computer Programming.* **1.** a special-purpose pointer value that is used to indicate the end of a linked list, or an empty list. Also, **nil pointer. 2.** in Lisp, an indication of the logical value False.

nilas *Hydrology.* a thin, elastic, gray-colored crust of ice formed on a calm sea.

Nile *Geography.* the world's longest river, flowing about 4100 miles northward from Lake Victoria in central Africa to the Mediterranean in Egypt.

nile *Nucleonics.* a dimensionless quantity expressing the unit of reactivity of a nuclear reactor; the number of niles for a particular reactor is expressed as $100(1 - k)$, where k is the effective multiplication factor.

Nilionidae *Invertebrate Zoology.* the false ladybird beetles, a family of coleopteran insects in the superfamily Tenebrionoidea.

nilpotent *Mathematics.* **1.** an element a of an algebraic object (such as a ring) is said to be nilpotent if $a^n = 0$ for some positive integer n; i.e., a multiplied by itself n times equals the additive identity of the object. For example, a square matrix or an operator may be nilpotent. The least integer n for which $a^n = 0$ is called the **index of nilpotence** of a. **2.** an ideal I of a ring R is said to be nilpotent if $I^n = 0$ for some integer n. Every nilpotent ideal is nil, but the converse is not true. **3.** a group G is said to be nilpotent if its ascending central series is of finite length; i.e., if $C_n(G) = G$ for some n.

nimbostratus *Meteorology.* a principal cloud type that is gray and often dark, appearing diffuse due to continuously falling rain, snow, or sleet but not accompanied by lightning, thunder, or hail; it is composed of suspended water droplets, sometimes supercooled, and falling raindrops or snow crystals or flakes, and occupies a layer of large horizontal and vertical extent.

nimbus *Astronomy.* a cloud or hazy light surrounding a bright object. *Meteorology.* a popular term for any cloud that produces precipitation; a rain cloud.

nimonic alloy *Materials.* a nickel-base superalloy also containing chromium, titanium, iron, and carbon; used in making turbine blades for jet engines.

N indicator see N SCOPE.

nines complement *Computer Programming.* the diminished 10's complement of a number x, found by subtracting every digit of x from 9. Also, COMPLEMENT-ON-9.

nine-track tape *Computer Technology.* magnetic tape on which nine bits can be recorded across the width of the tape, normally eight data bits and one parity bit.

ninhydrin *Organic Chemistry.* $C_9H_4O_3 \cdot H_2O$, white crystals or powder; loses water at 125°C, swells at 140°C, and melts at 240–245°C; soluble in water and alcohol; used to detect peptides, amines, amino acids, and amino sugars.

ninhydrin reaction *Biochemistry.* a test using a solution of ninhydrin to detect and measure the presence of amino acids, peptones, and peptides.

niningerite *Mineralogy.* $(Mg,Fe^{+2},Mn)S$, a cubic mineral having an undetermined hardness and a specific gravity of 3.21 to 3.68; found only in chondritic meteorites intergrown with troilite and nickel-iron.

niobic acid *Inorganic Chemistry.* $Nb_2O_5 \cdot xH_2O$, a white, water-insoluble substance; soluble in concentrated sulfuric acid; used in the analytical determination of niobium.

niobium *Chemistry.* a metallic element having the symbol Nb, the atomic number 41, an atomic weight of 92.9064, a melting point of 2468°C, and a boiling point of 4742°C; a slightly bluish-gray malleable metal that usually occurs with tantalum, from which it is separated by fractional distillation; used in steel alloys and in superconducting alloys with titanium. (Named for the Greek goddess *Niobe,* the daughter of King Tantulus, because of its association with tantulum.)

niobium carbide *Inorganic Chemistry.* NbC, black cubes or lavender-gray powder; insoluble in water; melts at 3500°C; used in special steels, carbide-tipped tools, and coating graphite for nuclear reactors.

NIOSH or **N.I.O.S.H.** National Institute of Occupational Safety and Health.

nip *Geology.* a small, low cliff produced at the high-water mark by wavelets. *Hydrology.* a point on the lobe of a meander where the crowding of the stream current toward the lobe erodes the bank.

Nipher shield *Engineering.* a conical copper shield for a rain gauge; used to prevent wind from affecting the rainfall catch, thereby improving the accuracy of the reading.

nippers *Mechanical Devices.* a type of pliers with sharp cutting edges used to cut wire.

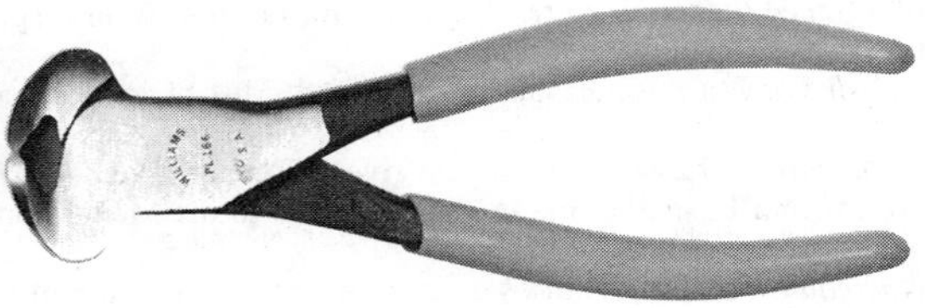

nippers

nipple *Anatomy.* the pigmented, protruding, cone-shaped structure at the center of each mammary gland, with about 20 tiny openings through which milk passes; it is surrounded by the darker-pigmented areola. *Mechanical Devices.* a short pipe or piece of tubing having threaded ends, a nipple-shaped projection, and an opening through which oil or grease is injected and regulated into machinery.

Nippotaeniidae *Invertebrate Zoology.* an order of segmented tapeworms, with an adhesive sucker on the head; endoparasites of freshwater fish.

Nirenberg, Marshall W. born 1927, American biochemist; awarded the Nobel Prize for explaining the genetic code.

Nishina, Yoshio 1890–1951, Japanese physicist; investigated cosmic rays.

Nissl bodies *Cell Biology.* distinct granules that are found in the cell body and dendrites of neurons and that are believed to be ribonucleoprotein particles involved in protein synthesis. Also, **Nissl substance.**

nit *Invertebrate Zoology.* a popular name for the egg or young of a louse or other ectoparasitic insect.

nit *Telecommunications.* a measure of information content; the information content of a message in nits is the negative of the natural logarithm of the probability of selecting that message from among all the messages picked. Also, NEPIT.

Nitelleae *Botany.* a group of delicate stoneworts belonging to the family Characeae, characterized by reproduction via oogonia, an antheridium covered by shield cells, and an entirely ecorticate thallus.

niter see POTASSIUM NITRATE.

niter cake see SODIUM BISULFATE.

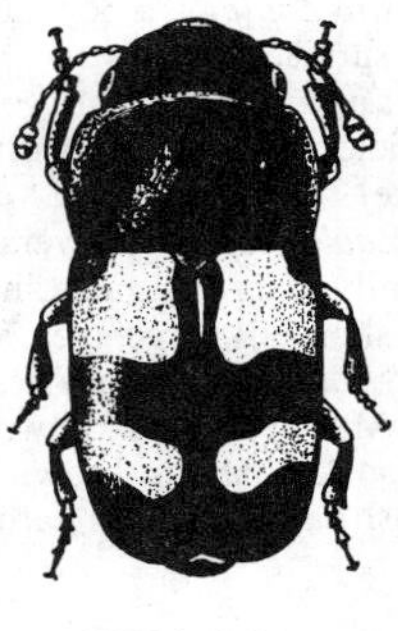

Nitidulidae

Nitidulidae *Invertebrate Zoology.* a large family of small beetles feeding on sap, coleopteran insects in the superfamily Cucujoidea, with antennae ending in a three-jointed club-shaped expansion.

nitinol *Materials.* any of a group of shape-memory alloys that are intermetallic compounds of nickel and titanium, having good fatigue properties.

nitrate *Chemistry.* **1.** any salt or ester of nitric acid, such as sodium nitrate, $NaNO_3$. **2.** any other compound that contains the NO_3^- group. *Materials.* a fertilizer consisting of sodium nitrate, $NaNO_3$, or potassium nitrate, KNO_3.

nitrate mineral *Mineralogy.* any of several generally rare minerals characterized by a fundamental anionic structure of NO_3^-, such as gerhardtite or nitromagnesite.

nitrate reductase *Enzymology.* an enzyme of the oxidoreductase class that catalyzes the reduction of nitrate to nitrite; occurs in certain oxidative bacteria that grow under anaerobic conditions. Also, **nitratase.**

nitrate reduction test *Microbiology.* an assay to determine whether a given bacterial strain is able to reduce nitrate, by examining the bacteriological culture medium for the presence of nitrite.

nitrate respiration *Biochemistry.* the reduction of nitrates to nitrites, nitrous oxide, or nitrogen under anaerobic conditions by facultative anaerobes.

nitrating acid *Inorganic Chemistry.* a mixture of concentrated sulfuric and nitric acids used in nitration. Also, MIXED ACID.

nitration *Organic Chemistry.* a reaction in which a hydrogen on a carbon atom is replaced by a nitro group ($-NO_2$) by treating with nitric acid, or a mixture, usually of nitric and sulfuric acids.

nitre see POTASSIUM NITRATE.

nitrene *Organic Chemistry.* an uncharged, electron-deficient organic group that contains a monocovalent nitrogen.

nitric *Chemistry.* **1.** of or relating to nitrogen. **2.** describing various compounds of nitrogen, especially those in which the element has a valence of five.

nitric acid *Inorganic Chemistry.* HNO_3, a toxic, corrosive, hygroscopic, colorless liquid; completely miscible with water; melts at –42°C and boils at 83°C; emits suffocating fumes and attacks most metals; a dangerous fire hazard in contact with organic substances. It is widely used in the manufacture of fertilizers and explosives, in metallurgy, photoengraving, and plastic manufacture, and for various other industrial purposes.

nitric anhydride see NITROGEN PENTOXIDE.

nitric oxide *Inorganic Chemistry.* NO, a colorless gas or blue liquid; slightly soluble in water; freezes at –163.6°C and boils at –151.8°C; noncombustible; used in the preparation of nitric acid, organic synthesis, and to bleach rayon.

nitridation *Materials Science.* in the production of metallic components, the formation of an iron-nitrogen compound as a wear-resistant surface that also resists thermal decomposition.

nitride *Inorganic Chemistry.* a compound made up of nitrogen and another element, usually a metal.

nitride nuclear fuel *Nucleonics.* fissionable nuclear fuel that is composed of a nitride of uranium or plutonium.

nitriding *Metallurgy.* a process of casehardening by diffusing nitrogen into a superficial layer of a ferrous part.

nitrification *Microbiology.* the process by which nitrogen in ammonia and organic compounds is oxidized to nitrites and nitrates by soil bacteria of the family Nitrobacteraceae.

nitrification inhibitor *Microbiology.* any substance that inhibits or blocks the nitrification process in microorganisms.

nitrifier *Microbiology.* a nitrifying microorganism.

nitrify *Microbiology.* to undergo or cause to undergo a process of nitrification.

nitrifying bacteria *Bacteriology.* a common term for bacteria of the family Nitrobacteraceae, which convert atmospheric nitrogen into various nitrogenous compounds.

nitrile *Organic Chemistry.* an organic compound possessing the cyanide group, $-C{\equiv}N$.

nitrile rubber *Materials.* a synthetic rubber made by copolymerizing butadiene and acrylonitrile; noted for its resistance to oil and fuel, and used in gaskets and hoses that carry oil and gasoline.

nitrilotriacetic acid *Organic Chemistry.* $N(CH_2CO_2H)_3$, a combustible, white, crystalline powder that is insoluble in water and most organic solvents; melts and decomposes at 246°C; used in commercial detergents.

nitrite *Chemistry.* any salt of nitrous acid, such as potassium nitrite, KNO_2.

nitrite reductase *Enzymology.* an enzyme that catalyzes the reduction of nitrite to ammonia in plants.

nitritoid *Chemistry.* **1.** resembling a nitrite. **2.** causing a reaction similar to that of a nitrite.

nitrituria *Medicine.* the presence of nitrites in the urine.

nitro *Chemistry.* **1.** containing the nitro ($-NO_2$) group. **2.** a shorter term for nitroglycerine.

nitroalkane *Organic Chemistry.* any alkane that contains the nitro ($-NO_2$) group.

***m*-nitroaniline** *Organic Chemistry.* $NO_2C_6H_4NH_2$, a yellow needlelike solid that is soluble in alcohol, ether, and acetone, and melts at 114°C.

***o*-nitroaniline** *Organic Chemistry.* $NO_2C_6H_4NH_2$, orange-red needles; soluble in alcohol and slightly soluble in water; melts at 72°C; an explosion risk and also toxic when absorbed by skin; used as a dye intermediate and in the synthesis of photographic antifogging agents.

nitroaromatic *Organic Chemistry.* an organic compound containing one or more benzene rings to which one or more nitro groups are attached.

Nitrobacter *Bacteriology.* a genus of Gram-negative bacteria of the family Nitrobacteraceae; flagellated, rod-shaped cells that oxidize nitrites to nitrates and occur in soil.

Nitrobacteraceae *Bacteriology.* a family of Gram-negative, soil bacteria that derive energy from oxidizing ammonia or nitrites; commonly referred to as nitrifying bacteria.

nitrobenzene *Organic Chemistry.* $C_6H_5NO_2$, greenish-yellow crystals or a yellow, oily liquid; slightly soluble in water and soluble in alcohol and ether; melts at 5.70°C and boils at 210.85°C; used in processing cellulose, and in the manufacture of aniline, benzidine, and other compounds. Also, OIL OF MIRBANE.

***o*-nitrobiphenyl** *Organic Chemistry.* $C_6H_5C_6H_4NO_2$, a combustible, light-yellow to reddish solid or liquid that is soluble in mineral spirits, pine oil, and turpentine, and insoluble in water; melts at 36–38°C and 165–170°C (13 torr) and boils at 330°C. It is toxic by ingestion and is used as a fungicide and wood preserver.

nitrobromoform see BROMOPICRIN.

nitrocarburizing *Metallurgy.* a process of casehardening by simultaneously diffusing carbon and nitrogen into a superficial layer of a ferrous part.

nitrocellulose filter *Molecular Biology.* a thin layer of cellulose ester used to blot or collect double-stranded nucleic acids; widely used in nucleic acid hybridization procedures.

nitrocellulose membrane *Biotechnology.* a selectively permeable membrane composed of nitrocellulose fibers and used to separate single- and double-stranded nucleic acids.

Nitrococcus *Bacteriology.* a genus of Gram-negative bacteria, of the family Nitrobacteraceae, that are nitrite-oxidizing, flagellated, coccoid cells found in marine habitats and reproduce by binary fission.

nitrocotton see CELLULOSE NITRATE.

nitro dye *Organic Chemistry.* any of a class of dyes whose molecules contain one or more nitro groups ($-NO_2$); the most important members are the nitrophenylamines, which provide shades of yellow, orange, and brown.

nitroethane *Organic Chemistry.* $C_2H_5NO_2$, a colorless liquid that is soluble in water and alcohol and boils at 114°C; used as a solvent and fuel additive.

nitro explosive *Organic Chemistry.* a class of organic compounds containing one or more nitro ($-NO_2$) or nitrate ($-ONO_2$) groups, including nitroglycerine and cellulose nitrate; widely used as explosives.

nitrogen *Chemistry.* a gaseous element having the symbol N, the atomic number 7, an atomic weight of 14.0067, a melting point of –209.9°C, and a boiling point of –195.5°C; a colorless, odorless, tasteless gas that makes up about four-fifths of the atmosphere; used in ammonia synthesis and as an inert gas, refrigerant, and fertilizer component. (From a French word for *niter,* or potassium nitrate.)

nitrogenase *Biochemistry.* one of two enzymes that catalyze the conversion of dinitrogen to ammonia and cause bacterial nitrogen fixation.

nitrogen balance *Geochemistry.* the net gain or loss of nitrogen in a soil.

nitrogen chloride see NITROGEN TRICHLORIDE.

nitrogen cycle *Ecology.* the continuous process by which nitrogen circulates among the air, soil, water, plants, and animals of the earth; nitrogen in the atmosphere is converted by bacteria into substances that green plants can absorb from the soil; animals eat these plants (or eat other animals that feed on the plants); the animals and plants die and decay; the nitrogenous substances in the decomposed organic matter return to the atmosphere and the soil.

nitrogen dating *Archaeology.* a dating method based on the gradual reduction of nitrogen in bone as collagen is broken down into amino acids and leached away.

nitrogen dioxide *Inorganic Chemistry.* NO_2, a brown gas or yellow liquid; decomposes in cold water; melts at –9.3°C and boils at –21.3°C; used in the production of nitric acid, as a catalyst, and as an oxidizer for rocket fuels.

nitrogen fixation *Bacteriology.* the conversion of atmospheric nitrogen into chemical compounds, such as ammonia, that can be used by green plants in the formation of proteins; this process is carried out by certain bacteria that are present on the nodules of the plants. *Biotechnology.* an industrial version of this natural process, carried out by bacterial action or by chemical means, used in producing fertilizers and other products.

nitrogen-fixing *Biology.* describing bacteria or other organisms that are involved in nitrogen fixation. Thus, **nitrogen-fixing bacteria.**

nitrogen fluoride see NITROGEN TRIFLUORIDE.

nitrogen monoxide see NITROUS OXIDE.

nitrogen mustard *Organic Chemistry.* any of several toxic compounds that are similar to mustard gas, but with nitrogen replacing the sulfur; used in medicine and as a military weapon.

nitrogen narcosis *Medicine.* a condition of confusion or unconsciousness caused by the formation of nitrogen bubbles in the blood and tissues (which impairs oxygenation); usually associated with too rapid a descent in deep-sea diving. Also popularly known as **the bends.**

nitrogenous [nī träj´ə nəs] *Chemistry.* containing nitrogen.

nitrogenous base *Biochemistry.* a basic compound, such as adenine, guanine, cytosine, or thymine, that contains nitrogen.

nitrogen pentoxide *Inorganic Chemistry.* N_2O_5, white, rhombic or hexagonal crystals, soluble in water and chloroform; melts at 30°C and decomposes at 47°C; used as a nitrating agent in chloroform solution. Also, NITRIC ANHYDRIDE.

nitrogen sequence see CARBON-NITROGEN-OXYGEN SEQUENCE.

nitrogen solution *Inorganic Chemistry.* a mixture of ammonium nitrate and ammonia water, directly applied to the soil as ferilizer or used in the manufacture of fertilizers.

nitrogen trichloride *Inorganic Chemistry.* NCl_3, a yellow oil or rhombic crystals, soluble in chloroform and insoluble in water; freezes at –40°C and boils at 71°C; toxic and a strong irritant, explosive when exposed to direct sunlight. Also, NITROGEN CHLORIDE.

nitrogen trifluoride *Inorganic Chemistry.* NF_3, a colorless, nonflammable gas, very slightly soluble in water; melts at –206.6°C and boils at –128.8°C; a severe explosion hazard; used in chemical synthesis and as a fuel oxidizer. Also, NITROGEN FLUORIDE.

nitrogen trioxide *Inorganic Chemistry.* N_2O_3, a red to brown gas or a blue solid or liquid, soluble in water and acid; melts at –102°C and decomposes at 3.5°C; toxic on inhalation and a strong irritant; used as a fuel oxidizer and to prepare alkali nitrites. Also, **nitrogen sesquioxide.**

nitrogen ylides *Organic Chemistry.* highly reactive nitrogen salts formed by the removal of a proton from a carbon adjacent to a heteroatom bearing a positive charge.

nitroglycerine *Organic Chemistry.* $CH_2NO_3CHNO_3CH_2NO_3$, a toxic, yellow, viscous liquid that is slightly soluble in water and soluble in alcohol; freezes at 13.1°C and explodes at 218°C; used as a high explosive, rocket fuel, and vasodilator.

nitroguanidine *Organic Chemistry.* $H_2NC(NH)NHNO_2$, long, thin, lustrous needles or small, thin, elongated plates that melt with decomposition at 239°C; used in explosives, especially flashless propellant powder.

nitroimidazole *Microbiology.* any of a number of synthetic antimicrobial substances that are effective against anaerobic organisms.

nitromagnesite *Mineralogy.* $Mg(NO_3)_2{\cdot}6H_2O$, a colorless or white, monoclinic mineral having a specific gravity of 1.58 and an undetermined hardness; found as an efflorescence in a limestone cavern in Kentucky.

nitrometer *Analytical Chemistry.* a glass apparatus used for the collection and measurement of nitrogen and other gases produced in chemical reactions. Also, AZOTOMETER.

nitromethane *Organic Chemistry.* CH_3NO_2, a colorless, explosive liquid that is soluble in water and alcohol; boils at 101°C and freezes at –29°C; used as a rocket fuel, gasoline additive, and solvent.

nitron *Organic Chemistry.* $C_{20}H_{16}N_4$, fine yellow needles that are insoluble in water, slightly soluble in alcohol, and soluble in acetone and chloroform; melts with decomposition at 189°C; used as a reagent for the detection of nitrate.

nitronium *Chemistry.* **1.** a nitryl ion. **2.** a salt containing this ion.

nitroparaffin *Organic Chemistry.* any of a series of compounds having the general formula $C_nH_{2n+1}NO_2$, with the nitro groups attached to a carbon atom via the nitrogen; they are colorless, pleasant-smelling liquids, slightly soluble in water; the lower members boil in the range 101 to 132°C and freeze in the range –18 to –104°C; used as solvents, inks, propellants, and fuel additives.

nitrophenide *Pharmacology.* $C_{12}H_8N_2O_4S_2$, a drug occurring in yellow rhomboid crystals; used in veterinary medicine to control coccidiosis infection, especially in poultry.

***o*-nitrophenol** *Organic Chemistry.* $NO_2C_6H_4OH$, yellow crystals soluble in hot water and alcohol; melts at 44-45°C; toxic by ingestion and used as an intermediate in organic synthesis.

***p*-nitrophenylhydrazine** *Organic Chemistry.* $C_6H_7N_3O_2$ orange-red crystals that are soluble in warm water and alcohol; deliquesces at 155°C; used as an analytical reagent.

nitrophosphate *Materials.* a fertilizer produced by the action of nitric acid (sometimes mixed with other acids) on phosphate rock; often supplemented with potassium salts to produce a complete fertilizer.

nitrophyte *Botany.* a plant that thrives in nitrogen-rich soils.

1-nitropropane *Organic Chemistry.* $CH_3CH_2CH_2NO_2$, a flammable colorless liquid; slightly soluble in water and miscible with organic solvents; boils at 132°; used as a rocket propellant, gasoline additive, and solvent.

2-nitropropane *Organic Chemistry.* $CH_3CHNO_2CH_3$, a flammable colorless liquid; slightly soluble in water and miscible with organic solvents; boils at 120°; used as a rocket propellant and gasoline additive, and as a solvent for vinyl and epoxy coatings.

nitrosamine *Oncology.* any of a number of compounds that are formed by combining nitrates with amines and that typically contain the group –N–N=O; some of these may be highly carcinogenic.

nitrosation *Organic Chemistry.* treatment of a compound with nitrous acid in order to introduce the –N=O group.

nitroso *Organic Chemistry.* indicating the presence of the radical –N=O in a compound.

Nitrosococcus *Bacteriology.* a genus of Gram-negative bacteria belonging in the family Nitrobacteraceae that oxidize ammonia to nitrite, contain intruded, lamellar membranes, and occur in soil or aquatic habitats.

Nitrosolobus *Bacteriology.* a genus of nitrifying bacteria that consist of Gram-negative, flagellated cells containing internal, vesicular membranes; found in soil.

Nitrosomonas *Bacteriology.* a genus of Gram-negative bacteria of the family Nitrobacteraceae that oxidize ammonia to nitrite and are composed of obligately autotrophic, rod-shaped cells possessing subpolar flagella.

nitrosourea *Oncology.* any of a number of alkylating drugs that are lipid-soluble and can cross the blood-brain barrier; used as antineoplastic agents in treating certain brain tumors.

Nitrosovibrio *Bacteriology.* a genus of Gram-negative bacteria that occur as rod-shaped, flagellated cells and reproduce by binary fission; found in the soil of the Hawaiian Islands.

Nitrospina *Bacteriology.* a genus of Gram-negative bacteria of the family Nitrobacteraceae that oxidize nitrite for energy, reproduce by binary fission, and occur in marine habitats.

nitrostarch *Organic Chemistry.* $C_{12}H_{12}(NO_2)_8O_{10}$, a flammable orange powder that is soluble in ether alcohol; used in explosives.

***m*-nitrotoluene** *Organic Chemistry.* $NO_2C_6H_4CH_3$, a yellow liquid that is soluble in alcohol, ether, and benzene, and boils at 232°C.

***o*-nitrotoluene** *Organic Chemistry.* $NO_2C_6H_4CH_3$, a combustible, yellow liquid that is insoluble in water and miscible with alcohol; boils at 220.4°C and freezes at –9.3°C; toxic by inhalation, ingestion, and skin absorption; used to produce toluidine and dyes.

***p*-nitrotoluene** *Organic Chemistry.* $NO_2C_6H_4CH_3$, toxic, combustible, yellow crystals; soluble in water and alcohol; melts at 51.7°C and boils at 238.3°C; used to produce synthetic dyes.

nitrourea *Organic Chemistry.* $NH_2CONHNO_2$, an explosive, white, crystalline powder that is slightly soluble in water and soluble in alcohol; melts at 158–159°C; used in explosives.

nitrous *Chemistry.* **1.** of or relating to nitrogen. **2.** describing various compounds of nitrogen, especially those in which the element has a valence of three.

nitrous acid *Inorganic Chemistry.* HNO_2, occurring only as a light blue solution that decomposes in water; used in organic synthesis and as a source of nitric oxide.

nitrous oxide *Inorganic Chemistry.* N_2O, a colorless gas or liquid, or cubic crystals; soluble in water and alcohol; melts at –102.4°C and boils at –89.5°C; narcotic in high concentrations; used as an anesthetic, aerosol propellant, and leak detector; popularly known as laughing gas. Also, NITROGEN MONOXIDE.

nitroxylene *Organic Chemistry.* $C_6H_3(CH_3)_2NO_2$, any of several isomers occurring as yellow liquid or crystalline needles that are insoluble in water and soluble in alcohol; used in organic synthesis and gelatin acceleration.

nitryl group *Chemistry.* the $-NO_2$ radical. Also, NITRONIUM.

nitryl halide *Inorganic Chemistry.* any compound composed of the nitryl group (NO_2) with an element from the halogen group (periodic table, Group VIIa), such as fluorine or chlorine.

Nitzschiaceae *Botany.* a family of marine, freshwater, and terrestrial diatoms belonging to the order Pennales, having straight, sigmoid raphe on both valves and usually living solitarily, but also in filamentous colonies.

nival *Ecology.* relating to or living in a snowy environment.

nival gradient *Hydrology.* the angle made by a nival surface with the horizontal.

nival surface *Hydrology.* the imaginary plane containing all the snowlines of the same geologic time period. Also, **nival plane.**

nivation *Hydrology.* the erosion of rock or soil beneath a snowbank or snow patch and along its fluctuating margins, due to frost action in conjunction with other processes such as chemical weathering and solifluction. Also, SNOW-PATCH EROSION.

nivation glacier *Hydrology.* a small, newly formed glacier representing the initial stage of glacier formation. Also, SNOWBANK GLACIER.

nivation hollow *Hydrology.* a small, shallow recess or depression produced by nivation and occupied intermittently by a snow patch or snowbank. Also, SNOW NICHE.

niveal *Hydrology.* of or relating to features and effects resulting from the action of snow and ice.

nivenite *Mineralogy.* a velvet-black variety of uraninite containing small amounts of the rare earths yttrium and cerium.

niveoglacial *Hydrology.* of, related to, or caused by the combined action of snow and ice.

niveolian *Hydrology.* of, related to, or caused by the simultaneous accumulation and intermixing of snow and airborne sand at the side of a gentle slope.

Nizymeniaceae *Botany.* a small family of northern Australian red algae of the order Gigartinales, characterized by profusely branched uniaxial thalli with similar gametangial and sporangial forms and obscure life histories.

N.K. an abbreviation for *Nomenklatur Kommission*, a committee of the Anatomical Society of Germany, which has revised (and given supplementary names to) the terminology of anatomy.

NK cell *Cell Biology.* a cell that is found in vertebrates and possesses a nonspecific cytolytic activity, lysing virus-infected cells. (An abbreviation for natural killer cell.)

n-key rollover *Computer Programming.* the ability of some terminals to store a sequence of key strokes and to transmit the sequence to the computer rather than sending each key stroke signal as it occurs.

NL not permitted. (From Latin *non licet.*)

NL or **N.L.** New Latin.

nl nanoliter.

N. Lat. or **N. lat.** north latitude.

N-level address *Computer Programming.* an indirect address that contains *N* reference addresses prior to reaching the desired memory location.

N-level logic *Electronics.* **1.** a logic circuit or operation that recognizes more than two distinct signal levels. **2.** a group of gates that are connected in such a fashion that no more than *N* gates appear in a series.

NLGI number *Materials Science.* a number indicating the relative consistency of a lubricating grease according to a system established by the National Lubricating Grease Institute.

N line *Spectroscopy.* a characteristic line in the X-ray spectrum of an atom, produced when an *N* electron is excited or when an *O*-shell electron falls into an *N* shell.

NLP natural language processing.

NLQ near-letter quality.

NM nautical mile.

nm nanometer.

NMA or **N.M.A.** National Medical Association.

NMR nuclear magnetic resonance.

N-Multistix *Chemistry.* the trade name for a reagent strip used in testing urine specimens for protein, glucose, ketones, bilirubin, occult blood, urobilinogen, and nitrite, and also to indicate urinary pH.

N-myc *Oncology.* a proto-oncogene that is found in neuroblastoma and small-cell carcinoma of the lung.

n.n. nomen novum; nomen nudum.

nn junction *Electronics.* the section of a semiconductor that serves as a transition zone between two n-type regions whose electrical properties differ.

n. nov. nomen novum.

No the chemical symbol for nobelium.

no. or **No.** north; northern; number. (From Latin *numero.*)

no-address instruction *Computer Programming.* an instruction whose address field is treated not as an address but as a constant to be used in the computation.

Noah's Dove *Astronomy.* another name for the constellation Columba.

no-atmospheric control *Space Technology.* any device designed to control a rocket or space vehicle outside the atmosphere or in areas where the atmosphere is so thin that the vehicle will not respond to aerodynamic controls.

Nobel, Alfred Bernhard 1833–1896, Swedish chemical engineer; invented dynamite; endowed the Nobel Prizes.

nobelium [nō bel′ē əm; nō bel′yəm] *Chemistry.* a very heavy radioactive element having the symbol No, the atomic number 102, and atomic weight 254; a synthetic element with nine short-lived isotopes, uncertain properties, and no known uses or compounds. (Named for the *Nobel* Institute for Physics in Stockholm, Sweden, where the existence of this element was first reported.)

Nobel Prize *Science.* an award for outstanding achievement in the fields of chemistry, physics, medicine or physiology, literature, and international peace; endowed by the inventor Alfred Nobel and presented annually since 1901. An additional award in economics was established by the Swedish Central Bank in 1969.

noble gas *Chemistry.* a gas that is unreactive (inert) or reactive only to a limited extent with other elements; six noble gases make up a group on the periodic table: helium, neon, argon, krypton, xenon, and radon. (So called because they do not mix readily with other elements; this was thought to be like the attitude of the nobility toward the common people.) Also, INERT GAS.

noble metal *Metallurgy.* any of several metallic elements, the electrochemical potential of which is much more positive than the potential of the standard hydrogen electrode; therefore, an element that resists oxidation.

noble potential *Physical Chemistry.* a potential that approaches or equals that of certain noble metals, such as gold, silver, or copper, which are chemically inert.

noble rot *Plant Pathology.* a condition that occurs on wine grapes when they become overripe on the vine and a fungus begins to grow; it produces the distinctive flavor of such wines as Sauterne and Tokay.

noble savage concept *Anthropology.* a term for the ethnocentric point of view in which tribal or nontechnological peoples are considered to be inherently more carefree, childlike, and naive (or more peaceful, compassionate, and sensitive) than people from industrialized cultures.

no-bottom sounding *Oceanography.* an unsuccessful attempt to measure water depth, especially using a weighted line, in which the ocean bottom is not reached.

no-break power *Electricity.* a backup power system in which load requirements of a device are fulfilled between the time the primary system fails and the auxiliary power is activated.

nocardamine *Microbiology.* an extracellular compound produced by certain microorganisms to bind to and subsequently take up ferric iron into the cell under iron-deficient conditions.

Nocardia *Bacteriology.* a genus of Gram-positive, aerobic bacteria that produce mycelia and have a distinctive cell wall composition; it includes some animal pathogens and some soil organisms.

Nocardiaceae *Bacteriology.* a family of aerobic bacteria of the order Actinomycetales that may be parasitic and/or pathogenic.

Nocardioides *Bacteriology.* a genus of aerobic bacteria that grow as branched substrate and aerial mycelia, reproduce by fragmentation into coccoid and rod-shaped cells, and are found in soil.

Nocardiopsis *Bacteriology.* a genus of actinomycete bacteria that form branching, substrate mycelia that undergo fragmentation and aerial mycelia producing spores; found in soil.

nocardiosis *Medicine.* an acute or chronic bacterial infection, usually of the lungs but with a tendency to spread to any organ of the body, especially to the brain; characterized by abscess formation. It is usually caused by *Nocardia asteroides* and sometimes by *N. farcinica* or *N. caviae*.

nociceptor *Physiology.* a pain receptor, usually a bare nerve ending without any organized end organ. Also, ALGESIRORECEPTOR.

nocodazole *Oncology.* $C_{14}H_{11}N_3O_3S$, an antineoplastic agent.

Noct. at night. (From Latin *nocte.*)

noctalbuminuria *Medicine.* the presence of abnormally high levels of albumin in the urine secreted during the night.

nocti- or **noct-** a combining form meaning "night."

Noctiliondae *Vertebrate Zoology.* the bulldog and harelipped bats, a family of the suborder Microchiroptera noted for feeding on insects and small fish they catch near the surface of streams and ponds; found in Central America and northern South America. Also, FISHERMAN BAT.

Noctilucaceae *Botany.* a family of marine dinoflagellates of the order Noctilucales that are nonphotosynthetic and phagotrophic, including highly buoyant species that are sometimes abundant enough to discolor the water.

Noctilucales *Botany.* an order of large flagellate algae of the class Dinophyceae, characterized by morphological modifications that enhance flotation and reduce flagellar activity.

noctilucent clouds *Meteorology.* rarely seen clouds of unknown composition that occur at great heights of between 75 and 90 kilometers and resemble thin cirrus clouds with a bluish or silverish color; they have been seen only at twilight during the summer.

noctiphobia see NYCTOPHOBIA.

Noct. maneq. at night and also in the morning. (From Latin *nocte maneque.*)

Noctuidae *Invertebrate Zoology.* a large family of medium-sized, usually dull-colored, night-flying moths in the superfamily Noctuoidea; the larvae (cutworms and armyworms) are destructive agricultural pests.

Noctuoidea *Invertebrate Zoology.* a large superfamily of medium to large moths, lepidopteran insects in the suborder Heteroneura, with reduced maxillary palpi.

nocturnal *Zoology.* describing an animal that is active mainly or exclusively at night, rather than during the day. *Botany.* describing a flower that opens at night and closes by day. (From the Latin word for night.)

nocturnal amblyopia see NIGHT BLINDNESS.

nocturnal emission *Physiology.* a release of seminal fluid during sleep.

nocturnal enuresis *Medicine.* nighttime bed-wetting; involuntary excretion of urine at night, especially during sleep.

nodal *Science.* relating to or resembling a node.

nodal line *Astronomy.* the line connecting the ascending and descending nodes of an orbit.

nodal point *Optics.* either of two points located on the axis of an optical system in such a way that a ray directed through one point emerges from the lens as if through the other point and in a parallel direction; if the same medium is on both sides of the lens, the nodal points coincide with the principal points. *Electricity.* in a transmission system, the junction points by which automatic switches and switching centers are nodal points in automated systems. *Oceanography.* points of zero displacement of water in amphidromic tidal systems, reduced from a line to a point by the Coriolis effect. Also, AMPHIDROMIC POINTS.

nodal progression *Astronomy.* a rotation of an orbit's nodal line in the same direction as the object is moving.

nodal recession *Astronomy.* a rotation of an orbit's nodal line in the direction opposite to the movement of object.

nodal rhythm *Physiology.* a heart rhythm produced when a pacemaker has been inserted in the atrioventricular node.

nodal tachycardia *Cardiology.* tachycardia in response to impulses arising in or near the atrioventricular node.

nodal tissue *Histology.* **1.** tissue from a lymph node. **2.** tissue from the sinoatrial node or atrioventricular node in the heart.

nodal zone *Oceanography.* the zone in which there is no tidal rise or fall in a system of standing waves, the direction of littoral transport differing on either side of the zone.

Nodaviridae *Virology.* a family of ssRNA-containing viruses that infect and replicate in insects.

Nodavirus *Virology.* the single genus of viruses of the family Nodaviridae; the type species, **Nodamura virus,** was isolated from mosquitoes and causes lethal infection in newborn mice.

node a point of juncture between parts, often forming a knot or protuberance; specific uses include: *Anatomy.* a small knot or lump of body tissue, either existing naturally or caused by disease. *Botany.* a joint in the stem of a plant, especially the joint from which a leaf starts to grow. *Geology.* a point along a fault marked by a change in the direction of apparent displacement. *Physics.* in a standing wave system, a point or locus of points that maintains zero amplitude. *Astronomy.* one of the two points in an orbit where the plane of the orbit intersects some reference plane. *Electricity.* a junction point in a circuit or other network. *Computer Technology.* a data entry point in a database management system. *Telecommunications.* the location at which transmitting or receiving equipment is connected to a communications network. *Mathematics.* **1.** a singular point on a curve. **2.** a vertex of a graph. By the duality principle, edges may also be viewed as nodes; in network applications, vertices are called nodes and edges are called branches. **3.** see CRUNODE.

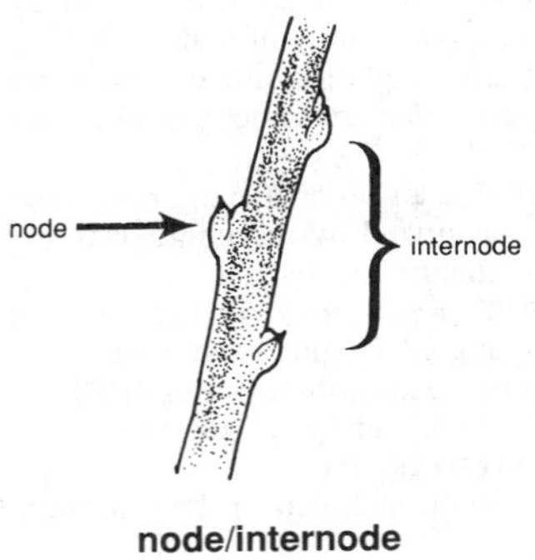

node/internode

node of Ranvier *Anatomy.* a periodic gap in the myelin sheath of a nerve fiber where the axon is incompletely covered by fingers of Schwann cells, typically occurring at 0.2- to 1.0-mm intervals in human nerves and acting as the sites for impulse propagation.

nodical month see DRACONIC MONTH.

Nodosariacea *Invertebrate Zoology.* a superfamily of foraminiferans in the suborder Rotaliina, with a radial calcite test, coiled, uncoiled, or spiral about the long axis.

nodose *Biology.* having many or noticeable protuberances; knobby, such as the **nodose ganglion** on the intestinal branch of the vagus nerve.

nodular *Science.* of, relating to, or occurring as a small lump or lumps.

nodular cast iron *Materials.* a cast iron treated so as to cause graphite to precipitate in spherical form during solidification, affording higher strength and ductility and thus combining the processing advantages of cast iron with the engineering advantages of steel. Also, DUCTILE (CAST) IRON, SPHERULITIC GRAPHITE CAST IRON.

nodular chert *Geology.* chert that occurs as dense, irregular, concretionary segregations.

Nodularia *Bacteriology.* a genus of cyanobacteria that grow as filaments in both freshwater and marine habitats and may cause diarrhea or vomiting in animals that drink water containing this bacteria.

nodular iron see AUSTEMPERED NODULAR IRON.

nodular nonsuppurative panniculitis see WEBER-CHRISTIAN DISEASE.

nodular powder *Metallurgy.* any metallic powder composed of round, but irregular particles.

nodule a small node; specific uses include: *Anatomy.* a small normal or pathological mass of closely packed tissue that can be distinguished from the surrounding area by touch. *Botany.* a knoblike swelling on the root of a legume that contains nitrogen-fixing bacteria. *Geology.* **1.** a small, hard mass or lump of a mineral or mineral aggregate whose composition generally contrasts with that of the surrounding sediment or rock matrix in which it is embedded. **2.** a concretionary lump of manganese, cobalt, iron, or nickel found on the deep-sea floor. **3.** a fragment of coarse-grained igneous rock that has been included in an extrusive rock.

nodules of the semilunar valves *Anatomy.* small fibrous nodes at the center of the free edge on each leaflet of the semilunar valves at the beginning of the pulmonary artery and aorta.

nodulizing *Engineering.* the process of consolidating fine mineral concentrates into lumps by kneading with a binder substance and occasionally by employing heat or chemical reactions.

nodulose *Biology.* having minute nodules or fine knobs.

Noeggerathiales *Paleontology.* an extinct order of vascular plants, tentatively placed with the Progymnospermopsida; Upper Carboniferous to Triassic.

Noether, Emmy [nō′thər] 1882–1935, German mathematician; noted for her work on noncommutative algebra.

Noetherian [nō′ə thēr′ē ən] *Mathematics.* relating to or based on the work of Emmy Noether.

Noetherian module *Mathematics.* a module that satisfies the ascending chain condition on submodules.

Noetherian operator *Mathematics.* a linear operator T between Banach spaces with closed range, such that $\alpha(T)$ and $\beta(T)$ are both finite, where $\alpha(T)$ is the dimension of the null space of T and $\beta(T)$ is the dimension of the null space of the adjoint of T. The difference $\chi(T) = \alpha(T) - \beta(T)$ is called the index of T.

Noetherian ring *Mathematics.* a ring R is said to be left (or right) Noetherian if one of the following equivalent conditions is satisfied: (a) R satisfies the maximum condition on left (or right) ideals; (b) every ideal of R is finitely generated; or (c) every prime ideal of R is finitely generated. A ring that is both left Noetherian and right Noetherian is said to be Noetherian.

Noether's theorem *Mechanics.* the theorem that establishes the identities of conserved quantities (if any) in a dynamical system, by examining the change in its Lagrangian resulting from a virtual displacement.

no-fire line *Military Science.* in artillery and naval gunfire support, a line short of which the supporting guns do not fire unless such fire is requested or approved by the commander of the supported troops; they may fire beyond the line at any time without danger to friendly troops.

nogalamycin *Microbiology.* an antibiotic produced by the bacterium *Streptomyces nogalater* and used against tumors.

no-go gauge *Engineering.* a gauge that does not fit a given part; used in conjunction with a go gauge to establish maximum and minimum dimension limits for that part.

Noguchi, Hideo 1876–1928, Japanese bacteriologist; produced pure cultures of syphilis spirochetes.

noise any undesired sound or disturbance; specific uses include: *Acoustics.* sound that consists of a random mix of different frequencies. *Electricity.* an unwanted, often random disturbance to a signal that tends to obscure the signal's information content; caused primarily by the random thermal motions of particles in the system. *Telecommunications.* **1.** any signal disturbance that interferes with the operation of a system. **2.** any random disturbance that obscures the clarity of a signal. Also, **noise distortion.**

noise analysis *Electricity.* the study of the sources of noise in a signal processing system, typically involving the determination of the frequency components responsible for the noise.

noise analyzer *Electronics.* an instrument that measures, displays, plots, or tabulates noise as a function of frequency.

noise control *Acoustical Engineering.* the prevention or reduction of noise, either at its source using such equipment as balancing machinery or at the listener's ear using methods such as shielding.

noise digit see NOISY DIGIT.

noise factor *Electronics.* the ratio of the signal-to-noise ratios at the output and the input of an amplifier or a system.

noise figure *Electronics.* the noise factor, expressed in decibels.

noise filter *Electronics.* a device that prevents spurious signals from passing through an alternating-current power line. *Acoustical Engineering.* a device added to a radio receiver to prevent spurious signals from entering.

noise generator *Acoustical Engineering.* a device that produces noise; sometimes used for calibrating sound measurement devices such as a real-time analyzer.

noise jammer *Electronics.* **1.** a device that produces signals that simulate electrical currents found in the atmosphere in order to convince an enemy that adverse communication conditions exist. **2.** a device used during World War II that transmitted white noise at the approximate frequency of an enemy's transmitter in order to obscure its signal.

noise killer *Telecommunications.* a device added to a communications circuit to prevent it from interfering with the operations of nearby circuits. Also, **noise suicide circuit.**

noiseless channel *Telecommunications.* a communications channel in which the effects of unwanted signals are negligible.

noise level *Acoustics.* a measure of the intensity of noise in a system, expressed in decibels relative to some specified reference.

noise limiter *Electronics.* a device that eliminates noise peaks that exceed the highest peak in the desired signal in order to limit the effects of atmospheric or man-made interference. Also, **noise silencer.**

noise-modulated jamming *Electronics.* random, unwanted signals that interfere with the ability of a radar receiver to pick up the desired signal.

noise pollution *Acoustics.* **1.** unwanted or harmful sounds in the environment. **2.** for a given locale, the average of the measured noise level over a period of time.

noise-power ratio *Electrical Engineering.* the ratio of the noise power at the output of a circuit to the noise power in the noise source.

noise rating *Acoustics.* **1.** a numerical representation of the level of noise that can be tolerated in a given situation. **2.** the evaluation of noise sound level in terms of loudness, perceived level, a noise-criterion curve, or other reference criteria.

noise reduction *Acoustical Engineering.* the use of electronic circuitry or mechanical means to reduce audible noise that is recorded or reproduced by a sound system, without affecting the desired sounds.

noise reduction coefficient *Acoustical Engineering.* the noise reduction capability, specified at 500 hertz or more, of an acoustic material (or a room lined with this material), determined by calculating an average of the Sabine absorption coefficients at 250, 500, 1000, and 2000 hertz.

noise suppression *Acoustical Engineering.* a method for minimizing surface noise when an audio tape or phonograph record is being played. *Electrical Engineering.* a method for minimizing or eliminating the effects of spurious electrical signals.

noise suppressor *Electronics.* a device that automatically shuts down the audio-frequency amplifier in a radio receiver when no signal is being received, to minimize background noise. Also, SQUELCH CIRCUIT.

noise temperature *Electronics.* given an ideal resistor whose resistance is that of the terminal pair in a system, the temperature at which the ideal resistor would generate the same noise power as that present at the system terminal.

noise tube *Electronics.* a type of gas-filled tube that generates white noise.

noise weighting *Electronics.* the process of calculating the effect of frequency-dependent noise in a network by taking into account the frequency response, giving the noise in each band a weight proportional to the response at that band.

noise word *Computer Programming.* in a keyword-in-context index, a common word such as "the" that should not be indexed.

noisy channel *Telecommunications.* a channel in which the effects of unwanted signals are highly prevalent.

noisy digit *Computer Technology.* the bit, usually 0, that is used to fill the rightmost position during a left shift operation; associated with floating-point arithmetic normalization. Also, NOISE DIGIT.

noisy mode *Computer Technology.* the use of a 1 bit instead of 0 as the noisy digit during normalization of a floating-point number.

Nolanaceae *Botany.* a dicotyledonous family of small South American shrubs and herbs of the order Solanales, characterized by internal phloem, glandular hairs, and an embryo with copious, oily endosperm.

no-load current *Electricity.* the flow of current into a network with an open-circuited output.

no-load loss *Electricity.* the power lost in a device such as a motor or transformer operating at rated voltage and frequency, but with the load disconnected.

noma *Medicine.* a severe gangrenous process involving the mucous membranes of the mouth or genitalia, most commonly seen in children with poor nutrition and hygiene. Also, GANGRENOUS STOMATITIS.

nomad *Anthropology.* a member of a tribe or people that practice nomadism as a way of life. Thus, **nomadic.**

nomadism *Anthropology.* a way of life in which there is no permanent residence site and the group moves from place to place according to the season, the available food supply, and other such factors; the route of travel often follows a traditional pattern.

Nomarski differential interference microscopy *Materials Science.* a microscopic system using the interference conditions generated by optical-path-length differences of two beams of coherent light to make visible detail that is not revealed by other light optical methods; used in both black-and-white and color contrast modes.

Nomarski microscope *Optics.* a microscope that measures the surface height profile and roughness of nonabsorbing reflective specimens, such as metallic surfaces.

nom. cons. nomen conservandum. Also, **nom. conserv.**

nom. dub. nomen dubium.

nomenclatural congress *Systematics.* a collection of rules, such as the International Code of Botanical Nomenclature Congress standards, governing the formation and application of scientific names to biological taxa.

nomenclatural type *Systematics.* the individual specimen with which a scientific name is associated; the taxon that determines the correct name of the higher taxon to which it belongs. Also, NOMENIFER, ONOMATOPHORE.

nomenclature *Systematics.* a classified system of names, as of organisms, anatomical structures, celestial objects, and so on.

nomen conservandum *Systematics.* in taxonomic literature, a notation used to indicate that a name, under provisions of some nomenclatural code, should not be changed. (A Latin term meaning "conserved name.")

nomen dubium *Systematics.* in taxonomic literature, a notation used to indicate a name that is invalid because it was not accompanied by an adequate description of the taxon to which it was applied. (A Latin term meaning "dubious name.")

nomenifer see NOMENCLATURAL TYPE.

nomen novum *Systematics.* in taxonomic literature, a notation used to indicate a proposed new name for a taxon. (A Latin term meaning "new name.")

nomen nudum *Systematics.* in taxonomic literature, a notation used to indicate that a name is not valid because of inadequate or insufficient material in the accompanying technical description. (A Latin term meaning "naked name.")

nomen rejiciendum *Systematics.* in taxonomic literature, a notation used to indicate that a name is no longer recognized under provisions of a nomenclatural code. (A Latin term meaning "rejected name.")

nomenspecies see TYPOLOGICAL SPECIES.

Nomina Anatomica *Anatomy.* the official body of anatomical nomenclature, as adopted and periodically revised by the International Congress of Anatomists.

nominal *Science.* in name but not necessarily in actuality; so called.
nominal band *Telecommunications.* in FAX systems, the frequency band of a signal wave equivalent to the width between zero frequency and maximum modulating frequency.
nominal bandwidth *Telecommunications.* **1.** in a given channel, the widest band of frequencies inclusive of guard bands. **2.** the space between the assigned frequency limits of a given channel.
nominal diameter *Geology.* the diameter calculated for a hypothetical sphere that would have the same volume as that calculated for a specific sedimentary particle.
nominal linewidth *Telecommunications.* in FAX transmissions, the average separation between the centers of adjacent recording or scanning lines.
nominal size *Design Engineering.* a generalized reference for identification purposes for the size of a particular part in which dimensions required for its fabrication are not precisely specified.
nominal speed see RATED SPEED.
nominal value *Electricity.* the rated value of resistance, capacitance, or impedance of a device under normal operating conditions as opposed to measured or actual value.
nominal weapon *Ordnance.* a nuclear weapon producing a yield of approximately 20 kilotons.
nom. nov. nomen novum.
nom. nud. nomen nudum.
nomograph *Mathematics.* **1.** a graphical method for exhibiting the value of a function of two independent variables. Two curves, indexed with the values of the independent variables, are given. A straight line is drawn between the two points on the curves corresponding to the independent variable values. This line intersects a third curve indexed by the value of the function. **2.** any similar or related graphical method.
nomothetic *Psychology.* relating to psychological studies or techniques that deal with people in general, as opposed to a single individual. Thus, **nomothetic goals, nomothetic emphasis.**
nom. rejic. nomen rejiciendum.
nonacosane *Organic Chemistry.* $CH_3(CH_2)_{27}CH_3$, rhombic crystals; insoluble in water and very soluble in alcohol; melts at 63.7°C and boils at 286°C (15 torr).
nonacoustic coupler *Computer Technology.* a modem built into a microcomputer or terminal that, when activated, communicates with another microcomputer over a telephone transmission line.
nonadaptive zone *Evolution.* a space representing inhospitable environmental conditions between adaptive zones, through which evolving organisms must pass during divergence or speciation.
nonadecane *Organic Chemistry.* $CH_3(CH_2)_{17}CH_3$, combustible leaflets; insoluble in water and soluble in alcohol; melts at 32°C and boils at 330°C; used in organic synthesis.
nonagon *Mathematics.* a polygon having nine sides.
nonambiguity *Telecommunications.* in coding, the property of identification without referring to preceding characters or to the spatial position of characters.
nonanal *Organic Chemistry.* $C_8H_{17}CHO$, a colorless, orange-rose smelling liquid; soluble in alcohol and mineral oil; boils at 190–192°C; used in perfumes and as a flavoring agent.
nonane *Organic Chemistry.* $CH_3(CH_2)_7CH_3$, a colorless liquid that boils at 150.2°C and freezes at –54°C; insoluble in water and soluble in alcohol; used in biodegradable detergents and organic synthesis.
nonanedioic acid see AZELAIC ACID.
non-A, non-B hepatitis *Pathology.* a clinical syndrome of acute viral hepatitis that occurs without the serologic markers of hepatitis A or B. It commonly follows blood transfusion or parenteral drug abuse and may be sporadic or epidemic. It is generally milder than hepatitis B, but a higher proportion of such infections become chronic and progress to cirrhosis.
nonanticipatory system see CAUSAL SYSTEM.
nonaqueous *Chemistry.* not composed of water.
nonaqueous solution *Chemistry.* a solution in which the solvent medium is not water.
nonartesian water *Hydrology.* groundwater that is not confined under pressure beneath relatively impermeable rock; it thus is not capable of rising if released, as artesian water would be.
nonatomic Boolean algebra *Mathematics.* a Boolean algebra in which no element $a \neq 1$ has the property that $ab = b$ for some b will imply $b = 0$. For example, a Boolean algebra of subsets of a set S is nonatomic if no subset A other than S itself has the property that $A \cap B = B$ for some subset $B \neq \emptyset$.
nonatomic measure *Mathematics* **1.** a real-valued measure λ defined on a sigma field A over a (topological) space X is said to be nonatomic if every set $E \in A$ with $|\lambda|(E) > 0$ contains a set $A \in A$ such that $0 < |\lambda|(A) < |\lambda|(E)$. $|\lambda|$ denotes the total variation measure of λ. The measure space (X,A,λ) is called a **nonatomic measure space. 2.** if (X,A,μ) is a measure space whose measure ring contains no atoms, then both the measure space X and its measure μ are said to be nonatomic.
nonbacterial thrombotic endocarditis see MARANTIC ENDOCARDITIS.
nonbanded coal *Geology.* a coal composed mainly of clarain or durain, without vitrain or bands of lustrous material.
nonbearing wall *Civil Engineering.* any wall not designed to support more than its own weight.
nonbenzenoid aromatic compound *Organic Chemistry.* an aromatic compound that does not contain benzene rings.
nonblackbody *Thermodynamics.* any real body; a body that reflects a portion of the electromagnetic radiation that is striking it or falling upon it.
nonblocking access *Telecommunications.* in telephony, the ability to connect any incoming line within the switching center to any outgoing line, provided that the outgoing line is unoccupied.
nonbonding orbital *Physical Chemistry.* a relationship between an electron and a nucleus in which the electron's energy remains constant as it draws closer to the nucleus so that neither bonding nor repulsion is expected to take place. Also, **nonbonding molecular orbital.**
non-bore-safe *Ordnance.* of or relating to a fuse that is not equipped with a safety device to prevent premature explosion of the projectile while still in the bore of the gun.
noncaking coal *Geology.* a hard or dull coal that does not cake or agglomerate when heated. Also, FREE-BURNING COAL.
noncalcic brown soil *Geology.* any of a group of zonal soils that develop in subhumid climates and are characterized by a slightly acidic, light-pink or reddish-brown A horizon overlying a light-brown or dull-red B horizon. Also, SHANTUNG SOIL.
noncapillary porosity *Geology.* the proportion of the total volume of large interstices in a rock or soil that do not hold water by capillarity.
noncentral force *Physics.* a force between two bodies that is not directed along the line connecting their centers of mass, as in some nuclear forces that depend in part on the spin orientations of the nucleons.
noncentral quadric *Mathematics.* a quadric surface possessing no points of symmetry, such as a paraboloid.
noncentrosymmetric structure *Crystallography.* a crystal structure that crystallizes in a space group with no center of symmetry.
noncholera vibrios *Bacteriology.* any of a group of vibrios that are morphologically and biochemically related to the cholera vibrios, but are not agglutinated by the cholera 01 antisera and thus do not cause cholera.
nonclastic *Geology.* **1.** exhibiting no evidence of being deposited mechanically or of being derived from preexisting rock. **2.** formed chemically or organically.
noncoding DNA *Molecular Biology.* a sequence of DNA contained in eukaryotic genomes that does not encode any genetic information and often consists of repetitive sequences.
noncognate ejecta see ACCIDENTAL EJECTA.
noncoherent scattering *Atomic Physics.* the scattering of photons from atoms resulting in noncoherent (out of phase) photons.
noncoincident demand *Electricity.* the total of the peak demands of all utilities in a given area, regardless of the occurrence times.
noncombat *Military Science.* relating to or designating vehicles or personnel not equipped or intended for combat service.
noncombatant *Military Science.* **1.** a person in a military unit who does not have an active combat role, such as a medical officer or a chaplain. **2.** anyone in a combat area who does not have military status, such as local civilians.
noncommissioned officer *Military Science.* an enlisted member of the armed forces who has been appointed to a position of leadership, but who is of lower rank than a commissioned officer or warrant officer; e.g., a sergeant, corporal, or petty officer.
noncompetitive inhibition *Biochemistry.* a reversible process in which a chemical substance alters the conformation of an enzyme molecule by binding outside the enzyme's substrate binding site.
noncomposite color picture signal *Telecommunications.* in color television transmission, a signal representing complete color picture information without the line- and field-synchronizing signals.
nonconductor see NONELECTROLYTE.

nonconformity *Geology.* **1.** an unconformity developed between overlying younger sedimentary rocks and underlying older igneous or metamorphic rocks that were exposed to erosion before being covered by the overlying sediments. Also, HETEROLITHIC UNCONFORMITY. **2.** see ANGULAR UNCONFORMITY.

nonconservative element *Oceanography.* an element, present in sea water in small quantities, that commonly undergoes temporary chemical changes resulting in variations in the total amount of the element at a given place and time.

nonconservative force *Mechanics.* a force for which the work done in displacing a particle from one point to another is dependent on the path taken by the particle in moving from the initial position to the final one; e.g., the force of friction. Thus, **nonconservative force field.**

nonconsumable electrode *Metallurgy.* in arc welding or arc melting, an electrode that does not melt during the operation; for instance, a tungsten or graphite electrode

noncoring bit *Engineering.* any drill bit that is designed to expel from the borehole all rock that it cuts away; used when a core sample is not needed or desired.

noncorrosive flux *Metallurgy.* in soldering or welding, a flux that does not corrode the part being joined

noncrossing rule *Physical Chemistry.* the principle that when two atoms in the same quantum state are plotted by relating potential energy to internuclear distance, the resulting curves will not cross, except when the states are asymmetrical.

noncrystalline solid *Materials Science.* an amorphous solid having a structure of atoms that is lacking in the long-range order of crystalline solids. Also, ANAMORPHOUS SOLID.

noncrystallographic symmetry *Crystallography.* symmetry within the asymmetric unit of a crystal structure. For example, the asymmetric unit of a crystalline protein may contain a dimer whose two subunits may have identical molecular structure, but, since they are not related by crystallographic symmetry, may have different environments. This noncrystallographic symmetry in proteins can be used as an aid in structure determination.

noncyclic terrace *Geology.* any of a series of stream terraces representing previous valley floors that were formed during periods when continued valley deepening was accompanied by lateral erosion.

nondegenerate *Mathematics.* **1.** an operator is nondegenerate if its nullity is zero; e.g., a bijection or isomorphism. **2.** an inner product f on a vector space V is nondegenerate if 0 is the only vector α in V, such that $f(\alpha,\beta) = 0$ for all β; equivalently, f is nondegenerate if and only if its associated linear operator T_f is nonsingular. In particular, if V is a finite dimensional vector space and f is bilinear, f is nondegenerate if and only if its matrix in any ordered basis for V is a nonsingular matrix. A nondegenerate form is sometimes called **nonsingular**.

nondegenerate amplifier *Electronics.* a device in which signals are amplified to ultrahigh or microwave frequencies by pumping them with alternating current whose frequency exceeds twice the signal frequency.

nondegenerative basic feasible solution *Mathematics.* any feasible solution to a linear programming problem in which all solution variables are positive nonzero.

nondelay fuse *Ordnance.* a fuse that functions quickly upon impact with the target; as the missile or projectile is slowed down by the target, inertia causes the firing pin to strike the primer.

nondepositional unconformity see PARACONFORMITY.

nondestructive breakdown *Electronics.* the breakdown that can occur between the gate and the channel of a field-effect transistor without causing the device to malfunction.

nondestructive testing *Materials Science.* any testing procedure that does not change the surface or interior of a component, so that it can be used in service, including such tests as radiography, ultrasonic testing, eddy current testing, liquid penetrant inspection, and electromagnetic testing. Also, **nondestructive evaluation.**

nondeterministic *Computer Programming.* describing a process that can do one of multiple things; which one it will do is not predetermined.

nondeterministic finite automaton *Computer Programming.* a finite automaton that has multiple state transitions from a single state for a given input symbol, or that has a null transition, not requiring an input symbol.

nondeterministic polynomial *Computer Programming.* a class of practical computational problems for which no effective algorithms have been developed. Solutions to such problems can be checked in polynomial time, but to find an unknown solution generally requires time that is an exponential function of problem size.

nondeviated absorption *Physics.* the absorption of radiation that does not exhibit any appreciable slowing of the waves.

nondigestible carbohydrates *Nutrition.* carbohydrates that cannot be broken down in the human digestive tract and hence provide no metabolizable energy.

nondimensionalizing *Mechanics.* see SCALING.

nondirective therapy see CLIENT-CENTERED THERAPY.

nondisjunction *Genetics.* the failure of homologous chromosomes to separate during meiosis, or the failure of sister chromatids to separate during mitosis.

nondisjunction mosaic *Genetics.* an organism composed of two distinct cell lines derived from a nondisjunction event in an earlier mitotic division.

nondissymmetric *Organic Chemistry.* describing a molecule that is superimposable upon its mirror image.

nondivergent flow *Oceanography.* the flow of a well-defined ocean current with distinct boundaries.

nonelectrolyte *Physical Chemistry.* any liquid or solid substance that resists the passage of an electric current, either in a solution or in its pure state, such as pure water, ethyl alcohol, or sucrose. Also, NONCONDUCTOR.

1-nonene see 1-NONYLENE.

nonequilibrium thermodynamics *Thermodynamics.* the study of irreversible processes and their rates in macroscopic systems. Also, IRREVERSIBLE THERMODYNAMICS.

nonequivalence gate see EXCLUSIVE-OR GATE.

nonerasable storage *Computer Technology.* any storage device or medium whose contents cannot be erased during processing, such as punched cards or microfilm.

nonesite *Petrology.* a porphyritic basalt consisting of phenocrysts of enstatite, labradorite, and augite embedded in a groundmass of plagioclase and augite.

nonessential amino acid *Nutrition.* an amino acid that is produced by an organism in sufficient quantities so that it need not be supplied by diet.

nonessential nutrient *Nutrition.* any nutrient that can be synthesized in sufficient quantities in the body and therefore does not need to be supplied in food.

noneuclidean geometry *Mathematics.* a geometry in which one or more of the axioms (usually the parallel postulate) of Euclidean geometry are omitted or modified. Also, **non-Euclidean geometry.**

nonevolutionary response *Ecology.* an adaptation made by an organism in response to changes in its environment.

nonexecutable statement *Computer Programming.* a program statement that provides information about the data or the way processing is to be done; does not result in a sequence of machine language instructions. Also, DECLARATION STATEMENT.

nonexpendable *Engineering.* relating to or describing a product, supply, tool, or piece of equipment that is not consumed or destroyed under normal operating conditions.

nonfaradaic path *Physical Chemistry.* a route followed by current traveling through an electrolytic solution, moved along by a continuous charging and discharging of stored energy (capacitance).

nonfat *Food Technology.* describing a food or food product that does not contain fat or fat solids.

nonfatal error *Computer Programming.* a software error that may result in a warning signal but does not terminate program execution or prevent a compiler from generating object code.

nonfat milk *Food Technology.* milk from which all or virtually all fat solids have been removed. Also, SKIM MILK.

nonfeasible method see GOAL-COORDINATION METHOD.

nonferrous *Metallurgy.* containing little or no iron.

nonferrous metal *Metallurgy.* any metal other than iron or steel.

nonflowing well *Engineering.* a well requiring a pump or other means to bring the liquid to the surface.

nonflow system see CLOSED SYSTEM.

nonfluent aphasia *Neurology.* a speech impairment caused by a brain lesion, in which an individual can write but cannot speak, except for a few ill-pronounced expletives. Also, BROCA APHASIA.

nonfragmenting projectile *Ordnance.* a smoke-producing projectile used in antiaircraft practice so that the location of the burst can be observed without a close burst destroying the target.

nongenetic reactivation *Virology.* a phenomenon by which an inactivated virus can produce viable progeny particles as the result of biochemical contributions from another virus.

nongonococcal urethritis *Medicine.* a urethritis without gonococcal infection, such as urethritis caused by *Chlamydia trachomatis.* Also, NONSPECIFIC URETHRITIS, SIMPLE URETHRITIS.

nongraded *Geology.* **1.** describing an unconsolidated sediment or cemented detrital rock consisting of particles of more than one size. Also, POORLY SORTED. **2.** describing a soil or unconsolidated sediment in which the constituent particles are essentially the same size.

nongranular leukocyte *Histology.* any white blood cell lacking prominent cytoplasmic granules, such as a monocyte or lymphocyte.

nongraphic character *Computer Programming.* a character that is sent to a display unit or a printer but does not produce a printable image, such as a letter-space or paragraph-return character.

nonheme iron *Biochemistry.* iron that is contained within a protein but is not bound within a prophyrin ring.

nonheme-iron protein *Biochemistry.* an iron-sulfur protein that contains nonheme iron complexed either with sulfur atoms derived from cysteine or with a mixture of inorganic and cysteine-derived sulfur atoms.

nonhistone protein *Biochemistry.* a eukaryotic DNA binding protein other than a histone, occurring in association with the DNA in a cell nucleus and believed to play a role in the regulation of gene activity. Also, **nonhistone chromosomal protein.**

nonholonomic constraint *Mechanics.* in classical mechanics, a constraint that is not holonomic; i.e., one that cannot be represented by generalized forces that are gradients of potential functions depending only on particle position.

nonholonomic system *Mechanics.* a mechanical system that is not holonomic; a system for which some or all of the constraint equations contain differentials dq_i of the generalized coordinates that cannot be integrated and therefore cannot be used to reduce the number of coordinates to equal the number of degrees of freedom of the system in order to completely describe the motion, such as a coin rolling on a tabletop.

nonhoming *Control Systems.* of a relay or device, not returning to its starting or home position.

nonhoming tuning system *Electronics.* a type of tuning system in which a motor rotating in the wrong direction will automatically right itself after it reaches the end of the dial.

nonhypergolic *Chemistry.* not igniting spontaneously on contact.

nonideal *Physical Chemistry.* describing a chemical system that does not conform to the behavior established for an ideal system, such as the equation of state for perfect gases or Raoult's law for ideal solution behavior. Thus, **nonideal gas, nonideal behavior.**

nonideal solution *Physical Chemistry.* a solution that shows a deviation from Raoult's law; i.e, the partial vapor pressure of its components is not exactly predictable from their mole fractions in the solution. Also, REAL SOLUTION.

Nonidet P-40 *Biochemistry.* a detergent, without ions, occurring in the polyoxyethylene *p-t*-octyl phenol series.

nonimage area *Graphic Arts.* the part of a printed page that receives no ink.

nonimpact printer *Computer Technology.* any of a class of printers that produce output without the printing element actually striking the paper, including electrophotographical, thermographic, ink-jet, and laser printers.

noninductive *Electricity.* having a negligible or zero inductance. Thus, **noninductive capacitor.**

noninductive resistor *Electricity.* a wire-wound resistor that is wound to have virtually no inductance.

noninductive winding *Electricity.* a winding in which the magnetic field of one section cancels the field of the section adjacent to it.

noninfectious *Medicine.* **1.** of a disease or condition, not infectious; not spread by contact, inhalation, and so on. **2.** of a person or other organism, not able to spread a disease.

noninteracting control *Control Systems.* in a multiple-input/multiple-output system, feedback control in which each input affects only one output.

noninverting amplifier *Acoustical Engineering.* an amplifier whose output signal is in phase with its input signal.

nonisolated system *Thermodynamics.* a system that interacts with its surroundings; i.e., heat transfer or work can cross the boundaries of the system.

nonlinear *Physics.* referring to a relationship between two quantities that are not directly proportional to each other. *Chaotic Dynamics.* of or relating to nonlinearity. *Mathematics.* of or relating to nonlinear programming.

nonlinear acoustics *Acoustics.* the study of nonlinear sonic and ultrasonic acoustic energy, using nonlinear differential equations to explain their behavior.

nonlinear amplifier *Electronics.* a device in which a change in the input signal does not result in a proportional change in the output signal.

nonlinear capacitor *Electricity.* a capacitor with a nonlinear mean-charge characteristic or a capacitance that varies with bias voltage.

nonlinear circuit *Electricity.* a network that cannot be described by linear differential equations in time. Also, NONLINEAR NETWORK.

nonlinear coil *Electromagnetism.* a coil whose core is easily saturated so that when little or no current passes through the coil, there is high impedance, and the converse is true with high current.

nonlinear control system *Control Systems.* a control system in which some or all of the outputs are not linear functions of the inputs.

nonlinear coupler *Electronics.* a device that produces an output signal whose frequency is a multiple of the input frequency wave, by coupling energy from the input circuit.

nonlinear crystal *Solid-State Physics.* any crystal that exhibits a response (such as magnetization, polarization, or strain) that is not linearly related to the externally applied agent (such as magnetic field, electric field, or stress).

nonlinear damping *Physics.* a condition in which the amplitude of a damped vibrational system does not decrease at a linear rate.

nonlinear detection *Electronics.* a process that extracts information from a radio-frequency signal by exploiting a nonlinear device or circuit.

nonlinear device *Electronics.* a device in which a change in the applied voltage does not proportionally change the current. Also, **nonlinear circuit component, nonlinear element.**

nonlinear dielectric *Electricity.* a dielectric whose dielectric constant varies with voltage.

nonlinear distortion *Electronics.* the distortion that occurs when the amplitude of an output signal is not directly proportional to that of the input signal.

nonlinear dynamics see CHAOS.

nonlinear elastic fracture mechanics *Materials Science.* the study of the growth of cracks and the problems associated with the accompanying plastic deformation that occurs in nonlinear elastic materials, such as ductile metals.

nonlinear equation *Mathematics.* any equation in *n* (independent) variables $x_1, x_2, \ldots, x_n = y$, where the coefficients a_i are constants.

nonlinear feedback control system *Control Systems.* a feedback control system in which the output signals are not linear functions of the input signals.

nonlinearity *Chaotic Dynamics.* a property of a dynamical system whereby the evolution of its variables depends on products of two or more of the present values of the variables. This feature is a necessary (though not sufficient) ingredient of systems that might be chaotic.

nonlinear material *Physics.* any substance whose response to an external influence is not proportional to the magnitude of the influence.

nonlinear molecule *Organic Chemistry.* a molecule that is branched or composed of rings.

nonlinear network see NONLINEAR CIRCUIT.

nonlinear optics *Optics.* the study of phenomena caused by the interaction of nonlinear materials with nearly monochromatic, directional, electromagnetic radiation, such as the light produced by high-power lasers.

nonlinear oscillator *Electronics.* a device that varies the frequency of an alternating current in response to an audio signal; commonly used in eavesdropping devices.

nonlinear planning *Artificial Intelligence.* a planning task in which subgoals interact and cannot be done in an arbitrary order; e.g., a task of painting the ladder and painting the ceiling.

nonlinear programming *Mathematics.* the branch of applied mathematics concerned with extrematizing functions of several variables, where constraints on the variables or the functions to be extrematized include at least one nonlinear function. If, in addition, any of these functions is not required to be differentiable, the process is known as **nondifferentiable programming.**

nonlinear reactance *Electronics.* **1.** the reaction of a coil when the voltage drop across it is not proportional to the rate of change in the current flowing through it. **2.** the reaction of a capacitor when the voltage drop across it is not proportional to the rate of change in its charge.

nonlinear refraction *Optics.* a change in the refractive index of a particular substance due to variation in the intensity of transmitted light.

nonlinear resistance *Electricity.* the behavior of an electric component that contradicts Ohm's law and produces a current that is not proportional to the voltage.

nonlinear spectroscopy *Spectroscopy.* the study of energy-level transitions using nonlinear effects such as ionization or multiphoton absorption.

nonlinear step-growth polymerization *Materials Science.* the formation of branched rather than linear chains by monomers with a functionality greater than two.

nonlinear system *Mathematics.* a system of equations that includes at least one nonlinear equation.

nonlinear taper *Electricity.* a nonuniform distribution of resistance throughout a potentiometer or rheostat element.

nonlinear vibration *Mechanics.* vibration in a system such that the restoring and damping forces are not proportional to displacement and velocity; e.g., an oscillating system in which a spring hardens or weakens with increased displacement.

nonliterate *Anthropology.* relating to or designating a society that does not have a written language. Also, PRELITERATE.

nonloaded Q *Electricity.* the *Q* value of an electric impedance without external coupling. Also, BASIC Q.

nonlocalized electron *Physics.* an electron whose orbital is not necessarily found between the two nuclei but is instead spread over several nuclei or over the volume of a crystal lattice.

nonmagnetic *Electromagnetism.* a property whereby a substance is not affected by magnetic fields or other magnetic phenomena.

nonmagnetic steel *Metallurgy.* any of various steels that are not attracted by a permanent magnet; for example, an austenitic stainless steel.

non-Mendelian *Genetics.* inheritance that results in abnormal segregation ratios when a mutant and a wild type are crossed; sometimes the genotype of only one parent is inherited and the genotype of the other is permanently lost; at other times the progeny of one parental type exceed those of the other.

nonmetal *Chemistry.* any chemical element that is not a metal or a metalloid; an element that is not capable of forming positive ions in an electrolytic solution.

nonmetallic *Chemistry.* **1.** of or relating to a nonmetal. **2.** not having the qualities or appearance of a metal. **3.** made of materials other than metal.

nonmetallic inclusion *Metallurgy.* in a metallic structure, nonmetallic particles that are foreign to the metal or alloy system, such as oxides or sulfides, which are at times intentionally added to improve properties, as in the case of dispersion-hardened products and mechanical alloys.

non-minimum-phase system *Control Systems.* a linear system whose transfer function has one or more poles with real parts that are nonzero or positive.

nonmonotonic logic *Artificial Intelligence.* a form of logic in which conclusions can be drawn based on assumptions of things that are typically true; later information could force such conclusions to be withdrawn.

nonmonotonic reasoning *Artificial Intelligence.* a reasoning process in which conclusions that are determined may be changed or "forgotten" based on later information.

non-Newtonian *Mechanics.* describing systems or processes that do not conform to the theories of Sir Isaac Newton; i.e., the speeds involved are too high, and/or the particles are too small, for Newton's laws of motion to apply.

non-Newtonian fluid *Fluid Mechanics.* a fluid whose shearing stress is found not to be proportional to the shearing rate; examples include Bingham plastic, pseudoplastic fluid, dilatant fluid, and thixotropic and viscoelastic fluids.

non-Newtonian fluid flow *Fluid Mechanics.* a fluid flow that does not exhibit characteristics of Newtonian flow; analysis of such flows is applied to problems of extrusion, fluid mixing, coating operation, and tubular flow.

non-Newtonian viscosity *Fluid Mechanics.* the viscosity of a fluid whose shearing stress and rate of strain are not proportional. Also, ANOMALOUS VISCOSITY.

nonnumeric character *Computer Programming.* any allowable character other than a numeral.

nonoccluded virus *Virology.* any of various insect-pathogenic viruses that do not produce proteinaceous occlusion bodies during infection.

nonohmic *Electricity.* describing a substance or circuit component that violates Ohm's law.

nonoic acid *Organic Chemistry.* $C_8H_{17}COOH$, any of several isomers having the same empirical formula; produced by the Fischer-Tropsch reaction.

nonopaque *Radiology.* not impervious to roentgen rays; radiolucent or transparent.

nonosteogenic fibroma *Medicine.* a small abnormality in the cortex of a bone, usually the lower femoral shaft of a child's bone, that is filled with fibrous tissue. Also, FIBROUS CORTICAL DEFECT.

nonoxidative thermal degradation *Materials Science.* the breaking of chemical bonds in polymers in the absence of oxygen, caused by energy acquired from heat, light, ionizing radiation, or chemical reaction; this process is significant when the availability of oxygen is limited.

nonoxynol-9 *Pharmacology.* $C_{33}H_{60}O_{10}$, the active ingredient in spermicides.

nonparalytic poliomyelitis *Medicine.* infection by poliomyelitis that does not lead to paralysis; characterized by muscle pain and stiffness, upper respiratory or gastrointestinal symptoms, and mild fever.

nonparametric *Statistics.* of a method, test, or the like, not requiring parameters or distributions.

nonparametric method *Statistics.* any of various inferential procedures whose conclusions do not rely on assumptions about the distribution of the population of interest. Also, DISTRIBUTION-FREE METHOD.

nonpareil *Graphic Arts.* **1.** a six-point slug or space. **2.** a former type size of about six points.

nonpenetrative *Geology.* of a deformational texture that affects only part of a rock.

nonpermissive cell *Cell Biology.* any cell in which a conditional mutation is lethal. *Virology.* any cell in which a given virus cannot replicate.

nonpersistent transmission *Virology.* a temporary surface infection of arthropod vectors that lasts for only seconds or minutes, after which the vector eliminates the virus.

nonpersistent war gas *Ordnance.* a chemical agent that, in an open space, is effective for 10 minutes or less at the point of dispersion.

nonplunging fold *Geology.* a fold whose axial surface is horizontal. Also, HORIZONTAL FOLD, LEVEL FOLD.

nonpolar *Physical Chemistry.* describing a substance that is composed of molecules with no permanent electric moment, having atoms that are bonded by shared electrons, nonelectrolytic in solution, and nonionizing in water.

nonpolar (covalent) bond *Physical Chemistry.* a bond in which the shared electron pair is equally attracted to both bonded atoms.

nonpolar dielectric *Materials Science.* any material that is a poor electric conductor and has a chemical structure that does not develop dipole orientational polarization when an electric field is applied; its total molecular moment is always zero.

nonpolarized relay see NEUTRAL RELAY.

nonpolar molecule *Physical Chemistry.* a diatomic molecule in which the positive and negative charges coincide instead of being separate.

nonprecision approach *Navigation.* any standard aircraft instrument approach procedure used when no electronic glideslope is provided.

nonprint code *Computer Technology.* a bit combination that inhibits line printing.

nonpriority interrupt *Computer Technology.* **1.** a request for central processing unit service in which no priority-interrupt scheme is active. **2.** a nonessential request for central processing unit service that can be postponed until all priority interrupts have been serviced.

nonprocedural language *Computer Programming.* any high-level programming language in which the user indicates the goals to be achieved rather than the specific steps required to achieve them.

nonproducer cell *Virology.* any cell that contains a viral genome but does not produce virus particles.

nonprotein nitrogen *Biochemistry.* nitrogen in the blood that is not a constituent of protein, such as the nitrogen associated with uric acid, creatine, and polypeptides.

nonprotic solvent *Chemistry.* a solvent that does not disassociate to produce protons.

nonpsychotic organic brain syndrome *Medicine.* a syndrome caused by transient or permanent dysfunction of the brain but in which psychotic symptoms are lacking.

nonrandom mating see ASSORTATIVE MATING.

non-rapid eye movement sleep see NREM SLEEP.

nonrecoverable error *Computer Programming.* a software error or hardware fault that causes processing to terminate prior to normal completion.

nonredundant system *Computer Technology.* a computer system designed with only the minimum number and variety of hardware components needed to accomplish its purpose. Also, SIMPLEX STRUCTURE.

nonreflective ink *Computer Technology.* any ink that can be recognized by an optical character reader.

nonrelativistic approximation *Physics.* an approximation in which any velocity of a physical entity is assumed to be much less than the speed of light.

nonrelativistic mechanics *Mechanics.* mechanics in which the particles under consideration move at speeds much slower than the speed of light. Similarly, **nonrelativistic kinematics.**

nonrelativistic particle *Physics.* a particle with a velocity that is low enough to disregard relativistic effects.

nonrelativistic quantum mechanics *Quantum Mechanics.* the theory of motion of a dynamical system whose motions are small compared with the speed of light, in which matter is neither created nor destroyed, and when the rest mass can be neglected.

nonremovable discontinuity *Mathematics.* a point in the domain of a function at which the function is discontinuous or undefined and where the function cannot be made continuous by redefining the function at that one point.

non-REM sleep see NREM SLEEP.

nonrenewable fuse unit *Electricity.* a fuse that cannot be restored for operation after the filament opens.

nonrenewable resource *Ecology.* a substance that is formed at a rate far slower than its rate of consumption and is therefore not able to be replenished within human time scales; e.g., fossil fuels. Also, FINITE RESOURCE.

nonrepro blue *Graphic Arts.* a light blue color that will not appear in a photograph; the color of pencils and printed lines commonly used in page layouts.

nonreproducible *Graphic Arts.* done so as not to reproduce photographically, such as lines drawn with a nonrepro blue pencil.

nonreproducing code *Computer Technology.* a control code that does not result in an output character, such as carriage and line-spacing control.

nonresident routine *Computer Programming.* any routine that is not stored in main memory and must be read from auxiliary memory prior to execution.

nonresonant antenna see APERIODIC ANTENNA.

nonresonant line *Electromagnetism.* a transmission line that is designed to suppress current and voltage standing waves and reflected waves.

non-return-to-zero *Computer Technology.* a method of writing to magnetic media in which the electrical current through the write mechanism does not drop to zero after each write operation.

nonreturn valve see CHECK VALVE.

nonrigid *Aviation.* describing a type of airship that is held in shape only by the pressure of the gas within, rather than by a supporting structure.

nonrotating disk see SEMICONDUCTOR DISK.

nonsecretor *Hematology.* an individual possessing A or B type blood whose saliva and other body secretions do not contain that particular A or B substance.

nonsegregated reservoir *Petroleum Engineering.* a solution-gas-drive oil reservoir in which oil-gas separation does not occur as a function of height or upward movement.

nonsegregating mixer *Mechanical Engineering.* a particulate-solid mixer, such as a barrel mixer, whose predominating mechanisms are diffusive or shear mixing; baffles, blades, or internal rotating devices are used to improve the mixing action.

nonselective mining *Mining Engineering.* a low-cost method of mining generally employing cheaper methods for large-scale operations.

nonsense codon *Molecular Biology.* a group of three nucleotides (UAG, UAA, and UGA) on an mRNA molecule that does not encode an amino acid and therefore causes termination of protein synthesis.

nonsense mutation *Molecular Biology.* a mutation that occurs when a codon specifying an amino acid is altered to a nonsense or chain-terminating codon.

nonsense suppressor *Genetics.* a gene coding for a mutant tRNA able to respond to one or more of the stop (nonsense) codons.

nonsequential computer *Computer Technology.* a computer that does not automatically proceed to the next instruction in the sequence, but must be directed to the location of each instruction in turn.

nonservo *Robotics.* describing any mechanism that is not a servomechanism.

nonservo robot *Robotics.* a robot with a manipulator that moves until it reaches a fixed stop.

nonshorting contact switch *Electricity.* a selector switch coupled with a movable contact that is narrower than the distance between two contacts; used to break an old circuit before a new circuit is completed. Also, **nonshorting switch.**

nonsingular *Mathematics.* invertible; e.g., a nonsingular matrix or nonsingular linear transformation.

nonsinusoidal waveform *Electricity.* any wave that is not a sine wave.

nonskid *Civil Engineering.* describing a concrete surface that has been treated with additives or indented while wet to roughen it.

nonsocial *Behavior.* see SOLITARY.

nonsorted polygon *Geology.* a patterned ground whose mesh is predominantly polygonal, and whose unsorted appearance results from the absence of border stones.

nonspecific *Medicine.* of a disease or condition, not having any single known cause. *Pharmacology.* of a drug or therapy, not directed against any specific cause of disease; having a general effect.

nonspecific immunity *Immunology.* a body's acquired or native resistance to infection or disease that is caused by factors other than specific antibodies or cells, such as genes or hormones.

nonspecific immunization *Immunology.* a natural resistance or immunity to a disease inherent in a species, not induced by antigens.

nonspecific urethritis see NONGONOCOCCAL URETHRITIS.

nonspherical nucleus *Nuclear Physics.* a deformed nucleus or a nucleus that, when at its lowest level of energy, appears to have an ellipsoidal shape.

nonstoichiometric compound see BERTHOLLIDE.

nonstoichiometry *Materials Science.* a compositional deviation in a phase from definite atom ratios.

nonstop *Transportation Engineering.* making no stops of any kind en route. Thus, **nonstop flight.**

nonstop computer *Computer Technology.* a highly redundant computer system that has virtually 100% reliability.

nonstorage camera tube *Electronics.* a television camera tube in which the picture signal is proportional, at any given instant, to the intensity of illumination on the area of the scene being broadcast.

nonstranded rope *Design Engineering.* a rope made of wire oriented in circular sheaths, as opposed to strands, with added strength, durability, and nonspinning properties, owing to the orientations of the sheaths in opposite directions.

nonstructural protein *Virology.* a protein that is coded by a viral genome intracellularly but is not incorporated into the mature virion.

nonsulfur vulcanization *Materials Science.* the introduction of a network of cross-links into an elastomer without the addition of sulfur to make the material strong and highly elastic.

nonsuppressible insulinlike activity see INSULINLIKE GROWTH FACTORS.

nonswappable program *Computer Programming.* a program of such high priority that it will not be temporarily unloaded from main memory so that another program may be executed.

nonsymmetrical aquifer *Petroleum Engineering.* an irregular (neither radial nor linear) water-bearing formation in an oil reservoir.

nonsymmetry see ASYMMETRY.

nonsynchronous *Electricity.* not related in frequency or speed to other quantities within a system.

nonsynchronous timer *Electronics.* a device placed at the receiving end of a communications link to reestablish the time relationship between pulses when no timing pulses are sent.

nonsynchronous transmission *Electronics.* a process that does not use a clock to control the spacing of signals that are transmitting data.

nonsynchronous vibrator *Electronics.* a device that interrupts a direct-current circuit at a frequency independent of other circuit factors and does not rectify the stepped-up alternating current that results.

nonsystematic joint *Geology.* a joint that is not part of a set or system of joints.

nontechnological *Anthropology* relating to or designating a society that does not employ the artifacts and techniques of contemporary industrial cultures. Also, PRETECHNOLOGICAL.

nontectonite *Petrology.* a rock in which the orientation and position of the fabric grains are not influenced by movement of adjacent grains, such as a rock formed by mechanical settling.

nonterminal symbol *Computer Science.* a symbol that names a phrase in a grammar.

nonterminal vertex *Mathematics.* **1.** in graph theory, a vertex of degree not equal to one. **2.** in a rooted tree, a vertex having at least one successor.

nonthermal decimetric emission *Electromagnetism.* radiation detected from the planet Jupiter having wavelengths greater than 4 centimeters; the range between 5 centimeters and 1 meter is known to have almost constant radiation flux.

nonthermal radiation *Physics.* the radiation from accelerated charged particles that are not in thermal equilibrium.

nontidal current *Oceanography.* a current caused by forces other than tides, such as ocean currents, rip currents, and longshore currents.

nontranscribed spacer *Molecular Biology.* any segment of DNA that connects two coding regions but that is not transcribed by RNA polymerase.

nontransferred arc *Metallurgy.* in arc cutting or welding, an arc that is not struck between an electrode and the workpiece.

nontrivial *Mathematics.* **1.** a subgroup (subring, subfield, subset, etc.) of a given group (ring, field, set, etc.) is said to be nontrivial if it equals something other than the entire group (ring, field, set, etc.), zero element, or empty set. **2.** a solution to an equation or problem is said to be nontrivial if it is nonzero.

nontronite *Mineralogy.* $Na_3Fe_2^{+3}(Si,Al)_4O_{10}(OH)_2 \cdot nH_2O$, a green or yellowish-green, monoclinic mineral of the smectite group, having a specific gravity of 2 to 3 and a hardness of 1 to 2 on the Mohs scale; found in fine-grained claylike form as an alteration product of volcanic glass, in weathered basaltic rocks, and in mineral veins.

nonuniform flow *Fluid Mechanics.* a fluid flow across a specified surface in which the instantaneous velocity can vary from point to point.

nonvenereal syphilis *Medicine.* a chronic inflammatory infection caused by an organism similar to *Treponema pallidum*; while it is found throughout the world, it usually affects children in arid regions; transmitted by direct, nonsexual contact and by the common use of eating and drinking utensils. Also, BEJEL, ENDEMIC SYPHILIS.

nonverbal communication *Linguistics.* a term for any of the methods by which people communicate other than spoken and written language, such as facial expression, gesture, or body movement or posture; this does not include formalized sign languages.

nonvibrio cholera *Pathology.* a disease that is clinically similar to cholera but that is caused by another agent such as *Escherichia coli, Salmonella,* or *Shigella.*

nonviscous neutral oil *Materials.* a neutral oil of viscosity less than 135 SSU at 37.7°C.

nonvolatile *Computer Technology.* describing a storage device that is designed to retain stored data when the power is turned off. Thus, **nonvolatile storage, nonvolatile memory.**

non-von Neumann architecture see DATA FLOW SYSTEM.

nonwetting *Materials Science.* describing a liquid metal that does not wet a specified surface.

nonwetting phase *Petroleum Engineering.* in a formation in which the reservoir fluid is two-phase (oil and water), the phase (oil) that does not wet the pore surfaces of the formation.

nonwetting sand *Geology.* sand consisting of angular particles of various sizes in a tightly packed lens that resists infiltration of water.

***n*-nonyl acetate** *Organic Chemistry.* $CH_3COO(CH_2)_8CH_3$, a colorless, pungent, combustible liquid; soluble in alcohol; boils at 208–212°C; used in making perfumes. Also, ACETATE C-9.

***n*-nonyl alcohol** *Organic Chemistry.* $CH_3(CH_2)_7CH_2OH$, a colorless combustible liquid with a floral odor; insoluble in water and soluble in alcohol; boils at 215°C and freezes at –5°C; used in perfumery and flavoring. Also, ALCOHOL C-9, OCTYL CARBINOL, PELARGONIC ALCOHOL.

nonylbenzene *Organic Chemistry.* $C_6H_5C_9H_{19}$, a combustible, aromatic, straw-colored liquid; boils at 245–252°C; used in manufacturing surface-active agents.

1-nonylene *Organic Chemistry.* C_9H_{18}, a combustible colorless liquid; insoluble in water and soluble in alcohol; boils at 149.9°C; used in organic synthesis, polymer gasoline, and as a lube-oil additive. Also, 1-NONENE.

nonyl phenol *Organic Chemistry.* $C_9H_{19}C_6H_4OH$, a combustible yellowish liquid; insoluble in water and soluble in alcohol; boils at 293°C and sets to glass below –10°C; used in resins, plasticizers, and lube-oil additives.

nonylphenoxyacetic acid *Organic Chemistry.* $C_9H_{19}C_6H_4OCH_2COOH$, a combustible, syrupy, yellowish liquid; insoluble in water, soluble in alkali, and miscible with organic solvents; used in making engine lubricants, fuels, and greases.

1-nonyne *Organic Chemistry.* $CH_3(CH_2)_6C{\equiv}CH$, a colorless liquid; insoluble in water and soluble in ether; boils at 150.8°C and freezes at –50°C.

noon *Astronomy.* the moment when the real or mean sun crosses an observer's meridian.

noon constant *Navigation.* a predetermined value that is added to a noon sight to determine the latitude.

noon sight *Navigation.* a measurement of the altitude of the sun at local apparent noon; used to determine latitude.

NOP or **NO OP** *Computer Programming.* **1.** a null instruction that causes no processing other than fetching the next instruction to be executed. **2.** anything that has no effect or fails to have the intended effect. (Short for "no operation.")

nopaline *Biochemistry.* an opine compound used by agrobacteria as a specific growth substance.

NOR *Mathematics.* a logic operator that operates on a finite set of statements; the value of the operator is true if all of the statements are false, and is false otherwise (i.e., if at least one statement is true).

nor- *Organic Chemistry.* a combining form indicating: **1.** a compound in which an Me group attached to a ring system has been replaced by H; used especially of steroids and terpenes. **2.** a compound in which one CH_2 group has been eliminated from a chain or ring; used especially of higher terpenes and steroids.

norbergite *Mineralogy.* $Mg_3(SiO_4)(F,OH)_2$, a yellow or tan orthorhombic mineral of the humite group, having a specific gravity of 3.18 and a hardness of 6 to 6.5 on the Mohs scale; found as crystals or grains in limestone or dolomite contact zones.

norbornane *Organic Chemistry.* C_7H_{12}, a solid that is insoluble in water and soluble in alcohol, ether, acetone, and benzene; sublimes at 87.5°C; contains two rings that share three carbon atoms.

NOR circuit *Electronics.* a logic circuit in which all of the inputs must be the same in order to produce the opposite condition, so that if a high voltage is applied to all the gates the voltage across the transistor will be low; the inverse of an OR circuit. Also, **NOR gate.**

Norden bombsight *Ordnance.* a synchronizing bombsight with gyroscopic stabilization, in wide use by United States and Allied forces in the World War II era. (Developed by Carl Lukas *Norden,* 1880–1965, Dutch inventor active in the U.S.)

Nordenfelt machine gun *Ordnance.* a Swedish 19th-century mechanically operated machine gun that included as many as ten stationary, parallel, horizontally mounted barrels.

nordenskiöldine *Mineralogy.* $CaSn^{+4}B_2O_6$, a colorless or yellow trigonal mineral occurring as tabular or lenslike crystals, having a specific gravity of 4.2 and a hardness of 5.5 to 6 on the Mohs scale; found in alkaline pegmatite and some tin deposits.

Nordenskjöld, Nils Adolf Erik [nôr´dən shēld´] 1832–1901, Swedish geologist and arctic explorer; first to sail through the Northeast Passage.

Nordenskjöld line *Meteorology.* a line that connects all the sites at which the mean temperature of the warmest month is equal to $51.4 - 0.1k$ (in °F), where k is the mean temperature of the coldest month.

Nordheim's rule *Solid-State Physics.* a rule stating that the residual resistivity of a binary alloy is proportional to the quantity $x(x - 1)$, where x is the mole fraction of one of the elements in the alloy, and $1 - x$ is the mole fraction of the other element.

nordihydroguaiaretic acid *Organic Chemistry.* $C_{18}H_{22}O_4$, a crystalline solid; melts at 185°C; slightly soluble in water and soluble in alcohol; used to retard rancidity in fats and oils; its use in food is now prohibited by the FDA.

nordmarkite *Petrology.* an alkali syenite of granitic or trachytoid texture containing microperthite, oligoclase, quartz, and biotite.

nor'easter *Meteorology.* a contraction for a northeaster or northeastern storm.

norepinephrine *Endocrinology.* a catecholamine neurotransmitter of the post-ganglionic sympathetic branch of the autonomic nervous system; also produced in small amounts as a hormone in the adrenal medulla. Also, **noradrenaline.**

norethindrone *Pharmacology.* $C_{20}H_{26}O_2$, a progestin occurring as a white to creamy white crystalline powder; used to treat abnormal uterine bleeding, amenorrhea, and endometriosis, and as a contraceptive, either alone or in combination with estrogen; administered orally.

Norian *Geology.* a European geologic stage of the Upper Triassic period, occurring after the Carnian and before the Rhaetian.

norite *Petrology.* a coarse-grained igneous rock consisting chiefly of basic plagioclase, with orthopyroxene the dominant mafic mineral over clinopyroxene; an orthopyroxene gabbro.

norm *Science.* a fixed or ideal standard. *Quantum Mechanics.* the square root of the integral over space of the product of a wavefunction and its complex conjugate. *Petrology.* the theoretical and somewhat arbitrary mineral composition of a rock expressed in terms of normative mineral molecules that has been derived from a chemical analysis according to a set procedure; the norm is used mainly for classification and comparison of igneous rocks. *Mathematics.* **1.** an absolute value or length. **2.** a non-negative real-valued function (denoted $\| \ \|$) defined on a vector (linear) space X such that for all $x, y \in X$: (a) $\|x\| = 0$ if and only if x is the zero vector; (b) $\|\alpha x\| = |\alpha|\ \|x\|$ for all real or complex numbers α (homogeneous); (c) $\|x + y\| \le \|x\| + \|y\|$ (triangle inequality). Also called a **vector norm.** Two norms $\| \ \|$ and $\| \ \|'$ on a (normed) space X are said to be equivalent if there exist constants $c_1 > 0$ and $c_2 > 0$ such that $c_1\|x\| \le \|x\|' \le c_2\|x\|$ for all $x \in X$. **3.** in particular, the norm (or modulus or length) of a vector x in Euclidean space (i.e., R^n) is given by

$$\| x \| = <x, x>^{1/2} = (\sum_{i=1}^{n} x_i^2)^{1/2},$$

where $<\ ,\ >$ is the usual dot product in R^n and x_i is the ith component of the vector x. $|x|$ is often used in place of $\|x\|$. **4.** let $f: R^p \to R^q$ be a bounded function on a subset D of p-dimensional space into q-dimensional space. Then the D-norm of f is the real number given by $\|f\|_D = \sup\{|f(x)| : x \in D\}$; that is, the supremum of the norms of all the vectors $f(x)$ as x ranges over D. When the set D is understood, it may not be referenced explicitly, and the notation $\|f\|$ is used instead of $\|f\|_D$.

norm- or **normo-** a prefix meaning "normal."

Norma *Astronomy.* the Level, an inconspicuous constellation that lies due south of Scorpius.

normal agreeing or conforming with a norm; as is natural, regular, expected, and so on; specific uses include: *Medicine.* free from infection, disease, or malformation. *Chemistry.* **1.** of an ionic compound, not containing replaceable hydrogen or hydroxide ions, as in a salt in which all available hydrogen has been replaced by a metal. **2.** of a solution, containing one gram-equivalent weight of solute per liter of solution. **3.** of a substance, consisting of atoms having the usual, accepted weight value. *Organic Chemistry.* of a hydrocarbon, consisting of a straight, unbranched chain of carbon atoms; often indicated by the prefix *n*-. *Meteorology.* the average value of a meteorological element over a fixed number of years that is considered standard for the country and the element; recommended international usage is to recalculate the normals at the end of every decade for the preceding thirty years. *Optics.* see NORMAL RAY. *Mathematics.* being perpendicular (so that a Riemannian manifold is implied). In particular, a vector normal to a hypersurface at a given point is perpendicular to every vector tangent to the hypersurface at that point. In the case of a differentiable curve embedded in a Riemannian n-manifold, $n-1$ normals to the curve can be defined. In 3-space, the tangent, principal normal, and binormal vectors form an orthogonal moving frame at each point of the curve.

normal acceleration *Mechanics.* the rate of change of velocity of a body in a direction normal to its velocity.

normal adjustment *Optics.* an image formed by an optical system when the distance from the observer to the image is proportional to the distance from the observer to the object, such as the image formed by a telescope of an object at infinity.

normal anticlinorium *Geology.* an anticlinorium in which the axial surfaces of the minor subsidiary folds converge downward.

normal axis *Mechanics.* an imaginary line perpendicular to the symmetry axis of a body in flight, or to the axis colinear with the body's velocity.

normal barometer *Engineering.* a highly accurate barometer acceptable for establishing pressure standards.

normal boiling point *Physical Chemistry.* the temperature at which a liquid substance boils at standard atmospheric pressure; e.g., 100°C for water.

normal chart *Meteorology.* any graphical representation of the distribution of the official normal values of a meteorological element. Also, **normal map.**

normal class *Mathematics.* a class of sets is said to be normal if it is closed under the formation of countable decreasing intersections and countable disjoint unions. For example, a sigma ring is a normal class.

normal component *Mathematics.* at a point on a hypersurface in Euclidean n-space, any n-vector can be expressed as the sum of a vector tangent to the hypersurface and a vector normal to the hypersurface, called the tangential component and normal component of the vector, respectively.

normal consciousness *Psychology.* the consciousness of the average person involved in ordinary daily activities, employing the usual faculties of attentiveness to environment and a normal level of balance between relaxation and arousal.

normal consolidation *Geology.* the consolidation of a sedimentary material that is in equilibrium with overburden pressure.

normal coordinate *Mechanics.* any of a set of coordinates that simplify the equations of motion in such a way that each equation involves only one coordinate.

normal curve see BELL CURVE.

normal cycle *Geology.* a cycle of erosion in which a region is reduced or lowered to base level mainly by running water, especially by the action of rivers. Also, FLUVIAL CYCLE.

normal density function see NORMAL DISTRIBUTION

normal derivative *Mathematics.* a directional derivative, in which the direction is taken normal to a hypersurface. Also denoted $\partial u/\partial n$, where u is the function being differentiated, and n is the (unit) normal vector.

normal dip see REGIONAL DIP.

normal dispersion *Geophysics.* an increase of seismic wave period with increasing time.

normal displacement see DIP SLIP.

normal distribution *Statistics.* a symmetric, bell-shaped probability distribution arising from the law of errors. It is uniquely identified by the specification of the expected value and the variance. This distribution is very widely used. Also, GAUSSIAN CURVE, NORMAL DENSITY FUNCTION.

normal dwarf *Medicine.* a term for a dwarf who is normally proportioned; i.e., a midget.

normal effort *Industrial Engineering.* a standard level of effort that a typical worker can readily sustain; a base level used for effort ratings.

normal electrode *Electricity.* an industry standard electrode used to measure electric potentials.

normal elemental time *Industrial Engineering.* the normal time a single basic elemental work motion requires for completion.

normal erosion

normal erosion *Geology.* erosion that is caused by running water and the weathering processes; normal erosion is primarily responsible for the present modification of the land surface. Also, GEOLOGIC EROSION.

normal extension *Mathematics.* an algebraic extension field F of a field K is said to be a normal extension field of F over K (or simply a normal extension) if one of the following equivalent conditions holds: (a) every irreducible polynomial in $K[x]$ (i.e., with coefficients in K) that has a root in F actually splits in $F[x]$; (b) F is a splitting field over K of some set of polynomials in $K[x]$; and (c) if K' is any algebraic closure of K that contains F, then any K-monomorphism from F to K' is actually a K-automorphism of F.

normal family *Mathematics.* let $C(G, \Omega)$ denote the space of continuous functions from an open subset G of the complex numbers into a complete metric space Ω. $C(G, \Omega)$ is itself a complete metric space using a standard construction. A normal family is a subset of $C(G, \Omega)$ with the property that every sequence in the family has a subsequence that converges to a function in $C(G, \Omega)$. This differs from the definition of sequential compactness only in the fact that the limiting function is not required to be a member of the normal family.

normal fault *Geology.* a fault in which the hanging wall has been displaced downward in relation to the footwall. Also, GRAVITY FAULT, NORMAL SLIP FAULT, SLUMP FAULT.

normal fluid *Physics.* the component of liquid helium II that behaves as an ordinary fluid possessing viscosity.

normal fold see SYMMETRICAL FOLD.

normal force *Mechanics.* the component of force that acts on a surface perpendicular to the surface, as opposed to the frictional force that acts parallel to the surface.

normal form *Computer Programming.* **1.** the form of a floating point number whose most significant bit is nonzero, that is, normalized. **2.** a logical representation of data that is designed in such a way as to reduce database maintenance problems by eliminating repeating groups and introducing independence of nonkey items. **3.** in general, a standardized representation of a kind of data that has desirable properties.

normal-form analysis *Statistics.* a preposterior analysis technique that links sequential acts or decisions into all possible strategies and considers the final outcome of each strategy.

normal freezing point *Physical Chemistry.* the temperature at which a liquid substance freezes at standard atmospheric pressure; e.g., 0°C for water.

normal frequency *Mechanics.* any frequency associated with a system's normal modes of vibration.

normal gravity *Geodesy.* a reference gravity field defined mathematically as a function of position, and closely approximating the earth's actual gravity field.

normal hemoglobin *Pathology.* a value used as a standard in measuring hemoglobin; the normal value is 14–16 grams per 100 cc of blood.

normal human plasma *Hematology.* sterile plasma obtained by pooling approximately equal amounts of the liquid portion of citrated whole blood from eight or more adult humans; used as a blood volume replenisher.

normal hydrostatic pressure *Hydrology.* the pressure at a given point in a porous stratum or well, approximately equal to the weight of a column of water which extends from that point to the surface.

normal impact *Mechanics.* the impact of a projectile where the trajectory and the surface it strikes are directly perpendicular.

normal impedance see FREE IMPEDANCE.

normal induction *Electromagnetism.* the maximum induction obtained in a magnetic substance subject to a magnetizing force that can vary between two limits; the normal induction represents the saturated magnetization of the material.

normality the fact of being normal; specific uses include: *Chemistry.* the concentration of a solution expressed in equivalents of solute per liter of solution.

normalization the process of making or becoming normal; specific uses include: *Computer Programming.* **1.** the process of scaling a floating-point number. **2.** the process of logically restructuring data files. *Metallurgy.* a simple heat treatment for steels that involves austenitizing followed by air cooling.

normalization factor *Quantum Mechanics.* the coefficient that must be multiplied with an eigenfunction in order to normalize the eigenfunction.

normalize to make or become normal; specific uses include: *Statistics.* to use a linear transformation that expresses a variable in terms of standard deviations from the mean. The resulting variable has mean 0 and variance 1. *Quantum Mechanics.* to multiply a wavefunction by the appropriate constant so that its norm becomes unity. *Computer Programming.* to carry out a normalization process. *Mathematics.* **1.** to multiply a vector by a scalar quantity so that the resulting vector has norm or length 1. Such a vector is said to be a unit vector or to be **normalized.** For example, given any vector x in an n-dimensional vector space over the complex numbers, $x/\langle x,x\rangle^{1/2} = x/\|x\|$ (where $\langle\ ,\ \rangle$ is the usual inner product and $\|\ \|$ is the usual vector norm) is a normalized vector pointing in the same direction as x. **2.** any process of rescaling a quantity so that a given integral or other functional of the quantity takes on a predetermined value. Thus, **normalized.**

normalized admittance *Electromagnetism.* the reciprocal of the value of the normalized impedance.

normalized coupling coefficient *Electromagnetism.* a measure, between the values of 0 and 1, representing mutual inductance between two inductively coupled circuits.

normalized current *Electromagnetism.* the current in a transmission line divided by the square root of the characteristic admittance.

normalized function *Mathematics.* a function with L^p norm equal to one for a given value of p, $1 \le p < \infty$. Also, **normal function.**

normalized impedance *Electromagnetism.* any impedance divided by the characteristic impedance associated with a waveguide or transmission line.

normalized structure factors *Crystallography.* the ratio of the value of the structure amplitude, $|F(hkl)|$, to its root-mean-square expectation value. It is denoted by $|E(hkl)|$, where $|E(hkl)| = |F(hkl)| / (\varepsilon f(hkl)_j)^{1/2}$.

normalized susceptance *Electromagnetism.* the susceptance of a waveguide element or transmission line element divided by the characteristic admittance (or equivalently, multiplied by the characteristic impedance).

normalized voltage *Electromagnetism.* any voltage divided by the square root of the characteristic impedance of a transmission line or waveguide.

normalizer *Mathematics.* let H be a subgroup of a group G. The set of all elements of G that fix H by conjugation is the normalizer of H in G; that is, $N_G(H) = \{g \in G : gHg^{-1} = H\}$.

normalizing selection *Evolution.* the process of natural selection in favor of the standard phenotype in a species and against any variation.

normal lens see STANDARD LENS.

normal magnetization curve *Electromagnetism.* a curve representing the magnetization of an originally unmagnetized sample as a function of the magnetizing force, which is increased from zero to some upper limit.

normal mass shift *Nuclear Physics.* the displacement among isotopes, caused by each having a different mass, that corresponds to the displacement that occurs when their mass is reduced.

normal mode see NATIVE MODE.

normal-mode helix *Electromagnetism.* a helical antenna with characteristic dimensions that are much less than the operating wavelength; the radiation pattern has a maximum intensity in a direction perpendicular to the helical axis.

normal mode of vibration *Mechanics.* a vibration of a coupled system that occurs in such a way that every part of the system vibrates with the same frequency, and the value of one normal coordinate oscillates while the values of all the other normal coordinates remain fixed.

normal-mode rejection ratio *Acoustical Engineering.* the ability of an amplifier to reject unwanted signals at a given frequency. *Electronics.* the frequency at which a digital direct-current voltmeter begins to reject the unwanted signals generated by an applied voltage.

normal moisture capacity see FIELD CAPACITY.

normal operation *Mechanical Engineering.* the operation of a boiler, pressure vessel, or other equipment, at or below the design environmental conditions of temperature and pressure.

normal operator *Mathematics.* an operator T that commutes with its (usually Hermitian) adjoint; also called a **normal transformation.** If T operates on a finite-dimensional inner product space and is represented by a complex $n \times n$ matrix A, then $A A^* = A^* A$, where A^* is the Hermitian adjoint of A. In this case A is called a normal matrix.

normal orientation *Computer Technology.* in optical character recognition, the parallel alignment of the line elements of the source document with the physical leading edge of the document.

normal pace *Industrial Engineering.* the rate of work output for a typical worker performing at normal (one hundred percent) productivity; the rate from which productivity incentives are computed.

normal pedal curve *Mathematics.* let O be a point not on a given curve γ. The normal pedal of γ with respect to O is the locus of the foot of the perpendicular (or normal) to the tangents at each point of γ.

normal performance *Industrial Engineering.* the amount of work that is expected of a trained operator working at an average pace.

normal pitch *Mechanical Engineering.* the distance between working faces of two adjacent gear teeth measured between the intersection of the line between the two faces.

normal-plate anemometer *Engineering.* an anemometer having a plate held perpendicular to the wind and attached from behind to a rigid spring; pressure on the plate causes it to move, and the movement is measured electronically to calculate the force and velocity of the wind.

normal polarity *Geophysics.* the natural remanent magnetism that is almost identical to the present ambient field.

normal pressure see STANDARD PRESSURE.

normal ray *Optics.* a ray of incident light that strikes a surface perpendicularly. Also, NORMAL.

normal reaction *Mechanics.* the perpendicular component of the force exerted by a surface on an object that contacts it.

normal saline *Physiology.* a solution composed of 0.85% salt and distilled water, equal to the salt content of blood serum and thereby able to maintain normal osmotic pressure in the body.

normal section *Mathematics.* the intersection of a surface S and a plane that cuts S and contains the normal to S at some point P of S.

normal series *Mathematics.* **1.** a subnormal series $G = G_0 > G_1 > \cdots > G_n$ of a group G such that G_i is normal in G for each i. (In general, a subnormal series is not necessarily a normal series.) **2.** a normal series for a module A is any chain of submodules $A = A_0 \supset A_1 \supset \cdots \supset A_n$.

normal slip fault see NORMAL FAULT.

normal soil *Geology.* a soil whose profile is more or less in equilibrium with the environment.

normal space *Mathematics.* a Hausdorff space with the property that any pair of disjoint closed sets can be contained in disjoint open neighborhoods. For example, an open subset of R^n is normal. Also, **normal topological space.**

normal spinel *Materials Science.* a crystal structure of oxides having the formula AB_2O_4, in which oxygen ions occupy the face-centered-cubic lattice sites, the A cations occupy the tetragonal and the B cations the octahedral interstitial sites; normal spinels are not ferrimagnetic.

normal spiral galaxy *Astronomy.* a spiral galaxy of type Sa, Sb, or Sc.

normal state see NUCLEAR GROUND STATE.

normal stress *Mechanics.* the component of stress on a surface perpendicular to the surface.

normal subgroup *Mathematics.* a subgroup H of a group G is said to be normal in G if $ghg^{-1} \in H$ whenever $g \in G$ and $h \in H$. Right and left cosets of a normal subgroup are identical, and the set of all cosets of a normal subgroup form a quotient or factor group. H is normal in G if and only if G is the normalizer of H. Also, **invariant subgroup.**

normal synclinorium *Geology.* a synclinorium in which the axial surfaces of the minor folds converge upward.

normal time *Industrial Engineering.* the average or standard time that a work process is expected to take when performed by a qualified operator without any delay.

normal transformation *Statistics.* a nonlinear transformation that converts a variable into a normally distributed variable.

normal twin *Crystallography.* twinned crystals in which the twin axis is perpendicular to a crystal face, the twin plane. This is the composition plane of the twin.

normal water *Oceanography.* a standard seawater preparation in which the chlorinity is between 19.3 and 19.5 parts per thousand, measured accurately to within 0.001 part per thousand; sometimes also called **Copenhagen water** because it is prepared by the Hydrographical Laboratories in Copenhagen.

normal water level *Cartography.* in surveying, the usual level and course of the water in a river, reservoir, lake, or pond, as defined by a shoreline of permanent land-type vegetation.

normal working area *Industrial Engineering.* the workspace that can be reached by the fingertips of the right or left hand when both elbows are pivoted at the edge of the workspace.

Norman *Architecture.* of or relating to English Romanesque architecture from the Norman Conquest (1066) to about 1180.

normative mineral see STANDARD MINERAL.

normed linear space *Mathematics.* a vector (linear) space endowed with a norm $\| \; \|$. A normed space V is also a metric space, with metric μ given by $\mu(x,y) = \|x - y\| = \|y - x\|$, for all $x, y \in V$. Also, **normed vector space.**

normoblast *Histology.* the smallest of the nucleated precursors of an erythrocyte. Also, METAKARYOCYTE.

normochromic *Cell Biology.* describing the normal color of erythrocytes or the normal hemoglobin content of red blood cells.

normocyte *Histology.* an erythrocyte of normal size and shape.

norm of a matrix *Mathematics.* a nonnegative, real-valued function (denoted $|||\;\;|||$) that is defined on the space M_n of $n \times n$ matrices with complex entries, such that, for all matrices $A, B \in M_n$ and all complex numbers a: (a) $|||A||| = 0$ if and only if A is the zero matrix; (b) $|||\alpha A||| = |\alpha| \, |||A|||$ for all real or complex numbers α (homogeneous); (c) $|||A + B||| \leq |||A||| + |||B|||$ (triangle inequality); and (d) $|||AB||| \leq |||A||| \; |||B|||$ (submultiplicative). Also called a **matrix norm.** A vector norm on matrices, that is, an ordinary norm satisfying axioms (a)–(c) but not necessarily (d), is also called a **generalized matrix norm.** A matrix norm satisfying axioms (b) and (c) but not necessarily (a) and (d) is called a **generalized matrix seminorm.** For example, the l_p norm that is defined for $A \in M_n$ by $|||A|||_p = (\sum_{i,j=1}^{n} |a_{ij}|^p)^{1/p}$ is a matrix norm, where $A = (a_{ij})$.

norm of reaction see RANGE OF REACTION.

normothermia *Physiology.* standard body temperature, or a temperature of the surroundings that neither excites nor depresses cell activity.

normovolemia *Hematology.* a condition of normal blood volume.

normovolemic *Hematology.* relating to or characterized by normovolemia; i.e., having a normal volume of blood circulating in the body.

Norplant *Medicine.* the trade name for a surgically implanted birth-control device that releases regular doses of the contraceptive progestin for an extended period.

Norris-Eyring reverberation *Acoustics.* an improvement on the Sabine equation for absorption, stating that $R_T = 0.049/S - \ln(1 - a)$, where R_T is the reverberation time for a 60-decibel decay, S is the surface area, and a is the absorption calculated with the Sabine coefficient.

Norrish, Ronald G. W. 1897–1978, British chemist; awarded the Nobel Prize for developing flash photolysis and flash spectroscopy.

norte [nôr´tā´] *Meteorology.* **1.** the north winter wind in Spain. **2.** a cold, strong northeasterly wind that blows in Mexico and on the shores of the Gulf of Mexico, resulting from an outbreak of cold air from the north.

north *Geodesy.* the primary reference direction relative to the earth, indicated by the positive direction of a line lying in a plane through the earth's axis of rotation. Also, ASTRONOMIC NORTH, COMPASS NORTH, MAGNETIC NORTH.

North America *Geography.* the continent occupying the northern part of the Western Hemisphere, from the Arctic Ocean to Central America, including the nations of Canada, the United States, and Mexico.

North American blastomycosis *Medicine.* a former term for blastomycosis, a chronic granulomatous and suppurative disease caused by *Blastomyces dermatitidis*, that starts as a respiratory infection and then spreads.

North American brown bear see GRIZZLY BEAR.

North America Nebula *Astronomy.* NGC 7000, a hydrogen emission nebula near Deneb in Cygnus whose shape resembles North America's.

North American high *Meteorology.* a weak area of high pressure that covers most of North America during the winter season. Also, **North American anticyclone.**

North Atlantic Current *Oceanography.* a group of broad currents that form a slower-moving extension of the Gulf Stream, extending from southeast of the Grand Banks, off Newfoundland, to the Norwegian Sea. Also, **North Atlantic Drift.**

North Cape Current *Oceanography.* an extension of the Norway Current that flows northeastward around North Cape, the northernmost part of Norway, producing unstable weather conditions as it enters the icy Barents Sea.

northeaster *Meteorology.* a northeast wind, especially a strong wind or gale.

northeast storm *Meteorology.* a frequently violent cyclonic storm system off the east coast of North America that occurs most often between September and April.

northeast trades *Meteorology.* the trade winds of the Northern Hemisphere.

North Equatorial Current *Oceanography.* the generic name given to the westward-flowing southern segment of the oceanic gyre in the North Pacific, Indian, and Atlantic Oceans; in the Indian Ocean it reverses its direction between May and September.

norther *Meteorology.* a northerly wind.

northerly turning error *Aviation.* an acceleration error in a magnetic compass, most evident during banked turns made from an initial north-south course.

northern anthracnose *Plant Pathology.* a disease of clover in North America, Europe, and Asia caused by the fungus *Kabatulla caulivora* and characterized by sunken linear brown lesions on the stems and petioles.

Northern blot *Molecular Biology.* an electroblotting technique for detecting a specific RNA molecule, in which RNA is transferred to a filter and is hybridized to radioactively labeled RNA or DNA.

Northern Hemisphere *Geography.* the half of the earth north of the equator.

northern lights *Astronomy.* a popular name for the aurora borealis. See AURORA BOREALIS.

north foehn *Meteorology.* a foehn condition that is sustained by a flow of wind from north to south across the Alps.
north frigid zone *Geography.* the area between the Arctic Circle (66°31'N) and the north pole. Also, ARCTIC ZONE.
north galactic spur *Astronomy.* part of a large shell of gas that is visible at X-ray and radio wavelengths and that is likely to be an old supernova remnant.
north geographic pole *Geography.* see NORTH POLE.
north geomagnetic pole *Geophysics.* see NORTH POLE.
North Island *Geography.* the northernmost of the two main islands of New Zealand.
north magnetic pole *Geophysics.* see NORTH POLE.
North Pacific Current *Oceanography.* a warm, eastward-flowing current in the Mid-Pacific between 25 and 45°N, part of the system (beginning with the Kuroshio Current and the Kuroshio Extension) that carries warm water toward the east from the Western Pacific south of Japan.
north point *Astronomy.* the point on the horizon directly below the north celestial pole.
north polar distance *Astronomy.* an archaic term referring to the angular distance between an object and the north pole, as measured along the shortest meridian between them.
North Polar Sequence *Astronomy.* a series of stars lying within 2° of the north celestial pole, whose carefully measured brightnesses were once used widely as photometric standards.
North Polar Spur see NORTH GALACTIC SPUR.
north pole or **North Pole** *Geography.* the northernmost point on the earth, near the middle of the Arctic Ocean. Also, NORTH GEOGRAPHIC POLE. *Geophysics.* the point on the surface of the earth toward which a magnetic needle points; the geomagnetic pole in the Northern Hemisphere. Also, NORTH MAGNETIC POLE, NORTH GEOMAGNETIC POLE. *Electromagnetism.* the pole of a dipole or magnet from which magnetic field lines are defined to emerge, as opposed to the south pole where magnetic field lines converge.
Northrup, John H. 1891–1987, American biochemist; awarded the Nobel Prize for preparing pure enzymes and virus proteins.
North Sea *Geography.* an arm of the North Atlantic east of Great Britain.
North Star *Astronomy.* **1.** another name for Polaris, a bright star now situated almost directly above the north pole. **2.** any other star located above the north pole at a given point in time.
north temperate zone *Geography.* the area between the Tropic of Cancer and the Arctic Circle.
north temperature zone *Meteorology.* the region of the earth's surface that stretches northward to the Arctic Circle.
northupite *Mineralogy.* $Na_3Mg(CO_3)_2Cl$, a colorless, brown, yellow, or gray, cubic mineral occurring as octahedral crystals, and having a specific gravity of 2.38 and a hardness of 3.5 to 4 on the Mohs scale; found in saline lake deposits.
north-upward presentation *Navigation.* a radarscope in which the ship's position is in the center of the screen, and gyro north is always at the top. Also, **north-upward plan position indicator**.
northwester *Meteorology.* a wind component of a northwesterly direction.
Norton's theorem *Electricity.* a theorem stating that a linear network with two output terminals can be represented as a single current source in parallel with a single impedance and with the terminals. The actual network can have any number of impedances and voltage and current sources, alhough the sources must be at the same frequency. The source and impedance, called an equivalent circuit, can be used to show the effect of the network on any device connected to the output terminals.
Norton transformation *Electrical Engineering.* a four-terminal network transformation of a series ladder element into a pi network in tandem with an ideal transformer.
Norwalk virus *Virology.* a small, nonenveloped virus that commonly causes acute gastroenteritis in humans via fecal-oral transmission.
Norway Current *Oceanography.* a branch of the North Atlantic Current that carries warm water north along the coast of Norway, ending in the North Cape Current. Also, **Norwegian Current.**
nor'wester *Meteorology.* a contraction for a northwester.
nos- or **noso-** a combining form meaning "disease," as in *nosophobia, nosotoxin.*
noscapine *Pharmacology.* $C_{22}H_{23}NO_7$, an nonanalgesic alkaloid derivative of opium, occurring as a white or whitish crystalline powder and administered orally to prevent and relieve coughing. Also available as a hydrochloride salt, **noscapine hydrochloride.**
nose *Anatomy.* the specialized structure on the face that serves as the organ of the sense of smell and as part of the respiratory system.
nose any of various features or structures regarded as analogous to the human nose; specific uses include: *Geology.* **1.** a short, plunging anticline that is without closure on one end. Also, STRUCTURAL NOSE, ANTICLINAL NOSE. **2.** a projecting, overhanging buttress of rock, or the projecting end of a hill, spur, ridge, or mountain. **3.** the central forward part of a parabolic dune. *Engineering.* the foremost, projecting portion of a vehicle, apparatus, or other object. *Fluid Mechanics.* the concentrated forward portion of a turbidity current.
nosean *Mineralogy.* $Na_8Al_6Si_6O_{24}(SO_4)\cdot H_2O$, a colorless to gray, blue, or brown cubic mineral of the sodalite group, having a hardness of 5.5 on the Mohs scale and a specific gravity of 2.30 to 2.40; found as grains only in volcanic bombs, phonolites, and related igneous rocks. Also, NOSELITE.
nose cone *Space Technology.* the cone-shaped leading section of a rocket or missile, consisting of a protective outer shell fit over one or more chambers that are designed to house equipment and instruments and a payload, either a satellite or a warhead.
nose fuse *Ordnance.* a fuse that is located in the foremost part of a projectile.
nose-heavy *Aviation.* of an aircraft or airframe, tending to sink at the nose; characterized by the downward inclination of the nose when the longitudinal control is released in any attitude of normal flight.
nose leaf *Vertebrate Zoology.* in some bats, a flap of skin resembling a leaf above the nostrils that is believed to have a tactile function.
noselite see NOSEAN.
Nosema *Invertebrate Zoology.* a genus of sporozoan protozoans that parasitize silkworms.
nose radius *Mechanical Engineering.* the radius of the rounded portion of the cutting edge of a tool.
nose spray *Ordnance.* fragments of a bursting projectile that are projected forward, as opposed to base spray, which is projected toward the rear, and side spray, which is projected toward the sides.
nosing *Building Engineering.* a projecting edge such as a part on the tread of a step that extends beyond a riser.
nosocomial *Medicine.* **1.** associated with a hospital. **2.** describing a disorder, unrelated to the patient's primary condition, that is acquired during a stay in the hospital.
nosocomial infection *Medicine.* an infection originating in the hospital; specifically, an infection that was neither present nor incubating prior to admittance to the hospital, but was acquired after hospitalization. The term usually refers to an infection developed by a patient, but may also apply to infections acquired by hospital personnel. Also, HOSPITAL-ACQUIRED INFECTION.
Nosodendridae *Invertebrate Zoology.* a small family of coleopteran insects commonly known as the wounded-tree beetles.
nosophobia *Psychology.* an irrational fear of disease, or an exaggerated concern over a mild ailment.
nosotoxin *Toxicology.* any toxin causing or associated with disease.
Nostoc *Bacteriology.* a genus of cyanobacteria, or blue-green algae, that form filamentous, gelatinous colonies and occur in a variety of habitats, including wet rocks and damp soil.
no-strike target list *Military Science.* a list of targets not to be destroyed because such destruction would hinder friendly military operations or friendly relations with indigenous personnel or governments.
nostril *Anatomy.* one of two openings of the nasal cavity, located on either side of the nasal septum.
nostrum *Pharmacology.* a term for a secret remedy, often patented, that is generally produced and sold by a nonmedically qualified source.
NOT *Computer Technology.* the logic operator that complements the single operand; for example, NOT A returns false if A is true and true if A is false.
Notacanthidae *Vertebrate Zoology.* the family of marine spiny eels, characterized by a dorsal fin composed of individual spines, a long body, a blunt snout, and a ventral mouth; found in moderate to deep waters in the Indian, Pacific, and Atlantic Oceans.
Notacanthiformes *Vertebrate Zoology.* an order of deep-sea eels of the class Actinopterygii, characterized by a long body with a tapering tail, connected abdominal fins, a reduced or absent caudal fin, and a protruding snout; found worldwide.
NOTAM *Telecommunications.* a code word used prior to giving essential information about flight conditions or operations; information is under the control of the National Flight Data Center. (An acronym for notice to airmen.)

Notaspidea *Invertebrate Zoology.* an order of gastropod mollusks in the subclass Opisthobranchia, with a small circular or conical shell.

notatin see GLUCOSE OXIDASE.

notch an angular cut or indentation; specific uses include: *Geology.* a deep, narrow V-shaped cut or hollow formed by undercutting near the high-water mark along the base of a sea cliff. *Electronics.* **1.** a rectangular depression that drops below the sweep line in some types of radar equipment. **2.** the point on a graph, especially the graph of a frequency response, where the line drops sharply before immediately returning to its original value. **3.** a drop in frequency response commonly seen in such frequency-rejection circuits as band-suppression filters. *Engineering.* a geometric (usually V-shaped) indentation or cut in the border or surface of an object.

notch antenna *Electromagnetism.* an antenna having a radiating surface with a slot whose purpose is to provide the radiation pattern.

notch filter *Electronics.* a device that sharply attenuates or rejects a given frequency band at both ends; commonly used in television transmitters to prevent sound interferences between channels.

notch graft *Botany.* a type of propagation in which the scion is inserted into a slit in the root stock.

notching *Mechanical Engineering.* the process of cutting various shapes from the end of a strip, blank, or part.

notching relay *Electricity.* a relay whose operation depends upon successive impulses from a predetermined number of individual impulses to complete its operation.

notch toughness *Materials Science.* the property of a material that represents its ability to withstand impact when notches or cracks are present; measured by the amount of energy required to break a standard-size specimen that contains a notch.

NOT circuit *Electronics.* a logic circuit in which the output is always the opposite of the input, so that if voltage applied to the gate is high the voltage across the output will be low, and vice versa. Also, NOT GATE, INVERTER CIRCUIT.

note *Acoustics.* **1.** a symbol denoting a musical tone of specific frequency and duration that is related to other notes in a common scale by either a whole step interval or a half step. **2.** the tone designated by this symbol, sounded on a musical instrument.

Noteridae *Invertebrate Zoology.* a small family of burrowing water beetles, coleopteran insects in the suborder Adephaga.

NOT function *Mathematics.* a logic operator such that the value of the operator is true if the given statements is false, and is false otherwise (i.e., if the statement is true). NOT A is written nA.

NOT gate *Electronics.* see NOT CIRCUIT. *Computer Technology.* an inverter circuit that computes the NOT function.

Notharctus *Paleontology.* a genus of extinct lemurlike primates in the infraorder Adapiformes of the suborder Prosimii, which lived in North America during the warm Eocene.

Notheiaceae *Botany.* a monospecific family of brown algae of the order Chordariales, characterized by a bushy appearance caused by the growth of new plants within the conceptacles of parental plants; occurs in New Zealand and Australia.

Nothosauria *Paleontology.* an order of aquatic diapsid reptiles in the superorder Sauropterygia, known principally from Middle Triassic strata in Europe and China; the nothosaurs ranged from 20 centimeters to 4 meters long and probably swam principally by trunk undulation; extant in the Lower to Upper Triassic.

Notidanoidei *Vertebrate Zoology.* the suborder of comb-toothed sharks of the order Selachii, having 6–7 gill slits, a dorsal fin set far back, and primitive jaw suspension; contains the single family Hexanchidae (cow sharks).

notifiable disease *Pathology.* any of certain specified diseases or infections that, if diagnosed or suspected, must be reported to federal, state, or local authorities.

no-till *Agronomy.* of or relating to the no-till technique of planting. Thus, **no-till farming**.

no-till system or **no-till method** *Agronomy.* of or relating to a technique of planting a field in which little or no tilling of the seedbed takes place beforehand, as when the topsoil is too thin or fragile for deeper plowing.

Notiomastodontinae *Paleontology.* a proposed subfamily of South American proboscideans in the suborder Euelephantoidea; extant in the Upper Pleistocene.

Notioprogonia *Paleontology.* an extinct suborder of primitive notoungulate mammals known from the Paleocene and Eocene of Asia and the Americas; they ranged from dog-size to bear-size.

notochord *Developmental Biology.* the rod-shaped body, composed of cells derived from the mesoblast, that is below the primitive groove of the embryo and is the center of development of the axial skeleton. Also, CHORDA DORSALIS.

notochordal canal *Developmental Biology.* a tunnel that extends from the primitive pit into the head process of some embryos. Also, CHORDAL CANAL.

notochordal plate see HEAD PROCESS.

Notodelphyidiformes *Invertebrate Zoology.* in some systems of classification, a tribe of parasitic nematodes in the Gnathostoma.

Notodelphyoida *Invertebrate Zoology.* an order of marine copepods that are commensal or parasitic in tunicates and mollusks.

Notodontidae *Invertebrate Zoology.* a family of nocturnal lepidopteran insects including the puss moths, in the superfamily Noctuidea; caterpillars are a harmful plant pest.

Notogea *Ecology.* another name for the Australian zoogeographical region.

Notommatidae *Invertebrate Zoology.* a large family of rotifers in the order Monogonota, having a cylindrical body and a slender posterior foot ending in two toes.

Notomyotina *Invertebrate Zoology.* a suborder of starfish, echinoderms in the order Phanerozonida, with a sucking disc at the tip of each tube foot.

Notonectidae *Invertebrate Zoology.* the waterboatmen or backswimmers, a family of carnivorous hemipteran insects with oarlike rear legs, that swim on their backs, breathing air from a bubble held by hairs on the abdomen.

Notonectoidea *Invertebrate Zoology.* a superfamily of hemipteran insects, true bugs in the subdivision Hydrocorisae.

notopodium *Invertebrate Zoology.* the upper lobe of the parapodium of a polychaete.

Notopteridae *Vertebrate Zoology.* the featherbacks, a small family of freshwater fish of the order Osteoglossiformes, found only in tropical Africa and Southeast Asia and distinguished by a long anal fin that runs nearly the entire length of the underside and is confluent with the tail fin.

Notoryctidae *Paleontology.* a monogeneric, still-living family of marsupial moles of Australia, incertae sedis, that are earless and almost sightless; there is no fossil record.

Notostigmata *Invertebrate Zoology.* a suborder of large leathery primitive omnivorous mites that live in leaf litter and beneath stones.

Notostigmophora *Invertebrate Zoology.* a subclass or suborder or centipedes in the class Chilopoda, having dorsal respiratory openings.

Notostraca *Invertebrate Zoology.* an order of branchiopods including the tadpole shrimps, having a flattened dorsal carapace, vestigial antennae, and up to 70 pairs of appendages.

Nototheniidae *Vertebrate Zoology.* the cod icefishes, a family of mostly bottom-dwelling fishes of the order Perciformes, found exclusively in Antarctic waters, and making up about two-thirds of the total Antarctic fish population. Also, ANTARCTIC COD, ANTARCTIC BLENNIES.

Notoungulata *Paleontology.* an order of hoofed mammals that achieved great diversity in South America in the Early Tertiary and survived in three families until the Middle Quaternary; the largest of its four suborders is the Toxodontia.

notum *Entomology.* the dorsal plate of the thoracic segments in insects.

nourishment *Nutrition.* the intake of food, or something that is taken as food. *Geology.* the replenishment of a beach by either natural or artificial means. *Hydrology.* see ACCUMULATION.

nova *Astronomy.* a white dwarf star in a binary system that brightens suddenly by about a dozen magnitudes when gaseous material accreted from its evolved companion ignites abruptly in a thermonuclear explosion.

novaculite *Geology.* a light-colored, dense, hard, fine-grained, even-textured siliceous sedimentary rock characterized by the dominance of microcrystalline quartz over chalcedony. Also, RAZOR STONE.

novaculitic chert *Geology.* a grayish chert that fragments into slightly rough, splintery pieces.

novalike symbiotic star *Astrophysics.* a binary star system that shows the same great outbursts in brightness as a nova.

novelty yarn *Textiles.* a yarn having interwoven irregularities.

novobiocin *Microbiology.* an antibiotic produced by the bacteria *Streptomyces niveus* and *S. spheroides*; it blocks DNA and RNA synthesis and is especially effective against staphylococci and other Gram-positive organisms; due to its high toxicity, its use is now generally limited to research.

Novocain [nō´və kān´] *Pharmacology.* a trade name for preparations of procaine hydrochloride, widely used as a local anesthetic in dentistry and minor surgery. Also, **novocain, novocaine.**

novolac resin *Materials.* a thermoplastic phenolformaldehyde resin obtained by the use of acid catalysts and excess phenol; used in molding and bonding materials, electrical insulation, clutch facings, and as a reinforcing agent.

Nowakowskiella *Mycology.* a genus of fungi belonging to the order Chytridiales that obtains its nutrients from dead organic matter and occurs in water and soil.

nowcasting *Meteorology.* **1.** a detailed description of the weather and forecasts for up to the next two hours. **2.** an area-specific forecast based on detailed observational data for up to the following twelve hours.

nowhere dense set *Mathematics.* a set A in a topological space X is said to be nowhere dense if the closure of A has an empty interior. Sets in X that are countable unions of nowhere dense sets are said to be of the first (Baire) category; all other sets are of the second category.

no-wind heading *Navigation.* the heading to be taken by an aircraft under conditions of no wind; it is equivalent to the desired course to be flown.

nox *Optics.* a unit of illumination used in measuring low luminance; equal to 10^{-3} lux.

noy *Acoustics.* a unit of measurement for perceived noise level.

nozzle *Mechanical Devices.* a contracting, tapering tube or vent used at the end of a pipe, tube, or hose to accelerate or direct the flow of a liquid. *Space Technology.* the part of a rocket thrust chamber in which the gases combined in the chamber are accelerated to high velocities as they exit the chamber.

nozzle blade *Aviation.* any of the blades or vanes in a nozzle diaphragm. Also, NOZZLE VANE.

nozzle-contraction-area ratio *Design Engineering.* the ratio between the nozzle cross-sectional area for gas flow at its inlet to that at its throat.

nozzle diaphragm *Aviation.* in a turbojet engine, a ring of radiating blades or vanes serving as nozzles through which gases pass from the combustion chambers to turn the turbine wheel. Also, NOZZLE RING.

nozzle-divergence loss factor *Fluid Mechanics.* a ratio of the momentum of the gas in a nozzle to that of a gas in an ideal nozzle.

nozzle efficiency *Mechanical Engineering.* the efficiency with which a nozzle converts potential energy into kinetic energy, given as a ratio between the actual change in kinetic energy and an ideal change at a specified pressure.

nozzle exit area *Design Engineering.* the cross section of a nozzle exit through which a measured amount of gas flows.

nozzle-expansion ratio *Design Engineering.* the ratio of the cross-sectional area for gas flow at the exit of a nozzle to the cross-sectional area that is available for gas flow at the throat.

nozzle-mix gas burner *Engineering.* a burner having a nozzle that combines air and fuel at the burner tile.

nozzle process *Nucleonics.* in the isotope separation of gaseous compounds, a process by which the gases are allowed to exhaust through a specially designed nozzle.

nozzle ring see NOZZLE DIAPHRAGM.

nozzle throat *Design Engineering.* the part of a nozzle with the smallest cross section.

nozzle throat area *Design Engineering.* the measured area of a nozzle throat, i.e., that portion with the smallest cross section.

nozzle thrust coefficient *Space Technology.* the amplification of thrust caused by gas expansion in a specific nozzle as it relates to the potential thrust exerted if the chamber pressure acted only over the throat area. Also, THRUST COEFFICIENT.

nozzle vane see NOZZLE BLADE.

NP nondeterministic polynomial; neuropsychiatry; neuropsychiatrist.

Np the chemical symbol for neptunium.

NP-complete problem *Computer Programming.* a class of computational problems for which no effective computer algorithm exists, such as the Clique and Knapsack problems. Known algorithms for such problems require time that is an exponential function of problem size.

NPIN transistor *Electronics.* a device that has a layer of high-purity (intrinsic) germanium wedged between its p-type base and n-type collector to increase its frequency range.

np junction *Electronics.* an area that serves as a transition between semiconductor material with the opposite electrical properties.

NPK or **N-P-K** *Agronomy.* the chemical symbols for nitrogen, phosphorus, and potassium, the major ingredients of complete fertilizers.

n-plus-one address instruction *Computer Programming.* an instruction that contains n data addresses along with the address of the next instruction.

NPN nonprotein nitrogen.

npn transistor *Electronics.* a junction that is formed by placing a thin layer of p-type semiconductor material between two layers of n-type semiconductor material, so that current will flow through both.

NPO nothing by mouth. (From Latin *nil per os.*)

NPO-body *Electricity.* any of various negative-positive-zero temperature-compensating dielectric materials having an ultrastable temperature coefficient and primarily used in ceramic capacitor applications.

NPP net primary production.

NPR noise-power ratio.

NPU net protein utilization.

NPV nuclear polyhedrosis virus.

N-RAS *Oncology.* a proto-oncogene that contains point mutations and is found in carcinoma of the genitourinary tract and thyroid, and in melanoma.

NRC or **N.R.C.** National Research Council; Nuclear Regulatory Commission.

nrdA gene *Genetics.* a gene that codes for ribonucleotide reductase. Also, DNAF GENE.

NREM sleep *Neurology.* a state of dreamless sleep characterized by slow brain waves of high voltage, slow and regular respiration, low and constant heart rate and blood pressure, and lower threshold for arousal, as opposed to REM sleep. Also, NON-REM SLEEP, NON-RAPID EYE MOVEMENT SLEEP.

NRM wind scale *Meteorology.* an adaptation of the Beaufort wind scale used by the U.S. Forest Service in the northern Rocky Mountains.

NRS nuclear reaction spectrometry.

NRT *Aviation.* the airport code for Narita, Tokyo, Japan.

NRZ non-return-to-zero.

ns nanosecond.

Ns nimbostratus.

NS new series; nuclear ship.

NSAID nonsteroidal anti-inflammatory drugs.

N scan see N DISPLAY.

N scope *Electronics.* a radarscope that produces an N display.

N shell *Atomic Physics.* the electron orbit of an atom, fourth closest to the nucleus, in which electrons whose principal quantum number is 4 occur.

NSP net secondary production.

N star *Astronomy.* a cool red star very much like an M-type star except that absorption bands of C2, CH, and CN replace the bands of TiO found in an M star.

nt or **nt.** newton; nit; nucleotide.

N-terminal see AMINO TERMINAL.

NTP or **N.T.P.** normal temperature and pressure.

NTS not to scale.

NTSB or **N.T.S.B.** National Transportation Safety Board.

n-tuple *Mathematics.* an ordered sequence of length n, where n is a positive integer; usually denoted by $(a_1, a_2, \ldots, a_n)$ or similar notation.

n-type conduction *Electronics.* the transmission of current in a semiconductor that arises from the action of electrons rather than holes.

n-type crystal rectifier *Electronics.* a crystal rectifier in which forward current flows when the semiconductor material is negative with respect to the metal.

n-type semiconductor *Electronics.* a semiconductor material in which there are more electrons than holes.

nubbin *Geology.* **1.** one of the bedrock knobs or small hills forming the last remnants of an eroded mountain crest or range in a desert region. **2.** a residual boulder that occurs on a desert dome or broad pediment.

Nubecula Major *Astronomy.* the Latin name for the Large Magellanic Cloud; seldom used in astronomical writing in English.

Nubecula Minor *Astronomy.* the Latin name for the Small Magellanic Cloud; seldom used in astronomical writing in English.

Nubian *Agriculture.* a breed of long-legged milk goat that is brown or black in color and has a prominent, curving nose. (From *Nubia*, a region in Africa.)

nu bodies *Cell Biology.* eukaryotic nucleosomes, complexes between 200 base pairs of DNA and a complex of histones that resemble beads on a string during interphase when viewed with an electron microscope.

nucellus *Botany.* the central mass of the ovary in which the embryo sac is formed.

nucha *Anatomy.* the nape of the neck.

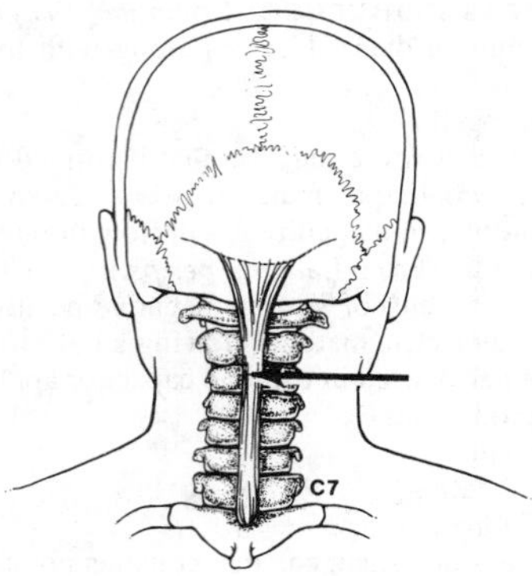

nuchal ligament

nuchal ligament *Anatomy.* the cordlike structure that supports the neck and attaches to the occipital bone superiorly and to all seven cervical vertebrae inferiorly.

nuchal organ *Invertebrate Zoology.* one of a pair of ciliated sensory pits on the head region of many polychaetes; they appear to be chemoreceptors and are important in the detection of food.

nuchal rigidity *Medicine.* the stiffness in the nape of the neck that is symptomatic of meningitis, often accompanied by pain and spasm on attempts to move the head.

nuchal tentacle *Invertebrate Zoology.* any of various filiform or thick, fleshy tentacles situated on the anterior segments of many annelids or the back of the prothorax of many insects.

nucle- or **nucleo-** a prefix meaning "nucleus."

nuclear [nook´lē ər] *Chemistry.* of or relating to an atomic nucleus. *Nucleonics.* of, relating to, or powered by nuclear energy. *Ordnance.* of or relating to nuclear weapons and their use in warfare.

nuclear absorption *Nuclear Physics.* the process by which a nucleus acquires energy.

nuclear adiabatic demagnetization *Physics.* the removal or reduction of the magnetic field due to the nuclear magnetic moments of a paramagnetic substance; the process is achieved at cryogenic temperatures and through the use of intense external magnetic fields.

nuclear airburst *Ordnance.* the explosion of a nuclear weapon in the air, at a height greater than the maximum radius of the fireball.

nuclear angular momentum see NUCLEAR SPIN.

nuclear area *Archaeology.* **1.** a location where large, complex societies occur at different times, such as the valley of central Mexico. **2.** the focus of activity in a site, such as a camp or village around which hunting or agricultural activity takes place.

nuclear atom *Chemistry.* the central atom of a molecule, to which other atoms or groups are attached.

nuclear battery *Nucleonics.* a battery that converts the energy of radioactive material to electrical energy.

nuclear binding energy see BINDING ENERGY.

nuclear boiler *Nucleonics.* a reactor whose primary coolant is water that is converted to steam from which energy may be drawn.

nuclear bomb see NUCLEAR WEAPON.

nuclear bonus effects *Ordnance.* desirable damage or casualties produced by the effects of friendly nuclear weapons, which cannot be accurately calculated in targeting or depended upon for a militarily significant result.

nuclear breeder *Nucleonics.* a reactor in which each generation produces more fissionable material than is used up in fission.

nuclear cage *Cell Biology.* a fibrous skeletal structure found within the eukaryotic nucleus that is associated with both the nuclear pores and loops of the chromosomal DNA.

nuclear capture *Nuclear Physics.* a process by which a particle, such as a neutron, proton, or electron, merges with a nucleus.

nuclear cataract *Medicine.* a cloudiness of the central area of the crystalline lens of the eye. Also, NUCLEAR SCLEROSIS.

nuclear chain reaction *Nucleonics.* a succession of fission processes in which each generation is produced by neutrons set free in a nuclear disruption of the preceding generation.

nuclear chemical engineering *Chemical Engineering.* the application of chemical engineering to such processes as the design of nuclear reactors, separating isotopes, and reclaiming radioactive fuel.

nuclear chemistry *Atomic Physics.* the branch of chemistry that studies reactions within the nucleus, such as fission and fusion.

nuclear cloud *Ordnance.* a term for the hot gases, smoke, dust, and other particles from a nuclear bomb and from its environment that are carried aloft in conjunction with the fireball produced by the detonation of a nuclear weapon.

nuclear collision *Nuclear Physics.* the bombardment of a nucleus by another nucleus or particles.

nuclear column *Ordnance.* a cylinder of water and spray through which the hot, high-pressure gases formed in the underwater explosion of a nuclear weapon are vented to the atmosphere; a column of dirt is formed in an underground explosion.

nuclear cross section *Nuclear Physics.* a quantity with the dimensions of an area that is equal to the fraction of the area of the beam perpendicular to the target that is effective in nuclear reactions.

nuclear damage assessment *Military Science.* the determination of the damage to the population, forces, and resources resulting from actual nuclear attack; **nuclear vulnerability assessment** is an estimation of the damage from a hypothetical attack.

nuclear decay mode *Nuclear Physics.* one of several reactions in which a nucleus undergoes radioactive decay, characterized by the isotope produced or the type of radiation emitted.

nuclear defense *Military Science.* the various methods, plans, and procedures that are involved in establishing and exercising defensive measures against the effects of an attack by nuclear weapons or radiological warfare agents.

nuclear delivery vehicle *Ordnance.* the part of a weapons system that provides the means of delivering a nuclear weapon to the target.

nuclear density *Nuclear Physics.* a quantity that establishes the mass per unit volume of a nucleus, based on a variety of experiments that reproduce the same figure.

nuclear device *Nucleonics.* any device that derives its power from nuclear (radioactive) sources.

nuclear-electric propulsion *Space Technology.* a propulsion system utilizing a nuclear reactor to produce electricity to be used either in an electric propulsion system or as a source of heat for a working fluid; used to power a **nuclear-electric rocket engine.**

nuclear emulsion *Nucleonics.* a photographic emulsion designed to record the tracks of an ionizing particle.

nuclear emulsion counter *Nucleonics.* an instrument that counts the number of tracks in a nuclear emulsion and thus can give a measure of the intensity of ionizing radiation.

nuclear energy *Nucleonics.* **1.** the energy released by a fission or fusion reaction. **2.** the binding energy of the system of particles forming an atomic nucleus.

nuclear engine *Nucleonics.* a thermodynamic engine that uses the heat released from fusion or fission processes to perform work on the working fluid.

nuclear engineering *Nucleonics.* a field of technology concerned with the handling, control, and application of nuclear materials and reactors to generate useful energy.

nuclear exoatmospheric burst *Ordnance.* the explosion of a nuclear weapon at an altitude above 120 kilometers, where atmospheric interaction is minimal.

nuclear explosion *Nucleonics.* an explosion resulting from fusion or fission.

nuclear family *Anthropology.* a family unit that consists solely of two parents and their children.

nuclear fission see FISSION.

nuclear force *Nuclear Physics.* the interaction that binds nucleons together in a nucleus, operating so that the strong nuclear force can overcome the repulsive (coulomb) force that separates bound nucleons by certain average distances.

nuclear fragments *Astrophysics.* nuclei produced by spallation of a heavy nucleus after collision with interstellar matter.

nuclear fuel *Nucleonics.* a fertile or fissionable material that acts as the primary source of energy in a nuclear reactor.

nuclear fuel cycle see REACTOR FUEL CYCLE.

nuclear fuel element *Nucleonics.* a sample of nuclear fuel material that is prepared for use in a reactor assembly.

nuclear fuel plate *Nucleonics.* a layer of nuclear fuel that is bound on either side by metallic cladding.

nuclear fuel reprocessing *Nucleonics.* the treatment of used nuclear fuel elements so as to recover and reuse the fissionable and fertile materials.

nuclear fusion *Nuclear Physics.* see FUSION.

nuclear geology see RADIOGEOLOGY.

nuclear ground state *Nuclear Physics.* a term used to describe a particle that is motionless and at its lowest level of energy. Also, GROUND STATE, NORMAL STATE.

nuclear gyroscope *Engineering.* a gyroscope utilizing the spin of electrons and atomic nuclei rather than a spinning mass.

nuclear heat *Nucleonics.* the amount of heat liberated from fission or fusion in a nuclear reactor.

Nuclearia *Invertebrate Zoology.* a genus of naked freshwater amoebae with filopodia that feed on algae.

nuclear induction *Physics.* the magnetic induction of a material achieved by the nonuniform population of available energy states in the nuclei of the medium, when it is placed in a magnetic field.

nuclear laser *Optics.* a laser in which gas molecules are excited by high-energy fission particles to produce constant surges of radiation.

nuclear magnetic moment *Nuclear Physics.* a measure of the strength of the magnetic force in a nucleus at a given point calculated to give the negative value of energy produced by the interaction of a nucleus with a magnetic field. Also, NEUTRON MAGNETIC MOMENT.

nuclear magnetic resonance *Physics.* a phenomenon in which transitions in the magnetic energy states of the nuclei of atoms are induced when the atoms are placed in a static magnetic field and subjected to an oscillatory magnetic field, perpendicular to the static field, and oscillating at some characteristic radio frequency. *Radiology.* the application of an external magnetic field to a solution in a constant radio-frequency field to ascertain the composition of organic compounds. This technique is used in magnetic resonance imaging to visualize soft tissues of the body by distinguishing between hydrogen atoms in different environments.

nuclear magnetic resonance flowmeter *Engineering.* a flowmeter equipped with a radio-frequency field superimposed on a magnetic field in order to resonate the nuclei of a flowing fluid; the velocity of the fluid is then determined by measuring the amount of decay of the resonance.

nuclear magnetic resonance spectrometer *Spectroscopy.* a device used to measure the characteristic energy absorbed and reemitted or dispersed by nuclei that are subjected to a static magnetic field and simultaneously irradiated with radio-frequency radiation.

nuclear magnetism *Physics.* the magnetism associated with the magnetic field generated by nuclei.

nuclear magneton *Nuclear Physics.* a unit of measure of the strength of the magnetic moment of baryons.

nuclear mass *Nuclear Physics.* the quantity of matter in a nucleus, which is less than the total mass of its nucleons by its binding energy divided by the square of the speed of light.

nuclear matrix *Cell Biology.* a fibrous skeletal structure found within the eukaryotic nucleus that is associated with both the nuclear pores and loops of the chromosomal DNA. Also, **nuclear cage.**

nuclear medicine *Medicine.* the clinical discipline dealing with the diagnostic, therapeutic, and investigative uses of radioactive isotopes, but not the therapeutic use of sealed radiation sources.

nuclear medicine imaging *Medicine.* a technique using radioactive tracers in the human body and recording their distribution as images.

nuclear membrane *Cell Biology.* an envelope surrounding the eukaryotic nucleus, consisting of two membranes traversed by a number of pores which permit communication between the nucleus and the cytoplasm.

nuclear molecule *Nuclear Physics.* a combination of two or more nuclei brought together in energetic collisions that overcome long-range electrostatic repulsive forces, so that molecular binding results from the attraction of short-range nuclear forces, as distinguished from chemical molecules held together by valence bonds.

nuclear moments *Nuclear Physics.* a collective term for electric moments generated by deviations of nuclear charge distribution from spherical symmetry, and magnetic (dipole and quadrupole) moments deriving from rotational motion and intrinsic spin of nucleons within the nucleus.

nuclear number see MASS NUMBER.

nuclear orientation *Nuclear Physics.* the alignment, with respect to an arbitrary axis, of a collection of separate nuclear spins, for which any direction is equally probable, so that the quantized axis of the nuclear spin system gives a net axis of polarization.

nuclear paramagnetism *Physics.* a form of paramagnetism associated with a substance due to the uneven population of magnetic moments of the nuclei, giving rise to a net magnetic moment.

nuclear physics *Physics.* the study of the properties and behavior of the nuclei of atoms. Thus, **nuclear physicist.**

Nuclear Physics

Situated at the center of the atom, the nucleus occupies just a very small fraction of the atomic volume yet contains most of its mass. Nuclear physics is concerned with the properties of all nuclei, their stability, their shapes, and the ways in which they can vibrate and rotate. The heaviest nuclei can split, a process referred to as fission.

To a very good approximation, nuclei consist of neutrons and protons. The proton is the nucleus of the hydrogen atom. The neutron, electrically neutral, together with the proton forms the nucleus of deuterium, a substance with the same chemical properties as hydrogen. The forces among these particles, referred to as the "strong nuclear interaction," hold the nucleus together. A very much weaker interaction also exists in nuclei and is responsible for β-radioactivity in which electrons and neutrinos are emitted from unstable nuclei. The wide variety of nuclei makes it possible to study these fundamental forces a variety of circumstances, leading to a deeper and more complete understanding of the structure of matter.

Radioactive nuclei play important roles in industry, and in other sciences such as archaeology, geology, chemistry, and biology. In medicine, radioactive nuclei are used in diagnosis and therapy. Neutrons are used in the oil exploration industry, while fission is the source of nuclear energy generated by nuclear reactors.

Nuclear reactions occur when beams of particles ("projectiles") produced by accelerators strike nuclei. These projectiles may be electrons, the more recently discovered pions, or other nuclei. Reactions can result in new, often unstable nuclei not found in nature. Reactions can induce relatively large-scale collective motions of the nucleus. Reactions can heat up a nucleus so that the individual neutrons and protons move more rapidly. Reactions therefore provide us with insights into the dependence of nuclear forces on energy and on the state of nuclear matter.

It is nuclear reactions that are responsible for the production of energy in the stars and for the production of the heavier elements when a star explodes.

At the present time, nuclear physicists are studying nuclear reactions resulting in the formation of dense and hot matter present an instant after the Big Bang. These studies will tell us much about the very early history of our universe. Nuclear physicists are studying the modifications induced in the state of the proton by its environment inside nuclei. These studies will reveal important aspects of the forces between quarks and gluons holding the proton together.

Herman Feshbach
Institute Professor Emeritus
Center for Theoretical Physics
Massachusetts Institute of Technology

nuclear polarization see NUCLEAR ORIENTATION.
nuclear polyhedrosis virus *Virology*. a subgroup of viruses of the genus *Baculovirus* that replicate and form occlusion bodies only in the host cell nucleus.
nuclear pore *Cell Biology*. one of a series of channels spanning the nuclear envelope of eukaryotic cells that allow transport of material between the nucleus and the cytoplasm.
nuclear potential *Nuclear Physics*. the level of energy that a nuclear particle may possess, based on its position in the field of a nucleus or in its relation to another particle.
nuclear potential energy *Nuclear Physics*. the average amount of energy found in the protons and neutrons of a nucleus that arise from the nuclear forces between them, excluding the electrostatic potential energy.
nuclear power *Nucleonics*. the power derived from fission or fusion processes in a nuclear reactor.
nuclear power plant *Nucleonics*. a facility for the commercial production of electrical energy using heat from the controlled fission chain reaction of a nuclear reactor, with appropriate shielding, containment structures, heat exchangers, and steam-driven turbine-generators.
nuclear propulsion *Naval Architecture*. a naval propulsion system employing a nuclear reactor as energy source. Existing nuclear propulsion systems use heat from the reactor to drive steam turbines.
nuclear protein *Oncology*. a polypeptide encoded by a family of genes (c-*myc*, c-*fos*, v-*fos*, v-*myc*, p53, adenovirus E1A), some of which activate transcriptional promoters or bind DNA.
nuclear quadrupole moment *Nuclear Physics*. a force arising from the deviation of shape from idealized sphericity of nuclear charge distribution on nonspherical (prolate and oblate) nuclei.
nuclear quadrupole resonance *Physics*. an effect similar to nuclear magnetic resonance whereby certain nuclei absorb energy from a radiofrequency field when the nuclei are placed in an inhomogeneous electric field.
nuclear radiation *Nuclear Physics*. a term used to identify particles, such as neutrons, electrons, or photons, that are released from the nucleus as by-products of radioactive decay and nuclear reactions.
nuclear radiation spectroscopy *Nuclear Physics*. the analysis of the energy spectra from nuclear reactions in order to elaborate on details of structure, energy partition, disintegration schemes, and the like.
nuclear radius *Nuclear Physics*. the distance from the center of a sphere to the radius at which the nucleus has decreased to half its central value; the unit of measure is the fermi, equal to 10^{-13} cm.
nuclear reaction *Nuclear Physics*. a reaction that involves a change in the nucleus of an atom, such as decay, fission, or fusion, as opposed to a chemical reaction that only affects the electron structure surrounding the the nucleus. Also, REACTION.
nuclear reaction spectrometry *Spectroscopy*. a spectroscopic technique in which the number of gamma rays produced when the solid surface of a sample is bombarded by a beam of ions is measured, in order to determine the concentration of an element as a function of its depth beneath the surface.
nuclear reactor *Nuclear Physics*. an assembly of fuel core, moderator, containment, control apparatus, and associated equipment used to create large-scale controlled nuclear reactions for the production of energy or nuclides or for research. Also, ATOMIC REACTOR, FISSION REACTOR, PILE.
nuclear relaxation *Physics*. the approach to the state of equilibrium in a system of nuclei having magnetic moments after an externally applied magnetic field is changed.
nuclear resonance *Nuclear Physics*. **1.** a state that is created when the nucleus collides with a bombarding particle and produces a burst of energy. **2.** a process in which a nucleus absorbs energy from radio-frequency fields at certain frequencies.
nuclear risk see DEGREE OF RISK.
nuclear safety line *Military Science*. a line that is used to delineate levels of protective measures and degrees of damage or risk to friendly troops, and/or to prescribe limits to which the effects of friendly weapons may be permitted to extend; it should follow well-defined topographical features.
nuclear sap see KARYOPLASM.
nuclear sclerosis see NUCLEAR CATARACT.
nuclear ship *Naval Architecture*. a surface ship powered by nuclear propulsion. Apart from some Soviet nuclear icebreakers, all present-day nuclear ships are combatant ships.
nuclear spallation see SPALLATION.
nuclear species see NUCLIDE.
nuclear spin *Nuclear Physics*. the total angular momentum of a nucleus measured as the vector sum of the angular momenta of all its component nucleons and its various components analogous to rotation around an axis. Also, NUCLEAR ANGULAR MOMENTUM.
nuclear stability *Nuclear Physics*. the resistance of an isotope to decay or fission as determined by the relative strength of nuclear forces that bind nucleons together and the repulsive (coulomb) force from the protons' electric charge; nuclides with magic numbers of protons and neutrons are especially stable.
nuclear submarine *Naval Architecture*. a submarine powered by nuclear propulsion, and therefore able to operate entirely submerged for months at a time.

nuclear submarine

nuclear superheating *Nucleonics*. the process of superheating steam that is produced in a nuclear reactor, either by recirculating the steam through the core of the reactor or by passing the steam through a second reactor.
nuclear support *Military Science*. the use of nuclear weapons against hostile forces in support of friendly air, land, and naval operations.
nuclear surface burst *Ordnance*. an explosion of a nuclear weapon at the surface of land or water, or above the surface at a height less than the maximum radius of the fireball.
nuclear thermionic converter *Nucleonics*. a device that has a nuclear reactor for a heat source and is capable of converting this heat into electrical energy.
nuclear time scale *Astrophysics*. the time a star takes to convert its usable hydrogen into helium; this may be measured in a few millions of years (large hot stars) or many billions (cooler stars).
nuclear transfer *Biotechnology*. a micromanipulation technique in which nuclei between cells are exchanged or transferred by slender glass needles. Also, **nuclear transplantation.**
nuclear transformation *Nuclear Physics*. a process in which one element is transformed into another, such as when uranium-238 is transformed into plutonium-239 by bombarding it with neutrons. Also, **nuclear transmutation.**
nuclear transplantation *Biotechnology*. a technique in which the nucleus of one cell is placed into another cell that already has a nucleus or in which the nucleus has been previously destroyed; the nuclei are exchanged using fine glass needles. Also, **nuclear transfer.**
nuclear triode detector *Electronics*. a device that can determine the location and energy level of ionizing radiation.
nuclear twin-probe gauge see PROFILING SNOW GAUGE.
nuclear underground burst *Ordnance*. the explosion of a nuclear weapon in which the center of detonation lies at a point beneath the surface of the ground. Also, **nuclear underwater burst.**
nuclear warfare *Military Science*. warfare involving the use of nuclear weapons.
nuclear warhead *Ordnance*. a warhead containing a charge that is designed to produce energy through a nuclear reaction; it may be nuclear fission, fusion, or both.

nuclear weapon *Ordnance.* a complete assembly that, upon completion of the prescribed arming, fusing, and firing sequence, is capable of producing the intended nuclear reaction and release of energy; such assemblies include implosion type, gun type, or thermonuclear type weapons. Thus, **nuclear weapon accident, nuclear weapon exercise, nuclear weapon maneuver.**

nuclear weapon degradation *Ordnance.* the degeneration of a nuclear warhead to such an extent that the anticipated nuclear yield is lessened significantly.

nuclear weapon employment time *Ordnance.* the time between the decision to fire a nuclear weapon and the delivery of the weapon to the target.

nuclear winter *Meteorology.* a hypothetical description of the devastation that would be caused by a global dust cloud following a large-scale nuclear detonation, characterized by worldwide darkness and extreme cold.

nuclear yield *Nucleonics.* a measure of the energy output of a nuclear reactor or nuclear explosion; a high yield generally refers to a larger cross section. *Ordnance.* specifically, the energy released in the detonation of a nuclear weapon, measured in terms of the kilotons or megatons of trinitrotoluene (TNT) required to produce the same energy release. Yields are categorized as follows: *very low,* less than 1 kiloton; *low,* 1 kiloton to 10 kilotons; *medium,* over 10 kilotons to 50 kilotons; *high,* over 50 kilotons to 500 kilotons; *very high,* over 500 kilotons. A metric ton of TNT releases 10^9 calories of energy.

nuclear Zeeman effect *Spectroscopy.* for an atomic species, the splitting of spectral lines that results from the interaction of the magnetic moment of the nucleus with an externally applied magnetic field.

nuclease *Enzymology.* any enzyme of the hydrolase class that splits the phosphodiester linkages in nucleic acids to form nucleotides or oligonucleotides.

nucleate *Nucleonics.* **1.** having a nucleus or in the form of a nucleus. **2.** to form into a nucleus.

nucleate boiling *Chemical Engineering.* a type of pool boiling in which the bubbles form at the liquid-solid interface and vapor escapes in jets or columns.

nucleating agent *Materials Science.* a substance, such as a colloidal silica, that forms nuclei for the growth of crystals in a polymer melt.

nucleation *Crystallograpy.* the condensation or aggregation process by which crystals are formed on a minute amount of substance that acts as a nucleus for subsequent crystalline growth; rain and snow are formed through this process.

nucleation and growth process *Metallurgy.* in any phase transformation and in recrystallization, the formation of discrete particles and their growth.

nucleation of crystals *Crystallography.* the action of a tiny seed crystal, dust particle, or other "nucleus" in starting a crystallization process; for example, in the seeding of a cloud for the production of ice crystals.

nuclei [nook´lē ī] the plural of nucleus.

nucleic acid [noo klā´ik] *Biochemistry.* a complex organic chain of nucleotides of DNA (deoxyribonucleic acid) or RNA (ribonucleic acid); found in chromosomes, mitochondria, ribosomes, bacteria, and viruses; i.e., in virtually all living cells.

nucleic acid hybridization *Biochemistry.* the pairing of complementary RNA and DNA strands to produce an RNA-DNA hybrid, or the partial pairing of DNA single strands from genetically different sources.

nuclein *Biochemistry.* a decomposition product of nucleoprotein, intermediate between native nucleoprotein and nucleic acid, that is a colorless, amorphous compound, soluble in dilute alkalis but insoluble in dilute acids.

nucleocapsid *Virology.* the viral protein-nucleic acid complex that is enclosed by a protein capsid.

nucleocytoplasmic ratio *Cell Biology.* the ratio of the volume of the nucleus to the volume of the cytoplasm of a cell.

nucleofuge *Organic Chemistry.* the group that leaves a substrate in a reaction.

nucleogenesis see NUCLEOSYNTHESIS.

nucleoid *Cell Biology.* a region in prokaryotic cells in which the DNA is found.

nucleokeratin *Neurology.* a variety of keratin found in the nervous system.

nucleolar *Cell Biology.* of or relating to the nucleolus.

Nucleolites *Paleontology.* a genus of atelostomate echinoids in the superorder Microstomata and order Cassiduloida; probably an infaunal deposit feeder; extant in the Upper Cretaceous.

nucleolonema *Cell Biology.* a dense region found in the center of the nucleolus of eukaryotic cells.

nucleolus *Cell Biology.* a discrete part of the eukaryotic nucleus that contains chromosomal clusters of ribosomal RNA genes and functions as the major site of ribosomal RNA synthesis and ribosome assembly.

nucleon *Nuclear Physics.* a proton or a neutron, particularly as a component of a nucleus; a general term for either proton or neutron.

nucleonics *Physics.* the study of the nature and behavior of nuclei and of the practical applications of this behavior.

Nucleonics

Nucleonics is a discipline in the physical sciences which concerns itself with the study of the nucleus of atoms and their constituents, the interaction of nuclei with other nuclei, and the interaction of nucleons, those components of the nucleus, with other nuclei and other nucleons. The field of nucleonics had its origin through early work by Roentgen, Curie, Rutherford, Chadwick, Hahn, Strassmann, Lawrence, and Fermi during the first third of the twentieth century. Protons and neutrons are typical nucleons which have provided important applications to date.

The energies involved in nuclear reactions range from a few electron volts to many billions of electron volts, in contrast to the energies involved in chemical reactions, which are limited to a few electron volts. The lifetimes of radioactive nuclei and nucleons range from picoseconds to many thousands of years. Because of the tremendous range of energies and lifetimes of the objects studied, workers in the field have grouped their studies into nuclear physics, medium energy physics, and particle or high-energy physics.

Today most applications of nucleonics are found in nuclear physics and a few have been found in medium energy physics. Primary civil applications range from nuclear reactors generating heat and electricity to the production of radioisotopes for medical research and treatment and as heat sources for space power systems. Defense applications have played an important role in submarine propulsion systems and nuclear warheads.

Today nuclear electric power plants supply twenty percent of the U.S. electrical supply and seventy percent in France. All major powers have nuclear reactors generating electricity, have nuclear warheads in their arsenals, and have nuclear submarines.

Currently nuclear fission, the splitting of uranium, plutonium, and thorium nuclei with neutrons, provides the major source of applications. Nuclear fusion joining nuclei, deuterium, and tritium is expected to make a major contribution for future world energy production.

Before the invention of accelerators, cosmic rays were the main source of nucleons for research; however, today, worldwide, particle accelerators provide the high-energy nucleons used to study the interaction of nucleons with nuclei and other nucleons. These studies are of paramount importance in the field of astrophysics and in attempting to understand the origin of the universe.

Harold M. Agnew
Former Director
Los Alamos National Laboratory

nucleonium *Atomic Physics.* a state in which a nucleus is bound to an antinucleus.
nucleophile *Physical Chemistry.* an ion or molecule that has a partial or complete negative charge, so that it can donate an electron pair to another atom. (Literally, "nucleus lover.")
nucleophilic *Physical Chemistry.* **1.** relating to a process in which electrons are donated to another atom or molecule. **2.** describing a substance that has electrons available for donation.
nucleophilic reagent *Physical Chemistry.* a molecule that forms a bond with another molecule by donating an electron pair during a chemical reaction.
nucleoplasm *Cell Biology.* the soluble phase of the eukaryotic nucleus, excluding the nucleolus. Also, KARYOLYMPH.
nucleoprotein *Biochemistry.* a compound, such as chromatin, that contains a simple basic protein, usually a histone or protamine, combined with a nucleic acid.
nucleosidase *Enzymology.* **1.** an enzyme that catalyzes the hydrolytic cleavage of an *N*-ribosylpurine to a ribose and a purine. **2.** any enzyme involved with the hydrolysis of a nucleoside.
nucleoside *Biochemistry.* a molecule in which a purine or pyrimidine base is bound to a pentose sugar.
nucleoside diphosphate sugar *Biochemistry.* a compound, such as uridine diphosphoglucose, that combines nucleoside diphosphate with a simple or complex sugar bonded to the 5'-diphosphate group.
nucleoside diphosphokinase *Enzymology.* an enzyme that catalyzes the transfer of a phosphate group from a nucleoside 5'-triphosphate to a nucleoside 5'-diphosphate.
nucleosome *Cell Biology.* a fundamental unit of eukaryotic chromatin, consisting of a histone protein core with a segment of DNA (typically 150–200 base pairs) wrapped around it, giving partially unfolded DNA a characteristic "beads-on-a-string" appearance.
nucleosynthesis *Astrophysics.* the creation of atomic nuclei in nuclear reactions at extremely high temperatures and pressures.
nucleotidase *Enzymology.* any enzyme of the phosphomonoesterase sub-subclass of the dydrolase class that catalyzes the reaction nucleotide + H_2O = nucleoside + orthophosphate.
nucleotide *Biochemistry.* the hydrolysis product of a nucleic acid, consisting of a purine or pyrimidine base combined with a ribose or deoxyribose sugar and with a phosphate group; a phosphate ester of a nucleoside.
nucleotide analog *Molecular Biology.* a synthetic molecule that is structurally similar to a specific nucleotide.
nucleotidyl *Enzymology.* the group formed when a hydroxyl group is removed from the phosphorus atom of a nucleotide.
nucleotidyltransferase *Enzymology.* any enzyme of the transferase group that catalyzes the transfer of a nucleotidyl group, such as DNA polymerase or RNA polymerase.
nucleotoxin *Toxicology.* any agent that is toxic to the cell nucleus.
nucleus [nook´lē əs] *Nuclear Physics.* the central region of an atom, made up of protons and neutrons (known collectively as nucleons). An atomic nucleus of atomic weight *A* and atomic number *Z* contains *A* nucleons, consisting of *Z* protons and *A* – *Z* neutrons. *Cell Biology.* a cellular organelle found in all living cells except mature erythrocytes and platelets and involved in processes such as growth, metabolism, and reproduction; it is bounded by a double membrane and contains the chromosomal DNA, which functions as the principal site of DNA and RNA synthesis. *Science.* generally, a central part around which other parts are concentrated or gathered. *Astronomy.* **1.** the loosely compacted mass consisting of ices and silicates and measuring a few kilometers in size, which forms the body of a comet. **2.** a small bright core in the head of a comet that is visible in a telescope. *Hydrology.* a small particle around which molecules of water may condense or freeze. *Computer Programming.* a set of programs in an operating system that carry out elementary functions such as memory management, security, and input/output control. (From a Latin word meaning "kernel" or "little nut.")
nuclide *Nuclear Physics.* a species of atom that is distinguished by the number of neutrons or protons it contains, and by the amount of energy in its nucleus. Also, NUCLEAR SPECIES, SPECIES.
Nuda *Invertebrate Zoology.* a class of comb jellies, ctenophores with no tentacles.
Nudaurelin β virus group see TETRAVIRIDAE.
nude *Science.* without clothes or covering; naked.
Nudechiniscidae *Invertebrate Zoology.* a family of heterotardigrades in the suborder Echiniscoidea, having uniform cuticles and terminal claws on the legs.
nudibranch *Invertebrate Zoology.* any organism that is a member of the order Nudibranchia.
Nudibranchia *Invertebrate Zoology.* an order of sea slugs and sea hares, gastropod mollusks in the class Opisthobranchia, having two pairs of tentacles and no mantle or shell.

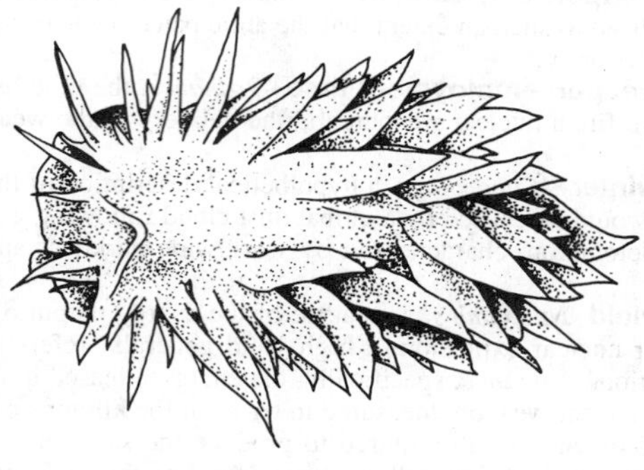

Nudibranchia

nudism *Psychology.* the practice of going nude, especially in public situations.
nudum *Invertebrate Zoology.* a general term referring to the absence or lack of some structure, particularly on the surface of an organism.
nuée ardente *Volcanology.* an avalanche of fiery ash enveloped in gas from a volcanic eruption. Also, GLOWING CLOUD, PELEAN CLOUD.
Nuevoleonian *Geology.* a North American Gulf Coast stage of the Lower Cretaceous period, occurring after the Durangoan and before the Trinitian.
nugget *Geology.* a large lump of gold, silver, or another precious metal found free in nature. *Metallurgy.* the weld metal in spot, seam, or projection welding.
nuisance minefield *Ordnance.* a minefield laid to delay and disorganize the enemy and to hinder their use of an area or route.
nuisance parameter *Statistics.* an unknown parameter whose value is not of interest to the investigator.
NUL null character; a character in ASCII.
null *Mathematics.* zero, empty, of size zero, or nonexistent.
null allele *Genetics.* an allele whose effect is either an absence of normal gene product at the molecular level or an absence of normal function at the phenotypic level.
nullary composition *Mathematics.* the selection of a particular element of a set, as in the construction of a choice function.
null balance *Electricity.* a condition in which no current is flowing.
null-balance recorder *Electrical Engineering.* an instrument used to measure electrical current or voltage, having a measuring circuit with a slide wire that adjusts itself to remain balanced with the current to be measured; a stylus attached to the slide wire records the adjustments.
null cell *Immunology.* a cell that lacks specific identifying surface markers for either T or B lymphocytes.
null character *Computer Programming.* **1.** a special character that is used as a filler with no effect on processing. **2.** in some programming languages, a figurative constant denoting the absence of any number or character.
null cone *Mathematics.* the set of null vectors at a given point x of a manifold; it forms a cone in the tangent space at x. Also called the **light cone** at x.
null-current circuit *Electronics.* a circuit used to measure current, in which the unknown current is opposed by a current generated by a voltage applied across an adjustable wire, so that the sum of the two currents always equals zero when measured by a direct-current amplifier.
null-current measurement *Electronics.* a current value that is determined by using a null-current circuit.
null detection *Electricity.* the use and manipulation of adjustable bridge circuit components in order to satisfy a zero-current requirement in a system or circuit.
null geodesic *Physics.* a curve in space-time that is constructed such that the infinitesimal interval between two neighboring points vanishes. *Mathematics.* a geodesic of length zero. This is possible only when the inner product on the tangent space (i.e., the metric) is indefinite; i.e., the space is pseudo-Riemannian. Also, ZERO GEODESIC.

null hypothesis *Statistics.* a statement about unknown parameters in the population that is to be assessed based on sample results.

nulli- a prefix meaning "none."

null indicator *Electrical Engineering.* a meter that indicates when there is no current or voltage in a circuit; used to show whether a bridge circuit is in balance. Also, **null detector.**

nullipara *Medicine.* a woman who has never borne a child.

nullisomy *Molecular Biology.* the absence of both members of a pair of chromosomes.

nullity *Mathematics.* the dimension (if it is finite) of the null space or kernel of a linear operator T on a linear space X; denoted nullity T. If both X and the image space are finite dimensional, then rank T + nullity $T = \dim X$.

null matrix *Mathematics.* any matrix with all entries equal to zero.

null method *Electrical Engineering.* a system for measuring current of unknown value by balancing out the measuring circuit to indicate zero deflection and then reading the balanced value controls. Also, ZERO METHOD.

null mutation *Genetics.* a mutation that produces neither a transcriptional nor a protein product.

null sequence *Mathematics.* any sequence that converges in some sense to zero; e.g., a sequence of numerical values that converges to zero, or a sequence of functions that converges almost everywhere to the zero function.

null set see EMPTY SET.

null space *Mathematics.* **1.** the null space of an operator T on a space X is the set of all elements x of X for which $Tx = 0$. If X is a linear space and T is a linear operator, then the null space of T is a subspace of X; in this case, the null space of T is usually referred to as the kernel of T. **2.** a vector space consisting of only the null vector.

null string *Computer Programming.* a string that contains no characters. Also, EMPTY STRING.

null vector *Physics.* a four-vector in the theory of special relativity in which the spatial part of the vector (in any Lorentz frame) has a magnitude equal to the product of the time component and the speed of light. *Mathematics.* **1.** a nonzero vector whose inner product with itself is zero. This is possible only when the inner product of the space is indefinite, as in Minkowski space. **2.** the (unique) vector that is the zero element of the additive group of a vector space. Also, ZERO VECTOR.

NUL lymphocyte *Immunology.* a white blood cell of the lymph system that appears to lack immunoglobulin or T lymphocyte antigens on its surface.

number *Mathematics.* **1.** any element of the complex numbers, such as a positive or negative integer, zero, a real or imaginary number, or the like. **2.** a cardinality.

number base *Computer Programming.* the radix of a number system; for example, 10 is the base of the decimal system, and 2 is the base of the binary system.

number-cruncher *Computer Technology.* an informal term for a computer that is designed to handle large scientific calculations involving much floating-point arithmetic. Thus, **number-crunching.**

number density *Physics.* a quantity describing the number of specified entities per unit volume, having units of m^{-3} or cm^{-3}.

number record printer *Telecommunications.* in a relay station, a printer that gives both the complete printed written record of channel numbers and the fixed routing line associated with each transmission.

number scale *Mathematics.* a representation of numerical quantities in linear order consisting of labeled points on a line.

number theory *Mathematics.* the branch of mathematics that studies the relationships between integers and their generalizations. **Elementary number theory** includes the study of divisibility, congruences, quadratic residues, arithmetic functions, prime numbers, etc. Other branches of number theory are **algebraic number theory** and **analytic number theory**.

numeral *Mathematics.* a written representation of a fixed numerical quantity; in particular, the digits 0, 1, 2, 3, 4, 5, 6, 7, 8, and 9.

numeration *Mathematics.* the process of listing a set of numbers in their usual order, from smallest to largest.

numeration system *Mathematics.* the representation of numerical quantities by the use of numerals, with each quantity being assigned a unique numeral or combination of numerals. Also, **number system.**

numerator *Mathematics.* the quantity a in a fraction a/b. The quantity b is the *denominator*.

numerical *Mathematics.* relating to or expressed in numbers; a real- or complex-valued function is sometimes called a **numerical function.**

numerical analysis *Mathematics.* the study of approximation methods (often involving computerized computations) for solutions to mathematical problems, based on two assumptions: (a) the formulas in question may themselves be approximations to more exact formulas, and (b) due to the fact that addition and multiplication are not, in general, associative in a digital computer, arithmetic operations give only approximate results. Numerical analysis thus includes the derivation of approximate expressions in solution to the problem at hand, and derivation of estimates for the error due to assumptions (a) and (b).

numerical aperture *Optics.* a measure specifying the resolving power of a microscope, calculated by multiplying the refractive index of the medium occupying the objective space by the sine of the angle between the most oblique ray entering the objective and the optical axis.

numerical control *Computer Programming.* **1.** the use of digital computer techniques in the control of machine tools such as lathes and milling machines, for example, by programming the positioning of machine work. Also, MACHINE-TOOL CONTROL. **2.** broadly, the use of numeric data in any automatic control process.

numerical display device *Electronics.* a device that displays the information as digits (rather than by analog means such as a pointer or deflection of a point of light).

numerical equation *Mathematics.* any equation with numbers for coefficients and constants.

numerical forecasting *Meteorology.* the process of forecasting the behavior of atmospheric disturbances based on the numerical solutions of the governing fundamental equations of hydrodynamics following observed initial conditions; usually performed on high-speed computing devices.

numerical indicator tube *Electronics.* an electron tube that can display numbers, and sometimes, letters and other symbols.

numerical radius *Mathematics.* given an $n \times n$ matrix A with complex entries, the real number $r(A) = \max\{\|z\| : z \in F(A)\}$; i.e., the maximum value of $\|z\|$ for z in the numerical range of A (where $\| \ \|$ denotes vector norm).

numerical range *Mathematics.* let A be a given $n \times n$ matrix with complex entries; let C^n denote the n-dimensional vector space over the complex numbers C; and let x^* denote the Hermitian adjoint of the vector x; i.e., the row vector that is the transpose of a column vector x, with each component replaced by its complex conjugate. Then the numerical range of A, denoted $F(A)$, is the set of all vectors z given by $z = x^*Ax$, where $x \in C^n$ and $x^*x = 1$. Also, FIELD OF VALUES.

numerical response *Ecology.* a change in the population size of a predatory species as a result of a change in the density of its prey.

numerical taxonomy *Systematics.* taxonomy that makes extensive use of numerical comparisons among taxa; often used as a synonym of phenetics.

numeric character set *Computer Programming.* the collection of all numerals.

numeric coding *Computer Programming.* a code system that includes only digits to represent data and instructions.

numeric data *Computer Programming.* data composed solely of numeric characters.

numeric keypad *Computer Technology.* an input device that is used for entering numeric characters; often a special area of a keyboard where number keys are clustered.

nunatak *Geology.* an isolated hill, knob, ridge, or peak of bedrock projecting prominently above the surface of a glacier and completely surrounded by glacial ice. (From an Inuit word for such a feature.)

nun buoy *Navigation.* a floating aid to navigation that is topped with a cone or truncated cone.

N unit *Optics.* a numerical expression used to simplify the value for the refractive index of the atmosphere; defined as $N = (n-1)10^6$, where n is the true index of refraction of the atmosphere.

nupercaine see DIBUCAINE.

nuptial flight *Invertebrate Zoology.* a mating flight during which sperm is passed from a male or males to a female, as in bees, between drones and a queen.

nurse *Medicine.* **1.** a person trained and licensed in the profession of nursing, concerned with the diagnosis and treatment of human responses to actual or potential health problems. **2.** to provide services that are essential to or helpful in the promotion, maintenance, and restoration of health and well-being. **3.** to breast-feed an infant. (Going back to a Latin word meaning to "feed" or "nourish.")

nurse cell *Histology.* in the ovary of certain animals, a type of cell that nourishes a developing egg cell.

nurse crop *Agriculture.* a crop that is planted with another crop in order to provide shade and prevent weed growth.

nurse graft *Botany.* a type of propagation in which the scion stays within the stock until the scion develops its own roots.

nursing *Medicine.* **1.** the work or profession of a nurse. **2.** the act of breast-feeding a baby.

Nusselt number *Physics.* a dimensionless quantity commonly used in heat transfer studies: the product of the rate of heat loss from a body and the characteristic dimension, divided by the product of the thermal conductivity and the temperature difference between the body and its surroundings. Thus, **Nusselt equation.**

nut *Botany.* a dry, hard-shelled, one-seeded fruit with a hard pericarp, often cultivated as a food crop.

nutant *Biology.* drooping or bent toward the body.

nutating antenna *Engineering.* an antenna in which a dipole or feed horn orbits the axis of a paraboloidal reflector without altering its polarization; used in conical-scan radar.

nutating-disk meter *Engineering.* a meter that measures the flow of a liquid by counting the number of times a disk in the path of the liquid is caused to wobble back and forth.

nutation *Mechanics.* the rhythmic variation of the angle between the precession axis and the body rotation axis, which can occur in a precessing body. *Botany.* the rhythmic growth pattern exhibited by the parts of some plants in response to differing growth rates.

nutational scanner *Optics.* a device in which an oscillating mirror and rotating prism systematically scan an image.

nutator *Engineering.* a drive mechanism used to move a radar beam in a circular, spiral, or conical path periodically.

nut coal *Mining Engineering.* a size of bituminous coal that will pass through a 2–3 inch mesh screen, but will not pass through a 0.75–1.5 inch mesh screen. Also, CHESTNUT COAL.

nut driver *Mechanical Devices.* a metal socket wrench that is mounted on a screwdriver-type handle, used to turn small nuts or self-tapping screws.

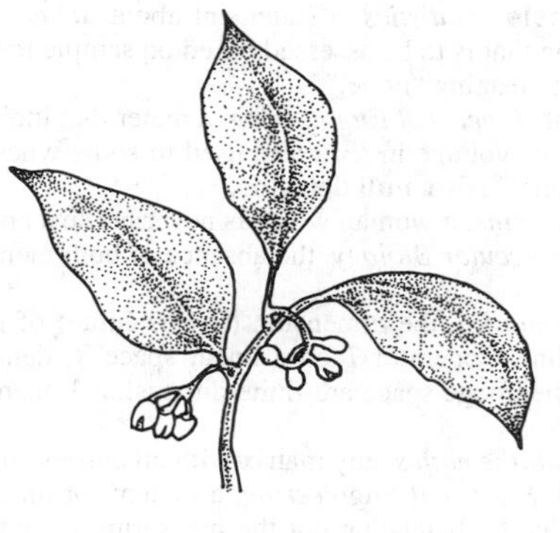

nutmeg

nutmeg *Botany.* the common name for *Myristica fragrans,* an evergreen tree of the family Myristicaceae, cultivated for its seeds, which are used as a spice and in the production of oil (**nutmeg oil**).

NutraSweet *Chemistry* a trade name for aspartame, $C_{14}H_{18}N_2O_5$, a widely used artficial sweetener. See ASPARTAME.

nutrient *Biology.* a substance or compound that provides nourishment; food. Also, **nutriment.**

nutrient cycle *Ecology.* the manner in which nutrients move through an ecosystem. Also, **nutrient cycling**.

nutrient density *Nutrition.* a general term for the concentration of the important nutrients in a particular food, in relation to its caloric value.

nutrification *Nutrition.* the addition of nutrients to a food at a level that significantly enhances the dietary value of the food.

nutrition *Biology.* the science of nourishment, including the study of nutrients that each organism must obtain from its environment in order to maintain life and reproduce.

Nutrition

Graham Lusk (1928) defined nutrition as "the sum of processes concerned in the growth, maintenance, and repair of the living body as a whole or of its constituent organs." This definition encompasses concepts stated by ancient scholars, such as Galen (130–200 AD), and by those who later were on the threshold of experimental science; e.g., Sanctorius (1743–1794), Lavoisier (1743–1794), and others in the 18th and 19th centuries.

Although the "processes" in Lusk's may be described in greater detail today due to more sophisticated state of scientific knowledge and technologic development, the processes continue to include ingestion and digestion of food, absorption, transport, metabolism and utilization (transformation, oxidation, degradation), storage, mobilization, and excretion of nutrients and their conversion products from foodstuffs and beverages. These processes differ with age of organisms, with species or variety, in disease, by genetic differences or environmental influence. Understanding such differences is part of the science of nutrition.

In 1795 Count Rumford first underscored the practical importance of "investigation of the science of nutrition" for planning of feeding of the poor. Much earlier than the introduction of the term "science of nutrition," the curative effect of particular foodstuffs on specific diseases (scurvy, night-blindness) was widely appreciated. Elaboration of such observations ultimately led to recognition of essential nutrients and to quantitation of their requirements.

That the processes, the sum of which comprise nutrition, occur in all living species has repeatedly been recognized (e.g., by Roget, 1836) and the science of nutrition is basic to all biological fields—e.g., agriculture, marine science, medicine, botany, microbiology, etc.—and to the multitude of substudies of each.

The complexity of nutrition science in each of the broad areas of biology can be appreciated by briefly considering the human. Here it is readily recognized that food use and availability are interrelated to or influenced by scientific and nonscientific considerations. These are as diverse as food composition, the application of food science and technology, and cultural aspects of foods, especially their religious significance, and socioeconomic, political, and environmental considerations.

The individual's reactions to all these are modulated by the physiological senses of taste and smell. Such is the complexity of factors that influence one's intake and, in turn, one's health. Although the science of nutrition includes knowledge of the roles of nutrients in maintaining health and the changes in food use patterns that are beneficial in management of disease (dietetics), this knowledge is not the sole determinant of food use.

Daily all persons are involved with the use of food but relatively few have a background in the science of nutrition. All, however, have attitudes impacted by their cultural background, socioeconomic position, and environment. The resulting individual differences account for much of the ever-present public conflict of beliefs and controversies regarding nutrition and foods. These controversies often are mistakenly attributed to the science of nutrition, when, in fact, they result from the general public lack of knowledge concerning that science and the dominant role played by nonscientific forces in molding food behavior, attitudes, and beliefs.

William J. Darby, M.D., Ph.D.
Professor of Biochemistry (Nutrition) Emeritus
Vanderbilt University School of Medicine

nutritional *Biology.* relating to or providing nutrition.

nutritional anemia *Medicine.* anemia due to an absence of a specific substance required for the synthesis of hemoglobin and for erythrocytic maturation, caused by malabsorption or poor dietary intake. Also, DEFICIENCY ANEMIA.

nutritional edema *Medicine.* a nutritional disorder characterized by anasarca and edema, caused by a long-term lack of protein or calories in the diet. Also, **nutritional dropsy**.

nutritional requirements *Nutrition.* the amount of calories and individual nutrients needed to sustain health and well-being and to meet physiological requirements for growth, pregnancy, and lactation.

nutritionist *Nutrition.* a professional who applies the knowledge of food and nutrition to the maintenance of health and the recovery from disease and trauma.

nutriture *Nutrition.* the condition of an individual's body in relation to overall nutrition or in regard to a particular nutrient. Also, **nutritional status.**

nut shell *Invertebrate Zoology.* the common name for marine bivalve mollusks of the genus *Nucula.*

Nuttalliellidae *Invertebrate Zoology.* a family of ticks in the suborder Ixodes, with a single rare African species.

nu value see ABBÉ NUMBER.

nux vomica *Botany.* **1.** an East Indian tree, *Strychnos nux-vomica,* of the logania family; the seed of its orangelike fruit contains strychnine. **2.** the strychnine-containing seed of this tree, formerly used in medicine as an emetic.

NW *Aviation.* the airline code for Northwest Airlines.

n-way selection statement *Computer Programming.* a control structure in which the group of statements to be executed is chosen from among multiple possible groups depending on the outcome of a set of Boolean expressions.

nyacyne see NEOMYCIN.

Nyasa, Lake *Geography.* a large lake (11,600 square miles) in the Great Rift Valley of southeastern Africa between Malawi, Mozambique, and Tanzania. Also, LAKE MALAWI.

nyatoh *Materials.* the resistant, medium textured wood of the Far East trees of the *Palaquium* and *Payena* species; used for building construction, furniture, and plywood.

nybble see NIBBLE.

NYC *Aviation.* the airline code for New York City.

nyct- or **nycto-** a prefix meaning "night."

Nyctaginaceae *Botany.* a dicotyledonous family of herbs, shrubs, and trees of the order Caryophyllales, characterized by ornamental flowers often having an involucre and a superior ovary with one ovule.

nyctalopia see NIGHT BLINDNESS.

Nycteridae *Vertebrate Zoology.* the slit-faced or hollow-faced bats, a monogeneric family of medium-small bats with very large ears, a well-developed noseleaf, and a deep concave spot between the eyes; found in woodland savannah, dry country, or dense forests of Africa, the Near East, Madagascar, Southeast Asia, and the East Indies.

Nyctibiidae *Vertebrate Zoology.* the potoos, a monogeneric family of moderately large nocturnal birds of the order Caprimulgiformes, having long wings, a large head with a small beak, and large downy patches on the sides and breast; found from Mexico to central South America and noted for laying only one egg atop a broken tree stub per clutch.

nyctinasty *Botany.* a movement in higher plants associated with changes in sunlight and temperature.

nyctophobia *Psychology.* a fear of night or darkness. Also, NOCTIPHOBIA.

Nyctribiidae *Invertebrate Zoology.* the bat tick flies, a family of dipteran insects in the subsection Acalyptratae.

N.Y.D. not yet diagnosed.

Nylander reagent *Chemistry.* a basic solution of bismuth subnitrate and Rochelle salt used to detect glucose in urine.

nylon *Materials Science.* **1.** any of various strong, elastic polyamide materials that can be shaped when heated into fibers, bristles, sheets, etc., for use in fabrics, coatings, brushes, and various other products. **2.** of or relating to such materials.

nymph [nimf] *Invertebrate Zoology.* **1.** the immature larval stage of various insects undergoing incomplete metamorphosis, generally similar in structure and function to the adult stage but lacking wings and mature genitalia. **2.** in the Acarina, the juvenile eight-legged form that precedes the adult stage.

nymphae [nim´fē] *Invertebrate Zoology.* the thickened processes on a bivalve shell where ligaments attach.

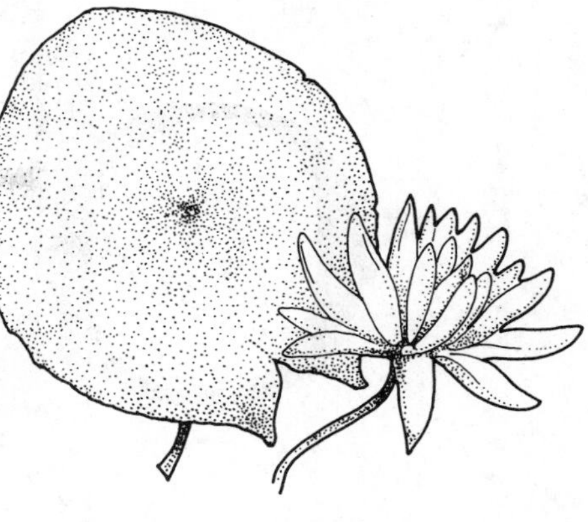

Nymphaeaceae (water lily)

Nymphaeaceae *Botany.* a family of dicotyledonous aquatic plants of the order Nymphaeales, characterized by long leaves with petioles, solitary bisexual flowers, and the presence of roots.

Nymphaeales *Botany.* an order of dicotyledonous aquatic plants of the subclass Magnoliidae, characterized by flowers with many parts, roots, and lack of cambium.

Nymphalidae *Invertebrate Zoology.* a family of bright colored four-footed butterflies, lepidopteran insects in the superfamily Papilionidea, having atrophied forelegs, spiny larvae, and angular pupae that hang upside down.

Nymphalinae *Invertebrate Zoology.* a subfamily of butterflies in the family Nymphalidae.

nymphochrysalis *Invertebrate Zoology.* a pupalike stage between larva and nymph in certain mites.

nymphomania *Psychology.* a condition in which a woman's need for sexual stimulation and gratification is so intense that it overshadows all other activities. Thus, **nymphomaniac.**

Nymphonidae *Invertebrate Zoology.* a family of marine arthropods in the subfamily Pycnogonida, having five-jointed palpi and ten-jointed ovigers.

nymphosis *Invertebrate Zoology.* in hemimetabolous insects, the process of developing into a nymph.

Nymphulinae *Invertebrate Zoology.* a subfamily of lepidopteran insects in the family Pyralididae; includes some aquatic species.

Nyquist contour *Control Systems.* a large semicircle used in constructing a Nyquist diagram.

Nyquist diagram *Control Systems.* a diagram in which a plot of the open-loop transfer function in the complex plane is made as the complex frequency is varied along the Nyquist contour; used to determine the stability of a control system.

Nyquist interval *Telecommunications.* the maximum time interval between regularly spaced instantaneous samples of a wave of specified bandwidth, for complete determination of the waveform of the signal.

Nyquist rate *Telecommunications.* **1.** with reference to the Nyquist interval, the maximum rate in which code events are unambiguously resolved in a channel for a specified range of frequencies. **2.** the Nyquist interval reciprocal.

Nyquist stability theorem *Control Systems.* a theorem stating that the net number of counterclockwise rotations about the origin of the complex plane, carried out by the value of the analytical function of a complex variable as its argument is varied around the Nyquist contour, is equal to the number of poles of the variable in the right half-plane minus the number of zeros in the right half-plane. Also, **Nyquist stability criterion.**

Nysa (asteroid) [nī´ sə] *Astronomy.* asteroid 44, discovered in 1857 and measuring 73 kilometers in diameter. The most reflective asteroid known, Nysa belongs to type E.

Nyssaceae *Botany.* a family of dicotyledonous trees and shrubs having alternate, simple leaves and small, mostly unisexual flowers in axillary or terminal heads.

nystagmus *Medicine.* a rhythmical movement of the eyeballs, either pendular or jerky.

nystatin *Microbiology.* an antibiotic produced by the bacterium *Streptomyces noursei*; used to treat a broad spectrum of fungi that cause infections of the mouth, intestines, skin, and vagina. Also, FUNGICIDIN.

nyxis *Surgery.* a puncture; paracentesis.

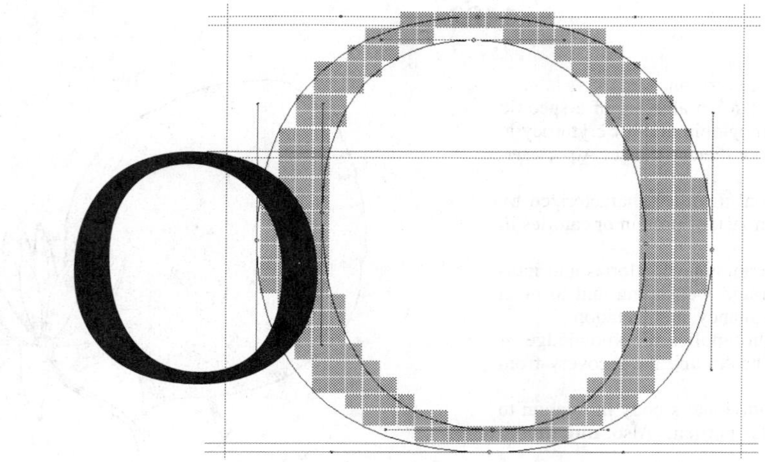

O the chemical symbol for oxygen.
o or **o.** ohm; order.
o- ortho-.
OA *Aviation.* the airline code for Olympic Airways.
OAF osteoclast-activating factors.
OAK *Aviation.* the airport code for Oakland, California.
oak *Botany.* any tree of the genus *Quercus* in the beech family Fagaceae, characterized by simple leaves and a rounded, thin-shelled nut (the acorn); growing exclusively in the Northern Hemisphere in as many as 600 different species. *Materials.* the very hard and durable wood of this tree, usually having a distinct grain and widely used for furniture.

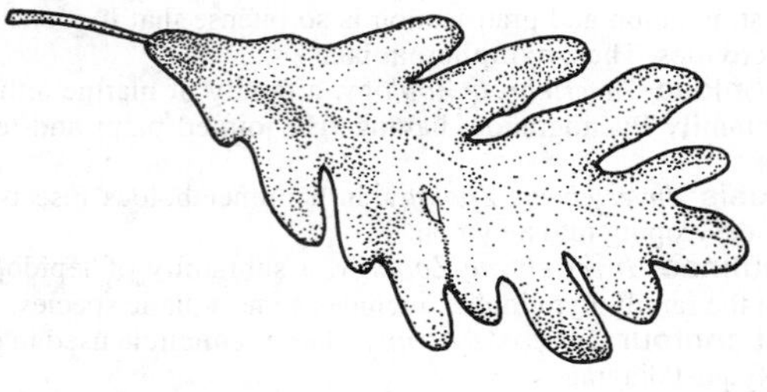

oak

oakmoss resin *Materials.* a resin obtained from the lichen, *Evernia pranastri,* that grows on oak and other trees; used in the manufacture of perfumes.
oakum *Materials.* a nautical caulking material composed of old hemp or jute ropes that have been unraveled and treated with tar.
oak wilt *Plant Pathology.* a disease of oak trees caused by the fungus *Chalara quercina*; characterized by drooping, shriveling, discoloration, and shedding of leaves from the crown downward and inward.
O antigen *Microbiology.* an antigen of certain Gram-negative bacteria, especially members of Enterobacteriacea, characterized by a long filament termed a flagellum that is used for locomotion; important in serologic classification.
OAP *Oncology.* an acronym for a chemotherapy regimen for cancer treatment that includes the use of the drugs Oncovin, ara-C, and prednisone.
oarfish *Vertebrate Zoology.* a very large, unusual fish of the family Regalecidae, having a thin, ribbonlike body that grows up to 30 feet long, antennaelike ventral fins, and a protrusible mouth; found in deepwater tropical and temperate seas and thought to be the basis for sea serpent fables.
oar weed *Botany.* a large brown alga, usually of the genus *Laminaria,* used as a source of iodine, fertilizer, and sometimes food.
oasis *plural,* **oases.** *Geography.* a fertile area in a desert, near a water source.
oat *Botany.* any plant of the genus *Avena* in the grass family Gramineae that is cultivated agriculturally for its seed, which is used in cereals and animal feed, and for its stalk, which is used for straw.
O attenuator *Electronics.* a device that reduces the amplitude of a signal without distorting it by passing it through a two-rung ladder circuit, in which resistances across the rungs are unequal, thus causing the impedances across the terminals to be unequal.
OB. or **ob.** obstetrics; obstetrical; obstetrician.
Ob *Geography.* a river in central Russia, flowing about 2300 miles north from its confluence with the Irtysh to the Arctic Ocean.
ob- a prefix meaning: inversely; the opposite of the usual position.
OB association *Astrophysics.* a loosely bound group of stars belonging to spectral classes O and B; found mainly in the disks of spiral galaxies.
obclavate *Biology.* inversely club-shaped, gradually narrowing near the distal end, as in a leaf.
obcompressed *Biology.* flattened in other than the usual direction, such as vertically or anteriorly rather than laterally.
obconic *Botany.* cone-shaped and attached at the pointed end rather than at the broad base. Also, **obconical.**
obcordate *Botany.* of a leaf, heart-shaped, with the pointed end serving as the base or spot of attachment.
obdiplostemonous *Botany.* having stamens arranged in two whorls, with those in the outer whorl located opposite the petals.
obduction *Medicine.* a seldom-used term for a medicolegal autopsy.
obeche *Materials.* the pale wood of the *Triplochitin scleroxylon* tree of West Africa, used for veneering.
obelisk *Architecture.* a tall, tapering, usually monolithic, squared pillar with a pyramidal tip. *Mathematics.* a three-dimensional solid that is a frustum of a regular, rectangular pyramid.
Obermayer's reagent *Chemistry.* a 4% solution of ferric chloride in dilute hydrochloric acid; used to test for indican (indoxyl sulfate) in urine.
Oberon *Astronomy.* the outermost satellite of Uranus; discovered by William Herschel in 1787, having a diameter of 1550 km, an orbital period of 13.5 days, and a distance from Uranus of 583,000 km. (From the name of the king of the fairies in medieval folklore.) Also, **Uranus IV.**
Oberth, Hermann born 1894, German physicist; a pioneer in the study of astronautics; proposed a liquid-propellant rocket in 1917.
obese [ō bēs´] *Anatomy.* excessively fat.
Obesumbacterium *Bacteriology.* a genus of Gram-negative, facultatively anaerobic bacteria of the family Enterobacteriaceae, occurring as slow-growing, rod-shaped cells that lack motility; a brewery contaminant.
ob-gyn obstetrician gynecologist; obstetrics and gynecology.
object *Optics.* **1.** the figure seen through or the image (real or virtual) formed by an optical system. **2.** the aggregation of points that acts as a source of light rays for an optical system. *Computer Science.* in an object-oriented programming system, a data structure containing data specific to the particular object and a pointer to the class to which the object belongs. *Mathematics.* **1.** a member of the class that, together with the sets of morphisms, makes up a category. **2.** anything that can be transformed from one set of coordinates to another.
object code *Computer Programming.* output from a compiler or assembler either in the form of machine-executable instructions or in relocatable form that requires link editing.
object computer *Computer Technology.* any computer that is in the process of executing an object program.

object function *Mathematics.* in an optimization problem, the function that is to be minimized or maximized subject to the given constraints. Also, **objective function.**

objective *Military Science.* the physical object of a military action; e.g., a definite tactical feature, the seizure or holding of which is essential to the commander's plan. Also, **objective area.** *Optics.* the mirror or lens in a telescope or microscope that first receives light rays from an object being viewed and forms the first or primary image; usually the optical element nearest the object. Also, **object glass, object lens.**

objective function *Robotics.* an equation that uses variables for errors, energy, and time to define scalar quantities and the trade-off relationships between these variables.

objective grating *Optics.* a diffraction grating placed over the objective of a telescope in order to form a diffraction pattern of a star in view, which can then be photographed and compared to other photographs in determining the relative magnitude of stars.

objective plane *Ordnance.* at the point of impact of a bomb or projectile, a plane that is tangent to the ground or coincides with the surface of the target.

objective prism *Optics.* a prism placed in front of the objective of a photographic telescope in order to form and record spectra of luminous objects in the field of view.

objective probability *Statistics.* any probability arising from long-term frequencies; often distinguished from subjective probability.

objective sign *Medicine.* a sign that can be seen, heard, or felt by a diagnostician. Also, PHYSICAL SIGN.

objective test *Psychology.* a test in which the correct answers are previously determined and no subjective judgements enter into the scoring.

object language *Computer Programming.* the intermediate or machine-readable language that is the output of a translating program.

object-oriented language *Computer Programming.* a programming language that focuses on a collection of data objects and the messages to which the objects respond, rather than on procedures. Objects are organized into classes of similar objects and may inherit behavior (procedures that implement messages) from classes and superclasses in a class hierarchy. Thus, **object-oriented programming.**

object permanence *Behavior.* the ability to understand that an object continues to exist even when it is not within immediate view; used as a measure of animal perception or of cognitive development in children. Also, **object constancy.**

object plane *Optics.* the plane, usually perpendicular to the optical axis, that contains the real or virtual image of an optical system.

object program *Computer Programming.* the output, in machine-language or relocatable form, of a translating program; a program that is capable of being executed by a given computer. Thus, **object program library.**

object-program library *Computer Programming.* a group of object-language programs that are maintained in storage and available for use; e.g., by combining them with application programs through link-editing.

object relations *Psychology.* an individual's emotional bonds with other people, activities, or things.

object relationship *Psychology.* the process by which an individual internalizes the external world.

object-relations theory *Psychology.* the principle that an individual's relationship with parents and others in early childhood will determine or significantly affect his or her later interpersonal relationships and sense of self-esteem.

object server *Computer Science.* a server that maintains a database of objects that can be accessed by one or more users over a network.

object space *Optics.* the region of space in front of and behind an optical system in which images, real or virtual, may be formed.

object time *Computer Programming.* the time during which an object program is executed.

obl oblique; oblong.

oblanceolate *Science.* of a leaf shape, tapering at the base so as to resemble an inverted spearhead.

oblate *Science.* of a spheroid body, flattened at the poles.

oblateness *Astronomy.* the degree to which the centrifugal force of rotation causes the shape of a planet to depart from that of a sphere; it is characterized by the ratio between the difference of the equatorial and polar radii to the equatorial radius.

oblate spheroid *Mathematics.* an ellipsoid of revolution in which the ellipse is rotated about its minor axis. Also, **oblate ellipsoid.** *Geodesy.* specifically, a planetary body, such as the earth, that has the shape of an ellipsoid of revolution and rotates about its semiminor axis.

obligate *Ecology.* restricted to a single mode of behavior or environmental condition, such as an obligate aerobe that is dependent on the presence of molecular oxygen to breathe. Also, **obligatory.**

obligate parasite *Ecology.* an organism that can exist only as a parasite.

obligate predator *Ecology.* a predator that is narrowly restricted as to its prey.

oblique [ō blēk′] *Science.* in an inclined position. *Anatomy.* **1.** of or relating to an anatomical structure that is sloping or inclined. **2.** of or relating to a plane or position that is neither transverse nor longitudinal, but rather is situated obliquely. *Botany.* having unequal or unlike sides, as on some leaves. *Mathematics.* **1.** of a polygon, having no right angle. **2.** of a cone, having an axis not perpendicular to the base.

oblique angle *Mathematics.* an angle that is not a right angle or an integral multiple of a right angle.

oblique axis *Graphic Arts.* a layout pattern in which a block of type is set as a tilted parallelogram or otherwise slanted away from the perpendicular.

oblique chart *Cartography.* a chart based on an oblique map projection.

oblique coordinates *Mathematics.* in an n-dimensional oblique-coordinate system, the coordinates of a point given by the directed distances of the point from the origin along the n axes of the coordinate system. The coordinates of a given point on the ith coordinate axis are zero in every position excepth the ith, and equal to the directed distance of the point from the origin in the ith position.

oblique coordinate system *Mathematics.* any n-dimensional coordinate system obtained from an n-dimensional Cartesian coordinate system by removing the requirement that the coordinate axes be mutually perpendicular.

oblique equator *Cartography.* in an oblique map projection, the great circle with a plane perpendicular to the axis of the projection.

oblique fault see DIAGONAL FAULT.

oblique fire *Military Science.* **1.** fire at a diagonal to the long side of the target. **2.** fire from a direction between the enemy's front and flank.

oblique fracture *Medicine.* a fracture that is at a diagonal angle, between the direction of, and at a right angle to, the axis of the bone.

oblique graticule *Cartography.* a network of lines on an oblique map projection representing meridians and parallels measured from an arbitrary reference line.

oblique-incidence transmission *Telecommunications.* in radio, the transmission of a wave obliquely up to the ionosphere and back down again.

oblique joint see DIAGONAL JOINT.

oblique lines *Mathematics.* lines that are neither parallel nor pependicular to a given line, to a given surface, or to each other.

oblique map projection *Cartography.* a map projection on which the vertical axis is inclined at an oblique angle to the line representing the geographic equator.

oblique Mercator map projection *Cartography.* a conformal, cylindrical map projection, developed by Mercator principles, in which points on the earth's surface are projected onto a cylinder tangent to an oblique great circle.

oblique meridian *Cartography.* a line on a map representing the great circle as perpendicular to an oblique equator; the oblique meridian of reference is the prime oblique meridian. Also, FICTICIOUS MERIDIAN.

oblique parallel *Cartography.* a circle or line on a map that connects all points of equal latitude parallel to the ficticious equator. Also, FICTICIOUS PARALLEL.

oblique pole *Cartography.* either of the two points on a map or globe 90° from the oblique equator, and through which the great circles representing the oblique meridians pass.

oblique rhumb line *Cartography.* **1.** a line on a map or globe that makes the same oblique angle with all the fictitious meridians it intersects. Also, FICTICIOUS RHUMB LINE. **2.** a line on a map or globe that forms the same oblique angle with all the meridians, real or ficticious, that it intersects.

oblique rotator theory *Astrophysics.* a model for magnetic stars with periodic variations in field strength which proposes that the magnetic axis lies at an angle to the rotation axis.

oblique section *Mathematics.* the curve of intersection made by a given plane as it intersects (cuts) a surface at a point P, making an angle $g \neq 0$ with the normal to the surface at P. By Meusnier's theorem, the curvature k at P of the oblique section thus formed is $k = \cos gk_n$, where k_n is the curvature of the normal section.

oblique shock wave *Fluid Mechanics.* a shock wave in a supersonic flow field wherein the wave is inclined at an oblique angle with respect to the direction of flow. Also, **oblique shock.**

oblique sphere *Astronomy.* the region of the celestial sphere that lies overhead for observers in the middle latitudes, where celestial objects appear to rise from the horizon at an oblique angle.

oblique triangle *Mathematics.* a plane triangle in which none of the angles is a right angle.

oblique type *Graphic Arts.* italic type. The term originally described only Electra Italic, but now is a synonym for *italic* in any typeface.

oblique visual range *Optics.* the greatest distance at which a given object can be seen along a line of sight tending toward the horizontal.

obliquity [ō blik′wə tē] *Astronomy.* the angle between the rotational axis of an object and its orbit.

obliquity factor *Optics.* a function proportional to the amplitudes of secondary waves traveling in various directions.

obliquity of the ecliptic *Astronomy.* the angle, approximately 23°27′, between the plane of the ecliptic and the celestial equator.

obliterated corner *Civil Engineering.* a survey corner at which previous survey marks have vanished, but which can be pinpointed from other marks or information.

obliteration *Medicine.* complete removal of an organ or other body part, whether by disease, degeneration, surgical procedure, irradiation, or otherwise.

obliterative appendicitis see APPENDICITIS OBLITERANS.

obliterative arterial disease *Cardiology.* a disease marked by complete closure of arterial blood vessels.

OBM ordnance bench mark.

Oboe *Navigation.* a British short-range navigation system having two radar transmitters that send out short bursts; these trigger a transponder on the mobile station, so that measurement of the distance from each of the transmitters yields a fix.

Obolellida *Paleontology.* some of the earliest brachiopods, a small, short-lived but widespread order of primitive brachiopods in the class Inarticulata; characterized by biconvex calcite shells; extant in the Lower and Middle Cambrian.

obovate *Biology.* inversely egg-shaped, such as a simple leaf shaped like a lengthwise section of an egg, with the petiole attached to the narrower or more pointed end.

Obovothyris *Paleontology.* an extinct genus of brachiopods in the order Terebratulida; characterized by a long-looped brachidium; extant in the Middle Jurassic.

obsequent *Geology.* of or related to a geological feature that does not resemble or coincide with the consequent feature from which it developed.

obsequent fault-line scarp *Geology.* a fault-line scarp that faces in a direction opposite the original fault scarp; or such a scarp in which the structurally upthrown block is topographically lower than the downthrown block.

observability *Control Systems.* the property of a system by which the initial values of all state variables can be determined through continuous observation of the output variables.

observable operator *Quantum Mechanics.* an Hermitian operator that yields the value of a physical observable quantity when applied to a state vector in a Hilbert space representing the states of a physical system.

observable quantity *Physics.* any physical quantity that is measurable.

observation *Science.* **1.** the process of obtaining information through the senses. **2.** an item of information obtained in this manner.

observational equation *Crystallography.* an equation expressing a measured value as some function of one or more unknown quantities. Observational equations are reduced to normal equations during the course of a least-squares refinement.

observational learning see IMITATION.

observational study see COMPARATIVE STUDY.

observation of fire *Military Science.* the process of watching fire in order to locate the points of impact or burst in relation to the target and to provide information used to correct firing data. Also, **observation post, observed fire, observed fire procedures.**

observation post *Military Science.* a forward position from which enemy activity can be observed and from which fire can be directed.

observation spillover *Control Systems.* unwanted sensor output from an active control system caused by changes in the control algorithm that omitted certain required modes.

observatory *Astronomy.* a building or facility for the observation of astronomical bodies or phenomena, usually having a powerful telescope.

observed altitude *Navigation.* the sextant altitude (of a body) after all corrections have been made.

observed angle *Cartography.* in surveying, an angle determined by direct instrumental observation, before any corrections are applied.

observed force see APPARENT FORCE.

observed gravity *Geodesy.* the value of gravity determined by direct instrumental observation at a station, before any corrections are applied.

observed latitude *Navigation.* latitude, as determined by observation of a body on or near the observer's meridian.

observed longitude *Navigation.* longitude, as determined by observation of a body on or near the prime vertical.

observed time see ACTUAL TIME.

observer *Control Systems.* a linear system that is driven by another linear system to produce an output that is a linear function of the state of the driving system. Also, STATE ESTIMATOR OR STATE OBSERVER.

observer bias *Psychology.* the influence of the observer's own experiences and attitudes on the results and evaluation of a study.

observer-target line *Ordnance.* an imaginary straight line from an observer or spotter to the target. Also, **observer-target range.**

observing angle *Military Science.* an angle at a target formed by the gun-target line and the observer-target line, representing the angular difference between the gun and the observer relative to the target.

obsession *Psychology.* a thought, idea, or impulse, generally of an upsetting nature, that preoccupies an individual and that cannot be suppressed.

obsessive *Psychology.* relating to or being an obsession.

obsessive-compulsive disorder *Psychology.* an anxiety disorder characterized by persistent and repetitive ideas and impulses and uncontrollable and repetitive acts and behaviors, which interfere with an individual's ability to function. Also, **obsessive-compulsive neurosis.**

obsessive-compulsive personality *Psychology.* a personality classification characterized by an extreme drive for perfection, persistent concern with conformity and orderliness, excessive rigidity, and an inability to relax.

obsessive disorder see OBSESSIVE-COMPULSIVE DISORDER.

obsidian *Geology.* a black or dark-colored volcanic glass characterized by conchoidal fracture, and usually composed of rhyolite composition. Also, HYALOPSITE, ICELAND AGATE, MOUNTAIN MAHOGANY. *Archaeology.* this volcanic glass, which was highly prized in prehistoric stone toolmaking, often used for projectile points and knife blades.

obsidian

obsidian hydration dating *Archaeology.* a method of dating in which the age of an obsidian artifact is established by measuring the thickness of its hydration rim (layer of water penetration) and comparing that to a known local rate of hydration.

obsidianite see TEKTITE.

obsolescence *Engineering.* the state of a device or system no longer being used or a method no longer practiced; usually due to technological or scientific improvements rather than to actual disintegration of the item or ineffectiveness of the method. Thus, **obsolescent.**

obsolete *Science.* no longer in active use. *Biology.* specifically, describing a part of an organism that becomes less distinct from generation to generation, such as the wing of a kiwi that has become reduced over many generations to a tiny structure no longer visible on the bird.

obstacle *Navigation.* any fixed object that must be avoided by a craft.
OB stars *Astronomy.* large, bright, hot stars that have short main-sequence lifetimes, often found in OB associations.
obstet obstetrical; obstetrics.
obstetric or **obstetrical** *Medicine.* of or relating to obstetrics.
obstetrical analgesia *Medicine.* the relief of pain during labor by administering opioids, by conduction analgesia, or by psychologic approaches.
obstetrical toad see MIDWIFE TOAD.
obstetric forceps *Medicine.* the forceps used for grasping and rotating the fetal head. Also, **obstetrical forceps.**
obstetrician [äb´stə trish´ən] *Medicine.* a specialist in obstetrics.
obstetrics [äb stet´riks] *Medicine.* the medical specialty dealing with pregnancy, labor, and childbirth.
obstipation *Medicine.* an obstruction of the intestine; severe constipation.
obstructed stream *Hydrology.* a stream whose valley has been blocked by an obstruction such as a landslide, glacial moraine, or lava flow.
obstruction *Medicine.* any blockage or clogging, usually caused by occlusion or stenosis.
obstruction marker *Navigation.* any navigational aid that marks the location of an obstruction to navigation. Thus, **obstruction beacon, obstruction buoy, obstruction light(s).**
obstruction moraine *Geology.* a moraine that formed where the movement of ice was obstructed.
obstructive atelectasis *Medicine.* an abnormal collapse of lung tissue caused by any factor, such as secretions, a foreign body, a tumor, or abnormal external pressure, which totally blocks the air passage and prevents air from traveling into the alveolar sacs and being absorbed into the bloodstream. Also, REABSORPTION, SECONDARY ABSORPTION, ACQUIRED ATELECTASIS.
obstructive emphysema *Medicine.* emphysema caused by partial obstruction of the air passages, which permits air to enter the alveoli but which resists expiration of the air.
obstructive hydrocephalus *Medicine.* hydrocephalus caused by a blocked ventricle. Also, NONCOMMUNICATING HYDROCEPHALUS.
obstructive jaundice *Medicine.* jaundice caused by an obstruction of the flow of bile into the duodenum. Also, MECHANICAL JAUNDICE.
obstructor *Ordnance.* in naval mine warfare, a device laid in order to obstruct or damage mechanical minesweeping equipment.
obtruncate *Biology.* to cut off the top or head of an organism, such as a tree.
obtund *Medicine.* to dull or blunt, usually in reference to sensation or pain.
obtundation *Neurology.* a condition of dulled sensibility.
obturating cup *Ordnance.* a cup that prevents the escape of propellant gases; the force of the pressure inverts the cup, thus activating a firing pin or similar mechanism.
obturation *Medicine.* **1.** the fact of closing or occluding. **2.** an intestinal obstruction or occlusion.
obturator *Medicine.* **1.** any structure that blocks an opening. **2.** a prosthesis used to close an opening in the hard palate, usually used to close a cleft palate. **3.** the stylus or removable plug often used during the insertion of a tubular instrument. *Ordnance.* **1.** an assembly used in guns firing separate-loading ammunition to prevent the escape of propellant gases around the breechblock; parts include the obturator pad and obturator spindle. **2.** a projectile component providing a similar function.
obturator artery *Anatomy.* a branch of the internal iliac artery that supplies blood to the pelvic bone, iliac muscle, and the obturator muscles.
obturator foramen *Anatomy.* a large opening in the hip bone bounded by the pubic bone and the ischium allowing passage of the obturator vessels and obturator nerve.
obturator membrane *Anatomy.* a fibrous sheath that almost closes the obturator foramen.
obturator nerve *Anatomy.* the nerve composed of branches from the second, third, and fourth lumbar nerves that serves the obturator externus and adductor muscles of the thigh and the articulations of the hip and knee.
obtuse *Mathematics.* see OBTUSE ANGLE. *Botany.* describing the blunt or rounded tip of a leaf with sides that form an angle of more than 90°.
obtuse angle *Mathematics.* an angle greater than 90° and less than 180°.
obtuse bisectrix *Crystallography.* an axis in a crystal bisecting an obtuse angle between optical axes.
obtuse triangle *Mathematics.* a plane triangle with one obtuse angle.
obumbrant *Zoology.* overhanging; projecting over some other part.
obv obverse.
obvallate *Biology.* surrounded by or as if by a wall.
obvolute *Biology.* overlapping.
OC officer candidate; on course; off center; on center.
Oc. or **oc.** ocean.
Occam's razor *Science.* a principle stating that entities must not be multiplied beyond what is necessary; often interpreted to mean that phenomena should be explained in terms of the simplest causes rather than the most complex. Also, LAW OF PARSIMONY. (From *William of Occam,* a medieval English philosopher who devised this principle.)
occasional fog signal *Navigation.* a fog signal, not commonly encountered, that sounds only under certain conditions.
occasional light *Navigation.* a lighted aid to navigation, not commonly encountered, that is exhibited only under certain conditions, such as to indicate a certain depth of tide.
occipital [äk sip´i təl] *Anatomy.* relating to the occiput or the occipital bone.
occipital arch *Invertebrate Zoology.* the part of the insect cranium that lies between the occipital suture and the postoccipital suture.
occipital artery *Anatomy.* an artery that supplies blood to the back of the head and scalp.
occipital bone *Anatomy.* the compound bone forming the lower back part of the skull.

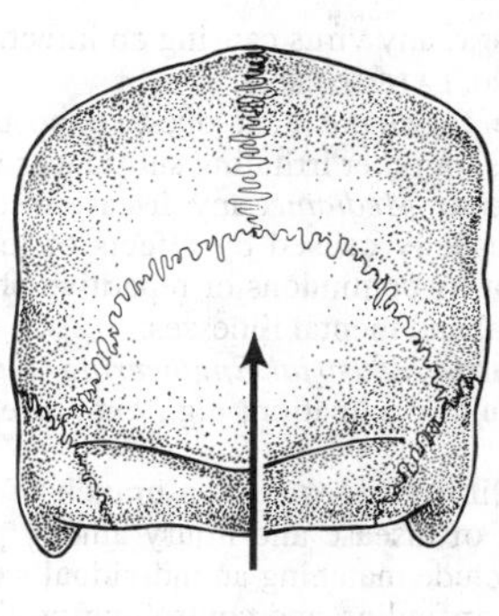

occipital bone

occipital chignon *Anthropology.* a rounded projection at the base of the brain case; a feature that occurs in the elongated skulls of Western European Neanderthals. (From its resemblance to the *chignon,* a French hairstyle.) Also, **occipital bun.**
occipital condyle *Anatomy.* one of two rounded projections of the occipital bone, lying on either side of the foramen magnum, that articulate with the atlas.
occipital crest external *Anatomy.* the ridge of bone extending from the external occipital protuberance to the foramen magnum.
occipital ganglion *Invertebrate Zoology.* one of the paired ganglia of the nervous system of an insect that lie just behind the brain.
occipital lobe *Anatomy.* the posterior lobe of either of the cerebral hemispheres covered by the occipital bone.
occipital pole *Anatomy.* the most posterior projection of either occipital lobe.
occipital protuberance *Anatomy.* a prominent tubercle of bone between the summit of the occipital bone and the foramen magnum that gives attachment to the nuchal ligament.
occipitofrontalis *Anatomy.* the muscle that moves the scalp, passing over the cranium and attached via a fibrous aponeurosis at one end above the frontal bone and at the other to the occipital bones.
occiput *Anatomy.* the back part of the head. *Invertebrate Zoology.* the dorsal or posterior head region of an insect.
occluded virus *Virology.* any insect-pathogenic virus that produces proteinaceous occlusion bodies containing virus particles.
occluding ligature *Surgery.* a ligature that cuts off blood flow to distal tissue.
occlusal disharmony *Medicine.* a condition in which the opposing occlusal surfaces of the teeth do not make accurate contact when the jaw is closed.

occlusion [ə kloo´zhən] the fact of closing or blocking off; specific uses include: *Anatomy*. an obstruction of a passage or vessel. *Medicine*. any contact between the incising or masticating surfaces of the upper and lower teeth. *Physiology*. the reduction in muscle tension when two or more afferent nerves sharing certain motor neurons are stimulated simultaneously, as opposed to the sum of the muscle tension when the nerves are stimulated separately. *Physics*. a process in which gas or liquid is trapped on the surface or within a solid mass.

occlusion bodies *Virology*. the large intracellular proteinaceous crystals in which virions are embedded.

occult [ə kult´] *Medicine*. describing a condition that is hidden from view or difficult to detect.

occultation *Astronomy*. **1.** a partial or complete eclipse of a star or planet by the moon. **2.** the passage of one celestial body in front of another, thereby hiding one body from view.

occult blood *Pathology*. a plasma occurring in feces in such minuscule amounts that its existence can only be detected by microscopic or spectroscopic examination.

occult hydrocephalus *Medicine*. hydrocephalus that produces no abnormal physical signs or symptoms.

occulting disk *Optics*. a small disk used in a telescope to block the view of a bright object in order to allow observation of a fainter one. Similarly, **occulting bar.**

occulting light *Navigation*. a lighted aid to navigation that eclipses at regular intervals so that the period of light is greater than the period of darkness.

occult mineral *Mineralogy*. a mineral that is expected to be present in a rock, perhaps from the evidence of chemical analysis, but that is not found there.

occult virus *Virology*. any virus causing an infection that does not produce symptoms. Also, LATENT VIRUS.

occupational acne *Medicine*. acne due to exposure to chlorinated hydrocarbons, tars, oils, or other irritating substances in a workplace.

occupational disease *Medicine*. any disease resulting from a specific job or workplace, usually caused by effects of long-term exposure to certain substances or by continuous or repetitive physical acts; this may include both physical and mental illnesses.

occupational hazard *Industrial Engineering*. a risk or danger that is inherent in a particular occupation; e.g., exposure to coal dust in coal mining.

occupational medicine *Medicine*. a branch of medicine concerned with the prevention of disease and injury among people at work; such prevention might include matching an individual's characteristics to certain kinds of work, or finding and controlling on-the-job hazards. Also, **industrial medicine.**

occupational neurosis *Psychology*. a neurotic reaction to one's own job, resulting in anxiety and physical symptoms that impede one's ability to continue to work.

occupational therapy *Medicine*. the use of avocational or vocational tasks as a form of therapy or rehabilitation.

occupation density *Transportation Engineering*. the available space per person in a station or vehicle.

occupation layer *Archaeology*. a layer in which an original deposit is preserved as it existed when the site was abandoned.

occupation span *Archaeology*. the time period during which a site is occupied.

occupation surface *Archaeology*. any surface used for human activities, such as a room floor, a stairway, or a walkway. Also, **occupation floor.**

occupied bandwidth *Telecommunications*. the frequency bandwidth such that, below its lowest and above its highest frequency range, the mean powers transmitted are each equal to 0.5% of the total mean power transmitted.

occupied territory *Military Science*. territory under the authority and effective control of a belligerent armed force; not applied to territory administered according to treaties and other civil agreements.

ocean *Geography*. **1.** the great body of sea water that covers two-thirds of the earth's surface and surrounds all of its dry land; the sea. **2.** any of the four main subdivisions of this great ocean: Pacific, Atlantic, Indian, and Arctic. The southern portions of the first three, which converge around Antarctica, are known collectively as the Antarctic Ocean.

ocean basin *Geology*. a depression on the surface of the lithosphere occupied by an ocean.

ocean circulation *Oceanography*. the large-scale movement of water in an ocean, usually in a closed, circular gyre.

ocean current *Oceanography*. a movement of ocean water that follows a more or less definite pattern, usually moving in a continuous flow but sometimes undergoing marked cyclical changes.

ocean floor *Geology*. the surface of the ocean basin.

ocean-floor spreading see SEA-FLOOR SPREADING.

Oceanian *Ecology*. the Polynesian zoogeographic subregion.

oceanic *Oceanography*. of or relating to the ocean: an *oceanic* species. *Geology*. of or relating to the areas of the ocean deeper than the littoral and neritic zones and beyond the epicontinental zone.

oceanic anticyclone see SUBTROPICAL HIGH.

oceanic basalt *Petrology*. the basalt rock occurring in the vicinity of oceanic island volcanoes.

oceanic climate see MARINE CLIMATE.

oceanic crust *Geology*. the thick mass of igneous rock that underlies the ocean basins, as distinguished from continental crust.

oceanic heat flow *Geophysics*. the average flow of heat from inside the earth through the oceanic crust, measured in microcalories per square centimeter per second.

oceanic high see SUBTROPICAL HIGH.

oceanic island *Geology*. an island that rises from the deep-sea floor.

oceanicity [ō´shə nis´ə tē] *Meteorology*. a description of the specific extent to which the climate of an area is influenced by the ocean.

oceanic rise *Geology*. a long, broad elevation of the ocean bottom.

oceanic trench *Geology*. see TRENCH, def. 3.

oceanic zone *Oceanography*. the biogeographic region of the open sea, as distinguished from that of the epicontinental zone.

oceanite *Petrology*. a picritic basalt with at least 50% olivine and a lesser amount of plagioclase.

oceanity see OCEANICITY.

oceanization *Geology*. a process by which continental crust is transformed into oceanic crust.

oceanodromous *Vertebrate Zoology*. of or relating to marine fish that are migratory.

oceanographer [ō´shə näg´rə fər] a person who specializes in or studies oceanography.

oceanographic or **oceanographical** relating to the scientific study of the ocean. Thus, **oceanographic survey, oceanographic research.**

oceanographic equator *Oceanography*. the zone of maximum surface water temperature (generally above 28°C) of the ocean, generally lying above the equator, but lying slightly below the equator during the summer in the Indian Ocean, the Western Pacific, and the Western Atlantic.

oceanographic model *Oceanography*. a hypothetical representation of a given type of marine environment.

oceanographic submersible *Naval Architecture*. a small submarine or submersible vessel designed for use in oceanographic research.

oceanographic submersible

Oceanography

Oceanography is the scientific study of that part of the earth that is covered with seawater. In the Soviet Union and some other countries it is called "Oceanology," a term that is preferable because it implies understanding of the ocean, where oceanography suggests only its description.

Oceanographers are concerned with much more than seawater. They are interested in the creatures—microbes, plants, and animals—that live in the sea, in the sediments on the sea floor and the rocks that underlie them, in the motions of the ocean waters, both waves and currents and the chaotic motions called turbulence, in the sandy beaches, rocky cliffs and other structures that form the boundaries of the sea, and in the exchange of matter and energy between the sea and the overlying air.

It used to be thought that the sea floor was a permanent feature of the earth's crust, unchanged for eons except for the slow accumulation of deep-sea sediments. Now we know that the sea floor is continuously renewed. There is no place in the sea that is much more than two hundred million years old—only about one-twentieth of the earth's lifetime.

The plants of the sea, on which all other life depends, live only in the upper water layers where bright sunlight penetrates. To maintain themselves in this sunlit zone, most ocean plants are microscopic in size, and the animals that eat them are not much larger. These animals are in turn eaten by larger animals, and eventually by fish that human beings can eat. But the number and weight of fishes is very small compared to the marine plants. The ocean can never be relied on to provide the major source of food for human beings.

Roger Revelle
Founder, Scripps Institution of Oceanography
Former Chancellor, University of California, San Diego

oceanography *Geography.* the scientific study of the sea, including its geography, geology, chemistry, physics, and biology.

oceanology **1.** the scientific study of the ocean; oceanography. **2.** the study of the ocean as it affects humans, including its economic, political, and military aspects. Thus, **oceanologist, oceanologic(al).**

Oceanospirillum *Bacteriology.* a genus of Gram-negative, aerobic bacteria that occur in marine habitats as motile cells with a respiratory metabolism.

ocean perch see REDFISH.

ocean station vessel *Naval Architecture.* a vessel designed to collect research data and transmit navigational information in one ocean area.

ocean sunfish *Vertebrate Zoology.* a brown and gray mola, *Mola mola,* of the family Molidae, characterized by a sharply truncated posterior half of the body behind the elongated dorsal and anal fins; found in tropical and temperate seas. Also, HEADFISH.

ocean thermal-energy conversion *Mechanical Engineering.* the conversion into electrical or other forms of energy of the energy arising from the difference between the warm, surface-ocean water and the cold, deep-ocean currents.

ocean tomography *Acoustical Engineering.* undersea imaging or remote-sensing in which arrays of transmitters and receivers detect differences in pulse travel times in order to determine temperature differences in the ocean areas through which the sounds travel.

Oceanus Procellarum see PROCELLARUM.

ocellar [ō sel′ər] *Petrology.* describing an igneous rock texture with phenocrysts of smaller crystals arranged radially or tangentially around the borders of larger euhedral crystals; of a rock having such a texture. *Zoology.* of or relating to an ocellus.

ocellate or **ocellated** *Zoology.* relating to or having ocelli, or eye-shaped marks.

ocellated turkey *Vertebrate Zoology.* a brilliantly colored wild turkey of Guatemala and the Yucatán, *Agriocharis ocellata,* having eye-shaped spots on its tail.

ocellation *Zoology.* **1.** the fact of having eyelike marks. **2.** any such mark or pattern.

ocellus [ō sel′əs] *plural,* **ocelli.** *Invertebrate Zoology.* **1.** a simple eye that is composed of a few sensory cells and a single lens; present in some insect larvae or adults and capable of sensing only light, darkness, and movement. **2.** an eye-spot pattern on the wing of a moth or butterfly, functioning as a means of intimidating predators. *Zoology.* any eyelike marking or spot, as on the tail feathers of a peacock.

ocelot [ä′sə lät; ō′sə lät] *Vertebrate Zoology.* a wild nocturnal and terrestrial cat, *Leopardus pardalis,* of the genus *Felis,* characterized by a tawny or grayish coat with black dots and stripes; found in humid tropical forests to dry scrub country from Arizona and Texas to northern Argentina. An endangered species in the U.S.

ocelot

OCG oral cholangiogram.

ocher [ō′kər] *Mineralogy.* a yellow, brown, or red earthy mineral consisting of a hydrated ferric oxide or other earthy, pulverulent metallic oxide; used as a pigment. Also, OCHRE.

ocher mutation *Genetics.* a mutation that results in an abnormally short polypeptide chain. Thus, **ocher mutant.**

ochlesis *Pathology.* any disease caused by overcrowding.

ochlophobia *Psychology.* an irrational fear of crowds or crowded places. Thus, **ochlophobic.**

Ochnaceae *Botany.* a family of dicotyledonous tropical evergreen trees and shrubs in the order Theales, which are characterized by anthers with terminal pores, simple leaves with parallel lateral veins, and often winged seeds.

Ochoa, Severo born 1905, Spanish-born American biochemist; awarded the Nobel Prize for synthesizing ribonucleic acid.

Ochoan *Geology.* a North American provincial of the uppermost Permian period, occurring after the Guadalupian and before the Lower Triassic period.

Ochotonidae *Vertebrate Zoology.* the pikas, a monogeneric family of the order Lagomorpha, characterized by short, rounded ears, short limbs, and no tail; found primarily in central Asia.

ochratoxin A *Biochemistry.* a mycotoxin that forms in certain species of *Penicillium* and *Aspergillus,* causing nephropathy in humans and livestock when ingested in contaminated food.

ochre see OCHER.

Ochrept *Geology.* a suborder of the soil order *Inceptisol* that develops in cold or temperate climates and that is characterized by a thin, light-colored surface layer overlying a cambic horizon.

Ochrobium *Bacteriology.* a genus of Gram-negative, rod-shaped or ellipsoidal iron bacteria of the familly Siderocapsaceae, encased in iron and surrounded by a transparent, horseshoe-shaped sheath.

ochroleucous *Biology.* pale ocher- or buff-colored.

Ochromonadaceae *Botany.* a family of flagellates of the order Ochromonadales, consisting almost exclusively of naked forms and often being very difficult to identify.

Ochromonadales *Botany.* an order belonging to the class Chrysophyceae that is distinguished by species that spend most of their lives as flagellate cells, living solitarily or in colonies, and generally free-living in freshwater phytoplankton.

ochronosis *Medicine.* an unusual discoloration, observed in certain patients with alkaptonuria, marked by pigmentation of cartilages and sometimes other tissues. (From Greek for "yellow disease.")

Ochrosphaeraceae *Botany.* a family of marine flagellate algae of the order Isochrysidales, characterized by a brief motile phase and a dominant coccoid phase, scaly cells, and the production of calcareous elements.

Ochteridae *Invertebrate Zoology.* the shore bugs, a family of hemipteran insects in the superfamily Ochteroidea that are black with a velvety sheen.

Ochteroidea *Invertebrate Zoology.* a superfamily of tropical and subtropical hemipteran insects in the subdivision Hydrocorisae, consisting of only one family, the Ochteridae.

ocotea oil *Materials.* an essential oil obtained by steam distillation from the wood of the Brazilian tree *Ocotea cymbarum*; used mainly as a source of safrole and in perfuming as a substitute for sassafras oil.

ocotillo [ō´kə tē´yo] *Botany.* a woody shrub, *Fouqueria splendens*, characterized by spines and a tight cluster of red flowers at the tip of each branch; found in arid regions in the southwestern U.S. and Mexico.

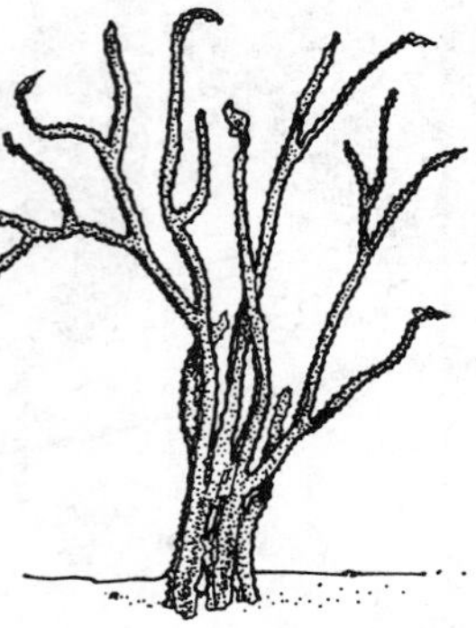

ocotillo

OCR optical character recognition; optical character reader.

ocrea *Botany.* a tubular sheath around a stem above a node formed by united stipules or leaf bases. Also, **ochrea.**

ocrylate *Surgery.* a tissue adhesive used in surgery.

oct. pint. (From Latin *octarius.*)

oct- or **octo-** a prefix meaning eight. Also, **octa.**

octadecanamide see STEARAMIDE.

***n*-octadecane** *Organic Chemistry.* $CH_3(CH_2)_{16}CH_3$, a combustible, colorless liquid; insoluble in water and soluble in alcohol, petroleum, and coal-tar hydrocarbons; melts at 28°C and boils at 318°C; used in solvents and organic synthesis.

***n*-octadecanoic acid** see STEAROLIC ALCOHOL.

octadecanol see STEARYL ACID.

1-octadecene *Organic Chemistry.* $C_{18}H_{36}$ or $CH_3(CH_2)_{15}CH{:}CH_2$, a combustible, colorless liquid; soluble in alcohol and insoluble in water; boils at 180°C (15 torr) and freezes at 17.5°C; used in organic synthesis and surfactants.

octadecenyl aldehyde *Organic Chemistry.* $C_{17}H_{33}CHO$, a flammable liquid that boils at 108–110°C; irritant to skin and mucous membranes; used in the manufacture of vulcanization accelerators, synthetic drying oils, and pesticides. Also, OLEYL ALDEHYDE.

9-octadecynoic acid see STEARYL ACID.

octafluorocyclobutane *Organic Chemistry.* C_4F_8, a nonflammable, colorless gas that is soluble in ether; freezes at –41°C and boils at –6°C; used as a refrigerant and heat transfer agent.

octafluoropropane *Organic Chemistry.* C_3F_8, a nonflammable, colorless gas that freezes at –183°C and boils at –36.7°C; derived from the electrofluorination of various organic compounds; used as a refrigerant and gaseous insulator, especially for radar waveguides.

octagon *Mathematics.* a polygon with eight sides.

octahedral cleavage *Crystallography.* the crystallographic cleavage associated with the four planes parallel to the surfaces of a regular octahedron.

octahedral coordination *Mineralogy.* the atomic structure of a mineral in which every cation is surrounded by six anions whose centers form the corners of a regular octahedron.

octahedral interstice *Materials Science.* sites between lattice points in a crystal structure, which when combined form an eight-sided polyhedron.

octahedral normal stress *Mechanics.* the normal component of the stress on the planes composing a regular octahedron whose vertices all lie on the principal stress axes; it is equal to the average of the three principal stresses at that point. Also, MEAN STRESS.

octahedral shear stress *Mechanics.* the tangential component of the stress on the planes composing a regular octahedron whose vertices all lie on the principal stress axes.

octahedrite *Geology.* a common iron meteorite, containing 6–18% nickel in the metal phase. *Mineralogy.* see ANATASE.

octahedron *Mathematics.* a polyhedron with eight faces.

octal *Computer Programming.* of or relating to a number system with a radix or base of 8.

octal base *Electronics.* a tube base that has eight uniformly spaced pins surrounding a central plastic pin or key, which assures correct insertion of the tube into a socket.

octal debugger *Computer Programming.* a debugging program that presents octal representations of address references.

octal loading program *Computer Programming.* a utility program used to enter program patches or data changes that are in octal representation.

octal number system *Mathematics.* the representation of numbers in base 8.

octamerous *Botany.* having eight members in each whorl; usually written 8-merous.

1-octanal see OCTYL ALDEHYDE.

octane *Organic Chemistry.* **1.** any of 18 chemical compounds that have the general formula C_8H_{18}; colorless, highly flammable liquids, some of which are found in petroleum. **2.** see OCTANE NUMBER.

***n*-octane** *Organic Chemistry.* $CH_3(CH_2)_6CH_3$, a flammable, colorless liquid; insoluble in water and soluble in alcohol and acetone; boils at 125.6°C and freezes at –56.8°C; used as a solvent, in organic synthesis, and in calibrations.

octanedioic acid see SUBERIC ACID.

octane number *Engineering.* a number indicating the antiknock value of a motor fuel, by giving the percentage of iso-octane in a blend of this mixture and normal heptane that manifests the same antiknock characteristics as the fuel sample in question.

octane requirement *Mechanical Engineering.* the fuel octane rating (expressed numerically) that is necessary for an internal-combustion engine to operate efficiently; based on the fuel characteristics and the engine design and compression ratio.

octane scale *Mechanical Engineering.* the standard range of numbers used to indicate the resistance of gasoline to detonation in an internal combustion engine, with normal heptane, producing the highest knock, rated 0; iso-octane, rated 100; and iso-octane +6 milliliters of TEL (tetraethyl lead), rated 120.3.

octanoic acid *Organic Chemistry.* $CH_3(CH_2)_6CO_2H$, a combustible, colorless, oily liquid with a slight odor and a rancid taste; slightly soluble in water and soluble in alcohol and ether; melts at 16°C and boils at 239.7°C (or 124°C at 10 torr); derived from the saponification and distillation of coconut oil; used in the manufacture of dyes, drugs, perfumes, and synthetic flavors. Also, CAPRYLIC ACID.

octanoic acid anhydride see CAPRYLIC ANHYDRIDE.

2-octanone see METHYL HEXYL KETONE.

Octans *Astronomy.* the Octant, a southern constellation containing the south celestial pole. Also, **Octantis.**

octant *Mathematics.* **1.** in plane geometry, one-eighth of a circle or a unit of plane angle equal to 45° (π/4 radians). **2.** one of the eight regions into which three-dimensional space is divided by the coordinate planes of a Cartesian coordinate system. *Navigation.* a double-reflecting, angle-measuring instrument that has an arc 90° in length. In modern usage, all such instruments are referred to as "sextants," regardless of the length of the arc.

octaphyllite *Mineralogy.* any of a group of trioctahedral mica minerals with a structure characterized by a ratio of eight cations for every ten oxygen and two hydroxyl ions.

octave *Physics.* a frequency range interval in which the ratio of the higher frequency to that of the lower frequency is 2:1. *Acoustics.* this interval used as the basis for the Pythagorean musical scales.

octave band *Electronics.* a band of frequencies in which the limits have the ratio of 2 to 1; an octave band extends from a lower frequency to twice the lower frequency.

octave-band analyzer *Acoustical Engineering.* a frequency analyzer that uses frequency increments of one octave.
octave frequency band *Physics.* any frequency band whose upper limit is twice that of the lower limit.
octavo *Graphic Arts.* a book with a trim size of approximately 6 by 9 inches that is printed on sheets folded to form 8 leaves or 16 pages.
1-octene *Organic Chemistry.* $CH_3(CH_2)_5CH{:}CH_2$, a flammable, colorless liquid that is soluble in alcohol and insoluble in water; freezes at –102.4°C and boils at 121.27°C. Also, 1-CAPRYLENE, 1-OCTYLENE.
2-octene *Organic Chemistry.* $CH_3(CH_2)_4CH{:}CHCH_3$, a flammable, colorless liquid occurring in *cis-* and *trans-* forms; soluble in alcohol and insoluble in water; the *cis* isomer boils at 124.6°C and freezes at –104°C; used in organic synthesis and in the manufacture of lubricants.
octet *Atomic Physics.* the eight electrons grouped in the outermost and *p* subshells of an atom or molecule, considered to be the most stable configuration for this shell. *Particle Physics.* a set of eight states of a single entity, not necessarily of one particle, made up of stable and metastable baryons.
octillion *Mathematics.* **1.** in American usage, the number 10^{27}. **2.** in British and German usage, the number 10^{48}.
Octocorallia *Invertebrate Zoology.* one of two subclasses of corals, colonial anthozoan coelenterates with eight branched tentacles, including soft corals, horny corals, and sea pens. Also, ALCYONARIA.
octode *Electronics.* an electron tube that has five auxiliary electrodes in addition to an anode, cathode, and a primary control electrode.
octoid *Design Engineering.* of or relating to a gear tooth form that resembles the involute form; used to generate the teeth in bevel gears.
octonary signaling *Telecommunications.* a form of signaling in which information is transmitted by the presence and absence or positive and negative variation of eight discrete levels of one parameter of the particular signaling medium.
octopamine *Endocrinology.* a catecholamine having little known biological activity, formed from tyramine and normally found within the nervous system only in small quantities.
octopamine hydrochloride *Pharmacology.* $C_8H_{11}NO_2{\cdot}HCl$, a drug formerly used to treat low blood pressure; the hydrochloric salt of octopamine, a biogenic amine.
octopine *Biochemistry.* a compound present in the muscles of a variety of marine invertebrates, such as the octopus, that resembles argenine and alanine.
Octopoda *Invertebrate Zoology.* an order of cephalopod mollusks, including the octopus and paper nautilus, in the subclass Dibranchia; characterized by a saclike body, eight arms with rows of suckers, and no internal shell.
Octopodidae *Invertebrate Zoology.* the octopuses, a family of cephalopod mollusks in the order Octopoda, characterized by a small saclike body, a large head, a strong beak, highly developed eyes, and sucker-bearing arms.
octopus *Invertebrate Zoology.* the common name for any member of the family Octopodidae, characterized by a soft, oval body and eight sucker-bearing arms; found mostly at the bottom of the sea.

octopus

oct-tree *Computer Science.* a method for representing three-dimensional space by successively breaking up a cubic region of space into eight cubes half its size, which are made its children in a tree structure.
octupole *Physics.* an arrangement of four electric dipoles or four magnetic dipoles arranged to give a system with zero net-dipole moment and zero net-quadrupole moment.
OCTV open-circuit television.
octyl- *Organic Chemistry.* a prefix used to denote the alkane group, C_8H_{17}–.
octyl aldehyde *Organic Chemistry.* $CH_3(CH_2)_6CHO$, a combustible, colorless liquid with a strong fruity odor; soluble in 70% alcohol and mineral oil and insoluble in glycerol; boils at 163°C; used to produce perfumes and flavors. Also, 1-OCTANAL, CAPRYLIC ALDEHYDE.
***m*-octyl bromide** see 1-BROMOOCTANE.
octyl carbinol see *n*-NONYL ALCOHOL.
***n*-octyl *n*-decyl adipate** *Organic Chemistry.* a combustible liquid with a boiling range of 250–254°C (5 torr); used as a low-temperature plasticizer. Also, NODA.
***n*-octyl *n*-decyl phthalate** *Organic Chemistry.* a clear combustible liquid with a boiling range of 232–267°C; used as a plasticizer for vinyl resins.
1-octylene see 1-OCTENE.
octyl formate *Organic Chemistry.* $C_8H_{17}OOCH$, a combustible, colorless liquid with a fruity odor; soluble in mineral oil and insoluble in glycerol; boils at 199°C; used as a flavoring agent.
octyl group see CAPRYL GROUP.
***n*-octyl mercaptan** *Organic Chemistry.* $CH_3(CH_2)_6CH_2SH$, a whitish combustible liquid that boils at 199°C and is soluble in alcohol; used in synthesis and as a polymerization conditioner.
octyl phenol *Organic Chemistry.* $C_8H_{17}C_6H_4OH$, combustible white flakes; insoluble in hot and cold water and soluble in alcohol; melts at 72–74°C and boils at 280–302°C; used in fungicides, dyestuffs, surfactants, plasticizers, and adhesives.
1-octyne *Organic Chemistry.* $CH_3(CH_2)_5CCH$, a colorless liquid; soluble in alcohol and ether; boils at 126°C. Also, CAPRILYDENE, HEXYLACETYLENE.
ocul- or **oculo-** a prefix meaning "eye."
ocular *Biology.* of or relating to the eye; eyelike. *Optics.* see EYEPIECE.
ocular lobe *Invertebrate Zoology.* a lobe in some beetles that projects from the thorax.
ocular prism *Optics.* a prism that angles the line of sight from a rangefinder into an eyepiece.
ocular skeleton *Vertebrate Zoology.* a rigid cartilaginous structure around the eyes of most submammalian vertebrates that includes a ring of intramembrous bones at the junction of the cornea and sclera.
oculist *Medicine.* a former term for an ophthalmologist or optometrist.
oculoglandular tularemia *Medicine.* an infectious disease that originates in rodents and rabbits and that can spread to humans by inoculation of the conjunctival sac or ocular contact with aerosols infected with *Francisella tularensis;* the symptoms include fever, chills, myalgia, headache, extremely painful conjunctivitis, photophobia, a mucopurulent discharge, local granulomatous lesions, and preauricular lymphadenopathy.
oculomotor *Physiology.* relating to or affecting the motion of the eye.
oculomotor nerve *Anatomy.* one of the two cranial nerves that control the actions of the eye muscles.
oculomotor nucleus *Anatomy.* the origin of the fibers of the oculomotor nerve, located in the rostral midbrain in the most ventral portion of the central gray matter.
oculomotor paralysis *Medicine.* a paralysis of the muscles that are controlled by the third cranial nerve.
oculonasal *Anatomy.* of or relating to the eyes and nose.
Oculosida *Invertebrate Zoology.* an order of protozoans in the subclass Radiolaria, with pores restricted to certain areas in the central capsule and an olive-colored material near the astropyle.
oculus *Science.* the Latin word for "eye."
OD right eye. (From Latin *oculus dexter.*)
OD or **O.D.** doctor of optometry; optical density; outside diameter; outside dimension; overdose; ordnance department; once a day.
O.D.A. right occipito-anterior, a position of the fetus. (From Latin *occipito-dextra anterior.*)
odaxetic *Neurology.* producing a biting or itching sensation.
odd-even nuclei *Nuclear Physics.* a group of 50 stable nuclei that have an odd number of protons and an even number of neutrons; unpaired protons and neutrons give half-integral (1/2, 3/2, 5/2, etc.) nuclear spin.

odd function *Mathematics.* a function f such that $f(-x) = -f(x)$ for all x in the domain of definition for f.

odd-leg calipers *Mechanical Devices.* a shoulder distance-measuring instrument with two hinged legs oriented in the same direction, one leg is pointed, and the other acts as a toe that runs against the edge of the workpiece.

odd number *Mathematics.* any integer that is not divisible by 2.

odd-odd nucleus *Nuclear Physics.* a group of 4 stable nuclei that have an odd number of protons and an odd number of neutrons; one proton and one neutron, each with one-half spin, remaining after pairing of the other protons and neutrons result in integral (l, 3) spin.

odd pages *Graphic Arts.* the recto (right-hand) pages in a book, numbered 1, 3, 5, 7, and so on.

odd parity *Computer Technology.* a binary representation in which the total number of 1's is odd; the parity or check bit is set to 1 if the basic code contains an even number of 1's, and to 0 if an odd number. *Quantum Mechanics.* see PARITY.

odd parity check *Computer Technology.* a simple error-detection scheme in which the total number of 1 bits in a code combination, including the extra parity bit, must be an odd number.

odd-pinnate *Botany.* of a compound leaf, having an extra unpaired leaflet at the tip of a pinnate petiole. Also, IMPARIPINNATE.

odds ratio *Statistics.* the ratio of the probability of the occurrence of an event to that of the occurrence of its complement.

odds-ratio estimator *Statistics.* in a 2×2 contingency table, the ratio between the two odd ratios in each row; can be regarded as a measure of the association.

odd term *Atomic Physics.* an atom or molecule in which the sum of all the electrons' angular-momentum quantum numbers produces an odd number.

Odiniidae *Invertebrate Zoology.* a family of flies, dipteran insects in the subsection Acalyptratae.

odinite *Petrology.* a greenish-gray lamprophyre occurring in dikes, consisting of phenocrysts of labradorite, augite, and sometimes hornblende within a groundmass of plagioclase and hornblende rods.

O display *Electronics.* a modified radar display that has an adjustable notch for measuring a target's distance.

Odium *Mycology.* an obsolete genus of mildew fungi believed to cause such diseases as dermatomycosis and candidosis.

Odobenidae *Vertebrate Zoology.* the walrus, a monospecific family in some classification systems placed in the suborder Pinnipedia, characterized by a massive, rounded body, reversible flippers, prominent whiskers above the upper lip, and two canine teeth forming tusks; found in Arctic waters.

odograph *Engineering.* an instrument installed in a vehicle that plots its route and mileage on a map.

odometer *Engineering.* **1.** an instrument for measuring the distance traveled by a vehicle or person. **2.** specifically, such an instrument in a motor vehicle to measure the number of miles or kilometers the vehicle has traveled.

Odonata *Invertebrate Zoology.* the dragonflies and damselflies, an order of predaceous hemimetabolous insects having slender bodies, biting mouthparts, large compound eyes, and two pairs of narrow, transparent, net-veined wings; nymphs are aquatic.

odont-, odonto- a prefix meaning "tooth."

odontalgia *Medicine.* pain in a tooth.

Odontaspididae *Vertebrate Zoology.* the sand tiger sharks, a family of harmless, sluggish, benthic sharks of the order Lamniformes, characterized by a flattened, conical head and a large, long mouth; found worldwide in tropical and warm-temperate waters.

odontectomy *Medicine.* the removal of a tooth by the reflection of a mucoperiosteal flap and excision of bone from around the root before pulling the tooth.

odontexesis *Medicine.* the scaling and polishing of the teeth.

odontoblast *Developmental Biology.* a cell in the dental papilla that secretes dentine in the developing tooth.

odontobothritis see ALVEOLITIS.

Odontoceta *Vertebrate Zoology.* a suborder of toothed cetaceans of the order Cetacea, including whales, dolphins, and porpoises, characterized by an asymmetric skull and a single blowhole.

Odontoceti *Vertebrate Zoology.* in some classification systems, a suborder of toothed whales that together with the baleen whales (Mysticeti) compose the single whale order Cetacea.

odontoclast *Histology.* a multinuclear cell that is active in reabsorbing the roots of deciduous teeth.

odontogenesis *Developmental Biology.* the development and formation of the teeth.

odontogenic *Histology.* **1.** of or relating to to the origin and development of teeth. **2.** having an origin in tissues associated with teeth.

odontogenic cyst *Medicine.* a cyst that develops from odontogenic epithelium.

odontogenic fibroma *Medicine.* a fibroma in the jaws that contains strands or islands of epithelial cells. Also, FIBROUS ODONTOMA.

Odontognathae *Paleontology.* a small superorder of toothed birds of the Middle and Upper Cretaceous in the subclass Neornithes; the superorder comprises six genera in two orders, the Hesperornithiformes and the Ichthyornithiformes, but is known from relatively few fossil specimens; the group is especially important as containing almost the only birds now represented in the fossil record between Ambiortus in the Lower Cretaceous and the more numerous birds of the Paleocene.

odontoid *Biology.* toothlike, such as the **odontoid peg,** a toothlike process that acts as a pivot for rotation of the head.

odontoid process *Anatomy.* the small toothlike process of the second cervical vertebra (the axis) around which the first vertebra (the atlas) revolves. Also, DENS.

odontology *Medicine.* another name for dentistry.

odontophobia *Psychology.* **1.** an irrational fear of teeth, especially the teeth of an animal. **2.** an irrational fear of having dental work performed on one's teeth.

odontophore *Invertebrate Zoology.* a structure supporting the radula in the mouth of most mollusks, except bivalves.

Odontophorinae *Vertebrate Zoology.* a subfamily of Phasianidae (in some schemes), including the 65 species of New World quail; distinguished by a resemblance to the Old World partridge but having a larger body, a strong, serrated bill, and no leg spurs.

Odontostomatida *Invertebrate Zoology.* an order of protozoans in the subclass Spirotrichia, with laterally compressed bodies and reduced ciliation.

odor *Neurology.* an emanation that is perceived by the sense of smell.

odorant *Neurology.* any substance capable of stimulating the sense of smell.

odoriferous homing *Electronics.* a system for navigating a submarine that employs ionized air produced by its exhaust gases.

odorize *Chemical Engineering.* to add scent, usually undesirable, to an otherwise odorless substance for purposes of detection.

odorography *Neurology.* a written description of odors or odorants.

odorous *Neurology.* yielding or diffusing an odor. Also, **odoriferous.**

Oe oersted.

Oecophoridae *Invertebrate Zoology.* a family of small inconspicuous moths, lepidopteran insects in the superfamily Tineoidea, with a comb of bristles on the antennae; larvae feed on leaves and flowers.

oedema see EDEMA.

Oedemeridae *Invertebrate Zoology.* a family of beetles, the false blister beetles, coleopteran insects in the superfamily Tenbrionoidea; soft-bodied, bright-colored adults frequent flowers; larvae feed on decaying wood, damaging damp timbers of wharves, bridges, and mines.

Oedipal or **oedipal** [ed′ə pəl] *Psychology.* relating to or resulting from the Oedipus complex. Also, **Oedipean.**

Oedipal complex see OEDIPUS COMPLEX.

Oedipal conflict *Psychology.* in Freudian theory, the conflict that a boy feels in early childhood when a desire to eliminate his father as a rival for his mother conflicts with fear of his father.

Oedipodiaceae *Botany.* a monotypic family of rare, dull-green alpine mosses of the order Splachnales, characterized by simple stems with terminal sporophytes, spreading leaves that form rosettes and become brittle when dry, and basal leaf cilia.

Oedipus complex [ed′ə pəs] *Psychology.* **1.** in psychoanalytic theory, a son's unconscious sexual love for his mother and his subsequent hostility and jealousy toward his father. **2.** a similar attraction of a daughter to her father and rivalry with her mother; more specifically called the ELECTRA COMPLEX. (From the Greek myth of *Oedipus,* who unwittingly killed his father and married his mother.)

Oedogoniales *Botany.* an order of freshwater algae in the division Chlorophyta, consisting of the single family Oedogoniaceae and characterized by simple or branched filaments with a basal holdfast cell.

Oegophiurida *Invertebrate Zoology.* an order of echinoderms in the subclass Ophiuroidea; the single living genus has few external skeletal plates.

Oegopsida *Invertebrate Zoology.* an order of cephalopod mollusks with no transparent corneal membrane over the eye.

O electron *Atomic Physics.* one of the electrons that is found in the *O* shell of an atom and displays the characteristics of the principal quantum number 5.
OEM original equipment manufacturer.
oenocyte *Invertebrate Zoology.* a specialized cell found in the fat body of insects; stores nutritive reserves and is important in certain phases of metabolism.
Oenothera *Botany.* the evening primrose, a genus of mostly North American biennial or annual herbs in the family Onagraceae, characterized by nocturnal yellow or white flowers.
Oepikellacea *Paleontology.* a superfamily of Paleozoic dimorphic ostracods in the extinct order Palaeocopida; characterized by biconvex shells without a sulcus.
Oerlikon *Ordnance.* **1.** a Swiss antiaircraft gun, including the twin-cannon 35-mm, 30-mm, and 20-mm guns and single-cannon 25-mm and 20-mm guns; used on ships and aircraft as well as on land. **2.** a Swiss 81-mm multiple-rocket system designed to launch two parallel banks of fifteen rockets each.
Oerskovia *Bacteriology.* a genus of actinomycete soil bacteria that form fragmenting mycelia and metabolize glucose either oxidatively or fermentatively.
Oersted, Hans C. 1777–1851, Danish physicist and chemist; a pioneer in the study of electromagnetism; produced the first aluminum.
Oersted or **oersted** *Electromagnetism.* a centimeter-gram-second unit that represents magnetic field intensity, equivalent to the intensity at the center of a single-turn circular (1-cm radius) coil with a current of $1/2\pi$ abamp; for example, the intensity of the earth's magnetic field is in the range of 0.25 to 0.70 oersted. (Named for Hans C. *Oersted.*)
OES Office of Endangered Species.
Oestridae *Invertebrate Zoology.* botflies, a family of two-winged, beelike flies in the subsection Calyptratae, having larvae parasitic in the nasal cavity of domestic animals and deer.
off-airways *Aviation.* any navigable airspace lying outside the boundaries of established airways.
off-carriage fire control *Ordnance.* a method of aiming a weapon with a sighting device, such as a director or aiming circle, that is not mounted on the weapon or its carriage. Also, **off-carriage fire control equipment.**
offense against the sine condition *Optics.* a measure of coma obtained by dividing the sagittal coma by the distance of the perpendicular, extending from the image point to the optical axis.
offensive grenade *Ordnance.* a grenade in a light, usually nonmetallic cover that is designed to inflict casualties through its blast and concussion effect rather than through fragmentation, thus allowing the thrower to continue advancing toward his target.
offensive mine countermeasures *Military Science.* measures intended to prevent the enemy from successfully laying mines.
offensive minefield *Ordnance.* a minefield laid in enemy territorial waters or waters under enemy control.
off-highway truck *Mining Engineering.* a truck whose size or weight bars it from use on public highways.
off-hook *Telecommunications.* **1.** in telephony, describing the condition that exists when a handset or receiver is taken off its switch. **2.** of or relating to the closed-loop or active state of a subscriber or PBX user's loop. **3.** one of two possible signaling states, as in ground connection versus battery connection, tone versus no tone.
off-hook service *Telecommunications.* in telephony, the automatic establishment of a connection between a caller and a receiver as a result of lifting the handset off its cradle.
office automation *Computer Technology.* any application of computers and communications technology designed to improve the productivity of clerical and managerial workers in an office environment.
offing *Navigation.* the deep part of the ocean that is visible from shore.
offlap *Geology.* an arrangement of sedimentary layers in a depositional basin in which each successively younger layer exposes a portion of the older layer beneath it. The updip edges of sedimentary units progress offshore. Also, REGRESSIVE OVERLAP.
off-line *Industrial Engineering.* **1.** of or relating to a work process that takes place off the main assembly line. **2.** of or relating to an electronic communication device that does not function in "real time" operation. *Computer Technology.* of equipment or operations, not in direct communication with a central processor. Thus, **off-line operation, off-line unit, off-line processing, off-line storage.** *Engineering.* of or relating to a borehole or other linear excavation that has deviated from the intended path. Also, **off line, offline.**
off-line analysis *Biotechnology.* the evaluation of a process that requires removal of a sample from the reaction to an analytical instrument for measurement; some inaccuracy in results can occur due to the disturbance of the reaction.
off-line cipher *Telecommunications.* an encryption that is not connected with any particular transmission system.
off-line equipment see AUXILIARY EQUIPMENT.
off-line mode *Computer Programming.* an operation that relates to, but is not in direct communication with, a central processor, such as reading tape input or printing output.
off-line programming *Design Engineering.* the process of developing a program for a machine remotely from the machine. *Robotics.* specifically, programming that is done away from a robot being programmed, using simulations and collections of data.
off-line programming control *Robotics.* a control in which the data for the path of motion has been developed at another location and stored in the robot's memory.
off-lining *Computer Technology.* the process of moving some operations, such as input and output functions, to peripheral equipment that is not in direct communication with a central processor.
offload *Computer Technology.* **1.** to move a job from one computer to a smaller or less-busy computer. **2.** to output data from the host computer to a peripheral device for printing or other final disposition.
off-page connector *Computer Programming.* in flow charting, a symbol used to link a sequence of logic from one page to its continuation on another page.
off-peak *Science.* not at a maximum; not occurring during a peak period.
offprint *Graphic Arts.* **1.** the reprint of an article that originally appeared as part of a larger publication. **2.** to reprint as a separate publication, as an article from a larger publication.
off-reef facies *Geology.* the inclined strata composed of reef detritus that is deposited along the seaward margin of a reef.
offset set at an angle or as a compensation or counterbalance; specific uses include: *Botany.* **1.** a short lateral shoot off an axillary bud that produces a daughter plant at its apex. **2.** a small bulb located near the base of a parent bulb. *Geology.* **1.** the horizontal displacement of a fault, measured perpendicular to the strike of the fault. **2.** a minor branch or spur of a mountain range. **3.** a level terrace on the side of a hill. *Mechanics.* the amount of strain between the initial linear portion of a stress-strain curve and a line parallel to that portion intersecting the curve at a given stress level; it provides a method of defining the yield stress. *Graphic Arts.* **1.** of ink or an image, transferred from one surface to another. **2.** see OFFSET PRINTING. *Computer Programming.* a linear displacement of the addresses compiled in a program relative to the base of the address space reserved for program execution. *Control Systems.* **1.** to substitute one value for another in order to improve the integrity or functioning of a system, as when a 4-mA signal is used to represent zero in a 4–20-mA system. **2.** the value used in such an operation. **3.** in a process control system, the difference between the desired and actual steady-state control points. *Mining Engineering.* a short crosscut or drift that is cut from a main gangway or level. *Naval Architecture.* a construction measurement made at an angle upward and outward from a vessel's baseline. *Ordnance.* the horizontal, forward distance that a missile travels after it hits the ground; measured from the center of the entry hole to the most forward point of the missile. *Cartography.* **1.** in surveying, a short line, extending from and perpendicular to a surveyed line, to a point or another line, thus locating the point or second line in relation to the surveyed line. **2.** in the construction of a map projection, a short extension of the meridians beyond the top and bottom latitude, made in order to place that latitude clearly in the body of the map rather than ambiguously on the border.
offset bombing *Military Science.* a bombing procedure that employs a reference or aiming point other than the actual target.
offset branching *Robotics.* the ability of a robot system to change direction at various points in the sequence of operations.
offset cab *Engineering.* a heavy equipment cab positioned off center in order to improve the visibility of the operator.
offset cutter *Mechanical Devices.* a cutting tool whose cutter is offset from the shank.
offset cylinder *Mechanical Engineering.* a reciprocating part with a crank that rotates around a center off the centerline.
offset deposit *Geology.* a mineral deposit formed partly by magmatic segregation and partly by hydrothermal solution injected into the country rock near the source rock.

offset distance *Ordnance.* the distance by which the desired ground zero or actual ground zero is offset from the center of an area target or from a point target.
offset drilling *Petroleum Engineering.* the drilling of a well on one tract of land to compensate for or prevent the loss of oil or gas due to the production of a well on adjacent land.
offset line *Engineering.* a line established parallel to and not far from a main survey line.
offset plotting *Ordnance.* a method for plotting firing data when each gun of a battery must operate with a different range and azimuth.
offset printing *Graphic Arts.* a widely used modern printing process in which a printing plate on one cylinder makes an impression on a second cylinder covered by a rubber blanket; the impression on the rubber is then transferred to paper carried by a third cylinder. Thus the printing plate does not make a direct impression on the paper as in earlier forms of lithography. Also, **offset lithography.**
offset ridge *Geology.* a discontinuous ridge of sedimentary rock whose strata have been displaced by faulting.
offset screwdriver *Mechanical Devices.* a double-headed tool consisting of a hexagonal or round steel bar with the ends bent at right angles and ending in a screwdriver tip; used for driving cross-head or slotted-head screws that are inaccessible to a standard screwdriver.
offset stream *Hydrology.* a stream that has been displaced laterally or vertically as a result of faulting.
offset yield strength *Mechanics.* the stress level at which a parallel line, at a specific offset to the initial linear portion of the stress-strain curve, intersects with the stress-strain curve.
offshore *Geology.* **1.** of or related to the comparatively flat, completely submerged zone that extends from the low-tide mark to the seaward edge of the continental shelf. **2.** the direction that is seaward from the shore. *Petroleum Engineering.* of or relating to a drilling or recovery operation in ocean waters in the relatively flat area between the outer edge of the shoreface and the edge of the continental shelf.
offshore bar see LONGSHORE BAR.
offshore drilling *Petroleum Engineering.* the drilling of wells on a submerged location, especially a continental shelf.

offshore drilling rig

offshore gas *Petroleum Engineering.* the natural gas extracted from continental shelves.
offshore slope *Geology.* the frontal slope below the outer edge of an offshore terrace.
offshore terrace *Geology.* a wave-built terrace of coarse sand and gravel that occurs in the offshore region.
offshore water *Oceanography.* the sea between the breaker zone and the edge of the continental shelf.
off-site facility *Chemical Engineering.* a process that is separate from a production facility, but is used to support the facility.
off soundings *Navigation.* water that is deeper than can be measured with an ordinary sounding lead, generally beyond the 100-fathom line.
off time *Metallurgy.* in a resistance welding cycle, a term for the time during which the electric current does not flow.
OFHC oxygen-free high-conductivity copper.
Of star *Astronomy.* an O star with a spectral class lower than O5; appears to be the largest and hottest star known.
Ogcocephalidae *Vertebrate Zoology.* the batfishes, a marine fish of the order Lophiiformes, characterized by a depressed body with a stout tail and winglike processes; found in all major oceans.
ogdosymmetric class *Crystallography.* a class of crystals having one-eighth the symmetry expected of a crystal in the same system.
ogee [ō´jē´] *Architecture.* a double curve resembling the letter S.
ogee arch *Architecture.* a pointed arch formed by a reversed pair of ogee curves. Also, KEEL ARCH.
ogee molding. a molding having the shape of an ogee.
ogive [ō´jīv´] *Architecture.* a pointed arch. *Geology.* one of a series of dark, curved structures convex downslope and repeating periodically down the surface of a glacier. *Ordnance.* the curved or tapered forward part of a projectile, including the nose; however, a nose fuse is not considered part of the ogive.
OH the chemical symbol for hydroxyl group; hydroxide ion; a hydroxide.
OH the laboratory designation for an Olduvai hominid.
OH/IR star *Astronomy.* a very cool, very large supergiant star that is losing mass rapidly; such stars show a strong infrared excess and an OH (hydroxyl) maser.
Ohm, Georg S. 1787–1854, German physicist; discovered Ohm's law of electric currents.
ohm [ōm] *Electricity.* a standard unit of electrical resistance, equal to the resistance of a circuit in which an electromotive force of one volt will maintain a current of one ampere; it has been defined as the resistance to a flow of steady current that will be offered by a thread of mercury 14.4521 grams in mass with a length of 1.063 meters and an unvarying cross-sectional area of one square millimeter, at a constant temperature of 0°C. (Named for Georg S. *Ohm.*)
ohmic *Electricity.* of or relating to ohms, or to a material that operates under the conditions stated in Ohm's law.
ohmic contact *Electricity.* a purely resistive contact between two surfaces or materials.
ohmic dissipation *Electronics.* a loss of energy in the form of heat when a current passes through an electrical resistance. Also, **ohmic loss.**
ohmic resistance *Electricity.* resistance to direct current as a result of properties of a particular material, circuit, or device that obeys Ohm's law.
ohmmeter *Engineering.* a meter that measures electrical resistance in a circuit, as expressed in ohms. Some types of ohmmeters also measure current flow.
Ohm's law *Electricity.* a law stating that the voltage across an element of a DC circuit is equal to the current through the element multiplied by the resistance of the element, in ohms; the current I (in amperes) is directly proportional to the total voltage E in the circuit, and inversely proportional to the total resistance R of the circuit; it is expressed in three forms: $E = IR$; $I = E/R$; $R = E/I$.
ohms per volt *Engineering.* a unit of measurement indicating the sensitivity of a meter; found by dividing the electrical resistance of the instrument, expressed in ohms, by the voltage.
Ohno's law *Genetics.* a law stating that the X-inactivation mechanism has preserved the ancestral X chromosome almost intact throughout the evolution of mammals; the inactivation of one X chromosome compensates for the different dosages of X-linked alleles in males and females, ensuring genic balance in both sexes.
-oic *Organic Chemistry.* a suffix designating a compound that contains a carboxyl group or a derivative of a carboxyl group.
-oid a suffix meaning "resembling."
oidiophore *Mycology.* a fungal filament or hypha that produces spores called oidia.
oidium *plural,* **oidia.** *Mycology.* a spore cell found in certain fungi belonging to the division Eumycota that is formed by the breaking up of hypha into cells.
oikocryst *Petrology.* an enclosing crystal in a poikilitic fabric.
Oikomonadidae *Invertebrate Zoology.* a family of protozoans in the order Kinetoplastida, having a single flagellum.
oikophobia *Psychology.* an irrational fear of one's own home or of being at home.

Con
& Infor
Tech

rs conducting gold-colored light.
chers.

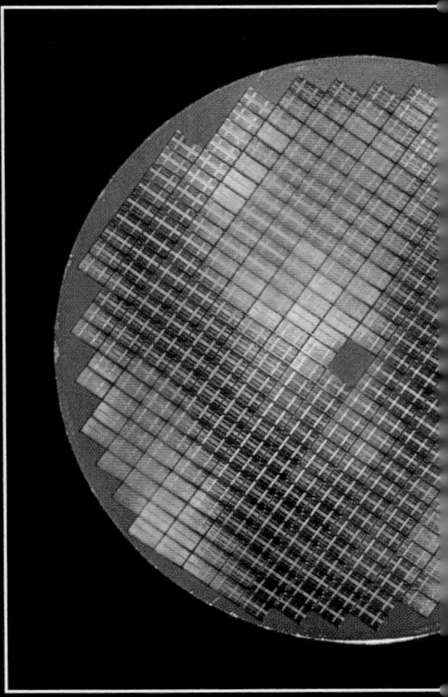

Wafer containing million-bit memory chip
Courtesy of International Business Machines C

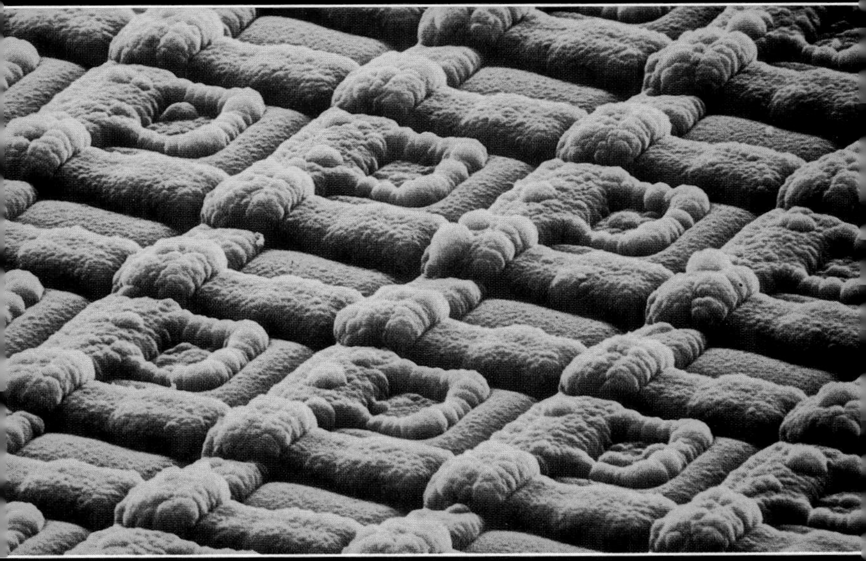

False-color scanning electron micrograph of the surface architecture of an integrated

Spectrum created by light diffracting off the numerous tiny pits in the surface of a compact disc.

n of the surface

oil *Geology.* another name for petroleum. See PETROLEUM. *Materials.* a general term for any of a wide variety of greasy, viscous, combustible substances that are liquid at room temperature or when slightly warmed and insoluble in water; such substances may be derived from animal, vegetable, or mineral sources. *Mechanical Engineering.* specifically, a petroleum product used as a lubricant in an internal-combustion engine. (Originally derived from the Greek word for the *olive* tree, because of the widespread use of olive oil in the ancient world.)

oil accumulation see OIL POOL.

oil-base mud *Petroleum Engineering.* drilling mud containing oil as the liquid component; used to drill through clay formations or in very deep wells in which bottom-hole temperatures may reach 300–400°F.

oil bath *Engineering.* **1.** a reservoir of oil in which moving parts of a machine or engine are immersed in order to provide them with constant lubrication. **2.** a vat of very hot oil used to temper steel. **3.** the pouring of oil on a device requiring lubrication.

oil beetle *Invertebrate Zoology.* any of the blister beetles of the genus *Meloe,* characterized by the emission of an oily fluid from its leg joints when disturbed.

oilbird *Vertebrate Zoology.* a nocturnal, fruit-eating bird, *Steatornis caripensis,* of the family Steatornithidae, distinguished by an oil that is derived from the fat of the young bird; found in South America. Also, GUACHARO.

oil break *Electricity.* any separation of contacts in oil.

oil buffer *Ordnance.* a device on certain automatic weapons that is designed to control firing speed and minimize the effect of recoil; used especially on 0.50-caliber guns.

oil burner *Mechanical Engineering.* any device that burns fuel oil, especially a home heating device of this type.

oil cake *Materials.* a cake of linseed, cottonseed, or soybean from which the oil has been extracted; used as food for livestock.

oil circuit breaker *Electricity.* a circuit breaker that is enveloped by an insulating oil to suppress the arc that is produced when a circuit is opened and to prevent damage to the contacts, as well as to provide cooling. Also, OIL SWITCH.

oil column *Geology.* in an oil-producing formation, the difference in elevation between the highest and lowest producing zones.

oil cooler *Mechanical Engineering.* **1.** a small air-cooled automotive radiator that cools the lubricant after return from the engine and before delivery to the oil tank. **2.** any heat exchanger or cooling unit used in an oil or hydrocarbon process.

oil cup *Engineering.* a vessel permanently affixed near a gear or other device needing lubrication, in order to provide a constant and uniform application of oil.

oil cut *Petroleum Engineering.* a term for oil mixed with drilling mud, extracted during exploration.

oil derrick see DERRICK.

oil dilution valve *Mechanical Engineering.* a valve that allows the dilution of lubricant by mixing it with fuel when the engine is off, facilitating cold weather starts.

oil extension *Materials Science.* the addition of hydrocarbon oils to rubber; commercially used to plasticize and soften the rubber and reduce its melt viscosity.

oilfield *Petroleum Engineering.* an area of land drilled for purposes of recovering petroleum, with boundaries corresponding to an oil pool or described by political or legal limits.

oil-field brine *Hydrology.* water having a high concentration of calcium and sodium salts; found in rocks during deep oil drilling.

oil-field emulsion *Petroleum Engineering.* a crude oil extracted from a well as an emulsion of oil and water.

oil-filled *Electricity.* describing a component or device that is filled with an insulating oil.

oil-filled cable *Electricity.* a cable that is impregnated with an insulating oil; usually includes longitudinal ducts with channels or reservoirs in order to maintain a positive oil pressure within the cable at all times.

oil filter *Engineering.* a device used in the oil lubrication system of an engine in order to remove grit, carbon, and other impurities from the oil passing through it.

oil floor *Geology.* the depth below which there is no economically useful oil accumulation in a sedimentary basin.

oil fogging *Engineering.* the spraying of a fine mist of oil in the gas stream of distribution or utilization equipment, in order to counteract the drying effects of gasoline on the equipment in the system.

oil furnace *Mechanical Engineering.* a combustion chamber that uses oil as the heat-producing fuel.

oil-gas process *Chemical Engineering.* a distillation process used to obtain a light gas-oil from petroleum.

oil gland see SEBACEOUS GLAND.

oil groove *Design Engineering.* one of the grooves in a bearing that collects and distributes oil.

oil hardening *Metallurgy.* the process of hardening a steel by quenching in an oil bath.

oil hole *Engineering.* a small hole through which oil is introduced to a bearing.

oil-hole drill *Mechanical Devices.* a twist drill that contains holes through which oil is released to lubricate the cutting edges.

oil horizon *Petroleum Engineering.* in a well, the upper oil surface or the stratum in which it is found.

oil-immersed *Electricity.* of a transformer, reactor, regulator, or other device, having the coils immersed in an insulating liquid such as oil or another viscous fluid.

oil-immersion objective see IMMERSION OBJECTIVE.

oil isoperms *Petroleum Engineering.* on an oil reservoir map, areas of equal oil permeability.

oilless bearing *Mechanical Engineering.* a self-lubricating bearing.

oil lift *Mechanical Engineering.* the hydrostatic lubrication of a journal bearing with oil at high pressure between the bottom of the journal and the bearing so that the shaft is raised and supported by an oil film.

oil of bitter almond see ALMOND OIL.

oil of cloves see CLOVE OIL.

oil of coriander see CORIANDER OIL.

oil of lavendar *Materials.* an oil distilled from the lavendar flowers, *Lavandula angustifolia* and *L. stoechas;* used in perfumes.

oil of mirbane see NITROBENZENE.

oil of peppermint see PEPPERMINT OIL.

oil of sage see SAGE OIL.

oil of sassafras see SAGE OIL.

oil of turpentine see TURPENTINE, def. 3.

oil of vitriol see SULFURIC ACID.

oil paint *Materials.* a commercial paint in which the base is an oil.

oil pool *Geology.* an economically useful accumulation of petroleum locally confined by subsurface geologic features.

oil pump *Mechanical Engineering.* a small auxiliary pump, driven from an engine crankshaft, that forces oil from the sump or oil tank to the bearings.

oil reclaiming *Engineering.* **1.** a continuous reprocessing and refiltering of oil through an oil lubrication system. **2.** an oil-purifying procedure in which oil is placed in settling tanks to remove solids and other undesired materials.

oil reservoir see OIL POOL.

oil ring *Mechanical Engineering.* a collar located in the lower part of a piston that prevents excess oil from being drawn above the piston during the suction stroke.

oil rock *Geology.* a stratum of rock that contains oil.

oil sand *Geology.* any porous stratum, unconsolidated, porous sand formation, or sandstone rock that contains petroleum or hydrocarbons, or from which oil can be obtained by drilling.

oil seal *Engineering.* **1.** a device that prevents the return of oil into one part of a system once it passes to another part. **2.** the use of oil as a seal to prevent the seepage of other fluids into a chamber.

oilseed *Materials.* any of several seeds, such as the castor bean, cottonseed, or sesame, from which an oil is expressed.

oil seep *Geology.* a naturally occurring emergence of liquid petroleum at the surface as a result of a slow, upward migration from its source through minute pores or fissures. Also, PETROLEUM SEEPAGE, SEEPAGE.

oil shale *Geology.* a finely layered brown, black, or gray kerogen-bearing shale from which liquid or gaseous hydrocarbons can be distilled. Also, KEROGEN SHALE.

oilstone *Materials.* a block of oiled, fine-grained stone that is used to put a keen edge on certain cutting tools.

oil switch see OIL CIRCUIT BREAKER.

oil tanker *Naval Architecture.* a vessel designed to carry petroleum oil products, especially crude oil. Very large oil tankers are by a wide margin the largest vessels ever built.

oil trap *Geology.* an undergound accumulation of petroleum that cannot escape from its reservoir rock. Also, TRAP.

oil varnish *Materials.* a flammable, organic protective coating that may be composed of a vegetable oil, such as linseed or tung, and solvent, or of a synthetic or natural resin and solvent; it is similar to a paint but does not contain a colorant.

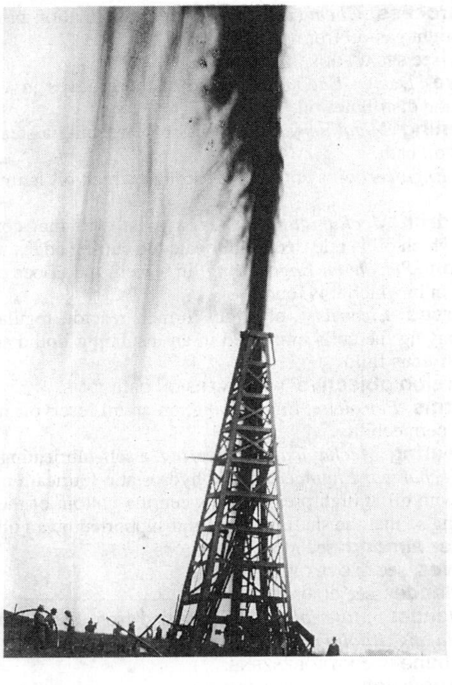

oil well

oil well *Petroleum Engineering.* a hole, usually vertical, bored into a reservoir to extract oil from porous formations.

O indicator see O SCOPE.

ointment *Pharmacology.* a semisolid preparation, usually containing medication, to be applied externally to the body.

oiticica oil *Materials.* a light-yellow drying oil expressed from the seeds of the oiticica tree, *Licania rigida,* of the rose family; used as a vehicle for paints and varnishes.

okaite *Petrology.* a feldspathoidal, ultramafic igneous rock consisting of hauyne and melilite and containing accessory biotite, perovskite, apatite, calcite, and opaque oxides.

okapi *Vertebrate Zoology.* a hoofed, even-toed ungulate mammal, *Okapia johnstoni,* of the family Bovidae, closely related to the giraffe, though smaller and having only a moderately long neck; characterized by a purplish, deep red, or near-black color, and horizontally striped rear limbs and buttocks; found in deep, dense forests of the Congo River Valley; unknown to zoologists until the 20th century.

okapi

Okazaki fragment *Molecular Biology.* a short segment of newly synthesized DNA found on the lagging strand of DNA during discontinuous DNA replication.

OKC *Aviation.* the airport code for Will Rogers World Airport, Oklahoma City, Oklahoma.

okenite *Mineralogy.* $Ca_{10}Si_{18}O_{46}\cdot 18H_2O$, a whitish triclinic mineral occurring as aggregates of acicular crystals or in compact to fibrous masses, having a specific gravity of 2.28 to 2.33 and a hardness of 4.5 to 5 on the Mohs scale; found in amygdules in basalt or related eruptive rocks.

Okhotsk, Sea of *Geography.* an arm of the western North Pacific between the Kamchatka Peninsula and the eastern Siberian mainland, covering approximately 582,000 square miles.

okra *Botany.* a tall annual plant, *Hibiscus esculentus,* of the family Malvaceae; native to Africa and now cultivated elsewhere, as in the southern United States, for its edible green pods which are used especially in soups. Also, GUMBO.

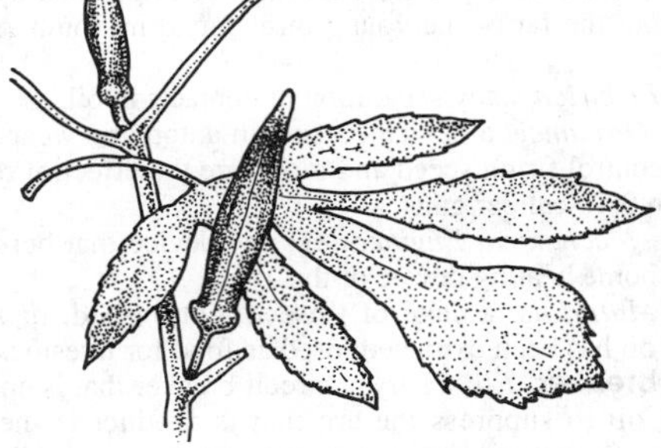

okra

OKT antibodies *Immunology.* commercially prepared monoclonal antibodies that define particular lymphocyte subsets.

okwen *Materials.* the moderately durable, mahoganylike wood of two African trees, *Brachystegia nigerica* and *B. kennedyi*; used in rough construction and temporary structures.

Ol. oil. (From Latin *oleum.*)

O.L. or **o.l.** left eye. (From Latin *oculus laevus.*)

-ol *Organic Chemistry.* a suffix designating that one or more hydroxyl groups are present in an organic compound.

O.L.A. left occipito-anterior, a position of the fetus. (From Latin *occipito-laeva anterior.*)

Olacaceae *Botany.* a family of terrestrial, mostly evergreen and tropical woody plants of the order Santales, with most species being hemiparasitic and attaching to the roots of other plants.

Olbers' paradox *Astronomy.* a paradox popularized by Heinrich Olbers in 1826, stating that if the universe consists of an infinite number of stars equally distributed through space, then every line of sight would come from a star and the night sky would be as bright as day. This is not observationally true; we receive no light from beyond a certain distance so that the night sky appears dark.

old age *Medicine.* an imprecise term for the later period of life. *Geology.* **1.** in the topographic development of a landscape or area, the last stage in the cycle of erosion, characterized by the reduction of the surface almost to base level and by landforms marked by simplicity of form and subdued relief. Also, TOPOGRAPHIC OLD AGE. **2.** in the development of a shore or coast, a stage that is characterized by a broad, wave-cut platform, a gently sloping sea cliff, and a coastal area approaching peneplanation. *Hydrology.* the developmental stage at which the efficiency of a stream is reduced, so that the load it receives is greater than it can carry, and aggradation becomes dominant.

Oldhaminidina *Paleontology.* a suborder of irregular articulate brachiopods in the order Strophomenida; extant in the Upper Carboniferous to Upper Triassic.

oldhamite *Mineralogy.* (Ca,Mn)S, a pale-brown, transparent cubic mineral, having a specific gravity of 2.58 and a hardness of 4 on the Mohs scale; found as spherules in meteorites.

Oldham's coupling see SLIDER COUPLING.

old ice *Hydrology.* a term for sea ice that has survived at least one summer's melt and shows smoother features than first-year ice.

old inflationary cosmology *Astrophysics.* an inflationary-universe cosmology that exposes a quantum-mechanical tunneling process at work; this process portends an oncoming symmetrically broken, but grand universe phase.

old lake *Hydrology.* **1.** a lake in an advanced stage of infilling by sediments or vegetation. **2.** a eutrophic or dystrophic lake. **3.** a lake whose shoreline is in an advanced stage of development.

oldland *Geology.* **1.** an extensive area of ancient crystalline rocks from which the materials of later sedimentary deposits were derived. **2.** a region of older land that supplied the material from which the coastal-plain strata were formed.

old mountain *Geology.* a mountain that was formed prior to the beginning of the Tertiary Period.

Oldowan culture *Anthropology.* the material culture present in Lower Pleistocene sites of early *Homo*; characterized by stone tools such as choppers and flakes; first found at Olduvai Gorge in East Africa.

Old Red Sandstone *Geology.* a thick sequence of nonmarine, predominantly red sandstones, conglomerates, and shales forming part of the Devonian System in Great Britain and northwestern Europe.

Oldshue-Rushton contractor *Biotechnology.* a liquid-liquid contractor that agitates its contents by means of vertical baffles attached to the walls of the apparatus.

old snow *Hydrology.* a term for deposited snow in which most of the original crystal shapes have been lost as a result of metamorphism. Also, FIRN SNOW.

Olduvai Gorge *Geography.* a canyon on the edge of the Great Rift Valley in Tanzania, East Africa, an important archaeological site at which several hominid fossils were found by the Leakey family.

Olduvai hominid *Anthropology.* any of the various hominid fossils found at Olduvai Gorge.

old workings *Mining Engineering.* a term for mines that have been abandoned, allowed to collapse, and in some instances sealed off.

Old World Europe, Africa, and Asia, as distinguished from the Americas.

Old World monkey see CATARRHINI.

Old World porcupine see PORCUPINE, def. 1.

old yellow enzyme see NADPH HYDROGENASE.

-ole a suffix meaning small.

ole- or **oleo-** a prefix meaning oil.

Oleaceae *Botany.* a family of woody plants in the order Scrophulariales, characterized by small, regular, perfect flowers having two stamens; includes the ash, privet, jasmine, olive, and lilac.

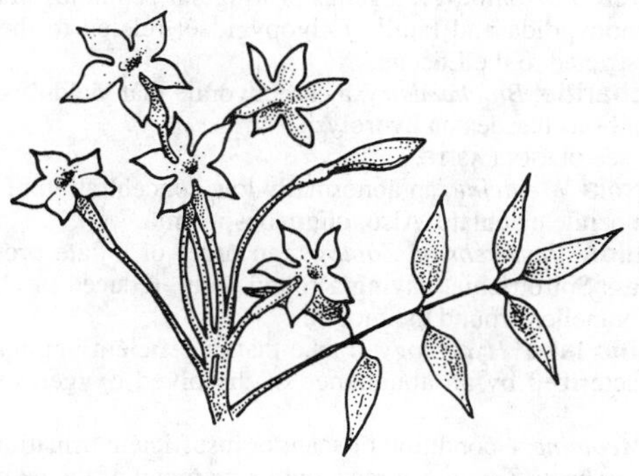

Oleaceae (jasmine)

oleander *Botany.* a poisonous shrub, *Nerium oleander,* of the dogbane family, characterized by evergreen leaves and clusters of pink, red, or white flowers; native to Eurasia, but now cultivated widely as an ornamental.

oleandomycin *Microbiology.* a macrolide antibiotic produced by the bacteria *Streptomyces antibioticus* that is similar to erythromycin and active against Gram-positive bacteria.

oleandrism *Toxicology.* poisoning due to ingestion or inhalation of the toxin contained in the oleander plant, *Nerium oleander*; symptoms may include gastrointestinal and cardiac disturbances.

oleanolic acid see CARYOPHYLLIN.

oleaster *Botany.* a shrub or small tree, *Elaeagnus angustifolia,* characterized by fragrant yellow flowers and an olivelike fruit; found in Eurasia. Also, RUSSIAN OLIVE.

oleate *Organic Chemistry.* a compound of an alkaloid or metal with oleic acid; used in externally applied medications and in soaps and paints.

olecranon *Anatomy.* the process on the proximal end of the ulna; the point of the elbow.

olefin *Organic Chemistry.* any of a class of unsaturated open-chain hydrocarbons containing one or more double bonds; obtained by cracking petroleum fractions at high temperatures; some used in the manufacture of fibers. Also, **olefine.**

α-olefin *Organic Chemistry.* an olefin that has a double bond between the two end or terminal carbons on the chain (α position).

olefin copolymer *Organic Chemistry.* a polymer containing two or more types of olefin monomers.

olefin fiber *Textiles.* any of a group of synthetic fibers formed from long-chain polymers of at least 85% ethylene, propylene, or other olefin units.

olefinic group *Materials Science.* a group of unsaturated hydrocarbons of the general formula C_nH_{2n} or the carbon-carbon double bond, including ethylene, propylene, and butene.

olefin resin *Organic Chemistry.* a polymeric material produced from olefin molecules, e.g., polyethylene from ethylene.

oleic acid *Organic Chemistry.* $CH_3(CH_2)_7CH{=}CH(CH_2)_7COOH$, a combustible, yellow-to-red oily liquid with a lardlike odor that darkens when it is exposed to air; insoluble in water and soluble in alcohol and ether; freezes at 4°C and boils at 286°C (100 torr) or at 194–195°C (1–2 torr); used in the manufacture of soap, ointments, and food-grade additives.

olein *Organic Chemistry.* $(C_{17}H_{33}COO)_3C_3H_5$, a combustible, oily, yellow triglyceride that is slightly soluble in alcohol; melts at 4–5°C; it is derived from refined natural oils and used in olive oil and textile lubricants.

Olenellidae *Paleontology.* a Cambrian family of trilobites in the order Redlichiida and the suborder Olenellina; characterized by a large, semicircular cephalon and a very small pygidium; Lower Cambrian.

Olenellus *Paleontology.* a genus of trilobites in the family Olenellidae; important in Lower Cambrian stratigraphy.

Olenus *Paleontology.* an early Paleozoic genus of trilobites in the order Ptychopariida and the stratigraphically important family Olenidae; widespread in the Upper Cambrian.

oleomargarine see MARGARINE.

oleometer *Engineering.* a device used to determine the weight and purity of an oil sample.

oleo oil *Materials.* a mixture of primarily olein and palmitin obtained from beef fat; used for making butterlike foods.

oleoresin *Materials.* the oily resinous sap of plants; most types are semisolid and tacky at room temperature and become soft and sticky at high temperatures.

oleoresinous varnish *Materials.* a varnish consisting of drying oils and resins that usually have been cooked.

oleo strut *Mechanical Engineering.* a shock absorber consisting of a telescoping structure that forces oil into an air chamber at a controlled rate, compressing the air and absorbing shock.

Olethreutidae *Invertebrate Zoology.* a family of small moths, including the codling moth and oriental peach moth, lepidopteran insects in the superfamily Tortricoidea, with a fringe of long hairs on the wings; larvae are serious plant pests.

oleum *Inorganic Chemistry.* a solution of SO_3 in concentrated H_2SO_4, called *fuming sulfuric acid.* (A Latin word meaning "oil.")

oleyl alcohol *Organic Chemistry.* $CH_3(CH_2)_7CH{=}CH(CH_2)_7CH_2OH$, a combustible, clear viscous liquid that boils at 195°C (8 torr) and freezes at 13–19°C; used in printing inks and as an antifoam agent.

oleyl aldehyde see OCTADECENYL ALDEHYDE.

olfaction *Physiology.* the means of smelling; the sensation of smell.

olfactology *Neurology.* the science of the sense of smell.

olfactometry *Neurology.* the determination of the degree of sensibility of the olfactory organ.

olfactory [ōl fak´tə rē] *Physiology.* of or relating to the sense of smell.

olfactory anesthesia see ANOSMIA.

olfactory aura *Medicine.* the disagreeable prodromal olfactory sensation preceding or symptomatic of an epileptic attack.

olfactory bulb *Vertebrate Zoology.* an outgrowth of the anterior portion of the endbrain that serves as the center for olfaction, consisting of a terminal swelling of olfactory nerve fiber endings; its size is proportionate to the acuteness of the organism's sense of smell; especially well-developed in fish and other lower vertebrates.

olfactory cell *Physiology.* one of a set of modified nerve cells distributed along the roof of the nasal cavity and specialized to produce the sensation of smell; one end extends into the nasal epithelium, where it branches into clusters of cilia, and the other end extends through the ethmoid bone into the olfactory bulb of the brain.

olfactory foramen *Anatomy.* an opening in the ethmoid bone through which the olfactory nerves pass.

olfactory gland *Physiology.* one of the numerous specialized glands located in the mucous membranes of the nose.

olfactory lobe *Vertebrate Zoology.* a modified formation projecting from the frontal lobe of each cerebral hemisphere that receives impulses from the olfactory apparatus and relays them to the diencephalon; well developed in most vertebrates, although very small in humans. Also, RHINENCEPHALON.

olfactory nerve *Anatomy.* any of a bundle of sensory nerves that conduct olfactory stimuli from the nasal cavity to the brain.

olfactory organ *Physiology.* a technical name for the nose.

olfactory pit *Developmental Biology.* the primordium of a nasal pit or cavity.

olfactory receptor *Physiology.* one of many cells in the nasal epithelium that are stimulated by substances capable of producing impulses in the olfactory passage, thus giving rise to the sense of smell.

olfactory region *Anatomy.* the upper third of the nasal cavities and nasal septum, lined with mucous membrane that contains olfactory sensory nerve endings.

olfactory stalk *Anatomy.* a structure that connects the olfactory bulb to the cerebrum.

olfactory tract *Anatomy.* a band of white matter containing olfactory nerves; projecting posteriorly from the olfactory bulb, passing under the surface of the frontal lobe of the cerebrum, and ending in olfactory regions of the brain, including the olfactory tubercle and the olfactory nucleus.

olfactory vesicle *Neurology.* a sac at the end of an olfactory cell from which the olfactory hairs grow. *Developmental Biology.* the vesicle in the embryo that later develops into the olfactory bulb and tract.

olibanum *Materials Science.* another name for frankincense or for an oil derived from frankincense.

olig- or **oligo-** a prefix meaning "small" or "few."

oligarchy [äl´i gär´kē] *Anthropology.* government by the few; a system in which a few people or an elite class rule the society.

oligemia *Medicine.* a deficiency of blood in the body. Also, HYPHEMIA, HYPOVOLEMIA.

oligoblast *Neurology.* a primitive macroglial cell of ectodermal origin that develops into an oligodendrocyte.

Oligocene [äl´i gə sēn´] *Geology.* **1.** the geologic epoch of the Tertiary period, extending from the end of the Eocene epoch to the beginning of the Miocene epoch. **2.** the rock strata formed during this epoch.

Oligochaeta *Invertebrate Zoology.* the earthworms, a class of hermaphroditic annelids having no larval stage, few chaetae, no parapodia, and a conspicuous girdle around the body that secretes mucus for copulation and to cocoon the eggs.

oligochaete *Invertebrate Zoology.* **1.** an organism belonging to the class Oligochaeta; an earthworm. **2.** of or relating to this class.

oligoclase *Mineralogy.* a member of the plagioclase feldspar isomorphous series (albite-anorthite), having a composition that ranges from $Ab_{90}An_{10}$ to $Ab_{70}An_{30}$; found in igneous rocks of intermediate to high silica content.

oligoclasite *Petrology.* a granular igneous rock, mainly composed of oligoclase. Also, OLIGOSITE.

oligocythemia *Medicine.* a former term for anemia.

oligodendrocyte *Neurology.* any cell of the oligodendroglia.

oligodendroglia *Neurology.* small non-neural supporting cells of the nervous system, having spheroid nuclei and delicate cytoplasmic branches, that envelop axons to form myelin sheaths. Also, OLIGOGLIA.

oligodendroglioma *Medicine.* an unusual, slow-growing glioma marked by many small, round or ovoid oligodendroglial cells distributed in a sparse fibrillary stroma.

oligodynamic *Microbiology.* effective in minute quantities; used especially of the toxic effect of heavy metal ions on microorganisms. Thus, **oligodynamic action, oligodynamic effect.** *Toxicology.* toxic in very small quantities; highly toxic.

oligoglia see OLIGODENDROGLIA.

Oligohymenophora *Invertebrate Zoology.* a class of the phylum Ciliophora; protozoans with a well-developed oral apparatus and compound ciliary organelles. Also, **Oligohymenophorea.**

oligolecithal *Developmental Biology.* describing eggs with a small amount of evenly distributed yolk, as the eggs of many invertebrates. Also, ISOLECITHAL, HOMOLECITHAL.

oligomenorrhea *Medicine.* a condition of markedly diminished menstrual flow.

oligomer see OLIGOMERIC PROTEIN.

Oligomera *Invertebrate Zoology.* a former subphylum of Vermes; worms with few or indistinct segments.

oligomeric protein *Biochemistry.* a protein containing two or more polypeptides that may be the same or different. Also, OLIGOMER.

oligomerous *Botany.* having only a few parts; usually applied to whorls that have fewer members than other whorls on the same flower.

oligomictic lake *Hydrology.* a lake that circulates only at irregular intervals, when abnormal cold spells occur.

oligomycin *Microbiology.* an antibiotic produced by a bacterium similar to *Streptomyces diatachromogenes* that acts against fungi.

oligonucleotide *Biochemistry.* the partial hydrolysis products of nucleic acids, consisting of a few nucleoside phosphate residues joined together.

oligonucleotide fingerprinting *Virology.* a process in which a print of an RNA virus strain is obtained by digesting the genomic RNA with an enzyme and subjecting the resulting oligonucleotides to two-dimensional separation.

oligonucleotide synthesizer *Biotechnology.* an automated device used for solid-phase chemical synthesis of linear nucleic acids that can be used as DNA probes, as synthetic genes, or for site-directed mutagenesis.

oligopelic *Geology.* of or related to a lake bottom deposit containing very little clay.

oligopeptide *Organic Chemistry.* a peptide containing not more than 10 amino acids.

oligophagous *Zoology.* feeding on only a few types of food.

oligophrenia *Pathology.* below normal mental development.

oligophyre *Petrology.* a light-colored diorite with oligoclase in both the phenocrysts and the groundmass.

Oligopithecus *Anthropology.* an early Oligocene catarrhine, found at the Fayum site in Egypt; *Oligopithecus savagei* is the only known species.

oligopneustic *Invertebrate Zoology.* of insects, having a reduced number of spiracles.

oligopod larva *Invertebrate Zoology.* a common form of larva of beetles and Neuroptera, predaceous, with well-developed legs and sense organs.

Oligopygida *Paleontology.* a genus of irregular echinoids in the extinct order Oligopygoida and family Oligopygidae; related to the clypeasteroids; restricted to the Eocene.

oligosaccharide *Biochemistry.* a carbohydrate that yields a small number of monosaccharides on hydrolysis.

oligosite see OLIGOCLASITE.

oligospermia *Medicine.* an abnormally low concentration of spermatozoa in the penile ejaculate. Also, **oligozoospermia.**

Oligotrichida *Invertebrate Zoology.* an order of ciliate protozoans in the subclass Spirotrichia, having a round body, reduced or absent cilia, and membranelles around the mouth.

oligotrophic lake *Hydrology.* a lake that is deficient in plant nutrients and characterized by an abundance of dissolved oxygen in its lower layer.

oliguria *Medicine.* a condition of scant or insufficient urination.

olingo *Vertebrate Zoology.* a raccoonlike mammal of the genus *Bassaricyo* that is nocturnal and eats fruit, characterized by large eyes and a long, ringed tail; found in tropical jungles from Nicaragua to Peru and Bolivia.

Oliniaceae *Botany.* a monogenetic family of dicotyledonous tanniferous shrubs and small trees of the order Myrtales, characterized by simple hairs and internal phloem next to the pith; native to tropical and southern Africa.

olipidium *Mycology.* in some classifications, a genus of mold fungi belonging to the family Olpidiaceae that is composed primarily of parasites of aquatic algae, microscopic aquatic plants, or moss and some plant pathogens.

olistolith *Geology.* an unrelated mass of rock occurring in an olistostrome.

olistostrome *Geology.* a sedimentary bed or layer accumulated as the result of sliding; generally without internal bedding and composed of intimately mixed heterogeneous materials.

oliva *Anatomy.* a rounded elevation, lateral to the upper part of each pyramid of the medulla oblongata; produced by an irregular mass of neurons, the nucleus olivaris caudalis, located just beneath its surface. Also, **olivary nucleus.**

olivaceaous *Biology.* **1.** resembling an olive. **2.** olive-colored.

olive *Botany.* any of the genus *Olea* in the family Oleaceae; especially *Olea europea*, an Old World evergreen tree or shrub widely cultivated since early human history for its berrylike fruit, the oil of the fruit, and the wood of the tree.

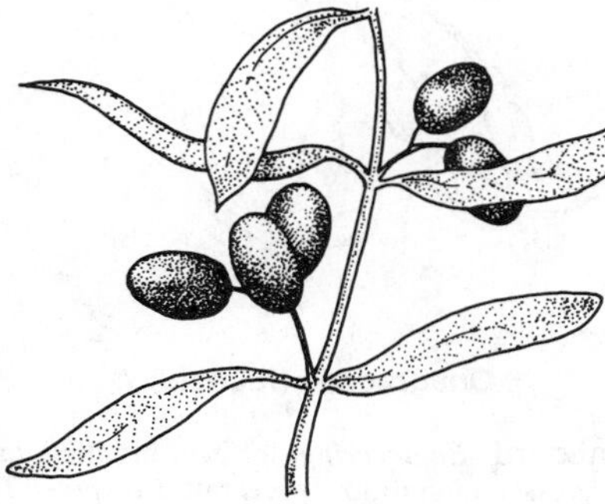

olive

olive family *Botany.* a popular name for the family Oleaceae.

olive-green mold *Mycology.* a fungal mold belonging to the species *Chaetomium olivaceum,* which is found on mushroom compost following pasteurization.

oliveiraite *Mineralogy.* $Zr_3Ti_2O_{10}\cdot 2H_2O$, an isotropic mineral consisting of a hydrated titanium zirconium oxide and considered a variety of zircon.

olive knot *Plant Pathology.* a disease of olive trees caused by the bacterium *Pseudomonas sevastonoi,* characterized by abnormal growths on the foliage, branches, and trunk. Also, **olive tubercle.**

olivenite *Mineralogy.* $Cu_2^{+2}(AsO_4)(OH)$, an olive-green to brown, vitreous, orthorhombic mineral, having a specific gravity of 4.37 to 4.46 and a hardness of 3 on the Mohs scale; found as acicular crystals and in globular or granular form in the oxidized zone of ore deposits.

olive oil *Materials.* a combustible, pale-yellow or greenish-yellow liquid that is expressed from the olive fruit and is soluble in ether and sparingly soluble in alcohol; used in foods, ointments, soap, lubricants, and cosmetics.

olive shell *Invertebrate Zoology.* a marine gastropod of the family Olividae, characterized by a highly colored, polished, elongated shell and a large mantle that surrounds the shell when extended; or the shell itself.

olivette *Electricity.* a large floodlight consisting of a single bulb, ranging from 500 to 1500 watts; often used for lighting stage entrances. Also, **olivet.**

Olividae *Invertebrate Zoology.* a family of burrowing marine gastropod snails in the order Neogastropoda, having cylindrical, glossy, usually brightly colored shells.

olivine *Mineralogy.* **1.** a general term used for members of the isomorphous series fayalite-forsterite (olivine group). **2.** a group name for orthorhombic silicates having the general formula $A^{+2}SiO_4$, where A^{+2} = Fe,Mg,Mn,Ni. Also, CHRYSOLITE.

olivine basalt *Petrology.* an igneous rock, or rock group, composed of basalt on the plane of critical silica saturation, containing olivine and diopside in the absence of nepheline and hypersthene.

olivine nephelinite *Petrology.* an extrusive igneous rock similar in composition to nephelinite except for the presence of olivine. Also, NEPHELINE BASALT, ANKARATRITE.

olivine sand *Materials.* a naturally occurring foundry sand used in place of more expensive silica sand.

ollenite *Petrology.* a variety of hornblende schist that contains abundant epidote, sphene, and rutile, with traces of garnet and other minerals.

Ollier-Thiersch graft *Surgery.* a very thin skin graft using the epidermis and a thin layer of dermis, often cut in long, broad strips. (Named for Léopold *Ollier,* 1830–1900, French surgeon, and Karl *Thiersch,* 1822–1895, German surgeon.) Also, THIERSCH'S GRAFT.

Olmstead, Frederick Law 1832–1903, American landscape architect; designed New York City's Central Park and other early urban parks.

-ology a suffix meaning "the study of" or "the knowledge of."

Ol. oliv. olive oil. (From Latin *oleum olivae.*)

O.L.P. left occipito-posterior, a position of the fetus. (From Latin *occipito-laeva posterior.*)

Olpidiaceae *Mycology.* a family of fungi belonging to the order Chytridiales that occur in water and soil; some species are parasites of algae, flowering plants, microscopic animals, or aquatic fungi, some live off decaying or dead organic matter, and some are pathogens.

Olpidiopsidaceae *Mycology.* a family of fungi belonging to the order Lagenidales that occur in freshwater and land environments; most are parasites of algae and some live off other oomycetes.

Olsen ductility test *Metallurgy.* one of several standardized tests to assess deep drawability.

O.L.T. left occipito-transverse, a position of the fetus. (From Latin *occipito-laeva transversa.*)

Olympus Mons *Astronomy.* the largest volcano on Mars, having a base diameter of 600 kilometers, a height of 27 kilometers, and an age of approximately 200 million years.

OMA *Aviation.* the airport code for Eppley Airfield, Omaha, Nebraska.

-oma *plural,* **-omas** or **-omata.** *Oncology.* a prefix denoting a tumor or neoplasm of the part or type indicated in the stem, as in *endometrioma, carcinoma.*

Oman, Gulf of *Geography.* an inlet of the Arabian Sea between southeastern Iran and the southeastern arm of the Arabian Peninsula.

Omar Khayyam [kī yam′] 11th century Persian mathematician, astronomer, and poet; reformed the Persian calendar; wrote the *Rubaiyat.*

omasum *Vertebrate Zoology.* the third stomach (of four) in ruminants, in which food chewed for the second time in the reticulum enters and is subjected to further churning before entering the fourth stomach (abomasum). Also, PSALTERIUM; MANYPLIES.

ombrometer see RAIN GAUGE.

ombroscope *Engineering.* an electrical or mechanical instrument that measures precipitation by means of heated, water-sensitive surfaces.

omega [ō mā′gə; ō mē′gə; ō meg′ə] **1.** the 24th and final letter in the Greek alphabet, written as Ω, ω. **2.** the last item of a group or series; the end.

Omega *Navigation.* a worldwide very low frequency (10–14 kHz) continuous-wave radio navigation system of medium accuracy in which eight stations transmit in succession on the same frequency.

Omega Centauri *Astronomy.* a large globular cluster in Centaurus, containing over 100,000 stars and stretching 65 light-years across; located about 16,000 light-years distant and appearing to the naked eye as a fuzzy 4th-magnitude star; designated **NGC 5139.**

omega hyperon *Particle Physics.* a negatively charged subatomic particle, the heaviest of the hyperons, with a mass of 1672 MeV and a spin of 3/2. Also, **omega particle, omega-minus particle.**

omega meson *Particle Physics.* a highly unstable and short-lived neutral vector meson with negative charge parity, G parity, and a mass of 784 MeV.

Omega Nebula *Astronomy.* a bright region of ionized hydrogen in the constellation of Sagittarius. Also, SWAN NEBULA.

omegatron *Electronics.* an instrument that photographs and identifies any radioactive gases that remain after an evacuation.

omental *Anatomy.* of or relating to the omentum.

omental foramen see EPIPLOIC FORAMEN.

omentum *Anatomy.* a fold of peritoneum extending from the stomach to adjacent organs of the abdominal cavity. The **greater omentum** (or **omentum major**) encloses the transverse colon, attaches to the stomach, and is suspended over the abdominal contents; the **lesser omentum** (or **omentum minor**) joins the lesser curvature of the stomach and the first part of the duodenum to the porta hepatis.

omicron the 15th letter of the Greek alphabet, written as O, o.

omicron *Microbiology.* a Gram-negative bacterial endosymbiont that is found in the cytoplasm of species of the aquatic ciliate *Euplotes.*

omission the fact of omitting or leaving out; specific uses include: *Geology.* the nonexposure or complete elimination of certain stratigraphic beds at the surface (or in a specified section) owing to disruption and displacement of the beds by faulting.

omission conditioning *Behavior.* a type of conditioning in which a reward that is constantly present is removed whenever a particular undesired response occurs. Also, **omission training.**

omission solid solution *Crystallography.* a crystal whose lattice sites are not completely occupied.

ommatidium *Entomology.* any of the numerous elongated cylindrical elements of the compound eye of insects and many crustaceans; includes a lens or cornea, a crystalline cone, and a sensory region.
ommatophore *Entomology.* a movable stalk bearing an eye.
omn. hor. every hour. (From Latin *omni hora.*)
omni- a combining form meaning "all."
omnibearing *Navigation.* characterizing a navigational aid that radiates or is detectable from all bearings. Thus, **omnibearing beacon.**
omnidirectional *Electronics.* of or relating to a device that radiates or receives signals in any direction.
omnidirectional antenna *Electromagnetism.* an antenna that radiates in all directions with equal intensity; also called a nondirectional antenna.
omnidirectional microphone *Acoustical Engineering.* a hydrophone that responds equally to sound waves arriving from all directions, thus having a spherical directivity pattern. Similarly, **omnidirectional hydrophone.**
omnigraph *Mechanical Engineering.* an automatic acetylene cutter that is controlled by and copies the motion of a mechanical pointer tracing a pattern. *Telecommunications.* a device that converts Morse code signals that have been punched onto tape into buzzer-produced audio signals.
omnimeter *Engineering.* a theodolite equipped with a microscope for the purpose of observing the angular motion of the telescope.
omnirange *Navigation.* a radio beacon that allows an aircraft to follow any radial to or from the station automatically. Also, **omnidirectional range.**
omnispective method *Evolution.* a method of classification in which intuitive and pragmatic considerations using phenotypic similarity form the criterion of relationship, and taking evolutionary history into consideration.
omnivore [äm´ni vôr´] *Zoology.* an organism that is omnivorous.
omnivorous [äm niv´ər əs] *Zoology.* **1.** feeding on both plants and animals. **2.** feeding on a wide variety of foods; not having a specific diet.
omnivory [äm niv´ə rē] *Zoology.* the fact of being omnivorous.
omn. man. every morning. (From Latin *omni mane.*)
omn. noct. every night. (From Latin *omni nocte.*)
omn. quar. hor. every quarter of an hour. (From Latin *omni quadrante hora.*)
omohyoid *Anatomy.* either of two muscles passing across the side of the neck from the hyoid bone to the scapula.
Omomyidae *Paleontology.* a tarsier-like family of primates that were widespread in the Eocene but became extinct in the early Miocene; probably ancestral to the modern genus *Tarsius.*
Omophronidae *Invertebrate Zoology.* the savage beetles, a small family of coleopteran insects in the suborder Adephaga.
Omo River Basin *Anthropology.* an important site in southern Ethiopia that contains australopithecines ranging from an early gracile form to a later super-robust form resembling *A. boisei* and *Homo erectus.*
omp *Microbiology.* an informal designation for any of several major outer membrane proteins of Gram-negative bacteria. (An acronym for outer membrane proteins.)
omphacite *Mineralogy.* a vitreous, green monoclinic mineral of the pyroxene group composing a solid solution of aegirine, jadeite, and augite (compositional range $JD2_{5-75}Aug_{25-75}AEG_{0-25}$), having a specific gravity of 3.29 to 3.39 and a hardness of 5 to 6 on the Mohs scale; granular to foliated in habit and found with garnet in eclogites and related rocks.
omphalectomy *Surgery.* the surgical excision of the umbilicus.
omphalic *Anatomy.* of or relating to the umbilicus (navel).
omphalitis *Medicine.* an inflammation of the umbilicus and surrounding parts.
omphalo- or **omphal-** a combining form denoting a relationship or resemblance to the umbilicus.
omphalocele *Medicine.* a protrusion, at birth, of part of the intestine through a large defect in the abdominal wall at the umbilicus.
omphaloma *Oncology.* a tumor of the umbilicus.
omphalomesenteric *Anatomy.* relating to the umbilicus and mesentery. *Developmental Biology.* see VITELLINE.
omphaloplasty *Surgery.* surgical repair of the umbilicus.
omphalotomy *Surgery.* the cutting of the umbilical cord.
omphalus *Anatomy.* a technical name for the umbilicus (navel). See UMBILICUS.
Omphralidae *Invertebrate Zoology.* a family of orthorrhaphous insects in the series Nematocera.
OMS ovonic memory switch.
o.n. every night. (From Latin *omni nocte.*)
Onagraceae *Botany.* a family of dicotyledonous shrubs and sometimes trees in the order Myrtales, characterized by internal phloem in the stem, an inferior ovary and many ovules, and twice as many stamens as petals. Several species within this family are cultivated for ornamentation, including the fuchsia and evening primrose.

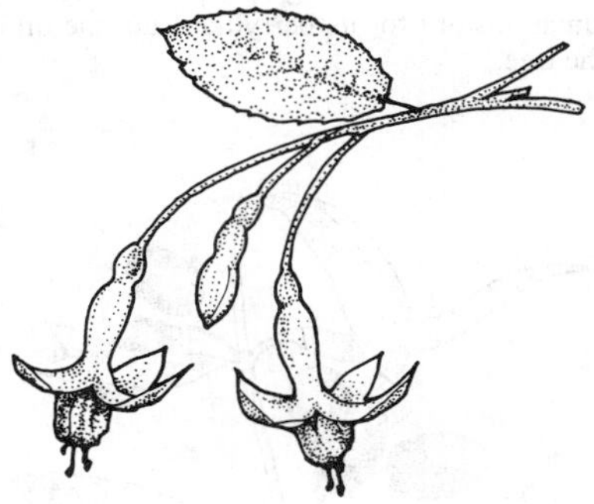

Onagraceae (fuchsia)

on-board or **onboard** *Engineering.* located, installed, or occurring in or on a vehicle, vessel, aircraft, or spacecraft. *Computer Technology.* **1.** of a controller or other printed circuit board, installed on a computer backplane. **2.** of a computer or other device, installed in an aircraft or other vehicle. Thus, **on-board computer.**
on-call target *Ordnance.* in artillery and naval gunfire support, a planned target that is not scheduled, but upon which fire is delivered when requested.
on-carriage equipment *Ordnance.* equipment that is mounted on the carriage of a weapon during transportation.
on-carriage fire control *Ordnance.* the method of aiming a weapon with a sighting device that is mounted on the weapon or its carriage.
once-through boiler *Mechanical Engineering.* a term for a boiler in which water does not recirculate.
onc gene *Genetics.* an oncogene in a retrovirus that is similar to proto-oncogenes found in vertebrate genomes.
Onchidiidae *Invertebrate Zoology.* a family of sluglike intertidal gastropod mollusks, in the subclass Pulmonata, with a posterior pulmonary sac and no shell.
oncho- a variant of the combining form *onco-.*
Onchocerca *Invertebrate Zoology.* a genus of nematode parasites of the superfamily Filarioidea.
onchocerciasis or **onchocercosis** *Medicine.* an infection with worms of the genus *Onchocerca* (usually *O. volvulus*), characterized by nodular swellings that form a fibrous cyst enveloping the coiled parasites; ocular complications may form, causing blindness in advanced cases as a result of the sensitization of the cornea to the microfilariae. Also, VOLVULOSIS.
onco- a combining form denoting: **1.** a tumor, swelling, or mass, as in *oncogene.* **2.** a barb or hook, as in *oncosphere.*
oncocyte *Histology.* an elongate columnar cell, having eosinophilic granular cytoplasm, found in salivary glands, certain endocrine glands, nasal mucosa, and other places.
oncocytoma *Medicine.* a tumor of the thyroid gland composed wholly or mainly of large cells (Hürtle cells); usually benign, it may occasionally become locally invasive or, rarely, metastasize.
oncogene [äng´kə jēn´] *Genetics.* a gene that induces cancer or other uncontrolled cell proliferation; a mutated or activated proto-oncogene that is associated with the development and proliferation of tumor cells.
oncogenesis *Oncology.* the production and causation of tumors.
oncogenic [äng´kə jen´ik´] *Oncology.* causing the formation of tumors. Also, **oncogenetic.**
oncogenic virus *Virology.* any virus that can cause cells to undergo unlimited proliferation; a cancer-causing virus.
oncogenous *Oncology.* developing from or occurring in a tumor.
oncolite *Geology.* a small, concentrically laminated, calcareous sedimentary structure formed by the accretion of successive, layered masses of gelatinous sheaths of blue-green algae; similar to, but smaller than, a stomatalite.
oncologist [äng käl´ə jist] *Oncology.* a specialist in the study and treatment of tumors.

Oncology

Oncology, from the Greek *onkos* ("bulk" or "mass"), means the study of tumors or cancers. The initial use of the word was as a substitute for the word "cancer" in order to mitigate patient anxiety since, in the public's mind, cancer implies a fatal malignant course. A tumor can be either benign or malignant. A benign tumor grows only at the site of origin and is easily dealt with. Malignant tumors can migrate from the primary site to other organs (metastasize) by invading tissue and gaining access to blood vessels.

This ability to metastasize is an important distinction between benign and malignant tumors, since most cancer deaths result from multiple secondary growths that destroy normal function in other vital organs. Understanding the importance and frequency of metastases has led to the development of systemic treatments that can reach circulating cells (e.g., chemotherapy).

At least three sets of genes are known to be instrumental in the transmutation of a normal cell to a benign growth to a malignant growth. Those that serve to restrain the capacity of cells to replicate are referred to as suppressor oncogenes and are inherited as recessive traits. A second set of genes, proto-oncogenes (the normal counterpart of an oncogene), is inherited as a dominant trait. Their products affect the ability of a cell to send or respond to signals that regulate growth. A third set of genes appears to govern migration of normal cells and the capacity of malignant cells to metastasize.

Understanding of the structure and function of oncogenes is just reaching practical application in the prevention, diagnosis, and treatment of cancer. The new biotechnology augurs well for further control of the disease in the foreseeable future.

Vincent T. DeVita, Jr.
Benno C. Schmidt Chair in Clinical Oncology
Attending Physician and Member
Program of Molecular Pharmacology and Therapeutics
Memorial Sloan-Kettering Cancer Center

oncology [äng käl´ə jē] the study of the physical, chemical, and biologic properties and features of tumors, including their causes, pathogenesis, and treatment. Thus, **oncologic, oncological.**

oncolysate *Oncology.* an antitumor agent.

oncolysis *Oncology.* the destruction or dissolution of tumor cells.

oncoma *Oncology.* a swelling or tumor.

oncomiracidium *Invertebrate Zoology.* a free-swimming, ciliated larval stage of a fluke.

oncosphere *Invertebrate Zoology.* the ciliated six-hooked larva hatched from the egg of many tapeworms.

Oncothecaceae *Botany.* a monospecific family of glabrous, tanniferous evergreen shrubs or small trees of the order Theales; native to New Caledonia.

oncotic pressure *Physiology.* the tension along a capillary wall caused by the difference in concentration of protein in the plasma and the interstitial fluid.

oncotomy *Surgery.* the surgical excision of a tumor or swelling.

oncotropic *Oncology.* having an affinity for tumor cells.

Oncovirinae *Virology.* a subfamily of viruses of the family Retroviridae that infect a wide range of animal species, including the feline leukemia virus and mouse mammary tumor virus.

oncovirus *Virology.* any virus that causes cancer.

ondometer *Electronics.* an apparatus that is designed to measure the frequency of the oscillations in high-frequency currents. Also, **ondograph.**

ondoscope *Electronics.* a glow-discharge tube used on an insulating rod to indicate the presence of high-frequency radiation, as the radiation ionizes the gas in its tube and causes it to glow.

-one *Organic Chemistry.* a suffix designating a ketone or a compound that is related or analogous to a ketone.

one-address code *Computer Programming.* a program segment using one-address instructions.

one-address instruction *Computer Programming.* an early instruction format that contained a single operand address along with an operation code.

one condition *Computer Technology.* the state of a memory element that represents a binary 1. Also, ONE STATE.

one day's supply *Military Science.* a unit or quantity of supply used to estimate the average daily expenditure under specific conditions; an ammunition day of supply may be expressed as the rounds of ammunition per weapon per day necessary to sustain operations.

one-digit subtractor see HALF-SUBTRACTOR.

one-dimensional array *Computer Programming.* a single row or column of data elements.

one-dimensional flow *Fluid Mechanics.* the hypothetical flow of fluid through a region in which all flow lines are parallel to some straight line.

one-dimensional lattice *Crystallography.* a crystal lattice model in which all the atoms or ions are arranged along a single line with periodic spacing.

one-ended tape Turing machine *Computer Technology.* an early abstract computing device that was altered so that the tape could be extended to the right but not to the left. (Named for Alan *Turing,* who invented this device in 1936.)

one-face-centered lattice *Crystallography.* a crystal lattice in which a central atom is located on one face of the lattice.

one gene-one enzyme hypothesis *Genetics.* a theory suggesting that a large class of genes exists in which each individual gene controls the synthesis or activity of a specific enzyme. Similarly, **one gene-one polypeptide hypothesis.**

one-group model *Nucleonics.* a model in which neutrons of all energies possess the same characteristics; used in the study of nuclear reactors.

one-hit kinetics *Virology.* the effect of concentration of a virus (in which a single particle can initiate infection) on the number of plaques or lesions induced. Also, SINGLE-HIT KINETICS.

oneirism *Medicine.* an abnormal dreamlike state of consciousness.

one-level code see ABSOLUTE PROGRAMMING.

one-level memory *Computer Technology.* computer memory in which all stored items are at the same level and are accessed by the same mechanism.

one-level subroutine *Computer Programming.* a subroutine that does not call other subroutines during execution.

one-look circuit *Ordnance.* a mine circuit that requires a single actuation by a given influence.

one-parameter group *Mathematics.* a one-dimensional Lie group.

one-parameter subgroup *Mathematics.* a one-parameter subgroup of a Lie group G is a differentiable curve $g(t)$, where g maps the real numbers into G, such that: (a) $g(t)g(s) = g(t + s)$ for all real numbers t, s; and (b) $g(0) = e$ (the identity element of G).

one-part code *Telecommunications.* a code in which text elements and their code groups are arranged in numerical, alphabetical, or some other systematic order.

one-particle exchange *Particle Physics.* the interaction between two elementary particles by the exchange (emission and absorption) of another elementary particle.

one-pass compiler *Computer Programming.* a compilation program that requires only one pass through the source program statements to produce a machine-language program.

one-pass operation *Computer Programming.* any operation of a compiler or assembler that generates a machine-language program with a single pass through the source language statements.

one-plus-one address instruction *Computer Programming.* an instruction that contains two address parts: one operand address and the address of the next instruction to be executed.

one-point compactification *Mathematics.* a compact topological space derived from the adjoining of a point z to a given topological space X not containing z; denoted X^+. The topology of X^+ consists of all the original open sets of X, the complement in X^+ of all the compact subsets of X, and X^+ itself. X is a topological subspace of X^+.

one-quadrant multiplier *Electronics.* a multiplier of an analog computer whose operation is limited to a single sign of both input variables.
one-sample test *Statistics.* any test based on a single specified sample.
one's complement *Mathematics.* given a number x expressed in base 2, the (base 2) number obtained by replacing 0's with 1's and 1's with 0's. For example, the one's complement of 110101_2 is 1010_2. *Computer Programming.* such a number used to represent a negative binary number; the diminished radix complement of the binary number system found by changing all the 1's in a number to 0's, and all the 0's to 1's, that is, by subtracting every bit from 1.
ones-complement code *Computer Programming.* a method of representing a negative binary number as its one's complement.
one-shot molding *Engineering.* the immediate production of urethane-plastic foam by mixing together various additives.
one-sided limit *Mathematics.* the value that a function approaches as the independent variable approaches a given number either from the left or from the right, but not from both sides. Known respectively as the **limit from the left** or **left-handed limit,** and **the limit from the right** or **right-handed limit.**
Onesquethawan *Geology.* a North American geological stage of the Lower and Middle Devonian period, occurring after the Deerparkian and before the Cazenovian.
one state see ONE CONDITION.
one-step growth curve *Virology.* a curve tracking a single cycle of a virus's replication in a synchronously infected cell culture.
one-step operation see SINGLE-STEP OPERATION.
one-tailed test *Statistics.* a test procedure that rejects the null hypothesis when the statistic is larger than a specified value, rather than outside an interval as in two-tailed tests; such a test arises, for example, when all values of the alternative hypothesis are smaller or larger than values in the null hypothesis.
one-time pad *Computer Programming.* a provably unbreakable data encryption technique using a key that is the same length as the message such that each ciphertext bit is the exclusive-or of the corresponding message and key bits.
one-to-many correspondence *Computer Programming.* a relationship between two or more entities in a database such that a "manager" may associate with one or more "subordinates," but a subordinate may answer to only one manager. Also, MANY-TO-ONE CORRESPONDENCE.
one-to-one *Mathematics.* a mapping (or function, transformation, etc.) f is said to be a one-to-one function on a set X if for every $y \in f(X)$, there is only one $x \in X$ such that $y = f(x)$. That is, no two elements of the set X have the same image under f. Also, **one-to-one correspondence, one-one.**
one-to-one assembler *Computer Programming.* an assembly program that produces one machine-language statement for each source-language statement. Also, **one-to-one translator.**
one-to-one function or **one-to-one mapping** *Mathematics.* see INJECTION.
O network *Electricity.* a four-branch closed-circuit impedance network that contains two series (upper and lower) and two shunt arms (input and output).
one-way analysis of variance *Statistics.* an analysis of variance that considers the effect of only one factor.
one-way communication *Telecommunications.* the broadcasting transmission of a message, either by sound and picture or by sound only, to a network of receiving devices such as radios or televisions over a particular set of frequencies or channels.
one-way incompatibility *Genetics.* an incompatibility between two plasmids that occurs when plasmid B is introduced into a cell containing plasmid A, but does not occur when plasmid A is introduced into a cell containing plasmid B.
one-way repeater *Telecommunications.* a repeater whose signals travel in one direction over transmission lines or in space.
one-way slab *Civil Engineering.* a concrete slab in which the structural action is essentially one-way, perpendicular to the supports of the slab.
one-way trunk *Telecommunications.* in a telephone network, a tie between two central offices that is used for calls originating at one of the offices only. Also, OUTGOING TRUNK.
one-year ice *Oceanography.* sea ice that is less than one year old.
ongioplasm *Cell Biology.* threadlike filaments of protoplasm that form the reticulum of fixed cells. *Neurology.* the granular material of an axon.
ongoing debugging *Computer Programming.* the corrective maintenance of software.
on grade *Civil Engineering.* on or at the same level or elevation.
on-hook *Telecommunications.* **1.** in telephony, describing the condition that exists when a handset or receiver is resting on its switch. **2.** of or relating to the open-loop or idle state of a subscriber or PBX line loop. **3.** one of two possible signaling states, as in ground connection versus battery connection, tone versus no tone.
onion *Botany.* any herbaceous plant of the genus *Allium* in the diverse family Liliaceae, especially *A. cepa,* a biennial plant native to Asia and now grown widely throughout the world for its pungent, edible bulb; common varieties include the yellow, white, and Bermuda onion.

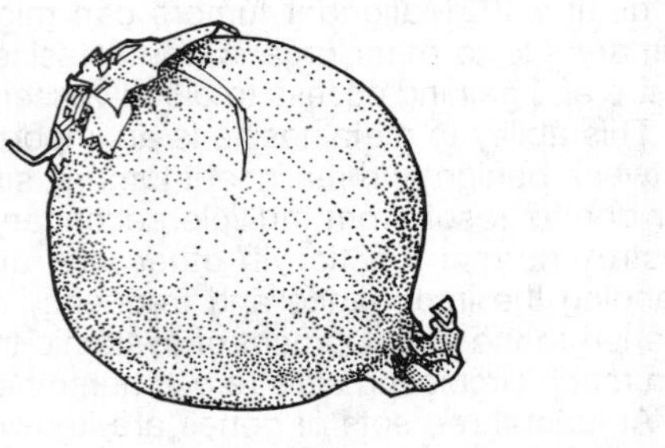

onion

onionskin paper *Materials.* a thin, lightweight, translucent, glazed paper having an undulating surface caused by a special heating and drying process; commonly used for airmail stationery.
onionskin weathering *Geology.* chemical exfoliation in which concentric shells of decayed rock that resemble the layers of an onion are loosened and separated from a fine-grained block of rock. Also, **onion weathering.**
onion smudge *Plant Pathology.* a disease of onion plants caused by the fungus *Colletotrichum circinans* and characterized by black rings or smutty spots on the bulb or neck. Also, **onion scab.**
onion smut *Plant Pathology.* a disease of onions caused by the fungus *Urocystis cepulae* and characterized by black patches on the leaves and scales.
Oniscoidea *Invertebrate Zoology.* a suborder of terrestrial and amphibious isopod crustaceans, including wood lice and pill bugs, having distinct abdominal segments that enable them to roll into a ball.
Oniscomorpha *Invertebrate Zoology.* the pill millipedes, arthropods in the class Diplopoda having flattened segments that enable them to roll into a ball.
onium *Chemistry.* a cation in which nitrogen has its maximum covalency, as in the ammonium ion NH_4^+. Such compounds include betaines, cholines, and amine oxides.
-onium *Chemistry.* a combining form used in the names of complex cations, as in *oxonium.*
onlap *Geology.* an arrangement of sedimentary layers within a depositional basin, in which the upper surface of each successively younger layer covers and extends beyond the margins of the older layer beneath it, as if deposited by a rising shoreline. Also, TRANSGRESSIVE OVERLAP.
onlay *Surgery.* a graft of tissue applied on the surface of a structure or organ.
on-line *Computer Technology.* describing equipment or operations in direct communication with and under control of a central processor. *Electronics.* describing any device that is connected to and capable of delivering the correct output to a system. *Engineering.* of a repair, test, or the like, performed without interrupting an operation. Thus, **on-line maintenance.** *Chemical Engineering.* see ON-STREAM. *Industrial Engineering.* of a factory or other facility, located on or near major routes or rail lines. *Transportation Engineering.* of airline flight equipment or personnel, in use or on active-duty status. Also, **on line, online.**
on-line analysis *Biotechnology.* the measurement of a process where sampling is done continuously via a sample line to the analytical device; this method allows for highly accurate results because it is accomplished without disturbing the reaction.
on-line analyzer *Mining Engineering.* a device used to investigate the material content at various stages in flotation or other mineral-processing flow sheets.
on-line central file *Computer Technology.* an organized file or a database that is maintained in a storage device under control of the central processor and is continuously available for real-time access.

on-line cipher *Telecommunications.* a type of encryption that is directly associated with the transmitting circuitry's particular transmission system and the reception circuits of its receiving stations.

on-line computer system *Computer Technology.* a computer system designed to provide a response within a short period of time.

on-line cryptographic operation *Telecommunications.* an operation in which messages are encrypted and simultaneously transmitted from one station, and then received and simultaneously decrypted by another station.

on-line database *Computer Programming.* a database that can be directly accessed by a user from a terminal.

on-line data reduction *Computer Programming.* the processing of data as it is received under direct control of the central processor.

on-line disk file *Computer Technology.* a magnetic disk storage device that is maintained under control of the central processor; used to expand the storage capacity of the computer.

on-line equipment *Computer Technology.* any device that operates in direct communication with and under the control of the central processor.

on-line inquiry *Computer Technology.* a customer service system that provides rapid response to requests for information by using a computer to rapidly access a repository of data. Also, **on-line query.**

on-line mode *Computer Technology.* the mode of operation in which the device is under control of the central processor.

on-line operation *Computer Technology.* any processing taking place under control of the central processor. *Telecommunications.* see ON-LINE CRYPTOGRAPHIC OPERATION.

on-line programming *Robotics.* programming in which the program is developed directly on the robot being programmed.

on-line real-time system *Computer Technology.* a system in which the computer communicates interactively with a user or monitoring device, processes input quickly, and returns output continuously; often used for continuous process control in industrial applications.

on-line search *Computer Programming.* the interactive search of a database such that response is prompt and search queries can be easily modified.

on-line secured system *Telecommunications.* any mixture of interconnected communications centers that have the capacity for on-line cryptographic operation.

on-line storage *Computer Technology.* any storage device under direct control of a central processor.

on-line typewriter *Computer Technology.* a typewriter input/output device under control of the central processor.

Onnes, Heike Kamerlingh 1853–1926, Dutch physicist; awarded the Nobel Prize for work with low temperatures and liquefying helium.

on-off control *Control Systems.* the control used to operate an on-off system.

on-off keying *Telecommunications.* a method of signaling in which a transmitter is turned on and off repeatedly.

on-off switch or **on/off switch** *Electricity.* a switch that turns a receiver or other electronic equipment on or off. Also, POWER-SWITCH.

on-off system *Control Systems.* a simple control system in which the device being controlled is either full on or full off, with no intermediate settings or operations.

on-off tests *Telecommunications.* a test performed on a receiver that has interference in order to determine the source of it by switching various suspected sources on and off while observing the subject receiver.

onomatophore see NOMENCLATURAL TYPE.

ONPG *ortho*-nitrophenyl-β-galactoside.

ONPG test *Bacteriology.* a test to detect activity of the enzyme β-D-galactosidase in bacteria, which, if present, hydrolyzes the substrate *o*-nitrophenyl-D-galactopyranoside to galactose and a colored indicator, *o*-nitrophenol.

Onsager, Lars 1903–1976, Norwegian-American chemist; awarded the Nobel Prize for work on irreversible thermodynamics.

Onsager equation *Physical Chemistry.* an equation that contrasts the conductance of an actual electrolytic solution, at a given concentration, to that of its pure solvent at zero concentration.

Onsager relations *Thermodynamics.* a set of relations claiming that for a system near thermodynamic equilibrium, the matrix of flux elements (such as heat conduction and heat diffusion), expressed as linear functions of various forces such as mass or temperature gradients (the conjugate affinities), is symmetric when appropriate definitions are selected for these fluxes and affinities. Also, **Onsager reciprocal relations.**

onshore *Geography.* situated on the shore or moving toward it from the sea.

on-stream *Chemical Engineering.* describing any component of a process that is actually in operation. Also, ON-LINE.

on-stream time *Chemical Engineering.* the amount of time during a process that a piece of equipment is actually in production.

ONT *Aviation.* the airport code for Ontario, California.

Ontario, Lake *Geography.* the smallest of the Great Lakes (7540 square miles), on the U.S.–Canada border between the state of New York and the province of Ontario.

on-the-fly printer *Computer Technology.* an impact printer whose type slugs do not stop moving while the impression is being made.

on-time performance *Transportation Engineering.* the proportion of an airline's flights that arrive by their scheduled time, or within some brief interval (normally defined as fifteen minutes) after their scheduled time.

onto *Mathematics.* a mapping (or function, transformation, etc.) f is said to map a set X onto a set Y if $f(X) = Y$; that is, if every $y \in Y$ has at least one preimage in X.

ontogeny *Developmental Biology.* the developmental history of an individual organism.

Onuphidae *Invertebrate Zoology.* a family of tube-dwelling polychaete annelids in the superfamily Eunicea having eyes, antennae, and gills that arise from the base of the dorsal cirrus.

on-vehicle materiel *Ordnance.* items that are issued with and carried on a military vehicle as part of a major combination; e.g., guns or spare parts.

onychectomy *Surgery.* an excision of a nail or nail bed. *Veterinary Medicine.* the removal of the claws.

onychia *Medicine.* an inflammation of the nail matrix. Also, **onychitis.**

onycho- or **onych-** a combining form meaning "nail" or "claw."

Onychodontidae *Paleontology.* a family of crossopterygian fishes tentatively placed in the suborder Onychodontiformes (Struniiformes); extant in the Devonian.

onychoma *Medicine.* a tumor of the nail or nail bed.

onychomalacia *Medicine.* any softening of the nails.

onychomycosis *Medicine.* a fungus infection of the nails, causing thickening, roughness, and splitting, usually caused by *Trichophyton rubrum* or *T. mentagrophytes.* Also, RINGWORM OF THE NAILS, TINEA UNGUIUM.

Onychopalpida *Invertebrate Zoology.* a suborder of mites in the order Acarina.

onychopathy *Medicine.* any disease of the nails.

onychophagia *Medicine.* the habit of biting the nails.

Onychophora *Invertebrate Zoology.* a phylum of small terrestrial caterpillarlike invertebrates having characteristics of both annelids and arthropods.

onychorrhexis *Medicine.* abnormal brittleness and splitting of the nails.

Onygenaceae *Mycology.* a family of fungi of the order Gymnascales, occurring in such animal parts as hoofs and horns.

O'nyong-nyong *Pathology.* an epidemic East African febrile disease that is caused by an alphavirus, carried by anopheline mosquitoes, and clinically similar to dengue. Also, **O'nyong-nyong fever.**

onyx *Mineralogy.* a banded, chalcedonic variety of quartz consisting of straight, parallel layers of different colors, primarily red, brown, yellow, white, and black; used in making jewelry.

onyx agate *Mineralogy.* a banded variety of agate featuring straight, parallel layers of white and various tones of gray.

onyx marble *Mineralogy.* a compact, usually translucent, variety of calcite or aragonite, similar in appearance to true onyx; used as decorative or architectural material. Also, MEXICAN ONYX, ORIENTAL ALABASTER.

onyx opal *Mineralogy.* a common variety of opal with straight, parallel markings.

oo- a combining form meaning "egg," as in *oocyte.* Also, **oö-.**

ooblast *Developmental Biology.* a primitive cell from which an ovum ultimately develops.

oocyesis *Medicine.* ovarian pregnancy.

oocyst *Invertebrate Zoology.* the enclosed zygote of various protozoans, especially the cyst formed around two conjugating gametes in certain sporozoa.

Oocystaceae *Botany.* a family of green algae of the order Chlorococcales containing all noncoenobial species that reproduce solely by autospores; characterized by usually solitary cells of varying size, shape, and ornamentation.

oocyte [ō´ə sīt´] *Developmental Biology.* a developing egg cell in either of two stages: The **primary oocyte** is derived from an oogonium by differentiation near the time of birth; it gives way to the **secondary oocyte** shortly before ovulation by a division that splits off the first polar body. If fertilized, the secondary oocyte divides into an ootid and the second polar body.

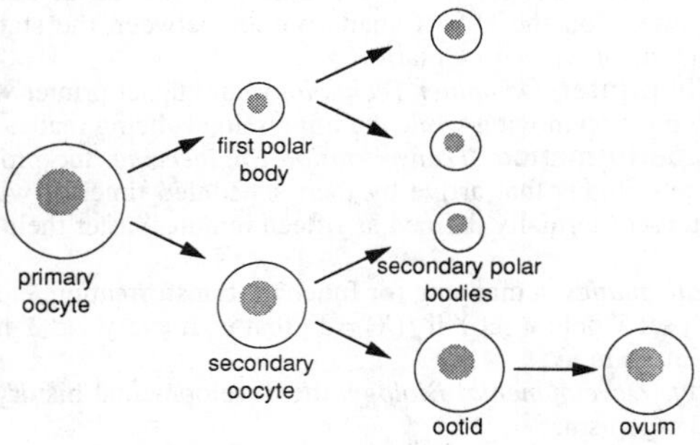

oocyte

Oodiniaceae *Botany.* a monogeneric family of nonphotosynthetic algae of the order Blastodiniales that are ectoparasitic on various marine and freshwater organisms.

ooecium see OVICELL.

oogamous [ō äg´ə məs] *Biology.* relating to or produced by oogamy.

oogamy [ō äg´ə mē] *Biology.* **1.** sexual reproduction characterized by the fusion of a small, motile sperm with a large, nonmotile egg, as in certain algae. **2.** broadly, the conjugation of any two dissimilar gametes; heterogamy.

oogenesis [ō´ə jen´ə sis] *Developmental Biology.* the proliferation, growth, and maturation of the ovum. Thus, **oogenetic.**

oogonium [ō´ə gō´nē əm] *Developmental Biology.* a female germ cell that divides several times before giving rise to a primary oocyte. *Botany.* in fungi and certain algae, a female sex organ in which large, nonmotile gametes are formed.

ookinesis *Developmental Biology.* the mitotic movements of the egg during maturation and fertilization.

ookinete *Invertebrate Zoology.* the motile zygote of certain protozoans, such as *Plasmodium.*

oolemma see ZONA PELLUCIDA.

oolicast *Petrology.* a small, subspherical opening in an oolitic rock resulting from the selective dissolution of an oolith while the matrix remains intact.

oolicastic porosity *Petrology.* the porosity of an oolitic rock that results from the removal of ooids and the formation of oolicasts.

oolite *Petrology.* a sedimentary rock, usually of limestone, clays, shales, or sandstones, consisting primarily of cemented ooliths; some oolites are used as building stone, iron ore, or bath clay. Also, EGGSTONE, ROESTONE.

oolith *Petrology.* a small, spherical or ellipsoidal accretionary body in sedimentary rock, usually calcium carbonate, that is 0.25–2.0 mm in diameter; may have a concentric or radial structure, with or without a nucleus; occurs in sedimentary rock as a result of inorganic precipitation or replacement. Also, **ooid.**

oolitic chert *Petrology.* chert consisting primarily of ooliths.

oolitic limestone *Petrology.* an even-textured variety of limestone consisting almost entirely of calcareous ooliths, with little or no interstitial material.

oology [ō äl´ə jē´] *Zoology.* the branch of zoology that deals with the study and collection of eggs, especially the study of the shape and coloration of birds' eggs. Thus, **oological.**

oomicrite *Petrology.* a limestone consisting of no less than 25% ooliths and no more than 25% intraclasts, with micrite in greater concentration than sparry-calcite cement.

oomicrudite *Petrology.* an oomicrite in which the ooliths exceed 1 millimeter in diameter.

Oomycetes *Mycology.* a class of fungi of the subdivision Mastigomycotina, found on land and in aquatic environments and including downy mildews and water molds, some of which infect crops; some spores have flagella located at both ends.

oophagous *Zoology.* feeding mainly or exclusively on eggs.

oophor- or **oophoro-** a prefix meaning "ovary."

oophoralgia *Medicine.* any ovarian pain.

oophorectomy *Surgery.* surgical removal of an ovary or ovaries.

oophoritis *Medicine.* the inflammation of an ovary. Also, OVARITIS.

oophorogenous *Medicine.* coming from the ovary.

oophorohysterectomy *Surgery.* the surgical removal of the uterus and ovaries. Also, OVARIOHYSTERECTOMY.

oophoroma *Medicine.* a tumor of the ovary.

oophoromalacia *Medicine.* the softening of an ovary.

oophoropathy *Medicine.* any disease of the ovaries.

oophoropexy *Surgery.* the surgical fixation or suspension of the ovary. Also, OVARIOPEXY.

oophororrhaphy *Surgery.* suture of an ovary.

oophorosalpingitis *Medicine.* the inflammation of an ovary and uterine tube.

oophyte *Botany.* in the alternation of generations of ferns, mosses, and liverworts, the generation or plant form that bears sexual organs.

ooplasm *Cell Biology.* the cytoplasm of an oocyte.

Oort, Jan Hendrick [ôrt] born 1900, Dutch astronomer; studied origin of comets; mapped structure of the galaxy.

Oort cloud *Astronomy.* a hypothesized cloud, roughly spherical in shape, containing millions of comets that have aphelia at distances of 30,000 to 100,000 astronomical units.

Oort's constants *Astrophysics.* two numbers used to describe a star's rotation velocity around the center of the Milky Way Galaxy: 15 kilometers per second (radial velocity) and –10 km/sec (tangential).

Oort's formulae *Astrophysics.* a series of formulas derived by Jan Oort to express the proper motions and radial velocities of stars (relative to the sun) produced by the differential rotation of the Milky Way Galaxy.

oospararenite *Petrology.* an oosparite in which the ooliths are the size of medium to coarse sand grains.

oosparite *Petrology.* a limestone consisting of no less than 25% ooliths and no more than 25% intraclasts, with the sparry-calcite cement more abundant than micrite.

oosparrudite *Petrology.* an oosparite in which the ooliths exceed 1 millimeter in diameter.

oosphere *Botany.* in the oogonium of fungi and certain algae, a female reproductive cell that becomes an oospore when fertilized.

oospore *Botany.* a thick-walled spore found within an oogonium that is formed from an oosphere after fertilization and from which a sporophyte develops.

Oosporidium *Mycology.* a genus of yeast fungi of the class Hyphomycetes, occurring on trees and characterized by vibrant fruiting bodies.

oostegite *Invertebrate Zoology.* any of several platelike structures on the coxae of the thoracic appendages of certain crustaceans; together they form a brood pouch for eggs and young.

Oosterhoff group *Astronomy.* either of two groups of globular clusters characterized as variable stars of two types, one having slightly weakened spectral lines of metals, and the other having very much weakened metallic lines.

ootheca *Entomology.* a batch of eggs encapsulated to form a packet or brood pouch, as in cockroach eggs.

ootid *Developmental Biology.* an egg cell following meiotic division.

ootype *Invertebrate Zoology.* a small chamber in the oviduct of flatworms that receives eggs, sperm, and yolk cells and passes the fertilized egg on to the uterus.

oovoid *Petrology.* a void in the center of an oolith, formed during incomplete replacement.

ooze *Geology.* **1.** an area of soft, muddy ground, such as a bog, that develops from the flow of a spring or stream. **2.** a soft, calcareous or siliceous, deep-sea deposit that is composed mainly of clay minerals with at least 30% skeletal remains of marine organisms. **3.** a soft, soupy mud or slime that typically overlies the bottom of a lake, river, or estuary. Thus, **oozy.**

oozoid *Invertebrate Zoology.* in tunicates, a zooid that develops from a fertilized egg and reproduces asexually by budding.

op. operation; opposite; opus.

OP or **O.P.** observation post; operation part.

opacifier *Materials.* a material such as a paint, glass, or enamel that is used to make a surface opaque, often to protect it from deterioration.

opacify *Science.* to make or become opaque.

opacite *Petrology.* a mass of microscopic, opaque grains found especially around hornblende phenocrysts in volcanic rocks, generally considered to consist primarily of magnetite dust.

opacity [ō pas´i tē] the quality or degree of being impenetrable to light, or sometimes to radiation, heat, and so on; specific uses include: *Optics.* the ratio of the light flux incident on a surface to the light flux transmitted by the surface; equivalent to the reciprocal of the transmission factor. *Physics.* the ability of a substance to block radiation, as measured by the radiation flux incident on its surface. *Materials Science.* the ratio of the reflectance of an enamel coating over black backing to the reflectance of a standard. *Graphic Arts.* the darkness of a given area of a film or plate, as measured by the amount or proportion of light absorbed by the emulsion. *Artificial Intelligence.* a term designating logic and procedures that are not included in the explanation facility because they are not relevant to the domain or to the reasoning process; examples are statistical computations and other mathematical algorithms.

opal *Mineralogy.* $SiO_2 \cdot nH_2O$, a colorless, white, yellow, red, brown, gray, or blue amorphous mineraloid, occurring generally in botryoidal or stalactic masses, having a specific gravity of 1.9 to 2.2 and a hardness of 5.5 to 6.5 on the Mohs scale; found as a low-temperature mineraloid in surface or near-surface deposits. Certain varieties of this mineral are widely used as gemstones.

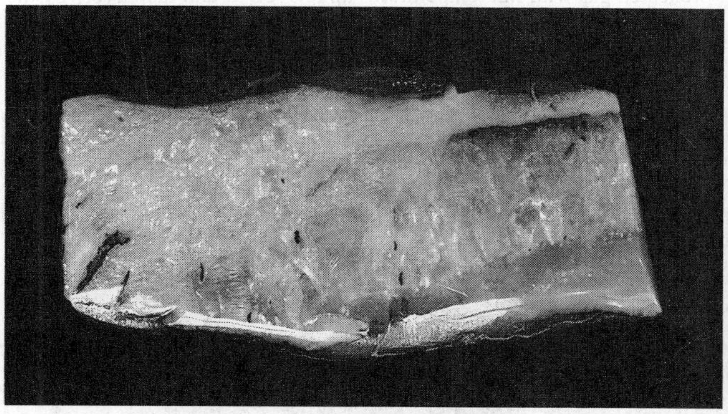

opal

opal agate *Petrology.* a banded variety of opal, agatelike in structure, displaying different shades of color within alternating layers of opal and chalcedony.

opal-CT *Petrology.* a crystalline variety of silica characterized by a poorly ordered fabric; often an intermediate phase in quartz chert formation.

opalescence [ō´pə les´əns] *Optics.* the milky iridescent coloration seen in certain minerals, such as opal, due to interference between surfaces of thin layers; also seen in some solutions due to reflection of light from fine, suspended particles.

opalescent [ō´pə les´ənt] *Optics.* having the quality of opalescence.

opalgia *Neurology.* a condition of facial neuralgia.

opal glass *Materials.* an opalescent or white glass made by adding fluorides to the glass mixture.

Opalinata *Invertebrate Zoology.* a superclass of protozoans in the subphylum Sarcomastigophora, having a flattened body covered with longitudinal rows of cilia; mainly gut commensals or endoparasites of frogs and toads.

opaline *Mineralogy.* **1.** any of a group of minerals that resemble or are related to opal. **2.** a variety of gypsum in earthy form or a magnesium-bearing variety of limestone.

opal mutation *Genetics.* a nonsense mutation that produces the mRNA codon UGA, which results in termination of a polypeptide chain.

opamp operational amplifier.

opaque [ō pāk´] *Science.* not transparent or translucent; characterized by opacity.

opaque attritus *Geology.* a constituent of coal formed by humic degradation that is impervious to light.

opaque medium *Optics.* **1.** a substance that does not transmit visible light. **2.** a substance that is impervious to light rays of specified wavelength or wavelengths.

opaque medium *Physics.* a medium that does not pass radiation of a particular frequency band.

opaque projector *Optics.* a projector that uses reflected light to project the image of an opaque object or picture onto a screen.

Oparin, Alexander I. 1894–1980, Russian biochemist; developed theory to explain that life on earth originated from chemical substances.

Oparin's theory *Biochemistry.* the concept that the chemical molecules necessary for life formed spontaneously in the early atmosphere of the earth; these molecules then combined to form more complex chemical substances, which eventually developed into living cells and then into independent organisms.

OPDAR *Engineering.* a laser system that is used to ascertain the elevation angle, azimuth angle, and slant range of a missile after firing. Also, optical radar. (An acronym derived from optical direction and ranging radar.)

Opegraphaceae *Mycology.* a family of fungi in the order Hysteriales that are lichenized or parasitic on lichens; found on the bark of trees in temperate zones and on firm leaves in tropical regions.

Opegraphales *Mycology.* a former name for an order of fungi now reclassified within the order Hysteriales.

open not closed; specifc uses include: *Physiology.* of a passage or vessel, not obstructed. *Medicine.* exposed to air; not covered by unbroken skin, e.g., an open wound. *Veterinary Medicine.* of a female, not pregnant. *Electricity.* of a circuit, discontinuous or interrupted. *Computer Programming.* **1.** to make a file or data set accessible for processing. **2.** a command used to access a file. *Graphic Arts.* of printed matter, widely spaced or leaded. *Linguistics.* of a class of items, readily admitting new members; e.g., English nouns.

open-angle glaucoma *Medicine.* a disease of the eye marked by increased intraocular pressure as a result of restricted outflow of the aqueous humor through the aqueous veins, Schlemm's canal, and the trabecular meshwork, excavation and degeneration of the optic disk, and nerve fiber bundle damage producing arcuate defects in the field of vision. Also, SIMPLE GLAUCOMA.

open-arc furnace *Metallurgy.* an arc furnace in which the arc is above the charge.

open architecture *Computer Technology.* the structure of a computer system to which add-ons, peripherals, and expansion facilities are easily appended.

open ball *Mathematics.* the open ball of radius r that is centered at a point x in a metric space X with distance function $d(x,y)$ is the set $\{y \in X : d(x,y) < r\}$; i.e., the set of all points within a fixed distance r from the point x. For example, if X is the real line, an open ball is an open interval; in general, if X is n-dimensional Euclidean space, an open ball is the interior of an n-sphere.

open bay *Geography.* an inlet of the sea that is exposed to the tides and surf, and usually much more broad than deep in penetration.

open-belt drive *Mechanical Devices.* a belt drive with parallel shafts that spin in the same direction.

open berth *Navigation.* a term for a shipping berth that is located in unprotected waters.

open-breech action *Ordnance.* in an automatic rifle, a system in which the breechblock is held open by a spring; when the spring is released by the trigger, the breechblock then moves forward and the cartridge is chambered; when the breechblock closes, a firing pin strikes the primer of the cartridge.

open bundle *Botany.* a vascular bundle in which a vascular cambium develops; a strandlike part of a plant's vascular system that contains cambium in addition to xylem and phloem.

open caisson *Civil Engineering.* a usually cylindrical caisson that is open at the top and bottom, constructed of steel, concrete, or masonry and serving as a portion of a foundation structure.

open-cast mining see OPEN-CUT MINING.

open-cell foam *Materials.* any foam material having interconnected cells.

open chain *Astronomy.* a rarely used term denoting a line of craters independently arranged without intersection.

open-circle DNA *Molecular Biology.* any circular DNA strand that lacks supercoiling and may contain a nick or gap in one strand. Also, **open-circular DNA.**

open circuit *Electricity.* **1.** a current path that is discontinuous or broken at one or more points such that it cannot conduct current or represent the voltage at its extremities. **2. open-circuit.** of or relating to a no-load condition, such as an open-circuit voltage of a battery.

open-circuited line *Electromagnetism.* a discontinuity in a microwave waveguide that reflects signals of infinite impedance.

open-circuit grinding *Mechanical Engineering.* a process in which material is passed through grinding equipment without being screened, classified, or otherwise checked for quality.

open-circuit impedance *Electricity.* in a transmission line, the driving point impedance of a line or four-terminal network in which the far end is left open.

open-circuit jack *Electricity.* a telephone jack in which a circuit break is effected until a plug that is connected to a closed external circuit is inserted.

open-circuit parameter *Electricity.* any of a set of four transmission equivalent-circuit parameters measured from one side of a network when the other side is open.

open-circuit potential *Physical Chemistry.* the potential of an electrode when there is no current flowing to or from the electrode.

open-circuit scuba *Engineering.* an elementary scuba diving system in which all gas that is exhaled is released immediately into the water.

open-circuit signaling *Telecommunications.* a form of signaling in which current does not flow while the circuit is in the idle state.

open-circuit voltage *Electricity.* **1.** the voltage at the terminals of a battery or other voltage source in a no-load condition. **2.** the voltage appearing at an electrode when there is no current flowing to or from the electrode.

open cluster *Astronomy.* a group of several hundred to several thousand stars lying in the plane of the Milky Way Galaxy, for example, the Pleiades. Also, **galactic cluster.**

open coast *Geography.* an unsheltered, harborless coast that is exposed to the weather and the sea.

open community *Ecology.* a term for a community that is readily available for colonization by immigrant species because some niches remain unoccupied.

open cut *Engineering.* **1.** a long, narrow trench on the surface of a mine working, from which ore is taken. Also, OPEN PIT. **2.** an unlined well or borehole.

open-cut mining *Mining Engineering.* the extrusion of ore from an open cut on the surface of a mine. Also, OPEN-CAST MINING.

open cycle *Thermodynamics.* a thermodynamic cycle that operates by allowing new mass to enter the system while also allowing the escape of exhaust.

open-cycle engine *Mechanical Engineering.* an engine in which the power fluid is used only once and replaced with fresh fluid.

open-cycle gas turbine *Mechanical Engineering.* a gas turbine prime mover in which hot products are expanded in the turbine element before being exhausted into the atmosphere.

open-cycle reactor system *Nucleonics.* in a nuclear reactor, a cooling system in which water is passed through the system only once and is then discarded without recirculation.

open-delta connection *Electricity.* two single-phase transformers that form two sides of a delta connection.

open die *Metallurgy.* in forging, a die that does not restrict the lateral flow of the stock. Thus, **open-die forging.**

open-ended *Computer Technology.* capable of being extended or expanded in size or capability without disrupting the original system. Thus, **open-ended system.**

open-ended spinning see ROTOR SPINNING.

open-ended wrench *Mechanical Devices.* a wrench with a fixed, open C-shaped jaw, designed to grasp the head of a nut or bolt from the side to tighten or loosen it.

open-end method *Mining Engineering.* a means of mining pillars so that no stump is left.

open fault *Geology.* a fault, or section of a fault, whose walls are separated along the fault surface.

open-field test *Behavior.* a type of experiment in which an animal is placed in a new and unknown space and aspects of its behavior are observed.

open file *Computer Programming.* a file that may be accessed for reading and writing.

open fire *Mining Engineering.* a term for a fire that occurs in a roadway or at the coal face of a mine.

open floor *Building Engineering.* a floor in which the joists are visible from the floor below.

open-flow capacity *Petroleum Engineering.* the rate of fluid flow from a gas well that is open to the atmosphere.

open-flow potential *Petroleum Engineering.* the rate of flow from a gas well that would result if only atmospheric pressure acted on the face of the wellbore; typically measured in thousands of cubic feet per day.

open-flow test *Petroleum Engineering.* analysis of flow rate and pressure drop while a gas well is open to the atmosphere; used in estimating the potential of a reservoir to deliver gas to the wellbore.

open-flow well *Petroleum Engineering.* a gas well that flows open to the atmosphere.

open fold *Geology.* a fold with moderately compressed limbs, such that the angle between the limbs is greater than 70°.

open form *Crystallography.* a crystal structure in which the crystal faces do not enclose a volume entirely.

open fracture see COMPOUND FRACTURE.

open-frame girder *Building Engineering.* a girder consisting of upper and lower booms connected by a vertical member and not braced by a diagonal member.

open-fuse cutout *Electricity.* an enclosed-fuse cutout in which the fuse holder and support are exposed.

open hawse *Naval Architecture.* the condition in which a vessel has two anchors out from the bow and the cables are not crossing.

open headstock *Mechanical Devices.* in a capstan lathe, a cross-sectionally pierced spindle that allows access to manipulate a bolt or pin.

open-hearth furnace *Metallurgy.* a reverberatory furnace having a shallow hearth and a low roof; once used extensively for steelmaking.

open-hearth process *Metallurgy.* **1.** a steelmaking process in which the charge is laid in a furnace on a shallow hearth and heated directly by preheated gas with flat temperatures up to 1750°C and radiatively by the furnace walls. Also, **open-hearth steelmaking. 2.** any liquid metal processing performed in an open-hearth furnace.

open-hearth process

open hole *Petroleum Engineering.* an uncased wellbore or the section of a drill hole below the casing.

open-hole completion *Petroleum Engineering.* oil well production that occurs in the absence of a casing or liner opposite the producing formation, resulting in the unrestricted flow of reservoir fluids into the wellbore.

open ice *Oceanography.* sea ice that is sufficiently broken up to allow navigation.

opening an open place; a hole or space; specific uses include: *Forestry.* an area within a forest that is thinly wooded in comparison with adjacent tracts. *Geography.* a break in a coastline or landform, especially between coastal islands or shoals. *Oceanography.* in sea ice, any break that reveals the water. *Anatomy.* an orifice or ostium.

opening die *Mechanical Engineering.* a screw-cutting die head that automatically clears the screw thread as it comes to the end.

opening snap *Cardiology.* a term for the sound caused by the movement of the mitral leaflet into the left ventricle at the beginning of the diastolic phase when blood begins to flow into the ventricle.

open interval *Mathematics.* a set of real numbers x satisfying $a < x < b$, where a and b are fixed real numbers; denoted by (a,b).

open jaws *Transportation Engineering.* a flight itinerary that resembles a round trip, except that the departure and arrival points are not identical (e.g., arrival at a different airport in the same metropolitan area as the departure airport).

open joint *Geology.* see FISSURE.

open lake *Hydrology.* **1.** any lake that has an effluent stream. **2.** any lake that is free of ice, emergent vegetation, or artificial obstructions.
open lead *Navigation.* a naturally occurring opening in an ice field that is wide enough to allow passage of a vessel.
open-link fuse *Electricity.* a fuse that is fabricated as a strip of material attached to an open terminal block.
open loop *Neurology.* a feedback path within the nervous system in which the input affects the output, but the output does not affect the input. *Electrical Engineering.* a signal path without feedback.
open-loop control system *Control Systems.* **1.** a control system in which the system outputs are solely a function of the system inputs. **2.** a system containing no feedback.
open-loop gain *Electrical Engineering.* the voltage, current, or power ratio of the output to the input for a system with no feedback. Often used to obtain parameters for calculation of the performance of a system where feedback will be used in normal operation.
open-loop robot *Robotics.* a robot that does not incorporate feedback in its control system.
open map *Mathematics.* a function between two topological spaces such that the image of any open set is an open set. Also, **open mapping**.
open mapping theorem *Mathematics.* **1.** the theorem that a bounded linear transformation (i.e., one that takes bounded sets to bounded sets) between Banach spaces is an open mapping. **2.** more generally, the theorem that a linear continuous surjective mapping between Fréchet spaces is an open mapping.
open mortise *Building Engineering.* a mortise made in the end of a structural member.
open pack ice *Oceanography.* floes of sea ice that cover 40–60% of the surface, seldom touching each other.
open-phase protection *Electricity.* the incorporation of an open-phase relay as an automatic device to interrupt power to a multiphase system if an open circuit occurs in any phase.
open-phase relay *Electricity.* a polyphase system in which a protective relay opens when one or more phases are open circuited.
open pit *Mining Engineering.* a broad, massive, often funnel-shaped excavation that is open to the air.

open-pit mine

open-pit mining *Mining Engineering.* a method of mining used to remove ores and minerals near the surface by first removing the waste or overburden and then breaking and loading the ore. The term is used to distinguish the surface mining of metalliferous ores from other types of surface mining, such as the strip mining of coal. Thus, **open-pit mine.**
open-pit quarry *Mining Engineering.* a quarry with an opening that is the full size of the excavation.
open plan *Building Engineering.* an interior building arrangement devoid of walls, partitions, or other distinct barriers.
open plug *Electricity.* a plug designed to hold telephone jack springs in an open position, with no external connections.
open population *Ecology.* a population that can freely interbreed with another population.
open porosity *Materials Science.* the condition of having spaces that can be filled by the addition of a liquid.
open range *Ecology.* an extensive, unrestricted area in which grazing animals can move about freely.
open reading frame *Molecular Biology.* a linear array of codon triplets in mRNA that encodes an amino acid sequence extending from a translation initiation codon to a termination codon.
open reduction *Surgery.* the surgical exposure of a fractured bone in order to realign it.
open resonator see BEAM RESONATOR.
open roadstead *Navigation.* an anchorage that offers little shelter from the sea.
open rock *Geology.* any stratum that is open or porous enough to hold a significant quantity of water, or to convey water along its bed.
open routine *Computer Programming.* **1.** a subroutine that appears as a code in each place it is used in a program, as distinguished from a closed routine, which is implemented as a single piece of code that can be called from different places in a program. **2.** a utility routine that changes the status of a file from closed to open.
open sand *Geology.* a relatively porous and permeable sand or sandstone formation that could provide good storage for oil.
open sea *Oceanography.* the main or central part of an ocean or sea, not enclosed between headlands nor within the jurisdiction of a nation.
open set *Mathematics.* **1.** any of the sets used to generate a topology on a space. **2.** a subset U of a metric space such that each point of U is an interior point of U.
open sheaf *Ordnance.* the lateral distribution of fire from two or more pieces so that adjoining points of impact or burst are separated by the maximum effective width of burst of the type of shell being used.
open shop *Computer Technology.* a computer center that functions such that the end users may operate computers and peripherals and enter programs and data without the assistance of the center staff.
open-side planer *Mechanical Devices.* a planer having one upright arm to allow machining of workpieces wider than its bed.
open-side tool block *Mechanical Devices.* a toolholder on a cutting machine having a T-clamp attached to a C-shaped block, with a screw-on gripper to hold tools. Also, HEAVY-DUTY TOOL BLOCK.
open sight *Ordnance.* any sight that is not a fully encircled aperture, especially a rear, notched sight.
open site *Archaeology.* a term for any site not located within a cave or rock shelter.
open-space structure *Geology.* in a carbonate sedimentary rock, a structure that develops by the filling or partial filling of cavities with internal sediments or cement.
open stope *Mining Engineering.* **1.** an unfilled cavity. **2.** an underground worksite, either unsupported or supported by timbers or pillars.
open-stope method *Mining Engineering.* a means of stoping generally involving no regular artificial supports, although a few props or cribs may be utilized to secure local patches of unstable ground.
open subroutine see IN-LINE PROCEDURE.
open system *Thermodynamics.* a thermodynamic system that allows both mass and energy to pass across the boundaries of the system. Also, NONFLOW SYSTEM. *Hydrology.* frozen ground in which additional groundwater is available through percolation or by capillary movement.
open-system architecture *Computer Technology.* the design of a data network for a distributed processing system that allows computers and peripheral devices from different vendors to communicate.
open-system interconnection *Computer Technology.* a seven-layer model that serves as a framework for establishing network architecture standards and protocols.
open timbering *Mining Engineering.* the technique of supporting a shaft or tunnel by placement of timber or steel supports 2 to 5 feet apart, with laggings and struts securing the intervening ground.
open traverse *Engineering.* in surveying, a traverse that begins from a station of known and established position but does not end close to the point of origin. Also, **open-end traverse.**
open tuberculosis *Medicine.* tuberculosis in which the bacilli are present in the excretions or secretions. Also, PULMONARY TUBERCULOSIS, TUBERCULOUS ULCERATION.
open tubular column see CAPILLARY COLUMN.
open universe *Astronomy.* a cosmological theory according to which the universe expands indefinitely because it lacks sufficient mass to halt the expansion by gravitational attraction.

open water *Oceanography.* in regions of ice, a large region of navigable water that is mostly free of floating ice, with no more than 10% of its surface covered by ice. *Hydrology.* **1.** water in a lake that does not freeze over in winter. **2.** lake water that is free of emergent vegetation or artificial obstructions.

open-web girder see LATTICE GIRDER.

open well *Civil Engineering.* an excavated well with a width or diameter large enough to allow a person to descend; usually at least 3 feet in diameter.

open wire *Electricity.* **1.** a wire that is unterminated. **2.** a wire that is supported above the earth's surface and is often left ungrounded.

open-wire carrier system *Telecommunications.* in carrier telephony, a system that uses an open-wire line.

open-wire circuit *Electricity.* an overhead line or a conductor that is broken or otherwise discontinued, thus impeding current flow.

open-wire line *Electricity.* a transmission line consisting of two straight, parallel wires that are held by bars of low-loss insulating material at regular intervals along the line in order to give a desired value of surge impedance; it acts as a pure resistance when properly terminated. Also, **open-wire feeder, open-wire transmission line.**

open-wire loop *Electricity.* a branch line that is connected to a main open-wire transmission line.

open workings *Mining Engineering.* any surface workings, such as a quarry or open-cast mine.

open-world assumption *Artificial Intelligence.* the assumption that if a proposition cannot be proved or disproved, its truth status is unknown; this is the normal assumption in mathematical logic.

open wound *Surgery.* a wound in which the tissues are exposed to the atmosphere.

opepe *Materials.* a strong, hard, yellow-brown wood of the *Nauclea diderrichii* tree of Africa; used in exposed areas for the construction of structures such as piers and bridges.

operable *Surgery.* able to be treated by surgery with a reasonable assurance of safety.

opera glasses *Optics.* a small, compact pair of binoculars, usually constructed on the Galilean principle; typically used for viewing an opera, play, sports event, or the like.

operand *Mathematics.* any quantity upon which an operation is performed. *Computer Programming.* **1.** the data upon which an operation is to be performed. **2.** the part of an instruction that contains the operand or the address of the operand.

operand fetch *Computer Programming.* a subtask in the processing of machine instructions in which the operand, located in memory, is moved to a register in the central processing unit.

operant *Engineering.* **1.** in operation; functioning or working. **2.** a person or thing that operates; an operator. *Behavior.* a unit of behavior as defined by its effect on the environment.

operant behavior *Behavior.* behavior defined in terms of the stimulus that proceeds from it rather than the stimulus that produces it.

operant chain *Behavior.* a series of behaviors that are built up through the process of shaping.

operant conditioning see INSTRUMENTAL CONDITIONING.

operate *Surgery.* **1.** to conduct a surgical procedure. **2.** an individual who has undergone an experimental surgical procedure, as distinguished from a control.

operate time *Electricity.* in a relay, the elapsed time from the instant when power is applied to the coil until the contacts are firmly opened or closed.

operating angle *Electronics.* the electrical angle of the input signal during which plate current is flowing in an amplifier or electron tube.

operating characteristic see LOAD CHARACTERISTIC.

operating characteristic curve *Statistics.* a graph representing the probability of accepting the null hypothesis as a function of the unknown parameter. *Industrial Engineering.* specifically, a graph representing the lot fraction or percentage defective relative to the probability of its acceptance by the sampling plan.

operating control *Transportation Engineering.* the part of a command and control system dealing with vehicle management.

operating fleet *Transportation Engineering.* for a transportation agency at any given time, the number of vehicles in use, either in motion or waiting in stations.

operating force *Military Science.* a force whose primary mission is to participate in combat or to provide integral support to combat forces.

operating handle *Ordnance.* the handle or bar that moves the operating lever of a gun.

operating instructions *Computer Technology.* in running a computer and peripheral equipment, the detailed instructions used for processing a specific job or sequence of jobs, as found in an operator's manual.

operating level *Industrial Engineering.* the normal level of supply needed to maintain operations between arrivals of successive shipments.

operating lever *Ordnance.* the lever that opens or closes the breech of a gun.

operating line *Chemical Engineering.* the line corresponding to compositions of passing streams in a graphical construction for a gas-liquid contactor or a distillation column.

operating part see OPERATION FIELD.

operating point *Electronics.* the point on a response curve that indicates the instantaneous electrode voltages or currents for the operating conditions under study.

operating position *Telecommunications.* in radio or telephone systems, the terminal of a communications channel overseen by an operator.

operating power *Electromagnetism.* the power that is actually supplied to a radio-transmitter antenna.

operating pressure *Engineering.* the pressure at which a system is functioning at any given time.

operating range *Robotics.* the range of a robot as defined by its working envelope and reach capability. Also, **operating space.** *Electronics.* the range of frequencies that will power a given reversible transducer.

operating ratio *Computer Technology.* the ratio of up time (time that a hardware device is reliably operating) to the total scheduled time.

operating stress *Materials Science.* the actual stress level that a material is subjected to during some mechanical operation.

operating system *Computer Technology.* the software, usually provided by the computer manufacturer, that controls the activities of a computer, such as job control and scheduling, allocation of system resources, input/output, and memory management. Also, EXECUTIVE SYSTEM, MASTER SCHEDULER.

operating time *Electronics.* the time period during which an electronic device is in operation.

operation *Surgery.* a surgical procedure; that is, any act performed with instruments or by the hands of a surgeon. *Computer Programming.* **1.** a defined action specified by a single computer instruction or program statement. **2.** the process of executing a defined action. **3.** a combination of an arithmetic or logical operator and one or more operands. *Military Science.* **1.** a military action or the execution of a strategic, tactical, service, training, or administrative military mission. **2.** also, **operations.** the process of combat, including movements, supply, attack, defense, and maneuvers necessary to gain the objective. **3.** of or relating to the planning or conduct of an operation. *Industrial Engineering.* a single industrial process or task. *Mathematics.* **1.** an **internal operation** on a set X is a mapping from $X \times X$ into X. **2.** an **external operation** on a set X is a mapping from $A \times X$ into X, where A is some set other than X. The elements of A are called operators on X.

operational *Engineering.* of a device or system, functioning and ready for use. *Military Science.* **1.** of or relating to the function and effectiveness of materiel and/or personnel. **2.** of or relating to the planning or conduct of an operation or a series of operations.

operational amplifier *Electronics.* an amplifier characterized by very high gain and input impedance, and very low output impedance. The response of a circuit incorporating an operational amplifier is established by the passive components external to the amplifier.

operational calculus *Mathematics.* the body of techniques used to transform problems in analysis into algebraic problems, often involving the solution of a polynomial equation. For example, the application of an integral transform, such as the Laplace transform, to an ordinary differential equation sometimes yields a more easily solvable equation. Also, **operational analysis.**

operationalize *Artificial Intelligence.* to convert general knowledge or principles (e.g.,"buy low, sell high") into executable decision procedures in terms of available data.

operational maintenance *Engineering.* the servicing and repair of equipment or a system that keeps it operational.

operational stage or **period** see CONCRETE OPERATIONAL STAGE.

operational taxonomic unit *Systematics.* in numerical taxonomy, a taxonomic group recognized provisionally for the purposes of a particular analysis.

operational unit *Geology.* an arbitrarily selected stratigraphic unit that is distinguished by certain objective criteria for use in analysis or for any other practical purpose. Also, PARASTRATIGRAPHIC UNIT.

operation analysis *Industrial Engineering.* the analysis of an industrial operation in terms of people, equipment, and processes in order to increase productivity, improve safety, reduce costs, etc.

operation analysis chart *Industrial Engineering.* a chart displaying the work movements required by a basic industrial process.

operation code *Computer Programming.* the mnemonic or binary code used in a machine instruction to identify the action to be performed. Also, INSTRUCTION CODE, OP CODE.

operation cycle *Computer Technology.* in computer control, the machine cycle or step during which the operand is read and the specified operation is executed.

operation decoder *Computer Technology.* in a computer control unit, a digital function that interprets the operation code field of a machine instruction and activates the circuitry to carry out an operation.

operation field *Computer Programming.* the portion of a machine instruction that contains the operation code. Also, OPERATING PART.

operation part *Computer Programming.* the portion of a computer instruction that designates the arithmetic or logical operation to be performed. Also, OPERATION FIELD.

operation process chart *Industrial Engineering.* a chart displaying all aspects of a manufacturing process and their sequential relationships, including methods, inspections, time allowances, and materials.

operation register *Computer Programming.* a special-purpose register that is used to hold the operation code of the next instruction to be executed; the code is subsequently decoded and executed.

operation sheet *Industrial Engineering.* a work plan for a specific work station, listing the operations that are to be performed and the materials, tools, and machinery that are to be used at that work station.

operations management *Industrial Engineering.* a field of management that involves the coordination of design, operation, and control functions in a facility that produces goods or services.

operations planning *Industrial Engineering.* the determination of methods for accomplishing specific tasks with available resources within a given time limit.

operations research *Mathematics.* the mathematical analysis of optimization problems, especially of dynamic systems subject to a given set of constraints. Applications include problems in linear and nonlinear systems, graph theory, game theory, and combinatorics. *Science.* any application of such mathematical analyses within a system in order to evaluate the effectiveness of the system in achieving identified goals.

operation time *Computer Technology.* the machine time required to complete the execution of a specific operation.

operation time chart see OPERATOR PROCESS CHART.

operative *Surgery.* of or relating to an operation. *Science.* in effect or operation.

operator *Engineering.* the person who observes and controls the working and sometimes the maintenance of a machine, device, or system. *Telecommunications.* specifically, the person who operates a telephone company switchboard. *Computer Programming.* **1.** the person who operates a computer. **2.** an arithmetic, logical, relational, or other symbol that identifies an action to be taken with one or more operands. *Artificial Intelligence.* in a state space search, an action that changes a state into a new state, e.g., a move in chess. *Mathematics.* **1.** a mapping, transformation, or function. **2.** an element of a set used to define an external operation on another set. *Genetics.* a sequence of DNA at which a repressor binds to prevent mRNA synthesis from the adjacent gene; it is characteristically composed of one or more palindromic sequences, and it controls gene transcription. Also, **OPERATOR SEQUENCE.** *Behavior.* the predator or prey object that an act of mimicry is intended to deceive.

operator hierarchy *Computer Programming.* the precedence rules that determine the order in which the operators used in an expression are executed.

operator interrupt *Computer Technology.* **1.** a pause in program execution that imparts some information to the operator and usually requires a response or other operator action. **2.** an interrupt caused by action of the operator.

operator process chart *Industrial Engineering.* a chart showing the relationships between the individual work elements and delays that are carried out by the right and left hands in the course of an operation. Also, OPERATION TIME CHART.

operator productivity *Industrial Engineering.* **1.** the work productivity of a machine operator in terms of output per worker hour. **2.** the ratio of standard time to actual performance time for a given task; a ratio of 1.00 signifies that the operator is meeting standard output. Also, **operator performance.**

operator sequence *Genetics.* see OPERATOR.

operator theory *Mathematics.* the study of (usually linear) operators, especially from the viewpoint of their spectra.

operator utilization *Industrial Engineering.* the percentage of total on-duty time that a machine operator is actually at work, operating or monitoring the machine or performing other tasks.

Opercularia *Invertebrate Zoology.* a genus of freshwater colonial ciliate protozoans, with an elongated vase-shaped body, that attach to other invertebrates or submerged objects.

Operculate *Mycology.* a series of fungi of the order Chytridiales, characterized by its distinct mode of discharging zoospores from its spore case, whereby a lid that is formed in the sporangial wall opens.

operculate ascus *Mycology.* in fungi of the class Pezizales, a type of spore sac that discharges its spores through an opening at its tip.

operculum *Anatomy.* a lid or flap covering an aperture, such as the tissue over an erupting tooth. *Vertebrate Zoology.* **1.** a lidlike body process, such as the fold of skin that grows over to enclose the gills during the development of a frog tadpole. **2.** see GILL COVER. *Invertebrate Zoology.* a lid or covering to a chamber, such as that which closes the opening of a gastropod shell when the animal is wholly inside.

operon *Genetics.* a unit consisting of adjacent cistrons that function coordinately under the control of an operator gene.

operon fusion *Molecular Biology.* the fusion of two operons in such a way that the structural genes of one operon are controlled by the regulatory sequences of the other; a process that is used in genetic engineering.

Opheliidae *Invertebrate Zoology.* a family of mud-ingesting polychaete annelids in the subclass Sedentaria, with a conical prostomium and relatively few body segments.

-ophenone *Organic Chemistry.* a suffix used to denote certain aromatic ketones, such as *acetophenone.*

Ophiacodontia *Paleontology.* a reptile suborder of the order Pelycosauria (ancient predatory forms of mammal-like reptiles), which led directly to the order Therapsida (the forerunners of mammals); extant from the Upper Carboniferous to the Upper Permian.

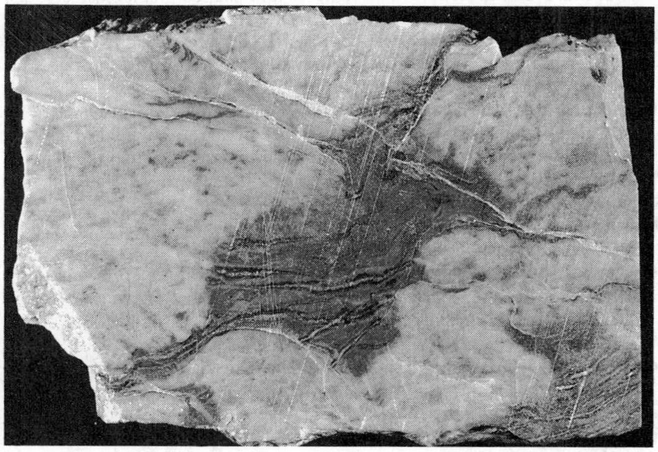

ophicalcite

ophicalcite *Mineralogy.* a recrystallized limestone that contains serpentine and calcite; it is formed by the metamorphosis of a siliceous dolomite.

ophidiasis see OPHIDISM.

ophidic *Vertebrate Zoology.* relating to, caused by, or derived from snakes.

ophidiophobia *Psychology.* an irrational fear of snakes.

ophidism *Toxicology.* poisoning by snake venom. Also, OPHIDIASIS, OPHIOTOXEMIA, OPHITOXEMIA.

Ophiidae *Vertebrate Zoology.* the cusk eels or brotulas, a family of elongate fishes of the order Gadiformes found in shallow coastal and in deep oceanic waters, as well as on coral reefs and in caves. Also, **Ophidiidae.**

Ophiocistioidea *Paleontology.* an extinct class of rare, free-living echinozoans characterized by a low, armless, dome-shaped test; extant in the Lower Ordovician to Middle Devonian.

Ophiocytaceae *Botany.* a monogeneric family of freshwater yellow-green algae of the order Mischococcales, characterized by elongated cylindrical multinucleate cells with multiple chloroplasts.

Ophioglossales *Botany.* the single order of ferns in the subdivision Ophioglossopsida, characterized by fleshy, subterranean stems, fleshy to membranous fronds, and massive sporangia.

Ophioglossidae *Botany.* the adder's tongue ferns, a small subclass of the class Polypodiopsida, characterized by the fertile spike opposed to the leaf blade of the sporophyte.

Ophioglossopsida *Botany.* one of three subdivisions in the fern division Filicophyta, containing the single order Ophioglossales and being usually terrestrial and sometimes epiphytic.

ophiolite *Petrology.* a group of mafic and ultramafic igneous rocks and pelagic sediments, commonly altered or metamorphosed, that were originally part of the oceanic lithosphere.

ophiolitic eclogite *Petrology.* any variety of eclogite associated with an ophiolite.

Ophiomyxidae *Invertebrate Zoology.* a family of brittle stars, echinoderms in the suborder Ophiomyxina.

Ophiomyxina *Invertebrate Zoology.* a suborder of primitive brittle stars, ophiuroid echinoderms in the order Phrynophiurida, with arms that move up and down as well as side to side; disc and arm plates are covered by a thick soft skin.

ophiopluteus *Invertebrate Zoology.* the larvae of many species of non-brooding ophiuroids, with four pairs of elongate arms bearing ciliated bands.

Ophiostomataceae *Mycology.* a family of fungi belonging to the order Eurotiales that occur on wood or tree bark; includes some species that are plant pathogens.

Ophiostomatales *Mycology.* a former term for an order of fungi belonging to the subdivision Ascomycotina, which has been reclassified within the order Eurotiales.

ophiotoxemia or **ophitoxemia** see OPHIDISM.

ophite *Petrology.* a diabase that has retained its embedded crystalline structure despite alteration of the pyroxene to uralite.

ophitic *Petrology.* describing an igneous rock texture that is characteristic of diabase or dolerite, which features lath-shaped plagioclase crystals partly or completely embedded in pyroxene crystals, commonly augite.

Ophiuchus *Astronomy.* the Serpent Bearer, a large constellation visible during summer in the Northern Hemisphere, straddling the celestial equator between Hercules and Scorpius.

Ophiuchus nebula *Astronomy.* a large, dark cloud of dust located in the constellation of Ophiuchus at a distance of about 800 light-years.

Ophiurida *Invertebrate Zoology.* the brittle stars, an order of echinoderms in the subclass Ophiuroidea, having dorsal arm shields and five unbranched arms capable of side to side movement only.

ophiuroid *Invertebrate Zoology.* belonging to, relating to, or resembling the subclass Ophiuroidea.

Ophiuroidea *Invertebrate Zoology.* a class of echinoderms including brittle stars, basket stars, and serpent stars, with long, narrow arms clearly set off from the central disc and a water vascular system that lacks ampullae.

ophryosis *Neurology.* spasm of the facial muscles causing movement of the eyebrows.

ophthalm- a combining form meaning "eye."

ophthalmalgia *Medicine.* any pain in the eye. Also, OPHTHALMODYNIA.

ophthalmectomy *Medicine.* the surgical removal of the entire eye without rupture.

ophthalmia *Medicine.* an extreme, often purulent, conjunctivitis.

ophthalmia neonatorum *Medicine.* a conjunctival inflammation occurring within the first 10 days of life; causes include *Neisseria gonorrhoeae*, *Staphylococcus*, *Streptococcus pneumoniae*, and *Chlamydia oculogenitalis*. Also, INFANTILE PURULENT CONJUNCTIVITIS.

ophthalmic [äf thal′mik; äp thal′mik] *Anatomy.* of or relating to the eye. *Pharmacology.* used in treatment of the eye. Also, **ophthalmologic, ophthalmological.**

ophthalmic nerve *Anatomy.* a sensory nerve, the first division of the fifth cranial (trigeminal) nerve, that supplies the eyeball, lacrimal glands, mucous lining of the eye, the nasal fossae, and the integument of the eyebrow, forehead, and upper eyelid.

ophthalmic zoster *Neurology.* an acute form of herpesvirus characterized by inflammation of sensor ganglia, erupting on or around the cornea. Also, HERPES OPHTHALMICUS.

ophthalmitis *Medicine.* an inflammation of the eye.

ophthalmo- a combining form meaning "eye."

ophthalmodynamometer *Medicine.* **1.** an instrument for determining the near point of convergence of the eyes. **2.** an instrument that measures the blood pressure in the retinal vessels.

ophthalmodynia see OPHTHALMALGIA.

ophthalmologist [äf′thə mäl′ə jist; äp′thə mäl′ə jist] *Medicine.* a physician who specializes in the diagnosis and medical and surgical treatment of diseases and defects of the eye and related structures.

ophthalmology [äf′thə mäl′ə jē; äp′thə mäl′ə jē] *Medicine.* the medical specialty dealing with the eye, including its physiology, diseases, and refractive errors.

ophthalmomalacia *Medicine.* the abnormal softening of the eyeball.

ophthalmometry *Medicine.* the measurement of the eye, especially by measuring the size of the images reflected from the cornea and lens to determine its refractive powers and defects.

ophthalmopathy *Medicine.* any disease of the eye.

ophthalmorrhagia *Medicine.* a hemorrhage from the eye.

ophthalmoscope *Optics.* an instrument, consisting of lenses, a light source, and a pierced concave mirror, that directs reflected light through the pupil; used by ophthalmologists to inspect the interior of the eye.

ophthalmoscopy *Medicine.* the examination of the interior of the eye using an ophthalmoscope.

ophthalmotoxin *Toxicology.* any poison that attacks the eye.

-opia a combining form denoting a condition or defect of the eye or of vision.

opianyl see MECONIN.

opiate *Pharmacology.* **1.** any medication containing or prepared from opium. **2.** broadly, any drug that induces sleep. Thus, **opiatic.**

Opiliaceae *Botany.* a family of terrestrial, mostly evergreen root-parasitic trees and shrubs of the order Santalales that are dicotyledonous and widespread in tropical and subtropical regions.

Opiliones see PHALANGIDA.

opine *Biochemistry.* a bacterial growth substance coded for by genes injected into the plant genome from *Agrobacterium,* a parasitic bacterium that causes crown gall disease in the plants it infects.

opine synthetase *Enzymology.* an enzyme-mediated system that uses cytoplasmic materials in plant cells to produce opines in those cells.

opioid *Pharmacology.* any synthetic narcotic having opiatelike activities, but not derived from opium.

opisometer *Engineering.* an instrument consisting of a thin wheel mounted on a threaded axle that is rolled along a line; used to measure the length of an irregular or curved line, as on a map.

Opisthandria see PENTAZONIA.

opisthaptor *Invertebrate Zoology.* in parasitic flatworms, a large posterior organ with hooks, anchors, or suckers that attach to the host. Also, **opisthohaptor.**

opisthobranch *Invertebrate Zoology.* **1.** any member of the subclass Opisthobranchia. **2.** of or relating to the opisthobranches.

Opisthobranchia *Invertebrate Zoology.* a subclass of mostly marine gastropod mollusks having two pairs of tentacles and a reduced or absent shell and mantle cavity; includes the sea slugs, sea butterflies, sea hares, and bubble shells.

Opisthocoela *Vertebrate Zoology.* in some classifications, an amphibian suborder of lower anurans, including the familes Discoglossidae (firebelly and midwife toads) and Rhynophrynidae (burrowing toads).

opisthocoelous *Anatomy.* of or relating to a vertebra with the centrum convex anteriorly and concave posteriorly.

Opisthocomidae *Vertebrate Zoology.* the hoatzin, a monotypic family of medium-sized gallinaceous birds constituting the order Opisthocomiformes, characterized by soft brown or buff plumage, a reddish head with a stiff crest, weak flight, and a nest consisting of a platform of sticks in a tree above water; its habitat is in rainforests of northern South America.

opisthognathous *Invertebrate Zoology.* having mouthparts ventral and posterior to the cranium. *Anatomy.* having a receding jaw.

opisthonephros *Vertebrate Zoology.* the functional adult kidney of lampreys, amphibians, and most fishes, consisting of a series of renal tubules leading to the opisthonephric duct.

Opisthopora *Invertebrate Zoology.* an order of oligochaetes with male gonopores that open posterior to the last pair of testes.

Opisthorchida *Invertebrate Zoology.* an order of flukes in the class Trematoda that are liver parasites in birds and mammals (including humans), with fish as an intermediate host; includes the Chinese liver fluke.

opisthosoma *Invertebrate Zoology.* the rear part of the body, especially in arachnids.

opisthotic *Anatomy.* of or relating to any anatomical structure that is located behind the ear, such as the temporal bone.

opisthotonos *Neurology.* a spasm of the neck and back muscles in which the head and heels are involuntarily thrown back and the body is arched forward.

opium *Pharmacology.* the air-dried, milky juice of the unripe seed pod of the opium poppy; it is highly addictive, but derivatives are used medicinally, chiefly as painkillers.

opium poppy *Botany.* a Eurasian poppy, *Papaver somniferum* or its variety *album,* cultivated as the source of opium.

opoka *Petrology.* a calcareous sedimentary rock that is porous and flinty, with conchoidal or irregular fracture; it consists of up to 90% fine-grained opaline silica hardened by organically derived silica.

opossum *Vertebrate Zoology.* **1.** a nocturnal marsupial, *Didelphis virginiana,* of the family Didelphidae, characterized by a ratlike skull and face and a long, scaly, mostly hairless prehensile tail; native to the eastern U.S. **2.** any of various related animals of the family Didelphidae, found throughout the New World and ranging in size from that of a mouse to that of a cat. (From a native American word for this animal; literally, "white animal;" recorded by Captain John Smith.)

opossum

opossum shrimp *Invertebrate Zoology.* a common name for small, shrimplike crustaceans of the order Mysidacea.

opp. opposed; opposite.

Oppenauer oxidation *Organic Chemistry.* a reaction in which a secondary alcohol is oxidized to the corresponding ketone.

Oppenheimer, J. Robert 1904–1967, American physicist; director of the Manhattan Project, which produced the first atomic bomb.

Oppenheimer-Phillips reaction *Nuclear Physics.* a type of stripping reaction that occurs in two stages: a deuteron approaching a nucleus breaks up into a proton and a neutron under the electrostatic repulsion of the positive charges that force the proton away, and the neutron is absorbed by the target nucleus, forming a compound nucleus that, if in an excited state, loses excess energy by gamma radiation.

Oppenheimer-Volkoff limit *Astrophysics.* the upper limit for the mass of a neutron star beyond which it collapses to become a black hole; equal to about 1.5–2 solar masses.

Oppenheim's inequality *Mathematics.* if A and B are positive, semidefinite $n \times n$ matrices, then

$$(\det A) \prod_{i=1}^{n} b_{ii} \leq \det (A \circ B);$$

that is, the product of the determinant of A and the elements of the main diagonal of B is less than or equal to the determinant of the Hadamard product of A and B.

opponent-colors theory *Optics.* the theory that some processes involved in color vision may produce two opposite outcomes.

opponent-process theory *Psychology.* the theory that a balance of motivation is necessary for the normal functioning of an individual, and that an overshift in one direction will produce an opponent process that brings the system back into balance; e.g., terror vs ecstasy.

opportunistic *Artificial Intelligence.* describing an expert system or program that is able to take advantage of data that may not always be available, but that provides valuable information leading to or constraining the solution when it is available. *Microbiology.* describing a usually harmless microorganism that becomes pathogenic under favorable conditions, as when the immune system of the host weakens to such an extent that it cannot defend itself against the microorganism.

opportunistic infection *Medicine.* an infection with an opportunistic microorganism; i.e., one that does not normally cause disease but that, under certain circumstances becomes pathogenic, usually because of the patient's weakened immune responses.

opportunistic microorganism *Microbiology.* see OPPORTUNISTIC.

opportunistic species *Ecology.* a species that is able to take advantage of temporary or local changes in a community.

opposed engine *Mechanical Engineering.* an engine with cylinders in the same plane but on opposite sides of the crankcase whose connecting rods work on a common crankshaft placed between the cylinders.

opposing signals *Transportation Engineering.* wayside signals controlling train movement in both directions on the same track.

opposing wind *Oceanography.* a wind blowing in the opposite direction to that in which the waves are traveling.

opposite facing or opposed to; specific uses include: *Botany.* situated on different or diametrically opposed sides of a stem (at the same level), as leaves when there are two on one node.

opposite tide *Oceanography.* a high tide at a point on the earth's surface exactly opposite the point at which a direct tide is occurring.

opposition *Physics.* a situation in which two quantities that vary periodically and have identical frequencies are out of phase with respect to each other by 180° or one half-cycle. *Astronomy.* the difference of 180° in longitude or 12h of right ascension between two celestial objects, most often the sun and a planet or asteroid orbiting beyond earth.

opposition effect *Astronomy.* a marked increase in brightness that occurs when a moon, asteroid, or planet lies at or close to opposition from the sun.

OPS outpatient service.

OPS 5 *Artificial Intelligence.* a knowledge-engineering language for developing rule-based expert systems.

-opsia a variant of the combining form *-opia.*

opsin *Biochemistry.* a protein that binds to retinal or vitamin A aldehyde to produce rhodopsin.

opsonic *Immunology.* of or relating to an opsonin.

opsonic action *Immunology.* a process whereby microorganisms or other cells become more susceptible to the engulfing action of phagocytes, caused by antibodies known as opsonins.

opsonic index *Immunology.* a measure of the relative content of the antibody opsonin in the blood of a person with an infectious disease, as compared to the opsonin content in normal blood.

opsonin *Immunology.* a substance, generally an antibody, that makes a cell or microorganism more susceptible to the engulfing action of phagocytes.

opsonization *Immunology.* a process in which an antigen is combined with the antibody opsonin, to make it more susceptible to the engulfing action of phagocytes.

-opsy a variant of the combining form *-opia.*

opt optical, optician, optics.

opt- or **opto-** a prefix meaning vision. Also, **optic-, optico-.**

optic *Biology.* of or relating either to the eye or to vision. *Optics.* **1.** any of the lenses, prisms, or mirrors of an optical instrument or system. **2.** a former term for an eyeglass or magnifying glass. **3.** see OPTICAL. (Going back to a Greek word meaning "to see.")

optical *Optics.* **1.** of or relating to light, vision, or optics. **2.** of or relating to techniques, devices, or systems that are designed to assist vision, utilize light, or employ the principles of optics. Also, OPTIC.

optical aberration see ABERRATION.

optical achromatism see VISUAL ACHROMATISM.

optical activity *Optics.* in optically active solutions and crystals, the rotation of the plane of polarized light as it passes through the substance; the angle of rotation is proportional to the thickness of the substance. Also, ROTATING POLARIZATION.

optical air mass *Geophysics.* the amount of atmosphere between an observer and a celestial object taken as a multiple of the zenith value; air mass is equal to 1 when the object is in the zenith and it increases as the secant of the object's zenith distance.

optical amplifier *Engineering.* a device that enhances the output of a signal without distortion by transforming the signal to light, amplifying the light, and then converting the light back to an electric signal.

optical analysis *Optics.* a method that employs the effects of transmitted light and the principles of optics to analyze the characteristics of a substance, such as its chemical composition or response to stress.

optical anisotropy *Optics.* the behavior of a substance or molecule whose effect on electromagnetic radiation varies with the direction of propagation; for example, the refractive index of a crystal of calcite varies with the direction of a ray of transmitted light.

optical anomaly *Physical Chemistry.* a phenomenon in which light passing through an organic compound does not refract in the manner expected from the compound's atomic structure.

optical aspherical lens see ASPHERICAL LENS.

optical axis *Anatomy.* a straight line from the center of the visual field to the fovea of the retina. *Optics.* the imaginary line that passes through the centers of curvatures of both optical surfaces of a single lens or the centers of all surfaces of a correctly centered lens system.

optical bar-code reader *Computer Technology.* any of a class of optical data-capture devices that read coded information from imprinted documents or labels; e.g., price codes in a supermarket or other store. Also, **optical bar-code scanner.**

optical bench *Engineering.* a rigid but movable bar or track equipped with mountings for optical components that is capable of precise longitudinal movement of one component relative to the others during experiments.

optical binary see OPTICAL DOUBLE STAR.

optical bistability *Optics.* the characteristic that causes a substance or optical device to have two stable states of transmission, one high and one low, depending on the intensity of the light passing through it, similar to a semiconductor in an electrical device.

optical branch *Solid-State Physics.* the branch in the dispersion relation corresponding to the optical mode of vibration of a crystal medium; the frequencies of this mode are higher than those of the acoustic mode and generally vary less with frequency.

optical calcite *Mineralogy.* a chemically pure, colorless variety of calcite whose crystals are transparent enough for optical uses, such as Nicol prisms.

optical center *Optics.* a point located on the axis of a lens wherein each ray passing through that point emerges from the lens on a path parallel to the corresponding incident ray.

optical character reader *Computer Technology.* an input device that scans a printed page, recognizes the characters by their shape, and converts the characters into character codes.

optical character recognition *Computer Technology.* a process by which machine-printed or hand-printed numerals, letters, and symbols may be converted into computer-processable information.

optical coating *Optics.* the process of coating an optical surface with transparent layers of evaporated calcium or magnesium fluoride to reduce reflectivity and provide a protective veneer; metallic substances are sometimes applied in cases where increased reflectivity is desired.

optical communication *Telecommunications.* the employment of electromagnetic waves in the area of the spectrum within or near the wavelengths of visible light for the transmission of signals.

optical comparator *Engineering.* a comparator equipped with optical devices capable of projecting the enlarged image of an object on a screen for analysis. Also, VISUAL COMPARATOR.

optical computer *Computer Technology.* a proposed very-high-speed computer that uses optical switches instead of transistors with fiber gates and laser light.

optical contact *Optics.* a contact between two surfaces so similar in form and close-fitting that they are separated by a distance appreciably less than a wavelength of light and require no bonding substance to adhere.

optical coupling *Electronics.* the coupling of signals from one electronic circuit to another by means of electromagnetic radiation, such as a light beam or light pipe; this provides isolation for the circuits using a short-length optical path.

optical crystal *Crystallography.* a crystal whose piezoelectric properties are used for ultraviolet and infrared experimental purposes; e.g., silver chloride, sodium chloride, and potassium iodide.

optical density *Optics.* a measure of the light absorption of a translucent medium, equivalent to the logarithm of the opacity, expressed as a fraction. Also, ABSORBANCE, DENSITY.

optical diffraction meter see LASER VELOCIMETER.

optical disk *Computer Technology.* a high-density storage medium that uses a laser to write information by burning a hole in the disk surface; information is read by a weaker laser beam. Also, LASER DISK.

optical-disk storage *Computer Technology.* computer storage composed of optical disks, which are generally not erasable. Also, **optical memory, optical storage.**

optical dispersion *Optics.* the separation of a beam of light into its constituent wavelengths or colors; caused by refraction that occurs when light passes through a transparent medium. Also, CHROMATIC DISPERSION, DISPERSION.

optical distance see OPTICAL PATH.

optical Doppler effect *Electromagnetism.* the observed shift in the frequency of light due to relative motion between the light source and the observer.

optical double star *Astronomy.* two stars that appear close together as viewed from earth, but are actually at great distances from each other; this effect is due to the alignment of both stars along the line of sight. Thus, **optical double galaxy.**

optical electronic reproducer see OPTICAL SOUND HEAD.

optical electronics see ELECTRO-OPTICS.

optical element *Optics.* a component in an optical instrument or system, such as a lens, prism, or mirror, that affects light traveling through the instrument; usually composed of a single material.

optical encoder *Electronics.* a device that converts information into digital data by interrupting light beams focused onto photoelectric devices.

optical exaltation *Physical Chemistry.* a phenomenon in which light passing through a substance exhibits a refraction that exceeds the calculated one; generally, all optical anomalies fall into this category.

optical fiber *Optics.* a transparent thread, such as fused silica, that transmits light; used in the technology of fiber optics. Also, LIGHT GUIDE.

optical-fiber sensor *Engineering.* an instrument that detects variations in the transmitted power or the rate of transmission of light in an optical fiber; used to measure such factors as temperature, pressure, voltage, rotation, and particle velocity.

optical figuring *Optics.* the process of grinding and polishing to shape optical elements or surfaces to specification.

optical finishing *Materials Science.* the forming of finished optical components, such as lenses and windows, from raw optical materials, including cutting, grinding, and polishing by abrasion.

optical flat *Optics.* **1.** a surface that is almost perfectly flat, to within 1/10 of a wavelength of light; used to transfer light in optical devices, such as telescopes. **2.** a glass or quartz disk, precisely polished flat on one or both sides; used in interferometry for testing the flatness of plane surfaces.

optical frequency *Physics.* a frequency of radiation in the optical region of the spectrum: the infrared, visible, and ultraviolet regions.

optical galaxy *Astrophysics.* a galaxy with a compact nucleus that resembles a star, due to its distance.

optical gauge *Engineering.* **1.** a gauge that magnifies the image of an object, thus allowing very sensitive measurements of minute details to be obtained. **2.** a gauge that determines the size of an object indirectly from its image.

optical glass *Materials.* any of several types of high-quality, color-free glass, such as flint or crown glass, that has specified refractive properties; used in lenses and other optical-system components.

optical harmonic *Solid-State Physics.* light that is generated in a transparent crystal by passing a high-power laser beam through it; the generated light has an integral multiple of the frequency of the incident laser light.

optical haze see TERRESTRIAL SCINTILLATION.

optical indicator *Engineering.* an indicator that magnifies the indicator diagram of the pressure in an engine cylinder and projects it on a glass screen, from which photographic recordings are made.

optical indicatrix *Crystallography.* a three-dimensional ellipsoid whose surface is defined by vectors from an origin having lengths proportional to the refractive indices in that direction.

optical instrument *Optics.* any system or device that affects light in an expressed way, such as forming an image or generating a given polarization or wavelength.

optical interference *Optics.* the effect occurring when two coherent light waves of the same frequency are combined. Also, INTERFERENCE, INTERFERENCE OF LIGHT.

optical isomerism *Physical Chemistry.* the existence of a pair of compounds whose molecular structures mirror each other, so that each form has an asymmetric atom and they rotate plane-polarized light in opposite directions. Thus, **optical isomer.**

optical lantern *Engineering.* an optical projection system designed to project color transparencies onto a screen. Also, SLIDE PROJECTOR.

optical length see OPTICAL PATH.

optical lens see LENS.

optical lever *Optics.* a device used to measure the slight angular displacements of a rotating object; light is beamed to a mirror that is located on the moving object, and the position of the reflected beam is noted.

optically active *Physical Chemistry.* describing a substance that causes optical activity, i.e., that rotates the plane of vibration of plane-polarized light. Thus, **optically active molecule.**

optically active polymer *Materials Science.* a biological or synthetic polymer that is capable of rotating the plane of polarization of light; usually associated with the presence of asymmetric intramolecular compensation due to the pressure of asymmetric carbon atoms.

optically effective atmosphere see EFFECTIVE ATMOSPHERE.

optically pumped laser *Optics.* a laser in which population inversion is activated by the absorption of light originating from an auxiliary light source.

optically pure *Organic Chemistry.* a term used to describe a sample of a pure chiral compound that is uncontaminated by its enantiomer.

optical mark reading *Computer Technology.* the process of optically interpreting a series of marks in specified positions on a piece of paper; may be either offline or under program control.

optical mask *Electronics.* a sheet of metal or other material that has an open pattern and is used to illuminate the photoresistive material on an integrated circuit.

optical measurement *Optics.* a measurement of the various characteristics of electromagnetic radiation given off by or reflected from a surface, or of such radiation passing through a medium.

optical meteor *Optics.* a phenomenon that occurs in the atmosphere and can be explained using principles of optics, such as a mirage, rainbow, or halo.

optical microscope *Optics.* an instrument that enlarges the image of a tiny specimen by passing visible light through a condenser and an objective lens. Also, LIGHT MICROSCOPE, PHOTON MICROSCOPE.

optical microscopy *Optics.* the use of an optical microscope.

optical mode *Solid-State Physics.* a vibrational mode in a crystal lattice in which neighboring atoms or molecules of different sublattices oscillate in opposing directions.

optical modulator *Telecommunications.* any device that impresses information onto a light beam; e.g., lasers or light-emitting diodes.

optical monochromator *Spectroscopy.* a monochromator used in spectroscopic instruments to select wavelengths in the visible, infrared, and ultraviolet regions of the electromagnetic spectrum.

optical null method *Spectroscopy.* an infrared spectroscopic technique in which the energy transmitted by a reference beam is adjusted to match that of the sample beam before recombination, e.g., by use of an attenuator.

optical oceanography *Oceanography.* the branch of oceanography that is concerned with the optical properties of seawater and of natural light in seawater.

optical parallax see PARALLAX.

optical parametric amplification see PARAMETRIC AMPLIFICATION.

optical parametric oscillator *Optics.* a device that pumps a nonlinear material with a laser to produce a coherent beam of light that can be tuned over a wide range of wavelengths.

optical path *Optics.* the distance of the path traversed in a medium by a ray of light, multiplied by the refractive index of that medium. Also, OPTICAL LENGTH, OPTICAL DISTANCE.

optical phase conjugation *Optics.* a phenomenon that reverses the direction of plane waves in a light beam by means of nonlinear optical effects; frequently used to replicate laser beams while eliminating previously introduced aberrations. Also, WAVEFRONT REVERSAL.

optical phenomena *Electromagnetism.* the physical phenomena involved with the generation, transmission, and detection of radiation in the infrared, visible, and ultraviolet ranges of the electromagnetic spectrum.

optical phonon *Solid-State Physics.* a quantized lattice vibration of the optical type.

optical printer see PROJECTION PRINTER.

optical projector or **optical projection system** *Optics.* see PROJECTOR, def. 1.

optical property *Electromagnetism.* any property that involves the interaction between light and a medium, including reflection, refraction, diffraction, and scattering.

optical proximity sensor *Engineering.* an optical device utilizing triangulation of reflected light to measure limited distances.

optical pulsar *Astronomy.* a pulsar that emits pulses of light at visible wavelengths.

optical pulse *Optics.* one of a series of regular flashes of light whose waveforms have short duration compared to the total time concerned.

optical-pulse compression *Optics.* the shortening of the duration of each optical pulse in a series by means of techniques similar to those used in chirp radar.

optical pumping *Optics.* a process used to trigger lasers, in which light of suitable frequency from an external source increases the energy levels of atoms within the laser from a lower quantum state to a higher state.

optical pyrometer *Engineering.* an instrument used to measure the temperature of a heated material by comparing the incandescent color of the surface with that of a test wire heated by electric current to a known temperature. Also, DISAPPEARING FILAMENT PYROMETER.

optical quality crystal *Materials Science.* a crystal that possesses good light transmission and refraction properties.

optical quenching *Optics.* a process in which internal or external forces reduce the intensity of emitted fluorescence, phosphorescence, and luminescence, as when molecules within a ray collide or when radiations of the same frequency traverse an oscillator in a different direction.

optical rangefinder *Engineering.* **1.** an instrument that determines the distance of a remote object by measuring the angle between two convergent beams of light on the object from disparate points of view. **2.** an instrument that measures the distance of a remote object using the time it takes a wave of sound, light, or electromagnetic energy to travel from the object to the instrument, as with sonar, laser, and radar systems.

optical reader *Computer Technology.* a class of data-capture devices that are used in the optical reading of marks, characters, or bar codes. Thus, **optical reading.**

optical recording *Engineering.* a recording made on photographic paper by a focused beam of light that shifts position in conjunction with a quantity being measured by a meter.

optical rectification *Optics.* an effect that occurs when one or more electromagnetic waves propagating in a nonlinear medium create a nonoscillating, second-order polarization, which in turn gives rise to an electrical voltage. *Photogrammetry.* a technique that corrects for a lack of parallelism between a subject and the camera during project printing.

optical reflectometer *Engineering.* an instrument used to measure the ratio of reflected flux to incident flux (reflectance) at wavelengths in or near the visible spectrum.

optical relay *Electronics.* an instrument that employs a light sensor to switch a circuit on and off as conditions change, as in an outdoor security light that switches on at night and off in daylight.

optical-rotary encoder *Robotics.* an encoder based on a revolving slotted disk that is placed between a light source and a detector.

optical rotation *Optics.* the rotation of the plane of polarization in a beam of light as it passes through an optically active substance; may be clockwise or counterclockwise.

optical rotatory dispersion *Optics.* a variation in optical rotation, caused by optically active substances as the wavelength is varied.

optical scanning *Optics.* the process of interpreting data in a bar-code or other visual form by means of a device that scans and identifies the data. Thus, **optical scanner.**

optical sound head *Electronics.* the component in a motion-picture projector that reproduces photographically recorded sound by means of light from an incandescent lamp focused onto the film's optical sound track and picked up by a photoelectric cell. Also, **optical electronic reproducer.**

optical sound recorder see PHOTOGRAPHIC SOUND RECORDER.

optical spectrograph *Spectroscopy.* an optical spectroscope in which the spectrum is recorded on photographic film.

optical spectrometer *Spectroscopy.* an optical spectroscope calibrated to measure either wavelengths of light or the refractive index of a prism.

optical spectroscope *Spectroscopy.* a device that produces an optical spectrum following the absorption or emission of radiant energy by a material sample. Thus, **optical spectroscopy.**

optical spectrum *Spectroscopy.* any spectrum having wavelengths anywhere in the ultraviolet, visible, or infrared regions of the electromagnetic spectrum.

optical square *Engineering.* in surveying, a hand-held instrument, using two mirrors arranged at a 45° angle, that works on the principle of the sextant to set out right angles.

optical storage see OPTICAL-DISK STORAGE.

optical superposition principle *Optics.* a principle stating that a resultant disturbance is the sum of two separate wave disturbances passing through each other unaffected; applicable only to linear phenomena, and not applicable when light intensity is exceedingly high.

optical surface *Optics.* the surface that lies between two media, such as air and glass, and reflects or refracts light.

optical system *Optics.* an assembly of elements such as mirrors, lenses, or prisms that affects light; frequently part of a system constructed to perform optical functions.

optical theorem *Quantum Mechanics.* a theorem relating the removal of flux from an incident beam that results from the destructive interference occurring between an incident wave and the forward-scattered amplitude in particle-scattering processes.

optical thickness *Optics.* the product of the physical thickness of a transparent optical medium multiplied by its refractive index.

optical thin film *Materials Science.* a coating of a single layer or multiple thin layers of amorphous, crystalline, or polymerized materials on surfaces of optical components such as lenses and mirrors; used to reduce reflections from optical parts, produce highly reflective surfaces, and protect components against abrasion and ambient moisture.

optical-to-optical interface device *Optics.* a device that transforms an image illuminated by noncoherent light into a coherently illuminated object that can be used as input for certain data processors.

optical tracking *Engineering.* the tracking of airplanes, missiles, or satellites using telescopes and ballistic cameras.

optical train *Optics.* a series of mirrors, lenses, and prisms through which light passes.

optical transition *Physics.* any energy transition occurring in an atom or molecule in which a photon is emitted or absorbed, having a frequency in the infrared, visible, or ultraviolet regions.

optical twinning *Crystallography.* the twinning of crystals so that one crystal is a mirror image of the other.

optical type font *Computer Technology.* a special font that is designed to be read easily both by humans and by optical character readers.

optical waveguide *Electromagnetism.* a waveguide that is designed to transmit electromagnetic radiation of the infrared, visible, or ultraviolet ranges, commonly using a dielectric material such as glass or plastic fiber.

optical window *Optics.* **1.** a piece of glass or other transparent material that admits light into an optical system and bars dust or moisture. **2.** a specific region in the electromagnetic spectrum that is singled out for a particular purpose; for example, an optical window through which visible and near-visible radiation will pass to the atmosphere is between 300 and 2000 nanometers.

optic angle *Crystallography.* an acute angle (generally designated $2V$) marking the angle between two optic axes in the XY plane, where X and Y represent the two smaller refractive indices n_x, n_y, in a biaxial indicatrix.

optic apraxia *Medicine.* an inability to draw or construct simple or complex shapes, from memory or from models. Also, GEOMETRIC APRAXIA, VISUAL APRAXIA, CONSTRUCTIONAL APRAXIA.

optic axis *Crystallography.* the direction in a birefringent crystal along which the ordinary and extraordinary rays travel at the same speed. Uniaxial crystals have one such axis; biaxial crystals have two.

optic canal *Anatomy.* a tubular passage in the lesser wing of the sphenoid bone through which the optic nerve and the ophthalmic artery pass.

optic capsule *Vertebrate Zoology.* one of a series of structures (fibrous in higher vertebrate embryos and cartilaginous in lower vertebrates) that surround the eye and are independent of any other structure, thus allowing unrestricted movement.

optic chiasma *Anatomy.* the location anterior to the hypothalmic infundibulum at which the optic nerve fibers from the medial half of each retina cross to the opposite side of the brain.

optic cup *Developmental Biology.* the primordium of the retina, formed by the collapse of the optic vesicle.

optic disk *Anatomy.* a small area on the retina where the fibers of optic nerve leave the eye; it is normally the only point on the retina that is insensitive to light. Also, BLIND SPOT.

optic gland *Invertebrate Zoology.* either of a pair of endocrine glands located near the brain of octopuses, cuttlefish, and squid.

optician *Engineering.* a person who makes and sells lenses, optical instruments, and eyeglasses.

optic lobe *Anatomy.* the superior colliculi, paired rounded eminences on the dorsal surface of the midbrain; with the auditory inferior colliculi they make up the corpora quadrigemina. *Vertebrate Zoology.* in lower vertebrates, one of a pair of prominences on the dorsal part of the midbrain serving mainly for processing visual information. Also, CORPORA BIGEMINA. *Invertebrate Zoology.* in certain arthropods, lateral lobes of the forebrain containing the visual centers of the compound eye.

optic nerve *Anatomy.* either of the paired, second cranial nerves that connect the retina to the brain and transmit neural impulses contributing to the sense of sight.

optic normal *Optics.* the axis positioned perpendicularly to the optic axis.

optics *Physics.* a branch of physics that studies the phenomena associated with electromagnetic radiation in the wavelength range of about 1 nm to 1 mm; particular emphasis is devoted to the visible, infrared, and ultraviolet regions.

Optics

Optics is the study of the generation, propagation, manipulation, and detection of light; the interaction of light with matter; and the use of light in a wide variety of scientific, industrial, commercial, and military devices and systems. Image formation has a special place because of the human visual system. Optics uses radiation from the ultraviolet, visible, and infrared regions of the electromagnetic spectrum.

Geometrical optics, the oldest but still an important field, means ray optics and rectilinear propagation. Traditionally it relates to natural light (incoherent) and the use of reflection and refraction that leads to devices such as prisms, lenses, and mirrors that are used in glasses, cameras, telescopes, binoculars, and microscopes.

Physical optics depends upon the wave nature of light and covers the phenomena of interference, diffraction, and polarization leading to such devices as diffraction gratings, polarizers, and interference filters; systems include interferometers and grating spectrometers. These wave phenomena are coherent light phenomena and have become of vital importance since the development of the laser—a convenient source of coherent light.

Quantum optics depends upon the recognition that light consists of photons. Thus, the phenomena include the photoelectric effect, photoconductivity, photoluminescence, nonlinear optics and, of course, the laser. The term photonics is sometimes used for the application of these and related phenomena.

Hybrid optical systems involve the interaction of light with sound (acousto-optics), with electric fields (electro-optics) and with magnetic fields (magneto-optics).

We use traditional optical devices almost every day and newer devices using fiber optical communications, laser disks, and laser scanning supermarket check-out systems are now commonplace. The future holds much in store for us all in the use of light, i.e., optics.

Brian J. Thompson
Professor of Optics and Provost
The University of Rochester

optic stalk *Developmental Biology.* the connection between the optic vesicle and the brain, serving at a later stage of development as the framework for the optic nerve.

optic tectum *Vertebrate Zoology.* a thick layer of gray matter composing the roof of the midbrain through which impulses from optic nerve fibers are relayed to the cerebral hemispheres; in lower vertebrates, it is characterized by the presence of the optic lobes; in mammals it is composed of two pairs of swellings, the corpora quadrigemina and the anterior colliculi.

optic tract *Anatomy.* bundles of optic nerve fibers posterior to the optic chiasma that carry impulses from the eyes to the thalamus and other portions of the brain.

optic vesicle *Developmental Biology.* a bulbous outgrowth developing from either side of the forebrain of the early embryo, from which the percipient parts of the eye are formed. Also, VESICULA OPHTHALMICA.

optimal allocation *Ecology.* the use of limited resources, especially energy and time, to best achieve certain goals.

optimal control *Control Systems.* a control method in which a system's response to a commanded input is judged to be optimal in terms of a specific performance criterion, given the dynamics of the process to be controlled and the limitations of measurement.

optimal control theory *Control Systems.* a theory based on the calculus of variations for dynamic systems in which an independent variable, such as time, is introduced and control variables are determined to maximize or minimize a measure of the performance of the system under specific constraints.

optimal feedback control *Control Systems.* a subfield of optimal control theory in which the current state of a system is used to determine the control variables.

optimal-foraging theory *Behavior.* a theory stating that an animal's behavior in obtaining food consists of actions that will produce the maximum intake for the minimum effort; e.g., large predators do not pursue small, highly elusive animals, because the energy loss from the chase would exceed any food intake that could be obtained by consuming the prey.

optimality theory *Ecology.* a theory in behavioral studies dealing with the limits and rewards of the best way to survive through natural selection.

optimal policy *Mathematics.* in a dynamic optimization problem, a sequence of decisions, called decision functions, changing the states of a system in a manner such that a given object function is minimized or maximized.

optimal programming *Control Systems.* a subfield of optimal control theory in which the control variables are determined as functions of time for a specific initial state of the system.

optimal proportions *Immunology.* the ratio to which antibodies and antigens are mixed in order to obtain the fastest reaction.

optimal regulator problem see LINEAR REGULATOR PROBLEM.

optimal smoother *Control Systems.* an algorithm that generates an estimate of a dynamic variable at a certain time, using all available data on past and projected performance.

optimal system *Mathematics.* a system in which the variables representing the various states are determined so that a given object function is minimized or maximized subject to the given constraints.

optimization *Mathematics.* the process of finding a solution to a problem that is represented by a mathematical model, subject to some given constraint. The desired solution will usually appear as a maximum or minimum.

optimization theory *Mathematics.* the body of techniques used to determine a specific solution in a defined set of possible alternatives that will best satisfy a selected object function or set of constraints; it includes linear, nonlinear, dynamic, and stochastic programming, control theory, and aspects of combinatorics. Also, MATHEMATICAL PROGRAMMING.

optimize *Computer Technology.* **1.** to design a program or system so as to minimize processing time or cost, or maximize throughput or efficiency. **2.** to make transformations to a program, e.g., a form of program produced by a compiler, to improve its efficiency.

optimized code *Computer Programming.* the machine-language product of a compiler that optimizes computer operations and memory for the target computer so that program execution will take the least amount of time and memory.

optimizer *Computer Programming.* a compiler or assembler that attempts to recognize and revise inefficient logic in order to optimize an object program.

optimizing control function *Control Systems.* the level in the functional decomposition of a control system that determines the required relationships among the system's variables in order to achieve an optimal or suboptimal performance, based on a given model of the plant and its environment.

optimum array current *Electromagnetism.* a current distribution in a broad side antenna array such that, for a specified side-lobe level, the beam width is as narrow or as small as possible.

optimum bunching *Electronics.* a bunching condition that produces the greatest output in a velocity-modulated tube, such as a klystron.

optimum charge *Ordnance.* a propelling charge that produces the maximum velocity at a given pressure.

optimum code *Computer Programming.* a computer code written so that it is efficient in some respect such as execution time, memory use, or cost.

optimum coupling see CRITICAL COUPLING.

optimum cure *Chemical Engineering.* a specified time and a definite temperature that will result in maximum desired properties, such as tensile strength or abrasion resistance.

optimum density *Transportation Engineering.* the traffic density at which a highway supports the greatest total volume of traffic; i.e., the density point above which congestion will cause traffic to slow.

optimum filter *Electronics.* a filter designed for the best possible discrimination between signals of a given set.

optimum flight *Space Technology.* the flight of an aircraft or spacecraft that is planned and navigated to achieve minimum flight time and minimum exposure to hazardous conditions.

optimum height *Ordnance.* the height of an explosion that will produce the maximum effect against a given target.

optimum height of burst *Ordnance.* the estimated height of burst at which a nuclear weapon of a given energy yield will produce a particular desired effect over the maximum possible area.

optimum moisture content *Geology.* the water content of a soil mass that has been compacted to its maximum dry weight by a specified force.

optimum programming *Computer Programming.* the practice of programming for maximum efficiency with respect to cost, execution time, storage requirements, or use of other system resources.

optimum reverberation time *Acoustics.* the ideal amount of time for sound to travel in an acoustic environment such as a concert hall or auditorium; normally less than one-third of a second for good listening conditions.

optimum speed *Transportation Engineering.* the average traffic speed at which a highway supports the greatest total volume of traffic. Also, CRITICAL SPEED.

optimum working frequency *Telecommunications.* at a preset time, the most potent frequency for ionospheric propagation of radio signals between two points.

optimum yield *Ecology.* the maximum number of members that can be obtained from a population without impairing its ability to produce a comparable number of replacement members. Thus, **optimum population size.**

optional pause instruction *Computer Programming.* an instruction that temporarily halts the execution of a program. Also, **optional halt instruction, optional stop instruction.**

option switch *Computer Technology.* a hardware switch or toggle that invokes an optional feature of a peripheral device, such as "near letter quality" printing. *Computer Programming.* a software value or parameter that replaces a default value to invoke an optional feature, such as selecting a different default disk drive.

opto- a combining form meaning "optic" or "vision," as *optoacoustic* or *optometrist.*

optoacoustic effect *Physics.* an effect in which a beam of light passing through a gaseous medium is capable of generating sound in the medium if the beam is periodically interrupted at some characteristic acoustic frequency.

optoacoustic spectroscopy *Analytical Chemistry.* the measurement of sound waves generated by the absorption of radiation by chemical substances, especially gases; the absorbed radiation is transformed into motion. Also, **optoacoustic detection method.**

optoelectronic *Electronics.* of or relating to a device that responds to optical power, emits optical radiation, or uses optical radiation for its own operation; any device that combines optic and electric parts, or that functions as an electrical-to-optical transducer (or vice versa). Thus, **optoelectronic device, optoelectronic element.**

optoelectronic amplifier *Engineering.* an amplifier that transforms electronic signals into optical signals, or vice versa, and then augments them.

optoelectronics *Electronics.* the branch of electronics that examines the technology of the coupling of functional electronic blocks with light beams; it often involves solid-state emitters and detectors of light. Also, ELECTRO-OPTICS.

optoelectronic scanner *Electronics.* a scanner that has an optical device, such as a mirror or lens, between its light source and its photoelectric device.

optoelectronic shutter *Engineering.* a camera shutter used to control or block a light beam by means of a Kerr cell.

optogalvanic spectroscopy *Spectroscopy.* a spectroscopic technique in which an arc or flame absorption spectrum is obtained by measuring the changes in voltage and current when a sample is irradiated by a laser.

optoisolator *Electronics.* a device that couples electric circuits with a light beam. Also, PHOTOCOUPLER.

optometer *Medicine.* any of several instruments for measuring the refractive error of an eye.

optometrist [äp täm′ə trist] *Medicine.* a person trained and licensed to practice optometry.

optometry [äp täm′ə trē] *Medicine.* **1.** the professional practice of primary eye and vision care, including the diagnosis, treatment, and prevention of disorders and the prescription of corrective lenses. **2.** use of an optometer.

optophone *Acoustical Engineering.* a device that recognizes printed text by means of a photoelectric cell and then produces sounds based on the text characters.

Opuntia *Botany.* a genus of cacti of the dicotyledonous xerophytic plant family, characterized by leafs that are reduced to spines, flowers borne singly directly on the stem, and fruit that is a berry; grown in California, Mexico, and some Mediterranean countries for its fruit. Also, PRICKLY PEAR.

Opuntia cactus

OPV oral polio vaccine.

oquasa *Vertebrate Zoology.* a brook trout, *Salvelinus aquassa,* that is small and dark blue in color; found especially in Maine.

OR *Mathematics.* a logic operator having the value false if all of its arguments are false, and true otherwise (i.e., if at least one of the arguments is true). *Computer Technology.* **1.** an electronic logic circuit that returns a binary 1 if any of the input signals is a 1, or a 0 if none of the input signals is a 1. **2.** an instruction that performs the OR function simultaneously on corresponding bits of two words. Also, DISJUNCTION.

OR or **O.R.** operating room; orienting response; operations research; organizational requirement.

or- or **oro-** a prefix meaning "mouth."

orach *Botany.* any plant of the genus *Atriplex,* especially *A. hortensis,* of the goosefoot family, cultivated for food. Also, **orache.**

oracle *Computer Science.* a procedure, usually imaginary, that can give a correct answer to a certain kind of question.

orad *Biology.* toward the mouth or the oral region.

oral *Biology.* relating to or involving the mouth. *Anatomy.* located in the mouth. *Pharmacology.* of a medication, taken through or applied in the mouth. *Psychology.* **1.** of or relating to the oral stage of psychosexual development. **2.** having or displaying oral character traits.

oral-aggressive type *Psychology.* a personality type that derives from insufficient satisfaction at the oral stage, characterized as hostile, pessimistic, and demanding.

oral arm *Invertebrate Zoology.* in a jellyfish, one of four armlike projections surrounding the mouth that bear nematocysts and are used for gathering food.

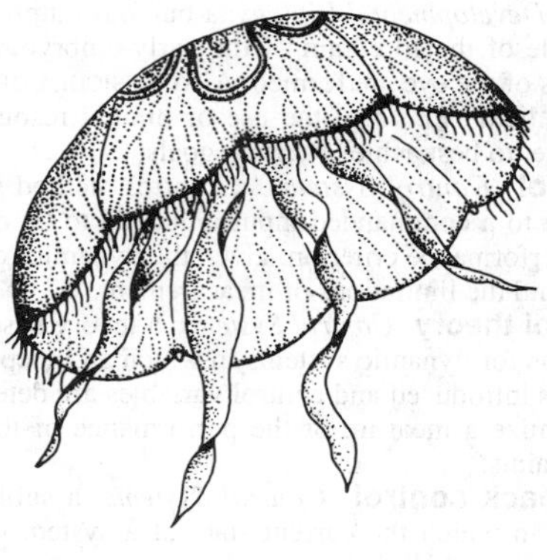

oral arms of a jellyfish

oral cavity *Anatomy.* the space in the mouth that contains the teeth, salivary ducts, and tongue.

oral character *Psychology.* a personality type that derives from unresolved conflicts during the oral stage of psychosexual development, considered to be manifested by extreme self-centeredness and by overindulgence in forms of oral gratification such as eating, smoking, and drinking. Also, **oral personality.**

oral contraceptive *Pharmacology.* any drug taken by mouth to prevent conception or pregnancy.

oral disk *Invertebrate Zoology.* **1.** the flattened upper end of a polyp with the mouth in its center and tentacles around the margins. **2.** the sucker-bearing mouth of some trematodes. **3.** the central part of a sea star's body.

oral eroticism *Psychology.* **1.** excessive pleasure derived from oral activities such as eating, smoking, or drinking, reflecting a fixation on the oral stage of psychosexual development. **2.** any excessive tendency to derive pleasure from oral activity.

oral-erotic stage see ORAL STAGE.

oral groove *Invertebrate Zoology.* a depressed region resembling a groove around the mouth of some protozoans.

oral herpes *Pathology.* a disease characterized by a cluster of small, transient blisters that occur chiefly at the edge of the lip or nostril, caused by herpes simplex virus type 1.

orally *Medicine.* by mouth.

oral-passive type *Psychology.* a personality type that derives from extensive satisfaction at the oral stage, characterized as lethargic, indifferent, and overdependent.

oral-receptive type *Psychology.* a personality type that derives from sufficient satisfaction at the oral stage, characterized as optimistic and friendly.

oral spear *Invertebrate Zoology.* in some nematodes, a spearlike protrusion used to puncture animal prey or plant cells.

oral stage *Psychology.* in psychoanalytic theory, the first stage of psychosexual development, lasting from birth to 18 months, during which the mouth is the focus of sexual gratification through the sensations associated with such activities as sucking and biting. Also, **oral period** or **phase.**

oral sucker *Invertebrate Zoology.* a suction device surrounding the mouth of certain invertebrates, such as flukes.

oral surgeon *Surgery.* a physician specializing in oral surgery.

oral surgery *Surgery.* the branch of medicine that deals with the surgical treatment of diseases, injuries, and defects of the mouth, jaws, and associated structures.

oral thermometer *Medicine.* an instrument designed for measuring temperature by mouth.

oral tradition *Anthropology.* the collected information of a culture in spoken form, including practical lore, moral lessons, history, tales, and mythology, that is passed from generation to generation. Also, **oral literature.**

orange *Botany.* **1.** any of various evergreen trees of the genus *Citrus* in the family Rutaceae, widely cultivated in the United States and elsewhere for their flavorful, nutrient-rich fruit. **2.** the fruit itself, which is globular with a leathery orange-colored rind and a sweet, juicy, membranous pulp. *Optics.* the color of this fruit; a color sensation that corresponds to radiation in the 597 to 622 nanometer wavelength region of the visible spectrum; perceived as a reddish yellow.

orange

orange blossom *Botany.* the flower of the orange tree of the genus *Citrus,* characterized by its white color and sweet odor; widely used in wreaths and bridal bouquets; the state flower of Florida.

orange mineral *Materials.* a red lead-oxide powder with low tinting strength, made by roasting lead carbonate or sublimed litharge in a furnace; used as a pigment in printing inks and primers.

orangeophil *Histology.* an acidophilic cell found in the anterior lobe of the adenohypophysis.

orange peel *Materials Science.* a term describing a pebbled film surface that has undulations similar to the skin of an orange.

orange-peel bucket *Mechanical Devices.* a grab bucket with at least three jaw segments that are hinged into one overlying support; so named because the jaws of the bucket resemble pieces of an orange peel.

orange rust *Plant Pathology.* a disease of blackberries and raspberries caused by the rust fungus *Gymnoconia interstitialis,* characterized by a powdery, orange mass of spores on the undersides of leaves and misshapen foliage.

orange sapphire *Mineralogy.* an orange-colored gem corundum.

orange spectrometer *Spectroscopy.* a β-ray spectrometer in which several modified double-focusing spectrometers having a common source and a common detector are used to obtain high transmission.

orange toner *Organic Chemistry.* $OHC_{10}H_8N_2C_6H_4SO_3Na$, a full-strength orange pigment used to produce printing inks.

orangewood *Botany.* the hard, fine-grained, creamy wood of the orange tree, used in inlaid carpentry work.

orangutan [ə rang´ə tan´] *Vertebrate Zoology.* the only true arboreal great ape, the large, long-armed anthropoid *Pongo pygmaeus* of the primate family Pongidae, characterized by a high forehead and a bulging snout; found in Sumatra and Borneo; an endangered species. (From pidgin Malay meaning "forest man.")

O-ray see ORDINARY RAY.

orb *Science.* **1.** a sphere or globe. **2.** another word for the eye or eyeball.

ORB omnidirectional radio beacon.

orbicular structure *Petrology.* a granular, igneous rock structure characterized by an abundance of spheroidal aggregates of megascopic crystals of varying minerals, arranged in concentric layers with or without a xenolithic nucleus.

orbicule *Geology.* a nearly spherical mass of rock, ranging from microscopic to several centimeters in diameter, and characterized by a distinct arrangement of concentric layers.

Orbiculoidea *Paleontology.* an extinct genus of inarticulate brachiopods in the suborder Acrotretidina, characterized by a functional pedicle and a phosphatic shell; Ordovician to Permian.

Orbiliaceae *Mycology.* a family of fungi belonging to the order Helotiales, consisting of some species that are lichens and some that are parasites of a wide variety of plants.

Orbiniidae *Invertebrate Zoology.* a family of sessile polychaete annelids of the subclass Sedentaria, characterized by a conical or globular prostomium, a weakly separated thorax and abdomen, and no appendages.

orbit *Physics.* **1.** a closed path that is the trajectory of a body acted upon by a force. **2.** a similar path that is not strictly closed. *Astronomy.* the path followed by celestial objects moving under gravity, such as the path of one celestial body around another celestial body (e.g., the earth around the sun or the moon around the earth) or the path of a spacecraft or artificial body around a celestial body (e.g., a communications satellite around the earth). *Oceanography.* the circular to elliptical path followed by a water particle affected by wave motion. *Anatomy.* either of the cavities in the skull containing the eyeballs. *Mathematics.* **1.** let G be a group that acts on a set S. The orbit of an element x of S is the subset $\{gx : g \in G\}$ of S. **2.** the orbit of an element a of a set S under a permutation θ is the subset of S consisting of all elements of the form $\theta^i a, i = 0, \pm 1, \pm 2, \ldots$. The orbits of the elements of S are the equivalence classes under the equivalence relation $a \sim b$ if and only if $b = \theta^i a$ for some integer i. If S is finite, then the ordered set $(a, \theta a, \theta^1 a, \ldots, \theta^{k-1} a)$ (where k is the least integer such that $\theta^k a = a$) consisting of the elements of the orbit of a, is called a cycle. (Going back to a Latin word meaning "wheel" or "circle.")

orbital *Space Technology.* of or relating to an orbit or orbiting body; used to form compounds such as **orbital period** and **orbital launch.** *Atomic Physics.* **1.** the volume of space within an atom in which there is a high probability of finding electrons at a given energy level. **2.** the space-dependent part of the Schrödinger wave function for an electron in an atom or molecule. *Physical Chemistry.* see ORBITAL THEORY.

orbital angular momentum *Mechanics.* the cross product of the position vector of a particle moving about an origin and its linear momentum. Also, ORBITAL MOMENT.

orbital current *Oceanography.* the flow of water that accompanies the orbital movement of the water particles in a wave.

orbital curve *Space Technology.* a track traced on the surface of a celestial body by a satellite having a different axis and period of rotation; used to chart displacement of the satellite's orbit due to rotation of the primary body.

orbital decay *Atomic Physics.* a loss of energy in an atom during which one of its electrons changes from a higher- to a lower-energy level. *Space Technology.* a decrease in the energy of an artificial satellite, which causes the orbital eccentricity and orbital semimajor axis to decrease.

orbital direction *Space Technology.* the path of a satellite relative to that of the body about which it orbits; measured as the angle of inclination to the path of the equator of the primary body.

orbital eccentricity *Space Technology.* the deviation of a body's orbit from a circle.

orbital electron *Atomic Physics.* an electron in one of the orbitals of an atom. Also, PLANETARY ELECTRON.

orbital-electron capture *Nuclear Physics.* an interaction during beta decay in which a nucleus captures an electron from the shell nearest the nucleus and releases a neutrino. Also, K-CAPTURE, K-ELECTRON CAPTURE.

orbital elements *Physics.* a group of seven parameters, semimajor axis, eccentricity, inclination, longitude of the ascending node, argument of periapsis, epoch, and period, that determine the characteristic motion of a body in a central force field following an inverse-square law.

orbital fossa *Invertebrate Zoology.* **1.** a depression or cavity from which the eyestalk of crustaceans arises. **2.** the region around the inner border of the compound eye of insects.

orbital magnetic moment *Quantum Mechanics.* the magnetic dipole moment associated with the motion of a charged particle about an origin, distinct from that due to the particle's spin.

orbital-metal plane see EYE-EAR PLANE.

orbital moment see ORBITAL ANGULAR MOMENTUM.

orbital motion *Physics.* the continuous motion of any body whose trajectory is a closed path about some point.

orbital period *Astronomy.* the time taken for one complete orbital circuit.

orbital plane *Mechanics.* a plane containing the orbit of a body or particle in a central force field, passing through the center of force.

orbital rendezvous *Space Technology.* **1.** the coming together of at least two orbiting objects with zero relative velocity at a predetermined time and location. **2.** the exact point at which such an event occurs in space.

orbital symmetry *Physical Chemistry.* the ability of specific molecular orbitals to be carried into themselves or their negatives by geometrical operations, such as a 180° rotation about an axis in the molecule's plane or reflection through the plane.

orbital theory *Physical Chemistry.* a description of the nature and behavior of electrons in an atom or molecule, based on the propositions that each individual electron moves in a definite wave pattern around a nucleus and that it is possible to predict accurately the region in which an electron with a specific energy level will be located.

orbital velocity *Astronomy.* the velocity an object must maintain to remain in a given orbit; this is constant in the case of a circular orbit, and varying in an elliptical, parabolic, or hyperbolic one.

orbite *Petrology.* a porphyritic gabbro with large phenocrysts of hornblende, or hornblende and plagioclase.

orbiting collision *Atomic Physics.* any interaction between an atom and an ion in which they approach each other closely and spend several orbital periods (of the atomic electrons) close together.

orbit point *Space Technology.* a point on or above the earth's surface that is used as a reference for determining or adjusting the orbit of a space vehicle.

Orbivirus *Virology.* a genus of viruses of the family Reoviridae that infect a wide range of vertebrates (including humans) and arthropods; transmitted by insects and ticks.

orca *Vertebrate Zoology.* the black and white toothed whale, *Orcinus orca,* popularly known as the killer whale; it grows to 30 feet in length and has teeth used only to capture its prey, which it swallows whole.

orcein *Cell Biology.* a cytological dye used for staining chromosomal preparations.

orchard *Agriculture.* an extensive area of fruit trees.

orchi- or **orch-** a prefix meaning testis. Also, **orchid-, orchido-, orchio-, orcho-.**

orchialgia *Medicine.* any pain in a testis. Also, **orchidalgia.**

orchid *Botany.* any member of the family Orchidaceae of the order Orchidales; a perennial often cultivated for its complex ornamental flower; found in temperate and tropical regions.

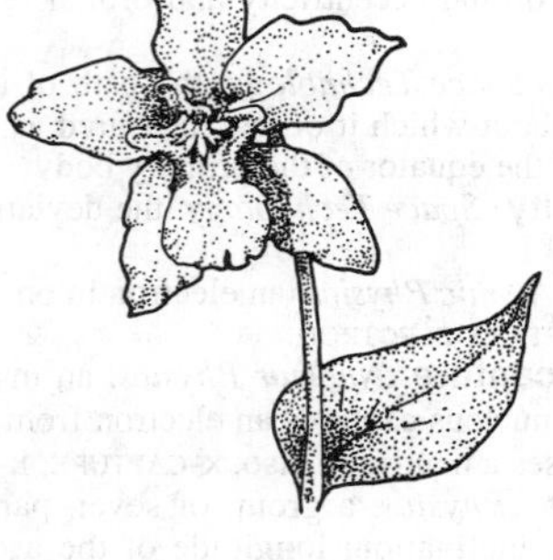

orchid

Orchidaceae *Botany.* the orchid family, a very large, cosmopolitan family of the order Orchidales that are monocotyledonous perennial herbs with showy, bilaterally symmetrical flowers, tuberous roots, and vertical rhizomes; usually found in tropical forests.

Orchidales *Botany.* an order of monocotyledonous mycotrophic herbs in the subclass Liliidae, having vessels usually confined to the roots, one or two stamens adnate to gynoecium, numerous tiny seeds, a minute, mostly undifferentiated embryo, and no endosperm.

orchidectomy *Surgery.* the surgical excision of one or both testicles. Also, TESTECTOMY, ORCHIECTOMY.

orchidic *Medicine.* of or relating to the testes.

orchiditis *Medicine.* an inflammation of a testicle.

orchidopexy *Surgery.* the surgical fixation of an undescended testicle in the scrotum. Also, **orchiopexy, orchidorrhaphy.**

orchidotomy *Surgery.* the incision and drainage of a testicle. Also, **orchiotomy.**

orchiectomy see ORCHIDECTOMY.

orchiepididymitis *Medicine.* the inflammation of a testicle and an epididymis.

orchil *Materials.* a violet coloring dye obtained from certain lichens, mainly *Rocella,* or any lichen that yields this dye.

orchiopathy *Medicine.* any disease of the testes.

orchiopexy *Surgery.* the surgical treatment of an undescended testicle, in which it is freed and implanted into the scrotum. Also, CRYPTORCHIDOPEXY.

orchitis *Medicine.* an inflammation of a testis, marked by pain, swelling, and a feeling of weight.

orcin *Organic Chemistry.* $CH_3C_6H_3(OH)_2$, a white crystalline prism that becomes red when exposed to air; has an intensely sweet, unpleasant taste and is soluble in water and alcohol; melts at 107°C and boils at 287–290°C; derived by the fermentation of lichens and used as a reagent in such carbohydrates as beet sugar, lignin, and pentoses. Also, ORCINOL, 5-METHYLRESORCINOL.

orcinol see ORCIN.

OR circuit *Electronics.* a logic circuit whose output is energized whenever one of its inputs is in a prescribed state, so that if a voltage is applied to at least one gate, a voltage will appear across the transistor. Also, OR GATE.

ORD *Aviation.* the airport code for O'Hare International, Chicago, Illinois.

order a system, sequence, or classification for ranking or arranging objects or items; specific uses include: *Systematics.* one of the principal or obligatory ranks in the taxonomic hierarchy, above family and below class. In botany, the order name ends in *-ales*; in zoology, it often ends in *-iformes*. *Chemistry.* a classification of chemical reactions described as first, second, third, and higher in accordance with the number of molecules that enter into the reaction. *Architecture.* **1.** a style of classical or Renaissance architecture as characterized by its type of column and entablature. **2.** a design and arrangement of columns and entablature. *Physics.* the approximate magnitude or range of magnitude of some quantity, usually expressed as the number of multiple factors of 10. Also, ORDER OF MAGNITUDE. *Mathematics.* a measurement of size or degree. For example, the order of a derivative or differential operator refers to how many times it is applied; i.e., D^n and $\partial^n f/\partial x_n$ both have order n. An $n \times n$ square matrix is said to be of order (or size) n, and a polynomial with n appearing as the largest exponent in any of its terms is said to be of order (or degree) n.

order-disorder transition *Solid-State Physics.* a transformation from a state of regularity (periodicity) to a state of random placement in the lattice structure of an alloy or a solid solution; the transformation occurs over a range of temperatures.

ordered array *Computer Programming.* a two-dimensional set of data elements in which the rows and columns are arranged in a specific sequence. *Statistics.* a display of a set of observations arranged in increasing or decreasing order to suit some specific purpose. Also, ARRAY.

ordered field *Mathematics.* a field F is said to be ordered if there is a subset P of F such that: (a) $P \cap (-P) = \Delta$ (where $-P$ denotes the subset of F containing the additive inverses of all the elements of P); (b) $P \cap \{0\} \cap (-P) = F$; and (c) if $a,b \in P$ then $a + b \in P$ and $ab \in P$. A linear ordering $<$ is induced on an ordered field by defining $a < b$ if $b - a \in P$. If $a < b$ and $a > 0$ implies that there is an integer n with $b \leq na$ (na means a added to itself n times) then F is **Archimedean ordered.**

ordered list *Computer Programming.* a collection of data elements that is structured as a list and arranged in a specific sequence.

ordered pair *Mathematics.* two elements x and y from a given set, where x is designated to be the first and y the second member of the pair; equivalently, an n-tuple with $n = 2$. Denoted (x,y). In general, $(x,y) \neq (y,x)$.

ordered ring *Mathematics.* a ring with an ordering defined on the underlying set of elements.

ordered search see BEST-FIRST SEARCH.

ordered solid solution *Metallurgy.* a solid solution in which the constituent atoms occupy preferred locations in the crystal lattice. Such a solution is said to have a superlattice.

ordering *Mathematics.* a reflexive, antisymmetric, transitive binary relation on a set X; usually denoted $\leq$. That is, for every $x,y,z \in X$, $x \leq x$; $x \leq y$ and $y \leq x$ imply that $x = y$; and $x \leq y$ and $y \leq z$ imply $x \leq z$. Also, **partial ordering** or **order.** If, in addition, either $x \leq y$ or $y \leq x$ holds for any $x,y \in X$, then the ordering $\leq$ is called a total or linear ordering, and X is called a chain or totally ordered set.

order of a differential equation *Mathematics.* an ordinary differential equation $F[x, y(x), y'(x), \ldots, y^r(x)] = 0$ for the determination of the function $y(x)$ is of order r if the rth derivative is the highest occurring in the equation.

order of a graph *Mathematics.* the number of vertices in a graph.

order of a group *Mathematics.* the cardinal number of the group; i.e., the number of elements in a group G; denoted $|G|$. G is said to be a finite group (or infinite group) if $|G|$ is finite (or infinite).

order of an element *Mathematics.* **1.** the order of an element a of a group G is the order of the cyclic subgroup of G that is generated by a. That is, if the cyclic subgroup $\langle a \rangle$ has finite order n, then a is said to be of order n; otherwise a is said to be of infinite order. **2.** the order of an element m of an R-module M is a left ideal $\lambda(m)$ of R generated by m; i.e., $\lambda(m) = \{r \in R : rm = 0\}$.

order of an elliptic function *Mathematics.* the number of poles, counting multiplicities, in a period parallelogram.

order of a pole *Mathematics.* if z_0 is a pole of a function $f(z)$ in a simply connected domain D, then the order of the pole z_0 is the smallest positive integer k such that the function $g(z) = \lim z \to z_0(z - z_0)^k f(z)$ is analytic at z_0.

order of a system of differential equations *Mathematics.* Let r_i be the order of the ith differential equation in a system of n ordinary differential equations. The number $r = \sum_{i=1}^{n} r_i$ is the order of the system. The general solution to the system contains r arbitrary constants.

order of a tensor *Mathematics.* the number of arguments of a tensor (viewed as a mixed multilinear functional); equivalently, the number of indices on the tensor components (as in the Ricci calculus).

order of a zero *Mathematics.* If z_0 is a zero of a single-valued function $f(z)$ in a simply connected domain D, then the order of the zero z_0 is the smallest positive integer k such that $f(z) = (z - z_0)^k g(z)$, where $g(z)$ is analytic at z_0 and $g(z_0) \neq 0$.

order of battle *Military Science.* the identification, strength, command structure, and disposition of a military force.

order of interference *Optics.* the difference in the number of wavelengths that occur in the paths of two interfering rays of light.

order of magnitude *Mathematics.* see ORDER.

order of phase transition *Thermodynamics.* a number that designates a classification of a change of phase; the order has a whole number value of one greater than the order of the derivative of a quantity with respect to the temperature for which there is a discontinuity at the onset of the change of phase; for example, if at the critical point, the density is not discontinuous through a phase change, but its first derivative with respect to the temperature is, then such a process is called a second-order phase transition.

order parameter *Physics.* a measure of the regularity in ordering of atoms in a binary crystal given by $P = (p_A - n_A)/n_B = (p_B - n_B)/n_A$, where subscripts A and B refer to the two different atoms, n refers to the relative numbers of the atoms ($n_A + n_B = 1$), and p is the relative probability of finding the correct atom at its expected lattice site.

order-preserving *Mathematics.* a function f is said to be order-preserving if $a \leq b$ implies that $f(a) \leq f(b)$ for every a and b in the domain of f.

order statistic *Statistics.* the rth-order statistic is the rth element of a sample after it has been arranged in increasing order. For example, the 1st-order statistic is the smallest observation; the $(n + 1)/2$ (n odd) is the median.

order strengthening *Materials Science.* a strengthening process resulting from a transformation of a disordered to an ordered structure.

ordinal number *Mathematics.* intuitively, a number such as first, second, twenty-seventh, etc., that specifies the ordering of the elements of a set, as opposed to cardinal numbers such as one, two, fifty-one, etc., that are used to count how many elements there are in a set. More generally, ordinal numbers are constructed as follows: an ordinal number is a well-ordered set α such that each element of α equals its initial segment; i.e., the initial segment $s(\xi) = \{\eta \in \alpha : \eta < \xi\} = \xi$ for all ξ in α. Used to extend the process of counting beyond the natural numbers. All natural numbers are ordinal numbers; the next ordinal number after all the natural numbers is the set w consisting of all the natural numbers. In general, if α is an ordinal number, then so is its successor set α^+. Also, **ordinal numeral.**

ordinal type *Computer Programming.* an ordered range of values, such as *integer*.

ordinary differential equation *Mathematics.* a differential equation in which the unknown functions depend on just one independent variable.

ordinary gear train *Mechanical Engineering.* a gear train in which the axes remain in a fixed position relative to the frame.

ordinary index *Optics.* the refractive index of the ordinary ray in an anisotropic crystal. Also, **ordinary refractive index.**

ordinary point *Mathematics.* **1.** a point on a curve that is not a node or a cusp is an ordinary point of the curve. **2.** the point $x = x_0$ is an ordinary point of the (ordinary) differential equation $d^n y/dx^n + a_1(x)d^{n-1}y/dx^{n-1} + a_2(x)d^{n-2}y/dx^{n-2} + \cdots + a_n(x)y = f(x)$ if the coefficient functions $a_k(x)$ and $f(x)$ are analytic at $x = x_0$. The point at infinity is an ordinary point if, after the substitution $z = 1/x$ is made, the point $z = 0$ is an ordinary point of the transformed ordinary differential equation.

ordinary ray *Optics.* one of two oppositely polarized divisions of an electromagnetic wave that are formed as the wave enters an anisotropic crystal; the ordinary ray obeys the normal laws of refraction, as distinguished from the extraordinary ray. Also, O-RAY.

ordinary tide *Oceanography.* a tide that has periods of between 12 and 24 hours.

ordinary wave component *Optics.* the component of electromagnetic radiation traversing an anisotropic uniaxial crystal that gives rise to the ordinary ray.

ordinate *Mathematics.* one of the two standard coordinate functions in the Cartesian plane; for a point P, the perpendicular distance from P to the x-axis. *Cartography.* on a map with a system of rectangular or oblique coordinates, the linear distance of a point from the horizontal line of reference (or x-axis, usually the equator), and parallel to the prime meridian (or y-axis); frequently called the y-coordinate.

ordination *Ecology.* the placing of species or communities along an environmental gradient.

ordnance *Engineering.* the weapons, explosives, chemicals, and ammunition used in warfare, as well as the equipment and supplies used to service any of these items.

Ordnance

In earlier times the word "ordnance" referred to the building of fortifications, and the acquisition, storage, distribution, and repair of military supplies, incorporating functions now divided among several branches of the army (artillery, engineers, and the organizations for supply and maintenance), and corresponding branches of the navy and the air force. The first systematic mapping of Britain was conducted by the Ordnance Survey. But in common modern usage the term is associated with guns, rockets, ammunition, and other explosive stores such as mines and demolition charges.

The history of guns and their projectiles began with the discovery of gunpowder, with significant advances coming with breech loading, mechanisms to absorb recoil, rifled barrels, automated ammunition handling, and fused shells. Specialized ammunition has been designed to achieve penetration of armor on ships or tanks, for the distribution of smoke or gas, and for illumination. Many types of fuses have been designed to cause the shells to explode at a chosen time or altitude, when close to a target, or after penetration. Many of these developments have also been applied to bombs dropped from aircraft.

While a gun accelerates its projectile to maximum velocity in a fraction of a second, leaving it to coast on its unguided ballistic trajectory thereafter, rocket or jet propulsion applies a gentler acceleration over a much longer time. This permits the mounting of a sophisticated guidance system in the missile, as well as making very long ranges possible. Guided missiles have displaced guns from many of their former roles, especially at longer ranges and against moving targets. Torpedoes, long effective against surface ships, are now preferred to depth charges as antisubmarine weapons.

George Lindsay
Senior Research Fellow
Canadian Institute of Strategic Studies

Ordovician *Geology.* **1.** a period of the Paleozoic era, occurring after the Cambrian and before the Silurian, dating from about 500 to 440 million years ago. **2.** the rock strata formed during this period.

ore *Geology.* **1.** any naturally occurring material that contains extractable minerals of economic value. **2.** the minerals thus extracted.

ore bed *Geology.* a layer of metal-rich rock occurring between rocks of sedimentary origin.

ore block *Mining Engineering.* a section of an ore vein that is bounded at top, bottom, and one or both ends, ready for excavation.

orebody or **ore body** *Geology.* a well-defined, solid, continuous mass of material containing enough ore to make extraction economically feasible.

ore bridge *Mining Engineering.* a large electric gantry crane used to transport, stockpile, and load ore.

ore cluster *Geology.* a group of interconnected, genetically related orebodies.

ore control *Geology.* any geologic feature that has influenced the formation and localization of ore.

Orectolobidae *Vertebrate Zoology.* the wobbegongs, a family of relatively small, bottom-dwelling sharks of the order Orectolobiformes, characterized by a very flattened, broad head with flaps of skin on the sides, rounded fins, and barbs in front of the nasal openings; found mainly in shallow tropical and subtropical waters of the Indo-Pacific and Red Sea, with one species in the Atlantic.

Orectolobiformes *Vertebrate Zoology.* the carpet sharks, an order of mostly marine benthic fishes of the superorder Galeomorphii, characterized by a short to blunt snout, lateral flaps with barbels on the nasoral grooves, and spineless dorsal fins; found in circumtropical zones.

ore deposit *Geology.* rocks that contain economically valuable minerals in sufficient quantities to be extracted profitably.

ore district *Geology.* a system composed of a combination of several ore deposits.

oreide bronze *Metallurgy.* the former designation of a copper-base alloy containing zinc and tin, having an undefined composition.

Orellan *Geology.* a North American geologic stage of the Middle Oligocene epoch, occurring after the Chadronian.

ore microscopy *Mineralogy.* the study of polished opaque ore mineral sections using a reflected light microscope.

orendite *Petrology.* a porphyritic leucite lamproite consisting of phenocrysts of phlogopite in a reddish-gray, nepheline-free groundmass of leucite, sanidine, phlogopite, amphibole, and diopside.

oreodont *Paleontology.* of or relating to the North American artiodactyls of the family Merycoidodontidae; the oreodonts were sheep-sized ruminants that first appeared in the Eocene and became very abundant in the Oligocene, but dwindled slowly down to extinction in the Pliocene.

Oreopithecidae *Paleontology.* an extinct family of apelike catarrhine primates; known from Italian sites of the Middle Miocene and perhaps from the Early Miocene of East Africa.

Oreopithecus *Anthropology.* a unique hominoid dating to the Miocene from coal deposits in Tuscany, Italy; about the size of a chimpanzee, it has some hominid features and may have been a separate branch in the evolution of a humanlike primate.

ore shoot *Geology.* a column of ore within an ore deposit that represents the richer or more valuable portion of the deposit. Also, SHOOT.

Oresme, Nicole 1325–1382, French prelate; a pioneer in the development of analytical geometry; worked on tax and coinage reform.

Orestes complex *Psychology.* the unconscious desire of a son to kill his mother. (From the Greek myth of *Orestes,* who killed his mother and her lover after they murdered his father.)

Ore's theorem *Mathematics.* a theorem stating that if G is a simple graph with $n \geq 3$ vertices such that the sum of the degrees of any two vertices is greater than or equal to n, then G is Hamiltonian.

org or **org.** organic.

organ *Anatomy.* a structure containing two or more different cell types that are organized to carry out a particular function of the body, such as the heart or kidneys. *Botany.* a grouping of tissues into a distinct structure that performs a specialized task, such as a leaf or stamen.

organ culture *Biotechnology.* the unlimited growth of isolated organs, usually embryonic in origin, in a sterile nutrient culture.

organelle *Cell Biology.* a discrete body found in the cytoplasm of a cell, defined by a surrounding membrane, and performing a specific function.

organic *Chemistry.* of or relating to any covalently bonded compound containing carbon atoms. *Biology.* relating to or involving an organism or organisms. *Medicine.* relating to or affecting an organ of the body. *Agronomy.* of or relating to organic farming or organic foods.

Organic *Architecture.* describing architecture based on natural forms; used in relation to Frank Lloyd Wright and his followers.

organic acid *Organic Chemistry.* an organic chemical compound containing one or more carboxyl radicals, such as acetic, formic, lactic, and all fatty acids.

organic bank *Geology.* a deposit of limestone that has been formed in place by organisms, usually under water.

organic bonded wheel *Mechanical Devices.* a grinding wheel covered with organically bonded abrasive particles.

organic brain syndrome *Medicine.* a group of psychological or behavioral signs resulting from one or more specific organic factors that affect brain function, such as may occur with cerebral arteriosclerosis or lead poisoning. Also, **organic mental disorder.**

organic chemistry *Chemistry.* a major branch of chemistry that deals with most compounds of carbon, including food and fuels.

Organic Chemistry

Organic chemistry is a branch of chemistry dealing with the structures, reactions, and synthesis of carbon compounds. Unlike most of the other elements, carbon forms strong bonds to other carbon atoms leading to chains and rings of infinite complexity. As implied by the definition "organic–derived from living organisms," the field was originally associated with the study of substances obtained from plant or animal sources supposedly created by a "vital force." During the nineteenth century, this definition was revised to accommodate the findings that organic compounds such as urea could be prepared in the laboratory from compounds which do not have their origin in living organisms.

The enormous breadth of the field is illustrated by the overlap of organic chemistry with disciplines ranging from cell biology to quantum mechanics. Plastics, dyes, synthetic fibers, and gasoline are organic compounds. Many organic compounds such as proteins and carbohydrates are intimately involved in controlling life processes. Medicinal agents, synthesized by organic chemists, are used in a wide range of therapeutic applications, as in cancer chemotherapy and the treatment of infectious diseases.

The development of new organic reagents, including organometallic species, has made possible the synthesis of highly functionalized, complex molecules such as vitamin B-12. Unusually reactive, transient species have been prepared to test the theories of bond-making and bond-breaking. Other molecules have been attached to proteins to probe the mechanisms of enzyme action.

Synthetic organic chemicals used as pesticides have led to improved food production and the control of malaria. However, problems have arisen as a result of the deleterious effects of some of these materials on humans, animals, birds, and plants. Solving these problems will provide important goals for the organic chemists of tomorrow.

Harry Wasserman
Eugene Higgins Professor Emeritus of Chemistry
Yale University

organic-cooled reactor *Nucleonics.* a nuclear reactor that uses an organic fluid such as a polyphenol as its coolant.

organic electrolyte cell *Electricity.* a wet cell that is composed of reactive metals, such as lithium, calcium, or magnesium, in conjunction with organic electrolytes, such as a lithium-cupric fluoride cell.

organic evolution *Evolution.* the process of phenotypic or morphological change in living organisms over many generations or through periods of geologic time.

organic farming *Agronomy.* a system of agriculture in which organic products and techniques are used, as, for example, when the soil is treated with manure rather than with chemical fertilizers or when plant pests are eliminated by introducing insects that prey on them rather than by spraying chemical insecticides.

organic fertilizer *Agronomy.* a term for any form of fertilizer that consists of natural animal and plant products, such as manure, compost, or bone meal.

organic foods *Nutrition.* generally, any foods that have been cultivated by the use of animal or vegetable fertilizers, rather than synthetic chemicals, including any vegetable or fruit grown in soil fertilized with natural compost rather than chemical fertilizers and not exposed to chemical pesticides or herbicides.

organic geochemistry *Geochemistry.* the branch of geochemistry dealing with naturally occurring carbon-based and biologically derived substances within the lithosphere, atmosphere, and hydrosphere.

organic lattice see GROWTH LATTICE.

organic mental disorder *Psychology.* a term for any behavioral disorder resulting from brain dysfunction or some other physiological cause.

organic-moderated reactor *Nucleonics.* a nuclear reactor that uses organic materials in order to moderate (slow down) the neutrons produced within the reactor.

organic pigment *Organic Chemistry.* any material with an organic chemical base used to add color to another material.

organic quantitative analysis *Analytical Chemistry.* the identification and measurement of the elements, functional groups, or molecules of an organic compound or mixture.

organic-reaction mechanism *Organic Chemistry.* the specific mechanism involved in producing a reaction within or between organic molecules.

organic reef *Geology.* a wave-resistant, ridgelike, sedimentary rock structure built by and composed primarily of the remains of marine sedentary or cement-binding organisms, especially corals and algae. Also, REEF.

organic rock *Petrology.* a sedimentary rock consisting principally of plant or animal remains.

organic salt *Organic Chemistry.* the reaction product of an organic acid and an inorganic or organic base.

organic soil *Agronomy.* soil consisting mainly of organic matter, such as peat.

organic solvent *Organic Chemistry.* a liquid organic-chemical compound that can dissolve solids, liquids, and gases.

organic texture *Geology.* a sedimentary rock texture that results from the activities of organisms.

organic weathering *Geology.* any biological process or alteration that contributes to the breakdown of rocks. Also, BIOLOGIC WEATHERING.

organism *Biology.* **1.** a living being; any form in which mutually interdependent parts maintain the various vital processes necessary for life to exist. **2.** a biological form considered as an entity, such as an animal, plant, or fungus.

organismic psychology *Psychology.* an approach to psychology that emphasizes the need to consider the organism as a unified biological entity, rejecting distinctions between mind and body.

organizational development *Industrial Engineering.* the use of behavioral science techniques to improve the effectiveness of an organization in solving problems, achieving goals, and coping with changes in the environment.

organizational psychology *Psychology.* the study of the psychology of organizations, dealing with effective relationships within organizations and the hierarchy and functioning of organizations.

organization chart *Industrial Engineering.* a chart showing the organization of an industrial enterprise, in terms of areas of responsibility and chains of reporting and command.

organizer *Developmental Biology.* a special group of cells on the dorsal side of the blastopore of vertebrate embryos that induces the dorsal axis of the embryo.

organizing pneumonia *Medicine.* pneumonia in which the lung signs fail to clear within the usual period; the healing period is characterized by organization and cicatrization of the exudate rather than by resolution and resorption.

organocopper reagent *Organic Chemistry.* any one of several organic reagents containing copper and often other metals; important in conjugate-addition reactions and in the displacement of leaving groups.

organ of Corti *Anatomy.* a spiral structure resting on the basilar membrane in the cochlear duct and containing hair cells that are the special sensory receptors for hearing; the hair cells are stimulated by sound vibrations and convert them into nerve impulses that are transmitted by the auditory nerve to the brain. Also, SPIRAL ORGAN.

organ of Leydig *Vertebrate Zoology.* either of the large groupings of lymphoid tissues along the length of the esophagus in certain fishes.

organ of Tömösvary *Invertebrate Zoology.* paired organs of possible olfactory function, found on the head of centipedes at the base of the antennae.

organogenesis *Developmental Biology.* the origin and development of organs during embryonic life.

organogenic *Geology.* of a rock or sediment containing organic material or derived from organic substances.

organoleptic *Physiology.* **1.** of or relating to a stimulus that makes an impression on a sense organ. **2.** of or relating to an anatomical part that is capable of receiving a sense impression.

organolite *Geology.* any rock composed primarily of organic material, particularly a rock derived from plant matter.

organoma *Oncology.* a tumor that is composed of organs or of specific portions of an organ.

organometallic compound *Organic Chemistry.* an organic compound that contains a metal linked to carbon; some are highly toxic or flammable and others are coordination compounds; often powerful catalysts that form useful coordination complexes. Also, METALLO-ORGANIC COMPOUND.

organopexy *Surgery.* a surgical operation to fix or support an organ, especially the uterus.

organophosphate *Organic Chemistry.* an organic chemical compound that contains a phosphate group; used as a fertilizer that provides phosphorus to deep-root systems.

organosol *Materials.* the colloidal dispersion of any insoluble material in an organic liquid, especially the finely divided dispersion of a synthetic resin in plasticizer.

organosulfur compound *Organic Chemistry.* an organic chemical compound that contains one or more sulfur atoms.

organotrophic *Microbiology.* describing a microorganism that derives energy from the oxidation of organic compounds.

organotropic *Microbiology.* describing microorganisms, often pathogens, that enter the body via the viscera or somatic tissue.

orgasm *Physiology.* the peak and culmination of sexual excitation, a series of strong, involuntary contractions of the muscles of the genitalia that is experienced as extremely pleasurable; usually accompanied in the male by ejaculation.

orgasmolepsy *Medicine.* **1.** a former term for an attack of epilepsy that occurs during orgasm. **2.** an attack of epilepsy that produces erotic sensations similar to orgasm; such attacks are caused by a discharge in the lobulus paracentralis or temporal lobe.

OR gate see OR CIRCUIT.

orgone *Psychology.* Wilhelm Reich's term for a supposed universal force that is the physical equivalent of the libido.

Oribatoidea *Invertebrate Zoology.* a superfamily of free-living mites with a leathery integument, including horny and beetle mites.

oriC *Genetics.* a genetic locus from which the genome of *E. coli* is replicated bidirectionally.

orient *Engineering.* **1.** to turn to the east or to fix in position in reference to the east. **2.** to set a map in alignment with the actual points on a compass or the landscape. *Optics.* the luster, sheen, and play of color seen on a gem-quality pearl; a pearl having such an appearance.

orientability of sound signal *Acoustics.* the property of a sound signal that enables the listener or other acoustic receiver to determine the relative direction or range of its source.

orientable manifold *Mathematics.* a differentiable manifold X is said to be orientable if there exists an atlas such that for any two charts (ϕ,U) and (ψ,V) of the atlas, the Jacobian determinant $(D\phi^i/D\psi^j)$ is positive at all points of $U \cap V$. A manifold defined in terms of such an atlas is said to be **oriented.**

oriental alabaster see ONYX MARBLE.

oriental amethyst *Mineralogy.* a purple variety of sapphire; or any amethyst of great beauty.

Oriental fruit moth *Invertebrate Zoology.* a moth, *Grapholitha molesta,* whose larvae infest and feed on twigs and fruit of peach and plum trees; introduced into the U.S. from the Orient.

Oriental poppy *Botany.* a poppy, *Papaver orientale,* characterized by bristly stems and leaves and scarlet, pink, or white flowers; native to Asia and cultivated as an ornamental.

Oriental region *Ecology.* an Asian zoogeographical region that includes the Indian, Ceylonese, Indo-Chinese, and Indo-Malayan subregions.

oriental topaz *Mineralogy.* a yellow variety of gem corundum.

orientation *Behavior.* a predisposition or frame of reference that affects an organism's reactions to certain stimuli, situations, or behaviors. *Psychology.* **1.** the awareness of the parameters of one's immediate environment; that is, the time, place, and other people present. **2.** a particular viewpoint, ideology, or world view. *Physics.* the relative direction of some vector or vector field with respect to another direction (reference direction). *Physical Chemistry.* **1.** the arrangement of atoms or radicals in a chemical compound. **2.** the positioning of atoms or radicals so that they are pointed in a definite direction rather than arranged at random. *Electromagnetism.* the physical positioning of an antenna that has directional properties. *Engineering.* the rotation of a map or instrument until the line of direction between any two of its points is parallel to the corresponding direction in nature. *Mathematics.* two coordinate systems $(x^1, \ldots, x^n)$ and $(y^1, \ldots, y^n)$ on an open set U of R^n are said to define the same orientation if the Jacobian determinant $D(x^i)/(Dy^j)$ is positive at all points of U. A chart (ϕ,U) on a manifold X defines an orientation of U by means of the orientation provided by the coordinates (that is, $\phi^i(x) = x^i$) on $\phi(U)$ in R^n).

orientation diagram *Geology.* a graphic representation used in structural petrology to show surfaces, lines, or points of structural discontinuity; a fabric diagram.

orientation effect *Physical Chemistry.* the effect of molecular interaction on the specific positioning of individual molecules.

orientation force *Physical Chemistry.* an attraction or repulsion arising from the interaction between the dipole moments of two molecules. Also, **orientation attraction.**

orientation matrix *Crystallography.* a matrix that defines the orientational relationship between two objects; used for defining the mutual orientations of the crystal, diffractometer circles, and detection system in a computer-controlled automatic diffractometer.

orientation polarization *Electricity.* a polarization effect resulting from molecules that are oriented so that permanent dipole moments arise from an asymmetric charged distribution. Also, DIPOLE POLARIZATION.

orientation vector *Mechanical Engineering.* a vector direction that indicates the orientation of a robot's moving part.

oriented *Geology.* of a specimen that is marked to show its precise original location and arrangement in space.

oriented core *Engineering.* a core removed from a borehole that is positioned in the same direction it had when still embedded in the ground.

oriented graph see DIRECTED GRAPH.

orienting response *Behavior.* the response of an organism to the sudden presentation of a new stimulus, including attention, arousal, and preparation for defense or escape. Also, **orienting reflex.**

orifice [ôr′i fis] *Science.* a small hole or opening.

orifice gas *Metallurgy.* in plasma arc welding or cutting, the gas that forms the plasma and exits from the torch orifice.

orifice gauge *Mechanical Engineering.* a flow gauge consisting of a thin orifice plate clamped between pipe flanges, with pressure takeoffs drilled into the adjacent pipes.

orifice meter *Engineering.* a gas or liquid flowmeter consisting in part of a disk with a specific hole that is placed transversely across a pipe; the difference in pressure between one side of the disk and the other is a measure of the rate of flow, as determined from standard tables.

orifice plate *Mechanical Devices.* a perforated disk that is bolted in a pipeline between two abutted flanges; used to measure the flow of oil or other liquid.

orig original.

origanum oil *Materials.* an essential oil used in pharmaceuticals and as a flavoring.

Origem Loop *Astronomy.* a large loop of gas, perhaps an ancient supernova remnant with a radius of about 200 light-years, that lies on the boundary of the constellations of Orion and Gemini.

origin the beginning or starting point; specific uses include: *Transportation Engineering.* the location from which a passenger's trip itinerary initially departs. *Cartography.* in surveying and map construction, the point from which angles and distances are measured. *Computer Technology.* the beginning address of a program segment, as determined by the linking loader when it reads the segment into memory; used to adjust all relative addresses in the segment and in external tables. *Genetics.* a sequence of DNA in which replication is initiated. *Mathematics.* the point in Cartesian space at which all the coordinates are zero.

original equipment *Computer Technology.* a term for equipment that is sold by one manufacturer to another for use in the latter's products; usually refers to complete components such as disk drives or control processors rather than parts such as circuit cards or memory units.

original interstice *Petrology.* an interstice created at the time that an associated enclosing rock is formed. Also, PRIMARY INTERSTICE.

original valley *Geology.* a valley formed by geologic processes, other than by the action of running water, that originate or occur above or below the earth's surface.

origin of coordinates *Cartography.* the point of intersection of the coordinate axes from which the coordinate system is built.

origin of replication *Molecular Biology.* the site on a sequence of DNA at which DNA replication begins.

oriole

oriole *Vertebrate Zoology.* any of various perching birds belonging to the New World family Icteridae or the Old World family Oriolidae, characterized by black plumage with colorful patterns, a long, slender body, and a short bill. Well-known North American species include the **orchard oriole**, *Icterus spurius*; **Scott's oriole**, *I. parisorum*; and the **northern oriole**, *I. galbula* (formerly classified as two separate species, the eastern Baltimore oriole and the western Bullock's oriole). (Going back to the Latin word for "gold.")

Orion [ō rī′ən] *Astronomy.* the Hunter, a large, bright constellation that lies on the celestial equator between Canis Major and Taurus, containing the stars Betelgeuse and Rigel; it is best seen in winter in the Northern Hemisphere. (Named for a famous hunter of Greek mythology.)

Orion

Orion arm *Astronomy.* the spiral arm of the Milky Way on whose inner edge the sun stands.

Orion association *Astronomy.* a large OB stellar association roughly centered on the Orion Nebula.

Orionids *Astronomy.* a meteor shower that peaks between October 18 and 20 with about 20 meteors per hour, and whose members appear to radiate from the constellation Orion.

Orion molecular cloud I *Astronomy.* a dense, molecular cloud lying behind the Orion Nebula; contains both the Becklin-Neugeubauer and Kleinmann-Low infrared sources.

Orion molecular cloud II *Astronomy.* an infrared and molecular emission complex located 12° northeast of the Trapezium, centered on an infrared cluster source.

Orion Nebula *Astronomy.* a cloud of ionized hydrogen lying about 1500 light-years from earth; barely visible to the unaided eye as a patch of light in the center of Orion's sword.

Orion spur *Astronomy.* a star-forming region that juts from one of the Milky Way spiral arms in the direction of the Orion arm toward the constellation Orion.

Orizaba *Geography.* an inactive volcano in southeastern Mexico; the highest peak in Mexico and third highest in North America. Also, CITLALTÉPETL.

ORL *Aviation.* the airport code for Orlando, Florida.

ornith ornithology.

Ornithella *Paleontology.* an extinct genus of articulate brachiopods in the order Terebratulida; characterized by a long and almost pentagonal loop; extant in the Middle Jurassic.

ornithine *Biochemistry.* $C_5H_{12}O_2N_2$, an amino acid present in the blood that is not incorporated into proteins; it is formed together with urea on hydrolysis of arginine by arginase in the urea cycle.

Ornithischia *Paleontology.* the "bird-hipped" dinosaurs, one of the two large orders of dinosaurs; characterized by a more advanced arrangement of pelvic bones than the Saurischia and possibly descended from an early saurischian; composed of several groups of herbivores, including the ankylosaurs, ceratopsians, ornithopods, and stegosaurs; the earliest known ornithischian is Pisanosaurus, known only from the Upper Triassic of South America; extant in the Triassic to Cretaceous.

ornithological [ôr′nə thə läj′i kəl] *Vertebrate Zoology.* of or relating to the scientific study of birds.

ornithologist [ôr′nə thäl′ə jist] *Vertebrate Zoology.* a person who specializes in the scientific study of birds.

ornithology [ôr′nə thäl′ə jē] *Vertebrate Zoology.* the branch of zoology that deals with the study of birds.

ornithophily *Botany.* pollination by the pollen carried by birds; common pollinators are hummingbirds and sunbirds.

ornithopod *Paleontology.* of or relating to the suborder Ornithopoda of ornithischian dinosaurs.

Ornithopoda *Paleontology.* a suborder of dinosaurs in the order Ornithischia, composed of virtually all the bipedal ornithischians (except the Pachycephalosauridae), including the large families Hypsilophodontidae, Iguanodontidae, and Hadrosauridae; extant in the Middle Triassic to Late Cretaceous.

Ornithorhynchidae *Vertebrate Zoology.* the duckbill platypus, a monospecific family of egg-laying, semiaquatic mammals in the order Prototheria, characterized by a furry, flattened tail, webbed feet, and a flattened rostrum that resembles a duck's bill; found in Tasmania and eastern Australia.

ornithosis *Medicine.* a disease of birds and fowl caused by *Chlamydia psittaci* and spread to humans by contact with these birds; symptoms are generally milder than psittacosis.

OR node *Artificial Intelligence.* a node in an AND/OR graph that is true if and only if its successor's is true.

Orobanchaceae *Botany.* a family of herbaceous root-parasites of the order Scrophulariales, characterized by a lack of chlorophyll, glandular hairs, reduced scalelike leaves, and numerous small seeds; native to the Northern Hemisphere.

orocline *Geology.* an orogenic belt having a horizontal curvature or sharp bend as a result of horizontal crustal deformation. Also, GEOFLEX.

orocratic *Geology.* of or relating to an interval of time characterized by widespread geologic deformation.

orogene see OROGENIC BELT.

orogenesis see OROGENY.

orogenetic *Geology.* formed as a result of mountain-building processes.

orogenic [ôr′ə jen′ik] *Geology.* of or relating to the processes by which great, elongate chains of mountains are formed.

orogenic belt *Geology.* a long, linear region in the earth's crust that has undergone folding or other crustal deformation during a particular orogenic cycle. Also, FOLDBELT, OROGENE.

orogenic cycle *Geology.* an interval of time during which a mobile belt evolves into a stabilized orogenic belt. Also, GEOTECTONIC CYCLE, TECTONIC CYCLE.

orogenic sediment *Geology.* a sediment formed during or as a result of crustal deformation and other mountain-building processes, or that is attributable to a region that has undergone such processes.

orogenic unconformity *Geology.* any locally produced, angular unconformity in a region affected by mountain-building movements.

orogeny [ə räj′i nē] *Geology.* **1.** the process by which mountains are developed. **2.** specifically, the process by which features such as thrusting, folding, and faulting are formed within mountain regions, as a result of intense deformation of strata within the earth's crust. Also, OROGENESIS.

orogeosyncline *Geology.* a geosyncline that later became an area of mountain building.

oroghydrography *Hydrology.* a branch of hydrography that deals with the relationship between mountains and drainage.

orograph *Engineering.* an instrument that records distance and elevation of land surfaces; used to make topographic maps.

orographic *Meteorology.* of, relating to, or caused by mountains. Thus, **orographic cloud.** *Geology.* a former term denoting features related to the structure and topography of mountains.

orographic ascent *Meteorology.* the upward displacement of air that is blowing over a mountain. Also, **orographic lifting.**

orographic desert see RAIN SHADOW DESERT.

orographic fog *Meteorology.* the supercooled air that condenses outward and is forced upward in the form of fog clouds as a result of a mountain in its path.

orographic rain *Meteorology.* the rain brought about by moist air or winds blowing over the rising slopes of mountains; caused by the cooling of the moist air after it is forced upward.

orography *Geography.* the geography of mountains.

Oromericidae *Paleontology.* an extinct family of the superfamily Cameloidea, composed of camel-like tylopod ruminants.

orometer *Engineering.* an aneroid barometer used for measuring and recording elevation above sea level.

oronasal *Anatomy.* of or relating to the mouth and nose.

oropesa sweep *Ordnance.* a method of naval minesweeping in which a length of sweep wire is towed by a single ship; lateral displacement is provided by an otter, while depth is controlled by a kite at the ship end and a float at the other end.

oropharyngeal *Anatomy.* of or relating to the oropharynx.

oropharynx *Anatomy.* the part of the pharynx between the soft palate and the upper edge of the epiglottis.

orophilous *Ecology.* living or thriving in subalpine regions. Thus, **orophile, orophily.**

orophyte *Ecology.* a plant that lives in subalpine regions. Thus, **orophytic.**

orotath *Geology.* an orogenic belt that has been substantially stretched lengthwise.

orotic acid *Biochemistry.* $C_5H_4N_2O_4$, an acid that, when mixed with 5-phosphoribosylpyrophosphate, produces inorganic pyrophosphate and orotidine 5'-phosphate.

orotracheal *Anatomy.* of or relating to the mouth and trachea.

orotracheal intubation *Medicine.* the insertion of a tube through the mouth into the trachea to serve as an airway.

Orowan strengthening *Materials Science.* a process in the precipitation strengthening of an alloy in which impenetrable precipitates are surrounded by dislocations.

Oroya fever *Medicine.* a specific, acute, febrile, endemic disease of the Peruvian Andes, caused by *Bartonella bacilliformis* and marked by high fever, rheumatic pains, progressive, severe anemia, and albuminuria. Also, CARRION'S DISEASE.

orphan *Graphic Arts.* the first line of a paragraph when it is printed alone at the bottom of a page or column; many word-processing systems can be set to automatically avoid such page or column breaks. *Genetics.* an isolated individual pseudogene that is related to members of a gene cluster; it can take on new functions following random mutations and appears to be evolutionarily important.

orphan drug *Pharmacology.* a drug that remains undeveloped or is otherwise neglected because of a limited potential for commercial gain.

orphan virus *Virology.* any virus that has not been classified as causing a disease.

orpiment *Mineralogy.* As_2S_3, a yellow monoclinic mineral occurring in foliated to granular masses or as short, prismatic crystals, having a specific gravity of 3.49 and a hardness of 1.5 to 2 on the Mohs scale; found as a hot spring deposit and in low-temperature veining.

Orr, John Boyd, Baron 1880–1971, Scottish physiologist; awarded the Nobel Prize for directing the United Nations Food and Agriculture Organization.

orrery *Astronomy.* a mechanical model solar system in which planets move at their true relative velocities around the sun.

orris *Materials.* an iris, *Iris germanica florentina,* whose fragrant rootstock is used in medicine and perfumes. Also, **orrice.**

Orsat analyzer *Analytical Chemistry.* an apparatus used to determine the composition of a mixture of gases by passing the mixture through a series of solvents, each of which absorbs only one of the gases.

Ortelius, Abraham 1527–1598, Flemish cartographer; he produced the first modern atlas in 1570.

orth- a prefix meaning "straight" or "upright."

orth. orthopedics; orthodontics.

Orthacea *Paleontology.* an early superfamily of articulate brachiopods in the extinct suborder Orthidina; derived from the Billingsellacea and characterized by impunctate shells; extant from the Early Cambrian to the end of the Permian.

Orthent *Geology.* a suborder of the soil order Entisol that develops on steep slopes; characterized by good drainage, a fine earth fraction, and an organic content that decreases with depth.

Ortheziinae *Invertebrate Zoology.* a subfamily of homopteran insects in the superfamily Coccoidea; all stages have abdominal spiracles; immature forms and adult females have a flat anal ring with setae and pores.

orthicon *Electronics.* a television camera that resembles an iconoscope, but has a greater light intensity and can be used in places with less ambient light.

Orthid *Geology.* a suborder of the soil order Aridisol that develops on relatively young land surfaces in desert regions; characterized by good drainage, a gray or brownish-gray color, and the absence of an argillic or natric horizon.

Orthida *Paleontology.* an order of brachiopods that includes the oldest known articulate families; extant from the Early Cambrian to the end of the Permian.

Orthidina *Paleontology.* a suborder of Paleozoic articulate brachiopods in the extinct order Orthida.

orthite see ALLANITE.

ortho- a prefix meaning "straight" or "upright."

ortho- *Organic Chemistry.* a prefix referring to the relative positions on a benzene ring of substituent atoms or radicals; when atoms are attached to adjoining carbon atoms, they are in the *ortho* position relative to one another.

ortho acid *Organic Chemistry.* an aromatic acid with a carboxyl group located in the *ortho* position.

orthoarsenic acid see ARSENIC ACID.

orthoaxis *Crystallography.* a diagonal axis in a monoclinic system at right angles to the vertical axis.

orthobaric density *Physics.* the density of liquid or of a saturated vapor (at a given temperature) that is in equilibrium with the liquid.

orthobituminous coal *Geology.* a bituminous coal that contains 87–89% carbon when analyzed on a dry, ash-free basis.

orthocenter *Mathematics.* the point inside a triangle at which the altitudes of the triangle intersect.

Orthoceras *Paleontology.* a genus of large nautiloids in the extinct order Orthocerida that are cone-shaped and several meters long; extant in the Lower Ordovician to Upper Triassic.

orthoceratite *Paleontology.* a fossil nautiloid of Orthoceras or a related genus.

orthocerid *Paleontology.* of or relating to the extinct nautiloid order Orthocerida; Ordovician to Triassic.

orthochem *Geochemistry.* a chemical precipitate formed within a sediment or depositional basin as a result of direct chemical action.

orthochorea *Neurology.* a form of random involuntary movement generally involving the hands and feet that occurs only when the person is standing up.

orthochromatic *Biology.* having normal staining characteristics, such as orthochromatic film, a photographic film that responds to colors in their normal brightness as seen by the eye.

orthochromatin *Biochemistry.* the common, stable form of chromatin.

orthochronological *Geology.* of or relating to orthochronology.

orthochronology *Geology.* a geochronology that is based on a standard biostratigraphical succession of significant faunas or floras, or on irreversible evolutionary processes.

orthoclase *Mineralogy.* $KAlSi_3O_8$, a colorless, white, cream-yellow, reddish or gray monoclinic mineral of the feldspar group, with partially ordered Al-Si arrangement, diomorphous with microcline, and having a specific gravity of 2.55 to 2.63 and a hardness of 6 to 6.5 on the Mohs scale; found as an essential constituent in granitic and syenitic rocks, and also occurs in some metamorphic and sedimentary rocks.

orthocone *Paleontology.* a cephalopod whose shell is a straight, tapering cone.

orthoconglomerate *Geology.* a conglomerate having an intact framework of gravel, pebbles, and coarse sand bound together by mineral cement and deposited by highly turbulent, ordinary water currents, such as surf.

orthocumulate *Petrology.* a cumulate consisting principally of at least one cumulus mineral and the crystallization products of an intercumulus liquid.

Orthod *Geology.* a suborder of the soil order *Spodosol,* containing aluminum, organic carbon, and iron.

orthodolomite *Petrology.* **1.** a primary dolomite or a dolomite formed by a sedimentary process. **2.** a tightly cemented dolomite rock with interlocking particles.

orthodontics [ôr′thə dän′tiks] *Medicine.* the branch of dentistry dealing with the correction and prevention of irregularities and malocclusion of the teeth.

orthodontist [ôr′thə dän′tist] *Medicine.* a dentist who specializes in orthodontics.

orthodrome *Navigation.* see GREAT CIRCLE.

orthodromic map projection see GNOMONIC MAP PROJECTION.

orthoferrites *Materials.* rare-earth magnetic oxides containing iron.

orthoferrosilite see FERROSILITE.

orthogenesis *Evolution.* a theory of evolutionary change stating that morphological changes in a species over time are the result of some internal or predestined impulse and not the result of natural selection or other external factors. Also, ARISTOGENESIS, ENTELECHY.

orthogeosyncline *Geology.* a geosyncline lying between continental and oceanic cratons, containing both volcanic and nonvolcanic belts. Also, GEOSYNCLINAL COUPLE, PRIMARY GEOSYNCLINE.

orthogneiss *Geology.* a coarse-grained, foliated, banded metamorphic rock, presumably formed by the metamorphism of igneous rock.

orthogonal [ôr thäg′ə nəl] *Mathematics.* perpendicular; normal; having an inner product equal to zero.

orthogonal antennas *Electromagnetism.* a pair of radar antennas, one receiving and one transmitting, that are arranged to detect polarization differences in transmitted and received signals.

orthogonal basis *Mathematics.* a basis of an inner product (Hilbert) space such that the inner product of any two nonequal members of the basis is zero. If, in addition, the inner product of any member of the basis with itself is 1, the basis is said to be orthonormal.

orthogonal complement *Mathematics.* if W is a subset of an inner product space V, the orthogonal complement of W, denoted W^T, is equal to all elements of V that are orthogonal to every element of W. If W is a subspace of V, then W^T is also a subspace of V.

orthogonal crystal *Crystallography.* a crystal in which all axes are at right angles to the others.

orthogonal functions *Mathematics.* two real-valued or complex-valued functions $f(x)$ and $g(x)$ on a space X whose inner product satisfies

$$<f,g> = \int_X f(x)g(x)w(x)dx = 0,$$

where $g(x)$ denotes the complex conjugate and $w(x)$ is a (positive) weight function.

orthogonal group *Mathematics.* the group of orthogonal transformations on an inner product space. On R^n with the usual dot product, this is the set of real orthogonal $n \times n$ matrices, with matrix multiplication as the group operation.

orthogonalization *Mathematics.* any process whereby a set of vectors is replaced by another set of mutually orthogonal vectors with the same span; for example, the Gram-Schmidt orthogonalization process. If, in addition, the members of the new set of orthogonal vectors are normalized, the process is referred to as orthonormalization.

orthogonal knowledge sources *Artificial Intelligence.* knowledge sources that involve different kinds of data or constraints and, therefore, are independent constraints on possible solutions.

orthogonally equivalent matrices *Mathematics.* two $n \times n$ complex matrices (or linear transformations) A and B are said to be orthogonally equivalent if there exists an $n \times n$ orthogonal matrix (or linear transformation) P such that $B = P^{-1}AP$.

orthogonal matrix *Mathematics.* a square matrix M whose inverse equals its transpose; i.e., $M^{T} = M^{-1}$. The determinant of an orthogonal matrix is ± 1; a real orthogonal matrix is unitary.

orthogonal polynomials *Mathematics.* a family of polynomials that are mutually (i.e., pairwise) orthogonal functions.

orthogonal projection *Mathematics.* **1.** let W be a subspace of an inner product space V and β be any vector in V. The orthogonal projection of β on W is a vector α of W such that $\alpha - \beta$ is orthogonal to every vector in W. By theorem, the orthogonal projection of an inner product space on a finite-dimensional subspace always exists. **2.** if every vector in V has an orthogonal projection on W, then the mapping that assigns each vector in V to its orthogonal projection on W is called the orthogonal projection of V on W.

orthogonal system *Mathematics.* a collection of real-valued or complex-valued mutually orthogonal functions f_α; i.e., functions satisfying

$$\int f_\alpha(x) f_\beta(x) dx = 0,$$

where $f_\beta(x)$ denotes the complex conjugate of $f_\beta(x)$ and $\alpha \neq \beta$. If, in addition,

$$\int f_\alpha(x) f_\alpha(x) dx = 1$$

for every n, the system and the set of functions are said to be orthonormal. Also, **orthogonal family.**

orthogonal trajectory *Mathematics.* a trajectory or curve that intersects and is normal to all curves (or surfaces) of a given family is said to be orthogonal with respect to the family.

orthogonal transformation *Mathematics.* a linear operator on a real inner product space that is an isometry; i.e., an orthogonal operator is an automorphism of the inner product space that preserves the inner product. Such operators are equal to their adjoint operators and are the elements of the orthogonal group. On a finite-dimensional space, orthogonal operators can be represented by orthogonal matrices.

orthogonal vectors *Mathematics.* perpendicular vectors. Two vectors u and v of a vector space are said to be orthogonal (also u is orthogonal to v, or v is orthogonal to u) if their inner product is zero; i.e., $< u, v > = 0$. A set of such vectors is an orthogonal set. A set of orthogonal vectors, none of which is the zero vector, is necessarily linearly independent.

orthograde *Anthropology.* in primate studies, the condition in which the body posture is upright or vertical, as in bipedalism.

orthographic *Linguistics.* of or relating to orthography.

orthographic chart *Cartography.* any chart based on an orthographic map projection.

orthographic map projection *Cartography.* a map projection in which lines from points on the earth's surface are projected from a point at infinity and are, therefore, perpendicular to a plane tangent to the surface of the earth, resulting in a perspective azimuthal projection.

orthographic projection *Mathematics.* a depiction of an object created by projecting its features onto a plane along lines perpendicular to the plane. *Crystallography.* the view of a crystal as seen by parallel projection. *Navigation.* a map created by projecting terrestrial points along parallel lines to a tangent plane. Its principal navigational use is for illustrating or graphically solving the navigational triangle.

orthography *Linguistics.* the study of spelling or of the accepted or correct spelling patterns of a particular language.

orthohelium *Atomic Physics.* a form of helium in which the spins of both electrons are parallel.

orthohexagonal axes *Crystallography.* the three axes that define the base of a right hexagonal crystal cell; each of these axes is perpendicular to the vertical axis of the cell.

orthohydrogen *Atomic Physics.* a form of hydrogen in which the spins of both nuclei in the molecule are parallel.

orthohydrous coal *Geology.* a coal that contains 5–6% hydrogen when analyzed on a dry, ash-free basis.

orthokeratology *Medicine.* a method of improving unaided vision by shaping the cornea with contact lenses.

orthokinesis *Biology.* the arbitrary movement of a cell or organism in response to a stimulus.

ortholignitous coal *Geology.* a coal that contains 75–80% carbon when analyzed on a dry, ash-free basis.

orthologous genes *Genetics.* homologous genes that have become differentiated in different species that are derived from a common ancestral species.

orthomagmatic stage *Geology.* the principal stage in the crystallization of silicates from magma, during which as much as 90% of the magma may crystallize. Also, ORTHOTECTIC STAGE.

orthometric correction *Geodesy.* a correction applied to overcome the effect of the orthometric error in the determination of elevation based on the measurement of gravity at various points.

orthometric elevation *Geodesy.* an elevation determined by the measurement of gravity, to which the orthometric correction has been applied.

orthometric error *Geodesy.* an error made in the measurement of the force of gravity at different elevations resulting from the spheroidal shape of the earth, centrifugal force, and the fact that level surfaces at different elevations are not exactly parallel.

orthometric height *Engineering.* the height above sea level as measured along a line directed to the earth's center of gravity.

orthomimic feldspars *Mineralogy.* minerals consisting of triclinic feldspars that, due to repeated twinning, appear to manifest a higher degree of symmetry with rectangular cleavages.

orthomolecular medicine *Medicine.* the practice of maintaining good health and treating disease by varying the concentration of substances normally present in and required by the human body, as by increasing the intake of ascorbic acid in the diet to control a respiratory infection.

orthomorphia *Surgery.* the surgical and mechanical correction of structural deformities.

orthomorphic *Cartography.* of or relating to a map or chart on which small shapes or areas are accurately represented.

orthomorphic chart see CONFORMAL CHART.

orthomorphic map projection see CONFORMAL MAP PROJECTION.

orthomorphic projection *Navigation.* any means for creating a map or chart that results in correct angular relationships.

Orthomyxoviridae *Virology.* a family of pleomorphic, enveloped RNA viruses that are mainly pathogens of the upper respiratory tract of humans, horses, pigs, and birds; transmitted by contact, aerosols, or water.

Orthonectida *Invertebrate Zoology.* an order or class of tiny wormlike mesozoans, endoparasites of marine invertebrates; generations alternate between asexual amoeboids and sexual free-swimming ciliates.

orthonormal basis *Mathematics.* a basis of an inner product (Hilbert) space such that: (a) the inner product of any member of the basis with itself is 1, and (b) the inner product of any two nonequal members of the basis is zero. It can be shown that a Hilbert space always admits an orthonormal basis, that two orthonormal bases have the same cardinality, and that any two Hilbert spaces with orthonormal bases of the same cardinality are isomorphic.

orthonormal coordinates *Mathematics.* in an inner product space, the coordinates of a vector with respect to an orthonormal basis.

orthonormal functions *Mathematics.* mutually orthogonal functions such that the inner product of each function with itself is 1; i.e., for each function $f(x)$ in question, $\int f(x)f(x)dx = 1$, where $f(x)$ indicates a complex conjugate.

orthonormal vectors *Mathematics.* two or more mutually orthogonal vectors, each of length 1. A set of pairwise orthonormal vectors is called an **orthonormal set.**

orthopedic *Medicine.* of or relating to the correction of deformities of the musculoskeletal system.

orthopedics *Medicine.* the medical specialty dealing with the preservation, restoration, and development of the function of the skeletal system, its articulations, and associated structures.

orthopedic surgeon *Medicine.* a surgeon who specializes in orthopedics. Also, **orthopedist, orthopod.**

Orthoperidae *Invertebrate Zoology.* the fungus beetles, a family of tiny coleopteran insects in the superfamily Cucujoidea.

orthophosphoric acid see PHOSPHORIC ACID.

orthophosphorous acid see PHOSPHOROUS ACID.

orthophotograph *Geology.* a photographic copy in which disturbances resulting from tilt and relief have been removed by using a perspective projection.

orthophyric *Petrology.* of or relating to the groundmass texture in certain holocrystalline porphyritic igneous rocks that contain feldspar crystals having quadratic or short, stumpy cross sections.

orthopnea *Pathology.* difficult breathing that is partially or wholly relieved by an erect sitting or standing position; associated with various cardiac or respiratory disorders.

orthopositronium *Particle Physics.* a sample of positronium in which the spins of the positron and the electron are parallel.

Orthopoxvirus *Virology.* a genus of serologically related and morphologically indistinguishable viruses of the subfamily Chordopoxvirinae that infect a wide range of mammals.

Orthopsidae *Paleontology.* a family of regular echinoids in the superorder Camarodonta and extinct order Orthopsida; extant in the Jurassic and Cretaceous.

Orthoptera *Invertebrate Zoology.* an order of large-headed hemimetabolous insects, including grasshoppers, crickets, locusts, and roaches; characterized by strong chewing mouth parts, compound eyes, and straight, narrow front wings that cover folded hind wings; some species are wingless, some have the hind legs enlarged for jumping, and many are harmful plant pests.

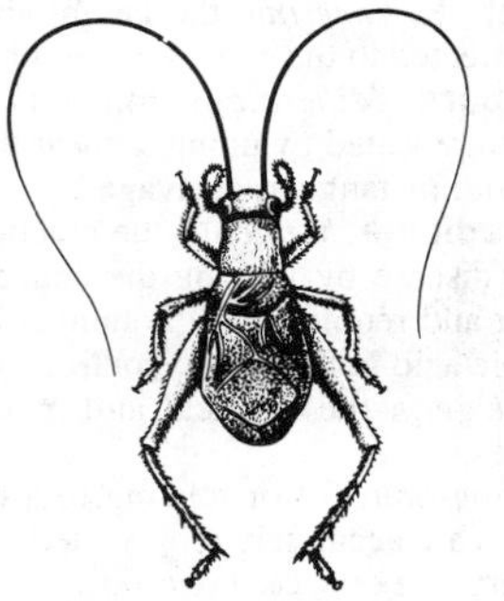

Orthoptera (bush cricket)

orthopteran *Invertebrate Zoology.* **1.** a member of the order Orthoptera. **2.** of or relating to this order.

orthopteron *Invertebrate Zoology.* a member of the order Orthoptera.

orthopterous *Invertebrate Zoology.* belonging to the order Orthoptera.

orthoptic *Mathematics.* given a point P on a (differentiable) plane curve γ, it is often possible to find tangents to γ that intersect the tangent at P at right angles. The locus of these intersections as P varies is called the orthoptic of γ.

orthopyroxene *Mineralogy.* any of several members of the enstatite-ferrosilite series (pyroxene group) that crystallize in the orthorhombic system, and generally contain no calcium and little or no aluminum. Also, **orthorhombic pyroxene.**

orthoquartzite *Petrology.* a sandstone that consists of 90–95% quartz and chert grains which are cemented with quartz deposited in crystallographic continuity with the detrital quartz. Also, SEDIMENTARY QUARTZITE.

orthoquartzitic conglomerate *Geology.* a light-colored orthoconglomerate composed of lithologically homogeneous quartzose residues, generally interbedded with pure quartz sandstone. Also, QUARTZ-PEBBLE CONGLOMERATE.

orthorhombic *Crystallography.* of or relating to a system of crystallization in which three unequal axes intersect at right angles.

orthorhombic pyroxene *Mineralogy.* any of several minerals of the enstatite-orthoferrosilite series crystallizing in the orthorhombic unit cell.

orthorhombic unit cell *Crystallography.* a unit cell in a lattice in which there are three mutually perpendicular twofold rotation axes (parallel to the three axes); as a result, while there are no restrictions on axial ratios, all interaxial angles are necessarily equal to 90° ($\alpha = \beta = \gamma = 90°$). Also, **orthorhombic lattice, orthorhombic system.**

Orthorrhapha *Invertebrate Zoology.* a large suborder of flies, one of two in the Diptera, with a pupa case that opens in a T-shaped cleft through which the adult escapes; includes mosquitos, gnats, horseflies, and crane flies.

orthoschist *Petrology.* a schist that is formed by the alteration of an igneous rock.

orthoscope *Optics.* an instrument that allows an ophthalmologist to examine the interior of the eye without inducing corneal refraction; a layer of water on the instrument prevents direct contact with the eye. Thus, **orthoscopy.**

orthoscopic *Medicine.* relating to, characterized by, or produced by normal vision.

orthoscopic eyepiece *Optics.* an eyepiece in which a planoconvex eye lens and a symmetrical triplet (three cemented elements) minimize distortion and spherical aberration.

orthoscopic system *Optics.* a system that minimizes distortion and spherical aberration. Also, RECTILINEAR SYSTEM.

orthoselection *Evolution.* a process of natural selection that acts continuously in the same direction over long periods of time.

orthosilicate *Materials Science.* any of a group of silicate structures based on a single silicate tetrahedral unit. Also, OLIVINES.

orthosite *Petrology.* a light-colored, coarsely granular igneous rock with orthoclase as the principal constituent.

orthostatic *Medicine.* describing or relating to an erect posture or position.

orthostatic hypotension *Medicine.* a drop in blood pressure associated with an upright position, usually occurring after a long period of standing still or upon assuming an erect position after sitting or lying in bed, often causing faintness, dizziness, and disturbed vision.

orthostratigraphy *Geology.* a standard classification of strata in which fossils are used to identify recognized biostratigraphic zones.

orthosymmetric crystal *Crystallography.* a crystal having the symmetry of an orthorhombic crystal.

orthotectic stage see ORTHOMAGMATIC STAGE.

orthotill *Geology.* an unstratified, unsorted deposit of glacial drift formed by immediate release from the transporting ice.

orthotomic system *Optics.* a system containing only those rays that may be intersected perpendicularly by an appropriate surface.

orthotonus *Neurology.* a spasmodic condition inducing the head, body, and limbs to be held rigid in a straight line. Also, **orthotonos.**

orthotopic graft *Surgery.* a tissue transplant grafted into its normal anatomical position.

Orthotrichaceae *Botany.* a family of large, dull-green, tuft-forming mosses of the order Orthotrichales, characterized by stems that are simple or have numerous erect branches, and by terminal sporophytes.

Orthotrichales *Botany.* a heterogeneous order of dull, tuft- or mat-forming mosses in the class Bryopsida that have stems with terminal sporophytes; found growing on rocks or tree trunks and branches.

orthotropic *Mechanics.* having significantly different properties along two or more orthogonal axes. Wood is a common example of an orthotropic material, since its properties, such as stiffness and Young's modulus, have different values depending on whether the direction is parallel or perpendicular to the grain of the wood.

orthotropic deck *Civil Engineering.* a floor or bridge deck notably stiffer in the direction of the span than in the direction perpendicular to the span.

orthotropic laminate *Materials Science.* a fiber-reinforced lamina with unidirectionally aligned fibers that make the material stronger and more resistant to stresses occurring in the longitudinal direction.

orthotropic plate *Civil Engineering.* a structural plate that has characteristics in one direction different from those in the other.

orthotropism *Botany.* a tendency to grow upward or downward in a vertical direction, in response to an orienting stimulus such as sunlight or gravity.

orthotropous *Botany.* having a straight ovule on a straight stalk so that the chalaza, hilum, and micropyle are in the same axial line.

orthotropy *Materials Science.* the existence of three perpendicular axes of symmetry in a material, so that a 180° rotation results in an identical structure.

orthotungstic acid see TUNGSTIC ACID.

orthovoltage *Radiology.* voltage that is in the range of 140–400 kilovolts; used in X-ray therapy.

Orthox *Geology.* a suborder of the soil order Oxisol that develops at low altitudes in humid tropical regions; characterized by a low to moderate organic-carbon content, a high water content throughout most of the year, and a mean annual soil temperature of at least 22°C.

ortolan *Vertebrate Zoology.* a small brown Old World bunting, *Emberiza hortulana,* of the family Fringillidae, often found in gardens; noted as a choice food.

Orussidae or **Oryssidae** *Invertebrate Zoology.* a small family of wood wasps, hymenopteran insects in the superfamily Siricoidea; larvae are parasitic on wood-boring insects.

orvietite *Petrology.* an extrusive rock consisting of plagioclase and sanidine in nearly equal amounts, with leucite, augite, minor biotite, and olivine, and trace amounts of apatite and opaque oxides; it is intermediate in composition between leucite phonolite and leucite tephrite.

ORY *Aviation.* the airport code for Orly, Paris, France.

oryx *Vertebrate Zoology.* a horselike antelope of the genus *Oryx* in the family Bovidae, having a stocky build, long, straight, slender horns, a light coat with dark markings, and a long tufted tail; found in the arid regions of Africa, Arabia, and Iraq.

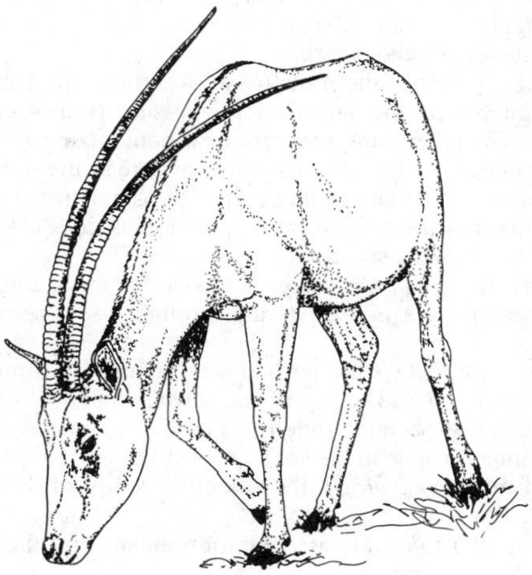

oryx

os *Anatomy.* **1.** *plural,* **ora.** the mouth or the opening of a hollow organ. **2.** *plural,* **ossa.** a term that, in combination with a qualifying adjective, is used to designate a particular bone or bony structure in Latinate medical terminology.

OS or **O.S.** left eye (from Latin *oculus sinister*); one side; ordnance survey.

Os the chemical symbol for osmium.

OSA or **O.S.A.** Optical Society of America.

Osage orange *Botany.* an ornamental dioecious North American thorny tree, *Maclura pomifera,* of the mulberry family Moraceae; characterized by its yellowish bark, milky sap, and aggregate green or yellow fruit.

Osagian or Osagean *Geology.* a North American provincial series of the Lower Mississippian period, occurring after the Kinderhookian and before the Meramecian.

osazone *Biochemistry.* an easy-to-isolate, high-melting solid monosaccharide derivative that contains two phenylhydrazine residues.

Osborn, Fairfield 1887–1969, American zoologist and conservationist; son of Henry Fairfield Osborn.

Osborn, Henry Fairfield 1857–1935, American paleontologist and author; made noted studies of early reptiles and mammals; headed the American Museum of Natural History.

osciducer *Electronics.* a transducer in which the frequency of an input signal varies proportionally from the center frequency of an oscillator. Also, **oscillating transducer.**

oscillate [äs′i lāt′] *Physics.* to carry on or produce a process of oscillation; to vary between one value or limit and another. (Going back to a Latin word meaning "to swing.")

oscillating conveyor *Mechanical Engineering.* a vibrating conveyor with a relatively low frequency and large amplitude of motion.

oscillating function *Mathematics.* a continuous function that is not monotone on any subinterval of a given interval is said to be oscillating on that interval. If the function is not monotone on any interval, it is said to be everywhere oscillating.

oscillating granulator *Mechanical Engineering.* a particle-reducing device with oscillating bars arranged cylindrically.

oscillating magnetic field *Electromagnetism.* a magnetic field that varies periodically in time, capable of inducing voltages and currents in conductive loops.

oscillating mine *Ordnance.* a hydrostatically controlled mine that maintains a preset depth independently of the rise and fall of the tide.

oscillating reaction *Physical Chemistry.* a reaction that proceeds in an oscillatory manner; i.e., in which concentrations oscillate with time or space rather than changing in a monotonic way.

oscillating screen *Mechanical Engineering.* a separation screen that oscillates a sifting screen parallel to another screen.

oscillating universe *Astronomy.* a variation on the closed universe cosmological model, in which the universe undergoes unending cycles of contraction and expansion.

oscillation [äs′i lā′shən] *Physics.* **1.** a periodic variation in the value of a physical quantity, especially a regular variation above and below some mean value; for example, the voltage of an alternating current, the pressure of a sound wave, or the position of a pendulum. **2.** a single movement or change of this type. *Robotics.* unwanted vibrations in a robot.

oscillation method *Crystallography.* a method in which a photograph of the X-ray diffraction pattern of a crystal is obtained by oscillating the crystal through a small angular range. Thus, **oscillation camera, oscillation photograph.**

oscillation of a function *Mathematics.* **1.** the oscillation of a real-valued function f on an interval is the difference between the maximum and minimum values of f on the interval (oscillation may be infinite). **2.** the oscillation of a real-valued function f at a point x is the limit as ε approaches 0 of the oscillation of f on the interval $[x - \varepsilon, x + \varepsilon]$. **3.** more generally, suppose f is a function from a topological space X into a metric space Y. For each $x \in X$, the oscillation of f is $\omega(x) = \inf\{\text{diam}(f(U)): U$ is a neighborhood of $x\}$; that is, the infimum of the diameters of images under f of neighborhoods of x. The function f is continuous at x if and only if $\omega(x) = 0$.

oscillator [äs′i lāt′ər] something that causes or displays oscillation; specific uses include: *Electronics.* a circuit or device that generates a periodic signal. *Physics.* a device that exhibits oscillation, as in the case of a mass-loaded spring that is released from a nonequilibrium position. *Quantum Mechanics.* a physical system that has a discrete set of possible energy levels or allowed values, between which intermediate energies never occur.

oscillator harmonic interference *Electronics.* the interference that appears in a superheterodyne receiver when the incoming signal interacts with the harmonics of the local oscillator.

Oscillatoria *Bacteriology.* a genus of cyanobacteria that grow as unbranched filaments in aquatic habitats or on damp rocks or soil, forming extensive mats; some species may contain gas vacuoles.

Oscillatoriales *Botany.* an order of blue-green algae that have filaments and are multicellular.

oscillator strength see F VALUE.

oscillatory circuit [äs′i lə tôr′ē] *Electricity.* a circuit in which inductance, capacitance, and resistance are so arranged that oscillations can be generated or sustained.

oscillatory current *Electricity.* a current that periodically reverses its flow, especially the flow in an LC tank circuit in which energy oscillates between an inductor and a capacitor.

oscillatory discharge *Electricity.* an electrical discharge generating an oscillating or alternating current that gradually decreases in amplitude.

oscillatory reaction *Chemistry.* a chemical reaction in which the variables or components of the system change in a periodic manner.

oscillatory shear *Fluid Mechanics.* low-amplitude oscillations in viscous fluids; used to study properties about the dynamic viscosity of the fluid.

oscillatory surge *Electricity.* a current or voltage surge containing positive and negative values.

oscillatory twinning *Crystallography.* parallel twinning that is repeated in a periodic manner.

oscillatory wave *Physics.* any wave that is periodic in time or whose source is an oscillator.

oscillistor *Electronics.* a bar of semiconductor material, such as germanium, that oscillates when subjected to a magnetic field and a direct current.

oscillogram *Engineering.* a cathode-ray tube display or permanent photographic record of the output data of an oscillograph.

oscillograph [ə sil′i graf′] *Engineering.* a sensitive electromechanical recorder that is able to provide a visual record of a waveform by recording values of an electrical quantity over time; sometimes equipped with a photographic recording system.

oscillographic polarography *Physical Chemistry.* a form of polarography using oscillographic scanning to measure the concentration of conductive elements in an electrolytic solution.

oscillometer *Cardiology.* an instrument used to measure oscillations, such as changes in the volume of the arteries.

oscillometric titration *Physical Chemistry.* a technique that uses variable radio frequencies to analyze changes in an electrolytic solution to which a titrant has been added.

oscillometry *Physical Chemistry.* a technique for measuring frequency in an electrode discharge in order to trace the progress of titrant in an electrolytic solution.

oscilloscope [ə sil′i skōp′] *Electronics.* an instrument that uses a cathode-ray tube or similar instrument to display fluctuating electrical signals on a fluorescent screen; common applications include the testing of electronic equipment and the monitoring of electrical impulses from the heart or brain.

oscilloscope tube *Electronics.* a cathode-ray tube that creates a visual representation of rapidly moving electric signals, which change in tandem with the signal's position.

Oscillospira *Bacteriology.* a genus of aerobic, motile, endospore-forming, rod-shaped bacteria of uncertain affiliation, formerly classified in the family Oscillospiraceae; found in the alimentary tract of herbivorous animals.

Oscillospiraceae *Bacteriology.* in former classifications, a family of bacteria of the order Caryophanales, containing the single genus *Oscillospira.*

oscin see SCOPOLINE.

Oscines *Vertebrate Zoology.* the largest suborder in the perching bird order Passeriformes, including most of the songbirds and about 45% of all recent birds; found worldwide.

O scope *Electronics.* a radarscope that produces an O display. Also, **O scan.**

osculating circle *Mathematics.* given a space curve *C*, a point *P* on *C*, and the tangent line *T* to *C* at *P*, the circle, if it exists, which is the limit as a point P' tends to *P* along *C* of the circles C' through P', tangent to *T* at *P*, and lying in the plane (if it exists) determined by *T* and P'. The radius of curvature of *C* at *P* is the radius of the osculating circle at *P*.

osculating orbit *Astronomy.* the orbit that a celestial body would follow if all perturbating forces of other celestial objects were suppressed and the body was subject only to the gravitational force of the sun.

osculating orbital elements *Astronomy.* the orbital elements that describe an osculating orbit.

osculating plane *Mathematics.* the osculating plane of a curve γ at a point *P* of γ is the plane spanned by the tangent and principal normal vectors at *P*. It may be viewed as the limit, as P' approaches *P* along γ, of planes that are tangent to γ at *P* and also contain P' of γ.

osculum *Invertebrate Zoology.* in sponges, the external opening of an excurrent canal.

-ose a suffix meaning "sugar."

Oseen's flow *Fluid Mechanics.* a fluid flow in which the fluid velocities are very low due to a large viscosity coefficient.

OSF or **O.S.F.** Office of Space Flight.

Osgood-Schlatter disease *Pathology.* an inflammation or partial separation of the tibial tubercle, usually resulting from overuse of the quadriceps muscle; characterized by swelling and tenderness that increase with exercise; most often seen in muscular, athletic adolescent boys. (Named for Robert Bayley *Osgood,* 1873–1956, American surgeon, and Carl *Schlatter,* 1864–1934, Swiss surgeon.)

OSHA or **O.S.H.A.** Occupational Safety and Health Administration.

O shell *Atomic Physics.* the electron orbit of an atom, fifth closest to the nucleus, in which electrons whose principal quantum number is 5 occur.

OSI Open System Interconnection.

-osis a suffix meaning: **1.** condition. **2.** disease. **3.** increase; formation.

OSL *Aviation.* the airport code for Oslo, Norway.

Osler, Sir William 1849–1919, Canadian physician and author; noted medical educator; promoted the relation between patient and physician.

Osler's disease see POLYCYTHEMIA VERA.

Osmeridae *Vertebrate Zoology.* the smelts, a family of marine and freshwater, elongate compressed fishes of the order Salmoniformes; prized for their sweet flavor; found in the northern Atlantic, Pacific, and Arctic oceans.

osmeterium *Entomology.* a fleshy eversible Y-shaped scent gland found at the anterior of some Lepidoptera larvae.

osmic *Chemistry.* relating to or containing osmium.

osmics *Neurology.* the science that relates to the olfactory organs and the sense of smell, as well as to odoriferous organs and substances.

osmium *Chemistry.* a metallic element having the symbol Os, the atomic number 76, an atomic weight of 190.2, a melting point of 2700°C, and a boiling point of 5300°C; it is a hard white metal of the platinum group that gives off fumes when heated in air. It is used as a hardener in alloys and as a catalyst. (Derived from a Greek word meaning "smell;" because of the powerful odor produced by oxidation of this element.)

osmium tetroxide *Organic Chemistry.* OsO_4, a colorless, pungent dimorphic compound with both crystalline and amorphous forms having a disagreeable odor; soluble in water and alcohol; boils at 130°C and melts at 40°C; derived by heating powdered osmium in air or by treating it with nitric acid, aqua regia, or chlorine; toxic by inhalation and a strong irritant to eyes and mucous membranes; used in microscope staining and photography. Also, **osmic acid.**

osmoceptor see OSMORECEPTOR.

osmolality *Chemistry.* the molality of a nonionizing substance in an ideal solution that has the same osmotic pressure as the solution itself.

osmolarity *Chemistry.* the molarity of a nonionizing substance in an ideal solution that has the same osmotic pressure as the solution itself.

osmole *Chemistry.* a unit equal to either the osmolality or osmolarity of a solution that has the same osmotic pressure as does a 1 mol/kg ideal solution of nonionizing substance.

osmometer *Analytical Chemistry.* a device for measuring the osmotic pressure exerted by a liquid passing through a semipermeable membrane.

osmometry *Physical Chemistry.* the measurement of osmotic pressure, used to determine the lowering of the solvent chemical potential when solute is added; especially suited to polymer solutions that require a membrane impermeable to the solute molecules.

osmonosology *Neurology.* the scientific study of disorders of the sense of smell.

osmophile *Microbiology.* any microorganism that thrives on or in media with high osmotic pressure.

osmophobia *Psychology.* an irrational fear of odors.

osmoreceptor *Neurology.* a specialized neuron or peripheral sense organ in the brain whose electrical activity is stimulated by changes in the extracellular fluid, chiefly an increase in sodium concentration. Also, OSMOCEPTOR. *Physiology.* a cell that reacts to stimuli produced by odors.

osmoregulatory mechanism *Physiology.* the process by which body cells or simple organisms maintain the osmotic pressure of the extracellular fluid.

osmosis [äz mō′sis; äs mō′sis] *Physical Chemistry.* the passage of a pure solvent through a semipermeable membrane, from a solution in which it has higher concentration to one in which it has lower concentration; this tends to equalize the concentration on either side of the membrane. For example, if a membrane will allow the passage of water molecules but not larger sugar molecules, the solvent (water) will tend to move from the region of higher concentration (pure water) to that of lower concentration (the solution of sugar and water). *Biology.* a manifestation of this process in living systems, such as the movement of water molecules in and out of cells past the cell wall.

osmotaxis *Biology.* the movement of cells as affected by the density of the liquid which contains them.

osmotic [äz mät′ik; äs mät′ik] of or relating to osmosis.

osmotic fragility *Physiology.* the susceptibility of a red blood cell to break apart when exposed to saline solutions of a lower osmotic pressure than that of the human cellular fluid.

osmotic pressure *Physical Chemistry.* **1.** the force exerted on a semipermeable membrane that is between a pure solvent and a solution, or between two solutions of differing concentration, caused by the tendency of solvent molecules to pass through the membrane into the more dilute solution. **2.** the amount of external pressure that must be applied in order to stop this force, so as to prevent the normal process of osmosis, or to oppose it, so as to bring about a process of reverse osmosis. Also, **osmotic gradient.**

osmotic shock *Physiology.* a rapid alteration in the osmotic pressure of a cell or virus, usually causing it to discharge its contents.

osmotrophy *Biochemistry.* a process in which nutrients are absorbed through the cell surface, common in bacteria and fungi.

Osmunda *Paleontology.* a still-living genus of large, eusporangiate ferns with massive, erect rhizomes; first appeared in the Permian.

Osmundaceae *Botany.* a cosmopolitan family of medium-sized, primitive, terrestrial ferns belonging to the order Filicales; characterized by a hairy rhizome, laterally flared leaf bases, and sporangia often borne on specialized fronds rather than in sori.

osotriazole see 1,2,3-TRIAZOLE.

osphradium *Invertebrate Zoology.* a patch of chemoreceptive sensory epithelium found on the posterior margin of the gill membranes in mollusks.

osphresiology *Neurology.* the study of the nature and associations of odors and odorants and of the sense of smell. Also, **osmology.**

Osphronemidae *Vertebrate Zoology.* the perciform family of "true" gourami, consisting of one species, *Osphronemus goramy*; characterized by a compact, oval body and a long filament extending from a ray on each pelvic fin; found in tropical fresh waters of Java, Sumatra, and Borneo.

osprey *Vertebrate Zoology.* a large, coastal fishing bird-of-prey, *Pandion haliaetus*, of the family Pandionidae, having a dark brown body, white underbelly and legs, a speckled neck, and spikes on the soles of the feet; found in tropical and temperate areas. Also, FISH HAWK.

osprey

osseous *Anatomy.* relating to or like bone; bony.

osseous tissue *Histology.* bone tissue.

osseus system *Anatomy.* the bones of the skeleton. Also, SKELETAL SYSTEM.

ossicle *Anatomy.* a small bone; especially the auditory bones of the middle ear, including the malleus, incus, and stapes. *Invertebrate Zoology.* a small bonelike structure, such as the calcareous toothlike structure in the gastric mill of crustaceans or the rods or plates that compose an echinoderm skeleton.

ossify *Physiology.* to produce bone or turn into bone.

ossifying fibroma *Medicine.* a benign tumor of bone derived from connective tissue, often occurring on vertebral column bones.

ossipite *Petrology.* a coarse-grained troctolite consisting of labradorite, olivine, magnetite, and minor clinopyroxene.

ostamer *Materials.* a commercial polyurethane rigid foam that is used to make bone fracture casts.

O star *Astrophysics.* an extremely hot, massive, blue star whose spectrum is characterized by lines of singly ionized helium; O stars have surface temperatures of 28,000 to over 50,000 kelvins.

Ostariophysi *Vertebrate Zoology.* a superorder of about 6000 teleost fishes, divided into two orders: Cypriniformes (the carps, minnow, and characins) and Siluriformes (the catfish).

oste- or **osteo-** a combining form denoting a relationship to a bone or to the bones.

ostealgia *Medicine.* pain in a bone.

ostectomy *Surgery.* the excision of a bone or a portion of a bone.

Osteichthyes *Vertebrate Zoology.* the class of bony fishes, having at least a partly ossified skeleton and a true gill cover; most species have a swim bladder.

osteitis *Medicine.* an inflammation of bone involving the haversian spaces, canals, and their branches; marked by tenderness and dull pain.

osteitis fibrosa cystica *Medicine.* an inflammatory degenerative condition, characterized by the formation of cysts and fibrous nodules on affected bones; caused by hyperfunction of the parathyroid gland.

osteoarthritis *Medicine.* a form of arthritis in which one or many joints undergo degenerative changes; marked by stiffness, tenderness to touch, and enlargment, which may be followed by deformity, subluxation, and synovial effusion. The most common form of arthritis, its cause is unknown, but emotional stress often aggravates the condition.

osteoarthropathy *Medicine.* a disorder affecting joints and bones, with periosteal and articular proliferation and severe pain; often found in diseases of the lungs.

osteoblast *Histology.* an immature bone cell that actively participates in bone formation.

osteoblastoma *Oncology.* a benign, painful, fairly vascular tumor of the bone that is characterized by the formation of primitive bone and fibrous tissue. Also, **osteoid osteoma.**

osteochondritis *Medicine.* an inflammation of both cartilage and bone, sometimes resulting in the splitting of pieces of cartilage into the joint.

osteochondroma *Oncology.* a benign tumor containing the elements of osteoma and chondroma, occurring on projecting adult bone capped by cartilage.

osteochondromatosis *Oncology.* a condition characterized by the presence of multiple osteochondromas.

osteochondromyxoma *Oncology.* a tumor that is part osteochondroma and part myxoma.

osteochondrosarcoma *Oncology.* a sarcoma that is composed of bone and cartilage tissue.

osteochondrosis *Medicine.* a disease involving ossification occurring chiefly during periods of rapid growth, characterized by necrosis followed by slow regeneration.

osteoclasis *Medicine.* the fracturing of a long bone without resorting to open operation, for the purpose of correcting a deformity.

osteoclast *Histology.* a large multinucleate cell that actively reabsorbs bone.

osteoclast-activating factors *Endocrinology.* a group of substances that may be produced by activated lymphocytes and may stimulate bone resorption.

osteoclastoma *Oncology.* a giant-cell tumor of the bone.

osteocyte *Histology.* a bone cell.

osteodermia *Medicine.* a condition characterized by the formation of bone-containing nodules on the skin. Also, OSTEOSIS CUTIS.

osteodynia *Medicine.* any pain in a bone.

osteodystrophy *Medicine.* defective bone formation, as in rickets or dwarfism, usually associated with inadequate diet or renal failure.

osteofibroma *Oncology.* a tumor composed of both bone and fibrous tissue.

osteofibrosis *Medicine.* bone fibrosis, usually involving red bone marrow.

osteogenesis *Physiology.* the production of bone.

osteogenesis imperfecta *Medicine.* a dominantly inheritable disease characterized by thin-walled, extremely fracture-prone bones deficient in osteoblasts, as well as malformed teeth and progressive deafness.

osteogenesis imperfecta congenita *Medicine.* a more severe form of osteogenesis imperfecta with the onset of multiple fractures in utero and autosomal recessive inheritance in some cases.

Osteoglossidae *Vertebrate Zoology.* the bonytongues, a small family of distinctive primitve fishes in the order Osteoglossiformes, having an elongated body, large bony scales, and a head protected by a hard bony plate; found in freshwater habitats of South America, Africa, Southeast Asia, and northern Australia.

Osteoglossiformes *Vertebrate Zoology.* an order of primitive freshwater teleost fishes that are exclusively tropical except for one North American species; includes the bonytongues, butterflyfish, featherbacks, and mooneyes.

osteoid *Histology.* newly synthesized bone matrix that has yet to be calcified.

osteolathyrism *Medicine.* a skeletal disorder in laboratory animals that are fed a diet containing the sweet pea or certain aminonitriles, resulting in pathologic changes in connective tissue.

Osteolepidae *Paleontology.* a family of primitive rhipidistian fishes in the extinct superfamily Osteolepidae; probably in the direct line of evolution of the amphibians; extant from the Devonian to Permian.

Osteolepiformes *Paleontology.* a suborder classification now generally abandoned in favor of a superfamily classification, Osteolepidoidea; extant from the Devonian to Permian.

Osteolepis *Paleontology.* a genus of crossopterygian fishes in the extinct superfamily Osteolepidoidea and family Osteolepidae; Devonian.

osteolipochondroma *Oncology.* an osteochondroma that contains fatty tissue.

osteolipoma *Oncology.* a lipoma containing bony tissue.

osteolith *Paleontology.* a fossil bone that has become completely mineralized.

osteologist *Anatomy.* a specialist in the study of bones.

osteology *Anatomy.* the scientific study of bones; the body of knowledge relating to the bones of humans and animals.

osteolysis *Medicine.* the softening, absorption, and dissolution of bony tissue, especially the removal or loss of calcium in bone.

osteoma *Oncology*. a benign tumor that is composed of bone tissue, usually appearing in late childhood or young adulthood.
osteomalacia *Medicine*. a condition marked by softening of the bones due to a decrease in the amount of available calcium and phosphorous, caused by dietary deficiencies or impaired function of one of the body organs involved in the absorption and metabolism of bone minerals.
osteometry *Anatomy*. the measurement of bones.
osteomyelitis [äs´tē ō´mī ə lī´tis] *Medicine*. an acute or chronic infection of bone tissue, usually occurring in childhood, most commonly caused by the bacterium *Staphylococcus aureus*.
osteon *Histology*. the histological unit of compact bone, consisting of concentric cylinders of osteocytes and bony salts surrounding a central canal containing blood vessels. Also, HAVERSION SYSTEM.
osteonal canal *Histology*. in compact bone, a microscopic canal that contains blood vessels and is surrounded by thin layers of bony salts and osteocytes. Also, HAVERSIAN CANAL.
osteonephropathy *Medicine*. any of a variety of syndromes involving changes in bone, usually accompanying renal disease.
osteopath [äs´tē ə path´] *Medicine*. a practitioner of osteopathy.
osteopathic [äs´tē ə path´ik] *Medicine*. of or relating to the practice of osteopathy.
osteopathy [äs´tē äp´ə thē] *Medicine*. a system of therapy based upon the idea that the body in correct adjustment has adequate nutrition, is in a favorable environment, and is capable of making its own remedies against infections and other toxic conditions. Osteopathy uses generally accepted physical, medicinal, and surgical methods of diagnosis and therapy, while placing chief emphasis on the importance of normal body mechanics and manipulative methods of detecting and correcting faulty structure.
osteopenia *Medicine*. a condition of reduced bone mass caused by the inability of the body to synthesize enough bone to keep pace with normal bone destruction.
osteopetrosis *Medicine*. a rare genetic disorder of the bone, characterized by abnormally dense bone, due to defective resorption of immature bone.
osteophony *Physiology*. the process of hearing sounds through bone.
osteoplasty *Medicine*. bone grafting or reconstructive surgery of the bone via plastic surgery.
osteopoikilosis *Medicine*. a hereditary condition that is asymptomatic, usually discovered accidentally on X-ray examination when ellipsoidal dense foci are seen in bones.
osteoporosis [äs´tē ō pə rō´sis] *Medicine*. a significant reduction in the amount of bone mass, leading to fractures after minimal trauma; occurring most often in postmenopausal women and in sedentary or immobilized individuals; it may lead to fractures after only minimal trauma, as of the hip, wrist, or spine, and can also lead to loss of stature and various deformities.
osteoradionecrosis *Radiology*. the destruction of bone tissue resulting from irradiation.
osteosarcoma *Oncology*. the most common malignant tumor of bone, affecting mainly the large long bones, characterized by pain, swelling, limitation of joint movement, and a high frequency of pathological fractures.
osteosclerosis *Medicine*. the hardening or abnormal density of bone.
osteoscope *Radiology*. a device that evaluates the performance of roentgen ray apparatus by making roentgenograms of a standard bone sample of the forearm.
osteosis cutis see OSTEODERMIA.
Osteostraci *Paleontology*. an extinct order of small, fishlike, jawless marine vertebrates characterized by a flat head armored with a bony shield; related to the modern lamprey; Silurian and Devonian.
osteotome *Medicine*. an instrument used to cut bone.
osteotomy *Medicine*. the surgical cutting of a bone.
osteotrophy *Medicine*. the nutrition of bone.
ostia *Entomology*. lateral openings in an insect body through which blood flows into the heart from the pericardial cavity.
ostiole *Biology*. a small orifice or pore, such as the small opening in the receptacle of algae, leading to the conseptacle.
ostium *Biology*. a mouth, entrance, or aperture, such as any of the lateral openings in the heart of an arthropod, any of the openings through which water enters the body of a sponge, or the opening at either end of the fallopian tubes.
ostium primum see FORAMEN PRIMUM.
ostomate *Surgery*. a patient who has undergone an enterostomy or ureterostomy.
Ostomidae *Invertebrate Zoology*. the bark-gnawing beetles, a mostly predaceous family of coleopteran insects belonging to the superfamily Cleroidea.
ostomy *Surgery*. an operation to create an artificial opening between two organs or viscera, or between an organ and the abdominal wall so that a body fluid, such as urine, may pass.
-ostomy a suffix meaning "the creation of an opening."
OSTP Office of Science and Technology Policy.
Ostracoda *Invertebrate Zoology*. a subclass of minute aquatic crustaceans, sometimes called seed shrimp, with a bivalve carapace and a reduced abdomen.
ostracoderm *Paleontology*. of or relating to several extinct jawless fishes of the Early Paleozoic, belonging to the Cephalaspidomorphi and Pteraspidomorphi.
Ostrea *Paleontology*. a genus of sessile bivalves in the superfamily Ostreacea that includes the common oyster; extant since the Cretaceous.
Ostreidae *Invertebrate Zoology*. the oysters, a family of sessile marine bivalve mollusks, with a lower valve that cements to the substrate, including both edible and inedible species.
Ostreobiaceae *Botany*. in some classification systems, a monogeneric family of green algae of the order Bryopsidales, having irregular siphonous filaments that are usually cylindrical or inflated; found in marine calcareous substrates.
ostreobium *Botany*. a genus of green algae, comprising the family Ostreobiaceae, that grow in marine calcareous substrates.
Ostreopsidaceae *Botany*. a family of unicellular biflagellates of the order Peridiniales that are photosynthetic, thecate, and epiphytic on red and brown algae or sometimes planktonic; some species are responsible for the ciguatera poisoning of tropical reef fish.
ostrich *Vertebrate Zoology*. an extremely large, flightless ratite bird of the genus *Struthio* in the family Struthionidae, reaching heights of ten feet, characterized by a long neck and long, strong legs; found in wide sandy plains and open savannah country in Africa. It is the tallest and heaviest living bird, up to 8 feet tall and over 300 pounds.

ostrich

ostropalean ascus *Mycology*. a kind of spore sac or ascus that discharges its spores through a narrow channel located in its cap.
Ostropales *Mycology*. an order of fungi belonging to the class Discomycetes, some forms obtain nutrients from dead organic matter, while others are parasites; found on wood, bark, and herb stems.
Ostwald, F. Wilhelm 1853–1932, German physical chemist; awarded the Nobel Prize for research on catalysis and speeds of chemical reactions.
Ostwald diagram *Analytical Chemistry*. a diagram plotting the percentage of carbon dioxide in a fuel against the percentage of oxygen in air.
Ostwald indicator theory *Chemistry*. a theory stating that indicators change color because they are weak acids or bases.

Ostwald law *Physical Chemistry.* a law that relates the degree of ionization to concentration in the case of a weak electrolyte, using the expression $K = \alpha^2C/1 - \alpha$, where K is the ionization constant, α is the degree of ionization, and C is the molar concentration of the electrolyte. Also, **Ostwald dilution law.**

Ostwald process *Chemical Engineering.* a commercial process used to produce nitric acid in which ammonia is heated with a catalyst forming nitric oxide that is oxidized to yield nitrogen dioxide, which in turn reacts with water to yield nitric acid and nitric oxide.

Ostwald ripening *Chemistry.* a phenomenon in which the smaller crystals in a solution reprecipitate onto larger crystals.

Ostwald's adsorption isotherm *Thermodynamics.* a statement that at constant temperature, the amount of adsorbed substance per unit amount of adsorbent dispersed in a gas or solution is proportional to the concentration of the adsorbent raised to some constant power.

Ostwald's rule *Chemistry.* a law stating that if a substance can exist in more than one modification, the least stable will form first and eventually change into the more stable.

Ostwald viscometer *Engineering.* an instrument that determines the relative viscosity of a liquid by drawing a test liquid into a capillary tube and noting the time it takes for the liquid to drop to a certain point, then comparing it with the time needed for a liquid of known viscosity to reach the same point.

osumilite *Mineralogy.* $(K,Na)(Fe^{+2},Mg)_2(Al,Fe^{+3})_3(Si,Al)_{12}O_{30}\cdot H_2O$, a dark-blue to black hexagonal mineral of the osumilite group occurring as short prismatic crystals, having a specific gravity of 2.64 and an undetermined hardness; found in volcanic rocks.

OT occupational therapy.

ot- or **oto-** *Anatomy.* a combining form denoting relationship to the ear.

otalgia *Medicine.* an earache or pain in the ear, often related to laryngeal, tonsillar, or nasopharyngeal causes.

Otariidae *Vertebrate Zoology.* the eared seals, fur seals, and sea lions, a family in the carnivore suborder Caniforma; characterized by long, slender bodies, short tails, and rudderlike rear fins; found in cold coastal waters in northeast Asia, western North America, South America, South Africa, southern Australia, New Zealand, and southern oceanic islands.

otavite *Mineralogy.* $CdCO_3$, a white, brown, or red trigonal mineral of the calcite group occurring as tiny rhombohedral crystals, having a specific gravity of 4.96 to 5.03 and an undetermined hardness; found with other secondary carbonates at Tsumeb, Namibia.

OTC over-the-counter (drug or prescription).

OTD organ tolerance dose.

other-directed *Psychology.* tending to be influenced by the beliefs and values of others rather than by one's own internal principles.

otic *Anatomy.* of or relating to the ear.

-otic a suffix meaning "of, relating to, or characterized by the preceding action or condition."

otic capsule *Developmental Biology.* the skeletal element enclosing the inner ear mechanism. Also, AUDITORY CAPSULE.

otic ganglion *Anatomy.* a parasympathetic ganglion in the infratemporal fossa that receives input from the glossopharyngeal nerve and innervates the parotid gland.

otic placode see AUDITORY PLACODE.

Otidiaceae *Mycology.* a family of fungi belonging to the order Pezizales that occur primarily in soil, and sometimes on decaying wood and charcoal.

otidium *Invertebrate Zoology.* the otocyst in mollusks.

Otitidae *Invertebrate Zoology.* a family of dipteran insects in the subsection Acalyptratae, with spotted or banded wings, metallic coloration, and larvae that usually feed on decaying matter.

otitis *Medicine.* an inflammation of the ear, which may be marked by pain, fever, hearing loss, and vertigo.

otitis externa *Medicine.* an inflammation of the external auditory canal, possibly caused by fungus or bacteria.

otoconium see OTOLITH.

otocyst *Invertebrate Zoology.* an organ of equilibrium consisting of a fluid-filled sac containing a statolith (in crustaceans commonly a grain of sand) which moves with respect to gravity and stimulates sensory cells. Also, STATOCYST. *Developmental Biology.* the otic capsule of the embryo.

otodynia *Medicine.* an earache or pain in the ear.

otolaryngologist *Medicine.* a physician who specializes in treatment of the ear, nose, and throat.

otolaryngoly *Medicine.* the branch of medicine concerned with treatment of the head and neck, including the ears, nose, and throat.

otolith *Anatomy.* one of a number of tiny calcium-containing granules in a gelatinous substance, adhering to the otolithic membrane in the utricle and saccule of the inner ear; movement of the membrane provides sensory information on the position and movement of the head in space. Also, OTOCONIUM, STATOCONIUM.

otologist *Medicine.* a specialist in otology.

otology *Medicine.* the branch of medicine that deals with the study, diagnosis, and treatment of diseases of the ear and related structures.

Otomat *Ordnance.* a French-Italian surface-to-surface cruise missile designed for ship-to-ship combat; it is powered by a turbojet engine, equipped with inertial guidance and active radar terminal homing, and delivers a 440-pound, semi-armor-piercing warhead at high subsonic speed and a maximum range of 37 to 50 miles.

-otomy a suffix meaning "a surgical incision."

otomycosis *Medicine.* a fungal infection of the external auditory canal, marked by pruritus and exudative inflammation.

otopathy *Medicine.* any disease of the ear.

otoplasty *Surgery.* plastic surgery of the ear to change size or correct deformities or defects.

otoporpa *Invertebrate Zoology.* a line of stinging cells on the umbrella of hydrozoan medusae.

otorrhagia *Medicine.* a hemorrhage from the ear.

otorrhea *Medicine.* a discharge from the ear.

otosclerosis *Medicine.* a disease of the bone enclosing the inner ear, causing fixation of the base of the stirrup in the oval window and resulting in progressively increasing deafness.

otoscope *Medicine.* an instrument designed for examination of the ear; it renders the tympanic membrane visible.

otoscopy *Medicine.* the examination of the inside of the ear.

ototoxic *Neurology.* poisonous to, or having an adverse effect on, the eighth nerve or related organs of hearing and balance. Thus, **ototoxicity.**

river otter

otter *Vertebrate Zoology.* any of various aquatic fish-eating mammals, primarily of the genus *Lutra* in the family Mustelidae, related to weasels, badgers, and skunks and characterized by a long, cylindrical body, flat head, webbed feet, and water-resistant fur; otters are found nearly worldwide, but are concentrated in the Northern Hemisphere. The **river otter,** *L. canadensis,* is a widely distributed North American species. *Naval Architecture.* an underwater board attached to a fishing trawl to keep the net open and to the side of the boat. *Ordnance.* a device used in naval minesweeping operations that, when towed, displaces itself sideways to a predetermined distance; it may be attached to a sweep wire or other sweeping device.

otter cat see JAGUARUNDI.

otter trawl *Naval Architecture.* a fishing trawl using underwater otters.

Otto, Nikolaus August 1832–1891, German inventor; developed an early internal-combustion engine; built the first four-cycle engine.

OTTO optical-to-optical interface device.

Otto cycle *Thermodynamics.* an ideal thermodynamic combustion cycle consisting of four processes: a compression at constant entropy; a constant-volume heat transfer to the system; an expansion at constant entropy; and a constant-volume heat transfer from the system. The thermal efficiency of the ideal Otto cycle increases with an increasing compression ratio. *Mechanical Engineering.* an actual version of this cycle, used in spark-ignition gasoline engines: first, an air-fuel mixture is drawn into the cylinder and compressed; second, heat transfer is added by the spark ignition of the fuel mixture in the cylinder; third, the gases produced by this combustion expand to move the piston downward for the power stroke; fourth, the burned gases in the cylinder are expelled.

Otto engine *Mechanical Engineering.* a four-stroke internal combustion engine in which two complete revolutions of the crankshaft correspond with the working cycle.

Otto-Lardillon method *Mechanics.* a numerical method to compute the trajectory of missiles with low velocities and quadrant angles of departure that may be high; it assumes that drag is proportional to velocity squared.

ottrelite *Mineralogy.* $(Mn^{+2},Fe^{+2},Mg)_2Al_4Si_2O_{10}(OH)_4$, a green, monoclinic and triclinic mineral having a specific gravity of 3.52 and a hardness of 6 to 7 on the Mohs scale; found as flakes in quartz veins in slates in Belgium.

OTU operational taxonomic unit.

otu mutation *Genetics.* a sex-linked gene in *Drosophila* whose different alleles cause several abnormalities in the ovary, all resulting in female sterility.

O-type backward-wave oscillator *Electronics.* a device that generates pulsating energy by producing a stream of electrons traveling in one direction, and microwaves traveling in the opposite direction, and then forcing the two to interact. Also, **O-type carcinotron.**

OU or **O.U.** both eyes. (From Latin *oculo utro.*)

ouabain *Organic Chemistry.* $C_{29}H_{44}O_{12}\cdot 8H_2O$, hygroscopic white plates; soluble in water and alcohol; loses water at 130°C and melts with decomposition at 190–200°C; used in medicine. Also, G-STROPHANTHIN.

ouachitite *Petrology.* a dark, olivine-free lamprophyre consisting of phenocrysts of biotite, some augite, and sometimes hornblende, with a glassy or analcime groundmass.

Ouchterlony test *Immunology.* a method used to analyze an antibody-antigen mixture, in which the components are placed in wells in a plate of gel (**Ouchterlony plate**). Also, **Ouchterlony technique.**

Oudeman law *Physical Chemistry.* a rule stating that the molecular rotations of the various salts of an acid or base decline at the same rate as the concentration of the solution drops to zero.

Oudin test *Immunology.* a method used to analyze an antibody-antigen mixture, in which an antigen solution is positioned above a column of antibody that has been incorporated in a gel. Also, **Oudin technique.**

Oughtred, William 1574–1660, British mathematician; invented the slide rule; introduced the signs of multiplication and proportion.

ounce *Metrology.* **1.** a unit of avoirdupois weight that is equal to 0.0625 pound. **2.** a similar unit of troy or apothecaries' weight that is equal to 0.0833 pound. **3.** see FLUID OUNCE.

ounce *Vertebrate Zology.* another name for the snow leopard, *Panthera uncia.*

ouncedal *Metrology.* a unit of force that is equal to the force that will produce an acceleration of one foot per second per second in a mass of one ounce.

-ous **1.** a combining form meaning: "possessing; having; full of." **2.** a combining form used in opposition to *-ic* to indicate an ion or acid having the lower of two oxidation states, as in *ferrous chloride.*

out *Artificial Intelligence.* in a truth maintenance system, a term describing a proposition that is currently not believed or supported.

outage *Electricity.* a failure of an electric power system or subsystem. *Petroleum Engineering.* the difference between rated capacity and actual content of a barrel, tank, or tank car.

outage method *Petroleum Engineering.* a technique for determining the volume of liquid in a tank by measuring the distance between the top of the tank and the top of the liquid. Also, **outage gauge.**

outage rate *Electrical Engineering.* the average number of outages per unit exposure time per component.

outboard *Naval Architecture.* **1.** away from a vessel's centerline. **2.** beyond the ends or sides of a vessel's hull. **3.** an outboard motor, or a boat powered by such a motor.

outboard motor *Naval Architecture.* an engine mounted externally on a vessel's stern, and usually fitted for easy removal; used typically on small pleasure craft. Also, **outboard engine.**

outboard profile *Naval Architecture.* a drawing of a vessel showing its external appearance as seen from one side (normally the starboard side).

outbreeding *Genetics.* a mating system in which matings between close relatives do not usually occur. *Agriculture.* the mating of unrelated animals of the same breed to eliminate unwanted characteristics.

outby *Mining Engineering.* nearer to or directed toward the entrance or shaft of a mine, thus farther from or directed away from the face.

outcrop *Geology.* the part of a stratum or other geologic formation or structure that is exposed at the earth's surface or that has been covered only by surficial deposits.

outcrop curvature *Geology.* see SETTLING.

outcrop map *Geology.* a geologic map that displays only the distribution and shape of outcrops, leaving all other areas blank.

outcrop water *Hydrology.* rain and surface water that seeps down through fault planes, old shafts, or surface drifts, or through porous and fissured rock that has been exposed at the surface.

outcrossing *Agriculture.* the mating of unrelated animals of the same breed in order to introduce a desired character to the strain that is otherwise absent.

outer automorphism *Mathematics.* an automorphism ν of a group G that cannot be written in the form $\nu(g) = hgh^{-1}$ for some $h \in G$; i.e., ν is not an inner automorphism. More precisely, an outer automorphism is an element of the quotient group of the automorphisms on G, modulo the subgroup of inner automorphisms on G.

outer bar *Geology.* a bar that is formed at the mouth of an ebb channel of an estuary by waves and currents.

outer beach *Geology.* an inland section of a beach that is generally dry, being reached by waves only during violent storms.

outer bottom *Naval Architecture.* the outermost (and lowermost) layer of plating on a double- or multibottom hull, the bottom shell actually in contact with the water.

outer bremsstrahlung *Physics.* a form of bremsstrahlung in which a fast charged particle is accelerated by the interaction with the nucleus of an atom and the energy loss appears as radiation of energies much higher than that of the ionization energy of the atom.

outer core *Geology.* the upper or outermost part of the earth's interior, extending from 2900 to 5100 kilometers below the surface.

outer-directed see OTHER-DIRECTED.

outer fix *Navigation.* a fix within the terminal area from which aircraft are cleared to the final approach fix or final approach course.

outer keel *Naval Architecture.* the vertical, external portion of a keel.

outer mantle see UPPER MANTLE.

outer marker *Navigation.* the marker beacon of an instrument landing system that is farthest from the runway (generally 4–7 miles). It indicates the point where an aircraft at the proper altitude will intercept the glide path.

outer measure *Mathematics.* a nonnegative, monotone, countably subadditive set function μ^* defined on the set of subsets of a given set X (or σ-ring R), such that $\mu^*(\emptyset) = 0$. If μ is a measure on a ring R and E is a member of a σ-ring containing R, then $\mu^*(E)$ may be viewed as the lower bound of sums of the type $\sum_{n=1}^{\infty} \mu(E_n)$, where E_n is a sequence of sets in R whose union contains E; these arise naturally in the attempt to extend measures from rings to larger classes of sets. They also are used to conveniently obtain measures when the family of subsets of X on which μ^* is defined is suitably restricted. Also, **Carathéodory outer measure.**

outer membrane *Cell Biology.* the outermost of two protein-containing phospholipid bilayers that surround mitochondria, chloroplasts, nuclei, and also Gram-negative bacteria, for example, and in the latter case providing an essential cell permeability barrier.

outer orbital complex *Physical Chemistry.* a system in which the electrons shared by molecules to form bonds have the same energy level as electrons not involved in the bonding process; found chiefly in metal molecules.

outer planet *Astronomy.* any of the planets beyond the asteroid belt; i.e., Jupiter, Saturn, Uranus, Neptune, or Pluto.

outface see DIP SLOPE.

outfall *Civil Engineering.* the site at which water is discharged from a conduit to a stream, lake, or ocean.

outflow *Engineering.* the flow of some product out of a process facility or vessel.

outflow cave *Geology.* a cave from which a stream emerges or is known to have emerged. Also, EFFLUENT CAVE.

outgassing *Astronomy.* the ejection of gaseous material from the interior of a planet.

outgoing trunk see ONE-WAY TRUNK.

out-group *Systematics.* any of a closely related group of taxa, such as two sister groups, that are used to establish the polarity of characters in the given taxa.

outgrowth *Meteorology.* the offshoot of a main cloud, such as the anvil of a cumulonimbus cloud, due to relative cloud and jetstream movements at different levels.

outlaw gene *Genetics.* a gene that has deleterious effects on other genes in the same genome, yet is favored by natural selection.

outlay graft *Surgery.* a modified inlay graft applied to correct an ectropion of the eyelid.

outlet *Electricity*. the point at which power is taken from a wiring system; a power-line termination that delivers a signal or operating power to equipment plugged into it. Also, RECEPTACLE. *Hydrology*. **1.** a stream or river that flows out of a lake or pond. **2.** the channel through which such a stream flows. **3.** the point at which a river or stream enters a large body of water, especially the sea.

outlet box *Electricity*. an enclosure in which terminations of electric wiring occur and allow connections to be made to electrical devices to receive power.

outlet glacier *Hydrology*. a stream of ice that flows outward from an ice sheet or ice cap through a mountain pass or valley, being forced to a channel or path by exposed rock.

outlet head *Hydrology*. the place where water from a lake flows into an effluent stream.

outlier *Geology*. an irregular area of younger, exposed rock that is surrounded by outcrops of stratigraphically older rocks. *Statistics*. any data point exibiting anomalous behavior.

outline map *Cartography*. a map that shows just enough geographic information to provide a basis for the additional information (e.g., cultural or political) placed on it.

outline processor *Computer Programming*. a text-editing system that allows the writer to impose a structural outline on the text, such as a subject-matter hierarchy.

out-of-line coding *Computer Programming*. a program statement that is not in the main flow of the program, such as a closed subroutine.

out of phase *Physics*. a condition in which two or more oscillating quantities having identical frequencies do not possess the same value at the same phase.

out-of-phase current *Electricity*. a reactive current in an AC circuit that is out of phase with voltage.

out-of-phase voltage *Electricity*. a voltage that is out of phase with the current.

out-of-range check *Computer Programming*. a memory protection feature designed to prevent a request for memory that is not within the range allocated to a program; if such a request is made, an exception is raised and the program terminated.

out-of-service jack *Electricity*. a test jack that cuts off current when a shorted plug is inserted.

outpatient *Medicine*. a patient who comes to a hospital or clinic for treatment, but who does not occupy a bed in the institution.

outplacement *Industrial Engineering*. a process in which an employee who has been or is about to be separated from a company is given various forms of counseling and assistance by the company to aid him or her in finding future employment elsewhere.

out-plant system *Computer Technology*. a distributed data processing system with remote terminals that communicate with a central processor.

output the product of a system or process; specific uses include: *Electronics*. **1.** the driving force, such as power, current, or voltage, that is delivered by a device. **2.** the terminals from which energy or information is emitted by a system or device. *Computer Technology*. **1.** the final result of computer processing. **2.** data that is transferred from main memory to a secondary storage device, or to a device such as a printer or display screen. **3.** to transfer data to a device in this manner. *Mechanical Engineering*. the total force produced by a machine.

output area *Computer Technology*. an area in main memory that is reserved for the results of processing. Also, **output block.**

output buffer *Computer Technology*. an area of main memory that is reserved for the temporary storage of output destined for a peripheral output device such as a printer, tape, or disk.

output bus driver *Electronics*. a device that intensifies a computer's output signal sufficiently to power another device, without overloading the supply line.

output capacitance *Electronics*. **1.** the capacitance in an *n*-terminal electron tube that is transferred between the output terminal and all other terminals connected to it, except for the input terminal. **2.** the capacitance of a device that appears at the output terminals.

output channel *Computer Technology*. a data channel that connects a central processing unit to peripheral devices, and over which data is transmitted for output.

output class *Computer Technology*. **1.** the type of output device, such as numeric/alphanumeric, graphical, or special (the latter including bar code and badge readers, point of sale terminals, and so on). **2.** the level of importance or priority among computer outputs competing for output devices.

output contact *Robotics*. any of various contact switches for grippers, tools, and other end effectors that are under the programmed control of a robotic system.

output device see OUTPUT UNIT.

output equipment *Computer Technology*. any device that enables a computer to communicate results of processing to the user or to copy data to off-line storage devices.

output exposure rate *Radiology*. the radiation exposure rate in a specified area per unit of time, expressed as roentgens per minute.

output gap *Electronics*. a device from which the usable power in the electron beam of a microwave tube can be extracted.

output impedance *Electronics*. the impedance calculated or measured at the terminals of a source, or at the output terminals of a transducer.

output indicator *Engineering*. a meter, connected to a radio receiver, that indicates variations in the strength of an output signal, but does not indicate the value of output.

output-limited *Engineering*. describing a device that must postpone its workings until an output operation ahead and in line has been completed.

output link *Telecommunications*. the final link in a communications chain.

output medium *Computer Technology*. the physical material upon which output data is presented or stored, such as paper, magnetic disk, or microfilm.

output meter *Electronics*. a meter that measures the output voltage of an oscillator or amplifier, given in volts or decibels.

output-meter adapter *Electronics*. a device that allows an output meter to be connected to the output end of a radio receiver during alignment.

output of the function see DEPENDENT VARIABLE.

output power see POWER OUTPUT.

output record *Computer Programming*. a unit of data that has been transmitted to a peripheral output device or that is being stored temporarily pending transmission to such a device.

output resistance *Electronics*. the real part of the output impedance.

output routine *Computer Programming*. a sequence of instructions that identify and organize output data, and direct and control transmission of the data to a peripheral output device.

output shaft *Mechanical Engineering*. a shaft that transmits power from a transmission or clutch.

output stage *Electronics*. the last stage of any electronic equipment or system; in a transmitter, this stage feeds the antenna.

output stream *Computer Technology*. **1.** the sequence of output files awaiting transmission to an output device. **2.** a stream (sequence of characters) that is written by a program.

output transformer *Electronics*. **1.** a transformer connected between the output of a transducer or amplifier and the load, usually to match the load impedance for maximum power transfer. **2.** a circuit that matches the load impedance to the output of the amplifier or transducer.

output tube *Electronics*. a tube that is specifically designed to be used in an amplifier's output stage.

output unit *Computer Technology*. any peripheral device capable of receiving output from a computer for human use and understanding or for off-line storage.

output variable See DEPENDENT VARIABLE.

output winding *Electromagnetism*. the winding of a saturable reactor that is connected directly to the load of a circuit.

outrigger *Naval Architecture*. **1.** an apparatus consisting of a framework that projects from the side of a canoe or other boat, supporting a float that provides stability for the craft. **2.** a boat equipped with such an apparatus. Thus, **outrigger canoe.** *Engineering*. any support, temporary or permanent, projecting outward from a principal structure and supported by a jack or block; used to increase stability.

outside air temperature see INDICATED AIR TEMPERATURE.

outside calipers *Mechanical Devices*. spring-action calipers with curved legs pointing inward; used to measure the outside dimensions of a workpiece.

outside diameter *Design Engineering*. the outer circumference of a pipe, including its wall thickness, as measured with a caliper tool.

outside extension *Telecommunications*. any telephone extension that is located on premises other than those connected to or containing the main station.

outside strake *Naval Architecture*. the outer strake of an in-and-out shell plating system; it requires a liner on the frame to make it flush to the inside strake on either side.

outwash *Geology.* **1.** sand and gravel that has been removed from a glacier by meltwater streams and laid down beyond the margin of the glacier. Also, GLACIAL OUTWASH, OVERWASH, OUTWASH DRIFT. **2.** soil material that has been transported down an incline by rainwater and deposited on more gently sloping land.
outwash apron see OUTWASH PLAIN.
outwash cone *Geology.* a steep-sided, cone-shaped deposit of glacial outwash that has been laid down by meltwater streams at the margin of a shrinking glacier. Also, WASH CONE.
outwash drift see OUTWASH, def. 1.
outwash fan *Geology.* a fan-shaped deposit of glacial outwash that has been laid down by meltwater streams in front of or beyond the terminal moraine of a glacier.
outwash plain *Geology.* a broad, extensive, flat or gently sloping sheet of glacial outwash formed in front of or beyond the terminal moraine of a glacier by the merging of outwash fans. Also, FRONTAL PLAIN, OUTWASH APRON, SANDUR, WASH PLAIN.

outwash plain

outwash terrace *Geology.* a flat-topped bank or benchlike deposit of outwash occurring along a valley, downstream from an outwash plain or terminal moraine. Also, FRONTAL TERRACE, OVERWASH TERRACE.
ouvarovite see UVAROVITE.
ova the plural of *ovum.*
oval *Mathematics.* any curve resembling an elongated circle or the longitudinal cross section of an egg.
ovalbumin *Biochemistry.* a glycoprotein found in egg white.
oval foramen see FORAMEN OVALE.

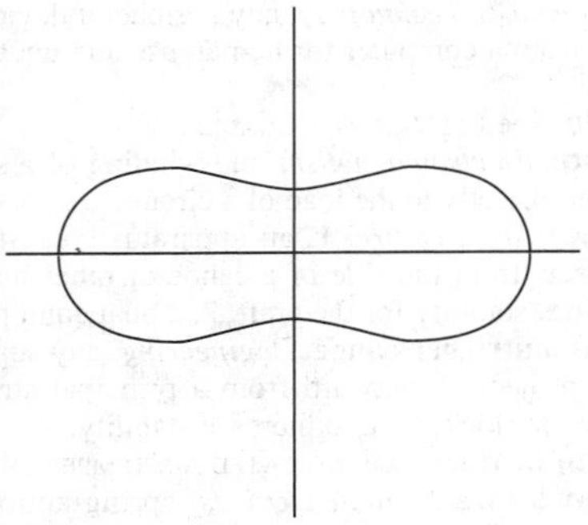

oval of Cassini

oval of Cassini *Mathematics.* the graph in the plane of the equation

$$(x^2 + y^2 + b^2)^2 - 4b^2 x^2 = k^2,$$

or in polar coordinates,

$$r^4 + b^4 - 2r^2b^2\cos 2\theta = k^2.$$

The values b and k are constants. When $b = k$, the curve is the lemniscate of Bernoulli. These curves are sections of a torus on planes parallel to the axis of the torus; equivalently, the locus of a point P, satisfying $F_1P \cdot F_2P = k^2$, where F_1P and F_2P represent the lengths of the line segments from the foci $F_1 = (-b, 0)$ and $F_2 = (b, 0)$, respectively.
oval window *Anatomy.* a membrane between the middle and inner ear, connecting the middle ear with the vestibular organs.
ovarian [ō vâr′ē ən] *Anatomy.* of or relating to the ovary. Thus, **ovarian artery, ovarian vein,** and so on. *Medicine.* affecting or involving the ovary. Thus, **ovarian cancer, ovarian cyst,** and so on.
ovarian agenesis *Medicine.* a chromosomal disorder resulting in the failure of development of the ovaries and in sexual infantilism.
ovarian dysmenorrhea *Medicine.* dysmenorrhea caused by disease of the ovaries.
ovarian follicle *Histology.* a blisterlike mass of cells on the surface of an ovary, containing a developing oocyte or ovum. Also, GRAAFIAN.
ovarian insufficiency *Medicine.* a deficiency of ovarian function, which can result in either oligomenorrhea, amenorrhea, or abnormal dysfunctional uterine bleeding.
ovarian ligament *Anatomy.* a fibrous band that connects the ovary to the uterus.
ovarian pregnancy *Medicine.* an ectopic pregnancy that occurs within an ovary.
ovarian seminoma see DYSGERMINOMA.
ovariectomy *Surgery.* the surgical removal or excision of one or both ovaries.
ovario- or **ovari-** a combining form meaning "ovary" or "ovarian."
ovariohysterectomy see OOPHOROHYSTERECTOMY.
ovariole *Entomology.* any of numerous tubule clusters that make up the insect ovary; the upper portion gives rise to immature eggs, which mature and are nourished by yolk from the lower portion.
ovariopexy see OOPHOROPEXY.
ovariosalpingectomy *Surgery.* the surgical removal of an ovary and uterine tube.
ovariotomy *Surgery.* the surgical removal of an ovary, or of an ovarian tumor.
ovariotubal *Anatomy.* relating to the ovary and the uterine tube.
ovaritis see OOPHORITIS.
ovarium *Anatomy.* another term for ovary, especially in technical use.
ovary *Anatomy.* the female gonad; either of two organs of the female reproductive tract in which ova form and which is responsible for the production of hormones that cause secondary sexual characteristics to develop. Also, SALPINX. *Botany.* the hollow structure at the lower part of the pistil in a flowering plant containing the ovules. (Derived from the Latin word for "egg.")

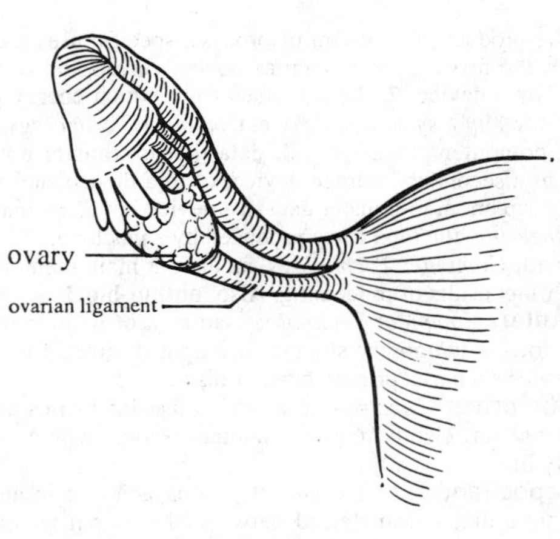

ovary

ovate *Science.* egg-shaped. *Botany.* of a leaf, having a shape like the longitudinal section of an egg, especially such a shape with the broader end at the base.
oven *Engineering.* a compartment in which substances are artificially heated for such purposes as baking, roasting, drying, or annealing. *Geology.* **1.** a sacklike pit or hollow in a rock formed by chemical weathering, and having an arched roof so that it resembles an oven. **2.** see SPOUTING HORN.

ovenbird

overcoupled circuit *Electronics.* a circuit in which two separate circuits are tuned to the same frequency and coupled closely together in order to obtain two response peaks.

overcritical binding *Atomic Physics.* a condition in which the energy of an atom is so vast that electrons and positron pairs are created spontaneously, predicted to occur if there are more than 173 protons in a nucleus.

overcritical electric field *Atomic Physics.* a condition in which an extremely strong electric field causes electron and positron pairs to appear spontaneously.

overcropping *Agronomy.* the exhaustion of the land by continuous cultivation and crop production. Also, **overbearing.**

overcuring *Chemical Engineering.* the process of allowing a material to exceed the necessary cure time, resulting in softness or brittleness. *Materials Science.* in vulcanization, a condition that occurs when too much sulfur has incorporated into rubber stock, resulting in a stiffer, harder, and less extrudable product than that at the optimum cure.

overcurrent *Electronics.* a current that is excessively high, generally caused by a short circuit.

overcut *Mining Engineering.* a cut driven along or near the top of a coal seam, sometimes used in thick seams or seams with sticky coal.

overcutting machine *Mining Engineering.* a coal-cutting machine, similar to a shortwall machine but with the cutter bar mounted at the top rather than at the bottom; used to cut a coal seam at a specified distance above the floor.

overdamping *Physics.* a condition in which a damped oscillatory system has damping that is greater than that required for critical damping.

overdeepening *Geology.* an erosive process whereby a glacier excessively deepens and widens an inherited preglacial valley to a level below that of the original surface.

overdominance *Genetics.* a condition in which heterozygotes express a more extreme phenotype than either homozygote. Also, SUPERDOMINANCE.

overdoor *Building Engineering.* situated or installed above a doorway. *Architecture.* a panel of ornamental woodwork set over a doorway.

overdose *Medicine.* **1.** an excessive, sometimes fatal, dose of medication. **2.** to administer an excessive dose of medication.

overdrafting *Hydrology.* the withdrawal of groundwater at a rate in excess of replenishment. *Metallurgy.* in rolling, the upward curving of the exiting product, caused by the differential speed of the rolls.

overdrilling *Engineering.* the act of drilling a run or length of borehole greater than the core-capacity length of the core barrel, causing loss of the core.

overdrive *Mechanical Engineering.* **1.** a method of reducing engine rpm in relation to speed, using a separate epicyclic gear unit. **2.** the gear unit used in this process. *Cardiology.* the artificial acceleration of the heartbeat by drugs or artificial pacemaker for the purpose of overcoming ectopic heart rhythm.

overdriven amplifier *Electronics.* an amplifier operating at levels that exceed those for which it was designed; usually producing a distorted output.

overdrive transmission *Mechanical Engineering.* a vehicle transmission system characterized by an overdrive gear unit.

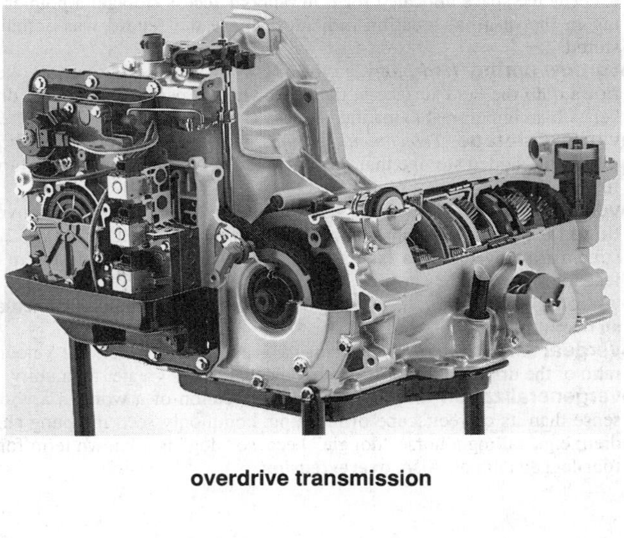

overdrive transmission

ovenbird *Vertebrate Zoology.* any member of the several species of North and South American birds of the family Furnariidae; known for its domed, ovenlike nest built on the ground.

overaging *Materials Science.* a process of aging for longer times or at greater temperatures than those at which the optimum mechanical properties are attained.

overall efficiency *Aviation.* the efficiency with which a rocket or jet engine converts the total heat energy of its fuel into effective propulsive energy; that is, the product of thermal efficiency and propulsive efficiency.

overall response *Electronics.* the ratio between a system's input and its output.

overarching weight *Mining Engineering.* the downward pressure resulting from the rocks above the active mine workings.

overarm *Mechanical Engineering.* an adjustable support at the end opposite the spindle of a milling cutter.

overbank deposit *Geology.* a fine-grained, clayey and silty sediment that is deposited on a floodplain by waters that overflow the stream channel.

overbending *Metallurgy.* in metal forming, bending beyond the designed amount in order to compensate for the elastic springback.

overbreak *Civil Engineering.* the amount of earth excavated beyond the neat lines of a cutting or tunnel. Also, **overbreakage.**

overbunching *Electronics.* a condition in which the fluctuation in electron velocity exceeds that which is required to generate energy in a velocity modulated tube, such as a klystron.

overburden *Geology.* **1.** the upper part of a sedimentary deposit that compresses and consolidates the underlying material. **2.** see REGOLITH. *Mining Engineering.* any barren rock material or soil that overlies a useful mineral deposit or coal seam, and that must be removed before mining. Also, BARING, TOP. *Metallurgy.* to introduce into a furnace excessive ore and flux in relation to fuel.

overburdened stream see OVERLOADED STREAM.

overcast *Meteorology.* of the sky, more than 95% covered by clouds. *Mining Engineering.* **1.** an enclosed airway made of concrete, tile, stone, or other incombustible material, allowing one air current to pass over another without interruption. **2.** to transport removed overburden from coal mined from surface mines to another site from which the coal has been mined.

overcast bombing *Military Science.* the process of bombing a target through a cloud layer, using radar or other equipment to aid in sighting.

overcoating *Building Engineering.* an added coat of paint or other protective material. Also, **overcoat.** *Engineering.* the process of extruding a plastic web past the edge of a substrate web.

overcompensation *Psychology.* the tendency to use more effort or react more strongly than is necessary in order to make up for a sense of weakness or deficiency in oneself; e.g., a person who feels himself to possess some unwanted character trait will adopt an opposing trait to an extreme degree.

overcompounded generator *Electricity.* a compound-wound generator whose series winding is proportioned such that the terminal voltage at a rated load is greater than that at no load.

overcompounding *Electricity.* **1.** as a load increases in a compound-wound generator, the use of sufficient series turns to raise voltage so that increased line drop will be compensated. **2.** the tendency of a motor to increase running speed as load resistance decreases.

overconsolidation *Geology.* the consolidation of sedimentary material beyond the normal level of consolidation for the existing overburden.

overcorrection *Behavior.* a behavior modification technique in which an enforced action serves as the punishment for a problem behavior and thereby corrects or undoes that behavior.

overdubbing *Acoustical Engineering.* to enhance recorded sound by integrating additional sounds or music into the basic soundtrack, usually by adding tracks in a multitrack recording. Thus, **overdubbed, overdubbing.**

overeating disease see PULPY KIDNEY DISEASE.

overexpose *Medicine.* to expose oneself excessively to the elements. *Graphic Arts.* to expose photosensitive film for too long or to too much light. Thus, **overexposed, overexposure.**

overexpression *Biotechnology.* a phenomenon in which heterozygotes may display greater gene expression of the phenotype than the same allele does in the phenotype of the homozygote.

overextension *Medicine.* extension beyond the normal limit, usually of a limb.

overfalls *Oceanography.* a turbulent, disturbed water surface caused by a conflict of strong currents, by winds opposing a current, or by a current passing over a submerged ridge or shoal.

overfire draft *Mechanical Engineering.* the air pressure in a boiler furnace when the main flame is burning.

overflight *Aviation.* a flight over a given area, especially a flight of military aircraft over foreign territory. *Transportation Engineering.* a commercial flight that does not stop at a scheduled intermediate landing point.

overflow *Science.* **1.** to spill over a rim or other boundary. **2.** something that spills over. *Hydrology.* **1.** of water, to flow out over the normal bank or channel onto the adjacent land. **2.** the amount of such flow. *Civil Engineering.* a device or structure over or through which excess water is allowed to flow. *Computer Programming.* a condition that occurs when the result of an operation exceeds the capacity of the system in which it occurs, as when a record or segment cannot be stored in the location assigned to it upon loading.

overflow bucket *Computer Technology.* an area of storage, often in a secondary storage device, where data can be stored when main memory is already full or in use.

overflow capacity *Engineering.* a measurement of the amount of liquid that would cause the overflow of a container.

overflow channel *Civil Engineering.* a channel or spillway structure used to carry overflow from a dam to the river behind it.

overflow continuous fermentation see CASCADE FERMENTATION.

overflow dam *Civil Engineering.* a dam across a stream, designed to allow water to overflow along its entire crest. Also, **overfall dam.**

overflow error *Computer Technology.* an error condition occurring when the value of the result exceeds the allowed range in length or magnitude.

overflow groove *Mechanical Engineering.* a groove provided in a plastics mold to allow for the escape of excess molten material.

overflow ice *Hydrology.* ice formed by water that rises through cracks in surface ice during high spring tides and then freezes.

overflow indicator *Computer Technology.* a single-bit flag that changes state, usually to 1, when an overflow occurs during an arithmetic operation. Also, **overflow check indicator.**

overflow pipe *Engineering.* a pipe placed in a container in order to keep the level of liquid at a specified height; excess liquid enters the upper open end of the pipe and is drained away.

overflow record *Computer Technology.* a data record or segment that exceeds the space allocated for it in random access storage; a pointer or tag in the original location indicates where the record was actually stored.

overflow spring *Hydrology.* a type of gravity spring whose water overflows onto the land surface at the edge where a permeable deposit dips beneath an impermeable mantle of rock.

overflow storage *Telecommunications.* in a store-and-forward switching center, added storage that prevents the loss of messages offered to a filled line store.

overflow stream *Hydrology.* **1.** a stream filled by water that has overflowed the banks of another stream or a river. Also, SPILL STREAM. **2.** a stream flowing out of a lake and carrying water to another stream or lake, or to the sea.

overfold *Geology.* a fold that has been tilted beyond the perpendicular; an overturned fold.

overgear *Mechanical Engineering.* a gear train in which the velocity ratio of the driven shaft to its driving mechanism is greater than unity.

overgeneralization *Psychology.* the application of a word in a wider sense than its correct scope of meaning, commonly seen in young children; e.g., calling a horse "doggie" because "dog" is a known term for a four-legged animal. Also, **overextension.**

overglaze *Materials Science.* a color or glaze applied to an existing ceramic glaze.

overgrafting *Surgery.* the process of placing a new skin graft on top of a previously healed graft; done for reinforcement and usually pertaining to split-thickness grafts.

overgrazing *Agronomy.* the exhaustion of the land by continuous grazing of animals.

overgrinding *Mining Engineering.* the grinding of an ore to a smaller particle size than that required to liberate it from other materials.

overgrowth *Biology.* excessive growth, as of a body part or a young plant. *Botany.* growth, as of brush or ground cover, over other plants. *Crystallography.* the growth of a crystal over the surface of another crystal of different composition.

overhand cut and fill *Mining Engineering.* a process of mining ore by which material is broken off from the roof of a drive and allowed to drop through chutes to a lower drive, from which it is removed.

overhand stope *Mining Engineering.* a stope in which the ore above the entryway is attached, so that ore that is broken off gravitates toward discharge chutes and the stope is self-draining.

overhand stoping *Mining Engineering.* a mining method, used widely in highly inclined deposits, in which ore is blasted from a series of ascending stepped benches, using both horizontal and vertical holes.

overhang *Architecture.* **1.** a projecting building element, such as a roof or second story. **2.** the extent that such an element projects beyond something below. *Medicine.* the extension of excessive filling material over the margin of a tooth cavity. *Cartography.* in aerial photography, any exposures taken before and after the boundary of the area to be surveyed, to assure complete coverage.

overhaul *Mechanical Engineering.* **1.** to disassemble, inspect, repair, then reassemble and test a piece of machinery. **2.** the process of overhauling an engine or other machinery. *Metallurgy.* to remove scale and other surface flaws from a cast part or mill product. *Transportation Engineering.* the overtaking of a vehicle or vessel by the load it is hauling, so that the line between the two slackens. Thus, **overhauling.**

Overhauser effect *Atomic Physics.* a phenomenon involving a substance with a nucleus of spin 1/2 and unpaired electrons when subjected to a radio frequency, matched to the electron spin resonance frequency, while in an external magnetic field; under such conditions the polarization of the nuclei is greatly enhanced.

overhead *Industrial Engineering.* the general, fixed costs involved in a manufacturing process or operation. *Computer Technology.* the time that a central processing unit spends on supervisory or administrative tasks rather than on directly productive tasks. *Chemical Engineering.* the vapor in a distillation column that reaches the top of the column and is condensed and separated; part is returned to the column and the remainder is removed as product.

overhead cableway *Mining Engineering.* a transportation system consisting of a strong overhead cable, suspended between towers, on which a car or traveler may run back and forth. From the car a bucket or pan may be lowered, loaded, then raised and locked to the car for transport.

overhead camshaft *Mechanical Engineering.* a camshaft that runs across the cylinder heads of an engine, opening the valves directly or through rockers. Also, **overhead cam.**

overhead fire *Military Science.* fire delivered above the heads of friendly troops.

overhead shovel *Mechanical Engineering.* a tractor loader that digs at one end, swings the bucket overhead, and dumps at the other end.

overhead-valve engine *Mechanical Engineering.* a vertical engine in which the inlet and exhaust valves operate in the cylinder head opposite the piston.

overheat *Science* **1.** to apply or undergo heat at a temperature sufficiently high to impair some property. **2.** of an engine or device, to malfunction because of such a level of heat.

overhit *Military Science.* to hit a target with more destructive force than necessary to accomplish the desired results.

overinclusion *Behavior.* the inability to eliminate from the behavioral repertoire inappropriate or inefficient responses that are associated with a particular stimulus.

overite *Mineralogy.* $CaMgAl(PO_4)_2(OH)\cdot 4H_2O$, a colorless to pale-green orthorhombic mineral of the overite group, occurring as minute platy crystals, and having a specific gravity of 2.53 and a hardness of 3.5 to 4 on the Mohs scale; found in variscite nodules.

overland flow *Hydrology.* the portion of surface runoff that flows downslope in thin sheets over the land surface toward stream channels. Also, UNCONCENTRATED FLOW.

overlap *Geology.* **1.** an arrangement of sedimentary layers in which each successively younger layer extends beyond or over the older, underlying layer, concealing its edges. **2.** in a fault, the horizontal component of separation, measured along the strike. **3.** the movement of an upcurrent section of a shore to a position extending seaward beyond a downcurrent section. *Cartography.* **1.** in aerial photography, any area included in two or more successive photographs; usually expressed as a percentage. **2.** any area on a map or chart that is included on an adjoining map or chart of the same series. *Metallurgy.* **1.** in welding, any undesirable weld metal that is deposited beyond the toe, face, or root. **2.** in resistance seam welding, the portion that is remelted in subsequent cycles. *Computer Programming.* to perform two or more tasks simultaneously, such as inputting data as the processor is computing.

overlap fault *Geology.* **1.** a fault in which the displaced layers are doubled back upon themselves. **2.** see THRUST FAULT.

overlap integral *Quantum Mechanics.* the integral over space of the product of one particle's wavefunction with another particle's complex conjugate wavefunction, which expresses the quantum interference (overlap) energy term that binds atoms to form molecules. Also, FRANK-CONDON FACTOR.

overlapped processing *Computer Programming.* a multiprogramming technique by which one program can be inputting, a second processing, and a third outputting all at the same time.

overlapping *Computer Programming.* **1.** a method of concurrent operation in which parts of two consecutive instructions are executed at the same time. **2.** a condition that occurs on a display screen when two or more windows appear to cross each other's borders.

overlapping code *Genetics.* the genetic code of overlapping genes, using the same sequence with a change in the reading frame or the same frame with a different start or stop point.

overlapping deletion *Genetics.* a compound deletion consisting of two separate deletions that overlap one another.

overlapping genes *Genetics.* two or more genes whose nucleotide coding sequences are partly or completely coextensive; found in certain bacteria and phages.

overlapping inversion *Genetics.* a compound chromosomal event in which an inversion overlaps or includes a segment that has already undergone a previous inversion.

overlapping orbitals *Atomic Physics.* two electron orbitals from different atoms in a molecule that enter a region of space where each has perceptible magnitude.

overlap radar *Engineering.* a radar system whose effective coverage in one sector overlaps with part of another sector.

overlay a layer that is superimposed on a substrate; specific uses include: *Engineering.* **1.** a surfacing mat laid on the top portion of a mat layer to provide a smooth finish to a plastic surfaces. **2.** a sheet of transparent medium placed over an existing map or drawing. *Graphic Arts.* the material contained on a single sheet in a color separation process. *Civil Engineering.* a layer of pavement added to an existing roadway that has deteriorated in order to prolong its life. *Computer Technology.* a section of computer code or data that is loaded into an area of main memory previously allocated to another segment or data set of the same program. Also, MEMORY OVERLAY.

overlay coating *Materials Science.* an oxidation-resistant, ductile coating applied by an electron-beam physical-vapor deposition process to protect advanced superalloys, as those used in jet engine turbine airfoils.

overlay transistor *Electronics.* a device that contains a large number of emitters linked together by means of diffusion and metallizing, increasing its current-handling capabilities and allowing it to produce high power amplification at exceedingly high frequencies.

overlearning *Behavior.* a type of learning in which practice continues even after the desired criterion of learning has been reached.

overline *Telecommunications.* in teletypewriter practice, the superimposition of one set of characters over another.

overload a load that is greater than can be accepted or handled; specific uses include: *Electronics.* a greater load than an amplifier or other electronic component or system is designed to carry; characterized by a waveform distortion or overheating. *Geology.* an amount of sediment that is in excess of that which a stream is capable of transporting, and therefore is deposited. *Psychology.* the fact of having too much information and stimuli to store in memory in the time available. *Computer Programming.* to carry out a process of overloading.

overload capacity *Electronics.* a defined level of current or power that a device can carry, usually only for a short time; generally higher than the rated load capacity.

overload current *Electronics.* a current that overwhelms a system's load capacity and may cause permanent damage.

overloaded stream *Hydrology.* a stream carrying more sediment than it can transport, so that its velocity is decreased and it is forced to deposit part of its load.

overloader *Mining Engineering.* a power-shovel loading machine utilizing a bucket that is dug into the material, usually sand or gravel, then retracted and swung over in a wide horizontal arc to a discharge point.

overloading *Computer Programming.* the assignment of multiple meanings to an operator, depending on the type of data to which it is applied; e.g., the symbol + could represent integer addition, floating-point addition, or matrix addition.

overload level *Electricity.* a threshold of current in excess of the rated output that may be safely applied to an electronic device or system.

overload protection *Electricity.* the use of a device, such as a circuit breaker or automatic limiter, that automatically disconnects a circuit when an excessive current or voltage condition is evident. Also, OVERCURRENT PROTECTION. *Robotics.* a safety feature on a robot that protects it from damage when excessive force is applied.

overload relay *Electricity.* a relay that opens a circuit in order to provide overload protection when the coil current rises above a predetermined value; actuated from excessive current or sometimes from an anomalous temperature or power value. Also, **overload release.**

overmatching plate *Ordnance.* an armored plate of a thickness greater than the diameter of the projectile.

overmatching projectile *Ordnance.* a projectile of a diameter greater than the thickness of the armored plate.

overmodulation *Electronics.* a state in which amplitude modulation is over 100%, resulting in distortion of the output signal.

overpass *Civil Engineering.* a bridge or viaduct used to provide clearance to traffic at a lower elevation.

overpoint *Chemistry.* in a distillation procedure, the temperature at which the first drop of distillate is collected.

overpotential see OVERVOLTAGE.

overpressure *Fluid Mechanics.* the amount of pressure that exceeds atmospheric pressure as the result of a blast explosion.

overprint *Graphic Arts.* to print additional copy, often in a different color, on a previously printed sheet. *Cartography.* to print or stamp additional information on an existing map or chart. *Geology.* the superposition of a new set of metamorphic structures on the original structures of a rock. Also, METAMORPHIC OVERPRINT, SUPERPRINT. *Geochemistry.* any disturbance of an isolated radioactive system that results in a gain or loss of radioactive or radiogenic isotopes, thus affecting the radiometric age that will be given the disturbed system.

overresponse *Neurology.* an abnormally intense response or reaction to a stimulus.

override *Mechanical Engineering.* **1.** to take control of an otherwise automatic mechanism or system to correct, supplement, or suspend its operation. **2.** an auxiliary device used to make such a correction.

overriding process control *Control Systems.* a priority system that allows one controller to override another when the process requires it.

overrun *Civil Engineering.* a cleared but unpaved area at the end of a runway used for extra landing roll for an airplane in an emergency. *Navigation.* to sail beyond a desired destination or turning point. *Graphic Arts.* **1.** to print additional copies of a book, magazine, etc. **2.** the result of such a process: an *overrun* of 5000 copies. *Computer Programming.* **1.** to load data to a register before previous data could be completely read. **2.** in unbuffered synchronous data transfer, a condition that occurs when an input/output activity exceeds the channel capacity.

overrunning clutch *Mechanical Engineering.* a coupling that transmits rotation in only one direction and that disconnects if the torque is reversed.

overshoot *Ordnance.* to go over or beyond the target. *Electricity.* a temporary response at the outset of an input signal that exceeds the steady-state, normal signal. *Electromagnetism.* the excessive transient effects brought on by a sudden change in voltage or current that usually decays rapidly to a steady-state condition. *Control Systems.* **1.** to carry a controlled variable or output beyond a final or desired value. **2.** the degree to which this occurs. *Robotics.* the amount by which a robotic end effector goes beyond the targeted position in response to a step change.

overshot *Mining Engineering.* a device employed to retrieve lost casings or drill pipes.

overshot wheel *Mechanical Engineering.* a vertical waterwheel turned on a horizontal shaft by water pouring over buckets attached to its circumference.

oversize chart *Cartography.* a chart on which the neatlines have been extended in an irregular fashion in order to include an area of importance on the chart and to avoid having to produce an inset or additional sheet.

oversize control screen *Mining Engineering.* a screen attached to machinery to prevent entry of coarse particles that might halt or hinder operation. Also, GUARD SCREEN, CHECK SCREEN.

oversize powder *Metallurgy.* the portion of a metallic powder that exceeds a specified particle size.

overspeed governor *Mechanical Engineering.* a governor that stops or slows a prime mover at excessive speeds.

overspeed protection *Transportation Engineering.* an automatic device designed to stop a vehicle that exceeds a preset speed.

overspin *Mechanics.* **1.** overstability caused by an excessive rate of spin of a spin-stabilized projectile: the nose of the projectile does not turn downward when it passes the summit of the path and begins to descend. **2.** see TOPSPIN.

oversquare engine *Mechanical Engineering.* an engine with the bore diameter larger than the stroke length.

overstability *Mechanics.* a condition in which the restoring forces acting on charged particles in a plasma are great enough to return the particles back to their equilibrium positions at speeds greater than their original outward speeds, thus causing an increasing oscillation. Also, **overstabilization.**

oversteepening *Geology.* an erosive process whereby an alpine glacier excessively steepens the sides of an inherited preglacial valley.

overstep *Geology.* **1.** an overlap in which the older strata of a complete sedimentary sequence are truncated by one or more younger strata. **2.** the progressive burial of the truncated edges of the strata underlying an unconformity. **3.** a stratum that has been deposited on the upturned edges of the underlying strata. **4.** see TRANSGRESSION.

overstory *Botany.* the layer of foliage forming a forest canopy.

overstory

overstrain *Medicine.* a condition of high fatigue brought about by activity; it is intermediate between fatigue and exhaustion.

overstress *Medicine.* excessive activity that brings about overstrain. *Materials Science.* to stress a material beyond the point of deformation. Thus, **overstressed, overstressing.**

overstrike *Computer Technology.* **1.** to print a character repeatedly in the same position in order to produce a boldface or enhanced effect. **2.** the capability of some printers to print two or more characters in the same position.

over-the-horizon radar *Electromagnetism.* a long-range radar system that is capable of seeking targets beyond the horizon by means of scattering the radar signal.

over-the-shoulder bombing *Military Science.* a method of loft bombing in which the bomb is released past the vertical in order that it may be thrown back to the target.

overthrow distortion *Telecommunications.* distortion that occurs when the maximum amplitude of the signal wavefront is greater than the steady state of amplitude of the signal wave.

overthrust

overthrust *Geology.* a large-scale, low-angle fault in which the hanging wall appears to have moved upward relative to the footwall. Also, OVERTHRUST FAULT, LOW-ANGLE THRUST.

overthrust fault see OVERTHRUST.

overthrust nappe *Geology.* the mass of rock forming the hanging wall of an overthrust. Also, **overthrust back, overthrust sheet, overthrust slice.**

overtide *Oceanography.* a shallow-water harmonic tidal constituent whose speed is a multiple of the speed of one of the basic constituents of the tide-producing force.

overtone *Acoustics.* a tone having a frequency that is an integral multiple of a fundamental frequency; there is an octave separation between a fundamental and its first overtone and two octaves of separation between the fundamental and its second overtone. *Mechanics.* in a vibrating structure or other vibrational system, any of those oscillatory modes that are higher than the lowest or fundamental frequency, and whose frequencies are integral multiples of the fundamental frequency.

overtone band *Spectroscopy.* a spectral band associated with vibrational transitions within a molecule in which the quantum number is a multiple of a given fundamental frequency.

overtone crystal *Electronics.* a quartz crystal that is ground in such a manner that it can perform at multiples of its base frequency or at two frequencies simultaneously, such as in a synthesizer.

overtrades *Meteorology.* a former name for Krakatao winds.

overtravel *Electrical Engineering.* the amount of continued movement of a responsive element once the input value falls below pickup. *Ordnance.* **1.** the rearward movement of a trigger beyond the point at which the gun is fired. **2.** in cocking a machine gun, the distance the firing notch moves beyond the sear notch.

overturn *Hydrology.* the circulation of water layers in a lake or sea, occurring when cooled surface water sinks and mixes with water at the bottom. *Geology.* see ISOCLINE.

overturned *Geology.* describing a fold or one of its limbs that has tilted so far beyond the perpendicular that the sequence of strata appears to be reversed. Also, INVERTED, REVERSED.

overvoltage *Electricity.* a voltage that is higher than the normal or predetermined limiting value for a system. Also, OVERPOTENTIAL. *Electronics.* the point at which an applied voltage surpasses the Geiger threshold in a radiation-counter tube.

overvoltage crowbar *Electricity.* a monitoring circuit to a power supply that prevents the output voltage from exceeding a preset value; in a failure condition, an overvoltage condition is prevented through a silicon-controlled rectifier (a low-resistance crowbar) placed across the output terminals.

overwash *Hydrology.* **1.** the part of a wave that flows over the highest part of a berm or other structure and does not flow back to the sea or lake. **2.** the water in restricted areas that flows over the low parts of barriers or spits, especially during high tides or storms. *Geology.* see OUTWASH, def. 1.

overwash mark *Geology.* a narrow, tongue-shaped sand ridge formed by an overwash on the landward side of a berm or other structure.

overwash plain see OUTWASH PLAIN.

overwash pool *Oceanography.* a tide pool that lies between a beach scarp and a berm, entered by the water only at high tide.

overwash terrace see OUTWASH TERRACE.

overweight *Medicine.* **1.** weighing more than normal. **2.** an excessive increase in adipose tissue **(obese overweight)** or in muscle and skeletal tissue **(muscular overweight)**.

overwind *Engineering.* **1.** the accidental hoisting of a mine shaft cage into or over the top of the head frame. **2.** to turn, wrap up, or coil something too tightly or too far, as with a winch or spring.

overwinding *Molecular Biology.* a term for the supercoiling or twisting of DNA upon itself.

overwinter *Botany.* to remain alive through the winter, such as tubers in the ground.

overwrite *Computer Programming.* to destroy stored or recorded information by moving new data to the same location or writing to the same output medium.

ovex *Organic Chemistry.* $ClC_6H_4OSO_2C_6H_4Cl$, a white crystal soluble in acetone and insoluble in water that melts at 86.5°C; toxic by ingestion and a strong irritant to skin; used as an insecticide and acaricide.

ovi- a combining form meaning "egg," as in *oviparous.*

ovicell *Invertebrate Zoology.* in many bryozoans, an external chamber for brooding newly hatched embryos. Also, OOECIUM.

ovicyst *Invertebrate Zoology.* in tunicates, a pouch in which the eggs develop.

oviducal gland *Invertebrate Zoology.* a secretory gland in the oviduct of certain invertebrates, often producing protective structures such as egg shells.

oviduct *Anatomy.* a tube that leads from the ovary to the uterus.

oviger *Invertebrate Zoology.* in sea spiders, a leg modified for carrying eggs, generally found in the male.

ovine *Immunology.* referring to a substance that is associated with or derived from sheep.

oviparous [ō vip´ə rəs] *Vertebrate Zoology.* laying eggs that develop and hatch outside the maternal body; common among birds, and among most bony fishes, amphibians, and reptiles.

oviposit *Zoology.* to lay or discharge eggs from the body, especially from a specialized organ of the body.

ovipositor *Entomology.* a specialized egg-laying organ formed from the last abdominal segments of insects and chelicerata; modified in worker bees and wasps to form the sting.

Ovis *Vertebrate Zoology.* a genus of the family Bovidae that includes the domestic sheep, and any of its wild relatives such as the bighorn, most of which have horns that form a lateral spiral and live in mountainous areas from western North America to western Asia.

ovisac *Invertebrate Zoology.* a structure or receptacle that receives and/or stores eggs.

ovo- a combining form meaning "egg," as in *ovoviviparous.*

ovoid *Biology.* oval or egg-shaped. *Botany.* see OVATE.

ovonic memory switch *Electronics.* a switch characterized by its ability, after moving from a highly resistive state to a conducting state, to stay in the conducting state until a current pulse reaches its highest resistive state.

ovonic threshold switch *Electronics.* a switch characterized by its ability, after moving from a highly resistive state to a conducting state, to return to a highly resistive state when current falls below a given value.

ovotesticular hermaphroditism *Medicine.* a rare form of hermaphroditism in which ovarian and testicular tissue are combined in the same gonad.

ovotestis *Medicine.* an abnormal gonad containing both testicular and ovarian tissue. *Invertebrate Zoology.* the reproductive glands of hermaphrodites that produce both eggs and sperm, usually not at the same time.

ovoviviparous *Vertebrate Zoology.* producing eggs that stay inside the maternal body for incubation and development and hatch just prior to or following extrustion; this is common among some elasmobranch fishes and many reptiles.

Ovshinshky effect *Electronics.* an element of thin-film solid-state devices that allows current to flow equally well in either a positive or negative direction.

ovulate [äv´yə lāt´; ōv´yə lāt´] *Physiology.* to discharge an egg from the ovum.

ovulation [äv´yə lā´shən; ōv´yə lā´shən] *Physiology.* the discharge of an ovum from a vesicular follicle of the ovary; this usually occurs on the 14th day after the first day of the last menstrual period.

ovule *Botany.* a structure in the ovary of a seed plant that encloses an embryo sac within a nucellus and develops into a seed after fertilization.

ovum *plural,* **ova.** *Cell Biology.* a haploid egg cell involved in sexual reproduction. *Architecture.* an egg-shaped ornament, as in an egg-and-dart molding.

Owen, Sir Richard 1804–1892, British anatomist and paleontologist; known for research on and naming of dinosaurs.

Owen bridge *Electronics.* a four-arm alternating current bridge that measures self-inductance in terms of capacitance and resistance and whose balance is independent of frequency.

Oweniidae *Invertebrate Zoology.* the bamboo worms, an order of polychaetes in the Sedentaria, with a body having a few long segments and a prostomium that usually ends in a frilled organ for collecting food.

owl *Vertebrate Zoology.* any nocturnal, predatory bird of the order Strigiformes, distinguished by a large head, an apparent absence of a neck, strong talons, and forward-positioned eyes; found in nearly all habitats worldwide (wherever prey is available) except Antarctica; the many familiar North American species include the **great horned owl,** *Bubo virginianus*; the **screech owl,** *Otus asio*; the **barn owl**, *Tyto alba*; the **barred owl**, *Strix varia*; and the **snowy owl,** *Nyctea scandiaca.*

owl

Owl Nebula *Astronomy.* a planetary nebula, in Ursa Major at about 2000 light-years' distance; its name derives from two large circular dark patches in an otherwise spherical shell, resembling owl eyes; designated **M 97.**

owner *Computer Programming.* in the data structure set concept, the record that defines the set, and to which members of the set are related and subordinated. Also, **owner record.**

OW unit see SABIN.

owyheeite *Mineralogy.* $Ag_2Pb_7(SbBi)_8S_{20}$, a silver-white to steel-gray, metallic, orthorhombic mineral occurring as fibrous masses or acicular crystals, and having a specific gravity of 6.22 to 6.51 and a hardness of 2.5 on the Mohs scale; found in hydrothermal veins.

ox *plural,* **oxen.** *Agriculture.* an adult castrated bovine male, used as a draft animal. *Vertebrate Zoology.* broadly, any member of the genus *Bos.*

ox- a variant of the combining forms *oxo-* and *oxy-.*

oxadiazon *Organic Chemistry.* $C_{15}H_{18}Cl_2N_2O_3$, a white solid that is slightly soluble in water; melts at 88–90°C; used as an herbicide.

oxalate *Organic Chemistry.* a salt or ester of oxalic acid that contains the radical $(COO)_2$.

oxalate plasma *Hematology.* blood plasma to which 1% ammonium oxalate has been added to prevent clotting.

oxalic acid *Organic Chemistry.* $(CO_2H)_2{\cdot}2H_2O$, a transparent, colorless dihydrate that melts at 101–102°C (the anhydrous form melts at 189–190°C); it occurs naturally in many plants and can be made by alkali extraction in sawdust; toxic by inhalation and ingestion and a strong irritant; used as an automobile radiator cleanser and in leather tanning and textile bleaching.

Oxalidaceae *Botany.* a diverse family of dicotyledonous trees, shrubs, and tuberous and bulbous herbs in the order Gerianales, characterized by regular, perfect flowers, seeds that have a basal aril, and capsular fruit.

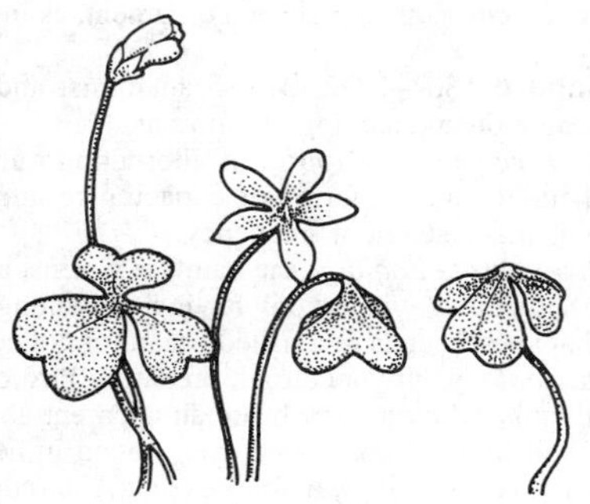

Oxalidaceae

oxalite see HUMBOLDTINE.
oxaloacetate *Biochemistry.* a metabolic intermediate, formed from aspartic acid by transamination, that couples with acetyl-CoA to form citrate.
oxalosis *Medicine.* a rare autosomal recessive metabolic error resulting in impaired glyoxylic acid metabolism, characterized by renal calculi, nephrocalcinosis, and renal insufficiency.
oxaluria *Medicine.* the excretion of an abnormally large amount of oxalates in the urine.
oxamide *Organic Chemistry.* $NH_2COCONH_2$, a white, odorless powder slightly soluble in water that melts at 419°C; used as a possible substitute for urea in fertilizers.
oxammite *Mineralogy.* $(NH_4)_2C_2O_4{\cdot}H_2O$, a yellowish-white to colorless orthorhombic mineral occurring as lamellar pulverulent masses, and having a specific gravity of about 1.5 and a hardness of 2.5 on the Mohs scale; found in guano deposits on the Guanape Islands, Peru.
oxammonium see HYDROXYLAMINE.
oxamyl *Organic Chemistry.* $C_7H_{13}N_3O_3S$, a white crystalline solid; melts at 100–102°C; used as an insecticide.
oxazine *Organic Chemistry.* any of several compounds used as dyes having the formula C_4H_5NO, and consisting of a ring containing four carbons, one oxygen, and one nitrogen.
oxazole *Organic Chemistry.* C_3H_3NO, a liquid that boils at 69°C; a five-membered heterocyclic compound used in organic synthesis.
oxbow *Agriculture.* a U-shaped piece of wood formerly used as part of an ox yoke. *Hydrology.* **1.** a looping stream meander having such an extreme bend that only a neck of land is left between the two parts of the stream. Also, OXBOW STREAM, HORSESHOE BEND. **2.** see OXBOW LAKE. *Geology.* **1.** the horseshoe-shaped abandoned channel of a former oxbow stream, left when it cuts a new channel across the narrow meander neck. Also, ABANDONED CHANNEL. **2.** in New England, land completely or partially enclosed within a bend of a stream.
oxbow lake *Hydrology.* a crescent-shaped lake formed when a stream cuts a new channel across the narrow neck of an oxbow, which later becomes separated from the new main stream by the deposition of silt. Also, OXBOW, HORSESHOE LAKE, MOAT, LOOP LAKE.
oxbow stream *Hydrology.* see OXBOW, def. 1.
Oxford *Agriculture.* a breed of mutton and medium wool sheep having a gray nose and ears, and no horns. (From *Oxford*, England, where the breed was developed.)
Oxfordian *Geology.* a European geologic stage of the Upper Jurassic period, occurring after the Callovian and before the Kimmeridgian. Also, DIVESIAN.
oxic *Ecology.* describing a soil layer from which most of the silica has been leached.
oxidant see OXIDIZER.
oxidase *Enzymology.* an enzyme of the oxidioreductase class that catalyzes the oxidation of a substrate with oxygen acting as the electron acceptor.
oxidase test *Bacteriology.* an assay used to identify aerobic bacteria by testing the ability of a given strain to transform certain amines into colored products.
oxidate *Geology.* a sediment composed of oxides and hydroxides of iron and manganese, crystallized from aqueous solution.
oxidation *Chemistry.* any reaction in which one or more electrons are removed from a species, thus increasing its valence (oxidation state); oxidation always occurs simultaneously with reduction, in which another species gains the electrons lost from the oxidized species. In an electrochemical cell, oxidation occurs at the anode. *Materials Science.* the deposit that forms on the surface of a metal as it oxidizes.
oxidation-fermentation test *Microbiology.* an assay that is used to determine whether a given bacterial strain carries out an oxidative or a fermentative mode of metabolism (or none at all) on a particular carbohydrate substrate.
oxidation number *Chemistry.* the number of electrons needed to restore an atom in a combined state to its elemental form.
oxidation pond *Civil Engineering.* a pond that holds partially treated waste water to allow algae, aquatic plants and microorganisms to decompose the organic waste.
oxidation potential *Physical Chemistry.* **1.** the difference in the amount of energy contained in an atom or an ion and the amount of energy remaining once an electron leaves its orbit. **2.** the energy that can be generated by an electrode when all the substances involved in the electrolytic process are in their standard states.
oxidation-reduction indicator *Analytical Chemistry.* a compound that changes color when it loses or gains electrons.
oxidation-reduction reaction *Chemistry.* any chemical change in which one species is oxidized (loses electrons) and another species is reduced (gains electrons).
oxidation state *Chemistry.* the condition of an atom as expressed by its oxidation number.
oxidative aging *Materials Science.* the aging of elastomers or other polymers in the presence of oxygen, primarily causing an increase in brittleness.
oxidative phosphorylation *Biochemistry.* the phosphorylation of ADP to ATP, or the creation of a pyrophosphate from an inorganic phosphate during a chemical reaction driven by a proton motive force powered by a respiratory chain.
oxide *Chemistry.* any binary compound of oxygen, especially with a metal.
oxide-coated cathode *Electronics.* a cathode or filament that has been coated with oxides of alkaline-earth metals, such as thorium oxide, to increase electron emission at low temperatures.
oxide dispersion strengthening *Materials Science.* a process of strengthening an alloy through internal oxidation, in which one of the components is oxidized within the matrix.
oxide-induced crack closure *Materials Science.* the effect of oxides within cracks to close them, decreasing crack growth or propagation.
oxide isolation *Electronics.* a method for isolating elements in an integrated circuit by inserting a layer of silicon oxide around each element.
oxide mineral *Mineralogy.* a mineral formed by the combination of one or more metallic elements with oxygen; for example, corundum, hematite, and spinel.
oxide nuclear fuel *Nucleonics.* fissionable nuclear fuel of the form UO_2 or PuO_2.
oxide passivation *Electronics.* a method for achieving stability in an integrated circuit by covering the surface with a layer of insulating oxide.
oxidite see SHALE BALL.
oxidize *Chemistry.* **1.** to convert a substance into an oxide; combine with oxygen. **2.** to undergo or cause to undergo oxidation; lose or remove electrons. Thus, **oxidized.**
oxidized hemoglobin see OXYHEMOGLOBIN.
oxidized shale see BURNT SHALE.
oxidized zone *Geology.* the part of an orebody or a mineral deposit in which the sulfides have been altered to carbonates and oxides by the action of surface waters bearing oxygen, soil acids, and carbon dioxide.
oxidizer *Chemistry.* an agent that oxidizes. *Space Technology.* specifically, a substance, usually containing oxygen, that supports the combustion reaction of a rocket fuel. Together, the fuel and oxidizer constitute a propellant. Also, OXIDANT.
oxidizing agent *Chemistry.* the species that gains electrons in an oxidation-reduction reaction and is itself reduced.
oxidizing atmosphere *Chemistry.* a space that is rich in oxygen, and where oxidation occurs.
oxidizing flame *Chemistry.* a gas flame in which the portion used contains excess oxygen.

oxidoreductase *Enzymology.* a class of enzymes that catalyze the reversible transfer of electrons from an oxidized substrate to a reduced substrate. Also, REDOX ENZYME.

oxime *Organic Chemistry.* any of several compounds containing the CH(=NOH) group and formed by a reaction between hydroxylamine and aldehydes and ketones. Also, **oxim.**

oximeter *Medicine.* **1.** an instrument that is used to measure oxygen in a controlled space, such as an oxygen tank or incubator. **2.** a photoelectric instrument used for measuring the degree of oxygen saturation in a fluid, such as blood.

oximetry *Medicine.* the calculation of the oxygen saturation level of hemoglobin in a blood sample by using an oximeter.

oxine *Organic Chemistry.* $HO(C_9H_6N)$, a needlelike solid that is soluble in alcohol, chloroform, acetone, and benzene; melts at 138–140°C; it is a compound that forms insoluble chelates with many metals; used extensively in inorganic quantitative analysis. Also, 8-HYDROXYQUINOLINE.

oxirane see EPOXIDE.

Oxisol *Geology.* an order of soils that develops in subtropical to tropical regions, and is characterized by the presence of free oxides of iron and aluminum, kaolin, and some quartz.

oxisuran *Oncology.* $C_8H_9NO_2S$, an antineoplastic drug.

oxo- *Organic Chemistry.* a prefix denoting the presence of oxygen in the keto group C=O; the approved formal nomenclature for *keto-* compounds.

oxoacid-lyase *Enzymology.* an enzyme of the lyase class that catalyzes the cleavage of an oxoacid. Also, **oxo-acid-lyase.**

oxodecenic acid *Entomology.* a pheromone secreted by honeybee queens which informs all members of the colony that the queen is present; when this pheromone is lacking, workers begin building queen cells and feeding young larvae royal jelly.

oxoferrite *Geology.* a variety of native iron containing some ferrous oxide in solid solution.

oxonium *Chemistry.* containing tetravalent basic oxygen. Thus, **oxonium compound.**

oxo process *Chemical Engineering.* a hydroformylation process used to convert alkenes to aldehydes in the presence of a catalyst; often continued to reduce the aldehydes to alcohols by catalytic hydrogenation.

oxy- a combining form: **1.** indicating the presence of oxygen in a compound. **2.** meaning "sharp," "quick," "sour," or "acidic."

oxyacanthine *Organic Chemistry.* $C_{37}H_{40}N_2O_6$, a needlelike alkaloid derived from the root of *Berberis vulgari;* soluble in alcohol, benzene, chloroform, and ether; melts at 216°C; used in medicine.

oxyacetylene cutting *Engineering.* the process of using an oxyacetylene torch to melt and cut through ferrous metal. Also ACETYLENE CUTTING.

oxyacetylene torch *Engineering.* an appliance using oxygen and acetylene to produce an extremely hot flame to braze, weld, or cut metal; exceeds the temperature of an oxyhydrogen flame. Also, ACETYLENE TORCH.

oxyacetylene welding *Metallurgy.* welding performed with the heat of an oxyacetylene torch.

Oxyaenidae *Paleontology.* a family of creodonts in the extinct suborder Hyaenodontia; the oxyaenids lived in Asia and North America in the Paleocene and Eocene; characterized by a broad head and a large, deep jaw, and similar to the mustelids in shape.

oxybenzone *Organic Chemistry.* $C_{14}H_{12}O_3$, colorless crystals; soluble in common organic solvents; melts at 65°C; used as a sunscreen in topical lotions.

oxybiotite *Mineralogy.* a phenocrystic biotite mineral whose composition is marked by increased levels of Fe(III).

oxycarboxin *Organic Chemistry.* $C_{12}H_{13}NO_4S$, off-white crystals that melt at 127.5–130°C; used to inhibit rust disease in greenhouse carnations.

oxycephaly *Medicine.* craniosynostosis in which there is premature closure of the lambdoid and coronal sutures, resulting in an abnormally high, peaked or conically shaped skull.

oxychloride cement *Materials Science.* a strong, extremely hard-setting cement, composed of a chemically combined oxide and chloride of magnesia; used in making composition floors.

oxycinesia *Neurology.* pain experienced during motion.

oxy-Cope rearrangement *Organic Chemistry.* a variation of the Cope rearrangement in which an enol is formed initially, and then tautomerized into its corresponding carbonyl compound.

oxyecoia *Neurology.* morbid acuteness of the sense of hearing.

oxyesthesia *Neurology.* morbid or abnormal acuteness of the senses.

oxygen *Chemistry.* a gaseous element having the symbol O, the atomic number 8, atomic weight 15.9994, a melting point of –218.4°C, and a boiling point of –182.962°C; a colorless, odorless, tasteless gas that is the most abundant element on earth, making up about 20% by volume of the atmosphere at sea level, about 50% of the material of the earth's surface, and about 90% of water. Oxygen is necessary for the life processes of nearly all living organisms and for most forms of combustion. It readily forms compounds with nearly all other elements except the inert gases, and it is used in blast furnaces, steel manufacture, chemical synthesis, and in resuscitation, and for many other industrial purposes. (A word coined by the French chemist Antoine Lavoisier, literally meaning "acid producer;" from the belief of the time that oxygen is a component of all acids.)

oxygen-18 see HEAVY OXYGEN.

oxygenase *Enzymology.* any enzyme that catalyzes the incorporation of oxygen from dioxygen into a single substrate.

oxygenate *Chemistry.* to combine or treat with oxygen.

oxygenated *Chemistry.* combined, treated, or saturated with oxygen.

oxygenated oil *Materials.* an oil that is treated, combined, or enriched with oxygen.

oxygenator *Medicine.* an apparatus that combines a substance with oxygen, either by chemical reaction or mechanically; one type is used in combination with pumps during open heart surgery to maintain circulation and oxygenate the blood.

oxygen burning *Nuclear Physics.* a process in which nuclei combine in stars, characterized by the fusion of two oxygen-16 nuclei at extreme temperatures and pressures.

oxygen cell see AERATION CELL.

oxygen corrosion *Metallurgy.* the reaction of a metal or alloy with oxygen.

oxygen cutting *Materials Science.* any of various processes in which material is heated to kindling temperature and cut with a stream of high-purity oxygen; used to cut materials to size and bevel or groove them for welding.

oxygen debt *Physiology.* the extra oxygen taken in by the body while recovering from exercise.

oxygen deficit *Geochemistry.* the difference between the actual quantity of dissolved oxygen in a sample of lake or seawater at a particular temperature and the saturation concentration at that temperature.

oxygen distribution *Oceanography.* the concentration of oxygen in seawater, which decreases from around 5 ml of oxygen per liter of seawater near the surface to practically nil at the greatest depths.

oxygen embrittlement *Metallurgy.* a weakening along grain boundaries in metals and alloys caused by the presence of oxygen, characterized by a reduction in ductility coupled with intergranular failures.

oxygen-flask method *Analytical Chemistry.* the burning of a substance in a closed flask containing oxygen in order to identify combustible elements by analysis of an alkali solution of the residue.

oxygen free radical *Biochemistry.* a monoatomic or diatomic species of oxygen containing an unpaired electron; for example, hydroxyl radical, superoxide radical.

oxygen-isotope fractionation *Geochemistry.* the difference in the ratio of oxygen-18 to oxygen-16 in oxygen-bearing minerals or other geologic materials, often used to determine temperature differences at the time of mineral formation.

oxygen mask *Engineering.* a face mask used to supply oxygen on demand to the wearer

oxygen minimum layer *Hydrology.* a subsurface layer containing little or no dissolved oxygen, especially when compared with the layers above and below.

oxygen point *Thermodynamics.* the boiling point of oxygen under a pressure of one standard atmosphere: –182.962°C.

oxygen ratio see ACIDITY COEFFICIENT.

oxygen steelmaking *Metallurgy.* any steelmaking process in which oxygen gas rather than nitrogen or hydrogen is used to remove carbon and phosphorus from pig iron. Thus, **oxygen (furnace) steel.**

oxygen toxicity *Physiology.* any distress caused by breathing oxygen under high pressure; symptoms may include lung and eye damage. Also, **oxygen poisoning.**

oxygen transfer *Biotechnology.* the transfer of oxygen from the air into a solution in a fermenter, usually expressed as a volumetric oxygen transfer coefficient.

oxygen transfer rate *Biotechnology.* the rate at which oxygen passes from the atmosphere into a solution, measured in relation to the rate of oxidation of a solution of sulfite to sulfate.

oxygen uptake rate *Biotechnology.* a measure of the average rate of oxygen consumption by a culture, dependent on cell concentration and expressed in volumetric terms.
oxygen yield *Biotechnology.* the amount of end products or biomass produced by an aerobic reaction, usually expressed as a rate of oxygen consumption for that culture.
oxygeusia *Neurology.* an abnormal acuteness of the sense of taste.
oxyheeite *Mineralogy.* $Pb_5Ag_2Sb_6S_{15}$, a silver-white to light-steel-gray orthorhombic mineral consisting of a sulfide of lead, silver, and antimony, occurring as masses or acicular needles, having a specific gravity of 6.22 to 6.51 and a hardness of 2.5 on the Mohs scale.
oxyhemoglobin *Biochemistry.* hemoglobin that contains bound O_2; formed from hemoglobin on exposure to alveolar gas in the lungs.
oxyhydrogen *Chemistry.* of, relating to, containing, or (usually) using a mixture of oxygen and hydrogen. Thus, **oxyhydrogen flame, oxyhydrogen torch, oxyhydrogen welding.**
oxykrinin see SECRETIN.
oxylophyte *Ecology.* a plant that grows in humus-rich soil regions. Thus, **oxylophytic.**
Oxymitraceae *Botany.* a monogeneric family of liverworts of the order Marchantiales, characterized by a persistent epidermis with simple stellate pores surrounded by thick-walled cells, thick ventral tissue, large ventral scales, and a capsule protected by an involucre.
Oxymonadida *Invertebrate Zoology.* an order of flagellate protozoans in the class Zoomastigophora; they are symbionts in the digestive tract of roaches and certain termites.
oxyneurine see BETAINE.
Oxynoticeras *Paleontology.* a genus of ammonoids in the late order Ammonitida and superfamily Psiloceratacеae; characterized by an involute oxycone and smooth, laterally compressed shell; Lower Jurassic.
oxyntic *Endocrinology.* secreting acid.
oxyntic cell see PARIETAL CELL.
oxyntic glands *Endocrinology.* the numerous, nearly straight, tubular glands in the mucosa of the fundus of the stomach that secrete mucus, pepsinogen, hydrochloric acid, and intrinsic factor.
oxyosmia *Neurology.* abnormal acuteness of the sense of smell. Also, **oxyosphresia.**
oxypathia or **oxypathy** see HYPERESTHESIA.
oxypetalous *Botany.* having sharply pointed petals.
oxyphile *Geochemistry.* see LITHOPHILE.
oxyphilia see EOSINOPHILIA.
oxyphilous *Botany.* living or thriving in acidic soil; acid-loving. Thus, **oxyphile, oxyphily.**
oxyphobous *Botany.* living or thriving in alkaline soil; not tolerant of acidic conditions. Thus, **oxyphobe.**
Oxyphotobacteria *Bacteriology.* a class of bacteria of the division Gracilicutes, consisting of Gram-negative aerobic organisms deriving energy from light and producing oxygen. It includes the blue-green bacteria (Cyanobacteria) and green prokaryotic algae (Prochlorophyta).
oxyphyte *Botany.* a plant that grows in an acidic environment. Thus, **orophytic.**
oxyquinoline see OXINE.
oxysphere see LITHOSPHERE.
Oxystomata *Invertebrate Zoology.* the true crabs, a subsection of marine burrowing crustaceans in the Brachyura, having a broad carapace and a triangular mouth.
oxytetracycline *Microbiology.* an antibiotic produced by the bacterium *Streptomyces rimosus* that inhibits protein synthesis and is effective against a variety of Gram-negative and Gram-positive microorganisms.
oxytocia *Medicine.* a process of rapid labor.
oxytocic *Medicine.* **1.** referring to or characterized by rapid labor. **2.** an agent that hastens childbirth by stimulating contractions of the myometrium.
oxytocin *Endocrinology.* a peptide hormone that is produced by the posterior lobe of the pituitary gland and induces contraction of the smooth muscle of the uterus and the myoepithelial cells of the mammary gland.
Oxytoxaceae *Botany.* a somewhat diverse family of thecate, marine, flagellate dinoflagellates of the order Peridiniales, occurring on temperate beaches and in tropical plankton.
Oxyuridae *Invertebrate Zoology.* the pinworms, a family of nematodes in the superfamily Oxyuroidea that are parasitic in the gut of vertebrates; the tail of the female ends in a sharp point. Also, **Oxyuroidea.**
Oxyurina *Invertebrate Zoology.* a suborder of parasitic nematodes in the order Ascaridida. Also, **Oxyurata.**
Oyashio Current *Oceanography.* a cold current that flows southwestward along the eastern side of the Kurile Islands, from about 52°N to about 44°N, east of Hokkaido, where it turns eastward and joins the Kuroshio Extension. Also, KURILE CURRENT.
oyster *Invertebrate Zoology.* the common name for sessile marine bivalve mollusks of the genera *Ostrea* and *Crassostrea,* family Ostreidae; adults have a rough, asymmetrical shell and a single adductor muscle.

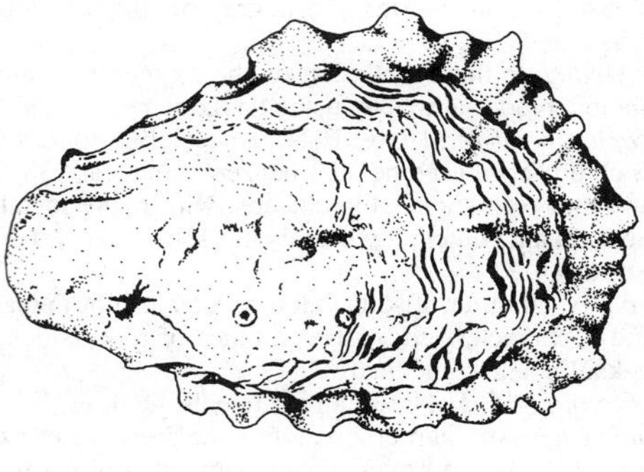

oyster

oystercatcher *Vertebrate Zoology.* any of several large, noisy shorebirds of the genus *Haematopus* of the family Haematopodidae, characterized by a white and black body, stout legs, and colorful feet and bill; noted for spearing shellfish during ebbtide when their shells are open.
oyster drill *Invertebrate Zoology.* the common name for marine snails that prey upon oysters and other shelled mollusks by drilling holes in the shell.
oyster leech *Invertebrate Zoology.* the common name for a flatworm in the class Turbellaria that feeds on oysters.
oystershell scale *Invertebrate Zoology.* a widely distributed homopteran scale insect, *Lepidosaphes ulmi,* that attacks and kills many plants, including fruit trees; the adults secrete scales that look like oyster shells.
Oz *Immunology.* an antigenic marker that distinguishes immunoglobulin human λ light chain subtypes.
oz. ounce; ounces. (From Italian *onza.*)
oz. ap. or **oz ap** ounce apothecary. Also, **oz. apoth.** or **oz apoth.**
Ozawainellidae *Paleontology.* an extinct family of Late Paleozoic foraminiferids in the superfamily Fusulinacea.
ozocerite or **ozokerite** *Geology.* a natural, brown to jet black, hydrocarbon-based paraffin wax occurring in irregular veins. Also, EARTH WAX, FOSSIL WAX, MINERAL WAX, NATIVE PARAFFIN.
ozone [ō´zōn´] *Chemistry.* O_3, a triatomic form of oxygen; a pungent, unstable blue gas that in the upper atmosphere forms a protective layer against excess ultraviolet radiation, and is also an ingredient of photochemical smog in the lower atmosphere; it is used in purification of drinking water and as an oxidizing agent.
ozone layer *Meteorology.* the layer of the upper atmosphere, from about 8 to 30 miles, where most atmospheric ozone is concentrated; it is now regarded as subject to depletion by industrial pollutants, such as fluorocarbons from aerosol sprays. Also, **ozonosphere.**
ozone resistance *Materials Science.* the resistance of a polymer to oxidative degradation by ozone.
ozone-resistance test *Materials Science.* a test for resistance to ozone attack in elastomers, in which samples are placed in an aluminum chamber and held at a specific stress with a constant temperature while a mixture of ozone in oxygen is passed over them. Cracks perpendicular to the direction of applied stress indicate degradation.
ozonide *Organic Chemistry.* an oily, thick compound based on a five-membered ring, consisting of two carbon atoms separated by one oxygen atom on one side of the ring and two oxygen atoms on the opposite side of the ring; occurring when ozone reacts with unsaturated compounds.
ozonization *Organic Chemistry.* the process of combining ozone with an unsaturated organic compound.
ozonolysis *Analytical Chemistry.* the use of ozone to break the unsaturated bonds in an organic compound in order to determine their positions by identifying the carbonyl compounds formed.
ozs. or **ozs** ounces.
oz. t. or **oz t** ounce troy.

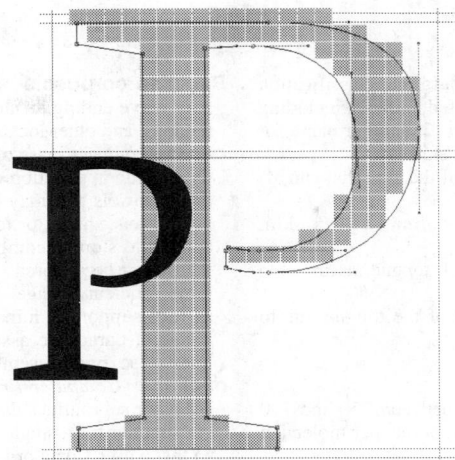

P

P the chemical symbol for phosphorus.
P or **p** parental generation. Also, **P., p., P_1.**
P or **P.** pressure; power; poise; position; pulse; presbyopia.
P or **P.** weight. (From French *poids.)*
P- *Aviation.* a former U.S. military designation for pursuit aircraft, as in *P-38*; later replaced by F- (fighter aircraft).
p or **p.** after. (From Latin *post.)*
p or **p.** part; pint; proton; per; past; pitch.
p- para-; peta-.
P_1 *Genetics.* the symbol for the immediate parents of the F_1 generation in a genetic cross.
^{32}P *Molecular Biology.* phosphorus-32, a radioactive isotope used to trace molecules, especially nucleic acids, within cells through its emission of beta radiation.
P-38 *Aviation.* a twin-engine pursuit (fighter) aircraft used in World War II; popularly known as the Lightning.
P-40 *Aviation.* a single-engine pursuit (fighter) aircraft widely used at the beginning of World War II; popularly known (in various modifications) as the Kittyhawk, Tomahawk, and Warhawk.
P-51 *Aviation.* an earlier designation of the F-51 fighter.
P53 *Oncology.* a tumor-suppressor gene that is found to be altered in astrocytoma and osteosarcoma, and in carcinoma of the breast, colon, and lung.
P-59 *Aviation.* an earlier designation of the F-59 jet fighter.
P-70 *Aviation.* an all-weather version of the A-20 attack bomber.
P-80 *Aviation.* an earlier designation of the F-80 jet fighter.
P430 *Biochemistry.* an iron-sulfur protein that serves as the first electron-carrier molecule in photosynthesis.
P450 *Biochemistry.* a cytochrome; a heme-containing protein that is a mono-oxygenase.
P680 *Biochemistry.* a type of chlorophyll that absorbs a maximum of 680 nanometers of light and serves as the reaction center for carriers transporting electrons to holes during photosynthesis.
P700 *Biochemistry.* a type of chlorophyll that absorbs a maximum of 700 nanometers of light and donates highly energized electrons for the reduction of NADP during photosynthesis.
PA posteroanterior.
PA or **P.A.** physician's assistant; psychiatric aide; power amplifier.
Pa the chemical symbol for protactinium.
Pa or **Pa.** pascal.
pA picoampere.
Paal-Knorr synthesis *Organic Chemistry.* a method of synthesizing a pyrrole from a 1,4-dicarbonyl compound by cyclization with ammonia or a primary amine in a sealed tube.
paar *Geology.* a depression produced by the separation of crustal blocks.
Paar turbidimeter *Analytical Chemistry.* an instrument used to measure the turbidity of a suspension of precipitate in a solution by visual occlusion of a light filament.
PABA [pä´bə; pa´bə] *Biochemistry. p*-aminobenzoic acid, a substance required by many organisms for the synthesis of folic acid; included in the B vitamin complex, although not an essential nutrient for humans. PABA absorbs ultraviolet light and is used as a topical sunscreen. Also, **PAB.**
PABX private automatic branch exchange.
Pac. Pacific.
paca *Vertebrate Zoology.* a generally solitary nocturnal rodent, *Cuniculus paca*, of the family Dasyproctidae, characterized by a white-spotted, large, powerful, almost tailless body and terrestrial habits, although generally found near water and a capable swimmer; native to Central and South America. Also, SPOTTED CAVY.

paca

Pacchionian bodies see ARACHNOIDAL GRANULATIONS.
pace *Military Science.* for ground forces, the speed of a column or element that is regulated to maintain a prescribed average speed. Also, **pace setter.**
pacemaker *Physiology.* **1.** any object, substance, or agent that influences the rate at which a process or reaction occurs, especially one that causes a regular rate to occur. **2.** also, **cardiac pacemaker.** specifically, the group of specialized cells located at the junction of the superior vena cava and the right atrium of the heart, which rhythmically initiate the heartbeat. *Medicine.* also, **artificial (cardiac) pacemaker.** an electronic device designed to produce impulses that will stimulate the heart muscle to contract at a certain regular rate; implanted beneath the skin or connected from the outside to treat certain cardiac conditions. *Biochemistry.* a substance whose rate of reaction sets the rate for a series of related reactions.
pachnolite *Mineralogy.* $NaCaAlF_6 \cdot H_2O$, a colorless to white monoclinic mineral, having a specific gravity of 2.98 and a hardness of 3 on the Mohs scale; found as prismatic crystals and granular masses as a secondary mineral in pegmatites.
Pachuca tank *Chemical Engineering.* a vertical, cylindrical tank used in the chemical treatment of ores, in which the pulp is reacted with suitable solvents for long periods, while the contents are agitated by compressed air.
pachy- a prefix meaning "thick."
pachycephalosaur *Paleontology.* of or relating to the "bone-headed lizards," ornithischian dinosaurs of the suborder Pachycephalosauria; extant from the Lower to Upper Cretaceous.
Pachycephalosauridae *Paleontology.* a family of small, bipedal ornithischian dinosaurs in the suborder Pachycephalosauria, which probably evolved from an unknown, very primitive ornithischian even before the divergence of the ornithopods.

pachyderm [pak´ə durm´] *Vertebrate Zoology.* **1.** a former classification for various large nonruminant, thick-skinned, hoofed animals, including the elephant, rhinoceros, hippopotamus, and others. **2.** another name for an elephant. (From a Greek phrase meaning "thick skin.")

pachyderma *Medicine.* an abnormal thickening of the skin that can affect the larynx, as well as external skin surface.

pachydermatous *Medicine.* unusually and abnormally thick skin, often caused by elephantiasis.

pachyglossal *Vertebrate Zoology.* having a thick tongue, as in some lizards.

pachyglossia *Medicine.* an abnormal thickness of the tongue due to local lymphangiectasis or to muscular hypertrophy.

pachygnathous *Medicine.* having a large jaw.

pachygyria see MACROGYRIA.

pachyman *Biochemistry.* a glucan that contains between 250 and 700 glucose residues with a number of β-linked branch points per molecule; present in the sclerotia of *Poria cocos.*

pachymeningitis *Neurology.* an inflammation of the dura mater, the outermost membrane covering the brain and spinal cord, producing symptoms resembling meningitis and sometimes resulting in spinal nerve radiculopathy.

pachymeninx see DURA MATER.

pachymeter *Engineering.* an instrument used to measure the thickness of an object or material.

pachynsis *Medicine.* a thickening, especially an abnormal thickening due to disease or injury.

Pachysolen *Mycology.* a genus of yeast fungi belonging to the order Endomycetales that are found in liquid mixes used for tanning.

pachytene *Cell Biology.* the third of five sequential steps composing the prophase I stage of meiosis, characterized by the presence of a synaptonemal complex between the fully paired homologous chromosomes. Genetic recombination is thought to occur at this stage. Also, **pachynema.**

Pachytichospora *Mycology.* a genus of yeast fungi belonging to the order Endomycetales; found in South Africa.

p-acid *Organic Chemistry.* a *para*-substituted benzoic acid.

Pacific *Geography.* the world's largest ocean, covering a third of the earth's surface (about 70,000,000 sq mi) from the Arctic to the Antarctic between the Americas and Asia and Australia.

Pacific barracuda *Vertebrate Zoology.* the barracuda *Sphyraena argentea*, characterized by a small, slender body and valued as a food fish; found in coastal seas from Alaska to lower California.

Pacific dogwood *Botany.* the dogwood tree, *Cornus nuttallii*, characterized by pointed, white or pinkish bracts and clustered scarlet fruits; found in North America.

Pacific Equatorial Countercurrent *Oceanography.* the equatorial countercurrent in the Pacific, which flows eastward between 3° and 10°N.

Pacific faunal region *Ecology.* the marine creatures living in the geographical region that extends from the offshore waters west of Central America, and running from the coast of South America up to the southern tip of California.

Pacific high *Meteorology.* a nearly permanent, subtropical, high-pressure system located in the north Pacific Ocean, at 30–45° latitude, or 1000 miles northeast of Hawaii. Also, HAWAIIAN HIGH.

Pacific North Equatorial Current *Oceanography.* a broad, slow current that flows westward entirely across the widest part of the Pacific between 10° and 20°N.

Pacific salmon *Vertebrate Zoology.* any salmon of the genus *Oncorhynchus,* especially the chinook salmon, *O. tshawtscha*; valued as a food fish.

Pacific series *Petrology.* one of the two large groups of igneous rocks, characterized by calcic and calc-alkalic rocks predominantly found in the circum-Pacific orogenic belt. Also, **Pacific suite.**

Pacific South Equatorial Current *Oceanography.* a very broad current that flows westward across the Pacific between 20°S and 3°N; sometimes called the Trade Drift in its lower reaches, where it is slower than near the equator.

Pacific temperate faunal region *Ecology.* the marine creatures who live in the geographical region that includes a narrow zone in the North Pacific Ocean, from Indochina to Alaska and along the east coast of Maine.

Pacific tree frog *Vertebrate Zoology.* a terrestrial frog, *Hyla regilla*, distinguished by the dark stripe along each side of its head; found in western North America.

Pacinian corpuscle *Anatomy.* a pressure-sensitive, encapsulated sensory nerve ending located in the skin of the fingers, and in mesenteries, tendons, and other locations.

pack a compact mass of material; specific uses include: *Surgery.* **1.** the therapeutic application of a material about a body or limb, such as wet or dry towels that may be hot or cold. **2.** the materials so applied, such as blankets, sheets, or towels. **3.** a tampon or other plug of soft material applied to stop bleeding. *Computer Programming.* to compress data so that it can be restored to its original form for later use, thus providing more efficient storage. *Mining Engineering.* waste rock or timber used as roof support in mines or used to fill excavations. Also, FILL. *Ordnance.* a parachute assembly unit containing the canopy and shroud lines. Also, **pack assembly.** *Oceanography.* see **pack ice.**

package *Computer Programming.* **1.** a term for a collection of programs or subroutines that are logically related and usually sold by a single supplier for a single all-inclusive price. **2.** a general program written so that a user's data organization will not impact the program. **3.** in Ada, a language construct that allows encapsulation of a set of related programs and data.

packaged forces *Military Science.* forces of varying size and composition that have been preselected for specific missions in order to facilitate planning and training.

packaged magnetron *Electronics.* a unit that contains a magnetron, its magnetic circuit, and its output matching device.

package dyeing *Materials Science.* a textile-dyeing process in which the yarn is wound onto a perforated steel tube or a compressible stainless-steel spring designed to allow inside-out or outside-in circulation of the dyebath.

package power reactor *Nuclear Physics.* a nuclear power plant scaled down so as to be easily transportable.

packaging density *Electronics.* **1.** the number of active devices per unit volume in a system or a subsystem. **2.** a computer's storage capacity, based on the number of information units in a specific segment of its magnetic medium.

pack artillery *Ordnance.* artillery weapons designed to be delivered by parachute or transported by animals; the weapons are transported in sections and reassembled in the field.

pack builder *Mining Engineering.* in anthracite and bituminous coal mining, a worker who fills or otherwise reinforces underground rooms, walls, or roofs to prevent caving. Also, **packer.**

pack carburizing *Metallurgy.* a process of case hardening, performed by diffusing carbon originating from a solid material packed around a part.

packed bed *Chemical Engineering.* a process vessel that consists of a vertical tower filled with particulate solids supported by a plate and a flow distributor; used in catalysis, ion exchange, distillation, and mixing.

packed bed reactor *Biotechnology.* a tubular bioreactor that is filled with a solid matrix; the medium enters and percolates through the matrix, during which the substrate is converted to product.

packed decimal *Computer Programming.* a data representation scheme in which two decimal digits are represented by one byte.

packed fluid *Petroleum Engineering.* fluid injected into the annulus between the tubing and the casing above a packer, so that pressure differences may be reduced across the packer and between the formation and the inside of the casing.

packed tower *Chemical Engineering.* an industrial mass-transfer device that consists of a vertical, cylindrical shell filled with packing supported by a plate and a flow distributor; used primarily for rectification, stripping, humidification, and distillation.

packet *Computer Science.* a part of a message sent over a packet-switching network, consisting of control, sequencing, and addressing information, and a section of message information. *Biology.* a somewhat cubical cluster of organisms formed as a result of cell division in three planes.

packet network *Computer Science.* a network of computers and communication links, over which communication is accomplished by packet switching.

packet radio *Computer Science.* a packet network, some of whose communication links are accomplished by radio transmission.

packet switching *Computer Science.* a method of network communication in which user messages are broken into short segments and packaged with control, addressing, and sequencing information to form packets, which are sent to their destination by transmitting them to successive nodes in the network.

packet transmission *Telecommunications.* the rapid transmission of packets of data over transmission lines; messages are stored in the fast-access core memory of high-speed computers.

pack hardening *Metallurgy.* any of several case hardening processes performed by packing a solid material around a part, and heating to a suitable temperature.

pack ice *Oceanography.* any free-floating sea ice that has been driven together into a more or less solid mass, from 50 meters to several kilometers across.

packing *Engineering.* any material that is used to cushion or protect packed goods, or to hold some item in place. *Surgery.* **1.** the application of gauze or other material to fill a wound or cavity. **2.** the material itself. *Acoustical Engineering.* the overcrowding of carbon particles in a carbon microphone, causing a reduction in the sensitivity of the device. *Geology.* the manner in which the solid particles or crystals in a sediment or sedimentary rock are arranged spatially. *Crystallography.* the atomic or ionic arrangement in a crystal that characterizes the crystal's structural features.

packing density *Electronics.* **1.** the amount of information a given storage medium, such as a tape, photograph, or magnetic drum, can hold. **2.** the number of elements within a given area of an integrated circuit. *Computer Programming.* the number of bits or storage cells per unit length, area, or volume on a storage medium; for example, the number of characters per track on a disk. *Geology.* the degree to which the grains of a sedimentary rock fill the gross volume of a rock, as contrasted with the spaces between the grains.

packing fraction *Physics.* a measure of the stability of an atomic nucleus, given by the ratio of the mass defect to the atomic number and expressed by the formula $(M_i - A)/A$, where M_i is the isotopic weight and A is the atomic number.

packing index *Crystallography.* the volume of space occupied by an atom or ion in a crystal, divided by the volume occupied by the unit cell.

packing proximity *Geology.* an approximate measure of the number of grains in a sedimentary rock that are in contact with neighboring grains.

packing radius *Crystallography.* one-half the distance between nearest-neighbor atom or ion centers in a crystal arrangement.

packing ratio *Molecular Biology.* the ratio of the calculated length of a helical DNA molecule to its length after compaction.

pack rat *Vertebrate Zoology.* a large rodent, *Neotoma cinerea,* of the family Cricetidae, characterized by a bushy tail and well-developed cheek pouches; known for its habit of carrying off small objects and for using sticks, bones, and feces to build nests that may reach six feet and may be inhabited for generations; native to the Rocky Mountains. Also, WOOD RAT.

pack rat

pack-rat midden *Archaeology.* a collection of artifacts or objects concealed at some point by a pack rat and remaining in an assemblage at that location.

pack rolling *Metallurgy.* the simultaneous hot rolling of two or more metallic sheets.

pack routine *Computer Programming.* a routine that compresses data and thus reduces the amount of storage required for a file.

packsand *Petrology.* a very fine-grained sandstone, loosely consolidated by a small amount of calcareous cement.

packstone *Petrology.* a sedimentary carbonate rock featuring an arrangement of granular material in a self-supporting framework, as well as some matrix of calcareous mud.

pack wall *Mining Engineering.* a wall of dry stone constructed alongside a roadway or in a waste area of a mine, designed to support the roof and to retain the packing material.

pactamycin *Microbiology.* an antibiotic produced by the bacterium *Streptomyces pactum* that acts against tumors by interfering with protein synthesis.

pad *Zoology.* any fleshy mass of tissue that cushions a weight-bearing part of the body, such as the underside of a dog's paw. *Anatomy.* any cushionlike mass of soft material, such as fat. *Engineering.* **1.** a layer of material used to protect an object. **2.** the spare metal projecting from a weld part of casting. *Space Technology.* see LAUNCHING PAD. *Electronics.* **1.** a fixed attenuator. **2.** to reduce the frequency in an inductance-capacitance circuit by greatly increasing the capacitance. **3.** see LAND. *Metallurgy.* refractory work on the base of a blast furnace.

pad deluge *Space Technology.* during a rocket launch, the spraying of water on the launching pad in order to reduce the temperature of parts of the pad or rocket.

padder *Electronics.* a device that is added to the oscillator-tuning circuit in a radio receiver to control calibration at the low-frequency end of the tuning range.

padding *Computer Programming.* the addition of dummy characters, words, or records to a data item, record, or block in order to achieve the required length.

paddle *Navigation.* a short pole with a flat blade, used to propel and steer a canoe or small boat. *Mechanical Devices.* any of various similar flat-bladed implements, such as the blades of a water wheel.

paddle-and-anvil *Archaeology.* describing a pottery-making method in which a wooden paddle and a stone or ceramic disk are used to smooth and shape a coiled pot. Also, **paddle-and-anvil technique.**

paddle blade *Aviation.* a broad, flat propeller blade with a squared tip.

paddlefish *Vertebrate Zoology.* either of two species of freshwater fish of the family Polyodontidae, characterized by a paddlelike snout, smooth skin, and an archaic cartiliginous skeleton; the **American paddlefish,** *Polyodon spathula*, lives in the Mississippi River and the Asian species lives in the Yangtze and other rivers of China.

paddlefish

paddlewheel *Mechanical Engineering.* a wheel at the side or stern of a ship, fitted with blades that enter the water more or less perpendicularly to propel the ship.

paddlewheel steamer *Naval Architecture.* a steam vessel propelled by rotating, horizontal-axis paddle wheels rather than by a propeller. A **sternwheeler** has a single paddle wheel astern; a **sidewheeler** has a pair of paddle wheels on either side amidships. Ocean-going paddlewheel steamers were usually sidewheelers.

paddy *Agriculture.* an irrigated rice field.

p-adic *Mathematics.* **1.** for a fixed prime p, a ***p*-adic integer** is a sequence $\{x_n\}_p$ of integers, $n = 0, 1, 2, 3, \ldots$, such that $x_n \equiv x_{n-1} (\mathrm{mod}\ p^n)$ for each $n \geq 1$. Two p-adic integers $\{x_n\}_p$ and $\{y_n\}_p$ are said to determine the same p-adic integer if and only if $x_{n-1} \equiv y_{n-1} (\mathrm{mod}\ p^n)$ for each $n \geq 1$. **2.** for a fixed prime p, a ***p*-adic number** is an expression of the form a/p^k, where a is a p-adic integer and k is a nonnegative integer. Two p-adic numbers a/p^k and b/p^m are said to determine the same p-adic number if and only if ap^m and bp^k are the same p-adic integer. **3.** for a fixed prime p, the set of p-adic numbers form a field called the ***p*-adic field.** Addition is defined by:

$$a/p^k + b/p^m = (ap^m + bp^k)/p^{(k+m)};$$

multiplication is defined by:

$$(a/p^k)(b/p^m) = ab/p^{(k+m)}.$$

padlock *Mechanical Devices.* a removable lock with a sliding, hinged, or pivoted shackle that can be passed through the eye or link and then secured to a hasp.

padouk *Materials.* a hard wood, colored with red and black stripes, from the *Pterocarpus* tree of India; used for furniture and cabinetry.

Paecilomyces *Mycology.* fungi formerly belonging to the genus *Penicillium*, now a genus belonging to the class Plectomycetes; occur in soils, food, and plant debris.

paedogamy *Invertebrate Zoology.* in ciliate protozoans, a process of self-fertilization involving fission and subsequent refusion of the nucleus. Also, AUTOGAMY.

paedogenesis *Invertebrate Zoology.* the generally parthenogenic production of offspring by an immature or larval animal; for example, the gall midge Miastor, whose larvae can give birth to twenty or thirty progeny. Also, PEDOGENESIS.

paedomorphosis *Evolution.* a phylogenetic evolutionary change in which the adults of a later species retain some of the characteristics of juveniles of the ancestral forms.

Paenungulata *Vertebrate Zoology.* in some classification systems, a division of eutherian mammals that includes the extinct orders of Embrithopoda, Pyrotheria, Dinocerata, and Pantodonta, and the living orders of Sirenia (manatees and dugongs), Proboscidea (elephants), and Hyracoidea (hyraxes).

Paeoniaceae *Botany.* the peonies, a temperate family of about 30 species in the order Dilleniales; dicotyledonous herbs and shrubs characterized by large cleft leaves, large perfect flowers with an intrastaminal disk, and seeds with abundant endosperm.

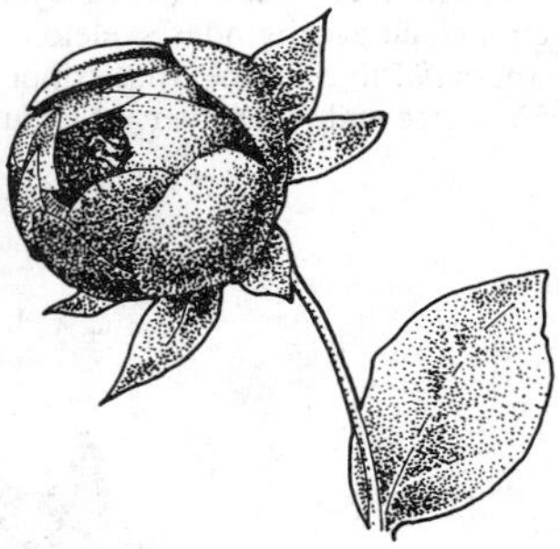

Paeoniaceae (peony)

paesa *Meteorology.* a violent north-to-northeast wind that blows over Lake Garda in Italy.

paesano *Meteorology.* a night breeze from the north, blowing down from the mountains over Lake Garda in Italy.

PAF platelet-activating factor.

page *Graphic Arts.* one side of a single sheet of paper in a book, magazine, newspaper, or the like. *Computer Programming.* **1.** in virtual storage systems, a fixed-length block of data, usually 512 to 4096 bytes, that is transferred as a unit between external storage and main storage. **2.** a block of data displayed at the same time on a display device. **3.** a printed paper form.

page boundary *Computer Programming.* in virtual storage systems, the address of the first word of a page of memory.

page data set *Computer Programming.* in virtual storage systems, a data set in external storage into which main storage pages are saved.

page eject see EJECT.

page fault *Computer Programming.* an interrupt that occurs when a program refers to a page that is not located in main storage, and thus must be retrieved from external storage.

page-in *Computer Technology.* the process of moving pages of programs or data from external storage to main storage. Conversely, **page-out.**

pagemaking program *Computer Science.* a software program used to create documents having the appearance of professionally printed pages.

page printer *Telecommunications.* **1.** a high-speed device that transposes messages onto a full-page format by printing them out one character at a time. **2.** an apparatus that is able to compose an entire page of text before it is printed.

pager *Telecommunications.* an electronic device that receives or sends a signal in order to notify its user about messages. Also, BEEPER.

page reader *Computer Technology.* in optical character recognition, a device that can read cut-form documents of various sizes under program control.

page skip *Computer Technology.* a control character sent to a printer that causes it to eject the current page and go to the top of the next page.

Paget, Sir James 1814–1899, English surgeon and pathologist; a founder of modern pathology; studied tumors and diseases of bones and joints; described Paget's disease.

page table *Computer Programming.* a table located in main storage that contains the actual addresses in main memory of pages in virtual memory.

pagetoid *Oncology.* resembling Paget's disease.

Paget's disease *Oncology.* a breast carcinoma marked by large cells with clear cytoplasm, involving the nipple or areola and the larger ducts.

paging *Computer Programming.* **1.** the transfer of pages of data between main storage and external storage. Also, **page turning. 2.** the display of pages of text or graphics on a screen.

paging rate *Computer Programming.* the number of page-ins and page-outs per unit of time.

paging system *Telecommunications.* any system that summons a person to the telephone; may be accomplished by an announcement over a loudspeaker or by means of a pager.

pagoda *Architecture.* **1.** a multitiered Buddhist shrine. **2.** a structure resembling such a shrine.

pagoda stone *Mineralogy.* **1.** a Chinese limestone that displays pagodalike designs of fossil orthoceratites when viewed in cross section. **2.** an agate characterized by pagodalike markings.

pagodite see AGALMATOLITE.

Paguridae *Invertebrate Zoology.* **1.** the hermit crabs, a family of small marine decapod crustaceans in the superfamily Paguridea that live in abandoned snail shells; the last pair of abdominal appendages is modified to hold the animal inside the shell. **2.** a superfamily of decapod crustaceans with an asymmetrical abdomen, including hermit crabs, land crabs, and stone crabs, that live in abandoned gastropod shells.

PAH *Biochemistry.* *p*-aminohippuric acid, the glycine amide of *p*-aminobenzoic acid, which is rapidly excreted, primarily by renal tubular secretion. The clearance of PAH administered by continuous infusion is considered to be the definitive measurement of effective renal plasma flow. Also, **PAHA.**

paha *Geology.* a low, elongated, glacial hill or rounded ridge composed primarily of drift, rock, or windblown sand, silt, or clay, and capped by a thick accumulation of loess.

pahoehoe [pä hō´ē hō´ē] *Volcanology.* a Hawaiian term for a basaltic lava flow typified by a smooth, wavy, or ropy surface and a glassy outer skin. Also, ROPY LAVA.

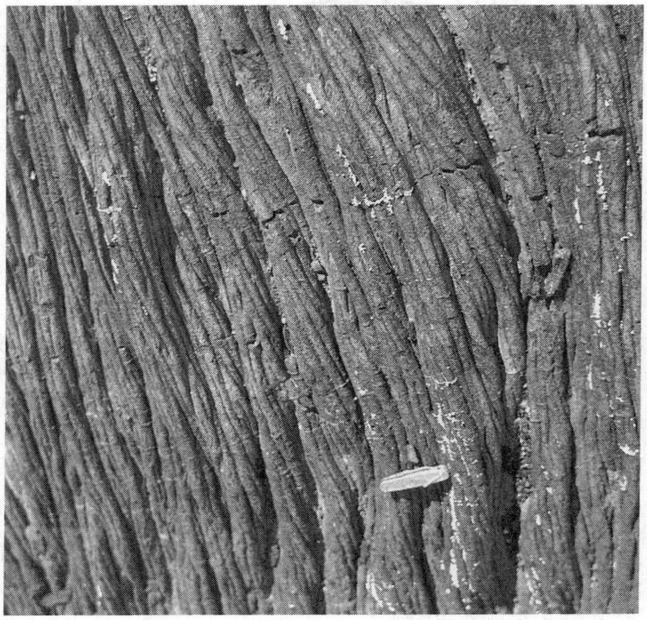

pahoehoe lava

paigeite see HULSITE.

pail *Mechanical Devices.* a nearly cylindrical plastic or metal container, usually with a handle, used to transmit or catch liquid. Also, BUCKET.

pain *Physiology.* a relatively localized sensation of discomfort, distress, or agony, resulting from the stimulation of specialized nerve endings.

pain spot *Physiology.* any of various spots on the skin where the sense of pain can be produced by a stimulus.

paint *Materials.* **1.** a coating or coloring mixture composed of pigment in a water, latex, or oil base. **2.** the pigment used in such a mixture. **3.** to apply such a mixture. *Electronics.* an informal term for the target image on a radarscope.

paintbrush *Mechanical Devices.* a brush for applying paint over a specified area, constructed of firmly bound bristles in a metal stock mounted on a handle. *Computer Technology.* on a computer graphics display, a patch of color that can be moved and used to "paint" the screen under program control or by use of an input device such as a mouse or joystick. Also, BRUSH.

painted butterfly *Invertebrate Zoology.* the butterfly *Vanessa cardui* characterized by brownish-black and orange wings and four eyespots on each hind wing; the larvae feed on thistles.

painted snipe *Vertebrate Zoology.* either of two snipelike birds of the family Rostratulidae distinguished by a female that is larger and more brightly colored than the male; found in South America and Old World tropics.

painted turtle *Vertebrate Zoology.* the freshwater turtle, *Chrysemys picta*, characterized by bright yellow markings on the head and neck and red markings on the margin of the carapace; found in the U.S.

painter *Meteorology.* a term for a frequent fog on the coast of Peru that often leaves a brownish deposit on exposed surfaces.

pain threshold *Physiology.* the point at which a stimulus of increasing intensity becomes painful.

painting *Computer Technology.* **1.** the process of filling in a selected graphic area with a color or pattern. **2.** the process of displaying graphics data or the movements of a graphical input device on a video screen.

paint roller *Mechanical Devices.* a device used to apply paint over large areas, consisting of a hollow sleeve of natural or synthetic fiber that slips onto an open wire cage which revolves about the end of a handle.

paint vehicle *Materials Science.* the fluid component of paint, acting as a carrier for the pigment.

pair two similar or corresponding things that are used together; specific uses include: *Electricity.* in electric transmission, two like conductors that are used to form an electric circuit. *Mechanical Engineering.* two components of a kinematic mechanism that are connected so as to constrain relative motion. Also, KINEMATIC PAIR.

pair bond(ing) *Behavior.* a bond between a male and a female of the same species.

paired-associate learning *Behavior.* a verbal learning technique in which a subject learns pairs of items and must respond with one half of the pair when presented with the other half.

paired cable *Electricity.* a cable composed of separate twisting pairs of conducting wires.

paired data *Statistics.* data arising from the measurement of the effect of, for example, two different treatments on the same experimental units.

paired terrace *Geology.* one of two equally elevated land surfaces that face each other from opposite sides of a stream valley and that represent remnants of the same floodplain or valley floor. Also, MATCHED TERRACE.

pairing a joining or coupling; specific uses include: *Electronics.* the meshing of two fields of a television picture into one frame, so that instead of being properly spaced, they appear in pairs; when this faulty interlace is serious, the lines of alternate fields can pair up and fall on one another, which cuts the vertical resolution in half.

pairing element *Mechanical Engineering.* either one of two parts of a mechanism that are connected to allow motion.

pairing energy *Nuclear Physics.* the energy gained by action of the attractive, short-range residual force (pairing force) between two like nucleons in an antisymmetric spin state.

pairing isomer *Nuclear Physics.* an instance of esoteric isomeric states resulting from different, very small-range motions of the nucleons that make up the nucleus.

pair potential approximation *Materials Science.* an equation that defines the total (cohesive) energy of a metal, based only on the nearest-neighbor interactions.

pair production *Nuclear Physics.* the phenomenon of relativistic materialization of energy that results in the simultaneous creation of a negative electron (negatron) and a positive electron (positron) when a photon is absorbed by a nucleus; the reverse of this process, annihilation, occurs when a positron collides with an electron and both disappear, their combined energy being converted into two photons of equal frequency.

paisanite see AILSYTE.

PAL *Robotics.* a programming language developed at Purdue University to describe end-effector positions, forces, torques, transformations, and primitives.

pal. paleontology; paleography.

Palade, George E. born 1912, Rumanian-born American physiologist; discovered ribosomes; awarded the Nobel Prize for the study of cell ultrastructure.

Palaeacanthaspidae *Paleontology.* an extinct family of primitive placoderms in the order Acanthothoraci; similar to the arthrodires and rhenanids; extant in the Lower Devonian.

Palaearctic region *Ecology.* the large zoogeographical region that consists of Europe, Africa north of the equator, Asia Minor, and Asia north of the Himalayas, and includes the European, Siberian, Mediterranean, and Manchurian subregions.

Palaecanthocephala *Invertebrate Zoology.* an order of spiny-headed worms in the phylum Acanthocephala, characterized by rows of hooks on the proboscis; intestinal parasites of fish and amphibians.

Palaechinoida *Paleontology.* a paraphyletic group of late Paleozoic regular echinoids in the subclass Periscoechinoidea; characterized by thick, tesselate plates; restricted to the Lower Carboniferous.

Palaemonidae *Invertebrate Zoology.* a family of shrimp, shallow-water decapods in the suborder Caridea, characterized by a long, compressed rostrum armed with teeth.

Palaeoanthropinae *Paleontology.* a subdivision of the family Hominidae that includes the Neanderthals and Rhodesian man.

Palaeocaridacea *Invertebrate Zoology.* an order of malacostracan crustaceans in the superorder Syncarida, characterized by the lack of a carapace and having exopodites on all appendages.

Palaeocaridae *Invertebrate Zoology.* a family of malacostracan crustaceans in the order Palaeocaridacea.

Palaeocharaceae *Paleontology.* an early family of charophytic algae, characterized by sinistrally coiled gyrogonites; extant in the Carboniferous.

Palaeoconcha *Paleontology.* a subclass or order of simple, smooth-hinged bivalves; mainly Paleozoic but in some classifications including some still-living forms. Also, CRYPTODONTA.

Palaeocopa or **Paleocopa** *Paleontology.* a subclass of copepod crustaceans known only from shells dredged from the south Pacific.

Palaeocopida *Paleontology.* an extinct order of ostracods, composing the suborders Beyrichicopina and Kloedenellocopina; generally sexually dimorphic; extant from the Ordovician to the Triassic, except for some recently discovered Tertiary and Holocene fossils that may be palaeocopids, and perhaps living representatives of the Holocene genus.

Palaeodonta *Vertebrate Zoology.* in some classification systems, a suborder of Artiodactyla that consists of several piglike families, including the hippopotamus and several extinct species.

Palaeognathae *Vertebrate Zoology.* the ratites, a superorder of birds in the subclass Neornithes in some classification systems, including the ostrich, cassowary, rhea, emu, kiwi, and their relatives; all are flightless (except for the tinamou) and are generally characterized by large size, strong legs, and a reduced breastbone.

Palaeoheterodonta *Invertebrate Zoology.* a subclass of bivalve mollusks that includes the freshwater oysters.

Palaeoisopus *Paleontology.* an extinct genus of arthropods; its probable modern descendant *Liphistius* is little changed; the fossil record is incomplete, but *Palaeoisopus* is probably related to the Pycnogonida, the modern sea spiders.

Palaeomastodontinae *Paleontology.* a subfamily of elephantoid proboscideans in the extinct family Gomphotheriidae; Oligocene.

Palaeomerycidae *Paleontology.* a family of pecorans in the superfamily Cervoidea, probably ancestral to modern deer and giraffes; wide holarctic distribution from the Oligocene to the Early Pliocene.

Palaeonemertini *Invertebrate Zoology.* the most primitive order of nemertean worms in the phylum Rhynchocoela, having no stylets on the proboscis and usually no eyes or cerebral organs.

Palaeonisciformes *Paleontology.* an extinct order of primitive actinopterygian fishes in the infraclass Chondrostei; relatively long-lived; extant from the Devonian to Cretaceous.

Palaeoniscoidei *Paleontology.* a suborder of early fishes belonging to the order Palaeonisciformes; several palaeoniscoid families independently evolved such advanced characteristics as more symmetrical tails and a smaller number of fin rays, providing a good example of mosaic evolution; extant from the Devonian to Cretaceous.

Palaeoniscum or **Palaeoniscus** *Paleontology.* a genus of primitive actinopterygian fishes in the extinct family Palaeoniscidae; extant in the Permian. Also, PALAEOTHRISSUM, GEOMICHTHYS.

Palaeopantopoda *Paleontology.* an extinct class or order of arthropods in the subphylum Pycnogonida; extant in the Devonian.

Palaeoparadoxidae *Paleontology.* a family of four-legged amphibians in the extinct order Desmostylia, of uncertain habits and affinities; known only from the Miocene and Pliocene of Eastern Asia and western North America.

Palaeophonus *Paleontology.* one of the oldest known arachnids, an extinct genus of scorpions of the Silurian; resembles some eurypterids but probably only as a result of convergent evolution.

Palaeopneustidae *Invertebrate Zoology.* a family of deep-sea heart urchins, echinoid echinoderms in the order Spatangoida, with an oval test and long spines.

Palaeopterygii *Vertebrate Zoology.* an equivalent name for the osteichthyan subclass Actinopterygii, which includes the ray-fin fishes.

Palaeoryctidae *Paleontology.* a family of eutherian mammals *incertae sedis,* but tentatively placed in the insectivore suborder Soricomorpha; holarctic distribution; extant in the Paleocene and Eocene.

Palaeosmilia *Paleontology.* a genus of solitary rugose corals in the extinct suborder Aulophyllum; characterized by many long septa, arranged radially; extant in the Devonian and Carboniferous.

Palaeospondylus *Paleontology.* a primitive fishlike vertebrate, known only from a one-inch fossil from the Permian of Scotland; *incertae sedis,* whether cyclostome, placoderm, or some other group.

Palaeotaxodonta *Invertebrate Zoology.* a subclass of bivalve mollusks characterized by equal shell valves and a row of short teeth along the hinge margin.

Palaeotheriidae *Paleontology.* a family of European perissodactyl hippomorphs in the superfamily Equoidea; the palaeotheres represent an early divergence from the line of descent of the Equidae, arising from Hyracotherium in the Eocene but becoming extinct in the Oligocene and leaving no known descendants; the type genus *Palaeotherium* was the largest equoid of the Eocene, growing to rhinoceros size.

Palaeotheriodont *Vertebrate Zoology.* of or relating to lophodont teeth that have longitudinal external tubercles united to the inner tubercles by transverse oblique crests.

palagonite *Geology.* a light brown to yellow altered basaltic glass that occurs in pillow lavas as amygdules or interstitial material.

palagonite tuff *Petrology.* a yellow or orange pyroclastic rock composed of angular fragments of altered basaltic glass.

palama *Vertebrate Zoology.* the webbing on the feet of aquatic birds.

palatal *Anatomy.* of or relating to the palate.

palatal sound *Linguistics.* the point of articulation in the formation of some consonant sounds, such as [sh], [z], [ch], [j], and [r], in which the tongue is pressed against the hard palate.

palate *Anatomy.* the roof of the mouth, consisting of an anterior hard palate and a posterior soft palate. *Physiology.* a popular term for the sense of taste.

palatine *Anatomy.* of or relating to the palate.

palatine bone *Anatomy.* either of two bones that together with the maxilla form the hard palate.

palatine canal *Anatomy.* a canal formed by a deep ridge on the posterior portion of the horizontal plate of the palate bone and its articulation with the maxillary bone, through which pass the descending palatine artery and the greater palatine nerve.

palatine gland *Anatomy.* any of the mucous glands of the soft palate and the posterior hard palate.

palatine process *Anatomy.* a thick, horizontal projection of the inner surface of the maxillary bone that forms part of the floor of the nostril and the roof of the mouth.

palatine suture *Anatomy.* the sagittal suture formed by the articulation of the palatine bones, or the suture between the palatine bones and the maxillary bones.

palatine tonsil *Anatomy.* either of two masses of lymphoid tissue in the lateral walls of the oral pharynx.

palatitis *Medicine.* an inflammation of the palate.

palatomaxillary arch *Anatomy.* the arch formed by the roof of the mouth from the teeth on one side to the teeth on the other side.

palatoplasty *Surgery.* plastic reconstruction of the palate, including repair of a congenital cleft palate.

palatoquadrate *Vertebrate Zoology.* of or relating to the cartilage forming the upper jaw or roof of the mouth of cartilaginous and bony fishes, and of most other nonmammalian vertebrates.

palau *Metallurgy.* a palladium-gold alloy, used as a platinum substitute in chemical analysis.

Palavasciaceae *Mycology.* a family of fungi belonging to the order Eccrinales characterized by multinucleate, thick-walled sporangiospores; occurs in the guts of marine isopods.

pale- a prefix meaning "old" or "ancient." Also, **palae-.**

palea *Botany.* **1.** one of the upper, scalelike, membranous bracts enclosing a grass flower. **2.** a bract at the base of an individual floret of some composite plants. **3.** the scales on certain fern stems. Also, **pale, palet.**

paleate *Botany.* covered with chaffy scales, used to describe some rhizomes.

paleic surface *Geology.* a smooth, preglacial erosion surface.

pale infarct see WHITE INFARCT.

paleo- a prefix meaning: "old" or "ancient." Also, **palaeo-.**

paleoagrostology *Paleontology.* the study of fossil grasses, a branch of paleobotany.

paleoalgology *Paleontology.* the study of fossil algae. Also, PALEOPHYCOLOGY.

paleoanthropology *Anthropology.* a branch of anthropology that involves the study of fossil remains to trace the development of certain physical characteristics. Thus, **paleoanthropologist, paleoanthropological.**

paleobiochemistry *Paleontology.* the organic chemistry of fossils; especially the study of chemical processes that occurred in extinct organisms.

paleobioclimatology *Paleontology.* the study of the long-range effects of climate on plants and animals in the geological past.

paleobiocoenosis *Paleontology.* a reconstruction of a taphonomic association of a functional community of extinct organisms.

paleobiogeography *Paleontology.* the study of the distribution of land and marine faunas and floras in the geological past.

paleobiology *Paleontology.* the study of life forms of the geological past; sometimes used as a synonym for paleontology, but better used to stress the part of paleontology that focuses on interpreting and reconstructing the life activities of fossil organisms. Thus, **paleobiologist, paleobiological.**

paleobotanic province *Geology.* an extensive area characterized and defined by similar plant fossils.

paleobotany *Paleontology.* the study of fossils of plants and their relationships to other organisms and to one another in the geological past. Thus, **paleobotanist, paleobotanical.**

paleoceanography *Oceanography.* the study of the history of ocean processes, especially the history of the life forms of ancient seas and their water conditions.

Paleocene [pā′lē ə sēn′] *Geology.* **1.** a geologic epoch of the Tertiary period, occurring after the Cretaceous period and before the Eocene. **2.** the rock strata formed during this epoch. Also, **Palaeocene.**

paleochannel *Geology.* a remnant of a stream channel in older rock that has been filled with or buried by the sediments of younger, overlying rock.

paleoclimate *Geology.* the climate of a designated interval of geologic time. Also, GEOLOGIC CLIMATE.

paleoclimatic sequence *Geology.* the chronology of changes in climate over geologic time, including the succession of alternating warm periods and ice ages.

paleoclimatology *Geology.* the study and analysis of climatic conditions throughout geologic time, including the interpretation of glacial deposits, fossils, and sedimentological data.

paleocrystic ice *Hydrology.* old sea ice, especially weathered polar ice that is at least 10 years old.

paleocurrent *Geology.* a current of water or wind that was active in ancient geologic time and whose direction can be inferred by the examination of sedimentary structures or rocks formed at that time.

paleodemography *Archaeology.* the study of ancient human populations and population changes.

paleodepth *Paleontology.* the depth at which a marine organism occurred in the geological past.

paleoecology *Paleontology.* the study of the relationships between organisms and their environments in the geological past, with special attention to fossil communities.

paleoenvironmental reconstruction *Archaeology.* the determination of the prehistoric environment of an archaeological site, using the methodologies of geology, botany, and zoology; for example, pollen analysis.
paleoequator *Geology.* the position of the equator for a specific interval of past geologic time, as determined by evidence such as fossil formations, coal deposits, and paleomagnetic measurements.
paleoethnobotany *Archaeology.* the study of plants used by ancient cultures.
paleofluminology *Geology.* the study of the formation, development, and processes of ancient stream systems.
Paleogene *Geology.* an interval of geologic time that incorporates the Paleocene, Eocene, and Oligocene epochs; equivalent to the Lower Tertiary period. Also, EOGENE.
paleogeographic event see PALEVENT.
paleogeographic stage see PALSTAGE.
paleogeography *Geology.* the study and description of the physical geography throughout the earth's geologic history.
paleogeologic map *Geology.* a map that displays the areal geology of a land surface as it appeared at some time in the geologic past, particularly such a map showing the surface immediately below a buried unconformity before it was covered by overlapping deposits.
paleogeology *Geology.* the study of the geological features, conditions, and events in a region during a past interval of geologic time.
paleogeomorphology *Geology.* the branch of geomorphology that involves the recognition of ancient erosion surfaces and the analysis of ancient topographies and topographic features that have either been removed by erosion or covered beneath the surface. Also, PALEOPHYSIOGRAPHY.
paleography *Anthropology.* the study of ancient forms of writing or of ancient written documents and inscriptions.
paleoherpetology *Paleontology.* the study of extinct snakes and their phylogeny.
paleoichnology *Paleontology.* the study of trace fossils, biogenic structures preserved in lithified sediments or volcanic ash or lava, as evidence of animal and plant activities.
Paleo-Indian *Archaeology.* **1.** one of the prehistoric people who migrated from Asia and settled throughout the Americas by about 10,000 BC. **2.** the earliest period in New World chronology, representing the time up to the development of agriculture and villages. **3.** also, **paleo-Indian.** of or relating to this culture or period.
paleoisotherm *Geology.* an imaginary line connecting points that were at the same temperature at some time in the geologic past.
paleokarst *Geology.* **1.** a karst rock or region that was subsequently buried under sediments. **2.** an area with karst topography that is no longer undergoing the processes that form karst terrain.
paleolatitude *Geology.* the latitude of a specific area on the surface of the earth at some former geologic time, expressed in relation to the paleoequator.
paleolimnology *Hydrology.* **1.** the study of the conditions and processes of lakes in the geologic past. **2.** the study of the history and sediments of existing lakes.
Paleolithic [pā´lē ə lith´ik] *Archaeology.* **1.** a period of cultural development defined by the use of stone tools, the oldest preserved form of technology; the time period extending from about 2.5 million years ago to about 10,000 years ago. **2.** of or relating to this period. Also, **paleolithic.**
paleolithologic map *Geology.* a map that depicts a land surface at a specific time in the geologic past with respect to lithological variations at a designated buried horizon or within a restricted area.
paleomagnetics *Geophysics.* the study of the earth's magnetic field through time, including its changes and the reasons for these changes, done mainly through analysis of remanent magnetization of rocks.
paleomagnetic stratigraphy *Geophysics.* the analysis of natural remanent magnetization to identify discrete stratigraphic units.
paleomagnetism see ARCHAEOMAGNETISM.
paleomalacology *Paleontology.* the study of extinct mollusks.
paleometeoritics *Geology.* the study of changes and differences in extraterrestrial debris as a function of time over extended segments of the geologic record.
paleomorphology *Paleontology.* the branch of paleontology concerned with the structure and form of extinct plants and animals.
paleomycology *Paleontology.* the study of fossil fungi, including mushrooms.
paleon. or **paleon** paleontology.

Paleontology

Paleontology is the study of past life forms by means of fossils and their traces. Fossils are the mineralized remains of these past life forms, ranging from microfossils to plant fossils and multi-cellular animals, both invertebrate and vertebrate. Different branches of paleonteology exist related both to the organisms studied and to the uses to which the information content of fossils are put.

Micropaleontology is the study of microorganisms, and is of particular importance in economic geology, for example, in locating oil deposits. Paleobotany is the study of plant remains, including leaves, fruit, spores, and pollen, among many others. Invertebrate paleontology is the study of invertebrate groups, including such groups as trilobites and ammonites, and vertebrate paleontology is the study of fish, amphibians, reptiles, birds, and mammals. As regards the information content of the subject, paleobiology attempts to interpret fossilized animals and plants as living entities, for example, their adaptations and way of life. Paleoecology aims to reconstruct the environment in which they lived, either directly from the fossils or from the associated fauna and flora and what that indicates by comparison with living ecologies.

Paleogeography looks at the distribution of fossil organisms in relation to past continental movements and reconstructs past patterns of distribution. An important facet of all these branches of paleontology is the study of how fossils accumulate and the processes that modify them before and after burial. This is known as the science of taphonomy, and its importance lies in the detection of biases in the fossil record. A combination of all these is incorporated in paleoanthropology, the study of human evolution. Fossilized human remains demonstrate the stages in human evolution, and study of paleoecology and paleogeography show the environments and distributrion in which human evolution occurred.

Peter Andrews
Department of Paleontology
Natural History Museum, London

paleontology *Biology.* the scientific study of life of the geologic past that involves the analysis of plant and animal fossils, including those of microscopic size, preserved in rocks; concerned with all aspects of the biology of ancient life forms.
paleopallium *Anatomy.* that portion of the cortex cerebri which, along with the archaeocortex, develops with the olfactory system, and which is older and less stratified than the neocortex. Also, **paleocortex.**
paleopalynology *Paleontology.* **1.** the study of fossil pollen and spores. **2.** the study of fossil palynomorphs, organic microfossils consisting of sporopollenin, chitin, or pseudochitin.
paleopathology *Anthropology.* the study of disease and medical biology of ancient human populations, especially as evidenced in skeletal remains or artifacts preserved from prehistoric or ancient times.
paleopedology *Geology.* the study of soils that existed during the geologic past, including the calculation of their ages.
paleophycology see PALEOALGOLOGY.
paleophysiography see PALEOGEOMORPHOLOGY.
Paleophytic *Paleontology.* **1.** a palynostratigraphic era lasting from approximately 500 million to 300 million years ago (Middle to Late Paleozoic). **2.** also, **paleophytic.** of or relating to this era.
paleoplain *Geology.* a broad, flat, or gently sloping degradational region that existed in the geologic past and is now buried beneath later deposits.

paleopole *Geology.* a magnetic or geographic pole of the earth that existed at some time in the geologic past.
Paleoptera *Invertebrate Zoology.* a subclass of primitive hemimetabolous insects including the Ephemoptera (mayflies) and Odonata (dragonflies).
paleosalinity *Geology.* the percentage of salt in a body of water at some time in the geologic past, as determined by the chemical analysis of sediment or formation water.
paleosere *Ecology.* an ecological succession that took place during the Paleozoic era.
paleoslope *Geology.* the inclination of an ancient land surface as it existed at some time in the geologic past.
paleosol *Geology.* an ancient soil or a buried soil horizon that was formed during the geologic past. Also, FOSSIL SOIL, BURIED SOIL.
paleosome *Geology.* in a composite rock or mineral deposit, a geometric constituent that appears to be of earlier origin than an associated younger constituent.
paleospecies see CHRONOSPECIES, def. 2.
paleostructure *Geology.* the geologic configuration of an area or sequence of rocks at some time in the geologic past.
paleotectonic map *Geology.* a map showing the geologic and tectonic features of an area as they existed at some time in the geologic past.
paleotemperature *Geology.* **1.** the average climatic temperature of a given region or at a given time in the geologic past. **2.** the temperature at which a geologic process took place in the geologic past.
paleothermal *Geology.* describing warm climates of the geologic past. Also, **paleothermic.**
paleothermometry *Geology.* the measurement or estimation of temperatures in the geologic past.
paleotopography *Geology.* the topography of a region at a specific time in the geologic past. Thus, **paleotopographical.**
paleotropical *Ecology.* belonging or relating to an area that stretches from Africa to Asia.
Paleotropical region *Ecology.* another name for the Ethiopian zoogeographical region.
paleovolcanology *Volcanology.* the study of volcanic activity during the geologic past.
Paleozoic [pā´lē ə zō´ik] *Geology.* **1.** a geologic era extending from the end of the Precambrian to the beginning of the Mesozoic, dating from about 600 to 230 million years ago. **2.** the rock strata formed during this era. Also, **Palaeozoic.**
paleozoology *Paleontology.* the study of fossil animals and their evolution. Thus, **paleozoological.**
palette *Graphic Arts.* the range of colors available (or used) for a particular artwork, publication, or the like. *Computer Technology.* the set of colors available for display on a video graphics monitor. *Geology.* a broad, flat, protruding ledge or sheet of calcite that represents a solutional remnant of the country rock in a cave. Also, SHIELD.
palevent *Geology.* a relatively sudden and brief geographic event that took place in the geologic past. Also, PALEOGEOGRAPHIC EVENT.
pale western cutworm *Invertebrate Zoology.* the larva of the noctuid moth *Agrotis orthogonia* that seriously damages grain, beet, potato, and other crops by feeding on the underground roots and stems; found in the western U.S. and Canada.
pali- or **palin-** a combining form meaning "again," as in *palingenesis.*
palikinesia *Neurology.* involuntary pathologic repetition of the same movement.
palila *Vertebrate Zoology.* an endangered species of honeycreeper, *Loxioides bailleui,* characterized by a thick, stubby bill, yellow head and breast, and gray back; found in Hawaii.
palilalia *Neurology.* a condition characterized by the repetition of a phrase or word with increasing rapidity.
palimpsest *Geology.* describing the structure or texture of a metamorphic rock in which remnants of an earlier structure or texture are preserved. *Hydrology.* of or relating to a type of drainage system in which a recent, anomalous drainage pattern has been superimposed on an older one. (From a Greek word meaning "rubbed again;" used to indicate a parchment or the like from which the writing has been erased to make room for another text.)
palindrome *Linguistics.* a word or group of words that reads the same backwards as forwards, such as Napoleon's supposed statment, "Able was I ere I saw Elba." *Genetics.* a nucleic acid sequence in which the base pairs read the same from each end on complementary strands, such as 5' AATGCGCATT 3' and 3' TTACGCGTAA 5'. Also, **palindromic sequence.** (From a Greek word meaning "recurring.")
palingenesis *Petrology.* the in situ formation of a new magma by melting or fusion of country rock or other rock material, without heat from an adjacent magma. *Biology.* the embryonic development that reproduces the ancestral features of the species in successive generations.
palinspastic map *Geology.* a geologic map that shows the features of a region in their original geographic positions, before the thrusting or folding of the crustal rocks took place.
Palinuridae *Invertebrate Zoology.* the spiny lobsters and crayfish, a family of large decapod crustaceans in the superfamily Scyllaridea, having no pincers on the legs and no appendages on the first abdominal segment.
palisade cell *Botany.* any of the chloroplast-containing elongated cells forming the layer below the upper epidermis of leaves.
Palisade disturbance *Geology.* a late Triassic episode of crustal deformation that supposedly produced a series of faultlike basins in the Appalachian area of eastern North America.
palisade mesophyll *Botany.* the layer of closely spaced palisade cells between the upper and lower epidermis of leaves that participates in photosynthesis. Also, **palisade, palisade layer, palisade parenchyma.**
palisades *Geology.* a series of rock cliffs that rise sharply from the margin of a stream, river, or lake, and commonly consist of basalt having a columnar structure.
Palladian *Architecture.* in the style of Andrea Palladio or his followers.
Palladian window *Architecture.* a window unit consisting of an arched central light flanked by two narrower rectangular lights.
Palladio, Andrea 1508–1580, Italian Renaissance architect and author; designed villas and palazzos near Vicenza.
palladium *Chemistry.* a metallic element having the symbol Pd, the atomic number 46, an atomic weight of 106.4, a melting point of 1552°C, and a boiling point of 2927°C; a silvery-white ductile metal. It is used as a base for dental alloys, electrical components, jewelry, and catalysts. (Named for the asteroid *Pallas.*)
pallanesthesia *Medicine.* the loss of the sense of vibration; insensibility to the vibrations of a tuning fork.
Pallas *Astronomy.* asteroid 2, discovered in 1802; it has a diameter of about 540 kilometers and belongs to type B.
pallasite *Geology.* **1.** a stony-iron meteorite consisting largely of glassy olivine crystals embedded in a nickel-iron matrix. **2.** an extremely basic meteoric or terrestrial rock containing more than 60% iron, if meteoric, or more iron oxides than silica, if terrestrial. Also, **pallas iron.**
Pallaviciniaceae *Botany.* a family of liverworts of the order Metzgeriales, characterized by a thallus of unistratose lamina having a thickened midrib and either prostrate rhizomes with erect bladeless stipes or closely prostrate thalli without stipes or rhizomes.
pallet *Engineering.* a tray or platform designed to be moved with a forklift, and on which stacked goods are placed. *Mechanical Engineering.* **1.** a lever, click, or pawl that regulates or drives a ratchet wheel. **2.** a disk or piston in a chain pump. *Building Engineering.* a flat piece of wood built into a mortar joint, providing a fixture to which joinery is nailed. *Horology.* in the escapement mechanism of a timepiece, one of two projections, attached to a pivot, that engages the teeth of the escape wheel, alternately stopping and releasing the wheel and transmitting its energy to the balance or pendulum. *Invertebrate Zoology.* one of a pair of shell-like plates on the siphon tubes of some bivalves.
palletized ship *Naval Architecture.* a ship fitted to carry and handle cargo on pallets.
pallet loading patterns *Design Engineering.* the resultant arrangement of products on a pallet in a predetermined manner by the use of an automatic or semiautomatic device that takes objects from a conveyor and positions them on a pallet.
pallet stone *Horology.* a pallet made out of a jewel or into which a jewel has been set to reduce wear and friction; usually referring to the pallet of a lever escapement.
palli- or **pallio-** a prefix meaning "cloak" or "mantle."
pallial artery *Invertebrate Zoology.* in lamellibranchs, one of a group of arteries that emanate from the aortic bulb and lead to the mantle.
pallial chamber see MANTLE CAVITY.
pallial line *Invertebrate Zoology.* in bivalves, the line of attachment between the mantle and the shell.
pallial nerve *Invertebrate Zoology.* in mollusks, one of a pair of large nerves that innervate the mantle.
pallial sinus *Invertebrate Zoology.* **1.** in bivalves, a sharp inward curve of the pallial line at the point of attachment of the siphonal retractor muscle. **2.** in brachiopods, any of the channels through which water flows in the mantle.

palliative [pal´ē ə tiv´] *Pharmacology.* **1.** of a medication or action, affording relief or reducing the severity of symptoms without curing the disease. **2.** an alleviating medicine or other agent.

pallium *Anatomy.* the cerebral cortex and associated white matter. *Invertebrate Zoology.* the mantle of a mollusk or brachiopod.

Pallopteridae *Invertebrate Zoology.* a family of dipteran flies in the subsection Acalyptratae.

pallor *Medicine.* an abnormal paleness, especially of the skin and mucous membranes.

palm *Anatomy.* the inner surface of the hand from the fingers to the wrist.

palm *Botany.* any of many species of the large family Arecaceae, most of which are slender monocotyledonous trees with an unbranched trunk and a germinal crown of large evergreen leaves; found worldwide in tropical or subtropical regions.

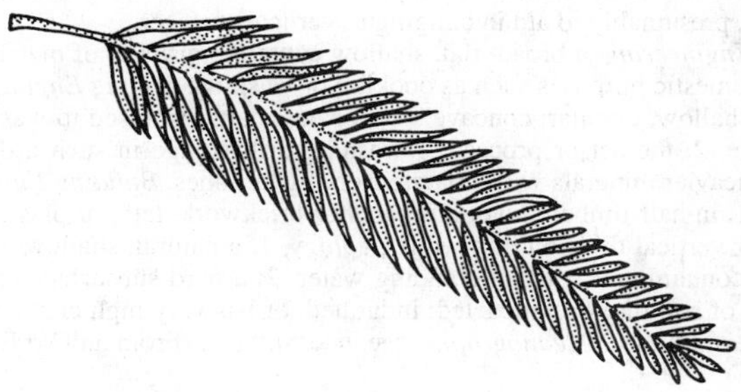

palm

Palmae see ARECACEAE.

Palmales see ARECALES.

palmar *Anatomy.* of or related to the palm.

palmar aponeurosis *Anatomy.* bundles of fibrous tissue radiating from the base of the palm to each finger.

Palmaria *Botany.* a genus of northern Atlantic, bladelike red algae of the family Palmariaceae, a commercial seaweed that is dried and eaten or fed to cattle.

Palmariaceae *Botany.* a family of red algae constituting the order Palmariales, characterized by multiaxial bladelike or tubular thalli and including the commercially important species *Palmaria palmata,* a seaweed used both for food and as a dietary supplement for cattle.

Palmariales *Botany.* an order in the red algae division Rhodophyta that is composed of the family Palmariaceae.

palmarosa oil *Materials.* an optically active, light yellow essential oil that consists primarily of geraniol; used in perfumes and flavoring.

palmar reflex *Physiology.* an inadvertent flexing of the fingers when the palm is lightly touched.

palmate *Zoology.* an organ or structure resembling a hand with extended fingers. *Vertebrate Zoology.* **1.** specifically, a feature of certain wading birds, such as ducks and geese, in which the feet have three front toes that are fully webbed with the hallux separate, resembling the palm of a hand. **2.** of a moose's antlers, having a broad, lobed, flat distal portion. *Botany.* shaped like a hand with the fingers outspread; having at least four equal divisions, lobes, or veins radiating from a common point; used to describe leaf structure or venation.

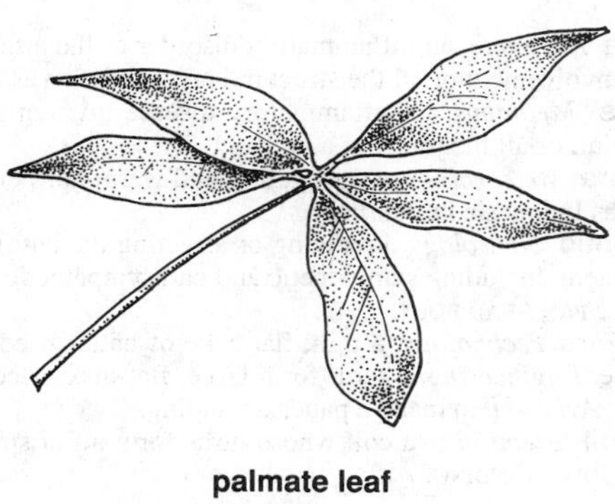

palmate leaf

palm butter *Materials.* a combustible, yellow-brown, buttery, edible solid at room temperature; used in soaps, pharmaceuticals, cosmetics, and as a softener in rubber processing.

palmchat *Vertebrate Zoology.* a small, noisy, olive-brown bird of the family Bombycillidae; noted for building an apartment-type, communal nest; usually occupies palms of the West Indian Islands.

palmella *Botany.* a genus of freshwater and terrestrial green algae of the family Palmellaceae, characteristically forming large masses of immobile cells in a gelatinous matrix.

palmelloid *Botany.* of or relating to a colony of nonmotile cells that divide and aggregate in a large gelatinous mass containing many generations, as the blue-green algae.

Palmellopsidaceae *Botany.* a family of the order Tetrasporales that encompasses those genera that form free or attached gelatinous colonies but lack pseudoflagella; native to various terrestrial, aubaerial, and freshwater habitats. Also, ASTEROCOCCACEAE.

Palmer Land *Geography.* a former name for the Antarctic Peninsula.

Palmer scan *Electronics.* a method by which a radar system scans both the azimuth and the elevation of a target area; a combination of a circular radar scan with a conical scan.

palmetto *Botany.* any of various palm trees of the genera *Sabal, Serenoa,* and *Thrinax,* characterized by fan-shaped leaves.

palmetto fiber *Botany.* a fiber from the young leaf stalks of a cabbage palm tree, *Sabal palmetto*; used in brushes, brooms, and weaving.

palmierite *Mineralogy.* $(K,Na)_2Pb(SO_4)_2$, a colorless to white trigonal mineral occurring as microscopic plates, having a specific gravity of 4.2 to 4.33 and an undetermined hardness; found in fumarole deposits on Mt. Vesuvius, Italy.

palmitate *Organic Chemistry.* an ester or salt of palmitic acid.

palmitic acid *Organic Chemistry.* $CH_3(CH_2)_{14}COOH$, combustible white crystals; insoluble in water and soluble in hot alcohol; melts at 62.9°C and boils at 215°C (15 torr); a saturated fatty acid derived from spermaceti and palm oil; used in soaps, waterproofing, and food additives. Also, HEXADECANOIC ACID.

palmitin *Organic Chemistry.* $C_{16}H_{98}O_6$, a white crystalline powder; soluble in water; prepared from glycerol and palmitic acid; used in the manufacture of soap.

palmitoleic acid *Organic Chemistry.* $C_{51}H_{30}O_2$, a combustible colorless liquid that is insoluble in water and soluble in alcohol; melts at 1.0°C and boils at 162°C (0.6 torr); an unsaturated fatty acid that is found in most fats, especially marine oils; used as a standard in chromatography.

palmityl alcohol see CETYL ALCOHOL.

palm nut *Botany.* the edible nut of the African oil palm that is cultivated as an important source of oil for margarine.

palm nut oil *Materials.* a yellow butterlike oil derived from the fruit of the oil palm; used as an edible fat and in making soap and candles. Also, **palm oil.**

Palmodictyaceae *Botany.* a family of green algae usually placed in the order Chlorococcales, characterized by spherical cells that live in a mucilaginous matrix or are attached to the tip of a dendroid stratified stalk and form filamentous colonies.

palmus *Neurology.* a rhythmic, spasmodic contraction of skeletal muscle, generally due to nerve irritation.

Palmyridae *Invertebrate Zoology.* a family of polychaete annelid worms in the division Errantia.

palometa *Vertebrate Zoology.* a pompano, *Trachinotus goodei,* characterized by long, tapering fins; found in tropical and temperate Atlantic seas.

palomino *Vertebrate Zoology.* a golden-colored horse with a white mane and tail, and often white markings on the face and legs; developed in the southwestern U.S. (A special use of Spanish meaning "resembling a dove.")

Palouse *Agriculture.* a medium-sized breed of hog having a white coat; crossbred for meat production from the Landrace and the Chester White breeds. (From the *Palouse* River in the western U.S.)

palouser *Meteorology.* a dust storm that occurs in the northwestern region of Labrador.

palp *Invertebrate Zoology.* any sensory appendage on or near the head.

palpable *Medicine.* **1.** perceptible to the touch or capable of being touched or palpated. **2.** of a condition, evident; plain.

palpable coordinate *Mechanics.* a generalized coordinate that explicitly appears in the Lagrangian of a system.

palpal organ *Invertebrate Zoology.* in male spiders, the tip of a pedipalp modified to convey sperm to the female.

palpate *Medicine.* to examine by touch; carry out a process of palpation. *Invertebrate Zoology.* relating to or having a palp or palps.

palpation *Medicine.* an examination by touch for the purposes of diagnosis, used to detect characteristics and conditions of local tissues or of underlying organs or tumors.

palpebra *Anatomy.* a technical name for the eyelid.

palpebral disk *Vertebrate Zoology.* a transparent, scaly disk that covers the eyelid of some lizards.

palpebral fissure *Anatomy.* the elliptical space between the margins of the eyelids when they are open.

palpebral fold see CONJUNCTIVAL FOLD.

Palpicornia *Invertebrate Zoology.* a superfamily of aquatic beetles, coleopteran insects in the suborder Polyphaga, including the moss beetle and water scavenger beetle.

palpiger *Invertebrate Zoology.* an exoskeletal plate on the jaw of an insect, bearing a labial palp.

Palpigradida *Invertebrate Zoology.* an order of tiny, whitish arachnids characterized by a long, segmented tail and no eyes; found in soil or under rocks in warm climates. Also PALPIGRADI, PALPIGRADA.

palpitate *Medicine.* to pulsate with abnormal rapidity; used specifically of a rapid increase in heartbeat, with or without irregularities in rhythm.

palpitation *Cardiology.* a patient's own sensation of an irregular or accelerated heartbeat.

palpocil *Invertebrate Zoology.* in some hydroids, a filamentous tactile process.

Palptores see PHALANGIDA.

palpus *plural,* **palpi.** *Invertebrate Zoology.* a sensory organ on or near the head; a palp.

palsa *Geology.* a mound of earth that is pushed up by or formed near the edge of a glacier; found in alpine and arctic areas.

palstage *Geology.* an interval of time in the geologic past when geographic conditions were more or less static, or were changing slowly and progressively with respect to factors such as sea level, surface relief, or distance from shore. Also, PALEOGEOGRAPHIC STAGE.

palsy [pôl´zē] *Medicine.* a general term for paralysis; specific forms of palsy include Bell's palsy, cerebral palsy, or Erb's palsy.

PAL system See PHASE-ALTERNATION LINE SYSTEM.

paludification *Ecology.* a process by which a peat bog impedes drainage and expands its size.

Palus Somnii *Astronomy.* the Marsh of Sleep, a dark area on the west side of the moon's Mare Tranquillitatis.

palustrine *Ecology.* of or relating to a marshy habitat. Also, **palustral.**

palygorskite *Mineralogy.* $(Mg,Al)_2Si_4O_{10}(OH){\cdot}4H_2O$, a soft but tough, white to gray monoclinic and orthorhombic mineral usually occurring in leatherlike sheets of minute interwoven fibers, having a specific gravity of 2.217 and an undetermined hardness; found in altered granitic and serpentine rocks, and in hydrothermal veins.

palynofacies *Paleontology.* generally, the organic residue in a sedimentary rock; more specifically, an assemblage of palynomorph taxa in a sediment.

palynology *Paleontology.* the branch of micropaleontology concerned with spores, pollen grains, and other palynomorphs, whether fossil or of the present.

palynomorph *Paleontology.* an independent, organic-walled microfossil, mainly from fossil pollen or spores but consisting of chitin and pseudochitin as well as sporopollenin, and thus partially originating from animals as well as plants.

palynostratigraphy *Paleontology.* the study of the formation and sequence of geological strata by analyzing pollen and other palynomorphs found in the strata; spores and pollen grains are especially important in correlating marine and nonmarine sediments, since they can be carried great distances by wind before being deposited.

palytoxin *Biochemistry.* $C_{129}H_{223}N_3O_{54}$, a toxin present in a number of *Palythoa* species, believed to be one of the most deadly substances known.

PAM pulse-amplitude modulation.

pamabrom *Pharmacology.* $C_{11}H_{18}BrN_5O_3$, a derivative of theophylline occurring as a fine white powder; formerly used to promote or increase the production of urine.

pamaquine naphthoate *Pharmacology.* $C_{42}H_{45}N_3O_7$, a drug that destroys the malarial parasite in the body; now largely replaced by primaquine, which is less toxic to humans.

Pamirs or **Pamir** *Geography.* a region of high mountains in the south central USSR and adjacent parts of northern Afghanistan, northern Pakistan, and western China.

Pampas [pam´pəz; päm´pəz] *Ecology.* usually, **the Pampas.** an extensive grassy plain in South America, occupying most of the interior of Argentina and adjacent areas, and similar to the prairies of central North America. Also, **Pampa, the Pampa.** (From a local Indian word.)

pampero *Meteorology.* a cold and dry southwesterly wind that blows across the Pampas of Argentina and Uruguay, following a deepening barometric depression; often accompanied by squalls, thundershowers, and a rapid decline of temperature; similar to a norther of the U.S. Plains region.

Pamphiliidae *Invertebrate Zoology.* a family of sawflies, the leafrollers or web-spinning sawflies, hymenopteran insects in the superfamily Megalodontoidea; larvae are leaf-eating tree pests that live in leaves stuck together with a web.

pampiniform *Anatomy.* shaped like a tendril.

pampiniform plexus *Anatomy.* a convoluted array of blood vessels formed by the union of veins from the testis and epididymis.

pamprodactylous *Vertebrate Zoology.* a condition in certain birds (especially common in some swifts) in which all four toes are directed forward, presumably to aid in clinging to vertical surfaces.

pan *Engineering.* a broad, flat, shallow container, usually of metal; used for domestic purposes such as cooking or washing. *Mining Engineering.* **1.** a shallow, circular, concave steel or porcelain dish used to wash drill sludge. **2.** the act or process of washing drill sludge in such a dish so that heavier minerals will separate from lighter ones. *Building Engineering.* **1.** in half-timbered work, a panel of brickwork, lath, or plaster. **2.** a major vertical division of a wall. *Geology.* **1.** a natural, shallow depression containing a body of standing water. **2.** a hard subsurface crust or layer of soil that is compacted, indurated, or has very high clay content. **3.** see SALT PAN. *Oceanography.* see PANCAKE ICE. (From an Old English word.)

pan *Telecommunications.* **1.** to move a motion-picture or television camera horizontally or vertically for the purpose of following a moving object, or to establish a wide view of a particular scene. **2.** a camera shot resulting from this. (Short for <u>pan</u>orama or <u>pan</u>oramic view.)

pan- a prefix meaning "all."

panacinar emphysema *Medicine.* a form of emphysema affecting all lung areas, with dilatation and atrophy of the alveoli and destruction of the vascular bed of the lung.

pan adapter see PANORAMIC ADAPTER.

Panadol *Pharmacology.* a brand name for acetaminophen, a widely marketed pain reliever. See ACETAMINOPHEN.

panagglutination *Immunology.* the nonspecific or broadly specific clumping together of certain antigens and antibodies in the blood.

panagglutinin *Immunology.* a nonspecific antibody that agglutinates with a wide variety of antigens in the blood.

Panagrolaimoidus *Invertebrate Zoology.* a genus of free-living, microscopic nematode worms.

Panama *Agriculture.* a large, hardy breed of white-faced, hornless sheep that produces large fleeces of white wool, crossbred in the United States from the Lincoln and Rambouillet breeds.

Panama, Gulf of *Geography.* a wide inlet of the Pacific formed by the bend in the southern coast of Panama.

Panama Canal *Geography.* a shipping canal across Panama, connecting the Atlantic and Pacific oceans.

Panama disease *Plant Pathology.* a disease of banana plants caused by the fungus *Fusarium oxysporum cubense*, which attacks the vascular system and causes foliage to wilt, yellow, and die.

pan-amalgamation process *Mining Engineering.* a method of gold and silver recovery in which crushed ore is mixed with salt, copper sulfate, and mercury; the released silver and gold amalgamize with the mercury.

panarteritis *Medicine.* an inflammatory disorder of the arteries, characterized by involvement of all the structural layers of the vessels.

panarthritis *Medicine.* an inflammatory disease involving all the tissues of a joint, or all the joints of a body.

panas oetara *Meteorology.* a strong, dry, and warm northerly wind that blows across Indonesia in February.

panautonomic *Neurology.* involving or affecting the entire autonomic nervous system, including sympathetic and parasympathetic systems.

pan bolt see PAN HEAD BOLT.

pancake *Food Technology.* a thin, flat cake of batter fried in a pan or on a griddle. *Engineering.* a term for a large, flat disk of concrete used for support. *Aviation.* to make a pancake landing.

pancake coil *Electricity.* a coil whose turns form a flat spiral arrangement. Also, DISK WINDING.

pancake engine *Mechanical Engineering.* a compact engine in which the cylinders are positioned one above the other, radially, horizontally, or side by side, as in an airplane wing.

pancake forging *Metallurgy.* any rough, flat forging.

pancake ice *Oceanography.* pieces of newly formed sea ice between 0.3 and 2.5 meters across, shaped by constant collisions in rough water into round, flat shapes with slightly raised edges.

pancake landing *Aviation.* an airplane landing in which the craft is leveled and stalled high above a surface, resulting in a steep, rapid descent with a forceful surface impact.

pancake theory *Astrophysics.* a theory proposing that very large clouds of gas condensed early in the universe's history and then were flattened by rotation to form the first superclusters of galaxies, and then the galaxies themselves.

pancarditis *Medicine.* a diffuse inflammation of the heart, involving the endocardium, pericardium, and myocardium.

Pancarida *Invertebrate Zoology.* a superorder of tiny malacostracan crustaceans having cylindrical bodies and brood pouches; found in underground waters, including salt springs and hot springs.

Panchaia *Astronomy.* a north temperate belt that lies to the east of Utopia and due north of Trivium Charontis on the surface of Mars; it appears as a dark region when viewed telescopically from the earth.

panclimax *Ecology.* an area including two or more fully evolved communities that have similar climates and wildlife.

pancreas [pang´krē əs] *Anatomy.* a large, elongated gland located behind the stomach, between the spleen and the duodenum. The endocrine part of the pancreas, consisting of small clusters of special cells called the islets of Langerhans, produces and secretes insulin and glucagon directly into the bloodstream; the exocrine part, consisting of secretory units called **pancreatic acini,** produces and secretes digestive enzymes directly into the duodenum.

pancreatectomy *Medicine.* the surgical removal of all or part of the pancreas.

pancreatic [pang´krē at´ik] *Anatomy.* of or relating to the pancreas. *Medicine.* involving or affecting the pancreas. Thus, **pancreatic cancer.**

pancreatic diarrhea *Medicine.* diarrhea caused by a deficiency of pancreatic digestive enzymes and characterized by the passage of stools having a high fat and nitrogen content.

pancreatic diverticulum *Developmental Biology.* one of two evaginations from the embryonic duodenum that later form the pancreas and its ducts.

pancreatic duct *Anatomy.* a tube through which pancreatic secretions pass from the pancreas to the small intestine. Also, WIRSUNG'S DUCT.

pancreatic islets *Endocrinology.* the islets of Langerhans, the irregular microscopic masses of cells that constitute the endocrine pancreas. Of the three cell types that form the islets, the α-cells produce glucagon, the β-cells produce insulin, and the δ-cells produce somatostatin.

pancreatic juice *Physiology.* a clear fluid product of the pancreas, carrying various enzymes.

pancreatic polypeptide *Endocrinology.* a peptide secreted by cells within the pancreas and duodenum; its function is not fully understood.

pancreatin *Biochemistry.* a mixture of enzymes, such as amylase, protease, and lipase, derived from the pancreatic juice of swine and beef cattle; used as a digestive aid in pancreatic deficiencies to allow the metabolization of starches, proteins, and fats.

pancreatitis *Medicine.* an acute or chronic inflammation of the pancreas, often associated with alcoholism or biliary tract disease, and marked by severe abdominal pain, nausea, and fever.

pancreatography *Radiology.* a radiologic study of the pancreas performed during surgical exploration, in which a water-soluble contrast medium is injected into the pancreatic duct and the film made while the abdomen is open.

pancreatolith *Medicine.* a pancreatic stone.

pancreatolithectomy *Surgery.* the surgical removal of a stone from the pancreas.

pancreatolithotomy *Surgery.* a surgical invasion of the pancreas for removal of a calculus. Also, PANCREOLITHOTOMY.

pancreatolysis *Medicine.* the destruction of pancreatic tissue.

pancreatopathy *Medicine.* any disease of the pancreas.

pancreatotomy *Surgery.* an incision of the pancreas. Also, **pancreatomy.**

pancreozymin see CHOLECYSTOKININ.

pancytopenia *Medicine.* an abnormal condition marked by a deficiency of all cell elements in the blood, including erythrocytes, white blood cells, and platelets.

panda

panda *Vertebrate Zoology.* **1.** a large, bearlike mammal, *Ailuropoda melanoleuca*, of the family Ailuropodidae, characterized by a heavy body covered with a thick, woolly coat that is white with black limbs, shoulders, ears, and patches around the eyes; found in mountainous forests near bamboo in China. Also, GIANT PANDA. **2.** a raccoonlike arboreal mammal, *Ailurus fulgens*, of the family Procyonidae, characterized by a long body and a bushy, ringed tail; found in mountainous forests near bamboo in Nepal, Bhutan, Sikkim, and China. Also, LESSER PANDA; RED PANDA; BEAR CAT. (From a native name used for the lesser panda in Nepal.)

Pandaceae *Botany.* a family of dicotyledonous dioecious trees and shrubs of the order Euphorbiales, characterized by simple distichous leaves arranged in flat sprays on lateral branches of limited growth; native to tropical Africa, Asia, and New Guinea.

Pandanaceae *Botany.* the family of trees, shrubs, and woody climbers composing the order Pandanales.

Pandanales *Botany.* a monofamilial order of monocotyledonous Old World trees, shrubs, and woody climbers in the subclass Arecidae, characterized by long and narrow parallel-veined leaves with spiny margins and often having stout prop roots.

Pandaridae *Invertebrate Zoology.* copepod crustaceans with a sucking mouth in the suborder Caligoida; ectoparasites of sharks.

pandemic *Medicine.* **1.** describing a widespread epidemic of a disease, occurring throughout the population of a country, a people, or the world. **2.** a disease of this type.

pandemonium model *Artificial Intelligence.* a model of human pattern recognition in which a collection of "demons" observe input data; a demon "squeaks" according to how well the data match its pattern; the demon that squeaks the loudest is accepted as recognizing the pattern.

Panderodontidae *Paleontology.* an important early family of Ordovician conodonts in the proposed order Panderodontida, whose teeth are generally simple, longitudinally grooved, and horn-shaped; Ordovician and Silurian.

Pandionidae *Vertebrate Zoology.* the osprey, a monotypic hawk family of the order Falconiformes; a moderately large, predatory bird characterized by plumage that is dark brown on top and white below; found worldwide, but most generally around coastlines, rivers, and lakes where it hunts for fish.

Pandora *Astronomy.* one of the two shepherd satellites of Saturn's F ring, about 100 kilometers in diameter; the other shepherd satellite is Prometheus.

Pandorae Fretum *Astronomy.* a dark feature in the southern tropics on Mars that connects Mare Serpentis and Mare Erythraeum and is an indicator of seasonal changes on the Martian surface.

pandurate *Botany.* of a leaf, fiddle-shaped.

p and v *Computer Science.* wait and signal; used in operations on semaphores. (From the Dutch phrase for "wait and signal.")

pane *Building Engineering.* a single plate of glass in a door or window. *Mechanical Devices.* **1.** one of the sides of a nut or bolt head. **2.** the narrow end of a hammer head. Also, PEEN.

panel *Building Engineering.* **1.** a distinct section or portion of a wall, ceiling, door, or other construction surface, usually a flat, rectangular area that is raised above or sunk below the surrounding area. **2.** to cover or provide an area with such sections or portions. *Materials.* a manufactured sheet of wood-based product that is available in standard sizes. *Engineering.* a flat surface or section containing instruments or controls for a device, machine, or vehicle, such as an aircraft. *Electronics.* a surface of this type serving as a switchboard or control board, including such elements as automatic overcurrent devices, switches for light, heat, or power control, and buses. Also, PANELBOARD. *Civil Engineering.* **1.** a covering that can be installed or removed as a unit. **2.** a slender, precast concrete member. **3.** the portion of a lattice girder between adjacent vertical struts. *Aviation.* one complete section of the wing of an aircraft, one section of a parachute canopy or aircraft skin, or a distinct, rectangular section of an aeronautical structure. *Mining Engineering.* a large portion of coal in the form of a rectangular block or pillar. *Computer Technology.* **1.** a predefined display image. **2.** see BOARD.

panelboard *Materials Science.* a compact paperboard used for paneling in walls, cabinets, and so on. *Electronics.* see PANEL.

panel code *Telecommunications.* a specified code that allows communication between ground units and aircraft.

panel coil see PLATE COIL.

panel cooling *Civil Engineering.* a cooling system utilizing floor, wall, or roof panels as cooling elements.

panel data see LONGITUDINAL DATA.

panel display *Electronics.* a method for presenting color-television pictures in which luminescent devices, such as light-emitting diodes, are controlled by signals transmitted by vertical and horizontal wires attached to their electrodes.

panel heating *Civil Engineering.* a heating system utilizing floor, wall, or roof panels containing electric coils or pipes that conduct hot air, water, or steam.

paneling *Building Engineering.* **1.** wood, plywood, or other material used for panels. **2.** a section or surface of panels.

panel length *Civil Engineering.* a set measurement for certain types of panels, particularly plywood.

panel point *Civil Engineering.* the point at which two members of a truss chord cross.

panel saw *Mechanical Devices.* a small handsaw with a long, tapering blade and filed teeth; used to finish cut panels, such as hardboard, plywood, or particle board.

panel strip *Building Engineering.* one of several wooden or metal strips laid on a wall or ceiling to conceal joints between composition material.

panel system *Building Engineering.* a wall in which individual factory-assembled units are anchored to the building frame and to one another.

panel truck *Mechanical Engineering.* a small, enclosed truck or van with side panels, typically used to carry light loads or as a delivery vehicle.

panel wall *Building Engineering.* a nonload-bearing partition between columns or piers.

panencephalitis *Neurology.* encephalitis, probably caused by a virus, that affects the gray and white matter of the brain simultaneously.

panendoscope *Medicine.* a cystoscope that permits visualization of the entire interior of the bladder.

panethite *Mineralogy.* $(Na,Ca,K)_2(Mg,Fe^{+2},Mn^{+2})_2(PO_4)_2$, an amber-colored, transparent monoclinic mineral, having a specific gravity of 2.9 to 3.0 and an undetermined hardness; found as minute grains only in the Dayton meteorite.

Paneth's (adsorption) rule *Physical Chemistry.* a rule stating that an element will be adsorbed on a solid substance with the opposite electrical charge, provided the resulting compound is relatively insoluble in the solvent.

panfan see PEDIPLAIN.

panga panga see WENGE.

Pangea or **Pangaea** *Geology.* a hypothetical protocontinent, consisting of all the continental crust of the earth that supposedly fragmented millions of years ago into the supercontinents of Laurasia and Gondwana from which the present-day continents were derived.

pangenesis *Cell Biology.* a theory of the inheritance of acquired characteristics developed by Charles Darwin and now considered obsolete; it postulated a hypothetical particle, a **pangene,** that arises from the somatic cells throughout the body, collects in the gametes, and transmits a blending of parental characteristics to the offspring.

pangenetic *Cell Biology.* of or relating to the theory of pangenesis.

pangolin [pan gō´lin; pang´gə lin] *Vertebrate Zoology.* a nocturnal, insectivorous mammal of the genus *Manis* and related genera of the order Pholidota, characterized by a long tail and body covered by large, armorlike, overlapping scales; some are terrestrial, some arboreal, and all are found in forests, thick bushes, or open savannah in Africa, India, Southeast Asia, and surrounding islands. Also, SCALY ANTEATER.

pangolin

panhead bolt *Mechanical Devices.* a rivet or bolt with a head shaped like a truncated cone. Also, **panhead rivet.**

panic *Psychology.* a sudden feeling of extreme fear, especially one that produces hysterical or irrational reactions. *Behavior.* flight behavior that is induced by fear. (From *Pan,* the ancient Greek god of woods and fields; it was thought that he had the power to affect people and animals in this way.)

panic attack *Behavior.* a period of sudden, intense anxiety that may last several hours but is usually shorter.

panic disorder *Psychology.* an anxiety state characterized by repeated and unpredictable panic attacks, which are usually short in duration but may be quite severe.

panicle *Botany.* **1.** a branched raceme with each branch also bearing a cluster of flowers, as on an oat plant. **2.** any branching flower cluster.

Panidae *Anthropology.* in most classification schemes, the family name for the two species of chimpanzee and the gorilla, placed within the superfamily Hominoidea.

panidiomorphic *Geology.* of an igneous rock or texture in which the constituent mineral crystals are bounded by their own crystal faces.

panmictic *Ecology.* of or relating to a population that breeds randomly.

panmictic unit see DEME.

panmixis *Ecology.* a completely random mating within a population.

Panmycin *Microbiology.* a trade name for the antibiotic tetracycline.

panniculectomy *Surgery.* a surgical procedure to remove superficial fat in the abdominal apron of an obese individual.

panniculitis *Medicine.* an inflammation of the panniculus adiposus marked by recurring cycles of fever.

panniculus *plural,* **panniculi.** *Anatomy.* a thin, sheathlike membrane or layer of tissue.

panniculus adiposus *Anatomy.* the layer of fat underlying the dermis.

panniculus carnosus *Anatomy.* a thin, sheathlike layer of muscular fibers just beneath the skin; well-developed in animals with a hairy coat, in man represented primarily by the platysma.

pannier see GABION.

panning technique *Biotechnology.* a technique that uses affinity chromatography to isolate or purify certain cell types.

Pannonian *Geology.* a European geologic stage of the Lower Pliocene epoch.

pannose *Biology.* having the texture or appearance of felt or woolen cloth.

pannus *Medicine.* **1.** an abnormal condition marked by vascularization and connective-tissue deposition beneath the cornea. **2.** an inflammatory overgrowth of tissues overlying the lining layer of synovial cells on the inside of a joint. *Meteorology.* a group of cloud fragments beneath a main cloud system, with a ragged or shredded aspect, either attached or separated from the main part of the cloud.

panophthalmitis *Medicine.* an inflammatory disorder affecting all the tissues of the eye, causing swelling and discomfort.

panoramic *Optics.* of or relating to a system or an instrument with a wide field of view, especially one that is capable of producing an image of 360° or close to 360°.

panoramic adapter *Electronics.* a device that can be attached to a search receiver to provide it with a visual presentation of the band of frequencies extending above and below the center frequency to which the search receiver is tuned.

panoramic camera *Photogrammetry.* a camera in which a complete or nearly complete 360° view of the terrain is obtained during a single exposure, either by means of a lens that revolves about an axis, or by a clockwork system that revolves the whole camera about an axis.
panoramic display *Electronics.* a device that presents the relative amplitudes of all signals received at various frequencies.
panoramic distortion *Photogrammetry.* a displacement of the image produced by a panoramic camera as a result of the cylindrical shape of the film and the scanning motion of the lens.
panoramic radar *Engineering.* nonscanning radar that transmits signals over a wide area in only one general direction.
panoramic receiver *Electronics.* a radio receiver that displays on a cathode-ray-tube screen the presence of all signals and their relative strengths across a wide frequency band.
panoramic sight *Ordnance.* a cannon sight that provides the gunner with a wide field of vision.
pan-pan-pan *Navigation.* the international voice-radio signal for an urgent message.
panphobia or **panophobia** *Psychology.* a fear of everything or all things. Also, **pantophobia.**
panplain *Geology.* **1.** an extensive plain formed by the merging of floodplains of adjacent streams. **2.** a level plain having a general inclination toward the sea.
panplanation *Geology.* a process or action by which a panplain is formed and develops.
pan-range *Electronics.* a radar detection method in which stationary targets appear as solid vertical deflections and moving targets appear as broken vertical deflections.
panspermia *Biology.* a 19th-century biogenentic theory proposing that the universe is full of minute germs or spores that develop upon finding a favorable environment.
pansporoblast *Invertebrate Zoology.* a sporont in some parasitic protozoans that produces two or more sporoblasts.
pansy *Botany.* a widely cultivated garden plant, *Viola tricolor,* having bright, showy flowers in a variety of colors.
pant- or **panto-** a prefix meaning "all" or "the whole."
pantagraph design *Robotics.* a type of robotic architecture in which all movement is in the vertical plane.
pantellerite *Petrology.* a green to black extrusive igneous rock, or rock group, essentially comprising a sodic quartz trachyte or peralkaline ryolite and containing diopside or acmite-augite, anorthoclase, quartz, and cossyrite phenocrysts in a glassy groundmass of acmite or feldspar.
Panthalassa *Geology.* the hypothetical proto-ocean that supposedly contained all the earth's oceans or oceanic crust, and that surrounded the protocontinent of Pangea.
pantheism *Anthropology.* a system of belief that identifies God or various gods as being present in all natural features, forces, and events.
pantheist *Anthropology.* **1.** relating to or following pantheism as a system of belief. **2.** a person or society that follows pantheism.
panther *Vertebrate Zoology.* a popular name for various large animals of the cat family, such as the mountain lion, *Felis concolor,* or the leopard, *Panthera pardus,* especially a leopard in its black color phase.
Panther *Military Science.* a popular name given to various military vehicles and aircraft, such as a German heavy tank of World War II, the U.S. F9F single-engine jet fighter, or a current U.S.-built light tactical helicopter.

Panther helicopter

panther fungus *Mycology.* a poisonous mushroom, *Amanita pantherina,* characterized by a brown cap covered with white cottony patches. Also, **panther amanita.**
panting *Physiology.* a process of heavy or labored breathing. *Naval Architecture.* **1.** a rhythmic pulsation of the forward shell plating, caused by pressure of seawater against the hull. **2.** a similar pulsation in the plating due to combustion in the boiler, usually indicating insufficient air.
panting beam *Naval Architecture.* a reinforcing girder intended to reduce panting of the shell plating. Similarly, **panting frame, panting stringer.**
Pantodonta *Paleontology.* an archaic order of piglike eutherian mammals in the superorder Paenungulata; early forms were dog-sized, but later forms, such as Barylambda, grew to 2–3 meters in length; holarctic distribution; extant in the Paleocene to Oligocene.
Pantodontidae *Vertebrate Zoology.* the butterflyfish, a monotypic family of surface-living freshwater fish of the order Osteoglossiformes, capable of jumping short distances and found in tropical West Africa.
pantograph *Engineering.* **1.** a device that is mounted on top of an electronic locomotive to transfer electricity from overhead wires. **2.** a graphical parallelogram-shaped device for copying an existing drawing or pattern.
pantography *Engineering.* the use of a system that sends and transcribes radar signals from an indicator to a remote receiver.
Pantolambdidae *Paleontology.* a two-genus family of short-limbed, sheep-sized ungulates in the extinct order Pantodonta; known only from the Upper Paleocene of North America.
Pantolambdodontidae *Paleontology.* a two-genus family of ungulates in the extinct order Pantodonta; known only from the Eocene of Eastern Asia.
Pantolambdoidea *Paleontology.* a proposed superfamily of ungulates in the order Pantodonta; extant in the Paleocene and Eocene.
Pantolestidae *Paleontology.* a family of semiaquatic protoeutherian mammals in the order Pantolesta that resembled seals and otters in some ways, although their exact affinities are unclear; holarctic distribution; extant in the Middle Paleocene to Lower Oligocene.
pantometer *Engineering.* an instrument used to measure angles, elevation, and distances.
pantophagous *Zoology.* eating a diet of both plant and animal matter; omnivorous.
pantophthalmidae *Invertebrate Zoology.* the wood-boring flies, a family of dipteran insects in the series Brachycera.
Pantopoda see PYCNOGONIDA.
pantothenate *Biochemistry.* a salt or ester of pantothenic acid.
pantothenic acid *Biochemistry.* $C_9H_{17}NO_5$, a vitamin constituent of coenzyme A, found in all human cells, although signs of deficiency do not appear in humans. Also, B_3.
Pantotheria *Paleontology.* a group of Jurassic mammals that are probably ancestral to the primitive marsupial and placental mammals; the pantotheres were shrewlike, egg-laying insectivores; affinities are uncertain because the fossil record is incomplete for this group.
pantropic or **pantropical** *Ecology.* found throughout the tropics; widespread in tropical regions. *Virology.* of a virus, capable of infecting a wide range of cells.
panuveitis *Medicine.* an inflammation affecting the entire uveal tract.
Panzerfaust *Ordnance.* a simple German hand-fired antitank rocket used during World War II, propelling a hollow charge bomb from a recoilless tube launcher.
panzootic *Virology.* of a virus, affecting many animals of different species.
Pap or **Pap.** Papanicolaou test or stain. Also, **pap.**
papagayo [päp′ə gī′yō] *Meteorology.* a violent, northeasterly fall wind that occurs along the Pacific coast of Nicaragua and Guatamala; characterized by the cold air mass of a norté which ascends over the Central American mountains and descends to the interior valley bringing fine, clear weather.
papain *Biochemistry.* an enzyme that breaks peptide bonds; derived from the latex of the papaya.
Papanicolaou, George [päp′ə nik′ə lou; päp′ə nēk′ə lou] 1883–1962; Greek-born American physiologist; developed the Papanicolaou test (Pap smear).
Papanicolaou stain *Chemistry.* any of a group of stains used to color exfoliated cells for diagnostic purposes, as in the Papanicolaou test (Pap smear).
Papanicolaou test see PAP SMEAR.

Papaveraceae *Botany.* a family of dicotyledonous herbs and softwooded shrubs, including the poppy, of the order Papaverales; characterized by alternate lobed or dissected leaves and large, perfect, regular flowers with many stamens; generally found in the Northern Hemisphere.

Papaverales *Botany.* an order of dicotyledonous herbs and softwood shrubs belonging to the subclass Magnoliidae, noted for producing isoquinoline alkaloids in lactifers or secretory cells and marked by often large and showy perfect flowers.

papaverine *Organic Chemistry.* $C_{20}H_{21}O_4N$, toxic, narcotic, white powdery crystals; insoluble in water, slightly soluble in alcohol, and soluble in chloroform; melts at 147°C; an alkaloid derived from opium or synthetically; used as a vasodilator in the treatment of hypertension.

papaya [pə pī´yə] *Botany.* **1.** a tropical American plant, *Carica papaya,* that is widely grown for its large, sweet, melonlike fruit. **2.** the fruit itself, eaten fresh or used as a source of juice.

paper *Materials.* **1.** a common material made of fibers such as wood pulp or rags that have been laid on a fine screen in liquid suspension; used as a medium for writing and printing, as an absorbent, for packaging or wrapping, and for many other purposes. **2.** a single sheet of such a material. **3.** of or relating to such a material. (From the *papyrus* plant, a noted source of writing material in ancient Egypt.)

paperback *Graphic Arts.* **1.** a book that is bound with a flexible paper cover, usually smaller and cheaper than a typical hardcover book and often a reprinting of the hardcover edition of a book. **2.** of or relating to such books.

paper birch *Botany.* the birch tree *Betula papyrifera,* characterized by a tough bark and yielding a valuable wood; it is found in North America, and is the state tree of New Hampshire. Also, CANOE BIRCH.

paperboard *Materials.* **1.** a thick, stiff cardboard composed of layers of paper or paper pulp that has been compressed. Also, PASTEBOARD. **2.** relating to or made of paperboard.

paper capacitor *Electricity.* a fixed capacitor consisting of two strips of metal foil that are separated by an insulating material and rolled together into a compact tube.

paper chromatography *Analytical Chemistry.* a separation technique in which a substance is placed at one end of a strip of paper that is then dipped in a solvent which carries each constituent to a different place along the paper; after drying, the spots are compared to standards.

paper cutter *Mechanical Devices.* a horizontal platform with adjustable guides on the end, a ruler perpendicular to the guides along the edge, and an attached vertical blade; used for cutting or trimming sheets of paper to required dimensions.

paper electrochromatography *Analytical Chemistry.* a separation technique in which a substance is placed on an electrolyte-impregnated strip of paper and a current is passed across the width of the strip, at right angles to the downward movement of the sample.

paper electrophoresis *Analytical Chemistry.* a separation technique in which a substance is placed on an electrolyte-impregnated strip of paper and a current is passed along the length of the strip in the direction of the sample flow.

paper machine *Mechanical Engineering.* a synchronized system of mechanical devices in which cellulose fibers are formed, pressed between rollers, dried, smoothed, wound on reels, cut into widths, and finally wound on rolls or cut into sheets.

paper-patched bullet *Ordnance.* an old type of lead bullet that was wrapped in thin paper to produce a tighter fit in a rifled barrel, thus improving accuracy; the paper also served as a gas seal and reduced lead fouling.

paper shale *Geology.* a variety of shale that, when weathered, splits easily into very thin, uniform layers that resemble sheets of paper.

paper spar *Geology.* a crystallized variety of calcite characterized by thin layers or paperlike plates.

paper tape *Computer Programming.* a strip of paper into which holes representing data are punched. Thus, **paper-tape unit, paper-tape reader, paper-tape code,** and so on.

paper-tape Turing machine *Computer Programming.* a variation of the Turing machine in which blank squares can be permanently imprinted with a symbol.

paper throw *Computer Technology.* the accelerated movement of paper through a printer more than one line space without printing.

paper wasp *Invertebrate Zoology.* the Vespidae, social wasps that build large underground or hanging spherical nests of chewed wood pulp resembling paper, with large numbers of cells into which their eggs are laid.

papier mâché [pā´pər mə shā´] *Materials.* paper or paper pulp that has been mixed with glue and other additives to form a material that is easily molded when damp, and that becomes hard and durable when dry; widely used to make art objects and other articles. (A French term literally meaning "chewed paper.")

Papilionidae *Invertebrate Zoology.* the swallowtails, a family of common butterflies, large, bright-colored lepidopteran insects in the superfamily Papilionoidea, having tail-like projections on the hindwings.

Papilionidae

Papilionoidea *Invertebrate Zoology.* butterflies, a superfamily of four-winged, diurnal, often bright-colored insects in the order Lepidoptera, having round-tipped clubbed antennae.

Papilionoideae *Botany.* a hardy cosmopolitan subfamily of the family Leguminosae that contains many of the familiar legumes, including the pea, and having butterflylike flowers with a large upper petal, two lateral petals, and two narrow lower petals.

papilla [pə pil´ə] *plural,* **papillae.** *Biology.* **1.** one of various small nipplelike protuberances that are concerned with the senses of touch, taste, and smell. **2.** a small vascular process at the root of the hair. **3.** a papule or pimple.

papillary [pap´ə lâr´ē; pə pil´ə rē] *Biology.* **1.** relating to or resembling a papilla or papillae. **2.** bearing or covered with papillae. Also, PAPILLATE.

papillary carcinoma *Oncology.* a type of carcinoma with fingerlike outgrowths, commonly seen in transitional-cell carcinoma of the urinary tract.

papillary muscle *Anatomy.* one of several cone-shaped muscles in the ventricles of the heart that connect the chordae tendineae of the atrioventricular valves to the ventricle wall.

papillary tumor see PAPILLOMA.

papillate see PAPILLARY.

papillectomy *Surgery.* the surgical excision of a papilla.

papilledema *Medicine.* edema of the optic disc, often due to raised intracranial pressure, as in malignant hypertension.

papilloma *Medicine.* a noncancerous tumor of the epithelium in which the epithelial cells grow outward from a surface to form a lobed or branching structure. Also, PAPILLARY TUMOR, VILLOMA.

papillomatosis *Medicine.* the widespread formation of papillomas.

papillomatous *Medicine.* describing or referring to a papilloma or papillomas.

Papillomavirus *Virology.* a genus of viruses of the family Papovaviridae through which small tumors, sometimes malignant but most often benign, can be introduced to a wide range of mammals including humans.

papillose *Medicine.* full of papillae.

papillotomy *Surgery.* a surgical incision into a papilla, as of the drodenal papilla.

Papovaviridae *Virology.* a family of DNA viruses that have small, nonenveloped icosahedral particles, share common antigens, replicate in the nuclei of host cells, and can in some cases induce tumors in a wide range of animals.

Papovavirus *Virology.* any virus of the family Papovaviridae.

Pappotherium *Paleontology.* a genus of early eutherians of the Lower Cretaceous of North America; *incertae sedis.*

pappus *Botany.* an often downy or bristly appendage on a seed that aids in the seed's dispersal, such as the bouyant parachutelike float of a dandelion seed.

Pappus of Alexandria c. 300 AD, Greek mathematician; wrote the *Collection,* a systematic account and extension of classical geometry.

Pappus' theorem *Mathematics.* **1.** a theorem of projective geometry: Let A, B, and C and A', B', and C' denote two sets of collinear points on lines l and l', respectively. Let P be the intersection of BC' and $B'C$, Q the intersection of CA' and $C'A$, and R the intersection of AB' and $A'B$. Then the points P, Q, and R are collinear. **2.** the area S of the surface of revolution formed by the rotation of a plane curve of length l about an axis in its plane that does not intersect it is $S = 2\pi\eta l$; that is, the product of the length of the curve and the circumference of the circle described by the centroid of the curve (where η is the distance of centroid of the curve to the axis). Also, **Pappus' (**or **Guldin's) first rule. 3.** the volume V of a solid formed by the rotation of a plane region of area F about an axis in its plane that does not intersect it is $V = F2\pi\eta$; that is, the product of the area and the circumference of the circle described by the centroid of the region (where η is the distance from the centroid of the region to the axis). Also, **Pappus' (**or **Guldin's) second rule.**

paprika [pap rē´kə] *Botany.* **1.** a variety of sweet pepper, *Capsicum annum,* of the family Solanaceae, cultivated for its long red fruit, which is ground up to make a widely used spice. **2.** the spice itself, a common feature of Eastern European cuisine.

Pap smear or **Pap test** *Pathology.* a widely used staining method of microscopic cytological examination, used to detect and diagnose cancers of the female genital and urinary tracts. Also, PAPANICOLAOU TEST. (Named for George *Pap*anicolaou.)

papula *Biology.* **1.** one of the small, ciliated projections of the body wall of an eichinoderm, serving for respiration and excretion. **2.** a small, somewhat pointed elevation of the skin caused by inflammation, accumulated secretion, or hypertrophy of tissue elements; a pimple.

papular acne see ACNE PAPULOSA.

papule [pap´yøol] *Medicine.* a small, solid, circumscribed elevation of the skin.

papulonecrotic *Medicine.* any papule formation that tends to result in central necrosis; applied especially to a variety of skin tuberculosis.

papyrus [pə pī´rəs] *Materials.* a writing material prepared from thin strips of an Egyptian aquatic plant, that have been laid together, soaked, pressed, and dried; used by the ancient Greeks, Romans, and Egyptians. *Botany.* the plant itself, *Cyperus papyrus*; thought to be common in ancient times and still found in the valley of the Upper Nile.

PAR *Aviation.* the airline code for the city of Paris, France; the principal airports are Charles de Gaulle (CDG), Orly (ORY), and Le Bourget (LBG).

par. or **par** paragraph; parallel.

par- or **para-** a prefix meaning: **1.** beside. **2.** closely related to.

par- or **paro-** a prefix meaning "to give birth."

para- *Organic Chemistry.* a prefix indicating the positions of substituted radicals in the 1,4-position of the benzene ring.

para-agglutination *Immunology.* the agglutination of a bacterium that is caused by contact of the bacterium with an antigen of a different bacterium.

para-anesthesia *Neurology.* a loss of feeling in the lower extremities. Also, PARANESTHESIA.

paraaortic body *Anatomy.* one of several small nodules of chromaffin tissue located along the abdominal aorta near the sympathetic ganglia. Also, PARAGANGLION.

para-appendicitis *Medicine.* an inflammation of the connective tissue adjacent to the appendix.

paraballoon *Electromagnetism.* an air-inflated radar antenna.

parabanic acid *Organic Chemistry.* $C_3H_2O_3N_2$, colorless plates that are soluble in water and decompose at 243°C; used in organic synthesis.

Parabasalia *Invertebrate Zoology.* a subclass of large zooflagellates possessing numerous flagella, a single nucleus, and numerous kinetosomes; found in the hindgut of roaches and termites.

parabasalidea *Invertebrate Zoology.* concentrations of Golgi material linked with one or two rootlets of flagella in flagellated protozoans. Also, KINETOSOMES.

Parabellum *Ordnance.* **1.** a German 7.92-mm, recoil-operated, air-cooled machine gun used during World Wars I and II. **2.** the German name for a 9-mm toggle-action automatic pistol sold in the U.S. under the trademark Luger; it was the official German Army sidearm from 1908–1938. **3.** a 9-mm cartridge used in many modern weapons, especially submachine guns; as distinguished from 9-mm Short cartridges.

parabiosis *Biology.* **1.** a reversible suspension of obvious vital activities (as by suitable drying of a rotifer). **2.** the experimental or natural union of two individuals with an exchange of blood. *Ecology.* a situation in which different species of social insects share the same nest without intermingling with each other. Thus, **parabiotic.**

parabiotic twins *Cell Biology.* two organisms that are physiologically connected, as in a union of their circulatory systems.

parabituminous coal *Geology.* a bituminous coal that contains 84–87% carbon when it is analyzed on a dry, ash-free basis.

parabola [pə rab´ə lə] *Mathematics.* **1.** the graph in the plane of the equation

$$Ax^2 + 2Bxy + Cy^2 + 2Dx + 2Ey + F = 0,$$

where $AC - B^2 = 0$, but not all of A, B, and C are zero. The axis of a parabola is often assumed to be parallel to the y-axis; the equation of the parabola then has the form $y = ax^2 + bx + c$ with $a \neq 0$. **2.** equivalently, the conic section formed by the intersection of a cone and a plane parallel to a generator of the cone. **3.** equivalently, the locus of points equidistant from a fixed point (focus) and a fixed line (directrix).

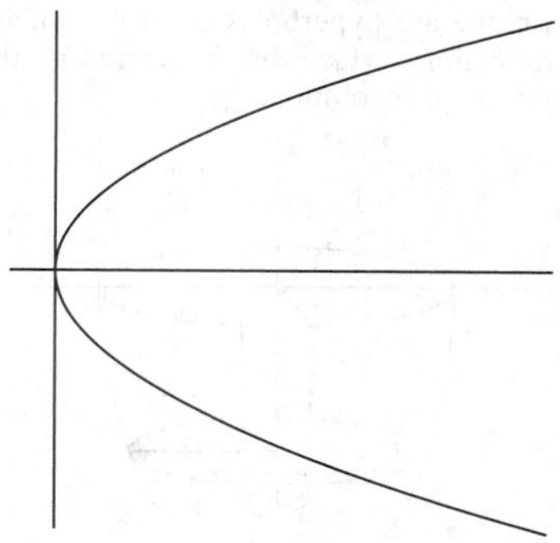

parabola

parabolic [pâr´ə bäl´ik] *Mathematics.* relating to or having the form of a parabola.

parabolic antenna *Electromagnetism.* an antenna having a reflector whose cross section is parabolic and is capable of concentrating the radiated power into a parallel beam.

parabolic differential equation *Mathematics.* a second-order differential equation of the form

$$\sum_{i,j=1}^{n} A_{ij}(\partial^2 u/\partial x_i \partial x_j) + \sum_{i=1}^{n} B_i(\partial u/\partial x_i) + Cu + D = 0$$

such that the following conditions hold: (a) the A_{ij}, B_i, C, and D are all differentiable functions of $(x_1, \ldots, x_n)$; (b) the quadratic form $\Sigma_{i,j=1}^{n} A_{ij}x_i x_j$ can be transformed to the form $\Sigma_{i=1}^{n} F_i x_i^2$, where the F_i have the same sign; and (c) that same transformation does not reduce all of the B_i to zero. The heat equation is a well-known example of a parabolic differential equation.

parabolic dune *Geology.* a low sand dune having a parabolic outline with horns that point windward.

parabolic flight *Space Technology.* a flight path following or describing a parabolic orbit; used to escape the gravitational pull of a celestial body.

parabolic growth *Materials Science.* a growth that is proportional to the square root of the time; many oxides whose growth depends on diffusion through the existing layer obey this relationship.

parabolic microphone *Acoustical Engineering.* a highly directional microphone that uses a parabolic reflector having the pickup at its focal point.

parabolic orbit *Astronomy.* an open (escape) orbit with an eccentricity of 1.0.

parabolic point of a surface *Mathematics.* a particular type of saddle point; a point on a surface at which the total curvature is zero.

parabolic reflector *Electromagnetism.* an antenna having a concave surface that is generated by translating a parabola normal to the plane in which it lies.

parabolic spiral *Mathematics.* the curve whose equation in polar coordinates is

$$(r - a)^2 = 4ak\theta,$$

where a and k are constants.

parabolic velocity *Astronomy.* the escape velocity from an orbit.

paraboloid [pə rab′ə loid] *Mathematics.* a solution surface of a general quadratic equation

$$Ax^2 + 2Bxy + 2Cxz + Dy^2 + 2Eyz + Fz^2 + 2Gx + 2Hy + 2Iz + J = 0$$

in three dimensions whose Hessian (in this case, a matrix of constants) has determinant zero (and hence a zero eigenvalue). If the Hessian has one zero eigenvalue and two nonzero eigenvalues of the same sign, the solution surface is an **elliptic paraboloid**. It is a bowl-shaped surface; the simplest example has the equation $z = x^2/a^2 + y^2/b^2$. Sections of the surface parallel to the z-axis are parabolas, and those parallel to the (x,y) plane are ellipses. If $a = b$, the surface is a **paraboloid of rotation** formed by rotating the parabola $z = x^2/a^2$ about its axis. If the Hessian has one zero eigenvalue and two eigenvalues of opposite sign, the solution surface is a **hyperbolic paraboloid.** It is a saddle-shaped surface; the simplest example has the equation $z = x^2/a^2 - y^2/b^2$. Sections of the surface parallel to the (y,z) and (x,z) planes are parabolas, and those parallel to the (x,y) plane are hyperbolas (or a pair of intersecting lines). *Engineering.* a reflecting surface that is formed by the revolution of a parabola about its axis of revolution.

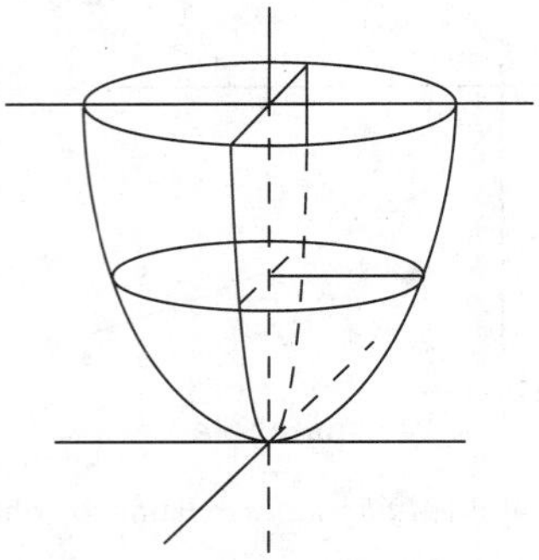

paraboloid

paraboloidal reflector *Electromagnetism.* a surface that is a portion of a paraboloid of revolution; used as an antenna reflector.

parabomb *Engineering.* a container equipped with a delayed-opening parachute.

paracaisson *Ordnance.* a container used to deliver artillery shells by airdrop; when assembled on the ground, it becomes a two-wheel, hand-drawn cart to transport the shells.

Paracelsus, Philippus 1493–1541, Swiss physician; pioneered the medical use of chemicals and specific remedies for specific diseases.

paracentesis *Surgery.* a puncture of the abdomen with a needle to aspirate peritoneal fluid. Also, ABDOMINOCENTESIS, CELIOCENTESIS, PERITONEOCENTESIS.

paracentric *Design Engineering.* a theft-deterrent fabrication built into a key and keyway system in which the ribs and grooves that project beyond the center are oriented longitudinally; generally used in pin-tumbler cylinder locks.

paracentric inversion *Genetics.* an inversion that does not include the centromere.

Paracentrotus lividus *Invertebrate Zoology.* a species of boring sea urchin that riddles rock walls with burrows; found along the coast of Europe.

parachor *Physics.* a quantity that is generally constant over a wide range of temperatures, given by the molecular weight of a liquid times the fourth root of the liquid's surface tension divided by the difference of the densities of the liquid and the vapor in equilibrium with it.

parachronology *Geology.* **1.** the correlation of stratigraphic units. **2.** a geochronology based on fossils that either supplement or take the place of biostratigraphically significant fossils.

parachute *Aviation.* **1.** an umbrellalike device of light fabric used to slow the descent of a person or object falling from a great height, as from an aircraft, by increasing air resistance. **2.** to descend or drop using such a device. *Engineering.* see PARACHUTE BRAKE. *Mining Engineering.* a type of safety catch for cages in mine shafts.

parachute brake *Engineering.* **1.** a parachute opened horizontally from the tail of an aircraft upon landing to slow its ground speed. **2.** a similar device used in auto racing as an aid in braking at the end of a race.

parachute deployment height *Military Science.* the height above the intended impact point at which a parachute or parachutes are fully deployed.

parachute flare *Ordnance.* a pyrotechnic device that produces intense illumination of brief duration; it is attached to a parachute and may be released from the air or fired from the surface.

parachute fragmentation bomb *Ordnance.* in low-level bombing, a fragmentation bomb that is dropped by parachute in order to allow the bombing plane to escape before the explosion.

parachute harness see HARNESS.

parachute-opening shock *Aviation.* the abrupt forces acting upon a parachute and its load at the moment it opens.

parachute troops see PARATROOPS.

parachute weather buoy *Engineering.* an automatic weather station equipped with a parachute and designed to be air dropped; used to transmit such information as wind speed, wind direction, barometric pressure, and air temperature.

paraclinal *Geology.* of or related to a stream or valley that is oriented parallel to the direction of the fold axes of a region.

paraclonal *Ecology.* describing a type of plant society whose inhabitants cannot regenerate if separated from their parents.

paracoagulation *Cell Biology.* the aggregation of a population of cells due to the addition of a staphylococcal clumping factor.

Paracoccidiodes *Mycology.* a genus of fungi belonging to the class Hyphomycetes; its only species, *P. brasiliensis*, occurs in South America, where it causes the disease paracoccidioidomycosis, which affects the lungs and other internal organs.

paracoccidioidal granuloma see SOUTH AMERICAN BLASTOMYCOSIS.

Paracoccus *Bacteriology.* a genus of Gram-negative bacteria of uncertain affiliation, composed of aerobic, nonmotile cocci or rod-shaped cells that are typically found in soil and oxidize a variety of carbon sources.

paracolitis *Medicine.* an inflammation of the outer coat of the colon.

paracolon bacteria *Microbiology.* a former name for certain nonlactose-fermenting intestinal bacteria; formerly classified in the genus *Paracolobactrum*, but now assigned to various other genera.

paracolpitis *Medicine.* inflammation of the tissues around the vagina.

paracompact *Mathematics.* a Hausdorff space X is said to be paracompact if every covering of X has a locally finite refinement. That is, there is a refinement of any covering of X such that, at every point $x \in X$, there exists a neighborhood of x that has a nonempty intersection with only a finite number of members of the refinement. All metric spaces and locally compact second countable Hausdorff spaces are paracompact. A manifold M must be paracompact in order for a partition of unity to exist on M.

paracondyloid *Vertebrate Zoology.* of or relating to the occipital bone on the outside of each condyle in some mammalian skulls.

paracone *Vertebrate Zoology.* a major cusp along the external border of an upper molar that functions in grinding and chewing food and corresponds to the paraconid of the lower molar; in certain mammals, it is the principal anterior, external cusp.

paraconformity *Geology.* an unconformity in which the strata above and below the break are parallel, and that exhibits little or no evidence of surface erosion or prolonged nondeposition. Also, NONDEPOSITIONAL UNCONFORMITY.

paraconglomerate *Geology.* sedimentary rocks containing sparse pebbles or cobbles, generally making up less than 10% of the total mass. Also, PARAGLOMERATE.

paraconid *Vertebrate Zoology.* a cusp of the primitive molar in the lower jaw that corresponds to the paracone of the upper molar; in certain mammals it is the anterior, internal cusp.

paracoquimbite *Mineralogy.* $Fe_2^{+3}(SO_4)_3 \cdot 9H_2O$, a pale-violet trigonal mineral, dimorphous with coquimbite, having a specific gravity of 2.11 and a hardness of 2.5 on the Mohs scale; found as rhombohedral crystals or in granular form with other sulfates.

paracrate *Engineering.* a rigid crate designed to shield materials being dropped by parachute.

paracrine *Endocrinology.* **1.** denoting a hormonal pathway characterized by the production of a biologically active substance that passes by diffusion within the extracellular space to a nearby cell where it initiates a response. **2.** of or relating to the secretion of a hormone by an organ that is not an endocrine gland.

Paracrinoidea *Paleontology.* a small group of echinoderms with bilateral symmetry and 2–4 crinoidlike arms; of uncertain position within the phylum; extant in the Middle Ordovician.

Paracryphiaceae *Botany*. a monospecific family of New Caledonian shrubs or trees of the order Theales, characterized by unicellular hairs, a dicotyledonous embryo, and simple toothed leaves.

paracrystalline aggregate *Molecular Biology*. a regular arrangement of molecules.

paracrystallinity *Materials Science*. a deformation in a crystalline structure that results in a broadening of X-ray diffraction lines.

Paracucumidae *Invertebrate Zoology*. a family of sea cucumbers, holothurian echinoderms in the order Dendrochirotida, having a simple calcareous ring.

paracystitis *Medicine*. an inflammation of the connective tissue and other structures surrounding the urinary bladder.

paradidymis *Anatomy*. a vestigial body composed of a few convoluted tubules in the anterior part of the spermatic cord near the epididymis.

paradigm [pâr´ə dīm´] *Science*. an example serving as a model or pattern. *Linguistics*. an example or display giving all the various possible forms of a word, such as the conjugation of a verb. *Artificial Intelligence*. the method or train of thought that a person follows in order to reason about a problem. Thus, **paradigmatic** [pâr´i də mat´ik].

Paradisaedae *Vertebrate Zoology*. the birds of paradise, a family of oscine birds distinguished by variable, very colorful plumage and elaborate plumes on the head, back, breast, tail, and wings; found in Australia, New Guinea, and adjacent islands.

paradox [pâr´ə däks´] *Science*. a thesis or argument that draws a seemingly contradictory conclusion from valid information.

paradoxical [pâr´ə däks´i kəl] *Science*. relating to or in the form of a paradox; seemingly self-contradictory. *Medicine*. describing various conditions characterized by a contradictory condition or process; e.g., in **paradoxical breathing,** a part of the lung deflates during inspiration and inflates during expiration.

paradoxical embolus *Medicine*. an embolus, usually a venous thrombus, that is transported to a peripheral arterial circulation through a cardiac septal defect with a right-to-left shunt.

Paradoxides *Paleontology*. a genus of large, early trilobites in the order Redlichiida, up to half a meter in length; widespread in the Middle Cambrian.

paraepilepsy see MINOR FOCAL EPILEPSY.

paraesophageal cyst *Medicine*. a bronchogenic cyst on the esophagus that is filled with mucoid material and epithelial cells.

par. aff. the part affected. (From Latin *pars affecta*.)

paraffin *Organic Chemistry*. any of a class of saturated aliphatic hydrocarbons having a straight or branched carbon chain and the general formula C_nH_{2n+2}, ranging in physical form from methane gases to waxy solids; found mainly in Pennsylvania and midcontinent petroleum. Also, ALKANE. *Materials*. see PARAFFIN WAX.

paraffin coal *Geology*. a soft, light-colored bituminous coal from which oil and paraffin may be produced.

paraffin dirt *Geology*. a rubbery clay soil found in the upper portion of a soil profile near gas seeps; believed to be formed by the biodegradation of natural gas.

paraffin distillate *Materials*. a distilled petroleum fraction that, when cooled, consists of a mixture of crystalline wax and oil.

paraffinicity *Organic Chemistry*. the paraffinic or saturated nature of crude petroleum or its products.

paraffin oil *Materials*. a combustible oil that is either pressed or dry-distilled from paraffin distillate; used in floor treatment and as a lubricant.

paraffin press *Engineering*. a filter press used to separate paraffin oil and paraffin wax from distillates during the petroleum refining process.

paraffin wax *Materials*. ordinary paraffin in its solid state, a combustible, translucent, tasteless, odorless, white hydrocarbon that is obtained from paraffin distillate; used in candles, paper coating, lubricants, cosmetics, polishes, foods, and as a protective sealant.

paraformaldehyde *Organic Chemistry*. $HO(CH_2O)_nH$, a polymer of formaldehyde in which *n* may range from 8 to 100; occurs as a combustible white solid that melts in the range 120–170°C; higher polymers are insoluble in water and all polymers are insoluble in alcohol; used as a disinfectant, fungicide, and component of adhesives and of contraceptive creams.

paraganglioma *Oncology*. a tumor of the tissue composing the paraganglia.

paraganglion *plural*, **paraganglia.** *Anatomy*. a collection of chromaffin cells occurring outside the adrenal medulla, usually near the sympathetic ganglia and in relation to the aorta and aortic branches; paraganglia secrete epinephrine or norepinephrine. Also, PARAAORTIC BODY.

paragenesis *Mineralogy*. a characteristic assemblage of minerals in a rock or an ore deposit, formed at the same time and apparently in equilibrium.

paragenetic sequence *Mineralogy*. the chronological order of crystallization of minerals or mineral assemblages in a rock or an ore deposit.

parageosyncline *Geology*. **1.** a geosyncline that lies within a craton or stable area. Also, INTRAGEOSYNCLINE. **2.** an oceanic depression of about the same age as the craton it borders. Also, MARGINAL GEOSYNCLINE.

parageusia *Neurology*. distortion of the sense of taste, usually involving an unpleasant taste in the mouth.

paraglomerate see PARACONGLOMERATE.

paragneiss *Geology*. a coarse-grained, banded, foliated metamorphic rock presumably formed by the metamorphism of sedimentary rock.

paragonimiasis *Medicine*. a pulmonary disease caused by infection with a lung fluke of the genus *Paragonimus*, characterized by hemoptysis, bronchitis, pain, diarrhea, and occasionally abdominal masses; occurring most commonly in Asia.

paragonite *Mineralogy*. $NaAl_2(Si_3Al)O_{10}(OH)_2$, a colorless to yellowish monoclinic mineral of the mica group occurring in metamorphic rock as compact masses and fine scales, having a specific gravity of 2.78 to 2.90 and a hardness of 2.5 on the Mohs scale.

paragranuloma *Medicine*. the most benign form of Hodgkin's disease, largely confined to the lymph nodes.

paragraph *Graphic Arts*. a distinct section of written or printed copy, usually set off by an indentation of the first line and generally having some unity of thought. *Computer Programming*. in COBOL, a set of one or more sentences constituting a logical processing unit and preceded by a paragraph name or header.

Paraguay River [pâr´ə gwā´; pâr´ə gwī´] *Geography*. a river that flows 1584 miles south from the Mato Grosso plateau in southwestern Brazil into the Paraná River.

parahopeite *Mineralogy*. $Zn_3(PO_4)_2{\cdot}4H_2O$, a colorless triclinic mineral, dimorphous with hopeite, occurring in fan-shaped aggregates of small tabular crystals, and having a specific gravity of 3.31 and a hardness of 3.5 on the Mohs scale; found as a secondary mineral in zinc-bearing ore deposits.

parahydrogen *Atomic Physics*. a state in which the nuclei of hydrogen molecules spin in opposite directions.

parainfluenza *Medicine*. a condition similar to influenza, caused by an myxovirus, and affecting primarily infants and young children.

parainfluenza virus *Virology*. any of a group of myxoviruses that have been isolated from patients with various upper-respiratory-tract diseases; classified as **parainfluenza 1 virus,** including the Sendai virus and hemadsorption type 2 virus; **parainfluenza 2 virus,** isolated from patients with acute laryngotracheobronchitis (croup); **parainfluenza 3 virus,** which causes bronchitis and pneumonia, especially in children; and **parainfluenza 4 virus,** associated with respiratory diseases in children.

parakeet *Vertebrate Zoology*. any of various small, colorful, and slender singing parrots in the family Psittacidae, characterized by long tails and a slightly hooked bill; popular as pets, they are found in tropical areas worldwide.

parakeet

parakeratosis *Pathology.* the preservation of nuclei in the epidermal cells of the stratum corneum, an abnormal condition found in psoriasis and exfoliative dermatitis.

parakinesia *Neurology.* an abnormality of motor function resulting in wild, spasmodic movements.

paralalia *Medicine.* any disturbance of speech, especially one characterized by distortion of sounds or the substitution of one sound for another.

paralaurionite *Mineralogy.* PbCl(OH), a colorless to white, soft monoclinic mineral, dimorphous with laurionite, occurring as lathlike to tabular crystals, and having a specific gravity of 6.15 and an undetermined hardness; found with laurionite and other secondary lead minerals.

paraldehyde *Organic Chemistry.* $C_6H_{12}O_3$, a colorless liquid trimer of acetaldehyde; soluble in water and miscible in most organic solvents and volatile oils; boils at 124.5°C and freezes at 12.6°C; used as a sedative, as a solvent, and in making rubber.

paraldol *Organic Chemistry.* $(CH_3CHOHCH_2CHO)_2$, white crystals that are soluble in water and boil at 90–100°C; used as a chemical intermediate and to make resins.

Paralepididae *Vertebrate Zoology.* the barracudinas, a cosmopolitan family of bathypelagic carnivorous fish of the order Myctophiformes, characterized by a body that is compressed anteriorly and round in posterior cross section.

paralexia *Neurology.* a form of alexia, characterized by the inability to comprehend written words and the substitution of meaningless combinations of words.

paralgesia *Neurology.* a painful form of paresthesia, characterized by abnormal, painful sensations without external cause.

paraliageosyncline *Geology.* a geosyncline that develops along a present-day continental margin.

paralic *Geology.* **1.** of or related to marine and continental deposits that were laid down on the landward side of a coast or in shallow water near the sea. **2.** of or related to coal deposits that occur along the margin of the sea.

paralimnion *Ecology.* the sum of all of the various factors that make up the portion of a lake in which vegetation flourishes. Thus, **paralimnetic.** *Hydrology.* the region along the shore of a lake, extending from the shoreline to the deepest limit of rooted vegetation.

parallactic angle *Astronomy.* an angle formed by a star in an astronomical triangle with vertices at the celestial pole or zenith and the subject star.

parallactic displacement *Astronomy.* the observed change in a star's position as a result of the motion of the earth in its orbit. Also, **parallactic motion.**

parallactic ellipse *Astronomy.* the apparent movement of a star's position caused by parallactic displacement. Also, **parallactic orbit.**

parallactic grid *Photogrammetry.* a rectangular grid marked on a transparent surface and introduced into the viewing system of a stereoscope to provide a floating system of orientation in the model in addition to, or instead of, floating marks.

parallactic inequality *Astronomy.* an irregularity in the motion of the moon as a result of the gravitational attraction of the sun. Also, **parallactic equation.**

parallactic motion *Astronomy.* the apparent motion of stars due to the earth's orbital motion.

parallax *Optics.* **1.** the apparent change in position of an object as viewed through an optical instrument, occurring when the observer shifts position. Also, OPTICAL PARALLAX. **2.** the angular degree of such a change. *Astronomy.* specifically, the angle between the two straight lines that join a celestial body to two different points of observation; e.g., two different points on the earth as it moves through space.

parallax bar see STEREOMETER.

parallax correction *Navigation.* a quantity used to adjust a sextant reading to its equivalent at the center of the earth.

parallax error *Optics.* an error that occurs when the observer's eye and the point on a scale do not lie in a line perpendicular to the scale; can be alleviated with a mirror.

parallax in altitude *Navigation.* the difference in altitude of a celestial body as measured on the surface of the earth and at the center of the earth; it varies with the altitude of the body and its distance from the earth.

parallax inequality *Oceanography.* the variation in tidal range or in the speed of tidal currents that is caused by the constantly changing distance between the earth and the moon.

parallax wedge *Photogrammetry.* a simple type of stereometer used to measure the heights of objects on pairs of stereo photographs.

parallel *Mathematics.* two lines l and m are said to be parallel if they lie in the same plane and if no point (except possibly the point at infinity) lies on both lines. Two planes are said to be parallel if they have no point in common. *Cartography.* **1.** a circle on the earth's surface that is parallel to the plane of the equator and connects all points of equal latitude. **2.** a line or circle on a map or chart that connects all points of equal latitude. *Geodesy.* **1.** a circle that is parallel to the primary great circle of a sphere or spheroid. **2.** a closed curve on an irregular spheroid that approximates a parallel on a sphere. *Physics.* **1.** describing any two vectors having identical directions. **2.** describing any two adjacent surfaces whose contour shapes are the same over a localized region. *Electricity.* a side-by-side connection in which the same voltage is applied to all components. *Computer Programming.* of or relating to data, programs, or events that are processed, transmitted, or occur simultaneously rather than in sequence.

parallel access *Computer Programming.* simultaneous access to each bit in a storage word or character.

parallel addition *Computer Programming.* addition performed simultaneously on all digits of the operands.

parallel algorithm *Computer Programming.* any algorithm in which computation sequences can be performed simultaneously.

parallel-axis theorem *Mechanics.* a theorem stating that the moment of inertia of a body about any axis is equal to its moment of inertia about a parallel axis through its center of mass, plus the product of its mass and the square of the perpendicular distance between the two parallel axes. Also, STEINER'S THEOREM.

parallel bedding see CONCORDANT BEDDING.

parallel buffer *Electronics.* a device, such as a magnetic core or flip-flop, that temporarily stores computer information in parallel, instead of series storing.

parallel by character *Computer Programming.* the simultaneous handling of all characters of a single computer word by using separate lines, channels, or storage cells.

parallel circuit *Electricity.* a circuit in which all the components are connected to each other in parallel.

parallel communications *Telecommunications.* **1.** a method of simultaneous transmission of information using different carrier frequencies on different communication channels. **2.** the simultaneous transmission of elements on a single path or wire.

parallel compensation see FEEDBACK COMPENSATION.

parallel computation *Computer Programming.* the simultaneous computations on various portions of a problem.

parallel computer *Computer Technology.* a computer that uses multiple arithmetic units to perform concurrent operations on data. Also, PARALLEL DIGITAL COMPUTER, PARALLEL PROCESSOR.

parallel conversion *Computer Programming.* the act of switching to a new computer system, during which time both the old and new systems are in use to reduce the occurrence of errors during conversion. Also, PARALLEL RUNNING.

parallel cousin *Anthropology.* in kinship terminology, the offspring of one's parental siblings of the same sex, such as the mother's sister's children or the father's brother's children.

parallel cut *Engineering.* a set of parallel holes used to form the initial cavity into which explosively loaded holes break during blasting. Also, BURN CUT.

parallel digital computer see PARALLEL COMPUTER.

parallel distributed processing see NEURAL NETWORK.

parallel dot matrix printer see LINE DOT MATRIX.

parallel drainage pattern *Hydrology.* a pattern of natural stream courses in which the streams and their tributaries are regularly spaced and flow parallel to one another over a large area.

parallelepiped [pâr´ə lel´ə pī´ped] *Mathematics.* a prism whose bases (and hence all faces) are parallelograms.

parallel evolution *Evolution.* **1.** the development of similar hereditary characters in species or other taxonomic groups not directly related in evolutionary descent. Also, **parallelism, parallel descent. 2.** the maintenance of a constant degree of divergence in the hereditary characters of two unrelated lineages.

parallel extinction *Optics.* the near total absorption of light moving through an anisotropic crystal at a direction parallel to the cleavage planes of that crystal.

parallel feed *Electronics.* the introduction of DC voltage into the electrodes of a device, in parallel with the AC circuit. Also, SHUNT FEED.

parallel firing *Engineering.* a method of electrically connecting detonators that are to be fired simultaneously.

parallel flow *Electricity.* **1.** the flow of electric current in a parallel circuit. **2.** a flow of electric current through a power system over signal paths other than the path contractually agreed to. Also, LOOP FLOW.

parallel fold see CONCENTRIC FOLD.

parallel grid *Radiology.* a radiographic grid in which the lead strips run parallel to each other along their longitudinal axis, rather than angled as in a focused grid.

parallel gripper *Robotics.* an end effector that has two opposing parallel fingers or jaws with which to grasp objects.

parallel impedance *Electricity.* any of a set of impedances that are connected across one pair of terminals.

paralleling reactor *Electromagnetism.* a reactive device that is used to correct the load division between transformers connected in parallel with unequal impedance voltages.

parallel input/output *Computer Technology.* the simultaneous transmission of data to and from a computer through multiple wires.

parallel intergrowth *Crystallography.* the interpenetrating growth of two crystals with at least one axis from each crystal parallel to an axis of the other.

parallel linkage *Mechanical Engineering.* a system of links that amplifies reciprocating motion.

parallel muscle *Anatomy.* a muscle composed of parallel fibers, such as the rectus abdominis muscle.

parallel of declination *Astronomy.* a circle on the celestial sphere connecting points of constant declination.

parallel of latitude *Astronomy.* a circle on the celestial sphere connecting points of equal latitude about a planet or star. *Navigation.* a line, parallel to the equator, that connects all points that are equally distant north or south from the equator.

parallelogram [pâr´ə lel´ə gram´] *Mathematics.* a quadrilateral satisfying one of the following equivalent conditions: (a) opposite sides are parallel; (b) opposite sides are equal; (c) opposite angles are equal; (d) two opposite sides are equal and parallel; or (e) the diagonals bisect each other.

parallelogram identity *Mathematics.* **1.** any vector norm $\| \ \|$ derived from an inner product (i.e., $\|x\| = <x, x> 1/2$) satisfies the parallelogram identity:

$$\|x\|^2 + \|y\|^2 = (1/2)(\|x + y\|^2 + \|x - y\|^2).$$

Conversely, if the parallelogram identity is satisfied, then the norm $\| \ \|$ is necessarily derived from an inner product $< , >$. **2.** in geometric terms, the sum of the squares of the lengths of the sides of a parallelogram equals the sum of the squares of the lengths of its diagonals. Also, POLAR IDENTITY.

parallel operation *Computer Programming.* any process in which two or more parts, usually similar, of an operation are performed simultaneously by individual units. *Electronics.* the practice of joining the outputs of two or more power sources, such as batteries, so that they power the same system or unit.

parallelotope *Mathematics.* **1.** a term for a parallelepiped in n dimensions. **2.** in particular, let H be the Hilbert space consisting of the set of sequences $x = (x_1, x_2, x_3, \ldots)$ of complex numbers where $\sum_{k=1}^{\infty} |x_k|^2$ is finite. Addition is defined by $x + y = (x_1 + y_1, x_2 + y_2, \ldots)$; scalar multiplication is defined by $ax = (ax_1, ax_2, \ldots)$; and an inner product $<x, y>$ is defined as $<x, y> = \sum_{k=1}^{\infty} x_k \bar{y}_k$, where x_k indicates a complex conjugate. Then the **Hilbert parallelotope** is the set of all x in H such that $|x_k| \leq 2^{-k}$ for each k.

parallel padding *Electricity.* a method of regulating a portion of the total current in each of two or more power supplies by employing a current-limiting or automatic-crossover-output characteristic in which each parallel supply adds to the total and pads or attentuates the output only when the load current demand exceeds the capability or limit setting of the original power supply.

parallel-plate capacitor *Electricity.* a capacitor consisting of two parallel metal plates with an adjoining dielectric filling between them.

parallel-plate laser *Optics.* a laser in which two small parallel plates, one reflecting and the other partially reflecting, force a light beam to bounce between them several times to build up a strong pulse before emerging.

parallel-plate oscillator *Electricity.* a push-pull, ultrahigh-frequency oscillator circuit composed of two parallel plates that act as the main frequency-determining elements.

parallel-plate waveguide *Electromagnetism.* two parallel conductive plates that are used to propagate circular, cylindrical electromagnetic waves having axes perpendicular to the plates.

parallel play *Behavior.* side-by-side play behavior without interaction, especially as observed in children of preschool age.

parallel processing *Psychology.* the simultaneous execution of more than one cognitive function. *Computer Programming.* the programming of a computer so that two operations are performed either simultaneously by different units or concurrently by the same unit. Also, **parallel programming.**

parallel processor *Computer Science.* **1.** see PARALLEL COMPUTER. **2.** an architecture in which multiple central processing units are connected, e.g., by shared memory or a communication mechanism.

parallel radio tap *Telecommunications.* in telephone tapping, a method in which a small radio transmitter is bridged across a target pair.

parallel rectifier *Electronics.* a device that is created when two or more rectifiers are connected to the same pair of terminals in a circuit in order to generate more current.

parallel representation *Computer Programming.* the representation of a digital variable so that each of the bits simultaneously appears on parallel bus lines.

parallel resonance *Electricity.* see ANTIRESONANCE.

parallel-resonant circuit *Electricity.* **1.** a resonant circuit in which the capacitor, inductor, and AC generator are connected in parallel in order to achieve a high impedance at the frequency to which the circuit is resonant. **2.** a resonant circuit in which the applied voltage is connected across a parallel circuit composed of a capacitor and an inductor. Also, ANTIRESONANCE CIRCUIT.

parallel resonant interstage *Electronics.* an intermediate stage that is created between two amplifier stages when their signals are coupled by a parallel inductance-capacitance circuit.

parallel ripple mark *Geology.* a ripple mark having a relatively straight crest and an asymmetric profile.

parallel roads *Geology.* a succession of horizontal beaches or wave-cut terraces that lie parallel to each other at different elevations on either side of a glacial valley.

parallel-rod oscillator *Electronics.* a device that generates alternating current at ultrahigh frequencies from the charging power supplied by two parallel rods in its circuit.

parallel rulers *Navigation.* an instrument used for transferring a line parallel to itself.

parallel running see PARALLEL CONVERSION.

parallels *Engineering.* the long, machined blocks used to rest, space, or support an object.

parallel sailing *Navigation.* a method for converting the distance sailed along a parallel of latitude into a difference of longitude.

parallel search storage *Computer Technology.* a storage system in which one or more parts of a storage location are searched simultaneously.

parallel series *Electricity.* **1.** a circuit in which two or more parallel circuits are connected in series. **2. parallel-series.** of or relating to such a series; thus, **parallel-series circuit, parallel-series capacitor.** Also, SERIES PARALLEL, SERIES-PARALLEL CIRCUIT.

parallel sheaf *Military Science.* in artillery and naval gunfire support, a sheaf in which the planes of fire of all pieces are parallel.

parallel shot *Engineering.* a test of lead, lag, polarity, and phase of the amplifier-to-oscilloscope circuits used in seismic prospecting when the amplifiers are connected in parallel.

parallel-slit interferometer *Optics.* an interferometer equipped with two parallel slits separated by an adjustable distance in order to fit over the objective of a telescope.

parallel sphere *Astronomy.* the apparent motion of celestial bodies as seen from the earth's poles; stars move parallel to the horizon.

parallel storage *Computer Technology.* a storage device in which all bits of a character or word are dealt with simultaneously.

parallel texture *Petrology.* a rock texture characterized by the orientation of tabular or prismatic crystals parallel to a plane or line.

parallel-T network *Electricity.* a resistance-capacitance network composed of separate T-networks with terminals connected in parallel that produces a null at one frequency; used most readily in capacitance measurements at radio frequencies. Also, TWIN-T FILTER.

parallel-T oscillator *Electricity.* a resistance-capacitance tuned oscillator having a negative feedback path in a parallel-T network; the oscillator frequency is determined by the null frequency of the network.

parallel transfer *Computer Technology.* the transfer of all bits of a character or word simultaneously.

parallel transmission *Computer Technology.* the transmission of all bits of a character or word simultaneously over separate channels.

parallel twin *Crystallography.* a twinned crystal in which the twin axis is a crystal edge lying in the composition plane.

parallel unconformity see DISCONFORMITY.

parallel-veined *Botany.* of or relating to a leaf that has veins running parallel to one another or nearly so, as distinguished from a leaf that is net-veined.

parallel wires *Electricity.* two conductors that are separated in parallel to each other; often used in the conduction of electricity across power transmission lines. Also, **parallel wire line.**

parallochthon *Geology.* rocks that were transported over intermediate distances and then deposited on or near an allochthonous mass during transit.

paralogous genes *Genetics.* homologous genes that have arisen through a gene duplication and that have evolved in parallel within the same organism.

paralutein cells *Histology.* certain epithelial cells of the corpus luteum.

paralysis [pə ral´ə sis] *Neurology.* a temporary or permanent loss of motor function, especially loss of sensation or voluntary motion, generally caused by disease, dysfunction, or injury of the central nervous system or of the peripheral neuromuscular system.

paralytic [pâr´ə lit´ik] *Neurology.* **1.** of or related to paralysis. **2.** a person affected with paralysis.

paralytic secretion *Physiology.* the constant exudate flowing from a gland after its secretory nerves have been severed.

paralytic spinal poliomyelitis *Medicine.* a form of paralysis involving principally the spinal cord; caused by one of the poliomyelitis viruses.

paralytic stroke *Neurology.* a sudden attack of paralysis, resulting from injury to the brain or spinal cord.

paralyzant *Neurology.* **1.** causing paralysis. **2.** an agent that causes paralysis.

paralyze [pâr´ə līz] *Neurology.* to affect with paralysis.

paramagnet *Electromagnetism.* any body or substance that is paramagnetic.

paramagnetic *Electromagnetism.* of a substance, having a magnetic permeability greater than that of a vacuum but significantly less than that of ferromagnetic materials such as iron; the permeability of a paramagnetic substance is considered to be independent of an external magnetizing force. Paramagnetic substances are magnetized parallel to the line of force in an applied magnetic field, in proportion to the strength of the field.

paramagnetic alloy *Metallurgy.* any alloy having a specific permeability greater than that of a vacuum and independent of magnetic field strength.

paramagnetic analytical method *Analytical Chemistry.* the analysis of a fluid mixture by measurement of the interaction of its constituents with a magnetic field.

paramagnetic crystal *Electromagnetism.* a crystal whose magnetic permeability is slightly greater than that of a vacuum but not greater than that of a ferromagnetic material.

paramagnetic Faraday effect *Optics.* the Faraday effect as observed in paramagnetic salts under frequencies at which splitting of the lower energy level occurs.

paramagnetic iron *Metallurgy.* iron that has been made paramagnetic by applying high pressure.

paramagnetic material see PARAMAGNETIC.

paramagnetic relaxation *Electromagnetism.* the establishment, over a period of time, of equilibrium exhibited by a paramagnetic substance when subjected to a change in an external magnetic field.

paramagnetic salt *Electromagnetism.* a salt that has paramagnetic properties and is used in adiabatic demagnetization.

paramagnetic spectrum *Spectroscopy.* any spectrum produced by the alignment of magnetic moments of paramagnetic materials (having unpaired electrons) in solution or in crystals, usually at microwave frequencies.

paramagnetic susceptibility *Electromagnetism.* the susceptibility of a paramagnetic material, generally positive and slightly greater than zero.

paramagnetism *Electromagnetism.* the fact of being paramagnetic; a weak, positive, attractive reaction to an applied magnetic field,

paramastigote *Cell Biology.* of or related to a cell form having an accessory flagellum in addition to a larger flagellum.

Parameciidae *Invertebrate Zoology.* a family of ciliate protozoans in the order Holotrichia, having an elongated body and a deep oral groove; found in decaying organic matter.

paramecin *Cell Biology.* a toxic factor produced by the kappa particle in killer strains of *Paramecium* and active on sensitive *Paramecium* cells.

Paramecium *Invertebrate Zoology.* a genus of pear- or slipper-shaped protozoans in the family Parameciidae, having uniform ciliation and two contractile vacuoles; found in stagnant water; often used in scientific experiments.

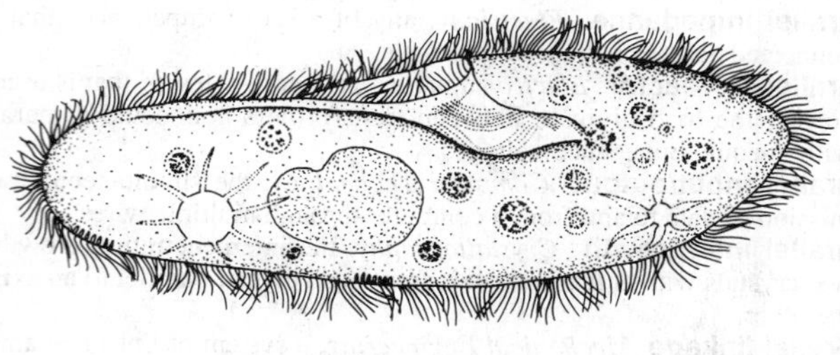

paramecium

paramecium [pâr´ə mē´shē əm; pâr´ə mē´sē əm] *plural,* **paramecia.** *Invertebrate Zoology.* a member of the genus *Paramecium.*

paramedian incision *Surgery.* a vertical incision in the abdominal wall parallel and lateral to the midline of the anterior abdominal wall.

paramedic *Medicine.* a person who acts in connection with the science or practice of medicine, especially a person trained in emergency medical procedures, such as in the military or in a fire or police department.

paramelaconite *Mineralogy.* $Cu_2^{+1}Cu_2^{+2}O_3$, a brilliant black tetragonal mineral occurring in prismatic crystals, having a specific gravity of 5.9 to 6.1 and a hardness of 4.5 on the Mohs scale; found with other secondary copper minerals.

Paramelina *Vertebrate Zoology.* in some classification systems, a marsupial order that includes the bandicoots.

paramere *Biology.* **1.** either of the halves of a bilaterally symmetrical animal, or of a segment or somite of such. **2.** any of several paired structures of an insect and especially of its ninth abdominal segment. **3.** one of a series of radiating parts or organs, as a ray of a starfish; an actinomere. *Invertebrate Zoology.* either of the paired lobes near the reproductive organs of some male insects.

paramesonephric duct *Developmental Biology.* either of the paired embryonic ducts, arising as a peritoneal pocket and extending caudally to the urogenital sinus, that differentiates as the vagina and uterus in the female, and degenerates to the appendix testis in the male. Also, MÜLLERIAN DUCT.

parameter [pə ram´ə tər] a variable quantity or value; specific uses include: *Physics.* a quantity that is used in a relationship involving several quantities and is held fixed, but in general may be allowed to have different values. *Mathematics.* **1.** let ϕ be a continuous mapping from R^k into an n-dimensional manifold M, where $k \leq n$. The k arguments of ϕ are called parameters. As they vary, they sweep out a k-dimensional surface in M. **2.** an independent variable that is not allowed to vary concurrently with other variables. *Statistics.* a quantity entering a probability distribution or a model. In classical inference, it is considered a fixed unknown constant; in Bayesian inference, it is considered a random variable. *Computer Programming.* **1.** data that can take different values at the execution time of a function, procedure, macroinstruction, or subroutine; a **formal parameter** is the parameter in the definition of the function, procedure, and so on; an **actual parameter** is provided by the calling program when the function is called. Also, ARGUMENT. **2.** a name whose value can be set to a constant at compile time. *Electricity.* **1.** an operating value, constant, or coefficient that is a dependent or independent variable such as a transistor-electrode current or voltage. **2.** a measure of system performance. **3.** the ratio of one coefficient to another when both are either fixed or variable.

parameterize or **parametrize** *Science.* to express something in terms of properties or characteristics.

parameterized curves *Mathematics.* a function from the real numbers R (or from a subset of the real numbers) into a manifold.

parameter tag *Computer Programming.* a constant that is used by multiple computer programs.

parameter word *Computer Programming.* a word that contains either the parameters or the addresses of the parameters to be passed to a subroutine.

paramethadione *Pharmacology.* $C_7H_{11}NO_3$, a drug occurring as a clear liquid, taken orally to control and prevent petit mal epilepsy.

parametric [păr´ə met´rik] *Science.* relating to or being a parameter or parameters.

parametric acoustic array *Acoustical Engineering.* a long array of transducers that uses differences in parameters, normally sound speed and particle velocity, to determine desired information, such as distance and direction, by producing a dual-frequency sound beam and using the resulting difference frequencies.

parametric amplification *Optics.* the amplification of a weak optical beam incident on a nonlinear crystal that occurs when an intense coherent pump beam interacts with the crystal. Also, OPTICAL PARAMETRIC AMPLIFICATION.

parametric amplifier *Electronics.* an amplifier whose operation is based on the use of a parametric device. Also, PARAMP, REACTANCE AMPLIFIER. *Optics.* a device that amplifies optical waves by means of interaction between a high-frequency laser beam and an optically nonlinear crystal.

parametric converter *Electronics.* a parametric device that changes an input signal at one frequency into an output signal at a different frequency.

parametric device *Electronics.* an electronic device whose operations are controlled by a time variation in one of its elements, generally a reactance.

parametric down-converter *Electronics.* a parametric device that changes the frequency of an output signal, so that it is lower than the input signal.

parametric equation *Mathematics.* an equation that includes an independent variable (i.e., a parameter) considered fixed in determining the solutions of the equation, so that when the parameter is changed, the solutions also change.

parametric excitation *Engineering.* a technique for initiating and maintaining oscillation within an electric or mechanical system by varying the energy produced by storage elements, such as capacitors, inductors, or springs.

parametric generation *Optics.* the conversion of one electromagnetic wave passing through a nonlinear material into two waves of lower frequency; the sum of the frequencies of the two resulting waves equals the frequency of the original wave.

parametric hydrology *Hydrology.* a branch of hydrology in which the relationships among the physical parameters involved in hydrologic events are studied, analyzed, and used to synthesize other hydrologic events.

parametric hypothesis *Statistics.* any hypothesis regarding unknown parameters in a population.

parametric mixing *Optics.* the mixing of electromagnetic waves from radiation incident on a nonlinear material that forms waves with linearly related frequencies.

parametric oscillator *Electronics.* a device that varies the amount of reactance in an energy-storage component to obtain oscillation. *Optics.* a device that employs a parametric amplifier to produce a coherent beam of light tunable over a wide range of wavelengths.

parametric programming *Mathematics.* a method of investigating the effects of proportional changes in the elements of a row or column of a matrix on an optimal linear programming solution. Also, SENSITIVITY ANALYSIS.

parametric up-converter *Electronics.* a parametric device that changes the frequency of an output signal, so that it is higher than the input signal.

parametrized voice-response system *Acoustical Engineering.* an electroacoustic system that converts a voice input into electroacoustical signals to be analyzed electronically for comparison with one or more predetermined characteristics, so that a voice within the established parameters will generate a certain response from the system.

parametron *Electronics.* a resonant circuit in which a signal oscillates at half the driving frequency; used as a digital computer element so that the oscillation represents a binary digit in the circuit.

paramilitary *Military Science.* **1.** of or relating to forces that are distinct from a nation's regular armed forces but resemble them in organization, equipment, training, or mission; e.g., the SS organization in Germany in the Hitler era. **2.** a member of such a force.

paramnesia *Neurology.* **1.** a memory disturbance in which a dream or fantasy is substituted for a memory of a real event. **2.** a state in which words are remembered, but are used without comprehension of their meaning.

paramo *Ecology.* a biological community that covers a wide area of the equatorial mountain peaks in the Western Hemisphere; made up essentially of grassland.

Paramoeba *Invertebrate Zoology.* a genus of flattened, naked marine amebae, one of the most numerous of its subclass, the Gymnamoeba.

paramorph *Mineralogy.* a type of pseudomorph having the same composition as the original crystal; for example, calcite after aragonite is a paramorph of aragonite.

paramorphism *Mineralogy.* the process by which a mineral undergoes a change in its internal structure such that a new mineral is formed, having the same composition and external form as the original mineral.

paramp see PARAMETRIC AMPLIFIER.

paramutation *Genetics.* a phenomenon in which one homologous allele in the heterozygous condition influences the phenotypic expression of another allele such that the latter behaves as if it were a mutant, even though no permanent change in its DNA sequence has occurred.

Paramyidae *Paleontology.* a family of the earliest known rodents in the infraorder Protrogomorpha and the superfamily Ischyromyoidea; the paramyids first appeared in the Upper Paleocene in North America, surviving into the Oligocene; probably ancestral to most later rodents and significant in early primate evolution as a competitor.

paramylon *Biochemistry.* a storage polysaccharide of algae that is made up of β(1–3)-glucan and is present as a discrete granule in the cytoplasm.

paramylum *Biochemistry.* a carbohydrate storage compound present in the euglenoids; chemically distinct from both starch and glycogen.

paramyoclonus *Neurology.* a condition characterized by shocklike contractions of muscles or portions of muscles.

paramyoclonus multiplex *Neurology.* a chronic but benign disorder characterized by sudden and frequent shocklike contractions of muscles or portions of muscles, especially of the trunk and legs, occurring from 10 to 50 times a minute and disappearing during sleep or motion; generally develops spontaneously but may follow episodes of trauma, infectious diseases, or poliomyelitis.

paramyosin *Biochemistry.* a protein component of muscle filament in invertebrates.

paramyotonia congenita *Medicine.* an inherited condition characterized by recurrent abnormal muscular weakness upon exposure to cold; transmitted as a dominant trait and considered to be a variety of the hyperkalemic form of periodic paralysis.

Paramyxea *Invertebrate Zoology.* a class of parasitic spore-forming protozoans of the phylum Ascetospora; includes only one species that lives in the cells of the pelagic larva of a polychaete annelid.

Paramyxoviridae *Virology.* a family of enveloped RNA viruses that enter the host cell by fusion of the viral envelope with the plasma membrane of the host cell and are transmitted in horizontal form; mainly airborne.

Paramyxovirus *Virology.* a group of viruses of the family Paramyxoviridae whose virions display both hemagglutinin and neuraminidase activities.

Paraná *Geography.* a river in central South America that flows about 2000 miles south-southwest from southeastern Brazil to the Rio de Plata in Argentina.

paranalgesia *Neurology.* an absence of sensitivity to pain in the lower extremities.

Paranaplasma *Bacteriology.* in former classifications, a genus of bacteria belonging to the taxonomic family Anaplasmataceae; species have been assigned to the genus *Anaplasma.*

paranasal *Anatomy.* located near or alongside the nasal cavities.

paranasal sinus *Anatomy.* one of several air-filled cavities in the maxilla, frontal, sphenoid, and ethmoid bones that are lined with mucous membrane and drain into the nasal cavity.

paranemic joint *Molecular Biology.* a DNA region in which complementary sequences are parallel rather than intertwined.

paranemic spiral *Molecular Biology.* two strands of nucleic acid that are coiled in opposite directions in such a way that they can be separated without uncoiling the strands.

paraneoplastic *Oncology.* of or relating to changes produced in tissues remote from a tumor or its metastases.

paranephritis *Medicine.* **1.** an inflammation of the adrenal gland. **2.** an inflammation of the connective tissue adjacent to the kidney.

paranephroma *Oncology.* a tumor of the adrenal gland.

paranesthesia *Medicine.* the anesthesia of only the lower half of the body, including the legs and lower trunk. Also, **para-anesthesia.**

paraneural *Anatomy.* situated or occurring beside or alongside a nerve.

paranoia [pâr′ə noi′yə] *Psychology.* **1.** a psychotic disorder that is characterized by delusions of persecution and sometimes hallucinations; the person often believes that this persecution occurs because of some special or unique gifts that he or she possesses. **2.** in popular use, any excessive, unfounded feeling of persecution or of suspicion about the motives of others.

paranoid [pâr′ə noid′] *Psychology.* **1.** relating to or affected with paranoia. **2.** a person affected with paranoia.

paranoid character *Psychology.* a personality type characterized by hypersensitivity and feelings of self-importance; individuals with this personality easily become suspicious of others and readily blame them or assign malevolent motives to their actions. Also, **paranoid personality.**

paranoid schizophrenia *Psychology.* a form of schizophrenia characterized by delusions of persecution or delusions of grandeur, or both; hallucinations are often present and the patient may fabricate his or her own reference framework to systematize these elements. Thus, **paranoid schizophrenic.**

paranoid state see PARANOIA.

paranomia *Neurology.* one of two different forms of aphasia: **myotactic paranomia,** which is characterized by an inability to name objects that are touched; or **visual paranomia,** which is characterized by an inability to name objects that are seen.

paranormal *Psychology.* relating to phenomena that have no conventional physical explanation, such as ESP or psychokinesis.

paranthelion [par′ ənt hē′ lē än′] *Astronomy.* a diffuse spot of light occurring in the sky at a distance of (usually) 120° from the sun, often showing rainbowlike colors. Also, MOCK SUN.

paranthropophytia *Ecology.* the destruction that takes place in an environment as a result of continuous or periodic disruptions, such as erosion caused by certain cultivation practices.

Paranyroca *Paleontology.* a genus of birds belonging to the extinct family Anatidae of the extinct order Anseriformes; known only from the Lower Miocene of North America.

Paraonidae *Invertebrate Zoology.* a family of small, slender polychaete annelid worms in the Sedentaria.

paraoperative *Surgery.* pertaining to the accessories essential to operative surgery; e.g., sterilization of instruments or care of gloves.

parapack *Engineering.* a package equipped with a parachute.

Paraparchitacea *Paleontology.* a superfamily of medium-sized ostracods in the extinct order Leperditicopida; Devonian to Permian.

paraparesis *Medicine.* a weakness or a slight degree of paralysis affecting only the legs and lower part of the body.

parapatric *Ecology.* of or relating to species that live in separate but adjoining habitats.

parapatric speciation *Evolution.* a process of speciation in which the ancestral population separates into two or more adjacent populations that do not interbreed.

parapertussis *Medicine.* an acute, highly communicable respiratory infection characterized by spasms of coughing, caused by the infecting agent *Bordetella parapertussis.*

parapet *Architecture.* **1.** a rampart raised above the main wall of a fortification. **2.** a low protective wall along the edge of a roof, balcony, or other structure. *Ordnance.* a protective earthwork, wall, or similar structure in front of a trench or gun emplacement.

paraphase amplifier *Electronics.* an amplifier that converts one input signal into two equally matched out-of-phase output signals.

paraphasia *Neurology.* partial aphasia characterized by the use of incorrect words or phrases, or of correct words in incorrect or senseless combinations.

paraphia *Neurology.* a disorder characterized by perversion of the sense of touch. Also, PSEUDAPHIA, PARAPSIA.

paraphilias *Psychology.* a group of psychosexual disorders involving sexual gratification through socially abnormal means, including such behaviors as fetishism, pedophilia, and exhibitionism.

paraphimosis *Medicine.* an abnormal constriction and retraction of the prepuce behind the glans penis; causes a painful swelling of the glans that can lead to gangrene if untreated.

paraphyletic *Systematics.* of or relating to a taxonomic group that includes a common ancestor and some, but not all, of it descendants.

paraphysis *Vertebrate Zoology.* a thin-walled sac that appears as a prominent fold of the tela choroidea in the most anterior part of the telencephalon and disappears in adult life; its function is unknown, but it may produce glycogen. *Botany.* any of the erect, sterile filaments often occurring in the reproductive structures of many fungi, algae, and ferns.

paraphysoid *Mycology.* **1.** a sterile filament of unknown function found in some fungi belonging to the subdivision Ascomycotina. **2.** a sterile structure forming the outer layer of the fruiting bodies of fungi belonging to the subdivision Basidiomycotina.

Paraphysomonas *Botany.* a genus of unicellular flagellates in the family Synuraceae, characterized as spochlorotic and nannoplanktonic.

Parapithecidae *Paleontology.* a family of primitive apes in the superfamily Parapithecoidea; known especially from the Middle Oligocene Fayum deposits of Egypt; one member of the family, Apidium, is probably ancestral to Oreopithecus.

paraplegia [pâr′ə plē′jē ə; pâr′ə plē′jə] *Neurology.* paralysis of the legs and lower part of the body, and possibly including the back and abdominal muscles; the most common cause is trauma to the spinal cord, but lesions may also result in paraplegia.

paraplegic [pâr′ə plē′jik] *Neurology.* relating to or affected by paraplegia. Also, **paraplectic.**

parapodium *Invertebrate Zoology.* **1.** in polychaetes, either of a pair of lateral projections, usually flat with bristles, that are used for locomotion or respiration. **2.** in some gastropod mollusks, either of the paired lobes extending from the foot, used in swimming.

parapositronium *Particle Physics.* a sample of positronium in which the spins of the positron and the electron are antiparallel.

Parapoxvirus *Virology.* a genus of viruses in the subfamily Chordopoxvirinae that infect ungulates.

parapraxis *Psychology.* a seeming error in speech or writing that actually reveals some unconscious or underlying truth.

paraprotein *Immunology.* an abnormal protein that occurs in the blood as a result of a pathological disturbance.

parapsia see PARAPHIA.

parapsychology *Psychology.* the attempted study of unexplainable phenomena such as ESP, telepathy, clairvoyance, psychokinesis, precognition, and poltergeist activity, using known psychological procedures and principles. Thus, **parapsychologist.**

Parapuzosia *Paleontology.* the largest known ammonoids, a genus of ammonitids in the superfamily Desmocerataceae; they occur worldwide, with fossil shells up to 1.7 meters (previously estimated about 30% larger); extant in the Upper Carboniferous.

paraquat *Organic Chemistry.* $[CH_3(C_5H_4N)_2CH_3]\cdot 2CH_3SO_4$, a highly toxic, yellow solid that is soluble in water. *Toxicology.* a contact herbicide and defoliant that is also toxic to mammals; contact causes skin irritation, cracking and shredding of the nails, and delayed healing of cuts and wounds; ingestion may lead to kidney or liver failure followed by pulmonary insufficiency.

pararammelsbergite *Mineralogy.* $NiAs_2$, a tin-white metallic mineral occurring in reniform masses or as tabular crystals, having a specific gravity of 7.12 to 7.25 and a hardness of 5 on the Mohs scale; found in mesothermal vein deposits.

pararectus incision *Surgery.* an oblique incision into the abdominal wall to one side of the midline.

para red *Organic Chemistry.* $C_{10}H_6(OH)NNC_6H_4NO_2$, a red pigment.

pararipple *Geology.* a large, generally symmetric ripple characterized by a gently sloping surface and no assortment of grains.

pararosaniline *Organic Chemistry.* $HOC(C_6H_4NH_2)_3$, combustible, colorless to red crystals; very slightly soluble in water and soluble in alcohol; melts at 205°C; used as a dye.

pararotavirus *Virology.* any of a group of viruses that lack the antigen specific to rotaviruses but are morphologically identical to rotaviruses. Also, GROUP B ROTAVIRUSES.

Parasaleniidae *Invertebrate Zoology.* a small family of sea urchins, echinoderms having an oval test and valves with end teeth.

paraschist *Petrology.* a schist formed by alteration of a sedimentary rock.

paraselene [par′ ə sə lēn′] *Astronomy.* a bright spot on a lunar halo formed in much the same manner as a paranthelion. Also, MOCK MOON.

paraselenic circle *Astronomy.* a halo around the moon at a distance of 22°, caused by refraction of moonlight by ice crystals in the earth's atmosphere.

Paraselloidea *Invertebrate Zoology.* a group of aquatic isopod crustaceans in the suborder Asellota.

Parasemionotidae *Paleontology.* a family of small neopterygian fishes in the extinct order Amiiformes; probably not ancestral to teleosts; extant in the Triassic.

parasexual cycle *Genetics.* a reproductive cycle that does not include meiosis, even though crossing over and recombination may occur in certain somatic cells.

parasheet *Aviation.* a simple type of parachute having the form of a regular polygon with rigging lines attached to the apexes; may be gathered (constrained by a hem cord at its periphery) or ungathered.

parasigmatism see LISP.

parasite *Biology.* an animal or plant living in or on an organism of another species (its host), obtaining from it part or all of its organic nutriment, and commonly exhibiting some degree of adaptive structural modification. The host is typically, but not always, harmed by the presence of the parasite; it never benefits from this presence. *Electricity.* an irregular electrical measurement due to a current in a circuit that is produced by an unintentional cause such as inequalities of temperature or of composition. *Computer Science.* see DEMON.

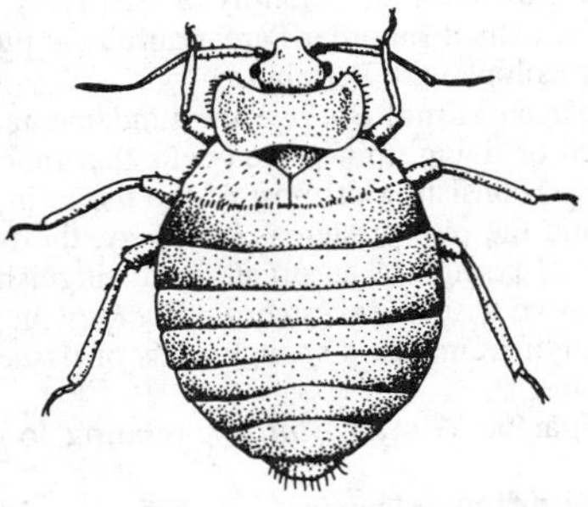

parasite (bedbug, *Cimex lectularius*)

parasite drag *Fluid Mechanics.* a drag produced by the combined effects of skin friction over an airfoil and extra pressure drag resulting from regions of flow separation over the surface (form drag).

parasitemia *Medicine.* the presence of parasites in the blood, especially malarial and other protozoan forms, and microfilariae.

parasitic [pâr´ə sit´ik] *Biology.* relating to or being a parasite; living as a parasite. *Electronics.* **1.** an undesired and inefficient signal in an electric circuit, generally one that is higher or lower than the desired frequency. **2.** an unavoidable property of a real component that causes its operation to deviate from that of an ideal component (e.g., parasitic resistance of an inductor or a capacitor, parasitic inductance of a resistance wire). *Volcanology.* describing a volcanic cone, crater, or lava flow occurring on the flank of and subsidiary to a larger cone. Thus, **parasitic cone, parasitic crater.**

Parasitica see ANOPLURA.

parasitic capacitance *Electricity.* the usually undesirable, inherent capacitance of a component; e.g., in a transformer, the capacitance between windings.

parasitic capture *Nucleonics.* a neutron absorption process that does not result in fission or the production of another element.

parasitic castration *Biology.* a restraint of function or development of gonads due to the infestation of a host with parasites.

parasitic current *Electricity.* current induced by a parasitic resistance, inductance, or capacitance.

parasitic element *Electromagnetism.* an antenna element that is driven indirectly by radiation from another antenna element.

parasitic inductance *Electricity.* the usually undesirable, inherent inductance in a component; e.g., the inductance of a wirewound resistor.

parasitic oscillation *Electronics.* the undesired signals that rise above or drop below the operating frequency of an oscillator or an amplifier, chiefly seen in those with vacuum tubes.

parasitic resistance *Electricity.* the usually undesirable, inherent resistance of a component; e.g., in a capacitor, the finite resistance of the dielectric.

parasitic suppressor *Electronics.* a device that inhibits parasitic high-frequency oscillations, usually consisting of a coil and resistor connected in parallel.

parasitism [pâr´ə sit´iz əm] *Biology.* a relationship between two species in which one, the parasite, benefits from the other, the host; it usually also involves some detriment to the host organism.

parasitoid *Biology.* of or relating to a parasite, especially one that practices parasitoidism.

parasitoidism *Biology.* a condition existing between various insect larvae and their hosts in which the larva feeds upon the living host tissues in an orderly sequence, such that the host is not killed until larval development is finished.

parasitoid predator *Entomology.* an insect that lays eggs on or near prey, with the emerging larvae living inside the host and feeding on host tissue; this differs from true parasitoidism in that a parasitoid always kills its host.

parasitology *Biology.* a branch of biology dealing with parasites and the effects of parasitism. Thus, **parasitologist.**

parasitophobia *Psychology.* an inordinate fear of parasites.

parasol ant *Invertebrate Zoology.* any of several tropical South American ants of the genus *Atta* that cut and chew bits of leaves and flowers into a mash that they use to cultivate a fungus garden. (So called because they carry leaves overhead in the manner of a parasol.)

paraspasm *Neurology.* a spasm that simultaneously affects the corresponding muscles on both sides of the body, especially the legs.

parasphenoid *Vertebrate Zoology.* in bony fishes and lower tetrapods, a dermal structure that is formed in the skin of the roof of the mouth and is slender in front and spread out posteriorly.

parasplenic *Anatomy.* located near or beside the spleen.

Parastacidae *Invertebrate Zoology.* a family of freshwater crayfish found in South America, Madagascar, New Guinea, Australia, and New Zealand.

parastate *Atomic Physics.* a state in which the nuclei of a diatomic molecule spin in the opposite direction.

parastratigraphic unit see OPERATIONAL UNIT.

parastratigraphy *Geology.* **1.** a supplemental classification of stratified rocks based on fossils other than those upon which the prevalent orthostratigraphy is based. **2.** a classification of strata based on an arbitrarily selected stratigraphic unit that is distinguished by certain objective criteria for use in analysis.

parastratotype *Geology.* an additional section in a region in which a stratotype was defined.

parastyle *Vertebrate Zoology.* a small cusp in front of the paracone of an upper molar.

Parasuchia see PHYTOSAURIA.

parasympathetic *Anatomy.* of or relating to the parasympathetic nervous system.

parasympathetic nervous system *Anatomy.* the part of the autonomic nervous system originating in the midbrain, hindbrain, and sacral region; responsible for such involuntary functions as slowing heart rate, modulating constriction of smooth muscles of the digestive tract and bronchioles, constriction of the pupils, and increasing glandular secretion, excluding the sweat glands.

parasympatholytic see ANTICHOLINERGIC.

parasympathomimetic *Pharmacology.* **1.** producing effects resembling those of stimulation of the parasympathetic nerve supply to a part of the body. **2.** any agent that produces effects resembling those caused by stimulation of the parasympathetic nerves.

parasystole *Cardiology.* an irregularity of heartbeat consisting of two overlapping rhythms caused by the stimulation of the heart by two foci (one in the sinoatrial node and one ectopic), each acting independently of the other.

paratacamite *Mineralogy.* $Cu_2^{+2}(OH)_3Cl$, a green to black trigonal mineral, trimorphous with atacamite and botallackite, occurring as rhombohedral crystals and in granular form, and having a specific gravity of 3.74 and a hardness of 3 on the Mohs scale; found as a secondary mineral in some copper-bearing deposits.

Parataeniellaceae *Mycology.* a family of fungi belonging to the order Eccrinales that occur in the hindgut of terrestrial isopods or in scarabid beetle larvae.

parataxic distortion *Psychology.* a distortion of perception and judgment, especially of interpersonal relationships, resulting from a tendency to perceive others according to a pattern determined by previous experiences.

parateny *Invertebrate Zoology.* the transmission of a parasite from one host to another without the occurrence of maturation or development in the parasite.

parathormone see PARATHYROID HORMONE.

Parathuramminacea *Paleontology.* an early superfamily of foraminiferids in the suborder Fusulinina; extant in the Lower Ordovician to Uppermost Carboniferous.

parathyrin see PARATHYROID HORMONE.

parathyroidectomy *Medicine.* the excision or surgical removal of the parathyroid glands.

parathyroid gland *Anatomy.* any of four small endocrine glands lying near the thyroid and producing hormones that regulate calcium and phosphorus metabolism.

parathyroid hormone *Endocrinology.* a polypeptide hormone that is secreted by the parathyroid glands and increases levels of blood calcium by stimulating the mobilization of calcium from bone and inhibiting the excretion of calcium by the kidneys. Also, PARATHORMONE, PARATHYRIN.
parathyroid insufficiency see HYPOPARATHYROIDISM.
parathyroidoma *Oncology.* a tumor of the parathyroid gland.
paratill *Geology.* an unstratified, unsorted deposit of glacial drift formed by ice-rafting in a marine or lake environment.
paratonia *Neurology.* a defect of muscle tone or tension in which a contraction persists and the limb involved is temporarily paralyzed in the position it has just assumed.
paratonic movement *Botany.* plant movement in response to an external stimulus such as gravity, sunlight, chemicals, heat, or electricity.
paratope *Immunology.* the area of an antibody molecule that makes contact with the antigenic determinant (epitope).
paratrichosis *Medicine.* any disorder in the growth of hair, especially the growth of hair in abnormal places.
paratroch *Invertebrate Zoology.* a ciliated band encircling the anus of certain coelemates, such as trochophores.
paratroops or **paratroopers** *Military Science.* an infantry unit trained to attack or land in combat areas by parachuting from airplanes. Also, PARACHUTE TROOPS.
paratype *Systematics.* any of the specimens in a type series, exclusive of the holotype.
paratyphoid fever *Medicine.* an acute infectious disease caused by the presence of the paratyphoid bacillus and similar to but milder than typhoid fever.
parauterine organ *Invertebrate Zoology.* a tough fibrous sac, arising from the tapeworm (cestode) mesenchyme, which connects with the uterus by forming around the uterine branches; receives and holds embryos.
parautochthonous *Geology.* of or related to a body of rock intermediate between autochthonous and allochthonous, which is displaced only a short distance from its place of origin.
paravaccinia virus *Virology.* a virus of the genus *Parapoxvirus* that infects cows, producing lesions similar to those of cowpox.
paravane *Engineering.* **1.** a hydrodynamic device that has sharp teeth-like structures on its front end and is towed on the end of a mine sweeping cable to cut the cables of anchored mines. **2.** an aerodynamic body attached to a cable and towed from an airplane in order to keep the line taut.
paravauxite *Mineralogy.* $Fe^{+2}Al_2(PO_4)_2(OH)_2 \cdot 8H_2O$, a colorless to greenish-white triclinic mineral of the paravauxite group occurring as small prismatic to tabular crystals, having a specific gravity of 2.36 and a hardness of 3 on the Mohs scale; found in pegmatites and the tin deposits of Llallagua, Bolivia.
paraventricular nucleus *Physiology.* a nucleus of nerve cells situated in the vicinity of the third ventricle in the hypothalamic region of the brain.
parawollastonite *Mineralogy.* $CaSiO_3$, a white or gray monoclinic mineral, dimorphous with wollastonite, consisting of a calcium metasilicate, and having a specific gravity of 2.913 and a hardness of 4.5 to 5 on the Mohs scale; found in igneous intrusions in limestones.
paraxial *Science.* lying along an axis.
paraxial optics see GAUSSIAN OPTICS.
paraxial rays *Optics.* the light rays whose paths are closest and nearly parallel to the axis of an optical system.
paraxial trajectory *Electricity.* in an axially symmetric electrical or magnetic field, the path that a charged particle takes when its distance from the field and its axis of symmetry and corresponding angle subtended between its axis have little effect on altering its trajectory.
paraxon *Neurology.* a subsidiary branch of an axon.
paraxonic *Vertebrate Zoology.* having the axis of the foot between the third and fourth digits, as in artiodactyls.
Parazoa *Invertebrate Zoology.* a branch of the animal subkingdom Metazoa that includes the single phylum Porifera (sponges), the only multicellular animal with no true digestive tract.
parazoan *Paleontology.* a type of multicellular animal that evolved into sponges in a line separate from all other animals, so that sponges are placed in the separate subkingdom Parazoa.
parchment *Materials.* **1.** an animal skin, especially that of a sheep or goat, that has been prepared for writing, drawing, or painting. **2.** a stiff, cream-colored paper that resembles parchment.
parchment paper *Materials.* a waterproof, grease-resistant paper produced by passing ordinary paper through sulfuric acid or zinc chloride.
PARCS *Ordnance.* an American radar designed to detect and analyze the approach of ICBMs. (An acronym for perimeter acquisition radar attack characterization system.)
Pardalotidae *Vertebrate Zoology.* in some classification systems, the diamond birds, a family of spotted tropical birds; placed in Dicaeidae in other systems.
Paré, Ambroise 1510–1590, French surgeon; introduced the use of artificial limbs; greatly improved the treatment of wounds.
paregoric [pâr´i gôr´ik] *Pharmacology.* a preparation of powdered opium, anise oil, benzoic acid, camphor, glycerin, and diluted alcohol, used to treat diarrhea by inhibiting contractions of the bowel muscles.
pareiasaur *Paleontology.* of or relating to the Permian cotylosaurs of the suborder Pareiasauroidea.
Pareiasauridae *Paleontology.* a family of massively built, herbivorous cotylosaurs in the extinct suborder Pareiasauroidea; probably descended from an early procolophonid; Permian.
parenchyma [pâr´ən kī´mə] *Botany.* the fundamental tissue of higher plants, composed of living, thin-walled cells that function in photosynthesis and storage, consisting of most of the tissue in leaves, roots, the pulp of fruits, and the pith of stems. *Histology.* the functional, usually secretory epithelial portion of an organ, as distinguished from the supporting and nutritive tissues. *Invertebrate Zoology.* in some simple animals such as Platyhelminthes, a spongy mass of tissue filling the space between the organs.
parenchymal [pâr´ən kī´məl] *Biology.* relating to or involving the parenchyma.
parenchymal jaundice *Medicine.* jaundice caused by destruction or functional impairment of the liver cells. Also, **parenchymatous jaundice.**
parenchymula *Invertebrate Zoology.* a larval stage of sponges in which the flagellated surface is flat and amoeboid cells are inside.
parent *Biology.* an organism that brings forth offspring. *Computer Programming.* **1.** the previous generation of an item or file that is required to create a new record. **2.** in data structures, a node on a tree that has a given node as one of its subtrees. *Nuclear Physics.* a radionuclide that yields a new nuclide, a daughter, when it disintegrates.
parental magma *Geology.* magma from which a particular igneous rock was formed or from which another body of magma was derived.
parental type *Genetics.* the genotype of parents in a genetic cross, especially with reference to genetic markers being studied in the cross.
parent clauses *Artificial Intelligence.* in resolution theorem proving, the two original clauses from which a new clause is derived by resolution.
parent compound *Chemistry.* **1.** a compound from which one or more molecules or compounds derive. Also, PARENT SUBSTANCE. **2.** see PARENT NAME.
parenteral *Medicine.* entering the body by some means other than the intestine.
parent figure *Psychology.* **1.** a person who emotionally represents one's father or mother and who is the object of the responses and attitudes, such as love or dependence, associated with the child-parent relationship. **2.** a mature individual with whom one identifies and who functions like a parent in providing protection, guidance, and love.
parenthesis-free notation see POLISH NOTATION, def. 1.
parenting *Psychology.* **1.** the rearing of children. **2.** the state of being a parent; parenthood. **3.** the techniques and methods required in rearing children.
parent log see TAFFRAIL LOG.
parent material *Geology.* unconsolidated mineral or organic matter from which a true soil develops.
parent name *Chemistry.* the part of the name of a compound that becomes the root of the name of its derivatives; e.g., *methane* is the parent name of *methanol.*
parent of a state *Quantum Mechanics.* the wavefunction for a system of *N* tightly bound particles alone, as related to the daughter state of a system in which there is one particle in addition to the *N* particles.
parent rock *Geology.* **1.** the mass of rock that gives rise to the parent material of a soil. **2.** see SOURCE ROCK.
parent substance see PARENT COMPOUND.
paresis [pə rē´sis] *Neurology.* partial or incomplete paralysis characterized by weakness and reduction in muscular power. *Medicine.* also, **general paresis.** a disease of the brain, syphilitic in origin, marked by progressive muscular weakness.
paresthesia *Neurology.* any sensation, such as tickling, pricking, burning, or tingling, that occurs without external stimulus.

Pareto distribution *Statistics.* distribution on the positive real line, given as $f(x) = (a/b)[b/(b+x)]^{a-1}$, $x > b$; widely used in reliability and in the analysis of income distribution.

Pareto optimality *Industrial Engineering.* the allocation of resources that results in the highest product output.

Pareto's law *Industrial Engineering.* a principle stating that, in most cases, a small fraction (15–20%) of the participants account for the major part of an activity (80–85%); for example, 20% of customers account for 80% of sales. Also, RULE OF 80–20.

parfocal eyepieces *Optics.* eyepieces that have common focal planes, enabling them to be interchanged on an instrument without being refocused.

parget *Building Engineering.* **1.** plaster, whitewash, or roughcast for coating a wall. **2.** plasterwork, especially in raised ornamental figures on walls.

parging *Civil Engineering.* a minimal coating of plaster or cement laid over masonry to seal it from moisture.

parhelic circle *Astronomy.* a whitish horizontal band that passes through the sun, either incomplete or extending around the horizon; caused by the reflection of the sun's rays from the vertical faces of ice prisms in the atmosphere. Also, **parheliacal ring.**

parhelion [pär hēl′ yən; pär hēl′ ē ən] *Astronomy.* a bright circular spot on a mock sun that is caused by refraction of light by cloud particles at high altitudes; usually two such spots occur at the same altitude above the horizon as the sun, but 22° on either side of it. Also, SUNDOG.

parhelium *Atomic Physics.* a state in which two electrons in helium spin in opposite directions. Also, **parahelium.**

Parian cement *Materials.* a hard plaster consisting of an intimate mixture of gypsum and borax that has been calcined and ground to powder.

Paridae *Vertebrate Zoology.* the Old World titmice and North American chickadees, a family of small, acrobatic, social songbirds of the passerine suborder Oscines; characterized by a large rounded head, fluffy underbelly, and short, strong legs.

paries [pâr′ ē ēz] *Biology.* the wall of a cavity or hollow organ in an animal or plant body.

parietal [pə rī′ə təl] *Anatomy.* **1.** of or related to the outer wall of an organ or cavity. **2.** of or related to the parietal bone. *Botany.* of or relating to a plant part that has a peripheral location; especially used to describe a placenta attached to an ovary's main wall rather than to the axis or a cross wall.

parietal bone *Anatomy.* either of a pair of bones that form the sides of the cranial vault.

parietal cell *Histology.* in the gastric glands of the stomach, a cell that secretes hydrochloric acid. Also, OXYNTIC CELL.

Parietales *Botany.* an order of plants in the Englerian system roughly equivalent to the order Violales.

parietal lobe *Anatomy.* the middle lobe of each cerebral hemisphere.

parietal pericardium *Anatomy.* the outer layer of the double serous membrane that surrounds the heart; it is surrounded by another layer, the fibrous pericardium.

parietal peritoneum *Anatomy.* the serous membrane that lines the abdominal cavity.

parietal pleura *Anatomy.* the serous membrane that lines the chest cavity.

paring chisel *Mechanical Devices.* a narrow, long-bladed chisel used to shave, cut, pare, or slice grooves for finishing woodwork.

paring gouge *Mechanical Devices.* a hand carving tool with a rounded blade of approximately 7 inches that is bevel-ground on its inside surface; used to cut curved shapes or outlines in woodwork.

pariobasidium *Mycology.* in certain fungi belonging to the subdivision Basidiomycotina, the portion of the fungal spore-bearing structure or basidium that contains the remains of the probasidium, in which fertilization occurs.

paripinnate *Botany.* of or relating to a leaf with an even number of leaflets and no terminal leaflet; even-pinnate.

Paris equation *Materials Science.* an equation describing crack growth in the intermediate range of the stress-intensity factor that is dominated by the striation growth process, expressed as $da/dN = C\Delta K^m$, where C and m are scaling constants, da/dN represents log-log plots of fatigue-crack growth rate per cycle, and ΔK represents the change in the stress-intensity factor.

Paris gun *Ordnance.* any of various German 210-mm guns having a 112-foot barrel and delivering an approximately 235-pound projectile at a range of almost 80 miles; used in the long-range bombardment of Paris during World War I.

parisite-(Ce) *Mineralogy.* $Ca(Ce,La)_2(CO_3)_3F_2$, a brownish-yellow trigonal mineral occurring in acute double hexagonal pyramids, having a specific gravity of 4.2 to 4.36 and a hardness of 4.5 on the Mohs scale; found in carbonaceous shales in Colombia.

parison *Materials Science.* a hot blob of soft or molten polymer that is partly blown or formed into a shape, before final shaping in a mold. *Engineering.* specifically, a piece of glass with an approximate shape in a preliminary forming process that is ready for final shaping. (From a French term meaning "to prepare.")

parison swell *Engineering.* the ratio of the cross-sectional area of a parison to the die opening.

parity *Mathematics.* the property of being either even or odd; i.e., two integers that are both even or both odd have the same parity. *Computer Technology.* a function that indicates whether the number of binary ones in a word is odd or even; used for error detection. *Particle Physics.* the property of an elementary particle or physical system that indicates whether or not its mirror image occurs in nature; it is designated as either even or odd, and defined by the symbol *P*. *Quantum Mechanics.* the property of a wavefunction that describes the behavior of a system whose physical coordinates are related by inversion about a center; if parity is even, the wavefunction is unchanged and if odd, the wavefunction is changed in sign only.

parity *Biology.* the condition or fact of having borne offspring.

parity bit *Computer Technology.* a binary digit that is appended to the digits of a logical component, such as a computer word, in order to maintain the sum of the digits in a condition of parity. The parity bit is generated when writing and checked when reading. Also, REDUNDANCY BIT.

parity error *Computer Technology.* a condition that occurs when the number of binary ones in a word is odd when expected to be even, or even when expected to be odd.

parity group *Crystallography.* a set of structure factors whose three indices are odd or even in an identical way. There are eight parity groups for three indices.

parity selection *Computer Technology.* a code or command to select whether even or odd parity is to be used.

parity-selection rules *Quantum Mechanics.* the rules governing the changes in parity of systems during transitions.

parity transformation *Telecommunications.* an alteration of the value of a transmitted character that delineates the number of one-bits.

Parkeriaceae *Botany.* the water ferns, a monogeneric family of aquatic ferns of the order Filicales, characterized by a hairy or naked short-creeping rhizome, small to medium fronds bearing sporangia along the margin, and netted veins.

parkerite *Mineralogy.* $Ni_3(Bi,Pb)_2S_2$, a bright bronze, metallic, monoclinic mineral occurring as minute grains, having a specific gravity of 8.4 to 8.5 and a hardness of 3 on the Mohs scale; found in some hydrothermal sulfide-arsenide assemblages.

parkerizing *Metallurgy.* an anticorrosion coating process based on phosphates.

Parker-Washburn boundary *Solid-State Physics.* a boundary within a crystal lattice consisting of a single array of dislocations; the boundary separates the crystal into two regions having different crystal axes.

Parkes process *Metallurgy.* one of several processes used to separate precious metals from lead.

parking apron *Civil Engineering.* a concrete or other hardened surface on which aircraft are parked.

parking brake see EMERGENCY BRAKE.

parking lot *Civil Engineering.* a ground-level area reserved for automobile storage.

parking orbit *Space Technology.* a temporary orbit of a spacecraft about a celestial body for some transient purpose, such as testing flight systems (often in earth orbit before interplanetary flight), assembling components, or awaiting conditions more favorable for movement.

Parkinson, James 1755–1824, English surgeon and paleontologist; described Parkinson's disease; wrote a major study of fossils.

Parkinsonia *Paleontology.* a genus of ammonoids in the order Ammonitida and superfamily Perisphinctaceae; laterally compressed and strongly ribbed; extant in the Middle Jurassic.

parkinsonism *Neurology.* any of a group of neurological disorders that are characterized by rhythmic muscle tremors, hypokinesia, muscular rigidity, and a slow, shuffling gait; this may develop in association with certain diseases, such as acute encephalitis and poliomyelitis, or certain drugs, especially derivatives of phenothiazine. (Named for James *Parkinson*.)

Parkinson's disease *Neurology.* a slowly progressive form of parkinsonism usually occurring late in life; characterized by masklike facial expression, tremor of resting muscles, slowing of voluntary movements, short accelerating gait, peculiar posture, and weakness of the muscles. Also, PARALYSIS AGITANS.

Parkinson's syndrome *Neurology.* a group of symptoms identical with parkinsonism; associated with encephalitis and also found among former professional boxers.

parkway *Civil Engineering.* a highway with a median strip and/or side strips of grass, shrubs, and trees.

Parmeliaceae *Botany.* a family of foliose shield lichens in the order Lecanorales.

Parmulariaceae *Mycology.* a family of fungi of the order Hysteriales that are primarily plant parasites, occurring in tropical regions.

Parnidae see DRYOPIDAE.

Parodiopsidaceae *Mycology.* a family of fungi belonging to the order Asterinales that form dark brown blotches on the surfaces of leaves, stems, and twigs; found exclusively in tropical regions.

parogenetic *Geology.* of or relating to material formed prior to the enclosing rock, especially a concretion that developed in a rock different from and older than its younger host.

paromomycin *Microbiology.* an antibiotic produced by the bacterium *Streptomyces rimosus* subsp. *paromomycinus*; active against a broad spectrum of Gram-positive, Gram-negative, and other bacteria, including certain intestinal bacteria. Also, CRESTOMYCIN.

paronychia *Medicine.* an inflammation of the folds of tissue surrounding the nail, with separation of the skin from the proximal portion of the nail, usually due to bacterial or fungal infection.

parosmia *Neurology.* any disorder of the sense of smell, especially a false sense of odors or the subjective perception of odors that do not exist. Also, **parosphresia.**

parotid *Anatomy.* situated or occurring near the ear.

parotid duct *Anatomy.* a duct beginning as small branches in the parotid salivary gland and opening into the mouth near the second molar of the upper jaw.

parotidectomy *Medicine.* the surgical removal of a parotid gland.

parotid gland *Anatomy.* the largest of the salivary glands, lying under the skin below the zygomatic arch and in front of the external ear on each side of the head.

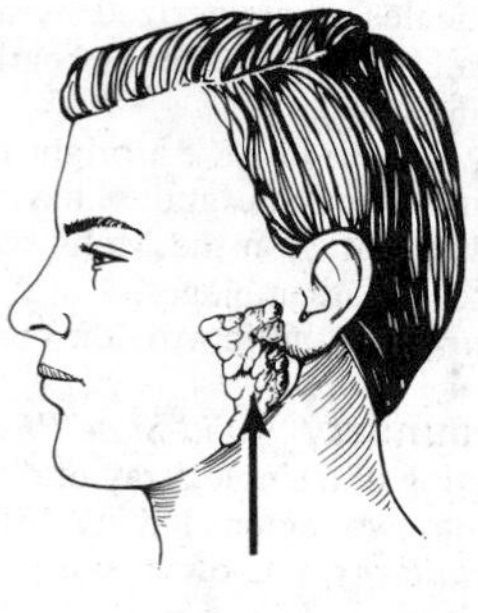

parotid gland

parotitis *Medicine.* an inflammation of the parotid (or salivary) glands, as in mumps.

parous *Medicine.* having given birth to one or more infants.

parovarian cyst *Medicine.* a cyst of mesonephric origin arising between the layers of the mesosalpinx, adjacent to the ovary.

paroxypropione *Pharmacology.* $C_9H_{10}O_2$, a soluble crystalline substance that inhibits the release of pituitary gonadotropic hormones. Also, *p*-HYDROXYPROPIOPHENONE.

paroxysm [pə räk´siz əm] *Medicine.* **1.** a periodic crisis in the progress of a disease; a sudden reappearance or increase in the intensity of symptoms. **2.** a spasm, convulsion, or seizure. **3.** in electroencephalography, a burst of electrical activity denoting cerebral dysrhythmia or epileptic discharges. *Psychology.* a sudden emotional outburst, as of crying or laughing.

paroxysmal eruption [pâr´ik siz´məl] see VULCANIAN-TYPE ERUPTION.

paroxysmal sleep *Neurology.* see NARCOLEPSY.

paroxysmal tachycardia *Cardiology.* a condition marked by sudden bursts of excessive heart rate followed by cessation of the heart's action; the rapid rate has its origin in an ectopic focus.

parrot *Vertebrate Zoology.* any of numerous birds of the family Psittacidae, including the cockatoo, lory, macaw, and parakeet; distinguished by a stout crooked beak, generally bright, colorful plumage, and often the ability to mimic a variety of sounds, including speech; found in lowland tropical and subtropical forests and woodland, open forest, and grassland of Central and South America, Africa, Madagascar, southern and southeastern Asia, Australasia, and Polynesia.

parrot

parrot disease or **parrot fever** see PSITTACOSIS.

parrotfish *Vertebrate Zoology.* any of various marine fishes, especially of the family Scaridae; so called because of their bright coloring and the shape of their jaws; found in tropical waters.

Parry arcs *Optics.* the faintly colored halos visible above and below the sun that arise from refraction caused by ice crystals with a preferred orientation.

pars anterior *Anatomy.* **1.** a subdivision of the anterior lobe of the pituitary gland. **2.** the interior part of the frontal convolution of the anterior lobe of the cerebellum, separated from the pars triangularis by the fissure of Sylvius. Also, **pars distalis.**

parsec *Astronomy.* a distance from which the earth's average distance from the sun would subtend 1 second of arc; equal to 206,265 astronomical units or about 3.26 light-years.

parser *Computer Science.* a program that determines how a given statement in a language could be derived from the grammar of the language, producing a parse tree or other data about the statement as output.

parser generator *Computer Science.* a program that constructs a parser from a specification of the grammar of a language and actions that are to be taken when phrases of the language are recognized.

parse tree *Computer Science.* a tree structure describing the derivation of a statement in a language according to the rules of a context-free grammar. The root of the tree is the "sentence" symbol, the leaves are the words or lexical units of the sentence, and each interior node represents a derivation step from the grammar; it may be annotated with additional information, as for compilation purposes.

parsettensite *Mineralogy.* a copper-red monoclinic, pseudohexagonal mineral having an approximate formula of $(K,Na,Ca)(Mn,Al)_7Si_8O_{20}(OH)_8{\cdot}2H_2O$, occurring in massive form, and having a specific gravity of 2.59 and a hardness of about 1.5 on the Mohs scale; found in manganese deposits in Switzerland.

Parseval's equation *Mathematics.* let $\{e_1, e_2, \ldots\}$ be a countable orthonormal system in a Hilbert space H. Parseval's equation for $x \in H$ is

$$\sum_{k=1}^{\infty} \langle x,e_k \rangle^2 = \|x\|^2.$$

In the usual application, an element x of H is equal to its Fourier series

$$\sum_{k=1}^{\infty} \langle x,e_k \rangle e_k$$

if and only if x satisfies Parseval's equation. (If the orthonormal system is complete, then the Fourier series of any $x \in H$ converges to x.) Also, **Parseval's identity.**

Parseval's theorem *Mathematics.* let the Fourier series of the real functions $f(x)$ and $g(x)$ be, respectively,

$$f(x) = a_0/2 + \sum_{k=1}^{\infty} (a_k \cos kx + b_k \sin kx) \text{ and}$$

$$g(x) = \alpha_0/2 + \sum_{k=1}^{\infty} (\alpha_k \cos kx + \beta_k \sin kx).$$

Then the integral of their product can be expressed in terms of the Fourier coefficients:

$$\int_{-\pi}^{\pi} f(x)g(x)dx = \pi[(a_0\alpha_0)/2 + \sum_{k=1}^{\infty} (a_k\alpha_k + b_k\beta_k)].$$

If $f(x) = g(x)$, the resulting identity is a particular case of Parseval's equation.

parsing *Computer Programming.* **1.** the breaking down of a complex entity into its elementary parts. **2.** the determination of how a given sentence (string of characters or tokens) could have been generated from a natural language. *Artificial Intelligence.* the process of ascribing structure to a linear sequence of words or symbols according to the rules of a grammar.

pars intermedia *Anatomy.* a subdivision of the anterior lobe of the pituitary gland.

parsley *Botany.* an edible annual or biennial herb of European origin, *Petroselinum crispum,* of the carrot family Apiaceae; widely used as a garnish and for flavoring.

pars nervosa see NEUROHYPOPHYSIS.

parsnip *Botany.* a biennial herb of Mediterranean origin, *Pastinaca sativa,* of the carrot family Apiaceae; cultivated for its thick, edible taproot.

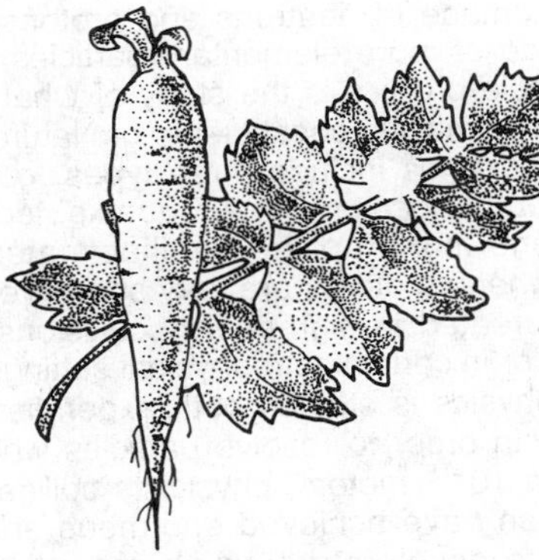

parsnip

parsnip yellow fleck virus group *Virology.* a genus of plant viruses with a narrow host range, transmitted mechanically or by aphids with a helper virus.

Parsons, Elsie Clews 1875–1941, American anthropologist; studied the culture of the Southwestern Indians.

parsonsite *Mineralogy.* $Pb_2(UO_2)(PO_4)_2 \cdot 2H_2O$, a pale-yellow to brown triclinic mineral occurring as prismatic crystals and fibrous masses or crusts, having a specific gravity of 5.37 to 5.72 and a hardness of 2.5 to 3 on the Mohs scale; rare and strongly radioactive; found with other uranyl phosphates.

Parson's steam turbine *Mechanical Engineering.* a reaction steam turbine, containing both stationary and moving blades of increasing size, in which the steam expands gradually from inlet pressure at the smallest section of blades to condenser pressure at the end. Also, **Parson's stage steam turbine.**

pars tuberalis *Anatomy.* a band of cells that extends from the anterior lobe of the hypophysis and wraps the infundibular stalk of the hypothalamus.

pars uterina placentae see MATERNAL PLACENTA.

part *Engineering.* an element of an assembly that has no value in and of itself, but that may be separated from the assembly.

part. particular.

part. aeq. equal parts. (From Latin *partes aequales.*) Also, **Part. aeq.**

part. dolent. painful parts. (From Latin *partes dolores.*) Also, **Part. dolent.**

parthenita *Invertebrate Zoology.* any of several immature stages of flukes.

parthenocarpy *Botany.* the development of fruit without fertilization, either spontaneously or artificially under cultivation.

parthenogenesis *Invertebrate Zoology.* unisexual reproduction in which young are produced by unfertilized females; common in rotifers, aphids, ants, bees, and wasps.

parthenomerogony *Developmental Biology.* the growth of a nucleated fragment of an unfertilized ovum.

parthenospore *Mycology.* see AZYGOSPORE.

partial *Science.* not total or general; incomplete.

partial carry *Computer Programming.* **1.** in parallel addition, the temporary storage of the carry bits rather than immediate propagation. **2.** the computer word in which the carry bits are stored.

partial cleavage see MEROBLASTIC.

partial common battery *Telecommunications.* in telephony, a system in which each individual telephone supplies the talking battery, and the switchboard supplies the supervisory and signaling battery.

partial condensation *Chemical Engineering.* a single contact separation process in which only a portion of a vapor stream is condensed.

partial correctness *Computer Science.* a condition of a program that is guaranteed to produce the correct result if it terminates.

partial correlation coefficient see COEFFICIENT OF PARTIAL CORRELATION.

partial derivative *Mathematics.* the ordinary derivative of a function of two or more variables taken with respect to one of the variables, while considering the other variables to be fixed. A function may be partially differentiable with respect to some, but not necessarily all, of its variables.

partial differential equation *Mathematics.* a differential equation involving partial derivatives of more than one variable, and in which the unknown function depends on more than one independent variable.

partial dislocation *Crystallography.* the edge of a dislocation in which a slip has occurred over a fraction of a lattice constant.

partial dominance *Genetics.* a relationship between two alleles that produces a phenotype intermediate between the phenotype of heterozygous individuals. Also, SEMIDOMINANCE.

partial-duration series *Geophysics.* a series, used primarily to determine return periods, that records all events within a given time exceeding certain set limits, such as droughts that exceed a certain duration.

partial eclipse *Astronomy.* a partial obscuration in which the eclipsing body does not completely overshadow or cover the body being eclipsed.

partial evaluation *Computer Science.* a program optimization technique in which a program is expanded (e.g., by unscrolling loops) and parts of the program that are constant at compile time are evaluated. The result may be a program that is larger, but faster and more suitable for parallel execution.

partial fractions *Mathematics.* a term for the technique of rewriting a given rational function $f(x)/g(x)$, where the degree of $f(x)$ is less than the degree of $g(x)$, as the sum of rational functions having the forms $A/p(x)^r$ and $(Bx + C)/q(x)^r$. The $p(x)$ and $q(x)$ are the distinct linear and quadratic irreducible factors of $g(x)$, respectively, and r takes the values from 1 to the multiplicities of $p(x)$ or $q(x)$ for each factor $p(x)$ or $q(x)$. Partial fractions are especially useful since they can be integrated by elementary methods.

partially mixed digester *Biotechnology.* a two-stage anaerobic digester with a lower solid unmixed layer and an upper mixed layer.

partially ordered set *Mathematics.* a set in which a partial order has been defined.

partially populated *Computer Technology.* describing an electronic circuit board that has room for additional components.

partially stabilized zirconia *Materials.* a tough, wear-resistant ceramic material that is used in metal-forming tools, dies, and bearings.

partial molal quantity *Physical Chemistry.* the portion of the total volume of a solution taken up by one of its elements, so that if one mole of ethanol is added to a large volume of water, the volume increases by 54.2 cm^3; conversely, if one mole of water is added to a large volume of ethanol, the volume increases by 14.1 cm^3.

partial molar volume *Physical Chemistry.* the portion of a solution's total volume related to one mole of one of its components.

partial node *Physics.* a point or locus of points in a standing wave system in which the amplitude of a characteristic quantity is a minimum through time, but not zero.

partial obscuration *Meteorology.* a sky-cover designation measured in tenths (between 0.1 and 0.9) to indicate that a portion of the sky is hidden as a result of surface-based obscuring phenomena; used for U.S. weather observation.

partial order *Mathematics.* a reflexive, antisymmetric, transitive binary relation on a set; usually denoted $\leq$. That is, a partial order in a set X is a relation $\leq$ in X such that for all x, y, and $z \in X$: (a) $x \leq x$; (b) if $x \leq y$ and $y \leq x$, then $x = y$; and (c) if $x \leq y$ and $y \leq z$, then $x \leq z$. Also, ORDERING, ORDER. If, in addition, for every x and y in X, either $x \leq y$ or $y \leq x$, then $\leq$ is called a total, simple, or linear order, and X is called a chain, or totally ordered set.

partial parasite see SEMIPARASITE.

partial pediment *Geology.* **1.** a broad, horizontal gravel-covered inter-stream bench or terrace. **2.** a broad, horizontal erosion surface formed by the merging of valley-restricted benches of about equal elevation and age that developed in neighboring valleys.

partial penetration *Ordnance.* the penetration of a projectile into a target in which neither the projectile nor light from the penetration of the projectile can be seen from the back of the target.

partial pluton *Geology.* the section of a composite intrusion representing the result of a single intrusive episode.

partial potential temperature *Meteorology.* the hypothetical temperature of dry elements within an air parcel under a partial pressure change to 1000 millibars.

partial pressure *Physics.* the pressure of an individual gas that additively contributes to the total pressure in a gas mixture.

partial pressure suit *Aviation.* a skintight space suit that, unlike a full pressure suit, does not enclose the entire body but does help counteract increased oxygen pressure in the lungs by covering and exerting pressure on most of the body.

partial racemization *Chemistry.* a racemization process that affects only a few asymmetric groups.

partial (schedule of) reinforcement see INTERMITTENT REINFORCEMENT.

partial-select output *Electronics.* the voltage generated by the unselected magnetic cells when a partial-read or partial-write pulse is applied to the magnetic core.

partial sum *Mathematics.* given an infinite series, the sum of a finite number of terms of the series, usually the first n terms for some n. Also, nTH PARTIAL SUM.

partial thermoremanent magnetization *Geophysics.* the remanent magnetization gained by rocks when observed within a range of temperatures, rather than throughout the entire cooling process.

partial tide see HARMONIC CONSTITUENT.

partial tone *Acoustics.* a nondivisible sinusoidal component of a tone.

partial veil *Mycology.* the portion of agarics or mushrooms extending from the stem to the margin of the mushroom cap. Also, INNER VEIL.

participant modeling *Behavior.* a behavior therapy technique in which the client observes as the therapist participates in a situation that produces a particular phobia and then the therapist guides the client through the same activity. Also, **participant observation.**

participant observer *Anthropology.* an observer who participates as fully as possible in the culture under study, including living among the people, speaking the language, and following the customs and dietary habits.

particle *Mechanics.* **1.** any finite object that may be considered to have mass and an observable position in space. **2.** see POINT PARTICLE. *Particle Physics.* a minute subdivision of matter such that it is a fundamental constituent so small that it cannot be further subdivided.

particle accelerator *Nucleonics.* a device designed to accelerate charged particles to high energies.

particle beam *Physics.* a very narrow, concentrated stream of particles.

particle board *Materials.* a grainless construction board of sawdust or wood particles with a resin binder.

particle detector *Nucleonics.* a sensitive device that can detect the presence of a passing particle by measuring the electrical disturbance it produces.

particle diameter *Geology.* an expression of particle size, equal to the length of a straight line through the center of a theoretically spherical sedimentary particle.

particle distribution function *Physics.* any function that gives the distribution of some property (such as velocity or energy) of a system of particles.

particle dynamics *Mechanics.* the study of the motion of particles as a result of external forces acting on them.

particle electrophoresis *Physical Chemistry.* the movement of electrically charged particles that are large enough to be observed directly, either with the naked eye or with the aid of a microscope.

particle emission *Nuclear Physics.* a type of radiation that does not generate electromagnetic rays.

particle energy *Physics.* the energy of a relativistic particle as given by the sum of the particle's kinetic energy, potential energy, and rest mass energy.

particle horizon *Astronomy.* the farthest distance from which electromagnetic signals can reach the earth, thus forming the boundary of the observable universe.

particle-induced X-ray emission *Analytical Chemistry.* the characteristic energy spectrum of X-rays that results after a sample to be analyzed has been bombarded with ions.

particle lens *Physics.* a device that produces an electric field or a magnetic field, or a combination of the two, that interacts with a charged particle beam, so as to focus the beam in a way similar to the way an optical lens focuses light.

particle mechanics *Mechanics.* the study of particles at rest or in motion.

particle physics a branch of physics that is concerned with subatomic particles, especially with the many unstable particles found by the use of particle accelerators and with the study of high-energy collisions.

Particle Physics

Our understanding of the structure of matter has proceeded through a series of divisions and subdivisions. Molecules consist of atoms, each of which is composed of electrons and a central core, the nucleus. A nucleus is made of neutrons and protons, which in turn are made of more elementary particles known as quarks. Particle physics is the study of what today we identify as the smallest constituents of Nature. Today's elementary particles include three types, *quarks, leptons,* and *gauge bosons.* Certain quarks, leptons, and gauge bosons make up the familiar material world around us, whereas others are very short-lived and are only briefly created in high-energy collisions either at accelerators or in certain astrophysical settings.

Particle physics is studied both experimentally and theoretically. In order to resolve particles whose sizes are less than 10^{-15} meters, physicists collide particles together which have achieved enormous energies by racing around accelerators which are miles across. Giant detectors, built and operated by teams of hundreds of physicists, are used to study the collisions. Since the structure of the particles ultimately determines the characteristics of the collisions, physicists learn about the particles themselves through these experiments.

Theoretical particle physicists have developed a mathematical framework for describing the interactions of elementary particles. This mathematical structure incorporates the general principles of quantum mechanics and special relativity and the specific dynamics of the known elementary particles. All of the predictions of this theory are consistent with experimental observation. Theoretical particle physicists are now seeking answers to ever more fundamental questions such as "Why do the elementary particles have the properties they have?" and "What is the relationship of particle physics to gravitational theory?"

Edward Farhi
Associate Professor of Physics
Massachusetts Institute of Technology

particle property *Particle Physics.* any of the qualities that characterize the behavior of an elementary particle, including mass, charge, parity, baryon number, spin, hypercharge, and isospin.

particle-scattering factor *Analytical Chemistry.* a factor used in the measurement of light scattering by macromolecules to compensate for the loss in scattered light intensity caused by destructive interference.

particle size *Geology.* the general dimensions of particles or mineral grains in a sediment or rock, assuming that the particles are spheres for the purposes of measurements. Also, GRAIN SIZE. *Metallurgy.* **1.** in powder metallurgy, the size of individual particles. **2.** in metallography, the size of discrete particles, such as precipitates or inclusions.

particle-size analysis *Geology.* a determination of the proportion or distribution of specific sizes of particles present in a sample of a rock, soil, or sediment. Also, SIZE ANALYSIS, SIZE-FREQUENCY ANALYSIS. *Engineering.* this process used in determining the proportions of particles of defined size fractions in a granular or powdered sample, or the result of the analysis.

particle-size distribution *Engineering.* the percentages by weight or number of each fraction into which a powder sample has been classified with respect to sieve number or particle size. *Agronomy.* specifically, the amounts of various soil separates in a soil sample determined by weight percentages.

particle track *Physics.* a trail indicating the path of a particle, such as a trail of small bubbles in a bubble chamber produced by the passage of an energetic particle through a superheated liquid.

particle velocity *Acoustics.* the instantaneous value of the distance traveled by a particle per unit time in a medium that is displaced from its equilibrium state by the passage of a sound wave.

particular solution *Mathematics.* a solution of the inhomogeneous linear equation $Lu = f$, where L is a linear operator and f is a forcing function. Boundary conditions may also be given. The term is used to emphasize the fact that the general solution is equal to any particular solution plus the set of homogeneous solutions.

particulate [pär tik´yə lāt´; pär tik´yə lit] *Materials Science.* a material composed of separate and distinct particles.

particulate inheritance *Genetics.* the Mendelian model of inheritance, according to which genetic information is transmitted from one generation to the next in distinct units; this overturned an earlier concept of blending inheritance, which held that the character of offspring was a mixture of contrasting parental traits.

particulate-mass analyzer *Engineering.* a device that measures dust concentrations in emissions from kilns, furnaces, and scrubbers.

particulate matter *Physics.* matter composed of particles that are not superficially bound together, such as sand.

parting a division or separation; specific uses include: *Geology.* **1.** a band or bed of waste material between mineral veins or beds. **2.** a thin sedimentary layer deposited along a surface of separation between thicker layers of dissimilar rock. **3.** a surface along which a hard rock separates easily or divides naturally into layers, as along a bedding plane. *Mineralogy.* the tendency of crystals to divide along planes that are structurally weak due to deformation, gliding, or twinning. *Metallurgy.* in precious metal refining, the separation of silver from gold.

parting cast *Geology.* a sand-filled tension crack that develops as a result of creep along the sea floor.

parting compound *Metallurgy.* in casting, a material used to facilitate the removal of the cast part.

parting lineation *Geology.* a sedimentary structure consisting of a series of small ridges and grooves oriented parallel to the direction of the current. Also, CURRENT LINEATION.

parting-plane lineation *Geology.* a series of small, linear, subparallel, shallow ridges and grooves of low relief on a laminated surface.

parting-step lineation *Geology.* a series of small, subparallel, steplike ridges formed where a parting surface cuts across several adjacent laminae.

parting stop *Building Engineering.* a thin strip of wood or zinc that separates the sashes in a double-hung window.

partition *Building Engineering.* an interior wall having a height of one story or less and dividing a building into rooms or sections. *Computer Programming.* **1.** a subunit of a larger entity. **2.** a fixed-size division of storage, e.g., into operating system and user parts. **3.** the area of storage allocated to a user program. **4.** to divide a data set into parts according to some condition.

partition chromatography *Analytical Chemistry.* a separation technique used for gas-liquid or liquid-liquid mixtures, based on the differing solubilities of the components.

partition coefficient *Analytical Chemistry.* a constant symbolizing the ratio of the concentration of a solute in the upper of two liquid phases in equilibrium to its concentration in the lower phase. Also, DISTRIBUTION COEFFICIENT. *Physical Chemistry.* see DISTRIBUTION COEFFICIENT.

partitioned data set *Computer Programming.* in direct-access storage, a data set that has been divided into partitions or members, each of which contains programs or data.

partition function *Physics.* a function from which all the thermodynamic properties of a system can be derived; given by the integral over phase space (or in the case of quantum statistical mechanics, the sum over all allowed states) of the quantity $\exp(-E/kT)$, where E is the energy, T is the absolute temperature, and k is the Boltzmann constant.

partitioning effect *Biotechnology.* the effect that an enzyme has on its surrounding area when it is immobilized on a solid support matrix.

partition law see DISTRIBUTION LAW.

partition noise *Electronics.* the unwanted signals that arise in an electron tube when the beam is split between two or more electrodes, such as between screen grid and anode in a pentode.

partition of a positive integer *Mathematics.* any representation of a positive integer n as the sum of positive integers.

partition of a set *Mathematics.* a partition of a set X is a collection of nonempty disjoint subsets of X whose union is X.

partition of unity *Mathematics.* let X be a topological space and $\{U_\alpha\}$ (where $\alpha \in I$, some index set) an open cover of X with the property that every point x of X belongs to only finitely many members of the open cover (that is, the cover is locally finite). Then a partition of unity on X, subordinate to the cover $\{U_\alpha\}$, is a collection of functions $\{f_\alpha\}$, $\alpha \in I$, such that: (a) f_α maps X to the real unit interval [0,1]; (b) for each $\alpha \in I$, $f_\alpha(x) = 0$ if x is not in U_α; and (c) $\sum_{\alpha \in I} f_\alpha(x) = 1$ for all $x \in X$. (This is a finite sum.) By theorem, if X is paracompact, it is always possible to find on it a partition of unity subordinate to any given locally finite covering. Partitions of unity are used in the theory of integration.

Partitivirus group *Virology.* a genus of viruses of the family Partitiviridae.

partiversal *Geology.* of or relating to a series of local beds that dip in various directions, ranging through approximately 180° and occurring near the end of a plunging anticline's axis.

partly cloudy *Meteorology.* **1.** the character of a day's weather when the average cloudiness has been from 0.4 to 0.7 for the 24-hour period; used in U.S. climatological observation. **2.** in general, the weather condition characterized by clouds that do not completely dull the sky, i.e., cloudiness between 0.2 to 0.6.

part of speech *Linguistics.* a syntactic category of words in a language, e.g., nouns or verbs.

parton *Particle Physics.* a hypothetical, pointlike particle that is a constituent of a hadron.

part programming *Control Systems.* **1.** manual programming that converts the planning and specification of the sequence of steps for a process into a written program. **2.** computer-assisted programming that converts written instructions into block instructions for a control system.

partridge *Vertebrate Zoology.* **1.** any of several medium-sized, gallinaceous game birds of the genera *Alectoris* and *Perdix* of the family Phasianidae; found in grasslands, shrubs, and semi-deserts of Europe, Asia, Africa, and Australia. **2.** see BOBWHITE.

partridge

partridge pea *Botany.* the plant *Cassia fasciculata* of the legume family, characterized by yellow flowers and feathery leaves that fold shut when touched; found in North America. (Probably so called because its fruit is eaten by partridges.)

partridgewood *Plant Pathology.* a rotten condition of wood of trees such as the oak, caused by the parasitic fungus *Xylobolus frustulatus.*

parts kit *Engineering.* a group of parts used for the repair or replacement of the worn or broken parts of a device.

parts layout *Design Engineering.* the layout of parts on stock.

parts list *Engineering.* a printed sheet that displays component parts of an item by name, part number, illustration, and price.

parts presenter *Robotics.* the equipment used to feed parts to machining and other robotic operations.

parturient [pär tůr′ē ənt] *Medicine.* **1.** of or relating to the process of childbirth. **2.** a woman who is giving birth or about to give birth.

parturifacient *Medicine.* **1.** a medication or process that triggers or induces labor or relieves the pain of labor in childbirth. **2.** inducing or accelerating labor.

parturiometer *Medicine.* a device used to determine the force of uterine contractions during childbirth.

parturition [pär′tə rish′ən] *Biology.* the act or process of giving birth. *Medicine.* specifically, in humans, the act of childbirth, including the stages of labor and the puerperium or postnatal period characterized by involution of the uterus.

part vic. in divided parts or doses. (From Latin *partes vicibus.*) Also, **Part. vic.**

party line *Telecommunications.* a central office line that serves two or more subscriber stations with a single pair of wires. *Computer Technology.* **1.** see DAISY CHAIN. **2.** a network to which multiple users are directly connected. *Civil Engineering.* a boundary line separating adjoining properties.

party-line bus *Computer Technology.* a single data communication line to which multiple units connect, with transaction priority controlled by closeness to the main processor.

party-line carrier *Telecommunications.* in telephony, a single-frequency system that directly transmits carrier energy to all other carrier terminals of the same channel.

party wall *Building Engineering.* a wall that provides joint service and access between two buildings.

Parulidae *Vertebrate Zoology.* the wood warblers, a family of small New World birds of the passerine suborder Oscines, characterized by often dramatically marked plumage.

parulis *Medicine.* an elevated nodule at the site of a fistula draining a chronic periapical abscess. Also, GUMBOIL.

parvafacies *Geology.* a subsection of any magnafacies that lies between designated time-stratigraphic planes or key beds traced across the magnafacies.

parvalbumin *Biochemistry.* a protein that has sequence homology with calmodulin, but only two calcium binding sites.

parvobacteria *Microbiology.* a general term for any bacterial strain of small, Gram-negative bacilli, such as species of *Pasteurella.*

Parvoviridae *Virology.* a family of unenveloped, isometric, ssDNA viruses with highly resistant virions, having a wide host range of vertebrates and a narrow host range of invertebrates.

Parvovirus *Virology.* a genus of the Parvoviridae family occurring primarily in cats, minks, and dogs, and frequently associated with arthritis, erythema infectiosum, and tumors.

parvule *Pharmacology.* a very small pill or granule.

parylene *Organic Chemistry.* a general term for thermoplastic film polymers produced from *p*-xylene and having the repeating structure $-(CH_2C_6H_4CH_2-)_n$, where n is approximately 5000; used as a pore-free coating and in ultrathin plastic films for capacitor dielectrics.

parylene capacitor *Electricity.* a highly stabilized fixed capacitor that utilizes parylene film as a dielectric and can be operated at temperatures as high as 170°C or as low as cryogenic temperatures.

PASA *Organic Chemistry.* $NH_2C_6H_3(OH)COOH$, *para*-aminosalicylic acid; a white or nearly white powder with a slight acetous odor; soluble in dilute sodium hydroxide and in dilute nitric acid, slightly soluble in ether, practically insoluble in benzene; melts at 150–151°C (with decomposition); used in medicine and in industrial preservatives. *Pharmacology.* an antibacterial preparation of this substance, used in the treatment of tuberculosis. Also, **PAS.**

pascal [pa skal′; pa skäl′] *Metrology.* the basic unit of pressure in the meter-kilogram-second system, equal to the pressure of a force of one newton per square meter. (From Blaise *Pascal.*)

Blaise Pascal

Pascal, Blaise [pa skal′; pa skäl′] 1623–1662, French mathematician, physicist, and philosopher; formulated Pascal's law; with Fermat, invented probability theory; studied the geometry of conics.

Pascal *Computer Programming.* a high-level, general-purpose programming language created by Nicholas Wirth and designed to support structured programming concepts.

Pascal distribution see NEGATIVE BINOMIAL DISTRIBUTION.

Pascal rules *Physical Chemistry.* the rules that govern the susceptibility of the orbitals in complex molecules to magnetic fields, based on the susceptibility of the constituent atoms and the nature of their bonds.

Pascal's law *Fluid Mechanics.* **1.** a law stating that a fluid in equilibrium contained in a vessel externally exerts a uniform isotropic pressure, i.e., equal intensity in all directions. **2.** a law stating that pressure differences between two points in a fluid in equilibrium while in a gravitational field are dependent only on the density and the difference in elevation.

Pascal's theorem *Mathematics.* the three points of intersection of pairs of opposite sides of a hexagon inscribed in a conic are collinear. Pascal's theorem is the dual of Brianchon's theorem.

Pascal's triangle *Mathematics.* a triangular array with the nth row consisting of the binomial coefficients $\binom{n}{k}$, $k = 0, 1, 2, \ldots, n$.

$$\begin{array}{c} 1 \\ 1 \quad 1 \\ 1 \quad 2 \quad 1 \\ 1 \quad 3 \quad 3 \quad 1 \\ 1 \quad 4 \quad 6 \quad 4 \quad 1 \\ 1 \quad 5 \quad 10 \quad 10 \quad 5 \quad 1 \\ 1 \quad 6 \quad 15 \quad 20 \quad 15 \quad 6 \quad 1 \\ \text{and so on.} \end{array}$$

Each number is equal to the sum of the two numbers to the right and left of it in the row above it. Sometimes the rows are arranged in a right triangle; i.e., left-justified. Also, BINOMIAL ARRAY.

Pasch, Moritz 1843–1930, German mathematician; recast the foundations of geometry.

Paschen-Back effect *Physics.* an effect related to the Zeeman effect in which spectral lines are split into multiplets with wide spreading when the external magnetic field is very strong, approaching a critical value; at even higher values, the splitting pattern returns to the normal Zeeman effect. (Named for Friedrich *Paschen,* 1865–1947, and Ernst *Back,* 1881–1959, German physicists.)

Paschen bodies *Pathology.* round bodies identified microscopically in the cytoplasm during infections of smallpox or vaccinia tissue; their exact role in infection is uncertain. (Named for Enrique *Paschen,* 1860–1936, German pathologist.)

Paschen-Runge mounting *Spectroscopy.* an arrangement in a spectrograph in which the diffraction grating and slit are fixed and the photographic plates are fastened to a fixed track along the corresponding Rowland circle.

Paschen series *Spectroscopy.* a set of spectral lines in the near infrared region of the hydrogen spectrum resulting from electron transitions from superior orbits to the third orbit. Also, RITZ-PASCHEN SERIES.

Paschen's law *Electronics.* a law stating that the voltage needed to generate a spark between two parallel-plate electrodes in a gas tube is proportional to the product of gas pressure and electrode spacing. Also, **Paschen's rule.**

pascoite *Mineralogy.* $Ca_3V_{10}^{+5}O_{28}\cdot 17H_2O$, an orange, water-soluble, monoclinic mineral, having a specific gravity of 1.87 to 2.45 and a hardness of about 2.5 on the Mohs scale; found as granular crusts on sandstone in uranium-vanadium districts of the western United States.

PASG pneumatic antishock garment.

Pasiphae *Astronomy.* the eighth moon of Jupiter, having a diameter of about 70 kilometers; discovered in 1908.

Pasla *Robotics.* a programming language developed by NEC for pick-and-place assembly robots.

pasmo *Plant Pathology.* a disease of flax caused by the fungus *Mycosphaerella linorum,* characterized by the formation of round brownish lesions on the cotyledons and lower leaves.

paspalism *Toxicology.* poisoning due to the ingestion of seeds of the grass *Paspalum scrobiculatum.*

paspalum staggers *Veterinary Medicine.* poisoning due to ingestion of paspalum grasses plagued with *Claviceps paspal*; results in tremors, incoordination, and paralysis.

PAS reaction see PERIODIC ACID-SCHIFF REACTION.

pass to move from one place to another; a movement from one place to another; specific uses include: *Geography.* **1.** a relatively low area allowing travel between two mountain peaks or slopes. **2.** a navigable channel connecting a river to the sea. *Aviation.* a brief, short run of an aircraft in flight, as by a fighter airplane maneuvering to attack enemy aircraft or by any aircraft surveying a potential landing area. *Space Technology.* a single orbit of an artifical satellite, beginning when it crosses the equator from the Southern Hemisphere to the Northern Hemisphere. *Computer Programming.* **1.** to scan or search data in a file for a given purpose. **2.** a single execution of a loop, a single cycle of processing, or a single scanning of a source program. *Mechanical Engineering.* **1.** the movement of combustion gases in a furnace. **2.** the motion of a cutting tool across the workpiece. *Mining Engineering.* a mine opening through which material is transported from a higher to a lower level. *Metallurgy.* in rolling, a single operation through a stand.

pass. passive.

passage a movement from one place to another, or the means of making such a movement; specific uses include: *Navigation.* the act of sailing from one place to another. *Surgery.* **1.** a channel, duct, gap, or other opening between organs or structures. **2.** an evacuation of the bowels or discharge of urine. **3.** the introduction of a catheter or probe into a body through a natural orifice. *Cell Biology.* the process of subculturing a tissue culture cell line. *Virology.* a laboratory procedure in which a host is infected with a virus or mixture of viruses, and the proceeds are subsequently recovered from that host after one infectious cycle and used to infect another host; this can be done to separate one virus from a mixture of viruses or to improve the production of a virus in a host.

passage bed *Geology.* a rock stratum that represents a transition in characteristics from rocks of one geological system to another or from the rocks above to the rocks below.

passage cell *Cell Biology.* a thin-walled plant cell thought to play a role in the passage of water and other substances between the cortex and the xylem.

passage grave *Archaeology.* a category of prehistoric tombs having a burial chamber and a separate passage into the tomb; it includes the earliest known megalithic graves, dating from about 5000 BC.

Passalidae *Invertebrate Zoology.* a family of shiny black or brown beetles, coleopteran insects in the superfamily Scarabaeoidea, having a short horn on the head; found in decaying wood.

passband *Electronics.* **1.** the full spectrum of frequencies that are transmitted with little loss by such devices as filters and amplifiers. **2.** the range of frequencies a given filter is designed to pass at low attenuation.

pass-by *Engineering.* a section of a single-track rail line with a double track. *Mining Engineering.* **1.** a passage around the working portion of a mining shaft. **2.** a turnout in an underground passageway, allowing cars to pass one another.

pass element *Electronics.* a device, such as a tube or power transistor, that is inserted into a DC power source to automatically increase its resistance when the output has to be lowered and to decrease its resistance when the output has to be raised.

passenger car *Engineering.* a railway car used to carry passengers rather than freight.

passenger mile *Transportation Engineering.* a measure of traffic carried, based on a single passenger carried a single mile. Thus, an airliner carrying 200 passengers on a 3000-mile flight provides 600,000 passenger miles of service.

passenger pigeon *Vertebrate Zoology.* a large pigeon, *Ectopistes migratorius,* of the family Columbidae. Present in huge numbers in eastern North America up through the Colonial era, it declined in population beginning in the mid-1800s and became extinct in 1914, one of the most significant extinctions of a species in historic times. (Named for its habit of frequent migration in huge flocks over long distances.)

passenger ship *Naval Architecture.* a ship designed primarily to carry passengers.

Passeres *Vertebrate Zoology.* an equivalent name for the suborder Oscines of the bird order Passeriformes.

passeriform *Vertebrate Zoology.* of or relating to the order Passeriformes.

Passeriformes *Vertebrate Zoology.* the largest order of birds, with over 5000 species, or about 60% of all birds, consisting of perching birds; characterized by a lack of the ambiens muscle, a well-developed vomer, and three toes in front with one behind.

passerine *Vertebrate Zoology.* **1.** of or relating to the order Passeriformes, the largest order of birds. **2.** a bird of this order.

Passifloraceae *Botany.* a family of dicotyledonous climbing vines or shrubs, including the passionflower, of the order Violales; characterized by regular flowers with an extra-staminal corona, vines with axillary tendrils, and arillate seeds with an oily endosperm.

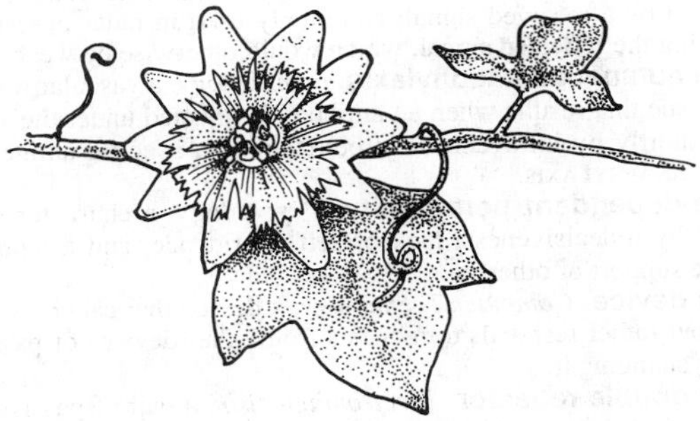

Passifloraceae (passionflower)

passing track *Engineering.* a railway sidetrack equipped with switches at both ends.

passionflower *Botany.* a climbing vine of the genus *Passiflora* of the family Passifloraceae that grows widely in warm regions of the Americas and elsewhere; noted for its large, showy flowers in various colors; the fruit of some species is edible. (Associated with the *Passion* of Jesus Christ; various parts of the flower were thought to symbolize different aspects of Christ's sufferings.)

passionfruit *Botany.* the edible fruit of the passionflower.

passivation *Electronics.* the practice of growing a thin oxide film on the surface of a semiconductor to protect exposed elements from environmental contaminates, thus ensuring the electrical stability of the device. *Metallurgy.* a major decrease in the reactivity of a metallic surface.

passive *Science.* **1.** not participating actively; inactive. **2.** influenced or acted upon by an external force or cause. *Medicine.* describing a condition that is harmful or unhealthy but not currently active. *Aviation.* of a vehicle or system, under external or preprogrammed guidance or control. *Military Science.* **1.** of or relating to military techniques, such as camouflage, deception, or dispersion, that do not utilize active weapons or involve initiative action. Thus, **passive (air) defense. 2.** of or relating to weapons, control systems, or surveillance techniques that operate on energy generated by the target or other external source while emitting no detectable energy of their own. Thus, **passive acoustic torpedo, passive homing guidance, passive detection, passive electronic countermeasures.**

passive accommodation *Robotics.* a change in the position or motion of a robot's manipulator due to external force.

passive-active cell *Metallurgy.* a corrosion cell created by adjacent active and passive areas.

passive agglutination test *Immunology.* a method used to detect antibodies that combine with soluble antigens.

passive-aggressive *Psychology.* displaying aggression in an indirect manner, without anger or confrontation.

passive-aggressive personality *Psychology.* a character disorder in which an individual displays hostility or aggression quietly and indirectly, by means of such behaviors as stubbornness, procrastination, pouting, forgetfulness, inefficiency, or noncompliance.

passive anaphylaxis see PASSIVE CUTANEOUS ANAPHYLAXIS.

passive AND gate *Engineering.* a fluid apparatus in which an output signal emerges only when both of two control signals appear simultaneously.

passive antenna *Electromagnetism.* an antenna that influences the directivity of an antenna system but is not directly connected to a transmitter or receiver.

passive antiroll system *Naval Architecture.* an antiroll system consisting of wing tanks joined by conduits, but with no pumping system; it is designed to slosh out of phase with the ship's roll, thus damping it.

passive armor *Ordnance.* armor that is designed to absorb the energy of shaped-charge ammunition.

passive clot *Hematology.* a clot formed in the sac of an aneurysm through which the blood has stopped circulating.

passive communications satellite *Space Technology.* a satellite that receives signals or reflects signals between stations, but does not transmit signals. Also, REFLECTOR SATELLITE, PASSIVE SATELLITE.

passive component *Electricity.* see PASSIVE ELEMENT.

passive congestion *Medicine.* hyperemia of a part as a result of impairment of the return of venous blood.

passive control *Mechanics.* see COMPLIANCE.

passive corner reflector *Electromagnetism.* a corner reflector that is energized by a received signal; commonly used in radar operations for improving the reflected signal, which would otherwise be weak.

passive cutaneous anaphylaxis *Immunology.* a vascular reaction in body tissue that results when an antibody is injected under the skin, followed shortly by intravenous introduction of a specific antigen. Also, PASSIVE ANAPHYLAXIS.

passive-dependent personality *Psychology.* a character disorder marked by indecisiveness, lack of self-confidence, and a tendency to seek the support of others.

passive device *Computer Technology.* a device that cannot initiate an action but rather responds to requests from other devices or passes data without changing it.

passive double reflector *Electromagnetism.* a pair of passive reflectors situated so as to reflect a microwave signal over mountains or ridges with little disturbance to the beamwidth.

passive earth pressure *Civil Engineering.* the resistance of a vertical earth face to the stress of alterations by a horizontal force due to active compression.

passive electronic countermeasures *Electronics.* a class of intelligence-gathering electronic devices that do not radiate energy, such as the surveillance equipment used to locate and examine electromagnetic radiation from radar and communications transmitters.

passive element *Electricity.* a device that is static in operation and that is not a source of energy, such as a conventional resistor, capacitor, inductor, diode, rectifier, or fuse. Also, PASSIVE COMPONENT. *Electromagnetism.* an antenna element that is energized by radio-frequency radiation from a neighboring element; a parasitic element.

passive filter *Electricity.* a filter having only passive components such as resistors, inductors, or capacitors, with no active elements such as transistors.

passive fold *Geology.* a fold in which the strata exert no structural control on the folding.

passive front see INACTIVE FRONT.

passive glacier *Hydrology.* a glacier whose movement is slow as a result of little accumulation and ablation.

passive guidance *Navigation.* a guidance system that relies on external inputs, such as detected radio or infrared signals, rather than on inputs obtained by an active onboard system, such as radar.

passive hemagglutination *Virology.* a test used to detect a virus-specific antibody by coating red blood cells with viral antigen; if the antigen is present, the cells will clump together (agglutinate).

passive homing *Navigation.* a homing system that is directed by an external input (e.g., an infrared source such as an aircraft engine) rather than by a target illuminated by an onboard or ground-based system such as radar.

passive immunity *Immunology.* a resistance to disease or infection acquired by an individual through the transfer of antibodies from an immune donor to a nonimmune recipient, or through the transfer of maternal antibodies to a fetus.

passive jamming *Electronics.* the use of reflectors to return unwanted or confusing signals back to the enemy radar system transmitting them.

passive junction *Electromagnetism.* a waveguide junction that does not have a source of energy.

passive margin *Geology.* a margin of a continent that lacks active plate boundaries at which the continental crust and oceanic crust meet.

passive metal *Metallurgy.* any metal or alloy, such as stainless steel, that is corrosion resistant when protected by a surface film.

passive method *Civil Engineering.* a method of erecting structures on permafrost ground, in which the permafrost is not disturbed; the structure base is insulated to prevent heat from the structure from melting the permafrost.

passive mine *Ordnance.* **1.** a mine whose anticountermining device has been operated, thus preventing the mine from being actuated. **2.** a mine that does not emit a signal to detect the presence of a target.

passive network *Electricity.* a network that has no energy source and no active gain elements.

passive permafrost *Geology.* a permafrost that was formed under earlier, colder conditions and that will not refreeze under present climatic conditions after it has been disturbed or destroyed. Also, FOSSIL PERMAFROST.

passive radar *Engineering.* **1.** the use of a radar system in the receive-only mode, so that the microwave electromagnetic energy is emitted from a target to be detected without revealing the presence or location of the radar system. **2.** a radar system that detects microwave energy from distant objects.

passive reflector *Electromagnetism.* a plane reflector that is used to reflect radar beams from places at which it would be impractical to set up transmitting equipment (such as at the top of a mountain or tower).

passive satellite see PASSIVE COMMUNICATIONS SATELLITE.

passive sensors *Ordnance.* a television camera set installed on fighter planes that enables aviators to positively identify aircraft at distances up to 10 times the range of the naked eye.

passive sensors (on an F-14 fighter plane)

passive smoking *Medicine.* a term for the effects of cigarette, cigar, or pipe smoke on people who do not smoke themselves, as when a nonsmoker is in an enclosed area for an extended period of time with another or others who are smoking; this is now regarded as injurious, particularly to those with existing respiratory problems.

passive solar system *Mechanical Engineering.* a solar heating or cooling system using natural heating, cooling, or evaporation rather than external sources of energy.

passive sonar *Acoustical Engineering.* a sonar system that receives acoustic signals from underwater objects in order to determine the direction and source of the signals, but does not transmit sounds; used by the military to detect the presence of enemy submarines.

passive system *Electronics.* a term for an electrical system that does not generate energy.

passive transducer *Electronics.* a device that transfers energy from one system to another without an internal source of power.

passive transport *Molecular Biology.* the movement of a molecule across a biological membrane without any expenditure of metabolic energy.

passivity the fact of being passive; specific uses include: *Chemistry.* in certain metals such as iron, a loss of normal chemical activity following a strong oxidizing agent or anodization.

passthrough *Computer Programming.* the process of a user's gaining access to another computer or network through the currently used computer or network.

password *Computer Programming.* a unique string of characters entered by a user to gain access to a system or to specific data within a system; used to prevent access to sensitive or confidential files and programs by unauthorized users.

past *Science.* the time before now; the time gone by. *Physics.* specifically, in relativity theory, the set of all events from which a signal could be emitted and travel at speeds that are not greater than the speed of light, so that the signal arrives at a contemporary event in question.

Past. *Pasteurella.*

past. pasteurize.

pasta [päs´tə] *Food Technology.* a collective term for various foods of Italian origin, consisting mainly of a mixture of flour and water, especially durum wheat flour; common types include spaghetti, macaroni, and noodles. (From the Italian word for "dough.")

paste *Materials.* **1.** a thick, creamy mixture of flour or starch and water that is used as an adhesive, especially for paper or similar lightweight materials. **2.** any material that is similar to paste in consistency. *Electricity.* a gelatinous electrolyte that is placed in dry batteries and electrolytic capacitors. *Metallurgy.* any thick mixture of metallic powder and a fluid, such as the mixture used for soldering or for magnetic particle inspection.

pasteboard see PAPERBOARD.

pastel *Materials.* **1.** a dried pastelike solid that is made from pigments ground with chalk and compounded with gum water. **2.** a chalklike crayon made from such paste. *Graphic Arts.* **1.** a picture made with such crayons, or the art of making such pictures. **2.** a soft, subdued shade of any color. *Botany.* the woad plant or the dye made from it.

Pasteur, Louis [päs tůr´; pas toor´] 1822–1895, French chemist and biologist; the founder of bacteriology; developed the germ theory of infection; invented the process of pasteurization; developed anthrax and rabies vaccines.

Louis Pasteur

Pasteur-Chamberland filter see CHAMBERLAND FILTER.

Pasteur effect *Microbiology.* in microorganisms capable of both fermentative and respiratory metabolism, a decrease in the rate of glucose utilization and the suppression of lactate accumulation when oxygen is present. Also, PASTEUR REACTION.

Pasteurella *Bacteriology.* a genus of Gram-negative, nonmotile, facultatively anaerobic, ovoid to rod-shaped bacteria of the family Pasteurellaceae; parasitic on and potentially pathogenic to humans and many animals.

Pasteurellaceae *Bacteriology.* a family of Gram-negative, nonmotile, facultatively anaerobic, coccoid to rod-shaped bacteria occurring as parasites in mammals and birds.

Pasteurella tularensis see FRANCISELLA TULARENSIS.

pasteurellosis *Medicine.* a general term for infections by microorganisms of the genus *Pasteurella.*

Pasteuria *Bacteriology.* a genus of budding, aquatic bacteria composed of nonmotile cells that have a cell wall lacking peptidoglycan; parasitic on freshwater crustaceans.

Pasteuriaceae *Bacteriology.* a former classification of a family of budding bacteria that are now assigned to the genera *Pasteuria* and *Planctomyces.*

Pasteurian *Science.* **1.** of or relating to Louis Pasteur. **2.** based on the research or theories of Louis Pasteur.

pasteurization [päs´chə rə zā´shən] *Food Technology.* a food-preservation process in which a mild heat treatment is applied for a specific period of time to kill or inactivate disease-causing organisms, without significantly altering the taste or quality of the food; used for milk and also for cheese, egg products, beer, fruit juices, canned meats, and various other products. A typical pasteurization treatment for milk is to expose it to a temperature of 145–150°F (62.8–65.5°C) for a period of at least 30 minutes. (Developed by Louis *Pasteur.*)

pasteurize [päs´chə rīz´] *Food Technology.* to carry out a process of pasteurization; subject a food to pasteurization.

pasteurizer *Engineering.* a device used to destroy disease-producing bacteria in a fluid, by keeping it at a specific elevated temperature for a specified period.

Pasteur pipet *Biotechnology.* a small-diameter, ungraduated glass-tube pipet with a constriction in the wider part for the insertion of a cotton plug.

Pasteur reaction see PASTEUR EFFECT.

Pasteur's salt solution *Analytical Chemistry.* a solution of potassium phosphate, calcium phosphate, magnesium sulfate, and ammonium tartrate in distilled water, used as a laboratory reagent.

pastille *Radiology.* a small paper disk coated with platinocyanide or barium whose green color turns brown upon exposure to X-rays; once used to estimate the quantity of X-rays administered and to test the intensity of ultraviolet radiation.

past of an event *Physics.* see PAST.

pastoral *Anthropology.* relating to or practicing pastoralism.

pastoralism *Anthropology.* a way of life in which domesticated herd animals provide the economic livelihood of a society.

Pastro *Robotics.* a programming language for the control of robots developed by Biomatic of Freiberg, Germany.

pasturage *Agriculture.* forage that is eaten directly by a grazing animal from a growing crop.

pasture *Agronomy.* an area of grassland that is suitable for the grazing of animals.

PA system see PUBLIC ADDRESS SYSTEM.

PAT paroxysmal atrial tachycardia.

pat. patient; patent.

patagium *Vertebrate Zoology.* **1.** in birds, a feathered web of skin that spans the angle in front of the elbow. **2.** a fold of skin in flying squirrels, flying lizards, and other arboreal gliding animals that encloses the limbs on both sides from neck to tail, enabling the animal to glide.

Patau syndrome *Genetics.* a syndrome caused by the presence of an extra chromosome 13, characterized by a cleft lip and cleft palate, as well as severe ocular, cerebral, and cardiovascular defects. Also, D_1 TRISOMY SYNDROME.

patch *Acoustical Engineering.* an interconnection between synthesizer channels so as to produce one unified sound. *Computer Programming.* **1.** to modify a program by inserting corrected machine code directly into the source code, object file, load file, or loaded program. **2.** the corrected machine code used to fix a program problem. Also, HARD PATCH. **3.** a small, ad-hoc correction to a program. *Electricity.* a temporary connection made between jacks or other terminations on a patch board.

patch board *Electricity.* a board or panel on which circuits are terminated in jacks or patch cords, so that various circuits can be temporarily connected as required; used in a wide variety of broadcast, communications, and computer applications.

patch bolt *Mechanical Devices.* a countersunk bolt with a square head that twists off when the bolt is driven in tightly, leaving a flush surface.

patch budding *Botany.* a grafting procedure in which a small rectangular area of bark bearing a bud is fitted into a corresponding opening in the bark of the stock.

patch-clamping *Biochemistry.* a technique for measuring the current flowing through individual ion channels of a cell, by means of a patch electrode pressed against the plasma membrane to form an electrically tight seal.

patch cord *Electricity.* a short cord that has a plug or a pair of clips on one end; used to connect two pieces of sound equipment, such as a phonograph and tape recorder or an amplifier and speaker. Also, ATTACHMENT CORD.

patchiness *Ecology.* a feature of a habitat within which significant spatial variation exists with respect to suitability for a species.

patch logging *Forestry.* a method of forest management in which patches of 40 to 200 acres are logged as single settings, then separated for as long as possible by living forest, in order to secure optimal seed dispersal for the patches and to avoid fire hazard.

patchouli oil *Materials.* the fragrant oil of the tropical Asian plant, *Pogostemon cablin,* that is used in perfumes.

patch reef *Geology.* **1.** a small, irregularly shaped, flat-topped subsection of an organic reef complex. **2.** a small, isolated, thick lens of limestone or dolomite rimmed by rocks of different characteristics.

patch test *Immunology.* a method used to diagnose a hypersensitivity of an individual, in which an allergy-producing substance is applied to unbroken skin and the skin is observed for irritation.

patefaction *Surgery.* the act of laying open.

patella *Anatomy.* the technical name for the kneecap, a flat triangular bone covering the knee, having a pointed apex that attaches to the patellar ligament.

Patellacea *Paleontology.* a still-extant superfamily of limpets in the order Archaeogastropoda and suborder Patellina; extant since the Silurian.

patellar *Anatomy.* relating to or involving the kneecap.

Patellariaceae *Mycology.* a family of fungi belonging to the order Hysteriales, which occurs on decaying organic matter or as a parasite of lichens.

patellectomy *Surgery.* the excision of the kneecap.

Patellidae *Invertebrate Zoology.* limpets, a family of intertidal gastropod mollusks in the order Archeogastropoda, characterized by a low conical shell that clings very strongly to a support when the limpet is disturbed.

patellofemoral *Anatomy.* of or related to the kneecap and the femur.

patent *Science.* an assignment by the government of various exclusive rights for the use and sale of a substance or process to its inventor or originator. *Medicine.* not closed; open or unobstructed; as a *patent* airway. *Botany.* expanded; spreading.

patent ductus arteriosus *Developmental Biology.* a cardiovascular handicap caused by the occasional failure of the ductus arteriosus to close postnatally, which can be surgically corrected.

patent foramen ovale *Medicine.* the persistence of the fetal foramen ovale after birth, often occurring in association with pentalogy of Fallot.

patent hammer see BUSH HAMMER.

patenting *Metallurgy.* in steel wiremaking, a thermal treatment consisting of heating above the transformation temperature, followed by controlled cooling.

patent log *Navigation.* any mechanical device for determining the distance traveled through water, especially a taffrail log.

patent medicine *Pharmacology.* any remedy that is protected by a patent or copyright, advertised to the public, and usually available without prescription.

Patera process *Metallurgy.* a former method used for the extraction of silver.

Paterinida *Paleontology.* an extinct order of inarticulate brachiopods with biconvex, phosphatic shells; of uncertain position within the phylum; extant from the Lower Cambrian to Middle Ordovician.

paternity *Genetics.* the fact of being a father; fatherhood.

paternity test *Genetics.* a method used in an investigation of the paternity of a child, in which genetic information from the blood groups of the mother, child, and supposed father is analyzed; this can definitively exclude the possibility that a certain individual is the father of the child, but cannot legally establish which individual is the father.

paternoite see KALIBORITE.

paternoster lake *Hydrology.* one of a series of small, circular lakes lying within rock basins in a glacial valley and connected by streams, rapids, or waterfalls, so that they resemble a rosary or a string of beads. Also, ROCK-BASIN LAKE, STEP LAKE, BEADED LAKE.

path a way or route that is followed; specific uses include: *Navigation.* the actual or intended course followed by a craft. *Robotics.* the motion of a robot along a line, a contour, or a series of points. *Computer Programming.* **1.** the logical sequence a computer takes in running a program. **2.** the route between two nodes in a network, or the route between a root and a node in a database. **3.** a route from one node to another in a graph. *Industrial Engineering.* in a PERT network, a series of activities and events. *Mathematics.* **1.** in graph theory, a finite alternating sequence of distinct vertices and edges

$$v_0, e_1, v_1, e_2, \ldots, v_{k-1}, e_k, v_k$$

in a graph such that v_{i-1} and v_i are the end vertices of the edge e_i, $1 \le i \le k$. Equivalently, a path is a trail with distinct vertices. **2.** a graph that consists solely of the edges and vertices listed in the sequence of a path. This graph is usually denoted P_n, where n is the number of vertices (= 1 + number of edges) of the path. **3.** a path in a region G of the real or complex plane is a continuous function γ: $[a,b] \to G$ for some real interval $[a,b]$. If the derivative $\gamma'(t)$ exists for each t in $[a,b]$ and γ' is continuous on $[a,b]$, then γ is said to be a differentiable path. **4.** more generally, a path from a point x_0 to a point x_1 of a topological space X is a continuous function γ: $[0,1] \to X$ such that $\gamma(0) = x_0$ and $\gamma(1) = x_1$. Also, ARC.

path- or **patho-** a prefix meaning "disease."

path. pathology; pathological.

path analysis *Genetics.* an analytical technique developed by Sewall Wright for studying both regular and irregular breeding systems.

path attenuation *Telecommunications.* the power loss between a transmitter and a receiver.

path coefficient *Telecommunications.* the ratio of power sent over a specified path to that sent over the most direct path.

path computation *Robotics.* the calculations used to specify the path of a robot or its end effector.

path control or **planning** see TRAJECTORY CONTROL.

path difference *Crystallography.* the difference in distance that two beams travel when scattered in the same direction from different points; as a result of such path differences, the beams may or may not be in phase.

pathergy *Immunology.* an abnormal reaction to an allergen, either subnormal or excessive.

pathfinder *Ordnance.* **1.** aircraft flown by highly skilled navigators to find and mark the target for a following bomber force. **2.** a radar signal that provides guidance to a target for missiles.

path length see SOFTWARE PATH LENGTH.

path of totality *Astronomy.* the area on the earth from which a solar eclipse is total.

pathogen *Medicine.* any virus, microorganism, or other substance that causes disease; an infecting agent.

pathogenesis *Medicine.* the source or development of a disease or disease process. *Cell Biology.* in particular, the cellular events and reactions occurring during the disease development.

pathogenic *Medicine.* giving rise to morbid tissue changes or to a pathological disorder or disease.

pathognomonic *Medicine.* a sign or symptom of a disease or pathological condition by which a diagnosis can be made.

pathol. pathology; pathological.

pathologic [path´ə läj´ik] *Medicine.* **1.** relating to or caused by some disease; pathological. **2.** of or relating to pathology.

pathological [path´ə läj´i kəl] *Medicine.* **1.** relating to or caused by a disease or morbid process. **2.** of or relating to the field of pathology. *Psychology.* of behavior, caused by or showing evidence of a mental disorder. *Mathematics.* exhibiting behavior that is not typical of its class; however, one generation's pathological examples often become the next generation's standard mathematics.

pathologic anatomy *Pathology.* the science or study of diseased, injured, or altered tissue and organs, usually by means of dissection.

pathologist [pa thäl´ə jist] *Medicine.* a physician who is specially trained and experienced in anatomical and experimental pathology.

Pathology

Pathology is mainly about limits—the range of normality and the point at which abnormality begins. In a general sense then, the discipline of pathology is of interest to all except the most ultraradical of thinkers who might hold that all things are normal, or the most ultraconservative who might hold that all things are abnormal. In a specific sense, pathology refers to the study of disease—plant pathology, animal pathology. When unqualified, the term is usually understood to mean the study of human disease and the establishment of disease diagnoses by laboratory analysis of tissue or body fluid samples. Often this examination involves the use of a microscope and/or electronic analytic equipment. Pathology has been regarded as the basis for medical practice. William Osler, one of the fathers of modern medicine, wrote "As is your pathology, so is your medicine."

As a discipline of clinical medicine, pathology encompasses four main specialties: anatomical pathology, hematologic pathology, medical microbiology, and medical biochemistry. Within these areas there are many subspecialties such as neuropathology, forensic pathology, pediatric pathology, virology, nuclear medicine, etc. As in other fields, many of the greatest strides in pathology have been made by those who have probed and extended the boundaries of the discipline. Thus pathology overlaps with molecular biology, genetics, immunology, hematology, basic microbiology, and many other areas.

Pathology is in essence a search for the truth about the existence and causes of specific diseases, but that truth is constantly changing as new scientific discoveries are made. The close link between the discipline and the forefront of medical research makes pathology an extremely dynamic field. Pathology offers unrivaled opportunity to explore specific disease-related issues in depth without some of the distractions which may occur in other branches of medicine. On the other hand there is also an important need in clinical medicine for broad-based general pathologists with a grasp of all four main areas of the discipline as they apply to the whole patient.

Kim Solez
Professor and Chairman of Pathology
University of Alberta

pathology [pa thäl´ə jē] *Medicine.* the branch of medical science that treats the essential nature of disease, especially the changes of structure and function in tissues and organs of the body that cause or are caused by disease.

pathophobia *Psychology.* an irrational fear of disease.

pathophysiology *Medicine.* the study of the cause of disordered function.

pathopsychology *Psychology.* the study of mental processes, particularly abnormal cognitive and perceptual processes, during the period of mental disorders.

pathotoxin *Plant Pathology.* a biologically produced chemical other than an enzyme that is an important factor in the inducement of a plant disease.

path plotting *Electromagnetism.* the plotting of the path followed by the microwave beam on a profile chart indicating the earth's curvature.

pathwise connected set see CONNECTED SET.

Patientia *Astronomy.* asteroid 451, belonging to type C and discovered in 1899; it has a diameter of 267 km.

patina [pa tē´nə; pat´i nə] *Geology.* **1.** broadly, any thin, colored film or layer formed on the surface of a rock or other material as a result of chemical weathering and long exposure. **2.** specifically, the greenish film that forms naturally on copper and bronze after long exposure to a moist environment. *Archaeology.* this surface layer, having a glazed or varnished appearance as formed on artifacts; thought to indicate great age. *Metallurgy.* a similar green coating on copper or on a copper-base alloy, formed by an intentional chemical treatment.

patio *Architecture.* an open-air paved courtyard within or adjacent to a house.

patio process *Metallurgy.* one of several methods for extracting silver from its ores.

patriarchal [pā´trē är´kəl] *Anthropology.* of or relating to a patriarchal society. Also, **patriarchic.**

patriarchal society *Anthropology.* a society in which the father is the supreme authority in the family, clan, or tribe; rules of patrilineal descent and patrilocal residence are usually followed. Also, **patriarchy.**

patrifocal family *Anthropology.* a family that consists of a father and his children.

patrilineal descent *Anthropology.* a descent system in which kinship is traced from the father's ancestors, children belong to the male's lineage, and members believe that they are descended from a common male ancestor.

patrilocal residence *Anthropology.* a practice in which a married couple resides with the man's family.

Patriot *Ordnance.* **1.** a surface-to-air missile used to intercept incoming enemy missiles; powered by a single-stage, solid-propellant rocket engine, with a launch weight of 2100 pounds and a length of 17 feet; it delivers a conventional warhead at a speed of Mach 3.5 and a range of 50 miles. **2.** a mobile weapons system consisting of a command station, a radar unit, and 8 to 16 launchers with four Patriot missiles each; an American antiaircraft system consisting of radar, computers, and surface-to-air missiles; officially designated **MIM-104.**

Patriot missile

patristic affinity *Evolution.* a degree of similarity between groups resulting from a common ancestry.

patroclinus *Genetics.* of or relating to progeny that resemble the father in a mating, rather than the mother.

patrol *Military Science.* a detachment of two or more persons sent to gather information or execute a mission; it is often a squadron or platoon and may be composed of ground, sea, or air forces.

patronite *Mineralogy.* a gray to black, soft monoclinic mineral with an approximate formula of VS_4, having a specific gravity of 2.82 and an undetermined hardness; mined in fine-grained masses as an ore of vanadium near Cerro de Pasco, Peru.

patronym *Anthropology.* a name that is derived from the name of a father or ancestor; e.g., Johnson, MacDonald, O'Neill. *Systematics.* a taxonomic or scientific name that is derived from the name of a person or persons; e.g., the plant genus *Bougainvillea,* named for the French explorer Louis Antoine de Bougainville. Also, **patronymic.**

pattern a plan, model, or form to be followed in some action or process; specific uses include: *Engineering.* **1.** an article or specimen made for the purpose of copying it to produce similar articles. **2.** specifically, a full-sized model used to render the mold in casting metal or thermoplastics. *Aviation.* a flight-path configuration followed or described by an aircraft. *Ordnance.* **1.** the distribution of a series of shots fired under the same conditions, either from a single gun or from a battery. **2.** of a shotgun, the distribution of pellets within a circle of given diameter from a single shot fired at a given distance.

pattern analysis *Computer Programming.* a phase of pattern recognition in which known information about the data is used to guide data gathering into patterns and pattern classes in order to discover the pattern present in the data.

pattern bombing *Military Science.* the technique of covering a target area with bombs uniformly distributed according to a plan.

patterned ground *Geology.* a ground surface feature generally found in periglacial areas, characterized by well-defined, approximately symmetrical forms, such as circles, polygons, steps, and stripes, produced by intensive frost action. Also, STRUCTURE GROUND, FROST-PATTERN SOIL.

pattern flood *Petroleum Engineering.* a waterflood of a petroleum reservoir in which injection wells are established in an areal pattern designed to sweep oil toward the boreholes of producing wells.

pattern formation *Chaotic Dynamics.* a description of nonlinear systems in terms of macroscopic variables (rather than microscopic particles) which are spatially nonuniform; nonlinear systems that might be chaotic. *Developmental Biology.* the generation and causes of the arrangements and symmetries of embryonic structures.

pattern generator *Electronics.* a device that generates signals for the test pattern used in servicing a television receiver.

pattern harmonization *Ordnance.* the process of adjusting the fixed guns of a fighter aircraft in order to produce the largest lethal pattern at a given range.

pattern laying *Military Science.* in land mine warfare, the laying of mines in a fixed relationship to each other.

pattern matching *Artificial Intelligence.* **1.** a technique used in natural language programs such as ELIZA that matches a statement against patterns, triggering an action or a response. **2.** a processing based upon matching a representation of a problem against a set of stored patterns, with actions to be taken when a pattern is matched.

pattern recognition *Psychology.* the ability to recognize a particular pattern, even if certain stimulus elements within the pattern are altered; a measure of childhood development. *Acoustical Engineering.* the ability to fill in missing parts of a sound pattern, based on previous exposure to the same pattern or similar patterns. *Computer Programming.* the automatic identification of shapes, forms, or configurations in a complex or noisy environment. *Robotics.* a robotic system's ability to recognize and identify figures, characters, forms, or patterns.

pattern-sensitive fault *Computer Programming.* a fault triggered by certain patterns in the data.

pattern shooting *Engineering.* in seismic prospecting, the firing of explosive charges arranged in a definite geometric pattern.

pattern wax *Materials.* a plastic material, easily molded at low temperatures and used to form patterns for the molding of prosthodontic devices, to form dental impressions, and as a dental adhesive.

Patterson function *Solid-State Physics.* a special function of three spatial coordinates used in the Patterson-Harker method of X-ray diffraction analysis, which is related to electron distribution.

Patterson-Harker method *Solid-State Physics.* a technique in X-ray diffraction analysis in which a Fourier series is constructed, using the squares of the absolute values of the structure factors, yielding a vectorial representation of the interatomic distances.

Patterson projection *Crystallography.* in crystal structure determinations by X-ray diffraction, a map made from the summation of a Fourier series that has the squares of the structure factor amplitudes as coefficients. Because these values of $|F|^2$ can be calculated from the diffraction intensities, the map can be computed directly. Ideally, the positions of the maxima in the map represent the endpoints of vectors between atoms, all referred to a common origin. Also, **Patterson map.**

Patterson superposition method *Crystallography.* see VECTOR SUPERPOSITION MAP.

Patterson vectors *Crystallography.* vectors between any two atoms in a crystal lattice that determine the crystal structure in the Patterson-Harker method of X-ray diffraction analysis.

Pattinson process *Metallurgy.* a process used to separate silver from lead.

patulin *Pharmacology.* $C_7H_6O_4$, a highly toxic antibacterial drug that is prepared from various fungi, especially of the genera *Asperigillus* and *Penicillium,* and formerly used as an antimicrobial.

patulous *Botany.* **1.** spreading widely from a center, as a calyx. **2.** of or relating to a plant that bears flowers loosely or in a dispersed manner.

paua *Vertebrate Zoology.* a large, edible abalone, *Haliotis iris,* whose shell is used in making jewelry; found in New Zealand.

Paucituberculata *Vertebrate Zoology.* the rat opossums, a monofamilial order of small, shrewlike marsupials having long narrow feet and a long tail; found in cool to temperate regions of South America.

Pauli, Wolfgang 1900–1958, Austrian physicist, living in the U.S.; awarded the Nobel Prize for the discovery of the Pauli exclusion principle; proposed the existence of the neutrino.

Wolfgang Pauli

Pauli anomalous moment term *Quantum Mechanics.* an extra term added to the Dirac equation to account for the anomalous magnetic moment of the particle. Also, PAULI TERM.

Pauli electron correlation *Quantum Mechanics.* the correlation of electron positions in space that results from the Pauli exclusion principle.

Pauli exclusion principle *Quantum Mechanics.* the principle stating that no two identical fermions may occupy the same quantum state simultaneously. Also, EXCLUSION PRINCIPLE, PAULI PRINCIPLE.

Pauli g-permanence rule *Atomic Physics.* a rule that governs the weak and strong forces in an atom: in *LS* coupling, the sum over J for specified L, S, and M_J of the weak-field g-factors is equal to the sum of the strong-field factors.

Pauli g-sum rule *Atomic Physics.* a rule claiming that the sum of the g-factors for energy levels consistent with the same J value is constant and is independent of the coupling scheme.

Paulinella *Invertebrate Zoology.* a genus of filose amoebae, having a shell and containing symbiotic blue-green algae.

Pauling, Linus born 1901, American chemist; awarded the Nobel Prize for his research on chemical bonds and molecular structure; awarded the Nobel Peace Prize for his role in the 1963 nuclear test ban treaty.

Pauling ionic radius *Materials Science.* the elective radius of the cations and anions whose ratios determine the coordination numbers.

paulingite *Mineralogy.* a colorless, cubic mineral of the zeolite group, with an approximate formula of $(K,Na)_2Ca(Si_{13}Al_4)O_{34}\cdot 13H_2O$, having an undetermined specific gravity and a hardness of about 5 on the Mohs scale; found as dodecahedral crystals in basalt vesicles.

Pauling rule *Solid-State Physics.* a rule that states the requirement for local electrical neutrality in a crystal-ion structure and determines the manner in which the coordination polyhedra are packed together in a complex structure.

Pauling scale see ELECTRONEGATIVITY SCALE.

Pauli principle see PAULI EXCLUSION PRINCIPLE.

Pauli spin matrices *Quantum Mechanics.* the angular-momentum matrices for spin ($\hbar$) given by the three 2×2 anticommuting matrices that represent the components of the spin operator.

Pauli spin space *Quantum Mechanics.* the two-dimensional vector space in which spin orientations are represented by vectors.

Pauli spin susceptibility *Solid-State Physics.* the susceptibility in a conductor associated with free electrons, which measures the tendency of these electrons to align their spins with an external magnetic field.

Pauli term see PAULI ANOMALOUS MOMENT TERM.

Pauli-Weisskopf equation *Quantum Mechanics.* the second-quantization form of the Klein-Gordon equation.

paunch *Anatomy.* a popular term for the belly or abdomen, especially a large, protruding belly. *Vertebrate Zoology.* another name for the rumen in cud-chewing animals.

paurometabolous metamorphosis *Invertebrate Zoology.* development through gradual metamorphosis; common among hemipterans (true bugs).

Pauropoda *Invertebrate Zoology.* a class of tiny terrestrial arthropods with cylindrical bodies, branched antennae, and nine to eleven pairs of legs; found in damp leaf litter.

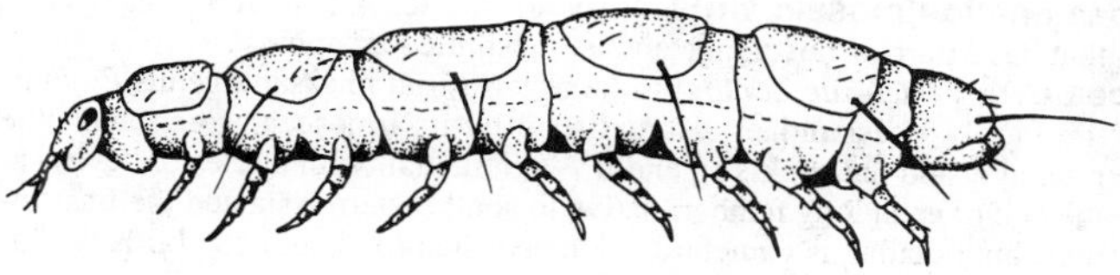

Pauropoda

pause *Acoustical Engineering.* a control on a tape recorder, compact disk player, or video tape recorder that allows movement to be stopped momentarily during recording or playback. Also, **pause control.**

Paussidae *Invertebrate Zoology.* flat-horn beetles, a family of small brown coleopteran insects belonging to the suborder Adephaga, having thick, broad-tipped antennae; found in ants' nests.

pave *Civil Engineering.* to cover a surface with pavement.

pavement *Building Engineering.* a hard floor or ground covering, generally of concrete, tile, or brick. *Civil Engineering.* **1.** the concrete surfacing of roads, streets, paths, and the like. **2.** a sidewalk. *Geology.* any bare rock surface that resembles a paved road surface in its texture, hardness, surface extent, horizontality, or close arrangement of its units.

paver *Mechanical Engineering.* **1.** any mobile machine that carries and lays material for paving. **2.** any brick, tile, stone, or other material that is used for paving.

Paveway *Ordnance.* any of three families of laser guidance units designed to transform a variety of standard unguided bombs into extremely accurate guided tactical weapons; the families are designated Paveway I, II, and III; the bombs are called **LGBs,** an acronym for Laser Guidance Bombs.

paviar *Building Engineering.* a hard brick material used in the construction of pavement.

pavilion *Architecture.* **1.** a tent or temporary structure, often ornamental, in a garden or recreation area. **2.** a wing of a building, usually having a specialized function, as in a hospital.

Pavlov, Ivan Petrovich [pav´läv´; pav´läf´] 1849–1936, Russian physiologist; awarded the Nobel Prize for his study of the digestive system; conducted noted early experiments in conditioned reflexes that influenced the development of behavioral psychology.

Pavlovaceae *Botany.* a family of primarily marine algae constituting the order Pavlovales.

Pavlovales *Botany.* a monofamilial order of algae of the class Chrysophyceae, characterized by either motile or palmelloid cells and a covering of electron-dense bodies, without scales.

Pavlovian [pav lōv´ ē ən; pav läv´ē ən] of or relating to I. P. Pavlov or his research and experiments, especially his famous technique of producing salivation in dogs by an artificial stimulus, such as the ringing of a bell, that was associated with food.

Pavlovian conditioning see CLASSICAL CONDITIONING.

Pavlov's pouch *Physiology.* a portion of the stomach of a dog, partitioned from the rest of the stomach and opened to the outside of the body; used to observe the flow of gastric juices.

Pavo *Astronomy.* the Peacock, a southern constellation, whose brightest star is 2nd magnitude, first introduced by Johann Bayer in 1603; it lies between Telescopium and the circumpolar Octans. (From the Latin word for peacock.)

pavonite *Mineralogy.* $(Ag,Cu)(Bi,Pb)_3S_5$, a lead-gray to tin-white metallic monoclinic mineral, having a specific gravity of 6.54 to 6.74 and a hardness of about 2 on the Mohs scale; found as small crystals and in massive form in hydrothermal veins with bismuthinite and alkinite.

Pavy's solution *Analytical Chemistry.* a modified Fehling's solution of copper sulfate, sodium potassium tartrate, potassium hydroxide, and ammonia; used in determining the concentration of sugars in solution by a colorimetric titration.

paw *Vertebrate Zoology.* **1.** an animal's foot. **2.** specifically, one of the feet of a quadruped mammal, usually having claws.

pawdite *Petrology.* a black or gray, fine-grained hypabyssal dike rock consisting of magnetite, titanite, biotite, hornblende, calcic plagioclase, and trace amounts of quartz.

pawl *Mechanical Engineering.* a pivoted catch, usually spring-controlled, that engages with a ratchet wheel to limit motion or to convert its own reciprocating motion to linear motion.

PAX private automatic exchange.

paxilla *Invertebrate Zoology.* in some types of starfish, one of a number of thick calcareous rods with tiny spines, found in place of the more usual spines.

Paxillaceae *Mycology.* a family of fungi belonging to the order Agaricales, occurring on soil and dead wood, and sometimes existing symbiotically with living plants.

Paxillosida *Invertebrate Zoology.* an order of starfish, echinoderms in the subclass Asteroidea, having large marginal plates, long spines, and pointed, suckerless feet.

Paxton, Joseph 1801–1865, English gardener and engineer; designed London's Crystal Palace (1851), a noted architectural innovation.

paydirt *Mining Engineering.* a poular term for gravel, sand, earth, or ore that can be mined profitably. Also, **pay ore.**

paying *Naval Architecture.* the process of completing the waterproofing of a wooden ship by pouring hot pitch into the seams, thus protecting the oakum caulking.

payload *Transportation Engineering.* the part of a vehicle's load that provides revenue, such as cargo or passengers. *Ordnance.* in a military missile, the warhead, its container, and its activating devices. *Aviation.* anything that a flight vehicle carries beyond what is required for its operation during flight.

payload-mass ratio *Aviation.* the ratio of the mass of a flight vehicle's payload to the total mass of the vehicle when loaded with the payload. *Space Technology.* specifically, the ratio of the mass of a rocket's propellant charge to the total mass of the rocket when charged with the propellant. Also, MASS RATIO.

payoff matrix *Mathematics.* in a finite, two-person zero sum game, the matrix $A = (a_{ij})$, where a_{ij} is the amount that the row player R obtains from the column player after independently selecting a row i and column j of the payoff matrix. If $a_{ij} < 0$, then R must pay $-a_{ij}$ to C.

pay sand *Petroleum Engineering.* the portion of oil-bearing or gas-bearing sand from which products may be recovered profitably.

pay streak *Mining Engineering.* **1.** an area where gold is concentrated in placer deposits. **2.** the portion of a vein of ore or minerals that is profitable to mine.

pay zone *Petroleum Engineering.* a geological formation from which oil and gas may be recovered profitably.

PB Pharmacopoeia Britannica; pulpobuccoaxial.

Pb the chemical symbol for lead. (From Latin plumbum.)

P band *Telecommunications.* a band of radio frequencies that ranges from 225 to 390 megahertz.

PBE Perlsucht Bacillen Emulsion.

PBI *Aviation.* the airport code for West Palm Beach, Florida.

PBI protein-bound iodine.

Pb-I-Pb junction lead-I-lead junction.

p-block elements *Chemistry.* elements belonging to the main groups III–VII and O with occupied *p* orbitals in their outer electronic shells.

P blood group *Immunology.* a system of red blood cell antigens that were originally characterized exclusively by their reaction to the immune rabbit serum known as anti-P; other closely related antigens are now included in this group.

PBO penicillin in beeswax.

pbr322 *Genetics.* a bacterial plasmid containing genes for resistance to the antibiotics ampicillin and tetracycline; specifically created for ease in cloning small DNA fragments of a few thousand nucleotide pairs or less.

P-branch *Spectroscopy.* in molecular spectra, a set of spectral lines associated with a decrease of one in the rotational quantum number, *J*.

Pb virus *Virology.* any of several viruses that have been isolated from the *Penicillium brevicompactum* (Pb) fungus.
PBX private branch exchange.
PBX access line *Electricity.* the interconnection between a private branch exchange (PBX) main and a switching center.
PC *Computer Science.* a shorter term for a personal computer, especially an IBM or IBM-compatible personal computer. See PERSONAL COMPUTER.
PC phosphocreatine; programmable controller; program counter.
P.C. or **p.c.** avoirdupois weight. (From Latin *pondus civile.*)
p.c. percent; percentage.
p.c. or **pc** after meals. (From Latin *post cibum.*)
PCB *Materials Science.* any of a group of toxic, chlorinated aromatic hydrocarbons used in a variety of commercial applications, including paints, inks, adhesives, electrical condensers, batteries, and lubricants; known to cause skin diseases and suspected of causing birth defects and cancer. (An acronym for polychlorinated biphenyl.)
PCc periscopic concave.
PCE pyrometric cone equivalent.
PCG phonocardiogram.
p chart *Industrial Engineering.* in quality control, a chart displaying the fraction of defectives, as represented by the letter *p.*
pCi picocurie.
PCM pulse-code modulation.
PCN personal communications network.
PCNB pentachloronitrobenzene.
PCP *Pharmacology.* $C_{17}H_{25}N{\cdot}HCl$, a potent analgesic and anesthetic used in veterinary medicine; abuse by humans may lead to serious psychological disturbances. Also, PHENCYCLIDINE HYDROCHLORIDE. *Organic Chemistry.* see PENTACHLOROPHENOL.
pcpt. perception.
PCR *Biotechnology.* polymerase chain reaction, a process for amplifying a DNA molecule by up to 10^6 to 10^9 fold; extremely important in biotechnology and in research.
pct. or **pct** percent; percentage.
PCV packed cell volume.
PCx periscopic convex.
P Cygni star *Astronomy.* a group of variable stars named for the prototype P Cygni, which has an irregular period and a spectrum characterized by broad emission lines with strong absorption on the violet side, originating from an expanding shell around the central star.
PD pulpodistal; interpupillary distance.
PD or **P.D.** potential difference; per diem. Also, **pd, p.d.**
Pd the chemical symbol for palladium.
pd or **p.d.** prism diopter; papilla diameter.
PDA predicted drift angle.
P-day *Military Science.* the precise time at which the production rate of an item equals the rate at which the item is required by the armed forces.
P_{dc} direct-current power.
PDGF platelet-derived growth factor.
PDM pulse-duration modulation.
p. dos. for a dose. (From Latin *pro dose.*)
PDP parallel distributed processing.
PDR *Physician's Desk Reference*; precision depth recorder.
PDT or **P.D.T.** Pacific Daylight Time.
PDX *Aviation.* the airport code for Portland, Oregon.
PE pentaerythritol; physical exam.
P.E. or PE probable error.
P.E. or **p.e.** printer's error. Also, **PE, pe.**
pea *Botany.* **1.** also, **garden pea.** an annual leguminous vine, *Pisum sativum* of the family Fabaceae, widely cultivated in cool climates for its edible globular seeds contained in dehiscent pods. **2.** also, **field pea.** a related plant, *Pisum arvense,* used in soups and as livestock fodder. **3.** the seed of either of these plants.
Peacekeeper *Ordnance.* a popular name for the M-X guided missile. See M-X.
peacetime force materiel requirement *Military Science.* the quantity of an item required to equip, provide a materiel pipeline, and sustain the peacetime force structure of the U.S. and its allies, and to support the scheduled establishment through normal appropriation and lead-time periods. Similarly, **peacetime force materiel assets.**
peacetime materiel consumption and losses *Military Science.* the quantity of an item consumed or worn beyond economical repair through normal appropriation and procurement lead-time periods.
peach *Botany.* **1.** a low-spreading, freely branching tree, *Prunus persica* of the family Rosaceae; widely cultivated in temperate zones for its edible fruit, a sweet, juicy, yellowish drupe with one large seed and a thin epicarp. **2.** the fruit of this plant.
peach leaf curl *Plant Pathology.* a disease of peaches caused by the fungus *Taphrina deformans,* characterized by an early puckering or crinkling and red to pale discoloration of young leaves, which eventually become thick and firm.
pea coal *Geology.* a size classification of anthracite coal, material that will pass through a 1 3/16-inch mesh screen, but not through a 9/16-inch mesh screen.
peacock *Vertebrate Zoology.* **1.** the male of the peafowl. **2.** loosely, any peafowl, male or female.
peacock blue *Organic Chemistry.* $HSOC_6H_4COH[C_6H_4N(C_2H_5)CH_2C_6H_4SO_3Na]_2$, a blue pigment used in inks, especially for multicolor printing.
peacock ore *Mineralogy.* any of various copper minerals, such as bornite or chalcopyrite, that exhibit a colorful and lustrous tarnish when exposed to air. Also, **peacock copper.**
pea enation mosaic virus *Virology.* a ssRNA-containing plant virus that has a narrow host range and is transmitted by aphids.
peafowl *Vertebrate Zoology.* a large terrestrial pheasant, genus *Pavo* of the family Phasianidae, distinguished by its iridescent green and blue plumage, and the cock's splendid 7-foot fan tail with circles or "eyes" at the end of each long feather; native to southeastern Asia and the East Indies, but popular as gamebirds, domesticated fowl, and zoo birds worldwide.

peafowl

peahen *Vertebrate Zoology.* a female peafowl.
pea iron ore see PEA ORE.
peak *Science.* **1.** the highest point or value. **2.** to reach the highest point. *Geology.* **1.** the conical or pointed top of a hill, mountain, mountain crest, or prominent summit. **2.** an individual mountain or hill taken as a whole, especially one that is isolated or has a pointed, conspicuous summit. *Meteorology.* an intersecting point between warm and cold fronts of a mature, extratropical, cyclonic weather circulation system.
peak-aged *Materials Science.* of or relating to certain alloys, such as those of aluminum and nickel, that have been age hardened so that their yield strength is a maximum.
peak-aging *Materials Science.* the process of aging a material to the point at which its yield strength is a maximum.
peak amplitude *Physics.* in a periodic quantity, the maximum value measured with respect to the zero value through one cycle.
peak attenuation *Telecommunications.* the reduction of response to a modulated wave that occurs on modulation crests.
peak cathode current *Electronics.* the highest value reached by the alternating current flowing through a cathode.
peak clipping *Acoustical Engineering.* the limiting of signal amplitudes during the processing of electroacoustic signals so that high peaks are clipped, thereby distorting sound when it is reproduced.
peak current *Analytical Chemistry.* in voltammetry, the maximum current rise due to formation of an increasing concentration gradient as the potential is shifted toward more negative values.
peak detector *Electronics.* a device that generates a DC output voltage whose value approximates the true peak value of the applied AC signal that it is tracking.

peak distortion *Telecommunications.* the highest amount of total signal distortion that is noted within a certain time span.

peak enthalpimetry *Analytical Chemistry.* an analytical method for measuring the heat content of a reagent at constant temperature when measured samples of a substance are added at given intervals.

peak envelope power *Electronics.* the average amount of power supplied to the antenna of a transmission line by a radio transmitter during one frequency cycle at the highest peak-to-peak values of the modulation envelope.

peaker *Electronics.* the fixed or adjustable inductance in a broadband amplifier that is used to resonate with stray and distributed capacitances to increase gain at high frequencies.

peak experience *Psychology.* Abraham Maslow's term for a moment of profound awe or ecstasy, felt so intensely that the individual may even seem unaware of space and time.

peak forward voltage *Electronics.* the highest instantaneous voltage applied to an electronic device in the direction with the least resistance to current flow.

peak gust *Meteorology.* the maximum instantaneous wind speed recorded at a weather station during a specified period, usually the 24-hour observational day.

peak-hour factor *Transportation Engineering.* the proportion by which peak hourly volume exceeds the average hourly volume.

peak hourly volume *Transportation Engineering.* the volume of traffic carried on a highway during the heaviest volume hour of the day.

peaking circuit *Electronics.* a device that can convert an input wave into a peaked waveform; commonly used to improve the high-frequency response in a broadband amplifier.

peaking network *Electronics.* a system that uses excess capacitance to increase the amplification of signals at the higher end of the frequency range.

peaking transformer *Electricity.* **1.** a transformer used to fire ignitrons and firetrons by rapid flux changes from one direction of saturation to the other, twice per cycle, which induces a highly peaked voltage pulse in a secondary winding; thus the number of ampere-turns in the primary transformer is high enough to produce many times the normal flux density values in the core. **2.** a transformer having a core designed to saturate relatively low values of primary current.

peak inverse anode voltage *Electronics.* the highest voltage applied to the anode of a circuit when current is flowing in the direction opposite to that for which the device was designed.

peak inverse voltage *Electronics.* the highest voltage applied, in the reverse direction, across the electrodes of a circuit.

peakless pumping *Mining Engineering.* a method of pumping a mine load so that the process is spread out over an entire day.

peak load *Electricity.* the maximum instantaneous or average load consumed or produced over a given time interval in an electronic system; AC power expressed as a product of peak voltage or peak current. Also, **peak power.** *Engineering.* the maximum quantity of material that can be transported by conveyor or other transport system.

peak overpressure *Ordnance.* the maximum value of overpressure at a given location; it is usually produced at the moment the shock or blast wave reaches the location.

peak period *Transportation Engineering.* a time of day (usually during the morning and evening commute hours) when traffic or demand for public transit is at its highest level.

peak period rate of flow *Transportation Engineering.* the rate of flow or volume of traffic carried on a highway during the period of maximum traffic.

peak plain *Geology.* a high-level plain formed by a series of accordant summits. Also, SUMMIT PLAIN.

peak power *Electromagnetism.* the maximum instantaneous power supplied by a radar pulse.

peak pressure *Ordnance.* the maximum pressure in the bore of a weapon produced by combustion of the propellant.

peak second algorithm *Telecommunications.* a technique for estimating the number of transmissions that will occur in a communication system during the busiest one-second interval of a timed study.

peak signal level *Electronics.* **1.** the highest instantaneous signal, power, or voltage allowed under specific operating conditions. **2.** the highest instantaneous signal, power, or voltage seen at any point in a facsimile transmission system, including auxiliary signals.

peak tank *Naval Architecture.* a tank below decks at the extreme bow or stern of a ship. Peak tanks are usually kept empty as voids, but sometimes provide potable water storage.

peak-to-peak amplitude *Physics.* the total range in amplitude of a periodic quantity through one cycle, measured as the maximum positive peak amplitude with respect to the negative peak amplitude.

peak-to-valley ratio *Telecommunications.* the ratio of the highest value of a modulated wave to its lowest value.

peak value *Electricity.* the maximum instantaneous value of a voltage, current, or power over a specified time interval. Also, CREST VALUE.

peak width *Analytical Chemistry.* the measurement of the time duration of a single rise and fall from a baseline on a graph of an analysis or process.

peak zone *Paleontology.* **1.** a stratigraphic zone dominated by an unusual abundance of one or more taxa. **2.** the stratigraphic zone in which a particular taxon reaches its greatest abundance.

Peano, Giuseppe 1858–1932, Italian mathematician; developed the rational number system from his five postulates for natural numbers; extended Boolean symbolic logic.

Peano axioms *Mathematics.* axioms that give rise to the construction of the natural numbers; at one time considered the fountainhead of all mathematical knowledge. In particular, let ω denote the set of all natural numbers (including zero). Then the axioms are (using the notion of successor sets): (a) $0 \in \omega$; i.e., 0 is a natural number, where $0 = \varnothing$ (sometimes this axiom is stated: there exists a natural number). (b) If $n \in \omega$, then $n^+ \in \omega$; i.e., every natural number has a successor that is a natural number, where $n^+ = n \cup \{n\}$. (c) If S is a subset of ω such that $0 \in S$ and $n \in S$ implies that $n^+ \in S$, then $S = \omega$; i.e., if a subset S of ω is a successor set, then $S = \omega$. (Axiom c is one version of the principle of mathematical induction.) (d) $n^+ \neq 0$ for all $n \in \omega$; i.e., 0 is not the successor of any natural number. (e) If $n^+ = m^+$, where $n,m \in \omega$, then $n = m$; i.e., no two natural numbers have the same successor.

peanut *Botany.* **1.** a low-growing annual legume, *Arachis hypogaea* of the family Fabaceae, widely cultivated in warm regions throughout the world for its edible seeds contained in underground pods. **2.** the pod or seed of this plant.

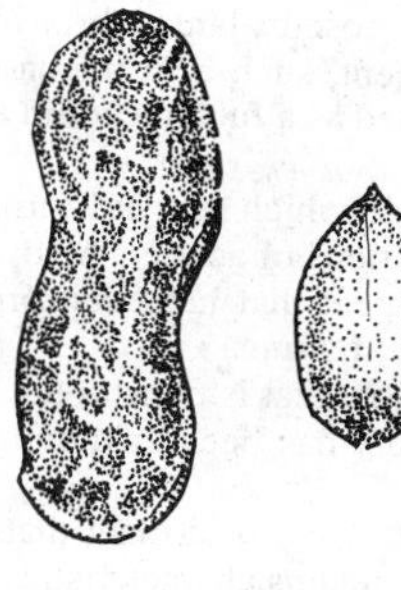

peanut

peanut oil *Materials.* a combustible, yellow to greenish-yellow nondrying oil that is expressed or extracted from peanuts; used in cooking, medicines, food, and as a substitute for olive oil and other edible oils. Also, ARACHIS OIL, GROUNDNUT OIL.

pea ore *Mineralogy.* a type of pisolitic limonite found as small, rounded grains or masses. Also, PEA IRON ORE.

pear *Botany.* **1.** any of several trees belonging to the genus *Pyrus* of the family Rosaceae, widely cultivated for their sweet, juicy fruit. The fruit is wider at the apical end, generally has a yellowish epicarp with white pulp, and has a core of several small seeds. **2.** the fruit of such trees. *Materials.* the fine-textured wood of the European timber tree, *Pyrus communis,* used for precision woodworking, laboratory equipment, and musical instruments.

pearceite *Mineralogy.* $Ag_{16}As_2S_{11}$, a metallic, black monoclinic mineral occurring as thin tabular crystals, having a specific gravity of 6.13 to 6.15 and a hardness of 3 on the Mohs scale; found in silver vein deposits of low to medium temperature.

pearl *Materials.* a smooth, hard, lustrous, spherical object composed of calcium or mineral aragonite and formed around a foreign irritant in certain mollusks.. *Pathology.* a small concretion of sputum noted in asthma sufferers. *Pharmacology.* a small, spherical hollow glass nodule containing a volatile inhalant such as amyl nitrite.

Pearl, Raymond 1879–1940, American biologist and statistician; a founder of biometrics.

pearl ash see POTASSIUM CARBONATE.

pearl essence *Materials Science.* a lustrous, silver-white substance that comes from the scales of certain fish or is synthetically derived and used as a pigment in lacquer and in the manufacture of simulated pearls.

pearlfish *Vertebrate Zoology.* any of several small fishes of the family Carapidae that live within pearl oysters, sea cucumbers, and starfish.

pearlite *Materials Science.* a two-phase lamellar microconstituent, consisting of ferrite and cementite, that forms in steels that are relatively slowly cooled or are isothermally transformed at relatively high temperatures. *Geology.* see PERLITE.

pearlite nose *Materials Science.* the portion of a TTT curve diagram for steels whose pearlite transformation rate is a minimum.

pearl-necklace lightning see CHAIN LIGHTNING.

pearl oyster *Invertebrate Zoology.* any of several marine bivalve mollusks of the family Pteriidae; some species form pearls of great value; found in waters of eastern Asia and the coasts of Panama and Baja California.

pearl polymerization *Materials Science.* a polymerization process in which a monomer or mixture of monomers is agitated continuously during polymerization; the resulting polymer is in a pearl shape that is separated from the suspending medium.

pearl spar *Mineralogy.* any of various crystalline carbonates, such as ankerite or dolomite, that exhibit a pearly luster.

pearly see NACREOUS.

Pearson, Karl 1857–1936, English mathematician; defined the standard deviation of a set of measurements; developed the chi-square test.

peasant *Anthropology.* **1.** a member of a contemporary peasant society. **2.** historically, a farmer or farm worker of low social status, especially in Europe, Latin America, or Asia.

peasant society *Anthropology.* a contemporary group of rural people living within a state or urban society who produce food primarily for their own subsistence and follow a traditional way of life.

pea-soup fog *Meteorology.* any fog of a particularly dense nature.

peasweep storm *Meteorology.* an early spring storm that occurs over the countries of Scotland and England.

peat *Geology.* a dark-brown or black mass of unconsolidated, semicarbonized, partially decomposed plant debris formed in an anaerobic, water-saturated environment, such as a marsh or bog. It is commonly used as fertilizer and is used as a fuel in certain areas, e.g., Ireland.

peat ball see LAKE BALL.

peat bog *Geology.* a bog in which peat has formed from the characteristic vegetation under conditions of acidity. Also, **peat bed, peat moor.**

peat breccia *Geology.* peat that has been broken up and later redeposited by water. Also, **peat slime.**

peat coal *Geology.* **1.** a coal that is transitional between peat and brown coal or lignite. **2.** peat that has been artificially carbonized for use as fuel.

peatmoss or **peat moss** *Botany.* a plant that partially decomposes in water, producing soil that retains characteristics of the plant.

peat-sapropel *Geology.* a degradation product that is transitional between peat and sapropel.

peat soil *Geology.* an acidic, humus-rich soil that contains a large amount of peat.

Peattie, Donald C. 1898–1964, American naturalist; the author of popular works on nature.

Peaucellier linkage *Mechanical Engineering.* a four-bar rhomboid mechanism that converts circular motion to linear motion. Also, **Peaucellier mechanism.**

pebble *Geology.* a small, roundish, waterworn or windworn rock fragment, larger than a granule and smaller than a cobble, generally ranging from 4 to 64 mm in diameter. Also, **pebblestone.** *Mineralogy.* a transparent and colorless quartz of low brilliance, with many uses in optics. Also, ROCK CRYSTAL.

pebble armor *Geology.* a desert armor composed chiefly of rounded pebbles.

pebble bed *Geology.* any conglomerate composed mainly of pebbles. Also, POPPLE ROCK.

pebble-bed reactor *Nucleonics.* a reactor whose fuel is composed of several small nuclear-fuel pellets stacked in the core.

pebble coal *Geology.* coal composed of rounded masses of coal cemented by coal material.

pebble conglomerate *Petrology.* a rock consisting of a consolidated mass of pebbles.

pebble dike *Geology.* **1.** any clastic dike composed mainly of pebbles. **2.** a tabular body containing sedimentary fragments in a matrix of igneous rock.

pebble heater *Chemical Engineering.* a type of regenerator made of small refractory particles enclosed in a brick-lined steel shell; used to temporarily store heat from one process so it may be used in another process.

pebble mill *Mechanical Engineering.* a horizontally mounted cylindrical pulverizing mill in which selected pebbles or large pieces of ore are used as grinding media; it has the advantage of not contaminating the material with iron.

pebble peat *Geology.* a thin peat formed by moss and algae in semiarid climates, accumulating under pebbles of quartz and chalcedony embedded in well-drained soil.

pebble phosphate *Geology.* a residual or transported sedimentary rock consisting of pebbles or concretions of phosphate minerals mixed with sand and clay.

pebble tool *Archaeology.* a simple form of stone tool, in which the original core is only slightly altered by striking off a few small flakes; the earliest known tool.

pebbly mudstone *Geology.* a delicately laminated mudstone consisting of thinly scattered pebbles embedded among distorted bedding planes.

pebbly sand *Geology.* an unconsolidated sedimentary deposit consisting of at least 75% sand and up to 25% pebbles.

pebbly sandstone *Geology.* **1.** a pebbly sand that has been consolidated. **2.** a sandstone consisting of 10–20% pebbles. **3.** in Scotland, a term used synonymously with conglomerate.

pebrine *Invertebrate Zoology.* a disease of silkworms caused by the sporozoan *Nosema bombycis*; epidemic in France and Italy in the 19th century.

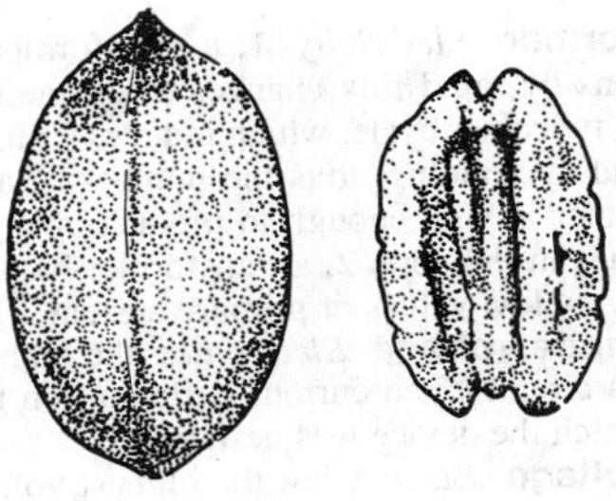

pecan

pecan *Botany.* **1.** a deciduous hickory tree, *Carya illinoinensis* of the family Juglandaceae, noted for its oblong, thin-shelled, edible nut and characterized by rough bark and hard but brittle wood. **2.** the nut of this tree.

peccary *Vertebrate Zoology.* a piglike mammal of the family Tayassuidae, characterized by a short, pointed head, a bristle-covered body, and long, slender legs with functional digits; found from the southwestern United States to Argentina.

peccary

peck *Metrology.* **1.** a traditional unit of dry measure that is used in the United States, equal to 8 quarts; equivalent to 8.81 liters. **2.** also, **imperial peck.** a similar unit of measure used in Great Britain, equivalent to 9.09 liters.

pecking order or **peck order** *Behavior.* **1.** a hierarchy of social status among the members of a flock of birds, demonstrated by the right of certain members to peck at or harass and dominate those lower in status without their retaliating. Thus, **peck(ing) right, peck(ing) dominance.** **2.** any similar status order within a group of animals of the same species. *Psychology.* in popular use, any clearly established hierarchy of authority or social status that is comparable to this, as in a business organization.

peckings *Building Engineering.* irregularly shaped, underburnt bricks; used for temporary work inside walls.

Peclet number [pə klā´] *Chemical Engineering.* a dimensionless parameter that is the product of the Reynolds and Prandtl numbers; used in forced convection for internal flow. *Fluid Mechanics.* see DIFFUSION NUMBER.

Pecopteris *Paleontology.* an extinct form genus of foliage that appears on the marattialean fern Psaronius, one seed fern, and several true ferns; extant in the Middle Carboniferous to Permian.

Pecora *Vertebrate Zoology.* an infraorder of the order Artiodactyla, composed of the high ruminants, including the deer, giraffe, okapi, cattle, antelope, sheep, and goat.

pecten *Vertebrate Zoology.* any of a number of comblike structures or organs that are found in animals. *Invertebrate Zoology.* a bivalve that belongs to the genus *Pecten.* (Going back to a Latin word meaning "comb" or "rake.")

Pecten *Invertebrate Zoology.* a genus of bivalves that includes most scallops.

pectic *Biochemistry.* relating to or containing pectin.

pectic acid *Biochemistry.* $C_{35}H_{50}O_{33}$, a uronic acid containing galactose and arabinose that is found in ripe fruits; used as an acidulant in pharmaceuticals.

pectic polysaccharides *Biochemistry.* polysaccharides, such as pectin, pectic acid, or pectates, extracted from plant walls using boiling water or dilute acids.

pectin *Biochemistry.* a homosaccharidic polymer of sugar acids of fruit that forms gels with sugar at the proper pH; used in pharmaceutical preparations and in the preparation of certain foods, such as jams and jellies.

Pectinariidae *Invertebrate Zoology.* the cone worms and trumpet worms, a family of polychaetes in the Sedentaria that construct conical tubes; parapodia are reduced, and heads have heavy hook-shaped setae for digging. Also, AMPHICTENIDAE.

pectinase *Enzymology.* an enzyme that is present in most plants; it catalyzes the hydrolysis of the polysaccharide pectin into sugars and galacturonic acid. It is used in biochemical research and in making jams and jellies.

pectinate *Biology.* shaped like a comb; having narrow, straight, close projections like the teeth of a comb.

Pectinatella *Invertebrate Zoology.* a genus of common freshwater bryozoans whose colony forms a gelatinous mass.

Pectinatus *Bacteriology.* a genus of Gram-negative bacteria of the family Bacteroidaceae, occurring in spoiled beer as curved, rod-shaped cells with flagella lining the concave side of each cell.

Pectinella *Invertebrate Zoology.* the equivalent name for *Sylncyclonema*, a genus of Pectinatella, containing several fossil species and a very rare living species found in deep waters of the Atlantic Ocean off Cuba and Antigua and the Pacific Ocean off Hawaii.

pectinesterase *Enzymology.* an enzyme that catalyzes the hydrolysis of pectin to pectic acid and methanol.

pectineus *Anatomy.* a flat quadrangular muscle located in the front and inner part of the thigh that flexes and adducts the thigh.

Pectinibranchia *Invertebrate Zoology.* a suborder of snails, including cowries, whelks, and cone shells; gastropod mollusks in the order Prosobranchiata that breathe with a comblike gill.

Pectinidae *Invertebrate Zoology.* a family of bivalve mollusks, including scallops, in the order Anisomyaria, having one central adductor muscle, a cylindrical foot, and no siphons; moves by clapping the valves to eject water rapidly from the mantle cavity.

pectization *Materials Science.* the process of changing into a gel or coagulate.

Pectobothridia *Invertebrate Zoology.* a subclass of parasitic flatworms in the class Trematoda, having hooks or suckers, or both.

pectolite

pectolite *Mineralogy.* $NaCa_2Si_3O_8(OH)$, a colorless, whitish, or grayish triclinic mineral with a vitreous to silky luster, occurring in aggregates of acicular crystals and fibrous masses, and having a specific gravity of 2.74 to 2.88 and a hardness of 4.5 to 5 on the Mohs scale; commonly found with zeolites in cavities in basalt.

pectoral *Anatomy.* of or related to the chest or breast.

pectoral fin *Vertebrate Zoology.* in fishes, one of a pair of fins located behind the gills and corresponding to the forelimbs of a quadruped.

pectoral girdle *Anatomy.* the bony arch supporting the arms, consisting of the scapulae and clavicles.

pectoralis major *Anatomy.* either of a pair of broad, thick, triangular muscles located in the upper part of the chest, arising out of the clavicle, sternum, and six upper ribs and inserting at the humerus of the arm; adducts, flexes, and rotates the arm medially.

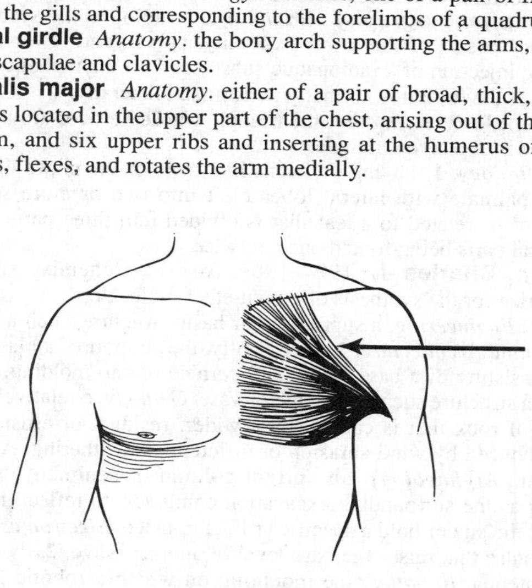

pectoralis major

pectoralis minor *Anatomy.* either of a pair of thin, flat, triangular muscles located on either side of the chest beneath the pectoralis major, arising from the third, fourth, and fifth ribs and inserted into the coracoid process of the scapula; draws the shoulder forward and downward.

peculiar galaxy *Astronomy.* any galaxy that does not fall into any of the three Hubble classes: elliptical, spiral, or irregular.

peculiar motion *Astronomy.* the real or true motion of a star with respect to the local standard of rest. Also, **peculiar velocity.**

peculiar part *Ordnance.* an ordnance part whose design and production are controlled by a single manufacturer.

peculiar star *Astronomy.* a star having a spectrum with peculiarities that does not fit into the usual spectral classification.

peculiar variable star *Astronomy.* a star characterized by an irregularly varying light period, such as novae, R Coronae Borealis, RR Tauri, and T Tauri stars.

ped *Agronomy.* a unit of soil structure, such as an aggregate, crumb, prism, block, or granule, that is formed by a natural process, as distinguished from a clod.

ped- or **pedo-** a prefix meaning: **1.** child. **2.** foot.

ped. pediatrics; pediatrician.

pedal *Mechanical Devices.* a foot lever pressed to activate another part within a mechanism. *Biology.* **1.** of or relating to the foot. **2.** the foot or base of a mollusk.

pedal coordinates *Mathematics.* let O be a fixed point not on a given curve γ. The pedal coordinates of a point P on γ are (r,p), where r is the distance from O to P and p is the perpendicular distance from O to the tangent to γ at P. Pedal coordinates are always given relative to some point O, known as the **pedal point**.

pedal curve *Mathematics.* let O be a fixed point not on a given curve γ. The pedal curve of γ with respect to O is the locus of the intersection of the envelope of tangents of γ with perpendiculars drawn to the tangents from O .

pedal disk *Invertebrate Zoology.* in sea anemones and anthozoan polyps, the flat base that attaches to the substrate.

pedal equation *Mathematics.* any equation describing a curve in terms of its pedal coordinates with respect to some fixed point O.

pedalfer *Agronomy.* a leached soil that develops in humid regions and is characterized by an accumulation or concentration of aluminum and iron oxides.

pedal ganglion *Invertebrate Zoology.* in mollusks, one of a pair of ganglia that supply nerves to the foot.

pedal gland *Invertebrate Zoology.* **1.** in snails and slugs, a gland in the foot that produces mucus to ease locomotion. **2.** in rotifers, a gland in the foot that secretes an adhesive substance.

Pedaliaceae *Botany.* a family of tropical terrestrial and sometimes aquatic herbs and shrubs of the order Scrophulariales, characterized by tricomes with a short stalk and a broad head composed of four or more mucilage-filled cells; includes sesame seed and unicorn plant species.

pedality *Agronomy.* the physical characteristics of a soil, as determined from the features of its constituent peds.

pedal lymphangiography *Radiology.* a radiographic study of the lymphatic vessels of the lower extremity, pelvis, and lower abdomen following injection of a radiopaque substance into a lymphatic vessel at the back of the foot; it is normally performed bilaterally.

pedal triangle *Mathematics.* the triangle whose vertices are the feet of the altitudes of a given triangle.

pedate *Biology.* **1.** having divisions like toes, or the claws of a bird's foot. **2.** palmate with lateral lobes cleft into two or more segments. *Botany.* of or related to a leaf that is divided into three parts, with the two lateral parts being forked once or twice.

Pedersen, Charles J. 1904–1989, American chemist; shared the Nobel Prize for the synthesis of crown-ether molecules.

pedestal *Engineering.* a supporting or basic structure, such as the base of a column. *Architecture.* **1.** specifically, the support of a classical column, consisting of a base, dado, and cornice or cap molding. **2.** a support for a structure such as a statue or vase. *Geology.* a relatively slender column of rock that is capped by a wider, residual or erosional rock mass produced by wind abrasion or differential weathering. Also, ROCK PEDESTAL. *Archaeology.* an upright column of sediment that is left standing as the surrounding excavation continues, to reflect the stratigraphy of the site or hold a specific artifact in place. *Electronics.* **1.** a flat-topped pulse that raises the base level of another wave. **2.** the base of a radar antenna. *Robotics.* the mounting on which a robotic arm is attached. (Going back to the Latin word for "foot," with the sense that it is the foot of a structure.)

pedestal design *Robotics.* a robot design centered on the vertical axis of a pedestal, confining the motion of the workpiece to a spherical envelope.

pedestal footprint *Robotics.* the amount of floor space taken up by the pedestal of a robot.

pedestal pile *Civil Engineering.* a cast-in-place concrete post formed by seating the pile in place and forcing the concrete to bulge out, creating a pedestal shape that holds the post in fixed place.

pedestal rock *Geology.* **1.** an isolated, residual or erosional rock mass supported by a rock pedestal. Also, **pedestal boulder.** **2.** see PERCHED BLOCK, def. 1.

pedestal sensor *Robotics.* a sensor that is mounted on the pedestal of a robot.

pedestal sight *Ordnance.* a sight mounted on a pedestal and used in the remote control of aircraft guns.

pedestrian survey or **pedestrian tactic** *Archaeology.* a method of examining a site in which surveyors, spaced at regular intervals, systematically walk over the area being investigated. Also, SURFACE SURVEY.

Pedetidae *Vertebrate Zoology.* the springhare, a monotypic family of large, nocturnal, semifossorial rodents of the suborder Sciuromorpha, characterized by long tails and large, harelike ears; native to South Africa. *Paleontology.* a small family of African rodents in the infraorder Myomorpha; the modern genus is little changed from the earliest fossils, which occur in the Miocene.

Pediastrum *Botany.* a common genus of free-floating freshwater green algae in the family Hydrodictyaceae, having biflagellate isogametes and flat, platelike colonies of two or more cells.

pediatric [pē´dē at´rik] *Medicine.* of or relating to the medical treatment of children. Thus, **pediatric nurse, pediatric surgery,** and so on.

pediatrician [pē´dē ə trish´ən] *Medicine.* a physician who specializes in the diagnosis and treatment of childhood diseases.

pediatrics [pē´dē at´riks] *Medicine.* the medical specialty concerned with the care of children and the diagnosis and treatment of childhood diseases.

pedicel [ped´ə səl] *Botany.* **1.** a small stalk bearing one flower only; either the main flower stalk or each secondary stalk in a branched inflorescence. **2.** the stem of a sporebearing or fruiting organ. *Zoology.* **1.** a stalklike fiber or support for certain internal organs. **2.** the second section of an insect antenna.

pedicellaria *Invertebrate Zoology.* minute pincers on the surface of some echinoderms; used to protect against ectoparasites.

pedicellate *Biology.* having or attached by a small stalk (a pedicel).

Pedicellinea *Invertebrate Zoology.* an order of colonial marine bryozoans with a basal stalk.

pedicle [ped´ə kəl] *Anatomy.* either of two short, thick pieces of bone projecting dorsally from the body of a vertebra, connecting the body and neural arch. *Invertebrate Zoology.* a short stalk composed of tough connective tissue that attaches a brachiopod to the substrate. Also, **peduncle.** *Surgery.* a part or structure that resembles a foot, stem, or narrow base, such as the base by which a nonsessile tumor might attach itself to normal tissue.

pedicle clamp see CLAMP FORCEPS.

pediculophobia *Psychology.* an irrational fear of infestation by lice.

pediculosis *Medicine.* a skin disease caused by infestation of lice and characterized by intense pruritus and cutaneous lesions.

Pediculus *Invertebrate Zoology.* a genus of blood-sucking lice dangerous to humans, serving as a vector for many major diseases, including epidemic relapsing fever and typhus rickettsia.

pedigree *Agriculture.* a chart of the family history of a purebred animal. *Genetics.* any diagram of the genetic history of an individual.

pedigree analysis *Genetics.* a diagram showing the ancestral relationships among individuals of a family over two or more generations.

Pedillidae *Invertebrate Zoology.* an order of beetles, coleopteran insects in the superfamily Tenebrionoidea.

pediment *Architecture.* **1.** in classical architecture, a framed gable usually containing a sculptured panel. **2.** a usually triangular ornament over a portico, door, or window. *Geology.* a broad, gently sloping, low-relief erosion surface, composed primarily of bare rock, that develops in an arid or semiarid region at the base of a receding mountain slope. Also, CONOPLAIN.

pedimentation *Geology.* the processes and actions involved in the formation and development of a pediment or pediments; the products of such processes and actions.

pediment dome see DESERT DOME.

pediment gap *Geology.* a broad opening formed by the enlargement of a pass joining two pediments on opposite sides of a mountain slope.

pediment pass *Geology.* a flat, narrow, rock-floored tongue that joins a pediment on one side of a mountain to a pediment on the other side.

Pedinellaceae *Botany.* the single family of the order Pedinellales, characterized by a radially symmetric cell, a ribbonlike flagellum surrounded by a crown of tentacles, and a posterior stalk.

Pedinellales *Botany.* a monofamilial order of the class Chrysophyceae, including freshwater and marine species that obtain nutrients from both photosynthesis and phagotopism.

Pedinidae *Invertebrate Zoology.* a family of sea urchins, echinoderms in the order Pedinoida, having a relatively flat, rigid test, solid primary spines, and hollow secondaries.

Pedinoida *Invertebrate Zoology.* an order of sea urchins in the superorder Diadematacea, consisting of a single family, the Pedinidae.

Pedinomonadales *Botany.* an order of green eukaryotic flagellates of the class Prasinophyceae, consisting of the single family Pedinomonadaceae; characterized by very small naked cells and a flagellum attached laterally or at one pole in a shallowly depressed area.

Pediococcus *Bacteriology.* a genus of Gram-positive bacteria of the family Streptococcus, occurring as nonmotile cocci in pairs or tetrads that reproduce by binary fission in two planes.
pediocratic *Geology.* of or relating to an interval of time during which little diastrophism occurs.
pedion *Crystallography.* a single face of a crystal with no symmetrically equivalent face.
Pedionomidae *Vertebrate Zoology.* the plains wanderers, a monotypic family of quail-like terrestrial birds characterized by short wings and tail and an unusual plumage of brown, buff, and black; found in deserts and open grasslands of central and southern Australia.
pedipalp *Invertebrate Zoology.* **1.** in arachnids, one of a pair of appendages on the second segment, modified for various purposes. **2.** a member of the Pedipalpida.
Pedipalpida *Invertebrate Zoology.* the whip scorpions, a group of arachnids usually divided into two orders, Uropygi and Amblypygi.

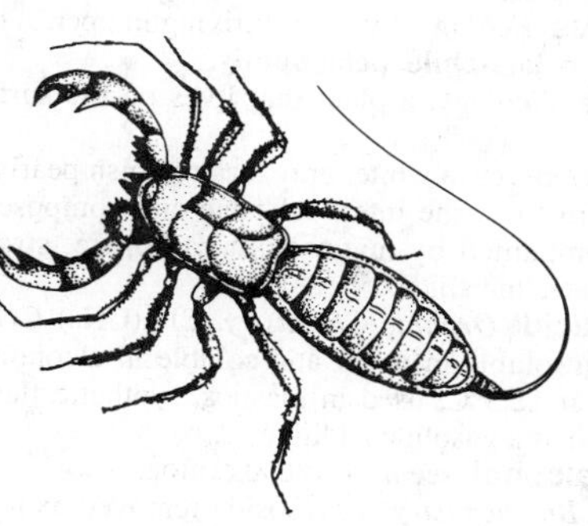

Pedipalpida

pediplain or **pediplane** *Geology.* a broad, thinly alluviated, rock-cut erosion surface produced in a desert region by the merging of two or more adjacent pediments. Also, DESERT PENEPLAIN, PANFAN.
pediplanation *Geology.* the processes involved in the formation and development of a pediplane or pediplanes.
pedo- a combining form meaning: **1.** soil, as in *pedogenics, pedography.* **2.** child, as in *pedodontics, pedology.*
pedocal *Geology.* an unleached soil that develops in arid or semiarid regions and that is characterized by an accumulation or concentration of carbonates, usually calcium carbonate.
pedodontics *Medicine.* the dental specialty concerned with the diagnosis and treatment of conditions of the teeth and mouth in children.
pedodontist *Medicine.* a dentist who specializes in the teeth and mouth conditions of children.
pedogenesis see SOIL GENESIS.
pedogenics *Geology.* a branch of soil science that involves the study of the origin, formation, and development of soil.
pedogeography *Geology.* a branch of geography that involves the study of the geographic distribution of soils.
pedography *Geology.* an aspect of soil science concerned with the systematic description of soils.
pedolith *Geology.* a surface formation that has undergone one or more soil-forming processes.
pedologic age *Geology.* the relative maturity of a soil profile.
pedologic unit *Geology.* a soil considered independently of its stratigraphic relationships.
pedology *Medicine.* the systematic study of the physiological and psychological development of children. *Geology* see SOIL SCIENCE.
pedometer *Engineering.* **1.** an instrument that measures the distance walked by counting the number of steps taken; the instrument is actuated at each step by the movement of a small weight balanced against a spring. **2.** an instrument used to measure the weight and length of a newborn baby.
Pedomicrobium *Bacteriology.* a genus of budding bacteria commonly coated with iron or manganese compounds; found in soil.
pedon *Geology.* the smallest unit or volume of soil representing or including all the horizons of a soil profile.
pedophile [ped´ə fīl´] *Psychology.* an adult who is sexually attracted to young children or who has sexual activity with young children.
pedophilia [ped´ə fēl´yə] *Psychology.* sexual attraction on the part of an adult for a child.
pedophobia *Psychology.* an irrational fear or dislike of children.
pedorelic *Geology.* of or relating to a soil feature that is derived from a preexisting soil horizon.
pedosphere *Geology.* the layer of the earth in which soil-forming processes take place.
pedoturbation *Archaeology.* any of the various processes by which soil surface is disturbed, such as the burrowing of animals.
peduncle *Anatomy.* a general term for collections of nerve fibers connecting various areas of the central nervous system. *Botany.* **1.** on a flower, the main stalk that supports an inflorescence. **2.** in certain thallophytes, a stalk supporting the fruiting body.
pedunculate *Biology.* having or growing on a small stalk or stem (peduncle).
peek *Computer Programming.* **1.** to examine the contents of a specified absolute-memory location from a high-level language. **2.** a command that allows the user to accomplish this, with the argument being the address of the memory location.
peeling *Materials Science.* the process of removing a layer of coating that has become detached because of poor adhesion. Also, **peel-back.**
peeling chisel *Mechanical Devices.* a paint chisel with a narrow, thin blade; used to peel paint from a surface.
peel test *Engineering.* a test used to determine adhesive strength by peeling apart bonded metal strips.
peel thrust *Geology.* a sheet of sedimentary rock that has been peeled from a sedimentary sequence, usually along a bedding plane.
peen *Mechanical Devices.* the rounded or wedge-shaped end of a hammer head opposite its face; used for bending, straightening, shaping, flattening, or cutting material.
peening *Metallurgy.* a process of cold-working the surface layer of a metallic part by hammering or shot blasting.
peephole mask *Computer Technology.* in a character recognition unit, one of a set of strategically placed points that theoretically render all input characters unique, regardless of style.
peephole optimization *Computer Science.* optimization, performed on generated code by a compiler, in which a linear pass is made over the code, examining a small region to determine if it can be improved; e.g., if a jump instruction to the next sequential location can be eliminated.
peep sight *Ordnance.* a rear sight with a small hole in which the front sight is centered during aiming.
peep slot *Ordnance.* in a combat vehicle, a slot that can be easily opened for viewing and closed for protection.
peg *Engineering.* **1.** a short, pointed, usually wooden pin or spike used to fasten pieces of wood or other materials together, or to close an opening. **2.** a projecting piece that serves as a support.
PEG pneumoencephalogram.
Pegasidae *Vertebrate Zoology.* the sea moths, a small family of little fishes of the order Gasterosteiformes, distinguished by a covering of armorlike bony plates, mothlike pectoral wings, and a long snout; found in warm waters of the Indo-Pacific from East Africa to Hawaii.
Pegasiformes *Vertebrate Zoology.* in some classification systems, an order of small fishes including the single family Pegasidae.
Pegasus *Astronomy.* the Flying Horse, a large but dim constellation in the Northern Hemisphere, visible in autumn.
pegging rammer *Metallurgy.* in casting, a tool used to ram sand in a mold.
peg graft *Botany.* a graft made by placing a wedge-shaped leafless section of a dormant plant into an opening in the stock and sealing it in with wax or a similar material.
pegmatite *Petrology.* a light-colored igneous rock or rock group from which feldspar, mica, or gemstones are derived; characterized by very coarse grains of interlocking crystals, with typically but not exclusively granitic composition. Also, GIANT GRANITE.
pegmatitic stage *Geology.* one of the successive stages in the normal crystallization of a magma that contains volatiles, during which the residual fluid is sufficiently enriched in volatile materials to allow the formation of pegmatites.
pegmatitization *Geology.* the process by which a pegmatite is formed or replaces another rock.
pegmatoid *Petrology.* an igneous rock similar to pegmatite in featuring a coarse-grained texture, but without the graphic intergrowths or typical granitic composition.
peg model *Geology.* a three-dimensional model used to illustrate the stratigraphic and structural conditions of subsurface geology, in which pegs of varying heights are mounted onto a flat surface to represent the contours of various strata.

I. M. Pei

Pei, I. M. [pā] born 1917, Chinese-American architect, designed the entrance pyramid at the Louvre (Paris) and the East Wing of the National Gallery of Art (Washington, D.C.).

Peierls-Nabarro stress *Materials Science.* the resolved shear stress required to move a dislocation on a slip plane in the slip direction. Also, **Peierls-Nabarro force.**

Peirce, Benjamin [purs] 1809–1880, American mathematician and astronomer; formulated the theory of linear associative algebras.

Peirce, Charles S. 1839–1914, American mathematician and philosopher; a pioneer in mathematical logic; formulated the philosophy of pragmatism.

Peirce-Smith furnace *Materials Science.* a nickel-smelting furnace used as a converter to remove iron and sulfur.

Peisidicidae *Invertebrate Zoology.* a family of polychaete annelid worms in the Errantia; found in mussel beds.

Pekilo process *Biotechnology.* a commercial process that cultures fungi of the species *Paecilomyces varioti* as animal feed.

Pekin or **Peking** *Agriculture.* the chief commercial breed of meat-producing duck in the U.S., having yellow skin, creamy white plumage, and orange shanks and feet. (From the city of *Peking* in China.)

Peking man *Anthropology.* a form of *Homo erectus* found in northern China, characterized by a massive brow ridge and a cranial capacity of 900–1200 cc; extant in the Pleistocene; one of the first hominid fossils of such antiquity to be found in association with stone tools. It has also been classified in the past as *Pithecanthropus* or *Sinanthropus.*

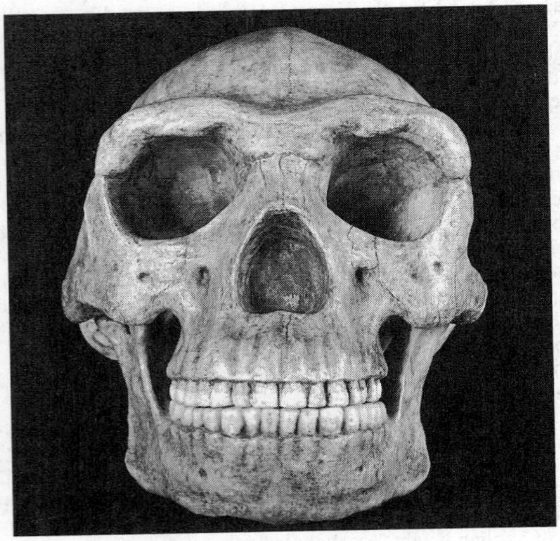

Peking man

PEL picture element.

pelage [pel′ij] *Vertebrate Zoology.* the fur or other soft surface covering of a mammal.

pelagic [pə laj′ik; pə lā′jik] *Oceanography.* of or relating to the open ocean, near the surface or in the middle depths, beyond the littoral zone and above the abyssal zone. *Hydrology.* relating to the deeper regions of a lake that are characterized by deposits of mud or ooze and by the absence of aquatic vegetation. *Ecology.* of or relating to aquatic organisms that live in the ocean, without direct dependence on the shore or bottom or on deep-sea sediment. (Going back to a Greek word meaning "the sea.")

pelagic limestone *Geology.* a fine-textured limestone formed in relatively deep water by the accumulation of calcareous tests of floating organisms.

pelagite see MANGANESE NODULE.

pelagochthonous *Geology.* of or relating to coal that is derived from a submerged forest or from driftwood.

pelagophilous *Ecology.* living or thriving in open surface waters of the ocean. Thus, **pelagophile, pelagophily.**

pelagophyte *Ecology.* a plant that lives on the surface waters of the ocean.

pelagosite *Geology.* a white, gray, or brownish pearly lustered, superficial crust formed in the intertidal zone and composed of calcium carbonate accompanied by magnesium carbonate, strontium carbonate, calcium sulfate, and silica.

pelargonic acid *Organic Chemistry.* $CH_3(CH_2)_7CO_2H$, a colorless to yellow oil; insoluble in water and soluble in alcohol; boils at 255.6°C and freezes at 12.5°C; used in plastics, synthetic flavors, and pharmaceuticals, and as a gasoline additive.

pelargonic alcohol see *n*-NONYL ALCOHOL.

pelargonin *Biochemistry.* a glycoside removed as red crystalline chloride from the dried petals of red pelargoniums or blue cornflowers.

peldon *Petrology.* a very hard, smooth, compact sandstone with conchoidal fractures, occurring in coal measures.

Peléan-type eruption *Volcanology.* a volcanic eruption characterized by extreme explosiveness and highly viscous magma. (From Mount Pelée, Martinique, which erupted disastrously in 1902.)

Pel-Ebstein fever *Medicine.* a recurrent fever, occasionally associated with malignant lymphoma or Hodgkin's disease. (Named for Pieter K. *Pel,* Dutch physician, 1852–1919, and Wilhelm *Ebstein,* 1836–1912, German physician.)

Pelecanidae *Vertebrate Zoology.* the pelicans, a monogeneric family of large marine and freshwater birds of the order Pelecaniformes, distinguished by a long, straight bill with an enormous gular pouch and a relatively long wingspan; found on or near coastal waters in eastern Europe, Africa, India, Sri Lanka, Southeast Asia, Australia, North America, and northern South America.

Pelecaniformes *Vertebrate Zoology.* an order of medium to large aquatic birds characterized by four webbed toes, a usually bare throat patch, and long wings; found worldwide, usually in coastal areas.

Pelecanoididae *Vertebrate Zoology.* the diving petrels, a monogeneric family of small, low-flying marine birds of the order Procellariiformes; found in all oceans, especially in southern seas and the Pacific Ocean.

Pelecinidae *Invertebrate Zoology.* a family of large parasitic wasps, hymenopteran insects in the superfamily Proctotrupoidea; the females have a long, tubelike abdomen.

P electron *Atomic Physics.* one of the electrons that is found in the *P* shell of an atom and displays the characteristics of the principal quantum number 6.

Pelecypoda see BIVALVIA.

P element *Molecular Biology.* a transposable element that is found in *Drosophila melanogaster* and that can integrate into the *Drosophila* genome, resulting in mutations.

Pele's hair [pā′lāz′] *Geology.* the naturally occurring filaments of volcanic glass blown out during quiet eruptions of fluid lava. (From the name of a Hawaiian goddess associated with a volcano where this occurs.) Also, CAPILLARY EJECTA, FILIFORM LAPILLI.

Pele's tears *Geology.* the small, tear-shaped, spherical, or cylindrical drops of volcanic glass that are usually trailed by pendants of Pele's hair.

Pelger nuclear anomaly *Medicine.* a hereditary anomaly of granulocytes that is characterized by small neutrophilic leukocytes in the peripheral blood with unusually coarse nuclear chromatin. (Named for Karel Pelger, 1885–1931, Dutch physician.) Also, **Pelger-Huët nuclear anomaly.**

pelican *Vertebrate Zoology.* a bird of the family Pelecanidae, having a relatively large, light body, a long wingspan, and a large gular pouch beneath a long bill; found near coasts and inland waters in eastern Europe, Africa, India, Sri Lanka, Southeast Asia, Australia, North America and northern South America. Two common North American species are the **brown pelican,** *Pelicanus occidentalis,* and the **American white pelican,** *P. erythrorhynchos.*

pelican

pelite *Geology.* **1.** a sediment or sedimentary rock composed of very fine clay-sized or mud-sized particles. **2.** a metamorphosed derivative of an argillaceous or aluminous rock or sediment. Also, PELYTE.

pelitic Geology. of or relating to a sedimentary or metamorphic rock composed of or derived from pelite.

pelitic hornfels *Petrology.* a fine-grained, nonfoliated metamorphic rock derived from pelite.

pelitic schist *Petrology.* a foliated crystalline metamorphic rock derived from pelite.

pellagra *Medicine.* a disease caused by a deficiency of niacin or by the inability to metabolize niacin, characterized by dermatitis, inflammation of mucous membranes, diarrhea, and psychic disturbances.

Pell equation *Mathematics.* the Diophantine equations $x^2 - Dy^2 = \pm 1$, where D is a positive integer that is not a perfect square. Also, FERMAT-PELL EQUATIONS.

pellet *Science.* a small rounded mass of a compressed substance such as food or medicine. *Pharmacology.* a small pill or granule, often a small ovoid mass of steroid hormones that is implanted beneath the skin to provide slow absorption. *Ordnance.* **1.** a small lead shot used in a shotgun charge. **2.** a small bullet. *Geology.* a small, fine-grained, spherical to elliptical aggregate composed almost entirely of clay-sized calcareous material, having no internal structure, and contained within a well-sorted carbonate rock. *Vertebrate Zoology.* **1.** in the feeding habits of owls, raptors, and other carnivorous birds, the clumps of undigestible matter regurgitated through the beak, containing fur, bones, feathers, and so on. **2.** fecal matter deposited by deer, elk, or lagomorphs.

Pelletier, Pierre 1788–1842, French chemist; with Joseph Caventous, isolated chlorophyll and discovered strychnine.

pelletierine tannate *Pharmacology.* $C_8H_{15}NO$, a mixture of the tannates of alkaloids obtained from pomegranate; formerly used to treat infestations of parasitic worms in the intestines. Also, PUNICINE TANNATE.

pelleting *Engineering.* a technique used to accelerate the solidification of an explosive charge by combining pellets of the explosive with the molten charge.

pelletization *Mining Engineering.* a method of working finely divided ore or coal so that it forms small, spherical aggregates about a half-inch in size. Also, **pelletizing.**

pellet mill *Mechanical Engineering.* a mechanical device that forces finely divided material into a roller, in which it is compacted by steam into a solid and cut into lozenges as it extrudes from the roller.

pelletron *Nucleonics.* an electrostatic particle accelerator whose charging system contains steel cylinders that are insulated at their ends with nylon so as to simulate a chain.

Pelliaceae *Botany.* a family of liverworts of the order Metzgeriales, characterized by organization in a stem and lateral leaf system or a thallose system, colorless or brownish rhizoids, and no ventral scales or appendages.

Pellicieraceae *Botany.* a monospecific family of dicotyledonous, glabrous mangrove trees of the order Theales, characterized by scattered raphide cells, branching stone cells, evergreen leaves, and a leathery, beaked, indehiscent fruit; native to Pacific Ocean shores from Costa Rica to Colombia.

pellicle *Invertebrate Zoology.* a translucent, hard, protective envelope of nonliving material covering many protozoans.

pellicularia disease *Plant Pathology.* a disease of coffee and other tropical plants caused by the fungus *Pellicularia koleroga*; characterized by a spotting of the leaves and the growth of filamentous strands of fungi over the leaves and twigs.

pellicular resin *Analytical Chemistry.* spheres or beads that are coated with a thin layer of ion-exchange resin; used in liquid chromatography.

pellicular water *Hydrology.* layers of water that adhere to and form a film on the surfaces of soil and rock particles in the zone of aeration. Also, ADHESIVE WATER, FILM WATER, SORPTION WATER.

pellis *Mycology.* in fungi, the tissue that is not part of the veils. Also CUTICLE.

pellotine *Organic Chemistry.* $C_{13}H_{19}O_3N$, colorless platelike crystals; slightly soluble in water and soluble in alcohol; melts at 111.5°C; found in the pellote (peyote) cactus, *Lophophora williamsi*, and used as a hypnotic.

Pelmatozoa *Invertebrate Zoology.* a subphylum of echinoderms, containing forms that are sessile during at least part of the life cycle; characterized by the presence of the mouth and anus on the upper surface and a skeleton of calcareous plates.

pelmatozoan *Paleontology.* of or relating to echinoderms of the subphylum Pelmatozoa, which were attached to the sea floor by a stem; although pelmatozoan niches are still abundantly occupied, the adoption of a more pelagic way of life by some echinoderms was significant in the evolution of animal life; extant from the Lower Cambrian to the present.

pelmicrite *Geology.* a limestone having a variable proportion of pellets and carbonate mud, especially one containing less than 25% each of intraclasts and ooliths; characterized by a volume ratio of pellets to fossils greater than 3:1, and a carbonate-mud matrix more abundant than its sparry-calcite cement.

Pelobacter *Bacteriology.* a genus of Gram-negative bacteria that occur as anaerobic, rod-shaped cells with a fermentative mode of metabolism.

Pelobatidae *Vertebrate Zoology.* the spadefoot or archaic toads, a family of lower anurans characterized by the absence of free ossified ribs; generally terrestrial and nocturnal and found in Europe, North Africa, North America, and southern Asia.

Pelobiontida *Invertebrate Zoology.* an order of naked, multinucleated amebas with only one pseudopod and no flagellated stages.

Pelodictyon *Bacteriology.* a genus of aquatic photosynthetic bacteria of the family Chlorobiaceae, composed of nonmotile, anaerobic, rod-shaped cells that contain gas vacuoles and reproduce by binary fission.

pelodite or **pellodite** *Geology.* a lithified rock flour composed of glacial pebbles embedded in a silt or clay matrix and formed by the redeposition of the fine fraction of a till.

Pelodytidae *Vertebrate Zoology.* the European spadefoot toads, a monogeneric family of lower anurans that are nocturnal and generally terrestrial; characterized by long hind legs, making them strong jumpers; found in western Europe and southwest Asia.

pelogloea *Geology.* a detrital slime derived from settled plankton and other organic material in natural waters.

Pelogonidae see OCHTERIDAE.

pelolithic see ARGILLACEOUS.

Pelomedusidae *Vertebrate Zoology.* the hidden-necked turtles, a small family of the order Testudines, distinguished by a habit of retracting the head by drawing it in sideways, sometimes leaving part of the head and neck exposed; found in fresh waters of tropical and subtropical areas of Africa, Madagascar, and South America east of the Andes.

Pelonemataceae *Bacteriology.* a family of cylindrical, gliding bacteria of uncertain classification; characterized by colorless cells forming filaments.

Pelopidae *Invertebrate Zoology.* a family of free-living mites in the order Sarcoptiformes, intermediate hosts of tapeworms.

Peloplocaceae *Bacteriology.* a former classification for a family of gliding bacteria now assigned to the family Pelonemataceae.

Peloponnesus *Geography.* a mountainous peninsula that forms the southern part of mainland Greece.

peloria *Botany.* an abnormal change in symmetry of structure in a flower that is normally irregular or bilaterally symmetrical.

Peloridiidae *Invertebrate Zoology.* a family of homopteran insects in the group Coleorrhyncha.
pelorus *Navigation.* a device for measuring the relative bearings of observed objects in degrees.
Pelosigma *Bacteriology.* a genus of nonmotile, Gram-negative bacteria occurring on mud in fresh or brackish water habitats as curved bundles of filamentous cells.
pelotherapy *Medicine.* the therapeutic use of mud, sand, or earth, especially in the treatment of skin diseases.
pelotons *Mycology.* clumps of filaments or hyphae found within the root cells of certain fungi living in the roots of plants without harming them.
pelphyte *Geology.* a deposit of fine, nonfibrous plant remains that forms on the bottoms of lakes.
pelsparite *Petrology.* a limestone containing intraclasts and oolites in percentages less than 25% each, having a volume ratio of fossils to pellets of less than 33.3% and more abundant sparry-calcite cement than micrite.
peltate *Botany.* **1.** of or relating to a leaf having the petiole attached to the lower or middle surface rather than at the base or end. **2.** of or relating to stalk parts having a similar type of attachment.

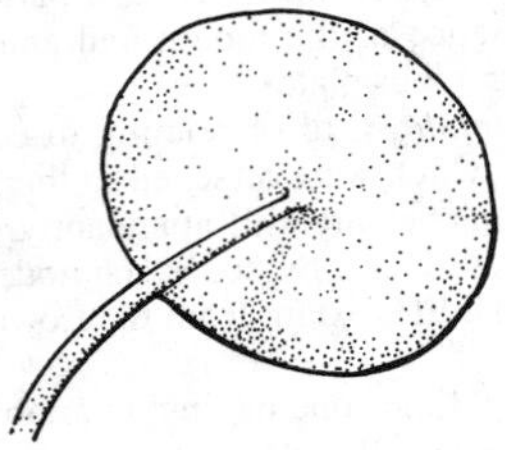

peltate leaf

Peltier, Jean [pel′tyā′; [pel′tyā′] 1785–1845, French physicist; discovered the Peltier effect in thermoelectricity.
Peltier coefficient *Physics.* a coefficient given by the the Peltier effect divided by the electric current passing through the junction.
Peltier effect *Physics.* the rate of heat being emitted or absorbed at a bimetal junction.
Peltier heat *Physics.* the heat gained or lost at a bimetal junction to the Peltier effect.
Peltigera *Botany.* the large type genus of foliaceous lichens of the family Peltigeraceae, characterized by shield- or tooth-shaped apothecia.
Peltigerales *Mycology.* in some classifications, an order of fungi belonging to the subdivision Ascomycotina, which includes certain lichens found on trees and mossy surfaces.
Pelton wheel *Mechanical Engineering.* an impulse hydraulic turbine in which a high-velocity jet of water strikes a series of cup-shaped buckets attached to the periphery of the rotor, producing a torque.
pelvic *Anatomy.* of or relating to the pelvis.
pelvic bone see HIP BONE.
pelvic fin *Vertebrate Zoology.* one of a pair of fins found on the ventral surface of most fishes, usually between the pectoral and anal fins; homologous to the hind legs of a quadruped.
pelvic girdle *Anatomy.* the hip bones that support the lower limbs.
pelvic index *Anatomy.* the ratio of the anterior-posterior diameter of the pelvis to its breadth (anterior posterior diameter × 100/breadth).
pelvic inflammatory disease *Medicine.* any pelvic infection of the upper female genital tract beyond the cervix.
pelvic presentation *Medicine.* see BREECH PRESENTATION.
pelvifixation *Surgery.* a surgical operation to secure an organ to the wall of the pelvic cavity.
pelvilithotomy see PYELOLITHOTOMY.
pelvimeter *Medicine.* an instrument for measuring the size of the pelvis.
pelvimetry *Medicine.* a measurement of the size and capacity of the pelvis.
pelvioscopy *Surgery.* **1.** the visual examination of the pelvis or associated organs. **2.** a pyeloscopy.
pelvis *Anatomy.* the ring of bones forming a bowl in which the digestive organs are housed and to which the thigh bones are attached. Also, **pelvic cavity.**
pelviscope *Medicine.* an endoscope used for examination of the pelvic organs of the female.
pelvisection *Medicine.* a cutting of the bones of the pelvis, as in pubiotomy.
pelycosaur *Paleontology.* **1.** of or relating to reptiles of the extinct order Pelycosauria. **2.** a reptile of this order.
Pelycosauria *Paleontology.* a Late Paleozoic order of mammal-like reptiles in the subclass Synapsida; ancestral to mammals; notable for large and complicated neural spines along the back that, in species like *Dimetrodon*, formed a sail of several square feet, possibly used to regulate body temperature; extant in the Carboniferous and Triassic.
pelyte see PELITE.
pemphigus *Medicine.* a distinctive group of skin and mucous membrane diseases, characterized by successive skin eruptions of crops of bullae.
pemphigus contageosus *Medicine.* an uncommon vesicular dermatitis endemic in tropical areas, chiefly affecting the armpits and groin.
pen *Engineering.* **1.** a small, fenced-off area used for confinement. **2.** a device to dam the water in a stream. **3.** a tubular device used for drawing or writing with ink.
Penaeaceae *Botany.* a family of tanniferous dicotyledonous evergreen shrubs of the order Myrtales, characterized by internal phloem; native to the Cape Province of South Africa.
Penaeidea *Invertebrate Zoology.* a section of decapod crustaceans in the suborder Natantia, consisting of prawns and shrimp with three pairs of pincer-bearing appendages.
pencil *Engineering.* a long, thin, rodlike instrument made of wood and having a center core of pointed graphite or crayon; commonly used for writing or drawing. *Mathematics.* in an n-dimensional manifold, an r-parameter family of one-dimensional curves, where $r \leq n - 1$; e.g., the lines in the plane passing through a given point, which could be a point at infinity, hence a pencil of parallel lines.
pencil beam *Electromagnetism.* a very narrow radar beam.
pencil-beam antenna *Electromagnetism.* a unidirectional antenna whose major lobe has a circular cross section.
pencil cave *Engineering.* hard, closely jointed shale that caves into wells in pencil-shaped fragments.
pencil cleavage *Geology.* the splitting or fracturing of a rock into long, slender pieces.
pencil follower *Computer Technology.* an input device that records the positions of an electronic pencil on an electronic tablet, thus enabling the user to convert a graphic image into digital form by tracing it on the tablet.
pencil gneiss *Geology.* a gneiss that splits into thin, roughly cylindrical aggregates of quartz-feldspar crystals. Also, STENGEL GNEISS.
pencil ore *Geology.* hard, fibrous masses of hematite that can be splintered or broken into thin rods.
pencil stone *Mineralogy.* a compact variety of pyrophyllite; formerly used to make slate pencils.
pencil tube *Electronics.* a long, thin tube that operates at ultrahigh frequencies; commonly used in oscillators or radio-frequency amplifiers.
pendant *Robotics.* a manually held device used to program a robot.
pendant cloud see TUBA.
pendant-drop method *Physics.* a technique used in measuring the surface tension of a liquid by measuring the elongation of a hanging drop suspended in the liquid.
pendant post *Building Engineering.* a post used to support a collar beam or other roof structure against a wall.
pendentive *Architecture.* a curved triangular surface that connects a dome with its square supporting structure.
pendent terrace *Geology.* a ribbon of sand that connects an isolated point of rock with a neighboring coast.
pendular water *Hydrology.* capillary water that forms rings around the points of contact between adjacent rock and soil particles in the zone of aeration.
pendulous gyroscope *Mechanics.* a gyroscope so mounted that its center of gravity is either above or below the intersection of two horizontal perpendicular axes about which the system can oscillate. Also, GYROSCOPIC PENDULUM; GYRO-PENDULUM.
pendulum [pen′jə ləm; pen′də ləm] *Physics.* any body that is free to swing under the influence of gravity; a simple practical example is a small weight suspended from the end of a string. *Horology.* specifically, a swinging lever, weighted at the lower end, that is used to regulate the movement of a clock. (From a Latin word meaning "to hang" or "to swing.")

pendulum anemometer *Engineering.* an anemometer that determines wind speed by measuring the angular deflection of a free-swinging plate.

pendulum clock *Horology.* any type of clock that uses a pendulum escapement mechanism to control its timekeeping movement.

pendulum day *Physics.* the amount of time a Foucault pendulum requires to make one complete rotation about the local vertical.

pendulum escapement *Horology.* in a clock, a type of escapement mechanism that relies on the regular oscillations of a pendulum to stop and release the escapewheel, thereby releasing the motion of the going train in steady, timekeeping impulses.

pendulum level *Engineering.* a leveling instrument that keeps the line of sight horizontal by means of a pendulum.

pendulum press *Mechanical Engineering.* a foot-operated punch press having a pendulumlike lever for applying power to the ram. Also, KICK PRESS.

pendulum saw *Mechanical Engineering.* a circular, crosscut saw mounted at the end of a swinging frame so that the weight of the saw and mounting are counterbalanced. Also, **pendulum crosscutting saw.**

pendulum scale *Engineering.* a device used to measure weight by balancing the load with the movement of one or more pendulums.

pendulum seismograph *Engineering.* a seismograph that gauges the relative motion between the earth and a pendulum; some of the instruments use optical magnification, electromagnetism, or electronic amplification to obtain a higher degree of amplification.

pendulum sextant *Navigation.* an artificial-horizon sextant using a pendulum as a horizontal reference.

penecontemporaneous *Geology.* relating to a process or resulting structure, feature, or mineral that occurred or formed immediately following deposition, but before the consolidation of the enclosing rock.

peneplain *Geology.* an extensive, nearly flat or level, almost featureless area of land assumed to represent the end-product of long-continued subaerial erosional processes. Also, BASE-LEVELED PENEPLAIN.

peneplanation *Geology.* the processes or actions involved in the formation and development of a peneplain.

penesaline *Ecology.* describing an environment that is intermediate between a normal marine environment and a saline environment.

penetrability the ability to penetrate; specific uses include: *Radiology.* the ability of roentgen rays to penetrate matter.

penetrance *Genetics.* the percentage of individuals having a specified genotype that show the expected phenotype under a defined set of environmental conditions.

penetrant *Materials Science.* a substance that increases liquid penetration into porous material or between contiguous surfaces. *Invertebrate Zoology.* in coelenterates, a type of nematocyst with barbs that inject a toxin. Also, STENOTELE.

penetrate to pass into or through; specific uses include: *Aviation.* to fly into a territory, airspace, or air defense. *Space Technology.* to enter an atmosphere.

penetrating excavation *Archaeology.* an excavating technique that exposes the vertical face of a site.

penetrating frequency *Electronics.* a limiting frequency below which radio waves are reflected by the ionosphere at vertical incidence, and above which radio waves penetrate the ionosphere layer. Also, CRITICAL FREQUENCY.

penetrating shower *Nuclear Physics.* a type of cosmic-ray shower that consists primarily of muons, which have enough power to penetrate up to 20 cm of lead.

penetrating wound *Surgery.* a wound caused by a sharp object, such as a nail, that extends into underlying tissues.

penetration the act or fact of passing into or through; specific uses include: *Ordnance.* the distance that a projectile enters into a target. *Military Science.* **1.** in land operations, an offensive that attempts to break through and disrupt the enemy's defense. **2.** in intelligence operations, the recruitment of agents within an organization or group for the purpose of acquiring information or influencing its activities. *Virology.* the step in the initiation of infection in which the virus particle penetrates the cell surface. *Metallurgy.* **1.** in welding, the distance from the initial surface of the base to the deepest fusion location. **2.** in casting, a defect caused by the penetration of molten metal between sand grains.

penetration aid *Ordnance.* any technique or device employed by offensive aerospace weapon systems, such as ballistic missiles or bombers, to increase the probability of penetrating hostile defenses; often designed to mislead enemy radar or divert defensive fire by simulating or masking aircraft or missile warheads. Also, **penaid.**

penetration ballistics *Mechanics.* the branch of terminal ballistics that deals with the motion and behavior of a missile before and after penetrating a target.

penetration depth *Physics.* the distance beneath the surface of a superconductor to which an external magnetic field is reduced by a factor of $1/e$ (approximately 0.37). *Electronics.* the thickness required in a hollow conductor in induction heating that would result in an effective resistance identical to that of the solid current, if a current were uniformly distributed over a cross section. *Electricity.* the nominal depth below the surface of a conductor at which current is concentrated by a skin effect such that the higher the frequency, the less the penetration.

penetration funnel *Geology.* a funnel-shaped impact crater that is formed by a small meteorite striking the earth at a relatively low velocity, and that contains nearly all the impacting mass within it.

penetration gland *Invertebrate Zoology.* in parasites such as flukes, a gland secreting an enzyme that breaks down the tissues of the host.

penetration number *Engineering.* a number indicating the relative density of a bituminous material, such as grease or wax.

penetration phosphors *Electronics.* a method for producing color images on a television screen in which phosphors from two different colors are applied to the screen in separate layers, so that a high-energy beam penetrates the first layer and excites the second, while a low-energy beam is stopped by the first layer and excites only this layer.

penetration probability *Quantum Mechanics.* the probability that a particle will tunnel through a potential barrier.

penetration rate *Mechanical Engineering.* the actual rate of penetration of a drilling tool.

penetration speed *Mechanical Engineering.* the speed at which a drill can penetrate through a rock or other material.

penetration test *Engineering.* a test to measure the relative density of unconsolidated sand or other substances at the base of a borehole.

penetrative *Geology.* of or relating to a deformation texture that is uniformly distributed throughout a rock, without notable discontinuities.

penetrometer *Engineering.* **1.** a device used to measure the consistency of a surface or solid by measuring the depth to which a standard needle pierces the sample. **2.** a device used to measure the penetrating power of a beam of electromagnetic radiation. *Radiology.* a device that measures the probing ability of a roentgen ray. Also, QUALIMETER.

Penex process *Chemical Engineering.* a petroleum-refinery process that isomerizes pentane and hexane, using platinum as a catalyst.

Penfield, Wilder Graves 1891–1976, Canadian neurologist; developed a surgical method for treating epilepsy.

penfieldite *Mineralogy.* $Pb_2Cl_3(OH)$, a colorless or white hexagonal mineral occurring in prismatic and tabular crystals, having a specific gravity of 5.82 and an undetermined hardness; found as a secondary mineral in lead slags at Laurium, Greece.

penguin *Vertebrate Zoology.* a flightless, upright, aquatic bird of the family Spheniscidae, distinguished by its excellence in swimming and diving; characterized by waterproof plumage that is generally blue-gray or blue-black above and white below, with small wings resembling flippers; found in large colonies on ice and rock on the islands and coasts of the Southern Hemisphere.

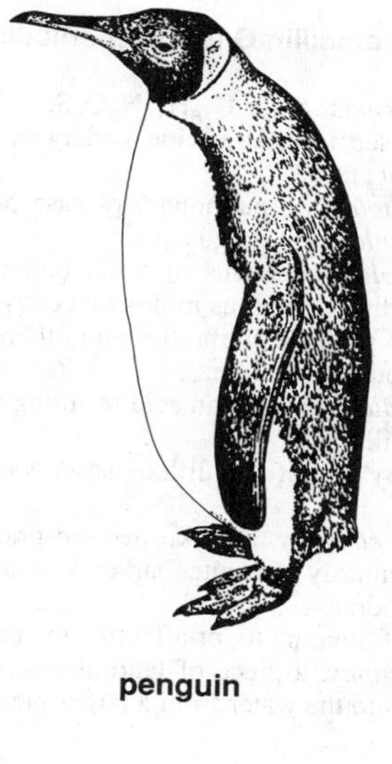

penguin

Penguin *Ordnance.* **1.** a Norwegian surface-to-surface missile designed for ship-to-ship combat; it is powered by a two-stage, solid-propellant motor equipped with inertial guidance and infrared homing and delivers a 264-pound (120 kg) high-explosive warhead at a maximum range of 11.5 to 17 miles (18.5–27 km). **2.** a similar air-to-surface missile with a single-stage motor and slightly smaller warhead (250 lb/113 kg).

Peniaceae *Botany.* a small family of placoderm desmids of the order Zygnematales, characterized by cylindrical or lunate cells composed of semicells that are not demarcated by a median constriction.

penicillamine *Pharmacology.* $C_5H_{11}NO_2S$, a drug produced by the degradation of penicillin; used to treat rheumatoid arthritis and lead poisoning, and to reduce blood copper levels in the treatment of liver degeneration.

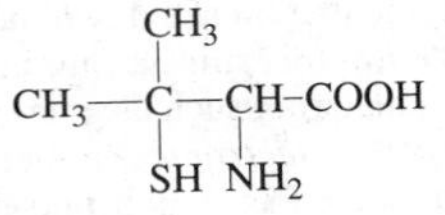

penicillamine

Penicillata *Invertebrate Zoology.* a subclass of minute millipedes having a soft integument bearing tufts of setae and 13–17 pairs of legs.

penicillate *Biology.* having a small, brushlike tuft of hair, as on a caterpillar.

penicillin [pen´i sil´ən] *Microbiology.* any of a large group of natural or semisynthetic antibiotics that are derived from strains of the fungi *Penicillum* or similar fungi; active against a wide range of Gram-positive, spheroid bacteria, including streptococci, as well as some Gram-negative spheroids, some spirochetes, and some fungi. Penicillin exerts a bacteriostatic and bacteriocidal effect by interfering with the synthesis of peptidoglycan.

penicillin acylase see PENICILLIN AMIDASE.

penicillin amidase *Enzymology.* an enzyme that catalyzes hydrolysis of nonpeptide amide bonds in penicillin, thus inactivating the antibiotic.

penicillinase *Enzymology.* an enzyme produced by certain bacteria that hydrolyzes a bond in penicillin and converts it to an inactive product, thereby increasing the bacteria's resistance to penicillin.

penicillin G *Microbiology.* the most widely used form of penicillin and the first of the penicillins developed for medicinal use; effective against a wide range of Gram-positive bacteria and certain Gram-negative coccoid bacteria; administered in combination with one of four salts: **penicillin G benzathine, penicillin G potassium, penicillin G procaine, penicillin G sodium.** Also, BENZYLPENICILLIN.

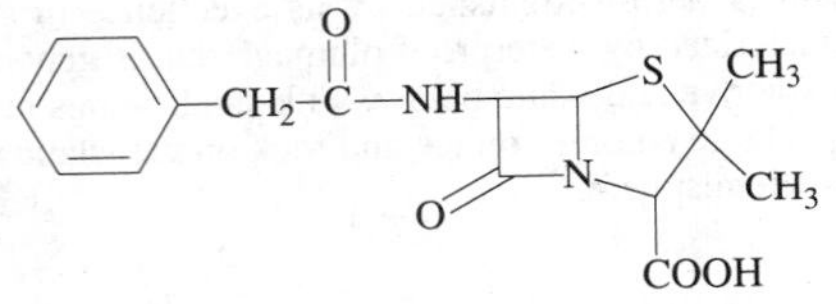

penicillin G (benzylpenicillin)

penicillin V *Pharmacology.* $C_{16}H_{18}N_2O_5S$, a form of the antibiotic penicillin that is used to treat a wide variety of infections. Also, PHENOXYMETHYLPENICILLIN.

penicilliosis *Pathology.* a pulmonary disease occurring as a result of infection with a *Penicillium* fungus.

Penicillium *Mycology.* a genus of fungi belonging to the class Hyphomycetes, including numerous molds that cause food to spoil; several species are used to produce antibiotics and others are used in food processing; found throughout nature.

penicilloic acid *Biochemistry.* an acid resulting from the breakdown of penicillin by penicillinases.

penicillus *Anatomy.* any of the tuftlike networks of arteries found in the spleen.

penikkavaarite *Petrology.* a dark, medium-grained, intrusive igneous rock consisting primarily of augite, barkevikite, and green hornblende in a feldspathic groundmass.

penile *Anatomy.* of, relating to, or affecting the penis.

peninsula *Geography.* a piece of land almost entirely surrounded by water, projecting into the water from a larger piece of land.

penis *Anatomy.* the male organ of copulation and of urinary excretion, consisting of a root, a body, and an extremity, or glans penis.

penis envy *Psychology.* **1.** in Freudian theory, the concept that young females feel incomplete without a penis and therefore wish for one; disputed by many recent theorists. **2.** the wish by a female for masculine attributes or advantages.

penitent *Hydrology.* referring to a jagged spike, consisting mainly of either ice or snow, that is produced by differential melting and evaporation as a result of exposure to intense solar radiation. *Geology.* **1.** see EARTH PILLAR. **2.** see NIEVE PENITENTE.

Penium *Botany.* a genus of placoderm desmids in the family Peniaceae.

Pennales *Botany.* an order of diatoms in the class Bacillariophyceae, characterized by isogamous or anisogamous reproduction with nonflagellate gametes, large chloroplasts, and an often circular form with radial markings.

pennant *Meteorology.* a graphic representation of windspeed on a synoptic weather chart, as indicated by a triangular flag pointing toward lower pressure from a wind-direction shaft.

pennantite *Mineralogy.* $Mn_5^{+2}Al(Si_3Al)O_{10}(OH)_8$, an orange, monoclinic mineral of the chlorite group, having a specific gravity of 3.06 and a hardness of 2 to 3 on the Mohs scale; found as tiny flakes in manganese ores.

pennate *Biology.* **1.** winged or feathered. **2.** resembling a feather, especially in having similar parts arranged on opposite sides of an axis like barbs on the rachis of a feather.

Pennatulacea *Invertebrate Zoology.* the sea pens or sea feathers, an order of sessile anthozoan coelenterates in the subclass Alcyonaria, characterized by a long fleshy stem, sometimes branched, with small polyps growing on the sides.

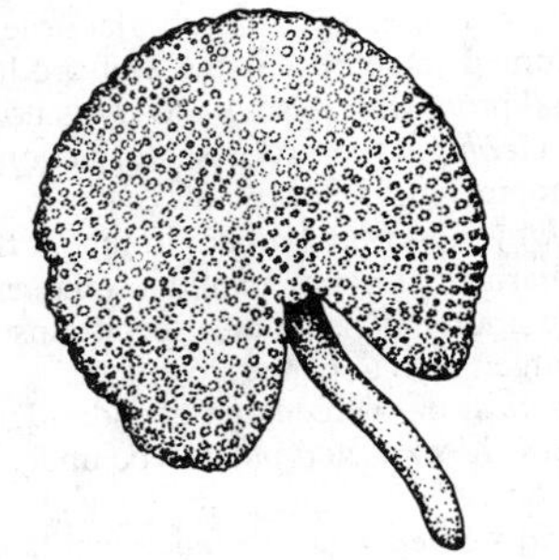

Pennatulacea

Pennellidae *Invertebrate Zoology.* a family of marine copepod crustaceans in the suborder Caligoida, ectoparasites of fish and whales.

Penniculina *Invertebrate Zoology.* protozoans in the order Hymenostomatida with brushlike ciliation in the cytopharynx.

Penning ionization *Atomic Physics.* a process in which ions are produced when atoms or molecules in a gas collide with metastable atoms.

Penning ion source *Nucleonics.* a device used to produce ions from gases at low pressures (approximately 10^{-3} torr); electrons are accelerated through the gas to cause ionization, and a magnetic field is used to selectively extract ions through a slit.

Penning trap *Engineering.* an instrument used to capture and isolate electrons by means of a magnetic field and the creation and superimposition of a parabolic electric potential.

penninite *Mineralogy.* a green or bluish pseudotrigonal variety of clinochlore; found in the Penninine Alps. Also, **pennine.**

Pennsylvanian *Geology.* **1.** a geologic period of the Upper Paleozoic era, after the Mississippian and before the Permian (from about 320 to 280 million years ago). **2.** the rocks formed during this time.

pennyroyal *Botany.* any of various plants of the mint family, Labiatae, having clusters of small purple flowers and leaves that yield a pungent oil, especially the **European pennyroyal,** *Mentha pulegium.*

pennyroyal oil *Materials.* a yellow to reddish essential oil extracted from the pennyroyal plant, used in flavoring, in medicine, and as an insect repellent.

pennyweight *Metrology.* a unit of troy weight equal to 24 grains or 0.5 troy ounce.

Penokean see ANIMIKEAN.

pen recorder *Engineering.* a device that continuously records electrical, mechanical, or pneumatic forces on a chart that moves past a recording stylus.

penroseite *Mineralogy.* (Ni,Co,Cu)Se_2, a metallic, lead-gray cubic mineral of the pyrite group, having a specific gravity of 6.58 to 6.9 and a hardness of 2.5 to 3 on the Mohs scale; found as reniform masses in hydrothermal veins.

Pensky-Martens tester *Chemical Engineering.* an apparatus used to determine the flash point of fuel oils, lubricating oils, asphalts, and other highly viscous petroleum products that require stirring during the testing procedure.

penstock *Engineering.* **1.** a sluice gate used to control a flow of water. **2.** a pressurized tube used to bring water to a waterwheel or turbine.

pent- or **penta-** a prefix meaning "five."

pentachloride *Chemistry.* any compound in which five chloride atoms are combined with one or more other elements.

pentachloroethane *Organic Chemistry.* $CHCl_2CCl_3$, a dense, toxic, colorless liquid that is insoluble in water; it boils at 159.1°C and freezes at –22°C; used as a decreasing solvent. Also, PENTALIN.

pentachloronitrobenzene *Organic Chemistry.* $C_6Cl_5NO_2$, off-white crystals; insoluble in water and soluble in alcohol; melts at 142–145°C; used as a fungicide and herbicide.

pentachlorophenol *Organic Chemistry.* C_6Cl_5OH, toxic white powder or crystals; slightly soluble in water and soluble in alcohol; melts at 190°C and boils with decomposition at 310°C; used as a fungicide, herbicide, bactericide, and algicide. Also, PCP.

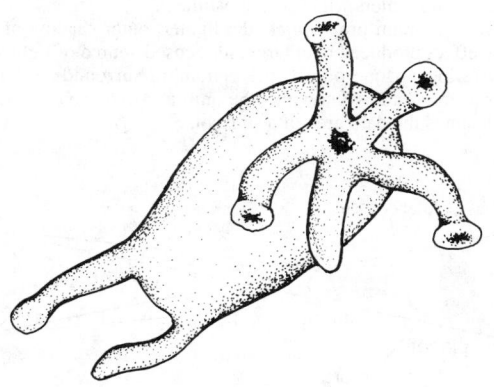

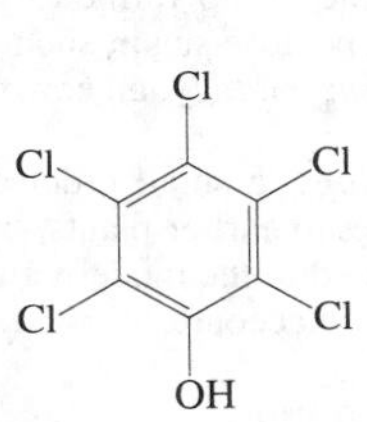

pentachlorophenol

pentacosane *Organic Chemistry.* $C_{25}H_{52}$, a hydrocarbon that is insoluble in water and melts at 54°C; derived from beeswax.

Pentacrinites *Paleontology.* an extinct genus of Mesozoic stemmed crinoids in the subclass Articulata; characterized by a long stem, star-shaped columnals, and repeatedly branching arms.

pentacrinoid *Invertebrate Zoology.* a sessile, stalked echinoderm larva with branching arms and an upward directed oral surface.

pentactinal *Zoology.* equipped with five rays, branches, or axes, as in a starfish.

pentactula *Invertebrate Zoology.* an echinoderm larva with five tentacles.

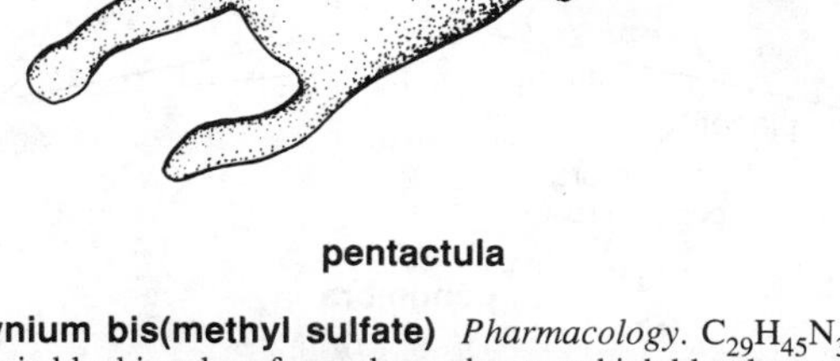

pentactula

pentacynium bis(methyl sulfate) *Pharmacology.* $C_{29}H_{45}N_3O_9S_2$, a ganglionic blocking drug formerly used to treat high blood pressure.

pentad a group or unit of five items; specific uses include: *Meteorology.* a period of five consecutive days that is used in climatological record-keeping; preferred to the seven-day week because it is an exact factor of the 365-day year.

pentadactyl *Vertebrate Zoology.* **1.** having five digits or fingerlike parts on the hand or foot. **2.** an organism of this type.

pentadactylism *Vertebrate Zoology.* the condition of having five digits on the hands or feet.

pentadecane *Organic Chemistry.* $CH_3(CH_2)_{13}CH_3$, a colorless liquid; insoluble in water and soluble in alcohol; boils at 270.5°C and freezes at 10°C; used as a chemical intermediate in organic synthesis.

pentadecanolide *Organic Chemistry.* $CH_2(CH_2)_{13}C(O)O$, a combustible, colorless, musky-smelling liquid that congeals to white crystals at a minimum temperature of 36°C; it is soluble in ethanol; used in perfumes.

pentadelphous *Botany.* of or relating to a flower having five sets of stamens, with the filaments more or less united within each set.

1,3-pentadiene see PIPERYLENE.

pentaerythritol *Organic Chemistry.* $C(CH_2OH)_4$, a white crystalline powder; soluble in water and slightly soluble in alcohol; melts at 262°C; used in explosives, plasticizers, varnishes, and pharmaceuticals.

pentaerythritol tetranitrate *Organic Chemistry.* $C(CH_2ONO_2)_4$, white crystals that are insoluble in water, slightly soluble in alcohol, and very soluble in acetone; melts at 139°C, decomposes above 150°C, and explodes at 210°C; used in explosives.

pentaerythritol tetrastearate *Organic Chemistry.* $C(CH_2OOCC_{17}H_{35})_4$, a hard, combustible, ivory-colored wax that softens at 67°C; used in polishes and textile finishes.

pentafluoride *Chemistry.* any compound in which five fluoride atoms are combined with one or more other elements.

pentagastrin *Endocrinology.* a synthetic pentapeptide derivative of gastrin, containing the four carboxy-terminal amino acids of gastrin and possessing full biological activity.

pentaglycerine see TRIMETHYLOLETHANE.

pentagon *Mathematics.* a polygon with five sides, usually convex.

pentahydrite *Mineralogy.* $MgSO_4{\cdot}5H_2O$, a white or light-blue triclinic mineral of the chalcanthite group occurring as granular masses, having a specific gravity of 1.90 and an undetermined hardness.

pentahydroxyhexoic acid see GLACTONIC ACID.

pentalin see PENTACHLOROETHANE.

pentalogy *Medicine.* five concurrent defects or symptoms, or a combination of five elements or factors.

pentalogy of Fallot *Medicine.* the defects included in the tetralogy of Fallot, associated with atrial septal defect or patent foramen ovale. (From Etienne-Louis Arthur *Fallot,* 1850–1911, French physician.)

pentamer *Virology.* in the virus capsid, a group of five-protein subunits that are arranged in a tight cluster to maximize contact and form the vertices of an icosahedron.

pentamerid *Paleontology.* of or relating to the early brachiopods of the extinct order Pentamerida.

Pentamerida *Paleontology.* an order of articulate brachiopods of the Early to Middle Paleozoic, characterized by impunctate, generally nonstrophic shells.

Pentameridina *Paleontology.* a Middle Paleozoic suborder of articulate brachiopods in the order Pentamerida, characterized by strongly biconvex shells.

pentamerous *Botany.* of or relating to a flower in which the members of each whorl are inserted in fives or multiples of five. Also, 5-MEROUS.

Pentamerus *Paleontology.* a genus of articulate brachiopods in the extinct suborder Pentameridina; extant in the Middle Paleozoic.

pentamethylmelamine *Oncology.* an active metabolic product of hexamethylmelamine under investigation as an antineoplastic agent.

pentanamide see VALERAMIDE.

pentandrous *Botany.* of or relating to a flower that has five stamens.

***n*-pentane** *Organic Chemistry.* $CH_3(CH_2)_3CH_3$, a highly flammable colorless liquid; insoluble in water and soluble in alcohol; boils at 36°C and freezes at –129.7°C; used as an anesthetic, a pesticide, and a blowing agent in making plastics. Also, AMYL HYDRIDE.

pentane candle *Optics.* a unit of luminous intensity equal to one-tenth that of a standard pentane lamp; equal to one candela.

1,5-pentanediol *Organic Chemistry.* $HOCH_2(CH_2)_3CH_2OH$, a combustible, viscous, colorless liquid; miscible with water and alcohol; boils at 240°C and freezes at –15.6°C; used as a hydraulic fuel, as an antifreeze, and in the manufacture of polyester.

pentane insoluble *Petroleum Engineering.* any of various substances such as resinous bitumens that may be separated from used lubricating oil by means of their insolubility in *n*-pentane (amyl hydride).

pentane lamp *Engineering.* a pentane-burning lamp formerly used as a standard of brightness.

pentanethiol see AMYL MERCAPTAN.

***n*-pentanoic acid** see VALERIC ACID.

1-pentanol *Organic Chemistry.* $CH_3(CH_2)_4OH$, a colorless liquid; miscible with alcohol; boils at 138°C and freezes at –79°C; used in pharmaceuticals and as a solvent. Also, AMYL ALCOHOL, PENTYL ALCOHOL.

2-pentanol see METHYL PROPYL CARBINOL.

3-pentanol *Organic Chemistry.* $CH_3CH_2CHOHCH_2CH_3$, a colorless liquid; slightly soluble in water and soluble in alcohol; boils at 115.6°C and freezes below –75°C; used as a solvent, as a flotation agent, and in pharmaceuticals. Also, DIETHYL CARBINOL.

2-pentanone *Organic Chemistry.* $CH_3COC_3H_7$, a flammable colorless liquid; slightly soluble in water and soluble in alcohol; boils at 101.7°C and freezes at –77.5°C; used as a solvent and flavoring. Also, METHYL PROPYL KETONE.

3-pentanone *Organic Chemistry.* $C_2H_5COC_2H_5$, a flammable colorless liquid; slightly soluble in water and soluble in alcohol; boils at 101°C and freezes at –42°C; used in medicine and in organic synthesis. Also, DIETHYL KETONE.

Pentaphragmataceae *Botany.* a family of dicotyledonous perennials, sometimes succulent herbs of the order Campanulales, containing a single genus, *Pentaphragma*; characterized by simple, spirally arranged leaves and the lack of alkaloids and a latex system; native to Southeast Asia and nearby Pacific islands.

Pentaphylacaceae *Botany.* a monogeneric family of evergreen shrubs and small trees belonging to the order Theales, characterized by the accumulation of aluminum and by horseshoe-shaped dicotyledonous embryo; native from southern China to the Malay Peninsula and Sumatra.

Pentastomida *Invertebrate Zoology.* the tongue worms, a small phylum of wormlike segmented animals, having no circulatory, respiratory, or excretory systems; parasites in the respiratory system of vertebrates; the larvae resemble mites. Also, LINGUALATIDA.

Pentatomidae *Invertebrate Zoology.* the shield bugs or stink bugs, a large family of hemipteran insects, many are bright colored and produce a foul smell for protection.

Pentatomoidea *Invertebrate Zoology.* a subfamily of hemipteran insects in the group Pentatomorpha, having sensory hairs and five-segmented antennae.

Pentatomorpha *Invertebrate Zoology.* a large group of hemipteran insects in the subdivision Geocorisae, having sensory hairs on the abdomen.

18-pentatriacontanone see STEARONE.

pentavalent *Chemistry.* any atom that can bond with five other atoms.

Pentazonia *Invertebrate Zoology.* a superorder of millipedes that can roll themselves into a ball; found in the southeastern United States and California. Also, OPISTHANDRIA.

1-pentene *Organic Chemistry.* $CH_3CH_2CH_2CH{=}CH_2$, a flammable colorless liquid; insoluble in water and soluble in alcohol; boils at 30°C and freezes at –165°C; derived from natural gasoline; used in high-octane motor fuel, pesticides, and organic synthesis.

2-pentene *Organic Chemistry.* $CH_3CH_2HC{=}CHCH_3$, a flammable colorless liquid available in *cis-* and *trans-*isomers; insoluble in water and soluble in alcohol; the *cis-*isomer boils at 37°C and freezes at –180°C, the *trans-*isomer boils at 36.4°C and freezes at –139°C; derived from natural gasoline; used in organic synthesis and as a polymerization inhibiter.

penthouse *Building Engineering.* **1.** an apartment or dwelling on the roof of a building. **2.** a specially designed, usually larger apartment on the top floor of a building. **3.** an enclosed space on a flat roof, used to enclose a stairway or an elevator or other apparatus.

pentice *Mining Engineering.* **1.** a rock pillar or heavy timber bulkhead at the bottom of a deep shaft of two or more compartments, through which the shaft may be sunk further. **2.** an overhead cover for workers at the bottom of a shaft or over a sinking shaft. **3.** a solid pillar of rock left at the bottom of a shaft for overhead protection during shaft sinking.

pentlandite *Mineralogy.* $(Fe,Ni)_9S_8$, a bronze-colored cubic mineral of the pentlandite group with a metallic luster, having a specific gravity of 4.6 to 5 and a hardness of 3.5 to 4 on the Mohs scale; found as granular masses in very basic rocks, it is the chief ore of nickel.

pentobarbital *Pharmacology.* $C_{11}H_{17}N_2O_3$, a short-acting to medium-acting barbituate, administered orally as a sedative and hypnotic. Also, **pentobarbitone.**

pentobarbital sodium *Pharmacology.* $C_{11}H_{17}N_2NaO_3$, a barbiturate drug used as a sedative, to prevent and treat convulsions, before administration of anesthesia in surgery, and as an adjunct to anesthesia.

pentode *Electronics.* a five-electrode electron tube containing an anode, a cathode, a control electrode, and two grids; this prevents the loss of secondary electrons that occurs in a tetrode, which has only one grid.

pentode transistor *Electronics.* a transistor that has four-point-contact electrodes and a body that serves as a base for three emitters and one collector.

pentolinium tartrate *Pharmacology.* $C_{23}H_{42}N_2O_{12}$, a ganglionic-blocking drug occurring as a white or cream-colored powder, formerly used to treat high blood pressure.

pentolite *Materials.* a dangerous high explosive consisting of equal parts of pentaerythritol tetranitrate and trinitrotoluene.

penton *Virology.* in an Adenovirus capsid, a complex structure comprising a viral polypeptide III base, one or two glycoprotein fibers, and a knob.

pentosan *Biochemistry.* any of a group of polysaccharides that form pentoses by hydrolysis; present in woody plant tissues, such as those in straw, cereal, or brans.

pentose *Biochemistry.* a monosaccharide, such as ribose, containing five carbon atoms in a molecule.

pentose phosphate pathway *Biochemistry.* a pathway through which animal tissues catabolize glucose to form NADPH and ribose 5-phosphate and to produce pentoses, especially the D-ribose. Also, PHOSPHOGLUCONATE PATHWAY.

pentosuria *Medicine.* a benign inheritable error of metabolism caused by a defect in the activity of the enzyme L-xylulose dehydrogenase, resulting in high levels of L-xylulose in the urine, but few other effects.

pentosyl *Enzymology.* the group formed by removing hydroxyl from the anomeric carbon of a pentose sugar, such as ribose.

pentosyltransferase *Enzymology.* an enzyme that catalyzes the transfer of a pentosyl group.

Pentoxylales *Paleontology.* a small order of gymnosperms that have no known affinities as yet to earlier plants, but that may be ancestral to the angiosperms and related to the modern ginkgo; extant in the Jurassic and possibly the Lower Cretaceous.

pentyl see AMYL.

pentyl alcohol see 1-PENTANOL.

pentylamine see *n*-AMYLAMINE.

pentylenetetrazol *Pharmacology.* $C_6H_{10}N_4$, a central stimulant used to produce convulsions in the evaluation of epilepsy; formerly used in shock treatment for mental disorders.

***p*-pentyloxyphenol** *Organic Chemistry.* $C_{11}H_{16}O_2$, a compound that melts at 49–50°C; used as a bactericide.

1-pentyne *Organic Chemistry.* $CH_3CH_2CH_2C{\equiv}CH$, a colorless liquid; insoluble in water and very soluble in alcohol; boils at 40.18°C and freezes at –90°C. Also, *n*-PROPYLACETYLENE.

2-pentyne *Organic Chemistry.* $CH_3CH_2C{\equiv}CCH_3$, a colorless liquid; insoluble in water and very soluble in alcohol; boils at 56.07°C and freezes at –101°C. Also, ETHYLMETHYL ACETYLENE.

penultimate-unit effect *Materials Science.* the effect of the unit before the final one on the reactivity of a chain radical; particularly pronounced in monomers with polar substituents.

penumbra [pə num´brə] *Optics.* the lighter, outer shadow of a double-shadow effect produced by a large, unfocused source of light shining on an interfering opaque object; the penumbra surrounds a dark, central portion of the shadow known as the umbra. *Astronomy.* the lighter area that surrounds the dark area of a sunspot.

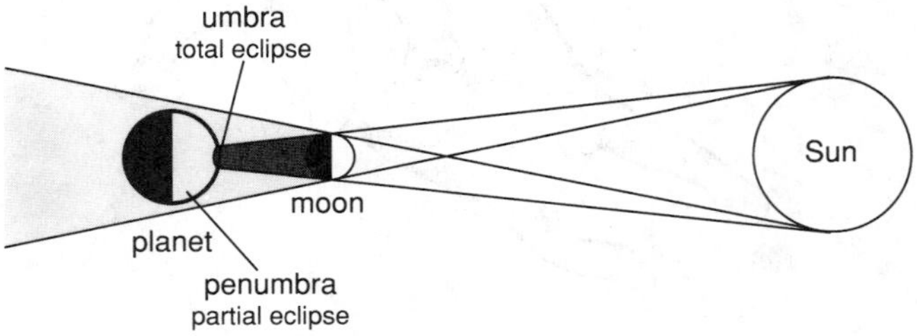

penumbra

penumbra field *Radiology.* the area of space treated by primary photons coming from only a portion of the radiation source.

penumbral *Optics.* relating to or having the form of a penumbra.

penumbral eclipse *Astronomy.* an eclipse in which the eclipsed object passes through the penumbral shadow.

penumbral waves *Astronomy.* waves observed in the light of H-alpha that move out from the center of a sunspot across the penumbra.

Penutian *Geology.* a North American geologic stage of the Lower Eocene epoch, occurring after the Bulitian and before the Ulatisian.

Penzias, Arno born 1931, American physicist; shared the Nobel Prize for studies of cosmic microwave radiation.

peony [pē´ə nē] *Botany.* any of various plants of the genus *Paeonia* in the family Ranunculaceae, especially the **common peony,** *P. officinalis,* or the **Chinese peony,** *P. albiflora,* widely cultivated for their large, showy flowers.

PEP peak envelope power.

peplomer *Virology.* on the surface of certain virus particles, a glycoprotein spike that serves as a virus receptor to bind the virion to the cell-surface receptors.

pepo *Botany.* a hard-walled fleshy berry with many seeds, characteristic of the gourd family, Cucurbitaceae, and found in the pumpkin, squash, melons, and cucumber.

pepper *Botany.* **1.** any of various plants belonging to the genus *Piper* in the family Piperaceae, especially *P. nigrum,* a tropical climbing shrub cultivated for its berries, from which the familiar spices **black pepper** and **white pepper** are made. Black pepper is made from the less ripe berries; the milder white pepper is made from riper berries with their outer coats removed. **2.** any of various plants of the genus *Capsicum* in the family Solanaceae, especially the many varieties of *C. annum,* whose fruit or seeds are eaten or ground up into condiments. **3.** the seeds or fruit of the plants of either of these genera.

peppermint *Botany.* an herb of the genus *Mentha* in the family Lamiaceae, cultivated for its aromatic oil and characterized by its dark green lanceolate leaves and spikes of small flowers.

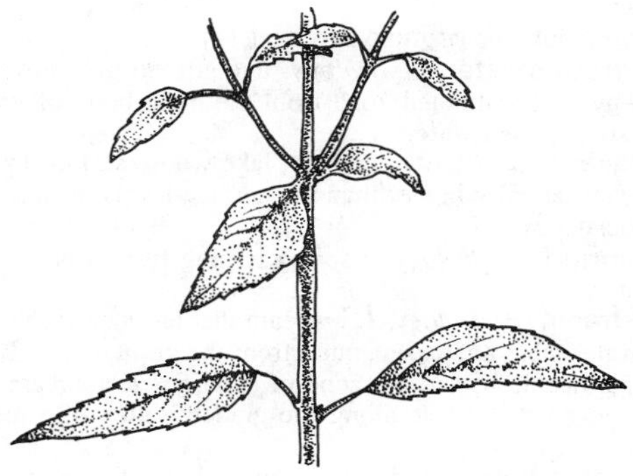

peppermint

peppermint camphor see MENTHOL.

peppermint oil *Materials.* an essential oil with a strong aromatic odor and taste; used as a flavoring and as a source of menthol.

PEPS permanent ethnographic probability sample.

pepsin *Enzymology.* an enzyme found in gastric juices that catalyzes the hydrolysis of proteins to form polypeptides.

pepsinogen *Biochemistry.* the inactive precursor of pepsin.

peptic *Physiology.* of or relating to the stomach, to digestive processes of the stomach, or to the digestive hormone pepsin.

peptic ulcer *Medicine.* a sharply circumscribed loss of tissue, involving chiefly the mucosa, submucosa, and muscular layer in areas of the digestive tract exposed to gastric juices containing acid and pepsin.

peptidase *Enzymology.* any enzyme that catalyzes the hydrolysis of peptide bonds. Also, **peptide hydrolase.**

peptide *Biochemistry.* any member of a class of compounds of low molecular weight that yield two or more amino acids on hydrolysis, and that form the constituent parts of proteins.

peptide bond *Organic Chemistry.* the –CO•NH– bond formed between the carboxyl group of one amino acid and the amino group of another amino acid; an amide linkage joining amino acids to form peptides.

peptide mapping *Biochemistry.* the distinctive two-dimensional pattern produced by a combination of peptides arising from a partial hydrolysis of a protein.

peptide synthetase *Enzymology.* any of a group of enzymes that catalyze the synthesis of peptide bonds.

peptidoglycan *Cell Biology.* an outer layer of the bacterial cell wall, composed of a complex of crosslinked oligosaccharides and peptides that confer rigidity to the cell.

peptidyltransferase *Enzymology.* an enzyme that catalyzes the transfer of a peptidyl group, especially important in protein synthesis.

peptization *Chemistry.* the dispersion of aggregates in solution by the addition of electrolytes to form a stable hydrophobic colloidal solution.

Peptococcaceae *Bacteriology.* a family of generally Gram-positive, spheroid, nonmotile, anaerobic bacteria occurring in soil and in the mouth, intestines, and respiratory tracts of animals and humans.

Peptococcus *Bacteriology.* a genus of Gram-positive bacteria of the family Peptococcaceae, occurring in the intestinal tract of animals as anaerobic cocci in pairs, tetrads, or groups and having a fermentative mode of metabolism.

peptone *Biochemistry.* a substance that contains proteoses and amino acids, taken from albumin, meat, or milk; used as a medium for bacterial growth.

peptonization *Biochemistry.* a process by which protein forms peptone.

Peptostreptococcus *Bacteriology.* a genus of Gram-positive, anaerobic bacteria of the family Peptococcaceae, found in the human intestinal tract as nonmotile, spherical cells in pairs or chains and having a fermentative metabolism.

peptotoxin *Toxicology.* any toxin found in or produced by a peptone or decomposing protein.

PER protein efficiency ratio.

per- a prefix meaning "throughout," "completely," "extremely."

per- *Chemistry.* a prefix denoting something extreme or complete, used in particular to indicate: **1.** an element in its highest oxidation state. **2.** the presence of the peroxy group, –O–O–. **3.** a complete addition or substitution.

Peracarida *Invertebrate Zoology.* a superorder of woodlice, sandhoppers, and shrimplike forms, crustaceans in the series Eumalacostraca; the females have a thoracic brood pouch.

peracetic acid *Organic Chemistry.* CH_3COOOH, a colorless liquid; soluble in water and alcohol; freezes at –30°C, boils at 105°C, and explodes at 110°C; used as a bleach, bactericide, and fungicide.

peracid *Chemistry.* an acid containing one or more pairs of peroxy groups, –O–O–.

peralcohol *Organic Chemistry.* a chemical compound containing the peroxy group, –O–O–, or the hydroperoxy group, –O–O–H.

peralkaline *Petrology.* of or relating to igneous rock in which the combined molecular proportion of sodium and potassium oxides exceeds that of aluminum oxide.

peraluminous *Petrology.* of or relating to igneous rock in which the molecular proportion of aluminum oxide exceeds that of sodium and potassium oxides combined.

Peramelidae *Vertebrate Zoology.* the short-haired bandicoots, a small family of insectivorous marsupials in the order Parmelina, having a long, pointed muzzle and powerful claws for digging; found in Tasmania, Australia, New Guinea, and nearby islands.

Peranema *Invertebrate Zoology.* a common genus of colorless protozoans with one anterior flagellum, found in stagnant water.

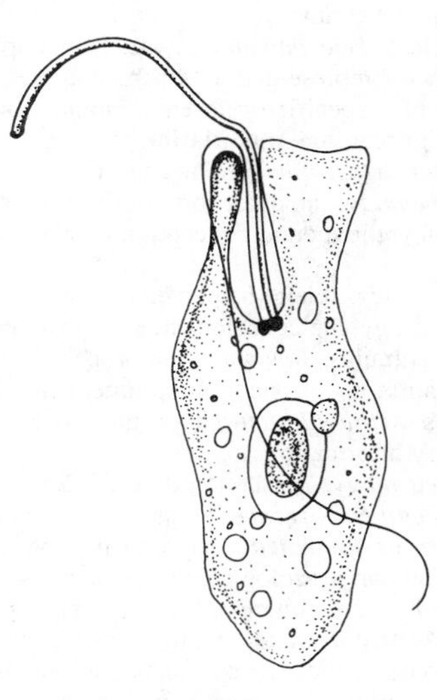

Peranema

perazine *Pharmacology.* $C_{20}H_{25}N_{35}$, a tranquilizer formerly used in the treatment of psychosis.
perbenzoic acid *Organic Chemistry.* C_6H_5COOOH, monoclinic plates that are insoluble in water and very soluble in alcohol; melts at 41–43°C and boils at 97–110°C (15 torr); used in chemical analysis and synthesis. Also, PEROXYBENZOIC ACID.
perbituminous coal *Geology.* a bituminous coal that contains more than 5.8% hydrogen when analyzed on a dry, ash-free basis.
percaine see DIBUCAINE.
per capita *Science.* per person; by or for each person. (From Latin; literally, "by head.")
per capita daily intake *Nutrition.* the daily consumption of nutrients per person; used as a statistical base for analyzing trends in human diets.
perceive *Physiology.* to be aware of through the senses; sense.
perceived noise decibel *Acoustics.* a unit of measure for perceived noise level.
perceived noise level *Acoustics.* **1.** the number of decibels of noise noticed by a listener, which, in a 50 hertz to 10 kilohertz range, can be determined by the formula PNL = 40.0 + 33.22 log $N(k)$, with $N(k)$ representing perceived noisiness in decibels. **2.** the average noise level of a given environment as relatively described by listeners and measured in units of perceived noise.
percent *Mathematics.* parts per one hundred; an expression of hundredths; denoted by %. For example, $n\%$ of a quantity x equals $xn/100$. Also, **per cent.**
percentage *Mathematics.* a statement of the rate or quantity per hundred, expressed in percent.
percentage differential relay *Electronics.* a differential relay operating at a current that is a fixed percentage of the current in the operating coils, rather than at a fixed current. Also, BIASED RELAY.
percentage log *Engineering.* a record in which the percentage of each type of rock found in a sample of cuttings is noted and plotted.
percentage map *Petroleum Engineering.* a contoured map showing the ratios of one component to the aggregate (e.g., sand to the sum of sand plus shale) by area.
percentage ripple *Electronics.* the ratio between the value of the alternating component in a unidirectional voltage to the value of its average voltage.
percent compaction *Engineering.* the ratio, expressed as a percentage, between the dry unit weight of a soil to the maximum unit weight achieved in a laboratory compaction test.
percent defective *Industrial Engineering.* the percentage of output items or lots that are found to be defective.
percent distortion *Telecommunications.* a percentage achieved by multiplying the ratio of the amplitude of a harmonic component to that of the fundamental component by 100.
percentile *Statistics.* the ith percentile is a value such that $i \times 100$ percent of the distribution is on its right. For example, the 50th percentile is the median.
percent make *Electronics.* **1.** the length of time a circuit remains closed during a pulse test, compared to the length of time the test signal is pulsating. **2.** the portion of time during which the springs in a telephone-dial pulse make contact.
percent modulation *Telecommunications.* **1.** in amplitude modulation, the modulation factor represented as a percentage. **2.** in angle modulation, the fraction of a specified reference modulation represented as a percentage. Also, **percentage modulation.**
percent of slope *Cartography.* see GRADIENT.
per centum *Mathematics.* an older form of the expression "percent."
percept *Psychology.* the mental conception of what is perceived by the physical senses.
perceptible *Physiology.* able to be perceived; evident to the senses.
perception *Physiology.* the conscious mental awareness and interpretation of a sensory stimulus. *Behavior.* the process of organizing and interpreting information about one's environment that has been acquired through the senses. *Artificial Intelligence.* the process of interpreting the meaning of sensory information.
perceptorium *Neurology.* see SENSORIUM, def. 2.
perceptron *Artificial Intelligence.* a class of networks implemented in electronic circuitry or simulated; proposed in 1958 as a cognitive machine to simulate human neurological and brain activity.
perceptual *Physiology.* relating to or based on perception.
perceptual constancy *Psychology.* the ability to distinguish the varying properties of objects such as size, shape, or color, despite the changing impression an individual may get from the objects.
perch *Vertebrate Zoology.* **1.** any of the many different members of the fish family Percidae; a common North American species is the **yellow perch,** *Perca flavescens.* **2.** a popular name for a group of unrelated species of deep-bodied fish of Percopsiformes, Centrarchidae, and Anabantoidei that generally resemble members of Percidae. **3.** a roosting place for a bird. *Navigation.* a pole driven into a rock or shoal as a navigational marker, with a topmark such as a ball or cage affixed to its top. *Metrology.* **1.** a unit of measure equal to 16.5 feet; a rod. **2.** a unit of area equal to 30.25 square yards; a square rod. **3.** a unit of measure for masonry, equal to 24.75 cubic feet.

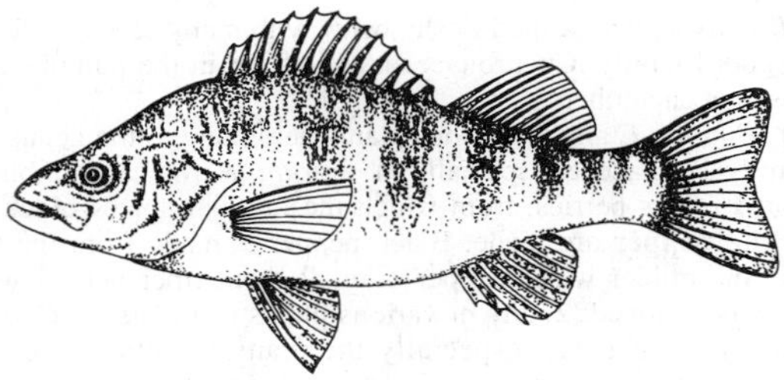

perch

perched aquifer *Hydrology.* an aquifer that contains perched groundwater.
perched block *Geology.* **1.** a large, detached rock fragment precariously or unstably positioned on the side of a hill. **2.** a rock that caps an earth pillar.
perched boulder see PERCHED BLOCK, def. 1.
perched groundwater *Hydrology.* unconfined groundwater that is separated by an unsaturated zone from the main body of groundwater below. Also, **perched water.**
perched lake *Hydrology.* a perennial lake whose surface level is at an elevation substantially higher than the surface levels of other bodies of water associated with it.
perched spring *Hydrology.* a spring issuing from a body of perched groundwater.
perched stream *Hydrology.* **1.** a stream that has a surface level higher than the water table and is separated from the groundwater lying below by an impermeable bed in the zone of aeration. **2.** a stream that flows onto an antecedent hillside along the graded valley of a higher-order stream into which it runs.
perched water table *Hydrology.* the surface of a body of perched groundwater. Also, APPARENT WATER TABLE.
Percheron [pur′shə rän′] *Agriculture.* a breed of short, heavily muscled draft horses that are usually dapple gray or black in color. (From *Perche,* a region in France where the breed originated.)
perching bed *Geology.* a body of rock of low permeability that supports a body of perched ground water.
perching bird *Vertebrate Zoology.* a term for any bird belonging to the order Passeriformes.
perchlorate *Inorganic Chemistry.* a salt of perchloric acid containing the $HClO_4$ group.
perchloric acid *Inorganic Chemistry.* $HClO_4$, a colorless, fuming liquid that freezes at –112°C and boils at 19°C; toxic and a strong irritant, detonated by shock or heat; used as a reagent in analytical chemistry, as a catalyst, in electropolishing, and for various other purposes.
perchloride *Chemistry.* the chloride of any element that has the highest amount of chlorine.
perchlorinate *Chemistry.* to combine with the maximum amount of chlorine.
perchloroethane see HEXACHLOROETHANE.
perchloroethylene *Organic Chemistry.* $Cl_2C{=}CCl_2$, a colorless liquid; insoluble in water and miscible with alcohol; boils at 121°C and freezes at –22.4°C; used as a dry-cleaning solvent, a degreasing agent for metals, and a component in the manufacture of fluorocarbons.
perchloromethane see CARBON TETRACHLORIDE.
perchloromethyl mercaptan *Organic Chemistry.* a trade name for trichloromethanesulfenyl chloride.
Percidae *Vertebrate Zoology.* the perches, a large family of the order Perciformes, generally characterized by an elongated, rounded body with spiny front dorsal fins; they are found strictly in freshwater habitats of the Northern Hemisphere.

Perciformes *Vertebrate Zoology.* a large order of perchlike fishes, ranging in size from very small to very large, usually having spiny fins, lacking an adipose fin, and found in marine and freshwater habitats worldwide; the largest vertebrate order.

Percoidei *Vertebrate Zoology.* a large suborder of perchlike fishes of the order Peciformes.

percolate *Science.* to undergo a process of percolation.

percolation *Physical Chemistry.* the gradual movement of a liquid through a porous medium. *Food Technology.* the process in which coffee is brewed in a percolator. *Hydrology.* the movement of water, under hydrostatic pressure or by the force of gravity, through the interstices in rock, soil, or other porous material. *Computer Programming.* **1.** the transfer of required data from secondary storage to main storage or from slower devices to higher-speed devices. **2.** in error recovery, the passing of control from a low-level recovery routine to a high-level recovery routine. **3.** in bubble sorting, the rising of lesser elements toward the top of the list. *Mining Engineering.* in the leaching treatment of minerals, a gentle flow of a solution through an ore bed to extract minerals. (Going back to a Latin word meaning "to filter.")

percolation filtration *Chemical Engineering.* a refinery process that percolates oils through an adsorbent material (e.g., clay) to remove impurities.

percolation leaching *Mining Engineering.* the selective extraction of a soluble mineral by introduction of a suitable solvent to seep into and through the material mass or pile containing the mineral.

percolation limit *Solid-State Physics.* **1.** the limit of the concentration of a magnetic element in a disordered crystalline alloy above which the material becomes ferromagnetic. **2.** the lower limit on the concentration of a foreign component needed to establish a continuous conduction path through an insulating matrix.

percolation network *Mathematics.* a (usually infinite) network whose edges have been randomly assigned values of 1 or 0 (called conducting or insulating links, respectively). The percolation problem for a given network is to determine the number or concentration of conducting links required for the network to become conducting.

percolation test *Civil Engineering.* a method of ascertaining the rate at which a soil can absorb waste fluids by calculating the drop in water level from a hole filled with water.

percolation zone *Hydrology.* on a glacier or ice sheet, the area where some degree of surface melting occurs, but where the snow layer is not completely soaked through or brought up to melting temperature, and the meltwater refreezes within the same layer.

percolator *Food Technology.* a device used for brewing coffee, in which boiling water bubbles up through a tube and then slowly filters downward through a perforated container holding ground coffee beans.

percolic *Anatomy.* around the colon.

Percomorphi *Vertebrate Zoology.* an equivalent name for the fish order Perciformes.

Percopsidae *Vertebrate Zoology.* the trout-perches, a small family of silvery, perchlike, foraging fishes composing the suborder Percopsoidei, characterized by a short length, rough scales, and small teeth; found in fresh waters of North America.

Percopsiformes *Vertebrate Zoology.* an order of small fishes intermediate between soft-rayed teleosts and spiny-rayed fishes, including the cavefishes, pirate perches, and trout-perches.

percurrent proliferation *Mycology.* in fungi, the growth of a germ tube through the top of a spore cell.

Percursariaceae *Botany.* a monospecific family of green algae of the order Ulvales, characterized by biseriate filaments that rise from a discoid base; often occurring in tangled masses in tide pools, salt marshes, and similar environments.

percussion *Mechanics.* the striking of one object against another, usually with significant force. *Medicine.* the act of striking a body part with sharp, short blows; the sound obtained is used as an aid in diagnosing the condition of the underlying part. *Ordnance.* **1.** the act of setting off an explosive charge with a sharp blow, such as that produced by a firing pin or the hammer of a gun. **2.** of or relating to weapons using percussion. Thus, **percussion firing mechanism, percussion lock, percussion long arms, percussion pistols, percussion system.** *Archaeology.* any process by which a stone core or mass is struck with a stone or other hard object to shape or sharpen it for use as a tool. Also, **percussion flaking.**

percussion bit *Mechanical Engineering.* a rock-drilling tool with chisel-like cutting edges that drills a hole by a chipping action when driven by impacts against a rock surface.

percussion cap *Ordnance.* a cap or cylinder containing a small high-explosive charge that can be detonated by a sharp blow. Also, **percussion primer.**

percussion drill *Mechanical Engineering.* a pneumatic drill in which a piston delivers hammer blows rapidly on a drill shank. Thus, **percussion drilling.**

percussion figure *Crystallography.* a fracture produced by a sharp blow to the surface of a crystal, characterized by radiating lines from the point of impact.

percussion mark *Geology.* a small, crescent-shaped or curved fracture produced on a hard, dense pebble by a sharp blow or violent impact.

percussion table see CONCUSSION TABLE.

percussion technique *Archaeology.* the use of percussion flaking in the making of stone tools. Also, **percussion method.**

percussion welding *Metallurgy.* a process of resistance welding performed by electrical sparking between the parts to be joined, followed by hammering.

percussor *Archaeology.* a hammerstone or similar object used to strike a stone in the process of percussion.

percutaneous *Medicine.* performed through the skin, as in a needle biopsy or an injection of a contrast medium in radiological examination.

percutaneous biopsy *Surgery.* a method for securing tissue for examination by insertion of a needle through the skin.

percutaneous nephrostomy tube *Surgery.* the surgical placement of a catheter into the kidney through the skin, providing diversion of urine.

percutaneous transluminal angioplasty *Surgery.* the dilatation of a blood vessel by inserting a balloon catheter through the skin and through the lumen of the vessel to the site of the narrowing, where the balloon is inflated; occasionally used to flatten or break an embolus. Also, BALLOON ANGIOPLASTY.

percylite *Mineralogy.* $PbCu^{+2}Cl_2(OH)_2$, a blue cubic mineral of dubious status occurring as small crystals, having a hardness of 2 to 2.5 on the Mohs scale and an undetermined specific gravity.

Perdicinae *Vertebrate Zoology.* a subfamily of Phasianidae (unless classified under subfamily Phasianinae) that includes the 130 species of Old world quail; distinguished from New World quail by their straight-edged bill, leg spurs, and short, stocky size.

peregrine falcon [per′i grin] *Vertebrate Zoology.* a small bird of prey, *Falco perigrinus* of the family Falconidae, noted for its rapid dives to grasp prey such as shorebirds and ducks; characterized by a blue-gray back with white neck and striped legs; found along coasts worldwide.

peregrine falcon

pereletok *Geology.* a frozen layer of ground above the permafrost that may remain frozen over several years. Also, INTERGELISOL.

perencephaly see PORENCEPHALIA.

perennate *Botany.* to survive from one growing season to the next or through a number of years, often with periods of reduced activity.

perennial *Science.* happening throughout the year or over a period of many years. *Botany.* **1.** a plant with underground vegetative parts that live more than two years, with the upper parts usually dying back seasonally in herbaceous plants and new growth occurring annually in woody plants. Also, **perennial plant. 2.** relating to or describing such a plant.

perennial lake *Hydrology.* a lake whose basin is filled with water throughout the year, and whose level fluctuates very little.
perennial spring *Hydrology.* a spring that flows continuously, rather than intermittently. Also, PERMANENT SPRING, LIVE SPRING.
perennial stream *Hydrology.* a stream whose bed lies below the water table, so that the stream flows continuously throughout the year. Also, PERMANENT STREAM, LIVE STREAM.
perezone *Geology.* a zone of deposition in which nonfossiliferous sediments accumulate between the low-tide mark and coastal lowlands undergoing erosion.
perf. perfect; perforated.
perfect *Science.* complete in all respects; flawless; exact. *Botany.* **1.** of or relating to a fungus that reproduces sexually or to the fungal stage in which sexual spores are produced. **2.** of or relating to a flower part that is complete, with all its constituent whorls present.
perfect binding *Graphic Arts.* a method of binding books or magazines using adhesive to attach trimmed pages to the spine. A book assembled in this way is **perfect bound.**
perfect cosmological principle *Astrophysics.* an assumption that for all observers, everywhere in space, and at all times, the universe appears the same on a large scale.
perfect crystal *Crystallography.* a single crystal in which the arrangement of atoms within each unit cell is uniform throughout.
perfect cube *Mathematics.* a quantity that is the product of three equal quantities in a group or ring; for example, $125 = 5^3$ and $x^3 + 3x^2 + 3x + 1 = (x + 1)^3$ are perfect cubes. Also, CUBE.
perfect dielectric see IDEAL DIELECTRIC.
perfect elastic collision see ELASTIC COLLISION.
perfect field *Mathematics.* a field F is said to be perfect if all finite extensions of F are separable.
perfect fifth *Acoustics.* an interval between complex tones whose frequency ratio of the fundamentals is 3 to 2, resulting in several equal frequencies (consonance) and few beat frequencies (dissonance).
perfect flower *Botany.* a flower having both pistils and stamens, as on most angiosperms.
perfect fractionation path *Physical Chemistry.* the line that represents inert crystals on a phase diagram.
perfect gas *Physical Chemistry.* see IDEAL GAS.
perfect graph *Mathematics.* a simple graph G such that, for every vertex-induced subgraph H of G, if m is the largest integer for which there is a subgraph of H isomorphic to the complete graph K_m, then m is at most the chromatic number of H. The strong perfect graph conjecture is that a graph G is perfect if and only if it is not a vertex-induced subgraph of G or the complement of G is an odd cycle with length greater than 3.
perfect inelastic collision see INELASTIC COLLISION.
perfecting press *Graphic Arts.* a rotary press that can print on both sides of a sheet during one pass through the press.
perfect lubrication *Engineering.* a condition in which the surfaces of two moving parts, usually made of metal, are completely enveloped in a film of liquid and do not touch each other.
perfectly diffuse radiator *Optics.* a surface that emits radiant energy in accordance with Lambert's cosine law.
perfectly diffuse reflector *Optics.* a surface that has a reflectance of unity and obeys Lambert's cosine law of emission for all directions of incident light. Also, UNIFORMLY DIFFUSE REFLECTOR.
perfectly inelastic collision *Physics.* a collision in which the colliding objects stick together.
perfectly mobile component *Physical Chemistry.* a component whose quantity within a system is dependent on external chemical factors, rather than on chemical factors within the system.
perfect matching *Mathematics.* in a graph, a matching that is incident with every vertex of the graph.
perfect number *Mathematics.* a positive integer that equals the sum of its divisors (other than itself); e.g., $6 = 1 + 2 + 3$ is a perfect number.
perfector press *Graphic Arts.* a flatbed press that can print on both sides of a sheet during one pass through the press.
perfect pitch *Acoustics.* a person's ability to perceive the exact pitch of a certain tone, and to reproduce the tone vocally. Also, ABSOLUTE PITCH.
perfect plasticity *Mechanics.* the idealization of zero workhardening for a solid; plastic deformation can thus continue without limit at a yield or limit stress.
perfect prognostic *Meteorology.* the observed atmospheric pressure pattern at the verifying time of the forecast of a meteorological element other than pressure; used in forecast studies.
perfect set *Mathematics.* a set A is a perfect set if it equals its derived set (i.e., the set of all the accumulation points of A).
perfect solution see IDEAL SOLUTION.
perfect square *Mathematics.* a quantity that is the product of two equal quantities in a group or ring; for example, $144 = 12^2$ and $x^4 - 2x^2 + 1 = (x^2 - 1)^2$ are perfect squares. Also, SQUARE.
perfect stage *Mycology.* a stage in the life cycle of certain fungi in which sexual spores are formed.
perfemic rock *Geology.* a class of igneous rocks in which the ratio of salic minerals to femic minerals is less than 1 to 7.
perfluorochemical *Organic Chemistry.* a hydrocarbon in which the hydrogen atoms have been replaced by fluorine atoms.
perfluoroelastomer *Materials.* a fully fluorinated elastomer that is resistant to solvents, bases, and jet fuels.
perfoliate *Botany.* of or related to a leaf whose base encircles a stem so that the stem appears to pass through the leaf. *Invertebrate Zoology.* of insect antennae, having platelike terminal joints surrounding the stalk.

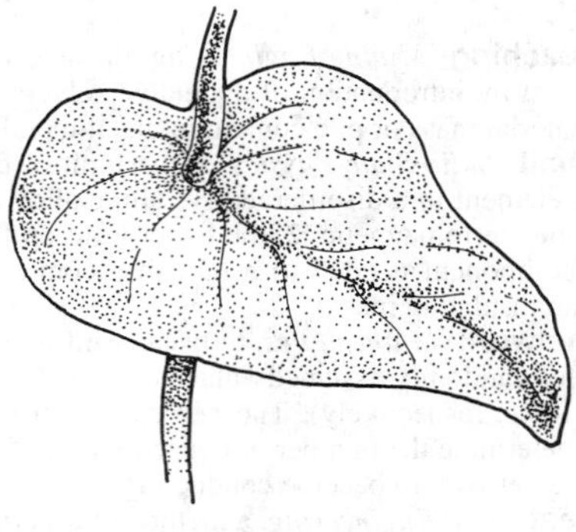

perfoliate leaf

perforate *Science.* to pierce or make a hole in something.
perforated brick *Building Engineering.* a clay brick with odd-sized vertical perforations to reduce weight, improve insulating properties, and reduce capillary attraction through a wall.
perforated completion *Petroleum Engineering.* a process of well completion in which the production casing or liner is punctured so that fluids may flow between the formation and the wellbore.
perforated crust *Hydrology.* a snow crust marked by pits and hollows formed as a result of ablation.
perforated-pipe distributor *Chemical Engineering.* a pipe perforated near one end; used to distribute vapors into a process liquid. Also, SPARGER.
perforated plate *Chemical Engineering.* a plate with holes that serves to disperse liquids into drops in a perforated-plate contacting tower. Also, SIEVE PLATE.
perforated-plate distributor *Chemical Engineering.* a perforated plate or screen that is placed across a channel to smooth out a nonuniform velocity profile of turbulent flow in a liquid-liquid extractor.
perforated-plate extractor *Chemical Engineering.* a liquid-liquid extraction vessel that uses perforated plates for enhanced extraction characteristics.
perforating see PERFORATION.
perforating wound *Surgery.* a wound that penetrates a viscus or body cavity.
perforation the process or result of perforating; specific uses include: *Surgery.* **1.** the surgical creation of a hole by boring or piercing. **2.** the hole so made. Also, FENESTRATION. *Ordnance.* the complete penetration of a projectile through an object. *Petroleum Engineering.* a procedure in which holes are made in the casing opposite the producing formation to allow oil or gas to flow into the well, or to allow water to be forced out into the formation, causing hydraulic fracturing.
perforation deposit *Geology.* an isolated kame formed by the accumulation of material in a vertical shaft piercing a glacier and leaving no outlet for water at the bottom.
perforator something that perforates; specifc uses include: *Telecommunications.* part of a local teletypewriter keyboard that punches coded holes into paper tape. Also, TAPE PERFORATOR.

perform to carry out some action; specific uses include: *Computer Programming.* **1.** to execute instructions in a computer. **2.** in COBOL, a verb that transfers control to a subroutine and executes it the specified number of times.

performance bond *Engineering.* **1.** a guaranteed bonding together of two surfaces. **2.** a bond signed by a contractor agreeing to perform to the specifications of a certain contract.

performance characteristic *Engineering.* any trait or quality present in a device, instrument, or machine when it is operational.

performance chart *Engineering.* a graphic representation of performance data.

performance curve *Engineering.* a graph showing the effectiveness of rotating equipment under a variety of operating conditions.

performance data *Engineering.* the information assessing the characteristics and effectiveness of a device, instrument, machine, or substance during use.

performance evaluation *Industrial Engineering.* the evaluation of a worker's productivity, reliability, and the like, in terms of set standards, for purposes of promotion, retention, etc. Also, **performance appraisal.**

performance index *Industrial Engineering.* the ratio between a worker's productivity and a defined standard productivity.

performance number *Engineering.* any of the number values over 100 in the octane scale that indicate the antiknock values of a motor (aviation) fuel.

performance rating *Industrial Engineering.* the process of evaluating a worker's performance in relation to normal performance. Also, EFFORT RATING, LEVELING.

performance sampling *Industrial Engineering.* the process of determining a worker's performance index over a short period by intermittent observation of his output.

perfory *Computer Programming.* the perforated strips that can be torn off the edges of continuous-form paper. Also, **perfs.**

perfume *Materials.* a blend of various substances, usually liquids, to produce a pleasant smell; perfumes typically consist of essential oils extracted from flower petals or other parts of plants; perfume materials are also derived from animals (e.g., musk, castor, ambergris) or produced synthetically. Perfumes are marketed as liquids and used as ingredients in soaps, lotions, creams, and cosmetics.

perfume oil *Materials.* any of various essential oils that are obtained from flowers, fruit, leaves, bark, roots, or other plant parts, or occasionally from animals, and used in perfume manufacture; plants whose oils are often used in perfumes include the cinnamon, gardenia, geranium, jasmine, lavender, rose, sandalwood, violet, and various citrus fruits such as the lemon and orange.

perfuse *Physical Chemistry.* to pass a fluid through or over something; cover or saturate with a fluid.

perfusion the act of perfusing; specific uses include: *Physiology.* the introduction of a fluid into an artery or vein. *Surgery.* **1.** the passage of a liquid over or through an organ or tissue. **2.** a liquid poured over or through an organ or tissue.

pergelation *Hydrology.* the formation of permafrost.

pergelic *Geology.* relating to a soil temperature regime characterized by a mean annual temperature of less than 0°C and the existence of permafrost.

pergelisol see PERMAFROST.

pergelisol table see PERMAFROST TABLE.

Pergidae *Invertebrate Zoology.* a family of sawflies, hymenopteran insects in the superfamily Tenthredinoidea.

perhumid climate *Meteorology.* one of the wettest of the climate types, with humidity index values of +100 or above.

perhydro- *Organic Chemistry.* a chemical prefix designating a fully hydrogenated hydrocarbon.

perhydrous coal *Geology.* a coal that contains more than 6% hydrogen when analyzed on a dry, ash-free basis.

peri- a prefix meaning "around."

periadenitis *Medicine.* an inflammation of tissues around a gland.

perianal *Anatomy.* located around the anus.

periangioma *Oncology.* a tumor surrounding a blood vessel.

perianth *Botany.* the set of structures enveloping the pistils and stamens, including the outer calyx of sepals and pistils or the inner corolla of petals.

periappendicitis *Medicine.* an inflammation of the tissues around the vermiform appendix, with decidual cells sometimes present in the peritoneum of the appendix.

periapsis *Astronomy.* the point at which an orbiting celestial body makes the closest approach to its primary body during its orbital revolution around that body.

periarteritis *Medicine.* an inflammation of the walls of an artery and the tissues around the artery, sometimes associated with a variety of systemic symptoms.

periastron *Astronomy.* the point at which one star in a binary system makes its closest approach to the other star.

periaxial encephalitis see SCHILDER'S DISEASE.

periaxonal *Neurology.* surrounding or encircling the axon of a nerve.

periblast *Developmental Biology.* **1.** a layer of cytoplasm that engulfs the blastodisc of an egg while the egg is participating in discoidal cleavage. **2.** the portion of the blastoderm of telolecithal eggs that consists of cells without complete cell membranes.

periblastula *Developmental Biology.* the blastula of a centrolecithal egg.

periblem *Botany.* in the histogen theory, the region of the apical meristem that produces the cortex.

periblinite *Geology.* a variety of provitrinite that consists of cortical tissue.

pericardial *Anatomy.* of or relating to the pericardium.

pericardial cavity *Anatomy.* the potential space between the parietal layer and the epicardium of the serous pericardium.

pericardial fluid *Physiology.* the liquid secretion of the inner layer of the pericardium.

pericarditis *Medicine.* an acute or chronic inflammation of the pericardium, sometimes characterized by the formation of adhesions and fibrous tissue.

pericardium *Anatomy.* the fibroserous sac surrounding the heart and the roots of the great vessels, consisting of an external layer of fibrous tissue and an inner serous layer; its base is attached to the central tendon of the diaphragm.

pericarp *Botany.* **1.** the part of a fruit formed by the ripening of the ovary wall, including the skin (epicarp), the often fleshy middle (mesocarp), and the membranous or stony inner layer (endocarp). **2.** in certain algae, a structure that holds the spores.

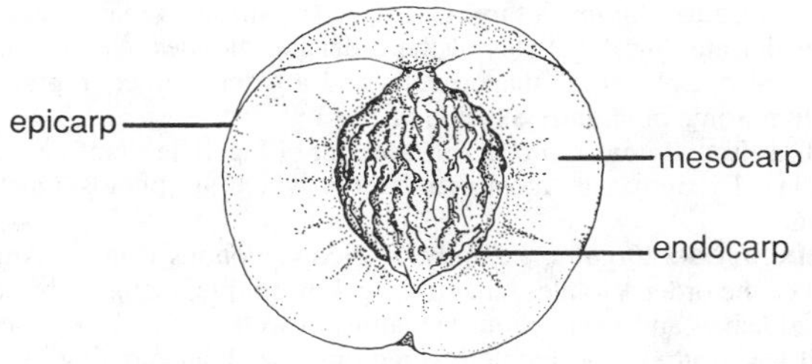

pericarp

pericenter *Physics.* a point along the orbit of a body that is closest to the center of gravitational force about which the body moves.

pericentric inversion *Genetics.* a chromosomal inversion that includes the centromere.

pericephalic *Anatomy.* surrounding the head.

pericerebral *Anatomy.* around the cerebrum.

perichaetium *Botany.* in mosses and liverworts, any of the leaves or bracts surrounding the sex organs, or the structure formed by this arrangement; a perichaetium surrounding the archegonium may be termed a perigynium, while one surrounding the antheridium may be termed a perigonium.

pericholangitis *Medicine.* an inflammatory reaction of the tissues that surround the bile ducts.

perichondrial *Anatomy.* of, relating to, or composed of perichondrium.

perichondrium *Anatomy.* a layer of dense, fibrous connective tissue that invests all cartilage except the articular cartilage of synovial joints.

perichrome *Neurology.* a nerve cell in which the chromophil substance, or stainable material, is arranged throughout the cytoplasm.

periclase *Mineralogy.* MgO, a cubic mineral of the periclase group occurring in granular form or as isometric crystals, having a specific gravity of 3.56 and a hardness of 5.5 on the Mohs scale; usually found in marbles.

periclinal *Geology.* of or relating to a pericline.

periclinal chimera *Genetics.* a plant composed of two genetically different tissues, one surrounding the other.
pericline *Geology.* any fold in which the beds dip radially inward toward or outward from a central point, forming a dome or basin. *Mineralogy.* a variety of albite characterized by elongation, and often by twinning, along the *b* axis.
pericline ripple mark *Geology.* a ripple mark forming an orthogonal pattern either parallel or transverse to the direction of the current, and having a wavelength less than 80 centimeters and an amplitude less than 30 centimeters.
periclinic twin law *Crystallography.* a twin law for feldspars in which the *b* axis of the unit cell is the twinning axis and the surface of the crystal is a rhombic section.
pericolic *Anatomy.* around the colon.
pericolitis *Medicine.* an inflammation of the tissue surrounding the ascending or descending colon, especially the peritoneal coat.
pericondensed polycyclic *Organic Chemistry.* of an aromatic compound in which three or more rings share common carbon atoms.
pericoronitis *Medicine.* an inflammatory reaction in the gingiva surrounding the crown of a partially erupted tooth.
pericranial *Anatomy.* of or relating to the pericranium.
pericranium *Anatomy.* the external periosteum of the skull.
pericycle *Botany.* a layer, consisting mainly of parenchyma cells, that forms the outer portion of the stele of a vascular plant.
pericystium *Anatomy.* **1.** the vascular tissue surrounding certain cysts. **2.** the tissues surrounding the gallbladder or urinary bladder.
pericyte *Histology.* a contractile cell that surrounds a capillary.
peridendritic *Neurology.* surrounding the dendrite of a nerve cell.
peridental see PERIODONTAL.
periderm *Botany.* a tissue, usually found in roots, that is bounded externally by the endodermis and internally by the phylum; a secondary protective tissue replacing the epidermis of the stem and root in plant axes exhibiting secondary growth, it is made up of cork, cork cambium, and phelloderm. *Developmental Biology.* a large-celled outer layer of the epidermis.
peridermioid *Mycology.* relating to a kind of cup-shaped fruiting body or aecium produced by fungal species belonging to the order Uredenales.
Peridiniaceae *Botany.* a family of widely distributed marine and freshwater thecate dinoflagellates of the order Peridiniales, characterized by spherical to polyhedral-shaped cells and a sulcus either limited to the cell hyposome or slightly extended.
Peridiniales *Botany.* an order of thecate, biflagellate dinoflagellates of the class Dinophyceae, most members of which are photosynthetic and marine.
Peridiscaceae *Botany.* a family of dicotyledonous tropical American trees of the order Violales, characterized by simple, leathery, basally triveined leaves and hypogynous apetalous flowers.
peridium *Botany.* the double-layered outer wall surrounding the spore-bearing organs of certain fungi.
peridot *Mineralogy.* **1.** a gem variety of forsterite, transparent to translucent and green in color. Also, EVENING EMERALD. **2.** a tourmaline with similar coloration.
peridotite *Petrology.* a dark, coarse-grained plutonic rock or rock group, consisting primarily of olivine, with or without mafic minerals such as amphiboles, pyroxenes, or micas. *Geology.* see UPPER MANTLE.
peridynamic loudspeaker *Acoustical Engineering.* a loudspeaker with a baffle that is designed to produce excellent bass response by minimizing the acoustic standing.
perifollicular *Histology.* of cells or material, surrounding a follicle.
perigean [per′ə jē′ən] *Astronomy.* of or relating to a perigee, especially the perigee of the moon. Also, **perigeal.**
perigean range *Oceanography.* the average of all monthly tide ranges occurring when the moon is at perigee.
perigean tidal currents *Oceanography.* tidal currents of increased velocity occurring monthly when the moon is at perigee.
perigean tide *Oceanography.* a tide of increased range occurring monthly when the moon is at perigee.
perigee [per′ə jē] *Astronomy.* the point at which an object that is in orbit around the earth, such as the moon or an artificial satellite, is nearest to the earth. (Derived from a Greek term meaning literally "near the earth.")
perigenic [per′ə jen′ik] *Geology.* describing a constituent of a rock or a mineral that is formed at the same time, but not at the same location, as the rock of which it is now a part.
periglacial *Geology.* of or relating to the geographical areas, climate, physical processes and conditions, and the resultant topography that are characteristic of the area bordering a glacier or ice sheet and are influenced by the cold temperature of the ice.

periglacial area

periglacial climate *Meteorology.* a climate that is characteristic of regions that border the outer perimeter of an ice cap or continental glacier, with extremely cold and dry winds that blow off the ice area; conducive to intense cyclonic activity.
periglial *Neurology.* situated around the glial cells in the brain.
perigonium *Botany.* see PERICHAETIUM. *Invertebrate Zoology.* a sac around the reproductive bud of a hydroid.
Perigordian *Archaeology.* describing a cultural tradition of western Europe in the Upper Paleolithic period. (From its identification with the *Perigord* region of southern France.)
perigynium see PERICHAETIUM.
perigynous *Botany.* of or relating to stamens, petals, or similar parts that are arranged on the edge of a cup-shaped structure (hypanthium) surrounding the pistil of a flower.
perihelion *Astronomy.* the point at which an object in orbit around the sun is nearest to it.
perihemal system *Invertebrate Zoology.* in echinoderms, a series of tubular canals containing the hemal vessels.
perihepatitis *Medicine.* a disease in which there is inflammation of the peritoneal capsule of the liver and of the tissues around the liver.
perikaryon *plural,* **perikarya.** *Cell Biology.* **1.** broadly, the cytoplasm of a cell body as distinguished from the nucleus and processes. **2.** specifically, the body of a nerve cell as distinguished from the nucleus, axon, and dendrites.
Perilampidae *Invertebrate Zoology.* a small family of wasps, hymenopteran insects in the superfamily Chalcidoidea, characterized by an arched thorax and spotted body; the larvae of these insects are parasitic in caterpillars.
perilemma *Cell Biology.* a membrane that surrounds the pellicle of certain ciliates.
perilla oil *Materials.* a combustible, light-yellow drying oil that is derived from the seeds of the mint plant, *Perilla ocimoides*; used in the Orient as a cooking oil, and elsewhere in the manufacture of varnish, ink, and artificial leather.
perilymph *Anatomy.* the fluid contained within the space that separates the membranous from the osseous labyrinth of the ear, surrounding and protecting the labyrinthine membrane. Also, **perilympha.**
perimagmatic *Geology.* of or relating to a hydrothermal mineral deposit that developed near its magmatic source.
perimedullary region *Botany.* the peripheral region of the pith.
perimeter [pə rim′i tər] *Science.* an outer edge, limit, or boundary. *Mathematics.* **1.** the boundary of a closed plane figure. **2.** the length of a closed curve bounding a plane figure. *Military Science.* a fortification or body of troops at the boundary of the territory controlled by a military force. *Optics.* an instrument used to determine the extent of a person's peripheral vision.

perimeter blasting *Mining Engineering.* a means of blasting in which outside rows of holes are loaded with very light continuous explosives that are fired simultaneously, so that shearing occurs from one hole to the next; this technique is used to minimize overbreak and leave clean-cut solid walls.

perimeter defense *Military Science.* a defense without an exposed flank, consisting of forces deployed along the perimeter of the defended area.

perimetrium *Anatomy.* the serous coat that covers the uterus.

perimetry *Medicine.* a determination of the limits and scope of the visual field.

perimysium *Anatomy.* the connective tissue demarcating a fascicle of skeletal muscle fibers. Also, INTERNAL PERIMYSIUM.

perinatal *Medicine.* occurring during or relating to a period before and after childbirth, generally defined as the twenty-eighth week of gestation through the seventh day after delivery.

perinatologist *Medicine.* a specialist in the branch of medicine dealing with perinatology.

perinatology *Medicine.* the branch of medicine that deals with the study of the fetus and infant during the perinatal period.

perineoplasty *Surgery.* plastic surgery of the perineum.

perineotomy *Surgery.* an incision into the perineum.

perineum *Anatomy.* **1.** the pelvic floor and associated structures occupying the pelvic outlet, bounded by the pubic arch and the coccyx. **2.** the region between the thighs, bounded in the male by the scrotum and anus and in the female by the vulva and anus.

perineural *Anatomy.* surrounding a nerve or nerves.

perineuritis *Neurology.* inflammation of the perineurium.

perineurium *Neurology.* the delicate connective-tissue sheath that surrounds each bundle of fibers in a peripheral nerve.

period a distinct and identifiable length of time; specific uses include: *Geology.* **1.** a unit of geologic time longer than an epoch, and representing an interval during which the rocks of a system were formed. **2.** informally, an interval of geologic time of any length, such as a glacial period. *Archaeology.* any specific interval of time in the archaeological record, such as the Upper Paleolithic period. *Physics.* the time elapsed for one complete oscillation or cycle. *Astronomy.* **1.** the interval of time between two phases of a periodic event. **2.** the interval between successive times of maximum brightness or minimum brightness of a variable star. *Nucleonics.* the amount of time required for the neutron flux density in a reactor to change exponentially by a factor of the natural logarithm base e (2.71828). *Chemistry.* a series of elements that form a horizontal row across the periodic table in order of increasing atomic number, beginning with an alkali metal and exhibiting a steady trend toward the last element, a noble gas. *Medicine.* a popular term for the menstrual period.

period-density relation *Astrophysics.* a relation between a star's period and its density after the star has undergone maximum expansion.

period doubling *Physics.* a transition from regular motion to chaotic motion. *Chaotic Dynamics.* **1.** a bifurcation that results in a destabilization of a periodic attractor in favor of an adjacent attractor with twice as many loops, causing the appearance of a frequency in the power spectrum which is one-half of a previous frequency. **2.** a route to chaos involving an infinite sequence of subharmonic bifurcations that accumulate at a value of the parameter beyond which chaos appears with an inverse period-doubling sequence that blurs residual, nearly periodic behavior.

period-doubling cascade *Chaotic Dynamics.* a sequence of period-doubling bifurcations that accumulate at a critical value of the parameter beyond which chaos is found.

periodic *Science.* occurring at regular intervals. *Chemistry.* of or relating to the periodic table of elements. *Chaotic Dynamics.* a behavior that repeats after an interval of time (the period). *Mathematics.* see PERIODIC FUNCTION.

periodic acid-Schiff reaction *Biochemistry.* a technique for detecting the presence of specific carbohydrates, such as cellulose or starch, in which the carbohydrates' hydroxyl groups are oxidized to aldehyde groups with periodic acid that reacts to Schiff's reagent by turning purple. Also, PAS REACTION.

periodic antenna *Electromagnetism.* an antenna whose input impedance is dependent on the frequency, such as a resonant antenna or an open-end wire antenna.

periodic comet *Astronomy.* a comet that returns to the inner solar system at regular intervals; usually applied to comets with orbital periods of less than 200 years.

periodic current *Oceanography.* a variable current whose values recur at more or less regular intervals.

periodic damping *Physics.* underdamping; the type of damping that is less than critical damping in a damped oscillatory system.

periodic duty *Electricity.* intermittent duty in which load conditions recur at regular time intervals.

periodic field focusing *Electronics.* a method for adjusting an electron beam so that it follows a trochoidal path and interacts with a focusing field at selected points.

periodic function *Mathematics.* **1.** a function f of a single real variable is said to be periodic with period T if whenever x is in the domain of f, then $x + T$ is also in the domain of f and $f(x + T) = f(x)$. **2.** a function f of n real variables is said to be multiply periodic if it is periodic in each of its variables separately. (For example, if $n = 2$, then f is said to be doubly periodic.) **3.** a function f of a complex variable z is said to be periodic with period c if $f(z + c) = f(z)$ for all z.

periodic kiln *Engineering.* a kiln used for an entire cycle of heating, cooling, and removal of product, in which the process is repeated for each batch.

periodic law *Chemistry.* a law stating that all properties of the elements that depend on atomic structure tend to change with increasing atomic number in a periodic way. *Nuclear Physics.* an early statement of this principle that governs classifying the elements according to the repetition of physical and chemical properties at definite intervals (periods); the law, when propounded in 1869, made allowance for elements that were not then known to exist but have since been established by confirmed discovery.

periodic line *Electricity.* a transmission line consisting of successively identical, similarly oriented sections, such as a loaded line having uniformly spaced loading coils.

periodic locust see SEVENTEEN-YEAR LOCUST.

periodic motion *Mechanics.* motion that repeats itself identically at regular time intervals.

periodic perturbation *Astronomy.* a small orbital oscillation in a celestial body due to gravitational attractive forces by other celestial bodies, such as planets. *Mathematics.* a perturbation function that is periodic.

periodic quantity *Physics.* any quantity that moves or oscillates in such a way as to repeat itself in a cyclic manner.

periodic spring *Hydrology.* a spring that rises and falls periodically, as a result of natural siphon action.

periodic table *Chemistry.* an arrangement of the chemical elements in order of increasing atomic numbers in a pattern that represents the periodic law; basic vertical groups consist of elements having similar properties, and elements having properties between those of adjacent neighbors form horizontal "periods."

periodic wave *Physics.* any wave whose displacement varies over time or space in a periodic manner.

period-luminosity relation *Astrophysics.* the relation between the period and the absolute magnitude of Cepheid variables; i.e., longer-period Cepheids are intrinsically brighter; used to determine distances.

period of vibration *Physics.* the amount of time required for a vibrational system to vibrate through one cycle.

periodontal *Anatomy.* **1.** relating to or occurring around a tooth. Also, PERIDENTAL. **2.** of relating to the periodontium. *Medicine.* of or relating to periodontics. Also, **periodontic.**

periodontics *Medicine.* the branch of dentistry that deals with the study and treatment of the periodontium.

periodontist *Medicine.* a dentist who specializes in periodontics.

periodontitis *Medicine.* an inflammatory reaction of the tissues surrounding a tooth, sometimes caused by the extension of gingival inflammation into the periodontium.

periodontium *Anatomy.* the tissues that invest or help to invest and support the teeth, including the gingivae, cementum, and alveolar and supporting bone.

periodontoclasia *Medicine.* a destructive or degenerative disease of the tissues supporting the teeth, including the alveolar bone, gingiva, and cementum.

periodontosis *Medicine.* a degenerative, noninflammatory condition of the tissues supporting the teeth, originating in one or more of the periodontal structures.

period-spectral relation *Astrophysics.* a correlation between the periods of Cepheid-type variable stars and their spectra, in which the stars that appear reddish in color have longer periods, and the stars that are bluish in color have higher temperatures and shorter periods.

perionychium *Anatomy.* the epidermis that forms the border around a nail.
periophthalmitis *Medicine.* an inflammation around the eye.
periosteum *Anatomy.* a fibrous connective tissue covering all bones of the body, and possessing bone-forming potentialities.
periostitis *Medicine.* a chronic inflammatory condition of the periosteum, marked by severe pain, tenderness, and swelling of the bone.
periostracum *Invertebrate Zoology.* the outer layer of the shells of mollusks and brachiopods, composed of a horny secretion called conchiolin.
periotic *Anatomy.* **1.** situated about the ear, especially the inner ear. **2.** the petrous and mastoid portions of the temporal bone.
peripancreatitis *Medicine.* an inflammation of tissues around the pancreas.
peripartum *Medicine.* relating to the mother during the last month of gestation or the first months following delivery.
peripediment *Geology.* the region of a pediplane extending across the younger rocks or alluvium of a basin that is beyond, but adjacent to, the region that developed on the older upland rocks.
peripheral [pə rif′ə rəl] *Science.* located at the outer limits of an area. *Anatomy.* relating to or situated away from the center of the body or an organ.
peripheral blood *Hematology.* a blood sample obtained from acral areas, or from the circulation remote from the heart, as from earlobe, fingertip, or heel.
peripheral buffer *Computer Technology.* a device that temporarily holds data being transferred between two devices with different transfer rates.
peripheral computer *Computer Technology.* a computer that performs auxiliary operations such as data storage conversions, sorting, and merging in support of a large data processing complex.
peripheral control unit *Computer Technology.* a device that links the central processing unit with a peripheral device.
peripheral depression see RING DEPRESSION.
peripheral equipment *Computer Technology.* any input/output devices or auxiliary storage units that are connected to a central processor by cables, such as visual display terminals, printers, and disk drives.
peripheral faults *Geology.* circular or arcuate faults that occur along the edges of an elevated or depressed region.
peripheral initiation *Ordnance.* a method of initiating detonation of a cylindrical explosive charge by simultaneously detonating all points on the periphery of the charge.
peripheral interface channel *Computer Technology.* a data path between peripheral equipment and the central processing unit.
peripheral isolate *Ecology.* a population of plants or animals that live in an area adjacent to their normal habitat.
peripheral-limited see INPUT/OUTPUT LIMITED.
peripheral male see SATELLITE MALE.
peripheral milling *Metallurgy.* a process of milling parallel to the axis of the cutting tool.
peripheral moraine see RECESSIONAL MORAINE
peripheral nervous system *Anatomy.* the nervous system outside of the brain and spinal cord.
peripheral operation *Computer Programming.* an operation that is not under direct control of the central processing unit. Also, **peripheral processing.**
peripheral processor *Computer Technology.* a small computer that performs input/output buffering, scheduling, and similar functions for a larger computer that actually processes the data.
peripheral sink see RIM SYNCLINE.
peripheral stream *Hydrology.* a stream that flows parallel to the edge of a glacier.
peripheral support computer see PERIPHERAL COMPUTER.
peripheral transfer *Computer Technology.* the transmission of data between two peripheral devices.
peripheral vision *Physiology.* vision produced by light falling on areas of the retina outside the macula.
periphery *Mathematics.* boundary.
periphlebitis *Medicine.* an inflammatory condition of the tissues around a vein, or of the external walls of a vein.
periphyses *Mycology.* in fungi, the hairlike filaments found lining a canal inside the fruiting body or ascocarp.
periphyton *Ecology.* a plant that lives on objects, such as the stems and leaves of plants; found at the bottom of a freshwater ecosystem. Also, **periphyte.**
periplasmic region *Microbiology.* **1.** in Gram-negative bacteria, the area between the outer membrane and the cytoplasmic membrane. **2.** in Gram-positive bacteria, the region between the cytoplasmic membrane and the cell wall.
periplasmic space *Cell Biology.* the region between the inner and outer membranes of a Gram-negative bacterial cell, or the region between the cell membrane and cell wall of a fungal or plant cell.
periplast *Histology.* the supporting and nutritive tissue in an animal organ.
periproct *Invertebrate Zoology.* in echinoderms, the plate or tough membrane surrounding the anal region.
peripteral *Architecture.* surrounded by a single row of columns.
Periptychidae *Paleontology.* a family of protoungulates in the extinct order Condylarthra; similar to modern tapirs, but with short legs and a long tail, and characterized by bunodont cheek teeth; known only from the Paleocene of North America.
Peripylina see PORULOSIDA.
perirenal insufflation *Radiology.* the introduction of air around the kidneys and adrenal glands to provide contrast for roentgenographic study.
perisarc *Invertebrate Zoology.* in some colonial hydrozoans, a translucent secreted exoskeleton surrounding each polyp or the connecting parts of the colony.
Periscelidae *Invertebrate Zoology.* a family of dipteran insects in the subsection Acalyptratae.
Perischoechinoidea *Invertebrate Zoology.* a largely extinct subclass of echinoderms with one extant genus of sea urchins, having very large spines, and sometimes club-shaped and bearing poison glands.
periscope [per′i skōp′] *Optics.* **1.** an instrument that contains an arrangement of prisms or mirrors enabling the viewer to see in a direction displaced, usually upward, from the line of sight; commonly used in submarines, tanks, and entrenchments. **2.** a lens that is thin, astigmatic, and contoured much like a meniscus lens. (Going back to a Greek term meaning "to look around.")

periscope

periscopic [per′i skäp′ik] *Optics.* relating to or involving the use of a periscope. Thus, **periscopic view.**
periscopic sextant *Navigation.* a bubble sextant that extends through the top of an aircraft so that celestial bodies can be seen.
periscopic sight *Ordnance.* a gunsight that raises the gunner's line of vision over an obstacle; similar to a periscope in design.
periseptal annulus *Microbiology.* a continuous zone of adhesion or fusion between the outer membrane and the cytoplasmic membrane in Gram-negative bacteria; formed around the circumference of the cell, one on each side of the future location of the cell division septum.
perisilic see SILICIC.
perisperm *Botany.* a nutritive tissue within certain seeds that do not contain an endosperm; derived from the nucleus and deposited outside the embryo sac.
Perissodactyla *Vertebrate Zoology.* the order of odd-toed hoofed ungulates; herbivorous, terrestrial mammals characterized by a large central digit on each foot; includes the horse, tapir, and rhinoceros.
peristalsis [per′i stôl′sis; per′i stäl′sis] *Physiology.* the wavelike muscle contractions that move food through the intestines.

peristaltic [per′i stôl′tik; per′i stäl′tik] *Physiology.* relating to or involving peristalsis.

peristaltic charge-coupled device *Electronics.* a device in which a voltage moves electrical charges through the circuit in much the same way contractions in the digestive tract move food through the stomach.

peristaltic pump *Biotechnology.* an apparatus used to transfer fluids with a flexible fluid-containing tube placed between rotating rollers that squeeze it intermittently.

Peristediidae *Vertebrate Zoology.* an equivalent name for Triglidae, a family of benthic marine fishes.

peristerite *Mineralogy.* an iridescent, whitish or bluish gem variety of sodic plagioclase, having sharp internal reflections of colors such as blue, green, and yellow.

peristomal pits *Invertebrate Zoology.* small pits found in the membrane surrounding the mouth in various invertebrates.

peristome *Botany.* a fringe of toothlike appendages around the opening of a moss capsule. *Invertebrate Zoology.* **1.** in many ciliated protozoans, a groove or depression leading to the mouth opening. **2.** in some freshwater and land snails, an outer lip surrounding the shell opening. **3.** in echinoderms, a thick, leathery membrane around the mouth. **4.** the first complete segment of most annelids, surrounding the mouth. Also, **peristomium.**

peristomial *Invertebrate Zoology.* in annelids, located on the peristome.

peristyle *Architecture.* a row of columns surrounding a temple or court.

perite *Geology.* a breccialike material that occurs in marine sedimentary rock, considered to be either a mixture of lava and sediment or a series of shallow intrusions of magma into wet sediment.

peritectic *Physical Chemistry.* **1.** describing a substance that has both a solid and a liquid component that forms when the liquid phase interacts with the solid phase during the cooling process; seen in compounds that have a different composition once they melt. **2.** a substance of this type.

peritectic point *Physical Chemistry.* the point at which a compound within a solid decomposes into a liquid with a different composition, occurring well before the melting point is reached.

peritectic reaction *Physical Chemistry.* a phase transformation in which, upon cooling, a liquid phase combines with a solid phase to produce a new solid phase.

peritectoid *Physical Chemistry.* **1.** describing a solid substance that, when cooled, reacts with another solid to form a third solid. **2.** a substance of this type.

peritectoid reaction *Physical Chemistry.* a phase transformation in which, upon cooling, two solid phases combine to produce a new solid phase.

perithecium *Mycology.* a roundish or flask-shaped fruiting body or ascocarp, having an opening at the top.

peritoneal [per′i tə nē′əl] *Anatomy.* of or related to the peritoneum.

peritoneal cavity *Anatomy.* either of the two sections of the abdominal cavity, the lesser peritoneal cavity or the greater peritoneal cavity.

peritoneocentesis see PARACENTESIS.

peritoneopexy *Surgery.* the surgical fixation of the uterus performed through the vagina.

peritoneoscope *Medicine.* an instrument used to examine the peritoneal cavity, often inserted through the abdominal wall.

peritoneoscopy *Surgery.* the examination of the structures and organs within the peritoneal cavity by means of an instrument introduced through the abdominal wall. Thus, **peritoneoscope.**

peritoneotomy *Surgery.* an incision into the peritoneal cavity.

peritoneovenous shunt *Surgery.* a prosthetic device used to divert fluid from the peritoneal cavity to a vein, as in the LeVeen peritoneovenous shunt.

peritoneum [per′i tə nē′əm] *Anatomy.* the serous membrane lining the abdominopelvic walls and investing the viscera.

peritonitis [per′i tə nī′tisl] *Medicine.* an inflammation of the peritoneum, marked by exudations in the peritoneum serum, abdominal pain, constipation, vomiting, and moderate fever.

peritonsillar *Anatomy.* located around a tonsil.

peritonsillar abscess *Medicine.* an abscess forming in acute tonsillitis around one or both faucial tonsils.

Peritrichia *Invertebrate Zoology.* an order of ciliate protozoans having a bell-shaped body and cilia spiraling around the mouth; most species are sessile and stalked; Also, PERITRICHA, PERITRICHIDA.

peritrichous *Invertebrate Zoology.* of protozoa, belonging to the Peritrichia. *Microbiology.* of bacteria or other microorganisms, having flagella distributed over the entire cell surface.

perityphlitis *Medicine.* appendicitis, or inflammation of the peritoneum surrounding the cecum.

perivitelline space *Cell Biology.* the fluid-filled region between the surface of the fertilized egg and the fertilization membrane.

Perkin reaction *Organic Chemistry.* a reaction in which aromatic, unsaturated carboxylic acids are formed by the condensation of aromatic aldehydes and acid anhydrides in the presence of an alkali salt of the acid. (Named for the English chemist Sir William Henry *Perkin*, 1838–1907.)

Perkinsea *Invertebrate Zoology.* a class of Apicomplexan protozoa containing a single species, *P. marinus*, a parasite of American oysters.

perleche *Medicine.* a bacterial or fungal infection of the lips, causing erosions and fissuring of the commissures; it can be caused by excessive drooling or overclosure of an edentulous mouth.

perlite *Geology.* a volcanic glass with rhyolitic composition whose texture consists of spheruloid cracks; generally has a higher water content than obsidian. *Materials.* this glass when heated and crushed to form a powder used in insulation, in acoustic tile, and as a lightweight aggregate in concrete.

perlitic *Petrology.* **1.** of or relating to a glassy igneous rock texture characterized by small spheruloids formed from cracks caused by contraction during cooling. **2.** of or relating to perlite.

Perls, Fritz 1893–1970, German psychiatrist; developed Gestalt therapy.

perlucidus *Meteorology.* a cloud variety of the species stratiformis, with transparent spaces between its elements that allow the sun, moon, blue sky, or clouds aloft to be seen; found only in the genera altocumulus and stratocumulus.

perm *Petroleum Engineering.* a unit of reservoir permeability.

Permafil *Materials.* the trade name for a mixture of monomers in which the liquid is completely polymerized and hardens without evaporation.

permafrost *Geology.* a permanently frozen soil, subsoil, or other ground deposit occurring in arctic regions or wherever the temperature remains below freezing (0°C) for two or more years. Also, PERGELISOL.

permafrost island *Geology.* a small, isolated patch of permafrost that is surrounded by unfrozen ground.

permafrost line *Geology.* a line on a map indicating the geographic border of the arctic permafrost.

permafrost table *Geology.* an irregular surface that represents the upper limit of permafrost in a particular region. Also, PERGELISOL TABLE.

permalloy *Metallurgy.* any of several iron-nickel-base alloys suitable for soft magnets. Also, PERMEABILITY ALLOY.

permanent *Mathematics.* a generalized matrix function, similar to det (determinant), except that in the sum used in forming the determinant, signum is replaced by +1. In particular, let A be a matrix with

$$\det A = \sum_{\sigma} (\operatorname{sgn} \sigma) \prod_{i=1}^{n} a_{i\sigma(i)},$$

where the sum runs over all $n!$ permutations σ of n items $[1, 2, \ldots, n]$ and sgn represents the signum function. The permanent of A, denoted by per$A$, is

$$\sum_{\sigma} \prod_{i=1}^{n} a_{i\sigma(i)}.$$

permanent anode *Metallurgy.* in an electrochemical process and in cathodic protection, an anode that does not dissolve.

permanent axis *Mechanics.* the axis about which a rigid body has the greatest moment of inertia, so that its rotational motion about this axis is stable.

permanent benchmark *Engineering.* a surveyor's mark that is easily recognizable, fashioned and located to withstand great wear; used as a standard reference point.

permanent current *Oceanography.* an ocean current whose flow is little changed by seasonal, tidal, or other factors.

permanent echo *Electromagnetism.* a signal reflected to a radar station from a fixed object.

permanent error *Computer Programming.* an error condition caused by malfunctioning hardware that cannot be easily diagnosed and corrected. Also, HARD ERROR.

permanent ethnographic probability sample *Anthropology.* a cross-cultural index that surveys a random sample of cultures for which there is extensive ethnographic information available and for which comparable statistical manipulations may be done.

permanent extinction *Hydrology.* the permanent drying up of a lake as a result of the destruction of the lake basin.

permanent fault *Computer Programming.* a computer error that consistently occurs when certain conditions exist.

permanent gas *Chemistry.* a gas that cannot be condensed or liquefied by pressure alone. *Thermodynamics.* a gas that is at a temperature far above its critical temperature.

permanent hardness *Chemistry.* water hardness that cannot be removed by boiling; a property of water that contains sulfates or chlorides, as distinguished from the temporary hardness of water that contains bicarbonates.

permanent ice foot *Hydrology.* a narrow strip of ice attached to a polar coast that does not melt completely in summer.

permanent magnet *Electromagnetism.* a ferromagnetic substance that has been subjected to a magnetic field strong enough to cause the material to retain its own magnetization indefinitely.

permanent-magnet focusing *Electronics.* an adjusting of the electron beam in a television picture tube by the magnetic field produced by permanent magnets mounted around the neck of the tube.

permanent-magnet generator *Electricity.* a generator in which the magnetic field is created by permanent magnets.

permanent-magnet loudspeaker *Acoustical Engineering.* an inductive loudspeaker in which a steady-state magnetic field is produced by permanent magnets.

permanent mold *Metallurgy.* in casting, a term for a mold that is used more than once.

permanent press *Textiles.* **1.** the process of applying a synthetic finish to fabrics and garments in order to make them retain desired creases and to impart shape retention and crease resistance. **2.** a fabric produced by such a process. **3. permanent-press.** of or relating to this process. Also, DURABLE PRESS.

permanent-press resin *Organic Chemistry.* a thermosetting resin used to impart crease resistance to textiles and fibers.

permanent set *Mechanics.* the plastic deformation of a body that remains after the applied load is removed. Also, PLASTIC DEFORMATION.

permanent-split capacitor motor *Electricity.* a capacitor motor that operates with the starting capacitor and auxilliary winding closed or operative in the circuit. Also, CAPACITOR START-RUN MOTOR.

permanent spring see PERENNIAL SPRING.

permanent storage *Computer Technology.* **1.** storage that cannot be modified. Also, FIXED STORAGE, READ-ONLY STORAGE. **2.** storage, such as magnetic tapes and diskettes, that does not lose its contents in the event of a loss of power.

permanent stream see PERENNIAL STREAM.

permanent tooth *Anatomy.* any of the thirty-two adult teeth, including replacements for the deciduous teeth.

permanent water *Hydrology.* a water source that stays constant throughout the year.

permanent wave *Fluid Mechanics.* a wave in a fluid whose streamline pattern remains constant in time in a coordinate system that moves with the wave.

permanganic acid *Chemistry.* $HMnO_4$, an acid known only in solution.

Permasyn motor *Electricity.* a synchronous motor that provides an equivalent DC field as a result of permanent magnets embedded in its squirrel-cage motor.

permatron *Electronics.* a thermionic gas tube in which conduction is controlled by an external magnetic field.

permeability [per′mē ə bil′ə tē] *Fluid Mechanics.* the capability of a porous substance or membrane to allow a fluid to filter through it. *Agronomy.* the ease with which water, air, or plant roots penetrate or pass through a soil horizon. *Engineering.* the relative ability of a rock or soil to conduct magnetic lines of force. *Electromagnetism.* a factor that is characteristic of the magnetic properties of a substance; given by the ratio of the magnetic flux induction B to the magnetizing force H, and symbolized by m; in most cases, B is parallel to H and m is a scalar quantity, otherwise m is a tensor.

permeability alloy see PERMALLOY.

permeability coefficient *Fluid Mechanics.* a quantity associated with a porous substance indicating its ability to allow fluid to pass through it; given by the rate of fluid flow through a unit cross section of the substance, subject to a unit pressure gradient while maintained at a specified temperature.

permeability number *Engineering.* a number used to indicate the relative ability of a substance to allow a fluid to permeate its surface.

permeability trap *Geology.* an oil trap formed by lateral variation of permeability within a reservoir bed.

permeability tuning *Electricity.* the process of tuning a resonant circuit by moving a ferrite core in or out of a coil, thus changing the effective permeability of the core and the inductance of the circuit.

permeable [per′mē ə bəl] *Science.* capable of being permeated. *Chemistry.* specifically, capable of being passed through by very small particles, such as ions.

permeable bed *Geology.* a porous reservoir formation through which oil, natural gas, or water can flow.

permeable membrane *Physical Chemistry.* a thin layer of natural or synthetic material that allows some substances, but not others, to pass through it; used in reverse osmosis.

permeameter *Engineering.* a device for measuring the permeability of soils or other materials, usually consisting of two reservoirs connected by a conduit containing the material being measured, as water is passed from one reservoir under varying conditions through the connecting conduit.

permeametry *Analytical Chemistry.* a method of measuring the average size of small particles in a gas or liquid by passing the mixture through a powder bed of known dimensions and recording the pressure drop and flow rate.

permeance *Electromagnetism.* the reciprocal of the reluctance of a magnetic circuit, symbolized by P and determined by the magnetic flux divided by the magnetomotive force.

permeant *Ecology.* an organism that habitually moves from community to community.

permeaplast *Cell Biology.* a cyanobacterial cell that has been exposed to agents causing spheroplast formation, usually to facilitate genetic transformation.

permease *Biochemistry.* a membrane protein that controls the passage of a substance through the membrane.

permeate *Science.* to pass, penetrate, or diffuse through.

permeation *Chemistry.* the diffusion or penetration of ions, atoms, or molecules through a permeable substance.

permeation gneiss *Petrology.* gneiss formed or altered by geochemically mobile materials passing through or into solid rock.

permeator *Chemical Engineering.* a membrane device used for separation that allows species to pass from one phase to another.

Permendur *Metallurgy.* an iron-cobalt alloy, at times with vanadium, that is suitable for soft magnets when high permeability at high field strength is required.

permenorm alloy *Metallurgy.* an iron-nickel alloy used in magnetic amplifiers and as a magnet core material.

Permian *Geology.* **1.** a geologic period of the Upper Paleozoic era, extending from the end of the Carboniferous period to the beginning of the Mesozoic era (from about 280 to 225 million years ago). **2.** the rocks formed during that time.

Permian extinction *Paleontology.* a period about 245 million years ago during which large numbers of marine invertebrate families and other species became extinct.

per mil or **per mill** *Science.* per thousand.

permineralization *Geology.* a process of fossilization by which additional mineral material is deposited in the pore spaces of original hard animal parts.

permissible dose *Radiology.* the suggested maximum amount of exposure to radiation over a specified time interval that an individual may safely endure and that is, therefore, allowable by current radiation protection guides.

permissible length *Naval Architecture.* a vessel's floodable length multiplied by its factor of subdivision.

permissible velocity *Civil Engineering.* the maximum safe speed at which water may flow through a channel, pipe, or other facility.

permissive action link *Ordnance.* a safety device that prohibits arming or launching a nuclear weapon system until a specified code or combination has been inserted.

permissive cell *Virology.* any cell in which a given virus can replicate, or in which a conditional mutation has no deleterious effects.

permissive host *Virology.* any organism or cell culture that permits the replication of a given virus, resulting in a productive infection.

permissive stop *Transportation Engineering.* a railroad signal at which trains are permitted to pause and then proceed at a slow speed through a "stop" indication, rather than waiting for the indication to change.

permissive temperature *Genetics.* the temperature range within which a given conditional lethal mutant can survive.

permittivity see DIELECTRIC CONSTANT.

permo-Carboniferous *Geology.* **1.** an interval of geologic time in which the Permian and Carboniferous periods are considered as a single unit. **2.** an interval of geologic time in which the Permian and Pennsylvanian periods are considered as a single unit. **3.** a geologic age and corresponding unit of rocks that is transitional between the Uppermost Pennsylvanian and the Lowermost Permian periods.

permselective membrane *Physical Chemistry.* a membrane that allows only one charge type of ions, either positive or negative, to pass through it.

permutation *Science.* a process of rearrangement or reorganization. *Mathematics.* a rearrangement of the elements of a given finite set; a one-to-one mapping of a set of finite cardinality onto itself. In particular, let $i_1, i_2, \ldots, i_r$ be r distinct elements of $I_n = \{1, 2, \ldots, n\}$. Then $(i_1 i_2 i_3 \cdots i_r)$ denotes the permutation (called a cycle of length r or r-cycle) that maps i_1 to i_2, i_2 to $i_3, \ldots, i_{r-1}$ to i_r, and i_r to i_1, and maps every other element of I_n to itself. A 2-cycle (k,l) is called a transposition, and a 1-cycle (k) is the **identity permutation.** Every nonidentity permutation in the permutation group on n objects is expressible as a product of disjoint cycles, and also as a product of (not necessarily disjoint) transpositions. A permutation is an **even** (or **odd**) **permutation** if the permutation can be written as the product of an even (or odd) number of transpositions.

permutation group *Mathematics.* the group of all bijections on a nonempty set S. The elements of the permutation group on a set S are called permutations. If S has finite cardinality n, then the permutation group on S is called the symmetric group on n objects and is denoted S_n.

permutation matrix *Mathematics.* a square matrix with exactly one entry in each row and column equal to +1, and all other entries equal to 0; so named because multiplication of a vector by a permutation matrix permutes the components of the vector. The determinant of a permutation matrix is ±1.

permutation modulation *Telecommunications.* a modulation scheme for sending digital information by means of band-limited signals in the presence of additive white gaussian noise.

permutation symbol *Mathematics.* the symbol ${}^{e}i_1 \cdots i_n$ defined by the following conditions: (a) ${}^{e}i_1 \cdots i_n = 0$ if any two of the indices $i_1, \ldots, i_n$ are equal; (b)${}^{e}i_1 \cdots i_n = +1$ if the indices are an even permutation of 1, 2, . . . , n; and (c) ${}^{e}i_1 \cdots i_n = -1$ if the indices are an odd permutation of 1, 2, . . . , n. Used in tensor calculus on an n-dimensional space; equal to the permutation tensor in the case of Euclidean space. Also called the **generalized Kroneker symbol** or **generalized Kroneker delta.**

permutation tensor *Mathematics.* a completely antisymmetric tensor field with number of arguments (indices) equal to the dimension of the underlying space. Its nonzero covariant (resp. contravariant) components are equal in magnitude to the square root of the determinant of the covariant (resp. contravariant) metric tensor, with sign equal to the sign of the permutation of the arguments. It coincides with the permutation symbol in the case of Euclidean space.

pernicious anemia *Medicine.* the progressive, megaloblastic anemia resulting from lack of vitamin B_{12}, sometimes accompanied by degeneration of the posterior and lateral columns of the spinal cord.

pernio *Medicine.* redness and swelling of the skin caused by excessive exposure to cold; chilblains.

perniosis *Medicine.* skin affections or chilblains on various parts of the body, usually caused by cold.

peroba *Materials.* the durable, fine-textured wood of the Brazilian tree, *Aspidosperma*; used for fine furnishings, veneers, and flooring.

perofskite structure *Materials Science.* a crystal structure found in perofskite, $CaTiO_3$, with oxygen ions at the centers of the faces of the unit cell, calcium ions at the corners, and titanium ions in the octaheral interstices.

Perognathae *Vertebrate Zoology.* a subfamily of pocket mice of the family Herteromyidae, containing the genera *Perognathus* and *Microdipodops*.

peronine see BENZYLMORPHINE HYDROCHLORIDE.

Peronospora *Mycology.* a genus of fungi belonging to the order Peronosporales, including species that cause downy mildews on plants such as onions, tobacco, beets, and spinach.

Peronosporaceae *Mycology.* a family of fungi belonging to the order Peronsporales that is composed of downy mildews, most of which are plant pathogens.

Peronosporales *Mycology.* an order of fungi belonging to the class Oomycetes, including plant pathogens and species that obtain nutrients from dead organic matter.

peroral *Surgery.* administered or performed through the mouth.

per os *Virology.* of virus transmission, through the mouth. (From *per os*, Latin for "by mouth.")

Perothopidae *Invertebrate Zoology.* a small family of beetles, coleopteran insects in the superfamily Elateroidea.

perovskite *Mineralogy.* $CaTiO_3$, an orthorhombic, pseudocubic mineral of the perovskite group, varying in color from yellow to grayish-black, characterized by an adamantine to submetallic luster, and having a specific gravity of 4.01 and a hardness of 5.5 on the Mohs scale; found in accessory amounts in basic alkaline rocks, talc schists, and metamorphosed limestones.

peroxidase *Enzymology.* an enzyme that oxidizes a substrate with hydrogen peroxide. See CATALASE.

peroxidase-antiperoxidase technique *Immunology.* an enzyme-labeling method used to identify antigens or antibodies in tissue sections, by joining specific immune complexes that are composed of horseradish peroxidase and rabbit antiperoxidase.

peroxide *Chemistry.* **1.** any compound that contains the peroxy group, –O–O–. These compounds are strong oxidizing agents and are used industrially as bleaches. **2.** a compound of this type used as a bleach, especially for the hair. **3.** a shorter term for hydrogen peroxide, H_2O_2.

peroxide number *Analytical Chemistry.* the number of millimoles of peroxide absorbed by 1000 grams of oil or fat; used to measure rancidity. Also, **peroxide value.**

peroxisome *Cell Biology.* an intracellular, membrane-bound organelle that contains oxidative enzymes and carries out oxidative reactions.

peroxybenzoic acid see PERBENZOIC ACID.

peroxy group *Chemistry.* a pair of oxygen atoms –O–O– that replace a single oxygen atom in certain compounds.

peroxysulfuric acid see CARO'S ACID.

perp. or **perp** perpendicular. Also, **perpend.**

perpendicular *Mathematics.* **1.** at right angles; meeting a given line or surface at a right angle. **2.** a straight line constructed from a given point so that it intersects a given line, curve, or surface at a right angle (i.e., at a right angle to the tangent at the point of intersection). **3.** a normal. **4.** two nonzero elements of an inner product space are said to be perpendicular, or normal, if their inner product is zero.

perpendicular axis theorem *Mechanics.* a theorem stating that the sum of the moments of inertia of a plane lamina about any two orthogonal axes in the plane is equal to its moment of inertia about a perpendicular axis passing through the intersection of the two orthogonal axes.

perpendiculars *Naval Architecture.* either of two vertical datum lines used to define a ship's submerged length. The forward perpendicular is at the junction of the waterline and the stem; the aft perpendicular is at the junction of the waterline and the aft end of the rudder post.

perpendicular slip *Geology.* the component of the net slip of a fault that is measured at right angles to the trace of the fault on any intersecting surface.

perpendicular slope *Geology.* a very steep slope or precipitous face.

Perpendicular style *Architecture.* the third phase (c. 1350–1550) of English Gothic architecture.

perpendicular throw *Geology.* the distance between two formerly adjacent points in a faulted bed, vein, or other surface, measured at right angles to the surface.

perpetual arrhythmia see CONTINUOUS ARRHYTHMIA.

perpetual frost climate *Meteorology.* the static climate of the ice cap regions of the world, in which the annual accumulation of snow and ice is never exceeded by ablation; characterized by a warmest-month mean temperature of less than 32°F. Also, ICE-CAP CLIMATE.

perpetual motion *Physics.* the concept, never demonstrated experimentally, that a machine can continue to operate for an indefinite time using only its own energy.

perpetual motion machine of the first kind *Physics.* **1.** a hypothetical device that, after being set into motion, would continue to operate indefinitely with no further external supply of energy to it. **2.** a hypothetical device whose output energy exceeds its input energy, thus violating the first law of thermodynamics (law of conservation of energy).

perpetual motion machine of the second kind *Physics.* a hypothetical device that operates by extracting heat from a source and completely converting it to other forms of energy, thus violating the second law of thermodynamics.

perpetual motion machine of the third kind *Physics.* a device in which a component is allowed to continue operation indefinitely without any external influences, such as a currrent in a superconductor.

perradial canals *Invertebrate Zoology.* in some jellyfish, radiating gastrovascular canals that join the circumferential canal.

Perret phase *Volcanology.* the stage of volcanic eruption characterized by the emission of large amounts of high-energy gas, often resulting in the enlargement of the volcanic conduit.

Perrin, Jean Baptiste 1870–1942, French physicist; awarded the Nobel Prize for calculating the value of Avogadro's number and proving the existence of atoms and molecules.

Perroncito's phenomenon or **Perroncito's spirals** see APPARATUS OF PERRONCITO.

Perron-Frobenius theorem *Mathematics.* if A is an irreducible $n \times n$ matrix with (real) nonnegative entries, then: (a) the spectral radius $\rho(A)$ is positive; (b) $\rho(A)$ is an eigenvalue of A; (c) there exists a vector $x \neq 0$ with n (real) nonnegative components such that $Ax = \rho(A)x$; (d) $\rho(A)$ is an algebraically (and hence geometrically) simple eigenvalue of A; (e) if the set $S = \{\lambda_n = \rho(A), \lambda_{n-1}, \ldots, \lambda_{n-k+1}\}$ of eigenvalues of modulus $\rho(A)$ has exactly k distinct elements, then each eigenvalue $\lambda_i \in S$ has algebraic multiplicity 1, and $S = \{e^{2\pi ih/k}\rho(A): h = 0, 1, \ldots, k-1\}$. That is, these maximum modulus eigenvalues are the kth roots of unity times $\rho(A)$. Also, if λ is any eigenvalue of A, then so is $e^{2\pi ih/k}\lambda$, for all $h = 0, 1, \ldots, k-1$; and (f) $\lim_{m\to\infty}[\rho(A)^{-1}A]^m = xy^T > 0$, where x and y are vectors with n (real) positive components satisfying $Ax = \rho(A)x$, $A^Ty = \rho(A)y$, and $x^Ty = 1$. In its original form, this theorem characterized only the eigenvalues and properties of matrices with positive entries; it was known as Perron's theorem. The name of Frobenius is associated with generalizations of Perron's results about positive matrices to nonnegative matrices.

Perron-Frobenius theory *Mathematics.* the study of nonnegative matrices (square matrices with real nonnegative entries) and their eigenvalues, especially applications of the Perron-Frobenius theorem.

perryite *Mineralogy.* $(Ni,Fe)_5(Si,P)_2$, a trigonal mineral found only in meteorites, having a specific gravity of 7.37 and an undetermined hardness.

persalic rock *Geology.* a class of igneous rocks in which the ratio of salic minerals to femic minerals is greater than 7 to 1.

persecution complex *Psychology.* a popular term for paranoia. See PARANOIA.

Perseids *Astronomy.* a major meteor shower that occurs annually between July 27 and August 17, with peak activity around August 12.

Perseus *Astronomy.* the Rescuer, a fairly conspicuous northern constellation lying in the Milky Way and best seen in the autumn.

Perseus

Perseus A *Astronomy.* an intense radio source in the constellation of Perseus, which is associated with Seyfert galaxy, NGC 1275.

Perseus arm *Astronomy.* one of several spiral arms of the Milky Way Galaxy located in the direction of Perseus at a distance of about 7000 light-years.

Perseus cluster *Astronomy.* a diffuse, irregular cluster of approximately 500 galaxies centered on Seyfert galaxy, NGC 1275.

Perseus OB1 *Astronomy.* a double stellar association which is visible to the unaided eye as a patchy, fuzzy light. Also, **H Persei, X Persei.**

Perseus OB2 *Astronomy.* an association of O and B type stars in a dense molecular cloud.

Perseus X-1 *Astronomy.* the strongest known X-ray source outside of the Milky Way Galaxy; it is centered on NGC 1275.

perseveration *Neurology.* the persistent repetition of a response after cessation of the causative stimuli; for example, the repetition of a correct answer to one question as the answer to succeeding questions.

Pershing *Ordnance.* a mobile, surface-to-surface, inertially guided missile that is designed to support ground forces by attacking long-range ground targets, having nuclear warhead capability and using solid propellant fuel. Pershing IA has a range of 450 miles; officially designated **MGM-31A.** Pershing II has a range of 1100 miles.

Persian Gulf *Geography.* an inlet of the Arabian Sea between southwestern Iran and Arabia.

Persian melon *Botany.* **1.** a variety of muskmelon, *Cucumus melo*, in the family Cucurbitaceae, having a fruit with dark green netted skin and firm, thick, orange flesh. **2.** the fruit of this plant.

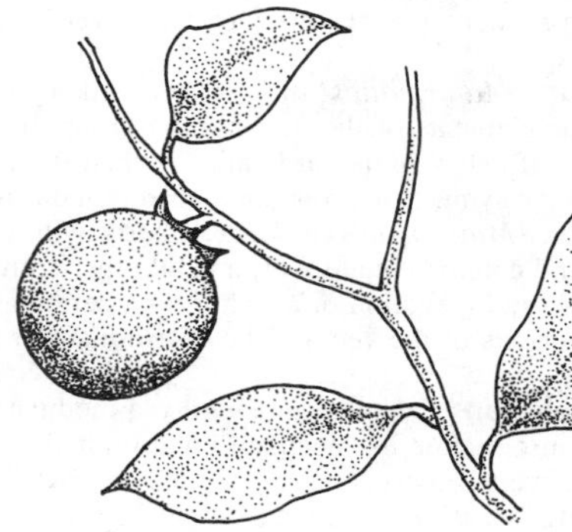

persimmon

persimmon *Botany.* **1.** any of the trees of the genus *Diospyros virginiana*, bearing astringent, plumlike fruit that is sweet and edible when ripe; found in North America. **2.** any of the trees of the genus *D. kaki*, bearing soft, orange fruit; found in China and Japan. **3.** the fruit of any of these trees.

persistence a condition of continued existence or occurrence; specific uses include: *Meteorology.* the ratio of the magnitude of the mean wind vector to the average windspeed without regard to direction; used to characterize the long-term nature of the wind at a given location. *Virology.* the ability of a virus to remain in a host for an extended period of time, whether or not it replicates to produce infectious progeny virions. *Electronics.* the length of time it takes for the phosphors on a cathode screen to lose their glow once the electron beam has swept past.

persistence forecast *Meteorology.* a weather forecast predicting that the future weather condition will be the same as the present one; used as a basis of comparison for measuring the skill of forecasts derived by other methods.

persistence hunting *Anthropology.* a method of hunting game by pursuing the prey, but not attempting to attack, until the prey finally stops or falls in exhaustion and can be killed.

persistence of vision *Physiology.* a visual image continuing briefly after visual stimulation; when looking at a neutral background, it may be either a positive image, retaining light and dark areas as originally perceived, or a negative image, in which light and dark are reversed.

persistent *Botany.* of or relating to a leaf that withers but remains attached to the stalk during the winter. *Zoology.* having continuity without change in function or structure, as the gills of a fish.

persistent current *Physics.* an electric current magnetically induced in a superconductor that continues without any reduction.

persistent-image device *Electronics.* an optoelectronic amplifier that is able to sustain an image for an exact length of time by incorporating the properties of light-sensitive and light-emitting devices.

persistent infection *Virology.* a viral infection in which the virus does not cause lysis, but instead reaches a state of equilibrium with the host.

persistent slip bands *Materials Science.* the characteristic surface deformations that are an early manifestation of fatigue in metal which has been plastically deformed, causing an increase in photoelectric emission; visible under ultraviolet illumination.

persistent train *Astronomy.* a faintly glowing streak from a meteor that persists for more than one second.

persistent transmission *Virology.* a process involving a vector and a virus in which the vector assumes a long acquisition feed, followed often by a latent period of nontransmission and always by an extended transmission period.

persistent vegetative state *Neurology.* a condition of profound nonresponsiveness in the wakeful state caused by brain damage; characterized by a nonfunctioning cerebral cortex, the absence of any discernible response to external stimuli, and the inability to communicate.

persistent virus infection *Virology.* a process of infection similar to circulative transmission in which the virus attains equilibrium with the host.

persistent war gas *Ordnance.* a war gas that is effective at the point of dispersion for more than ten minutes.

persistron *Electronics.* a panel that combines the properties of electroluminescence and photoconductivity to generate a persistent image when a pulsed signal is received.

personal communications network *Telecommunications.* a proposed type of wireless network in which phone numbers are assigned to persons rather than to telephones.

personal computer *Computer Technology.* a general term for a relatively inexpensive and compact computer designed for individual home or business use, typically for such applications as word processing, database reference, and financial calculations.

personal computer

personal construct theory *Psychology.* George Kelly's theory that personality is determined by how an individual construes or interprets his or her environment.

personal equation *Science.* **1.** a variation that is due to the idiosyncracies of an individual. **2.** an allowance that is made for this variation.

personal error *Navigation.* a known, consistent error in observations that can be compensated for in reducing sights.

personal identification code *Computer Programming.* a unique six-digit number encoded on a magnetic strip on a plastic card that identifies a user and allows access to a special-purpose computer system.

personalistic probability see SUBJECTIVE PROBABILITY.

personality *Psychology.* the specific pattern of psychological traits and attributes that characterize a particular individual; generally considered to be determined by a variety of forces, both hereditary and societal, and referring to both the apparent and inferable aspects of the individual.

personality development *Psychology.* the development over time of an individual's characteristic traits and behavior patterns.

personality disorder *Psychology.* a disorder marked by pathological trends in personality structure, most often manifested by abnormal actions or behaviors.

personality test *Psychology.* any device or technique used to assess personality or measure personality characteristics and traits.

personality theory *Psychology.* any of various theoretical approaches that attempt to offer an explanation for human behavior.

personal motivation see SECONDARY MOTIVATION.

personal rapid transit *Transportation Engineering.* a transport system in which small, self-propelled vehicles (usually seating two to six passengers with exclusive use of the vehicle) operate under automatic control on an exclusive guideway.

personal space *Psychology.* the physical area immediately around a person, in which the uninvited presence of another or others is regarded as an intrusion.

personal trait *Psychology.* a personality trait that is specific to an individual and can be used to characterize this individual, as distinguished from a common trait that characterizes a group. Also, **personal disposition.**

person-centered therapy see CLIENT-CENTERED THERAPY.

personnel *Military Science.* individuals who are required, in either a military or civilian capacity, to accomplish an assigned mission.

personnel carrier *Ordnance.* a motor vehicle used to transport troops and their equipment.

personnel monitoring *Nucleonics.* the process of monitoring an individual's exposure to radioactive contamination by means of a dosimeter.

personnel reaction time *Military Science.* the time required by personnel to take prescribed protective measures after receiving a nuclear-strike warning.

personology *Psychology.* **1.** the study of personality. **2.** the belief that all behavior should be studied in terms of its relationship with personality.

perspective *Graphic Arts.* a method of representing three-dimensional objects on a two-dimensional plane or curved surface.

perspective axis see AXIS OF HOMOLOGY.

perspective chart *Cartography.* any chart based on a perspective map projection.

perspective grid *Photogrammetry.* a network of lines drawn or superimposed on a single photograph that represents the perspective of a larger system of lines on the ground or datum plane.

perspective map projection *Cartography.* a map projection developed by projecting straight lines from a point of projection through the points on the sphere to be mapped onto a plane tangent to the sphere; the point of projection lies at an arbitrary point on the diameter of the sphere that passes through the point of tangency between the sphere and the plane of projection.

perspective projection *Navigation.* a system of creating a map by projecting elements on the surface of the earth from a single point (including infinity).

perspectivity *Mathematics.* in projective geometry, a one-to-one correspondence between points on two distinct lines having the property that the lines joining corresponding points are concurrent (i.e., intersect in a single point, which may be the point at infinity). The composition of a finite sequence of perspectivities is called a **projectivity**.

perspiration *Physiology.* the release of fluid by the sweat glands of the skin.

perspire *Physiology.* to secrete a fluid from the sweat glands of the skin, usually as the result of extreme heat or strenuous exertion.

Perssoniellaceae *Botany.* a monospecific family of liverworts belonging to the order Jungermanniales; plants are large, dorsiventral, flattened, and sparsely branched, and grow close to the ground, with purplish, unicellular, scattered rhizoids; found only in the mountains of southwestern New Caledonia.

Pers sunshine recorder *Engineering.* a sunshine recorder for which the sun furnishes the time scale.

persuader *Electronics.* an element in a storage tube that drives secondary electron emissions toward the dynodes in an electron multiplier.

persulfuric acid see CARO'S ACID.

PERT *Industrial Engineering.* a management technique developed in 1958 by the U.S. Navy for calculating the most likely completion time for a project, using pessimistic, optimistic, and most-likely estimated times for the completion of each segment of the project. (An acronym for Program Evaluation and Review Technique.)

pert. pertaining.

perthite *Geology.* an intergrowth of potassium and sodium feldspar in which the potassium-rich phase serves as the host for the sodium-rich phase.

perthitic *Geology.* relating to a texture produced by the intergrowth of sodium and potassium feldspars in which the sodium feldspar occurs as small strings, blebs, or films.

perthosite *Petrology.* a leucocratic syenite consisting almost entirely of perthite, with less than 3% mafic minerals.

pertinency factor *Computer Programming.* in information retrieval, the ratio of the number of relevant documents retrieved to the total number of documents retrieved.

pertitanic acid see TITANIUM TRIOXIDE.

perturbation [per´tər bā´shən] *Physics.* an influence on a system that only disturbs the system slightly. *Astronomy.* a deviation in the motion of any celestial object as a result of the gravitational pull of other bodies. *Medicine.* the process of blowing air into the uterine tubes to force out secretions or other substances, thereby rendering the tubes unobstructed. *Mathematics.* given a function f of one or more independent variables, a perturbation is another function of one or more of the variables that produces small changes in the values of f.

perturbation equation *Physics.* an equation, typically a differential equation, that governs the behavior of a perturbation.

perturbation motion *Physics.* in a mechanical system, the particular motion of a perturbation through the system, not to be confused with the motion of the system with a perturbation superimposed on it.

perturbation quantity *Physics.* any quantity in a perturbed system with which the perturbation is associated.

perturbation theory *Mathematics.* the study of solutions to partial and ordinary differential equations, based on the assumption that perturbations in the given conditions of a problem cause only small changes in the solution. Applications include finding approximate solutions to classical problems from physics and quantum mechanics.

perturbed field *Electrical Engineering.* a field is regarded at a point as perturbed if the magnitude or direction varies when an object is introduced into the vicinity.

Pertusaria *Botany.* a large genus of crustose lichens of the family Partusariaceae, characterized by fruiting bodies in knoblike structures; one of the sources of litmus.

Pertusariales *Mycology.* in some classifications, an order of fungi belonging to the subdivision Ascomycotina that includes certain lichens found on the ground and on trees and rocks.

pertussis *Medicine.* an acute, highly communicable respiratory disease that is characterized in its typical form by paroxysms of coughing, expulsion of clear sticky mucus, and often by vomiting. Also, WHOOPING COUGH.

pertussis toxin *Toxicology.* a toxin produced by the bacterium *Bordetella pertussis* that causes whooping cough.

Peru balsam *Materials.* a dark, aromatic, molasseslike liquid with a bitter taste, obtained from the Central American leguminous tree, *Myroxylon pereirae;* used in medicine, perfumes, and in the manufacture of chocolate. Also, BLACK BALSAM, CHINA OIL, INDIAN BALSAM.

Peru Current *Oceanography.* a cold, broad current that flows northward along the west coast of South America, from about 40° to 20°S. Also, HUMBOLDT CURRENT.

Perutz, Max F. born 1914, Austrian-born English molecular biologist; shared the Nobel Prize for the X-ray analysis of hemoglobin.

peruviol see NEROLIDOL.

perveance *Electronics.* the space-charge-limited cathode current of a diode divided by 3/2 the power of the anode voltage in an electron tube.

perversion *Psychology.* a change in behavior to what is abnormal or unnatural.

pervious *Science.* permitting a passage through; permeable.

pes- a prefix meaning foot.

pessary *Medicine.* a cup-shaped instrument placed in the vagina and used either to support the uterus or as a contraceptive device. *Pharmacology.* any medication in the form of a suppository to be inserted in the vagina.

pessulus *Vertebrate Zoology.* a bony or cartilaginous bar in the lower end of a bird's windpipe that crosses the windpipe dorsoventrally at the division into bronchi.

pest *Biology.* a subjective term describing any organism that is regarded as harmful, irritating, or offensive to humans, either directly or indirectly through its effect on animals and plants; the term is typically applied to rats and other rodents, to insects that transmit disease or destroy crops, and to pathogenic fungi and bacteria. (From the Latin word for "plague.")

pest control *Ecology.* any of various techniques to manage pest populations through biological or chemical agents.

peste des petits ruminants *Veterinary Medicine.* an acute or subacute viral disease of goats and sheep that is caused by *Morbillivirus*; characterized by fever, inappetence, congestion, and lesions of the mucous membranes of the mouth and gastrointestinal tract. Also, PSEUDORINDERPEST.

pesticide *Toxicology.* any poison, organic or inorganic, that is used to destroy pests of any sort, including insecticides, fungicides, rodenticides, herbicides.

pestilence *Medicine.* any infectious epidemic or virulent contagious disease.

pestilent *Medicine.* tending to produce pestilence.

pestis minor *Pathology.* an ambulatory form of the plague.

Pestivirus *Virology.* a genus of the Togaviridae family, often transmitted transplacentally, resulting in mucosal disease of cattle, swine fever, and diseases of sheep.

pestle *Science.* a club-shaped instrument that is used to pound or crush into powder substances that are placed in a mortar.

PET positron emission tomography.

pet. petroleum.

peta- a prefix meaning one-quadrillionth.

petal *Botany.* one of the leaf-shaped parts of the corolla of a flower, usually brightly colored and conspicuous.

Petalichtyida *Paleontology.* a small extinct order of flattened placoderms, characterized by two shields of armor and scimitarlike lateral spines; similar in some ways to arthrodires but probably benthonic; extant in the Devonian.

petalite *Mineralogy.* $LiAlSi_4O_{10}$, a colorless, white, gray, or pink monoclinic mineral, usually occurring in cleavable masses, and having a specific gravity of 2.3 to 2.5 and a hardness of 6 to 6.5 on the Mohs scale; found in granite pegmatites.

Petalodontida *Paleontology.* an order of primitive fishes in the subclass Chondrichthyes, characterized by flat-crowned teeth in a pavement dentition (embedded in the palate), probably used to crush mollusks; extant in the Carboniferous and Permian.

petaloid *Botany.* having the form of or otherwise resembling petals, as characteristic of certain bracts or sepals.

Petaluridae *Invertebrate Zoology.* a family of dragonflies in the suborder Anisoptera.

petasma *Invertebrate Zoology.* in decapod crustaceans, a membranous modification of the first abdominal appendage.

petechia *Medicine.* a nonraised, perfectly round, purplish-red pinpoint spot, usually caused by submucous or intradermal hemorrhage.

Peter of Abano 1250–1318, Italian scholar; translated classical scientific works; an innovative physiological and astronomical theorist.

Petersen graph *Mathematics.* in graph theory, a cubic graph on ten vertices; i.e., a graph, each of whose ten vertices has exactly three edges incident on it.

Petersen's theorem *Mathematics.* a theorem postulating that every cubic graph with no cut edges has a perfect matching.

petiole [pet´ē ōl] *Botany.* the usually slender stalk attaching a leaf to the stem. *Entomology.* **1.** the thin stalk in the Hymenoptera that attaches the thorax to the abdomen. **2.** any stem- or stalklike structure or process.

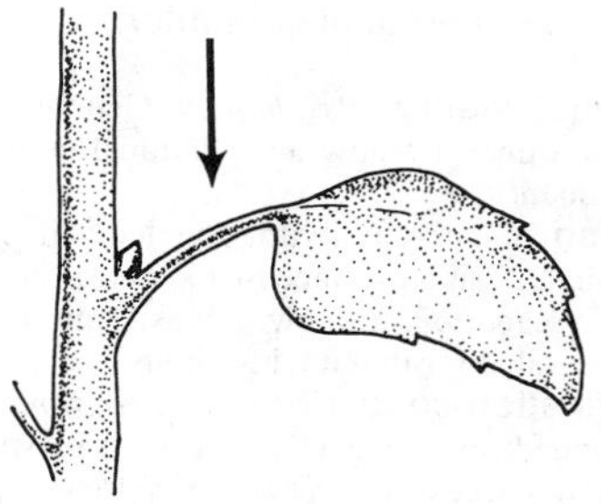

petiole

petiolule *Botany.* the small or partial stalk of a leaflet on a compound leaf.

Petit, Alexis [pə tē´] 1791–1820, French physical chemist; with Dulong, formulated the law of specific heats.

petite mutant *Genetics.* a mutation that occurs in certain strains of yeast, characterized by defective mitochondrial function.

petitgrain oil *Materials.* an essential oil used in perfumes and flavoring.

petit mal [pə tē´ mäl´] *Medicine.* an epileptic seizure in which sudden transient lapses of consciousness occur, often many times in succession; during the seizure the patient loses facial expression and ceases all voluntary motor activity for a brief period. Also, ABSENCE SEIZURE.

petit mal epilepsy *Medicine.* a form of epilepsy characterized by sudden transient lapses of the conscious state. Symptoms often include smacking of the lips, a blank stare, blinking of the eyes, and sometimes twitching of facial muscles or hands. Also, ABSENCE ATTACK SEIZURE.

petrel *Vertebrate Zoology.* any of several small, black-and-white seabirds of the families Procellariidae, Hydrobatidae, and Pelecanoididae; noted for their habit of flying close above the surface of the water to pick up food.

petri dish [pē´trē] *Microbiology.* a circular, shallow glass or plastic dish with a larger dish fitting over it to serve as a lid; used for tissue cultures and for growing microorganisms on solid culture. (Named for Julius Richard *Petri,* 1852–1921, German bacteriologist.) Also, CULTURE DISH.

Petrie, Sir Flinders 1853–1942, English archaeologist; studied Stonehenge; a founder of Egyptology, excavated Memphis and Thebes.

petrifaction *Geology.* a process of fossilization by which dissolved inorganic matter replaces the original organic matter, producing a stony substance. Also, **petrification.**

petrified wood *Geology.* a wood that has been permineralized by silica, in such a manner that the original shape and structure of the wood are preserved. Also, SILICIFIED WOOD, WOODSTONE.

Petriidae *Invertebrate Zoology.* a small family of beetles, coleopteran insects in the superfamily Tenebrionoidea.

Petrocelidaceae *Botany.* a family of red algae belonging to the order Gigartinales, characterized by prostrate, discoid thalli and predominantly intercalary tetrasporangia.

petrochemical *Organic Chemistry.* any of various organic compounds that are derived directly or indirectly from petroleum or natural gas. Also, PETROLEUM CHEMICAL.

petrochemistry *Organic Chemistry.* the chemistry and reactions of materials derived from petroleum or natural gas.

Petroforge machine *Materials Science.* a machine in which high velocities supply the power for deformation processes; used in high-energy-rate forming, chiefly in closed-die forging operations and for powder compaction.

petrogenesis *Petrology.* the branch of petrology dealing with the origins of rocks, especially igneous rocks. Also, **petrogeny.**

petrogenic grid *Petrology.* a diagram whose coordinates are parameters of the rock-forming environment, such as temperature and pressure, and on which equilibrium curves are plotted to indicate the limits of the stability fields of certain minerals and mineral assemblages.

petroglyph *Archaeology.* a design carved or chipped into a rock surface. Also, **petrogram.**

petrographer *Geology.* a person who studies petrography.

petrographic facies *Geology.* facies distinguished mainly on the basis of composition and appearance, without consideration of form, boundaries, or relationship to one another. Also, **petrofacies.**

petrographic microscope *Optics.* a polarizing microscope used primarily by mineralogists and crystallographers to analyze and identify unknown materials.

petrographic period *Geology.* the interval of geologic time over which a rock association extends.

petrographic province *Geology.* a broad region in which the igneous rocks have similar petrographic characteristics, indicating that they were formed from a common magma or in the same period of igneous activity. Also, MAGMA PROVINCE, IGNEOUS PROVINCE, COMAGMATIC REGION.

petrography *Geology.* any systematic classification of igneous and metamorphic rocks on the basis of mineralogical and textural relationships, especially by means of microscopic examination.

petrol [pet´rəl] a British term for gasoline.

petrol. petrology; petroleum.

Petrolacosaurus *Paleontology.* the earliest known diapsid, a genus of North American araeoscelid reptiles in the family Petrolacosauridae; extant in the Carboniferous.

petrolatum [pet´rə lat´əm] *Materials.* a translucent, oily, semisolid, amorphous, yellowish or whitish mass obtained from petroleum and used as a lubricant, as a rust preventive, and in cosmetics and medicine.

petroleum [pə trō´ lē əm] *Geology.* a naturally occurring liquid mixture of complex hydrocarbon compounds that yields combustible fuels, petrochemicals, and lubricants upon distillation; usually found in deposits beneath the earth's surface and thought to have originated from plant and animal remains of the geologic past. It is by far the most widely used fuel source in the industrialized world and is also used in many industrial products, such as plastics, synthetic fibers, and drugs.

petroleum benzin *Materials.* a special grade of ligroin.

petroleum chemical see PETROCHEMICAL.

petroleum coke see COKE.

petroleum engineering *Engineering.* a branch of engineering that includes the recovery, production, distribution, and storage of oil, gas, and liquefiable hydrocarbons, as well as exploration for these products. Thus, **petroleum engineer.**

Petroleum Engineering

Petroleum Engineering is a scientific discipline devoted to the development and application of technology related to the economic production of energy through wellbores. In 1901, Anthony Lucas, an Austrian mining engineer, drilled the discovery well for the Spindletop field in east Texas, ushering in the modern oil industry. Lucas was probably the first petroleum engineer, since he realized the need for understanding petroleum reservoirs and methods for producing them, although at that time, most engineers were concerned only with improving drilling methods and keeping wells flowing at high rates. The first formal degree was conferred by the University of Pittsburgh in 1915.

During the 1920s, the use of geophysics led to the discovery of a number of large fields which were drilled on extremely close spacing and produced at maximum rates because oil production was still governed by the "law of capture"—an 1875 judicial opinion that oil and gas were like wild game and belonged to the person who reduced them to possession.

By the early 1930s, this approach caused massive overproduction and enormous waste which, coupled with limited demand during the Depression, led to very low oil prices and the virtual collapse of the U.S. petroleum industry. Faced with economic chaos, the Texas Railroad Commission passed regulations limiting oil production to market demand, deeming initial oil-in-place to be an asset not subject to "capture." Petroleum engineering then became the medium through which sound conservation policies were devised and implemented.

During the 1930s and 1940s, publications on reservoir mechanics and phase behavior laid a sound technical foundation for reservoir engineering. Later, integrated development of production systems for giant overseas fields, deep drilling, offshore and arctic operations, advances in the physics/chemistry of well logging, fast computers for reservoir simulation, improved recovery methods, and use of seismic data resulted in the explosive expansion of the technologies employed by petroleum engineers.

This broadening of interests ended in specialization, i.e., drilling, production, or reservoir engineering, as well as formation evaluation, reservoir management, and petroleum economics. This expansion will continue as new concepts and tools become available and are used by the petroleum engineer to solve upcoming problems associated with environment and safety, operations in frontier areas, enhanced recovery from existing fields, cost cutting, and improved productivity.

J. E. Warren
Director
Frontier Resources International, Inc.

petroleum ether *Materials.* a low-boiling, flammable hydrocarbon mixture produced by the fractional distillation of petroleum; used as a solvent.

petroleum geology *Geology.* the branch of economic geology that is concerned with the origin, migration, accumulation, and occurrence of hydrocarbon fuels, and exploration for these fuels.

petroleum isomerization process *Chemical Engineering.* a petroleum-refinery process that converts straight chain compounds into branched isomers with the use of a catalyst, in order to obtain higher octane gasolines and other products.

petroleum microbiology *Microbiology.* a branch of microbiology that deals with the relationship between microbes and the formation and industrial uses of petroleum.

petroleum processing *Chemical Engineering.* the process of separating and purifying crude oil. Also, **petroleum refining.**

petroleum trap *Geology.* a stable underground geological or physical formation, usually consisting of sand or porous rock surrounded by impervious rock or clay, in which liquid or gaseous hydrocarbons are trapped.

petroleum wax *Materials.* a solid hydrocarbon derived from petroleum and produced by solvent dewaxing.

petroliferous *Geology.* of or relating to any geologic feature or formation that contains petroleum.

petrologen see KEROGEN.

petrologist [pə träl′ ə jist] *Geology.* a person who studies petrology or is an expert in petrology.

petrology [pə träl′ ə jē] *Geology.* the scientific study of rocks.

Petrology

A branch of geology or of what is now more broadly referred to as the earth sciences, petrology encompasses a body of information that extends far beyond its simple definition. Curiosity about rocks, and thought respecting their origins, can be traced back to Theophrastus (c. 315 BC) and Pliny the Elder (c. 75 AD); and "scientific" studies were reported as early as the mid-1700s. But detailed work on rock structures and compositions, work that went beyond mere description, had to await tools that modern physics and chemistry brought to hand; and what is now called "petrology" includes topics in crystallography, mineralogy, and paleontology, and strays as far into inorganic and physical chemistry as into physics.

This diversity of subject matter arises from the fact that rocks are complex mixtures of minerals, and that their structures and mineral compositions are influenced by their origins and thermal histories. An initial focus on igneous rocks was thus soon broadened to cover metamorphosed schists and gneisses that derived from them. And after that, it was inevitable that sedimentary rocks, which told so much about geological history, came to be included; that rock compositions began to be quantitatively expressed by making use of the optical properties of their consitutent minerals; and that the mineral constituents themselves began to be defined in terms of crystallographic habit and chemical composition.

But in the past 50 years or so, the study of inorganic rocks has also prompted development of an organic counterpart, with the principles of petrographic analysis being applied to coals and cokes. Such analyses reflect important aspects of coal quality and, in conjunction with paleobotanical studies, offer unique information about ancient earth environments.

Norbert Berkowitz
Emeritus Professor of Fuel Science
University of Alberta

petromictic *Geology.* describing a sediment or deposit composed of an assortment of metastable rock fragments. Also, **petromict.**

petromorph *Geology.* in a cave, a secondary mineral deposit or cave formation that is exposed at the surface by erosion of the surrounding limestone.

Petromyzonidae *Vertebrate Zoology.* the lampreys, a family of freshwater eel-like fishes of the class Agnatha, characterized by a sucking mouth with rasping teeth, and lacking a jaw, scales, and paired fins; found in cooler waters of the Northern and Southern Hemispheres.

Petromyzontiformes *Vertebrate Zoology.* the lampreys, an order of eel-like fishes belonging to the class Agnatha and containing the single family Petromyzonidae.

petrophysics *Geology.* the branch of petroleum geology concerned with the study of the physical properties of reservoir rocks.

petrosal nerve *Anatomy.* either of two nerves serving the face and forming the sphenopalatine ganglion: the motor root derived from the facial nerve through the superficial petrosal nerve and the sympathetic root from the carotid plexus through the deep petrosal nerve, the two joining to form the Vidian nerve.

petrosal process *Anatomy.* a projection of the sphenoid bone of the skull.

Petrosaviaceae *Botany.* a family of small mycotrophic, nonchlorophyll-containing herbs belonging to the order Triuridales, characterized by perfect flowers, reduced leaves, and numerous seeds; native to Asian forests.

petticoat insulator *Electricity.* a bell-shaped insulator having a hollow, outward flaring in its lower part, which increases the length of a surface-leakage path and keeps part of it dry at all times; used for high-voltage insulation.

petty officer *Military Science.* in the U.S. Navy or Coast Guard, a noncommissioned officer ranking below a chief petty officer; there are three grades: first class, second class, and third class.

petzite *Mineralogy.* Ag_3AuTe_2, a steel-gray to iron-black metallic cubic mineral having a specific gravity of 8.7 to 9.4 and a hardness of 2.5 to 3 on the Mohs scale; found with other tellurides in vein deposits.

Petzval condition *Optics.* a condition that must be met by at least two lenses in an optical system in order to eliminate curvature of field.

Petzval curvature *Optics.* **1.** the curvature of the image surface of an optical system corrected for spherical aberration, coma, and astigmatism. **2.** the curvature of the surface of an optical system that matches the total of all the optical surfaces in the system.

Petzval lens *Optics.* a lens system that has two widely separated pairs of achromatic lenses; used for portrait photography, motion picture projection, and other applications in which loss of clarity at the edges of an image are acceptable.

Petzval surface *Optics.* the paraboloidal surface on which a sharp image formed by a doublet lens free from astigmatism may be found.

peuroseite *Mineralogy.* $(Ni,Cu,Pb)Se_2$, a gray mineral consisting of a selenide of nickel, copper, and lead that occurs in columnar masses.

pewee *Vertebrate Zoology.* any of several small flycatchers of the genus *Contopus*, including the wood pewee and the phoebe; characterized by an olive-colored or gray back and a call that sounds somewhat like its name; found in the New World.

pewee

pewter *Metallurgy.* any of several alloys based on tin and lead, now used mainly for decorative objects; widely used for houseware in colonial America.
pex- or **pexo-** a prefix meaning "surgical fixation."
pexicyst *Cell Biology.* a type of extrusion on certain protozoans that adheres to the prey upon its discharge from the cell.
pexis *Surgery.* the fixation of organs and structures, either by natural means or by surgical intervention.
Peyer's patch *Histology.* aggregated nodules of lymphoid tissue located in the small intestine, especially the ileum.
peyote [pā ō´tē] *Pharmacology.* another name for the mescal cactus or hallucinogenic mescal buttons. See MESCAL.
Peyronie's disease *Medicine.* a disease of unknown origin that results in induration of the corpora cavernosa of the penis, producing a fibrous chordee. (Named after Francois de *Peyronie,* 1678–1747, French surgeon.)
Peyssonneliaceae *Botany.* a family of red algae belonging to the order Cryptonemiales, characterized by compact crustose thalli. Formerly, SQUAMARIACEAE.
Peziza *Mycology.* a genus of fungi belonging to the order Pezizales, characterized by light-brown, cup-shaped fruiting bodies.
Pezizaceae *Mycology.* a family of fungi belonging to the order Pezizales which occurs on soil or decaying plant debris.
Pezizales *Mycology.* an order of fungi belonging to the subdivision Ascomycotina, characterized by a cup-shaped athecium; it obtains nutrients from dead organic matter. Also, CUP FUNGI.
pezograph see REGMAGLYPT.
PF or **P.F.** power factor. Also, **pf, p.f.**
pF or **pf** picofarad.
Pfaff, Johann [pfäf; fäf] 1765–1826, German mathematician; developed Pfaffian differential equations; worked on the combinatorial theory of series.
Pfaffian differential equation *Mathematics.* a first-order linear differential equation of the form

$$P(x, y, z)dx + Q(x, y, z)dy + R(x, y, z)dz = 0,$$

where the functions P, Q, and R are continuously differentiable. More generally, exterior differential systems made up of 1-forms are called **Pfaff systems.**
Pfannenstiel's incision *Surgery.* a curved, transverse abdominal incision, just above the pubic symphysis, made through the skin down to the external sheath of the rectus muscles, which are separated from each other in the midline, after which the peritoneum is opened vertically; used especially during pelvic operations. (Named for Hermann *Pfannenstiel,* 1862–1909, German gynecologist.)
Pfeiffer phenomenon *Immunology.* the sudden rupture of the bacterium *Vibrio cholerae* that occurs when it is incubated with a specific antibody and complement.
PFM pulse-frequency modulation.
PFTE polytetrafluoroethylene.
pfu plaque-forming unit.
Pfund series *Spectroscopy.* a set of spectral lines in the far-infrared region of the hydrogen spectrum.
PG *Pharmacopoeia Germanica*; prostaglandin.
pg or **pg.** page; picogram.
PGA pteroylglutamic acid (folic acid).
p-group *Mathematics.* a group in which every element has order equal to a power of some fixed prime p. If a p-group H is a subgroup of some group G, then H is said to be a p-subgroup of G. By theorem, a finite group G is a p-group if and only if the order $|G|$ of G is a power of p.
PH Pharmacopeia.
PH or **P.H.** public health; past history.
Ph or **ph** phenol.
pH *Chemistry.* a symbol for the logarithm of the reciprocal of the hydrogen-ion concentration of an aqueous solution, used to express its acidity or alkalinity. At 25°C (77°F), a neutral solution, such as pure water, has a pH of 7; a pH under 7 indicates that the solution is acidic, and a pH over 7 indicates that the solution is alkaline. (An abbreviation for potential of Hydrogen.)
ph phase; phot.
PH-30 *Biochemistry.* a protein, found on the surface of sperm, that controls the fusion of sperm with an egg, which is the crucial event in fertilization.
PHA phytohemagglutinin.
PHA or **P.H.A.** Public Health Agency.
phac- or **phaco-** a prefix meaning: **1.** lens. **2.** a mole or freckle.
phacellite see KALIOPHILITE.
Phacelocarpaceae *Botany.* a monogeneric family of red algae of the order Gigartinales, characterized by flat or terete, profusely branched, uniaxial thalli, similar gametangial and sporangial forms, and superficially formed spermatangia.
Phacidiaceae *Mycology.* a family of fungi belonging to the order Phacidiales that occur on the bark and leaves of higher plants and are sometime pathogenic, causing such diseases as blight and canker.
Phacidiales *Mycology.* an order of fungi belonging to the class Discomycetes, composed of plant parasites and characterized by a cup-shaped spore case, formed within vegetative hyphae, that ruptures to expose the hymenium at maturity.
phacoidal structure see FLASER STRUCTURE.
phacolite *Mineralogy.* a type of chabazite with colorless, lenticular crystals caused by twinning.
phacolith *Geology.* a minor, concordant, usually granitic, lens-shaped body with a concave base that has been intruded into folded sedimentary strata.
phacopid *Paleontology.* **1.** of or relating to trilobites of the Middle Paleozoic order Phacopida, which were widespread and abundant during the Devonian but died out toward the end of the period. **2.** an organism of this type.
Phacops *Paleontology.* a genus of large trilobites in the order Phacopida; extant in the Silurian and Devonian.
Phacotaceae *Botany.* a family of solitary flagellates of the order Volvocales, characterized by a noncellulosic wall-like structure; restricted to nonmarine habitats and usually planktonic.
Phacus *Invertebrate Zoology.* a genus of green protozoans with a single flagellum and an asymmetrical, flattened body that cannot change shape to any appreciable extent.

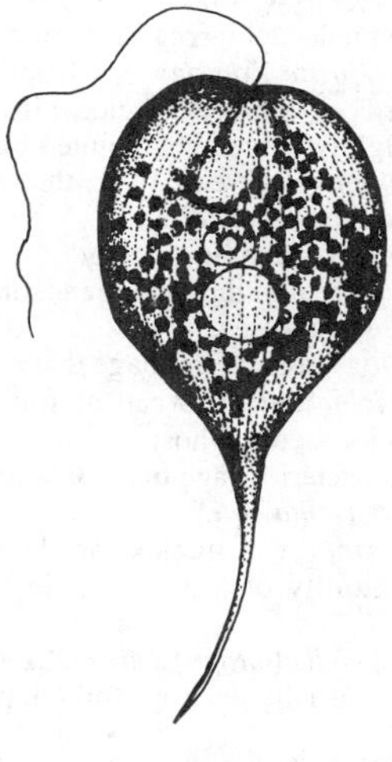

Phacus

Phaedra complex [fē´drə] *Psychology.* the sexual love of a mother for her son. (From the Greek myth of *Phaedra,* who fell in love with her stepson Hippolytus.)
Phaenocephalidae *Invertebrate Zoology.* a family of Japanese beetles, coleopteran insects in the superfamily Cucujoidea.
phaeo- a prefix meaning "dusky." Also, PHEO-.
Phaeocoleosporae *Mycology.* a dark spore found in certain imperfect fungi.
Phaeocystaceae *Botany.* a family of marine algae of the order Prymnesiales, characterized by a dominant phase of aflagellate cells embedded in mucilage, sometimes in such large quantities as to give an oily appearance to the sea.
Phaeodermatiaceae *Botany.* a monogeneric family belonging to the order Phaeothamniales, characterized by a parenchymatous thallus and reproduction by means of peculiar uniflagellate swarmers.
Phaeodictyosporae *Mycology.* a dark spore found in certain imperfect fungi that has both vertical and horizontal cell walls or septa, resembling a brick wall.

Phaeodidymae *Mycology.* a dark, two-celled spore found in certain imperfect fungi.

Phaeodorina see TRIPYLINA.

Phaeohelicosporae *Mycology.* a dark, spiral-shaped spore found in certain imperfect fungi.

phaeohyphomycosis *Pathology.* a usually opportunistic infection with the *Dematiacious* fungi.

Phaeophragmiae *Mycology.* a dark, multicellular spore found in certain imperfect fungi.

Phaeophyceae *Botany.* the brown algae, a large class in the division Chromophycota (or in some systems, the division Phaeophyta), whose members are eukaryotic, multicellular, autotrophic, marine, and usually benthic organisms, ranging from microscopic filamentous tufts to the giant kelps.

Phaeophyta *Botany.* in some systems of classification, a division of the kingdom Plantae, containing a single class, Phaeophyceae, the brown algae.

Phaeoplacaceae *Botany.* a monogeneric family of algae in the order Phaeothamniales, characterized by a cushionlike thallus that grows epiphytically on water plants.

Phaeosphaeriaceae *Mycology.* a family of fungi that belong to the order Pleosporales, composed of species that live off decaying organic matter or are parasites of plants or other fungi.

Phaeosporeae *Mycology.* a dark, unicellular spore found in certain imperfect fungi.

Phaeostaurosporae *Mycology.* a dark, fork- or star-shaped spore found in certain imperfect fungi.

Phaeothamniaceae *Botany.* a monogeneric family of freshwater algae in the order Phaeothamniales, characterized by cells that produce four swarmers each.

Phaeothamniales *Botany.* an order of algae belonging to the class Chrysophyceae and containing species at the highest level of organization in the class, characterized by a filamentous or parenchymatous thallus, with cell division that is a modified autospore formation.

Phaeotrichiaceae *Mycology.* a family of fungi that belong to the order Pleosporales and live on decaying vegetable material or dung.

Phaethontidae *Vertebrate Zoology.* the tropic birds, a monogeneric family of marine birds of the order Pelecaniformes, characterized by long tailfeathers, a slightly curved and pointed bill, and no gular sac.

Phaffia *Mycology.* a genus of yeast fungi that belong to the class Hyphomycetes and reproduce by budding.

phag- or **phago-** a prefix meaning "eating."

phage [fāj] *Virology.* any virus that has been isolated from a prokaryote. Most phages are bacterial viruses.

phage lambda *Virology.* a bacteriophage that contains double-stranded DNA, infects *Escherichia coli,* and can be found in either the lytic or the lysogenic form in its bacterial host.

phage T4 *Virology.* a bacteriophage that contains linear double-stranded DNA and infects *Escherichia coli.*

phage typing *Bacteriology.* a method of classifying bacterial strains based on the susceptibility of a given strain to a range of bacteriophages.

phagocyte [fag´ə sīt´] *Cell Biology.* a specialized cell, such as a macrophage, that ingests and usually destroys foreign particulate matter or microorganisms.

phagocytic [fag´ə sit´ik] *Cell Biology.* of or relating to phagocytes or the ingestive action of phagocytes.

phagocytin *Biochemistry.* a bactericidal substance found in neutrophilic leukocytes.

phagocytosis *Cell Biology.* the process by which certain phagocytes can ingest extracellular particles by engulfing them, functioning either in mammals as a defense mechanism against infection by microorganisms or in many protozoans as a means of taking up food particles.

phagolysosome *Cell Biology.* a large vesicle found in phagocytes, formed by the fusion of a phagosome and a lysosome, in which ingested material is degraded.

phagomania *Medicine.* an abnormal craving for food; obsessive preoccupation with eating.

phagomaniac *Medicine.* a person who suffers from the condition of phagomania.

phagosome *Cell Biology.* a membrane-bound vesicle formed in phagocytes by the pinching off of a region of the plasma membrane to enclose recently ingested material.

phagotrophy *Biochemistry.* the ingestion of nutrients in particulate form, as by certain protozoa.

Phalacridae *Invertebrate Zoology.* the shining flower beetles, a family of coleopteran insects in the superfamily Cucujoidea.

Phalacrocleptes *Invertebrate Zoology.* a highly specialized type of parasitic ciliate protozoan that lacks cilia, but undergoes classical ciliate binary fission and conjugation.

Phalacrocoracidae *Vertebrate Zoology.* the cormorants, a family of medium-sized aquatic birds of the order Pelecaniformes, characterized by a small, usually colored gular sac and dark plumage; found nearly worldwide, except at high latitudes.

Phalaenidae see NOCTUIDAE.

phalangectomy *Surgery.* the removal of a phalanx of a finger or toe.

phalanger [fə lanj´ər] *Vertebrate Zoology.* any of various arboreal, tree-climbing marsupials of the family Phalangeridae, characterized by a long bushy tail and foxlike ears; found in Australia and New Guinea. (So called because of the webbed structure of the second and third digits of the hind feet.)

phalanger

Phalangeridae *Vertebrate Zoology.* the brush-tailed opossums, cuscuses, and phalangers, a family of nocturnal, arboreal marsupials of the order Diprotodontia, characterized by long prehensile tails, soft, dense fur, and a snout that is either blunt or elongated; they are found in Australia, Tasmania, New Guinea, and islands ranging from the Celebes to the Solomons.

phalanges [fə lan´jēz] *Anatomy.* the bones of the fingers or toes.

Phalangida *Invertebrate Zoology.* the daddy longlegs and harvestmen, an order of arachnids that resemble spiders but have no silk gland; they are characterized by very long legs and a segmented abdomen. Also, OPILIONES, PALPTORES.

phalangitis *Medicine.* an inflammation of one or more phalanges (the bones of the fingers or toes).

phalanx [fā´langks] *plural,* **phalanges** or **phalanxes.** *Military Science.* **1.** originally, a formation used by ancient Greek armies, in which a large group of infantry soldiers armed with spears stood tightly massed together. **2.** any similar close or massed formation of troops. *Anatomy.* **1.** any of the bones of the fingers or toes. **2.** any one of a set of plates arranged in rows making up the reticular membranes of the organ of Corti. *Botany.* a filament-united stamen bundle in plants with multiple stamen groups.

Phalanx *Ordnance.* a close-in weapons system providing automatic, autonomous terminal defense against antiship cruise missiles; includes search and track radars, weapons control, and a 20-mm M61 gun that fires subcaliber penetrators.

phalarope *Vertebrate Zoology.* any of three species of small, aquatic birds of the family Phalaropodidae, characterized by lobate toes; the females are larger and more brightly colored than the males, which brood the eggs.

Phalaropodidae *Vertebrate Zoology.* the Phalaropes, a monogeneric family of small aquatic birds of the order Charadriiformes; related to the sandpiper and plover but distinguished by a longer neck and a long, slender bill; found in arctic and subarctic regions of North America and Eurasia and in equatorial regions in winter.

phall- or **phallo-** a prefix meaning "penis."

Phallaceae *Mycology.* a family of fungi belonging to the order Phallales, which is composed of stinkhorns; found in tropical to warm regions.

Phallales *Mycology.* an order of fungi that of the class Gasteromycetes that is composed of the stinkhorns; characterized by an unpleasant odor and by a fruiting body receptacle that expands when its outer layers break; found in rotting wood and soil in Australia and certain tropical regions.

phallic character *Psychology.* a personality type that derives from unresolved conflicts during the phallic stage of psychosexual development; considered to be manifested by such traits as boastfulness, vanity, and aggression. Also, **phallic personality.**

phallic stage *Psychology.* in psychoanalytic theory, the third stage of psychosexual development, from 3 to 7 years of age, characterized by a focus on the child's own genitals as the source of sexual gratification. Also, **phallic period** or **phase.**

Phallostethidae *Vertebrate Zoology.* a family of small, fully scaled, slender fishes of the order Atheriniformes, whose males are distinguished by an unusual, asymmetrical copulatory organ; these fishes are found in fresh and brackish waters of the Philippines, Malaysia, and Thailand.

Phallostethiformes *Vertebrate Zoology.* in some classification systems, an order of small fishes composed of the single family Phallostethidae.

phallotoxin *Biochemistry.* one of a class of poisons found in certain fungi, such as *Amanita phallosides,* with symptoms, including vomiting and extreme fatigue, that appear within hours after ingestion; considered to be less toxic than amatoxins, but may be fatal.

phallus [fal′əs] *Developmental Biology.* the rudiment of embryonic or fetal tissue formed from the genital folds, becoming the penis in the male and the clitoris in the female. *Anatomy.* the penis itself.

phaner- or **phanero-** a prefix meaning "visible."

phanerite *Petrology.* any igneous rock with a microscopically observable crystalline texture.

phaneritic *Petrology.* of or relating to a largely crystalline, igneous rock texture that is visible to the naked eye. Also, COARSE-GRAINED PHENOCRYSTALLINE.

phanerophyte *Botany.* a type of perennial tree or shrub that bears its shoots high in the air, usually defined as more than 25 centimeters above ground level.

Phanerorhynchidae *Paleontology.* a single-genus family of actinopterygian fishes belonging to the extinct infraclass Chondrosteia and order Phanerorhynchiformes; extant in the Upper Carboniferous.

Phanerozoic *Geology.* the part of post-Precambrian time for which the corresponding rocks show abundant evidence of life.

Phanerozonida *Invertebrate Zoology.* an order of starfish, echinoderms in the class Asteroidea, characterized by gills on the aboral upper surface and large skeletal plates around the edge of the arms and central disk. Also, **Phanerozonia.**

Phanodermatidae *Invertebrate Zoology.* a family of free-living, marine nematode roundworms in the superfamily Enoploidea.

phanotron *Electronics.* a hot-cathode gas diode that uses an arc discharge in mercury vapor or an inert gas.

phantasm [fan taz′əm] *Neurology.* an illusion or mental image of something that does not exist, unprovoked by external stimuli. Also, PHANTOM.

phantastron *Electronics.* a monostable pentode circuit that produces sharp pulses at an adjustable and accurately timed interval when triggered by a signal; used in radar systems for gating and sweep-delay functions.

phantasy see FANTASY.

phantom [fan′təm] a ghostlike image; an apparition; specific uses include: *Neurology.* **1.** of or related to a nonexistent source of sensation; applied specifically to a limb that has been amputated, but is the apparent source of pain. Thus, **phantom limb. 2.** see PHANTASM. *Nucleonics.* a collection of materials used to simulate human tissue in testing radiation absorption and backscattering. *Geology.* a bed or member of a sedimentary sequence that is missing from a given stratigraphic section but is characteristically present in other sequences of similar age. *Petrology.* a crystal with a discernible outline of a former period of crystallization of a fossil, or of some partially obliterated rock structure bounded by inclusions, bubbles, or foreign material.

phantom bottom see DEEP-SCATTERING LAYER.

phantom circuit *Telecommunications.* a circuit that originates from center taps on two side circuits, with no additional wire lines.

phantom-circuit loading coil *Electricity.* a loading coil that can introduce an inductance into a phantom circuit with a minimum of inductance into the constituent side circuits.

phantom-circuit repeating coil *Electricity.* a repeating coil that is utilized at the terminal end of a phantom circuit.

phantom crystal *Crystallography.* a crystal in whose interior an earlier stage of crystallization is outlined.

phantom group *Electricity.* a group of four open-wire conductors from which a phantom circuit is derived. *Telecommunications.* a phantom circuit and its two constituent side circuits.

phantom horizon *Geology.* on a seismic reflection profile, a line drawn parallel to the nearest actual dip segment at all points along the profile.

phantom material *Nucleonics.* any material, such as beeswax or masonite, used in constructing a phantom for testing radiation and its effects.

phantom midge see CHAOBIRADAE.

phantom network *Materials Science.* a theory proposing that an elastomer consists of a network of Gaussian chains connected in an arbitrary manner; the physical effect of the chains is assumed to be confined exclusively to the junction to which they are attached.

phantom repeating coil *Electricity.* a repeating coil in a side circuit or a phantom circuit.

phantom signals *Electronics.* signals of unknown origin appearing on a radar screen that do not correspond to the target; generally the cause is unknown, although they may result from faulty circuitry, interference, or jamming.

phantom tumor see PSEUDONEOPLASM.

Phar. or **phar.** pharmacopoeia; pharmacy; pharmacist; pharmaceutical; pharmacology.

PharB or **Phar.B.** Bachelor of Pharmacy. (From Latin *Pharmaciae Baccalaureus.)*

PharC or **Phar.C.** Pharmaceutical Chemist.

PharD or **Phar.D.** Pharmaceutical Doctor.

Pharetronida *Invertebrate Zoology.* an order of sponges in the subclass Calcina, with a calcareous skeleton.

PharG or **Phar.G.** Graduate in Pharmacy.

PharM or **Phar.M.** Master of Pharmacy. (From Latin *Pharmaciae Magister.)*

Pharm. or **pharm.** pharmacist; pharmacy; pharmaceutical; pharmacology.

pharmaceutical [fär′mə soot′i kəl] *Pharmacology.* **1.** of or relating to pharmacy or to drugs. Also, **pharmaceutic. 2.** any medicinal drug.

pharmaceutical chemistry *Chemical Engineering.* the chemistry of drugs, medicines, and other pharmaceutical products.

pharmacist [fär′mə sist] *Pharmacology.* a person who is trained and licensed to prepare and sell or dispense drugs and compounds and to prepare prescriptions.

pharmaco- a prefix meaning "drug" or "medicine," as in *pharmacometrics, pharmacotherapy.*

pharmacodiagnosis *Pharmacology.* the use of drugs in diagnosing disease.

pharmacodynamics *Pharmacology.* the study of the physiological and biochemical effects of drugs, the mechanism of their actions, and the correlation of their actions with their chemical structure.

pharmacoendocrinology *Pharmacology.* the study of the influence of drugs on the secretion of the endocrine glands.

pharmacogenetics *Genetics.* a branch of biochemical genetics that is concerned with the relationship between genetic factors and variations in responses to drugs.

pharmacognosy *Pharmacology.* the branch of pharmacology dealing with study of natural drugs, including their botanical sources, constituents, and biological, biochemical, and economic characteristics. Also, **pharmacognostics.**

pharmacography *Pharmacology.* a written account or description of a drug.

pharmacokinetics *Pharmacology.* the way drugs in the body are absorbed, distributed, localized in tissues, changed, and excreted over a period of time.

pharmacolite *Mineralogy.* $CaHAsO_4 \cdot 2H_2O$, a colorless to white monoclinic mineral occurring in botryoidal or fibrous form, having a specific gravity of 2.68 to 2.72 and a hardness of 2 to 2.5 on the Mohs scale.

pharmacologic [fär′mə kə läj′ik] *Pharmacology.* of or relating to pharmacology or to the properties and reactions of drugs.

pharmacologic pyrogen *Pharmacology.* any drug that produces fever.

pharmacologist [fär′mə käl′ə jist] *Medicine.* a person who is a specialist in pharmacology.

Pharmacology

Pharmacology is the science of drugs (medicines), their discovery and uses—the general aspects of the how and why of drugs. It stands in a central place in the biomedical sciences, with no clearcut separation from biochemistry, physiology, pathology, or microbiology, or from medicine in general. It is involved with understanding the mechanism of action of drugs and provides the knowledge for their rational use.

By-products of pharmacological studies are the leads provided for the invention of new drugs that avoid the shortcomings of the old. Pharmacology is a relatively young science. It is particularly important to the use of chemotherapeutic agents in medical practice. In return, pharmacological knowledge is significantly enhanced by observations of chemotherapy.

In recent times pharmacologists have been concerned not only with the actions of whole drugs, but with the specific segments of these drugs. Pharmacology is becoming more and more molecular in its outlook. A knowledge of pharmacology is an essential element in medical practice, and is a basis for the discovery of new medicines.

The multiplication of useful antiparasitic drugs since 1940 has been an unmatched phenomenon. In 1909, Paul Ehrlich brought forward arsphenamine (Ehrlich's 606; Salvarsan). The next useful antiparasitic drug was sulfanilamide in 1938. It was hailed as the fourth useful synthetic agent. It was also the taking-off point for a myriad of synthetic useful agents, and also for the antimetabolite theory. This arose from studies of its reversal by *p*-aminobenzoic acid and constituents of DNA (thymine, purines, cytosine, and uracil). The modern Goodman and Gilman *The Pharmacological Basis of Therapeutics* pays tribute to these beginnings. Current progress characterizes in detail the cell receptors for functional drugs and the synthesis of compounds that fit these cell receptors. Thus progress in pharmacology encompasses a sizable element of cell biology, sophisticated chemistry, and sophisticated cell reactions.

George H. Hitchings
The Wellcome Research Laboratories
Burroughs Wellcome Company

pharmacology [fär′mə käl′i jē] *Medicine.* the science that deals with the origin, nature, chemistry, effects, and uses of drugs.

pharmacometrics *Pharmacology.* the quantitative evaluation of drug activity and chemical identification of the active components.

pharmacophobia *Psychology.* an irrational fear of medicine or medical treatment.

pharmacopoeia [fär′mə kə pē′ə] *Pharmacology.* an authoritative treatise on drugs and their preparations; a list of drugs and their uses. The first official U.S. Pharmacopeia (abbr. USP) was published in 1820 and listed 217 drugs; new editions appear every five to ten years.

pharmacosiderite *Mineralogy.* $KFe_4^{+3}(AsO_4)_3(OH)_4 \cdot 6\text{–}7H_2O$, a cubic mineral occurring in green, yellowish-green, or brown cubic crystals, having a specific gravity of 2.798 and a hardness of 2.5 on the Mohs scale; found in oxidized arsenical ores. Also, CUBE ORE, IRON SINTER.

pharmacotherapeutics *Pharmacology.* the study of the uses of drugs in the treatment of disease.

pharmacotherapy *Pharmacology.* the use of drugs or medications in the treatment of disease.

pharmacy [fär′mə sē] *Pharmacology.* **1.** the branch of health sciences dealing with the preparation, dispensing, and proper administration of drugs. **2.** a place at which drugs are prepared or dispensed.

pharyngeal [fə rinj′ē əl; fâr in jē′əl] *Anatomy.* of, relating to, or involving the throat.

pharyngeal bursa *Developmental Biology.* an inconstant blind sac on the caudal part of the pharyngeal tonsil, representing persistence of an embryonic communication between the anterior tip of the notochord and the roof of the pharynx. Also, LUSCHKA'S BURSA, TORNWALDT'S CYST.

pharyngeal cleft *Developmental Biology.* an ectodermal groove on the surface of the embryo lying between branchial arches.

pharyngeal pouch *Developmental Biology.* a lateral diverticulum of the pharynx that meets a corresponding groove in the ectoderm, which may open to the outside as a gill slit.

pharyngeal reflex *Medicine.* a normal neural reflex, elicited when the soft palate or posterior pharynx is touched, in which the palate elevates, the tongue retracts, and the pharyngeal muscles contract. Also, GAG REFLEX.

pharyngeal tooth *Vertebrate Zoology.* one of a set of teeth, located in the pharynx of many fishes, that aids in feeding.

pharyngitis [fâr′ən jī′tis] *Medicine.* an inflammation of the throat or pharynx, often accompanied by pain, dryness, congestion of the mucous membrane, and fever.

Pharyngobdellae *Invertebrate Zoology.* a suborder of aquatic and terrestrial leeches with a nonprotrusible, sucking pharynx and no teeth or jaws.

pharyngoconjunctival fever *Pathology.* an epidemic viral infection, marked by a combination of pharyngitis and conjunctivitis, caused, in part, by an inflammation of the adenoids; common in school-age children.

pharyngodynia *Medicine.* any pain in the throat; sore throat.

pharyngology *Medicine.* the branch of medical science dealing with the study of the pharynx or throat and the diagnosis and treatment of its diseases.

pharyngomycosis *Medicine.* any fungal infection of the pharynx.

pharyngopathy *Medicine.* any disease of the pharynx.

pharyngoplasty *Surgery.* a plastic operation on the pharynx.

pharyngoscope *Medicine.* an instrument that is used to examine the pharynx.

pharynx [fâr′ingks] *Anatomy.* the musculomembranous passage between the mouth and posterior nares rostrally and the larynx and esophagus caudally. Also, THROAT.

phas- or **phaso-** a prefix meaning "speech."

phase a particular stage or aspect of something; specific uses include: *Archaeology.* an interval of time in the archaeological record, especially a relatively limited time within a specific locality or region. *Chemistry.* the distinct condition of a substance, as a solid, liquid, or gas; a quantity of matter that is homogeneous throughout its environment. In each phase, the substance may exist at various pressures and temperatures. *Materials Science.* a portion of a system that is uniform both in its chemical composition and in its physical properties and that is separated from other homogeneous portions of the system by boundary surfaces. *Metallurgy.* specifically, a physically homogeneous portion of any alloy system. *Physics.* **1.** a particular stage or point in a sequence through which time has advanced, measured from some arbitrary starting point. **2.** the relative angular displacement of one sinusoidal quantity with respect to a reference angle or to another sinusoidally varying quantity of the same frequency. **3.** the argument of a harmonic function, as in $y = Ae^{i}(wt - kx)$, where $(wt - kx)$ is the phase. *Astronomy.* the varying illuminated area of the moon and planets caused by the relative positions of the earth, the sun, and the illuminated body. *Cartography.* in surveying, the apparent displacement of an object as a result of one side of the object being more strongly illuminated than the other, which may result in errors in measurement. *Mathematics.* **1.** a constant added to the argument of a trigonometric function. **2.** a former term for the argument of a complex number.

phase advancer *Electricity.* a phase modifier that supplies leading reactive volt-amperes to an electronic system to which it is connected.

phase-alternation line system *Telecommunications.* a color TV system used in eastern Europe, in which the relative phases of the chrominance signal are reversed in alternate lines for the purpose of improving picture color and minimizing phase errors. Also, PAL SYSTEM.

phase angle *Physics.* the relative angular displacement between a periodic quantity and a reference angle or between two sinusoidally varying quantities of identical frequencies. Also, PHASE DIFFERENCE. *Electricity.* the angle of a steady-state electrical quantity, such as voltage, current, or impedance, described as a phasor. *Astronomy.* the angle formed by the earth and sun when viewed from the moon or another object.

phase-balance relay *Electricity.* a relay that operates by a difference between two quantities characterizing different phases of a polyphase circuit.

phase boundary *Physical Chemistry.* the interface between two or more substance phases, such as the interface between a gas and a liquid or that between two immiscible liquids.

phase change *Physical Chemistry.* the transition of a substance from one phase to another by melting, freezing, boiling, condensing, or sublimation. *Physical Chemistry.* a shift in the phase angle between two sinusoidal quantities by some mechanism that can effectively delay one quantity relative to the other.

phase coherence *Physics.* a condition of coherence, having a phase difference that does not change with time, between two or more waves.

phase comparator *Electrical Engineering.* a device that receives two radio signals of the same frequency and then produces two video outputs proportional to the sine and cosine of the phase difference of the inputs.

phase-comparison relaying *Electricity.* a system in which two terminals are monitored every half cycle so that transmitted signals appear as a continuous signal received when no fault exists between terminals, or as a periodic signal that indicates a fault does exist; used to detect faults in an electric power device.

phase conductor *Electricity.* in a polyphase circuit, any conductor other than the neutral conductor.

phase conjugate system *Optics.* a system that uses geometry or interferometry to rectify aberrations in wavefronts.

phase conjugation see OPTICAL PHASE CONJUGATION.

phase constant *Electromagnetism.* a hypothetical part of the propagation constant for a plane wave of specified frequency traveling through a transmission line or medium; expressed in radians per unit length, it is the constant of proportionality between the phase lag of a field component (or voltage or current) and the distance through which the wave propagates.

phase-contrast imaging *Materials Science.* a process in which the image of a specimen is produced from refractive light; used in the study of the structural features in solid opaque polymers in electron microscopy when electron beams traveling through adjacent regions of a material are out of phase.

phase-contrast microscope *Optics.* a microscope that detects phase variations of light passing through an object and translates them into corresponding variations of light intensity to form an image of an object; used to examine the surface structure of transparent objects.

phase control *Electronics.* **1.** an element that changes the phase angle at which the AC line initiates conduction in a gas tube, such as a thyratron or ignitron. Also, PHASE-SHIFT CONTROL. **2.** see HUE CONTROL.

phase converter *Electricity.* a converter that changes the number of phases in an AC power system without affecting the frequency.

phase correction *Telecommunications.* a method by which synchronous telegraph mechanisms are kept in more or less correct phase relationship.

phase corrector *Acoustical Engineering.* a network that is designed to adjust phase differences during signal processing.

phase crossover *Control Systems.* the point on the plot of a loop ratio at which the phase angle is 180°.

phased array *Electromagnetism.* an array of antenna radiating elements, each of which can be driven with an adjustable phase to steer the resultant beam in a desired direction.

phased-array radar *Ordnance.* a phased array that can detect and track targets over a wide sector without mechanical movement of the antenna.

phase defect *Astronomy.* the angular extent of the illuminated portion of the disk of the moon or a planet compared to its full disk. *Electrical Engineering.* the phase differential between the actual current in a real capacitor and that of a hypothetical capacitor, which has an equivalent ideal loss-free flow.

phase delay *Telecommunications.* the delay that a single unmodulated frequency incurs in passing through a channel, usually measured in the number of cycles of the frequency being transmitted.

phase detector *Electronics.* a device that delivers a DC voltage proportional to the phase deviation of a receiver oscillator to keep it synchronous with the transmitted signal; used in television circuits. Also, **phase discriminator.**

phase deviation *Telecommunications.* the peak difference between an instantaneous angle of a phase-modulated wave and the angle of the sine-wave carrier.

phase diagram *Physical Chemistry.* a graphic representation of the relationships between the physical state of a substance and its pressure and temperature; e.g., a diagram showing the change in phase of a sample of carbon dioxide or water as pressure is decreased at constant temperature. *Materials Science.* a graphic representation of the phases present in a materials system at various temperatures and compositions; sometimes at various pressures, but usually at atmospheric pressure. *Metallurgy.* a representation of temperature and composition boundaries of the phases present in an alloy system. Also, CONSTITUTION DIAGRAM.

phase difference *Physics.* see PHASE ANGLE.

phase distortion *Telecommunications.* distortion that arises when the relative phases of the singular components of an input signal are different from those of the output signal. Also, **phase-frequency distortion.**

phase equalizer *Electronics.* a network that compensates for phase-frequency distortion for a selected range of frequencies.

phase equilibria *Physical Chemistry.* the equilibrium relationships of a substance undergoing a phase change, such as from solid to liquid, under various conditions of temperature and pressure.

phase excursion *Telecommunications.* the difference between the instantaneous angle of a modulated wave and the angle of the carrier, as it occurs in angle modulation.

phase factor *Solid-State Physics.* a multiplier of the structure factor that cannot be determined experimentally and must be estimated or guessed; provides information on the relative position of atoms constituting a unit cell. *Electronics.* see POWER FACTOR.

phase front *Physics.* a contour in a field of waves whose surface is constructed by points having a common phase.

phase function *Optics.* the angular distribution of light reflected from an object that is illuminated by rays from a given direction.

phase generator *Electronics.* an instrument that accepts and passes single-phase input signals at a specified frequency range, or produces its own signals, in which the phase of the output signal can be varied by tuning calibrated dials on the generator.

phase inequality *Oceanography.* the variation in tidal range or in the speed of tidal currents that is caused by changes in the phase of the moon.

phase inversion *Electronics.* a condition in which the output of a circuit produces a phase difference of 180° between two waves with the same shape and frequency.

phase inverter *Electronics.* **1.** a device that converts the phase of a signal by 180°; commonly used to power a push-pull amplifier stage or to change the polarity of a pulse. **2.** a device that produces two output signals that are 180° out-of-phase with each other.

phase jitter *Electronics.* a condition that causes inconsistency in the timing of the rising edge or trailing edge of pulses during transmission or processing.

phase lag *Physics.* the amount of negative angular displacement that exists between two sinusoidally varying quantities. *Oceanography.* the angular retardation of the maximum of one of the harmonic constituents of the observed tide behind the corresponding maximum of the same constituent of the hypothetical equilibrium tide. Also, EPOCH, TIDAL EPOCH.

phase line *Military Science.* a line used to control and coordinate military operations; usually a terrain feature extending across the zone of action.

phase lock *Electronics.* a technique of adjusting the phase of an oscillator signal so that it will follow the phase of a reference signal.

phase-locked communication *Telecommunications.* a system in which oscillators at both the transmitter and receiver are held in phase, and messages are passed on by phase shifts of the modulating signal.

phase-locked loop *Electronics.* a circuit that locks the phase of an oscillator to the phase of a reference signal.

phase-locked system *Engineering.* a radar system containing an immobile local oscillator that determines target data by calculating the phase shift of the echo.

phase margin *Control Systems.* in a stable system, the difference between 180° and the phase of the loop ratio at the gain-crossover frequency.

phase matching *Optics.* a technique used in a multiwave nonlinear optical process to increase the distance over which energy transfers between waves.

phase meter *Engineering.* a device that measures the phase angle (difference) between two alternating currents or electromotive forces.

phase modifier *Electricity.* a device that changes the phase of the current in an electronic system to which it is connected.

phase modulation *Telecommunications.* a special type of modulation in which the angle relative to the unmodulated carrier is varied in accordance with the instantaneous value of the communicated message.

phase-modulation detector *Electronics.* a device that extracts data from a phase-modulated carrier signal.

phase-modulation transmitter *Electronics.* a radio transmitter that broadcasts intelligence over a phase-modulated signal.

phase modulator *Electronics.* a device that varies the phase of carrier signal in accordance with variations in voltage level.

phase of compiler *Computer Science.* a major section of the compilation process, generally involving examination of the entire program, e.g., syntax analysis, optimization, or code generation.

phaseolin *Biochemistry.* $C_{20}H_{18}O_4$, a phytoalexin produced in response to stress by plants of the genus *Phaseolus,* such as French or kidney beans.

phaseolotoxin *Biochemistry.* a poison generated by the bacteria *Pseudomonas syringae* that inhibits the enzyme which catalyzes the production of citrulline from ornithine and carbamoyl phosphate.

phase plane *Mathematics.* in the theory of ordinary differential equations, the plane spanned by the dependent variable x and its derivative dx/dt with respect to the independent variable t; used in the qualitative theory of ordinary differential equations; for example, closed curves represent periodic solutions.

phase-plane analysis *Control Systems.* a method of system analysis in which some quantity characterizing the system's position is plotted as a function of position for certain values of initial conditions.

phase plate *Optics.* an instrument that changes the relative phase of the components in polarized light; used in phase-contrast microscopy.

phase portrait *Control Systems.* a graph showing some quantity that characterizes the system as a function of position for certain values of initial conditions.

phase-probability relationships *Crystallography.* equations representing the probability that the phase of a Bragg reflection in X-ray diffraction studies has a certain value. Such equations form the basis of phase determination in direct methods.

phase problem *Crystallography.* the problem of determining the phase angle to be associated with each structure factor, so that an electron-density map may be calculated from a Fourier series with structure factors (including both amplitude and phase) as coefficients. The measured intensities of diffracted beams give only the squares of the amplitudes; the phases cannot normally be determined experimentally. Phases may, however, be calculated for any postulated structure and combined with the experimentally determined amplitudes to give an electron-density map. They may also be measured by the method of isomorphous replacement.

phaser *Electromagnetism.* a microwave device that can shift the phase of an electromagnetic wave in a waveguide. *Telecommunications.* in FAX systems, a device that adjusts equipment so that the recorded elemental area has the same relation to the record sheet as the corresponding transmitted elemental area has to the subject copy in the direction of the scanning line.

phase resonance *Physics.* a resonance condition in which the driving frequency and the response frequency are identical.

phase response *Electronics.* in an amplifier or a two-port network with a sinusoidal input, the difference between the phase at the output and the phase at the input, as a function of frequency.

phase reversal *Physics.* an angular shift in phase by 180° or one half-cycle.

phase-reversal modulation *Telecommunications.* in data transmission, a type of pulse modulation in which the reversal of the signal phase is used to distinguish between the two binary states employed.

phase rule *Materials Science.* a rule that governs variations in phase diagrams, expressed as $F = C - P + 2$, in which F represents the degree of freedom, C the number of components, and P the number of phases in equilibrium. At constant pressure, the phase rule is $F = C - P + 1$.

phase rule see GIBBS PHASE RULE.

phase separation *Materials Science.* a condition in which two polymers or a polymer and solvent are no longer miscible and separate into two phases; usually caused by temperature change.

phase separation point see CLOUD POINT.

phase sequence see POSITIVE PHASE SEQUENCE.

phase-sequence relay *Electricity.* in a polyphase system, a relay, device, or relay circuit that is actuated when voltages reach the maximum positive amplitude in a predetermined phase sequence. Also, **phase-sequence indicator, phase-rotation relay.**

phase shift *Physics.* a change in the angular phase of a sinusoidally varying quantity with respect to a reference angle or another sinusoidal quantity of the same frequency. *Electronics.* specifically, variations between corresponding points on input and output signal waveshapes, expressed as degrees of lead or lag.

phase-shift circuit *Electronics.* a system that shifts the phase of a voltage with respect to a reference voltage at the same frequency.

phase-shift control see PHASE CONTROL, def. 1.

phase-shift discriminator *Electronics.* a device that generates an output signal that is proportional to the phase difference between two input signals.

phase shifter *Electricity.* a circuit or device whose output voltage or current may be adjusted to have some desired phase relation with the input voltage.

phase-shifting transformer *Electricity.* a transformer that is connected across a polyphase circuit and provides voltages of the proper phase for energizing varmeters, var hour meters, or other instruments; it produces differences in the phase angle for two circuits, which become the basis for energizing the instruments. Also, **phasing transformer.**

phase-shift keying *Telecommunications.* a type of phase modulation in which the phase of the carrier is discretely varied in reference to the phase of the previous signal element.

phase-shift oscillator *Electronics.* an oscillator composed of an amplifier and a circuit that shifts the phase of a signal by 180°, connected between the output and the input of the amplifier.

phases of the moon *Astronomy.* the apparent changes in the size and shape of the moon caused by the differing amounts of sunlight reflected by the moon toward the earth during the moon's orbit; the eight phases are the new moon, waxing crescent, first quarter, waxing gibbous, full moon, waning gibbous, last quarter, and waning crescent.

phases of the moon

phase solubility *Physical Chemistry.* the range of solubilities among the solid, liquid, and gaseous constituents in a given solvent.

phase-solubility analysis *Analytical Chemistry.* the analysis of components in a solid substance by comparing the weight of the original sample with the weight of the sample dissolved in a solvent; based on the solubility of the components in the solvent.

phase space *Mathematics.* an even-dimensional space, with one set of coordinate functions that are the dependent variables in a system of ordinary differential equations, and a complementary set of coordinate functions equal to the (generalized) derivatives of the dependent variables; a generalization of the concept of phase plane. *Physics.* an abstract Euclidean space used to describe a system having n degrees of freedom; the space will have $2n$ dimensions: one set of n dimensions for the generalized coordinates and another set for the corresponding momenta.

phase splitter *Electricity.* **1.** a circuit that generates, from a single AC input signal, two equal-amplitude output signals that are 180° apart in phase. **2.** a device that produces, from a single input wave, two or more output waves that differ in phase.

phase stability *Nucleonics.* in a synchrotron, the stability of the motion of a charged particle, which is maintained by accelerating the particle slightly earlier than the peak of the accelerating potential.
phase titration *Analytical Chemistry.* an analytical technique that involves the titration of a mixture of two miscible liquids with a liquid that is miscible in only one of them, and the use of a ternary (three-phase) diagram to determine the endpoint.
phase-transfer catalysis *Organic Chemistry.* in a two-phase organic water system, a process in which a catalyst enhances the rate of transfer of water-soluble reactant across an interface to the organic phase.
phase transformation *Physical Chemistry.* the changing of a substance's phase by freezing, melting, boiling, condensing, or sublimation. *Metallurgy.* specifically, any of several solid-state reactions that change the phases present in a metallic microstructure, usually affecting properties. *Chaotic Dynamics.* an abrupt change from one physical state to another. Also, **phase transition.**
phase undervoltage relay *Electricity.* a relay that is tripped when one phase voltage in a polyphase circuit is reduced.
phase variation *Genetics.* an alteration in the type of flagella produced by a bacterium.
phase velocity *Physics.* in wave propagation, the velocity of a surface of constant phase; this would be the speed of the wave.
phase winding *Electricity.* a winding on the armature of an AC generator that delivers voltages differing in phase.
Phasianidae *Vertebrate Zoology.* the pheasants, quails, and partridges, a family of gallinaceous game and domesticated birds of the order Galliformes, generally plump and with tail lengths varying from very short to very long; found worldwide in tropical and temperate regions, but concentrated in the Old World.
phasic arrhythmia *Cardiology.* an irregularity of the heartbeat marked by its speeding up during inhalation and slowing down during exhalation; it does not require treatment and is most noticeable in the young and old. Also, SINUS ARRHYTHMIA.
phasing line *Electronics.* the portion of the length of a facsimile scanning line that is used for the phasing signal.
phasing signal *Electronics.* a signal that corrects the picture position along the scanning line in a facsimile system.
phasitron *Electronics.* an electron tube that generates a frequency-modulated audio signal from the field effects induced by a magnetic coil attached to its glass envelope.
Phasmatidae *Invertebrate Zoology.* a family of insects with slender cylindrical bodies, in the order Phasmida. Also, **Phasmidae.**
phasmid *Invertebrate Zoology.* **1.** in nematodes, one of a pair of posterior sensory organs. **2.** of insects, belonging to the order Phasmida.
Phasmida *Invertebrate Zoology.* an order of large nocturnal insects, the stick insects and leaf insects, that mimic twigs or leaves for camouflage. Also, **Phasmatodia, Phasmoidea.**
Phasmidea *Invertebrate Zoology.* a class of nematodes with phasmids and porelike amphids. Also, **Phasmidia.**
phasor *Physics.* a vector representation of a sinusoidally varying quantity in which the vector rotates about the origin in the *x–y* plane, with the vector's length indicating the amplitude of the quantity, the *x*-component giving the instantaneous real value of the quantity, and the angle with the *x*-axis representing the phase (angular displacement) of the quantity. *Solid-State Physics.* the low-energy excitation of valence electrons in a conductor by a slowly varying phase modulation of the charge-density wave. *Electricity.* a polar (magnitude and angle) representation of a complex (real and imaginary) number; generally used to represent steady-state electrical quantities, such as voltage, current, and impedance at a single frequency.
PhB or **Ph.B.** Bachelor of Pharmacy; *Pharmacopoeia Britannica.*
PhD or **Ph.D.** Doctor of Philosophy.
Phe phenylalanine.
pheasant [fez´ənt] *Vertebrate Zoology.* **1.** any of several large gallinaceous birds of the family Phasianidae, characterized by long pointed tail feathers and brilliantly colored feathers in the male; native to Asia, but now established in many parts of Europe and America. **2.** any one of various similar birds, such as the ruffed grouse of the southern U.S.
pH electrode *Analytical Chemistry.* a glass electrode used in pH meters as a hydrogen ion sensor.
α-phellandrene *Organic Chemistry.* $C_{10}H_{16}$, a combustible, toxic, colorless monoterpene occurring as *d-* and *l-*optical isomers; insoluble in water and soluble in ether; both boil at 174°C; used in flavorings and in perfumes.
phellem *Botany.* the outer layer of cells of the periderm. Also, CORK.
phelloderm *Botany.* a layer of parenchymous tissue often containing chlorophyll and formed by the cork cambium.
phellogen *Botany.* the meristemic portion of the peridium that forms cork on its outer surface and phelloderm internally. Also, CORK CAMBIUM.
phen- or **pheno-** a prefix meaning "showing" or "displaying."
phenacaine hydrochloride *Pharmacology.* $C_{18}H_{22}N_2O_2·HCl$, a local anesthetic applied topically, especially in the conjunctiva of the eye.
phenacemide *Pharmacology.* $C_9H_{10}O_2$, a toxic drug formerly used to alleviate and prevent convulsions in the treatment of epilepsy and to manage seizures with more than one cause.
phenacetin *Pharmacology.* $CH_3CONHC_6H_4OC_2H_5$, odorless, bitter-tasting white crystals or powder, soluble in alcohol, chloroform, and ether and slightly soluble in water; melts at 135°C; used in medicine as an analgesic. Also, ACETOPHENETIDIN, ACETPHENETIDIN.
Phenacodontidae *Paleontology.* a family of medium-sized herbivorous mammals in the extinct order Condylarthra, ancestral to horses and other perissodactyls, and probably to proboscideans and other later mammals; the earlier genera were small, but some later genera were as large as a modern pony; holarctic distribution, Paleocene and Eocene.
Phenacodus *Paleontology.* a genus of condylarths in the family Phenacodontidae; similar to *Hyracotherium*; intermediate between early carnivores and early herbivores; extant in the Lower Eocene.
phenakite *Mineralogy.* Be_2SiO_4, a vitreous, trigonal mineral, ranging from colorless to yellow, pink, or brown, and having a specific gravity of 2.93 to 3.0 and a hardness of 7.5 to 8 on the Mohs scale; found as prismatic crystals in alpine veins, greisens, and granite pegmatites. Also, **phenacite.**
phenanthrene *Organic Chemistry.* $C_{14}H_{10}$, carcinogenic colorless crystals; insoluble in water and soluble in alcohol; melts at 100°C and boils at 340°; used in explosives and in the synthesis of dyes and drugs.

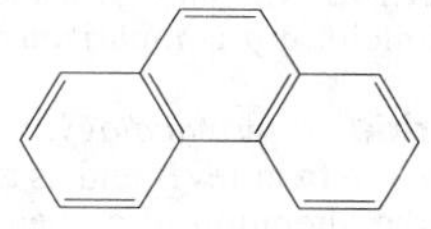

phenanthrene

phenanthridine *Biochemistry.* an intercalating agent that has antitumor and typanocidal properties.
phenanthroline *Organic Chemistry.* $C_{12}H_8N_2$, any of several heterocyclic compounds derived from phenanthrene by substituting two N atoms for two CH groups in the ring.
phenanthroline indicator *Analytical Chemistry.* a red-colored reagent used as an indicator in oxidation-reduction reactions, especially for ferrous compounds.
phenarsazine chloride *Organic Chemistry.* see ADAMSITE.
phenazine *Organic Chemistry.* $C_6H_4N_2C_6H_4$, combustible yellow crystals; very slightly soluble in water and partially soluble in alcohol; melts at 170°C and boils above 360°C; used in dye-making, in organic synthesis, and as a larvicide.
phencyclidine hydrochloride see PCP.
phene [fēn] *Genetics.* a phenotypic character that is controlled by genetic factors.
phenethyl acetate *Organic Chemistry.* $C_6H_5CH_2CH_2OOCCH_3$, a combustible colorless liquid that is soluble in alcohol and in most fixed oils; boils at 238-239°C; used in perfumes and as a reagent. Also, 2-PHENYLETHYL ACETATE.
phenethyl alcohol *Organic Chemistry.* $C_6H_5CH_2CH_2OH$, a combustible, colorless, floral-smelling liquid; slightly soluble in water and soluble in alcohol; boils at 219°C and freezes at –27°C; used in soaps, in flavorings, and as a preservative. Also, 2-PHENYLETHYL ALCOHOL.
phenethyl isobutyrate *Organic Chemistry.* $(CH_3)_2CHCOOC_2H_4C_6H_5$, a combustible, colorless, pleasant-smelling liquid; soluble in alcohol and ether; used in perfumes and flavorings. Also, 2-PHENYLETHYL ISOBUTYRATE.
phenetic [fə net´ik] *Systematics.* of or relating to general or overall similarity. *Evolution.* of or relating to overall similarity based on traits selected without regard to evolutionary history, including features arising from common ancestry, parallel evolution, and convergent evolution. *Genetics.* of or relating to observable characteristics or aspects of a species that share a common ancestry.

phenetic classification *Systematics.* a classification based on general or overall similarity.

phenetic method *Evolution.* a method of phenetic classification.

phenetidine *Organic Chemistry.* $NH_2C_6H_4OC_2H_5$, a combustible, colorless, oily liquid that turns brown when exposed to light or air; insoluble in water and soluble in alcohol; ***o*-phenetidine** boils at 228–230°C and solidifies at about –20°C, ***p*-phenetidine** boils at 253–255°C and solidifies at 2–4°C; used to manufacture dyes and pharmaceuticals and as a laboratory reagent.

phenetole *Organic Chemistry.* $C_6H_5OC_2H_5$, a combustible colorless liquid that is insoluble in water and soluble in alcohol; it boils at 172°C and melts at –30°C.

Phengodidae *Invertebrate Zoology.* the fire beetles, a family of coleopteran insects in the superfamily Cantharoidea.

phengophobia *Psychology.* an irrational fear of daylight.

phenindione *Pharmacology.* $C_{15}H_{10}O_2$, a fast-acting drug administered orally to prevent blood clotting; toxicity limits its use.

pheniramine maleate *Pharmacology.* $C_{16}H_{20}N_2O_4{\cdot}C_4H_4O_4$, an oral antihistamine that is used to treat allergies. Also, PROPHENPYRIDAMINE MALEATE.

phenobarbital [fē′nō bär′bə tôl′] *Organic Chemistry.* $C_{12}H_{12}O_3N_2$, a white crystalline powder occurring in three forms: a stable form that melts at 174°C and two unstable forms that melt at 156–157°C and 166–167°C, respectively; all three forms produce salts with alkali metals and with complex metallic derivatives. Also, **phenobarbitone, phenylethylbarbituric acid.** *Pharmacology.* this substance in the form of a long-acting barbiturate drug used as a sedative, to induce sleep, and to treat convulsions; abuse may lead to addiction.

phenoclast *Petrology.* a fragment in a sediment or sedimentary rock, larger than and easily distinguished from the fine-grained matrix in which it is embedded.

phenoclastic rock *Petrology.* a clastic rock containing phenoclasts of various shapes and sizes.

pheno-coefficient method *Chemistry.* a method used to determine the efficiency of a germicide by comparison with phenol under controlled conditions.

phenocoll hydrochloride *Pharmacology.* $C_{10}H_{14}N_2O_2{\cdot}HCl$, a crystalline drug formerly used to treat fever and as a painkiller.

phenocopy *Genetics.* the alteration of a phenotype by either nutrition or exposure to environmental stress during development.

phenocryst *Geology.* a relatively large, prominent crystal embedded in a finer-grained matrix or groundmass of igneous rock. Also, PROPHYRITIC CRYSTAL.

phenogenetics *Genetics.* a branch of developmental genetics dealing with the genetics of phenotypic expression.

phenogram *Systematics.* a branching diagram depicting phenetic relationships among taxa.

phenol [fē′nôl] *Organic Chemistry.* **1.** C_6H_5OH, a toxic, white crystalline mass; soluble in water and alcohol; melts at 43°C and boils at 181.75°C; used in solvents, resins, and weed killers. Also, CARBOLIC ACID. **2.** a class of aromatic compounds characterized by the attachment of at least one hydroxy group to the benzene ring; includes phenol, resorcinol, the cresols, naphthols, and xylenols.

phenolate process *Chemical Engineering.* an industrial absorption process that combines sodium phenolate and caustic soda to purify fuel gas.

phenol coefficient *Analytical Chemistry.* a value used to describe the effectiveness of a disinfectant; the coefficient indicates the highest dilution of the disinfectant that sterilizes a given bacteria culture under standardized conditions, in which phenol efficacy is set at one.

phenol extraction *Chemical Engineering.* a petroleum-refinery solvent-extraction process used to improve the lubricating, stability, and flow properties of lubricating oils by removing asphalt and sulfur compounds with molten phenol.

phenol-formaldehyde resin *Organic Chemistry.* a thermosetting resin used as a molding material for mechanical and electrical parts.

phenol-furfural resin *Organic Chemistry.* a phenolic resin used in injection molding because it hardens only after curing conditions are reached.

phenolic resin *Materials.* a strong thermosetting resin that resists electricity, water, and acids; produced by condensing a phenol with an aldehyde; used as molding (compression, transfer, or injection) or in laminates and protective coatings.

phenological *Meteorology.* of or relating to phenology, the study of periodic biological phenomena that affect climate.

phenology *Meteorology.* the study of periodic biological phenomena in relation to climate, particularly seasonal changes. Phenological events, or stages of plant growth, serve as bases for the analysis of local seasons and climatic zones.

phenolphthalein *Organic Chemistry.* a pale yellow powder with an approximate formula of $(C_6H_4OH)_2C_2O_2C_6H_4$; loses color in neutral or acid solution and turns pink to deep red in the presence of small to moderate amounts of alkali; insoluble in water and soluble in alcohol; melts at 261°C; used as an acid-base indicator, laxative, and reagent.

phenol poisoning see CARBOLISM.

phenol process *Chemical Engineering.* a petroleum-refinery, single-solvent process that uses phenol to purify oil.

***p*-phenolsulfonic acid** *Organic Chemistry.* $HOC_6H_4SO_3H$, a yellow liquid that turns brown on exposure to air; soluble in water and alcohol; used in water analysis, pharmaceuticals, and dye manufacturing.

phenolsulfonphthalein *Organic Chemistry.* a red crystalline powder with an approximate formula of $(C_6H_4OH)_2COSO_2C_6H_4$; slightly soluble in water and alcohol; used as an acid-base indicator and medical reagent. Also, **phenol red.**

phenomenological psychology *Psychology.* an approach to psychology that focuses on the individual's immediate experiences and his or her own perception of those experiences. Also, **phenomenological approach** or **perspective, phenomenology.**

phenon *Systematics.* a group of similar organisms.

phenoplast *Petrology.* a large rock fragment occurring within a sedimentary rock that was plastic at the time of incorporation into the groundmass.

phenothiazine *Organic Chemistry.* $C_{12}H_9NS$, toxic green or green-yellow crystals; slightly soluble in water and alcohol; melts at 185°C and boils at 371°C; used as an insecticide and in pharmaceutical manufacture. Also, THIODIPHENYLAMINE.

phenotole *Organic Chemistry.* $C_6H_5OC_2H_5$, a combustible colorless liquid that is insoluble in water and soluble in alcohol; boils at 172°C.

phenotype [fē′nō tīp′] *Genetics.* the appearance or other characteristics of an organism, resulting from the interaction of its genetic constitution with the environment, as opposed to its underlying hereditary determinants, or genotype.

phenotypic *Genetics.* of, relating to, or characteristic of a phenotype or an individual organism.

phenotypic lag *Genetics.* the delay of expression of a newly acquired mutant character.

phenotypic mixing *Virology.* a condition in which a progeny that is simultaneously infected by different viruses contains structural proteins or other components derived from both viruses.

phenotypic mutant *Genetics.* a mutant that is defined by the characteristic phenotype it exhibits.

phenotypic plasticity *Ecology.* the ability of an individual organism to alter its physiology in response to changes in environmental conditions.

phenotypic suppression *Genetics.* any occurrence that causes one phenotype not to appear in the progeny of a cross.

phenotypic variance *Genetics.* the variance among individuals with respect to some phenotypic trait or traits; usually affected by environmental influences.

phenoxyacetic acid *Organic Chemistry.* $C_6H_5OCH_2COOH$, a partial combustible powder; soluble in water and alcohol; melts at 98°C and boils with decomposition at 285°C; it is used in pharmaceuticals, pesticides, dyes, and as a precurser in fermenting antibiotics, especially penicillin V.

phenoxybenzamine hydrochloride *Organic Chemistry.* $C_{18}H_{22}O$ $NCl{\cdot}HCl$, white crystals; slightly soluble in water; melts at 139°C; used in medicine.

2-phenoxyethanol *Organic Chemistry.* $C_6H_5OCH_2CH_2OH$, a combustible, colorless, faintly aromatic liquid; partially soluble in water; boils at 244.9°C and freezes at 14°C; used as a perfume fixative, insect repellent, and topical anesthetic. Also, ETHYLENE GLYCOL MONOPHENYL ETHER.

phenoxymethylpenicillin see PENICILLIN V.

phenoxypropanediol *Organic Chemistry.* $C_9H_{12}O_3$, white crystals; soluble in water and alcohol; melts at 53°C and boils at 315°C; used as a plasticizer and an ingredient in resins and lacquers.

phenoxy resin *Organic Chemistry.* a thermoplastic polyether resin that has a high molecular weight and the basic molecular structure $-[OC_6H_4C(CH_3)_2C_6H_4OCH_2CH(OH)CH_2]_n-$, where *n* is approximately 100; used for injection molding, coatings, and adhesives.

phentolamine hydrochloride *Organic Chemistry.* $C_{17}H_{19}ON_2{\cdot}HCl$, white crystals; soluble in water; melts at 240°C; used in medicine.

phenyl [fen´əl; fēn´əl] *Organic Chemistry.* C_6H_5–, a group composed of a benzene ring from which a hydrogen has been removed.

phenylacetaldehyde *Organic Chemistry.* $C_6H_5CH_2CHO$, a combustible colorless low-melting solid with an odor similar to hyacinth; slightly soluble in water and soluble in alcohol; boils at 193–194°C and melts at 33–34°C; used in perfumes and flavorings. Also, α-TOLUIC ALDEHYDE.

phenylacetic acid *Organic Chemistry.* $C_6H_5CH_2COOH$, combustible, white platelike crystals with a floral odor; soluble in hot water and alcohol; melts at 76–78°C and boils at 265°C; used in perfumes, flavorings, and in the manufacture of penicillin G. Also, α-TOLUIC ACID.

phenylalanine *Biochemistry.* $C_9H_{11}O_2N$, one of the twenty amino acids that make up proteins; it is essential for optimal growth in infants and for nitrogen equilibrium in adults. Also, β-PHENYLALANINE, α-AMINOHYDROCINNAMIC ACID.

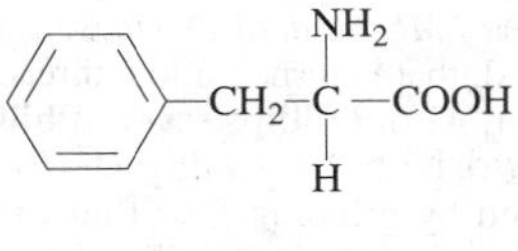

phenylalanine

phenylalanine deaminase test *Microbiology.* an evaluation of the enzymatic ability of a microorganism to convert phenylalanine into phenylpyruvic acid.

phenylalanine hydroxylase *Enzymology.* an enzyme that catalyzes the conversion of phenylalanine to tyrosine by the addition of hydroxyl groups to phenylalanine; the genetic disease phenylketonuria results from a lack of this enzyme.

***N*-phenylanthranilic acid** *Organic Chemistry.* $(C_6H_5NH)C_6H_4CO$ OH, crystals that are soluble in hot alcohol; decomposes at 183–184°C; used to detect vanadium in steel.

phenylbenzene see BIPHENYL.

phenyl bromide see BROMOBENZENE.

phenylbutazone *Organic Chemistry.* $C_{19}H_{20}N_2O_2$, an aromatic, white to yellow bitter-tasting powder; very slightly soluble in water and very soluble in ether; melts at 103–107°C; used as an anti-inflammatory drug.

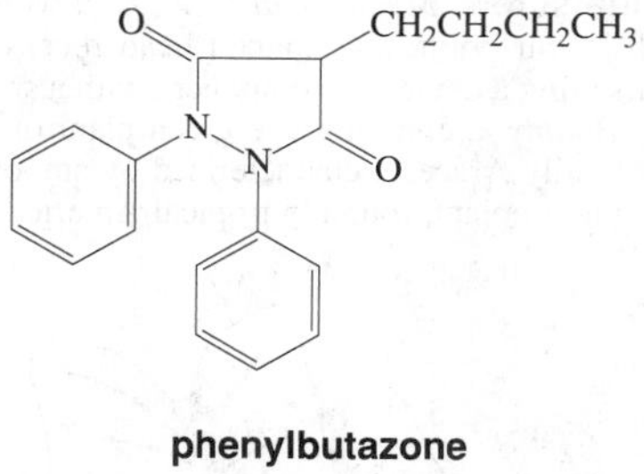

phenylbutazone

phenyl cyanide see BENZONITRILE.

phenylcyclohexane *Organic Chemistry.* $C_6H_5C_6H_{11}$, a combustible, colorless oily liquid that is insoluble in water and very soluble in alcohol; boils at 237.5°C and freezes at 5°C; it is used as a solvent and laboratory reagent.

phenyldichloroarsine *Organic Chemistry.* $C_6H_5AsCl_2$, a toxic liquid; insoluble in water and soluble in alcohol; boils at 255–257°C and freezes at –20°C; used as a poison gas.

phenyl diglycol carbonate *Organic Chemistry.* $(C_6H_5OOCOCH_2CH_2)_2O$, a combustible, colorless solid; insoluble in water and soluble in most organic solvents; melts at 40°C; used as a plasticizer.

phenylenediamine *Organic Chemistry.* $C_6H_4(NH_2)_2$, any of three isomers that are derived from benzene and occur as toxic colorless crystals; ***m*-phenylenediamine** is very soluble in water, soluble in alcohol, melts at 65°C, and boils at 282–287°C; ***o*-phenylenediamine** is soluble in water, very soluble in alcohol, melts at 102–104°C, and boils at 252–258°C; ***p*-phenylenediamine** is slightly soluble in water, soluble in alcohol, melts at 147°C, and boils at 267°; used as a dye, reagent, and photographic developing agent.

phenylephrine hydrochloride *Pharmacology.* $C_9H_{13}NO_2{\cdot}HCl$, a vasoconstrictor that is used to dilate the pupil of the eye, to maintain blood pressure during anesthesia, to treat low blood pressure due to shock, to prolong spinal anesthesia, to treat rapid heartbeat, and as a decongestant.

phenyl ether see DIPHENYL OXIDE.

2-phenylethyl acetate see PHENETHYL ACETATE.

2-phenylethyl alcohol see PHENETHYL ALCOHOL.

phenylethylene see STYRENE.

2-phenylethyl isobutyrate see PHENETHYL ISOBUTYRATE.

phenyl fluoride Organic Chemistry. C_6H_5F, a colorless, flammable liquid with a benzene odor; miscible with alcohol and ether and insoluble in water; freezes at –40°C and boils at 84.9°C; used as a reagent and in insecticides. Also, FLUOROBENZENE.

phenylformic acid see BENZOIC ACID.

***N*-phenylglycine** *Organic Chemistry.* $C_6H_5NHCH_2COOH$, water-soluble crystals that melt at 127–128°C; used in manufacturing indigo dye.

phenylglyoxylonitril oxime *O,O*-diethyl phosphorothioate *Organic Chemistry.* $(H_5C_2O)_2PSONCCNC_6H_5$, a yellow liquid; soluble in water at a ratio of 7 parts to 1 million; boils at 102°C (0.01 torr); used as an insecticide in stored products.

phenylhydrazine *Organic Chemistry.* $C_6H_5NHNH_2$, a toxic, combustible, colorless liquid that is slightly soluble in water and soluble in alcohol; it boils at 243.5°C and freezes at 19.35°C; used in analytical chemistry and organic synthesis.

phenyl isothiocyanate see PHENYL MUSTARD OIL.

phenylketonuria *Medicine.* an autosomal recessive trait resulting in an inability to metabolize phenylalanine, causing mental retardation, neurologic manifestations, light pigmentation, and eczema, unless the disorder is detected early and the infant receives a diet low in phenylalanine.

phenyl mercaptan see THIOPHENOL.

phenylmercuric acetate *Organic Chemistry.* $C_6H_5HgOCOCH_3$, toxic white prisms; slightly soluble in water and soluble in alcohol; melts at 148–150°C; used as an antiseptic, fungicide, and herbicide.

phenylmercuric chloride *Organic Chemistry.* C_6H_5HgCl, toxic white crystals; insoluble in water and slightly soluble in hot alcohol; melts at 251°C; used as an antiseptic, fungicide, and germicide.

phenylmercuric hydroxide *Organic Chemistry.* C_6H_5HgOH, toxic whitish crystals; slightly soluble in water and soluble in alcohol; melts at 197–205°C; used as a fungicide, germicide, alcohol denaturant, and as the primary compound in organic mercury derivatives.

phenylmercuric oleate *Organic Chemistry.* $C_6H_5HgOOC(CH_2)_7CH{=}CHC_8H_{17}$, a toxic, white, crystalline powder; insoluble in water and soluble in organic solvents; melts at 45°C; used as an antimildew agent in paints and as a fungicide and germicide.

phenylmercuric propionate *Organic Chemistry.* $C_6H_5HgOCOCH_2CH_3$, a toxic, whitish, waxlike powder that melts at 65–70°C; used in paints as a fungicide and bactericide.

phenylmercuriethanolammonium acetate *Organic Chemistry.* $[(HOC_2H_4)NH_2(C_6H_5Hg)]OOCCH_3$, toxic, white water-soluble crystals; used as an insecticide and fungicide.

phenylmethane see TOLUENE.

phenylmethyl acetate see BENZYL ACETATE.

***N*-phenylmorpholine** *Organic Chemistry.* $C_6H_5N(CH_2CH_2)_2O$, a toxic, combustible, white solid; soluble in water; melts at 57°C and boils at 268°C; used as an insecticide, in photography, and in dyestuffs.

phenyl mustard oil *Organic Chemistry.* C_6H_5NCS, a toxic, combustible, yellow or colorless liquid with a strong unpleasant odor; insoluble in water and soluble in alcohol; boils at 221°C and freezes at –21°C; used in medicine and organic synthesis. Also, PHENYL ISOTHIOCYANATE.

phenylphenol *Organic Chemistry.* $C_6H_5C_6H_4OH$, one of two isomers occurring as toxic, combustible, whitish crystals; insoluble in water and soluble in alcohol; ***o*-phenylphenol** melts at 56–58°C and boils at 275°C; ***p*-phenylphenol** melts at 164–165°C and boils at 305–308°C; used in making rubber and as a fungicide and dye intermediate. Also, XENOL.

***N*-phenylpiperazine** *Organic Chemistry.* $C_6H_5N(CH_2CH_2)_2NH$, a combustible yellow oil; insoluble in water and soluble in alcohol; boils at 286.5°C and freezes at 18.8°C; used in manufacturing pharmaceuticals and synthetic fibers.

phenylpropane see PROPYLBENZENE.

1-phenyl-1-propanol *Organic Chemistry.* $C_6H_5CH(OH)CH_2CH_3$, an oily liquid that is insoluble in water and soluble in alcohol; boils at 213–215°C; used in medicine and perfumes.

phenylpropanolamine hydrochloride *Pharmacology.* $C_9H_{13}NO \cdot HCl$, a drug that constricts the blood vessels; used as a decongestant and to dilate the air passages in treating allergies; it has also been used as a central nervous system stimulant and an appetite depressant.

3-phenylpropionic acid see HYDROCINNAMIC ACID.

phenylpropyl acetate *Organic Chemistry.* $C_6H_5CH_2CH_2CH_2OOCCH_3$, white combustible crystals; soluble in alcohol; used in perfumes, flavorings, and as a laboratory reagent.

phenylpropyl alcohol *Organic Chemistry.* $C_6H_5CH_2CH_2CH_2OH$, a combustible, colorless floral-smelling liquid; insoluble in water and soluble in alcohol; boils at 235°C or 119°C at 12 torr; used in perfumes, flavorings, and as a laboratory reagent.

phenylpropyl aldehyde *Organic Chemistry.* $C_6H_5CH_2CH_2CHO$, a combustible, colorless, floral-smelling liquid; soluble in alcohol; boils at 202–205°C; used in perfumes and flavorings.

phenylpropyl chloride *Organic Chemistry.* $C_6H_5CH_2CH_2CH_2Cl$, a combustible, colorless to pale-yellow liquid that boils at 219–220°C; used in organic synthesis and as a laboratory reagent.

1-phenyl-3-pyrazolidinone *Organic Chemistry.* $C_6H_5NNHC(O)CH_2CH_2$, combustible, water-soluble crystals that melt at 121°C; used as a photographic developer and laboratory reagent.

phenylpyruvic acid *Biochemistry.* $C_6H_5CH_2$–CO–COOH, a metabolite produced by transamination of phenylalanine; present in the urine of phenylketonurics.

phenyl salicylate *Materials.* $C_6H_4(OH)CO_2C_6H_5$, a crystalline ester that is used as a stabilizer in plastics and medicines. Also, SALOL.

phenylthiourea *Organic Chemistry.* $C_6H_5NHCSNH_2$, crystals that melt at 148–150°C; used in human genetics studies because they are perceived either as tasteless or as having a bitter taste, depending on the heredity of the taster.

phenyl tolyl ketone *Organic Chemistry.* $C_{14}H_{12}O$, one of two isomers that are insoluble in water and soluble in alcohol; **phenyl *o*-tolyl ketone** occurs as an oily liquid that freezes below −18°C and boils at 309°C; **phenyl *p*-tolyl ketone** occurs as colorless crystals that melt at 59°C and boil at 326°C; used as a fixative in perfumes.

phenytoin *Pharmacology.* $C_{15}N_{12}N_2O_2$, a drug used to prevent and treat convulsions in grand mal epilepsy; it has also been used to treat irregular heartbeat. Also, DIPHENYLHYDANTOIN.

pheo- see PHAEO-.

pheochromoblast *Histology.* a cell that develops into a pheochromocyte.

pheochromocytoma *Medicine.* a usually benign, encapsulated, vascular tumor of the adrenal medulla or sympathetic paraganglia; the cardinal symptom is chronic or intermittent hypertension.

pheresis *Hematology.* a procedure in which blood is withdrawn, one or more components are selectively removed, and the remainder of the blood is reinfused into the donor; used to treat conditions in the donor or to obtain blood elements for other treatments or research.

pheromone *Physiology.* a substance released by the body that causes a predictable reaction by another individual of the same species. *Entomology.* an odorous substance secreted by insects; it often serves as a specific attractant, social communicator, or sexual stimulant.

pH gradient *Biochemistry.* the rate of change of hydrogen ion concentration as a function of time.

phi [fī] the 21st letter of the Greek alphabet, written as Φ or ϕ.

Phialophora *Mycology.* a genus of fungi belonging to the class Hyphomycetes; some species cause chromyosis, a human skin disease.

phialospore *Mycology.* a fungal spore formed on a hyphal branch, either singly or in a chain.

phi function see EULER PHI FUNCTION.

phi grade scale *Geology.* a logarithmic scale in which the diameter value of the Wentworth scale is replaced by the negative logarithm to the base 2 of the particle diameter (in millimeters), so that conventional statistical methods can be applied to sedimentary data. Also, **phi scale.**

phil- or **philo-** a prefix meaning "attraction."

Philadelphia chromosome *Pathology.* an aberration in an acrocentric chromosome, marked by a foreshortening in the length of the long-arm; it is found primarily in the megakaryocytic, granulocytic, and erythrocytic cells but is not present in the lymphoid structure of most chronic leukemia patients.

Philippine mahogany *Botany.* **1.** any of several trees of the genus *Shorea* and related genera, characterized by a reddish or brown wood that is used as lumber and in furniture. **2.** the wood of these trees.

Philippines *Geography.* a group of about 7000 islands in the Malay Archipelago north of Indonesia.

Philips hot-air engine *Mechanical Engineering.* a compact hot-air engine that uses only one cylinder and piston to operate at 3000 rpm with a hot-chamber temperature of 12,000°F, a maximum pressure of 50 atmospheres, and a mean effective pressure of 14 atmospheres.

Philips ionization gauge *Electronics.* an ionization gage that is designed to increase the length of the electron path in a vacuum tube by applying a high voltage across its two electrodes and then deflecting the electron beam with a powerful magnetic field, thus increasing the odds that the electrons and gas molecules in the tube will collide.

philipstadite *Mineralogy.* $[Ca_2(Fe,Mg)_5(Si,Al)_8O_{22}(OH)_2]$, a mineral consisting of a basic calcium, iron, magnesium, and aluminum silicate; it crystallizes in the monoclinic system and is a member of the amphibole group.

Phillips and Johnson tube *Biotechnology.* an apparatus that monitors oxygen flow in a fermenter without disturbing the reactions that are occurring.

Phillipsastrea *Paleontology.* a genus of colonial, astraeoid rugose corals in the extinct superfamily Zaphrenticae and family Phillipsastraeidae; extant in the Devonian.

Phillips head screw *Mechanical Devices.* a screw, having a cross-shaped slot in its head, that is turned into a threaded or drilled hole with a Phillips screwdriver. Also, **Phillips screw, Phillips head.**

Phillips head screwdriver *Mechanical Devices.* a screwdriver with a cross-head tip formed by grinding four flutes in the end of a pointed blade; used to loosen or tighten a Phillips head screw. Also, **Phillips screwdriver.** (From a trade name.)

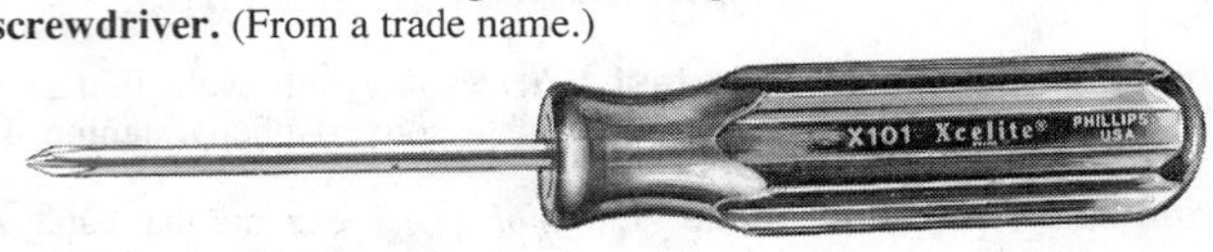

Phillips head screwdriver

Phillipsia *Paleontology.* a genus of small isopygous trilobites in the order Proetida; one of the latest known trilobites; Carboniferous.

Phillipsiellaceae *Mycology.* a family of fungi belonging to the order Hysteriales that occur inconspicuously in warm to temperate regions on the surfaces of decaying or living plant leaves.

phillipsite *Mineralogy.* $(K,Na,Ca)_{1-2}(Si,Al)_8O_{16} \cdot 6H_2O$, a white or reddish monoclinic mineral of the zeolite group occurring as twinned crystals and in spherulites, having a specific gravity of 2.2 and a hardness of 4 to 4.5 on the Mohs scale; a principal constituent of the red-clay sediments of the Pacific Ocean, it is also found in cavities in basalt.

Phillips machine screw *Mechanical Devices.* a wood or metal holding screw with a four-point, star-shaped head recess in various head forms; used for setting a device or component with a special tool.

philodendron *Botany.* a climbing evergreen plant of the genus *Philodendron* of the family Araceae, characterized by smooth, shiny, leaves; often grown as a houseplant; found in tropical America.

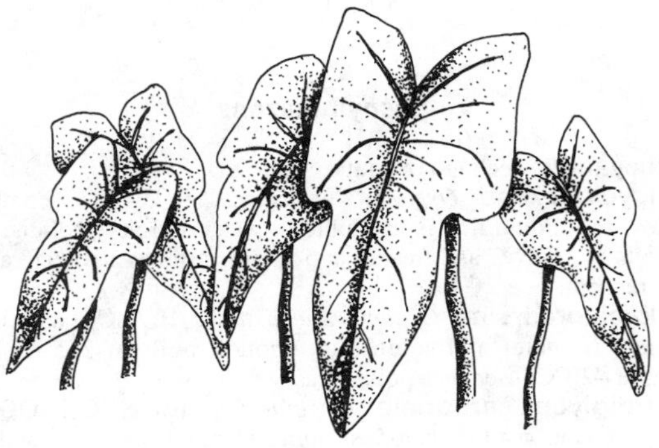

philodendron

Philolaus c. 480–400 BC, Greek Pythagorean philosopher; the first to propose a theory of the earth's movement around a "central fire."

philologist [fil äl′ə jist] *Linguistics.* an older term for a linguist.

philology [fil äl′ə jē] *Linguistics.* **1.** the study of written language, especially the study of literary texts. **2.** an older term for language study in general, now replaced by *linguistics.*

Philomiragia *Bacteriology.* a genus of bacteria, certain species of which are pathogenic in muskrats; formerly classified within the genus *Yersinia.*

Philomycidae *Invertebrate Zoology.* a family of slugs in the subclass Pulmonata, having no shell or gills, and a mantle cavity that acts as a respiratory organ.

philopatric *Ecology.* of or relating to species or groups that remain in or habitually return to their native regions or territories. Thus, **philopatry.**

Philopteridae *Invertebrate Zoology.* a family of biting lice in the order Mallophaga that are parasites on birds.

philtrum *Anatomy.* the vertical groove from the upper lip to the septum of the nose.

Philydraceae *Botany.* a family of erect, perennial, tanniferous herbs of the order Liliales, characterized by parallel-veined leaves, a rhizome, tuber, or thickened stem base, and no secondary growth; native to Australia.

phi meson *Particle Physics.* a neutral vector meson that has negative charge parity, G parity, and a mass of 1016 MeV; on decaying, it usually produces two kaons and three pions.

phimosis *Medicine.* an obstruction or constriction of the preputial orifice so that the prepuce cannot be retracted back over the glans.

Phiomyidae *Paleontology.* a family of Eocene rodents in the superfamily Thryonomyoidea; the phiomyids survived into the Miocene.

PHL *Aviation.* the airport code for Philadelphia International, Philadelphia, Pennsylvania.

phleb- or **phlebo-** a prefix meaning "vein."

phlebectomy *Surgery.* the removal of a vein or a segment of a vein.

phlebite *Petrology.* any roughly banded or veined variety of metamorphic rock or migmatite.

phlebitis [flə bī´tis] *Medicine.* the inflammation of a vein, marked by infiltration of the coats of the vein and accompanied by the formation of a thrombus; characterized by edema, stiffness, and pain in the affected part. Also, THROMBOPHLEBITIS.

phleboclysis *Surgery.* the injection of fluids into a vein.

phlebogram *Radiology.* **1.** an X-ray image of veins that have been injected with a contrast medium. **2.** a tracing of the venous pulse made with a phlebograph or sphymograph. Also, VENOGRAM.

phlebograph *Radiology.* an instrument used to record the venous pulse.

phlebography *Radiology.* **1.** the process of making a phlebogram. **2.** a description of the veins. Also, VENOGRAPHY.

phlebolith *Medicine.* a vein stone; a concretion or calculus in a vein.

phleboplasty *Surgery.* plastic operation for repair of a vein.

phlebosclerosis *Medicine.* the fibrous thickening or hardening of the wall of the veins.

phlebostasis *Medicine.* **1.** a retardation of the flow of blood in a vein. **2.** the temporary separation of a portion of the blood from the general circulation by applying a tourniquet on an extremity.

phlebothrombosis *Medicine.* the presence of a clot in a vein, not associated with inflammation of the wall of the vein.

phlebotomic *Zoology.* describing an organism, especially an insect, that sucks blood.

phlebotomy *Medicine.* the practice of drawing blood as a therapeutic measure, through venesection or the incision of a vein.

Phlebovirus *Virology.* a genus of viruses of the Bunyaviridae family that infect a wide host range of vertebrates and are transmitted mainly by phlebotomic sandflies.

phleger corer *Oceanography.* an instrument used to extract core samples from the ocean bottom by means of a weighted cylindrical assembly, having a tailfin and check valve that keeps water from entering the upper section and allows the core sample to be washed out as the corer is lifted.

phlegm [flem] *Physiology.* **1.** a viscous, mucous secretion, such as that which is produced by the mucous linings of the respiratory system and discharged through the mouth. **2.** in classical and medieval thought, one of the four humors of the body, regarded as producing calmness or lethargy.

phlegmasia alba dolens *Medicine.* an inflammation of the femoral vein, characterized by swelling of the leg, usually without redness; sometimes seen after childbirth or after a severe febrile illness.

phlegmasia cerulea dolens *Medicine.* an acute fulminating form of deep venous thrombosis characterized by reactive arterial spasms, pronounced edema of the affected extremity, severe cyanosis, purpuric areas, petechiae, severe pain accompanied by swelling, cyanosis, and edema of the extremity followed by circulatory collapse and shock.

phlegmatic [fleg mat´ik; flə mat´ik] *Psychology.* **1.** one of the four basic personality types defined in ancient and medieval thought, described as being calm or lethargic; said to result from the predominance of phlegm over the other bodily humors. **2.** of or relating to this personality type.

phlegmatic type see PHLEGMATIC, def. 1.

phlegmon *Medicine.* **1.** a spreading, diffuse inflammatory reaction to streptococcal infection, forming a suppurative or gangrenous lesion that may extend into the deep subcutaneous tissues and muscles. **2.** a solid, swollen, inflamed mass of pancreatic tissue occurring as a complication of pancreatitis.

Phleogenaceae *Mycology.* a family of fungi of the order Auriculariaceae, found on decaying plant matter in tropical to temperate regions.

phleomycin *Microbiology.* an antibacterial antibiotic produced by the bacterium *Streptomyces verticillatus* that is also used against tumors.

Phloeidae *Invertebrate Zoology.* the bark bugs, a family of hemipteran insects in the superfamily Pentatomoidea.

phloem [flō´əm] *Botany.* a food-conducting tissue that is found in vascular plants, composed mainly of sieve elements, parenchyma cells, fibers, and sclereids.

phloem necrosis *Plant Pathology.* a diseased condition of a plant in which the phloem turns brown and decays.

phloem ray *Botany.* a vascular ray or phloem plate between two medullary rays. Also, BAST RAY.

phlogopite *Mineralogy.* $KMg_3Si_3AlO_{10}(F,OH)_2$, a yellow-brown to copper-colored monoclinic mineral of the mica group, having a specific gravity of 2.76 to 2.90 and a hardness of 2 to 2.5 on the Mohs scale; found as prismatic crystals in ultrabasic rocks and metamorphosed limestones.

phlorhizin *Organic Chemistry.* $C_{21}H_{24}O_{10}{\cdot}2H_2O$, a bitter-tasting glycoside occurring as white needles; slightly soluble in water and soluble in alcohol; melts at 110°C; extracted from the root bark of apple, cherry, plum, and pear trees; used in medicine to produce glycosuria. Also, **phloridzin, phlorizin.**

phloroglucinol *Organic Chemistry.* **1.** $C_6H_3(OH)_3{\cdot}2H_2O$, odorless white-to-yellow crystals; slightly soluble in water and soluble in alcohol; melts at 218–221°C, depending on the heating intensity; used in the manufacture of pharmaceuticals, as a decalcifying agent for bones, and as a preservative for cut flowers. **2.** see 1,3,5-TRIHYDROXYBENZENE. *Botany.* a stain used to detect the distribution of lignin in woody tissue; generally acidified before application. Also, **phloroglucin.**

Phlyctidiaceae *Mycology.* a family of fungi belonging to the order Chytridiales that are parasitic to freshwater algae and microscopic animals; some species are plant pathogens.

pH measurement *Analytical Chemistry.* measurement of the activity of hydrogen ions in a solution using an indicator solution or a pH meter.

pH meter *Analytical Chemistry.* a millivolt meter that measures the potential difference between the pH electrode and the reference electrode in terms of pH value of the sample solution in which they are immersed.

phobia [fō´bē ə] *Psychology.* an obsessive, persistent fear of a certain object or situation that does not arouse similar fearful feelings in the average person.

-phobia *Psychology.* a combining form meaning "fear;" used to form nouns such as *claustrophobia.*

phobic [fō´bik] *Psychology.* **1.** relating to or affected with a phobia or phobias. **2.** a person affected with a phobia.

-phobic *Psychology.* a combining form meaning "fear;" used to form adjectives such as *claustrophobic.*

phobic disorder *Psychology.* a neurotic disorder marked by a morbid, persistent fear of some object or situation, in which the fear is significantly greater than the situation warrants. Also, **phobic neurosis.**

phobic reaction *Psychology.* any of various physiological manifestations of a phobic disorder, such as perspiring, breathlessness, rapid heartbeat, trembling, or nausea.

phobophobia *Psychology.* an irrational fear of being afraid, or of developing a phobia.

Phobos *Astronomy.* the larger of the two satellites of Mars, having a potato shape and numerous craters; measuring about 27 by 22 by 20 km.

Phocaenidae *Vertebrate Zoology.* the porpoises, a family of small, beakless marine mammals of the order Odontoceta, distinct from the dolphins; found in coastal waters throughout the Northern Hemisphere.

Phocidae *Vertebrate Zoology.* the earless seals or hair seals, a family of mostly marine mammals of the suborder Caniformia, characterized by a torpedo-shaped body and small front flippers, which give them excellent swimming ability, but make land dwelling difficult; found on ice fronts and coastlines in polar and temperate oceans.

phocomelia *Medicine.* a rare developmental anomaly in which the upper portion of one or more limbs is missing, so that the hands or feet or both are attached to the body by short, irregularly shaped stumps, resembling the flippers of a seal; the condition is seen primarily as a side effect of the drug thalidomide taken during early pregnancy. *Genetics.* a similar syndrome that is inherited as an autosomal recessive; in homozygous individuals, the hands and feet are attached directly to the trunk by a single bone. (From a Greek word meaning "seal.")

Phodilidae *Vertebrate Zoology.* in some classification systems, a monospecific family of bay owls in the order Strigiformes.

phoebe [fē´bē] *Vertebrate Zoology.* a perching bird of the genus *Sayornis* in the family Tyrannidae, characterized by a black to brownish gray body, a lighter underbelly, and a low crest on the head; its name is onomatopoetic for its call; found in eastern North America.

Phoebe *Astronomy.* Saturn IX, a moon of Saturn discovered in 1898, it has a diameter of about 230 km.

phoenicochroite *Mineralogy.* $Pb_2(CrO_4)O$, a red monoclinic mineral occurring in polycrystalline masses, having a specific gravity of 7.01 and a hardness of 2.5 on the Mohs scale; found in altered quartz-galena veins with other secondary minerals.

Phoenicopteridae *Vertebrate Zoology.* the flamingos, a family of large, long-necked, wading birds composing the order Phoenicopteriformes; distinguished by pink plumage, a large, crooked bill, and very thin, long legs with webbed toes; found in freshwater and marine shallow lakes and lagoons in South America, the Caribbean, Eurasia, and Africa.

Phoenicopteriformes *Vertebrate Zoology.* the monofamilial order of birds comprising the family Phoenicopteridae.

Phoeniculidae *Vertebrate Zoology.* the woodhoopoes, a monogeneric family of arboreal and climbing birds of the order Coraciiformes; characterized by scarlet legs and a long, slender scarlet bill, a long tail, and dark, multicolored glossy plumage; found in forests and wooded savannahs of sub-Saharan Africa.

Phoenix [fē´niks] *Astronomy.* the Firebird, a southern constellation that contains one 2nd-magnitude and two 3rd-magnitude stars, lying between Hydrus and Sculptor. *Ordnance.* a long-range air-to-air missile equipped with radar electronic guidance and homing; it has a top speed of Mach 5 and a range of 130 nautical miles; officially designated **AIM-54.** (From a famous bird in Greek mythology that upon its death was able to create itself anew from its own funeral ashes.)

Pholadidae *Invertebrate Zoology.* piddocks, a family of wood-boring marine clams, bivalve mollusks in the subclass Eulamellibranchia.

Pholadomya *Paleontology.* a genus of burrowing bivalves in the subclass Anomalodesmata and the order Pholadomyoida; distinguished by their wide posterior gape; extant from the Triassic to the present.

Pholidae *Vertebrate Zoology.* the gunnels, a family of marine fish of the order Perciformes, characterized by a slender, slimy, compressed body; found in the North Pacific and Atlantic Oceans. Also, **Pholididae.**

Pholidophoridae *Paleontology.* a family of ray-finned teleost fishes in the extinct order Pholidophoriformes; extant in the Triassic and Jurassic.

Pholidophoriformes *Paleontology.* the earliest teleosts, an extinct order of small neopterygian fishes in the division Teleosti; extant in the Triassic and Jurassic.

Pholidota *Vertebrate Zoology.* the pangolins or scaly anteaters, a monofamilial order of mammals covered with hard, epidermal scales and having a long, sticky tongue; native to Africa and southeast Asia.

Phomaceae see SPHAEROPSIDACEAE.

Phomales see SPHAEROPSIDALES.

phon [fän] *Acoustics.* a unit of loudness level, equal in decibels (referenced to 0.0002 microbars) to the mean sound-pressure level of a 1000-hertz tone presented to listeners facing the source.

phon- a combining form meaning "sound" or "voice," as in *phonic.*

phon. phonetic; phonetics.

phonation *Physiology.* the mechanism of producing sounds with the vocal cords.

phone *Telecommunications.* a shorter word for telephone. See TELEPHONE.

-phone a combining form meaning "sound" or "voice," as in *telephone.*

phoneme [fō´nēm´] *Linguistics.* the most basic unit of sound in a language; i.e., the smallest difference in sound that distinguishes one word from another. Thus in English the sounds [p] and [b] are phonemes that distinguish the word *pat* from the word *bat,* but the slightly varying sounds of the letter "p" in *pot* and *spot* are not phonemes because no two English words are distinguished from each other by this variation alone. English is typically described as having 44 or 45 phonemes.

phonemic [fə nē´mik] *Linguistics.* **1.** relating to phonemes or phonemics. **2.** relating to or used in a phonemic system of pronunciation.

phonemics *Linguistics.* the study of phonemes, or the description of a language according to its phonemes.

phonemic synthesizer *Acoustical Engineering.* a voice synthesizer that creates basic phonic sounds to simulate speech.

phonemic system *Linguistics.* a pronunciation system in which each unique meaningful sound of a language is represented by one and only one symbol; e.g., the letters "a" in *bake,* "ay" in *day,* "ea" in *steak,* or "ai" in *mail* all are represented by [ā].

phone patch *Electronics.* a device that is used to temporarily join an amateur or citizen-band transceiver to a telephone system.

phone plug *Electricity.* a standard, rod-shaped, two-contact plug having a .75-inch-diameter shank and an insulated ball-shaped tip; originally designed for patching telephone circuits, but now widely used in other electronic applications.

phonetic [fə net´ik] *Linguistics.* **1.** of or relating to phonetics. **2.** of or relating to the phonetic spelling of words.

phonetic alphabet *Linguistics.* **1.** a set of letters or similar symbols used to represent the sounds of a language or of various different languages, usually one in which each unique meaningful sound (phoneme) has a unique symbol. **2.** specifically, the International Phonetic Alphabet (IPA). See IPA. *Telecommunications.* in radio and telephone transmission, a list of symbols that has a separate representation for every distinguishable sound of human speech.

phonetics [fə net´iks] *Linguistics.* **1.** the study of the sounds that make up the human system of speech, including such aspects as the physical process by which sounds are produced by the speaker, the way in which sounds are perceived by the listener, and the relationship between sounds and the letters and other symbols used to represent sounds. **2.** a description or analysis of the sound system of a particular language.

phonetic search *Computer Programming.* an information retrieval technique that produces the initial request and any information of similar phonetic sound.

phonetic spelling *Linguistics.* **1.** a spelling of a word that corresponds closely or exactly to the way the word is pronounced; e.g., in Italian and Spanish most words have phonetic spellings because there are few silent letters and vowels have a single consistent pronunciation. **2.** an altered spelling of a word that corresponds more closely to the way the word is pronounced; e.g., "nite" for *night* or "wuz" for *was.*

phonic [fän´ik] *Linguistics.* of or relating to speech sounds. *Physiology.* of or relating to the voice.

phonic motor *Electricity.* a simple synchronous motor that operates under low power provided by electronic oscillators; the output frequency can be measured by a revolution counter. Also, **phonic wheel, phonic drum.**

phonics [fän´iks] *Linguistics.* a method of teaching reading that emphasizes the relationship of letters to the speech sounds they represent.

phonism *Neurology.* see AUDITORY SYNESTHESIA.

phono- a combining form meaning "sound" or "voice," as in *phonograph.*

phonoangicgraphy *Cardiology.* a graphic recording of the sounds that are produced by the flow of blood through an artery; used as a diagnostic technique for determining the extent to which blood vessels have narrowed.

phonocardiograph *Medicine.* an apparatus used to provide a continuous recording of beat-to-beat changes in heart rate, including irregular heart sounds and murmurs.

phonocatheterization *Cardiology.* the insertion of a microphone into the heart or major vascular system by catheter, in order to hear or record cardiac sounds.

phonoelectrocardioscope *Medicine.* an instrument incorporating a double-beam, cathode-ray oscilloscope with a fluorescent screen, permitting the simultaneous, direct visual recording of two phenomena, such as the phonocardiogram and the electrocardiogram.

phonograph *Acoustical Engineering.* an instrument designed to reproduce sound from records, consisting of a turntable, an arm with a stylus for playback, and one or more channels including a voltage amplifier and loudspeaker. Also, RECORD PLAYER.

phonographic *Acoustical Engineering.* of or relating to a phonograph or to sound reproduced in this way.

phonograph pickup *Acoustical Engineering.* a transducer that is located at the end of a turntable arm and converts variations in grooves on a record, as sensed by a pickup needle, into a signal for processing by a phonograph.

phonograph record *Acoustical Engineering.* a vinylite or shellac disk with grooves representing a recorded sound, used for reproducing the sound as a phonograph needle travels through it.

phono jack *Electronics.* a receptacle that accepts a phono plug to an amplifier in an audio system.

phonolite *Petrology.* a light-colored, extrusive igneous rock or rock group, consisting primarily of alkali feldspar and feldspathoids; the extrusive equivalent of nepheline syenite.

phonon *Solid-State Physics.* a quantum of vibration excited by the acoustic mode of a crystal lattice; the vibration is usually thermally excited.

phonon-electron interaction *Solid-State Physics.* an interaction between a thermal vibration in a crystal lattice and an electron, resulting in a change in the phonon wave vector and the electron momentum.

phonon emission *Solid-State Physics.* the generation of phonons in response to electron scattering within the crystal lattice, to anharmonic lattice forces due to the interaction with other phonons, or to X-ray or high-energy particle bombardment.

phonon wind *Solid-State Physics.* a stream of nonthermal phonons that propels charge carriers through a crystal.

phonophobia *Psychology.* **1.** the fear of one's own voice, or of speaking. **2.** an irrational fear of any noise or sound.

phono plug *Electronics.* a connector used on wires that carry audio signals; used to interconnect components of an audio system.

phonopsia *Neurology.* the condition in which visualization of color is caused by acoustic stimulation.

phonoreception *Physiology.* the means by which a nerve cell receives sound stimuli.

phonoscope *Cardiology.* an instrument for monitoring the movements of the heart photographically, by recording the movements of the diaphragm as it is affected by the heart's sounds.

phonotelemeter *Engineering.* an instrument that approximates the distance of firearms in action by measuring the interval between the flash and the arrival of the sound from the discharge.

phony disease *Plant Pathology.* a disease of peaches and other fruit trees, caused by the virus *Nanus mirabilis*; characterized by abnormally dark green leaves, dwarfing, and an eventual cessation of fruit production.

phony mine *Ordnance.* an object used to simulate a mine or provide false signals to mine detectors. Similarly, **phony minefield.**

phorate *Organic Chemistry.* $C_7H_{17}O_2PS_3$, a toxic clear liquid; insoluble in water and miscible with carbon tetrachloride; boils at 118–120°C (0.8 torr); used as an insecticide.

phorbol ester *Biochemistry.* a highly carcinogenic polycyclic compound derived from croton oil that can activate protein kinase C by mimicking diacylglycerol. Also, TUMOR PROMOTERS.

-phore *Biology.* a suffix meaning "a thing or part that carries or bears; a carrier."

phoresy *Ecology.* a type of dispersal in which one animal clutches on to the body of a larger animal of another species and rides along with it for some distance before leaving.

Phoridae *Invertebrate Zoology.* hump-backed flies, a family of small dipteran insects in the series Aschiza, having short antennae; adults frequent plants, and larvae are found in decaying organic matter.

Phormidium *Bacteriology.* a genus of cyanobacteria that forms mats of filamentous cells on the sediments of marine lagoons or salt-marsh habitats.

phorogenesis *Geology.* the shifting or slipping of the crust with respect to the mantle.

Phoronida *Invertebrate Zoology.* a small phylum of small, wormlike marine animals, having tentacles around the mouth, that live in a chitinous tube in shallow coastal waters.

phorophyte *Ecology.* a plant that provides physical support for another plant growing on it, but does not provide nourishment; commonly found in tropical rain forests.

phose [fōz] *Neurology.* any subjective sensation, especially of light or color.

phosgene see CARBONYL CHLORIDE.

phosgenite *Mineralogy.* $Pb_2(CO_3)Cl_2$, a colorless to white, yellow, gray, or brown tetragonal mineral with adamantine luster, occurring as prismatic to tabular crystals, and having a specific gravity of 6.133 and a hardness of 2 to 3 on the Mohs scale; often found in association with cerussite in the oxidized zones of lead deposits.

phosis *Neurology.* the production of subjective sensations, especially of light or color.

phosph- *Chemistry.* a combining form indicating the presence of phosphorus or phosphoric acid.

phosphagen *Biochemistry.* a substance, such as phosphocreatine, that can phosphorylate ADP and thus act as a reserve of ATP.

phosphatase *Enzymology.* any enzyme that catalyzes the hydrolysis of an ester of phosphoric acid.

phosphate [fäs′fāt′] *Chemistry.* **1.** any salt of phosphoric acid containing the PO_4^{3-} group. **2.** any compound containing the phosphate group. *Mineralogy.* any mineral compound containing the anionic tetrahedral PO_4^{3-} group.

phosphate anion *Inorganic Chemistry.* PO_4^{3-}, a negatively charged ion derived from phosphoric acid.

phosphate buffer *Analytical Chemistry.* a standard laboratory solution that is prepared from KH_2PO_4 and Na_2HPO_4; used as a pH reference solution.

phosphate cement *Materials.* an inorganic adhesive produced by mixing phosphoric acid with oxides or silicates; used in dental work.

phosphate coating *Metallurgy.* anticorrosion coating consisting of a phosphate.

phosphate desulfurization *Chemical Engineering.* a petroleum-refining process that uses tripotassium phosphate to remove hydrogen sulfide from gases.

phosphate glass *Materials.* a transparent glass composed primarily of phosphorus pentoxide; resistant to fluorine chemicals but not to water.

phosphate group *Chemistry.* the phosphate functional group PO_4^{3-}.

phosphate recovery process *Mining Engineering.* a method for recovering phosphate from low-grade, phosphorus-bearing shale; established by the U.S. Bureau of Mines.

phosphate rock *Materials Science.* a natural rock that consists largely of calcium phosphate; used as source of elemental phosphorus and as a raw material in the manufacture of phosphate fertilizers, phosphoric acid, and animal feed. Also, PHOSPHORITE.

phosphate surveying *Archaeology.* a technique of analyzing soil samples for phosphates; high concentrations may indicate the location of human settlements.

phosphatic *Chemistry.* of, relating to, or containing a phosphate or phosphoric acid.

phosphatic nodule *Geology.* a black, gray, or brown earthy mass or pebble of variable size and shape, having a hard, shiny surface; found in marine strata or forming along the sea floor; containing or cemented by calcium phosphate.

phosphatidic acid *Biochemistry.* a compound, containing no head alcohol, from which phosphoglycerides are derived.

phosphating *Metallurgy.* forming a phosphate coating on a metallic part. Also, **phosphatizing.**

phosphazene see POLYPHOSPHAZENE.

phosphide *Inorganic Chemistry.* any compound composed only of phosphorus and another element.

phosphine *Inorganic Chemistry.* PH_3, a colorless, poisonous, inflammable gas or colorless liquid that is slightly soluble in cold water and soluble in alcohol; melts at –133.5°C and boils at –87.4°C. It ignites spontaneously, is toxic on inhalation, and is a strong irritant; used in organic preparations and as a condensation catalyst. Also, HYDROGEN PHOSPHIDE.

phosphinic acid *Organic Chemistry.* an organic derivative of hypophosphorous acid that contains the radical $-H_2PO_2$ or $=HPO_2$.

phosphite *Inorganic Chemistry.* any salt of phosphorous acid containing the PO_3^{3-} ion.

phosphite ester *Inorganic Chemistry.* a compound of the type $P(OR)_3$, where *R* is an aryl or alkyl group.

phospho- *Chemistry.* a combining form indicating the presence of phosphorus or phosphoric acid.

phosphocreatine *Biochemistry.* $C_4H_{10}N_3O_5P$, a compound, found in vertebrate muscle tissue, that is hydrolyzed under anaerobic conditions to provide energy for muscle contractions. Also, PC.

phosphodiesterase *Biochemistry.* an enzyme that catalyzes the hydrolytic cleavage of a phosphodiester bond, as in the conversion of cyclic AMP to 5′-adenylate.

phosphodiester bond *Biochemistry.* either of two ester bonds present in compounds that are obtained by esterification of two hydroxyl groups of the same phosphoric acid molecule.

phosphoenolpyruvic acid *Biochemistry.* $CH_2{=}O(OPO_3H_2)COOH$, a phosphate produced from 2-phosphoglyceric acid that gives a phospho group to ADP in the Embden-Meyerhof pathway and to hexoses in anaerobic bacteria.

phosphoferrite *Mineralogy.* $(Fe^{+2},Mn^{+2})_3(PO_4)_2 \cdot 3H_2O$, a colorless, pale green, or reddish-brown orthorhombic mineral occurring as octahedral crystals and in granular form, having a specific gravity of 3.1 to 3.29 and a hardness of 3 to 3.5 on the Mohs scale; found as an alteration mineral in granite pegmatites.

phosphoglucoisomerase *Enzymology.* an enzyme that catalyzes the conversion of glucose 6-phosphate to fructose 6-phosphate.

phosphoglucomutase *Enzymology.* an enzyme that catalyzes the interconversion of glucose 1-phosphate and glucose 6-phosphate.

phosphogluconate pathway see PENTOSE PHOSPHATE PATHWAY.

phosphoglyceride *Biochemistry.* a component in membranes that is produced when glycerol forms ester bonds with two molecules of phosphatidic acid.

phospholan *Organic Chemistry.* $C_6H_{14}O_3PNS$, a colorless to yellow solid that melts at 37–45°C; used to kill mites in cotton crops.

phospholipase *Enzymology.* any enzyme that catalyzes the hydrolysis of a phospholipid.

phospholipid *Biochemistry.* a lipid with phosphorus in the form of phosphoric acid, functioning as a structural component in membranes. Also, **phospholipin.**

phosphomolybdic acid *Inorganic Chemistry.* $H_3PMo_{12}O_{40} \cdot H_2O$, yellowish crystals that are soluble in water and alcohol and that melt at 78–90°C; a strong oxidizing agent in aqueous solution; used as a reagent and catalyst. Also, 12-MOLYBDOPHOSPHORIC ACID.

phosphomonoesterase *Enzymology.* **1.** any enzyme that catalyzes the hydrolysis of an ester (R–O–PO(OH)$_2$) to orthophosphate and the alcohol R–OH. **2.** see ALKALINE PHOSPHATASE.

phosphonecrosis see PHOSPHORUS NECROSIS.

phosphonic acid *Organic Chemistry.* any of various acids having the general formula $AOP(OH)_2$, where A is an organic radical; for example, phenylphosphonic acid, $C_6H_5OP(OH)_2$.

phosphonitrile *Materials.* a plastic that is derived from chlorophosphontitrile; noted for high elasticity and temperature resistance.

phosphophyllite *Mineralogy.* $Zn_2(Fe^{+2},Mn^{+2})(PO_4)_2 \cdot 4H_2O$, a colorless to pale blue-green monoclinic mineral, having a specific gravity of 3.13 and a hardness of 3 to 3.5 on the Mohs scale; found as prismatic to tabular crystals in granite pegmatite.

phosphoprotein *Biochemistry.* a protein made up of certain phospho groups, including serine and threonine phosphates.

phosphoprotein phosphatase *Enzymology.* an enzyme that removes phosphate from a phosphorylated protein.

phosphor [fäs´fôr´] *Physics.* a material that is capable of phosphorescence; i.e., that can absorb any of various types of energy and emit visible light after the energy source is removed.

phosphoramidite method *Biotechnology.* a nonaqueous solid-phase technique for synthesizing oligonucleotides; the growing chain is anchored to a solid support and then reacted with the 5'-end of the next nucleotide to be polymerized.

phosphor bronze *Metallurgy.* any of several copper-tin alloys containing up to 0.35% phosphorus.

phosphor dot *Electronics.* any of the tiny dots of phosphor material that represents one of the primary colors on the screen of a television picture tube; each dot is grouped with two others on the screen.

phosphoresce [fäs´fə res´] *Physics.* to exhibit phosphorescence.

phosphorescence [fäs´fə res´əns] *Physics.* **1.** an emission of light that follows the absorption of energy by a substance from sources such as visible light, infrared light, ultraviolet light, electricity, cathode rays, or X-rays, and that continues for a relatively long time after the energy supply has ceased. **2.** any continuing emission of light that occurs in the absence of significant heat, such as the bioluminescence of fireflies.

phosphorescent [fäs´fə res´ənt] *Physics.* exhibiting phosphorescence.

phosphoric acid *Inorganic Chemistry.* H_3PO_4, a colorless liquid or rhombic crystals; very soluble in cold water; deliquescent; melts at 29.32°C; it is toxic and an irritant, and is corrosive to ferrous metals; used in fertilizers, pharmaceuticals, ceramics, food, and carbonated beverages. Also, ORTHOPHOSPHORIC ACID.

phosphoric acid polymerization *Chemical Engineering.* a refinery process that uses phosphoric acid to catalyze polymerization of unsaturated hydrocarbons into compounds that boil in the gasoline range.

phosphoric anhydride *Inorganic Chemistry.* P_4O_{10}, a white powder, very deliquescent; melts at 580–585°C and sublimes at 300°C. It reacts violently with water to produce heat; a fire hazard and corrosive to skin; used as a surfactant and dehydrating agent.

phosphoric diesterhydrolase *Enzymology.* an enzyme that catalyzes the hydrolysis of a diester of phosphoric acid (R–O–PO_2^-–O–R').

phosphoric monoester hydrolase *Enzymology.* an enzyme that catalyzes the hydrolysis of a monoester of phosphoric acid.

phosphorimetry *Analytical Chemistry.* an analytical procedure in which a test substance is irradiated with ultraviolet light and the amount of phosphorescence emitted is measured.

phosphorism *Toxicology.* poisoning due to ingestion or inhalation of phosphorus; symptoms may include toothache and jaw problems, loss of appetite, weakness, and anemia. Also, **phosphorus poisoning.**

phosphorite *Petrology.* a sedimentary rock characterized by an abundance of phosphate minerals.

phosphorite see PHOSPHATE ROCK.

phosphorized copper *Metallurgy.* copper that has been deoxidized by adding phosphorus; electrical conductivity is impaired if the residual phosphorus exceeds 0.01%.

phosphorogen *Physics.* a substance that allows the phenomenon of phosphorescence when it is combined with another substance, as manganese combined with zinc sulfide.

phosphorolysis *Biochemistry.* the breaking of a bond within a molecule by reaction with phosphoric acid.

phosphorous acid *Inorganic Chemistry.* H_3PO_3, a colorless, vitreous, deliquescent mass; soluble in alcohol; melts at 70°C and decomposes at 200°C; used in testing for mercury, and as a reducing agent. Also, ORTHOPHOSPHOROUS ACID.

phosphorous salt see SODIUM AMMONIUM (HYDROGEN) PHOSPHATE.

phosphorrösslerite *Mineralogy.* $MgHPO_4 \cdot 7H_2O$, a yellowish, monoclinic mineral occurring as small crystals, having a specific gravity of 1.73 and a hardness of 2.5 on the Mohs scale. Also, **phosphor rösslerite, phosphorroesslerite.**

phosphor tin *Metallurgy.* a master alloy of tin and phosphorus, used to make bronzes.

Phosphorus *Astronomy.* the ancient Greek name for the planet Venus when seen as a morning star.

phosphorus [fäs´fər əs] *Chemistry.* a widely occurring nonmetallic element having the symbol P, the atomic number 15, an atomic weight of 30.9738, a melting point of 44.1°C (white form), and a boiling point of 280°C (white form); it occurs in three main allotropic forms: **white** (or **yellow**) **phosphorus, red phosphorus,** and **black phosphorus.** It is used in matches, fertilizers, pesticides, and various other industrial products. Phosphorus is an essential element of the human diet and is the main component of bones and teeth; it is involved in some form in virtually all processes of metabolism. (From *Phosphorus,* a bright star; going back to the Greek word for "light;" referring to the property of white phosphorus of igniting in air at ordinary temperatures.)

phosphorus necrosis *Toxicology.* ulceration, tissue death, and bone damage in the jaw of an individual chronically exposed to yellow phosphorus. Also, PHOSPHONECROSIS, PHOSSY JAW.

phosphorus nitride *Inorganic Chemistry.* P_3N_5, an amorphous white solid; insoluble in cold water and soluble in organic solvent; decomposes at 800°C; used in doping semiconductors.

phosphorus-nitrogen ratio *Oceanography.* the almost constant ratio of phosphorus to nitrogen in seawater and in plankton, by weight about 7:1 and by atom about 15:1.

phosphorus oxide *Inorganic Chemistry.* any compound containing phosphorus and oxygen only, such as P_4O_{10}, phosphoric anhydride.

phosphorus oxychloride *Inorganic Chemistry.* $POCl_3$, a colorless fuming liquid; decomposes in water, alcohol, and acid; melts at 2°C and boils at 105.3°C; toxic and a strong irritant; used in gasoline additives, hydraulic fluids, and as a catalyst. Also, PHOSPHORYL CHLORIDE.

phosphorus pentabromide *Inorganic Chemistry.* PBr_5, a yellow rhombic mass; decomposes in water and at 106°C; hygroscopic; used as a brominating agent. Also, **phosphoric bromide.**

phosphorus pentachloride *Inorganic Chemistry.* PCl_5, a yellowish to white tetragonal, fuming, crystalline mass; decomposes in water and acid; sublimes at 162°C; used as a catalyst and a chlorinating and dehydrating agent. Also, **phosphoric chloride.**

phosphorus pentasulfide *Inorganic Chemistry.* P_2S_5, a toxic, gray to yellow crystal; insoluble in cold water and decomposes in hot water, deliquescent; melts at 286–290°C and boils at 514°C. It is a dangerous fire hazard, ignited by friction; used in insecticides, safety matches, and ignition compounds. Also, THIOPHOSPHORIC ANHYDRIDE.

phosphorus sesquisulfide *Inorganic Chemistry.* P_4S_3, a toxic, yellow rhombic crystalline substance; insoluble in cold water and decomposes in hot water; melts at 174°C and boils at 408°C; an irritant and a dangerous fire hazard; used in matches and organic synthesis. Also, TETRAPHOSPHORUS TRISULFIDE.

phosphorus thiochloride *Inorganic Chemistry.* $PSCl_3$, a colorless fuming liquid that decomposes in water; melts at –35°C and boils at 125°C; used in insecticides and oil additives.

phosphorus tribromide *Inorganic Chemistry.* PBr_3, a colorless, fuming, corrosive liquid with a penetrating odor; used in analysis, synthesis, and as a catalyst.

phosphorus trichloride *Inorganic Chemistry.* PCl_3, a colorless, fuming, corrosive liquid that is soluble in ether and decomposes in water; melts at –112°C and boils at 75.5°C; used as an additive and catalyst. Also, **phosphorus chloride.**

phosphorus triiodide *Inorganic Chemistry.* PI_3, red, hexagonal, deliquescent crystals that decompose in water; melts at 61°C; flammable and an irritant; used in organic synthesis.

phosphorus trisulfide *Inorganic Chemistry.* P_2S_3 or P_4S_6, grayish-yellow crystals that are soluble in alcohol and decompose in water; melts at 290°C and boils at 490°C; a dangerous fire hazard; used as an organic reagent. Also, TETRAPHOSPHORUS HEXASULFIDE, THIOPHOSPHORUS ANHYDRIDE.

phosphorylase *Enzymology.* an enzyme that catalyzes phosphorolysis.

phosphorylation *Organic Chemistry.* a process in which a phosphate group is combined with an organic molecule; occurs naturally in cellular metabolism.

phosphoryl chloride see PHOSPHORUS OXYCHLORIDE.

phosphoserine *Biochemistry.* $(HO)_2PO{-}CH_2{-}CH({-}NH_2){-}COOH$, an amino acid that forms when the phosphorylation of the hydroxyl group of serine is stimulated, frequently as the result of hormonal action.

phosphosiderite *Mineralogy.* $Fe^{+3}PO_4{\cdot}2H_2O$, a reddish-violet to yellow monoclinic mineral, dimorphous with strengite, occurring as tabular crystals and fibrous botryoidal crusts, and having a specific gravity of 2.76 and a hardness of 3.5 to 4 on the Mohs scale; found in pegmatites.

phosphotransacetylase *Enzymology.* an enzyme that transfers an acetyl group between coenzyme A and orthophosphate.

phosphotransferase *Enzymology.* an enzyme that catalyzes the movement of a phosphate group from a donor molecule to an acceptor.

phosphotungstic acid *Inorganic Chemistry.* $H_3PW_{12}O_{40}{\cdot}xH_2O$, yellow to green triclinic crystals; soluble in water and alcohol and decomposes at its melting point; a strong irritant; used as a reagent and catalyst. Also, PHOSPHOWOLFRAMIC ACID, TUNGSTOPHOSPHORIC ACID.

phosphotungstic pigment *Organic Chemistry.* a green or blue pigment manufactured by precipitating basic dyestuffs with solutions of phosphotungstic acid or phosphomolybdic acid; used in inks, paints, and enamels. Also, TUNGSTEN LAKE.

phosphowolframic acid see PHOSPHOTUNGSTIC ACID.

phosphuranylite *Mineralogy.* $Ca(UO_2)_3(PO_4)_2(OH)_2{\cdot}6H_2O$, a deep yellow, strongly radioactive, orthorhombic mineral occurring as powdery aggregates, having a specific gravity of about 4.1 and a hardness of about 2.5 on the Mohs scale; found as a secondary mineral in pegmatites and sandstone uranium deposits.

phossy jaw see PHOSPHORUS NECROSIS.

phosvitin *Biochemistry.* a phosphoprotein isolated from vitellin in egg yolk. Also, **phosphovitellin.**

phot *Optics.* a unit used for very high levels of illumination, equal to 1 lumen per square centimeter or 10,000 lux. Also, CENTIMETER CANDLE.

phot. photograph; photography; photographic.

photic *Biology.* of or relating to light, especially as it affects living organisms.

photic zone *Ecology.* the section of an aquatic system in which there is enough sunlight for photosynthesis to occur.

Photidae *Invertebrate Zoology.* a family of sandhoppers, marine amphipods in the suborder Gammaridea, having a laterally compressed body and no carapace.

photism *Neurology.* a subjective sensation of light or color induced by a stimulus of another sense, such as hearing, taste, smell, or touch.

photo- or **phot-** a combining form meaning "light."

photoablation *Surgery.* the removal of tissue through volatilization using laser-generated ultraviolet radiation.

photoacoustic spectroscopy *Spectroscopy.* a spectroscopic technique for obtaining spectra of opaque samples, such as powders, in which a sample is exposed to acoustic-frequency-modulated light to produce a photoacoustic signal whose magnitude is related to the ultraviolet or infrared absorption coefficient of the sample.

photoaddition *Physical Chemistry.* a process that uses radiant energy to produce a single molecule from two unsaturated molecules.

photoallergic *Immunology.* relating to, characterized by, or producing photoallergy.

photoallergy *Immunology.* a delayed immunologic type of photosensitivity involving radiant energy and a chemical substance to which the individual was previously sensitized.

photoautotroph *Biology.* an autotroph that derives its energy and nourishment from light.

photoautotrophic *Biology.* autotrophic and obtaining energy from light.

photobacterium *Bacteriology.* **1.** any bacterium producing luminescent substances. **2.** an individual organism of the genus *Photobacterium.*

Photobacterium *Bacteriology.* a genus of Gram-negative, motile, facultatively anaerobic, spheroid or rod-shaped bacteria of the family Vibrionaceae, possessing polar flagella and distinguished by a requirement for high concentrations of salt. Two species, *P. phosphoreum* and *P. leiognathi,* are bioluminescent, occurring in sea water, in the alimentary tract of certain fishes, and on the luminous organs of certain fishes and cephalopods.

photobase *Cartography.* in aerial photography, the distance between the principal points of two adjacent photographs. Also, BASELINE.

photobiological hydrogen production *Biotechnology.* any of a wide variety of biological processes that use light and a biocatalyst to produce hydrogen and oxygen.

photobiology *Biology.* a branch of biology dealing with the effect of light and other forms of radiant energy on living organisms and biological processes. Thus, **photobiologist.**

photobiotic *Biology.* living or thriving only in light.

photocatalysis *Physical Chemistry.* a spontaneous reaction that is promoted by or speeded up by radiant energy.

photocatalyst *Physical Chemistry.* a substance that facilitates the utilization of sunlight, as chlorophyll does in photosynthesis.

photocathode *Electronics.* an electrode that emits electrons when exposed to electromagnetic radiation such as light; found in phototubes, television camera tubes, and other devices that are sensitive to light.

photocell *Electronics.* a device that generates electrical energy, usually as a voltage or current, from light energy. Also, PHOTOELECTRIC CELL, ELECTRIC EYE.

photocell relay *Electronics.* a switching circuit that is activated by a signal sent by a photocell when it absorbs or does not absorb light.

photochemical *Physical Chemistry.* relating to or involving the chemical effect of light.

photochemical cell see BECQUEREL CELL.

photochemical decomposition see PHOTOLYSIS.

photochemical oxidant *Physical Chemistry.* a substance that becomes an oxidant through the absorption of light or another form of radiant energy.

photochemical reaction *Physical Chemistry.* a chemical change brought about by the interaction of the molecules of a substance with light; e.g., photosynthesis, photography.

photochemical smog *Meteorology.* the resulting chemical pollutants in existence in the atmosphere from the burning of hydrocarbons and the nitrogen oxides that are released in the presence of sunlight.

photochemistry *Physical Chemistry.* the branch of chemistry concerned with chemical reactions caused or influenced by the effect of light. Thus, **photochemist.**

photochromic *Physical Chemistry.* relating to, characterized by, or producing photochromism.

photochromic compound *Organic Chemistry.* an organic compound that undergoes a reversible color change when exposed to a suitable wavelength of light.

photochromic glass *Materials.* a chemically treated glass that darkens or changes color when exposed to light.

photochromic reaction *Physical Chemistry.* any reaction that involves the change of color of a substance due to incident radiation.

photochromism *Physical Chemistry.* a color change, usually darkening, that certain organisms undergo when exposed to light of a particular wavelength.

photochromogen *Microbiology.* any microorganism in which the synthesis of carotenoid pigments and development of yellow or orange color pigmentation requires exposure of the cells to light; the best-known examples are bacteria of the genus *Mycobacteria,* some species of which are pathogenic to humans.

photoclinometer *Engineering.* a type of surveying instrument that photographically measures the distance to which a well deviates from the vertical.

photoclinometry *Geology.* a method for obtaining information about the slope of a feature from an image brightness distribution.

photocoagulator *Medicine.* an instrument that uses a controlled intense beam of light, such as a xenon arch light or argon laser, to treat retinal detachment and destruction of abnormal retinal vessels.

photocomposer *Graphic Arts.* a machine designed to arrange and reproduce multiple images; used in making composite negatives and printing plates.

photocomposition *Graphic Arts.* any of several processes of producing and arranging type photographically, such as phototypesetting.

photocompositor *Graphic Arts.* **1.** a machine designed to produce and arrange type by photographic means. **2.** a person who operates a phototypesetting machine.

photoconduction *Solid-State Physics.* an effect, exhibited in many substances, in which the electrical conductivity is increased when the substance is exposed to electromagnetic radiation, usually in the visible region of the spectrum. Thus, **photoconductive.**

photoconductive cell *Electronics.* a device in which conductance is proportional to the amount of light striking the cell; used to measure or detect electromagnetic radiation.

photoconductive device *Electronics.* a device that changes the level of an electric signal when light alters its conductivity.

photoconductive film *Electronics.* a film of material that conducts electricity more easily when exposed to light.

photoconductive gain factor *Electronics.* the ratio of the number of electrons flowing, per second, through a cube of semiconducting material to the number of photons absorbed, per second, by this material.

photoconductive meter *Electronics.* a device in which a battery provides power through a photoconductive cell attached to a milliammeter; used to measure light intensity.

photoconductivity *Solid-State Physics.* the property or degree of increase in electrical conductivity of some nonmetallic solids when exposed to light.

photoconductivity gain *Electronics.* the ratio between the number of electrons flowing through a photoconductive circuit to the number of electrons generated by light.

photoconductor *Solid-State Physics.* any nonmetallic solid that exhibits an increase in electrical conduction as the result of being exposed to electromagnetic radiation.

photocontour map *Cartography.* a topographic map on which aerial photographs are used to show planimetric details in their proper positions.

photocontrol index map *Photogrammetry.* a map or photoindex on which points used for ground control are depicted and identified.

photocooperation *Entomology.* a mutualistic relationship in which two separate species prosper; e.g., when ants and aphids forage on the same shrubs, the ants stimulate aphids to secrete honeydew directly into their mouths, while the ants provide protection to the aphids from natural predators.

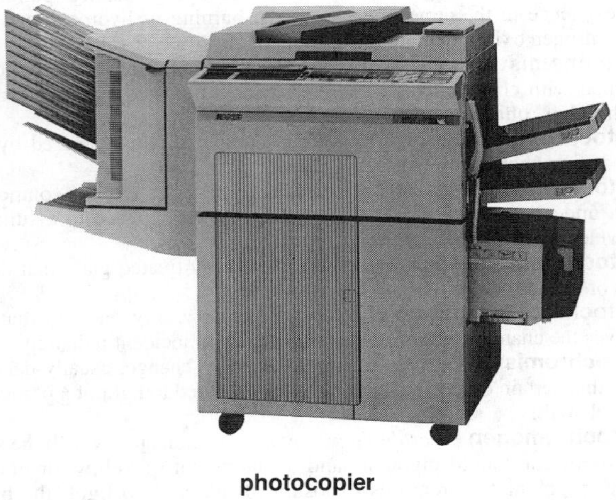

photocopier

photocopier *Graphic Arts.* a reprographic machine designed to make photographic copies, usually by an electrostatic printing process. Also, **photocopy machine, photocopying machine.**

photocopy *Graphic Arts.* **1.** a photographic reproduction, especially one produced by electrostatic printing. **2.** to make such a reproduction.

photocoupler see OPTOISOLATOR.

photocurrent *Physical Chemistry.* an electric current produced when radiant energy, such as ultraviolet light, strikes an electrode.

Photocyta *Bacteriology.* a proposed taxonomic group of bacteria that includes the halobacteria and the eubacteria, based on their common three-dimensional ribosomal structure

photodecomposition see PHOTOLYSIS.

photodegradation *Physical Chemistry.* any process in which radiant energy decomposes a substance.

photodelineation *Photogrammetry.* the selection and enhancement of significant surface features by graphic means directly on an aerial photograph.

photodermatitis *Medicine.* any abnormal skin reaction in which light is an important causative factor.

photodetachment *Physical Chemistry.* a process in which a photon of light is absorbed and an electron is detached.

photodetector *Electronics.* a device that senses and measures light. Also, PHOTOSENSOR.

photodichroic material *Optics.* a material containing photosensitive molecules that, when exposed to plane polarized light, will show variation in color depending on the orientation of the molecules.

photodiffusion effect see DEMBER EFFECT.

photodimerization *Physical Chemistry.* a process in which an unexcited molecule gains atoms from a highly charged unsaturated molecule of the same species as a result of radiant energy.

photodiode *Electronics.* a diode in which current flow is regulated by light intensity. Also, **photoconductor diode.**

photodisintegration *Nuclear Physics.* the decomposition of a nucleus that takes place after it absorbs gamma rays.

photodisruption *Surgery.* the disruption of tissues by laser-generated rapid molecular ionization.

photodissociation *Physical Chemistry.* a process in which atoms are removed from a molecule by the absorption of a quantum of radiant energy.

photodosimetry *Nucleonics.* a method of dosimetry in which photographic film is used to record the cumulative dose of ionizing radiation.

photodraft *Mechanical Devices.* the photographic equivalent of a design layout on sheet metal containing an emulsion coating; used in tool fabrication operations.

photodynamic *Optics.* powerful in the light; said especially of the action of fluorescent substances in the light.

photodynamic effect *Microbiology.* a phenomenon in which certain microorganisms, especially bacteria, are damaged or killed when treated with certain fluorescent dyes and subsequently exposed to light under aerobic conditions.

photodynamics *Biology.* the study of the effect of light on living organisms.

photodynamic therapy *Oncology.* an antitumor light-radiation procedure in which an intravenously injected substance concentrates selectively in metabolically active tumor tissue where, upon exposure to red laser light of a specific wavelength, it is converted into molecules toxic to the tumor tissue. Also, PHOTORADIATION.

photoecology *Ecology.* the use of photography, especially aerial photography, in ecological studies.

photoelasticity *Optics.* a technique in which double refraction is induced in materials subject to mechanical stress, such as plastic, so that disturbance patterns can be analyzed and corrected. Thus, **photoelastic, photoelastic effect.**

photoelectret *Solid-State Physics.* an electret produced from a photoconductor in an electric field in the absence of light.

photoelectric *Electronics.* of or relating to the electrical effects of light, including the emission of electrons, the generation of a voltage, or a change in resistance.

photoelectric absorption *Electronics.* an emission of electrons caused by illumination.

photoelectric cell see PHOTOCELL.

photoelectric colorimeter *Engineering.* an instrument that gives a reading for a quantitative determination of color, by means of a phototube or photocell, various color filters, an amplifier, and a meter.

photoelectric colorimetry *Analytical Chemistry.* the determination of the concentration of a colored solution using the tri-stimulus values of three primary light filter-photocell combinations.

photoelectric constant *Electronics.* the ratio between the frequency of the radiation causing electron flow and the voltage equaling the radiation absorbed by the electron; this corresponds to Planck's constant divided by the electron charge.

photoelectric control *Control Systems.* any control device or system based on changes caused by a photosensitive device absorbing light or other forms of electromagnetic radiation; used especially to control the rate of operation of a piece of equipment. Thus, **photoelectric control device, photoelectric control system, photoelectric cutoff control.**

photoelectric counter *Electronics.* a device in a photocell or phototube that tallies the number of times an object intercepts the light beam shining on it.

photoelectric densitometer *Engineering.* a dust-sampling instrument that measures the density or opacity of a thin material by shooting a beam of light through the material and measuring the light transmitted with a photocell and meter.

photoelectric device *Electronics.* a device that generates an electrical signal when exposed to infrared, visible, or ultraviolet radiation.

photoelectric door opener *Control Systems.* a control system that uses a photoelectric cell as part of a device that opens and closes a power-driven door.

photoelectric effect *Physics.* a phenomenon in which electrons are liberated from a material when exposed to electromagnetic radiation of sufficient frequency.

photoelectric flame-failure detector *Control Systems.* a device that uses a photoelectric cell to detect when a fuel-consuming flame is extinguished in order to cut off the flow of fuel.

photoelectric fluorometer *Engineering.* a device that uses a photoelectric cell to determine the amount of fluorescence in a chemical sample that has been subjected to ultraviolet or visible light.

photoelectric imaging *Graphic Arts.* a technology used to scan an image, convert it to an electronic signal, and reconvert it to a corresponding light pattern on a photosensitive surface; used, for example, in color separation, platemaking, and facsimile (FAX) transmission.

photoelectric infrared radiation see NEAR-INFRARED RADIATION.

photoelectric intrusion detector *Electronics.* a burglar-alarm system that is activated when an intruder interrupts the beam of light illuminating a phototube.

photoelectricity *Electronics.* the electricity produced when light or other forms of electromagnetic radiation strike certain materials such as cesium, selenium, and silicon.

photoelectric lighting control *Electronics.* a means of varying the light intensity in a given area by employing a photoelectric relay.

photoelectric liquid-level indicator *Engineering.* a device used to determine the liquid level in a tank or process vessel in which the rising liquid level obstructs a light beam of a photoelectric system.

photoelectric loop control see LOOP CONTROL.

photoelectric magnitude *Astronomy.* the magnitude of a celestial object when measured with a photoelectric photometer attached to a telescope.

photoelectric photometer *Engineering.* a device using a photocell, phototransmitter, or phototube to measure luminous intensity.

photoelectric photometry *Optics.* the use of photoelectric sensors and devices to measure intensity of light. *Astronomy.* specifically, the study of the brightnesses of celestial objects by means of photocathode instruments and standardized filters to isolate the objects' brightnesses in specific regions of the spectrum.

photoelectric plethysmograph *Medicine.* an apparatus for measuring and recording ear opacity by using a tiny phototube and lamp clipped to the ear to measure the fullness of blood vessels; also worn by aircraft pilots during high-altitude flights to indicate the need for more oxygen.

photoelectric pyrometer *Engineering.* a device for measuring high temperatures by means of an electric current generated by a photoelectric cell when it receives focused radiation from a furnace or other hot body.

photoelectric reader *Computer Technology.* any device that can read punches or marks by photoelectric means.

photoelectric reflectometer *Engineering.* an electronic device that uses a photocell or phototube to measure the reflectance of light or radiant energy from substances.

photoelectric register control *Control Systems.* a register control that uses photoelectric equipment to detect changes in the amount of light reflected from a moving surface containing register marks, a certain design, or defects.

photoelectric relay *Electronics.* a device whose ability to open and close an electric circuit is controlled by the interaction of light, a photocell, and an amplifier; this forms the basic unit in a photoelectric alarm system. Also, LIGHT RELAY.

photoelectric sorter *Control Systems.* a photoelectric control system that can sort objects by their color, size, shape, or other light-altering characteristics.

photoelectric transmissometer *Engineering.* an electronic device used to determine visibility on an airport runway, by measuring the obscuring effect of clouds or fog on a beam of light that is projected on a photocell.

photoelectric tube see PHOTOTUBE.

photoelectric turbidimeter *Engineering.* an instrument that measures the degree to which a solution is muddied, by projecting the light from a photocell through the solution to reveal loss of light intensity.

photoelectrochemistry *Physical Chemistry.* the study of electrochemical reactions that are affected or promoted by light.

photoelectrolysis *Physical Chemistry.* a process that uses radiant energy rather than electrical energy to produce a chemical change in an electrolytic solution.

photoelectromagnetic effect *Electronics.* the current that results when radiation strikes the flat surface of an intermetallic semiconductor placed in a magnetic field parallel to its surface.

photoelectron *Electronics.* an electron that is emitted when light or other forms of electromagnetic radiation strike certain materials, such as cesium, selenium, or silicon.

photoelectron spectroscopy *Spectroscopy.* a branch of electron spectroscopy that studies the energy of electrons emitted when a substance is bombarded by ultraviolet or X-ray radiation.

photoelectrosynthesis *Physical Chemistry.* the production of products in a reaction driven in a nonspontaneous direction by light in an electrochemical cell, usually at a semiconductor electrode.

photoemission *Electronics.* the ejection of electrons from a material, usually a solid, when it is exposed to electromagnetic radiation and other forms of light.

photoemission threshold *Electronics.* the amount of radiant energy needed to eject an electron from certain materials, such as cesium, selenium, or silicon.

photoemissive cell *Electronics.* a device that senses light or other forms of radiation by measuring the electrons emitted from the surface of a photocathode.

photoemissive tube photometer *Electronics.* an instrument using electronic amplifications and a tube made of photoemissive material to obtain highly accurate measurements of luminous intensities (emission of electrons).

photoemissivity *Electronics.* the quality of a material that causes it to emit electrons when exposed to light and other forms of electomagnetic radiation.

photoemitter *Solid-State Physics.* a solid material that when illuminated with electromagnetic radiation emits electrons, as in the photoelectric effect.

photoengraving *Graphic Arts.* **1.** a process for acid-etching a photographic image onto a metal relief printing plate. **2.** a plate produced by this process. Thus, **photoengraver.**

photoengraving zinc *Metallurgy.* a hardened zinc-base alloy, containing iron, cadmium, lead, magnesium, and aluminum; suitable for photoengraving plates.

photoferroelectric effect *Solid-State Physics.* an effect in which certain ferroelectric materials experience changes in the electric field within the substance and the ferroelectric remanent polarization when exposed to light having an energy close to the band gap energy of the material.

photofinishing *Graphic Arts.* the processing of photographs for commercial uses.

photofission *Nuclear Physics.* a type of fission that takes place in nuclei of both uranium and thorium when bombarded with photons; so called because the fission is induced by radiation.

photoflash bomb *Engineering.* an aerial bomb that creates an explosion to produce a brilliant flash of light for aerial night photography.

photoflash unit *Graphic Arts.* a portable light source added to cameras to trip the flash; it generally consists of a capacitor-discharge power source, a flash tube, a battery to charge the capacity, and occasionally a high-voltage pulse generator. Thus, **photoflash bulb, photoflash lamp.**

photoflood lamp *Electricity.* a tungsten lamp that runs at an excessive voltage to give a brilliant illumination for a few hours duration in its short life; used in television and photographic applications.

photog. photographer; photography.

photogalvanic cell *Physical Chemistry.* an electrochemical cell in which an electric current flow is produced by irradiation with light.

Photogrammetry

Strictly speaking, photogrammetry means the measurement of images on a photograph. More generally, however, photogrammetry refers to the science, art, and technology of using photographs primarily to construct maps but also for other purposes, such as evaluation of timber stands, classification of soils, interpretation of geology, acquisition of military intelligence, and exploration of space. The practice of photogrammetry involves both the interpretation of features on photographs and the measurement of images on them.

Photogrammetry may be categorized into three types—aerial photogrammetry, terrestrial photogrammetry, and space photogrammetry. In the case of aerial photogrammetry, photographs of an area are taken by a precision camera normally mounted on an airplane flying over the area, with the camera axis vertical or near vertical. With terrestrial photogrammetry, the camera is located on or near the ground at a fixed, known location; and the camera axis is essentially horizontal. Space photogrammetry is used to explore the extraterrestrial, with the camera located on the earth or in a spacecraft or perhaps on a heavenly body.

Photogrammetry could not, of course, be developed and practiced until photography itself was invented by Louis Daguerre of France in 1839. Some ten years later, Col. Aimé Laussedat of the French Army Corps of Engineers directed the first experiments in using photogrammetry for map preparation. Photogrammetry was first introduced in the United States in 1894 for mapping of the Canada-Alaska Territory border by the U.S. Coast and Geodetic Survey.

Prior to the Wright Brothers' invention of the airplane in 1902, aerial photogrammetry was performed using kites and balloons with obviously limited scope and usefulness. Accordingly, most photogrammetry up until then was done using terrestrial photographs. Beginning in 1913, aerial photographs for mapping purposes were taken using airplanes. Aerial photogrammetry was used during World War I, primarily for reconnaissance, and in World War II, Korea, Vietnam, and the Persian Gulf War for map preparation, reconnaissance, and intelligence.

Space photogrammetry was not possible until space flight was perfected, beginning in the late 1950s. Cameras mounted on orbiting satellites provided vastly improved methods for surveying and mapping the earth's surface, as well as for other applications such as weather forecasting. With later space vehicles, photogrammetry provided the first detailed images of the moon and of Venus, Mars, and other planets.

Clearly, the old adage, "The sky is the limit," is no longer valid. Who knows what the limit will be in the future as additional advances in photogrammetry are made?

Jack B. Evett
Professor and Associate Dean
College of Engineering
University of North Carolina at Charlotte

photogalvanism *Physical Chemistry.* the conversion of physical or chemical energy into electrical energy in a electrochemical cell, aided by radiant energy.

photogenic *Biology.* emitting or producing light, such as certain bacteria. *Medicine.* caused or produced by light, such as certain skin conditions or photogenic epilepsy. *Graphic Arts.* providing an attractive or suitable subject for photography.

photogenic epilepsy *Neurology.* a form of reflex epilepsy in which seizures are induced by intermittent flashes of light.

photogeologic anomaly *Geology.* any systematic deviation of a photogeologic factor, such as topography, vegetation, or drainage, from the expected norm for a given area.

photogeologic map *Geology.* a map based on information derived from the geologic interpretation of a group or series of aerial photographs.

photogeology *Geology.* the geologic interpretation, identification, and recording of land features and structures through the use of aerial photography. Thus, **photogeologic, photogeological.**

photogeomorphology *Geology.* the study of landforms through the use of aerial photography.

photoglow tube *Electronics.* a device in which the operating voltage is raised to a point that ionization and flow discharge take place when a specified illumination is attained; used to open and close connections in a circuit.

photoglycine see GLYCIN.

photogoniometer *Engineering.* an instrument for obtaining the direction of a ray from the nodal point of a camera lens to the image of any point on the photograph, by measuring the horizontal and vertical angles with reference to two perpendicular planes.

photogram *Photogrammetry.* an image made on a photosensitive material by initiating direct contact with the object and then exposing the emulsion to light.

photogrammetric *Photogrammetry.* relating to, using, or produced by means of photogrammetry. Thus, **photogrammetric survey.**

photogrammetric control *Photogrammetry.* horizontal control established by means of phototriangulation.

photogrammetric triangulation see PHOTOTRIANGULATION.

photogrammetry *Cartography.* the technique of using aerial photographs in conjunction with stereoscopic viewing equipment to develop accurate maps and charts.

photograph *Graphic Arts.* a positive or negative image produced through the process of photography.

photograph coordinates *Photogrammetry.* a system of rectangular or polar coordinates that is superimposed on a photograph in order to define positions on that photograph.

photographic *Graphic Arts.* relating to, used in, or produced by photography.

photographic barograph *Meteorology.* a mercury barometer in which the upper or lower meniscus level is measured photographically.

photographic emulsion *Graphic Arts.* see EMULSION.

photographic field *Optics.* the area viewed by a camera; determined by focal length of the lens and the film format being used.

photographic film *Graphic Arts.* see FILM.

photographic fixing *Graphic Arts.* see FIXATION.

photographic interpretation see PHOTOINTERPRETATION.

photographic magnitude *Astronomy.* the magnitude of a celestial object as measured on a photographic emulsion; because emulsions are more sensitive to the blue region of the spectrum, a blue star measures brighter than a red star of the same apparent visual brightness.

photographic meteor *Astronomy.* a meteor detected by photography and studied to determine its origin, relative velocity, size, and orbit.

photographic negative *Graphic Arts.* see NEGATIVE.

photographic photometry *Spectroscopy.* a photometric technique in which a comparator-densitometer is used to measure emulsion density in order to analyze a photographed spectrum.

photographic positive *Graphic Arts.* see POSITIVE.

photographic proof *Graphic Arts.* see PROOF.

photographic recording *Telecommunications.* a form of facsimile recording in which a photosensitive surface is exposed to a signal-controlled source of light.

photographic sound recorder *Electronics.* a device that generates a modulated light beam and moves a light-sensitive device in tandem with the beam to produce a photographic recording of sound signals. Thus, **photographic sound recording.**

photographic sound reproducer *Electronics.* a device that uses a light beam to convert a photographic sound recording back to audio signals.

photographic surveying see PHOTOTOPOGRAPHY.

photographic zenith tube *Optics.* a photographic telescope of fixed or limited movement that is used primarily to calculate the time at which a star crosses the zenith and to determine latitude variation.

photograph meridian *Optics.* the image of any horzontial line in the object space of a photograph that is parallel to the principal plane.

photograph nadir (point) *Optics.* the point where a vertical line running through the perspective center of a camera lens penetrates the plane of the photograph. Also, NADIR POINT.

photograph parallel *Optics.* the image of any horzontial line in the object space of a photograph that is perpendicular to the principal plane.

photography *Graphic Arts.* the art and process of reproducing images on a photosensitive surface using light or another form of energy.

photogravure *Graphic Arts.* an intaglio platemaking process using resin granules instead of the usual screen.

photoheliograph *Astronomy.* a telescope used exclusively for photographing the disk of the sun.

photoheterotroph *Biology.* any organism that uses light as an energy source, but that also derives carbon from organic compounds.

photohomolysis *Physical Chemistry.* a process in which electromagnetic radiation severs a covalent bond existing between two atoms so that each atom then has one electron from the bond.

photoinduced *Physical Chemistry.* induced by light.

photoinduction *Biochemistry.* any process in which light furthers the growth of certain plants or fungi.

photoinhibition *Biochemistry.* **1.** damage to the light-gathering process in photosynthesis caused by excess light energy trapped by a chloroplast. **2.** the negative effect of light on spore production in certain fungi.

photoinitiation *Materials Science.* the formation of light-generated radicals which initiate polymerization.

photointerpretation *Photogrammetry.* the identification and characterization of objects in photographs, particularly for topographic studies. Also, PHOTOGRAPHIC INTERPRETATION.

photoionization *Physical Chemistry.* a process in which a photon of light absorbs electrons from an atom or a molecule. Also, ATOMIC PHOTOELECTRIC EFFECT.

photoisland grid *Electronics.* the photosensitive surface of a television camera tube from which electrons are emitted.

photoisolator see OPTOISOLATOR.

photoisomer *Physical Chemistry.* a molecule whose structure is altered by radiant energy. Thus, **photoisomerization.**

photojunction battery *Nucleonics.* a nuclear battery in which a radioactive substance irradiates a phosphor, which in turn emits light; the light is converted into electrical energy by means of a silicon junction.

photolithography *Graphic Arts.* a type of lithography using photographic processes to produce an image on the surface of the material used to make a printed impression, originally a lithographic stone but now usually a metal printing plate. *Computer Technology.* a process by which the pattern of an integrated circuit is imprinted on a silicon wafer coated with a photoresist mask containing the pattern on it and spinning ultraviolet light through it.

photology *Optics.* the science concerned with the use of light or heat to record images.

photoluminescence *Physics.* the luminescent emission of electromagnetic radiation from a material after it has absorbed another form of electromagnetic radiation, particularly infrared or ultraviolet radiation.

photolysis [fō täl´ə sis] *Physical Chemistry.* **1.** the decomposition of a substance into simpler units as a result of its absorbing light; e.g., the separation of hydrogen from hydrogen sulfide in water. **2.** any process in which radiant energy produces a chemical change. Also, PHOTOCHEMICAL DECOMPOSITION, PHOTODECOMPOSITION.

photolytic [fō´tō lit´ik] *Physical Chemistry.* relating to or involving a process of photolysis.

photom. photometry; photometric.

photoma *Neurology.* the subjective sensation of seeing flashes of light or color, with no objective basis.

photomacrograph see MACROPHOTOGRAPH.

photomagnetic effect *Physics.* the paramagnetism exhibited in certain substances when they are in a phosphorescent state.

photomagnetoelectric effect *Electromagnetism.* an effect in which a voltage is generated across a sample of semiconductor material when it is properly positioned in a magnetic field and one face is illuminated.

photomap *Cartography.* a map based on the simple reproduction of an aerial photograph, without reference to any source maps or surveys but sometimes including an overprint emphasizing certain planimetric details or supplying supplemental information.

photomask *Electronics.* a material, such as film or glass, upon which the etching pattern for a semiconductor device or an integrated circuit is drawn.

photomechanical *Graphic Arts.* **1.** relating to or describing any process using photographic methods to create or treat a printing surface, as in photolithography, photoengraving, or rotogravure. **2.** a machine that etches relief printing plates based on signals transmitted from a scanning device.

photomechanical plate *Graphic Arts.* a printing plate made by exposing a photographic negative or positive onto an emulsion-coated surface and used in a mechanical printing process such as lithography.

photomechanochemistry *Physical Chemistry.* the branch of science concerned with the conversion of chemical energy into mechanical energy through the effect of radiant energy.

photometer *Optics.* an instrument that contrasts the luminous intensities of a light source against a standard reference source.

photometric [fō´tō met´rik] *Optics.* relating to or used in a process of photometry. Also, **photometrical.**

photometric binary *Astronomy.* an eclipsing binary variable star discovered by the fluctuations in its lightcurve as detected by photometric instruments.

photometric parallax *Astronomy.* a rough distance to a star determined by deriving its intrinsic brightness from its photoelectric colors and comparing this with its apparent brightness in the sky.

photometric titration *Analytical Chemistry.* a titration in which the titrant and the sample form a metal complex that can be measured by a change in light absorbance of the titrated solution.

photometry [fō täm´ə trē] *Optics.* **1.** the measurement of the intensity, distribution, color, absorption factor, and spectral distribution of visible light, and occasionally of near-ultraviolet and near-infrared light. **2.** the science of making such measurements. Thus, **photometrist.** *Astronomy.* the use of photometric instruments and techniques to make a precise, quantitative determination of the amount of electromagnetic energy that is received from a celestial object.

photomicrograph see MICROPHOTOGRAPH.

photomicroscope *Optics.* an instrument combining a microscope and a camera, used to make microphotographs.

photomorphogenesis *Botany.* the control of plant growth, development, and differentiation by the duration and nature of light, independent of photosynthesis.

photomosaic *Graphic Arts.* **1.** a work of art that is created by assembling several photographs. **2.** a composite photograph. *Photogrammetry.* see AERIAL MOSAIC.

photomultiplier cell *Electronics.* a device combining a photodetector to sense light and an electron multiplier to amplify the resultant current.

photon [fō´tän´] *Quantum Mechanics.* the quantum of the electromagnetic field that manifests itself by absorption or emission only in multiple quantum units of energy $E = h\nu$, where h is Planck's constant and ν is the frequency of the electromagnetic wave; a unique massless particle that carries the electromagnetic force. *Optics.* the amount of light falling on the retina of the eye when the pupil has an aperture of 1 square millimeter and the illuminated surface outside the eye has a brightness of 1 candela per square meter.

photon activation analysis *Physics.* the chemical analysis of a substance by spectroscopic means, wherein a gamma ray photon is absorbed by a stable nuclide to produce a radioactive nuclide.

photon coupled isolator *Electronics.* a device that couples an infrared emitter diode to a photon detector over an abbreviated shielded light path, in order to achieve exceedingly high circuit isolation.

photon coupling *Electronics.* the process in which two circuits are coupled by a beam of light.

photon curve *Petroleum Engineering.* a plot of gamma-radiation (photon) scatter versus depth during wellbore radioactive logging; used in detecting density variations at different well depths.

photonegative *Electronics.* having negative photoconductivity, a property in certain materials that causes them to exhibit less conductivity and more resistance when exposed to light; occasionally seen in selenium. *Biology.* exhibiting an aversion to light; photophobic.

photon emission spectrum *Physics.* a spectrum indicating the intensities of optical (infrared, visible, and ultraviolet) frequencies emitted by a material.

photoneutron *Nuclear Physics.* a type of neutron produced in a monoenergetic beam that is released from a nucleus during a photonuclear reaction.

photon flux *Optics.* the number of photons in an incident beam of light received by a surface over a unit of time.

photon gas *Physics.* an electromagnetic field that is treated as if it were composed of a system of photons.

photon microscope see OPTICAL MICROSCOPE.

photon theory *Quantum Mechanics.* Einstein's theory of photoemission that treats light as being composed of particles having energy proportional to the frequency of the light.

photonuclear reaction *Nuclear Physics.* a type of nuclear reaction that occurs when a photon collides with a nucleus.

photooxidation *Chemistry.* oxidation that is induced by light. *Materials Science.* specifically, light-induced oxidation in polymers that results in photodegradation.

photopathy *Biology.* any pathological effect that is produced or stimulated by light. Thus, **photopathic.**

photoperceptive *Zoology.* describing an organism that is able to perceive light. Thus, **photoperception.**

photoperiod *Biology.* the amount of time per day that an organism is exposed to light.

photoperiodic *Biology.* of or relating to a photoperiod or to photoperiodism.

photoperiodism *Biology.* the physiological and behavioral response of an organism to seasonal or daily changes in light, especially as demonstrated by changes in vital processes such as plant flowering. Also, **photoperiodicity.**

photophilic *Biology.* thriving in strong light or requiring abundant light for complete and normal development.

photophobia *Medicine.* an abnormal visual intolerance of light, as seen in albinism, measles, and Rocky Mountain spotted fever. *Psychology.* an irrational fear of light.

photophobic *Biology.* shunning or avoiding light; exhibiting negative phototropism. *Botany.* growing best in reduced illumination. *Medicine.* exhibiting physical or psychological photophobia.

photophore gland *Vertebrate Zoology.* a luminous phosphorescent organ formed from modified mucous cells and used by deep-sea elasmobranchs and teleosts in bioluminescence.

photophoresis *Physics.* a phenomenon whereby unidirectional motion is imparted to a system of very fine particles of solid or liquid while either suspended in a gas or falling through a vacuum when an intense beam of light is fired at it.

photophosphorylation *Biochemistry.* the production of ATP from ADP in combination with the electron transport system in an organism dependent on photosynthesis.

photophygous *Biology.* preferring or thriving in shade.

photopia *Physiology.* vision in bright light; day vision. Also, **photopic vision.**

photopic *Physiology.* **1.** of the eye, adapted to bright light. **2.** of or relating to photopia.

photopigment *Biochemistry.* the light-sensitive chemicals in the rods and cones of the eye.

photopolymer *Materials Science.* any of a class of polymers whose physical properties are permanently altered by exposure to light.

photopositive *Electronics.* having positive photoconductivity, a property in certain materials that causes them to exhibit less resistance and more conductivity when exposed to light; observed in selenium.

photoradiation see PHOTODYNAMIC THERAPY.

photoradiometer *Radiology.* an apparatus that measures the quantity of roentgen rays penetrating a given surface.

photoreactivating enzyme *Enzymology.* an enzyme that, when exposed to visible light, is induced to repair DNA that has previously been damaged by exposure to ultraviolet light; an enzyme that cleaves the thymine dimer formed as a result of UV damage.

photoreactivation *Molecular Biology.* a process by which DNA that is damaged by exposure to ultraviolet light recovers when exposed to visible light in the presence of a photoreactivating enzyme. Also, LIGHT REPAIR. *Biotechnology.* any laboratory DNA repair process that requires the presence of light.

photoreactive chlorophyll *Biochemistry.* a molecule that gathers light and converts it into energy.

photorealism *Computer Technology.* any of various techniques intended to provide computer graphics that are comparable to photography in their fidelity to actual images in nature.

photoreception *Physiology.* the process of detecting radiant energy, especially of wavelengths between 370 and 760 nm, the range of visible light.

photoreceptor *Physiology.* a receptor sensitive to light; a rod or cone.

photoreduction *Biochemistry.* a process in a photosynthetic organism that decreases the amount of light-dependent substances, as when NAD^+ or carbon dioxide is reduced during the Calvin cycle.

photoresist *Materials.* a light-sensitive polymer used in photolithography.

photorespiration *Biochemistry.* a process in which an organism takes in oxygen and releases carbon dioxide in the presence of light, occurring during photosynthesis in which there is a low concentration of carbon dioxide and intensive levels of light.

photoscanner *Engineering.* a device that photographs gamma rays as they move through tissue from an injected radioactive substance; used to study the distribution of a radioactive isotope or a radiopaque dye in a body organ or part.

photoscope *Radiology.* a fluoroscope with a manually held display. Also, SKIASCOPY.

photoscopy *Radiology.* an examination of the body with a photoscope.

photosensitive *Biology.* readily stimulated by radiant energy.

photosensitive emulsion *Graphic Arts.* see EMULSION.

photosensitivity *Biology.* the fact of being readily stimulated by light. *Medicine.* any abnormal response to exposure to light, especially of the skin, resulting in accelerated burning and blistering.

photosensitize *Materials Science.* to make a material sensitive to photons by applying a photosensitive emulsion.

photosensor see PHOTODETECTOR.

photosphere *Astronomy.* the outermost visible layer of the sun, consisting of a shallow layer of strongly ionized gases.

photostabilization *Materials Science.* the incorporation of stabilizers, such as hindered amines, organonickel compounds, benzoates, phosphites, and ultraviolet absorbers, to prevent photodegradation in polymers to be used in light exposure applications.

photostat *Graphic Arts.* an inexpensive form of photographic print used in printing to reproduce line art. Also, STAT.

photostethoscope *Medicine.* an instrument that is used to record the heartbeat of a fetus by transforming amplified sounds into light pulsations.

photosynthate *Biochemistry.* a carbohydrate formed during photosynthesis.

photosynthesis [fō´tō sin´thə sis] *Biochemistry.* **1.** the fundamental chemical process in which green plants (and blue-green algae) utilize the energy of sunlight or other light to convert carbon dioxide and water into carbohydrates, with the green pigment chlorophyll acting as the energy converter. This process releases oxygen and is the chief source of atmospheric oxygen. Photosynthesis is often described as the most important chemical reaction on earth; it provides green plants with their complete energy requirement, and most other living organisms obtain their own nutrients from these plants, either directly or indirectly. **2.** a similar process occurring in certain bacteria, in which oxygen is not released. *Chemistry.* any chemical combination caused by the action of light.

photosynthetic [fō´tō sin thet´ik] *Biochemistry.* relating to or taking part in a process of photosynthesis. Thus, **photosynthetic bacteria.**

photosynthetic unit *Biochemistry.* a group of photosynthetic pigment molecules that supply light to one reaction center in photosystem I or II.

photosystem I *Biochemistry.* a series of reactions that occur during the light phase of photosynthesis, in which a pigment system stimulated by light waves of less than 700 nanometers shifts the light energy to energy carriers, such as NADPH, to be used by carbon dioxide.

photosystem II *Biochemistry.* a series of reactions that occur during the light phase of photosynthesis in which the pigment system is stimulated by light waves of less then 685 nanometers, used in the photolysis or splitting of water molecules.

phototaxis *Biology.* movement of an organism toward or away from a source of light. Thus, **phototactic, phototactically.**

phototheodolite *Engineering.* an optical-tracking instrument made up of a theodolite mounted on a camera that takes a series of photographs from several different points; used in terrestrial photogrammetry.

phototherapy *Medicine.* the use of light rays to treat disease, especially diseases of the skin.

photothermoelasticity *Optics.* the change in optical properties produced in a homogeneous isotropic medium subjected to mechanical stress induced by regular changes in temperature.

phototimer *Radiology.* a device that limits the exposure interval by terminating exposure when the amount of incident radiation reaches a preset quantity.

phototonus *Biology.* **1.** the normal sensitivity to light in an organism or part of an organism. **2.** an irritability exhibited by cytoplasm when exposed to light of a certain intensity.

phototopography *Engineering.* a method of surveying in which measurement points are derived from photographs taken from designated ground or aerial points. Also, PHOTOGRAPHIC SURVEYING.

phototoxicity *Toxicology.* photosensitivity, usually to the skin, induced by a poison; the symptoms may include dermal irritation and lesions. Thus, **phototoxic.**

phototoxin *Biochemistry.* a plant that causes an allergic reaction when touched or eaten by a person or animal subsequently exposed to light.

phototransistor *Electronics.* a device that amplifies an input signal, whose current has been enhanced by illumination in order to produce an output signal greater than that an equivalent photodiode could produce.

phototriangulation *Cartography.* a surveying method that uses photogrammetry to extend horizontal or vertical control points, so that the perspective principles of overlapping photographs are related into a spatial solution. Also, AEROTRIANGULATION, AERIAL TRIANGULATION, PHOTOGRAMMETRIC TRIANGULATION.

phototronic photocell see PHOTOVOLTAIC CELL.

phototroph *Biology.* any organism utilizing light as its principal source of energy.

phototrophic bacteria *Microbiology.* any bacteria, such as purple and green sulfur aquatic bacteria, that derives energy from light by the process of photosynthesis. Also, PHOTOBACTERIA.

phototropism *Botany.* plant movement or orientation in response to the location of light. Also, HELIOTROPISM. *Solid-State Physics.* the reversible phenomenon of color change in a solid as the result of exposure to light.

phototube *Electronics.* an electron tube that converts light energy into electrical energy in direct proportion to the intensity of light absorbed by the photoemissive material on its cathode.

phototube cathode *Electronics.* the terminal in a phototube that is coated with photoemissive material and has a negative polarity.

phototube current meter *Engineering.* an instrument that determines the velocity of water currents, by means of a perforated disk that rotates with the current, and through which a beam of light is directed and reflects from a mirror onto a phototube.

phototube relay *Electronics.* a device that uses a light beam to open and close the circuits of mechanical devices, such as counters and safety controls.

phototypesetter *Graphic Arts.* **1.** a machine designed to create lines of type by projecting type characters onto film or photographic paper. **2.** a person who operates such a machine.

phototypesetting *Graphic Arts.* the use of a phototypesetter to create reproducible type characters.

photovaporization *Surgery.* the vaporization by laser-generated radiation of intracellular and extracellular fluids, providing an incision with cauterization of adjacent vessels.

photovaristor *Electronics.* a varistor in which the relationship of voltage to current is altered by the illumination of a light-sensitive material, such as cadmium sulfide or lead telluride; used as a semiconductor.

photoviscoelasticity *Optics.* the change in optical properties produced in viscoelastic material subjected to mechanical stress.

photovisual magnitude *Astronomy.* a photographic determination of the magnitude of a celestial object through a combination of filter and emulsion that approximates the spectral sensitivity of the human eye.

photovoltaic *Electronics.* of or relating to the photovoltaic effect.

photovoltaic cell *Electronics.* a device that produces an electric current when exposed to light; principally constructed with silicon, selenium, or germanium. Also, PHOTOTRONIC PHOTOCELL.

photovoltaic effect *Electronics.* the production of a voltage in a semiconductor material, such as selenium, when light or other forms of electromagnetic radiation strike it.

photovoltaics *Electromagnetism.* a field of semiconductor technology dealing with the direct conversion of electromagnetic radiation, as sunlight, into electricity.

photox cell *Electronics.* a photovoltaic cell that produces a voltage between a copper base and a film of cuprous oxide when it is exposed to light or other forms of electromagnetic radiation.

photronic cell *Electronics.* a photovoltaic cell that produces a voltage across a layer of selenium when it is exposed to light or other forms of electromagnetic radiation.

Phoxichilidiidae *Invertebrate Zoology.* a family of sea spiders, marine arthropods in the subphylum Pycnogonida, having four pairs of walking legs, rudimentary or no pedipalpi, and ovigerous legs found in males only.

Phoxocephalidae *Invertebrate Zoology.* a family of marine amphipod crustaceans in the suborder Gammaridea, having a fusiform body, burrowing appendages, and a rostrum that forms a broad head; males are smaller than females.

Phractolamidae *Vertebrate Zoology.* a monotypic family of fish of the order Gonorynchiformes, characterized by a slender gray body and a protractile mouth containing only two teeth; found in weedy, muddy water of the Congo and Niger rivers of West Africa.

Phragmidiaceae *Mycology.* a family of fungi belonging to the order Uredenales, plant parasites that primarily occur on the Rosoideae of the northern temperate regions.

Phragmites *Botany.* a genus of tall grasses having plumed heads, especially the common reed *P. australis* (or *P. communis*); found in marshy areas.

Phragmobasidiomycetes *Mycology.* a class of fungi belonging to the subdivision Basidiomycotina, which is characterized by the horizontal and vertical cell walls of its fruiting body or basidium.

phragmocone *Invertebrate Zoology.* the conical, chambered portion of the internal shell of some cephalopods, especially cuttlefish. Also, **phragmacone.**

phragmoid *Botany.* of or relating to a plant having septae positioned perpendicularly to the axis, as the conidia of certain fungi.

Phragmonemataceae *Botany.* a family of red algae belonging to the order Porphyridiales, distinguished by a thalli consisting of linear cell aggregations.

phragmoplast *Cell Biology.* a cylindrical microtubular structure assembled between two daughter nuclei in a dividing plant cell to transport precursor material to a developing cell wall.

phragmosome *Cell Biology.* a region of differentiated cytoplasm that forms in the equatorial region of a dividing plant cell providing a site for phragmoplast assembly.

phragmosporae *Mycology.* a spore containing two or more horizontal cell walls; found in certain imperfect fungi.

phreatic *Hydrology.* of or relating to ground water. *Volcanology.* of or relating to the explosion or eruption of nonincandescent material, such as steam or mud, from a volcano.

phreatic cycle *Hydrology.* the time period during which the water table rises and falls. Also, CYCLE OF FLUCTUATION.

phreatic gas *Geology.* a gas generated by contact between atmospheric or surface water and ascending magma.

phreatic surface *Hydrology.* see WATER TABLE.

phreatic water *Hydrology.* **1.** a former term for water occurring in the upper part of the zone of saturation. **2.** see GROUNDWATER.

phreatic water discharge see GROUNDWATER DISCHARGE.

phreatic zone *Geology.* a zone of groundwater saturation. *Hydrology.* see ZONE OF SATURATION.

Phreatoicidae *Invertebrate Zoology.* a family of freshwater isopod crustaceans in the suborder Phreaoicoidea, having a laterally compressed body and the first thoracic segment fused with the head.

phreatomagmatic *Geology.* describing a volcanic explosion caused by contact between magma and groundwater or shallow ocean water, in which both magmatic gases and steam are expelled.

phreatophyte *Ecology.* a deep-rooted plant that obtains its moisture from the water table or the layer of soil just above it.

phren- or **phreno-** a prefix meaning **1.** diaphragm. **2.** mind.

phrenectomy *Medicine.* the surgical removal of all or part of the diaphragm.

-phrenia *Medicine.* a combining form used in the names of mental disorders, as *schizophrenia.*

phrenic *Anatomy.* of or relating to the diaphragm. *Psychology.* of or relating to the mind.

phrenic nerve *Anatomy.* the nerve that transmits impulses to and from the diaphragm.

phrenitis *Medicine.* an inflammation of the diaphragm.

phrenodynia *Medicine.* any pain in the diaphragm.

phrenogastric *Anatomy.* of or relating to the diaphragm and the stomach.

phrenology [frə näl´ə jē] *Psychology.* an earlier theory that linked mental functions to particular parts of the skull and asserted that the size and shape of the skull indicated the strength or weakness of a particular mental faculty. Thus, **phrenologist, phrenological.**

phrenophobia *Psychology.* an irrational fear of thinking, or of losing one's mind.
phrenoplegia *Medicine.* a paralysis of the diaphragm.
phrenoptosis *Medicine.* the downward displacement of the diaphragm.
phrenospasm *Neurology.* a spasm of the diaphragm.
phrynoderma *Medicine.* a follicular hyperkeratosis frequently accompanied by mild neuritic and eye symptoms; seen in East Indian laborers fed on a diet of maize meal, probably caused by a vitamin A deficiency.
Phrynophiurida *Invertebrate Zoology.* an order of basket stars, echinoderms in the subclass Ophiuroidea, having disk and arms covered by skin, no dorsal armshields, and long, narrow branched arms that move both up and down and from side to side.
PHS or **P.H.S.** Public Health Service.
pH standard *Analytical Chemistry.* any of a number of solutions of known pH value that are available from the U.S. National Institute of Standards and Technology for use in calibrating a pH meter; the original pH standard is pure water, with pH 7.0, which is defined as neutral.
phthalate [thal´ăt] *Organic Chemistry.* a salt or ester of phthalic acid containing the radical $C_6H_4(COO)_2$=; used for buffers, for standard solutions, and in vacuum pumps.
phthalate buffer *Analytical Chemistry.* a standard laboratory solution of potassium hydrogen phthalate, $KHC_8H_4O_4$; used as a pH reference solution.
phthalate ester *Organic Chemistry.* a plasticizer made by the action of alcohol on phthalic anhydride.
phthalazine *Organic Chemistry.* $C_6H_4CHN_2CH$, pale yellow crystals; soluble in alcohol; melts at 91°C.
phthalein *Chemistry.* any of a group of compounds formed by treating phthalic anhydride with phenols; used to derive certain important dyes.
phthalic acid [thal´ik] *Organic Chemistry.* $C_6H_4(COOH)_2$, colorless crystals; slightly soluble in water and soluble in alcohol; melts (and decomposes) at about 230°C; used in dyes, in perfumes, and as a reagent.
***p*-phthalic acid** see TEREPHTHALIC ACID.
phthalic anhydride *Organic Chemistry.* $C_6H_4(CO)_2O$, combustible white crystals; soluble in hot water and alcohol; melts at 131°C and boils at 295°C, with sublimation below the boiling point; used in plasticizers, polyesters, dyes, resins, and insect repellents.
pthalin *Chemistry.* any of a group of compounds obtained by the reduction of the phthaleins.
phthalocyanine pigments *Organic Chemistry.* $(C_6H_4)C_2N$, a group of lightfast organic pigments with a basic structure of four isoindole groups joined by four nitrogen atoms; used in enamels, plastics, inks, and rubber goods.
phthalonitrile *Organic Chemistry.* $C_6H_4(CN)_2$, toxic, combustible, buff-colored crystals; insoluble in water and soluble in acetone and benzene; melts at 138°C; used as an insecticide and in organic synthesis.
phthiocol *Biochemistry.* $C_{11}H_8O_3$, a yellow crystalline substance that is produced by the human tubercle bacillus, *Mycobacterium tuberculosis*, having certain properties of vitamin K.
phthisis *Medicine.* **1.** the progressive wasting away of the body or any of its parts. **2.** tuberculosis, especially of the lungs.
phugoid *Aviation.* of or relating to variations in the slow longitudinal motion of the center of mass of an aircraft in flight.
phugoid oscillation *Aviation.* a long-period longitudinal oscillation of an airplane's pitch axis, with shallow climbing and diving movements about the median flightpath and little change in the angle of attack.
phyc- or **phyco-** a prefix meaning "seaweed," "algae."
Phycitinae *Invertebrate Zoology.* a subfamily of snout moths, lepidopteran insects in the family Pyralididae, having narrow wings striped in gray.
phycobilin *Biochemistry.* a photosynthetic pigment found in both red and blue-green algae.
phycobiliproteins *Biochemistry.* the subunits of phycobilisomes, each with different spectral properties, that collect light energy from a variety of wavelengths and then transfer it to the phycobilisome, thus allowing it to absorb light from a broad spectrum.
phycobilisome *Biochemistry.* a small, spherical particle composed mainly of phycobilin that is found on the photosynthetic components of red algae and cyanobacteria.
phycobiont *Botany.* the algal component of a lichen or other symbiotic association.
phycocyanin *Biochemistry.* a photosynthetic protein that has a bound phycobilin; found in blue-green algae.
phycocyanobilin *Biochemistry.* $C_{31}H_{38}O_2N_4$, a phycobilin that contains an ethylidene side chain and one asymmetric carbon atom.
phycoerythrin *Biochemistry.* a photosynthetic protein that has a bound phycobilin; found in red algae.
phycoerythrobilin *Biochemistry.* a phycobilin that contains seven conjugated double bonds, an ethylidine side chain, and two asymmetric carbon atoms.
Phycolepidoziaceae *Botany.* a monotypic family of liverworts in the order Jungermanniales, characterized by microscopic, filamentous, green, creeping plants with a high level of reduction of both the gametophyte and the sporophyte; known only from Dominica.
phycology *Botany.* the branch of botany dealing with algae.
Phycomycetes *Mycology.* a former term for a class of fungi belonging to the order Eumycetes, characterized by its lack of cell walls. Also, LOWER FUNGI.
Phycopeltis *Botany.* a genus of green algae belonging to the order Trentepohliales, forming a monostromatic layer on the cuticle of leaves.
phycoplast *Cell Biology.* a structure that develops in the cleavage plane between two daughter cells in certain species of green algae.
Phycosecidae *Invertebrate Zoology.* a family of beetles, coleopteran insects in the superfamily Cucujoidea.
phycovirus *Virology.* any virus that infects a form of blue-green algae. Also, CYANOPHAGE.
phyl- or **phylo-** a prefix meaning: **1.** race. **2.** kind.
phylactic *Medicine.* referring to or producing phylaxis.
Phylactolaemata *Invertebrate Zoology.* a class of freshwater colonial bryozoans with a horseshoe-shaped lophophore.
phylaxis *Medicine.* protection against infection; specifically, the bodily defense against infection.
phyletic [fī let´ik] *Evolution.* of or relating to a line of direct descent, or a course of evolution.
phyletic evolution *Evolution.* **1.** a sequence of changes that occur more or less uniformly throughout the geographical range of a species over an extended period of time. **2.** see ANAGENESIS, def. 2.
phyletic extinction *Evolution.* evolution in which a species becomes extinct but is replaced by another species that is directly descended from it, so that the species lineage continues. Also, PSEUDOEXTINCTION.
phyletic gradualism *Evolution.* a model in which directional evolutionary change proceeds at a slow and steady pace, resulting in a steady increase in biological diversity. Also, GRADUALISM.
phyletic lineage *Evolution.* an ancestor-descendant sequence between two successive branching points of a phylogenetic tree.
phyletic speciation *Evolution.* a type of evolution in which one species gradually evolves away from another species and eventually replaces it. Also, SUCCESSFUL SPECIATION.
phyll- or **phyllo-** a prefix meaning "leaf."
Phyllachoraceae *Mycology.* a family of fungi belonging to the order Xylariales that occur on grasses or other plants and that sometimes are pathogenic, causing tar spot disease.
phyllary *Botany.* one of the bracts forming a whorl at the base of an inflorescence of a composite plant.
phyllite *Petrology.* an argillaceous metamorphic rock, intermediate in grade between slate and mica schist, featuring cleavage surfaces with a silky sheen due to the presence of minute mica crystals.

phyllite

Phyllobacterium *Bacteriology.* a genus of Gram-negative aerobic bacteria of the family Rhizobiaceae, composed of rod-shaped cells that are distinguished by growth in star-shaped groups in a liquid culture medium.

Phyllobothrioidea see TETRAPHYLLIDEA.

phyllobranchiate gill *Invertebrate Zoology.* in crustaceans, a gill formed of flat, leaflike plates arranged like book pages. Also, LAMELLAR GILL.

Phylloceras *Paleontology.* the type genus of smooth-shelled ammonoids in the order Phylloceratida; Jurassic and Cretaceous.

phylloceratid *Paleontology.* of or relating to the ammonoid order Phylloceratida; extant from the Triassic to the Cretaceous.

phylloclade *Botany.* a flattened stem or branch resembling a leaf and performing a photosynthetic function, as on a cactus.

phyllode *Botany.* a flattened, sometimes expanded petiole that functions as a photosynthetic organ in the absence of a leaf, as is found in many acacias.

Phyllodocidae *Invertebrate Zoology.* a family of free-swimming polychaete annelids, having long, narrow bodies, reduced parapodia, and flattened, leaflike cirri.

Phyllodrepaniaceae *Botany.* a family of small, dark-green mosses of the order Dicranales that form cushions on decaying wood in lowland tropical forests; characterized by sparingly branched stems with terminal sporophytes, complanate assymetric leaves that occur in four rows, and smooth erect capsules.

phyllody *Plant Pathology.* the abnormal development of leaves from parts of a flower, generally caused by a virus.

phyllofacies *Geology.* facies distinguished on the basis of stratification characteristics.

phyllogenesis *Botany.* the process in which plants develop leaves.

phyllogenetic *Botany.* of or relating to the development of leaves.

Phyllogoniaceae *Botany.* a family of shiny tropical and subtropical mosses in the order Isobryales, noted for forming mats on or hanging from trees.

Phyllograptus *Paleontology.* a scandent graptolite in the order Graptoloidea; notable for its oval leaf shape; Lower Ordovician.

phylloid *Botany.* of or relating to leaves; leaflike.

Phyllolepida *Paleontology.* an order of relatively small, flattened placoderms; the known fossils do not disclose the position of eyes or mouth; related to the arthrodires; extant in the Upper Devonian.

phyllome *Botany.* **1.** a leaf of a plant or any leaflike part such as a sepal, petal, or stamen. **2.** the leaves of a plant as a whole; foliage.

phyllomorphic stage *Geology.* the most advanced geochemical stage of diagenesis, characterized by the replacement of clays by micas, feldspars, and chlorites.

phyllonite *Petrology.* a metamorphic rock macroscopically resembling phyllite but formed by mylonization of coarser rocks followed, or accompanied, by recrystallization and growth of new minerals.

phyllophagous *Zoology.* describing an organism that feeds on leaves.

Phyllophoraceae *Botany.* a family of red algae belonging to the order Gigartinales, characterized by erect, encrusting, or parasitic thalli with erect fronds, and spermatangia occurring in patches.

phyllophore *Botany.* the terminal bud of a stem, especially the stem of a palm.

Phyllophoridae *Invertebrate Zoology.* a family of sea cucumbers, holothurians in the order Dendrochirotida, having a brown or gray U-shaped body, a naked body wall, tiny ossicles, and tentacles arranged in two circles.

phylloplane *Botany.* the surface of a leaf, used especially in reference to a leaf as a habitat for microorganisms.

phyllosilicate *Mineralogy.* a group of silicate minerals in which the SiO_4 tetrahedra are linked together in flat, infinite sheets; generally flaky in habit with one prominent cleavage, soft, and low in specific gravity; includes mica and clay minerals.

phyllosoma *Invertebrate Zoology.* a free-swimming, flat, transparent larval form of marine crayfish and spiny lobsters.

phyllospondylous *Vertebrate Zoology.* describing vertebrae that have a hypocentrum but no pleurocentra and a neural arch that encloses the notochord, forming a transverse structure that supports the ribs; used especially in reference to larval labyrinthodont amphibians.

Phyllosporaceae *Botany.* a family of brown algae of the order Fucales, characterized by axes up to 10 meters long that have pinnate branches and are attached by a discoid or conical holdfast, and by flattened, leaflike laterals; native to Australia and New Zealand.

Phyllostictales see SPHAEROPSIDALES.

Phyllostomatidae *Vertebrate Zoology.* the New World leaf-nosed bats and vampire bats, a diverse family of bats of the suborder Microchiroptera, characterized by a flap of skin tissue above the nostril called a nose leaf; found in tropical and subtropical regions of Central and South America and in the southwestern U.S. Also, **Phyllostomidae.**

phyllotaxy *Botany.* the arrangement or distribution of leaves on a stem. Also, **phyllotaxis.**

Phyllothalliaceae *Botany.* a monogeneric family of liverworts of the order Metzgeriales, characterized by leafy, robust, fleshy plants with colorless rhizoids scattered on ventral surfaces, especially at the nodes; native to Tierra del Fuego and New Zealand.

phyllovitrinite see PROVITRINITE.

phylloxera [fil´ik sâr´ə; fi läks´ə rə] *Invertebrate Zoology.* any of several plant lice of the genus *Phylloxera* in the subfamily Phylloxerinae, especially the **grape phylloxera,** *P. vitifoliae,* a serious grapevine pest of Europe and the western U.S.; it virtually destroyed the vineyards of Western Europe in the mid to late 1800s, until the vines were saved by grafting European plants onto the rootstocks of vines native to the eastern U.S., which are resistant to the grape phylloxera.

Phylloxerinae *Invertebrate Zoology.* a subfamily of plant lice, aphid-like homopteran insects; the genus *Phylloxera* includes serious plant pests.

phylogeny [fə läj´ə nē] *Evolution.* the evolutionary history of a species or other taxonomic group. Also, **phylogenesis.**

phylogenetic [fī´lō jə net´ik] *Evolution.* of or relating to the evolutionary relationships within and between groups. Also, **phylogenic.**

phylogenetic classification see NATURAL CLASSIFICATION.

phylogenetic tree *Genetics.* a branching diagram that indicates evolutionary relationships in a group of organisms by placing ancestral groups closer to the "trunk" of the tree and derived descendants at the ends of branches.

phylogeography *Ecology.* the study of the geographic distribution of taxonomically related groups.

phylum [fī´ləm] *plural,* **phyla** [fī´lə]. *Systematics.* one of the primary or obligatory taxonomic ranks (in all kingdoms except Plantae, in which it is replaced by the division) below kingdom and above class. (A coined Modern Latin word; derived from a Greek word meaning "tribe.")

phyma *Medicine.* any skin tumor or cutaneous tubercle, especially one produced by exudation into the subcutaneous tissue or the corium.

Phymatidae *Invertebrate Zoology.* ambush bugs, a family of predatory hemipteran insects, having a chirping organ and prehensile forelegs; most active during summer when they hunt their insect prey around flowers.

phymatosis *Medicine.* the growth of small nodules (phymas) on the skin.

Phymosomatidae *Invertebrate Zoology.* a family of echinoderms with a rigid test in the order Phymosomatoida.

Phymosomatoida *Invertebrate Zoology.* an order of echinoderms with a sculptured test, interambulacral plates each having one spine and tubercle, and no large, angular plates in the periproct.

phys. physical; physics; physician; physiological; physiology.

Physa *Invertebrate Zoology.* a common genus of air-breathing freshwater snails with a short spiral shell.

Physalopteridae *Invertebrate Zoology.* a family of nematodes in the superfamily Spiruroidea, parasites in the respiratory and digestive systems of vertebrates.

Physaraceae *Mycology.* a family of slime mold fungi belonging to the order Physarales.

Physarales *Mycology.* an order of fungi of the class Myxomycetes.

physi- or **physio-** a prefix meaning "nature."

physiatrics *Medicine.* the branch of medical science that focuses on the diagnosis, treatment, and prevention of disease with physical means rather than with drugs, such as by massage, infrared or ultraviolet light, electrotherapy, hydrotherapy, heat, or exercise.

physiatrist *Medicine.* a physician who specializes in physiatrics.

physical *Science.* **1.** of or relating to matter or the material world. **2.** of or relating to physics or another physical science. *Medicine.* **1.** of or relating to the body. **2.** see PHYSICAL EXAMINATION.

physical adsorption *Physical Chemistry.* a relatively weak process of adsorption in which the adsorbate molecules are held to the adsorbent surface by secondary forces, rather than by chemical bonding as they are in chemisorption. Also, PHYSISORPTION.

physical anthropology see BIOLOGICAL ANTHROPOLOGY.

physical change *Chemistry.* a change of matter from one form to another without an accompanying change in its chemical properties.

Physical Chemistry

Physical chemistry is the branch of chemistry that deals with the application of physical principles to the study of chemical systems and phenomena. It is concerned with the physical properties of chemical compounds, their structure and chemical bonding, and their energetics, mechanisms, and rates of reactions. The main branches of this field include thermodynamics, kinetics, surface chemistry, electrochemistry, spectroscopy, and theoretical chemistry (e.g., quantum chemistry and statistical mechanics).

Physical chemistry began in the 1800s with studies of the kinetic theory of gases, which involved prediction of the properties of gases in terms of the behavior of molecules, and chemical thermodynamics, which related the heat, energy, and chemical changes that occur in reactions. Early physical chemistry was also concerned with the properties of solutions and electrochemistry. The development of quantum mechanics led to interest in the nature of bonding in chemical compounds and computational approaches to the determination of their structure and properties.

More recent work has been concerned with understanding the mechanisms and rates of reactions of atoms and small molecules prepared in known quantum states (e.g., in molecular beams) and correlations of these with theoretical predictions. Very fast reactions are also studied using laser pulses of picosecond or femtosecond duration. These latter studies are sometimes included in the area of chemical physics, although the distinction from physical chemistry is not clear.

Experimental physical chemistry has greatly benefitted in recent years from the development of sophisticated instrumentation, such as ultrahigh vacuum systems for the study of surfaces and synchronotron light sources for spectroscopic investigations. Advances in theoretical chemistry have followed from improvements in the size and speed of digital computers. Related areas include physical organic chemistry, which is mainly concerned with reactions of organic species, and biophysical chemistry, dealing with biological systems.

Allen J. Bard
Hackerman/Welch Regents Chair
The University of Texas at Austin

physical chemistry *Chemistry.* the study of chemical phenomena through the application of physical laws and concepts in order to describe mathematically observable chemical behavior.

physical climate *Meteorology.* the actual climate of a given locality as distinguished from the solar or mathematical climate.

physical climatology *Meteorology.* a branch of climatology that deals with the explanation of climate, as distinguished from the facts of it or its presentation.

physical compatibility *Engineering.* a characteristic of two or more substances that enables them to be used together successfully.

physical constant *Physics.* any physical quantity whose value is fixed, such as the speed of light through a vacuum or the mass of an electron.

physical-constraint gripper *Robotics.* a gripper that restrains an object by grasping and holding it.

physical data independence *Computer Programming.* an attribute of a file structure in which the physical structure of the data can be modified without requiring a change in the logical file structure.

physical deaeration *Materials Science.* a method of achieving limited corrosion protection involving reduction of the dissolved oxidant concentration.

physical device table *Computer Technology.* a table used by the operating system that contains information about the physical characteristics of the computer system's peripheral devices, such as device type, status, and addresses of input/output routines.

physical disability *Medicine.* a popular or legal term for any permanent physical impairment or malfunction that prevents a person from carrying on the full range of physical activities associated with a normal individual.

physical double galaxy *Astronomy.* two galaxies that orbit a common center of mass.

physical double star *Astronomy.* two stars that orbit a common center of mass.

physical electronics *Electronics.* the branch of electronics that is concerned with the physical phenomena central to electronics, such as discharges, field emissions, and conduction.

physical examination *Medicine.* an examination and investigation of a person's body in order to determine his or her general state of physical health.

physical exfoliation *Geology.* exfoliation that results from the actions of physical forces, such as the expansion of freezing water or the release of overburden pressure.

physical file structure see FILE ORGANIZATION.

physical fitness *Medicine.* **1.** a general term for a combination of qualities enabling an individual to carry out his or her daily activities with alertness and efficiency, and without any undue fatigue, discomfort, or pain. **2.** the ability to carry out strenuous physical exercise or to perform well in various tests of agility, strength, and stamina.

physical geography *Geography.* a branch of geography concerned with the natural features and inanimate phenomena of the earth's surface. Also, PHYSIOGRAPHY.

physical geology *Geology.* the branch of geology concerned with the processes and forces affecting the morphology of the earth or its constituent materials, such as rocks, minerals, and magmas.

physical law *Physics.* any law in physics that describes specific phenomena and applies to all cases in which these phenomena occur, such as the law of conservation of momentum in mechanics.

physical layer *Computer Technology.* layer one of the seven-layer open system interconnection (OSI) model for network architectures; the physical communication circuit layer concerned with transmitting data bits.

physical libration *Astronomy.* see LIBRATION.

physical map *Molecular Biology.* a map indicating the order of DNA sequences on a chromosome or chromosome segment, produced by nucleotide sequencing and heteroduplex assays.

physical measurement *Physics.* any quantitative measurement, direct or indirect, of a physical quantity, such as length or electric current.

physical medicine *Medicine.* the use of physical therapy procedures to help physically diseased or injured patients return to a useful life.

physical metallurgy *Metallurgy.* the art and science of the structure, properties, and solid-state transformations of a metal or alloy.

physical meteorology *Meteorology.* the branch of meteorology that studies optical, electrical, acoustical, and thermodynamic atmospheric phenomena, as well as the laws of radiation and the elucidation of clouds and precipitation.

physical oceanography *Oceanography.* the study of the physical characteristics and movements of the ocean, and the relationships between these factors, the atmosphere, and the sea floor.

physical optics *Optics.* a branch of physics that considers light to be a wave phenomenon, as distinguished from geometrical optics, which views light as a series of rays traveling in straight lines.

physical pendulum *Physics.* an apparatus consisting of a body that is allowed to rotate freely about a horizontal axis on which it is pivoted, as distinguished from a simple pendulum.

physical property *Chemistry.* a characteristic of a pure substance that does not involve a chemical change, such as its density, color, or hardness.

physical realizability *Control Systems.* the ability or inability to construct a network based on some transfer function.

physical record *Computer Technology.* a unit of data on a storage medium separated from other related data by some indicator, such as an interrecord gap on magnetic tape; may contain one or more logical records.

physical residue *Geology.* an accumulation of rock debris that results from physical weathering processes occurring in a place.
physical science *Science.* a science that is concerned with the nature and behavior of nonliving matter; e.g., physics, geology, meteorology, astronomy, and some areas of chemistry.
physical sign see OBJECTIVE SIGN.
physical stratigraphy *Geology.* a classification based on the physical aspects of rocks.
physical symbol system hypothesis *Artificial Intelligence.* a hypothesis proposing that representation and processing of symbolic structures are necessary and sufficient to account for intelligence.
physical system see CAUSAL SYSTEM.
physical testing *Engineering.* the determination of the inherent properties of a material with the aid of mechanical instruments and observations.
physical theory *Physics.* any statement that attempts to explain physical phenomena by describing them as the the result of some underlying assumption.
physical therapist *Medicine.* a person who is professionally trained and certified in the techniques of physical therapy.
physical therapy *Medicine.* **1.** a specialized school of medicine that deals with the treatment of disease and injury by the application of physical agents, such as massage, manipulation, therapeutic exercises, cold, heat, hydrotherapy, electrical stimulation, and light. **2.** any specific program of physical exercise intended to promote recovery from disease or injury.
physical time *Geology.* geologic time determined by measuring a physical phenomenon or process, such as the radioactive decay of elements.
physical weathering see MECHANICAL WEATHERING.
physician *Medicine.* **1.** a licensed practitioner of medicine, who has completed an approved course of study and graduated from a college of medicine or osteopathy, who has passed the National Board Examination, and who has been certified by the appropriate board of specialty. **2.** specifically, such a practitioner other than a surgeon.
physician's assistant *Medicine.* a health professional who has been trained and certified in certain aspects of the practice of medicine in order to carry out various medical procedures under the supervision of a physician. Also, **physician assistant, physican's associate.**
physicist [fiz′i sist] *Physics.* **1.** a scientist who specializes in or studies physics. **2.** anyone who applies physics, in either an experimental or a theoretical way.
physics *Science.* the scientific study of matter, energy, motion, and force. (From a Greek term meaning "the science of nature.")
physiogenic disease *Plant Pathology.* a plant disease caused by an unfavorable environmental or physical factor such as temperature extremes, light or water excess or deficiency, or injury.
physiognomy [fiz′ē äg′nə mē] Anatomy. a person's general physical form; the shape and appearance of the face and body. *Psychology.* the judging or defining of personality and mental attitudes through the study of the face and other physical features.
physiographic diagram *Geology.* a small-scale map on which landforms are depicted using a standardized set of simplified pictorial symbols representing an aerial view of such landforms from an angle of about 45°. Also, LANDFORM MAP, MORPHOGRAPHIC MAP.
physiographic feature *Geology.* a prominent physiographic form or a noticeable part of such a landform.
physiographic form *Geology.* a landform as considered in relation to its origin, cause, or history.
physiographic province *Geology.* a region whose parts exhibit similar geologic structures and climate, and whose pattern of topographic relief differs significantly from that of adjacent regions, indicating a unified geomorphic history.
physiography *Geography.* **1.** see PHYSICAL GEOGRAPHY. **2.** see GEOMORPHOLOGY.
physiol. physiology; physiological; physiologist.
physiological [fiz′ē ə läj′i kəl] *Science.* of or relating to the normal functioning of an organism.
physiological antidote *Toxicology.* any substance that counteracts the effects of a poison by producing an opposite biological effect.
physiological motivation see PRIMARY MOTIVATION.
physiological psychology *Psychology.* the study of behavior through its relationship to physiological processes.
physiological race *Mycology.* a group of organisms sharing a common physiological characteristic, such as host preference.
physiological tetanus *Neurology.* see TETANUS.
physiologic cup see OPTIC CUP.
physiologic dead space *Physiology.* the portion of the respiratory tract that includes the anatomical dead space as well as the space in the alveoli which is occupied by air that does not participate in oxygen-carbon dioxide exchange.
physiologic diplopia *Physiology.* the double vision produced naturally for all objects seen outside a fixed point of concentrated vision.
physiologic race see CRYPTIC SPECIES.
physiologic tremor *Physiology.* a normal, slight body quivering, usually occurring at the rate of 10 to 12 per second; may be observed in outstretched hands.
physiologist [fiz′ē äl′ ə jəst] *Biology.* a specialist in physiology.

Physics

Physics is the description of mass, energy, and their interaction, as experienced in our universe. As an inductive science, physics expresses many phenomena in terms of a few (perhaps, eventually one) general principles. As a deductive science, physics tests its principles by comparing their application to the outcomes of both controlled and serendipitous observations.

Physics is an old science. The physics of most phenomena on a human scale and directly tangible to the human senses was well elucidated more than a century ago. Today's physics focuses on effects far removed from direct human experience. These remote frontiers and examples of them include the very large (the "Big Bang" expansion at the origin of the universe); the very small (quarks, subcomponents of the constituents of the nuclei of atoms); the very long (the $>10^{31}$ year proton lifetime); the very short (the 10^{-23} s lifetime of a subnuclear meson's decaying via the strong interaction); the very hot (the behavior of the sun's core at a temperature of 15,000,000° kelvin); and the very cold (the behavior of paramagnetic salts at a few millidegrees above absolute zero). Remarkably, some of these apparently obscure phenomena have an enormous effect on human society via the now familiar technology in fields such as electronics, telecommunications, aeronautics, and nuclear energy.

The future course of physics is not known, but, historically, physics has progressed via an interlocking of theory, experiment, and technology. Maxwell's prediction of electromagnetic waves was tested by Hertz's experiments. The electroweak theory predictions of a new kind of neutrino interaction and the existence of W and Z particles were separately verified by large international teams of physicists during the 1970s and 1980s. Conversely, the Michaelson-Morley experiment inspired Einstein's theory of special relativity. The discovery of charge conjugation-parity (CP) violation in K meson decay well preceded theoretical understanding. Technology based on physics developed 50 years ago or more underlies much of today's experimental physics. The implications of today's physics for a new generation of technology are yet to be discovered.

Marvin L. Marshak
Professor of Physics
University of Minnesota

Physiology

Physiology is the science of how living organisms function, from the smallest forms of life to multicellular organisms such as plants or animals. Since function is obligatorily coupled to structure, this interrelationship presents a principal focus at all levels of biological organization. Approaches that stem from physics and chemistry provide the format for analysis of cellular and subcellular processes in the service of function. Thus where general physiology may search for fundamental processes that may be common to all cells, comparative physiology and mammalian physiology define the functioning of different systems and different organisms at varying levels of complexity.

Principles that apply at one level have counterparts at another; for instance, understanding the mechanisms that govern the flow of electrical current across cell membranes underlies the basis for electrical signaling in the nervous system. Claude Bernard formulated the principle of the constancy of the internal environment (called homeostasis) that is illustrated by such integrated and controlled activities as the adaptive responses of respiratory, contractile, and hormonal systems and the set-points of blood pressure and body temperature. While it is clear that analysis of functional activity at the cellular and subcellular level is becoming increasingly more quantitative and molecular, it is also evident that there are functions that are only understandable at the integrated or whole animal level, such as perception, consciousness, or exercise.

Physiology deals with the experimental analysis of normal processes in normal systems in contrast to pathophysiology that subserves medicine as attended by injury and disease. There is a clear reciprocity of fundamental insight that emerges from comparisons of normal and abnormal processes. The study of physiology lends an understanding of life processes and provides us with profound opportunities to fulfill the precept "Think of all, hold on to the beautiful."

Joseph F. Hoffman
Higgins Professor of Cellular and Molecular Biology
Yale University School of Medicine

physiology [fiz′ē äl′ ə jē] *Biology.* **1.** a branch of biology dealing with the functions of living organisms or their parts. **2.** the organic processes or functions in an organism or in any of its parts or of a particular bodily process.

physisorption see PHYSICAL ADSORPTION.

Physodermataceae *Mycology.* a family of fungi belonging to the order Chytridiales that occur on land and in water; composed of parasites of vascular plants.

physohydrometra *Medicine.* the presence of gas and fluid within the uterus.

physometra *Medicine.* the presence of air or gas in the uterine cavity resulting in its distention.

Physopoda see THYSANOPTERA.

Physosomata *Invertebrate Zoology.* a superfamily of amphipod crustaceans in the suborder Hyperiidea, having small or no eyes, and some species with inflated bodies.

physostigmine *Organic Chemistry.* $C_{15}H_{21}O_2N_3$, an alkaloid occurring as toxic, colorless-to-pink crystals; slightly soluble in water and soluble in alcohol; unstable forms melt at 86–87°C and stable forms melt at 105–106°C; derived from the dried ripe seed of the African climbing plant *Physostigma venenosum*; also available as a salicylate, $C_{15}H_{21}O_2N_3{\cdot}C_7H_6O_3$, and sulfate, $(C_{15}H_{21}O_2N_3)_2{\cdot}H_2SO_4$; all three forms are used in medicine to produce miosis and decrease of intraocular pressure in glaucoma and to reverse effects produced by overdosage of anticholinergic drugs. Also, ESERINE.

physostigminism *Toxicology.* poisoning that is due to the ingestion of physostigmine; symptoms may include weakness, severe respiratory and gastrointestinal disturbances, and convulsions, possibly leading to death.

phyt- or **phyto-** a prefix meaning "plant."

Phytalmiidae *Invertebrate Zoology.* a family of dipteran insects in the subsection Acalyptratae.

phytal zone *Ecology.* the section of a shallow lake in which rooted vegetation grows.

Phytamastigophorea *Invertebrate Zoology.* a class of free-living, plantlike, flagellate protozoans in the subclass Sarcomastigophorea that are green or colorless and have only one or two flagella.

phytase *Enzymology.* an enzyme in plants that catalyzes the hydrolysis of phytic acid to inositol and phosphoric acid.

phyteral *Geology.* morphologically recognizable forms of vegetal matter (as distinguished from macerals) that occur in a coal mass.

phytic acid *Organic Chemistry.* $C_6H_6[OPO(OH)_2]_6$, an odorless, acidic-tasting, white to yellow liquid that is soluble in water and alcohol; it occurs naturally in many cereal grains and is derived from corn steep liquor; used to treat hard water, to chelate heavy metals in fat and oil processing, and as a nutrient.

phytin *Organic Chemistry.* the calcium-magnesium salt of phytic acid; a form of inositol found in plants and used as a dietary supplement.

phytoalexin *Biochemistry.* any of a group of compounds formed in plants in response to fungal infection, physical damage, chemical injury, or a pathogenic process that inhibit or destroy the invading agent.

phytobiont *Botany.* of or related to organisms that generally live on or within plants. Also, **phytobiontic.**

phytochorology see PLANT GEOGRAPHY.

phytochrome *Biochemistry.* a bluish conjugated protein whose response to relative periods of light and darkness regulates the flowering cycle in plants.

phytoclimatology *Meteorology.* the study of the climate of a confined space within an open area, such as in plant or wooded communities.

phytocoenosis *Ecology.* **1.** a collection of plants that are distinct from one another, yet share the same habitat. **2.** the total plant population of a given habitat. Also, **phytocenoses.**

phytocollite *Geology.* a black, gelatinous, nitrogenous mass of humus that occurs beneath or within peat deposits.

Phytodiniaceae *Botany.* a family of freshwater, unicellular flagellates composing the order Phytodiniales.

Phytodiniales *Botany.* an order of freshwater, unicellular flagellates of the class Dinophyceae, having a nonmotile vegetative state that is enclosed in a cell wall and attached to other algae and aquatic plants.

phytogenesis *Biology.* the origin and development of plants.

phytogenic *Ecology.* caused by or consisting of plants. Thus, **phytogenic dam, phytogenic dune**.

phytogeocoenosis *Ecology.* all the plants of a community together with their physical environment.

phytogeography see PLANT GEOGRAPHY.

phytoglycogen *Biochemistry.* a polysaccharide that resembles glycogen and is present in certain algae.

phytohemagglutinin *Hematology.* a lectin isolated from the red kidney bean, *Phaseolus vulgaris;* it is a hemagglutinin that agglutinates mammalian erythrocytes and a mitogen that stimulates predominantly T lymphocytes. Also, **phytolectin.**

phytol *Organic Chemistry.* $C_{20}H_{40}O$, a combustible, odorless liquid that is insoluble in water and soluble in most organic solvents; boils at 202–204°C (10 torr); obtained by the decomposition of chlorophyll; used in the synthesis of vitamins E and K.

Phytolaccaceae *Botany.* a loosely knit family of dicotyledonous herbs or woody plants of the order Caryophyllales that usually have anomalous secondary growth and are glabrous, saponiferous, and somewhat succulent; found in tropical and subtropical America.

phytolith *Paleontology.* a small biogenic structure formed by a living plant and preserved in rock; only recently studied.

phytology *Science.* a former term for botany.

Phytomastigina *Invertebrate Zoology.* a subclass of flagellate protozoans in the class Mastigophora; some species are green and capable of photosynthesis, some are colorless but similar in structure.

phytometer *Engineering.* a device used to measure physiological responses to various environmental factors on a group of plants grown under controlled conditions.

Phytomonadida *Invertebrate Zoology.* an order of green, spherical, flagellate protozoans in the subclass Phytomastigina, which feed as plants; similar to algae and classified as algae in some systems. Also, VOLVOCINA, VOLVOCIDA, PHYTOMONADINA.

phytonadione *Organic Chemistry.* $C_{31}H_{46}O_2$, a yellow, viscous, odorless liquid; insoluble in water, slightly soluble in alcohol, and soluble in benzene, chloroform, and vegetable oils; stable in air and decomposes in sunlight; used as a food supplement. Also, VITAMIN K_1.

phytoncide *Plant Pathology.* any substance that prevents the growth of an attacking organism, and thus gives a plant resistance to disease or infection.

phytone *Biochemistry.* a peptone made from plant protein.

phytoparasite *Ecology.* a parasitic plant.

phytopathogen *Pathology.* an organism that causes diseases in plants.

phytopathology *Botany.* the study of plant diseases, especially as they relate to parasites.

phytophagous *Zoology.* feeding mainly or exclusively on plants.

Phytophthora *Mycology.* a genus of fungi of the order Peronosporales, causing diseases in plants such as chestnuts, apples, and pears.

phytoplankton *Ecology.* a type of plant plankton, such as algae, that is the basic food source in many aquatic and marine ecosystems.

Phytoreovirus group *Virology.* either of two plant viruses of the Reoviridae family, having a wide host range of plants and transmitted by insect vectors, particularly the cicadellid leafhopper.

Phytosauria *Paleontology.* an order of thecodonts in the superorder Archosauria, widespread in the northern continents in the Triassic; about 2 m long, the phytosaurs resembled crocodiles but were not ancestral to them. Also, PARASUCHIA.

Phytoseiidae *Invertebrate Zoology.* a family of mites in the suborder Mesostigmata.

phytosociology *Ecology.* the science or study of vegetation that examines all aspects of plant life including its organization, interdependence, development, and geographical distribution.

phytotoxin *Biochemistry.* an exotoxin produced by certain species of high plants, notably abrin, ricin, crotin, and robin. *Toxicology.* any toxin produced by a plant.

phytotron *Botany.* a large room or area in which plants are grown under controlled environmental factors for the purpose of scientific comparison and evaluation. Also, BIOTRON.

pi [pī] Science. the 16th letter of the Greek alphabet, written as Π or π. *Mathematics.* for any circle, the constant irrational number that is the ratio of the circumference to the diameter; denoted by π, and approximately equal to 3.14159265.

PI programmed instruction; present illness; protamine insulin.

pia arachnoid *Vertebrate Zoology.* in reptiles and amphibians, a tough, protective inner covering surrounding the central nervous system; it is made from neural crest cells, is vascular, and has contact with the brain.

Piaget, Jean [pē´ä zhä´] 1896–1980, Swiss psychologist; formulated a noted description of childhood cognitive development.

Jean Piaget

pia mater *Neurology.* the innermost of the three meninges of the brain and spinal cord, consisting of a plexus of blood vessels contained in a fine areolar tissue.

pian bois *Medicine.* a form of cutaneous leishmaniasis of the New World that occurs in the forests of the Guianas and northern Brazil; caused by *Leishmania braziliensis guyanesis*, it is transmitted primarily by *Lutzomyia umbratilis*, and is characterized by multiple, widespread, deep skin ulcers with nodular lymphatic metastases. Also, FOREST YAWS.

piano wire *Metallurgy.* a thin carbon-steel wire that is cold drawn to very high tensile strength.

piassava *Botany.* **1.** the *Leopoldina piassaba* or the *Attalea funifera* palm of South America, from which a coarse woody fiber is obtained. **2.** the fiber from these trees, used in making brooms, mat, and the like.

PIAT *Ordnance.* a British antitank weapon used during World War II; it fired a 3-pound, rocket-propelled, hollow-charge grenade capable of penetrating 4 inches of armor. (An acronym for Projector Infantry Anti Tank.)

pi attenuator *Electricity.* a set of resistors organized and arranged in a pi network that serves as an attenuator.

piazza [pē ät´sə] *Architecture.* a public square surrounded by buildings. Also, PLAZA.

pibal see PILOT-BALLOON OBSERVATION.

pi bond *Physical Chemistry.* a type of covalent bond in which there is an overlap of orbitals (i.e., a region of greater electron density) both above and below the bond axis of the two atoms, and there is a plane of zero electron density along the bond axis itself. Thus, **pi bonding.**

PIC personal identification code.

pica [pi´kə] *Graphic Arts.* **1.** a printer's unit of measurement equal to 12 points (about 1/6 inch); commonly used in page layout, especially to measure the length of lines or the size of artwork. **2.** a type size equal to 12 points. **3.** a common typewriter type face of this size.

pica *Medicine.* the compulsive eating of nonnutritive items, such as dirt, gravel, clay, hair, or flaking paint or plaster, as food.

Picard, Charles [pē kär´; pi kärd´] 1856–1941, French mathematician; developed Picard's theorem for approximation of differential equations.

Picard, Jean 1620–1682, French astronomer; greatly improved measurements of the earth's circumference and meridian.

Picard method *Mathematics.* an iterative method for solving differential equations. The differential equation $y' = f(x,y)$ with initial condition $y(x_0) = y_0$ is equivalent to the integral equation

$$y(x) = y_0 + \int_{x_0}^{x} f(t,y(t))dt.$$

Then let

$$Ty = y_0 + \int_{x_0}^{x} f(t,y(t))dt.$$

Under certain continuity conditions, the iterative sequence

$$Ty_n = y_{n+1}(x) = y_0 + \int_{x_0}^{x} f(t,y_n(t))dt$$

converges uniformly to the solution $y(x)$. Also, **Picard-Lindelöf iteration method.**

Picard's big theorem *Mathematics.* let f be an analytic function with an essential singularity at $z = a$. Then in each neighborhood of a, f assumes each complex number, with one possible exception, an infinite number of times. Also, **Picard's great** or **second theorem.**

Picard's little theorem *Mathematics.* Let f be an entire function such that there exist complex numbers a and b with $f(z) \neq a$ and $f(z) \neq b$ for all z. Then f is a constant. Also, **Picard's first theorem.**

Picatinny test *Engineering.* a method used to determine the sensitivity of high explosives, in which the explosives are subjected to various degrees of impact.

Piccard, Auguste [pē kär´; pi kärd´] 1884–1962, Swiss physicist; invented the airtight balloon gondola and the bathyscaph; explored the stratosphere and sea depths. His son **Jacques Piccard** (born 1922), Swiss-Belgian oceanographer, explored the Marianas Trench.

Pichia *Mycology.* a large genus of yeast fungi belonging to the order Endomycetales; found throughout nature in trees, the tunnels of wood-boring beetles, grains, flour, wine, feces, and skin.

Picidae *Vertebrate Zoology.* the woodpeckers, a large family of birds in the order Piciformes, characterized by straight, pointed bills adapted for pecking wood, relatively large heads, long tongues, and often unusual plumage of striking color; found in tropical, subtropical, and deciduous forests worldwide, except Madagascar, New Zealand, Australia, and surrounding islands.

Piciformes *Vertebrate Zoology.* an order of hole-nesting land birds containing the families of jacamars, puffbirds, barbets, honeyguides, toucans, and woodpeckers.

Picinae *Vertebrate Zoology.* the subfamily of true woodpeckers of the family Picidae composed of 23 genera and about 170 species, including the **olive-backed woodpecker,** *Dinopium rafflesi*, and the **great spotted woodpecker,** *Picoides major*.

pick *Mechanical Devices.* a heavy iron or steel tool with a pointed tip at one or both ends and a handle fastened to the midpoint of its head; used for hacking through such hard surfaces as concrete or stone, or for chipping softer materials, such as asphalt or compacted soil. Also, **pickax.** *Mining Engineering.* **1.** to clean the walls of an excavation. **2.** to remove adhering material, such as dust, from coal. *Textiles.* **1.** a single filling yarn in a piece of fabric. **2.** in determining the count of a cloth, the number of warp or weft threads per inch. *Computer Technology.* to select an item on a display screen using a light pen or mouse.

pick-and-place robot *Robotics.* a simple robot with only one function, to move objects from one place to another.

pickerel *Vertebrate Zoology.* **1.** any of various small species of pike found in eastern North America, such as the **chain pickerel,** *Esox niger*, a popular food and game fish. **2.** a general name for members of the pike family, especially the walleye.

pickerel frog *Vertebrate Zoology.* the meadow frog, *Rana palustris*, which is similar to the leopard frog but has squarish dark spots on the back; found in eastern North America.

pickerel frog

pickerelweed *Botany.* any plant of the genus *Pontederia,* especially *P. cordata,* characterized by spikes of blue flowers; found in shallow fresh water in eastern North America.

Pickering, Edward Charles 1846–1919, American astronomer; a pioneer in stellar photometry; first to discover a double star (Mizar).

pickeringite *Mineralogy.* $MgAl_2(SO_4)_4{\cdot}22H_2O$, a colorless to white to faintly colored monoclinic mineral of the halotrichite group occurring in aggregates of acicular crystals, having a specific gravity of 1.79 and a hardness of 1.5 on the Mohs scale; found in weathered aluminous and pyrite-bearing rocks.

Pickering series *Spectroscopy.* in the spectra of very hot O-type stars, a set of spectral lines associated with singly ionized helium.

picket *Military Science.* a soldier or detachment of soldiers forward of a position, acting as sentinels to warn against an enemy advance.

picket bullet *Ordnance.* a conical, flat-nosed bullet. Also, SUGARLOAF BULLET.

picket ship *Naval Architecture.* a warship, normally an escort type, stationed at the edge of a formation or in a particular area, as a radar and observation picket to detect approaching enemy aircraft.

pick hammer *Mechanical Devices.* a hammer with a pointed tip at one end of the head and a blunt face at its opposite end.

picking *Mining Engineering.* **1.** the extraction of waste rock or other valueless material, or any specially selected mineral, from ore between mine and mill. **2.** the selective mining of high-grade ore. *Graphic Arts.* an unwanted removal of fiber or coating from the surface of paper during printing, usually due to a defect in the paper or the use of tacky ink.

pickle *Engineering.* to carry out a process of pickling. *Food Technology.* a cucumber (or, rarely, another food) that has been subjected to this process.

pickle liquor *Metallurgy.* an acid bath that has been used for pickling metallic materials.

pickle patch *Metallurgy.* a portion of scale that has not been removed after pickling.

pickle stain *Metallurgy.* a stain caused by pickling, when rinsing is inadequate.

pickling *Food Technology.* a method of preserving and seasoning foods by soaking them in brine or in an acidic solution, often with **pickling spices** such as allspice, bay leaf, and black pepper. *Chemical Engineering.* a treatment in the chrome tanning process in which hides and skins are conditioned with salt and acid solutions. *Metallurgy.* the process of removing scale or other surface compounds by immersion in a suitable aggressive liquid; sometimes electrochemically assisted.

pickling acid *Chemistry.* an inorganic acid, such as sulfuric, phosphoric, or hydrochloric acid, that is used to remove oxides, scale, and other impurities from metals.

pickoff *Electronics.* **1.** a device that converts mechanical movements into proportional electric signals. **2.** to monitor characteristics in a circuit, such as voltage or current, without interfering with its operation. **3.** a device that monitors linear or angular displacement in a circuit. *Mechanical Engineering.* an automatic device that removes the finished part from a press die after it has been stripped.

Pick's disease *Medicine.* **1.** a rare, progressive, degenerative disease of the brain very similar in clinical symptoms to Alzheimer's disease; it affects mainly the frontal and temporal lobes of the brain and produces slow disintegration of intellect, personality, and emotion. (Named for Arnold *Pick,* 1867–1926, Czech neurologist.) **2.** a former term for pericardial pseudocirrhosis of the liver; a congestive cirrhosis resulting from restrictive pericarditis. (Named for Friedel *Pick,* 1867–1926, Czech physician.)

pickup *Electricity.* **1.** a transducer used to receive the signal required for any equipment operation by the utilization of an active conversion from sound, light, or another measurable quantity into a corresponding electrical signal, for example, in a microphone, telephone, or television camera. **2.** the minimum of current, voltage, power, and so on at which a relay will complete its intended function. **3.** any device that serves as a sensor of a signal or a quantity, such as a temperature sensor or vibration detector. *Aviation.* in an automatic pilot, an electronic device that detects deviation from the desired flight path. *Metallurgy.* in metal working, an accumulation of material on a working tool. *Nuclear Physics.* a type of nuclear reaction in which the particle bombarding the nucleus takes a nucleon from the nucleus and binds the nucleon to itself.

pickup voltage *Electricity.* the lowest value at which the contacts of a relay will open or close Also, **pickup value, pickup current, pickup power.**

pico- *Electricity.* a combining form that represents 10^{-12}, or one-trillionth. Formerly called *micromicro-*.

picoammeter *Engineering.* an instrument used to measure small electric currents.

picoampere *Electricity.* a unit of electric current equal to one-trillionth of an ampere.

pico-carmine *Cell Biology.* a cytological dye that is used to stain the cytoplasm of a cell yellow.

picocurie *Nucleonics.* a unit of radioactivity equal to one-trillionth of a curie.

picodnavirus *Virology.* a former term for an autonomous virus of the Parvoviridae family.

picofarad *Electricity.* a unit of capacitance equal to one-trillionth of a farad.

picogram *Metrology.* a unit of mass equal to one-trillionth of a gram. Also, **picogramme.**

picohenry *Electricity.* a unit of inductance equal to one-trillionth of a henry.

picoline *Organic Chemistry.* $C_5H_4N(CH_3)$, one of three isomers occurring as a combustible colorless liquid; soluble in water and alcohol; **α-picoline** boils at 129°C and freezes at –69.9°C; **β-picoline** boils at 143.5°C and freezes at –18.3°C; **γ-picoline** boils at 144.9°C and freezes at 3.8°C; used in making pharmaceuticals and rubber.

picolinic acid *Organic Chemistry.* $C_5H_4N{\cdot}COOH$, white needles that are slightly soluble in water and soluble in alcohol; melts at 136–137°C and sublimes before boiling; it is used as a chemical reagent. Also, 2-PYRIDINECARBOXYLIC ACID.

picomole *Chemistry.* a measure of the amount of a substance equal to one-trillionth of a mole.
picoplankton *Ecology.* plankton in the size range 0.2–2.0 μm.
Picornaviridae *Virology.* a family of small, ssRNA-containing viruses that are marked by virus-coded polymerase, cytoplasmic replication, and a narrow host range.
picornavirus *Virology.* any virus of the family Picornaviridae.
picosecond *Metrology.* a unit of time equal to one-trillionth of a second.
picotite *Mineralogy.* $(Mg,Fe^{+2})(Al,Cr)_2O_4$, a dark-brown chromian variety of spinel, commonly found in peridotites.
Picotte, Susan La Flesche 1865–1915, American physician; the first American Indian woman to become a medical doctor.
picowatt *Metrology.* a unit of power equal to one-trillionth of a watt.
picr- or **picro-** a prefix meaning "bitter."
picramic acid *Organic Chemistry.* $C_6H_5N_3O_5$, flammable, explosive, red crystals; slightly soluble in water and soluble in alcohol; melts at 169–170°C; used in dye manufacture.
picric acid *Organic Chemistry.* $C_6H_2(NO_2)_3OH$, bitter-tasting yellow crystals; soluble in water and alcohol; melts at 122°C and explodes about 300°C; used in explosives, matches, and electric batteries and for other industrial purposes.
picrite *Petrology.* a dark, fine- to medium-grained igneous rock with abundant olivine and smaller amounts of pyroxene, hornblende, and plagioclase.
Picrodendraceae see EUPHORBIACEAE.
picromerite *Mineralogy.* $K_2Mg(SO_4)_2{\cdot}6H_2O$, a colorless to white, water-soluble, monoclinic mineral of the picromerite group, occurring as crystalline encrustations, and having a specific gravity of 2.028 and a hardness of 2.5 on the Mohs scale; found with other evaporite minerals in oceanic salt deposits.
picropharmacolite *Mineralogy.* $H_2Ca_4Mg(AsO_4)_4{\cdot}11H_2O$, a colorless to white triclinic mineral occurring in botryoidal forms and as tiny acicular crystals, having a specific gravity of 2.62 and an undetermined hardness.
picrotoxin *Biochemistry.* $C_{30}H_{34}O_{13}$, a glycoside obtained from the seed of *Anamirta cocculus*; occurring as flexible, shining, prismatic crystals or as a microcrystalline powder that stimulates the respiratory center of the medulla. Also, COCCULIN.
pictograph *Linguistics.* **1.** a sign or symbol used to represent a word or idea. **2.** specifically, such a symbol used in a prealphabetic writing system. Thus, **pictographic.** *Graphic Arts.* any pictorial image that conveys an idea, such as the familiar "no smoking" symbol. *Archaeology.* a design or drawing painted on a rock surface. Also, PETROGRAPH.
pictomap *Photogrammetry.* a reproduction of a photomosaic in which the photographic images have been converted, through the use of masking techniques and overlays, into interpretable colors and symbols.
Pictor *Astronomy.* the Painter's Easel, a faint southern constellation between Dorado and Carina; it has no bright stars.
picture *Graphic Arts.* **1.** a positive or negative photographic image. **2.** any visual representation, especially a photograph, that is reproduced in a printed work. *Telecommunications.* the image seen on a television receiver. *Computer Programming.* **1.** in COBOL, a description of data elements using rules concerning numeric or alphanumeric data, location of decimal points, and length. **2.** a digitally represented graphic image.
picture carrier *Telecommunications.* a carrier frequency that is used for transmitting color information. Also, LUMINANCE CARRIER.
picture compression *Computer Programming.* **1.** the representation of a digital picture by short codes for frequently occurring colors or blocks of colors, and longer codes for infrequent occurrences, thus reducing redundancy. **2.** in general, any system for reducing the number of bits required to represent a digital picture.
picture element *Electronics.* **1.** the portion of the copy seen by a facsimile scanner at any given moment. **2.** the portion of a television scanning line being explored at any given moment during the scanning process. Also, ELEMENTAL AREA, PIXEL.
picture frequency *Electronics.* **1.** the number of complete pictures scanned per second in a television system. **2.** see FRAME FREQUENCY. *Telecommunications.* in FAX technology, the frequencies that result from scanning a subject copy.
picture grammar *Computer Programming.* a set of formal rules for the description of picture structure.
picture plane *Photogrammetry.* the plane occupied by a photographic film, on which light rays from the object can be focused to form an image.
picture segmentation *Computer Programming.* the division of a complex picture into smaller segments and the subsequent description of the whole in terms of its parts, their relationships, and their properties.
picture signal *Telecommunications.* in television or FAX transmission, the signal that results from the scanning process.
picture transmission *Telecommunications.* any transmission of a photograph with special regard for tone reproduction.
picture tube *Electronics.* a cathode-ray tube that displays an image in a television receiver by changing the intensity of its electron beam as the beam scans the illuminated portion of the tube's fluorescent screen.
picture-tube brightener *Electronics.* a small step-up transformer that is placed between the socket and base of a picture tube to compensate for the loss of image brightness that normally occurs as the tube ages.
PID pelvic inflammatory disease.
pidgin [pij′in] *Linguistics.* **1.** a minimal second language that is a combination of the vocabulary and pronunciation patterns of two or more languages, created when groups speaking different languages have a need to communicate, as for trade or negotiations; grammatically, it usually is a simplified form of one of the languages. **2.** loosely, any simplified or abridged form of a language used by nonnative speakers. (Said to be from a mispronunciation of the word *business*.)
piebaldism *Medicine.* a condition in which the hair or skin is spotted or blotched partly white and partly brown, as in part albinism and vitiligo.
piece dyeing *Textiles.* the process of dyeing a whole piece of fabric as it is taken from the loom, as distinguished from dyeing the yarns before weaving or knitting.
piece goods *Textiles.* fabric manufactured in long lengths and standard widths, sold in retail stores in a cut length specified by a customer.
piece mark *Engineering.* an identification mark for a part of a machine or structure, found on a drawing and sometimes on the part itself, that indicates the position or order in which the part belongs in the assembly. Also, ASSEMBLY MARK.
piecemeal stoping *Geology.* magmatic stoping in which only isolated blocks of roof rock are detached and engulfed by magma.
piece part *Design Engineering.* an individual item or workpiece; a piece of manufacturing stock.
piece rate *Industrial Engineering.* a worker compensation scheme in which workers are paid a fixed rate per "piece" or unit of output.
piece root grafting *Botany.* a grafting process in which pieces of cut seedling roots are used as stalk.
piecewise *Mathematics.* relating to any property of a function of a single real variable that holds on a set of (convex) intervals whose union is the domain of the function; e.g., **piecewise differentiable, piecewise linear, piecewise monotone,** etc.
piecewise-linear system *Control Systems.* a system in which the input quantities can be divided into a finite number of intervals and, within each interval, the output quantity is a linear function of the input quantity.
piecework *Industrial Engineering.* work that is paid at a piece rate, rather than at an hourly rate or day wage.
piedmont *Geology.* **1.** an area, plain, or other feature located at the base of a mountain or mountain range. **2.** relating to any feature lying or formed at the base of a mountain or mountain range.

piedmont

piedmont angle *Geology.* the sharp change of slope between a lowlands and uplands, such as between a hill and a plain.

piedmont benchland *Geology.* one of a series or a system of piedmont steps. Also, **piedmont stairway, piedmont treppe.**

piedmont bulb see EXPANDED FOOT.

piedmont glacier *Hydrology.* a thick sheet of ice formed at the base of a mountain range by the spreading out and joining of ice streams from higher elevations.

piedmont gravel *Geology.* a coarse gravel transported from high ground by mountain torrents and spread out on relatively flat ground, where the velocity of the water has decreased.

piedmontite see PIEMONTITE.

piedmont lake *Hydrology.* a body of water that occupies a basin hollowed out by a piedmont glacier.

piedmont plateau *Geology.* a plateau situated between the mountains and the plains or the ocean.

piedmont scarp *Geology.* a small, low cliff formed in alluvium on a piedmont slope at the base of a steep mountain range, more or less parallel to it. Also, SCARPLET.

piedmont step *Geology.* in a piedmont area, an extensive or regional terracelike bench that slopes outward or down-valley, usually in a series at different levels. Also, **piedmont bench, piedmont flat.**

Piedraia *Mycology.* a genus of fungi belonging to the order Myriangiales; its only species, *P. hortae,* is found on hair and causes the hair disease **black piedra.**

Piedraiaceae *Mycology.* a family of fungi belonging to the order Myriangiales, which forms hard, dark masses at the base of hair in humans, causing black piedra.

pi electron *Physical Chemistry.* an electron that takes part in pi bonding.

piemontite *Mineralogy.* $Ca_2(Al,Mn^{+3},Fe^{+3})_3(SiO_4)_3(OH)$, a reddish-brown or purplish-red monoclinic mineral of the epidote group occurring as prismatic or acicular crystals and in massive form, having a specific gravity of 3.45 to 3.52 and a hardness of 6 on the Mohs scale; found in schist and volcanic rocks. Also, PIEDMONTITE.

pier *Architecture.* **1.** a masonry mass that supports an arch, beam, or lintel. **2.** a thickened section of wall, as between two windows. *Civil Engineering.* a structure constructed on posts extending out over the water; used as a landing place for ships. *Building Engineering.* a pillar or post on which a door is hung.

Pierce, George Washington 1872–1956, American physicist; developed the Pierce oscillator; with Arthur Edwin Kennelly, discovered motional impedance.

Pierce gun *Electrical Engineering.* an electron gun that is designed with the electrodes shaped so that the potential at the edges of the beam balances the space charge and accelerates the electronics through the desired aperture.

piercement see DIAPIR.

Pierce oscillator *Electronics.* an oscillator that has a piezoelectric crystal, such as quartz, connected between the input and output terminals of the tube or transistor, instead of a coil or capacitor.

Pierce's disease *Plant Pathology.* a virus disease of grapes, characterized by dwarfing, wilting, mottling between leaf veins, premature defoliation and fruit ripening, and death of the vine.

piercing *Materials Science.* a method for producing metal tubing by using a mandrel between two tapered rolls; the inlet taper compresses a heated billet that rotates, and the cross section continuously changes from round to oval, creating a cavity in the center of the billet as it moves over the mandrel.

piercing fold see DIAPIR.

piercing gripper *Robotics.* an end effector that pierces a workpiece in order to handle it.

Pieridae *Invertebrate Zoology.* a family of common butterflies, composed of whites, sulfurs, and orange-tips; lepidopteran insects in the superfamily Papilionoidea, including the cabbage white whose larvae are a serious plant pest.

Piesmatidae *Invertebrate Zoology.* a family of leaf bugs, hemipteran insects in the group Pentatomorpha.

Piesmidae *Invertebrate Zoology.* the ash-gray leaf bugs, a family of hemipteran insects in the superfamily Lygaeoidea; plant eaters that transmit a viral plant disease.

piezo- a prefix meaning "pressure."

piezocardiogram *Cardiology.* a graphic tracing of variation in pressure caused by the heart's pulsation, often recorded through the esophagus.

piezochemistry *Chemistry.* the study of chemical reactions that occur at very high pressures, such as under the earth's crust.

piezocrystallization *Geology.* the crystallization of a magma under the influence of direct pressure.

piezoelectric [pī ē´zō i lek´trik; pī ā´zō i lek´trik] *Solid-State Physics.* describing the ability of a solid to generate a voltage when subjected to a mechanical stress, or the ability to generate a mechanical force when subjected to a voltage. *Acoustical Engineering.* describing a sound device whose operation is based on the energy produced by the change in dimensions of a piezoelectric material, such as quartz crystals. Thus, **piezoelectric microphone, piezoelectric transducer.**

piezoelectric ceramic *Materials.* any ceramic that exhibits piezoelectric properties.

piezoelectric crystal *Solid-State Physics.* a crystal that exhibits the piezoelectric effect; such a crystal is used in electro-audio devices such as crystal microphones, speakers, and phonograph pickups.

piezoelectric detector *Engineering.* an instrument used to measure seismic activity; made up of a stack of piezoelectric crystals with intervening metal foils that collect charges produced on the crystal faces when the crystals are strained by pressure from an inertial mass mounted above the stack.

piezoelectric effect *Solid-State Physics.* an electromechanical effect by which mechanical forces acting upon a ferroelectric material can produce an electrical response, and electrical forces can produce a mechanical response.

piezoelectric element *Electronics.* a crystal, such as quartz, that produces an electric voltage when it is twisted or squeezed, and, conversely, that twists, bends, expands, or contracts when a voltage is applied to it; commonly used as a transducer to convert mechanical or acoustical signals into electric signals or to regulate frequency in a crystal oscillator.

piezoelectric gauge *Engineering.* an instrument used to measure blast pressures resulting from explosions and pressures created in firearms, using a piezoelectric substance that produces a voltage when under pressure.

piezoelectric hysteresis *Solid-State Physics.* the hysteretic behavior of a piezoelectric crystal in which the electric polarization depends on the stress history as well as on the mechanical stress applied to the crystal.

piezoelectricity [pī ē´zō i lek tris´i tē; pī ā´zō i lek tris´i tē] *Solid-State Physics.* a voltage or electric field in a dielectric crystal produced by the piezoelectric effect; this can occur in certain nonmetallic materials, such as quartz crystals or ceramics and plastics.

piezoelectric polymer see PIEZOPOLYMER.

piezoelectric semiconductor *Solid-State Physics.* a semiconductor that exhibits the piezoelectric effect, such as quartz or barium titanate.

piezoelectric transducer *Electronics.* a device that uses the interaction between an electric charge and the deformation of a piezoelectric crystal to convert mechanical or acoustical signals into electrical ones, especially such a device used in a microphone. Also, CRYSTAL TRANSDUCER.

piezoelectric vibrator *Solid-State Physics.* a sample of piezoelectric material that is attached to electrodes and mounted near some other vibrating element; the piezoelectric sample is used to excite resonant frequencies in the second material.

piezogene *Geology.* describing the formation of minerals largely under the influence of pressure.

piezoglypt see REGMAGLYPT.

piezomagnetism *Solid-State Physics.* magnetism resulting from mechanical stress or the reverse effect.

piezometer *Engineering.* **1.** a device used to measure fluid pressure, particularly a device inside a vessel that contains gas or liquid. **2.** a device that measures the compressibility of materials, especially a device that measures the variation in volume of a material that is subjected to hydrostatic pressure.

piezometric surface see POTENTIOMETRIC SURFACE.

piezooptical effect *Optics.* an alteration in the refractive index of transparent material, induced by mechanical stress.

piezopolymer *Organic Chemistry.* a polymer film that can convert heat and pressure into electricity. Also, PIEZOELECTRIC POLYMER.

piezotransistor accelerometer *Engineering.* a device used to measure acceleration, by recording the current variation across the *pn* junction of a transistor that has received a concentrated force from a seismic mass.

piezotropic *Fluid Mechanics.* of or relating to piezotropy.

piezotropy *Fluid Mechanics.* a behavior exhibited by fluids that show a functional dependence of the form $dr/dt = f(dp/dt)$, where r is the fluid density, p is the fluid pressure, t is the time, and f (the coefficient of piezotropy) is a function of thermodynamic variables.

PIF prolactin inhibitory factor.

pi filter *Electronics.* a filter constructed as a pi network.

pig *Vertebrate Zoology.* **1.** an omnivorous mammal of the family Suidae, having a compact body with thick skin covered with bristles, and a protruding, flattened snout surrounded by tusks (in certain species); native to forests and woodlands of the Old World and domesticated worldwide. Also, HOG, SWINE. **2.** specifically, the common domesticated animal *Sus scrofa. Agriculture.* a term for a young animal of this type, usually less than 10 weeks old.

pig any of various materials or devices thought to resemble the common animal in some way, as in shape or appearance; specific uses include: *Engineering.* a brush, blade, or swab that is forced through a pipe or duct to clean it. *Nucleonics.* a heavily shielded lead container used to store radioactive materials. *Metallurgy.* **1.** a casting that is eventually remelted, and not used as such. **2.** an oblong mass of metal that has been run while still molten into a mold of sand or the like. **3.** the mold for such a mass of metal.

pig *Electronics.* an ion source that functions on the same principles as a Philips ionization gauge. (An acronym for Philips ionization gauge.)

pig bed *Metallurgy.* a bed of sand used for molding pigs, into which molten metal is poured.

pigeon *Vertebrate Zoology.* any of several birds belonging to the family Columbidae, characterized by a plump, large-breasted body covered with dense, soft feathers and an ability to produce a milklike liquid in the crop for feeding its young; found worldwide except in high northern latitudes, Antarctica, and Madagascar. Pigeons have often been domesticated, as for racing or carrying messages. The type often found in urban locales of North America is the rock dove, *Columba livia.*

pigeonhole principle *Mathematics.* if n objects are distributed over m places, with $n > m$, then some place receives at least two objects.

pigeonite *Mineralogy.* $(Mg,Fe^{+2},Ca)(Mg,Fe^{+2})Si_2O_6$, a brown to black monoclinic mineral of the pyroxene group occurring as short prismatic crystals and small grains, having a specific gravity of 3.30 to 3.46 and a hardness of 6 on the Mohs scale; found in basic igneous rocks and the groundmass of intermediate volcanics.

pigeon milk *Physiology.* the regurgitated liquid, having a milklike appearance, that the adult pigeon feeds the young.

pigfish *Vertebrate Zoology.* a grunt of the species *Orthopristis chrysoptera*; found off the Atlantic coast of the southern United States. (From the characteristic sound it makes, thought to resemble a pig.)

piggery *Agriculture.* a place in which pigs are bred and housed.

piggyback plant *Botany.* the plant *Tolmiea menziesii* of the saxifrage family that produces a new plant at the base of its leaves; native to western North America; it is popular as a houseplant. Also, **pickaback plant.**

piggyback twistor *Electronics.* a device that employs a thin, narrow tape of magnetic material wrapped around a fine copper conductor to store information, while a similar material is wrapped, piggyback style, on top of the first to detect stored information.

pig iron *Metallurgy.* high carbon iron produced in a blast furnace.

pig lead *Metallurgy.* lead molded in pigs.

pigment *Biochemistry.* any normal or abnormal coloring matter of the body. *Chemistry.* a natural or synthetic substance that is used in suspension to impart color to another substance; distinguished from a dye, which is used in solution. *Materials.* specifically, such a coloring substance mixed with water, oil, or other liquid in order to create paint.

pigmentation *Physiology.* **1.** coloration of an organ or organism due to pigment deposits. **2.** changes in hue of the skin or tissue, either by normal or by disease process; caused by pigment deposits.

pigment cell *Cell Biology.* a specialized cell containing granules that impart color or pigment, such as a melanocyte.

pigmented paper *Graphic Arts.* a lightly coated printing paper that is nearly as smooth as a fully coated paper but less expensive. Also, FILM-COATED PAPER, MATTE-COATED PAPER.

pignola see PINE NUT.

pigtail *Electricity.* a term for a short, flexible wire used to connect a stationary terminal with a terminal that has a limited range of motion such as those on relay armatures; it is usually stranded and braided and contains a long lead or leads that are used for mounting.

pigtail splice *Electricity.* a splice made by tightly twisting together the bared ends of parallel conductors.

pika *Vertebrate Zoology.* a small herbivorous mammal of the family Ochotonidae, closely related to rabbits and hares, but distinguished by short, rounded ears, a compact body with short limbs, and no visible tail; found in central and northern Asia and North America.

pike *Geology.* the summit or top of a mountain or hill that is peaked or pointed.

pike *Vertebrate Zoology.* **1.** any one of five species of predatory fish of the family Esocidae, characterized by elongate bodies with pointed snouts and caudal and anal fins toward the rear; found in fresh water in temperate Northern Hemisphere regions; noted as a sport fish. The **northern pike,** *Esox lucius,* is widely distributed in eastern and northern North America. **2.** any of various somewhat similar fish.

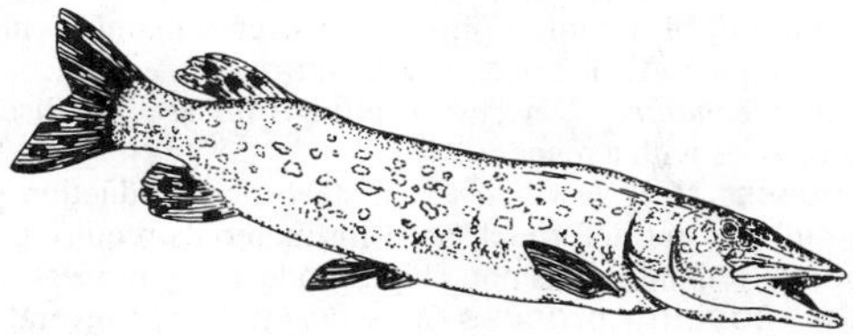

northern pike

pike perch *Vertebrate Zoology.* a popular name for the walleye or walleyed pike (actually a perch), *Stizostedion vitreum vitreum.*

pil- or **pilo-** a prefix meaning "hair."

Pilacraceae *Mycology.* in some classifications, a family of fungi belonging to the subdivision Basidiomycotina.

Pilaira *Mycology.* a genus of fungi belonging to the order Mucorales that are found on dung.

Pilargidae *Invertebrate Zoology.* a family of small polychaete annelid worms in the Errantia.

pilaster *Architecture.* a rectangular column that protrudes from and beyond a wall surface, often having a base and capital.

pilchard *Vertebrate Zoology.* a small ocean fish of the genus *Sardina* or *Sardinops,* especially *Sardina pilchardus,* the commercial sardine of Europe.

pilchard oil *Materials.* a combustible, pale-yellow oil that is expressed from the pilchard fish, a member of the herring family; used in soap and paints.

pile *Textiles.* raised loops, cut loops, and other erect fibers deliberately woven or knitted into fabric to form a uniform, even surface, such as on velvet, terry cloth, imitation fur, and carpets. *Engineering.* a columnar timber, steel, or reinforced concrete stake that is sunk into the ground to support a load or resist lateral pressure. *Nucleonics.* see NUCLEAR REACTOR.

pileate *Biology.* having a pileus. *Mycology.* specifically, describing fungi that have a pileus or mushroom cap.

pileated woodpecker *Vertebrate Zoology.* a large woodpecker, *Dryocopus pileatus,* that is characterized by black-and-white plumage and a prominent red crest.

pile beacon *Navigation.* a fixed aid to navigation, consisting of a single pile with some sort of identification.

pile bent *Civil Engineering.* a multiple of posts or piles situated at a crosswise angle to the length of a structure, held together for stability by capping or bracing.

pile cap *Civil Engineering.* **1.** a structural device fastened over the top of a pile to consign a load equally to multiple piles or posts. Also, GIRDER CAP, RIDER CAP. **2.** the footing that sits atop the pile bent.

pile dike *Civil Engineering.* a dike that is formed by a line of braced or lashed piles.

piledriver *Mechanical Engineering.* a framed construction for driving piles, in which a heavy weight is raised and dropped on a pile head or in which a steam hammer drives the pile.

pile extractor *Mechanical Engineering.* a sheet-piling extractor that operates in a manner similar to a pile driver, except that the force of the blow is upward rather than down.

pile fabric *Materials.* a three-dimensional fabric that is usually machine produced and used for carpeting.

pile foundation *Civil Engineering.* the structural arrangement of piles that consigns loads to bed rock or to ground well below the surface.

pile hammer *Mechanical Engineering.* the heavy weight or ram of a piledriver.

pile helmet *Civil Engineering.* a steel cap that is used to cover and protect the head of a concrete pile during driving. Also, HELMET.

pile lighthouse *Navigation.* a lighthouse built on a platform supported by screwpiles. Also, SCREWPILE LIGHTHOUSE.

pile shoe *Civil Engineering.* a cast-iron device fitted to the foot of a pile to protect it during the driving process and to assist in stability and breaking ground.

pileum *Vertebrate Zoology.* a bird's head from the bill to the nape.

pileup *Electronics.* the elements, such as contact arms, assemblies, or springs, mounted on a relay with a layer of insulation between each.

pileus *Mycology.* the cap of a mushroom of the subdivision Basidiomycotina, characterized by its umbrellalike shape. *Invertebrate Zoology.* the umbrella or bell of a jellyfish. *Meteorology.* an accessory cloud with a thin aspect, in the form of a cap, hood, or scarf, found above or attached to the top of a cumuliform cloud; occurs mainly with cumulus and cumulonimbus. Also, CAP CLOUD, SCARF CLOUD.

Pilger mill *Mechanical Devices.* a rolling mill used in discontinuous milling processes with a mandrel.

Pilger process *Materials Science.* a steel-tube production process in which the tube is hot deformed; used for materials requiring high-flow stresses for deformation, it is not a high-productivity process.

Pilger tube-reducing process *Metallurgy.* one of several processes by which the diameter and wall thickness of a tube are reduced by rolling.

Pilidae *Invertebrate Zoology.* a family of amphibious freshwater snails in the order Pectinibranchia, having both gill and pulmonary sac.

pilidium *Invertebrate Zoology.* a type of larva found in nemerteans, having flaplike lobes that grow on either side of the mouth and resembling a helmet.

Pilifera *Vertebrate Zoology.* a collective name for all vertebrates with hair, i.e., mammals.

piliferous layer *Botany.* the hair-bearing absorbing region of the root epidermis.

Pilimelia *Bacteriology.* a genus of bacteria of the family Actinoplanaceae, order Actinomycetales, consisting of soil organisms that possess spherical or cylindrical sporangia containing rod-shaped spores.

pilin *Biochemistry.* the protein that composes bacterial pili.

piling *Graphic Arts.* a term for an accumulation of ink or pigment on a printing plate, roller, or blanket.

pill *Pharmacology.* a small, usually soluble medicated mass that is shaped to facilitate swallowing; along with medication, it contains a filler and an excipient substance that facilitates working it into the desired pillular form. *Electromagnetism.* the termination of a microwave stripline.

pillar *Civil Engineering.* a post or column. *Mining Engineering.* a section of ore or coal supporting the overlying strata or hanging wall within a mine. *Geology.* **1.** any naturally occurring large rock formation that is shaped like a pillar. **2.** a columnar-shaped body of rock between adjacent joints.

pillar bolt *Mechanical Devices.* a bolt projecting from a part similar to a stud for added support.

pillar buoy *Navigation.* a floating aid to navigation with a tall structure rising from the center.

pillar crane *Mechanical Engineering.* a crane that is rotated about a fixed pillar so that the load is moved horizontally along the periphery of a circle.

pillar drill *Mechanical Devices.* a drilling device with a spindle and table supported by a slidable pillar bracket and base assembly.

pillar file *Mechanical Devices.* a double-cut, slender, flat steel file; used to finish narrow openings such as slots and keyways.

pillaring *Ordnance.* a rapidly rising column of smoke that may be produced by a misfunctioning white phosphorous bomb or projectile; it is a negative result because it does not obscure a large enough area.

pillar line *Mining Engineering.* an airflow that has definitely passed through an inaccessible, abandoned area or has ventilated a pillar area, regardless of the amount, or lack, of methane in the air.

pillar press *Mechanical Engineering.* a punch press framed by two upright columns through which the driving shaft passes as the slide moves between them.

pillbox *Ordnance.* a small, low fortification that houses machine guns or similar weapons; usually made of concrete, steel, or filled sandbags.

pillbox antenna *Electromagnetism.* an antenna having a cylindrical parabolic reflector that is closed by parallel end surfaces spaced so that only one mode can be excited; the antenna is fed by a radiating element along the focal line.

pilling *Textiles.* a defect in fabric caused when fibers are rubbed or pulled out of yarns and entangle with intact fibers, forming soft, fuzzy balls on the fabric surface.

Pilling-Bedworth ratio *Materials Science.* the relative specific volumes of oxide and metal when an oxide forms on the surface of a metal, expressed as the ratio, Wd/Dw, where W is the molecular weight of the oxide, d is the density of the metal, D is the density of the oxide, and w is the molecular weight of the metal; predicts the type of oxide growth.

pillow breccia *Petrology.* a volcanic deposit containing pillows and fragments of lava in a matrix of palagonite tuff.

pillow distortion see PINCUSHION DISTORTION.

pillow lava *Volcanology.* bulbous masses of lava, resembling a pile of pillows, that flow from an underwater eruption; its form results from the chilling of magma in water. Also, ELLIPSOIDAL LAVA.

pillow structure *Geology.* a primary sedimentary structure that resembles a pillow. *Petrology.* a structure of certain extrusive igneous rocks or lavas characterized by agglomeration of rounded masses in the shape of pillows, bolsters, or filled sacks; the structure is attributed to subaqueous deposition.

pilmer *Meteorology.* a heavy rainshower that occurs in England.

Pilobolaceae *Mycology.* a family of fungi belonging to the order Mucorales, characterized by a distinctive cell wall called a **trophocyst;** found primarily on dung, and also in riverbanks and algae.

Pilobolus *Mycology.* a genus of fungi belonging to the order Mucorales, characterized by a forcible ejection of ripe sporangium.

pilocarpine *Organic Chemistry.* $C_{11}H_{16}N_2O_2$, an alkaloid occurring as colorless, water-soluble needles that melt at 34°C; derived from jaborandi, the dried leaflets of the tropical American tree, *Pilocarpus pennatifolius*; used to constrict pupils and relieve intraocular pressure in the treatment of glaucoma.

pilocystic *Oncology.* of or related to a hollow or cystlike structure containing hairs; a term used to describe certain dermoid tumors.

pilomotor reflex *Physiology.* a tensing of the smooth muscles of the skin attached to integumentary hairs, when stimulated by light, by touch, or by a drop in temperature; causes an erection of the hairs commonly termed gooseflesh.

pilose *Medicine.* hairy; covered with hair.

pilosebaceous *Anatomy.* relating to the hair follicles and sebaceous glands.

pilot *Aviation.* **1.** to operate, control, or guide a flight vehicle from within the vehicle; to fly an aircraft. **2.** a person or persons performing this function, especially someone who is legally qualified and licensed to do so. *Navigation.* a specially qualified navigator with extensive local knowledge, used for conducting ships in harbors, bays, and other inshore waters. *Computer Programming.* of or relating to a new or experimental program, project, or device. *Mechanical Engineering.* a guide for positioning two adjacent parts; usually consists of a projection on one part that fits into a recess on the other part. *Telecommunications.* a signal wave, usually a single frequency, sent over a system for the purpose of supervision, synchronization, or reference.

PILOT *Computer Programming.* a special-purpose programming language designed for writing computer-aided instruction programs. (An acronym for programmed inquiry, learning, or teaching.)

pilotage *Navigation.* the directing of a vessel in proximity to the land, using primarily geographic reference points.

pilotaxitic *Geology.* of or relating to the texture of the groundmass of a holocrystalline igneous rock, consisting of lath-shaped microlites of feldspar arranged in a glass-free, felted mesh.

pilot balloon *Engineering.* a small rubber balloon that is tracked by a theodolite after being released from the ground; used to determine the direction and speed of air currents at high altitudes.

pilot-balloon observation *Meteorology.* a method of winds-aloft observation that examines wind speeds and directions in the atmosphere above a weather reporting station; determined by readings of the elevation and azimuth angles of a theodolite, along with visual tracking of a pilot balloon. Also, PIBAL.

pilot bit *Mechanical Devices.* a noncoring, diamond-set drill bit in the center of a main coring-bit hole saw that protrudes beyond its face or cylindrical blade; used to anchor the bit or saw in position before cutting.

pilot boat *Navigation.* a small, maneuverable vessel used to carry pilots out to incoming ships.

pilot cell *Electricity.* a cell of a battery used to take readings indicative of the overall state of the battery, since temperature, voltage, and specific gravity are assumed to be consistent within the entire battery.

pilot channel *Civil Engineering.* a small tunnel or shaft driven before a larger tunnel is built to afford access for the larger project.

pilot chart *Navigation.* a chart, usually small-scale, that provides information on ocean currents, normal weather, and other topics of interest to the navigator.

pilot drill *Mechanical Engineering.* a small drill used to start a hole to keep a larger drill true to center.

pilot electrode see IGNITOR, def. 1.

pilotfish *Vertebrate Zoology.* a small marine fish, *Naucrates ductor* of the family Carangidae, often seen swimming near sharks and other much larger fish. (So called because of an earlier belief that it guided the larger fish.)

pilot hole *Engineering.* a small hole drilled prior to a borehole to serve as a guide for the drilling of the larger hole.

pilothouse *Naval Architecture.* a raised, enclosed housing around a ship's steering or pilotage position. Also, WHEELHOUSE.

pilot lamp *Electricity.* a small incandescent or neon light that indicates that a device is operating.

pilotless aircraft *Aviation.* an aircraft, such as a drone, that is designed to fly without a human pilot. Also, ROBOT AIRCRAFT, RPV.

pilot light *Engineering.* a small permanent flame used to ignite a gas burner.

pilot-line operation *Industrial Engineering.* the initial production-line manufacturing of a new product or item, usually on a small production scale, during which production methods are analyzed, adjusted, and streamlined.

pilot model *Industrial Engineering.* the initial model of a new product being brought into production, usually a simple, basic model suited to further refinement or alteration as production experience is gained.

pilot parachute *Aviation.* a small, usually spring-operated auxiliary parachute attached to the apex of a main parachute canopy and used to facilitate its opening.

pilot plant *Industrial Engineering.* the first plant designed to produce a new product, or to employ a new technology; usually built on a small scale and used particularly when further refinement or development is anticipated.

pilot production *Petroleum Engineering.* the extraction of a limited amount of gas or oil for purposes of determining reservoir and product characteristics before engaging in full recovery operations.

pilot protein *Biochemistry.* a protein that assists in the transfer of DNA from a donor cell to its recipient during bacterial mating.

pilot report *Meteorology.* an in-flight weather report by an aircraft pilot or crew member that includes data on the location and extent of reported weather phenomena, time of observation, phenomenon description, and phenomenon altitude. Also, **pirep.**

pilot rules *Navigation.* federal regulations governing the conduct of vessels on inland waters.

pilot's discretion *Transportation Engineering.* an air traffic control authorization given a pilot to climb or descend to a new altitude according to the pilot's judgment.

pilot station *Naval Architecture.* the position on a ship's bridge from which piloting is done. *Navigation.* a location at sea where incoming ships commonly pick up pilots.

pilot streamer *Geophysics.* a theoretical, nonluminous lightning streamer that lays down a weak ionization path, which is followed by the stepped leader in cloud-to-ground discharges.

pilot system *Computer Programming.* a system for evaluating new procedures by creating a realistic model to process a representative sample of real data.

pilot test *Computer Programming.* a test of a computer system under the exact conditions for which the system was designed.

pilot tone *Telecommunications.* an unmodulated tone of a specified frequency, sent to control an alarm or to trigger some automatic control device.

pilot tunnel *Engineering.* a small tunnel created in the preparatory phase of a blasting operation prior to mining or tunnel building; used to obtain information about the ground and to create a free face.

pilot-venturi flow element *Engineering.* an instrument used to measure the velocity of fluid flow, consisting of a venturi flume, rather than Pitot tubes.

pilot waters *Navigation.* **1.** areas in which the services of a pilot are required or desirable. **2.** inshore waters in which navigation is by piloting.

pilsner *Food Technology.* a dry, golden beer brewed in the city of Plzen (Pilsen), Czechoslovakia. Other beers brewed in this style are called **pilseners** or **pils**.

Piltdown man *Anthropology.* the name given to a supposed discovery of a subspecies of *Homo sapiens*, proposed as a "missing link" between apes and human. Purportedly a fossil from the Early Pleistocene found in 1912 at Piltdown in Sussex, England, it actually consisted of the jaw of a modern ape and the skull of a medieval human; it was definitively shown to be a hoax by sophisticated testing in the 1950s.

pilus *plural,* **pili.** *Anatomy.* any of the filamentous appendages of the skin, consisting of modified epidermal tissue; a hair. *Cell Biology.* a short, hairlike appendage extending from the surface of certain bacterial cells, composed mainly of the protein pilin; involved in a variety of functions including bacterial conjugation and bacteriophage adhesion. Also, FIMBRIA.

pimaricin *Organic Chemistry.* $C_{33}H_{47}NO_{13}$, crystals that are soluble in water and organic solvents; decompose at around 200°C; used in medicine to kill the fungus *Candida albicans.* Also, TENNECETIN.

pimelic acid *Organic Chemistry.* $HO_2C(CH_2)_5CO_2H$, combustible crystals that melt at 105°C; slightly soluble in water and soluble in alcohol; used in polymers, plasticizers, and biochemical research.

pimelic ketone see CYCLOHEXANONE.

pimento *Botany.* **1.** a type of pepper tree, *Capsicum annuum* of the family Solanaceae, cultivated for its thick-fleshed, sweet, red fruit. **2.** this fruit. Also, **pimiento. 3.** see ALLSPICE.

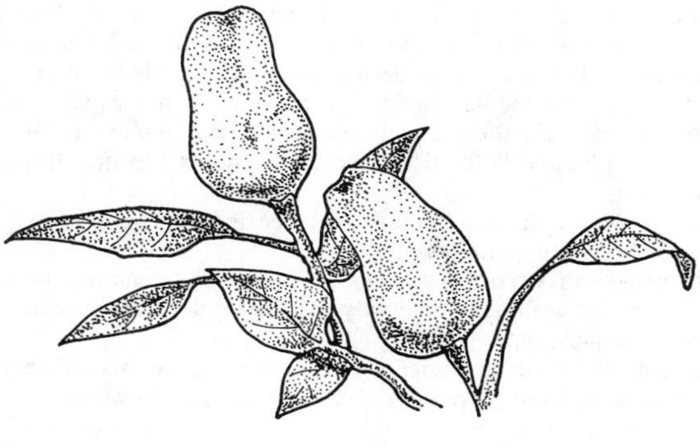

pimento

pi meson see PION.

pi mode *Electronics.* a magnetron design in which the phases of anode openings, connected across from where electrons interact with an alternating magnetic field, differ by pi radians.

pimpernel *Botany.* any plant of the genus *Anagallis,* of the primrose family, having scarlet or white flowers that close in threatening weather, especially the **scarlet pimpernel,** *A. arvensis.*

pimple *Medicine.* a papule or pustule, usually of the face, neck, or upper trunk; common in acne vulgaris.

pimple mound *Geology.* a low, flattened, roughly circular or elliptical dome composed of a sandy loam that is coarser than and entirely distinct from the surrounding soil; found on the coast of east Texas and southwestern Louisiana.

pimple plain *Geology.* a plain characterized by the presence of numerous, conspicuous pimple mounds.

pi-mu atom see PIONIUM.

pin *Mechanical Devices.* a short, narrow piece of wire with a point at one end and a head at the other, used as a fastener. *Mechanical Engineering.* **1.** a cylindrical piece of wood or metal used to fasten two parts together or to support one part that is suspended from another, allowing freedom of motion. **2.** a small axle about which a lever oscillates. **3.** a spindle about which a pulley is supported. *Surgery.* a long, slender metal rod used to fix the ends of fractured bones. *Electronics.* a terminal on a tube, a plug, or a connector. *Robotics.* a cylindrical fastener that provides one degree of freedom when two parts are joined together.

PIN see PIN CONSTRUCTION.

piña [pēn´yä] *Materials.* the silky fiber obtained from the large leaves of pineapple plants. Also, PINEAPPLE FIBER. (From the Spanish word for pineapple.)

Pinaceae *Botany.* the pine family, mostly evergreen and resiniferous trees and shrubs of the class Pinatae, characterized by spirally arranged needlelike leaves, and by staminate and pistillate cones with spirally arranged scales; includes pines, cedars, and larches.

pinaceous *Botany.* of, relating to, or resembling the family Pinaceae.

pinacocyte *Invertebrate Zoology.* a large flat cell that forms part of the outer surface of certain sponges.

pinacoderm *Invertebrate Zoology.* a surface epithelium composed of pinacocytes.

pinacoid *Crystallography.* a pair of opposite parallel faces of a crystal. If the faces are perpendicular to a chosen axis, they are named accordingly as base, front, and side pinacoids.

pinacoidal *Crystallography.* relating to or characterized by pinacoids.

pinakiolite *Mineralogy.* $(Mg,Mn^{+2})_2(Mn^{+3},Sb^{+3})BO_5$, a black, monoclinic mineral having a metallic luster, occurring in tabular crystals, and having a specific gravity of 3.88 and a hardness of 6 on the Mohs scale; polymorphous with fredrikssonite, orthopinakiolite, and takeuchiite.

Pinales *Botany.* in some classifications, an order of gymnospermous woody trees and shrubs in the subdivision Pinicae, including fir, pine, spruce, yew, cypress, and redwoods. Also, CONIFERALES.

Pinatae *Botany.* in some classifications, a class of the subdivision Pinicae, containing various families of gymnosperms; characterized by either deciduous or evergreen leaves, branches with short spur shoots and long shoots, small scaly leaves or needlelike leaves, and either a cone or drupelike structure. Also, PINOPSIDA.

pinboard *Computer Technology.* see BOARD.

pincers *Mechanical Devices.* an iron or steel gripping tool with two short handles and two grasping jaws joined by a pivot; used for pulling nails or brads. *Medicine.* see FORCEPS. *Invertebrate Zoology.* any grasping structure that opens and closes with a scissorlike motion, especially the modified appendage of a decapod crustacean. Also, CHELAE. *Vertebrate Zoology.* the median deciduous incisor teeth in horses.

pincer tongs *Mechanical Devices.* a pair of tongs having V-shaped jaws and a globular body; used, especially by blacksmiths, to grip thin rods or bolts.

pinch *Geology.* a marked thinning or squeezing together of a rock layer, or a narrow part of an orebody.

pinch-and-swell structure *Geology.* a structural condition in which a vein is pinched at frequent intervals, leaving thick parts between, resembling a string of pearls.

pinch bar *Mechanical Devices.* a metal bar similar to a crowbar, with one clawed end used for pulling spikes or rolling large wheels.

pinch device *Physics.* any plasma device designed to confine the plasma by means of the pinch effect; two common types are the **q-pinch device** and the **z-pinch device.**

pinch effect *Electricity.* the constriction of ionized gas in the center of an electron tube through which heavy current is passed. A vertical motion in the tip of a pickup stylus in disk reproduction media caused by variations of the angle between the groove walls. Also, RHEOSTRICTION.

pin cherry see BIRD CHERRY.

pinch graft *Surgery.* a small split-thickness skin graft.

pinch-off blade *Engineering.* in a blow molding process, a mechanism that seals off the parison before blowing, thus allowing the flash to be cooled and removed easily.

pinch-off voltage *Electronics.* in a field-effect transistor, the voltage needed to block current flow between the source and the drain.

pinch pass *Metallurgy.* in sheet rolling, a pass that causes very limited reduction of thickness.

pinch resistor *Electronics.* a device in which a P-type material is layered over an N-type material to greatly increase the device's resistance value.

pinch trimming *Metallurgy.* a process for trimming the edge of a tube.

pinch-tube process *Engineering.* a process in plastics blow molding in which a tube is lowered between mold halves and pinched off when the mold closes.

PIN construction *Electronics.* a semiconductor structure that has low-resistant P-type and N-type regions with a high-resistant instrinsic region sandwiched in between. (An acronym for positive-instrinsic-negative construction.)

pincushion cactus *Botany.* any of various rounded, low-growing spiny cacti of the genus *Mammillaria.*

pincushion distortion *Optics.* a distortion in an image formed by an uncorrected lens that causes the straight lines of a square to appear curved and concave; the opposite of a barrel distortion. Also, PILLOW DISTORTION.

pin diode *Electronics.* a device in which the silicon substrate is doped with P-type material on one side and N-type material on the other, producing a lightly doped layer between the two which acts as a dielectric barrier; commonly used as a variable microwave attenuator.

pine

pine *Botany.* a tree of the genus *Pinus* in the family Pinaceae, including various coniferous, evergreen trees with elongated needles in fascicles on spur shoots; growing naturally only in the Northern Hemisphere; includes many species native to North America. *Materials.* the soft to hard, durable, light-colored wood of such a tree, widely used for furniture, cabinetry, and construction.

pineal *Biology.* shaped like a pinecone. *Anatomy.* of or relating to the pineal body.

pineal body *Anatomy.* a small, flattened organ lying between the superior colliculi and secreting the hormone melatonin, which is thought to regulate reproductive function. Also, **pineal gland.**

pineal hormone *Endocrinology.* any hormone produced by the pineal body, including melatonin and possibly vasotocin.

pinealoblastoma *Oncology.* a pinealoma of poorly differentiated pineal cells. Also, **pineoblastoma.**

pinealocyte *Neurology.* the principal cell constituting the pineal body, characterized by light-staining cytoplasm, prominent nucleoli, and a large multilobar nucleus.

pinealoma *Oncology.* a rare tumor of the pineal body composed of large epithelial cells and resulting in hydrocephalus, a type of visual paralysis, early puberty, and walking disorders. Also, **pinealocytoma.**

pinealopathy *Neurology.* any disease or disorder of the pineal body.

pineapple

pineapple *Botany.* **1.** a tropical monocotyledonous herb, *Ananas comosus,* of the family Bromeliaceae; characterized by spiky swordlike leaves and widely cultivated in tropical regions for its large, multiple fruit that develops from a fleshy inflorescence and is covered by scaly bracts and crowned by a tuft of sharp leaves. **2.** the fruit of this plant.

pineapple fiber see PIÑA.

pinecone *Botany.* the cone or strobilus of any pine tree.

pinecone fish *Vertebrate Zoology.* a small marine fish of the family Monocentrididae, characterized by plated scales having a pinecone pattern, jagged fin spines, and bioluminescent organs under the jaw that house light-producing symbiotic bacteria; widely distributed in the Indian and Pacific Oceans.

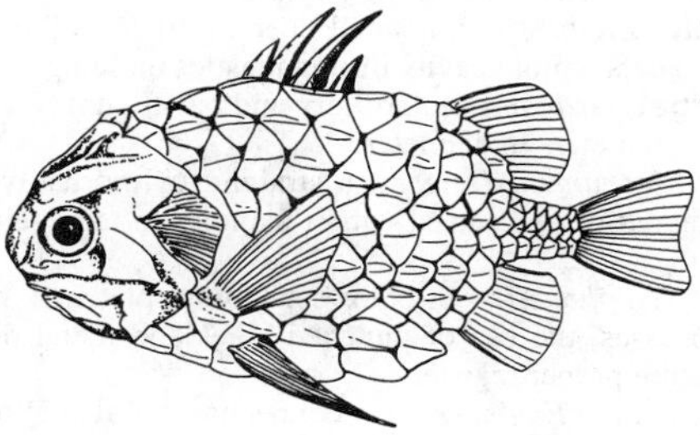

pinecone fish

pinene *Organic Chemistry.* $C_{10}H_{16}$, either of two isomers occurring as colorless transparent liquids that are insoluble in water and soluble in alcohol; both isomers are terpene hydrocarbons derived from sulfate wood turpentine; **α-pinene** boils at 155–160°C and is used as a solvent, flavoring, and odorant; **β-pinene** boils at 165–166°C and is used in polyterpene resins, perfumes, and flavorings.

pinene hydrochloride see BORNYL CHLORIDE.

pine nut *Botany.* an edible seed borne in the pinecones of various species of *Pinus*, having a thin hard shell and a fleshy, nutty-tasting kernel. Also, PIGNOLA.

pine oil *Materials.* a combustible, colorless to light amber oil having a strong piny odor; extracted and fractionated from the wood of the tree *Pinus palustris* and used as an odorant, disinfectant, penetrant, wetting agent, and reagent.

piner *Meteorology.* a strong breeze that blows from the north or northeast in England.

pine tar *Materials.* a viscid, blackish-brown liquid with a turpentine-like odor; distilled from pine wood and used in paints, roofing, soaps, and medicines.

pine tar pitch *Materials.* the residue after distillation of the volatile oils from pine tar.

pine-tree array *Electromagnetism.* an antenna array constructed of a vertical curtain of horizontal radiating elements with reflectors behind them.

pi network or **pi-network** *Electricity.* a network formed from a pi section. *Electronics.* a network with three branches, one connected between input and output, and the other two connected across the input and output terminals, respectively. (Named for the resemblance of the schematic to the Greek letter Π.)

pinfeather *Vertebrate Zoology.* a young, not fully developed feather just emerging from a bird's skin and still covered by a horny sheath that is later cast off.

pin-feed printer *Computer Technology.* a printer that uses pins or bumps to feed perforated-edge paper into the carriage.

ping *Acoustical Engineering.* the sonic or ultrasonic signal generated by a sonar device.

pinger *Acoustical Engineering.* a simple active transmitter device consisting of a frequency oscillator, an amplifier, and a projector that transmits a signal for uses such as tracking a submarine or signaling distress.

pingo *Hydrology.* in an arctic permafrost region, a small hill or large cone-shaped mound of soil-covered ice that is a localized upwarp of the land surface caused by the hydrostatic pressure of groundwater.

pingo ice *Hydrology.* relatively clear ice occurring in a permafrost region.

pingo remnant *Geology.* a rimmed depression formed when the crest of a pingo ruptures, exposing the ice core, which melts and results in the partial or total collapse of the crest.

ping-pong *Computer Programming.* a technique in which a virtually endless file can be processed by alternating continuously between two tape drives or other physical storage units. (From its resemblance to the rapid back-and-forth play in the game of ping-pong, or table tennis.)

ping-pong effect *Acoustical Engineering.* the shifting of sound back and forth between two stereo channels.

pinguecula *Anatomy.* a small, yellow, mass of connective tissue on the conjunctiva of the eye, usually on the nasal sclerocorneal junction; seen in elderly people. Also, **pinguicula.**

pinguite see NONTRONITE.

pinhole *Science.* any small hole or opening, like that made by a pin. *Materials Science.* any undesirable small void formed in a material, especially in a metallic product such as an electroplated coating or a foil.

pinhole camera *Optics.* a camera that has a tiny, sharp-edged opening as its aperture instead of a lens, so that light reflected from a subject through the hole will form a properly focused image inverted from top to bottom.

pinhole chert *Petrology.* a chert that contains weathered pebbles with minute holes or pores.

pinhole detector *Engineering.* an instrument used to find tiny holes or defects in moving material with the use of a photoelectric device.

pinic acid *Organic Chemistry.* $C_9H_{14}O_4$, white prisms; slightly soluble in water; melts at 135–136°C and boils at 214–216°C (10 torr); used in plasticizers and lubricants.

Pinicae *Botany.* in some classifications, a large subdivision of the division Pinophyta, including woody evergreen or deciduous shrubs and trees that are monoecious or dioecious; most have a simple trunk with excurrent branches, simple needlelike or scalelike leaves, and wood with resin canals. Also, CONIFEROPHYTA.

piniform *Science.* shaped like a pine cone; conical or cone-shaped.

pinion [pin´yən] *Mechanical Engineering.* the smaller of a pair of high-ratio toothed spur wheels. *Vertebrate Zoology.* **1.** the end of a bird's wing including the carpus, metacarpus, and phalanges. **2.** a bird's flight feather.

pinite *Mineralogy.* **1.** any of various fine-grained gray, green, or brown alteration products derived from minerals such as cordierite, nepheline, spodumene, and feldspar. **2.** a variety of muscovite that is used in ceramics.

pin jack *Electricity.* a small jack designed for use with a pin plug, so that the plug can be inserted quickly.

pin joint *Mechanical Devices.* a swivel joint held together by a removable pin to allow independent freedom of motion such as between a door and its frame.

pin junction *Electronics.* a device that has three distinct semiconducting regions, one that has P-type impurities, one that is electrically pure (intrinsic), and one that has N-type impurities.

pink *Botany.* any of various plants of the genus *Dianthus,* in the family Caryophyllaceae. *Graphic Arts.* a color perceived as pale red or reddish white.

pink disease *Plant Pathology.* a bark disease affecting rubber, cacao, citrus, coffee, and other trees; caused by the fungus *Corticum salmonicolor* and characterized by a pink covering of hyphae on the stems and branches.

pinkeye *Medicine.* an acute, contagious conjunctivitis of humans and various domestic animals.

pink noise *Acoustics.* noise that has a nearly constant loudness at all frequencies within its frequency range.

pink root *Plant Pathology.* a disease of onions and garlic caused by fungi of the genera *Phoma* and *Fusarium*; characterized by a pink or reddish discoloration and rotting of roots.

pink rot *Plant Pathology.* **1.** a disease of potatoes caused by the fungus *Phytophthora erythroseptica* and characterized by a pink-colored wet area of decay on the cut surfaces of the tubers. **2.** a disease of apples caused by the fungus *Tricothecium roseum,* resulting in the breakdown and decay of the plant tissue. **3.** a disease of the celery plant caused by the fungus *Sclerotinia sclerotiorum* and resulting in a watery soft breakdown and decay of the plant tissue.

pink snow see RED SNOW.

pin metal *Metallurgy.* the popular name of a copper-base alloy containing 37.5 percent zinc; officially designated **UNS C27400.**

pinna *Anatomy.* the external ear; the projecting part of the ear lying outside the head. (A Latin word meaning "wing.") Also, AURICLE.

pinnacle *Geology.* **1.** a high, tapering, or pointed tower or spire-shaped pillar of rock, found alone at the crest of a mountain or hill. **2.** a mountain or hill having a pointed summit. **3.** see REEF PINNACLE. *Architecture.* a small, upright ornamental structure crowning a projecting member such as a tower or spire.

pinnacled iceberg *Oceanography.* an iceberg weathered in such a way that pinnacles and spires are formed.

pinnacle reef see REEF PINNACLE.

pinnal *Anatomy.* of or relating to the pinna.

pinnate *Botany.* describing compound leaves that have more than three leaflets arranged in two rows on one stalk, in a manner similar to that of a feather.

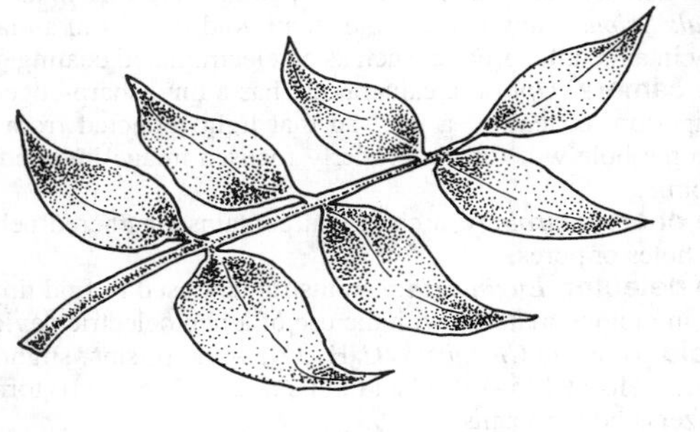

pinnate leaf

pinnate drainage pattern *Hydrology.* a pattern of natural stream courses in which the main stream is joined at acute angles by many closely spaced subparallel tributaries.

pinnate joint *Geology.* see FEATHER JOINT.

pinnate muscle *Anatomy.* any muscle whose fascicles are attached along a central tendon, like plumes of a feather, to an elongate tendon that is approximately as long as the muscle.

pinnatifid *Botany.* describing a leaf that is divided in a pinnate fashion, with narrow, acute divisions extending about halfway down to the midrib.

pinning *Solid-State Physics.* the accumulation of impurities at or near dislocations that prevents further motion of dislocations and results in hardening of the solid .

Pinnipedia *Vertebrate Zoology.* in some classifications, an order of aquatic mammals including the sea lions, fur seals, walruses, and earless seals.

pinnoite *Mineralogy.* $MgB_2O_4 \cdot 3H_2O$, a yellow, tetragonal mineral occurring in fibrous nodular masses and granular form, having a specific gravity of 2.27 to 2.29 and a hardness of 3.5 on the Mohs scale; found in salt deposits.

Pinnotheridae *Invertebrate Zoology.* a family of tiny decapod crustaceans, the pea-, oyster-, or mussel crabs, in the subsection Brachygnatha, having small eye stalks and a circular, membranous carapace; commensal in polychaete tubes and burrows and the mantle cavity of bivalves and snails.

pinnulate *Biology.* having pinnules.

pinnule *Biology.* a secondary branch of a plumelike organ, such as a leaf. *Invertebrate Zoology.* any of a row of slender, flexible appendages along each side of the arms of crinoids. Also, **pinnula.**

pinocyte *Cell Biology.* a cell that exhibits pinocytosis.

pinocytosis *Cell Biology.* the nonspecific uptake of extracellular fluid by a cell via small vesicles derived from the plasma membrane; thought to be a method of active transport across the cell membrane.

pinocytotic *Cell Biology.* relating to a pinocyte or to pinocytosis. Also, **pinocytic.**

pinolite *Petrology.* a metamorphic rock consisting of crystals and granular aggregates of magnesite in a groundmass of phyllite or talc.

piñon [pin yōn´] *Botany.* **1.** any of several pine trees of the southwestern U.S. bearing edible, nutlike seeds. Also, PINYON PINE. **2.** the seed of such a tree. Also, **piñon nut.**

Pinophyta *Botany.* in some classifications, a division of the plant kingdom containing the gymnosperms, seed-bearing plants having true roots, stems, and leaves but not having seeds enclosed in carpels or a double fertilization process. Also, GYMNOSPERMAE.

Pinopsida see PINATAE.

pinosome *Cell Biology.* any of the small, membrane-bound cytoplasmic vesicles that are pinched off of the plasma membrane and containing fluid taken up by a cell during pinocytosis. Also, **pinocytotic vesicle.**

pinosylvin *Botany.* a toxic compound produced in the heartwood of *Pinus sylvestris* (Scotch pine) that protects the wood against insects, fungi, and various chemicals.

Pinot Blanc or **pinot blanc** [pē nō´ blängk´] *Botany.* an acidic green grape grown in Alsace, Italy, and California. *Food Technology.* a light, fruity white wine made mostly or entirely with this grape. Known in Italy as **pinot bianco.**

Pinot Noir or **pinot noir** [pē nō´ nwär´] *Botany.* a sweet, mild red grape grown in Burgundy; used to make wine and blended with other varieties to make Champagne. *Food Technology.* a wine made entirely of Pinot Noir grapes.

pin plug *Electricity.* a plug that is approximately as thick as an ordinary pin; designed for quick insertion and removal, and used with a pin jack.

pinpoint *Navigation.* a position in air navigation determined by identifying a feature directly beneath the aircraft.

pinpoint gate *Engineering.* a small opening (0.76 millimeter) through which resin enters a mold cavity during plastics molding.

pinpoint target *Ordnance.* in artillery and naval gunfire support, a target less than 50 meters in diameter.

pin punch *Mechanical Devices.* a straight, cylindrically shaped tool having a squared-off end; used with a hammer to tap against a pin and remove it.

pin register *Graphic Arts.* in color separation, platemaking, and other printing processes, the use of punched holes and metal pins to secure copy and ensure proper register.

pin rod *Mechanical Engineering.* a connecting rod that allows two parts to function as one.

pint *Metrology.* **1.** a traditional unit of capacity used in the United States, equal to 16 fluid ounces or one-half quart; equivalent to 0.473 liter. **2.** a similar unit of measure used in Great Britain, equivalent to 0.568 liter. Also, IMPERIAL PINT. **3.** see DRY PINT.

pinta *Medicine.* an infectious skin disease caused by the bacterium *Treponema carateum,* believed to be transmitted by person-to-person contact and characterized by the presence of a rash or spots on the skin which may be white, coffee-colored, blue, red, or violet; the disease occurs in parts of tropical America and responds to penicillin therapy. (Going back to a Latin word meaning "to paint.")

pintadoite *Mineralogy.* $Ca_2V_2^{+5}O_7 \cdot 9H_2O$, an inadequately described green mineral that is slowly soluble in water, occurring as an efflorescence on sandstone.

pin timbering *Mining Engineering.* a means of supporting mine roofs either by driving bolts into strong, firm structures to support lower, weaker layers or by binding several weak layers into a beam strong enough to support itself across the work place.

pintle *Mechanical Devices.* **1.** an upright pivot pin or bolt on which another part turns. **2.** a cast metal base for a wooden post.

pintle center *Ordnance.* the theoretical center of a weapon; used in computing firing data.

pintle chain *Mechanical Devices.* a sprocket wheel chain with its links connected by pintles.

pintle valve *Mechanical Devices.* a short extension of a needle valve that initiates regulation of fluid through the valve.

pinto bean *Botany.* a variety of *Phaseolus vulgaris* having mottled seeds, widely grown in the southern U.S.

pinulus *Invertebrate Zoology.* a five-rayed sponge spicule having one longer spiny ray that sometimes projects from the sponge.

pinus see PINEAL BODY.

pin vise *Mechanical Devices.* a manually operated vise fitted with a collet or chuck, used to secure fine materials such as a timepiece mechanism or small wires, taps, or drills.

pinwheel *Horology.* in a striking or chiming clock, a wheel that has regularly spaced projecting pins that engage the hammers, causing them to strike.

pinwheel inclusions *Virology.* protoplasmic bodies that appear as virus-coded protein sheets in plants infected with potyviruses.

pinworm *Invertebrate Zoology.* a nematode worm, *Enterobius vermicularis,* of the order Oxyuroidea, that is parasitic in the gut of humans and animals; females lay their eggs around the anus, causing itching.

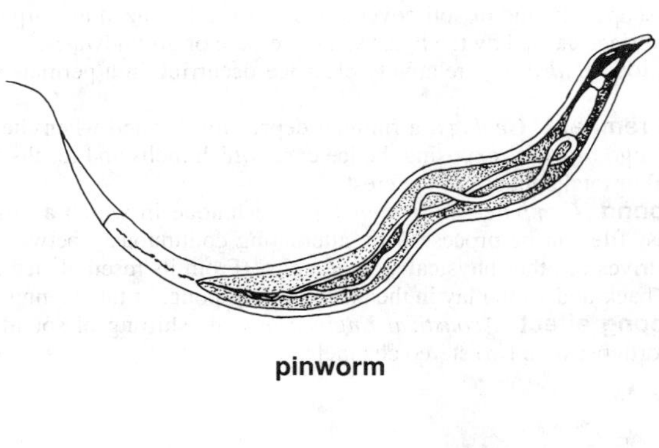

pinworm

pin wrench *Mechanical Devices.* a wrench having two projecting pins to fasten or loosen a round nut with two corresponding face holes.

pinyon or **pinyon pine** see PIÑON.

PIOCS *Computer Programming.* procedures and routines that supervise and control channel programs. (An acronym for Physical Input/Output Control System.)

pion *Particle Physics.* a short-lived meson that is primarily responsible for the nuclear force and that exists as a positive or negative particle with mass 273.2 times the electron mass or as a neutral particle with mass 264.2 times the electron mass. Also, PI MESON.

pion condensate *Nuclear Physics.* a condition resulting from interaction of a large number of pairs of particles, each pair made up of a positive pion and a negative pion, with the nucleons causing them to form a coherent state.

pioneer *Science.* a person who is among the first in a given area of scientific theory, practice, or research. *Ecology.* an organism that establishes itself in a barren region and thus begins an ecological cycle in which many other species will emerge. *Military Science.* a member of a field engineer unit.

Pioneer *Space Technology.* a series of U.S. deep-space probes launched between 1958 and 1973; included the first successful flybys of Jupiter and Venus.

pioneer tunnel *Mining Engineering.* a small tunnel located parallel to and ahead of a main tunnel, from which crosscuts are driven to the path to be traversed by the main tunnel.

pionemia see LIPEMIA.

pionium *Particle Physics.* an exotic atom that is similar in structure to the hydrogen atom but with the proton replaced by a pion and the electron replaced by a muon. Pionium is unique among atoms that have been observed in the laboratory in that all of its constituents are unstable particles not found in ordinary matter. Also, PI-MU ATOM.

pionnotes *Mycology.* in fungi of the species *Fusarium,* a mass of spores that looks fatty or greasy.

Piophilidae *Invertebrate Zoology.* the skipper flies, a family of small dipteran insects in the subsection Acalyptratae whose larvae (found in cheese, pork products, fungi, feces, and carrion) jump into the air when disturbed.

piotine see SAPONITE.

pip see BLIP.

pipage *Engineering.* **1.** the conveyance of liquid through a pipe. **2.** the pipes used in a given operation.

pipe *Mechanical Devices.* a long, hollow metal or plastic tube used to carry fluids, gases, or particle solids or to provide structural support for a building or water channel. *Geology.* **1.** a vertical, elongated, cylinder-shaped orebody. Also, NECK, ORE CHIMNEY. **2.** a tubular cavity in calcareous rocks that is often filled with sand and gravel. *Volcanology.* a vertical conduit leading from the magma chamber below a volcano to the earth's surface. Also, BRECCIA PIPE. *Botany.* a hollow stem of a plant. *Computer Programming.* **1.** the channel or medium through which data passes when being transferred from one program or task to another, as in Unix. **2.** a sequence of processing units in a pipelined architecture. *Metallurgy.* an axial defect in a casting or wrought product.

pipe amygdule *Geology.* an elongated or tubular amygdule sometimes found near or at the base of a lava flow.

pipe bit *Mechanical Devices.* a bit used for anchoring pipe lengths to bedrock.

pipe clamp *Mechanical Devices.* **1.** a casing-type clamp used to hoist or support pipe sections. **2.** a clamp consisting of a sliding, screw adjusted jaw that runs along a pipe length to a specified position; used to hold large boards or frames together while gluing.

pipe clay or **pipeclay** *Materials.* any white to grayish-white, highly elastic clay, especially such a clay suitable for making tobacco pipes. Also, BALL CLAY, CUTTY CLAY. *Geology.* a mass of fine clay that forms the bedrock surface upon which the gravel of old river beds often rests.

pipe coupling *Mechanical Devices.* a female-threaded collaring device used as a bridge.

pipe culvert *Civil Engineering.* a section of pipe that permits water to flow under a roadway or other embankment.

pipe cutter *Mechanical Devices.* a hand tool used to cut pipe by grasping it with three sharp-edged wheels that are forced inward by screw pressure and cut into the pipe as the tool rotates.

pipe die *Mechanical Engineering.* a screw plate used to cut threads on a pipe.

pipe elbow meter *Engineering.* a device that can be adjusted to take measurements of the flow of material around a bend in a pipe.

pipefish *Vertebrate Zoology.* any long, thin marine or freshwater fish of the family Syngnathidae, having a tubular snout and covered with bony plates.

pipe fitting *Mechanical Devices.* any object used to join pipe lengths, such as a coupling, an elbow, or a nipple. *Engineering.* the process of installing or repairing a pipe system.

pipe flow *Engineering.* the movement of fluid through a closed conduit.

pipe hanger *Mechanical Devices.* a bracket, clip, ring, or loop used to suspend pipes on ceilings or beams; sometimes used with an adjustable rod and ring.

pipe harrow *Agriculture.* a type of harrow having several iron pipes fitted with spikes to break up the earth.

pipe laying *Engineering.* the process by which pipe is put into a trench to create a piping system.

pipeline *Engineering.* a line of pipe with pumping or compressing machinery and apparatus for conveying a fluid.

pipelined architecture *Computer Technology.* a computer architecture in which a sequence of processing units is arranged so that each unit passes its results to the next and the units operate in parallel, as in an assembly line.

pipeline mixer *Mechanical Engineering.* an electric mixer designed to attach to a pipeline and mix fluids without interrupting flow through the line; used extensively in food processing.

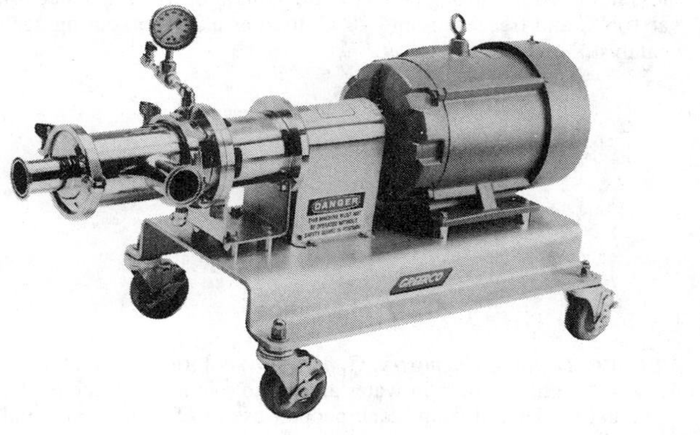

pipeline mixer

pipelining *Computer Programming.* a procedure in which execution of the next instruction or sequence of instructions begins before the previous instruction or sequence of instructions has completed, thus increasing processing speed.

pipe nail *Mechanical Devices.* a broad, flat-headed wrought-iron nail used as a support to the underside of a pipe core.

pipe pile *Civil Engineering.* a cylindrical steel pile, 6 to 30 inches in diameter, used for underpinning a structure; usually driven with both ends open for access and then hollowed and filled with concrete in situ.

pipe plug *Mechanical Devices.* a plug with male threads and a hexagonal, square, or round head; used to seal the female ends of pipe fittings or valves.

Piperaceae *Botany.* the pepper family, dicotyledonous aromatic herbs and woody plants or vines in the order Piperales, characterized by alternate leaves, a solitary ovule, and usually spherical ethereal oil cells.

piperaceous *Botany.* relating or belonging to the family Piperaceae.

piperacillin sodium *Pharmacology.* $C_{23}H_{26}N_5NaO_7S$, a semisynthetic broad-spectrum antibacterial.

Piperales *Botany.* an order of dicotyledonous herbs, shrubs, and trees in the class Magnoliopsida, having reduced flowers usually borne in dense fleshy spikes or racemes and spherical ethereal oil cells.

piperazine *Organic Chemistry.* $C_4H_{10}N_2$, combustible, colorless, glassy needles that absorb carbon dioxide from the air; soluble in water and alcohol; melts at 104–107°C and boils at 145°C; used as an insecticide, corrosion inhibitor, and medicinal accelerator.

piperazine dihydrochloride *Organic Chemistry.* $C_4H_{10}N_2 \cdot 2HCl$, white needles that are soluble in water; used in insecticides, fibers, and pharmaceuticals.

piperazine hexahydrate *Organic Chemistry.* $C_4H_{10}N_2 \cdot 6H_2O$, white crystals that are soluble in water and alcohol; melts at 44°C and boils at 125–130°C; used in pharmaceuticals, in insecticides, and in medicine as an anthelmintic.

Piper Cub

Piper Cub *Aviation.* the trade name for a single-wing, single-engine, two-passenger airplane built by the Piper Aircraft Corporation.

piperidine *Organic Chemistry.* $(CH_2)_5NH$, a toxic, combustible, colorless liquid with a strong pepper odor; soluble in water and alcohol; boils at 106°C and freezes around –8°C; used as a solvent, curing agent, and catalyst.

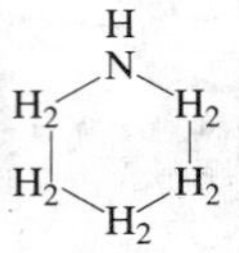

piperidine

piperine *Organic Chemistry.* $C_{17}H_{19}NO_3$, colorless monoclinic crystals that are slighty soluble in water and soluble in alcohol; melts at 130°C; an alkaloid derived from black pepper; used to flavor brandy and as an insecticide.

piperism *Toxicology.* poisoning due to excessive ingestion of black pepper, the dried fruit of *Piper nigrum*; characterized by vomiting and nervous system disorders.

pipernoid *Geology.* describing the appearance and texture of certain extrusive igneous rocks, characterized by parallel bands of dark patches and stringers within in a light-colored groundmass.

piperocaine hydrochloride *Organic Chemistry.* $C_{16}H_{23}NO_2 \cdot HCl$, a white, bitter-tasting crystalline powder that is soluble in water and alcohol; melts at 172–175°C; used in medicine as a local anesthetic.

pipe rock *Petrology.* a marine sandstone that is rich in scolites.

piperonal *Organic Chemistry.* $C_8H_6O_3$, combustible white crystals that are slightly soluble in water and soluble in alcohol; melts at 35.5–37°C and boils at 263°C; used in perfumes, suntan preparations, and mosquito repellants.

piperylene *Organic Chemistry.* CH_2=CHCH=$CHCH_3$, a highly flammable colorless liquid that occurs in *cis* and *trans* forms; the *cis* form freezes at –141°C and boils at 44°C, and the *trans* form freezes at –87°C and boils at 42°C; both forms are insoluble in water and soluble in alcohol; used in polymers and as a chemical intermediate.

pipe scale *Engineering.* any residual material, such as rust, that sticks to the inside of pipes and lessens its ability to carry heat or raises the pressure drop for fluid flow.

pipe shrinkage *Materials Science.* a void at the surface of a casting, caused by volume contraction that occurs during solidification.

pipe still *Petroleum Engineering.* an apparatus used to distill crude petroleum, consisting of coiled tubes enclosed in a furnace.

pipestone *Petrology.* a pink or mottled pink-and-white argillaceous stone; used by Native Americans for carving tobacco pipes. Also, CATLINITE.

pipe stopper *Mechanical Devices.* an expanding drain plug, used to close the outlet of drain pipes for testing.

pipe tap *Engineering.* a small, threaded hole in a pipe wall used to obtain samples from a pipe or to connect controlling or measurement devices.

pipe thread *Design Engineering.* see THREAD.

pipe tongs *Mechanical Devices.* an adjustable wrench that is used to grip, tighten, and secure pipe of varying sizes by squeezing its handles together.

pipe-to-soil potential *Electricity.* the electric charge potential that is effected by a subterranean pipe and the chemical constituents of the soil surrounding it, as a result of electrolytic action.

pipette [pi pet´] *Chemistry.* **1.** a small laboratory device used for dispensing a known volume of liquid or gas, typically consisting of a narrow tube into which fluid is drawn up and held by closing off the upper opening. **2.** to measure or transfer a liquid or gas by means of a pipette. Also, **pipet.**

pipe vesicle *Geology.* a slender vertical cavity that extends upward from the base of a lava flow.

pipe vise *Mechanical Device.* a vise with curved jaws lined with teeth; used for gripping pipe during cutting, reaming, and threading operations.

pipe wrench *Mechanical Devices.* a wrench used to grasp and rotate a pipe in one direction by means of two serrated jaws.

Pipidae *Vertebrate Zoology.* the tongueless frogs, a family of lower anurans of small to moderately large size, having ribs and adapted to aquatic life; found in Africa south of the Sahara as well as in South America.

piping *Engineering.* an arrangement of pipes set up to convey fluid. *Hydrology.* subsurface erosion by percolation that results in the caving in and forming of narrow tunnels or pipes through which soluble or granular materials are removed from the soil. Also, TUNNEL EROSION. *Textiles.* a cord of material used as an ornamental trimming on clothing, upholstered furniture, etc.

pipistrelle *Vertebrate Zoology.* any of numerous common insectivorous brown bats of the genus *Pipistrellus* of the family Vespertilionidae, especially *P. pipistrellus*; found in barns, church spires, and other large buildings, especially in Europe and Asia.

pipit *Vertebrate Zoology.* any of various small ground-dwelling songbirds of the genus *Anthus,* in the family Motacillidae, occurring nearly worldwide in grasslands and along shores; resembling the lark in its appearance, habits, and ability to sing during flight; a widespread North American species is the **water pipit**, *A. spinoletta*.

pipobroman *Oncology.* $C_{10}H_{16}Br_2N_2O_2$, an antineoplastic agent occurring as a white crystalline powder; used in the treatment of polycythemia vera and certain forms of leukemia.

pi point *Electricity.* the frequency at which the insertion phase shift of an electronic device is either 180° or an integral multiple of 180°.

piposulfan *Oncology.* $C_{12}H_{22}N_2O_8S_2$, an antineoplastic agent that is chemically related to pipobroman.

pipper *Optics.* a tiny aperture located in the reticle of an optical sight to aid in aiming.

pipper image *Optics.* the light projected through the pipper.

pippin *Botany.* **1.** any of several varieties of round green apples. **2.** another term for a seed.

Pipridae *Vertebrate Zoology.* the manakins, a family of small, brightly colored birds of the order Passeriformes, having a small, stocky build and a short, broad, slightly curved bill; found in tropical forests of southern Mexico to Peru.

piptoblast *Invertebrate Zoology.* a sessoblast that has no solid attachment to objects but rather lies free in the colonies.

Piptocephalidaceae *Mycology.* a family of fungi of the order Zoopagales that occur in soil, dung, and plant debris and are parasitic to Mucorales.

Pipunculidae see DORILAIDAE.

piqué [pi kā´; pē kā´] *Textiles.* a cotton, cotton blend, or rayon fabric having lengthwise cords or ribs and sometimes a honeycomb weave. (A French word meaning "quilted.")

piqûre [pi kyůr´; pē kyůr´] *Surgery.* a puncture, especially Bernard's (diabetic) puncture.

piranha [pə rän´ə; pə rän´yə] *Vertebrate Zoology.* any of several small, carnivorous, freshwater South American fishes of the genus *Serrasalmus* in the family Serrasalmidae; a voracious predator with saw-edged teeth, it attacks in great numbers and a group is able to devour a large animal in minutes.

Pirani gauge *Physics.* a device that is used to measure the thermal conductivity of a gas over the range of pressures of about 10^{-3} to 1 mm Hg.

piricularin *Toxicology.* a substance that is toxic to the rice plant and to the fungus *Pyricularia oryzae,* which produces piricularin as well as a copper-binding protein that inhibits its toxic properties.

piriform *Science.* pear-shaped. *Mathematics.* the graph in the plane of the equation $y^2 = ax^3 - bx^4$, where a and b are constants.

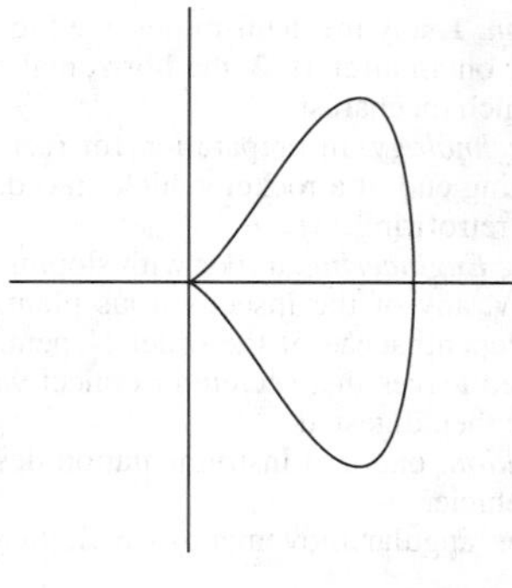

piriform

piriformis *Anatomy.* a muscle that extends from the anterior surface of the sacrum to the femur and rotates the thigh laterally.

pirimiphos-ethyl *Organic Chemistry.* $C_{13}H_{24}N_3O_3PS$, a straw-colored liquid that decomposes at 130°C; kills soil insects in vegetable crops.

pirimiphos-methyl *Organic Chemistry.* $C_{11}H_{20}N_3O_3PS$, a straw-colored liquid that decomposes around 100°C; used as a pesticide.

Piromonas *Mycology.* a genus of fungi occurring in rumen.

Piroplasmia *Invertebrate Zoology.* a subclass of protozoans in the class Sporozoea, having no cilia, flagella, or spores; disease-causing parasites in the blood of domestic animals, they are transmitted by ticks.

piroplasmid *Invertebrate Zoology.* **1.** any protozoan of the subclass Piroplasmia. Also, **piroplasm. 2.** relating or belonging to the subclass Piroplasmia.

Piroplasmida *Invertebrate Zoology.* the single order of protozoa of the subclass Piroplasmia.

piroplasmosis see BABESIOSIS.

pirssonite *Mineralogy.* $Na_2Ca(CO_3)_2 \cdot 2H_2O$, a colorless to white orthorhombic mineral, occurring as prismatic to tabular crystals, having a specific gravity of 2.352 and a hardness of 3 to 3.5 on the Mohs scale; found in saline lake deposits.

Piry virus *Virology.* virus of the Rhabdoviridae family that infects humans with fever, anthralgia, myalgia, and disorders of the abdomen; antigenically related to VSV.

pisanite *Mineralogy.* $(Fe^{+2},Cu)SO_4 \cdot 7H_2O$, a cuprian variety of melanterite.

Pisces [pī´sēz] *Vertebrate Zoology.* the superclass of fishes, or vertebrates adapted to life in water and generally possessing fins and scales; all are cold-blooded with two-chambered hearts; includes the classes Chondrichthyes, Osteichthyes, Agnatha, and Placodermi. *Astronomy.* the Fishes, an extensive but faint northern constellation which lies in the zodiac between Aquarius and Aries, and is seen best in the autumn. (From the Latin word for fish.)

Piscis Austrinus *Astronomy.* the Southern Fish, a small constellation located in the southern hemisphere whose brightest star is 1st-magnitude Fomalhaut. Also, **Piscis Australis.**

Piscis Volans see VOLANS.

piscivorous *Zoology.* feeding mainly or exclusively on fish. Thus, **piscivory.**

pi section *Electricity.* a passive filter or tuner section having three impedences that are usually represented on a schematic in a shape resembling the Greek letter Π.

pi section filter *Electricity.* a pi section that is employed as a low-pass or a high-pass filter, depending upon the position of the capacitors in the circuit.

pisiform *Anatomy.* resembling a pea in shape and size. Also, **pisiformis.**

Pisionidae *Invertebrate Zoology.* a family of crawling polychaete worms in the Errantia, with two pairs of peristomial cirri and no prostomial antennae.

pisolite *Petrology.* a sedimentary rock consisting of a concretion of pisoliths.

pisolith *Petrology.* a small (pea-sized) rounded or ellipsoidal accretionary body, usually of calcium carbonate, found in sedimentary rocks, especially limestones and dolomites.

Pisolithus *Mycology.* a genus of fungi of the order Sclerodermatales that live in the roots of various trees without harming them.

pisolitic *Petrology.* of, relating to, or resembling the texture of pisoliths or pisolite.

pisolitic tuff *Petrology.* of or relating to a tuff consisting of accretionary lapilli or pisoliths.

pisosparite *Petrology.* a limestone consisting of at least 25% pisoliths and no more than 25% intraclasts, with sparry-calcite cement more abundant than micrite.

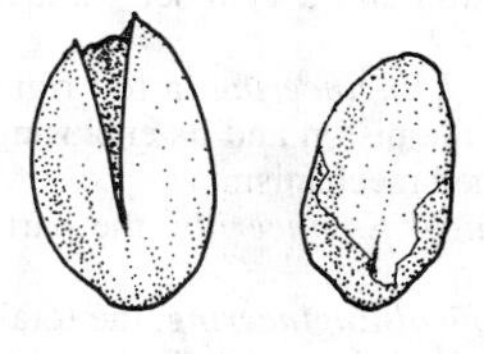

pistachio

pistachio *Botany.* **1.** a small dioecious evergreen tree, *Pistacia vera,* of the family Anacardiaceae, having drupaceous fruit that contains an edible greenish seed in a tough shell. **2.** the edible seed of this tree, called a **pistachio nut.**

pistil *Botany.* the seed-producing part of a flower, including the ovary, style, and stigma.

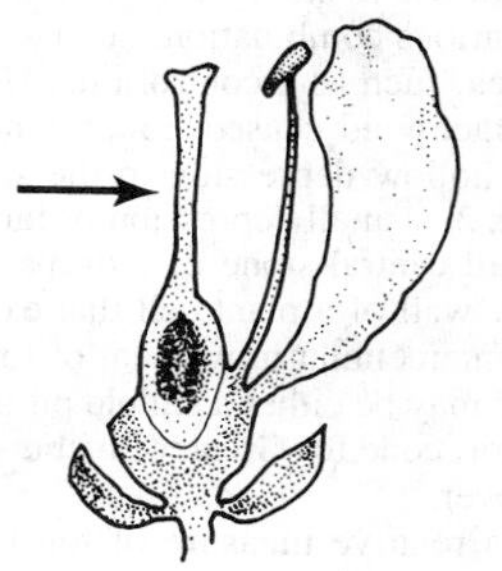

pistil

pistillate *Botany.* **1.** having a pistil or pistils. **2.** having a pistil but no stamens.

pistol *Ordnance.* **1.** a handgun, usually having a short barrel; it may be single shot, repeating, semiautomatic, or automatic. Unlike a revolver, a pistol has a chamber that is an integral part of the barrel. **2.** of or relating to a pistol. Thus, **pistol belt, pistol holster, pistol lanyard,** and so on. **3.** the firing mechanism of an antisubmarine depth charge.

piston *Mechanical Engineering.* the working part of a pump, hydraulic cylinder, or engine that moves back and forth in the cylinder to control the passage of fluid. *Electromagnetism.* a movable plunger used in a waveguide to provide variable impedance for impedance matching purposes.

piston attenuator *Electromagnetism.* a microwave device used in waveguides, in which an attenuator introduces a variable amount of impedance by moving an element along the longitudinal axis of the guide.

piston blower *Mechanical Engineering.* a piston-operated, positive-displacement air compressor.

piston corer *Mechanical Engineering.* a corer, consisting of a steel tube with a lead attached to its upper end, used to recover distorted vertical sections of sediment.

piston displacement *Mechanical Engineering.* the amount of air displaced by a piston in a cylinder from the bottom to the top of its stroke; it is equal to the distance the piston travels times the internal cross section of the cylinder.

piston drill *Mechanical Engineering.* a heavy drill mounted on a horizontal bar or on a short horizontal arm fastened to a vertical column. Also, RECIPROCATING DRILL.

piston engine *Mechanical Engineering.* an engine in which the periodic compression of the working fuel causes the oscillation of a piston in a cylinder. Also, DISPLACEMENT ENGINE, RECIPROCATING ENGINE.

piston flow *Fluid Mechanics.* a flow pattern that develops during upflow of a liquid-gas mixture in a vertical pipe; the pattern consists of large plugs of gas that churn their way up the pipe at moderate gas velocities. Also, SLUG FLOW.
piston head *Mechanical Engineering.* the point above the top ring of a piston.
piston meter *Mechanical Engineering.* a device that measures the flow rate in a piece of equipment by recording the piston's movement in relation to the liquid's buoyant force. Also, **piston-type area meter.**
piston pin see GUDGEON PIN.
piston ring *Mechanical Engineering.* a split metal ring used for sealing the gap between a piston and a cylinder wall, thus inhibiting leakage. Also, PACKING RING.
piston rod *Mechanical Engineering.* a rod rigidly coupled to a piston that moves parallel to the piston and its enclosing cylinder, transmitting movement to an attached mechanism.
piston skirt *Mechanical Engineering.* the part of a piston below the head bore.
piston speed *Mechanical Engineering.* the total distance a piston travels during a given period of time, usually expressed in feet per minute.
piston valve *Mechanical Engineering.* a cylindrical steam engine slide valve for emission and exhaust of steam.
piston viscometer *Mechanical Engineering.* an instrument that measures the viscosity of a liquid by recording the rate at which a piston moves through the liquid.
pit a hole or cavity; specific uses include: *Mining Engineering.* **1.** a general term for a coal mine. **2.** any mine, quarry, or excavation site that is worked by open cut to extract valuable material. *Archaeology.* **1.** an area dug out by humans for use in storage, food preparation, refuse disposal, or the like; used in various combinations such as **cooking pit** or **firepit.** **2.** any excavation area, such as a control pit. *Metallurgy.* in a metallic product, a small surface void caused, for instance, by localized corrosion. *Medicine.* **1.** a hollow depression in the surface of a structure or organ. **2.** a pockmark. **3.** a small depression or fault in dental enamel.
pit *Botany.* **1.** the hard central stone of a drupaceous fruit. **2.** a minute cavity in a secondary wall of a plant cell that exposes the primary wall and functions in the intercellular movement of food and water; it is one-half of a **pit-pair** and may be either a simple pit or a bordered pit.
PIT *Aviation.* the airport code for Greater Pittsburgh, Pennsylvania.
PIT plasma iron turnover.
pitch *Acoustics.* **1.** a relative measure of the frequency of a specific sound as judged by several listeners; in music, pitch is identified by a letter (A, B, C, D, E, F, or G) and is indicated by the vertical placement of notes on a music staff. **2.** the relative frequency (high or low) of a sound as perceived by a listener. *Mechanics.* **1.** the angular displacement of a body about a transverse horizontal axis parallel to the lateral axis of the body. **2.** to move about such a horizontal axis. *Aviation.* a deviation from level horizontal flight. *Robotics.* a vertical wrist movement in a robotic arm. *Architecture.* the slope of a roof, expressed as a ratio of rise to span. *Geology.* see PLUNGE. *Design Engineering.* the distance between successive elements arranged similarly between two points on a surface or part, such as the grooves that separate tracks of a disk recording or the threads of a screw. *Graphic Arts.* on a typewriter or in a given printing font, a unit of width expressing the number of characters that will fit in a one-inch space, as in 12-pitch type.
pitch *Materials.* **1.** the black, viscous residue resulting from the distillation of various tars or petroleum and used for caulking, paving, roofing, waterproofing, etc. **2.** any of various bituminous substances that are used for similar purposes; for example, asphalt. *Botany.* the resinous sap of pine trees.
pitch acceleration *Mechanics.* the angular acceleration of a body about its horizontal axis.
pitch attitude *Aviation.* the attitude of a flight vehicle as defined by the relationship between the longitudinal body axis of the vehicle and a horizontal reference line or plane.
pitch axis *Mechanics.* the horizontal axis about which a vehicle pitches. Also, PITCHING AXIS.
pitchblende *Mineralogy.* a massive to banded botryoidal variety of uraninite, brown to black in color and pitchy to dull in luster, having a specific gravity of 6.5 to 9 and a hardness of 5.5 on the Mohs scale; found in sulfide-bearing hydrothermal veins; the most important ore of uranium. Also, URANITE, PITCH ORE.
pitch canker *Plant Pathology.* a disease of pine trees caused by the fungus *Gibberella fujikuroi.*
pitch chain see SPROCKET CHAIN.
pitch circle *Design Engineering.* in a toothed gear wheel, an imaginary circle along which the tooth pitch is measured concentrically with the gear axis, defined at its thickest point on its teeth.
pitch coal see BITUMINOUS LIGNITE.
pitch control *Aviation.* **1.** any mechanism designed to control the pitch of a propeller or rotor on an aircraft. **2.** the horizontal control of an aircraft achieved using such mechanisms.
pitchdown *Space Technology.* in preparation for reentry, a downward movement of the leading end of a rocket vehicle in order to compensate for the lift incident to retrofiring.
pitched roof *Building Engineering.* a roof with sloping surfaces.
pitcher plant *Botany.* any of the insectivorous plants of the families Sarraceniaceae and Nepenthaceae of the order Nepenthales; named for the urn- or vase-shaped leaves that secrete or collect water in which insects drowned and are then digested.
pitch indicator *Aviation.* onboard instrumentation designed to display the pitch of a flight vehicle.
pitching *Aviation.* the angular movement of a flight vehicle about its lateral axis.
pitching axis see PITCH AXIS.
pitching chisel *Mechanical Devices.* a chisel with a squared, flat cutting edge that is used on the face of a soft stone such as sandstone and struck with a hammer to form an edge. Also, **pitching tool.**
pitching fold see PLUNGING FOLD.
pitching moment *Mechanics.* a moment about the pitch axis.
pitching piece see APRON PIECE.
pitch line see CAM PROFILE.
pitch mining *Mining Engineering.* the process of recovering coal from beds with steep slopes.
pitch opal *Mineralogy.* a yellowish to brownish variety of common opal of inferior quality, with a luster similar to that of pitch.
pitch ore see PITCHBLENDE.
pitchout *Aviation.* an abrupt turn from a straight line of flight, as in breaking formation to enter a landing pattern.
pitchover *Aviation.* the depression of an airplane nose effected by deflection of the aft elevator. *Space Technology.* **1.** the programmed movement of a rocket as it starts from the vertical, and then turns to point in the direction of its trajectory. **2.** the location in space at which this sequence occurs.
pitchstone *Geology.* a type of volcanic glass characterized by a waxy, dull, resinous, pitchy luster and containing a higher percentage of water than obsidian. Also, FLUOLITE.
pit crater *Volcanology.* see SINK.
pit furnace *Metallurgy.* in steelmaking, a furnace used for tempering.
pith *Botany.* **1.** in most vascular plants, the strand of spongy tissue that is located in the center of the stem and is used for storage. Also, MEDULLA. **2.** a similar spongy substance occurring in another plant part, as in the lining of the skin of an orange.
pithecanthropoid *Paleontology.* relating to or resembling the genus *Pithecanthropus* or one of its members.
Pithecanthropus *Paleontology.* the name first given to several hominids (including Atlanthropus, Peking man, and Solo man), all of whom are now generally classified as *Homo erectus.*

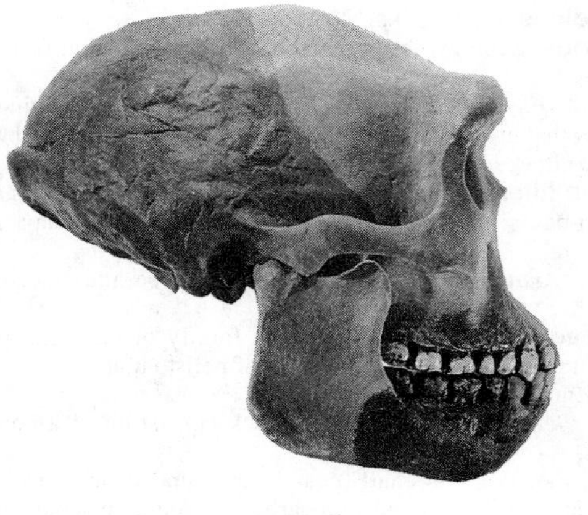

Pithecanthropus

pithecoid *Zoology.* resembling an ape; apelike.
pi theorem see BUCKINGHAM'S PI THEOREM.
pithouse *Archaeology.* a dwelling with an underground floor and aboveground earthen or brush walls supported by interior posts; found in the U.S. Southwest.
pith ray see MEDULLARY RAY.
pitman *Mining Engineering.* a person who works in or in the vicinity of a pit. *Mechanical Engineering.* the extension on a pumping unit that joins the crank with the walking beam, thus converting rotary motion to reciprocating motion.
pitometer *Engineering.* an instrument that measures the flow of gas or liquid by means of two pitot tubes, one turned upstream and the other turned downstream.
pitometer log *Engineering.* a record of variations in liquid flow measured by a pitometer.
Pitot, Henri [pē tō′] 1695–1771, French physicist; inventor of the pitot tube.
pitot pressure *Fluid Mechanics.* the pressure measured at the open end of a pitot tube.
pitot tube or **Pitot tube** *Engineering.* an instrument consisting of two concentric tubes bent into an L shape, with the inner tube open at the end directed upstream, and the outer tube connected to a pressure-indicating device; used to measure the velocity of flow. Also, IMPACT TUBE.
pitot-tube anemometer *Engineering.* an instrument consisting of a tube with one end facing into the direction of airflow and a registering device on the other end that measures wind velocity.
pitressin *Biochemistry.* a polypeptide, stored in the posterior lobe of the pituitary and secreted by the hypothalamus, that regulates hydration in the body.
pit-run gravel see BANK GRAVEL.
pit sampling *Mining Engineering.* a method of evaluating shallow alluvial deposits or ore dumps by digging small, untimbered pits and extracting samples for analysis. Thus, **pit sample.**
pit saw *Mechanical Devices.* a handsaw used by two workers to cut logs; one stands on or above the piece being sawed and the other below it, usually in a pit. Also, CLEARING SAW.
pit slope *Mining Engineering.* the angle of the wall of an open pit or cut measured against an imaginary plane along the crests of the berms or extending from the slope crest to slope toe.
pitta *Vertebrate Zoology.* a ground-foraging bird of the family Pittidae, characterized by a stub tail, a stout body, and brilliant plumage marked by sharply contrasting colors; found in Australia and southern Asia.

pitta

pitted keratolysis *Pathology.* a bacterial infection of the soles of the feet, causing round depressions in the weight-bearing areas.
pitted outwash plain *Geology.* an outwash plain whose surface is characterized by numerous shallow, irregular depressions, such as kettles. Also, **pitted plain.**
pitted pebble *Geology.* a pebble characterized by marked concavities that either are not related to the texture of the rock in which it appears or are formed by differential weathering.
pitticite *Mineralogy.* an amorphous mineraloid consisting of hydrous ferric arsenate-sulfate, occurring in gray, brown, and yellow reniform crusts and masses, having a specific gravity of about 2.2 to 2.5 and a hardness of about 2 to 3 on the Mohs scale; found as a recent deposit from mine or spring waters. Also, **pittizite.**
Pittidae *Vertebrate Zoology.* the pittas, a family of small plump birds of the order Passeriformes characterized by colorful plumage of blues, greens, and yellows and having a very short tail, short rounded wings, long legs, and a strong bill; found in evergreen and deciduous forests, bamboo jungles, mangroves, and wooded ravines of south and central Africa, southern Asia, Australia, and the Solomon Islands.
pitting *Mining Engineering.* **1.** the process of digging or sinking a pit. **2.** the sinking of small pits or shafts to expose material for valuation and testing. *Metallurgy.* the process of forming localized, small but sharp, cavities on a metallic surface by corrosion or by nonuniform coating. *Medicine.* **1.** the formation of a small depression, usually by scarring. **2.** the removal of elements such as iron granules from erythrocytes by the spleen, without damage to the cells. **3.** an indentation of the skin remaining for a few minutes after removal of firm finger-pressure, distinguishing fluid edema from myxedema.
pitting edema *Medicine.* a form of edema that remains temporarily after pressure is applied to the skin.
pitting potential *Metallurgy.* the voltage above which pits are formed in chemical or electrochemical dissolution.
Pittosporaceae *Botany.* a family of dicotyledonous trees or shrubs of the order Rosales, often climbing and distinguished by unitegmic tenuinucellar ovules, well-developed secretory canals, and unique chemical features; best developed in Australia.
pi-T transformation see Y-DELTA TRANSFORMATION.
pituicyte *Neurology.* a modified branched neuroglial cell in the posterior lobe of the pituitary gland.
pituitary [pi too′i târ′ē] *Physiology.* **1.** of or relating to the pituitary gland or the hormones produced by it. **2.** see PITUITARY GLAND.
pituitary dwarfism *Genetics.* any of a group of human hereditary growth disorders associated with the inheritance of an autosomal recessive gene that either reduces the production of growth hormone by the pituitary gland or interferes with the growth hormone receptors on cells.
pituitary gland *Anatomy.* a glandular body located at the base of the brain and attached by a stalk to the hypothalamus, from which it receives important neural and vascular outflow; the anterior portion secretes several hormones under the control of the hypothalamus, and the posterior part (neurohypophysis) stores and releases hormones synthesized in the hypothalamus. Also, HYPOPHYSIS.
pituitary gonadotropin *Endocrinology.* either of the two glycoprotein hormones (luteinizing hormone and follicle-stimulating hormone) that are produced by the anterior pituitary and are required for the maintenance of normal gonadal function.
pit viper *Vertebrate Zoology.* any of various venomous snakes of the family Crotalidae, including the rattlesnake, copperhead, water moccasin, and others having a heat-sensing organ (pit) between the eye and nostril and hollow perforated fangs; found primarily in the Americas.
Pityaceae *Paleontology.* a family of Pteridospermous plants once erroneously thought to belong to the order Cordaitales; known only as petrifactions of branches and wood; extant in the Lower Carboniferous.
pityriasis [pit′i rī′ə sis] *Medicine.* any of a group of skin diseases marked by the formation of dry scaling patches of skin. Common types are **pityriasis alba,** characterized by the presence of round or oval hypopigmented patches on the cheeks of children and adolescents, and **pityriasis rosea,** in which a solitary "herald plaque" is followed by smaller pink lesions that spread as a rash over the trunk and other unexposed parts of the body.
Pitzer equation *Physical Chemistry.* a formula that estimates how much heat organic and simple inorganic compounds generate when they vaporize.
PIV peak inverse voltage.
pivot *Mechanical Engineering.* a short, pointed shaft that forms the center and fulcrum on which something balances, oscillates, or turns. *Horology.* specifically, in a timepiece, the finely turned, hard, and highly polished end of an arbor that, when inserted in a bearing hole in the plate or bridge, allows the arbor to revolve freely as an axle. A pivot mounted in the plate is known as a **lower pivot**; a pivot mounted in a bridge is known as an **upper pivot.** *Mathematics.* **1.** in linear programming, the positive entry a_{ij} in a tableau that is used with Gaussian operations to convert column j to a column of the identity matrix having a 1 in row i. **2.** the act of converting the column in this manner.

pivotal *Engineering.* of, relating to, or serving as a pivot.
pivotal fault see ROTARY FAULT.
pivot bearing see STEP BEARING.
pivot bridge *Civil Engineering.* a swing bridge that rotates to allow passage.
pivoting *Mathematics.* in Gaussian elimination, the interchanging of rows (**partial pivoting**) or rows and columns (**full pivoting**) so as to place the element with largest absolute value on the diagonal.
pivoting point *Naval Architecture.* the point along a vessel's center line around which it pivots when turning; usually near the bridge or pilot house.
pivot joint *Anatomy.* a joint formed by the articulation of a post with an embracing ring, as in the proximal articulation of the radius and ulna.
PIXE proton-induced X-ray emission.
pixel [piks´əl] *Computer Technology.* the smallest element of a display image, corresponding to a single displayed spot or color triad on a display, or to a single input spot from a camera. (A word coined from the phrase pic̲ture e̲lement.) Also, PICTURE ELEMENT.
pixmap *Computer Technology.* a representation of a graphical image in which each pixel is individually specified as a color or gray-scale value.
pk. or **pk** park; peak; peck (dry measure).
Pk. or **Pk** park; peak (in place names).
pK *Chemistry.* a measurement of the degree of completeness of a reversible reaction, defined as the negative logarithm of the equilibrium constant *K*; used, for example, to describe the extent of disassociation of a weak acid.
PK psychokinesis.
pkg. or **pkg** package; per kilogram.
PKU phenylketonuria.
pl. place; plural; plate.
PL pulpolingual; partial loss.
PL/1 *Computer Programming.* a high-level programming language designed to accommodate a number of applications, including scientific computations and business and systems programming.
PLA programmable logic array; pulpolinguoaxial, pulpolabial.
placanticline *Geology.* a generally asymmetric, gentle upward arch of the continental platform that lacks a typical outline and corresponds to the western plains-type fold.
place *Mathematics.* **1.** in a positional notation system, a position corresponding to a given power of the base. Also, COLUMN. **2.** let F and K be fields and let $\{F,\infty\}$ denote the extension of F by the adjoining of ∞, with the usual algebraic rules for ∞. A place from K into F is a homomorphism ϕ: $K \rightarrow \{F,\infty\}$ such that $\phi(1) = 1$; ϕ is said to be F-valued. The elements of K that are not mapped into ∞ are called finite and form a valuation ring of K; the others are called infinite.
placebo [plə sē´bō] *Medicine.* any intentionally noneffective medical treatment, especially an inactive or inert substance prescribed to replace medication that is desired by a patient but which cannot be given or which would be inappropriate. Placebos are also given to control groups used in experimental research to compare the results with those of the experimental drug. (From a Latin phrase meaning, "I will please.")
placebo effect *Psychology.* the tendency of any form of therapy or treatment to have at least some success, because of factors such as the patient's desire to improve or his confidence in the treatment and the therapist. Also, **placebo response.**
placeholder *Mathematics.* a symbol in an expression that may be replaced by any element of the set. *Computer Programming.* an area in memory reserved for data to be entered later.
placenta [plə sen´tə] *Developmental Biology.* a structure formed by fusion of the chorion with the wall of the uterus that serves to attach the embryo to the uterine wall and to exchange nutrients, wastes, and gases between the maternal blood and the embryonic blood. *Botany.* **1.** in a flowering plant, the tissue of the ovary where ovules form and remain attached while they mature, usually on the enlarged or modified margins of carpillary leaves. **2.** a sporangia-bearing structure in ferns.
placenta accreta *Medicine.* an abnormal condition in which part or all of the placenta adheres to the uterine wall.
placental [plə sen´təl] *Biology.* of or relating to a plant or animal placenta. *Vertebrate Zoology.* of or relating to a placental mammal.
placental barrier *Developmental Biology.* the placental separation of fetal from maternal blood and blood-borne materials of greater than molecular size.
placental insufficiency *Medicine.* a dysfunctional placenta adversely affecting the fetal environment.
placental lactogen see CHORIONIC SOMATOMAMMOTROPIN.
placental mammal *Vertebrate Zoology.* a mammal whose unborn young receive nourishment in the uterus by means of a placenta; i.e., any mammal except a marsupial or monotreme.
placental transfusion *Medicine.* the return to a newborn infant, through the umbilical vessels, of some of the blood contained in the fetal placenta.
placenta previa *Medicine.* the presence of a placenta in the lower segment of the uterus, partially or completely covering the opening of the cervix.
placentate *Botany.* having a placenta.
placentation *Developmental Biology.* the process of placenta formation and the result, especially with respect to taxonomically relevant aspects of structure. *Botany.* the arrangement of the placenta in a plant ovary and of the ovules on the placenta.
placer *Geology.* a surficial mineral deposit, especially such a deposit containing gold or other valuable metal that has been concentrated mechanically from weathered debris. Also, ORE OF SEDIMENTATION.
placer dredge *Mining Engineering.* a dredge used to mine metals from placer deposits, consisting of a chain of closely connected buckets that pass over an idler tumbler and an upper or driving tumbler; the system is mounted on a structural-steel ladder carrying a series of rollers.
placer mine *Mining Engineering.* any mine extracting gold from sand or gravel. Also, GRAVEL MINE.
placer mining *Mining Engineering.* the recovery of gold or heavy minerals from placer deposits by the use of running water.
placic horizon *Geology.* a black to dark red, thin layer of soil that is commonly cemented with iron and is almost impermeable to water.
placid *Invertebrate Zoology.* a cuticular plate found on the neck of kinorhynchs that is adapted for closing over the retracted head.
Placida *Invertebrate Zoology.* a genus of gastropod mollusks that lacks a shell; found in the Mediterranean Ocean and eastern Atlantic Ocean.
placode *Developmental Biology.* an ectodermal thickening in the early embryo, from which sense organs or other ectodermal derivates often develop .
Placodermi *Paleontology.* an extinct class of primitive jawed fishes characterized by bony plates of armor covering their head and flanks; the placoderms arose in the Silurian and flourished in the Devonian, replacing the jawless ostracoderms.
Placodontia *Paleontology.* an order of highly specialized, mollusk-eating diapsid marine reptiles; generally considered sauropterygians but still of uncertain affinities; the placodonts ranged from 1 to 3 meters in length; extant in the Lower to Upper Triassic.
Placothuriidae *Invertebrate Zoology.* a family of sea cucumbers in the order Dendrochirotida, with overlapping plates covering the body wall and paired processes on the calcareous ring.
Placozoa *Invertebrate Zoology.* a phylum of Metazoa containing two species of small (2–3 mm), flattened, amebalike marine organisms; the body lacks symmetry, has no organs, and has no muscular or nervous system; they feed on algae and protozoans by extracellular predigestion.
plage [pläzh] *Astronomy.* a bright spot in the solar chromosphere, appearing in the vicinity of a sunspot; formerly called a bright flocculus. (A French word meaning "beach.")
Plaggept *Geology.* a suborder of the soil order Inceptisol, characterized by having very thick surface horizons of mineral and organic matter; formed by long, continued manuring and mixing.
plagiaplite *Petrology.* a dioritic aplite consisting of plagioclase with or without green hornblende, with traces of quartz, biotite, and muscovite.
Plagiaulacidae *Paleontology.* a family of early multituberculate mammals in the extinct suborder Plagiaulacoidea; Jurassic and Cretaceous.
Plagiaulacoidea *Paleontology.* an extinct suborder of mammals belonging to the extinct Mesozoic order Multituberculata; it includes the families Arginbaataridae and Paulchoffatidae; Jurassic and Cretaceous.
plagio- a combining form meaning "oblique."
plagiocephaly *Medicine.* an asymmetrical and twisted condition of the head, resulting from irregular closure of the cranial sutures.
Plagiochilaceae *Botany.* a family of terrestrial and epiphytic liverworts of the order Jungermanniales; characterized by erect, diagonally erect, or prostrate plants without underleaves or bracteoles, ranging in color from brown to yellow to bright green; most species are native to the Southern Hemisphere.
plagioclase *Mineralogy.* a series of important, rock-forming triclinic silicates of the feldspar group with the general formula (Na,Ca)Al (Al,Si)Si_2O_8, and ranging from albite, $NaAlSi_3O_8$, through the intermediate members oligoclase, andesine, labradorite, and bytownite, to anorthite, $CaAl_2Si_2O_8$.

plagioclimax *Ecology.* a community whose natural development stops as a result of human activity.
plagiodont *Vertebrate Zoology.* describing a snake that has obliquely set or converging sets of palatal teeth.
plagiogravitropism *Physics.* a response of phenomena that are partially influenced by gravity and occur at oblique angles to the direct gravitational action.
Plagiogyriaceae *Botany.* a monogeneric family of medium-sized terrestrial ferns of the order Filicales that grow in high elevations of wet tropical areas; characterized by a hairy creeping or erect rhizome, pinnatifid or pinnate fronds, and sporangia borne on special fertile fronds.
plagiohedral *Crystallography.* of or relating to a crystal structure having obliquely arranged spiral faces.
plagionite *Mineralogy.* $Pb_5Sb_8S_{17}$, a lead-gray or blackish, monoclinic mineral with metallic luster, occurring as thick tabular crystals and in massive form, having a specific gravity of 5.54 and a hardness of 2.5 on the Mohs scale; found with other sulfosalts in vein deposits.
Plagiorchiida *Invertebrate Zoology.* an order of trematode flukes, parasitic on many vertebrates.
Plagiosauroidea *Paleontology.* a superfamily of labyrinthodont amphibians in the extinct order Temnospondyli; characterized by a very short skull (6 inches in Plagiosaurus); extant in the Triassic.
plagiosere *Ecology.* a community whose development has been changed as a result of human activity or other external intervention.
Plagiotheciaceae *Botany.* a family of mostly slender, shiny yellowish mosses of the order Hypnobryales that form mats on humus, soil, logs, tree bases, and rocks; characterized by simple or branching stems that are prostrate or ascending, lateral sporophytes that arise from the lower branch portions, and mostly complanate leaves.
plagiotropism *Botany.* the process by which plant parts turn toward a position that is oblique or perpendicular to the orienting stimulus. Thus, **plagiotropic.**
plague *Medicine.* a severe, acute, communicable disease that is caused by a bacterium, *Pasteurella pestis*; it occurs both endemically and epidemically worldwide and has caused mass deaths at various times in history. Primarily a disease of rats and other rodents, it is transmitted to humans by flea bites, by contact with or ingestion of infected animals, or by human-to-human infection. The most common forms are bubonic plague, pulmonic plague, and septicemic plague.
plague count *Biotechnology.* a technique used to determine the number of phage particles or infected bacterial cells in a suspension; the sample is diluted and used to inoculate an agar plate, and the number of cultures produced are then counted.
plain *Geology.* **1.** any extensive area of relatively level, smooth, flat, or gently rolling land at low elevations, without noticeable hills, mountains, or valleys. **2.** a prairie or any broad, extensive area of level or rolling land that is generally treeless and marked by shrubby vegetation. (From the Latin word for "flat.")
plain-carbon steel *Materials.* a steel that contains no alloying elements other than carbon. Also, WROUGHT CARBON STEEL.
plain concrete *Civil Engineering.* a term for concrete that contains no reinforcement to counteract load-bearing forces, usually used to encompass light steel to prevent shrinkage or cracking.
plain film *Radiology.* a term for a radiograph taken without the use of radiopaque material, usually prior to examination with a contrast medium.
plain-laid *Design Engineering.* describing rope in which the strands are twisted together opposite to the twist in the strands.
plain milling cutter *Mechanical Devices.* a cylindrical cutter with teeth along its periphery; used for milling plain, flat surfaces. Also, SLAB CUTTER.
plain of denudation *Geology.* a relatively flat surface area that has been almost or completely eroded to sea level.
plain of lateral planation *Geology.* an extensive, smooth area at the base of a mountain or an escarpment formed by the merging of the widened floodplains of neighboring streams moving laterally against their banks.
plain of marine denudation *Geology.* **1.** a level or nearly level surface formed by the gradual advancement of ocean waves upon the land. **2.** a hypothetical area representing such a surface subsequent to uplift and partial subaerial erosion. Also, **plain of submarine denudation.**
plain of marine erosion *Geology.* a hypothetical platform that represents a level surface of unlimited width below sea level produced by the beveling action of marine erosion over a prolonged period. Also, SEA PLAIN, MARINE PLAIN.
plain sail *Naval Architecture.* any or all of a vessel's ordinary working sails.
plain-sawing or **plainsawing** *Forestry.* the process of cutting wood so that the width of the board is approximately tangential to the growth rings and the thickness is approximately radial.
plains-type fold *Geology.* an anticlinal or domelike feature of the continental platform, having no typical outline or related synclinal feature.
plain text *Telecommunications.* words that are not in code; words in English or another natural language.
plain tract *Hydrology.* the lower section of a stream, having a low gradient and a wide flood plain.
plain weave *Textiles.* the simplest of the three types of weave (twill and satin being the other two), having filling threads passing alternately under and over successive warp threads, and alternating each row; used in about 80% of all woven fabric.
Plaisancian or **Plaisanzian** *Geology.* a European geologic stage of the Lower Pliocene epoch, occurring after the Pontian of the Miocene epoch and before the Astian; after the Zanclean and before the Calabrian. Also, PIACENTIAN, PIACENZIAN.
plaiting *Geology.* in some schists, a texture that results when relict bedding planes intersect with well-developed cleavage planes. Also, GAUFRAGE.
plait point *Chemistry.* the point at which two partially miscible liquids become completely miscible following the addition of another liquid that is completely miscible in both of them.
plakula *Invertebrate Zoology.* a larval stage of some tunicates, a flattened hollow mass of cells.
plakula theory *Invertebrate Zoology.* the hypothesis advanced in 1884 by Otto Butschli that the earliest true metazoan may have been a plakula.
plan a proposed method or system for doing something; specific uses include: *Architecture.* a drawing of the horizontal section of a building, showing its chief architectural and design elements.
planar *Science.* flat or level. *Mathematics.* relating to, lying in, or embedded in a geometric plane.
planar array *Electronics.* a group of ultrasonic transducers that are mounted onto a single sheet so that they will fit into the hull of a sonar-carrying vessel.
planar-array antenna *Electromagnetism.* an antenna array in which all the centers of the radiating elements lie in the same plane.
planar cross-bedding *Geology.* cross-bedding whose base lies on a surface that has been flattened and leveled by erosion.
planar device *Electronics.* a device in which all the contacts to all the *pn* junctions terminate on the same geometric plane.
planar diode *Electronics.* a diode constructed as a planar device.
planar flow structure see PLATY FLOW STRUCTURE.
planar graph *Mathematics.* a graph that can be drawn on a plane so that its edges intersect only at their end vertices; i.e., a graph that can be embedded on a plane. By theorem, for every simple planar graph there exists a **planar embedding** in which all the edges of the graph can be drawn as straight line segments. Also by theorem, a 3-connected planar graph has essentially only one plane embedding, but a planar graph of connectivity at most two may have several distinct plane embeddings. Kuratowski's theorem characterizes all planar graphs.
planar growth *Materials Science.* the growth of a smooth solid-liquid interface during solidification.
Planaria *Invertebrate Zoology.* a genus of tiny free-living freshwater flatworms of the order Tricladida.
planarian [plə nâr´ē ən] *Invertebrate Zoology.* any free-living flatworm of the turbellarian order Tricladida, living in fresh water or on land and used extensively in regeneration studies.

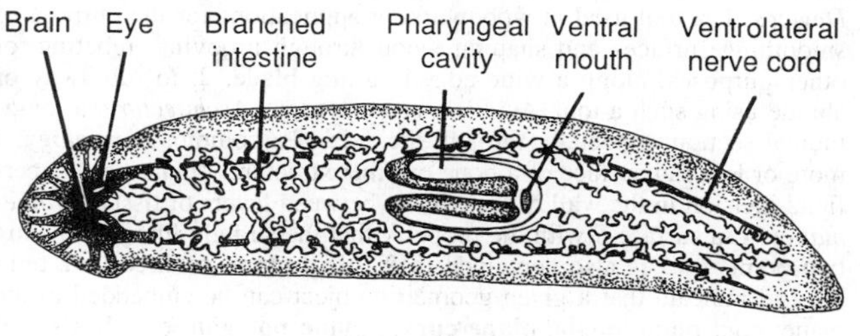

planarian

planar linkage *Mechanical Engineering.* a linkage with motion restricted to only two dimensions.

planar photodiode *Electronics.* a photodiode constructed as a planar device.

planar process *Engineering.* a silicon transistor and integrated circuit manufacturing process in which an oxide layer is grown on a silicon substrate, and a series of etching and diffusion steps are used to produce the active regions and junctions inside the substrate.

planar transistor *Electronics.* a transistor constructed as a planar device.

planate *Science.* having a plane or flat surface. *Geology.* specifically, describing a surface that has been leveled or flattened as a result of erosion processes.

planation *Geology.* the processes by which a surface is flattened or leveled by the erosive action of meandering streams, waves, water currents, wind, or glaciers.

planation stream piracy *Hydrology.* the diversion of water from the upper part of a smaller stream to the channel of a larger meandering stream, as a result of the leveling of the interstream area by the lateral erosion of the meandering stream.

planchet *Radiology.* a glass or metal receptacle on which radioactive samples are mounted for the measurement of radioactivity. *Materials.* a flat piece of metal that is stamped as a coin; a coin blank.

Planck, Max [plängk] 1858–1947, German physicist; formulated a radiation law that became the foundation of quantum theory.

planck *Physics.* a unit of action given as the product of an energy of 1 joule and a time of 1 second.

Planck era *Astronomy.* the first period in the early universe (10^{-43} second after the Big Bang), when the strong force appeared.

Planckian locus *Optics.* locus points on a chromaticity diagram that includes the chromaticities of blackbody radiators.

Planck length *Physics.* the distance over which quantum fluctuations are theorized to become significant: $(Gh/2\pi c^3)^{1/2} = 1.616 \times 10^{-35}$ m, where G is the gravitational constant, h is Planck's constant, and c is the speed of light in a vacuum.

Planck mass *Physics.* the mass of a particle having a reduced Compton wavelength that is equal to the Planck length, as given by $(hc/2\pi G)^{1/2} = 21.77 \times 10^{-6}$ grams, where h is Planck's constant, c is the speed of light in a vacuum, and G is the gravitational constant.

Planck oscillator *Quantum Mechanics.* an oscillator that absorbs or emits radiation only in units of Planck's constant times the frequency of the oscillator.

Planck's constant *Quantum Mechanics.* a fundamental physical constant, having the dimensions of angular momentum and action and the value $6.6260755 \times 10^{-34}$ joule-second in SI units.

Planck's law *Quantum Mechanics.* a fundamental law stating that radiation is composed of indivisible units of energy, or quanta, whose energies are proportional to the frequency of the radiation: $E = h\nu$, where h is Planck's constant and ν is the frequency of the oscillator absorbing or emitting energy.

Planck's radiation formula *Physics.* a formula relating the intensity of radiation from a blackbody to the wavelength λ and the absolute temperature T of the body:

$$E_\lambda d\lambda = (hc^3/\lambda^5) \cdot d\lambda/(e^{hc/\lambda kT} - 1),$$

where h is Planck's constant, c is the speed of light, k is the Boltzmann constant, and $E_\lambda d\lambda$ is the intensity of radiation between wavelengths and $\lambda + d\lambda$. Also, **Planck distribution law.**

Planctomyces *Bacteriology.* a genus of budding, appendaged bacteria that attach themselves to a surface with long slender stalks, occurring in fresh water.

plane a flat or level surface or device; specific uses include: *Mechanical Devices.* **1.** a tool used in woodworking applications for the purpose of smoothing surfaces and shaping wood through grooving, rebating, or other purposes, along a wide-edged cutting blade. **2.** to rub away or abrade using such a tool. *Aviation.* see AIRPLANE. *Architecture.* a longitudinal section, especially through the axis of a column. *Physiology.* a more or less flat surface of a bone or other structure. *Medicine.* a superficial incision in the wall of a cavity or between layers of tissue. *Mathematics.* **1.** a surface S with the property that the straight line joining any two points of S is also contained in S. Also, **plane surface. 2.** a term used to indicate that a given geometric object can be embedded in the plane; e.g., **plane angle, plane curve, plane polygon,** etc. **3.** a set of points and lines, together with rules for incidence and axioms in a particular geometric structure; e.g., **affine plane, projective plane.**

plane atmospheric wave *Meteorology.* an atmospheric wave that is depicted by using two-dimensional rectangular Cartesian coordinates, as opposed to using a wave on the spherical earth.

plane bed *Geology.* a bed of sedimentary rock having elevations or depressions less than the largest size of the rock material.

plane correction *Engineering.* an adjustment made to surveying data to provide a common reference plane.

plane-dendritic crystal *Crystallography.* a crystal of ice exhibiting hexagonal symmetry with dendritic structure, and having its long dimension perpendicular to the principal axis. Also, **plane dendrite.**

plane earth *Electromagnetism.* the earth when it is considered to have a plane surface as opposed to a spherical surface; used in calculations of ground waves.

plane-earth attenuation *Electromagnetism.* the attenuation of a ground wave in an imperfectly conducting plane earth, as distinguished from a ground wave in a perfectly conducting plane.

plane geometry *Mathematics.* the geometric study of figures whose parts lie in a single plane, such as lines and polygons.

plane graph *Mathematics.* a planar graph that has been drawn on the plane.

plane group *Crystallography.* any of 17 groups of symmetry elements that produce regularly repeating patterns in two dimensions. Also, PLANE SYMMETRY GROUP. *Mathematics.* **1.** the three-parameter group of isometries of the Euclidean plane. **2.** a subgroup of such isometries.

plane jet *Hydrology.* a flow pattern in which the water flowing into another body of water spreads out in the shape of a parabola whose dimensions are related to the distance downstream from the mouth.

plane lamina *Mechanics.* any body whose mass is concentrated on a single plane; e.g., a flat plate.

plane mirror *Optics.* a mirror on a plane that forms an image of an object so that the mirror surface is perpendicular to and bisects the line joining all corresponding object-image points.

plane of departure *Mechanics.* the plane containing the path of a projectile as it leaves the muzzle of a gun.

plane of fire *Mechanics.* the plane that contains a gun and its target; i.e., the vertical plane that contains the line of sight.

plane of incidence *Physics.* a geometric plane defined by the propagation direction of a wave incident on a surface and the line normal to the surface at the point of reflection or refraction.

plane of maximum shear stress *Mechanics.* at a given point, either of the two orthogonal planes across which the shear stress is maximum at that point. They lie at 45° to the minimum and maximum principal stress axes and parallel to the intermediate principal stress axis.

plane of mirror symmetry *Crystallography.* an imaginary plane that divides a crystal into two parts, each of which is a mirror image of the other. Also, PLANE OF REFLECTION.

plane of polarization *Electromagnetism.* a plane that contains the electric field vector and the direction of propagation in an electromagnetic wave.

plane of reflection *Optics.* a plane that includes both the direction of radiation reflected from a surface and the perpendicular of that surface. *Crystallography.* see PLANE OF MIRROR SYMMETRY. *Mathematics.* see PLANE OF SYMMETRY.

plane of saturation see WATER TABLE.

plane of site *Ordnance.* a horizontal plane defined by the gun-target line and a horizontal line perpendicular to the gun-target line at the muzzle of the gun.

plane of symmetry *Optics.* an imaginary plane dividing a body into two portions that are mirror images of each other. *Crystallography.* see MIRROR PLANE. *Mathematics.* a plane with respect to which a geometric configuration in space is symmetric; i.e., for every point in the configuration, there is a unique corresponding point such that the line joining the points is bisected by and perpendicular to the plane. Also, PLANE OF REFLECTION, MIRROR PLANE.

plane of yaw *Mechanics.* a plane determined by the tangent to the projectile trajectory and the axis of the projectile.

plane-parallel resonator *Optics.* a beam resonator incorporating two plane mirrors positioned perpendicularly to the axis of the beam.

plane-parallel texture *Petrology.* a rock texture in which the constituents are aligned parallel to a plane as opposed to a line.

plane Poiseuille flow *Fluid Mechanics.* the flow of a fluid through a narrow slot in which a pressure gradient and the volumetric flow rate are measured together to determine the viscosity of the fluid.

plane polarization *Crystallography.* the process of affecting electromagnetic radiation so that the electric vectors of the waves are confined to a single plane. *Optics.* see LINEAR POLARIZATION.

plane-polarized light *Optics.* light that has been polarized so that there is a single direction of vibration. *Crystallography.* electromagnetic radiation in which the electric vectors of the waves are confined to a single plane.

plane-polarized wave *Electromagnetism.* an electromagnetic wave in an isotropic homogeneous medium in which the electric field vector remains in a fixed plane that also contains the direction of propagation.

plan equation *Mechanical Engineering.* a mathematical representation of horsepower, taking into account the mean effective pressure, the length of piston stroke, the net area of the piston, and the number of cycles over a specified time.

planer *Mechanical Engineering.* a mechanical plane; i.e., a power machine used to reduce and smooth the rough surface of wood or other materials. *Civil Engineering.* a machine fitted with milling cutters that is used to smooth a previously softened road surface. *Mining Engineering.* a fixed-blade instrument used in continuous longwall mining of narrow seams of friable coal. *Graphic Arts.* in hot-metal typesetting, a smooth block of hard wood used to level the type in a chase.

planer saw *Mechanical Engineering.* a circular saw designed for cutting across wood grain.

plane sailing *Navigation.* a method of solving problems involving courses, distances, and so on (but not including difference of longitude), in which the portion of the earth traversed is considered as a flat plane.

plane strain *Mechanics.* a state of strain where there exists a direction in which no displacement occurs, and along which strain is independent of distance.

plane stress *Mechanics.* a state of stress where there exists a direction in which no displacement occurs, and along which stress is independent of distance.

plane surveying *Engineering.* any topographical surveying that makes no corrections for the curvature of the earth's surface.

plane symmetry group *Crystallography.* see PLANE GROUP.

planet *Astronomy.* **1.** any of the nine primary celestial objects that orbit the sun: Mercury, Venus, Earth, Mars, Jupiter, Saturn, Uranus, Neptune, and Pluto. **2.** a similar body orbiting another star. (From a Greek word meaning "wanderer," because it appeared to the ancients that the planets changed their position significantly in relation to the seemingly fixed stars.)

plane table *Cartography.* an instrument used for plotting survey lines directly from field observations, made up of a drawing board supported by a tripod and fitted with a measuring device that is pointed at the observed area.

plane-table method *Mining Engineering.* a means of measuring the area of a mine roadway using a plane table; from a central point on the drawing board, the distances to the perimeter of the roadway are measured along various offsets and plotted to scale on the paper along the correct lines.

planetarium *Astronomy.* **1.** a special projector used to display the nighttime sky on a hemispherical dome; primarily used for instruction in astronomy and related subjects. **2.** an auditorium equipped with such a device.

planetary *Astronomy.* relating to or resembling a planet. *Mechanical Engineering.* of or relating to a planet gear train or similar mechanism.

planetary aberration *Optics.* an apparent displacement of a planet or other object moving through the solar system, caused by the length of time it takes for an object's light to reach the earth.

planetary alignment *Astronomy.* a condition occurring when several planets lie relatively close together as seen from the sun.

planetary atmosphere *Astronomy.* the gases gravitationally attached to a planet.

planetary boundary layer *Meteorology.* the atmospheric layer below the free atmosphere that extends from the earth's surface to the geostrophic wind level; this includes the surface boundary layer and the Ekman layer.

planetary circulation see GENERAL CIRCULATION.

planetary drive *Robotics.* a gear-reduction arrangement that consists of a central gear that meshes with an epicyclic train of gears.

planetary electron see ORBITAL ELECTRON.

planetary gear train *Mechanical Engineering.* an assembly of meshed gears consisting of a central (or "sun") gear, a coaxial internal or ring gear, and one or more intermediate pinions supported on a revolving carrier.

planetary geology *Geology.* the application of the principles and techniques of geology to the scientific study of planets and their satellites. Also, **planetary geoscience.**

planetary nebula *Astronomy.* a transient shell of highly ionized gas thrown off by a star as it evolves into a white dwarf. (So called because such gaseous shells, seen through a telescope, often roughly resemble the planets Uranus and Neptune.)

planetary nebula symbiotic star see SUBDWARF SYMBIOTIC STAR.

planetary orbit *Astronomy.* the orbit that a planet follows as it moves about the sun.

planetary perturbation *Astronomy.* the irregularity in a planet's orbit due to the gravitational forces of other celestial bodies.

planetary physics *Astrophysics.* a detailed study of the planets of our solar system from physical and chemical standpoints to determine their atmospheric constitution, composition, and magnetospheres.

planetary precession *Astronomy.* a small portion of the precession of the equinoxes that comes from perturbations of the planets on the body of the earth.

planetary system *Astronomy.* all the celestial bodies that revolve around the sun, including planets, moons, asteroids, comets, and dust. Also, SOLAR SYSTEM.

planetary vorticity effect *Geophysics.* the effect that differences in the earth's vorticity at various latitudes have on the relative vorticities of flows having a meridional element.

planetary wave *Meteorology.* see LONG WAVE.

planetary wind *Meteorology.* a system of wind that resides in the earth's atmosphere due to the forces of solar radiation and the rotation of the earth.

planetesimal *Astronomy.* a small body that is the precursor of a planet in the early stages of solar system development, or the fragmented result of a cataclysmic collision between a planet and another celestial body.

planetesimal theory *Astronomy.* a theory that the solar system formed from the collisional aggregation of a great many planetesimals that condensed from the solar nebula.

planet gear *Mechanical Engineering.* any gearwheel whose axis describes a circular path around that of another wheel.

planetographic coordinates *Astronomy.* **1.** the position of a point on the surface of a planet determined by latitude from the planet's equator and by longitude from a reference meridian. **2.** the position of an object in the sky as if seen from a planet's center.

planetography *Astronomy.* **1.** a description of the physical features of a planet. **2.** see PLANETOLOGY.

planetoid see ASTEROID.

planetology *Astronomy.* the branch of astronomy dealing with planets and satellites, including their surface features, structures, chemical compositions, and atmospheres. Also, PLANETOGRAPHY.

plane trigonometry *Mathematics.* trigonometry of plane figures, as opposed to spherical or spatial trigonometry.

Planet X *Astronomy.* **1.** a hypothetical tenth planet in our planetary system beyond the orbit of Pluto. **2.** any planet hypothesized to be orbiting beyond the farthest known planet in a system.

plane wave *Physics.* a wave whose phase fronts are a plane; sometimes approximated by a spherical wave when at large distances from the source.

plane-wave initiation *Ordnance.* the method of initiating detonation of an explosive charge by simultaneously detonating all points of the rear surface of the charge with a flat detonation wave.

planform *Aviation.* the shape or outline of an object, especially a wing, as seen from above. Also, PROJECTED PLANFORM.

plani- a combining form meaning "flat" or "plane."

planidium *Invertebrate Zoology.* a legless larval stage of some parasitic dipteran insects.

planimeter *Engineering.* an instrument that mechanically measures the area of a plane surface by tracing its boundary lines.

planimetric map *Cartography.* a map that shows only horizontal position of surface details, rather than surface relief or topography.

planimetry *Cartography.* **1.** in surveying, the theories and techniques of accurate measurement of horizontal distances and locations. **2.** collectively, those elements of a map that represent the surface details without reference to elevation or surface relief.

planing *Engineering.* the action of smoothing or shaping a wood, metal, or plastic surface with a hand or motor tool.

planing hull *Naval Architecture.* a hull designed with a step or sharp chine for efficient hydroplaning at high speed.

Planipennia *Invertebrate Zoology.* a suborder of insects, including the antlions and lacewings, in the order Neuroptera, with two membranous wings; predatory larvae have mandibles adapted for piercing and sucking.

planishing *Mechanical Engineering.* the process of giving a smooth finish to metal surfaces by a rapid series of overlapping, light hammer-like blows, or by rolling in a planishing mill.

planishing hammer *Mechanical Devices.* a hammer with convex faces used to smooth and shape sheet metal.

planisphere *Astronomy.* **1.** a circular star map drawn for a given latitude, having a rotating mask that shows which constellations are visible at any time and date in the year. **2.** a map in which the celestial sphere is projected onto a plane surface, especially in a polar projection, and to which a grid system is added to aid in location of celestial objects.

planitia *Astronomy.* the official term used in naming a broad flat area on a planet or moon.

plank *Forestry.* **1.** a long, flat piece of wood that is thicker than a board. **2.** lumber formed into such pieces.

planking *Building Engineering.* **1.** an arrangement or collection of planks, as in a floor. **2.** the process of laying planks. *Naval Architecture.* the hull shell of a wooden vessel; planking runs longitudinally, as opposed to the mainly transverse structural framing.

plankter *Biology.* an individual organism within an aggregation of plankton.

plankton [plangk´tən] *Biology.* a collective term for the wide variety of plant and animal organisms, often microscopic in size, that float or drift freely in water because they have little or no ability to determine their own movement; found worldwide in both aquatic and marine environments and representing the basic level of many feeding relationships. (Going back to the Greek word for "wanderer.")

planktonic [plangk tän´ik] *Biology.* relating to, being, or characteristic of plankton.

planning *Industrial Engineering.* the management function that involves the defining of goals for an organization and the determination of tasks and resources that are necessary to obtain those goals.

planning by abstraction *Computer Programming.* a problem-solving method in which a problem is simplified and solved, and the solution is then used as a guide in solving the original problem.

Planobispora *Bacteriology.* a genus of actinomycete bacteria that are found in soil as branching substrate and spore-bearing aerial mycelia.

planoblast *Invertebrate Zoology.* a free-swimming medusa form of hydroids.

planoclastic rock *Petrology.* a clastic rock whose constituents are of uniform size.

Planococcus *Bacteriology.* a genus of Gram-positive bacteria that occur as aerobic, flagellated cocci in pairs and tetrads and possess a yellow-brown pigmentation.

planoconcave lens *Optics.* a lens that has one plane surface and one concave surface.

planoconformity *Geology.* the relationship among conformable rock layers that are nearly uniform in thickness and noticeably parallel throughout.

planoconvex lens *Optics.* a lens that has one plane surface and one convex surface.

planoconvex spotlight *Electricity.* a brightly illuminated light that can be used as a spotlight, having a power output of 100–2000 watts.

planocylindrical lens *Optics.* a lens that has one plane surface and one surface shaped like a portion of a cylinder; used to correct astigmatism in the eye and to produce astigmatism.

planography *Graphic Arts.* printing on a flat surface; i.e., any printing process in which the printing and nonprinting areas are on the same plane, such as offset lithography. Also, **planographic printing.**

Planomonospora *Bacteriology.* a genus of soil bacteria of the order Actinomycetales that form branching substrates and aerial mycelia, the latter possessing sporangia, each containing one spore.

Planonephraceae *Botany.* in some classifications, an equivalent name for Hemiselmidaceae.

Planosol *Geology.* a group of intrazonal, hydromorphic soils that develop in warm, humid to subhumid climates, having a leached surface layer above a clay pan or hardpan.

planospiral *Invertebrate Zoology.* of foraminiferans and gastropods, having the shell coiled in one plane.

plan position indicator *Electronics.* an instrument that displays signals from various targets at points on its screen that correspond to points on a circular map of the area under surveillance, with the radar's antenna at the center of the map.

plansifter *Food Technology.* an apparatus consisting of a group of sieves mounted together so that the material being sifted is divided into a number of different sizes; used in flour milling.

plant *Botany.* **1.** any member of the kingdom Plantae, generally characterized by the ability to produce food by photosynthesis, thick cell walls containing cellulose, a lack of the power of locomotion, and a relatively open growth pattern. **2.** a small growth of this type, such as a vegetable or herb, as opposed to a larger shrub or tree. *Industrial Engineering.* an engineering or production facility; a factory, electric power station, or the like. *Geophysics.* the manner in which a geophone is placed on or in the earth. *Computer Programming.* an instruction that is formed by the execution of the program and is stored in a location for use at a later time.

Plantae *Botany.* the plant kingdom; in most systems, it includes all extant and extinct eukaryotic multicellular organisms that have cellulose cell walls, are capable of producing their own food, are not capable of spontaneous movement, and lack obvious nervous or sensory organs, including vascular plants, mosses, liverworts, and hornworts; some classification systems may also include fungi, algae, bacteria, and blue-green algae as plants.

Plantaginaceae *Botany.* the single family in the order Plantaginales.

Plantaginales *Botany.* a monofamilial order in the class Asteridae, containing herbs and some shrubs with simple spirally arranged leaves and small, mostly perfect, wind-pollinated flowers.

Plantago *Botany.* a genus of herbs of the family Plantaginaceae, including three species whose seeds are used as a cathartic.

plantain *Botany.* **1.** a plant, *Musa paradisiaca,* of the family Plantaginaceae, whose cooked fruit is a staple food in many tropical regions. **2.** any herb of the genus *Plantago.*

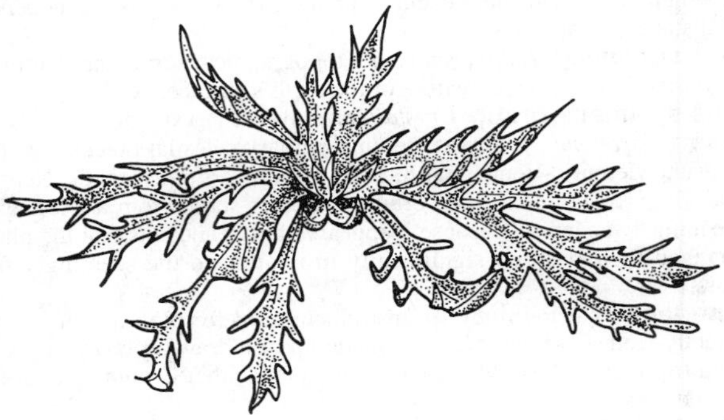

plantain

plantalgia *Medicine.* pain in the sole of the foot.

planta pedis *Anatomy.* the undersurface, or sole, of the foot.

plantar *Anatomy.* relating to or located on the sole of the foot.

plantaris *Anatomy.* a small dorsal calf muscle between the gastrocnemius and the soleus that flexes the foot.

plantar reflex *Physiology.* a reaction to touch on the soles of the feet, normally resulting in flexing of the toes.

plantar wart *Medicine.* a painful viral wart occurring on pressure-bearing areas of the sole of the foot.

plantation *Agriculture.* a large-scale farm in a tropical or subtropical area, usually devoted to the production of a single crop, such as cotton or sugar. *Surgery.* the surgical insertion or application of tissues, organs, or devices into a body, including implantation, replantation, and transplantation.

plant casein see AVENIN.

plant cost *Graphic Arts.* any factor in print production that is not affected by the size of an edition or press run, including payments to writers and artists, illustration and design costs, typesetting, platemaking, and the like. *Industrial Engineering.* see FIXED COST.

plant decomposition *Control Systems.* a method of partitioning a large-scale control system into a number of subsystems.

Plante cell *Electricity.* a lead-acid cell in which the active material forms on plates by electrochemical means over repeated charging and discharging intervals, rather than by the application of a prepared paste.

planter *Agriculture.* **1.** a machine that plants seeds in the ground in a single row. **2.** a person who owns or manages a plantation.

plant fermentation *Biochemistry.* a process in which carbohydrates are partially broken down without consuming molecular oxygen.

plant fiber *Textiles.* any textile fiber produced by or derived from plants, such as cotton, flax, or hemp.

plant geography *Botany.* the branch of botanical science concerned with the spatial distribution of plants. Also, GEOGRAPHICAL BOTANY, PHYTOCHOROLOGY, PHYTOGEOGRAPHY.
plant growth substance *Biotechnology.* an organic regulatory substance that at low concentrations promotes, inhibits, or alters plant growth. Also, **plant growth regulator.**
plant hormone *Biochemistry.* any chemical substance produced in minute amounts in one part of a plant and transported to another part of the plant, where it has a specific effect. Also, PHYTOHORMONE.
plantigrade *Vertebrate Zoology.* describing a foot posture in which the entire foot is in contact with the ground, as opposed to digitigrade; common in humans, bears, and many insectivores.
plant key *Botany.* a method of plant identification through analysis of contrasting characters; used to subdivide a particular plant group into branches.
plant kingdom *Botany.* a primary division of living things consisting of all plants, especially as used in contrast to the animal and mineral kingdoms. See PLANTAE. Also, VEGETABLE KINGDOM.
plant pathology *Botany.* the branch of botanical science concerned with the diseases of plants. Thus, **plant pathologist.**

Plant Pathology

Plant pathology is the study of disease and its control in plants. Plant pathology should more properly be called plant medicine, as it is for plants what medicine is for humans and veterinary medicine is for animals. Plant pathology studies the microbes and the abiotic (environmental) factors that cause disease in plants, the mechanisms and dynamics of disease development, and the methods of preventing and controlling disease.

Plant diseases have been affecting the survival and welfare of humans since prehistoric times. Crop failures caused by blasting and mildew were considered, along with human diseases and war, among the great scourges of mankind. Plant diseases have caused shortages of food of varying severity throughout history and in many developing countries they continue to do so to this date. Besides reducing available food, plant diseases commonly reduce incomes for farmers, increase food prices for consumers, may wipe out entire plant species (e.g., the American chestnut), make foodstuffs unfit to eat, and result in pollution of the environment with pesticides.

Plant pathology as a science began about 1850 AD when first fungi and later bacteria, viruses, and nematodes were shown to cause plant diseases. Since then, great improvements have been made in detection and diagnostic techniques, in the physiology, epidemiology, and genetics of disease development, and in cultural, biological, and chemical measures for controlling plant diseases. The increasing world population and improving standards of living demand ever greater amounts of food products, to be produced by genetically uniform crop plants cultivated at high densities but with reduced application of pesticides. These conditions increase plant diseases and the need for their control. Genetic engineering and biotechnology already contribute to plant disease control. The challenge to plant pathology is just beginning to take on a worldwide significance as the race intensifies to produce more, and more wholesome, food while still safeguarding our environment.

George N. Agrios
Professor and Chairman
Department of Plant Pathology
The University of Florida

plant physiology *Botany.* the branch of botanical science concerned with the life processes of plants.
plant protease concentrate see BROMELAIN.
plant reovirus group *Virology.* a term sometimes used to refer to the Phytoreovirus and Fijivirus groups, collectively.
plant rhabdovirus group *Virology.* a group of plant viruses of the family Rhabdoviridae that have bacilliform particles and can replicate in insect vectors.
plant sociology see PHYTOSOCIOLOGY.
plant tissue culture *Biotechnology.* a technique used to grow undifferentiated plant cells on synthetic medium for experimentation or commercial purposes.
plant toxicant *Toxicology.* any poison occurring naturally in a plant.
plantula *Entomology.* a small, cushionlike, sometimes adhesive structure on the undersurface of tarsal segments in certain insects.
plant virus *Virology.* a virus that can infect and replicate in plant cells; one of three major types of viruses, with animal viruses and phages.
planula *Invertebrate Zoology.* the ciliated, free-swimming larva of a coelenterate.
Planuloidea *Invertebrate Zoology.* a former name for Mesozoa.
plaque [plak] *Medicine.* **1.** any patch or flat area on a body part or surface. **2.** also, **dental plaque.** a soft, thin film of food debris, mucin, and dead epithelial cells deposited on the teeth, providing a medium for bacteria growth. *Virology.* an identifiable area in a cell monolayer or bacterial growth layer due to lysogenic or growth-inhibiting infection by a virus. *Veterinary Medicine.* a disease caused by *Yeroinia pestis,* affecting mainly the lymphatics and lungs, found in cats and dogs and now recognized as hazardous to humans. *Biotechnology.* an area of clearing in a bacterial lawn on an agar plate, caused by viral lysis of the bacteria cells.
plaque assay *Virology.* a method of determining the concentration of infective particles by calculating the number of plaques induced on a lawn of cells or bacteria in a viral solution.
plaque-forming units *Virology.* in a plaque assay, the concentration of plaques per unit of virus suspension.
plaque mutant *Virology.* a mutant virus producing plaques that are different in size or appearance from those produced by the wild-type virus. Also, **plaque-type mutant.**
plaque neutralization test *Virology.* a process in which a virus is identified or the concentration of a specific antibody in antiserum is measured by means of a plaque assay. Also, **plaque reduction test.**
plaque picking *Virology.* the selection of a virus from an individual plaque arising from progeny of a single virus precursor.
plas- or **plaso-** a prefix meaning "formation," "development."
plash *Hydrology.* a puddle, pool, or small pond of standing water that forms after a flood or heavy rain, or from melting snow.
plasm- or **plasmo-** a prefix meaning "the basic substance of a cell."
plasma [plaz´mə] *Histology.* the fluid portion of the blood or of lymph. *Physics.* a gas consisting of ions, electrons, and neutral particles; the behavior of the gas is dominated by the electromagnetic interaction between the charged particles. *Geology.* a portion of a soil material that has the potential to be, or that has been, moved, reorganized, or concentrated by soil-forming processes. *Mineralogy.* a green, translucent or semitranslucent variety of chalcedonic quartz, sometimes having white or yellow spots. (From a Greek word meaning "to mold or shape.")
plasma accelerator *Physics.* a device that is capable of producing a high-velocity plasma stream by means of magnetic field configurations.
plasma-arc cutting *Metallurgy.* a method used in cutting metals, in which an area is melted with an arc, and the metal is then cut by high-velocity, high-temperature ionized gas. Also, **plasma-arc welding.**
plasma cathode *Electronics.* a cathode in which ionized gas is used as a source of electrons instead of a solid material.
plasma cloud *Astrophysics.* a moving cloud of ionized gases caught in the solar wind.
plasmacyte *Histology.* an agranular, basophilic leukocyte that is active in antibody production. Also, **plasma cell.**
plasmacytoma *Oncology.* a tumor composed of plasmacytes. Also, PLASMOMA.
plasma diode *Electronics.* **1.** a tube that generates energy when heat is applied to a space between two electrodes which is filled with cesium vapor. **2.** a tube in which an ionized gas generates electricity that conducts in only one direction.
plasma display *Electronics.* a system for presenting data that produces one element of a dot-matrix display from the interactions of conductors and gas deposited onto a glass plate.

plasma engine *Space Technology.* an electrical-reaction engine in which magnetically accelerated plasma is used as the propellant. Also, **plasma jet.**

plasmagel *Cell Biology.* a firm, viscous phase of the cell cytoplasm, usually located at the outer extremity of a pseudopodium, that can reversibly transform into plasmasol.

plasmagene *Cell Biology.* any self-replicating gene located in the cytoplasm of a cell rather than in the nucleus, including genetic material found within endosymbionts or organelles such as mitochondria.

plasma generator *Electronics.* any device, such as a plasma accelerator, plasma engine, or plasma torch, that generates a stream of high-velocity electrons and positive ions.

plasma gun *Electronics.* a machine that converts neutral gases into plasma by exposing them to intense thermionic flux.

plasma heating *Physics.* the increase of random thermal motion in a plasma by a variety of methods, such as ohmic heating, adiabatic compression, ion-wave or electron-wave heating, beam injection, and magnetic pumping.

plasma instability *Physics.* a sudden deformation in the quasi-static velocity or position distribution in a plasma as well as an abrupt change in the associated electromagnetic field.

plasma jet *Space Technology.* **1.** a plasma engine. **2.** the jetstream ejected by such an engine.

plasmalogen *Biochemistry.* one of a group of glycerol-based phospholipids in which the aliphatic side chains are not connected by ester linkages; a high percentage are found in the central nervous system.

plasma mantle *Geophysics.* a layer of plasma found immediately below the magnetopause, having a tailward flow of speeds from 100 to 200 km per second, and showing a decrease of density, temperature, and speed with the increase of depth inside the magnetosphere.

plasma membrane *Cell Biology.* a protein-containing lipid bilayer that surrounds a cell, defining the interface between the cell and its environment and providing a semipermeable barrier to the entry of molecules into the cell. Also, CELL MEMBRANE.

plasmapause *Geophysics.* the outermost point of the plasmasphere at which plasma density decreases sharply by a factor of 100 and more.

plasmapheresis *Medicine.* the removal of plasma from withdrawn blood, and reinjection of the packed cells; the procedure may be done as a therapeutic measure or to collect plasma components.

plasma physics *Physics.* a branch of physics that studies the properties and interactions of highly ionized gases.

plasma protein *Hematology.* any one of the hundreds of different proteins present in blood plasma, including carrier proteins, fibrinogen and other coagulation factors, immunoglobulins, enzyme inhibitors, and many other types of proteins.

plasma radiation *Physics.* the electromagnetic radiation emitted from a plasma, most of which is due to the free-free electron transitions.

plasma rocket *Space Technology.* a rocket propelled by a plasma engine. Also, ELECTROMAGNETIC ROCKET.

plasma sheath *Electronics.* an envelope of ionized gas that engulfs a spacecraft, moving through the atmosphere at hypersonic speed, and interferes with the transmission and reception of radio signals.

plasma sheet *Geophysics.* an area of hot plasma that starts at about 30,000 miles and reaches the orbital path of the moon; composed of particles whose thermal energies typically are 2 to 4 keV.

plasmasol *Cell Biology.* a fluid, soluble phase of the cell cytoplasm that can transform from and into plasmagel.

plasmasphere *Geophysics.* an area of relatively dense plasma surrounding the earth, extending 2 to 6 earth radii and composed mostly of protons and electrons having thermal energies not more than several eV.

plasma spraying *Materials Science.* a process that forms a dense, pore-free, polymer powder coating, using a torch that produces and controls a high-velocity inert gas stream (plasma) at temperatures in the 2500–8000 K range.

plasma tail *Astronomy.* the portion of a comet's tail composed of ionized gas.

plasmatherapy *Hematology.* the therapeutic use of blood plasma.

plasmatic *Hematology.* relating to or like plasma.

plasmatocytes *Entomology.* the most active white blood cells in insects; they can spread over surfaces of parasites forming a membranous capsule or become granular.

plasma torch *Engineering.* a torch used for cutting metals, in which a plasma gas is forced through an arc in a water-cooled tube, with consequent ionization and very high temperatures.

plasmatorrhexis *Cell Biology.* the bursting of a cell due to pressure exerted from within.

plasmatron *Electronics.* a gas-discharge tube in which a neutral gas, usually helium, is ionized so that it will conduct electricity across electrodes and, under certain conditions, amplify microwave frequencies.

Plasmaviridae *Virology.* a family of pleomorphic mycoplasma viruses having a circular, supercoiled dsDNA genome.

Plasmavirus *Virology.* the single genus of the family Plasmaviridae.

plasma wave *Physics.* a disturbance that propagates through a plasma, initiated by a localized displacement of charged particles in the plasma.

plasmic *Hematology.* **1.** see PLASMATIC. **2.** rich in protoplasm.

plasmid *Genetics.* a small, closed entity of double-stranded DNA forming an extrachromosomal self-replicating genetic element in many bacteria and some eukaryotes, often carrying genetic sequences that give the host cell a survival advantage such as resistance to antibiotics; widely used in genetic engineering as a cloning vector.

Plasmids

Plasmid: the etymology is -plasm, "living tissue," from *cytoplasm*; and -id, "diminutive," by analogy with *chromatic* and *plastic*. This term was defined in 1952 to embrace any extra-chromosomal hereditary determinant. This was intended to defuse the operationally vacuous dispute as to whether certain entities were "viruses" or "symbionts," on the one hand, or "genes" on the other. In fact, *plasmid* overlaps *virus* and *symbiont*: when either of the latter is inherited by the progeny of a given cell, and has a phenotypic effect, it may best be thought of as a plasmid.

Indeed, some bacteriophages (e.g., *Escherichia coli* P1) and many animal viruses are vertically transmitted in just this way. On the other hand, plasmids may move in and out of the chromosome (e.g., *E. coli* l bacteriophage, or the F plasmid), in which case they can be described as episomes. In 1990, 739 articles published had "plasmid" in their title, and the journal *Plasmid* is devoted to these entities.

Another fruit of this redefinition has been the reexamination of the origin of mitochondria and chloroplasts—it is now generally accepted that these have come from symbiotic bacteria or cyanophytes.

The first plasmids to be extensively studied were the F(ertility) and RTF (resistance transfer factors) in *E. coli.* These are fairly small circular DNA molecules, and plasmid is often used in this narrower sense. Such plasmids have been indispensable workpieces in biotechnology. On the other hand, some megaplasmids are a large fraction of the normal chromosome; and the boundary between plasmid and chromosome segment depends on how one chooses to define the chromosome (e.g., minute and double minute chromosomes). Plasmids have been found to inhabit mitochondria in some fungi, and bits of DNA in the nucleus that were not regularly segregated like chromosomes might be considered intranuclear plasmids. While most plasmids are circular, some are linear (as indeed are some bacterial chromosomes).

Joshua Lederberg
Professor and President-Emeritus
The Rockefeller University
Nobel Laureate in Physiology or Medicine

plasmid rescue *Genetics.* the beneficial transformation of a bacterium by its incorporation of plasmid genes into the genome.
plasmin *Biochemistry.* an enzyme, produced by plasminogen, that removes fibrin clots from the blood stream.
plasminogen *Biochemistry.* a plasma protein that changes into plasmin when hydrolyzed by an Arg–Val bond. Also, PROFIBRINOLYSIN, PROPLASMIN.
plasminogen activator see UROKINASE.
plasmodesma *Cell Biology.* a narrow cytoplasmic channel connecting adjacent plant cells through the intervening cell wall.
plasmodial *Microbiology.* of or relating to the genus *Plasmodium.*
Plasmodiidae *Invertebrate Zoology.* a family of protozoans in the order Haemosporina, parasites in the blood of vertebrates, transmitted by mosquitos; includes the malaria parasite.
Plasmodiophoraceae *Mycology.* the single family of fungi belonging to the order Plasmodiophorales, which is composed of species that are pathogenic to plants, causing galls.
Plasmodiophorales *Mycology.* the single order of fungi belonging to the class Plasmodiophoromycetes, which parasitize algae, aquatic plants, and aquatic fungi; they are found in aquatic or semiaquatic environments.
Plasmodiophorida *Invertebrate Zoology.* an order of slime molds in the protozoan subclass Mycetozoa, endoparasites of plant roots; sometimes classified as fungi.
Plasmodiophoromycetes *Mycology.* a class of fungi that is currently listed within the division Myxomycota, but whose taxonomic association is not yet clear; they cause diseases in certain aquatic fungi, algae, and plants.
Plasmodium *Microbiology.* the malarial parasites, a genus of coccoidan protozoa belonging to the suborder Haemosporina of the order Eucoccidiida and parasitic in the erythrocytes of mammals. Several species, including *P. falciparum* and *P. malaria,* cause malaria in humans.

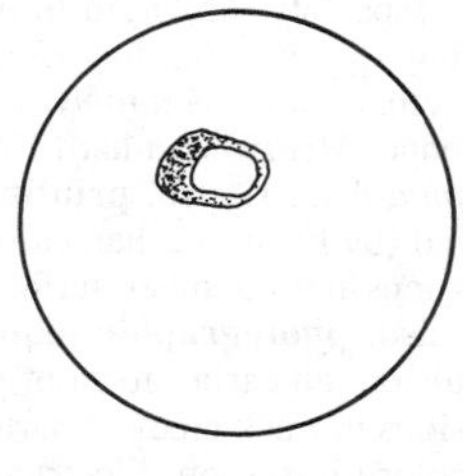

Plasmodium

plasmodium *Microbiology.* **1.** a motile mass of protoplasm varying in size and form and containing numerous nuclei. **2.** a parasite of the genus *Plasmodium.*
Plasmodroma *Invertebrate Zoology.* in some classification systems, a subphylum of the phylum Protozoa, including Mastigophora, Sarcodina, and Sporozoa.
plasmogamy *Invertebrate Zoology.* the joining of individual cells without fusion of nuclei in the formation of a plasmodium, especially in slime molds.
plasmoid *Physics.* an isolated aggregate of charged and neutral particles that is capable of maintaining its identity for periods much longer than the time between collisions of the constituent particles.
plasmolysis *Physiology.* **1.** broadly, the dissolution of cell parts. **2.** specifically, the contraction or shrinking of a plant cell membrane away from the cell wall after osmotic loss of water.
plasmoma see PLASMACYTOMA.
plasmon *Genetics.* the sum of extrachromosomal hereditary factors in a given cell, including the mitochondria, the protoplasts, and any plasmids or other genetic factors. *Solid-State Physics.* a quantum of excitation of a longitudinal mode-wave character in the electron gas of a metal.
Plasmopara *Mycology.* a genus of fungi belonging to the order Peronosporales and causing downy mildew in such plants as grapes.
plasmotomy *Invertebrate Zoology.* the division of a multinucleate plasmodium into separate individuals without fission of the nuclei, especially in slime molds.
plast- or **plasto-** a prefix meaning "repair."
plaster *Materials.* **1.** a pasty mixture, generally of lime or gypsum with water and sand, that is used as a coating on walls and ceilings. **2.** see PLASTER OF PARIS. **3.** a pasty mixture, usually containing herbs or other medicinal substances, that is spread on the body, either directly or after being saturated into a base material such as cloth. **4.** to apply any of these materials.
plasterboard *Materials.* a material consisting of paper-covered sheets of gypsum and felt; used for insulating or covering walls, and as a lath.
plasterboard nail *Mechanical Devices.* a nail with a barbed or ringed shank and a larger head; used for securing plasterboard to framing.
plaster coat *Building Engineering.* a thin layer of plaster that covers walls throughout buildings or other structures.
plaster conglomerate *Geology.* a conglomerate made up entirely of boulders derived from a partially uncovered monadnock, forming a wedgelike mass on the side of the monadnock.
plaster ground *Building Engineering.* a wood piece fabricated to limit the thickness of a plaster coat applied to a wall; used primarily around windows and doors and above the floor strip on a wall.
plaster of Paris *Inorganic Chemistry.* $(CaSO_4)_2 \cdot H_2O$, a white noncombustible powder; slightly soluble in water and soluble in acid; loses 1.5 water at 128°C and becomes anhydrous at 163°C; used in tile, plaster, paints, and paper, and in metallurgy. *Materials.* a commercial form of this powder, composed primarily of the hemihydrate of calcium sulfate, made by calcining gypsum until it is partially dehydrated; the powder is mixed with water to form a paste that sets to form casts, art objects, or building materials. (Originally prepared from the gypsum of *Paris,* France.) Also, DRIED GYPSUM, CALCINED GYPSUM.
plaster shooting *Engineering.* a surface blasting process in which a charge of gelignite, primed with a safety fuse and detonator, is placed in close contact with a rock or boulder and covered completely with stiff, damp clay.
plaster slab *Building Engineering.* a perforated block made from plaster of Paris; used in partition construction.
plaster slap *Building Engineering.* a mixture of plaster of Paris and coarse sand that is used for coarse plastering, as between studs in a wall.
plastic *Materials Science.* **1.** any of various synthetic or organic materials that can be molded or shaped, generally when heated, and then hardened into a desired form; for example, polymers, resins, and cellulose derivatives. Plastics are used in virtually all areas of industrial technology; among the major applications are packages and containers, construction materials, paints, adhesives, pipelines, automobile parts, electronic components, textiles, medical and scientific devices, and various consumer items such as toys and houseware. **2.** relating to or composed of such a material. Thus, **plastic cement, plastic clay, plastic film, plastic paint, plastic wood**, and so on. *Chemistry.* describing any substance or material that is capable of being shaped or molded, with or without the application of heat. *Mechanics.* able to be deformed in any direction and to retain this deformed shape without rupture. *Surgery.* serving to shape tissues or restore a lost part, as in plastic surgery.
plastic anisotropy *Materials Science.* variations in directions in the resistance of a material to plastic deformation.
plasticate *Engineering.* to make a material malleable with a heating or kneading device. Also, PLASTIFY.
plastic bonding *Engineering.* any process used to join plastics; usually involving heating, adhesives, pressure, solvents, or radio frequency.
plastic bronze *Metallurgy.* a copper-base alloy containing 30% lead; used for bearings.
plastic clay *Materials.* a brick earth composed of silica and alumina combined with a small percentage of lime, magnesia, soda, or other salts. Also, FOUL CLAY.
plastic clot *Hematology.* a clot formed from the intima of an artery at the point of ligation, permanently obstructing the artery.
plastic deformation *Materials Science.* the permanent deformation arising from the relative displacement of atoms or molecules. *Geology.* a metamorphic process involving the recrystallization of previously existing minerals under pressure and stress into new minerals. *Mechanics.* see PERMANENT SET.
plastic equilibrium *Geology.* a condition of stress within a soil mass or a portion of a soil mass in which extensive deformation results in the mobilization of its ultimate resistance.
plastic-film capacitor *Electricity.* a capacitor made by rolling alternate layers of aluminum foil and plastic dielectric into a compact tube much like a paper capacitor; the foiled strips are then staggered to project alternately from the row ends, and leads are soldered to the foil. Also, FILM CAPACITOR.

plastic flow *Physics.* a phenomenon in which solids may undergo a variety of extensive, irreversible deformations after the applied stress reaches a critical value.

Plasticine *Materials.* the trade name for a synthetic substitute for modeling clay.

plastic instability *Materials Science.* severe localization of the plastic deformation in materials under tensile stresses (necking); because of the decreased cross-sectional area, the true stress increases tend to fracture them.

plasticity *Mechanics.* the fact of being plastic; the property of a body by which it undergoes plastic deformation when the applied stress exceeds a certain value, known as the yield value.

plasticity index *Geology.* the range of water content at which a soil behaves plastically, equal to the percentage difference between the liquid limit and the plastic limit.

plasticization *Materials Science.* a process used to lower glass transition temperature; it improves the flexibility of certain polymers, allowing them to remain flexible well below the glass transition temperature of unplasticized materials.

plasticize *Engineering.* to make a material malleable by mixing it with a plasticizer, or by applying heat.

plasticizer *Engineering.* a material, usually organic, that is capable of imparting flexibility to nonplastic material or improving the flexibility of ceramic mixtures. *Materials Science.* see WATER REDUCER.

plasticlast *Geology.* an intraclast made up of calcareous mud that was fragmented while still soft.

plastic limit *Geology.* the minimum amount of water mixed with a given sediment or soil that enables the soil to be rolled into a thin thread, without breaking the thread.

plasticorder *Engineering.* an instrument that measures the temperature, viscosity, and shear-rate relationships of a plastic substance to determine its eventual behavior.

plasticoviscosity *Mechanics.* the property of a material whose rate of plastic deformation when subjected to stresses exceeding the yield stress is a linear function of the applied stresses.

plastic paint *Materials.* a thick-texture paint that can be worked to a patterned finish.

plastic plate *Electronics.* a collection of dielectric materials used as a base for a semiconductor device. *Graphic Arts.* **1.** a direct offset-printing plate made on paper or other material coated with a thermoplastic vinyl resin; used for short, medium-quality press runs. **2.** a molded printing plate made by pouring thermosetting plastic powder into a plastic mold; used for long press runs of type and simple line art.

plastics *Materials Science.* materials of high molecular weight that consist primarily of synthetic polymers or condensates, which can be shaped by flow into objects of diverse shapes and sizes.

plastic strain *Materials Science.* a permanent displacement of material, as in slip or twinning; the displacement remains after the stress has been removed.

plastic surgeon *Surgery.* a specialist in plastic surgery.

plastic surgery *Surgery.* a surgical procedure to repair, remodel, or restore defective or injured tissue or body parts or to improve their shape or appearance. Also, **plastic operation.**

Plastic Wood *Materials.* the trade name for a compound that is used to patch and fill woodwork.

plastic wrap *Materials.* a thin, transparent sheet of plastic that can cling to other substances; used to wrap and store food and for microwave cooking.

plastic zone *Geology.* in an explosion crater, a region bordering the rupture zone at an increased distance from the shock site, having less fracturing than and only small permanent deformations in comparison to the rupture zone. *Materials Science.* a heavily plastically deformed region in a material, generally adjacent to the tip of a crack that can cause crack-tip blunting.

plastid *Cell Biology.* any of a number of membrane-bound organelles found in plant cells and performing a specific function for the cell, such as a photosynthetic chloroplast.

plastify see PLASTICATE.

plastisol *Materials.* a dispersion of resin in a plasticizer that gels when heated.

plastocyanin *Biochemistry.* a blue, copper protein that transports electrons in chloroplast membranes from photosystem I to photosystem II during photosynthesis.

plastogene *Cell Biology.* a gene located in a plastid of a plant cell rather than in the nucleus.

plastoglobuli *Biochemistry.* a group of globules, found in plastids, that contain principally lipid.

plastome *Molecular Biology.* the genetic complement of a plastid.

plastometer *Engineering.* **1.** an instrument used to measure the viscosity or flexibility of a material. **2.** an instrument that measures the flow characteristics of a thermoplastic resin as it moves through an orifice at a certain pressure and temperature. **3.** a machine for determining the stress and strain properties of metals at high temperatures and at various rates of strain.

plastron *Vertebrate Zoology.* the ventral portion of the shell of a turtle or tortoise, composed of four bony plates and covered with epidermal scales. *Invertebrate Zoology.* **1.** in some adult aquatic insects, a thin layer of gases held in place by extremely fine, charged hairs on the body surface, to allow for respiration when the insect is submerged. **2.** in heart urchins, a modified plate on the underside. **3.** in spiders, a ventral plate on the cephalothorax.

plat *Cartography.* a plan drawn to scale that shows the boundaries and subdivisions of a piece of land, intended for use during development or sale of that land, and not necessarily showing other planimetric, relief, or cultural detail.

Plata see RIO DE PLATA.

Platacidae *Vertebrate Zoology.* an equivalent name for Ogcocephalidae, a family of marine fishes commonly known as batfishes.

Platanaceae *Botany.* a family of monoecious dicotyledonous trees in the order Hamamelidales, having simple deciduous leaves, large and conspicuous stipules, and small, densely clustered fruit.

Platanistidae *Vertebrate Zoology.* the freshwater or river dolphins, a family of toothed whales of the order Odontoceta, characterized by a long, slender rostrum, and a prominent bulge on the forehead; found in southern Asia and South America.

Plataspidae *Invertebrate Zoology.* shiny, oval, true bugs, a family of hemipteran insects in the superfamily Pentatomoidea.

platband *Architecture.* **1.** a flat structural member, as a lintel or flat band. **2.** a shallow molding having a flat face.

plate any of various thin, flat objects or devices, such as the large, circular dish on which food is served; specific uses include: *Metallurgy.* a thin, flat piece of metal, especially one used to provide support or protection of a surface, fitting, or joint. *Building Engineering.* a base member, as of a partition or other frame. *Anatomy.* a flat structure or layer, such as a thin layer of bone. *Medicine.* a hard fitting to which artificial teeth are attached. *Graphic Arts.* **1.** also, **printing plate.** a surface that can be etched or engraved (by hand, mechanically, or photographically) and from which impressions onto another surface can be made for the purpose of printing. **2.** also, **photographic plate.** a glass plate coated with photosensitive emulsion; an earlier form of photographic film, still used with nuclear emulsions to track charged partricles. **3.** an illustration in a book, especially a color illustration. *Electricity.* **1.** the electrode in a cell, battery, or tube toward which current flows or to which electrons are attracted; an anode. **2.** of or relating to an anode. Thus, **plate current, plate efficiency, plate modulation, plate neutralization, plate power, plate saturation,** and so on. *Horology.* the solid base onto which are mounted the wheels, pinions, springs, and screws that compose the movement of a timepiece. *Geology.* **1.** one of several large, mobile blocks of continental or oceanic crust, together with some portion of the asthenosphere, that move as a single, nearly rigid unit. **2.** a hard, smooth, thin, flat fragment of rock or stone, such as a flagstone.

plate amalgamation *Metallurgy.* a metallic plate used for amalgamating gold with mercury.

plateau *plural,* **plateaus** or **plateaux.** *Geology.* an extensive, nearly level land area, with at least one steep side, that is higher than the surrounding area and usually at least 2000 feet above sea level. *Electronics.* the point in a response curve where an increase in the independent variable no longer affects the dependent variables. (From the French word for "plate.")

plateau basalt *Geology.* a large, extensive basaltic lava flow or series of flows from fissure eruptions that accumulate to form a plateau. Also, FLOOD BASALT.

plateau characteristic *Electronics.* a relationship between two variables in which the dependent variable reaches a value that does not change with further increase of the independent variable.

plateau glacier *Hydrology.* a glacier formed on a mountain plateau, usually overflowing the edges in hanging glaciers.

plateau gravel *Geology.* a sheet or patch of surficial gravel on a plateau or other region above the height at which stream-terrace gravel is usually found.

plateau level *Petroleum Engineering.* the highest level of production attained by an oil field.
plateau mesa see MESA.
plateau mountain *Geology.* a mountain formed by gradual differential erosion and sculpturing of a plateau, rather than by uplift of crustal deformation.
plateau plain *Geology.* an extensive plain bordered by escarpments from which a sublevel summit region rises.
plateau problem *Mathematics.* the problem of finding a minimal surface with a given (nonplanar) curve as boundary.
plateau ring structure *Astronomy.* a former term for a circular structure with an elevated floor, located on the moon.
plate battery *Electricity.* a battery that supplies the DC plate battery of an electron tube. Also, B BATTERY.
plate-bearing test *Engineering.* a method used to estimate the load-bearing capacity of a soil by adding weight to a 1 square foot plate at the site.
plate budding *Botany.* in grafting, the insertion of a rectangular cutting containing a bud underneath a flap of bark on the stock, covering all exposed wood on the stock.
plate burning *Graphic Arts.* the use of a photographic process to create an impression on a printing plate.
plate cam *Mechanical Engineering.* a flat open cam for sliding movement.
plate center *Astronomy.* the point on a photographic plate that corresponds to the optical axis of a telescope.
plate circuit see ANODE CIRCUIT.
plate coil *Mechanical Engineering.* a coil constructed of two metal sheets fastened together, in which one or both of the sheets are embossed to form a passage through which heat flows. Also, PANEL COIL.
plate constants *Astronomy.* numbers that transform the measured coordinates of a star's image on a plate into standard coordinates by means of mathematical formulas.
plate conveyor *Mechanical Engineering.* a conveyor with a series of trough-shaped steel plates joined together with a slight overlap to form an articulated band.
plate crystal *Hydrology.* an ice crystal that exhibits typical hexagonal symmetry, but has relatively little thickness parallel to its principal axis.
plate current see ANODE CURRENT.
plate-current shift *Electricity.* a change in the DC plate current of an rf amplifier during amplitude modulation.
plate cylinder *Graphic Arts.* on a rotary printing press, the metal cylinder around which the printing plate is wrapped.
plated circuit *Electronics.* a printed circuit that is made by etching a conductive pattern onto an insulating base.
plate dissipation see ANODE DISSIPATION.
platedog *Graphic Arts.* on a rotary press, the heavy metal plate on which plates are locked into position for printing.
plated wire memory *Computer Technology.* an early storage device designed to store data on wires coated with magnetic film; replaced by semiconductor technology.
plate efficiency see ANODE EFFICIENCY.
plate-fin exchanger *Mechanical Engineering.* a layered heat-transfer device in which each layer consists of a fin between flat metal sheets sealed off on two sides by bars to form a passage for the flow of fluids.
plate floor *Building Engineering.* a reinforced-concrete floor, often used in office buildings because of its flexibility in altering the shape and size of partition rooms.
plate girder *Civil Engineering.* a girder formed by a web plate running between parallel angle pieces.
plate girder bridge *Civil Engineering.* a support structure made functional by a series of interlocking plate girders.
plateglass *Materials Science.* a high-quality, pressed glass that has been ground and polished to remove defects; used in large windows, mirrors, and other applications.
plate-grid capacitance *Electricity.* the internal capacitance between the plate and the control grid of an electron tube. Also, INTERELECTRODE CAPACITANCE, FEEDBACK CAPACITANCE.
plate iceplate impedance see ANODE IMPEDANCE.
plate input power see ANODE INPUT POWER.
platelet [plāt′lit] *Histology.* a cytoplasmic fragment that occurs in the blood of vertebrates and is associated with blood clotting. Also, THROMBOCYTE. *Hydrology.* one of the small ice crystals that join together to form a layer of floating ice and serve as seed crystals for the further thickening of the ice cover.
platelet-activating factor *Endocrinology.* an ether lipid autacoid that is released from the membrane phospholipids of platelets, certain cells of the hemopoietic system, and vascular endothelium; its physiological actions include vasodilation, bronchoconstriction, decreasing of renal blood flow, induction of platelet aggregation, and activation of neutrophils.
platelet-derived growth factor *Endocrinology.* a growth factor, originally purified from platelets, that is produced by the association of two polypeptides (PDGF-A and PDGF-B) into a dimeric configuration; a trophic factor for cells of mesodermal origin and neuroectodermal origin, including smooth muscle, fibroblasts, and glia. *Hematology.* this substance, contained in the alpha granules of platelets, whose action contributes to the repair of damaged vascular walls.
plate-load impedance see ANODE IMPEDANCE.
platemaking *Graphic Arts.* any process used to produce a printing plate.
plate modulation see ANODE MODULATION.
platen *Engineering.* a part of a machine with a flat surface that transmits pressure, and to which dies, fixtures, backups, or electrode holders are attached. *Graphic Arts.* **1.** the flat surface upon which paper rests as it comes into contact with the ink-bearing surface of a press, especially a platen press. **2.** a hard roll in a typewriter that acts as a backing against which paper is pressed when it is struck by the type keys. *Mechanical Engineering.* a flat surface for exchanging heat in a boiler or heat exchanger.
plate neutralization see ANODE NEUTRALIZATION.
platen press *Graphic Arts.* a printing press that squeezes together two flat surfaces, one bearing the paper and the other an inked form.
plate proof *Graphic Arts.* an impression taken from a finished plate, used to check type and art before a reproduction proof (repro) is made.
plate pulse modulation see ANODE PULSE MODULATION.
plate resistance see ANODE RESISTANCE.
plate saturation see ANODE SATURATION.
plate scale *Astronomy.* the degrees, minutes, or seconds of arc per millimeter on a photographic plate.
plate spacing *Electricity.* the difference between plates in a fixed or variable capacitor; in a fixed capacitor, it is the same thickness as the dielectric thickness. Also, CAPACITOR AIR GAP.
plate tectonics *Geology.* a modern geological theory of tectonic activity according to which the earth's crust is divided into a small number of large, rigid plates whose independent movements relative to one another cause deformation, volcanism, and seismic activity along their margins.

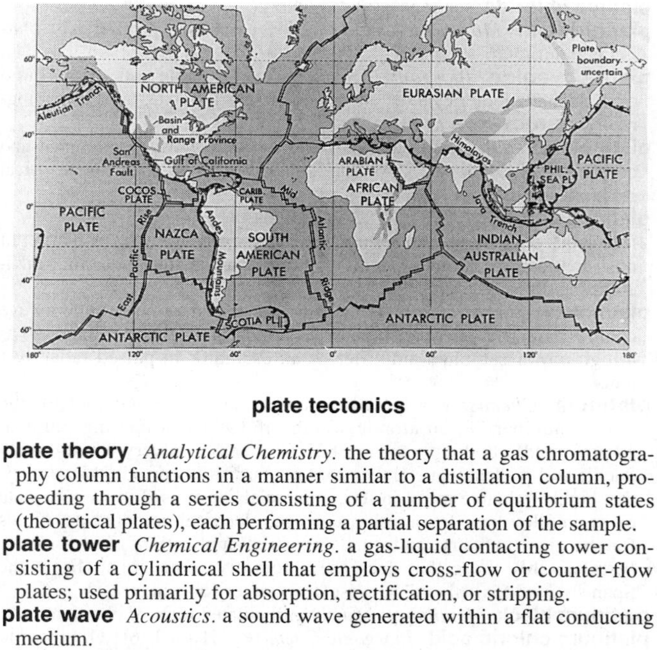

plate tectonics

plate theory *Analytical Chemistry.* the theory that a gas chromatography column functions in a manner similar to a distillation column, proceeding through a series consisting of a number of equilibrium states (theoretical plates), each performing a partial separation of the sample.
plate tower *Chemical Engineering.* a gas-liquid contacting tower consisting of a cylindrical shell that employs cross-flow or counter-flow plates; used primarily for absorption, rectification, or stripping.
plate wave *Acoustics.* a sound wave generated within a flat conducting medium.
plate winding *Electricity.* **1.** an inductor that is connected in series between the plate of a vacuum tube and the positive power-supply voltage. **2.** in a plate-circuit output transformer, the primary winding.

platform a horizontal surface raised above the level of the surrounding area; specific uses include: *Ordnance.* **1.** a solid, flat support for an artillery piece; it may be temporary or permanent. **2.** a metal support at the base of some guns upon which the crew stands while operating the gun. see STRIKING PLATFORM. *Geology.* **1.** any level or approximately level surface or flat, elevated ground, such as a terrace or a plateau. **2.** that part of a continental craton consisting of stable basement rock covered by nearly horizontal or flat-lying sedimentary strata. **3.** a level or gently sloping erosion surface that extends underwater from the shore out to the sea or a lake. *Mining Engineering.* a wooden floor on the side of a gangway, situated at the bottom of an inclined seam, to which coal flows by gravity and from which the material may then be shoveled into mine cars.

platform balance *Engineering.* a weighing machine consisting of a flat plate mounted on a balance beam that holds the material to be weighed.

platform beach *Geology.* a sharply curved ridge or bar of sand and gravel that develops on a wave-cut platform.

platform blowing *Engineering.* a method of shaping large plastic parts using a blow-molding process in which the parts being molded do not sag.

platform conveyor *Mechanical Engineering.* a conveyor with continuously moving steel or hardwood platforms on which the loads are placed.

platform deck *Naval Architecture.* a partial deck below the lowest complete deck on a ship.

platform facies see SHELF FACIES.

platform frame *Building Engineering.* a building method in which floor platforms are framed independently to a one-story height in a multistory building, with each story having a box sill. Also, WESTERN FRAME.

platform gantry *Building Engineering.* a gantry fabricated to support a platform upon which a scaffold is built to handle materials.

platform mobility *Robotics.* the ability of a platform on which a robot is mounted to move horizontally.

platform reef *Geology.* a relatively small organic reef characterized by a flat upper surface.

platina *Metallurgy.* a zinc-base alloy containing 25% copper.

plating *Metallurgy.* the process of coating an object with a metallic material by deposition. *Biotechnology.* a technique used to grow pure cultures of microorganisms by streaking a petri dish with a sterile loop dipped into a suspension containing the microorganisms.

plating rack *Metallurgy.* the rack that holds parts being plated.

platinic *Chemistry.* of or containing platinum, especially in the tetravalent state.

platinic chloride see PLATINUM CHLORIDE.

platiniridium *Materials.* a natural alloy composed primarily of platinum and iridium.

platinocyanide *Inorganic Chemistry.* any double salt of platinum cyanide and another cyanide, such as $BaPt(CN)_4 \cdot 4H_2O$; used in photography. Also, CYANOPLATINITE.

platinoid *Metallurgy.* an alloy having a high electrical resistance and containing 62% copper, 22% zinc, and 15% nickel; used for electrical resistors and thermocouples.

platinous chloride see PLATINUM DICHLORIDE.

platinosis *Toxicology.* poisoning due to the exposure to or inhalation of dust containing platinum salts; symptoms may include breathing difficulties and skin allergies.

platinotron *Electronics.* a tube that produces and amplifies microwave energy, requiring a permanent magnet, as does a magnetron; it is used as a high-power saturated amplifier or an oscillator in pulsed radar systems.

platinum *Chemistry.* a metallic element having the symbol Pt, the atomic number 78, an atomic weight of 195.09, a melting point of 1769°C, and a boiling point of 3827°C; a heavy, silvery-white, highly ductile and malleable metal that does not tarnish or corrode. *Metallurgy.* this highly valuable metal in its commercial form; used in its pure state or as an alloy base for jewelry, coinage, electrical components, molds for the glass industry, and other industrial applications, and for catalysts, notably those that decrease automotive emissions. (From the Spanish word for "silver;" because of its silvery appearance.)

platinum black *Metallurgy.* a finely particulated form of platinum.

platinum chloric acid *Inorganic Chemistry.* $H_2PtCl_6 \cdot 6H_2O$, brown to red hygroscopic crystals; soluble in water and absolute alcohol, and very soluble in ether; melts at 60°C and decomposes above 115°C. Also, CHLOROPLATINIC ACID.

platinum chloride *Inorganic Chemistry.* $PtCl_4$, brown to red crystals; very soluble in water and slightly soluble in alcohol; decomposes at 370°C; used for electroplating, for microscopy, and as an analytical reagent. Also, PLATINUM TETRACHLORIDE, PLATINIC CHLORIDE.

platinum dichloride *Inorganic Chemistry.* $PtCl_2$, olive green, hexagonal crystals, very slightly soluble in water and insoluble in alcohol, decomposes at 581°C; it is used to make platinum salts. Also, PLATINOUS CHLORIDE.

platinum electrode *Physical Chemistry.* a platinum wire electrode that is used as a reference to measure the amount of electrical charge in an electrolytic solution.

platinum iodide *Inorganic Chemistry.* PtI_2, a black powder; insoluble in water and acid; decomposes at 360°C.

platinum-iridium alloy *Metallurgy.* any of several platinum-base alloys containing up to 30% iridium; used for jewelry, chemical equipment, thermocouples.

platinum metal *Chemistry.* any of six metals belonging to group VIII B of the Periodic Table, ruthenium, osmium, rhodium, iridium, palladium, and platinum, that occur naturally as alloys. Also, **platinum family, platinum group.**

platinum oxide *Inorganic Chemistry.* any compound composed of platinum and oxygen only; for example, platinous oxide, PtO, or platinic oxide, PtO_2.

platinum resistance thermometer *Metrology.* a highly accurate thermometer, composed partly of platinum, that provides the basis of the International Practical Temperature Scale of 1968 from 259.35° to 630.74°C; used in industrial thermometers in the 0° to 650°C range. Also, CALLENDER THERMOMETER.

platinum-rhodium alloy *Metallurgy.* any of several platinum-base alloys containing 3 to 40% rhodium; used for high-temperature applications, catalysts, and thermocouples.

platinum sponge *Metallurgy.* a very porous platinum material used as a catalyst.

platinum sulfate *Inorganic Chemistry.* $Pt(SO_4)_2 \cdot H_2O$, toxic, hygroscopic, yellow plates that are soluble in water and alcohol; used in microtesting.

platinum tetrachloride see PLATINUM CHLORIDE.

Plato 427–347 BC, Greek philosopher; founded the Academy; established the central role of mathematics in scientific education; applied mathematics to astronomy.

platomaxillary arch *Anatomy.* the palate and maxillary bones.

Platonic [plə tän´ik] *Science.* of, relating to, or characteristic of Plato or his philosophy.

Platonic graph *Mathematics.* the graphs formed by the vertices and edges of the five regular (Platonic) solids.

Platonic solid *Mathematics.* any of the five polyhedra whose faces are congruent polygons; i.e., a tetrahedron, cube, octahedron, dodecahedron, or icosahedron. By theorem, there are no others.

Platonic year see GREAT YEAR.

Platonism [plāt´ə niz´əm] *Science.* the philosophy and writings of Plato. *Mathematics.* specifically, a philosophical view of mathematics in which the external reality of mathematical objects is asserted.

platoon *Ordnance.* an army unit immediately subordinate to a company; may consist of several sections or squads.

Platte *Geography.* a river in the central U.S. formed by the confluence of the **North Platte** and **South Platte,** both rising in northern Colorado; flows eastward from central Nebraska to join the Missouri.

platte *Geology.* a hard knob of resistant rock that occurs in a glacial valley or emerging from an existing glacier, and frequently produces a split at the glacier's snout.

plattnerite *Mineralogy.* PbO_2, an iron-black, metallic to adamantine, tetragonal mineral of the rutile group occurring in fibrous masses and thin prismatic crystals, having a specific gravity of 9.42 to 9.56 and a hardness of 5.5 on the Mohs scale; occurs as an alteration product of other lead minerals.

platy *Geology.* **1.** of or relating to a sedimentary particle that is three times as long as it is thick. **2.** referring to a sandstone or limestone that tends to split into layers ranging from 2 mm to 10 mm in thickness.

platy- a prefix meaning "broad," "flat."

Platyasterida *Invertebrate Zoology.* a mostly extinct order of primitive starfish in the subclass Asteroidea, having rows of elongated skeletal plates on flat, leaflike arms and no suckers on the feet.

Platybelodon *Paleontology.* a genus of shovel-tusked proboscideans in the extinct family Gomphotheriidae; Platybelodon was a widespread bunodont mastodon; extant in the Miocene.

platycelous *Vertebrate Zoology.* describing a vertebra that is flat or ventrally concave and dorsally convex.

platycephalic *Anthropology.* having a flattened head, as in certain Neanderthal specimens.

Platycephalidae *Vertebrate Zoology.* the flatheads, a fairly large family of marine and brackish water fishes composing the suborder Platycephaloidei in the order Scorpaeniformes, having a broad, flat head covered with small, dense scales and spiny, bony ridges; found in the Indo-Pacific and tropical east Atlantic Oceans, where they bury themselves in the sandy, muddy seafloor.

Platyceratacea *Paleontology.* an extinct superfamily of globose gastropods in the suborder Trochina; extant in the Late Paleozoic.

Platycopa *Invertebrate Zoology.* an order of marine ostracod crustaceans with two pairs of antennae, two pairs of thoracic appendages, and no heart or eyes. Also, **Platycopina.**

Platyctenea *Invertebrate Zoology.* an order of tentacled, creeping comb jellies in the phylum Ctenophora, having a body flattened along the oral-aboral axis.

platy fish *Vertebrate Zoology.* any of several small, yellow-gray freshwater fishes of the genus *Xiphophorus*; popular in home aquariums. Also, MOON PLATY.

platy flow structure *Petrology.* an igneous rock structure that is characterized by tabular sheets indicative of stratification during magma intrusions; formed by contraction during cooling. Also, LINEAR FLOW STRUCTURE, PLANAR FLOW STRUCTURE.

Platygasteridae *Invertebrate Zoology.* parasites in gall flies, a family of hymenopteran insects in the superfamily Proctotrupoidea, having a long ovipositor; females lay the eggs inside the host, where the larvae develop.

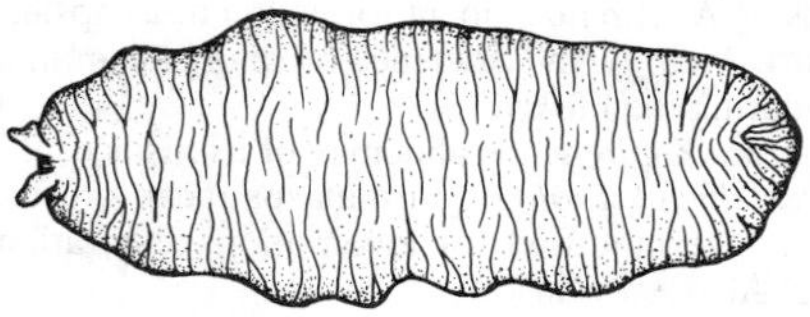

Platyhelminthes

Platyhelminthes *Invertebrate Zoology.* the phylum of flatworms, including the two parasitic classes of Trematoda (flukes) and Cestoda (tapeworms), and the free-living class Turbellaria; dorsoventrally flattened and lacking respiratory and circulatory systems.

platykurtic distribution *Statistics.* a frequency distribution that is less heavily concentrated about the mean than the normal distribution, and thus is wide and rather flat about the mode.

platykurtosis *Statistics.* the state of having a platykurtic distribution.

platymyarian *Invertebrate Zoology.* of certain nematodes, having flat muscle cells with supporting fibrils and contractile fibers in a basal zone next to the epidermis.

platynite *Mineralogy.* $PbBi_2(Se,S)_3$, a metallic, iron-black trigonal mineral occurring in thin plates resembling graphite, having a specific gravity of 7.98 and a hardness of 2 to 3 on the Mohs scale.

Platypodidae *Invertebrate Zoology.* the ambrosia beetles and pinhole borers, a family of coleopteran insects belonging to the superfamily Curculionoidea, having a long, cylindrical body and a snout on the head; a tree pest that burrows into wood.

platypus [plat′i pu̇s′] *Vertebrate Zoology.* an egg-laying aquatic mammal, *Ornithorhynchus anatinus* of the family Ornithorhynchidae, characterized by even, dense fur, no ear pinnae, webbed feet with hollow, sharp claws in the rear feet capable of injecting poison, and a toothless, broad bill; found in freshwater lakes, streams, and lagoons of eastern Australia and Tasmania. Also, DUCK-BILLED PLATYPUS. (Going back to a Greek phrase meaning "flat-footed.")

platypus

platyrrhine [plat′i rin] *Anthropology.* having a flat nose; used especially to describe the monkeys of the New World, which evolved separately from the Old World primates.

platyrrhine monkey *Vertebrate Zoology.* any primate of the family Cebidae from tropical America; species are chiefly herbivorous and arboreal and often have a prehensile tail; includes the howler monkeys, squirrel monkeys, capuchin monkeys, woolly monkeys, and sakis. Also, NEW WORLD MONKEY.

Platyrrhini *Paleontology.* an infraorder of primates in the order Anthropoidea, composing the New World monkeys; the New and Old World monkeys diverged in the Oligocene, but have remained alike because of parallel evolution.

platysma *Anatomy.* either of the broad plates of muscle fibers immediately beneath the superficial fascia in the side of the neck that depress the mandible and draw the lower lip downward.

Platysomidae *Paleontology.* a family of deep-bodied fishes in the extinct suborder Platysomoidei; related to the palaeoniscids; characterized by elongate dorsal and anal fins; extant from Carboniferous to Triassic.

platyspondylia *Medicine.* a congenital bone disorder characterized by abnormalities in the size and shape of the tubular and flat bones, vertebrae, and skull.

Platysternidae *Vertebrate Zoology.* the big-headed turtle, a monospecific family characterized by a proportionately large head that cannot be retracted, a long flat shell, a long tail, and skin covered with hard scales; found in freshwater habitats of southeast Asia.

Platyzomataceae *Botany.* a monotypic family of terrestrial ferns of the order Filicales, characterized by very slender fronds, a short, creeping rhizome clothed with stiff golden hairs, and sporangia attached individually on the veins rather than in sori.

play see PLAY BEHAVIOR.

playa [plī′ə] *Geology.* **1.** in an arid or semiarid region, a low, flat, dried up, barren area at the bottom of a closed lake basin from which rainwater quickly evaporates, often leaving behind deposits of salt. **2.** a small, sandy area at the mouth of a stream or along the shore of a bay. **3.** a flat, alluvial area along a coast, as distinguished from a beach.

playa

playa lake *Hydrology.* in an arid or semiarid region, a shallow lake that temporarily covers or fills a playa during the wet season, or a temporary lake that leaves or forms a playa upon evaporation.

playback *Acoustical Engineering.* the reproduction of sound from a previously recorded medium such as a tape or disk.

playback accuracy *Control Systems.* the accuracy with which an automatic control system executes a given position command. *Robotics.* specifically, a measure of the difference between the place a robot or its end effector is supposed to be and the place it actually is.

playback experiment *Molecular Biology.* a procedure by which a DNA strand is recovered from an RNA-DNA hybrid and is shown to contain nonrepetitive DNA through reassociation kinetic experiments.

playback head *Electronics.* the element in a tape recorder that converts the audio pattern formed on the tape's magnetic field into corresponding electrical signals. Also, REPRODUCE HEAD.

playback robot *Robotics*. a robot that is led through a sequence of movements by an operator and is able to repeat them continuously.

play behavior *Behavior*. behavior consisting of erratic or spontaneous movements that are not directed toward an immediate biological purpose, usually occurring in young animals; thought to be a form of rehearsal or preparation for adult activities such as hunting.

Playfair, John 1748–1819, Scottish mathematician; reformulated Euclid's parallel postulate; popularized Hutton's geological theories.

Playfair's law *Geology*. a statement based on John Playfair's observation that streams cut their own valleys, which are proportional in size to their streams, and that the stream junctions in such valleys are accordant in level.

play therapy *Psychology*. a treatment approach used with children that involves play activities and materials; based on the theory that a child can express thoughts and feelings more directly in play than in a formal conversation or interview.

plaza *Architecture*. **1.** a large, open public square surrounded by buildings; a piazza. **2.** any urban open area. **3.** another term for a shopping mall.

pleasure-pain principle *Psychology*. a principle stating that behavior is motivated by the seeking of pleasure and avoidance of pain.

pleasure principle *Psychology*. in psychoanalytic theory, the internal force that motivates an individual to satisfy libidinal impulses.

pleated cartridge *Mechanical Devices*. a convoluted filter cartridge appearing as an accordion.

Plecoptera *Invertebrate Zoology*. the stoneflies, a primitive order of hemimetabolous insects that spend most of their lives as predacious aquatic larvae; short-lived adults are clumsy fliers, having four membranous wings, a soft body, and long slender cerci.

plectane *Invertebrate Zoology*. one of a set of striated cuticular plates that give support to genital papillae in certain nematode worms.

Plectascales *Mycology*. a former term for the fungi belonging to the subdivision Ascomycotina whose spore-sacs or asci are unevenly distributed in the spore-bearing structures.

plectenchyma *Botany*. a parenchymatous tissue formed by tubular cells or twisted filaments, especially in lichens and fungi.

Plectoidea *Invertebrate Zoology*. a superfamily of free-living nematodes with spiral amphids.

Plectomycetes *Mycology*. a class of fungi belonging to the subdivision Ascomycotina, characterized by ascospores that are generally one-celled and asci that are scattered throughout the spore case.

Plectonema *Bacteriology*. a genus of cyanobacteria that are photosynthetic and filamentous; some species are able to fix nitrogen under microaerobic conditions.

plectonemic spiral *Molecular Biology*. a spiral that consists of two strands of nucleic acid which are coiled in the same direction, as in a DNA double helix, and which cannot be separated unless the strands are uncoiled.

plectostele *Botany*. a protostele in which the xylem is divided into plates that alternate with phloem across the stele.

Plectrovirus *Virology*. a genus of bacteriophages of the family Inoviridae, having short, bullet-shaped virions and a circular ssDNA genome.

pleg- or **plego-** a prefix meaning "paralysis."

-plegia a combining form meaning "paralysis;" as *quadriplegia*.

Pleiades *Astronomy*. a small, bright star cluster about 400 light-years away in the constellation of Taurus. Also, M 45, THE SEVEN SISTERS.

Pleidae *Invertebrate Zoology*. a family of very small backswimmers, hemipteran insects in the superfamily Pleoidea that feed on water fleas; they have a highly arched back and regressed hindwings.

pleio- or **pleo-** a prefix meaning "more than one." Also, PLEI-, PLIO-.

pleiomorphism *Biology*. **1.** the occurrence of more than a single distinct form in the life cycle of some organisms, such as rusts. **2.** the ability to change shape.

pleiotaxy *Botany*. the state of having more than the usual number of floral whorls.

pleiotropic *Genetics*. of a gene or mutation, having multiple effects.

pleiotropy *Genetics*. a phenomenon in which a single gene affects a range of phenotypic characteristics in an organism. Also, **pleiotrophy.**

Pleistocene [plī es´tə sēn; plīs´tə sēn] *Geology*. the geologic epoch of the Quaternary period extending from the end of the Pliocene to the beginning of the Holocene, and the rocks formed during that time.

Pleistocene extinction *Evolution*. a mass extinction of species from approximately 2 million to 10,000 years ago.

plench *Space Technology*. a tool combining pliers and a wrench that is used by astronauts.

plenishing nail *Mechanical Devices*. a short nail with a small head used to attach pieces of furniture together.

plenum [plen´əm] *Astronomy*. space that is filled with matter, as distinguished from a pure vacuum. *Engineering*. a system of ventilation in which air is forced into an enclosed space so that the outward air pressure is greater than the inward air pressure, and the direction of leakage is outward instead of inward.

plenum chamber *Engineering*. a compartment in which the interior air pressure is higher than the exterior air pressure; air is forced into this chamber for slow distribution through ducts.

plenum system *Mechanical Engineering*. an air conditioning system in which the air propelled into a building is maintained at a higher pressure than the atmosphere.

pleochroic halo *Optics*. a darkly colored ring or group of such rings found in certain minerals around small radioactive inclusions; caused by alpha particle irradiation.

pleochroism *Optics*. a characteristic of a transparent crystal causing it to exhibit different colors when white light travels through it from different directions. Also, POLYCHROISM.

pleocytosis *Medicine*. the presence of an abnormally large number of cells in the cerebrospinal fluid.

pleodont *Vertebrate Zoology*. having solid teeth, as some reptiles.

Pleoidea *Invertebrate Zoology*. a superfamily of aquatic true bugs, hemipteran insects in the subdivision Hydrocorisae.

Pleomassariaceae *Mycology*. a family of fungi belonging to the order Pleosporales, which occur in cool to warm temperate regions and live off decaying branches of woody plants.

pleomorphic *Biology*. able to change shape, as certain protozoa and bacteria.

pleomorphic phages *Virology*. phages of variable sizes and shapes that contain dsDNA with no evident formation of a capsid.

pleomorphism *Biology*. the existence of an organism in more than one distinct form during the life cycle.

pleon *Invertebrate Zoology*. the abdomen of a crustacean.

pleopod *Invertebrate Zoology*. in malacostracan crustaceans, an abdominal appendage modified for locomotion, respiration, or carrying eggs or young. Also, SWIMMERET.

Pleosporales *Botany*. an order of lichenized fungi in the class Loculoascomycetes. *Mycology*. an order of fungi belonging to the class Loculoascomycetes, which is characterized by a distinct, down-growth of hyphae in the locule prior to ascus formation; members are either parasites or live off nonliving organic matter.

plerocercoid *Invertebrate Zoology*. the wormlike, final larval stage of tapeworms that develops in an intermediate host.

plerome *Botany*. in the histogen theory, the innermost region of the apical meristem which gives rise to all primary tissues internal to the cortex.

plerotic water see GROUNDWATER.

Plesiadapiformes *Paleontology*. a suborder including perhaps the oldest primates; they arose in the Cretaceous but did not diversify greatly until the Late Paleocene, dying out at the end of the Eocene; they were very unlike modern primates and recent fossil evidence links some or all groups of **plesiadapiforms** to the order Dermoptera.

Plesiocidaroida *Paleontology*. an extinct order of echinoids in the subclass Euechinoidea, but exact affinities are uncertain; one of the early euechinoids that began to spread adaptively in the Upper Triassic after the class had almost become extinct at the end of the Permian.

Plesiomonas *Bacteriology*. a genus of Gram-negative bacteria of the family Vibrionaceae, occurring as facultatively anaerobic, polarly flagellated, rod-shaped cells in aquatic habitats or in the intestinal tract of certain mammals.

plesiomorph *Evolution*. an ancestral or primitive hereditary character retained in later forms of an organism.

plesiomorphic *Evolution*. of or relating to a plesiomorph.

plesiosaur *Paleontology*. any of the large marine reptiles of the extinct order Plesiosauria; the plesiosaurs represent a continuation of the aquatic adaptation of their cousins, the Nothosauria, but may not have descended directly from them; extant in the Triassic to Cretaceous.

Plesiosauria *Paleontology*. an order of reptiles in the extinct superorder Sauropterygia; the plesiosaurs, which grew as long as 13 meters (Kronosaurus) and had relatively large heads and long necks and tails, represent an almost complete adaptive return to the sea, returning only seldom to land; extant in the Triassic to Cretaceous.

Plesiosaurus *Paleontology*. a genus of marine reptiles in the family Plesiosauridae; extant in the Lower Jurassic.

plesiotype *Systematics.* a specimen that has been annotated by someone other than the original author yet determined to belong within the same classification as the type specimen.

plesiotypic state *Evolution.* the ancestral character state in a transformation series. Also, **pleisiotypic state.**

Plessy's green *Inorganic Chemistry.* $CrPO_4$, a green pigment; insoluble in water and soluble in acids; used as paint pigment and as a catalyst. Also, ARNAUDON'S GREEN, CHROMIC PHOSPHATE.

Plethodontidae *Vertebrate Zoology.* the lungless salamanders, a large family of the suborder Ambystomatoidea, characterized by adults that, lacking lungs or gills, obtain oxygen through the skin or the mouth, which are richly supplied with blood vessels; found in North America, northern South America, and southern Europe.

plethysmograph *Medicine.* an instrument for determining and recording variations in the volume of an organ, part, or limb and in the volume of blood present or passing through it.

Pleuorchloridellaceae *Botany.* a monogeneric family of freshwater yellow-green algae of the order Heterogloeales, characterized by spherical, unattached, solitary cells without surrounding mucilage.

pleur- or **pleuro-** a prefix meaning "side."

pleura [plůr´ə] *plural,* **pleurae.** *Anatomy.* a delicate serous membrane lining the thoracic cavity and investing the lungs.

Pleuracanthodii *Paleontology.* a former name for the order Xenacanthida, a diverse group of generally freshwater sharks; most species were characterized by a postcranial spine; Carboniferous and Permian.

pleural *Anatomy.* of or related to the pleura.

pleural cavity *Anatomy.* the space between the parietal and visceral layers of the pleura.

pleural effusion *Medicine.* the presence of liquid in the pleural space.

pleurapophysis *Anatomy.* a rib or the process on cervical or lumbar vertebra that is homologous to a rib.

pleurisy [plůr´ə sē] *Medicine.* an inflammation of the pleura, which may be either acute or chronic, characterized by dyspnea and stabbing pain; common causes include bronchial carcinoma, pneumonia, and tuberculosis; it may result in permanent adhesions between the pleura and adjacent surfaces.

pleuritis pleurobranchia *Invertebrate Zoology.* in malacostracan crustaceans, a gill on the body just above a basal leg segment.

Pleurocapsa group *Microbiology.* a large group of cyanobacteria, or blue-green algae, that form branching filaments; some members exhibit spherical vegetative cells and some possess elongated cells.

pleurocarpous *Botany.* of a moss that has the archegonia, capsule, and stalk on a short sidebranch or alongside a stem rather than at the top of the main branch; as distinguished from acrocarpous.

pleurocentesis *Medicine.* the surgical puncture of the chest wall for aspiration of fluid. Also, THORACENTESIS.

Pleuroceridae *Invertebrate Zoology.* a family of common freshwater snails, gastropod mollusks in the order Pectinibranchia, having an elongated spiral shell.

Pleurochloridaceae *Botany.* a family of freshwater or marine yellow-green algae in the order Mischococcales, characterized by solitary, free-living, usually uninucleate cells.

pleurococcus *Botany.* a genus of usually terrestrial green algae of the family Protococcaceae that forms thin, filmy colonies on damp rocks, tree bark, and similar environments.

Pleurocoelea see TECTIBRANCHIA.

Pleurodictyum *Paleontology.* a widespread genus of colonial tabulate corals in the extinct suborder Favositina and family Micheliniidae; almost always found with the commensal orbiculous worm Hicetus at the center of the colony; Silurian and Devonian.

Pleurodira *Vertebrate Zoology.* the side-necked turtles, a suborder of Testudines characterized by species that withdraw their heads laterally into the carapace.

pleurodontia *Vertebrate Zoology.* a primitive tooth arrangement common to lizards in which the tooth sits on a ledge of the inner jaw and is attached by the side and base.

pleurodynia *Medicine.* paroxysmal pain in the intercostal muscles, caused by muscular rheumatism or irritation of pleural surfaces.

pleurogenous *Anatomy.* relating to or originating from the pleura. *Botany.* occurring on the side of a structure or body.

pleurolophocercous cercaria *Invertebrate Zoology.* an immature fluke with a long tail, protrusible oral sucker, and pigmented eyespots.

Pleuromastigaceae *Botany.* a family of flagellates usually placed in the order Cryptomonadales, characterized by a single flagellum, brownish chloroplasts, and compressed cells.

Pleuromastigophora *Invertebrate Zoology.* an arthropod subclass of class Chilopoda (centipedes), characterized by the presence of spiracles that run along each side of the body; includes the orders Scolopendromorpha, Geophilomorpha, and Lithobiomorpha.

Pleuromeiaceae *Paleontology.* a family of Mesozoic lycophytes in the extinct order Pleuromeiales; generally about 1 meter high.

Pleuromeiales see LYCOPODIOPHYTA.

pleuron *Invertebrate Zoology.* **1.** one of the small plates on either side of the body segments of most arthropods. **2.** in crustaceans, a lateral downward extension of the exoskeleton.

Pleuronectidae *Vertebrate Zoology.* the right-eyed flounders, a cosmopolitan family of flatfishes of the order Pleuronectiformes, characterized by the placement of both eyes on the right side of the head; most species are important food fishes.

Pleuronectiformes *Vertebrate Zoology.* the flatfishes, an order including the flounders and halibuts, characterized by a laterally compressed, asymmetrical deep body, with both eyes on one side of the head; found mainly in marine waters.

Pleuronematina *Invertebrate Zoology.* a suborder of ciliated protozoans in the order Hymenostomatida, having undulating membranes near the mouth and sensory caudal bristles.

pleuroperitoneal cavity *Vertebrate Zoology.* in all pulmonate vertebrates but mammals, a body cavity containing both the lungs and the abdominal viscera.

Pleurophascaceae *Botany.* a monotypic moss family of the order Dicranales that forms robust loose tufts with creeping stems, erect branches, and lateral sporophytes; found in peaty substrates of upland swamps in New Zealand and Tasmania.

pleuropneumonia *Medicine.* pleurisy complicated by inflammation of the lungs. *Veterinary Medicine.* an infectious disease of cattle and other mammals caused by organisms of the *Mycoplasma* group, characterized by combined inflammation of the pleura and the lungs.

pleuropneumonialike organism *Microbiology.* a former classification for a group of microorganisms that lack a true cell wall and resemble the bacterium *Mycoplasma mycoides,* the causative agent of pleuropneumonia in cattle; now classified among certain species of the bacterial genus *Mycoplasma.* Also, PPLO.

Pleurotomaria *Paleontology.* a genus of Paleozoic gastropods in the superfamily Pleurotomariacea; characterized by its asymmetrical trochiform shell; Jurassic and Cretaceous.

Pleurotomariacea *Paleontology.* a superfamily of aspidobranch gastropods in the order Archaeogastropoda that arose in the Upper Cambrian from bellerophontid stock; still extant.

Pleurotus *Mycology.* a genus of fungi belonging to the family Tricholomataceae that are found on wood; some species, such as *P. ostreatus* (oyster fungus), are edible, while others are toxic.

Pleuroziaceae *Botany.* a widespread family of liverworts of the order Jungermanniales, characterized by bilateral growth by means of apical cells, purplish pigmentation, sporadic lateral-intercalary branches, rigid stems, and a lack of asexual reproduction.

Pleuroziopsidaceae *Botany.* a monotypic family of large, dendroid mosses of the order Hypnobryales that form loose mats on soil, humus, and logs; characterized by unique branch lamellae and secondary stems bearing branched fronds; found in moist forests of coastal western North America and eastern Asia.

plexectomy *Surgery.* the surgical excision of all or part of a plexus.

Plexiglas *Materials.* a trade name for a transparent methyl methacrylate that is lightweight, resistant to weathering, and formed in sheets and rods.

plexitis *Neurology.* inflammation of a nerve plexus.

plexopathy *Neurology.* any disease or disorder of a plexus, especially of a nerve plexus.

plexus *Anatomy.* an interlacing network of nerves, blood vessels, or lymphatic vessels. *Geology.* the region of a subglacial deposit that surrounds a giant's kettle. *Evolution.* anastomosing lineages.

pli pounds per linear inch.

pliability *Materials Science.* a property of soft plastics that gives the material increased flexibility, ductility, resistance to strain and stress, and an ability to cling to surfaces.

plica *Biology.* a fold, as of skin or of a leaf.

plication *Surgery.* **1.** a procedure for folding or tucking tissues. **2.** a fold. *Geology.* an episode of intense folding or crumpling that takes place on a small scale.

Pliensbachian *Geology.* a European geologic stage of the Lower Jurassic period, occurring after the Sinemurian and before the Toarcian.

pliers *Mechanical Devices.* a small pincers with broad, long, flat-roughened jaws; used to hold small objects or to bend and cut wire.

Plimsoll mark *Naval Architecture.* any of several lines on the side of a merchant ship that indicate the allowable load levels for various seasons and conditions. Also, **Plimsoll line.** (Named for the English reformer and legislator Samuel *Plimsoll,* who successfully lobbied for the adoption of this safety measure by Britain in 1875.)

Plinian eruption *Volcanology.* an explosive eruption characterized by a tall, sustained eruption column that yields fallout of pumice and ash.

Plinian-type eruption see VULCANIAN-TYPE ERUPTION.

plinth *Architecture.* **1.** the lowest member of a pedestal. **2.** the projecting base of an exterior wall. *Geology.* the lower or outer surface of a seif dune that lies outside the slip-face boundaries and has never been subjected to sand avalanches.

plinthite *Geology.* a highly weathered soil material composed of clay and quartz mixed with other diluents, containing large amounts of aluminum and iron oxides and very small amounts of humus.

Pliny the Elder [plin´ē] 23–79 AD, Roman naturalist; wrote an influential 37-volume *Natural History.*

Pliny the Younger c. 61–113 AD, Roman statesman and scholar, the nephew of Pliny the Elder; wrote a detailed account of the eruption of Vesuvius.

Pliocene *Geology.* the geologic epoch of the Tertiary period extending from the end of the Miocene to the beginning of the Pleistocene, and the rocks formed during that time.

Pliohippus *Paleontology.* a one-toed North American horse descended from Merychippus and ancestral to Equus; extant in the Middle to Upper Pliocene; characterized by strongly hysodont molar teeth.

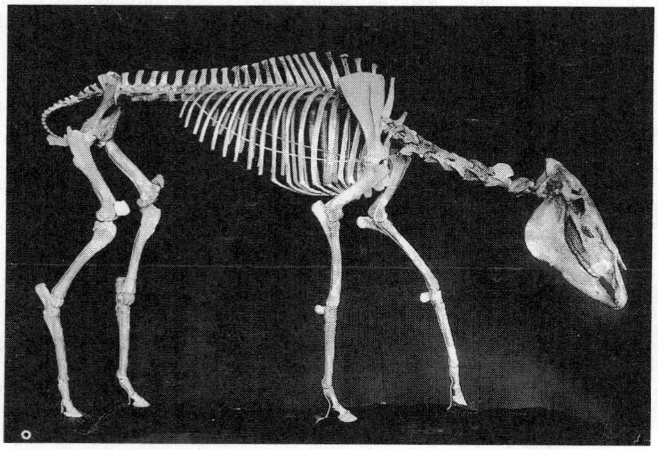

Pliohippus

Pliohyracidae *Paleontology.* an extinct family of coneys in the order Hyracoidea; predominantly African but also found in a limited range in Europe and Asia; extant in the Eocene to Miocene.

pliomagmatic zone see EUGEOSYNCLINE.

Pliopithecoidea *Paleontology.* a superfamily of Oligocene and Miocene primates in the infraorder Catarrhini, known from a few remains found at East and North African sites; may include *Dendropithecus* and *Propliopithecus.*

Pliopithecus *Anthropology.* a fossil hominoid from the Miocene and Pliocene that somewhat resembles the modern gibbon.

pliothermic *Geology.* of or relating to an interval of geologic time during which the climate was warmer than average.

pliotron *Electronics.* any hot-cathode vacuum tube with at least one grid.

plissé [pli sā´] *Textiles.* a thin cotton fabric treated with a chemical causing shrinkage to produce a crinkled effect, generally in a lengthwise stripe pattern. The same effect is sometimes achieved by weaving with yarns having different degrees of shrinkage. (From a French term meaning "to pleat.")

Plocamiaceae *Botany.* a family of red algae belonging to the order Gigartinales, characterized by erect, abundantly branched, uniaxial thalli with similar sporangial and gametangial forms and separately formed spermatangia.

Ploceidae *Vertebrate Zoology.* the weaver finches, a family of small to medium-sized gregarious birds of the passerine suborder Oscines, known for their large complex nests; found in Africa and Asia.

ploidy *Genetics.* the number of complete chromosome sets that are present in a cell or organism; the root of terms such as *haploid* (having one chromosome set), *diploid* (two chromosome sets), *triploid* (three sets), and *polyploid* (many sets); also used with specific numbers, as in *16-ploid.*

Ploima *Invertebrate Zoology.* a major order of rotifers belonging to the class Monogononta, containing a diverse group of predatory and filter-feeding aquatic species.

Plokiophilidae *Invertebrate Zoology.* a family of predatory bugs, hemipteran insects in the superfamily Cimicoidea, that live in spider webs.

plomb *Surgery.* any inert material that is inserted into a body cavity for therapeutic purposes.

plombage *Surgery.* a surgical procedure for filling a cavity or space in a body with plomb.

plot *Civil Engineering.* **1.** in surveying, a detailed map of an area made from precise measurements taken from field notes. **2.** an area of land. Also, LOT.

Plotosidae *Vertebrate Zoology.* the sea catfish eels, a family of mainly marine fish of the order Siluriformes, characterized by a long eel-like body lacking scales; generally found in large schools near reefs in the Indo-Pacific basin.

plotter *Engineering.* an automatically controlled writing instrument that draws a dependent variable on a display board in relation to one or more independent variables. *Computer Technology.* specifically, an output device used with digital computers that produces graphic or pictorial representations of computer data on hard copy by drawing with one or more automatically controlled pens that move in small discrete steps. Also, DIGITAL PLOTTER, ELECTROMECHANICAL PLOTTER, INCREMENTAL PLOTTER. *Navigation.* a device to assist in the plotting of courses or bearings on a chart or plotting sheet. It usually consists of a protractor and one or more fixed or moveable straightedges.

plotting chart *Navigation.* a chart used for solving navigational problems, such as dead reckoning, lines of position, or great-circle courses, by plotting.

plotting sheet *Navigation.* a blank chart, showing only an unspecified grid and one or more compass roses; used to track a vessel's position when out of sight of land.

plough the British spelling of PLOW.

Plough *Astronomy.* the British term for the Big Dipper, the seven brightest stars in the constellation of Ursa Major.

plover *Vertebrate Zoology.* any of the numerous shorebirds of the family Charadriidae, characterized by a short tail, a plump breast, thin legs, and a short, hard-tipped bill; found worldwide.

plover

plow *Agriculture.* **1.** a farm implement that is used to cut, break, or turn the soil to prepare it for planting or other agricultural purposes. **2.** to make a furrow or turn over a layer of soil. **3.** to cultivate, or turn over weeds. *Engineering.* **1.** a grooving or shaping plane with blades and an adjustable guide. It can make a single groove in wood, or a completely grooved surface. **2.** an indentation cut parallel to the grain in a piece of wood. *Mining Engineering.* **1.** a continuous coal mining machine with cutters that bite into the longwall face along which it is operated. **2.** a V-shaped belt scraper used to remove coal and debris from the return belt of a conveyor. Also, PLOUGH.

plow depth *Agronomy.* the depth to which a certain type of plow bottom penetrates the soil.
plow marks *Archaeology.* marks left in buried soil indicating that the land has been plowed at some remote time, giving evidence of ancient agricultural activity.
plowshare *Agriculture.* the cutting edge of a plow that tears the furrow slice loose from the soil. Also, SHARE. *Hydrology.* a wedge-shaped feature that develops on the surface of snow as a result of further ablation of foam crust.
Plowshare *Nucleonics.* a program of the U.S. Atomic Energy Commission (replaced by the Nuclear Regulatory Commission) for the research and development of peaceful uses of nuclear explosives.
plow sole *Geology.* a highly compacted, subsurface soil layer having a higher bulk density and a lower porosity than the soil above or below it, created by the effects of repeated plowing to the same depth. *Agronomy.* see HARDPAN. Also, **plow pan.**
plow steel *Metallurgy.* steel that contains from 0.5 to 0.95% carbon.
plow wind *Meteorology.* a weather phenomenon, consisting of strong, straight line winds, that occurs in the midwestern U.S. and is associated with squall lines and thunderstorms; damage is usually confined to a narrow zone like that due to tornadoes.
plow zone *Archaeology.* the top layer of the soil to the depth at which a plow will penetrate and disturb archaeological deposits.
PLSS portable life-support system.
Plücker, Julius [ploo′kər] 1801–1868, German mathematician; major contributions in analytical geometry, especially in abridged notation.
plucking *Geology.* a process of glacial erosion by which fairly large fragments of bedrock that have been weakened along joints or fissure planes by the action of freezing water are loosened, pried off, and carried away as the glacier advances.
plug *Science.* **1.** a piece of material that is used to fill or stop up a hole. **2.** a small piece of material that is removed from a larger object. **3.** to fill an opening by inserting a plug. *Geology.* **1.** a roughly cylindrical, vertical mass of solidified magmatic material representing the conduit of an inactive volcano. **2.** a lava-filled crater that remains after the surrounding material has been worn away by erosion. **3.** a mass of sediment that fills part of an abandoned, cutoff stream channel. *Electricity.* a male connector that can be inserted into a jack, outlet, receptacle, or socket. *Metallurgy.* **1.** the mandrel used in tube working. **2.** the protruding portion of a die. *Mining Engineering.* a watertight seal in a mine shaft.
plug-and-feather method *Mining Engineering.* a method used in quarrying to reduce the size of large masses of stone by making a row of shallow holes along the line where a break is desired, then splitting the rocks with a wedge and a hammer. Thus, **plug-and-feather hole.**
plugboard *Computer Technology.* see BOARD, def. 2.
plug-compatible hardware *Computer Technology.* a piece of computer hardware that can be connected to and used by a computer from a different manufacturer.
plug dome *Geology.* an upheaval of consolidated lava from the conduit of a volcano that forms a steep-sided mound above and around the volcanic vent.
plug flow digester *Biotechnology.* an anaerobic digester that consists of a long tube or elongated, covered pit into which material to be digested is introduced.
plug forming *Engineering.* a plastics manufacturing process in which a plug, under heat, vacuum, or pressure, gives an initial shape to a part before forming is completed.
plug fuse *Electricity.* a fuse that is used in a standard screw-based socket.
plugging *Engineering.* the placing of a solid barrier in a process flow system, such as a pipe or a reactor, in order to stop the flow. *Electricity.* a method used to brake an electric motor by reversing its connections, in order to provide torque in the opposite direction.
plughole *Mining Engineering.* **1.** a passageway left open to maintain normal ventilation while an old portion of a mine is sealed. **2.** a hole used to hold an explosive charge or for a bolt.
plug-in unit *Electricity.* a component or subassembly whose pins or contacts are oriented so that all connections can be made simultaneously when the unit is pressed into a suitable socket.
plug meter *Engineering.* a flow meter having a tapered plug that measures the flow rate according to the height the plug must be raised to permit fluid flow.
plug reef *Geology.* a small, triangular reef with a seaward-pointing apex that grows through openings between linear shelf-edge reefs.
plug valve *Mechanical Devices.* a valve fitted with a plug that rotates 90° to open or close an area, allowing fluid to flow. Also, **plug cock.**
plum *Botany.* **1.** any of various shrubs and small trees of the genus *Prunus*, family Rosaceae, that bear smooth-skinned, reddish to purple drupaceous stone fruit. **2.** the fruit itself.
plum *Geology.* a clast that is enclosed by a matrix of different material, particularly a pebble occurring in a conglomerate.
plumage *Vertebrate Zoology.* the feather covering of birds, sometimes colored for puposes of camouflage, courtship display, etc., and divided in certain stages of a bird's growth to indicate age or maturity, such as natal plumage, juvenal plumage, first winter plumage, first nuptial plumage, and the like.
Plumatellina *Invertebrate Zoology.* an order of freshwater colonial bryozoans in the class Phylactolaemata, having U-shaped bodies.
plumb *Engineering.* **1.** of or relating to the position of an object or structure that is vertical, as determined by a plumb bob. **2.** to establish or test for such a position.
Plumbaginaceae *Botany.* the single family in the order Plumbaginales.
Plumbaginales *Botany.* a monofamilial order in the class Caryophyllales, composed of dicotyledonous shrubs, lianas, and perennial herbs having chalk glands that exude water and calcium salts, pentamerous flowers, and a compound ovary containing a single basal ovule.
plumb bob *Mechanical Devices.* a pear-shaped or globular, pointed, tapering weight that is attached to a plumb line to measure its vertical center.

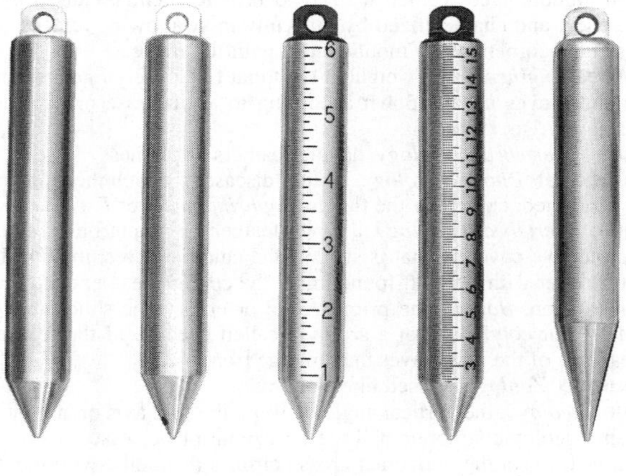

plumb bobs

plumb bond *Civil Engineering.* a masonry bond in which the vertical joints line up.
plumber *Building Engineering.* a person who installs and repairs the piping and fixtures used in the distribution of water in a building.
plumbing *Building Engineering.* **1.** the system of pipes and other apparatus for conveying water and liquid waste from a building. **2.** the work or business of installing and servicing such a system. *Civil Engineering.* a method used to find a vertical line or to ensure that an object is set straight. *Electromagnetism.* a collective term encompassing coaxial lines, waveguide devices, junctions, bends, and other types of microwave transmission hardware.
plumbism see LEAD POISONING.
plumb level *Mechanical Devices.* a level consisting of a piece of hardwood or metal fitted with curved glass vials that are oriented horizontally and vertically to a frame; the vials are filled with liquid, and an air bubble is left to aid in the determination of the horizontal plane and highest vertical point. Also, **plumb rule.**
plumb line *Mechanical Devices.* a length of line with a plumb bob that hangs free; used to determine verticality of structures. *Engineering.* a cord with a plumb bob used to center a survey instrument over a reference point. *Geophysics.* a line of force in a geopotential field to which the direction of gravity is everywhere tangential.
plum blotch *Plant Pathology.* a disease of plums caused by the fungus *Phyllosticta congesta,* characterized by small, gray or brown angular spots on leaves and fruit.
plumboferrite *Mineralogy.* $PbFe_4^{+3}O_7$, a black, opaque, trigonal mineral occurring in cleavable masses and thick tabular crystals, having a specific gravity of 6.07 and a hardness of 5 on the Mohs scale.

plumbogummite *Mineralogy.* $PbAl_3(PO_4)_2(OH)_5 \cdot H_2O$, a white to yellowish gray, translucent, trigonal mineral of the crandallite group occurring in massive form and as tiny prismatic crystals, having a specific gravity of 4.014 and a hardness of 4.5 to 5 on the Mohs scale; found as a secondary mineral in the oxidized zones of lead-bearing deposits.

plumbojarosite *Mineralogy.* $PbFe_6^{+3}(SO_4)_4(OH)_{12}$, a soft brown trigonal mineral of the alunite group, with a specific gravity of 3.64 to 3.66 and an undetermined hardness; found as secondary crusts and masses in the oxidation zones of lead deposits.

plumbous oxide see LEAD MONOXIDE.

plumbous sulfide see LEAD SULFIDE.

plumbum *Chemistry.* the Latin name for lead, from which the symbol Pb was derived.

plume *Botany.* a featherlike appendage on a fruit or seed that aids in wind dispersal. *Vertebrate Zoology.* the conspicuous feather or feathers of a bird. *Geology.* a deep-seated upwelling of magma within the earth's mantle. Also, MANTLE PLUME.

plume structure *Geology.* a ridgelike, plume-shaped tracing on the surface of a master joint, generally lying parallel to the upper and lower surfaces of the constituent rock unit. Also, **plumose structure.**

plumicome *Biology.* a supporting structure with plumelike tufts, found in many invertebrates, such as sponges.

plummer block *Mechanical Devices.* a box-form casing holding a metal load that bears the journal of a shaft with a horizontal split to reduce wear. Also, BEARING BLOCK, PILLOW BLOCK.

Plummer-Vinson syndrome *Medicine.* a rare disorder usually occurring in middle-aged women with hypochromic anemia, caused by iron deficiency, and characterized by difficulty in swallowing, cracks or fissures at the corners of the mouth, and a painful tongue.

plummet *Engineering.* a weighted float that fits loosely into a rotameter tube and moves up and down according to an increase or decrease of fluid flow. Also, FLOAT.

plumose *Vertebrate Zoology.* having feathers or plumes.

plum pocket *Plant Pathology.* a plum disease in which the stone of the fruit is aborted; caused by the fungus *Taphrina pruni* or *T. communia.*

plumule *Vertebrate Zoology.* a down feather; in young birds, a part of the protective covering that is shed as the chick matures; in adult birds, a soft feather with no shaft found under the contour feathers and providing insulation. *Botany.* the primary bud or embryonic shoot above the cotyledon or cotyledons of a seed; so called because of the featherlike appearance of the new leaves that emerge from it.

plumulose *Zoology.* shaped like a plumule.

plunge *Geology.* the vertical angle between the fold axis or any inclined line in a geologic structure and the horizontal plane. Also, PITCH. *Engineering.* **1.** to set the horizontal cross hair of a theodolite with the direction of a grade. **2.** see TRANSIT.

plunge angle see ANGLE OF DEPRESSION.

plunge basin *Geology.* a deep, extensive cavity or hollow formed in a stream bed at the foot of a waterfall or cataract by the action of the falling water.

plunge point *Oceanography.* **1.** the point at which a plunging wave curls over and falls. **2.** the final point at which a wave may be said to break before it rushes up the beach.

plunge pool *Hydrology.* **1.** a plunge basin. **2.** a deep, circular lake that occupies such a plunge basin after the stream has been diverted or the waterfall no longer exists. Also, WATERFALL LAKE.

plunger *Mechanical Devices.* a device consisting of a broomsticklike handle with a suction cup-shaped piece of rubber on its end; used for clearing plumbing traps and waste outlets. In popular use, informally called a **plumber's helper** or **plumber's friend.** *Engineering.* **1.** a sliding device in a piece of equipment that moves by or against fluid pressure. **2.** the long rod or piston of a reciprocating pump. **3.** see FORCE PLUG. *Electromagnetism.* a piston device used to introduce impedance to a waveguide by inserting a metal cylinder to a certain depth.

plunger pump *Mechanical Engineering.* a reciprocating pump in which a solid piston displaces the fluid; used for moving water or pulps.

plunging breaker *Oceanography.* a breaking wave that steepens gradually, but breaks very rapidly with a crash of water.

plunging cliff *Geology.* a steep, overhanging face of rock that directly borders deep sea water, and whose base lies considerably below the water level.

plunging fold *Geology.* a fold that angles steeply in relation to the horizontal plane. Also, PITCHING FOLD.

pluri- a combining form meaning: **1.** several. **2.** more.

pluriglandular see MULTIGLANDULAR.

plurilocular sporangium *Botany.* a multicelled sporangium divided by septa into several compartments, as found in some brown algae.

pluripotent *Biology.* generally, capable of maturing or developing in any of several ways. *Cell Biology.* specifically, possessing the capacity to differentiate along a variety of pathways; undifferentiated.

plurivorous *Biology.* living on several hosts.

plus *Mathematics.* **1.** the mathematical symbol + ; used to indicate addition or an operation that is analogous to addition. **2.** greater than zero; positive.

plus angle see ANGLE OF ELEVATION.

plush *Textiles.* a fabric made of silk, cotton, or wool, whose pile is more than 1/8 inch high, giving it a velvety feel and appearance.

plus-minus sequencing *Molecular Biology.* the Sanger-Coulson enzymatic method of determining the sequence of a specific segment of DNA.

plus-90 orientation *Computer Programming.* in optical character recognition, the specific position indicating that line elements on a document are perpendicular to the leading edge of the reader.

plus sieve *Metallurgy.* a term for the portion of a metallic powder or other powder that does not pass through a specified sieve size.

plus strands *Molecular Biology.* strands of DNA or RNA that have the same sequence polarity.

plus zone *Computer Programming.* the bit positions in a data representation code that are used to designate the algebraic plus sign.

pluteus *Invertebrate Zoology.* a free-swimming ciliated larva of ophiuroids and echinoids, having long armlike appendages.

Pluto *Astronomy.* the farthest known planet from the sun; discovered by Clyde Tombaugh in 1930 and having a diameter of 2302 kilometers, a single moon (Charon), and a year equal to 248 earth years. (Named for the ancient Greek and Roman god of the underworld.)

plutology *Geology.* the scientific study of the earth's interior.

pluton *Geology.* **1.** a body of igneous rock formed underground by the consolidation of magma or by the metasomatic replacement of an older rock. **2.** an intrusion of igneous rock within or between other rock formations.

Pluton *Ordnance.* a French tactical surface-to-surface missile designed to deliver a 10–15 kiloton nuclear warhead; powered by a dual-thrust, solid-propellant rocket motor, equipped with inertial guidance, and having a maximum range of 62 miles (100 km).

plutonic *Geology.* of or relating to pluton. Also, ABYSSAL, PLUTONIAN.

plutonic breccia *Geology.* a breccia that consists of angular fragments of older rocks surrounded by younger, plutonic rock.

plutonic metamorphism *Geology.* the regional metamorphism that takes place at high temperatures and pressures deep within the earth's crust, generally accompanied by strong deformation.

plutonic rock *Geology.* a medium- to coarse-grained rock formed deep within the earth's crust by crystallization of magma or by chemical alteration.

plutonic water *Hydrology.* water contained in or derived from magma at considerable depths.

plutonism *Geology.* **1.** all the processes and phenomena related to and associated with the formation of plutons. **2.** the theory that the earth was formed by the solidification of a molten mass. *Toxicology.* poisoning caused by exposure to radiation from plutonium; in laboratory animals, symptoms include liver damage and graying of hair.

plutonium [plü tō′nē əm] *Chemistry.* a radioactive metallic element that has the symbol Pu, the atomic number 94, an atomic weight (for its most stable isotope) of 244, a melting point of 640°C, and a boiling point of 3230°C; it occurs as fifteen isotopes (having mass numbers from 232 to 246) and in six allotropic forms. (Named for the planet *Pluto*; in the periodic table it is next to neptunium, and Neptune is the closest planet to Pluto.)

plutonium-238 *Nuclear Physics.* a radioactive isotope of plutonium with a half-life of 87.74 years; formed by decay of neptunium and used in radioisotopic thermoelectric generators.

plutonium-239 *Nuclear Physics.* a radioactive isotope of plutonium with a half-life of 24,100 years; created by irradiating uranium-238 in a reactor and used as a fuel in fast reactors and nuclear weapons.

plutonium bomb *Ordnance.* an atomic fission bomb using plutonium-238.

plutonium oxide *Inorganic Chemistry.* PuO_2, yellowish-green cubic crystals; slightly soluble in acid; highly toxic, dangerously radioactive, and a carcinogen; used to prepare plutonium halides.

plutonium reactor *Nucleonics.* a reactor that operates on the transuranic element plutonium.

pluvial *Meteorology.* **1.** of or relating to precipitation, particularly an abundant amount. **2.** describing an interval of geologic time characterized by a large amount of precipitation; usually applied to periods of heavy rainfall in the lower latitudes. *Ecology.* describing a time period or climate characterized by heavy rainfall. *Geology.* describing a geologic episode, process, deposit, or feature that results from or is affected by the action of rain. (From the Latin word for rain.)

pluvial lake *Hydrology.* a lake formed during a period of increased rainfall, particularly such a lake formed during the glacial advances in the Pleistocene epoch that is now either extinct or exists as a remnant.

pluviilignosa *Ecology.* a plant community that grows in the equatorial lowlands; characterized by vegetation that flowers, fruits, and leafs all year long.

pluviofluvial *Hydrology.* of or relating to a geologic process or feature that results from or is affected by the combined action of rainwater and streams.

pluviograph see RECORDING RAIN GAUGE.

pluviometer *Meteorology.* a rain gauge.

pluviometric coefficient *Meteorology.* the ratio of the monthly normal precipitation to 1/12 of the annual normal precipitation for any month at a given weather station.

pluviophilous *Ecology.* living or thriving under conditions of heavy rainfall. Thus, **pluviophile, pluviophily.**

pluviophobous *Ecology.* not tolerant of heavy rainfall. Thus, **pluviophobe.**

ply *Materials Science.* one of the sheets of veneer that are glued together to form plywood. *Textiles.* the number of single fiber strands twisted together to form a yarn, such as three-ply yarn. *Artificial Intelligence.* **1.** in a game-playing search, one level of the search tree, representing one move by one player. **2.** hyphenated with a number, the number of moves ahead that a program or search is able to examine before running out of time, e.g., 8-ply.

plymetal *Materials.* a bonded composite of dissimilar metals in sheet form.

Plymouth Rock *Agriculture.* a breed of chicken, characterized by its brown color and brown eggs; bred also for meat production. (From *Plymouth Rock,* in Massachusetts, near the location where the breed was developed.)

plywood *Materials.* thin sheets of wood glued together, with the grain of each consecutive piece positioned at a right angle to the preceding one to give strength and prevent warping; widely used in construction.

ply yarn *Textiles.* a yarn made by twisting together two or more single strands.

PM permanent magnet; phase modulation; pulpomesial.

P.M. post mortem. (Latin "after death.")

P.M. or **PM** afternoon. (From Latin *post meridiem.*) Also, **p.m., pm.**

Pm the chemical symbol for promethium.

PMD postmortem dump.

PMH previous medical history; past medical history.

PMI point of maximal impulse.

PMM pentamethylmelamine.

PMN polymorphonuclear granulocyte.

PMR proportionate mortality ratio.

PMS *Medicine.* a shorter name for premenstrual syndrome. See PREMENSTRUAL SYNDROME.

PMS notation *Computer Technology.* a concise graphical method of describing the physical structure of a computer system. (An acronym for processor-memory-switch notation.)

PN or **P.N.** Practical Nurse.

pNa *Chemistry.* the negative logarithm of the activity or concentration of a sodium ion in a solution.

pneo- a combining form meaning "breath," "breathing."

pneudraulic *Engineering.* of or relating to a mechanism involving both pneumatic and hydraulic action.

pneumat- or **pneumato-** a combining form meaning: **1.** breath; air. **2.** lungs.

pneumatic [noo mat´ik] *Engineering.* **1.** set in motion by air. **2.** operated by air pressure or by compressed air. (From a Greek word meaning "breath" or "wind.")

pneumatic atomizer *Mechanical Engineering.* an atomizer in which compressed air is used to produce moisture drops in the diameter range of 5 to 100 micrometers.

pneumatic control valve *Mechanical Engineering.* a valve in which compressed air forced against a diaphragm is opposed by a spring to regulate fluid flow.

pneumatic conveyor *Mechanical Engineering.* a system by which air velocity created by the expansion of compressed air through nozzles conveys loose material, such as sand, grain, or catalyst, through tubes. Also, AIR CONVEYOR.

pneumatic deception device *Ordnance.* an inflatable dummy that is intended to deceive the enemy regarding the strength or location of forces; e.g., a dummy tank or weapon.

pneumatic drill *Mechanical Engineering.* a hard-rock drill operated by a reciprocating piston, hammer action, or turbo drive. Thus, **pneumatic drilling.**

pneumatic drive *Robotics.* a motor or actuator that is driven by compressed air.

pneumatic duct *Vertebrate Zoology.* the duct that joins the air bladder to the alimentary canal of a physostomous fish.

pneumatic hammer *Mechanical Engineering.* a hammer in which compressed air is used to produce the impacting blows.

pneumatic hoist see AIR HOIST.

pneumatic injection *Mining Engineering.* a method developed by the U.S. Bureau of Mines for fighting underground coal fires, in which incombustible material, such as rock wool or dry sand, is injected through boreholes drilled from the surface into underground passageways.

pneumatic loudspeaker *Acoustical Engineering.* a loudspeaker in which sound is produced by modulating an air stream with an acoustic signal.

pneumatic riveter *Mechanical Engineering.* a high-speed riveting machine with a rapidly reciprocating piston driven by compressed air.

pneumatics *Fluid Mechanics.* the study of gas behavior in static, closed systems.

pneumatic weighing system *Engineering.* a weighing system in which the load is detected by a nozzle and balanced by regulating the air pressure in an opposing capsule.

pneumatocele *Medicine.* **1.** a hernial protrusion of lung tissue, as through a congenital fissure of the chest. **2.** a usually benign, thin-walled, air-containing cyst of the lung, as in staphylococcal pneumonia. **3.** a tumor or sac that contains gas, especially a gaseous swelling of the scrotum.

pneumatocodon *Invertebrate Zoology.* the external wall of a medusoid float (pneumatophore) of the order Siphonophora, e.g., composed of epidermis.

pneumatocyst *Botany.* a cavity in a pneumatophore.

pneumatogenic *Geology.* of or relating to a mineral or rock deposit that was formed by a gaseous agent.

pneumatolysis *Geology.* the alteration of rock or the crystallization of minerals as a result of the actions of gaseous substances given off by solidifying magma.

pneumatolytic *Geology.* of or relating to any deposit or process formed or influenced by gaseous emanations from solidifying magma.

pneumatolytic metamorphism *Petrology.* contact metamorphism in which rock composition is altered by the chemical activity of magmatic gases.

pneumatolytic stage *Geology.* a stage in the cooling and crystallization of a magma, during which the solid and gaseous phases are in equilibrium.

pneumatophore *Botany.* a channel in the specialized upward-growing roots of some swamp plants that apparently facilitates gas exchange for the submerged roots. *Invertebrate Zoology.* a buoyant apical float filled with air, formed from the first, bladderlike polyp in colonial hydrozoans of the order Siphonophora, such as the Portuguese man-of-war.

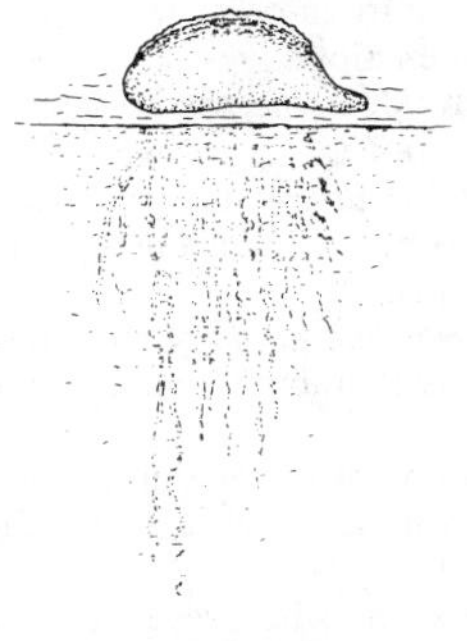

pneumatophore

pneumatosaccus *Invertebrate Zoology.* the internal wall of a pneumatophore, composed of gastrodermis, with a lumen divided into a distal air sac and an epidermal gas gland.

pneumatosis *Medicine.* the abnormal accumulation of gas anywhere in the body.

pneumaturia *Medicine.* the passage of urine charged with air or gas.

pneumectomy see PNEUMONECTOMY.

pneumo- a combining form meaning: **1.** breath; air. **2.** lungs.

pneumoangiography *Medicine.* the examination of the pulmonary and bronchial blood vessels by contrast radiography.

pneumobacillus *Bacteriology.* a bacterium, *Klebsiella penumoniae,* causing a type of pneumonia and other diseases of the respiratory tract.

pneumocardial *Anatomy.* of or relating to the lungs and the heart.

pneumocentesis *Medicine.* the surgical puncture of a lung.

pneumococcus [noo´mō käk´əs] *Bacteriology.* a bacterium, *Diplococcus penumoniae,* that causes lobar pneumonia and that is associated with such other diseases as pericarditis and meningitis.

pneumoconiosis *Medicine.* any disease of the lungs caused by chronic inhalation of irritating chemicals or dust, and characterized by permanent deposition of substantial amounts of particulate matter in the lungs. It may range from relatively harmless forms of anthracosis or siderosis to the destructive fibrosis of silicosis.

pneumocystis carinii pneumonia *Medicine.* an infection of the lungs associated with *Pneumocystis carinii*, occurring in debilitated premature infants and in patients with impaired immune systems, such as AIDS victims. Also, **pneumocystic pneumonia.**

pneumoencephalography *Medicine.* the radiographic visualization of the fluid-containing structures of the brain after cerebrospinal fluid has been withdrawn and replaced with air, oxygen, or helium.

pneumoenteritis *Medicine.* an inflammation of both the lung and intestines.

pneumography *Medicine.* **1.** an anatomical description of the lungs. **2.** a graphic record of respiratory movements. **3.** an X-ray of a part after injection of a gas.

pneumohemothorax *Medicine.* the presence of gas or air in the pleural cavity, sometimes deliberately introduced for the purposes of medical examination and testing.

pneumolithiasis *Medicine.* the presence of concretions or calculi in the lungs.

pneumomalacia *Medicine.* an abnormal softening of lung tissue.

pneumomelanosis *Medicine.* the blackening of the lung caused by inhaled coal dust or soot, a form of pneumoconiosis.

pneumomycosis *Medicine.* any inflammation of the lungs that occurs as a result of a fungal rather than a bacterial infection.

pneumon- or **pneumono-** a combining form meaning: **1.** breath; air. **2.** lungs.

pneumonectomy *Medicine.* the surgical excision of lung tissue, especially of an entire lung. Also, PNEUMECTOMY.

pneumonia [nů mōn´yə; noo mōn´yə] *Medicine.* an inflammation and consolidation of lung tissue, usually caused by inhaled pneumococci of the species *Diplococcus pneumoniae*, but also caused by other bacteria, viruses, rickettsiae, and fungi; historically a major cause of death but now effectively treated with antibiotics.

pneumonic [nů män´ik; noo män´ik] *Medicine.* **1.** of, relating to, or affecting the lungs. **2.** relating to or affected by pneumonia.

pneumonic plague *Medicine.* an acute, febrile infectious disease with a high mortality rate in which there is extensive involvement of the lungs and the sputum is loaded with causative organisms.

pneumonitis *Medicine.* an inflammation of the lungs.

pneumonolysis *Medicine.* a division of the tissue attaching the lung to the wall of the chest cavity in order to permit collapse of the lung; formerly used to treat tuberculosis.

pneumopericardium *Medicine.* the presence of air or gas in the cavity of the pericardium.

pneumoperitoneum *Medicine.* the presence of air or gas in the peritoneal cavity; it may occur spontaneously, or may be deliberately introduced for X-ray examination.

pneumostome *Invertebrate Zoology.* in terrestrial and aquatic snails of the gastropod subclass Pulmonata, an opening into the mantle cavity used for respiration.

pneumotectic *Geology.* of or relating to the process and products of the consolidation of a magma that are influenced by the gaseous constituents of the magma.

pneumothorax *Medicine.* the presence of air or gas in the pleural space.

pneumotropic *Biology.* turning toward or having an affinity for lung tissues; used especially in reference to infective agents.

Pneumovirus *Virology.* a genus of viruses of the family Paramyxoviridae that cause respiratory ailments in humans and animals.

pnicogen *Chemistry.* any of the elements of group Va of the periodic table: nitrogen, phosphorus, arsenic, antimony, and bismuth.

pnictide *Chemistry.* any compound that contains the species X^{3-}, with X as nitrogen, phosphorus, arsenic, antimony, or bismuth.

pnigophobia *Psychology.* an irrational fear of choking or suffocating.

pnip transistor *Electronics.* a device in which a layer of germanium is placed between the base and the collector to produce a signal with a high-frequency range.

pn junction *Electronics.* the boundary separating two regions of semiconductor material with opposing types of impurities.

pnpn diode *Electronics.* a device consisting of four alternate layers of P- and N-semiconductor material, with the ability to block in one direction only, when the positive end is positively biased.

pnp transistor *Electronics.* a bipolar junction transistor in which an N-type base is placed between a P-type emitter and a P-type collector.

PNS *Aviation.* the airport code for Pensacola, Florida.

PNS peripheral nervous system.

PO or **P.O.** by mouth. (From Latin *per os.*) Also, **po, p.o.**

Po the chemical symbol for polonium.

Poaceae *Botany.* see GRAMINEAE.

poacher *Vertebrate Zoology.* any of various slender, marine fishes of the family Agonidae, characterized by a body covered with bony plates; found in deep waters of the North Pacific. Also, SEA POACHER.

Poales see CYPERALES.

Pobeda Peak *Geography.* the highest peak of the Tian Shan mountains (24,406 feet), on the border of western China.

pocill. in prescriptions, a little cup. (From Latin, *pocillum.*)

pock *Medicine.* a pustule or skin lesion formed as a result of an eruptive disease, such as smallpox.

pock assay *Virology.* a method of determining the concentration of infective particles in a cell culture by calculating the number of pocks induced by the virus on the allantoic membrane of a chicken egg.

Pockels cell *Optics.* a device containing an electro-optical crystal that aligns the electric field along the direction of an incident beam of light; used to modulate laser beams.

Pockels effect *Optics.* an electro-optical effect in which the refractive properties of piezoelectric crystals are altered by an applied electric field; the amount of change is proportional to the intensity of the field.

Pockel's readout optical modulator *Electronics.* a device that stores data in the form of images and employs a blue laser to write and a red laser to read or process data. Also, PROM.

pocket *Mining Engineering.* **1.** a small cavity, patch, or localized enrichment of an ore or mineral. **2.** a bin for ore storage. *Anatomy.* any saclike cavity in the body.

pocket beach *Geology.* **1.** a small, narrow, crescent-shaped beach formed in an enclosed or sheltered area along the coast and displaying well-sorted sands. **2.** see BAYHEAD BEACH.

pocket chisel *Mechanical Devices.* a woodworking chisel with a medium length blade.

pocket gopher *Vertebrate Zoology.* any of many burrowing rodents of the family Geomyidae, characterized by large, external, fur-lined cheek pouches; found in North and Central America.

pocketing *Robotics.* the use of numerical control equipment to cut and remove metallic materials.

pocket mouse *Vertebrate Zoology.* a nocturnal burrowing rodent of the family Heteromyidae that is very small (up to 5 inches) and has a long tail and external, fur-lined cheek pouches; found in parts of western North America.

pocket valley *Geology.* a valley whose head is surrounded by steep walls, and from whose base underground water flows as a spring.

pod *Botany.* **1.** a dry dehiscent legume. **2.** the bivalve case that contains the seeds of some plants, such as beans or peas. *Aviation.* a self-contained, usually detachable container, compartment, or housing on an aircraft; may be within the craft (as with an engine pod) or streamlined and carried externally (as with a wing pod). *Mechanical Devices.* **1.** a straight groove or channel in the barrel of an auger. **2.** a bit socket in a brace. *Geology.* a former term used to describe an orebody having an elongated, lenticular shape.

POD postoperative day.

pod- or **podo-** a combining form meaning "foot."

podalgia *Medicine.* a pain in the foot.

podalic *Medicine.* relating to the feet.

Podargidae *Vertebrate Zoology.* the frogmouths, a bird family of the order Caprimulgiromes, characterized by a large, flat, broad, hooked bill with a wide gape; found in southeast Asia, Australia, Indonesia, and Sri Lanka.

Podaxaceae *Mycology.* the single family of fungi belonging to the order Podaxales, occurring primarily on ant nests in sandy soil.

Podaxales *Mycology.* an order of fungi belonging to the class Gasteromycetes, characterized by crumbly spore tissue and an appearance similar to that of fungi of the order Agaricales.

Podbielniak extractor see ALFA LAVAL CONTRACTOR.

pod blight *Plant Pathology.* a destructive disease affecting plants of the legume family, caused by various fungi of the genus *Diaporthe*.

podetium *Botany.* on certain lichens, a shrublike or stalklike outgrowth of the thallus that bears the apothecium or fruiting body.

podiatrist [pō dī´ə trist] *Medicine.* a health-care professional specializing in care of the feet. Also, CHIROPODIST.

podiatry [pō dī´ə trē] *Medicine.* the care of the feet, especially the diagnosis and treatment of foot disorders. Also, CHIROPODY.

Podicipedidae *Vertebrate Zoology.* the grebes, a family of carnivorous freshwater diving birds composing the family Podicipediformes, characterized by lobed toes, colorful plumage used in courtship, strong flying ability, and clumsy land behavior display; found in freshwater lakes and marshes nearly worldwide.

Podicipediformes *Vertebrate Zoology.* the order containing the single family Podicipedidae, the grebes. Also, **Podicipitiformes.**

podite *Invertebrate Zoology.* a segment of an arthropod appendage; usually used with a prefix to describe its specific position on the leg, such as basipodite, propodite. Also, PODOMERE.

podium [pō´dē əm] *Architecture.* **1.** a base or platform upon which a building rests. **2.** a raised platform, as for a speaker or conductor on a stage. (Going back to a Greek term meaning "a little foot.")

pod nacelle *Aviation.* an engine nacelle carried under an airplane wing.

podo *Materials.* the soft wood of the East African podo tree; used for construction, packing crates, or furniture.

podobranch *Invertebrate Zoology.* in malacostracan crustaceans, a gill attached to the basal segment of a leg.

Podocarpaceae *Botany.* a family of evergreen, resiniferous trees and shrubs of the order Pinatae, characterized by scalelike or needlelike leaves, staminate cones with spirally or oppositely arranged scales, reduced pistillate cones, and two cotyledons; native mainly to the Southern Hemisphere.

Podocopa *Invertebrate Zoology.* a large subclass of small aquatic ostracod crustaceans, with leglike second antennae, three pairs of thoracic appendages, and cylindrical, well-developed posterior rami.

Podocopida *Invertebrate Zoology.* an order of ostracod crustaceans (the "modern" ostrocods), having valves with no anterior notch and a valve surface that is sometimes sculptured, and lacking compound eyes and a heart; mostly marine and freshwater, with a few terrestrial forms.

podocyst *Invertebrate Zoology.* a contractile, vesiclelike extension of the embryonic foot in gastropod mollusks.

pododermatitis *Veterinary Medicine.* see FOOT ROT.

pododynia *Neurology.* a pain in the heel and sole of the foot.

Podogona *Invertebrate Zoology.* an order of tropical arachnids with an elongated body, segmented abdomen, and posterior retractile tubercle. Also, RICINULEI.

Podolampaceae *Botany.* a family of tropical, marine, thecate dinoflagellates of the order Peridiniales, characterized by unique cingulate plates lacking transverse grooves.

podomere see PODITE.

podophyllin *Pharmacology.* the main constituent of podophyllum resin; used as a topical caustic on genital warts.

podophyllotoxin *Pharmacology.* $C_{22}H_{22}O_8$, a highly toxic substance that has antitumor properties and inhibits DNA synthesis; the main active ingredient of podophyllin.

podophyllotoxin *Toxicology.* $C_{22}H_{22}O_8$, a highly toxic poison found in podophyllum, having antineoplastic properties.

podophyllum *Pharmacology.* the dried seed and roots of the mandrake root, *Podophyllum peltatum,* used to obtain podophyllum resin and podophyllotoxin.

podophyllum resin *Pharmacology.* a powdered mixture of resins obtained from podophyllum; commonly used as a topical caustic agent.

Podospora *Mycology.* a genus of fungi belonging to the order Sordariales that are found on plant debris, dung, and seeds.

Podostemaceae *Botany.* the single family of the order Podostemales.

Podostemales *Botany.* a monofamilial order of the subclass Rosidae, containing tropical aquatic herbs that grow mostly submerged in fast, stony-bedded rivers but produce aerial flowers and fruit during times of low water.

Podoviridae *Virology.* a family of bacteriophages having isometric or elongated heads, short noncontractile tails, and linear dsDNA genomes; includes the T7 phage group.

pod rot *Plant Pathology.* a disease of cacao plants caused by the fungus *Monilia roreri* and characterized by lesions on the pods.

Podzol *Geology.* a group of zonal soils that develop in cool to temperate, moist climates; characterized by surficial mats of organic matter and thin layers of organic minerals overlying a highly leached, whitish-gray horizon and a dark brown illuvial horizon.

podzolization *Geology.* **1.** the processes by which the acidity of a soil increases as a result of the depletion of bases, forming surface horizons that have been leached of clay. **2.** the processes by which a Podzol develops.

POE port of entry; port of embarkation.

poecil- or **poecilo-** a combining form meaning "variegated."

Poeciliidae *Vertebrate Zoology.* the live-bearers, a large family of colorful small fishes of the order Antheriniformes; distinguished by the presence of a gonopodium in males; found in freshwater and away from open seas in many habitats throughout the Americas.

Poecilosclerida *Invertebrate Zoology.* the largest order of sponges in the class Demospongiae, having a skeleton composed of several different kinds of spicules, usually joined by spongin.

Poecilostomatoida *Invertebrate Zoology.* an order of copepod crustaceans that are parasitic on marine fishes.

Poeobiidae *Invertebrate Zoology.* a family of aberrant free-swimming polychaete worms, having a transparent body with no segments, appendages, or setae.

Poggendorff, Johann Christian 1796–1877, German physicist and scientific editor; invented the galvanometer.

pogonip *Meteorology.* an ice fog that forms in the mountain valleys of the western U.S.

pogonochore *Botany.* a plant that produces plumed disseminules.

Pogonophora *Invertebrate Zoology.* a phylum of deep-sea beard worms, having a long, threadlike body in a chitinous, sessile, cylindrical tube and a beardlike bunch of tentacles used in respiration and feeding.

Pogson ratio *Astronomy.* the ratio in brightness between successive stellar magnitudes; approximately 2.512 (the fifth root of 100). (Proposed by N. *Pogson* in 1856.) Also, **Pogson scale.**

poikil- or **poikilo-** a combining form meaning "variegated," "irregular."

poikilitic *Petrology.* of or relating to an igneous rock texture in which small crystals of one mineral are scattered irregularly and without common orientation in larger crystals of another mineral.

poikiloblast *Geology.* a large crystal produced by recrystallization during metamorphism and containing many inclusions of small grains.

poikiloblastic *Petrology.* of or relating to a metamorphic rock texture resembling the poikilitic texture of igneous rocks, with small grains of one constituent lying within larger xenoblasts. Also, SIEVE TEXTURE.

poikilocarynosis *Oncology.* the various cell types and cell formations occurring in Bowen's disease.

poikilocytosis *Medicine.* the presence of abnormally shaped red blood cells, or **poikilocytes,** in the peripheral blood. Also, **poikilocythemia.**

poikilophitic *Geology.* relating to the texture of certain igneous rocks in which thin, narrow crystals of feldspar are completely included in large, anhedral pyroxene crystals.

poikilotherm *Biology.* an animal that exhibits poikilothermy; a "cold-blooded" animal.

poikilothermy *Biology.* the exhibition of body temperature that varies with environmental temperature.

poikilotope *Geology.* in a sedimentary rock having a poikilotopic fabric, a large crystal that encases smaller crystals of a different mineral.

poikilotopic *Geology.* of or relating to the fabric of a recrystallized carbonate rock or a chemically precipitated sediment in which the larger of the multisized crystals enclose smaller crystals of a different mineral. Also, **poikilocrystallic.**

Poikilovirus *Virology.* a proposed genus of the family Herpesviridae.

Poincaré, Henri [pwan ka rā´] 1854–1912, French mathematician; made major contributions to the theory of functions, differential equations, and celestial mechanics.

Poincaré conjecture *Mathematics.* the assertion that a compact, simply connected three-dimensional manifold without boundary is homeomorphic to the three-dimensional sphere.

Poincaré electron *Electromagnetism.* a classical model of an electron in which the electron is held together by nonelectromagnetic forces and therefore has no self-stress; the model is unstable and consequently has infinite self-energy as a point electron.

Poincaré map *Mechanics.* see POINCARÉ SURFACE OF SECTION.

Poincaré recurrence theorem *Mathematics.* **1.** let $T: R^n \to R^n$ be a volume-preserving homomorphism on Euclidean n-space. Then for almost every point x of R^n, T is recurrent on any open set containing x; that is, any open set U containing a given point x also contains infinitely many points of the set $\{T^k(x) : k = 1, 2, 3, \ldots\}$. **2.** more generally, if X is a measure space with finite measure, then any measure-preserving transformation on X is recurrent.

Poincaré's theorem *Mathematics.* curl grad $a = 0$; div curl $a = 0$. $d(\omega \wedge \Theta) = d\omega \wedge \Theta + (-1)^r \omega \wedge d\Theta$, where r is the degree of Θ.

Poincaré surface of section *Mechanics.* a method of showing the characteristics of a trajectory without examining its complete time development, by sampling the trajectory periodically and plotting the rate of change of a quantity under study against the value of that quantity at the beginning of each period. Also, POINCARÉ MAP, SURFACE OF SECTION.

poinciana *Botany.* any of several tropical trees of the legume family, especially *Caesalpina pulcherrima* or *Peltophorum pterocarpum,* having showy scarlet, yellow, or orange flowers. (Named for M. de *Poinci,* 17th-century governor of the French Antilles.)

poinsettia [poin set´ē ə; poin set´ə] *Botany.* the plant *Euphorbia pulcherrima* of the spurge family, having large scarlet or white leaves and a small greenish-yellow flower; native to Mexico and Central America; often used as a Christmas flower. (Named for Joel Roberts *Poinsett,* 1799–1851, U.S. minister to Mexico, where he discovered the plant.)

Poinsot, Louis [pwan sō´] 1777–1859, French physicist and mathematician; formulated the concept of torque and theory of static couples.

Poinsot ellipsoid see INERTIA ELLIPSOID.

Poinsot motion *Mechanics.* the motion of a torque-free rotating rigid body in space, in general whose angular velocity vector precesses regularly about the constant angular momentum vector.

Poinsot's method *Mechanics.* a method of describing Poinsot motion, using a geometric construction in which the inertia ellipsoid rolls without slip on the invariable plane.

Poinsot's spiral *Mathematics.* the graph in the plane of the equation in polar coordinates $a = r \cosh n\theta$ or $a = r \sinh n\theta$, where n is an integer and a is any constant.

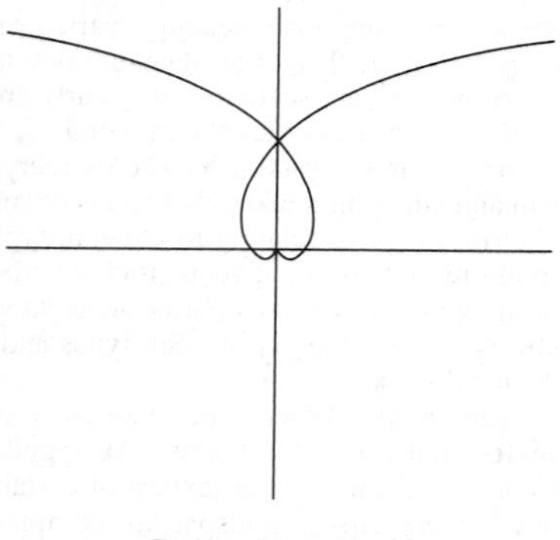

Poinsot's spiral

point a location in space; specific uses include: *Mathematics.* **1.** an element of a topological space. **2.** an element of a geometry that indicates position but not distance, direction, or size. **3.** the mathematical symbol, usually denoted by a period, that separates the integral part of a quantity from the fractional part of the quantity represented by a positional notation; e.g., decimal point in base ten notation, binary point in base two notation. *Geography.* a sharp, usually low piece of land projecting into the sea. *Transportation Engineering.* any location at which a landing or takeoff is made. *Navigation.* **1.** an older division of the compass equal to 1/32 of the circle, or 11.25°. **2.** a specific place on the earth. *Ordnance.* **1.** to aim a firearm or lay a gun or other weapon on a target. **2.** the tip or front end of a projectile. *Military Science.* the individual or group that precedes an advancing force. *Robotics.* the position of a gripper or end effector within a coordinate system or relative to a workpiece. *Graphic Arts.* **1.** the basic unit of type measurement, equal to 0.01384 inch (1/72 inch). **2.** in typesetting, a punctuation mark, especially a period.

point angle *Design Engineering.* an angle that forms the point or edge of a cutting tool.

point at infinity see IDEAL POINT.

point bar *Geology.* one of a series of low, curved ridges of sand and gravel enclosed by a growing meander and formed by the gradual addition of accretions as the channel migrates toward the outer bank. Also, MEANDER BAR.

point-bearing pile *Building Engineering.* a pile that depends for support on the soil or rock beneath its foot.

point-blank range *Mechanics.* a distance between the target and the firing point of a projectile, which is so short that any deflection of the trajectory from a straight line may be totally ignored.

point charge *Electricity.* an electric charge assumed to exist at a single point, thus having neither area nor volume.

point contact *Electronics.* the pressure contact between the surface of a semiconductor and a metallic point.

point-contact diode *Electronics.* a diode in which a fine wire, known as a cat whisker, is pressed against a semiconductor surface; used in microwave applications.

point-contact rectifier *Electricity.* a rectifier having a diode in which stray capacitance is minimized by using a point contact as a junction, thus allowing current to flow away from the junction in a radial fashion; used in high-frequency applications.

point-contact transistor *Electronics.* a device that has two or more point contacts adjacent to one another on the surface of a semiconductor and is capable of high-current amplification.

point defect *Crystallography.* a defect occurring in a crystal at a specific lattice site as opposed to affecting a larger (planar) area.

point diagram *Petrology.* a fabric diagram on which the preferred orientations of fabric elements, such as crystallographic axes of quartz grains, are plotted as points. Also, SCATTER DIAGRAM.

pointed arch *Architecture.* an arch with a point at its apex.

pointer *Engineering.* a needle or hand on a dial that moves to indicate measurement or direction. *Ordnance.* the person in an artillery crew who aims the weapon. *Computer Programming.* a data element containing the address or location of an item of data. *Mechanical Devices.* a tool used to remove old mortar from brick joints before new mortar is inserted.

Pointers *Astronomy.* the two outermost stars of the Big Dipper's bowl, known as Alpha and Beta Ursae Majoris; a line through them passes near Polaris, the pole star.

point estimate *Statistics.* a single value that is used to estimate a population parameter.

point fire *Military Science.* fire from more than one gun concentrated on a small target.

point fuse *Ordnance.* a fuse employed in the front end of a projectile or warhead.

point group *Crystallography.* a group of symmetry operations that leave unmoved at least one point within the object to which they apply. Symmetry elements include simple rotation and rotatory-inversion axes; the latter include the center of symmetry and the mirror plane. Since one point remains invariant, all rotation axes must pass through this point and all mirror planes must contain it. A point group is used to describe isolated objects, such as single molecules or real crystals.

point harmonization *Ordnance.* the process of adjusting the fixed guns of a fighter aircraft so that they will fire at a single point at a given range.

pointing *Ordnance.* the process of setting an artillery piece at a given elevation and direction. *Building Engineering.* **1.** the process of completing or finishing off the joints in a masonry structure. **2.** the mortar used as a finishing touch to stone work. *Materials Science.* to dress the surface of a stone with a pointed tool. *Metallurgy.* the process of reducing the size of an end portion of a rod, wire, tube, or pipe.

pointing trowel *Mechanical Devices.* a masonry trowel, typically with a five-inch, triangularly shaped blade, used to finish mortar joints and for small patching jobs.

point man *Military Science.* the lead soldier of an infantry patrol during combat operations.

point-mode display *Computer Technology.* a mode in which a point can be established on a display device by means of two computer words, one indicating vertical location, the other horizontal location.

point-mode servoing *Robotics.* a method for moving a robot's end effector to a specific location.

point mutation *Molecular Biology.* a mutation consisting of an alteration in a single nucleotide of a nucleic acid.

point of application *Mechanics.* the position on a rigid body at which a force is applied.

point of arrival *Navigation.* the position a craft is assumed to reach after following certain courses for certain distances from a specific starting place.

point of call *Transportation Engineering.* any airport at which a flight makes a scheduled landing.

point of contraflexure *Mechanics.* a point across which the bending moment on a flexural member changes sign; the bending moment at that point is zero. Also, POINT OF INFLECTION, POINT OF REFLECTION.

point of control *Industrial Engineering.* the number or percentage of defectives in a "borderline" lot, which stands an equal chance of acceptance or rejection by quality control.

point of departure *Navigation.* the initial point of embarkation of a voyage.

point of destination *Navigation.* the final point of arrival of a voyage.

point of fall *Ordnance.* the point on the falling section of a trajectory that is level with the muzzle of the gun. Also, LEVEL POINT.

point of impact *Ordnance.* **1.** the point at which a projectile, bomb, or re-entry vehicle impacts or is expected to impact. Similarly, **point of burst. 2.** the point in the dropped zone at which the first parachutist or air-dropped cargo item lands or is expected to land.

point of inflection *Mathematics.* let f be a function that is differentiable in a neighborhood of a point x_0 of its domain. f is said to have a point of inflection at x_0 if and only if f' has a local extremum at x_0. If the graph of f is a plane curve, then $(x_0, f(x_0))$ is a point at which the curve changes from concave to convex, relative to some fixed line, and the tangent of f intersects the graph at x_0. Also, INFLECTION POINT, INFLECTIVE POINT. *Mechanics.* see POINT OF CONTRAFLEXURE.

point of intersection *Civil Engineering.* the point at which the tangent sections at either end of a curved section of road would meet if extended.

point of maximal impulse *Cardiology.* the point on the chest at which the strongest impulse from the left ventricle is felt, normally occurring inside the left mamillary line.

point of no return *Navigation.* the point in a flight or voyage after which it is easier or safer to continue to the destination than to return to the starting point.

point-of-origin system *Computer Technology.* a computer system in which data collection occurs at its source or origin.

point of reflection see POINT OF CONTRAFLEXURE.

point-of-sale terminal *Computer Technology.* a terminal that records sales data at the time and place a sale is made and then inputs the data directly into a central computer.

point of switch *Civil Engineering.* in a railroad turnout, the point at which wheels actually pass from one track to the other.

point of tangency *Civil Engineering.* the point at which a curved section of roadway meets a tangent (straight) section.

point of the compass see COMPASS POINT.

point of zero charge see ZERO CHARGE POTENTIAL.

point particle *Mechanics.* an idealized object that has mass and possibly charge or other attributes, but which is considered to have zero volume. Also, MATERIAL PARTICLE.

point processing *Robotics.* the processing of one program using information from another program or routine.

point provenience *Archaeology.* the location (provenience) of a specific object at an exact point on a site.

point rainfall *Meteorology.* **1.** the rainfall measured in a rain gauge during a specific time interval, which is often one storm. **2.** an estimate of the amount of rainfall that might have been measured at a given point.

points *Agriculture.* the important features of an animal that are noted in the judging of livestock. *Electronics.* see BREAKER POINTS.

point set topology *Mathematics.* the theory of topological spaces based on general axioms about sets of points, as distinguished from topological spaces that are constructed from geometrical spaces by removing one or more of their properties.

point source *Physics.* a source (of radiation, mass, or charge) whose dimensions are small compared to the distance to the observation point. *Astronomy.* any light source sufficiently distant that it has no detectable angular extent.

point-source light *Electricity.* a type of lamp with a concentrated radiating element.

point-source method *Radiology.* the administration of radiation to the urinary bladder wall, using a small point source of radiation centered within a Foley catheter bag containing a contrast medium solution.

point spectrum *Mathematics.* let $T: X \to X$ be a linear operator on a complex Banach space X, I the identity operator on X, λ a complex number, and $T_\lambda = \lambda I - T$. Then λ belongs to the point spectrum of T if T_λ^{-1} does not exist. If λ does not belong to the point spectrum of T, then it belongs to one of the following sets: the resolvent set of T, the residual spectrum of T, or the continuous spectrum of T.

point system *Industrial Engineering.* a system of evaluating a job for compensation purposes by assigning point values to various features of the job, e.g., decision-making required, physical arduousness, working conditions, etc.; the total point value determines the payment for the job.

point target *Ordnance.* **1.** a conventional target of such small dimension that it requires the accurate placement of ordnance in order to neutralize or destroy it. **2.** a nuclear target in which the ratio of the radius of damage to the target radius is equal to or greater than five. *Electromagnetism.* an object that returns a target signal by reflection from a relatively simple discrete surface.

point-to-point communication *Telecommunications.* a direct radio communication between two terminal installations.

point-to-point control *Robotics.* a control in which a limited number of points along a robot's path are specified, and the remaining points are calculated by the system. Thus, **point-to-point robot, point-to-point programming.**

point-transfer device *Photogrammetry.* a stereoscopic device used to mark corresponding image points on overlapping photographs.

point transposition *Electricity.* an ordered interchange of position within an open-wire line and its phases; it occurs so that effects of mutual capacitance and inductance with consequent interference are minimized or balanced without material distortion of the normal wire configuration at a distance comparable to the wire separation.

pointwise convergence *Mathematics.* a sequence $\{f_1, f_2, f_3, \ldots\}$ of functions as defined on a set S is said to **converge pointwise** to a function f if the sequence $\{f_1(x), f_2(x), f_3(x), \ldots\}$ converges to $f(x)$ for every x in S.

poise [pwäz] *Fluid Mechanics.* the unit of dynamic viscosity in the centimeter-gram-second system of units, equivalent to the viscosity of a fluid exhibiting tangential resistance of one dyne between two fluid layers of one square centimeter each, separated by one centimeter, while having a differential velocity of one centimeter per second. (Named for Jean-Louis-Marie Poiseuille.)

poised mine *Ordnance.* a naval mine in which the ship counter has been run down to "one," thus setting the mine to detonate at the next actuation.

poised stream *Hydrology.* a stream that neither deposits nor erodes sediment.

Poiseuille, Jean-Louis-Marie [pwä zwē´] 1799–1869, French physician and physiologist; studied blood circulation; discovered Poiseuille's law.

poiseuille *Fluid Mechanics.* the unit of dynamic viscosity in the meter-kilogram-second system of units, equivalent to the viscosity of a fluid exhibiting tangential resistance of one newton between two fluid layers of one square meter each, separated by one meter, while having a differential velocity of one meter per second; the unit is equal to 10 poise.

Poiseuille flow *Fluid Mechanics.* the steady flow of fluid in a pipe with circular cross section and infinite length, caused by a pressure gradient along the axis of the pipe.

Poiseuille's law *Fluid Mechanics.* a law stating that the volumetric flow in a tube of circular cross section is given by the quantity $\pi r^4 Dp/8lh$, where r is the radius of the tube, Dp is the pressure difference across the ends of the tube, l is the tube length, and h is the dynamic viscosity.

poison *Toxicology.* **1.** any substance that has the inherent ability to kill, injure, or damage living organisms, especially a substance that is destructive in relatively small doses. **2.** to administer such a substance. *Chemistry.* a trace impurity that reduces or eliminates catalytic activity by being adsorbed at the active sites, such as carbon monoxide, which affects the formation of ammonia from hydrogen and nitrogen gases. *Electronics.* a material that inhibits the flow of electrons from the surface of a cathode. *Atomic Physics.* a substance that limits phosphorescence in an irradiated material. *Nucleonics.* a material that can absorb neutrons without experiencing fission, thus lowering the reactivity of a nuclear reactor. (Derived from the Latin word for "drink," from the idea of a poisonous drink.)

poison bean *Botany.* the plant *Glottidium vesicaria* of the legume family, having yellow flowers and pods containing highly poisonous seeds; found in the southeastern U.S.

poison gas *Materials.* any of various toxic gases, such as phosgene or chlorine, especially those used in chemical warfare.

poison gland *Vertebrate Zoology.* a gland in the spines of fins or tails of certain fishes, in the hind claws of the platypus, and as an oral gland in certain snakes, lizards, and mammals, such as shrews and bats, that secretes a poisonous mucuslike substance; used in self-defense, predation, etc.

poison hemlock *Botany.* the biennial herb, *Conium maculatum*, of the family Apiaceae, having finely divided leaves and white flowers; it contains the alkaloid coniine in its leaves and fruits.

poisoning *Toxicology.* the harmful effect on a living organism of ingestion, inhalation, or contact with a poison.

poison ivy *Botany.* the climbing, aborescent or shrubby plant, *Rhus radicans*, of the family Anacardiaceae, characterized by glossy trifoliolate leaves, greenish flowers, and white berries producing an oil that is highly irritating to human skin.

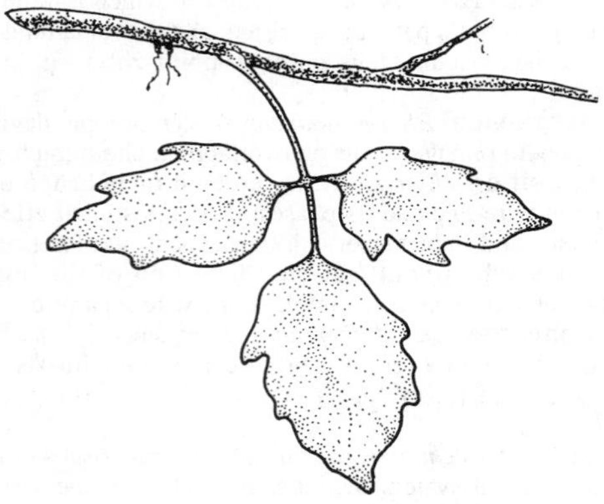

poison ivy

poison oak *Botany.* **1.** a plant, *Rhus diversiloba*, of the family Anacardiaceae, resembling poison ivy and producing a similar irritaing oil. **2.** broadly, any of several types of poison ivy or poison sumac plants.

poisonous *Toxicology.* causing a harmful effect when ingested, inhaled, or in contact with a living organism.

poison sumac *Botany.* a tall bush, *Rhus vernix*, of the family Anacardiaceae, having pinnately compound leaves and drooping clusters of white berries that produce a highly irritating oil. Also, **poison dogwood.**

Poisson, Siméon Denis [pwä sōn´] 1781–1840, French mathematician and physicist; formulated the Poisson distribution and Poisson's equation.

Poisson arrivals *Transportation Engineering.* arrivals in a queue as modeled by use of a Poisson distribution rule.

Poisson-Boltzmann equation *Physical Chemistry.* an approximate equation governing the distribution of ions near a charged particle or surface; generally most valid at low concentrations.

Poisson bracket *Mathematics.* let f and g be continuously differentiable functions of the variables $(x_1, \ldots, x_n, p_1, \ldots, p_n)$. The Poisson bracket (f,g) is then defined by

$$(f,g) = \sum_{k=1}^{n} (\partial f/\partial p_k \partial g/\partial x_k - \partial g/\partial p_k \partial f/\partial x_k).$$

The Poisson bracket is antisymmetric and satisfies the Jacobi identity, and the Poisson bracket of a function with itself or a constant is zero. In addition, $(f_1+f_2,g) = (f_1,g) + (f_2,g)$; $(f_1f_2,g) = f_2(f_1,g) + f_1(f_2,g)$. The latter identity is analogous to the product rule of differential calculus.

Poisson constant *Physics.* the quotient of the universal gas constant R divided by the specific heat at constant pressure C_p for a specified gas, denoted by k.

Poisson curve *Transportation Engineering.* the probability curve formed by a Poisson distribution.

Poisson distribution *Statistics.* the distribution of the count X of the number of randomly occurring events in a fixed interval of time. It has a wide variety of applications as a distribution for the occurrence of rare events. *Transportation Engineering.* this probability distribution in which $f(x) = k^x e^{-k}/x!$, where $x = 0, 1, 2, \ldots$; commonly used to generate trips in a traffic model.

Poisson integral formula *Mathematics.* an explicit formula for solution of the Dirichlet problem $\Delta u = u_{xx} + u_{yy} = 0$ on the interior of the unit circle in terms of given boundary values f; i.e., in polar coordinates,

$$u(r, \theta) = (1/2\pi)\int_0^{2\pi} f(t)(1 - r^2)\,[1 - 2r\cos(\theta - t) + r^2]^{-1} dt.$$

This solution is valid if the boundary data are continuous.

Poisson number *Mechanics.* the reciprocal of the Poisson ratio.

Poisson process *Statistics.* a random process describing the total number of randomly occurring events over time; times between successive events are exponential random variables. The number of events in a fixed time interval is a **Poisson random variable.**

Poisson ratio *Mechanics.* the absolute ratio of the transverse strain to the axial strain of a long specimen under an axial tensile or compressive stress at its ends.

Poisson relation *Geophysics.* for a body longitudinally stressed within its elastic limits, the ratio of unit elongation (c) to unit lateral contraction (s).

Poisson's equation *Mathematics.* Laplace's equation with sources; i.e., $\Delta u = -f(x_1, \ldots, x_n)$, where Δ is the Laplacian operator. For example, if f is mass density, the solution u is the gravitational potential to which it gives rise.

Poisson summation formula *Mathematics.* suppose that, for a given function f, the series $\sum_{k=-\infty}^{\infty} f(2\pi k + t)$ will converge uniformly to a function of bounded variation. Then

$$\sum_{k=-\infty}^{\infty} f(2\pi k) = \sum_{k=-\infty}^{\infty} \left(\int_{-\infty}^{\infty} f(x)e^{-ikx}dx\right).$$

Poisson transform *Mathematics.* given a real function f, the integral transform

$$F(x) = (1/2\pi)\int_0^{\infty} t(x^2 + t^2)^{-1} f(t)\, dt.$$

Pojetaia *Paleontology.* an early genus of pelecypod mollusks in the subclass Anomalodesmata; found in Lower Cambrian strata in Australia.

poke *Computer Programming.* **1.** to write information into a specified absolute memory location from a high-level language. **2.** a command that allows the user to accomplish this.

pokeweed *Botany.* a tall herb, *Phytolacca americana*, characterized by juicy purple berries and a purple root used in medicine, and young edible shoots; found in North America. Also, **poke-root.**

pokeweed mitogen *Immunology.* a substance extracted from the pokeweed plant that stimulates mitosis in lymphocytes and activates T lymphocytes; composed of five (Pa 1–5) mitogens.

poky mutant *Genetics.* a mutant strain of the *Neurospora*, having a slow rate of growth caused by a deficiency in its respiratory chain.

pol *Molecular Biology.* any gene that encodes a DNA polymerase enzyme.

pol I *Enzymology.* one type of RNA polymerase that catalyzes the production of the large fragment of ribosomal RNA (rRNA). Also, RNA POLYMERASE I.

pol II *Enzymology.* one type of RNA polymerase that catalyzes the production of both messenger RNA (mRNA) and heterogeneous RNA (hnRNA). Also, RNA POLYMERASE II.

pol III *Enzymology.* one type of RNA polymerase that catalyzes the production of the small fragment of ribosomal RNA (5S rRNA) and transfer RNA (tRNA). Also, RNA POLYMERASE III.

polacke [pə läk´ē] *Meteorology.* a cold, dry, northeasterly katabatic wind occurring over the region of Bohemia as air descends from the Sudeten mountains near Poland.

Poland or **Poland China** *Agriculture.* a breed of large hog that is black with white tips on the tail, feet, and head; developed in Ohio.

polar *Geography.* of or relating to the North or South Pole, or the regions surrounding them. *Science.* of or relating to any system or region having opposing points or tendencies, such as a magnet, an electrolytic cell, and so on. *Mathematics.* **1.** let Γ be a conic in the plane and P a point external to Γ from which tangents to Γ are drawn. The polar of P with respect to Γ is the line l joining the two points of tangency, and P is the pole of l with respect to Γ. **2.** if P lies on the conic, then l is the tangent to Γ at P. If P lies inside the conic, then draw any two lines m and n through P, each intersecting Γ in two points, and find their poles M and N. Then the polar l of P is the line MN. Again P is the pole of l.

polar air *Meteorology.* an air mass with weather characteristics that are developed over high latitudes, such as continental polar air and maritime polar air.

polar anticyclone see ARCTIC HIGH.

polar automatic weather station *Meteorology.* an automatic, unmanned weather station used in frigid and polar climates to measure meteorological elements and transmit the data by radio; the equipment is designed to function on ice or slush surfaces and is encased in a sled-like structure with pontoons for stability.

polar axis *Crystallography.* an axis in a crystal that exhibits different properties at its two ends, having directionality (like an arrow). These properties include face development and charge accumulation, and they may be used to define the directionality of the axis. *Astronomy.* **1.** the diameter of a sphere that connects its two poles. **2.** the axis, parallel to the earth's axis, about which an equatorially mounted telescope revolves to keep a celestial object in the field of view. *Mathematics.* the directed line relative to which the angular coordinate of a polar coordinate system is measured. In the usual system, this is the positive x-axis (or the positive real axis).

polar bear *Vertebrate Zoology.* the large, white bear, *Ursus maritimus*, of the family Ursidae, related to the brown bear of the same genus, but distinguished by a longer body, a long, slender head, white fur, hair-covered soles, and a strictly carnivorous diet, mainly seals; it often reaches a length of 9 feet and a weight of 1000 pounds; found on arctic coasts, islands, and sea ice of Eurasia and northern North America.

polar bear

polar body *Cell Biology.* one of the two small cells produced and discarded during each of the two meiotic divisions that yield the haploid ovum.

polar bond *Physical Chemistry.* a bond between two atoms in which electrons are shared unequally; i.e., one atom acquires a partial negative charge and the other a partial positive charge, and the entire molecule remains neutral. Also, **polar bonding, polar covalent bond.**

polar cap *Hydrology.* an ice sheet that is centered at one of the earth's poles. *Astronomy.* seasonal or permanent deposits of snow, ice, or frozen material in the polar regions of another planet, e.g., Mars.

polar capsule *Invertebrate Zoology.* a specialized cell in the spores of some Sporozoa, with a coiled thread that can be shot out, probably an anchoring device.

polar chart *Cartography.* any chart that is based on a polar map projection.

polar circle *Cartography.* either the Arctic Circle (**north polar circle**) or the Antarctic Circle (**south polar circle**), drawn on maps to indicate the lowest latitude in either hemisphere at which the sun does not set during the summer solstice.

polar climate *Meteorology.* the climate of a polar region, usually one that is too cold to allow the growth of trees. Also, ARCTIC CLIMATE, SNOW CLIMATE.

polar compound *Chemistry.* a compound that has an electric dipole moment and interacts with other polar molecules through electrostatic forces.

polar continental air *Meteorology.* a shallow air mass originating over land or frozen ocean areas in the polar latitudes, characterized by cold surface temperatures, low moisture content, and strong stability in the lower layers.

polar convergence *Oceanography.* the line along which cold polar water moving toward the equator meets and sinks below warmer subpolar water.

polar coordinates *Mathematics.* **1.** a coordinate system for the real plane, in which the position of a point $P = (x, y)$ in the plane is described in terms of the radius r of the circle about the origin on which P lies and the angle θ, measured counterclockwise around from the positive x-axis, of a directed ray from the origin on which P lies. The coordinates of P are then (r, θ), and P is said to be represented in polar form. If P is not the origin, then $r = (x^2 + y^2)^{1/2}$, and $\tan \theta = y/x$. Generally, $0 \le \theta < 2\pi$. **2.** a coordinate system for the complex plane, in which the position of a complex number $z \in C$ in the plane is described in terms of the radius r of the circle about the origin on which z lies and the angle θ, measured counterclockwise around from the real axis, of a directed ray from the origin on which z lies. The coordinates of z are then (r, θ), and the notation $z = re^{i\theta}$ is used, where $e^{i\theta} = \cos \theta + i \sin \theta$. z is said to be represented in polar form. If, in rectangular coordinates, $z = a + ib$, then (for $z \neq 0$) $r = |z| = (a^2 + b^2)^{1/2}$, and $\tan \theta = b/a$. Generally $0 \le \theta < 2\pi$. *Ordnance.* in artillery and naval gunfire support, the direction, distance, and vertical correction from the observer or spotter position to the target.

polar crystal *Crystallography.* a crystal with ionic bonding between atoms.

polar cyclone see POLAR VORTEX.

polar day *Astronomy.* a 24-hour period in which the sun is always above the horizon for observers within the Arctic or Antarctic Circles.

polar desert *Geography.* an area, especially in the polar regions, having little or no vegetation due to extremely cold temperatures.

polar diagram *Physics.* any diagram that uses polar coordinates to describe the magnitude of some quantity in terms of the distance from the origin and the relative angle about the origin (with respect to the x-axis).

polar distance *Astronomy.* the angular distance of an object from a celestial pole.

polar easterlies *Meteorology.* the shallow and diffuse body of easterly winds that are located north of the Aleutian and Icelandic lows in the mean of the Northern Hemisphere.

polar-easterlies index *Meteorology.* a measurement of the strength of easterly winds between the latitudes of 55°N and 70°N; determined from the average sea-level pressure differences between these latitudes, and expressed as the east-to-west wind component in both meters and tenths of meters per second.

polar electrojet *Geophysics.* a narrow belt of intense electric current flowing through the lower ionosphere in the polar regions.

polar fiber *Cell Biology.* any of the long microtubules that occur in the spindle during cell mitosis and extend from the spindle poles to the equator.

polar firn *Hydrology.* firn that is formed at low temperatures where no liquid water is present and no melting takes place. Also, DRY FIRN.

polar fox see ARCTIC FOX.

polar front *Meteorology.* the major transitional region of a semipermanent and semicontinuous front separating polar air masses at the high latitudes from the relatively warmer easterly subtropical air masses and southwesterly tropical air masses; very susceptible to cyclogenesis.

polar-front theory *Meteorology.* a theory stating that a polar front, which separates air masses of tropical and polar origin, gives rise to cyclonic disturbances that intensify and travel along the front.

polar glacier *Hydrology.* a glacier that has a subfreezing temperature throughout its mass and that experiences no melting even in summer.

polar granule *Entomology.* a small particle or grain at the pole of a bacterial cell.

polar grid *Cartography.* a grid system used on maps of polar regions, having the origin at the pole, the x-axis along the 0° and 180° meridians, and the y-axis along the 90°E and 90°W meridians.

polar group *Biochemistry.* a water-soluble chemical group, such as a hydrophilic side chain of an amino acid. *Chemistry.* a group having a large dipole moment, such as –OH or –COOH. *Materials Science.* any structure in which a strong attractive force shifts some electrons to cause an unbalanced distribution.

polar high see ARCTIC HIGH.

polar ice *Geophysics.* any sea ice that is more than one year old and more than 3 meters thick. Also, ARCTIC PACK.

polarimeter *Optics.* an instrument used to measure the rotation of the plane of polarization by materials placed within it.

polarimetric analysis *Analytical Chemistry.* the measurement of the number of degrees and direction of rotation of a plane of polarized light as it passes through an optically active compound, determined by the use of prisms.

polarimetry *Optics.* the measurement of the rotation of the plane of polarization of radiant energy; used as a basis for chemical analysis.

Polaris *Astronomy.* Alpha (α) Ursae Minoris, the brightest star in Ursa Minor; it lies now less than 1° from the north celestial pole. Also, POLE STAR. *Ordnance.* a strategic ballistic missile designed to be launched from a submerged or surfaced submarine; it uses solid-propellant fuel, is equipped with inertial guidance, and carries a nuclear warhead; it has a range of 1200 to 2500 nautical miles, depending on the model; officially designated **UGM-27.**

Polaris missile

polariscope *Optics.* an instrument used to view stress patterns in doubly refracting materials that are placed between the polarizer and analyzer of the instrument.

Polaris correction *Navigation.* a quantity that, when applied to the observed altitude of Polaris, yields the latitude.

polariton *Solid-State Physics.* a quantum of interaction between a phonon and a photon.

polarity the fact of being polar; specific uses include: *Physics.* **1.** a physical property of some systems by which there exist two points having opposite characteristics, such as an electric dipole. **2.** the characteristic of an alternating quantity indicating whether it is in its positive half-cycle or its negative half-cycle. *Electricity.* **1.** a characteristic of the poles of a magnet (north or south), or of the terminals of a battery (positive or negative.) **2.** the direction in which a direct current flows. *Telecommunications.* in a television picture signal, the sense of the potential of a portion of a dark area of a scene relative to that of a portion of a light area. *Mathematics.* a property of a geometric configuration that opposite parts are distinguishable from one another. *Evolution.* the direction of evolution or change within a transformation series. *Biochemistry.* **1.** the difference between the 5' and 3' terminals of nucleic acids. **2.** a mutation within a gene to inhibit the translation of distal genes from the same operon.

polarity effect *Electronics.* a condition in which the magnitude of the breakdown voltage between two electrodes depends on the polarity of the voltage between the electrodes.

polarity epoch *Geophysics.* an era in the earth's history in which the magnetic field has one predominant polarity.

polarity event *Geophysics.* a relatively short period of time (100,000 years or less) within a polarity epoch in which the polarity of the earth's magnetic field is opposite to the predominant polarity of that epoch.

polarity gradient *Genetics.* a measure of the ability of a polarity mutation in one gene to affect the expression of other genes in an operon; usually dependent on the distance of the mutant codon from the next chain initiation signal.

polarity mutation *Genetics.* a mutation in one gene that affects the rate of expression of one or more genes that are adjacent to it on a chromosome. Also, **polar mutation.**

polarizability *Materials Science.* a proportionality factor that measures the average dipole moment per unit of local field strength. *Electricity.* the property of a molecule or an atom whose electron cloud can be readily deformed by an external electron field, thus inducing an electric dipole moment.

polarization *Physics.* a phenomenon associated with wave propagation in certain media, by which the displacement vector of the wave disturbance varies in time in a definite way. *Electricity.* **1.** the separation of the positive and negative charges of an atom or molecule by an external force, typically an electric field. **2.** the direction of electric lines in a radio wave. **3.** an increase in the internal resistance of a battery cell, which shortens the cell's useful life as a result of an active chemical change occurring within it. **4.** a vector quantity equal to the electric dipole moment per unit volume of a material. Also, DIELECTRIC POLARIZATION, ELECTRIC POLARIZATION. *Optics.* a state in which rays of light or other radiation traveling in different directions exhibit different properties. *Crystallography.* see PLANE POLARIZATION.

polarization charge see BOUND CHARGE.

polarization diversity *Telecommunications.* **1.** a system of reception and transmission that lessens the effects of selective fading of the vertical and horizontal components of a signal. **2.** any method of diversity reception and transmission by which the same information signal is received and transmitted simultaneously on right-angled polarized waves.

polarization ellipse *Physics.* an ellipse that is traced out by the tip of the time-varying displacement vector of a polarized wave, when measured at a fixed point in space.

polarization error *Navigation.* a fluctuation in the direction of radio waves at sunrise and sunset due to rapid changes in the ionosphere.

polarization factor *Crystallography.* a factor that takes into account the reduction in intensity on X-ray scattering as a result of polarization on reflection.

polarization fading *Telecommunications.* fading that takes place as a result of changes in the direction of polarization of the received signal with respect to the polarization of receiving antennae; it occurs because of fluctuations in electron density along the propagation path.

polarization isocline *Meteorology.* a locus of all points at which the inclination to the vertical of the plane of polarization of the diffuse sky radiation has the same value.

polarization of atoms *Crystallography.* the separation of charges in an atom in response to some stimulus.

polarization of light *Crystallography.* a condition in which the electric vector of electromagnetic radiation is no longer equally distributed in the plane normal to the direction of the propagation of light.

polarization potential *Physical Chemistry.* the voltage that opposes the direct, useful voltage in an electrolytic cell. *Electromagnetism.* either the electric field vector or magnetic field vector from which an electric scalar potential or a magnetic vector potential, respectively, can be derived. Also, HERTZ VECTOR.

polarization spectroscopy *Spectroscopy.* a saturation spectroscopic technique in which a laser is used to polarize certain molecules of a sample that are then detected through their induction of polarization in a probe beam.

polarize *Science.* to cause to undergo a process of polarization.

polarized *Science.* exhibiting or affected by polarization.

polarized electrode *Physical Chemistry.* an electrode that has sufficient electrolytic activity to be identified as an anode or a cathode.

polarized electrolytic capacitor see ELECTROLYTIC CAPACITOR.

polarized electromagnetic radiation *Electromagnetism.* an electromagnetic wave whose electric field vector, together with the direction of propagation, lies in a fixed plane rather than in one of random orientation.

polarized ion source *Electronics.* a device that generates a beam of ions with the same alignment and spin.

polarized light *Optics.* light whose waves vibrate transversely on a single plane, as distinguished from ordinary light whose waves vibrate in all directions.

polarized meter *Engineering.* an instrument used to measure the direction or voltage of a current, by means of a zero-center scale and an indicating needle that moves in relation to the polarity of the current's voltage or direction.

polarized neutrons *Physics.* an aggregate of neutrons whose spins have a collective net direction rather than a random distribution of orientations (which would typically result in no net direction).

polarized plug *Electricity.* an electrical plug that can be inserted in only one position; designed for use with a polarized receptacle.

polarized receptacle *Electricity.* a protected receptacle, such as a socket, that can receive a plug only in one position, thus helping to prevent incorrect connections.

polarized relay *Electricity.* a relay in which movement of the armature depends upon the direction of the energizing current or voltage in the relay coil. Also, **polar relay.**

polarized-vane ammeter *Engineering.* an instrument in which an electric current moves through a small coil, distorting the field of a permanent magnet, and a vane moves in relation to the distorted field to provide a rough measurement of the current.

polarizer *Optics.* a device, such as the Nicol prism, that absorbs all vibrations from transmitted light except those within a single plane, so that the natural light emerges as polarized light.

polarizing angle see BREWSTER ANGLE.

polarizing filter *Optics.* a device that transmits only one plane of polarized light; used in photography to control reflections, and in photomicrography to distinguish crystals.

polarizing microscope *Optics.* a microscope that contains a polarizer and analyzer to examine birefringent substances in linearly polarized light.

polar keying *Telecommunications.* in telegraphy, a signal in which current flows in only one direction for marking, and in the other for spacing.

polar lake *Hydrology.* a lake whose surface temperature never rises above 4°C.

polar lights see AURORA BOREALIS, AURORA AUSTRALIS.

polar lobe *Cell Biology.* in some mollusks, a protrusion that is found at one end of the cell prior to the first cleavage and remains associated with only one of the daughter cells.

polar low see POLAR VORTEX.

polar map projection *Cartography.* any map projection that has a geographic pole at its center.

polar maritime air *Meteorology.* a polar air mass that initially possesses characteristics similar to those of continental polar air, having low surface temperature, low moisture content, and great stability in the lower layers that, after passing over warmer water, becomes unstable while maintaining a moderately low temperature and a high moisture content.

polar membrane *Cell Biology.* a multilayered membrane, found at the cell poles of certain helical and vibrioid bacteria that is attached to the inner face of the cytoplasmic membrane.

polar meteorology *Meteorology.* the study of meteorological principles that considers atmospheric conditions in both northern and southern high latitudes or polar ice cap regions.

polar migration see POLAR WANDERING.

polar modulation *Telecommunications.* a form of amplitude modulation in which the positive and negative excursions of the carrier are modulated by two separate, distinct signals.

polar molecule *Physical Chemistry.* a molecule in which a pair of opposite charges of equal magnitude are permanently separated at a specific distance from each other.

polar navigation *Navigation.* techniques of position finding and course determination used in polar regions.

polar night *Astronomy.* a 24-hour period in which the sun is never above the horizon for observers within the Arctic or Antarctic Circles.

polar nucleus *Botany.* either of two usually haploid nuclei in an embryo sac of a seed plant that fuses with a gamete to form the endosperm nucleus.

polarogram *Analytical Chemistry.* the graphic representation of an electroanalytical technique in which the current at a dropping mercury electrode is plotted against the electrode potential.

polarographic analysis *Analytical Chemistry.* the measurement of the current at an electrode in an electrolysis cell as a function of the applied potential, i.e., a current-voltage relationship; techniques include alternating-current polarography, square-valve polarography, and so on. Also, POLAROGRAPHY.

polarographic cell *Analytical Chemistry.* a flask, cup, or jar containing reference and indicator electrodes immersed in an electrolyte solution; used to hold a test solution for polarography.

polarography see POLAROGRAPHIC ANALYSIS.

Polaroid *Optics.* a trade name for a thin sheet of colorless plastic that is treated with an iodine solution, creating thin, parallel chains of polymeric molecules containing conductive iodine atoms; it produces polarized light from unpolarized light by dichroism; used to reduce glare in optical and lighting devices.

Polaroid camera

Polaroid camera *Optics.* a brand of camera with a special film that contains its own developing and printing agents, producing a finished positive print within a minute or two after the photograph is taken; originally developed by Edwin H. Land in 1948. Also, LAND CAMERA.

polaron *Solid-State Physics.* in a crystal lattice, an electron surrounded by a cloud of phonons; the combination produces local deformation in the crystal lattice due to electron-ion interaction.

polar orbit *Space Technology.* a space vehicle orbit that passes over or near the earth's poles.

polar orthographic map projection *Cartography.* a map projection having the plane of projection tangent to the earth's sphere at the axis of rotation (the pole) and parallel to the equator; parallels are circles of true scale, and meridians are straight lines radiating from the origin.

polar outbreak *Meteorology.* the movement of a cold air mass from its source region, usually a vigorous thrust of cold polar air moving in an equatorward direction.

polar radiation pattern *Acoustical Engineering.* a plot of the acoustic response of an acoustic-to-electroacoustic transducer in all directions on a specific plane.

polar regions *Geography.* the areas roughly coterminous with the northern and southern frigid zones.

polar resolution *Computer Programming.* the process of determining the magnitude of a vector and the angle it makes with the x-axis.

polar symmetry *Crystallography.* the symmetry associated with a crystal axis.

polar telescope *Optics.* a telescope equipped with a rotating mirror system that allows an individual to view celestial objects through a fixed eyepiece.

polar timing diagram *Mechanical Engineering.* a diagram showing the events of an engine cycle relative to the crankshaft position.

polar transmission *Telecommunications.* a method of signaling, using three states: a mark, a space, or the absence of space, indicating a no-signal condition. Since there is zero voltage between signals, the process is sometimes referred to as **polarized return to zero.**

polar trough *Meteorology.* an atmospheric wave trough that has sufficient amplitude to reach the tropics in the upper air as it travels in the circumpolar westerlies; at the surface, it is reflected as a trough in the tropical easterlies, and is characterized by westerly winds at higher elevations.

polar variation *Geophysics.* a small change or wobble in the earth's axis of rotation.

polar vector *Mathematics.* a vector field in R^3 that transforms properly (i.e., as multiplied by the Jacobian matrix) under coordinate changes, regardless of the sign of the Jacobian determinant.

polar vortex *Meteorology.* the large-scale cyclonic circulation at the mid- and upper-tropospheric layer centered mainly over the polar regions; characterized by a two-center vortex, one near Baffin Island and the other over northeast Siberia. Also, POLAR CYCLONE, POLAR LOW.

polar wandering *Geology.* the movement of the earth's rotation poles and magnetic poles over geologic time. Also, CHANDLER MOTION, POLAR MIGRATION.

polC gene see DNAE GENE.

polder *Civil Engineering.* **1.** the reclamation, encompassing, and draining of lands taken from the sea by means of dikes, as in the Netherlands. **2.** an area of land reclaimed in this way.

pole *Mechanics.* the point at which the axis of rotation or symmetry of a body passes through its surface. *Astronomy.* **1.** either of two such points on the surface of the earth; the North Pole or South Pole. **2.** either of two similar points in the heavens about which the stars seem to revolve. *Physics.* one of two points, parts, or regions that have opposing qualities or tendencies, such as the ends of a magnet, the electrodes of an eltrolytic cell, or the terminals of a battery. *Biology.* either end of the axis of a nucleus, cell, organ, or complete organism, about which the parts seem to be symmetrically arranged. *Cell Biology.* **1.** specifically, either end of the spindle formed in a cell during mitosis. **2.** the point in a cell where a process or extension originates. *Electricity.* an output terminal as part of a switch, or two output terminals as part of a double-pole switch. *Optics.* **1.** either of the points on the extremities of the lens axis at the position where the axis intersects with the lens surfaces. **2.** the geometric center of a convex or concave mirror. *Crystallography.* an orientation direction perpendicular to a face of a crystal.

pole *Mathematics.* **1.** let $f(z)$ be a function of a complex variable that is analytic in a simply connected domain D, except at a point z_0. If there exists a positive integer k such that $(z - z_0)^k f(z)$ is analytic at z_0, then z_0 is said to be a pole of $f(z)$. The smallest such positive integer k is called the order of the pole. **2.** let Γ be a conic in the plane and l a line intersecting Γ in at most two points, at which points tangents to Γ are drawn. The point P external to Γ at which the tangents intersect is the pole of l with respect to Γ, and l is the polar of P with respect to Γ. If l is tangent to the conic, then P is the point of tangency to Γ. If l lies outside the conic, then choose any two points M and N on l and find their polars m and n. Then the pole P for l is the point of intersection of the lines m and n. Again, l is the polar for P.

pole blight *Plant Pathology.* a fatal disease of white pines, characterized by stunting of new growth, yellowing or shortening of needles, and excess resin flow.

polecat *Vertebrate Zoology.* **1.** a small, carnivorous, dark-brown mammal, *Mustela putorius*, of the weasel family, that ejects a fetid fluid when disturbed; found in Europe. **2.** another name for the North American skunk.

polecat

pole cell *Invertebrate Zooology.* a cleavage cell of various insect embryos.

pole changing control *Electromagnetism.* a method of obtaining two or more running speeds of a three-phase motor by changing the number of magnetic poles.

pole-dipole array *Engineering.* an arrangement of electrodes used in a drill logging or surveying process, in which one current electrode is set at infinity while others are guided across the structure to be studied.

pole dominance *Particle Physics.* a scattering amplitude property in which, in the complex analysis of the energy and scattering angle, the dominating terms of the Laurent series near a pole are those terms with negative powers.

pole face *Electromagnetism.* the smooth face of a permanent magnet or a magnetic core that faces an air gap.

pole-face winding *Electromagnetism.* winding in the pole face of a motor or generator; used to neutralize the cross-magnetizing armature reaction under the pole faces.

pole figure *Metallurgy.* a graphic representation of the preferential orientation of a polycrystalline metal or alloy.

Polemoniaceae *Botany.* a family of dicotyledonous plants in the order Solanaceae, characterized by regular flowers with three carpels, gland-tipped hairs, a continuous xylem ring, and no internal phloem.

Polemoniales see SOLANALES.

pole piece *Electromagnetism.* a piece of ferromagnetic material at the end of an electromagnet or permanent magnet, whose shape controls the magnetic flux distribution.

pole-pole array *Engineering.* an arrangement of electrodes, used in a logging or lateral search process, in which one current electrode and one potential electrode are moved in proximity across the structure to be studied.

pole-positioning *Control Systems.* in linear control theory, a design technique that relies on the proper choice of a linear feedback law to position any or all of a system's closed loop poles.

polestar *Astronomy.* see POLARIS.

polestar recorder *Engineering.* a device used to determine the degree of cloudiness during night and early morning darkness; it is made up of a camera that keeps the polestar (Polaris) in constant view, recording the clouds that come between it and the star.

Pole-tek test *Materials Science.* a commercial testing process for nondestructive evaluation of wood that uses sound velocity to determine the presence of decay by comparison with the velocity of a standard, decay-free pole.

pole tide *Oceanography.* a tide caused by the Chandler wobble of the earth, a nutation that has a period of 428 days and a theoretical amplitude of 6 mm.

pole-zero configuration *Control Systems.* a method of analyzing a system for stability, natural motion, frequency response, and transient response by plotting the poles and zeros of its transfer function in the complex plane.

polhode *Mechanics.* for a rotating rigid body of arbitrary shape subjected to no forces other than its own weight and the reaction of the support at its center of mass, the curve traced by the tip of its angular velocity vector with respect to a frame of reference coinciding with its principal axes of inertia.

Polhode cone see BODY CONE.

poli- or **polio-** a combining form meaning "gray," especially the gray matter of the brain.

polian vesicle *Invertebrate Zoology.* an elongate muscular sac suspended from the ring canal in echinoids and holothurians; used in maintaining pressure in the organism's water vascular system.

poling *Electricity.* the deliberate adjustment of an electromagnetic field polarity, especially in wire-line applications in which transpositions are used between sections of wire or lengths of cable to allow the opposition of residual cross-talk couplings in individual sections or lengths. *Metallurgy.* a reducing step in the refining of copper, formerly effected by immersing a green pole into the molten crude copper; currently effected by using a reducing gas such as propane.

polio *Medicine.* a shorter name for the disease poliomyelitis. See POLIOMYELITIS.

polioencephalitis *Neurology.* inflammation of the gray matter of the brain and brain stem.

polioencephalomyelitis *Medicine.* an inflammatory disease of the gray matter of the brain and spinal cord.

poliomyelitis [pō´lē ō mī´ə lī´təs; pō´lē ō mī lī´təs] *Medicine.* an acute viral disease, occurring sporadically and in epidemics, characterized in the minor illness by fever, sore throat, headache, stiff neck and back, and vomiting. The major illness is characterized by involvement of the central nervous system, with possible paralysis and atrophy of muscles or muscle groups resulting in permanent deformity. The disease is now largely controlled by vaccines. Also, POLIO, INFANTILE PARALYSIS.

poliomyelitis vaccine *Immunology.* an oral Sabin vaccine or killed Salk vaccine that contains the three types of poliomyelitis virus.

poliomyelopathy *Neurology.* any disease of the gray matter of the spinal cord.

poliovirus *Virology.* a picornavirus of the genus *E. enterovirus* that causes poliomyelitis in humans; three serotypes are known.

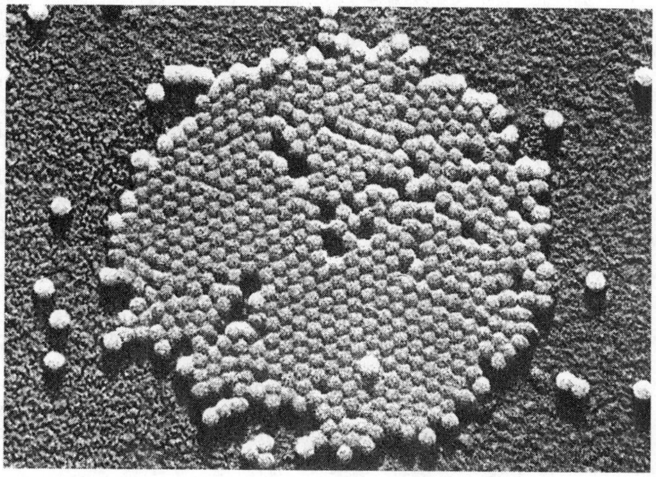

poliovirus

polish *Materials.* **1.** to make a surface smooth and lustrous, usually by applying friction. **2.** a substance used to augment the effect of friction in such a procedure; it may provide color and protection as well as luster to the surface.

polishing *Mechanical Engineering.* the process of using abrasive materials to smooth, polish, and brighten a surface, such as metal, rock, or wood. *Chemical Engineering.* any chemical process that is used to remove small amounts of impurities or to improve color as a final measure, as in food or petroleum products. *Food Technology.* the third and final stage of rice milling, in which hulled and pearled rice is spun in cones that are lined with leather or sheepskin. The fully processed rice is called **polished rice.** Also, BRUSHING. *Textiles.* a finishing process for yarn or fabric in which starch, wax, resin, or another material is applied, generally with rollers, to impart a smooth, glossy surface.

polishing roll *Mechanical Engineering.* a roll or series of rolls on a plastic mold with highly polished, chrome-plated surfaces; used to produce a smooth surface on a plastic sheet as it is extruded.

polishing wheel *Mechanical Devices.* a wheel covered with a soft material, such as leather or canvas; used for polishing machine parts and for fine-finishing tools.

Polish notation *Computer Programming.* **1.** a method of forming a mathematical expression in which there are no parentheses and each operator is either unary or binary. Also, LUKASIEWICZ NOTATION, PARENTHESES FREE NOTATION. **2.** a form of this method in which the operator precedes its operands. Also, PREFIX NOTATION. **3.** a form in which the operator follows its operands. For example, the expression $(a+b)*c$ would be written $ab+c*$.

political anthropology *Anthropology.* a branch of anthropology that focuses on the political organization of a particular culture, or the comparison of political systems cross-culturally.

political geography *Geography.* a branch of geography dealing with the effect of geography on politics, especially on national boundaries and relations between states.

poll *Science.* a sampling of the opinion of a limited group of people on a specifc issue, for the purpose of analysis or prediction. *Agriculture.* to remove the horns of cattle or other animals.

pollard *Botany.* a tree cut back almost to the trunk to produce a thick mass of branches. *Vertebrate Zoology.* an animal, such as a stag or sheep, having no horns.

pollarding *Forestry.* the process of cutting back the crown of a tree in order to produce a dense head of shoots beyond the reach of animals.

polled *Agriculture.* **1.** describing cattle or other domestic animals that lack horns. **2. Polled.** describing a breed or variety without horns, such as the Polled Shorthorn breed of cattle.

pollen *Botany.* the microspores of seed-producing plants, tiny granules that contain mature or immature male gametophytes; released from the anthers of flowers to fertilize the pistils.

pollen analysis *Archaeology.* the study of pollen grains in soil samples from an archaeological site; provides information on ancient human use of plants and plant resources. Thus, **pollen diagram, pollen zone.**

pollen count *Botany.* an estimation of the number of pollen grains landing on a given area, usually one square centimeter, during a specified time.

pollen grain *Botany.* a single granule of pollen.

pollen mother cell *Botany.* a usually diploid cell in a pollen sac that undergoes meiotic division to form a tetrad of pollen grains; a microsporocyte of seed plants.

pollen-restoring gene *Genetics.* a gene that allows a plant to produce normal microspores, even in the presence of a cytoplasmic male sterility factor.

pollen sac *Botany.* the chamber in the anther of a seed plant in which pollen grains are formed.

pollen tube *Botany.* a tubular growth on the wall of a pollen grain that enters the embryo sac, providing a passage for male reproductive cells to move through to the ovule.

pollination *Botany.* the transfer of pollen from the anthers to the stigmas in angiosperms or to the micropyle in gymnosperms, for the purpose of fertilization; usually by means of insects, wind, birds, bats, or water.

pollination by bees

polling *Telecommunications.* **1.** a process that interrogates each terminal on a shared communications line to determine which terminal or terminals have messages waiting to be transmitted over a common bus. **2.** a process by which a processor allows an external unit or units to feed it with information selectively in turn.

polling list *Telecommunications.* in a time-sharing system, a roster of transmitting devices that are sequentially scanned.

pollinium *Botany.* an agglutinated or coherent mass of pollen grains that is transported, usually by an insect, in pollination, as in plants of the milkweed and orchid families.

pollucite *Mineralogy.* $(Cs,Na)_2Al_2Si_4O_{12}\cdot H_2O$, a transparent, colorless to white cubic mineral of the zeolite group, occurring in fine-grained masses and rarely as cubic crystals, having a specific gravity of 2.936 and a hardness of 6.5 to 7 on the Mohs scale; found in granite pegmatites and used as a gemstone.

pollutant *Ecology.* any substance or agent that causes pollution.

pollute *Ecology.* to cause pollution in something; contaminate. Thus, **polluted.**

pollution *Ecology.* any alteration of the natural environment producing a condition that is harmful to living organisms. Pollution may occur naturally, as when an erupting volcano emits sulfur dioxide, but the term usually refers to the negative effect of human activities; e.g., automobile exhaust emissions, oil spills, the dumping of industrial wastes in the water supply, the overuse of pesticides and chemical fertilizers, improper disposal of solid wastes, and so on. (Ultimately derived from a word meaning "mud" or "dirt.")

Pollux *Astronomy.* Beta (β) Geminorum, a first-magnitude star in the constellation of Gemini. (From the Greek legend of Castor and *Pollux,* two devoted friends; Castor is the other bright star in this constellation.)

polonium *Chemistry.* a radioactive metallic element having the symbol Po, the atomic number 84, an atomic weight of 210, a melting point of 250°C, and a boiling point of 962°C; a very hazardous member of the uranium decay series that has no stable isotopes, occurs naturally in uranium ores, and is prepared artificially by neutron bombardment of bismuth; used as a source of alpha particles and neutrons, and as a moisture determinant and static dissipater. (From the Medieval Latin name for *Poland,* the birthplace of its codiscoverer, Marie Sklodowska Curie.)

polonium-210 *Nuclear Physics.* a radioactive isotope of polonium with a half-life of 138.4 days; found in pitchblende and radium-lead residues, it decays by alpha-ray emission and is useful, because of its alpha activity, in neutron sources.

Polonyi, John born 1929, Canadian chemist; awarded the Nobel Prize for the study of the reaction dynamics of chemiluminescence.

poly [päl´ē] *Textiles.* a shorter name for polyster fabrics or garments. See POLYESTER. *Pathology.* see POLYMORPHONUCLEAR LEUKOCYTE.

poly- a combining form meaning: **1.** many, much. **2.** polymerous.

poly. polytechnic; polygon.

polyA or **poly-A** polyadenylate, polyadenylation.

polyacetal *Materials.* a hard, tough, slippery plastic that is derived from the polymerization of formaldehyde; used as a metal substitute. Also, ACETAL RESIN.

polyacid *Chemistry.* **1.** having more than one replaceable hydrogen atom. **2.** an acid having more than one replaceable hydrogen atom. **3.** of a base or basic radical, capable of neutralizing several molecules of an acid radical.

polyacrylamide *Organic Chemistry.* $(CH_2CHCONH_2)_x$, a white, solid, water-soluble polymer based on acrylamide; used as a thickening or suspending agent, adhesive additive, or food additive, and in biochemistry in the study of biological compounds. Also, **polyacrylamide gel.**

polyacrylamide gel electrophoresis *Physical Chemistry.* in a polyacrylamide gel, the migration of electrically charged particles (ions) under an applied electric potential.

polyacrylate *Organic Chemistry.* a polymeric ester or salt of acrylic acid.

polyacrylic acid *Organic Chemistry.* $(H_2C–CHCOOH)_x$, an acrylic or acrylate water-soluble resin that is formed by the polymerization of acrylic acid; used as a suspending agent, and in adhesives, paints, and hydraulic fluids.

polyacrylic fiber *Organic Chemistry.* a continuous-strand fiber extruded from an acrylate resin.

polyacrylonitrile *Organic Chemistry.* a polymer of acrylonitrile, used in manufacturing various synthetic fibers such as Orlon and Dynel.

polyadelphous *Botany.* of a flower, having stamens joined by their filaments into three or more separate bundles..

polyadenoma *Oncology.* adenomas occurring in many glands.

polyadenomatosis *Oncology.* the condition of many adenomas occurring in one body part.

polyadenopathy *Pathology.* any disease affecting several glands at once.

polyadenylate *Molecular Biology.* a polymer of adenylic acid.

polyadenylation *Molecular Biology.* a process by which a sequence of adenylic acid residues is added to the 3' end of many eukaryotic mRNA molecules immediately after transcription.

polyalcohol *Organic Chemistry.* a polyhydric alcohol.

polyalgesia *Neurology.* a disorder in which a single painful stimulus produces the sensation of multiple painful stimuli. Thus, **polyalgesic.**

polyalgorithm *Mathematics.* a collection of algorithms designed to solve a given class of problems, with rules for implementing them.

polyallomer *Organic Chemistry.* a copolymer of propene and other alkenes having a uniform crystalline structure and a mixed chemical composition, designed to provide physical characteristics intermediate between those of its components but having a better balance than a blended polymer; used in film, sheeting, wire cables, and vacuum-formed, injection-molded products.

polyamide *Organic Chemistry.* a natural or synthetic polymer in which the structural units are linked by amide groupings (–CONH–); **natural polyamides** occur in soybean, peanut, and corn proteins and are used to manufacture plastics, textiles, and adhesives; **synthetic polyamides** are used to manufacture textiles, especially various types of nylon.

polyamine *Chemistry.* a compound that contains more than one amino group.

polyandrous *Anthropology.* of, relating to, or practicing polyandry. *Botany.* relating to or exhibiting polyandry. Also, **polyandric.**

polyandry *Anthropology.* the cultural practice in which one woman is married to two or more husbands at the same time. *Botany.* the fact of having a large number of stamens.

Polyangiaceae *Microbiology.* a family of cylindrical gliding bacteria of the order Myxobacterales that occur in decaying organic matter and in soils.

poly-A polymerase *Enzymology.* an enzyme found in the nucleus that catalyzes the addition of several adenine nucleotides to the 3' end of a primary transcript of messenger RNA.

polyargyrite *Mineralogy.* $Ag_{24}Sb_2S_{15}$, a gray to black opaque mineral of metallic luster consisting of a sulfide of antimony and silver.

polyarteritis *Medicine.* multiple inflammatory and destructive arterial lesions.

polyarteritis nodosa *Medicine.* a widespread inflammation of small or medium-sized arteries with formation of nodules within the walls of those vessels; symptoms are varied and related to the tissues and organs supplied by the affected arteries.

polyarthritis *Medicine.* the inflammation of several joints at the same time. Thus, **polyarthritic.** *Veterinary Medicine.* see JOINT-ILL.

poly-A tail *Molecular Biology.* a sequence of adenylate residues added to the 3' end of certain eukaryotic mRNA molecules; thought to function to increase the stability of these molecules.

polyatomic *Chemistry.* containing more than two atoms.

polyaxon *Neurology.* a nerve cell characterized by horizontal dendrites that give off multiple axons or branches. *Invertebrate Zoology.* a sponge spicule having three or more axes radiating from a central point. Thus, **polyaxonic.**

polybasic *Chemistry.* of an acid, disassociating in solution to yield two or more H^+ ions per molecule.

polybasite *Mineralogy.* $(Ag,Cu)_{16}Sb_2S_{11}$, an iron-black monoclinic pseudohexagonal mineral with metallic luster, occurring as thin tabular crystals and in massive form, having a specific gravity of 6.1 to 6.30 and a hardness of 2 to 3 on the Mohs scale; found in silver vein deposits of low to moderate temperature.

polybenzimidazole *Materials.* a structural high-temperature adhesive polymer.

polybenzoxale *Materials.* a structural adhesive polymer that yields useful bonds above 260°C and has good high-temperature aging characteristics.

polyblennia *Medicine.* the secretion of excessive quantities of mucus.

Polybrachiidae *Invertebrate Zoology.* a family of deep-sea sedentary marine animals in the class Thecanephria, having 2 to 200 tentacles.

polybutadiene *Organic Chemistry.* a synthetic, thermoplastic, combustible, rubberlike polymer made from butadiene.

polybutylene *Organic Chemistry.* any of various combustible thermoplastic polymers of isobutene, butene-1, or butene-2; used in rubber products, lube-oil additives, and adhesives. Also, **polybutene.**

polycarbonate *Organic Chemistry.* $[COOC_6H_5C(CH_3)_2C_6H_5O]_x$, a synthetic thermosplastic polymer of carbonic acid derived by the condensation of bisphenols with a phosgene or phosgene derivative; transparent, nontoxic, noncorrosive, weather-resistant, and stain-resistant; used in molded products, nonbreakable windows, and household appliances.

polycarboxylic *Organic Chemistry.* a combining form indicating the presence of two or more carboxyl (COOH) groups.

polycentric having many centers; specific uses include: *Cell Biology.* of a cell, having multiple centrosomes. *Genetics.* of a chromosome, having several chromomeres. *Biology.* of an organism, having more than one center of development or differentiation.

Polychaeta *Invertebrate Zoology.* a class of mostly marine worms in the phylum Annelida, with anterior tentacles and palps and most segments bearing parapodia with bristles; free-swimming or sessile in tubes or burrows; often brightly colored, most 5–10 centimeters in length.

polychlorinated biphenyl see PCB.

polychloroprene *Materials Science.* see NEOPRENE.

polychlorotrifluoroethylene *Materials.* a thermoplastic fluoropolymer that is tough and resistant to heat and chemicals.

polychondritis *Medicine.* inflammation involving many cartilages of the body.

polychroism see PLEOCHROISM.

polychromatic *Optics.* exhibiting many colors.

polychromatic radiation *Electromagnetism.* electromagnetic radiation having a range of frequencies rather than a narrow band.

polychromatophilia *Cell Biology.* the quality of being stainable with various stains or tints; affinity for various stains. Also, **polychromatia.**

polychromatophilic *Cell Biology.* relating to or displaying polychromatophilia.

polycistronic *Molecular Biology.* describing any mRNA molecule that encodes more than one polypeptide chain. Thus, **polycistronic mRNA, polycistronic messenger.**

Polycladida *Invertebrate Zoology.* an order of free-swimming flatworms in the class Turbellaria, often brightly colored, with flattened, leaflike bodies and many eyes, testes, and ovaries; most have tentacles.

polyclimax *Ecology.* an ecological community that maintains its stability through a number of environmental factors, including topography, fire, and interactions among the various species.

polyclimax theory *Ecology.* the theory that over time and within a particular climatic region there will be numerous climax communities, probably due to changes in soil condition.

polyclinal fold *Geology.* any of a group of adjacent folds having similar surface axes randomly oriented.

polyclonal *Molecular Biology.* describing a population of heterogeneous antibodies derived from multiple clones, each of which is specific for one of a number of determinants found on an antigen.

polyclonal activator *Immunology.* a substance that activates many clones of lymphocytes.

polyclonal antiserum *Immunology.* a serum that contains many different antibodies against a specific antigen.

polyclonal mixed cryoglobulin *Biochemistry.* a cryoglobulin that contains heterogeneous immunoglobin and occasionally extra serum proteins, so that it belongs to more than one class.

polyclonia *Neurology.* a disease characterized by multiple, continuous, and uninterrupted clonic spasms that are similar to but distinct from chorea and tic.

polycondensation *Organic Chemistry.* a process of polymerization in which recurring structural units are formed from simpler molecules by eliminating a simple substance such as water.

polyconic chart *Cartography.* any chart that is based on a polyconic map projection.

polyconic (map) projection *Cartography.* a map projection, neither conformal nor equal area, on which the central geographic meridian is represented by a straight line, the other meridians are represented by curved lines, and the parallels are arcs of nonconcentric circles which have centers lying on the line that represents the central meridian and radii determined by the lengths of the elements of cones that are tangent along the parallels.

polycoria *Medicine.* the deposit of reserve material in an organ or tissue so as to produce enlargement. *Zoology.* the existence of more than one pupil in an eye.

polycotyledon *Botany.* a plant having more than two cotyledons within the seed, such as most cone-bearing trees.

polycrase-(Y) *Mineralogy.* $(Y,Ca,Ce,U,Th)(Ti,Nb,Ta)_2O_6$, a black orthorhombic mineral, occurring as aggregates of prismatic crystals, having a specific gravity of 4.30 to 5.87 and a hardness of 5.5 to 6.5 on the Mohs scale; found in granite pegmatites.

polycrystal *Materials Science.* the attribute of a structure that is composed of many crystals.

polycrystalline *Materials Science.* of, having, or relating to a rock or metal composed of aggregates of variously oriented crystals.

Polyctenidae *Invertebrate Zoology.* the bat bugs, a family of hemipteran insects in the superfamily Cimicoidea having short, flattened bodies, no eyes or wings, and short antennae; they are ectoparasites on bats.

polycyclic *Biology.* having many whorls or turns, as a shell. *Chemistry.* of a molecule, containing three or more closed atomic rings.

polycyesis *Medicine.* a multiple pregnancy.

polycystic *Medicine.* containing or consisting of many cysts.

polycystic kidney *Medicine.* an abnormal condition characterized by enlargement of the kidneys with the formation of many cysts, resulting in kidney failure and death; the disease occurs in three forms, congenital, childhood, and adult.

polycystic ovary syndrome see STEIN-LEVENTHAL SYNDROME.

Polycystina *Invertebrate Zoology.* a class of Radiolarian protozoans having a siliceous skeleton and a perforated capsular membrane.

polycystoma *Medicine.* a condition characterized by the proliferation of cysts in the breast.

polycythemia *Hematology.* any abnormal increase in the total red cell mass of the blood, especially polycythemia vera.

polycythemia vera *Hematology.* a disorder that is characterized by an absolute increase in red blood cell mass and total blood volume, often associated with leukocytosis and thrombocytosis. Also, **polycythemia rubra vera.**

polydactylic *Medicine.* having more than the normal number of fingers or toes. Also, **polydactylous.**

polydactyly *Medicine.* the developmental anomaly of having more than the normal number of fingers or toes. Also, **polydactylism.**

polydemic *Ecology.* of or relating to a species that occurs in several different geographic areas.

Polydesmida *Invertebrate Zoology.* the largest order of millipedes, with 19–22 segments, containing nearly 30 families worldwide. Also, **Polydesmoidea.**

polydiene elastomer *Materials.* a synthetic rubber that is produced by polymerization of diene rubber; used widely in manufacturing.

polydipsia *Medicine.* an excessive or insatiable thirst, as is symptomatic of diabetes.

polydisperse *Chemistry.* of a colloidal system, composed of particles of many sizes.

polydispersity *Chemistry.* the tendency of colloidal particle size to vary within a dispersion.

Polydnaviridae *Virology.* a family of viruses infecting certain parasitic hymenopteran insects, having multiple supercoiled dsDNA genomes and replicating in the nuclei of host cells.

Polydnavirus *Virology.* a genus of viruses belonging to the family Polydnaviridae.

Polydolopidae *Paleontology.* an extinct family of rodentlike South American marsupials in the suborder Caenolestoidea; extant in the Paleocene and Eocene.

polydymite *Mineralogy.* $NiNi_2S_4$, a gray, opaque, metallic, cubic mineral of the linnaeite group, occurring as octahedral crystals and in massive form, having a specific gravity of 4.5 to 4.8 and a hardness of 4.5 to 5.5 on the Mohs scale; found in vein deposits with other sulfides.

polyelectrolyte *Organic Chemistry.* a polymer having high molecular weight and producing large chain-type ions in solutions; may be natural, such as protein, or synthetic, such as salts of polyacrylic acid; used in industry for various forms of water and soil treatment.

polyembryony *Zoology.* the production of more than one embryo from a single egg or ovule.

polyene *Organic Chemistry.* any unsaturated aliphatic or alicyclic compound having more than four carbon atoms in the chain and at least two double bonds.

polyester [päl´ē es´tər; päl´ē es´tər] *Textiles.* any manmade fiber filament made of at least 85% polyester resin; characterized by crease-resistance, quick-drying and shape-retaining properties, high strength, and durability. Also, **polyester fiber.** Thus, **polyester fabric.**

polyester resin *Organic Chemistry.* any of a group of synthetic resins, such as Dacron or Mylar, that are formed by the polycondensation of carboxylic acids with dihydroxyl alcohols; when catalyzed, such resins cure or harden at room temperature under little or no pressure; characterized by strength and resistence to moisture and chemicals; widely used in reinforced plastics, textiles, magnetic tapes, automotive parts, and structural applications.

polyesthesia *Neurology.* a disorder of sensation in which a single stimulus is perceived as several stimuli.

polyesthetic *Physiology.* relating to or affecting several senses or sensations.

polyestradiol phosphate *Oncology.* a polymer of estradiol phosphate used in treating prostrate cancer.

polyestrous *Physiology.* describing an animal that completes two or more fertile cycles during a mating season.

polyether *Materials Science.* a commercially produced polymer made from cyclic ether, used as a functional material or as an intermediate for further processing.

polyetherimide *Materials Science.* an amorphous thermoplastic resin that is inherently flame resistant; used for aircraft panels and seat component parts.

polyether ketone *Materials Science.* an engineering thermoplastic that has a high melting point and excellent abrasion and radiation resistance, and that is used for wire and cable coverings and insulation and in composites.

polyether resin *Organic Chemistry.* a polymer that contains the linkage $(–CH_2–CHR–O–)_x$.

polyethersulfone *Materials.* an engineering thermoplastic that is strongly resistant to high temperatures and hydrolysis.

polyethylene [päl′ē eth′ə lēn] *Materials.* a hard-to-soft, ductile, easily molded thermoplastic that is chemical resistant and has good insulating properties; it has many industrial uses, primarily in the form of packaged film for food products and garment bags, and in pipe, electrical insulation, and molded products.

polyethylene glycol *Organic Chemistry.* any of various combustible polymers formed by condensation of ethylene glycol and having the general formula $HOCH_2(CH_2OCH_2)_nCH_2OH$ or $H(OCH_2CH_2)_nOH$; forms vary from clear, colorless, viscous liquids to waxy solids; generally soluble with water and alcohol; used in making cosmetics, pharmaceuticals, and rubber, and as plasticizers, binders, and food additives.

polyethylene glycol induced fusion *Biotechnology.* the fusion of cells in a tissue culture caused by the dissolving of cell membranes by a hydrocarbon solvent.

polyethylene terephthalate *Organic Chemistry.* $(–C_{10}H_8O_4–)_x$, a thermoplastic polyester derived from ethylene glycol; melts at 265°C, resists combustion, and is self-extinguishing; used to make films, fibers, and recording tapes.

polyfluoroethylepropylene *Materials.* a continuous plastic-coating material that is applied by a plasma-spray process and is strongly resistant to high temperatures.

polyfoam *Materials.* the material used in foam mattresses, consisting of bubbles of air or carbon dioxide in a polymer (often polyurethane) matrix.

polyforming *Chemical Engineering.* a petroleum-refinery process to manufacture gasoline from petroleum gases, in which the gases containing unsaturated hydrocarbons are heated and injected into the middle of the cracking furnace. Also, **polyform process.**

Polygala *Botany.* the milkworts, a genus of plants of the family Polygalaceae containing many species including *P. senega,* the senega snakeroot of North America.

Polygalaceae *Botany.* a family of herbaceous or woody dicotyledonous plants in the order Polygalales, having an extrafloral nectary, a closed ring of xylem, a bicarpellate pistil, and monadelphous stamens.

polygalactia *Medicine.* an excessive secretion of milk.

polygalacturonase *Enzymology.* an enzyme that catalyzes the hydrolysis of bonds in polymers of galacturonic acid, such as pectate.

Polygalales *Botany.* an order of woody and herbaceous dicotyledonous plants of the subclass Rosidae, characterized by a continuous xylem cylinder, simple leaves, and hypogynous perfect flowers.

polygamous [pə lig′ə məs] *Anthropology.* relating to or practicing polygamy. *Vertebrate Zoology.* of an animal, having two or more mates at a time. *Botany.* having both perfect and imperfect flowers on the same plant or on different plants of the same species. Also, **polygamic.**

polygamy [pə lig′ə mē] *Anthropology.* **1.** the fact of being married to two or more persons at the same time. **2.** specifically, a cultural practice that permits a man to have more than one wife at a time, or less often, a woman to have more than one husband. Thus, **polygamist.**

polygene *Genetics.* any of a class of genes that combine to control a quantitative character such as height or skin color in humans. *Geology.* **1.** an igneous rock made up of two or more minerals. Also, POLYMERE. **2.** see POLYGENETIC, def. 2. Thus, **polygenic.**

polygenesis *Anthropology.* the concept that humans evolved from several independent sets of ancestors, rather than from a single pair. *Biology.* the theory that a species or other group developed from more than one source or in more than one place at a time.

polygenetic *Geology.* **1.** describing a material that was formed as a result of more than one process, derived from more than one source, or that originated and developed at various places and times. **2.** referring to a substance that is composed of or consists of more than one type of material. Also, POLYGENE.

polygenetic soil *Agronomy.* soil formed by two or more different and contrasting processes. All the horizons are not genetically related.

polygenic trait *Genetics.* a trait that is determined by many genes, each of which has only a partial effect on the expression of the trait.

polygeosyncline *Geology.* a geosynclinal-geoanticlinal zone along the margin of a continent that receives sediments from the borderland on its oceanic side.

Polyglactin *Surgery.* the trade name for semicrystalline polymer materials used as absorbable suture materials.

polyglycol *Organic Chemistry.* any dihydroxy ether formed by dehydration of two or more glycol molecules.

polyglycol distearate *Organic Chemistry.* $C_{17}H_{35}COO(CH_2CH_2O)_xOCC_{17}H_{35}$, a soft, combustible, off-white solid; melts at 43°C; slightly soluble in alcohol; used in polishing pastes and as a resin plasticizer.

Polygnathidae *Paleontology.* a family of conodonts belonging to the order Ozarkodinida that arose in the Early Devonian and became extinct in the Middle Carboniferous.

polygon [päl′ē gän′] *Mathematics.* a simple closed curve in the plane, made up of line segments.

Polygonaceae *Botany.* the single family comprising the order Polygonales.

polygonal [pə lig′ə nəl] *Mathematics.* relating to or having the form of a polygon.

Polygonales *Botany.* a monofamilial order of dicotyledonous plants in the subclass Caryophyllidae, characterized by stems that are often swollen at the nodes, simple and usually alternate leaves, a unilocular ovary, and well-developed endosperm.

polygonal ground *Geology.* a type of patterned ground characterized by polygonal arrangements of rock, soil, and vegetation, and formed on a flat or gently sloping surface by frost action. Also, CELLULAR SOIL.

polygonal karst *Geology.* a karst whose surface is completely divided into a polygonal network.

polygonal method *Mining Engineering.* a method of computing ore reserves by assuming that the area of influence of each drill hole extends halfway to adjacent drill holes.

polygonal ring structure *Astronomy.* a lunar crater whose walls approximate the shape of a polygon, often a hexagon.

polygonization *Solid-State Physics.* the alignment of metastable dislocations on the edges of plastically bent crystals during annealing so as to accommodate the formation of polygonal domains.

polygonized structure *Materials Science.* a subgrain structure produced in the early stages of annealing. The subgrain boundaries are a network of dislocations rearranged during heating.

polygraph see LIE DETECTOR.

polygynous *Anthropology.* of, relating to, or practicing polygyny. *Botany.* relating to or exhibiting polygyny.

polygyny *Anthropology.* the cultural practice in which one man is married to two or more wives at the same time. *Botany.* the fact of having many pistils or styles.

polygyria *Neurology.* a condition in which the brain has an excessive number of convolutions.

polyhalite *Mineralogy.* $K_2Ca_2Mg(SO_4)_4{\cdot}2H_2O$, a colorless, white, gray, or pink triclinic mineral, massive to fibrous in habit, having a specific gravity of 2.78 and a hardness of 3.5 on the Mohs scale; found in sedimentary evaporite deposits; a source of potash for fertilizer.

polyhead *Virology.* a phage head that is elongated due to mutations.

polyhedral *Science.* having many faces or sides. *Mathematics.* relating to or having the form of a polyhedron.

polyhedral angle *Mathematics.* an angle formed by a vertex of a polyhedron and all of the faces sharing that vertex.

polyhedrin *Virology.* a matrix protein constituting occlusion bodies that are produced during cytoplasmic polyhedrosis and nuclear polyhedrosis; it forms a protective crystal around the virions.

polyhedron [päl′ē hē′drən] *plural,* **polyhedra.** *Mathematics.* a solid bounded by planes. The faces of a polyhedron are polygons. *Virology.* an inclusion body that transmits viruses and is produced in the cells of certain virus-infected insects; it contains virions embedded in a matrix of polyhedrin.

polyhedrosis *Virology.* a viral disease of invertebrates, especially insect larvae, characterized by disintegration of tissues and the formation of polyhedral occlusion bodies. Also, **polyhedral disease.**

polyhidrosis see HYPERHIDROSIS.

polyhydramnios *Medicine.* an excess of amniotic fluid.

polyhydrazide *Materials.* a strong synthetic fiber material that is used for tire yarn.

polyhydric *Organic Chemistry.* denoting a compound that contains more than two hydroxyl groups. Thus, **polyhydric alcohol, polyhydric phenol.**

Polyhymenophora *Invertebrate Zoology.* a diverse class of often large, commonly free-living ciliated protozoans formerly known as the Spirotricha. Also, **Polyhymenophorea.**

Polyideaeceae *Botany.* a family of red algae of the order Gigartinales characterized by branched, compact, cylindrical, or flattened thalli arising from a spreading prostrate base.

polyimide *Organic Chemistry.* any of a group of high polymers containing an imide group (–CONHCO–) in the polymer chain; characterized by high tensile strength, high temperature stability, and resistance to friction, wear, and radiation; used in aerospace vehicles, semiconductors, magnetic tapes, and internal combustion engines.

polyisoprene *Organic Chemistry.* $(–C_5H_8–)_x$, a thermoplastic polymer that is the major component of natural rubber; also produced synthetically.

polykaryocyte *Cell Biology.* a giant cell containing several nuclei.

Polykrikaceae *Botany.* a monogeneric family of nonthecate, marine dinoflagellates of the order Gymnodiniales, containing organisms that are pseudocolonial or have a series of energids with two or more nuclei and flagellar pairs.

polylactic resin *Organic Chemistry.* a soft, elastic resin derived from heating lactic acid with a fatty oil or oils; used to produce water-resistant coatings.

polyliner *Engineering.* a vertically ribbed, perforated insert that conforms to the inside of a cylinder in an injection-molding apparatus.

polylinker *Molecular Biology.* a linker DNA molecule containing multiple restriction sites.

polylysogeny *Virology.* a condition in which bacterial strains are lysogenic for several different phages.

polymer [päl´ə mər] *Organic Chemistry.* a large molecule formed by the union of at least five identical monomers; it may be natural, such as cellulose or DNA, or synthetic, such as nylon or polyethylene; polymers usually contain many more than five monomers, and some may contain hundreds or thousands of monomers in each chain.

Polymera *Invertebrate Zoology.* a division of animals with clearly defined linear segments; in some classifications the same as Annelida.

polymer accelerator *Materials Science.* an organic additive used to speed the vulcanization of rubber, usually in conjunction with other process activators.

polymerase *Enzymology.* any enzyme that catalyzes the formation of a polymer from monomers.

polymerase chain reaction *Biotechnology.* a technique in which repeated cycles of DNA synthesis are carried out to produce a large number of a specific DNA sequence.

polymer crazing *Materials Science.* a defect in plastic articles consisting of distinct surface cracks or minute frostlike internal cracks, resulting from stresses within the article that exceed the cohesive strength of the plastic structural units.

polymer crystalite *Materials Science.* an ordered structure arrangement within some polymers.

polymere *Geology.* see POLYGENE, def. 1.

polymeric *Organic Chemistry.* of or relating to a polymer. *Chemistry.* of two or more compounds, having the same elements combined in the same proportions but having different molecular weights.

polymeric molybdate see ISOPOLYMOLYBDATE.

polymerism *Chemistry.* the state of being polymeric. *Botany.* the state of being polymerous.

polymerization *Organic Chemistry.* the formation of a polymer, a chemical process in which a series of simple structural units (monomers) combine to form large, chainlike molecules (macromolecules).

polymer melt *Materials Science.* a high-temperature rubbery state in polymers in which the chain segments are thermally agitated into liquid-like Brownian motions, and at which the polymers are processed by extrusion or molding processes.

polymerous *Botany.* having many members in each whorl.

polymetamorphic diaphthoresis *Geology.* the changes during the second phase of metamorphism that took place in response to lower-grade metamorphic conditions than those to which the rock was previously adjusted.

polymetamorphism *Geology.* polyphase or multiple metamorphism in which two or more metamorphic events in succession have left evidence of their occurrence upon the same rocks. Also, SUPERIMPOSED METAMORPHISM.

polymethyl methacrylate *Organic Chemistry.* a thermoplastic polymer derived from methyl methacrylate, used in lighting fixtures and optical instruments.

polymicrogyria *Neurology.* a malformation of the brain characterized by the development of numerous small convolutions. Also, MICROGYRIA.

polymictic *Petrology.* describing clastic sedimentary rock composed of more than one rock type or mineral species.

polymictic lake *Hydrology.* a lake that has no lasting thermal stratification.

polymignite *Mineralogy.* $(Ca,Fe,Y,Th)(Nb,Ti,Ta,Zr)O_4$, a rare black, orthorhombic mineral that is weakly to moderately radioactive, having a specific gravity of 4.77 to 4.85 and a hardness of 6.5 on the Mohs scale; it is found as prismatic crystals in pegmatites in Norway. Also, **polymignyte.**

polymorph *Biology.* any of two or more forms of a single organism, such as different castes of certain ants. *Systematics.* any of the two or more forms in a polymorphic group. *Crystallography.* any of several crystals of the same substance that occur naturally in many forms. Also, **polymorphic modification.** *Histology.* see GRANULOCYTE.

polymorphic *Science.* **1.** occurring in many forms. **2.** appearing in various forms in different stages of development. *Systematics.* of or relating to morphological variations occurring within a single taxonomic group. Also, **polymorphous.**

polymorphic system *Computer Technology.* referring to a computer system in which major parts or units are held in a common pool, assigned to executing programs based on need, and returned to the pool when they are no longer needed.

polymorphic type *Computer Programming.* an abstract data type, such as a linked list, that could be implemented in different ways or could take multiple forms, such as a linked list of integers or a linked list of reals using a similar record format.

polymorphism the state or condition of being polymorphic; specific uses include: *Biology.* **1.** the occurrence of different forms of individuals in a single species. **2.** the occurrence of different structural forms in a single individual at different periods in the life cycle. *Genetics.* the occurrence of two or more different phenotypes in a population. *Crystallography.* the property of crystallizing in two or more forms with distinct structures, as opposed to dimorphism if only two forms occur.

polymorphonuclear cells *Immunology.* leukocytes with lobulated nuclei that take part in inflammatory reactions; major subpopulations are neutrophils, eosinolphils, basophils. Also, **polymorphonuclear leukocytes.**

polymorphonuclear granulocyte *Immunology.* a cell that is associated with bone marrow and has a multibodied nucleus and granular cytoplasm in its mature form. Also, NEUTROPHIL.

polymyarian *Invertebrate Zoology.* of nematode worms, having many muscle cells in each quadrant of a cross section.

polymyos *Medicine.* the simultaneous inflammation of many muscles.

polymyositis *Medicine.* a chronic, progressive disease of skeletal muscle, characterized by symmetrical weakness of the limb girdles, neck, and pharynx, and usually resulting in deformity; sometimes associated with malignancy.

polymyxin *Microbiology.* any of five antibiotics (designated A, B, C, D, E) produced by strains of the soil bacterium *Bacillus polymyxa* and acting against various Gram-negative bacteria.

polyneme hypothesis *Molecular Biology.* the hypothesis that a new chromatid is composed of more than one double-stranded DNA molecule.

Polynemidae *Vertebrate Zoology.* the threadfins, the single family of carnivorous fishes in the suborder Polynemoidei.

Polynemoidei *Vertebrate Zoology.* a suborder of carnivorous fishes of the order Perciformes, distinguished by a pectoral fin divided in two parts, with the bottom half resembling threads that act as feelers; found in murky marine waters in tropical seas worldwide.

Polynesia *Geography.* the region of Oceania containing those Pacific island groups east of Melanesia and Micronesia, and stretching from the Hawaiian Islands south to New Zealand.

Polynesian subregion *Ecology.* a distinct zoogeographical region that includes the oceanic islands of Polynesia.

polyneuritic *Neurology.* of or relating to polyneuritis.

polyneuritis *Neurology.* simultaneous inflammation of multiple spinal nerves, characterized by paralysis, pain, and wasting of muscles.

polyneuropathy *Neurology.* any disease that affects multiple nerves simultaneously.

Polynoidae *Invertebrate Zoology.* a family of scale worms, crawling polychaetes in the Errantia, with a dorso-ventrally flattened body and rows of scales on the back.

polynomial [päl´i nō´mē əl] *Mathematics.* **1.** a formal power series in one or more indeterminates with only finitely many nonzero coefficients. If the coefficients are required to be elements of a particular algebraic object *A* (ring, field, and so on), the polynomial is said to be a polynomial over *A*. **2.** a function constructed using only addition, subtraction, and multiplication.

polynomial time *Computer Programming.* a term characterizing the complexity of a given algorithm in which the time to solve a problem of size *n* increases with *n* no more rapidly than a polynomial in *n* increases with *n*.

polynuclear *Science.* having several or many nuclei. Also, **polynucleate, polynucleated.**

polynuclear hydrocarbon *Organic Chemistry.* a hydrocarbon molecule with two or more closed rings.
polynucleotide *Biochemistry.* a series of nucleotides, such as DNA and RNA, in which the 3' of the pentose of one is linked by a phosphodiester group to the 5' position of the pentose of the other.
polynucleotide kinase *Enzymology.* an enzyme that catalyzes the addition of phosphate to the 5' hydroxyl end of a nucleic acid molecule produced by an endonuclease.
polynucleotide ligase *Enzymology.* an enzyme that catalyzes the connection of adjacent 5' and 3' ends of nucleic acid molecules through the formation of phosphodiester bonds.
polynucleotide phosphorylase *Enzymology.* an enzyme that catalyzes the cleavage of a polynucleotide by phosphate.
polynya [pə lin´yə] *Oceanography.* a large expanse of water that is surrounded by ice; usually found near the shore, often at the mouth of a large river. (From a Russian word meaning "emptiness" or "openness.") Also, **polyn'yá.**
polyodontia *Physiology.* the presence of supernumerary teeth.
Polyodontidae *Vertebrate Zoology.* the paddlefishes, a family of sturgeonlike fishes of the order Acipenseriformes, characterized by an elongate, paddlelike snout; found in the Mississippi and Yangtze Rivers.
polyolefin *Organic Chemistry.* a resin made by polymerizing olefin.
polyoma *Oncology.* a tumor caused by a virus having a broad host range; originally isolated from parotid gland tumors of mice inoculated with Gross leukemia virus.
Polyomavirus *Virology.* a genus of viruses of the family Papovaviridae that occur commonly and usually without symptoms in their natural hosts (primarily humans, rabbits, mice, hamsters, and monkeys) but are highly oncogenic when injected into laboratory animals.
polyopia *Medicine.* the formation of more than one image of an object on the retina; multiple vision. Also, **polyopsia, polyopy.**
Polyopisthocotylea *Invertebrate Zoology.* an order of flukes, trematodes in the subclass Monogenea, parasites with posterior suckers or clamps.
polyoxyalkylene resin *Organic Chemistry.* a condensation polymer produced from an oxyalkene.
polyoxyl (8) stearate *Organic Chemistry.* a soft, waxy or pasty, cream-colored solid, soluble in acetone, ether, and ethanol; it is used as an emulsifier in bakery products. Also, **polyoxyethylene (8) stearate.**
polyoxymethylene *Materials.* a polymer of formaldehyde with polar groups, having excellent mechanical and high-temperature properties.

polyp

polyp [päl´ip] *Medicine.* an abnormal mass of tissue that projects outward from a mucous membrane surface, especially the nose. *Invertebrate Zoology.* the sessile form of coelenterates, with a hollow, more or less cylindrical body and tentacles around the mouth; it may be colonial, as in the case of corals, or solitary, as with hydras.
polyparaphenylene *Materials.* an electrically conductive polymer formed by melt processing.
polypectomy *Surgery.* the surgical removal of a polyp.
polypentadiene *Materials.* a synthetic polydiene elastomer.
polypeptide *Biochemistry.* any of the class of compounds made up of a single chain of amino acid residues linked by peptide bonds, having a larger molecular weight than a peptide but less than a protein.
polypetalous *Botany.* having a flower or corolla of distinct, rather than fused, petals. Also, CHORIPETALOUS.
Polyphaga *Invertebrate Zoology.* a suborder including around 90% of all beetles, insects in the order Coleoptera, with no cross-veins on hindwings; larvae have no legs or legs with only one claw.
polyphagous *Zoology.* feeding on a wide variety of foods.
polypharmacy *Medicine.* **1.** the administration of several drugs together. **2.** an instance of excessive medication.
polyphase *Electricity.* in AC power-line supply circuitry, a group of (usually three) carrying currents having the same frequency and uniformly spaced phase differences. Thus, **polyphase circuit, polyphase transformer, polyphase meter, polyphase rectifier,** and so on.
polyphase structure *Materials Science.* the structure of a material consisting of several phases.
polyphenism *Genetics.* the occurrence in a population of several phenotypes that are the result of environmental influences rather than differences in genotypes.
polyphenol *Organic Chemistry.* any polymeric phenol.
polyphenol oxidase *Enzymology.* any of a group of enzymes that act on diols to form semiquinones.
polyphenyl *Organic Chemistry.* any of a group of direct colors used as a dye.
polyphenylene oxide *Organic Chemistry.* a polyether resin of 2,6-dimethylphenol.
polyphenylene sulfide *Materials.* a high-strength, highly elastic, chemically and environmentally resistant plastic that can be rendered electrically conductive by incorporating up to 30% carbon black.
polyphobia *Psychology.* an irrational fear of many different things.
polyphonic *Acoustical Engineering.* capable of producing two or more tones at a time.
polyphosphate *Materials.* a metal corrosion inhibitor chemical that limits the crystallization of minerals to laminar growth on a metal surface.
polyphosphazene *Organic Chemistry.* any compound having a ring or chain with alternating phophorus and nitrogen atoms, with two substituents on each phosphorus atom; such compounds include cyclic trimers, cyclic tetramers, and high polymers. Also, PHOSPHAZENE.
polyphosphoric acid *Inorganic Chemistry.* $H_{n+2}P_nO_{3n+1}$ (for $n > 1$), any of a series of strong acids, viscous water-white liquid; soluble in water; hygroscopic; toxic on ingestion; strong irritant; used as a dehydrating or catalytic agent and in metal treatment.
polyphyletic *Genetics.* of a species or other taxonomic group, displaying hereditary characters from two or more distinct ancestral lineages as a result of convergent evolution. *Systematics.* of a group of organisms, formally classified together but differing in ancestry. *Hematology.* arising or descending from more than one cell type.
polyphyletism *Hematology.* the theory that the various corpuscles and cells of the blood have their origin from two or more distinct varieties of primordial (mother) cells.
polyphyllous *Botany.* **1.** having separate, distinct leaves not fused to each other. **2.** having many leaves.
polyphyodont *Vertebrate Zoology.* **1.** a kind of dentition found in some mammals, fish, and reptiles in which teeth are replaced continuously. **2.** of, belonging, or pertaining to a polyphyodont vertebrate.
polypide *Invertebrate Zoology.* the soft portion of a bryozoan individual that can withdraw inside the zooecium.
polypivaloactone *Materials.* a polyester, hard-fiber elastomer having a high degree of elasticity, recovery, and extension.
Polyplacophora *Invertebrate Zoology.* chitons, marine mollusks in the class Amphineura, with a dorso-ventrally compressed oval body and a shell of eight overlapping plates on the back. Also, LORICATA.
polyplanetism *Biology.* the condition of having two or more motile stages with resting stages in between.
polyplegia *Neurology.* paralysis affecting several muscles simultaneously.
polyploid *Genetics.* of a cell, nucleus, or organism, having more than two haploid sets of chromosomes; more than diploid; having a ploidy greater than two.
polyploid virus *Virology.* a virus whose particles contain a variable number of genomes.
polyploidy *Genetics.* the state of being polyploid.

Polypodiaceae *Botany.* a large family of tropical, epiphytic and terrestrial ferns of the order Polypodiales (or Filicales), characterized by moundlike exindusiate sori on lower frond surfaces, a scaled creeping rhizome, and bilateral golden spores.

Polypodiales *Botany.* in some systems of classification, a large order in the subdivision Filicopsida, containing the true ferns, characterized by a small sporangia with a definite number of spores.

Polypodiophyta *Botany.* in some systems of classification, a division of the plant kingdom containing the ferns, flowerless vascular plants that reproduce by spores and have creeping, erect, or trunklike stems, wiry to fleshy roots, and sporangia borne in clusters on the leaves.

Polypodiopsida *Botany.* in some systems of classification, a class of ferns in the division Polypodiophyta, characterized by large, spirally arranged compound leaves with groups of sporangia on their undermargins. Also, **Polypodiatae.**

Polyporaceae *Mycology.* a family of fungi belonging to the order Agaricales, occurring on wood and including species that cause timber rot as well as some edible wild mushrooms.

polypore *Mycology.* any member of the genus *Polyporus,* especially *P. sulfphureus,* which forms large, brightly colored growths on old or fallen trees. Also, PORE FUNGUS.

Polyporus *Mycology.* a genus of fungi of the family Polyporaceae, found primarily on wood and characterized by its sometimes very large, bracket-shaped or mushroom-shaped fruiting bodies; certain species cause rot diseases in deciduous trees.

polypotome *Surgery.* a cutting instrument used for removing polyps.

polypotrite *Surgery.* a surgical instrument for crushing polyps.

polypropylene *Organic Chemistry.* $(-C_3H_6-)_n$, a translucent white solid; insoluble in cold organic solvents and softened by hot solvents; melts at 168–171°C; a synthetic, crystalline, thermoplastic polymer used to make molded articles, fibers, film, and toys.

polypropylene glycol *Organic Chemistry.* $CH_3CHOH(CH_2OCHCH_3)_xCH_2OH$, a polymeric material used as a solvent and in hydraulic fluids.

polyprotein *Biochemistry.* a large protein that is formed by the continuous translation of adjacent genes on polycistronic mRNA, then cleaved to form multiple functional proteins.

Polyprotodonta *Paleontology.* in some classifications, a suborder of marsupials that arose in the Cretaceous and includes several living groups such as the Didelphoidea and Dasyuroidea; some authorities have now dropped the term and elevated each of these two groups to suborder level.

Polypteridae *Vertebrate Zoology.* the bichirs and reedfish, the single family of the order Polypteriformes.

Polypteriformes *Vertebrate Zoology.* an order of primitive bony fishes, characterized by a snakelike body covered with hard ganoid scales, fanlike pectoral fins, a dorsal fin divided into a series of spines attached to a membrane, and a swim bladder capable of breathing atmospheric oxygen; found in tropical freshwater rivers and lakes of Africa.

polypyrrole *Materials.* a black, highly conductive film material that is produced by electrochemical polymerization.

polyquinoxaline *Materials.* a high-temperature structural adhesive.

polyribosome *Cell Biology.* a functional unit of protein synthesis composed of several ribosomes simultaneously translating a single molecule of mRNA. Also, **polysome.**

polyrod antenna *Electromagnetism.* a microwave antenna constructed from parallel dielectric rods (typically of polystyrene) that when stimulated at one end radiate from the opposite end.

polysaccharide *Biochemistry.* any of a group of carbohydrates composed of long chains of simple sugars; e.g., starch, cellulose, insulin, or glycogen.

polysaprobic *Ecology.* describing a polluted habitat, such as a lake, that has a high concentration of decomposing organic matter and an absence or minimal concentration of free oxygen.

polysepalous *Botany.* having distinct and separate sepals, not fused to each other. Also, CHORISEPALOUS.

polyserositis *Medicine.* a chronic inflammation of serous membranes with fibrous thickening of the serosa and constrictive pericarditis.

polysiloxane *Organic Chemistry.* $(-R_2SiO-)_n$, a straight-chain compound that consists of silicon atoms single-bonded to oxygen and arranged so that each silicon atom is linked with two oxygen atoms.

polysomy *Genetics.* a condition in which all chromosomes are present, and some but not all are present in greater than the diploid number; usually expressed with a precise indication of the number and the chromosome, such as trisomy 21.

polyspermy *Physiology.* the fertilization of an ovum by multiple spermatozoa.

Polysphondylium *Mycology.* a genus of fungi of the class Dictyosteliomycetes occurring as cellular slime molds of varied colors.

polyspore *Botany.* in certain red algae, any of twelve to sixteen asexual spores.

polystele *Botany.* a stele with vascular units in the parenchyma.

polystely *Botany.* a condition in which vascular tissues exist in two or more steles in a stem and interconnect only occasionally.

Polystigmatales *Mycology.* in some classifications, an order of fungi of the subdivision Ascomycotina; most are plant parasites.

Polystomatoidea *Invertebrate Zoology.* a superfamily of flukes, trematode flatworms in the subclass Monogenea, having posterior suckers and hooks; parasitic on vertebrates, especially fishes, amphibians, and reptiles.

Polystylifera *Invertebrate Zoology.* a suborder of proboscis worms belonging to the order Hoplonemertini, having many short stylets attached to a sickle-shaped base.

polystyrene *Organic Chemistry.* $(-C_6H_5CHCH_2-)_n$, a combustible, transparent polymerized styrene of high strength and impact resistance that is an excellent electrical and thermal insulator; widely used for such purposes as packaging, injection molding, insulation, and lamination.

polystyrene acrylonitrile resin *Materials Science.* a tough thermoplastic copolymer with excellent resistance to acids, bases, and a few solvents.

polysulfide *Chemistry.* any sulfide whose molecules contain two or more sulfur atoms.

polysulfide elastomer *Materials.* an elastomer with excellent oil and gasoline resistance that is used for such items as gaskets, sealants, and hoses for automobiles, boats, and aircraft. Also, THIOKOL.

polysulfide rubber *Organic Chemistry.* a synthetic polymer made by the reaction of sodium polysulfide with an organic dichloride.

polysulfide treating *Chemical Engineering.* a petroleum-refining process that removes sulfur dioxide and sulfur trioxide from sulfuric acid sludges remaining from petroleum treatments.

polysulfone *Materials.* a high-strength plastic with high temperature, oil, solvent, and corrosion resistance that has one of the highest service temperatures of any melt-processable thermoplastic.

polysynaptic *Neurology.* of or relating to the interaction of multiple neurons. Also, MULTISYNAPTIC.

polysynthetic twinning *Crystallography.* twinning composed of three or more multiple crystals in which the twin planes are all parallel.

polytechnic or **polytechnical** *Science.* of or relating to a variety of applied sciences or technical subjects. Thus, **polytechnic institute.**

polytef *Surgery.* a polymer of tetrafluoroethylene used in prostheses and in the fixation or augmentation of bones.

polytene *Genetics.* a giant chromosome with a characteristic banding consisting of many identical chromatids parallel to one another; found in certain protozoans, dipteran insects, and flowering plants.

polytenization *Genetics.* a process in which successive replications of a chromosome set are generated without separation of the replicas.

polyterpene resin *Organic Chemistry.* an amber-colored thermoplastic resin or viscous liquid from polymerization of turpentine; soluble in most organic solvents; used in paints, polishes, and rubber plasticizers.

polytetrafluoroethylene *Materials.* a fluoropolymer that is resistant to chemicals and heat; used for electrical insulation, chemically inert gaskets and parts, and low-friction applications.

polytetrahydrofuran *Materials.* a strong, highly abrasive-resistant polyether derived from tetrahydrofuran; used as an intermediate for polyurethane elastomer or in the manufacturing of spandex fibers.

polytheism *Anthropology.* the worship of several or numerous supernatural beings who are of varying importance and often independent of each other. Thus, **polytheistic.**

polythene *Organic Chemistry.* the British name for polyethylene.

polythetic *Systematics.* describing a group that cannot be defined on the basis of any single character, but only on a combination of characters.

polythetic classification *Systematics.* a method of classification that uses a large number of characters to determine whether an organism is to be included in the taxon; none of the characters alone can determine classification.

Polytrichales *Botany.* an order of perennial mosses having rigid, simple stems that arise from a horizontal underground rhizome.

Polytrichidae *Botany.* a distinctive subclass of mosses of the class Bryopsida, characterized by dark-green very tall or sometimes matted plants with simple, sparsely branched stems and crisped or curled leaves.

polytrophia *Medicine.* excessive nutrition. *Bacteriology.* deriving nourishment from many sources. Also, **polytrophy.**

polytrophic *Medicine.* relating to or characterized by polytrophia. *Bacteriology.* relating to or exhibiting polytrophia.

polytropic atmosphere *Meteorology.* a model atmosphere constituted by hydrostatic equilibrium with a constant nonzero lapse rate.

polytropic compression curve *Physics.* a graphical representation of the pressure of a gas as a function of its volume, the curve being one of a family of curves for various values of n in the equation $PV^n = k$, where k is a constant.

polytropic process *Thermodynamics.* a process in which a sample of gas is compressed or expanded while keeping the quantity pV^n held at a constant value; if the process is at constant pressure, then $n = 0$; if the process is at constant temperature for an ideal gas, then $n = 1$; if the process is at constant entropy for an ideal gas, then n is given by the ratio of specific heat at constant pressure to that at constant volume; and if it is at constant volume, then n is infinitely large.

polytype *Crystallography.* any form of a crystal that is capable of occurring naturally in more than one form.

polytypic having or occurring as several or many types; specific uses include: *Systematics.* of a taxon, containing two or more subordinate taxa, such as a family that contains two or more genera or a genus that contains two or more species. *Crystallography.* of or relating to a polytype or polytypism.

polytypic species *Evolution.* a species that is composed of several subspecies or geographical variants.

polytypism *Crystallography.* a special form of polymorphism that is one-dimensional and is shown by certain close-packed and layered structures; the variation occurs because identical layers of structure may be stacked in different sequences.

polyunsaturated [päl′ē un sach′ə rā′tid] *Chemistry.* containing two or more double or triple bonds per molecule.

polyunsaturated acid *Organic Chemistry.* a fatty acid having two or more double bonds per molecule, such as linoleic and linolenic acids.

polyunsaturated fat *Nutrition.* any of a class of animal or vegetable fats or oils containing multiple double or triple molecular bonds that are unsaturated by the incorporation of hydrogen; such fats and oils are associated with a low level of cholesterol in the blood.

polyurethane foam *Materials.* a flexible or rigid cellular substance that is created through the reaction of a polyester with a diisocyanate; used primarily for insulation or padding.

polyurethane resin *Organic Chemistry.* a resin resulting from the reaction of diisocyanates with a phenol, amine, or hydroxylic or carboxylic compound to make a polymer with free isocyanate groups.

polyurethane rubber *Organic Chemistry.* a synthetic polyurethane resin elastomer made by the reaction of a diisocyanate with a polyester.

polyuria *Pathology.* the passage of an excessive amount of urine, a characteristic of diabetes.

polyvalent *Chemistry.* having more than one valence. Also, MULTIVALENT.

polyvalent number *Computer Programming.* a number made up of several figures, each of which represents a characteristic of the item being described.

polyvalent vaccine *Immunology.* a vaccine that contains protective antigens made up of several strains of a single species of pathogen.

polyvinyl [päl′ē vī′nəl] *Materials.* any polymerization product of a monomeric vinyl compound, such as vinyl chloride.

polyvinyl acetal *Materials.* a thermoplastic resin that is made by condensing polyvinyl alcohol with an aldehyde; primarily made into sheets and used for coatings, adhesives, or molded products.

polyvinyl acetate *Organic Chemistry.* $(-H_2CCHOOCCH_3-)_x$, a combustible, colorless, transparent solid; insoluble in water, gasoline, oils, and fats and soluble in alcohol; resistant to weathering; a thermoplastic high polymer used in adhesives, film, inks, and latex paints.

polyvinyl alcohol *Organic Chemistry.* $(-CH_2CHOH-)_x$, a combustible, white to cream-colored powder; soluble in water; decomposes at 200°C; a synthetic polymer used in adhesives and as an emulsifying agent, thickener, and stabilizer.

polyvinyl butyral *Materials.* a tough, flexible, transparent, vinyl resin that is made by reacting polyvinyl alcohol with butyraldehyde; used as the interlayer in safety glass and in textile coating. Also, BUTVAR.

polyvinyl carbazole *Materials.* a brown thermoplastic resin that softens at 150°C, and has excellent dielectric properties, good heat and chemical stability, but poor mechanical strength; used as a paper-capacitor impregnant and as a substitute for electrical mica.

polyvinyl chloride *Organic Chemistry.* $(-H_2CCHCl-)_x$, white powder or colorless granules available as film, sheet, fiber, and foam that is resistant to weathering and moisture and has good dielectric properties; it decomposes at 148°C, evolving toxic fumes; a polymer of vinyl chloride that is widely used for many industrial purposes, as in food packaging films, piping and conduits, siding, and for other construction uses.

polyvinyl chloride acetate *Organic Chemistry.* $(C_2H_4O{\cdot}C_2H_3Cl)_x$, a copolymer of vinyl chloride and vinyl acetate that is used for wire coverings and protective garments.

polyvinyl dichloride *Organic Chemistry.* a combustible but self-extinguishing polymer that has high strength and superior chemical resistance over a wide temperature range; used for pipes and fittings that carry hot, corrosive materials.

polyvinyl (ethyl) ether *Organic Chemistry.* $[-CH(OC_2H5)CH_2-]_n$, a rubbery solid that is colorless when pure, insoluble in water and soluble in almost all organic solvents; used for pressure-sensitive tape.

polyvinyl fluoride *Organic Chemistry.* $(-H_2CCHF-)_n$, a vinyl fluoride polymer in film form, characterized by superior resistance to weather, having high strength and low permeability to air and water; it evolves toxic fumes on heating; used for packaging and in electrical equipment.

polyvinyl formate resin *Organic Chemistry.* $(-CH_2-CHOOCH-)_x$, a clear resin used in plastics.

polyvinylidene chloride *Organic Chemistry.* $(-CH_2-CCl_2-)$, a thermoplastic polymer that is tasteless, odorless, abrasion resistant, impermeable to flavor, highly inert to chemical attack, and has low vapor transmission; used in the packaging of food products, pipes, upholstery, and latex coatings. Also, SARAN.

polyvinylidene fluoride *Organic Chemistry.* $(-H_2C-CF_2-)$, combustible, self-extinguishing and nondripping crystals that are resistant to electricity, acids, alkalies, and oxidizers; a fluorocarbon polymer available in powder, pellets, solution, and dispersion; used as insulation, tank linings, and sealant, and in protective paints and coatings, diodes, and soldered joints.

polyvinylidene resin see VINYLIDENE RESIN.

polyvinyl isobutyl ether *Organic Chemistry.* $[-CH_2CHOCH_2CH(CH_3)_2-]_n$, a combustible, white, opaque elastomer or viscous liquid, insoluble in water, ethanol, and acetone and soluble in most other organic solvents; used in adhesives, waxes, and lubricating oils.

polyvinyl methyl ether *Organic Chemistry.* $(-CH_2CHOCH_3-)_n$, a combustible, colorless liquid, soluble in water at 32°C and in most organic solvents; used in pressure-sensitive adhesives, printing inks, and as a protective colloid in emulsions.

polyvinyl resin *Organic Chemistry.* any resin or polymer derived by polymerization or copolymerization of vinyl monomers $(CH_2{=}CH-)$.

polzenite *Petrology.* a lamprophyre rock, or rock group, containing olivine and melilite.

Pomacentridae *Vertebrate Zoology.* the damselfishes, a large family of small, brightly colored fishes of the order Perciformes, characterized by a deep, laterally compressed body with a single nostril on each side; found worldwide in tropical and temperate regions, especially where reef-building corals are found.

pomaceous *Botany.* relating to or characteristic of pomes.

Pomadasyidae *Vertebrate Zoology.* the grunts, a large family of marine fishes of the order Perciformes, characterized by slender bodies with a spiny dorsal fin continuous with a rayed one; the popular name comes from the practice of grinding its teeth together, with the sound amplified by the swim bladder; found in shallow tropical waters, usually near reefs, distributed worldwide. Also, HAEMULIDAE.

Pomatomidae *Vertebrate Zoology.* the bluefishes, a family of predatory schooling fishes of the order Perciformes, characterized by a long, stout body that is blue on top, with silvery sides and white underneath, a low spiny dorsal fin followed by a rayed one, and a forked tail; found worldwide in tropical waters except the east Pacific.

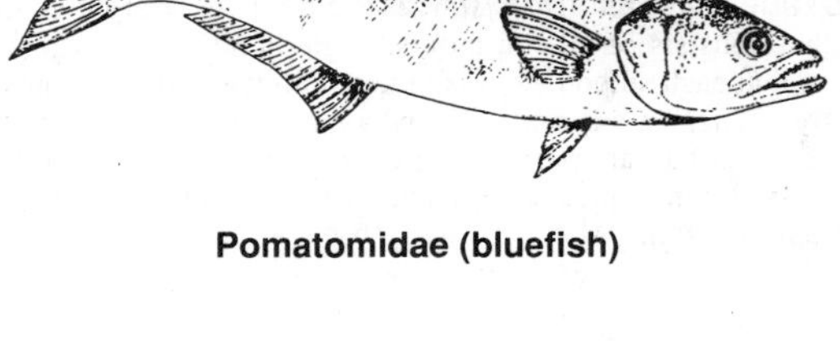

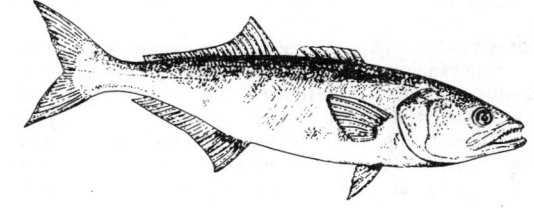

Pomatomidae (bluefish)

pome *Botany.* a type of simple fruit that is characteristic of the rose family, in which a core containing the seeds is surrounded by a fleshy receptacle, as in an apple, pear, or quince.

pomegranate [päm´ə gran´it] *Botany.* **1.** a small deciduous tree, *Punica granatum* of the family Punicaceae, cultivated for its red pomelike fruit having a thick skin, several cells, and tangy edible pulp and seeds. **2.** the fruit of this tree.

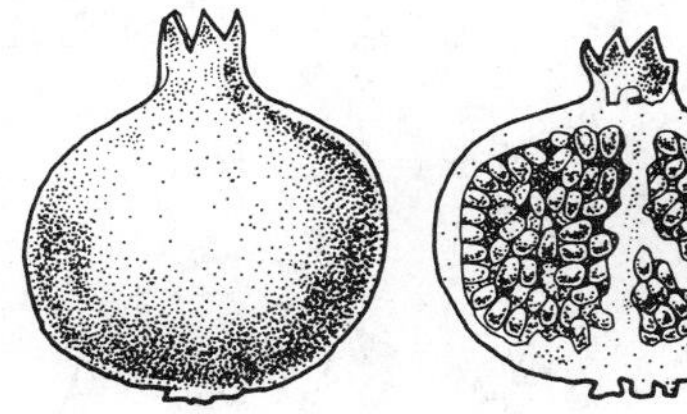

pomegranate

Pomeranchuk theorem *Particle Physics.* a theorem regarding the equivalence of limits of the cross sections of both a particle and its antiparticle when such particles are projected toward a target particle at sufficiently high energies.

Pomeron *Particle Physics.* a singularity in the complex representation of the scattering amplitude (Regge pole) located at +1.

pomology *Agriculture.* the science of growing fruits.

POMP *Oncology.* an acronym for a cancer chemotherapy regimen consisting of prednisone, Oncovin, methotrexate, and 6-mercaptopurine.

Pompilidae *Invertebrate Zoology.* a family of wasps in the superfamily Pompiloidea, having long bristle-covered legs, an iridescent body, and dark wings; they paralyze spiders and store them in their nests for food.

Pompiloidea *Invertebrate Zoology.* a superfamily of wasps, hymenopteran insects with slender bodies and long legs, hunters of spiders and crickets.

pom-pom *Ordnance.* a type of shipborne light antiaircraft cannon.

PONA analysis *Analytical Chemistry.* an analytical procedure to determine the amounts of paraffins, olefins, naphthenes, and aromatics in gasolines. (An acronym for paraffins, olefins, naphthenes, aromatics.)

poncelet *Physics.* a unit of power equivalent to 980.655 watts.

Ponchon-Savarit method *Chemical Engineering.* a graphical method used to solve binary and ternary extraction systems by plotting mass balances on enthalpy-concentration diagrams.

pond *Geography.* **1.** a natural body of standing fresh water occupying a surface depression, regarded as smaller than a lake and larger than a pool. **2.** a similar artificial body of water, as on a farm for water supply.

pond. by weight. (From Latin *pondere.*)

pondage *Hydrology.* water whose natural flow is regulated by holding it in a reservoir for short periods of time.

pondage land *Geology.* the area of land where water is stored or held as dead water during flooding. Also, FLOOD FRINGE.

ponded stream *Hydrology.* a stream in which a pond has formed because the normal streamflow has been interrupted.

ponderosa pine *Botany.* a tall western North American timber pine, *Pinus ponderosa* of the family Pinaceae, having long, clustered dark green needles and yellowish bark.

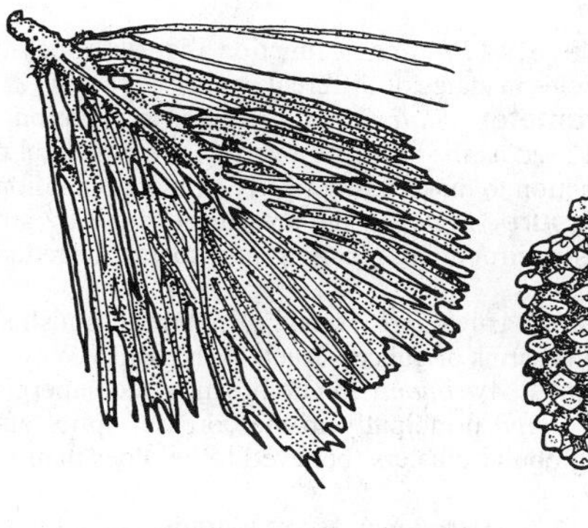

ponderosa pine

ponding *Hydrology.* **1.** the natural formation of a pond in a stream whose normal streamflow has been interrupted. **2.** the artificial containment of water in a stream to form a pond. *Building Engineering.* the accumulation of rainwater on a flat roof because of poor drainage. *Civil Engineering.* a technique of curing concrete by purposely flooding a dammed area.

ponente *Meteorology.* a westerly wind that blows on the Mediterranean coast of France, the northern Roussillon region, and Corsica.

Ponerinae *Invertebrate Zoology.* a subfamily of hunting ants, hymenopteran insects in the family Formicidae, having slender bodies, large mandibles, a strong sting, and a sound-producing organ.

pongid *Anthropology.* any of the four living apes, including the chimpanzee, gorilla, orangutan, and gibbon.

Pongidae *Vertebrate Zoology.* the great apes or anthropoid apes, a family of Primates characterized by massive bodies with forelimbs longer than the hindlimbs and capable of bipedalism, but generally moving on all fours; includes the orangutan, chimpanzee, and gorilla.

poniente *Meteorology.* a westerly wind that blows in the Straits of Gibraltar.

pono- or **pon-** a combining form meaning "pain" or "hard work."

ponograph *Medicine.* an instrument for estimating and recording sensitivity to pain.

pons *Science.* the Latin word for bridge. *Anatomy.* the section of the brain between the medulla and the midbrain. Also, BRIDGE OF VAROLIUS.

Pons, Jean Louis 1761–1831, French astronomer; discovered Encke's comet and 36 other comets.

pons asinorum *Mathematics.* the nickname for one of the first geometric propositions presented by Euclid: if two sides of a triangle are equal, the angles opposite these sides are also equal. *Science.* any problem that a beginner must solve in order to gain a basic understanding of a subject. (From a Latin phrase meaning "the donkey's bridge.")

Pontederiaceae *Botany.* a widespread family of aquatic and semiaquatic, tanniferous, monocotyledonous herbs of the order Liliales, characterized by vessels usually confined to the roots and expanded floating or emergent blades; includes the common water hyacinth and the pickerelweed.

Pontian *Geology.* a European geologic stage of the Upper Miocene epoch, occurring after the Pannonian and before the Kimmerian of the Pliocene epoch.

pontic *Geology.* of or relating to sediments or facies deposited in relatively deep, still water. *Medicine.* an artificial tooth on a fixed partial denture that replaces a lost natural tooth and restores its function.

pontine flexure *Developmental Biology.* a bending in the hindbrain of the embryo. Also, BASICRANIAL FLEXURE.

pontobulbar *Neurology.* of or relating to the pons and the medulla oblongata.

Pontodoridae *Invertebrate Zoology.* a family of carnivorous polychaetes containing a single species, *Pontodora pelagica,* a small slender worm having a transparent body, few segments, and a short, eversible proboscis.

pontoon *Naval Architecture.* **1.** a vessel used as a float to support part of a load, e.g., a pontoon bridge. **2.** a flat-bottom boat; may be fitted for wrecking and lifting operations or used to carry goods and passengers within a harbor. **3.** a lifeboat with internal subdivisions instead of flotation tanks. *Aviation.* a hollow, watertight structure attached to a seaplane or similar aircraft for the purpose of giving it the ability to take off and land on water with stability and buoyancy. Also, FLOAT.

pontoon bridge *Civil Engineering.* a bridge supported by hollow, cylindrical floats anchored to the bottom of the body of water, such as Lake Washington Bridge near Seattle.

Pontosphaeraceae *Botany.* a family of flagellate marine algae of the order Coccosphaerales, distinguished by ellipsoidal coccoliths usually having a raised margin.

pony *Agriculture.* a small horse, usually one that is less than 14 hands (about 5 feet) high when full-grown.

ponyfish see SLIPMOUTH.

pool *Hydrology.* **1.** a small, natural body of standing water, usually fresh, regarded as smaller than a pond. **2.** a small, still, deep reach of a stream. **3.** any small collection of water on a surface, such as a rain puddle. *Civil Engineering.* an artificially confined body of water above a dam or the closed gates of a lock. *Building Engineering.* see SWIMMING POOL. *Medicine.* **1.** a common group of blood donors from whom blood plasma is collected. **2.** to mix plasma from various donors. *Hematology.* an accumulation of blood in any part of the body due to retardation of the venous circulation.

pool cathode *Electronics.* an electrode that has a pool of mercury from which electrons are ejected.
pooled plasma *Hematology.* a mixture of plasma from several donors.
Poole-Frenkel effect *Materials Science.* an electronic conduction involving a field-induced thermal excitation of electrons from traps into the conduction band of a dielectric thin film. Since the traps are intrinsic, the conduction process is a bulk effect and not controlled by the electrode material.
pool reactor *Nucleonics.* a nuclear reactor whose core is suspended in a pool of water that acts as a coolant, moderator, reflector, and radiation shield.
pool spring *Hydrology.* a pool of water that is created and fed by an underground spring.
pool tube *Electronics.* any tube, such as an excitron, ignitron, or mercury-arc rectifier, that has a pool of mercury as its cathode. Also, **pool-cathode tube.**
poop *Naval Architecture.* a relatively small, raised partial deck at the stern of a ship above the quarterdeck.
poop deck *Naval Architecture.* a small weather deck at the top of a poop.
poor rate *Horology.* the rating given to a timepiece with an irregular error, or with an error of more than 10 seconds fast or slow per day.
poorwill or **poor-will** *Vertebrate Zoology.* a nocturnal nightjar, *Palaenoptilus nuttallii,* of the family Caprimulgidae, the only bird known to hibernate (often on the same rock crevice each year); characterized by large eyes and a very large mouth, a plump body, and a two-syllable call.
pop *Computer Programming.* to remove an item from the top of a stack or pushdown list. *Mining Engineering.* a short blasthole used to reduce larger sections of rock or to trim a working face. Also, **pop hole, pop shot.**
pop. population.
popcorn noise *Electronics.* the heat-generated noise that resembles the sound of popcorn popping and appears at random intervals in semiconductor devices, especially operational amplifiers.
poplar *Botany.* any of various slender, quick-growing trees of the genus *Populus,* family Salicaceae; characterized by simple, alternate leaves, flowers and fruit borne in catkins, and scaly buds on a bitter bark. *Materials.* the fine-grained, yellowish wood of these trees; used for paper and paneling.

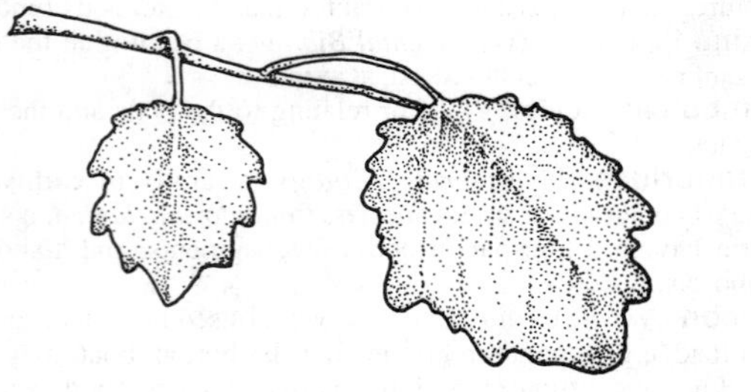

poplar

poplin *Textiles.* one of a group of durable, plain weave fabrics with fine cross ribs due to warp threads that are finer than the filling threads; made from silk, cotton, wool, manmade fibers, or blends.
popliteal artery *Anatomy.* a large artery located behind the knee.
popliteal nerve *Anatomy.* the nerve that serves the muscles behind the the knee.
popliteal space *Anatomy.* a space between the muscles in the back of the leg that is widest at the knee and deepest above the articular end of the femur.
popliteal vein *Anatomy.* a large vein draining the muscles in the back of the leg.
popliteus *Anatomy.* a muscle, extending from the femur to the tibia, that flexes and rotates the leg.
pop-off valve see PRESSURE-RELIEF VALVE.
poppet *Mining Engineering.* the frame of a pulley or the headgear above a mine shaft. *Civil Engineering.* a framework that supports the bow or stern of a vessel for launching. *Mechanical Devices.* a spring-loaded ball that engages a groove, cut, or notch in a wheel, such as the movable headstock of a lathe, with a revolving mandrel. Also, **poppet head.**
poppet valve *Mechanical Engineering.* a conical-shaped, heat-resistant steel disk that engages a conical seat, and is used in internal-combustion and steam engines to control inlet and exhaust.
popping pressure *Mechanical Engineering.* the inlet pressure at which a safety valve disk opens.
popple rock see PEBBLE BED.
poppy *Botany.* any of many species of the genus *Papaver,* family Papaveraceae, characterized by milky juice, large, showy flowers, and capsular fruit; *P. somniferum* is the source of opium.

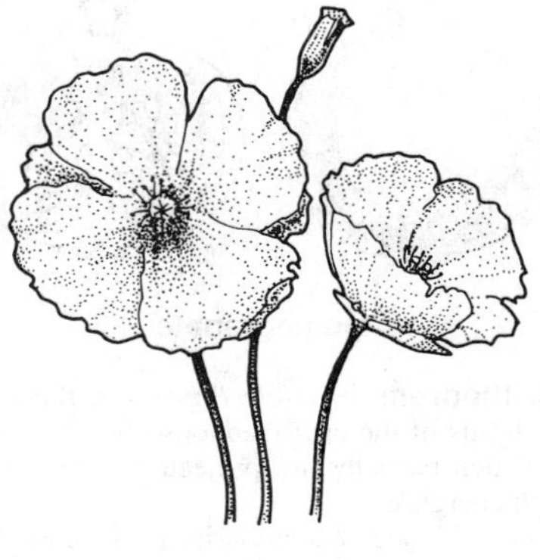

poppy

populate *Computer Technology.* **1.** to add electronic components to a circuit board. **2.** to fill a database in a new system with relevant operational data.
population *Ecology.* any group of organisms of the same species living in a specific geographic area. *Statistics.* all the items, quantities, or values being evaluated in a statistical study from which samples are taken. *Archaeology.* the total set of items or materials to be studied in a given category, as all the sites in a region or all the artifacts at a site. A **population parameter** is a single specific characteristic of a population.
population bottleneck *Ecology.* a sharp decline in population that causes the gene pool to shrink.
population covariance *Statistics.* a measure of association between two populations.
population cycle *Ecology.* the periodic fluctuation in the density of a given species.
population density *Ecology.* see DENSITY.
population dispersal *Biology.* the process by which groups of living organisms expand the space or range within which they live.
population dispersion *Biology.* the spatial distribution at any particular moment of the individuals of a species of plant or animal.
population ecology *Ecology.* the study of the various factors that affect the size of a particular population over a period of time.
population genetics *Genetics.* a branch of genetics that deals with the relative occurrence of different alleles and genes in populations.
population geography *Geography.* the branch of geography dealing with the relationships between geography and population patterns, including birth and death rates and migration.
population inversion *Atomic Physics.* a condition in which the number of atoms at higher energy levels exceed those at lower energy levels.
population mean *Statistics.* the expected value of an arbitrary element of a population.
population of levels *Physics.* a function that describes the relative numbers of particles in states of different energy.
population parameter *Statistics.* for a given population, a quantity that is fixed and used as the value of a variable in a general distribution or frequency function to make it descriptive of that population.
population pressure *Ecology.* the pressure exerted by an increasing population on its environment, resulting in dispersal or reduction of the population.
Population I stars *Astronomy.* young, very bright, bluish stars; found principally in spiral arms of galaxies such as the Milky Way.
Population II stars *Astronomy.* predominantly red supergiant stars of low luminosity; found principally in the cores of spiral and elliptical galaxies and in globular clusters; believed to be older than Population I stars.
Population III stars *Astronomy.* as yet unfound stars, hypothesized to have formed before Population II stars.

Mat
& Su

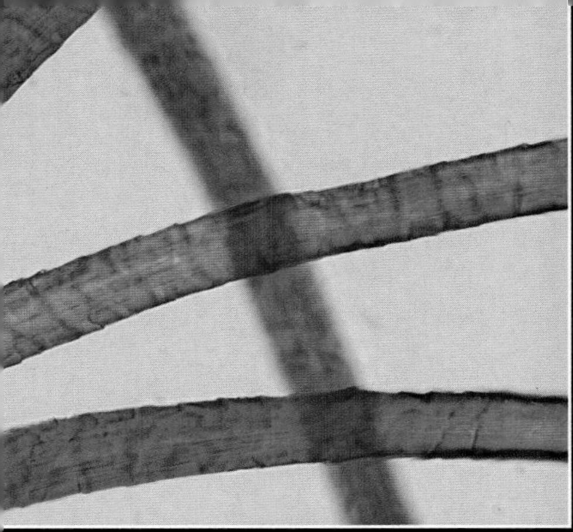

revealing scales on the shafts.
Photos.

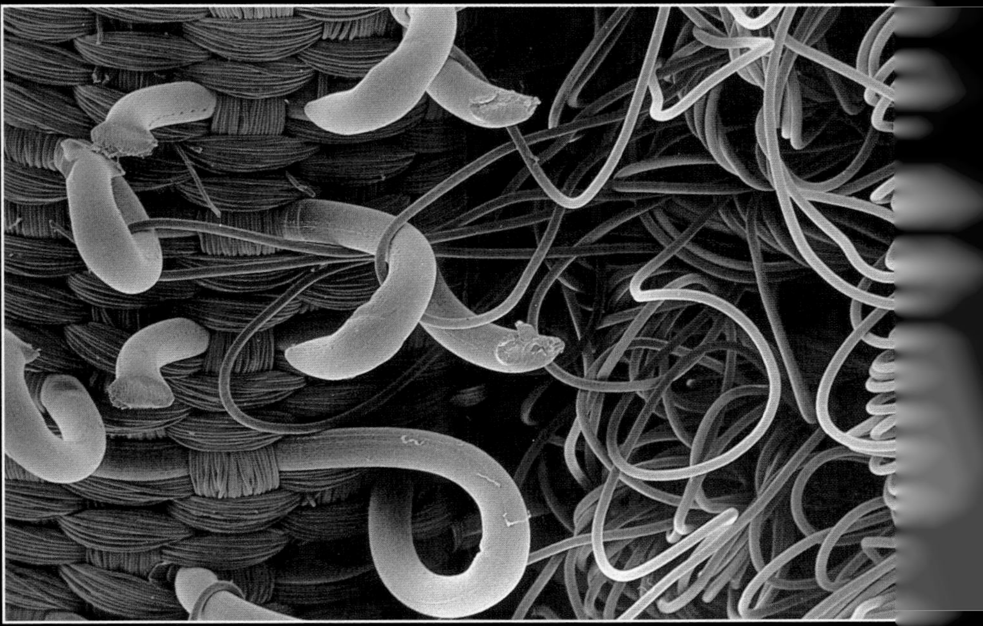

False-color scanning micrograph of Velcro being pulled apart. SPL/Photo Researchers.

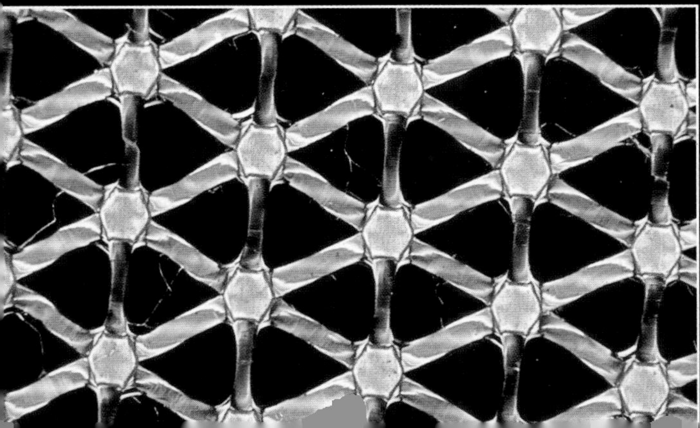

High-temperature superconductor magnetically levitating a small magnet. Liquid nitrogen used in the superconductor causes the glowing vapor. D. Parker/SPL/Photo Researchers.

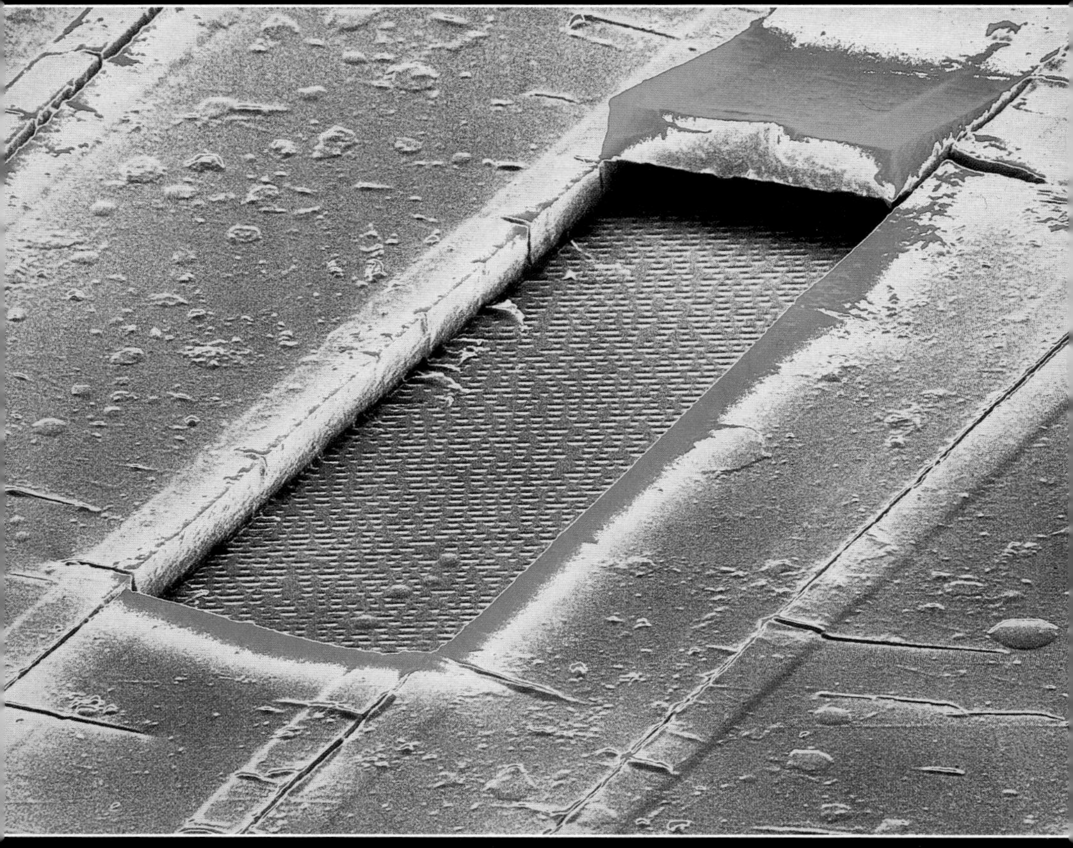

alse-color scanning electron micrograph of the surface of a compact disc cracked to show the layer of epressions below that represent the digitized musical signals read by a laser.

pop-up menu *Computer Programming.* a list of options that is displayed for selection by the user. The menu appears on the screen ("pops up") when a selection is needed and disappears after a selection has been made.

pop valve *Mechanical Devices.* a spring-loaded safety valve.

p/o ratio *Biochemistry.* the ratio of moles of ATP synthesized to moles of oxygen atoms reduced to water during oxidative phosphorylation.

p orbital *Atomic Physics.* the orbital of an atom in which the electron has an angular momentum quantum number 1.

porcelain *Materials.* a hard, white, translucent, nonporous ceramic material composed primarily of kaolin, feldspar, and quartz fired first at a low temperature and then fired again at a very high temperature.

porcelain capacitor *Electricity.* a ceramic-dielectric-based fixed capacitor whose dielectric is made of porcelain or a related substance; it is molecularly fused to alternate layers of fine silver electrodes and forms a monolithic unit that requires no case or hermetic seal.

porcelain enamel see VITREOUS ENAMEL.

porcelain jasper *Mineralogy.* a hard, naturally baked, impure red clay that was formerly considered to be a variety of jasper.

porcelaneous *Geology.* describing rock that has the appearance of unglazed porcelain.

porcelaneous chert *Petrology.* a hard variety of chert with a smooth fracture surface and an opaque to subtranslucent china-white appearance, resembling chinaware or glazed porcelain.

Porcellanasteridae *Invertebrate Zoology.* a family of deep-sea starfish in the suborder Paxillosina, having a large disc, high, narrow marginal plates, and no intestine.

Porcellanidae *Invertebrate Zoology.* the porcelain crabs, a family of decapod crustaceans in the group Anomura, having a flat, calcified carapace, a broad abdomen, and a well-developed tail fan.

porcellanite *Petrology.* a hard, dense siliceous rock, such as impure chert, characterized by the texture, luster, and general appearance of unglazed porcelain.

porcine *Science.* **1.** relating to or like a pig. **2.** relating to a substance or process that is derived from or associated with pigs.

porcine polyoencephalitis see TESCHEN DISEASE.

porcupine *Vertebrate Zoology.* **1.** any of the herbivorous, nocturnal, terrestrial rodents of the family Hystricidae; characterized by long, sharp spines or quills covering the back and side; found in desert, savannah, and forest regions of southern Europe, southern Asia, the East Indies, and Africa. Also, OLD WORLD PORCUPINE. **2.** any of the generally nocturnal, arboreal rodents of the family Erethizontidae, characterized by quills and a prehensile tail and feet adapted to climbing trees; found from the arctic coasts of North America to northern Argentina. Also, NEW WORLD PORCUPINE. (From a Middle French word meaning literally "thorn pig.")

porcupine

porcupine boiler *Mechanical Engineering.* a boiler having dead-end tubes projecting from a vertical shell.

porcupinefish *Vertebrate Zoology.* a fish of the family Diodontidae, named for its covering of sharp spines and ability to inflate its body when attacked; found in shallow, tropical waters worldwide.

pore *Biology.* a small, more or less round aperture by which matter passes through a wall or membrane, as in the skin or a leaf, for perspiration, absorption, and so on. *Metallurgy.* any small void in a metallic material, electroplated coating, or powder compact. *Astronomy.* a tiny, short-lived dark spot in the sun's photosphere. *Geology.* **1.** a small hole or passageway in a rock or soil. **2.** see INTERSTICE.

pore canal see OSTIUM.

pore complex *Cell Biology.* the combination of protein subunits that form a nuclear pore in eucaryotic cells.

pore compressibility *Geology.* the fractional change in the pore volume of a reservoir rock per unit change in pressure upon that rock.

pore diameter *Design Engineering.* the average diameter of the openings of a mesh or other porous material.

pore diffusion *Fluid Mechanics.* the movement of fluids through porous solids and membranes, such as reverse osmosis, zeolite adsorption, dialysis, and reactant diffusion into porous catalysts.

pore fungus *Mycology.* a common term for fungi belonging to the families Boletaceae and Polyporacea, characterized by pores or tubes, rather than gills, that encase their spores.

pore ice *Hydrology.* ice that more or less fills the pore spaces in the ground as a result of the freezing of interstitial water in place, with no additional water being present.

Porellaceae *Botany.* a cosmopolitan family of large, showy liverworts of the order Jungermanniales, characterized by complicate-bilobed leaves with a unique, noninflated ventral lobe and by rhizoids at the bases of underleaves.

porencephalia or **porencephaly** *Neurology.* **1.** the presence of cysts or cavities in the brain cortex communicating through a pore with the arachnoid space. **2.** a congenital defect characterized by one or more cavities or cysts within one cerebral hemisphere, whether or not they communicate with the arachnoid space, marked by a cystic outpouching of the ventricles. Also, PERENCEPHALY.

pore pressure or **pore-water pressure** see NEUTRAL STRESS.

pore-size distribution *Geology.* the variations in pore size exhibited by each type of rock in reservoir structures.

pore space *Geology.* the total volume of all the open spaces in a rock or soil. Also, **pore volume.**

pore water see INTERSTITIAL WATER.

porfiromycin *Oncology.* $C_{16}H_{20}N_4O_5$, an antineoplastic antibiotic derived from *Streptomyces ardus.*

porgy *Vertebrate Zoology.* any of the food fish, *Pagrus pagrus* of the family Sparidae; found in the Mediterranean and off the Atlantic coasts of Europe and North America.

Poria *Mycology.* a genus of fungi of the family Polyporaceae, which is characterized by inverted fruiting bodies and the cause of certain rots; found on wood.

poriaz *Meteorology.* violent northeasterly winds that blow on the Black Sea near the Bosporus in the Soviet Union.

Porifera *Invertebrate Zoology.* the phylum of sponges, the most primitive of all multicellular organisms, having no true tissues or organs; filter feeders with pores and canals through which water is propelled by ciliary beating; all are sessile and almost all are marine.

porin *Microbiology.* any of several major proteins found in the outer membrane of Gram-negative bacteria that, grouped as dimers or trimers, form transmembrane channels for the entry of certain molecules into the cell.

pork *Agriculture.* the meat of pigs.

porker *Agriculture.* a young pig that is being fattened for market.

Porlezzina *Meteorology.* an easterly wind from the Gulf of Porlezza that blows on Lake Lugano, bordering both Italy and Switzerland.

Porocharaceae *Paleontology.* a family of charophytic algae, ancestral to the three post-Paleozoic families: Raskyellaceae, Characeae (still extant), and Clavatoraceae; extant from the Carboniferous to Cretaceous.

porocyte *Invertebrate Zoology.* a large tube-shaped cell in some sponges through which water can flow.

porogamy *Botany.* the entry of the pollen tube into the micropyle of an ovule in a seed plant.

porosimeter *Engineering.* an instrument used to measure the porosity of a rock sample by comparing the bulk volume of the sample with the aggregate volume of the pore spaces between the grains.

porosis *Medicine.* **1.** the formation of the callus in the repair of a fractured bone. **2.** the formation of cavities, vacuoles, or pores.

porosity *Physics.* the ratio of the total amount of void space in a material (due to poses, small channels, and so on) to the bulk volume occupied by the material. *Metallurgy.* the amount of small cavities, usually undesirable, but at times intentional, present in a metallic product. *Geology.* see TOTAL POROSITY.

porous *Materials.* of or relating to a material that contains pores; able to be permeated by air, water, and so on.

porous bearing *Design Engineering.* a bearing fabricated from a powdery metal substance that has been impregnated with oil by a vacuum treatment.

porous metals *Metallurgy.* a metallic material that is intentionally porous; often used for filtration.

porous mold *Engineering.* a plastics mold created from a mixture of diffuse material, such as powdered metal, that has many tiny holes for the passage of liquids or air.

Poroxylaceae *Paleontology.* a monogeneric family of pteridospermous plants, once erroneously thought to belong to the Cordaitales, but now assigned to the extinct order Callistophytales; Carboniferous.

porpezite *Mineralogy.* a natural alloy of gold with 5–10% palladium. Also, PALLADIUM GOLD.

porphin *Biochemistry.* a porphyrin that does not have a substituted form.

porphobilinogen *Biochemistry.* $C_{10}H_{14}O_4N_2$, a derivative of pyrrole that forms during the degradation of hemoglobin; trace amounts are naturally present in urine.

porphrite see PORPHYRY.

Porphyra *Botany.* a genus of red algae of the family Bangiaceae, having gelatinous red or purple fronds; cultivated as a food source in China and Japan.

porphyria *Medicine.* a metabolic disorder affecting blood pigment, marked by a significant increase in the formation and excretion of porphyrins.

Porphyridiaceae *Botany.* a family of marine and freshwater red algae of the order Porphyridiales, characterized by spherical or ovoid cells and single-celled thalli.

Porphyridiales *Botany.* an order of the red algae division Rhodophyta, characterized by a very low level of organization, single-celled thalli, and essentially unknown life histories.

Porphyridium *Botany.* a widespread red algae genus of the family Porphyridiaceae.

porphyrin *Biochemistry.* any of a group of compounds containing the porphin structure of four pyrrole rings connected by methine bridges in a cyclic configuration, to which a variety of side chains are attached; usually metalled, e.g., with iron to form heme.

porphyrin test *Microbiology.* a test determining whether or not a *Haemophilus* strain requires a particular growth factor, termed the X factor.

porphyrinuria *Medicine.* the presence of an excess of phorphyrin in the urine.

porphyritic *Geology.* **1.** an igneous rock texture in which large phenocrysts are embedded within a fine crystalline and/or glassy groundmass. **2.** of, relating to, or resembling porphyry.

porphyroblast *Geology.* a large, pseudoporphyritic crystal formed by metamorphism.

porphyroblastic *Geology.* of or relating to a texture of recrystallized metamorphic rocks in which large crystals are distributed throughout a finer-grained crystalloblastic groundmass.

porphyroclastic structure see MORTAR STRUCTURE.

porphyrogranulitic *Petrology.* of or relating to an ophitic texture with large phenocrysts of feldspar and either augite or olivine set in a matrix of lath-shaped feldspar crystals and small, irregular crystals of augite.

porphyroid *Petrology.* **1.** a blastoporphyritic, or sometimes porphyroblastic, metamorphic rock of igneous origin. **2.** a porphyrylike rock of feldspathic, metasedimentary constituency.

porphyroskelic *Geology.* of or relating to the fabric of a soil whose plasma consists of a dense matrix in which skeleton grains are embedded in a fine-grained groundmass.

porphyrotope *Geology.* a large crystal set in a finer-grained matrix within a sedimentary rock exhibiting a porphyrotopic fabric.

porphyrotopic *Geology.* describing the fabric of a recrystallized carbonate rock or a chemically precipitated sediment in which the larger of the multisized crystals are enclosed in a finer-grained matrix. Also, **porphyrocrystallic.**

porphyry

porphyry *Petrology.* any igneous rock with numerous large phenocrysts embedded in a fine-grained groundmass. Also, PORPHRITE.

porphyryl *Hematology.* a name for hemin from which iron has been removed.

porpoise *Vertebrate Zoology.* a marine, air-breathing mammal of the family Phocoenidae that is actually a beakless whale; characterized by a long, torpedo-shaped body, a horizontal tail, and a blowhole on the highest part of the head; found in northern coastal zones and along portions of the South American and southeast Asian coastlines.

porpoise

Porro-Koppe principle *Photogrammetry.* a principle stating that camera lens distortion may be eliminated by viewing a photograph through an optical system that has the same distortion characteristics as the lens used to make the original exposure.

Porro prism *Optics.* a 45°–90°–45° reflecting prism whose surfaces form the 90° angle reflecting the light beam through a total angle of 180°.

Porro-prism erecting system *Optics.* a perpendicular arrangement of two Porro prisms that adjusts an inverted image, so that its orientation is the same as that of the object imaged; used in prism binoculars and some telescopes.

port *Geography.* **1.** the commercial area within a harbor, containing facilities for taking on and discharging passengers and cargo. **2.** a city or town having such facilities. *Naval Architecture.* **1.** the left side of a vessel, when facing toward the bow. **2.** an opening or entryway in a vessel's side. *Computer Technology.* **1.** an input/output connection between a central processing unit and a peripheral device. **2.** a connector on a device to which cables for other devices can be attached. **3.** an input/output channel of a central processing unit to which a peripheral device can be attached. *Engineering.* an opening, generally valve-controlled, through which fluid enters or leaves an engine. *Acoustical Engineering.* an opening in a bass reflex-type loudspeaker, used to improve the bass response characteristics of the speaker enclosure. *Nucleonics.* an opening to the core of a nuclear reactor; used for inserting objects to be irradiated, and sometimes for allowing a beam of radiation to emerge. *Electricity.* **1.** an opening in a waveguide component through which energy is fed or withdrawn and sometimes measured. **2.** an entrance or an exit for a network, usually consisting of a pair of terminals that are fed into it or withdrawn from it. (From the Latin word for "gate.")

port *Food Technology.* a heavy, sweet fortified wine; most ports are classified as **ruby port** (very sweet and dark) or **tawny port** (milder and paler). (Named for the Douro Valley near Oporto, *Portugal,* where it was first made.)

Porta, Giambattista Della 1535–1615, Italian biologist; wrote major studies of optics and of human and plant physiognomies.

portability *Engineering.* the fact of being portable; the ability to be conveniently carried and transported. *Computer Programming.* the capability of a program or data to be run on another type of computer without modification.

portable *Engineering.* of, relating to, or describing any object that can be conveniently carried and transported.

portable audio terminal *Computer Technology.* a small, light-weight terminal with a keyboard that can communicate with a host system via an ordinary telephone receiver.

portable computer *Computer Technology.* any computer that is intended to be conveniently moved or carried about, especially a laptop computer.

portable data terminal *Computer Technology.* a small, light-weight terminal used to collect data and transfer it to the host computer or a storage medium.

portable genetic element *Molecular Biology.* any movable piece of nucleic acid, such as a plasmid or a transposon, that can insert into and remove itself from other segments of nucleic acid.

porta hepatis *Anatomy.* the fissure between the right and left lobes of the liver, containing the hepatic arteries, the hepatic portal vein, and the hepatic ducts. Also, HILUS OF THE LIVER, PORTAL.

port-aiguille *Surgery.* an apparatus used to hold needles for surgery.

portal *Anatomy.* **1.** an entry or gateway of nerves or vessels into an organ. Also, **porta. 2.** see PORTA HEPATIS. *Architecture.* a door, gate, or entranceway, especially when on a grand or monumental scale. *Mining Engineering.* **1.** a mine entrance. **2.** the rock face at which the digging of a tunnel is begun. *Engineering.* a skeletal structure made up of two uprights joined by a horizontal at the top.

portal circulation *Physiology.* any blood vessels that connect with the capillary systems of two organs before the blood flows back to the heart.

portal cirrhosis *Medicine.* a disease of the liver, frequently associated with chronic alcoholism, in which fatty change of liver cells leads to enlargement of the liver and later to contraction of the liver. Also, LAENNEC'S CIRRHOSIS.

portal crane *Mechanical Engineering.* a jib crane carried on a four-legged portal, and which runs along rails.

portal hypertension *Medicine.* an increase in venous pressure in the portal venous system, caused by compression or by occlusion in the portal or hepatic vascular system, and resulting from conditions such as cirrhosis of the liver, heart failure, and portal vein thrombosis.

portal system *Anatomy.* a system of veins conveying blood from the intestines and stomach to the liver.

portal vein *Anatomy.* a large vein conveying blood from the stomach and other abdominal organs to the liver.

Porter, George born 1920, English chemist; shared the Nobel Prize with Ronald Norrish for work in high-speed chemical reactions.

Porter, Rodney Robert 1917–1985, American biochemist; shared the Nobel Prize with Gerald Edelman for analyzing the chemical structure of antibodies.

porter *Food Technology.* a dark, full-bodied bittersweet British-style ale, made with well-roasted barley.

Porterfield *Geology.* a North American geologic stage of the Middle Ordovician period, occurring after the Ashby and before the Wilderness.

Portevin-Le Chatelier effect *Physics.* the formation of a repeating pattern of deformation on a specimen subjected to a uniformly increasing stress.

porthole *Engineering.* an opening in the side of a ship or aircraft, usually covered with glass. *Mechanical Devices.* **1.** an opening for the intake or exhaust of a fluid, such as in a valve seat or valve face. **2.** the opening through which a bit or core barrel passes, discharging the drilling medium.

porthole die *Metallurgy.* in extrusion of hollows, a die capable of manufacturing intricate products.

portico *Architecture.* a colonnaded entrance porch.

Portland cement or **portland cement** *Materials.* a widely used hydraulic cement made of pulverized limestone and clay or shale, used in the construction industry. (So named from its resemblance to stone from the English Isle of *Portland,* a place noted for its limestone quarries.)

Portlandian *Geology.* a European geologic stage of the Uppermost Jurassic period, occurring after the Kimmeridgian and before the Berriasian of the Cretaceous period.

portlandite *Mineralogy.* $Ca(OH)_2$, a colorless, transparent trigonal mineral occurring as minute, hexagonal plates, having a specific gravity of 2.23 and a hardness of 2 on the Mohs scale; found in the chalk-dolerite contact zone at Scawt Hill, Ireland, and in Portland cement, from which it derives its name.

portligature *Surgery.* a surgical instrument facilitating application of a ligature or suture in a deep wound.

port of entry *Civil Engineering.* **1.** a place that provides customs and excise facilities for the collection of duties on imported goods. **2.** a place of entry or only available opening.

port operations service *Telecommunications.* a nautical communications service between ship stations or between ship stations and coast stations that receives and transmits messages dealing with the movement and safety of ships and their occupants.

Portuguese man-of-war *Invertebrate Zoology.* the common name for *Physalia pelagica,* a bright-colored floating oceanic hydrozoan, having an air-filled chamber above the sea surface and dangling tentacles up to 50 feet long; nematocysts produce a toxin dangerous to humans.

Portuguese man-of-war

portulaca *Botany.* any of several plants of the genus *Portulaca,* especially *P. grandiflora,* characterized by thick, fleshy leaves and showy, brightly colored flowers; widely cultivated.

Portulacaceae *Botany.* a family of dicotyledonous herbs and shrubs belonging to the order Caryophyllales, characterized by simple, often succulent leaves, generally perfect and hypogynous flowers, and an instability of chromosome numbers.

Portunidae *Invertebrate Zoology.* a large family of true crabs in the section Brachyura, having a broad carapace and the rear pair of legs flattened into powerful swimming paddles.

port-wine stain see NEVUS FLAMMEUS.

Porulosida *Invertebrate Zoology.* an order of protozoans in the subclass Radiolaria, with many pores in the central capsule.

POS *Aviation.* the airport code for Port of Spain, Trinidad.

PoS terminal see POINT OF SALE TERMINAL.

pos. positive; position; positron.

Poseidon [pə sī′dən] *Ordnance.* a two-stage, solid-propellant, submarine-launched, strategic ballistic missile; it is equipped with inertial guidance, nuclear warheads, and a maneuverable bus that can carry up to 14 re-entry bodies which can be directed at up to 14 separate targets; officially designated **UGM-73.**

poset *Mathematics.* a partially ordered set.

Posidoniaceae *Botany.* a monogenetic family of glabrous, rhizomatous marine herbs of the order Naiadales, characterized by poorly developed xylem; they typically grow submerged or on tidal rocks in Mediterranean and Australian waters.

posigrade rocket *Space Technology.* an auxiliary rocket designed to fire in the direction of flight for added thrust, as when two stages are separating.

posistor *Electronics.* a thermistor that exhibits a greater resistance as the temperature rises.

position a certain location in space; specific uses include: *Cartography.* **1.** the location of a point on the surface of the earth, the spheroid, or the geoid (two-dimensional position) or in space (three-dimensional position). **2.** the coordinates that define the location of a point. *Industrial Engineering.* in work motion studies, an elemental motion in which an object is placed at a specific location.

position accuracy and repeatability *Robotics.* a measure of a robot's ability to repeat the same movement with no discernible error over a period of time.

positional astronomy *Astronomy.* the science that determines the positions of celestial bodies.

positional information *Cell Biology.* instructions from which it is believed that cells detect their relative spatial positions within an organism, thus influencing or determining their subsequent course of differentiation.

positional isomer *Chemistry.* any of two or more molecules that are composed of the same number and types of atoms but differ in the position of a certain atom or group.

positional notation *Mathematics.* any system of representing numerical quantities by a sequence of digits so that the relative position of each digit as well as its numeric value determines its significance. For example, the representation of numbers by the radix or base system is a form of positional notation.

positional parameter *Computer Programming.* a parameter that must appear in a specified position relative to other parameters, whether or not other parameters are designated.

position angle *Astronomy.* the angle, measured from due north toward east, at which a companion star in a binary or multiple system lies relative to its primary star. *Navigation.* the angle of the navigational triangle at which any observed body is located. Also, PARALLACTIC ANGLE.

position defense *Military Science.* a defensive strategy in which the bulk of a defending force is deployed in selected tactical areas where the decisive battle is to be fought; the defense is based on the ability of these forces to maintain their positions and to control the terrain between them.

position effect *Genetics.* a change in a gene's expression that occurs when its position on a chromosome changes due to a crossover or a chromosomal mutation.

positioning *Mechanical Engineering.* a tooling function in which a workpiece is manipulated as needed to accommodate working tools.

positioning error *Robotics.* the degree by which a robot's ability to move to a specified point in space is in error when compared against absolute accuracy or repeatability.

positioning time *Mechanical Engineering.* the time needed to move a machining tool between a pair of coordinates. *Computer Technology.* the amount of time it takes a transducer to be positioned at the beginning of the data to be read or written; on a magnetic disk it is equal to seek time plus latency.

position operator *Quantum Mechanics.* the operator corresponding to a position in Hilbert space that has a continuous spectrum of possible values.

position-plotting sheet *Navigation.* a piece of paper marked only with longitude and latitude lines; used for plotting courses and position when land, visual aids to navigation, and depth of water are not important, as when well out at sea.

position representation *Quantum Mechanics.* a representation in which state functions are eigenfunctions of the position operator.

position sensor *Engineering.* an instrument that measures a position, immediately transforming the information so that it may be transmitted. Also, **position transducer.**

position telemetering *Engineering.* a system of voltage measurement transmitting by placing a device in a bridge circuit that creates relative magnitudes of electrical amounts or phase relationships.

position vector *Mathematics.* given a point P in Euclidean n-space, the vector with length equal to the distance from the origin to P, and orientation in the direction from the origin to P. Also, RADIUS VECTOR.

positive being in a condition or direction opposite to one that is termed negative; specific uses include: *Mathematics.* greater than zero in some sense; e.g., an element of the real numbers is positive if it lies to the right of zero in the natural ordering. *Electricity.* having the type of electric charge in which electrons are minimally present, as that developed on a glass object that has been rubbed with silk. *Chemistry.* describing an element or compound that tends to lose electrons. *Physics.* describing the part of an electric cell toward which the current flows. *Mechanics.* moving or tending in the direction considered as that of increase or progress, e.g., as with a clockwise motion. *Graphic Arts.* a photographic image on which the normal tonal values are present; i.e., light and dark tones correspond to those of the original subject. Thus, **positive image.** *Medicine.* describing a test, investigation, or the like in which the sign or condition under study is actually found to be present. *Biology.* directed toward the source of a stimulus.

positive acceleration *Mechanics.* **1.** any instance of acceleration that is in the same direction as the velocity. **2.** specifically, an increase in velocity; a process of speeding up.

positive afterimage *Physiology.* a visual image persisting after viewing an object, in which the light and dark fields appear the same as in the original image.

positive angle *Mathematics.* an angle subtended by rotating a ray in the counterclockwise direction.

positive axis *Mathematics.* the section of an axis in a Cartesian coordinate system corresponding to positive coordinate values. *Meteorology.* a locus of maximum streamline curvature within an easterly atmospheric wave; used in tropical synoptic wave analysis.

positive bias *Electronics.* a condition in which a vacuum tube's control grid attracts more electrons than the cathode.

positive birefringence *Optics.* double refraction in which the velocity of the ordinary ray surpasses the velocity of the extraordinary ray.

positive charge *Electricity.* an electric charge about a material body, characterized by a field having fewer electrons than normal.

positive clutch *Mechanical Engineering.* a clutch that transmits torque without slip.

positive column *Electronics.* a glow that appears between the Faraday dark space and the anode in a glow-discharge or cold-cathode tube. Also, POSITIVE GLOW.

positive conditioning see POSITIVE REINFORCEMENT.

positive control *Molecular Biology.* a type of bacterial gene regulation in which an activator binds to DNA to increase both the efficiency of RNA polymerase binding to the DNA and the efficiency of transcription. *Transportation Engineering.* the separation of all air traffic, in altitude and horizontal position, by air traffic control.

positive crystal *Optics.* a crystal in which the refractive index of the extraordinary ray surpasses the refractive index of the ordinary ray.

positive definite *Mathematics.* **1.** an $n \times n$ nonsingular Hermitian matrix (with complex entries) A is said to be a positive definite matrix if $x^*Ax > 0$ for all nonzero vectors $x \in C^n$, where x^* denotes Hermitian conjugate and C^n denotes the n-fold Cartesian product of the complex plane with itself. If the requirement $x^*Ax > 0$ is weakened to $x^*Ax \geq 0$, A is said to be **positive semidefinite.** The terms **negative definite** and **negative semidefinite** can be defined for A by reversing the corresponding inequalities or by saying that $-A$ is positive definite or semidefinite, respectively. **2.** a complex-valued function f of a real variable is a **positive definite function** if the $n \times n$ matrix with (i,j)th entry $f(x_i - x_j)$ is positive semidefinite for all choices of n points in the real numbers R, and for all n. If f is a positive definite function, then: (a) $f(-x) = \bar{f}(x)$ for all $x \in R$, where $\bar{f}(x)$ is the complex conjugate of $f(x)$; (b) $f(0) \geq 0$; (c) f is a bounded function and $|f(x)| \leq f(0)$ for all $x \in R$; and (d) if f is continuous at 0, then it is continuous everywhere.

positive dihedral *Aviation.* an upward inclination of a wing or other support surface. (The term *dihedral* alone usually refers to positive dihedral; negative dihedral is called **anhedral.**)

positive-displacement compressor *Mechanical Engineering.* a compressor in which the fluid pressure increases as the volume of the closed space is decreased.

positive-displacement meter *Engineering.* an instrument that measures fluid volume by enclosing a designated volume of flow and then passing it downstream; the measurement is given by the number of times the operation is performed.

positive-displacement pump *Mechanical Engineering.* a pump in which the pressure of a measured quantity of entrapped liquid rises, as in a reciprocating piston cylinder or rotary vane.

positive distortion see BARREL DISTORTION.

positive draft *Mechanical Engineering.* a pressure in a furnace or the gas passage of a steam-generating unit that is greater than atmospheric pressure.

positive electricity *Electricity.* the effects existing in a body associated with a deficiency of electrons, as the electricity appearing on a glass object rubbed with silk.

positive electrode see ANODE.

positive electron see POSITRON.

positive element *Chemistry.* any of the elements in the upper left of the periodic table, which can yield electrons to other elements and become positively charged. *Geology.* see POSITIVE LANDFORM.

positive estuary *Hydrology.* an estuary whose water has been measurably diluted of seawater by land drainage.

positive feedback *Science.* a kind of feedback that disturbs or prevents equilibrium between the input and output in a system or process. *Control Systems.* feedback that is fed back in phase with the input of a system so as to increase the amplification of the system. Also, REGENERATIVE FEEDBACK, REGENERATION. *Psychology.* in popular use, a response to one's behavior that supports or praises the pattern of behavior.

positive g *Mechanics.* **1.** an instance of acceleration that is in the same direction as that of gravity. **2.** see POSITIVE ACCELERATION, def. 1. *Space Technology.* specifically, the force exerted on the human body in a gravitational field or during acceleration so that the force of inertia acts in a head-to-foot direction.

positive gene control *Genetics.* an intensification of DNA transcription through the binding of expressor molecules to promoter sites on the chromosome.

positive interference *Genetics.* an interaction between homologous chromosomes such that a crossover at one locus between them reduces the likelihood of another crossover in the vicinity of the first.

positive ion *Chemistry.* an atom or group of atoms that have lost one or more electrons, thus carrying a positive charge. *Physics.* any positively charged particle.

positive-ion sheath *Electronics.* a layer of positively charged ionized gas that forms around the wires of a grid in a gas-filled tube, inhibiting the flow of current.

positive landform *Geology.* an upstanding topographical form that has risen with respect to the adjacent land or that has been formed by an excess of material. Also, POSITIVE ELEMENT.

positive lens see CONVERGING LENS.

positive logic *Electronics.* digital logic in which the more positive voltage represents the binary digit 1 and the less positive level represents the binary digit 0.

positive matrix *Mathematics.* an $n \times n$ square matrix is a positive matrix if all its entries are positive real numbers.

positive meniscus lens see CONVERGING MENISCUS LENS.

positive modulation *Electronics.* a process by which a television signal is varied to link an increase in brightness to an increase in transmitted power.

positive mold *Engineering.* a device, used in plastics molding, that traps all molding resin when the mold is shut.

positive movement *Geology.* **1.** an uplifting of the earth's crust in relation to an adjacent area. **2.** the creation of a positive shoreline.

positive phase sequence *Electricity.* the prescribed order in which phase voltages of a polyphase system attain maximum values. Also, PHASE SEQUENCE.

positive-positive conflict see APPROACH-APPROACH CONFLICT.

positive ray *Electronics.* a stream of charged atoms that arises in the anode and has the ability to attract electrons. Also, CANAL RAY.

positive real function *Mathematics.* a function f of a complex variable is said to be a positive real function if: (a) $f(z)$ is real and positive for z real; and (b) the real part of $f(z)$ is nonnegative if the real part of z is nonnegative.

positive reinforcement *Behavior.* **1.** an event or reward that follows a behavioral response and increases the likelihood that the behavior will be repeated. **2.** the presentation of such an event or reward. Thus, **positive reinforcer.**

positive sanction *Anthropology.* a form of reward or recognition conferred by a society on a member for adhering to social norms or standards; this may be formal, such as a ceremonial honor, or informal, such as praise.

positive segregations *Materials Science.* the macro-segregations of solute elements, such as A-segregations (pencil-like) and V-segregations (downward pointing), forming in the upper two-thirds of an ingot during the solidification process of steel.

positive-sense genome *Virology.* a genome that is homologous with the viral mRNA.

positive-sense strand *Virology.* **1.** the strand that functions as the messenger (mRNA) in RNA. **2.** in DNA, the strand that has the same sequence as the mRNA. Also, **positive strand.**

positive set *Mathematics.* the set P of positive elements of an ordered field.

positive shoreline see SHORELINE OF SUBMERGENCE.

positive-strand virus *Virology.* a virus with a positive-sense genome.

positive supercoiling see DNA WRITHING.

positive temperature coefficient *Thermodynamics.* a coefficient relating a characteristic change in the quantity of a substance or system, such as the change in resistance, volume, or length, to a change in temperature.

positive temperature gradient *Acoustics.* a temperature that rises above a negative gradient in a region where temperature increases as depth increases, causing the occurrence of a shadow zone, sound channel, or similar phenomenon due to the bending of sound away from higher velocities and toward lower velocities, as described in Snell's law.

positive terminal *Electricity.* the terminal of a device with a voltage greater than the device common or the negative terminal.

positive transfer *Behavior.* the application of one's previous learning experience to present learning tasks; e.g., employing one's familiarity with a typewriter in learning to operate a word processor.

positive transmission *Telecommunications.* in television transmissions, a way of sending signals so that increased initial light intensity also promulgates an increase in the transmitted power.

positive zero *Computer Programming.* a zero value with a positive sign.

positron *Particle Physics.* the antiparticle of the electron, having the same mass and spin as the electron, but opposite charge and magnetic moment. Also, POSITIVE ELECTRON.

positron camera *Engineering.* an instrument that detects positrons emitted by short-lived radioisotopes used as tracers in the human body.

positron emission *Nuclear Physics.* a type of beta decay in which a nucleus emits a positron and a neutrino.

positron emission spectroscopy *Spectroscopy.* a spectroscopic technique in which a low, monoenergetic positron beam is used to bombard a solid surface, and the energy of emitted positrons is measured to determine energy lost by adsorption of surface molecules.

positron emission tomography *Radiology.* tomography produced by the detection of gamma rays emitted from tissues after administration of a natural biochemical substance into which positron-emitting isotopes have been incorporated; the paths of the gamma rays are interpreted by a computer, and the resulting tomogram represents local concentrations of the isotope-containing substance; used for the in vivo study of human metabolic and physiological functions.

positronium *Particle Physics.* the simplest atom, consisting of an electron bound with a positron; formed in a collision between a positron and a gas atom, resulting in the capture of an atomic electron by the positron.

positronium velocity spectroscopy *Spectroscopy.* a spectroscopic technique in which a low, monoenergetic positron beam is used to bombard a solid surface, and the velocity of emitted positronium atoms is measured to determine the energy and momentum of electrons near the surface.

posology *Pharmacology.* the branch of pharmacology concerned with the dosage in which medicines should be administered.

poss. possible.

possible capacity *Transportation Engineering.* the maximum capacity of a section of highway under existing road and traffic conditions.

possible ore *Mining Engineering.* ore whose existence is reasonably possible, considering geologic-mineralogic relationships and the extent of ore bodies previously developed.

possible reserves *Petroleum Engineering.* primary petroleum reserves that may be present despite the absence of data to prove them.

post *Civil Engineering.* a structural member placed vertically as support. *Mining Engineering.* **1.** a support between the roof and floor of a coal seam, used with certain mining machines and augers. **2.** a pillar of ore or coal.

post *Computer Programming.* **1.** to enter data into a record. **2.** to notify the operating system of an event. **3.** to perform a transaction by making the appropriate modifications to a database, as in posting a payment to a customer's account.

Post, Emil 1897–1954, American mathematician; independently formulated the truth table; extended two-valued logic to *n*-valued systems.
post- a combining form meaning "after," "behind."
postaccelerating electrode see INTENSIFIER ELECTRODE.
postacceleration *Electronics.* the acceleration of electrons in a tube that takes place after the beam has been deflected.
postalbumin *Hematology.* a serum protein that has an electrophoretic mobility between albumin and alpha-globulin at pH 8.6.
post-and-beam construction *Building Engineering.* a type of wall construction in which beams, rather than studs, serve to support the larger posts.
post-and-lintel *Architecture.* of or relating to a system of construction based on vertical supports (posts) and horizontal beams (lintels).
postbasic stare *Neurology.* a facial appearance, often seen in children with posterior basic meningitis, in which the eyeballs roll downward and the upper eyelids retract.
post-boost vehicle *Ordnance.* the port of a missile able to launch and direct reentry vehicles on separate trajectories.
post brake *Mechanical Engineering.* a brake that consists of two upright posts mounted on either side of the drum and that operates on brake paths bolted to the drum cheeks; sometimes fitted on a steam winder or haulage.
postbranchial body see ULTIMOBRANCHIAL BODY.
postcard stock *Graphic Arts.* a sturdy paper, often coated on one side, used in the production of postcards.
postcentral gyrus *Anatomy.* the ridge of gray matter immediately posterior to the central sulcus of the cerebral hemisphere.
postcentral sulcus *Anatomy.* a sulcus on the superolateral surface of the cerebrum, separating the postcentral gyrus from the remainder of the parietal lobe.
Post-Classic or **Postclassic** *Archaeology.* the final pre-Columbian period in New World cultural history, following the Classic; characterized by metalworking, complex urban societies, militarism, and secularism.
postcollision interaction *Atomic Physics.* the interaction that occurs between the atomic and subatomic particles that are products of a collision reaction.
postcondition *Artificial Intelligence.* a fact or set of facts that will be true after an action has been taken.
postconventional level *Psychology.* in Lawrence Kohlberg's analysis of moral development, the highest level, in which moral values are developed and sustained on individual rather than collective principles. This level itself is divided into **stage five,** in which the individual focuses on balancing his or her beliefs with those of society, and **stage six,** in which the individual's own moral principles prevail. Also, **postconventional morality.**
postcranial *Biology.* located posterior to the head.
postcure bonding *Engineering.* a high-temperature process used after parts have been steam pressed under pressure to increase the heat-resistance of their adhesive bonds.
postdiastolic *Cardiology.* after diastole.
post-edit *Computer Programming.* to edit a previous computation's results.
postemphasis *Acoustical Engineering.* see DEEMPHASIS.
postencephalitic parkinsonism *Medicine.* a group of nervous conditions, such as those of Parkinson's disease, following encephalitis.
postequalization see DEEMPHASIS.
poster- or **postero-** a combining form meaning "behind."
posterior *Zoology.* the back or dorsal side of an animal.
posterior chamber *Anatomy.* a narrow space between the iris of the eye and the lens, the suspensory ligament of the lens and the ciliary body.
posterior distribution *Statistics.* the distribution of a quantity of interest, conditional on data.
posterior pituitary hormone *Endocrinology.* any of the hormones that are produced in the posterior lobe of the pituitary, such as vasopressin, oxytocin, or their precursor, neurophysin.
posterization *Graphic Arts.* a color separation process in which the intermediate gray tones of a continuous-tone image are converted into two or three tones.
posteroanterior *Anatomy.* from back to front.
posteroexternal *Anatomy.* situated on the outside of a posterior part.
posterointernal *Anatomy.* situated within and to the rear.
posterolateral *Anatomy.* situated on the side and toward the posterior aspect.
posteromedial *Anatomy.* situated in the middle of the back side.
posterosuperior *Anatomy.* situated behind and above.
poster paint *Materials Science.* a brightly colored, opaque, water-based paint with a gum binder; used on posters.
postesophageal *Anatomy.* behind the esophagus.
postfix notation see REVERSE POLISH NOTATION.
postforming *Engineering.* a high-temperature process that shapes or bonds flexible thermoset laminates before the laminate becomes set.
post-Freudian see NEO-FREUDIAN.
postglacial *Geology.* **1.** describing the interval of geologic time that began with the total disappearance of continental glaciers in the middle latitudes or from a specified area. **2. Postglacial.** the Holocene.
postheating *Metallurgy.* in welding, heating after the joining is completed.
posthitis *Medicine.* an inflammation of the prepuce, or foreskin of the glans penis.
posthole *Civil Engineering.* a cavity ready made, usually by an auger, in which to seat a vertical stake.
posthumous structure *Geology.* a structural feature in overlying rock strata that exhibits or is similar to the structure of the older, generally more deformed, underlying rocks.
posthypnotic amnesia *Psychology.* a hypnotic subject's inability to remember afterward what occurred during the hypnotic state.
posthypnotic suggestion *Psychology.* a suggestion made to a subject during hypnosis that the subject will behave in a particular way after being awakened from the hypnotic state; the subject usually has no recollection of the command.
postignition *Chemistry.* surface ignition that follows the normal spark.
postincrement *Computer Science.* a CPU feature in which the contents of an index register are automatically increased by a fixed amount after its use, thus pointing to the next data in an array.
postindexed *Computer Programming.* of or relating to an effective address obtained after adding the contents of a specified index register to the given indirect (preindexed) address.
postinfectious encephalitis *Medicine.* encephalitis that occurs as a complication of another infection, such as chicken pox or measles; characterized by headache, neck pain, fever, nausea, and vomiting, with the possible occurrence of neurological disturbances.
postirradiation syndrome *Radiology.* the signs and symptoms, including hemorrhage, anemia, and malnutrition, experienced by animals or humans following massive exposure to radiation.
postmagmatic *Geology.* relating to events or reactions that occurred after the major portion of a magma was crystallized.
postmating isolation *Evolution.* a condition in which intrinsic factors effective after mating prevent interbreeding between two or more populations.
postmature *Medicine.* relating to the birth of an over-term infant.
post meridian or **post meridiem** *Science.* after noon; of, or relating to, the portion of the day that occurs between the hours of noon and the following midnight.
post mold *Archaeology.* the outline of a deteriorated post, indicating the former location of some structure.
postmortem [pōst môrt´əm] *Medicine.* **1.** relating to or occurring after death. **2.** an examination of the body after death; an autopsy. *Science.* any analysis of an operation or activity after it has ended. *Computer Programming.* an analysis of the data or of storage locations after a program has abnormally ended.
postmortem clot *Hematology.* a blood clot formed in the heart or in a large blood vessel after death.
postmortem dump *Computer Programming.* a listing of the contents of registers and storage locations obtained after a program has ended.
postmortem routine *Computer Programming.* a routine used in debugging or diagnostics that lists the contents of registers and storage locations. Also, **postmorten program.**
postnasal [pōst nāz´əl] *Medicine.* located or occurring behind the nose or in the nasopharynx, as a flow of mucus.
postnasal drip *Medicine.* a flow of mucus from the back of the nasal cavity onto the pharyngeal surface, usually caused by a cold or allergies.
postnatal [pōst nāt´əl] *Medicine.* occurring after birth.
postnecrotic cirrhosis *Medicine.* a form of cirrhosis characterized by widespread necrosis and collapse of the reticular hepatic framework, with the formation of large scars of connective tissue and regenerating nodules of liver tissue.
postnova *Astronomy.* a nova that has faded to the brightness it had before its explosion.

postobsequent stream *Hydrology.* a stream that follows the direction of the underlying strata, although the stream developed independently of the original relief and after the opposite-moving stream into which it flows.

postoperative *Surgery.* subsequent to a surgical procedure.

postorder *Computer Science.* an order of visiting trees, in which the children of a node are examined first, in left-to-right order, followed by examination of the node itself.

postorogenic *Geology.* relating to geologic events or processes that took place after a period of intense mountain-building activity, or the rocks or features formed after such a period.

postorogenic phase *Geology.* the last phase in an orogenic cycle, during which the present mountainous landscapes were produced.

postpartum *Medicine.* after childbirth.

postprandial *Medicine.* after a meal.

postprecipitation *Chemistry.* the resultant precipitation of a chemically different species that forms on the surface of an initial precipitate, often a common ion.

postprocessor *Computer Programming.* **1.** a program that performs final computations or data organization to put the results of another program into the desired form. **2.** a program that converts the data produced by an emulator into the format that is required by the emulated system. *Robotics.* a program that converts graphical data to a form that can be used by numerical control or other computer equipment.

post sight *Ordnance.* a front sight that projects upward from the barrel of a gun.

postsphygmic *Cardiology.* after a pulse wave.

post-strike reconnaissance *Military Science.* a mission to obtain information that can be used to measure the results of an attack.

postsynaptic *Neurology.* situated or occurring after a synapse.

post-tensioning *Engineering.* the application of a load to prestressed wires in a prestressed concrete structure.

post-term birth *Medicine.* the birth of an infant occurring after the normal term.

post-transcription *Biochemistry.* of or relating to principal RNA transcripts before they leave the nucleus as mature mRNA.

post-transcriptional processing *Biochemistry.* a process during which enzymes modify certain RNA polymerases, such as ribosomal RNA in prokaryotic and eukaryotic cells, before they begin their various functions.

post-translation *Biochemistry.* of or relating to peptide bonds that have been formed in a protein.

post-translational modification *Biochemistry.* **1.** an alteration to a polypeptide chain after it has been synthesized. **2.** a process in which enzymes alter a polypeptide chain following its translation from its mRNA.

post-traumatic stress disorder *Psychology.* an anxiety disorder associated with an extremely stressful event, such as military combat, and characterized by such symptoms as painfully vivid recollections of the trauma, sleep disturbances, and an overall sense of numbness and detachment from others. Also, TRAUMATIC NEUROSIS.

post-tuning drift *Electronics.* the increase in frequency that arises when there is a drop in the varactor temperature of a frequency-agile source, such as the fast-tuning oscillators used in jamming equipment.

posture *Physiology.* the bearing or carriage of the body.

postuterine *Anatomy.* behind the uterus.

postvulcanization bonding *Materials Science.* a process used in bonding metallic and elastomeric components, such as rubber shock and noise mounts, in which elastomers are joined by an adhesive following vulcanization of the component parts, resulting in bonds that perform under severe environmental conditions.

postweld heat treatment *Materials Science.* any of various thermal treatments of a component following welding to reduce the residual stresses inherent in welding, prevent postweld cracking, change mechanical properties, and improve corrosion resistance.

pot. potential.

pot. a drink. (From Latin *potus.*)

potable [pōt´ə bəl] *Science.* of a liquid, suitable for drinking.

potable water *Hydrology.* water that is safe and suitable to drink; drinking water.

potamic *Science.* of or relating to rivers.

Potamogalinae *Vertebrate Zoology.* a mammalian subfamily of the family Tenrecidae, composed of the genera, *Potamogale,* the giant African water or otter shrew, and *Micropotamogale*, the dwarf African water or otter shrew.

potamogenic rock *Petrology.* a sedimentary rock precipitated from river water.

Potamogetonaceae *Botany.* a large family of moncotyledonous floating or submerged freshwater herbs of the order Najadales, having creeping rhizomes, erect, leafy branches, and flowers in spikes or racemes.

Potamogetonales *Botany.* in some classification systems, an order of monocotyledonous, flowering, aquatic and semiaquatic plants of the subclass Alismatidae.

potamology *Hydrology.* the scientific study of rivers.

potamoplankton *Biology.* plankton found in rivers.

potarite *Mineralogy.* PdHg, a silver-white, opaque, metallic, tetragonal mineral occurring as small grains or nuggets, having a specific gravity of 14.88 and a hardness of 3.5 on the Mohs scale; found in diamond-bearing placer deposits in Guyana.

potash *Inorganic Chemistry.* **1.** a commercial form of potassium carbonate, K_2CO_3. **2.** a popular name for various other compounds that contain potassium, such as potassium hydroxide, KOH, or potassium chloride, KCl. (Originally "pot ashes;" it was derived from wood ashes and produced in large iron pots.)

potash alum see KALINITE.

potash bentonite see POTASSIUM BENTONITE.

potash blue *Inorganic Chemistry.* a pigment produced by oxidizing ferrous ferrocyanide made from a mixture of potassium ferrocyanide and ferrous sulfate; used to make carbon paper.

potash feldspar see POTASSIUM FELDSPAR.

potash lake *Hydrology.* a salt lake containing a high concentration of dissolved potassium salts.

potassic *Petrology.* of or relating to a rock or mineral rich in potassium.

potassium [pə tas´ē əm] *Chemistry.* a metallic element having the symbol K, the atomic number 19, an atomic weight of 39.098, a melting point of 63°C, and a boiling point of 770°C. It is a soft, silver-white, extremely reactive alkali metal that is fairly abundant in the earth's crust; it is essential for plant growth and for human and animal nutrition and is used extensively in its compound form as a fertilizer component, as a laboratory reagent, in heat-exchange alloys, and for various other industrial purposes. (A word coined by its discoverer, the English chemist Sir Humphry Davy; from *potash.*)

potassium-40 *Nuclear Physics.* a radioactive isotope of potassium with a half-life of billions of years.

potassium-42 *Nuclear Physics.* a radioactive isotope of potassium with a half-life of 12.4 hours; used as a radiotracer in medicine.

potassium acetate *Organic Chemistry.* $KC_2H_3O_2$, a combustible, white crystalline powder; soluble in water and alcohol and insoluble in ether; melts at 292°C; used in medicine, as a dehydrating agent, and in crystal glass manufacture.

potassium acid carbonate see POTASSIUM BICARBONATE.

potassium acid saccharate *Organic Chemistry.* $HOOC(CHOH)_4COOK$, a combustible, off-white powder; slightly soluble in cold water and soluble in hot water; used in rubber, soaps, and detergents.

potassium alginate *Organic Chemistry.* $(C_6H_7O_6K)_n$, a hydrophilic colloid; soluble in water and insoluble in alcohol; used as a food thickener and stabilizer.

potassium aluminate *Inorganic Chemistry.* $K_2Al_2O_4{\cdot}3H_2O$, colorless crystals; very soluble in water and insoluble in alcohol; used in dying, printing, and paper manufacture.

potassium aluminum fluoride *Inorganic Chemistry.* K_3AlF_6, a toxic, white powder; slightly soluble in water; used as an insecticide.

potassium aluminum sulfate *Inorganic Chemistry.* $KAl(SO_4)_2{\cdot}12H_2O$, colorless, cubic, octagonal or monoclinic crystals; very soluble in water and insoluble in alcohol; melts at 92.5°C and boils at 200°C; used in dyeing, manufacture of paper, purification of water, and as a food additive. Also, **potassium alum.**

potassium antimonate *Inorganic Chemistry.* $KSbO_3$, white crystals that are soluble in water. Also, POTASSIUM STIBNATE.

potassium argentocyanide *Inorganic Chemistry.* $KAg(CN)_2$, white light-sensitive crystals; soluble in water and alcohol and insoluble in acids; highly toxic; used in silver plating and as an antiseptic. Also, SILVER POTASSIUM CYANIDE, POTASSIUM CYANOARGENATE.

potassium-argon dating *Geology.* the determination of the age, in years, of a mineral or rock by measuring the ratio of radioactive potassium-40 to argon-40 present in the sample. Also, K-AR DATING. *Archaeology.* this same method of dating used to establish the antiquity of human or humanoid remains, such as those found in volcanic deposits in East Africa.

potassium arsenate *Inorganic Chemistry.* K_3AsO_4, toxic, colorless needles; very soluble in water and insoluble in alcohol; melts at 288°C; deliquescent; used in insecticides, in preserving hides, and in printing textiles. Also, MACQUER'S SALT.

potassium arsenite *Inorganic Chemistry.* $KH(AsO_2)_2 \cdot H_2O$, a white powder; soluble in water and slightly soluble in alcohol; hygroscopic; toxic on ingestion and inhalation; used in silvering mirrors. Also, POTASSIUM METAARSENITE.

potassium bentonite *Geology.* a potassium-bearing clay produced by the alteration of volcanic ash. Also, POTASH BENTONITE.

potassium bicarbonate *Inorganic Chemistry.* $KHCO_3$, colorless monoclinic crystals or white powder; soluble in water and insoluble in alcohol; decomposes at 100–200°C; used in baking powders, soft drinks, medicines, and fire extinguishing agents. Also, POTASSIUM ACID CARBONATE, BAKING SODA.

potassium bifluoride *Inorganic Chemistry.* KHF_2, colorless cubic crystals; very soluble in water and insoluble in alcohol; deliquescent; decomposes at 225°C; used in etching glass, in solder flux, and as a catalyst and electrolyte. Also, FREMY'S SALT.

potassium binoxalate *Organic Chemistry.* $KHC_2O_4 \cdot H_2O$, toxic white crystals with a bitter taste; soluble in water and insoluble in alcohol; decomposes when heated; used for stain removal, in photography, and as a laboratory reagent. Also, SALT OF SORREL.

potassium biphthalate *Organic Chemistry.* $HOOCC_6H_4COOK$, colorless crystals; soluble in water; used as a buffer in pH determinations. Also, **potassium hydrogen phthalate.**

potassium bismuth tartrate *Organic Chemistry.* $C_4H_2O_9Bi_3K \cdot 4H_2O$, an odorless white powder; soluble in water; used in medicines (formerly, in treating syphilis).

potassium bisulfate *Inorganic Chemistry.* $KHSO_4$, colorless rhombic crystals; soluble in water and insoluble in alcohol; deliquescent; melts at 214°C and decomposes at its boiling point; used in wine making and as a reagent. Also, ACID POTASSIUM SULFATE.

potassium bisulfite *Inorganic Chemistry.* $KHSO_3$, colorless crystals; soluble in water and insoluble in alcohol; decomposes at 190°C; used in reduction of organic compounds, and as an antiseptic and a preservative in foods.

potassium bitartrate *Organic Chemistry.* $KHC_4H_4O_6$, a flammable, white crystalline or powdery solid; soluble in boiling water and insoluble in alcohol; used in baking powder and medicines, and as a food additive. Also, CREAM OF TARTAR.

potassium borofluoride see POTASSIUM FLUOROBORATE.

potassium borohydride *Inorganic Chemistry.* KBH_4, white cubic crystals; very soluble in water and slightly soluble in alcohol; decomposes at 500°C; toxic on ingestion and a dangerous fire hazard; used as a source of hydrogen, a reducing agent, and in plastics manufacture.

potassium bromate *Inorganic Chemistry.* $KBrO_3$, colorless trigonal crystals; soluble in water and slightly soluble in alcohol; decomposes at 370°C; a strong oxidant and dangerous fire hazard in contact with organic substances; used as a reagent and food additive.

potassium bromide *Inorganic Chemistry.* KBr, colorless cubic crystals; soluble in water and alcohol; slightly hygroscopic; melts at 734°C and boils at 1435°C; toxic on ingestion and inhalation; used in photography, and in spectroscopy and infrared transmission.

potassium cadmium iodide see CADMIUM POTASSIUM IODIDE.

potassium carbonate *Inorganic Chemistry.* K_2CO_3, colorless monoclinic crystals; soluble in water and insoluble in alcohol; hygroscopic; melts at 891°C and decomposes at boiling point; an irritant in solution; used in optical and color TV tubes, and as a food additive. Also, POTASH, PEARL ASH, SALT OF TARTAR.

potassium chlorate *Inorganic Chemistry.* $KClO_3$, colorless monoclinic crystals; soluble in water and alcohol; melts at 356°C and decomposes at 400°C; used as an oxidizing agent, in explosives, and in textile printing.

potassium chloride *Inorganic Chemistry.* KCl, colorless cubic crystals; soluble in water and slightly soluble in alcohol; melts at 770°C and sublimes at 1500°C; used as a fertilizer and salt substitute, and in pharmaceuticals and spectroscopy.

potassium chloroplatinate *Inorganic Chemistry.* $K_2[PtCl_6]$, toxic, yellow cubic crystals; soluble in water and insoluble in alcohol; decomposes at 250°C; used in photography and as a laboratory reagent.

potassium chromate *Inorganic Chemistry.* K_2CrO_4, toxic, yellow rhombic crystals; soluble in water and insoluble in alcohol; melts at 968.3°C; used as a reagent, and in chromate pigments and inks.

potassium chromium sulfate see CHROMIUM POTASSIUM SULFATE.

potassium citrate *Organic Chemistry.* $K_3C_6H_5O_7 \cdot H_2O$, a colorless or white, odorless, crystalline or powdery solid with a saline taste; soluble in water and insoluble in alcohol; decomposes at 230°C; used as an antacid and a food stabilizer and buffer.

potassium cobaltinitrite see COBALT POTASSIUM NITRITE.

potassium cyanate *Inorganic Chemistry.* KCNO, toxic, colorless tetragonal crystals; soluble in water and insoluble in alcohol; decomposes at 700–900°C; used in medicine, in the manufacture of organic chemicals, and as an herbicide.

potassium cyanide *Inorganic Chemistry.* KCN, very toxic, colorless cubes or white granules; soluble in water and alcohol; deliquescent; melts at 634.5°C; used in the extraction of gold and silver, in insecticides, and in electroplating.

potassium cyanoargenate see POTASSIUM ARGENTOCYANIDE.

potassium cyanoaurite see POTASSIUM GOLD CYANIDE.

potassium dichromate *Inorganic Chemistry.* $K_2Cr_2O_7$, red monoclinic or triclinic crystals; soluble in water and insoluble in alcohol; decomposes at 500°C; a strong oxidant and dangerous fire hazard; used as an oxidizing agent, in explosives, and in dyeing and printing.

potassium feldspar *Mineralogy.* alkali feldspar that includes the molecule $KAlSi_3O_8$; specifically, microcline, orthoclase, and sanidine. Also, POTASH FELDSPAR.

potassium ferricyanide *Inorganic Chemistry.* $K_3Fe(CN)_6$, red monoclinic crystals; soluble in water and insoluble in alcohol; decomposes at melting point; used in tempering steel, electroplating, and as a reagent. Also, RED PRUSSIATE OF POTASH.

potassium ferrocyanide *Inorganic Chemistry.* $K_4Fe(CN)_6 \cdot 3H_2O$, lemon-yellow monoclinic crystals; soluble in water and insoluble in alcohol; decomposes at its boiling point; used in tempering steel, dyeing, explosives, and engraving. Also, YELLOW PRUSSIATE OF POTASH.

potassium fluoride *Inorganic Chemistry.* KF or $KF \cdot 2H_2O$, colorless cubic crystals; very soluble in water and insoluble in alcohol; deliquescent; melts at 858°C and boils at 1505°C; used as an insecticide and preservative, and in glass etching and fluorination.

potassium fluoroborate *Inorganic Chemistry.* KBF_4, toxic, colorless rhombic or cubic crystals; soluble in water and slightly soluble in alcohol; decomposes at 350°C; used in chemical research. Also, POTASSIUM BOROFLUORIDE.

potassium fluorosilicate *Inorganic Chemistry.* K_2SiF_6, colorless cubic or hexagonal crystals; slightly soluble in water and very slightly soluble in alcohol; decomposes at melting point; used in metallurgy, ceramics, and as an insecticide. Also, POTASSIUM SILICOFLUORIDE.

potassium gluconate *Organic Chemistry.* $KC_6H_{11}O_7$, an odorless, fine, white crystalline powder; soluble in water and insoluble in alcohol; decomposes at 180°C; used in vitamin tablets.

potassium glutamate *Organic Chemistry.* $KOOC(CH_2)_2CH(NH_2)COOH \cdot H_2O$, an odorless, free-flowing white powder; soluble in water and slightly soluble in alcohol; used as a flavor enhancer and a salt substitute.

potassium glycerophosphate *Organic Chemistry.* $K_2C_3H_5O_2 \cdot H_2PO_4 \cdot 3H_2O$, a yellow liquid with an acid taste; miscible with water and soluble in alcohol; used as a food additive and nutritional supplement.

potassium gold chloride *Inorganic Chemistry.* $K[AuCl_4] \cdot 2H_2O$, yellow monoclinic crystals; soluble in water, alcohol, and acids; decomposes at 357°C; used in photography, medicine, and painting glass and porcelain. Also, **potassium chloroaurate.**

potassium gold cyanide *Inorganic Chemistry.* $K[Au(CN)_2]$, colorless rhombic crystals; soluble in water and slightly soluble in alcohol; used in electrogilding. Also, POTASSIUM CYANOAURITE.

potassium hydroxide *Inorganic Chemistry.* KOH, white rhombic crystals; soluble in water and very soluble in alcohol; deliquescent; boils at 1320–1324°C; toxic and highly corrosive; used in soap manufacture and bleaching, and as a food additive and paint remover. Also, **potassium hydrate.**

potassium hypophosphite *Inorganic Chemistry.* KH_2PO_2, white crystals or powder; soluble in water and alcohol; very deliquescent; decomposes on heating; a fire risk and may explode in contact with strong oxidants.

potassium iodate *Inorganic Chemistry.* KIO_3, colorless monoclinic crystals; soluble in water and insoluble in alcohol; melts at 560°C and decomposes above 100°C; used as a reagent and in medicine.

potassium iodide *Inorganic Chemistry.* KI, colorless or white cubic crystals or granules; soluble in water and slightly soluble in alcohol; melts at 681°C and boils at 1330°C; used as a reagent, in photographic emulsions, spectroscopy, and infrared transmission.

potassium linoleate *Organic Chemistry.* $C_{17}H_{31}COOK$, a light tan paste that is soluble in water; used as an emulsifying agent.

potassium manganate *Inorganic Chemistry.* K_2MnO_4, green rhombic crystals that are soluble in water; decomposes at 190°C; a strong oxidant and dangerous fire hazard; used in bleaching, batteries, and photography.

potassium metaarsenite see POTASSIUM ARSENITE.

potassium metabisulfite *Inorganic Chemistry.* $K_2S_2O_5$, colorless monoclinic plates; slightly soluble in cold water and alcohol; decomposes at 190°C; used as an antiseptic, a reagent, and in wine making and engraving. Also, POTASSIUM PYROSULFITE.

potassium nitrate *Inorganic Chemistry.* KNO_3, colorless rhombic or trigonal crystals; soluble in water and insoluble in alcohol; melts at 334°C and decomposes at 400°C; a strong oxidant and dangerous fire hazard; used in explosives, glass manufacture, and curing foods. Also, NITER, SALTPETER.

potassium nitrite *Inorganic Chemistry.* KNO_2, white to yellowish prisms; soluble in water and hot alcohol; deliquescent; melts at 440°C and decomposes at boiling point; a strong oxidant and fire hazard on heating; used in chemical analysis and as a food additive.

potassium oxalate *Organic Chemistry.* $K_2C_2O_4 \cdot H_2O$, toxic, odorless, colorless crystals; soluble in water and efflorescent in warm dry air; decomposes upon heating; used in analytical chemisty, in photography, and as a bleach.

potassium oxide *Inorganic Chemistry.* K_2O, colorless cubic crystals; very soluble in water and alcohol; hygroscopic; decomposes at 350°C; used as a reagent and intermediate.

potassium percarbonate *Inorganic Chemistry.* $K_2C_2O_6 \cdot H_2O$, a granular white mass that is soluble in cold water; melts at 200–300°C; a strong oxidant and fire hazard; used in analysis, photography, and textile printing.

potassium perchlorate *Inorganic Chemistry.* $KClO_4$, colorless rhombic crystals; soluble in water and very slightly soluble in alcohol; decomposes at 400°C; a strong oxidant and fire hazard; used in explosives and photography, and as an oxidizer.

potassium permanganate *Inorganic Chemistry.* $KMnO_4$, purple rhombic crystals; soluble in water and decomposes in alcohol; decomposes below 240°C; a strong oxidant and fire hazard; used as a disinfectant, in medicine, and in air and water purification.

potassium peroxide *Inorganic Chemistry.* K_2O_2, white, deliquescent, amorphous crystals; melts at 490°C and decomposes at boiling point; an irritant, strong oxidant, and fire hazard; used as an oxidizing and bleaching agent, and in oxygen masks.

potassium persulfate *Inorganic Chemistry.* $K_2S_2O_8$, colorless triclinic crystals; soluble in water and insoluble in alcohol; decomposes below 100°C; a strong oxidant and fire hazard; used in bleaching and desizing textiles, and as an oxidizing agent and antiseptic. Also, **potassium peroxydisulfate.**

potassium phosphate *Inorganic Chemistry.* any of the orthophosphates of potassium: KH_2PO_4 (monobasic), K_2HPO_4 (dibasic), or K_3PO_4 (tribasic).

potassium polymetaphosphate *Inorganic Chemistry.* $(KPO_3)_n$, a polymeric white powder; insoluble in water; used as a fat emulsifier and moisture retainer in food.

potassium pyrophosphate *Inorganic Chemistry.* $K_4P_2O_7 \cdot 3H_2O$, colorless deliquescent crystals; soluble in water and insoluble in alcohol; used in soaps and detergents, and as a peptizing and dispersing agent. Also, TETRAPOTASSIUM PYROPHOSPHATE.

potassium pyrosulfite see POTASSIUM METABISULFITE.

potassium silicate *Inorganic Chemistry.* K_2SiO_3, colorless crystals; soluble in water and insoluble in alcohol; melts at 976°C; used in manufacture of glass and detergents, and as a catalyst. Also, SOLUBLE POTASH GLASS.

potassium silicofluoride see POTASSIUM FLUOROSILICATE.

potassium sodium ferricyanide *Inorganic Chemistry.* $K_2NaFe(CN)_6$, orange-red monoclinic crystals; soluble in water and decomposes at melting point; used in photography.

potassium sodium tartrate see ROCHELLE SALTS.

potassium sorbate *Organic Chemistry.* $C_6H_7KO_2$, a white powder that is soluble in water and decomposes at 270°C; used as a bacteriostat and preservative in foods.

potassium stannate *Inorganic Chemistry.* $K_2SnO_3 \cdot 3H_2O$, colorless crystals; soluble in water and insoluble in alcohol; highly toxic; used in dyeing, printing textiles, and tin plating.

potassium stibnate see POTASSIUM ANTIMONATE.

potassium sulfate *Inorganic Chemistry.* K_2SO_4, colorless rhombic or hexagonal crystals; soluble in water and insoluble in alcohol; used as a reagent, in medicine, and in the manufacture of glass.

potassium sulfide *Inorganic Chemistry.* K_2S, yellow to brown cubic crystals; soluble in water and alcohol; deliquescent; melts at 840°C; a dangerous fire hazard; used as a chemical reagent, a depilatory, and in medicine.

potassium sulfite *Inorganic Chemistry.* $K_2SO_3 \cdot 2H_2O$, white to yellowish hexagonal crystals; soluble in water and slightly soluble in alcohol; decomposes at melting point; used in photography, medicine, and as a preservative.

potassium tetraiodocadmate see CADMIUM POTASSIUM IODIDE.

potassium thiocyanate *Inorganic Chemistry.* KNCS, toxic, colorless rhombic prisms; soluble in water and alcohol; deliquescent; melts at 173.2°C and decomposes at 500°C; used as a reagent, and in textiles, photography, and medicine.

potassium undecylenate *Organic Chemistry.* $CH_2{=}CH(CH_2)_8COOK$, a toxic, fine white powder; soluble in water and decomposes at 250°C; used to inhibit fungal and bacterial growth in pharmaceuticals and cosmetics.

potassium xanthate *Organic Chemistry.* KC_2H_5OCSS, toxic, colorless or pale yellow crystals; soluble in water and alcohol and insoluble in ether; used as an analytical reagent and fungicide.

potato *Botany.* **1.** any plant of approximately 150 species of the genus *Solanum*, an herbaceous annual with underground lateral stems that produce a short, thick, edible tuber with nodes (eyes) and internodes; originally cultivated by the Indians of South America, probably in the Andes valleys, and introduced to Europe in the 1500s. **2.** the tuber itself, the world's most widely grown vegetable; a diet staple in Europe and North America. Also, **white potato, Irish potato.**

potatobug or **potato bug** *Invertebrate Zoology.* a yellow-and-black beetle, *Leptinotarsa decemlineata,* that feeds on the leaves of potato plants; a noxious pest in potato-growing regions throughout the U.S. Also, **potato beetle.**

potato stone *Mineralogy.* a geode resembling a potato in shape, particularly such a geode made up of hard silicified limestone having an inner lining of quartz crystals.

potato virus X group see POTEXVIRUS GROUP.

pot core *Electromagnetism.* a magnetic core, typically powdered iron or ferrite material that is shaped in the form of a pot having a center post; a magnetic lid is used to close a magnetic circuit between the center post and the pot's rim, while the windings are about the center post and are totally enclosed.

pot die forming *Mechanical Engineering.* the process of forming sheet or plate metal by applying internal pressure that causes the workpiece to assume the contour of a hollow die.

pot earth see POTTER'S EARTH.

potential *Science.* **1.** having latent power or ability. **2.** possible rather than actual; not now in effect. *Mathematics.* a function or set of functions whose spatial derivatives can form a vector field, as in the case of the electric field and its associated electric potential: $E = -\text{grad}(f)$. *Physics.* the amount of work per unit mass (or charge) that is required to move a mass (or charge) through a gravitational (or electrostatic) field, from an infinite distance to the point at which the potential is to be evaluated. *Electricity.* see ELECTRIC POTENTIAL. (From a Latin word meaning "having power.")

potential barrier *Physics.* a region in a potential field in which a particle would encounter opposition, causing a decrease in its kinetic energy. Also, **potential hill.**

potential density *Physics.* the mass density of a compressible fluid after it is adiabatically expanded or compressed to a pressure of 1 bar (10^5 newtons per square meter).

potential difference *Electricity.* the voltage difference in potential between two points in a circuit.

potential divider see VOLTAGE DIVIDER.

potential drop *Fluid Mechanics.* the difference in pressure between two barometric surfaces or lines in a fluid. *Electricity.* see VOLTAGE DROP.

potential energy *Mechanics.* the stored capacity of a body or system to do work by virtue of its configuration and position.

potential energy of deformation see STRAIN ENERGY.

potential evaporation see EVAPORATIVE POWER.

potential evapotranspiration *Hydrology.* the ideal maximum amount of water that will be lost from a particular area by evapotranspiration, assuming sufficient water is available through precipitation.

potential flow *Fluid Mechanics.* an irrotational fluid flow wherein the fluid velocity can be written as the gradient of a scalar function.

potential gradient see VOLTAGE GRADIENT.

potential index of refraction *Meteorology.* an atmospheric index of refraction devised so that it has no height variation in an atmosphere with no gain or loss of heat. Also, **potential refractive index.**

potential instability see CONVECTIVE INSTABILITY.

potentiality *Science.* the fact of having potential, or something that has this quality.

potential of zero charge see ZERO CHARGE POTENTIAL.

potential ore *Mining Engineering.* **1.** a term for ore presumed to exist but not yet discovered. **2.** ore known to exist but whose recovery is not yet economically practical.

potential scattering *Quantum Mechanics.* the scattering from a nucleus or stationary charge in which the nuclear and particle wavefunctions do not significantly overlap, enabling the problem to be defined merely in terms of a nuclear potential function.

potential temperature *Thermodynamics.* the temperature of a compressible fluid that ideally would result if it were compressed or expanded with no gain or loss of heat to some standard pressure, typically one bar (10^6 dynes/cm^2).

potential theory *Mathematics.* the study of functions associated with Laplace's equation, especially of harmonic functions.

potential transformer see VOLTAGE TRANSFORMER, def. 2.

potential transformer phase angle *Electricity.* the measured angle between the primary voltage vector and the secondary voltage vector reversed; it is positive when the secondary voltage vector leads the primary voltage vector.

potential vorticity *Fluid Mechanics.* the product of the static stability and the absolute vorticity; conservative in adiabatic flow as in the quantity given by $(h/q)(\partial q/\partial p)$, where h is the absolute vorticity of a fluid element, q is the potential temperature, and p is the pressure.

potential well *Physics.* a region in a field where a particle may be trapped because its kinetic energy decreases to zero as its potential energy increases to a maximum.

potentiometer [pō´ten shē äm´ə tər] *Electrical Engineering.* **1.** an instrument that measures electromotive force or potential difference by comparing a part of the voltage to be measured against a known electromotive force, then using the law of fall of potential to compute the final measurement. **2.** a continually adjusting resistor having a sliding contact, which is typically mounted on a rotating shaft; used primarily as a voltage divider. Also, TRIMMER POTENTIOMETER.

potentiometric *Electrical Engineering.* relating to or determined by the use of a potentiometer.

potentiometric cell *Analytical Chemistry.* a container or vessel that holds the reference and indicator electrodes and an electrolyte solution that is being potentiometrically titrated.

potentiometric controller *Control Systems.* a controller that amplifies an error signal in order to maintain a constant load on an input circuit.

potentiometric map *Hydrology.* a map on which contour lines and other symbols are used to represent the potentiometric surface of an aquifer. Also, PRESSURE-SURFACE MAP.

potentiometric model study *Petroleum Engineering.* the use of an analogic electrical-resistance model and Darcy's law to predict conditions in gas-condensate oil reservoirs.

potentiometric surface *Hydrology.* an imaginary surface defined by the level to which water will rise in a well under its full pressure head, and representing the total static level of groundwater. Also, PIEZOMETRIC SURFACE, PRESSURE SURFACE.

potentiometric titration *Analytical Chemistry.* a titration at zero (or constant) current in which the end point is determined by reading electrode-potential variations due to changes in the concentration of the potential-determining ion.

potentiostat *Engineering.* a device used for the direct study of corroding metals in which both anodic and cathodic reactions are taking place simultaneously.

Potexvirus group *Virology.* a genus of plant viruses in which the virions are flexuous, rod-shaped filaments that contain linear single-stranded RNA; the host range is narrow, and the viruses are easily transmitted mechanically and by contact. (A name derived from *potato virus X group.*)

pot furnace *Engineering.* **1.** a furnace for melting glass in pots. **2.** any of the small, vertical furnaces used to smelt batches of enamel in a crucible.

pothole a hole or depression in a surface; specific uses include: *Civil Engineering.* a rough, irregular hole in a pavement surface, caused by the effects of weathering or heavy use. *Geology.* **1.** any pit, hole, or hollow in the earth. **2.** a vertical or nearly vertical columnar opening in limestone. Also, AVEN, CENOTE. **3.** a shaftlike cave having an opening toward the surface. *Hydrology.* **1.** a rounded, steep-sided depression in a coastal marsh that contains water at or below the level at low tide. Also, ROTTEN SPOT. **2.** a large, shallow depression that occurs between dunes on a prairie and often contains an intermittent pond or marsh. **3.** see MOULIN. *Archaeology.* a depression or pit left by a pothunter.

pothole

pothunter or **pot-hunter** *Archaeology.* a person who collects archaeological objects or excavates sites in an unscientific manner for personal gain, and whose actions result in the destruction of surrounding data.

Potier diagram *Electricity.* a vector diagram showing phase relationships between current and voltage in an AC circuit containing reactance.

potification *Engineering.* the process of making water fit to drink.

potlatch *Anthropology.* an activity of certain American Indian groups of the Pacific Northwest, in which great quantities of food and possessions were given away to guests to enhance the prestige of a host chief.

potometer *Engineering.* a device that measures the transpiration rate of a plant, consisting of a small, closed, water-filled vessel in which the cut end of the plant is immersed.

potomology *Civil Engineering.* the study of the behavior of flow in river channels; used to predict the effect on flow of structures such as bridge piers.

potoo *Vertebrate Zoology.* a nocturnal, insectivorous bird of the family Nyctibiidae, found in tropical America and characterized by a rounded head, a large mouth, and mottled brown plumage with powder down patches on the sides and breast; noted for laying a single egg on top of a broken tree stub.

pot planting *Forestry.* the setting out for growing of young trees in pot-shaped containers (made of such materials as concrete, peat, or polythene) in which they have been raised.

pot plunger *Engineering.* a device that forces melted plastic into the closed cavity of a transfer mold.

potrero *Geology.* an elongated beach ridge consisting of a series of accretionary dune ridges encircled by mud flats and isolated from the coast by a lagoon and barrier island.

potsherd *Archaeology.* any pottery fragment that has archaeological significance. Also, SHERD.

potted circuit *Electricity.* a circuit that has been embedded in a potting compound, usually wax or resin, to protect it from effects such as corrosion, vibration, condensation, and tampering.

potted line *Electricity.* any of a series of straight-line circuits that together equal a pulse-forming network; they are immersed in oil and are enclosed in a metal container.

Potter-Bucky diaphragm or **grid** *Radiology.* a moving grid of thin lead strips on an X-ray table that reduces the amount of secondary radiation reaching the film, thus increasing detail and contrast.

Potter-Elvejhem homogenizer *Microbiology.* an apparatus designed to homogenize, or break up, eukaryotic cells.

potter's earth or **potter's clay** *Materials.* a plastic clay that is free of iron and other impurities and is used by potters. Also, POT EARTH.

potter's wheel *Engineering.* a rotating circular table mounted on a vertical spindle on which clay is shaped by hand before it is fired.

pottery *Materials.* **1.** ceramic articles made of clay that has been fired. Also, CLAYWARE. **2.** the art of making such articles. **3.** a place where such articles are made.

Pottiaceae *Botany.* a family of dull, often xerophytic mosses of the order Pottiales that form tufts and cushions on rocks, trees, and soil in temperate and polar regions; most species are characterized by erect stems, terminal sporophytes, and contorted leaves.

Pottiales *Botany.* a cosmopolitan order of mosses in the class Bryopsida, characterized by stems with terminal sporophytes, erect and little-branched gametophytes, and broad-to-narrow lanceolate-to-lingulate leaves.

potting *Electronics.* a process by which components are embedded into insulating material held in a container that becomes the outer surface of the unit.

potto *Vertebrate Zoology.* a short-tailed, arboreal African primate of the family Lorisidae, especially *Perodicticus potto*; characterized by large eyes, an elongated snout, thick fur, an opposable thumb and big toe, and four spines that extend from the neck; it eats fruits, leaves, and insects. Also, BUSH BEAR, TREE BEAR.

potto

Pott's disease *Medicine.* an inflammation of one or more of the vertebrae; tuberculosis of the spine. (First described by Sir Percival *Pott,* 1714–1778, English surgeon.) Also, SPONDYLITIS.

Pott's fracture *Medicine.* a fracture of the lower part of the fibula, often accompanied by serious injury to the lower tibial articulation.

Potyvirus group *Virology.* a genus of plant viruses in which the virions are flexuous filaments that contain linear ssRNA; they infect a wide range of plants and are easily transmitted mechanically, by aphids through nonpersistent transmission, or by whiteflies, fungi, or mites. (A name derived from *potato virus Y group.*)

pouch *Anatomy.* a pocketlike space or sac, as of the peritoneum. *Zoology.* any baglike or pocketlike part, such as the pocket for marsupial young or the sac beneath the bill of a pelican.

poudrage *Surgery.* the application of a powder to surfaces, e.g., pleural surfaces, to promote adhesion.

poult *Agriculture.* **1.** a young turkey, especially one that is less than eight weeks old. **2.** the young of chicken or other fowls.

poultice *Medicine.* a preparation of absorbent substances wetted with oily or watery fluids, generally heated and sometimes medicated, applied to the skin to soothe, relax, or stimulate an irritated or aching part of the body.

poultry *Agriculture.* a collective term for any domestic fowl, especially chickens, that are raised primarily for meat or egg production. (From French; going back to the Latin word for chicken.)

pounce *Materials Science.* a fine powder, often of charcoal, used in transferring a design through a perforated pattern.

pound *Metrology.* **1.** a unit of avoirdupois weight equal to 16 ounces, equivalent to 0.45 kilogram. Also, **pound mass. 2.** a unit of troy or apothecaries' weight equal to 12 ounces, or 0.37 kilogram. (Going back to a Latin word meaning "weight.")

poundal *Metrology.* a unit of force equal to the amount of force needed to give a mass of one pound an acceleration of one foot per second per second.

pound-foot see FOOT POUND.

pound force *Metrology.* in the foot-pound-second system, a unit of force equal to the force that produces an acceleration equal to the acceleration of gravity when acting on a mass of one pound.

pounds per square foot *Metrology.* a unit of measure equal to the amount of pressure applied by a mass of one pound acting on an area of one square foot.

pounds per square inch *Metrology.* a unit of measure equal to the amount of pressure applied by a mass of one pound acting on an area of one square inch.

pounds per square inch differential *Engineering.* a measurement indicating the pressure difference between two points in a fluid-flow system.

pounds per square inch gauge *Mechanics.* gauge pressure (the difference between total pressure and ambient pressure) expressed in the unit of pounds per square inch.

Poupart's ligament see INGUINAL LIGAMENT.

Pourbaix diagram *Chemical Engineering.* a pH-potential diagram used extensively in corrosion studies.

pour-plate *Microbiology.* describing a procedure, used as a counting method or a method of growing pure cultures, in which a liquid inoculum is added to molten medium in a petri dish, and the combined ingredients are then mixed by swirling before the plate is incubated. Thus, **pour-plate culture.**

pour point *Fluid Mechanics.* the minimum temperature at which a liquid will flow.

Pourtalesiidae *Invertebrate Zoology.* a family of deep-sea echinoids, the bottle urchins, in the order Holasteroida.

pour test *Engineering.* a procedure in which a fluid is chilled under certain conditions to determine the lowest temperature at which it can flow.

poverty of movement *Neurology.* relative immobility, usually as a result of parkinsonism.

powder *Chemistry.* a substance made up of an aggregation of small particles. *Materials.* **1.** any dry material in a fine, granulated state. **2.** to reduce material to such a state. *Hydrology.* see POWDER SNOW.

powder avalanche see DRY-SNOW AVALANCHE.

powder bag *Ordnance.* a bag that holds the propellant charge used with separate-loading ammunition; once made of a special silk called cartridge silk, it is now usually made of a cotton fabric called cartridge cloth; the bag must be completely consumed in the explosion.

powder blast see MUZZLE BLAST.

powder clutch *Mechanical Engineering.* an electromagnetic disk clutch in which the space between the clutch members is filled with dry, finely divided magnetic particles.

powder diffraction *Crystallography.* diffraction by a crystalline powder (a mass of randomly oriented microcrystals) consisting of lines or rings rather than separate diffraction spots. The diffraction pattern obtained is that which is expected for a set of randomly oriented crystals. Also, **powder method.**

powder diffraction camera *Crystallography.* a hollow, film-lined cylindrical device in which monochromatic X-rays enter through a slot on the side and diffract from a powdered sample of crystalline material along the axis of the cylinder so that the diffraction pattern is received on the film. Also, **powder camera.**

powder flowmeter *Engineering.* an instrument for measuring the flow rate of a metal powder.

powder forging *Materials Science.* a process in which powder is compacted to a preform shape, sintered, cooled and then reheated, or cooled from the sintering temperature to the forging temperature, and then forged.

powderkeg *Engineering.* a small metal container that holds gunpowder or blasting powder.

powder lubricant *Metallurgy.* in powder metallurgy, any additional agent that helps in the removal of the green compact.

powder metal *Metallurgy.* any particulate metal or alloy.

powder metallurgy *Metallurgy.* the science and techniques of fabricating metallic components by a combination of compression and heating, with no or only partial melting.

powder method *Solid-State Physics.* an X-ray diffraction technique in which a monochromatic X-ray beam is aimed at a powder sample of crystallites having random orientations; the scattered radiation is detected by use of film or a moveable counter.

powder mine *Mining Engineering.* a mining excavation that is filled with powder for the purpose of rock blasting.

powder-moisture test *Engineering.* a method used to calculate the amount of moisture in a propellant by drying it under specified conditions.

powder molding *Engineering.* a method used in plastics molding, in which polyethylene powder is melted against the heated interior of a mold.

powder pattern *Crystallography.* the diffraction pattern produced by all possible orientations in a powder diffraction camera. *Electromagnetism.* a pattern formed by very fine powders or colloidal suspensions when applied to the surface of a magnetic body; the pattern can indicate the magnetic domains of the body in detail.

powder processing *Materials Science.* any of a wide variety of processes involving powders, including powder metallurgy, polymer powder coatings, and ceramic production.

powder rolling see ROLL COMPACTING.

powder snow *Hydrology.* dry, loose, fallen snow that was deposited under low temperature conditions and that has not been compacted in any way. Also, POWDER.

powder train *Engineering.* **1.** a device placed in fuses to produce time action. **2.** a line of explosive devices set down for destruction by burning.

powdery mildew *Plant Pathology.* a plant disease caused by the powdery mildew fungus of the family Erysiphaceae (or sometimes by an imperfect fungus of the genus *Oidium*) producing white powdery growths on leaves, stems, buds, and flowers.

powdery mildew fungus *Mycology.* any of various fungi of the family Erysiphaceae that produce large amounts of powdery spores on hosts, causing powdery mildew disease.

powdery scab *Plant Pathology.* a disease of potato tubers caused by the fungus *Spongospora subterranea* and characterized by rough, nodular, corky areas that may burst and release masses of spores.

Powell, Cecil Frank 1903–1969, English physicist; awarded the Nobel Prize for advances in photographing cell nuclei and studies of mesons.

Powell, John Wesley 1834–1902, American geologist; explored the Colorado River; studied the formation of canyons.

powellite *Mineralogy.* $CaMoO_4$, a transparent yellow, brown, gray, or blue tetragonal mineral, isomorphous with scheelite, having a specific gravity of 4.23 and a hardness of 3.5 to 4 on the Mohs scale, and occurring as pyramidal crystals and in massive to foliated form; found in the oxidized zones of ore deposits as a secondary mineral and minor ore of molybdenum.

power the ability to do or act; being able to cause some change or activity; specific uses include: *Engineering.* **1.** any type of physical energy, as of electricity, water, and so on. **2.** energy produced by a machine, as distinguished from hand labor. **3.** relating to or driven by electrical or mechanical energy. *Physics.* the time rate at which work is performed (energy is transformed) by a system. *Optics.* **1.** the reciprocal of the focal length in a lens or mirror. **2.** the magnifying power of a telescope or microscope. *Mathematics.* **1.** another term for exponent; e.g., the expression x^b can be read as "x raised to the power b" or "x to the bth power." **2.** the power of a point P in the plane of a circle with respect to the circle is the value of the product $PA \cdot PB$, where A and B are the points of intersection with the circle of any secant through P, and the lengths PA and PB are directed. *Statistics.* in testing, the ability of a test statistic to distinguish between alternative hypotheses, as measured by the power function.

power amplification *Electronics.* see POWER GAIN.

power amplifier *Electronics.* a circuit that increases the power delivered to one or more audio speakers, usually the last stage of a multistage amplifier circuit.

power attenuation see POWER LOSS.

power axiom *Mathematics.* an axiom stating that the power set of any set is itself a set.

power bandwidth *Telecommunications.* in an audio amplifier, the frequency range for which half its rated power is obtainable at rated distortion.

power brake *Mechanical Engineering.* an automotive brake operated by pressure from a power source, such as a compressed-air reservoir, proportionately to a smaller amount of pressure on the brake pedal.

power breeder *Nucleonics.* a nuclear reactor designed to provide power while producing more fissionable material than it uses.

power-cable logging *Forestry.* any logging operation that involves the use of power-operated cables for hauling.

power car *Mechanical Engineering.* a railroad car that furnishes heat and electric power to a train. *Aviation.* **1.** on an airship, especially a rigid airship, a nacelle or car housing an engine. Also, ENGINE CAR. **2.** on an airplane, a wing pod housing a jet engine or other power plant.

power circuit *Electricity.* the portion of the wiring of an electrical installation that is used to supply power to various motor-driven electrical devices through power sockets or receptacles.

power-control rod *Nucleonics.* a neutron-absorbing rod that is used in a reactor to provide fine control over the reactivity.

power cord see LINE CORD.

power cylinder *Control Systems.* a pneumatic or hydraulic actuator that consists of a piston in a cylinder.

power dam *Civil Engineering.* a barrier built across a river to generate electricity in the form of hydroelectric power.

power density *Electromagnetism.* a measure of the amount of radiated electromagnetic power traversing a unit area, usually expressed in units of watts per square centimeter. *Nucleonics.* the power per unit volume generated within the core of a nuclear reactor.

power detection *Acoustical Engineering.* a process by which an audio signal is supplied directly from a receiver to a loudspeaker or a recorder.

power detector *Acoustical Engineering.* a device that extracts sounds from powerful radio signals with little distortion.

power divider *Electromagnetism.* a device that allows for variable power distribution at a branch point in a waveguide.

power drill *Mechanical Engineering.* a drill driven by steam, compressed air, or electricity.

power drive *Robotics.* a servomechanism that uses any source of power, such as direct current, pneumatic, hydraulic, or other forces, to drive a robot.

power-driven *Mechanical Engineering.* describing a component or piece of equipment that is moved, rotated, or operated by electrical or mechanical energy. Thus, **power-driven fan.**

power efficiency *Statistics.* a method for comparing hypothesis tests using the sample sizes for two tests having the same power (i.e., two tests that will detect a false null hypothesis with the same probability).

power excursion *Nucleonics.* an abrupt increase in the reactivity of a nuclear reactor resulting in a sudden increase in power output.

power factor *Electricity.* **1.** the ratio of active power to apparent power. For sinusoidal voltage and current, the power factor is $\cos(\theta)$ when θ is the phase difference between the voltage and current. **2.** the ratio of total watts to the root-mean-square (RMS) volt-amperes of an AC circuit. Also, PHASE FACTOR.

power flow *Electromagnetism.* the rate of electromagnetic energy transport across a specified unit area by an electromagnetic field.

Powerforming *Chemical Engineering.* a petroleum-refinery process with fixed-beds that reforms petroleum naphtha catalytically to produce high-octane motor gasoline.

power frequency *Electricity.* **1.** the power-line frequency for AC power distribution, which is 60 Hz in the U.S. **2.** the frequency of an alternating current generator, or of a DC-to-AC inverter.

power function *Statistics.* in testing, the probability of rejecting the null hypothesis as a function of the values of the unknown parameter. Good tests have high power function for values of the parameter in the region constituting the alternative hypothesis.

power gain *Electronics.* the ratio of the increased output power delivered by a transducer to the power delivered to its input. Also, POWER AMPLIFICATION. *Electromagnetism.* a measure of an antenna's ability to transmit a signal in a specified direction, given by 4π times the ratio of the intensity of radiation to the total power delivered to the antenna.

power generator *Electricity.* a device used to produce electrical impulses through a system such as an electric generator or a magnetohydrodynamic, thermionic, or thermoelectric power generator.

power grip *Anthropology.* the ability of a primate to grasp objects in the hand with the fingers wrapped around it and the thumb reinforcing the fingers.

powerhouse see GENERATING STATION.

power-law fluid *Fluid Mechanics.* a fluid whose shear stress measured at some point is proportional to the shear rate at the same point raised to some power, such as pseudoplastic and dilatant fluids.

power-law profile *Meteorology.* a formula used for the determination of wind variation with height in the surface boundary layer.

power level *Electricity.* the amount of power that is transmitted beyond a point in an electric system, usually expressed in decibels, indicating the ratio to a given reference power (such as one watt or one milliwatt). *Nucleonics.* the power output of a nuclear reactor, usually expressed in watts.

power line *Electricity.* a set of two or more wires that conduct electric power from one location to another.

power-line carrier *Electricity.* a carrier frequency, generally below 600 kHz, that is used to transmit control signals between stations over power lines without interfering with the line's normal transmission.

power-line interference *Telecommunications.* unwanted signals that are caused by radiation from electrical power transmission lines.

power-line monitor *Electronics.* a device that continuously tracks the voltage level across a power line.

power-line transposition *Electricity.* an ordered interchange of position within an open-wire line and its phases which is implemented so that effects of mutual capacitance and inductance with consequent interference are minimized or balanced without material distortion of the normal wire configuration at a distance comparable to the wire separation.

power loss *Electronics.* the ratio of the input power that is absorbed by a transducer's circuit to the decreased output power it delivers, usually dissipated as heat and expressed in decibels. Also, POWER ATTENUATION.

power output *Electronics.* the energy that is delivered by a power amplifier to another component, such as an audio speaker; measured in watts. Also, OUTPUT POWER.

power pack *Electronics.* a unit that converts the voltage of a power line or battery into energy at voltage levels suitable for a given device, as in a battery-operated radio, toy, or the like.

power package *Mechanical Engineering.* an automotive engine and its accessories installed and removed as a complete unit.

power plant *Electrical Engineering.* a plant, including all the equipment and buildings, in which some form of energy is converted to electric energy, such as a hydroelectric or steam generating station.

power projection *Military Science.* a manifestation of sea power in which naval forces use their strength to influence operations or international relations on land.

power rating *Electricity.* **1.** the power required by an electronic device for normal operations, as specified by the manufacturer. **2.** the specified power output of a generator or an amplifier.

power ratio *Electromagnetism.* for a waveguide that is improperly terminated, the ratio of maximum power to minimum power.

power reactor *Nucleonics.* a nuclear reactor that is used to provide useful power as distinguished from a research reactor.

power relay *Electricity.* a heavy-duty relay that can switch significant amounts of power.

power resistor *Electricity.* a heavy-duty resistor designed to carry large currents without overheating; may be cooled by air convection, air blasts, or water.

power saw

power saw *Mechanical Engineering.* a power-driven working saw, such as a bench saw, scroll saw, or circular saw.

power semiconductor *Electronics.* a device that employs the electrical properties of a semiconductor material to deliver large amounts of power to its load.

power series *Mathematics.* any finite or infinite series of the form

$$a_0 + a_1x + \cdots + a_nx^n + \cdots,$$

where the coefficients a_n may be constants, functions, matrices, and so on; denoted by $\sum_{n=0}^{\infty} a_nx^n$, by $\sum_n a_nx^n$, or by $\sum a_nx^n$. If x is regarded merely as a place holder or indeterminate without regard to convergence, the power series is called a **formal power series.**

power set *Mathematics.* for a given set X, the (unique) set consisting of all the subsets of X; denoted by 2^X or $P(X)$. The power set of a finite set of cardinality n has 2^n elements; hence the notation.

power shovel *Mechanical Engineering.* a power-driven excavating and loading shovel, consisting of a bucket at the end of a radial arm suspended from a boom.

power station see GENERATING STATION.

power steering *Mechanical Engineering.* an automotive steering system in which power from the engine is applied to augment the driver's mechanical turning of the steering wheel.

power stroke *Mechanical Engineering.* the pressure supplied to a piston by expanding steam or gases.

power supply *Electricity.* **1.** a generator or transformer-rectifier-filter arrangement that produces the power needed to operate electrical equipment. **2.** a power reserve that is stored in a power line or in batteries. *Electronics.* a circuit that converts AC power to one or more DC voltages required by electronic equipment.

power supply circuit *Electricity.* a network of electronics that is used to convert current from direct to alternating.

power switch see ON-OFF SWITCH.

power switchboard *Electricity.* the part of a switch box that contains one or more panels with switches mounted for the purpose of control, circuit-protection measurement, and equipment regulation; it can also carry main power switching and interruption devices with associated connections.

power switching *Electricity.* a tranference from one high-level power-supply circuit output to another.

power train *Mechanical Engineering.* all the moving parts that connect an engine to the point at which work is accomplished. Also, DRIVE TRAIN.

power transfer theorem see MAXIMUM POWER TRANSFER THEOREM.

power transformer *Electricity.* a transformer, usually iron core, that supplies power to electrical equipment.

power transistor *Electronics.* a transistor that delivers and dissipates large amounts of power without malfunctioning; commonly used in audio and switching circuits.

power-transmission line *Electricity.* an electric power line that provides large amounts of power at a high voltage to remote locations. Also, ELECTRIC MAIN.

power-transmission tower *Electricity.* a reinforced steel tower that supports a high-voltage power-transmission line; designed to maintain a large space between conductors (and between each conductor and the earth's surface) in order to prevent corona discharge.

power traverse *Ordnance.* the process of using a powered mechanism to turn a gun or change the firing direction.

power tube *Electronics.* a tube that handles more current than an average electron tube, so that it is able to generate far more energy; used in the last stages of audio-frequency or radio-frequency amplifiers.

power typing *Computer Programming.* in word processing, the high-speed entry of text to be proofread and corrected at a later time.

power winding *Electricity.* the output winding through which a controlled current flows in a magnetic amplifier or a saturable reactor.

pox *Medicine.* **1.** any disease that is marked by skin eruptions containing pus, especially one caused by a virus, such as chicken pox, cowpox, or smallpox. **2.** a former term for syphilis.

Poxviridae *Virology.* a family of viruses that infect mammals, birds, and insects, having large, structurally complex virions that are brick-shaped or ovoid and contain dsDNA. Also, **poxvirus group.**

Poynting, John Henry 1852–1914, English physicist; contributions to electromagnetics, including theorem on constant of gravitation.

Poynting-Robertson effect *Astrophysics.* the spiraling of small particles of solar system dust into the sun as a result of solar radiation impinging on their leading hemispheres.

Poynting theorem *Electromagnetism.* a theorem stating that the rate of electromagnetic energy loss in a specified region of space is equal to the sum of the dissipation rate (heat loss) and the outward rate of electromagnetic energy loss flowing across the boundary of the region.

Poynting vector *Electromagnetism.* a vector whose magnitude is proportional to the power per unit area present at a given point and time and whose direction indicates the direction of the energy flow.

pozzolana or **pozzolan** *Materials.* a finely ground, burnt clay, shale, or siliceous tuff or ash used in making cement because of its ability to harden underwater when mixed with lime. Also, **pozzuolana.**

pp. or **pp** pages.

p.p. or **pp** after meals (postprandial). Also, **PP.**

PP or **P.P.** near point of accommodation. (From Latin *punctum proximum.*) Also, **pp, p.p.**

ppb or **ppb.** parts per billion. Also, **p.p.b., P.P.B.**

pp60c-src *Biochemistry.* the protein kinase encoded by the c-*src* gene.

PPD or **P.P.D.** purified protein derivative.

ppg picopicogram.

ppGpp *Biochemistry.* guanosine 5'-diphosphate, 3'-diphosphate, one of two unusual nucleotides known as magic spots, which regulate metabolism in microorganisms; the other is pppGpp.

P1 phage *Virology.* a tailed phage that replicates efficiently in cells but does not integrate into the host chromosome.

PPI plan position indicator.

pp junction *Electronics.* the boundary on a semiconductor substrate that lies between two p-type regions, with different concentrations of impurities.

PPLO see PLEUROPNEUMONIALIKE ORGANISM.

ppm pulse per minute.

ppm or **ppm.** parts per million. Also, **p.p.m., P.P.M.**

PPM pulse-position modulation.

PPO blend *Materials.* a polyphenylene oxide plastic alloyed with polystyrene to improve processability and to lower the tensile strength and heat deflection temperature.

p power-line carrier *Electricity.* a carrier frequency that is generally below the level of 600 kHz; it is used to transmit control signals between stations over power lines without interfering with the line's normal transmission of power.

pppGpp *Biochemistry.* guanosine 5'-triphosphate, one of two unusual nucleotides known as magic spots, which regulate metabolism in microorganisms; the other is ppGpp.

PPR pestes des petits ruminants.

P-protein *Botany.* a phloem protein; a proteinaceous substance found in phloem of seed plants, especially in sieve tube members.

pps pulse per second.

ppt. or **Ppt.** precipitate.

pptn. precipitation.

PQ permeability quotient.

PQQ pyrroloquinoline quinone.

pr. pair.

Pr the chemical symbol for praseodymium.

Pr. propyl; presbyopia.

PR or **P.R.** far point of accommodation. (Latin *punctum remotum.*)

practical astronomy *Astronomy.* a branch of astronomy that focuses on determining the apparent positions and motions of celestial objects; applications include nautical astronomy and geodesy.

practical capacity *Transportation Engineering.* the maximum capacity of a section of highway under existing traffic conditions, with due regard to safety. This differs from possible capacity by the emphasis on safety.

practical system *Metrology.* another name for the MKSA (meter-kilogram-second-ampere) system of units.

practice ammunition *Ordnance.* ammunition used in training or target practice; it contains a propelling charge and either an inert filler or spotting charge. Similarly, **practice mine.**

pradolina see URSTROMTHAL.

Praesepe *Astronomy.* an open cluster in the constellation Cancer that is faintly visible to the unaided eye as a hazy patch. Also, THE BEEHIVE, THE MANGER, M 44.

pragma *Computer Programming.* **1.** a statement in a programming language that conveys information about a particular implementation or optimization suggestions from the programmer and can be ignored if not applicable or implemented. **2.** statements in a programming language that are directives to the compiler, not part of the executable code.

pragmatics *Artificial Intelligence.* aspects of a natural language that are arbitrarily dependent on a particular language, culture, or facts about the world. *Computer Programming.* the analysis of the relationships between characters or groups of characters and the way they will be interpreted and used by the receiver.

prairie *Ecology.* an extensive tract of level or rolling grassland, usually without trees except along river banks; characteristic of most of the central United States and Canada prior to agricultural development in the 1800s. *Botany.* the continuous tall-grass cover that grows in such grasslands in humid continental-hot summer and subtropical humid climates.

prairie chicken *Vertebrate Zoology.* a grouse, *Typhanuchus cupido,* of the family Tetraonidae, characterized by elaborate mating rituals on established dance grounds, including a display of erect black head feathers and inflated, orange neck pouches; found on the American plains. Also, **prairie fowl, prairie grouse.**

prairie climate see SUBHUMID CLIMATE.

prairie dog *Vertebrate Zoology.* any of several diurnal, burrowing rodents of the genus *Cynomys* in the family Sciuridae, having a reddish-gray coat covering a stout, heavy body; found in open plains and plateaus in western North America and Mexico.

prairie dog

prairie soil *Geology.* a group of zonal soils that develop in temperate, humid regions and are characterized by a dark to grayish-brown A horizon that grades through a brown to yellow-brown B horizon into lighter-colored parent material. Also, BRUNIZEM.

prairie wolf *Vertebrate Zoology.* a popular name for the coyote, especially in older use. See COYOTE.

prand. meal; meals; eating. (From Latin *prandium.*)

Prandtl, Ludwig [prän´təl] 1875–1953, German physicist; introduced concepts of boundary layer and "mixing length" in fluid dynamics.

Prandtl-Meyer relation *Fluid Mechanics.* the relation for two-dimensional supersonic flow between the inclination, I, of the flow to a boundary and the Mach number, M; a supersonic flow has a definite value of v, the Prandtl-Meyer parameter.

Prandtl mixing length see MIXING LENGTH.

Prandtl number *Fluid Mechanics.* a dimensionless quantity used in momentum and heat transfer calculations, given by the kinematic viscosity of a fluid divided by the molecular diffusivity.

prase *Mineralogy.* a dull, light- or grayish-green variety of granular microcrystalline quartz, found with jasper.

praseodymium [pras´ē ō dī´ mē əm; praz´ē ō dī´mē əm] *Chemistry.* a rare-earth (lenthanide) metallic element that has the symbol Pr, the atomic number 59, an atomic weight of 140.9, a melting point of 930°C, and a boiling point of 3200°C; it is a yellow, easily tarnished metal that is used in thermoelectric materials, glazes, glasses, and lasers. (From a Greek word meaning "leek-green," because it occurred in the greenish fraction of didymium.)

prasinite *Petrology.* a greenschist containing roughly equal amounts of hornblende, chlorite, and epidote.

Prasinophyceae *Botany.* a class of marine, brackish, and freshwater flagellates in the division Chlorophycota, characterized by a flagellar root system and often having scale-covered cells and flagella.

Prasinovolvocales *Botany.* an order of green algae that have lateral appendages in the flagellum.

Prasiolales *Botany.* an order of subaerial green algae of the class Chlorophyceae, characterized by multicellular thalli containing uninucleate cells with an axile stellate chloroplast having a central pyrenoid.

prasopal *Mineralogy.* a green variety of common opal, colored by chromium. Also, **prase opal.**

p. rat. aetat. *Medicine.* in proportion to age. (From Latin *pro ratione aetatis.*) Also, **P. rat. aetat.**

pratinicole *Vertebrate Zoology.* any of several birds belonging to the genus *Glareola,* having a short bill, a forked tail, and long, narrow, pointed wings. *Ecology.* any pratinicolous organism.

pratinicolous *Ecology.* describing a type of organism that lives in meadows or low grass regions.

Pratt truss *Civil Engineering.* a particular arrangement of roof support made by means of compressed vertical members counterbalanced by diagonal members in tension. Also, N-TRUSS.

Prausnitz-Kustner reaction *Immunology.* an immediate hypersensitive skin reaction that occurs when an individual is tested positive for the presence of reaginic antibodies in the serum.

prawn *Invertebrate Zoology.* any of various shrimplike crustaceans, especially of the genera *Penaeus* and *Palaemon.*

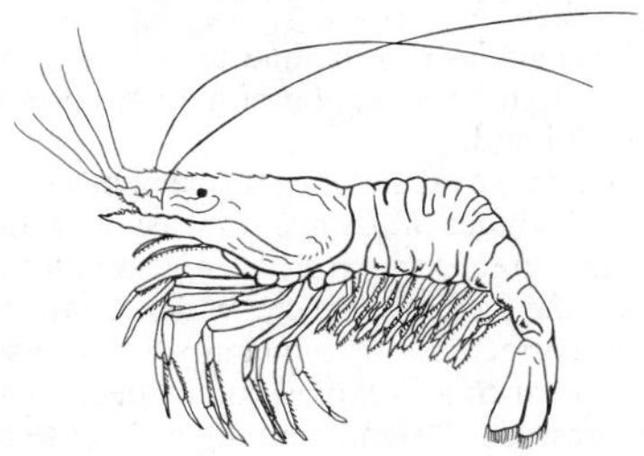

prawn

praxis *Psychology.* Jean Piaget's term for a child's system of coordinated and deliberate movements acquired during the sensory-motor stage of development. (A Greek word meaning "action.")

pre- a prefix meaning "before" or "in front of."

preadapt *Evolution.* to undergo preadaptation.

preadaptation *Evolution.* a condition in which an organism possesses characters or traits that would enable it to adapt easily to a given potential change in its environment, so that such a change would not be a threat to the survival of the species. Also, **preadaption.**

prealloyed powder *Materials Science.* a metal powder produced by atomizing a molten alloy by impingement with a fluid stream of gas or liquid; the fluid disperses the liquid alloy into small droplets that solidify into powder particles.

prealpine facies *Geology.* in neritic areas, geosynclinal facies distinguished by thick deposits of limestone and coarse terrigenous sediments, generally overlain by flysch.

preamble *Telecommunications.* the beginning section of a message that includes information such as the message number, date/time, or address.

preamplifier *Acoustical Engineering.* a low-impedance amplifier that amplifies an electronic signal produced by the impact of acoustic waves on a transducer, so that the generated signal can be processed further. *Electronics.* a circuit that increases the strength of a signal from a low-level source, so that it can be amplified further without appreciable degradation; used primarily for radio-frequency or microwave amplification. Also, **preamp.**

preanesthesia *Medicine.* preliminary anesthesia; light anesthesia or narcosis that is induced by medication before the administration of general anesthesia. Thus, **preanesthetic.**

prearranged fire *Military Science.* fire that is formally planned and executed against targets or target areas of known location; it is usually planned well in advance and executed at a predetermined time or during a predetermined period of time.

preassault operation *Military Science.* an operation conducted in the objective area prior to the actual assault; it includes reconnaissance, minesweeping, bombardment, bombing, underwater demolition, and destruction of beach obstacles.

preassembled *Engineering.* of, relating to, or describing an object that has been put together beforehand.

preataxic *Neurology.* occurring prior to the onset of ataxia.

preaxial *Science.* situated or occurring before an axis. *Anatomy.* in front of the body axis; denoting the lateral (radial) aspect of the upper limb and the medial (tibial) aspect of the lower limb.

prebiotic *Ecology.* denoting the period before the existence of organic life on earth.

prebreaker *Mechanical Engineering.* a device for breaking down bulk materials before feeding them to a grinder or crusher.

prec. preceding; preceded.

Precambrian *Geology.* **1.** the era that includes all geologic time from the formation of the earth to the beginning of the Paleozoic era (from about 4.6 billion to 570 million years ago). **2.** of or relating to this era, or to the rocks formed during that time.

precancer *Oncology.* a condition that tends eventually to become malignant.

precancerous *Oncology.* of or relating to any lesion that is not yet malignant but that in time may be, and that is interpreted to be a precursor to cancer and its resultant pathological process.

precast concrete *Materials.* a concrete that is cast and partly matured before it is set into position.

precava *Anatomy.* the superior vena cava.

precedence relation *Computer Programming.* in computer languages, the definition of the order in which operators will be executed in a mathematical expression. For example, in the expression $A + B * C$, the multiplication * takes precedence over + and is performed first.

preceding limb *Astronomy.* the western edge of a celestial object.

preceding spot *Astronomy.* the western of the two sunspots in a pair.

precentral *Anatomy.* situated in front of the center of an organ or part.

precentral gyrus *Anatomy.* the ridge of gray matter immediately anterior to the central sulcus of the cerebral hemisphere.

precession *Mechanics.* **1.** the angular motion of the spin axis of a body about an axis fixed in space. **2.** see PRECESSION RATE. *Astronomy.* specifically, a slow, periodic movement of the earth's rotation axis, caused by gravitational tugs from other celestial objects in the solar system. *Navigation.* a change in the rotational axis of a gyroscope or other object that is acted upon by a torque.

precession camera *Crystallography.* a device that produces a direct photographic representation of the reciprocal lattice of a crystal by rotating the crystal and the photographic film simultaneously.

precession constant *Astronomy.* the average annual movement of the equinoxes westward: 50.2564 seconds of arc.

precession frequency *Mechanics.* the frequency with which the spin axis of a precessing body rotates about the angular momentum axis.

precession in declination *Astronomy.* the change in an object's declination caused by precession.

precession in right ascension *Astronomy.* the change in an object's right ascension caused by precession.

precession method *Crystallography.* the use of a specific camera for recording the X-ray diffraction pattern of a crystal, resulting in a photograph that is an undistorted magnified image of a given layer of the reciprocal lattice. The necessary camera and crystal motion involve the precession of one crystal axis about the direction of the direct beam, and the film is continuously maintained in the plane perpendicular to this precessing axis.

precession of the equinoxes *Astronomy.* the earlier occurrence of the equinoxes in each sidereal year, caused by the precession of the axis of rotation of the earth and the resulting gradual changes in the coordinates of the stars.

precession of the nodes *Astronomy.* a gradual shift in the orbital planes of a binary system.

precession photograph *Crystallography.* a photograph of the X-ray diffraction pattern of a crystal, produced with a precession camera.

precession rate *Mechanics.* the rate or frequency with which the angular velocity vector of a precession body orbits its angular momentum vector.

prechlorination *Civil Engineering.* in water treatment, the chlorination of water prior to filtration.

precious metal *Metallurgy.* silver, gold, or any of the platinum group metals.

precious stone *Mineralogy.* a genuine gemstone, especially one held in high regard and of high commercial value due to qualities such as rarity, durability, and beauty.

precipice *Geology.* a very steep, tilted, vertical, or overhanging wall or rock surface.

precipitable *Chemistry.* capable of being precipitated.

precipitable water *Meteorology.* the total atmospheric water vapor collected by a vertical column of unit cross-sectional area between any two specified levels; often expressed in terms of the height to which the water substance would stand if it were completely condensed and collected in a vessel of the same unit cross section. Also, **precipitable water vapor.**

precipitant *Chemistry.* any substance that when added to solution causes a precipitate to form.

precipitate *Chemistry.* **1.** to cause a solid substance to settle out of a liquid or gaseous solution. **2.** to be deposited out of a solution, either by gravity or by a chemical reaction produced by the addition of a reagent. **3.** a substance, usually small crystalline particles, precipitated from a solution. *Meteorology.* to fall to the earth's surface as a condensed form of water.

precipitation *Chemistry.* **1.** the process of depositing a substance from a solution, by the action of gravity or by a chemical reaction. **2.** the sum of particles that are deposited in such a process. *Meteorology.* **1.** any form of water particles, such as frozen water in snow or ice crystals, or liquid water in raindrops or drizzle, that falls from clouds in the atmosphere and reaches the earth's surface. **2.** an amount, usually expressed in inches of liquid water depth, of the water substance that has fallen at a given point over a specified period of time; measured by a rain gauge. *Immunology.* a reaction between a soluble antibody and a soluble antigen, resulting in the formation of a substance (known as a precipitate) that separates, in solid particles, from a liquid. (Originally meaning "to throw headlong" or "to hurl downward.")

precipitation area *Meteorology.* **1.** a region over which precipitation is falling, represented on a synoptic surface weather chart. **2.** the region from which a precipitation echo is received.

precipitation attenuation *Electromagnetism.* the attenuation of transmitted radio waves in the atmosphere due to the presence of precipitation.

precipitation ceiling *Meteorology.* a ceiling classification applied when the ceiling value is the vertical visibility upward into precipitation; necessary when precipitation obscures the cloud base and prevents a height determination.

precipitation cell *Meteorology.* an element of a precipitation area over which the precipitation is more or less continuous.

precipitation clutter suppression *Electronics.* a technique that reduces interference caused by rain in a radar system.

precipitation current *Meteorology.* the downward moving electrical charge, from cloud region to the earth, that occurs in a fall of electrically charged rain or other hydrometeors.

precipitation echo *Meteorology.* a radar echo that is returned by precipitation.

precipitation electricity *Geophysics.* **1.** the study of the nature and distribution of electrical charges carried by particles of precipitation such as raindrops and snowflakes. **2.** the electric charge carried by particles of precipitation.

precipitation-evaporation ratio *Meteorology.* for a given locality and month, an empirical expression used for classifying climates numerically on the basis of precipitation and evaporation. Also, **precipitation-effectiveness ratio.**

precipitation facies *Geology.* facies that exhibit evidence of depositional conditions, primarily sedimentary structures, such as cross-bedding and ripple marks, and primary components, such as fossils.

precipitation gauge *Meteorology.* any of various instruments that measure precipitation in a designated area.

precipitation-generating element *Meteorology.* a small volume of supercooled cloud droplets in which ice crystals form and then grow much more rapidly than in a lower, larger cloud mass.

precipitation hardening see AGE HARDENING.

precipitation intensity *Meteorology.* the rate of precipitation, usually expressed in inches per hour.

precipitation inversion *Meteorology.* a decrease in precipitation that occurs along with increasing elevation of ground above sea level; found in mountainous regions.

precipitation noise *Electronics.* unwanted signals that arise when another antenna, or a conductor near an antenna, rapidly charges and discharges energy into the atmosphere.

precipitation number *Analytical Chemistry.* a standard that is used by the petrochemical industry to determine the amount of asphalt in a petroleum-lubricating oil.

precipitation physics *Meteorology.* the study of the formation and precipitation of solid and liquid hydrometeors released from clouds.

precipitation static *Telecommunications.* in a receiver, a type of interference experienced during rainstorms, snowstorms, or duststorms; caused by the impact of charged particles against the antenna or by induction fields created by nearby corona discharges.

precipitation station *Meteorology.* a station limited to the collection of precipitation observations.

precipitation titration *Analytical Chemistry.* a volumetric titration during which a precipitate forms; the end point can be determined by potentiometry.

precipitation trail see VIRGA.

precipitation trajectory *Meteorology.* an echo that is observed on range-height indicator scopes and represents the height-range pattern of snowfall as made by isolated precipitation-generating elements of a few miles in diameter.

precipitator see ELECTROSTATIC PRECIPITATOR.

precipitin *Immunology.* an antibody that reacts with an antigen to form a precipitate.

precipitinogen *Immunology.* a soluble antigen that stimulates the formation of a precipitate.

precipitin reaction *Immunology.* any process in which antigen and antibodies react together near their equivalence point, often forming cross-linked precipitates; or if the reaction occurs in a supporting medium like agar gel, the reactants form precipitin arcs that can be used to identify antigens and antibodies.

precipitin test *Immunology.* a method used to detect a specific antibody-antigen reaction, by observation of the presence or absence of the formation of a precipitate.

precision the accuracy, exactness, or refinement of a measurement or tool; specific uses include: *Statistics.* a measure of the accuracy of a measurement subject to random error, or in general of the variability of a random variable; the reciprocal of the variance. *Mathematics.* the degree to which the correctness of a quantity is expressed, such as the number of places to which π is expressed; the degree of refinement of a number, often expressed as "significant digits;" not to be confused with *accuracy,* which is the degree of correctness of a number. *Robotics.* the measurement of accuracy and repeatability in a robotic system.

precision adjustment *Ordnance.* the process of adjusting a single weapon in a battery in order to place its center of impact on the target.

precision approach *Navigation.* a standard instrument approach procedure in which an electronic glideslope/glidepath is provided. Also, **precision approach procedure.**

precision approach radar *Navigation.* radar equipment that allows an air traffic control facility to display azimuth, elevation, and range of aircraft so that the controller can issue instructions to the pilot of an approaching aircraft.

precision attribute *Computer Programming.* a set of numbers that denotes the length of a numeric string and base point normalization.

precision bombing *Military Science.* in general, any bombing directed at a specific point target; specifically, horizontal bombing directed at such a target and using precision instruments and equipment.

precision casting *Metallurgy.* **1.** any casting process, such as the lost wax process, that produces a part requiring little or no further metal removal. **2.** a casting made by such a process.

precision depth recorder *Engineering.* a recorder that creates a graphic display of a sonic depth trace on electrosensitive paper. Also, **precision graphic recorder.**

precision fire *Military Science.* fire in which the center of impact is placed accurately on a small target.

precision forging *Materials Science.* any metal-forging process that produces a part requiring little or no further metal removal.

precision grinding *Mechanical Engineering.* the process of machine grinding materials to specified dimensions and low tolerance.

precision grip *Anthropology.* the ability of a primate to hold an object between the fingers and the thumb; highly developed in humans to provide a great range of delicate movements for tool use.

precision net *Electricity.* an artificial line that is designed and adjusted to provide an accurate balance for a loop and subscriber set or line impedance over a four-wire terminating set or other device utilizing a hybrid coil.

precision sweep *Electronics.* a technique by which the movement of the electron beam across a radar screen is accelerated for a given portion of time, to increase resolution and range accuracy.

Pre-Classic or **Preclassic** see FORMATIVE.

precoat *Metallurgy.* in investment casting, a fine refractory material applied to the pattern prior to the application of the final ceramic coat.

precoat filter *Engineering.* a filter that removes solids from a liquid-solid mixture after the inner surface of the filter medium has been precoated with built-up solids or filter aid.

precoating *Engineering.* a process in which an inert material is applied to a filter medium before suspended solids are filtered from a solid-liquid mixture.

precocial *Zoology.* describing a species in which the young are at a relatively advanced stage of development when born.

precocious *Biology.* appearing, developing, or maturing earlier than is usual. *Botany.* of a plant, producing flowers or fruits ahead of the usual time, such as before leaves appear.

precognition *Psychology.* purported knowledge of the future, obtained through some paranormal means.

pre-Columbian *Anthropology.* relating to the history or culture of the Americas before the arrival of the expedition of Christopher Columbus in 1492 and the subsequent European colonization.

pre-Columbian sculpture

precombustion chamber *Mechanical Engineering.* a small chamber adjacent to the cylinder head of some compression ignition engines into which the oil fuel is injected at the end of the compression stroke.

precompact set *Mathematics.* a subset S of a metric space with the property that for each $\varepsilon > 0$, there exists a finite set of points of S such that the union of open balls of diameter ε centered at those points completely covers S. Also, TOTALLY BOUNDED SET.

precompiler *Computer Programming.* **1.** a program that processes source code of another program prior to compilation; may convert programmer statements into valid source code and enforce programming standards. **2.** a program designed to detect errors in a program and provide corrections.

precomputation *Navigation.* a method for determining a line of position quickly after taking a sight by figuring the computed altitude (and its presumed observed altitude) before the sight and comparing it with the altitude actually observed.

precomputed altitude *Navigation.* the expected altitude of a celestial body, calculated before the sight and with sextant altitude corrections applied with reversed sign.

precondition *Artificial Intelligence.* a set of conditions, often expressed as a predicate calculus formula, that must be satisfied before a rule or set of codes can be executed. *Control Systems.* a fact or set of facts that must be true before an action can be taken; e.g., before a robot can pick up an object, its hand must be empty. *Computer Science.* the left-hand side of an if-then rule.

preconditioning *Behavior.* the presentation of two stimuli consecutively without reinforcement in order to condition the subject to the second stimulus.

preconduction current *Electronics.* the electrons that are ejected from the anode in a gas tube before conduction takes place.

preconscious *Psychology.* **1.** in psychoanalytic theory, all the contents of the mind not immediately at a conscious level, but which can readily be brought into consciousness. **2.** of or relating to such material.

preconsolidation pressure *Geology.* the maximum effective stress or pressure exerted on a soil or unconsolidated sediment by overlying material that results in compaction.

preconventional level *Psychology.* in Lawrence Kohlberg's analysis of moral development, the first level, characterized by obedience, acceptance, and the evaluation of behavior solely in terms of its consequences, without any concept of right and wrong. This level is itself divided into **stage one,** marked by a concern with reward and punishment, and **stage two,** reflecting interest in self-gratification but lacking awareness of the needs of others. Also, **preconventional morality.**

precool *Mechanical Engineering.* to reduce the temperature of a working fluid before its use by a machine. *Food Technology.* to refrigerate a food, such as meat or produce, before shipping. Thus, **precooler.**

precursor something that goes before; specific uses include: *Materials Science.* a polymer or polymer-bonded ceramic fiber which upon heating carbonizes to produce graphite fibers or coverts into ceramics fibers.

precursor mRNA *Molecular Biology.* the primary RNA product of gene transcription, which is initially found in the eukaryotic nucleus as heterogeneous RNA (hnRNA) and lacks any of the later modifications to the RNA molecules. Also, PRE-MRNA.

precursor sweeping *Military Science.* the sweeping of an area by relatively safe means in order to reduce the risk of subsequent mine countermeasure operations.

predation *Ecology.* the killing and eating of an animal of one species by an individual of the same or different species.

predation pressure *Ecology.* the effect that predators' consumption has on a prey population.

predator [pred′ə tər] *Ecology.* **1.** an animal that kills other animals for food. **2.** any organism that kills and consumes other organisms.

predator-prey model *Chaotic Dynamics.* dynamical models of the population dynamics of interrelated species. Nonlinear interactions can lead to chaotic behavior.

predatory [pred′ə tôr′ē] *Ecology.* relating to or characterisitic of a predator. Thus, **predatory behavior.**

predazzite *Petrology.* a recrystallized limestone marble resembling pencatite but containing more calcite than brucite.

predecessor something that goes before; specific uses include: *Mathematics.* **1.** if x and y are elements of a partially ordered set X, with $x \leq y$ and $x \neq y$, then x is said to be a predecessor of (or less than) y. **2.** in a directed graph, if there is an arc (u,v), u is a predecessor of v and v is a successor of u.

predecrement *Computer Science.* a CPU feature in which the contents of an index register are automatically decreased by a fixed amount before its use, thus pointing to the next data to be processed.

predefined function *Computer Programming.* a commonly used mathematical function that has been defined and included in the utility libraries of the programming language for use by the programmer.

predetection combining *Electronics.* a technique that compensates for fading in a radio signal by combining several signals.

predetermined time system *Industrial Engineering.* a system providing times for specific work elements under varying conditions.

prediastolic *Cardiology.* before or at the beginning of the diastole.

predicate *Linguistics.* in English and many other languages, one of the two basic constituents of a sentence, along with the subject; the predicate has a verb that describes an action or state of being of the subject. *Mathematics.* a mathematical function of one or more variables; when values from appropriate domains are assigned to the variables, the result is a statement to which a value of "true" or "false" may be assigned. Also, **predicative expression.** *Artificial Intelligence.* in Lisp, a function that tests a condition and returns a value of T (true) or nil (false).

predicate calculus *Mathematics.* the mathematical representation of logical statements using the concepts of statement calculus and the additional notions of terms, predicates, and quantifiers. *Artificial Intelligence.* a form of mathematical logic that includes quantified variables, using the quantifiers "for all" and "there exists;" more powerful than propositional calculus. The statement "All men are mortal" would be represented $\forall_x$ [man(x) -> mortal(x)] in predicate calculus. Also, FIRST-ORDER LOGIC, FIRST-ORDER PREDICATE CALCULUS.

predictability the fact of being predictable; specific uses include: *Navigation.* a measure of how accurately a navigator can determine position in terms of geographical coordinates. *Chaotic Dynamics.* the possibility of forecasting the long-range future from the knowledge of the present state of the system; viewed as a key feature of systems that are termed deterministic or dynamical. For chaotic behavior, accurate longer-range forecasting requires increasing accuracy.

predicted firing *Ordnance.* the act of firing at a moving target based on predictions of the target's position at the moment of impact with the projectile. Thus, **predicted fire.**

predicted-wave signaling *Telecommunications.* a type of system that optimizes detection in the presence of severe noise by using mechanical resonator filters and other detector circuits to process previously ascertained information on the arrival and completion times of each pulse.

prediction any declaration or estimate regarding the future; specific uses include: *Statistics.* an inference regarding future observations rather than features of a population. *Meteorology.* an appraisal of future weather conditions; a forecast.

predissociation *Physical Chemistry.* the breakup of a highly energized molecule after it absorbs electromagnetic radiation.

prednisolone *Pharmacology.* $C_{21}H_{28}O_5$, a glucocorticoid, derived from cortisol, used as an anti-inflammatory drug. Also, METACORTANDRALONE.

prednisone *Pharmacology.* $C_{21}H_{26}O_5$, a synthetic derivative of the glucocorticoid cortisone that is administered orally to treat inflammation. Also, DELTACORTISONE.

predozzite *Petrology.* a variety of limestone containing abundant periclase and brucite.

preeclampsia *Medicine.* a toxemia of pregnancy generally accompanied by hypertension, proteinuria, and edema, usually occurring during the latter half of pregnancy.

preedit *Computer Programming.* to edit input data prior to processing.

preemphasis *Electronics.* the process of highlighting a particular band of frequencies by amplifying those frequencies with respect to others; used especially to allow higher audio frequencies to override distortion in FM sound broadcasting. Also, ACCENTUATION.

preemphasis network *Electronics.* a circuit added to a system that favors one range of frequencies over another. Also, EMPHASIZER.

preemptive strike *Military Science.* **1.** an attack initiated on the basis of an expectation that an enemy attack is imminent. **2.** an attack intended to prevent a possible enemy attack.Also, **preemptive attack.**

preening *Behavior.* the act of grooming and maintaining plumage, performed by birds.

preexcitation *Cardiology.* the premature excitation of part of the ventricle occurring in Wolff-Parkinson-White Syndrome, caused by cardiac impulses transmitted via a pathway which does not possess the normal physiologic delay of the atrioventricular node. In an electrocardiograph, it is characterized by a short PR interval and a long QRS interval.

pre-exponential factor *Physical Chemistry.* in the Arrhenius equation, the constant factor preceding the exponential that expresses the relationship between absolute temperature and the rate of a chemical reaction. Also, FREQUENCY FACTOR.

preferred datum *Geodesy.* a geodetic datum selected, because of its accuracy or the area it covers, as the basis for combining several adjacent or overlapping data into a single datum. Also, MAJOR DATUM.

preferred orientation *Materials Science.* an alignment of grains, inclusions, or other microstructural features in a particular direction or plane in a material. *Petrology.* specifically, in structural geology, any nonrandom systematic orientation of linear or planar fabric elements, such as that exhibited by a rock with parallel mica plates.

preferred values *Electronics.* a system that standardizes range values for resistors and capacitors and minimizes the number of different sizes that must be kept in stock. Also, **preferred numbers, preferred values of components.**

prefilter *Engineering.* a filter used in the preliminary phase of a filtering process to remove large solids from a liquid before it is put through a separator-filter.

prefiring *Materials Science.* any of several processes, including calcination, prereaction, dead burning, and fusion or melting, that subject raw ceramic materials to thermal transformation to prepare them for further processing.

prefix notation see POLISH NOTATION, def. 2.

preflight check *Transportation Engineering.* a thorough check of an aircraft and its systems, particularly controls, fuel, and aircraft weight, performed before the flight.

preform *Materials Science.* the production of pellets, tablets, or other shapes designed to facilitate filling of cavities with plastic molding powders. *Acoustical Engineering.* a small slab of thermoplastic material, such as a shellac or vinylite, that is placed between two stampers for pressing a record disk. *Archaeology.* a piece of stone that has been prepared for tool-making but not yet developed to its final form.

preformation *Developmental Biology.* a theory of early embryologists stating that a homunculus exists inside an egg or sperm, and development consists of an increase in the size of this miniature adult. An alternate theory to epigenesis.

preforming *Metallurgy.* any preliminary forming operation, such as the initial pressing of metal powder.

prefrontal *Anatomy.* of or relating to the front of the head or anterior to the frontal lobes of the brain. *Vertebrate Zoology.* **1.** of or relating to a bone in the skull of some vertebrates that is anterior and lateral to the frontal bone. **2.** of or relating to a plate or scale in some reptiles and fishes that is in the center or on each side of the head and anterior to the frontal scale.

prefrontal lobotomy *Surgery.* a psychosurgical procedure in which the connecting fibers are cut to separate the frontal lobes from the rest of the brain.

prefrontal squall line *Meteorology.* a squall line or instability line located in the warm air sector of a wave cyclone, usually 50–300 miles in advance of the cold front and moving in nearly the same direction as the cold front.

Pregl, Fritz [prāg´əl] 1869–1930, Austrian chemist; awarded the Nobel Prize for microanalysis of organic compounds.

preglacial *Geology.* **1.** relating to any interval of geologic time that preceded a period of glaciation, specifically the time immediately preceding the Pleistocene epoch. **2.** of or relating to any material that underlies glacial deposits.

Pregl procedure *Analytical Chemistry.* a technique used to analyze very small test samples; it involves the thermal decomposition of the sample and subsequent oxidation of the products for analysis.

pregnancy *Medicine.* **1.** the condition of a human female for the duration of time, approximately 266 days, between conception and childbirth; absolute signs of pregnancy are fetal movements, sounds of the fetal heart, and demonstration of the fetus by X-ray or ultrasound. **2.** this time period itself.

pregnane *Endocrinology.* the fully saturated hydrocarbon ring structure that is the parent compound for the synthesis of the progestins, the glucocorticoids, and the mineralocorticoids.

pregnanediol *Biochemistry.* $C_{21}H_{36}O_2$, a substance, found in the urine, resulting from the breakdown of progesterone.

pregnant *Medicine.* in the condition of pregnancy; expecting a child.

pregnenolone *Biochemistry.* $C_{21}H_{32}O_2$, a compound formed from cholesterol in the steroid hormone biosynthetic pathway. Also, 3-HYDROXYPREGN-5-EN-20-ONE.

preheater *Mechanical Engineering.* an exchanger in which a material, substance, or fluid is heated prior to further thermal or mechanical treatment.

preheat roll *Engineering.* a heated roll, used in plastic-extrusion coating, that fits between the pressure and unwind rolls to heat the substrate prior to coating.

prehensile *Vertebrate Zoology.* describing a vertebrate tail that can be used as a grasping organ by curling ventrally (dorsally in a species of porcupine), as in suspending from tree branches.

pre-Hilbert space *Mathematics.* a linear space endowed with an inner product; that is, a normed vector space. A complete pre-Hilbert space is a Hilbert space.

prehis. prehistoric; prehistory.

prehistoric *Anthropology.* relating to or from the time in the past before written records and other such formal historic data existed; variously defined but generally regarded as any time more than 5000 to 7000 years ago. *Biology.* describing an organism that existed in this time period, especially one that is no longer extant today.

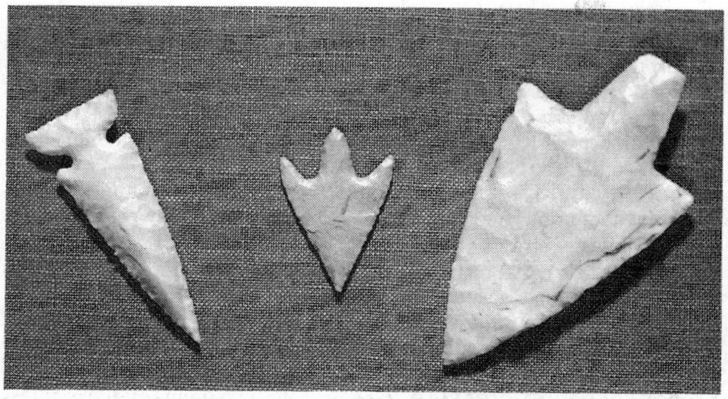

prehistoric arrow points

prehistory *Archaeology.* **1.** the study of cultures before written history, or of more recent cultures lacking formal historical records. **2.** the period of time before written history.

prehnite *Mineralogy.* $Ca_2Al_2Si_3O_{10}(OH)_2$, a green to grayish-white or colorless orthorhombic mineral occurring as granular masses and crystal aggregates, having a specific gravity of 2.90 to 2.95 and a hardness of 6 to 6.5 on the Mohs scale; found in cavities in basic igneous rocks.

prehormone see PROHORMONE.

preignition *Mechanical Engineering.* the ignition of the charge in a gasoline engine cylinder before normal ignition by the spark, caused by overheated plug points.

preimage *Mathematics.* see INVERSE IMAGE.

pre-Imbrian *Astronomy.* the period of lunar geological history before 3.85 billion years ago; the date of the impact that created the Imbrium basin on the moon.

preindexed *Computer Programming.* of or relating to an indirect address prior to adding the contents of a specified index register to determine the effective (postindexed) address.

preinitiation *Ordnance.* the initiation of the fission chain reaction in the active material of a nuclear weapon prior to attaining the designed or maximum compression or degree of assembly.

prejudice *Psychology.* **1.** a persistently held, often negative, judgment about people, objects, concepts, or groups, based on stereotypical presumptions about what is being judged. **2.** any attitude formed without sufficient information or evidence.

prelaunch survivability *Military Science.* the likelihood that a delivery or launch vehicle will survive an enemy attack under an established condition of warning.

prelim or **prelim.** preliminary.

preliminary elevation *Cartography.* in surveying, an elevation determined after field measurements are completed and corrected for variations in index, level, rod, and temperature, but which may be subject to future corrections.

preliminary stage see PRESTAGE.

preliminary waves *Geophysics.* the waves of an earthquake that appear on the seismograph before the surface waves, due to their high speeds in the earth's interior; includes P-waves (first preliminary waves) and S-waves (second preliminary waves).

preliterate *Anthropology.* **1.** occurring at a time before the development of written language. **2.** see NONLITERATE.

Prelog, Vladimir born 1906, Serbian-born Swiss chemist; awarded the Nobel Prize for work in the stereochemistry of organic molecules and reactions.

prelogging *Forestry.* the removal of trees that occurs before a major logging, involving those trees that would be lost, damaged, or cause obstruction during the major logging.

pre-main-sequence star *Astronomy.* a phase in the evolution of a star before its hydrogen ignites, in which the star's energy is generated by gravitational contraction and mass accretion.

premating isolation *Evolution.* a condition in which extrinsic factors, such as geographical separation, effective before mating or fertilization prevent the interbreeding between two or more populations.

premature *Medicine.* **1.** relating to or involved in a premature birth. Thus, **premature infant, premature labor. 2.** describing any event or condition that occurs earlier than is normal or expected.

premature birth *Medicine.* the birth of an infant occurring before the expected time, usually during a period before thirty-seven weeks of gestation.

premature chromatin condensation *Molecular Biology.* the condensation of chromosomes earlier than is usual in an interphase nucleus following introduction of the nucleus into another cell in which the nucleus is entering mitosis.

premature ejaculation *Medicine.* a male psychosexual disorder in which ejaculation occurs shortly after the commencement of sexual intercourse.

premaxilla *Anatomy.* an element derived from the median nasal processes in the embryo that later fuses with the maxilla.

premenstrual syndrome *Medicine.* a syndrome that may be experienced several days before the onset of menstrual flow, characterized by bloating, swelling, emotional changes, headache, food cravings, breast sensitivity, constipation, and fatigue. Also, PMS.

premise *Science.* any statemant or assumption that is established beforehand as the basis for an argument or theory. *Artificial Intelligence.* specifically, the "if" part of a production rule, composed of a Boolean combination of one or more clauses that examine data values or facts.

premix *Engineering.* **1.** the materials completely mixed before molding begins that will eventually form the finished product in a plastics mold. **2.** a premixing of anything, such as concrete

premix gas burner *Engineering.* a fuel burner having a combustion chamber in which the fuel and air are mixed prior to ignition.

premolar *Anatomy.* any one of the eight permanent teeth anterior to the molars and posterior to the canine teeth.

premorse *Botany.* of a plant part, such as certain roots, having an abruptly truncated end as if broken or bitten off.

pre-mRNA see PRECURSOR mRNA.

premunition *Immunology.* a state of protective immunity that is maintained by the persistence of a small number of specific pathogenic organisms in the body.

prenat. prenatal.

prenatal *Biology.* existing or occurring prior to birth. *Medicine.* **1.** describing a condition of the fetus existing prior to birth. **2.** of or relating to care or treatment of the mother or the fetus prior to birth.

prenderol see 2,2-DIETHYL-1,3-PROPANEDIOL.

prenex normal form *Artificial Intelligence.* a representation of formulas in first-order predicate calculus, in which all quantifiers are at the front of the formula.

prenova *Astronomy.* the physical state of an evolved binary star system just before its outburst as a nova.

preoperational stage or **period** *Psychology.* the second of Jean Piaget's four stages of cognitive development, between ages 2 and 7, in which the child develops language skills, begins to think in symbolic or imaginative terms, and can classify objects, but still has difficulty with certain fundamental concepts.

preoperative *Surgery.* preceding a surgical procedure.

preorder *Computer Science.* an order of visiting trees, in which a node is examined first, followed by recursive examination of its children, in left-to-right order, in the same fashion.

preorogenic *Geology.* of or relating to geologic events or processes that took place before a period of intense mountain-building activity, or rocks or features formed before such a period.

preorogenic phase *Geology.* the first phase of an orogenic cycle, during which geosynclines are formed.

preparation fire *Military Science.* fire delivered on a target preparatory to an assault.

prepared conditioning *Behavior.* the belief that certain common fears or aversions stem from evolutionary experience; e.g., threatening features of the prehistoric environment such as snakes, heights, and open spaces are common phobias, but dangerous modern phenomena such as guns and electrical wires are not.

preparedness *Behavior.* a genetically determined capacity or readiness for learning.

preplanned *Military Science.* of or relating to military support that is planned in advance of operations.

preplastication *Engineering.* a process in which injection-molding powders are melted in a separate compartment before they are put in an injection cylinder.

preplumbed system *Electromagnetism.* a system of fixed nontunable waveguides or coaxial transmission lines.

prepolymer *Organic Chemistry.* **1.** a plastic or resin intermediate whose molecular weight is between that of the original monomer and that of the final, cured resin. **2.** a reaction intermediate of a monomeric isocyanate and a polyol.

prepolymer molding *Materials Science.* a process in which polyol reacts with isocyanate to produce a liquid prepolymer that can be pumped and mixed with a second blend containing more polyol, catalyst, or blowing agent to produce urethane foam.

pre-position *Industrial Engineering.* in work motion studies, an elemental motion in which an object is placed in its proper orientation at a specific location.

pre-positioned *Ordnance.* describing a bomb, mine, or similar weapon that has been placed at the desired point of detonation and equipped with a remote or internal actuation mechanism. *Military Science.* of or relating to military units, equipment, and supplies that are placed near the point of planned use to reduce reaction time and ensure timely support of a specific force during the initial phases of an operation.

preposterior analysis *Statistics.* an analysis undertaken before sampling, usually in order to assess the value of experimentation or to choose the best experiment.

prepotential *Cardiology.* a slow depolarization of the cell prior to depolarization spike.

prepreg *Materials Science.* a fibrous composite material having unidirectional fibers embedded in a resin matrix in sheets or strips ready for forming. (Short for pre-impregnation.)

prepress *Graphic Arts.* of or relating to those stages in the production of printed material that precede the actual printing.

prepress proof *Graphic Arts.* any of a variety of simple photographic or photomechanical proofs, such as bluelines, used to check imposition; often serves as an inexpensive substitute for press proofs.

preprocessor *Computer Programming.* **1.** a program designed to prepare or organize data for processing. **2.** a program that scans a program for preprocessor statements that are executed prior to compilation. **3.** a program that converts data to the format required by an emulator. **4.** a precompiler.

preprogrammed robot *Robotics.* any robot that must follow a built-in program without the ability to adapt to changes in its environment.

preprogramming *Computer Programming.* the prerecording of instructions or procedures into a machine, such as a manufacturing robot.

preproinsulin *Endocrinology.* the precursor molecule for insulin as it is released from the ribosome of the pancreatic β-cell, consisting of a single-chain 110-amino-acid polypeptide from which the 23-amino-acid leader sequence is cleaved to form proinsulin.

preproparathyrin *Endocrinology.* the precursor molecule for parathyroid hormone as it is released from the ribosome of the cells of the parathyroid gland, including the leader sequence that is cleaved to form proparathyrin. Also, **preproparathyroid hormone.**

preprosthetic *Surgery.* preceding insertion of a prosthesis.

prepsychotic *Psychology.* relating to or characterized by behavior that leads to the determination that an individual is likely to become psychotic.

prepuce *Anatomy.* a fold of skin, especially those covering the glans penis in the male and the clitoris of the female. Also, FORESKIN.

prepupa *Entomology.* **1.** a quiescent stage between the larval period and the true pupal period. **2.** the final nymphal instar of thrips.

prepupal period *Entomology.* the stage before the pupa in the development of a holometabolous insect; a stage in which feeding usually ceases and sometimes a cocoon is produced.

prereaction *Materials Science.* a process in which premixed ceramic raw materials are heated to a temperature sufficient to allow reaction to take place between phases to prepare the material for further processing.

preread head *Computer Technology.* a read head located adjacent to another read head that reads the data prior to the second read head reading the same data.

prereduced pellets *Materials Science.* iron ore that has been prepared for market by sizing and concentration to yield a pellet containing 87–92% Fe; used in an electric furnace in steel production.

pres. present.

presacral insufflation *Radiology.* a radiographic procedure in which gas, usually carbon dioxide, is injected around the kidneys through a needle inserted into the retrorectal space to provide contrast for imaging of the kidneys, adrenal glands, and neighboring structures.

presaturation dispersal *Ecology.* a condition in which a large number of individuals emigrate from a population that is at low density; usually these individuals tend to be fit and able to survive elsewhere.

presby- a combining form meaning "of or relating to old age."

presbycusis [prez´bi kyoo´sis] *Medicine.* the natural loss of hearing with age, which is normal within certain limits. Also, **presbyacusis.**

presbyophrenia *Psychology.* a form of senile psychosis characterized by severe memory defects but a relatively normal level of behavior and alertness.

presbyopia *Medicine.* the farsightedness that may occur as a consequence of aging, resulting from the loss of elasticity of the crystalline lens.

prescaler *Electronics.* **1.** a circuit that extends the frequency range of an input signal by a fraction. **2.** a circuit that increases the frequency limit of a counter by dividing the input frequency by an exact amount, generally 10 or 100.

prescribed burning *Forestry.* a fire-prevention process in which small fires are set in litter on the forest floor in order to reduce existing fuel and confine fires to a certain area.

prescribed load *Ordnance.* the combination of primer, powder, and bullet recommended to achieve a specific result for a specific weapon.

prescribed nuclear load *Ordnance.* a specified quantity of nuclear weapons to be carried by a delivery unit; it may vary depending on the tactical situation, the nuclear logistical situation, and the capability of the unit to transport and utilize the load.

prescribed nuclear stockage *Ordnance.* a specified quantity of nuclear weapons, components of nuclear weapons, and warhead test equipment to be stocked in special ammunition supply points or other logistical supply points.

prescription *Medicine.* **1.** an order, usually in written form, by a physician to an authorized person to provide a medicine or some other form of medical treatment to a patient named on the order. **2.** the medicine or medical treatment so prescribed.

prescription drug *Medicine.* an official description designating a drug that is available only through an authorized prescription from a physician, as opposed to a drug (**over-the-counter drug**) such as aspirin that may be purchased by the general public without restriction.

prescutum *Invertebrate Zoology.* the anterior of three cuticular plates on the thoracic segments of an insect.

presection suture *Surgery.* a surgical stitch or series of stitches performed on a tissue prior to incision.

preselection *Computer Programming.* a technique in which data is read from the next input tape into a buffer before the current data has been fully processed.

preselector *Electronics.* in a radio or television receiver, a narrow bandpass filter that suppresses undesired frequencies before reaching the main amplification stages. *Telecommunications.* in telephone exchange operations, a device that selects a particular switch automatically before it takes control of an idle trunk.

presence *Acoustical Engineering.* **1.** a sensation that the original sound source of a recorded sample is in the same room as the listener due to the high fidelity of the playback unit. **2.** the prominence of a voice or instrument in a sound recording and reproduction system, achieved by boosting the sound levels in the 2- to 8-kilohertz range.

presenile *Medicine.* relating to or exhibiting the characteristics of premature old age. Thus, **presenility.**

presensitized plate *Graphic Arts.* a printing plate that has been coated with photosensitive emulsion by the manufacturer and thus arrives at the printing plant ready to receive an impression.

presentation *Behavior.* any behavior pattern in which an animal displays a certain body part or feature to another member of the same species. Also, **presenting.** *Medicine.* see LIE. *Electronics.* see RADAR DISPLAY.

presenting part *Medicine.* the portion of the fetus that is touched by the examining finger through the cervix during labor, or the portion of the fetus that is in advance during birth.

present position *Ordnance.* the position of the target at the instant the gun is fired. Similarly, **present elevation, present range.**

preservative *Food Technology.* any chemical additive that prevents or retards spoilage; the most commonly used are sodium benzoate, sulfur dioxide, nitrates, and nitrites.

preserve *Food Technology.* **1.** to prepare and package a food to keep it from spoiling, as by canning or pickling. **2.** also, **preserves.** any food product made from fruit preserved with sugar. **3.** also, **preserves.** a thick fruit syrup containing large pieces of fruit or whole fruits.

preset see INITIALIZE, def. 1.

preset guidance *Engineering.* a guidance system set into a missile before launching that remains unaltered.

preset parameter *Computer Programming.* a parameter whose value is bound when the program is designed, coded, or compiled.

preset tool *Mechanical Engineering.* a tool used to set an initial parameter controlling another device.

presheaf *Mathematics.* a presheaf F over a topological space X is an assignment of a set $F(U)$ to each open set U of X, together with a collection of mappings (restriction homomorphisms) $r_V^U\colon F(U) \to F(V)$ between the sets of F such that: (a) r_U^U is the identity map on U; and (b) if U,V, and W are sets in X such that $U \ldots V \ldots W$ (i.e., U contains V and V contains W), then $r_W^U = r_W^V \circ r_V^U$. The sets $F(U)$ are called sections of F over U. Morphisms can be defined between presheaves on X.

preshrunk *Textiles.* describing a fabric or garment that has had a shrinking process applied to it in a factory to minimize the degree of shrinkage later on.

presintering *Metallurgy.* in powder metallurgy, a preliminary sintering operation performed at a temperature lower than the final sintering temperature.

presoma *Invertebrate Zoology.* the anterior portion of the body of an invertebrate with no clearly defined head.

presort *Computer Programming.* **1.** the first step in some sorts in which the data records are divided into strings that equal or exceed some minimum length. **2.** the off-line sorting of data prior to processing.

direct gravure printing press

press *Graphic Arts.* **1.** a machine used to transfer images from inked type or plates onto paper; a printing press. **2.** a building housing one or more such machines. **3.** a publisher or a printing company. *Mechanical Engineering.* a machine that applies pressure to a workpiece by means of a tool, usually in such cold-working operations as cutting, bending, drawing, and squeezing. *Agriculture.* see PRESS WHEEL.

pressboard *Materials.* a strong, highly glazed board made from organic fibers, such as wood chips, or particles pressed to form a board; used for construction.

press bonding *Engineering.* a process in which a platen press or other device applies pressure to a material or structure to create a bond.

press brake *Materials.* a power press used primarily for producing long bends, as in corrugating or seaming, but also used for embossing, trimming, and punching.

press-brake forming *Materials Science.* a metal-forming process in which relatively long, narrow pieces of sheet metal or plates are bent.

press camera *Optics.* a camera that contains interchangeable lenses and takes sheet films in formats between 2.5 × 3.5 and 4 × 5 inches; formerly in wide use by photojournalists.

pressed brick *Materials Science.* a high-quality brick that is molded under pressure to produce a smooth face; used in exposed surface work.

pressed density *Metallurgy.* in powder metallurgy, the density of a green compact.

pressed glass *Materials.* a molded glass that is shaped or patterned by the action of a plunger thrust into the mold while the glass is molten.

pressed loading *Engineering.* a process in which bulk material is put under pressure to reduce its volume for loading.

press fit *Engineering.* a tight fitting of structural parts made by a press. Also, FORCE FIT.

press forging *Metallurgy.* the process of forging a metallic stock in a press.

pressing a process in which something is pressed into shape; specific uses include: *Metallurgy.* **1.** in powder metallurgy, the forming of a compact by pressure. **2.** in metalworking, the shallow drawing of a flat mill product. *Acoustical Engineering.* the stamping of a heated preform during the manufacture of a record

pressolution see PRESSURE SOLUTION.

pressolved *Geology.* relating to a sedimentary bed or rock in which the constituent grains have been subjected to pressure solution.

pressor *Physiology.* a substance that tends to cause an increase in blood pressure.

press polish *Engineering.* a glossy finish on plastic sheet stock, produced by rubbing it against smooth metal under heat and pressure.

press proof *Graphic Arts.* an actual printed sheet taken from a press after it has run through all the stations; used to check color, tone, and imposition.

press run *Graphic Arts.* the running of a printing press for a specific job, or the quantity that is run.

press slide *Mechanical Engineering.* the reciprocating member of a power press on which the punch and upper die are fastened.

press-to-talk switch *Electronics.* a switch on a microphone that must be pressed when the microphone is to pick up and transmit sounds.

pressure *Physics.* the force that is exerted per unit area. *Mechanics.* **1.** specifically, the condition of a fluid, such as air or water, in which it exerts a perpendicular force per unit area on an area element of a surface in contact with it that has the same value at this point, regardless of the orientation of the surface. **2.** a measurement of this.

pressure altimeter *Meteorology.* an instrument that precisely measures changes in atmospheric pressure along with the altitude changes of an aircraft. Also, BAROMETRIC ALTIMETER.

pressure altitude *Meteorology.* the altitude that corresponds to a given value of atmospheric pressure according to the International Civil Aeronautical Organization standard atmosphere.

pressure angle *Mechanical Engineering.* the angle in a toothed gearing between the common tangent to the pitch circles of two teeth at their point of contact and the common normal at that point.

pressure bar *Mechanical Engineering.* a bar that holds the edge of a metal sheet during such press operations as punching, stamping, or forming.

pressure breccia see TECTONIC BRECCIA.

pressure broadening *Spectroscopy.* collision broadening in which the widening of a spectral line results from an increase in collision rate brought on by an increase in applied pressure.

pressure cable *Electricity.* a paper-insulated power cable that is operated under a hydrostatic pressure greater than atmospheric pressure through the use of a gas (usually nitrogen) contained in an outer steel pipe or an outer reinforced lead sheath.

pressure capacitor *Electricity.* a capacitor whose dielectric is pressurized gas.

pressure carburetor see INJECTION CARBURETOR.

pressure casting *Metallurgy.* **1.** any casting process in which pressure is applied to the molten metal. **2.** the resulting product.

pressure center *Meteorology.* the central point of an atmospheric high or low.

pressure chamber *Engineering.* an enclosure used to perform tests on equipment, in which the pressure of the environment is controlled.

pressure-change chart *Meteorology.* a chart indicating the changes in atmospheric pressure on a constant-height surface over a given period of time.

pressure coefficient *Thermodynamics.* a quantity given by the fractional change in the pressure of a gas sample divided by the fractional change in the temperature under specified conditions, such as constant volume.

pressure cone see SHATTER CONE.

pressure contact *Electricity.* the contact made by pressing together two conducting surfaces, thus allowing electricity to flow in a circuit.

pressure-containing member *Mechanical Engineering.* the part of a pressure-relieving device in direct contact with the pressurized medium in a vessel being protected.

pressure contour *Meteorology.* a line that joins equal points of barometric pressure or the point at which a constant pressure surface meets a plane parallel to sea level.

pressure cooker or **pressurecooker** *Engineering.* an airtight cooking vessel that preserves or cooks foods quickly by means of superheated steam under pressure.

pressure cycle fermenter *Biotechnology.* a highly effective tower fermenter whose contents are circulated through an exterior limb of a tower and reinjected under pressure at the bottom of the tower.

pressure decline *Petroleum Engineering.* a drop in reservoir pressure due to pressure drawdown from gas or oil production. Also, **pressure depletion.**

pressure deflection *Engineering.* the reactive movement of a primary sensing element in a gauge under pressure by a fluid being measured.

pressure diffusion *Materials Science.* a bonding process combining pressure and diffusion at an elevated temperature.

pressure drawdown *Petroleum Engineering.* a drop in reservoir pressure caused by gas or oil production.

pressure drop *Fluid Mechanics.* the difference in pressure between two points in a fluid.

pressure-drop manometer *Engineering.* a U-shaped instrument in which each of its two opened ends connects with a different spot in a flow system to measure a change in system pressure occurring between the two points.

pressure dye test *Engineering.* a test in which a liquid-dye-filled vessel is pressurized under water to reveal any points of leakage.

pressure effect *Spectroscopy.* any effect on a spectral line that results from a change in pressure.

pressure elements *Engineering.* the components of a pressure-measurement gauge directly affected by changes in gas or liquid pressure that cause the indicating element on the gauge to alter its position.
pressure-enthalpy chart *Physics.* a graph of the pressure of a gas as a function of its enthalpy, for various values of various parameters such as temperature, entropy, and specific volume. Also, ENTHALPY-PRESSURE CHART.
pressure-fall center *Meteorology.* a point of maximum decrease in atmospheric pressure over a specified period of time.
pressure filtration *Materials Science.* the process of die pressing a slurry while the liquid is removed by filtration; commonly used in paper manufacture and in special cases for ceramics. Also, WET PRESSING.
pressure flaking *Archaeology.* the technique of producing a tool by pressing a hard object against a stone core or mass to remove flakes.
pressure force *Fluid Mechanics.* the force within a fluid mass arising from a difference in pressure. *Meteorology.* see PRESSURE-GRADIENT FORCE.
pressure forming *Engineering.* a plastics-manufacturing process in which heat and pressure are used to force a plastic sheet against a mold surface.
pressure fringe see PRESSURE SHADOW.
pressure gauge *Engineering.* any instrument that measures the force per unit area exerted by a confined fluid. *Ordnance.* specifically, an instrument that determines the pressure in the bore or chamber of a gun when the charge explodes.
pressure gradient *Fluid Mechanics.* the rate of change of pressure of a fluid over a region of space while measured at a fixed time. *Meteorology.* the rate of atmospheric decrease by units of horizontal distance along the line in which the pressure decreases most rapidly.
pressure-gradient force *Meteorology.* the force due to differences of pressures within the atmosphere. Also, PRESSURE FORCE.
pressure head *Fluid Mechanics.* see HEAD.
pressure hydrophone *Acoustical Engineering.* a hydrophone having an output that corresponds to the instantaneous sound-pressure level of underwater sound that strikes its reactive element.
pressure ice *Oceanography.* pack ice that exhibits differential movement in different parts of the pack, caused by contact with land or by varying atmospheric or ocean conditions, and resulting in shifting and deformation of the ice; the typical forms are hummocks, pressure ridges, rafted ice, ropak, tented ice, and weathered ice.
pressure ice foot *Oceanography.* a narrow strip of ice formed along a polar coast by the freezing together of pressure ice.
pressure ionization *Meteorology.* an atmospheric pressure effect on the molecular or electric field charges of ions in which electrons are gained or lost.
pressure jump *Meteorology.* a sudden, sharp increase in barometric pressure, occurring along an atmospheric front and preceding a storm.
pressure jump line *Meteorology.* a graphical representation of an atmospheric condition in an atmosphere conducive to the formation of thunderstorms, in which a sudden rise in pressure is plotted on a line and is followed by a higher pressure level than that which preceded the jump.
pressure melting *Physics.* the melting of a substance, e.g., ice, by applying pressure to its surface. Also, REGELATION.
pressure melting temperature *Physics.* the temperature at which a substance melts when it is subjected to abnormal pressures.
pressure microphone *Acoustical Engineering.* a microphone whose diaphragm vibrates in accordance with the instantaneous pressure acting upon it.
pressure mine *Ordnance.* a land mine whose fuse responds to the direct pressure of a vehicle or a person's foot, or a naval mine whose fuse responds to the hydrodynamic pressure field caused by a moving ship or submarine.
pressure pad *Engineering.* a steel-reinforced component that fits into the face of a plastic mold to aid in the absorption of the closing pressure. *Acoustical Engineering.* a small piece of cloth or felt in a tape recorder that holds the tape against the heads.
pressure pan *Geology.* a firmly compacted, subsurface soil layer that is higher in bulk density and lower in total porosity than the soil above or below it, and which formed as a result of artificially applied pressure, such as tillage. Also, TRAFFIC PAN.
pressure pattern *Meteorology.* the general geometric characteristics of atmospheric pressure distribution, as plotted by isobars on a constant-height weather chart.
pressure pickup see PRESSURE TRANSDUCER.
pressure plate *Mechanical Engineering.* the plate in an automobile disk clutch that presses against the flywheel.
pressure-plate anemometer *Engineering.* a device that measures wind speed by determining the drag that the wind exerts on a solid body.
pressure plateau *Geology.* a broad, uplifted section of a thick lava flow that has been elevated by the intrusion from below of new lava that does not reach the surface.
pressure point *Physiology.* **1.** a small area of the skin that is very sensitive to pressure. **2.** a place at which a large blood vessel is close to the surface of the skin, so that pressure may be applied to this point to control bleeding from an artery.
pressure process *Chemical Engineering.* a timber treatment that forces a preservative, such as zinc oxide and creosote, into the cells of the wood in order to prevent decay.
pressure radius *Petroleum Engineering.* the effective radius of the zone of higher reservoir pressure encircling a water-injection well.
pressure rating *Engineering.* the allowable internal operating pressure of a tank, vessel, or piping that is used to transport liquids or gases.
pressure-regulating device *Engineering.* a valve that controls the pressure in a process system by adjusting its position between full-open and full-closed.
pressure regulator *Engineering.* a balanced valve that maintains the pressure in a system within an acceptable range.
pressure release *Geophysics.* a release of pent-up energy from within rock formations that, through erosion or other means, have rid themselves of some superincumbant pressure.
pressure-release jointing *Geology.* a process by which the confining pressure on a once deeply buried rock is released when the rock is brought closer to the surface by erosion, resulting in the exfoliation of the rock.
pressure relief *Engineering.* the elimination of excess pressure in a pressurized system by means of a component that automatically allows the escape of liquid or gas from the system.
pressure-relief device *Mechanical Engineering.* a device that relieves pressure that has risen beyond a specified limit and recloses upon return to normal operating conditions.
pressure-relief valve *Mechanical Engineering.* a safety valve used on pressure vessels or pump discharge lines to release pressures that exceed a preset limit. Also, POP-OFF VALVE, SAFETY VALVE.
pressure-retaining member *Mechanical Engineering.* the part of a pressure relieving device loaded by the restrained pressurized fluid.
pressure ridge *Geology.* **1.** a seismic feature produced by transverse shock waves within the earth and the shortening of an area of land. **2.** an elongated uplift of the solidifying crust of a lava flow. *Hydrology.* a ridge of glacier ice formed by horizontal pressure associated with the flow of the glacier.
pressure ring *Mining Engineering.* a ring around a large excavation site showing signs of pressure, such as distortion of openings, shear cracks, and minor slabbing of rocks.
pressure-rise center *Meteorology.* a graphical representation of maximum positive pressure tendency on a synoptic weather chart; a point of maximum increase in atmospheric pressure over a given period of time.
pressure roll *Engineering.* a roll in a plastics-extrusion molding apparatus that acts with the chill roll to put pressure on the substrate and the molten extruded web.
pressure seal *Engineering.* a tight closure which pressure-proofs the contacting surfaces of two parts that move in relation to each other.
pressure-sensitive *Materials.* **1.** describing an adhesive that bonds most efficiently when a light pressure is applied. **2.** describing an item, such as a label, employing this type of adhesive.
pressure shadow *Petrology.* an area adjoining a porphyroblast or detrital grain distinguished from the groundmass, composed of aggregates of new grains growing on opposed sides of the host, thereby creating an elongated structure with a sense of movement and direction. Also, PRESSURE FRINGE, STRAIN SHADOW.
pressure shift *Spectroscopy.* an increase in the wavelength at which the intensity of a spectral line is at a maximum resulting from an increase in pressure.
pressure solution *Petrology.* a solution in a sedimentary rock, occurring preferentially at grain boundary surfaces where the external pressure is greater than the hydraulic pressure of the interstitial fluid. Also, PRESSOLUTION.
pressure-stabilized *Aviation.* of a nonrigid airship or other membranous structure, relying on internal pressure for structural stability.

pressure still *Chemical Engineering.* a petroleum-refinery apparatus used to crack oils by reducing the amount of unsaturated hydrocarbons formed and preventing straight distillation by raising the vaporization point of the oil above the temperature required to crack it.

pressure storage *Engineering.* the trapping and enclosing of a volatile liquid or liquefied gas under pressure to keep it from evaporating.

pressure suit *Aviation.* a garment designed to exert pressure upon the body, thus maintaining normal respiratory and circulatory functions under low-pressure conditions (as in space or at high altitudes) without ambient pressurization. Full-pressure suits covering the entire body are often distinguished from partial pressure suits covering part of the body.

pressure surface see POTENTIOMETRIC SURFACE.

pressure-surface map see POTENTIOMETRIC MAP.

pressure switch *Electricity.* a switch that automatically turns on or off in response to changes of pressure in a system.

pressure system *Meteorology.* a cyclonic scale feature of atmospheric circulation, used to denote a high or a low. *Engineering.* a network of equipment operating at an internal pressure that is higher than the external atmospheric pressure.

pressure tank *Chemical Engineering.* any storage tank that operates at pressures above atmospheric pressure.

pressure tap *Engineering.* a small opening in the side of a pressurized, fluid-filled pipe or vessel to which a pressure-measuring instrument is connected in order to determine static pressure. Also, PIEZOMETER OPENING, STATIC PRESSURE TAP.

pressure technique or **pressure method** *Archaeology.* the use of pressure flaking in the making of stone tools.

pressure tendency *Meteorology.* **1.** the character and amount of atmospheric pressure change for a specified period that ends at the time of observation. Also, BAROMETRIC TENDENCY. **2.** see TENDENCY.

pressure tensor *Physics.* a tensor used commonly in fluid mechanics and magnetohydrodynamics in which the diagonal elements equal the pressures.

pressure thrust *Space Technology.* a measure of rocket thrust, calculated by multiplying the cross-sectional area of the exhaust stream at the nozzle exit by the difference between the exhaust pressure and the ambient pressure.

pressure topography see HEIGHT PATTERN.

pressure transducer *Electronics.* a device that converts the pressure in a gas or a liquid into current or voltage for measurement. Also, PRESSURE PICKUP.

pressure-travel curve *Mechanics.* a plot of the pressure against the travel of a projectile within the bore of a weapon.

pressure treating *Chemical Engineering.* any treating process in which a chemical or chemicals are applied at a pressure above atmospheric pressure.

pressure tube *Hydrology.* a narrow, deep, cylindrical hole formed in a glacier when a stone that has become warmer than the surrounding ice sinks.

pressure-tube anemometer *Engineering.* an instrument that measures wind speed by calculating the pressure difference between wind blowing into a tube and wind blowing across a tube; the resultant pressure difference is proportional to the square of the wind speed.

pressure tube reactor *Nucleonics.* a reactor whose fuel elements are contained inside tubes that also contain coolant at high pressure; the tube assembly is contained in a tank of moderating fluid at low pressure.

pressure tunnel *Civil Engineering.* a tunnel through which water flows under a pressure head; any free water surface would be above the roof line of the tunnel.

pressure vessel *Engineering.* a container for fluids, often of steel or aluminum, that can withstand pressures above or below atmospheric pressures.

pressure viscosity *Fluid Mechanics.* the property of some petroleum-based lubricating oils to show an increase in viscosity when subjected to an increase in pressure.

pressure water reactor *Nucleonics.* a reactor whose uranium fuel elements are cooled and moderated by water under high pressure to keep it from boiling.

pressure wave *Meteorology.* **1.** a short-period oscillation of pressure such as that associated with the transmission of sound through the atmosphere; usually recorded on sensitive microbarographs. **2.** a wave that exists in the variation of atmospheric pressure on any scale, usually excluding normal diurnal and seasonal trends.

pressure welding *Metallurgy.* welding aided by pressure applied to the joining parts.

pressurization *Engineering.* **1.** a process in which inert gas or dry air above the pressure of the atmosphere is put inside a radar system or coaxial line, to prevent erosion caused by moisture and to inhibit high-voltage breakdown at high altitudes. **2.** a process in which near normal atmospheric pressure is achieved in an enclosed space having high or low external pressure, as in an aircraft cabin. **3.** low pressurization that is used for covered athletic arena "domes" and airship bags.

pressurize *Engineering.* **1.** to keep atmospheric pressure at a normal level in an enclosed space that has high or low external pressure. **2.** to apply pressure to a structure. *Aviation.* specifically, to produce and maintain within a flight vehicle an air pressure higher than the ambient atmospheric pressure in order to compensate for low pressures at high altitudes.

pressurized *Aviation.* containing air pressure higher than the ambient atmospheric pressure; subject to pressurization; used to form compounds such as **pressurized cabin, pressurized capsule, pressurized cockpit, pressurized flight,** and so on.

pressurized blast furnace *Engineering.* a smelting furnace operated under pressure to reduce iron ore to pig iron, in which the off-gas line is constricted to allow more air to pass through the furnace at a faster rate, increasing the smelting rate.

pressurized suit *Aviation.* a garment, similar to a pressure suit, that must be inflated before it will exert pressure on the body.

press wheel *Agriculture.* a heavy roller that is dragged behind a plow to compress the seedbed. Also, FURROW PRESS.

presswork *Graphic Arts.* the actual printing of a piece, as distinguished from the other stages in the printing process (composition, makeup, and binding).

prestage *Space Technology.* **1.** the stage at which a partial flow of propellants into the thrust chamber of a liquid rocket is ignited prior to ignition of the full flow. **2.** the flow itself. Also, PRELIMINARY STAGE.

prester *Meteorology.* a whirlwind or waterspout that is accompanied by lightning and occurs in the Mediterranean and Greece.

prestige suggestion *Psychology.* an attempt to influence an individual's point of view by associating a desired response with prestigious persons.

prestore *Computer Programming.* to store data required by a program prior to the execution of the program.

prestress *Engineering.* to apply an internal force, such as tension wire or cable, to a structure so that it will withstand its working load more effectively or with less deflection.

prestressed concrete *Materials Science.* a concrete to which an internal compressive stress has been applied by means of wires or rods, so that a tensile stress equal to the compressive one can be applied in service; the net stress is then zero.

prestrike reconnaissance *Military Science.* reconnaissance undertaken to obtain complete information about known targets for later use by a strike force.

presumptive test *Biotechnology.* a test used to detect fecal contamination of water; the water sample is inoculated into tubes of lactate broth, in which the production of gas indicates that the water has been contaminated.

presuppression *Geophysics.* in seismic prospecting, the suppression of the first events of a seismic shot until the appearance of the first strong arrivals.

presynaptic *Neurology.* situated or occurring before a synapse.

presystolic *Cardiology.* before or at the beginning of the systole.

pretakeoff check *Transportation Engineering.* a three-hour warning given a crew to prepare for takeoff.

pretechnological *Anthropology.* **1.** occurring at a time before the development of basic industrial techniques, such as metalworking for tools and implements and the use of wheeled vehicles and simple machines. **2.** see NONTECHNOLOGICAL.

pretensioning *Engineering.* a process in which steel reinforcements in a concrete beam are subjected to tension before the concrete hardens. Also, HOYER METHOD OF PRESTRESSING.

preterm *Medicine.* **1.** occurring earlier in pregnancy than expected; premature. **2.** a baby born prematurely, especially one that is born before the 37th week of pregnancy and weighs less than $5^1/_2$ lb.

pretrigger *Electronics.* a signal that activates a device before a transmitted signal arrives.

pretty-print *Artificial Intelligence.* to print an expression using indentation to show the nesting of substructures.

pretty-printer *Artificial Intelligence.* a function that will pretty-print a given expression.

prevailing current *Oceanography.* the flow most frequently observed at a place during a specified period of time.
prevailing visibility *Meteorology.* the greatest horizontal visibility that is equaled or exceeded throughout half of the horizon circle; in the case of quickly changing conditions, it is the average of the prevailing visibility while the observation is being taken.
prevailing westerlies *Meteorology.* the dominating west to east motion of the atmosphere on both northern- and southern-hemispheric poleward sides of the subtropic pressure belts; the westerly movement of the general circulation at mid-latitude.
prevailing wind *Meteorology.* the most frequently observed wind direction during a given period, such as a day, month, season, or year. Also, **prevailing wind direction.**
preventive maintenance *Industrial Engineering.* scheduled routine maintenance that is done at regular intervals rather than in response to a malfunction, in order to keep machinery in efficient working order and to slow down its deterioration.
preventive medicine *Medicine.* a general term for the effort to prevent disease from occurring in unaffected individuals, as opposed to the treatment of individuals who are already affected; activities that can be considered part of preventive medicine include immunization, the use of antiseptics, yearly physical examinations, the restriction of hazardous substances in the environment, the monitoring of dietary intake, and counseling programs to promote good health practices.
preventive war *Military Science.* a war initiated in the belief that military conflict, while not imminent, is inevitable, and that to delay would involve greater risk. Similarly, **preventive attack.**
previewing *Computer Programming.* in optical character recognition, the attempt to gain information, such as ink density or relative position, about characters on an incoming source document.
previous element coding *Telecommunications.* in digital television transmission, a method of coding in which each picture element depends upon the similarity of the preceding element.
previtamin H see CAROTENE.
previtrain *Geology.* the woody lenses occurring in lignite that correspond to vitrain in higher-ranked coal.
Prévost, Pierre [prā vō´] 1751–1839, Swiss philosopher and physicist; noted for his work in magnetism and heat; formulated Prevost's theory, a law of exchanges.
Prevost's theory *Thermodynamics.* a theory stating that a body is continuously exchanging heat with its environment, radiating energy which is independent of its environment, and increasing or decreasing its temperature depending upon whether it absorbs more radiation than it emits.
prewashing *Textiles.* a finishing technique in which a fabric or garment is washed to give it an intentionally soft or faded look.
preweld interval *Metallurgy.* in resistance spot welding, the time from the end of pressure application to the current application.
prewhitening filter see WHITENING FILTER.
prewithdrawal demolition target *Ordnance.* a target that is prepared for demolition preliminary to a withdrawal and may be demolished at the discretion of the officer responsible for the demolition operation.
PRF prolactin releasing factor.
prf. proof.
PRG *Aviation.* the airport code for Prague, Czechoslovakia.
Priabonian *Geology.* a European geologic stage of the upper Eocene epoch, thought to consist of the Auversian and the Bartonian.
Priacanthidae *Vertebrate Zoology.* the bigeyes or catalufas, a family of nocturnal, predatory marine fishes of the order Perciformes, characterized by very large eyes and a bright red color; found in tropical and subtropical seas worldwide.
priapism *Medicine.* the persistent abnormal erection of the penis, usually without sexual desire, accompanied by pain and tenderness; it may result from diseases and injuries of the spinal cord, from urinary calculi, or from acute leukemia. (From *Priapus*, the Greek god of male procreative power.)
priapitis *Medicine.* an inflammation of the penis.
Priapulida *Invertebrate Zoology.* a small phylum of marine worms that are cylindrical and ringed, having many spines and wartlike appendages on the proboscis, abdomen, and tail; mud-dwelling predators. Also, **Priapuloidea.**
Pribnow box *Molecular Biology.* in prokaryotic DNA, a sequence of nucleotides having a large number of A-T pairs and a consensus sequence of TATAAT; precedes the transcription start site and functions as the site in the promoter to which a subunit of RNA polymerase binds.
priceite *Mineralogy.* a white triclinic mineral with the approximate formula $Ca_4B_{10}O_{19}\cdot 7H_2O$, occurring as irregular masses and nodules, characterized by an earthy to porcelainous luster, and having a specific gravity of 2.42 and a hardness of 3 to 3.5 on the Mohs scale.
Price-Jones curve *Hematology.* a frequency distribution curve of erythrocyte diameters in a peripheral blood smear, estimated visually with the aid of an optical micrometer.
Price meter *Oceanography.* an instrument used to measure ocean currents, made up of six rotating cone-shaped cups attached to a vertical axis and kept in position by a tail vane and heavy weight; each rotation of the cups transmits a signal through headphones.
prickleback *Vertebrate Zoology.* any of various blennioid fishes of the family Stichaeidae, characterized by spiny rays in the dorsal fin; found in cold waters.
prickly ash *Botany.* a shrub or small tree, *Zanthoxyluum americanum*, of the rue family; it is characterized by aromatic leaves and prickly branches; the bark is used in medicine.
prickly heat see MILIARIA.
prickly pear *Botany.* any of various cactus plants of the genus *Opuntia*, having flattened, spiny stem joints, yellow, orange, or reddish flowers, and pear-shaped, often edible, fruit; or the fruit of this cactus.

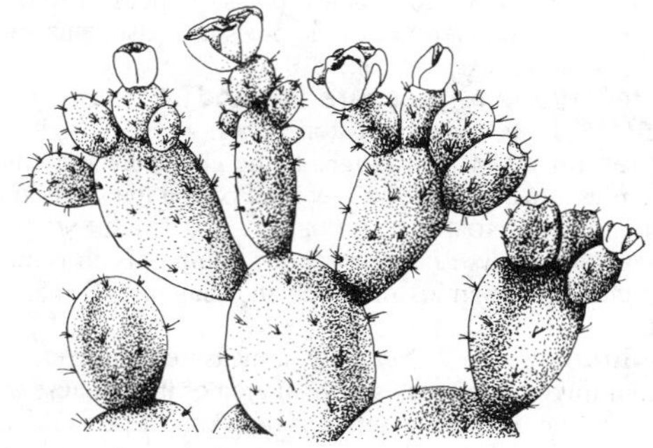

prickly pear

prick punch *Mechanical Devices.* a short, cylindrical steel tool with a long beveled point ground to a 30–60° angle; used to accurately mark a hole centerline of a workpiece before drilling, or to indent the final cutting lines on metalwork. Also, LAYOUT PUNCH.
Priestley, Joseph 1733–1804, English chemist; discovered oxygen (independently of Karl Scheele), ammonia, hydrochloric acid, carbon monoxide, nitrous oxide, and sulfur dioxide.
Prigogine, Ilya born 1917, Belgian chemist; awarded the Nobel Prize for his work on the theory of dissipative structures in nonequilibrium thermodynamics.
primacy effect *Psychology.* the tendency for the first items presented in a series to be better remembered or more influential than those presented later.
primal scream *Psychology.* the cathartic scream that is characteristic of the primal therapy process.
primal therapy *Psychology.* a form of psychotherapy in which the subject relives earlier traumatic experiences and reacts with violent and spontaneous screaming, crying, or the like, in order to release repressed anger or pain.
primaquine *Pharmacology.* $C_{15}H_{21}N_3O$, a drug that is effective against certain types of malaria; used often in the form of the yellow crystals **primaquine phosphate** and sometimes in combination with other antimalarial drugs.
primary being first in time, sequence, or importance; specific uses include: *Chemistry.* of a substance, occurring in the first or simplest form, such as a primary alcohol. *Geology.* relating to a comparatively newly formed coast or shoreline whose features have been produced mainly by nonmarine agencies. *Mineralogy.* **1.** describing the minerals or texture in a rock at the time of formation. **2.** of or relating to a mineral deposit that has not been affected by supergene enrichment. *Astronomy.* **1.** the most massive body in a system of celestial bodies that revolve about a local common center of gravitational forces. **2.** the star that appears most luminous in a multiple or binary system. *Metallurgy.* any metal produced by extraction from a natural source, such as an ore, or from sea water. *Graphic Arts.* any of the four colors that are used in color printing: yellow, magenta (red), cyan (blue), and black.

primary *Mathematics.* **1.** a proper ideal Q of a ring R is said to be a primary ideal if the following conditions hold for any $a,b \in Q$: (a) $ab \in Q$; and (b) if a is not in Q then $b^m \in Q$, for some integer $m > 0$. Every prime ideal is primary. **2.** suppose R is a commutative ring with identity and that A is a proper submodule of an R-module B. A is then said to be a **primary submodule** of B provided that $r \in R$, b is not in A, and $rb \in A$ imply that r^mB is contained in A, for some integer $m > 0$. A primary submodule A of a module B is said to belong to a prime ideal P (where P is a prime ideal of the underlying ring R) or to be a P-primary submodule of B if P is of the form $P = \{r \in R : r^mB$ is contained in A, for some $m > 0\}$.

primary accent *Linguistics.* **1.** in a word with more than one accented syllable, the strongest accent, indicated in pronunciation by a heavier mark. In the word "primate," the first syllable has the primary accent and the second syllable has a secondary accent: [prī´māt´]. Also, PRIMARY STRESS. **2.** the symbol indicating such an accent.

primary air *Mechanical Engineering.* the air that flows up and through the fire beds to supply the oxygen necessary for the combustion of fuel in a furnace.

primary alcohol *Organic Chemistry.* an organic compound in which the $-CH_2OH$ moiety attached either to one alkyl group and two hydrogen atoms or (in methanol) to three hydrogen atoms; distinguished from secondary alcohols, which have two alkyl groups and one hydrogen, and from tertiary alcohols, which have three alkyl groups.

primary amine *Organic Chemistry.* an amine in which only one of the hydrogen atoms of ammonia is replaced by an alkyl group.

primary arc *Geology.* **1.** a curved section of elongated mountain zones that represent major areas of very recent tectonic activity. **2.** see INTERNIDES.

primary basalt *Petrology.* any basaltic rock or magma derived by partial melting in the mantle and not affected by subsequent magmatic differentiation.

primary battery *Electricity.* a battery that consists of primary cells.

primary beam see USEFUL BEAM.

primary bench mark see TIDAL BENCH MARK.

primary body *Astronomy.* see PRIMARY.

primary breaker *Mechanical Engineering.* a machine, such as a gyratory or jaw crusher, that reduces the size of coarse ore from blasting operations. Also, **primary crusher.**

primary calcium phosphate see MONOBASIC CALCIUM PHOSPHATE.

primary carbon atom *Organic Chemistry.* a carbon atom that is singly bonded to only one other carbon atom.

primary care *Medicine.* medical care that is basic to the diagnosis and treatment of a disease or condition, usually the initial step in seeking specialized care.

primary cell *Electricity.* an electrochemical cell that is self-initiating insofar as the chemical energy of its constituents is changed to electrical energy when current is permitted to flow; it cannot be recharged electrically because of the irreversibility of the chemical reaction that occurs within it, but it can be used as a subgrade standard cell for calibration purposes.

primary cell culture *Biotechnology.* a freshly isolated culture of cells that are derived from a particular organ, tissue, or an organism's blood; these cultures usually have a low growth fraction but are representative of the cell types from which they were derived; subculturing of the primary culture creates a secondary culture.

primary circuit *Electricity.* **1.** the circuit that is associated with the primary winding of a transformer. **2.** an input circuit.

primary circulation *Meteorology.* the prevailing atmospheric circulation on a planetary scale that must exist in response to radiation differences with latitude, the rotation of the earth, and the particular distribution of land and oceans; required from the viewpoint of conservation of energy.

primary clay see RESIDUAL CLAY.

primary closure see PRIMARY SUTURE.

primary coil see PRIMARY WINDING.

primary colors *Optics.* **1.** the three colors, red, yellow, and blue, from which all other colors originate. **2.** the three spectral colors, red-orange, green, and blue-violet, which when mixed in various proportions provide an adequate match to all other colors. *Graphic Arts.* the specific values of red, yellow, and blue that are used in color printing.

primary constituent *Materials Science.* a portion of a solid that forms at a temperature above that of an invariant reaction such as a eutectic and is still present in the microstructure after the invariant reaction is completed.

primary consumer *Ecology.* an organism that feeds on plant products and that thus represents the first stage of consumption in a food chain; e.g., rabbits or field mice.

primary control program *Computer Programming.* the operating system program that is responsible for job scheduling and other basic operating system functions.

primary crater *Geology.* **1.** a surface depression formed directly by the high-velocity impact of a meteorite or other projectile. **2.** see TRUE CRATER.

primary creep *Mechanics.* the initial stage in the creep process, characterized by a high and nonconstant strain rate.

primary culture *Cell Biology.* the initial culture of a cell or tissue taken directly from an organism.

primary current flow *Electronics.* a current that produces the electrical activities in a semiconductor device. Also, PRIMARY FLOW.

primary cyclone *Meteorology.* any cyclone, particularly a frontal cyclone, within whose circulation one or more secondary cyclones have developed. Also, PRIMARY LOW.

primary dip *Geology.* the slight dip of a bedded deposit assumed at the time of its deposition. Also, ORIGINAL DIP, DEPOSITIONAL DIP.

primary drilling *Engineering.* a preparatory phase in which holes are drilled in a solid rock ledge before it is blasted.

primary electron *Electronics.* **1.** the electron with the greater energy following the collision of two electrons. **2.** an electron that bombards atoms to produce additional electrons. **3.** an electron that is emitted directly by a material rather than as a result of a collision.

primary emission *Electronics.* the stream of electrons arising from a primary cause, such as the heating of the cathode in an electron tube, and not from a secondary effect, such as a bombardment by other electrons.

primary energy *Engineering.* an energy source occurring in nature, such as coal or solar heat, before it is converted to a usable form.

primary excavation *Civil Engineering.* the breaking up or turning of soil that has not previously been disturbed.

primary extinction *Solid-State Physics.* in X-ray diffraction, a reduction of intensity of the strongest reflections as compared to the weaker reflections when the specimen is a very perfect crystal. *Crystallography.* see EXTINCTION.

primary fabric see APPOSITION FABRIC.

primary fault *Electricity.* the first breakdown that occurs in the insulation of a conductor in an electric circuit; usually followed by a flow of power or current through the system.

primary flake *Archaeology.* a flake of stone from which smaller flakes are removed, either to refine the initial flake or for separate use as tools.

primary flat joint *Geology.* a nearly horizontal joint plane occurring in igneous rocks.

primary flow see PRIMARY CURRENT FLOW.

primary flow structure *Petrology.* parallel features in igneous rocks developed concurrently with the original formation of enclosing rock, resulting from flow effects, such as gas bubbles drawn out in lava or layers of xenoliths parallel to the preferred flow direction.

primary focus *Optics.* the focal line with the greatest image distance in an astigmatic system. Also, MERIDIONAL FOCUS, TANGENTIAL FOCUS.

primary frequency *Telecommunications.* a frequency that is designated for normal usage.

primary front *Meteorology.* the principal and usually original front in a frontal system in which secondary fronts exist.

primary fuel cell *Electricity.* a fuel cell in which the fuel and oxidant burn continuously.

primary geosyncline see ORTHOGEOSYNCLINE.

primary gneiss *Petrology.* a rock exhibiting metamorphic characteristics, such as foliation, lineation, or other planar or linear structures, but considered to be igneous because of a lack of observable granulation or recrystallization.

primary gneissic banding *Petrology.* a banding of igneous rocks of heterogeneous composition, resulting from either admixture of two, partially miscible magmas or intimate admixture of magma with country rock, into which it has been injected along planes of bedding or foliation.

primary great circle *Cartography.* a great circle used as the origin in a system of spherical coordinates, such as the equator or the prime meridian.

primary growth *Botany.* the growth that occurs as a result of cell division at apical meristems and that gives rise to primary tissue; as distinguished from secondary growth.

primary immune response *Immunology.* an immunological reaction to the first encounter with an antigen; it is normally small, but sensitizes or primes the individual.
primary index *Computer Technology.* an index that contains the values of primary keys, in sequential order, and pointers to the corresponding records. Also, GROSS INDEX.
primary institution *Anthropology.* any of the fundamental institutions that identify a culture, such as its subsistence methods, group organization, or marriage customs.
primary instrument *Engineering.* an apparatus from which measurements can be determined without the use of a reference instrument.
primary interstice see ORIGINAL INTERSTICE.
primary key *Computer Technology.* in databases, a key that uniquely defines a record and can be used to locate the record or sort the entire database. Also, MAJOR SORT KEY.
primary knock-on *Materials Science.* a term for the energy distribution that arises when radiation passes through a solid.
primary lateral sclerosis *Medicine.* a degenerative disease of the lateral motor tracts of the spinal cord accompanied by spastic weakness of the extremities.
primary lesion *Medicine.* **1.** the first distinguishable sign of a skin disease, such as a pustule, papule, or vesicle. **2.** a chancre, as in syphilis or tuberculosis.
primary letters *Graphic Arts.* in a given font, those lowercase letters having no ascenders or descenders; in this typeface (Times Roman) they are a, c, e, m, n, o, r, s, u, v, w, x, and z.
primary lights *Optics.* the set of three lights used in the tristimulus colorimetric analysis of solutions.
primary low see PRIMARY CYCLONE.
primary magma *Geology.* molten rock material originating below the crust of the earth.
primary memory see SHORT-TERM MEMORY.
primary meristem *Botany.* the meristem that is derived directly from embryonic tissue and that produces the epidermis, vascular tissue, and cortex.
primary metal *Metallurgy.* metal that is derived directly from ore rather than from scrap.
primary metal production *Materials Science.* any system for producing metals directly from ores.
primary motivation *Psychology.* an unlearned, universal motivation that is derived from basic biological needs, such as hunger or thirst.
primary nondisjunction *Molecular Biology.* nondisjunction of the sex chromosomes in diploid organisms.
primary oocyte *Cell Biology.* one stage in a developing vertebrate egg, derived from an oogonium, which will subsequently yield a secondary oocyte and the first polar body.
primary optical area *Graphic Arts.* the point, usually near the upper left-hand corner, on which a reader's eye tends first to fall on a printed page; an important compositional factor, especially in advertising.
primary optic axis *Optics.* the axis in a crystal along which all light rays travel at the same velocity.
primary orogeny *Geology.* the mountain-building processes involving deformation, regional metamorphism, and granitization, typical of that part of an orogenic belt farthest from the craton.
primary phase *Thermodynamics.* the crystalline phase that is capable of existing in equilibrium with a particular liquid.
primary phloem *Botany.* phloem tissue derived from the apical meristem and consisting of protophloem and metaphloem.
primary photocurrent *Electronics.* a photocurrent arising from nonohmic contacts that are unable to replenish charge carriers passing out of the opposite contact and whose maximum gain is unity.
primary pit field *Botany.* a thin area in a plant cell's primary wall in which one or more pits may develop if a secondary wall is formed
primary porosity *Geology.* the natural porosity in reservoir sediments or rocks that developed during the final stages of sedimentation or that was present in the constituent particles at the time of deposition.
primary power cable *Electricity.* the cable that connects an outside power generating station to a main facility and associated metering devices.
primary-process thinking *Psychology.* in psychoanalytic theory, the irrational and emotionally determined mental processes associated with the id.
primary producer *Ecology.* an organism that is the basic food source in an ecosystem; usually a green plant, which obtains its food from simple substances, such as carbon dioxide and inorganic nitrogen.
primary production *Ecology.* the rate at which self-feeding organisms, such as green plants, store energy as carbohydrates to be consumed by other organisms.
primary productivity *Ecology.* the productive capabilities of self-feeding organisms (primary producers).
primary radar *Engineering.* radar in which the target reflects a portion of the transmitted energy back to the transmitter. Also, PRIMARY SURVEILLANCE RADAR.
primary radiation *Physics.* the radiation that arrives at an observation point directly from its source with no intervening interaction.
primary rainbow *Optics.* the rainbow produced when light undergoes one internal reflection in water droplets; commonly seen after a rain shower.
primary ray *Radiology.* rays emitted directly from a source, such as an X-ray tube or a radioactive substance, without interaction with matter. *Botany.* see MEDULLARY RAY.
primary recovery *Petroleum Engineering.* the production of a reservoir requiring only natural-energy recovery methods, such as the use of a water drive.
primary refuse *Archaeology.* unwanted objects or materials that are located directly at the site where they were used, rather than having been transported from somewhere else.
primary register *Computer Technology.* a general-purpose register that can be utilized directly by computer programs.
primary reinforcer see UNCONDITIONED REINFORCER.
primary relay *Electricity.* the relay that sends the initial impulse in a circuit, which then processes a sequence of operations.
primary rock *Petrology.* rock composed of newly formed particles that have never been constituents of other rocks and are not the result of alteration or replacement, such as limestone precipitated out of a solution.
primary root *Botany.* the first plant root that develops from the embryo itself, and from which secondary and tertiary roots branch out.
primary scattering *Physics.* scattering by which scattered radiation is detected after suffering only one scattering collision.
primary sedimentary structure *Geology.* a sedimentary structure that formed in place at or near the time of deposition, but before the final consolidation of the rock in which it is found.
primary service area *Telecommunications.* a region in which the ground wave of a broadcast station is not hampered by interference or fading.
primary sewage sludge *Biotechnology.* the part of sewage that has been water precipitated and is no longer raw, but is not yet fit to use.
primary sexual characteristic *Anatomy.* any of the body structures directly concerned in reproduction, as the testes, ovaries, and external genitalia. Also, **primary sex character.**
primary skip zone *Electromagnetism.* an area surrounding a radio transmitter that is sufficiently beyond the ground-wave range but not past the skip-distance; radio reception is not reliable in this zone.
primary sodium phosphate see MONOBASIC SODIUM PHOSPHATE.
primary solid solution *Materials Science.* a constituent of alloys that forms when atoms of one element are distributed throughout the lattice of a second element.
primary speciation *Evolution.* the division of a single evolutionary line into two or more separate evolutionary lines.
primary standard *Science.* an agreed-upon standard for a unit of measurement that is maintained in a national laboratory.
primary station see MAIN-SCHEME STATION.
primary storage see INTERNAL STORAGE.
primary stratification *Geology.* the arrangement or disposition of sedimentary strata that developed when the sediments were initially deposited.
primary stratigraphic trap *Geology.* a structure produced by the deposition of clastic materials or by chemical deposition, in which petroleum or natural gas is trapped.
primary stress *Mechanics.* a nonself-limiting stress component due to an applied load in a body under a condition of equilibrium. *Linguistics.* see PRIMARY ACCENT, def. 1.
primary structure *Aviation.* any of the assemblage of structural members or elements that carry most of an aircraft's critical loads, including the wings, fuselage, empennage, landing gear, and associated devices, fittings, and braces. *Space Technology.* on a missile, any structural member or element whose failure would imperil the missile or abort a mission. *Biochemistry.* the first level of organization of a polypeptide or polynucleotide chain that is the amino acid sequence of the protein and the nucleotide sequence of the nucleic acid.

primary succession *Ecology.* **1.** the first succession on a previously uninhabited site, such as terrain laid down by a volcano. **2.** plant succession that takes place when an area is transformed from barren earth or water into a mature and stable plant community.

primary surveillance radar see PRIMARY RADAR.

primary suture *Surgery.* the prompt surgical closure of a wound. Also, PRIMARY CLOSURE.

primary syphilis *Medicine.* the earliest stage of syphilis, from the manifestation of the chancre on the genitalia or the oral cavity to the appearance of the eruption.

primary target *Ordnance.* a target at the highest level of attack priority.

primary tectonite *Petrology.* a tectonite characterized by a depositional fabric.

primary tillage *Agronomy.* the original plowing of a field at the beginning of the growing season. Also, **primary till.**

primary tissue *Botany.* any plant tissue formed during primary growth *Histology.* any of the following types of animal tissue: epithelium, connective tissue, muscle, and nervous tissue.

primary trait or **primary disposition** see CENTRAL TRAIT.

primary transcript *Molecular Biology.* the initial, newly synthesized RNA molecule, prior to excision of the introns.

primary treatment *Civil Engineering.* the key method for pollutant removal from sewage by means of sedimentation.

primary type *Evolution.* the prototypic species from which other species evolve.

primary voltage *Electricity.* the voltage that flows across the primary winding of a transformer. Also, TRANSFORMER INPUT VOLTAGE.

primary wall *Botany.* the older part of a cell wall, usually formed when the cell is expanding and characteristically richer in pectins than the secondary wall.

primary wave *Telecommunications.* a radio wave that moves via a direct path.

primary weapon *Ordnance.* the principal type of weapon used by a combat unit.

primary winding *Electricity.* the input winding of a transformer. Also, PRIMARY COIL.

primary xylem *Botany.* xylem tissue derived from the apical meristem and consisting of protoxylem and metaxylem.

primase *Enzymology.* an enzyme that initiates production of the precursor fragments of primer RNA on the lagging strand during discontinuous DNA replication. The primer RNA is then elongated by DNA polymerase III to produce Okazaki fragments. Also, RNA PRIMASE.

primate (baboon)

primate *Vertebrate Zoology.* a member of the order Primates.

Primates [prī´mə tēz´] *Vertebrate Zoology.* an order of omnivorous mammals, consisting of three suborders: Anthropoidea, including humans, great apes, gibbons, and Old World and New World monkeys; Prosimii, including lemurs, loris, and the like; and Tarsioidea, including tarsiers; members of this order are characterized by highly developed brains, forward directed eyes, use of the hands, and varied locomotion, and particularly by complex, flexible behavior involving a high level of social interaction and adaptability.

primatology *Vertebrate Zoology.* the study of primates, especially of the behavior of nonhuman primates. Thus, **primatologist.**

primavera *Botany.* a tall tree, *Cybistax donnell-smithii,* of the bignonia family, having bright yellow flowers. *Materials.* the hard, light wood of such a tree, used for fine furniture and interior decoration. Also, WHITE MAHOGANY.

prime *Science.* of or relating to a principal object, location, or person. *Engineering.* **1.** to prepare an explosive for firing, as when a detonator is placed in a cartridge or charge. **2.** to lay a preparatory substance, such as a penetrant primer, on a wood surface, as before painting. *Mathematics.* **1.** also, **prime number.** a positive integer greater than 1 that can be evenly divided only by 1 and itself. The primes are 2 (the only even prime), 3, 5, 7, 11, 17, ... **2.** thus, an element p of a commutative ring R with identity is prime if: (a) p is not a unit, and (b) if p is a divisor of ab then p divides either a or b, or both. Every prime element of a ring is also irreducible and, in a unique factorization domain, every irreducible element is prime. For example, an element of a given polynomial ring $R[x]$ is said to be a prime polynomial if it cannot be expressed as the product of two polynomials, neither of which is of degree zero. **3.** a **prime factor** is a factor of a given quantity that is prime. **4.** a proper ideal P in a ring R is a **prime ideal** if for any ideals A and B of R, if AB is contained in P, then either A or B, or both, are contained in P. Every prime ideal is also primary, and if R is a commutative ring with identity, then every maximal ideal is also prime. **5.** the **prime radical** of a ring R is the intersection of all prime ideals of R. If R has no prime ideals, then the prime radical of R is R itself. If the prime radical of R is the zero ideal, then R is said to be **semiprime.** **6.** a ring R is a **prime ring** if the zero ideal is a prime ideal; that is, if I and J are ideals such that $IJ = 0$, then at least one of I or J is the zero ideal. **7.** the intersection P of all subfields of a given field F is itself a field with no proper subfields and is known as the **prime subfield** of F. If the characteristic of F is a prime p, the P is isomorphic to the integers modulo p; if the characteristic of F is zero, then P is isomorphic to the field of rational numbers.

prime coat *Materials.* see PRIMER.

prime contractor *Engineering.* a person or group that makes a direct agreement to perform the work of a project and that subsequently may delegate specific tasks to subcontractors.

primed *Immunology.* describing an individual who has been exposed to a particular antigen and has undergone an initial immune response to it.

primed charge *Ordnance.* a charge ready in all aspects for ignition.

prime fictitious meridian *Navigation.* the meridian used as the reference point for measuring the longitude of a fictitious graticule.

prime field *Mathematics.* a field that contains no proper subset that is itself a field.

prime focus *Optics.* the point on which light is focused to form an image; in a reflecting telescope it is at the open end, along the central axis of the primary mirror.

prime grid meridian *Navigation.* the meridian used as the reference point for measuring the longitude of a grid system.

prime meridian *Cartography.* the meridian used as the origin in a system of grid or spherical coordinates, and designated as 0°; almost universally considered the meridian that passes through the Royal Observatory at Greenwich, England.

prime movement *Astronomy.* an ancient astronomical concept that depicted stars as fastened to an outermost crystal sphere with a rotation from east to west once in 24 hours.

prime mover *Mechanical Engineering.* **1.** an engine or device by which a natural source of energy is converted into mechanical power. **2.** the initial agent that puts a machine in motion, such as electricity.

prime number theorem *Mathematics.* a theorem from analytic number theory concerning the density of the set of prime numbers. Let $\pi(x)$ be the number of primes less than or equal to x. Then $\pi(x) \sim x/\log x$; that is, $\lim_{x \to \infty} \pi(x) \log x/x = 1$. As x approaches infinity, there are approximately $x/\log x$ primes less than or equal to x.

prime oblique meridian *Navigation.* the meridian used as the reference point for measuring longitude on an oblique projection system.

primer [prī´ mər] *Materials.* any paint or similar material that is used to provide an initial coating for a surface in preparation for a finishing coating. Also, PRIME COAT. *Engineering.* a device that ignites an explosive charge and is itself ignited by electricity, friction, or percussion. *Molecular Biology.* **1.** see PRIMER DNA. **2.** see PRIMER RNA.

primer cup *Engineering.* a small, open, metal container used for loading a primer.

primer detonator *Engineering.* an enclosed, metal device that contains a primer, detonator, and at times, an intervening delay charge.

primer DNA *Molecular Biology.* a short single-stranded DNA fragment that is required to initiate polymerization of new DNA nucleotides.

prime register *Computer Technology.* in a central processing unit with duplicate general-purpose registers, one of the inactive registers.

primer pocket *Ordnance.* an opening in the base of a cartridge case or shell that is designed to hold the primer or primer cap.

primer RNA *Molecular Biology.* a piece of RNA that is thought to initiate the formation of Okazaki fragments.

primer seat *Ordnance.* in a gun using separate-loading ammunition, a chamber in the breech mechanism into which the primer is placed.

prime transverse meridian *Navigation.* the meridian that is used as the reference point for measuring longitude on a transverse projection system.

primeval atom [prī mē´vəl] *Astrophysics.* a term coined by cosmologist Georges Lemaître to denote the hypothetical tightly packed mass of original nuclear particles in existence at the beginning of time.

primeval nebula *Astronomy.* a former term for the original nebula of dusty gas out of which the sun and planets condensed; now called the **solar nebula.**

prime vertical *Astronomy.* the great circle passing through the zenith and the west and east points on the horizon.

prime vertical circle *Cartography.* the circle passing through the east and west points on the horizon and passing through the normal at the location of the observer.

prime vertical plane *Cartography.* the plane perpendicular to the meridian plane and containing the normal at the location of the observer; the intersection of the prime vertical plane and the circle of the horizon are the east and west points of the particular system considered.

primi- a combining form meaning "first," as in *primigravida, primiparous.*

primigravida *Medicine.* a woman who is pregnant for the first time.

primine *Botany.* the outer integument of an ovule.

priming *Mechanical Engineering.* **1.** the process of charging a quantity of fuel or fluid to a carburetor, cylinder, or pump, to facilitate starting the vehicle or machine. **2.** the delivery of steam containing water in suspension by a boiler in an ebullition or frothing phase. *Materials.* the process of applying primer to a surface, as before painting it. *Molecular Biology.* the process of initiating nucleic acid polymerization from the 3' end of a strand of DNA.

priming of the tides *Oceanography.* the periodic speeding up of the time of occurrence of high and low water (caused by changes in the relative positions of the moon and sun), resulting in the arrival of high water before the moon reaches its zenith.

priming pump *Mechanical Engineering.* a fuel pump that supplies fuel to an injector or carburetor in starting an engine.

primipara *Medicine.* a woman who has had one pregnancy resulting in a viable infant. Thus, **primiparity, primiparous.**

primite *Invertebrate Zoology.* the anterior member of two gregarine protozoans joined in syzygy.

Primitiopsacea *Paleontology.* a small Paleozoic superfamily of dimorphic ostracods belonging to the extinct order Palaeocopida and suborder Beyrichicopina.

Primitiopsis *Paleontology.* a Paleozoic genus of ostracods in the extinct superfamily Primitiopsacea.

primitive *Anthropology.* a former term used to describe a nonliterate, nontechnological culture or society, such as those that obtain food by hunting and gathering and use tools made from organic materials or stone rather than metal; now generally avoided as negative and judgmental. *Biology.* **1.** not specialized or differentiated; not fully developed. **2.** at an embryonic or other early stage of development. *Evolution.* of or relating to hereditary characters or traits found in ancestral forms of organisms, usually less specialized or complex than those in later related organisms. *Artificial Intelligence.* a basic attribute of a domain; an element of meaning. Also, SEMANTIC PRIMITIVE. *Computer Technology.* **1.** of or relating to a fundamental unit that cannot be divided, such as a single letter, symbol, or program element. **2.** of or relating to a basic operation that cannot be interrupted or divided. *Mathematics.* **1.** a ring R is a left (or right) **primitive ring** if there exists a simple faithful left (or right) R-module. By theorem, a simple ring R with identity is primitive, and a commutative ring R is primitive if and only if R is also a field. **2.** an ideal P of a (not necessarily commutative) ring R is said to be a left (or right) primitive ideal if the quotient ring R/P is a left (or right) primitive ring. By theorem, an ideal P of a ring R is left (or right) primitive if and only if P is the left (or right) annihilator of a simple left (or right) R-module. **3.** a square $n \times n$ matrix A with nonnegative (real) entries is said to be a **primitive matrix** if it is irreducible and has only one eigenvalue of maximum modulus. **4.** an antiderivative of a function.

primitive abstract data type *Computer Technology.* a data type, including the associated operations, that is typically implemented in a high-level language, such as integers, reals, and pointers.

primitive act *Artificial Intelligence.* in conceptual dependency, one of a small number of abstract actions, such as physical transfer, in terms of which natural language verbs are analyzed.

primitive cell *Crystallography.* the unit cell of the smallest possible volume for a given space lattice, as opposed to centered or other nonprimitive cells.

primitive circle *Mathematics.* the stereographic projection of the great circle (of a given projected sphere) whose plane is parallel to the plane of projection. Also, EQUATOR.

primitive culture or **primitive society** *Anthropology.* see PRIMITIVE.

primitive element *Mathematics.* **1.** given an extension field F of a field K, an element u of F such that u and K generate F (so thus F must be a simple extension of K). **2.** the **primitive element theorem** states that: (a) if F is separable over K, then F is a simple extension of K, and (b) more generally, F is a simple extension of K if and only if there are only finitely many intermediate fields (extension fields of K which are proper subfields of F).

primitive equations *Fluid Mechanics.* the Eulerian equations of fluid motion in which the primary dependent variables are the components of the fluid velocity.

primitive gut see ARCHENTERON.

primitive interatrial foramen see FORAMEN PRIMUM.

primitive knot see HENSON'S NODE.

primitive lattice *Crystallography.* a crystal lattice with lattice points only at the origins of the unit cell.

primitive period *Acoustics.* the smallest increment of a sound wave that repeats itself; for a perfect sinusoidal standing wave, it is one wavelength from peak to peak or from valley to valley, and for sound it is measured from compression to compression or from rarefaction to rarefaction. *Mathematics.* **1.** the minimal period of a simply periodic function. **2.** any of the periods a_k of a multiply periodic function f such that any period of f is of the form $\sum n_k a_k$, where the n_k are integers.

primitive polynomial *Mathematics.* **1.** a nonzero polynomial with integer coefficients, such that those coefficients have 1 as their greatest common divisor. **2.** more generally, a nonzero polynomial over a unique factorization domain (e.g., the integers) whose coefficients have a unit (i.e., an element possessing a multiplicative inverse) as their greatest common divisor.

primitive property *Thermodynamics.* a property in a system that can be determined or measured by performing a standardized experiment on the system.

primitive Pythagorean triple *Mathematics.* a Pythagorean triple (a,b,c) such that a, b, and c have no common factor.

primitive root of unity *Mathematics.* **1.** for a positive integer m, a positive integer a such that: (a) a is a root of unity (modulo m), and (b) a belongs to the exponent $\phi(m)$ modulo m, where $\phi(m)$ is the Euler phi function of m. That is, $a^{\phi(m)} \equiv 1(\text{mod } m)$ and $\phi(m)$ is the least positive integer k for which $a^k \equiv 1(\text{mod } m)$. By theorem, the only integers m that have primitive roots are 2, 4, p^n, and $2p^n$, where p is an odd prime and n is any positive integer. **2.** more generally, a is a primitive root of unity in a field F if: (a) a is an nth root of unity for some n, and (b) there is no $k < n$ such that a is a kth root of unity.

primitive streak *Developmental Biology.* a thickening along the midline of some early vertebrate embryos, indicating the prospective of invagination.

primitive subproblem *Artificial Intelligence.* a subproblem that is simple enough to be considered directly solvable or executable.

primitive translation *Crystallography.* a translation operation along any one the three axes of a crystal that leaves the crystal unchanged.

primitive water *Hydrology.* water that has been entrapped in the earth's interior since its formation.

primodium see ANLAGE.

primordial black hole [prī môr´dē əl] *Astrophysics.* a hypothetical black hole created during the highly compressed stages of the universe immediately following the Big Bang.

primordial germ cells *Cell Biology.* cells that are the progenitors of the egg or sperm cells; they migrate to the developing gonads during early embryonic development, undergo meiosis, and differentiate into gametes.

primordial gut see ARCHENTERON.

primordial solar nebula see SOLAR NEBULA.

primordium see ANLAGE.

primosome *Molecular Biology.* a multiprotein complex that binds to DNA and is required for the priming step that occurs prior to synthesis of each Okazaki fragment.

primrose *Botany.* any of various plants belonging to the genus *Primula* of the family Primulaceae; the **common primrose,** *P. vulgaris*, is a popular garden flower.

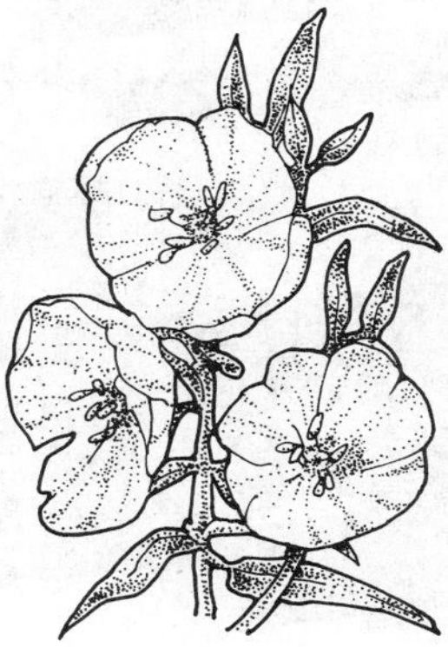

primrose

Primulaceae *Botany.* a family of dicotyledonous herbs and shrubs of the order Primulales, characterized by perfect, mostly hypogynous flowers with stamens opposite the petals, a unilocular ovary, and a dry capsular fruit with several or many seeds; it includes the cyclamen and the primrose.

Primulales *Botany.* a large order of dicotyledonous herbs and woody plants in the subclass Dilleniidae, characterized by sepals joined in a calyx, petals joined in a sympetalous corona, and stamens opposite the corolla lobes.

princess tree see ROBINA.

principal axes of thermal ellipsoids *Crystallography.* three mutually perpendicular directions, along two of which the amplitude of vibration of an atom, represented by an ellipsoid, is at a maximum and at a minimum. Each axis is characterized by an amplitude and a direction.

principal axis *Mechanics.* one of the three axes in a rigid body with regard to which the tensor of inertia is diagonalized. Also, **principal axis of inertia.** *Optics.* the line that passes through a lens surface at the center of curvature. *Acoustical Engineering.* the direction of maximum response of a transducer, expressed in terms of angular coordinates.

principal axis of strain *Mechanics.* an axis normal to one of the mutually perpendicular three planes at a point on which there is no shear strain. For a linear elastic material, it is the same as the principal axis of stress.

principal axis of stress *Mechanics.* an axis normal to one of the mutually perpendicular three planes at a point on which there is no shear stress.

principal branch *Mathematics.* a particular restriction of a multivalued analytic function to a region of the complex plane, with a choice of value (the principal value) at each point so that the resulting single-valued function is continuous.

principal component analysis *Statistics.* a technique for the analysis of a multidimensional data set based on the identification of the orthogonal linear combination of the variables that accounts for the majority of the variability in the data.

principal curvatures *Mathematics.* at a given point P on a surface, the absolute maximum and minimum values of the curvature of the normal sections at P. The reciprocals of the principal curvatures are the **principal radii.**

principal-distance error *Photogrammetry.* an instrument error, especially significant in relief modeling, that results from improper calibration of the aerial camera, diapositive printer, or projector.

principal E plane *Electromagnetism.* a plane that is defined by the electric field vector and the direction of propagation.

principal focus see FOCAL POINT.

principal function *Mechanics.* the time integral of the kinetic energy less the potential energy of a system; it is stationary with respect to small variations of the system's behavior. Also, HAMILTON'S PRINCIPAL FUNCTION.

principal H plane *Electromagnetism.* a plane defined by the magnetic field vector and the direction of propagation; it is perpendicular to the principal E plane.

principal ideal *Mathematics.* an ideal of a ring R that is generated by a single element r of R; denoted (r). A ring R with the property that every ideal is principal is called a **principal ideal ring.** If R is also an integral domain, it is called a **principal ideal domain.**

principal item *Engineering.* a term for an object that is of major interest due to its economic value, complexity, rarity, or strategic importance, and is therefore studied and observed in detail.

principal lobe *Physics.* in a radiation field (commonly from a dipole antenna), the region that carries most of the radiation away from the dipole, found to be at 90° to the dipole axis and cylindrically symmetrical about the axis, having a lobe-shaped cross section in a radial plane containing the dipole axis.

principal meridian *Civil Engineering.* the central strip dividing the two opposing flows of traffic on a freeway. *Cartography.* in surveying, the line extending north and south along the surface of the earth that corresponds to the astronomical meridian, which passes through the initial point of the survey.

principal moment of inertia *Mechanics.* the moment of inertia of a rigid body about a principal axis.

principal moments *Physical Chemistry.* the moments of inertia calculated with repect to the three principal axes of a rigid molecule.

principal normal *Mathematics.* at a point P on a given space curve γ, a unit vector in the osculating plane and perpendicular to γ (i.e., perpendicular to the tangent of γ at P). Also, **principal normal vector.** The principal normal and tangent vectors span the osculating plane of γ at P.

principal plane *Optics.* a plane that passes through the principal points of a lens and is positioned perpendicularly to the optical axis. *Cartography.* in aerial photography, the truly vertical plane that passes through the internal perspective center of an accidentally tilted or deliberately oblique photograph.

principal plane of stress *Mechanics.* one of the three mutually perpendicular planes at a point across which the shearing stress at that point vanishes.

principal point *Cartography.* in aerial photography, the point on the plane of a photograph intersected by a perpendicular line from the interior perspective center. *Optics.* the point at which the principal planes and optical axis intersect on a lens. Also, **principal points of a lens.**

principal-point radial *Cartography.* a line or direction with its origin at a principal point on a photograph.

principal quantum number *Atomic Physics.* a quantum number whose value determines the energy and orbital of an electron in an atom and serves to specify its wave function, derived from Bohr's theory of a one-electron atom

principal rafter see COMMON RAFTER.

principal ray *Optics.* the ray of an oblique narrow beam of light, having a small angle of convergence or divergence, that passes through the center of the pupils.

principal section *Optics.* the plane that contains both the optic axis and a given ray of light in an uniaxial crystal.

principal series *Spectroscopy.* for alkali elements, a set of spectral lines representing a change in total orbital angular momentum accompanying a transition from the p state to the lowest s state (ground state).

principal station see MAIN-SCHEME STATION.

principal strain *Mechanics.* the change in length of a material line segment lying on one of the principal axes of strain with respect to its original length.

principal stress *Mechanics.* the normal stress on a principal plane of stress; the maximum and minimum principal stresses are the algebraically largest and smallest normal stresses to be found at a point.

principal value *Mathematics.* **1.** the value of a multivalued analytic function corresponding to the principal branch of the function; for example, principal values are defined for the arc-trigonometric functions. **2.** a Cauchy principal value.

principal vertical circle *Cartography.* a circle passing through the north and south points of the horizon and including the normal at the point of the observer; it coincides with the celestial meridian.

principal vertical plane *Cartography.* the plane passing through the north and south points on the horizon and containing the normal at the location of the observer; it coincides with the plane of the celestial meridian.

principle *Science.* a law of science that states or explains how a device or a physical process works.

principle of action and reaction *Mechanics.* another term for Newton's third law of motion, which states that for every action there must be an equal and opposite reaction.
principle of constant proportions *Oceanography.* the relationship to one another of the ions of conservative elements, such as sodium, chloride, sulfate, magnesium, calcium, and potassium, whose ratios are almost invariable in open ocean water, regardless of variations in total salinity. Also, MARCET'S PRINCIPLE, LAW or RULE OF CONSTANT PROPORTIONS.
principle of covariance *Physics.* a principle of special relativity stating that the laws of physics have the same mathematical forms in all inertial frames of reference.
principle of duality see DUALITY PRINCIPLE.
principle of equivalence see EQUIVALENCE PRINCIPLE.
principle of insufficient reason *Statistics.* the historically controversial principle stating that if no reason can be found to consider any of a set of outcomes of a random variable more likely to occur than the others, then the outcomes should be assigned equal probability.
principle of least action see MAUPERTUIS' PRINCIPLE.
principle of least constraint see GAUSS' PRINCIPLE.
principle of least energy see LEAST-ENERGY PRINCIPLE.
principle of least time see FERMAT'S PRINCIPLE.
principle of optimality *Control Systems.* the principle stating that in an optimal system, any portion of the optimal state trajectory is optimal between the states that are joined by it.
principle of radial displacement *Photogrammetry.* the principle that points vertically above one another on the ground appear to be displaced radially with respect to one another on a photograph, the center of which is the principal point or nadir.
principle of reciprocity see RECIPROCITY THEOREM.
principle of stationary time see FERMAT'S PRINCIPLE.
principle of superposition *Physics.* a principle applied to linear phenomena, stating that the response to a sum of stimuli equals the sum of the responses to the individual stimuli. Also, SUPERPOSITION PRINCIPLE.
principle of the maximum *Mathematics.* a nonconstant analytic function f defined on a domain D cannot assume its maximum modulus on the interior of D. If D is closed, bounded, and simply connected, the maximum-modulus theorem states that the maximum value of $|f|$ occurs on the boundary of D. Usually called the MAXIMUM PRINCIPLE.
principle of the minimum *Mathematics.* a nonconstant, nonzero analytic function f defined on a domain D cannot assume its minimum modulus on the interior of D. If D is closed, bounded, and simply connected, the maximum-modulus theorem states that the maximum value of $1/|f|$, and hence the minimum value of $|f|$, occurs on the boundary of D. Also, MINIMUM PRINCIPLE.
principle of virtual work *Mechanics.* the principle stating that in order for a condition of equilibrium to exist for a conservative mechanical system or elastic structure, the total virtual work of all external forces acting on it must be zero for any admissible virtual displacement.
Prinsiaceae *Botany.* a family of marine flagellate algae of the order Isochrysidales, characterized by coccolith-bearing cells with a rudimentary haptonema and without microtubules.
print *Graphic Arts.* **1.** to produce text copy, artwork, and so on, on paper or another medium, by the application of inked type, plates, blocks, or the like. **2.** a collective term for the process, product, industry, and craft of producing text and artwork in this manner. **3.** a developed photographic image, usually a positive made from a negative. **4.** a lithographic reproduction of a work of art or photograph.
printability *Graphic Arts.* the degree to which a paper surface will reproduce a desired image; factors include the smoothness, cushion, coating coverage, and ink receptivity of the paper.
print driver *Computer Technology.* a program that controls the operations of a printer, such as buffering the output, printing a line of output, advancing the form, and tabulating.
printed circuit *Electronics.* **1.** a circuit in which components are connected by conductive strips, painted or etched onto an insulating board. **2.** the board itself; a printed circuit board.
printed circuit board *Design Engineering.* a plastic card or rectangular device onto which various chemical elements and substrates are laid so that wiring can be applied; instrinsic to certain electronic and computerized devices including robots, and constituting the control electronics and motor drives in the power control system.
printed-wiring armature *Electricity.* an armature whose conductors consist of etched metal lines on both sides of a thin insulating disk; used with variable high-speed motors and servomotors.
printer *Graphic Arts.* **1.** a person or business whose work is to prepare or produce printed copy. **2.** a machine designed to produce typed copy through electronic impulses, now often generated by a computer or word processor. **3.** any of the four color positives used in four-color printing.

computer printer

printer's error *Graphic Arts.* in proofreading, an error not appearing in the manuscript or previous galleys; distinguished from author's alterations in billing for corrections. Also, PE.
printer's ink see PRINTING INK.
printer's palsy *Toxicology.* poisoning due to exposure to the antimony found in printers' materials; symptoms may include inflammation of the nerves and paralysis.
printer's rule or **ruler** see TYPE GAUGE.
print head *Computer Technology.* the mechanism in a printer that forms a character and prints it on the paper. Also, **printing element.**
printing *Graphic Arts.* any of the various methods of creating printed material, such as letterpress or lithography. *Textiles.* the process of applying color or colors to fabric or yarns in definite patterns, using engraved blocks, rollers, screens, etc.
printing calculator *Computer Technology.* a calculator that displays data on paper rather than, or in conjunction with, a visible display.
printing ink *Materials Science.* a mixture of carbon black or other pigments in a vehicle of mineral oil or linseed oil that flows smoothly, dries quickly, and has a consistency allowing it to hold enough color to make printed matter legible; used to transfer the image on a press plate to the printing surface.
printing press *Graphic Arts.* a usually power-driven machine designed to transfer images from inked type or plates onto paper or other substrate.
printing process *Graphic Arts.* the entire process of producing a printed work; commonly divided into four stages: composition, makeup, presswork, and binding.
printing telegraphy *Telecommunications.* in telegraphy, a system by which received signals are automatically recorded in printed characters.
printmaking *Graphic Arts.* the use of printing methods and technology to produce original works of art or reproductions.
print member *Computer Technology.* the component of a printer that determines the form of a character to be printed, such as a print chain or print wheel.
printout *Computer Technology.* the paper copy of the output from a computer system.
printout paper *Graphic Arts.* the paper, often in continuous folded and perforated sheets, designed to receive the output of a computer or word processor.
print position *Computer Technology.* any location on a form where a character may be printed.
print queue *Computer Technology.* a sequence of files, ordered by arrival time, size, priority number, or some other method, waiting to be printed by the operating system.
print server *Computer Technology.* a computer connected to one or more printers that can receive print jobs from other computers via a network and print them.

print-through *Electronics.* the process by which an audio signal is transferred from one layer of magnetic tape to the next on a tape-recorder reel.

print train *Computer Technology.* **1.** a revolving carrier in a chain printer on which the type slugs are mounted to make impressions on the paper. **2.** the electronic character set in a laser printer that serves a similar function.

print wheel see DAISY WHEEL.

prion *Biochemistry.* an infectious particle that does not contain DNA or RNA, but consists only of a hydrophobic protein; believed to be the tiniest infectious particle.

Prioniodidae *Paleontology.* a former term for a group of conodonts, superseded by the term Prioniodontidae.

Prioniodinidae *Paleontology.* a family of conodonts in the order Prioniodinida, having composite teeth; they arose from the family Chirognathidae at the end of the Ordovician; they were highly diversified and widespread by the Late Devonian, declining in the Permian, and extinct in the Middle Triassic.

Prioniodontidae *Paleontology.* a family of conodonts with composite teeth in the class Conodonti and the order Prioniodontida; known only from the Early Ordovician.

prionodont *Vertebrate Zoology.* having simple, similar teeth in a saw-like row.

Prionodontaceae *Botany.* a small tropical family of robust, light-green, glossy mosses of the order Isobryales that form loose mats on tree bark and moist humus; characterized by creeping stems, ascending branched, secondary stems, lateral sporophytes, and lanceolate leaves that taper to a fine point.

priorite *Mineralogy.* $(Y,Ca,Fe,Th)(Ti,Nb)_2(O,OH)_6$, an orthorhombic, black or dark-brown to yellow mineral massive in habit or occurring as prismatic crystals, having a specific gravity of 4.95 and a hardness of 5 to 6 on the Mohs scale; found in feldspathoidal pegmatites and as a detrital mineral in placers. Also, AESCHYNITE-(Y).

priority the fact of being more important or of taking precedence; specific uses include: *Computer Technology.* a feature that allows different tasks, data elements, and so on, to be ranked in order of importance for processing. *Military Science.* an indication of the relative importance of targets, missions, and other operational tasks; it is not an exclusive or final designation of the order in which these tasks are accomplished.

priority processing *Computer Technology.* **1.** a method used by operating systems in which the order of jobs selected for execution is determined by a system of priorities. **2.** a method of handling interrupts in which an interrupt service routine can be interrupted only by a device of higher priority. The user program has the lowest priority.

priority queue *Computer Technology.* a data structure in which each inserted element has a priority number, and each removal or access takes the earliest of the elements with highest priority.

Prioropidae *Vertebrate Zoology.* an equivalent name for Laniidae, the bird family of shrikes.

prisere *Ecology.* see PRIMARY SUCCESSION, def. 2.

prism [priz´əm] *Optics.* an element bounded by two polished plane surfaces that deviates or disperses a beam of light; commonly a wedge-shaped piece of glass. *Crystallography.* a crystal having faces parallel to the vertical axis and intersecting the horizontal axes. *Geology.* an elongated, narrow, wedge-shaped sedimentary structure for which the ratio of width to thickness is greater than 5:1, but less than 50:1. Also, CRYSTAL FORM. *Mathematics.* a prismatoid with congruent polygons as bases; the lateral faces are all parallelograms. If the lateral faces of a prism are perpendicular to the bases, the prism is a **right prism;** if, in addition, the bases are regular (congruent) polygons, it is a **regular prism.**

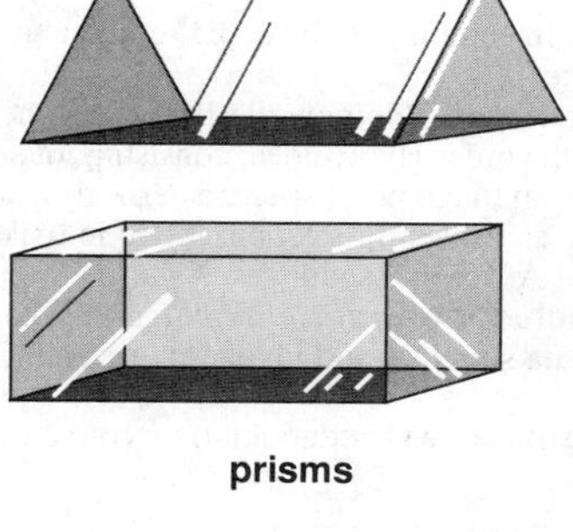

prisms

prismatic [priz mat´ik] *Science.* of or relating to prisms.

prismatic astrolabe *Engineering.* an instrument that determines the exact astronomical position of stars at a designated time, using a pan of mercury to form an artificial horizon and a prism positioned in front of a horizontal telescope.

prismatic coefficient see LONGITUDINAL COEFFICIENT.

prismatic compass *Engineering.* a hand-held surveyor's compass that contains a prism positioned so that the compass can be read while the sight is being taken.

prismatic error *Optics.* a deviation or distortion that results when an optical element, such as a mirror, has unparallel faces. *Navigation.* an error in the reading of a sextant due to lack of parallelism of the two surfaces of an optical element such as a mirror or shade glass.

prismatic joint *Robotics.* a joint consisting of two nested links that slide into or along side each other.

prismatic layer *Invertebrate Zoology.* the middle layer of a mollusk shell, composed of polygonal prisms of calcium carbonate between the periostracum and the nacre.

prismatic plane *Crystallography.* in noncubic crystals, any plane parallel to the principal axis.

prismatic surface *Mathematics.* given a fixed line and a broken line lying in a plane not containing the fixed line, the surface formed by the union of all lines parallel to the fixed line that intersect the broken line.

prismatoid *Mathematics.* a polyhedron with the property that each of its vertices lies on one of two parallel planes, so that it has two parallel faces (bases). Triangles, trapezoids, or parallelograms form the other faces.

prism binoculars *Optics.* binoculars that employ two astronomical telescopes, with total reflecting prisms, to erect images, thus allowing shorter tube lengths than would be found in other telescopes of equal power. Also, **prismatic binoculars.**

prism crack *Geology.* a mud crack forming polygonal patterns on the surface of a drying mud puddle that causes the sediment to break into prisms which stand perpendicular to the bedding.

prism diopter *Optics.* a unit of measure for the deviating power of a prism, which is expressed by multiplying the tangent of the angle of deviation of a ray of light by 100; used primarily for narrow-angle prisms.

prism joint *Robotics.* a robotic articulation in which a sliding motion constitutes the only degree of freedom.

prism level *Engineering.* a surveying level with a prism positioned so that the level bubble can be determined by the surveyor without looking away from the eyepiece.

prismoid *Mathematics.* a prismatoid with an even number of vertices, half on each parallel face (base), and with parallelograms or trapezoids as the other faces. If the bases are congruent, the result is a prism.

prism spectrograph *Spectroscopy.* a spectrograph in which a prism is used to separate incident light into component wavelengths; used in the analysis of materials by flame photometry.

pristane *Organic Chemistry.* $C_{19}H_{40}$, a colorless, combustible liquid derived from shark liver and herring oil; soluble in most organic solvents; boils at 296°C and freezes at –100°C; used as a precision lubricant. Also, 2,6,10,14-TETRAMETHYLPENTADECANE.

Pristidae *Vertebrate Zoology.* the sawfishes, a family of bottom-dwelling rays composing the order Pristiformes, distinguished by a long snout with sharp pointed teeth of uniform length protruding from either side and gills on the ventral surface; found in all tropical seas and in Lake Nicaragua in Central America.

Pristiformes *Vertebrate Zoology.* the sawfishes, an order comprising the family Pristidae.

Pristiophoridae *Vertebrate Zoology.* the sawsharks, a family of marine sharks composing the order Pristiophoriformes, characterized by a long snout with alternating long and short teeth that are sharp and pointed, barbels on the underside of the snout, and small pectoral fins behind gill slits; found off the coasts of Australia, Korea, Japan, and South Africa.

Pristiophoriformes *Vertebrate Zoology.* the sawsharks, the marine fish order comprising the family Pristiophoridae.

privacy system *Telecommunications.* a radio system, used in overseas telephone transmissions, that scrambles an outgoing signal to such a degree that it is undecipherable to a listener who does not have access to an unscrambling device. Also, **privacy equipment, privacy transformation.**

private *Military Science.* the lowest military rank in most armies; in the U.S. Army, an enlisted soldier of one of the three lowest grades; in the U.S. Marine Corps, an enlisted soldier of one of the two lowest grades; **private first class** is the highest grade of private in both services.

private data *Computer Technology.* a set of data open only to a single user or to a single program module.

private exchange *Telecommunications.* a telephone exchange that serves only one organization and does not have the capability to connect to a public telephone system. Thus, **private branch exchange, private automatic exchange.**

private line *Telecommunications.* a point-to-point telephone line that serves only one user. *Computer Technology.* see DEDICATED LINE.

private stream *Hydrology.* a stream that diverts all or part of the drainage of another stream.

privileged directions *Optics.* the two perpendicular directions of vibration of a plane polarized light beam incident on a uniaxial crystal plate that emerges as a single beam of plane polarized light.

privileged instruction *Computer Technology.* a machine instruction that can be executed only when the computer is in a special state, called **privileged mode** or **supervisor mode,** and is usually available to the operating system but not to other users.

PRL prolactin.

p.r.n. or **prn** as needed. (From Latin *pro re nata.*) Also, **PRN.**

pro- a combining form meaning "before" or "in front of."

proactive inhibition *Psychology.* the interference of material learned earlier or the recall of similar material learned later.

proamnion *Developmental Biology.* the part of the embryonic area at the front and side of the head that remains without mesoderm for some time.

Proanura *Paleontology.* a primitive order of amphibians of the Triassic, ancestral to the modern order Anura; significant especially because one of the larger gaps in the fossil record is that between the Early Paleozoic proanurans and the diversified Cenozoic forms; only a handful of later Paleozoic and Mesozoic taxa are known.

prob. probable; problem.

probabilistic automaton *Computer Technology.* a device, with a finite number of internal states, that scans words over a finite alphabet, with the next state determined according to probabilistic functions of the current input word. Also, STOCHASTIC AUTOMATON.

probabilistic fracture mechanics *Materials Science.* a statistical technique for predicting the failure rate of structures in actual use, by calculating failure probabilities as a means of comparing the reliability of the structure against some predetermined limit, indicating the parameters to which the failure probabilities are sensitive, and estimating the gains in reliability that may follow from improvements in design, manufacture, or inspection.

probabilistic sequential machine *Computer Technology.* a probabilistic automaton that can print output words over a finite alphabet, with both its scanning and printing functions determined according to probability. Also, STOCHASTIC SEQUENTIAL MACHINE.

probability *Statistics.* the measure of the uncertainty associated with events of unknown outcome. Formally, the size of a set. Interpretations of probability include physical properties of phenomena, subjective states of uncertainty, and logical relations between sentences.

probability current density *Quantum Mechanics.* a vector whose component normal to a plane gives the probability per unit time that a particle will pass through the plane; the quantum mechanical analogue to current density in which probability is assumed to flow through space time.

probability curve *Statistics.* a curve that describes the distribution of probability over the values of a random variable.

probability density *Quantum Mechanics.* the probability of finding a single particle with a wavefunction ψ at a given location x at a given time t, expressed as $P(x,t) = |\psi(x,t)|^2$.

probability density function *Statistics.* a function that describes the probability distribution of a continuous random variable by representing probabilities of intervals as areas under the density curve; the derivative of the cumulative distribution function. Also, DENSITY FUNCTION.

probability distribution *Statistics.* a distribution of all possible values of a random variable with an indication of their probabilities.

probability forecast *Meteorology.* a weather forecast of the probability of occurrence of one or more of a mutually exclusive set of meteorological contingencies, as distinguished from a series of categorical statements.

probability of damage *Military Science.* the likelihood that damage will occur to a target, expressed as a percentage or a decimal.

probability of detection *Military Science.* the likelihood that a search object will be detected under given conditions, if it is in the area searched.

probability relationships *Crystallography.* equations representing the probability that a phase has a certain value; they form the basis of phase determination in direct methods.

probability space *Mathematics.* a measure space Ω with sigma algebra F and measure P such that $P(\Omega) = 1$. Ω is the sample space, and its elements ω are sample points. The elements A of F are called events; P is a probability measure; and $P(A)$ is called the probability of the event A. Denoted (Ω, F, P).

probability theory *Mathematics.* the area of mathematics that uses the ideas of limits and measure theory to analyze the likelihood of events occurring within a given sample space, given various constraints.

probable *Military Science.* describing a damage assessment of an enemy aircraft seen to break off combat in circumstances that indicate it must be a loss, although it is not actually seen to explode or crash. Thus, **probably destroyed.**

probable error *Statistics.* a quantity equal to 0.6745 times the standard deviation; a normally distributed population would have half of its elements within one probable error of the mean.

probable error deflection *Ordnance.* an error in deflection that is exceeded as often as not. Similarly, **probable error height of burst, probable error range.**

probable maximum precipitation *Meteorology.* a theoretical representation for the greatest depth of precipitation for a given duration that is physically possible over a particular area at a certain time of year.

probable ore *Mining Engineering.* **1.** ore whose existence is reasonably but not absolutely assured. **2.** ore that is partially exposed, usually on two or three sides, but not fully blocked out.

Proboscidea *Vertebrate Zoology.* the elephants, a monofamilial order of large mammals, characterized by a conspicuous long nose that is muscular and flexible, large fan-shaped ears, leathery skin, and large upper incisors growing into tusks; found in Africa and in southern Asia.

proboscis [prō bäs′is; prō bäs′kis] *Vertebrate Zoology.* the trunk of an elephant. *Invertebrate Zoology.* an anterior protuberance, variously modified as, for example, a sucking organ in leeches or a sensory organ in nemertine worms. *Anatomy.* a humorous name for the human nose, especially a large nose.

Proca equations *Quantum Mechanics.* a set of equations, developed by analogy with Maxwell's equations, describing spin one particles having mass, and of which there are four equivalent sets.

procaine *Pharmacology.* $C_{13}H_{20}N_2O_2$, a compound used mainly as a local and spinal anesthetic.

procaine amide *Pharmacology.* $C_{12}H_{21}N_2O_2$, white crystals used in the treatment of cardiac arrhythmias.

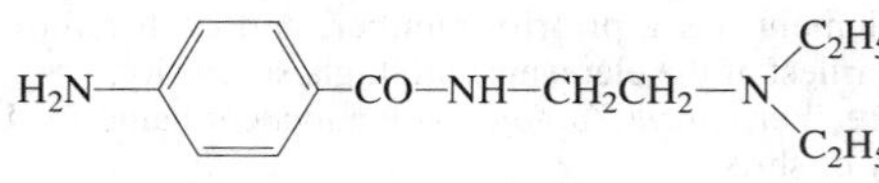

procaine amide

procaine hydrochloride see ETHYL *p*-AMINOBENZOATE HYDROCHLORIDE.

procaine penicillin G *Organic Chemistry.* $C_{16}H_{18}N_2O_4S \cdot C_{13}H_{20}N_2O_2 \cdot H_2O$, stable white crystals that are slightly soluble in alcohol; added to animal feed as an antibiotic.

procambium *Botany.* the undifferentiated tissue of the apical meristem from which primary vascular tissues and cambium develop.

procapsid *Virology.* a capsid precursor that does not contain nucleic acid, formed during virion assembly.

procarbazine hydrochloride *Oncology.* $C_{12}H_{19}N_3O \cdot HCl$, an antineoplastic drug used mainly in the treatment of Hodgkin's disease and other lymphomas.

procarp *Botany.* in certain red algae, the reproductive structure of a female gametophyte.

Procaviidae *Vertebrate Zoology.* the hyraxes, coneys, or dassies, the single family of the order Hyracoidea, consisting of small mammals that resemble rodents and lagomorphs; characterized by short rounded ears, short sturdy legs, and soles for traction; found in rocky areas, scrub, and open grassland of Africa and southwest Asia.

procedural attachment *Artificial Intelligence.* the ability to attach a procedure to a data structure and invoke it, e.g., an if-needed method in a frame system.

procedural language see PROCEDURE-ORIENTED LANGUAGE.

procedural memory *Psychology.* the part of an individual's unconscious memory containing information about how to do something, such as riding a bicycle.

procedural representation *Computer Technology.* the representation of objects by procedures in an appropriate programming language rather than by tables of characters or other static structures.

procedural semantics *Artificial Intelligence.* representation of the meanings of words by procedures that construct the meanings, as distinguished from a declarative, dictionarylike representation.

procedure *Computer Technology.* a sequence of program instructions that collectively accomplish some task, may be referenced at multiple points in a program, and may have one or more input and output parameters; generally used in reference to high-level languages.

procedure division *Computer Technology.* one of the main components of a COBOL program that contains the statements specifying the actions to be carried out in solving a problem.

procedure-oriented language *Computer Technology.* a high-level, machine-independent language that enables the user to describe the steps for the solution of a problem via a set of algorithms or procedures; examples include FORTRAN, COBOL, and PL/1. Also, PROCEDURAL LANGUAGE.

procedure turn *Navigation.* **1.** a prescribed maneuver used when an aircraft must reverse direction to begin its approach. **2.** a maneuver in which an airplane turns away from its track and turns in the opposite direction in order to fly along the reciprocal of its original track.

proceed-to-select signal *Telecommunications.* a signal that confirms the receipt of a call request signal and, in turn, indicates that selection signals may be transmitted.

proceed-to-transmit signal *Telecommunications.* a signal that confirms the receipt of a call request signal, and, in turn, indicates that a data receiver or teleprinting is attached to the circuit, ready to receive the transmission.

Procellarian *Geology.* **1.** referring to the lithologic map units and topographic forms of the lunar surface that are part of or are related closely to the maria. **2.** of or relating to the interval of geologic time during which such features developed.

Procellariidae *Vertebrate Zoology.* the shearwaters, fulmars, and petrels, a family of oceanic birds of the order Procellariiformes, characterized by unusual tubelike nostrils; generally resembling albatrosses and having a large wingspan for gliding; found in oceans worldwide.

Procellariiformes *Vertebrate Zoology.* an order of marine birds including the albatross, petrel, shearwater, and fulmar, characterized by tubelike nostrils on their beaks; found in oceans worldwide.

procephalon [prō sef´ə län] *Invertebrate Zoology.* the anterior part of an insect's head, in front of the segment bearing the mandibles.

proceptive behavior see SOLICITING.

proceratops [prō sâr´ə täps´] *Paleontology.* a hornless, plant-eating ceratopsian dinosaur of the Cretaceous; characterized by a bony frill on the back of the head spreading out over the neck.

proceratops eggs

procercoid *Invertebrate Zoology.* a six-hooked larval stage in some tapeworms that develops in the body cavity of copepods, the usual intermediate host.

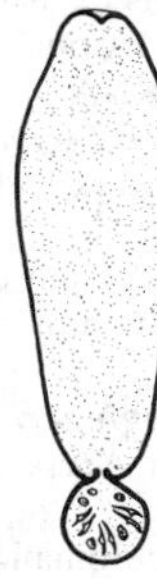

procercoid

process to perform a series of activities, or the series itself; specific uses include: *Computer Technology.* **1.** to perform operations on data in a computer. **2.** in multiprogramming, a program that is in a state of execution or would be executing if all of its required resources were available. **3.** a systematic procedure designed to perform some action. *Engineering.* a continuous or periodic series of actions organized and conducted to achieve an end result, such as a chemical manufacturing process. *Mathematics.* a sequence of random variables defined on a common probability space. Also, STOCHASTIC PROCESS. *Anatomy.* a projection or an outgrowth of a larger structure, particularly of bone.

process analyzer *Chemical Engineering.* an instrument used to monitor and control an industrial process that quantitatively measures properties such as composition, pH, moisture, and temperature.

process annealing *Materials Science.* the process of heating steel sheet or wire between cold-working operations to slightly below the eutectoid temperature and then cooling it; used to improve workability by softening the material.

process-bound see CPU-BOUND.

process box *Robotics.* a method of programming robotic control functions in which there is only one point of entry and one point of exit.

process camera *Optics.* a camera that reproduces, enlarges, or reduces flat copy; used exclusively for optical printing and animation.

process camera see COPY CAMERA.

process chart *Industrial Engineering.* a chart relating the actual movements and operations involved in a specific process.

process color *Graphic Arts.* **1.** a color produced by overprinting two or more color-separated plates. Thus, **process blue, process red, process yellow. 2.** any of the three primaries (yellow, magenta, or cyan) used to produce such a color. **3.** see PROCESS PRINTING.

process control *Computer Technology.* the use of computers to control industrial processes such as continuous-flow chemical processing.

process control chart *Industrial Engineering.* a process chart on which actual progress made toward completing a process can be checked against scheduled goals. Also, GANTT CHART.

process control system *Control Systems.* a system that automatically controls continuous processes such as chemical manufacturing, power generation, or water flow.

process description *Computer Science.* see TASK DESCRIPTOR.

process dynamics *Engineering.* the correspondent responses that occur among the units of a process system.

processed cheese *Food Technology.* a food product made by mixing and melting together various natural cheeses, usually with emulsifiers and preservatives added; the melted mixture is extruded into sheets and sliced or poured into block molds.

processed gene *Genetics.* in a eukaryotic genome, a pseudogene that lacks introns and contains a poly(A) sequence (consisting of several adenine nucleotides) at its 3' end. Also, RETROGENE.

processed meat *Food Technology.* any meat product that is commercially cured and preserved, such as ham, bacon, or salami.

process engineering *Engineering.* the conception, design, development, and implementation of a production process.

process furnace *Chemical Engineering.* an industrial device that conveys heat through circulation of combustion gases; used for processes such as the heat treating of metals, drying, and oxidation.

process heat reactor *Nucleonics.* a nuclear reactor whose generated heat is used for manufacturing and industrial processes.

processing *Computer Science.* a general term for the amending, deleting, or rearranging of data to convert it from one format to another, or to translate it into an intelligible form. *Graphic Arts.* the conversion of a latent photographic image into a visible image. *Molecular Biology.* a series of modification steps, such as excising introns, by which a primary RNA transcript is converted into its final functional product, which is ribosomal, messenger, or transfer RNA.

processing program *Computer Technology.* any application or support program that accomplishes work and that is not involved in control of the system.

processive enzyme *Enzymology.* an enzyme that remains bound to and continues to act on a substrate molecule through several repetitious catalytic events.

process layout *Industrial Engineering.* a physical shop layout in which machines are grouped by the type of function or process that they perform.

process lens *Optics.* **1.** an apochromatic lens used in the production of transmutation products such as plutonium. **2.** the objective lens in a cinematic projector; projects the images recorded on film onto the screen.

process metallurgy *Metallurgy.* the art and science of extracting, separating, and refining metals and alloys.

processor *Computer Technology.* **1.** a device that interprets and executes instructions. **2.** a large program, such as a language translator.

processor complex *Computer Technology.* the central part of a large computer system consisting of multiple computers working together.

processor-limited see CPU-BOUND.

processor status word *Computer Technology.* a machine register that contains information on the current state of the CPU, such as condition codes, instruction address, and processor mode. Also, STATUS WORD.

process printing *Graphic Arts.* full-color reproduction using color separation.

process schizophrenia *Psychology.* a chronic schizophrenic condition that develops gradually, appears to be the result of organic rather than environmental causes, and has a low likelihood of recovery.

process simulation *Computer Technology.* the use of programs to simulate manufacturing or other processes.

process time *Industrial Engineering.* **1.** the whole elapsed time that a given production process is determined to take; thus the minimum time that can be scheduled for it. **2.** the time required for completion of the machine portion of a work cycle.

processus maxillaris see MAXILLARY PROCESS.

process variable *Chemical Engineering.* any physical parameter in an industrial process that is changing or being changed over time.

prochiral *Organic Chemistry.* of or relating to the site in a molecule attached to two constitutionally identical ligands.

prochirality *Organic Chemistry.* the property exhibited by a prochiral molecule of having at least one pair of features that can be distinguished only by reference to a chiral object.

prochiral molecule *Organic Chemistry.* an organic compound that lacks chiral centers but contains one or more carbon atoms to which are attached two identical and two different substituents.

Prochloron *Invertebrate Zoology.* a genus of symbiotic algae found only in the tunic or cloacal region of colonial tunicates of the family Didemnidae.

procoagulant *Biochemistry.* any agent that functions as precursor to an active blood coagulation factor, such as prothrombin.

Procoela *Vertebrate Zoology.* the toads, tree frogs, and related species, an amphibian suborder of higher anurans, characterized by vertebrae that are concave in front and by sacral vertebrae joined to the urostyle by a double condyle.

procoelous *Vertebrate Zoology.* belonging or relating to a kind of ball-and-socket vertebral arrangement common in certain reptiles and salamanders, in which the anterior end of the centrum is concave and the posterior end is convex, so that the posterior bulge of each vertebral element fits into the hollow of the anterior portion of the vertebra behind it.

procollagen *Biochemistry.* the precursor molecule of collagen, synthesized in the fibroblast, osteoblast, and so on, and cleaved to form collagen extracellularly.

Procolophonia *Paleontology.* an extinct suborder of small, short-jawed cotylosaurian reptiles in the subclass Anapsida; includes the superfamily Procolophonoidea; extant in the Permian and Triassic.

Proconodontidae *Paleontology.* a family of conodonts in the class Cavidonti and the order Proconodontida; extant in the Upper Cambrian and Ordovician.

Proconodontus *Paleontology.* a genus of conodonts in the order Proconodontida and family Proconodontidae; the oldest known conodont; ancestral to the Belodellida, the second order in the class Cavidonti; extant in the Upper Cambrian and Ordovician.

Proconsul *Paleontology.* a genus of primates known only in East Africa in the Miocene and related to the dryopithecines; Proconsul was a sexually dimorphic, fruit-eating, arboreal quadruped about the size of a modern baboon; probably ancestral to the major groups of living hominoids.

proct- or **procto-** a combining form meaning "anus," "rectum."

proctatresia see IMPERFORATE ANUS.

proctiger *Invertebrate Zoology.* in insects, the reduced final abdominal segment bearing the anus.

proctitis *Medicine.* an inflammation of the rectum.

proctocele see RECTOCELE.

proctodaeum *Developmental Biology.* a posterior invagination into the embryonic ectoderm that is the progenitor of the anus. Also, **proctodeum.**

proctologist *Medicine.* a person who specializes in or practices proctology.

proctology *Medicine.* the branch of medical science that deals with the study of the anatomy, functions, and diseases of the rectum and anus.

proctopexy *Surgery.* the surgical fixation of the rectum to some other part.

proctoplasty *Surgery.* plastic surgery of the rectum. Also, RECTOPLASTY.

proctor compaction test *Agronomy.* a test to determine soil density.

proctorrhaphy *Surgery.* the surgical repair of the rectum.

proctoscope *Medicine.* an instrument used to examine the rectum and anal canal. Also, RECTOSCOPE.

proctosigmoidectomy *Surgery.* the surgical removal of the rectum and sigmoid colon, generally accompanied by the formation of an abdominal colostomy.

Proctotrupidae *Invertebrate Zoology.* a family of wasps, hymenopteran insects in the superfamily Proctotrupoidea, parasites of the larvae of beetles and dipteran flies.

Proctotrupoidea *Invertebrate Zoology.* a superfamily of black and brown wasps, hymenopteran insects in the suborder Apocrita, endoparasites of insect, spider, and centipede larvae.

procumbent *Botany.* of or related to a plant or stem that is lying loosely on the ground but not sending down roots.

procurement *Military Science.* the process of obtaining personnel, services, supplies, and equipment. *Archaeology.* any process by which materials are obtained from the environment, such as the collecting of plant foods.

procuticle *Invertebrate Zoology.* in arthropods, a soft, pliant material that is secreted by the epidermis after molting and hardens into the sclerites of the cuticle.

Procyon [prō´ sē än´] *Astronomy.* Alpha (α) Canis Minoris, the 1st-magnitude star in the Lesser Dog.

Procyonidae *Vertebrate Zoology.* the raccoons, red panda, coati, and their allies, a family of carnivorous mammals characterized by a short, broad face, rounded or pointed ears, and a usually banded tail; found in temperate and tropical parts of North and South America and the Himalayas of east Asia.

prod cast *Geology.* the surface impression of a prod mark that appears as a short ridge rising downcurrent and ending abruptly. Also, IMPACT CAST.

prodelta *Geology.* the gently sloping area of a delta beyond the delta front that lies completely underwater.

prodelta clay *Geology.* **1.** very fine sand, silt, or clay carried by a river and deposited on the floor of a sea or lake beyond the main body of the delta. **2.** the material from which bottomset beds develop.

Prodinoceras *Paleontology.* one of the earliest uintatheres, a genus of large ungulates in the order Dinocerata and the family Uintatheriidae; known only from the Upper Paleocene of Mongolia.

prod mark *Geology.* a short tool mark running parallel to the current that was produced by an object which plowed into the bottom of a stream and was then lifted up. Also, IMPACT MARK, SLICKENSLIDE.

prodrome *Medicine.* an early or premonitory sign of a disease, even before the usual symptoms appear.

produce [prō´doos´] *Agriculture.* agricultural products such as fruits or vegetables, as distinguished from grains and staple crops.

producer *Chemistry.* an apparatus that produces fuel gas by passing air over hot coke. *Ecology.* see PRIMARY PRODUCER.

producer gas *Materials.* a highly flammable, toxic gas obtained by burning coal or coke with a restricted supply of air, or by passing air and steam through a bed of incandescent fuel so that the carbon dioxide formed is converted into carbon monoxide; a cheap gas with a low caloric content that is used where transportation is not required.

producer's risk *Industrial Engineering.* in quality control, the statistical likelihood that a given sampling process will lead to rejection of a lot that actually should meet acceptance standards.

producing reserves *Petroleum Engineering.* developed petroleum reserves that are to be recovered from wells in the area of a reservoir in which full secondary-recovery operations are being undertaken.

product anything that is produced; specific uses include: *Chemistry.* a compound formed by a chemical reaction. *Mathematics.* **1.** the result obtained when one quantity is multiplied by another or others; e.g., in the equation $ab = c$, c is the product of the factors a and b. **2.** in general, the product of two elements of an algebraic object (such as a ring, field, etc.) is the multiplication of those elements relative to the operation in the object analogous to multiplication of real numbers. **3.** see CARTESIAN PRODUCT. **4.** see INNER PRODUCT. **5.** see DIRECT PRODUCT.

product detector *Electronics.* a device that detects the modulation of a carrier by analog multiplication of the incoming signal and a locally generated signal. Also, **product demodulator.**

Productidina *Paleontology.* a suborder of articulate brachiopods in the extinct order Strophomenida; spines protrude from the shell in most families; extant from the Early Devonian to the end of the Permian.

production *Engineering.* the total output resulting from a manufacturing process. *Computer Technology.* **1.** the completion of useful work by a computer system, not including developing and testing new programs. **2.** a rule in a formal language grammar that describes how parts of a string may be replaced by other strings. **3.** a production rule. *Ecology.* the assimilation of nutrients into biomass and energy.

production capacity *Design Engineering.* the efficiency peak of product output rate without changing specifications for a given product element type, plant facility, and equipment.

production cell *Robotics.* a flexible unit that takes in raw material and puts out a finished product.

production control *Industrial Engineering.* the management of the production process from the time an order is received until its disposition is complete, to ensure that goods will be produced on time and at the lowest possible cost.

production engineering *Industrial Engineering.* the branch of industrial engineering that deals with the design and operation of productive processes and production facilities.

production memory *Artificial Intelligence.* in a production system, the memory that represents the rules used in analyzing and responding to a problem; analogous to human long-term memory.

production model *Industrial Engineering.* a standard model of a product, assembled or manufactured under normal factory conditions.

production program *Computer Technology.* **1.** an application program that solves problems or produces similar useful results; as contrasted with programs that are designed for support or housekeeping, or programs still under development. **2.** a proprietary program intended for internal use in a business.

production rate *Design Engineering.* an accounting of the quantity of assemblies, subassemblies, or parts produced over a given period of time, such as a month or a year.

production reactor *Nucleonics.* a nuclear reactor that is used to produce transmutation products such as plutonium.

production rule *Artificial Intelligence.* a method of representing knowledge by stating a condition and an action; an "if <condition> then <action>" or "if <condition> then <conclusion>" statement.

production rule shell see RULE-BASED SHELL.

production system *Industrial Engineering.* a general term for any system designed to produce products or goods for consumption. *Computer Technology.* a fully tested and proven computer system that is operational and ready for productive work. *Artificial Intelligence.* a forward-chaining, rule-based system, in which the contents of a working memory are compared to a set of rules, which are identified, and one or more selected to be fired or executed.

production tape *Acoustical Engineering.* a tape that is released for commercial use after all processing in the studio has been completed, including recording, editing, overdubbing, and splicing. Also, **production master.**

productive infection *Virology.* the formation of complete particles in a cell during viral infection.

productive time *Industrial Engineering.* the part of a worker's time that is spent in advancing a product toward completion, as opposed to personal time or maintenance time.

productivity the fact of producing or being able to produce; specific uses include: *Petroleum Engineering.* the capacity of a well to produce liquid or gaseous hydrocarbons. *Industrial Engineering.* **1.** the relative output of work by workers or machines, typically measured in output per man-hour or machine-hour. **2.** the output of goods and services of an organization relative to its inputs (labor, capital, materials, and so on).

product layout *Industrial Engineering.* a work flow layout in which machines are grouped according to the sequence of steps in the production of a specific product.

product measure *Mathematics.* let $M = (X,A,m)$ and $M' = (X',A',m')$ be two sigma finite measure spaces. The sigma field $A \otimes A'$ (of subsets of the direct product $X \otimes X'$) is the smallest sigma field that contains all sets of the form $A \times A'$, where $A \in A$ and $A' \in A'$. A unique measure $m \otimes m'$ exists on $A \otimes A'$ such that, for all $A \in A$ and $A' \in A'$, $(m \otimes m')(A \times A') = m(A)m'(A')$. The measure space $M \times M' = (X \times X', A \otimes A', m \otimes m')$ is called the **product measure space** of X and X', with product measure $m \otimes m'$.

product model see MULTIPLICATIVE MODEL.

product-moment coefficient see COEFFICIENT OF CORRELATION.

product-moment correlation *Statistics.* see CORRELATION.

product of inertia *Mechanics.* relative to two orthogonal axes through a body, the integral of the product of these two coordinates of a differential quantity (such as a differential area or a differential mass) over the differential quantity.

product recovery membrane fermentation *Biotechnology.* an experimental fermentation process that allows continuous ethanol recovery from a fermenter broth; ethanol diffuses across a membrane and is carried away while the sugar substrate is retained until completely utilized.

product topology *Mathematics.* let $X_1, \ldots, X_n$ be a finite collection of topological spaces with the topologies $U_1, \ldots, U_n$, respectively. A base for product topology on the Cartesian product $\mathrm{X}_{k=1}^{n} X_k$ is the set of n-tuples $\{(U_1, \ldots, U_n) : U_k \in U_k \text{ for } k = 1, \ldots, n\}$; that is, the open sets of the product topology are constructed from the Cartesian products of the open sets of the original topologies. The product topology is the coarsest topology for which the projection maps are continuous.

Productus *Paleontology.* a genus of medium-sized articulate brachiopods in the extinct suborder Productidina, about 4 cm across; known only from the Carboniferous of Eurasia.

product water *Chemical Engineering.* the effluent water stream from a water purification operation.

proenkephalin *Endocrinology.* a polypeptide precursor of enkephalin that is cleaved to release six copies of met-enkephalin and a single copy of leu-enkephalin .

proenzyme see ZYMOGEN.

proestrus *Physiology.* the period of added follicular activity prior to estrus in mammals.

proetid *Paleontology.* a trilobite of the order Proetida; the proetids were the only trilobites to survive the Devonian.

Proetida *Paleontology.* an order of isopygous trilobites characterized by a large and vaulted glabella and a narrow rostral plate; extant in the Ordovician to Permian.

proeutectic constituent *Materials Science.* the primary constituent in a eutectic system.

proeutectoid constituent *Materials Science.* the primary constituent in a eutectoid system.

profile an outline or contour, or a representation of this; specific uses include: *Archaeology.* **1.** a vertical wall or face of an excavation pit. **2.** a record or representation of this, including color, soil type, features, and content. Also, SECTION. *Cartography.* in surveying, a cross section of the surface of the earth, and frequently the underlying strata, taken along a given line. *Geology.* **1.** the configuration produced when the plane of a vertical section intersects the ground surface. **2.** see PROFILE SECTION. **3.** in seismic prospecting, the information obtained from a shot point by one line of receivers. *Geophysics.* a record of the variations in a property or of a series of events represented graphically in which the property is usually the ordinate and compared with another property such as time or distance. *Hydrology.* a vertical section of a body of surface water or of any potentiometric surface, such as a water table. *Petrology.* in structural petrology, a cross section of a homoaxial region of cylindrical folds drawn perpendicular to the fold axes. Also, TECTONIC PROFILE. *Aviation.* see AIRFOIL PROFILE.

profile drag *Aviation.* the part of wing drag resulting from the shape of the airfoil and surface friction.

profile grinding see FORM GRINDING.

profile line *Geology.* the upper line on a diagram showing the intersection of a vertical plane with the ground surface.

profile of equilibrium *Geology.* the slope of the floor of a body of water, taken in a vertical plane, having a gradient that is sufficient to maintain a balance between deposition of sediment and erosion by waves and currents. *Hydrology.* the longitudinal profile of a graded stream or a stream having a gradient that is sufficient to maintain a balance between the amount of material the stream can transport and the amount of material supplied. Also, EQUILIBRIUM PROFILE, GRADED PROFILE.

profile refinement *Crystallography.* refinement of an equation representing the profile of a powder diffraction pattern with that experimentally observed. The equation takes into account the positional and vibration parameters of each atom in the unit cell and these parameters are refined by a least-squares procedure. Also, RIETVELD PROFILE REFINEMENT, RIETVELD ANALYSIS.

profile section *Geology.* a graphic representation showing the slope and configuration of the ground surface along a given line as it would appear if cut by a vertical plane. Also, PROFILE.

profile thickness *Aviation.* the greatest distance between the upper and lower contours of an airfoil profile, measured perpendicular to the mean line of the profile; the maximum thickness of a wing, or other surface.

profilin *Biochemistry.* an actin-binding protein that forms a complex with G-actin and renders it incompetent to nucleate F-actin formation.

profiling *Engineering.* **1.** the action of shaping an outline of a mold or other object with the use of a cutting tool. **2.** any operation that produces an irregular contour on a workpiece, usually by means of a tracer or template-controlled duplication equipment. **3.** an automated contouring using computer and sensor and numerical control manufacturing.

profiling machine *Mechanical Engineering.* a modified milling machine using a tracer or template-controlled duplicating equipment to produce an irregular contour on a workpiece.

profiling snow gauge *Hydrology.* a radioactive gauge used for measuring the water equivalent and density-depth distribution of a snow cover. Also, NUCLEAR TWIN-PROBE GAUGE.

profilograph *Engineering.* a device that moves over a surface, creating a graphic recording of the surface texture.

profilometer *Engineering.* an electrical instrument that measures the roughness of a surface, by means of a coil in an electrical field attached to an arm that passes over the surface and transmits a current proportional to the roughness of the surface to the indicating element.

proflavine sulfate *Organic Chemistry.* $C_{13}H_{11}N_3 \cdot H_2SO_4$, red-brown crystals, soluble (with fluorescence) in water; used as an antiseptic.

profound mental retardation *Psychology.* the most serious level of mental retardation, characterized by an IQ under 25; those affected have very limited motor coordination and speech ability and must be aided and supervised in all activities throughout their lives.

profunda *Anatomy.* relating to any deeply lying veins or arteries, such as the arteries of the upper arm.

profundal zone *Ecology.* the section of a freshwater ecosystem where little sunlight penetrates. Generally it is found in deep-water systems such as lakes; in ponds and other shallow systems it is usually absent.

progametangium *Mycology.* an extension branch of fungi of the order Mucorales that produces the sex cells, gametangium, and a suspensor cell (usually a zygospore).

Proganosaurus *Paleontology.* a genus of anapsid reptiles in the order Chelonia and family Chelonidae; extant in the Eocene.

progastrin *Endocrinology.* the 34-amino acid polypeptide precursor of gastrin that is cleaved to form the mature 17-amino-acid polypeptide. Also, BIG GASTRIN.

progeny [präj´ə nē] *Biology.* the offspring of a plant or animal; descendants.

progeny testing *Genetics.* a determination of the genotype of a parent through controlled observations of the progeny.

progeria *Medicine.* a condition of premature senility or aging in young children who exhibit the physical characteristics normally associated with the aged. Also, HUTCHINSON-GILFORD DISEASE.

progestational [prō´jes tā´shə nəl] *Medicine.* **1.** describing a phase of the menstrual cycle, just prior to menstruation, that is conducive to the implantation and growth of the fertilized ovum. **2.** referring to progesterone or to a pharmaceutical preparation with similar properties.

progestational hormone *Endocrinology.* one of the hormones, especially progesterone, that support the induction and maintenance of the secretory phase of the uterine endometrium during the menstrual cycle.

progesterone [prō jes´tə rōn] *Endocrinology.* $C_{21}H_{30}O_2$, the steroid hormone that is secreted by the corpus luteum of the ovary and stimulates changes in the endometrium of the uterus in preparation for the implantation of the fertilized egg; produced by the placenta in large quantities during pregnancy.

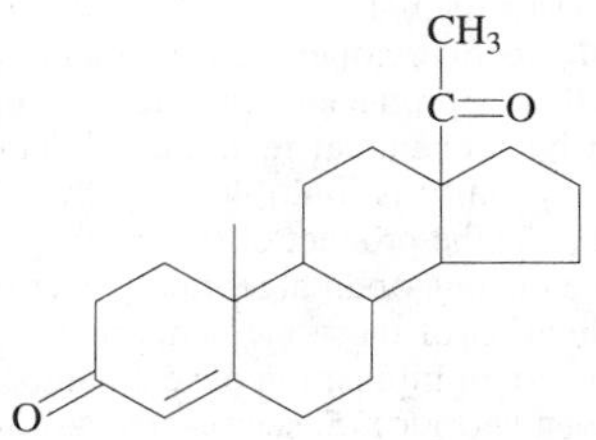

progesterone

progestin *Endocrinology.* any of the steroid hormones, especially progesterone, that prepare the uterus for implantation and pregnancy. Also, **progestogen.**

proglacial *Hydrology.* referring to the area directly in front of or just beyond the outer margin of a glacier or ice sheet, and the deposits or other features formed by or derived from the glacier ice.

proglottid *Invertebrate Zoology.* one of the body segments of an adult tapeworm; each contains hermaphroditic reproductive organs and breaks off, when full of eggs, to be excreted from the host.

proglucagon *Endocrinology.* the precursor polypeptide of glucagon that is cleaved to form the mature hormone.

prognathism [präg´nə thiz´əm] *Anthropology.* the condition of having protruding jaws, especially as indicated by a gnathic index over 103.

prognathous [präg nā´thəs] *Anthropology.* showing prognathism; having protruding jaws. Also, **prognathic.**

prognosis [präg nō´sis] *Medicine.* a prediction of the probable course or outcome of a disease, injury, or developmental abnormality, based on such factors as the patient's condition and medical history and the usual course of the disease or abnormality.

prognostic [präg näs´tik] *Medicine.* relating to or acting as a prognosis.

prognostic chart *Meteorology.* a chart that shows weather characteristics such as the expected pressure pattern of a given synoptic chart at a specified future time.

prognostic equation *Meteorology.* a mathematical expression that governs the system and contains a time derivative of a known quantity which is used to predict the value of that same quantity at a later time or date when other factors in the equation are known, such as the vorticity equation.

progonoma *Oncology.* a tumor caused by misplacement of tissue resulting from fetal atavism to a stage that does not occur in the present life history of the species but that did occur in ancestral forms of the species.

progradation *Geology.* the seaward growth of a shoreline or coastline by deposition of river-borne sediments or by the accumulation of material thrown up by waves or transported by longshore drifting.

prograde metamorphism *Geology.* the metamorphic alterations of a rock that result from an increase in the pressure and temperature to which the rock had previously adjusted.

prograde motion *Astronomy.* **1.** rotation from west to east around an axis. **2.** the orbital movement of a planet in the same direction as the sun's rotation. Also, **prograde orbit.**

prograding shoreline *Geology.* a shoreline in the process of being built seaward or lakeward by deposition and accumulation.

program *Computer Technology.* **1.** an explicit specification of definitions and sequences of instructions in a particular programming language, written to produce a specified result. **2.** to design, code, and test such specifications. *Telecommunications.* **1.** a sequence of signals, either audio or video, transmitted for information or for entertainment purposes. **2.** a set of instructions written into the memory of a program-controlled switching system to achieve some function or functions. *Aviation.* the preset flight path and operations to be followed by a guided missile in flight. (Going back to a Greek term meaning literally "public writing.")

program analysis *Computer Programming.* **1.** the methodical investigation of a problem in order to determine the functions to be employed by the associated computer program. **2.** the analysis of a given program to determine its behavior; for example, its computational complexity. Also, ANALYSIS.

program block *Computer Technology.* a section of a computer program in a problem-oriented language that functions in some sense as an independent program.

program check *Computer Technology.* a system in which errors occurring in a program are detected by an input/output channel.

program compatibility *Computer Technology.* a measure of the degree to which a program that compiles, executes, and produces results on one computer will compile, execute, and produce identical results on another computer.

program control *Control Systems.* any process whose set point is varied periodically in order to make the process variable vary in a predetermined manner.

program counter *Computer Technology.* in a central processing unit, an incrementing register that contains the address of the next instruction to be executed. Also, CONTROL COUNTER or REGISTER, INSTRUCTION COUNTER or REGISTER, SEQUENCE COUNTER or REGISTER.

program design *Computer Technology.* the phase of program development in which the hardware and software resources are determined and the problem-solving algorithms are specified and written.

program element *Computer Technology.* a part of a computer system that executes an instruction sequence in a program.

program generator *Computer Technology.* a program that creates other programs automatically.

program level *Acoustical Engineering.* the volume level of a magnetic recording, based on display by a peak program meter of peak sound levels that last for a certain duration.

program library *Computer Technology.* an organized collection of available programs and routines stored for later use.

program listing *Computer Technology.* a printout produced by a compiler or assembler showing the source program, error messages, symbol cross references, and other relevant information about the object program.

program logic *Computer Technology.* the control structure of a computer program, including branches, subroutine calls, and other sequences.

programmable *Computer Technology.* **1.** of a computer device, able to be programmed. **2.** specifically, describing a device that can be programmed by the end user after purchase from the manufacturer.

programmable calculator *Computer Technology.* an electronic calculator that is capable of executing small programs, which are either stored or entered via the keys.

programmable controller *Control Systems.* a control device that uses computer logic and programmable memory to control industrial applications. Also, **programmable logic controller.**

programmable counter *Electronics.* a counter that divides an input frequency by a number, *n*, programmed into decades of synchronous down counters, which with additional logic will give the equivalent divide-by-*n* counter system.

programmable function key *Computer Technology.* a key on a terminal or personal computer keyboard that can be assigned a function or operation by a computer program.

programmable logic array see FIELD-PROGRAMMABLE LOGIC ARRAY.

programmable manipulator *Robotics.* any device controlled by a programmed set of instructions that can be stored in memory and modified or replaced.

programmable power supply *Electricity.* a power supply subject to digital control signals that change its voltage; provisions can also be made for manual selection of a range of preset output voltages.

programmable read-only memory see PROM.

program maintenance *Computer Technology.* the on-going process of modifying a program to meet changing needs, to keep it free of errors, and to take advantage of new technologies.

programmed cell death *Cell Biology.* a hypothesis proposing that cells are programmed to die after dividing a certain fixed number of times.

programmed check *Computer Technology.* an error-detection procedure performed by a program rather than automatically by the hardware.

programmed dump *Computer Technology.* a memory dump forced by the deliberate execution of an instruction at a particular point in a program.

programmed halt *Computer Technology.* the termination of a program as the result of the execution of a program stop instruction. Also, **programmed stop.**

programmed learning *Behavior.* a learning method consisting of materials presented in a predetermined order with provision for students to check their answers and to proceed at their own pace, usually involving the use of teaching machines, computers, or workbooks.

programmed logic array *Electronics.* a circuit that is built as a programmable logic array in which the programming connections are permanent rather than alterable by the user.

programmed marginal check *Computer Technology.* the variation of voltage levels by a computer program to test computer components for preventive maintenance.

programmed operators *Computer Technology.* computer operations that are actually implemented as subprograms or procedures in a programming language.

programmed timer see CYCLE TIMER.

programmed turn *Space Technology.* a preset, electronically controlled pitchover of a rocket or guided missile.

programmer *Computer Technology.* a person who participates in one or more of the program development phases: specification, design, coding, testing, and maintenance.

programmer-defined macroinstruction *Computer Technology.* a segment of code that can be defined by the programmer and referenced by a mnemonic code within the program; when the macro is referenced, it is replaced by its definition, perhaps with parameter substitutions.

programming *Computer Technology.* **1.** a general term for the process of designing, coding, and testing programs for carrying out operations on a computer. **2.** the process of converting a stated problem into a machine-sensible form.

programming environment *Computer Science.* an integrated set of interactive tools to aid the programming process, including program editors, compilers, debugging aids, and the like.

programming language *Computer Technology.* any language used by a programmer to write a computer program.

programming panel *Control Systems.* a device that can be used to edit a program or to insert a program into a programmable controller and monitor it.

programming scan time *Robotics.* the amount of time it takes for a processor to execute a complete set of instructions one time.

programming unit *Computer Programming.* a hand-held device with which an operator can program and store the instructions for a computer device, such as a robot, to carry out one or more tasks.

program mode *Computer Science.* a mode of central-processing unit operation in which user programs are run; certain operations, such as input/output, that are reserved for the operating system are prohibited.

program monitor *Telecommunications.* a monitor that is used to observe the quality of a radio or television broadcast.

program parameter *Computer Technology.* a parameter for a subprogram that can have a different value each time the subprogram is called.

program register see INSTRUCTION REGISTER.

program-sensitive fault *Computer Technology.* a hardware fault that occurs only during a specific combination or sequence of program instructions.

program specification *Computer Technology.* a document that describes the task to be performed by a computer program, including the input to be provided, the processing required, and the output expected.

program star *Astronomy.* a term for a star whose properties are currently under study.

program step *Computer Technology.* a single operation in a program, typically one instruction.

program stop *Computer Technology.* an instruction that causes the program to terminate execution under certain conditions. Also, HALT INSTRUCTION, STOP INSTRUCTION.

program storage *Computer Technology.* the portion of main memory reserved for the storage of programs, functions, and other routines.

program tape *Computer Technology.* a magnetic tape containing a stored copy of a program to facilitate loading into a computer's memory for execution.

program test *Computer Technology.* a technique of checking a program for errors by running it on a set of data or sample problems with known results.

program translation *Computer Technology.* the process of converting a program written in one language into another language.

program unit *Computer Technology.* a main program or subprogram.

progression *Mathematics.* a (countable) sequence of mathematical objects, each of which is determined from its predecessors according to some rule; e.g., arithmetic progression or geometric progression.

progressive *Medicine.* describing a disease that increases in scope or severity over time.

progressive bonding *Engineering.* a process in which a resin adhesive is prepared by the application of heat followed by the application of pressure.

progressive die *Metallurgy.* in metal working, a die used for more than one operation.

progressive forming *Metallurgy.* a process of sequentially performing more than one forming operation, either with a progressive die or with a set of separate dies.

progressive lateral sclerosis *Medicine.* a disease marked by degeneration of the lateral motor tracts of the spinal cord, resulting in progressive atrophy of the muscles as well as exaggerated reflexes.

progressive metamorphism *Geology.* a change in the degree of metamorphic alteration of a body of rock from a lower to a higher grade.

progressive muscular dystrophy *Medicine.* a disease in which there is progressive atrophy of the muscles, with its origin in the muscle and not in the spinal centers. Also, ERB'S ATROPHY.

progressive overflow *Computer Programming.* the storage of overflow records in the next available consecutive location.

progressive proofs *Graphic Arts.* in process printing, a series of proofs in which each color is printed separately in the order it will be printed, after which further proofs are run in progressive combination to show the addition of each color.

progressive relaxation (training) *Behavior.* a training technique in which a subject learns to relax muscle groups, beginning with the muscles that are easiest to control and eventually developing the ability to relax the entire body, in order to minimize the body's reaction to stressful situations. Also, RELAXATION TRAINING, RELAXATION THERAPY.

progressive sand wave *Geology.* a sand wave whose migration is caused by high water velocities and shallow depths.

progressive scanning *Telecommunications.* in television, a rectilinear scanning process in which the distance from center-to-center of successively scanned lines is equal to the nominal linewidth. Also, SEQUENTIAL SCANNING.

progressive shrinkage *Textiles.* an incremental reduction in the length or width of a fabric with each washing or dry cleaning.

progressive sorting *Geology.* a natural sorting of sedimentary particles in a body of water by which the average grain size decreases systematically in a downcurrent direction.

progressive wave *Meteorology.* a wave or wavelike disturbance that moves relative to the earth's surface.

progs [prägz] *Graphic Arts.* an informal expression for progressive proofs.

proguanil hydrochloride *Pharmacology.* $C_{11}H_{17}Cl_2N_5$, a folic acid antagonist used to prevent and treat malaria, but rarely prescribed in the U.S. because of resistance developed by the malarial parasite. Also, CHLOROGUANIDE HYDROCHLORIDE.

progymnosperm *Paleontology.* of or relating to the ancestral gymnosperms in the class Progymnospermopsida.

Progymnospermopsida *Paleontology.* an extinct class of plants, generally accepted since its proposal in 1960 as the earliest group of gymnosperms; includes the associated genera *Callixylon* and *Archaeopteris*; extant in the Devonian and Carboniferous.

prohaptor *Invertebrate Zoology.* the anterior adhesive organ of a fluke.

prohead *Virology.* a phage head precursor that does not contain DNA, formed during the assembly of a bacteriophage.

prohormone *Endocrinology.* a precursor of a hormone. Also, PREHORMONE.

proinsulin *Endocrinology.* the precursor polypeptide of insulin that is cleaved at two sites to form the mature dimeric insulin peptide and the inactive C-peptide fragment.

Projapygidae *Invertebrate Zoology.* a family of bristletails, primitive wingless insects in the order Diplura.

project *Engineering.* a planned undertaking in a research or development field that is geared toward the accomplishment of a singular goal, such as the production of an item or the solution to a problem. *Computer Science.* in a relational database system, an operation that forms a new relation, whose tuples are formed from a subset of the fields of the tuples in an existing relation, with duplicates removed.

projected planform see PLANFORM.

project engineering *Engineering.* **1.** the designing, planning, and supervision of the construction of a manufacturing or municipal facility. **2.** the planning and carrying out of a specific project, such as a research project.

projectile *Mechanics.* any object that is projected by an external force and that continues to travel by its own inertia. *Ordnance.* specifically, such an object fired from a gun as a weapon of war; e.g., a bullet, artillery shell, or grenade; the term is also applied to rockets and guided missiles.

projectile ogive *Ordnance.* a hollow, metallic cone that covers the front end of a projectile in order to reduce air resistance.

projectile point *Archaeology.* a stone, bone, or wooden point that is attached to a weapon such as an arrow, dart, or spear.

projection *Psychology.* **1.** the attributing to another person of a characteristic or impulse that one finds emotionally unacceptable in oneself, as when an unfaithful husband accuses his wife of infidelity. **2.** the fact of unwittingly attributing one's own beliefs and values to others. **3.** the perceiving of events around oneself in terms of one's own needs and wishes. *Neurology.* the connection of parts of the brain or spinal chord, through projection fibers, with other parts of the central nervous system. *Cartography.* **1.** the process of systematically extending lines or planes from features on a spherical surface so that they intersect with a plane surface, thus transferring the features on the sphere to the plane. **2.** a map or chart produced by a systematic application of the process of projection. Also, PROJECTION MAP. **3.** in surveying, the process of extending a surveyed line beyond the points that define its position. *Mathematics.* **1.** let $\prod_{i \in I} A_i$ be the direct product of a family of algebraic objects (such as rings, modules, etc.), indexed over a set I, with a typical element in the product denoted $\{a_i\}_{i \in I}$. For each $k \in I$, the **canonical projection** π_k: $\prod_{i \in I} A_i \to A_k$ is given by $\pi_k(\{a_i\}_{i \in I}) = a_k$. For example, if $\prod_{i \in I} A_i$ is Euclidean n-space, then π_k maps each point to its kth coordinate. **2.** see ORTHOGONAL PROJECTION. **3.** an idempotent linear map. **4.** the map from a bundle to its base space.

projection area *Anatomy.* any one of several specific areas of the cerebral cortex that receive sensory input.

projection cathode-ray tube *Electronics.* a tube that produces extremely luminous but relatively small images, which are projected, through lenses, onto a larger screen.

projection chamber see PROJECTION SPARK CHAMBER.

projection display *Electronics.* any system that employs lenses to project images, such as those generated by a cathode-ray tube, onto a larger screen.

projection drawing see DESCRIPTIVE GEOMETRY.

projection fibers *Anatomy.* the nerve fibers that conduct sensory impulses to the sensory areas of the cerebral cortex.

projection map *Cartography.* see PROJECTION, def. 2.

projection microradiography *Physics.* a technique in X-ray imaging in which an electron beam is focused extremely narrowly and is thus able to generate X-rays from an effective point source; a target specimen is placed very close to the X-ray source and far away from the recording medium for magnification.

projection microscope *Optics.* an X-ray microscope using projection microradiography or contact microradiography to magnify the image.

projection operator *Quantum Mechanics.* an operator that projects a state vector onto a single axis, or that maps one state into another.

projection plan position indicator *Electronics.* an instrument in which radar signals are displayed as magenta-colored arcs against a plotting surface. Also, **projection PPI.**

projection platemaking *Graphic Arts.* a method in which printing plates are made by projecting a microfilm image (often computer-generated) onto a photosensitive plate; used in printing telephone directories, catalogs, and other low-quality texts.

projection printer *Optics.* a system that duplicates a film image essentially by rephotographing it. Also, OPTICAL PRINTER.

projection printing *Graphic Arts.* a process of photographic englargement in which the image is projected onto a photosensitive surface and the projected image is developed.

projection slide *Graphic Arts.* a color transparency that is mounted in a small frame of paper or plastic and is designed to be viewed with the aid of a projector.

projection spark chamber *Nucleonics.* a spark chamber designed so that a particle passing through the gap perpendicular to the electric field produces electrons whose streamers provide information about only the projection of the particle track perpendicular to the field. Also, PROJECTION CHAMBER.

projection thermography *Engineering.* a method used to measure surface temperature in which heat radiation creates a pattern on a screen of luminescent material in an optical system, and the pattern is then interpreted to determine the surface temperature.

projection welding *Metallurgy.* resistance welding performed at a selected location, such as a projection.

projective curvature tensor *Mathematics* a four-argument tensor field P on a manifold M that vanishes identically if M is projectively flat, that is, if M has the same geodesics (with possibly different parametrizations) as it does when it is given a flat (pseudo-) Riemannian metric. P is invariant under projective mappings of M into itself, coincides with the Riemann curvature tensor in Ricci-flat ($R_{ij} = 0$) spaces, and has trace zero ($P^{is}_{ks} = 0$). In component form,

$$P^{ij}_{km} = R^{ij}_{km} - (R^{i}_{k}\,\delta^{j}_{m} - R^{j}_{k}\,\delta^{i}_{m})/(n-1),$$

where n is the dimension of M. The projective curvature tensor is due to Weyl, but the term Weyl tensor usually refers to the conformal curvature tensor.

projective geometry *Mathematics.* the area of mathematics that has its origins in the study of perspective drawing, and includes the study of geometric properties invariant under projections, algebraic representations of projective planes, collineations, projective groups, and some topics from non-Euclidean geometry.

projective group *Mathematics.* the group of collineations of a projective space or any of various groups of transformations arising in the study of projective geometry. For example, the set of all projectivities on a line for which two given fixed points are double points forms a projective group. The set of collineations on the plane is a projective group with a large number of different types of subgroups.

projective limit see INVERSE LIMIT.

projective line *Mathematics.* the image of a circle under the stereographic projection; equivalently, a line whose endpoints at infinity have been identified as being the same point.

projective module *Mathematics.* A module P over a ring R is said to be projective if, given any diagram of R-module homomorphisms,

$$\begin{array}{c} P \\ \downarrow_f \\ A \rightarrow B \rightarrow 0 \\ g \end{array}$$

with the bottom row exact (i.e., g is an epimorphism), there exists an R-homomorphism $h\colon P \rightarrow A$ such that the following diagram commutes ($gh = f$):

$$\begin{array}{c} P \\ {}_h\!\downarrow \quad \downarrow_f \\ A \rightarrow B \rightarrow 0 \\ g \end{array}$$

projective plane *Mathematics.* a projective plane (P,L,I) is a set of points P and a set of lines L together with an incidence relation I between the points and lines that satisfies the following axioms: (a) There are at least four points, no three of which are collinear. (b) Every pair of distinct lines is incident with a unique point. (c) Every pair of distinct points is incident with a unique line. By theorem, in a projective plane, the set P has cardinality ≥ 7.

projective space *Mathematics.* a projective space (P,L,I) is a set of points P and a set of lines L together with an incidence relation I between the points and lines that satisfies the following axioms: (a) There exists at least one line and a point not on the line. (b) Every pair of distinct lines l_1 and l_2 is incident with a unique point, denoted $l_1 \vee l_2$. (c) Every pair of distinct points P and Q is incident with a unique line, denoted $P \vee Q$. (d) If P Q, R, and T are distinct points such that there exists a point $M \in P$ with $(P \vee Q) \vee (R \vee T) = M$ (that is, M is the intersection of the line joining P and Q with the line joining R and T), then there exists another point $N \in P$ such that $(P \vee R) \vee (Q \vee T) = N$ (that is, N is the intersection of the line joining P and R with the line joining Q and T). A projective plane is also a projective space.

projective technique *Psychology.* any test or procedure that presents relatively ambiguous stimulus materials to an individual in order to gain insight into his or her personality; the individual's response projects aspects of personality into the task. Also, **projective test.**

projectivity *Mathematics.* in projective geometry, the composition of a finite sequence of perspectivities.

projector *Optics.* a two-lens system that projects the magnified image of a transparency or opaque matter onto a screen. Also, OPTICAL PROJECTOR, OPTICAL PROJECTION SYSTEM. *Acoustical Engineering.* a horn that is used to direct acoustic energy in a specified direction. *Ordnance.* **1.** any mechanism or piece of equipment that launches a projectile; for example, a gun or grenade launcher. **2.** specifically: an unrifled tube used to launch flares, grenades, and mortar shells; a rack used to launch target rockets; a gun or mortar used to launch antisubmarine projectiles.

prokaryote *Cell Biology.* any organism of the kingdom Procaryotae, including the bacteria and cyanobacteria; characterized by the lack of a defined nucleus, and the possession of a single double-stranded DNA molecule and a very small range of organelles; almost all such organisms are unicellular and have a true cell wall containing peptidoglycan.

Prokhorov, Alexander born 1914, Soviet physicist; shared the Nobel Prize for the development of masers.

Prolacertiformes *Paleontology.* an extinct suborder of early lepidosaurian reptiles in the order Eosuchia; recent authorities prefer to classify this group as an order, Protorosauria; extant in the Triassic.

prolactin *Endocrinology.* a protein hormone that is produced by the lactotrophs of the anterior lobe of the pituitary gland and that stimulates milk secretion by the mammary glands and encourages activity in the corpus luteum. Also, LACTOGENIC HORMONE.

prolactin inhibitory factor *Endocrinology.* a hypothalamic factor, possibly dopamine, that inhibits the release of prolactin from the adenohypophysis. Also, **prolactin release-inhibiting factor**.

prolactinoma *Oncology.* a tumor of the pituitary gland that secretes prolactin.

prolactin-releasing factor *Endocrinology.* a hypothalamic factor, possibly thyrotropin-releasing hormone, that stimulates the release of prolactin from the adenohypophysis.

prolamellar body *Cell Biology.* a system of membranes found within an etioplast that, upon exposure to light, develop rapidly into the normal chloroplast membrane stacks.

prolamin *Biochemistry.* a simple protein occurring in plants that is soluble only in strong alcohol solutions.

prolapse *Medicine.* the slipping or falling down of a part or organ, such as the protrusion of a portion of the iris through a wound in the cornea or the slipping of the uterus into the vagina.

prolapsed bedding *Geology.* a bedding marked by a series of flat folds with nearly horizontal axial planes contained wholly within a bed whose boundaries have not been disturbed.

prolapse of the cord *Medicine.* the early expulsion of the umbilical cord during labor, prior to the delivery of the fetus; this may result in fetal death.

prolapse of the uterus *Medicine.* the falling down or descent of the uterus in such a way that the cervix is within, partially outside, or completely outside the vaginal orifice.

prolate *Mathematics.* elongated in the direction of a line joining the poles; opposite of oblate.

prolate cycloid *Mathematics.* a trochoid in which the radius of the circle is less than the distance from the center of the rolling circle to the point describing the curve.

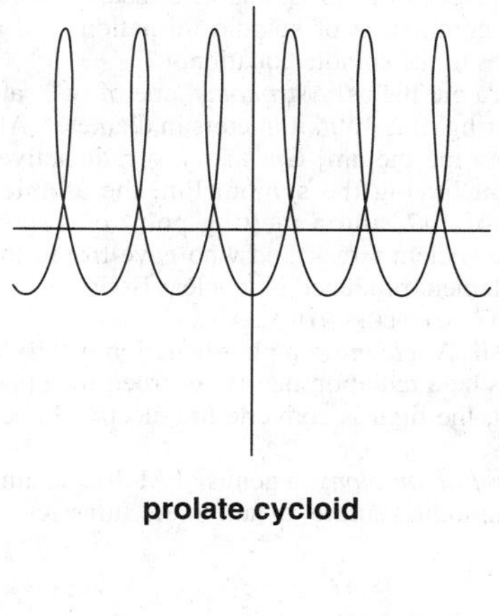

prolate cycloid

prolate spheroid *Geodesy.* a spheroid that has the shape of an ellipsoid of revolution and rotates about its semimajor axis.

prolecanitid *Paleontology.* of or relating to the long-ranging ammonoids of the order Prolecanitida; extant in the Devonian to Triassic.

Prolecithophora *Invertebrate Zoology.* an order of freshwater and marine turbellarian flatworms.

proleg *Invertebrate Zoology.* one of several short, fleshy appendages found in the immature stages of some insects, such as caterpillars; used in walking or crawling.

proliferation *Military Science.* a term for the process by which one nation after another gains possession of nuclear weapons or the right to determine the use of such weapons, with each nation potentially able to launch a nuclear attack upon another nation.

proline *Biochemistry.* $C_5H_9NO_2$, the only one of the twenty protein-forming amino acids that contains an imino from a secondary amino group, a primary component of collagen. Also, 2-CARBOXYLIC ACID.

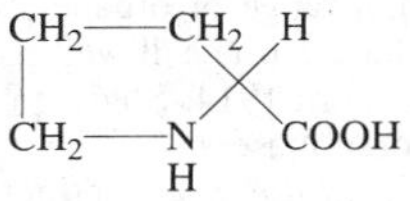

proline

proline racemase *Enzymology.* an enzyme that catalyzes the interconversion of optical isomers of proline.

prolinuria *Medicine.* the presence of proline in the urine.

PROLOG *Artificial Intelligence.* a high-level nonprocedural programming language built on the principles of predicate calculus; that is, a program is expressed as a set of facts and a set of rules for deriving new facts. (An acronym for programmed logic.)

prolongation *Cartography.* **1.** in surveying, an extension of a line beyond the last surveyed segment of that line, in the same direction as the segment. **2.** the extension of a line tangent to a curve at the point of extension.

prolonge *Ordnance.* a rope with a hook or loop that is used to move a gun carriage or vehicle into position.

proluvium *Geology.* a complex, breakable deltaic sediment that accumulates at the base of a slope as a result of intermittent torrential washing of fragmented material.

prolyl racemase *Enzymology.* an enzyme that catalyzes the interconversion of optical isomers of prolyl, the group produced when a hydroxyl group is removed from the 1 carbon in proline.

PROM [präm] *Computer Technology.* an integrated memory circuit that can be programmed once by the user and generally cannot be altered thereafter. (An acronym for programmable read-only memory.)

promastigote *Cell Biology.* an elongate or pear-shaped cell form with a central nucleus and, at the anterior end, a kinetoplast and a basal body from which arises a single, long, slender flagellum; adopted by certain parasites, such as trypanosomes, at certain stages in their life cycle.

promazine hydrochloride *Organic Chemistry.* $C_{17}H_{20}N_2S{\cdot}HCl$, whitish hygroscopic crystals or powder, turning blue or pink on prolonged exposure to air; soluble in water and alcohol, insoluble in benzene; melts at 172–182°C; used as a tranquilizer and a food additive.

promenade deck *Naval Architecture.* an upper, weather deck on a passenger vessel, designed for passengers to walk along. Also, HURRICANE DECK.

prometaphase *Cell Biology.* a transition period between the prophase and metaphase stages of mitosis, characterized by disruption of the nuclear envelope, completion of spindle formation, and the migration of the chromosomes to the spindle equation.

Prometheus [prə mē´thē əs] *Astronomy.* one of two satellites that shepherd Saturn's F ring; it is 150 kilometers in diameter. Also, **1980 S27.**

promethium [prə mē´thē əm] *Chemistry.* a radioactive rare-earth (lanthanide) element having the symbol Pm, the atomic number 61, an atomic weight of 147, and a melting point of 1160°C. (Named for *Prometheus,* the ancient Greek god who gave fire to the human race; it is an artificial element produced by nuclear fission.)

promethium-147 see FLORENTIUM.

promethium cell *Nucleonics.* a photojunction battery in which promethium-147 emits beta radiation that is absorbed by a phosphor which, in turn, emits light; the light is converted to electrical energy by means of silicon junctions.

Promicroceras *Paleontology.* a genus of Mesozoic ammonoids characterized by an ammonitic suture; extant in the Jurassic.

Promicromonospora *Bacteriology.* a genus of bacteria of the order Actinomycetales, occurring in soil as yellow mycelia and reproducing by fragmentation.

prominence *Astrophysics.* an eruption of relatively cool, incandescent gas from the sun's surface that can be seen in the corona in a solar eclipse or by observing strong spectral lines in its emission spectrum.

promiscuity *Behavior.* a mating system in which the male and female do not form lasting pair bonds; one male may mate with several females, or one female with several males.

promiscuous DNA *Molecular Biology.* pieces of DNA that have been transposed to a different location, such as from an organelle to the nucleus.

promitochondrion *Cell Biology.* an organelle that is similar to a mitochondrion but is unable to carry out electron transport; found in certain facultatively fermenting organisms, including yeast.

promitosis *Oncology.* a form of cell division in tumors, with mitosislike division in the nucleolus or karyosome, and amitosislike division in the remainder of the cell.

promontory *Geology.* **1.** a high, prominent area of land or rocky cliff that projects out into a body of water beyond the shoreline. Also, COBB, REACH, NESS, NOOK. **2.** a cape of any elevation having a bold termination. **3.** a bluff or prominent hill that juts out into or overlooks a lowland.

promontory

promoter *Chemistry.* a substance that is added in small amounts to a catalyst to improve the latter's activity, selectivity, or longevity. *Molecular Biology.* a nucleotide sequence on a DNA molecule that, when attached to an RNA polymerase molecule, will initiate transcription of a specific structural gene.

promoter down mutation *Molecular Biology.* a mutation that decreases the efficiency of transcription initiation or inactivates the promoter, preventing mRNA synthesis.

promoter gene *Molecular Biology.* a gene that encodes a protein that functions to activate expression of other genes.

promoter up mutation *Molecular Biology.* a mutation that increases the efficiency of transcription initiation or increases binding of RNA polymerase to the promoter.

PROM programmer *Computer Technology.* a device used to program PROMs or reprogram EPROMs by electrically rewriting selected memory cells. Also, **PROM burner.**

prompt *Computer Technology.* a message, symbol, or audible signal generated by the processor to request input from a user.

prompt critical *Nucleonics.* a critical condition of a nuclear reactor that is due to the prompt neutrons alone; any contribution of delayed neutrons would thus render the system supercritical.

prompt neutron *Nuclear Physics.* a neutron released in a primary nuclear fission process with almost zero delay (probably within 10^{-12} seconds).

prompt radiation *Nuclear Physics.* the burst of radiation produced by nuclear fission, including prompt neutrons and instantaneous gamma rays.

promycelium *Mycology.* a short filament produced in the germination of a spore that bears small spores and then dies.

promyelocyte *Histology.* a young myelocyte as it develops from a myeloblast.
pronase or **Pronase** *Biochemistry.* a compound containing a variety of fungal proteinases that degrade proteins into amino acids.
pronate *Anatomy.* to rotate the forearm so that the palms of the hands are downward and backward. *Robotics.* to turn a robot around so that the back or protected side is exposed and facing up.
pronation *Robotics.* the movement of a manipulator into a position in which the back or protected side is facing up.
pronator *Physiology.* a muscle that turns a body part prone, such as the muscle that turns the palm downward.
prone *Science.* in a position with the face downward; lying face down.
pronephros *Developmental Biology.* the most anterior and precocious of three types of kidney in vertebrates.
pronghorn *Vertebrate Zoology.* an antelopelike diurnal or nocturnal ruminant mammal, *Antilocapra americana,* composing the family Antilocapridae; the males and most females have two pronglike horns of bone material on their heads covered with a sheath of skin that is shed once a year; found in the grasslands and deserts of western North America.

pronghorn

prong reef *Geology.* a wall reef characterized by irregular buttresses that develop perpendicular to its axis in the leeward and seaward directions.
Prono *Robotics.* a programming language used in geometric modeling and machining.
pronograde *Anthropology.* of primates, characterized by a posture in which the body is parallel to the ground; moving on all four limbs.
pronogradism *Anthropology.* see QUADRUPEDALISM.
pronormoblast *Histology.* a young erythrocyte prior to the disintegration of its nucleus. Also, MEGALOBLAST OF SABIN.
pronotum *Entomology.* the dorsal sclerite of the prothorax of an insect.
pronucleus *Cell Biology.* a haploid nucleus of a gamete involved in fertilization; may be found in an ovum, sperm cell, or pollen tube.
proof *Graphic Arts.* a sample of a printed page produced by a compositor or printer and used to check the accuracy of a printed text, the tonal values of process colors, and the like; various kinds of proofs are produced at different stages of the printing process, including galley proofs, prepress proofs, press proofs, and reproduction proofs. *Engineering.* a copy of a mold impression with the use of a cast. *Food Technology.* a measure of the alcoholic content of a liquid, especially an alcoholic beverage, equal to 0.5% by volume. *Mathematics.* within a given axiomatic theory, a finite sequence of statements such that each statement is either an axiom or may be deduced from previous statements by logical rules of inference. Also, FORMAL PROOF. A theorem or provable statement is the last statement of some proof, and an axiom may be viewed as a one-step proof.
proof cartridge *Ordnance.* a cartridge used in test firing; it is usually loaded to produce a breech pressure approximately 25% greater than that produced by the heaviest available cartridge for that weapon. Also, **proof charge.**
proof firing *Ordnance.* the process of firing a weapon, usually with overloaded cartridges, rounds, or shells, in order to test the serviceability of the weapon or its mount. Also, **proof fire.**
proof load *Engineering.* a predetermined load to which a specimen or structure is submitted before being accepted for use.
proof mark *Ordnance.* a mark on a weapon to indicate that it has been inspected or proof fired. Also, **proofmark.**
proofread *Graphic Arts.* to read proofs in order to detect and mark errors.
proofreader *Graphic Arts.* a person who reads proofs in order to detect and mark errors. *Molecular Biology.* a mechanism that maintains the correct base sequence of nucleic acids during replication, transcription, or translation.
proof resilience *Mechanics.* the tensile strength necessary to stretch an elastomer from zero elongation to its breaking point, expressed in units of foot-pounds per cubic inch of the original dimension.
proof stress *Mechanics.* **1.** the specified stress to be applied to a body to assess its ability to support service loads. **2.** see YIELD STRESS.
proof testing *Materials Science.* the process of exposing a sample of a component to conditions similar to or exceeding those that will be encountered during its service life; used to determine the suitability of a component for its application.
pro-opiomelanocortin *Endocrinology.* the polypeptide precursor molecule that can be cleaved to produce corticotropin, β- and γ-lipotropin, β-endorphin, and α-, β-, and γ-melanocyte-stimulating hormone.
prop *Engineering.* any rigid support, such as a pole, beam, stick, or rod. *Aviation.* an informal designation for propeller-driven aircraft.
prop. property; proposition; propaganda.
propadiene see ALLENE.
propaedeutic *Science.* preliminary; introductory to some science or area of inquiry.
propaedeutic stratigraphy see PROSTRATIGRAPHY.
propaganda *Military Science.* any form of communication that is designed to influence the opinions, emotions, attitudes, or behavior of any group in order to benefit the sponsor of the communications, either directly or indirectly; **black propaganda** purports to emanate from a source other than the true one; **gray propaganda** does not specifically identify its source; **white propaganda** acknowledges its true source.
propagate *Science.* **1.** to reproduce or cause to reproduce. **2.** to produce (a signal, sound, or other effect) at a distance, as by electromagnetic waves or compression waves. Thus, **propagation, propagative.**
propagated blast *Engineering.* a blast commonly used for ditching, and consisting of a number of unprimed explosive charges and one primed hole, with each charge detonated by the explosion of the adjacent one, transmitting the shock through wet soil.
propagated error *Computer Technology.* an error that occurs in one operation and causes resultant errors in subsequent operations.
propagation anomaly *Physics.* a change in the characteristics of propagation, typically due to some resonance condition in the medium.
propagation constant *Electromagnetism.* a complex quantity that measures the properties of propagation for a given medium or transmission line at a specified frequency; the real part is the attenuation constant (nepers per unit length), and the imaginary part is the phase constant (radians per unit length).
propagation delay *Electronics.* the time it takes for a signal to travel through operating circuits, generally measured in nanoseconds.
propagation forecasting *Meteorology.* a forecast method that utilizes the known or predicted vertical distribution of the index of refraction over a given area to indicate the wave motion performance of radar or microwave radio systems that operate in that area.
propagation loss *Telecommunications.* attenuation of signals passing between two points on a transmission path.
propagation mode *Electromagnetism.* a characterization of the field pattern for electromagnetic waves propagating through a medium with respect to their direction of propagation; examples include transverse electric (TE), transverse magnetic (TM), and transverse electric/magnetic (TEM).
propagation path *Telecommunications.* a path between transmitter and receiver that many include ionospheric scatter, direct tropospheric scatter, E-layer skip, F1-layer and F2-layer skip, and echo.
propagation time delay *Telecommunications.* on a given transmission path, the time it takes for a wave to travel between two points.
propagation velocity *Electromagnetism.* the velocity of electromagnetic propagation through a given medium.

propagator *Quantum Mechanics.* a function of two spatial coordinates that yields the state for all time after zero when multiplied by the state of a system at time zero and integrated over space.

propagule *Ecology.* the minimum number of individuals of a species required to colonize a habitable island. *Botany.* **1.** a seed, bud, or other offshoot capable of developing into a new and fully independent plant. **2.** a reproductive structure of brown algae. Also, **propagulum.**

Propalticidae *Invertebrate Zoology.* a family of beetles, coleopteran insects in the superfamily Cucujoidea, having flattened bodies and dense hairs covering the dorsal surface; found under the bark of dead trees.

propanal see PROPIONALDEHYDE.

propane [prō´pān´] *Organic Chemistry.* $CH_3CH_2CH_3$, a colorless gas with a natural gas odor; soluble in alcohol and slightly soluble in water; boils at –42.5°C; freezes at –187.7°C; a dangerous fire risk; widely used in industrial and household fuel, as a solvent, refrigerant, and aerosol propellant, and in organic sysnthesis.

propane fractionation *Chemical Engineering.* a petroleum-refinery stabilization process that removes propane as an impurity from gasoline. Also, DEPROPANIZING PROCESS.

1-propanethiol see *n*-PROPYL MERCAPTAN.

propanil *Organic Chemistry.* $C_9H_9C_{12}NO$, an herbicide used for weed control near food crops.

propanoic acid see PROPIONIC ACID.

propanol see PROPYL ALCOHOL.

propanol-2 see ISOPROPYL ALCOHOL.

2-propanone see ACETONE.

proparathyrin *Endocrinology.* the immediate precursor polypeptide from which the mature parathyrin is cleaved. Also, **proparathyroid hormone.**

propargyl alcohol *Organic Chemistry.* $HC{\equiv}CCH_2OH$, a colorless flammable liquid with a geraniumlike odor; boils at 114°C and freezes at –53°C; soluble in water and alcohol; used as a chemical intermediate, solvent stabilizer, and corrosion inhibitor.

propargyl bromide *Organic Chemistry.* $HC{\equiv}CCH_2Br$, a flammable liquid that has a sharp odor; boils at 88–90°C; used as a soil fumigant and a chemical intermediate.

propargyl chloride *Organic Chemistry.* $HC{\equiv}CCH_2Cl$, a flammable liquid; insoluble in water and miscible with benzene and ethanol; boils at 58°C; used as a chemical intermediate and a soil fumigant.

propellant *Space Technology.* **1.** the agent providing thrust for a rocket, consisting of a fuel, an oxidizer, or usually a mixture of a fuel, an oxidizer, and an additive. **2.** a constituent ingredient in a propellant mixture, such as the fuel or oxidizer. *Ordnance.* the rapidly burning charge that propels a bullet, shell, rocket, or missile. *Materials.* a compressed inert gas that is used to dispense the contents of an aerosol container when pressure is released.

propellant mass ratio *Space Technology.* the ratio of the mass of the propellant in a rocket to the total mass of the rocket including propellant. Also, **propellant mass fraction**.

propeller *Mechanical Engineering.* **1.** an assembly of radiating blades around a revolving hub that produces thrust or power. **2.** relating to or driven by such an assembly. Thus, **propeller screw, propeller aircraft. 3.** any device or apparatus that causes an object to move. **4.** a wind-driven device that provides mechanical energy, as in a windmill.

propeller blade *Mechanical Devices.* any of two or more wing-shaped blades attached to a propeller boss and positioned so that each forms part of a helical surface; used to propel ships and aircraft.

propeller boss *Mechanical Devices.* the hub of a propeller to which two or more blades are attached. Also, **propeller hub.**

propeller cavitation *Fluid Mechanics.* the formation of gas bubbles on the surface of a propeller's blades where the motion of the propeller in a liquid causes the local pressure to fall below the vapor pressure of the liquid.

propeller fan *Mechanical Engineering.* a fan consisting of an impeller or rotor carrying several blades of airscrew form, usually driven by a direct-coupled motor.

propeller horsepower *Mechanical Engineering.* a measure of the horsepower actually delivered to a craft's propellers.

propeller racing *Naval Architecture.* the acceleration of a propeller due to lessened resistance when the ship's pitching and heaving lift it largely clear of the water.

propeller shaft *Mechanical Engineering.* the main drive shaft that transmits power from an engine to the propeller.

2-propenoic acid see ACRYLIC ACID.

4-propenylanisole see ANETHOLE.

propenyl guaethol *Organic Chemistry.* $C_{11}H_{14}O_2$, a free-flowing white powder that has an odor and taste similar to vanilla; melts at 85–86°C; very soluble in fats and very slightly soluble in water; used for artificial vanilla flavoring and as a flavor enhancer.

proper *Mathematics.* **1.** if a set A is a subset of a set B but $A \neq B$ (i.e., there is at least one element of B that is not in A) then A is a **proper subset** of B. **2.** an algebraic object A of a given class, such as a group, ring, module, etc., is a **proper subobject** (e.g., proper subgroup, subring, submodule, etc.) of another object B of the same class if: (a) it is a member of that same class, (b) its elements are a subset of the set of elements of B, and (c) there exists at least one element of B which is not an element of A.

proper character *Mathematics.* see CHARACTER.

proper class *Mathematics.* a class that is not a set.

properdin *Immunology.* a normal serum protein that has the ability to kill bacteria and viruses in the presence of complement and magnesium ions.

proper fraction *Mathematics.* **1.** a rational number expressed in the form a/b, whose numerator a has absolute value less than that of its denominator b. **2.** in general, the quotient $p(x)/q(x)$, where $p(x)$ and $q(x)$ are polynomials and the degree of $p(x)$ is less than the degree of $q(x)$.

proper function see EIGENFUNCTION.

proper Lorentz transformation *Physics.* a Lorentz transformation in which the determinant of the transformation matrix is +1.

proper motion *Astronomy.* the very small observed motion of a star in right ascension and declination, usually expressed in fractions of a second of arc per year.

proper prefix *Computer Science.* a string that is a prefix of another string and shorter than it.

proper time *Physics.* in general relativity, the integral over the distance ds along a trajectory, in which ds is defined in four-space by the following relation: $ds^2 = g_{mn}(x)dx^m dx^n$, where g_{mn} is the metric of the manifold.

property *Science.* a characteristic quality, capability, function, or the like.

proper value see EIGENVALUE.

proper vector see EIGENVECTOR.

prophage [prō´fāj´] *Virology.* a phage genome that is replicated in the host cell and maintained by integration into the host chromosome or as an extrachromosomal plasmid.

propham *Organic Chemistry.* $C_{10}H_{13}NO_2$, a brownish solid that melts at 87–88°C; used as an herbicide.

prophase [prō´fāz´] *Cell Biology.* the initial stage of both mitosis and meiosis, characterized by condensation of the chromosomes.

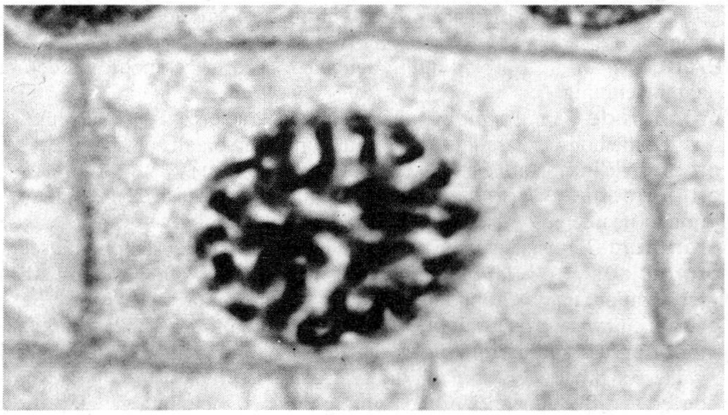

prophase

prophenpyridamine maleate see PHENIRAMINE MALEATE.

prophylactic [prō´fi lak´tik] *Medicine.* **1.** tending to prevent disease; relating to prophylaxis. **2.** an agent that tends to prevent disease. **3.** in popular use, a condom.

prophylaxis [prō´fi lak´sis] *Medicine.* any measures or treatment taken to prevent disease; preventive treatment.

prophyritic crystal see PHENOCRYST.

propidium *Biochemistry.* an agent inserted between adjacent base pairs in dsDNA or in regions of ssDNA that attacks certain protozoa known to cause sleeping sickness.

propiodal *Pharmacology.* $C_9H_{24}I_2N_2O$, a white crystalline powder, formerly used as a source of iodine.

β-propiolactone *Organic Chemistry.* $C_3H_4O_2$, a reactive liquid that boils and then decomposes at 162°C or 61°C (20 torr); used as a chemical intermediate.

propionaldehyde *Organic Chemistry.* C_2H_5CHO, a flammable white liquid with a suffocating odor; soluble in water; boils at 48.8°C; used as a disinfectant, preservative, and in the manufacture of plastics. Also, PROPANAL, PROPYL ALDEHYDE.

propionate *Organic Chemistry.* an ester or a salt of propionic acid, such as sodium propionate (CH_3CH_2COONa).

Propionibacteriaceae *Bacteriology.* a family of Gram-positive, rod-shaped or filamentous, anaerobic but aerotolerant bacteria occurring on the skin and in the intestinal and respiratory tracts; closely related to the actinomycetes.

propionic acid *Organic Chemistry.* CH_3CH_2COOH, a colorless, oily liquid with a rancid odor; soluble in water and alcohol; boils at 140.1°C; used in pharmaceuticals, perfumes, esters, herbicides, and flavorings, and as a preservative. Also, PROPANOIC ACID.

propionic anhydride *Organic Chemistry.* $(CH_3CH_2CO)_2O$, a colorless liquid with a pungent odor; soluble in alcohol and ether; decomposed by water; boils at 167–169°C; a strong irritant to tissues; used as an esterifying agent, and in pharmaceuticals and dyestuffs.

Propionigenium *Bacteriology.* a genus of anaerobic, asporogenous bacteria, one species of which metabolizes succinate to propionate.

Propionispira *Bacteriology.* a genus of Gram-negative, obligately anaerobic bacteria occurring in certain trees as motile, curved rods or spiral filaments, able to fix nitrogen.

propjet see TURBOJET.

proplasmin see PLASMINOGEN.

propleuron *Invertebrate Zoology.* a lateral plate on the side of the first segment of an insect's thorax.

Propliopithecus *Anthropology.* a fossil hominoid from the Fayum site in Egypt that dates from the Middle Oligocene and shows a combination of hominoid and cercopithcoid features; it may be ancestral to the Old World primates.

propodite *Invertebrate Zoology.* the sixth segment of the leg of a malacostracan crustacean. Also, **propodus.**

propodium *Invertebrate Zoology.* the front part of the foot of a mollusk.

proportion *Science.* the relationship of one part to another or to the whole. *Mathematics.* see RATIO.

proportional *Science.* relating to or based on a proportion; relative. *Mathematics.* having the same or a constant ratio.

proportional band *Acoustics.* a frequency analysis band that uses proportional rather than equal bandwidths for frequency analysis, as in a 1/3-octave spectral analyzer.

proportional counter *Nucleonics.* a device that operates in the proportional region to count radiation events, particularly α- and β-particles and X-rays.

proportional dividers *Mechanical Devices.* a compass with two slotted, double-ended legs with pointed tips joined by an adjustable pivot and screw assembly to allow regulation of the width of the object to which the points are touched, thereby providing a means of proportional distance measurements. Also, **proportional compass.**

proportional (elastic) limit *Mechanics.* the maximum stress beyond which a solid's strain deviates from proportionality with stress; it coincides with the elastic limit for some substances, notably metals. Also, LIMIT OF PROPORTIONALITY.

proportional navigation *Navigation.* a method of homing navigation in which the missile turn rate is directly proportional to the turn rate in space of the line of sight. Also, **proportional guidance**.

proportional parts *Mathematics.* assuming a finite set of numbers $\{x_1, \ldots, x_n\}$, any set of numbers of the form $\{ax_1, \ldots, ax_n\}$ is then said to contain the parts proportional to the original set; i.e., ax_k and x_k are proportional parts.

proportional reducer *Graphic Arts.* a chemical solution used to reduce the density and contrast in a photographic negative in proportion to the amount of silver present in a given part of the image.

proportional region *Nucleonics.* a range of voltages applied to a radiation counter tube where the charge collected in an isolated count is proportional to the amount of charge liberated by the initial ionizing event.

proportional sensitivity *Neurology.* the quantitative relationship between the intensity of a stimulus and the magnitude of the corresponding response.

proportional spacing *Graphic Arts.* any type system, such as Linotype, in which the space occupied by a given letter of the alphabet is proportional to its width, as opposed to the uniform spacing of a standard typewriter.

proportioning probe *Engineering.* a sensing device used in leak testing that can regulate the air-tracer gas ration while keeping the amount of flow transmitted to the testing device constant.

proportioning reactor *Electromagnetism.* a saturable-core reactor in which an increase in the input current from zero to the rated value produces an output current that increases proportionally from cutoff to the full load value.

proposition *Mathematics.* **1.** a statement to be proved; a theorem or problem. **2.** in the theory of statement calculus, a statement or assertion that has the property of being either true or false.

propositional calculus *Mathematics.* see STATEMENT CALCULUS.

propositional database *Artificial Intelligence.* a database of facts, stored in the form of logical formulae, that allows retrieval of facts that match (unify with) a query predicate.

propositus *Genetics.* in a human pedigree, the affected individual who first came to the attention of the clinician performing a pedigree study. Also, INDEX CASE, PROBAND.

propping agent *Petroleum Engineering.* a granular material introduced into drilling fluid during hydraulic fracturing so that fractures will be propped open and the fluid may be withdrawn more easily. Also, **proppant.**

propranolol *Pharmacology.* $C_{16}H_{21}NO_2$, a beta-adrenergic blocking agent that decreases cardiac rate and output and reduces blood pressure; effective in the prophylaxis of migraine.

proprietary *Pharmacology.* describing a medicinal substance that is protected against free competition by patent, trademark, copyright, or other means. Thus, **proprietary chemical, proprietary drug, proprietary medicine.** *Computer Science.* describing a program that is owned by an individual or software vendor, is usually copyrighted, and whose use requires a fee.

proprioception *Physiology.* the conscious perception of body position, balance, or movement, or of the location of a body part. Thus, **proprioceptive.**

proprioceptor *Physiology.* any sensory nerve terminal that gives information concerning movements and position of the body. *Robotics.* a transducer that transmits feedback signals concerning the position of a robot arm or end effector.

prop root *Botany.* a root that grows downward into the soil from above ground level to brace or give added support to the plant, as on a mangrove or a corn plant.

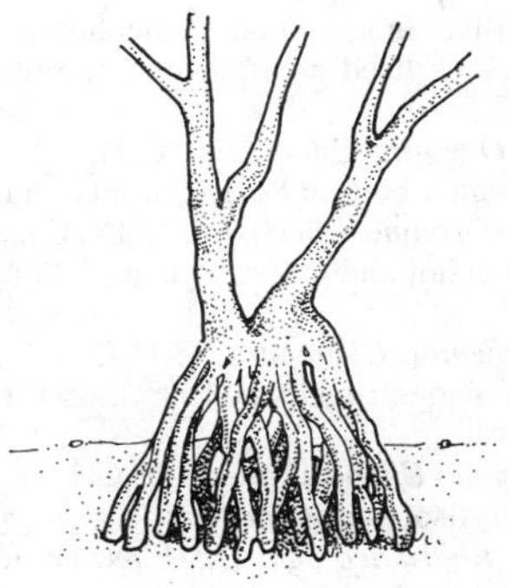

prop root

propterygium *Vertebrate Zoology.* the anterior of three principal basal cartilages in some paired fish fins, such as those of sharks and rays.

proptosis *Medicine.* the forward displacement or bulging of any organ, as the protrusion of the eyeball.

propulsion *Mechanics.* the process of applying a force, pushing ambient fluid out the propulsor at a velocity higher than the ambient from the rear end of the propulsor on an object, to cause it to move forward. Thus, **propulsive.**

propulsion system *Mechanical Engineering.* a system that produces a required change in the momentum of a vehicle by changing the velocity of the fluid or propellant passing through the propulsive device or engine.

propulsive coefficient *Naval Architecture.* the ratio of a vessel's effective horsepower (ehp) to its shaft horsepower (shp).
propyl [prō′pil] *Organic Chemistry.* being or containing a propyl group.
propyl- a combining form indicating the presence of a propyl group.
***n*-propyl acetate** *Organic Chemistry.* $C_3H_7OOCCH_3$, a flammable, colorless liquid with a pleasant odor; slightly soluble in water; miscible with alcohols; boils at 101–102°C; used as a solvent and a flavoring agent, and in perfumes, laquers, plastics, and resins.
propylacetone see METHYL BUTYL KETONE.
***n*-propyl acetylene** see 1-PENTYNE.
propyl alcohol *Organic Chemistry.* $CH_3CH_2CH_2OH$, a flammable, colorless liquid; soluble in water and alcohol; boils at 97.2°C; used as a solvent, antiseptic, and chemical intermediate, and in brake fluids. Also, PROPANOL.
propyl aldehyde see PROPIONALDEHYDE.
***n*-propylamine** *Organic Chemistry.* $C_3H_7NH_2$, a highly flammable, colorless liquid; soluble in water and alcohol; boils at 47.8°C; used as a sedative and a laboratory reagent.
propylbenzene *Organic Chemistry.* $C_6H_5C_3H_7$, a colorless liquid; soluble in alcohol and ether; sparingly soluble in water; boils at 159°C; used as a solvent and in dyeing textiles. Also, PHENYLPROPANE.
propylene [prō′pə lēn′] *Organic Chemistry.* $CH_3CH{=}CH_2$, a highly flammable, colorless gas; soluble in alcohol and ether, and slightly soluble in water; boils at −47.7°C; used in the manufacture of plastics and isopropyl alcohol.
propylene dichloride *Organic Chemistry.* $CH_3CHClCH_2Cl$, a colorless, stable liquid with a chloroformlike odor; insoluble in water; miscible with most common solvents; boils at 96.3°C; used as a solvent, dry-cleaning fluid, metal degreaser, and fumigant.
propylene glycol *Organic Chemistry.* $CH_3CHOHCH_2OH$, a colorless, viscous, hygroscopic, stable liquid; miscible with water and alcohol; boils at 187.3°C; used as an antifreeze, solvent, lubricant, plasticizer, preservative, and bactericide, and in suntan lotions and pharmaceuticals.
propylene glycol alginate *Organic Chemistry.* $C_9H_{14}O_7$, a white powder; soluble in water; used as a stabilizer, thickener, food additive, and emulsifier.
propylene glycol monomethyl ether *Organic Chemistry.* $C_4H_{10}O_2$, a colorless liquid; soluble in water and ether; boils at 120.1°C; used as a solvent for acrylics, dyes, and inks.
propylene glycol monoricinoleate *Organic Chemistry.* $C_{21}H_{30}O_4$, a pale yellow, oily liquid with a mild odor; insoluble in water; used as a plasticizer and lubricant.
propyleneimine *Organic Chemistry.* C_3H_7N, a flammable, colorless liquid; soluble in water; boils at 66–67°C; thought to be a carcinogen; used as an intermediate in organic synthesis. Also, 2-METHYLAZIRIDINE.
propylene oxide *Organic Chemistry.* C_3H_6O, a flammable, colorless liquid with an etherlike odor; soluble in alcohol and partially soluble in water; boils at 33.9°C; used as a solvent, fumigant, and in urethane foams.
propyl formate *Organic Chemistry.* $C_4H_8O_2$, a flammable liquid; slightly soluble in water; boils at 81.3°C; used as a flavoring.
***n*-propyl furoate** *Organic Chemistry.* $C_8H_{10}O_3$, a colorless, fragrant liquid; soluble in alcohol and ether; boils at 210.9°C; used as a flavoring.
propyl gallate *Organic Chemistry.* $C_3H_7OOCC_6H_2(OH)_3$, colorless crystals; soluble in alcohol; melts at 150°C; used to retard rancidity in cooking fats and oils.
propyl group *Organic Chemistry.* the $CH_3CH_2CH_2$– radical that is derived from propane. Also, **propyl radical.**
propylhexedrine *Pharmacology.* $C_{10}H_{21}N$, an adrenergic compound occurring as a clear, colorless liquid; inhaled as a decongestant.
propyliodone *Organic Chemistry.* $C_{10}H_{11}O_3NI_2$, white crystals; practically insoluble in water; melts at 186–187°C; used in medicine as a radiopaque medium.
propylite *Petrology.* a hydrothermally altered variety of andesite that resembles greenstone and consists of calcite, chlorite, epidote, serpentine, quartz, pyrite, and iron oxides.
***n*-propyl mercaptan** *Organic Chemistry.* C_3H_7SH, a liquid that boils at 67–73°C; used as an herbicide. Also, 1-PROPANETHIOL.
***n*-propyl nitrate** *Organic Chemistry.* $C_3H_7NO_3$, a whitish liquid that boils at 110°C; insoluble in water and soluble in alcohol and ether; used as a rocket fuel.
propylparaben *Organic Chemistry.* $C_{10}H_{12}O_3$, colorless crystals or white powder; slightly soluble in boiling water; melts at 95–98°C; used in medicine and as a food preservative and fungicide.
propylthiopyrophosphate *Organic Chemistry.* $C_{12}H_{28}P_2S_2O$, a yellowish liquid used as an insecticide.
propylthiouracil *Pharmacology.* $C_7H_{10}N_2OS$. a drug used to prepare patients for thyroid surgery and to maintain patients with hyperthyroidism who are poor surgical risks.
propyne see ALLYLENE.
Prorastomidae *Paleontology.* a single-genus family of sirenians known only from Jamaican fossils of the Middle Eocene; the skull of *Prorastomus* displays a more primitive structure than that of any other known Sirenian.
prorenin *Endocrinology.* the precursor of the enzyme renin.
Prorhynchidae *Invertebrate Zoology.* a family of common free-living flatworms in the class Turbellaria, having the mouth at the anterior end of the body, strong body musculature, and a male reproductive system opening into the buccal cavity.
Prorichthofenia *Paleontology.* a genus of aberrant strophomenid brachiopods characterized by the rootlike spines on their pedicle valve; extant in the Permian.
Prorocentraceae *Botany.* a family of unicellular flagellates comprising the order Prorocentrales.
Prorocentrales *Botany.* a monofamilial order of unicellular, thecate flagellates of the class Dinophyceae, usually occurring in marine waters and characterized by a lack of flagellar grooves, a theca composed of a left and right plate, and 8 to 14 minute plates.
Prorocentrum *Invertebrate Zoology.* a genus of marine dinoflagellate protozoans having an anterior spine; one species produces a shellfish toxin, and some species cause "red tides."
Prorodon *Invertebrate Zoology.* a genus of aquatic ciliated protozoans in the class Kinetofragminophorea, having uniformly ciliated, cylindrical or ovoid bodies.

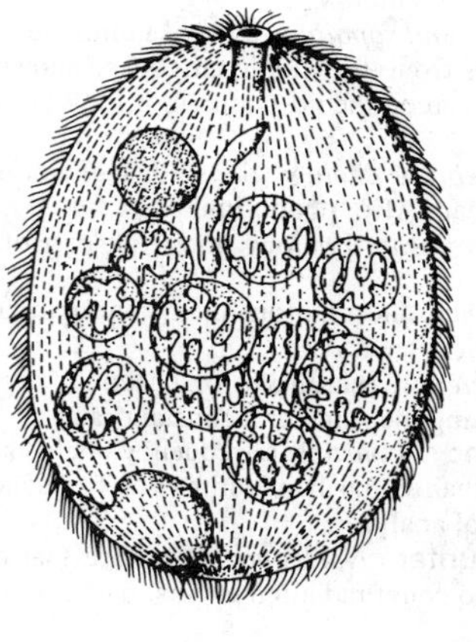

Prorodon

prorubricyte *Histology.* a basophilic normoblast.
Prosauropoda *Paleontology.* an infraorder of saurischian dinosaurs that arose from the theropods in the Triassic and developed along a different line from that of the predatory carnosaurs, leading to the quadrupedal herbivorous sauropods in the Jurassic; extant in the Triassic to Cretaceous.
prosector *Anatomy.* a person who dissects anatomical subjects for demonstration.
prosencephalon see FOREBRAIN.
prosenchyma *Botany.* an older term for the tissue of the woody and bast portions of plants, consisting of long, narrow, pointed cells. Thus, **prosenchymatous.**
Proseriata *Invertebrate Zoology.* an order of small, mostly marine turbellarian flatworms.
prosimian *Vertebrate Zoology.* **1.** a primitive living primate. **2.** an early ancestral primate.
proso- a combining form meaning "forward" or "anterior."
Prosobranchia or **Prosobranchiata** *Invertebrate Zoology.* a subclass of gastropod mollusks having two tentacles, two eyes, and a well-developed spiral or cone shell; includes winkles, whelks, and limpets. Also, STREPTONEURA.
prosodus *Invertebrate Zoology.* in sponges, a small channel between an incurrent canal and a flagellated cavity.

prosoma *Invertebrate Zoology.* the anterior part of the body in some invertebrates with no clear division between head and thorax, especially the cephalothorax of arthropods.

prosopalgia see TRIGEMINAL NEURALGIA.

prosopite *Mineralogy.* $CaAl_2(F,OH)_8$, a colorless to grayish-white monoclinic mineral occurring as small tabular crystals and in granular form, having a specific gravity of 2.894 and a hardness of 4.5 on the Mohs scale; found in cryolite deposits.

prosoplegia *Neurology.* a condition of paralysis of the face.

Prosopora *Invertebrate Zoology.* a suborder of freshwater annelid worms in the class Oligochaeta that easily break into separate segments, each of which can regenerate.

prosopospasm *Neurology.* a spasm of the facial muscles.

prosopyle *Invertebrate Zoology.* **1.** the tiny channel through a porocyte cell in ascon sponges. **2.** a small pore connecting an incurrent canal to a radial canal in syncon and lycon sponges.

prospect *Mining Engineering.* **1.** a property whose mineral value is believed to be economically significant but is yet to be proved. **2.** to carry on the process of prospecting.

prospecting *Mining Engineering.* the process of searching for minerals or oil by examining the surface or by drilling.

prospector *Mining Engineering.* a person who explores for valuable minerals or evaluates mineral discoveries.

prostacyclin *Endocrinology.* a prostaglandin, PGI_2, synthesized by endothelial cells lining the cardiovascular system; it is the most potent known inhibitor of platelet aggregation, as well as a powerful vasodilator, and thus is a physiologic antagonist of thromboxane A_2. Also, **prostaglandin I_2.**

prostaglandin *Endocrinology.* any of the family of cyclooxygenase metabolites of arachidonic acid. They are autacoid hormones, are widely distributed in many tissues, and exert diverse biological effects, including vasodilation, smooth muscle contraction or relaxation, and regulation of renal function.

prostate [präs´tāt] *Anatomy.* **1.** a muscular gland surrounding the urethra below the bladder; it contributes to the seminal fluid a secretion containing acid phosphatase, citric acid, and proteolytic enzymes, which account for the liquefaction of the coagulated semen. Also, **prostate gland, prostrata. 2.** of or relating to this gland. *Medicine.* affecting or involving this gland. Also, **prostatic.**

prostatectomy *Medicine.* the surgical removal of part or all of the prostate.

prostatitis *Medicine.* an inflammation of the prostate.

prostheca *plural,* **prosthecae.** *Microbiology.* an appendage of a bacterial cell that forms a narrow extension and is enclosed by the cell wall; in some bacteria, such as *Rhodomicrobium,* it has a reproductive function, while in others, such as *Caulobacter,* it has an adhesive function.

prosthecate bacteria *Microbiology.* bacteria that form one or more prosthecae during their life cycle.

Prosthecomicrobium *Bacteriology.* a genus of coccoid aquatic bacteria that possess prosthecae and divide by budding.

prosthesis [präs thē´sis] *plural,* **prostheses** [präs the´sēz]. *Medicine.* an artificial substitute for a missing part of the body, such as a hand, leg, eye, or tooth.

prosthetic [präs thet´ik] *Science.* serving as a substitute. *Medicine.* of or relating to prosthesis.

prosthetic group *Biochemistry.* a class of nonamino acid compounds that are tightly bound to proteins and facilitate enzyme activity.

prosthetics *Medicine.* the science and study of prostheses, including their design and application.

prosthion-basion line *Anthropology.* in anthropometry, the point of bone found between the upper central incisor teeth (prosthion) and the anterior margin of the foramen magnum in the median plane (basion).

prosthodontics *Medicine.* the branch of dentistry that deals with the replacement of teeth or their associated parts with artificial devices to restore and maintain oral functioning. Also, DENTAL PROSTHETICS.

Prostigmata *Invertebrate Zoology.* a suborder of mites in the order Acarina having one pair of spiracles near the mouth parts; some are parasites of plants and animals, and some are free-living predators.

prostomium *Invertebrate Zoology.* in annelid worms, an incomplete segment anterior to the mouth, modified for various purposes.

prostratigraphy *Geology.* a preliminary stage in the analysis of a rock unit, involving lithologic and paleontologic studies without consideration of the time factor. Also, PROPAEDEUTIC STRATIGRAPHY.

prostyle *Architecture.* an open portico of columns at the front of a building.

prot- a combining form meaning "first," "foremost," or "earliest."

protactinium *Chemistry.* a radioactive element having the symbol Pa, the atomic number 91, an atomic weight of 231.036, and a melting point around 1500°C; **protactinium-231,** the most stable of 13 isotopes with a half-life of about 33,000 years, is a hard, white, highly toxic metal produced by the disintegration of uranium and too rare for commercial use. (Literally, *proto-actinium.*)

protactinium-ionium age method *Geology.* the determination of the age, in years, of certain deep-sea sediments by measuring the ratio of radioactive protactinium-231 to ionium (thorium-230) present in the sediment under study.

protagon *Neurology.* a fatty substance found chiefly in the white matter of the brain, consisting of a mixture of cerebrosides and sphingomyelin.

protalus rampart *Geology.* a curved ridge of angular boulders and other coarse debris formed along the downslope margin of an existing or melted snowbank. Also, **protalus.**

protamine *Biochemistry.* any of a group of low-molecular-weight proteins, made up of about 30 arginine residues; found in salmon sperm DNA.

protandry *Physiology.* a form of hermaphroditism in which the male gonad matures before the female gonad. *Biology.* a form of dichogamy in which the male organs mature before the female organs are receptive.

protanomaly *Medicine.* a form of color blindness in which there is a diminished ability to differentiate between red, orange, yellow, and green.

protanopia *Medicine.* a form of color blindness in which there is inability to differentiate between red, orange, yellow, and green. Also, RED BLINDNESS.

protargol *Microbiology.* a silver-containing proteinaceous substance used as an antiseptic and for the staining of protozoan flagella.

Proteae *Bacteriology.* in some systems of classification, a tribe of Gram-negative, rod-shaped, motile, facultatively anaerobic bacteria of the family Enterobacteriaceae; known to ferment dextrose.

Protean behavior [prō´tē ən] *Behavior.* a behavior pattern in which an animal engages in some abrupt change in appearance or movement in order to deceive or elude a predator. (From *Proteus,* an ancient Greek god who could change his form at will.)

protease *Enzymology.* any enzyme, such as pepsin or trypsin, that catalyzes the hydrolysis of a protein during the first stage of its degradation to a simpler substance. Also, PROTEINASE, PROTOLYTIC ENZYME.

protectant *Pharmacology.* any agent, such as a sunscreen, that affords defense against a deleterious influence. *Plant Pathology.* any pesticide that is sprayed on a plant before it comes in contact with a pathogen.

protected location *Computer Technology.* a memory location whose contents are protected against accidental or intentional alteration, or unauthorized access.

protection code *Computer Technology.* the part of a task descriptor that lists the authorizations it has for various actions.

protection key *Computer Technology.* a bit string associated with the current program that must match the key of each block of memory that it references; if the keys do not match, then memory access is denied.

protectite *Petrology.* a rock produced by crystallization of a primary magma.

protective *Science.* affording defense or immunity. *Pharmacology.* see PROTECTANT.

protective barrier *Radiology.* a shield of lead, concrete, plastic, or other radiation-absorbing material that provides adequate body protection against radiation hazards.

protective capacitor *Electricity.* a power-line bypass capacitor, used to protect the line against overvoltage.

protective coating *Materials Science.* any surface coating that protects a component from wear, corrosion, or oxidation. Similarly, **protective finish.**

protective coloration *Zoology.* markings and coloring that serve as protective camouflage.

protective grounding *Electricity.* the use of a grounding system to protect humans from dangerous electric currents; consists of a neutral conductor as part of a secondary power-distribution system.

protective relay *Electricity.* a relay that protects equipment.

protective resistance *Electricity.* a resistance that is used in series with a gas tube or other tubular device to limit current flow to a safe level.

protector *Electricity.* **1.** a circuit breaker or fuse that acts to prevent electric equipment. **2.** a connection device, such as a safety ground or a ground-fault interrupter, that protects an operator from electric shock.

protector block *Electricity.* a spark-gap arrester used to ground voltages in excess of 350 volts; it consists of a usually rectangular piece of carbon with an insulated metal insert (or porcelain, with a carbon insert) forming a gap that breaks down and provides a path to ground.

protector gap *Electricity.* a spark gap arrester consisting of two carbon blocks separated by an air gap.

protector tube *Electronics.* a device, such as a glow-discharge tube or a cold-cathode tube, that is placed between electrodes to prevent extremely high voltages.

protectoscope *Optics.* an instrument that permits a soldier in a tank to see outdoors through the ports of the vehicle without being exposed to hostile fire; similar to the periscope used on a submarine.

Proteeae *Bacteriology.* in some classifications, a tribe of Gram-negative, rod-shaped, motile, facultatively anaerobic bacteria of the family Enterobacteriaceae, known to ferment dextrose.

proteid *Biochemistry.* another term for a protein. See PROTEIN.

Proteida *Vertebrate Zoology.* in some classification systems, a suborder coextensive with the family Proteidae.

Proteidae *Vertebrate Zoology.* the mudpuppies and olms, an aquatic salamander family in the suborder Salamandroidea, characterized by long bodies with relatively small limbs, feathery external gills, and incomplete metamorphosis; found in lakes and streams of eastern North America and southern Europe.

protein *Biochemistry.* any of a group of complex organic compounds, consisting essentially of combinations of amino acids in peptide linkages, that contain carbon, hydrogen, oxygen, nitrogen, and usually sulfur. Widely distributed in plants and animals, proteins are the principal constituent of the protoplasm of all cells and are essential to life. (Going back to a Greek word meaning "first" or "primary;" because of the fundamental role of proteins in sustaining life.)

protein A *Biochemistry.* a protein, present in the cell walls of many strains of *Staphylococcus aureus,* that binds to the fragment of a variety of immunoglobulins; used to collect antigen and antibody complexes.

proteinaceous [prō´ti nā´shəs] *Biochemistry.* relating to or composed of proteins.

proteinase see PROTEASE.

protein biosynthesis *Molecular Biology.* a process by which a given sequence of mRNA information is used to produce a specific chain of amino acids to yield a protein.

protein blotting *Biotechnology.* a method for identifying proteins in a complex mixture; separated proteins are transferred from a gel medium to a protein-binding nitrocellulose sheet for analysis.

protein-bound iodine *Biochemistry.* iodine that binds to proteins in the blood.

protein C see FACTOR XIV.

protein coat see CAPSID.

protein determination *Biotechnology.* any of a variety of techniques that use spectrophotometric devices or amino acid analyzers to measure the amount of protein in a given sample.

protein efficiency ratio *Nutrition.* a method for determining the nutritive value of a protein, determined by measuring the gain in weight of young rats that are fed a diet which contains 10% of the particular protein.

protein engineering *Biotechnology.* a technology by which proteins can be isolated and studied by manipulation of the genes that encode them.

protein equivalent *Nutrition.* the corresponding nutritional value of a protein based on the amino acid composition of the protein, the digestibility of the protein, and the availability of the digested products.

protein fibers *Textiles.* a class of fibers encompassing both natural animal fibers, such as silk and wool, and fibers derived from protein-containing substances such as corn zein, peanuts, and soy beans.

protein folding *Biochemistry.* a process in which a protein spontaneously forms into its correct tertiary structure and is held by cystine bonds and by attractive forces between atoms.

protein g *Biochemistry.* a protein, present in the cell walls of the group *G Staphylococci,* that resembles protein A in binding only at the Fc section of the immunoglobulin G molecule, but differs by binding to a wider portion of the subclass of this molecule.

protein hormone receptors *Endocrinology.* specific molecules that are found on the cell surface and are necessary for the action of such hormones as insulin and growth hormone.

protein kinase *Enzymology.* an enzyme that phosphorylates proteins, specifically to the amino acids serine, threonine, and tyrosine.

protein-lysine 6-oxidase see LYSYL OXIDASE.

protein quality *Nutrition.* the relative nutritional value of a protein based on the amino acid composition of the protein, the digestibility of the protein, and the availability of the digested products; the relative biological value defined in various terms, including the fraction of absorbed nitrogen that is retained in the body, the growth rate of young animals as a function of the dietary level of the protein, and the minimal protein concentration required to establish nitrogen balance in adults.

protein S *Hematology.* a vitamin K-dependent plasma protein that inhibits blood clotting by serving as a cofactor for activated factor XIV (protein C).

protein structure *Molecular Biology.* any of the several parameters by which protein conformation is evaluated, such as the amino acid sequence, interactions with the polypeptide chain, and subunit association.

protein synthesis *Molecular Biology.* a process by which a given sequence of mRNA information produces a specific chain of amino acids to yield a protein.

proteinuria *Medicine.* the presence of excess protein, usually albumin, in the urine; when persistent, it is usually a sign of renal disease or renal complications of another disease, but proteinuria may also result from fever or heavy exercise.

Proteocephaloidea *Invertebrate Zoology.* an order of tapeworms in the subclass Cestoda having four suckers; parasites of fishes, amphibians, and reptiles.

proteoglycan *Biochemistry.* a protein, present in connective tissue and cartilage, that is glycosylated to a variety of polysaccharide chains, frequently sulfated.

proteolysin *Enzymology.* an enzyme that catalyzes the hydrolysis of peptide bonds in proteins. Also, **proteolytic enzyme.**

proteolysis *Biochemistry.* the hydrolysis of peptide bonds in proteins.

Proteomyxida *Invertebrate Zoology.* the only order in the subclass Proteomyxidia.

Proteomyxidia *Invertebrate Zoology.* a subclass of protozoans in the class Actinopodea, having filopodia or reticulopodia and no skeleton or shell; parasites of algae and other aquatic plants.

proteoplast *Cell Biology.* a plant cell plastid that stores proteins.

proteose *Biochemistry.* a secondary protein derivative or a mixture of split products formed by a hydrolytic cleavage of the protein molecule more complete than that which occurs with the primary protein derivatives, but not so complete as that which forms amino acids.

protero- a combining form meaning "earlier" or "former."

Proteromonadida *Invertebrate Zoology.* an order of flagellate protozoans in the class Zoomastigophora having one or two pairs of flagella and a single mitochondrion.

Proteromonas *Invertebrate Zoology.* a genus of protozoa in the order Proteromonadida that are common parasites in the intestines of amphibians, reptiles, and some mammals.

Proterosuchia *Paleontology.* the earliest known archosaurians, a suborder of reptiles in the extinct order thecodontia; probably ancestral to the thecodonts and dinosaurs; extant in the Permian and Triassic.

Proterotheriidae *Paleontology.* a family of horselike South American ungulates in the order Litopterna; the proterotheres arose in the Paleocene, and they became extinct in the Pliocene at about the time that true horses immigrated to South America.

Proterozoic *Geology.* **1.** the geologic time representing the more recent division of the Precambrian period, characterized by unmetamorphosed sedimentary rock. **2.** the oldest era of earth history, corresponding to the entire Precambrian period.

Proteutheria *Vertebrate Zoology.* an equivalent name for the family Tupaiidae, the tree shrews.

prothoracic gland *Entomology.* an endocrine gland on the prothorax of insects that secretes molting hormones.

prothorax *Entomology.* the first segment of an insect's thorax, containing the first pair of legs.

prothrombin *Biochemistry.* a protein found in plasma that produces thrombin when triggered by prothrombiase. Also, FACTOR II, THROMBINOGEN.

prothrombin time *Pathology.* the clotting or coagulation period of blood in lab reagents, normally 10–12 seconds, and the screening for unusual action of factors involved in coagulation.

prothymosin α *Endocrinology.* the precursor peptide for the thymic hormone thymosin α_1; unlike many prohormones, prothymosin α is believed to be biologically active and to play an immunoregulatory role along with other thymic factors.

proticity *Biochemistry.* the movement of protons from high-to-low potential energy in the proton circuit of an oxidative phosphorylation.

protist *Biology.* any member of the kingdom Protista.
Protista *Biology.* one of five kingdoms in a commonly used system of classification; it includes eukaryotic unicellular organisms such as protozoa, paramecia, bacteria, yeasts, slime molds, and unicellular algae.
protium see LIGHT HYDROGEN.
proto- or **prot-** a combining form meaning "first," "foremost," or "earliest."
protoactinium see PROTACTINIUM.
Protoalcyonaria *Invertebrate Zoology.* an order of deep-water, permanently solitary, monomorphic octocorals.
Protobranchia *Invertebrate Zoology.* a subclass of primitive bivalve mollusks with gills consisting of two rows of filaments hanging down from the mantle. Also, **Protobranchiata.**
Protoceratidae *Paleontology.* an extinct family of North American selenodont artiodactyls, classified by some in the suborder Tylopoda but really still *incertae sedis*; characterized by long limbs and up to three pairs of horns, from the nose to the back of the head; extant in the Eocene to Pliocene.
protocerebrum *Entomology.* the first set of supra-oesophageal ganglia that comprise the front of the arthropod "brain" of an insect and innervate the compound eyes.
Protochordata *Invertebrate Zoology.* a subphylum of animals of phylum Chordata with gill slits, a hollow dorsal nervous system, and a notochord, but no heart and no brain.
protoclastic *Petrology.* of or relating to igneous rocks exhibiting granulation and deformation of earlier-formed crystals of constituent minerals, because of the differential flow of the magma before consolidation.
Protococcidiida *Invertebrate Zoology.* an order of primitive parasitic protozoans of phylum Apicomplexa, of which merogamy is absent; parasitic to marine annelids, although pathogenicity is not known.
protocol [prōt′i kôl; prōt′i kōl] a formal or customary procedure; specific uses include: *Medicine.* **1.** an explicit, detailed plan of an experiment or treatment. **2.** the original notes made on a necropsy, experiment, or instance of disease. *Computer Programming.* **1.** a set of rules to be followed in communication with a computer device or program. **2.** see COMMUNICATION PROTOCOL. **3.** in an object-oriented system, the interface to a class, i.e., the set of messages understood by members of the class. *Artificial Intelligence.* a transcript of what a human subject says when asked to "think aloud" while solving a problem.
protocol converter *Computer Technology.* a device that translates the data transmission code and protocol of one computer or peripheral device to that of another computer or device, thereby enabling communication between otherwise incompatable devices.
protoconch *Invertebrate Zoology.* the oldest part of a gastropod shell, representing the area where shell development started in the larva.
protoconodonts *Paleontology.* an early group of animals that have until recently been considered primitive conodonts with very small and weakly phosphatized teeth; some authorities would now classify them together with paraconodonts but separate from the larger group known as euconodonts, the "true conodonts;" extant from the Upper Precambrian to Lower Ordovician.
Protocucujidae *Invertebrate Zoology.* a small family of tiny beetles, coleopteran insects in the superfamily Cucujoidea, with an elongate body, and a prognathous head bearing protuberant eyes.
protoculture *Anthropology.* the simplest aspects of culture noted in the nonhuman primates, such as toolmaking and other innovative learned behaviors.
protoderm see DERMATOGEN.
protodiastolic *Cardiology.* relating to early diastole, immediately following the second heart sound.
protodolomite *Mineralogy.* a crystalline carbonate of calcium and magnesium with a disordered structure, differentiated from the mineral dolomite in that the metallic ions are found in the same crystallographic layers rather than in alternating layers.
Protodonata *Paleontology.* an extinct class of large-winged, predaceous insects; a Carboniferous offshoot of the Odonata, the class that includes the modern dragonflies; achieved a wingspan of up to two feet; extant from the Carboniferous to Triassic.
Protodrilidae *Invertebrate Zoology.* a family of polychaetes in the Archiannelida having a small, thin body that is dorso-ventrally flattened, a heavily ciliated mouth, and no parapodia; found in marine and brackish waters.
protoenstatite *Mineralogy.* a modified, unstable form of $Mg_2Si_2O_6$, artificially produced by thermal decomposition and grinding of talc; formerly known as metatalc.
Protoeumalacostraca *Paleontology.* a name given to certain putative eumalacostracan crustaceans of the early Devonian that seem ancestral to the more advanced shrimplike animals that appeared in the great eumalacostracan radiation of the late Devonian and are still widely represented today.
protofilament *Cell Biology.* any of thirteen filaments contained in a single microtubule structure.
protogalaxy *Astronomy.* an immense concentration of gas and dust from which a galaxy is formed.
Protogastraceae *Mycology.* a family of rare fungi of the order Protogastrales, occurring exclusively on the roots of plants in Maine.
Protogastrales *Mycology.* an order of fungi of the class Gasteromycetes, characterized by having a double peridium and a small fruiting body with one hymenial cavity.
protogenic *Chemistry.* of an acidic solvent, capable of donating protons.
protogyny *Physiology.* a form of hermaphroditism in which the female gonad matures before the male gonad.
protohistory *Archaeology.* **1.** a transitional time period between prehistory and recorded history, for which both archaeological and historical data are employed. **2.** a branch of study concerned with this period. Thus, **protohistoric(al), protohistorian.**
protointraclast *Geology.* a component of limestone that was formed during a premature attempt at resedimentation while the limestone was still in a loose, viscous, or plastic state.
protolanguage *Linguistics.* an ancestral language that is reconstructed by comparing elements in known related languages and creating a hypothetical form of their common parent language.
protolith *Petrology.* the unmetamorphosed rock from which a given metamorphic rock is derived. Thus, **protolithic.**
protolytic enzyme see PROTEASE.
protomerite *Cell Biology.* the smaller, anterior portion of the cell of a gregarine.
protometer *Cell Biology.* a sensor that is theorized to exist in a cell to detect and respond to alterations in proton motive force.
Protomonadida *Invertebrate Zoology.* an order of protozoans in the class Mastigophora, having one or two flagella; both free-living and parasitic forms. Also, **Protomonadina.**
Protomycetaceae *Mycology.* the only family of fungi of the order Protomycetales.
Protomycetales *Mycology.* an order of fungi of the class Hemiascomycetes that is parasitic and pathogenic to flowering plants, causing lesions and galls.
protomylonite *Petrology.* a mylonitic rock formed from contact-metamorphic rock, characterized by granulation and flowage resulting from overthrusts controlled by the contact surfaces between the intrusion and the country rock.
Protomyzostomidae *Invertebrate Zoology.* a small family of hermaphroditic polychaetes with an elongate-oval, dorso-ventrally flattened body, five pairs of lateral organs, and no lateral cirri; endoparasites of basket stars.
proton [prō′tän] *Physics.* a stable elementary particle found in the nuclei of ordinary matter having a positive charge of approximately 1.602×10^{-19} coulombs, a mass of 1.007593 amu or 1.67239×10^{-24} grams, and a spin quantum number of 1/2.
proton accelerator *Nucleonics.* a device designed to accelerate protons to high energies.
proton acid see BRÖNSTED ACID.
protonate *Chemistry.* to provide a proton (to an atom, molecule, etc.).
proton binding energy *Physics.* the minimum amount of energy required to extract a proton from a nucleus.
proton bridge *Materials Science.* an atomic bond characteristic of hydrogen bonding, in which the hydrogen ion simultaneously attracts two different electronegative atoms, forming a bridge between them.
proton capture *Nuclear Physics.* a type of reaction in which a proton merges with a nucleus.
proton channel *Cell Biology.* a passage in the inner mitochondrial membrane that facilitates the movement of protons out of the mitochondrial matrix during oxidative phosphorylation.
proton-electron-proton reaction *Nuclear Physics.* a type of nuclear reaction in which two protons and an electron combine to form a deuteron and a neutrino.
protonephridium *Invertebrate Zoology.* an excretory organ found in most flatworms, rotifers, and certain polychaetes, used in discharging water and waste products.

protoneuron *Neurology.* the first neuron in a peripheral reflex arc. *Invertebrate Zoology.* in low metazoa, a unit of the nerve net that lacks polarization.

proton gradient *Biochemistry.* the gradient of hydrogen ions set up across a membrane by active pumping of protons from one side to the other.

protonic acid see BRÖNSTED ACID.

proton-induced X-ray emission *Spectroscopy.* an analytical method in which a sample is bombarded with protons and the energy spectrum of the resulting X-rays is used to identify the elements present in the sample.

protonium *Atomic Physics.* a state in which a proton and an antiproton are bound together.

proton magnetometer *Electromagnetism.* a sensitive device designed to measure the frequency of the proton resonance in ordinary water.

proton microscope *Electronics.* an instrument that uses protons instead of electrons to form the image of minute objects for viewing.

proton motive force *Biochemistry.* the electrochemical gradient generated by a membrane potential and a protein gradient across a membrane.

proton number see ATOMIC NUMBER.

proton-proton chain *Nuclear Physics.* a group of reactions in which energy is released through a series of fusion reactions that convert hydrogen into helium; considered to be an important source of energy production in hydrogen-rich stars.

proton-proton reaction *Nuclear Physics.* the primary or commencing reaction in a proton-proton chain.

proton pump *Biochemistry.* a term for any action that furthers the proton transport process.

proton-recoil counter *Nucleonics.* a device that detects energetic neutrons by observing the recoil of protons.

proton resonance *Spectroscopy.* the absorption of energy at certain characteristic frequencies by protons that are subjected simultaneously to a static magnetic field and an alternating magnetic field.

proton-scattering microscope *Solid-State Physics.* a microscope in which protons are discharged from a cold cathode, then accelerated and focused onto a crystal in an evacuated chamber; reflections from the crystal surface are recorded as they strike a fluorescent screen.

proton-stability constant *Physical Chemistry.* the reciprocal of the dissociation constant for a weak base in solution.

proton storage ring *Nucleonics.* a device designed to store beams of high-energy protons by means of evacuated circular chambers and magnetic fields.

proton synchrotron *Nucleonics.* a device designed to accelerate protons in a circular orbit of constant radius by means of a time-varying magnetic field.

proton translocators *Biochemistry.* a group of ionophores that enhance the permeability of protons in a lipid bilayer of membrane.

proton transport *Biochemistry.* the movement of protons across a membrane that is mediated by membrane proteins such as bacteriorhodopsin.

proton vector magnetometer *Electromagnetism.* a proton magnetometer that permits measurement of horizontal intensity or vertical intensity as well as total intensity.

Protoodiniaceae *Botany.* a monospecific family of marine algae of the order Blastodiniales, living as ectoparasites on the hydromedusa of certain hydrozoans.

proto-oncogene *Oncology.* a normal cellular gene from which an oncogene can arise as a result of mutation or recombination.

Protophallaceae *Mycology.* a family of fungi whose classification is in transition, occurring in tropical to suptropical regions.

protophilic *Chemistry.* of a basic solvent, able to combine with a hydrogen ion; acting as a proton acceptor.

Protophyta *Botany.* in some taxonomic systems, the lowest division of the plant kingdom, containing the algae and variously defined to include also the blue-green algae, yeasts, fungi, lichens, bacteria, and viruses.

protophyte *Botany.* **1.** any member of the division Protophyta. **2.** any simple plant.

protoplanet *Astronomy.* a planet in the early stage of its formation, when it has accreted nearly all its present mass.

protoplasm [prōt′ə plaz′əm] *Cell Biology.* a general term for the basic material of all plant and animal cells, composed mainly of nucleic acids, proteins, lipids, carbohydrates, and inorganic salts; this includes the cytoplasm, nucleoplasm, and their contents. (From a Greek phrase meaning "first form" or "first material.")

protoplasmic [prōt′ə plaz′mik] *Cell Biology.* relating to or consisting of protoplasm.

protoplast [prōt′ə plast] *Cell Biology.* a plant cell from which the cell wall has been enzymatically removed.

protoplast fusion *Molecular Biology.* an in vitro genetic transfer process in which a hybrid protoplast is formed by fusing protoplasts from different strains, species, or genera.

protopodite *Invertebrate Zoology.* the basal part of the biramous crustacean leg, to which the endopodite and the exopodite are attached; consists of the coxopodite and the basopodite. Also, **protopod.**

Protopteridales *Paleontology.* an extinct order of fernlike plants that produced sporangia at the edges of their leaves; formerly thought to be a primitive fern, but affinities now considered uncertain; may be related to the Aneurophytales; extant in the Devonian.

protoquartzite *Petrology.* a well-sorted variety of sandstone intermediate in composition between orthoquartzite and subgraywacke, with 75–95% quartz and chert, less than 15% detrital clay matrix, and 5–25% unstable materials. Also, QUARTZOSE SUBGRAYWACKE.

Protoraphidaceae *Botany.* a family of marine diatoms of the order Pennales, distinguished by the lack of a raphe and by a transverse or oblique row of labiate processes at each pole.

protore *Mining Engineering.* a primary mineral deposit whose economic viability depends upon enrichment or upon some combination of technological advancement and market demand.

Protosiphonaceae *Botany.* a monospecific family of green algae of the order Chlorococcales, characterized by single cells, often occurring in extensive patches, that have a bulbous upper portion and a colorless rhizoid penetrating moist soil.

Protosirenidae *Paleontology.* a monogeneric family of sirenians of the Middle Eocene; widely distributed in tropical to temperate waters, the **protosirens** are generally considered to be ancestral to the modern dugongs and manatees.

Protospondyli see SEMIONOTIFORMES.

protostar *Astronomy.* a contracting gas cloud in the embryonic stage of star formation before the onset of thermonuclear reactions in its interior.

protostele *Botany.* the simplest type of stele with a solid core of xylem surrounded by phloem, such as in most roots and the first-formed portion of some primitive stems.

protostoma see BLASTOPORE.

protostomate *Invertebrate Zoology.* relating to, belonging to, or resembling Protostomia.

protostome *Invertebrate Zoology.* any member of Protostomia.

Protostomia *Invertebrate Zoology.* a series of the Eucoelomata, including the mollusks, annelids, and arthropods, all having a mouth arising from the blastopore. Also, ECTEROCOELIA.

protostratigraphy see PROSTRATIGRAPHY.

Protosuchia *Paleontology.* an extinct suborder of medium-sized archosaurian reptiles in the order Crocodilia; extant in the Upper Triassic and Lower Jurassic.

protothecosis *Pathology.* a disease that is produced by *Prototheca zopfri* and may be introduced into the body by trauma or implantation in the tissue.

Prototheria *Vertebrate Zoology.* a mammalian subclass containing the single order Monotremata, or egg-laying mammals. (All other mammals are placed in the subclasses Allotheria or Theria.) The prototherians arose in the Late Triassic and diversified widely in the Jurassic and Cretaceous; they declined thereafter, and are represented today only by the platypus and the echidnas, which are found in Australia.

prototroch *Invertebrate Zoology.* a ring of ciliated cells located above the mouth in trochophore larvae.

prototroph *Biology.* **1.** a microorganism, usually a microbial mutant, that has the same requirements for growth as the original strain. **2.** an organism that is able to synthesize all of its metabolites from inorganic material and a carbon source such as glucose; it usually has no nutritional requirement. Thus, **prototrophic.**

prototropy *Organic Chemistry.* the interconversion of structural isomers by transfer of a proton.

prototype [prōt′ə tīp′] an early or original form; specific uses include: *Biology.* a primitive or ancestral form of an organism. *Systematics.* see TYPE SPECIES. *Engineering.* a full-scale model of a structure or piece of equipment, used in evaluating form, design, fit, and performance. *Psychology.* the characteristic model or example on which a concept is based.

protoxylem *Botany.* the first xylem to be formed from the procambium, usually lying closest to the pith and maturing as the organ elongates.

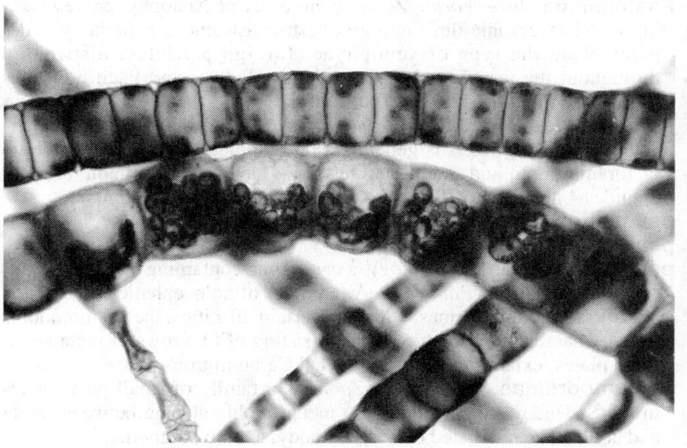

Protozoa

Protozoa [prō´tə zō´ə] *Biology.* a phylum or subkingdom including all single-celled animals with membrane-bound organelles; they may be aquatic or parasitic, with or without a test, solitary or colonial, sessile or free-swimming, moving by cilia, flagella, or pseudopodia. (From a Greek phrase meaning "first animal.")

protozoa *Biology.* a plural of *protozoan.*

protozoan [prō´tə zō´ən] *plural,* **protozoa** or **protozoans.** *Biology.* **1.** an individual organism of the phylum or subkingdom Protozoa, such as an ameba or a paramecium. Also, PROTOZOON. **2.** of or relating to the Protozoa.

protozoology *Biology.* the scientific study of Protozoa.

protozoon see PROTOZOAN.

protractor *Engineering.* an instrument for measuring angles on drawings or for drawing lines at specific angles; usually consists of a half or full circle, of a suitable material such as plastic, marked in degrees. *Surgery.* a surgical instrument used to extricate bits of bone, a bullet, or other foreign matter from a wound.

protrude *Science.* to project or cause to project; thrust forward.

protrusion *Science.* the fact of protruding, or a thing that protrudes.

protuberance *Physiology.* **1.** a projecting part or prominence. **2.** a swelling or bulge. Thus, **protuberant.**

Protura *Invertebrate Zoology.* an order of tiny wingless insects without eyes or antennae and with piercing mouthparts for consuming fungal cells; found in soil and decayed organic matter.

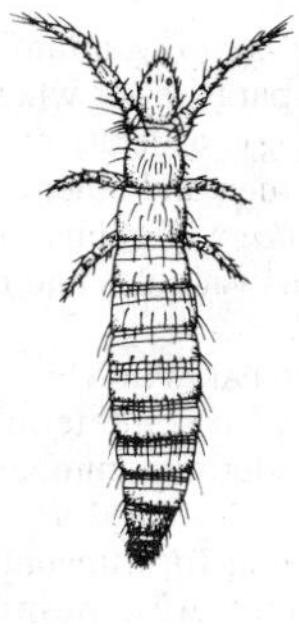

Protura

protyle *Science.* a theoretical substance that was formerly thought to be the source of all actual chemical elements. Also, **protyl.**

proud flesh *Surgery.* the proliferation of exuberant granulation tissue at the site of a surface wound or ulcer.

Proust, Joseph Louis 1754–1826, French chemist; proposed and confirmed the law of definite proportions (or constant composition).

proustite *Mineralogy.* Ag_3AsS_3, a scarlet red trigonal mineral, dimorphous with xanthoconite, massive in habit or as prismatic or rhombohedral crystals, having a specific gravity of 5.55 to 5.64 and a hardness of 2 to 2.5 on the Mohs scale; found in low-temperature vein deposits. Also, LIGHT-RED SILVER ORE, LIGHT RUBY SILVER, RUBY SILVER.

Prout, William 1785–1850, English chemist; formulated Prout's hypothesis; discovered stomach acids; classified foods into groups.

Prout's hypothesis *Physical Chemistry.* the earlier conception that all atoms are composed of hydrogen atoms, or of some other universal substance.

provenance *Geology.* the original place or area, including the associated rocks, from which the constituent particles of a sedimentary rock or facies are derived. Also, PROVENIENCE.

provenience *Archaeology.* the recording of the location where an artifact or feature was found. *Geology.* see PROVENANCE.

proven reserves *Petroleum Engineering.* oil reserves that have been discovered and proved recoverable by production at commercial flow rates. Also, **proved reserves.**

proventriculus *Entomology.* a section of the insect foregut; in an insect that eats solid food, it is modified for grinding and shredding; in sucking insects, it consists of a valve opening into the midgut; in bees and certain beetles, it permits fluids but not solid food to enter the midgut, as when separating pollen from nectar.

Providencia *Bacteriology.* a genus of Gram-negative bacteria of the tribe Proteeae, occurring as motile, rod-shaped cells and typically associated with human urinary tract infections.

province *Oceanography.* a general term for a division of the ocean based on the physical characteristics of the ocean environment itself, as distinguished from life-zone divisions such as littoral, neritic, or pelagic. *Geology.* see PHYSIOGRAPHIC PROVINCE.

provincial series *Geology.* a time-stratigraphic series of rocks associated only with a particular region and corresponding to a major division of geologic time.

proving ground *Ordnance.* a site where weapons, military vehicles, and so on are tested. *Science.* any place or context in which something is tested, such as a theory or a piece of equipment.

proving ring *Mechanical Devices.* a calibration test ring with an elastic shell used for determining the accuracy of various machines by the application of a force along its variable diameter.

proving stand see TEST STAND.

provirus *Virology.* a viral genome that has become integrated into the chromosomal DNA of the host cell or into a plasmid.

provisional ligature *Surgery.* a suture applied during the initial stages of an operation, and removed before completion of the procedure.

provisional unit *Military Science.* personnel and equipment that are temporarily organized for a limited period of time to accomplish a specific mission.

provitamin *Biochemistry.* a substance that produces vitamins by a natural reaction, as when 7-dehydrocholesterol forms vitamin D_3 in the skin in the presence of sunlight.

provitamin A see CAROTENE.

provitrain *Geology.* a vitrain in which some plant structures are visible under a microscope. Also, TELAIN.

provitrinite *Geology.* a variety of vitrinite in which plant cell structure is microscopically visible. Also, PHYLLOVITRINITE.

prow *Naval Architecture.* the bow or front part of a vessel.

prox. proximate.

proxemics *Anthropology.* the study of how body language and the space between individuals expresses certain cultural standards for social interaction.

Proxima Centauri *Astronomy.* a red dwarf star belonging to the Alpha Centauri triple star system; at a distance of only 4.22 light-years, it is the sun's closest neighboring star.

proximal *Anatomy.* **1.** situated closer to any point of reference. **2.** the point of attachment of a muscle or limb closest to the tubule. *Robotics.* near the base or pedestal and away from the end effector of a robot.

proximal clot *Hematology.* a clot formed in a blood vessel proximal to a ligature.

proximal convoluted tubule *Anatomy.* any of the twisted and coiled tubes leading from the glomeruli within the kidneys.

proximal factor *Behavior.* a mechanical feature, especially a physiological feature, that causes a behavior.

proximate analysis *Chemical Engineering.* an arbitrary and empirical test for analyzing compounds in a mixture, allowing predictions to be made of the fuel's behavior in a furnace.

proximate factor *Behavior.* an immediate environmental change that elicits a normal response in certain species, e.g., the onset of winter as a cue for hibernation.

proximity effect *Electricity.* a redistribution of current in one conductor caused by the nearby location of another current-carrying conductor.

proximity fuse *Ordnance.* a fuse that initiates detonation by remotely sensing the presence, distance, and possibly the direction of a target or its environment; it may function by means of a signal generated by the fuse or emitted by the target, or by sensing a change in a natural field surrounding the target. Also, INFLUENCE FUSE, VT FUSE.

proximity sensor *Robotics.* a sensor that uses light, sound, magnetic fields, or other forces to detect or measure the distance to an object without touching it. Also, NONCONTACT SENSOR.

proximoceptor *Physiology.* a specialized sense organ able to detect stimuli close to the body.

proxytype see NEOTYPE.

prozone *Immunology.* an occurrence in which a visible agglutination reaction does not occur in an antibody-antigen mixture, due to the presence of excess antibodies.

PRR pulse repetition rate.

prs. pairs; preconscious.

prudent limit of endurance *Aviation.* the time period in which an aircraft may remain airborne and still retain a safe margin of fuel.

prune *Agriculture.* to cut living or dead parts, such as twigs, branches, or roots, from a plant in order to increase fruit or flower production, stimulate growth, or alter form. *Artificial Intelligence.* to discard certain nodes from a search tree once it has been determined that they are not part of the solution or do not look promising. *Food Technology.* any of several varieties of plum that can be dried without spoiling.

Prunellidae *Vertebrate Zoology.* the accentors or hedge sparrows, a monogeneric family of small birds of the psserine suborder Oscines, characterized by creeping habits and a solitary lifestyle; found in Eurasian forests and meadows.

pruning shears *Mechanical Devices.* shears with long, sharp blades used to trim branches of plants.

pruritus *Medicine.* **1.** an uncomfortable sensation of the skin that provokes the desire to rub or scratch; itching. **2.** any of various conditions marked by itching, especially when specified by a modifying term, as in **pruritus senilis, pruritus hiemalis.**

pruritus vulvae *Medicine.* severe itching of external female genitalia.

Prussian blue *Inorganic Chemistry.* dark blue crystals of ferriferrocyanide; used as a pigment. Also, IRON BLUE, BLUE IRON FERROCYANIDE.

prussiate *Toxicology.* an older term for cyanide and its compounds.

prussic acid see HYDROCYANIC ACID.

pry bar *Mechanical Devices.* a flat, thin steel bar, approximately twelve inches long, with beveled notches in both ends, one slightly curved and one bent at a 90° angle; used for removing nails and spikes and for prying off wood.

Prymnesiaceae *Botany.* a family of marine and freshwater algae of the order Prymnesiales, characterized by a dominant motile phase and a short or long haptonema that may coil in some species; one genus, *Prymnesium,* contains species that produce a toxin lethal to fish.

Prymnesiales *Botany.* an order of marine and freshwater algae of the class Chrysophyceae, characterized by cells that have either a motile or nonmotile phase dominant in their life history and that are covered with unmineralized organic scales.

Prymnesiophyceae *Botany.* a class of mostly unicellular and marine flagellate algae of the division Chromophycota, characterized by cells that are usually covered with one or more layers of organic scales.

Prymnesium *Botany.* a genus of flagellate algae of the family Prymnesiaceae, noted for producing a toxin that can kill fish; particularly troublesome in some brackish fish ponds in Israel.

Pryophacaceae *Botany.* a small family of thecate, photosynthetic, marine dinoflagellates of the order Peridiniales.

pryrrolidine *Organic Chemistry.* C_4H_9N, a yellowish liquid that boils at 87°C; used in pharmaceuticals, insecticides, and fungicides.

ps picosecond; picoseconds.

p.s. or **ps** per second.

PS power supply; power steering.

PS3 *Bacteriology.* a strain of thermophilic bacteria of uncertain affiliation.

psalterium see OMASUM.

psamm- or **psammo-** a combining form denoting a resemblance or relationship to sand.

Psamment *Geology.* a suborder of the soil order Entisol, characterized by having the texture of a loamy, fine or coarse sand, and containing less than 35% coarse fragments.

Psammetidae *Invertebrate Zoology.* a primitive family of large protozoans of the class Xenophyophorea, characterized by a strong test with a haphazard arrangement of xenophyae.

Psamminida *Invertebrate Zoology.* an order of Xenophyophores lacking linnelae (organic threads) in the test; consistency of the body is dependent on the type of xenophyae (foreign particles) distributed throughout the test; found in the Atlantic, Indian, and Pacific Oceans and in the Antarctic; includes four families.

Psamminidae *Invertebrate Zoology.* a family of xenophyophores characterized by a test that is generally solid (sometimes fragile), made up of foraminiferan and radiolarian tests and having little organic cement; includes 4 genera and 10 species, with diameters of 5–50 mm.

psammite see ARENITE.

psammitic see ARENACEOUS.

psammocarcinoma *Oncology.* a carcinoma containing calcium.

Psammodontidae *Paleontology.* a family of holocephalic fishes in the order Psammodontiformes; still of uncertain affinities, the psammodonts have a characteristic tooth pattern consisting of two rows of rectangular tooth plates; extant in the Devonian and Carboniferous.

Psammodrilidae *Invertebrate Zoology.* a family of small polychaetes in the Sedentaria, with little or no metamorphic change between larval and adult stage, a ciliated cylindrical body, and no parapodia.

psammoma *Oncology.* a tumor derived from fibrous tissue of the meninges and occurring along the blood vessels of the membranes of the brain and skull; characterized by the formation of psammoma bodies. Also, SAND TUMOR.

psammoma bodies *Pathology.* oval-shaped, microscopic, concentric sandlike substances that are found in the skin and connective tissue tumors and are linked to chronic inflammations.

psammon *Ecology.* a community made up of microscopic plants and animals that live in-between the grains of sand along sea shores and lake-shore areas.

psammophilic *Ecology.* describing organisms that live or thrive in sandy soil. Also, **psammophilous.**

psammophyte *Ecology.* a plant that lives or thrives in sandy soil.

psammosarcoma *Oncology.* a sarcoma containing a sandy deposit.

psammosere *Ecology.* an ecological succession that begins in an environment of barren sand.

psec. or **psec** picosecond.

Pselaphidae *Invertebrate Zoology.* the ant-loving beetles, a family of tiny red or yellowish coleopteran insects in the superfamily Staphylinoidea having elongated bodies and large maxillary palps; found in ants' nests.

psellism see STUTTERING.

psephicity *Geology.* a numerical value for expressing the roundability of a sand or pebble-sized mineral fragment, equal to the ratio of specific gravity to hardness in air, or the ratio of one less than the specific gravity to hardness in water.

psephyte *Geology.* a sediment or sedimentary rock equivalent to a rudite, and made up of large, coarse fragments set in a variable matrix. Also, **psephite**. *Hydrology.* a lake-bottom deposit made up primarily of coarse, fibrous plant remains.

Pseudaliidae *Invertebrate Zoology.* a family of nematode roundworms in the order Strongyloidea; parasites of whales.

Pseudallescheria *Mycology.* a genus of fungi of the order Microascales that cause diseases in man and other animals, such as cattle.

Pseudanabaena *Bacteriology.* a genus of filamentous photosynthetic cyanobacteria containing gas vacuoles and occurring in aquatic environments.

pseudaphia *Neurology.* see PARAPHIA.

pseudergates *Entomology.* in some termites, a juvenile worker form capable of undergoing an indefinite number of moults without developing into adulthood.

pseudo- [soo´dō] a combining form meaning "false," often denoting a close or deceptive resemblance. Also, **pesud-.**

pseudoadiabat *Meteorology.* a line representing a pseudoadiabatic expansion (rising air within precipitating clouds) of an air parcel, utilizing approximate computations.

pseudoadiabatic chart see STÜVE CHART.

pseudoadiabatic expansion *Geophysics.* the process whereby a condensed water substance is removed from a system and treated by the thermodynamics of an open system.

pseudoallele *Genetics.* any of two or more genes that behave as alleles of the same gene when put to the allelism test, but can be separated by crossing over. Thus, **pseudoallelic, pseudoallelism.**

pseudoallochem *Geology.* a material that resembles an allochem, but was formed in place within a calcareous sediment by a secondary process, such as recrystallization, rather than by direct deposition.

Pseudoanemoniaceae *Botany.* a small family of red algae of the order Cryptonemiales, characterized by erect gametangial thalli composed of a short axis that develops into a whorl of axes where spermatangia are formed.

pseudoangina *Cardiology.* symptoms of angina, such as precordial pain, fatigue, and lassitude, without the presence of organic heart disease. Also, FALSE ANGINA.

pseudoaquatic *Biology.* indigenous to moist regions but not actually aquatic.

Pseudoborniales *Paleontology.* an extinct order of sphenopsid plants of the Devonian.

pseudobranch *Invertebrate Zoology.* a gill-like structure on the side of the foot in pulmonate gastropod snails, usually with no respiratory function.

pseudobreccia *Petrology.* a partly and irregularly dolomitized variety of limestone with a mottled appearance similar to that of breccia, or with a weathered surface that appears fragmental.

pseudobrookite *Mineralogy.* $(Fe^{+3},Fe^{+2})_2(Ti,Fe^{+3})O_5$, a brown or black orthorhombic mineral that resembles brookite, occurring as tabular crystals, having a specific gravity of 4.39 and a hardness of 6 on the Mohs scale; found in cavities of some volcanic rocks.

Pseudocaedibacter *Bacteriology.* a genus of Gram-negative bacteria of uncertain affiliation that grow as endosymbionts in certain strains of *Paramecium* and that may be capable of conferring the killer trait on a host.

pseudocannel coal *Geology.* a variety of cannel coal composed largely of humic matter.

pseudocapillitum *Mycology.* in fungi of the order Myxomycetes, a thready part of the fruiting body where the spores are located.

pseudocarp *Botany.* **1.** broadly, any fruit with parts in addition to the mature ovary and its contents, such as an apple or pineapple. **2.** a specific type of pseudocarp having a number of achenes borne on the surface of an enlarged, fleshy receptacle, such as the strawberry. Also, ACCESSORY FRUIT, FALSE FRUIT.

pseudocatalase *Biochemistry.* a substance that triggers the degradation of H_2O_2 into water and O_2 in certain lactic acid bacteria, such as strains of *Lactobacillus, Streptococcus,* and *Veillonella.*

pseudocentrum *Vertebrate Zoology.* the main body of a vertebra, formed by fusion of the dorsal or dorsal and ventral arcualia, as in tailed amphibians.

pseudocholinesterase see CHOLINESTERASE.

pseudochromesthesia *Neurology.* **1.** a form of synesthesia in which certain sounds induce a sensation of a distinct visual color. **2.** see CHROMATIC AUDITION.

pseudoclonus *Neurology.* a clonic muscular contraction that is very brief, despite continued stimulus.

Pseudococcidae see ERIOCOCCIDAE.

pseudocoel *Invertebrate Zoology.* in certain bilateral metazoans such as rotifers and roundworms, a type of body cavity derived from the blastocoel of the embryo; there is no peritoneum in the cavity, and mesoderm may line the inner surface.

Pseudocoelomata *Invertebrate Zoology.* a group of phyla containing animals with a pseudocoel, including Entoprocta, Aschelminthes, and Acanthocephala.

pseudocoelomate see ASCHELMINTHES.

pseudocol *Geology.* a surface feature represented by a narrowing of a stream valley diverted by a glacial ponding, and produced by the cutting through of a cover of glacial drift and the subsequent uncovering of a former col.

pseudoconcretion *Geology.* a rounded sedimentary structure that resembles a concretion but did not originate from the orderly precipitation of mineral matter in the pores of a sediment.

pseudoconditioning *Behavior.* the elicitation of a response to a previously neutral stimulus when that stimulus is presented following a series of conditioned stimuli.

pseudoconglomerate *Geology.* a rock that closely resembles a normal sedimentary conglomerate.

pseudo cross-bedding *Geology.* **1.** an inclined bedding having foreset beds that seem to dip into the current, and that form by deposition in response to ripple-mark movement. **2.** a formation that resembles cross-bedding but was formed by the slumping and sliding of a semiconsolidated mass of sediments without distortion, rather than by deposition.

pseudocrystal *Crystallography.* a substance composed of crystalline grains that are slightly worn and solidly compacted by other mineral accretions so as to strongly resemble a true crystal.

pseudocumene *Organic Chemistry.* C_9H_{12}, a liquid that boils at 168.89°C and freezes at –43.91°C; soluble in alcohol, benzene, and ether; used in the production of perfumes and dyes. Also, **pseudocumol.**

Pseudocycnidae *Invertebrate Zoology.* a family of copepod crustaceans in the suborder Caligoida; ectoparasites on fish gills.

pseudocyesis *Medicine.* a false pregnancy; the presence of signs of pregnancy, such as an enlarged abdomen or absence of the menses, without conception.

pseudodeficiency rickets see VITAMIN D-RESISTANT RICKETS.

pseudodiffusion *Geology.* the mixing together of thin, successively deposited layers of slowly accumulated marine sediments by water motion or the activities of subsurface organisms.

Pseudoditrichaceae *Botany.* a monotypic family of tiny mosses usually placed in the order Funariales, characterized by erect stems with terminal sporophytes and spreading, falcate leaves; species grow loosely on calcareous soil.

pseudodominance *Genetics.* the expression of a recessive allele in the absence (by deletion) of the dominant allele.

pseudoephedrine *Pharmacology.* $C_{10}H_{15}NO$, a stereoisomer of ephedrine having less pressor action and central stimulant effects than ephedrine.

pseudoephedrine hydrochloride *Pharmacology.* $C_{10}H_{15}NO{\cdot}HCl$, the hydrochloride salt of ephedrine, occurring as a white powder or fine crystals; widely used as a nasal decongestant and bronchodilator, administered orally.

pseudoepithecium *Mycology.* in fungi, a certain layer on the surface of the spore-bearing structure (known as the apothecium).

pseudoextinction *Ecology.* see PHYLETIC EXTINCTION.

pseudofault *Geology.* a rock feature that resembles a fault but was produced by weathering along joint, shrinkage, or bedding planes, rather than by displacement.

pseudofeces *Invertebrate Zoology.* the debris rejected by the filtering mechanism in bivalve mollusks that settles on the mantle and is periodically flushed out by briskly pulling the valves together.

pseudofibrous peat *Geology.* a pliable, incoherent peat that has a fibrous texture.

pseudofront *Meteorology.* a small-scale frontal system between a mass of rain-cooled air from the thunderstorm clouds and the warm surrounding air.

pseudoganglion *Neurology.* a thickening of a nerve trunk, resembling a ganglion.

pseudogene *Genetics.* a gene that is very similar to a known gene at a different locus, but is made nonfunctional by additions or deletions to its structure that prevent normal transcription or translation.

pseudoglanders *Veterinary Medicine.* an infectious fungal disease of equines arising from *Corynebacterium pseudotuberculosis,* characterized by nodules on the lymph nodes; often mistaken for cutaneous glanders. Also, CASEOUS LYMPHADENITIS, EQUINE ULCERATIVE LYMPHANGITIS.

pseudogley *Geology.* a dense, highly compacted, silty soil that has alternately been subjected to waterlogging and rapid drying.

pseudo-Goldstone bosons *Particle Physics.* bosons that accompany the breakdown of accidental symmetries in certain gage theories regarding weak and electromagnetic interactions.

pseudogradational bedding *Geology.* in a metamorphosed sedimentary rock, a structure whose original textural gradation, from coarse at the base to finer at the top, seems to be reversed.

pseudohemophilia see VON WILLEBRAND'S DISEASE.

pseudohemoptysis *Pathology.* spitting of blood that comes from some source other than the lungs or bronchial tubes.

pseudohermaphroditism *Physiology.* a condition in which an individual has sexual organs of only one sex, but has either genital openings or external tissue exhibiting one or more traits of the opposite sex.

pseudohypertrophic muscular dystrophy see DUCHENNE MUSCULAR DYSTROPHY.

pseudoinfarction *Cardiology.* the simulation of a heart attack, especially as seen in an electrocardiogram, caused by other cardiopathies.

pseudoinstruction *Computer Technology.* a symbolic instruction that is not a member of the machine instruction set for the computer and may actually be implemented as a procedure consisting of normal instructions, or may be an instruction to the assembler such as a request to allocate a block of data storage. Also, QUASI-INSTRUCTION.

pseudoionone *Organic Chemistry.* $C_{13}H_{20}O$, a yellowish combustible liquid; soluble in alcohol, benzene, and ether; boils at 143–145°C (12 torr); used in perfumery and cosmetics.

pseudokame see RESIDUAL KAME.

pseudo-Kaposi sarcoma *Oncology.* a dermatitis condition combined with underlying vascular fistula, which closely resembles Kaposi's sarcoma though it is not a sarcoma.

pseudokarst *Geology.* an irregular topography that resembles karst but was not formed by the dissolution of limestone, especially a rough-surfaced lava field in which ceilings of lava tubes have collapsed.

pseudolanguage *Computer Programming.* **1.** a type of language in which instructions are written with symbols indicating specific operation codes and addresses, requiring a compiler to translate it to machine language before it can be implemented. **2.** a language that is similar to a programming language but may not be implemented, used to describe algorithms for expository purposes. Also, **pseudocode.**

Pseudolepicoleaceae *Botany.* a family of small, brownish liverworts of the order Jungermanniales characterized by irregular, variable branching and rhizoids arising from underleaf bases; native chiefly to south temperate and subarctic regions.

pseudoleucite *Mineralogy.* a pseudomorph after leucite, comprising a mixture of analcime, nepheline, and orthoclase.

pseudoleukemia *Oncology.* a former term for a group of conditions accompanied by symptoms, especially swelling of the lymph glands, that resemble leukemia but are not accompanied by leukemic blood findings; includes Hodgkin's disease, multiple myeloma, and tuberculosis and syphilis of the lymph glands.

pseudolysogeny *Virology.* a phenomenon in which lysogeny appears to occur between the genome of a temperate bacteriophage and its host but is not truly established.

pseudomalachite *Mineralogy.* $Cu_5^{+2}(PO_4)_2(OH)_4$, a dark-green, monoclinic mineral that resembles malachite, trimorphous with ludjibaite and reichenbachite, occurring generally in botryoidal or massive form, having a specific gravity of 4.08 to 4.35 and a hardness of 4.5 to 5 on the Mohs scale; found in the oxidized zones of copper deposits.

pseudomalignancy *Oncology.* any of a group of tumors that appear to be malignant under microscopic examination but are benign in clinical behavior; the group includes **pseudolymphoma, pseudomelanoma,** and **pseudosarcoma.**

Pseudomonadaceae *Bacteriology.* a family of Gram-negative, straight or curved rod-shaped, motile, aerobic bacteria with polar flagella that occur in fresh water, salt water, and soil.

Pseudomonadales *Bacteriology.* in some classification systems, an order of bacteria having polar flagella. See EUBACTERIALES.

Pseudomonadineae *Bacteriology.* a former classification for a suborder of the order Pseudomonadales, characterized by having pigments.

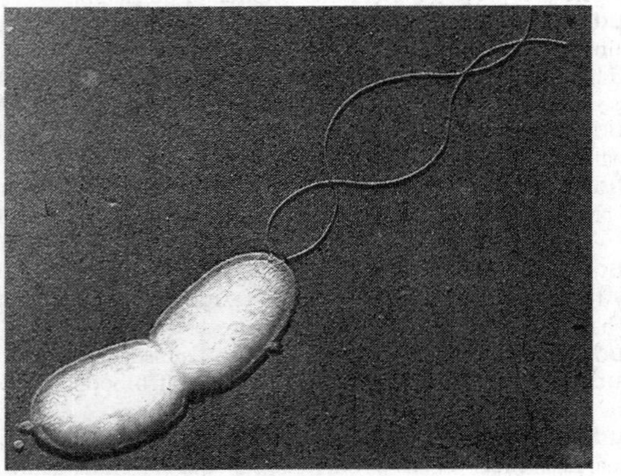

Pseudomonas fluorescens

Pseudomonas *Bacteriology.* a genus of Gram-negative aerobic bacteria of the family Pseudomonadaceae, occurring widely as curved or straight rods in soil and water; some species are pathogenic for animals or plants. The fluorescent species *P. fluorescens* is an environmental contaminant and occasionally an opportunistic pathogen for humans, causing infections of wounds, the urinary tract, and the blood stream.

pseudomorph *Mineralogy.* a mineral whose outward crystal form is the same as that of another species of mineral, which it has replaced through substitution, incrustation, or alteration. Also, FALSE FORM.

pseudomorphine *Organic Chemistry.* $C_{34}H_{36}N_2O_8 \cdot 3H_2O$, a compound occurring in opium and prepared by the oxidation of morphine.

pseudomotor *Neurology.* relating to or producing abnormal movements.

pseudomountain *Geology.* a mountain formed by differential erosion, rather than by uplift, due to porphyroblast formation in the fine-grained portions.

pseudomucinous cystadenoma *Medicine.* a benign cystic ovarian tumor lined with epithelial cells and filled with mucinous material.

pseudomurein *Cell Biology.* a cell wall polymer that occurs in certain archaebacteria and is resistant to certain antibiotics and to the enzyme lysozyme.

pseudoneoplasm *Oncology.* a cell formation that resembles a tumor but regresses to normal morphology. Also, PHANTOM TUMOR.

pseudoneuroma *Neurology.* a tumor or mass on a nerve that resembles a neuroma, generally forming at the end of an amputation stump.

Pseudonocardia *Bacteriology.* in former classifications, a genus of bacteria of the family Nocardiacea occurring in manure and soil as substrate and aerial mycelia bearing cylindrical spores; species have been reassigned to other genera.

pseudonodule *Geology.* a primary sedimentary structure consisting of a spherical mass of sandstone encased in shale or mudstone. Also, SAND ROLL.

pseudonoise see pseudorandom noise.

pseudo-oolith *Geology.* a round or spherical particle in a sedimentary rock that resembles an oolith but is different in origin and lacks the radial or concentric structure of an oolith. Also, FALSE OOLITH.

pseudo-operation *Computer Technology.* **1.** an operation that is not physically implemented in hardware as part of a computer's operational repertoire. **2.** see PSEUDOINSTRUCTION.

pseudoparalysis *Neurology.* a momentary interruption or dysfunction of motor skills, usually brought on by pain, but without actual paralysis.

pseudoparaphysis *Mycology.* in fungi, sterile threads attached to the roof and bottom of the spore-bearing structure or ascocarp.

pseudoparasite *Microbiology.* an organism that resembles but is not necessarily a parasite; it may be a commensal or a temporary parasite.

pseudoparenchyma *Botany.* interwoven fungal hypha in a thallophyte that make up false tissue and resemble the parenchyma of higher plants.

pseudopericardial *Cardiology.* apparently, but not actually, arising from the pericardium.

pseudoperiodic motion see CONDITIONALLY PERIODIC MOTION.

Pseudophyllidea *Invertebrate Zoology.* an order of tapeworms in the subclass Cestoda; both larvae and adults are endoparasites of fish, birds, and invertebrates.

pseudoplasmodium *Microbiology.* a multicellular aggregation of ameboid cells found in certain slime molds.

pseudoplastic fluid *Fluid Mechanics.* a fluid in which the viscosity instantaneously decreases with an increase in the rate of shear; the shear stress increases with shear rate taken to a fractional power of less than unity.

pseudoplasticity *Materials Science.* an instantaneous decrease in the apparent viscosity of a substance with increase in the rate of shear.

pseudopodium *Cell Biology.* a retractible, usually temporary cellular extension formed by certain ameboid organisms, composed of membrane-bounded cytoplasm and functioning in food ingestion and locomotion. Also, **pseudopod.** *Botany.* in certain mosses, a thin leafless branch of the gametophyte that often bears gemmae.

pseudoporphyritic *Petrology.* describing an igneous rock texture in which larger crystals are embedded in a macrocrystalline groundmass but were partly formed after rock consolidation.

pseudoposematic see PSEUDOSEMATIC.

pseudopotential theory *Materials Science.* a theory that explains aspects of the bonding process in metals, such as the effect of crystal structure on the cohesive energy.

pseudorabies see AUJESZKI'S DISEASE.

pseudorandom noise *Telecommunications.* a noise that is produced by a definite process but passes at least one of the standard tests for statistical randomness.

pseudorandom numbers *Mathematics.* a sequence of numbers from some set (usually the integers from 1 to *n* or the real numbers between 0 and 1) such that the sequence is generated by a deterministic formula or algorithm but the numbers chosen appear to be random in that they pass tests devised to check that the numbers were selected from the set with equal likelihood and that the selections were independent of one another. Pseudorandom numbers are used in situations where random numbers are needed.

pseudorecombinant *Virology.* describing a virus produced through the in vitro technique of combining nucleic acid of similar viruses with multipartite genomes; used for mapping genetic coding in the nucleic acid of plant viruses.

pseudoreduced compressibility *Chemistry.* the compressibility of a gaseous system, calculated at reduced conditions using pseudoreduced properties of the system.

pseudoreduced properties *Chemistry.* the reduced-state relationships of chemical systems calculated using the pseudocritical properties.

pseudo-Riemannian manifold *Mathematics.* a manifold whose metric tensor is not positive definite, but maintains its signature over the entire manifold.

pseudorinderpest see PESTES DES PETITS RUMINANTS.

pseudo ripple mark *Geology.* a feature on a bedding plane that resembles a ripple mark but was produced by lateral pressure resulting from slumping or by local, small-scale tectonic deformation, rather than by the action of winds, waves, or currents.

pseudorutile see ARIZONITE.

pseudoscalar *Physics.* describing any scalar quantity that suffers a sign change under a space or time reflection.

pseudoscalar coupling *Particle Physics.* a conjectured interaction between a pion and a nucleon wherein the energy results from the product of the pseudoscalar field of the pion and a bilinear pseudoscalar function of the nucleon fields.

pseudoscalar meson *Particle Physics.* a semistable meson with zero spin and odd parity, such as the pion, kaon, and eta meson. Also, **pseudoscalar particle.**

pseudosclerosis *Neurology.* any of a group of degenerative disorders of the brain that are characterized by progressive paralysis and rigidity but lack the lesions associated with sclerosis.

pseudoscopic image *Photogrammetry.* a stereoscopic image or model in which the usual impressions of elevation and depression are reversed.

Pseudoscorpionida *Invertebrate Zoology.* the false scorpions, an order of tiny arachnids that superficially resemble scorpions but lack the tail and sting; the body is flattened and eyes are lacking or poorly developed.

pseudosematic *Ecology.* describing a trait or characteristic that serves to deceive a predator or prey object. Also, PSEUDOPOSEMATIC.

pseudosolarization see SABATTIER EFFECT.

pseudosolution *Physical Chemistry.* any solution, such as a colloid solution, that does not act according to the usual physical laws of solutions.

Pseudosphaeriaceae *Mycology.* a family of fungi of the order Dothideales that either live off nonliving organic matter or are parasitic to leaves, plant tissue, tree bark, or legumes.

Pseudosporidae *Invertebrate Zoology.* a family of amoeboid protozoans, parasites in algae and flagellated protozoans.

pseudostatic spontaneous potential *Petroleum Engineering.* the peak spontaneous potential current theoretically measurable in a downhole, mud-column log in shaly sand.

pseudostratified epithelium *Histology.* a type of epithelium consisting of a single layer of cells, some of which reach all the way from the basement membrane to the surface.

Pseudosuchia *Paleontology.* the "paracrocodiles," a suborder of archosaurs in the order Thecodontia; may be related to the dinosaurs; some were bipedal and some endothermic; some would reclassify the pseudosuchians in the order Crocodilia and various thecodont suborders; extant in the Triassic to Cretaceous.

pseudosymmetry *Crystallography.* the resemblance of a crystal to a form of symmetry higher than that which is characteristic of the mineral, usually as a result of twinning.

pseudotachylite *Petrology.* a black vein rock superficially similar to tachylite but produced by extreme mylonitization and/or partial melting.

pseudotensor *Physics.* **1.** a quantity that can be transformed as a tensor under a rotation but is transformed as a tensor accompanied by a change in sign under a space inversion. **2.** a quantity that transforms as a tensor under the Lorentz transformations, but suffers a change in sign under space or time reflections.

Pseudothelphusidae *Invertebrate Zoology.* a family of New World freshwater crabs, decapod crustaceans in the section Brachyura, having a flat carapace and a biramous mandibular palp.

pseudotillite *Geology.* a rock that resembles tillite but was either formed on land by the flow of nonglacial mud or deposited by an underwater turbidity flow.

pseudotoxin *Toxicology.* a poisonous extract from leaves of *Atropa belladonna* that causes blurred vision and hallucinations.

Pseudotriakidae *Vertebrate Zoology.* the false cat sharks, a monogeneric family of rare, harmless, deep-bottom sharks of the order Carcharhiniformes characterized by a long, low first dorsal fin.

pseudotuberculosis *Medicine.* an infection occurring in animals such as rodents, sheep, and birds; characterized by nodules similar to those of tuberculosis; it occurs in humans, though rarely, from infection by *Pasteurella pseudotuberculosis.*

pseudotype *Virology.* a virus particle that contains the genome of one virus within the capsid of another, resulting from a mixed infection.

pseudounconformity *Geology.* a relationship among sedimentary strata that seems unconformable, but is characterized by an excess accumulation of sediment.

pseudovalve *Cardiology.* an unusual formation on the parietal endocardium of the left ventricle, especially with insufficiency of the aortic valves.

pseudovector *Physics.* **1.** a quantity that transforms as a vector under a rotation in space, but that undergoes a sign change under spatial inversion. **2.** a quantity that, under a Lorentz transformation, transforms as a vector, but that under a spatial or time reflection undergoes a sign change.

pseudovector coupling *Particle Physics.* a conjectured interaction between a particle and a nucleon wherein the energy of interaction is derived from a bilinear pseudovector of the nucleon fields.

pseudovector meson *Particle Physics.* a meson particle having spin 1, and positive parity, and may be described by a pseudovector field.

pseudovirion *Virology.* a virion that contains only host-cell nucleic acid.

pseudovitrinite *Geology.* a coal maceral that superficially resembles vitrinite, but is higher in reflectance, has a slitted structure, as well as remnants of cellular structures, rare fracture patterns, higher relief, and little or no inclusions of pyrite.

pseudovitrinoid *Geology.* a pseudovitrinite that occurs in bituminous coals.

pseudovolcano *Geology.* a large crater or circular depression thought to be the result of meteoric impact, cauldron subsidence, or phreatic explosion, rather than the result of recent volcanic activity.

pseudowavellite see CRANDALLITE.

psf or **p.s.f.** pounds per square foot.

P shell *Atomic Physics.* the electron orbit of an atom, farthest from the nucleus, in which electrons whose principal quantum number is 6 occur.

psi [sī] the 23rd letter of the Greek alphabet (Ψ, ψ).

psi *Psychology.* the purported mental processes and energies of parapsychology, involving phenomena that cannot be explained in terms of natural laws.

psi or **p.s.i.** pounds per square inch.

psia or **p.s.i.a.** pounds per square inch absolute.

psid or **p.s.i.d.** pounds per square inch differential.

psi function *Mathematics.* the special function of a complex variable that is the derivative of the logarithm of the gamma function; i.e.,

$$\psi(z) = d\,[\ln \Gamma(z)]/dz = \Gamma'(z)/\Gamma(z).$$

psig or **p.s.i.g.** pounds per square inch gauge.

psilate *Botany.* lacking ornamentation, usually used in reference to pollen.

Psilidae *Invertebrate Zoology.* the rust flies, a family of dipteran insects in the subsection Acalyptratae, having a slender brown body and a head with long eye-stalks; larvae are pests of crops.

Psilocerataceae *Paleontology.* a superfamily of ammonoids in the order Ammonitida; origin uncertain, whether from Ceratitina or Phylloceratida; extant in the Lower Jurassic.

psilocin *Pharmacology.* a hallucenogenic substance closely related to psilocybin.

psilocybin *Pharmacology.* $C_{13}H_{18}O_3N_2P_2$, a hallucinogenic crystalline compound having indole characteristics, isolated from the mushroom *Psilocybe mexicana.*

psilomelane *Mineralogy.* **1.** a general term for hard, massive, manganese oxides that have not been specifically identified. **2.** see ROMANECHITE.

Psilophytales *Paleontology.* an order of dichotomously branched plants; the type genus *Psilophyton* has been thoroughly studied; they are the oldest known terrestrial plants having a vascular structure; extant in the Devonian.

psilopsid *Botany.* a member of the Psilopsida.

Psilopsida *Botany.* in some classification systems, a subdivision or class of simple vascular plants including fossil and living forms, characterized by the lack of true roots, and often by the lack of leaves; the living forms are generally considered the oldest of all vascular plants.

Psilorhynchidae *Vertebrate Zoology.* a monogeneric family of small, loachlike fish of the suborder Cyprinoidei, closely related to the homalopterid loaches, but without barbs and having a spindle-shaped body with a flat snout and belly; found in mountain streams in parts of India.

psilosos *Medicine.* the falling out or loss of hair.

Psilotales *Botany.* in some classification systems, a monofamilial order of the subdivision Psilopsida, including two genera, *Psilotum* and *Tmesipteris*; characterized by the lack of true roots and often the lack of leaves, and having the most simple construction of all living vascular plants.

Psilotatae *Botany.* a class of the division Psilotophyta.

Psilotophyta *Botany.* in some classification systems, a division of very simple vascular plants including fossil forms and nine living species thought to be representative of the first land plants.

Psilotopsida *Botany.* in some classification systems, the subdivision of Psilotophyta containing nine living species.

psi particle or **psi-prime particle** see J PARTICLE.

P site *Molecular Biology.* one of two tRNA binding sites on the ribosome; it holds the tRNA molecule that has just discharged its amino acid to the end of the polypeptide chain.

Psittacidae *Vertebrate Zoology.* the bird family of parrots, lories, macaws, and cockatoos; a diverse family of the order Pisttaciformes characterized by a short, deep, strongly hooked bill, a large fleshy tongue, and often colorful plumage; found primarily in lowland tropical and subtropical forests and woodlands of South America, Central America, Africa, southern Asia, Australia, and Polynesia.

Psittaciformes *Vertebrate Zoology.* the order of parrots and their allies, including the single family Psittacidae.

psittacosis [sit´ə kō´ sis] *Medicine.* an infectious viral disease of birds that is sometimes transmitted to man, especially by birds such as parrots or parakeets, and which manifests itself as pneumonia with symptoms of headache, nausea, constipation, sore throat, chills, and fever. Also, PARROT DISEASE, PARROT FEVER.

PSK phase-shift keying.

PSN public switched network.

psoas *plural,* **psoai.** *Anatomy.* either of the two muscles connecting the spinal column to the thighbone.

Psocoptera *Invertebrate Zoology.* the booklice or barklice, an order of lice having a small, soft body, large head, biting mouthparts, long antennae, and sometimes membranous wings; some are pests in grains and book paste. Also, CORRODENTIA.

Psolidae *Invertebrate Zoology.* the sea cucumbers, a family of holothurians in the order Dendrochirotida having armorlike overlapping plates covering the body, a soft ventral sole, and dorsal mouth and anus.

Psophiidae *Vertebrate Zoology.* the trumpeters, a monogeneric family of tailless wading birds of the order Gruiformes, characterized by long legs and neck, a fowl-like bill, and a mostly black plumage that is velvety on the head; found in tropical rain forests of northeastern South America.

psophometer *Engineering.* an instrument that measures the noise voltage in electrical circuits, and whose reading equals one-half of the true noise voltage in the circuit.

psoriasis [sə rī´ə sis] *Medicine.* a common, chronic inheritable skin disease characterized by red patches covered with silvery scales, occurring primarily on elbows, knees, scalp, and trunk.

psoroptic mange *Veterinary Medicine.* hair loss or mange in cattle, horse, sheep, and other animals as caused by scab mites of the genus *Psoroptes.*

psorosis *Plant Pathology.* a virus disease of citrus trees characterized by scaly and brittle bark, chlorosis in the leaves, retarded growth, and gummy exudants. Also, SCALY BARK.

PSP phenolsulfonphtalein; pseudostatic spontaneous potential.

PSP *Aviation.* the airport code for Palm Springs, California.

p-spot *Astronomy.* the leading or preceding spot of a sunspot pair.

PSRO Professional Standards Review Organization.

PST or **P.S.T.** Pacific Standard Time.

^{32}P suicide *Molecular Biology.* the radioactive disintegration of an unstable isotope of phosphorus that is incorporated into a bacteriophage nucleic acid sequence, causing breaks in the nucleic acid molecules.

PSW processor status word.

psych. psychology; psychological; psychiatry; psychiatric.

psych- a combining form meaning "psyche" or "mind."

psychanopsia *Neurology.* a form of aphasia characterized by the inability to recognize what is seen. Also, PSYCHIC BLINDNESS.

psychataxia *Neurology.* a mental condition characterized by the inability to maintain focus, attention, or concentration.

psyche [sīk´ē] *Psychology.* **1.** the mind in its totality, considered separately from the physical organism. **2.** all of one's self, including the soul or spirit. (From *Psyche,* the Greek goddess who personified the soul.)

Psyche *Astronomy.* asteroid 16, discovered in 1852; it has a diameter of 264 kilometers and belongs to type M.

psychedelic [sīk´ə del´ik] *Pharmacology.* of or relating to a group of drugs that produce hallucinations and altered states of consciousness, such as LSD or mescaline. *Psychology.* relating to altered states of consciousness, especially those produced by such drugs.

psychiatric [sī´kē at´rik] *Psychology.* **1.** of or relating to the practice of psychiatry. **2.** relating to or being a mental or emotional disorder.

psychiatric genetics *Genetics.* the study of abnormal genetic patterns to determine how they correlate with abnormal behavior patterns.

psychiatric social worker *Psychology.* a social work professional who is specifically trained to work with patients and families in clinical mental health settings.

psychiatrist [sə kī´ə trist] *Medicine.* a physician who specializes in psychiatry; a person who studies, diagnoses, and treats mental and emotional disorders.

psychiatry [sə kī´ə trē] *Medicine.* the branch of medicine concerned with the study, diagnosis, treatment, and prevention of mental and emotional disorders.

psychic blindness see PSYCHANOPSIA.

Psychidae *Invertebrate Zoology.* the bagworm moths, a large family of lepidopteran insects in the superfamily Tineoidea; larvae build a baglike case in which the legless, wingless adult females live after metamorphosis; male adults are large, hairy moths.

psycho- [sī´kō] a combining form meaning "psyche" or "mind."

psychoacoustics *Acoustics.* the study of auditory detection and discrimination as it relates to behavior and other psychological aspects of response.

psychoactive *Psychology.* relating to or being a substance that has a significant effect on the mind or behavior. Thus, **psychoactive drug.**

psychoanal. psychoanalysis; psychoanalytical.

psychoanalysis [sī´kō ə nal´ə sis] *Psychology.* **1.** a theory of human personality developed by Sigmund Freud, based on the division of the human mind into the ego, id, and superego, and focusing on the influence of unconscious motivation, conflict, and symbolism. **2.** a group of techniques for the study and treatment of abnormal behavior, derived from Freud's theory of personality, including free association, dream analysis, and interpretation of resistance and defense mechanisms.

psychoanalyst [sī´kō an´ə list] *Psychology.* a psychiatrist or clinical psychologist who has been specially trained in the theory and practice of psychoanalysis, and who applies these techniques as a means of treating patients.

psychoanalytic or **psychoanalytical** *Psychology.* of or relating to the theories or techniques of psychoanalysis.

psychoanalytic psychotherapy *Psychology.* a form of therapy based on the principles of psychoanalysis, but briefer and less intense than classical psychoanalysis and involving less emphasis on childhood.

psychobiology *Psychology.* **1.** the study of the relationship between biological processes and behavior. **2.** an approach to psychology that emphasizes the integration of the individual's environmental, psychological, and biological factors. Thus, **psychobiologist.**

psychodiagnosis *Psychology.* any procedure used to assess psychological abnormalities or mental disorders.

Psychodidae *Invertebrate Zoology.* the moth flies and sandflies, a family of dipteran insects having a small hairy body and hairy wings.

psychodrama *Psychology.* a therapeutic technique developed by J. L. Moreno in which the individual acts out certain conflict situations in front of a group, in the hope of gaining insight into the nature of the conflict.

psychodynamics *Psychology.* the study of the motivational forces that affect the development and behavior of an individual, such as drives, emotions, and biological needs. Also, **psychodynamic theory, psychodynamic approach.**

psychogalvanic reflex *Physiology.* a variance in electrical resistance over the skin, resulting from an emotional state or from focused attention.

psychogalvanometer *Engineering.* a device that measures mental reactions by recording the effect of the direct application of charged electrodes to the skin.
psychogenic *Medicine.* **1.** having its origin in the mind. **2.** referring to a symptom, illness, or condition produced as a result of psychological rather than organic causes.
psychohistory *Psychology.* the use of psychoanalytic theories or techniques to interpret the actions of historical figures.
psychoimmunology *Medicine.* the study of the effect of psychological factors on the body's immune system.
psychokinesis [sī´kō ki nē´sis] *Psychology.* the supposed ability to use mental powers to move or change the shape of objects or otherwise influence external events. Thus, **psychokinetic** [sī´kō ki net´ik].
psychol. psychology; psychologist.
psycholinguistics *Linguistics.* the branch of linguistics dealing with the study of the psychological factors involved in language and communication and their effects on individual and group interaction.
psychological [sī´kə läj´i kəl] **1.** of or relating to the science of psychology. **2.** relating to or involving mental processes.
psychological anthropology *Anthropology.* a branch of anthropology studying the effect of cultural influences on individual personality.
psychological dependence *Psychology.* a strong craving for a psychoactive substance, such as nicotine, marijuana, or caffeine, that is based on the pleasurable feelings the substance provides, rather than on a physiological need.
psychological determinism *Psychology.* the doctrine that an individual's every action has either a conscious or unconscious cause, leaving no human behavior to chance.
psychological operations *Military Science.* planned operations to influence the emotions, motives, objective reasoning, and behavior of foreign governments, organizations, groups, and individuals. Also, **psychological warfare.**
psychologist [sī käl´ə jist] *Psychology.* a specialist in psychology; an individual who is professionally trained in the study of behavior.
psychology [sī käl´ə jē] the scientific study of the mind and behavior.
psychometrics *Psychology.* any of various efforts to measure aspects of behavioral functioning, particularly through the use of standardized tests.
psychomotor *Psychology.* relating to control of the muscles by the nervous system.
psychomotor epilepsy *Neurology.* an epileptic seizure marked by impaired consciousness and amnesia during the episode, often accompanied by coordinated but abnormal movements of the arms and legs. Also, TEMPORAL-LOBE EPILEPSY.
psychoneuroimmunology see PSYCHOIMMUNOLOGY.
psychoneurosis see NEUROSIS.
psychopath see SOCIOPATH.
psychopathic personality see SOCIOPATHIC PERSONALITY.
psychopathology *Psychology.* the scientific study of mental disorders, including their nature, causes, and processes.
psychopharmacology *Psychology.* **1.** the study of the effects of drugs on psychological processes. **2.** the therapeutic use of drugs, such as antidepressants, in the treatment of mental disorders.
psychophysics *Psychology.* a branch of psychology that deals with the relationships between sensory stimuli and behavior, in order to explain and predict the functioning of sensory systems.
psychophysiology *Psychology.* a branch of psychology that deals with the interrelationship of psychology and physiology.
psychosexual *Psychology.* **1.** relating to the relationship between sexual and psychological processes. **2.** relating to Freud's theories of sexuality and sexual development.
psychosexual role see SEX ROLE.
psychosexual stages (of development) *Psychology.* in psychoanalytic theory, the five stages of sexual development, each identified by a specific area of gratification; the individual's adult personality can be affected if development is impeded or unresolved at any specific stage.
psychosis [sī kō´sis] *plural,* **psychoses.** *Psychology.* a severe mental disorder characterized by delusions, a loss of contact with reality, and other extreme behavioral disturbances; generally requiring institutionalization or special treatment.

Psychology

While the roots of psychology date back to the ancient Greek philosophers (Socrates, Plato, Aristotle) who posed fundamental questions about the human mind, psychology as a science was born in the latter part of the 19th century with the idea that mind and behavior—like the movement of the planets or the reaction of chemicals—could be the subject of scientific analysis.

The concern of psychology—how and why organisms do what they do—encompasses a broad spectrum of phenomena: learning, cognition, perception, intelligence, personality (including the development of normal and abnormal behavior), emotion, motivation, social behavior, and behavioral genetics (the extent to which individual differences are inherited or shaped by the environment).

These issues have been studied from a number of different perspectives, but three major approaches predominate. The *neurobiological approach* seeks to specify the neural and endocrine events that underlie behavior and mental processes (e.g., the changes that occur in nerve-cell connections as a result of learning; the brain structures involved in consolidating memories; the array of hormones released during emotional arousal).

The *behavioral approach* studies those activities of an organism that can be observed and measured directly. It seeks to determine the relationship between stimuli and responses without conjecturing about intervening biological processes (e.g., the stimuli that elicit aggressive responses; the schedule of reinforcement that produces the most efficient learning; the speed of response as a function of the intensity of visual signals).

The *cognitive approach* attempts to understand the nature of human thought and intelligence. Using an information-processing analysis, cognitive psychologists develop models (and, in turn, theories) of such mental processes as perceiving, remembering, reasoning, decision making, and problem solving (e.g., models of the way that visual information is processed and patterns recognized; the way verbal information is represented in memory, stored, and later retrieved; the sequence of mental operations used in problem solving). The models, which may or may not involve conjectures about neural mechanisms, are usually formulated as computing machines of some sort. Their implications can be explored by simulating the model on a computer and then comparing its predictions against observed behavior.

These three approaches—neurobiological, behavioral, and cognitive—are not mutually exclusive; all have proved valuable in studying and understanding various psychological phenomena. The work of psychologists taking the neurobiological approach overlaps with the neurosciences; that of cognitive psychologists overlaps with cognitive science, linguistics, and computer science.

In addition to numerous specializations focused on basic research in psychology are more applied areas which include clinical and counseling, industrial, engineering, educational, environmental, and health psychology.

Richard C. Atkinson
Professor of Psychology and Chancellor
University of California, San Diego

psychosocial stages (of development) *Psychology.* Erik Erikson's eight stages of human development, each of which focuses on a specific crisis that must be resolved in order for healthy adjustment and transition to the next stage to take place.

psychosomatic [sī´kō si mat´ik] *Medicine.* relating to a physical condition or physical symptoms caused by or intensified by psychological factors. Thus, **psychosomatic medicine, psychosomatic disorder.** *Psychology.* relating to processes or conditions that involve both the mind and the body.

psychosomatograph *Engineering.* a device that measures and creates a graphic display of muscular action currents or physical movements of a person being tested for mental-physical coordination.

psychosurgery [sī´kō sur´jə rē] *Surgery.* the use of brain surgery in the treatment of mental disorders, as in a lobotomy. Thus, **psychosurgeon, psychosurgical.**

psychotaxis *Psychology.* a defense mechanism consisting of the automatic mental and behavioral adjustments that are made by individuals in order to maintain a pleasant state of mind and avoid conflict.

psychotherapist [sī´kō ther´ə pist] *Psychology.* a person who is trained and licensed to practice psychotherapy.

psychotherapy [sī´kō ther´ə pē] *Psychology.* a general term for the treatment of abnormal behavior or mental disorder by psychological means, usually, but not exclusively, through patient interaction with a trained therapist. Also, **psychotherapeutics.**

psychotic [sī kät´ik] *Psychology.* **1.** relating to a psychosis or a behavior pattern symptomatic of psychosis. **2.** a person who exhibits such behavior.

psychoticism *Psychology.* a factor that is used to distinguish normal, schizophrenic, and manic-depressive individuals from each other; those ranking high in this characteristic are generally aggressive, cold, impulsive, and creative.

psychotogen *Psychology.* a drug or other substance that produces a psychotic reaction.

psychotomimetic *Psychology.* **1.** a drug producing effects that resemble psychotic behavior, such as mescaline or LSD. **2.** of or relating to such a drug.

psychotropic *Psychology.* **1.** any drug that produces a mind-altering or behavior-altering state, such as sedatives, stimulants, psychedelics, and narcotics. **2.** of or relating to such a drug.

psychro- a combining form meaning "cold."

Psychrobacter *Bacteriology.* a proposed genus to be included in the family Neisseriaceae, composed of psychrotrophic coccobacilli.

psychroesthesia *Neurology.* a subjective sensation of coldness in a part of the body that feels warm to the touch.

psychrometer *Engineering.* an instrument used to measure the moisture content of the atmosphere, composed of two thermometers, one having a wet bulb and the other a dry bulb.

psychrometric *Engineering.* relating to or based on information from a psychrometer.

psychrometric chart *Thermodynamics.* a graph whose vertical scale is absolute humidity and whose horizontal scale is temperature; each of the curves on the chart is composed of points representing different conditions of a gas-vapor system.

psychrometric formula *Thermodynamics.* a formula that relates vapor pressure to the readings of a psychrometer and a barometer.

psychrometric ratio *Thermodynamics.* a quantity used in humidity and saturation calculations given by the ratio of the heat transfer coefficient to the product of the mass-transfer coefficient and the humid heat in a gas-vapor system.

psychrometry *Engineering.* the study of the physical laws that influence air and water mixture.

psychrophile *Ecology.* an organism that lives or thrives at low temperatures. *Microbiology.* specifically, a cold-loving bacterium.

psychrophilic *Ecology.* of or relating to organisms that live or thrive at low temperatures. Also, **psychrophilous.**

psychrophobia *Psychology.* an abnormal fear of cold.

psychrophyte *Ecology.* an organism that lives or thrives in cold climates. Thus, **psychrophytic.**

psychrotroph *Microbiology.* a microorganism that can grow at low temperatures but grows optimally between 15° and 20°C. Thus, **psychrotrophic.**

Psyllidae *Invertebrate Zoology.* the jumping plant lice, a family of tiny homopteran insects in the series Sternorrhyncha, having hindlegs modified for jumping, a short three-segmented beak, and a body resembling that of a cicada.

PSYOP psychological operations.

PSZ partially stabilized zirconia.

Pt the chemical symbol for platinum.

pt or **pt.** patient.

PT or **P.T.** physical therapy; physical training; Pacific time.

Pt. Point; Port (in place names).

pt. or **pt** part; pint; pints; point; port.

PTA plasma thromboplastin antecedent.

Ptaeroxylaceae *Botany.* a family of trees and shrubs in the order Sapindales, characterized by pinnate leaves, arranged spirally or opposite, with oblique leaflets; found in the tropics, Southern Africa, and Madagascar.

ptarmic *Medicine.* relating to or producing spasmodic sneezing.

ptarmigan [tär´mi gən] *Vertebrate Zoology.* a gamebird of the genus *Lagopus* in the grouse family Tetraonidae; characterized by dark plumage that becomes white in winter, feathered feet, a plump body, and short, rounded wings; found in forest, prairie, and tundra of North America and northern Asia and Europe.

ptarmigan

PT boat *Naval Architecture.* a small, fast, easily maneuverable fighting ship of the U.S. Navy used for patrol duty and also to carry torpedoes for combat; ranging in length from 70 to 90 feet and over 50 long tons in displacement and capable of speeds of more than 50 knots; used especially in the Pacific campaign of World War II. (An initialism for patrol torpedo boat.) Also, MOTOR TORPEDO BOAT, MOSQUITO BOAT.

PTC plasma thromoboplastin component; phenylthiocarbamide.

PTD posttuning drift.

Pteranodon *Paleontology.* a genus of large flying reptiles in the extinct subclass Archosauria and suborder Pterodactyloidea; their wingspan may have extended as much as 25 feet; extant in the Late Cretaceous.

Pteraspis *Paleontology.* a genus of heterostracan jawless fish whose shield projects forward beyond the mouth and forms a medial dorsal spine; the rear half of the body is not armored except by scales; extant in the Lower Devonian.

Pterasteridae *Invertebrate Zoology.* a family of large deepwater starfish in the order Spinulosida that practice brood care and have a fin-like membrane edging the arms.

Pterichthyodes *Paleontology.* a genus of placoderms in the subclass Antiarcha; distinguished by a pair of unusual jointed appendages extending below and behind the head; known only from the Middle Devonian of Scotland.

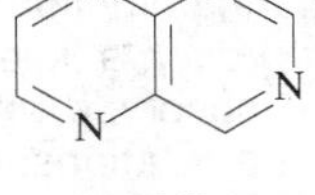

pteridine

pteridine *Organic Chemistry.* yellow plates derived from pigments in the wings of butterflies; soluble in water; characterized by a fused pyrazine and pyrimidine ring system.

Pteridophyta *Botany.* in some classification systems, a division containing all nonseed-bearing vascular plants.

Pteridospermales *Botany.* an order of extinct gymnosperms of the Devonian and Triassic periods, characterized by fernlike fronds, secondary stem thickening, and apparent seed formation. Also, CYCADOFILICALES, SEED FERNS.

Pteridospermaphyta *Paleontology.* the seed ferns, an extinct division of gymnospermous plants that often resembled tree ferns but bore seeds and other pollen-bearing structures on the fronds; formerly called Cycadofilicales; Upper Devonian to Jurassic. Also, PTERIDOSPERMAE.

Pteriidae *Invertebrate Zoology.* a family of bivalve mollusks including pearl oysters and wing shells, having valves of unequal size; found in warm sea shallows. Also, AVICULIDAE.

pterinophore *Cell Biology.* a set of color pigments found in vertebrates, plants, and insects.

Pterobranchia *Invertebrate Zoology.* a class of little-known, tiny, sessile, colonial tentacled worms in the Hemichordata that live in branching secreted tubes in deep waters.

Pterobryaceae *Botany.* a large family of generally glossy, dendroid or frondose, tropical and subtropical mosses of the order Isobryales that form tufts and mats on trees and rocks; characterized by creeping and deeply colored primary stems with reduced scale leaves, erect or ascending, frequently branched secondary stems, and lateral sporophytes.

pterochore *Botany.* a plant that produces winged disseminules.

Pterocladiophilaceae *Botany.* a monospecific family of red algae in the order Cryoptonemiales, having a small, cushion-shaped thalli without pigmentation; parasitic on members of the family Gelidiaceae.

Pteroclidae *Vertebrate Zoology.* the sandgrouse; a bird family of the order Columbiformes, characterized by a partridgelike body with a dovelike head, dull-colored plumage, and a short bill and legs; found in deserts, semideserts, dry grasssland, arid savannah, and bushveld of Europe, southern Asia, and Africa. Also, **pteroclodidae.**

pterodactyl [ter´ə dak´təl] *Paleontology.* **1.** one of the flying or gliding archosaurian reptiles of the suborder Pterodactyloidea. **2.** of or relating to this suborder.

Pterodactyloidea *Paleontology.* a suborder of archosaurs in the extinct order Pterosauria; relatively small (sparrow- to heron-sized) and probably inefficient flyers; extant in the Jurassic and Cretaceous.

Pterodactylus *Paleontology.* the type genus of flying reptiles of the suborder Pterodactyloidea; extant in the Jurassic and Cretaceous.

pteroic acid *Biochemistry.* $C_{14}H_{12}N_6O_3$, a derivative of folic acid that forms during the hydrolysis of pteroylglutamic acids, such as folic acid.

Pteromalidae *Invertebrate Zoology.* a large family of hymenopteran insects, including the jewel wasps, in the superfamily Chalcidoidea; tiny, metallic parasites of insect pests.

Pteromedusae *Invertebrate Zoology.* a suborder of hydrozoans in the order Trachylina, having four swimming lobes on the bipyramidal body.

Pterophoridae *Invertebrate Zoology.* the plume moths, a family of lepidopteran insects in the superfamily Pyralidoidea, having a slender, delicate body and wings divided into fringed lobes; its larvae are crop pests.

Pteropoda *Invertebrate Zoology.* the sea butterflies, a suborder of gastropod mollusks in the order Opisthibranchia; small free swimmers with the side of the foot modified into fins.

Pteropodidae *Vertebrate Zoology.* the Old World fruit bats, a family composing the suborder Megachiroptera; characterized by large eyes, large, funnel-shaped ears, a sturdy thumb, and a large claw; among the largest of all bats; found in tropical and subtropical regions of Asia, Africa, and Australia, and in islands of the Pacific.

pteropod ooze *Geology.* a calcareous deep-sea deposit consisting of at least 45% skeletal remains of marine organisms, especially pteropods.

Pteropsida *Botany.* a large subdivision or class of vascular plants, characterized by leaves believed to have originated as branched stem systems, having parenchymatous leaf gaps in the stele; depending upon the system of classification, it may include only the ferns or the ferns, gymnosperms, and angiosperms.

pterosaur [ter´ə sôr´] *Paleontology.* **1.** one of the flying reptiles of the order Pterosauria. **2.** of or relating to this order.

Pterosauria *Paleontology.* an order of archosaurs distinguished by the leathery membranes stretched between elongated fingers that enabled them to glide or soar; the wing span ranged from a few inches to over 40 feet; they flourished worldwide during the Jurassic but were dying out by the Late Cretaceous.

Pterosidomorphi *Vertebrate Zoology.* a subdivision of ostracoderms (jawless fishes) containing the major genera *Pteraspis* and *Palaeodus.*

Pterospermaceae *Botany.* a family of gold-brown marine algae belonging to the order Chrysosphaerales, characterized by autospores surrounded by a wall.

pterostigma *Invertebrate Zoology.* a thick opaque spot on the front edge of the wings of some insects.

pterostilbene *Biochemistry.* a phytoalexin secreted by the grape vine.

pterygium *Medicine.* an abnormal triangular growth of mucous membrane on the conjunctiva from the inner canthus to the border of the cornea or beyond, with the apex of the growth pointed toward the pupil.

pterygoid *Anatomy.* shaped like a wing.

pterygoid bone *Vertebrate Zoology.* in the skull of most lower vertebrates, one of four large, medially situated bones of the palatal complex that is located on the lower surface of the palatal cartilage.

pterygoid plate *Anatomy.* either of the sections of a pterygoid process.

pterygoid process *Anatomy.* either of two processes descending from the point of juncture of the main body and the great wing of the sphenoid bone of the skull.

pterygopalatine fossa *Anatomy.* a small space between the pterygoid process, the maxilla, and palatine bones that houses the pterygopalatine nerve ganglion.

pterygoquadrate *Developmental Biology.* relating to the first branchial arch of the upper jaw.

Pterygota *Invertebrate Zoology.* a subclass of the class Insecta, including the more advanced orders, having wings or having had wings at an earlier evolutionary stage.

pterygota *Materials.* the pale, light wood of the African trees *Pterygota bequaertii* and *P. macrocarpa*; used for plywood and furniture.

Pterygotus *Paleontology.* a widespread genus of large sea scorpions in the extinct order Eurypterida and family Pterygotidae; characterized by a terminal tail fin in place of the usual spine; some of the pterygotids grew to lengths of 6 feet; they were widespread in the Silurian but became extinct in the Devonian.

PTH parathyroid hormone.

Ptilidiaceae *Botany.* a monogeneric family of medium-sized, prostrate, brownish-red liverworts of the order Jungermanniales; having asymmetrical lobed leaves with an incubous orientation; found in arctic regions.

Ptiliidae *Invertebrate Zoology.* the feather-winged beetles, a family of tiny coleopteran insects in the superfamily Staphylinoidea, having a flattened body and hindwings fringed with long hairs; found among fungus spores or ant colonies. Also, TRICHOPTERYGIDAE.

Ptilodontoidea *Paleontology.* a suborder of multituberculate mammals in the subclass Allotheria; the ptilodontoids arose in the Triassic, diversified in the Late Cretaceous, and became extinct in the Oligocene.

Ptinidae *Invertebrate Zoology.* the spider beetles, a family of coleopteran insects in the superfamily Bostrichoidea, having a tiny oval or cylindrical body; they are pests in foods.

PTM pulse-time modulation.

Ptolemaic system [täl´ə mā´ik] *Astronomy.* a theory developed by Ptolemy about 150 AD, which held that a motionless earth is the center of the universe with the sun, moon, and planets revolving around it in eccentric circles and epicycles, while the fixed stars are attached to an outer sphere concentric with the earth; generally accepted in the European community until the development of the Copernican theory.

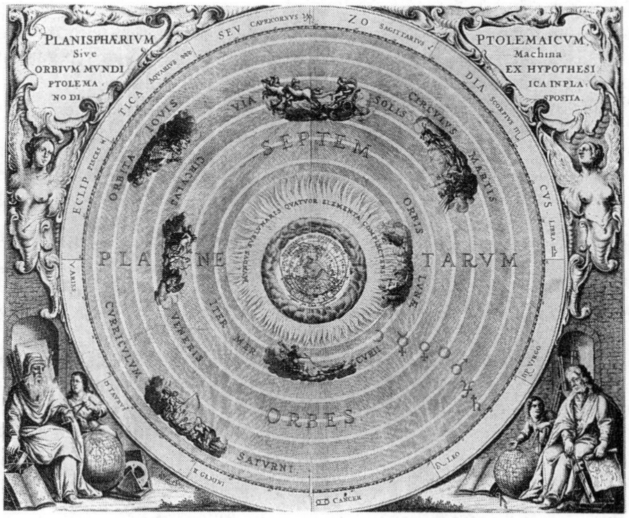

Ptolemaic system

Ptolemy [täl′ə mē] c. 100–165 AD, Greco-Egyptian astronomer, geographer, and mathematician; proposed the Ptolemaic system.
ptomaine [tō′mān; tō mān′] *Toxicology.* any of a group of highly toxic alkaloids produced by bacterial action. *Biochemistry.* a more general term for all alkaloid compounds formed in decaying animal or vegetable matter by the bacterial decomposition of proteins, including both toxic compounds and others that are nontoxic.
ptomainemia *Toxicology.* the presence of one or more ptomaines in the blood.
ptomaine poisoning *Toxicology.* an older term for various types of food poisoning associated with the eating of ptomaines.
ptomainotoxism *Toxicology.* poisoning by a ptomaine.
ptosis *Medicine.* **1.** the prolapse of an organ or part. **2.** the drooping of the upper eyelid, from paralysis of the third nerve or from sympathetic innervation. Also, BLEPHAROPTOSIS.
pts. or **pts** pints.
PTT partial thromboplastin time.
PTY *Aviation.* the airport code for Panama City, Panama.
ptyalagogue see SIALAGOGUE.
ptyalin *Biochemistry.* an enzyme present in saliva that triggers the hydrolysis of starch into sugars, such as dextrin, matos, and glucose, and the hydrolysis of sucrose to glucose and fructose. Also, PTYLASE, SALIVARY AMYLASE, SALIVARY DIASTASE.
ptyalism see SALIVATION.
Ptychodactiaria *Invertebrate Zoology.* an order of sea anemones, anthozoan coelenterates with no cilia on the mesenteric filaments; found in polar waters.
Ptychodus *Paleontology.* a widespread genus of early sharks in the extinct family Ptychodontidae; noted for its flattened, shell-crushing teeth; Cretaceous.
Ptychomitriaceae *Botany.* a mostly tropical, monogeneric family of dark- or olive-green mosses of the order Grimmiales that form mats or cushions on rocks and tree trunks; characterized by frequently branched, erect-ascending stems and a haplolepideous peristome of 16 papillose teeth that are divided into filiform prongs.
Ptychomniaceae *Botany.* a family of shiny, yellowish-green mosses of the order Isobryales that grow on humus, tree trunks, and branches; distinguished by an eight-ribbed capsule.
Ptyctodontida *Paleontology.* an order of arthrodires in the extinct class Placodermi, having considerably weaker armor than other placoderms, but at least remnants of all the major plates; distinguished by a short skull and shoulder girdle, and the pelvic claspers of the males; possibly ancestral to sharks and rays; Devonian and Lowermost Carboniferous.
ptygma *Geology.* the appearance of disharmonic folds of pegmatitic material in granitic or metamorphic rocks, such as migmatite or gneiss. Also, **ptygmatic fold.**
p-type crystal rectifier *Electronics.* a crystal rectifier in which current flows when the semiconductor material is positive with respect to the metal.
p-type semiconductor *Electronics.* an extrinsic semiconductor having a low concentration of electrons and a high hole density, so that the majority carriers are holes.
p-type silicon *Electronics.* silicon that has been doped with more of the acceptor-type atoms, such as boron, than of the donor-type atoms, such as phosphorus, so that its majority carriers are holes.
PU peptic ulcer; propellant utilization.
Pu the chemical symbol for plutonium.
puberty [pyoo′bər tē] *Physiology.* the time of life when secondary sex changes occur in primates and humans, marked by the onset of menses in females and sperm production in males.
puberulent *Botany.* covered with fine, short hairs or down.
puberulic acid *Biochemistry.* $(HO)_3(C_7H_2O)COOH$, an antibiotic isolated from *Penicillium puberulum* and other *Penicillium* species that attacks Gram-positive bacteria.
pubes *Anatomy.* **1.** the hairs covering the pubic region. **2.** the pubic region.
pubescence *Botany.* a soft down that covers the surface of certain plants. *Anatomy.* the arrival of puberty.
pubic [pyoo′bik] *Anatomy.* of or relating to the pubis.
pubic arch *Anatomy.* the arch formed by the inferior borders of the pubic bones.
pubic crest *Anatomy.* a raised area on the rostral border of the superior ramus of the pubic bone.
pubic symphysis *Anatomy.* the ventral (fibrocartilagenous) articulation of the pubic bones.
pubis *Anatomy.* either of the two bones forming the frontal arch of the pelvis. Also, OS PUBIS.
public-address system *Acoustical Engineering.* a system that amplifies the sound of a performer, speaker, or the like, in order to project the sound above the level of the ambient noise. Also, PA SYSTEM.
public data *Computer Technology.* data to which all users or program modules have read access.
public health *Medicine.* the branch of medicine dealing with safeguarding and improving community health through organized community effort, involving prevention of disease, control of communicable disease, and health education.
pubofemoral *Anatomy.* relating to the pubis and the femur.
Puccinia *Mycology.* a genus of rust fungi belonging to the class Uredenales that causes diseases in plants and cereals, such as barley and oats.
Pucciniaceae *Mycology.* the largest family of fungi belonging to the order Uredenales; composed of species that are primarily pathogenic to wheat and other cereal plants; found worldwide.
Pucciniastraceae *Mycology.* a family of fungi belonging to the order Uredenales, which is composed of fern rusts that occur on fern fronds or the leaves of seed plants.
puccoon *Botany.* **1.** any of various plants, such as the bloodroot and plants of the genus *Lithospermun* of the borage family, that yield a red dye. **2.** the dye itself.
pucellas *Mechanical Devices.* a tool similar to tongs or shears used to handle and shape molten glass.
pucherite *Mineralogy.* $BiVO_4$, a reddish-brown orthorhombic mineral occurring as tabular or acicular crystals, having a specific gravity of 6.25 to 6.57 and a hardness of 4 on the Mohs scale; a minor ore of vanadium.
puddingstone *Geology.* a type of conglomerate in which rounded, differently colored pebbles provide a sharp contrast to the abundant finegrained matrix or cement, so that in cross section the rock resembles an old-fashioned plum or raisin pudding; found especially in Great Britain.
puddle *Engineering.* to saturate soil in order to settle the dirt of an area for agricultural or construction purposes.
puddling *Metallurgy.* a former process for manufacturing wrought iron from pig iron.
pudendum *Anatomy.* the external genital organs, especially those of the female. Also, **pudenda.** (Going back to a Latin term meaning "to be ashamed.")
pudu *Vertebrate Zoology.* a small, reddish deer of the family Cervidae; the smallest deer of the Americas at 12 to 13 inches tall; having a thick, dark-brown coat, short legs, and small, upright antlers; found in Chile.
pueblo [pweb′lō] *Architecture.* an ancient dwelling of the Indians of the Southwest U.S., typically consisting of compounds of multistoried, contiguous houses with courtyards and large circular rooms; some continue to be occupied.

pueblo

Pueblo *Archaeology.* a stage in various chronologies of the Southwestern U.S., typically spanning the time period from 700 AD to the 1700s.
puelche [pwel´chā] *Meteorology.* an easterly wind that has crossed the Andes mountains on the South American west coast.
Puercan *Geology.* a North American continental stage of the Lower Paleocene, occurring before the Dragonian.
puerpera *Medicine.* a term for a woman who has just given birth.
puerperal *Medicine.* **1.** of or relating to a woman who has just given birth. **2.** of or relating to childbirth.
puerperalism *Medicine.* any disorder related to childbirth.
puerperal sepsis *Medicine.* an infection acquired by the mother during the time following childbirth.
puerperium *Medicine.* the time following childbirth during which the physical changes of pregnancy resolve, usually about six weeks.
Puerto Rican cherry see ACEROLA.
puff *Meteorology.* a local light wind that causes a rippling effect on the surface of a body of water. *Molecular Biology.* a decondensed region of a polytene chromosome that serves as a site of active RNA transcription; especially evident in certain *Drosophila* larvae cells.
puff adder *Vertebrate Zoology.* a large, thick-bodied viper, *Bitis arietans,* or a similar viper, *B. inornata,* that inflates its body and hisses when disturbed; found in Africa.
puffball *Botany.* the fruiting body of some fungi, especially *Lycoperdon* and allied genera, which retains spores until they are fully mature, then releases them as puffs of fine dust.
puffbird *Vertebrate Zoology.* any of several tropical American birds of the Bucconidae family that are related to the barbets; characterized by a large head with feathers often fluffed out.
puff cone see MUD CONE.
puffer *Vertebrate Zoology.* any of various fishes of the family Theraponidae that can inflate the body with air or water until it appears globe-shaped, with the spines in the skin erect; several species contain the nerve poison tetrodotoxin.
puffin *Vertebrate Zoology.* an auklike diving bird of the genera *Fratercula* and *Lunda* of the family Alcidae, having a broad, brightly colored bill, a short tail, a stout body, and webbed feet; found on the coasts of the Northern Hemisphere.

puffin

pug mill *Mechanical Engineering.* a machine in which blades are used to mix and temper a plastic material.
Pugnax *Paleontology.* an extinct genus of articulate brachiopods in the order Rhynchonellida; characterized by strong central ribs with a smooth shell surface on both sides; Devonian and Carboniferous.
pulaskite *Petrology.* a light-colored, feldspathoid-bearing variety of alkali syenite, having a granular or trachytoid texture and consisting primarily of alkali feldspar, soda pyroxene, arfyedsonite, and nepheline.
Pulfrich refractometer *Optics.* a refractometer in which a prism or block of high refractive index and known angle measures the refractive index of solids or liquids.
pull-apart *Geology.* a sedimentary structure formed prior to compaction that resembles boudinage; composed of beds that have been stretched and torn apart into relatively short slabs.
pull crack *Metallurgy.* in a casting, an undesirable fracture caused by the residual stresses generated upon solidification.
pull-down menu *Computer Technology.* a set of options displayed across the top of a video screen, so that when one option (such as Style) is selected by means of a function key or mouse, a set of suboptions (such as Font, Size, Type Style, and Character) are displayed below the original option.
puller see COME-ALONG.
pullet *Agriculture.* a young hen, especially one that is between 5 and 18 months old.
pullet disease see BLUE COMB.
pulley *Mechanical Devices.* one of the simple machines, typically consisting of a metal wheel fixed to a shaft, with a rope, belt, band, or the like passing over the wheel to transmit motion or energy.
pulley stile *Building Engineering.* the vertical sides of a double-hung window, having a pulley over which the sash cord passes. Also, HANGING STILE.
pulling *Electronics.* a condition in which an oscillator changes frequency under the influence of another source; frequently seen in mixers where two oscillators share the same circuit.
pull-in torque *Mechanical Engineering.* the greatest steady torque at which a motor will attain normal speed after accelerating from a standstill.
pullorum disease *Veterinary Medicine.* a fatal egg-borne disease affecting chicks within the first two or three weeks after hatching, caused by *Salmonella pullorum* and manifested by incoordination, drowsiness, constant cheeping, and diarrhea with whitish fecal pasting around the vent. Also, BACILLARY WHITE DIARRHEA.
pullout *Materials Science.* a fracture or operative failure mode of fiber composites in which, under certain loading conditions, fibers do not break but debond and are pulled out of the matrix.
pull-out torque *Mechanical Engineering.* the greatest torque at which a motor will operate without a sudden loss of speed.
pull strength *Mechanics.* in tensile testing, a measure of the bond strength in pounds per square inch.
pullulan *Biochemistry.* a glucan that is principally composed of maltotriose units joined by alpha glucosidic bonds, synthesized by *Aureobasidium pullulans.*
pullulanase *Enzymology.* an enzyme that catalyzes the hydrolysis of α-1,6-glucosidic bonds at branch points in an α-1,4-linked chain.
pullulation *Biology.* **1.** reproduction by budding, as in yeast cells. **2.** the process of sprouting or germination, as in a seed.
pulm- or **pulmo-** a combining form meaning "lungs." Also, **pulmon-, pulmono-.**
pulmoaortic canal see DUCTUS ARTERIOSUS.
pulmonary [pul´mə nâr´ē] *Anatomy.* relating to or affecting the lungs.
pulmonary adenomatosis see JAAGSIEKTE.
pulmonary alveolus *Anatomy.* one of a cluster of small outpocketings at the end of a bronchiole through which the exchange of carbon dioxide and oxygen between alveolar air and capillary blood takes place.
pulmonary anthrax *Medicine.* a form of internal anthrax caused by the inhalation of dust particles contaminated with anthrax bacillus; characterized by chills, back and leg pains, rapid respiration, cough, fever, and extreme prostration. Also, WOOL SORTERS' DISEASE.
pulmonary artery *Anatomy.* one of the vessels conveying blood from the right ventricle of the heart to the lungs.

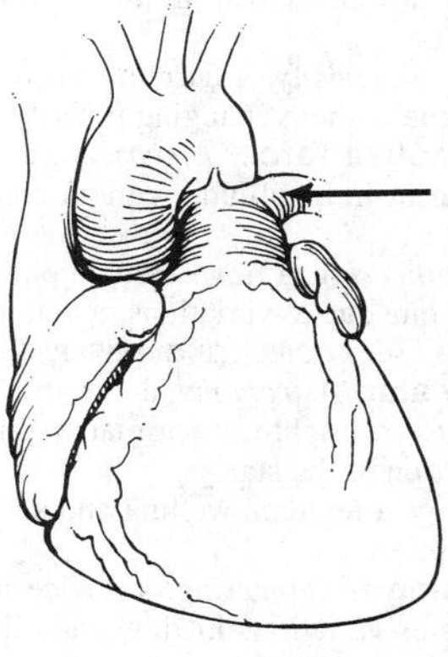

pulmonary artery

pulmonary aspergillosis *Medicine.* a disease of the lungs resulting from infection with the fungus *Aspergillus fumigatus.*

pulmonary circulation *Physiology.* the flow of blood through the network of blood vessels between the right ventricle and the left atrium of the heart, including the pulmonary capillaries, veins, and arteries.

pulmonary edema *Medicine.* the collection of fluid in the air sacs and interstitial tissue of the lungs, usually resulting from ventricular failure or mitral stenosis.

pulmonary hamartoma see ADENOCHONDROMA.

pulmonary incompetence *Cardiology.* faulty closure of the pulmonic valve, permitting regurgitation of blood into the right ventricle during diastole.

pulmonary plexus *Anatomy.* either of the networks of sympathetic and parasympathetic nerves that join one another and enter each lung at the hilum and run with the bronchi.

pulmonary stenosis *Medicine.* a narrowing of the opening into the pulmonary artery from the right ventricle.

pulmonary tuberculosis see OPEN TUBERCULOSIS.

pulmonary valve *Anatomy.* a one-way, semilunar valve at the exit of the pulmonary artery from the right ventricle of the heart.

pulmonary vein *Anatomy.* any of the veins conveying oxygenated blood from the lungs to the left atrium of the heart.

Pulmonata *Invertebrate Zoology.* a subclass of terrestrial and freshwater snails and slugs, gastropod mollusks with no gills; the mantle cavity is modified as a lunglike respiratory sac.

pulmonic *Anatomy.* relating to the lungs or the pulmonary artery.

pulp *Anatomy.* any soft, moist, pithlike matter, such as that beneath the enamel and dentine of a tooth or that contained within the spleen. *Botany.* **1.** the spongy, succulent tissue within a fruit or vegetable. **2.** the soft pith at the interior of a plant stem. *Materials.* a soft, moist, slightly cohering mass of material used to make paper; it is generally composed of ground-up wood, linen, or rags.

pulpboard *Materials Science.* a board made from wood pulp.

pulp chamber *Anatomy.* the section of the internal tooth containing pulp. Also, **pulp cavity.**

pulper *Mechanical Engineering.* a machine that converts wood or other materials to pulp.

pulping *Engineering.* a mechanical or chemical process that disintegrates wood fibers to form pulp. Also, DEFIBRATION.

pulp molding *Engineering.* a plastics-manufacturing process in which a resin-impregnated pulp substance is shaped with a vacuum apparatus and then oven-cured and molded.

pulpotomy *Surgery.* the removal by surgery of a section of the pulp structure of a tooth, usually the coronal portion.

pulpwood *Materials Science.* spruce or other soft wood suitable for making paper.

pulpy kidney disease *Veterinary Medicine.* a disease of lambs or young sheep; an acute and usually fatal toxemia caused by the bacterium *Clostridium perfringens*, type D. It may be precipitated by overfeeding or a high-carbohydrate diet. Rapid postmortem autolysis of the kidneys has given rise to the name "pulpy kidney." Also, ENTEROTOXEMIA, OVEREATING DISEASE.

pulpy peat see SEDIMENTARY PEAT.

pulsar [pul´sär] *Astrophysics.* a rapidly spinning neutron star that periodically beams flashes of radio (and sometimes optical) energy.

pulsatance *Physics.* the angular velocity of some periodic quantity, expressed in radians per seconds.

pulsate *Mechanics.* to move with an alternating pattern of increasing and decreasing motion.

pulsating current *Electricity.* a current that repeatedly increases and decreases in amplitude without changing polarity.

pulsating electromotive force *Electricity.* a voltage that repeatedly increases and decreases in amplitude without changing sign. Also, **pulsating voltage.**

pulsating flow *Engineering.* a flow within a piping system that is irregular in pressure and rate due to variations in compressors and pumps.

pulsating universe see OSCILLATING UNIVERSE.

pulsating variable star *Astronomy.* a variable star, such as Cepheids or Mira, that changes in brightness, primarily due to an alternating expansion and contraction of the star.

pulsation *Physiology.* a regular swelling and shrinking motion, such as that of the heart muscle.

pulsation dampening *Engineering.* a device in a piping system that regulates or eliminates variations in flow caused by components of the system, such as compressors and pumps.

pulse *Physics.* an abrupt change in a quantity, characterized by a rise and a decline, typically occurring over a short time interval in comparison to the time scale of interest. *Electricity.* **1.** a sharp variation of a current or voltage having a normally constant value. **2.** of or relating to such a variation. *Telecommunications.* **1.** a shift in the energy potential at a point for a relatively short period of time. **2.** a single impulse of a telephone dial or similar signal. *Physiology.* **1.** the rhythmic pressure of the blood against the walls of a vessel, particularly an artery. **2.** a stimulus of short duration, such as light, sound, or an electric current.

pulse altimeter *Engineering.* an instrument that measures the altitude of an aircraft, in which radar signals are transmitted in short pulses, and the time delay between the leading edge of the pulse and the pulse returned from the ground is measured to calculate altitude.

pulse amplifier *Electricity.* an amplifier designed to amplify electric pulses without changing their waveform.

pulse amplitude *Physics.* the magnitude of a pulse measured with respect to the normally constant value.

pulse analyzer *Electronics.* an instrument that examines a pulse to establish its characteristics, such as repetition rates and amplitude, and then displays the results on a cathode-ray tube screen.

pulse chase experiment *Biotechnology.* a technique that follows the metabolism of a compound by tracing the movement of a radioactive label as it moves with the compound.

pulse circuit *Electronics.* a circuit that generates or processes pulses of current or voltage.

pulse code *Telecommunications.* a code made up of various combinations of pulses, such as the Baudot code or Morse code.

pulse column *Chemical Engineering.* a separation or extraction process column in which a rapid motion of relatively short amplitude is pulsed through the column contents to give improved extraction rates.

pulse compression *Electronics.* a filtering technique that prevents signals which do not correspond to the transmitted signal from entering a circuit.

pulse-compression radar *Telecommunications.* a radar system that linearly spreads its transmitted signal to reduce the peak power handled by the transmitter; the receiver uses a linear filter so that the signal is transformed to a short pulse for the radar display.

pulse counter *Electronics.* a device that records the total number of pulses received over a given time span.

pulse decay time *Telecommunications.* the time needed for the pulse amplitude to descend from 90% to 10% of the peak value. Also, FALL TIME.

pulse-delay network *Electronics.* a network whose output is a pulse of the same shape as the input pulse, but delayed by a fixed or variable time period.

pulse demoder *Telecommunications.* a circuit that responds only to pulse signals that have the specified spacing between pulses for which the device is adapted. Also, CONSTANT-DELAY DISCRIMINATOR.

pulsed-field gel electrophoresis *Molecular Biology.* a gel electrophoretic procedure used to separate DNA molecules based on frequently altering the direction of the electric field.

pulse dialing *Telecommunications.* a system of telephone dialing in which electrical pulses that correspond to the digits in a number called are generated by manipulating a rotary dial or push buttons, as distinguished from tone dialing.

pulse discriminator *Electronics.* a device that responds only to pulses with a given characteristic, such as those at a specific amplitude.

pulsed laser *Optics.* a laser that emits short pulses of coherent radiation in fixed intervals, rather than a continuous stream of energy; used for ranging and tracking or for situations in which brief but high output power is desired.

pulsed light *Optics.* a light beam whose output is modulated so as to transmit short, regular bursts of light; similar to radar pulses.

pulsed-light ceilometer *Engineering.* a pulse radar device that determines cloud altitude, using visible light rather than radio waves. Also, **pulsed-light cloud-height indicator.**

pulse-Doppler radar *Engineering.* a pulse radar in which the Doppler effect is used to determine the speed at which the target is moving.

pulsed oscillator *Electronics.* an oscillator whose output is a stream of pulsed oscillating signals separated by periods of zero output, rather than a continuous sinusoidal signal.

pulsed reactor *Nucleonics.* a research reactor that is designed to operate in a pulsed mode in which short surges of neutron flux density are produced with intensities much greater than the tolerance level of steady-state operation.

pulse droop *Electronics.* a distortion in a pulse's shape, in which, characteristically, an incline appears on top of a normally flat-top rectangular pulse.

pulsed ruby laser *Optics.* a laser in which a synthetic ruby is activated by an electronic flashtube to produce a high-power pulse of brief duration.

pulse duration *Telecommunications.* **1.** the time interval between the first and last instants at which the instantaneous value is a specified fraction, often 50%, of the peak pulse amplitude. **2.** in radar transmission, the measurement of the pulse transmission time in microseconds. Also, PULSE LENGTH, PULSE WIDTH.

pulse-duration modulation *Telecommunications.* a type of modulation by which the duration of a pulse varies in accordance with a characteristic of the modulating signal. Thus, **pulse-length modulation, pulse-width modulation.**

pulsed video thermography *Engineering.* a process used to determine defective areas of an object, in which heat is applied for a short time to the material to be tested, and an infrared detector system reveals hot or cold areas adjacent to the defective areas.

pulse form *Physics.* the shape of a pulse when plotted as a function of time or position.

pulse-forming network *Electronics.* a network that converts electricity generated by a charging source into rectangular wave-shaped pulses; used in microwave and radar equipment.

pulse-frequency modulation *Telecommunications.* a type of modulation by which the pulse-repetition frequency of a carrier varies in accordance with a characteristic of the modulating signal. Also, PULSE INTERVAL MODULATION.

pulse-frequency spectrum see PULSE SPECTRUM.

pulse generator *Electricity.* a piece of electronic equipment that is capable of generating a pulsed signal, that is, an output that rapidly changes polarity.

pulse group see PULSE TRAIN.

pulse hardening *Metallurgy.* a process in which very rapid heating is used to case-harden a ferrous alloy; the energy derives from a brief, but intense, pulse of high-frequency current.

pulse height *Electronics.* the amplitude of a pulse, measured in volts.

pulse-height analyzer *Nucleonics.* an electronic device that accumulates (counts) pulses and sorts them according to their height, storing a counter in an appropriate channel.

pulse-height discriminator *Electronics.* a circuit that accepts only those pulses having a specific amplitude. Also, AMPLITUDE DISCRIMINATOR.

pulse-height selector *Electronics.* a circuit that accepts input pulses whose amplitude lies between two given values. Also, AMPLITUDE SELECTOR, DIFFRACTIONAL.

pulse-height spectrum *Physics.* a spectrum of wavelengths and pulse heights (strengths) obtained in activation analysis.

pulse integrator *Electronics.* a circuit that extends the duration of a pulse.

pulse interleaving *Telecommunications.* a process of combining independent pulse trains on the basis of time-division multiplex for transmission along a common path. Also, **pulse interlacing.**

pulse interval see PULSE SPACING.

pulse-interval modulation see PULSE-FREQUENCY MODULATION.

pulsejet engine *Space Technology.* a compressorless jet engine in which intermittent combustion produces a series of explosions occurring at the approximate resonance frequency of the engine to provide thrust. An early form of pulsejet engine was used to propel the German V-1 rockets of World War II. Also, AEROPULSE ENGINE.

pulse jitter *Telecommunications.* a random or systematic variation of pulse spacing in a pulse train.

pulse labeling *Molecular Biology.* a short exposure of an experimental system, such as a population of cells, to a radioactive material that is incorporated into cell constituents, which is then traced as a function of time.

pulse length see PULSE DURATION.

pulse-link repeater *Electronics.* a device that transfers signals from the input and output leads of one telephone circuit to their counterparts in another telephone circuit.

pulse-modulated radar *Engineering.* a radar system in which signals are transmitted in a pattern of discrete pulses.

pulse modulation *Telecommunications.* a system of modulation in which the amplitude, duration, position, or presence of pulses is controlled so as to represent a message that is to be communicated.

pulse period *Telecommunications.* in telephony, the time needed for a single opening and closing of a rotary dial. Also, IMPULSE PERIOD.

pulse pressure *Physiology.* in blood pressure, the contrast between the maximum systolic pressure and the minimum diastolic pressure.

pulser *Chemical Engineering.* a reciprocating plunger or piston pump that produces a rapid reciprocating motion throughout the continuous phase of a pulse column. *Electronics.* a device that generates high-energy, short-lived electrical pulses for radar and microwave equipment.

pulse radar *Engineering.* a radar system in which pulses are transmitted between relatively long intervals, and an active component system receives echoes in the interval following each pulse.

pulse radiolysis *Physical Chemistry.* a technique for analyzing fast chemical reactions, in which a sample of a compound is subjected to a quick pulse of ionizing radiation, and then examined to determine the effects of the radiation.

pulse rate *Physiology.* the number of rhythmic pulsations that can be felt per minute in a peripheral artery.

pulse regeneration *Electronics.* the process by which a pulse is restored to its original shape, following distortion caused by moving through a circuit or a given medium, such as gas.

pulse repeater *Electronics.* a network of devices that receives and retransmits pulses from one circuit to the next; commonly used in telephone signaling systems.

pulse repetition rate *Electricity.* the frequency at which a signal pulses. Also, **pulse recurrence rate, pulse repetition frequency.**

pulse-shape discrimination *Biotechnology.* the process of distinguishing different types of ionizing particles based on the amplitude and lifetime of their signals.

pulse shaper *Electronics.* **1.** a device that changes the shape of a pulse, such as from a sine wave to a square wave. **2.** a device that changes the characteristics of a pulse so that it can perform other functions, before it is transmitted to another circuit.

pulse spacing *Physics.* in a train of pulses, the time that elapses between corresponding points in successive pulses. Also, PULSE INTERVAL.

pulse spectrum *Physics.* the frequency spectrum of a pulse indicating the frequency components and their relative magnitudes. Also, PULSE-FREQUENCY SPECTRUM.

pulse stretcher *Electronics.* **1.** a circuit that increases the duration of a pulse, generally to improve its sound quality. **2.** a circuit that produces a long pulse when triggered by a short pulse; mainly used in computers.

pulse synthesizer *Electronics.* a circuit that generates pulses to replace those absent from a sequence as a result of interference.

pulse-time-modulated radiosonde *Engineering.* a radio transmitter that is carried aloft and transmits meteorological data in pulses phased at specific intervals. Also, TIME-INTERVAL RADIOSONDE.

pulse-tracking system *Engineering.* a radar system in which the reflection of a high-energy pulse directed toward a target is used to obtain information about the speed, direction, and range of the target.

pulse train *Physics.* a continuous series of pulses, each having identical characteristics and constant pulse spacing. Also, IMPULSE TRAIN, PULSE GROUP.

pulse transformer *Electronics.* a transformer whose frequency response is wide enough to transmit pulses without distorting their shape.

pulse transmitter *Electronics.* **1.** a device that transmits sound waves at power levels well above average. **2.** a device that generates and transmits pulses for long-range transmissions; commonly used in telemetering and pilot wire devices.

pulse wave *Physiology.* the rhythmic movement of a vein or artery as perceived by touch or as recorded graphically.

pulse width see PULSE DURATION.

pulse-width discriminator *Electronics.* a device that passes or responds to pulses whose width is between specified limits.

pulsing *Military Science.* in naval mine warfare, a method of operating magnetic and acoustic sweeps in which the sweep is energized by current that varies or is intermittent in accordance with a predetermined schedule.

pulsometer *Mechanical Engineering.* a steam pump without a piston, in which an automatic ball valve admits steam alternately to a pair of chambers, forcing out water sucked in by steam condensation after the previous stroke.

pultrusion *Materials Science.* a process for making certain products, such as rods and electrical insulators, from glass-reinforced plastics by passing continuous bundles of glass fiber impregnated with liquid resin through an oven at the rate of about 0.5 m per minute at about 140°C.

pulv. powder. (From Latin *pulvis.*)

pulverite *Petrology.* a nonclastic sedimentary rock consisting of an aggregate of silt or clay, and having a texture similar to that of a lutite of clastic origin.
pulverize *Science.* to reduce to dust or powder.
pulverizer *Mechanical Engineering.* a device used to break down material to dust or powder, as by pounding or grinding.
pulverulite *Petrology.* a limestone containing an uncemented, granular, friable calcium carbonate mass.
pulvillus *Entomology.* a cushioned pad between the pair of claws on the tarsus of the leg of many insects.
pulvinus *Botany.* **1.** a cushionlike swelling made up of thin-walled cells at the base of a leaf or petiole that functions in turgor movements of the plant. **2.** a thickened region in the stem or leaf sheath of grasses, usually containing an intercalary meristem.
puma [pyoo´mə; poo´mə] *Vertebrate Zoology.* another name for the mountain lion, a wild cat of the Americas. See MOUNTAIN LION.
pumice [pum´is; pyoo´mis] *Geology.* a highly vesicular, glassy, volcanic rock, compositionally similar to rhyolite and often light enough to float on water. Also, **pumicite, pumice stone.** *Materials.* this naturally occurring, highly porous, material, containing mostly silica and some alumina, soda, and potash; often used as an abrasive in polishing and in soaps, concrete, cement, and plaster.
pumice fall *Volcanology.* a showerlike descent of pumice from an eruption cloud.
pumiceous *Geology.* **1.** of or relating to the texture of a pyroclastic rock characterized by an abundance of small cavities and a spongy, frothy appearance. **2.** a rock exhibiting such a texture.
pumilith *Geology.* a deposit of volcanic ash that has been consolidated or turned to stone.
pump *Fluid Mechanics.* **1.** a device that converts mechanical energy into fluid energy by any of various technologies, typically by suction or compression; used to move water, air, or other fluids into, through, or out of a system. **2.** to move a fluid by means of such a device. *Physiology.* **1.** a physical mechanism or process that is comparable to a mechanical pump. **2.** to move a liquid or gas within the body by means of this, as in the heart. *Physics.* **1.** an energy source that increases the number of highly energized electrons in an electron stream. **2.** to increase the energy level in an atom by bombarding it with electomagnetic radiation; applied to such devices as lasers, masers, and parametric amplifiers. *Electronics.* a signal that periodically changes the reactance of the varactor in a parametric amplifier.
pumpability test *Engineering.* a test that determines the lowest temperature at which a petroleum fuel oil may be effectively pumped.
pump-action *Ordnance.* describing a shotgun or other weapon with a sliding mechanism that is moved back and forth manually to clear, reload, and cock the weapon. Also, SLIDE-ACTION, TROMBONE-ACTION.
pumped hydroelectric storage *Electricity.* a form of energy storage in which extra electrical energy produced at low-demand times is used to pump water into a reservoir; the water is released at high-demand times to drive hydroelectric generators.
pumpellyite *Mineralogy.* **1.** a general name for three monoclinic, green to brown or pink, hydrous, calcium-aluminum silicate minerals containing iron, magnesium, or manganese; for example, pumpellyite-(Mg), having the formula $Ca_2MgAl_2(SiO_4)(Si_2O_7)(OH)_2 \cdot H_2O$, occurring in acicular to fibrous form, and having a specific gravity of 3.18 to 3.34 and a hardness of 5 to 6 on the Mohs scale; found widespread in rocks of diverse composition and origin. **2.** a mineral group name.
pumpellyite-prehnite-quartz facies *Petrology.* the set of metamorphic mineral assemblages consisting of metagraywackes containing albite, quartz, prehnite, pumpellyite, chlorite, and sphene, formed under low-temperature and moderate-pressure conditions.
pump gun *Ordnance.* a weapon using pump-action firing.
pumping *Fluid Mechanics.* the removal of fluids from a reservoir or vessel by externally applied mechanical means, such as a pump, compressor, or the like. *Physics.* in maser or laser technology, the process of irradiating an active medium so that the atoms or molecules in the medium will make a transition to the excited state, by absorbing the energy of the radiation.
pumping loss *Mechanical Engineering.* the power consumed when fresh air is sucked into a cylinder to purge it of exhaust gas.
pumping pressure *Petroleum Engineering.* the pressure necessary to pump acid, water, or gas into a pressurized oil reservoir.
pumping radiation *Physics.* the electromagnetic radiation used to pump the particles in a maser or laser medium from the ground state to the excited state.
pumpkin *Botany.* any of several species of the genus *Cucurbita* of the gourd family Cucurbitaceae, especially the edible fruit of the decumbent vine *C. pepo*; characterized by large-lobed leaves, somewhat prickly hairy vines, yellow flowers, and a large globular fruit with a hard orange rind and a hollow core filled with seeds.
pump oscillator *Electronics.* a device that generates high-frequency alternating current for masers and parametric amplifiers.
puna *Ecology.* a basin located along the western side of the higher Andes, characterized by tufted grasses and large, widely spaced, plants.
punch *Mechanical Devices.* **1.** a short, cylindrical steel device with a square, tapered head and a knurled or hexagonal area for gripping; used to mark a hole in metal or wood by striking with a hammer, and for driving out headless bushings or rivets. **2.** a tool used to force metal into a die. *Archaeology.* a pointed tool, usually of bone, stone, or wood, used to perforate a material such as hide or shell. *Surgery.* a surgical instrument used to indent, perforate, or cut a segment of tissue or to remove foreign matter. *Computer Technology.* **1.** a hole in a computer card or paper tape, used in combinations to represent information. **2.** to code information on cards or paper tape.
punch biopsy *Surgery.* the removal for examination of a small amount of living tissue by means of a punch.
punch card or **punched card** *Computer Technology.* a stiff, fixed-size card in which holes are punched to record data in a coded format; now generally being replaced by ordinary paper output but earlier a standard medium for computer data. Also, TABULATING CARD.

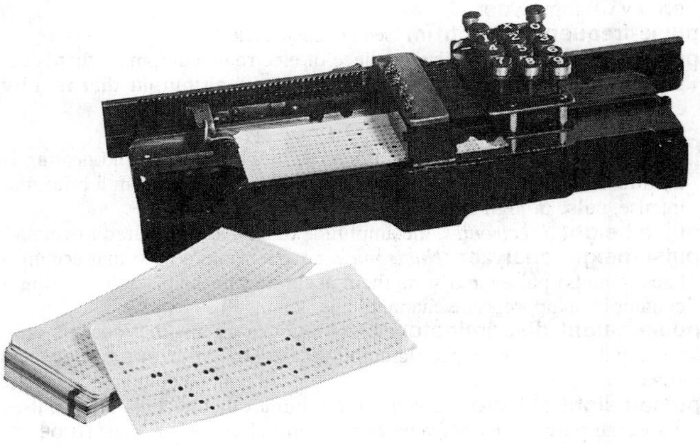

punch card

punchdrunk encephalopathy or **punchdrunk syndrome** *Neurology.* a condition occurring in professional boxers from the cumulative effects of repeated injuries, including a state of forgetfulness, slowness in thinking, slurred speech, and slow, uncertain movements, especially of the legs. Also, DEMENTIA PUGILISTICA.
punched-card reader see CARD READER.
punched-card reproducer see CARD REPRODUCER.
punched-card sorter see CARD SORTER.
punched-plate screen *Engineering.* a device used to sort size classifications of fragmented solids, made up of a flat plate with openings that vary in size and shape according to the sorting being done.
punched tape see PAPER TAPE.
punching *Engineering.* **1.** a process of machine shaping in which mating die sections determine the shape of the final product. **2.** a fragment removed from a metal plate or other material by a punch.
punch press *Mechanical Engineering.* a power-driven press in which metal sheets are cut or shaped in dies, by means of pressure or impact.
punch tape see PAPER TAPE.
punch-through *Electronics.* a condition in which a high voltage at a transistor's collector spreads across the entire base region, causing a breakdown.
Punctariaceae *Botany.* a family of membranous, tubular, and saccate brown algae belonging to the order Dictyosiphonales; characterized by a parenchymatous thallus that is often differentiated into a large-celled medulla and a small-celled cortex.
punctate *Biology.* full of minute points; dotted.
punctiform *Biology.* marked by dots or appearing as dots.

punctuated equilibrium *Evolution.* a model of evolutionary change in which long periods of relative evolutionary stability are interrupted (or "punctuated") by brief epochs of drastic evolutionary change, including mass extinctions and rapid speciation. Also, **punctuationalism.**

punctuation bit *Computer Technology.* a bit used as a delimiter for variable-length words, items, or records.

punctum remotum see FAR POINT.

puncture *Engineering.* a hole made with or as if with a sharp, pointed object. *Medicine.* see PUNCTURE WOUND. *Electricity.* a disruptive discharge of current through the body of an insulating material or other solid dielectric.

punctured neighborhood *Mathematics.* in topology, the set $N - \{x_0\}$, where N is a neighborhood of x_0.

puncture voltage *Electricity.* the voltage at which an electrical puncture occurs.

puncture wound *Medicine.* a relatively deep and narrow injury caused by a sharp object such as a nail, knife, wood splinter, or glass fragment.

pungent *Biology.* **1.** ending in sharp or piercing points. **2.** acrid to the sense of taste or smell.

Punicaceae *Botany.* a monogeneric family of dicotyledonous shrubs or small trees of the order Myrtales, often thorny, tanniferous, and producing pyridine alkaloids; native from the Balkan region to northern India.

punicine tannate see PELLETIERINE TANNATE.

punish *Behavior.* to carry out an act of punishment. *Artificial Intelligence.* to decrease the importance of a rule or network connection that is implicated in an incorrect decision, often by reducing its numerical weight.

punishment *Behavior.* **1.** an aversive stimulus that is applied following an undesired response. Also, **punisher. 2.** the act of applying such an aversive stimulus, or of withdrawing a rewarding stimulus following an undesired response.

punishment training *Behavior.* a type of training in which a particular response elicits a negative reinforcement.

Punnett square *Genetics.* a graphic checkerboard used to determine the types of zygotes, genotypes, and phenotypes resulting from a particular genetic cross.

punt *Naval Architecture.* **1.** a small, rectangular flat-bottomed boat used for general work in a harbor. **2.** a similar boat used for pleasure on inland rivers and waterways, often propelled with a pole.

pupa [pyoo´pə] *Invertebrate Zoology.* in insects with complete metamorphosis, an immobile, nonfeeding stage between the larva and the adult; the organism is usually enclosed in a cocoon or a cuticular case, and it undergoes extensive anatomical changes.

puparium *Entomology.* a hard barrel-shaped casing enclosing the pupa of certain members of the Diptera order.

pupate *Entomology.* **1.** of an insect larva, to metamorphose into a pupa. **2.** of or relating to the pupa stage.

pupfish *Vertebrate Zoology.* any of several rare species of desert topminnow (or killifish) of the family Cyprinodontidae, known for their ability to survive extreme temperatures; characterized by hard-cased eggs that stay dormant in desert mud until enough rain creates a pool for them to hatch; found primarily in fresh water throughout the tropics and in New World temperate zones.

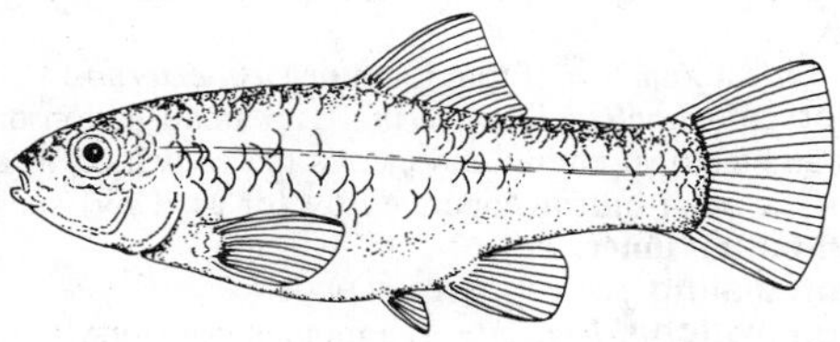

pupfish

pupil *Anatomy.* the opening in the center of the iris that controls the amount of light entering the eye. (Going back to a Latin term meaning literally "a little doll;" from the fact that one can see a tiny reflection of oneself when looking into another person's pupils.)

pupillary *Anatomy.* relating to the pupil of the eye.

pupillary reflex *Physiology.* **1.** the closing of the pupil when exposed to bright light. **2.** the opening of the pupil in reaction to pain.

Pupin, Michael 1858–1935, Serbian-born American physicist and engineer; improved telephone, telegraph, and radio networks.

Pupipara *Invertebrate Zoology.* a group of three dipteran fly families with larvae ready to pupate immediately after hatching.

pupiparous *Invertebrate Zoology.* describing an insect that bears fully developed larvae ready to pupate.

Puppis *Astronomy.* the Poop Deck, a large southern constellation in the Milky Way that was part of the old constellation of Argo Navis (Jason's Ship).

Puppis A *Astronomy.* a radio-bright supernova remnant that also emits X-rays.

Purbeckian *Geology.* in Great Britain, a geologic stage of the uppermost Jurassic period, occurring after the Bononian and before the Cretaceous period, equivalent to the European Upper Portlandian stage.

Purcell, Edward Mills born 1912, American physicist; awarded the Nobel Prize for developing a method of measuring magnetic fields in atomic nuclei.

pure without foreign, inappropriate, or inferior matter; in its true state; specific uses include: *Chemistry.* describing a substance that is free from contamination, that contains no detectable extraneous materials, and that has standard physical and chemical properties. *Genetics.* of a strain or breed, not mixed with any other. *Science.* describing the theoretical or abstract aspects of some area of science, without regard for practical applications.

purebred *Genetics.* descended from an inbred strain. *Agriculture.* **1.** of an animal, belonging to one of the recognized breeds of livestock. **2.** an animal bred from two purebred parents.

purebred quadrivalent *Genetics.* the association of four homologous chromosomes at meiosis in an inbred line of cells or organisms.

pure coal see VITRAIN.

pure culture see AXENIC CULTURE.

pure flutter *Cardiology.* atrial flutter in which there is regularity of rhythm.

pure forest *Forestry.* a forest composed primarily of one species, generally having at least 80% trees of the same species.

pure geometry *Mathematics.* the axiomatic study of geometry, as opposed to the study of geometric objects occurring in other contexts.

pure imaginary number *Mathematics.* a complex number of the form bi, where b is real.

pure line *Genetics.* a homozygous strain produced by inbreeding.

pure mathematics *Mathematics.* the study of mathematical theories and structure without regard to practical applications.

pure motion *Mechanics.* the movement of a system in which the paths of all the particles of the system remain parallel to their original direction; i.e., perfect linear motion.

pure projective geometry *Mathematics.* the axiomatic study of invariance of geometric systems with respect to a notion of projection.

pure research see BASIC RESEARCH.

pure rotation *Mechanics.* the motion of a rigid body whose orientation in space changes, but whose center of mass remains fixed.

pure shear *Mechanics.* a state of stress when there is only shear stress on a material surface particularly oriented at a point but no normal stresses.

pure state *Quantum Mechanics.* a single state described by a unique wavefunction that completely describes the system.

pure tone see SIMPLE TONE.

pure translation see TRANSLATIONAL MOTION.

pure Trojan group *Astronomy.* the total of all the Trojan asteroids that orbit the sun at Jupiter's distance, 60° ahead and behind the planet.

purga *Meteorology.* a violent blizzardlike storm that occurs in the tundra regions of northern Siberia in the winter.

purgative *Pharmacology.* **1.** causing the evacuation of the bowels; cathartic. **2.** a cathartic, especially one that stimulates peristalsis.

purge *Science.* to clean thoroughly, removing impurities or other unwanted material. *Medicine.* **1.** to induce an evacuation of the bowels by cathartics. **2.** a purgative medication. *Computer Technology.* to delete files or other information from secondary storage in a computer system to make room for new information. *Engineering.* to replace substances in a container with inert substances, usually so that explosive mixtures will not form. *Space Technology.* to remove residual fuel or oxygen from the tanks or lines of a rocket after a test firing. Thus, **purging.**

purge date *Computer Technology.* a date specifying when an associated file or set of data may be erased to provide space for other information.

purified cotton *Surgery.* purified, bleached, and sterilized cotton, used as a surgical dressing. Also, ABSORBENT COTTON.

puriform *Medicine.* resembling pus.

purify *Chemistry.* to remove extraneous materials from a substance, as by fractionation of a solution. *Engineering.* to clear an area or object of all undesirable matter. *Computer Technology.* to remove errors from data. Thus, **purified, purification.**

purine *Biochemistry.* an organic nitrogenous base, consisting of a purine ring fused to an imidazole ring, that, when substituted, becomes adenine or guanine.

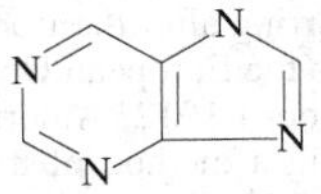

purine

purinemia *Hematology.* the presence of purine bases in the blood. Thus, **purinemic.**

purity *Chemistry.* the quality or degree of being pure. *Optics.* the degree to which a color has been diluted with white light, or has been mixed with other colors.

purity coil *Electronics.* a coil that controls the color purity in a television picture tube by producing a magnetic field that directs an electron beam to its assigned color phosphor.

purity control *Electronics.* a device that varies the amount of current flowing through a purity coil in a television picture tube.

purity magnet *Electronics.* an element that controls the color purity in a television picture tube by producing a magnetic field that directs an electron beam to its assigned color phosphor; sometimes used in place of a purity coil.

purity of state *Physics.* the certainty of a system existing in a particular quantum state.

purity plate *Microbiology.* a petri plate inoculated with a given culture of microorganisms and incubated to determine the presence or absence of contamination.

Purkinje, Johannes [per kin´jē] 1787–1869, Czech naturalist; isolated and named protoplasm; discovered Purkinje's phenomenon.

Purkinje cell *Histology.* a flask-shaped cell forming a single layer in the cerebellar cortex.

Purkinje fiber *Histology.* a specialized cardiac muscle cell that conducts impulses through the walls of the ventricles.

Purkinje's network *Cardiology.* conductive tissue beneath the endocardium. Also, RAMI SUBENDOCARDIALES.

Purkinje's phenomenon *Physiology.* the phenomenon in which the region of maximum visual acuity shifts from red-yellow to blue-green as the intensity of illumination decreases and the eye becomes scotopic. Also, **Purkinje's effect, Purkinje's shift.**

purl *Hydrology.* **1.** a swirling or eddying stream or rivulet that rapidly moves around obstructions. **2.** a popular term for any stream that makes a soft, murmuring sound. *Textiles.* **1.** a knitting stitch that creates horizontal ridges across the fabric. **2.** a looped edge on a border.

purlin *Building Engineering.* a horizontal timber in a roof frame that supports the roofing material or the common rafters. Also, **purline.**

puromycin *Microbiology.* an antibiotic produced by the bacterium *Streptomyces alboniger* that inhibits protein synthesis and that has been used against tumors.

purple *Optics.* any color between red and blue. *Chemistry.* a substance of this color used as a dye or indicator.

purple bacteria *Microbiology.* photosynthetic bacteria that reduce carbon dioxide in the presence of sulfur compounds; the **purple sulfur bacteria** are classified in the family Chromatiaceae, and the **purple nonsulfur bacteria** are classified in the family Rhodospirillaceae.

purple blende see KERMESITE.

purple blotch *Plant Pathology.* a disease of onions, garlic, and shallots caused by the fungus *Alternaria porri* and characterized by small white irregular spots that enlarge into large purple blotches.

purple boundary *Optics.* the line that connects the ends of the spectrum locus on a chromaticity diagram, forming a loop in which coordinates of all real colors fall.

purple grackle *Vertebrate Zoology.* the eastern variety of the North American common grackle, *Quiscalus quiscula,* having an iridescent purple back.

purpleheart *Materials.* a strong, durable wood with a fine grain from the genus *Peltogyene,* used for furniture and decorative veneers.

purple light *Geophysics.* a luminous glow observable in the eastern sky just before sunrise and in the western sky just after sunset; caused by the joint effects of two types of scattered light: red-rich light from a turbid layer of the middle atmosphere and blue-rich light scattered from higher regions of the atmosphere.

purple membrane *Microbiology.* any of several specialized regions of the cytoplasmic membrane of certain strains of *Halobacterium,* composed mainly of the purple pigment bacteriorhodopsin and facilitating the synthesis of ATP energy by illuminated cells.

purple nonsulfur bacteria see PURPLE BACTERIA.

purple of Cassius see GOLD STANNATE.

purple plague *Electronics.* a condition that develops when gold and aluminum come in contact with each other in a semiconductor device, causing it to malfunction and produce a purple color.

purple sage *Botany.* an herb, *Salvia leucophylla,* of the mint family having spiky purple flowers and silvery leaves.

purple salt see POTASSIUM PERMANGANATE.

purple sulfur bacteria see PURPLE BACTERIA.

purple-top *Plant Pathology.* a virus disease of potato plants characterized by a purplish or yellowish discoloration of the shoot tips, abnormal enlargement of axillary branches, and wilting.

purposivism *Behavior.* the theory that behavior results from the interaction of goals, or purposes, with certain stimulus conditions.

purpura *Medicine.* a small hemorrhage in the skin, mucous membrane, or serosal surface, appearing as purple patches on the skin due to spontaneous bleeding in the subcutaneous tissues. Thus, **purpuric.**

purpurin *Organic Chemistry.* $C_{14}H_5O_2(OH)_3$, red to orange crystals that melt (in vacuo) at 256°C; soluble in alcohol, ether, and hot water; used as a stain for microscopy and a dye for cotton.

purpurite *Mineralogy.* $Mn^{+3}PO_4$, a deep-red or purple, orthorhombic mineral, isomorphous with heterosite, occurring in small, irregular masses, having a specific gravity of 3.3 and a hardness of 4 to 4.5 on the Mohs scale; found as an alteration mineral in granite pegmatites.

purpurogallin *Organic Chemistry.* $C_{11}H_8O_5$, red crystals that decompose at 274–275°C; used as an antioxidant.

purpuroidea *Paleontology.* a genus of Mesozoic prosobranchiate gastropods in the order Mesogastropoda; characterized by a thick, robust shell, a wide aperture, and a spiral series of shoulder tubercles; extant in the Jurassic and Cretaceous.

purring tremor see BRUISSEMENT.

purse seine *Naval Architecture.* a type of fishing net used to capture fish that swim well clear of the bottom. The net surrounds a school of fish on all sides, and then the lower edges are pulled together ("pursed") to prevent escape downward. Thus, **purse seiner.**

pursuit *Military Science.* an offensive operation designed to catch or cut off a hostile force that is trying to escape, with the aim of destroying it.

pursuit course *Ordnance.* a course pointing at all times directly at the object being pursued.

pursuit curve see TRACTRIX.

purulent *Medicine.* **1.** containing pus or consisting of pus. **2.** forming pus. Thus, **purulence.**

Purús [pü roos´] *Geography.* a river rising in the Peruvian Andes and flowing about 2000 miles northeast into the Amazon.

pus *Medicine.* a viscous, yellowish fluid product of inflammation, consisting of a liquid (**liquor puris**) containing dead white blood cells (leukocytes).

p. us. ext. for external use. (From Latin *pro usu externo.*)

push button *Mechanical Engineering.* **1.** a small button or knob that completes an electrical circuit and causes the activation of a device or component. **2. push-button.** operated by (or as if by) such a button. Thus, **push-button tuner.**

push-button dialing see TOUCH-TONE DIALING.

push-button switch *Electricity.* **1.** a momentary-contact switch that is actuated only when pressed down, and resets itself when released. **2.** a maintained-contact switch that remains actuated even when released; such a push button must be pressed again to change states.

push development *Graphic Arts.* a technique for increasing the effective speed of a film by deliberately underexposing it and then extending its normal development time.

push-down automaton *Computer Technology.* a finite-state automaton with an auxiliary stack of symbols that are distinct from the symbols in the input string.

push-down stack or **list** *Computer Technology.* a set of memory locations used for temporary storage of a sequence of elements, in which the elements are retrieved in a last-in first-out order. Also, CELLAR.

push drill *Mechanical Devices.* a ratchet-type hand drill that rotates a drill point bit when pushed down against a surface; used for making small holes in soft materials.

pusher kiln *Materials Science.* a kiln used in firing advanced ceramics, in which the objects to be fired are set on ceramic pushing trays and conveyed through the kiln, with each tray pushing the tray ahead of it.

pushing *Electricity.* an abrupt change in the resonant frequency of a circuit due to an applied voltage variation.

push moraine *Geology.* a wide, smooth, arcuate ridge of accumulated glacial drift that has been mechanically pushed along by an advancing glacier or ice sheet. Also, THRUST MORAINE, UPSETTED MORAINE.

push-pull *Electronics.* of an electronic device, having components with balanced signals that are opposite in phase (180° out of phase with each other). Thus, **push-pull amplifier, push-pull oscillator, push-pull transistor.**

push-pull currents see BALANCED CURRENTS.

push-pull electret transducer *Electronics.* a device that converts energy by passing it through a specially treated foil placed between two electrodes, causing the electrodes to pull in opposite directions proportional to the voltage applied.

push-pull sound track *Acoustical Engineering.* a two-channel sound track in which the modulation is 180° out of phase between the two tracks.

push-pull transformer *Electronics.* a transformer with two secondary windings wound in such a way that the voltages on the secondary windings are respectively in phase and 180° out of phase with the voltage on the primary winding.

push-pull voltages see BALANCED VOLTAGES.

push-push amplifier *Electronics.* a circuit in which power is amplified by two devices in which current and voltage have the same magnitude and polarity; commonly used as a frequency multiplier.

push-ridge moraine see PUSH MORAINE.

push rod *Mechanical Engineering.* a rod through which the tappet of an overhead valve engine operates the rocker arm to open and close the valves.

push-to-talk switch see PRESS-TO-TALK SWITCH.

push-up stack or **push-up list** see QUEUE.

puss caterpillar *Invertebrate Zoology.* the larva of various *Megalopyge* species; it has stinging hairs associated with poison-secreting glands and can cause blisters, pain, swelling, and paralysis if handled.

pustular [pus´chə lər] *Medicine.* relating to, of the nature of, or characterized by the presence of pustules. Also, **pustulous.**

pustule [pus´chůl] *Medicine.* a small, circumscribed elevation of the skin containing pus.

pustulina *Paleontology.* a genus of small, planoconvex brachiopods in the extinct superfamily Spiriferacea; extant in the Middle Devonian.

put *Computer Technology.* an instruction that causes a data record to be written to an output file or device.

putrefaction *Biochemistry.* the act or fact of putrefying.

putrefy *Biochemistry.* to decompose, with the production of a fetid substance; a term applied especially to the decomposition of proteins and other organic substances, with the production of foul-smelling compounds such as ammonia, hydrogen sulfide, and mercaptans.

putrescent *Biochemistry.* relating to or being a product of putrefaction.

putty *Materials.* **1.** a doughy adhesive material used for sealing. **2.** a creamy cement of lime and water that is mixed with plaster of Paris and sand for use as a finished plaster coating.

putty knife *Mechanical Devices.* a broad, flexible-bladed knife having a squared-off end; used to apply glazing compound or putty to a surface.

putty powder *Materials.* a soft abrasive composed of tin oxide.

puttyroot *Botany.* an orchid, *Aplectrum hyemale,* characterized by a thin, naked rootstock that produces a leafless stalk with yellow-brown flowers.

puy *Volcanology.* a small remnant volcanic cone formed during a relatively short period of activity.

puzzle box *Behavior.* any chamber or container used for investigation and experimentation in animal learning; e.g., a maze or a Skinner box.

puzzolana or **puzzolan** see POZZOLANA.

PV polyvinyl.

PVA polyvinyl acetate; polyvinyl alcohol.

PVAc polyvinyl acetate.

p-value *Statistics.* in testing, the probability of observing a value of the test as discrepant from the null hypothesis as the one actually observed. Small *p*-values are taken to be evidence that the null hypothesis should be rejected.

PVC polyvinyl chloride; premature ventricular contractions.

PVD *Aviation.* the airport code for Providence, Rhode Island.

PVDC polyvinyl dichloride.

p-V diagram *Thermodynamics.* a plot that serves to illustrate a thermodynamic cycle or process in which pressure is plotted on the vertical and volume is plotted on the horizontal.

PVE polyvinyl ether ether

PVI polyvinyl isobutyl ether.

PVM polyvinyl methyl ether.

PVP polyvinyl pyrrolidone.

PVR *Aviation.* the airport code for Puerto Vallarta, Mexico.

PVT pressure, volume, and temperature.

pW picowatt.

P wave see IRROTATIONAL STRAIN.

P-wave see COMPRESSIONAL WAVE.

PWM *Aviation.* the airport code for Portland, Maine.

pwr or **pwr.** power.

PX physical exam.

Px patient; patients.

px pneumothorax.

py- a combining form meaning "pus," as in *pyarthrosis.*

pyarthrosis *Medicine.* the formation of pus in a joint cavity.

pycnidiospore *Mycology.* in certain fungi, a kind of spore produced by a fruiting body called a pycnidium.

pycnidium *Mycology.* in certain fungi, a hollow fruiting body that produces pycnidiospores.

pycniospore *Mycology.* in certain fungi, a spore produced by a fruiting body called a pycnium.

pycnite *Mineralogy.* a form of topaz that occurs in massive columnar aggregates.

pycnium *Mycology.* the flask-shaped fruiting body of rust fungi.

pycno- or **pycn-** a combining form meaning "thick" or "dense."

pycnocline *Geophysics.* **1.** a gradient marking vertical changes in density. **2.** a water layer in lakes, seas, or oceans that shows an abrupt change in density with increasing depth. Thus, **pycnoclinal.**

Pycnodontiformes *Paleontology.* an extinct order of laterally compressed neopterygian fishes, the **pycnodonts,** which flourished in the Jurassic and Cretaceous; characterized by their disk shape and knobby, flattened teeth on the palate and lower jaw.

Pycnogonida *Invertebrate Zoology.* the sea spiders, a class of marine arthropods with small bodies and very long legs, usually four pairs. Also, **Pycnogonum.**

Pycnogonidae *Invertebrate Zoology.* a family of cold water sea spiders, arthropods in the class Pycnogonida, with no chelifores or palpi; the males have legs modified for carrying eggs.

pycnometer *Engineering.* a device for determining the specific gravity of small quantities of oil or other liquids. Also, PYKNOMETER.

pycnometry *Physics.* the liquid density measurements made by weighing the liquid in a container of known volume.

Pycnonotidae *Vertebrate Zoology.* the bulbuls, a family of fruit-eating birds of the passerine suborder Oscines, characterized by bristles on the nape, thick plumage on the rump, and sometimes boldly patterned body plumage; found in Africa, southern Asia, Japan, and the Phillippines.

pycnosis see PYKNOSIS.

pyelolithotomy *Surgery.* the surgical removal of a calculus from the pelvis of the kidney. Also, PELVILITHOTOMY.

pyelonephritis *Medicine.* an inflammation of the kidney, especially the renal pelvis, due to bacterial infection. Also, **pyelitis.**

pyeloplasty *Surgery.* a plastic operation on the kidney pelvis.

pyelostomy *Surgery.* the formation of an opening into the kidney pelvis for diversion of urine from the ureter.

pyemia *Medicine.* a disease due to the presence of pus-producing microorganisms in the blood, marked by the formation of multiple abscesses.

Pygaster *Paleontology.* a genus of Mesozoic sea urchins in the extinct family Pygasteridae; extant in the Jurassic and Cretaceous.

Pygasteridae *Paleontology.* a family of irregular euechinoids in the extinct order Pygasteroida; extant in the Jurassic and Cretaceous.

Pygasteroida *Paleontology.* an extinct order of asymmetrical irregular euechinoids in the superorder Eognathostomata (formerly placed in the Diadematacea); extant in the Jurassic and Cretaceous.

pygidium *Invertebrate Zoology.* the rear part of the body, specifically: in annelids, the terminal part of the body bearing the anus; in some scale insects, an abdominal section formed by fused segments; in some crustaceans, the caudal portion; in some insects, the last body segment.

Pygmalion effect [pig māl´yən] *Psychology.* the tendency to influence others to behave in a way that will confirm one's prior evaluation of them. (From the Greek myth of *Pygmalion,* a sculptor who fell in love with a statue he had created, which then came to life.)

PYG medium *Microbiology.* a bacteriological medium containing peptone, yeast extract, and glucose, used to culture anaerobic microorganisms.

pygmy rattlesnake *Vertebrate Zoology.* a North American snake belonging to the genus *Sistrurus* of the family Crotalidae, distinguished from other rattlers by its plated head shield, small rattles, and weak venom; the best known in the U.S. is the **Massasauga,** *S. catenatus.*

Pygopodidae *Vertebrate Zoology.* the flap-footed lizards or snake lizards, a family in the suborder Sauria, having a snakelike appearance except for small flaplike appendages located about one-third of their length from the head; found in Australia and New Guinea.

pykno- a combining form meaning "thick" or "dense."

pyknolepsy *Neurology.* a form of epilepsy characterized by extremely frequent episodes of petit mal.

pyknometer see PYCNOMETER.

pyknosis *Pathology.* a thickening or condensing, often seen in the dissolution or breakdown of a cell when the nucleus diminishes in mass and the chromatin reforms to a thick, amorphous composition.

pylephlebitis *Medicine.* an inflammation of the portal vein orbit branches.

pylon *Aviation.* **1.** a streamlined, fitted strut or other device used to attach integral systems or accessories such as external tanks, bombs, and surveillance pods to an aircraft. **2.** a surface reference object used by pilots when performing aerial maneuvers or during aerial races.

pyloric [pī lôr´ik] *Anatomy.* relating to the pylorus or the surrounding area of the stomach.

pyloric cecum *Vertebrate Zoology.* in certain fishes (particularly ray-finned fishes), a tube-shaped pouch that is located externally but penetrates the intestinal wall; it functions in absorbing food. *Invertebrate Zoology.* **1.** in insects, any of a series of long diverticula that increase the absorptive area of the midgut epithelium. **2.** in starfishes, either of two digestive pouches in each arm.

pyloric sphincter *Anatomy.* a circular muscle surrounding the base of the stomach that controls the flow of chyme into the small intestine.

pyloric stenosis *Medicine.* a narrowing of the pyloric orifice of the stomach, restricting the flow of food into the small intestine; it may occur as a congenital defect in newborns or, in adults, secondary to an ulcer or fibrosis. Also, **pyloristenosis.**

pyloromyotomy *Surgery.* an incision of the muscles of the pylorus, as in the operation for congenital pyloric stenosis.

pyloroplasty *Surgery.* plastic surgery to relieve pyloric obstruction or to accelerate gastric emptying, expecially after truncal vagotomy in the treatment of a duodenal ulcer.

pylorospasm *Medicine.* a spasm of the pylorus; a condition in adults generally related to severe gastritis or a gastric ulcer.

pylorotomy *Surgery.* the surgical incision of the pylorus.

pylorus [pī lôr´əs] *Anatomy.* the distal aperture of the stomach surrounded by a strong band of circular muscle, through which materials pass into the duodenum.

pyo- a combining form meaning "pus," as in *pyorrhea.*

pyocephalus *Neurology.* a brain abscess characterized by the presence of purulent fluid in the cerebral ventricles.

pyocin *Microbiology.* a bacteriocin that is produced by certain species of *Pseudomonas* and that acts as an antimicrobial agent against closely related organisms.

pyocyanin *Microbiology.* a blue-green crystalline antibiotic produced by the bacterium *Pseudomonas aeruginosa*; it is effective against numerous bacteria and fungi.

pyoderma *Medicine.* a skin disease marked by lesions producing pus.

pyogenesis *Medicine.* the formation of pus.

pyomyoma *Oncology.* a leiomyoma that excretes pus.

pyonephritis *Medicine.* any inflammation of a kidney accompanied by the formation of pus.

pyonephrosis *Pathology.* the suppurative destruction of the kidney, with total or almost complete loss of renal function.

pyophylactic *Medicine.* serving to defend against purulent infection.

pyorrhea [pī´ə rē´ə] *Medicine.* **1.** a pus-containing discharge. **2.** an inflammatory reaction of the tissues surrounding a tooth.

pyr- *Science.* a combining form meaning "fire" or "heat." *Chemistry.* a combining form indicating formation by heat, particularly an acid derived by the loss of one molecule of water following a heating process.

PYR *Microbiology.* a substrate, *N,N*-dimethylaminocinnamaldehyde, used to identify certain bacteria based upon their characteristic hydrolysis of this compound.

pyracetic acid see PYROLIGNEOUS ACID.

Pyralididae *Invertebrate Zoology.* a large family of snout moths, lepidopteran insects in the superfamily Pyralidoidea, with a slender body and long legs. Also, **Pyralidae**.

Pyralidoidea *Invertebrate Zoology.* the snout moths, a superfamily of lepidopteran insects having a slender body and long legs; the larvae are serious plant pests.

pyramid *Architecture.* an ancient Egyptian tomb having a square base and four sloping triangular sides that meet at the apex. *Science.* **1.** anything having this shape. **2.** anything having sides that taper to meet at the apex. *Crystallography.* a crystal form having three, four, six, eight, or twelve faces that intersect at a common point on the vertical axis. *Mathematics.* a prismatoid with a single point substituted for one of the bases; the other base is a polygon and the lateral surfaces are triangles with a common vertex. A pyramid with an *n*-gon (polygon with *n* sides) as base is called an ***n*-sided pyramid.** A tetrahedron is a pyramid with a triangle as base. A **truncated pyramid** is a pyramid cut by a plane parallel to the base, so that the resulting object has two similar bases. Thus, **pyramidal, pyramidical, pyramidic.**

pyramidal cleavage *Crystallography.* the cleavage of a pyramidal crystal, occurring parallel to one of the natural faces of the crystal.

pyramidal crystal *Crystallography.* any crystal having the form of a pyramid.

pyramidal horn *Engineering.* a horn whose sides flare out to form a pyramid shape.

pyramidal iceberg see PINNACLED ICEBERG.

pyramidal peak *Geology.* see HORN, def. 1.

pyramidal surface *Mathematics.* given a fixed point and a broken line lying in a plane not containing the point, the surface formed by the union of all rays from the point which intersect the broken line.

pyramidal tracts *Anatomy.* the system of fibers arising from pyramidal cells in the cerebral cortex and descending into the spinal cord to the motor neurons. Also, **pyramidal system.**

Pyramidellidae *Invertebrate Zoology.* a family of gastropod mollusks in the order Tectibranchia, having a small shell and chitinous jaws and stylets; they are blood-sucking ectoparasites of marine invertebrates.

pyramid of biomass *Ecology.* a representation of the relationships of all organisms living within a community, in which each species is placed at a level equivalent to its nutritional output.

pyramid of energy *Ecology.* a representation of energy relationships within a community, in which each species is placed at a level equivalent to its energy output.

pyramid of numbers *Ecology.* a numeric representation of the individual organisms nourished at each feeding level of an ecosystem's food chain.

pyramid pebble see DREIKANTER.

pyramid sight *Ordnance.* a triangular or pyramid-shaped front sight that projects above the barrel of a firearm.

Pyramimonadaceae *Botany.* a family of green algae of the order Pyramimonadales, characterized by two or four flagella of equal length usually attached at the anterior end of the cell, and having at least three layers of organic scales on the cell body and two layers of differently structured scales on the flagella.

Pyramimonadales *Botany.* an order of flagellates of the class Prasinophyceae, having two to four or more flagella and one to several layers and types of organic scales on both the cell and the flagella.

pyran *Organic Chemistry.* C_5H_6O, either of two cyclic compounds in which the ring consists of five carbon atoms and one oxygen atom.

pyranometer *Engineering.* a device that measures solar radiation by comparing its heating effect on two metallic strips with the effect produced by an electric current on the same strips. Also, SOLARIMETER.

pyranose *Organic Chemistry.* the six-membered ring form of a sugar; i.e., a hexose in which the oxygen ring bridges carbon atoms 1 and 5 (in the aldoses) or 2 and 6 (in the ketoses).

pyrargyrite *Mineralogy.* Ag_3SbS_3, a deep-red trigonal mineral occurring in massive form or as prismatic crystals, having a specific gravity of 5.82 to 5.85 and a hardness of 2.5 on the Mohs scale; found in low-temperature silver veins and an important ore of silver. Also, DARK-RED SILVER ORE, DARK RUBY SILVER, RUBY SILVER.

Pyraustinae *Invertebrate Zoology.* the web worms and cornborers, a subfamily of lepidopteran insects in the family Pyralididae whose large larvae are serious plant pests.

pyrazinamide *Pharmacology.* $C_5H_5N_3O$, an antibacterial drug that is derived from nicotinic acid, used in the treatment of tuberculosis to inhibit the growth of the tubercular bacteria.

pyrazofurin *Oncology.* $C_9H_{13}N_3O_6$, an antineoplastic agent.

pyrenoid *Botany.* a small, colorless protein structure found within the chloroplast of certain algae, associated with the formation and storage of starch.

pyrenolichen *Botany.* any lichen of the order Pyrenulales.

Pyrenomycetes *Botany.* in some systems of classification, a class of one-celled or two-celled fungi having unitunicate asci.

Pyrenomycetes *Mycology.* a class of fungi of the subdivision Ascomycotina, characterized by a single-walled ascus and by an ascocarp that is initiated by a coiled hypha which instigates its ascus-producing development system.

Pyrenophoraceae *Mycology.* a family of fungi of the order Pleosporales that are parasitic to plant leaves and stems or crop grasses.

Pyrenulaceae *Botany.* a family of crustose lichens in the order Pyrenulales, most often found on tree bark in the tropics.

Pyrenulales *Botany.* an order of lichens in the class Ascolichenes (or Pyrenomycetes), containing over 1500 species distinguished by perithecia having true paraphyses and unitunicate asci.

pyrethrum *Materials Science.* a toxic, nonvolatile hydrocarbon of the kerosene type that is derived by extraction of chrysanthemum flowers and used as a natural insecticide.

Pyrex *Materials.* a trade name for a borosilicate glass with a low coefficient of thermal expansion, and high chemical, heat shock, and thermal resistance; widely used for laboratory and pharmaceutical apparatus, and also for electrochemical equipment, domestic cooking ware, and fiber manufacture.

pyrexia *Medicine.* a fever or febrile condition; an abnormal elevation of body temperature.

Pyrgotidae *Invertebrate Zoology.* a family of flies, dipteran insects in the subsection Acalyptratae, medium or large with long wings and a large rounded head; the larvae are endoparasites in beetles.

pyrheliometer *Engineering.* an instrument used to measure the total solar radiation striking a surface.

Pyricularia *Mycology.* a genus of fungi of the class Hyphomycetes that cause a rice disease called rice blast.

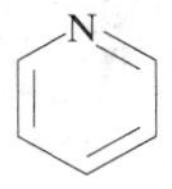

pyridine

pyridine *Organic Chemistry.* $N(CH)_5$, a colorless to pale yellow liquid; boils at 116°C and freezes at –42°C; soluble in water or ether; used as a denaturant and solvent.

2-pyridinecarboxylic acid see PICOLINIC ACID.

pyridine-linked dehydrogenase *Enzymology.* a dehydrogenase that requires a pyridine nucleotide (NAD^+ or $NADP^+$) as a coenzyme.

pyridostigmine bromide *Pharmacology.* $C_9H_{13}BrN_2O_2$, a drug used orally or injected to treat myasthenia gravis and as an antidote for certain muscle-relaxing drugs.

pyridoxal *Biochemistry.* one of the forms of vitamin B_6; chemical name 2-methyl-3-hydroxy-4-formyl-5-hydroxymethylpyridine.

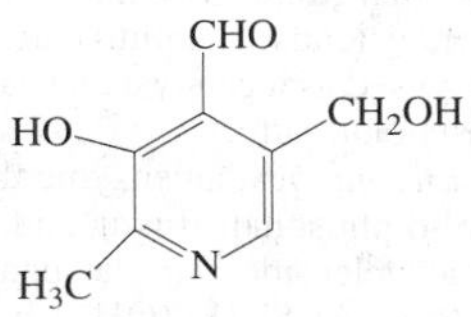

pyridoxal

pyridoxal phosphate *Enzymology.* a coenzyme that plays a major role in amino acid metabolism.

pyridoxine *Biochemistry.* one of the forms of vitamin B_6; chemical name 5-hydroxy-6-methyl-3,4-pyridinedimethanol.

pyridoxine hydrochloride *Pharmacology.* the hydrochloric salt of pyridoxine; used in the prevention and treatment of vitamin B_6 deficiency.

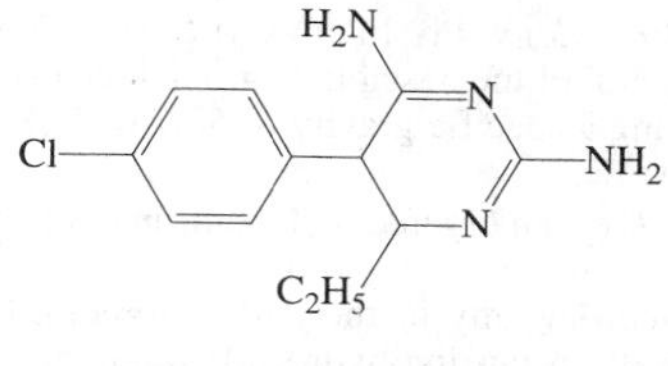

pyrimethamine

pyrimethamine *Pharmacology.* $C_{12}H_{13}ClN_4$, a folic acid antagonist occurring as a white crystalline powder; administered orally as an antimalarial drug.

pyrimidine *Biochemistry.* $C_4H_4N_2$, an organic nitrogenous base consisting of a heterocyclic compound that, when substituted, becomes uracil, cytosine, or thymine; it is also the parent compound of the barbituates.

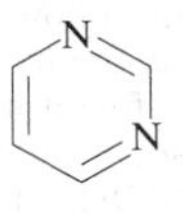

pyrimidine

pyrimidinetrione see BARBITURIC ACID.

pyrite *Mineralogy.* FeS_2, a brass-yellow, opaque, metallic, cubic mineral of the pyrite group, dimorphous with marcasite, having a specific gravity of 5.00 to 5.02 and a hardness of 6 to 6.5 on the Mohs scale; occurs as cubes, octahedrons, and pyritohedrons, and in granular form; found in rocks of all types, the most abundant and widespread sulfide mineral. It is popularly known as "fool's gold" because of its superficial resemblance to the precious metal. Also, IRON PYRITE, MUNDIC.

pyrite

pyrite roasting *Mining Engineering.* a process by which iron pyrite is heated in the presence of air to produce iron oxide sinter, which is used in steel mills, and elemental sulfur.

pyritization *Geology.* the process in which pyrite is introduced into any variety of rock, or replaces another mineral in a rock or other material, such as the hard parts of a plant or animal.

pyritohedron *Crystallography.* a hemihedral isometric form bounded by twelve equal but not equilateral pentagons. Also, **pyritoid.**

pyro- *Science.* a combining form meaning "fire" or "heat." *Chemistry.* a combining form indicating formation by heat, particularly an acid derived by the loss of one molecule of water following a heating process.

pyroaurite *Mineralogy.* $Mg_6Fe_2^{+3}(CO_3)(OH)_{16}\cdot 4H_2O$, a colorless, yellow-white, or green trigonal mineral of the hydrotalcite group, dimorphous with sjögrenite, occurring as tabular crystals, having a specific gravity of 2.12 and a hardness of 2.5 on the Mohs scale; found in serpentine rock.

pyrobelonite *Mineralogy.* $PbMn^{+2}(VO_4)(OH)$, a fire-red, transparent, orthorhombic mineral of the descloizite group, occurring as minute acicular crystals, having a specific gravity of 5.37 to 5.58 and a hardness of 3.5 on the Mohs scale.

pyrobiolite *Petrology.* an organic rock featuring volcanically altered organic remains.

pyrobitumen *Geology.* any of the dark-colored, comparatively hard, nonvolatile, naturally occurring hydrocarbon complexes that are generally associated with mineral matter, and which yield bitumens upon heating. Also, **pyritobitumen.**

pyroborate see BORAX.

pyrocatechuic acid see CATECHOL.

pyrocellulose see GUNCOTTON.

pyroceram *Materials Science.* a white, opaque ceramic that is nonporous and shock-resistant; it is used for coatings and for mechanical and electrical parts.

pyrochlore *Mineralogy.* $(Ca,Na)_2Nb_2O_6(OH,F)$, a cubic mineral of the pyrochlore group, ranging from yellowish-brown or reddish-brown to black in color, having a specific gravity of 4.45 to 4.48 and a hardness of 5 to 5.5 on the Mohs scale; found in pegmatites, carbonatites, and some alkalic igneous rocks. Also, PYRRHITE.

pyrochroite *Mineralogy.* $Mn^{+2}(OH)_2$, a trigonal mineral of the brucite group, occurring in foliated form or as tabular crystals, having a specific gravity of 3.25 and a hardness of 2.5 on the Mohs scale; characterized by a white to pale blue or green color and pearly luster when fresh, becoming dark brown to black and dull upon exposure.

pyroclast *Volcanology.* an individual particle or fragment of clastic rock material of any size that is formed by volcanic explosion or ejected from a volcanic vent.

pyroclastic *Volcanology.* of or relating to pyroclasts.

pyroclastic flow *Volcanology.* an ash flow that takes place at any temperature.

pyroclastic ground surge *Volcanology.* the comparatively thin layer of rock of varying thickness found around a volcanic vent.

pyroclastic rock *Petrology.* a rock containing fragments of volcanic products ejected explosively or aerially from volcanoes.

pyroconductivity *Solid-State Physics.* electrical conductivity in a solid that is normally nonconductive, achieved through the application of high temperatures up to the fusion point.

Pyrocystaceae *Botany.* a monogeneric family of unicellular marine flagellates that composes the order Pyrocystales.

Pyrocystales *Botany.* a monofamilial order of marine flagellates of the class Dinophyceae, containing pelagic, free-living, unicellular, and photosynthetic organisms, characterized by large, highly vacuolate, nonmotile vegetative cells.

Pyrodictium *Bacteriology.* a genus of filamentous, thermophilic bacteria of the order Thermoproteales that is halotolerant; originally isolated from an undersea volcanic area.

pyroelectric effect *Crystallography.* the development of a small potential difference across certain crystals as the result of a temperature change; observed in crystals such as ferroelectric barium titanate, cane sugar, or tourmaline. Also, **pyroelectricity.**

pyrogallate *Organic Chemistry.* any salt or ester of pyrogallol.

pyrogallol *Organic Chemistry.* $C_6H_3(OH)_3$, lustrous white crystals that melt at 133°C and boil at 309°C, turning gray on exposure to light; derived by autoclaving gallic acid with water; used in photography, engraving, dyes, and synthetic drugs. Also, **pyrogallic acid.**

pyrogen *Biochemistry.* any substance that induces fever.

pyrogenesis *Geology.* **1.** the intrusion and extrusion of a magma. **2.** the products derived from a magma.

pyrogenetic *Chemistry.* formed by heat. *Geology.* of or relating to pyrogenesis. *Biochemistry.* of or relating to a pyrogen.

pyrogenic distillation *Chemical Engineering.* the thermal decomposition of hydrocarbons in a petroleum refinery at high temperatures, high pressures, or both.

Pyrolaceae *Botany.* a family of mycotrophic perennial herbs or half-shrubs of the order Ericales, characterized by creeping rhizomes and poorly developed secondary tissues, often evergreen, and usually growing in acid soils in cool regions of the Northern Hemisphere

pyroligenous *Chemical Engineering.* relating to or produced by the destructive distillation of wood. Also, **pyrolignic.**

pyroligneous acid *Materials.* a crude acetic acid that is made by destructively distilling wood or pine tar; used in smoking meats. Also, PYRACETIC ACID, WOOD VINEGAR.

pyrolithic acid see CYANURIC ACID.

pyrolusite *Mineralogy.* $Mn^{+4}O_2$, an iron-black or dark-gray, tetragonal mineral of the rutile group, occurring as prismatic crystals and in massive, fibrous, and granular form, having a specific gravity of 5.06 and a hardness of 2 to 6.5 on the Mohs scale; widely found as an alteration product of other manganese minerals, and the chief ore of manganese.

pyrolysis *Chemistry.* a process of decomposition by heat. Thus, **pyrolytic.**

pyromagma *Geology.* a highly mobile magma that is oversaturated with gases and that lies at a shallower depth than a hypomagma.

pyromania [pī′rō mā′nē ə] *Psychology.* an irrational urge to set destructive fires and watch them burn.

pyromaniac [pī′rō mā′nē ak] *Psychology.* a person who is affected by pyromania.

pyromellitic acid *Organic Chemistry.* $C_6H_2(COOH)_4$, colorless prisms that are slightly soluble in water; melts at 275–276°C; used in the production of waxes, lubricants, and polyesters.

pyromellitic dianhydride *Organic Chemistry.* $C_6H_2(C_2O_3)_2$, a white powder that melts at 286°C; used for curing epoxy resins.

pyrometamorphism *Petrology.* contact metamorphism occurring at temperatures near the melting points of the constituent minerals in contact with magma.

pyrometric cone *Materials Science.* a triangular piece of ceramic that bends or melts to indicate that a certain temperature has been reached.

pyrometry *Engineering.* a field of technology concerned with the measurement of high temperatures. Also, **pyrometrics.**

pyromorphite *Mineralogy.* $Pb_5(PO_4)_3Cl$, a green, yellow, brown, or gray hexagonal mineral of the apatite group, having a specific gravity of 7.04 and a hardness of 3.5 to 4 on the Mohs scale; occurring as prismatic crystals and in botryoidal form in the oxidized zones of lead-bearing deposits. Also, GREEN LEAD ORE.

pyromucic acid see FUROIC ACID.

pyron *Physics.* a unit of power flux, equal to 1 international table calorie uniformly traversing across an area of 1 square centimeter in 1 minute; equivalent to 697.8 watts per square meter.

pyron detector *Electronics.* a device that converts alternating current to pulsating current by passing it through metallic points; generally made of iron pyrites and copper.

Pyronemataceae *Mycology.* a family of fungi of the order Pezizales occurring on burned or sterilized surfaces or on dung.

pyrope *Mineralogy.* $Mg_3Al_2(SiO_4)_3$, a fire-red cubic mineral of the garnet group, having a specific gravity of 3.5 to 3.8 and a hardness of 6.5 to 7.5 on the Mohs scale; found as rounded grains in peridotites, serpentinites, and their detrital deposits, and used as a gemstone when transparent.

pyrophanite *Mineralogy.* $Mn^{+2}TiO_3$, a blood-red, finely crystalline to scaly, trigonal mineral, having a specific gravity of 4.54 and a hardness of 5 to 6 on the Mohs scale; found in manganese ores.

pyrophobia *Psychology.* an irrational fear of fire. Thus, **pyrophobic.**

pyrophoric *Chemistry.* capable of igniting spontaneously in air. Thus, **pyrophoric propellant, pyrophoric alloy.**

pyrophosphatase *Enzymology.* any enzyme that catalyzes the hydrolysis of pyrophosphate.

pyrophosphate *Biochemistry.* either of two phosphate groups joined by esterification.

pyrophosphate cleavage *Biochemistry.* a cleavage of the linkage between the two phosphate groups of pyrophosphate-containing compounds; it ensures irreversibility of many synthetic steps involving nucleotide triphosphates.

pyrophosphoric acid *Inorganic Chemistry.* $H_4P_2O_7$, a viscous liquid that is soluble in water; it tends to solidify at room temperature to a solid that melts at 61°C; used as a catalyst and as a stabilizer for organic peroxides. Also, DIPHOSPHORIC ACID.

pyrophosphorylase *Enzymology.* an enzyme that catalyzes the formation of a nucleoside 5′-diphosphate sugar and pyrophosphate from a sugar 1-phosphate and a nucleoside 5′-triphosphate.

pyrophyllite *Mineralogy.* $Al_2Si_4O_{10}(OH)_2$, a white, green, gray, or brown, monoclinic and triclinic mineral, occurring in foliated to fibrous masses resembling talc, characterized by a pearly to greasy luster, and having a specific gravity of 2.65 to 2.90 and a hardness of 1 to 2 on the Mohs scale; found in metamorphic rocks and also in quartz veins and granites.

pyroschist *Petrology.* a schist or shale with sufficient carbon to burn with a bright flame or to yield volatile hydrocarbons upon heating.

pyrosilicate *Materials Science.* any of a group of silicate structures based on a pair of silicate tetrahedral units sharing one corner only.

pyrosmalite *Mineralogy.* any member of the series ferropyrosmalite-manganpyrosmalite, having the general formula $(Fe^{+2},Mn^{+2})_8Si_6O_{15}(OH,Cl)_{10}$; brown, gray, or green trigonal minerals occurring as thick hexagonal prisms or in massive form, having a specific gravity of 3.06 to 3.19 and a hardness of 4 to 4.5 on the Mohs scale; found in iron mines.

Pyrosomida *Invertebrate Zoology.* an order of tunicates belonging to the class Thaliacea, with tubelike swimming colonies, up to 10 meters in length, of zooids surrounded by a common tunic; many are bioluminescent. Also, PYROSOMATIDA.

pyrosphere *Geology.* the zone of layer of the earth below the lithosphere, consisting partly of magma. Also, BARYSPHERE, MAGMOSPHERE.

pyrostibite see KERMESITE.

pyrostilpnite *Mineralogy.* Ag_3SbS_3, a hyacinth-red, monoclinic mineral, dimorphous with pyrargyrite, having a specific gravity of 5.94 and a hardness of 2 on the Mohs scale; found as aggregates of tabular crystals in low-temperature hydrothermal vein deposits.

pyrotechnic *Materials.* relating to or resembling pyrotechnics.

pyrotechnics *Materials.* **1.** a collective term for various chemicals for producing smoke or light; military uses include signaling, illuminating, and screening; nonmilitary uses include flares and fireworks. **2.** the art of making or using such materials. Also, **pyrotechny.**

Pyrotheria *Paleontology.* an extinct order of large South American ungulates; includes two small families, the Protheriidae and the Colombitheriidae; extant from the Eocene or Paleocene to Oligocene.

Pyrotheriidae *Paleontology.* a family of elephantlike ungulates in the extinct order Pyrotheria; characterized by long, columnar limbs; extant in the Eocene and Oligocene.

pyroxene *Mineralogy.* a group name for orthorhombic or monoclinic rock-forming silicate minerals characterized by the general formula ABZ_2O_6; where A = Ca, Fe^{+2}, Li, Mg, Mn^{+2}, Na, Zn; B = Al, Cr^{+3}, Fe^{+2}, Fe^{+3}, Mg, Mn^{+2}, Sc, Ti, V^{+3}; and Z=Al, Si; colors include white, yellow, green, brown, and greenish black; specific gravity ranges from 3.2 to 4.0 and hardness ranges from 5 to 7 on the Mohs scale; found as stout prismatic crystals and in massive form in igneous and high-temperature metamorphic rocks rich in magnesium and iron.

pyroxenite *Petrology.* a dark-colored, coarse-grained, igneous rock formed by crystallization of magma and consisting primarily of pyroxene, with traces of hornblende, biotite, or olivine.

pyroxenoids *Mineralogy.* any of a number of triclinic silicate minerals having a Si:O ratio of 1:3 as do the pyroxenes, but whose structures consist of single twisted SiO_3 chains of variable repeat distance, instead of the pyroxene structure of single straight SiO_3 chains of regular repeat distance; for example, rhodonite, wollastonite-1T, and pectolite.

pyroxylin cement *Materials.* a nitrocellulose compound used in manufacturing artificial silk, leather, and oilcloth. Also, **pyroxyline.**

pyrrhite see PYROCHLORE.

Pyrrhocoridae *Invertebrate Zoology.* a family of brightly colored hemipteran insects, the red bugs, fire bugs, or cotton stainers, in the superfamily Pyrrhocoroidea, that suck plant juices; they are plant pests, especially on cotton.

Pyrrhocoroidea *Invertebrate Zoology.* a superfamily of true bugs, hemipteran insects in the group Pentatomorpha.

Pyrrhophyta *Botany.* in some systems of classification, a small division of unicellular flagellate algae characterized by an absence of cell walls and by golden-brown or yellowish green plastids.

pyrrhotite *Mineralogy.* $Fe_{1-x}S$, where x = 0–0.17, a metallic bronze-yellow to brown, monoclinic and hexagonal magnetic mineral occurring as granular aggregates and tabular crystals, having a specific gravity of 4.53 to 4.77 and a hardness of 3.5 to 4.5 on the Mohs scale; found mainly in basic igneous rocks.

pyrrobutamine phosphate *Pharmacology.* $C_{20}H_{22}ClN{\cdot}2H_3PO_4$, a long-acting antihistaminic drug, occurring as a white crystalline powder and administered orally.

pyrrole *Organic Chemistry.* $(CH)_4NH$, a colorless oil (when pure) that boils at 130°C, freezes at –24°C, and polymerizes in light; a very weakly basic, cyclic substance that is obtained in the destructive distillation of bone oil and used in the manufacture of pharmaceuticals.

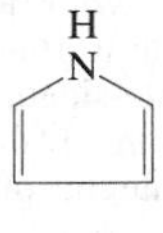

pyrrole

pyrrolidine *Organic Chemistry.* C_4H_9N, a yellowish liquid that boils at 89°C; used in pharmaceuticals, insecticides, and fungicides.

2-pyrrolidone *Organic Chemistry.* C_4H_7ON, a combustible, pale-yellow liquid that boils at 245°C and is soluble in water, ethanol, ethyl ether, chloroform, and benzene; used as a plasticizer and a polymer solvent, and in insecticides and inks.

pyrrone *Organic Chemistry.* an aromatic polyimidazopyrrolone that is synthesized from dianhydrides and tetramines; stable to 482°C and temperature-resistant to 600°C; used in film, adhesives, and laminates.

Pyrrophyta *Botany.* a division of lower unicellular and biflagellate yellowish-green to golden brown algae that form starch, starchy compounds, or oil as food reserves; it includes the dinoflagellates and the cryptomonads.

pyruvate *Biochemistry.* a salt, ester, or anionic form of pyruvic acid; the end product of glycolysis.

pyruvate carboxylase *Enzymology.* an enzyme of the ligase class that catalyzes the irreversible reaction ATP + pyruvate + HCO_3 = ADP + orthophosphate + oxaloacetate, a reaction necessary for gluconeogenesis from lactate or amino acids forming pyruvate; its deficiency causes severe psychomotor retardation and lactic acidosis in infants.

pyruvate decarboxylase *Enzymology.* an enzyme involved in the production of ethanol and CO_2 from flucose; occurs in yeast.

pyruvate dehydrogenase *Enzymology.* a multienzyme complex that removes carbon dioxide from a pyruvate molecule forming acetyl CoA.

pyruvic acid *Biochemistry.* $C_3H_4O_3$, an intermediate in metabolism, 2-oxopropanoic acid, that is produced during glycolysis and converted to coenzyme A, which is required for the Krebs cycle.

Pythagoras [pə thag´ə rəs] c. 582–507 BC, Greek philosopher and mathematician; formulated the Pythagorean theorem.

Pythagorean [pə thag´ə rē´ən] of or relating to Pythagoras or his philosophical or mathematical thought.

Pythagorean scale *Acoustics.* a musical scale based upon a sequence of perfect fifths, with intervals having a 3/2 frequency ratio. Also, TEMPERED SCALE.

Pythagorean theorem *Mathematics.* a right triangle with hypotenuse of length c and sides of length a and b satisfies the relation $a^2 + b^2 = c^2$.

Pythagorean triple *Mathematics.* three positive integers a, b, and c satisfying $a^2 + b^2 = c^2$; also called **Pythagorean numbers.** A triangle with the sides forming a Pythagorean triple must be a right triangle.

Pythiaceae *Mycology.* a family of fungi of the order Peronsporales, living in marine environments as parasites of algae and mollusks.

Pythidae *Invertebrate Zoology.* a family of beetles with long, narrow bodies that are parallel-sided and ranging from 3 to 20 mm in length, with a head that tapers slightly in back; some species are covered with hairs; they are found in rotten trees (under the bark) on which they feed, mainly in the Arctic, but also in Australia, New Zealand, and Chile. There are 16 genera and about 50 species.

Pythium *Mycology.* a genus of fungi belonging to the order Peronosporale, including over 100 species, some of which are parasites of plants or animals, while others obtain nutrients from dead organic matter.

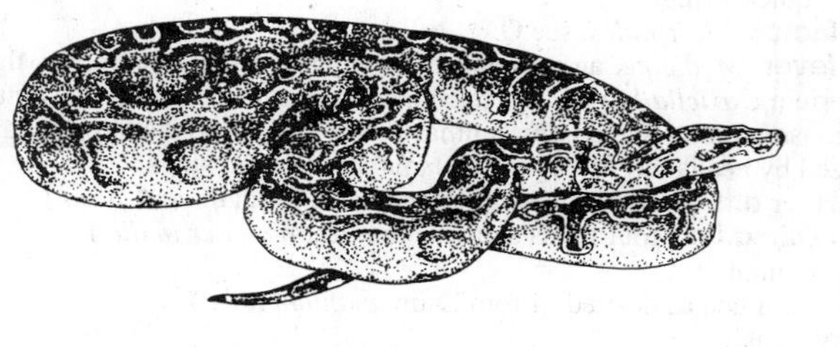

python

python *Vertebrate Zoology.* any member of the subfamily Pythoninae, nonvenomous, predatory, constricting, egg-laying snakes with a colorful patterned layer of scales.

Pythoninae *Vertebrate Zoology.* the pythons, a subfamily of egg-laying snakes of the family Boidae including 20 species, found mainly in tropical parts of the Old World in or near water and in trees.

pyuria *Medicine.* the presence of pus in the urine.

pyxidium *Botany.* a seed vessel that bursts open into a top and bottom part by means of a transverse circular split, with the top coming off like a lid. Also, **pyxis.**

Pyxis *Astronomy.* the Compass, a faint constellation of the southern celestial hemisphere; it contains no star brighter than 4th magnitude.

Q *Electronics.* the ratio of reactance to resistance; used to figure the merit of a capacitor, inductor, or LC circuit. *Thermodynamics.* a unit of heat energy that is equivalent to 10^{18} British thermal units.

q *Genetics.* the symbol for the long arm of a chromosome.

q quartile; query; quintal.

Q10 *Chemistry.* the ratio of the rate of a reaction to the rate of the reaction at 10°C lower. Also, TEMPERATURE COEFFICIENT.

Q band *Electromagnetism.* a radio-frequency band with a frequency range of 36 to 46 gigahertz.

QBE query by example.

Qβ phage *Virology.* an ssRNA phage used extensively to study RNA phage and bacterial cell function; it is specific for enterobacteria containing the F plasmid.

Q-branch *Spectroscopy.* in molecular spectra, a set of spectral lines associated with a change in the vibrational quantum number without a corresponding change in the rotational quantum number, *J*.

QCD quantum chromodynamics.

q-cycle *Biochemistry.* a theoretical pathway first suggested in the respiratory loop model to explain certain phenomena, such as the extrusion of protons at complex III in the mitochondrial electron transport chain.

Q.D. or **q.d.** daily. (From Latin *quaque die.*)

q.e. which is. (From Latin *quod est.*)

QED quantum electrodynamics.

Q.E.D. which was to be shown or demonstrated. (From Latin *quod erat demonstrandum.*)

Q.E.F. which was to be done. (From Latin *quod erat faciendum.*)

Q.E.I. which was to be found out. (From Latin *quod erat inveniendum.*)

QF *Aviation.* the airline code for Qantas Airways Ltd.

QF quick firing.

Q factor *Electronics.* see Q.

Q fever *Medicine.* an influenzalike bacterial disease caused by the bacterium *Coxiella burneti,* an organism propagated in sheep and cattle and transmitted to humans by animal contact and by inhalation; characterized by headache, fever, and inflammation of the lungs.

Q.H. or **q.h.** each hour; every hour. (From Latin *quaque hora.*)

Q.I.D.or **q.i.d.** four times a day. (From Latin *quater in die.*)

ql. quintal.

q.l. as much as desired. (From Latin *quantum libet.*)

qlty. quality.

QM quartermaster.

Q.M. or **q.m.** every morning. (From Latin *quoque matutino.*)

Q machine *Physics.* a plasma device that generates a highly ionized, confined plasma by means of thermionic emission and surface ionization.

Q-machine plasma *Physics.* a magnetically confined plasma column produced by the thermionic emission of electrons from a heated tungsten plate which is bombarded with a beam of cesium.

Q magnitude *Astronomy.* the magnitude of a celestial object determined at a wavelength of 19.5 micrometers in the thermal infrared.

Q multiplier *Electronics.* a circuit that can increase the range (the *Q*) of an adjustable voltage, because of its high selectivity to a particular frequency.

q.n.s. quantity not sufficient.

Q.O.D. or **q.o.d.** every other day.

Q.P. or **q.p.** as much as you please. (From Latin *quantum placet.*) Also, **Q.P.L., q.pl.**

qq. questions.

Q.Q.H. or **q.q.h.** every four hours. (From Latin *quaque quarta hora.*)

Qq.hor. every hour. (From Latin *quaque hora.*)

qq. v. which (plural) see. (From Latin *quae vide.*)

qr. quarter.

QRS complex *Medicine.* the pattern recorded by an electrocardiogram in the process of registering the ventricular contraction of the heart; the downward deflection represents a **Q wave,** the upward deflection an **R wave,** and the subsequent downward deflection an **S wave.**

Q.S. or **q.s.** sufficient quantity. (From Latin *quantum satis.*)

Q signal *Telecommunications.* in radiotelegraphy, a three-letter code, beginning with Q, that is used to represent a service-oriented sentence; for example, QRN means "I am troubled by static."

QSO quasi-stellar object.

QSS quasi-stellar radio source.

q.suff. as much as suffices. (From Latin *quantum sufficit.*)

Q switch *Optics.* any of various devices that cause a crystal laser to produce a high-energy pulse of extremely short duration.

Q-switched laser *Optics.* a laser that uses a Q-switch to prevent lasing action until all the impurity atoms reach the metastable state; used to generate a short-energy pulse of very high intensity. Also, **Q-spoiled laser.**

qt. quantity; quart.

qtd. quartered.

qty. quantity.

qu. or **ques.** question.

quad *Cartography.* an area on a sphere that is bounded by meridians and parallels having the same number of degrees of arc between them, such as a 10° × 10° quad. *Thermodynamics.* a unit of energy equivalent to 10^{15} British thermal units or approximately 1.055×10^{18} joules. *Electricity.* **1.** a unit consisting of four separately insulated conductors, usually twisted together in pairs. **2.** to arrange in such a configuration. Thus, **quadded, quadding.** *Electronics.* a circuit design in which devices, such as transistors, are paired so that if one fails the other will continue to function in the circuit. *Graphic Arts.* in metal typesetting, a blank used to fill a space, as, for example, between words.

quad. quadrant.

quadded redundancy *Electronics.* a redundancy scheme applied to logic gates in which each gate is quadruplicated and each such gate will have twice as many inputs as the nonredundant gate replaced; the outputs of the stages are connected in such a way that errors made in earlier stages can be corrected in later stages.

quad density *Computer Technology.* a floppy disk formatting scheme that allows for the storage of up to four times the capacity of a single-density disk; for example, double-sided, double-density disks.

quadergy *Electrical Engineering.* **1.** the product of the reactive power and the time interval. **2.** the integral of reactive power with respect to time.

quad in-line *Electronics.* an integrated circuit that is characterized by having room for 48 pins per unit, arranged in four rows.

quadr- or **quadri-** a combining form meaning "four."

quadrangle *Mathematics.* a plane polygon with four vertices; a quadrilateral. A complete quadrangle is the figure formed by four points, called vertices, in the same plane, no three of which lie on the same line, and the six lines, called sides, determined by pairs of these vertices. *Architecture.* a rectangular courtyard or lawn enclosed and defined by surrounding buildings. Also, **quad.** *Cartography.* a rectangular area shown on a map or plat, usually bounded by specific meridians and parallels. Thus, **quadrangle map.**

quadrant a group or arrangement of four parts; specific uses include; *Mathematics.* **1.** an arc of 90°, bounded by a circle, or the region of the circle bounded by two radii and an arc of 90°. **2.** any of four regions into which a plane is divided by rectangular coordinate axes. **3.** more generally, any of four parts into which a geometric configuration is divided by two perpendicular lines. *Physiology.* any one of four corresponding parts or quarters, as of the abdominal surface or of the visual field in perimetry. *Mechanical Engineering.* an instrument that changes horizontal reciprocating motion to vertical up-and-down motion. *Cartography.* a surveying instrument consisting of a graduated arc of 90° and a sighting device or alidade; used to take azimuths and sometimes altitudes. *Navigation.* **1.** a double-reflecting, angle-measuring instrument that has an arc 180° in length. In modern usage, all such instruments are referred to as "sextants," regardless of the length of the arc. **2.** a quarter of a circle, centered on a navigational aid. Quadrants cover the area between cardinal headings and are named for the intercardinal heading they contain. *Naval Architecture.* a fitting on a rudder head that is engaged by the steering gear, serving the same purpose as a tiller. *Electromagnetism.* a unit of electrical induction equivalent to 1.00049 H. Also, INTERNATIONAL HENRY, SECOHM.

quadrantal corrector *Navigation.* either of two soft iron spheres or cylinders placed on each side of a magnetic compass to correct for quadrantal deviation.

quadrantal deviation *Navigation.* the deflection of a magnetic compass that changes sign (easterly or westerly) with each 90° change in vessel heading; this is due to induced magnetism in the horizontal soft iron of the vessel.

quadrantal error *Navigation.* **1.** an error in the reading of a gyro compass that changes sign (easterly or westerly) with each 90° change in vessel heading. **2.** a false indication of the direction of a radio bearing due to large metal structures on the vessel; usually maximum for bearings broad on the bow and broad on the quarter.

quadrantal sphere *Navigation.* a sphere that is used in a quadrantal corrector.

quadrant angle of fall *Mechanics.* the angle at the altitude of firing between a projectile's line of fall and the horizontal.

quadrant elevation *Ordnance.* the angle between the horizontal base relative to the trajectory and the bore axis of a gun immediately before firing; it is the sum of the elevation, the angle of sight, and the complementary angle of sight.

Quadrantids *Astronomy.* a meteor shower occurring annually around January 31, whose origin appears near the joint border of Boötes and Hercules.

quadrant mount *Ordnance.* a device for holding the gunner's quadrant while the gun's elevation is being adjusted.

quadrant sight *Ordnance.* a sight that is used in adjusting the elevation of a gun.

quadraphonic [kwä´drə fän´ik] *Acoustical Engineering.* describing a sound system that employs four separate channels for the processing of electroacoustic signals, thereby theoretically improving the spatial stereophonic qualities of reproduced sound. Also, QUADRASONIC.

quadraphonics [kwä´drə fän´iks] *Acoustical Engineering.* a method of sound reproduction through the use of a quadraphonic sound system. Also, **quadraphony.**

quadrasonic see QUADRAPHONIC.

quadrate bone *Vertebrate Zoology.* in the skulls of tetrapods, birds, and certain fishes, a posterior jaw element that forms a surface for the joint of the lower jaw.

quadratic *Mathematics.* any expression of degree two. A homogeneous quadratic is a quadric.

quadratic equation *Mathematics.* **1.** a polynomial equation of the form $ax^2 + bx + c = 0$, where a is nonzero; the coefficients a, b, and c are generally taken to be integers or real numbers. **2.** in general, any second-degree polynomial equation.

quadratic form *Mathematics.* a bilinear symmetric mapping of a vector space V into its field of scalars. Usually expressed as

$$f = \sum_{i=1}^{n} \sum_{j=1}^{n} a_{ij} x_i x_j,$$

where the x_i form a basis for the dual vector space V^* and a_{ij} are scalars. The matrix $A = (a_{ij})$ is called the matrix of the quadratic form. A quadratic form is often written as $f = XAX^T$; X^T denotes the transpose of the row matrix $X = (x_1, \ldots, x_n)$. Also, **quadratic.**

quadratic formula *Mathematics.* the formula for the roots of the quadratic equation $ax^2 + bx + c = 0$ in terms of its coefficients; the roots are $[-b + \sqrt{(b^2 - 4ac)}]/2a$ and $[-b - \sqrt{(b^2 - 4ac)}]/2a$. The expression $b^2 - 4ac$ is the discriminant; the quadratic equation has two real roots, one real root, or two imaginary roots according to whether the discriminant is positive, zero, or negative, respectively.

quadratic interaction criterion *Materials Science.* an improvement of the Tsai-Hill criterion that takes into account stress interactions when predicting failure of a fiber-reinforced composite under loading conditions prevalent in service use.

quadratic programming *Mathematics.* the collection of techniques for determining the extremals of a system of quadratic inequalities.

quadratic reciprocity law *Mathematics.* the theorem stating that if p and q are distinct odd primes, then

$$(p|q)(q|p) = (-1)^{[(p-1)/2][(q-1)/2]},$$

where $(p|q)$ and $(q|p)$ are Legendre symbols.

quadratic residue *Mathematics.* let a and m be relatively prime positive integers. a is said to be a quadratic residue modulo m if there exists some integer x such that $x^2 \equiv a(\text{mod } m)$; i.e., x^2 is congruent to a modulo m. If a is not a quadratic residue modulo m, then a is a **quadratic nonresidue** modulo m. There exist exactly $(p - 1)/2$ quadratic residues and $(p - 1)/2$ quadratic nonresidues modulo p, where p is an odd prime.

quadratic Stark effect *Atomic Physics.* a phenomenon in which a split in an atom's spectral lines is observed by an amount that is proportional to the square of the externally applied electric field.

quadratic Zeeman effect *Atomic Physics.* a phenomenon in which a split in an atom's spectral lines is observed by an amount that is proportional to the square of the externally applied magnetic field.

quadratojugal *Vertebrate Zoology.* a small membrane bone connecting the jugal and quadrate bones on each side of the skull of amphibians and other lower vertebrates.

quadratrix of Hippias *Mathematics.* the graph in the plane of the equation $y = x \tan(\pi y/2)$.

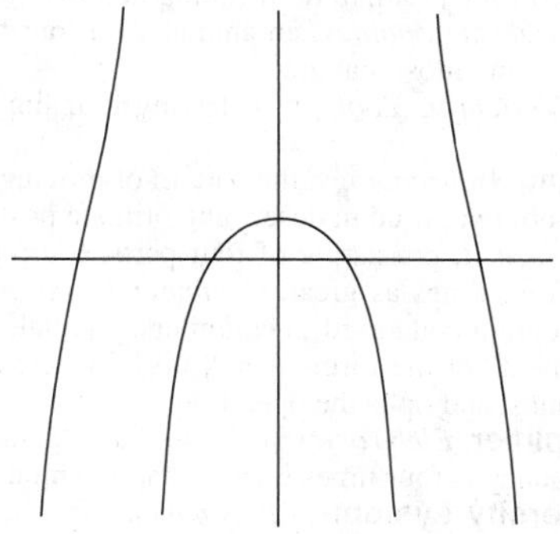

quadratrix of Hippias

quadrature *Astronomy.* the moment when the earth and the sun lie 90° apart as seen from a planet or the moon. *Mathematics.* **1.** a determining of the definite or indefinite integral of a function. The term quadrature is used in preference to integration when "integration" is likely to refer to finding the solutions of a differential equation. **2.** the process of determining a square equal in area to a given area. The quadrature of a circle is known as squaring the circle and is impossible to accomplish with ruler and compass. **3.** a **numerical quadrature** is the approximate numerical computation of a one-dimensional definite integral $\int_a^b f(x)dx$, given some information concerning individual values of the integrand and of its general properties. Standard computation formulas include the Gaussian quadrature and Simpson's formula.

quadrature amplifier *Electronics.* a circuit that produces two signals at the same frequency and whose phase angles differ by 90°; used in the testing of instruments, transmitters, and color-television receivers.

quadrature component *Electricity.* the component of a sinusoidal waveform that is either leading or lagging a reference waveform by 90°; this applies to voltage current or power.

quadrature current see REACTIVE CURRENT.

quadrature modulation *Telecommunications.* the modulation of two carrier components that are 90° apart in phase, using separate modulating functions.

quadric *Mathematics.* **1.** any homogeneous expression of degree two. **2.** a quadric surface.

quadric cone *Mathematics.* a quadric surface that satisfies an equation in standard form: $x^2/a^2 + y^2/b^2 - z^2/c^2 = 0$. Also, REAL QUADRIC CONE.

quadriceps *Anatomy.* a muscle having four heads.

quadriceps femoris *Anatomy.* the large muscle on the front of the thigh, having four heads: rectus femoris, vastus lateralis, vastus medialis, and vastus intermedius.

quadric surface *Mathematics.* any solution surface of the general quadratic equation

$$Ax^2 + 2Bxy + 2Cxz + Dy^2 + 2Eyz + Fz^2 + 2Gx + 2Hy + 2Iz + J = 0.$$

There are seventeen distinct quadric surfaces; they are various types of cones, ellipsoids, hyperboloids, and paraboloids.

quadricycle *Mechanical Engineering.* **1.** a four-wheeled cycle that is propelled by pedals. **2.** a four-wheel vehicle with a two-wheeled axle, a two-wheeled front carriage, and a bicycle seat for the driver.

quadridentate ligand *Chemistry.* a ligand having four points of attachment.

quadrigeminal bodies see CORPORA QUADRIGEMINA.

quadrilateral *Mathematics.* a plane polygon with four sides; a quadrangle. A **complete quadrilateral** is the figure formed by four lines, called sides, in the same plane, no three of which are concurrent, and the six points, called vertices, determined by pairs of sides.

quadrillion *Mathematics.* **1.** in American usage, the number 10^{15} or 1,000,000,000,000,000. **2.** in British usage, the number 10^{24}.

quadriplegia [kwä´drə plā´jē ə] *Medicine.* a condition of paralysis involving all four limbs. Also, TETRAPLEGIA.

quadriplegic [kwä´drə plā´jik] *Medicine.* a person affected by paralysis involving all four limbs. Also, TETRAPLEGIC.

quadrivalent *Molecular Biology.* of or relating to an association of four homologous chromosomes at meiosis.

quadrulus *Cell Biology.* an arrangement of four rows of cilia that spiral along the buccal cavity of paramecia.

quadrumanous *Biology.* having all four feet adapted for use as hands, as do monkeys. (From a Latin term meaning literally "four hands.")

quadruped *Vertebrate Zoology.* an animal with four feet, such as reptiles, amphibians, and most mammals.

quadrupedal *Vertebrate Zoology.* **1.** having four limbs. **2.** moving on all four limbs.

quadrupedalism *Anthropology.* the pattern of moving on four limbs as distinguished from two; used in describing primate posture.

quadruple *Science.* **1.** consisting of four parts; fourfold. **2.** to increase by four; make four times as great. *Computer Programming.* a form of intermediate program code used in compilers, equivalent to a small "assignment statement" of the form "$R = X$ op Y" where R is the result, X and Y are operands, and op is the operation.

quadruple amplifier *Electronics.* a circuit that generates an output signal whose frequency is four times greater than the input signal.

quadruple-diversity system *Telecommunications.* a receiving system that utilizes a space-diversity technique and a frequency-diversity technique simultaneously.

quadruple point *Physical Chemistry.* the temperature at which reactions of four separate phases take place at the same rate or are in equilibrium, as may occur when a saturated solution contains an excessive amount of solute.

quadruplet [kwä droop´lit] *Biology.* any of four children who are born at one birth, usually arising from multiple ovulation and statistically rare unless ovarian follicle stimulants have been used in the treatment of infertility.

quadruple thread *Design Engineering.* a multiple thread with four individual helices equally spaced around the circumference of the threaded object; the lead is approximately four times the pitch.

quadruplex circuit *Telecommunications.* a telegraph circuit that is able to carry two messages in each direction simultaneously.

quadrupole *Electromagnetism.* a characteristic distribution of charges or magnetic poles formed by placing two electric or magnetic dipoles in close proximity to each other, usually in an alternating-pole arrangement. *Nuclear Physics.* **1.** a focusing magnet used in particle accelerators, consisting of two sets of opposing poles; each pair exerts focusing forces in the vertical plane and defocusing forces in the horizontal plane. When the poles are reversed in alternate quadrupoles, a beam of positive particles is constrained to a neutral magnetic field at the central symmetry axis. Also, **quadrupole lens. 2.** a nuclear configuration composed of two dipoles of equal but opposite moment.

quadrupole field *Electromagnetism.* a field due to four charges of equal magnitude located in the same region of space.

quadrupole lens *Electromagnetism.* a device used to focus charged particle beams, providing a quadrupole field (either electric or magnetic) that is produced by an alternating-polarity circular arrangement of poles.

quadrupole moment *Electromagnetism.* a quantity that mathematically characterizes a quadrupole distribution of charges or magnetization.

quadrupole spectrometer *Analytical Chemistry.* a type of mass spectrometer with four parallel cylindrical rods in which an alternating potential superimposed on a steady potential between pairs of rods filters out all ions except those of a certain mass.

quadtree *Computer Science.* a method of representing two-dimensional space by successively breaking a square of space into four squares half its size, which are made its children in a tree data structure.

quad word *Computer Technology.* a unit of storage with a length four times that of the standard word for a given computer.

quagmire [kwag´mīr´] *Ecology.* an area of wet or swampy ground.

quahog [kwô´hôg´] *Invertebrate Zoology.* an edible clam, *Venus mercenaria,* having a thick shell; found in waters along the Atlantic coast.

quail *Vertebrate Zoology.* a cosmopolitan game bird that belongs to the family Phasianidae; it is characterized by a small bill, a short, plump body with generally colorful or heavily marked plumage, long, pointed wings, and a short tail; found in forests, woodland scrub, grasslands, and tundra; common North American species include the **California quail,** *Lophortyx californicus,* and the **bobwhite,** *Colinus virginianus.*

quail

Quail *Ordnance.* an air-launched decoy missile carried internally on the B-52 bomber; used to deceive enemy radar, interceptor aircraft, and air defense missiles; officially designated **ADM-20.**

quailbrush *Botany.* a silvery gray shrub, *Atriplex lentiformis breweri,* of the goosefoot family, that is salt tolerant; it is found in southern California.

quake sheet *Geology.* a well-defined bed that resembles a slump sheet but was formed by an earthquake and that results in the production of a load cast having no horizontal slip.

quaking bog *Geology.* a water-saturated or floating peat bog that is unstable when walked upon.

qual. qualitative; quality.

qualified name *Computer Technology.* a data set name that is further described by a specification of the class to which it belongs or by association with additional names.

qualifier *Computer Technology.* a name that is associated with another name in order to provide additional information and to distinguish the latter from others with the same name.
qualimeter see PENETROMETER.
qualitative analysis *Analytical Chemistry.* the identification of a gas, liquid, or solid or of the components of a mixture, based on the physical and chemical properties of the sample.
qualitative simulation *Artificial Intelligence.* the simulation of a physical system described by differential equations whose form is known, but whose coefficients are unknown or known only approximately; it can provide the set of possible behaviors of the system.
quality *Science.* an essential or distinctive characteristic or property of a thing. *Thermodynamics.* the quality, X, that represents the fraction of mass in the vapor phase and that is used to determine specific properties such as enthalpy, entropy, and specific volume for a wet mixture, where X is numerically equal to one minus the moisture (M); $X = (1 - M)$.
quality control *Industrial Engineering.* a general term for the process of testing an industrial output (or input) for conformity to design specifications, including defining testing procedures, minimum quality standards, and the like.
quality engineering *Design Engineering.* a manufacturing division or department concerned with the integrity of parts and assemblies from an engineering standpoint, according to verification of design and production requirements and specifications.
quality factor *Nucleonics.* a factor that when multiplied by the absorbed dose of ionizing radiation gives a quantity expressible on a common scale for all ionizing radiations.
quality of snow *Meteorology.* the amount of ice in a snow sample, expressed as a percentage of the weight of the sample.
quantal *Physics.* of or relating to quanta or quantum mechanics.
quantal response analysis *Statistics.* an analysis of the dependence of one or more categorical responses on a set of predictors.
quantasome *Cell Biology.* a small particle found on thylakoid membranes in plant chloroplasts, postulated to be the functional unit of photosynthesis.
quantic *Mathematics.* a former term for a homogeneous polynomial expression in two or more variables.
quantification *Science.* the fact or process of quantifying.
quantifier *Mathematics.* a prefix operator in a logical expression that binds variables by specifying their quantity in some sense. In particular, a **universal quantifier** is a phrase such as "for every" and "for all;" written symbolically as $\forall$. An **existential quantifier** is a phrase such as "there exists;" written symbolically as $\exists$.
quantify *Science.* to express as a number or amount.
quantile *Statistics.* any of the values in a data set that divide the distribution into a given number of equal proportions; for example, a quartile is one of three quantiles that divide a frequency into four equal parts. Also, FRACTILE.
quantimeter *Radiology.* a device that measures the amount of roentgen rays produced by an X-ray tube.
quantitative analysis *Analytical Chemistry.* the determination of the precise percentage composition of a sample.
quantitative character *Genetics.* any characteristic whose phenotype is expressed quantitatively, such as height, weight, and skin color in humans.
quantitative genetics *Genetics.* the statistical analysis of the inheritance of genetically determined characteristics.
quantitative geomorphology *Geology.* a mathematical analysis of the descriptive parameters of landform geometry and geomorphic processes.
quantitative inheritance *Genetics.* the occurrence of phenotypes that show a continuous and quantitative distribution.
quantitative microanalysis *Materials Science.* a technique in which an electron microprobe analyzer is used to provide quantitative data on the elemental composition of a small portion of a material.
quantity *Mathematics.* **1.** the property of magnitude. **2.** something having magnitude, size, extent, or amount. **3.** a figure or symbol representing this.
quantization *Science.* the process of quantizing. *Quantum Mechanics.* the restriction of the magnitude of some physical quantity to be a member of a discrete set of values; these are typically integral multiples of some fundamental unit, or quantum. *Telecommunications.* **1.** a process by which the continuous range of values of a given signal is separated into nonoverlapping subranges, each of which is represented by a discrete value within the subrange called the **quantization level**; whenever the signal value falls within a given subrange, the output has the corresponding discrete value. **2.** of, related to, or resulting from this process. Thus, **quantization distortion.**
quantize *Science.* to limit the size of a variable to a particular set of values that are multiples or increments of a definite unit.
quantized frequency modulation *Telecommunications.* a type of frequency modulation that employs quantization. Similarly, **quantized pulse modulation.**
quantized vortex *Physics.* **1.** the quantized circulation of the superfluid component of helium II about a thin cylindrical region of the nonsuperfluid (normal) component. **2.** the quantized magnetic flux associated with the superconductive component of current in a superconductor about a thin cylindrical region of nonsuperconductive core.
quantizer *Electronics.* a device that converts an analog signal into digital form, by converting the analog value to the closest one of a set of predefined discrete values.
quantum [kwän´təm] *plural,* **quanta.** *Quantum Mechanics.* **1.** the fundamental unit of electromagnetic energy that is absorbed in integral multiples of $E = h\nu$, where h is Planck's constant, or, equivalently, $E = \hbar\omega$, where $\omega = 2\pi\nu$. **2.** of the possible values of a wave, one of the subranges specified by quantization, and represented by a specific value within the subrange. (From a Latin word meaning "how much.")
quantum acoustics *Acoustics.* the study of properties of sound waves in a medium due to the quantum-mechanical nature of that medium; such properties usually appear at low temperatures, short wavelengths, or high frequencies by the kinetic energy, and are shown by the equation $E = h\nu - E_B$, where h represents Planck's constant, ν represents the frequency of sound waves, and E_B represents the binding energy of the molecules.
quantum barrier penetration see TUNNELING.
quantum chemistry *Physical Chemistry.* the branch of chemistry that relates the laws of quantum mechanics to various chemical phenomena.
quantum chromodynamics *Particle Physics.* a theory that describes the strong interactions among quarks and seeks to explain why quarks combine in certain configurations to form the observed patterns of elementary particles; the strong forces are explained in terms of color charge replacing electric charge.
quantum defect *Atomic Physics.* the difference between the effective quantum number and the principal quantum number of an atomic energy level.
quantum detector *Physics.* a device that is capable of detecting a specified quantized amount of electromagnetic energy, but is incapable of detecting any amount less than the specified level; examples are Geiger counters and photoelectric cells.
quantum efficiency see QUANTUM YIELD.
quantum electrodynamics *Particle Physics.* a quantum field theory that describes the interactions of electrically charged particles through the electromagnetic field; the particle carrying the electromagnetic force is the photon; this includes the weak electromagnetic interaction characterized by the dimensionless coupling constant $\alpha \approx 1/137$. Also, QUANTUM THEORY OF LIGHT.
quantum electronics *Electronics.* the branch of electronics that examines the relationship of energy to matter, such as the movement of atoms and the characteristics of crystals, as explained by quantum physics.
quantum evolution *Evolution.* an extremely rapid evolutionary change within a single lineage, resulting from a sudden and drastic change in an organism's environment, or from migration of the organism to an environment radically different from its original habitat.
quantum field theory *Quantum Mechanics.* a theory in which a physical field is considered as a collection of particles and forces, and observable properties of an interacting system are expressed as finite quantities rather than as state vectors.
quantum gravity *Quantum Mechanics.* the quantum field theory version of Einstein's gravitation theory.
quantum hydrodynamics *Physics.* a field devoted to the study of superfluid mechanics, such as the fountain effect.
quantum hypothesis *Quantum Mechanics.* the formulation of ideas by Max Planck that became the basis for modern quantum mechanics, particularly the assumptions that a physical system has a discrete set of possible energy values between which intermediate values are not allowed, and that absorption and emission of radiation occurs in quanta of radiant energy.
quantum-mechanical operator *Quantum Mechanics.* a linear, Hermitian operator whose operation on a state yields some physical quantity or its expected value.

Quantum Mechanics

Quantum mechanics is a theory that predicts the outcomes of physical experiments in terms of probabilities. Such a theory is necessary to explain the behavior of elementary particles such as the electron. Electrons have no internal parts that make one electron different from another, yet identical experiments performed with the same initial conditions on different electrons lead to a variety of results. Thecries of so-called "hidden variables," developed in an effort to explain these different results without resorting to quantum-mechanical probabilities, have all led to predictions that disagree with observations. The failures of "hidden variable" theories seem paradoxical, but all paradoxes thus far have been resolved in favor of quantum mechanics.

The probabilistic nature of quantum mechanics is linked to the Heisenberg Uncertainty Relation, which specifies an unavoidable lower limit in the precision with which one can simultaneously measure the position and the momentum of any object. In spite of persistent attempts, nobody has ever conceived of an instrument that could overcome the Heisenberg limit.

Theoretically, the position and momentum of any particle can be described by the particle's quantum-mechanical wave function, in which the wavelength is inversely proportional to the momentum of the particle. According to classical wave theory, accuracy in measuring any wavelength is improved if the wave is spread out over a longer distance; but accuracy in measuring the position is improved when the wave is concentrated over a smaller distance, because the particle can be anywhere in the region where the wave is not zero. These conflicting requirements prevent the simultaneous measurement of both variables in any given direction with unlimited accuracy.

Representation of a particle by a wave is facilitated by the Schrodinger wave equation, which can be used to derive the wave as a function of position and time for any particle whose potential energy is known at each position. The Schrodinger equation, combined with the application of known physical laws for the potential energy, has provided explanations for an impressive variety of previously mysterious phenomena involving atomic spectra, the behavior of particles in magnetic fields, alpha particle emission by atomic nuclei, the temperature dependence of the conductivity of metals, the tunneling of electrons through insulating barriers, superfluid behavior of liquid helium, and many others. Quantum mechanical calculations have described in precise detail the existence of previously undreamed of phenomena, and thus provided the basis for the invention and development of immensely valuable devices (e.g., the laser, the SQUID, and the "atomic clock").

John D. McGervey
Professor of Physics
Case Western Reserve University
Author, *Introduction to Modern Physics*

quantum mechanics *Physics.* a modern field in physics that is based on the premise that energy and momentum are quantized and that, at the atomic and subatomic levels, the effects of quantization are significant; this theory apparently supersedes classical theory. Also, QUANTUM THEORY.

quantum noise *Electrical Engineering.* the noise associated with the random fluctuations in the arrival or spontaneous emission of a quantum of energy.

quantum number *Quantum Mechanics.* a number designating which member of an allowed set of magnitudes a quantized physical quantity possesses, usually an integer or half-integer; this number is usually an eigenvalue of the operator associated with the quantity.

quantum solid *Solid-State Physics.* a solid in which the atoms possess relatively large motion in the quantum ground state even at the absolute zero of temperature, as the result of weak interaction and small mass.

quantum speciation *Genetics.* the rapid evolution of a new species from a small peripheral population of the ancestral species through genetic drift and the founder effect; this usually involves a small number of mutations with significant phenotypic expression. Also, SALTATIONAL SPECIATION.

quantum state *Quantum Mechanics.* the condition of a quantum mechanical system, described by a wavefunction or state vector in some function space.

quantum statistics *Physics.* the treatment of statistical mechanics in which the boson or fermion nature of the particles is taken into account.

quantum theory see QUANTUM MECHANICS.

quantum theory of heat capacity *Physics.* a theory in thermodynamics that accounts for the decrease of specific heats of substances at low temperatures near absolute zero.

quantum theory of light see QUANTUM ELECTRODYNAMICS.

quantum theory of radiation *Quantum Mechanics.* the theory describing electromagnetic radiation as a dynamical field composed of quanta or photons, and the interaction of these fields with matter.

quantum theory of spectra *Quantum Mechanics.* a theory that explains atomic spectra as the result of the quantization of the bound electrons' energies and the quantization of all degrees of freedom in bound systems with spectral lines corresponding to transitions between stationary states.

quantum theory of valence *Physical Chemistry.* a theory that uses the principles of quantum mechanics to explain various properties and characteristics of molecules, such as the stability of a chemical bond.

quantum-wave equation *Quantum Mechanics.* any partial differential equation governing the temporal behavior of wavefunctions such as the Schrödinger equation.

quantum yield *Physical Chemistry.* the ratio of the total number of moles of light-absorbing substances formed in a system to the intensity of the absorbed light. *Electronics.* the average number of electrons for each absorbed quantum of light that is emitted from a cathode in a phototube. Also, QUANTUM EFFICIENCY.

quaquaversal *Geology.* describing rock strata and geologic structures that dip away from a central point in all directions.

quar. or **quart.** quarter; quarterly.

quarantine [kwôr´ən tēn´] *Medicine.* **1.** the isolation of a person who is affected with a communicable disease, or of an apparently well person who has been exposed to such a disease, in an attempt to control the spread of the disease. **2.** the detention, for a specified period of time, of vessels, vehicles, or travelers coming from places where epidemic disease exists or is suspected to exist. (From an Italian term meaning "forty days;" from the historic practice in which a ship coming into port was confined in isolation for a period of forty days for this reason.)

quarantine anchorage *Civil Engineering.* an anchorage reserved for vessels under health quarantine.

quarantine buoy *Navigation.* a buoy that marks a quarantine anchorage area.

quark [kwärk] *Particle Physics.* any of a group of elementary particles that make up the hadrons; quarks are acted on by the strong, electroweak, and gravitational forces. These types (flavors) of quarks are identified: up (u) and down (d) (the two types found in ordinary matter); strange (s); charm or charmed (c); bottom or beauty (b); and top or truth (t). As many as eighteen types of quarks are considered possible, according to the theory of quantum chromodynamics. (A term coined by the American physicist Murray Gell-Mann; said to be from a word used by the Irish author James Joyce in his novel *Finnegan's Wake.*)

quark-gluon plasma *Nuclear Physics.* a theoretical state of nuclear particles in which the quarks and gluons merge to form an unstructured group.

quarkonium *Particle Physics.* an atom made up of a quark and an antiquark; assembled by colliding electrons and positrons at 10 GeV or higher, causing photons to sometimes form as quark-antiquark pairs; types of quarkoniums include charmoniums and bottomoniums.

quark star *Astronomy.* a hypothesized star consisting of degenerate quarks, matter which is denser than that of a neutron star.

quarry *Mining Engineering.* a pit or excavation from which slate, stone, or the like is obtained by blasting or cutting.

quarry face *Mining Engineering.* in quarrying, the freshly split face of ashlar, squared off for the joints only and used most commonly for massive work.

quarrying machine *Mechanical Engineering.* a machine that drills boreholes or cuts tunnels into native rock.

quart *Metrology.* **1.** a traditional unit of capacity used in the United States, equal to 32 fluid ounces or one-eighth gallon; equivalent to 0.946 liter. **2.** a similar unit of measure used in Great Britain, equivalent to 1.14 liters. Also, IMPERIAL QUART. **3.** see DRY QUART.

quartan *Medicine.* a fever with 72 hours between attacks. Thus, **quartan malaria.**

quarter one-fourth of a larger amount; one of four equal parts; specific uses include: *Metrology.* any of various units of measure equal to one-fourth of a larger standard unit, such as: (a) a unit of weight equal to one-fourth of a ton; (b) a unit of weight equal to one-fourth of a hundredweight; (c) a measue of grain equal to 25 pounds. *Naval Architecture.* the section of a ship to either side of the stern. As a position angle, "starboard quarter" or "port quarter" extends roughly from the stern to 45° away from the stern.

quarterdeck *Naval Architecture.* **1.** a partial deck above the main deck of a sailing ship, extending from aft of amidships to the stern, and below the poop. **2.** the section of the upper deck aft of the main mast, or aft of the point at which the main mast would be.

quarter hard *Metallurgy.* describing a metallic product that is harder than dead soft, but softer than full hard.

quartering sea *Navigation.* a sea that approaches a vessel from a bearing of approximately 135° or 225° relative.

quarter section *Cartography.* in surveying, one of the four quadrants of a section, measuring approximately one-half mile on each side and containing 160 acres; usually identified as the northeast, southeast, southwest, or northwest quarter.

quarter-square multiplier *Electronics.* an analog computer component that performs high-accuracy multiplication by implementing a number of diode-resistor networks.

quarter-turn drive *Mechanical Engineering.* a belt drive between pulleys having axes at right angles.

quarter-wave *Electromagnetism.* a distance whose length is one-quarter of a wavelength.

quarter-wave antenna *Electromagnetism.* an antenna whose electrical length is one-quarter of the wavelength of the operating frequency.

quarter-wave attenuator *Electromagnetism.* a transmission-line attenuating device in which two parallel grids of dissipative material are placed an odd number of quarter-wavelengths apart in the line; reflections from one grid nullify reflected waves from the other grid.

quarter-wave plate *Optics.* a plate of doubly refracting crystal that introduces a phase difference of one-quarter cycle between the ordinary and extraordinary rays passing through it perpendicularly; used to measure refraction and to convert polarized light.

quarter-wave stub *Electromagnetism.* a stub of one-quarter of a wavelength of the fundamental frequency that, when shorted at the far end, provides high impedance to the odd harmonics (including the fundamental) and low impedance to the even harmonics. Also, **quarter-wave line, quarter-wave transmission line.**

quarter-wave termination *Electromagnetism.* a terminating device consisting of a metal end plate and a dissipative grid positioned one-quarter of a wavelength before the end; the design is nonreflective.

quarter-wave transformer *Electromagnetism.* an impedance matching device consisting of a one-quarter wavelength section of transmission line. Also, **quarter-wave matching section.**

quartic *Mathematics.* any (usually homogeneous) expression of degree four. A quartic equation is any polynomial equation of degree four (not necessarily homogeneous). Also, BIQUADRATIC.

quartic curve *Mathematics.* the graph in the plane of a general quartic equation in two variables. Quartic curves include conchoids, cardiods, Cassinian ovals, and lemniscates.

quartile *Statistics.* one of the three numbers that divide a distribution into four equal parts.

quartz [kwôrts] *Mineralogy.* SiO_2, a transparent to translucent trigonal mineral with a vitreous luster, commonly white or colorless but also occurring in a variety of colors, massive in habit or as hexagonal prismatic crystals, and having a specific gravity of 2.65 and a hardness of 7 on the Mohs scale; the most abundant and widely distributed of all minerals.

quartz andesite see DACITE.

quartzarenite *Petrology.* a sandstone containing more than 95% quartz framework grains separated chiefly by cement rather than by matrix; essentially an orthoquartzite.

quartz basalt *Petrology.* a basalt that contains small amounts of quartz or another silica mineral in the groundmass.

quartz clock *Horology.* any timepiece with a movement controlled by the oscillations of a quartz crystal subjected to an electrical field. Also, **quartz crystal clock.**

quartz crystal *Mineralogy.* a transparent, colorless quartz of low brilliance, often occurring in distinct crystals; used in optics and electronics. Also, ROCK CRYSTAL. *Electronics.* an industrial form of this material, either natural or produced synthetically; quartz crystals vibrate at a specific frequency when a voltage is applied to them; they are widely used to control the frequency of an oscillating system.

quartz crystals

quartz-crystal filter *Electronics.* a filter that includes one or more quartz-crystal resonators as elements.

quartz-crystal resonator *Electronics.* a resonant circuit using a quartz crystal as the element that determines the frequency; used as components of a quartz-crystal filter, or to control the frequency of an oscillator. Also, CRYSTAL RESONATOR.

quartz delay line *Electronics.* a device that employs quartz to delay the transmission of audio signals.

quartz diorite *Petrology.* a plutonic rock or rock group similar to diorite in composition but containing quartz as 5% to 20% of the light colored minerals. Also, TONALITE.

quartz-fiber electroscope *Electronics.* an instrument in which an electrical charge changes the angle of a gold-plated quartz fiber suspended in a glass jar; used to measure voltage.

quartz-flooded limestone *Petrology.* a limestone with an abundance of quartz particles suddenly appearing as a result of nearby wind or water action, but rapidly decreasing in an upward direction and completely disappearing, all within a few centimeters.

quartz glass see FUSED QUARTZ.

quartz graywacke *Petrology.* a variety of graywacke with abundant grains of quartz and chert and less than 10% each of feldspar and rock fragments.

quartz-iodine lamp *Electronics.* a lamp in which a tungsten filament is surrounded by iodine gas in a quartz envelope, whose properties allow the lamp to generate more heat and pass ultraviolet rays.

quartzite *Petrology.* **1.** a quartz rock derived from metamorphosed quartz sandstone or chert. **2.** a sedimentary rock consisting of pure quartz sandstone cemented with silica, so that the rock breaks across the sand grains.

quartzitic sandstone *Petrology.* sandstone containing 100% quartz grains cemented together with silica.

quartz latite see RHYODACITE.

quartz mine *Mining Engineering.* a mine in which the deposit of a valuable material, such as gold, is found in veins or fissures rather than in placers; quartz is the principal accessory mineral is these deposits.

quartz monzonite *Petrology.* a granitic plutonic rock that is composed of major quartz, plagioclase, and orthoclase with small amounts of biotite and hornblende and traces of apatite, zircon, and opaque oxides; the ratio of alkali feldspar to total feldspar is between 35 and 65%. Also, ADAMELLITE.

quartz movement *Horology.* a very accurate electronic movement in which the natural frequency of vibrations of a quartz crystal is used to regulate the operation of the timepiece.

quartz oscillator *Electronics.* a circuit that generates alternating current at a frequency determined by vibration of a quartz crystal resonator.

quartzose *Geology.* describing any substance, especially sediment or rock, that is composed primarily of quartz.

quartzose arkose *Petrology.* a sandstone with 50–85% quartz, chert, and metamorphic quartzite, 15–25% feldspars and feldspathic crystalline rock fragments, and 0–25% micas and micaceous metamorphic rock fragments.

quartzose chert *Petrology.* a vitreous, shiny, and sparkly chert composed of a heterogeneous mixture of pyramids, prisms, and faces of quartz that are visible only under high magnification, and combined with chert in which the secondary quartz is largely anhedral.

quartzose graywacke *Petrology.* **1.** a sandstone with 50–85% quartz, chert, and metamorphic quartzite, 15–25% micas and micaceous metamorphic rock fragments, and 0–25% feldspars and feldspathic crystalline rock fragments. **2.** a graywacke that has lost its micaceous constituents due to abrasion, and is thus similar to an orthoquartzite.

quartzose sandstone *Petrology.* a well-sorted sandstone with at least 95% clear quartz grains and at most 5% groundmass. Also, QUARTZ SANDSTONE.

quartzose shale *Petrology.* a green or gray shale consisting mainly of rounded silt-size quartz grains, associated with very mature sandstones.

quartzose subgraywacke see PROTOQUARTZITE.

quartz-pebble conglomerate see ORTHOQUARTZITIC CONGLOMERATE.

quartz plate see CRYSTAL PLATE.

quartz porphyry *Petrology.* a medium-grained porphyritic igneous rock of granitic composition, composed of quartz and alkali feldspar phenocrysts set in a microcrystalline or cryptocrystalline groundmass. Also, GRANITE PORPHYRY.

quartz resonator see QUARTZ-CRYSTAL RESONATOR.

quartz sandstone see QUARTZOSE SANDSTONE.

quartz schist *Petrology.* a schist with foliation due principally to the presence of streaks and lenticles of nongranular quartz.

quartz strain gauge *Electronics.* a device that measures small deformations of a substance by measuring the voltage produced when a quartz element is attached to it.

quartz syenite *Petrology.* a quartz-bearing plutonic rock, or rock group, intermediate in composition between true syenite and granite; 5–20% of the light-colored minerals are quartz.

quartz watch *Horology.* any watch with a movement controlled by the oscillations of a quartz crystal subjected to an electrical field.

quartz wedge *Optics.* a piece of quartz whose thin wedge shape allows any thickness of quartz to be superimposed onto a biaxial mineral sample being viewed under a polarizing microscope; used to determine the sign of double refraction of the sample from the interference figures produced and viewed in convergent light.

quas see KAUS.

quasar [kwā´zär] *Astronomy.* a compact quasi-stellar radio source or visible object; characterized by spectra with high redshifts, implying great distances and consequent high rates of energy production; thought to be the most distant and most luminous objects in the universe. (From quasi-stellar radio source, the first type of quasar discovered.)

quasi- a combining form meaning: **1.** almost. **2.** seemingly. **3.** resembling.

quasi-acoustical holography *Acoustical Engineering.* the assemblage of sonograms made in the B mode into a three-dimensional image volume that creates a realistic three-dimensional facsimile of an object.

quasi-atom *Atomic Physics.* a system in which the nuclei of two colliding atoms briefly overlap, so that they seem to be one atom with a combined atomic number.

quasi-cratonic *Geology.* describing a region of oceanic crust at the margin of a continent that is believed to have been continental material which stretched and foundered during expansion. Also, SEMICRATONIC.

quasi-empiricism *Mathematics.* an approach to mathematics that views it as a living, growing subject in which proofs and counterexamples are used to generate a constantly changing understanding of mathematical problems and their solutions.

quasi-equilibrium *Geology.* in a stream cross section, a condition of balance or grade in which a state of near equilibrium tends to occur in a reach of the stream contemporaneously with the establishment of a relatively smooth longitudinal profile.

quasi-equivalence theory *Virology.* a theory that explains the absence of strict icosahedral symmetry in the arrangement of chemical subunits in a viral capsid, suggesting that the bonds between identical subunits are not precisely the same throughout the particle, but are deformed in slightly different ways in different parts of the structure.

quasi-free electron theory *Solid-State Physics.* a modification applied to the free electron theory by which a conduction electron in a metal is described by an effective mass different from its free mass; this modification accounts for the presence of the periodic variation in the electrical potential throughout the crystal.

quasi-hydrostatic approximation *Meteorology.* the application of the hydrostatic equation as the vertical equation of motion, resulting in the inference that the vertical accelerations are small without constraining them to be zero.

quasi-instruction see PSEUDOINSTRUCTION.

quasi-molecule *Atomic Physics.* the structure produced when two colliding atoms come close enough for their electron clouds to overlap without actually forming a stable molecule.

quasi-parallel execution *Computer Technology.* the processing of a set of coroutines by a serial processor in which the order of execution is arbitrary and each coroutine is executed independently of the rest in order to simulate parallelism.

quasi-particle *Physics.* in a system of entities with characteristic features of particles such as mass, momentum, and energy, an entity that carries these features without existing as a true free particle; an example is a phonon oscillation in a crystal lattice.

quasi-periodic *Mathematics.* **1.** a function $f(z)$ is said to be quasi-periodic if there exist a period ω and a constant η such that $f(z + \omega) = f(z) + \eta$ for all z in the domain of f. For example, $f(x) = x + \sin x$ is quasi-periodic with $\omega = \eta = 2\pi$. If $\eta = 0$, then f is periodic. **2.** alternatively, a function $f(z)$ is said to be quasi-periodic if there exist a period ω and constants η and c such that $f(z + \omega) = ce^{\eta z}f(z)$ for all z in the domain of f. The theta functions of Jacobi and the sigma and zeta functions of Weierstrass are well-known examples. If $\eta = 0$ and $c = 1$, then f is periodic. *Chaotic Dynamics.* nonperiodic behavior that is the combination of a finite number (two or more) of periodic signals with incommensurate frequencies.

quasi-periodicity *Chaotic Dynamics.* **1.** behavior that is the sum of two or more incommensurate periodic oscillations. **2.** a route to chaos involving a sequence of Hopf bifurcations, the first two generating incommensurate frequencies and a 2-torus attractor. Thereafter the two oscillations may lock to a periodic attractor that period doubles to chaos, or a third incommensurate frequency generated by a third Hopf bifurcation may lead (with high probability) directly to chaos.

quasi-periodic motion *Physics.* the motion of a body, usually represented in phase space, in which the motion is close to being cyclic.

quasi-random code generator *Telecommunications.* in the design and evaluation of wide-band communications links, a high-speed, pulse-code-modulation information source that provides an approximately random sequence for testing.

quasi-reflection *Optics.* the strong return of light that occurs when its rays strike suspensoids, such as dust particles, whose diameters exceed the light's wavelength.

quasi-reversible *Physical Chemistry.* describing a process or reaction that approaches reversibility, but that shows small deviations from an equilibrium path. Thus, **quasi-reversibility.**

quasi-square wave *Electricity.* a slightly nonsquare waveform that is close enough to square for all practical purposes.

quasi-stable elementary particle *Particle Physics.* an older term for an elementary particle that does not decay upon strong interactions and that has a lifetime of at least 10^{-20} seconds.

quasi-static process see REVERSIBLE PROCESS.

quasi-stationary front *Meteorology.* a frontal system that is stationary or nearly stationary; generally, a front that moves at a speed of less than five knots. Also, STATIONARY FRONT.

quassia *Botany.* **1.** a small tree, *Quassia amara,* characterized by pinnate leaves, bright red flowers, and a bitter-tasting wood. **2.** the wood of such a tree. *Pharmacology.* a bitter drug prepared from the wood of such trees, used as a tonic and a purge, as an insecticide, and as a substitute for hops.

quater in prescriptions, four times. (From Latin *quattuor.*)

Quaternary [kwät′ər nâr′ē; kwə ter′nə rē] *Geophysics.* the second period of the Cenozoic age, beginning three million years ago and extending to the present. *Geology.* relating to the rocks and deposits formed during this time.

quaternary *Metallurgy.* of an alloy, having four main constituents.

quaternary ammonium base *Organic Chemistry.* $R_4N^+OH^-$ or R_4NOH, an ammonium hydroxide in which the ammonium hydrogen has been replaced by an alkyl or other organic group.

quaternary ammonium compound *Organic Chemistry.* any of a group of ammonia salts in which the nitrogen atom is bonded to four organic groups; all such salts are surface-active compounds; they are useful as antiseptics and disinfectants because of their ability to disrupt bacterial cell membranes.

quaternary ammonium salt *Organic Chemistry.* $R_4N^+A^-$, a nitrogen compound in which a central nitrogen is covalently bonded to four alkyl or other organic groups and ionically bonded to one acid.

quaternary carbon atom *Organic Chemistry.* a carbon atom that is single-bonded to four other carbon atoms.

quaternary phase equilibria *Physical Chemistry.* the thermodynamic relationship among four nonreactive components that exhibit varying degrees of mutual solubility; for example, two solvents, each composed of a feed liquid and a solute.

quaternary phosphonium hydroxide *Organic Chemistry.* a phosphorus compound in which a central phosphorus is bonded to four organic groups, such as R_4POH.

quaternary signaling *Telecommunications.* in electrical communications, a mode in which data are transmitted by the plus and minus variations of four discrete levels of a single parameter of the signaling medium.

quaternary structure *Biochemistry.* the fourth level of organization within a protein molecule, representing the orientation and bonding between its polypeptide chains.

quaternary system *Physical Chemistry.* a relationship among four elements, such as four phases or four components, in which chemical reactions are taking place under the same conditions and at the same rate.

quaternion *Mathematics.* a generalization of complex numbers developed by the Irish mathematician Sir William Rowan Hamilton. The set of quaternions forms a skew (i.e., noncommutative) field as well as a vector space of dimension four over the real numbers. The basis elements of the vector space are usually written 1, i, j, and k. They satisfy the following rules: (a) $i^2 = j^2 = k^2 = -1$; and (b) $ij = -ji = k$, $jk = -kj = i$, $ki = -ik = j$. A typical quaternion has the form $a + bi + cj + dk$, where a, b, c, and d are real numbers. A quaternion is sometimes called a **hypercomplex number**.

quay [kē, kā, kwā] *Civil Engineering.* a wharf or landing place constructed along the edge of a body of water.

quebrachitol *Organic Chemistry.* $(HO)_5C_6H_6OCH_3$, colorless crystals that melt at 191°C; a monomethyl ether of L-*chiro*-inositol that occurs in many plants, as in the latex of the rubber tree, *Hevea braziliensis.*

quebracho [kā brä′chō] *Materials.* a combustible substance obtained from two tropical American trees, *Aspidosperma quebracho* and *Quebracho lorentzi,* used in tanning, dyeing, ore flotation, and flavoring.

queen *Invertebrate Zoology.* in the social colonies of bees, wasps, ants, and termites, the mature, fertile female that lays eggs.

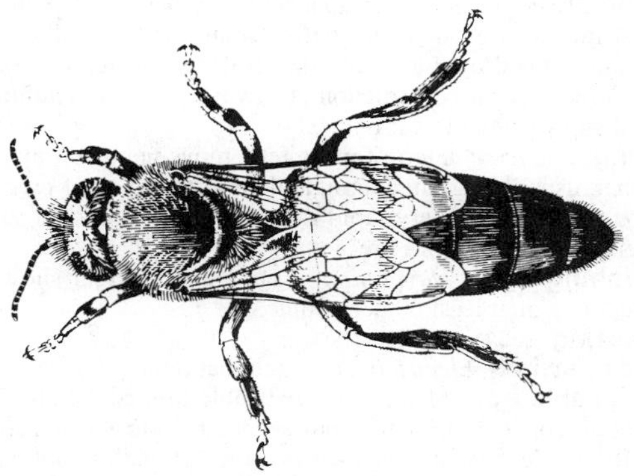

queen bee

queenfish *Vertebrate Zoology.* a silver and blue drum, *Seriphus politus,* found along the coast of California.

queen palm *Botany.* a feather palm, *Arecastrum romanzoffianum,* having large leaves from 7 to 12 feet and hanging clusters of small fruit; found in South America.

Queensland tick typhus *Medicine.* an acute infectious and contagious disease caused by *Rickettsia australis,* transmitted by ticks of marsupials and wild rodents of Queensland, Australia.

queenstownite see DARWIN GLASS.

quelea *Vertebrate Zoology.* a red-billed weaverbird, *Quelea quelea,* whose large flocks destroy grain crops; found in Africa.

quellung reaction *Microbiology.* a phenomenon in which in the outer covering of the bacterial cell wall, the capsule, becomes more visible (i.e., looks darker) when a certain antiserum is applied to it. Also, **quellung phenomenon.** (From a German word meaning "swelling," though little actual swelling takes place.)

quench *Science.* to cool suddenly by plunging in liquid. *Electronics.* to terminate by applying a voltage, as the flow of electrons in a vacuum tube.

quench aging *Metallurgy.* a process of aging that occurs while a supersaturate solid solution is rapidly cooled.

quench and tempered steel *Materials.* any of the easily welded, treated, low-carbon steels with a yield strength of 80,000 to 125,000 psi and lower ductility than HSLA steel.

quench anneal *Metallurgy.* to anneal an austenitic ferrous alloy by solution heat treating followed by rapid cooling.

quenchant *Materials Science.* a medium used to provide rapid cooling following heat treatment; water and brine are quenchants used when rapid cooling is required; oil, molten salt, and still air provide slower cooling rates.

quench correction *Biotechnology.* a technique to correct for quenching in the observable counts per minutes that are produced by a scintillation counter; the true value is given in terms of disintegrations per minute.

quenched spark gap *Electricity.* a spark gap that is capable of rapid deionization.

quenched water *Oceanography.* an area of turbulent water in which air bubbles absorb and scatter sound waves, decreasing the efficiency of sound propagation through the water and interfering with sonar technology.

quench frequency *Electronics.* the number of times per second that a circuit ceases and resumes oscillating.

quench hardening *Metallurgy.* a hardening of many ferrous alloys and a few nonferrous alloys, occurring after rapid cooling; generally caused by a martensitic transformation.

quenching *Mechanical Engineering.* the process of quickly cooling a material, such as steel, by immersion in a cold liquid or gas. *Electronics.* **1.** the process by which a discharge is terminated in a gas tube. **2.** a process by which a heated element, such as gas or metal, is rapidly cooled. **3.** a process by which alternating current is applied to an electrode to prevent continuous oscillation in a superregenerative receiver. *Solid-State Physics.* a reduction in the intensity of luminescent radiation from a crystal due to the dissipation of energy in forms other than radiation; this effect arises from the presence of defects and impurities. *Atomic Physics.* a phenomenon in which an intense electric field, such as a crystal field, causes an electron to reduce its average angular momentum. *Biotechnology.* a process that reduces the number of photons impinging on a photomultiplier of a scintillation counter, resulting in a counting rate lower than the true disintegration rate. *Immunology.* an occurrence during the process of immunofluorescence, by which bound antigens absorb light from the fluorescence of an antibody that has been excited by ultraviolet light.

quenching frequency *Electronics.* the point at which oscillation is suppressed in a superregenerative receiver.

quenching oscillator *Electronics.* a circuit that produces the alternating current which is used to suppress oscillation in a superregenerative receiver.

quenching stress *Metallurgy.* the residual stresses induced by rapid cooling.

quenite *Petrology.* a dark, fine-grained hypabyssal rock consisting primarily of anorthite and chrome diopside, with less olivine and little bronzite.

quenselite *Mineralogy.* $PbMn^{+3}O_2(OH)$, a pitch-black, opaque monoclinic mineral occurring as tabular crystals, having a specific gravity of 6.84 and a hardness of 2.5 on the Mohs scale; found in Sweden.

quenstedtite *Mineralogy.* $Fe_2^{+3}(SO_4)_3 \cdot 10H_2O$, a violet, transparent triclinic mineral occurring in aggregates of tiny crystals, having a specific gravity of 2.147 and a hardness of 2.5 on the Mohs scale; found with other secondary iron sulfates.

querbalken *Microbiology.* crossbands found in the prosthecae of certain bacterial species of *Asticcacaulis* and *Caulobacter.*

quercetin *Biochemistry.* $C_{15}H_{10}O_7$, a yellow, crystalline, flavonoid pigment that is found in oak bark, lemon juice, asparagus, and other plants; used to reduce abnormal capillary fragility. Also, MELETIN, SOPHORETIN.

quercitol *Organic Chemistry.* $C_6H_7(OH)_5$, colorless crystals that are soluble in water and that melt (decompose) at 234–235°C; a deoxyinositol found in acorns and oak bark. Also, **quercite.**

***d*-quercitol** *Biochemistry.* $C_{20}H_{28}ClNO_8P_2$, a substance derived from acorns.

quercitrin *Organic Chemistry.* $C_{21}H_{26}O_{11}$, yellow crystals that are insoluble in water and soluble in alcohol, ether, acetone, chloroform, and benzene; melts at 176–179°C. Also, **quercimelin, quercitroside.**

quern [kwurn] *Archaeology.* an ancient grinding stone for grain; the earliest form, the **saddle quern,** typically consisted of a concave base used with a handstone for rubbing seeds into flour; this was later replaced by a **rotary quern,** with one stone turning inside another.

query *Computer Programming.* **1.** a program instruction that requests information from a database. **2.** any request by a user for information from a computer, such as the status of a program, the time, or a list of active users. **3.** to ask for information that meets certain conditions, such as the last record entered or all records starting with a certain letter.

query by example *Computer Programming.* to request information from a database by specifying the search conditions in a sample record.

query language *Computer Programming.* **1.** a language in which queries to a database can be specified. **2.** a high-level language designed to allow queries of databases without the knowledge of any codes or key words.

query program *Computer Programming.* a program designed to allow users or programs to interrogate a database and then store or print out the retrieved information.

question-answering system *Computer Programming.* **1.** a system that can answer relatively free-form questions, such as questions in a natural language subset, by deduction and retrieval from a database. **2.** a system in which direct answers are provided to submitted queries, rather than a collection of references that may contain the desired answer.

Quetelet, (Lambert) Adolphe [kā´tə lā] 1796–1874, Belgian mathematician; carried out the first statistical breakdown of a national census.

Quetelet index see BODY MASS INDEX.

quetzal [ket säl´] *Vertebrate Zoology.* a rare, brilliantly colored arboreal forest bird, *Pharomachrus mocino*, of the family Trogonidae, characterized by a plump body, a very long, plumed tail in the male, and a bushy, feathered crown; native to Central America.

quetzal

Quetzalcoatlus *Paleontology.* a genus of large pterosaurs in the extinct order Pterosauria and suborder Pterodactyloidea; the vulturelike Quetzalcoatlus achieved a wing span of 12 meters; known only from a few bones found in Texas; extant in the Upper Cretaceous.

queue [kyoo] a file or line of people, objects, or items waiting their turn; specific uses include: *Transportation Engineering.* a line of individuals or vehicles waiting for service or for clearance to proceed. *Industrial Engineering.* a group of jobs waiting to be processed in a shop. *Computer Technology.* **1.** a line or list of items in a computer system that are waiting to be processed. **2.** a data storage structure in which new items are put at the end of a list and items to be processed are removed from the front of the list, so that the order of precessing is first-in, first-out. Also, PUSH-UP LIST, PUSH-UP STACK.

queued access method *Computer Programming.* a collection of procedures that employ queues of data blocks to organize and synchronize the transfer of data between the computer and the various input/output devices.

queue discipline *Transportation Engineering.* the sequence in which members of a queue are served. In most traffic conditions this is first come, first served. In some cases, however, priority traffic may enjoy a "head of the line" privilege, as in a carpool lane on a highway on-ramp.

queue-driven system *Computer Programming.* a computer system that uses queues to control various processing tasks.

queueing configuration *Transportation Engineering.* the arrangement of a queuing system, having a simple single queue, several independent lines separately serviced, or one line serviced by several exit points.

queueing model *Industrial Engineering.* for planning purposes, a simulation of an actual situation involving jobs or customers waiting in a queue to be processed.

queueing service rate *Transportation Engineering.* the rate at which vehicles are able to exit from the front of a queue.

queueing system *Transportation Engineering.* an arrangement of vehicles in a queue to pass a service point or traffic obstruction.

queueing theory *Mathematics.* a branch of discrete mathematics that solves optimization problems involving limited resources satisfying random demands; e.g., customers waiting for service.

queue length *Transportation Engineering.* the number of vehicles waiting in a queue at any given time.

quibinary *Computer Technology.* of or relating to a binary-coded decimal code in which each decimal digit is represented by seven bits. Also, BIQUINARY.

quick *Geology.* **1.** describing a sediment that becomes soft or loose when mixed with water and flows easily under a load or gravitational force. **2.** describing a soil in which a decrease in intergranular pressure causes the water to flow upward, significantly reducing the soil's bearing capacity. **3.** describing a highly porous soil with a tendency to absorb heat readily. *Mining Engineering.* describing a deposit of material that is economically valuable or productive.

quick-break fuse *Electricity.* a thermal protective device that opens rapidly to disconnect a circuit when current becomes excessive.

quick-break switch *Electricity.* a switch that abruptly breaks a circuit.

quick-change gearbox *Mechanical Engineering.* an arrangement of gears on a machine tool set up so that the speed relationship between gears can be quickly varied.

quick clay *Geology.* a clay that loses practically all its shear strength and flows plastically upon being disturbed.

quick dipping *Materials Science.* a coating process in which an object is dipped into a large tank of coating material (typically paint), removed, and drained over a trough that collects and returns drainage to the tank; used to coat construction steelwork, paint agricultural equipment, and apply primer to car bodies.

quickening *Medicine.* a term for the fetal movements felt by an expectant mother, usually beginning around the fourth month of pregnancy.

quick fire *Ordnance.* a very rapid rate of fire that is used against moving targets; faster than rapid fire.

quick-flashing light *Navigation.* a navigational light showing brief flashes at a rate of at least 60 per minute.

quick freezing see FLASH FREEZING.

quick-make switch *Electricity.* a switch that rapidly connects a circuit.

quick malleable iron *Metallurgy.* malleable iron containing 2.2% carbon, 1.5% silicon, 0.3–0.6% manganese, and 0.75 to 1% copper

quick return *Mechanical Engineering.* a reciprocating motion in operating the tool of a shaping machine in which the return is made more rapidly than the cutting stroke, in order to reduce idling time.

quicksand *Geology.* a mass or thick bed of fine sand made up of smooth, loose grains, generally saturated with upward-flowing water, resulting in a soft, semiliquid, highly mobile and unstable mass that yields under pressure and that readily swallows heavy objects touching its surface. Also, RUNNING SAND.

quicksilver *Chemistry.* a popular name for mercury.

quicksort *Computer Science.* a fast-sorting algorithm in which a partitioning key value is selected from the existing file, the file is scanned from the left until a key value greater than the partition value is found, and from the right until a key value less than the partition value is found; the records are exchanged and the scans are continued. When the scans from the left and the right meet, the process is repeated recursively on the left and right subfiles around the meeting point.

quickstone *Petrology.* a consolidated rock that flowed under the influence of gravity before lithification; a lithified quick sediment.

quick time *Military Science.* a rapid marching rate in which 120 30-inch paces are taken each minute.

quiesce [kwī es´] *Computer Programming.* to reject new jobs in a multiprogramming system while continuing to process current jobs, usually in preparation for taking the system down.

quiescent [kwī es´ənt] *Science.* not active; dormant or at rest. *Electronics.* a condition in which a circuit is operating but does not have signals applied to it.

quiescent center *Botany.* a region in the apical meristem of many roots in which cells do not divide or divide much slower than surrounding cells.

quiescent culture *Cell Biology.* a population of cells that are not rapidly proliferating.

quiescent period *Telecommunications.* the period between pulse transmissions.

quiescent point *Electronics.* the set of voltages and currents in a circuit or device when no external signal is being applied to it.

quiescent prominence *Astronomy.* sheets of relatively cool solar material that hang in the corona for days or months.

quiescent push-pull *Electronics.* a push-pull amplifier designed so that current flows only when a signal is being processed.

quiescent value *Electronics.* the level of current or voltage in an electrode when no signals are present.

quiet battery *Electronics.* a device that generates energy with so little interference that it may be used for voice transmissions.

quieting sensitivity *Electronics.* the minimum amplitude that an incoming radio signal must have to meet the signal-to-noise specifications of a given FM receiver.

quiet reach see STILL WATER.

quiet sleep see NREM SLEEP.

quiet sun *Astronomy.* the months or years during the minimum of the sunspot cycle, at which time the sun displays few spots, prominences, or flares, and little radio activity.

quiet sun noise *Astrophysics.* an electromagnetic noise from the sun during times of sunspot minimum.

quiet tuning *Electronics.* a technique for operating a radio receiver in which no signal is sent unless it is tuned to the exact frequency of the incoming radio signal.

Quiinaceae *Botany.* a family of Amazonian trees, shrubs, and vines of the order Theales; characterized by mucilage channels, leaf veins sheathed by thick-walled fibers, and wooly seeds without endosperm.

quill *Vertebrate Zoology.* **1.** in birds, the hollow portion of the base of a feather that is embedded in a follicle of the skin and filled more or less with remains of mesodermal tissue from the feather's development. Also, CALAMUS. **2.** in porcupines, hedgehogs, and spiny anteaters, a loosely attached barblike spine that serves as a defense. *Mechanical Devices.* a hollow shaft surrounding another shaft; used in various devices.

quill drive *Mechanical Engineering.* a hollow, nonrotating shaft into which another shaft is inserted and rotated under power, providing axial movement, as in a drilling machine spindle.

quilted surface *Geology.* a topography characterized by uniformly convex hills separating relatively narrow valleys.

quinacrine hydrochloride *Organic Chemistry.* $C_{23}H_{30}ClN_3O{\cdot}2HCl{\cdot}2H_2O$, a bitter, yellow crystalline powder; soluble in water, alcohol, and methanol; melts at 248°C. *Pharmacology.* this substance in the form of an antiparasitic drug, taken orally to treat malaria, giardiasis, and tapeworm. Also, CHINACRIN HYDROCHLORIDE, MEPACRINE HYDROCHLORIDE.

quinaldine *Organic Chemistry.* $C_9H_6NCH_3$, a colorless oily liquid that is soluble in alcohol, ether, and chloroform; boils at 246°C; used as an antimalarial drug.

quinalizarin *Organic Chemistry.* $C_{14}H_8O_6$, red crystals that are soluble in alkaline solutions, acetic acid, and sulfuric acid; melts above 275°C; used to dye cottons and as a reagent for beryllium.

quinaphthol *Pharmacology.* a drug used as an antiseptic in the intestines. Also, CHINAPHTHOL.

quinary *Science.* relating to or consisting of five items or members.

quinary code *Computer Technology.* a binary-coded decimal code that uses combinations of five bits to represent digits.

quince *Botany.* **1.** either of two small trees, *Cydonia oblonga or C. sinensis,* of the rose family, bearing a hard, yellowish pear-shaped fruit; native to Asia. **2.** the fruit of such a tree, used in preserves and jellies.

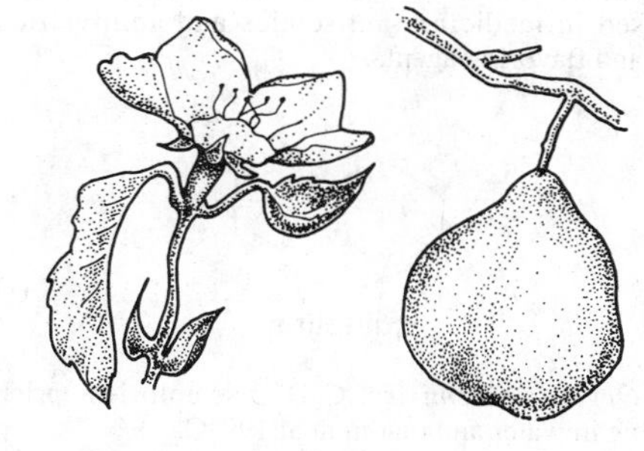

quince

quincunx *Botany.* an overlapping arrangement of five petals or leaves in which two are exterior, two are interior, and one is partly exterior and partly interior.

quincunx planting *Forestry.* a pattern of planting in which four young trees are set out to form the corners of a square and a fifth tree is planted at its center.

quinestrol *Pharmacology.* $C_{25}H_{32}O_2$, a synthetic estrogen; used in oral contraceptives.

quinhydrone *Organic Chemistry.* $C_{12}H_{10}O_4$, a green powder; soluble in alcohol, ether, ammonia, and hot water; melts (sublimes) at 171°C; used in solution to coat electrodes for pH determination.

quinhydrone electrode *Analytical Chemistry.* a platinum wire in a standard solution of quinhydrone, $C_6H_4O_2{\cdot}C_6H_4(OH)_2$, used as a reversible electrode standard for pH determinations.

quinic acid *Organic Chemistry.* $C_6H_7(OH)_4COOH{\cdot}H_2O$, a white crystalline solid; soluble in alcohol, water, and acetic acid; melts at 162°C; found in many plants and used as a medicine. Also, KINIC ACID.

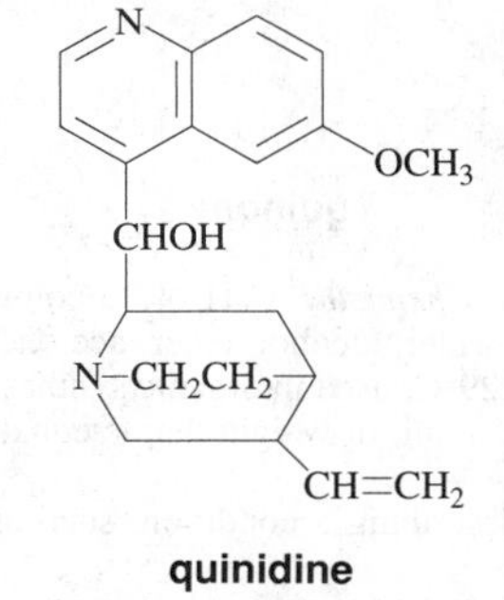

quinidine

quinidine *Organic Chemistry.* $C_{20}H_{24}N_2O_2$, a crystalline alkaloid; soluble in alcohol, benzene, and chloroform; melts at 174°C; an extract from cinchona bark, used in making pharmaceuticals. Also, **ß-quinine.**

quinine [kwī´nīn´] *Organic Chemistry.* $C_{20}H_{24}N_2O_2$, a white, bitter, crystalline alkaloid that is soluble in alcohol, ether, chloroform, and pyrimidine; melts at 177°C; obtained from cinchona bark and used as an antimalarial and antipyretic.

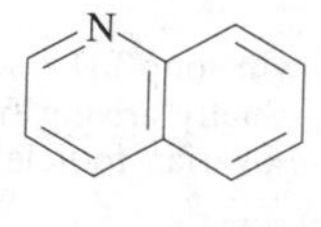

quinine

quinine water *Food Technology.* carbonated water that contains lemon, lime, sweetener, and a small amount of quinine; used as a mixer. Also, TONIC, TONIC WATER.

quininism see CINCHONISM.

quinizarin *Organic Chemistry.* $C_{14}H_6O_2(OH)_2$, orange crystals that are soluble in alcohol, ether, and acetic acid; melts at 200–203°C; an important isomer of alizarin, used in making dyes.

quinoidine *Organic Chemistry.* a thick material containing several alkaloid compounds that remain in solution after the extraction of crystalline alkaloids from cinchona bark; a medicine soluble in alcohol, chloroform, and dilute acids.

quinoline *Organic Chemistry.* C_9H_7N, a yellowish aromatic liquid; soluble in alcohol, ether, acetone, and benzene; boils at 238°C and freezes at –15°C; used in medicine (antiseptics and antipyretics) and as a preservative and flavoring agent.

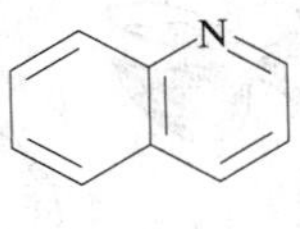

quinoline

quinolone *Organic Chemistry.* C_9H_7ON, colorless prisms that are slightly soluble in water and that melt at 199°C.

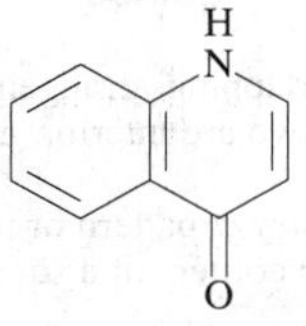

quinolone (4-quinolone)

quinone *Organic Chemistry.* $OC(CH{=}CH)_2CO$, combustible, irritating yellow crystals; soluble in alcohol, ether, and alkalis; melts at 116°C; used as an oxidizing agent, in photography, and in making dyes and hydroquinone.

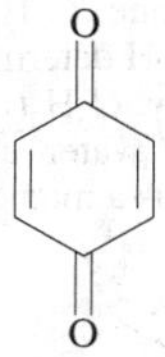

+

$H^+ + e$

quinone

quinoxaline *Organic Chemistry.* $C_8H_6N_2$, a colorless crystalline powder that is soluble in water, alcohol, ether, acetone, and benzene; melts at 30°C and boils at 229°C; used in organic synthesis.

quintal *Metrology.* **1.** a unit of weight that is equal to 100 kilograms. **2.** see HUNDREDWEIGHT.

quintan *Medicine.* describing a condition, such as a fever, that recurs every fifth day.

quintan fever or **quintana fever** see TRENCH FEVER.

quintant *Navigation.* a double-reflecting, angle-measuring instrument that has an arc 144° in length. In modern usage, all such instruments are referred to as "sextants," regardless of the length of the arc.

quintic *Mathematics.* any expression of degree five. A quintic equation is any polynomial equation of degree five (not necessarily homogeneous).

quintillion *Mathematics.* **1.** in American usage, the number 10^{18}, or 1,000,000,000,000,000,000. **2.** in British and German usage, the number 10^{30}.

quintuplet [kwin tup´lit; kwin toop´lit] *Biology.* one of five children who are born at one birth, usually arising from multiple ovulation and statistically very rare unless ovarian follicle stimulants have been used in the treatment of infertility.

quira see MACAWOOD.

quirk *Building Engineering.* **1.** a narrow groove adjacent to a bead or molding and sunk flush into the face of the work. **2.** a V-shaped groove in the finishing coat of plaster where it abuts the return on a door or window.

quirk bead *Building Engineering.* **1.** a bead with a quirk on one side, as on the edge of a board. **2.** a bead that lies flush with the adjacent surface and that is separated from it by a quirk on each side. **3.** a corner bead with the quirks at either side at right angles to each other. **4.** a bead with a quirk on its face.

quit *Computer Programming.* to exit the program that is currently in use.

Q unit *Thermodynamics.* 10^{18} British thermal units or 1.055×10^{21} joules; a unit of heat energy commonly applied to fuel reserves.

quoin *Building Engineering.* one of the members forming a solid exterior angle or corner of a building. *Graphic Arts.* a wedge made of wood or iron used to secure type in a chase.

quoin post *Civil Engineering.* a heel post.

quotation *Artificial Intelligence.* see QUOTE.

quote *Artificial Intelligence.* **1.** in Lisp, a pseudofunction indicating that a symbol itself, rather than the value of the symbol, is to be used as a function argument. **2.** to use a single-quote character as an abbreviation for this. **3.** to specify a function argument using this. Also, QUOTATION.

Quotid. daily. (From Latin *quotidie.*)

quotient [kwō´shənt] *Mathematics.* **1.** the result of performing the arithmetic operation of division; in particular, the quantity $q = a/b$, where a and b are given quantities and $b \neq 0$. **2.** a quotient field, group, ring, etc.

quotient field *Mathematics.* the smallest field containing a given integral domain R; constructed by forming the ring of quotients of R by the set S of its nonzero elements. For example, the quotient field of the integers is the field of rational numbers.

quotient group *Mathematics.* let N be a normal subgroup of a group G. The collection of all the distinct left (or right) cosets of N in G is called the quotient group of G by N (or G mod N); denoted G/N (or $N\backslash G$). The group operation in G/N is given by $(aN)(bN) = (ab)N$, where the usual rules for cosets are observed. Also, FACTOR GROUP.

quotient ring *Mathematics.* **1.** let U be an ideal of a ring R. The collection of all the distinct cosets of U in R, obtained by considering U as an additive subgroup of R, is called the quotient ring of R by U (or R mod U); denoted R/U. The ring operations in R/U are given by $(rU) + (sU) = (r + s)U$ and $(rU)\,(sU) = (rsU)$, where the usual rules for cosets are observed. Also, FACTOR RING, RESIDUE CLASS RING. **2.** a ring of quotients.

quotient rule *Mathematics.* if f and g are differentiable functions of a real variable, then their quotient f/g is also differentiable, with the derivative given by the quotient rule, at each x such that $g(x) \neq 0$:

$$[f(x)/g(x)]' = (f'(x)g(x) - g'(x)f(x))/(g(x))^2.$$

quotient set *Mathematics.* if r is an equivalence relation on a set X, then the collection of equivalence classes into which X is partitioned by r is the quotient set of X; denoted X/r and read "X modulo r." Also, **quotient class of X.**

quotient space *Mathematics.* **1.** let M be a closed linear subspace of a normed linear space E. The quotient space E/M is the set of objects of the form $x + M$, where $x \in E$. Linear operations are defined by: $(x + M) + (y + M) = (x + y) + M$ and $a(x + M) = (ax) + M$. The **quotient norm** on E/M is defined by the rule: $\|x + M\| = \inf\{\|x + m\| : m \in M\}$; it can be shown to be a norm on E/M. In addition, if E is a Banach space, then so is E/M, and if M and E/M are both complete, then so is E. **2.** a topological space on which a quotient topology has been defined.

quotient topology *Mathematics.* let $f\colon X \to Y$ be a map from a topological space X onto the set Y. The quotient topology on Y is defined by specifying that a subset U of Y is open if and only if the preimage of U in X is open; that is, $f^{-1}(U)$ is an open set in the topology on X. f is called the **quotient map**, and the quotient topology is the weakest topology for which the quotient map is continuous.

q.v. **1.** which see; used to refer from one part of a text to another. (From Latin *quod vide.*) **2.** as much as you please. (From Latin *quantum vis.*)

Q-value *Nuclear Physics.* a measure of the energy balance in a given system or process, so that for nuclear decay it represents the amount of energy released; in a nuclear reaction it is the difference in total kinetic energy between the particles released and the ones captured.

qwerty [kwer´tē] *Computer Technology.* the usual arrangement of letters for typewriters and computer keyboards, in which Q, W, E, R, T, and Y are the first six letters from left to right in the top row of letter keys.

Qy. or **qy.** query.

R rare earth; ratio; roentgen; Rydberg constant.
R *Behavior*. response.
R *Chemistry*. a symbol for an organic hydrocarbon radical.
r radius.
r or **R** Rankine scale; Réaumur; resistance.
r *Statistics*. the symbol for the correlation coefficient.
R.A. right ascension.
Ra the chemical symbol for radium.
ra range.
RAA remote-access admittance.
Raabe's convergence test [räb´əz] *Mathematics*. let $\{x_n\}$ be a sequence of nonzero elements of R^p. If there are a real number $a > 1$ and a natural number N such that $|x_{n+1}|/|x_n| \leq 1 - a/n - 1/n$ for $n \geq N$, then the series $\sum_{n=1}^{\infty} x_n$ is absolutely convergent. If there are a real number $a \leq 1$ and a natural number N such that $|x_{n+1}|/|x_n| \geq 1 - a/n$ for $n \geq N$, then the series $\sum_{n=1}^{\infty} x_n$ is not absolutely convergent.
rabal *Meteorology*. a method of determining the speed and direction of winds in the atmosphere above a weather station, using a theodolite for the visual tracking of a radiosonde balloon.
rabbet *Engineering*. **1.** a shoulder or recess cut into the edge of a piece of wood; used to receive the edge of a similar piece. **2.** to cut a groove or a channel into a part. **3.** a strip added to a part that forms a stop or a seal. **4.** a joint formed by fitting two members together.
rabbit *Vertebrate Zoology*. any of numerous members of the family Leporidae in the order Lagomorpha, distinguished by long hindlimbs, long ear pinnae, rapid reproduction capabilities, and naked young; found worldwide; a widespread North American species is the **Eastern cottontail,** *Sylvilagus floridanus*.
rabbit any of various devices or objects thought of as analogous to the common animal, as in their motion; specific uses include: *Nucleonics*. a small container that is hydraulically or pneumatically driven through a tube traversing the core of a nuclear reactor; used to expose specimens to radiation or neutron flux. *Petroleum Engineering*. a small plug driven through flow lines to clear them or check for obstructions.
rabbit ears *Electronics*. a small, V-shaped indoor television antenna consisting of a pair of telescoping aerials. *Metallurgy*. a V-shaped die used in metal working when folds or wrinkles are desirable.
rabbit fever see TULAREMIA.
rabble *Metallurgy*. a hoelike tool used to stir molten metals or alloys.
rabbling *Metallurgy*. the process of stirring molten metal, ore, or other such material, using a rabble to remove slag from the surface and draw the material into a ball.
Rabi, Isador Isaac [rä´bē] 1898–1988, American physicist; awarded the Nobel Prize for measuring the magnetic movements of disturbed atomic nuclei.
rabid [rab´id] *Virology*. affected with rabies.
rabies [rā´bēz] *Pathology*. an acute, usually fatal disease, a form of viral encephalitis, that is transmitted through the saliva of an infected animal (usually a dog, cat, skunk, or raccoon); the symptoms include paresthesia and a burning sensation, hyperexcitability, agitation, delirium, hallucinations, hypersalivation, fear of fluids, convulsions, paralysis, and coma. Also, HYDROPHOBIA.
rabies virus *Virology*. a virus of the genus *Lyssavirus* in the family Rhabdoviridae that can infect most mammals and that causes the disease rabies; transmitted in the saliva of a rabid animal.
Racah coefficient *Quantum Mechanics*. a coefficient used in the coupling of angular momenta.
raccoon *Vertebrate Zoology*. any of various omnivorous, nocturnal, arboreal North American carnivores in the genus *Procyon* of the mammalian family Procyonicae, especially the **northern raccoon,** *P. lotor,* which has a black band across the forehead and eyes and a bushy ringed tail and is found throughout most of North America.

raccoon

race *Biology*. an interbreeding subgroup of a species whose individuals are geographically, physiologically, or chromosomally distinct from other members of the species. *Anthropology*. **1.** a geographical variation in the human population, identified by a range of genetic characteristics such as hair and skin color, eye color and shape, facial features, body build, and blood group; now a disputed term that is not regarded as technically precise. **2.** any of three such population groups into which humans have traditionally been classified; i.e., Caucasoid, Mongoloid, or Negroid. *Oceanography*. a very rapid current, or the narrow channel through which it flows. *Civil Engineering*. a channel that conducts water to or from the place where it performs work. *Design Engineering*. either of a concentric pair of steel rings in a ball or roller bearing, designated the **inner race** and **outer race**.
race condition *Electricity*. in a control circuit, an ambiguous condition occurring when one flip-flop changes before a second has had sufficient time to latch.
racemase *Enzymology*. an enzyme that catalyzes the racemization of an optically active substance such as lactic acid.
racemate *Chemistry*. a mixture of dextrorotatory and levorotatory isomers of the same compound in equal amounts; each of the isomer's rotations cancels the other, resulting in a mixture that does not rotate plane polarized light either left or right, i.e., a mixture that does not show optical activity. Also, **racemic mixture.**
raceme *Botany*. an inflorescence in which the individual flowers are borne on stalks that spring from a long central stem. Thus, **racemic.** *Chemistry*. see RACEMATE.

racemic acid *Organic Chemistry.* $C_2H_4O_2(COOH)_2 \cdot HOH$, a colorless crystalline solid; soluble in water and alcohol; melts at 206°C; used in organic synthesis. Also, INACTIVE TARTARIC ACID.

racemization *Chemistry.* the transformation of optically active substances into optically inactive substances or mixtures. Also, **racemation.**

racemose *Botany.* relating to, characteristic of, or arranged in racemes. *Anatomy.* any anatomical structure that is divided into clusters resembling a bunch of grapes on its stem.

racetrack *Nucleonics.* a system of several isotope separators that are assembled in the form of an oval racetrack and that operate on a common magnetic field.

raceway *Electricity.* a term for a channel that is designed and used solely for holding wires, cables, or bus bars.

rachi- or **rachio-** a combining form meaning "spine." Also, RHACHI- or RHACHIO-.

Rachiglossa *Invertebrate Zoology.* a group of predatory snails, gastropod mollusks in the suborder Pectinibranchia, having a rachiglossate radula and often large, ornate shells.

rachiglossate radula *Invertebrate Zoology.* in carnivorous marine snails, a modified feeding organ (radula) having up to three cone-shaped teeth in each row, contained within an extendable proboscis.

rachiochysis *Medicine.* an effusion or leakage of fluid within the spinal column.

rachiomyelitis *Medicine.* an inflammation of the spinal cord structures.

rachis *Botany.* any of various axial structures, especially the axis of a raceme or other elongated inflorescence. *Anatomy.* another term for the spinal column.

rachischisis *Medicine.* any of several types of spina bifida, in which the spinal meninges and spinal column are exposed on the surface.

rachitis see RICKETS.

Rachycentridae *Vertebrate Zoology.* the cobia, a monotypic family of game and food fish of the order Perciformes; characterized by a long, pointed head with a projecting lower jaw and three dark stripes on the side of the body; found in the Atlantic and Indo-West Pacific Oceans.

racial group *Anthropology.* see RACE.

racial unconscious see COLLECTIVE UNCONSCIOUS.

raciation *Evolution.* the process of evolutionary change that produces slight morphological or phenotypic variations, or new races, within the population of an existing species.

racism *Psychology.* the attitude or belief that one's own race or ethnic group is superior to another or others, or that some other race or group is inferior to others.

racist *Psychology.* **1.** a person who has an attitude of racism. **2.** relating to or showing an attitude of racism.

rack any of numerous types of metal or wooden frameworks; specific uses include: *Mechanical Engineering.* a straight length of toothed gearing designed to mesh with the teeth of a pinion. *Aviation.* a framework used for the placement of avionics equipment within an aircraft. *Civil Engineering.* a meshed metal barrier used to stop waterborne garbage from entering a waterway. *Mining Engineering.* an inclined trough used to wash and separate ore. *Mechanical Devices.* see RELAY RACK.

rack and pinion *Mechanical Engineering.* a gear arrangement for converting rotary motion into linear motion, or vice versa, consisting of a straight, toothed bar (the rack) that meshes with a small gear wheel (the pinion).

rack-and-pinion steering *Mechanical Engineering.* a vehicle steering mechanism in which a pinion gear at the end of the steering column engages with a rack, moving it to either the right or the left in order to transmit steering movements.

racket mycelium *Mycology.* in fungi that are skin pathogens, hyphal masses whose tips resemble swellings and thus look like tennis rackets. Also, RACQUET MYCELIUM.

racking *Civil Engineering.* a stepped finish used between progressive layers of an unfinished masonry wall. *Metallurgy.* in electroplating, the process of placing parts on a conductive rack. *Petroleum Engineering.* the orderly placement of pipe in the derrick after the pipe has been hoisted from the wellbore during oil drilling.

rack railway *Civil Engineering.* a railway used in mountain regions, having an intermediate cogged rail that is set between the running rails and that engages with cogwheels on the locomotive. Thus, **rack rail, rack locomotive.**

rack saw *Mechanical Devices.* any of various wide-toothed saws.

rack spur *Mechanical Devices.* a spur gear with an infinitely large pitch diameter.

racon *Navigation.* a radar beacon that, when triggered by a craft's radar, returns a distinctive signal giving both bearing and range information.

Racopilaceae *Botany.* a family of slender dark-green mosses of the order Hypnobryales that form flat mats on tree bases, moist soil, and stones; characterized by creeping stems with irregular to subpinnate branching and dimorphous leaves occurring in three apparent rows; native to the tropics and the Southern Hemisphere.

racquet mycelium see RACKET MYCELIUM.

rad *Nucleonics.* a unit of an absorbed dose of radiation equivalent to 100 ergs per gram. (An acronym for radiation absorbed dose.)

rad or **rad.** radian; radical; radius; radix.

radappertization *Food Technology.* the sterilization of food by irradiation in doses sufficient to reduce the number of organisms below detectable levels.

radar *Engineering.* a system that uses reflected electromagnetic radiation to determine the velocity and location of a targeted object; widely used in such applications as aircraft and ship navigation, military reconnaissance, automobile speed checks, and weather observations. *Military Science.* of or relating to military techniques that involve the use of such a system. Thus, **radar bombing, radar reconnaissance, radar intelligence**, and so on. (An acronym for radio detection and ranging.)

radar-absorbing material *Materials.* material that reduces the intensity of a radar signal by attenuating the return echo of the radar waves.

radar altimeter *Navigation.* a device that determines the height of an aircraft above the ground by downward-looking radar.

radar antenna *Electromagnetism.* an antenna system that radiates a concentrated directional beam and also receives the reflected signal after echoing from a target.

radar antenna

radar antijamming *Electronics.* the measures taken to prevent an enemy from disrupting a radar system.

radar approach *Aviation.* an aircraft instrument approach procedure that uses radar, especially precision-approach radar.

radar approach control *Transportation Engineering.* a control system that employs precision-approach radar information to provide speed, course, and altitude for the positive control of aircraft approaching an airport.

radar astronomy *Astronomy.* the branch of astronomy that utilizes radar techniques to map the surfaces of planetary and celestial bodies such as the moon, the asteroids, and the planet Venus, for the purpose of determining topography, surface properties, and periods of rotation.

radar attenuation *Electromagnetism.* the ratio of transmitted power to received power in a radar beam.

radar beacon *Navigation.* a navigational aid that consists of a transmitter operating in the radar-frequency band. There are two general types: *racon,* which gives both bearing and range information, and *ramark,* which gives bearing information only.

radar beam *Electromagnetism.* a beam of radio-frequency energy in space bounded by the locus of points where a radar signal is decreased to one-half of the value about the center of the beam.

radar bearing *Navigation.* a bearing that has been obtained by radar.

radar bombsight *Ordnance.* an airborne radar system that determines a target's relative velocity and position, and releases the bombs.

radar camera *Optics.* a camera that is designed to photograph images appearing on a radarscope. Also, **radarscope camera.**

radar camouflage *Military Science.* the use of radar-absorbent or radar-reflecting materials to alter the radar-echoing characteristics of an object's surface.

radar cell *Electromagnetism.* a volume in space bounded by a radial length of one radar pulse by one radar beam width.

radar climatology *Meteorology.* the analysis of statistics in time and space of radar weather echoes.

radar command guidance *Military Science.* a missile guidance radar system that determines missile and target positions, computes flight path corrections, and transmits the corrections to the missile.

radar conspicuous object *Electromagnetism.* a term for any object that returns a strong radar echo.

radar constant *Electronics.* any radar function to which equations are applied, such as beam width, pulse length, or noise level.

radar contact *Engineering.* the perception and identification of an echo on a radar screen.

radar controller *Navigation.* any controller on the ground who uses radar information to give directions to aircraft, such as an air controller (tactical military) or air traffic controller (civilian).

radar countermeasure *Military Science.* any technique used to interfere with an enemy's radar system, such as jamming, interception, evasion, or deception.

radar coverage *Electronics.* the range within which a radar system can detect an object.

radar-coverage indicator *Transportation Engineering.* a device that determines the distance for which an aircraft should be tracked, and provides a measure of station range for quality control.

radar cross section see ECHO AREA.

radar danning *Military Science.* in naval mine warfare, a method of navigating by using radar to maintain a safe distance from a line of dan buoys.

radar data filtering *Electronics.* a program that directs a computer to delete certain radar data and alert the crew to the deletion.

radar deception *Military Science.* see ELECTRONIC WARFARE.

radar decoy *Ordnance.* an object that is used to deceive enemy radar; it has the same or similar radar reflective characteristics as the target.

radar display *Electronics.* the representation of targets on a radar screen, usually produced on a cathode-ray tube screen. Also, RADAR PRESENTATION.

radar dome see RADOME.

radar echo *Electronics.* a blip on a radar screen that indicates the presence of an object or a weather phenomenon.

radar equation *Electromagnetism.* an equation relating the transmitted power, received power, and radar antenna gains to the range and area of a target.

radar fence *Space Technology.* see FENCE.

radar fire *Military Science.* gunfire directed at a target that is being tracked by radar.

radar fix *Navigation.* a position determined by radar, by itself or in conjunction with other information.

radar-frequency band *Electromagnetism.* the range of frequencies of microwave radiation in which radar operates.

radargrammetry *Photogrammetry.* the theories and techniques of obtaining reliable measurements by means of radar rather than by visible light waves.

radar homing *Military Science.* the use of a missile-borne radar guidance system to guide a missile to a target.

radar horizon *Navigation.* the point at which a radar beam is tangent to the surface of the earth. Due to refraction of the radar rays in the atmosphere, this is about 15% greater than the distance to the visible horizon.

radar intelligence item *Electronics.* an object that is picked up by radar, but that cannot be identified at the moment of its appearance.

radar jamming *Military Science.* a technique in which electromagnetic radiation is used to disable an enemy's radar system.

radar marker *Engineering.* a fixed object that continuously emits a radar signal, and that is used as a reference point on a radar display.

radar meteorological observation *Meteorology.* a study of radar echoes in terms of orientation, coverage, intensity, height, movement, and unique characteristics, to be utilized in the determination of severe weather, such as hurricanes, tornadoes, and thunderstorms.

radar meteorology *Meteorology.* a branch of meteorology that studies the scattering of radio radar waves from all types of atmospheric phenomena.

radar mile *Electromagnetism.* the time required for a radar signal to travel one statute mile to a target and return: 10.75 microseconds.

radar nautical mile *Electromagnetism.* the time required for a radar signal to travel one nautical mile to a target and return: 12.37 microseconds.

radar netting *Engineering.* a network of several radar stations relaying to a single station to furnish integrated target information.

radar picket *Military Science.* a mobile radar unit placed at a distance from a protected area to furnish an early-warning system and to increase radar range.

radar presentation see RADAR DISPLAY.

radar pulse *Electromagnetism.* a short duration of radio-frequency energy, typically in the range of microseconds, emitted by a radar transmitter.

radar range *Electromagnetism.* the maximum distance for which a radar system is effectively capable of detecting objects.

radar-range equation *Electromagnetism.* an equation that relates the range of a radar system to variables such as the radar antenna gain, transmitted power, target cross-sectional area, and minimum detectable signal.

radar receiver *Electronics.* a device that converts signals reflected from a radar target into a form that can be displayed on a radarscope.

radar receiver-transmitter *Electronics.* a component that produces the energy beamed at a radar target and converts the signal reflected back from the target for display on the radarscope.

radar reflection *Electromagnetism.* the reflection of a radar signal off an object; the portion of the transmitted signal that reflects from the cross-sectional area of the target.

radar reflectivity *Electromagnetism.* a ratio of the reflected radar electromagnetic intensity from a target to the transmitted intensity.

radar reflector *Electromagnetism.* a reflector mounted on a target object and used intentionally to reflect radar signals back to the radar receiver, or to deflect the signals away from the receiver.

radar relay *Electronics.* a radar system that transmits a video from the tracking station to a distant receiver unit.

radar repeater *Electronics.* a device that reproduces signals displayed by a radarscope on screens at remote locations.

radar return *Navigation.* a portion of a radar signal that is reflected back to the transmitting antenna. Also, ECHO.

radar scanning *Engineering.* the process of passing a radar beam through space in a specified pattern in an effort to locate a target.

radarscope *Electronics.* a cathode-ray tube that displays signals reflected back from a radar target.

radarscope camera see RADAR CAMERA.

radarscope overlay *Engineering.* a transparent overlay placed on a radar screen for comparison and identification purposes.

radar set *Engineering.* a complete assembly of radar equipment, including transmitter, antenna, receiver, and indicator.

radar shadow *Electromagnetism.* an area that is blind to radar because of an obstruction such as a reflective target or an absorbing medium.

radar silence *Military Science.* a period in which the radar transmission of electronic signals is prohibited on some or all frequencies.

radarsonde *Engineering.* **1.** a system for measuring and transmitting meteorological data from a weather balloon triggered from a secondary radar source. **2.** a radar system that measures the range, elevation, and azimuth of radar placed in an airborne radiosonde.

radar station *Engineering.* any site from which a radar set operates.

radar storm detection *Meteorology.* the use of radar in determining stormy conditions as a result of their radar-detectable characteristics, such as liquid or frozen water drops that reflect radar echoes.

radar surveying *Cartography.* an airborne surveying process in which radar is used to measure the distance between two ground radio beacons placed along a baseline in difficult or hard-to-reach terrain.

radar target *Electromagnetism.* any object that is detected and tracked on a radar screen.

radar telescope *Electromagnetism.* a large radar antenna used to receive electromagnetic radiation from beyond the earth's atmosphere.

radar threshold limit *Electromagnetism.* a point in space relative to the focal point of an antenna, at which an initial detection limit criterion can be satisfied.

radar tracking *Engineering.* the process of following the position and velocity of a targeted object with a radar system.

radar transmitter *Electronics.* the section of a radar system in which radio-frequency energy is generated and keyed.

radar transponder see RADAR BEACON.

radar triangulation *Engineering.* the process of locating a target using two separate radar stations to form a triangle with the target.

radar upper band see UPPER BRIGHT BAND.

radar volume *Electromagnetism.* the volumetric range of a radar that is capable of scanning in three dimensions.

radar weather observation see RADAR METEOROLOGICAL OBSERVATION.

radar wind *Meteorology.* the wind speed and direction as determined by a radar tracking device (such as a balloon, radio transmitter, or radar reflector) that discloses information about the winds as the radar tracking process continues on the ground.

radar wind system *Engineering.* a radar system that measures the range, elevation, and azimuth of a balloon-borne object, giving an indirect measure of air-wind data.

Rad CM radar countermeasure.

radechon *Electronics.* a tube that stores electrical charges on a sheet of mica placed between a continuous metal backing plate and a fine-mesh screen; commonly used in communications systems to improve the signal-to-noise ratio, to carry out signal comparison, or to convert a signal's time base. Also, BARRIER-GRID STORAGE TUBE.

radectomy *Medicine.* the partial or complete excision of the root of a tooth.

Rademacher function *Mathematics.* let X be the closed unit interval, i.e., [0,1], with Lebesgue measure. For every positive integer n, define the Rademacher function by $f_n(x) = +1$ or $f_n(x) = -1$, according to whether k is odd or even, where k is the integer for which $(k-1)/2^n \leq x < k/2^n$.

Radfoam process *Textiles.* in nonwoven fabric production, a wet-laid process for handling long-fibered stocks in which the fibers are suspended in foam rather than water; this yields fabrics with high levels of bulk, softness, and porosity.

radiability *Radiology.* the property of being penetrated by radiation, especially roentgen rays.

radiac *Nucleonics.* the process of detecting, measuring, and analyzing nuclear radiation. (An acronym for radioactivity detection, identification, and computation.)

radiac set *Nucleonics.* a complete system capable of detecting, identifying, and computing the intensities of nuclear radiation. Also, **radiac instrument, radiacmeter.**

radiad *Medicine.* toward a radius or radial side.

radial *Science.* **1.** of or relating to a radius. **2.** radiating from a central point. *Anatomy.* relating to a radius of the body, such as the radius of the forearm or the radial aspect of the arm. *Design Engineering.* see RADIAL TIRE. *Navigation.* a magnetic bearing from a vortac radio beacon. *Photogrammetry.* in aerial photography, any line or direction from the radial center of the photograph to any other given point. The radial center is presumed to be the principal point of the photograph unless otherwise stated.

radial acceleration see CENTRIPETAL ACCELERATION.

radial aerotriangulation *Cartography.* an aerotriangulation in which horizontal control extension is accomplished by a combination of resection and intersection using directions of images from the radial centers of overlapping photographs.

radial artery *Anatomy.* an artery in the forearm, arising from the brachial artery and supplying the lateral forearm, wrist, and hand.

radial aspect *Anatomy.* the lateral aspect of the arm, as distinguished from the medial or ulnar aspect.

radial assumption *Cartography.* an assumption made by the viewer of an aerial photograph containing both tilt displacement and relief displacement regarding its radial center: if relief is the major consideration of the viewer, the photograph nadir point should be used as the radial center; if tilt is the major consideration, the isocenter should be used.

radial band pressure *Ordnance.* the pressure exerted on the rotating band by the walls of the gun tube, and hence against the projectile wall at the band seat, due to the engraving of the band by the gun rifling.

radial-beam tube *Electronics.* a tube in which an externally applied magnetic field rotates an electron beam around the axis of the tube; used for high-speed switching.

radial bearing *Mechanical Engineering.* a bearing in which the movement of the load transmitted is radial to the axis of rotation.

radial canal *Invertebrate Zoology.* **1.** in medusae, digestive canals that extend from the stomach to the ring canal at the margin of the umbrella. **2.** in sponges, any of numerous canals lined with choanocytes between the main cavity and the surface. **3.** in a starfish's water-vascular system, a canal extending from the central oral ring canal to the tip of each arm.

radial cell *Cell Biology.* a type of glial cell, found especially in the developing nervous system, that is thought to guide the movements of neurons to their ultimate destinations.

radial center *Photogrammetry.* the point on an aerial photograph from which the radials or directions are drawn or measured to the images of the features on the photograph; depending on the type of information required, the radial center may be the principal point, isocenter, or photograph nadir.

radial chromatography *Analytical Chemistry.* a separation technique involving a round disk of absorbent paper with a strip cut from the edge to the center that is used as a wick to dip into a solvent; the solvent climbs the wick, touches the sample, and elutes in concentric rings.

radial cleavage *Developmental Biology.* embryonic cleavage in which all divisions are either parallel or at right angles to the longitudinal axis of a fertilized egg.

radial distortion *Photogrammetry.* the linear displacement of image points toward or away from the center of the image field, resulting from the fact that objects at different angular distances from the camera lens undergo different magnifications.

radial distribution *Mathematics.* let $f(r, \theta, \phi)$ be a real-valued function of spherical coordinates. The radial distribution of f is the function $F(r)$ equal to the average of the values of f on the surface of the sphere Σ of radius r, centered at the origin; i.e.,

$$F(r) = 1/(4\pi r^2) \int_\Sigma f(r, \theta, \phi)\, r \sin\theta \, d\theta d\phi.$$

radial distribution function *Physical Chemistry.* a function that plots the density, relative to the average density, of molecules at a given distance from some particular molecule; used to describe the molecular structure of liquids.

radial Doppler effect *Electromagnetism.* the part of the optical Doppler effect that is dependent on the direction of the relative velocity between the source and the observer.

radial drainage pattern *Hydrology.* a pattern of natural stream courses in which the streams diverge outward from a high central area. Also, CENTRIFUGAL DRAINAGE PATTERN.

radial draw forming *Mechanical Engineering.* a process for shaping metals by gradually applying tangential stretch and radial compression simultaneously.

radial drill *Mechanical Engineering.* a drilling machine in which the drill head is encased in a toolhead and saddle and moves along a horizontally projecting arm that rotates around a vertical pillar.

radial drilling *Engineering.* the process of drilling several holes in a single plane radiating from a common point.

radial engine *Mechanical Engineering.* an internal-combustion engine in which the cylinders are arranged like spokes of a wheel at regular intervals around the crankshaft; used formerly as an air-cooled aircraft engine.

radial error *Ordnance.* the distance between the actual point of impact and the desired point of impact, measured on a theoretical plane perpendicular to the flight of the projectile.

radial faults *Geology.* a system of faults that diverge outward from a central point.

radial flow *Engineering.* **1.** a working fluid flowing primarily along the radii of rotation. **2.** a spokelike flow of reservoir hydrocarbons toward a focal area.

radial force *Mechanical Engineering.* in machining, a force that is exerted on a cutting tool in a direction opposite to that of the depth of cut.

radial forging *Metallurgy.* a metalworking process in which a workpiece is fed into and through reciprocating dies whose closure is controlled to produce the desired product contour; depending on the number of dies and the rotation, the contour may be circular, square, rectangular, hexagonal, and so on.

radial grating *Electromagnetism.* a grating of wires arranged radially on a circular frame, used in a circular waveguide to allow transverse electric waves to pass while blocking the zero-order transverse magnetic waves.

radial keratotomy *Surgery.* an operation to correct myopia in which a series of incisions is made in the cornea from its outer edge toward its center in spokelike (radial) fashion.

radial lead *Electricity.* a lead or pigtail that is attached perpendicularly to the axis of a component such as a resistor or capacitor, rather than axially from an end.

radial load *Mechanical Engineering.* a load that is oriented perpendicularly to the bearing axis.

radial locating *Mechanical Engineering.* one of three locating problems (the others being concentric locating and plane locating) used to maintain an efficient working relationship between a workpiece, a cutter, and the body of a machine tool.
radial motion *Mechanics.* a motion away from or toward a central axis or point. *Astronomy.* see RADIAL VELOCITY.
radial nerve *Anatomy.* a branch of the brachial plexus that innervates skin and muscles on the dorsal side of the arm.
radial network *Transportation Engineering.* a system in which most lines pass through a central hub.
radial of a curve *Mathematics.* let γ be a space curve and O a fixed point not on γ. For each point P of γ, draw the line segment having O as one endpoint and having length and orientation equal to the radius of curvature of γ at P. The locus of endpoints (other than O) of such line segments is the radial of γ with respect to O.
radial rake *Mechanical Engineering.* the angle between the cutter tooth surface and a radial line that passes through the cutting edge in a plane that is perpendicular to the cutter axis.
radial saw *Mechanical Engineering.* a circular power saw in which the blade is suspended from a traverse head that is mounted on a rotatable arm.
radial search *Navigation.* a search conducted outward along one or more radial vectors from a given point.
radial selector *Engineering.* a device that can be set to any desired omnibearing to control a course-line-deviation indicator.
radial stress *Mechanics.* the component of stress toward or away from the central axis of a curved member.
radial symmetry *Science.* a configuration of a structure in which similar parts radiate from the central axis in a regular pattern and in which the object or organism can be divided into two similar halves along any of several planes that pass through its center. *Biology.* specifically, such a pattern as the basic body plane of an organism.
radial tire *Design Engineering.* an automotive tire whose cords run across its width while an extra layer of fabric is laid up around the circumference between the plies and the thread. Also, **radial-ply tire.**

radial tire

radial velocity *Mechanics.* the component of velocity away from or toward a central axis or point. *Astronomy.* specifically, a star's relative velocity directly toward or away from the earth, as determined by measuring the Doppler shifts of its spectral lines.
radial wave equation *Physics.* the second-order ordinary differential equation for the radial dependence of a wavefunction, which results from applying the separation of variables technique to the solution of a two-dimensional or three-dimensional partial differential wave equation, such as the Schrodinger and Helmholtz equations.
radian *Mathematics.* a unit of angular measurement: one revolution = 2π radians = 360°. One radian = $(360/2\pi)°$. It measures an arc of length equal to the radius of the circle.
radiance *Optics.* the flux density or radiant energy per unit area of a radiating surface. Also, STERADIANCY.
radian frequency see ANGULAR FREQUENCY.
radian length *Physics.* for a sinusoidally varying quantity, the distance given by the wavelength divided by 2π, thus giving the distance spanned by the angle of one radian.
radiant *Physics.* **1.** emitting radiation or heat. **2.** transmitted by radiation. *Optics.* transmitting rays of light. *Astronomy.* a point on the celestial sphere from which a shower of parallel-moving meteors appears to radiate outward due to perspective effects.
radiant density *Physics.* the instantaneous energy density (energy per unit volume) in a medium through which electromagnetic energy is passing.
radiant efficiency *Optics.* the ratio of the radiant flux emitted by a source to the power consumed by that source.
radiant emittance *Electromagnetism.* the radiance flux per unit area emerging from a radiant surface. Also, **radiancy, radiant exitance.**
radiant energy *Physics.* see RADIATION.
radiant exposure *Optics.* the intensity of light incident on a surface integrated over the time of exposure. Also, EXPOSURE.
radiant flux *Optics.* the rate of a flow of radiant energy, usually measured in watts per hour.
radiant flux density *Electromagnetism.* the electromagnetic radiation power per unit area flowing across a boundary.
radiant heating *Mechanical Engineering.* **1.** a heating system that radiates heat from a surface to its surroundings. **2.** any of various means of heating objects or persons by radiation without heating the intervening air.
radiant intensity *Electromagnetism.* the electromagnetic radiation power per unit solid angle about the direction under consideration; expressed in units of watts per steradian.
radiant power *Electromagnetism.* **1.** the time-rate of electromagnetic energy passage across a unit surface. **2.** the rate at which electromagnetic energy is emitted from a source.
radiant quantities *Optics.* those physical quantities, such as radiance, that can be measured and that are not influenced by the response of the human eye.
radiant reflectance *Electromagnetism.* the ratio of electromagnetic power reflected from an object to that which is incident on the object.
radiant superheater *Mechanical Engineering.* a part of a steam-generating unit in which steam is heated above its saturation temperature by the transfer of radiant heat from combustion products to the steam.
radiant transmittance *Electromagnetism.* the ratio of electromagnetic power transmitted through an object to that which is incident on the object.
radiant-type boiler *Mechanical Engineering.* a water-tube boiler in which the tubes are arranged around the outer surface of the combustion area of the furnace.
Radiata *Invertebrate Zoology.* in some classifications, a major grouping of radially symmetrical animals including Porifera, Coelenterata, and Ctenophora.
radiate *Physics.* to give off rays, as of light or heat; emit radiation.
radiated interference *Telecommunications.* interference that is transmitted through the atmosphere, based on the laws of electromagnetic wave propagation.
radiated power *Electromagnetism.* the total electromagnetic power radiated by a transmitting antenna.
radiating curtain *Electromagnetism.* a vertical plane of dipole elements, each positioned so as to reinforce the other's signal and placed a quarter of a wavelength in front of a vertical reflecting curtain.
radiating element *Electromagnetism.* the basic component of a transmitting antenna that radiates and receives radio-frequency energy.
radiating guide *Electromagnetism.* a waveguide that can radiate microwave energy into open space via slots, horns, gaps, and so on.
radiating power see EMISSIVE POWER.
radiating scattering *Physics.* the scattering of radiation through the interaction with matter that alters the course of the original path.
radiation *Physics.* **1.** the emission of waves or particles from a source and the propagation of these waves or particles through a medium; when used without a modifier, the term usually refers to electromagnetic radiation. **2.** by extension, a stream of particles such as electrons, neutrons, protons, or alpha particles. **3.** the complete process by which waves or particles are emitted, pass through a medium, and are absorbed by another body. **4.** the energy transferred by such processes. *Anatomy.* a structure made up of divergent elements, such as one of the fiber tracts in the brain. *Evolution.* the process by which a group of species diverge from a common ancestral form, resulting in an overall increase in biological diversity. *Cartography.* in surveying, the process of locating the position of an unknown point by determining its direction (either the azimuth or bearing) and distance from a known point. (From a Latin word meaning "ray" or "beam.")

21-cm radiation *Astronomy.* the 1420-megahertz radio signal emitted by neutral hydrogen in its hyperfine transition; the radiation is used to map the location of hydrogen gas in galaxies.

radiation accident *Nucleonics.* any incident in which radioactivity unintentionally escapes to expose individuals or the environment.

radiational cooling *Meteorology.* the rapid dissipation of heat of the earth's surface and adjacent air that begins just after dusk and whenever the earth's surface suffers a net loss of heat because of terrestrial radiation; the mechanism by which a dramatic temperature fall occurs in valley areas surrounding hills or mountains.

radiational inversion *Meteorology.* a land surface inversion resulting from rapid radiational cooling of air just above ground level, usually occurring on cold winter nights.

radiation angle *Electromagnetism.* an angle measured with respect to a horizontal or vertical reference indicating the direction of electromagnetic energy propagation from a transmitting antenna.

radiation area *Nucleonics.* a region where the radiation dose is at least five millirem over a period of one hour.

radiation barrier *Radiology.* any protective barrier against ionizing radiation.

radiation biochemistry *Biochemistry.* the study of the effects of ionizing radiation on living things; it includes the study of substances having the ability to protect cells and body tissue against the harmful effects of ionizing radiation.

radiation biology see RADIOBIOLOGY.

radiation-bounded nebula *Astronomy.* an emission nebula whose central star is not hot enough to ionize the entire cloud of gas.

radiation budget *Geophysics.* the amount of radiation accepted and emitted from a given area within a given time.

radiation burn *Medicine.* a burn that results from excessive exposure to X rays, radium, sunlight, or other radiant energy.

radiation catalysis *Chemistry.* an acceleration of the rate of a chemical reaction through the application of energy in the form of electromagnetic waves.

radiation cataract see IRRADIATION CATARACT.

radiation characteristic *Telecommunications.* any identifying feature of a radiating signal, such as frequency.

radiation chart *Geophysics.* a chart that, given the vertical distribution of temperature and humidity, aids in the calculation of the upward and downward fluxes of radiation.

radiation chemistry *Chemistry.* a branch of chemistry dealing with high-energy radiation interaction.

radiation chimera *Genetics.* a chimera that is produced by destroying or damaging the genome or developing tissues of an organism with ionizing radiation as a technique for studying the importance of cell lineage and development.

radiation cooling *Electronics.* any process of losing heat other than by conduction or convection.

radiation correction see COOLING CORRECTION.

radiation corrosion *Metallurgy.* the corrosion of metals or alloys exposed to high fluxes of ionizing radiation, such as neutron radiation.

radiation counter *Nucleonics.* any instrument designed to count ionizing events as a measure of the intensity of radiation.

radiation counter tube see COUNTER TUBE.

radiation damage *Nucleonics.* any destructive effect brought about by radiation exposure, including decomposition, dissociation, biological changes, and changes in mechanical, electrical, and thermal properties.

radiation damping *Electromagnetism.* the process of damping a system by absorbing radiant energy from it. *Quantum Mechanics.* in quantum electrodynamics, damping due to the virtual interaction of particles with the zero point field.

radiation decontamination *Nucleonics.* the removal of radioactive materials from a site where it is not desired.

radiation dermatitis see RADIODERMATITIS.

radiation dose *Nucleonics.* the total amount of ionizing radiation absorbed by a body.

radiation dose rate *Nucleonics.* the total amount of ionizing radiation absorbed by a body per unit time.

radiation effects *Cell Biology.* mutations in cells that alter their genetic structure, interfere with their division, or kill them.

radiation efficiency *Electromagnetism.* the ratio of radiated antenna power to the total power supplied to the antenna.

radiation era *Astrophysics.* a period following the Big Bang that lasted between 10 and 100 seconds, when the temperature dropped to a billion kelvins and during which radiation came to dominate the universe.

radiation exposure state *Military Science.* the health condition of a military unit based on the cumulative radiation doses received; expressed as a symbol that indicates the potential for future operations and the degree of risk if the unit is exposed to additional nuclear radiation.

radiation field *Electromagnetism.* the portion of an electromagnetic field that is radiated away from an antenna and outward into space, as distinguished from the induction field that remains in the vicinity of the antenna.

radiation filter *Electromagnetism.* a medium that can selectively block certain frequency ranges while being transparent to others.

radiation fog *Meteorology.* a type of fog that occurs when radiational cooling of the earth's surface at night lowers the air temperature near the ground to or below its dew point, particularly in valleys surrounded by hills or coastal mountains; it can persist overnight to just before sunrise.

radiation gauge *Nucleonics.* a device that measures the intensity and quantity of ionizing radiation.

radiation genetics *Genetics.* the study of the effects of radiation on genes and patterns of inheritance.

radiation hardening *Engineering.* the process of improving the ability of a device or material to withstand nuclear radiation.

radiation hazard *Medicine.* the destructive effect of radioactive substances on living cells and tissues, with specific regard to radiation intensity or accumulated radiation dose.

radiation hygiene *Radiology.* the discipline involving human protection from radiation injury. Also, RADIOLOGICAL HEALTH.

radiation illness see RADIATION SICKNESS.

radiation impedance see RADIATION RESISTANCE.

radiation intelligence *Military Science.* intelligence derived from noninformational electromagnetic energy unintentionally emanated by foreign devices, equipment, and systems, excluding energy generated by the detonation of nuclear or atomic weapons.

radiation intensity *Nucleonics.* the perpendicular component of radiant energy flux density passing across a specified unit surface area in a unit amount of time. *Electromagnetism.* the electromagnetic power per unit solid angle for a particular direction under consideration.

radiation ionization *Physics.* any ionization process that is achieved by the interaction of electromagnetic radiation and a neutral atom or molecule.

radiation law *Physics.* any of four physical laws that collectively describe radiation phenomena in a blackbody: Planck's law, Kirchhoff's law, Wien's law, and the Stefan-Boltzmann law.

radiation length *Nucleonics.* the mean path distance that a relativistic charged particle travels through a specified sample of matter after its radiation energy is reduced by a factor of $1/e$ (approximately 0.368), where e is the base of the natural logarithm.

radiationless transition *Physics.* a transition from one energy state to another in which the energy that would normally be emitted or absorbed as radiation is instead received or given up by a nearby particle or system.

radiation loss *Mechanical Engineering.* the amount of heat that escapes from a boiler into the atmosphere due to conduction, radiation, and convection.

radiation microbiology *Microbiology.* the branch of microbiology that deals with the reaction of microorganisms to radiation.

radiation monitoring *Nucleonics.* the continuous measurement of radiation present in a given area.

radiation necrosis *Radiology.* the death of tissue following exposure to injurious doses of radiation. Also, RADIONECROSIS.

radiation noise see ELECTROMAGNETIC NOISE.

radiation oscillator see PLANCK OSCILLATOR.

radiation oven *Engineering.* an oven that utilizes tungsten filament infrared lamps with reflectors to create the high-temperature environment needed to bake surface coatings.

radiation pattern *Electromagnetism.* a pattern indicating the directional characteristics of a radiation field associated with an antenna.

radiation physics *Physics.* a branch of physics that is concerned with the study of radiation ionization and the interaction between radiation and matter.

radiation preservation *Food Technology.* the use of ionizing radiation or other radiation to attack microoganisms and sterilize food products.

radiation pressure *Electromagnetism.* a physical pressure imposed by an electromagnetic field on an object. *Acoustics.* the force per unit area exerted by a sound wave on a surface.

radiation protection *Nucleonics.* the use of certain materials to provide a barrier against exposure to radiation.

radiation protection guide *Nucleonics.* a table of maximum permissible dosages of radiation established by the Federal Radiation Council.

radiation pyrometer *Engineering.* an instrument that determines the temperature of a body by measuring the thermal radiation emitted.

radiation quality *Physics.* the frequency spectrum associated with a radiation field that passes through a specified substance of a given depth.

radiation resistance *Acoustics.* the resistance of a surface area to an acoustic wave, described for a projector radiating sound into water by the equation $R_r = \rho c A$, where ρ represents the equilibrium resistance, c represents the sound velocity, and A represents the surface area.

radiation safety *Nucleonics.* the methods, materials, and regulations used to protect personnel who handle or are exposed to radioactive materials.

radiation scattering *Physics.* the scattering of radiation through its interaction with matter.

radiation shield *Engineering.* a wall placed between a radiant energy source and something that is radiation-sensitive, to prevent the body or object (or person) from becoming irradiated.

radiation sickness *Radiology.* **1.** the nausea and vomiting resulting from therapeutic irradiation. **2.** a toxic condition resulting from exposure to or ingestion of over 1 gray of ionizing radiation, with symptoms of **acute radiation syndrome** ranging from white cell depression to rapid death. Also, RADIATION ILLNESS, ROENTGEN INTOXICATION, X-RAY SICKNESS.

radiation softening *Materials Science.* a condition that occurs when some alloys are exposed to radiation; additional precipitation may occur, existing precipitates may coarsen, and dislocation structures introduced by cold working may recover, resulting in a softening of the material.

radiation source *Nucleonics.* **1.** a sealed capsule that contains an artificial radioisotope; used in radiography, teletherapy, nuclear batteries, gauges, and so on. **2.** any material that is radioactive.

radiation standards *Nucleonics.* a list of procedures and regulations used in the handling and applications of radioactive materials.

radiation therapy see RADIOTHERAPY.

radiation thermocouple *Electricity.* an infrared detector arranged in such a way that radiation falls on half of its junctions and causes several thermocouples connected in series to increase in temperature, thus generating a voltage.

radiation thermometer see RADIATION PYROMETER.

radiation vacuum gauge *Engineering.* a reduced-pressure device in which gas ionization from an alpha source fluctuates with differences in the density of the gas being measured.

radiation warning symbol *Nucleonics.* an easily recognizable symbol identifying an area that may contain radioactive materials.

radiation zone see FRAUNHOFER REGION.

radiative *Physics.* emitting radiation.

radiative braking *Astronomy.* a measurable reduction of a star's rotational velocity due to the emission of electromagnetic radiation.

radiative capture *Nuclear Physics.* a capture reaction in which a neutron is absorbed by the target nucleus and the excess energy in the resulting compound nucleus is emitted as gamma radiation.

radiative collision *Physics.* a collision between two charged particles resulting in a partial conversion of kinetic energy to electromagnetic radiation.

radiative correction *Quantum Mechanics.* the adjustment to the theoretical value of a physical quantity when the field theory leading to the calculation is quantized.

radiative diffusivity *Meteorology.* a property intrinsic to a given layer of the atmosphere that governs the rate at which the layer will warm or cool as a result of the transfer of infrared radiation; it is dependent on the temperature and water-vapor content of the layer and the pressure within the layer.

radiative equilibrium *Astrophysics.* the condition in a star in which each volume of stellar gas radiates, absorbs, and reradiates equal amounts of energy per second.

radiative recombination *Physics.* a recombination process by which electromagnetic radiation is emitted due to the lowering of the potential energies of the two oppositely charged particles.

radiative transfer *Physics.* the transmission of heat by electromagnetic radiation. Also, **radiative transport.**

radiative transition *Quantum Mechanics.* a transition between quantum states that results in the emission of radiation.

radiator something that radiates; specific uses include: *Mechanical Engineering.* **1.** a heating apparatus, typically a coil of tubes through which hot water or steam passes. **2.** an apparatus used to cool circulating fluids, typically a series or coil of thin-walled tubes that are exposed to air or another fluid; commonly used in water-cooled engines. *Physics.* any body that emits radiation (of energy, particles, or waves). *Electromagnetism.* specifically, a body that is capable of emitting electromagnetic radiation, either by inherent properties or by stimulation from an external source such as excitation or ionization. *Acoustics.* the source of an acoustic sound wave; i.e., a vibrating object that disturbs the medium around it, thereby producing acoustic waves at the same frequency as the frequency of vibration.

radiator temperature drop *Mechanical Engineering.* a measurement of the difference in temperature of a coolant liquid in an internal-combustion engine that occurs between the time it enters the radiator and the time it leaves.

radiatus *Meteorology.* a cloud variety whose elements are arranged in parallel bands that seem to converge toward one point on the horizon or, when the bands are crossing the entire sky, toward two opposite points; this variety occurs in cirrus, altocumulus, altostratus, and stratocumulus clouds.

radical *Organic Chemistry.* **1.** an electrically neutral organic group possessing one or more unpaired electrons. **2.** an ionic group having one or more charges, either negative or positive. *Surgery.* directed to and addressing the cause or source of a disease or dysfunction. *Mathematics.* **1.** a given root of a quantity; symbolized by $\surd$ or $\sqrt{\ }$. For example, $a^{1/n}$ is written $\sqrt[n]{a}$. **2.** the radical of an ideal I in a commutative ring R is the intersection of all prime ideals of R that contain I; denoted Rad I. If the set of prime ideals containing I is empty, then Rad I is defined to be R. By theorem, Rad $I = \{r \in R : r^n \in I \text{ for some } n > 0\}$. **3.** a Jacobson radical. (From the Latin word for "root.")

radical axis *Mathematics.* given a pair of intersecting circles, the line joining the two points of intersection, or the tangent at a single point of intersection. The radical axis of two nonconcentric coplanar circles is the locus of all points having equal powers with respect to the two circles. By theorem, it is a straight line perpendicular to the line of centers of the circles.

radical coupling *Materials Science.* a stepwise polymerization mechanism in which the reaction requires an oxidizing agent; for example, the commercial preparation of polymers containing acetylene units.

radical equation see IRRATIONAL EQUATION.

radical mastectomy *Surgery.* the surgical removal of the entire breast, pectoral muscles, axillary lymph nodes, and related skin and tissue; a treatment for breast cancer. Also, HALSTED'S OPERATION.

radical operation *Surgery.* a surgical procedure involving extensive tissue resection to eliminate and prevent recurrence of a disease. Also, **radical surgery.**

radical polymerization *Materials Science.* a common method of polymerization of unsaturated monomers; typically it involves a chain reaction with radical transfer.

radical ring *Mathematics.* a ring that equals its Jacobson radical.

radical sign *Mathematics.* the symbol $\surd$ or $\sqrt{\ }$.

radicidation *Food Technology.* the treatment of food by ionizing radiation in doses sufficient to kill all non-spore-forming bacteria.

radiciform *Biology.* shaped like a root. *Medicine.* shaped like the root of a tooth.

radicle *Botany.* the embryonic, or primary, root that grows out of the hypocotyl of seed plants. *Anatomy.* any of the smallest branches of a vessel or nerve.

radicofunctional name *Organic Chemistry.* a name designating an organic compound, consisting of two key words; the first word refers to the parent compound or groups involved, the second to the functional group; e.g., benzyl chloride.

radicotomy see RHIZOTOMY.

radicular *Botany.* of or relating to a root or radicle. *Anatomy.* of or relating to a radicle.

radicular cyst *Medicine.* a cyst with a wall of fibrous connective tissue at the root of a tooth, resulting from chronic infection of a granuloma.

radiculitis [rə dik′yə lī′tis] *Neurology.* an inflammation of a spinal nerve root, especially the portion of the root that lies between the spinal cord and the intervertebral canal.

radiculoneuropathy *Neurology.* any disease of the spinal nerve roots and nerves.

radiculopathy *Neurology.* any disease of the roots of spinal nerves.

radiculotomy see RHIZOTOMY.

radio *Telecommunications.* **1.** a system of communicating over a short or long distance by means of modulating and radiating electromagnetic waves. **2.** the process or business of broadcasting news and entertainment programs by such a a system. **3.** relating to, used in, or sent by such a system. Thus, **radio broadcast, radio station**, and so on. *Electronics.* **1.** a receiver or transmitter that converts electromagnetic waves into understandable sounds. **2.** of or relating to such a device. Thus, **radio antenna, radio dial**, and so on.

radio- a combining form meaning: **1.** radioactive, as in *radiocarbon.* **2.** radiant energy, particularly of the radio-frequency range of the electromagnetic spectrum. **3.** radio waves, as in *radioacoustics.*

radioacoustic ranging *Engineering.* a method for determining a sea vessel's location through the use of sound waves. Also, **radioacoustic position finding, radioacoustic sound ranging.**

radioacoustics *Telecommunications.* the study of the production, reproduction, and transmission of sounds transmitted between two points by radiotelephony.

radioactinium *Nuclear Physics.* a member of the actinium series that exists as the thorium isotope with a half-life of 18.5 days; it is both radioactive and toxic, and is found in a number of minerals.

radioactive *Nuclear Physics.* describing a material having an unstable nucleus that decomposes and emits radiation; displaying radioactivity.

radioactive balloon method *Radiology.* a method of applying radiation therapy to the urinary bladder wall by inserting a Foley catheter bag into the bladder and filling it with a radioactive solution.

radioactive carbon dating see RADIOCARBON DATING.

radioactive chain see DECAY SERIES.

radioactive clock *Nuclear Physics.* an isotope, such as potassium-40 or carbon-14, that is exceedingly long-lived and spontaneously decays into a stable element at a constant rate; used to determine absolute geological age.

radioactive cloud *Nucleonics.* a portion of the atmosphere containing masses of air made radioactive by a nuclear explosion or radiation accident. Also, ATOMIC CLOUD.

radioactive cobalt see COBALT-60.

radioactive contaminant *Nucleonics.* a radioactive material that, by a radiation accident, is allowed to spread over an area and subsequently renders the area unsafe or unusable.

radioactive dating see RADIOMETRIC DATING.

radioactive debris *Nucleonics.* radioactive matter that is dispersed over a region by a nuclear explosion.

radioactive decay *Nuclear Physics.* **1.** the disintegration of an unstable nucleus into one or more different isotopes, occurring spontaneously and accompanied by the emission of alpha or beta particles, or gamma radiation. **2.** a decrease in the quantity of radioactive material over time. Also, RADIOACTIVITY, RADIOACTIVE DISINTEGRATION or TRANSFORMATION.

radioactive decay series see DECAY SERIES.

radioactive disintegration see RADIOACTIVE DECAY.

radioactive element *Nuclear Physics.* any element having isotopes that emit various forms of radiation while spontaneously decomposing into different nuclides. Also, RADIOELEMENT.

radioactive emanation see ACTINON.

radioactive equilibrium *Nuclear Physics.* the condition in which the rate of decay of the parent isotope parallels the rate of decay of every intermediate daughter isotope.

radioactive fallout see FALLOUT.

radioactive half-life see HALF-LIFE.

radioactive heat *Thermodynamics.* heat that is generated within the body of a radioactive substance as the result of absorption of the radiation occurring inside the substance.

radioactive isotope *Nuclear Physics.* an isotope of an element that is radioactive, such as tritium, which is an isotope of hydrogen. Also, RADIOISOTOPE.

radioactive label *Biotechnology.* a radioactive isotope that is incorporated into a compound in order to follow the progress of the compound through a series of reactions. Thus, **radioactive labeling.**

radioactive material *Nucleonics.* any substance that spontaneously emits ionizing radiation.

radioactive metal *Thermodynamics.* a metallic element, such as actinium, radium, or uranium, that emits radiation spontaneously and continuously and that has an appreciable ability to penetrate most other materials.

radioactive mineral *Mineralogy.* any mineral containing uranium or thorium as a principal component, such as uraninite, pitchblende, thorianite, and carnotite.

radioactive salt *Nucleonics.* a salt having at least one radioactive constituent, such as radium bromide.

radioactive series see DECAY SERIES.

radioactive snow gauge *Meteorology.* a device that continuously records the water equivalent of snow on a surface as a function of time, by using radioactive salt encased in a lead-shielded collimeter and a Geiger counter to measure the amount of radiation depleted by the snow.

radioactive source *Nucleonics.* a sample of radioactive material that is used as a source of ionizing radiation.

radioactive standard *Nucleonics.* a sample of material that, at a given time, contains a known amount and type of radioactive isotope; used to calibrate instruments that measure radiation.

radioactive sterilization *Nucleonics.* the use of ionizing radiation to destroy microorganisms in a body or tissue.

radioactive tracer *Nucleonics.* a substance that is doped with radioactive material and injected into a physical or biological system so that its distribution can later be observed by radiation detection means. Also, RADIOTRACER.

radioactive transformation see RADIOACTIVE DECAY.

radioactive waste *Nucleonics.* unusable waste material (solid, liquid, or gaseous) that is radioactive; produced in nuclear reactors, the mining of radioactive materials, industrial and medical practices, and research.

radioactive-waste disposal *Nucleonics.* the handling and disposal of radioactive waste, usually by burial or dilution.

radioactivity *Nuclear Physics.* **1.** the property of unstable nuclei in certain atoms to spontaneously emit ionizing radiation in the form of alpha or beta particles and sometimes gamma rays during measurably delayed nuclear transitions, the length of delay being measured as the nuclide's half-life. **2.** the radiation thus emitted. **3.** see RADIOACTIVE DECAY.

radioactivity analysis see ACTIVATION ANALYSIS.

radioactivity log *Engineering.* a record of the radioactivity found in an oil-well borehole.

radio aid to navigation *Electronics.* any process or technique that applies the knowledge of how radio waves travel through space to obtain navigational information.

radioallergosorbent test *Immunology.* a radioimmunoassay technique that measures specific IgE antibodies to a variety of allergens; used as an alternative to skin tests to determine sensitivity to suspected allergens.

radio altimeter *Engineering.* an instrument that uses earth-reflected radio waves to determine altitude.

radio approach aid *Transportation Engineering.* any ground-based radio system that aids aircraft in making a landing approach.

radioassay *Analytical Chemistry.* a procedure, such as radioimmunoassay, that measures the radiation intensity of radioactive substances.

radio astronomy *Astronomy.* a branch of astronomy devoted to making observations at radio wavelengths.

radio atmometer *Engineering.* an instrument that measures the relationship between sunlight and evaporation in plant foliage.

radio attenuation *Electromagnetism.* for one-way propagation, the ratio of powers measured in the transmission lines of a pair of transmitting and receiving antennas.

radio aurora *Telecommunications.* the process of modifying the ionosphere with high-frequency radio waves in order to improve long-distance communication.

radioautograph or **radioautogram** *Engineering.* an image produced by autoradiography.

radioautography see AUTORADIOGRAPHY.

radio autopilot coupler *Engineering.* the equipment that allows an electronic navigational signal to operate an autopilot.

radio B battery *Electricity.* a B-type battery commonly used in a radio set, consisting of 15–30 cells connected in series.

radio beacon *Navigation.* a radio transmitter that broadcasts a characteristic signal, allowing craft with the proper receivers to determine its direction or distance.

radio beam *Electromagnetism.* a concentrated beam of radio-frequency energy used for communication and radar applications.

radio bearing *Navigation.* the bearing of a radio transmitter as determined by a receiver capable of radio direction finding.

radiobiology *Biology.* the branch of biology concerned with the effects of all types of radiation, especially ionizing radiation, on living organisms and tissues. Thus, **radiobiological, radiobiologist.** *Radiology.* the use of radioactive tracers to study metabolic processes. Also, RADIATION BIOLOGY.

radio bomb *Ordnance.* an electronic bomb that is set off by radio waves reflected from the target.

radio broadcasting *Telecommunications.* a collective term for the radio transmission of program material that is intended for general reception by the public.

radiocarbon *Nuclear Physics.* another name for carbon-14, a radioactive isotope of carbon with a mass number of 14 and a half-life typically described as 5568 years; used in dating ancient organic materials.

radiocarbon age *Archaeology.* the most likely statistical age of an object dated by radiocarbon; such a date comes from the laboratory expressed in **radiocarbon years** before present (calculated as the age before 1950 AD) with a plus or minus factor; e.g., 3618 ± 120.

radiocarbon dating *Archaeology.* another name for carbon-14 dating, the technique of deriving an approximate age from organic material, such as bone, charcoal, or shell, by measuring the loss of radiocarbon (carbon-14) that begins dissipation at death. Also, RADIOACTIVE CARBON DATING.

radiocarcinogenesis *Oncology.* the formation of cancer by exposure to ionizing radiation.

radiocardiogram or **radiocardiograph** *Cardiology.* a graphic record produced by radiocardiography. Also, RADIOELECTROCARDIOGRAM.

radiocardiography *Cardiology.* the graphic tracing of the change in concentration over time in a selected chamber of the heart of an introduced radioactive isotope. Also, RADIOELECTROCARDIOGRAPHY.

radiocesium see CESIUM-137.

radiochemical analysis see RADIOMETRIC ANALYSIS.

radiochemistry *Chemistry.* the branch of chemistry that studies the properties, applications, and relationships of radioactive elements. Thus, **radiochemical, radiochemist.**

radiochroism *Radiology.* the capacity of a substance to assimilate certain radioactive and roentgen rays.

radiochronology *Geology.* a method of absolute age dating based on the ratio between radioactive elements and their radioactive decay products present in a sample.

radio climatology *Meteorology.* the study of seasonal and regional variations in the transmission of radio energy through the atmosphere.

radio communication *Telecommunications.* a general term for any system of communication via electromagnetic waves at radio frequencies, including radiotelegraph, radiotelephone, and radio facsimile.

radio compass see AUTOMATIC DIRECTION FINDER.

radio control *Electricity.* the use of radio waves to control a remote apparatus. Similarly, **radio command.**

radiocurable *Radiology.* of a disease, curable by radiation therapy.

radio deception *Military Science.* the employment of radio to deceive an enemy, including the utilization of deceptive headings or enemy call signals.

radiodense *Radiology.* describing a material that is not penetrable by roentgen rays at the commonly used diagnostic energy range; radiodense areas appear light or white on exposed film. Also, RADIOPAQUE.

radiodensity *Radiology.* the property or degree of being radiodense. Also, RADIOPACITY.

radiodermatitis *Radiology.* an inflammation of the skin due to excessive exposure to radiation, particularly X rays and gamma rays.

radio detection and location *Space Technology.* an electronic system used to detect, locate, and predict future satellite positions.

radio detection and ranging see RADAR.

radio direction finder *Navigation.* a radio receiver that is capable of determining the direction from which a radio signal is arriving. Thus, **radio direction finding.**

radio duct *Geophysics.* a shallow, nearly horizontal layer in the atmosphere that, through vertical temperature and moisture gradients, produces an effect by which radio waves can become trapped or propagated anomalously.

radio echo see RADAR RETURN.

radio echo observation *Engineering.* a method of determining the distance of an object by measuring the elapsed time between transmission and reception of the reflected radio pulse.

radioecology *Ecology.* the branch of ecology that studies the effects of radioactive materials on organisms in an ecosystem and the pathways by which such materials are dispersed through the inorganic environment. It deals especially with radiation generated by nonnatural sources, such as nuclear power plants or nuclear weapons.

radioelectrocardiogram see RADIOCARDIOGRAM.

radioelectrocardiography see RADIOCARDIOGRAPHY.

radioelement see RADIOACTIVE ELEMENT.

radio emission *Electromagnetism.* the emission of electromagnetic energy having frequencies in the radio-frequency range of the spectrum.

radio energy *Electromagnetism.* electromagnetic energy having wavelengths in the radio-frequency range.

radio engineering *Engineering.* **1.** the study of radio waves. **2.** the design, development, and manufacture of radio equipment.

radio-facsimile system *Telecommunications.* a facsimile (FAX) system that transmits signals via radio rather than by wire.

radio fadeout *Telecommunications.* a condition in which incoming radio signals in the ionosphere fade away due to a sudden and unusual change in ionization. Also, **radio blackout.**

radio-field intensity *Electromagnetism.* the magnitude of an electric or magnetic field as measured at a specified location where radio-frequency radiation is passing; usually measured in the direction of maximum radiation.

radio field-to-noise ratio *Electromagnetism.* for a given location, the ratio of the radio-field intensity to the field strength of the electromagnetic interference.

Radiofilaceae *Botany.* a poorly described family of brackish and freshwater green algae of uncertain affiliation, characterized by unbranched filaments of spherical cells in a gelatinous sheath.

radio fix *Navigation.* the process of determining position by means of radio bearings. *Telecommunications.* the location of a radio transmitter by obtaining cross bearings on the transmitter from two or more listening stations.

radio frequency *Telecommunications.* a frequency that is useful for radio transmission, usually between 10 kHz and 300,000 MHz.

radio-frequency alternator *Electricity.* a rotating generator that produces radio-frequency power at frequencies above power-line values; used for high-frequency heating applications.

radio-frequency amplifier *Electricity.* **1.** a channel in which an incoming radio-frequency signal is amplified in a superheterodyne circuit. **2.** any amplifier of radio-frequency signals.

radio-frequency bandwidth *Telecommunications.* a band of frequencies constituting 99% of the total amount of radiated power, extended to incorporate any discrete frequency in which the power is equal to or greater than 0.25% of the total radiated power.

radio-frequency cable *Electromagnetism.* a transmission line designed to operate at radio frequencies.

radio-frequency choke *Electricity.* a low-inductance coil used to impede the flow of radio-frequency currents; its core is generally air or pulverized iron.

radio-frequency component *Telecommunications.* a part of a signal or wave consisting of radio-frequency alternations and not incorporating its audio rate of change in amplitude frequency.

radio-frequency current *Electricity.* **1.** the intensity of a generated RF signal, with units of microamps. **2.** an alternating current having a frequency above 10,000 Hz.

radio-frequency filter *Electricity.* an electric filter that is capable of attenuating signals at undesirable radio frequencies or enhancing signals at desired radio frequencies.

radio-frequency generator *Electricity.* a power source consisting of an oscillator, an amplifier, a power supply, and associated controlling equipment; commonly used in industrial and dielectric heaters, and capable of supplying sufficient radio energy at the required frequency for induction or dielectric heating purposes.

radio-frequency head *Engineering.* a unit containing a radar transmitter and part of a radar receiver.

radio-frequency heating see ELECTRONIC HEATING.

radio-frequency interference *Telecommunications.* **1.** equipment-induced noises that interfere with radio reception. **2.** the intrusion of unwanted electromagnetic noise or undesired signals into a submarine cable system.

radio-frequency measurement *Electricity.* the use of various devices and techniques to measure signal parameters at frequencies outside the normal audible range; examples include a digital counting or scaling device that operates over a given time period and calculates the total number of events, and an electronic circuit that produces a direct current proportional to the frequency of its input signal.

radio-frequency oscillator *Electronics.* a device that generates alternating current at radio frequencies.

radio-frequency preheating *Engineering.* the preheating of plastic-molding materials with an intense radio frequency to aid in molding operations and decrease molding time.

radio-frequency pulse see RADIO PULSE.

radio-frequency reactor *Electronics.* a device that opposes the flow of high-frequency alternating current, so that direct current can pass through a circuit.

radio-frequency sensor *Control Systems.* a device that uses radio signals to determine and transmit the position of an object to a robotic system.

radio-frequency shielding *Electromagnetism.* the process of shielding radio-frequency energy by means of conductive enclosures that isolate a particular component.

radio-frequency signal generator *Electronics.* an instrument that produces the radio frequencies needed to test and service electrical equipment, especially radios and televisions.

radio-frequency spectrometer *Spectroscopy.* a device used to measure the intensity of radio-frequency radiation emitted or absorbed by atomic or molecular species.

radio-frequency spectrum see RADIO SPECTRUM.

radio-frequency SQUID *Electronics.* a device that conducts high-frequency radiation across a hairlike gap in a superconducting loop.

radio-frequency transformer *Electromagnetism.* a transformer that is designed to transfer radio-frequency energy by means of a magnetic field, typically having an air core or a core of low-eddy current material.

radio-frequency transmission line *Electromagnetism.* a transmission line designed for operation at radio frequencies. Also, **radio-frequency line.**

radio galaxy *Astronomy.* a radio source, associated with an optical galaxy, that has an energy output in the range of 10^{35} to 10^{38} watts; many show signs of explosive activity in the center and have two jets extending from either side.

radiogenic *Nuclear Physics.* of or relating to material produced by radioactive decay.

radiogenic isotope *Nuclear Physics.* an isotope that is created by the decomposition of a radioactive isotope, but that may or may not be radioactive itself.

radiogenic lead *Nuclear Physics.* a stable isotope that is a by-product of uranium and thorium decay, a process that has taken place in rocks and minerals since the earth first formed.

radiogeology *Geochemistry.* the study of the distribution of radioactive elements and their isotopes in the earth's crust and their effects on geologic phenomena. Also, ISOTOPE GEOLOGY, ISOTOPE GEOCHEMISTRY, NUCLEAR GEOLOGY.

radioglaciology *Geophysics.* the analysis of glaciers and glacier ice, especially for determining ice depth through the use of radar.

radiogold *Oncology.* any of several radioactive isotopes of gold, especially gold-198; used as a diagnostic scintiscanning agent and as a therapeutic anticancer agent.

radiogold seed *Radiology.* a small radioactive isotope of gold about 2.5 mm long and 0.8 mm thick, permanently implanted into tissue for interstitial radiation therapy.

radiogoniometer *Electronics.* an instrument that measures the phase difference between two radio signals in a radio direction finder.

radiogoniometry *Engineering.* the process of locating a radio transmitter using directional bearings obtained from a radiogoniometer.

radiogram *Telecommunications.* a message that is transmitted by radio waves, as by radiotelegraphy.

radiographic sensitivity *Nucleonics.* a measure of the quality of a radiograph, expressed as the ratio of the minimum discontinuity detectable on the film to the base thickness of the film.

radiography *Radiology.* the process of registering on film the images produced by X-ray examination.

radio guard *Telecommunications.* a radio station, aircraft, or ship that is assigned to listen for and record transmissions, and to manipulate traffic on a specified frequency for a designated unit.

radio guidance *Electronics.* the use of radio signals to control the flight path of a missile from the ground.

radio hole *Geophysics.* the loss or fading of radio transmission due to the anomalous refraction of radio waves.

radio homing aid *Navigation.* a ground-based radio system that provides a homing signal to aircraft.

radio horizon *Telecommunications.* the set of points at which direct rays from an antenna become tangential with the earth's surface.

radioimmunity *Radiology.* a condition of decreased sensitivity to radiation, sometimes resulting from repeated exposure to radiation.

radioimmunoassay *Immunology.* a number of different, sensitive techniques for measuring antigen or antibody titers with the use of radioactively labeled reagents.

radioimmunosorbent test *Immunology.* a method used to measure immunoglobulin molecules that contain epsilon heavy chains (known as IgE antibodies) occurring in serum.

radio inertial guidance system *Engineering.* a radio command guidance system characterized by an inertial system, used for partial guidance in the event of radio guidance failure, or for furnishing current for accurate radar guidance information.

radio intelligence *Telecommunications.* information about an enemy or potential enemy that is obtained by intercepting radio messages.

radio interception *Telecommunications.* the act of tuning in on a radio transmission that was not originally intended to be monitored by the listener.

radiointerferometer *Engineering.* a radiotelescope that measures and records minute angular distances, as tiny as one second of arc, from interference separating celestial radio waves.

radioiodine *Nuclear Physics.* a radioactive isotope of iodine, such as the isotope iodine-131 with a half-life of 8.04 days; used in medicine to trace the size and activity of the thyroid gland.

radioiron *Chemistry.* a radioactive isotope of iron, having an atomic weight of 59 and a half-life of 46 days; used primarily as a tracer in biochemistry.

radioisotope see RADIOACTIVE ISOTOPE.

radioisotope assay *Analytical Chemistry.* the separation and measurement of a radioactive tracer for the purpose of determining the distribution of the substance to which it was attached.

radioisotope battery see NUCLEAR BATTERY.

radioisotope thermoelectric generator *Nucleonics.* a device designed to convert nuclear energy to electrical energy by means of a voltage produced in a thermocouple circuit heated by radioactive emission.

radio landing aid *Navigation.* a precision, ground-based radio system that provides guidance data to landing aircraft.

Radiolaria *Invertebrate Zoology.* a subclass of marine protozoans in the Sarcodina; most are spherical with radiating filopodia, a spiny siliceous exoskeleton, and a perforated central capsule.

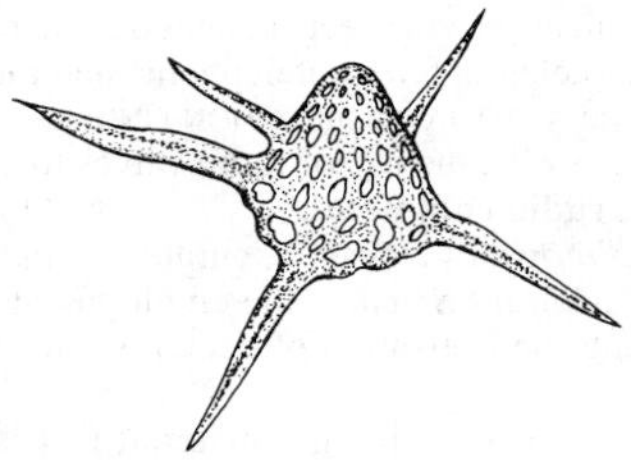

Radiolaria

radiolarian *Paleontology.* an actinopod that belongs to the subclass Radiolaria, appearing as a siliceous skeleton with a range from the Cambrian to the present; found in marine pelagic environments.

radiolarian chert *Geology.* a hard, dense, microcrystalline or cryptocrystalline quartz rock composed mainly of consolidated, homogeneous, siliceous skeletal remains of Radiolaria.

radiolarian earth *Geology.* an unconsolidated, siliceous deposit formed from the skeletal remains of Radiolaria.

radiolarian ooze *Geology.* a soft, fine-grained, deep-sea siliceous sediment composed of at least 30% skeletal remains of Radiolaria.

radiolarite *Geology.* **1.** the hard, fine-grained, consolidated, homogeneous equivalent of radiolarian earth. **2.** radiolarian ooze that has been hardened or consolidated by pressure, cementation, or heat. **3.** a fossil shell of a radiolarian.

radiole *Entomology.* in filter-feeding polychaetes (serpulid and sabellid fanworms), a paired feeding structure near the mouth that collects detritus and plankton from the surrounding water.

radiolesion *Radiology.* a sore arising from exposure to radiation.

radio line of position *Navigation.* a line of position determined by radio information, most typically a radial bearing.

radiolitic *Petrology.* **1.** of or relating to an igneous rock texture characterized by radial, fanlike aggregates of needles, resembling sectors of spherulites. **2.** of or relating to limestone whose constituents radiate from central points and are bound together by cement that forms less than 50% of the rock.

Radiology

Radiology may be defined as the branch of medicine that utilizes electromagnetic radiation, including radio waves, gamma waves, X-rays, and high-frequency sound waves, in the diagnosis and treatment of most diseases. Radiology, indeed, is an extraordinary medical discipline affecting virtually all human organ systems. The rapid advances in radiology, particularly over the last several decades, are unprecedented. In order to give a sense of the importance of the specialty of radiology, the various subdisciplines of the specialty are here briefly described.

Three major areas exist: diagnostic radiology, nuclear radiology, and radiation oncology. Of additional significance are the fields of radiological physics and radiobiology, which are ancillary sciences of major importance to radiology.

Diagnostic radiology is extraordinary in the scope of the modalities utilized in the contributions to the understanding of normal function and the abnormalities inherent in various disease states. The development and introduction of a number of image-producing modalities, using different types of radiant and other energies in radiological diagnosis, includes plain roentgenograms, which still constitute almost 70% of the studies in diagnostic radiology (e.g., the chest, the musculoskeletal system, the gastrointestinal tract). Recently developed is cross-sectional imaging obtained by the utilization of computerized tomography (CT) and magnetic resonance (MRI). In addition, beams of high-frequency sound waves—called ultrasonography (US)—are used in the evaluation of the gallbladder and biliary tract and in the diagnosis of various congenital entities in the fetus in utero as well as in many other vital studies.

The development and increasing utilization of CT, MRI, and US as parts of diagnostic radiology have revolutionalized the practice of medicine. The impact of these three modalities has been truly spectacular. As a result of these technical advances, well-defined subspecialty sections in diagnostic radiology have developed. These subdisciplines include specialization in virtually all organ systems (e.g., chest, musculoskeletal, neurological, gastrointestinal, genitourinary, and cardiovascular) and also in interventional pediatrics, obstetrics and gynecology, and mammography. This subspecialization in diagnostic radiology has contributed significantly to the acquisition of additional diagnostic information with all its implications for increasing medical knowledge and for a highly significant improvement in patient care.

Of increasing importance is the use of nuclear radiology as an important tool of diagnostic radiology. This subspecialty deals with the injection of radioisotopic material for diagnostic scanning of various organ systems (e.g., heart, lungs, kidney, thyroid, musculoskeletal) for the detection of many disease states. Here, too, advances have been unique, particularly with the development of single photon emission computed tomography (SPECT) and positron emission tomography (PET).

Radiation oncology encompasses the use of high-energy ionizing radiation to accomplish the destruction of malignant (neoplastic) tissues. The patient may be treated with an external beam of X rays produced by a linear accelerator or may have a radioisotope implanted directly into a malignant tumor. Increasingly, radiation therapy is combined with various chemotherapeutic agents, resulting in increasing success in the treatment of different neoplastic disorders (e.g., acute leukemia, certain metastatic tumors).

Both radiological physics and radiobiology have become meaningfully interrelated with the major areas of diagnostic radiology, nuclear radiology, and radiation oncology. These two basic scientific fields make it possible to provide radiologists with the technical information that will lead to the important decisions they must make in the many subspecialty areas listed above. Science and clinical activities are thus joined with a felicitous result in patient care.

Harold G. Jacobson, M.D.
Emeritus Professor and Chairman
Distinguished University Professor
Department of Radiology
Albert Einstein College of Medicine
Montefiore Medical Center, Bronx, N.Y.

radiolocation *Engineering.* a technique for determining the position and velocity of an object by radar, in which it is assumed that radio waves propagate at a constant velocity and are bounded by lines.

radio log *Telecommunications.* a log, usually maintained by a radio operator, of transmitted and received radio messages and information.

radiological *Nucleonics.* of or relating to nuclear energy and radioactivity. *Radiology.* of or relating to radiology. Also, **radiologic.**

radiological agent *Nucleonics.* any source of ionizing radiation.

radiological defense *Military Science.* defensive measures taken against the radiation hazards resulting from the employment of nuclear and radiological weapons.

radiological dose *Nucleonics.* the total amount of radiation absorbed by a body upon exposure to an ionizing radiation source.

radiological operation *Military Science.* the use of radioactive materials or radiation-producing devices to cause casualties or restrict the use of terrain; this includes the intentional employment of fallout from nuclear weapons, but not the initial destructive effect of a nuclear explosion. Also, **radiological warfare.**

radiological survey *Nucleonics.* an evaluation of the radiation levels and distribution in an area known to be contaminated with radioactivity.

radiological technician *Radiology.* a person who has received special training in the procedures necessary to perform X-ray examinations. Also, **radiological technologist.**

radiologist *Radiology.* a physician whose specialty is the use of radiant energy in the diagnosis and treatment of disease.

radiology *Medicine.* the science that deals with the use of radiant energy, such as X rays, radium, and radioactive isotopes, in the diagnosis and treatment of disease.

radiolucency *Radiology.* the property of being radiolucent. Also, RADIOPARENCY, RADIOTRANSPARENCY, ROENTGENLUCENCY.

radiolucent *Electromagnetism.* describing the property of being transparent to radio waves (and waves of higher frequencies). *Radiology.* allowing the passage of roentgen rays or other forms of radiant energy with little attenuation; radiolucent areas show up as dark spaces on exposed film. Also, RADIOPARENT, RADIOTRANSPARENT, ROENTGENLUCENT.

radioluminescence *Physics.* a luminescence that results from the irradiation of X rays or gamma rays on a body.

radiolysis *Physical Chemistry.* the breakdown of molecules that results from radiation, as when water, used to cool a nuclear reactor core, separates into hydrogen and oxygen.

radio magnetic indicator *Navigation.* an aircraft radio receiver that is coupled with a gyro compass so that it will indicate the direction of a selected navigational aid and its bearing, relative to the heading of the aircraft.

radio mast *Engineering.* a structure used to raise a radio antenna.

radio meteor *Astronomy.* a meteor detected by bouncing a radar signal off the train of ionized air left in its wake.

radio meteorology *Meteorology.* a branch of meteorology that examines the propagation of radio energy through the atmosphere and the use of radio and radar equipment for meteorological purposes.

radiometer *Electronics.* an instrument that detects and measures the intensity of radiant energy, such as X rays and microwaves.

radio meterograph *Telecommunications.* an apparatus set up for the automatic radio transmission of the indications of meterological equipment.

radiometric age *Geology.* the age of a fossil, mineral, rock, or geologic event given in years and determined by measuring the relative abundance of radioactive elements and their decay products.

radiometric analysis *Analytical Chemistry.* a measure of the disintegration rate of a radioactive compound; compared to a specific activity standard for the purpose of identifying the compound. Also, RADIOCHEMICAL ANALYSIS.

radiometric dating *Nucleonics.* a method of determining the approximate age of certain objects based on the ratio of a radioisotope concentration to that of a stable isotope. Also, RADIOACTIVE DATING.

radiometric magnitude *Astronomy.* a magnitude determined for a celestial body that includes all the radiation received at all wavelengths.

radiometrics *Archaeology.* any of the various methods of dating based on measurement of the known rate of decay of radioactive isotopes as they form stable elements.

radiometric titration *Analytical Chemistry.* a titration in which a radioactive indicator is used to trace the transfer of a substance between two liquid phases in equilibrium.

radiometry *Physics.* the detection and measurement of radiation fields, their spectral intensities, and energies.

radiomimetic *Chemistry.* having or producing radiationlike effects (similar to those of ionizing radiation) on biological organisms. Thus, **radiomimetic substances.**

radiomimetic activity *Biology.* the effects of certain chemicals, such as nitrogen mustard, urethane, and fluorinated pyrimidines, that imitate the physiological action of X rays, used in treating cancer by suppressing new cell growth.

radio mirage *Electromagnetism.* the detection of radar targets at phenomenally long range due to radio ducting.

radiomutation *Genetics.* a mutation that occurs as a result of a radiation dose.

Radiomycetaceae *Mycology.* a family of fungi belonging to the order Mucorales, occurring primarily in soil and dung in temperate regions.

radio nebula *Astrophysics.* a rapidly moving cloud of gas that emits at radio wavelengths by synchrotron processes or thermally in collisions with interstellar particles.

radionecrosis *Pathology.* the destruction of tissue caused by profuse radiation with X rays or gamma rays.

radio network *Telecommunications.* a group of radio stations that broadcast the same program simultaneously or otherwise communicate with one another.

radioneuritis *Neurology.* an inflammation of the nerves due to exposure to radiation.

radio noise *Electromagnetism.* electromagnetic noise having radio frequencies.

radionuclide *Nuclear Physics.* a nuclide containing isotopes that decay and emit radiation.

radiopacity see RADIODENSITY.

radio-paging system *Telecommunications.* a service that transmits a signal, usually brief messages or a beep tone, by radio from a public network telephone to a portable receiving device. Also, **radio-paging service.**

radiopaque [rā′dē ə pāk′] *Radiology.* impenetrable by X rays or by any form of radiant energy; radiodense.

radioparency see RADIOLUCENCY.

radioparent see RADIOLUCENT.

radiopasteurization *Engineering.* a process of pasteurization through the use of low-energy radiation.

radiophare see RADIO BEACON.

radiophases *Materials Science.* in radioactive waste management, the radioactive waste that contains radionuclides and is therefore continuously changing; as distinguished from encapsulants, the nonradioactive part of the waste.

radiophotoluminescence *Physics.* a luminescent phenomenon exhibited by certain minerals as the result of irradiation with beta or gamma rays and then exposure to light.

radiophylaxis *Radiology.* the mitigating effect of the reaction to a small dose of radiation, subsequent to large doses of radiation.

radio pill *Electronics.* a device that monitors a given physiological activity in an animal, such as the pH value of its stomach acid.

radio position finding *Engineering.* the process of locating a radio transmitter using two or more radio direction finders.

radiopotentiation *Radiology.* the ability of a drug or other agent to heighten the desired effect of radiation.

radio proximity fuse *Ordnance.* a fuse equipped with a radio transmitter and radar antennae that receives radio waves reflected from a target and actuates the fuse at a preset distance from the target.

radio pulse *Electromagnetism.* an intense, short duration of radio-frequency electromagnetic energy. *Telecommunications.* a radio-frequency carrier amplitude-modulated by a pulse. Also, RADIO-FREQUENCY PULSE.

radio range *Telecommunications.* a radio outlet that sends signals which, when received by the proper equipment, provide navigational information. Also, RANGE.

radio range station *Navigation.* a radio aid to navigation that provides course guidance, such as a VOR station.

radioreaction *Radiology.* a bodily reaction, particularly of the skin, to radiation.

radio receiver *Electronics.* see RADIO.

radio recognition *Telecommunications.* the determination by radio means of the friendly character, enemy character, or the individuality of another radio station; used mostly in military communications.

radio recombination line *Spectroscopy.* a spectral line in the radio-frequency range, resulting from an energy-level transition in an atom or ion having a principal quantum number greater than 50.

radio relay system *Telecommunications.* a point-to-point radio transmission system in which intermediate radio stations or radio repeaters receive, amplify, and retransmit radio signals. Also, RELAY SYSTEM.

radio repeater *Telecommunications.* a repeater that serves as an intermediate station in the transmission of radio programs or radio communications signals from one fixed station to another.

radioresistance *Radiology.* the resistance of tissue or cells to the injurious effects of radiation.

radio scan see SCANNING RADIO.

radiosensibility *Radiology.* the property of being radiosensible.

radiosensible *Radiology.* of or relating to skin, tumor tissue, and so on that is receptive to radiant energy.

radiosensitive *Radiology.* of or relating to tissues or organisms that are easily affected or destroyed by radiation.

radiosensitivity *Radiology.* the property of being radiosensitive.

radiosensitizer *Radiology.* a chemotherapeutic agent that is used to enhance the effect of radiation therapy.

radio shielding *Electricity.* the use of a metallic shroud over wiring and ignition devices in order to reduce electronic interference against radio communications.

radio-signal reporting code *Telecommunications.* a code used for reporting the quality of radiotelephone or radiotelegraph transmission. Also, SIGNAL REPORTING CODE.

radio silence *Telecommunications.* a period span during which all or some radio equipment is kept inoperative.

radiosodium *Chemistry.* the radioactive isotope of sodium, having an atomic mass of 24 and a half-life of 14.9 hours; used as a tracer in biochemistry.

radiosonde [rā′dē ə sänd′] *Meteorology.* a balloon-borne instrument that simultaneously measures and transmits meteorological data; it consists of transducers for measuring pressure, temperature, and humidity; a modulator for converting the output of the transducers to a quantity that controls a property of the radio-frequency signal; a selector switch that determines the sequence of the transmittal; and a transmitter to generate the radio-frequency carrier.

radiosonde commutator *Electricity.* a device that transmits temperature and humidity signals in a radiosonde; it consists of a series of alternate electrically conducting and insulating strips that are scanned by a contact for electronic impulses which indicate temperature and humidity values.

radiosonde observation *Meteorology.* an evaluation of temperature, relative humidity, and pressure aloft from radio signals received from a radiosonde; the height of each significant pressure level of the observation is computed from these data.

radiosonde-radio-wind system *Engineering.* a system consisting of radiosonde and radiosonde ground equipment, used to measure and transmit meteorological data from the upper atmosphere, including pressure, temperature, and humidity, as well as wind vectors.

radio sonobuoy see SONOBUOY.

radio source *Astronomy.* any celestial object that emits radio-frequency radiation; no nature or origin is implied.

radio source counts *Astronomy.* the number of radio sources detected by an antenna in a given area of sky with a brightness greater than some specified limit.

radio spectrum *Telecommunications.* the total range of frequencies in which useful radio waves can be promulgated; it extends from the audio range to about 300,000 megahertz. Also, RADIO-FREQUENCY SPECTRUM.

radio spectrum allocation *Telecommunications.* the specification of the frequencies of the radio spectrum that can be utilized by various radio systems.

radio star *Astronomy.* a star characterized by an unusually strong emission at radio wavelengths; this includes pulsars, flare stars, some types of infrared stars, and some types of X-ray stars.

radio storm *Astronomy.* strong radio-frequency radiation emanating from the sun in conjunction with solar flare eruptions or other forms of solar activity for periods of hours or days.

radio sun *Astrophysics.* the sun studied by means of its radio-frequency emission.

radio tail object *Astronomy.* an extragalactic object featuring a strong radio-emitting jet or tail.

radiotelegraphy *Telecommunications.* the transmission of telegraphic codes via radio waves, as distinguished from wire lines.

radiotelemetry *Telecommunications.* the presentation of automatically detected or measured information at a remote location using radio-frequency electromagnetic radiation as the means of transmission from the original source.

radio telephone *Telecommunications.* **1.** a radio transmitter and receiver used in conjunction for two-way voice communications. **2.** of or related to telephony over radio channels.

radiotelephony *Telecommunications.* the transmission of speech via modulated radio waves, without interconnecting wires.

radio telescope *Engineering.* a radio receiver capable of amplifying, recording, and determining the direction of radio waves.

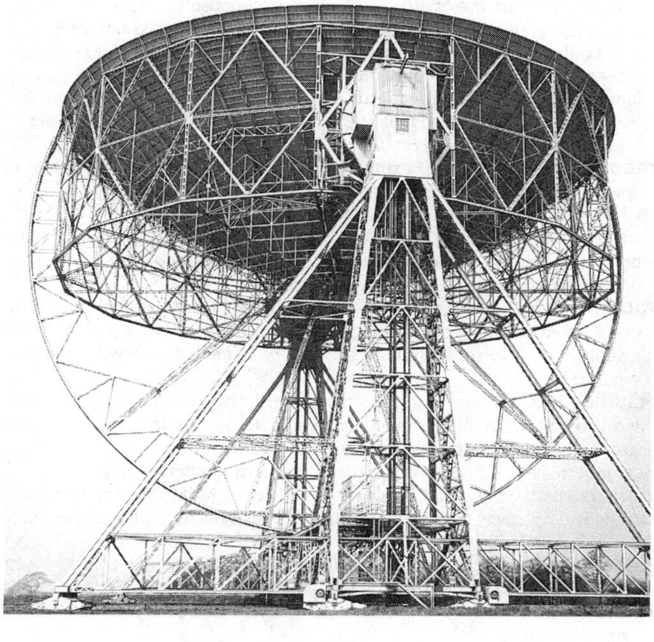

radio telescope

radioteletype *Telecommunications.* a teletypewriter/teleprinter combination that operates over a radio channel.

radioteletypewriter *Telecommunications.* a teletypewriter and its associated equipment that operates over a radio channel rather than over wires.

radiotherapeutics *Radiology.* a field of medicine that is concerned with the use of ionizing radiation for therapeutic purposes.

radiotherapy *Radiology.* the treatment of disease by ionizing radiation. Also, RADIATION THERAPY.

radiothorium *Nuclear Physics.* a common term for the isotope of thorium that has a mass number of 228.

radio time signal *Telecommunications.* a time signal that is sent via radio broadcast.

radio tower *Telecommunications.* **1.** a freestanding or guyed tower on which an antenna is attached to increase the range of radio transmission or reception. **2.** a tall tower that serves as an antenna.

radiotoxic *Radiology.* causing radiation sickness.

radiotracer see RADIOACTIVE TRACER.

radio tracking *Engineering.* a process in which a radio or radar beam locks onto a target and continuously monitors its range.

radio transmission *Telecommunications.* the process of sending signals through space at radio frequencies by means of radiated electromagnetic waves.

radio transmitter *Electricity.* a complete set of equipment that generates radio-frequency power and modifies it with communications signals, then delivers signals to an antenna to be radiated into space as electromagnetic waves.

radiotransparency see RADIOLUCENCY.

radiotransparent see RADIOLUCENT.

radio transponder *Electronics.* a transponder capable of receiving and transmitting radio waves.

radiotropic *Radiology.* influenced by radiation.

radio tube *Electronics.* a vacuum tube used in a radio receiving set.

radio wave *Electromagnetism.* an electromagnetic wave whose frequency is in the radio-frequency range.

radio wavefront distortion *Electromagnetism.* a change in the direction of the propagation of a radio wave.

radio-wave propagation *Electromagnetism.* the transfer of radio-frequency energy through space by the propagation of electromagnetic radiation.

radio window *Geophysics.* an interval of radio frequencies, ranging from about 6 to 30,000 megahertz, that are not blocked by the earth's atmosphere.

radish *Botany.* **1.** the small, pungent, red, black, or white edible root of the plant, *Raphanus sativus,* of the mustard family; usually eaten raw in salads and as a relish. **2.** this plant itself.

radish

radist *Navigation.* a system for direction-finding and position-finding by radio. It involves finding the phase differences between two continuous-wave signals. Also, RAYDIST.

radium *Chemistry.* a radioactive element having the symbol Ra, the atomic number 88, an atomic weight of 226.0, a melting point of 700°C, and 14 radioactive isotopes; the only usable isotope, radium-226, is a highly toxic, luminescent, white solid derived from uranium ores and used in medical treatment, as a source of neutrons, and in industrial radiography. (From the Latin word for *ray,* because it emits rays.)

radium age *Nucleonics.* the approximate age of a mineral, based on the concentration of radium atoms originally present and currently present, and the time when equilibrium is established with ionium.

radium-beam therapy *Radiology.* the treatment by radiation emitted from a source located at a distance from the body. Also, TELERADIUM TREATMENT.

radium bromide *Inorganic Chemistry.* $RaBr_2$, colorless to yellowish crystals; soluble in water or alcohol; melts at 728°C and sublimes at 900°C; highly radioactive and toxic; used in medicine and physics.

radium carbonate *Inorganic Chemistry.* $RaCO_3$, a white or light brown powder that is insoluble in water; highly radioactive and toxic.

radium cell *Nucleonics.* a tubelike capsule having thin walls containing radium; used in tracking experiments.

radium chloride *Inorganic Chemistry.* $RaCl_2$, yellowish-white or brownish crystals that are soluble in water and alcohol; melts at 1000°C; highly radioactive and toxic; used in medicine and physics.

radium necrosis *Toxicology.* a destruction of tissue, primarily bone, due to exposure to radium.

radium needle *Nucleonics.* a needlelike radium cell used for insertion into tissue; usually made of a gold or platinum alloy.

radium plaque *Nucleonics.* a surface over which radium is distributed; shielding is low in one direction to allow for the transmission of beta and gamma rays.

radium sulfate *Inorganic Chemistry.* $RaSO_4$, colorless to white crystals that are insoluble in water and acids; highly radioactive and toxic.

radium therapy *Radiology.* the treatment of disease by the use of a radioactive substance or radiant energy, especially for the destruction of malignant tissues.

radius *Mathematics.* **1.** a line segment joining the center of a circle (or sphere) to the circle (or sphere). **2.** the length of that line segment. *Anatomy.* the shorter, lateral bone of the forearm.

radius cutter *Mechanical Engineering.* a milling cutter having teeth designed to cut a radius on a workpiece.

radius of convergence *Mathematics.* the radius of a circle, the circle of convergence, in the complex plane inside of which a given power series converges and outside of which it diverges. The series may converge at none, some, or all points on the circle of convergence. Let $\sum_{n=1}^{\infty} a_n(z-a)^n$ be a power series of real or complex terms. If the sequence $\{|a_n|^{1/n}\}$ is bounded, set $\rho = \lim \sup(|a_n|^{1/n})$; otherwise, set $\rho = +\infty$. Then the radius of convergence R for the power series is: (a) $R = 0$ if $\rho = +\infty$; (b) $R = 1/\rho$ if $0 < \rho < +\infty$; or (c) $R = +\infty$ if $\rho = 0$. If $|z - a| < R$, the series converges absolutely; if $|z - a| > R$, the series diverges; and if $0 < r < R$, the series converges uniformly on the disk $|z - a| \leq r$. If the coefficients a_n and a are all real numbers, it is common to refer to the interval of convergence $(a - R, a + R)$.

radius of curvature *Mathematics.* at a given point of a curve, the radius of the osculating circle of curvature at that point.

radius of damage *Ordnance.* in a nuclear explosion, the distance from ground zero at which there is a 50% probability of achieving the desired damage.

radius of gyration *Materials Science.* the radius at which it may be assumed that the entire mass of an object is concentrated for the purposes of rotational kinematics.

radius of integration *Ordnance.* the distance from ground zero that defines the area within which the effects of a nuclear detonation and conventional weapons are to be integrated.

radius of protection *Engineering.* the radius of a circle within which a lightning rod prevents lightning from striking.

radius of rupture *Ordnance.* in an underground explosion, the maximum distance from the center of the charge at which destruction will occur.

radius of safety *Ordnance.* in a nuclear detonation, the horizontal distance from ground zero beyond which the effects on friendly troops are acceptable.

radius of visibility *Navigation.* the radius of a circle within which an object can be seen under specified conditions.

radius ratio *Physical Chemistry.* the ratio of the radius of a cation to the radius of an anion when anions surround a cation in a crystal lattice structure.

radius rod *Engineering.* a rod with a marking point at one end so that, as it is swung around, it marks out a circle or a part of a circle with a radius equal to the length of the rod; generally utilized in restricting the movement of a moving part.

radius vector *Astronomy.* a straight line connecting the sun and an object traveling around it in an elliptical orbit. *Mathematics.* see POSITION VECTOR.

radix *Mathematics.* see BASE, def. 1.

radix approximation *Mathematics.* the result of rounding applied to the representation of a given number in radix notation.

radix complement *Mathematics.* suppose a number has been represented in radix notation (base b), with n places to the left of the radix point and m places to the right of the radix point. The radix complement of the given numeral is that numeral which, when added to the given number, results in the number represented as 1 followed by n zeros to the left of the radix point and all zeros to the right of the radix point. The **radix-minus-one complement** is one less than the radix complement; i.e., the numeral with n digits to the left of the radix point and m digits to the right all equal to $b - 1$. Also, TRUE COMPLEMENT.

radix notation *Mathematics.* the positional representation of a numerical quantity, in which digits appearing in successive positions indicate coefficients of successive integral powers (or place values) of the radix or base. The place values usually decrease from left to right. The **radix point** is a dot (or comma, in Europe) written between the digits to indicate the transition from nonnegative to negative place value; e.g., a decimal point is a radix point for base ten.

radix sort *Computer Science.* a sorting algorithm that can be used when the keys can be viewed as numbers composed of successive digits (or words composed of letters). A file is sorted into bins based on the value of the least-significant key digit, the bins are appended, and the result is sorted on the next most significant digit, and so forth. Also, BIN SORT.

radix transformation *Computer Programming.* the process of converting numbers from one number system to another, as from decimal to binary.

radome *Electromagnetism.* a protective plastic housing for a radar antenna, having material that is transparent to radio-frequency waves. Also, RADAR DOME.

radon *Chemistry.* a gaseous radioactive element having the symbol Rn, the atomic number 86, an atomic weight of 222, a melting point of –71°C, a boiling point of –62°C, and 18 radioactive isotopes; it is an extremely toxic, colorless gas; it can be condensed to a transparent liquid and to an opaque, glowing solid; it is derived from the radioactive decay of radium and is used in cancer treatment, as a tracer in leak detection, and in radiography. (From the word *radium,* the substance from which it is derived.).

radon-219 *Nuclear Physics.* an isotope of radon that forms part of the actinium series and has a half-life of 4 seconds.

radon-220 *Nuclear Physics.* an isotope of radon that forms part of the thorium series and has a half-life of 56 seconds.

radon-222 *Nuclear Physics.* an isotope of radon that forms part of the uranium series and has a half-life of 3.82 days.

radon breath analysis *Medicine.* the determination of the volume and composition of specific substances in respired gases.

Radon-Nikodyn theorem *Mathematics.* suppose (X, A, μ) is a totally sigma finite measure space and ν is a sigma finite signed measure on A that is absolutely continuous with respect to μ. Then there exists a finite-valued measurable function f on X such that $\nu(E) = \int_E f d\mu$ for every measurable set E. The function f is unique in the sense that if there exists another function g so that $\nu(E) = \int_E g d\mu$, then $f = g$ except on a set of measure zero with respect to μ.

radon seed *Radiology.* a small gold tube filled with radon gas that is detectable radiographically; used in interstitial radiation therapy.

rad/s radians per second. Also, **rad/sec.**

Radstockian *Geology.* a European geologic stage of the Upper Carboniferous period, occurring after the Staffordian and before the Stephanian, and equivalent to the Upper Westphalian.

radula *plural,* **radulae.** *Invertebrate Zoology.* a tonguelike feeding organ in most mollusks, having numerous rows of chitinous teeth for grinding food; in some mollusks it may be extended and used as a boring organ.

Radulaceae *Botany.* a monogeneric family of medium to large liverworts of the order Jungermanniales, characterized by irregularly pinnate to bipinnate branching, rhizoids only on the leaf lobuli, and a total lack of underleaves; native to tropical and southern temperate zones.

radurization *Food Technology.* a low-level radiation treatment designed to enhance the keeping properties of food by reducing the number of spoilage organisms.

rafaelite *Petrology.* a nepheline-free hypabyssal syenite composed of orthoclase, analcime, and calcic plagioclase.

raffinase *Enzymology.* $C_{18}H_{32}O_{16}$, an enzyme that catalyzes the hydrolysis of raffinose.

raffinate *Chemical Engineering.* the portion of an oil that is not dissolved during the solvent refining of lubricating oil.

raffinose *Organic Chemistry.* $C_{18}H_{32}O_{16}$, a combustible white powdery solid that is soluble in water and very slightly soluble in alcohol; melts at 118–119°C; derived by the hydrolysis of cottonseed meal from sugar beet concentrates; it breaks down to fructose, glucose, and galactose on hydrolysis; it is used in bacteriology and the preparation of other saccharides. Also, MELITOSE, MELITRIOSE.

Rafflesiaceae *Botany.* a family of dicotyledonous tropical and subtropical root-parasites of the order Rafflesiales, having a lack of chlorophyll, a dissected and filamentous vegetative body, and fleshy flowering shoots that rise endogenously from the host; one species, *Rafflesia arnoldii,* has the largest flower in the world, up to three feet in diameter.

rafos *Navigation.* a hyperbolic position-determining system that uses underwater transmission of sound; the sound signals are produced at shore stations and detected at the vessel by microphones lowered to a specific depth. (From "sofar" spelled backward.)

raft *Engineering.* **1.** a floating structure made of timber, logs, or casks. **2.** a floating platform inflated with air or made of bouyant material. *Geology.* **1.** see FLOAT COAL. **2.** a rock fragment that is found drifting freely and more or less vertically in magma. *Hydrology.* an accumulation or jam of floating logs or debris formed naturally in a stream.

rafted ice *Oceanography.* the most extreme stage of pressure ice, characterized by ice floes stacked on one another. Also, **raft ice.**

rafter *Building Engineering.* a roof-supporting beam immediately beneath the roofing material.

rafter dam *Civil Engineering.* a dam formed by horizontal timbers that meet at midstream, resembling roof rafters turned on their side.

raft foundation *Civil Engineering.* a structural foundation slab composed of uninterrupted reinforced concrete that is larger in size than the structure it supports.

rafting *Geology.* the transporting of rock or soil by floating ice or by organic material, such as seaweed. *Hydrology.* a deformation of floating ice in which one floe overrides another, as a result of extreme pressure in an ice pack.

raft lake *Hydrology.* a temporary body of water that is confined along a stream by an accumulation of floating logs or other debris.

rag *Petrology.* a British term for a hard, coarse rock that develops a rough, irregular surface upon weathering, such as a flaggy sandstone. Also, **ragstone.**

rag bolt see BARB BOLT.

rag-content paper *Graphic Arts.* a high-quality paper containing at least 25% cotton or linen fiber (**rag**) by weight. Also, **rag paper.**

ragfish *Vertebrate Zoology.* a deep-sea fish of the family Icosteidae, having an extremely flexible body due to its soft, cartilaginous body; found in the waters of the North Pacific.

ragged *Graphic Arts.* of lines of type, having even word spacing and uneven lengths; unjustified. Copy having an unjustified right-hand margin is **ragged right;** copy having an unjustified left margin (rare in text but fairly common in advertising) is **ragged left.**

raggiatura *Meteorology.* brief land squalls that descend with great force from valleys and canyons in the highlands of Italy.

raglanite *Petrology.* a nepheline diorite that contains oligoclase, nepheline, and corundum, with traces of mica, calcite, magnetite, and apatite.

ragweed *Botany.* any of several composite plants of the genus *Ambrosia,* whose airborne pollen is the most common cause of hay fever, especially the **common ragweed,** *A. artemisiifolia,* and the **giant ragweed,** *A. trifida,* both of which are widespread North American species.

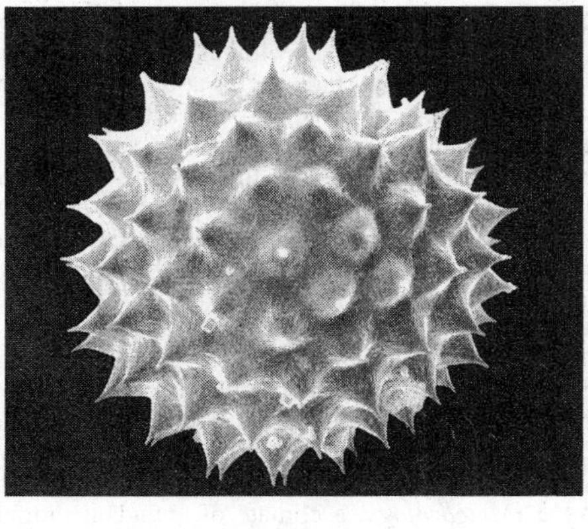

ragweed

Rahnella *Bacteriology.* a genus of Gram-negative, facultatively anaerobic, rod-shaped bacteria of the family Enterobacteriaceae, occurring in freshwater habitats.

raid *Military Science.* a small-scale operation involving swift penetration of hostile territory and planned withdrawal upon completion of the assigned mission. *Anthropology.* the practice of carrying out a short-term, carefully organized attack on another group in order to gain captives, food, herd animals, or other possessions. Also, **raiding.**

rail *Engineering.* **1.** a steel bar laid upon crossties to provide track for flange-wheeled vehicles. **2.** a bar placed between posts, used as a barrier. **3.** the chain or inner surface of a crawler. *Mechanical Engineering.* a high-pressure fitting on some internal-combustion engines that receives exhaust gases from several cylinders.

rail *Vertebrate Zoology.* any of numerous cosmopolitan marsh birds of the family Rallidae; characterized by a plump body, a long, pointed bill, short legs, a short tail, usually very long toes, and a loud call; found worldwide.

rail

rail anchor *Civil Engineering.* an anchor that holds a railroad track longitudinally in position.

rail capacity *Civil Engineering.* the maximum number of trains that can pass in two-way traffic along a section of railroad line.

rail clip *Civil Engineering.* **1.** the base plate that supports a rail on crossties. **2.** a fastener that holds a crane or derrick securely on a track to prevent tipping. **3.** an attachment that holds a detector bar on a track.

rail-fence jammer see CONTINUOUS-WAVE JAMMER.

railhead *Civil Engineering.* **1.** the closest port of access to a railroad from which transportation, loading, and disembarking may begin. **2.** the farthest point to which railroad track has been laid.

railing *Civil Engineering.* **1.** a bannister or support for a staircase. **2.** a balustrade. **3.** a wooden edge surrounding a plywood panel. *Electronics.* an image, resembling a fence railing, that appears on a radar screen when it is jammed with more than 50 kilohertz of energy.

rail joint *Civil Engineering.* a rigid connection between sections of railroad track.

Raillietiellidae see PENTASTOMIDA.

rail rapid transit *Transportation Engineering.* a high-volume, intraurban transit service with electric vehicles running on duorail lines. Also, HEAVY RAIL TRANSIT.

railroad *Civil Engineering.* a permanent road laid with iron or steel tracks forming one or more continuous lines, used to move linked cars or carriages pulled by locomotive engines from one place to another for the transportation of passengers and goods. *Transportation Engineering.* a collective term for the entire system of transportation using this road, including the tracks, cars, engines, workers, stations and other buildings, and all other fixed and movable property involved in such an enterprise.

railroad engineering *Transportation Engineering.* the planning, design, development, operation, economics, and use of rail transportation. Also, **railway engineering.**

railroad jack *Mechanical Engineering.* **1.** a portable mechanism used to lift heavy objects. **2.** a hoisting mechanism operated by electric motors to lift a locomotive off its wheels while it is undergoing repairs. **3.** a hydraulic-lift mechanism operated by hand lever or motor.

railroad thermit *Metallurgy.* a mixture of particulated aluminum and red iron oxide to which nickel and other alloy elements are added.

rail transit *Transportation Engineering.* an urban rail service; the term may refer to light rail or rail rapid transit, but generally not to mainline commuter service or automated guideway transit.

railway *Transportation Engineering.* another name for a railroad, especially one that operates over short distances.

railway dry-dock *Civil Engineering.* a dry-dock facility with a sloping track on which vessels are drawn from the water. Also, MARINE RAILWAY.

rain *Meteorology.* precipitation in the form of water drops usually having diameters greater than 0.5 mm, except when the drops are widely scattered and may be smaller; the intensity of rainfall at any given time and place is classified as very light, light, moderate, or heavy.

rain and snow mixed *Meteorology.* precipitation that consists of both rain and wet snow, usually occurring when the temperature of the air layer near the ground is just slightly above freezing.

rain attenuation *Telecommunications.* the weakening of radio waves after they have passed through an area in which there is rainfall.

rainbow *Optics.* an arc that displays concrete bands of colors and appears opposite the sun when solar rays are refracted and reflected in raindrops, spray, or mist. *Petroleum Engineering.* the iridescence that appears on water due to a thin film of crude oil; such iridescence on drilling fluid circulating in a well indicates contamination or contact with fresh oil. *Electronics.* a technique that guards against deceptive radar signals produced by reflective devices such as decoys and chaffs.

rainbow cactus *Botany.* a stiff, erect cactus, *Echinocereus pectinatus rigidissimus,* characterized by a cylindrical body with numerous intelocking spines and pink flowers; found in Arizona and Mexico.

rainbow granite *Petrology.* a granite featuring pink, yellowish, or reddish mottling on a dark-green or black background or dark mottling on a pink background.

rainbow roof *Architecture.* a pitched roof whose slopes are slightly convex.

rainbow snake *Vertebrate Zoology.* a burrowing snake, *Farancia erytrogramma,* having a red and yellow belly, red and black stripes along the sides, and a sharp-tipped tail; found in the southeastern U.S.

rainbow trout *Vertebrate Zoology.* a large trout, *Salmo gairdnerii,* characterized by a bright pinkish coloring; native to streams from Baja California to Alaska, and introduced to other areas as a game and food fish. A rainbow trout that has entered or returned from the sea is called a **steelhead.**

rain cast see RAIN PRINT.

rain cloud *Meteorology.* a cloud of any type from which rain falls.

rain crust *Hydrology.* a snow crust that is characterized by a dimpled surface, formed when surface snow is melted or dampened by rain and then refreezes.

raindrop *Meteorology.* an elliptical or spherical drop of water with a diameter greater than 0.5 mm that falls from a cloud through the atmosphere.

raindrop impression or **raindrop imprint** see RAIN PRINT.

rain erosion *Ecology.* the transfer of material down a hillslope as a result of rainfall; normally consisting of rain splash, which is the detachment and down-slope transfer of small soil particles by raindrop impact, and soil wash, which is the down-slope movement of material by surface water flow.

rain erosion

Rainey's corpuscle *Invertebrate Zoology.* the crescent-shaped spore of an encysted protozoan in the phylum Sarcodina. Also, MIESCHER'S CORPUSCLE.

rainfall *Meteorology.* the amount of precipitation of any kind, especially the amount that is measured by a rain gauge, taking into account a small amount of direct condensation.

rainfall frequency *Meteorology.* the number of times during a specified number of years that precipitation of a certain magnitude occurs, or is predicted to occur, at a given weather station.

rainfall penetration *Hydrology.* the depth to which water from a particular rainfall has been able to infiltrate the surface of the soil.

rainfall regime *Meteorology.* the character of seasonal rainfall distribution at any place, including equatorial, tropical, monsoonal, oceanic, or continental westerlies, or Mediterranean rainfall.

rain forest or **rainforest** *Ecology.* **1.** an area characterized by tall, broad-leaved, densely growing, primarily evergreen trees and a moist climate, having a dry season that is brief or absent; commonly found in the tropics, subtropics, and some sections of the temperate zone. **2.** see TROPICAL RAIN FOREST. **3.** see TEMPERATE RAIN FOREST.

rain gauge *Engineering.* an instrument that collects and measures rainfall, expressed in inches or centimeters of depth. Also, OMBROMETER.

raininess *Meteorology.* **1.** in general, the quantitative character of rainfall for a place. **2.** an estimate of the amount of precipitation during a given past century, before the beginning of instrumental observations; based on historical records of floods and droughts, as well as the availability of these records.

rainmaker *Meteorology.* a person who is able or who purports to be able to induce an increased amount of rainfall.

rainmaking *Meteorology.* a collective term for all activities or artificial methods, both scientific and nonscientific, that are used to increase the amount of precipitation released from a cloud, such as cloud seeding.

rain pillar *Geology.* an upward-projecting column of soil or soft rock that is capped and protected by pebbles or concretions, formed by differential erosion caused by falling rain.

rain print *Geology.* a small, shallow, circular pit with a slightly raised rim, formed by the impact of a raindrop falling in soft sand, clay, or mud. Also, RAIN CAST, RAINDROP IMPRINT, RAINDROP IMPRESSION.

rain shadow *Meteorology.* the geographical region on the lee side of a mountain or mountain range that receives less precipitation than the windward side.

rain shadow desert *Ecology.* a section of arid land lying in a rain shadow. Also, OROGRAPHIC DESERT.

rain splash *Agronomy.* the movement of soil particles by the impact of rain; a cause of erosion.

rainsquall *Meteorology.* a brief weather event in advance of a thunderstorm that arises as a result of a strong convective atmosphere, in which cumulus clouds are present and become the catalyst for a squall of rain blowing outward from the thunderstorm. Also, THUNDERSQUALL.

rain stage *Meteorology.* an ideal thermodynamic process of water condensation from moist air that rises without heat transfer at temperatures above the freezing point; this stage begins at the condensation level.

rain track *Meteorology.* any visible pattern that consistently reflects higher than normal rainfall in a given area.

rainwash or **rain wash** *Geology.* **1.** the washing away or movement of loose soil and other surface material by ground rainwater before the water has been channeled into well-defined streams. **2.** the material transported or washed away by rainwater. **3.** the rainwater that removes loose surface material in this manner.

rainwater *Meteorology.* another term for rain. See RAIN. *Hydrology.* surface water that has fallen to the earth as rain, but that does not contain dissolved material from the soil and is thus relatively soft.

Rainwater, L. James 1917–1986, American physicist; he shared the Nobel Prize for the proposal that atomic nuclei are asymmetrical spheroids.

rainy climate *Meteorology.* a climate of abundant rainfall; generally used to describe tree climates, not polar climates.

ainy season *Meteorology.* an annually recurring period of one or more months, during which precipitation is at a maximum for a region; generally used to describe such a period as a primary recurring feature of the tropics and subtropics. Also, WET SEASON.

raised beach *Geology.* a beach that has been elevated above the high-water mark and separated from the present beach by local crustal movements or by the lowering of the sea level.

raised bog *Ecology.* an area made up of moist, spongy, usually domed-shaped soil that forms on lake sediments, uniform clay substrates, and occasionally on valley-bog surfaces; commonly found in central Ireland.

raised flooring *Civil Engineering.* a floor made of panels that can be taken up to give access to the area immediately beneath it.

raisin *Food Technology.* a sweet variety of grape that has been dried by the sun or artificial means; used as a snack and in cooking and baking.

Rajakaruna engine *Mechanical Engineering.* a rotary engine having a combustion chamber with sides connected at their ends by pin-joints.

Rajidae *Vertebrate Zoology.* the skates, a worldwide family of marine, cartilaginous rays of the order Rajiformes, characterized by a protracted snout and broadly rounded pectoral fins.

Rajiformes *Vertebrate Zoology.* a worldwide order of chondrichthian fishes, including skates, bats, and rays.

rake *Mechanical Devices.* a combing tool with curved or straight tines, used for gathering dead plant material or cultivating soil. *Mechanical Engineering.* the angle, measured in degrees, formed by the leading face of a cutting tool and the surface behind the cutting edge. *Naval Architecture.* an angle away from the vertical of an upright fitting such as a mast, funnel, or cutwater. Masts and funnels frequently have a rake aft, while cutwaters often have a forward rake. *Building Engineering.* the exterior finish and trim applied parallel to the sloping end walls of a gabled roof. *Geology.* the inclination of an ore shoot or other linear geologic structure from the horizontal, as measured in the plane of the associated veins, faults, or foliation. *Military Science.* to fire across a target in a sweeping manner.

raked joint *Civil Engineering.* a masonry wall joint in which the still pliable mortar has been scooped out to a specific depth.

raking bond *Building Engineering.* a bond used to strengthen heavy-load-carrying footings with diagonal courses across the wall, which successively cross one another. Also, DIAGONAL BOND.

raking flashing *Building Engineering.* a flashing used on the side walls of a chimney projecting from an inclined roof.

rale *Medicine.* an abnormal sound from the pulmonary airway on auscultation of the chest, associated with diseases of the bronchi or lungs.

Ralfsiaceae *Botany.* a family of brown algae of the order Ectocarpales, characterized by prostrate, crustose thalli with filaments either united or embedded in mucilage; they characteristically form spots or patches on rock substrate.

Rallidae *Vertebrate Zoology.* the rails, coots, and gallinules, a diverse family of chickenlike, ground-nesting, woodland marsh birds of the order Gruiformes; found worldwide.

ralstonite *Mineralogy.* $Na_xMg_xAl_{2-x}(F,OH)_6 \cdot H_2O$, a colorless or white cubic mineral occurring as octahedral or cubic crystals, having a specific gravity of 2.56 to 2.64 and a hardness of 4.5 on the Mohs scale; found with cryolite.

ram *Vertebrate Zoology.* **1.** a male sheep. **2.** the male of certain other animals, such as a goat or antelope. *Engineering.* a guided piece in a machine that exerts pressure, or drives or forces a material by impact. *Mechanical Engineering.* **1.** the plunger of a pump. **2.** the moving weight in a pile-driving hammer. **3.** the side of a shaping machine on which the cutting tool is mounted. *Hydrology.* an underwater projection from an ice wall, iceberg, floe, or ice front, usually caused by melting and erosion of the unsubmerged part. Also, APRON, SPUR. *Aviation.* the forward movement through space of an air duct, scoop, inlet, or tube.

RAM [ram] *Computer Technology.* an acronym for random-access memory. See RANDOM-ACCESS MEMORY.

r.a.m. relative atomic mass. Also, **RAM.**

ramalina *Botany.* a genus of fructiose lichens belonging to the family Usneaceae, characterized by a thallus with flattened, tufted, dichotomously branched lobes; several species are sources of dye or perfume.

Raman, Sir Chandrasekhara Venkata [rä´mən] 1888–1970, Indian physicist; awarded the Nobel Prize for his work in light diffusion and the discovery of the Raman effect.

Raman effect *Optics.* the scattering of light passing through a transparent medium, caused by the light's interaction with the vibrational or rotational energy of the medium's molecules. Also, **Raman scattering.**

Raman-Rayleigh ratio *Optics.* the ratio of the light beams passing through a transparent medium (Raman) to the light beams passing at a horizontal angle to the medium (Rayleigh).

Raman spectrophotometry *Spectroscopy.* the photometric measurement of spectra produced by light scattered from a sample, usually at right angles to an exciting beam from a quartz mercury lamp. Raman lines observed are either Stokes or anti-Stokes; this measures vibrational motions and is thus complementary to infrared spectroscopy.

Raman spectroscopy *Spectroscopy.* the study of radiant energy scattered inelastically when a sample is irradiated with an intense beam of monochromatic light, usually from a laser.

Raman spectrum *Spectroscopy.* a diagram, graph, or any other display indicating the intensity of Raman scattered light with respect to its frequency.

Ramanujan's sums *Mathematics.* let m be a positive integer greater than 1, and let k be any positive integer. Ramanujan's sums are sums of the form $c_m(k) = \sum_h e^{2\pi i hk/m}$, where h runs through the complete set of reduced residues modulo m. (Since $e^{2\pi i x}$ is periodic, it does not matter which complete set of reduced residues is used.) If m and n are relatively prime, then $c_m(k)c_n(k) = c_{mn}(k)$ for all integers k; i.e., Ramanujan's sums are multiplicative functions of their indices. Also, $c_m(1) = \mu(m)$ for all m, where $\mu(m)$ is the Möbius function of m; and $c_p(k) = -1$ for every prime p and every integer k not a multiple of p. This is used in the theory of representation of integers as sums of squares.

Ramapithecinae *Anthropology.* a proposed subfamily of hominoids that may include Ramapithecus and Kenyapithecus; most authorities now consider the latter a hominid, however, and not closely related to Ramapithecus.

Ramapithecus *Anthropology.* an extinct group of homoid primates in the family Pongidae that is known only from fossil remains, inhabiting woodland areas from East Africa to Asia from about 15 million to 8 million years ago; it was once thought to be the earliest-known direct ancestor of humans and considered a separate genus, but it is now generally identified as an ape related to the orangutan and classified under the genus *Sivapithecus.*

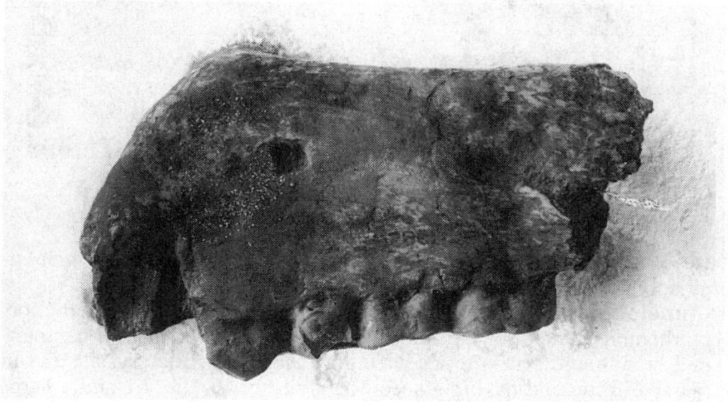

Ramapithecus

ramark *Navigation.* a radar beacon that transmits a signal either continuously or at intervals; developed by the U.S. Coast Guard. (An acronym for radar marker.)

ramate *Biology.* having branches.

rambla *Geology.* the dry bed of a transitory stream or any dry ravine, especially in the western U.S.

Rambouillet [ram´bü lā´] *Agriculture.* a breed of sheep raised for wool and mutton; it was developed from the smaller Merino breed. (From *Rambouillet*, France.)

ramdohrite *Mineralogy.* $Ag_3Pb_6Sb_{11}S_{24}$, an opaque, metallic, dark-gray monoclinic mineral occurring as twinned prismatic to lance-shaped crystals, having a specific gravity of 5.43 to 5.44 and a hardness of 2 to 3 on the Mohs scale; found in hydrothermal veining at Chocaya, Bolivia.

ram effect *Mechanical Engineering.* a compressing effect in a jet engine or in the manifold of a piston engine, due to the forward motion of an air scoop or inlet through the air.

Ramelli, Agostino 1531–1590, Italian inventor; designed pumps and many other machines.

ramentum *plural,* **ramenta.** *Botany.* a thin, membranous scale on the shoots or leaves of certain ferns.

ramet *Ecology.* an offshoot or module formed by vegetative growth in some plants and modular invertebrates that is actually or potentially independent physiologically, e.g., the runners of the strawberry or the polyps on a colonial hydroid.

ram-fed injection molding *Materials Science.* a high-pressure injection molding process that utilizes a ram to force the molten plastic into a mold.

rami *Invertebrate Zoology.* the plural of *ramus*; projecting parts or elongated processes of body parts, primarily of appendages.

ramie *Botany.* see RHEA.

ramiform *Botany.* **1.** having branches; branched. **2.** having the form of a branch; branchlike.

ramin *Materials.* the pale, fine-textured wood of an East Asian tree, *Gonystylus bancanus;* used for interior design, flooring, furniture, and precision instruments. Also, MELAWIS.

ramjet or **ram jet** *Aviation.* **1.** a ramjet engine. **2.** an aircraft or missile that is equipped or propelled by such an engine.

ramjet engine *Aviation.* a compressorless jet engine that is essentially a tube or duct with both ends open. Air is admitted at one end, compressed by the forward movement of the flight vehicle, heated by the combustion of fuels, and expelled at the other end, producing thrust. Such an engine cannot operate under static conditions.

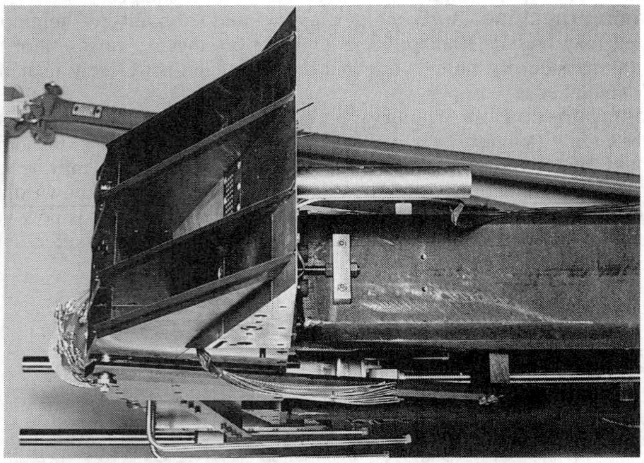

ramjet engine

ramjet missile *Space Technology.* a missile propelled by a ramjet engine.

rammelsbergite *Mineralogy.* $NiAs_2$, a white or pinkish, metallic, orthorhombic mineral of the löllingite group, trimorphous with pararammelsbergite and krutovite, occurring as short prismatic crystals and in massive form, and having a specific gravity of 6.97 to 7.1 and a hardness of 5.5 to 6 on the Mohs scale. Also, WHITE NICKEL.

rammer *Ordnance.* **1.** a hand-operated or powered tool that is used to seat a projectile in a gun or firearm. **2.** a tool used to remove a live projectile from the bore of a gun.

ramming *Metallurgy.* in sand casting, the process of packing sand around a pattern in a mold.

ramoff *Metallurgy.* in a sand casting, a flaw caused by improper ramming of the sand.

Ramon flocculation test *Immunology.* a method used to achieve a neutralized mixture of a specific toxin and antitoxin, in which the antitoxin is added in increments to a constant amount of the toxin until a zone of flocculation occurs.

Ramon titration *Immunology.* a method used to determine the amount of antibody that is needed to form a precipitate with a constant volume of a specific antigen.

Ramón y Cajal, Santiago 1852–1934, Spanish histologist; awarded the Nobel Prize for isolating the neuron and discovering the structure of the nervous system.

ramose *Biology.* having many lateral divisions or branches.

ramp *Engineering.* a walkway laid upon supports to form an inclined plane. *Aviation.* **1.** an inclined structure used to guide or launch an aircraft or rocket missile. **2.** in general, any paved or prepared area around a hangar or runway; an apron. *Electrical Engineering.* **1.** a change in output from one value to another that occurs at a predetermined linear rate. **2.** a voltage or current that varies at a constant rate.

rampant *Architecture.* having one impost or abutment higher than the other. Thus, **rampant arch, rampant vault.**

rampart *Military Science.* historically, a bank or mound of earth used as a defensive structure, as around a fort or castle. *Geology.* **1.** a narrow ridge or wall of boulders, gravel, or reef rubble that has been built up by waves along the seaward edge of a reef flat. **2.** a ridge or wall of unconsolidated material along a beach, formed by strong currents and wave action. **3.** a crescent or ring of pyroclasts that have been deposited around the top of a volcano.

rampart wall *Geology.* a steep, erosional remnant of porous, permeable, poorly cemented limestone that has formed along the seaward margin of a terrace.

ramp generator *Electronics.* a circuit that generates a voltage that increases linearly during one cycle of a sweep, and then drops to zero at the start of the next cycle.

Ramphastidae *Vertebrate Zoology.* the toucans, a family of medium to large, insect-eating and fruit-eating birds of the order Perciformes; having a very large, brightly colored bill; native to the forests of Central and South America.

Rampichthyidae *Vertebrate Zoology.* the knifefish family, certain nocturnal and herbivorous species, indigenous to the Americas, that have bladelike bodies and emit electrical discharges.

ramp sight *Ordnance.* a metal sight that is mounted on a ramp and adjusted by moving it back and forth along the incline.

ramp valley *Geology.* a valley or trough formed between high-angle thrust faults. Also, **ramp trough.**

ramp weight *Aviation.* the weight of an aircraft that is loaded to operating levels with fuel and payload.

ram rocket *Space Technology.* **1.** a hybrid propulsion system for unmanned vehicles using rocket and ramjet propulsions, in which the vehicle is launched by a rocket. The nozzle is jettisoned at the conclusion of the rocket burn, an air inlet is opened or extended, and the ramjet operation begins. **2.** the rocket component in such a system. **3.** a vehicle propelled by such a system.

Ramsauer effect *Atomic Physics.* a condition in which there is a sharp decline in the electron-scattering cross section of the atoms in a noble gas for electrons with energy below 25 electron volts.

Ramsay, Sir William 1852–1916, Scottish chemist; awarded the Nobel Prize for the discovery and analysis of argon (with Rayleigh) and other inert gases.

Ramsay-Shields-Eötvös equation *Thermodynamics.* a modification to the Eötvös rule which states that the molar surface energy of a liquid is proportional to the quantity $T_c - T - 6$ K, where T is the temperature and T_c is the critical temperature; this rule holds provided that T is not too close to T_c.

Ramsay-Young rule *Thermodynamics.* a rule stating that the ratio of the boiling points of two liquids of similar chemical character is approximately constant, independent of the vapor pressure at which they are measured.

ramsdellite *Mineralogy.* $Mn^{+4}O_2$, an opaque, metallic, gray-black orthorhombic mineral trimorphous with akhtenskite and pyrolusite, having a specific gravity of 4.37 to 4.83 and a hardness of 2 to 4 on the Mohs scale; found in massive and finely crystalline form in various manganese deposits.

Ramsden, Jesse 1735–1800, English instrument maker; invented divided circles, equatorial mounting, and other astronomical instruments.

Ramsden circle see EXIT PUPIL.

Ramsden eyepiece *Optics.* an eyepiece having two planoconvex lenses that face inward and that are separated by a distance equal to their shared focal length.

Ramsey, Norman born 1915, American physicist; awarded the Nobel Prize for developing the separated oscillatory fields technique.

Ramsey fringes *Physics.* the fringes or oscillations present in the number of transitions in a molecular beam with particle speed v that passes through two radio-frequency fields separated by a distance L; found to be a function of the quantity (L/v).

Ramsey's theorem *Mathematics.* given any positive integers s and t, the theorem that there is a smallest integer $r(s,t)$ such that every graph with $r(s,t)$ or more vertices contains either a subgraph isomorphic to K_s or a set of t independent vertices. The numbers $r(s,t)$ are called **Ramsey numbers.**

ramsonde *Engineering.* a cone-tipped rod that is used to measure the hardness of snow.

ram travel *Engineering.* the distance moved by an injector ram when filling a mold.

ramus *plural,* **rami.** *Anatomy.* **1.** a branch of an anatomical structure. **2.** a primary division of a blood vessel or nerve. *Vertebrate Zoology.* **1.** the vertical portion of the mammalian mandible. **2.** a barb of a feather. *Invertebrate Zoology.* see RAMI.

Ranavirus *Virology.* a genus of viruses of the family Iridoviridae that infects amphibians.

ranch *Agriculture.* **1.** a farm that is devoted to the rearing of horses, cattle, or sheep, especially a large farm on which cattle are raised for food. **2.** a farm that is devoted to the raising of a particular animal or crop.

Rancholabrean *Geology.* in southern California, a geologic stage of the Upper Pleistocene epoch, occurring after the Irvingtonian.

rancidity *Food Technology.* the rank, stale smell or taste of foods in which fats have spoiled due to hydrolysis or oxidation.

R and D research and development. Also, **R & D.**

randannite *Mineralogy.* a dark variety of diatomaceous earth specific to the Puy-de-Dôme region of France.

randkluft *Hydrology.* a crevasse at the head of a mountain glacier that separates moving ice and snow from the surrounding wall of the valley when no ice apron is present.

random access *Computer Technology.* a method of storage access in which the access time to any item of data is constant and independent of its address and order in storage. *Telecommunications.* **1.** in a recorded medium, the ability to select and immediately play any portion of the program, regardless of its sequential location; this capability exists for a videodisk or compact disk, but not for a videotape or audio tape, where access is sequential only. **2.** the ability of a television remote-control tuner to select any channel available, rather than moving sequentially from one channel to the next number above or below it.

random-access *Computer Technology.* describing a device that utilizes a method of storage access in which the ability to gain access to a particular item is not affected by its location in relation to other items. Thus, **random-access programming, random-access file,** and so on. *Telecommunications.* describing a television set, videodisk player, or other such device that allows the random selection of individual program segments, channels, or the like.

random-access discrete address *Telecommunications.* a radio communication system, employing pulse modulation and a broadcast carrier, in which users share a single wide band as compared to a system in which each user has an individual narrow band.

random-access input/output *Computer Programming.* an input/output control capability that allows the random processing of records stored on a direct-access device, in which seeks are overlapped with other processing and are issued in an order that minimizes the average seek time.

random-access memory *Computer Technology.* RAM, a storage device that allows data to be directly written to, or directly read from, any memory location without concern for its relation to other memory locations. Also, **random-access storage.**

random coil *Physical Chemistry.* an irregularly shaped polymer that appears in solutions. Also, CYCLIC COIL.

random error *Statistics.* a random variable included in a variety of statatistical models to summarize discrepancies between the data and the model that are due to unpredictable sources or to sampling variation. Also, SAMPLING ERROR, STOCHASTIC ERROR.

random forecast *Meteorology.* a forecast that utilizes one given set of meteorological contingencies on a random basis; thus used as a standard to compare forecast skill with another forecast method.

random function *Mathematics.* any mapping having the form $f\colon R\# \times (\Omega, F, P) \rightarrow R\#$, where $f(x,\omega) = Y_x(\omega)$, and Y_x is a random variable depending on x; that is, a mapping of the Cartesian product of an interval in the extended real numbers $R\#$ and a given probability space into $R\#$ such that, for fixed x, the mapping f is a random variable on (Ω, F, P).

randomization *Statistics.* in the design of experiments, a random assignment of units to the different experimental conditions.

randomized block design *Statistics.* an experimental design in which units are arranged into homogeneous groups in order to eliminate differences between groups and increase experimental accuracy.

randomized group design see COMPLETELY RANDOMIZED DESIGN.

randomized test *Statistics.* in hypothesis testing, the introduction of an auxiliary random device in order to decide whether to accept or reject the null hypothesis.

randomizing *Computer Technology.* a technique for locating an indirectly addressed file by reducing the range of keys with which it is associated to successively smaller ranges of addresses, until the desired address is found.

randomizing scheme *Computer Technology.* a storage-management technique in which records are distributed among several file storage locations so that accesses to the files are distributed equally and waiting times are equalized.

random line *Cartography.* in surveying, a line run as nearly as possible from one survey station toward a second station that cannot be seen from the first; the true position of the second station can then be determined by an offset from the random line to that station. Also, **random traverse.**

random mating *Genetics.* a mating system in which each female gamete has an equal chance of being fertilized by each male gamete in the population.

random minelaying *Military Science.* in land mine warfare, the laying of mines without a pattern.

random number *Mathematics.* any of a sequence of numbers from some set (usually the integers from 1 to n or the real numbers between 0 and 1) having the property that every number in the sequence has the same chance of being any number in the set and different numbers in the sequence are independent, in that the probability of a number in the sequence is not influenced by the other numbers in the sequence. In practice, pseudorandom numbers are often used to serve the purposes of random numbers.

random number generator *Computer Programming.* a program or hardware unit that produces pseudorandom numbers according to some algorithm and a given starting key.

random-ordered sample *Statistics.* a set in which every possible order of the elements is equally probable.

random processing *Computer Programming.* the processing of data records without regard to their sequence in a file.

random pulsing *Telecommunications.* a continuous, varying pulse-repetition rate, promulgated by continuous frequency change or noise modulation.

random sample *Statistics.* a sample consisting of independent and identically distributed random variables.

random scission *Materials Science.* the breaking of chemical bonds in the molecular chain during the degradation of certain polymers; an initial break, occurring at a weak link in the polymer chain, produces free radicals at the ends of the chain fragments; however, intramolecular and intermolecular chain transfers produce a random distribution of radical sites, and subsequent scission results in a random distribution of molecular weights in the material.

random variable *Statistics.* a quantity having one and only one numerical value, based on the outcome of an experiment; real-valued function defined over a sample space. Also, STOCHASTIC VARIABLE. *Mathematics.* a real-valued function $X(\omega)$ defined on a probability space (Ω, F, P) such that the preimage of every Borel set in the real numbers R is a member of F; equivalently, $X(\omega)$ is a measurable function on the measurable space (Ω, F) for all $\omega \in \Omega$. The argument ω is often omitted and the random variable written as X. A sequence of random variables defined on a common probability space is called a **countable (stochastic) process.**

random vector *Mathematics.* an n-tuple of random variables is called an n-dimensional random vector.

random vibration *Mechanics.* irregular oscillations of a structure due to excitations by many irregularly timed pulses with different characteristics, such as wavelength and amplitude.

random walk *Materials Science.* **1.** a diffusion process. **2.** a stage in the development of the random coil model for polymer chains; it explained the eutrophic origin of rubber elastic forces by postulating free rotation of the carbon-carbon bonds. *Mathematics.* **1.** a sequence s_n of directed line segments in n-dimensional Euclidean space, joined end-to-end, such that the length and direction of s_n are independent of s_k for all $k < n$. **2.** more generally, let $S_0, S_1, S_2, \ldots$ be the Markov process of successive sums of independent, identically distributed random variables $X_1, X_2, \ldots$ defined on a common probability space; that is, $S_n = \sum_{k=1}^{\infty} X_k$, where $S_0 = 0$. Any Markov process having the same distribution as $S_0 + x, S_1 + x, S_2 + x, \ldots$ is then called a random walk; such a process may be viewed as "starting from x." (Note that x is a real number and that S_n is real-valued.)

random winding *Electricity.* a coil winding in which successive turns are wound haphazardly rather than in regular layers.

random X-inactivation *Genetics.* in the homogametic sex of a species, the process by which the second of two sex chromosomes (the X chromosome in human females) is inactivated more or less randomly in all somatic cells, thereby eliminating the double dose of alleles.

R and R *Military Science.* **1.** rest and recuperation; a period of leave given to a unit recently withdrawn from combat. **2.** any recuperative or restorative period. Also, **R & R.**

Raney nickel *Organic Chemistry.* a pyrophoric, dark gray powder or crystals derived by leaching aluminum from an alloy of aluminum-nickel with caustic soda solution. *Metallurgy.* a form of this very fine nickel powder used in industry, mainly as a catalyst.

rang *Petrology.* a subdivision of an order in the CIPW igneous rock classification system.

range the extent or scope of action; or the limits between which variation is possible; specific uses include: *Engineering.* the maximum operating distance of a system or the operating limits of an instrument. *Physics.* the maximum distance between two bodies or particles over which the forces acting between them can be considered appreciable. *Military Science.* **1.** the distance between a weapon or tracking station and a target. **2.** the distance limiting the operation of an aircraft, ship, vehicle, weapon, and so on. **3.** an area that is set aside for target practice. *Geography.* **1.** a row or chain of mountains. **2.** an open stretch of uncultivated land on which animals, especially cattle, can graze or forage for food. *Ecology.* **1.** the general region in which a certain type of animal or plant is typically found; e.g., the *range* of the gray wolf is much more limited today than in the past. **2.** the area throughout which a specific animal moves in search of food during the course of a day. *Navigation.* **1.** two or more objects that, when lined up, mark a recommended course or other specified bearing. **2.** the difference in height between successive high and low tides. *Robotics.* **1.** the maximum distance that a robot can move its arm or wrist. **2.** the envelope that defines the movement of a robotic arm or wrist. *Cartography.* a series of contiguous sections within a township situated north and south of each other. *Telecommunications.* see RADIO RANGE. *Statistics.* the difference between the highest and lowest values in a sample or distribution; a measure of dispersion. *Mathematics.* given a function or mapping f from a set X to a set Y, the subset of Y consisting of those elements that have a preimage in X; that is, the set $\{y \in Y : f(x) = y \text{ for some } x \in X\}$. Also, CODOMAIN.

range adjustment *Ordnance.* an adjustment to the firing data of a weapon in order to correct the actual range of fire.

range-amplitude display *Electronics.* the representation of targets on a radar screen in which the range scale is based on time.

range angle *Ordnance.* in aerial bombing, the angle formed by the aircraft-target line and the vertical aircraft-ground line at the instant of bomb release. Also, DROPPING ANGLE.

range arithmetic *Computer Programming.* a former term for interval arithmetic.

range attenuation *Electromagnetism.* a decrease in power density caused by the divergence of flux lines with distance.

range calibration *Engineering.* the process of tuning or adjusting a radar set, so that it will indicate the correct range.

range calibrator *Electronics.* **1.** a device with which a radar set can be adjusted to give the correct range. **2.** a device to determine how far a signal can travel and remain intelligible.

range check *Computer Programming.* a data-validation process to ensure that numerical data falls within specified limits.

range coding *Engineering.* a technique for coding radar signals from transponder beacons to register as a series of illuminated rectangles on a radarscope.

range comprehension *Electronics.* the range of values in a frequency-modulated sonar system.

range control *Aviation.* on a radar indicator, a control used to alter the display range. *Space Technology.* control over the range of a guided missile.

range control chart *Aviation.* a chart used in flight to plot fuel consumption in order to determine whether the desired range may be attained.

range correction *Ordnance.* an adjustment to firing data in order to compensate for changes in the range due to nonstandard conditions of weather, ammunition, or materiel. Also, **range-correction board.**

range delay *Electromagnetism.* a control permitting an operator to present on the radarscope only those echoes from targets lying beyond a certain distance from the radar.

range determination *Ordnance.* the process of determining the distance between a weapon and its target; methods include visual estimation, firing the weapon, using range-finding instruments, and plotting. Also, **range estimation.**

range deviation *Ordnance.* the distance by which a projectile falls short of its target or strikes beyond the target.

range disk *Ordnance.* a graduated disk connected to the elevation mechanism of a gun; used in setting the range. Also, **range drum.**

range dispersion diagram *Ordnance.* a chart that indicates the expected distribution of shots fired using the same firing data within eight successive range areas.

range error *Ordnance.* the difference in range between the point of impact of a specific projectile and the center of impact of a group of projectiles fired with the same firing data.

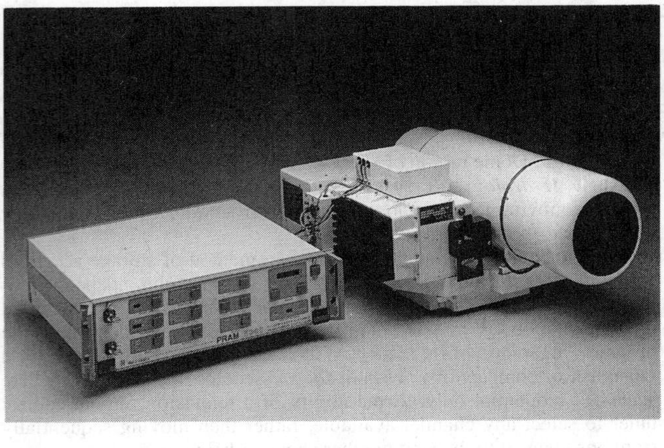

laser rangefinder

rangefinder any of various devices that determine the distance from an observer to a particular object; specific uses include: *Electronics.* an electronic device that determines how far away an object is by noting how long it takes for a radio wave or laser to travel to and from the object. *Ordnance.* a device that determines the distance from a weapon or tracking station to the target. *Optics.* an optical device that measures the distance of a remote object, especially to assist in setting the focus of a camera lens. *Telecommunications.* a movable, calibrated unit of the receiving mechanism of a teletypewriter through which the selecting interval can be situated with regard to the start signal.

range gate capture *Electronics.* a technique that generates false target signals that cause a radar system to leave the real target and follow a false one.

range gating *Electronics.* a process in which a radar system selects signals from a target that fall within a very narrow range.

range height indicator *Engineering.* a radar device that simultaneously determines the height and range of a targeted object.

range-imaging sensor *Robotics.* a device that measures the distance from a robot's end effector to the various parts of an object, in order to form an image of the object.

range ladder *Ordnance.* a method of adjusting the range of naval gunfire by firing an initial shot that is substantially in front of or beyond the target, then making small, equal adjustments on succeeding shots until the proper range is determined.

range lights *Navigation.* two or more lights that, when lined up, mark a recommended course or other specified bearing.

range line *Cartography.* in surveying, the north or south boundary of a township or township section. Also, TOWNSHIP LINE.

range mark see DISTANCE MARK.

range of a loop *Computer Programming.* all of the instructions in an interactive loop, beginning with the loop control statement and ending with the last instruction in the loop.

range of motion *Medicine.* any body action involving the muscles, joints, and natural directional movements, such as abduction, extension, flexion, pronation, and rotation.

range of reaction *Genetics.* the range of all possible phenotypes that may develop, by interaction with various environments, from a given genotype. Also, NORM OF REACTION.

range of tide see TIDE RANGE.

range of visibility *Navigation.* the farthest distance at which a light or an object can be seen under normal conditions.

range probable error *Ordnance.* the median error in range for a specific weapon; provided in the weapon's firing tables and used as a measure of accuracy.

range rake *Ordnance.* a T-shaped device used by a flank observer to calculate an angular measurement of range deviation; the distance between pegs set in the crossbar of the T subtends a specific angle at the base.

range rate *Electronics.* the rate at which a tracked target's distance is changing with respect to time.

range recorder *Engineering.* an instrument for recording the distance versus time relationship of an object. *Acoustical Engineering.* a display system that is designed to record on paper the range and bearing information processed by a sonar system.

range ring *Electronics.* a circular mark on a radar screen that can be adjusted to indicate the distance from the radar set to the target; used on plan position indicators, it corresponds to a range step on a type M indicator.

range rod *Engineering.* a long rod used for sighting points or lines, or for making a ground point in surveying. Also, **range pole.**

Ranger program *Space Technology.* a series of U.S. lunar probes, launched during 1961–1965, that before impact provided the first clear closeup views of the moon's surface.

range scale *Ordnance.* **1.** a graduated scale on the sight or mount of a gun that indicates the gun's elevation. **2.** a table in the firing data that indicates elevation settings to produce various ranges with standard charges. **3.** a scale on the side of a plotting board on which the range of a moving target is recorded for calculating the firing data.

range sensing *Engineering.* a measurement of the distance between an object and a robot's end effector.

range space *Mathematics.* any set of which the range is a subset.

range step *Electronics.* the vertical displacement of a signal on an M-indicator radar set; used to measure the distance from the set to the target.

range strobe *Electromagnetism.* an index mark that is displayable on various types of radar indicators to assist in determining the exact range of a target.

range surveillance *Engineering.* the electronic surveillance of a given area.

range sweep *Electronics.* a sweep that measures the distance between a radar station and the target.

range table *Ordnance.* the part of a weapon's firing table that indicates the correct elevations for various ranges under various conditions.

range-tracking element *Electronics.* a component in a radar set that measures a target's range and time by opening a range gate to accept a signal from a moving target just before it arrives.

range wind *Military Science.* the true horizontal wind in the vertical plane defined by a weapon's line of fire.

range zero *Electronics.* an alignment of a radar sweep in which there is no distance between the station and the target.

range zone *Geology.* a body of strata that represents the total geographic and stratigraphic extent of occurrence of a particular fossil form. Also, ACROZONE, ZONITE.

ranging *Military Science.* **1.** the process of establishing target distance; types of ranging include echo, intermittent, manual, navigational, explosive echo, optical, and radar. **2.** the process of locating an enemy gun by watching its flash or listening to its repeat. **3.** the systematic, wide-scale scouting of an area, especially by aircraft.

ranging oscillator *Electronics.* a device that generates the alternating current used by a cathode-ray screen to indicate a target's range.

Ranidae *Vertebrate Zoology.* the riparian frogs, a family of the order Anura containing about 600 species of moderate-sized to large frogs; nearly worldwide in distribution.

rank the relative position of a given item in some scale or order; specific uses include: *Geology.* **1.** a method of classifying coal according to the percentage of carbon that it contains, with anthracite being the highest in rank and lignite the lowest. **2.** see METAMORPHIC GRADE. *Behavior.* the position of an individual animal within a dominance hierarchy. *Anthropology.* see STATUS. *Mathematics.* **1.** the rank of an $n \times m$ matrix M is the number of linearly independent rows (or columns) of M. Also, **row rank, column rank.** If M is a matrix over a division ring (e.g., the real or complex numbers), then by theorem, the row rank of M equals the column rank of M. **2.** the rank of a linear transformation is the dimension of the image space; equivalently, the rank of the matrix that represents the transformation. **3.** the rank of a system of homogeneous linear equations is the rank of the matrix that represents the system. **4.** suppose $f: R^n \to R^p$ consists of p functions in n variables and that the Jacobian matrix of f (i.e., the derivative of f) exists. Then the rank of f is the rank of the Jacobian matrix. **5.** the rank of a free Abelian group F is the cardinal number of any basis X of F; denoted $|X|$. By theorem, any two bases of F have the same cardinality. **6.** let f be an entire function with zeros $\{a_1, a_2, \ldots\}$ repeated according to multiplicity and arranged so that $|a_1| \le |a_2| \le \cdots$. Then f is said to be of finite rank if there exists an integer p such that

$$\sum_{n=1}^{\infty} |a_n|^{-(p+1)} < \infty.$$

If p is the smallest such integer, then f is said to be of rank p. If f is not of finite rank, then f is said to be of infinite rank. **7.** the rank of a matroid is the number of elements in any base of the matroid.

rank correlation *Statistics.* in nonparametric statistics, a measure of the degree of correlation between two sets of rankings.

rank estimator *Naval Architecture.* a measurement of a ship's seakeeping capability, derived from its dimensions, waterplane, and prismatic coefficients.

Rankine, William [rang'kin] 1820–1872, Scottish engineer and physicist; a founder of thermodynamics, developed important descriptions of waves and elasticity.

Rankine body *Fluid Mechanics.* any of the shapes obtained from analysis of source-sink combinations with uniform flow; flow past a cylinder (uniform flow and a doublet) can be regarded as a special case of flow past a Rankine body.

Rankine cycle *Thermodynamics.* an ideal thermodynamic cycle that consists of four processes: heat transfer to the system at constant pressure; an expansion at constant entropy; a constant-pressure heat transfer from the system; and a compression at constant entropy; used as a standard of efficiency. Also, STEAM CYCLE.

Rankine efficiency *Mechanical Engineering.* a measurement of the operating efficiency of a steam plant or engine as compared to an ideal system operating on the Rankine cycle under designated steam pressure and temperature conditions.

Rankine-Hugoniot equations *Fluid Mechanics.* a set of equations that describe the conservation of mass, momentum, and energy flux across a discontinuity or shock; they relate the pressure, density, and enthalpy of a fluid to the velocity of a shock wave, before and after the wave passes through the fluid. Also, **Rankine-Hugoniot relations.**

Rankine temperature scale *Thermodynamics.* an absolute scale of temperature, in which zero value occurs at absolute zero and the ice point of water (at one standard atmosphere) occurs at 459.67°R; the degree increments are identical to those of the Fahrenheit temperature scale.

Rankine vortex *Fluid Mechanics.* a rectilinear vortex of finite circular section with a vertical axis, subject to gravity and having a free surface.

ranking method *Industrial Engineering.* a rating system for evaluation of jobs or individual workers, in which each is ranked in order from first (most important, or best) to last (least important, or worst).

rankinite *Mineralogy.* $Ca_3Si_2O_7$, a colorless or white transparent monoclinic mineral occurring in massive and granular form, having a specific gravity of 2.94 to 3 and a hardness of 5.5 on the Mohs scale.

rank mimicry *Behavior.* imitation by an animal of lesser status of the behavior of a higher-status individual.

rank order see DOMINANCE HIERARCHY.

rank society *Anthropology.* a society in which a position of high status or prestige is available only by inheritance to members of certain classes.

rank-sum test see MANN-WHITNEY U TEST.

Ranney well *Civil Engineering.* a well in which the central shaft is fed by horizontal perforated pipes; suited to thin, shallow aquifers.

ransomite *Mineralogy.* $Cu^{+2}Fe_2^{+3}(SO_4)_4 \cdot 6H_2O$, a transparent, sky-blue monoclinic mineral, having a specific gravity of 2.632 and a hardness of 2.5 on the Mohs scale; found as tufts of acicular crystals in the fire zones of mines at Jerome and Bisbee, Arizona.

ranula *Medicine.* a cystic tumor located on the undersurface of the tongue or the floor of the mouth.

Ranunculaceae *Botany.* the buttercup family, containing some 50 genera and over 1800 species, most of which are herbaceous; characterized by divided or lobed leaves, usually arising from the base or in an alternate arrangement, and flowers that may be solitary or borne in a raceme or byme; found worldwide, but mostly in temperate and cold climates.

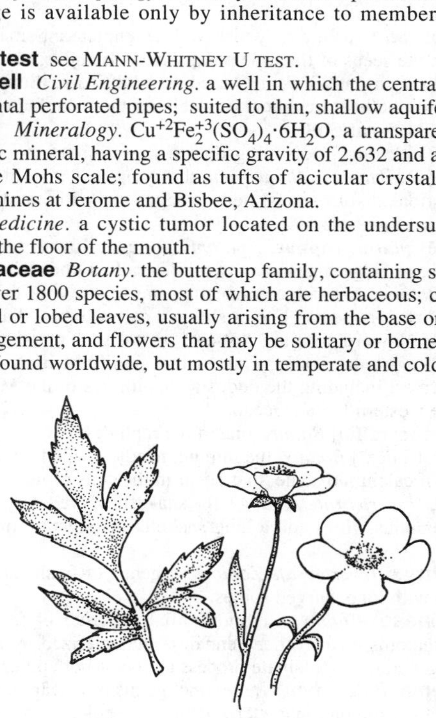

Ranunculaceae (buttercup)

raob [rä′äb′] *Meteorology.* a shorter term for a radiosonde observation. (From radiosonde observation.)

Raoult, Francois Marie [rä ool′] 1830–1901, French chemist; working with solvents, formulated Raoult's law and a general law of freezing.

Raoult's law *Physical Chemistry.* a law stating that the partial vapor pressure of a substance in equilibrium with a solution is equal to the mole fraction of that substance times the vapor pressure of the pure substance; it applies only to an ideal liquid solution when the vapor is an ideal gas.

rapakivi *Petrology.* a granite or quartz monzonite containing large, rounded crystals of orthoclase mantled with plagioclase. Also, WIBORGITE.

rapakivi texture *Petrology.* a texture of igneous and metamorphic rocks characterized by large phenocrysts of potassium feldspar mantled with sodium feldspar and set within a finer-grained matrix of quartz and colored minerals; first described in Finnish granites.

Rapateaceae *Botany.* a family of monocotyledonous, perennial, rhizomatous herbs of the order Commelinales, characterized by vessels in all vegetative organs and basal leaves with a folded, open sheath and a parallel-veined blade; native to tropical South America.

rape *Botany.* a plant, *Brassica napus,* of the family Cruciferae (mustard family), having leaves that are used as food for hogs and sheep and that yield rapeseed oil.

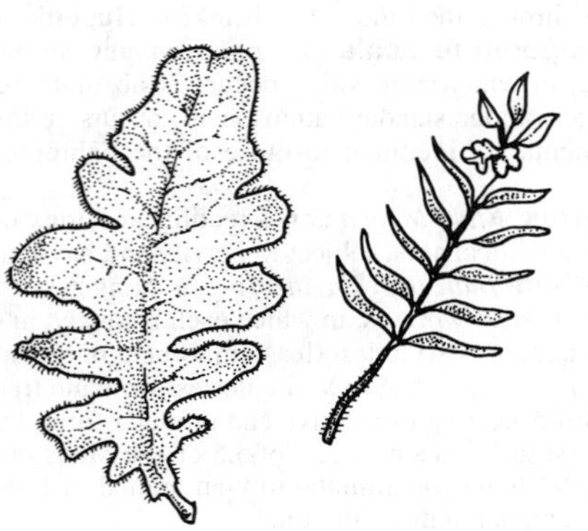

rape

rapeseed oil *Materials.* a brownish-yellow, unpleasant-smelling oil obtained from the seeds of the rape plant, *Brassica napus*; soluble in alcohol and solidifying at 20°C; it is used as a lubricant, illuminant, and a component in the heat treatment of steel and the manufacture of rubber substitutes. Also, **rape oil.**

raphania *Toxicology.* poisoning due to ingestion of the seeds of the wild radish, *Raphanus raphanistrum*; the symptoms may include severe gastrointestinal disturbances, hemorrhage, nephritis, and respiratory failure. Also, RHAPHANIA.

raphe [rā′fē] *plural,* **raphae.** *Anatomy.* a general term relating to the line, furrow, or ridge that unites two bilaterally symmetrical parts of a structure. *Botany.* **1.** a ridge connecting the hilum with the chalaza in certain ovules. **2.** a median line or slit on a cell wall of a diatom. (Going back to a Greek word meaning "to sew" or "to stitch together.")

Raphidae *Vertebrate Zoology.* an extinct family of birds of the order Columbiformes, including the dodos and solitaries of the Mascarene Islands of the western Indian Ocean.

raphide [raf′id; rā′fid] *Botany.* one of the raphides.

raphides [raf′i dēz′] *Botany.* the minute, needle-shaped crystals, usually composed of calcium oxalate, that occur in the cells of many plants.

Raphididae *Invertebrate Zoology.* the snakeflies, predatory insects with four membranous wings and a long snakelike elongation of the prothorax.

Raphidiophrys *Invertebrate Zoology.* a genus of freshwater heliozoan protozoans with long, curved spines.

Raphidoidineae *Botany.* a monofamilial suborder of freshwater and terrestrial diatoms of the order Pennales, characterized by a duo-slitted raphe and generally by a labiate process at one or both poles.

Raphidonema *Paleontology.* an extinct genus of calcareous sponges in the order Pharetronida; characterized by a vaselike shape and frequent twinned growth; extant in the Triassic to Cretaceous.

Raphidophyceae *Botany.* a class of mostly freshwater cellular flagellates belonging to the division Chromophyta, containing the single order Raphidomonadales (or Chloromonadales); characterized by two heterodynamic flagella attached to the cell's anterior groove, with one flagellum pointed to the anterior and having hairs, and the other flagellum pointed to the posterior and lacking hairs.

rapid-access loop *Computer Technology.* a small portion of storage that can be accessed more quickly than the remaining portion.

rapid eye movement see REM.

rapid fire *Ordnance.* a rate of fire faster than slow or deliberate fire, but slower than quick fire; in rapid-fire handgun competition, the shooter must fire five shots in ten seconds.

rapid flow *Hydrology.* water flowing at a rate greater than the velocity of a long surface wave in still water. Also, SHOOTING FLOW, SUPERCRITICAL FLOW.

rapidly reassociating DNA *Molecular Biology.* repetitive sequences of DNA that hybridize rapidly due to the ease with which the strands find complementary sequences.

rapid quenching *Metallurgy.* the process of cooling a molten alloy at a very high rate, usually more than 100,000°C per second; the resulting structure is either amorphous or microcrystalline.

rapids *Hydrology.* **1.** a part of a stream or river in which the current moves significantly faster than usual, as from a narrowing or steepening of the channel or the presence of obstructions; it does not have the definite break in slope of a waterfall. **2.** a swift, turbulent current flowing through such an area.

rapid sequence camera *Graphic Arts.* a camera that feeds film automatically and enables a number of photographs to be taken in rapid succession with a single depression of the shutter button.

rapid solidification *Materials Science.* a processing method used to produce metal with enhanced mechanical performance properties; the metal is cooled very rapidly, causing it to solidify into a noncrystalline or microcrystalline structure with no massive second-phase particles.

rapid storage *Computer Technology.* storage that has a relatively short access time. Also, **rapid memory.**

rapid traverse *Mechanical Engineering.* a quick movement of the table of a machine tool that brings a workpiece up to the cutting mechanism and returns the mechanism to the starting position after the cut.

Rapier *Ordnance.* a British close-range, anti-aircraft missile system; it includes a four-round launcher, an optical tracker, and a power unit; the missiles are powered by a solid-propellant motor and carry a high-explosive warhead at speeds greater than Mach 2.

raptor *Vertebrate Zoology.* **1.** a bird of prey. **2.** specifically, a member of the order Strigiformes or Falconiformes, which hunts by seizing prey from the air and carrying it off; characterized by hook-tipped beaks and talons.

raptorial *Zoology.* adapted for seizing or tearing prey; a term used especially of carnivorous birds.

raptorial legs *Entomology.* legs adapted for catching prey either with pincerlike claws and spines or bristles or with a jack-knife conformation, as in the praying mantis.

RAR radioaccoustic ranging.

Rarden gun *Ordnance.* a British 30-mm automatic cannon mounted on or transported by military vehicles; it has a range of 10,200 meters.

rare bases *Biochemistry.* the modified forms of purine and pyrimidine nucleoside phosphates that are found in transfer RNA molecules.

rare earth *Chemistry.* **1.** the oxide of a rare-earth element. **2.** another term for a rare-earth element.

rare-earth alloy *Metallurgy.* any alloy containing appreciable amounts of one or more rare-earth elements; for instance, mischmetal.

rare-earth element *Chemistry.* a member of a series of fifteen chemically related metallic elements having atomic numbers ranging from 57 to 71, placed in a special row of the periodic table. This group consists of lanthanum, cerium, praseodymium, neodymium, promethium, samarium, europium, gadolinium, terbium, dysprosium, holmium, erbium, thulium, ytterbium, and lutetium. (They are not earths and are not literally rare; they are so called because they were associated with other more familiar substances known as *common earths.*)

rare-earth magnet *Electromagnetism.* a magnet that is manufactured with a rare-earth element, such as a rare-earth cobalt magnet; it can have as much as ten times the coercive force as a typical magnet.

rare-earth metal see RARE-EARTH ELEMENT.

rare-earth mineral *Mineralogy.* any mineral whose composition includes a high percentage of rare-earth elements such as ytterbium and cerium.

rare-earth salt *Inorganic Chemistry.* any of various salts of the rare-earth elements (the lanthanide series), especially a mixture derived from the mineral monazite.

rarefaction *Acoustics.* an area of expansion in a sound wave; the portion of a three-dimensional sound wave in which the medium compression due to acoustic energy is minimum and that, if measured as a standing waving of pressure versus time, would be a point at which signal amplitude is minimum.

rarefaction wave *Fluid Mechanics.* an expansion wave; the pressure jump across such a wave is less than zero.

rarefied gas *Fluid Mechanics.* a very low density gas.

rare genotype advantage *Genetics.* a hypothesis proposing that certain genotypes are advantageous only when they are rare in a population; such selective advantages are progressively lost as the genotype becomes more numerous.

rare species *Biology.* any species that is found infrequently, especially one that is absent or minimally present in a sampling of a given area. *Ecology.* a species of animal or plant that is not in immediate danger of becoming extinct, but that is the subject of protective regulations or conservation measures because of its limited population and distribution.

Rarita-Schwinger equation *Quantum Mechanics.* a partial differential equation describing the behavior of isolated relativistic spin 3/2 particles and their antiparticles.

RAS *Oncology.* a family of genes (c-Ha-*ras*l, v-Ha-*ras*, Ns [Gs]) that encode for GTPases, enzymes that are located at the plasma membrane or in cytoplasm; first identified as transforming genes in Harvey sarcoma virus (H-*ras* or Ha-*ras*) and Kirsten sarcoma virus (K-*ras* or Ki-*ras*).

Rascal *Ordnance.* a guided missile formerly used by the U.S. Air Force; it was propelled by a liquid rocket engine, had a maximum speed of Mach 2.5, and a range of over 100 miles.

raschel knit *Textiles.* a versatile type of warp knitting resembling tricot, done by a special machine.

Raschig process *Chemical Engineering.* a process for producing phenol by the hydrolysis of chlorobenzene, produced by the chlorination of benzene with hydrochloric acid and air.

Raschig ring *Chemical Engineering.* a type of column packing that consists of a hollow, steel, or ceramic cylinder open at both ends.

rash *Medicine.* a general term for any localized or generalized skin eruption.

Ra-Shalom *Astronomy.* asteroid 2100; discovered in 1978, it is an Aten-type astroid with a year shorter than the earth's and a diameter of 3 to 4 km.

rashing *Mining Engineering.* a term for soft, scaly or flaky earth or rock found immediately underneath a coal seam, often containing carbonaceous matter.

rasorial *Zoology.* having feet adapted for scratching the ground; a term used especially of birds.

rasorite see KERNITE.

rasp *Mechanical Devices.* a flat, round, or half-round, coarse file with cutting points used to shape and finish wood, soft metal, leather, and bone.

raspatory *Surgery.* a filelike surgical instrument used to abrade firm materials, such as bone. Also, XYSTER.

raspberry *Botany.* **1.** the fruit of any of various shrubs of the genus *Rubus* of the rose family, consisting of small, juicy, usually red, black, or pale yellow drupelets forming a cone-shaped cap. **2.** any shrub bearing this fruit.

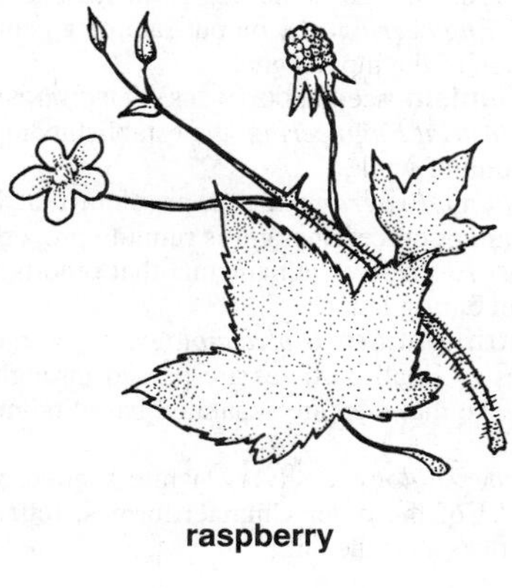

raspberry

raspite *Mineralogy.* $PbWO_4$, a brownish-yellow to gray monoclinic mineral with an adamantine luster occurring as tabular crystals, having a specific gravity of 8.465 and a hardness of 2.5 to 3 on the Mohs scale; dimorphous with stolzite.

Rassenkreis *Evolution.* a polytypic species.

R association *Astronomy.* a young stellar association of hot stars with nearby reflection nebulosity.

Rast camphor method *Organic Chemistry.* a method for determining the molecular weights of unknown substances by noting the effect the unknown molecule has on lowering the freezing point of camphor.

raster *Electronics.* **1.** the track produced on a cathode-ray tube by the scanning of the electron beam, as in a television set or computer display screen. **2.** the pattern produced by an array of such tracks.

raster graphics *Computer Technology.* a method of producing a graphic or pictorial image within a display space by using an array of picture elements arranged in rows and columns that are illuminated by a succession of line-by-line sweeps across the display screen.

raster scanning *Electronics.* the movement of a radar antenna in which the signal's elevation oscillates horizontally, so that its trace resembles the one taken by an electron beam across a television screen.

Rast method *Analytical Chemistry.* a method of determining the molecular weight of organic compounds that uses camphor in the melting point depression.

Rastrites *Paleontology.* a genus of graptolites in the order Graptoloidea and family Monograptidae; characterized by a uniserial rhabdosome and long, isolated theca, which are hooked at the tip; Lower Silurian.

rat *Vertebrate Zoology.* **1.** any long-tailed rodent of various genera of the family Muridae and other families of the suborder Myomorpha, distinguished from mice by their larger size and dental features; found worldwide and noted throughout history as destructive pests and as carriers of disease. Among the species commonly found in North America are the **Norway** or **brown rat**, *Rattus norvegicus*; the **black** or **roof rat**, *R. rattus*; and the **cotton rat**, *Sigmodon hispidus.* **2.** any of various rodents similar in size or appearance to Muridae, such as the kangaroo rat.

cotton rat

RAT remote-association test; rocket-assisted torpedo.

ratbite fever *Medicine.* either of two diseases that are transmitted by the bite of infected rats, typically caused in North America by *Streptobacillus moniliformis* (**Haverfill fever**) and in Asia by *Spirillum minus* (**sodoku**); the symptoms include fever, headache, malaise, nausea, vomiting, and rash.

ratchet *Mechanical Devices.* **1.** a device that consists of a cylindrically shaped bar or wheel with teeth on its inner surface that mesh with a loose-pivoted cog or pawl, allowing movement only in one direction at one time. **2.** to move incrementally with or as if with such a device. **3.** see CLICKWHEEL.

ratchet brace *Mechanical Devices.* **1.** a device with a ratchet-driven chuck, used in carpentry when a tight area allows only a small turn of a screw or bolt. Also, **ratchet tool.** **2.** a lever with a ratchet-driven chuck used for manually drilling holes with a square-threaded feed screw.

ratchet coupling *Mechanical Engineering.* a machine part that joins two shafts by the use of a ratchet system, so that the driven shaft can move in only a single direction and can overrun the driving shaft.

ratchet drill *Mechanical Devices.* a hand drill with a square tapered shank used with a ratchet brace, wheel, and pawl.

ratcheting *Mechanical Devices.* of a device, having or operating with a ratchet. Thus, **ratcheting wrench, ratcheting jack.**

ratcheting box wrench

ratchet screwdriver *Mechanical Devices.* a screwdriver having a ratchet mechanism that allows it to be turned while maintaining a firm grip on the handle; used for driving and removing screws.

ratchet wheel see CLICKWHEEL.

rat distillate *Chemical Engineering.* a term for gasoline and other fuels as they come from the condenser during refining, before impurities have been removed by processing.

rate *Science.* **1.** a quantity that is measured in relation to a unit of time, such as miles per hour or interest per year. **2.** to make such a measurement of a quantity, or to appraise something on the basis of such a measurement. **3.** a proportion that does not include a unit of time, such as an infant mortality rate. (Ultimately derived from a Latin word meaning "to judge.")

rate action *Control Systems.* see DERIVATIVE ACTION.

rate climb *Aviation.* a climb at a constant rate; a steady climb.

rate constant *Physical Chemistry.* the proportionality constant, typically expressed as k, that relates the speed of a chemical reaction rate to some function of reactant concentrations; it is usually a strong function of temperature. Also, REACTION CONSTANT.

rate control *Control Systems.* a control in which the rate or velocity of motion is determined by the position of the controller.

rated appraised or classified according to a rating of some kind; specific uses include: *Electricity.* specified as an operating limit or range, as for a given device under normal or given conditions. Thus, **rated current, rated frequency range, rated output, rated voltage,** and so on. *Aviation.* of an aircraft or piece of equipment, specified as the best or maximum for safe and efficient performance; used to form compounds such as **rated altitude** and **rated engine speed.**

rated capacity *Mechanical Engineering.* **1.** any limit, for safety or other reasons, on volume, pressure, temperature, or the like for vessels, pumps, transfer lines, heaters, or other equipment. **2.** the output of any piece of equipment operating in conformance with a particular criterion; e.g., the maximum holding ability of a boiler, determined by the pounds of steam delivered per hour under designated pressure and temperature conditions.

rated engine speed *Mechanical Engineering.* the maximum speed at which an engine motor can rotate to maintain a reliable and efficient performance level.

rate descent *Aviation.* a descent at a constant rate; a steady descent.

rate-determining step *Chemistry.* in a multistep reaction, the particular step that is the slowest and that therefore determines the rate of the entire reaction.

rated flow *Engineering.* **1.** the flow rate at which a vessel or pipe is designed to operate. **2.** the normal operating rate.

rated horsepower *Mechanical Engineering.* the maximum allowable amount of power that can be produced continuously by an engine or turbine motor.

rated load *Mechanical Engineering.* **1.** the maximum load that a machine or vehicle can carry. **2.** the power output that an engine is designed to produce.

rated relieving capacity *Mechanical Engineering.* the relieving capacity for which a pressure release such as a valve is vented, as measured by a given code or standard.

rated speed *Computer Technology.* the maximum speed or data rate of a device or facility, without allowance for delaying functions such as checking or tabbing. Also, NOMINAL SPEED.

rate effect *Electronics.* the tendency for a semiconductor device to generate large amounts of energy when it receives a surge of voltage.

rate feedback *Electronics.* a signal proportional to the rate of change of the output returned and combined with the input signal.

rate-grown transistor *Electronics.* a device whose substrate contains alternate layers of p-type and n-type material; formed by simultaneously placing two impurities, such as gallium and antimony, into a molten semiconductor material, and then abruptly raising and lowering the melt's temperature.

rate gyroscope *Mechanical Engineering.* a gyroscope suspended in a single gimbal and restrained by a spring, having bearings that serve as its output axis; it produces a couple that is proportional to the rate of rotation.

rate integrating gyroscope *Mechanical Engineering.* a gyroscope that can rotate about only one orthogonal axis, having mainly viscous restraint of its spin axis about its output axis; the angular displacement of a gimbal produces an output signal that is proportional to the integral of the angular rate of the base around the input axis.

rate meter or **ratemeter** *Nucleonics.* a radiation detector that shows instantaneous radiation intensity or the rate of radioactive emissions.

rate of approach *Aviation.* **1.** the rate at which an aircraft executes a series of movements (such as descent, circling, or entering a landing pattern) in preparation for landing. **2.** the rate at which an aircraft approaches some other object or location.

rate-of-change map *Geology.* a stratigraphic map showing the rate at which the structure, thickness, or composition of a given stratigraphic unit changes over time.

rate of change of acceleration *Mechanics.* the rate at which acceleration changes with a variable, such as time or position.

rate of climb *Aviation.* the rate of an aircraft's gain in altitude, usually expressed in feet per minute.

rate-of-climb indicator *Aviation.* a flight instrument used to indicate the rate at which an aircraft is increasing or decreasing altitude; it operates by differential air pressure. Also, VERTICAL SPEED INDICATOR.

rate of closure *Aviation.* the rate at which two aircraft flying toward each other close the distance between themselves; equal to the sum of the speeds of the two aircraft.

rate of departure *Aviation.* **1.** the rate at which an aircraft is leaving some designated flight path or location. **2.** the rate at which two aircraft flying away from each other increase the distance between themselves; equal to the sum of the speeds of the two aircraft.

rate of descent *Aviation.* **1.** the rate at which an aircraft is descending, as shown by a rate-of-climb indicator. **2.** the rate at which an object is falling, such as a parachute and its load.

rate of detonation *Ordnance.* the rate at which the detonation of an explosive progresses; usually expressed as yards or meters per second.

rate of fire *Ordnance.* the number of rounds fired per weapon per minute.

rate of flow *Fluid Mechanics.* see FLOW RATE.

rate of march *Military Science.* the average number of miles or kilometers to be traveled in a given period of time, including all ordered halts; it is expressed in miles or kilometers per hour.

rate of reaction *Chemistry.* the velocity of a chemical reaction, expressed in terms of either the rate of formation of products or the rate of disappearance of reactants.

rate of return *Aviation.* the rate at which a flight vehicle moves toward its base, especially military aircraft after a mission or a reentering spacecraft.

rate of rise *Engineering.* the rate of pressure increase that occurs during leak testing of a vessel.

rate of sedimentation *Geology.* the quantity of sediment accumulated in an aquatic environment in a given time interval, generally represented as thickness of accumulation over time.

rate of strain see STRAIN RATE.

rate of strain hardening *Metallurgy.* in a process of plastic deformation, the change of true stress relative to the true strain.

rate receiver *Electromagnetism.* a receiver that reads a signal emitted by a launched missile and gives the rate of the missile's speed.

rate response *Engineering.* the output rate of a central system, measured as a function of the input signal.

rate servomechanism see VELOCITY SERVOMECHANISM.

rate setting *Industrial Engineering.* the establishment of the amount of pay or time per unit of work.

rate test *Computer Programming.* a problem that is used to check whether a program or processing unit is running properly.

rate transmitter *Electronics.* an antenna that reports a missile's speed to ground control during launch.

rate zonal centrifugation *Biotechnology.* a separation technique in which molecules or small particles are passed through gradients of increasing density, so that they are separated based on molecular size and conformation.

ratfish *Vertebrate Zoology.* a silvery, white-spotted cartilaginous fish, *Hydrolagus collie* of the order Chimaeriformes, found in the northern Pacific Ocean; it has a ratlike tail.

rat flea *Invertebrate Zoology.* any of various fleas, such as *Xenopsylla cheopis*, the carrier of bubonic plague, that infest rats and sometimes also migrate to humans or other hosts.

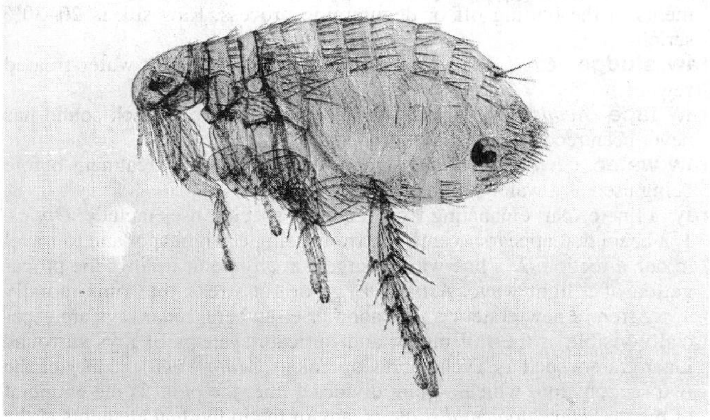

rat flea

rat guard *Naval Architecture.* a shallow cone fitting around a mooring line or other ship-to-shore line, used to prevent rats from moving freely along the line to board a pierside vessel. Also, **ratcatcher.**

rathite *Mineralogy.* $(Pb,Tl)_3As_5S_{10}$, a dark-gray, metallic, monoclinic mineral occurring as twinned, prismatic crystals, having a specific gravity of 5.37 and a hardness of 3 on the Mohs scale; found in crystalline dolomite at Valais, Switzerland.

Rathke, Martin Heinrich 1793–1860, German anatomist; discovered gill slits and arches in bird and mammal embryos and Rathke's pouch in vertebrate embryos; described lancet fish.

Rathke's pouch or **Rathke's pocket** *Developmental Biology.* a diverticulum from the embryonic buccal cavity, from which the anterior lobe of the pituitary gland develops. Also, CRANIOPHARYNGEAL POUCH.

rathole *Mining Engineering.* an auxiliary hole resembling a rat's hole, drilled alongside a main borehole and used for access or storage.

ratiaria larva *Invertebrate Zoology.* the second, hourglass-shaped, free-swimming larva of the siphonophore jellyfish *Velella.*

rating *Science.* the process of appraising and classifying items according to grade or rank. *Engineering.* the operational limit of a device when performing under specific conditions.

rating factor *Industrial Engineering.* in a time study, the worker's actual pace as compared to a normal pace for the operation being studied.

rating nut *Horology.* a nut on a pendulum rod used to adjust the accuracy of a timepiece by raising or lowering the weight or bob, thus changing the period of the swing of the pendulum.

ratio *Mathematics.* a statement of the relative size of two quantities (numbers, functions, and so on), expressed as a quotient. The ratio of *A* to *B* is written as *A*/*B* or *A*:*B*. The **inverse ratio** (or **reciprocal ratio**) of these same terms *A* to *B* is *B*/*A* or *B*:*A*.

ratio arm circuit *Electronics.* the portion of a resistance measuring circuit, known as a Wheatstone bridge, that can be adjusted to various resistance ratios.

ratio control system *Control Systems.* a control system in which the ratio of two process variables is kept constant despite changes in the variables themselves.

ratio-delay study *Industrial Engineering.* a set of random observations of work activities used in estimating the amount of time devoted to specific elements of the work. Also, WORK SAMPLING.

ratio detector *Electronics.* an FM detector circuit that extracts audio signals from a FM radio wave and rejects AM signals.

ratio deviation see MODULATION INDEX.

ratio map *Geology.* a stratigraphic map showing the ratio of thickness of one rock type to that of another in a given stratigraphic unit.

ratio meter *Engineering.* a meter whose deflection is proportional to the ratio of the currents passing through its coils, thus measuring the quotient of two electrical quantities.

rational *Behavior.* **1.** of behavior, based on reason; considered to be an appropriate response to the situation at hand. **2.** of an individual, possessing the power of reason; able to solve problems and make decisions. *Mathematics.* relating to a rational number or a rational function.

rational-emotive therapy *Behavior.* a form of cognitive behavior therapy based on the theory that problems are caused by irrational thinking; designed to help the subject modify such thinking.

rational expression see ALGEBRAIC EXPRESSION.

rational formula *Hydrology.* an equation used to calculate the maximum rate of discharge from a particular drainage basin, equal to the product of the rate of rainfall, the area of the drainage basin, and a runoff coefficient that depends on the characteristics of the basin.

rational function *Mathematics.* a function having the form

$$h(x_1, \ldots, x_n) = f(x_1, \ldots, x_n)/g(x_1, \ldots, x_n),$$

where $f(x_1, \ldots, x_n)$ and $g(x_1, \ldots, x_n)$ are polynomials in the n variables $x_1, \ldots, x_n$; i.e., a function that is the quotient of two polynomials.

rational horizon see CELESTIAL HORIZON.

rational indices law *Materials Science.* a law stating that the indices of natural growth or cleavage planes are small numbers.

rationalization *Psychology.* **1.** a defense mechanism in which an individual offers seemingly logical, but actually questionable, reasons to justify his or her unacceptable behavior, the real motives for which are being repressed. **2.** the general attribution of logical yet questionable reasons to account for one's inappropriate behavior or feelings.

rationalize *Psychology.* to offer a logical yet questionable reason for feelings or behavior. *Mathematics.* to simplify an expression or fraction by eliminating radicals.

rationalized unit *Electricity.* any of certain systems of electrical units for which the factor of 4π is removed from field equations and appears in explicit expressions. In Coulomb's law, this factor is absent from more widely used rationships; in Heaviside-Lorentz units, this is done directly with a modification to the unit charge and unit pole; in MKSA units, it is performed indirectly by modifying values of the permittivity and permeability of free space.

rational number *Mathematics.* **1.** any number that can be represented in the form a/b, where a and b are integers and $b \neq 0$. **2.** equivalently, an element of the ring of quotients of the integers.

ratio of expansion *Mechanical Engineering.* a comparison of the volume of steam in an engine cylinder when the piston has ended its stroke to the volume of steam in the cylinder when the piston is in the cutoff position; the ratio of expansion is the reciprocal of the cutoff.

ratio of reduction *Engineering.* the ratio of maximum-sized stones before they enter a crusher to the size of the crushed stone.

ratio of rise *Oceanography.* the ratio of the height of a tide at one place to its height at another place.

ratio of specific heats *Physical Chemistry.* the ratio between the specific heat of a substance at a constant pressure and its specific heat at a constant volume.

ratio of transformation *Electricity.* in a transformer, the ratio of output voltage to input voltage (or, equivalently, of input current to output current), usually equal to the ratio of turns on the primary winding to turns on the secondary winding. For a current transformer, this ratio is inverted.

ratio scale *Statistics.* a graphical system in which displayed values correspond to proportions, and in which there is a true zero point or origin.

ratio test see CAUCHY RATIO TEST.

ratite *Vertebrate Zoology.* **1.** having a flat breastbone. **2.** any bird of the superorder Palaeognathae, such as an ostrich or kiwi, characterized by a ratite (flat) breastbone.

ratline *Naval Architecture.* on a sailing vessel, any of the rope rungs that form a ladder between the shrouds, allowing seamen to climb the rigging. *Textiles.* the three-stranded, tarred hemp from which such rope is made. Also, **ratline stuff, ratlin.** *Military Science.* a procedure for moving personnel or materiel by clandestine means across a denied border.

rato or **RATO** [rā´tō´] *Aviation.* an auxiliary rocket propulsion system used to provide added thrust for an air or space vehicle during takeoff. (An acronym for rocket-assisted takeoff.)

rato unit *Aviation.* an apparatus designed for rocket-assisted takeoff, usually consisting of a rocket, a propellant, and an ignition system.

rat race *Electronics.* see HYBRID RING.

rat-tail *Veterinary Medicine.* a swollen condition of the hair papillae on a horse's leg, caused by a bacterial infection due to poor hygiene. *Metallurgy.* in a sand casting, a flaw caused by the expansion of the sand.

rat-tail cactus *Botany.* a Mexican cactus, *Aporocactus flagilliformis,* whose slender, cylindrical stem resembles a rat's tail.

rat-tail file *Mechanical Devices.* a slender file with a round, tapered end, used for enlarging holes and for various detail work.

rattlesnake *Vertebrate Zoology.* any snake of the genera *Crotalus* and *Sistrurus*; venomous snakes of North and South America distinguished by a tail modified into a series of horny, loose-fitting structures that produce a buzzing sound when vibrated. Also, **rattler.**

rattlesnake

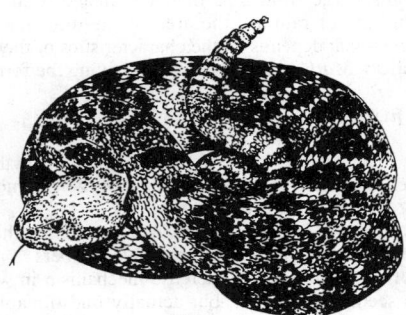

rattlesnake ore *Geology.* a mottled ore of carnotite and vanoxite that is gray, black, and yellow in color and resembles the spots of a rattlesnake. Also, KLAPPERSTEIN.

rattle stone or **rattlestone** *Geology.* a concretion of concentric layers of differing composition in which the central part has become detached from the outer part, so that it produces a rattling sound when shaken.

rat typhus see MURINE TYPHUS.

rauhaugite *Petrology.* a carbonatite containing ankerite or dolomite.

Rauracian *Geology.* in Great Britain, a geologic substage of the Upper Jurassic period, occurring after the Argovian and before the Sequanian.

Rauscher leukemia virus *Virology.* a virus complex of the family Retroviridae, subfamily Oncovirinae, genus Type C oncovirus, consisting of a replication-competent murine leukemia virus and a replication-defective SFFV, that causes lymphoid leukemia in mice.

raushelback rotor *Oceanography.* a free-turning propeller used to measure the flow rate of ocean currents.

Rauwolfia *Botany.* a genus of apocynaceous tropical trees containing over 100 species, many of which provide medicinal alkaloids such as reserpine.

ravelly ground *Geology.* a term for rock that breaks into small pieces when drilled and that tends to partly cave into the drill hole or to bind the drill string when it is pulled.

raven *Vertebrate Zoology.* a large, glossy black bird belonging to the genus *Corvus* in the family Corvidae, known worldwide; characterized by a large bill, a shaggy feathered neck, a wedged tail, a harsh croaking voice, and the ability to mimic human speech.

Raveneliaceae *Mycology.* a family of fungi of the order Uredenales occurring on plant leaves and stems in tropical to subtropical regions.

ravine *Geology.* a deep, narrow, steep-sided hollow that is smaller than a canyon and larger than a gully, and is usually excavated by running water.

ravinement *Geology.* **1.** the formation of a ravine or ravines. **2.** see GULLY EROSION. **3.** an irregular junction that marks a break in sedimentation in shallow-water marine deposits.

raw data *Science.* rough or unprocessed information that is not in its final form.

rawhide *Materials.* **1.** an untanned animal skin, especially a cattle skin. **2.** a whip or rope made of this material.

raw humus see MOR.

rawin *Meteorology.* a method of determining wind speeds and directions in the atmosphere above a weather station by tracking a balloon-borne radar target, responder, or radiosonde transmitter with either radar or a radio direction-finder.

rawinsonde *Meteorology.* a method of upper-air observation that consists of an analysis of the wind speed and direction, temperature, pressure, and relative humidity aloft by means of a balloon-borne radiosonde tracked by a radar or radio direction-finder.

rawin target *Meteorology.* a device that is airborne by balloon and tracked by radar to determine the direction and velocity of the wind.

raw map *Geophysics.* a seismic map having time as the *Z* coordinate.

raw material *Materials.* material that has not yet been processed and incorporated into a finished good in a production or manufacturing process.

raw score *Statistics.* the untreated score or measurement actually observed in a test or experiment.

raw sewage *Civil Engineering.* untreated effluance or waste matter.

raw silk *Textiles.* the fiber of a silkworm as it is reeled from the cocoon, before the gummy coating (sericin) has been removed from the filaments in the boiling off or degumming process. Raw silk is 20–30% sericin.

raw sludge *Civil Engineering.* the result of partially water-treated sewage.

raw tape *Acoustical Engineering.* magnetic tape on which sound has never been recorded. Also, VIRGIN TAPE.

raw water *Civil Engineering.* any water that requires treatment before being used as a water supply.

ray a line or part emanating from a center; specific uses include: *Optics.* **1.** a beam that appears to emanate from a single bright spot and to travel in one direction. **2.** a line whose tangent at any point follows the propagation of a light wave. *Astronomy.* a bright streak that runs radially away from a new crater on the moon or elsewhere; lunar rays are especially visible at the full moon, and intricate systems of rays surround lunar craters such as Tycho and Copernicus. *Mathematics.* either of the two sections into which a point divides a line; the point is the endpoint of the ray. *Developmental Biology.* any of the individual elements at the distal end of an early embryo, foretelling development of the metatarsal or metacarpal bones and the phalanges of the digits. *Vertebrate Zoology.* **1.** any of the arms or branches of a radiate animal such as a starfish. **2.** any of the bony or cartilaginous rods in the fins of a fish.

ray *Vertebrate Zoology.* any of numerous cartilaginous fishes of the order Rajiformes, such as the stingray and the electric ray, having a flat body, eyes on the upper surface, and a slender, whiplike tail.

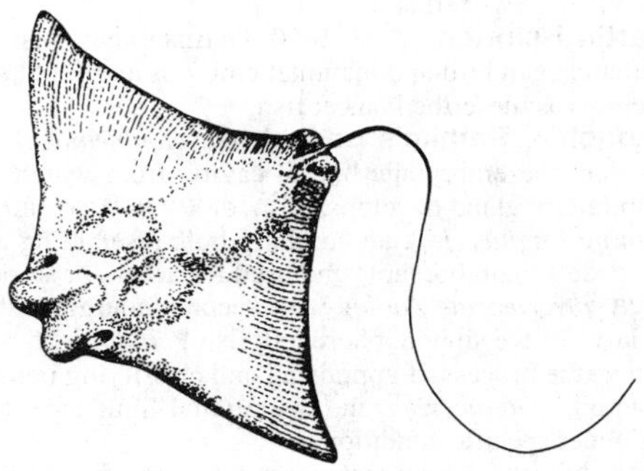

ray

Ray (Wray), John 1628–1705, English naturalist; made fundamental contributions to plant classification; the first to record the distinction between monocotyledons and dicotyledons.

ray acoustics *Acoustics.* the study of acoustics based upon the theory that sound travels in straight paths through a homogeneous medium. Also, GEOMETRICAL ACOUSTICS.

ray crater *Astronomy.* see RAY.

Raydist *Navigation.* a system for distance-finding and position-finding by radio. It involves comparing the phase differences between two continuous-wave signals. Also, RADIST.

ray ellipsoid see FRESNEL ELLIPSOID.

rayfin fish *Vertebrate Zoology.* any fish belonging to the subclass Actinopterygii of the class Osteichthyes, with over 18,000 species found worldwide.

ray fungus see ACTINOMYCES.

ray initial *Botany.* a meristematic cell in the vascular cambium that gives rise to ray cells of the secondary xylem and secondary phloem.

Raykin fender *Civil Engineering.* a fender used to protect ships alongside a pier or wharf, consisting of connected steel plates, each bonded sandwich fashion to layers of rubber.

rayl [rāl´] *Acoustics.* the meter-kilogram-second unit for specific acoustic impedance at a given surface, defined as the complex ratio of the sound pressure averaged over the surface to the volume velocity through the medium; it is expressed in units of N-sec/m^3.

Rayleigh, Baron (John William Strutt) [rā´lē] 1842–1919, English physicist; developed theories of sound and light; awarded the Nobel Prize for the discovery of argon.

rayleigh *Optics.* a unit of flux used to measure the luminous intensity of the night sky and of the aurora; equal to 10^6 photons per 4π steradians per cm^2 vertical column per second. (Named for Baron *Rayleigh.*)

Rayleigh atmosphere *Meteorology.* an ideal atmosphere that consists of only those particules that are smaller than about one-tenth the wavelength of all radiation occurring on that atmosphere.

Rayleigh balance *Electromagnetism.* an apparatus used to assign the value of one ampere of electrical current by means of balancing forces between interacting coils and gravitation.

Rayleigh-Bernard convection *Fluid Mechanics.* incompressible convection between fixed boundaries for viscous fluids in which buoyancy results from temperature perturbations.

Rayleigh criterion *Optics.* the criterion that the images of two point objects are resolved when the principal maximum of the diffraction pattern for one point lies on the first minimum of the diffraction pattern for the other.

Rayleigh cycle *Electromagnetism.* the portion of a magnetization cycle that is part of the initial magnetization from zero to the upward bend.

Rayleigh disk *Acoustics.* a device for measuring sound velocity with a thin disk set at an angle of 45° to a sound beam, producing a torque on the disk from which the velocity may be derived.

Rayleigh interferometer *Optics.* an instrument that splits a single light beam into two, then recombines them with a collimating lens in a tube with a sample gas or liquid, thus producing a shift in the interference fringes; used to determine the refractive index of the sample. Also, **Rayleigh refractometer.**

Rayleigh-Jeans law *Physics.* a relation between the differential intensity dE_λ of blackbody radiation within a narrow range of wavelengths; it is equal to $2\pi ckT\ \lambda^{-4}d\lambda$, where c is the speed of light, k is the Boltzmann constant, T is the absolute temperature, and λ is the wavelength; the law is only experimentally valid for long wavelengths and leads to the ultraviolet catastrophe for short wavelengths.

Rayleigh line *Spectroscopy.* in Raman spectroscopy, a line whose frequency (wavelength) is equal to that of the incident monochromatic light.

Rayleigh number *Thermodynamics.* a dimensionless parameter, arising in free convection flow problems, that can be expressed as a product of two other dimensionless parameters, the Grashof number and the Prandtl number; the Rayleigh number is usually applied in calculating the convective heat transfer coefficient for a given flowing fluid.

Rayleigh number 2 *Thermodynamics.* a dimensionless quantity, Ra_2, used in the study of natural convection, given by the product of the Grashof number and the Prandtl number.

Rayleigh number 3 *Thermodynamics.* a dimensionless quantity, Ra_3, used in the study of combined natural and forced heat convection, given by product of the Rayleigh number 2, the Nusselt number, and the tube diameter, divided by the entry length of the tube.

Rayleigh prism *Optics.* a system of prisms that creates a greater dispersion of light than a single lens.

Rayleigh ratio *Optics.* the ratio of incident light to the light dispersed by Rayleigh scattering; used in photometry and refractometry.

Rayleigh reciprocity theorem *Electromagnetism.* a theorem stating that the effective heights and lengths, radiation resistances, and antenna patterns of an antenna are the same whether it is transmitting or receiving.

Rayleigh relation *Physics.* a ratio expressing the energy distribution of optically heterogeneous systems in which a portion of the beam of light is scattered and the intensity of the beam is reduced.

Rayleigh-Ritz method *Mathematics.* a method for estimating the smallest eigenvalue λ of a self-adjoint linear equation $L\psi = \lambda\psi$ by considering the equivalent variational (minimization) problem and applying conveniently chosen trial functions to the variational integral; especially in the case in which the trial functions are linear combinations of a complete (and possibly orthonormal) family.

Rayleigh scattering *Electromagnetism.* the scattering of electromagnetic radiation by particles having characteristic dimensions that are much smaller than the radiation wavelength. *Optics.* specifically, the scattering of light by fine dust particles as it passes through a transparent medium.

Rayleigh's dissipation function *Mechanics.* a function found in the equations of motion of a system undergoing small oscillations with friction; it represents the frictional force that is proportional to the velocity but opposite in direction, and it is given by a positive definite quadratic form in the time derivatives of the coordinates. Also, DISSIPATION FUNCTION.

Rayleigh-Taylor instability *Fluid Mechanics.* the instability of superimposed fluids of differing densities across an interface, driven by gravity, that leads to mixing if the higher density layer is superior.

Rayleigh wave see SURFACE WAVE.

Raymond concrete pile *Civil Engineering.* a pile formed by driving a thin steel shell into the soil with a mandrel, then filling the shell with concrete.

Raynaud's phenomenon [rā nōz´] *Medicine.* an intermittent bilateral blanching or bluish discoloration of the fingers, toes, ears, and nose, accompanied by severe pallor, numbness, or pain; commonly brought on by emotional stress or by exposure to cold. When this condition persists with no evidence of an underlying disease cause, it is called **Raynaud's disease.** (Named for Maurice *Raynaud,* 1834–1881, French physician.)

rayon *Textiles.* any of various semisynthetic textile fibers composed of regenerated cellulose derived from trees, cotton, and woody plants, and extruded through a spinneret. Rayons are characterized by a high dye acceptance, superior strength and durability, and good draping qualities. (Supposedly formed from the words ray, because of the shiny appearance of the fabric, and cotton, for which it is a substitute.)

ray path *Telecommunications.* a straight path between the transmitting and receiving sites.

ray surface *Optics.* a geometrical construction for determining the velocities and directions of rays vibrating in an anisotropic crystal.

ray system *Astronomy.* a term for the "splatter" pattern of bright material surrounding large fresh craters.

ray tracing *Optics.* a technique that relies on Snell's law and trigonometric functions to trace the paths of rays traversing or passing through an optical element or system. *Computer Science.* the use of this technique in the production of realistic computer-generated images; a large number of simulated rays of light from a light source are followed as they are reflected and refracted by simulated objects, ultimately resulting in an image of the simulated scene.

razon *Ordnance.* a radio-controlled glide bomb that can be guided in range and azimuth during flight; as opposed to an azon, which can be guided in azimuth only.

razorback *Vertebrate Zoology.* a wild or semiwild hog of the southern United States, having a narrowed, ridgelike back; a descendant of escaped domestic hogs rather than a distinct species. *Geology.* a sharp, narrow ridge that resembles the back of a razorback hog.

razor clam *Invertebrate Zoology.* any bivalve mollusk of the family Solenidae having a long, slightly curved shell resembling an old-fashioned straight razor, especially the **Pacific razor clam,** *Siliqua patula.*

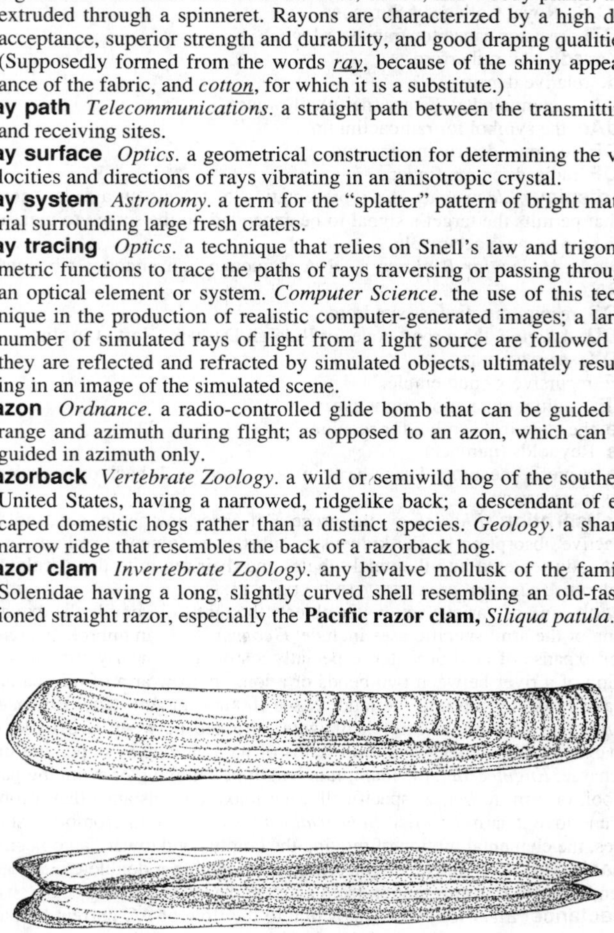

razor clam

razor stone see NOVACULITE.

Rb the chemical symbol for rubidium.

RB1 *Oncology.* a tumor-suppressor gene found to be altered in retinoblastoma, osteosarcoma, and carcinoma of the breast, bladder, and lung.

RBC or **R.B.C.** red blood cell; red blood count.

RBE relative biological effectiveness.

R body *Cell Biology.* a structure present within certain bacterial endosymbionts of protozoans that is thought to confer the killer capability upon that protozoan.

R-branch *Spectroscopy.* in molecular spectra, a set of spectral lines associated with an increase of one in the rotational quantum number, *J.*

RBS 70 *Ordnance.* a Swedish surface-to-air or air-to-air tube-launched missile powered by a solid-propulsion boost/sustain motor and equipped with laser-beam guidance; it delivers a prefragmented 1-kg warhead at supersonic speed and a range of 5 km.

RC resistance-capacitance.

RCC remote center compliance; resistance-capacitance coupled.
RC circuit *Electricity.* a circuit that has both resistance and capacitance arranged in series or in parallel. Also, **RC network.**
RC constant *Electricity.* the time constant (in seconds) for an RC circuit to charge or discharge; equal to the product of the resistance (R) times the capacitance (C).
RCM radar countermeasures.
R.C.N. Royal College of Nursing.
r colony see ROUGH COLONY.
R Coronae Borealis star *Astronomy.* a type of variable star that abruptly declines as much as four magnitudes in brightness and regains its normal brightness slowly; the return to full brightness is thought to be the result of a clearing of dust from the star's atmosphere.
RC oscillator resistance capacitance oscillator.
R.C.P. Royal College of Physicians.
R.C.S. Royal College of Surgeons.
RCTL resistor-capacitor-transistor logic.
rd or **rd.** rod; rutherford.
r.d. relative density.
RDA recommended dietary (daily) allowance.
RdAc the symbol for radioactinium.
RDE receptor-destroying enzyme.
RDF radio-direction finder.
R display *Electronics.* the representation of targets on a radar screen that permits the target's signal to be expanded on the screen to measure the range more accurately.
rDNA *Molecular Biology.* a DNA sequence that encodes ribosomal RNA.
RDS respiratory distress syndrome.
RDU *Aviation.* the airport code for Raleigh/Durham, North Carolina.
RDX see CYCLONITE.
RE recursively enumerable.
R.E. radium emanation; right eye.
Re the chemical symbol for rhenium.
Re Reynolds (number).
re- a prefix meaning: **1.** again, as in *regeneration.* **2.** back or backward, as in *recession.*
reabsorption *Physiology.* the process of absorbing again; e.g., the selective absorption by the kidneys of substances (glucose, sodium, etc.) already secreted into the renal tubules and their return to the circulating blood. *Medicine.* see OBSTRUCTIVE ATELECTASIS.
reach any of various actions or objects thought to resemble the extending of the arm; specific uses include: *Geography.* **1.** an unbroken stretch or expanse of land or water, especially a straight or nearly straight section of a river between two bends or a long, narrow arm of the sea or a lake extending up into the land. **2.** see HARBOR REACH. *Civil Engineering.* **1.** a continuous stretch of water situated between two locks or canals. **2.** a channel's length in terms of its area, depth, slope, and discharge. *Engineering.* the maximum length of a moving or swinging part, tool, or arm. *Robotics.* specifically, the maximum distance that a robot can move its arm or wrist. *Industrial Engineering.* in micromotion studies, the elemental motion of moving the empty hand toward an object.
reach rod *Mechanical Engineering.* a rod having a double eye at each end, used to transmit motion from a reversing rod to a lifting shaft.
reactance *Electricity.* the opposition to the flow of AC current by pure inductance or capacitance of the circuit; expressed in ohms.
reactance amplifier see PARAMETRIC AMPLIFIER.
reactance drop *Electricity.* a voltage drop across a circuit due to inductance or capacitance.
reactance frequency multiplier *Electronics.* a frequency multiplier that generates harmonics of a sinusoidal source by utilizing a nonlinear reactor.
reactance relay *Electricity.* an impedance relay that is activated when the reactance of a circuit connected to it drops below a predetermined value.
reactance tube *Electronics.* a device, such as a transistor or vacuum tube, in which reactance is produced by placing the collector current or anode current in phase with the base current or grid current.
reactance-tube modulator *Electronics.* a circuit that uses a reactance tube controlled by the modulating signal to modulate the frequency of the carrier.
reactant *Chemistry.* any substance or molecule that participates in a chemical reaction.
reactant ratio *Space Technology.* in a rocket engine, the ratio of oxidizer to fuel expressed in terms of mass flow.
reaction an opposite action or counteraction; specific uses include: *Mechanics.* the equal force or torque offered by a body in opposition to a force or torque applied to it, according to Newton's third law of motion. *Nuclear Physics.* see NUCLEAR REACTION. *Chemistry.* see CHEMICAL REACTION. *Physiology.* any response to physical stimulation, such as a reflex reaction. *Psychology.* the mental and emotional response that is elicited in response to a given situation. *Control Systems.* see POSITIVE FEEDBACK.
reaction bonding see REACTION SINTERING.
reaction boundary or **reaction curve** see REACTION LINE.
reaction center *Biochemistry.* a site in the chloroplast that receives the energy trapped by chlorophyll and accessory pigments and initiates the electron transfer process.
reaction chain see ACTION CHAIN.
reaction constant see RATE CONSTANT.
reaction engine *Aviation.* an engine that produces thrust by means of its reaction against some product that is ejected from it, such as a stream of gases of combustion released as exhaust.
reaction fin *Naval Architecture.* a stationary multifinned device mounted ahead of a vessel's rotating propeller and designed to feed water into the propeller so that the water is moving in a direction opposite to the propeller's rotation, thus stabilizing the propeller's performance.
reaction formation *Psychology.* a defense mechanism in which an individual behaves in a way directly opposite from an underlying impulse, thereby denying his real feelings.
reaction hot pressing *Materials Science.* in ceramics processing, the densification of two or more powders which undergo a reaction at or near the pressing temperature under the simultaneous application of heat and pressure.
reaction injection molding *Materials Science.* in the production of polymer components, a process in which monomers are mixed by impinging streams just prior to mold entry, and the polymer forms in the mold under low molding pressures; used primarily for polyurethane products, especially in the automotive industry for parts such as bumpers, fender extensions, and spoilers.
reaction kinetics see CHEMICAL KINETICS.
reaction line *Physical Chemistry.* the boundary along which the crystalline phase of a substance reacts with the liquid phase as the substance cools, to form a second crystalline phase, seen in substances formed from three elements. Also, REACTION BOUNDARY, REACTION CURVE.
reaction motor *Space Technology.* a solid-fuel rocket motor. *Aviation.* loosely, any reaction engine.
reaction pair *Mineralogy.* any two minerals, one of which is formed by reaction with liquid at the expense of the other; especially, adjacent minerals in the discontinuous reaction series.
reaction principle *Mineralogy.* the assertion that the principal igneous rock-forming minerals may be grouped into two series, termed reaction series, the members of which have a reaction relation to one another.
reaction propulsion *Space Technology.* propulsion in reaction to rearward projection of a substance; based on Newton's third law of motion, stating that for every reaction there is an equal and opposite reaction. Such reaction is achieved by means of a reaction engine.
reaction rim *Petrology.* the outer zone of a composite crystal in an igneous rock; it is formed by reaction of the core mineral with the surrounding magma. Also, CORONA.
reaction series *Mineralogy.* any series of minerals whose members are related in such a way that each member is derived from the preceding member by reaction with the melt; such series are classified as either continuous (plagioclase feldspars) or discontinuous (ferromagnesian minerals). Also, BOWEN'S REACTION SERIES.
reaction sintering *Materials Science.* a process in which a compacted powder is heated to a temperature where sintering takes place simultaneously with a chemical reaction; for example silicon powder heated in the presence of nitrogen to produce silicon nitride. Also, REACTION BONDING.
reaction-specific energy *Behavior.* the energy that motivates the occurrence of a specific behavior.
reaction time *Physiology.* the time elapsing between the application of a stimulus and the reaction to that stimulus. Also, LATENT PERIOD.
reaction turbine *Mechanical Engineering.* a turbine having nozzles or jets mounted on a moving wheel so that torque is created from the gradual decrease of steam pressure from inlet to exhaust; characterized by an acceptably high stage efficiency over a broad range of the ratio of blade speed to jet speed.

reaction wheel *Mechanical Engineering.* a device that can store angular momentum; used in spacecraft to supply the rotational force needed to maintain a desired orientation.
reaction wood *Botany.* abnormal wood formed in tilted stems; this includes compression wood, which forms on the undersides of tilted conifers, and tension wood, which usually forms on the upper sides of tilted stems of broadleaved trees.
reaction zone *Chemical Engineering.* the zone of heating in a kiln in which the charge is burned, decomposed, reduced, and oxidized.
reactivation the fact of becoming active again; specific uses include: *Virology.* the activation of a latent virus or a defective virus as a result of coinfecting the viruses and reassorting their genomes. *Immunology.* the restoration of activity to immune serum that has had its activity destroyed.
reactive *Science.* **1.** tending to react; volatile or unstable. **2.** relating to or characterized by reaction. *Electricity.* being inductive or capacitive, as opposed to resistive.
reactive bond *Chemistry.* a chemical bond between two atoms that is easily broken and invaded by other atoms.
reactive calcination *Materials Science.* a process of heating inorganic material to remove volatile components in a manner that creates solid-state reactions between oxides, resulting in the formation of spinels; the process involves solid-state diffusion of metallic ions and is sometimes accompanied by solid-state or gaseous transport of oxygen.
reactive component *Electricity.* in the phasor representation of an AC circuit, the component of voltage or current that does not add power.
reactive current *Electricity.* the component of alternating current that is in quadrature with the voltage. Also, IDLE CURRENT, INACTIVE CURRENT, WATTLESS CURRENT, QUADRATURE CURRENT.
reactive dye *Materials Science.* a dye that is fixed by reacting chemically with the fiber molecules; used especially in dyeing cellulose-based textiles.
reactive factor *Electricity.* the proportion of reactive volt-amps to the total available volt-amps in a system.
reactive hemolysis *Immunology.* the destruction of red blood cells that is caused by a stable complex of activated C5 and C6 components of a complement.
reactive intermediate *Chemistry.* an unstable product that is formed during a middle step in a composite reaction.
reactive ion etching *Electronics.* a technique in which an electric field causes energized ions to strike the surface of a semiconductor substrate at a perpendicular angle; used in integrated circuit fabrication.
reactive load *Electricity.* a load device that is capacitive or inductive, not resistive.
reactive muffler *Acoustical Engineering.* a muffler that dampens sound by reflecting it back toward its source.
reactive power *Electricity.* power, measured in volt-amps-reactive (vars), that cannot do work. For sinusoidal voltage and current, reactive power is $I{\cdot}V{\cdot}\sin\theta$, where θ is the phase difference between the voltage and current.
reactive voltage *Electricity.* the component of a phasor that represents voltage of an AC circuit in quadrature with current in the circuit.
reactive volt-ampere meter see VARMETER.
reactivity the fact of reacting or taking part in a reaction; specific uses include: *Chemistry.* **1.** the capacity to combine chemically with another atom, molecule, or radical. **2.** a kinetic property of a chemical species, expressed in terms of the size of the rate constant for a specified reaction, relative to another species. *Nucleonics.* a measure of the departure of the state of a nuclear reactor from the critical state where the reactor is just self-sustaining, expressed as a percentage of the multiplication constant k, in units of dollars, cents, or hours; positive values of the reactivity correspond to supercritical states, and negative values to subcritical states.
reactivity ratio *Materials Science.* in copolymerization reactions, the ratio of the reaction of each species with its own monomer.
reactor something that causes or that is the medium for a reaction; specific uses include: *Nuclear Physics.* a shorter term for a nuclear reactor. See NUCLEAR REACTOR. *Chemical Engineering.* any process equipment used to carry out a chemical reaction. *Electricity.* see ELECTRIC REACTOR.
reactor fuel cycle *Nucleonics.* a cycle that includes the preparation and use of fuel elements in a nuclear reactor followed by the recovery and reprocessing of the radioactive by-products into new useful fissionable fuel elements. Also, NUCLEAR FUEL CYCLE.
reactor-grade uranium *Nucleonics.* uranium that is enriched to about 3.5% uranium-235.
reactor period *Nucleonics.* the amount of time required for neutron flux density to exponentially change by a factor of e (about 2.71828), the base of the natural logarithm.
reactor physics *Nucleonics.* a branch of physics concerned with the processes that occur within the core of a nuclear reactor.
reactor vessel *Nucleonics.* a large container built around a nuclear reactor core to prevent the escape of radioactive materials or radiation.
read to carry out an electronic process that is in some way comparable to the human activity of understanding written or printed text; specific uses include: *Computer Technology.* to obtain data from an input device or storage medium without changing the contents. *Electronics.* to translate magnetic spots, characters, or punched holes, into electrical impulses. *Telecommunications.* in radio or telephone communication, the ability to understand clearly.
Read diode *Electronics.* a diode that converts energy with an extremely high frequency and power exceeding hundreds of thousands of volts.
reader an electronic device that analyzes data in a manner comparable to human reading; specific uses include: *Computer Technology.* any device that transcribes data from an input medium and enters it into the computer. *Graphic Arts.* **1.** an electronic device designed to scan the perforated holes in a tape and transmit corresponding signals to a typesetting machine. **2.** see OPTICAL CHARACTER RECOGNITION UNIT.
reader-interpreter *Computer Technology.* a service routine that reads the input stream, recognizes control information, and interprets it by executing the appropriate programs.
read error *Computer Technology.* a condition in which data from a storage medium or input device is not read correctly for reasons such as equipment failure or damage to the recording surface.
read-eval-print loop see LISP INTERPRETER.
read head *Computer Technology.* a sensor used to read data from a medium such as a tape, disk, or drum.
read-in *Computer Technology.* to acquire or interpret source data and transmit the data to internal storage.
readiness time *Industrial Engineering.* the time required for a system to start its intended function, measured from the time it is unassembled to the time it begins operating properly, including warm-up time, but excluding maintenance time.
reading *Engineering.* **1.** the data displayed by an instrument. **2.** the observation or recording of such data.
reading frame *Molecular Biology.* the correct way, of three possible ways, to translate a given mRNA sequence to produce a functional protein based on the position of the initiation codon.
reading frame shift *Molecular Biology.* a shift in the manner in which the nucleotides are grouped into codons for translation, resulting from a mutation, and producing a nonfunctional sequence of amino acids.
reading microscope see CATHETOMETER.
reading point see BREAKPOINT.
reading rate *Computer Technology.* a measure of the speed at which an input device can read units of data during a specified unit of time.
reading station *Computer Technology.* the part of a reader that physically reads the data on a data input medium. Also, **read station.**
reading telescope see CATHETOMETER.
read-only *Computer Technology.* relating to or employing read-only memory (ROM).
read-only memory *Computer Technology.* ROM, a memory device containing data that can be read but cannot be altered; the memory contents are either permanent, built into the device during its manufacture, or semipermanent, alterable only under special conditions. Also, READ-ONLY STORAGE, FIXED MEMORY.
read-only storage see READ-ONLY MEMORY.
readout *Computer Technology.* the representation of data in a visual form such as a display, printout, plot, or photograph.
readout station *Telecommunications.* a receiving or recording radio station at which information is received as a transmitter reads the information out.
read protection *Computer Programming.* a protection code used with a file of data to prevent unauthorized users from reading the data.
read screen *Computer Technology.* the transparent component of an optical character reader, through which the document that is to be read appears.
readthrough or **read-through** *Genetics.* **1.** transcription that occurs beyond a normal terminator sequence in DNA when the RNA polymerase fails to recognize the stop codon. **2.** translation beyond the stop codon of a messenger RNA molecule.
read time see ACCESS TIME, def. 2.

read-while-writing *Computer Technology.* the reading of a record or set of records into storage simultaneously with the writing of another record or set of records from storage.

read/write channel see DATA CHANNEL.

read/write check indicator *Computer Technology.* a device on some computers that indicates whether an error occurred during reading or writing and instructs the computer to take specific corrective actions.

read/write head *Computer Technology.* an electromagnetic device capable of reading, writing, and erasing data encoded as polarized bit patterns on magnetic disk, tape, or drum.

read/write memory *Computer Technology.* any memory device that allows both reading and writing.

ready queue see CENTRAL-PROCESSOR QUEUE.

ready-to-receive signal *Telecommunications.* a signal returned to a facsimile transmitter that gives the message that a facsimile receiver can accept the transmission.

reafforestation see REFORESTATION.

reagent [rē ā´jənt] *Chemistry.* any substance that reacts or participates in a chemical reaction. *Analytical Chemistry.* specifically, a substance that is used in a reaction for the purpose of testing, analyzing, or detecting other substances; types of reagents include precipitants, solvents, oxidizers, and reducers.

reagent chemical *Analytical Chemistry.* a high-purity (grade) laboratory chemical used in analytical techniques and reactions where impurities must be absent or at known concentrations.

reagent solution *Chemistry.* an aqueous solution, usually 10% concentration, of any reagent that is used in the laboratory.

reagin [rē´ə jin] *Immunology.* **1.** an antibody that readily sensitizes the skin and that occurs in allergic reactions such as asthma. Also, **reaginic antibody. 2.** an antibody that occurs in association with syphilis. Also, WASSERMAN REAGIN.

real address see DIRECT ADDRESS.

real axis *Mathematics.* the horizontal axis in the Argand diagram of the complex plane; all complex numbers of the form $a + 0i$, where a is real.

real data type *Computer Programming.* a data type whose values represent approximations to the real numbers, i.e., the floating point types.

real fluid flow *Fluid Mechanics.* a flow that takes into consideration the energy lost by the flowing fluid through friction with the boundaries restricting the fluid's motion; real fluid flow in a piping system, for example, would take into consideration the fluid head loss due to the fluid friction encountered in the piping elements.

real force *Mechanics.* a force that can be traced to the effect of an actual physical origin, as distinguished from a fictitious force that is postulated to account for some observed effect involving rotation of a reference frame.

real function *Mathematics.* **1.** a function of one or more real variables that takes on real values. **2.** an analytic function whose power series expansion about some real number has only real coefficients.

realgar *Mineralogy.* AsS, an orange-red, monoclinic mineral with a resinous luster, occurring in massive form and as short prismatic crystals, having a specific gravity of 3.56 and a hardness of 1.5 to 2 on the Mohs scale; found in low-temperature veins with other arsenic-antimony minerals.

real gas *Thermodynamics.* a gas whose properties deviate from those of the hypothetical ideal gas due to interactions occurring between the gas molecules. Also, IMPERFECT GAS.

real image *Optics.* the reproduction of an object formed by the convergence of light rays that have traversed a lens or other image-forming device.

reality principle *Psychology.* in psychoanalytic theory, an individual's awareness of the societal demands of the environment and subsequent regulation of behavior in order to meet those demands.

realizability *Control Systems.* a property of the transfer function of a network that consists only of resistances, capacitances, inductances, and ideal transformers.

reallification *Mathematics.* the process of restricting complex parameters or variables to real values only; for example, vector components in a vector space or the parameters in a Lie group. The properties of the reallified object can often be derived from the complex form. Also, REAL RESTRICTION.

real number *Mathematics.* a number representing a point on the real line; an element of the completion of the rational numbers. The real numbers, denoted R, form a field, and any complete ordered field is isomorphic to the field of real numbers. Constructions of the real numbers include the method of Dedekind cuts and that of Cauchy sequences.

real orthogonal group *Mathematics.* the group of $n \times n$ orthogonal matrices with real entries. The subgroup of matrices with determinant equal to 1 is the **special real orthogonal group** or **real unimodular group.**

real part *Mathematics.* the value of a for a complex number $a + bi$, where a and b are real numbers; b is the imaginary part.

real power see ACTIVE POWER.

real restriction see REALLIFICATION.

real solution see NONIDEAL SOLUTION.

real source *Acoustics.* a body that produces sound waves in a medium composed of a different material.

real-space averaging *Crystallography.* a computational method for improvement of phases found for Bragg reflections in the X-ray diffraction pattern of a crystal, applied whenever there are two or more identical chemical units in the crystallographic asymmetric unit. In an initial electron density map, the electron densities of the identical units are averaged. A new set of phases is then found by Fourier inversion of this averaged structure, and with these a new map is synthesized with the observed $|F|$ values.

real storage *Computer Technology.* a hardware device that serves as the main storage in a virtual storage system. Also, **real memory.**

real time *Science.* the actual time during which a physical process takes place.

real-time *Computer Technology.* **1.** describing an operating mode in which the input and processing of data occurs concurrently with an external activity and in time to affect that activity or subject to strict constraints on the elapsed time required to complete processing, as in the updating of an airline reservation system upon the input of reservation data. **2.** relating to real time.

real-time analyzer *Acoustics.* an electronic device that measures frequency as a function of sound pressure level or time, by dividing the selected frequency range by a device-specific number of equal frequency filters and making the measurements over each of the ranges simultaneously.

real-time clock *Computer Technology.* **1.** a clock in a computer that keeps track of date and time. **2.** an interval timer that allows a computer to monitor elapsed time between events or to request that the clock generate an interrupt after a certain period of time. Thus, **real-time clock queue.**

real-time control *Computer Technology.* a processing or controlling mechanism that occurs simultaneously with the operation or process in progress.

real-time control system *Computer Technology.* a control system that processes input events fast enough to control a physical system directly as it operates, as in an aircraft autopilot.

real-time operating system *Computer Technology.* a scheduling system that provides fast response to input signals and transactions, often maintaining several levels of priority and processing higher-priority tasks first.

real-time processing *Computer Technology.* a mode of computer operation in which data processing and transmission occurs concurrently with a transaction that is immediately affected by the results of the processing. Also, **real-time operation.**

real-time program *Computer Programming.* a program that runs concurrently with an external activity that imposes time constraints upon the processing of data. Also, **real-time system.**

real variable *Mathematics.* a variable whose domain of definition is a subset of the real numbers.

ream *Engineering.* to enlarge or clear out an already existing hole, especially by means of a reamer.

ream *Materials Science.* a standard quantity of paper sheets; usually contains 500 sheets, but may vary from 480 to 516 sheets.

reamed extrusion ingot *Metallurgy.* stock used for tube extrusion, consisting of a shell that has been machined internally.

reamer *Mechanical Devices.* a rotating finishing tool that consists of a metal fluted cone with sharpened edges and a handle for turning; used with a bit for removing burrs from the ends of various kinds of metal pipe after cutting, or for finishing and truing a cored or drilled hole. *Medicine.* a dental drill having a spiral blade; used especially to enlarge root canals.

reaming bit *Mechanical Devices.* a bit used with a reamer to enlarge and smooth a bore hole.

reannealing *Molecular Biology.* the process of renaturing complementary single-stranded DNA molecules to yield duplex molecules; usually refers to strands that were originally separated prior to melting.

reaper

reaper *Agriculture.* any of several early types of machines used for cutting (or reaping) grain, the best known of which was patented by Cyrus McCormick. Also, **reaping machine.**

rear area *Military Science.* the area in the rear of the combat and forward areas; for any particular command, it extends forward from the rear boundary of the command to the rear boundary of the next lower level of command; it is used primarily for the performance of combat service support functions. Also, ARMY SERVICE AREA.

rear echelon *Military Science.* the elements of a force that are not required in the objective area.

rear guard *Military Science.* a security detachment that protects the rear of a column from a hostile attack; during a withdrawal, the rear guard delays the enemy by armed resistance, destroying bridges and blocking roads.

rearming *Ordnance.* **1.** the process of replenishing the stores of ammunition, bombs, or other armament items for an aircraft, ship, tank, or armored vehicle. **2.** a process of resetting the fuse on a bomb or projectile so that it will detonate at the desired time.

rear projection *Electronics.* **1.** a television system in which the picture is projected onto a glass screen so that it can be seen from the opposite side of the screen; rear projection is the standard television technology for all but certain large-screen sets. **2. rear-projection.** relating to or using such a system. **3.** a technique used in motion-picture filming in a studio, in which the actors stand in front of a screen on which some kind of background action or scenery appears, to give the illusion of filming on location.

rearrangement reaction *Chemistry.* a chemical reaction in which the atoms of a single compound recombine to form a new compound with the same molecular weight as the original compound but different chemical properties. *Nuclear Physics.* a nuclear reaction that causes neutrons and protons to exchange nuclei.

rear response *Acoustical Engineering.* the relative directivity of a transducer between 90° and 270° with respect to the direction of maximum response, centered on 0°.

rear sight *Ordnance.* the sight mounted nearest the breech end of a weapon; it is usually used in conjunction with the front sight to establish a true line of sight to the target or aiming point.

rear suspension *Mechanical Engineering.* an arrangement of parts (e.g., hangers, springs, and shock absorbers) connecting the rear axle of an automobile to the chassis frame.

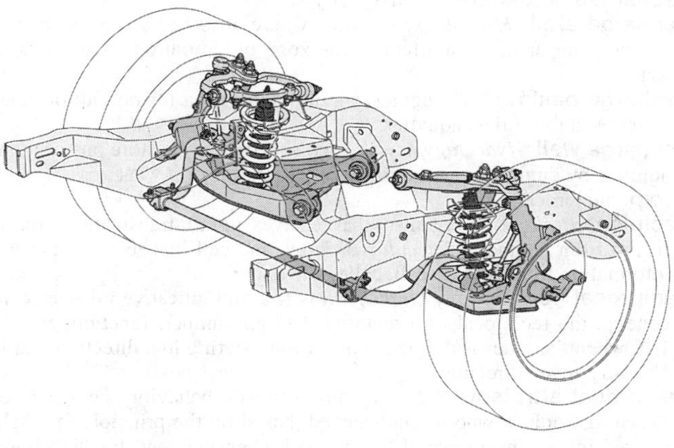

rear suspension

reasonableness *Computer Programming.* a term for the absence of gross errors in data or results, achieved by testing the values against specified criteria.

reasoning under uncertainty *Artificial Intelligence.* reasoning about situations, e.g., in medical diagnosis, in which good or complete data is not available, and for which decisions must be made based on available data and knowledge of likelihoods of the various possibilities. Also, UNCERTAIN REASONING.

reason maintenance system see TRUTH MAINTENANCE SYSTEM.

reassociation *Molecular Biology.* the reannealing of complementary single-stranded DNA into double-stranded DNA; the term usually refers to strands that were originally separated by melting.

reassociation kinetics *Molecular Biology.* a measurement of the rate at which complementary DNA strands reassociate to form duplexes.

reassortant *Virology.* a hybrid virus that contains genome segments from different parents, resulting from the coinfection of a cell with different strains of a particular virus.

reassortment *Virology.* the production of a hybrid virus that contains parts derived from the genomes of two different viruses in a mixed infection.

reattribution therapy *Psychology.* therapy that encourages the subject to reevaluate his explanation of his own behavior and that of others, so that he will recognize the influence of external circumstances or, conversely, become aware of internal motives.

Réaumur, René de [rā´ə myůr´] 1683–1757, French physicist and naturalist; invented the Réaumur temperature scale; his extensive writings include works on geometry, metallurgy, entomology, and botany.

Réaumur temperature scale *Thermodynamics.* a temperature scale in which the freezing point of water at one standard atmosphere occurs at 0°R and the boiling point occurs at 80°R.

rebat *Meteorology.* a daytime lake breeze blowing across Lake Geneva, Switzerland.

reboiler *Chemical Engineering.* a process heating device that receives liquid from the bottom stage of a fractioning tower, vaporizes it with steam or other hot process fluid, and returns it to the column.

reboot *Computer Technology.* to restart the operating system, often after an error has occurred. Also, REINITIALIZE.

rebound *Neurology.* a reflex response in which the sudden withdrawal of a stimulus is followed by movement in the intended direction and then a jerk, or rebound, in the opposite direction. *Materials Science.* the ability of a material to retract rapidly to roughly its original size and shape; one of the properties of a typical elastomer. *Geology.* the upward adjustment of the crust to maintain mass-density equilibrium when a mass that was loading it is removed, as by the melting of glacial ice.

reboyo *Meteorology.* a Brazilian coastal storm from the southwest that occurs during the rainy season and persists for a full day.

rebreather *Engineering.* a closed-loop oxygen-delivery system that includes a gas mask and oxygen supply.

rec. fresh. (From Latin *recens.*)

recalescence *Materials Science.* the increase in temperature of a solidifying liquid that is growing dendritically; caused by the transfer of the latent heat of fusion into the undercooled liquid. *Metallurgy.* spontaneous heating of gamma iron as it is transformed to alpha upon cooling.

recall *Psychology.* the process of bringing previously learned information or past experiences from memory back into one's consciousness. *Telecommunications.* **1.** in telephony, the process of bringing an operator into an established circuit by operating the switch hook of the subscriber's set. **2.** a flashing signal sent to an attendant's switchboard.

recall factor *Computer Programming.* in an information retrieval system, a comparison of the numbers of documents found to the total number searched.

recapitulation *Biology.* a theory stating that, as an organism develops, it passes through a sequence of embryonic or juvenile stages that resemble the evolutionary forms of its successive ancestors; this is now regarded as a gross oversimplification.

recA protein *Biochemistry.* a protein encoded by a recA gene in *Escherichia coli* and recA-like gene products in other bacteria that triggers the joining of a single-stranded piece of DNA with its complementary sequence.

recarburize *Metallurgy.* **1.** to restore the original carbon content of a decarburized alloy. **2.** to increase the carbon content of molten cast iron or steel.

recast *Cartography.* to change the horizontal datum of a map to a different one by changing the appropriate geographic values of the map's graticule.

recBCD pathway *Molecular Biology.* a mechanism for general recombination in *Escherichia coli* and related bacteria following bacterial conjugation.

receding leg *Military Science.* in tracking a moving target, the part of the target's course during which the slant distance increases with the target's movement.

received power *Electromagnetism.* the total power that is picked up by a receiving antenna and delivered to the amplifying circuit for signal processing.

receive-only *Telecommunications.* of or related to a teleprinter that can receive and print signals, but not transmit them.

receiver a device or apparatus that receives; specific uses include: *Electronics.* any of various devices that convert electrical energy into images or sounds; for example, a television set or a radio. *Acoustical Engineering.* specifically, a component of a sound system that combines the functions of a preamplifier, power amplifier, and tuner; current models also have controls to receive sound from other sources such as a tape deck, CD player, TV set, or VCR. *Telecommunications.* the part of a telephone that is held to the ear to hear. *Chemical Engineering.* a vessel used to collect and hold a process stream. *Mechanical Engineering.* an instrument that balances the flow of air from a compressor as it moves into a pipeline and clears the air of moisture and oil. *Ordnance.* the metal part of a firearm that includes the breech action and firing mechanism; the basic component to which the barrel and stock are attached.

receiver bandwidth *Electronics.* the difference of frequencies of the half-power points on a receiver's response curve.

receiver gating *Electronics.* the practice of applying a voltage to a receiver at the point in the operating cycle at which reception is required.

receiver incremental tuning *Electronics.* the feature that permits the receiver of a transceiver to be tuned within three kilohertz of the transmitter frequency.

receiver noise threshold *Electronics.* the level an unwanted signal must reach before it will be picked up by a receiver.

receiver radiation *Electromagnetism.* the radiation due to the oscillator circuit of a receiving system; this radiation is responsible for some of the noise present in the processed signal.

receiving antenna *Electromagnetism.* an antenna used to convert electromagnetic waves to radio-frequency currents.

receiving area *Electromagnetism.* a factor given by the product of the antenna gain and the square of the wavelength divided by 4π; this factor multiplied by the power density gives the total received power of a receiving antenna.

receiving controller *Transportation Engineering.* the controller at a new station to which an aircraft is assigned as it passes from one ground control station's control zone to another.

receiving house *Chemical Engineering.* a building in a petroleum refinery in which process streams are physically observed.

receiving loop loss *Telecommunications.* in telephony, the section of the repetition equivalent which is assignable to the subscriber line, station set, and battery supply circuit that are located on the receiver's end.

receiving station *Mechanical Engineering.* the section of a conveyor system that is designed for the storage and loading of bulk material.

receiving tube *Electronics.* a vacuum tube that operates at low energy levels; commonly used in radio receivers, computers, and sensitive control equipment.

recemented glacier see GLACIER REMANIÉ.

recency effect *Psychology.* the principle that the most recently presented items or experiences will most likely be remembered best.

recent *Evolution.* describing species or other taxonomic groups that still exist, as oppposed to those that are extinct.

Recent see HOLOCENE.

receptacle *Botany.* **1.** in angiosperms, the typically expanded portion of the shoot to which the floral units are attached; it may enlarge to become part of the fruit as in the common strawberry, *Fragaria.* **2.** in ferns and nonvascular plants, the point on a leaf or thallus that typically bears the reproductive structures such as sporangia, gemmae, or reproductive organs. *Electricity.* see OUTLET.

reception *Telecommunications.* **1.** the conversion into useful information of modulated electric signals or electromagnetic waves transmitted through the air or via wires or cables. **2.** the act of receiving, listening to, or watching signals that carry information. **3.** the quality or fidelity with which such signals are received.

receptive character *Psychology.* a personality type that is characterized by passivity and excessive dependence on the support and guidance of others.

receptor *Physiology.* a specialized sensory nerve structure (exteroceptor, interoceptor, or proprioceptor) at the peripheral end of a sensory neuron that responds to specific types of stimuli. *Cell Biology.* a site in a cell, usually on a membrane that combines with a chemical to specifically alter cell function.

receptor-destroying enzyme *Enzymology.* an enzymatic component of a virion that destroys the surface receptors of the host cell.

receptor recycling *Molecular Biology.* the movement of a membrane receptor back out to the surface of the cell after delivery of its ligand to the interior of the cell.

receptosome see ENDOSOME.

recess *Engineering.* a depression or groove on the surface of an object. *Geology.* **1.** any surface indentation. **2.** an area in which the axial traces of folds concave outward, toward the edge of the folded belt.

recessed tube wall *Mechanical Engineering.* a wall of a boiler furnace having niches that allow for the support of waterwall tubes so they will be partly in contact with radiant combustion gases.

recession the process or result of moving back or retreating; specific uses include: *Geology.* **1.** a backward movement or retreat of a cliff or steep slope as it undergoes a process of weathering or erosion. **2.** the landward movement of a shoreline or beach undergoing erosion. Also, RETROGRESSION. **3.** the withdrawal of a body of water that exposes formerly submerged areas and produces a movement of the shoreline away from higher land. *Hydrology.* **1.** a decrease in the length or overall volume of a glacier, usually as the result of melting. Also, RETREAT, GLACIAL RETREAT, GLACIAL RECESSION. **2.** the gradual retreat of a waterfall upstream.

recessional moraine *Geology.* a ridgelike accumulation of glacial drift deposited during a temporary halt in the final retreat of a glacier or during a slight readvance during a period of general recession. Also, PERIPHERAL MORAINE, STADIAL MORAINE.

recession curve *Hydrology.* a hydrograph that shows the decrease in the rate of runoff after a rainfall or the melting of snow.

recession of galaxies *Astronomy.* an observed relationship indicating that the greater the distance to a galaxy, the greater its velocity of recession (as shown by its Doppler-generated redshift).

recessive *Genetics.* **1.** describing an allele that is phenotypically expressed only when homozygous; when present in the heterozygous condition, it is masked by the phenotypic expression of a dominant allele. **2.** a trait or character determined by such an allele.

recessive lethal *Genetics.* describing an allele that is lethal only when it is homozygous in a cell.

recessiveness *Genetics.* the property of an allele that is obscured in the phenotype of a heterozygote by a dominant allele; or the property of a phenotype that is obscured by a dominant phenotype.

recessive oncogene see TUMOR-SUPPRESSOR GENE.

recf pathway *Molecular Biology.* a recA-dependent pathway for recombination in *Escherichia coli* that is independent of the recBCD mechanism.

rechamber *Ordnance.* to cut out or alter the chamber of a firearm in order to use larger cartridges.

recharge *Electricity.* to charge again with electricity; e.g., an automobile battery. *Hydrology.* **1.** the process by which water is absorbed and added to the zone of saturation, thus replenishing it. **2.** the amount of water so added. Also, INCREMENT, INTAKE.

rechargeable battery see STORAGE BATTERY.

recharge area *Hydrology.* any area where absorbed water eventually reaches one or more aquifers in the zone of saturation. Also, INTAKE AREA.

recharge basin *Civil Engineering.* a holding area for rainfall or other water used to refill an aquafier.

recharge well *Hydrology.* a well used to artificially restore one or more aquifers by pumping water into them. Also, INJECTION WELL.

recip. reciprocal.

recipient *Biology.* an organism that receives a graft, transplant, or transfusion from another. *Molecular Biology.* any cell that receives genetic information from another cell, called the donor.

reciprocal [rē sip´rə kəl] *Mathematics.* **1.** a multiplicative inverse. **2.** in general, the reciprocal of a quantity A (e.g., number, function, etc.) is $1/A$, when the latter is defined. *Navigation.* bearing in a direction that is 180° to a given direction.

reciprocal altruism *Behavior.* any altruistic behavior occurring between individuals who are not related, based on the principle that help offered in one instance will be returned when the need for help is reversed.

reciprocal cross *Genetics.* a genetic cross of two genotypes in which sex influences the outcome; for example, the cross of a male horse with a female donkey results in a hinny, while the cross of a female horse with a male donkey results in a mule.

reciprocal determinism *Behavior.* a theory of behavior maintaining that a person's thoughts, his or her behavior, and the situations in which he or she is involved all interact and affect one another.

reciprocal ferrite switch *Electromagnetism.* a waveguide device that can switch an input signal to one of two output waveguides by means of a Faraday rotator, which can be activated by an external magnetic field.

reciprocal genes see COMPLEMENTARY GENES.

reciprocal hybrid *Genetics.* a hybrid progeny resulting from a reciprocal cross.

reciprocal impedance *Electricity.* either of two impedances, Z_1 and Z_2, that are so related with respect to an impedance Z (invariably a resistance) as to satisfy the equation $Z_1Z_2 = Z^2$.

reciprocal inhibition *Physiology.* a condition in which the action of one group of a pair of antagonistic muscles is inhibited by the excitation of the other group, as when the biceps is prevented from contracting and bending the arm at the elbow while the the triceps is extending the arm; a result of reciprocal innervation. *Psychology.* the inhibition of an anxiety-provoking response by the practice of deep muscle relaxation.

reciprocal innervation *Physiology.* the innervation of muscles around the joints, where the motor centers are connected in pairs in such a way that when one is excited, the motor center of the corresponding antagonist is inhibited.

reciprocal junction *Electromagnetism.* a multiple waveguide junction in which the transmission coefficient for one port to a neighboring port is identical to the transmission coefficient for the neighboring port to the original port.

reciprocal lattice *Crystallography.* the lattice with axes a^*, b^*, c^*, related to the crystal lattice or direct lattice (with axes a, b, c) in such way that a^* is perpendicular to b and c; b^* is perpendicular to a and c; and c^* is perpendicular to a and b. a^*, b^*, and c^* are related to a, b, and c, as, for example, by

$$a^* = (b \times c)/(a \cdot b \times c).$$

Rows of points (zone axes) in the direct lattice are normal to nets (planes) of the reciprocal lattice, and vice versa. The repeat distance between points in a particular row of the reciprocal lattice is inversely proportional to the interplanar spacing between the nets of the crystal lattice that are normal to this row of points. The same relation holds between the spacing in rows of the crystal lattice and the spacings of planes of the reciprocal lattice.

reciprocal laying *Ordnance.* a method of laying two guns so that they fire on parallel planes; the guns sight each other and then swing outward through supplementary angles until their planes of fire are parallel.

reciprocal leveling *Cartography.* in surveying, the method of trigonometric leveling in which vertical angles are measured at both ends of the line, thus eliminating errors.

reciprocal observation *Cartography.* in surveying, one of a pair of observations taken forward and backward along the same line, such as a back azimuth.

reciprocal raid *Anthropology.* the practice of carrying out a retaliatory raid by an injured party.

reciprocal recombination *Genetics.* the production of new reciprocal-linkage arrangements in gametes during meiotic recombination; e.g., production of combinants Ab and aB from parental types AB and ab.

reciprocal space *Crystallography.* a mathematical construct used to predict positions at which the diffraction pattern of a crystal will occur. *Solid-State Physics.* see WAVE-VECTOR SPACE.

reciprocal strain ellipsoid *Mechanics.* a mathematical representation for the state of strain at a point; it is an ellipsoid of a certain shape and orientation which, under homogeneous strain, is transformed into a set of orthogonal diameters of a sphere.

reciprocal transducer *Electronics.* an electrical transducer in which the ratio of the short circuit current at one pair of terminals to the voltage at the other pair is the same in either direction.

reciprocal translocation see BALANCED TRANSLOCATION.

reciprocal vectors *Crystallography.* the vectors that form the primitive translations in the reciprocal lattice of a crystal.

reciprocal velocity region *Nuclear Physics.* an area in which the probability that a neutron will be captured by a given element is inversely proportional to the neutron's velocity.

reciprocating beam conveyor *Mechanical Engineering.* a system consisting of parallel beams that move back and forth, with tilting pushers that move objects along a conveyor.

reciprocating compressor *Mechanical Engineering.* a machine that compresses gases, composed of one or several cylinders; each cylinder contains a piston that is moved by a crankshaft through a connecting rod.

reciprocating drill see PISTON DRILL.

reciprocating engine see PISTON ENGINE.

reciprocating flight conveyor *Mechanical Engineering.* a system consisting of beams that move back and forth, with hinged flights that move materials along a conveyor trough.

reciprocating-plate extractor *Chemical Engineering.* a mechanically agitated liquid-liquid extraction tower in which both liquids are sparged in separately onto plates that are moved by a motor. Also, **reciprocating-plate column.**

reciprocating-plate feeder *Mechanical Engineering.* a tray that shakes back and forth to feed process units with granular or pulverized material.

reciprocating saw *Mechanical Devices.* a portable, heavy-duty saw that can be fitted with various forward- and backward-moving blades; used for rough cutting wood and metal.

reciprocating-screw molding machine *Materials Science.* an apparatus used in polymer processing in which pellets of a plastic are fed, compacted, melted, and then forced into a mold cavity by means of a plunger.

reciprocating screen *Mechanical Engineering.* a horizontal sieve used in solids classification that is shaken back and forth by an eccentric gear.

reciprocation *Science.* the complementary interaction of two distinct entities. *Electronics.* a process by which a reciprocal impedance (or network) is derived from a given impedance (or network).

reciprocity [res´i präs´i tē] *Anthropology.* a system in which the exchange of goods is conducted within the groups of a society or community according to a prescribed ritual. *Mathematics.* **1.** a quadratic reciprocity. See QUADRATIC RECIPROCITY LAW. **2.** a relationship between pairs of dissimilar mathematical objects that remains true when the objects are interchanged, as in Frobenius reciprocity; a more general concept than duality.

reciprocity calibration *Acoustical Engineering.* the calibration of a transducer, based on reference values established by the National Bureau of Standards, for use in reciprocity calibrations of other comparable transducers.

reciprocity law *Graphic Arts.* in photography, the formulation that exposure equals intensity (the amount of light) × time (the time that the light is allowed to act on a photographic emulsion).

reciprocity theorem *Physics.* a general principle stating that the behavior of a system is unchanged by reciprocating the input and the output. *Acoustics.* a theory stating that a transducer can be calibrated and then used for reciprocity calibrations of other comparable transducers, with the transducers being used in the calibration as a receiver for one portion of the calibration and a source for another portion of the calibration. *Electricity.* a theorem stating that the exchange of electromotive forces at any given point in a network and the current produced at any other point generates the same current from the same electromagnetic force. *Electromagnetism.* a theorem stating that for a pair of loop antennas, the ratio of the current in the first antenna to the voltage in the second antenna (when the first is used as the transmitter) is equal to the ratio of the current in the second to the voltage in the first (when the second is used as the transmitter).

reciprocity theorem report *Acoustics.* a report-form listing of standard reference values established by the National Bureau of Standards for a transducer, such as a microphone, so that the transducer can be calibrated and then used for reciprocity calibrations of other comparable transducers.

recirculating-ball steering *Mechanical Engineering.* a mechanism in which steering movements are transmitted by steel balls placed between a worm gear and a nut.

recirculator *Engineering.* an underwater breathing device that delivers oxygen to the user until the supply is depleted.

reckless DNA degradation *Genetics.* the rapid extensive degradation of DNA that occurs in certain rec-mutants of *Escherichia coli* upon exposure to DNA-damaging treatment, such as UV radiation.

reckoning *Navigation.* any means of determining the position of a craft, as in the phrase "dead reckoning."

reclaim *Science.* to restore something to a condition that is similar to its original one. *Ecology.* specifically, to restore wasteland, desert, submerged land, or other previously unusable land to a condition where it can be used as an agricultural site, a wildlife habitat, or for some other purpose.

reclaimed rubber *Materials.* elastomer scrap sold in the form of pellets that are inexpensive and easy to process, often used as fillers. Similarly, **reclaimed oil.**

reclaimer *Computer Programming.* see GARBAGE COLLECTOR.

reclaim rinse *Metallurgy.* a rinse that is used to remove and recover a processing solution carried by a metallic part.

reclamation the act or fact of reclaiming; specific uses include: *Civil Engineering.* a process of extensive drainage of areas of low-lying land from the sea or other marshy lands for potential practical use. Also, ACCRETION. *Archaeology.* any of various processes by which artifacts move from an archaeological context to an active status, as when a later society makes use of objects deposited earlier.

reclinate *Botany.* of or related to a plant or plant part, especially a leaf, that is turned or bent downward.

reclined fold see RECUMBENT FOLD.

reclosing relay *Electricity.* a programmed relay that functions automatically to close a circuit breaker after it has operated to open the circuit.

rec-mutant *Genetics.* a group of radiation-sensitive mutations that are characterized by defective recombination.

recognin *Oncology.* any of a group of protein fragments produced from cancer cells that are capable of recognizing specific cells, including astrocytin and malignin.

recognition *Psychology.* positive acknowledgment of an achievement. *Behavior.* **1.** the ability to discriminate between known and unknown stimuli. Also, RETENTIVE DISCRIMINATION. **2.** see SPECIES RECOGNITION. *Computer Programming.* the process of sensing or identifying an input.

recognition differential *Acoustics.* the detection threshold; the minimum detectable signal-to-noise ratio, giving a measure of a sonar system's effectiveness.

recognition gate see DECODING GATE.

recognition signal *Military Science.* any prearranged signal by which individuals or units may identify one another.

recognize-act cycle *Artificial Intelligence.* the cycle of operation in a forward-chaining production system, in which rules whose left-hand sides are satisfied by the data are identified (recognize), one or more of the enabled rules is selected (conflict resolution), and the selected rules are executed (act).

recognizer *Computer Science.* a program or abstract device that can read a string of symbols and decide whether the string is a member of a particular language.

recoil to move backward or away; specific uses include: *Mechanics.* **1.** the backward force exerted against its support by a gun upon firing, equal in magnitude and opposite in direction to the force that is applied to the projectile. Also, GUN REACTION. **2.** of a gun, to move backward in such a manner. *Physics.* of a particle or system, to be set in motion or undergo a change of motion as the result of a collision.

recoil adapter *Ordnance.* a device fastened between a gun and its mount to adapt it for mounting and to absorb the shock of recoil; used especially on antiaircraft guns.

recoil booster *Ordnance.* a part of a machine gun that is designed to trap some of the propellant gas from the barrel and maintain proper recoil action when the weapon is fired from a nonstandard position.

recoil brake *Ordnance.* the component of the recoil system that absorbs the recoil energy and prevents further rearward movement of the recoiling parts.

recoil click *Horology.* in a spring-powered timepiece, a spring-loaded pawl that prevents the mainspring from unwinding (except to power the going train) and from being overwound, by allowing it to recoil slightly against the click mechanism.

recoil electron *Physics.* an electron that is set into motion as the result of a collision.

recoil escapement see ANCHOR ESCAPEMENT.

recoiling parts *Ordnance.* the parts of a weapon that move during recoil; usually includes the tube, breech housing, breechblock assembly, and recoil mechanism.

recoil ion spectroscopy *Atomic Physics.* a technique for analyzing highly energized atomic states in which relatively light atoms in a gaseous substance are bombarded with heavy, highly ionized, fast-moving projectiles.

recoilless *Ordnance.* relating to or designating a weapon that is designed to counteract or eliminate recoil; most recoilless weapons employ high-velocity gas ports.

recoilless ammunition *Ordnance.* ammunition designed to be used in recoilless weapons; the propellant gases are released in a manner and quantity appropriate to the recoilless effect.

recoilless gun *Ordnance.* a lightweight launcher-type artillery weapon with a smooth bore and open breech; the propellant is muzzle-loaded and the weapon is fired electrically or by mechanical remote control.

recoilless rifle *Ordnance.* **1.** a weapon consisting of a light recoilless artillery tube and a lightweight tripod mount; the smaller 57-mm version can be fired from a shoulder mount. **2.** a similar, heavier weapon that may be fired from a ground mount or from a vehicle; capable of destroying tanks.

recoil mechanism *Ordnance.* **1.** in an artillery gun or similar weapon, a shock-absorbing mechanism that gradually decreases the energy of the recoil; it may be spring-powered, hydraulic, or pneumatic. **2.** in an automatic, semiautomatic, or selfloading weapon, a mechanism that rejects the empty cartridge, reloads, and sets the weapon in a ready position. Also, **recoil-operated.**

recoil particle *Physics.* a particle that is set into motion as the result of a collision.

recoil pit *Ordnance.* a hole dug behind the breech of a gun to allow space for the breech during recoil.

recoil velocity *Ordnance.* the velocity of the recoiling parts of a weapon during recoil. Also, **recoiling mass.**

recombinant *Genetics.* **1.** a cell or organism that is produced as a result of the independent assortment and crossing over of chromosomes during meiosis. **2.** of or relating to such a cell or organism.

recombinant DNA *Molecular Biology.* a hybrid DNA molecule that is created by the in vitro combination of DNA from different sources.

recombinant DNA technology *Genetics.* any of various techniques for separating and recombining segments of DNA or genes.

recombination the process or result of combining again; specific uses include: *Genetics.* **1.** the regrouping of genes in an offspring caused by the independent segregation and crossing over of chromosomes during meiosis. **2.** the exchange of portions between two DNA molecules. *Physics.* the coming together and subsequent neutralization of particles having equal and opposite charges. *Materials Science.* at the atomic level of conducting materials, the combination of a free electron and the hole; the hole captures the electron, producing a neutral atom.

recombination coefficient *Electronics.* the rate at which positive ions recombine with negative ions, divided by the product of the number of positive ions and the number of negative ions per unit volume.

recombination energy *Physics.* the amount of energy that is released when two charged particles (of equal and opposite charge) recombine.

recombination frequency *Genetics.* the frequency with which recombinant types appear in the progeny of parents in a test cross; progeny that are not recombinant are of the parental type; geneticists use the recombination frequency to determine the degree of linkage exhibited in a cross.

recombination radiation *Solid-State Physics.* the electromagnetic radiation released when a conduction electron recombines with a hole in the valence band of a semiconductor.

recombination repair *Molecular Biology.* a process by which aberrations in a sequence of DNA can be repaired.

recommended daily allowance *Nutrition.* another term for recommended dietary allowance, especially in popular use.

recommended dietary allowance *Nutrition.* the daily intake of calories and nutrients recommended by the Food and Nutrition Board of the National Research Council as necessary for the maintenance of normal good health.

recomposed granite *Petrology.* an arkose containing minimally reworked or decomposed residue in a consolidated feldspathic mass such that, upon cementation, the rock resembles granite with a less even grain and a greater percentage of quartz. Also, RECONSTRUCTED GRANITE.

recomposed rock *Petrology.* a rock formed in situ by the cementation of fragments that have broken off due to surface weathering.

recon *Genetics.* the smallest unit of DNA capable of recombination.

reconditioned carrier reception *Electronics.* a method by which distortions, such as amplitude variations, are removed from a radio signal to provide more clarity.

reconditioning *Engineering.* the restoration of an object to its working condition. *Behavior.* the reestablishment of a previously extinguished conditioned response by reintroduction of the unconditioned stimulus.

reconnaissance [rē kän′ə səns] *Military Science.* a mission undertaken to obtain information about the activities and resources of an enemy, or the geophysical characteristics of a particular area; it may employ visual observation or other detection methods. *Engineering.* an investigation to obtain information on a given area, such as its weather, geography, and demography. (From a French term meaning "to recognize.")

reconnaissance aircraft *Aviation.* any aircraft that is designed, modified, or used to carry out aerial reconnaissance.

reconnaissance by fire *Military Science.* a reconnaissance method in which fire is placed on a suspected enemy position to cause disclosure of the position by enemy movement or return of fire.

reconnaissance drone *Aviation.* an unmanned vehicle, either remotely piloted or computer guided, used to obtain information of a military or nonmilitary nature.

reconnaissance in force *Military Science.* an offensive operation designed to discover or test an enemy's strength, or to obtain other information.

reconnaissance map *Cartography.* a map based on the reconnaissance of a military target or enemy, and on the meteorological, hydrographic, or geographic characteristics in a particular area.

reconnaissance spacecraft *Space Technology.* an earth satellite, or another type of spacecraft operating in near-earth areas, used to obtain information regarding enemy resources and activities.

reconnaissance survey *Cartography.* a rapidly executed and relatively inexpensive survey of an area intended for a more thorough survey at a later date.

reconnection *Astronomy.* the rejoining of solar magnetic lines of force severed by the annihilation of the magnetic field across a neutral zone.

reconsequent see RESEQUENT.

reconsequent stream see RESEQUENT STREAM.

reconstitute *Science.* to restore something to its former state, form, or condition.

reconstituted *Materials Science.* describing a material that has been reconstructed; liquids are generally reconstituted by adding water to a dry solid from which the water was originally evaporated; solids may be reconstituted by combining scraps with a binding substance. Thus, **reconstituted orange juice, reconstituted mica.** Similarly, **reconstructed coal, reconstructed stone.**

reconstitution *Materials Science.* the act or fact of creating a reconstituted substance. *Food Technology.* specifically, a process in which water is added to a dehydrated or concentrated food product in order to restore it to its former state. *Computer Programming.* the restoration of a file to an earlier version, often following a dump and restart procedure.

reconstitution site *Military Science.* a location chosen by the command authority surviving an enemy attack at which a damaged or destroyed headquarters can be reformed, from survivors of the attack or other predesignated replacement personnel.

reconstructed glacier see GLACIER REMANIÉ.

reconstructed granite see RECOMPOSED GRANITE.

reconstruction *Solid-State Physics.* the migration of atoms and molecules within a crystal in order to achieve a more stable state of equilibrium in the lattice structure.

reconstruction conjecture *Mathematics.* let G and H be a finite undirected graph on n vertices and suppose that there is a bijection α from $V(G)$ to $V(H)$ such that $G - v$ is isomorphic to $H - \alpha(v)$ for every $v \in V(G)$. The reconstruction conjecture states that, if $n \geq 3$, then G and H are isomorphic. Any graph G for which the reconstruction conjecture is true is called **reconstructible.**

reconstructive surgery *Surgery.* a plastic surgery procedure for restoring bodily form or function impaired by disease or injury.

reconstructive transformation *Materials Science.* a polymorph transformation or crystalline modification in which interatomic bonds are broken and structural relationships are completely altered.

record *Science.* any of various items that are created or established to provide a permanent source of information or other material, as in a computer. *Acoustical Engineering.* the medium on which music or other material is preserved for playing on a phonograph.

record changer *Acoustical Engineering.* an armlike holder on a turntable that allows records to be stacked together above the turntable for automatic sequential drop down and playback.

recorder *Science.* a device or system that records data.

record gap *Computer Technology.* a former term for interblock gap.

recording *Science.* **1.** the process of preserving something in a reproducible form. **2.** a product of such a process.

recording balance *Analytical Chemistry.* a special balance equipped with an electromagnetic- or servomotor-driven apparatus to record weight.

recording density *Computer Technology.* a measure of the number of bits that can be recorded per unit surface of a floppy disk or per unit length of a magnetic tape.

recording head *Electronics.* a device that converts sound in the form of electrical energy into magnetic energy, for storage on a tape or disk.

recording instrument *Engineering.* a device that makes a permanent record of its measurements.

recording lamp *Electronics.* a device that generates light whose intensity varies with the audio-frequency signal sent through it; used in motion picture sound-recording systems.

recording level *Electronics.* the signal level necessary for recording sound waves.

recording noise *Electronics.* the unwanted signals that are generated during the recording processes from amplifiers and other components.

recording rain gauge *Engineering.* a rain gauge that records the amount of rainfall as a function of time. Also, PLUVIOGRAPH.

recording storage tube *Electronics.* a device that stores an image as electrical energy for transmission as an electrical signal.

recording thermometer *Meteorology.* an electronic instrument that produces a continuous graphic record of the temperature and sometimes also the barometric pressure at a given location.

recording thermometer

recording trunk *Telecommunications.* a trunk that extends from a local, centralized office or restricted branch exchange to a toll office; used for communication with toll operators only.

recording unit *Archaeology.* a specific location or feature that is defined as an entity for the purpose of recording archaeological data.

record layout *Computer Programming.* a description of the contents and physical arrangement of the elements in a data record. Also, **record format.**

record length *Computer Programming.* the number of characters, bytes, or words in a record. Also, **record size.**

record locking *Computer Programming.* in a multiuser environment, the limiting of access to a record for some users while another user is updating the record.

record mark *Computer Technology.* a special character used to limit the number of characters in a data transfer or to indicate the end of a block of records on a tape.

record observation *Meteorology.* a weather observation that is taken at regular and equal intervals, such as hourly, and usually on the hour.

record-playback robot *Robotics.* a robot that stores data on critical points along a path as it is moved under operator control, then plays them back in order to perform a task.

record player see PHONOGRAPH.

record separator *Computer Technology.* a character that indicates the end of one record and the beginning of the next.

record storage mark *Computer Technology.* a character that indicates the length of a record that is being read.

recover *Science.* to carry out or undergo a recovery process.

recoverable item *Ordnance.* an item that normally is not consumed in its use and that thus may be returned for repair or disposal.

recovery a return to an earlier desired state; a restoration or reclamation; specific uses include: *Medicine.* **1.** a return to health after an illness. **2.** the process by which this occurs. *Computer Programming.* **1.** the restoration of data from a backup file. **2.** the return of a system to an operational state following some type of failure. **3.** the resolution by a specified data station of an error condition that occurs during data transfer. *Mining Engineering.* **1.** the process of extracting valuable material from a mine, well, or deposit. **2.** the amount or percentage of valuable material contained within an extracted mass of material. *Hydrology.* **1.** a rise in the static water level of a well that occurs when the flow from that well or a nearby well ceases. Also, GROUNDWATER RECOVERY. **2.** any process through which groundwater is removed from a source. Also, GROUNDWATER WITHDRAWAL. **3.** the amount of groundwater thus removed. *Archaeology.* the act or process of obtaining artifacts from a site for the purpose of deriving archaeological data. *Materials Science.* a low-temperature annealing heat treatment designed to increase the electrical conductivity and to eliminate residual stresses introduced during plastic deformation, without substantially reducing the strength of the cold-worked material. *Metallurgy.* **1.** in thermal treating, the alleviation of work hardening without recrystallization. **2.** in ore processing, the fraction of a wanted element that is recovered. *Aviation.* **1.** the process of returning a vehicle in flight to equilibrium after some nonequilibrious maneuver, such as a spin or dive. **2.** a conversion of kinetic energy to potential energy, such as the deceleration of air in a ramjet engine duct. *Military Science.* **1.** in air operations, the phase of a mission that involves the return of an aircraft to the base. **2.** in amphibious reconnaissance, the physical extraction of landed forces or their link-up with friendly forces. *Space Technology.* the retrieval of all or part of a satellite or other space vehicle after launch, especially of a satellite or section containing data or instrumentation after a mission.

recovery area *Space Technology.* the location or area where a space vehicle recovery is planned or actually occurs.

recovery capsule *Space Technology.* a capsule that is designed to survive reentry and be recovered.

recovery factor *Petroleum Engineering.* in a petroleum reservoir, the ratio of recoverable oil to the whole of the oil in place.

recovery package *Space Technology.* a package that is attached to a main reentry body and contains devices such as a radio transmitter or balloons which serve to locate the reentry body after impact.

recovery party *Ordnance.* a unit that attempts to recover disabled ordnance materiel from designated recovery points and transport it to the repair facility.

recovery room *Medicine.* a specially equipped hospital room to which a surgical patient is taken after surgery while still under anesthesia, where the vital signs can be closely observed as the patient recovers consciousness.

recovery routine *Computer Programming.* a routine that is used to isolate and correct errors that occur during processing.

recovery ship *Space Technology.* a ship assigned to pick up a space vehicle or capsule following an at sea landing, a method particularly associated with early American manned space flights.

recovery system *Computer Programming.* a program that records the progress of processing activities to allow the reconstruction of a run in the event of a complete crash.

recovery temperature see ADIABATIC RECOVERY TEMPERATURE.

recovery theory *Archaeology.* the various principles that archaeologists employ in the process of recovery, such as where and how to search for sites and how to excavate those sites.

recovery time *Nucleonics.* the minimum time required by a radiation counter tube to record two consecutive ionizing events at full magnitude. *Electronics.* the minimum time required for a circuit to respond to a change in a signal's amplitude; for the control electrode in a gas tube to begin functioning following a break in the current flow; for a communications tube to return to normal energy levels following the reception or transmission of a signal; or for a radar system to regain at least half of its sensitivity following the transmission of an electrical pulse, in order to receive the signal returned from the target.

recovery unit *Archaeology.* a defined area from which archaeological materials are recovered, such as an excavation pit.

recovery vehicle *Mechanical Engineering.* a vehicle equipped with a hoist, boom, or winch for the purpose of retrieving other vehicles.

recrudescent typhus see BRILL'S DISEASE.

recruitment *Physiology.* an increase in the action of a reflex when the initiating stimulus is prolonged without a change in intensity, due to the activation of additional motor neurons.

recrystallization *Crystallography.* a technique of repeated crystallizations; used to produce newer, purer crystals by dissolving and reforming existing crystals by evaporation, sublimation, or the cooling of a melt. *Metallurgy.* specifically, the production of new metal grains through a process of recrystallization annealing. *Geology.* the solid-state metamorphic formation of new crystalline mineral grains in a rock.

recrystallization annealing *Metallurgy.* a medium-temperature annealing heat treatment in which new grains are nucleated and grow at the expense of the deformed ones designed to eliminate all of the effects of the strain hardening produced during cold working.

recrystallization breccia see PSEUDOBRECCIA.

recrystallization flow *Geology.* a rock deformation in which solution and redeposition, solid diffusion, or local melting causes a rearrangement of molecular structure.

recrystallization temperature *Crystallography.* the minimum temperature at which full recrystallization occurs within a specified time.

recrystallization texture *Crystallography.* a preferential grain orientation often observed in recrystallized structures; it may cause property directionality.

rect. rectangle.

rect. or **Rect.** rectified. (From Latin *rectificatus.*)

rect- a combining form meaning "right" or "straight."

rectangle *Mathematics.* a quadrilateral with four right angles; equivalently, a parallelogram, all of whose angles are equal (i.e., right angles).

rectangular *Mathematics.* relating to or having the form of a rectangle.

rectangular cavity *Electromagnetism.* a hollow resonant cavity whose interior is bounded by a rectangular parallelepiped.

rectangular chart *Cartography.* any chart based on a rectangular map projection.

rectangular coordinate system see CARTESIAN COORDINATE SYSTEM.

rectangular cross ripple mark *Geology.* a ripple mark resulting from the concurrent or successive interference of two sets of waves that intersect at right angles and enclose a rectangular pit.

rectangular distribution *Statistics.* see UNIFORM DISTRIBUTION.

rectangular drainage pattern *Hydrology.* a pattern of natural streams and their tributaries that displays many right-angle bends and sections that are approximately equal in length. Also, LATTICE DRAINAGE PATTERN.

rectangular game see MATRIX GAME.

rectangular hyperbola *Mathematics.* a hyperbola having equal major and minor axes; equivalently, a hyperbola having perpendicular asymptotes. Also, EQUIANGULAR HYPERBOLA.

rectangular map projection *Cartography.* a map projection in which the points on the surface of the earth are projected onto a cylinder that is tangent to the earth and the parallels are evenly spaced.

rectangular parallelepiped *Mathematics.* a right prism having rectangles as bases. Also, **rectangular solid.**

rectangular polyconic map projection see MODIFIED POLYCONIC MAP PROJECTION.

rectangular pulse *Electronics.* a signal characterized by an amplitude that goes from zero to another value for a brief period, and then drops to zero.

rectangular scanning *Electronics.* two-dimensional scanning in which a beam is perpendicularly superimposed on another beam, with the former moving slowly and the later moving rapidly.

rectangular search *Navigation.* a search procedure in which each searching craft or group is assigned a specific rectangular search area.

rectangular surveys *Cartography.* a method of dividing a parcel of land into smaller units that are surveyed individually; the parcel to be surveyed is divided east to west by a base line, which is intersected perpendicularly by a principal meridian to establish the initial point, from which the partitions are then further divided into townships of equal size, each containing 36 sections of land.

rectangular wave *Electronics.* a signal whose level alternates between two fixed values. Also, **rectangular wave train.**

rectangular waveguide *Electromagnetism.* a waveguide whose transverse cross section is rectangular.

rectangular weir *Civil Engineering.* a rectangular notched device used on waterways to regulate and measure water height and upstream flow.

rectenna *Electronics.* a device that converts microwave energy into direct current energy.

recti- a combining form meaning "right" or "straight."

Recticornia *Invertebrate Zoology.* a family of amphipod crustaceans in the superfamily Genuina, characterized by straight first antennae on the anterior margin of the head.

rectifiable curve *Mathematics.* a curve (path) given by a continuous function γ: $[a,b] \to C$ (i.e., a complex-valued function of a real interval) that is of bounded variation. Also, **rectifiable path.**

rectification a process of adjusting, correcting, or setting right; specific uses include: *Chemistry.* the redistillation of a liquid in order to purify it. *Electricity.* the process of changing an alternating current to a direct current. *Geology.* **1.** an alignment made to correct a deviation of stream channel or bank. **2.** the simplification and straightening of an irregular, notched shoreline by the erosion of headlands and offshore islands, and by the deposition of erosional waste or of sediment from neighboring rivers. *Cartography.* the process of projecting a tilted or oblique photograph onto a horizontal reference plane by mathematical, graphic, or photographic methods; unlike the process of transformation, rectification does not remove relief displacement.

rectification distillation *Chemical Engineering.* the separation of components of a liquid mixture with the use of a rectifying column to purify the light product.

rectification factor *Electricity.* the rate of change in a DC current caused by a change in amplitude of the AC voltage applied to the electrode.

rectified current *Electricity.* a current that has been changed from alternating current to direct current; the product of rectification.

rectifier *Electricity.* a device through which current can flow only in one direction; often used alone or in sets to convert AC current into DC pulsating current.

rectifier filter *Electricity.* a circuit that eliminates voltage fluctuations caused by converting an alternating current to a direct current.

rectifier instrument *Electronics.* a device in which a direct current meter combines with a rectifier to measure alternating currents or voltages as direct voltage.

rectifier stack *Electronics.* a device that converts current from alternating to direct by passing it through a stack of pressurized disks that are commonly made of copper-oxide or selenium.

rectifier transformer *Electronics.* a device that employs one wire to supply alternating current energy to a circuit and another to supply energy to the electrodes of the circuit's rectifier.

rectifying column *Chemical Engineering.* **1.** a complex distillation operation in which the rectifying section is operated separately from the stripping section, usually at a different pressure, to save energy. **2.** the lower part of a distillation column, below the feed stream.

rectilinear [rek′tə lin′ē ər] *Mathematics.* **1.** composed of line segments. **2.** bounded by lines or line segments. **3.** moving in or forming a straight line. Also, **rectilineal.**

rectilinear-Cartesian robot *Robotics.* a continuous-path, extended-reach robot that uses a bridge and trolley construction to give it a large, rectangular work envelope.

rectilinear evolution *Evolution.* an evolutionary change in a single line of descent that follows a clear direction for a long period of time, with no consideration of how that direction is maintained.

rectilinear generators *Mathematics.* any set of straight lines that generate a ruled surface.

rectilinear motion *Mechanics.* **1.** for particles, motion along a straight line. **2.** for a rigid body, motion in which all particles of the body move the same distance along parallel straight line paths.

rectilinear scanning *Electronics.* a technique of moving a marker, such as a radar antenna or an electronic beam, across an area in a repetitive pattern of straight, narrow, parallel strips.

rectilinear shoreline *Geology.* a long shoreline that forms a relatively straight line.

rectilinear system see ORTHOSCOPIC SYSTEM.

recto *Graphic Arts.* a right-hand, odd-numbered page in a book or other printed text; the opposite of a *verso* (left) page. (From the Latin word for right.)

recto- a combining form meaning "rectum."

rectocele *Medicine.* the protrusion of the rectum into the vagina. Also, PROCTOCELE.

rectoplasty *Surgery.* plastic surgery performed on the rectum. Also, PROCTOPLASTY.

rectorite *Geology.* a white, clay-mineral mixture having a regular interstratification of two layers of mica and one or more layers of water.

rectorrhaphy *Surgery.* surgical repair of the rectum. Also, PROCTORRHAPHY.

rectoscope see PROCTOSCOPE.

rectourethral *Anatomy.* relating to the rectum and urethra.

rectovaginal *Anatomy.* relating to the rectum and vagina.

rectrix *Vertebrate Zoology.* a tail feather in a bird, important in controlling flight direction.

rectum *Anatomy.* the terminus of the large intestine, beginning at the sigmoid colon and ending at the anus.

rectus *Anatomy.* a general term referring to a straight structure, as a muscle.

rectus abdominis *Anatomy.* the segmented abdominal muscle, extending from the pubis to the xiphoid process.

rectus femoris *Anatomy.* the portion of the quadriceps femoris muscle that extends along the ventral aspect of the thigh, from the anterior inferior iliac spine to the patella.

recumbent fold *Geology.* an overturned anticlinal or synclinal fold having a nearly horizontal axial plane. Also, RECLINED FOLD.

recuperability *Telecommunications.* the ability to continue to function after a loss of the primary communications facility resulting from a natural disaster or enemy attack.

recuperation *Medicine.* the recovery of health and strength.

recuperative air heater *Engineering.* an air heater in which the heat-transferring metal parts remain stationary to form a boundary between the heating and cooling fluids.

recuperator *Engineering.* a device that transfers heat from combustion products to a cool air current entering through a series of thin-walled ducts.

recurrence *Medicine.* the return of symptoms after a remission. Thus, **recurrent.**

recurrence interval *Hydrology.* **1.** the average time interval between floods of a given size, either as a yearly maximum in an annual flood series or independently of any other time period in a partial duration series. **2.** the average time period between actual occurrences of any hydrologic event of a given magnitude.

recurrence rate see REPETITION RATE.

recurrent folding *Geology.* a type of folding characterized by thinning formations at the crest, and caused by periodic deformation or subsidence. Also, REVIVED FOLDING.

recurrent function *Mathematics.* a function T on a topological space X is said to be recurrent at a point x if any open set containing x also contains infinitely many points of the form $T^k(x)$, where $k = 1, 2, 3, \ldots$. Also, **recursive function.**

recurrent nova *Astronomy.* a cataclysmic variable star that erupts like a nova, but does not brighten as much as a true nova and returns to its pre-eruption brightness faster than a true nova.

recurring demand *Military Science.* a request by an authorized requisitioner to satisfy a materiel requirement that is expected to recur periodically. Also, **recurring issue.**

recursion *Chaotic Dynamics.* the repetitive application of simple rules beginning from a particular initial condition. *Computer Programming.* **1.** a process of defining a program, function, routine, or procedure in terms of itself. **2.** a repetition of the same operations, in which the outcome of one operation becomes the input of the next. **3.** an instance of a call by a procedure to itself as a subprocedure.

recursion formula *Mathematics.* a formula that permits computation of the nth term a_n of a sequence if a_k is known for $k < n$, if boundary (initial) conditions are given; also called a **recursion relation**. Recursion formulas in two or more indices are defined similarly. A sequence for which a recursion formula is known is sometimes called a **recursion sequence**, or is said to be **recursive**.

recursive *Computer Programming.* **1.** relating to or involving recursion. **2.** defined in terms of itself.

recursive algorithms *Computer Programming.* the program instruction set that contains repeating steps of a process as a direct consequence of the steps that precede it.

recursive descent *Computer Programming.* a method of writing a parser in which a grammar rule is written as a procedure that recognizes that phrase, calling subroutines as needed for subphrases and producing a parse tree or other data structure as output.

recursively enumerable language *Computer Programming.* a language whose sentences can be enumerated by a recursive program, i.e., any language described by a formal grammar.

recursive procedure *Computer Programming.* a repetitive procedure that calls itself, using results of the previous call, until a solution is reached. Similarly, **recursive routine, recursive subroutine.**

recurvature *Meteorology.* a direction change in the motion of severe tropical cyclones, from westward and poleward to eastward and poleward; frequently occurs as the storm moves into the middle latitudes.

recurved *Science.* curving inward, upward, or backward.

recurved spit *Geography.* a spit whose end has been turned landward by currents or inflowing tides.

Recurvirostridae *Vertebrate Zoology.* the avocets and stilts, a family of wading birds of the order Charadriiformes, having very long legs and long, slender bills; found worldwide in tropical and temperate zones.

recycle *Engineering.* **1.** to carry out a recycling process. **2.** to reuse an item or material. Thus, **recycled, recyclable.** *Ecology.* to participate in an organized recycling program. *Electronics.* to return to a previous condition. *Space Technology.* **1.** in a countdown, to stop the count and return to some earlier point. **2.** to recheck a rocket or other object.

recycle base *Military Science.* a base at which aircraft are serviced between missions.

recycle mixing *Chemical Engineering.* the blending of fresh and recycled stock before entry into a process vessel.

recycle ratio *Chemical Engineering.* the ratio in an industrial process of previously treated stock to fresh feed.

recycle stock *Chemical Engineering.* the portion of the charge in a petroleum-refinery cracking process that is recirculated into the process.

recycling *Engineering.* any process of recovering or extracting valuable or useful materials from waste or scrap. *Ecology.* specifically, the reuse of specific consumer or industrial items in order to conserve scarce materials, reduce pollution and littering, and so on. Organized municipal recycling programs typically recycle aluminum cans, glass and plastic containers, and newsprint or other paper. *Nucleonics.* the use of fissionable reactor fuel after it has already been used and chemically reprocessed.

red *Optics.* the color sensation that corresponds to radiation in the 622–770 nm wavelength region of the visible spectrum.

red algae *Botany.* the common name for algae within the division Rhodophyta, in which the chlorophyll is masked by a red or purplish pigment.

Red Angus *Agriculture.* a breed of beef cattle developed from red calves born to Aberdeen Angus cattle.

red antimony see KERMESITE.

red arsenic see REALGAR.

red bed or **redbed** *Geology.* a continental sedimentary deposit that is composed mainly of sandstone, siltstone, and shale; its red color is caused by the presence of ferric oxide. Also, RED ROCK.

red blindness see PROTANOPIA.

red blood cells *Hematology.* a popular term for the erythrocytes, the disk-shaped, nonnucleated cells containing hemoglobin and serving to transport oxygen from the lungs to various parts of the body. Also, **red cells, red corpuscles, red blood corpuscles.**

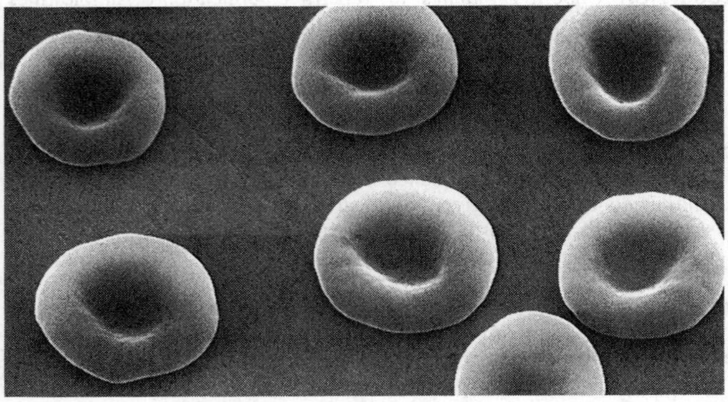

red blood cells

red brass *Metallurgy.* a copper-base alloy nominally containing 15% zinc; designated **UNS C23000.**

red clay *Geology.* a fine-grained, bright to reddish brown or chocolate colored deep-sea marine deposit formed by the slow accumulation of insoluble windblown particles, meteoric and volcanic dust, pumice, and other debris at depths greater than 3500 meters. Also, BROWN CLAY.

red cobalt see ERYTHRITE.

red copper ore see CUPRITE.

red core see RED STELE.

red deer *Vertebrate Zoology.* a common Eurasian deer, *Cervus elaphus,* of the family Cervidae, having a reddish summer coat and being similar in appearance to the elk, only smaller.

Reddish-Brown Lateritic soil *Geology.* a group of zonal, lateritic soils that develop in warm to temperate climates from a mottled red parent material and are characterized by having a reddish-brown surface horizon over a B horizon of red clay.

Reddish-Brown soil *Geology.* a group of zonal soils that develop in warm, temperate to tropical semiarid climates and are characterized by having a reddish, light-brown surface horizon, below which is a heavier, more reddish horizon and a light-colored accumulation of calcium carbonate.

red dog see BURNT SHALE.

red dwarf star *Astronomy.* a dim red star with a diameter about half that of the sun and a surface temperature between 2000 and 3000 kelvins; of spectral type M, they are probably the most numerous class of star in the universe.

red dwarf white star *Astronomy.* a star that is approximately 10,000 times fainter than the sun with a surface temperature less than 4000 kelvins.

red earth *Geology.* the leached, red, deep, clayey soil characteristic of a tropical climate. Also, RED LOAM.

redefine *Computer Programming.* in COBOL, to use the same storage area for different data items during program execution.

redeployment *Military Science.* the transfer of a unit, an individual, or supplies deployed in one area to another location, within the same area, in another area, or in the zone of interior for the purpose of further employment.

redeposition *Geology.* the formation of a new accumulation by the solution and reprecipitation of mineral matter or by the deposition of sediment that has eroded and been transported from its place of origin.

redesignated site *Military Science.* a facility surviving an attack that may be redesignated as the command center to carry on the functions of an incapacitated headquarters or facility.

Redeye *Ordnance.* a lightweight artillery weapon that can be carried and shoulder-fired; it is used for low-altitude air defense of troops in the forward combat area; officially designated **FIM-43.**

Redfield, William 1789–1857, American oceanographer; proposed theories of ocean currents and of rotary motion of storms.

redfish *Vertebrate Zoology.* a rockfish, *Sebastes marinus,* of the North Atlantic that is commonly caught for food. Also, OCEAN PERCH.

red fox *Vertebrate Zoology.* a medium-sized fox, *Vulpes vulpes,* having an orange-red to reddish-brown coat.

red giant star *Astronomy.* a large reddish or orange star in a late phase of its evolution; usually of type K or type M, it has a surface temperature of between 2000 to 3500 kelvins and a radius perhaps 100 times the star's original size.

red giant tip *Astronomy.* a region on the Hertzsprung-Russell diagram that identifies the upper tip of the red giant branch; it represents the helium or carbon flash point where density and temperature of the core have become so high that an ash ignition occurs and serves as the fuel for a new series of nuclear reactions.

red-green-blue *Computer Technology.* **1.** the three primary colors used in producing a color image on a color display. **2.** describes a specification of a color as three binary numbers representing the strengths of red, green, and blue required to produce the color.

red gum see EUCALYPTUS.

red-hardness *Metallurgy.* the attribute of a steel that does not soften when heated to red color; for example, certain high-speed tool steels.

redhorse fish *Vertebrate Zoology.* any of several freshwater suckers of the genus *Moxostoma,* in the family Catostomidae, having reddish fins.

Redi, Francesco 1621–1697, Italian physician; proved that worms were not spontaneously generated in carrion.

redia *Invertebrate Zoology.* a cylindrical larval stage of parasitic trematode flatworms, with an oral sucker and rudimentary appendages; it embeds in the tissues of its snail host.

redifferentiation *Physiology.* the return of dedifferentiated tissue to its original condition, or to a similar condition.

redingtonite *Mineralogy.* $(Fe^{+2},Mg,Ni)(Cr,Al)_2(SO_4)_4 \cdot 22H_2O$, a pale-purple monoclinic mineral of the halotrichite group, occurring in fibrous masses, and having a specific gravity of 1.761 and an undetermined hardness.

red. in pulv. reduce to a powder. (From Latin *redactus in pulverem.*)

redintegration *Surgery.* the restoration of a lost or injured part.

redirected behavior *Behavior.* an action that is deflected from the object that aroused the behavior and directed at an irrelevant substitute object. Also, **redirection.**

red iron ore see HEMATITE.

redistribution *Electronics.* the alteration of electrical charges on the storage surface of a tube by electrons migrating from another area of the same surface; mainly seen in television tubes and other charge-storage tubes. *Anthropology.* the accumulation of goods for a specific purpose, such as a seasonal feast or rite of passage, at which time the goods are given to the participants in fair proportions.

red lead *Materials.* **1.** Pb_3O_4, a red or orange, poisonous, water-insoluble powder that is used as a base for protective paint, in lead glass, in storage battery plates, and as a lute in pipe fitting. Also, **red lead oxide. 2.** a bright-orange, anti-corrosive, priming paint that includes this substance.

red leaf *Plant Pathology.* any of various nonparasitic diseases characterized by the appearance of red discoloration on the foliage.

red leg *Veterinary Medicine.* a disease occurring in captive amphibians; characterized by emaciation and ulcers of the skin, nose, and toes, as well as hemorrhaging of the legs, abdomen, and skeletal muscles.

Redler conveyor *Mechanical Engineering.* a conveyor in which material such as coal or cement is dragged through a duct by U-shaped or skeletonized impellers along an endless chain.

red loam see RED EARTH.

red magnetism *Geophysics.* the magnetism characteristic of the southern pole of a magnet or of the earth's south magnetic pole.

red mercuric iodide see MERCURIC IODIDE.

red metal *Metallurgy.* **1.** a popular name for copper. **2.** a copper matte.

red mud *Geology.* a terrigenous mud that contains up to 25% calcium carbonate, whose red color is caused by the presence of ferric oxide. *Metallurgy.* an iron-rich waste stream formed during the purification of bauxite, an aluminum ore.

red nucleus *Neurology.* a mass of reticular fibers in the gray matter of the roof of the mesencephalon in vertebrates.

redox *Chemistry.* **1.** a shorter term for oxidation-reduction. **2.** relating to or producing the processes of oxidation and reduction.

redox cell *Electricity.* a liquid electrolyte cell that is designed to convert the energy of the reactants to electrical energy.

redox enzyme see OXIDOREDUCTASE.

redox initiator *Materials Science.* a polymerization initiator composed of a mixture of peroxide and a reducing agent.

redox-mediated sensor *Biotechnology.* a sensor that uses a transducer by which electrons generated by an enzyme are transferred to an electrode surface by a mediator such as ferrocene.

redoxomorphic stage *Geochemistry.* in the postdepositional transformation of sediment, the earliest geochemical stage involving changes in minerals mainly as a result of reduction and oxidation reactions.

redox potential *Physical Chemistry.* the amount of voltage released by an inert electrode immersed in a redox system.

redox potentiometry *Analytical Chemistry.* the use of electrode probes to measure solution potential in oxidation or reduction reactions.

redox system *Chemistry.* a system in which oxidation and reduction take place.

redox titration *Analytical Chemistry.* a titration that involves the transfer of electrons from reductant to oxidant.

red panda see PANDA, def. 2.

red pepper *Botany.* **1.** a pepper plant, *Capsicum annuum longum,* whose red pods are widely used as a piquant spice or condiment. **2.** the ripe fruit of the sweet pepper plant, *Capsicum annuum grossum,* eaten as a vegetable. See PEPPER.

red phosphorus *Chemistry.* an allotrope of phosphorus, a violet-red amorphous powder that is obtained by prolonged heating of white phosphorus, and is much less reactive than white phosphorus; used as an additive to semiconductors and safety matches, and in fertilizers.

Red Poll *Agriculture.* a breed of medium-sized dual-purpose cattle originating in England, having a red coat and no horns. Also, **Red Polled.**

red precipitate see MERCURIC OXIDE, def. 1.

red prussiate of soda see SODIUM FERRICYANIDE.

red queen hypothesis *Evolution.* a theory stating that each evolutionary advance by one species represents a deterioration of the environment of other species, so that each species must evolve as rapidly as possible merely to survive. (Taken from the famous statement by the Red Queen in *Alice in Wonderland,* "Now, here, you see, it takes all the running you can do just to stay in the same place.")

redressement *Medicine.* **1.** a second application of a wound dressing. **2.** the correction of a deformity.

red ring *Plant Pathology.* a disease of tomato plants caused by the insect *Cyrtopeltis varians* and characterized by rust-colored markings on the stems and petioles.

red ring disease *Plant Pathology.* a disease of coconut palms caused by the nematode *Aphelenchoides cocophilus* and characterized by red rings in the cross section of the stem.

red rock see RED BED.

red rot *Plant Pathology.* any of various fungus diseases of sugarcane, sisal, and various evergreen and deciduous trees; characterized by red patches on leaves and stems.

red rust *Plant Pathology.* a disease of tea, citrus, and certain other subtropical plants, caused by the green alga *Cephaleuros virescens* and characterized by reddish-brown discoloration of the leaves and twigs.

Red Sea *Geography.* a long, narrow sea between Arabia and northeastern Africa; connected to the Mediterranean Sea by the Suez Canal.

red seaweed see RED ALGAE.

red sector *Navigation.* a portion of the circle of visibility of a light in which the light appears red. It indicates a danger area.

redshift or **red shift** *Astrophysics.* a Doppler shift of observed spectral lines toward longer wavelengths caused by the recession of the object; believed to occur because of the movement of celestial objects outward at increasing rates of speed, and providing the basis for theories suggesting that the universe is constantly expanding.

Red Sindhi *Agriculture.* a breed of red dairy cattle similar to the Brahman and able to withstand very warm climates. (From the province of *Sind,* in Pakistan.)

red snapper *Vertebrate Zoology.* any of several large, reddish-colored snappers of the genus *Lutjanus,* such as *L. campechanus* of the Gulf of Mexico; widely used as a food fish.

red snow *Hydrology.* a term for an arctic or alpine snow surface colored by the presence of microscopic red or pink algae. Also, PINK SNOW.

Red Spot see GREAT RED SPOT.

red spruce *Botany.* a spruce tree, *Picea rubens,* of eastern North America, having reddish-brown bark and cones; its light, soft wood is commonly used to make boxes.

red squirrel *Vertebrate Zoology.* a reddish North American squirrel, *Tamiasciurus hudsonicus.*

red star see RED GIANT STAR, RED DWARF STAR.

red stele *Plant Pathology.* a disease of the strawberry plant caused by the fungus *Phytophthora fragariae,* characterized by red, enlarged plant roots, followed by the dwarfing, wilting, and death of the plant. Also, RED CORE.

redstone *Petrology.* any reddish sedimentary rock, especially red bed.

Redstone *Ordnance.* a mobile, surface-to-surface guided missile with nuclear-warhead capability and a range of approximately 350 miles; it is designed to support ground operations as a supplement to long-range artillery; officially designated **PGM-11.**

Redstone missile

red stripe *Plant Pathology.* **1.** a disease of timber caused by the fungus *Polyporus vaporarius* that causes decay and is characterized by reddish linear markings. **2.** a disease of sugarcane caused by the bacterium *Xanthomonas rubrilineans* and characterized by red linear markings on leaves and destruction of the vascular system.

red tetrazolium see TRIPHENYLTETRAZOLIUM CHLORIDE.

red thermit *Metallurgy.* a mixture of particulated aluminum and red iron oxide that gives off heat upon ignition.

red thread *Plant Pathology.* a disease of turf grasses caused by the fungus *Cortecium fuciforme* and characterized by red stromata intertwined with pinkish hyphal threads.

red tide *Ecology.* a population explosion of certain marine dinoflagellates, occurring especially off the coasts of California and Florida and resulting in a reddish or reddish-brown color of the water; red tides kill many marine animals and render mollusks extremely toxic to humans because of the water-soluble poisonous alkaloid that the dinoflagellates produce.

redtop grass *Botany.* any of several bentgrass species of the genus *Agrostis,* characterized by reddish panicles; commonly cultivated for lawns and pastures.

reduce to decrease in size, weight, number, and so on; specific uses include: *Chemistry.* to submit a substance to a process of chemical reduction. *Surgery.* to realign or restore to a normal position, as in reducing a fracture. *Ordnance.* to clear a stoppage in a weapon.

reduced charge *Ordnance.* **1.** the smaller of the two propelling charges available for naval guns. **2.** a charge employing a reduced quantity of propellant to fire a gun at a shorter range.

reduced equation of state *Physics.* a general equation of state that relates the reduced pressure, the reduced volume, and the reduced temperature.

reduced form *Mathematics.* a fraction a/b is said to be expressed in reduced form if a and b are relatively prime.

reduced frequency see STROUHAL NUMBER.

reduced instruction set computer *Computer Technology.* a microprocessor with a small set of built-in, frequently used, basic instructions for rapid execution of jobs.

reduced mass *Mechanics.* for a system of two bodies with masses m_1 and m_2 exerting equal and opposite forces on one another, the quotient $m = m_1m_2/(m_1 + m_2)$. This makes it possible to describe a two-body system by the single body of mass m.

reduced nickel *Metallurgy.* a form of nickel that is produced by a hydrometallurgical process, such as the hydrogen reduction of a nickel salt solution.

reduced-order controller *Control Systems.* a control algorithm that ignores certain modes of the structure to be controlled in order to compute control commands more rapidly.

reduced pressure *Thermodynamics.* the value of the pressure of a substance in a given state divided by the value of that same pressure at the critical point. *Meteorology.* the value of atmospheric pressure at mean sea level or another specified level, as reduced from station pressure or actual pressure.

reduced proper motion *Astronomy.* a quantity similar to the absolute magnitude of a star, but based on its proper motion.

reduced residue *Mathematics.* **1.** a residue class induced by a congruence of the form $x \equiv a \pmod{m}$, where m is a positive integer greater than 1 and a is any integer relatively prime to m, is called a **reduced residue class** or **prime residue class. 2.** a **complete set of reduced residues** (mod m) is a set $S = \{r_i\}$ such that: (a) $i \neq j$ implies that $r_i \neq r_j$ (mod m); that is, no two elements of S are congruent (modulo m); (b) $r \in S$ implies that r is relatively prime to m; and (c) any integer that is relatively prime to m is congruent to some element of S. In addition, if $0 < r \leq m$ for all $r \in S$, then S is called a **reduced set of least positive residues.** If $-m/2 < r \leq m/2$ for all $r \in S$, then S is called a **reduced set of least residues.**

reduced states see CORRESPONDING STATES.

reduced telemetry *Telecommunications.* basic telemetry data that is converted into a useful form.

reduced temperature *Thermodynamics.* the value of the temperature of a substance in a given state divided by the value of that same temperature at the critical point.

reduced tillage or **reduced till** *Agronomy.* any of various methods by which plowing of the soil is carried out at a shallow depth.

reduced value *Thermodynamics.* the actual quantitative value of a property such as pressure or temperature, divided by the value of the same property at its critical point. Also, **reduced property.**

reduced viscosity *Engineering.* the relationship between specific viscosity and concentration in plastics processing.

reduced volume *Thermodynamics.* the value of the volume of a substance in a given state divided by the value of that same volume at the critical point.

reduced word *Mathematics.* let $X = \text{U}_{i \in I} G_i$, where $\{G_i\}$ is a family of mutually disjoint groups. A reduced word on X is a word $(a_1, a_2, \ldots)$ such that the following conditions apply: (a) no $a_i \in X$ is the identity element of its group G_i; (b) for all $i,j \geq 1$, a_i and a_{i+1} are not members of the same group G_k; and (c) if $a_k = 1$, then $a_i = 1$ for all $i \geq k$. The set of all reduced words on X is denoted by $\prod^*_{i \in I} G_i$. A group is formed by $\prod^*_{i \in I} G_i$, called the free product of the family $\{G_i\}$.

reducer *Chemistry.* **1.** a substance that adds hydrogen to an element or compound. **2.** a substance that adds electrons to an element or compound, thereby reducing it. Also, **reducing agent.** *Graphic Arts.* **1.** any chemical solution used to remove silver from a photographic negative or print, thus decreasing the density of the image and altering the contrast. Also, **reducing agent. 2.** any substance, usually a solvent, used to thin the consistency of ink on a printing press. *Engineering.* a threaded fitting that is larger at one end than at the other. Also, **reducer design.** *Biology.* an organism that is responsible for degrading or decomposing organic matter.

reducible configuration *Mathematics.* a plane graph C with boundary B of the infinite region such that the 4-colorability of any planar graph G containing C can be deduced from the 4-colorability of a graph derived from $G - E(C)$. Specifically, every possible coloring of B is extended to a coloring of C by using Kempe chains. The four-color theorem proof is completed by providing a large list of reducible configurations and showing that every planar graph contains at least one of them.

reducible polynomial *Mathematics.* a polynomial $f(x)$ of degree $n \geq 1$ with coefficients in some ring R is said to be reducible over R if it can be written in the form $f(x) = g(x)h(x)$, where $h(x)$ and $g(x)$ are also polynomials over R and neither is of degree zero. If $f(x)$ is not reducible, it is said to be irreducible; constants are neither reducible nor irreducible.

reducing atmosphere *Chemistry.* an atmosphere of hydrogen around a reaction; an enclosed space from which air has been displaced by a reducing gas such as hydrogen or carbon monoxide, thereby providing electrons.

reducing coupling *Engineering.* a coupling used to join a smaller pipe to a larger pipe.

reducing enzyme *Enzymology.* **1.** an enzyme that catalyzes the addition of hydrogen atoms to a substrate. **2.** an enzyme that catalyzes the addition of electrons to a substrate.

reducing flame *Chemistry.* a gas flame that lacks enough oxygen for full combustion, and contains enough extra fuel to act as a reducing agent.

reducing glass *Optics.* a lens that reduces the apparent size of an object viewed through it; used by painters to achieve the perception of distance from their work.

reducing series *Evolution.* the trend in plant evolution toward reduced size and simpler morphology.

reducing sugar *Organic Chemistry.* any sugar, including monosaccharides and most disaccharides, that easily reduces alkaline solutions of metallic salts, such as copper and silver.

reducing valve *Mechanical Devices.* **1.** a valve used to reduce fluid pressure in a supply line. **2.** a valve fitted with a lever to regulate steam pressure in the piping of a boiler system.

reductant *Chemistry.* the electron donor in an oxidation-reduction (redox) reaction. *Metallurgy.* any reducing material used in a smelting or refining process.

reductase *Enzymology.* **1.** an enzyme of the oxidoreductase class that catalyzes reactions in which metabolites are reduced. **2.** any enzyme that acts as a reducing agent.

reductio ad absurdum *Artificial Intelligence.* literally, reduction to absurdity, the demonstration that a proposition is false by the absurdity or incongruity of the conclusion to which it would lead. *Mathematics.* see INDIRECT PROOF.

reduction the process of reducing; specific uses include: *Chemistry.* a chemical change that involves a gain of electrons, either by a removal of oxygen or an addition of hydrogen, or simply by the addition of electrons. *Surgery.* the restoration to a normal location or position, as in the correction of a hernia or the setting of a fracture. *Geology.* the lowering of a land surface as a result of erosion. *Computer Programming.* see DATA REDUCTION.

reduction cell *Chemistry.* an electrolytic cell that produces a free metal through electrolysis.

reduction gear *Mechanical Engineering.* a gear that slows down the output speed of a mechanism.

reduction index *Geology.* the rate of abrasive wear of a sedimentary particle in the course of transportation, expressed as the quotient of the difference between the mean weight of the particle before and after transport and the product of the mean weight before transport and the distance traveled.

reduction in strength *Computer Programming.* an optimization in which an operator is changed to a less-expensive operator; e.g., $x * 2$ becomes $x + x$.

reductionism *Science.* the theory that complex phenomena are best explained in terms of their simplest elements.

reductionist *Science.* **1.** relating to or applying the principle of reductionism. **2.** a person who applies or advocates this principle.

reduction mammaplasty *Surgery.* plastic surgery of the breast for purposes of reducing size by excision of tissue.

reduction of area *Metallurgy.* in a tensile test, the percentage decrease in cross-sectional area, determined from the original to the minimum post-test dimensions.

reduction of star places *Astronomy.* the determination of mean star position when calculated from observations of apparent position.

reduction of tides *Oceanography.* the manipulation of tidal data to obtain mean values for the various tidal constituents. Similarly, **reduction of tidal current.**

reduction potential *Physical Chemistry.* the decrease in voltage that occurs when a positively charged ion drops to a neutral or less positively charged state, or when a neutral atom becomes a negatively charged ion.

reduction ratio *Engineering.* **1.** the ratio of feed size to product size in a milling operation. **2.** the reduction in speed obtained through gearing. **3.** the ratio of current that is input to a transformer to the output current.

reduction roller *Mechanical Engineering.* any of a set of tightly fitted rollers designed to reduce the size of a material; used in flour production, metal forming, and various other applications.

reduction sphere *Geology.* a white, leached, spherical body formed in a reddish or brownish sandstone, usually surrounding an organic nucleus or pebble.

reduction step *Computer Programming.* in shift-reduce parsing, the reduction of items at the top of the stack to a phrase that encompasses those items.

reduction to center *Cartography.* **1.** in surveying, the process of applying corrections to a direction observed at an eccentric station to reduce it to what it would be if there were no eccentricity. **2.** the value of the corrections applied during the reduction-to-center process.

reduction to the ellipsoid *Cartography.* in surveying, the process of calculating the length of a line lying on an ellipsoid of reference with the same longitudes and latitudes at its end points as those of a given line lying on or at an average elevation above the geoid.

reduction to the geoid *Cartography.* in surveying, the process of calculating the length of a line having all points at zero elevation and having the same longitudes and latitudes for its end points as does a corresponding line lying at a constant but nonzero elevation. Also, **reduction to sea level, reduction to mean sea level.**

reduction to the meridian *Navigation.* the process of determining the meridian altitude of a body from an observation of the body near the meridian.

reduction to the sun *Astronomy.* a correction applied to the observed radial velocity of an object to compensate for the earth's orbital motion.

reductive tricarboxylic acid cycle *Biochemistry.* a cyclic reaction by which acetyl CoA is oxidized to carbon dioxide, providing reducing equivalents, such as NADH.

reductone *Organic Chemistry.* HOCH:C(OH)CHO in the enol form, a solid that is soluble in water and alcohol; melts at 161°C; used in organic reactions and as a reducing agent in alkaline solutions.

redundancy the fact of repeating or duplicating; specific uses include: *Genetics.* the existence of several identical copies of the same gene or nucleotide sequence in a chromosome or genome. *Computer Technology.* **1.** the duplication of system components as a means of checking for accuracy of results or improving reliability. **2.** the repetition of data among several files. *Telecommunications.* that part of the gross information content of a message that can be dropped without any significant loss of the original information.

redundancy bit see PARITY BIT.

redundant excessive or unnecessary; specific uses include: *Military Science.* of or relating to an attack on a target that is inoperative or useless due to previous attacks. Thus, **redundant bombing, redundant target.** *Building Engineering.* describing a member that is not necessary to support the load of the structure.

redundant character *Computer Programming.* a character added to a group of characters to detect a computer or data recording malfunction. Also, **redundancy check character.**

redundant code *Telecommunications.* any code that uses more elements than are needed to represent the transmitted information.

redundant digit *Computer Programming.* a binary-coded decimal digit with an extra check bit.

Reduviidae *Invertebrate Zoology.* the single family of insects of the group Reduvioidea; most species are predators on insects; some suck the blood of mammals and transmit disease-causing endoparasites.

Reduvioidea *Invertebrate Zoology.* the assassin bugs, a group of hemipteran insects in the subdivision Geocorisae having a three-segmented proboscis.

reduzate *Geology.* a sediment, such as coal or black shale, that is rich in organic carbon and iron sulfide as a result of being accumulated under reducing conditions.

red water *Ecology.* see RED TIDE.

redwater fever see BABESIASIS.

redwing *Vertebrate Zoology.* a European thrush, *Turdus iliacus,* having maroon feathers on its wings and flanks.

red-winged blackbird *Vertebrate Zoology.* a blackbird, *Agelaius phoeniceus,* of North America; the male has bright scarlet patches on the bend of its wings. Also, **redwing blackbird.**

redwood *Botany.* **1.** a coniferous tree, *Sequoia sempervirens,* that is found in coastal areas from central California to southern Oregon; noted for its great height (200 to 300 feet); it is characterized by needlelike leaves, fire-resistant bark, and soft, light, long-lasting, reddish-brown wood. Also, **California redwood, coast redwood. 2.** any of various other trees that yield a reddish wood or produce a red dye, such as *Brosimum rubescens* of Brazil or *Chukrasia tabularis* of India. *Materials.* the wood of the tree *Sequoia sempervirens*; characterized by high strength and resistance to decay, making it valuable in construction.

California redwoods

redwood viscometer *Engineering.* a device that determines the viscosity of petroleum oils by measuring the time it takes for an amount of liquid to pass through an orifice.

Red-Yellow Podzolic soil *Geology.* a group of zonal, acidic soils that develop in warm, temperate, or tropical humid climates and that are characterized by having a leached, red to yellowish-red to bright yellowish-brown surface layer over a subsoil containing clay, as well as aluminum and iron oxides.

red zinc ore see ZINCITE.

reed *Botany.* **1.** any of various tall, stout grasslike plants, especially *Phragmites australis* or *Arundo donax.* **2.** the stalk of such a plant. *Engineering.* a thin blade, leaf, or strip used as a vibrator, relay, or oscillator. *Textiles.* a comblike device on a loom that holds the warp threads apart and pushes each new filling thread against those already woven. *Acoustical Engineering.* of or relating to a reed horn.

Reed, Walter 1851–1902, American army surgeon; headed the commission that confirmed Finlay's theory on the cause of yellow fever.

Reed and Prince screwdriver *Mechanical Devices.* an industrial variation of a Phillips head screwdriver with a special sharp crosstip. Also, **Reed & Prince screwdriver.**

Reed cells see REED-STERNBERG CELLS.

reed contact *Electronics.* a miniature hermetically sealed device used to switch electrical voltages and currents; made of two solid metal reeds sealed in a glass vial containing an inert or reducing atmosphere; the reeds are aligned so that their tips slightly overlap and therefore make contact when they are magnetically drawn together. Also, DRY-REED CONTACT.

reed horn *Acoustical Engineering.* a musical instrument that produces sound by vibration of a flexible mouthpiece and amplification by a resonant pipe that can be changed in effective length by the selection of various key holes that have fixed diameters.

reeding *Metallurgy.* the corrugating on a metal, as on the rim of a coin.

reedmergnerite *Mineralogy.* $NaBSi_3O_8$, a colorless, transparent, triclinic mineral of the feldspar group, occurring as small prismatic crystals, having a specific gravity of 2.77 and a hardness of 6 to 6.5 on the Mohs scale; found in unmetamorphosed dolomitic oil shales in Utah.

reed relay *Electromagnetism.* a relay device in which two magnetic reeds are sealed inside of a glass tube and are actuated to contact each other when sufficient current flows through a coil wound about the glass tube or about an auxiliary ferrite core structure upon which the glass tube is mounted.

Reed-Sternberg cells *Pathology.* giant histiocytic cells whose nuclei are enclosed in abundant amphophilic cytoplasm and contain prominent nucleoli; the common histologic characteristic of Hodgkin's disease. Also, DOROTHY REED'S CELLS, REED CELLS, HODGKIN'S CELLS.

reed taper *Mechanical Devices.* a standard taper used on a hollow lathe spindle.

reef *Geology.* **1.** any offshore rock mass, narrow ridge, or chain of rocks that lies at or near the surface of the water. **2.** see ORGANIC REEF. *Mining Engineering.* a lode, vein, or major deposit of ore. *Surgery.* a folding or tucking of tissue performed during surgery.

reef breccia *Petrology.* a rock produced by consolidation of limestone fragments that have broken off from a reef due to the effects of waves, winds, and/or tides.

reef cap *Geology.* a deposit of fossil-reef material covering an island or mountain.

reef cluster *Geology.* a group of reefs composed entirely or partially of contemporaneous growth, found within a geologic province or circumscribed area. Also, HERMATOPELAGO.

reef complex *Geology.* the reef core together with the heterogenous and contiguous fragmentary material derived from the core by abrasion.

reef conglomerate see REEF TALUS.

reef core *Geology.* in an organic reef, the centrally located solid rock mass constructed in place by the reef-building organisms.

reef detritus *Geology.* any debris derived from the erosion of an organic reef. Also, **reef debris.**

reef flank *Geology.* the part of a reef surrounding, interfingering, and locally overlying the reef core.

reef flat *Geology.* the summit of a reef, or a flat, stony expanse of coral fragments and sand that is generally dry at low tide.

reef front *Geology.* the upper part of the outer (seaward) slope of a reef, extending from above the dwindle-point of abundant coral and algae to the reef edge.

reef-front terrace *Geology.* a shelflike or benchlike eroded surface that slopes seaward. Also, REEF TERRACE.

reef knoll *Geology.* **1.** a small, prominent, rounded hill of coralline limestone, often having a mushroom-shaped top. **2.** a present-day reef in the shape of a small, rounded hill. Also, KNOLL.

reef limestone *Petrology.* a limestone containing the remains of sedentary organisms, such as corals and sponges, and sediment-binding organic constituents, such as calcareous algae. Also, CORAL ROCK.

reef milk *Geology.* a very fine-grained matrix formed on the landward side of a reef, composed mainly of white, opaque microcrystalline calcite that has been derived by abrasion of the reef core and flank.

reef patch *Geology.* a coral growth formed independently on a shelf less than 70 meters deep, often in the lagoon of an atoll or of a barrier reef.

reef pinnacle *Geology.* a small, isolated spire or column of rock or coral that often rises close to the surface in the lagoon of an atoll. Also, PINNACLE, PINNACLE REEF, CORAL PINNACLE.

reef rock *Petrology.* a hard, massive, unstratified rock consisting of carbonate sand, shale, and the calcareous remains of sedentary organisms, all cemented by calcium carbonate.

reef segment *Geology.* any part of an organic reef that lies between gaps, passes, or channels.

reef slope *Geology.* the face or flank of a reef that rises from the sea floor.

reef talus *Geology.* a massive, stratified incline consisting of reef detritus that has been deposited along the seaward margin of an organic reef. Also, REEF CONGLOMERATE.

reef terrace see REEF-FRONT TERRACE.

reef tufa *Geology.* prismatic, fibrous calcite deposited directly from supersaturated water upon the internal sediment of a reef knoll.

reef wall *Geology.* **1.** a wall-like growth composed of living coral and the skeletal remains of reef-building organisms that acts as a partial barrier between adjacent environments. **2.** the vertical cliff beneath the surface and extending below most reef and coral growths.

reel *Mechanical Devices.* a revolving frame or spool for winding thread, yarn, rope, film, tape, or other materials.

reel and bead see BEAD AND REEL.

reeling *Textiles.* the process of unwinding raw silk filaments from several cocoons and joining them into a single thread held together by the gum that naturally coats each filament.

reel number *Computer Technology.* a sequential number assigned to a tape reel when a file extends over more than one reel of magnetic tape. Also, **reel sequence number.**

reel-to-reel tape *Acoustical Engineering.* audio recording tape designed to be fed from one reel to another through the recording heads, as distinguished from cassette tape.

reemergence *Hydrology.* see RESURGENCE, def. 2.

reentrant *Engineering.* describing an object that has one or more sections pointing inward. *Computer Programming.* describing a program or routine that can be shared by several tasks concurrently, but of which only one copy is stored in memory. Thus, **reenterable.** *Geography.* a major inlet on a coastline. *Cardiology.* of or relating to reentry.

reentrant angle *Crystallography.* an angle extending inward, as the angle between twin crystals.

reentrant pathway *Cardiology.* the pathway taken by the return of an impulse to the site of the recently activated heart muscle once it has ceased to be refractory; occurs in reciprocal heart rhythms.

reentrant program *Computer Programming.* in a multiprocessing environment, a program that can be interrupted, entered again by another task, and then reentered at the point of interruption for further processing. Also, **reenterable program.**

reentrant winding *Electricity.* an armature winding whose ends are joined to make a closed circuit.

reentry *Space Technology.* **1.** the descent of a spacecraft or other object into the sensible atmosphere from higher altitudes; an instance of this process. **2.** of, relating to, or occuring during this process. Thus, **reentry body, reentry heating, reentry slowdown.** *Cardiology.* a theoretical mechanism in which the normal sinus impulse which has entered the heart's area of low responsiveness "reenters" the area of normal responsiveness, stimulating that area and eliciting a premature contraction.

reentry angle *Space Technology.* the angle of an object's trajectory relative to the horizon as it reenters the sensible atmosphere after flight above it.

reentry nose cone *Space Technology.* a nose cone designed especially for safe reentry and protected by an outer shield.

reentry point *Computer Programming.* the address or label of the instruction in a main program at which that program is reentered after a subroutine or an interrupting program has been executed.

reentry system *Computer Technology.* a character recognition system in which the data that are input for reading are printed by a computer that is associated with the data reader. *Ordnance.* the components of a ballistic missile or space vehicle that direct one or more reentry vehicles on a terminal trajectory toward a selected target.

reentry trajectory *Space Technology.* that portion of a reentering object's trajectory starting at reentry and ending at the surface or some target.

reentry vehicle

reentry vehicle *Space Technology.* any vehicle designed to leave the sensible atmosphere and then reenter it to land on earth. This term may refer to either orbital or space vehicles as well as boostglide vehicles.
reentry window *Space Technology.* an interval of time during which conditions are favorable for reentry.
reepithelialization *Medicine.* the regrowth or placement by surgical means of epithelial tissue over a denuded area.
reevesite *Mineralogy.* $Ni_6Fe_2^{+3}(CO_3)(OH)_{16}\cdot 4H_2O$, a yellow to greenish-yellow trigonal mineral of the hydrotalcite group, found as fine-grained aggregates in meteorites, having a specific gravity of 2.80 to 2.88 and an undetermined hardness.
ref. reference; refined, refining; reformed, reformation.
reference acoustic pressure *Acoustics.* a sound pressure to which 0 decibels of a sound pressure level is referenced; the most common reference level is 1 micropascal (formerly the reference was 0.0002 dyne per square centimeter).
reference angle *Electromagnetism.* the angle of incidence of a radar beam onto a target, when measured with respect to the normal to the surface of the target.
reference block *Computer Programming.* a block in a numerical control program containing information that allows the program to resume after an interruption.
reference burst see COLOR BURST.
reference craft *Navigation.* the craft used as the origin of the frame of reference in a relative plot; it is usually but not always the craft aboard which the plot is being made.
referenced *Computer Programming.* of a variable, having its value read within a sequence of code.
reference datum *Geodesy.* a geometric or numerical quantity, or set of such quantities, that is applied to a point, plane, or curved surface and used as a base or reference from which further quantities are measured.
reference delusion see DELUSIONS OF REFERENCE.
reference dimension *Design Engineering.* a dimension in which no tolerance is applied; used for information purposes only on a design drawing, not to govern machining operations.
reference dipole *Electromagnetism.* a straight, half-wave dipole antenna element that is tuned for a specific operating frequency and is used as a standard unit for comparing other antenna elements.
reference direction *Navigation.* a direction (usually north or the craft's head) used as a starting point in measuring the angular distance to some other direction.
reference electrode *Physical Chemistry.* an electrode that generates consistent levels of energy, without polarization, such as a hydrogen electrode or a calomel electrode; used as a standard to measure the potential of other electrodes.
reference ellipsoid see REFERENCE SPHEROID.
reference frame see FRAME OF REFERENCE.
reference frequency *Telecommunications.* a frequency that has a fixed and specified position in the frequency spectrum with respect to the assigned frequency.
reference group *Psychology.* a group that an individual belongs to or identifies with, and which he or she uses as a guide to define behavior and develop values and goals.
reference level *Engineering.* the level of a quantity to which others of the same quantity are compared. *Acoustical Engineering.* a quantitatively measured sound characteristic such as sound pressure level or volume that is equivalent to 0 decibels on a specific decibel reference scale. *Oceanography.* a stratum of the ocean, at a specified depth, that can be used as a datum against which the values of scalar phenomena at other depths can be measured, such as water pressure, current movement, height of tide, and sound transmission.
reference line *Navigation.* a line relative to which a position is determined. *Military Science.* in artillery or naval gunfire support, a convenient and readily identifiable line used by the observer or spotter as the line to which spots will be related; it is one of three types of spotting lines.
reference listing see COMPILER LISTING.
reference locality *Geology.* any area that contains a reference section.
reference lot *Industrial Engineering.* a standard lot to which production lots are compared.
reference mark *Electronics.* an indicator on a printed circuit that gives the scale dimensions and the edges of its circuit board.
reference noise *Electronics.* the magnitude that an unwanted signal must reach before it can be read by a noise-meter, generally at 10^{-12} watt at 1000 hertz.
reference plane see DATUM PLANE.
reference point see DATUM POINT.
reference protein *Nutrition.* a theoretical standard against which other foods can be compared in terms of ideal amino acid balance; the hen egg and human milk products, established by nutritionists and biologists as having the most perfect balance of amino acids in a natural protein substance, are used as a basis for comparison.
reference range *Engineering.* the aerial range established by a radar coverage indicator.
reference record *Computer Programming.* a record that is output by the compiler, listing the operations and positions for a routine as well as its segmentation and allocation to storage.
reference rounds *Ordnance.* standard ammunition rounds that are fired for comparison during ammunition tests.
reference section *Geology.* a rock section or group of sections used to supplement, or sometimes replace, the type section in order to correlate a certain part of the geologic column.
reference seisomometer *Engineering.* a seismic detector used in seismic prospecting to record and compare successive explosions.
reference spheroid *Geodesy.* a mathematical figure differing slightly from a perfect sphere and closely approaching the geoid in shape and size, used as a surface of reference in geodetic surveys. Also, REFERENCE ELLIPSOID, SPHERIOD OF REFERENCE.
reference station *Oceanography.* a tidal station whose tide or tidal current constants have been calculated from observations over a considerable period of time, and which is used as a standard of comparison for simultaneous observations at subordinate stations; tide predictions are given for certain reference stations in tide tables, making it possible to calculate corresponding predictions for other places.
reference strain *Microbiology.* a strain, other than the type strain of a species, that is used to identify or define any product produced by that strain, such as a particular antigen.
reference supply *Electronics.* a device, such as a Zener diode, that supplies voltage at a stable and constant level; used to measure varying voltages in analog computers and other devices.
reference system see FRAME OF REFERENCE.
reference time *Computer Programming.* a moment in time to which the time measurements in a digital computer are referenced.
reference tone *Engineering.* a stable tone of known frequency used as a frequency reference in multitrack recording.
reference vehicle *Navigation.* a vehicle used as the origin of the frame of reference in a relative plot.
reference voltage *Electricity.* a voltage against which any current or voltage can be measured.
reference volume *Acoustics.* a unit of audio intensity that is used for expressing the power level of a complex sound wave.
reference white *Telecommunications.* in television transmission, the level at the point of observation that corresponds to the preset maximum excursion of the picture signal in the white direction. Also, **reference white level.**

referencing *Cartography.* in surveying, the process of establishing a station's position and preventing loss of that location in the future by measuring the horizontal distances and directions from that station to nearby permanent landmarks.

referential ambiguity *Artificial Intelligence.* in natural language understanding, the apparent possibility that a word, such as a pronoun, might refer to one of several other words or concepts.

referent identification *Artificial Intelligence.* the process of determining to what object(s) in a program's model of the world a phrase in natural language refers.

referred pain *Neurology.* any pain that is felt in an area other than the site of its origin; i.e., pain that originates in one part of the body but is perceived as coming from another area. Also, SYNALGIA, TELALGIA.

refine *Chemical Engineering.* **1.** to remove impurities from a substance or free the substance from foreign matter; remove unwanted components. Products that are refined include petroleum, metals, lubricatiing oils, and food products, such as sugar. **2.** to separate a mixture into its component parts.

refined *Chemical Engineering.* describing a substance that has been subjected to a process of refining; i.e, that has had impurities or foreign matter removed.

refinement *Chemical Engineering.* the act of refining; the removal of impurities or foreign matter. *Mathematics.* **1.** the covering $V = \{V_i\}$ of a topological space X is said to be a **refinement of a covering** $U = \{U_i\}$ of X if for every V_i there exists a U_i such that V_i is contained in U_i. **2.** let $G = G_0 > G_1 > \cdots > G_n$ be a subnormal series. A **one-step refinement** of this series is any series of the form

$$G = G_0 > G_1 > \cdots > G_i > N > G_{i+1} > \cdots > G_n \quad \text{or}$$
$$G = G_0 > G_1 > \cdots > G_n > N,$$

where N is a normal subgroup of G_i and (if $i < n$) G_{i+1} is normal in N. A **refinement of a subnormal series** S is any subnormal series obtained from S by a finite sequence of one-step refinements. A refinement of S is said to be a **proper refinement** if its length is larger than the length of S. **3.** a **refinement of a normal series** of modules is a normal series obtained by inserting a finite number of additional submodules between the given submodules. A **proper refinement** is one with length larger than that of the original series. By theorem, a composition series has no proper refinements and is therefore equivalent to any of its refinements. **4.** a grid N' is said to be a **refinement of a grid** N if N' is obtained by adding one or more lines to those forming N.

refinement of a crystal structure *Crystallography.* a process of improving the parameters of an approximate (trial) atomic arrangement in a crystal structure until the best fit of calculated structure factor amplitudes to those observed is obtained. The process usually requires many successive stages.

refinery *Chemical Engineering.* an industrial facility for refining a material such as petroleum, sugar, or metals.

refining temperature *Metallurgy.* the temperature at which a metal refining operation is performed.

reflect *Physics.* to cast back a ray of light, heat, or sound; be involved in a process of reflection. Thus, **reflected.**

reflectance *Computer Technology.* in optical character recognition, the relative value of the ink color or characters, compared to the background. *Physics.* see REFLECTIVITY. *Electricity.* see REFLECTION FACTOR.

reflectance spectrophotometry *Spectroscopy.* a spectroscopic technique for the measurement of the ratio of radiant flux reflected from a light-diffusing sample to the radiant flux reflected by a particular standard.

reflected binary code see GRAY CODE.

reflected impedance *Electricity.* the effect of an impedance on one winding, when the actual impedance occurs on another winding. Generally, the value of the actual impedance must be scaled by the turns ratio of the two windings. For example, the effect of load resistance (connected to the secondary winding) on the primary winding. Also, REFLECTED RESISTANCE.

reflected ray *Physics.* a ray of light, sound, or heat after reflection.

reflected resistance see REFLECTED IMPEDANCE.

reflected wave *Physics.* any wave that is reflected from a surface or discontinuity between two media.

reflecting *Physics.* acting to reflect; involved in a process of reflection.

reflecting antenna *Electromagnetism.* an antenna that uses a reflector to provide a directive and concentrated beam.

reflecting curtain *Electromagnetism.* a vertical array of half-wave dipole antenna elements lying in a common plane and placed one-quarter of a wavelength behind a radiating curtain for high gain.

reflecting electrode *Electronics.* an element that reverses the direction of the electron stream in a microwave oscillator tube.

reflecting goniometer *Crystallography.* a device for measuring the angle between crystal faces by measuring the angle through which a crystal has to be rotated from a position at which one face reflects a narrow beam of light into a stationary detector to a position at which a second face reflects.

reflecting grating *Electromagnetism.* a grating of conductive wires placed in a waveguide to allow certain modes of propagation to pass while blocking other modes.

reflecting microscope *Optics.* a microscope whose reflecting objective consists of mirrors rather than a lens, thus enabling it to be used with infrared and ultraviolet radiation as well as visible light.

reflecting prism *Optics.* a prism that has several polished surfaces so it can either transmit or reflect light; used in place of mirrors and designed to minimize dispersion.

reflecting sign *Civil Engineering.* a symbol that bounces back an image when caught in certain light; used in traffic management.

reflecting spectrograph *Optics.* a solar spectrograph in which long-focus concave mirrors serve as the collimator and camera element.

reflecting telescope *Optics.* a telescope that forms an image of a distant object by focusing its light with a concave paraboloidal mirror, then magnifying the image with an eyepiece.

reflection *Physics.* the turning back of a ray of light, sound, or heat when it strikes a surface that it does not penetrate. *Optics.* **1.** specifically, the return of a light ray from a surface with no change in the ray's wavelength; the return may be specular, diffuse, or mixed. **2.** an image produced by this process. *Mathematics.* **1.** the mirror image of a plane (or space) geometric configuration across a line (or plane). **2.** the function $f(-x)$, given a real function $f(x)$.

reflection-absorption spectroscopy *Spectroscopy.* a spectrographic technique in which the absorption spectrum is studied after it is reflected from some reflecting surface.

reflection altimeter see RADIO ALTIMETER.

reflection angle see ANGLE OF REFLECTION.

reflection coefficient *Physics.* a dimensionless ratio given by the amplitude of a reflected wave divided by the amplitude of the incident wave.

reflection density *Optics.* the logarithm of the ratio comparing the luminance of a perfectly diffuse reflector (a surface that reflects 100% of incident light) to the luminance of the surface in question.

reflection diffraction *Physics.* a technique in electron diffraction analysis whereby an electron beam is directed at a grazing angle (angle of incidence almost 90°) to the surface of the sample.

reflection factor *Electricity.* **1.** the ratio of the load current ideally delivered by a source when the source and load impedances are matched to a load with an impedance that is not matched. Also, MISMATCH FACTOR, TRANSMISSION FACTOR. **2.** the ratio of total flux that is reflected from a surface to the incident flux. Also, REFLECTANCE. *Physics.* the ratio of electrons reflected to electrons that enter a reflector space such as a reflex klystron.

reflection goniometer *Crystallography.* an instrument that measures the angles between crystal faces through reflection of parallel light beams from successive faces.

reflection HEED *Physics.* a technique that is used in electron diffraction analysis to determine the lateral arrangement of atoms in the topmost layers of the surface of a crystal. (An acronym for high-energy electron diffraction.)

reflection law *Physics.* a law stating that in the phenomenon of reflection, the angle of incidence is equal to the angle of reflection direction.

reflection lobe *Electromagnetism.* a lobe of radiation formed by the constructive superposition of signals that reflect from the surface of the earth.

reflection loss *Electricity.* **1.** the portion of transmission loss that occurs as a result of the reflection. **2.** the ratio of incident-minus-reflected power to incident power at a discontinuity, expressed in decibels.

reflection nebula *Astronomy.* an extensive cloud of interstellar dust that reflects light from nearby stars.

reflection plane see PLANE OF REFLECTION.

reflection profile *Engineering.* a seismic profile in which the spread geometry is designed to enhance reflected energy.

reflection seismology see REFLECTION SHOOTING.

reflection shooting *Petroleum Engineering.* a seismic prospecting operation in which the reflective properties of sound waves are used to determine the depth and size of an underground oil pool. Also, REFLECTION SEISMOLOGY.

reflection spectrum *Physics.* the frequency or wavelength spectrum of radiation after having suffered a reflection from some surface.

reflection survey *Petroleum Engineering.* the detection of the presence and depth of underground formations using reflection shooting.

reflection twin *Crystallography.* either of two twin crystals related by reflection across a common plane.

reflective *Optics.* serving to reflect light; not transparent or translucent. *Graphic Arts.* specifically, describing print material of this type, such as a positive photographic print, a line drawing on artist's board, or text copy on repro paper, as opposed to film negatives of the same material.

reflective coating *Optics.* a thin coating, such as evaporated silver or a double layer of dielectric materials, that is applied to a surface, usually glass, to increase its reflective properties over a wide or narrow band of wavelengths. Also, MIRROR COATING.

reflective code see GRAY CODE.

reflective spot *Computer Technology.* a piece of metallic foil at each end of a magnetic tape that reflects light to indicate to the tape drive the beginning and the end of the recording surface.

reflectivity *Physics.* the ratio of energies of the reflected wave to that of the incident wave; the portion of incident radiation reflected by a surface of discontinuity. Also, REFLECTANCE.

reflectometer *Engineering.* a photometer used to measure the optical reflectance of a surface. *Electromagnetism.* **1.** a device that measures microwave power in a waveguide traveling in both directions and thus determines the reflection coefficient of the guide. **2.** a device that measures the reflectance of radiation from a reflecting surface.

reflector something that reflects; specific uses include: *Engineering.* a smooth metal surface, wire grating, or array of elements whose purpose is to reflect radiation in a desired direction. *Transportation Engineering.* a surface or object that reflects the light of a vehicle's headlights as an aid to location in the dark; placed along lane boundaries or on obstructions, worn by cyclists or by pedestrians, and so on. *Electromagnetism.* an element in an antenna that changes the direction of radiation received by another antenna. *Electronics.* a cathode that changes the direction of an electron stream. *Acoustical Engineering.* a device that directs distant audio signals to a microphone. *Graphic Arts.* a gray or white card that is used in photography to reflect light into a shadow area of the subject. *Nucleonics.* a layer of water, beryllium, or graphite that surrounds the core of a reactor so as to reduce the amount of escaping neutrons. *Geophysics.* any surface, horizon, or interface that reflects incident seismic waves.

reflector characteristic *Electronics.* a graph displaying the energy generated by the voltage that appears between the reflector electrode and the cathode in a reflex klystron.

reflector microphone *Acoustical Engineering.* a microphone that has a highly directive response pattern due to a parabolic reflector around the pickup, located at the focal point.

reflector plate *Optics.* a transparent mirror that reflects an image to the eye, mounted on gun sites and bomb sites.

reflector satellite see PASSIVE COMMUNICATIONS SATELLITE.

reflector telescope see REFLECTING TELESCOPE.

reflector voltage *Electronics.* the voltage that appears between the reflector electrode and the cathode in a reflex klystron.

reflex *Neurology.* **1.** any reflected action or movement; the sum total of a given involutary activity. **2.** see REFLEX ACTION. **3.** see REFLEX ARC. *Optics.* **1.** reflected or reflecting. **2.** a reflected image of an object; reflection.

reflex action *Neurology.* an involuntary, practically immediate movement in response to a stimulus applied to a peripheral organ and transmitted to the nervous centers of the brain or spinal cord.

reflex angle *Mathematics.* an angle that is greater than 180° but less than 360°.

reflex arc *Neurology.* the pathway traversed by an impulse during a reflex action, extending from a receptor to an effector, usually via some part of the central nervous system.

reflex baffle *Acoustical Engineering.* a loudspeaker baffle that redirects some of the sound at the rear of the diaphragm in order to reinforce sound at the front.

reflex bladder *Medicine.* a urinary bladder that only functions in response to simple reflex action through the sacral cord. Also, SPASTIC BLADDER.

reflex bunching *Electronics.* the expansion and contraction an electron stream experiences when it is forced to change directions by a wave of direct current energy in a microwave tube.

reflex camera *Optics.* a camera in which a mirror reflects the image to be photographed onto a ground-glass screen, thus improving the focus and composition of the photograph.

reflex circuit *Electronics.* a circuit that amplifies a radio signal twice, before and after the intelligence it is carrying is extracted.

reflexed *Botany.* describing a plant or plant part, especially a leaf, that is bent or twisted downward or backward.

reflex epilepsy *Neurology.* an epileptic seizure that is triggered by sensory stimulation such as music, a sudden noise, or objects of touch or sight.

reflexive *Mathematics.* **1.** given a set A, a relation R on $A \times A$ is said to be reflexive if $(a, a) \in R$; that is, if a is related to itself, for all a in A. R is an equivalence relation if it is reflexive, symmetric, and transitive. **2.** a normed linear space is said to be reflexive if it is equal to its second dual; i.e., if the dual of the dual is isometrically isomorphic to the original space. Every reflexive normed linear space is a Banach space, and every L^p space $(1 < p < \infty)$ is reflexive.

reflex klystron *Electronics.* a tube that generates ultrahigh frequencies by using an electrode to channel the electron stream back into the chamber where it was first modulated, instead of channeling it into a second chamber. Also, **reflex oscillator.**

reflexogenic *Neurology.* stimulating relex action. Also, **reflexogenous.**

reflexology *Neurology.* the scientific study of motor behavior in terms of simple and complex reflexes.

reflex sight *Optics.* in an optical or computing system, a device that reflects an image onto a reflector plate so that it can be superimposed onto the target by a human eye.

reflex time *Neurology.* the elapsed time between the stimulation of a sensory organ and the response of a muscle group involved in the reflex action.

reflowing *Engineering.* a process of melting and resolidifying a surface coating.

reflux *Physiology.* a backward or return flow, as of the stomach contents into the esophagus, or of urine from the bladder into a ureter. *Chemical Engineering.* a process by which vapor from the top of a distillation column is condensed and sent back to the column to provide a contacting liquid.

reflux condenser *Chemical Engineering.* a process device connected to the top of a distillation column that condenses the vapor drawn off the top of the column.

reflux ratio *Chemical Engineering.* the ratio of the amount of overhead vapor condensed and returned to an industrial distillation column to the amount of vapor drawn off as product.

refolding *Geology.* a process whereby folds of one generation undergo strain by a force of different orientation.

refoliation *Geology.* a foliation that follows and is oriented differently from an earlier foliation.

reforestation *Forestry.* the reestablishment and development of a timber crop on forest land. Also, REAFFORESTATION.

reform *Petroleum Engineering.* to subject to a reforming process.

reformat *Computer Programming.* **1.** to change the arrangement and/or organization of data. **2.** to reinitialize a disk.

reformate *Materials Science.* the product of a reforming process.

Reformatskii reaction *Organic Chemistry.* a condensation reaction that produces β-hydroxy esters by adding zinc to a mixture of an α-halogen ester, usually the α-bromo ester, with an aldehyde or ketone in ether or hydrocarbon solution.

reforming *Petroleum Engineering.* a petroleum-refinery cracking process that converts straight-run gasoline or naphtha to a branched structure, resulting in a higher octane rating.

refract *Mechanics.* to cause to deviate; subject to a process of refraction. *Optics.* to ascertain errors of ocular refraction; determine a refractive condition.

refracted ray *Physics.* any ray of electromagnetic radiation that undergoes refraction; the beginning point of the ray is the point of refraction.

refracted wave *Physics.* the portion of a wave that suffers a change in direction as the result of encountering a discontinuity in the propagation medium. Also, TRANSMITTED WAVE.

refracting prism *Optics.* a prism used to cause a beam of light to deviate from a straight course; because the amount of deviation is a function of the wavelength of the light, the prism can be used to break a beam of white light into a spectrum.

refracting sphere *Optics.* a transparent sphere made of material having a refractive index different from the medium surrounding it, so that it refracts light which passes through.

refracting telescope *Optics.* a telescope that gathers light from a distant object, passes it through a converging or compound lens to form an image of the object, then uses an eyepiece to magnify the image; the first type of telescope invented. Also, REFRACTOR TELESCOPE.

refraction *Physics.* a change in the direction of propagation when a wave passes from one medium to another of different density. *Medicine.* the determination of the refractive errors of the eye and their correction by lenses.

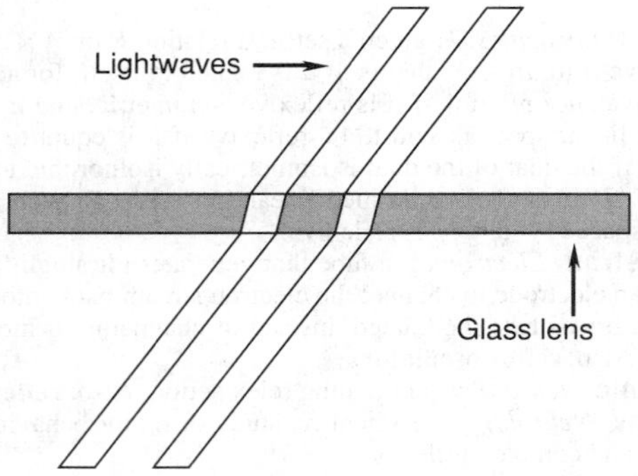

refraction

refractionation *Materials Science.* a method of fractionating or precipitating a polymer that relies on refining each initial fraction by reprecipitation; this usually achieves significantly better separations than bulk fractionation.

refraction coefficient *Oceanography.* the square root of the ratio of the spacing between adjacent orthogonals in a given area for a specific wave period and direction; it measures the way that refraction increases the length of the wave crest and thereby diminishes wave height.

refraction correction *Navigation.* an adjustment in the sextant altitude of a celestial body due to the curvature of light as it passes through the atmosphere. It is of importance only for bodies near the horizon or observed from a high altitude in an airplane.

refraction diagram *Oceanography.* a drawing that shows the position of wave crests in a given ocean area for a specific wave period and direction.

refraction error *Optics.* an error in the observed altitude of a body caused by the atmospheric bending of the light from the body when it is observed close to the horizon.

refraction loss *Electromagnetism.* a loss in a wave passing through a medium due to nonuniformities in the medium.

refraction of water waves *Oceanography.* the change in the direction of movement of waves in shoaling water that is caused by the differential effect of curving bottom contours.

refraction process *Engineering.* a seismic survey in which the distance between sensors is much larger than the depth being measured.

refraction profile *Engineering.* a seismic profile in which the spread geometry is designed so as to enhance refracted energy.

refractive *Physics.* **1.** having the power to refract. Also, **refractile. 2.** relating to a process of refraction.

refractive index *Optics.* the ratio of the phase velocity of electromagnetic radiation in a vacuum (or, for practical purposes, in air) to that in a transparent medium. Also, INDEX OF REFRACTION.

refractive shooting *Petroleum Engineering.* a technique used in seismic prospecting, in which the refractive properties of sound waves are used to determine the location of underground deposits.

refractivity *Physics.* the quality or degree of being refractive; the power to refract. *Electromagnetism.* a measure of the ability of a material to refract electromagnetic waves; a quantity expressed by $(c/v) - 1$, where c is the velocity of light in a vacuum and v is the velocity of the wave in a medium.

refractometer *Optics.* a device that measures the refractive index of a liquid or solid.

refractometry *Optics.* the measurement of the refractive index of a given substance; often used in chemical analysis.

refractor telescope see REFRACTING TELESCOPE.

refractory *Materials.* describing a material that has a high softening point and a very high melting point. Thus, **refractory alloy, refractory carbide, refractory cement, refractory enamel, refractory metal, refractory hard metal, refractory oxide.**

refractory-lined firebox boiler *Mechanical Engineering.* a horizontal fire-tube boiler having the front section of its shell lying over the refractory furnace, with the rear of the shell housing the first-pass tubes, and the upper section of the shell housing the second-pass tubes.

refractory period *Physiology.* a brief period of time following the stimulation of a nerve, during which the nerve cannot be restimulated. *Behavior.* a brief period of time following the initial movement of a set of similar movements in which a second movement cannot be initiated even if it is not antagonistic to the first. Also, **refractory phase.**

refrangible *Physics.* having the ability to be refracted.

refresh *Surgery.* to denude a wound of epithelium in order to enhance tissue repair. *Computer Technology.* **1.** to restore the digital information on a cathode-ray tube or television monitor at regular intervals to produce the continuous appearance of images. **2.** to restore stored data values in a dynamic memory.

refrigerant *Materials.* any material whose properties (e.g., low vaporization temperature) make it suitable for use in refrigeration; the substance may be a liquid (such as ammonia or Freon) or a solid (such as ice or solid carbon dioxide). *Pharmacology.* **1.** having the ability to relieve fever and thirst. **2.** any cooling remedy, such as an acidulous drink or an evaporating lotion.

refrigerated truck *Mechanical Engineering.* a truck whose interior is designed to operate as a refrigerator, used to transport perishable or frozen items.

refrigeration *Mechanical Engineering.* the process of cooling an area or object to a temperature that is significantly lower than the temperature of the surrounding environment. *Food Technology.* the use of this process to preserve foods. *Medicine.* the therapeutic application of or exposure to low temperatures.

refrigeration condenser *Mechanical Engineering.* a component of a mechanical refrigerator plant, in which a volatile agent is cooled under pressure and liquefied, discharging its heat to the environment.

refrigeration cycle *Thermodynamics.* any thermodynamic cycle in which there is a heat transfer from a cold source; in such a cycle, power is drawn from an external source and the heat transfer from the cycle exceeds the heat transfer from the cold source by the amount of work done to operate the cycle.

refrigeration system *Mechanical Engineering.* a closed-flow system used to extract heat from a controlled space by compressing, condensing, and finally expanding a refrigerant so that cooling is produced at a lower temperature and heat is rejected at a higher temperature.

refrigerator *Mechanical Engineering.* an enclosed, insulated appliance or area that is used to keep food or other perishable items cool.

refrigerator car *Mechanical Engineering.* an insulated freight car designed for use as a refrigerator.

Refsum's disease *Medicine.* a rare recessive hereditary disorder of lipid metabolism, caused by accumulation of phytanic acid, and marked by retinitis pigmentosa, incoordination of voluntary muscular action, chronic polyneuritis, and cardiac damage.

Refugian *Geology.* a North American geologic stage of the Eocene and Oligocene epochs, occurring after the Fresnian and before the Zemorrian.

refugium *plural,* **refugia.** *Ecology.* a section that remains stable within an area undergoing extreme, usually climatic, change, thus providing native plants and animals with a sanctuary.

refuse *Archaeology.* any materials or remains left behind or discarded by humans.

refusion *Surgery.* the return to circulation of blood after temporary removal or blockage.

refutation completeness *Artificial Intelligence.* the ability of a proof procedure, such as resolution, eventually to derive a contradiction if the set of axioms is contradictory.

refutation graph *Artificial Intelligence.* a data structure, in the form of a tree or directed acyclic graph, that shows the facts, axioms, and intermediate conclusions from which a contradiction was derived in a resolution proof.

reg *Geology.* **1.** a broad, thin sheet of coarse, closely packed wind-polished gravel and small stones lying on an alluvial soil in a desert plain from which fine sand has been removed by the wind. Also, GRAVEL DESERT. **2.** a stony desert of the plains, such as the Algerian Sahara.

R.E.G. radioencephalogram.

Regalecidae *Vertebrate Zoology.* the oarfishes, a monogeneric family of ribbon-shaped fish belonging to the order Lampriformes, found in all oceans.

regelation *Hydrology.* a two-stage process in which ice (or any other substance that expands upon freezing) melts under applied pressure, and then the resulting meltwater refreezes upon the release of that pressure; seen in the flow of glaciers or in the formation of a snowball by pressure.

regenerant *Chemistry.* a chemical solution used to reactivate an ion-exchange bed.

regenerate *Biology.* to undergo a process of regeneration; grow new tissues or organs. *Chemical Engineering.* to collect and purify energy or materials for reuse in another process in order to save money and resources. *Electronics.* **1.** to restore an electrical signal to its original shape. **2.** to restore data to its original form in a charge-storage tube, in order to prevent fading and decay effects. Thus, **regenerated.**

regenerated fiber *Textiles.* **1.** a semisynthetic fiber produced by extracting a natural substance, such as protein or cellulose, from a material, then extruding that substance through a spinneret to form a fiber. **2.** fiber produced by dissolving waste silk or wool fibers, then "regenerating" them through extrusion.

regenerated flow control *Hydrology.* the control of glacial drainage as a result of the modification of morainal features by the readvance of a glacier whose flow had previously stopped.

regenerated glacier *Hydrology.* **1.** a glacier that begins to readvance after it had previously stopped flowing. **2.** see GLACIER REMANIÉ.

regeneration *Biology.* the growth of new tissues or organs to replace those lost or damaged by injury; this process is very variable in animals, being greatest in the lower animals, where a complete organism can sometimes be regenerated from a few cells, and least in the mammals, where it is limited to wound-healing and the regrowth of peripheral nerve fibers; it is a common occurrence in plants. *Computer Technology.* the process of restoring a computer storage element when its ability to store information deteriorates. *Control Systems.* see POSITIVE FEEDBACK. *Electronics.* see REGENERATIVE FEEDBACK. *Nucleonics.* the treatment of used nuclear fuel elements so that they may be used again in the reactor.

regenerative *Science.* relating to or involved in a process of regeneration.

regenerative air heater *Mechanical Engineering.* an air heater in which the heat-transmitting surfaces are exposed in an alternating pattern to the heat-surrendering gases and to the air.

regenerative amplifier *Electronics.* a circuit that increases its power and selectivity by returning the amplified signal back to the circuit without inducing oscillation.

regenerative braking *Electricity.* the application of power to a motor so that the motor acts like a generator. This converts the rotational energy into electrical energy, thus braking the motor and producing electric power.

regenerative chaff *Military Science.* an electronic deception technique in which clouds of chaff are sequentially ejected by a missile or decoy.

regenerative cooling *Mechanical Engineering.* a process to cool gases in which compressed gas is cooled by expansion through a nozzle, and the cooled expanded gas passes through a heat exchanger to further cool incoming compressed gas.

regenerative cycle *Mechanical Engineering.* a cycle in a steam engine that uses heat that would ordinarily be lost, by extracting exhaust steam from the turbine in stages and using it to heat the feedwater.

regenerative detector *Electronics.* a circuit in which an audio signal is fed back into the radio receiver in order to increase the sensitivity and amplification of the carrier signal.

regenerative divider *Electronics.* a circuit that reduces an incoming signal's frequency by modulating, amplifying, and recycling a portion of it.

regenerative engine *Space Technology.* a liquid rocket engine whose chamber is cooled by the circulation of the fuel, the oxidizer, or both.

regenerative feedback *Electronics.* a process in which part of an amplified signal is returned to the input in the same phase, without inducing oscillation, in order to increase amplification. *Control Systems.* see POSITIVE FEEDBACK.

regenerative fuel cell *Electricity.* a fuel cell in which the reaction product can be recycled to derive more reactants.

regenerative pump *Mechanical Engineering.* a rotating-vane instrument that creates high liquid heads at low volumes with the use of mechanical impulse and centrifugal force. Also, TURBINE PUMP.

regenerative reactor *Nucleonics.* a reactor that simultaneously produces useful energy and fissionable material.

regenerative read *Computer Programming.* a destructive read operation in which the data taken from a storage location is automatically written back into that location.

regenerative receiver *Electronics.* a radio receiver in which regeneration is used to enhance the amplification that takes place when the audio component is being extracted from an incoming signal.

regenerative repeater *Telecommunications.* **1.** a repeater in which the pulse signals are amplified, reshaped, retimed, and then retransmitted. **2.** a repeater used exclusively for digital transmission.

regenerative storage see CYCLIC STORAGE.

regenerative track *Computer Technology.* a track used in conjunction with read/write heads, arranged on a magnetic drum or disk so that signals are regenerated during each revolution to retain recorded data.

regenerator *Chemical Engineering.* a process device that returns the system to its full capacity.

Regge, Tullio [reg´ə; rā´gā´] born 1931, Italian physicist; provided an early description of quantum-mechanical scattering.

Regge family *Particle Physics.* a theoretical grouping of elementary particles that have the same Regge trajectory.

Reggeism see REGGE THEORY.

Regge pole *Particle Physics.* a theoretical model used to describe the scattering of elementary particles at high energies; such poles are identified as singularities in the complex plane of angular momentum.

Regge recurrence *Particle Physics.* a hadron from a sequence of hadrons arranged in sequence by increasing spin number and increasing mass.

Regge theory *Particle Physics.* an approach to the relationship between the asymptotic behavior of high-energy scattering amplitudes of elementary particles (hadrons) and their resonances through the use of Regge poles. Also, REGGEISM.

Regge trajectory *Particle Physics.* the path followed by a Regge pole in the complex plane as the center of mass is varied.

regime *Geology.* a regular or systematic pattern of action or occurrence, or a condition having widespread influence. *Hydrology.* **1.** in a stream channel, the reaching, over a period of years, of a state of equilibrium between erosion and deposition. **2.** the condition of a stream with respect to its average flow rate, determined by measuring the volume of water passing different cross sections in a given period of time.

regimen *Medicine.* a strictly regulated scheme of diet, exercise, or other activity designed to preserve or restore health or achieve some other result, such as weight loss. *Hydrology.* **1.** the flow characteristics of a stream, such as its volume, velocity, and capacity to transport sediment. **2.** the analysis of the total amount of water related to a lake, including losses and gains, over a given period of time. **3.** the total amount of water related to a particular drainage basin. **4.** see BALANCE.

regiment *Military Science.* a unit of ground forces that usually consists of two or more battalions or battle groups, a headquarters unit, and supporting units; it is larger than a battalion but smaller than a brigade.

regimental landing team *Military Science.* a task organization consisting of an infantry regiment reinforced by those elements that are required for initiation of its combat function ashore.

Regiomontanus (Johann Müller) 1436–1476, German mathematician and astronomer; wrote the first systematic treatise on trigonometry.

region an area having more or less definite boundaries; specific uses include: *Geography.* a portion of the earth's surface that is distinguished from adjacent lands by one or more characteristics or features. Depending on the criteria used to distinguish them, such regions may be geological, climatic, economic, linguistic, political, and so on. *Ecology.* such a division based on climate and geographical features, as well as on the plants and animals native to the area. *Anatomy.* a general term for certain sites on the surface area of the body within defined boundaries, such as the abdominal region or lumbar region. *Computer Technology.* **1.** a group of memory addresses that are relative to the same reference address. **2.** a group of machine addresses that refer to a base address. **3.** in multiprogramming, the internal storage space that is allocated to a task. *Mathematics.* **1.** a connected open set. **2.** a region of a planar graph is an area of the plane that is bounded by edges and contains neither edges nor vertices.

regional address *Computer Programming.* the address of a machine instruction within a series or region of consecutive related addresses.

regional anatomy *Anatomy.* the study of the structures of specific regions of the body, such as the thoracic cavity, abdominal, head, or limbs.

regional anesthesia *Medicine.* the loss of sensation of a region of the body produced by an injection of anesthetic solution into or near the nerve trunks involved in the operative field. Also, **regional block anesthesia.**

regional center *Telecommunications.* a long-distance telephone office of the highest rank in the routing of telephone calls.

regional character species *Ecology.* a plant species that grows only within a particular plant community.

regional dip *Geology.* the nearly uniform, usually low-angled inclination of sedimentary beds over a wide area. Also, NORMAL DIP.

regional enteritis *Medicine.* a chronic recurrent disease, generally of young adults, in which inflammation of portions of the small intestine is accompanied by abdominal pain, diarrhea, fever, and weight loss. Also, CROHN'S DISEASE, TERMINAL ILEITIS, REGIONAL ENTEROCOLITIS, REGIONAL ILEITIS.

regional enterocolitis see REGIONAL ENTERITIS.

regional forecast see AREA FORECAST.

regional geography *Geography.* the geographical study of particular portions of the earth's surface.

regional geology *Geology.* the geology of any relatively large region with respect to the spatial distribution and position of stratigraphic units, structural features, and surface forms.

regional gravity *Geodesy.* the measured effects on the observed gravity anomalies at a given point resulting from density irregularities at much greater depths than the structures being surveyed at that point.

regional ileitis see REGIONAL ENTERITIS.

regional metamorphism *Geology.* metamorphism affecting rocks over an extensive region, rather than over a restricted or local area.

regional metasomatism *Geology.* metasomatism affecting rocks over an extensive area, whereby the introduced material may be the result of partial fusion from deep-seated magmatic sources.

regional migration *Petroleum Engineering.* the movement of oil or gas horizontally through a formation in response to artificial pressure differences resulting from extraction at well sites.

regional slope *Geology.* the generally uniform inclination of rock layers or the surface of the land over a wide area.

regional slope deposit *Geology.* a sedimentary deposit that has been distributed as a thin sheet over a regional slope.

regional snowline *Hydrology.* the level above which the accumulation of snow is greater than its ablation year after year, as averaged over a large area.

regional survey *Archaeology.* a broad survey that includes the total environmental setting surrounding an archaeological site.

regional system *Archaeology.* a system of time divisions such as those used in the Americas, based on major technological or social changes that produced regional cultures rather than local ones.

regional unconformity *Geology.* a continuous unconformity extending throughout an extensive region and usually representing a relatively long period of time.

region of escape see EXOSPHERE.

regioselective *Organic Chemistry.* relating to the preference of a chemical reaction for a certain position or structural isomer, producing a greater yield of the product associated with the position or isomer. Also, **regiospecific.**

register *Graphic Arts.* **1.** the proper and consistent alignment of printed text, especially on facing pages. **2.** the proper alignment of multiple (superimposed) impressions, especially in four-color printing; impressions that are not correctly aligned are said to be **out of register.** *Computer Technology.* a special-purpose storage location within the central processing unit, in which data and intermediate results are temporarily stored for use in arithmetic and logical operations. *Telecommunications.* **1.** the part of an automatic switching telephone system that receives impulses and that controls subsequent switching operations. **2.** to accurately match two or more patterns, such as the three images in color television. *Mechanical Engineering.* a lid or sliding plate in a heating device that controls the air flow in the combustion process. *Military Science.* to adjust fire in order to determine firing data corrections.

register breadth *Naval Architecture.* the breadth of a hull used for the computation of register tonnage; normally the maximum molded beam.

register capacity *Computer Technology.* **1.** the range of values for numerical quantities that can be held in a register. **2.** the number of registers in a central processing unit.

register circuit *Electronics.* a circuit whose memory elements store bits of coded information that can be retrieved from the circuit in that code or in another one.

register control *Control Systems.* a method of automatically controlling the position of a mark or design as part of a process.

registered nurse *Medicine.* a graduate nurse who has been legally authorized (**registered**) to practice after examination by a state board of nurse examiners or similar regulatory authority and is legally entitled to use the designation **R.N.**

register length *Computer Technology.* the number of characters and bits that a register can store.

register mark *Graphic Arts.* a small mark, often a circled cross, that is drawn on copy or printed on layout sheets to ensure proper registration of plates. *Mechanical Engineering.* a reference point used to keep accurate alignment on the working surface of sheet metal or other machinable material.

register ton *Naval Architecture.* a unit of cubic capacity for a vessel, equal to 100 cubic feet; equivalent to 2.83 cubic meters.

register tonnage *Naval Architecture.* the cubic capacity of a ship, as expressed in register tons. **Gross register tonnage** is the total enclosed space of the vessel, less certain minor exemptions; **net register tonnage** exempts all crew and engineering spaces.

registration *Graphic Arts.* the process of placing copy for multiple impressions in the proper alignment, or the accuracy with which this placement is achieved. *Psychology.* the first stage in learning and memory, when stimuli impressions are made on the central nervous system.

registration fire *Military Science.* fire delivered to obtain accurate data for use in delivering accurate fire on subsequent targets.

registration mark *Computer Technology.* in plotter operation, a mark that permits alignment of the plotter pen to ensure accurate placement of the plotted data.

registration point *Military Science.* a terrain feature or other designated point on which fire is adjusted for the purpose of determining firing data corrections. Also, CHECKPOINT.

regma *Botany.* a dry fruit composed of at least three carpels that separate from the axis when mature, as in geraniums.

regmagenesis *Geology.* the production of global strike-slip faults by earth forces. Also, RHEGMAGENESIS.

regmaglypt *Geology.* any small, well-defined indentation or pit on the surface of a meteorite. Also, PIEZOGLYPT, PEZOGRAPH.

Regmatodontaceae *Botany.* a monogeneric family of tropical, slender, shiny, olive-green mosses of the order Isobryales that form dense mats on tree trunks; characterized by creeping branched stems, lateral sporophytes, and a distinct diplolepideous peristome of 16 blunt teeth.

regolith *Geology.* the entire layer or mantle of loose, unconsolidated material, including soils, sediments, and rock fragments, that overlies bedrock and forms the surface of the land. Also, RHEGOLITH, MANTLE ROCK, OVERBURDEN.

Regosol *Geology.* a group of azonal soils that develops from deep, unconsolidated deposits of rock or soft minerals and is characterized by having no definite genetic horizons.

regradation *Hydrology.* the formation of a new longitudinal profile by a graded stream.

regrade *Military Science.* to determine that specific classified information requires a higher or lower degree of protection against unauthorized disclosure than currently provided, and to change the classification designation to reflect such higher or lower degree. Also, RECLASSIFY.

regression a process of going back; specific uses include: *Evolution.* a reversal in the direction of evolution to a less-developed form or to an average type. *Psychology.* a defense mechanism that is characterized by a return to an earlier, usually less adequate stage of behavior as a way of dealing with a current problem or conflict. *Oceanography.* a withdrawal of the sea from the land, sometimes in conjunction with a rising landmass. *Geology.* the land area resulting from such withdrawal. *Hydrology.* the theory that some rivers have their sources on the rainier sides of mountain ranges and gradually erode their heads backward until the ranges are cut through. *Biochemistry.* see CATABOLISM. *Statistics.* see LINEAR REGRESSION.

regression coefficient *Statistics.* the coefficient describing the effect of an independent variable in a linear regression equation.

regression conglomerate *Geology.* a coarse sedimentary deposit that was formed during a retreat of the sea.

regression line *Statistics.* in a bivariate regression model, a line describing the expected behavior of the dependent variable as a function of the independent variable. Also, GALTONIAN CURVE.

regression of nodes *Astronomy.* the slow westward motion of the nodes of the moon's orbit as a result of perturbations by the earth and the sun; one cycle takes approximately 18.6 years.

regressive *Science.* involved with, causing, or carrying out a process of regression.

regressive overlap see OFFLAP.

regressive reef *Geology.* any of a series of nearshore reefs that have been built upon basin deposits during the lowering of the sea level or the rising of a landmass.

regressive ripple *Geology.* an asymmetric ripple mark formed by a current and oriented in a direction that is opposite to the general movement of current flow; for example, in the lee of an obstruction.

regressive sediment *Geology.* sediment deposited during the retreat of water from land or during the rising of land, exhibiting an offlap arrangement.

regressive shock therapy *Psychology.* electroconvulsive treatment in which the patient is first shocked into a convulsive period, and then is maintained by means of a steady flow of mild electrical current in a state resembling sleep; a severe treatment for schizophrenics that is now rarely used.

regroup airfield *Military Science.* a military or civilian airfield at which aircraft reassemble after H-hour to rearm and resume armed alert, to conduct further combat missions, or for overseas deployment.

regula falsi *Mathematics.* a method of approximating an unknown quantity by first making an estimate and then using known properties of the quantity to improve the estimate.

regular being normal or as expected; usual or consistent; specific uses include: *Physiology.* occurring at proper or fixed intervals, as a *regular* heartbeat. *Botany.* describing flowers that are radially symmetrical or identical, or that can be divided into equal parts; for example, the sunflower has *regular* symmetry. *Electromagnetism.* having defined direction; not scattered or diffused. *Military Science.* regarding or belonging to the standing army of a state. *Mathematics.* **1.** a nonzero element r of a ring R is said to be regular if r is neither a left nor right zero divisor. If R has an identity, this means that r has a left and right multiplicative inverse. **2.** a polygon is said to be regular if all its sides are of the same length; equivalently, if all its angles are congruent. **3.** a polyhedron is said to be regular if all its faces are congruent. There exist exactly five regular polyhedra: tetrahedron, octahedron, cube, dodecahedron, and icosahedron; they are the Platonic solids. **4.** a graph G is said to be regular if each of its vertices has the same degree; if every vertex has degree r, then G is said to be regular of degree r.

regular connective tissue *Histology.* a dense connective tissue whose collagenous fibers are arranged in regular patterns.

regular element *Industrial Engineering.* a basic work element that is a regular part of the work cycle. Also, REPETITIVE ELEMENT.

regular embedding *Mathematics.* an embedding from a topological space Y into a topological space X that is also a homeomorphism. An embedding is a homeomorphism if and only if the topology of the image of Y coincides with the topology induced by X.

regular expression *Computer Programming.* an algebraic expression that denotes a regular language.

regular grammar *Computer Programming.* a grammar that denotes a regular language; its productions can only have on the right-hand side either a terminal string or a terminal string followed by a single nonterminal.

Regularia *Invertebrate Zoology.* a grouping of echinoid echinoderms, spherical sea urchins with mouth and anus at opposite poles.

regular ideal *Mathematics.* a left (or right) ideal I in a ring R is said to be regular if there exists an element e of R such that $r - re \in I$ (or $r - er \in I$) for all $r \in R$. Every ideal in a ring with identity is regular.

regularity the fact of being regular; specific uses include: *Materials Science.* a characteristic of the molecular structure of polymers in which the molecules can be described by only one species of atoms or atom groups linked in a single sequential pattern.

regularization *Quantum Mechanics.* the formal introduction of extra fields into a quantized field theory, whose masses will be allowed to approach infinity, in order to remove ambiguities in calculating certain integrals.

regular language *Computer Programming.* a computer language that is described by a regular grammar, or recognizable by a finite automaton; e.g., a simple item such as a variable name or a number in a programming language.

regular lay *Design Engineering.* the lay of a rope in which wires in a strand are twisted opposite to the direction of the strand itself.

regular precession *Mechanics.* the precession of a rotating body without nutation, in other words, in such a way that the angle between the precession and the body rotation axes remains constant.

regular rayon *Materials.* a synthetic fiber made from processed cellulose and having a low wet modulus and high degree of shrinkage when washed; widely used in the textile trade due to its high water-holding capacity, uniformity, biodegradability, and cleanliness.

regular reflection see SPECULAR REFLECTION.

regular representation *Mathematics.* the representation of a group afforded by the action of the group on its own group ring.

regular solution *Physical Chemistry.* a solution in which the components are randomly distributed as if the solution were ideal, but that is nonideal in that the components have different interaction energies.

regular topological space *Mathematics.* a topological space X that satisfies the first and third separation axioms: i.e., given two distinct points x and y, there is an open set that contains y but not x; and given a closed set F and a point x not in F, there are disjoint open sets U_1 and U_2 such that $x \in U_1$ and F is contained in U_2. Also, **regular space.**

regular variable star *Astronomy.* any variable star whose light fluctuates in a cyclical manner.

regular ventilating circuit *Mining Engineering.* all places within a mine through which the flow of air occurs naturally and without the assistance of a blower fan or ventilation tubing.

regulate *Engineering.* to monitor a system or device continuously and adjust it as necessary to maintain or achieve desired results. Thus, **regulated, regulating, regulative, regulatory.**

regulated circuit *Electricity.* a circuit in which voltage or current flow is continuously regulated.

regulated item *Ordnance.* an item that is closely supervised in distribution due to its scarcity, cost, or highly technical or hazardous nature.

regulated power supply *Electricity.* a power supply that regulates voltage or current through variations in line voltage, output load, and ambient temperature.

regulating reservoir *Civil Engineering.* a source of stored water that can be increased or decreased to allow for changes in water flow.

regulating rod *Nucleonics.* a control rod that can be inserted into the core of a reactor much more rapidly than a shim rod but only provides fine control over reactivity.

regulating station *Military Science.* a command agency established to control all movements of personnel and supplies into or out of a given area.

regulating system see AUTOMATIC CONTROLLER.

regulating transformer *Electricity.* a transformer that is used for voltage, phase angle, or both in a regulated circuit; controls the output within specified limits and compensates for fluctuations of load and input voltage and phase angle (when applicable) over a specified range.

regulating winding *Electricity.* a winding of a main power unit in which taps are altered to control voltage or phase angle of the regulated circuit through the series unit.

regulation a process of controlling or correcting; specific uses include: *Biology.* the adaptation of an organism's form or behavior to changed conditions in its environment. *Control Systems.* a process in which a quantity (such as speed, temperature, or voltage) is held constant by an electrical or electromechanical system that automatically corrects errors through a feedback loop. *Electricity.* the change of one of the controlled or regulated output parameters that result from the change of one or more of the unit's variables within specification limits between no-load and full-load capacities in a transformer, generator, or other source. *Electronics.* the difference between the maximum and minimum voltage drops across a given anode-current range in a gas tube.

regulation of constant-current transformer *Electricity.* the maximum allowed departure of a secondary current from its rated load; it is expressed in a percentage of the rated secondary current with rated primary voltage and frequency applied.

regulative egg *Developmental Biology.* an egg in which the cells of the fertilized ovum retain equipotentiality during cleavage stages, a process known as **regulative development.**

regulator a person or thing that regulates; specific uses include: *Control Systems.* a device that can vary the quantity of something according to a set plan or hold it to a predetermined value. *Horology.* **1.** in a timepiece with a balance escapement, a device used to adjust the timekeeping rate by shifting the position of two pins which control the working length of the hairspring, thereby changing its period of oscillation. Also, INDEX. **2.** a precise, weight-driven timepiece controlled by a compensation pendulum; the dial usually has a long minute hand in the center, with separate, smaller dials for the second and hour hands above and below the center. *Mining Engineering.* an opening in a door or wall in a mine's return airway, used to increase resistance and reduce the volume of air flowing.

regulator gene *Genetics.* a genetic unit that regulates or suppresses the activity of one or more structural genes. Also, **regulatory gene.**

regulator pin *Horology.* either of two pins that control the working length of the hairspring on the regulator of a timepiece. Also, INDEX PIN.

regulator problem see LINEAR REGULATOR PROBLEM.

regulatory control function *Control Systems.* the level of a large-scale control system that carries out the decisions of the optimizing controller through the input of set points, desired trajectories, or targets. Also, DIRECT CONTROL FUNCTION.

regulatory enzyme *Enzymology.* any enzyme that regulates a metabolic process and whose catalytic activity can be modified.

regulatory protein *Biochemistry.* a protein that controls cellular or physiological functions; for example, insulin, which regulates sugar metabolism.

regulatory response *Ecology.* the reversible cybernetic reaction of an organism, either behavioral or physiological, to changes in its environment.

regulon *Genetics.* a disjunct group of genes that are controlled by the same regulator gene.

regulus *Metallurgy.* crude metal that is formed below the slag cover during smelting or refining.

Regulus *Astronomy.* Alpha Leonis, a white triple star of spectral type B8 that lies at a distance of approximately 67 light-years and has a visual magnitude of 1.3. Also, LITTLE RULER. *Ordnance.* a surface-to-surface, jet-powered guided missile with nuclear-warhead capability; it is designed to be launched from a cruiser or surfaced submarine; officially designated **RGM-6/15.** (From Latin; literally, "the little king.")

Regur *Geology.* a dark-colored, calcareous, intrazonal soil having a high clay content and formed from rocks low in silica. Also, BLACK COTTON SOIL.

regurgitate *Medicine.* to carry on or cause a process of regurgitation.

regurgitation *Medicine.* a backward flowing, such as the return from the stomach back to the mouth of food that has been swallowed, or the backflow of blood through a defective heart valve.

rehabilitate *Medicine.* to carry on or cause a process of rehabilitation.

rehabilitation *Medicine.* **1.** the restoration of normal form or function after disease, injury, or addiction. **2.** the restoration of an ill or injured patient to self-sufficiency or to gainful employment at his or her highest attainable skill. *Military Science.* the processing of units or individuals recently withdrawn from combat during which personnel are rested, reinforced, and provided special facilities and training, materiel is reconditioned or replaced, and the unit is generally prepared for employment in future operations.

rehabilitation engineering *Engineering.* the research and development of devices that enable handicapped persons to function independently in society.

rehash *Computer Science.* **1.** in a hash table storage scheme, to calculate a new hash value for an item when the previous hash value caused a collision with an existing item. **2.** the algorithm used to calculate the new hash value.

Rehbinder effect *Materials Science.* an effect exerted on the mechanical properties of solids by a gaseous or liquid medium; may facilitate the plastic flow of materials or decrease their strength and plasticity; often occurs in conjunction with other environmental effects such as dissolution, corrosion, and hydrogen embrittlement.

Rehbock weir formula *Fluid Mechanics.* a formula that relates the dimensions of a sharp-edged weir bounded by parallel walls to a certain dimensionless parameter alpha; alpha can then be used, along with the weir dimensions, to calculate the weir volumetric flow rate.

rehearsal *Behavior.* the act of practicing a behavior.

rehearsal buffer *Behavior.* a process that prevents rapid loss of items that are in short-term memory.

reheating *Thermodynamics.* the process of passing a fluid that has been partially expanded at constant entropy back to a superheater before subjecting it to further expansion; used in pneumatic systems for operating power tools and in heat-treating processes.

reheat test *Materials Science.* a test that evaluates a refractory material for permanent shrinkage or expansion; the dimensions of test bricks are measured, and the bricks are then heated in a furnace at a specified rate to a temperature which is sustained for a specific time; after they have cooled, the bricks are measured again.

Rehfuss tube [rā´fəs] *Medicine.* a stomach tube with a graduated syringe, designed to remove the stomach contents after a special meal for analysis of gastric secretion in the **Rehfuss test.** (Named for Martin Emil *Rehfuss,* 1887–1964, American physician.)

Reich, Ferdinand [rīk] 1799–1882, German chemist; with Hieronymus Theodor Richter, discovered indium.

Reich, Wilhelm 1897–1957, Austrian psychoanalyst, lived in the U.S.; formulated theory of character armor; developed orgone therapy.

Reichenbach's lamellae *Geology.* thin, platy inclusions of foreign minerals occurring in iron meteorites.

Reichert-Meissl number *Analytical Chemistry.* a measure of the number of milliliters of 0.1 N base required to neutralize the water-soluble volatile obtained from 5 grams of fat.

Reich process *Chemical Engineering.* an industrial process that recovers and purifies carbon dioxide from a fermentation process through oxidation of the organic impurities and chemical dehydration.

Reichstein, Tadeus born 1907, Polish-born Swiss organic chemist; awarded the Nobel Prize for isolating and analyzing the hormones of the adrenal cortex.

Reid equation *Petroleum Engineering.* an equation relating flow rate in a gas well to pitot-tube readings over various impact pressures.

Reid mechanism see ELASTIC REBOUND THEORY.

Reid vapor pressure *Petroleum Engineering.* the vapor pressure of gasoline at a temperature of 100°F as measured in a test bomb.

Reighardiidae *Invertebrate Zoology.* a family of tongue worms in the phylum Pentastomida, having a cuticle covered with tiny spines and a rounded, lobeless posterior; endoparasites in the nasal passages of vertebrates.

Reil's ribbon see MEDIAL LEMNISCUS.

reimbursed time *Computer Technology.* a term for machine time that is lent or leased to an organization and is usually reimbursed.

Reimer-Tiemann reaction *Organic Chemistry.* a reaction that produces phenolic aldehydes by heating a phenol with chloroform in an alkaline solution.

Reinartz crystal oscillator *Electronics.* a circuit in which alternating current is generated by coupling the crystal current with the current produced by a resonant circuit tuned to half the crystal frequency.

reincarnation *Psychology.* the belief that the souls of dead people are revived and then inhabit the bodies of later generations of people or of other life forms.

reindeer *Vertebrate Zoology.* a large Old World northern deer, *Rangifer tarandus tarandus,* of the deer family (Cervidae), having large hooves with very large dewclaws and large antlers in both sexes. Reindeer are closely related to and identified with the North American caribou, and are often classified in the same species.

reindeer

Reinecke's salt *Analytical Chemistry.* a precipitating agent for organic bases and amines, formed by fusion of ammonium thiocyanate with ammonium dichromate.

reinfection *Medicine.* **1.** a second infection by the same pathogenic agent, occurring either during the primary infection or after recovery. **2.** a second infection of the same organ by a different pathogen.

reinforced beam *Civil Engineering.* a concrete beam that is centrally strengthened by internal steel rods to resist tensile or sheer stresses.

reinforced brickwork *Civil Engineering.* masonry in which reinforcers, such as thin rods, steel-wire mesh, or expanded metal, have been sunk into the joints as a brace.

reinforced clostridial medium *Microbiology.* a medium that is supplemented with blood, egg yolk, or milk and used to culture species of *Clostridium.*

reinforced column *Civil Engineering.* a column, usually concrete, into which steel rods have been placed internally to act as a brace against sheer forces.

reinforced concrete *Civil Engineering.* concrete in which steel bars or wires (at least 0.6% by volume) are embedded to increase tensile load-bearing capacity.

reinforced plastic *Materials.* any plastic strengthened with a fibrous material such as asbestos, cloth, glass fiber, or metal fiber; used in automobile components and mechanical devices.

reinforcement *Behavior.* **1.** an event (reward or punishment) that follows a response and increases or decreases the likelihood that it will recur. Also, **reinforcer. 2.** the act of applying such an event or reward. *Civil Engineering.* any material, such as steel rods or wire mesh, that is embedded in concrete or masonry to strengthen its tensile strength or load-bearing capacity.

reinforcement conditioning see OPERANT CONDITIONING.

reinforcement effect *Behavior.* the principle that behavior which is rewarded will tend to recur, while behavior which is punished will not.

reinforcement planting *Forestry.* the process of restocking or filling in failed areas in a crop or stand by additional planting or sowing.

reinforcements *Military Science.* additional personnel or materiel supplied to a military unit.

reinforcement schedule see SCHEDULE OF REINFORCEMENT.

reinforcement value *Behavior.* a measure of the degree to which the attaining of a goal provides reinforcement.

reinforcing bar *Civil Engineering.* a rigid steel member used in concrete or masonry to resist lateral forces or shear stresses.

reinfusion *Medicine.* the infusion of a body fluid that has been withdrawn from the same patient; e.g., the reinfusion of ascitic fluid after ultrafiltration.

reinitialize see REBOOT.

reinnervation *Neurology.* the restoration of nerve control of a paralyzed muscle or organ, either spontaneously or by nerve grafting.

reinsertion of carrier *Electronics.* in a radio receiver, the process of recombining two sidebands of the received signal in order to reconstruct a replica of the unmodulated carrier signal.

reintegration *Biology.* a restoration of integration following a condition of disruption. *Psychology.* a restoration of harmonious mental functioning following a disintegration of the personality in mental illness.

reinversion *Surgery.* the restoration to normal position of an inverted organ, especially a uterus.

Reissert synthesis *Organic Chemistry.* a synthesis of indoles that occurs under basic conditions starting with diethyl oxalate and *o*-nitrotoluene or a substituted derivative.

Reissner-Nordstrom solution *Physics.* a solution in general relativity theory that describes a charged black hole that is not rotating.

Reissner's membrane *Anatomy.* an epithelial membrane of the inner ear that separates the cochlear duct from the scala vestibuli. Also, VESTIBULAR MEMBRANE.

Reiter's syndrome *Medicine.* an arthritic disorder affecting adult males, marked by a triad of symptoms: urethritis, conjunctivitis, and arthritis; the origin is unknown, and the disease follows a self-limited but relapsing course. Also, **Reiter's disease.** (Identified by Hans *Reiter,* 1881–1961, German physician.)

rejected recharge *Hydrology.* water that reaches the water table, but is then released because a full or overflowing aquifer cannot hold it.

rejection *Immunology.* an immune response against a grafted tissue or transplanted organ, causing failure of the graft or organ to survive.

rejection band *Electromagnetism.* a band of frequencies whose upper limit is the cutoff frequency of a uniconductor waveguide. *Physics.* any band of frequencies whose intensities are characteristically reduced or eliminated by some filtering process.

rejection region *Statistics.* the range of values of the sample statistic that will cause the null hypothesis to be rejected. Also, CRITICAL REGION.

rejuvenated fault scarp *Geology.* a fault scarp that has been freshened by renewed movement along an old fault line after partial dissection or erosion had taken place. Also, REVIVED FAULT SCARP.

rejuvenated stream *Hydrology.* a mature stream in which renewed erosional activity restores the stream to a more youthful stage of development. Also, REVIVED STREAM.

rejuvenated water *Hydrology.* water that has been returned to the earth's water supply as a result of compaction and metamorphism.

rejuvenation *Physiology.* a renewal of strength and vigor. Also, **rejuvenescence.** *Geology.* **1.** the restoration of youthful features of a landscape or landform in an area that had been worn down close to base level. **2.** the renewal of any geologic process. Also, REVIVAL. *Hydrology.* **1.** the stimulation of a stream, either by uplift or by a drop in sea level, to renew erosive activity. **2.** the restoration of youthful vigor in a mature stream.

rejuvenation head see KNICK.

rel *Electromagnetism.* the centimeter-gram-second unit of magnetic reluctance, equivalent to one ampere-turn per magnetic line of force.

rel. relative; relation.

relapse *Medicine.* the return of a disease after its apparent or partial cessation.

relapsing fever *Medicine.* any of a group of infectious diseases caused by the spiral microorganism of the genus *Borrelia* and transmitted by lice and ticks; it is characterized by episodes of fever, chills, headache, muscular pain, and nausea. More commonly seen in Asia, South America, and Africa than in North America, it occurs often during wars and famines. Also, AFRICAN TICK FEVER, FAMINE FEVER.

relation *Science.* the condition or state of one entity when considered in connection or comparison with another. Thus, **relational.** *Computer Programming.* **1.** in assembler programming, the comparison of two expressions to determine how the value of one expression relates to the value of the other. **2.** in a relational database model, a table of tuples (records) and attributes (data element values). *Mathematics.* a relation between two sets X and Y is a subset R of $X \times Y$. x is said to be R-related to y if $(x, y) \in R$.

relational algebraic language *Computer Programming.* a high-level language used to devise and characterize logical relations between entities in a relational database.

relational calculus language *Computer Programming.* a high-level language that uses predicate calculus to define logical relations in relational databases; examples are PROLOG and SQL.

relational database *Computer Programming.* a database management system in which tuples of related data values are stored and used in data retrieval operations. Also, **relational data structure, relational system.**

relational difference *Archaeology.* a method of defining variation in artifacts according to which other artifacts they are found with.

relational dimension *Archaeology.* a characteristic of an artifact based on the other artifacts it is found in association with.

relational expression *Computer Programming.* an expression that describes the relationship between two terms, consisting of an arithmetic expression followed by a relational operator, such as ">" (greater than), followed by another arithmetic expression; the expression may be either true or false.

relational operator *Computer Programming.* **1.** an operator that makes a comparison between two operands, such as *greater than* or *less than* and returns true or false. **2.** an operator that combines various relations into one. **3.** in COBOL, a reserved word, group of reserved words, or group of relational characters.

relative *Science.* not absolute; comparative or conditional. *Navigation.* **1.** of a bearing or other direction, measured as an angle clockwise from the head of the craft. **2.** of a speed with respect to a craft.

relative abundance *Ecology.* the ratio of the number of members of a single group in a given area to all the members of all the groups found in this area. *Nucleonics.* see ABUNDANCE RATIO. *Physical Chemistry.* see ISOTOPIC ABUNDANCE.

relative address *Computer Technology.* a memory address to which a reference or base address must be added in order to obtain the effective address of the operand or instruction. Also, FLOATING ADDRESS.

relative age *Geology.* the age of a geologic feature, structure, or event expressed in relation to other features, structures, or events, rather than in years.

relative atomic mass see ATOMIC WEIGHT.

relative attenuation *Electronics.* in an electrical transducer or filter, the ratio of the attenuation at a given frequency to the minimum attenuation in the frequency band of use.

relative azimuth *Navigation.* the horizontal direction of a celestial point, measured from the head of the craft.

relative bandwidth *Electronics.* the ratio, in an electric filter, of the bandwidth under consideration to the specified reference bandwidth, such as one in which the attenuation between frequencies is three decibels.

relative bearing *Navigation.* the horizontal direction to another object, measured from the head of the craft.

relative biological effectiveness *Nucleonics.* a quantity equal to the inverse ratio of the absorbed ionizing radiation in a tissue to that of a reference amount that is required to produce a similar effect.

relative braking distance *Transportation Engineering.* the distance maintained between two vehicles (or trains) moving in the same direction on the same line; equal to the braking distance of the trailing vehicle minus that of the lead vehicle plus a designated margin of safety.

relative chronology *Geology.* the study of geologic time where the time-order is based on superposition or on fossil content, rather than on age in years.

relative coding *Computer Programming.* the process of writing machine instructions using relative or symbolic addresses.

relative compaction *Engineering.* the ratio of the dry density of soil measured in situ divided by the maximum density of the soil measured in a standardized compaction test.

relative complement *Mathematics.* let A be a subset of a set X. The relative complement of a subset S of A consists of those elements of X that are in A but not in S; denoted $A\backslash S$ or $A - S$.

relative contour see THICKNESS LINE.

relative current *Oceanography.* a current that is a function of the dynamic slope of an isobaric surface, the current being determined from an assumed layer of no motion.

relative dating *Geology.* the proper chronological placement of a geologic feature, structure, or event in the geologic time scale in relation to another or others, without reference to its age in years. *Archaeology.* any of various methods of dating based on establishing how old an artifact or feature is relative to some other item or event; examples include stratigraphy, seriation, and cross dating.

relative density see SPECIFIC GRAVITY.

relative direction *Navigation.* any direction measured from the head of the craft.

relative divergence see DEVELOPMENT INDEX.

relative error *Mathematics.* the error in a quantity, divided by the quantity; often expressed as a percentage.

relative error of closure *Cartography.* in surveying, a value used to express the degree of accuracy of a survey, determined by dividing the total error of closure by the total length of the traverse; usually expressed as a fraction with a numerator of one, e.g., 1/1245.

relative erythrocytosis *Medicine.* a relative increase in the number of red blood cells as a consequence of loss of the fluid portion of the blood. Also, **relative polycythemia.**

relative force *Engineering.* the ratio of the force of an experimental propellant divided by the force of a standardized propellant determined under the same test conditions.

relative frequency *Statistics.* a count of the occurrences of an event or phenomenon divided by the number of observations. *Ecology.* specifically, a method for determining how common a given species is to an area by calculating the ratio of its frequency to the sum of the frequency of all the other species present.

relative fugacity *Physical Chemistry.* the ratio of an idealized partial pressure (fugacity) in a given gas or solution to the standard idealized partial pressure.

relative gain *Electromagnetism.* the gain in a specified direction compared to a reference antenna, the latter being an isolated lossless half-wave dipole antenna whose gain is measured in an equatorial plane containing the specified direction.

relative gain array *Control Systems.* an analytical device that uses an array to compare single-loop control to multivariable control in process control multivariable applications.

relative geologic time see RELATIVE TIME.

relative gravity *Geodesy.* the difference between a measurement of the force of gravity at the point of observation and the measurement at an established gravity station where the value is taken as standard.

relative gravity instrument *Engineering.* an instrument that measures the difference in the acceleration of gravity between two test locations.

relative humidity *Meteorology.* the ratio of the actual vapor pressure of the air to the saturation vapor pressure; usually expressed as a percentage. Also, HUMIDITY.

relative hypsography see THICKNESS PATTERN.

relative interference effect *Acoustical Engineering.* the decibel ratio of the amplitude of a signal of a specified frequency to that of a single-frequency signal that has equal interference characteristics as the reference signal.

relative isohypse see THICKNESS LINE.

relatively closed set *Mathematics.* let A be a subset of a topological space X. A subset S of A is relatively closed in A if the relative complement $A\backslash S$ of S in A is an element of the relative topology U_A.

relatively compact *Mathematics.* a subset A of a topological space X is said to be relatively compact if the closure of A is compact.

relatively compact set see CONDITIONALLY COMPACT SET.

relatively open set *Mathematics.* let A be a subset of a topological space X. A subset S of A is relatively open in A if S is an element of the relative topology U_A; that is, S is the intersection of A with some open set of U.

relatively prime *Mathematics.* **1.** the integers $a_1, \ldots, a_n$ are said to be relatively prime if their greatest common divisor is 1; this is denoted by $(a_1, \ldots, a_n) = 1$. **2.** more generally, the elements $r_1, \ldots, r_n$ of a ring R with (multiplicative) identity are said to be relatively prime if their greatest common divisor is the multiplicative identity. For example, in the ring of polynomials over the integers, $3x^2 + 4x - 5$ and $2x + 1$ are relatively prime.

relative magnetometer *Engineering.* a magnetometer that measures magnetic field strength at a test point based on the magnetic field of a calibrated reference point.

relative maximum see LOCAL MAXIMUM.

relative minimum see LOCAL MINIMUM.

relative molecular mass *Chemistry.* the ratio of the average mass per molecule of a naturally occurring form of an element (or compound) to 1/12 of the mass of an atom of nuclide carbon-12. Also, MOLECULAR WEIGHT, MOLECULAR MASS.

relative momentum *Mechanics.* the momentum of a body in an inertial reference frame that is fixed to another body's center of mass. Also, APPARENT MOMENTUM.

relative motion *Mechanics.* the motion of a body in an inertial reference frame that is fixed to another body's center of mass. Also, APPARENT MOTION.

relative movement *Navigation.* the motion of another craft or object from the point of view of one's own craft.

relative movement line *Navigation.* a line drawn on a maneuvering board that indicates the movement, over time, of another vessel in reference to one's own.

relative orbit *Astronomy.* the apparent orbit of the fainter component in a binary star system when measured against the brighter component.

relative permeability *Geology.* the ratio between the ability of a rock to conduct a particular fluid at partial saturation and its ability to conduct the fluid at 100% saturation. *Materials Science.* the ratio of the magnetic permeability of a material to the permeability of free space.

relative permittivity *Materials Science.* the ratio of the permittivity of a material to the permittivity of a vacuum; important in dielectric materials.

relative plot *Navigation.* the plotting, on a maneuvering board, of the positions and movements of other vessels in order to predict possible risk of collision.

relative position *Navigation.* the position of a target craft relative to a reference craft.

relative power gain *Electromagnetism.* for a transmitting and receiving antenna pair, the ratio of the power measured at the receiver input terminals to the power measured at the output terminals of the transmitter.

relative pressure response *Acoustical Engineering.* the electromagnetic energy generated by an acoustic receiver transducer, such as a microphone, in response to an acoustic reference level under stated conditions.

relative quickness *Ordnance.* the ratio of the quickness of a test propellant to the quickness of a standard propellant, measured under the same conditions.

relative refractive index *Optics.* the ratio of the refractive index in one medium to that in another medium.

relative relief see LOCAL RELIEF.

relative resistance *Electricity.* a ratio of resistance between a known material and a subject material of equal dimensions and temperature.

relative response *Electronics.* the difference between the response of a device under a specific condition and the response of that device under a reference condition, generally expressed in decibels.

relative roughness factor *Fluid Mechanics.* a property of piping material and piping diameter that, along with the Reynolds number of the flow field, permits the estimation of a friction factor.

relative scatter intensity *Optics.* the ratio of the intensity of light scattered in a given direction to the intensity of light scattered in the direction of the incident beam.

relative search *Navigation.* a search in which search areas are assigned with respect to a moving point rather than a fixed position.

relative sector search *Navigation.* a search made by radial sectors relative to a moving point.

relative speed *Transportation Engineering.* the speed of one vehicle relative to that of other vehicles nearby. In a head-on approach, this is equal to the sum of the individual speeds.

relative square search *Navigation.* a square made by square or rectangular search areas relative to a moving point.

relative stability test *Analytical Chemistry.* a test used to determine the amount of water pollution in a sample by checking for the absence of oxygen, using methylene blue.

relative sunspot number see WOLF NUMBER.

relative tilt *Photogrammetry.* **1.** the amount of tilt in the plane of a photograph with reference to another, and not necessarily horizontal, plane, such as the preceding or subsequent photograph in a series. **2.** the difference between the photograph perpendicular and an arbitrary reference direction, such as the photograph perpendicular of the preceding or subsequent photograph in a series.

relative time *Geology.* geologic time as determined by the placing of events in order of their occurrence, especially when this placement is based on paleontological evidence or on layered rock sequences. Also, RELATIVE GEOLOGIC TIME.

relative topography *Meteorology.* see THICKNESS PATTERN. *Mathematics.* let X be a topological space with topology U. The relative topology on a subset A of X is given by $U_A = \{A \cap U : U \in U\}$. The sets $A \cap U$ are, by definition, the open sets in the relative topology U_A.

relative transmission level see TRANSMISSION LEVEL.

relative transmitting response *Acoustical Engineering.* the acoustic energy, expressed in decibels, transmitted by an acoustic transmitter transducer, such as a loudspeaker, in response to an electromagnetic reference level signal under stated conditions.

relative velocity *Mechanics.* the velocity of a body in an inertial reference frame that is fixed to another body's center of mass. Also, APPARENT VELOCITY.

relative viscosity *Materials Science.* the ratio of the kinematic viscosity of a specific solution of a polymer to the kinematic viscosity of the solvent.

relative volatility *Chemistry.* the volatility of a material relative to one whose volatility is taken as unity.

relative wind *Navigation.* the wind velocity relative to a reference craft; of particular importance in carrier flight operations, for which the relative wind over the carrier deck is crucial.

relativistic *Physics.* relating or subject to the special or general theory of relativity.

relativistic beam *Physics.* a particle beam whose velocity is comparable to the speed of light.

relativistic electrodynamics *Electromagnetism.* a branch of electromagnetism that involves the study of electromagnetic phenomena and interactions among particles that travel at velocities comparable to the speed of light.

relativistic kinematics *Particle Physics.* the study of particle motion using relativistic theory and disregarding the causes of motion.

relativistic mass *Particle Physics.* the mass of a relativistic particle, which is significantly greater than the particle's rest mass if the velocity is greater than about one-tenth of the speed of light.

relativistic mechanics *Physics.* the study of mechanics in association with the special theory of relativity.

relativistic particle *Particle Physics.* a particle that has a velocity which is comparable to the speed of light and that may be appropriately treated with the theory of relativity.

relativistic quantum theory *Quantum Mechanics.* a quantum theory of particles that is consistent with the principles of special relativity, specifically including the rest mass of the particle.

relativistic theory *Physics.* any theory that is derived from the theory of general or special relativity.

relativistic velocity *Physics.* the velocity of a body on the order of, but not exceeding, the speed of light in a vacuum; such velocities range from about 3×10^7 to 3×10^8 meters per second.

relativity *Science.* the fact of being relative; the condition of a quantity or quality that is affected by something else. *Physics.* **1.** see GALILEAN RELATIVITY. **2.** see RELATIVITY THEORY.

relativity theory *Physics.* a theory introduced by Albert Einstein, concerned with physical laws (such as those that describe mass, space, motion, and time) as they are formulated by different observers in uniform relative motion with respect to each other. The **special relativity theory** (1905) postulates that all the laws of nature have the same form in all inertial frames of reference. It also postulates that the speed of light is a constant, independent of any relative motion between the light source and the observer. Among the implications of these postulates are the following: (a) the maximum attainable velocity in the universe is that of light in a vacuum; (b) the mass of an object increases with its velocity; (c) mass and energy can be considered different aspects of the same quantity ($E = mc^2$); (d) a moving stick appears to contract along the direction in which it moves; (e) a moving clock appears to run slower than an identical clock at rest; (f) time must be described in terms of the motion of some system, or the relative motion of two systems, rather than in absolute terms. The **general relativity theory** (1915) postulates that in the neighborhood of a given point, it is not possible to distinguish between a gravitational field and an accelerated frame of reference; i.e., gravitational mass and inertial mass are completely equivalent.

relaxation the fact of relaxing; specific uses include: *Physiology.* **1.** a lessening of tension. **2.** an easing of pain. *Physics.* the approach to a steady state or equilibrium of a system after the system has undergone a sudden change due to some external influence. *Mechanics.* **1.** the phenomenon that stress decreases due to creep in a material whose strain is held fixed. **2.** a decrease in elastic resistance in an elastic material under an applied stress beyond the yield stress. *Materials Science.* a decrease in stress over long time intervals resulting from plastic flow. *Volcanology.* the release of tension in magma following the initial explosions.

relaxation circuit *Electronics.* a circuit that combines an energy storage device with another device that has a bistable characteristic. The output of such a circuit is usually a nonsinusoidal periodic waveform.

relaxation complex *Biochemistry.* a group of three proteins bound to some *E. coli* supercoiled plasmids that convert supercoiled DNA to a nicked open circle.

relaxation inverter *Electronics.* a device that employs a relaxation oscillator to convert direct current into alternating current.

relaxation method *Mathematics.* a method of approximating solutions to a system of linear equations in which errors from an initial approximation are regarded as constraints to be minimized, i.e., relaxed, by successive approximations so that a desired degree of accuracy is attained.

relaxation oscillations *Physics.* a cyclic pattern in which a displacement is increased linearly to a maximum value and then dropped back to zero amplitude; a sawtooth waveform.

relaxation oscillator *Electronics.* a circuit that generates a nonsinusoidal waveform, such as a sawtooth or rectangular form, by slowly charging and then rapidly discharging a capacitor and inductor through a resistor.

relaxation suture *Surgery.* a suture used for the immediate closure of a wound and formed so that it can later be loosened to relieve undue tissue strain.

relaxation test *Engineering.* a test for the measurement of the decrease of stress with time on a material kept under constant total stress.

relaxation therapy or **relaxation training** see PROGRESSIVE RELAXATION TRAINING.

relaxation time *Physics.* in any of several physical systems, the *e*-folding time that is characteristic of a system when it is disturbed suddenly and then allowed to return to its steady-state condition; i.e., the average length of time that is required for a system to move toward its equilibrium state. *Solid-State Physics.* specifically, the average length of time an electron travels through a metal before it changes its momentum upon scattering.

relaxed control *Genetics.* the ability of some plasmids to continue replicating after bacteria cease dividing, resulting in a bacterial cell containing hundreds of plasmids.

relaxed DNA *Genetics.* a molecule of DNA that is not supercoiled.

relaxed mutant *Molecular Biology.* a bacterial mutant that continues to synthesize RNA in a growth medium lacking certain amino acids or other nutrients.

relaxin *Endocrinology.* a polypeptide hormone that is secreted during pregnancy by the placenta and the corpus luteum of the ovary; inhibits uterine contactility, softens the cervix, and relaxes the ligaments of the pubic symphysis.

relaxing enzyme *Enzymology.* any enzyme, such as a helicase, that causes an unwinding of the DNA helix prior to replication.

relay *Electricity.* a device that is activated by variations in an electric circuit and that upon such activation makes or breaks one or more connections in the same or another circuit. *Telecommunications.* **1.** any transmission system that passes on a signal from one communication link to another. **2.** on multihop radio systems, any station that acts as an intermediary. *Space Technology.* one of an early series of U.S. low-altitude communications satellites.

relay center *Telecommunications.* a switching center from which messages are automatically routed.

relay contact *Electricity.* either of a pair of contacts that operate to open or close depending upon the motion of the armature within the relay.

relay control system *Control Systems.* a control system in which the controller reacts only when the error signal has reached a predetermined value.

relay rack *Mechanical Devices.* a standard steel or metal rack that holds electronic components on 19-inch panels positioned at various levels in a frame assembly.

relay satellite *Space Technology.* see RELAY.

relay selector *Electricity.* a relay that registers digits and holds a circuit for a magnetic impulse-counter selector.

relay station see REPEATER STATION.

relay system *Electricity.* an assembly of linked relays. *Telecommunications.* see RADIO RELAY SYSTEM.

relearning *Behavior.* the process of learning a behavior or acquiring information that was learned previously but is believed to have been forgotten.

release a process of letting go, or a device that does this; specific uses include: *Mechanical Engineering.* a device that is designed to hold or free a mechanism as required. *Industrial Engineering.* in micromotion studies, the elemental motion of relinquishing control of an object. *Ordnance.* the intentional separation of a free-fall aircraft store, such as a bomb, from its suspension equipment for purposes of employment of the store; the term is also used in relation to other air-delivered weapons such as rockets and missiles.

release adiabat *Mechanics.* a curve in state space describing the relaxation of a material after being instantaneously released from a high pressure state.

release altitude *Aviation.* an aircraft's altitude at the time it releases something such as a bomb, glider, or parachutist.

released mineral *Mineralogy.* a mineral formed during the crystallization of a magma as the result of an earlier phase's failure to react with the liquid.

release factor *Biochemistry.* a component of the specialized transport system involved in the transport of cobalamin (vitamin B_{12}) across the wall of the intestine.

release fracture *Geology.* a fracture that forms as a result of the relief of stress in one particular direction.

release-inhibiting factor *Endocrinology.* a substance, such as somatostatin, that is produced by the hypothalamus and that acts on the anterior pituitary to inhibit the release of a specific hormone. Also, INHIBITING FACTOR.

releaser *Behavior.* any behavior pattern or physical feature that produces a certain response from another animal; e.g., vocal sounds, odor, or coloration.

releaser stimulus *Behavior.* any stimulus that sets off instinctive behaviors in an animal.

release therapy *Psychology.* psychotherapy that is based on the subject's release of emotional tension through the expression of submerged hostility or frustration.

releasing hormone *Endocrinology.* a substance that is produced by the hypothalamus and acts on the anterior pituitary to stimulate the release of a specific hormone. Also, **releasing factor.**

releasing mechanism see INNATE RELEASING MECHANISM.

reliability *Engineering.* the probability that a product will be operational after a period of usage or over a specified time period, based on testing of the product under a prescribed operation and operating environment. *Statistics.* **1.** the analysis of the life distribution of systems subject to random failures. **2.** the probability that a system will accomplish a given task.

reliability engineering *Industrial Engineering.* the branch of engineering that deals with the causes of product failures and seeks to eliminate such failures.

relic *Archaeology.* an object surviving from an earlier culture, especially a valuable or symbolic object. *Geology.* **1.** a landform that remains despite decomposition, or a landform that has been left behind after most of its substance has disappeared. **2.** a trace of a particle found in a sedimentary rock, such as a fossil fragment. **3.** see RELICT, def. 2. (Going back to a Latin word meaning "remains.")

relict *Ecology.* **1.** a group of organisms representing the surviving remnants of a population that was formerly more widespread in a certain area. **2.** a remnant species from an earlier geological period. Also, **relict species.** *Geology.* **1.** describing a topographic feature that remains after parts have been removed or have disappeared. Also, **relicted. 2.** a mineral, structure, or feature of a rock that was formed earlier and persists despite processes that tend to destroy it. Also, RELIC.

relict dike *Geology.* a tabular body within a granitized mass that represents a dike placed in position before granitization occurred, which was relatively resistant to the granitization process.

relict glacier *Hydrology.* the remains or part of the remains of a larger, older glacier.

reliction *Hydrology.* **1.** a slow, gradual withdrawal of the water in a lake, stream, or sea that permanently exposes the bottom and leaves it uncovered. **2.** the land left so exposed.

relict lake *Hydrology.* **1.** a lake remaining in an area that was once covered by the sea or a larger lake. **2.** a lake that has become separated from the sea or another original body of water.

relict mineral *Mineralogy.* a mineral of an earlier rock that has survived in a later rock despite destructive processes or unfavorable conditions.

relict permafrost *Hydrology.* a permafrost that was formed sometime in the past and that still persists in areas in which it could not form in the present.

relict sediment *Geology.* a sediment that was originally deposited in equilibrium with its environment but is unrelated to its present environment, although it has not been buried by later sediments.

relict soil *Geology.* a soil that was formed on a preexisting landscape but has not been buried under younger sediments.

relict texture *Geology.* an original texture of a mineral deposit that has been preserved after partial or total replacement or alteration.

relief *Geology.* **1.** the shape or configuration of part of the earth's surface as it is evidenced by variations in altitude and slope or by irregularities of the land. Also, TOPOGRAPHIC RELIEF. **2.** the collective variations in elevation of a land surface. **3.** for a particular region, the difference in elevation between its hilltops or mountain summits and its lowlands or valleys. *Crystallography.* the apparently rough surface of a crystal when viewed in thin section under a microscope due to difference in index of refraction. *Cartography.* collectively, the surface irregularities of a specific land area, represented graphically by a number of methods (contours, shading, hachures, and so on) on maps, and according to scale on three-dimensional models. *Mechanical Engineering.* **1.** a slight change in the dimensions of a machine part that allows for clearance around a cutting edge. **2.** a passage in a tailstock center, cut so that a tool can be moved nearer to the center of the work. *Architecture.* see RELIEF SCULPTURE.

relief angle *Mechanical Engineering.* the angle between a relieved surface and a tangential plane at a cutting surface.

relief displacement *Cartography.* in aerial photography, an effect in which points vertically above each other on the ground appear to be displaced radially from the principal point of the photograph, toward or away from the photographic nadir.

relief frame *Mechanical Engineering.* a frame or ring in a large steam engine that lies between the rear of a slide valve and the inside of a steam-chest cover to prevent the access of steam to the greater part of the valve, thus reducing pressure on the valve and materially reducing friction.

relief hole *Engineering.* a blast hole fired to break up the ground in tunneling, after the cut holes and before the lifter holes are fired.

relief limonite *Mineralogy.* a minutely craggy, cavernous, porous, sulfide-derived limonite that occurs in the location originally occupied by the grain of sulfide.

relief map *Cartography.* a type of map that shows the topography of the earth in terms of its elevation above or below a specified datum, usually mean sea level, by means of such graphic techniques as contour lines, hachures, shading, or tinting.

relief model *Cartography.* any type of three-dimensional representation of the topography of a specific geographic area.

relief printing see LETTERPRESS.

relief sculpture *Architecture.* the projection of a figure or part from its background; in a **bas-relief** the figures stand out from their background less than one-half their suggested thickness, and in a **high relief**, they project from the background more than one-half their implied thickness. Also, RELIEF.

relief sculpture

relief valve see BLOWOFF VALVE, def. 1.

relief well *Petroleum Engineering.* a directional well drilled to intersect a blowing (out-of-control) well and flood it with water or heavy drilling fluid so as to bring it under control. Also, KILLER WELL. *Civil Engineering.* a hole bored at the base of an earth dam to relieve naturally occurring porous water pressure.

relieving *Metallurgy.* a process in which an embossed metal surface is rubbed with an abrasive material to expose its base-metal color. *Mechanical Engineering.* **1.** a process of lessening the pressure in a closed space, such as a tank or boiler, to maintain a desirable level. **2.** the process of interrupting a bearing surface in order to improve alignment or lubrication. **3.** the process of milling or grinding some of the material behind the cutting edge of a tool in order to supply clearance.

relieving anode *Electronics.* the auxiliary anode in a pool-cathode tube that offers an alternative conducting path, thus reducing the current flowing to the principal anode.

relieving platform *Civil Engineering.* a jetty situated adjacent to a retaining wall; used as a dock or holding point before lowering heavy equipment or transportation down to the wall.

reline *Ordnance.* to replace a worn gun liner in order to give it the ballistic characteristics of a new gun; a replaceable gun liner is called a **removal liner** or **removable liner.**

relish *Engineering.* the projection of the shoulder of the tenon in a mortise-and-tenon wood joint.

Relizian or **Relizean** *Geology.* a North American geologic stage of the Miocene epoch, occurring after the Saucesian and before the Luisian.

relocatable code *Computer Programming.* a machine-language code having instructions with relative addresses that are modified as the program is relocated.

relocatable program *Computer Programming.* a computer program that is written so that it can be stored and executed in any part of memory.

relocate *Computer Programming.* to modify the instructions in a program to permit the loading and execution of the program in a particular memory area.

relocating loader *Computer Programming.* a loader program that collects the relocatable program units to be loaded, assigns memory space to them, relocates addresses, resolves external references, and loads the programs into memory.

relocation factor *Computer Programming.* the base address of the memory area assigned to a program unit. Adding this factor to a relative address in a relocatable program yields an absolute address.

relocation hardware see DYNAMIC ADDRESS TRANSLATOR.

relocation register see INDEX REGISTER.

reluctance *Electromagnetism.* a measure of the opposition presented to the magnetic flux in a magnetic circuit; it is analogous to the resistance in an electric circuit and is expressed by the ratio of the magnetomotive force to the magnetic flux.

reluctance motor *Electricity.* a synchronous motor that derives torque from variations in reluctance with rotor position, rather than from permanent magnet or DC excitation.

reluctance pressure transducer *Engineering.* a transducer that detects changes in pressure by utilizing the corresponding change in magnetic reluctance measurable in a magnetic circuit within the device.

reluctivity *Physics.* the reciprocal of the magnetic permeability associated with a medium.

rem *Nucleonics.* a unit of ionizing radiation equivalent to the dosage that produces the same amount of damage as does one roentgen of high-voltage X-rays; 100 rem = 1 sievert. (An acronym for roentgen equivalent, man.)

REM [rem] *Physiology.* rapid eye movement, the series of pronounced, jerky eye movements that identify REM sleep.

remainder *Mathematics.* **1.** the quantity left after one number is divided by another an exact number of times; i.e., given a quantity (e.g., an integer or a polynomial) m and a (nonzero) quantity q, then a quantity r is the remainder of m when it is divided by q if there is a quantity s such that $r = m - qs$ and $0 \leq r < q$ if m is an integer, or the degree of r is less than the degree of q if m is a polynomial (the degree of 0 is taken here to be 0). **2.** if $\sum_{k=1}^{\infty} a_k$ is a convergent infinite series, then a remainder of the series is

$$\sum_{k=1}^{\infty} a_k - \sum_{k=1}^{n} a_k = \sum_{k=n+1}^{\infty} a_k$$

for each $n > 1$; i.e., the sum of the terms remaining after the nth partial sum is computed.

remainder formula *Mathematics.* any of several expressions for computing the remainder of a convergent infinite series.

remainder theorem *Mathematics.* **1.** let $f(x)$ be an element of the polynomial ring $R[x]$, where R is a ring with identity. Then for any element c of R, there exists a unique $q(x)$ in $R[x]$ such that $f(x) = q(x)(x - c) + f(c)$. That is, the remainder when $f(x)$ is divided by $x - c$ is $f(c)$. This is the basis for using synthetic division (also known as Horner's method) to evaluate polynomials rapidly. **2.** see CHINESE REMAINDER THEOREM.

remaining velocity *Mechanics.* the velocity at any point along the path of fire of a projectile.

Remak, Robert 1815–1865, German biologist; a pioneer in embryology; with Johannes Peter Müller, distinguished the ectoderm, mesoderm, and endoderm.

Remak's ganglion *Anatomy.* a sympathetic ganglion located in the wall of the right atrium, between the superior vena cava and the coronary sinus.

remalloy *Materials.* a permanently magnetized alloy composed of iron, molybdenum, and cobalt.

remanence *Electromagnetism.* a residual magnetic flux density that remains in a magnetic circuit after the magnetomotive force is removed. Thus, **remanent.** *Acoustical Engineering.* specifically, the net magnetization that remains on a magnetic recording material after it has been saturated and the field has been removed.

remanent magnetization *Geophysics.* the permanent magnetization orientation of a rock, established at the time of the rock's formation and conforming either to local ambient magnetic fields or to the orientation of the earth's magnetic field at the time.

remedial operation *Chemical Engineering.* a corrective process performed so that an operation or product can meet set standards or conditions.

remineralization *Physiology.* the restoration of minerals to a tissue, such as bone.

Remington *Ordnance.* any of the small-arms weapons manufactured by the Remington Arms Company; among the most popular were the revolver introduced in 1859 and the single-shot, rolling-block military rifle developed in 1865 and used widely from 1870 to 1900.

Remipedia *Invertebrate Zoology.* a class of crustaceans that resembles centipedes; found in caves on some Atlantic islands.

remission *Medicine.* **1.** a partial or complete decrease or subsidence of the clinical and subjective characteristics of a chronic or malignant disease; it may be spontaneous or the result of therapy. **2.** the period during which such a decrease or subsidence occurs.

remittent fever *Medicine.* diurnal variations of a fever with exacerbations and remissions but not a return to normal.

remodulator *Electronics.* a device that converts the modulation of a signal from one form to another without loss of intelligence, as when FM is converted to AM.

remolded soil *Geology.* a soil whose natural internal structure has been altered or disturbed by handling, so that the soil lacks shear strength and is more compressible.

remolding index *Geology.* the ratio between the modulus of deformation of an undisturbed soil and that of a remolded soil.

remora *Vertebrate Zoology.* any of several marine fish of the family Echeneididae that characteristically attaches to a host (usually a shark) by a sucker disk on the top of its head; the sucker is an extension of the dorsal fin; found throughout the world's temperate and tropic seas. Also, SUCKERFISH.

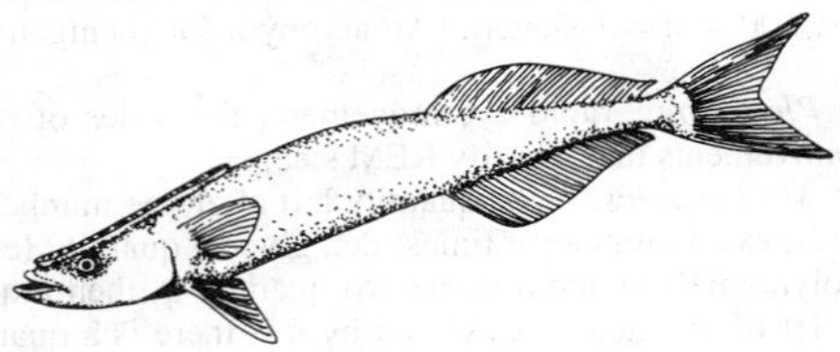

remora

remote *Electronics.* **1.** describing a system, program, or device that is accessed by means of remote control. **2.** a remote-control device. See REMOTE CONTROL.

remote access *Computer Technology.* the ability to communicate with a distant data processing facility through a data link.

remote-access admittance *Robotics.* a special piece of hardware with built-in sensors and actuators that can be programmed to rotate, translate, or tilt an object in the final stages of assembly tasks.

remote-association test *Psychology.* a test of creativity in which the subject is asked to provide a fourth word that links three other seemingly unrelated words.

remote calculator *Computer Technology.* a keyboard device that communicates with a central processor through a data link, enabling the computer to be used for calculating solutions to problems presented from remote locations.

remote-center compliance *Robotics.* a device that allows a part being gripped by a robot to rotate around the tip of the end effector and make assembly easier to accomplish.

remote computing system *Computer Technology.* a hardware configuration in which the users communicate with a central processor from individual remote consoles that can be operated simultaneously.

remote computing system exchange *Computer Technology.* a device that handles simultaneous transmissions between a central processor and a number of remote consoles.

remote computing system log *Computer Technology.* a record of events that occur while remote consoles are in communication with a central processor, including such information as the amount of traffic for an individual terminal or the types of errors or volumes of data transmitted.

remote console *Computer Technology.* a user interface unit in a location distant from the central processor, equipped with facilities to exchange data with the central processor.

remote control *Control Systems.* the control of equipment from a significant distance away, by means of wires or by radio, sound, light, or other such means. *Electronics.* a small device that is used to control the operation of a machine, such as a television set or a VCR, from a distance. Also, REMOTE.

remote data station *Computer Technology.* a terminal unit in a location distant from the central processor that acts as a data collection point, sending data to and receiving data from the central processor. Also, **remote data terminal.**

remote debugging *Computer Technology.* the process of testing and correcting a program from a remote terminal.

remote delivery *Military Science.* in mine warfare, the delivery of mines to a target area by means other than direct emplacement; the exact location of such mines may not be known.

remote gun control *Ordnance.* the process of laying and adjusting a gun using a remote control system.

remote-indicating compass *Navigation.* a compass that has one or more repeaters giving readings at different locations.

remote indicator *Electronics.* in radar, a device that receives signals from a target indicating distance and direction, and then transmits its own signal, with the same information, to a radar system located some distance away. Also, REPEATER.

remote inquiry *Computer Technology.* the interrogation of a central processor by remote users.

remote job entry *Computer Technology.* the submission of jobs to a central computing system from a remote input device.

remotely piloted vehicle *Aviation.* an unmanned aircraft that is directed from either the ground or another aircraft; generally used for such missions as reconnaissance and targeting.

remote Mach number *Aviation.* the Mach number of a moving body taken as a whole, as distinguished from the local Mach number of a specific section of the body.

remote manipulator *Engineering.* a mechanical or robotic device that allows remote handling of a material; often used in hostile environments, such as space or a contaminated enclosure.

remote pickup *Telecommunications.* the process of recording television or radio program material at a remote location and transmitting it directly to a centralized broadcast studio.

remote procedure call *Computer Science.* a call to a procedure that is implemented on another computer or server to which the computer of the calling program is connected via a network.

remote sensing *Engineering.* the gathering and analysis of data from an object physically removed from the sensing equipment, as in satellite and aerial photography or subsurface detection instruments. *Electricity.* the mechanism or process by which a power supply senses the voltage at a load in order to assess lead drop and maintain load regulation.

remote sensing imagery *Photogrammetry.* any photographic or electronic image obtained at a remote station or platform, such as an airplane or satellite, including images produced by various types of photography, radar, or microwave recordings.

remote subscriber *Telecommunications.* a subscriber to a network that does not have direct access to the switching center, but can gain access through a distant facility such as a line concentrator or carrier multiplex system.

remote terminal *Computer Technology.* **1.** a terminal located away from the site of a central processor, used for data collection and transmission but having no direct operating control over the central processor. **2.** a terminal or workstation connected to a computer via a network.

remote testing *Computer Technology.* a method for monitoring the rate at which work is processed by a computer system; programs, test data, and instructions are supplied to individual computer operators, who record events that occur during test runs and convey the information to programmers for use in organizing work flow.

remote-unit effect *Materials Science.* in a polymer, the effect on the reactivity of the chain radical of any unit other than the last two on the chain; often observed in copolymers.

removable discontinuity *Mathematics.* a (real) function f has a removable discontinuity at a point $x = a$ if: (a) f is defined in a neighborhood of a; (b) f is discontinuous at a; and (c) $\lim_{x \to a} f(x)$ exists and equals some finite number A. The function f^*, obtained from f by redefining $f(a) = A$, is then continuous at a.

removable medium *Computer Technology.* any storage medium that is not fixed to its read/write device, such as a magnetic tape cassette or floppy disk.

removable singularity *Mathematics.* a (complex) function f has a removable singularity at a point $z = a$ if there is a function g that is: (a) analytic at $z = a$, and (b) equal to f in some region $0 < |z - a| < R$. In particular, the Laurent expansion of f in a neighborhood of a contains no negative powers of $(z - a)$; the Laurent series becomes the Taylor series and f may be made analytic by redefining $f(a) = \lim_{z \to a} f(z)$.

REM rebound *Neurology.* a phenomenon by which a subject denied REM sleep will compensate by increasing the amount of REM sleep in the next sleep cycle.

REM sleep *Neurology.* a state of sleep during which dreaming frequently occurs; it is characterized by rapid eye movements, fast low-voltage brain waves, mild involuntary muscle jerks, irregular heart rate and respiration, and a higher threshold of arousal; typically it lasts from 5 to 20 minutes, occurs at about 90-minute intervals, and, in adults, occupies approximately 20% of overall sleep time. Also, ACTIVE SLEEP, DESYNCHRONIZED SLEEP, DREAMING SLEEP, D SLEEP, FAST WAVE SLEEP, PARADOXICAL SLEEP, RAPID EYE MOVEMENT SLEEP.

Renaissance [ren´ə säns´] *Science.* a period in European history in which there was a renewed awareness of the culture and thought of ancient Greece and Rome, and a general emergence of interest in the physical world, the human individual, and human society. It dates roughly from the year 1350 or 1400 to the year 1600, and marks the transition from the medieval world to the modern world; it is noted as an era of great accomplishment in the fine arts, architecture, science, literature, and scholarship. *Architecture.* the rebirth of classical style and forms of architecture that occurred during this cultural period, originating in Italy in the 1400s and spreading to the rest of Europe in the following century. (From a Medieval French term meaning "rebirth;" literally, "to be born again.")

renal [rēn´əl] *Anatomy.* of or relating to the kidney. *Medicine.* involving or affecting the kidneys.

renal artery *Anatomy.* the large artery branching from the abdominal aorta and conducting blood into the kidney through the renal hilus.

renal calculus *Medicine.* an abnormal stony concretion in the kidney.

renal-cell carcinoma *Medicine.* a malignant tumor of epithelial cells of the kidney occurring in middle-aged to older persons of either sex, though most common in males. Also, HYPERNEPHROMA.

renal corpuscle *Anatomy.* the portion of a nephron that consists of the glomerulus and the Bowman capsule.

renal dwarfism *Medicine.* dwarfism that results from chronic renal disorders in children, as congenital kidney malformations, chronic nephritis, and Fanconi's syndrome.

renal failure *Medicine.* the inability of the kidneys to excrete wastes, concentrate urine, and conserve electrolytes; this condition may be either chronic or acute.

renal medulla *Anatomy.* the innermost portion of the body of a kidney, lying between the cortex and the renal pelvis.

renal papilla *Anatomy.* a cone-shaped projection into one of the minor calyces of the kidney, having 10 to 20 openings of collecting tubules.

renal pelvis *Anatomy.* the large, funnel-shaped structure of the kidney from which urine passes into the ureter.

renal pyramid *Anatomy.* one of several pyramidal-shaped masses of collecting tubules located in the medulla of the kidney.

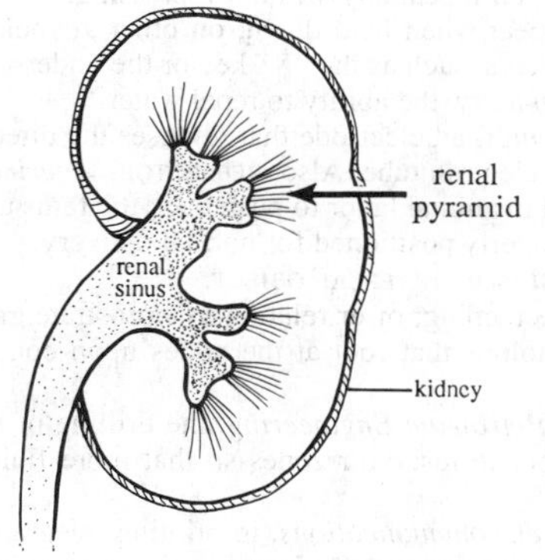

renal pyramid

renal rickets *Medicine.* a metabolic bone disease occurring in children as a result of renal insufficiency.

renal threshold *Physiology.* the level of concentration of a substance in the plasma at which it begins to pass through the kidneys into the urine.

renal tubular acidosis *Medicine.* a renal tubular defect marked by inability to excrete acid urine, low plasma bicarbonate, high plasma chloride concentrations, and hypokalemia; muscle weakness and renal stones may be complicating factors.

renal tubule *Anatomy.* one of several types of canals, lined with epithelium, that produce and convey urine to the calyces.

renal vein *Anatomy.* the large vein that drains blood from the kidney.

renaming rule *Mathematics.* in a predicate H, any variable x represents its domain of variability and can be renamed to a variable not occurring in H according to the following rules: (a) a variable x that is bound in H is renamed immediately after the quantifier and in the whole range of the quantifier; and (b) a variable x that occurs freely in parts of H is renamed in all those parts of H where it occurs freely. These two cases are called **bound renaming** and **free renaming**, respectively.

renardite *Mineralogy.* $Pb(UO_2)_4(PO_4)_2(OH)_4 \cdot 7H_2O$, a rare, strongly radioactive, yellow orthorhombic mineral, having a specific gravity of 4.35 and a hardness of 3.5 on the Mohs scale; found as crusts and lamellar masses with other secondary uranium minerals; may be the same as dewindtite.

renaturation *Molecular Biology.* the return of a protein or nucleic acid to its native conformation after it has been unfolded, generally achieved by placing it in a mild denaturing solution, such as dilute urea solutions for proteins.

renaturation kinetics *Molecular Biology.* a measurement of the reformation of a protein or nucleic acid molecule to its native conformation as a function of time.

rendering *Food Technology.* the separation of animal fat, usually by melting; used to obtain products such as lard, fish oil, and whale oil. *Graphic Arts.* the depiction of some actual object in artistic or graphical form. *Architecture.* specifically, the representation of a proposed building or other architectural feature in a perspective and scaled view. *Computer Science.* the generation of a graphical image from a model of a three-dimensional object or scene that appears to be a visual image of the object or scene.

rendezvous [rän´dá voo´] *Military Science.* **1.** a prearranged meeting from which an action or phase of an operation is begun, or to which to return after completion. **2.** in land warfare, an easily found terrain location at which visitors to a unit, headquarters, or facility are met and escorted to their destination. *Aviation.* **1.** the meeting of two objects at a preplanned time and place at matched velocities (zero relative velocity). **2.** the place where such a meeting occurs or is planned to occur. (From French; literally, "present yourself.")

rendezvous area *Military Science.* in an amphibious operation, the location in which the landing craft and amphibious vehicles meet to form waves after being loaded and before moving to the line of departure.

Rendoll *Geology.* a suborder of the soil order Mollisol that develops under humid, temperate conditions in highly calcareous parent materials; characterized by a thick, dark upper horizon grading to a pale lower horizon.

Rendu-Osler-Weber disease see HEREDITARY HEMORRHAGIC TELANGIECTASIA.

Rendzina *Geology.* a group of intrazonal soils that develops in humid to semiarid climates from highly calcareous parent material; characterized by a dark-brown to black surface horizon over a light-gray or yellow, soft calcareous horizon.

renette *Invertebrate Zoology.* in marine trematodes, a gland or pair of glands believed to function in osmoregulation, located ventrally in the pseudocoel near the pharynx.

renewable energy *Science.* any naturally occurring source of energy that is theoretically inexhaustible, such as solar, wind, tidal, and hydroelectric power, as opposed to sources such as fossil fuels, which have finite supply levels that cannot be replenished except over a geologic time scale.

renewable resource *Chemical Engineering.* a natural resource that can be replenished by natural means at rates comparable to its rate of consumption, such as food production by photosynthesis.

renewed consequent stream see RESEQUENT STREAM.

rengas *Materials.* the very hard wood of the East Asian trees of the *Gluta* and *Melanorrhoea* species; used in cabinets and veneers.

Renibacterium *Bacteriology.* a genus of aerobic bacteria of the order Actinomycetales, typically occurring in pairs of rod-shaped cells or coccobacilli and causing kidney disease in certain fish.

reniform *Science.* kidney-shaped.

renin *Enzymology.* an enzyme that is formed by the kidney and catalyzes cleavage of the leucine-leucine bond in angiotensin to generate angiotensin-1.

renin substrate see ANGIOTENSINOGEN.

rennet *Vertebrate Zoology.* the lining membrane of the fourth stomach in calves and other ruminants, noted specifically for the production of the enzyme rennin. *Food Technology.* an extract of the rennet membrane used to curdle milk in making cheese, junket, and the like.

rennin see CHYMOSIN.

Renninger effect see DOUBLE REFLECTION.

renninogen *Biochemistry.* a zymogen, present in the gastric glands, that converts to rennin upon secretion.

Renn-Walz process *Metallurgy.* one of several processes to reclaim iron from a waste stream in lead or zinc smelting.

reno- or **ren-** a prefix meaning "the kidney;" "involving or affecting the kidney."

renography *Medicine.* an X-ray examination of the kidney following the injection of a radiopaque substance.

renopathy see NEPHROPATHY.

renormalizability *Quantum Mechanics.* the property possessed by some quantum field theories that allows any infinities that arise in the theory to be eliminated by renormalizing fundamental quantities.

renormalization *Quantum Mechanics.* a procedure applied to certain quantum field theories in which the observable values of quantities are substituted for their predicted values, in order to remove infinities that arise in the theories.

renormalization group methods *Physics.* the statistical methods for substances that are near their critical points, in which the canonical ensemble is generalized by dividing the substance into cells and forming a new ensemble consistent with the thermodynamic variables of each of the cells.

renormalization transformation *Mathematics.* any transformation involving a change of scale.

renovascular *Medicine.* of or relating to the blood vessels of the kidneys.

renovation *Military Science.* the process of restoring weapons, ammunition, or other materiel to a condition as close as possible to the original condition.

rensselaerite *Petrology.* a compact, fibrous variety of yellow, white, or black talc, pseudomorphous after pyroxene, occurring in Canada and northern New York; it is harder than talc and takes a good polish.

reorder cycle *Industrial Engineering.* the time between a resupply order and delivery. Also, **reorder lead time.**

reorder point *Industrial Engineering.* the inventory level at which it is appropriate and economical to reorder an inventory item.

Reoviridae *Virology.* a family of viruses in which the particles are unenveloped and icosahedral with a wide range of hosts, including vertebrates, insects, and plants. (From respiratory enteric orphan viruses.)

reoxygenation *Radiology.* a phenomenon in which previously radioresistant tumor cells become more radiosensitive by coming closer to capillaries after destruction of other tumor cells due to previous irradiation.

rep *Nucleonics.* a former unit of ionizing radiation equivalent to the dosage of 93 ergs absorbed per gram of tissue. (An acronym for roentgen equivalent, physical.)

rep *Textiles.* a plain weave fabric having distinct rounded, crosswise ribbing, made of almost any kind of fiber. Also, REPP.

Rep. let it be repeated. (From Latin *repetatur.*)

repaint *Computer Science.* to regenerate the information in a displayed window by erasing the contents of the window and then redrawing the information in it.

repair *Engineering.* to restore a faulty product to operating condition. *Medicine.* to restore damaged tissue to a healthy state by natural or surgical processes.

repairability *Industrial Engineering.* the degree to which a machine or product is conveniently designed for restoration to service after failure.

repair cycle *Engineering.* the period of time in which a product is removed from operation, repaired, and returned to operation.

repair enzyme *Enzymology.* a DNA polymerase that catalyzes the replacement of damaged and missing segments of single strands in double-stranded DNA.

repair nuclease *Enzymology.* an enzyme that contributes to the repair of damaged DNA by removing the damaged segment, allowing DNA polymerase and DNA ligase to insert a correct replacement segment.

repair synthesis *Molecular Biology.* the recognition, enzymatic removal, and replacement of damaged DNA.

repand *Botany.* describing a plant or plant part, especially a leaf, that has wavy or undulating margins.

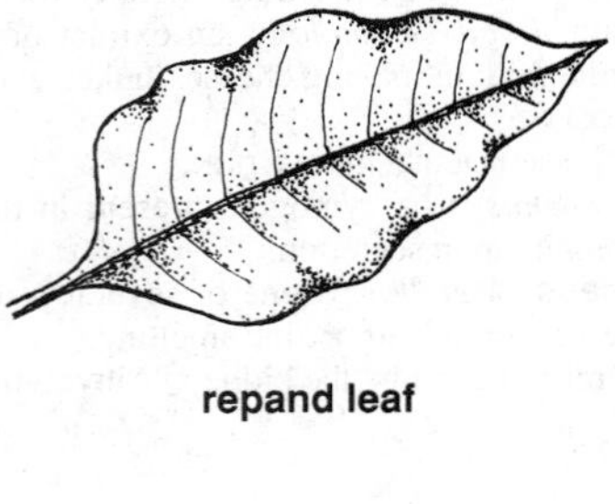

repand leaf

repatency *Surgery.* a surgical procedure to reestablish flow in a blocked part or vessel.

repeat see REPEAT KEY.

repeatability *Robotics.* the ability of a robot to return to the same position again and again.

repeat accuracy *Robotics.* the amount of variation in the actual position of a robotic manipulator from one cycle to the next when it is sent to the same position again and again.

repeated root see DOUBLE ROOT.

repeated twinning *Crystallography.* the periodic formation of a twin structure made up of many similar individual crystals.

repeater something that repeats; specific uses include: *Ordnance.* a firearm that can deliver several shots without reloading; the term is usually applied to rifles and carbines. *Horology.* a type of timepiece that has the ability, controlled by a button or a pull-cord, to repeat the last hour struck. *Electricity.* a well-situated automatic retransmission station that serves to extend the range of other stations. *Telecommunications.* **1.** a combination of equipment that receives one-way or two-way communication signals and delivers corresponding signals that are amplified and/or reshaped. Thus, **repeater station. 2.** a switch by which dialed telephone pulses are repeated from the central office to a remote office. *Electronics.* see REMOTE INDICATOR.

repeater compass see COMPASS REPEATER.

repeater jammer *Electronics.* a device that intercepts and modifies an enemy's radar signal so that it sends back false data.

repeater signal *Transportation Engineering.* an advance signal warning that the next stop signal is at danger.

repeat glass *Optics.* an instrument in which four lenses form four identical images of the same subject; used to view the repetition of a single pattern, as, for example, in textile design.

repeating coil *Electricity.* a transformer having a 1:1 ratio; used to prevent conductive coupling between two sections of a telephone line.

repeating decimal *Mathematics.* a decimal representation of a real number whoses fractional part is composed of a finite sequence of digits that repeat indefinitely, or that terminate after a finite number of digits. Repeating decimals correspond to rational numbers.

repeat key *Computer Technology.* a key that can be held down in order to transmit a signal repeatedly without repressing. On some keyboards, any key will repeat when held down; on other keyboards, only certain keys are repeat keys, such as the "X" key or the underscore.

repellency *Chemistry.* the ability to repel water.

repeller *Electronics.* an electrode that reverses the direction of the electron beam in an electron tube. Also, REFLECTOR. *Veterinary Medicine.* an instrument used in animal labor to push back the fetus until the head and limbs can be properly positioned for normal delivery.

repelling object see UNIVERSAL OBJECT.

repent *Botany.* creeping; of or relating to a prostrate growth form, or to plants having stolons that root at the nodes upon contact with the soil surface.

reperforation *Petroleum Engineering.* the drilling of new holes in oil-well tubing opposite reservoir zones so that more fluid will drain into the wellbore.

reperforator *Telecommunications.* in printing telegraph systems, a device that punches a paper strip in accordance with incoming signals, thus permitting retransmission of the signals as printed messages when the paper strip is later passed through a printing telegraph machine. Also, TAPE REPERFORATOR.

repertoire see BEHAVIOR REPERTOIRE.

repetition *Geology.* the reproduction of certain stratigraphic layers at the surface or in a specified section of rock as a result of disruption and displacement caused by faulting or intense folding.

repetition equivalent *Telecommunications.* in a telephone linkup, a measure of the grade of transmission encountered by the users, including such factors as volume, distortion, and noise.

repetition frequency see REPETITION RATE.

repetition instruction *Computer Programming.* an instruction that causes an operation or group of operations to execute repeatedly, either a certain number of times or until a certain condition is met.

repetition rate *Telecommunications.* the rate at which regularly recurring signals, such as radar pulses, are repeated. Also, RECURRENCE RATE, REPETITION FREQUENCY.

repetitive addressing *Computer Programming.* an addressing system in which instructions can be written without designating an operand address, using the address affected by the previously executed instruction in the operation of the next instruction.

repetitive DNA *Molecular Biology.* any repeated sequence of DNA; such sequences constitute a significant fraction of the eukaryotic genome and include varying degrees of repetition.

repetitive element see REGULAR ELEMENT.

repetitive manufacturing *Industrial Engineering.* a type of manufacturing in which standard items are produced in large volume in discrete units, such as refrigerators or automobiles, as opposed to continuous manufacturing of products such as fluids, power, or the like.

repetitive timing *Industrial Engineering.* in time-motion studies, a timing technique based upon repeated (but not continual) trials of a work process.

Repettian *Geology.* a North American geologic stage of the Lower Pliocene epoch, occurring after the Delmontian and before the Venturian.

repi *Hydrology.* any lake, pond, or other body of standing water associated with a sink or subsidence of a land surface.

replacement a process of replacing; specific uses include: *Geology.* **1.** the process by which a new, partly or wholly chemically different mineral grows in the body of an old mineral or mineral aggregate as a result of simultaneous capillary solution and deposition. **2.** the gradual movement of the ocean either toward or away from land areas. *Paleontology.* the process in which inorganic matter is substituted for the organic constituents of an organism during fossilization.

replacement bit *Mechanical Devices.* see RESET BIT.

replacement demand *Engineering.* the economic or operational demand for a replacement part of a product.

replacement deposit *Mineralogy.* a mineral deposit that results from the replacement of a mineral or mineral aggregate by another mineral of differing composition.

replacement dike *Geology.* a dike produced by the gradual conversion of wall rock by solution along fractures or permeable zones.

replacement factor *Engineering.* the predicted percentage ratio of products that will require replacement to the total number of products in use during a given period of time.

replacement series *Ecology.* a method of studying the competition between mixed populations of two plant species by planting the species in various proportions to each other while keeping the overall density constant.

replacement site *Genetics.* any position within a gene at which a point mutation alters the encoded amino acid sequence.

replacement study *Industrial Engineering.* a study carried out to determine if a plant or item of equipment is due for replacement, or can economically be kept in operation.

replacement texture *Geology.* a rock texture showing mineral replacement.

replacement transfusion see EXCHANGE TRANSFUSION.

replacement vector *Genetics.* a lambda bacteriophage that has been produced in the laboratory to receive foreign DNA in recombinant DNA experiments.

replacement vein *Geology.* a vein that has been transformed by mineral replacement. Also, SUBSTITUTION VEIN.

replantation *Surgery.* the replacement of a body part to the location from which it was lost or removed, such as an amputated limb. Also, REIMPLANTATION.

replenisher *Pharmacology.* an agent that restores what has been lost, used up, or lacking. *Ordnance.* a cylinder in the recoil mechanism that allows for oil expansion due to the heat created during firing and then returns the oil to the recoil system when the weapon cools; it contains recoil oil and a spring-activated piston.

replenishment *Geology.* the stage in the development of a cavern during which the presence of air in the passages allows the formation of mineral deposits by the action of water. *Hydrology.* the process by which water is absorbed and added to the zone of saturation; recharge.

repletion *Nutrition.* the condition of being full, or the act of making full.

replica *Science.* an exact copy or model. *Engineering.* a film applied to a test surface and then removed for inspection under an electron microscope in order to indirectly determine the characteristics of the surface.

replica grating *Optics.* a diffraction grating in which thermosetting plastic is cast over the original; used to avoid the costly process of making original gratings.

replica master *Control Systems.* a teleoperator control device that is structurally similar to the slave unit that it controls. *Robotics.* a machine that is similar to a robot but is operated by a human, whose movements are duplicated by another robot.

replica plating *Microbiology.* the process of printing bacteria from one plate to another, usually in searching for mutants.

replicase *Enzymology.* an RNA polymerase that catalyzes the synthesis of RNA from the ribonucleoside 5-triphosphates, using RNA as a template. Also, RNA REPLICASE.

replicating fork *Genetics.* a Y-shaped region of a chromosome that is a growing point in DNA replication.

replication *Science.* the repetition of an experiment in order to ensure accuracy. *Genetics.* the duplication of a DNA molecule during which both strands are used as templates. *Virology.* the duplication of DNA or RNA viral genomes. *Analytical Chemistry.* the formation of a reverse image of a surface by means of an impression on a receptive material that is thin enough to be used in electron microscopic techniques.

replication bubble *Genetics.* an eye-shaped DNA structure formed by the separation of two strands of DNA at the origin of DNA replication.

replication origin *Genetics.* a specific sequence of DNA at which DNA synthesis is initiated.

replicative *Genetics.* characterized by or capable of replication.

replicative form *Virology.* an intracellular form of viral nucleic acid consisting of a viral strand paired with a synthesized complementary strand, produced during replication.

replicative intermediate *Genetics.* see RI.

replicombinase *Genetics.* a transposase that catalyzes replicative recombination.

replicon *Genetics.* a length of DNA that behaves as an autonomous unit during DNA replication; any region of DNA that contains an origin of replication and is thus capable of being replicated under appropriate conditions.

replicon fusion *Genetics.* a tranposon-mediated reaction that joins two complete replicating systems; in the process, a replicon containing a transposon becomes fused with a replicon that lacks the element.

replisome *Enzymology.* a complex of DNA polymerase and other enzymes that are active in DNA replication; identified at the replication fork of bacterial DNA.

replum *Botany.* in mustards (Brassicaceae), the site of attachment of the ovules to the placenta that later develops into a septum or partition separating the seeds within the mature fruit.

reply *Telecommunications.* any radio-frequency signal that is transmitted in response to an interrogation. Also, RESPONSE.

report *Computer Programming.* any document that is output by a processor.

reporter gene *Genetics.* a gene used to help identify or locate another gene.

report generator *Computer Programming.* a program that is designed to gather specified data from files and organize it according to a preset format for printout or display.

reporting point *Aviation.* a geographical point to which an aircraft position report can be related; e.g., so many miles on a given bearing from some point.

reporting time interval *Telecommunications.* the elapsed time for the transmission of data from the originating terminal to the end receiver.

report of survey *Military Science.* an official, standardized report concerning the loss, damage, or destruction of property.

report program *Computer Programming.* a program that is designed to generate custom reports from the data stored in one or more files.

report program generator *Computer Programming.* a general-purpose, business-oriented programming language that is designed to generate reports according to requirements specified by the user.

repose period *Volcanology.* an interval of gaseous emissions between volcanic outbursts.

repositioning *Surgery.* the restoration of a body structure to its original and normal location.

repoussage *Graphic Arts.* the process of hammering out an etching or photoengraving plate from the back, often to clean up depressions made during a previous printing.

repoussé *Graphic Arts.* a technique for making an image on a relief printing plate by making impressions on the back side of the plate.

repp *Textiles.* see REP.

Reppe process *Chemical Engineering.* the high-pressure vinylation, ethynylation, and polymerization reactions of acetylene.

Rep protein *Biochemistry.* a protein, encoded for plasmid, that triggers its replication. *Enzymology.* an enzyme that acts as a helicase and causes unwinding of the DNA helix; produced by the *rep* mutant strain of *Escherichia coli.* (An acronym for repetitive extragenic palindromic protein.)

representation *Artificial Intelligence*. any formalized technique for presenting facts and the rules for manipulating these facts, such as predicate calculus, production rules, semantic networks, and frames. *Mathematics*. a linear action of a group G on a vector space V is called a representation of G. V is called a **representation space** or **representation module**. The representation of G can be extended to a representation of the group ring R_G (called a **regular representation**), so V is a module over the ring R_G. For such rings, the languages of representations and modules are equivalent.

representation hypothesis *Artificial Intelligence*. the hypothesis that intelligence consists primarily of a symbolic representation of the world, manipulation of that representation by reasoning processes, and connection of it to perception and action.

representation theory *Mathematics*. the study of representations of groups, rings, and algebras over fields of various characteristics. *Quantum Mechanics*. the selection of a Hilbert space of eigenfunctions in which to represent the wave function of a system, and the appropriate assignment of matrices to operators and vectors to states.

representative pattern *Cartography*. a feature selected from a dense group of similar features to be included on a map as a characteristic of the entire group.

representative sample *Engineering*. a sample that is large enough and of such average composition as to allow the results of its testing and valuation to represent a specific volume of the body or system of which it is a part.

repressible enzyme *Enzymology*. **1.** an enzyme whose rate of formation is decreased by an increased concentration of one or more by-products, a process serving as a control mechanism in certain bacterial and mammalian metabolic systems. **2.** an enzyme that is produced only in the absence of a by-product.

repressing *Metallurgy*. in powder metallurgy, applying pressure to a compact that has already been sintered.

repression *Psychology*. **1.** the most important defense mechanism, by which anxiety-provoking thoughts or impulses are prevented from becoming conscious. **2.** any method of dealing with anxiety by blocking out undesirable thoughts. *Biochemistry*. the prevention of the biosynthesis of glucose within an organism. *Genetics*. the inhibition of gene transcription by a repressor.

repressor *Genetics*. a molecule, usually a protein, made by a regulator gene that bonds to an operator locus and blocks transcription of that operon.

repressor gene see REGULATOR GENE.

repressuring *Petroleum Engineering*. the process of injecting pressurized water or gas into a reservoir to increase oil recovery.

repro [rē´prō] *Graphic Arts*. **1.** a reproduction proof. **2.** the collection of such proofs to be used in a printed text.

reproduce head see PLAYBACK HEAD.

reproduction *Biology*. the process of producing offspring, which serves to perpetuate or increase the size of a species; achieved either sexually, when two individuals are involved in reproducing new individuals, or asexually, when one individual organism reproduces new individuals. *Science*. see REPLICATION.

reproduction paper *Graphic Arts*. high-quality paper that is coated on one side and used in printing reproduction proofs.

reproduction proof *Graphic Arts*. in printing, a final corrected proof to be used for camera copy.

reproduction ratio *Graphic Arts*. the amount of reduction in reproducing copy, expressed as the measure (usually width) of an original image divided by the corresponding measure on the reproduction.

reproduction scale *Cartography*. the scale at which a map in production is intended to be compiled and published.

reproduction speed *Telecommunications*. in facsimile transmission, the area of copy recorded per unit time.

reproductive *Biology*. relating to, involved in, or serving the process of producing offspring.

reproductive behavior *Behavior*. any behavior that has to do with the breeding and rearing of offspring, such as courtship, nest building, or maternal behavior.

reproductive curve *Ecology*. a representation of the average number of offspring produced for individuals for all ages in a population.

reproductive effort *Biology*. the amount of energy that an organism uses in reproduction.

reproductive isolating mechanism *Genetics*. any biological property of an organism that interferes with its interbreeding with organisms of other species.

reproductive isolation *Ecology*. the inability of a population to reproduce with other populations of the same or related species for physiological (rather than geographic) reasons.

reproductive potential *Biology*. the number of offspring a female of a given age can be expected to produce.

reproductive success *Biology*. the number of surviving offspring produced by an individual at a given time.

reproductive system *Biology*. the system of internal and external genital organs that accomplish reproduction of a species.

reproductive value *Biology*. the number of offspring produced by a given female during a lifetime.

reprographics *Graphic Arts*. any of several processes employing mechanical means to copy an original, such as facsimile or xerography.

REP sequence *Genetics*. a palindrome that occurs in the chromosomes of *Escherichia coli* and *Salmonella typhmurium*; thought to function in the regulation of transcription or in nucleoid structure. (An acronym for repetitive extragenic palindromic sequence.)

reptant see REPENT.

Reptantia *Invertebrate Zoology*. a suborder of decapod crustaceans, including lobsters, crayfish, and crabs, in which one pair of appendages is chelate and the rest are adapted for crawling; none has a laterally compressed body.

reptation *Materials Science*. a model of dynamic macromolecular structure introduced to explain flow properties of polymers by postulating wriggling, snakelike motions of a polymer chain inside a tube formed by its molecular neighbors.

reptilase *Hematology*. an enzyme derived from Russell's viper venom, used to determine blood clotting time.

reptile *Vertebrate Zoology*. any cold-blooded vertebrate animal belonging to the class Reptilia, including extinct dinosaurs as well as crocodilians, turtles, lizards, snakes, amphisbaenids, and tuatara. (Going back to a Latin word meaning "to creep" or "to crawl.")

Reptilia *Vertebrate Zoology*. the reptiles, a class of aquatic or terrestrial, cold-blooded vertebrates having bodies covered with horny scales or plates, breathing by means of lungs, and usually laying eggs outside the body.

repulsion *Mechanics*. any force that tends to push two bodies farther apart, such as the coulomb repulsion between like electrical charges; the opposite of attraction. Also, **repulsive force.** *Genetics*. the occurrence of opposite chromosomes in a double heterozygote of the two mutant alleles of interest.

repulsion-induction motor *Electricity*. a single-phase alternating-current motor that is repulsion starting and inductance running; used where a large starting torque is needed, as in an electric fan. Also, **repulsion-start induction motor.**

repulsion motor *Electricity*. a motor in which the attraction of opposite magnetic poles and repulsion of like poles produces rotational force.

request/grant logic *Computer Programming*. in an interrupt-servicing program, a routine that receives requests for service and grants access, often in accordance with a specified priority scheme.

request repeat system *Telecommunications*. in data transmission, a method of control in which the receiving terminal is able to detect a transmission error and automatically sends a request for retransmission.

request stop *Transportation Engineering*. a place where a transit vehicle or train will stop, but only upon passenger request. Also, DEMAND STOP.

required supply rate *Ordnance*. the estimated amount of ammunition required to sustain operations of a specific force for a specific period; it is expressed in rounds per weapon per day for ammunition fired by weapons, and in appropriate units of measure per day for other ammunition items.

required thickness *Design Engineering*. the wall thickness of boilers or other pressure vessels, calculated by standardized formulas before corrosion allowance is added.

requisition *Military Science*. a demand or request for personnel, supplies, or services that are authorized but not available without specific request.

requisitioning objective *Military Science*. the maximum quantities of materiel to be maintained on hand and on order to sustain current operations; it includes the operating level, the safety level, and the order and shipping time or procurement lead time.

reradiation *Physics*. radiation that is emitted as a result of a previous absorption of radiation. *Telecommunications*. the unwanted local radiation of signals within a receiver that cause interference or pinpoint its location.

rerailer *Transportation Engineering.* either of a pair of devices that channels railroad locomotives or cars back onto the tracks.

rering locked in *Telecommunications.* on magnetic lines, a universal cord circuit feature by which the calling or called party can resignal the operator, causing the supervisory lamps of the cord circuit to stay lit until the operator responds.

rerun *Computer Programming.* **1.** to repeat the execution of a program, usually following an error condition or a machine failure. **2.** the process of re-executing a program. *Telecommunications.* in radio or television, the complete repetition of a transmission. *Chemical Engineering.* to redistill a liquid material because of an improper separation.

rerun point *Computer Programming.* see CHECKPOINT.

rerun routine *Computer Programming.* a recovery routine whose instructions reconstitute a routine from the last rerun point in the event of a computer malfunction or a coding or operating mistake. Also, ROLLBACK ROUTINE.

RES reticuloendothelial system.

RESA or **R.E.S.A.** Research Society of America.

resawing *Engineering.* the process of cutting lumber to the final thickness of a timber product, such as planks, boards, or slabs.

rescinnamine *Pharmacology.* $C_{35}H_{42}N_2O_9$, an alkaloid drug obtained from the plant species *Rauwolfia,* formerly used as a tranquilizer and to treat high blood pressure.

rescue *Virology.* the activation of a defective virus from a latent stage either by recombination or by complementation of the defective functions. *Molecular Biology.* see MARKER RESCUE.

rescue dump *Computer Programming.* the recording of the entire contents of memory on a backing storage for use in restart and recovery.

rescue grass *Botany.* a brome grass, *Bromus unioloides* or *B. catharticus,* characterized by clusters of flattened spikelets and grown for forage; found in South America.

rescue ship *Naval Architecture.* a ship utilized primarily for rescuing personnel and salvaging equipment in cases of sea disasters.

rescue/salvage ship

resealing pressure *Mechanical Engineering.* the inlet pressure at which leakage in a mechanism stops after shutting a pressure relief valve.

research *Science.* scientific investigation that is performed in order to discover new information or to develop or improve products and technology.

research facility *Design Engineering.* a building in which the research and development of high technology occurs through the use of computer workstations, engineering design workstations, and working production models of products under research.

research octane number *Engineering.* a rating of motor-vehicle fuel that predicts the amount of knocking that will occur in an engine; a higher number indicates less engine knocking.

research reactor *Nucleonics.* a nuclear reactor that is used primarily for experimental research and training purposes.

research rocket *Space Technology.* a rocket vehicle designed to collect data, especially about itself or its own equipment through in-flight tests.

research ship *Naval Architecture.* a ship fitted for and employed in oceanographic or other scientific research.

research vehicle *Space Technology.* a rocket vehicle designed to collect data, especially about an atmosphere through which it flies.

reseau [rā zō´] *Photogrammetry.* a grid plate; a network of fine lines etched on a glass plate, used for calibration of plotting instruments and film distortion. *Astronomy.* specifically, such a plate used as an aid in determining the positions of stars. *Meteorology.* a worldwide network of weather reporting stations that have been chosen to represent the meteorology of the globe.

resect *Surgery.* to remove all or part of an organ or tissue.

resection *Surgery.* the removal of all or part of an organ or tissue. *Cartography.* **1.** in surveying, the process of determining the unknown horizontal position of a survey station by observing directions from that station to stations of known position. **2.** the line or direction observed from the unknown station during the resection process.

resection station *Cartography.* in surveying, a station of previously unknown position located through use of the resection process.

resectoscope *Medicine.* an instrument with a wide-angle telescope and an electrically activated wire loop, used for removing tissue of the prostate gland through the urethra.

Resedaceae *Botany.* a family within the order Capparidales consisting of primarily Mediterranean annual or perennial herbs or shrubs, characterized by fruit that is an open capsule, berry, or series of radially arranged carpels; thought to be related to the mustards because some members produce mustard-oil-type compounds.

resedimentation *Geology.* **1.** the redeposition of sedimentary material. **2.** the mechanical deposition of material in cavities of a later age. **3.** the process of underwater, downslope movement of sediment under the influence of gravity.

resequent *Geology.* **1.** describing a physiographic feature that resembles or agrees with some feature of consequent origin, but was formed at a later time. **2.** describing a valley, stream, or drainage system that follows an earlier pattern, but on a newer, lower surface. Also, RECONSEQUENT.

resequent fault-line scarp *Geology.* an erosional fault scarp along a fault line that faces in the same direction as the original fault scarp, or one in which the downthrown block is topographically lower than the upthrown block.

resequent stream *Hydrology.* a stream whose course follows the same direction as an earlier, original stream, but developed later and at a lower level than the original stream. Also, RECONSEQUENT STREAM, RENEWED CONSEQUENT STREAM.

reserpine *Pharmacology.* $C_{33}H_{40}N_2O_9$, an alkaloid drug that is obtained from the plant species *Rauwolfia*; used in the treatment of high blood pressure.

reserve *Military Science.* **1.** a portion of a body of troops that is kept to the rear or withheld from action at the beginning of an engagement to be available for a decisive movement. **2.** members of the military who are not in active service but may be called to active duty. **3.** a military unit composed of reserve personnel. *Aviation.* **1.** or **reserves.** aircraft in a fleet that are in excess of current active needs and kept out of service until needed. **2.** the fuel supply specified to remain in an aircraft's tanks at the completion of a flight as a margin of safety. **3.** any accumulated supplies, equipment, or personnel in excess of current needs. Used to form compounds such as **air reserve, reserve buoyancy,** and **reserve force.** *Computer Programming.* in a multiprogramming system, to allocate a memory area or a peripheral device to the operation of a particular program.

reserve battery *Electricity.* a secondary cell that is used as a storage battery.

reserve bouyancy *Naval Architecture.* the difference between a vessel's full-load displacement and the maximum displacement at which the vessel can float; a measure of how much flooding the vessel can safely endure without sinking, equivalent to the watertight space above the vessel's waterline.

reserve capacity *Engineering.* the capacity of a battery to keep a vehicle operating if the charging system fails; measured in minutes.

reserve cell *Histology.* **1.** a cell that does not stain readily. **2.** a small, undifferentiated cell located at the base of the columnar epithelium; e.g., in the respiratory passages.

reserved demolition target *Ordnance.* a demolition target that may not be demolished without authorization from a specific level of command because of its importance to a tactical or strategic plan, or its inherent structural importance, or because its demolition must be executed in the face of the enemy.

reserved minerals *Mining Engineering.* a term for valuable minerals owned by the government, for which applicants may apply for rights to mine and prospect.

reserved word *Computer Programming.* a word that has a specific meaning or value in a computer operating system or language translator and is therefore unavailable for other use by the programmer. Also, KEYWORD.

reserve force *Cardiology.* the amount of cardiac energy available to satisfy the body's circulatory requirements in activity above the basal metabolic level.

Reserve Officers Training Corps *Military Science.* a group of students at certain colleges and universities who are given instruction toward becoming officers in the armed forces. Also, ROTC.

reserve parachute *Aviation.* an auxiliary parachute, usually having a slightly reduced diameter and worn in a chest pack, for use if the main parachute fails to open.

reserves *Engineering.* the amount of workable minerals, petroleum, or other such useful or valuable material that is calculated to lie within prescribed boundaries. *Military Science.* see RESERVE. *Ordnance.* see RESERVE SUPPLIES.

reserves-decline relationship *Petroleum Engineering.* the relation between the fall in production rate over time to the remaining reserves in a petroleum reservoir.

reserve supplies *Ordnance.* supplies accumulated in excess of immediate needs to ensure continuity of an adequate supply. Also, RESERVES.

reservoir [rez´ər vôr´; rez´ər vwär´] *Civil Engineering.* **1.** a tank, receptacle, or other repository that is used to hold and store water or another fluid. **2.** specifically, a large area, natural or artificial, that holds water for a municipal water supply or for such other purposes as irrigation or recreation. *Geology.* any pool of crude oil or natural gas. *Petroleum Engineering.* a porous, permeable, subsurface sedimentary rock formation in which crude oil or natural gas accumulates. *Hydrology.* any natural or artificial storage place from which water may be withdrawn as needed. *Biology.* a place in an animal or plant in which fluid is collected or retained. *Immunology.* an organism carrying a virus or germ to which it is immune.

reservoir cycling *Petroleum Engineering.* the process of repressurizing an oil reserve by injecting dry gas into it.

reservoir drive mechanism *Petroleum Engineering.* the means by which oil and gas are propelled through a porous reservoir, such as water drive and gas drive. Also, OIL-WELL DRIVE.

reservoir dynamics *Petroleum Engineering.* the properties and performance of fluid flow in a petroleum reservoir.

reservoir fluid *Geology.* any fluid, such as natural gas, oil, or water, that is trapped beneath the surface by a reservoir formation.

reservoir pressure *Geology.* the pressure on fluids in a reservoir formation that causes flow to the bore hole. Also, BOTTOM-HOLE PRESSURE.

reservoir rock *Geology.* any porous or permeable rock, such as sandstone, that yields natural gas or oil.

reset *Computer Technology.* **1.** to set a circuit or program to its initial state. **2.** to set a flag circuit or variable to zero.

reset action *Control Systems.* a floating action in which the speed of the final control element is proportional to the extent of proportional-position action.

reset bit *Mechanical Devices.* a drill bit made from diamonds previously used on another bit that are reset on the crown of a new bit blank.

reset condition *Electronics.* **1.** a condition in a computer in which all of the binary elements are set at zero. **2.** a condition (set of currents, voltages, or signal level) defined for a device or circuit, considered a starting or rest condition to which the device can be brought by applying a special signal.

reset cycle *Computer Programming.* the return of a cycle index counter to its initial or preselected condition.

reset mode *Computer Technology.* in analog computers, a state during which initial conditions are applied to the system.

reset pulse *Electronics.* **1.** a signal that resets the magnetic cell in a computer's storage section. **2.** an electrical signal that resets a computer's counter to zero or to some other predetermined value. **3.** a pulse that returns a device or circuit to its defined reset condition.

reset rate *Engineering.* the number of times per minute that an output signal from a control system is changed as the result of a changing feedback response.

reshabar *Meteorology.* a strong, dry, bora-type, northeast wind that blows down mountain ranges in southern Kurdistan in Iran.

residence time *Chemical Engineering.* the amount of time a unit of mass remains in a volume. Also, **residence.** *Nucleonics.* the amount of time that radioactive debris remains in the atmosphere following a nuclear explosion or accident; usually expressed as a half-life.

resident routine *Computer Programming.* a routine that is loaded into main storage and left there at all times.

residual left over; remaining; specific uses include: *Geology.* **1.** describing a mineral deposit formed by either mechanical or chemical concentration. **2.** describing a physiographic feature that represents a part or trace of a once greater mass and that remains standing after the surrounding areas have been eroded. **3.** describing any material that is left after the weathering of rock in place. *Petroleum Engineering.* any oil left over after a refining process, such as the heavy black oils used in heating plants and ship boilers. *Medicine.* remaining in an organ or part following normal discharge or expulsion, as **residual air.**

residual anticline *Geology.* in salt tectonics, a local structural high between two adjacent rim synclines. Also, RESIDUAL DOME.

residual body *Cell Biology.* an intracellular lysosome that contains indigestible material.

residual charge *Electricity.* the excess charge on capacitor plates after an initial discharge due to dielectric absorption.

residual clay *Geology.* any very finely divided clay material that was formed in place by the weathering of rock. Also, PRIMARY CLAY.

residual compaction *Geology.* the difference between the amount of compaction that will occur for a given increase in applied stress and the amount of compaction that has occurred at a given time.

residual current *Electronics.* **1.** the current produced by the velocity of the electrons ejected from a heated cathode. **2.** the vector sum of the currents flowing through several wires in an electric supply circuit.

residual deposit *Geology.* see RESIDUE.

residual deviation *Geodesy.* a magnetic deviation that is still recorded by a compass even after compensation or adjustments are made.

residual dome see RESIDUAL ANTICLINE.

residual elements *Metallurgy.* harmless or undesirable impurities present in metals and alloys in small quantities.

residual error *Statistics.* the difference between the value of a quantity in a series of observations that have been corrected for systematic errors and the value of the same quantity obtained from the determination of the mean, or another adjustment, of that series.

residual error ratio *Physics.* the difference between the optimum result obtained in practice and the exact result predicted by theory.

residual field *Electromagnetism.* a magnetic field that remains in the core of a coil after the current ceases.

residual flux *Electricity.* the induced flux that remains in a magnetic material after the magnetizing force has been reduced to zero. Also, RESIDUAL INDUCTION.

residual flux density *Electromagnetism.* the magnetic induction present in a sample of ferromagnetic material subject to a cyclic magnetizing force, measured when the magnetizing force is zero. Also, **residual magnetic induction.**

residual forces *Military Science.* unexpended forces that have been deliberately withheld from utilization, but have an immediate combat potential for continued military operations.

residual free gas *Petroleum Engineering.* free gas, occurring in a gas cap of a depleted petroleum reservoir, that is in equilibrium with residual liquid hydrocarbons.

residual induction see RESIDUAL FLUX.

residual intensity *Spectroscopy.* for a spectral line, the ratio of the intensity of radiation at some wavelength to the intensity in the adjacent continuum.

residual ionization *Physics.* the ionization in a gas contained in a closed vessel, which is due to cosmic rays passing through the gas.

residual kame *Geology.* a mound or ridge of sand or gravel that was formed by the wearing away of deposits of glaciofluvial material in glacial lakes or on the flanks of hills of till. Also, PSEUDOKAME.

residual liquid or **residual magma** see REST MAGMA.

residual magnetism *Electricity.* the amount of magnetism still remaining in the core of an electromagnet when the current through the coil ceases.

residual map *Geology.* a stratigraphic map showing small-scale variations of a given stratigraphic unit.

residual material *Geology.* the unconsolidated parent material from which soils are formed, developed in place from the weathering of underlying rock.

residual method *Metallurgy.* a method of magnetic particle inspection in which particles are supplied to a specimen after the magnetic field is removed.

residual mineral *Geology.* a mineral that has been concentrated in place by the leaching of rock and by weathering.

residual mode *Control Systems.* a characteristic mode of motion of a structure that is deliberately ignored in the algorithm for an active control system when the model is being refined.

residual radiation *Nucleonics.* after a nuclear explosion, remaining radioactivity caused by the presence of fissionable by-products and materials made radioactive by neutron bombardment. Also, RESTRAHLEN. *Optics.* the nearly monochromatic infrared radiation remaining after light (or other radiation) is reflected several times by substances such as quartz or fluorite crystals.

residual range *Electricity.* the distance over which a particle can still produce ionization after it has lost some of its energy in passing through matter.

residual resistance *Solid-State Physics.* the electrical resistance of a metal whose temperature is lowered to almost absolute zero; the resistance existing under this condition is caused by lattice imperfections in the metal.

residuals *Statistics.* in regression analysis, the deviations of the actual values of the dependent variable from the values predicted based on the estimated model.

residual sediment see RESISTATE.

residual set *Mathematics.* a subset of a topological space whose complement is a countable union of nowhere dense sets.

residual shrinkage *Textiles.* the percentage of shrinkage that will still occur in a fabric or garment that has been preshrunk.

residual spectrum *Mathematics.* let $T: X \to X$ be a linear operator on a complex Banach space X, I the identity operator on X, λ a complex number, and $T_\lambda = \lambda I - T$. Then λ belongs to the residual spectrum of T if (a) T_λ^{-1} exists, and (b) the domain of T_λ is not dense in X. If λ does not belong to the residual spectrum of T, then it belongs to one of the following sets: the resolvent set of T, the point spectrum of T, or the continuous spectrum of T.

residual stress *Materials Science.* any stress in a material resulting from nonuniform plastic deformation caused by the more deformed portion pulling on the less deformed one and vice versa; later release of such stresses may cause distortions, and the removal of these stresses is a reason for recovery heat treatments.

residual swelling *Geology.* the difference between the prefreezing level of the ground and the level reached by settling after a complete thaw.

residual valley *Geology.* a dividing trough between uplifted mountains.

residual vibration *Design Engineering.* vibration levels that occur in a machine or process after all known or identifiable sources of vibration are eliminated.

residual voltage *Electricity.* the vector sum of the voltages to ground of the multiple phase wires of an electric supply circuit.

residue something that is left over; specific uses include: *Agronomy.* plant parts that remain on the soil surface after harvesting. *Geology.* the accumulation of insoluble rock debris left in place after almost all the soluble constituents have been removed by weathering. Also, RESIDUAL DEPOSIT. *Chemical Engineering.* any original material remaining after a chemical process or reaction is complete. Also, RESIDUUM. *Mathematics.* **1.** see QUADRATIC RESIDUE. **2.** more generally, an element of a quotient ring; equivalently, a coset of an ideal in a ring. **3.** a set A of integers is called a **complete set of residues** (modulo m) if no two of them are congruent (modulo m), and if every integer is congruent to a member of A (modulo m). For example, each of the sets $\{0, 1, \ldots, m-1\}$ and $\{-3, -2, \ldots, m-4\}$ forms a complete set of residues (modulo m). **4.** let $f(z)$ be analytic in some domain D. If $z_0 \in D$ is a pole or an essential singularity of $f(z)$, then the coefficient a_{-1} of the Laurent expansion at $z = z_0$ is called the residue of $f(z)$ at z_0.

residue class *Mathematics.* **1.** the solution set of a congruence relation. **2.** in particular, the equivalence class induced by the congruence relation $x \equiv a$ (modulo m); i.e., the set of integers congruent to some integer a (modulo m).

residue system *Computer Programming.* a number system in which each digit position has a different radix and all radix pairs are relatively prime.

residue theorem *Mathematics.* a theorem stating that if the function f is analytic in the region D, except for a finite number of isolated singularities $a_1, \ldots, a_n$, and if γ is a simple closed rectifiable curve in D that does not pass through any of the singularities a_i, then the line integral of f around γ is equal to $2\pi i$ times the sum of the residues of f at each of the singularities a_i.

residuum *Chemical Engineering.* see RESIDUE.

resilience *Mechanics.* **1.** the ability of a strained body to recover its size and shape after deformation, especially when the strain is produced by compressive stress. Also, ELASTIC RESILIENCE. **2.** the recoverable potential energy stored in an elastic body when stressed within its elastic limit. *Computer Technology.* the ability of a system to continue to run in the event of a component failure. *Textiles.* also, **resiliency.** the qualities of a fabric that enable it to recover its original shape and size after release from wrinkling or crushing; some fabrics, such as wool and silk, have natural resiliency, and others can be chemically treated to impart a certain amount of resiliency.

resilin *Entomology.* a rubbery protein substance found in some insect wing muscles or in leaping muscles like those of the grasshopper; when catch mechanisms are released, energy that is stored and released by resilin helps to project the insect forward.

resin [rez′in] *Organic Chemistry.* any one of several solid or semisolid natural or synthetic organic products, usually translucent polymers that do not conduct electricity; used in plastics, textiles, paints, and varnishes.

resin-anchored bolt *Engineering.* a roof-bolting technique in which a bolt is anchored in the resin placed at the back of the hole in a glass cartridge that ruptures when the bolt is inserted.

resin-bonded composite *Materials.* a material composed of a range of filler types held together by synthetic resin; used for such objects as insulation, molded products, and tools. Similarly, **resin-bonded plywood.**

resin duct *Botany.* any of a series of longitudinal canals developed in the secondary vascular system of many gymnosperms for the secretion of resin. Also, **resin canal.**

resin-in-pulp ion exchange *Chemical Engineering.* an ion-exchange process that utilizes an anion-exchange resin and a heavy slurry of ground uranium ore in acid-leach liquor to produce ore.

resinite *Geology.* a coal maceral of the exinite group consisting of elliptical or spindle-shaped bodies containing resinous compounds.

resinography *Chemistry.* the study of the properties and treatment of resins, polymers, and plastics.

resinoid *Organic Chemistry.* a thermosetting synthetic resin, in either its fusible state or its final infusible state.

resin opal *Mineralogy.* a honey to ocherous-yellow form of common opal, characterized by a resinous luster.

resinous *Materials Science.* of or relating to resin or to materials that contain or resemble resin. Thus, **resinous cement.**

resinous coal *Geology.* a usually younger coal whose attritus contains a relatively large percentage of resinous material.

resinous luster *Mineralogy.* the reflection of light from the surface of a fractured mineral or rock that resembles resin in appearance.

resin roof bolting *Mining Engineering.* the use of a bonding resin to secure metal roof bolts into rock holes.

resin tin see ROSIN TIN.

resist *Materials Science.* a protective coating that prevents a particular reaction. *Graphic Arts.* specifically, a solution of bichromated gum that is applied to a deep-etch offset plate to keep the developing solution from etching nonimage areas. *Metallurgy.* **1.** a coating applied to a portion of a material for preventing chemical attack, electrodeposition, or vapor phase deposition. **2.** in brazing, a material that selectively prevents the flow of the filler.

resistance the fact of opposing or acting against; specific uses include: *Psychology.* **1.** an individual's opposition, usually on an unconscious level, to revealing certain psychological material during the course of treatment by an analyst. **2.** in Hans Selye's analysis of reaction to stress, the second stage, in which most bodily functions gradually return to normal levels. Also, **resistance stage.** *Virology.* the degree to which a host can limit the effects of an infection, ranging from tolerance, in which symptoms are suppressed, to hypersensitivity, in which only a few cells surrounding the infected cell are affected, to immunity, in which the virus does not multiply due to a lack of susceptible cells. *Mechanics.* the ratio of the frictional forces to the speed of a system under damped harmonic motion. Also, MECHANICAL RESISTANCE. *Electricity.* **1.** the ratio of applied electromotive force to the resulting current in a circuit; measured in ohms and following Ohm's law. Resistance opposes the flow of current, generates heat, controls electron flow, and helps supply the correct voltage to a device, depending on the material used, the length and cross-sectional area of the conductor, and the temperature. **2.** the opposition presented to the flow of direct current by a material or device; measured in ohms, kilohms, or megohms. **3.** in an AC circuit, the real component of impedance.

resistance box *Electricity.* a laboratory device used to connect resistors in various combinations so that the resistance can be varied in (decade) steps.

resistance brazing *Metallurgy.* a process of brazing in which the heat is generated by a current flowing in the part.

resistance-capacitance circuit see RC CIRCUIT.

resistance-capacitance constant see RC CONSTANT.

resistance-capacitance coupled amplifier *Electronics.* an amplifier in which the output of one stage is coupled to the following stage by a resistor-capacitor network. Also, **resistance-coupled amplifier.**

resistance-capacitance network *Electricity.* any combination of components connected to form an RC circuit.

resistance-capacitance oscillator *Electronics.* a circuit that generates an alternating current at a frequency determined by a network of resistors and capacitors.

resistance coefficient *Fluid Mechanics.* **1.** in engineering applications, the resistance coefficient (λ) is equal to the diameter of a given pipe, divided by the length of pipe in which the pressure decreases by an amount equal to the dynamic pressure corresponding to the mean velocity. **2.** in physics literature, the resistance coefficient is based on the radius of the pipe; therefore, the corresponding physics coefficient of resistance (λ) is half of that used in engineering applications.

resistance coupling *Electronics.* a method of coupling individual circuits so that their currents pass through the same resistor. Also, RESISTIVE COUPLING.

resistance drop *Electricity.* the voltage across a resistor.

resistance element *Electricity.* a resistor, or any device, such as a capacitor, generator, or semiconductor, that may be connected directly to any other device.

resistance factor see RESISTANCE PLASMID.

resistance furnace *Engineering.* an electric furnace in which heat is generated by electric resistance.

resistance grounding *Electricity.* an electrical grounding in which lines are connected to ground through a resistor.

resistance heating *Materials Science.* a rise in temperature resulting from electric current flow in materials of high resistivity.

resistance lamp *Electricity.* a common incandescent lamp having a resistance element (a tungsten filament) that is heated to a high temperature until it gives off light.

resistance loss *Electricity.* the amount of power lost in a circuit due to the resistance of the wiring and other conductors.

resistance magnetometer *Engineering.* a magnetometer that uses the known change in resistance of a material occurring while the material is immersed in a magnetic field.

resistance material *Electricity.* any material, such as carbon, having such high resistivity that it can be used to make a discrete resistor.

resistance measurement *Electricity.* the process of determining the value of a resistance by passing a small amount of current through the resistance and reading the value in ohms directly from the scale of the ohmmeter.

resistance meter *Engineering.* an instrument that measures electrical resistance in conductive material.

resistance methanometer *Engineering.* a catalytic methanometer in which platinum may be used as the filament that heats the detecting element and acts simultaneously as a resistance-type thermometer.

resistance plasmid *Molecular Biology.* a bacterial plasmid possessing genes that confer resistance to certain drugs upon the host cell. Also, RESISTANCE FACTOR, R PLASMID.

resistance start motor *Electricity.* a squirrel-cage type motor that has resistors inserted in each phase, which gradually cut out as the motor comes up to speed.

resistance strain gauge *Electronics.* a device that measures expansions and contractions in an object under stress, by attaching it to a metal strip whose resistance changes when flexed.

resistance thermometer *Engineering.* a thermometer that uses a metal or semiconductor material with an accurate function of electrical resistance to temperature. Also, **resistance pyrometer.**

resistance transfer factor *Genetics.* the component of a resistance plasmid that enables the plasmid DNA containing genes for resistance to antibiotics to be transferred between bacterial cells.

resistance welding *Metallurgy.* welding in which the heat is generated by a current flowing in the part. Thus, **resistance seam welding.**

resistance wire *Metallurgy.* any alloy wire that has high electrical resistivity and is therefore suitable for making heating elements or electrical resistors.

resistate *Geology.* a sediment composed of chemically resistant minerals that are rich in weathering residues. Also, RESIDUAL SEDIMENT.

resist dyeing *Textiles.* a process in which yarns are treated with a dye-resistant substance, then woven with untreated yarn into a cloth; when the cloth is dyed, only the untreated yarns take the dye, producing a contrasting effect.

resisting moment *Mechanics.* an internal moment produced by a structural member to oppose the external bending moment.

resistive coupling see RESISTANCE COUPLING.

resistive flowmeter *Engineering.* a flowmeter that uses a manometer consisting of a conductive fluid whose level rises and falls with flow rate to produce a variable resistance, proportional to the flow rate.

resistive load *Electricity.* a terminating impedance that is nonreactive, so that the load current is in phase with the electromotive source.

resistive unbalance *Electricity.* an imbalance of current or voltage in a transmission line caused by a difference in the resistance in each of its two wires.

resistivity *Materials Science.* an electrical property of materials, representing the inherent ability of a material to resist current flow; expressed as $r = R(A/l)$, where r is resistivity, A is the area of a conductor, l is its length, and R is its resistance.

resistivity factor *Geochemistry.* see FORMATION FACTOR.

resistivity index *Petroleum Engineering.* the ratio of the actual electrical resistivity of a rock at a certain level of water saturation to the resistivity of the rock itself; used in electrical well logging.

resistivity method *Engineering.* a mineral exploration method in which measurements are made directly by recording the ratio of voltage to current when a current is forced to flow through a section of ground to be tested.

resistivity surveying *Archaeology.* a survey technique that measures the electrical resistance of the ground; higher levels of resistance are registered over stone structures and lower levels over ditches or pits.

resistor *Electricity.* a component in an electric circuit that provides opposition to or limits the current flow according to Ohm's law. Also, **resistor element.**

resistor bulb *Engineering.* an instrument that measures temperature through a conductor with an applied electrical potential from the change in current resulting from a change in resistance brought on by a temperature change.

resistor-capacitor-transistor logic *Electronics.* a logic circuit in which capacitors are used to enhance switching speeds.

resistor color code *Electricity.* a system established by the Electronic Industries Association for marking the value in ohms of a resistor with bands of color. The first band represents the first digit, the second band the second digit, the third the number of zeros to follow the first two digits, and the fourth the percentage of tolerance. The colors and their values are black 0, brown 1, red 2, orange 3, yellow 4, green 5, blue 6, violet 7, gray 8, white 9; in the fourth band, silver represents 10% tolerance, gold 5%, and no color 20%.

resistor core *Electricity.* an insulating core of porcelain, cement, phenolic materials, or plain pressed paper, around which resistance wire is wound.

resistor furnace *Engineering.* a furnace that develops heat by the flow of current through affixed resistive elements.

resistor network *Electricity.* a network composed entirely of resistors.

resistor oven *Engineering.* an oven heated by the flow of electrical current through resistors to produce temperatures reaching 800°F.

resistor termination *Electronics.* a pad of thick film material that overlaps the resistor in the section of a circuit that transforms electrons into energy.

resistor-transistor logic *Electronics.* a logic circuit in which the logic function is performed by resistors, and transistors are used to invert the output.

resist printing *Textiles.* a printing method in which a dye-resistant substance is printed on a fabric before the fabric is dyed; the fabric is washed after dyeing to remove the resist material, leaving a white printed pattern.

resite see C STAGE THERMOSETTING RESIN.

resnatron *Electronics.* a tube that oscillates and amplifies ultrahigh and microwave frequencies.

resojet (engine) *Aviation.* a pulsejet engine that operates at resonance frequency. Also, RESONANT JET, AERORESONATOR.

resole *Materials.* a thermosetting resin in its unformed and uncured state that is fusible and soluble in organic solvents. Also, A-STAGE RESIN.

resolution the extent to which two or more entities can be separated or distinguished from one another; specific uses include: *Physics.* **1.** the degree to which closely related quantities can be discriminated. **2.** the difference between the maximum and the minimum amounts of energy at which an instrument will respond to a beam of particles. **3.** the representation of a vector in terms of its components. *Materials Science.* the smallest distance between points or lines in a material that can be individually distinguished. *Geodesy.* the shortest distance over which differences in the force of gravity, such as those caused by two or more close but separate disturbing bodies, can be determined. Also, **resolution limit.** *Optics.* the ability of a lens or lens system to form images with fine detail; quantitatively expressed by the smallest angular or linear separation of two object points that are reproduced or perceived as separate entities. *Spectroscopy.* the ability of a spectroscopic instrument to separate or distinguish between different wavelengths of radiation. *Electronics.* the least possible increment of change in an electronic signal. *Electromagnetism.* the ability of a radar system to distinguish between two targets in close proximity. *Organic Chemistry.* the process of separating a racemic mixture into its two component optical isomers or enantiomers. *Artificial Intelligence.* a sound and complete method for obtaining proofs by contradiction. The axioms and the negation of the desired conclusion are converted to a set of clauses in conjunctive normal form, pairs of clauses from the set are resolved, and the resolvents are added to the set. If the empty clause is produced, the theorem is proved. *Control Systems.* the smallest amount of error that can be determined and acted upon by an automatic control system.

resolution cell *Electromagnetism.* a conical volume of space associated with a radar whose diameter is given by the product of the slant range and the beamwidth, and whose length is the pulse length.

resolution chart *Optics.* a chart having patterns of lines spaced progressively closer together; used to determine the resolving power of a lens or lens system. *Telecommunications.* see TEST PATTERN.

resolution error *Computer Technology.* any error caused by a computer's inability to represent very small changes in a variable.

resolution in azimuth *Engineering.* the minimum angle of separation between two radar objects at the same range that can be differentiated by radar equipment.

resolution in range *Engineering.* the minimum difference in range between two radar objects at the same azimuth that can be differentiated.

resolution of a crystal structure *Crystallography.* the process of distinguishing individual parts of an object when examining it with radiation. Most X-ray structures of small molecules are determined to a resolution of 0.8–1.0 Å. At this resolution each atom is fairly distinct. For macromolecules, the resolution is not as high.

resolution of a vector *Mathematics.* given a vector v, the determination of a set of vectors parallel to the coordinate axes (or to any other given set of axes) whose vector sum equals v.

resolution of the identity *Mathematics.* let M be a sigma algebra in a set Ω, and $B(H)$ the space of bounded linear operators on a Hilbert space H. A resolution of the identity on M is a mapping $E: M \to B(H)$ with the following properties: (a) $E(\varnothing) = 0$ and $E(\Omega) = I$, where 0 and I are the zero and identity operators, respectively. (b) For each $\omega \in \Omega$, $E(\omega)$ is a self-adjoint projection. (c) For every $\omega', \omega'' \in \Omega$, $E(\omega' \cap \omega'') = E(\omega')E(\omega'')$. If, in addition, $\omega' \cap \omega'' = \varnothing$, then $E(\omega' \cup \omega'') = E(\omega') + E(\omega'')$. (d) For every $x, y \in H$, the set function $E_{x,y}$ is a complex measure on M, where $E_{x,y}(\omega) = (E(\omega)x,y)$. Used in studying the spectra of linear operators.

resolution of transposition *Molecular Biology.* the final step that results in the integration of a transposon into a piece of DNA.

resolution reading *Optics.* the number of lines per millimeter that an optical system or photographic system is capable of resolving, as determined with a resolution chart.

resolution wedge *Telecommunications.* on a test pattern, a cluster of gradually converging lines that are used to gauge resolution.

resolvase *Enzymology.* an enzyme that catalyzes the site-specific recombination between two transposons which are direct copy repeats in a cointegrate structure.

resolve to carry on a process of resolution; specific uses include: *Artificial Intelligence.* in the resolution method, to form a new clause from a pair of given clauses; given two clauses that have exactly one pair of unifiable predicates whose signs are opposite, a new clause is formed from the remaining literals of both clauses, with substitution of variables from the unification of the complementary literals. *Chemistry.* to separate a racemic mixture into optically active components. *Optics.* to separate the individual parts of an image.

resolve ambiguity see DISAMBIGUATE.

resolve motion-rate control *Robotics.* a method of controlling the velocity vectors of the end points of a manipulator by using the angular velocities of the joints.

resolvent *Artificial Intelligence.* the result obtained by resolving two clauses.

resolvent kernel *Mathematics.* a function $R(x, y, \lambda)$ appearing in the integrand of the solution to the general integral equation

$$f(x) = g(x) + \lambda \int_a^b K(x, y) f(y) dy.$$

The solution $f(x)$ is actually a series, found by subdividing the interval $a < x < b$, replacing the integral by a sum of integrals over the subintervals, solving the resulting algebraic equations for f, and then passing to the limit of infinitely many subdivisions. The result is a solution of the given integral equation, having the form

$$f(x) = g(x) + \lambda \int_a^b R(x, y, \lambda) g(y) dy.$$

A solution involving a resolvent kernel is also called a Fredholm solution.

resolvent set *Mathematics.* let $T: X \to X$ be a linear operator on a complex Banach space X, I the identity operator on X, λ a complex number, and $T_\lambda = \lambda I - T$. Then λ belongs to the resolvent set of T if: (a) T_λ^{-1} exists, (b) the domain of T_λ is dense in X, and (c) T_λ^{-1} is continuous. The operator T_λ^{-1} is called the resolvent of T at the point λ. If λ does not belong to the resolvent of T, then it belongs to one of the following sets: the residual spectrum of T, the point spectrum of T, or the continuous spectrum of T; i.e., λ is an element of the spectrum of T.

resolver *Electronics.* **1.** a device that separates a quantity into distinct parts. **2.** a device that accepts the angular position of an object, such as the rotor on an electric machine, and then converts it into two electric signals, proportional to the sine and cosine of the angle. *Electricity.* a device that converts signals into analog or digital signals proportional to two or three components of a vector.

resolving power a measure of the power of resolution; specific uses include: *Optics.* a measure of the ability of an optical instrument to form separate images of two object points spaced closely together. *Spectroscopy.* a measure of the ability of a spectroscopic instrument to separate or distinguish between different wavelengths (frequencies) of radiation. *Electromagnetism.* a measure of a radar system's ability to distinguish between two targets in close proximity; this is expressed as the minimum angular or radial separation between the two targets for which the system can detect them as separate objects.

resolving time *Engineering.* the minimum possible time elapsed between samples taken by an instrument.

resonance *Physics.* a phenomenon exhibited by a freely oscillating system in which the system's response has a large amplitude when driven by an externally applied oscillation at a frequency equal to the natural oscillation frequency of the system. *Quantum Mechanics.* the process by which the wave amplitude and probability are transferred between two degenerate states, in a manner analogous to the energy transfer between two harmonic oscillators. *Electricity.* an effect in which the impedance of a network becomes very small over a narrow frequency range; occurs in LC circuits where the inductive and capacitive reactances are equal. *Chemistry.* see MESOMERISM.

resonance absorption *Quantum Mechanics.* the absorption of radiation by an atom at a frequency corresponding to some transition between stationary states.

resonance bridge *Electricity.* a bridge used to measure inductance and capacitance of an element by determining at what frequency the element causes the bridge to resonate.

resonance curve *Electricity.* a Cartesian plot of voltage or impedance against frequency; used to show the resonant properties of a circuit.

resonance energy *Physics.* the kinetic energy associated with a particle that is preferentially absorbed or scattered because of the presence of a characteristic resonance level in a compound nucleus.

resonance fluorescence *Nuclear Physics.* the resonant scattering from a nucleus of an atom. *Atomic Physics.* see RESONANCE RADIATION.

resonance frequency *Physics.* the frequency at which the response of an oscillating system is a maximum when driven by an external driving force. *Quantum Mechanics.* the frequency at which resonance absorption occurs; the difference between some pair of atomic energy levels divided by Planck's constant. *Electricity.* the frequency at which a resonant circuit resonates.

resonance hybrid *Chemistry.* an ideal hybrid used to describe certain compounds that cannot be accurately represented by any single structure; the actual structure is a weighted sum of the approximate structures.

resonance integral see OVERLAP INTEGRAL.

resonance ionization spectroscopy *Spectroscopy.* a spectroscopic technique for detecting single atoms or molecules in a gas by using a laser beam at a resonant wavelength to excite and remove electrons, which are then ionized in an electric field.

resonance level *Quantum Mechanics.* an unstable state formed preferentially at certain energies in collisions between two particles; at these energies the cross sections reach peaks.

resonance line *Spectroscopy.* for a particular atom, the spectral line corresponding to the longest wavelength capable of producing a transition between the ground state and an excited state.

resonance luminescence see RESONANCE RADIATION.

resonance method *Electricity.* a means of calculating the impedance of a circuit element, by which the resonance frequency of a resonant circuit containing the element is measured. *Metallurgy.* an ultrasonic testing method used to determine the thickness of a metal by tuning to the resonant frequency that in turn reveals the thickness of the metal.

resonance radiation *Atomic Physics.* the glow generated from a gas or a vapor when the energy levels of highly energized atoms are raised by incident photons; the light produced does not necessarily bear the same frequency as that of the incident photons. Also, RESONANCE FLUORESCENCE, RESONANCE LUMINESCENCE.

resonance reaction *Nuclear Physics.* a type of nuclear reaction that occurs when the energy of the incident particle is at or near a characteristic value.

resonance spectrum *Spectroscopy.* an emission spectrum produced by irradiating a sample with light at a specific frequency or set of frequencies.

resonance structure *Organic Chemistry.* any one of two or more possible configurations of the same compound that have identical geometry but different arrangements of their electrons; the actual structure of the molecule is approximately the weighted average of its resonance structures.

resonance transformer *Electricity.* a high-voltage transformer in which the secondary circuit is tuned to the power-supply frequency. *Electronics.* a device in which electrons are energized by two voltages of equal magnitude but of opposite polarities.

resonance trough *Meteorology.* a large-scale atmospheric pressure trough that forms at an appropriate wavelength away from a dominant trough.

resonance vibration *Mechanics.* forced vibration at a natural frequency of the system; characterized by a large amplitude of vibration.

resonant *Electricity.* of or relating to resonance.

resonant antenna *Electromagnetism.* an antenna that radiates a high amount of energy at a particular frequency, due to the establishment of standing wave patterns in the antenna.

resonant capacitor *Electricity.* a tubular inductor that is wound to have capacitance in series with its inductance.

resonant cavity maser *Physics.* a maser whose paramagnetic active medium is placed in a resonant cavity.

resonant-chamber switch *Electromagnetism.* a waveguide device that switches between two or more waveguide branches in which tuned resonant cavities serve the purpose of the switches; the process of detuning a cavity does not transmit energy into its associated waveguide branch.

resonant circuit *Electricity.* a circuit that contains both inductance and capacitance of such values as to give resonance at a certain operating frequency; this frequency can be changed by altering the value of inductance or capacitance.

resonant detector *Physics.* a detector of electromagnetic radiation that is designed to detect radiation of only selected frequencies producing a resonance condition in the detector.

resonant diaphragm *Electromagnetism.* a waveguide diaphragm that is designed to introduce no impedance at a particular frequency.

resonant frequency see RESONANCE FREQUENCY.

resonant helix *Electromagnetism.* in certain transmission lines and resonant cavities, an inner helical conductor that carries currents having the same frequency as the rest of the line or cavity.

resonant iris *Electromagnetism.* in a circular waveguide, a window whose dimensions can produce resonant effects in the guide.

resonant jet see RESOJET (ENGINE).

resonant line *Electromagnetism.* a transmission line whose values of distributed inductance and capacitance define a resonance condition for the line when driven at a certain frequency.

resonant-line oscillator *Electronics.* an oscillator in which the frequency is determined by the length of a section of transmission line.

resonant-line tuner *Electronics.* a tuner that incorporates one or more sections of transmission line as reactances.

resonant Raman effect *Atomic Physics.* a process in which a photon with energy exactly matched to the transition energy between two atomic energy levels promotes an atomic electron to an excited state, which decays in the same step.

resonant-reed relay *Electricity.* a relay designed for high-speed or frequent switching, consisting of a metal reed and coil to which an AC voltage is applied to actuate the relay, generally at its resonant frequency.

resonant resistance *Electricity.* the value of resistance at a resonance frequency.

resonant scattering *Quantum Mechanics.* the absorption and reemission of radiation by an atom or molecule at a frequency corresponding to some atomic transition.

resonant voltage step-up *Electricity.* the capability of an inductor and a capacitor in a series-resonant circuit to supply a voltage several times greater than the input voltage.

resonant wavelength *Electromagnetism.* a combination of capacitive and inductive diaphragms used in a waveguide, designed so that certain frequencies are allowed to pass while others are reflected back.

resonate *Electricity.* to be resonant or to produce resonance.

resonating cavity *Electromagnetism.* a short section of a waveguide that is terminated at one or both ends to provide an enclosure that can support resonant standing waves.

resonator *Physics.* any body or system that exhibits a resonant condition at a characteristic frequency.

resonator grid *Electronics.* an electrode that couples electrons from an electromagnetic field with those from an electron beam at the same frequency; used in velocity-modulated tubes.

resonator wavemeter *Electromagnetism.* a circuit that is used to determine the wavelength or frequency of a wave by achieving a resonance condition.

resorb *Physical Chemistry.* to absorb or adsorb again.

resorbed reef *Geology.* a circumscribed reef having many isolated reef patches that are closely distributed about the main reef mass.

resorcinism *Toxicology.* poisoning due to excessive exposure to resorcinol; symptoms may include severe gastrointestinal disturbances, neuropathy, convulsions, and death.

resorcinol *Organic Chemistry.* $C_6H_4(HO)_2$, combustible, white needles; soluble in water, alcohol, and ether; melts at 111°C and boils at 281°C; used in dyes, resins, and pharmaceuticals. Also, **resorcin.**

resorcinol acetate *Organic Chemistry.* $HOC_6H_4O_2CCH_3$, a combustible, viscous, pale-yellow or amber liquid; slightly soluble in water and soluble in alcohol and most organic solvents; boils (with decomposition) at 283°C; used in medicine and cosmetics. Also, **resorcinol monoacetate.**

resorcinol diglycidyl ether *Organic Chemistry.* $C_{12}H_{14}O_2$, a combustible, straw-yellow liquid; miscible with most organic resins; boils at 172°C (0.8 torr); used in epoxy resins.

resorcinol-formaldehyde resin *Organic Chemistry.* a resin composed of phenol and formaldehyde; soluble in water, alcohol, and ketones; used in making rapid-setting glues for wood products.

β-resorcyclic acid *Organic Chemistry.* $(OH)_2C_6H_3COOH$, a white solid; soluble in water, alcohol, and ether; melts at 210–220°C; used in dyes and pharmaceuticals.

resorption *Physical Chemistry.* the absorption or adsorption of material that is released after previously being absorbed or adsorbed in the same medium. *Petrology.* the process of complete or partial refusion or solution of previously crystallized minerals, by and in a magma.

resource *Science.* an available supply of a material or mineral; any time that is or can be put to use.

resource allocation *Industrial Engineering.* in multiproject scheduling, the process of distributing resources (labor, capital, and so on) among several mutually independent projects in order to minimize delays or bottlenecks affecting any of them. Also, **resource deployment.**

resource-holding potential *Behavior.* a measure of the ability of a male animal to obtain and defend some territory or resource.

resource partitioning *Ecology.* the patterns by which potentially competing species use different essential resources.

respiration *Physiology.* the exchange of gases between the body and the atmosphere; the external and internal processes of breathing. The external, mechanical process involves the muscular activity of the lungs, bringing in oxygen that is absorbed by the blood (inhalation) and taking away waste gases from the blood (exhalation). The internal process allows the interchange of oxygen and carbon dioxide within the body cells, in which oxidation of food nutrients provides energy for cell activity and produces the waste materials of carbon dioxide and water.

respirator *Engineering.* **1.** a device for maintaining artificial respiration. **2.** a device for protecting the respiratory tract with or without equipment supplying oxygen or air.

respiratory *Physiology.* of or relating to respiration. *Anatomy.* of or relating to the respiratory system.

respiratory acidosis *Medicine.* a form of acidosis arising from some respiratory disturbance, such as lung disease.

respiratory arrest *Medicine.* a loss of effective respiratory function which results in the cessation of breathing.

respiratory arrhythmia *Cardiology.* a normal irregularity of the heartbeat caused by the speeding up of the beat during inhalation and its slowing down during exhalation; it occurs commonly in children and older people. Also, SINUS ARRHYTHMIA.

respiratory center *Physiology.* a group of nerve cells located in the medulla oblongata that control and regulate breathing.

respiratory chain *Biochemistry.* the series of proteins that transport electrons from substrates to molecular oxygen in aerobic cells. *Molecular Biology.* see ELECTRON TRANSPORT CHAIN.

respiratory dead space *Physiology.* the volume of the air passages from the nose down to the level at which interchanges of carbon dioxide and oxygen do not take place.

respiratory distress syndrome (of newborn) *Medicine.* a disease of unknown cause occurring most often in infants born prematurely or of diabetic mothers, or in infants delivered by cesarean section; characterized by difficulty in breathing, bluish discoloration of the skin due to insufficient oxygen in the blood, fragile alveoli, and the inability of pulmonary tissues to expand during inspiration.

respiratory epithelium *Histology.* the pseudostratified, ciliated, columnar epithelium that lines the respiratory passages.

respiratory metabolism see AEROBIC METABOLISM.

respiratory minute volume *Physiology.* the volume of gas moved in and out of the respiratory passages per minute.

respiratory quotient *Physiology.* the ratio of the volume of carbon dioxide exhaled to the volume of oxygen inhaled.

respiratory syncytial virus *Virology.* a virus of the genus *Pneumovirus* in the family Paramyxoviridae that resembles the influenza virus and replicates in human cell lines, causing acute respiratory infection, especially in infants and children; also infects mammals, such as cows and chimpanzees.

respiratory system *Anatomy.* a collective term for the system of organs, including the trachea, bronchial tubes, and lungs, that accomplish the exchange of oxygen and carbon dioxide between an organism and its environment.

respiratory tree *Anatomy.* the branching system of bronchial tubes in the lungs that direct air into and out of the lungs. *Invertebrate Zoology.* a branched tubular organ in the cloaca of holothurians (sea cucumbers), thought to function in respiration.

respirometer *Engineering.* a device used to monitor and record aspects of respiration, such as rate and volume.

respirometry *Biology.* the study or measurement of respiration.

respond *Behavior.* to carry out an action following a stimulus.

respondent behavior see RESPONSE.

responder *Electronics.* the component in a radar system that returns signals used to detect, identify, and locate objects. Also, **responsor.**

responder beacon *Electronics.* the radar beacon in a transponder that emits the signals of the responder.

response *Behavior.* the behavior that follows a stimulus. *Telecommunications.* see REPLY.

response area *Industrial Engineering.* in a queuing situation, a sector for which one particular unit is primarily responsible, such as the area to which a given rescue unit is assigned.

response bias *Psychology.* any tendency to make a response that is independent of the experimental or functional conditions; e.g., answering "no" to all questions.

response characteristic *Control Systems.* the representation of system response as the function of an independent variable such as frequency.

response cost *Behavior.* a punishment procedure that involves the contingent removal of a reinforcement following inappropriate or problem behavior.

response fatigue see BEHAVIORAL FATIGUE.

response latency see LATENCY.

response prevention *Behavior.* a form of treatment for fear and avoidance behavior that entails the prevention of avoidance responses during exposure to anxiety-provoking stimuli. Also, FLOODING.

response set *Psychology.* **1.** in reaction-time experiments or similar situations, an individual's concentration on the muscular phase of the process rather than the perceptual phase, which focuses on the stimulus. **2.** a readiness to respond in a particular manner.

response time the time between an occurrence and a reaction produced by the occurrence; specific uses include: *Electricity.* the time between an event and the response of an instrument or circuit to it; this is important in switching circuits and measurement meters. *Computer Technology.* the elapsed time between a computer event, such as data input, and the system's reaction to it. *Control Systems.* the amount of time required for a control operation to begin after the order has been given.

responsor see RESPONDER.

rest and recuperation see R AND R.

restart *Aviation.* to start a rocket or other engine after it has been cut off. *Computer Technology.* **1.** a resumption of computer operations after a temporary halt. **2.** to perform such a resumption. **3.** to start a program again from a checkpoint. Also, REBOOT.

restart capability *Space Technology.* the ability of a rocket, rocket motor, or rocket stage to start again after it has been cut off.

restart point see CHECKPOINT.

rest density *Physics.* the local density of a fluid at rest in a Lorentz frame of reference.

rest energy *Physics.* the energy that is equivalent to the mass of a particle at rest in an inertial frame of reference; equal to the rest mass times the speed of light squared.

restenosis *Cardiology.* that recurrent stenosis of a heart valve that remains after corrective surgery.

rest force *Cardiology.* the amount of heart power necessary to maintain circulation when the body is resting.

rest frame *Physics.* the Lorentz frame of reference in which the total momentum of the system vanishes.

rest hardening *Geology.* the increase, over time, in the strength of a clay following its deposition, remolding, or alteration under shear stress.

restharrow *Botany.* a shrub, *Ononis spinosa*, of the legume family, characterized by pink flowers and tough roots that hinder a plow or harrow; found in Europe.

restiform body see INFERIOR CEREBELLA PEDUNCLE.

resting cell *Cell Biology.* any cell not undergoing mitosis, such as a terminally differentiated cell that is usually arrested in a G_0 or resting stage of its cell cycle.

resting metabolic rate *Nutrition.* the rate at which energy is expended for the maintenance of respiration, circulation, peristalsis, body temperature, and other vegetative functions of the body.

resting potential *Physiology.* the electrical potential across the membrane of a nerve cell or muscle cell when an action potential is not occurring.

resting spore *Biology.* a thick-walled spore that remains dormant for long periods before germination, withstanding adverse conditions.

Restionaceae *Botany.* a monocot family within the order Restionales, consisting of perennial arid-climate herbs with a tufted appearance; characterized by seeds that typically contain mealy endosperm and fruit that appears as a loculicidal capsule, small nut, or achene; found in the Southern Hemisphere.

Restionales *Botany.* an order of monocots containing five families: Flagellariaceae, Joinvilleaceae, Restionaceae, Centrolepidaceae, and Hydatellaceae.

restitution *Photogrammetry.* the correction of the positions of object points, which are inaccurate as a result of both tilt and relief displacement in aerial photographs, to their true or map positions.

restitution nucleus *Cell Biology.* a nucleus in which the chromosomes fail to separate during meiotic or mitotic division, due to a nuclear spindle malfunction and resulting in a diploid or tetraploid nucleus, respectively.

rest magma *Geology.* the still-molten part of a magma that is left after the crystallization of many of its minerals during a long series of differentiations. Also, RESIDUAL LIQUID, RESIDUAL MAGMA.

rest mass *Physics.* the mass of a particle that is at rest in a Lorentz frame.

restoration *Biology.* the return to a previous state or condition, especially a condition of vitality or health. *Nutrition.* the addition of nutrients to a food in order to replace those lost in processing.

restore *Computer Technology.* **1.** to reset a counter, register, switch, or indicator to its initial condition or previous value. **2.** to periodically regenerate a charge in a volatile storage medium.

restorer pulses *Electronics.* the signals that restore energy lost when capacitance from two circuits are coupled in a computer's flip-flop component.

restoring force *Mechanics.* a single force or a system of forces that is able to act in an equal and opposite way to another force or forces, in order to restore a displaced system to the equilibrium position.

restrahlen see RESIDUAL RADIATION.

Restrahlen absorption *Materials Science.* the first of two minor absorption peaks that occur in the microstructure of ionic solids at lower frequencies of radiation; caused by an interaction of the incident electromagnetic radiation with the ions themselves.

restrainer *Graphic Arts.* in a developing solution, a chemical, often potassium bromide, that prevents overly rapid development.

restraint of loads *Engineering.* any of a number of techniques for fastening down or securing cargo to prevent movement during transport.

restricted *Science.* confined or limited.

restricted airspace *Aviation.* the airspace over a restricted area.

restricted area *Military Science.* **1.** an area in which special restrictive measures are employed to prevent or minimize interference between friendly forces; may be land, sea, or air. **2.** an area under military jurisdiction in which special security measures are employed to prevent unauthorized entry. *Navigation.* a defined portion of airspace within which the flight of aircraft may be limited or forbidden.

restricted basin *Geology.* a depression in the sea floor that is often oxygen-depleted as a result of topographically limited water circulation. Also, SILLED BASIN, BARRED BASIN.

restricted gate *Engineering.* the small opening used in transfer or injection molding between the cavity and the injection runner that is designed to snap off upon ejection of the mold.

restricted internal rotation *Physical Chemistry.* the rotation that is inhibited by the presence of π-bonds or delocalized electrons.

restricted propellant *Space Technology.* a solid propellant in which the surface area available for burning is limited by an inhibitor.

restricted proper motion *Astronomy.* a correction applied to proper motion for precession, nutation, aberration, and the like, taking into account the angular movement of a star relative to surrounding stars over a given period of time, usually a year.

restricted waters *Navigation.* a defined portion of navigable waters within which the operation of vessels may be limited or forbidden.

restriction *Mathematics.* If $f: Y \to Z$ is a function (or map) between the sets Y and Z, and the set X is contained in Y, then the function $g: X \to Z$ may be defined by $g(x) = f(x)$ for each $x \in X$. Then the function g is the restriction of f to X, and f is an extension of g to Y. It is customary to write $g = f|X$. *Molecular Biology.* a mechanism by which a bacterial strain degrades foreign DNA by cleaving it at specific sequences of nucleotides.

restriction analysis *Molecular Biology.* the mapping of specific sites on a sequence of DNA, using restriction enzymes that cleave the DNA at specific sites.

restriction endonuclease *Enzymology.* any of several enzymes that destroy foreign DNA molecules by cleavage at specific sites. Also, DNA RESTRICTION ENDONUCLEASE, RESTRICTION ENZYME.

restriction enzyme see RESTRICTION ENDONUCLEASE.

restriction fragments *Molecular Biology.* a series of DNA fragments that result from the cleavage of a double-stranded DNA molecule at specific sites by restriction enzymes.

restriction map *Molecular Biology.* a map of the specific oligonucleotide locations at which each of a number of restriction enzymes would cleave the DNA.

restriction mapping *Molecular Biology.* the process by which each of a number of restriction endonuclease recognizes and cleaves a specific sequence of DNA, yielding a known, arranged order of these restriction sites on a segment of DNA.

restriction modification *Molecular Biology.* the process by which a prokaryotic cell can degrade foreign DNA using its restriction enzymes, while preventing degradation of its own DNA by modifying it with a methylating enzyme.

restriction-modification methylase *Molecular Biology.* any of the enzymes that methylate DNA at a specific site in certain bacteria, protecting the DNA against cleavage by the restriction endonuclease that recognizes that site.

restriction site *Molecular Biology.* the location on a double-stranded DNA at which a restriction enzyme recognizes and cleaves a specific sequence.

restrictive condition *Genetics.* any environmental condition that allows a mutant phenotype to be expressed.

resue *Mining Engineering.* **1.** to remove barren rock by mining or stripping so that a narrow but rich vein of material is exposed and may be extracted in a clean condition. **2.** to open up a stope in the wall rock rather than in the vein.

resultant *Mathematics.* **1.** let

$$f(x) = a_0x^n + a_1x^{n-1} + \cdots + a_{n-1}x + a_n \quad \text{and}$$

$$g(x) = b_0x^m + b_1x^{m-1} + \cdots + b_{m-1}x + b_m$$

be polynomials over a ring R, such that both a_n and b_m are nonzero. The resultant $R(f,g)$ of f and g is the determinant of the $(m + n) \times (m + n)$ matrix (r_{ij}) (and an element of R) defined as follows: if $1 \le i \le m$, then

$$r_{ij} = a_{j-i} \text{ for } 0 \le j - i \le n, \text{ and } r_{ij} = 0 \text{ otherwise;}$$

and if $m + 1 \le i \le m + n$, then

$$r_{ij} = b_{j-i-m} \text{ for } 0 \le j - i - m \le n, \text{ and } r_{ij} = 0 \text{ otherwise.}$$

The matrix may be visualized as follows, where all the blank spaces are filled in with zeros:

$$\begin{array}{cccccccccl}
a_0 & a_1 & \cdots & a_{n-1} & a_n & & & & & \\
 & a_0 & a_1 & \cdots & a_{n-1} & a_n & & & & \\
 & & \cdots & \cdots & \cdots & \cdots & & & & (m \text{ rows}) \\
 & & & a_0 & a_1 & \cdots & a_{n-1} & a_n & & \\
b_0 & b_1 & \cdots & b_{m-1} & b_m & & & & & \\
 & b_0 & b_1 & \cdots & b_{m-1} & b_m & & & & \\
 & & \cdots & \cdots & \cdots & \cdots & & & & (n \text{ rows}) \\
 & & & b_0 & b_1 & \cdots & b_{m-1} & b_m & &
\end{array}$$

Two polynomials f and g are relatively prime if and only if their resultant is nonzero. **2.** in addition, the resultant may be viewed as a function of the $m + n$ polynomial coefficients, or simply of $m + n$ variables, in which case it can be shown to be homogeneous of degree m in its first set of m variables and of degree n in its second set of n variables. **3.** the sum of two or more vectors.

resultant force *Mechanics.* a single composite force having the same net effect as all the actual forces that act simultaneously at the same point. Also, COMPOSITION OF FORCES.

resultant rake *Mechanical Engineering.* the angle between the cutting tooth surface and an axial plane through the tooth point, taken in a plane perpendicular to the cutting edge.

resultant wind *Meteorology.* the vectorial average of all wind directions and speeds for a specified level at a specified place for a certain period, such as a month.

resupinate *Botany.* of or relating to a plant or plant part that appears to be reversed or twisted; this is characteristic of orchids in which the flower appears to be twisted 180° on the main shoot. *Mycology.* of or relating to certain fungi in which the hymenial layer develops on the upward side.

resurgence *Hydrology.* **1.** the place where an underground stream appears at the surface to become a surface stream. Also, RISE, RISING, EMERGENCE. **2.** the reappearance of a sinking stream from a cave. Also, REEMERGENCE.

resurgent *Geology.* **1.** describing magmatic water or gases that originate on the surface, in the atmosphere, or in the country rock of the magma. **2.** describing pyroclastics formed from fragments of the volcanic cone or earlier lavas. Also, ACCESSORY.

resurgent cauldron *Geology.* a volcanic subsidence structure in which the cauldron block has been uplifted following subsidence, and usually forms a structural dome.
resurgent ejecta see ACCESSORY EJECTA.
resurrected *Geology.* describing a previously buried geologic feature that has been exposed by erosion and thus restored to its previous status in the existing relief. Also, EXHUMED. *Hydrology.* describing a stream that follows an earlier drainage system after a brief period of submergence has slightly masked the old course with a thin film of sediments.
resurvey *Cartography.* a survey in which the lines of an earlier, original survey are retraced and the stations are recovered and used for control purposes. A **dependent resurvey** is conducted in order to restore a missing comer; an **independent resurvey** is conducted to supersede the original survey and establish new boundaries and subdivisions.
resuscitate [rē sus´i tāt] *Medicine.* to carry out a process of resuscitation; restore to life or consciousness.
resuscitation [rē sus´i tā´shən] *Medicine.* the restoration to life or consciousness of a person who is apparently dead, including such measures as artificial respiration or cardiac massage.
resuscitator [rē sus´i tā´tər] *Engineering.* a system that supplies and pumps oxygen to a person who is unable to breath properly.
ret *Chemistry.* to break down fibers by the action of enzymes.
RET *Oncology.* a proto-oncogene that shows rearrangement and is found in carcinoma of the thyroid.
retained austenite *Materials Science.* austenite that is not transformed into martensite during quenching.
retained water *Hydrology.* water that remains in rock or soil after gravity water has drained out.
retainer *Engineering.* any device that serves to keep a part in place. *Medicine.* **1.** an orthodontic device for maintaining the teeth and jaws in a desired postion. **2.** any clasp, attachment, or other device designed to fix and stabilize a prosthetic appliance.
retainer plate *Engineering.* the mounting plate that holds removable mold parts during molding.
retainer wall *Engineering.* a barrier constructed beyond a primary storage tank to keep stored liquids from escaping a containment area upon failure of the storage tank. *Civil Engineering.* see RETAINING WALL.
retaining ring *Mechanical Devices.* **1.** a metal ring within a ball bearing mechanism that maintains the proper distribution of the balls for the casing. **2.** a clip of spring steel forming a shoulder inside a shaft or reaming shell to prevent a core lifter from entering a core barrel or to locate a pair of mating parts in an axial direction. Also, CIRCLIP.
retaining valve *Mechanical Devices.* an additional valve embedded in a pumping system; used to inhibit water from running back between strokes in deep wells.
retaining wall *Civil Engineering.* a wall built to hold in place a mass of earth, often at the edge of a terrace or excavation.
retard *Science.* to slow the development of; hinder or restrict. *Mechanical Engineering.* in tuning an internal-combustion engine, a delay adjustment made to the setting of the distributor so that the ignition spark for each cylinder is generated in sequence.
retardant something that delays, diminishes, or restricts; specific uses include: *Chemistry.* a substance that is capable of reducing the speed of a chemical reaction.
retardation a process or instance of hindrance or delay; specific uses include: *Chemistry.* the slowing down or stopping of a chemical reaction. *Medicine.* **1.** subnormal development or maturation. **2.** see MENTAL RETARDATION. *Optics.* the difference between the optical path taken by light that penetrates an object and the path taken by light that does not. *Navigation.* the time delay between the meridian passage of the moon and local high water at any given point. *Oceanography.* the amount of time by which corresponding tidal phases occur later from one day to the next, averaging around 50 minutes.
retardation coil *Electromagnetism.* a high-inductance coil that allows low frequency and direct currents to pass while blocking audio-frequency currents; used in telephone communications.
retardation plate or **retardation sheet** *Optics.* an optical element whose two principal axes resolve an incident polarized beam into two mutually perpendicular polarized beams, which recombine to form a particular single polarized beam; may produce full-wave, half-wave, or quarter-wave retardations. Also, WAVE PLATE.
retarded *Psychology.* suffering from mental retardation; a term now generally regarded as pejorative and not in clinical use.
retarded field *Electromagnetism.* the electric or magnetic field that is derived from the retarded potentials.
retarded potentials *Electromagnetism.* the electric scalar potential and the magnetic vector potential occurring at a time t and at a point in space derived from the state of charges and currents that existed at earlier times.
retarder *Chemical Engineering.* a substance that delays or prevents any of various processes, such as the setting of cement or the vulcanization of rubber. *Mechanical Engineering.* **1.** a power-operated braking device placed along the classification tracks of a humping yard to control the speed of railroad cars. **2.** an elongated piece of material that is used to stimulate the activity of hot gases that flow within a fire-tube boiler.
retarding basin *Civil Engineering.* a basin in the course of a stream or channel that serves to reduce the peak stream flow.
retarding conveyor *Mechanical Engineering.* any conveyor that inhibits the movement of its load; used for heavy loads moving at an angle that tends to speed up the conveying medium.
retarding-field oscillator *Electronics.* a circuit that shifts direct current through a grid lying between the cathode and the anode to convert it to alternating current; used in microwave generators.
retarding potential *Physics.* a potential that causes slowing in the motion of a charged particle passing through it.
retard transmitter *Electronics.* a device in a transmitter that delays the transmission of a signal after actuation.
rete [rē´tē] *plural,* **retia** [rē´shē ə] *Anatomy.* a net or network, as of arteries or veins.
rete cord see MEDULLARY CORD, def. 2.
rete mirabile *Vertebrate Zoology.* a circulatory structure found in fish kidneys and swim bladders, consisting of a small artery or vein that splits into a mass of smaller vessels which then reunite into a similar vessel. (From Latin; literally, "a marvelous network.")
retene *Organic Chemistry.* $C_{18}H_{18}$, plates that are soluble in benzene and ligroin; melts at 100°C; used in organic synthesis. Also, 7-ISOPROPYL-1-METHYL PHENANTHRENE.
retention *Physiology.* **1.** the act or process of keeping in possession, or of holding in place or position. **2.** the holding back of matter that would normally be excreted as waste. *Psychology.* the ability to recall and recognize previously learned behavior or experience at a time when it is not being actively performed.
retention cyst *Medicine.* a cyst resulting from obstruction of the flow of secretion in the excretory duct of a gland.
retention defect *Neurology.* a disorder affecting short-term memory of names, numbers, or events.
retention index *Analytical Chemistry.* the relationship of the retention volume with an arbitrary assigned value to the substance being tested during gas chromatography.
retention period *Computer Technology.* the length of time that data is to be preserved before it can be overwritten.
retention suture *Surgery.* one or more closing surgical stitches, of very strong material and encompassing large amounts of tissue with each stitch, performed to reinforce and relieve tension from the primary suture line and to prevent postoperative wound disruption.
retention time *Analytical Chemistry.* the time at which a gas chromatogram peak reaches its maximum. *Electronics.* the maximum amount of time that can elapse between writing and reading from a storage tube.
retention volume *Analytical Chemistry.* the product of retention time and flow rate of solvent in gas chromatography.
retentive discrimination see RECOGNITION.
retentivity *Electromagnetism.* the residual flux density resulting from the saturation induction of a magnetic material.
rete testis *Anatomy.* the network of tubules in the mediastinum testis that conducts sperm from the straight seminiferous tubules into the efferent ductules.
retgersite *Mineralogy.* $NiSO_4 \cdot 6H_2O$, a blue or emerald-green, water-soluble, tetragonal mineral, dimorphous with nickel-hexahydrite, occurring as thick tabular crystals and in fibrous form, having a specific gravity of 2.04 to 2.07 and a hardness of 2.5 on the Mohs scale; found as a secondary mineral in nickel-bearing deposits.
Retgers' law *Solid-State Physics.* a law stating that properties of an isomorphous crystalline mixture are continuous functions of the composition percentages.
rethrolone *Organic Chemistry.* the five-member ring portion of a pyrethrin.
reticle *Optics.* a system of marks located in the focal plane of an optical instrument or on the light shield of a cathode-ray oscilloscope, used to aid in sighting, aligning, or measuring. Also, GRATICULE, RETICULE.

reticul- a combining form meaning "network," denoting a relationship to a reticulum or reticular structure.
reticular *Science.* relating to or resembling a net or network. *Anatomy.* of or relating to a reticulum.
reticular cell *Histology.* see RETICULOCYTE.
reticular degeneration *Pathology.* a deterioration or alteration in the tissue to a netlike degeneration of a lower form.
reticular density *Mathematics.* the number of lattice points in a given square unit of measurement.
reticular fiber *Histology.* any of a group of delicate, branching fibers that form a latticework within tissues, as in the red marrow, spleen, and lymph nodes.
reticular formation *Anatomy.* a network of nerve cells and fibers that extends from the spinal cord through the brain stem to the diencephalon; it is thought to regulate wakefulness and sleepiness, and other reflexes.
Reticulariaceae *Mycology.* the plasmodial slime molds, a family of fungi of the order Liceales.
reticular system see RETICULOENDOTHELIAL SYSTEM.
reticular tissue *Histology.* loose connective tissue that contains principally reticular fibers.

reticulate leaf

reticulate *Biology.* having a two-dimensional network pattern of fibers, veins, or lines; e.g., the nucleus in a cell or the small veins in a leaf. *Geology.* **1.** of a vein or lode, having a netlike texture. **2.** describing a rock texture in which partially altered crystals form a network that encloses remnants of the original, unaltered mineral. Also, **reticulated.** *Materials.* having a network of fine lines. Thus, **reticulated glass.**
reticulated bar *Geology.* any of a group of slightly submerged sandbars forming two sets, diagonal to the shoreline in a crisscross pattern.
reticulate evolution *Evolution.* an evolutionary change that results from the repeated recombination of genes between previously separated populations.
reticulate python *Vertebrate Zoology.* a python, *Python reticulatus,* of southeastern Asia and the East Indies; considered the longest snake in the world (up to 35 feet in length).
reticulation *Graphic Arts.* a weblike pattern of lines that may appear on the surface of a photographic negative as a result of extreme temperature changes during processing.
reticule see RETICLE.
reticulin *Biochemistry.* a protein that exhibits distinctively shaped fibers under electron microscopy; high concentrations are found in the spleen.
reticulite see THREAD-LACE SCORIA.
reticulo- a combining form meaning "network," denoting a relationship to a reticulum or reticular structure.
Reticuloceras *Paleontology.* a genus of involute ammonoids in the superfamily Gastrioceratacеae; an important zone fossil for the Carboniferous (along with its cousins Goniatites and Gastrioceras).
reticulocyte *Hematology.* a large, immature erythrocyte. *Histology.* a cell found in reticular tissue.
reticulocyte response *Hematology.* an increase in the formulation of reticulocytes in response to a bone marrow stimulus, such as that provided by administration of a hematinic agent.
reticuloendothelial granulomatosis *Medicine.* any of a group of rare diseases characterized by excessive formation of phagocytic cells, including Letterer-Siwe's disease, Hand-Schuller-Christian disease, and eosinophilic granuloma.
reticuloendothelial system *Anatomy.* the system of more or less strongly phagocytic cells (macrophages, histiocytes, and microglia) distributed throughout the body, but primarily found in lymph nodes and in blood and lymph sinuses in the liver, spleen, and bone marrow; functions as a defense system.
reticulohistiocytoma *Oncology.* a proliferation of granulomas composed of lipid-laden histiocytes and multinucleated giant cells with pale eosinophilic cytoplasm having a ground-glass appearance. Also, **reticulohistiocytic granuloma.**
reticuloma *Oncology.* histiocytic malignant lymphoma.
reticulopodia *Invertebrate Zoology.* a type of branched, interconnected pseudopodia found in the protozoan order Foraminifera.
reticulospinal tract *Anatomy.* a system of fibers extending from the reticular formation in the pons and medulla oblongata through the ventral and lateral portions of the spinal cord; it influences various reflex mechanisms including muscle tone, as well as respiratory and cardiovascular activities.
reticulum any network or netlike structure or system; specific uses include: *Cell Biology.* a protoplasmic network in cells; e.g., the flattened double membrane sheets of the endoplasmic reticulum. *Histology.* see RETICULAR TISSUE. *Vertebrate Zoology.* the second compartment in the ruminant stomach, where vast numbers of microorganisms digest and ferment swallowed food, breaking down protein, polysaccharides, fats, and cellulose.
Reticulum *Astronomy.* the Net, a faint constellation of the Southern Hemisphere that lies near the bright star Achernar.
reticulum cell sarcoma of the brain see MICROGLIOMATOSIS.
Reticulum system *Astronomy.* a probable dwarf galaxy located near the Large Magellanic Cloud and likely a member of the Local Group.
retina [ret′i nə] *Anatomy.* the inner lining of the eye, which receives images formed by the lens and transmits them through the optic nerve to the brain. *Computer Technology.* the scanning unit in an optical character recognition system. Also, **retina character reader.**
retinaculum *Anatomy.* a structure that retains an organ or tissue in place. *Surgery.* an instrument used to retract tissues during surgery. *Invertebrate Zoology.* **1.** in Collembola (springtails), an appendage on the third abdominal segment that holds the furcula in place. **2.** in Cirripedia (barnacles), a small hooked process that retains the egg sac. *Entomology.* in certain lepidopterans, a clasp on the forewing that catches and holds the frenulum of the hindwing.
retinal *Anatomy.* of or relating to the retina. *Biochemistry.* an aldehyde present in visual pigments whose molecule has 20 carbon atoms, a six-member ring, and a sequence of conjugated double bonds that fashion a chromophore.
retinal astigmatism *Medicine.* any defect of vision caused by irregularity in the curvature of the cornea or the lens, preventing light rays from focusing at a single point on the retina.
retinal illuminance *Optics.* the luminous flux received per unit area of the retina; measured in trolands.
retinalite *Mineralogy.* a honey-yellow to light-green variety of massive serpentine having a resinous or waxy luster.
retinal pigment see RHODOPSIN.
retinal retinitis see VASCULAR RETINOPATHY.
retinal rods *Anatomy.* the rod-shaped photosensitive cells of the retina, serving night vision and detection of motion.
retinasphalt *Mineralogy.* a light-brown variety of retinite; generally found with lignite.
retinene *Biochemistry.* the chief carotenoid pigment in the retina, which turns yellow when exposed to light.
retinite *Mineralogy.* any of a large group of fossil resins characterized by amberlike appearance but lacking succinic acid, found generally in brown coals and peat.
retinitis *Medicine.* an inflammation of the retina.
retinitis pigmentosa *Genetics.* a degenerative and atrophic hereditary human disease of the retina, inherited as an autosomal recessive gene; homozygous individuals suffer progressive blindness, with an onset of night blindness that is generally followed by migration of pigment and progressively reduced peripheral vision.
retinoblastoma *Oncology.* a congenital malignant neoplasm of the retina composed of embryonic retinal cells, generally occurring in the early years of childhood.
retinocerebral angiomatosis see VON HIPPEL-LINDAU DISEASE.
retinochoroiditis *Medicine.* an inflammation of the retina and the choroid.
retinoid *Anatomy.* resembling the retina. *Biochemistry.* a class of compounds that have molecules containing 20 carbon atoms and are related to retinal compounds and others associated with vitamin A activity.
retinol see VITAMIN A.
retinopathy *Medicine.* **1.** an inflammation of the retina. **2.** any degenerative noninflammatory disease of the retina.

retinoschisis *Medicine.* a degenerative condition resulting in splitting of the retina or separation of the retinal layers.

retinula *Entomology.* the light-sensitive component of the ommatidium or compound eye; consists of radially arranged cells possessing a photoreceptor component or rhabdomere.

retire *Navigation.* to compensate for the movement of a craft between observations by moving later observations back along the course line to the earlier observation. *Military Science.* to carry on an operation of retirement.

retired line of position *Navigation.* a line of position that has been moved back along the course line to correspond to an earlier time.

retirement *Military Science.* an operation in which a force moves away from an enemy with which it is not in direct contact.

retort *Chemical Engineering.* **1.** a glass vessel, typically a bulb with a long, downward-pointing neck; used to distill or decompose substances, usually solid or semisolid substances, by heating. **2.** a cylindrical refractory chamber used to heat coal or ore.

Retortamonadida *Invertebrate Zoology.* an order of flagellated parasitic protozoans in the class Zoomastigophora.

retort pouch *Food Technology.* a lightweight, flexible container or pouch in which foods are heated and sterilized; often made of laminated film or aluminum film.

retouch *Graphic Arts.* to treat a photographic negative or print to correct defects, improve tonal qualities, or make other desired alterations; a variety of materials and techniques are used, including pencils, dyes, and airbrush. *Archaeology.* **1.** to thin, sharpen, straighten, or otherwise refine an existing stone tool for further use. **2.** the process of altering a tool in this way.

retouch color *Graphic Arts.* an aniline dye used in retouching photographs.

retrace blanking *Electronics.* the practice of blanking out a picture tube while the electron beam scans across a television or an oscilloscope screen from the end of one line to the start of the next line, to obscure trace lines.

retrace interval *Electronics.* the time it takes an electron beam to go from the end of one line to the start of the next line during its trace of an oscilloscope or television screen. Also, **retrace time, return time.**

retrace line *Electronics.* the path an electron beam follows as it travels from one field to another on a television or oscilloscope screen. Also, **return line.**

retracker *Transportation Engineering.* see RERAILER.

retract *Mathematics.* a subset R of a topological space X is said to be a retract on X if there exists a continuous map ρ from X to R that fixes R; i.e., $\rho(x) = x$ for all $x \in R$.

retractor *Anatomy.* any retractile muscle. *Surgery.* an instrument for separating and holding back tissues to expose a site for surgery.

retral *Biology.* at or toward the back; posterior.

retransmission unit *Electronics.* the device that controls the transmission of a signal through a two-way communication system.

retreat *Military Science.* **1.** a withdrawal of a force that is in direct contact with the enemy. **2.** a signal for lowering the flag in the evening, or a ceremony held as the flag is lowered. *Mining Engineering.* mine workings directed opposite to advance work, the completion of which will allow the area to be finished and abandoned.

retreater *Engineering.* a mercurial thermometer that displays an incorrect higher value due to a decreased constriction in the mercury column.

retreat gun see EVENING GUN.

retrieval *Computer Programming.* the process of retrieving data or files. *Psychology.* the process of recovering information from long-term memory; analogous to information retrieval in computers.

retrieval request *Computer Programming.* a request to the computer operator to retrieve a file.

retrieve *Computer Programming.* **1.** to extract data from a file or files. **2.** to move a file from archival storage to on-line storage.

retro- a prefix meaning "backward" or "behind."

retroaction *Physiology.* action in a reversed direction; reaction.

retroactive inhibition *Psychology.* the interference that current learning places upon the recall of similar material already learned.

retroactive refit see RETROFIT.

retrobuccal *Anatomy.* relating to the back part of the mouth.

retrocardiac *Anatomy.* situated or occurring behind the heart.

retrocecal *Anatomy.* situated or occurring behind the cecum.

retrocerebral gland *Entomology.* a group of neurosecretory cells, located behind the insect brain, that play a role in the process of metamorphosis, growth, and reproduction.

retrocession *Medicine.* an abnormal dropping backward of the entire uterus.

retrocolic *Anatomy.* situated or occurring behind the large intestine.

retrodirective *Optics.* of a mirror, reflector, and so on, causing reflected rays to return along paths that are parallel to the corresponding incident rays.

retrofire *Space Technology.* **1.** to ignite a retrorocket. **2.** the action of a retrorocket when ignited.

retrofire time *Space Technology.* the time period during which retrorockets are being fired or are programmed for firing in order to slow the velocity of a reentry capsule.

retrofit *Engineering.* to modify an item after purchase to correct an improper design, to replace a faulty component, or to make new improvements to the older model. *Building Engineering.* to install new fixtures, pipes, and so on, in a previously constructed building.

retroflexion *Physiology.* the bending of an organ so that its top is turned backward, especially the bending backward of the body of the uterus forming an angle with respect to the cervix.

retrogene see PROCESSED GENE.

retrogradation *Chemistry.* a process of physical and chemical changes in aqueous solutions or gels brought on by aging, resulting in simpler molecular forms. *Geology.* the retreat of a shoreline or coastline caused by wave erosion.

retrograde *Mechanics.* moving backward or against the usual direction of flow. *Biology.* degenerating, deteriorating, or catabolic.

retrograde amnesia *Medicine.* a loss of memory of events that occurred prior to a particular time in a person's life, especially prior to a traumatic or stressful event that caused the amnesia.

retrograde condensation *Organic Chemistry.* the condensation of the vapor phase of a hydrocarbon in contact with its liquid phase that occurs when either the pressure decreases at constant temperature or the temperature increases at constant pressure.

retrograde evaporation *Organic Chemistry.* the evaporation of the liquid phase of a hydrocarbon in contact with its vapor phase that occurs when pressure increases at constant temperature or the temperature decreases at constant pressure.

retrograde metamorphism *Petrology.* the formation of higher-grade metamorphic minerals into lower-grade metamorphic minerals in response to a change in physical condition, such as a lower temperature. Also, DIAPHTHORESIS, RETROGRESSIVE METAMORPHISM.

retrograde motion *Astronomy.* **1.** with planets in their orbits, motion westward. Also, **retrograde orbit. 2.** in rotation, clockwise movement as viewed from above the object's north pole.

retrograde movement *Military Science.* any movement of a command to the rear or away from the enemy; it may be forced or voluntary and may be classified as withdrawal, retirement, or delaying action. Similarly, **retrograde personnel.**

retrograde reservoir *Geology.* a reservoir in which hydrocarbons are initially in the gaseous state, change to the liquid state as pressure is reduced and the bubble-point line is passed, and then return to the gaseous state upon further reduction of pressure.

retrograde vernier *Cartography.* on surveying instruments, a type of sliding auxiliary scale that aids in reading fractional parts of the smallest division on the primary scale; its divisions are larger, and its numbers run in a direction opposite to that of numbers on the main scale.

retrograde wave *Meteorology.* an atmospheric wave that moves in a direction opposite to that of the flow in which the wave is embedded.

retrograding shoreline *Geology.* a shoreline that is being cut back toward the land as a result of wave erosion. Also, ABRASION SHORELINE.

retrogression *Medicine.* **1.** the degeneration or atrophy of an organ or tissue. **2.** a return to an earlier or less complex condition. *Psychology.* the process of dealing with a conflict by returning to behavior appropriate to an earlier period of development. *Meteorology.* the movement of an atmospheric wave or pressure system in a direction opposite to that of the main flow in which it is embedded. *Geology.* see RECESSION. *Astronomy.* see RETROGRADE MOTION.

retrogressive metamorphism see RETROGRADE METAMORPHISM.

retrogressive metamorphosis *Invertebrate Zoology.* a change of form from an apparently more developed stage to a more primitive stage, as when free-swimming tunicate larvae become sessile adults. Also, **retromorphosis.**

retroid viruses *Virology.* a former term for viruses similar to members of the family Retroviridae, in which replication occurs through reverse transcription of RNA genomes.

retroinfection *Medicine.* the infection of the mother by the fetus.

retrojection *Surgery.* the irrigation of a body cavity by injection of a fluid.

retrolental fibroplasia *Medicine.* blindness resulting from the administration of excessive concentrations of oxygen to premature infants, causing the formation of fibrous tissue behind the lens of the eye.

retromammary *Anatomy.* situated or occurring behind the breast.

retronasal *Anatomy.* situated or occurring behind the nose.

retroperitoneal *Anatomy.* situated or occurring behind the peritoneum.

retroreflection *Physics.* a process of reflection in which the reflected rays and incident rays are antiparallel.

retroreflector *Physics.* any device that is designed for retroreflection.

retroregulation *Molecular Biology.* a method of gene regulation in which the expression of a gene is influenced by a segment of DNA lying downstream of the gene.

retrorocket *Space Technology.* on a space vehicle, a rocket designed to produce thrust opposed to forward motion; used to slow forward motion or to separate a fallaway section or companion body from the remaining body in flight. Also, BRAKING ROCKET.

retrorse *Biology.* bent or directed downward or backward.

retrostalsis *Physiology.* reversed or backward peristalsis in the alimentary canal, sometimes associated with gastric discomforts such as heartburn.

retrosternal *Anatomy.* situated or occurring behind the breastbone.

retrothrust *Space Technology.* thrust in a direction opposite to the forward motion of a rocket or other object.

retrotransposon *Genetics.* a transposable element in which transposition involves a process of reverse transcription with an RNA intermediate similar to that of a retrovirus.

retroversion *Medicine.* a tipping or turning backward of an organ, as of the uterus.

Retroviridae *Virology.* a family of animal viruses having spherical enveloped particles that contain ssRNA, in which the genome replicates through reverse transcription, including the subfamilies Lentivirinae, Oncovirinae, and Spumavirinae; the viruses occur widely among vertebrates, and transmission is either vertical or horizontal.

retrovirus *Virology.* any virus of the family Retroviridae.

retry *Computer Technology.* **1.** to resend the current block of data a specified number of times, or until it is entered correctly or accepted. **2.** to retry an input/output operation that previously generated an error.

retting *Materials Science.* in the processing of natural fibers, a biological process that removes gums and stem tissue from soft fibers; carried out in water, dew, or snow through the action of certain groups of bacteria and fungi. *Chemical Engineering.* a synthetic process that is comparable to this.

return something that goes back to an earlier position or state; specific uses include: *Geophysics.* a surface seismic wave that has traveled from its epicenter to the observation station along the long arc (greater than 180 degrees) and then returned to the station. *Architecture.* the continuation of a molding, cornice, or other member in a different direction, usually at a right angle. *Design Engineering.* an attachment to a desk, usually at right angles, forming a table for a typewriter or computer. *Computer Programming.* **1.** an instruction at the end of a subroutine that causes control to be passed back to the calling program. **2.** to initiate or carry out such an instruction. **3.** see CARRIAGE RETURN. *Electronics.* **1.** see TARGET, def. 1. **2.** see ECHO.

return address *Computer Programming.* an address in a calling program, often transmitted to a subroutine, to which control returns after the subroutine has been executed.

return bead *Building Engineering.* see QUIRK BEAD.

return bend *Mechanical Devices.* a U-shaped pipe fitting that connects two parallel pipes such that fluid or gas flowing through one pipe returns in the opposite direction through the other pipe.

return code *Computer Programming.* any code that has some effect on subsequent instructions.

return connecting rod *Mechanical Engineering.* a connecting rod having its crankpin end on the same side of the crosshead as the engine cylinder.

return difference *Control Systems.* the difference between 1 and the value of a loop transmittance.

return flow *Hydrology.* water used for irrigation that is not lost by evapotranspiration, but is returned either to its source or to another body of ground or surface water. Also, WASTE WATER, RETURN WATER.

return-flow burner *Mechanical Engineering.* a device in a boiler furnace that regulates the amount of oil to be burned through an atomizer by the amount of oil that is recirculated to the storage area.

return idler *Mechanical Engineering.* a rolling or idling component of a conveyor belt system that fits under the cover plates, and upon which the conveyor belt moves when the material it has carried is unloaded.

return interval *Electronics.* see RETRACE INTERVAL.

return key *Computer Technology.* **1.** a key on some keyboards that causes transfer to the processor of a previously entered command or series of commands. **2.** a key that repositions the cursor to the beginning of the next display line.

return line *Electronics.* see RETRACE LINE.

return loss *Telecommunications.* a ratio, expressed in decibels, of the reflected incident upon a discontinuity to the power thrown back from the discontinuity. *Electronics.* the measure of the dissimilarity between two impedances.

return nosing *Building Engineering.* the mitered overhang of a tread in a stair.

return streamer *Geophysics.* the violently luminous final stroke of a cloud-to-ground discharge in which the streamer ascends from the ground to its cloud base. Also, **return stroke.**

return time see RETRACE INTERVAL.

return wall *Building Engineering.* an interior wall of about the same height as the outside wall of a building.

return water *Petroleum Engineering.* the reinjection of salt water recovered with oil into a reservoir during a water flood. *Hydrology.* see RETURN FLOW.

return wire *Electricity.* the wire through which current must pass to complete a circuit.

retuse *Botany.* of a leaf, having a rounded apex that is notched in the center.

retzian *Mineralogy.* any of three rare, weakly radioactive, orthorhombic arsenates of general formula $A_2B(AsO_4)(OH)_4$, where A=Mn^{+2}, Mg and B=Ce, La, Nd; occurring as brown to reddish-brown prismatic crystals, having a specific gravity of 4.15 to 4.49 and a hardness of 3 to 4 on the Mohs scale; found at the Sterling Hill mine, Ogdensburg, New Jersey.

reusable *Computer Programming.* of a routine, able to be used for two or more tasks.

Reuss model *Materials Science.* a model used to predict the anisotropy of crystalline regimes in a polymer; yields an elastic constant.

revascularization *Medicine.* the restoration of blood supply, either naturally (e.g., after a wound) or surgically (e.g., by means of a vascular graft or prosthesis).

reveal *Building Engineering.* a jamb between a window or door frame and the outer surface of a wall.

reveille [rev′ə lē] *Military Science.* a signal, usually a bugle call, used to wake personnel at the beginning of the day; a formation held shortly after the signal. (From a French word meaning "to awaken.")

reveille gun *Military Science.* the firing of a gun at the first note of reveille or at sunrise. Also, MORNING GUN.

reverberant sound field *Acoustics.* the field produced in an enclosed area from sound waves that combine with oncoming and reflected waves.

reverberate *Acoustics.* of sound, to persist after the sound source has ceased.

reverberation *Acoustics.* the persistence of sound after cessation of its source.

reverberation chamber *Acoustics.* a room having very little sound absorption and therefore multiple echoes in all directions, so that if a sound is measured at a particular point it will appear to come equally from all directions.

reverberation time *Acoustics.* a quantitative measure of how quickly sound decays by 60 decibels in a specific environment; given by the equation $R_T = 0.16\ V/(A + \alpha V)$, where R_T is the reverberation time, V is the volume of the enclosure, A is the total absorption, and α represents the absorption coefficient in air.

reverberatory furnace *Metallurgy.* a furnace with a shallow hearth and a roof that deflects the flame and radiates heat toward the hearth or the surface of the charge.

Reverdin's needle *Surgery.* a surgical needle with an eye that may be opened or closed by a movable slide. (Named for Jacques Louis *Reverdin,* 1842–1929, Swiss surgeon.)

reversal film *Graphic Arts.* photographic film (e.g., a transparency) that produces a positive image from a positive subject with one camera exposure.

reversal of dip *Geology.* a local change in the direction of the regional dip near a fault, such that the beds curve in a direction opposite to the drag folds of the fault.

reversal spectrum *Spectroscopy.* the dark-line spectrum that is observed when an intense white light is passed through a luminous gas, and that corresponds to the bright-line emission spectrum of the gas.

reversal speed *Aviation.* the lowest airspeed at which control reversal is evident.

reversal temperature *Spectroscopy.* in the spectroscopic analysis of light from a blackbody source that is passed through a luminous gas, the temperature of the blackbody source associated with the disappearance of a given spectral line appearing as a bright line at lower temperatures and a dark line at higher temperatures.

reverse *Mechanical Engineering.* **1.** to change the running direction of a machine or mechanism. **2.** the mechanism or setting used to accomplish this change. *Graphic Arts.* **1.** a reversed image, especially white letters printed on a black background. **2.** to make such an image.

reverse bias *Electronics.* a voltage applied to a semiconductor P-N junction to make the P side negative with respect to the N side.

reverse billowing *Materials Science.* a method of blow-forming a superplastic alloy, in which the alloy sheet is initially bulged away from the female cavity, and then pressure is reversed to blow the sheet into the die to produce the desired shape.

reverse Brayton cycle *Thermodynamics.* a refrigeration cycle involving these processes: fluid is compressed at constant entropy and then cooled by reversible constant-pressure cooling; the high-pressure fluid expands reversibly in the engine and exhausts at low temperature; the cooled fluid passes through the cold storage chamber and picks up heat at constant pressure, then returns to the suction side of compressor. Also, DENSE-AIR REFRIGERATION CYCLE.

reverse Carnot cycle *Thermodynamics.* an ideal thermodynamic cycle consisting of four processes: an expansion at constant entropy; a constant-temperature expansion; a compression at constant entropy; and a constant-temperature compression. These processes are reversed and occur in the reverse order in the Carnot cycle.

reverse cell *Meteorology.* a circulating fluid system in which the circulation in a vertical plane is thermally indirect; that is, its cooler air rises relative to its warmer air.

reverse circulation drilling *Mining Engineering.* a method of drilling in which a substance (bit coolant, cuttings-removal liquid, drilling fluid, mud, air, or gas) is introduced into a borehole outside the drill rods and forced upward through the drill rods.

reverse code dictionary *Computer Programming.* a listing in alphabetic or alphanumeric code order in which the user can look up the corresponding English words or terms.

reverse current *Electronics.* the direct current that flows when a reverse bias is applied to a semiconductor device.

reverse-current protection *Electricity.* the process of protecting a device from damage caused by current flowing opposite to the normal direction, for example, with a series diode.

reverse-current relay *Electricity.* in a DC circuit, a relay that indicates when current is flowing opposite to the normal direction.

reverse curve *Mathematics.* a curve composed of two arcs joined end to end with their centers on opposite sides of the curve. Also, S-CURVE.

reversed *Geology.* see OVERTURNED.

reversed arc *Geology.* an island arc that is concave toward the open sea.

reversed consequent stream *Hydrology.* a stream whose course follows a direction inconsistent with the slope and configuration of the geologic structure of the area.

reverse deionization *Chemistry.* a process that removes all ions from a solution through the use of an anion-exchange unit followed by a cation-exchange unit.

reversed image *Graphic Arts.* a printed image created by the reverse process. *Optics.* see INVERTED IMAGE.

reverse-direction flow *Computer Programming.* in flowcharting, a flow in a direction other than the standard left-to-right or top-to-bottom direction.

reversed magnetic field *Geophysics.* see REVERSED POLARITY.

reversed polarity *Geophysics.* the remanent polarity of rocks that is opposite to the current polarity of the earth's magnetic field, evidence that the earth experienced a geomagnetic reversal at some point in its history.

reverse drawing *Metallurgy.* a process of drawing in the opposite direction of the previous drawing pass.

reversed stream *Hydrology.* a stream whose direction of flow has been reversed as a result of glacial action, landsliding, capture, or gradual tilting of the area.

reversed tide *Oceanography.* a tide that is out of phase with the apparent motions of the principal attracting body, so that low tide occurs directly beneath that body and is accompanied by another low tide on the opposite side of the earth; the opposite of a direct tide.

reverse electron transport *Biochemistry.* the transport of electrons powered by proton motive force in an "uphill" flow; critical to photosynthesis in bacteria from the *Rhodospirillinea* family.

reverse fault see THRUST FAULT.

reverse feedback see NEGATIVE FEEDBACK.

reverse flange *Engineering.* a projecting flange formed from shrinking rather than stretching.

reverse flow *Transportation Engineering.* a traffic flow along a highway in the direction opposite to that of the primary traffic flow; e.g., outbound traffic during a morning inbound rush hour.

reverse-flowage fold *Geology.* a fold deformation in which flow has thickened the crests of the anticlines and thinned the troughs of the synclines.

reverse graft *Botany.* a technique used to dwarf a plant by inserting the scion in a reversed position to the stock.

reverse gyrase *Enzymology.* an enzyme that catalyzes the winding of the DNA molecule.

reverse key *Electricity.* a key used to reverse the polarity of a circuit.

reverse lay *Design Engineering.* the lay of a wire rope with strands alternating in a left and right pattern.

reverse mutation *Genetics.* a second mutation in a mutant gene that restores the nucleotide sequence to that of the wild type and allows the production of a functional protein. Also, REVERSION.

reverse osmosis *Chemical Engineering.* the application of external pressure to oppose the natural process of osmosis. For example, water would normally move from a region of higher concentration (such as pure fresh water) into one of lower concentration (such as a solution of water and salt). Assuming that a membrane will allow the passage of water molecules but block larger salt molecules, reverse osmosis then would cause water to move out of the salt solution. This process provides a means of desalinizing seawater.

reverse passive anaphylaxis *Immunology.* a process by which an antibody combines with a cell-surface antigen, resulting in an immediate hypersensitive response in an individual.

reverse passive hemagglutination *Virology.* a test for the presence of a virus antigen, in which red blood cells are coated with virus-specific antibody; the cells agglutinate when a virus antigen is present.

reverse-phase chromatography *Analytical Chemistry.* chromatography in which hydrophobic carbon chains exchange with hydrophilic groups in the support matrix in order to identify and separate proteins.

reverse-phase high-performance liquid chromatography *Analytical Chemistry.* a separation technique in which the inert column material is impregnated with a nonpolar liquid (stationary phase) and a polar solvent is used as the mobile phase; used for high-molecular-weight compounds.

reverse-phase partition chromatography *Analytical Chemistry.* a technique of paper chromatography in which a low-polarity material such as a grease is spread on paper and then a high-polarity phase such as an acid or organic solvent is allowed to flow over it to separate the sample.

reverse pinocytosis see EMIOCYTOSIS.

reverse pitch *Mechanical Engineering.* a propeller-blade pitch that creates a thrust in a direction that is opposite to that normally obtained.

reverse polarity *Metallurgy.* in direct-current arc welding, the polarity that makes the electrode positive.

reverse Polish notation *Computer Programming.* a form of Polish notation in which the operators follow the operands; for example, $A + B$ is represented as $AB+$. Also, POSTFIX NOTATION.

reverse-printout typewriter *Engineering.* an electronic typewriter capable of printing both to the right and to the left, so as to avoid a carriage return.

reverse process *Graphic Arts.* in platemaking, reversing the tonal values of copy, usually by making a positive print of a photographic negative, then shooting another negative.

reverse saddle *Geology.* a mineral deposit that follows the bedding plane and is associated with the trough of a synclinal fold. Also, TROUGH REEF.

reverse selection *Molecular Biology.* the process of reversing the trend of a previous selection pressure through artificial selection.

reverse sexual dimorphism *Ecology.* a phenomenon in which mature females are larger than males.

reverse similar fold *Geology.* a fold having strata that are thickened on the limbs and thinned on the axes.
reverse slip fault see THRUST FAULT.
reverse slope *Geology.* a hill that descends away from a ridge. *Military Science.* any slope that descends away from the enemy.
reverse stratification *Anthropology.* the disturbance of a site so that earlier materials lie above later ones; caused by natural processes such as erosion or by activities of later occupants.
reverse transcriptase *Enzymology.* an RNA-dependent DNA polymerase, found in viruses, that catalyzes the synthesis of DNA from deoxyribonucleoside 5'-triphosphates, using RNA as a template.
reverse transcription *Genetics.* the process of synthesizing DNA from an RNA template using reverse transcriptase.
reverse triiodothyronine *Endocrinology.* 3,3',5-triiodothyronine, a thyroid hormone that has very low potency and is apparently produced by 5-monodeiodination of thyroxine during times of calorigenic deprivation, in order to reduce the level of metabolic activity in the body. Also, **reverse T_3.**
reverse video *Computer Technology.* the display of dark characters on a light background on a cathode-ray-tube screen.
reverse voltage *Electricity.* the maximum instantaneous voltage that occurs in a circuit with a polarity opposite that of normal conduction, as with diodes.
reversibility *Science.* the tendency of natural processes to return to equilibrium. *Physical Chemistry.* the fact of being reversible; a situation in which change from a given state does not proceed directly forward to a permanent new state. *Thermodynamics.* the condition of a thermodynamic process that can be reversed and that will restore the system in question to its original state. *Psychology.* the conceptual ability to recognize that certain processes can, if reversed in order, be restored to their original state; e.g., pouring milk from a bottle to a glass and then back again.
reversibility principle *Optics.* the principle that a beam of light reflected directly back onto itself will retrace in the reverse direction the path or paths it traversed before being reflected. *Physics.* see MICROSCOPIC REVERSIBILITY.
reversible *Science.* capable of being reversed; i.e., of going through a series of changes in either direction, forward or backward. *Physical Chemistry.* **1.** relating to a process or change that can be reversed and that, once initiated, immediately thereafter proceeds in the opposite direction as the conditions of the reaction system change. **2.** describing a colloidal system that can be restored to its original state after it has been transformed, as in the restoration of evaporated egg white by the addition of water.
reversible booster *Electricity.* a booster that has the ability to add and subtract from the voltage of a circuit.
reversible change see REVERSIBLE PROCESS.
reversible chemical reaction *Chemistry.* a chemical reaction, usually incomplete, that is capable of proceeding in either direction, by a change in conditions.
reversible counter *Computer Technology.* a device, such as a register, in which the stored value can be incremented or decremented according to a specified control signal.
reversible cycle *Thermodynamics.* an ideal thermodynamic cycle in which both the system in question and all its surroundings are restored to their original state; that is, there is no net heat transfer and no net work.
reversible drill *Mechanical Devices.* an electric drill with a spiral fluted bit that can be operated in either a forward drilling-in motion or a reverse drilling-out motion, by the flip of a switch.
reversible electrode *Physical Chemistry.* an electrode that derives its potential from reversible ionic charges.
reversible engine see REVERSIBLE HEAT ENGINE.
reversible heat engine *Thermodynamics.* an ideal heat engine that is able to operate without the constraint of friction and other resistances.
reversible motor see REVERSING MOTOR.
reversible path *Thermodynamics.* a path that a system follows, in which at any point on the path, the direction of motion can be reversed by any infinitesimal change in the thermodynamic conditions.
reversible-pitch propeller *Aviation.* a propeller having a blade that may be adjusted so that its pitch is reduced to and beyond the zero value, to the negative pitch range.
reversible process *Thermodynamics.* an ideal thermodynamic process that can be exactly repeated in the reverse direction by means of an infinitesimal change in the external conditions; the entropy change for the system and surroundings is zero for such a process.
reversible reaction *Physical Chemistry.* a chemical reaction that reverses its direction as it continues, so that it eventually reaches a point at which forward and reverse reactions occur simultaneously with equal speed; i.e., a state of chemical equilibrium has been reached before all of the reactants have been converted to products.
reversible steering gear *Mechanical Engineering.* a vehicle steering gear in which road shock and irregular wheel movements are transmitted through the system and are felt in the steering control.
reversible transducer *Electronics.* a device that can simultaneously transfer power from one source to another in either direction.
reversible transit circle *Engineering.* a transit circle capable of 180° rotation for evaluation and elimination of instrument errors.
reversing current *Oceanography.* a tidal current that alternately flows in more or less opposite directions, with a slack water period at each reversal of direction.
reversing dune *Geology.* a dune that tends to develop in height, but shows limited migration as a result of seasonal shifts in the dominant wind direction.
reversing layer *Astrophysics.* the cool, lower layer of the solar chromosphere, directly above the photosphere.
reversing mill *Metallurgy.* any mill in which the stock is passed sequentially in opposite directions.
reversing motor *Electricity.* a motor whose direction of rotation can be reversed while running at full speed, for example, by changing electric connections. Also, REVERSIBLE MOTOR.
reversing switch *Electricity.* a switch used to reverse the connections of one part of a circuit.
reversing thermometer *Engineering.* a mercury thermometer that retains a temperature reading until inversion of the thermometer.
reversing water bottle see NANSEN BOTTLE.
reversion *Biology.* a return in some degree to an earlier ancestral type or average condition not exhibited by parents or for several generations, such as a return by an animal from domestication to a feral state; e.g., the razorback hog of North America. *Genetics.* see REVERSE MUTATION. *Materials Science.* the tendency for rubber to become soft and tacky with notable loss of strength after prolonged heating.
revertant *Genetics.* a gene derived by reverse mutation or an organism carrying such a gene.
revet-crag *Geology.* one of a series of narrow, pointed ridges of eroded layers inclined against a mountain spur.
revetment *Archaeology.* a retaining wall that supports an earthwork structure. *Ordnance.* **1.** a wall, faced with concrete, stone, or similar material, that provides protection against explosions. **2.** a wall made of earth or of sandbags that provides protection against bombing or similar attacks. *Civil Engineering.* a protective sheltering for soil or surface bedrock to deter erosion by weather or water.
revitalization *Anthropology.* a social activity with religious overtones that has the goal of restoring past power and infusing new hope to a society; often seen in smaller societies being engulfed and acculturated by a dominant group; for example, the ghost dance and cargo cult movements. Also, **revitalization movement, revitalization activity.**
revival *Geology.* see REJUVENATION, def. 2.
revive *Medicine.* to restore to consciousness or life. *Surgery.* to refresh diseased surfaces in order to promote union. *Chemistry.* to restore or reduce a metal or other substance to a natural or uncombined state.
revived fault scarp see REJUVENATED FAULT SCARP.
revived folding see RECURRENT FOLDING.
revived stream see REJUVENATED STREAM.
revolute *Science.* of a leaf, machine part, and so on, curled or rolled backward at the tip or margins.

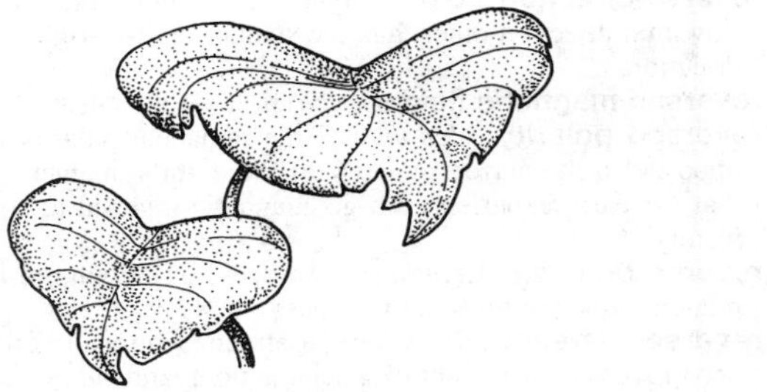

revolute leaves

revolute-coordinate robot see JOINTED-ARM ROBOT.
revolute joint *Design Engineering.* a pipe joint in which the ends of the pipes are curled downward and backward. *Robotics.* a joint based on a pin and having only one degree of freedom.
revolution a movement around; specific uses include: *Mechanics.* **1.** the motion of a body around a closed path, relative to some point that is internal or external to the moving body. **2.** see ROTATION. *Astronomy.* **1.** the orbiting of one celestial body around another. **2.** a single complete course of such an orbit. *Geology.* a period of major crustal deformation when folds and faults are formed.
revolution counter *Mechanical Engineering.* a device that records the number of revolutions of a motor or other revolving machinery. Also, AUTOMATIC COUNTER.
revolutions per minute *Mechanics.* a unit for describing angular velocity, based on one full revolution in one minute; this equals 0.1047198 radian per second.
revolutions per second *Mechanics.* a unit for describing angular velocity, based on one full revolution in one second; this equals 6.283185 radians per second.
revolution table *Naval Architecture.* a table giving the ship speed corresponding to any given propeller rotation rate.
revolve *Mechanics.* to carry on a process of revolution; move in a circular or curving path such as an ellipse. Thus, **revolving.**
revolver *Ordnance.* a handgun with a rotating cylinder that contains several firing chambers, thus allowing repeated fire with successive pulls of the trigger. Thus, **revolver action.** *Mechanics.* any system or object that revolves. *Navigation.* an indeterminate position that results from an attempted three-point fix by horizontal angles when the three observed objects and the vessel are all on or near the circumference of a circle.
revolving-block engine *Mechanical Engineering.* a type of engine in which the piston action is combined with the rotation of the engine block.
revolving door *Building Engineering.* an entrance door that consists of four rigid leaves that revolve around a central pivot.
revolving shovel *Mechanical Engineering.* a mechanical digging apparatus designed so that its base sits on a vertical pivot, enabling it to swing in any direction.
revolving storm *Meteorology.* a cyclonic or similar storm in which the wind revolves about a central low-pressure area.
revulsion *Medicine.* a method of treatment by which blood is drawn from one area to another, as in counterirritation.
reward *Behavior.* **1.** a stimulus, stimulus-object, situation, or verbal statement that is presented upon completion of a successful performance of a task and that tends to increase the probability of the behavior involved. **2.** to make such a presentation. *Artificial Intelligence.* to increase the importance of a rule or network connection that helped make a correct decision, often by increasing its numerical weight.
reward training *Behavior.* a method of learning in which correct responses are rewarded.
rewind *Electronics.* **1.** a control on a tape recorder or a videocassette recorder that returns the tape to its starting position (or any earlier position), usually at a high speed. **2.** to move a tape backward to its starting point or any earlier position.
rewire *Building Engineering.* to install new wiring.
reworked *Geology.* describing any geologic material, such as a sediment or rock fragment, that has been transported naturally from its place of origin and has been incorporated into a younger deposit.
rewrite *Computer Programming.* **1.** to replace an expression by an equivalent expression; erase and reset. **2.** to return data to an area of storage after reading from that location. **3.** to restore a storage device to its state prior to reading.
rewrite rule *Computer Programming.* a rule that describes a transformation of an expression, e.g., $-(x-y) \Rightarrow (y-x)$.
Reye's syndrome [rāz] *Medicine.* an acute, often fatal illness of children characterized by encephalopathy, hepatitis, fatty accumulation in the liver and renal tubules, fever, vomiting, and loss of consciousness progressing to coma and convulsions. (First described by Ralph Douglas Kenneth *Reye,* 1912–1978, Australian pathologist.)
reyn *Fluid Mechanics.* a unit of dynamic viscosity; 1 poise = 1 dyne sec/cm^2 = 0.0000145 reyn. (Named in honor of the physicist Osborne Reynolds.)
Reynier's isolator *Engineering.* a steel enclosure for providing a germ-free environment for animals; the isolator includes electricity, purified air exchange, glove box, and viewing port.
Reynolds, Osborne 1842–1912, British physicist; in hydrodynamics, discovered Reynolds number and Reynolds stresses.
Reynolds analogy *Chemical Engineering.* a relationship in turbulent heat or mass transfer, showing the similarity between the mechanisms for the transfer of energy, momentum, and mass.
Reynolds criterion *Fluid Mechanics.* a Reynolds number above which a fluid flow is considered to be turbulent, and below which the flow is considered to be laminar for a given flow condition in a pipe, or over a flat plate; the Reynolds criterion for fluid flowing in a pipe is about 2300.
Reynolds effect *Meteorology.* a process of drop growth in clouds that involves the net evaporation from cloud drops warmer than others and net condensation on the cooler drops.
Reynolds equation *Fluid Mechanics.* an equation composed of the characteristic parameters of a particular flow field used to calculate a Reynolds number for that flow field.
Reynolds model *Oceanography.* a hydrodynamic laboratory model of ocean currents that is concerned especially with inertial and frictional forces, and that is based mainly on calculations involving the Reynolds number.
Reynolds number *Fluid Mechanics.* a dimensionless parameter that expresses the ratio of inertia forces to viscous forces in a flow field, by which the state of the flow is determined; a high Reynolds number implies that the flow is turbulent, while a low Reynolds number implies that the flow is laminar.
Reynolds stress *Fluid Mechanics.* an apparent stress that must be added to the stress caused by the mean velocity gradients to make up for the exchange of momentum between adjacent layers of fluid, due to velocity fluctuations in turbulent flow.
Reynolds stress tensor *Fluid Mechanics.* a tensor that is associated with Reynolds stresses; tensors involve magnitude, direction, and orientation.
rezbanyite *Mineralogy.* $Pb_3Cu_2Bi_{10}S_{19}$, a metallic, gray orthorhombic mineral of questionable status, having a specific gravity of 6.24 and a hardness of 2.5 on the Mohs scale, and occurring in massive to granular form; may be a mixture.
RF rimfire.
RF or **rf** radio frequency; releasing factor.
Rf the chemical symbol for rutherfordium.
rf rangefinder; reducing flame; rapid-fire.
R factor *Genetics.* a bacterial plasmid that confers resistance to one or more antibiotics to the host bacterial cell. Also, R PLASMID, RESIDUAL R.
R-factor *Crystallography.* see DISCREPANCY INDEX.
ρ factor see RHO FACTOR.
RFI radio-frequency interference.
RF power supply radio-frequency power supply.
RF reactor radio-frequency reactor.
RF resistance see HIGH-FREQUENCY RESISTANCE.
RF signal generator radio-frequency signal generator.
RF SQUID *Electronics.* see RADIO-FREQUENCY SQUID.
RG *Aviation.* the airline code for Varig, S.A.
R galaxy *Astronomy.* any rotationally symmetrical galaxy lacking spiral or elliptical structure; formerly called D galaxy.
RGB red-green-blue.
R group *Biochemistry.* the characteristic by which the 20 amino acids present in proteins differ from each other; these side chains vary in size, structure, electric charge, and solubility in water.
RGU system *Astronomy.* the standardized set of red, green, and ultraviolet filters used with a photometer to characterize a star's spectrum.
Rh *Genetics.* see RH FACTOR.
Rh the chemical symbol for rhodium.
r.h. or **RH** relative humidity.
Rhabdiasoidea *Invertebrate Zoology.* a superfamily of parasitic nematode roundworms with smooth cuticles and no pharyngeal bulb.
Rhabdinopora *Paleontology.* a genus of planktic graptolites in the family Anisograptidae; extant in the Lower Ordovician.
rhabdion *Invertebrate Zoology.* any of the sclerotized segments lining the mouth cavity in nematode roundworms.
rhabdite *Invertebrate Zoology.* **1.** in many turbellarian flatworm species, one of the small crystalline rods secreted by cells in the epidermis. **2.** in some insects, the pair of appendages that unite to form the sharp-edged ovipositor or sting. *Mineralogy.* see SCHREIBERSITE.
Rhabditia *Invertebrate Zoology.* a subclass of nematode roundworms belonging to the class Phasmidea, having a mouth stylet and a tripartite esophagus.

Rhabditidia *Invertebrate Zoology.* an order of nematode roundworms belonging to the subclass Rhabditia, parasites of humans and domestic animals.

rhabditiform *Invertebrate Zoology.* a rodlike body resembling rhabdite, found singly or in bundles within or between the epidermal cells of turbellarian flatworms.

rhabditiform larva *Invertebrate Zoology.* a free-living larval stage of some parasitic and many soil-dwelling nematodes, having a central and a posterior bulb in the esophagus.

Rhabditoidea *Invertebrate Zoology.* a superfamily of parasitic nematode roundworms in the order Rhabditidia.

Rhabdocoela *Invertebrate Zoology.* a former order of aquatic turbellarian flatworms, now divided into three orders: Catenulida, Macrostomida, and Neorhabdocoela.

rhabdocyst *Cell Biology.* a type of extrusome that occurs in certain ciliated organisms.

Rhabdodendraceae *Botany.* a monogeneric, dicotyledonous family of tropical South American shrubs of the order Rosales, often having unusual secondary growth and parenchyma with resin-filled scattered secretory cavities.

rhabdolith *Botany.* a spine or projecting formation that is found on coccoliths, the fossilized remains of golden-yellow algae within the order Chrysophyta.

rhabdom *Invertebrate Zoology.* a group of four, minute, pigmented, rodlike structures (rhabdomeres) in the center of the retinula of a compound eye of arthropods. Also, **rhabdome.**

rhabdomere *Invertebrate Zoology.* one of the rods making up a rhabdom in the compound eye of an arthropod.

Rhabdomonadales *Botany.* an order of euglenoids with a single emergent flagellum and rigid, colorless cells.

rhabdomyoma *Medicine.* a benign tumor that is derived from striated muscle.

rhabdomyosarcoma *Oncology.* a highly malignant tumor composed of striated muscle derived from primitive cells and occurring in three forms: **pleomorphic rhabdomyosarcoma** generally affects the extremities of adults; **alveolar rhabdomyosarcoma** affects the extremities, trunk, and orbital region of adolescents; **embryonal rhabdomyosarcoma** affects the head, neck, lower genitourinary tract, and extremities of infants and children.

Rhabdoniaceae *Botany.* a family of red algae belonging to the order Gigartinales, characterized by erect, multibranched thalli of uniaxial construction, and by similar gametangial and sporangial forms; some species are used as food in the Orient.

rhabdophane *Mineralogy.* **1.** any of three weakly radioactive, brown, pink, or yellowish-white hexagonal minerals of the rhabdophane group having the general formula $A(PO_4)\cdot H_2O$, where A = Ce, La, Nd, occurring as aggregates of acicular crystals, and having a specific gravity of 3.94 to 4.0 and a hardness of 3.5 on the Mohs scale. **2.** a mineral group name.

Rhabdophorina *Invertebrate Zoology.* a suborder of aquatic ciliated protozoans in the order Gymnostomatida that paralyze and ingest other protozoans.

rhabdosarcoma *Oncology.* a sarcoma that is composed of muscle fiber tissue.

rhabdosome *Paleontology.* a colonial graptolite that develops from a single individual.

Rhabdosphaeraceae *Botany.* a family of flagellate marine algae belonging to the order Coccosphaerales, distinguished by the presence of rhabdoliths.

Rhabdoviridae *Virology.* a large family of enveloped, bullet-shaped, ssRNA-containing viruses with a wide host range, including vertebrates, invertebrates, arthropods, and plants, and an equally wide range of vectors.

rhabdovirus *Virology.* one of the Rhabdoviridae family of viruses.

Rhabdoweisiaceae *Botany.* a family of small, erect mosses of the order Dicranales that form rich green cushions on acidic rocks and in moist alpine crevices; characterized by terminal sporophytes, numerous branches, and leaves that crisp when dry.

rhabdus *Invertebrate Zoology.* a simple sponge spicule with a single axis.

rhachi- or **rhachio-** see RACHI-, RACHIO-.

Rhachitheciaceae *Botany.* a family of small epiphytic mosses of the order Orthotrichales, characterized by erect stems with terminal sporophytes and irregular branching, and by blunt, oblong leaves with a single costa.

Rhachitomi *Paleontology.* a seldom-used grouping of Late Paleozoic labyrinthodont amphibians in the order Temnospondyli; most authorities now prefer to subdivide the Temnospondyli into a dozen superfamilies; the most important of these are the Eryopoidea, Trematosauroidea, and Plagiosauroidea.

rhachitomous *Vertebrate Zoology.* relating to or having vertebrae with centra consisting of separate, unfused bony elements.

Rhacophoridae *Vertebrate Zoology.* the Old World tree frogs, a family of amphibians belonging to the suborder Neobatrachia, characterized by distinct disks on the digits; they are found in Asia, Africa, Indonesia, and Australia.

Rhacopilaceae *Botany.* a family of mosses in which only dwarf males are known.

Rhadinovirus *Virology.* a genus of viruses of the family Herpesviridae, subfamily Gammaherpesvirinae.

Rhaetian *Geology.* a European geologic stage of the Uppermost Triassic period, occurring after the Norian and before the Hettangian of the Jurassic period. Also, **Rhaetic.**

Rhagionidae *Invertebrate Zoology.* the snipe flies, a family of dipteran insects that are usually brownish or gray with spotted wings; most larvae are predatory, and some adults bite humans.

rhagon *Invertebrate Zoology.* a sponge with a large central cavity surrounded by flagellated chambers.

Rhamnaceae *Botany.* a large family of dicotyledonous, primarily tropical trees and shrubs belonging to the order Rhamnales, characterized by simple, pinnately veined leaves, free stamens, flowers with stamens oppposite the petals, and a drupe type of fruit; the family includes many ornamental plants, such as the lilac.

Rhamnales *Botany.* an order of dicotyledonous plants of the subclass Rosidae, characterized by a single set of stamens opposite the petals, a generally well-developed intrastamenal disk, and two or more locules in the ovary; the order includes three families: Rhamnaceae, Leeaceae, and Vitaceae.

rhamnose *Organic Chemistry.* $C_6H_{12}O_5$, white needles that are very soluble in water and soluble in absolute alcohol; the β-anomer form melts at 122°C; it is the most common, naturally occurring deoxysugar, and is used in synthetic sweetener research. *Biochemistry.* this deoxysugar in its natural form, occurring in poison sumac and in glycoside combination in flower pigments.

rhamphoid *Biology.* shaped like a beak.

Rhamphorhynchoidea *Paleontology.* a suborder of flying reptiles in the extinct subclass Archosauria and order Pterosauria; the earliest known pterosaurs; extant in the Triassic and Jurassic.

Rhamphorhynchus *Paleontology.* a genus of fish-eating pterosaurs in the suborder Rhamphorhynchoidea, having a wing span of about 3 feet; characterized by a very long tail with a steering fin at the tip and an extremely elongated finger on each side, from which their leathery wings were suspended; known from well-preserved specimens of the Upper Jurassic from Germany and Africa.

rhamphotheca *Vertebrate Zoology.* the horny covering of a bird's beak.

rhaphania see RAPHANIA.

Rhaphidiophyta see CHLOROMONADIDA.

rhapidosomes *Microbiology.* rod-shaped or cylindrical structures that have been found in a number of bacterial species; variously considered as antimicrobial agents, bacteriophage tails, membranous particles, or locomotory apparatuses.

Rhazes 865–925, Arabic physician; wrote the *Liber continens,* a compendium and expansion of classical medical knowledge.

Rh blocking serum *Immunology.* a type of blood serum that combines with Rh-positive blood and prevents the activity of anti-Rh sera that may be introduced later.

Rh blocking test *Immunology.* a method used to detect the presence of Rh antibodies in plasma.

rhe *Fluid Mechanics.* in the centimeter-gram-second system, the unit of measure for fluidity (the reciprocal of velocity); rhe = 1/poise. (From the Greek word for "stream.")

Rhea [rē´ə] *Astronomy.* one of the moons of Saturn, discovered in 1672; it is 1530 kilometers in diameter and orbits Saturn once every 4.52 days. (Named for the Greek mythological figure *Rhea,* a Titan and the sister and wife of Cronus.) Also, SATURN V.

rhea [rē´ə] *Botany.* **1.** *Boehmeria nivea,* a shrub belonging to the nettle family, Urticaceae, that is cultivated for its flaxlike fiber used in high-quality papers and fabrics. **2.** the fiber of this plant. Also, CHINA GRASS, RAMIE.

rhea *Vertebrate Zoology.* a large, flightless bird belonging to the order Rheiformes, characterized by long, powerful legs with three toes and plumose feathers over most of the body; the heaviest of all New World birds; found in South America.

rhea

rhebok *Vertebrate Zoology.* a large, deerlike antelope, *Pelea capreolus,* having pale-gray, curly fur and straight horns; found in South Africa.

rhegmagenesis see REGMAGENESIS.

rhegolith see REGOLITH.

rheid *Geology.* **1.** a substance showing flow structure. **2.** of or relating to such a substance.

Rheidae *Vertebrate Zoology.* the family of large flightless rheas, composing the order Rheiformes and native to South America.

rheid fold *Geology.* a fold whose layers have deformed by flow as if they were fluid. Also, RHEOMORPHIC FOLD.

rheidity *Geology.* the capacity of material to flow within the earth.

Rheiformes *Vertebrate Zoology.* the order of large flightless birds containing the single family Rheidae.

Rheinberg illumination *Optics.* a technique in which colored light is projected through a specimen by means of color filters fitted over the condenser; this is a modification of the dark-field method. Also, OPTICAL STAINING.

Rheinmetall *Ordnance.* one of several German 20-mm and 30-mm automatic cannons employed as aircraft, antiaircraft, and antitank weapons during World War II.

Rheintochter *Ordnance.* a German antiaircraft rocket used during World War II.

Rhenanida *Paleontology.* an order of skatelike placoderms that consists of only a few genera, notably *Gemuendina*; characterized by a broad head, terminal mouth with pointed teeth, and sharply tapering body; extant in the Devonian; the rhenanids were worldwide and have been thoroughly studied.

rhenium [rē´nē əm] *Chemistry.* a metallic element that has the symbol Re, the atomic number 75, an atomic weight of 186.2, a melting point of 3180°C (one of the highest of any element), and a boiling point of about 5400°C; it is a silvery-white solid or gray to black powder, with the widest range of valences of any element; it is derived from molybdenum or copper ores and is used in tungsten and molybdenum alloys and in electrical components and filaments. (From the Latin name for the *Rhine* river of Germany; so named by its discoverers, the German scientists Walter Nodack, Ida Tacke, and Otto Berg.)

rhenium halide *Inorganic Chemistry.* any compound consisting of the element rhenium with an element from the halogen group (periodic table Group VIIa), such as rhenium pentachloride, $ReCl_5$.

rhenium heptasulfide *Inorganic Chemistry.* Re_2S_7, a brownish-black powder; insoluble in water and ignites on heating in air; decomposed by heat; used as a catalyst.

rheo- a combining form meaning "flow."

rheobase *Physiology.* the minimum electrical current strength needed to excite a given tissue when the stimulus duration is very long.

rheocasting *Metallurgy.* casting a metallic alloy that is partially melted and partially solid.

rheoelectroencephalograph see ELECTROENCEPHALOGRAPH.

rheogoniometry *Mechanics.* rheological tests performed to determine the various stress and shear actions on both Newtonian and non-Newtonian fluids.

rheoignimbrite *Geology.* a rock on the slope of a volcanic crater that was formed by deposition and consolidation of ash flows and volcanic gaseous materials, and that developed secondary flowage as a result of high temperatures.

rheological *Mechanics.* of or relating to the science of rheology.

rheology *Mechanics.* the science of the flow and deformation of matter, generally organic and inorganic liquids with suspended particles.

rheomorphic *Petrology.* relating to or characterized by a process of rheomorphism.

rheomorphic fold see RHEID FOLD.

rheomorphic intrusion *Petrology.* the injection of mobilized or rheomorphic country rock into the igneous intrusion that caused the rheomorphism; the intrusion often resembles the metamorphosed country rock.

rheomorphism *Petrology.* the mobilization of a rock by partial or complete fusion, accompanied and sometimes promoted by the addition of new material due to diffusion.

rheooptical properties *Materials Science.* the flow-dependent optical properties of polymers subjected to stresses; these properties are studied using birefringence, infrared dichroism, Raman and fluorescence polarization, and light and neutron scattering.

rheopectic *Materials Science.* a phenomenon in which viscosity increases with time under a constantly applied stress.

rheopectic fluid *Fluid Mechanics.* a fluid that shows an increase in apparent viscosity with time.

rheopexy *Physical Chemistry.* a property that causes certain slow-gelling, thixotropic sols to gel more quickly when the containing vessel is shaken gently.

rheophile *Ecology.* an organism that flourishes in free-flowing waterways.

rheophilous *Ecology.* living or thriving in free-flowing water. Thus, **rheophile, rheophilic, rheophily.**

rheophilous bog *Ecology.* an area of damp soil made up of organic debris and water that drains in from another source.

rheophobous *Ecology.* not tolerant of free-flowing water. Also, **rheophobic.**

rheophyte *Ecology.* a plant that lives or thrives in free-flowing water.

rheoplankton *Ecology.* plankton living in flowing water.

rheoreceptor *Invertebrate Zoology.* a specialized sensory cell of fishes and aquatic amphibians that detects the movement or flow of air or water past the animal's body. Also, **rheotactic receptors.**

rheostat *Electricity.* a variable resistor having a resistance element and a continuously adjustable contact arm.

rheostatic braking *Engineering.* the process of stopping the rotation of a motor by using the kinetic energy of the motor and load to generate electrical energy for subsequent dissipation in a braking rheostat.

rheostatic control *Electricity.* the process of controlling resistance in a circuit containing a rheostat or potentiometer.

rheostriction see PINCH EFFECT.

rheotaxial growth *Engineering.* a process used in the formation of solid-state devices in which a chemical vapor is deposited on a liquid layer with high carrier mobility.

rheotaxis *Biology.* a directed movement response of a motile organism to a water or air current, either into the current (positive rheotaxis) or with the current (negative rheotaxis).

rheotron see BETATRON.

rheotropic brittleness *Metallurgy.* a type of ductility impairment that may be alleviated by mild prestressing.

rheotropism *Biology.* an orientation response of an organism to a water or air current.

rhesus see RHESUS MONKEY.

rhesus blood group system *Immunology.* a group of red blood cell antigens that were originally defined by their reaction to rabbit or guinea pig serum that had been injected with the immunized blood of rhesus monkeys; the group now includes red blood cell antigens that react to certain human antisera.

rhesus factor see RH FACTOR.

rhesus monkey [rē´səs] *Vertebrate Zoology.* a small, yellowish-brown macaque, *Macaca mulatta*, of the monkey family Cercopithecidae, having a short tail; noted for its active, playful disposition; it is often used in biological research; native to India. Also, **rhesus macaque**.

rhesus monkey

Rheticus, Georg Joachim 1514–1576, German astronomer and mathematician; assisted Copernicus; wrote trigonometric tables.

rheum *Medicine.* a thin, watery discharge of the mucus membranes, especially during a cold.

rheumatic [roo mat´ik] *Medicine.* relating to or affected by rheumatism.

rheumatic arteritis *Medicine.* a generalized inflammation of arteries and arterioles due to rheumatic fever; it may lead to fibrosis and constriction of the arteries.

rheumatic carditis *Medicine.* an inflammation of the heart occurring in connection with rheumatic fever; it may lead to scarring of the valves with stenosis.

rheumatic encephalopathy *Medicine.* an inflammatory condition of the brain associated with rheumatic fever.

rheumatic endocarditis *Medicine.* an inflammation of the endocardium or lining membrane of the heart occurring in rheumatic fever, with special involvement of the valves.

rheumatic fever *Medicine.* the very sudden onset of a disease that follows an infection with type A streptococci; marked by abdominal and joint pain, fever, various types of cardiac involvement, and skin disorders; it usually occurs in childhood and may cause permanent heart damage. Also, ACUTE RHEUMATIC FEVER.

rheumatic heart disease *Medicine.* damage to the heart muscle or heart valves caused by rheumatic fever.

rheumatic pneumonia *Medicine.* an inflammation of the lungs in acute rheumatic fever.

rheumatic valvulitis *Cardiology.* inflammation of a heart valve caused by rheumatic fever.

rheumatism [roo´mə tiz´əm] *Medicine.* a general term for any of various disorders marked by inflammation, degeneration, or metabolic derangement of the connective tissues, including the muscles, tendons, joints, bones, and nerves. The symptoms include pain, stiffness, and limitation of movement.

rheumatoid [roo´mə toid] *Medicine.* **1.** relating to or affected by rheumatism. **2.** resembling rheumatism.

rheumatoid arthritis *Medicine.* a chronic systemic disease of unknown cause, affecting the connective tissue; it is especially common in women and is characterized by pain, stiffness, and joint deformity. Also, ARTHRITIS OBLITERANS.

rheumatoid nodules *Medicine.* the subcutaneous knobs or swellings, usually occurring over bony prominences or joints, seen in patients with rheumatoid arthritis.

rheumatoid spondylitis *Medicine.* a chronic arthritis of unknown cause, usually occurring in young men, affecting the vertebrae and sacroiliac joints, progressively leading to fusion and deformity.

rhexis *Pathology.* the rupture of a blood vessel, organ, or cell.

rhexistasy *Geology.* the mechanical breaking up and subsequent transport of surface residual materials, such as old soils.

Rh factor *Genetics.* an antigen carried on the erythrocytes of humans, labeled Rh^+ (Rh positive), who possess an autosomal dominant allele on chromosome 1 that codes for this antigen; those who are homozygous recessive, labeled Rh^- (Rh negative), lack the antigen and produce antibodies that destroy Rh^+ erythrocytes. An Rh^- woman who is pregnant with an Rh^+ fetus may produce sufficient antibodies to decimate the fetus' blood cells and precipitate a spontaneous abortion; an anti-Rh gamma globulin can be administered to the pregnant woman to prevent Rh hemolytic disease. Also, RHESUS FACTOR. (So named because it was first discovered in rhesus monkeys.)

RHI range height indicator.

rhigolene *Chemistry.* a volatile petroleum distillate, formerly used to produce local anesthesia by freezing.

rhin- or **rhino-** a combining form meaning "nose."

rhinal *Anatomy.* relating to the nose.

Rhincodontidae *Vertebrate Zoology.* an alternate name for the circumtropical shark family Rhiniodontidae, which contains the whale shark.

Rhine *Geography.* a river that flows northward for approximately 850 miles from southeastern Switzerland through western Germany and The Netherlands to the North Sea.

rhinencephalon *Anatomy.* a collective term for the parts of the brain formerly believed to be involved with olfaction, including the olfactory bulbs, olfactory tracts and striae, and the hippocampus and amygdala in the limbic system. *Vertebrate Zoology.* see OLFACTORY LOBE.

rhinestone *Materials.* an artificial gem composed of paste and often cut to resemble a diamond or other gemstone.

Rhine wine *Food Technology.* any wine from the Rhine Valley of western Germany, such as Kabinett, Spätlese, and Liebfraumilch.

rhinitis *Medicine.* an inflammation of the mucous membrane of the nasal passages with profuse discharge of mucus; common examples occur as head colds and seasonal allergies.

Rhinobatidae *Vertebrate Zoology.* a family of small, circumtropical sharklike fishes of the order Rhinobatiformes, including the guitarfish and fiddlerfish.

Rhinobatiformes *Vertebrate Zoology.* an order of marine tropical and warm temperate fish of the superorder Batoidea; found worldwide, usually in shallow waters.

rhinoceros *Vertebrate Zoology.* any of five species of large, herbivorous, thick-skinned mammals of the family Rhinocerotidae, characterized by one or two medial, conical horns composed entirely of agglutinated horny fibers with no internal bony support; the three Asian species belong to the genus *Rhinoceros,* and the two African species belong to the genus *Diceros.* All five species are endangered.

rhinoceros

rhinoceros beetle *Invertebrate Zoology.* any of several large scarab beetles, especially of the genus *Dynastes,* characterized by a male that has one or more prominent horns on the head and prothorax; found in tropical regions.

Rhinocerotidae *Vertebrate Zoology.* a mammalian family of the order Perissodactyla, characterized by a large body with short, stocky legs, and a large head with one or two fibrous conical, boneless horns; the family includes the modern rhinoceroses and more than two dozen extinct genera, including the largest known land mammal, *Baluchitherium,* which stood 15 feet high at the shoulder; extant in the Eocene to the present.

rhinocheiloplasty *Surgery.* plastic reconstructive surgery of the nose and lips.

Rhinochimaeridae *Vertebrate Zoology.* a family of small to medium-sized, deep-sea, cartilaginous fish of the order Holocephalii, found in the Atlantic, Pacific, and Indian Oceans.

Rhinocryptidae *Vertebrate Zoology.* the tapaculos, a family of small birds of the passerine suborder Furnarii, characterized by short, rounded wings, a compact, plump body, and poor flying but excellent running skills; found in neotropical forests and grasslands of South and Central America.

rhinogenous *Medicine.* originating in the nose.

rhinolaryngology *Medicine.* the branch of medical science that deals with the nose and the larynx and their diseases.

rhinolaryngoscope *Medicine.* a tubular instrument with combined light and telescopic systems; used in examining the interior of the nose and larynx.

rhinological *Medicine.* of or relating to the science of rhinology.

rhinology *Medicine.* the branch of medical science that deals with the nose and its diseases.

Rhinolophidae *Vertebrate Zoology.* the Old World leaf-nosed or horseshoe bats of the suborder Microchiroptera, characterized by a complex, fleshy muzzle that sometimes forms horseshoelike ridges; found in Africa, Asia, Australia, and Europe.

rhinopharyngitis *Medicine.* an inflammation of the mucous membrane of the nose and pharynx.

rhinophore *Invertebrate Zoology.* the modified cephalic tentacles in some mollusks in the subclass Opistobranchia, such as sea hares and nudibranchs; thought to serve in chemoreception.

rhinophycomycosis *Pathology.* a fungal infection caused by *Entomorphora coronata* and marked by the growth of copious polyps in the subcutaneous tissues of the nose and nasal sinuses; in chronic cases blindness may occur, followed by cerebral interference.

rhinoplastic *Surgery.* relating to or involved with plastic surgery or reconstruction of the nose.

rhinoplasty *Surgery.* plastic surgery or reconstruction of the nose.

Rhinopomatidae *Vertebrate Zoology.* the mouse-tailed bats, a family of insectivorous bats of the suborder Microchiroptera, characterized by a long, slender tail; native to northern and central Africa and south central Asia.

Rhinopteridae *Vertebrate Zoology.* a monogeneric family of cartilaginous bat rays of the order Myliobatiformes; found in shallow tropical and temperate waters worldwide.

rhinorrhea *Medicine.* a profuse watery discharge from the nose. Also, **rhinorroea.**

rhinoscleroma *Medicine.* a chronic infectious disease of the nose and adjacent areas marked by the growth of nodules; caused by *Klebsiella rhinoscleromatis*; occurs in Egypt, Eastern Europe, and Central and South America.

rhinoscope *Medicine.* an instrument for examining the interior of the nose. Also, NASOSCOPE.

rhinoscopy *Medicine.* the examination of the interior of the nose.

rhinosporidiosis *Pathology.* a chronic, localized fungal infection of the mucocutaneous tissues caused by *Rhinosporidium seeberi*; characterized by the development of polyps, tumors, papillomas, or wartlike lesions; endemic in parts of East Asia, also found in many temperate and tropical regions worldwide.

Rhinosporidium *Mycology.* a genus of fungi whose taxonomic affiliation is still undecided, but is frequently classified as belonging to the class Chytridiomycetes; it contains the species *R. seeberi*, which causes rhinosporidiosis.

Rhinotermitidae *Invertebrate Zoology.* a family of subterranean and damp-wood termites in the order Isoptera, having a fontanelle (a depressed pale spot between the eyes).

rhinotheca *Vertebrate Zoology.* the horny covering of the upper part of a bird's beak.

Rhinovirus *Virology.* a genus of viruses of the family Picornaviridae that cause upper respiratory infections in humans and other mammals; the common cold viruses.

rhipicephalus *Invertebrate Zoology.* a large and widely distributed genus of ticks that are parasitic on many mammals and some birds; vectors for several serious diseases including Rocky Mountain spotted fever.

Rhipiceridae *Invertebrate Zoology.* a family of rare cedar beetles, coleopteran insects in the superfamily Elateroidea.

Rhipidiaceae *Mycology.* a family of fungi belonging to the order Leptomitales, which is composed of water molds that live on fruits or twigs in warm to tropical climates.

Rhipidistia *Paleontology.* a suborder of extinct, lobe-finned fishes.

rhipidistian *Paleontology.* a member of the suborder Rhipidistia in the extinct order Crossopterygii, early fishes including the Holoptychidae and Osteolepidae; they are ancestral to vertebrates and were characterized by the union of the basal bones of the medial fins; extant from the Devonian to the Permian.

rhipidium *Botany.* of or relating to a fan-shaped cyme in which all the branches develop in one plane.

Rhipiphoridae *Invertebrate Zoology.* a family of beetles, coleopteran insects in the superfamily Meloida; larvae parasitize wasps, bees, and cockroaches and undergo hypermetamorphosis.

rhiz- or **rhizo-** a combining form meaning "root."

rhizanthous *Botany.* of or relating to a plant that flowers or appears to flower from the root.

rhizautoicous *Botany.* in mosses, of or related to the connection of an antherial branch to an archegonial branch by rhizoids.

rhizic water see SOIL WATER.

Rhizidiaceae *Mycology.* a family of fungi that belong to the order Chytridiales, composed of parasites of freshwater algae and soil fungi.

Rhizidiomyces *Mycology.* a genus of fungi belonging to the class Hyphochytriomycetes, composed of unicellular organisms found in soil and in water as parasites of certain algae and water molds.

Rhizidiomycetaceae *Mycology.* a family of fungi belonging to the order Hypochytriales; they are parasites on aquatic fungi and nonliving organic matter.

Rhizidiovirus group *Virology.* a genus of fungal viruses that primarily infect developing sporangia through vertical transmission.

rhizine *Mycology.* the rootlike hyphal strands that anchor a lichen to its substrate and also function in absorption.

Rhizobiaceae *Bacteriology.* a family of Gram-negative, rod-shaped, motile, aerobic bacteria without endospores, occurring in the roots of certain plants.

Rhizobium *Bacteriology.* a genus of Gram-negative bacteria belonging to the family Rhizobiaceae, occurring as aerobic, nitrogen-fixing, rod-shaped cells that are symbiotically associated with certain leguminous plants.

rhizocarpous *Botany.* of or relating to plants that have perennial root systems but whose above-ground vegetative portion dies back annually. Also, **rhizocarpic.**

Rhizocephala *Invertebrate Zoology.* an order of highly modified barnacles in the class Cirripedia that parasitize decapod crustaceans such as crabs and hermit crabs; the adults have saclike bodies without appendages, digestive tract, or shell, but with a rootlike structure that grows into the host's tissues.

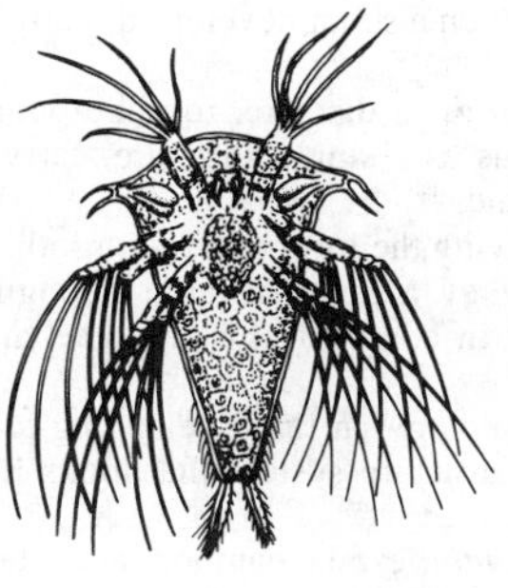

Rhizocephala

rhizocephalous *Invertebrate Zoology.* relating to the Rhizocephala.

Rhizochloridaceae *Botany.* a monogeneric family of yellow-green algae of the order Rhizochloridales, characterized by free-living, solitary, amoeboid cells.

Rhizochloridales *Botany.* an order of yellow-green algae of the class Xanthophyceae, characterized by naked, free-living or attached ameboid cells without flagella; found in freshwater, brackish, and marine environments.

Rhizoclonium *Botany.* a genus of green algae belonging to the family Cladophoraceae, characterized by stratified walls and short, rhizoidal branchlets.

rhizoconcretion *Geology.* a mass that has formed around the root of a living plant. Also, ROOT CAST.

rhizoctol see METHYLARSINIC SULFIDE.

Rhizoctonia *Mycology.* a genus of imperfect fungi belonging to the order Agonomycetales, characterized by a lack of conidia.

rhizodermis see EPIBLEM.

Rhizodontidae *Paleontology.* a family of fishes belonging to the extinct superfamily Osteolepidoidea; they were extant in the Devonian and Carboniferous.

rhizogenic *Botany.* of or relating to a plant or plant part that produces roots. Also, RHIZOGENOUS.

rhizogenous see RHIZOGENIC.

Rhizogoniaceae *Botany.* a family of lustrous mosses of the order Bryales that form tufts or mats in rock crevices and on tree bases, rotting logs, and humus; characterized by tomentose stems, sporophytes that are lateral or near the base of the stems, and a well-developed bryoid peristome; found in tropical and subtropical regions of the Southern Hemisphere.

rhizoid [rī´zoid´] *Botany.* **1.** a rootlike structure in nonvascular plants, such as mosses, liverworts, and fern prothalli, that lacks the vascular tissue of a true root but provides similar functions such as attachment to a substrate and absorption. **2.** having the shape or form of a root; rootlike. Also, **rhizoidal.**

Rhizomastigida *Invertebrate Zoology.* an order of chiefly freshwater, microscopic, ameboid protozoans in the class Zoomastigophorea, having one or two flagella. Also, **Rhizomastigina.**

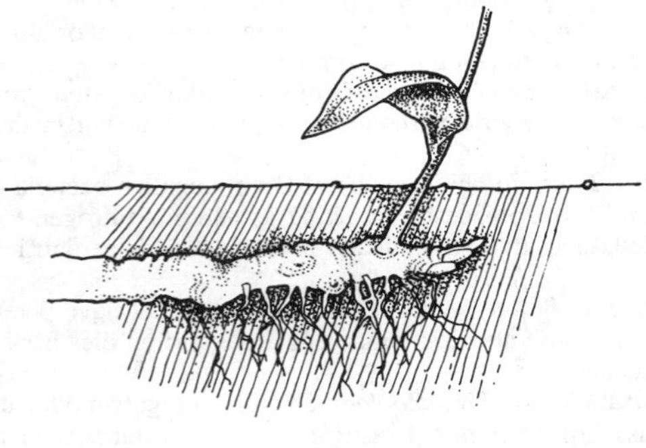

rhizome

rhizome [rī´zōm´] *Botany.* an underground, rootlike stem, usually growing horizontally, that bears buds in the axils of highly reduced leaves and that provides a mechanism for asexual reproduction through a process of cloning in which a shoot develops directly from the underground portion.

rhizomorph *Mycology.* a discrete, rootlike bundle of somatic fungal hyphae that functions as a single structure, carrying the fungus across nonfavorable substrates.

rhizomorphous having the form of a root; rootlike.

Rhizomucor *Mycology.* a genus of fungi belonging to the order Mucorales, which occurs in compost and produces an abundance of aerial mycelia.

rhizomycelium *Mycology.* in fungi belonging to the family Cladochytriaceae, a rootlike structure system that looks like the hyphal masses called mycelia.

Rhizomyidae *Paleontology.* the bamboo rats, a family of burrowing rodents belonging to the infraorder Myomorpha and the superfamily Spalacoidea; they first appeared in the Miocene; six species still survive in Africa and Asia.

Rhizophagidae *Invertebrate Zoology.* a family of root-eating beetles, tiny coleopteran insects belonging to the superfamily Cucujoidea. Also, MONOTOMIDAE.

rhizophagous *Ecology.* feeding on roots; root-eating.

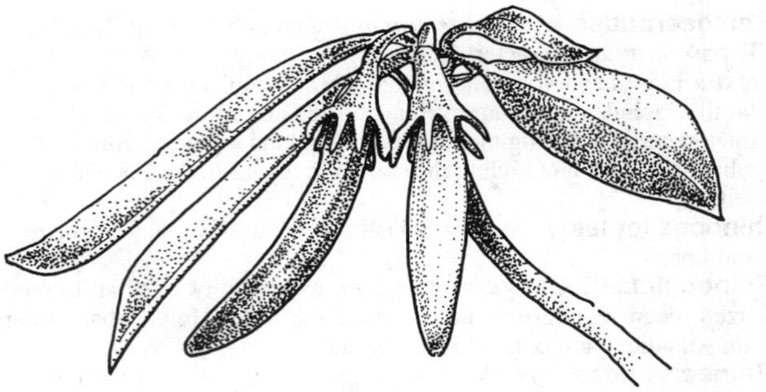

Rhizophoraceae

Rhizophoraceae *Botany.* the mangrove family; a family of dicotyledonous, primarily woody plants, trees, and shrubs belonging to the order Rhizophorales; characterized by upward-growing roots containing pneumatophores, channels that apparently facilitate gas exchange for the submerged root system; typically occurring in tropical aquatic or wetland habitats.

Rhizophorales *Botany.* an order of the subdivision Rosidae that includes a single family, the Rhizophoraceae.

rhizophore *Botany.* any of the leafless branches that develop in certain club mosses of the genus *Seleginella*, arising from the stem at forking points and growing downward toward the soil where they may function as roots.

Rhizophydium *Mycology.* a genus of fungi belonging to the order Chytridiales; some species obtain nutrients from dead organic matter, while others are aquatic or terrestrial plant parasites.

Rhizophyllidaceae *Botany.* a family of red algae of the order Gigartinales, characterized by strap-shaped, prostrate or erect, branched thalli with similar gametangial and sporangial forms.

rhizoplane *Botany.* the external surface of roots, including adherant soil particles and debris.

rhizoplast *Cell Biology.* a cytoplasmic fibril that connects the nucleus with the parabasal body or the blepharoplast in certain ciliated or flagellated cells.

rhizopod [rī´zə päd´] *Invertebrate Zoology.* **1.** a member of the class Rhizopodea. **2.** of or relating to this class.

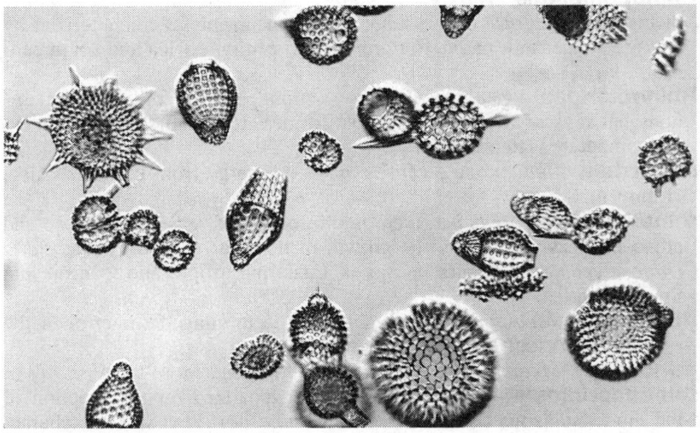

rhizopods

Rhizopodea *Invertebrate Zoology.* a class of protozoans in the superclass Sarcodina, having various types of pseudopodia, including lobopodia, filopodia, and rhizopodia, but lacking cilia and flagella.

rhizopodium *Invertebrate Zoology.* a branching, filamentous pseudopodium that forms an anastomosing network; it is found chiefly in Foraminifera.

Rhizopus *Mycology.* a genus of fungi belonging to the order Mucorales, found in organic matter such as soil and fruits and parasitic on vegetables and fruits; causes the disease mucormycosis, which affects the nose and sinuses.

Rhizosoleniaceae *Botany.* a family of diatoms constituting the suborder Rhizosoleniineae.

Rhizosoleniineae *Botany.* a monofamilial suborder of mostly marine planktonic diatoms in the order Centrales, characterized by a strongly developed pervalvar axis and a valve structure radiating from a raised point near the center of the valve.

rhizosphere *Geology.* a soil region in which increased microbiological activity results from the presence of plant roots. Also, ROOT ZONE.

Rhizostomeae *Invertebrate Zoology.* an order of jellyfish, coelenterates in the class Scyphozoa, having fused arms and vast numbers of tiny mouths; medusae have a tall, narrow, lobed umbrella with no tentacles.

rhizotomy *Surgery.* the surgical division of a root, such as the roots of spinal nerves within the spinal canal, usually performed to alleviate pain. Also, RADICOTOMY, RADICULOTOMY.

Rh negative *Genetics.* see RH FACTOR.

rho [rō] the 17th letter of the Greek alphabet, written as P or ρ.

rhod- or **rhodo-** a combining form meaning "rose."

rhodamine B *Organic Chemistry.* $C_{28}H_{31}ClN_2O_3$, green crystals or reddish-violet powder; soluble in water and alcohol, slightly soluble in acids; melts at 165°C; used as a textile dye, tissue stain, and reagent in organic syntheses. Also, TETRAMETHYL RHODAMINE.

rhodanese *Enzymology.* an enzyme, produced in the liver, that catalyzes the transfer of sulfur from the thiosulfate ion ($S_2O_3^{2-}$) to the cyanide ion (CN^-). Also, THIOSULFATE SULFURTRANSFERASE.

Rhodanian orogeny *Geology.* a brief, late Miocene period of crustal deformation and mountain-building activity.

rhodanic acid see THIOCYANIC ACID.

rhodanine *Organic Chemistry.* $C_3H_3NOS_2$, a toxic, prismatic yellow solid; soluble in hot water and ethyl ether; decomposes, often violently, at a melting point of 166°C; used in organic synthesis and as a laboratory reagent.

Rhode Island Red *Agriculture.* a breed of chicken that lays brown eggs and is characterized by a long, flat back, red plumage, yellow skin, and yellow shanks tinged with red.

Rhodesian man *Paleontology.* the name given to the remains of several hominids found near Kabwe, Zambia (formerly Northern Rhodesia), in 1921; the cranial capacity of the single skull is 1280 cc, and the remains in general are close to Neanderthal morphology, with a few features that are closer to *Homo erectus* and *H. sapiens*; the dating is uncertain, and estimates range from 125,000 to 600,000 years ago.

Rhodesian trypanosomiasis *Medicine.* an East African infectious disease caused by *Trypanosoma rhodesiense* and transmitted by the tsetse fly; characterized by the presence of parasites in the circulating blood, inflammation of the lymph nodes, accumulation of fluid in the tissues, and inflammation of the heart muscle. Also, SLEEPING SICKNESS.

Rhodininae *Invertebrate Zoology.* a subfamily of mud-ingesting bamboo worms, annelids in the family Maldanidae.

rhodite *Mineralogy.* a natural alloy composed of gold and rhodium. Also, **rhodium gold.**

rhodium *Chemistry.* a metallic element having the symbol Rh, the atomic number 45, an atomic weight of 102.9, a melting point of 1970°C and a boiling point of 3730°C; a grayish-white, ductile metal that is obtained from platimun ores and gold gravels; used in high temperature alloys with platinum or palladium and in electrical components and optical instruments. (From the Greek word for "rose;" because of its color in solution.)

rhodium acetate *Organic Chemistry.* $[(CH_3CO_2)_2Rh]_2$, a catalyst used for reactions of diazo compounds.

rhodium chloride *Inorganic Chemistry.* $RhCl_3$, a toxic reddish-brown powder; insoluble in water; decomposes at 450–500°C and sublimes at 800°C; used to make rhodium fluoride. Also, **rhodium trichloride.**

rhodium fluoride *Inorganic Chemistry.* RhF_3, red rhombic crystals that are insoluble in water and acids; sublimes above 600°C. Also, **rhodium trifluoride.**

rhodizite *Mineralogy.* $(K,Cs)Al_4Be_4(B,Be)_{12}O_{28}$, a colorless, white, or yellow cubic mineral occurring as dodecahedral or tetrahedral crystals, having a specific gravity of 3.36 to 3.44 and a hardness of 8 to 8.5 on the Mohs scale; found only in pegmatites.

Rhodobacter *Bacteriology.* a genus of bacteria proposed to include certain motile and nonmotile species that divide by binary fission; they would be reassigned from the genus *Rhodopseudomonas.*

Rhodobacteriineae *Bacteriology.* in former classifications, a suborder of the order Pseudomonadales containing all bacteria that use light to obtain energy except the genus *Rhodomicrobium.*

Rhodochaetaceae *Botany.* a monospecific family belonging to the order Rhodochaetales, having a branched, filamentous thallus of elongated cells with pit connections.

Rhodochaetales *Botany.* an order belonging to the red algae division Rhodophyta, composed of the monospecific family Rhodochaetaceae.

rhodochrosite *Mineralogy.* $Mn^{+2}CO_3$, a pink to red or brown trigonal mineral of the calcite group, forming two series with calcite and with siderite, found in massive form or as rhombohedral crystals, and having a specific gravity of 3.7 and a hardness of 3.5 to 4 on the Mohs scale.

Rhodochytriaceae *Botany.* an algae family of the order Chlorococcales, known to be parasitic on various terrestrial flowering plants; characterized by flask-shaped cells having branched rhizoids on the basal portion, no chloroplast, and a bright-red protoplast.

Rhodococcus *Bacteriology.* a genus of aerobic, nonmotile bacteria of the order Actinomycetales, occurring in soil or in aquatic environments; some species are animal pathogens.

Rhodocyclus *Bacteriology.* a proposed genus of the family Rhodospirillaceae, composed of spiral-shaped photosynthetic bacteria.

rhododendron [rō´də den´drən] *Botany.* any of the evergreen or deciduous shrubs or trees of the genus *Rhododendron* of the heath family, characterized by beautiful, large clusters of pink, purple, or white flowers and oval or oblong leaves.

rhododendron

rhodolite *Mineralogy.* a rose- or violet-red variety of garnet with a brilliant luster, intermediate in composition between pyrope and almandine.

Rhodomelaceae *Botany.* the largest family of the red algae division Rhodophyta, belonging to the order Ceramiales and occurring mostly in marine habitats; characterized by multibranched, uniaxial thalli and spermatangia occurring in large quantities on special axes.

Rhodomicrobium *Bacteriology.* a genus of Gram-negative bacteria belonging to the family Rhodospirillaceae of the order Rhodospirillales, composed of ovoid pigmented cells found in soil.

rhodonite *Mineralogy.* $(Mn^{+3},Fe^{+2},Mg,Ca)SiO_3$, a pink to brownish-red triclinic mineral occurring in massive form and as tabular crystals, having a specific gravity of 3.57 to 3.76 and a hardness of 5.5 to 6.5 on the Mohs scale; found in manganese-bearing ore deposits.

Rhodophyceae *Botany.* a class of red algae belonging to the division Rhodophyta.

Rhodophyllaceae *Mycology.* a family of fungi belonging to the order Agaricales, which is composed of gill mushrooms that primarily live off decaying organic matter.

Rhodophyllidaceae *Botany.* a family of red algae of the order Gigartinales, characterized by flattened, branched, uniaxial thalli, gametangial and sporangial thalli of similar form and morphology, and superficially formed and scattered spermatangia.

Rhodophyta *Botany.* a division of eukaryotic red algae, primarily marine and characterized by the presence of photosynthetic and accessory phycobilin pigments; exhibiting reproductive and morphological diversity and lacking flagellate stages.

Rhodopila *Bacteriology.* a proposed genus of bacteria, including some species formerly classified in the genus Rhodopseudomonas.

rhodoplast *Botany.* a chromatophore within the red algae group.

Rhodopseudomonas *Bacteriology.* a genus of purple bacteria composed of flagellated cocci or rod-shaped cells that are capable of carrying out photosynthesis.

rhodopsin *Biochemistry.* the retinal, visual pigment that allows night vision. Also, RETINAL PIGMENT, VISUAL PURPLE.

rhodora *Botany.* a low deciduous shrub, *Rhododendron canadense,* of the heath family, characterized by purplish-rose flowers that appear before the leaves; found in North America.

Rhodospirillaceae *Bacteriology.* the purple nonsulfur bacteria, a family of motile, anaerobic photosynthetic bacteria of the suborder Rhodospirillineae that are found in various aquatic habitats and in moist soils. Also, ATHIORHODACEAE.

Rhodospirillales *Bacteriology.* the single order of bacteria that obtain energy from light; cells are Gram-negative, anaerobic, usually aquatic, and either spherical, rod-shaped, or vibrio-shaped.

Rhodospirillineae *Bacteriology.* the purple bacteria, a suborder of the order Rhodospirillales.

Rhodotorula *Mycology.* a genus of yeast fungi belonging to the class Hyphomycetes; some of its species are common environmental contaminants, and others cause infections in humans.

rhodotoxin *Toxicology.* any of several toxins found in the leaves and flowers of various species of *Rhododendron*, *Kalmia*, and *Leucothoe*.

rhodoxanthin *Organic Chemistry.* $C_{40}H_{50}O_2$, a yellow ketonic carotenoid pigment that is slightly soluble in alcohol and soluble in benzene and chloroform; first isolated from the red berries of the Irish yew (*Taxus baccata*), and used in the food, drug, and cosmetic industries.

Rhodymeniaceae *Botany.* a family of marine, red algae of the order Rhodymeniales, characterized by stalked, flattened, or tubular thalli that are usually multibranched and multiaxial.

Rhodymeniales *Botany.* an order belonging to the red algae division Rhodophyta, characterized by crustose and discoid or erect and frondose thalli and by superficial spermatangia.

rho factor *Molecular Biology.* a hexameric protein found in prokaryotes that is required for certain transcription terminations in bacteria. Also, ρ FACTOR.

rho gene *Genetics.* a gene that encodes the rho factor.

Rhoipteleaceae *Botany.* a family of dicotyledonous trees of the order Jugalandales, consisting of a single genus with a single species; an aromatic, wind-pollinated deciduous tree with pinnate leaves, small flowers, and a samaroid fruit; found in China and Vietnam.

rhombencephalon *Developmental Biology.* the posterior part of the three primary vesicles in the vertebrate embryo, later dividing into the metencephalon and myelencephalon. Also, HINDBRAIN.

rhombic *Science.* having the shape of a rhombus.

rhombic antenna *Electromagnetism.* a horizontal antenna composed of four conductors arranged in a rhombus and fed at one apex; the antenna pattern produced lies in a horizontal plane.

rhombic dodecahedron *Crystallography.* a geometric form having twelve faces, each described by a parallelogram.

rhombic sulfur *Chemistry.* a stable allotrope of sulfur, in the form of yellow rhombic-shaped crystals.

Rhombifera *Paleontology.* a class of brachiole-bearing echinoderms in the subphylum Blastozoa, distinguished by the respiratory rhombs in adjoining thecal plates; formerly classified with the diploporans as a class Cystoidea.

rhombochasm *Geology.* a parallel-sided break in the continental crust that is infilled by the underlying oceanic crust as a result of spreading and separation.

rhomboclase *Mineralogy.* $HFe^{+3}(SO_4)_2 \cdot 4H_2O$, a colorless, white, yellow, or gray orthorhombic mineral occurring as platy crystals, having a specific gravity of 2.23 and a hardness of 2 on the Mohs scale; found with other secondary sulfates.

rhombogen *Invertebrate Zoology.* a stage in some parasitic mesozoans; produced by an adult nematogen, it develops into a ciliated, free-swimming larva.

rhombohedral *Crystallography.* having a six-sided form.

rhombohedral system *Crystallography.* another way of describing hexagonal axes (which are described by Miller-Bravais indices) in terms of Miller indices. When a trigonal unit cell is described in terms of rhombohedral axes with a threefold axis of rotation or inversion, $a = b = c$ and $\alpha = \beta = \gamma$ (less than 120° but not 90°). When it is described in terms of hexagonal axes (with c unique) $a = b \neq c$ and $\alpha = \beta = 90°$ and $\gamma = 120°$. Also, **rhombohedral lattice.**

rhombohedral unit cell *Crystallography.* a unit cell in which there is a threefold rotation axis along one body diagonal of the unit cell. This symmetry requirement makes all three axial lengths necessarily the same and all three interaxial angles necessarily equal, although the value is not restricted ($a = b = c$, $\alpha = \beta = \gamma$).

rhombohedron *Crystallography.* a solid figure, such as a crystal, having six sides, each of which is a rhombus (diamond-shaped). There are two ways that rhombohedral axes can be oriented with respect to hexagonal axes. These are referred to as obverse and reverse orientations.

rhomboid *Mathematics.* a parallelogram with unequal adjacent sides.

rhomboidal prism *Optics.* a prism in which four parallel sides and two parallel ends slanted at 45° angles deflect light from its path as it enters the prism, without affecting the light's form.

rhomboid current ripple see RHOMBOID RIPPLE MARK.

rhomboideus *Anatomy.* either of two muscles of the back, the **rhomboideus major** and **rhomboideus minor,** that support the scapula and act to draw it slightly upward.

rhomboid ripple mark *Geology.* an asymmetric ripple mark formed by currents of water moving uniformly over a sandy surface, characterized by a netlike pattern of diamond-shaped tongues of sand, each having two acute angles: one pointing downcurrent and the other pointing upcurrent. Also, RHOMBOID CURRENT RIPPLE, OVERHANGING RIPPLE.

Rhombozoa *Invertebrate Zoology.* a class of the phylum Mesozoa, ciliated, multicellular organisms with a two-layered body that lacks skeletal, muscular, nervous, digestive, respiratory, and excretory systems; parasites in cephalopod mollusks; sometimes considered equivalent to Mesozoa.

rhomb-porphyry *Petrology.* a porphyritic variety of alkaline syenite that contains augite, sparse olivine, and anorthoclase or potassium oligoclase in phenocrysts of rhombohedral cross section, set within an alkali feldspar groundmass.

rhombus *Mathematics.* a parallelogram with all sides equal.

rho meson *Particle Physics.* a very short-lived unstable meson with mass 1490 times the mass of an electron; plays an essential role in the electrical structure of nucleons.

Rhône wine *Food Technology.* any wine from the Rhône Valley of south-central France, such as Côtes du Rhône, Châteauneuf-du-Pape, Hermitage, St. Joseph, or Gigondas.

rhoophilous *Ecology.* living or thriving in creeks. Thus, **rhoophile, rhoophily.**

rhoophyte *Ecology.* a plant that lives or thrives in creeks.

Rhopalidae *Invertebrate Zoology.* a family of true bugs, hemipteran insects in the superfamily Coreoidea that feed chiefly on grasses and occasionally on certain trees, such as the box elder.

rhopalium *Invertebrate Zoology.* a small, clublike sense organ of coelenterate medusae.

Rhopalocera *Invertebrate Zoology.* a former classification of Lepidoptera consisting of the butterflies, as distinct from Heterocera, the moths; replaced by classification based on wing venation.

rhopalocercous cercaria *Invertebrate Zoology.* a larval form of flukes, with the tail as wide as or wider than the body.

Rhopalodia *Botany.* a genus of green algae belonging to the family Epithemiaceae.

Rhopalodinidae *Invertebrate Zoology.* a family of sea cucumbers, holothurian echinoderms in the order Dactylochirotida, with a flask-shaped body and adjacent mouth and anus.

Rhopalosomatidae *Invertebrate Zoology.* the cricket wasps, a family of hymenopteran insects in the superfamily Scolioidea, with parasitic larvae that live and feed in sacs attached to the abdomen of crickets.

rho protein *Molecular Biology.* a cell protein that is required for rho-dependent termination of transcription in bacteria.

rhoptries *Invertebrate Zoology.* electron-opaque, elongate bodies found in the apical complex of the infective bodies of parasitic protozoans of the phylum Apicomplexa.

rhourd *Geology.* a sand dune formed by the intersection of other dunes and resembling a pyramid in shape.

RHP *Behavior.* resource-holding potential.

rhubarb *Botany.* any of several species of the genus *Rheum* of the buckwheat family (Polygonaceae), especially *R. officinale*, cultivated for its medicinal rhizome, and *R. rhabarbarum,* cultivated for its edible, fleshy leafstalks.

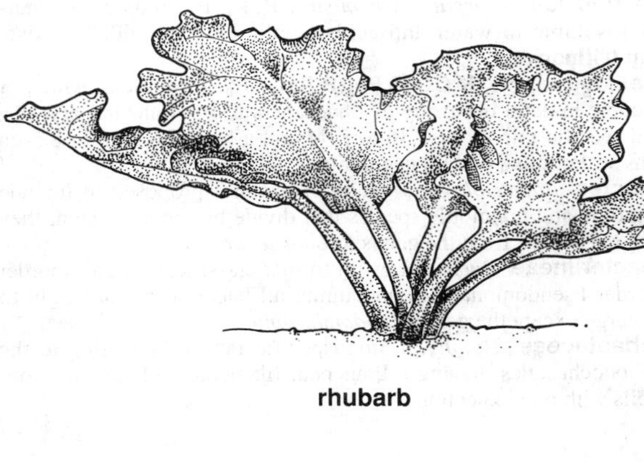

rhubarb

rhumb bearing *Cartography.* in surveying, the direction, usually expressed as angular distance from a reference direction, of a rhumb line that passes through two terrestrial points; usually measured from 0° at the reference direction clockwise to 360°.

rhumb line *Cartography.* a line or direction on the surface of the earth that follows a single compass bearing and makes the same angle with all meridians, resulting in a curve spiraling toward the poles; shown as a diagonal straight line on a Mercator projection. Also, MERCATOR LINE.

rhumb-line course *Navigation.* the direction of the rhumb line from departure to destination.

rhumb-line sailing *Navigation.* the process and techniques of plotting and sailing a course along a rhumb line.

Rhus *Botany.* a genus of vines and shrubs of the family Anacardiaceae, containing toxic species producing a highly allergic oleoresin mixture called urushiol; includes poison ivy, poison oak, and poison sumac. Also, TOXICODENDRON.

rhyncho- a combing form meaning "beak" or "snout."

Rhynchobatidae *Vertebrate Zoology.* a family of small tropical marine fishes of the order Rhinobatiformes, characterized by a broad, blunt snout and a stout tail.

Rhynchobdellae *Invertebrate Zoology.* an order of aquatic leeches in the class Hirudinea, having an eversible proboscis and blood with no hemoglobin; ectoparasites of fish and mollusks. Also, **Rhynchobdellidae.**

Rhynchocephalia *Vertebrate Zoology.* the tuatura, an order of primitive, lizardlike reptiles found on the islands off the coast of New Zealand.

rhynchocoel *Invertebrate Zoology.* the dorsal cavity that houses the proboscis in nemertine worms of the phylum Rhynchocoela; filled with a fluid that, when compressed by muscles, everts the proboscis through a pore.

Rhynchocoela *Invertebrate Zoology.* a phylum of unsegmented thread or ribbon worms having complete digestive and vascular systems and an eversible proboscis. Also, NEMERTEA, NEMERTINA.

rhynchodaeum *Invertebrate Zoology.* a short, anterior dorsal cavity in nemertine worms, situated between the rhynchocoel and the proboscis pore.

Rhynchodina *Invertebrate Zoology.* a suborder of ciliated protozoans, in the order Thigmotrichida, having sparse or absent cilia and a sucker in place of a cytostome; found on the gills of aquatic invertebrates.

Rhynchonella *Paleontology.* an extinct genus of sharply beaked articulate brachiopods in the order Rhynchonellidae and superfamily Rhynchonellacea; characterized by impunctate shells and the absence of spondylia; extant in the Jurassic and Cretaceous.

Rhynchonellidae *Invertebrate Zoology.* a family or order of brachiopods having a triangular shell, a sharp, hooklike beak, and tentacles; found in shallow water attached to the substrate.

rhynchophorous *Zoology.* having a beak.

rhynchosaur *Paleontology.* of or relating to the reptiles of the extinct family Rhynchosauridae.

Rhynchosauridae *Paleontology.* an extinct family of large, herbivorous reptiles in the order Rhynchocephalia; extant in the Triassic.

rhynchosporium *Plant Pathology.* a fungus that causes barley leaf blotch and from which the toxin rhynchosporiside is derived.

Rhyniophyta *Paleontology.* an extinct division of primitive, leafless plants with sporangia at the tips of branches; the simplest of all known vascular plants, first known from the genera *Rhynia* and *Horneophyton* of the Rhynie chert assemblage; extant in the Silurian and Devonian.

Rhyniopsida *Paleontology.* an extinct class of primitive plants in the division Rhyniophyta; includes some psilophytes; extant from the Middle Silurian to Middle Devonian.

Rhynochetidae *Vertebrate Zoology.* the kagu, a monotypic family of flightless cranelike birds of the order Gruiformes, characterized by a long bushy crest; found in the forests of New Caledonia.

rhyodacite *Petrology.* an extrusive porphyritic igneous rock, or rock group, intermediate in composition between rhyolite and dacite, with phenocrysts of quartz, plagioclase, and biotite in a fine-grained to glassy matrix of alkali feldspar and silica minerals; the extrusive equivalent of granodiorite. Also, DELLENITE, QUARTZ LATITE, TOSCANITE.

rhyolite *Petrology.* a light-colored, silica-rich volcanic or hypabyssal rock, or rock group, normally having small phenocrysts of quartz and alkali feldspar in a glassy to cryptocrystalline groundmass; the extrusive equivalent of granite. Also, LIPARITE.

rhyolitic glass *Volcanology.* volcanic glass having a chemical composition similar to that of rhyolite. Also, OBSIDIAN.

rhyolitic lava *Volcanology.* a highly viscous, silica-rich lava.

rhyolitic magma *Volcanology.* a silica-rich magma that, if cooled at or near the surface, will produce a rhyolite.

rhyolitic tuff *Geology.* a tuff formed from fragments of rhyolitic lava.

Rhyssa *Invertebrate Zoology.* a genus of ichneumon flies, hymenopteran insects that parasitize horntail wasps.

rhythmeur *Radiology.* an apparatus that produces rhythmic breaks in the current of an X-ray machine.

rhythmic accumulations *Geology.* regular patterns of ripples and cusps formed in sediment on the beach or ocean floor by currents and waves.

rhythmic crystallization *Petrology.* a process, exhibited in igneous rocks, by which different minerals crystallize in concentric layers, resulting in an orbicular structure.

rhythmicity *Cardiology.* the ability of the heart muscle to beat regularly without artificial stimulation.

rhythmic layering *Geology.* easily observable layering in an igneous intrusion that shows repeating zones of varying composition.

rhythmic sedimentation *Geology.* a consistently repeated sequence of two or more rock units which, through sedimentary succession, form a recognizable pattern and indicate a recurrence of the same sequence of conditions.

rhythmic stratification *Geology.* a repeating pattern occurring in layers of sediment.

rhythmic succession *Geology.* a succession of rock units that exhibit continual, repeated changes in lithology.

rhythmite *Geology.* any of the units formed as part of a rhythmic succession or by rhythmic sedimentation. Also, **rhythmic unit.**

rhythm method *Medicine.* a popular term for a method of preventing conception in which coitus is restricted to a "safe" period, avoiding the days just prior to and after the time of ovulation.

rhytid- or **rhytido-** a combining form meaning "wrinkle."

rhytidectomy *Surgery.* the surgical excision of skin to eliminate wrinkles.

Rhytidiaceae *Botany.* a monotypic family of large, robust mosses of the order Hypnobryales that form loose mats on humus or calcareous soil; characterized by a single costa, rugose leaves, and a hypnaceous peristome, and native to arctic-alpine and boreal regions of the Northern Hemisphere.

rhytidoplasty *Surgery.* plastic surgery involving the removal of excess skin to reduce or eliminate wrinkling.

Rhytisma *Mycology.* a genus of fungi belonging to the order Phacidiales; the species *R. acerinum* causes the tree disease tar spot.

Rhytismataceae *Mycology.* a family of fungi belonging to the order Phacidales, characterized by ascospores that are enclosed in gelatinous membranes; parasitic on and pathogenic to plants.

Rhytismatales *Mycology.* a former term for an order of fungi belonging to the subdivision Ascomycotina; it has been reclassified as the family Rhytismataceae.

rhyzoid *Botany.* an outgrowth from a fungus, alga, bryophyte, or pteridophyte gametophyte that anchors to a substrate and possibly serves in absorption.

RI *Genetics.* a partially double-stranded structure that is formed as an intermediate during replication of certain ssRNA viral genomes. (An acronym for replicative intermediate.)

RI refractive index.

ria *Geography.* **1.** a wedge-shaped or funnel-shaped indentation of a rocky coastline, formed by the submergence of an estuary or river mouth. **2.** a river mouth.

RIA radioimmunoassay.

ria coast *Geography.* a coastline, such as the Atlantic coast of North America, where rias alternate with long rocky promontories.

rib *Anatomy.* a member of the twelve pairs of bones that extend from the thoracic vertebrae to the sternum and form the walls of the thorax.

rib any of various structures or features thought of as analogous to the human rib; specific uses include: *Botany.* **1.** the main, often thickened central vein in a leaf. **2.** a thickened area of vascular tissue or protrusion. *Geology.* a rock layer or dike that forms a small ridge on a steep mountainside. *Architecture.* **1.** any of the raised moldings separating the panels of a Gothic vault. **2.** any curved, projecting structural member, as of a wooden sailing ship. *Mining Engineering.* **1.** the side of a pillar or an entry wall. **2.** a solid pillar or barrier of coal or ore left to stand as support. **3.** a thin stratum of material, such as stone, in a seam of coal. *Aviation.* in an airfoil, a chordwise (transverse) structural member that gives the airfoil its shape and carries its load from the skin to the spars.

rib and furrow *Geology.* sets of small, transverse, arched cross-lamination markings in stratified rock that are confined to long, narrow, grooves oriented parallel to the current flow and separated from one another by very narrow continual ridges. Also, **rib-and-furrow structure.**

riband jasper see RIBBON JASPER.

rib arch *Civil Engineering.* an arch formed of two or more parallel arched ribs.

ribavirin *Virology.* 1-D-ribofuranosyl-1,2,4-triazole-3-carboxamide, an analogue of guanosine that is active against both DNA and RNA viruses.

ribbed-clamp coupling *Mechanical Devices.* a rigid coupling with a longitudinal split to allow a shim to separate its halves when used in a borehole shaft.

ribbed moraine *Geology.* one of a group of locally branching, smoothly rounded arched ridges curving in the downstream direction of a glacier, but curving upstream adjacent to eskers.

ribbed smoke sheet *Materials.* a flat, ribbed piece of rubber which is made by passing coagulated latex through grooved rollers; has a variety of manufacturing uses.

ribbed vault *Architecture.* a vault supported by diagonal ribs. Also, RIB VAULT.

ribble see RIPPLE TILL.

ribbon *Textiles.* a narrow woven strip of fabric made of various fibers, having selvage, cord finished, or fused edges, and used for trimming. *Building Engineering.* a horizontal wood piece nailed into studs to support the ends of floor joists. *Petrology.* a particular set of parallel streaks or bands in a rock or mineral. *Mathematics.* a plane figure generated by a line segment intersected at its midpoint by a curve and moving perpendicular to the curve.

ribbon banding *Petrology.* a series of thin parallel strata of contrasting colors, occurring in the bedding of a sedimentary rock and giving the rock an appearance resembling bands of ribbons.

ribbon bomb *Volcanology.* an elongated, flattened volcanic bomb derived from ropes of lava.

ribbon cable *Electricity.* a flat, flexible multiconductor cable, made of specially treated flexible wire.

ribbon conductor *Electricity.* wire that is used to form a ribbon cable.

ribbon diagram *Geology.* a drawing in perspective of a continuous geologic cross section along a curved or winding line.

ribbonfish *Vertebrate Zoology.* any of the thin, ribbonlike, deep-bodied marine fishes of the families Trachipteridae, Regalecidae, and Lophotidae, usually fragile, deepwater species; many, such as the oarfish, are quite large.

ribbon gut *Surgery.* a broad, absorbable band of animal intestine used to support and reinforce suture lines.

ribbon jasper *Mineralogy.* banded jasper having parallel, ribbonlike stripes of alternating colors or shades of color. Also, RIBAND JASPER.

ribbon lightning *Geophysics.* the common streak lightning that, under the influence of strong winds blowing at right angles to an observer's line of sight, appears to spread out across the sky.

ribbon microphone *Acoustical Engineering.* a microphone that produces a figure-eight-shaped response pattern, due to the ribbon shape of its magnetic pickup.

ribbon mixer *Mechanical Engineering.* an agitator in which materials are mixed by a long, revolving metal spiral.

ribbon parachute *Aviation.* a parachute whose canopy consists of an arrangement of closely spaced ribbons held in place by equally spaced tapes. The spacing makes the parachute highly porous, thus increasing stability and reducing opening shock.

ribbon reef *Geology.* within the Great Barrier Reef off the coast of Australia, a linear reef of variable length whose extremities are curved inwardly, forming a festoon along the precipitous edge of the continental shelf.

ribbon rock *Petrology.* a rock whose appearance is marked by a succession of thin layers of varying composition or color.

ribbon slate *Petrology.* slate formed by incomplete metamorphism manifested as residual bedding planes that cut across the cleavage surface, with characteristic ribbons of various colors.

ribbon snake *Vertebrate Zoology.* a long-tailed garter snake of the genus *Thamnophis proximus* or *T. sauritus,* characterized by a brownish body and yellow or orange stripes; found in eastern and central North America.

ribbon vein see BANDED VEIN.

ribbon windows *Architecture.* a series of windows set in a continuous band across the facade of a building.

ribbon worm *Invertebrate Zoology.* any of various thin, unsegmented marine worms of the phylum Nemertea, varying in length from an inch to over 80 feet; able to contract and stretch to an extreme extent.

rib cage *Anatomy.* the barrel-shaped enclosure formed by the ribs and their connecting bones.

ribi cell fractionator *Biotechnology.* an apparatus that ruptures cells by high shear forces exerted by hydraulic pressure.

rib marks *Materials Science.* marks found on the surface of broken glass in the form of raised arcs perpendicular to the direction in which the fracture occurred.

riboflavin [rī´bə flā´vən] *Biochemistry.* $C_{17}H_{20}N_4O_6$, the heat-stable factor of the vitamin B complex; a water-soluble vitamin that is essential in the human diet, it serves as a component of two coenzymes or prosthetic groups of flavoproteins, which function as hydrogen carriers in oxidation-reduction processes. Also, LACTOFLAVIN, VITAMIN B_2.

riboflavin 5'-phosphate *Biochemistry.* the phosphoric acid ester of riboflavin.

ribofuranosyladenine see ADENOSINE.

ribofuranosylcytosine see CYTIDINE.

ribonuclease *Enzymology.* an enzyme that catalyzes the hydrolysis of phosphate ester linkages in ribonucleic acid. Also, RNASE.

ribonucleic acid [rī´bō noo klā´ik] *Biochemistry.* RNA, a natural polymer that is present in all living cells and that is important in protein synthesis. See RNA.

ribonucleic acid polymerase *Biochemistry.* an enzyme that triggers the formation of ribonucleic acid from ribonucleoside 5'-triphosphate with a strand of DNA or RNA serving as a template.

ribonucleic acid processing *Biochemistry.* a term for modifications, such as splicing, cleavage, capping, and addition of poly-A tails, of primary RNA transcripts.

ribonucleoprotein *Biochemistry.* one of a group of conjugated proteins that have molecules of ribonucleic acid as their prosthetic group.

ribonucleoside *Biochemistry.* a nucleoside in which the purine or pyrimidine base is combined with ribose.

ribonucleotide *Biochemistry.* a nucleotide in which the purine or pyrimidine base is combined with ribose.

ribophage *Virology.* a bacteriophage characterized by the existence of an RNA genome.

ribophorins *Biochemistry.* glycoproteins of the endoplasmic reticulum that interact with ribosomes during cotranslational insertion of membrane or secreted proteins.

riboprine *Oncology.* $C_{43}H_{58}N_2O_{13}$, an antineoplastic agent.

ribose *Biochemistry.* $C_5H_{10}O_5$, a pentose sugar present in the structural components of ribonucleic acid, riboflavin, and other nucleotides and nucleosides.

riboside *Biochemistry.* a glycoside that contains ribose as the sugar component.

ribosomal biosynthesis *Biotechnology.* the assembly of ribosomal particles from RNA and protein components, coordinated in eukaryotes and prokaryotes so that neither excess protein nor excess nucleic acids accumulate.

ribosomal cistron *Molecular Biology.* gene coding for ribosomal RNA.

ribosomal protein *Molecular Biology.* any of a number of proteins that, together with RNA molecules, compose a ribosome.

ribosomal ribonucleic acid *Biochemistry.* an organelle that accounts for most of the RNA in the cell and that apparently is responsible for ribosome function. Also, **ribosomal RNA.**

ribosomal RNA genes *Genetics.* a series of tandem repeating genes found on the chromosomes of eukaryotes that code for the three units of ribonucleic acid composing ribosomal RNA (rRNA).

ribosome *Cell Biology.* the functional unit of protein synthesis, composed of a cellular complex of RNA and protein molecules and organized into two subunits.

ribosome binding sequence *Molecular Biology.* the Shine-Dalgarno sequence of four to seven nucleotides that occurs in the leader section of an mRNA molecule and functions to properly orient the mRNA with the ribosome for protein synthesis.

ribovirus *Virology.* a virus that has an RNA genome.

ribozyme *Enzymology.* a ribonucleic acid molecule that can catalyze specific biochemical reactions especially in the processing of some RNA.

rib pillar *Mining Engineering.* a pillar with a large length-to-width ratio.

rib rifling *Ordnance.* a type of rifling in which the lands and grooves are of equal width.

ribulose *Biochemistry.* $C_5H_{10}O_5$, the ketose analogue of ribose. Also, 2-ARABOKETOSE.

ribulose 1,5-diphosphate *Biochemistry.* $C_5H_{12}O_{11}P_2$, the pentose sugar that reacts with carbon dioxide in photosynthesis; synthesized from arabinose by isomerization with pyridine.

ribut *Meteorology.* a brief squall that occurs in relatively calm winds from May to November in Malaya.

rib vault see RIBBED VAULT.

rib weave *Textiles.* a variation of plain weave that forms ribs running either lengthwise or crosswise by having two or more successive picks weave alike or by using coarser yarn for the rib than for the ground.

RIC *Aviation.* the airport code for Richmond, Virginia.

Riccati equation *Mathematics.* any first-order differential equation of the form $y' = A(x) + B(x)y + C(x)y^2$; any second-order linear differential equation can be transformed into a Riccati equation.

Ricciaceae *Botany.* a family of terrestrial and floating aquatic liverworts of the order Marchantiales, characterized by closely branched rosette-forming plants, absent or rudimentary thallus pores, and sporophytes consisting only of capsules scattered within the thallus tissue.

Ricci calculus *Mathematics.* the rules of arithmetic for tensor fields expressed in component form; formerly called **tensor calculus** or **absolute differential calculus,** prior to the development of other symbolic methods for manipulating tensor fields.

Ricci-Curbastro, Gregorio 1853–1925, Italian mathematician; developed tensor analysis.

Ricci equations *Mathematics.* a set of formulas for the difference between covariant second derivatives of tensor quantities when the order of differentiation is reversed. The difference depends only on the (undifferentiated) original tensor field and the curvature tensor. Also, **Ricci identities.**

Ricci's lemma *Mathematics.* the statement that the covariant derivative of the metric tensor field on a Riemannian manifold vanishes.

Ricci tensor *Mathematics.* a contraction of the curvature tensor whose components are, by definition, $R_{ik} = R^j_{ikj}$, where it can be demonstrated that $R_{ik} = R_{ki}$. The Ricci tensor appears in the Einstein equations of general relativity theory.

rice *Botany.* **1.** an annual marsh grass, *Oryza sativa*, widely cultivated in warm climates for its starchy, grainlike seeds, which are used for food in many countries, particularly in Asia. **2.** the seeds of this plant.

rice bird see BOBOLINK.

rice coal *Geology.* a size classification of anthracite coal that will pass through a 5/16-inch mesh screen, but not through a 3/16-inch mesh screen.

rice grains *Astronomy.* a granulated appearance to the sun's photosphere caused by small, constantly moving convection cells.

Rice neutralization *Electronics.* a process in which a voltage is applied to the grid of a vacuum tube in order to eliminate feedback.

Rice neutralizing circuit *Electronics.* a circuit that eliminates the capacitance between the electrodes of an amplifier tube; used in the transmission and reception of radio signals.

rice paper *Materials.* **1.** a thin paper made from rice grass straw. **2.** a similar paper made from the pith of various plants, especially *Tetrapanax papyriferus*, a small Asian tree of the ginseng family.

rice rat *Vertebrate Zoology.* a rat of the genus *Oryzomys,* characterized by an extremely long tail, especially *O. palustris*; found in the southern U.S., Mexico, and Central America.

Rice's bromine solution *Analytical Chemistry.* a standard laboratory reagent used in tests for urea; contains bromine and sodium bromide in aqueous solution.

Richards, Dickinson Woodruff 1895–1973, American physician; with Cournand, shared the Nobel Prize for the technique of heart catheterization.

Richards, Theodore 1868–1928, American chemist; discovered the radioactive isotope of lead; awarded the Nobel Prize for determining atomic weights.

Richards box *Crystallography.* a method for building a model of the three-dimensional structure of a crystalline protein. A half-silvered mirror is held at 45° to the electron-density sections and also at 45° to the model that is being built. The images of the model and of the electron-density maps appear superposed when viewed through the half-silvered mirror.

Richardson, Henry Hobson 1838–1886, American architect; revived the Romanesque style and designed Trinity Church, Boston.

Richardson, Lewis Fry 1881–1953, English meteorologist; pioneer in weather forecasting; introduced the Richardson number.

Richardson, Owen W. 1879–1959, English physicist; awarded the Nobel Prize for work in thermionics and the electron theory of matter.

Richardson automatic scale *Engineering.* a machine that measures the weight of items on a moving conveyor belt.

Richardson equation *Electronics.* an expression for the relationship between the density of the electrons ejected from a heated conductor and its temperature. Also, **Richardson-Dushman equation.**

Richardson number *Fluid Mechanics.* a number that is a ratio of two expressions, arrived at through studying the limit of stability of shearing flow in a stably stratified fluid, by equating the kinetic energy involved in raising an incremental parcel of fluid with the product of the applied force and the directed distance in the line of motion of the given incremental parcel.

Richardson plot *Electronics.* a graph that shows the relationship between the current density of the electrons emitted by a heated conductor and the temperature of the conductor; generally the relationship produces a straight line.

richellite *Mineralogy.* a yellowish-brown to reddish-brown, compact massive to foliated, amorphous mineral with an approximate formula of $Ca_3Fe^{+3}_{10}(PO_4)_8(OH,F)_{12}\cdot nH_2O$, having a specific gravity of about 2 and a hardness of 2 to 3 on the Mohs scale; found in Richelle, Belgium.

Richer, Jean 1630–1696, French astronomer; discovered that the magnetic force of gravity varies with location on the globe.

Richet, Charles Robert 1850–1935, French physiologist; awarded the Nobel Prize for the discovery of anaphylaxis.

rich mixture *Chemistry.* an air-fuel mixture that contains a concentration of combustible fuel that exceeds the optimum level for such a mixture with the given amount of air.

Richmondian *Geology.* a North American geologic stage of the Upper Ordovician period, occurring after the Maysvillian and before the Lower Silurian period.

Richter, Burton born 1931, American physicist; shared the Nobel Prize with Samuel Ting for their independent discoveries of the psi (or J) particle.

Richter, Charles F. 1900–1985, American seismologist; developed the Richter scale for measuring the intensity of an earthquake.

Richter, Hieronymus Theodor 1824–1898, German chemist; with Ferdinand Reich, discovered indium.

Richter, Jeremias 1762–1807, German chemist; formulated the Law of Equivalent Proportions; anticipated the Law of Definite Proportions.

richterite *Mineralogy.* $Na_2Ca(Mg,Fe^{+2})_5Si_8O_{22}(OH)_2$, $Mg/(Mg+Fe^{+2})=$ 0.5–1.0, a brown, yellow, green, or rose-red monoclinic mineral of the amphibole group occurring as long prismatic crystals, having a specific gravity of 2.97 to 3.13 and a hardness of 5 to 6 on the Mohs scale; found in contact metasomatic deposits and some alkalic igneous rocks.

Richter scale *Geophysics.* a logarithmic scale ranging from 1 to 9 that expresses the magnitude (**Richter magnitude**) of an earthquake based on a measurement of the amount of energy dispersed during the event. (Named for Charles F. *Richter.*)

Richthofenia *Paleontology.* a Paleozoic genus of conical articulate brachiopods in the extinct order Strophomenida and suborder Productidina; similar in appearance to cup-shaped corals; extant in the Permian.

ricin *Materials.* a white, poisonous protein powder derived from the bean of the castor-oil plant, *Ricinus communis.*

Ricinidae *Invertebrate Zoology.* a family of bird lice in the order Mallophaga, ectoparasites of birds.

ricinism *Toxicology.* poisoning due to ingestion of seeds of the castor-oil plant, *Ricinus communis*; symptoms may include severe gastrointestinal disturbances, hemorrhage, and death.

ricinoleic acid *Organic Chemistry.* $CH_3(CH_2)_5CH(OH)CH_2CH{:}CH(CH_2)_7COOH$, a combustible, colorless to yellow liquid; insoluble in water and soluble in most organic solvents; melts at 5.5°C and boils at 227°C (10 torr); derived from the saponifiction of castor oil; used in making soaps and textile finishing.

ricinoleyl alcohol *Organic Chemistry.* $CH_3(CH_2)_5CH(OH)CH_2CH{:}CH(CH_2)_7CH_2OH$, a combustible, colorless liquid; boils at 178°C (0.5 torr); used in organic synthesis, protective coatings, polyesters, pharmaceuticals, and plasticizers.

Ricinulei *Invertebrate Zoology.* a small order of tiny, ground-dwelling arachnids, having a hard oval body and an anterior dorsal hood that can be lowered to cover the mouthparts. Also, **Ricinuleida.**

rickardite *Mineralogy.* Cu_7Te_5, a deep-purple, opaque, metallic, orthorhombic, pseudotetragonal mineral occurring in compact masses, having a specific gravity of 7.53 and a hardness of 3.5 on the Mohs scale; found in gold-quartz veins and copper sulfide ores.

Rickerian curve *Ecology.* a general relationship between the stock and recruitment in fisheries in which the number of recruits is an exponential function of the parental stock.

rickets *Medicine.* a deficiency disease of early childhood in which a lack of vitamin D results in defective bone growth. Also, RACHITIS.

ricketsiosis *Medicine.* any disease caused by *Rickettsia,* such as Rocky Mountain spotted fever, typhus, and rickettsialpox.

Rickettsia [ri ket′sē ə] *Bacteriology.* **1.** a genus of Gram-negative bacteria of the family Rickettsiaceae that occur in the guts of lice, fleas, ticks, and mites, by which they are transmitted to humans and other animals. **2.** any scotobacterium of the order Rickettsiales.

Rickettsiaceae *Bacteriology.* a family of bacteria of the order Rickettsiales that are parasites or pathogens of arthropods, by which they are transmitted to other animals, including humans.

Rickettsiales *Bacteriology.* an order of Gram-negative, spheroid or rod-shaped bacteria that are parasites and pathogens of humans and other animals, and thrive within the cells of the host.

rickettsialpox *Medicine.* a mild and short-lasting mite-borne disease caused by *Rickettsia akari* and transmitted by mice; characterized by a skin lesion, rash, fever, backache, and headache.

Rickettsieae *Bacteriology.* a tribe of bacteria of the family Rickettsiaceae that are parasites of certain invertebrates and vertebrates and pathogens of humans. The pathogenic species may be classified in four main groups: I, causing what is termed classic typhus; II, causing spotted fever; III, causing scrub typhus; and IV, causing miscellaneous diseases such as Q fever and trench fever. Also, **Rickettsiae.**

Rickettsiella *Bacteriology.* a genus of Gram-negative bacteria of the family Rickettsiaceae, occurring as intracellular, rod-shaped parasites in arthopods but not parasitic mammals.

ricolettaite *Petrology.* a dark, coarse-grained syenite-gabbro composed of calcic plagioclase, olivine, biotite, and augite.

ricotta [ri kät′ə] *Food Technology.* a white unripened whey cheese, similar in appearance to cottage cheese.

rictus *Vertebrate Zoology.* the posterior corner of the mouth.

RID radial immunodiffusion.

riddle *Mechanical Devices.* a coarse sieve used in foundries for removing foreign matter from granular materials.

Rideal-Walker test *Microbiology.* a test used to determine the dilution of a phenol-containing disinfectant that causes the same rate of death to a given organism as does a standard concentration of phenol.

rideau *Geology.* **1.** a slightly elevated portion of the ground. **2.** any small ridge or mound of earth.

rider *Graphic Arts.* any of a set of rigid plastic or metal rollers used in combination with soft rollers to break down or distribute ink on a press. Also, **rider roller.** *Naval Architecture.* an auxiliary member that follows and supports a framing member, especially along the top of a keel. *Mining Engineering.* a crossbeam that slides between the guides in a sinking shaft, used to guide and steady the hoppit as it moves up and down the shaft. It is carried by the hoppit but not attached to it.

ridge *Geology.* **1.** a long, narrow, usually sharp-crested and steep-sided elevation of the earth's surface. **2.** the narrow, elongated crest of a hill or mountain. **3.** a long, steep-sided elevation of the deep sea floor. Also, SUBMARINE RIDGE. *Agronomy.* the raised area of soil thrown up when a plow creates a furrow. *Architecture.* the horizontal line along the apex of a sloping roof. *Meteorology.* an elongated, wedge-shaped area of relatively high atmospheric pressure, associated with an area of maximum of anticyclonic curvature of wind flow. *Anatomy.* see CRISTA.

ridge board *Building Engineering.* a horizontal timber mounted onedge at the peak of a roof.

ridge cap *Building Engineering.* a wood or metal cap that runs along the ridge line of a roof to protect the intersection of sloping roof surfaces.

ridge course *Building Engineering.* the top course of slate or tile on a roof.

ridged ice *Oceanography.* rough and irregular pack ice that has been thickened by extensive ridging, the ridges occasionally reaching heights of 30 meters and lengths of several miles.

ridge fault *Geology.* a fault structure consisting of a set of two faults bordering a horst.

ridge height *Agronomy.* the depth of a furrow, measured by the distance from the bottom of the channel to the top of the ridge above it. Also, CHANNEL DEPTH.

ridge pole *Building Engineering.* the uppermost horizontal supporting member in a roof, to which the common rafters are fastened along the ridge.

ridge roll *Building Engineering.* a rounded section formed into a ridge shape to which lead or zinc flashing is secured to cover the tops of ridge courses, thus sealing the roof.

ridge roof *Building Engineering.* a roof whose common rafters meet in an apex.

ridge-top trench *Geology.* a trench formed at or near the top of a high, steep-sided mountain ridge by the slow, gradual displacement of a rock slab along shear surfaces parallel to the side slope of the ridge.

ridge waveguide *Electromagnetism.* a waveguide whose longitudinal ridge protrudes into the guide so as to increase the transmission bandwidth by lowering the cutoff frequency.

ridging *Oceanography.* the formation of pressure ridges in pack ice. *Surgery.* in plastic surgery, a visible line or ridge at the margin of a area that has been surgically planed.

ridgling *Agriculture.* a male animal, especially a horse, whose testicles do not descend into the scrotum. Also, **ridgeling.**

riding correction *Crystallography.* the riding motion of one atom on another. In this model two bonded atoms, A and B, one of which, B, is much lighter than the other. Atom B then "rides" on atom A in such .a way that the translational motion of B contains all the motion of A plus an additional motion that is not correlated with that of A. Analyses of thermal parameters lead to corrections for this effect in crystal structures. It is most common for hydrogen atoms riding on heavier atoms.

riebeckite *Mineralogy.* $Na_2(Fe^{+2},Mg)_3Fe_2^{+3}Si_8O_{22}(OH)_2$, $Mg/(Mg+Fe^{+2})=0–0.49$, $Fe^{+3}/(Fe^{+3}Al)=0.7–1.0$, a blue or black monoclinic mineral of the amphibole group, occurring as long prismatic crystals and in fibrous or granular form, having a specific gravity of 3.2 to 3.382 and a hardness of 5 on the Mohs scale; found in felsic igneous rocks, banded ironstones, and schists.

Riecke's principle *Mineralogy.* the assertion that mineral grains under stress dissolve more readily than unstressed grains, and that solution and recrystallization occur most readily at points of maximum and minimum stress.

Riedel's disease [rē′dəlz] *Medicine.* a chronic inflammation of the thyroid with a stony hardening of fibrous tissue adhering to the adjacent structures. Also, **Riedel's struma, Riedel's thyroiditis.** (Named after Bernhard *Reidel,* German surgeon, 1846–1916.)

riedenite *Petrology.* an igneous rock consisting of large, tabular crystals of biotite within a granular groundmass containing nosean, biotite, pyroxene, and trace amounts of sphene and apatite.

Riefler clock *Horology.* a pendulum clock using an escapement invented by Sigmund Riefler in which the impulse is transmitted by a suspension spring to a pendulum made of a nickle steel rod with a weight resting on an aluminum tube.

riegel *Geology.* a low, lateral ridge or bar of bedrock on the floor of a glacial valley. Also, ROCK BAR, THRESHOLD, VERROU.

Riegler's reagent see SODIUM NAPHTHIONATE.

Riegler's test *Analytical Chemistry.* any of three tests, for albumin or dextrose in the urine and for hydrochloric acid in the gastric juices. (Named after Emanuel *Riegler,* 1854–1929, German chemist.)

Rieke diagram *Electronics.* a chart that depicts the power constraints exhibited by microwave devices such as klystrons and magnetrons.

Riellaceae *Botany.* a monogeneric family of erect aquatic liverworts of the order Sphaerocarpales, characterized by a thallus differentiated into a stemlike axis and a dorsal, sometimes broadly lobed, unilateral wing; native to North Africa and South America.

Riemann, (Georg) Bernhard [rē′män] 1826–1866, German mathematician; devised Riemann surfaces; developed Riemannian geometry.

Riemann curvature tensor *Mathematics.* the form the curvature tensor field takes on a Riemannian manifold. Although the curvature tensor in general is defined via the connection (covariant differentiation), the curvature tensor on a Riemannian manifold may be computed directly from the metric tensor field, since the connection itself depends only on the metric tensor. The Riemann curvature tensor exhibits a permutation symmetry not possessed by the general curvature tensor. A famous theorem of Riemann states that if the curvature tensor is known, the metric is also determined. Also, **Riemann-Christoffel tensor.**

Riemann function *Mathematics.* a function useful in solving a linear hyperbolic second-order partial differential equation $Lu = g$, with Cauchy conditions along a noncharacteristic initial curve γ. The Riemann function v is a solution, if it exists, of the adjoint partial differential equation $L\tilde{\ }v = 0$ whose first derivatives meet certain conditions along the characteristic curves of L. The Riemann function differs from a Green's function in that it is a regular function and does not depend on the initial curve γ.

Riemann hypothesis *Mathematics.* the assertion that all nontrivial zeros of the zeta function are at points $s = 1/2 + it$ inside the critical strip.

Riemannian geometry *Mathematics.* **1.** the geometry of Riemannian manifolds; i.e., those having an infinitesimal squared distance ds that is everywhere positive. **2.** the non-Euclidean geometry in which the parallel postulate has been replaced by the postulate: Every two lines intersect. Riemannian plane geometry can be visualized as the geometry on the surface of a sphere in which "lines" are taken to be great circle arcs.

Riemannian manifold *Mathematics.* a smooth manifold X together with a continuous 2-covariant tensor field g, called the metric tensor, such that: (a) g is symmetric, and (b) for each $x \in X$, the bilinear form g_x on the tangent space is positive definite. (If g_x fails to be positive definite but is nondegenerate, the manifold is said to be pseudo-Riemannian.) Also, **Riemann manifold, Riemann space.**

Riemann integral *Mathematics.* suppose the interval $[a,b]$ on the real line has been partitioned with points x_i such that

$$a = x_0 < x_1 < \cdots < x_n = b.$$

The Riemann integral of a bounded real-valued function $f(x)$ on $[a,b]$ is then defined to be

$$\lim_{n \to \infty} \sum_{i=1}^{n} f(a_i)(x_i - x_{i-1}),$$

where $x_{i-1} \leq a_i \leq x_i$ for each i. When this limit exists and is finite, it is unique and is denoted $\int_a^b f(x)dx$; otherwise, the Riemann integral of $f(x)$ on (a,b) is said to be undefined. Intuitively, the Riemann integral of a bounded real-valued function f of a single real variable is defined as the limit, when it exists, of the sum of the areas of narrow rectangles approximating more and more closely the area of the plane under the graph of $f(x)$. Riemann integrals in n-dimensional space are evaluated in terms of iterated integrals.

Riemann invariants *Fluid Mechanics.* the conserved quantities in fluid flow; the flow characteristics.

Riemann-Lebesgue lemma *Mathematics.* the theorem that the coefficients in the Fourier series of an integrable (i.e., L^1) function have limit zero.

Riemann mapping theorem *Mathematics.* the theorem that any simply connected proper open subset of the complex numbers ¢ can be mapped analytically and one-to-one to the unit disc $\{z : |z| < 1\}$.

Riemann method *Mathematics.* a method for solving hyperbolic second-order partial differential equations based on the use of the Riemann functions.

Riemann sphere *Mathematics.* a sphere whose stereographic projection is the finite complex plane ¢. Usually taken to be tangent to the complex plane at the origin, the Riemann sphere is analytically diffeomorphic to the one-point compactification of C.

Riemann-Stieltjes integral *Mathematics.* a generalization of the Riemann integral that is obtained by replacing the quanties $\Delta x_k = x_k - x_{k-1}$ in the Riemann sums (whose limit defines the integral) with the quantities $\Delta g(x_k) = g(x_k) - g(x_{k-1})$.

Riemann surface *Mathematics.* the Riemann surface for a multivalued function $W(z)$ of a complex variable are the regions of the complex plane (the domain of W) corresponding to the branches of $W(z)$. They may be visualized as copies of the complex plane, called **Riemann sheets,** stacked above each other and joined along the branch cuts, on which $W(z)$ is single-valued.

Riemann zeta function *Mathematics.* the function of a complex variable $s = \sigma + it$, given by $\zeta(s) = \sum_{n=1}^{\infty} n^{-s}$. The series converges uniformly on every compact subset of the half-plane $\sigma > 1$, and ζ is holomorphic there. For $\sigma > 1$, $\zeta(s) = s \int_1^{\infty} [x]x^{-1-s}dx$, where $[x]$ is the greatest integer function of x. The critical strip of $\zeta(s)$ is the region $0 < \sigma < 1$; it is so named because relatively little is known about the behavior of $\zeta(s)$ in this region. The trivial roots of $\zeta(s)$ are the negative even integers; all other (nontrivial) roots are inside the critical strip.

rieske protein *Biochemistry.* one of a group of proteins, containing iron and sulfur atoms, that are found in mitochondria and a wide variety of bacteria, generally carrying an E_m value within a 150 to 330 mV range.

Riesling [rēs´ling; rēz´ling] *Botany.* a green wine grape widely grown in warm regions of France, Germany, California, and Australia. The chief varieties are **Johannisberg Riesling, Rhine Riesling,** and **White Riesling.** *Food Technology.* a crisp, fruity, acidic white wine made entirely or mostly with this grape.

Riesz, Frigyes 1880–1956, Hungarian mathematician; pioneer in functional analysis, independently formulated Riesz-Fischer theorem.

Riesz representation theorem *Mathematics.* **1.** let X be a locally compact Hausdorff space. Every nonnegative linear functional on the space of functions on X with compact support can be written as a countably additive integral with respect to a measure depending only on the functional. **2.** let X be an arbitrary measure space. The dual of $L^p(X)$ is then isomorphic to $L^q(X)$, where $L^p(X)$ is the L^p space on X and $1/p + 1/q = 1$.

Riesz-Schauder theorem *Mathematics.* the spectrum of a compact operator has no accumulation points, with the possible exception of zero.

Riesz theorem *Mathematics.* for $1 \leq p \leq \infty$, $L^p(X)$ is a complete normed space, i.e., a Banach space. The case $p = 1$ is known as the **Riesz-Fischer theorem.**

Rietveld analysis or **Rietveld profile refinement** see PROFILE REFINEMENT.

rifamycin *Microbiology.* an antibiotic produced by the bacterium *Streptomyces mediterranei* that is effective against both Gram-negative and Gram-positive bacteria, including the bacterium that causes tuberculosis, *Mycobacterium tuberculosis.* Formerly knwon as **rifomycin.**

riffle *Hydrology.* **1.** a shallow area extending across a stream bed, over which water rushes quickly and is broken into waves by obstructions under the water. **2.** the water flowing over such an area. *Mining Engineering.* the ridged lining of a sluice, usually consisting of transverse bars or slats arranged to catch heavy minerals.

riffler *Mechanical Devices.* a small flat, triangular, round, or half-round double-ended rasp with fine-toothed and curved tips; used for finishing details in wood or metal, including enlarging holes.

rifle *Ordnance.* a long-barreled, shoulder-supported firearm with longitudinal grooves cut into the surface of its bore; in modern rifles the grooves follow a helical pattern to impart a rotary motion to the projectile, thus increasing accuracy and range. *Mechanical Devices.* **1.** a drill core with spiral grooves, used to cut similarly shaped surfaces in the bore of a pipe. **2.** a wood strip covered with emery, used to sharpen scythes. *Engineering.* a borehole that has a spiral groove.

rifle bracket *Ordnance.* a metal clamp, usually on a motor vehicle, for holding a rifle in an easily accessible position.

rifle grenade *Ordnance.* a grenade designed to be fired from a launcher attached to the muzzle of a rifle or carbine. Similarly, **rifle-grenade cartridge, rifle-grenade launcher.**

rifle-grenade ogive *Ordnance.* a hollow metal protective shell over the front end of a rifle grenade that allows the grenade to be reused after firing.

rifle range *Ordnance.* see TARGET RANGE.

rifling *Ordnance.* **1.** the process of cutting spiral grooves into the bore of a gun to provide for the spinning action of a projectile around its long axis. **2.** the system of spiral grooves in such a gun.

rift *Geology.* **1.** a narrow, high passage in a cave. **2.** a narrow fissure or other opening in a rock, caused by cracking or splitting. *Hydrology.* a rocky or shallow place in a stream.

rift-block mountain *Geology.* a mountain or uplifted block of crustal material that is bounded by faults along its length.

rift lake or **rift-valley lake** see SAG POND.

rift saw *Mechanical Devices.* **1.** a circular saw used for cutting wood radially from a log. **2.** a circular saw with at least four toothed projections to saw cants into flooring strips.

rift valley *Geology.* **1.** a valley that has developed between two normal faults. Also, **rift trough.** **2.** the deep central crevice in the crest of the mid-oceanic ridge. Also, MID-OCEAN RIFT, CENTRAL VALLEY, MEDIAN RIFT VALLEY.

Rift Valley fever *Medicine.* a disease caused by a mosquito-borne virus found on domestic animals or on African wild game, characterized by fever, headache, sensitivity to light, muscle pain, loss of appetite, and reduction of leukocytes in the blood.

rig *Naval Architecture.* **1.** the arrangement of a sailing vessel's masts, spars, and sails. A ship with **full rig** or **ship rig** had at least three masts with square sails. **2.** to set up a ship's rigging. **3.** to set up or prepare a vessel or fitting for use or service. *Aviation.* **1.** to assemble and align the major components of an aircraft, particularly its airfoils and other surfaces. **2.** to fit out an aircraft with control cables and other devices. **3.** to prepare and pack a parachute or its harness. *Mechanical Engineering.* **1.** a combination of trucking equipment, such as a tractor-trailer. **2.** broadly, any truck, trailer, or the like. *Mining Engineering.* the arrangement of equipment necessary for a given drilling operation.

Rigel [rī´gəl; rī´jəl] *Astronomy.* Beta (β) Orionis, a blue-white star of magnitude 0.0 and spectral type B; it lies about 900 light-years away.

rigging *Naval Architecture.* the lines and gear associated with a vessel's sailing rig. **Standing rigging** refers to lines that are fixed (except for tightening or adjustment), while **running rigging** refers to lines that are hauled in or let out during normal operation. *Aviation.* **1.** the assembly and alignment of aircraft components. **2.** the system of cables, wires, lines, fittings, and so on with which an aircraft is rigged. **3.** the shroud lines attached to a parachute.

Righi, Augusto 1850–1920, Italian physicist; independently discovered Righi-Leduc effect; developed the electrical oscillator.

Righi experiment *Optics.* an experiment generating effects in a light beam that are similar to beats between sounds whose frequencies vary slightly.

Righi-Leduc effect *Physics.* an effect whereby a temperature gradient is produced in a conductor perpendicular to an applied magnetic field and another (applied) temperature gradient, which is also perpendicular to the magnetic field.

right-and-left-hand chart *Industrial Engineering.* see SIMO CHART.

right angle *Mathematics.* an angle of 90°.

right-angle prism *Optics.* a prism that deflects a beam of light from right to left or from up to down through a right angle; found in astronomical telescopes, projectors, rangefinders, and other optical instruments.

right ascension *Astronomy.* the celestial coordinate analogous to longitude on the earth; it is measured in hours, minutes, and seconds of time eastward from the vernal equinox.

right ascension circle *Astronomy.* **1.** in equatorial coordinates, any great circle that passes through both celestial poles and intersects the celestial equator at right angles. **2.** the calibrated dial on an equatorial mounting's polar axis that shows which right ascension the telescope is pointing toward.

right associative operator *Computer Science.* an operator in an arithmetic expression such that if there are two adjacent occurrences of the operator, the right one should be done first.

right-brain *Psychology.* of or relating to the thought processes associated with the right hemisphere of the brain, such as generalization, imagination, and creativity.

right-continuous function *Mathematics.* a real function f is said to be right-continuous at a point c if $\lim_{x \to c+} f(x) = f(c)$; i.e., $f(x)$ approaches $f(c)$ from above, or from the right, or for $x > c$. A left-continuous function is similarly defined.

right-cut tool *Mechanical Devices.* a lathing tool with its cutting edge on the right side when viewed from the tool's end.

right-eyed flounder *Vertebrate Zoology.* any of various flatfishes of the family Pleuronectidae having both eyes on the right side of the head.

right-hand *Mechanical Engineering.* of or relating to those tools or parts (such as drills, cutters, or screw threads) that are designed to rotate clockwise or to the right as they cut. *Building Engineering.* of a door, having the hinges on the right when viewed from the exterior of the building, room, etc., to which the doorway leads.

right-hand cutting tool *Mechanical Devices.* a cutting tool in which material is ground to the appropriate angle on its left-hand side with a machining process occurring from right to left.

right-hand derivative *Mathematics.* a real function f is said to have a right-hand derivative at a point c if $\lim_{x \to c+}[f(x) - f(c)]/[x - c]$ exists and is finite; i.e., the limit exists as x approaches c from above, or for $x > c$. The left-hand derivative of f is similarly defined.

right-handed *Neurology.* using the right hand more frequently and with greater dexterity than the left. Thus, **right-handedness.** *Mechanical Engineering.* rotating clockwise. *Crystallography.* a crystal structure characterized by that which is a mirror image of a left-handed structure.

right-handed coordinate system *Mathematics.* **1.** a system visualized in Euclidean space by extending the right thumb and forefinger at a right angle and the right middle finger at right angles to both of them. Then the positive x, y, and z axes are taken in the directions of the thumb, forefinger, and middle finger, respectively. It is so named because a screw with a right-handed thread placed along the z axis will advance in the positive z direction when it is turned so that the positive x axis rotates directly into the positive y axis. **2.** more generally, on a Riemannian n-manifold, where the coordinate functions have gradient vectors, the coordinate system is right-handed if the determinant of the gradients (in the given order) is positive.

right-hand helicity *Quantum Mechanics.* the property of a particle such that its spin and angular momentum are parallel.

right-hand polarization *Electromagnetism.* for a circularly or elliptically polarized electromagnetic wave, the polarization of the field vector at a fixed point in space that is observed to rotate in the right-hand rule sense.

right-hand rule *Electromagnetism.* **1.** a rule stating that if the thumb of a person's right hand points along the direction of the current in a conductor, the fingers would curl about the conductor in the direction of the magnetic field circulation produced by the current. **2.** a rule stating that for current passing through a region of magnetic field that is perpendicular to the current direction, if the thumb and index finger of the right hand are extended perpendicularly to each other and the thumb is aligned with the current while the index finger indicates the magnetic field direction, then the magnetic field force acting on the charges is given by the direction outwardly normal to the palm of the right hand.

right-hand screw *Mechanical Engineering.* a screw whose threads allow coupling only by turning in a clockwise direction.

right-hand taper *Electricity.* a taper having more resistance in the clockwise half of the operating range of a rheostat than in the counterclockwise half.

right heart bypass *Cardiology.* surgery that diverts the blood flow away from the right atrium and right ventricles so that it passes directly to the pulmonary arteries.

righting arm *Naval Architecture.* in a heeling ship, the horizontal distance between the vessel's metacenter and its center of gravity; the moment arm that acts to bring it back to a full upright position.

righting moment *Naval Architecture.* the force acting on a heeling ship to bring it back to a full upright position, equivalent to the righting arm multiplied by the vessel's displacement (weight).

right-justified *Graphic Arts.* describing spacing within a line of type that is arranged or adjusted so that the last character is in a desired position, usually flush right.

right-justify *Computer Programming.* to place data items so that they occupy consecutive positions, ending at the rightmost position of a field or display ground.

right-laid *Design Engineering.* a lay of a rope or cable in which the individual wires or fibers twist to the right (clockwise).

right-lateral fault *Geology.* a strike-slip fault in which the relative displacement appears to be offset to the right when viewed across the fault plane. Also, **right-lateral slip fault, right-slip fault.**

right of way *Transportation Engineering.* the right granted to a vehicle to proceed ahead of another; e.g., the vehicle on the right (in the U.S.) at a four-way stop sign. *Civil Engineering.* a deeded portion of land granting easement for a specific purpose, such as overhead power lines or garage access. Also, **right-of-way.**

right parasternal impulse *Cardiology.* cardiac impulses originating from the right sternal border.

right-reading *Graphic Arts.* of photographic paper or film, reading normally (from left to right) when viewed from the emulsion side.

right rudder *Navigation.* displacing the rudder to the right so that the craft makes a turn to the right.

right section *Mathematics.* a section of a given surface by a plane normal to the surface.

right sphere *Astronomy.* the name given to the celestial sphere as observed from the earth's equator.

right strophoid *Mathematics.* the strophoid of a line L with respect to points A and B, where A is the foot of the perpendicular from the pole B to the line.

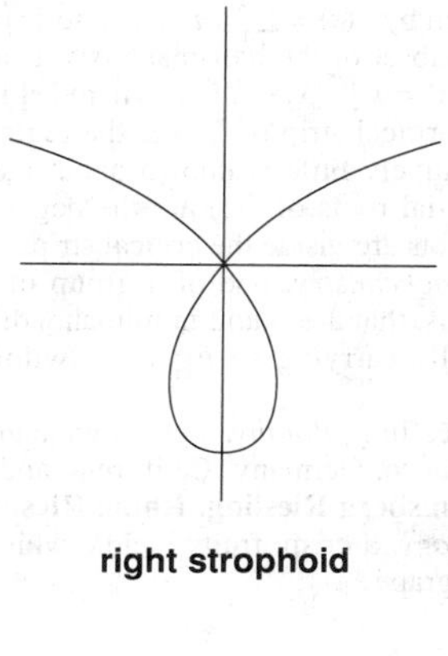

right strophoid

right triangle *Mathematics.* a plane triangle that has one right angle.
right way up *Geology.* a term for the original order of deposition in an upward succession of layers. Also, **right side up.**
right whale *Vertebrate Zoology.* any of several large whalebone whales of the genus *Balaena* in the family Balaenidae; found throughout the polar seas, though greatly reduced in number due to hunting.
rigid *Mechanics.* able to resist deformation so as to retain an existing shape and size. *Medicine.* exhibiting or characterized by rigidity; stiff or inflexible. *Aviation.* **1.** of a lighter-than-air craft, supported by an internal framework. **2.** of a rotor, fixed at its root.
rigid arch *Building Engineering.* a continuous arch fabricated without hinges or joints and attached to the abutments.
rigid arm conveyor see ARM CONVEYOR.
rigid body *Mechanics.* **1.** a perfect body that does not change its shape or size regardless of the force applied to it; i.e., the relative position of its component particles is absolutely fixed in position relative to one another. **2.** an actual body whose behavior approaches that of an ideal rigid body, such as a steel beam. Thus, **rigid-body motion, rigid-body dynamics.**
rigid copper coaxial line *Electromagnetism.* a coaxial cable whose central conductor and outer conductor are formed by joining rigid pieces of copper.
rigid coupling *Mechanical Engineering.* a device that joins shafts so that their axes are directly in line.
rigid frame *Building Engineering.* a structural steel skeleton in which beams and columns are rigidly connected without the use of hinges, so that angles formed by adjoining members do not change.
rigidity *Mechanics.* the fact of being rigid; resistance to deformation. *Medicine.* stiffness or inflexibility, especially that which is abnormal or morbid. *Astrophysics.* the momentum per unit of charge in cosmic ray studies; for relativistic particles it is proportional to the energy per nucleon.
rigidizer *Engineering.* a supporting structure designed to eliminate unwanted vibrations.
rigid pavement *Civil Engineering.* a roadway or airstrip foundation constructed of concrete slabs and made to withstand and distribute heavy loads.
rigor [rig′ər] *Medicine.* **1.** a condition of rigidity or stiffness. **2.** a chill, especially one associated with a fever.
rigor mortis [rig′ər môr′tis] *Pathology.* the stiffening of a dead body that takes place about one to seven hours after the time of death due to the hardening of the muscular tissue in reaction to the congealing of mysinogen and paramyosinogen; it dissipates in five or six days, at the onset of decomposition. (From Latin; literally, "stiffness of death.")
Riley reaction *Organic Chemistry.* a common method for preparing compounds containing two adjacent carbonyl groups; it involves the oxidation of aldehydes or ketones with selenium dioxide.
rill *Hydrology.* **1.** a small, trickling stream of water. **2.** the channel that is eroded by such a stream. *Geology.* **1.** a small, temporary, troughlike hollow that carries wave water from a beach to the sea or a lake, especially following an outgoing tide. **2.** a small channel in the wall, floor, or ceiling of a cave that was formed by circulating water. **3.** any small channel resulting from rill erosion. *Astronomy.* see RILLE.

rills

rille *Astronomy.* a long, narrow, trenchlike valley on the surface of the moon; it may be straight or winding. Also, RILL.
rille crater *Astronomy.* a lunar crater that forms part of a rille.
rillenstein *Geology.* a pattern of tiny grooves formed by solution on the surface of a soluble rock.
rill erosion *Geology.* the formation of many, minute, closely spaced channels as a result of an uneven removal of surface soil by fine streams of running water. Also, **rill wash, rilling, rillwork.**
rill flow *Hydrology.* surface runoff that flows in very small, irregular channels.
rill mark *Geology.* **1.** a small, branching channel formed in beach mud or sand by a rill or by a retreating tide, often found on the sides of an obstruction such as a shell or a half-buried pebble. **2.** a small, downsloping, branching channel formed by a streamlet that emerges on a sand or mud flat.
rillstone see VENTIFACT.
rim *Design Engineering.* **1.** the outer part of a wheel connected by spokes to a hub. **2.** the raised or projected outer edge or border of a device or part. *Metallurgy.* on an ingot, an outer layer of metal whose composition is different from the composition of the center. *Geology.* see RIMROCK, def. 2.
RIM reaction injection molding; reproductive isolating mechanism.
rima *Geology.* a long, narrow opening or fissure.
rim blight *Plant Pathology.* a disease of tea plants caused by members of the fungus genus *Cladosporium,* characterized by yellowing and eventual browning of leaf borders.
rim cement *Geology.* in detrital sediments, a thin deposit of calcium carbonate that forms on the surface of a grain having the same composition.
rim clutch *Mechanical Engineering.* a frictional contact clutch having areas on its surface that exert internal or external pressure to the rim.
rim drive *Acoustical Engineering.* a mechanism for driving a phonograph turntable by means of a rubber-coated wheel on the inside of the turntable.
rime *Hydrology.* an opaque, granular deposit of ice crystals formed when droplets of supercooled fog come in contact with a vertical surface whose temperature is below freezing. Also, HARD RIME.
rime fog see ICE FOG.
rimfire or **rim-fire** *Ordnance.* **1.** relating to or describing a cartridge in which the priming mixture is located in the rim of the base. **2.** describing a firearm using such cartridges.
rim gypsum *Geochemistry.* gypsum occurring in thin films between anhydrite crystals.
rimless *Ordnance.* describing a type of cartridge in which the base is equal in diameter to the body of the case; such cartridges are center-fire.
rimmed *Ordnance.* describing a type of cartridge in which the base is larger in diameter than the body of the case; it may be center-fire or rimfire.
rimmed kettle *Geology.* in a moraine, a deep, bowl-shaped hole or depression having raised edges.
rimmed solution pool *Geology.* a pool in rock that has a hardened rim formed by the deposition of lime during evaporation.
rimmed steel *Metallurgy.* a low-carbon steel that is not fully deoxidized; thus, evolving carbon monoxide during solidification.
rimming wall *Geology.* a steep, ridgelike erosional remnant consisting of continuous layers of porous, permeable, and poorly cemented detrital limestones.
rimpylite *Mineralogy.* one of a group of hornblendes characterized by green and brown color and containing large amounts of $[(Al,Fe)_2O_3]$.
rim ridge *Geology.* a small ridge of till that delimits the edge of a moraine plateau.
rimrock or **rim rock** *Geology.* **1.** a ledge or cliff overlooking lower ground and formed by the outcropping of a horizontal layer of resistant rock on an elevated area. **2.** a cliff or vertical face of an outcrop of rock in a canyon wall. Also, RIM. **3.** bedrock that forms or rises above the edge of a gravel deposit or placer.
rimsherd *Archaeology.* a fragment of the rim of a pottery artifact.
rimstone *Geology.* a thin, crustlike, calcium-containing deposit that forms a ring around an overflowing basin or pool of water.
rim syncline *Geology.* in salt tectonics, a local depression that forms a border around a salt dome. Also, PERIPHERAL SINK.
rimu *Materials.* the reddish brown wood of a New Zealand tree, *Dacrydium cupressinum;* used in making furniture and construction. Also, IMOU PINE.
rimy *Hydrology.* covered with rime.

rincon [rin´kän´; rin´kōn´] *Geology.* especially in the southwestern United States, a term for: **1.** a square-cut recess in a cliff. **2.** an angular indentation in the borders of a mesa or plateau. **3.** a small, secluded valley. **4.** a bend in a stream.

rind *Botany.* the thickened outermost layer of a fruit or fruiting body.

rinderpest *Veterinary Medicine.* an epidemic, often fatal viral disease of cattle, sheep, and goats, manifested by fever, diarrhea, ulcerations of the mucous membranes, and by a reddish, watery discharge from the mouth, nose, and eyes. Also, CATTLE PLAGUE.

ring a circular band or formation; specific uses include: *Mechanical Devices.* a heavy, welded circular piece of steel; used in various chain assemblies with other connectors or with spring-loaded snaps. *Organic Chemistry.* any closed loop of bonded atoms in a molecule. *Geology.* see RING STRUCTURE. *Design Engineering.* a member of a chain link, through whose center compression or tension is applied, producing normal force on radial sections. *Oceanography.* a detached meander or eddy of a western boundary current, grown too large to continue moving with the current. *Computer Programming.* **1.** see CIRCULAR LIST. **2.** a list structure in which the list and all sublists are circularly linked. *Mathematics.* a ring R is a set X together with two internal operations $(x, y) \rightarrow x + y$ and $(x, y) \rightarrow xy$, called respectively (ring) addition and (ring) multiplication, such that: (a) X is an Abelian group under addition; and (b) multiplication is associative and distributive with respect to addition. If, in addition, there exists an element $1 \in X$ such that $1x = x1 = x$ for all $x \in X$, then R is called a **ring with identity**. If multiplication in R is commutative, R is called a **commutative ring** or an **Abelian ring**.

ring-and-bead sight *Ordnance.* a gunsight system in which the rear sight is a ring and the front seat is a bead or post.

ring architecture *Computer Science.* see TOKEN RING.

ring auger *Mechanical Devices.* an auger with a ring or eye at its upper shank end, which is turned when a bar is inserted.

ring barking see BAND GIRDLE.

ring-billed gull *Vertebrate Zoology.* a gull, *Larus delawarensis,* characterized by a black ring around the bill; found in North America.

ring bivalent *Molecular Biology.* in a meiotic chromosome, a configuration caused by terminal chiasma in both arms.

ringbolt *Mechanical Devices.* an eyebolt with a ring through its eye.

ringbone *Veterinary Medicine.* an abnormal bony growth on the pastern or coronet of a horse, often resulting in lameness.

ring canal *Invertebrate Zoology.* **1.** in echinoderms, a circular canal near the mouth, part of the water vascular system. **2.** in free-swimming coelenterate medusae, a circular part of the gastrovascular system around the edge of the umbrella.

ring chromosome *Molecular Biology.* **1.** an abnormal chromosome in which both arms are joined, forming a ring structure. **2.** an association of metacentric tetrads of chromosomes, forming a temporary ring.

ring circuit *Electromagnetism.* a waveguide hybrid T configuration of a ring having several branches.

ring complex *Geology.* an association of ring-shaped igneous intrusions, such as ring dikes and cone sheets.

ring compound *Chemistry.* a compound whose structure contains a closed chain or ring of atoms; a cyclic compound.

ring counter *Electronics.* a device in which a voltage pulse repeatedly circulates around a loop of circuits.

ring current *Geophysics.* a westward-moving electric current that is believed to circle the earth during geomagnetic storms, causing a significant, worldwide decrease in the horizontal component of the earth's magnetic field in low latitudes.

ring data structure *Computer Programming.* data that is organized in storage by a chain of pointers, with the last pointer directed back to the chain's beginning.

ring depression *Geology.* in a cryptoexplosion structure, a round, depressed area that surrounds the centrally uplifted region. Also, RING SYNCLINE, PERIPHERAL DEPRESSION.

ring dike *Geology.* a dike with a curved or roughly circular outcrop that is either vertical or inclined away from the center of the arc. Also, **ring-fracture intrusion.**

ring DNA see CIRCULAR DNA.

ring dove *Vertebrate Zoology.* **1.** a wood dove of Europe, characterized by a whitish patch on each side of the neck. **2.** a dove, *Streptopelia risoria,* characterized by a black half ring about its neck; found in Europe and Asia.

ringdown *Telecommunications.* in telephone switching, a method of signaling an operator in which the telephone ringing current is transmitted over the line to operate a drop or a self-locking relay and lamp.

ringed spherulites *Materials Science.* a spiraled or ringed arrangement of lamellae in the morphology of some crystalline polymers; often associated with a periodic twisting of the lamella.

ringent *Botany.* of or related to a bilabiate corolla that opens widely upon reaching anthesis, such as in the inflorescence of a mint or snapdragon flower. *Zoology.* having gaping mouthlike parts, as in the shells of certain bivalves.

Ringer's solution *Chemistry.* a solution of 0.86 gram of sodium chloride, 0.03 gram of potassium chloride, and 0.033 gram of calcium chloride in boiled and purified water; used as a salt solution.

ring fault *Geology.* **1.** a steep-sided, cylindrical-shaped fault pattern associated with cauldron subsidence. Also, **ring fracture. 2.** a fault that borders a rift valley.

ring fissure *Geology.* a roughly circular crack on a playa that is produced by drying and forms around a point source.

ring-fracture stoping *Geology.* a large scale intrusion of magma associated with cauldron subsidence that involves the downward sinking of pieces of country rock.

ring galaxy *Astronomy.* a galaxy characterized by a bright elliptical ring with clumps of ionized hydrogen clouds on its periphery; thought to have been involved in a galaxy/galaxy collision.

ring gate *Civil Engineering.* a gate used to regulate flow into a morning-glory type spillway without retarding the passage of ice and the like. *Metallurgy.* a gate with a widened opening containing a centered disk that prevents molten metal from falling in a direct vertical stream.

ring gauge *Mechanical Devices.* a steel ring with a particular inside tolerance used to verify the outside diameter of a cylindrical object. Also, **ring gage.**

ring gear *Mechanical Engineering.* a ring-shaped gear in an automobile differential that is moved by the propeller shaft pinion and transmits power through the differential to the line axle.

ringing *Acoustics.* a repetitious and undesired sound output by a sound system during the time it takes for the filter circuit to discharge. *Control Systems.* an oscillation that occurs in the output of a system after a sudden change in the input.

ringing time *Engineering.* the period of time that a piezoelectric crystal continues to oscillate after no ultrasonic waves are produced, measured in microseconds.

ring isomerism *Organic Chemistry.* a geometrical isomerism in certain cyclic alkenes in which bond lengths and bond angles prevent the existence of the *trans* isomer of the molecule.

ringite *Geology.* an igneous rock formed by the mixing of silica- and carbonate-containing magmas.

ring jewel *Mechanical Devices.* a pivot-bearing jewel used in a gyroscope or timepiece.

ring laser gyro *Engineering.* an integrating rate gyroscope in which two laser beams move in opposite courses over a ring-shaped path created by three or more mirrors, to measure rotation without the aid of a spinning mass, so that its performance is not affected by accelerations; used in missile guidance systems.

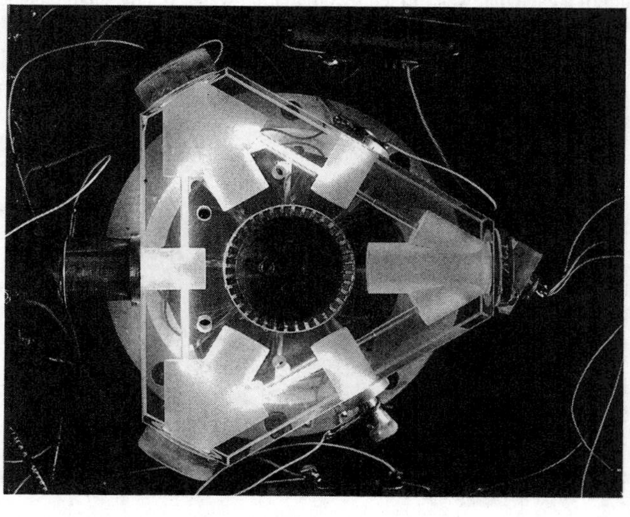

ring laser gyro

ringlock nail *Mechanical Devices.* a nail ringed with grooves to provide an extra means of securing a device under a load.

ring micrometer *Optics.* a ring set in the focal plane of a telescope in order to gauge differences in right ascension and declination.

ring modulator *Acoustical Engineering.* an electronic device that generates sum and difference frequencies from two or more frequencies.

Ring Nebula *Astronomy.* an ellipsoidal planetary nebula in Lyra, so named because in telescopes and astrophotos it appears like a smoke ring. Also, **M 57.**

ring of quotients *Mathematics.* **1.** let S be a multiplicative subset of a commutative ring R. Then define an equivalence relation $\sim$ on the set $R \times S$ by the rule $(r_1,s_1) \sim (r_2,s_2)$ if and only if $s'(r_1s_2 - r_2s_1) = 0$ for some $s' \in S$. The collection of all distinct equivalence classes under $\sim$ is called the ring of quotients of R by S; denoted $S^{-1}R$. The notation r/s is sometimes used instead of (r,s). Also, **ring of fractions, quotient ring. 2.** if R is an integral domain, then $S^{-1}R$ is a field, called the quotient field. If R is a ring with identity and P is a prime ideal of R, then $S = R\backslash P$ (i.e., the elements of R that are not in P) is a multiplicative subset of R, and the ring of quotients $S^{-1}R$ is called the localization of R at P.

ring of sets *Mathematics.* a collection of sets closed under union and symmetric difference.

ring-oil *Mechanical Engineering.* the process of oiling a machine part, usually a bearing, by moving the oil to the part that needs oiling using a ring that lies upon and turns with a journal and dips into a container of the lubricant.

ring-opening polymerization *Materials Science.* a reaction in which the seven-carbon ring structure is opened; used in many reactions in the polymerization of caprolactam.

ring plain *Astronomy.* a former term for a large lunar crater. Also, RINGWALL STRUCTURE.

ring-porous *Botany.* of or related to wood that has larger or more numerous vessels in the spring than in the summer, so that the earlier wood appears as a ring of small holes in cross section.

ring power-transmission line *Electricity.* a power-transmission line that connects several load or distribution points in a circular configuration.

ring-roller mill *Mechanical Engineering.* a mill that crushes or pulverizes material by means of rollers that move the material and press it against the sides of a revolving bowl. Also, ROLLER MILL.

ring rot *Plant Pathology.* **1.** an infectious disease of potatoes caused by the bacterium *Corynebacterium sepedonicum*, characterized by decay in the vascular ring within the tuber. **2.** a disease of sweet potatoes caused by the fungus *Rhizopus stolonifer*; characterized by rings of dry rot around the root.

ring scission *Materials Science.* in polymerization of complex monomers, the breaking of a tight C–O–C ring in order to rearrange the bonds and join the molecules.

ring sight *Ordnance.* a sight in the shape of a ring or a series of concentric rings; it may be mechanical or optical.

ring slide *Biotechnology.* a large, thick slide with fixed, raised rings from 1 to 3 centimeters in diameter on its face; often used in serological tests.

rings of Saturn

rings of Saturn *Astronomy.* the seven shining, thin, flat rings around the planet Saturn, made up of particles, forming numerous thin ringlets that orbit the planet.

rings of stars *Astronomy.* illusory rings of stars occasionally seen on photographic plates.

ring spinning *Textiles.* a spinning system using a special frame that drafts the roving, twists the yarn, and winds it on a bobbin simultaneously and continually, generally producing a stronger yarn than other types of spinning.

ring spot *Plant Pathology.* any of several plant diseases caused by fungi or viruses, characterized by darkened, depressed lines and rings in leaves, stems, and fruit.

ring stress *Mining Engineering.* a zone of higher than pre-existing stress in rock; such zones surround all development excavations.

ring structure *Geology.* any secondary ring-shaped surface structure of the earth.

ring syncline see RING DEPRESSION.

ring system *Organic Chemistry.* the arbitrary designation of certain four-, five-, and six-member compounds as closed ring molecules, including combinations of carbon, nitrogen, oxygen, sulfur, and other elements.

ringtail *Vertebrate Zoology.* another name for the cacomistle, a slender, nocturnal mammal of the Southwest U.S. and Mexico that is related to the racoons but distinguished by its long, bushy black and white ringed tail.

ringtail (cacomistle)

ring test *Immunology.* a method used to detect the presence of an antigen-antibody reaction, in which a soluble antigen is layered with a corresponding soluble antibody in a test tube; a band of white precipitate becomes visible at the interface of the two substances to indicate a positive reaction.

ring theory *Mathematics.* the study of rings and their homomorphisms.

ring time *Electronics.* **1.** the length of time it takes for a pulse of energy transmitted to an echo box to dissipate; measured in microseconds and used to determine a radar system's performance capabilities. **2.** the period of time in which an alternating current progressively decreases its amplitude and stops oscillating.

ringwall structure see RING PLAIN.

ringworm *Medicine.* a general name for various fungal infections of the skin, hair, or nails marked by ring-shaped or oval itchy lesions. Also, TINEA.

ringworm corrosion *Materials Science.* selective galvanic corrosion occurring in plain-carbon steel piping near the end of the pipe that has been especially heated to forge the flange portion.

ringworm of the nails see ONYCHOMYCOSIS.

rinkite *Mineralogy.* $(Na,Ca,Ce)_3Ti(Si_2O_7)_2OF_3$, a rare, reddish-brown, yellowish-green, or yellow, weakly radioactive, monoclinic mineral, having a specific gravity of 2.93 to 3.5 and a hardness of 5 on the Mohs scale; found as crystals and in massive form in nepheline syenites and their pegmatites. Also, **rinkolite.**

rinneite *Mineralogy.* $K_3NaFe^{+2}Cl_6$, a colorless, violet, rose, or yellow trigonal mineral occurring in coarse, granular masses, having a specific gravity of 2.347 and a hardness of 3 on the Mohs scale; found in saline deposits.

rinse *Graphic Arts.* in photography, a brief, clean-water bath used to decrease the carryover of solutions between stages of the processing cycle.

RIO *Aviation.* the airport code for Rio de Janeiro, Brazil.

Rio de Plata *Geography.* the wide estuary of the Paraná and Uruguay rivers, between eastern Argentina and southern Uruguay. Also, PLATA.

Riodininae *Invertebrate Zoology.* a subfamily of small, delicate, bright-colored butterflies, lepidopteran insects in the family Lycaenidae.

Rio Grande [rē´ō grand´; rē´ō grän´dā] *Geography.* a river that rises in the San Juan Mountains of southwestern Colorado and flows 1885 miles to the Gulf of Mexico.

Riojasaurus *Paleontology.* a genus of early herbivorous prosauropod dinosaurs in the family Melanosauridae; up to 10 meters long; transitional between plateosaurs and sauropods; the only South American representative of the family, but other melanosaurs have been found in Europe and southern Africa; extant in the Upper Triassic.

riometer *Engineering.* a meter that measures the level of cosmic radiation from space as an indicator of the changes in the ionosphere, by assuming the radiation is a constant value and the absorption of the radiation by the ionosphere fluctuates.

Rio Negro *Geography.* a river that flows 1400 miles from eastern Colombia into the Amazon.

riot-control operations *Military Science.* the employment of riot-control agents and special tactics, formations, and equipment in the control of violent disorders.

riot gun *Ordnance.* a short-barrelled shotgun used in guard duty or riot control; it usually has a 20-inch cylinder bored barrel.

rip *Oceanography.* a turbulent agitation of the water's surface, usually caused by a conflict between currents, between currents and the wind, or between a strong current and an irregular bottom. *Hydrology.* see RIPPLE, def. 1. *Engineering.* to saw wood along the direction of the grain.

riparian [ri pâr´ē ən] *Geology.* relating to or situated on the bank of a river or stream. Thus, **riparian land.** *Biology.* living or found on the banks of rivers or streams. *Agriculture.* relating to or holding riparian rights.

riparian right *Agriculture.* a right held by the owner of riverfront land, such as the right to reasonable use of the river water for irrigation. Thus, **riparian owner.**

rip channel *Geology.* a deep channel cut on the shore by a rip current.

rip cord *Aviation.* **1.** a cord or cable that holds a parachute pack closed and, when pulled during free-fall, allows the pack and parachute to open. **2.** a cord or cable used to open a rip panel.

rip current *Oceanography.* a seaward flow of water that has been piled up on shore by waves and wind, consisting of feeder currents, the neck, and the head.

ripe *Botany.* of or related to a plant or plant part, especially a fruit or grain, that is fully developed or mature. *Forestry.* of or relating to timber in a forest area that is ready to be cut for commercial value or other forest management goals. *Geology.* describing peat in an advanced state of decomposition.

ripe snow *Hydrology.* a term for wet, coarse, crystalline snow that will produce meltwater runoff with additional melting.

ripidolite *Mineralogy.* a ferroan variety of clinochlore.

rip panel *Aviation.* a patch or panel on an aerostat envelope (such as a dirigible or free balloon), usually near the top, which can be ripped open for emergency deflation.

ripper *Mechanical Devices.* a long bar or thin steel blade used to break up solid material, such as rock, pavement, or roofing.

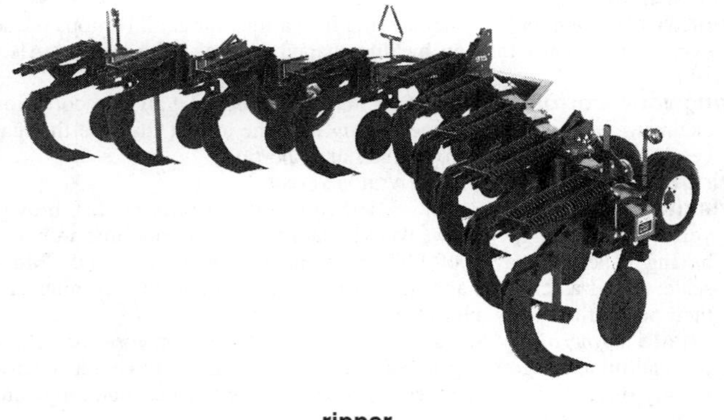

ripper

ripping bar *Mechanical Devices.* a high-carbon steel bar, from one to several feet in length, with one end bent into a forked hook and the other into a slightly angled chisel; used for tasks such as heavy prying and wrecking, providing leverage for lifting heavy objects, or as a nail puller.

ripping chisel *Mechanical Devices.* a hexagonal steel rod with a wide chisel end, a beveled notch, and a teardrop-shaped nail slot for removing nails and cleaning mortises.

ripping face support *Mining Engineering.* a timber or steel structure providing support at the ripping lip.

ripping lip *Mining Engineering.* during roadway enlargement, the end of the enlarged section and the site at which the work is being conducted.

ripping punch *Mechanical Devices.* a punch-press tool with a rectangular cutting edge; used to crosscut metal plates.

ripple a slight disturbance of a surface; specific uses include: *Hydrology.* **1.** a small disturbance of the water's surface, as by a breeze or by the distant movement of a craft within the water. Also, RIP. **2.** a shallow expanse of running water in a stream, broken into small waves by a rocky or uneven bottom. *Fluid Mechanics.* surface waves of short wavelength in deep fluid dominated by surface tension. *Geology.* a tiny ridge of sand on the bedding surface of a sediment, particularly a ripple mark. Also, SEDIMENTARY RIPPLE. *Electricity.* the presence of an alternating current component in a direct-current signal. *Graphic Arts.* the finish of a paper. *Textiles.* a comblike device used to remove seeds or capsules from flax, hemp, and other fibrous materials used for making fabric.

ripple bedding *Geology.* **1.** a bedding surface covered by ripple marks. **2.** the small-scale rippling characteristic of sand that is deposited rapidly.

ripple biscuit *Geology.* a bedding structure produced by lens-shaped lamination of sand in a bay or lagoon.

ripple current *Electrical Engineering.* the AC component of the current superimposed on a pulse or DC background current.

ripple drift *Geology.* a small-scale, cross-lamination pattern produced as sediment is deposited by migrating ripples.

rippled sandstone *Geology.* a formation of sandstone that is ripple-marked.

rippled sandstone

ripple filter *Electronics.* a filter circuit that passes the direct current that is received from a rectifier, while reducing the attendant alternating current.

ripple index *Geology.* on a ripple-marked surface, the ratio of the horizontal distance between two crests or troughs of a ripple to the vertical distance between the crest of the ripple and the trough of an adjacent ripple. Also, RIPPLE-MARK INDEX, VERTICAL FORM INDEX.

ripple lamina *Geology.* a sedimentary structure formed internally, rather than on the surface, by the action of waves or currents on sand or silt. Also, **ripple lamination.**

ripple load cast *Geology.* a ripple mark whose load cast is characterized by an accentuated trough and crest, as well as by oversteepening of its original deposition layers. Also, LOAD-CASTED RIPPLE.

ripple mark or **ripple-mark** *Geology.* **1.** a pattern of undulating symmetrical or asymmetrical ridges and grooves formed by wave action or by water or air currents moving over the surface of loose, unconsolidated sedimentary material. **2.** one of the small ridges produced on such a surface. **3.** one of a series of parallel ridges and grooves formed by wind moving over the surface of snow.

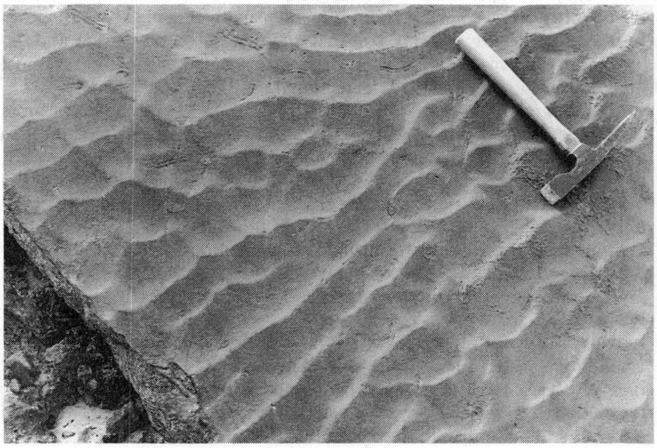

ripple mark

ripple-mark index see RIPPLE INDEX.

ripple quantity *Physics.* the oscillating component of a quantity superimposed on a relatively large constant value.

ripple scour *Geology.* a shallow, linear trough exhibiting crosswise ripple marks.

ripple sort see BUBBLE SORT.

ripple symmetry index *Geology.* the ratio of the horizontal length of the upcurrent side of a ripple mark to the horizontal length of the downcurrent side.

ripple tank *Physics.* a device consisting of a shallow tray filled with a liquid and a mechanism that generates surface waves for the purpose of demonstrating wave patterns, such as interference phenomena.

ripple till *Geology.* a sheet of glacial till exhibiting low, winding, smooth-topped ridges lying at right angles to the direction of ice movement and grouped into narrow parallel belts. Also, RIBBLE.

ripple voltage *Electrical Engineering.* the small amount of AC voltage fluctuation remaining at the output of a DC power supply; usually eliminated by use of regulated power supplies.

riprap *Civil Engineering.* a lightweight stone covering used to protect soil or surface bedrock from erosion by water or the elements.

ripsaw *Mechanical Devices.* a slightly tapered, coarse-toothed saw with about 3 teeth per inch; used to cut wood in the direction of the grain.

RISC reduced instruction set computer.

rise *Science.* **1.** to ascend or increase in value. **2.** the amount or rate at which this occurs. *Astronomy.* the path that an ascending celestial body takes when crossing the visible horizon. *Geology.* **1.** the top portion of a landform that is higher than the surrounding area. **2.** an upward slope in land. **3.** a long, broad, smooth elevation in the deep-sea floor. Also, SWELL. *Hydrology.* see RESURGENCE, def. 1.

rise of tide *Oceanography.* the height of a tide above the tidal datum.

rise pit *Geology.* a surface cavity or hollow through which an underground stream emerges.

riser something that rises or that causes a rising movement; specific uses include: *Geology.* one of the vertical or steeply sloping faces of a steplike landform. *Civil Engineering.* the vertical board or rise between stairs. *Chemical Engineering.* the part of the bubble cap assembly on a distillation column that serves to connect the cap to the tray and to guide the rising vapor through the cap during operation. *Materials Science.* an extra reservoir of liquid metal connected to a casting; as the riser solidifies after the casting, it can provide liquid metal to compensate for shrinkage in the casting. *Petroleum Engineering.* **1.** in an offshore drilling operation, the piping that extends from the platform to the hole, through which drilling is conducted. **2.** piping through which gas or liquid may flow upward. *Metallurgy.* in casting, molten metal stored above the mold cavity to assure proper feed during solidification.

riser plate *Civil Engineering.* a steel plate forming the vertical face of a stairway.

rise time *Control Systems.* the time it takes for the energy generated by a system to change from approximately 10% of its final value to 90% of its final value. *Electronics.* **1.** the time it takes for a signal to rise from zero or a reference level to its highest amplitude. **2.** the time it takes for the pointer on an electric instrument to reach 90% of its final value after power is applied.

rishitin *Biochemistry.* a phytoalexin bred by potato tubers.

rising dune see CLIMBING DUNE.

rising hinge *Building Engineering.* a hinge with a loose leaf that rises on its center pin when opened, causing the attached door to close automatically.

rising limb *Hydrology.* a part of a hydrograph that indicates an increase in stream flow as a result of rainfall or snowmelt.

rising mine *Ordnance.* a naval mine of positive buoyancy that is released from a sinker by a ship influence or timing device; it may fire by contact, hydrostatic pressure, or other means.

rising-sun magnetron *Electronics.* a tube in which two resonators, whose currents operate at different frequencies, are arranged alternately around the perimeter of the anode, so that they resemble sun rays; used to generate AC power at microwave frequencies.

rising tide see FLOOD TIDE.

risk *Statistics.* **1.** in classical decision theory, a function of the unknown parameter describing the expected loss associated with a decision rule. **2.** in Bayesian decision theory, the expected loss of a decision rule, in which expectations are taken over all unknown quantities.

risk aversion *Ecology.* the preference of a tactic that minimizes the variance of rewards.

risk proneness *Ecology.* the preference of a tactic that offers a chance of high rewards due to the high variance in payoff.

risk-sensitive foraging *Ecology.* a foraging situation in which a choice has to be made between a certain payoff and a gamble with the same mean value.

Risley prism system *Optics.* an arrangement of two thin, usually identical prisms, which facilitates simultaneous rotation in like or opposite directions; used in ophthalmology to test ocular convergence.

Riss *Geology.* **1.** a European geologic stage of the Pleistocene epoch, occurring after the Mindel and before the Würm. **2.** in the Alps, the third glacial stage of the Pleistocene epoch.

Rissoacea *Paleontology.* an extinct superfamily of small gastropods in the order Caenogastropoda; extant in the Cenozoic.

Rissoellaceae *Botany.* a monospecific family of red algae of the order Gigartinales, having flat, multiaxial thalli with dichotomous branching and marginal proliferations; found in the Mediterranean.

Riss-Würm *Geology.* in the Alps, the third interglacial stage of the Pleistocene epoch, occurring after the Riss glacial stage and before the Würm.

Ritchey, George Wills 1864–1945, American astronomer; invented and constructed telescopes and telescopic devices.

Ritchey-Chretien optics *Optics.* a modification of the Cassegrain telescope having a hyperbolic primary and secondary mirror and a wide field; free of spherical aberration and coma, but subject to astigmatism.

Ritchie wedge *Optics.* a photometer in which two white diffusing surfaces arranged at a 90° angle form a wedge shape, such that a light source illuminating one surface can be compared to a standard source illuminating the other surface.

rite of passage *Anthropology.* any ceremony that celebrates an individual's passing to a new phase of life, performed in circumstances such as birth, puberty, adulthood, marriage, parenthood, and death.

Rittenhouse, David 1732–1796, American astronomer and instrument maker; introduced spider lines in telescope focus; measured state boundaries and the earth's distance from the sun.

Ritter, Karl 1779–1859, German geographer; founder of comparative geography.

Ritter reaction *Organic Chemistry.* a reaction that forms amides by adding alkenes or tertiary alcohols to nitriles in an acidic solution.

Rittinger's law *Mechanical Engineering.* a law stating that the energy consumed in a crushing or grinding operation is directly proportional to the area of the new surface produced.

ritual *Psychology.* a compulsively repeated series of actions carried out as a defense against anxiety.

ritualization *Behavior.* the process of transforming a behavior into one that communicates more effectively, usually by repeating and exaggerating the original behavior.

ritualized fighting *Behavior.* fighting that is restricted in certain ways and that is not directed toward the death of the opponent, but is intended to establish dominance in some form; occurs among members of the same species competing for mates, territory, food supply, and so on.

Ritz formula *Atomic Physics.* an equation that is used to analyze the spectra of atoms.

Ritz method see RAYLEIGH-RITZ METHOD.

Ritz-Paschen series see PASCHEN SERIES.

Ritz's combination principle *Spectroscopy.* an empirical rule stating that the sums and differences of the frequencies of spectral lines are usually equal to other observed frequencies. Also, COMBINATION PRINCIPLE.

river *Hydrology.* a natural stream of fresh water that is larger than a brook or creek, has a permanent or seasonal flow, and moves in a definite channel toward a sea, lake, or another river. *Geography.* the channel or course permanently or seasonally occupied by such a stream.

river bar *Geology.* a ridge or mound of alluvium accumulated along the banks, in the channel, or at the mouth of a river.

river basin *Geology.* the entire tract of land drained by a river and its tributaries.

riverbed or **river bed** *Geology.* a channel through which a river now flows or once flowed.

riverboat *Naval Architecture.* any vessel designed for operation on rivers, including towboats and barges; typically characterized by shallow draft, a flat bottom, and relatively broad beam.

river bottom *Geology.* the low-lying alluvial land through which a river flows.

river breathing *Hydrology.* a term for the rising and falling of the water level of a river.

river buoy *Navigation.* buoy used to mark the channel in a river.

river-deposition coast *Geology.* the coast of a delta bordered by lowlands and characterized by seaward bulges in the form of lobes that are crossed by river distributaries.

river dolphin *Vertebrate Zoology.* a toothed, freshwater whale of the family Platanistidae; characterized by reduced eyes, a bulging forehead, and a long rostrum (snout); native to South America and Asia.

river drift *Geology.* rock material that is deposited by a river in one location after being transported from another location.

river end *Hydrology.* the lowest point of a river that has no outlet to the sea, located where the water disappears as a result of evaporation or percolation.

river engineering *Civil Engineering.* the field of engineering as it applies to the flow of rivers or the control of such flow.

river flat see ALLUVIAL FLAT.

river forecast *Hydrology.* a prediction of expected river stage or discharge, based on hydrologic and meteorologic factors.

river gauge *Engineering.* a gauge used to measure the height of a body of water. Also, STREAM GAUGE.

river ice *Hydrology.* ice that is formed on or carried by a river.

riverine *Geology.* found in or near a river, as an ore deposit. *Biology.* living alongside or near a river.

riverine operations *Military Science.* operations conducted by forces organized to cope with and exploit the unique combination of land and water characteristics of a riverine area; such operations are designed to locate and destroy hostile forces and to achieve or maintain control of the riverine area; they may combine land, naval, and air operations as appropriate.

river mining *Mining Engineering.* the mining or excavation of an existing river, either by dredging or by deflecting its course.

river morphology *Hydrology.* the study of a river channel and the network of tributaries within the river basin, in which an aerial view and the shape of a given cross section within a limited reach are examined at several points along the channel. Also, CHANNEL MORPHOLOGY, FLUVIOMORPHOLOGY, STREAM MORPHOLOGY.

river otter *Vertebrate Zoology.* an otter, *Lutra candensis*, characterized by brown and silver fur; native to lakes and streams in the U.S. and Canada.

river pattern see CHANNEL PATTERN.

river patterns *Materials Science.* any of the branching radial ridges that appear on the smooth surface of a fracture at a slight deviation from the original plane of fracture; a distinct feature of a fracture surface that can be used to locate the fracture origin.

riverplain valley *Geography.* a valley in the final stage of development, with a broad floodplain filled with alluvium.

river profile see STREAM PROFILE.

river run gravel *Geology.* natural gravel found in deposits that have been acted upon by running water.

Rivers, W. H. R. 1864–1922, English anthropologist; applied geneological method and psychoanalytic theory to ethnography.

river system *Hydrology.* a natural system consisting of a river together with all its tributaries. Also, WATER SYSTEM.

river tide *Oceanography.* a tide resembling an ocean tide and occurring in rivers that empty directly into the sea.

river valley *Geography.* a valley cut and usually occupied by a river.

riverwash *Geology.* **1.** any soil material that has been carried and deposited by a river. **2.** an alluvial deposit in a riverbed or flood channel that is subject to erosion and deposition during recurring floods.

rivet *Mechanical Devices.* a headed pin or bolt used to join two or more workpieces by passing the shank through a hole in each piece and then hammering down the protruded end to form a second head.

riveter *Mechanical Devices.* a device with two plierslike handles and a nose that flattens out a flanged rivet when fastening it to an item, creating a sandwich bond; used for repairing thin metal, canvas, or leather items, especially when only one side of the material is accessible.

riveting *Metallurgy.* the joining of two or more members of a structure by the insertion and stamping of metal rivets.

riveting die see SETTING PUNCH.

riveting hammer *Mechanical Devices.* a narrow hammer with a corresponding pane used to drive rivets and beat metal.

rivet pitch *Engineering.* the spacing between rivets in metalworking.

rivet set *Mechanical Devices.* a steel punch with a hollowed or cupped face used to close up and form rivet tails.

rivet tongs *Mechanical Devices.* long, narrow-jawed tongs used to grasp red-hot rivets and place them into holes.

Riviera *Geography.* **1.** usually, **the Riviera.** the Mediterranean coast of southern France and northwestern Italy. **2. riviera.** by extension, any strand or shoreline region, especially when utilized as a resort area.

riving *Geology.* the cracking, splitting off, or fracturing of a rock, especially by the action of frost.

riving knife see FROE.

rivulet *Hydrology.* a small stream, brook, or river.

rivulose *Botany.* having a dendritic or highly branching pattern.

RJE remote job entry.

RK galaxy *Astronomy.* a ring galaxy that contains a single dominant knot or condensation in its ring system.

R Leonis *Astronomy.* a long-period (or Mira-type) variable star that ranges from 4th magnitude to 11th and back again in a period that averages 312 days.

rII locus *Molecular Biology.* a segment of the bacteriophage T4 chromosome.

R loop *Molecular Biology.* a region on a duplex DNA strand in which an RNA sequence has paired with one DNA strand and has displaced the second strand, causing it to loop out.

RMA or **R.M.A.** Rubber Manufacturers Association.

R magnitude *Astronomy.* stellar brightness measured at 6800 Ångstroms.

R meson *Particle Physics.* a meson resonance that has experimentally been resolved into three peaks labeled R_1, R_2, and R_3 with respective masses of 1640, 1700, and 1750 MeV.

R meter *Nucleonics.* an instrument that indicates the intensity in roentgens of X-rays and gamma rays. Also, ROENTGEN METER.

r.m.m. relative molecular mass.

R Monocerotis *Astronomy.* an irregular variable star located at the wide end of the cone-shaped nebula NGC 2261.

RMP pathway *Biochemistry.* a type of cyclic metabolic pathway through which certain bacteria take up formaldehyde.

RMR resting metabolic rate.

RMS reason maintenance system.

rms or **RMS** root-mean-square.

RMS sound pressure see EFFECTIVE SOUND PRESSURE.

RMV respiratory minute volume.

RN or **R.N.** Registered Nurse.

Rn radon.

RNA *Biochemistry.* a linear, usually single-stranded polymer of ribonucleotides, each containing the sugar ribose in association with a phosphate group and one of four nitrogenous bases: adenine, guanine, cytosine, or uracil. RNA is found in all living cells; in prokaryotic and eukaryotic cells, it encodes the information needed to synthesize proteins (i.e., it copies "instructions" that it receives from DNA); in certain viruses, it serves as the genome. (An abbreviation for ribonucleic acid.)

RNA-coding triplet *Molecular Biology.* a group of three nucleotides, called a codon, that specifies a particular amino acid.

RNA-driven hybridization *Molecular Biology.* a hybridization technique used to form DNA-RNA hybrids in which excess RNA molecules are added to ensure the hybridization of all single-stranded complementary DNA to RNA sequences.

RNA ligase *Enzymology.* an enzyme that catalyzes the connection of RNA molecules by forming phosphodiester linkages that join the loose 5' and 3' ends of the two molecules.

RNA polymerase *Enzymology.* an enzyme that catalyzes the synthesis of RNA polymers from individual ribonucleoside 5'-triphosphates using DNA as a template. Also, DNA-DIRECTED RNA POLYMERASE, RIBONUCLEIC ACID POLYMERASE.

RNA polymerase I see POL I.

RNA polymerase II see POL II.

RNA polymerase III see POL III.

RNA primase see PRIMASE.

RNA replicase see REPLICASE.

RNase see RIBONUCLEASE.

RNasin *Biochemistry.* an inhibitor of RNase.

RNA splicing *Molecular Biology.* a processing procedure in which two exons, or coding sequences, are joined together in a eukaryotic mRNA after excision of the intervening introns; used to produce a mature RNA molecule. Also, ALTERNATIVE SPLICING.

RN galaxy *Astronomy.* a ring galaxy with an off-center nucleus.

RNO *Aviation.* the airport code for Reno, Nevada.

RO receive-only.

ROA *Aviation.* the airport code for Roanoke, Virginia.

roach *Vertebrate Zoology.* **1.** a shorter name for the cockroach, an insect belonging to the family Blattidae. See COCKROACH. **2.** a freshwater sunfish of the genus *Lepomis*; found in North America.

road *Civil Engineering.* any stable strip of land used as a travel surface, especially a paved or smoothed surface used by motor vehicles. *Mining Engineering.* any passageway or tunnel in a mine. *Geology.* **1.** one of a series of erosional terraces formed in a glacial valley as the water level dropped in an ice-dammed lake. **2.** see ROADSTEAD. (From an earlier English word referring to a trip on horseback; related to the word *ride*.)

roadbed *Civil Engineering.* **1.** the part of a road used most frequently by vehicular traffic. **2.** the foundation of a railway.

road grade *Civil Engineering.* the camber or rise of a roadway, usually for drainage purposes.

road octane *Engineering.* the measured antiknock rating of a motor fuel determined under normal driving conditions or using a chassis dynamometer.

roadrunner *Vertebrate Zoology.* a large, long-tailed, ground-dwelling cuckoo of the genus *Geococcyx* of the family Cuculidae, known for its running speed; characterized by speckled brown plumage, often colorful cheek patches, and a crested head; ranges from California to Mexico and Texas. Also, CHAPARRAL COCK.

roadrunner

roadstead *Geology.* a sheltered area of water near the shore with good holding ground for anchors.

road test *Engineering.* an automobile test given under normal driving conditions to evaluate the performance of automobiles, fuels, or accessory products.

roadway *Civil Engineering.* the main part of a road, as opposed to the outer edge or sidewalk.

roan antelope *Vertebrate Zoology.* a large, light-colored animal of the Bovidae family that can become dangerous when wounded; characterized by a horn formed around a single bony core that is kept as long as the animal lives; found from South Africa to Ethiopia.

roan antelope

roaring *Veterinary Medicine.* a disease of horses characterized by loud or rough breathing, caused by respiratory obstruction or vocal chord paralysis.

roaring forties *Meteorology.* the stormy ocean regions occurring between 40° and 50° latitude, usually in the Southern Hemisphere, distinguished by a continuous belt of ocean with strong prevailing westerly winds.

roaring sand *Geology.* a clean, dry, desert dune sand that emits a low roaring sound when disturbed.

roast *Food Technology.* to cook meat or other food by long exposure to dry heat, either in an oven or by direct exposure to a fire. *Metallurgy.* to heat an ore or ore concentrate prior to smelting.

roaster *Agriculture.* **1.** a chicken that is suitable for roasting, especially one that weighs over four pounds. **2.** any domestic animal, such as a pig or rabbit, that is roasted whole. *Engineering.* a device for the industrial roasting or heating of raw materials such as ores, or for the removal of sulfur.

roasting regeneration *Chemical Engineering.* the process of heating a substance to an elevated temperature to remove impurities by volatization.

robalo see SNOOK.

Robbins, Frederick Chapman born 1916, American physician; with John Franklin Enders, shared the Nobel Prize for discovering tissue cultivation of polio virus.

Robertinacea *Invertebrate Zoology.* a superfamily of deep-sea foraminiferans in the suborder Rotaliina.

Robert of Chester c. 1110–1160, English monastic scholar; translated the *Algebra* of al-Kwarizimi; devised astronomical tables.

Roberts' linkage *Mechanical Engineering.* a device used in the early 19th century to create straight metal guides for the slides in a metal planner.

Robertsonian change *Genetics.* a chromosomal mutation due to centric fusion or centric fission.

Robertson-Seymour theorem *Mathematics.* if $G_1, G_2, G_3, \ldots$ is a countable sequence of graphs and if G_1 is planar, then there exist positive integers i and j with $i < j$ such that G_1 is isomorphic to a minor of G_j.

Robertson-Walker solutions *Physics.* solutions to three classes of models for the universe in general relativity corresponding to flat, open, or closed universes.

Roberval, Gilles 1602–1675, French mathematician; anticipated the discoveries of infinitesimal calculus; proposed a theory of gravitation.

robin *Vertebrate Zoology.* any of several small, red-breasted or yellow-breasted thrushlike birds of the family Turdidae; common in Europe, North America, and Australia. The familiar **American robin**, *Turdus migratorius*, is characterized by a brownish-gray plumage and chestnut-red breast and abdomen.

robin

robina *Materials.* a coarse, strong, attractive wood of the European tree, *Robina pseudoacacia*; used in making furniture. Also, EMPRESS TREE, PRINCESS TREE.

Robin Hood's wind *Meteorology.* a term used in Britain for a raw and penetrating weather condition that is characterized by near freezing temperatures and saturated air.

Robin law *Physics.* a law stating that when a system in chemical or physical equilibrium is subjected to an increase in pressure, the system will react so as to decrease the volume; equivalently, a decrease in pressure will cause an increase in volume.

Robins-Messiter system *Mechanical Engineering.* a stacking conveyor system having one or two wing conveyors that receive material from a conveyor belt.

Robinson, Sir Robert 1886–1975, English organic chemist; awarded the Nobel Prize for his analyses of plant substances.

Robinson annelation reaction *Organic Chemistry.* the formation of conjugated cyclohexenones by reaction of cyclohexanones with methyl vinyl ketone or its equivalent, followed by intramolecular aldol condensation.

robinsonite *Mineralogy.* $Pb_4Sb_6S_{13}$, an opaque, metallic, bluish-gray triclinic mineral occurring as thin prismatic crystals and in massive form, having a specific gravity of 5.6 to 5.75 and a hardness of 2.5 to 3 on the Mohs scale; usually found as a primary mineral in mercury mines.

Robitzsch actinograph *Engineering.* a pyranometer that measures light (radiant) intensity from the difference in light energy absorbed on both black and white surfaces of the device, causing a temperature difference and motion of the indicator.

roble *Materials.* the wood of various North and South American trees of the oak and beech families, especially the California white oak, *Quercus lobata*.

robot *Robotics.* any mechanical device that can be programmed to perform a number of tasks involving manipulation and movement under automatic control. Because of its use in science fiction, the term *robot* suggests a machine that has a humanlike appearance or that operates with humanlike capacities; in actuality modern industrial robots have very little physical resemblance to humans.

robot aircraft see PILOTLESS AIRCRAFT.

robot assembly *Design Engineering.* a small, lightweight robot that fits and assembles parts or components with speed and accuracy, using its ability to grasp, position, assemble, and adjust workpieces to design specifications.

robot bomb see FLYING BOMB.

robot deburring *Design Engineering.* a process developed using a robot to remove burrs and other blemishes.

robot design *Design Engineering.* **1.** the geometric, kinematic, electrical, and mechanical features of a robot. **2.** the process of creating a robot.

Robotics

Around 1923, the word "robot" came into general use, originating from Karel Capek's play R.U.R. (Rosaum's Universal Robots). Isaac Asimov, together with John Campbell, began to formulate the "Three Laws of Robots" that were published in 1942 in a story called "Runaround." It was in this story that the word "robotics" was coined.

As we moved from fiction to science in the 1960s, robotics became the science of designing, building, and applying robots. The Robotic Industries Association has defined an industrial robot as "a programmable multifunctional machine designed to both manipulate and transport parts, tools, or specialized implements through variable programmed paths for the performance of specific manufacturing tasks."

Manufacturers are investing increasing amounts of money in new forms of automation, including robots, to increase productivity, reduce product cost, and improve product quality and reliability in order to retain their competitive position. Robots are being applied successfully to accomplish these objectives.

Robots have always excelled in applications in which freeing humans from repetitious tasks or dangerous operating environments is a goal. Advances in visual, tactile, acoustic, and magnetic sensing capabilities have provided robots with the ability to perform many of the basic manufacturing functions, including visual inspection and intricate welding operations. A new generation of robots is being applied in areas such as special finishing, laser and water-jet cutting, multistep assembly, and detail inspection. Clearly, robot technology is important now and in the future as one of the tools manufacturers can use to their advantage to enhance their competitive position in the global market.

V. Daniel Hunt
President
Technology Research Corporation

robotics *Industrial Engineering.* the branch of industrial engineering that deals with the design, manufacture, operation, and use of programmable multifunction machines, or industrial robots. *Artificial Intelligence.* any use of machines to perform manual tasks without human intervention.

robot machining cell *Design Engineering.* a cell of machines serviced by a robot for loading and unloading.

robot manipulator *Design Engineering.* any of the manipulative functions that are incorporated into robot design, such as a paint sprayer or a simple gripper used to clamp and unclamp parts.

robot motion *Design Engineering.* the motions developed by a robot element.

robot programming *Design Engineering.* computer programming within a robotic device in which numerical control techniques form the basis for operation.

robot programming language *Robotics.* any computer language that can be used to send instructions to or control a robot.

robot system *Robotics.* a group of devices that form a network to control the work of a robot.

robust *Anthropology.* heavily or powerfully built; a term used to describe a hominid having a build of this type. *Artificial Intelligence.* describing a program or reasoning technique that is able to work despite difficulties such as unexpected developments or missing, ambiguous, or incorrect inputs. *Statistics.* a term that is applied to a procedure that is relatively insensitive to any violations of the assumptions on which it is based or to procedures that are themselves based on weaker assumptions.

robusta *Botany.* **1.** another name for the coffee tree *Coffea canephora,* cultivated in warm regions of the Old World and native to western tropical Africa. **2.** the seeds of this plant, from which coffee is made.

robust Bayesian inference *Statistics.* a Bayesian inference that is not very sensitive to the *a priori* probabilities or to the specification of the statistical model.

robust estimator *Statistics.* an estimator that is not very sensitive to the presence of anomalous values in the sample.

robustness *Acoustical Engineering.* the durability of a microphone; that is, its ability to withstand rough treatment away from the studio.

ROC *Aviation.* the airport code for Rochester, New York.

rocambole *Botany.* a European plant, *Allium scorodoprasum,* of the amaryllis family that is closely related to the leek; used as a flavoring.

Roccella *Botany.* orchella weed, a genus of lichenized fungi of the family Roccellaceae that is the source of a purple dye.

Roccellaceae *Mycology.* a family of fungi belonging to the order Hysteriales that occur on rocks in warm coastal regions.

Rochalimaea *Bacteriology.* a genus of Gram-negative bacteria of the family Rickettsiaceae, usually occurring as small extracellular coccobacilli in an arthropod host.

Roche limit *Astronomy.* the distance from the center of a planet at which another celestial body (such as a satellite) would be broken apart by tidal forces. (Named for Edouard *Roche,* 1820–1883, French astronomer, who first calculated it.)

Rochelle salt *Organic Chemistry.* $KNaC_4H_4O_6 \cdot 4H_2O$, transparent, colorless, efflorescent crystals or white powder, soluble in water and insoluble in alcohol; melts at 70–80°C. *Materials.* a commercial preparation of this material; used in medicines, mirrors, baking powder, and piezoelectric equipment. Also, POTASSIUM SODIUM TARTRATE.

Roche lobes *Mechanics.* for two mutually gravitating bodies rotating about a common center, two imaginary tear-shaped regions surrounding each of the bodies, whose points meet at the center of rotation, and which remain stationary in a coordinate system that is rotating with the bodies.

roche moutonnée *Geology.* a small, protruding, glacially sculpted knob of bedrock with a gently inclined, smoothly rounded, and striated upstream side and a steep, rough, rugged downstream side. Also, SHEEPBACK, SHEEPBACK ROCK.

Rochon polarizing prism *Optics.* a prism in which two quartz or calcite wedges transmit the ordinary ray, without change, and deflect the extraordinary ray; used to produce plane-polarized light.

rock *Petrology.* any naturally formed aggregate of one or more minerals, consolidated or not, with some degree of mineralogic and chemical constancy; in popular use the term is usually restricted to those aggregates that are hard, compact, and coherent.

rockair *Space Technology.* a solid-propellant research rocket that carries equipment for making observations about or from high altitudes; the rocket is fired from an aircraft in flight while the aircraft is climbing.

rock alignment *Archaeology.* a large design created by an arrangement of rocks, typicaly found across desert plains; usually the complete image can be discerned only from high above.

rock art *Archaeology.* any of various images or designs produced on rock by prehistoric peoples, usually by painting or carving.

rock asphalt see ASPHALT ROCK.

rock association *Petrology.* a group of chemically and petrographically related igneous rocks occurring within a petrographic province, the chemical data of which may be plotted as smooth curves on variation diagrams. Also, ROCK KINDRED.

rock avalanche *Geology.* a rapid downslope flowage of rock fragments, during which the fragments may become pulverized, typically resulting from large rockfalls and rockslides. Also, STURZSTROM.

rock awash *Oceanography.* a rock that is exposed at any stage of the tide between mean high water and the sounding datum.

rock bar see RIEGEL.

rock-basin lake see PATERNOSTER LAKE.

rock bass *Vertebraate Zoology.* the game fish *Ambloplites repestris* of the sunfish family, characterized by an olive-green body about 9 inches long and a dark red iris; found in freshwater streams of the eastern U.S.

rock bench see STRUCTURAL BENCH.

rock bit *Engineering.* any one of several types of roller or drag-type bits with jagged teeth; used on rotary-type drills for drilling large-size holes in soft to medium-hard rocks, particularly in the petroleum industry.

rockbolt or **rock bolt** *Engineering.* a bar, usually constructed of steel, that is inserted into predrilled holes in rock and secured for ground control. Also, ROOF BOLT.

rock bolting *Engineering.* the process of using rock bolts to provide reinforcement or suspension of highly fissured rock.

rockbridgeite *Mineralogy.* $(Fe^{+2},Mn^{+2})Fe_4^{+3}(PO_4)_3(OH)_5$, a dark-green orthorhombic mineral occurring as botryoidal crusts, fibrous masses, and tiny, elongated prismatic crystals, forming a series with frondelite, and having a specific gravity of 3.3 to 3.49 and a hardness of 3.5 to 4.5 on the Mohs scale; found in granite pegmatites and in a limonite deposit in Rockbridge County, Virginia.

rock bulk compressibility *Geology.* the fractional change in volume of the bulk volume of a rock per unit change in pressure.

rock bump *Mining Engineering.* the sudden release of large vertical stresses, due to the weight of rocks over a coal seam, or of large lateral stresses, due to structural or tectonic folds, or thrusts.

rockburst *Mining Engineering.* the sudden, violent, and hazardous failure of rock strained beyond its elastic limit, occurring around or within a mining operation.

rock candy *Food Technology.* large, transparent, hydrated crystals of sugarcane.

rock cave see SHELTER CAVE.

rock channeler *Mechanical Engineering.* a quarrying machine that cuts an artificial seam into a mass of stone.

rock cleavage *Petrology.* the tendency or capacity of a rock to split along a particular surface.

rock crab *Vertebrate Zoology.* any of several crabs that live along rocky beaches, especially those of the genus *Cancer,* such as *C. irroratus,* which is found on the eastern coast of North America and which has rear legs that are modified for running.

rock creep *Geology.* a slow, downslope flowage of rock materials.

rock crystal see QUARTZ CRYSTAL.

rock cycle *Geology.* the sequences through which earth materials may pass when subjected to geological processes.

rock-defended terrace *Geology.* **1.** a river terrace protected from undermining by a ledge or outcrop of resistant rock at its base. **2.** a marine terrace protected from wave erosion by a mass of resistant rock at the base of the wave-cut cliff that is formed in the overlying sediments. Also, ROCK-PERCHED TERRACE.

rock desert *Geology.* an upland desert area that has been swept clean of sand and dust by wind, leaving the bedrock completely exposed or covered by a thin layer of coarse rock fragments. Also, ROCKY DESERT.

rock drill *Mechanical Engineering.* a drill, generally having bits detachable from the drill stem, that is designed to bore relatively short holes through rock for blasting purposes; powered either by electricity, compressed air, or steam.

rock drumlin *Geology.* a smooth, streamlined, glacially modeled hill with a core of bedrock usually covered by a thin layer of till. Also, FALSE DRUMLIN, DRUMLINOID.

rock dust *Mining Engineering.* a finely powdered limestone that is sprayed on mine surfaces in order to blanket highly explosive coal dust.

rock duster *Mining Engineering.* a machine that blows rock dust over the interior surfaces of a coal mine in an effort to prevent coal dust explosions. Also, **rock dust distributor.**

rock element *Petrology.* a coherent, intact, basic rock constituent featuring quantifiable or qualifiable physical, mechanical, and petrographic properties.

rocker *Civil Engineering.* a support that permits both sliding and pivoting motions to allow for structural expansion or contraction. *Mining Engineering.* a small digging bucket mounted on two rocker arms, used in gold mining. *Graphic Arts.* a small steel plate having one curved and one toothed edge for roughening a copperplate to create a mezzotint. *Ordnance.* a movable support on a field gun carriage that permits elevation adjustment without changing the angle of position setting.

rocker arm *Mechanical Engineering.* a center-pivoted lever in an internal combustion engine that transmits motion from a cam or a push rod to a valve stem.

rocker bearing *Civil Engineering.* a pivoting bridge support.

rocker bent *Civil Engineering.* a bent hinged at one or both ends to allow for structural expansion or contraction.

rocker cam *Mechanical Engineering.* a cam, frequently found in a rockshaft, having a rocking movement.

rocker panel *Engineering.* the automobile paneling that covers the midsection of the sides of the passenger compartment.

rocket *Space Technology.* **1.** a shorter term for a rocket engine. See ROCKET ENGINE. **2.** a vehicle, projectile, or device propelled by a rocket engine. *Ordnance.* of or relating to weapons using rockets or rocket propulsion. Thus, **rocket ammunition, rocket bomb, rocket missile.**

rocket airplane *Aviation.* **1.** an airplane that utilizes rocket propulsion as its primary or only source of motive force. **2.** an aircraft that carries and fires rocket ammunition.

rocket antenna *Electromagnetism.* a transmitting antenna mounted on an unmanned rocket for tracking the rocket.

rocket assist *Aviation.* **1.** the use of rocket motors to provide extra thrust for an aircraft during takeoff or flight. **2.** the motor or motors that provide this thrust.

rocket-assisted takeoff see RATO.

rocket-assisted torpedo *Ordnance.* an underwater weapon that is shot into the air by rocket power, dropped into the water by parachute, and guided toward its target by a homing device.

rocket astronomy *Astronomy.* a branch of astronomy in which rockets flying on ballistic paths are used to lift instruments above the atmosphere in order to collect astronomical data for study.

rocket bomb *Ordnance.* any rocket-propelled missile launched from the ground.

rocket chamber see COMBUSTION CHAMBER.

rocket electrophoresis *Immunology.* a method used to measure antigens, by which antigens are introduced into a charged medium containing an antibody, and the rocketlike trails that the antigens create are observed.

rocket engine *Space Technology.* a reaction engine that is entirely self-contained; i.e., that contains within itself everything necessary for the combustion of its fuel, and that therefore does not require an external medium of air for combustion (thus it can operate in outer space). The simplest rocket engine consists of a combustion chamber and a nozzle; a more complex type may have more than one combustion chamber. The two main types are defined by the type of fuel used: **liquid-propellant** or **solid-propellant**. (In strict usage, the latter is distinguished as a rocket *motor* rather than a rocket engine.)

rocket fuel *Materials Science.* any of various liquid or solid substances that are used as fuel in rocket engines, such as liquid hydrogen or gasoline; characterized by the potential for extremely rapid, controlled combustion and subsequent production of large volumes of gas at high pressure and temperature. In popular usage, this term often refers to both the fuel and the oxidizer used in a rocket propellant.

rocket glider *Aviation.* a glider vehicle that is carried to altitude by rocket propulsion, then released into gliding flight.

rocket igniter *Space Technology.* an electrical, chemical, or mechanical device used to initiate combustion of a rocket propellant.

rocket immunoelectrophoresis see LAURELL ROCKET TEST.

rocket launcher *Space Technology.* a structure or device for launching a rocket; may be mobile or stationary, fixed in place on the ground or mounted on an aircraft or ship.

rocket lightning *Geophysics.* a rare form of lightning that moves across the sky slowly enough to be seen moving by the naked eye; it is thought to be caused by a low electric field strength at the tip of its leader. (So called because it seems to move at the speed of a rocket.)

rocket missile *Space Technology.* any missile that uses rocket propulsion in flight.

rocket motor *Space Technology.* **1.** another term for a rocket engine. **2.** in strict usage, a simple solid-fuel rocket propulsion device consisting of a single combustion chamber and nozzle.

rocket nose section *Space Technology.* the streamlined leading section of a rocket: in a rocket weapon, the nose section contains a warhead or other ordnance; in a satellite vehicle, the nose section may be ejected from the rocket to become the satellite; in a space vehicle, the nose section is also the nose cone.

rocket nozzle *Space Technology.* the exhaust nozzle of a rocket engine.

rocket ogive *Ordnance.* a hollow, conical metal case over the front end of a rocket warhead that is designed to reduce air resistance.

rocket projectile *Space Technology.* **1.** a missile or other projectile propelled by a rocket. **2.** an object that is launched like a projectile but continues in motion by a combination of inertia and rocket propulsion.

rocket propellant *Materials Science.* a collective term for the combination of materials utilized for power by a rocket engine, i.e., a rocket fuel and an oxidizer such as liquid oxygen or nitroglycerin.

rocket propulsion *Space Technology.* the generation of thrust by means of a rearward ejection of combustion gases, as from a rocket engine; the use of a rocket engine to produce reaction propulsion.

rocket ramjet *Space Technology.* a ramjet engine fitted at its duct with a rocket; used to increase speed to a point at which the ramjet may operate efficiently. Also, DUCTED ROCKET.

rocketry *Space Technology.* the science and study of rockets, including all steps from theory to application; the activity of building and launching rockets.

rocketship *Space Technology.* a spacecraft or other flight vehicle using rocket propulsion for its principal or only source of motive power.

rocket sled *Space Technology.* a rail-mounted sled that is accelerated to high speeds by means of a mounted rocket engine; used especially for experimentation in gravity tolerance and crash-survival techniques.

rocketsonde see METEOROLOGICAL ROCKET.

rocket staging *Space Technology.* the separation during flight of a stage or half-stage from the body of a main rocket vehicle, thus freeing the separated stages to decelerate or enter into a separate flightpath.

rocket thrust *Space Technology.* the propulsive force of a rocket engine, usually expressed in pounds.

rocket tube *Space Technology.* an apparatus composed of a combustion chamber and nozzle. Several rocket tubes may be placed together for increased thrust.

rocket vehicle *Space Technology.* a vehicle powered by a rocket engine and may be used to carry a crew and cargo into space, a satellite into orbit, or a missile toward its target.

Rockeye *Ordnance.* a 500-lb. cluster bomb designed for use against heavy tanks and other armored targets.

rock fabric see FABRIC.

rock failure *Geology.* a fracture of a rock that has been stressed beyond the maximum differential stress it can sustain under deformation conditions.

rockfall or **rock fall** *Geology.* **1.** the relatively free-falling, freshly detached segments of bedrock from cliffs, mountainsides, or other very steep slopes. **2.** the mass of fallen or moving rock material transported by such a landslide. Also, STURZSTROM.

rockfall

rock fan *Geology.* an eroded, fan-shaped bedrock surface having its apex where an intermittent mountain stream emerges onto a piedmont slope, and occupying the area at which a pediment meets the mountain slope.

rock-fill *Civil Engineering.* any pieces of broken rock or stone used as infill.

rock-fill dam *Civil Engineering.* an earth dam or barrier constructed of small, closely compacted stones or rocks.

rock-floor robbing *Geology.* a form of erosion in which sheets of moving water remove crumbling debris from rock surfaces in desert mountains.

rock flour *Geology.* a finely ground powder of rock particles, usually formed by the abrasive action of a glacier on underlying rock. Also, GLACIER MEAL, GLACIAL FLOUR.

rock flowage *Geology.* see FLOW.

rock flower *Botany.* any of the early flowering shrubs of the genus *Crossosoma*, having thick, narrow leaves and solitary flowers; found in arid regions of the southwestern U.S.

rockforming *Geology.* the minerals, such as quartz, feldspar, and calcite, that occur as the dominant constituents of rocks and determine their classification.

rock glacier *Geology.* an accumulation of boulders and finer material cemented by ice below the surface that moves slowly downhill under its own weight. Also, TALUS GLACIER.

rock-glacier creep *Geology.* the downslope movement of tongues of rock glaciers.

rock gypsum *Mineralogy.* massive, coarsely crystalline to fine-grained gypsum occurring in sedimentary beds.

rocking bar *Horology.* the plate or bar of a keyless watch that carries the winding mechanism.

rocking furnace *Mechanical Engineering.* a cylindrical melting furnace that rolls back and forth on a geared cradle.

rocking pier *Civil Engineering.* a hinged bridge pier that allows for structural expansion or contraction.

rocking stone *Geology.* a finely perched boulder or stone that can be moved backward and forward slightly, using little force. Also, ROGGAN.

rocking valve *Mechanical Engineering.* a steam-engine valve that rotates in its seat to control fluid flow.

rock island see MEANDER CORE, def. 1.

rock kindred see ROCK ASSOCIATION.

rock magnetism *Geophysics.* the occurrence of remanent magnetism possessed by rocks containing iron-oxide minerals, or the study of remanent magnetism's characteristics.

rock meal see DIATOMACEOUS EARTH.

rock mechanics *Geophysics.* a branch of both mechanics and geology that studies the response of rocks to the various forces in their environment.

rock milk *Mineralogy.* a soft, white, plastic, calcareous deposit found coating limestone cave walls.

rockoon *Space Technology.* a hybrid flight vehicle composed of a rocket (especially a small, solid-fuel sounding rocket) that is carried aloft by a large plastic balloon and then fired. (A combination of <u>rock</u>et and ball<u>oon</u>.)

rock pedestal *Geology.* see PEDESTAL.

rock pediment *Geology.* a broad, flat or gently inclined, low-relief erosional surface developed on bedrock at the foot of a receding mountain slope.

rock-perched terrace see ROCK-DEFENDED TERRACE.

rock permeability *Geology.* the ability of a rock to receive, hold, or pass fluids such as oil, gas, or water.

rock pillar *Geology.* **1.** a column of rock formed by differential weathering or erosion along a plane of weakness. **2.** a columnar structure in a cave that is formed of residual bedrock rather than minerals deposited by the action of water.

rock pool *Geology.* a pool of water deposited by an ebbing tide along a rocky shoreline.

rock pressure see GROUND PRESSURE.

rock river *Geology.* a long, narrow rock stream.

rock salt see HALITE.

rock saw *Mining Engineering.* a mechanical mining instrument, consisting of a moving steel band or blade and a slurry of abrasive materials, such as diamonds; used for cutting narrow slots or channels into large blocks of material to facilitate removal.

rockshaft *Engineering.* an oscillating shaft. *Mining Engineering.* a shaft through which rock can be brought into a mine.

rock shell *Invertebrate Zoology.* the common name for a mollusk of genus *Murex*, having a heavy, often colorful shell, rows of spines, and an irregular pattern of teeth on the radula; carnivores of bivalves and the source of ancient royal purple dye.

rock shelter or **rockshelter** *Geology.* a cave with a roof of overlying rock that extends beyond the sides of the cave. *Archaeology.* such a shallow cave or cliff overhang used by humans to provide shelter from the elements; so utilized by hunters and gatherers into historic times.

rock silk *Mineralogy.* a variety of asbestos characterized by its silky texture.

rockslide or **rock slide** *Geology.* **1.** a landslide consisting of a sudden, rapid, downward movement of segments of freshly detached bedrock slipping over a bedding, joint, fault surface, or other such surface of weakness. **2.** the mass of rock material moving in or moved by such a landslide. Also, **rock slip.**

rock stack *Geology.* see STACK, def. 2.

rock step see KNICK.

rock-stratigraphic unit *Geology.* a rock subdivision based on homogeneous lithological characteristics and identified by its observable physical features or by the dominance of a certain rock type or combination of types. Also, LITHOSTRATIGRAPHIC UNIT, GEOLITH.

rock stream *Geology.* **1.** a streamlike mass of rocks and rock fragments that are moving or have moved slowly downslope. Also, STONE RIVER. **2.** see BLOCK STREAM.

rock stripes see STONE STRIPES.

rock system *Geophysics.* all natural structures and related forces that are capable of affecting an area of the earth's crust that will become part of an engineering structure.

rock terrace *Geology.* a stream terrace formed on the side of a valley, and composed of strong bedrock that remains after weaker beds above and below have eroded.

rock type *Petrology.* **1.** one of the three major rock classifications: igneous, sedimentary, and metamorphic. **2.** a particular rock with a specific group of characteristics, such as basalt.

Rockwell hardness *Engineering.* a unit of metal hardness according to the Rockwell hardness test.

Rockwell hardness test *Engineering.* a method of determining the relative hardness of metals and case-hardened materials, by measuring the depth of the penetration of a steel ball on softer metals or of a conical diamond point on harder metals.

rock wood see MOUNTAIN WOOD.

rocky desert see ROCK DESERT.

Rocky Mountains or **Rockies** *Geography.* the main mountain system of western North America, extending from northern Alaska to northern Mexico.

Rocky Mountain goat *Vertebrate Zoology.* an antelopelike wild goat, *Oreamnos americanus*, characterized by long, white hair; found in the mountainous regions of western North America.

Rocky Mountain spotted fever *Medicine.* an acute, infectious, and sometimes fatal tick-borne disease caused by *Rickettsia rickettsii*; characterized by fever, bone and muscle pain, headache, and rash. The disease occurs only in North and South America.

Rocky Mountain subregion *Ecology.* a distinct zoogeographical region of North America that includes the Rocky Mountains and other adjacent areas.

rocky point effect *Electronics.* the violent discharges that appear briefly between electrodes in high-voltage transmitting tubes.

Rococo [rə kō´kō] *Architecture.* the final phase of the Baroque period, characterized by ornate, romantic interior decoration in light colors.

rod *Mechanical Devices.* **1.** a slender, cylindrically shaped metal or wood bar, usually threaded along part or all of its length, used with nuts and washers for fastening, bracing, and mounting. **2.** a bar with a slotted, screwed, or tapered end to receive a drill bit. *Geology.* an elongated sedimentary particle whose width to length ratio is less than 2:3 and whose thickness to width ratio is greater than 2:3. Also, ROLLER. *Nucleonics.* a long, narrow body of material that is inserted into a nuclear reactor, the extent of insertion providing finite control over the reactor's reactivity; the rod may contain fuel elements, a neutron absorbing material, or fertile material. *Bacteriology.* a term used to describe the shape of one of the basic types of bacteria (bacilli). *Histology.* a rod-shaped cell in the retina that is sensitive to dim light. Also, **rod cell.**

rod *Metrology.* **1.** a unit of length equal to 16.5 feet, or 5.03 meters. **2.** see SQUARE ROD.

rod bit *Mechanical Devices.* a drill bit that fits into a reaming shell, which is then threaded onto a drill rod.

rod bolt *Mechanical Devices.* a long, double-headed bolt.

rod chisel *Mechanical Devices.* a blacksmith's chisel held by flexible, elastic handles. Also, **rod cutter.**

rod coupling *Mechanical Devices.* a double-threaded pin coupling used to connect two drill rods.

rodding *Petrology.* a linear structure in metamorphic rocks in which the stronger parts, such as vein quartz or quartz pebbles, have been formed in the shape of parallel rods.

rodent *Vertebrate Zoology.* a member of the large worldwide mammalian order Rodentia, characterized by chisel-like, ever-growing incisor teeth.

Rodentia *Vertebrate Zoology.* the largest order of mammals including rats, mice, hamsters, squirrels, chipmunks, gophers, porcupines, woodchucks, and beavers; found worldwide except in Antarctica and a few oceanic islands.

rodenticide *Materials.* any of various toxic substances used to kill rodents.

rodent ulcer *Oncology.* a slowly developing ulceration of a basal cell carcinoma of the skin, usually of the face, that works inward, destroying deeper tissue and bone.

rod gap *Electricity.* a spark-gap arrester, typically consisting of two 1/2-square-inch rods: one grounded and the other connected to the line conductor.

rodingite *Petrology.* a dense, medium-grained to coarse-grained, buff to pink gabbrolike rock, containing calcium, with grossular and diallage as essential minerals.

rod level *Engineering.* a level, such as a spirit level or glass bubble, attached to a survey rod for correct vertical orientation of the rod.

rod mill *Mechanical Engineering.* a machine that pulverizes material by means of long, heavy iron rods. *Metallurgy.* **1.** in metal working, a hot mill used to shape rods. **2.** in metal finishing, equipment in which a rod abrades the surface of a part.

rod pump *Petroleum Engineering.* a sucker-rod pump that can be placed in or removed from tubing in an oil well without disturbing the tubing. Also, INSERT PUMP.

Rodrigues' formula *Mathematics.* an equation for the *n*th function (or term) of some special class of functions (or sequence) in terms of derivatives of a polynomial. For example, Rodrigues' formula for the Legendre polynomials is $P_n(x) = 1/(2^n n!)d^n(x^2 - 1)^n/dx^n$.

rod saw *Mechanical Devices.* a thin, round, sectioned rod that is covered with tiny teeth or tungsten carbide particles and tensioned in a hacksaw frame; used especially to cut intricate curves in metal, ceramics, and plastics.

rod-shaped *Bacteriology.* a basic classification of bacteria types, including those organisms (bacilli) that have roughly the form of a rod.

rod string *Mechanical Engineering.* the drill rods coupled to form a connecting link between the core barrel and bit in a borehole and the drilling mechanism at the top of the borehole.

rod thermistor *Electronics.* a device that provides high heat-controlled resistance over a long period of time with little loss of power.

rod weeder *Agriculture.* a farm machine that uses a slowly rotating rod to pull weeds from beneath the soil surface.

roe *Vertebrate Zoology.* **1.** the mass of eggs contained within the ovarian membrane of the female fish. **2.** the sperm of the male fish. **3.** see ROE DEER.

roedderite *Mineralogy.* $(Na,K)_2(Mg,Fe^{+2})_5Si_{12}O_{30}$, a colorless, hexagonal mineral of the osumilite group, forming a series with eifelite, occurring as imperfect platy crystals, and having a specific gravity of about 2.6 and an undetermined hardness; found almost exclusively in meteorites.

roe deer *Vertebrate Zoology.* a small, agile deer, *Capreolus capreolus*, characterized by a male with forked, three-pointed antlers; found in Europe and Asia. Also, ROE.

roemerite see RÖMERITE.

Roentgen, Wilhelm Konrad [rent´gən] 1845–1923, German physicist; awarded the Nobel Prize for the discovery of X-rays (which are also known as Roentgen rays).

roentgen *Nucleonics.* a unit of ionizing-radiation dosage of X-rays or gamma rays, equivalent to the amount that produces one electrostatic unit of charge from secondary ionization (either positive or negative) per 0.001293 gram of air at atmospheric pressure (2.58×10^{-4} coulombs per kilogram of air). Also, RÖNTGEN.

roentgen intoxication see RADIATION SICKNESS.

roentgen meter see R METER.

roentgenography *Physics.* radiography using X-rays.

roentgenologist see RADIOLOGIST.

roentgenology see RADIOLOGY.

roentgenolucent see RADIOLUCENT.

roentgenoluminescence *Physics.* any luminescent effect brought on by X-rays.

roentgenopaque see RADIODENSE.

roentgenoparent see RADIOLUCENT.

roentgenotherapy see RADIOTHERAPY.

roentgen-per-hour-at-one-meter *Nucleonics.* a unit of X-ray and gamma-ray source strength, equivalent to the rate of radiation dosage of one roentgen per hour received at a distance of one meter from the source in air.

roentgen rays *Radiology.* another term for X-rays. (From their discoverer, Wilhelm Konrad *Roentgen.*) See X-RAYS.

roesslerite see RÖSSLERITE.

roestelioid *Mycology.* a long, tubelike fruiting body or aecium, such as that found in fungi belonging to the genus *Gymnosporangium.*

roestone see OOLITE.

rofla *Geology.* an extremely narrow, winding gorge often formed by streams of water from a melting glacier.

ROFOR *Meteorology.* an international code word used to indicate a route forecast along a specific air route.

ROFOT *Meteorology.* an international code word used to indicate a route forecast, with units in the English system.

Rogallo wing *Aviation.* **1.** a flexible, delta-wing plan in which three rigid members are shaped in the form of an arrowhead and joined by a flexible fabric, which inflates upward under flight loads; originally limited to paragliders, but now found on some powered aircraft. **2.** a flight vehicle having such a wing plan.

rogenstein or **roggenstein** *Geology.* a sedimentary rock made up of small, rounded, accretionary bodies that are joined by clayey cement.

Rogers, Carl R. 1902–1987, American psychologist; introduced nondirective therapy.

Roget, Peter Mark [rō zhā´] 1779–1869, English lexicographer and physician; wrote and lectured on applied physics and mathematics.

roggan see ROCKING STONE.

rogue *Genetics.* a phenotypic variant usually displaying inferior characteristics that occurs in a population of standardized individuals. *Biology.* an individual that varies from the standard; usually used in reference to a plant. *Plant Pathology.* to remove plants by roguing.

roguing *Plant Pathology.* a mechanical process of removing or destroying diseased plants or plant parts from a seed crop in order to reduce the possibility of the spread of infection.

Rohrbach solution *Materials.* a toxic, clear yellow liquid derived from mercuric barium iodine; used to detect alkaloids and to separate minerals by their specific gravity.

Rohrer, Heinrich born 1933, Swiss physicist; with Binning, awarded the Nobel Prize for the development of the scanning tunneling microscope.

roil *Hydrology.* a small, turbulent, swiftly flowing section of a stream.

roily *Hydrology.* **1.** turbulent or swirling. **2.** muddy or sediment-filled. Thus, **roily water.**

rolamite mechanism *Mechanical Engineering.* a basic device consisting of two rollers bordered by two parallel planes and enclosed by a stationary S-shaped band under tension, so the rollers turn with almost no sliding friction. (A term coined by its inventor, the American engineer Donald F. Wilkes.)

Roland see U.S. ROLAND.

role *Behavior.* the position of an individual within a group, and the behavior that is associated with that position.

role indicator *Computer Programming.* a code that is associated with a key word to identify its role in a sentence, such as noun, verb, or adjective.

role model *Psychology.* a popular term for a person who serves as a model or example for another or others, especially a prominent person whose achievements can be emulated by a younger person.

role-playing *Behavior.* **1.** the act of behaving according to a role that is not one's own. **2.** the enactment of a role in a contrived situation in order to understand the role more clearly. *Artificial Intelligence.* a question-and-answer technique for acquiring knowledge from an expert.

role reversal *Behavior.* a form of therapy in which the therapist acts out the role of the client and the client assumes the role of the other person in the interaction.

roll *Aviation.* the rotational or oscillatory motion of an aircraft, rocket, or missile about a longitudinal axis passing through its body. *Cartography.* in aerial photography, a rotation of the camera about either the photographic *x*-axis or the exterior *x*-axis. *Mechanical Engineering.* a cylindrical body set in bearings that is used to shape, crush, flatten, move, or print material. *Mining Engineering.* a thinning of a coal seam caused by the protrusion of shale, siltstone, or sandstone from the roof into the seam, sometimes replacing the seam entirely. *Geology.* a sedimentary structure produced by deformation involving underwater slump or vertical collapse.

roll acceleration *Mechanics.* the angular acceleration of an aircraft, rocket, or missile about a longitudinal axis passing through its body.

roll axis *Mechanics.* the longitudinal axis through the body about which an aircraft, rocket, or missile rolls. Also, **rolling axis.**

rollback *Military Science.* the progressive destruction or neutralization of opposing defenses, beginning at the periphery and working inward, to permit deeper penetration of succeeding defense positions.

roll bonding *Materials Science.* a process commonly used for producing metallic laminar composites, such as claddings and bimetallics, in which sheets of the two materials are layered and then compressed by rollers; the pressure breaks up the oxides at the surface of the metals, bringing the surfaces into atom-to-atom contact and permitting them to weld.
roll cloud *Meteorology.* a popular name for an arcus. See ARCUS.
roll compacting *Materials Science.* a process for compressing powder particles, in which loose powder in a hopper feeds between two rotating rollers; the compacted powder strip emerging from the rolls is passed through a sintering furnace and subsequently hot-rolled or cold-rolled. Also, POWDER ROLLING.
roll control *Engineering.* a system to control the movement about the longitudinal axis of a rocket.
roll crusher *Mechanical Engineering.* a reduction crusher, having one or two toothed or corrugated rollers, into which material is fed from above and discharged at the bottom.
rolled glass *Materials.* a sheet of glass formed by the extrusion of molten glass between rollers.
rolled gold *Metallurgy.* gold cladded to a substrate in proportions smaller than the minimum specified for gold-filled.
rolled joint *Engineering.* a joint formed by expansion of a tube set in a tube sheet hole.
roller *Mechanical Devices.* a revolving cylinder over or on which an object is placed so it can be moved about freely, such as rollers mounted on drawers. *Oceanography.* a long, massive wave that retains its form while traveling a great distance, producing large breakers on exposed coasts. *Geology.* see ROD.

roller

roller analyzer *Engineering.* an instrument that separates and records different particle sizes down to the near-micrometer level.
roller bearing *Mechanical Engineering.* a shaft-bearing, having parallel or tapered steel rollers enclosed by outer and inner rings or races, designed for heavy loads.
roller bottle apparatus *Biotechnology.* a system used for the propagation of specialized cell cultures, in which the culture is placed on a set of revolving rollers that keep the particles in suspension.
roller cam follower *Mechanical Engineering.* a follower having a shaft with a rotating wheel at one end.
roller chain *Mechanical Engineering.* a driving or transmission chain having links of pin-mounted rollers that connect sideplates.
roller coating *Engineering.* the use of rollers to apply paint or other coatings to ensure even distribution over a surface.
roller cone bit *Engineering.* a rotary boring bit consisting of two to four cone-shaped, toothed rollers that are turned by the rotation of the drill rods; used in hard rock in deep drilling operations. Also, **roller bit.**
roller conveyor *Mechanical Engineering.* a conveyor having a horizontal track made up of a series of closely spaced rollers over which material is moved by gravity or propulsion.
roller drawing *Materials Science.* in textile production, a drawing-down method which uses rollers to spin yarn from the sliver; typically, two or more sets of rollers are used, each set faster than the previous one and each set spaced apart slightly more than the maximum fiber lengths of the assembly; as it is drawn through the rollers, the sliver is elongated and attenuated.
roller drying *Chemical Engineering.* a type of web drying in which the material to be dried makes a sinusoidal path around rollers while heat is supplied externally by blowing air.
roller gate *Civil Engineering.* a crest gate at the entrance of a dam spillway that operates by means of large cogs that connect with sharply sloping racks, along which the gate travels as it is being opened.
roller leveling *Mechanical Engineering.* a process of leveling in which the material is fed through a series of small-diameter staggered rolls.
roller mill see RING-ROLLER MILL.
roller printing *Textiles.* the process of printing fabric with engraved rollers or cylinders. Also, DIRECT PRINTING.
roller pulverizer *Mechanical Engineering.* a machine having rotating rollers that reduces solid material to very small particles.
roller stripping *Graphic Arts.* during a press run, a failure of the ink to stick to the metal ink rollers.
Rolle's theorem *Mathematics.* a special case of the mean value theorem of differential calculus; let f be continuous on the closed interval $[a,b]$ and differentiable on the open interval (a,b). If $f(a) = f(b)$, then there exists a point x_0, with $a < x_0 < b$, at which the derivative of f is zero; i.e., $f(x_0) = 0$.
roll feeder see DRUM FEEDER.
roll film *Graphic Arts.* photographic film that has an opaque paper backing and that is wound onto a spool in a continuous roll, so that more than one photograph can be taken without changing film.
roll forging *Metallurgy.* forging with a rotary die of special configuration.
roll forming *Metallurgy.* in metal forming, a process in which a strip of flat-rolled metal of appropriate width is progressively shaped or bent to a desired cross section by a number of roll stands; may be used for almost any metal with sufficient ductility to be bent cold.
roll-in *Computer Technology.* the process of bringing an operation into main memory during activation of that operation.
rolling *Mechanics.* relative motion between two bodies in contact along a line or point, without any slipping. *Metallurgy.* fabricating a metallic product by compressing it between rolls. *Naval Architecture.* a pendulumlike oscillation of a vessel around its longitudinal axis, as a vessel heels first to one side, then the other under the influence of wind and wave action and the hull's stability characteristics. Excessive rolling is a sign of poor stability.
rolling barrage see CREEPING BARRAGE.
rolling beach *Geology.* the part of an accumulation of boulders and pebbles at the base of a sea cliff that is being ground into finer particles.
rolling circle *Molecular Biology.* the conformation of a circular DNA molecule during its replication.
rolling circle mechanism *Molecular Biology.* a mechanism for the replication of an unnicked circular DNA molecule.
rolling colter *Agriculture.* a flat, steel disk suspended from the beam of a plow, used to separate the furrow slice from the furrow wall. Also, DISK COLTER.
rolling contact *Mechanics.* the point or line of contact between a surface and a body that rolls along on that surface.
rolling contact bearing *Mechanical Engineering.* a machine bearing having rolling parts that fit between an inner and an outer ring.
rolling door *Engineering.* a door set on wheels either horizontally or vertically along a set of tracks to assist opening and closing.
rolling friction *Mechanics.* the force of friction that exists between a rolling body and the surface over which it is rolling.
rolling radius *Design Engineering.* in an automotive vehicle, the distance from the center of an axle to the ground.
rolling texture *Materials Science.* a characteristic of metals and metal alloys that have been subjected to a process of deformation by rolling, so that a certain crystal plane is preferentially parallel to the rolled surface and a certain crystallographic direction is preferentially parallel to the rolling direction in most grains.
rolling transposition *Electricity.* transposition in which the conductors of an open-wire circuit are rotated in a helical manner.
roll mill *Mechanical Engineering.* a sequential arrangement of mills used to crush and grind material.
roll-off *Electronics.* **1.** the gradual rise in attenuation over a range of frequencies. **2.** the attenuation that increases with a rise in frequency. Also, SLOPE.
roll orebody *Geology.* a body of uranium or vanadium having an S- or C-shaped cross section within sedimentary rock.
roll-out *Computer Technology.* the process of removing an operation from main memory and recording its contents to an auxiliary device.

roll straightening *Engineering.* a process of straightening metal stock of various shapes by passing it through a series of staggered rolls that are usually in horizontal and vertical planes.

roll threading *Mechanical Engineering.* the process of making a threaded metal workpiece by rolling the metal between grooved circular rolls or grooved straight lines.

roll-tube technique *Microbiology.* a process in which a test tube containing bacteria is rolled during incubation; used in growing pure cultures of aerobic and anaerobic bacteria.

roll welding *Metallurgy.* solid-state welding in which pressure is applied by rolls.

rolock wall *Building Engineering.* a type of hollow consisting of a bull stretcher and header; used to impede rainwater into a structure. Also, ROWLOCK WALL.

ROM [räm] *Computer Technology.* the portion of memory that can be read from, but not written to. (An acronym for read-only memory.)

ROM *Aviation.* the airline code for the city of Rome, Italy.

ROMable code *Computer Programming.* a program made to be stored permanently in read-only memory.

roman *Graphic Arts.* **1.** usually, **Roman.** a class of typefaces having serifs and a marked difference between the thick and thin strokes; e.g., Times Roman or Garamond. **2.** any typeface in its conventional form, as opposed to its other forms such as italic, boldface, bold italic, and so on.

Romanche trench *Geography.* a deep trench in the mid-Atlantic ridge near the equator.

romanechite *Mineralogy.* $(Ba,H_2O)(Mn^{+4},Mn^{+3})_5O_{10}$, a black to gray, opaque, monoclinic mineral occurring in massive, botryoidal, and earthy form, having a specific gravity of 4.71 and a hardness of 5 to 6 on the Mohs scale; derived from weathered manganese carbonates and silicates; found in residual and carbonate replacement deposits. Also, PSILOMELANE.

Romanesque [rō´mə nesk´] *Architecture.* a style of architecture based on classic Roman forms and characterized by heavy masonry construction; it was prevalent in Western Europe from about 1000 to 1150 AD for churches, monasteries, and other religious buildings.

Romano or **romano** *Food Technology.* a hard Italian cheese that is similar to Parmesan but with less sharpness and usually more moisture.

Romanowsky's stain *Hematology.* the prototype of the many eosin-methylene blue stains used for blood smears and malarial parasites, including Giemsa stain, Leishman's stain, and Wright's stain. Also, **Romanovsky's stain.**

Romberg's sign *Neurology.* a sign of sensory ataxia in which the body cannot maintain equilibrium when standing with feet together and eyes closed, associated with diseases of the posterior columns of the spinal cord.

roméite *Mineralogy.* $(Ca,Fe^{+2},Mn^{+2},Na)_2(Sb,Ti)_2O_6(O,OH,F)$, a pale-yellow to brown cubic mineral of the stibiconite group occurring as small octahedral crystals, having a specific gravity of 4.7 to 5.4 and a hardness of 5.5 to 6.5 on the Mohs scale.

Romeldale *Agriculture.* a breed of medium-sized, hornless, white-faced sheep that produce both wool and meat; developed in the U.S. from the Romney, Rambouillet, and Corriedale breeds.

römerite *Mineralogy.* $Fe^{+2}Fe_2^{+3}(SO_4)_4{\cdot}14H_2O$, a rust-brown to yellow triclinic mineral occurring as thick tabular crystals and in granular form, having a specific gravity of 2.18 and a hardness of 3 to 3.5 on the Mohs scale; found with other secondary sulfates. Also, ROEMERITE.

Romer method *Optics.* a method, based on delays observed in successive eclipses of the innermost satellites of Jupiter, that was used to calculate the speed of light in the seventeenth century.

ROMET *Meteorology.* an international code word used to indicate a route forecast, measured in metric units.

Romney *Agriculture.* a breed of hardy, long-wool, mutton-type sheep that are hornless, white-faced, and have a thick forelock. Also, ROMNEY MARSH. (From *Romney* Marsh, an area of pasture in England.)

Ronchi test *Optics.* a test for determining the quality of a curved mirror using a transmission grating with 40–200 lines per inch; more effective than the Foucault knife-edge test.

rondada *Meteorology.* a Spanish nautical term for wind that shifts diurnally from northwest through north, east, south, and west.

rongeur *Surgery.* a surgical device used to cut bone, cartilage, and other tough tissues.

rongstockite *Geology.* a medium-grained to fine-grained plutonic rock composed mainly of zoned plagioclase and orthoclase.

röntgen see ROENTGEN.

Röntgen rays *Crystallography.* see X-RAYS.

rood *Metrology.* **1.** a historic unit of land area, equal to one-quarter acre. **2.** any of several other historic measures varying from 5 1/2 to 8 yards. **3.** see ROD.

roof *Architecture.* the cover of a building, including the framework and the materials. *Building Engineering.* a frame that supports the upper covering of a house or other building. *Geology.* the country rock that extends along the upper surface of an igneous intrusion. *Mining Engineering.* the rock, commonly shale and usually carbonaceous, found immediately above a coal seam, orebody, or other tabular deposit.

roof beam *Building Engineering.* in roofing, a load-bearing member.

roof board *Building Engineering.* a board that is fixed to common rafters as sheathing or undercovering to a roof; it serves as a foundation for shingles, slate, or other roofing materials. Also, ROOFERS.

roof-deck *Building Engineering.* a portion of flat roofing used for various human activities such as gardening or sun bathing.

roofed dike *Geology.* a dike having an upward termination.

roof filter *Electronics.* a circuit that blocks high-frequency interference caused by outside sources in a telephone system.

roof foundering *Geology.* the collapse of the roof of a magma chamber, usually following the eruption of a large quantity of magma from below. Also, **roof collapse.**

roofing *Building Engineering.* the application of roof covering materials. *Materials.* **1.** any of various materials used in roof contruction, such as tar paper, tin, or slate. **2.** describing such materials. Thus, **roofing putty, roofing slate.**

roofing nail *Building Engineering.* a short nail with a broad head; used for asphalt layup applications as shingles.

roof line *Building Engineering.* **1.** the contouring aspect to the top of a roof. **2.** the point at which a wall meets a roof.

roof pendant *Geology.* a downward projection or sag of country rock into the upper part of an igneous intrusion. Also, PENDANT.

roof prism *Photogrammetry.* a prism having two surfaces inclined at 90° to each other, through which an image may be reversed left-to-right, but not top-to-bottom (reverted). *Optics.* see AMICI PRISM.

roof truss *Building Engineering.* timbers fastened together to form a truss, adding strength to the support of a roof deck or framing.

rook *Vertebrate Zoology.* a crow, *Corvus frugilegus,* that often nests in large flocks in trees near buildings; found in Europe.

rookery *Zoology.* a breeding ground, used primarily for seals and birds.

rook polynomial *Mathematics.* consider a plane configuration of n squares of equal sides, each of which shares at least one side with another square. For a given configuration, let r_k denote the number of ways of choosing k squares so that no designated square is in the same horizontal row or vertical column as any other designated square. The ordinary generating function

$$R(x) = \sum_{k=0}^{n} r_k x^k$$

of the sequence is called the rook polynomial of the configuration; so named because of the particular case of placing nontaking rooks on a chessboard. The exact values of r_k depend on the plane configuration.

room *Building Engineering.* the space within a building or structure separated by walls or partitions from other spaces or rooms. *Geology.* an open area in a cave, or an expanded part of a cave passage. *Mining Engineering.* **1.** an area abutting an entry or airway in which coal is mined. **2.** a wide working space within a flat mine.

room acoustics *Acoustics.* the study of the behavior of sound in enclosed areas or rooms, including studies involving the use of echo chambers, reverberant rooms, live-end-dead-end rooms, and large rooms, such as concert halls and churches.

room constant *Acoustical Engineering.* a measure of the absorptivity of a room given by the formula $R = S/(1-\gamma)$, where S is the boundary area of the room and γ is the average absorption coefficient of the room; used in the equation for computing critical distance.

room noise *Acoustical Engineering.* the ambient noise in any enclosed location.

rooseveltite *Mineralogy.* $BiAsO_4$, a gray monoclinic mineral of the monazite group occurring as thin botryoidal crusts, having a specific gravity of 6.86 and a hardness of 4 to 4.5 on the Mohs scale; found in cassiterite veinlets in felsic lava flows.

rooster *Vertebrate Zoology.* the adult male of certain fowl, especially chickens.

roostertail *Hydrology.* water or spray occurring at the intersection of two crossing waves, taking on a form that resembles a plume of feathers.

root *Botany.* in vascular plants, the underground organ that serves to anchor the plant in the soil and to absorb and store moisture and nutrients from the soil; it typically develops from the radicle and lacks nodes and internodes. *Design Engineering.* the base or bottom of a screw thread. *Geology.* **1.** the lower margin of an orebody. **2.** the part of a fold nappe that was connected to the source, or basement rock, on which the nappe originally rested. **3.** a low-density, subsurface, downward projection of crustal material that acts as isostatic compensation for a surplus of mass at the surface. Also, DOWNWARD BULGE. *Civil Engineering.* **1.** the part at which a weld fuses with the base metal. **2.** the part at which a dam runs into the ground as it joins the bank. *Mathematics.* **1.** a number or quantity that, when raised to some exponent, equals a given number or quantity. Also, RADIX. **2.** a root of a single-valued function equation $f(x) = 0$ is a zero of the function; i.e., an element z_0 in the domain D of f at which $f(z_0) = 0$. If, in addition, D is a simply connected domain and f is analytic in D, and if k is the smallest positive integer such that $f(z) = (z - z_0)^k g(z)$, where $g(z)$ is analytic at z_0 and $g(z_0) \neq 0$, then z_0 is said to be a root (or zero) of order k. *Computer Programming.* the primary node in a tree structure. The root node has no parents, and all other nodes of the tree are descendants of the root. *Linguistics.* see ROOT WORD.

root canal *Anatomy.* the narrow pulp cavity extending through the root of a tooth. *Medicine.* a shorter term for root canal therapy.

root canal therapy *Medicine.* a procedure for treatment of disease of the pulp of a tooth, involving removal of the affected tissue and its replacement with a filler material.

root cap *Botany.* the parenchymatous shield that protects the meristematic region of a root.

root cast *Geology.* **1.** a nearly vertical, slender, tubelike sedimentary structure formed by the filling in of an opening left by a plant root. **2.** see RHIZOCONCRETION.

root circle *Design Engineering.* a hypothetical circle at the base of the tooth spaces of a gear.

root clay see UNDERCLAY.

root collar *Botany.* the swelling at the base of a tree trunk at which the roots begin.

root crack *Metallurgy.* in welding, a fracture occurring at the root of the weld.

rooted tree *Mathematics.* a directed tree that has a root.

rooter *Agriculture.* a towed plow with tines or teeth, used to break up the soil.

root fillet *Design Engineering.* the rounded corner at the angle of a gear tooth flank and the bottom land.

root hair *Botany.* a single-celled, short-lived strand that develops as an outgrowth of the primary root epidermis and acts to increase the surface area for absorption.

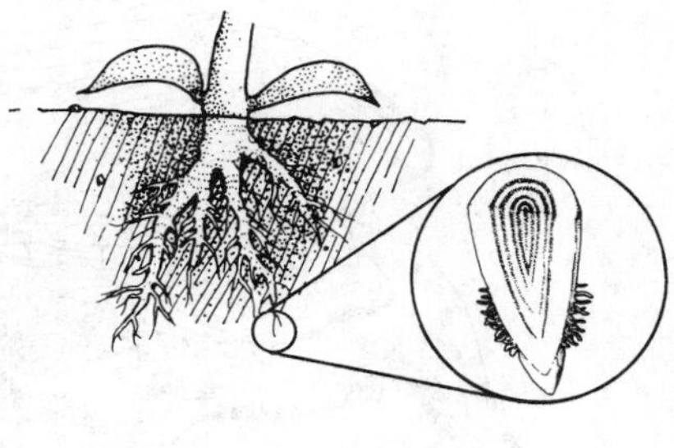

root hair

rooting *Botany.* the production of roots by a plant.

root knot *Plant Pathology.* any of various plant diseases caused by root-knot nematodes, characterized by abnormal, large swelled growths on the roots.

root locus *Electrical Engineering.* a plot of the solutions of the equation $KG(s)H(s) + 1 = 0$ in the complex plane for all values of s. $KG(s)H(s)$ is the loop gain.

root locus plot *Control Systems.* a plot in the complex plane of values in which the loop transfer function of a feedback control system becomes a negative number.

root-mean-square *Electrical Engineering.* the square root of the average value of the squares of related values.

root-mean-square error see STANDARD DEVIATION.

root-mean-square sound pressure see EFFECTIVE SOUND PRESSURE.

root-mean-square value *Physics.* the square root of the average (mean) value of the square of a quantity. *Statistics.* the square root of the arithmetic mean of the squares of the deviation of observed values from their arithmetic mean.

root of an equation *Mathematics.* any number, quantity, or value that satisfies the given equation.

root of joint *Metallurgy.* the location at which the parts to be joined are closest to each other.

root of unity *Mathematics.* let K be a field with multiplicative identity 1_K and n a positive integer. An element a of K is then said to be an ***n*th root of unity** in K if $a^n = 1_K$; that is, a is a root of $x^n - 1_K$ in the polynomial ring $K[x]$. The set of all nth roots of unity forms a cyclic multiplicative subgroup (of order at most n) of the multiplicative group of nonzero elements of K. a is a **primitive *n*th root of unity** in K if a is an nth root of unity and there is no $k < n$ such that a is a kth root of unity.

root of weld *Metallurgy.* the location at which the weld bead intersects the base near the root of joint.

root opening *Metallurgy.* in welding, the distance between the parts near the root of joint.

root pass *Metallurgy.* in welding, the first pass performed near the root of joint.

root penetration *Metallurgy.* the depth that a weld extends into the root of joint.

root rot *Plant Pathology.* **1.** a phase of plant disease in which the roots discolor and decay. **2.** a disease so characterized.

Roots blower *Mechanical Engineering.* a gas compressor, consisting of a pair of hourglass-shaped members that interlock and rotate together within an enclosure to deliver large volumes of gas at relatively low pressure ratios to a process or pipeline.

root segment *Computer Programming.* **1.** a program segment that remains in storage during the execution of an overlay program. **2.** the highest segment in the hierarchy of a database.

root sheath *Geology.* a hollow rhizoconcretion.

rootstock *Botany.* **1.** the underground rhizome system of a vascular plant. **2.** see STOCK, def. 1.

root-sum-square value *Physics.* the square root of the sum of the squares of a distribution of quantities.

root symbol *Computer Programming.* in formal language grammar, a metavariable such as S that is the starting point for a substitution process. Also, **root component.**

root system *Mathematics.* a set Φ of vectors in Euclidean R^n is called a root system in R^n if: (a) Φ is finite, spans R^n, and does not contain the zero vector; (b) if $\alpha \in \Phi$, then the only multiples of α in Φ are $\pm\alpha$; (c) a reflection through a hyperplane normal to any element of Φ leaves Φ invariant; and (d) $2 < \alpha,\beta > / < \alpha,\alpha >$ is an integer for all $\alpha,\beta \in \Phi$, where $< , >$ is the usual inner product in R^n.

root test see CAUCHY ROOT TEST.

root vertex *Mathematics.* a vertex v in a directed graph G such that there are directed paths from v to all the remaining vertices of G. If G is a tree, there is at most one root vertex; if there is a root vertex in a tree G, it has indegree zero and every other vertex of G has indegree one.

root word *Linguistics.* the basic word of which an inflected word is a variant; for example, the verb *open* is the root word for *opened, opening, opens, opener, reopen,* and so on.

rootworm *Invertebrate Zoology.* the common name for any insect larva or nematode worm that feeds on or parasitizes the roots of plants.

root zone *Geology.* **1.** see RHIZOSPHERE. **2.** the source, or part of the basement rock to which a fold nappe was originally connected. **3.** the region in the crust from which low-angle thrust faults originate.

ropak *Oceanography.* an extreme form of ridging in which slabs of heavy sea ice have been forced into a vertical position in pressure ice, often rising 8 meters above the surrounding ice.

rope *Materials.* a strong cord composed of fiber or wire strands that have been twisted or braided together. *Ordnance.* an element of chaff consisting of a long roll of metallic foil or wire; **rope chaff** is designed for broad, low-frequency responses. *Food Technology.* gummy yellowish patches appearing on white bread, caused by masses of *Bacillus* that survive baking and form spores when the bread is stored under damp conditions. The damaged product is called **ropy bread.**

rope drive *Mechanical Engineering.* an arrangement of ropes that run at high speed through pulleys to transmit power over distances.

rope-lay conductor *Electricity.* a cable made by laying rope, that is, by twisting together conventional three-strand rope. The term "lay" designates the number of turns of strands per unit length of rope; a soft lay has high ultimate strength, while a hard lay has high abrasive resistance.

rope sheave *Mechanical Devices.* a grooved wheel consisting of cast steel or heat-treated alloys to receive a rope bight; used in hauling overhead hoists, cranes, and the like. Also, ROPE WHEEL.

rope socket *Mechanical Devices.* a steel device with a tapering hole that is attached in pairs to the end of a wire cable in order for a load to be fastened to the assembly.

ropeway *Engineering.* a tram system consisting of towers that support a line or double line of ropes, usually wire, along which articles of moderate weight may be transported by gravity or power; used for hauling ore over mountainous terrain or ferrying supplies across a river.

ropiness *Food Technology.* a type of food spoilage in which long stringy threads of polysaccharides are formed by bacteria; this occurs in milk and carbonated beverages as well as in bread.

Roproniidae *Invertebrate Zoology.* a family of tiny parasitic wasps, hymenopteran insects in the superfamily Proctotrupoidea, having 14-segment antennae.

ropy lava see PAHOEHOE.

Roquefort [rōk´fərt] *Food Technology.* **1.** a semisoft cheese that is made with sheep's milk and ripened in caves near the town of Roquefort, France, by blue mold, *Penicillium roqueforti*, growing in its curd. **2.** also, **roquefort.** a similar strong-tasting cheese having a bluish mold, that is made elsewhere.

rorqual [rôr´kwəl] *Vertebrate Zoology.* the slender baleen whales of the family Belaenopteridae, including the blue whale, finback whale, sei whale, Bryde's whale, minke whale, and humpback whale; all are swift swimmers, have a deeply fluted belly and throat, and summer in cold polar waters and then migrate to warmer waters during the winter.

Rorschach test [rôr´shäk; rôr´shak] *Psychology.* a projective approach to personality assessment, in which a subject responds to a series of ten ambiguous ink blots; the responses are theorized to indicate significant elements of the subject's personality. (Developed by the Swiss psychiatrist Hermann *Rorschach,* 1884–1922.) Also, INKBLOT TEST.

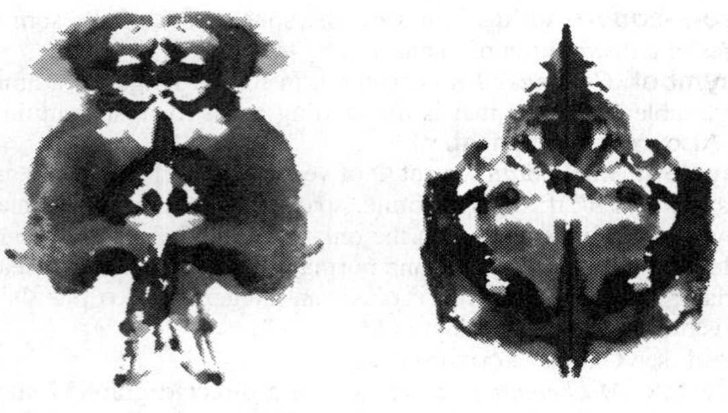

Rorschach test

ROS *Oncology.* a proto-oncogene found in astrocytoma; the exact damage or abnormality of this gene is unknown.

Rosa and Dorsey method *Electromagnetism.* a method by which the speed of light is determined based on measurements of the capacitance of a capacitor in electromagnetic units, as well as the capacitance in electrostatic units as derived from the dimensions of the capacitor.

rosacea *Medicine.* an acnelike, chronic, inflammatory disorder of middle age, marked by skin lesions and dilated capillaries of the face. Also, ACNE ROSACEA.

Rosaceae *Botany.* a dicotyledonous family of the order Rosales consisting of approximately 120 genera and 3400 species of temperate and warm-climate trees, shrubs, and herbs; characterized by a fleshy cuplike receptacle called a **hypanthium,** such as the hip developed by a rose, with seeds lacking endosperm; the family includes many species that produce edible fruit such as apples, cherries, almonds, plums, peaches, apricots, strawberries, pears, and raspberries.

Rosales *Botany.* a large order of dicotyledonous plants in the subclass Rosidae, containing over 20 families including Rosaceae.

rosaniline *Organic Chemistry.* $HOC(C_6H_4NH_2)_2C_6H_3(CH_3)NH_2$, reddish-brown crystals; slightly soluble in water and soluble in alcohol and acids; melts at 186°C; used as a dye and fungicide. Also, FUCHSIN.

rosasite *Mineralogy.* $(Cu^{+2},Zn)_2(CO_3)(OH)_2$, a green to sky-blue monoclinic mineral of the rosasite group, having a specific gravity of 4.0 to 4.2 and a hardness of about 4.5 on the Mohs scale; found in the oxidized zones of zinc-copper-lead ore deposits.

roscherite *Mineralogy.* $Ca(Mn^{+2},Fe^{+2})_2Be_3(PO_4)_3(OH)_3 \cdot 2H_2O$, an olive-green to brown or red monoclinic and triclinic mineral occurring as tabular to prismatic crystals and fibrous masses, having a specific gravity of 2.934 and a hardness of 4.5 on the Mohs scale; found in pegmatites.

roscoelite *Mineralogy.* $K(V^{+3},Al,Mg)_2(AlSi_3)O_{10}(OH)_2$, a rare, weakly radioactive, monoclinic mineral of the mica group occurring in minute scales or flakes, having a specific gravity of 2.97 and a hardness of 2.5 on the Mohs scale; found in uranium-vanadium mines, and with native gold and tellurides.

rose *Botany.* **1.** any erect, climbing, or trailing shrub of the genus *Rosa,* family Rosaceae, characterized by prickly stems and showy flowers of generally red, pink, white, or yellow color; cultivated in many varieties and considered the world's most popular garden flower. **2.** the flower of any such plant. **3.** any of various plants having flowers that are superficially similar to the rose, or the flower of such plants; used especially in compound terms such as **desert rose** *(Adenium* spp.) or **Christmas rose** *(Hellebrous niger).*

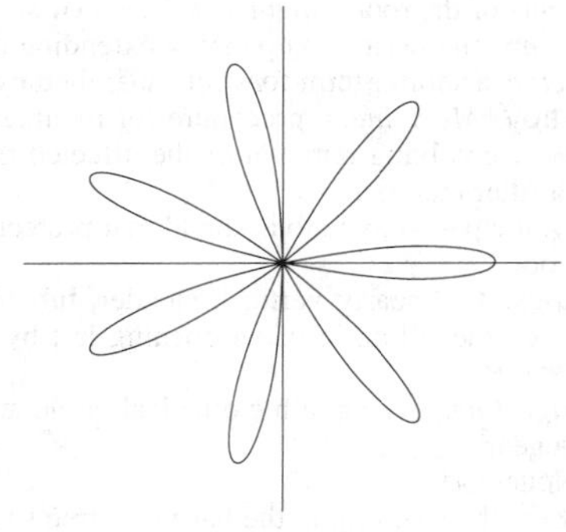

rose

rose *Mathematics.* the graph in the plane of the equation in polar coordinates $r = a \cos n\theta$ (or $r = a \sin n\theta$). If n is an even (odd) integer, the rose has $2n$ (resp. n) loops or leaves, evenly distributed about the origin, intersecting only at the origin, and mutually tangent at the origin when n is even.

roseate spoonbill *Vertebrate Zoology.* a spoonbill, *Ajaia ajaja,* characterized by rose-colored plumage and a bare head; found in tropical regions of the Americas.

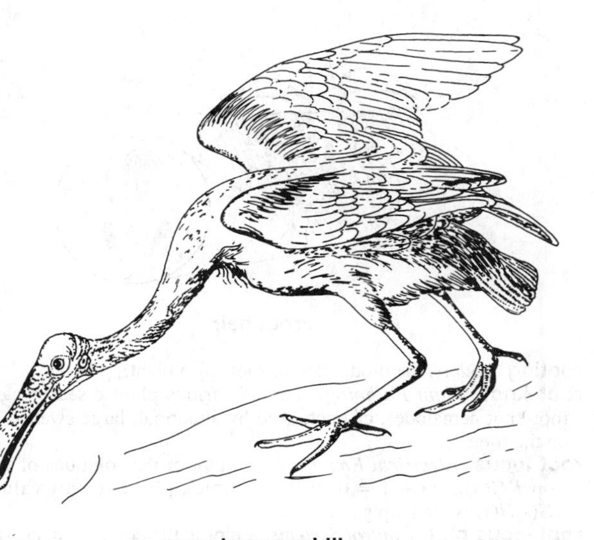

roseate spoonbill

rose bit *Mechanical Devices.* a hardened steel or alloy, cylindrically shaped, noncoring boring tool with a serrated face; used to finish drilled holes or to cut bits, casing, and other metal objects lost in a borehole.

rose diagram *Geology.* a graph consisting of radiating rays that represent quantities or values in several directions of bearing.

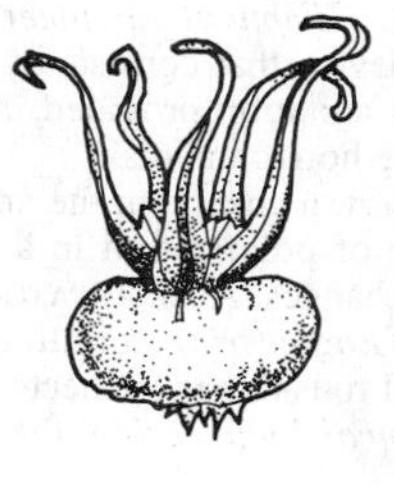

rose hip

rose hip *Botany.* the fleshy, enlarged, cuplike receptacle of *Rosa* species; morphologically considered a hypanthium.

roselite *Mineralogy.* $Ca_2(Co^{+2},Mg)(AsO_4)_2 \cdot 2H_2O$, a pink or rose-red monoclinic mineral of the roselite group occurring as prismatic or tabular crystals, having a specific gravity of 3.69 and a hardness of 3.5 on the Mohs scale; forms a series with wendwilsonite.

rosemary *Botany.* an evergreen shrub of the mint family, *Rosmarinus officinalis,* characterized by pale-blue tubular flowers and rough, narrow leaves that are used as a seasoning, for medicinal purposes, and in perfumes; native to the Mediterranean region.

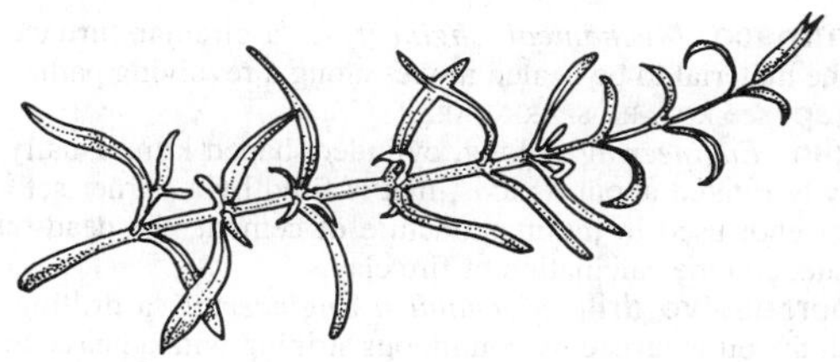

rosemary

Rosenmund reaction *Organic Chemistry.* a reaction that produces an aldehyde by catalytic hydrogenation of an acid chloride.

rose of Sharon *Botany.* **1.** a shrub, *Hibiscus syriacus,* of the mallow family, characterized by showy white, pink, or purple flowers; widely cultivated. Also, **Althea**. **2.** a St.-John's-wort, *Hypericum calycinum,* characterized by evergreen foliage and yellow flowers; probably the plant mentioned in the Bible. Also, **Aaron's beard**.

rose oil *Materials.* a volatile, fragrant, sweet liguid occurring in various colors and obtained by distilling fresh roses; used in flavorings and perfumes. Also, **rose flower oil.**

roseola [rō´zē ō´lə; rō´zē´ə lə] *Medicine.* **1.** any rose-colored rash. **2.** specifically, the rash of rubella (German measles). See RUBELLA.

roseola infantum *Medicine.* see EXANTHEMA SUBITUM.

roseola typhosa *Medicine.* the rose-colored eruption of typhus or typhoid fever.

rose opal *Mineralogy.* a pink to red variety of opaque common opal.

rose quartz *Mineralogy.* a rose-pink variety of crystalline quartz, usually massive in habit; used as a gemstone or an ornamental stone. Also, BOHEMIAN RUBY.

rose reamer *Mechanical Devices.* a heavy-duty machine reamer that cuts through material at its beveled end only.

Rose's metal *Metallurgy.* a low-melting, bismuth-base alloy containing tin and lead.

rosette *Science.* a circular cluster or arrangement around a central axis, resembling a rose. *Botany.* **1.** a cluster of leaves that develops at the base of a plant prior to the flowering stage, thought to be for accumulation of reserves for the flowering and fruiting cycle. **2.** the cluster of four cells that develop in the embryo of a pine seed. *Plant Pathology.* a symptom of various plant diseases, in which stems are shortened and produce a clustering of leaves in the form of a rosette; caused by fungi, viruses, or nutrition deficiencies, often a zinc deficiency. *Metallurgy.* in a metallic microstructure, any rounded constituent containing a spiral design. *Mineralogy.* a symmetrical, rose-shaped, crystalline aggregate formed by barite, marcasite, or pyrite in sedimentary rock.

Rosette Nebula *Astronomy.* a bright ionized hydrogen region in Monoceros surrounding the young open cluster NGC 2244.

rosetting *Immunology.* a method used to detect or isolate cells by mixing them with other cells to which they bind.

Rose-Waaler test *Immunology.* a method used to determine the presence of the rheumatoid factor in individuals, in which an antibody-antigen complex (in the presence of complement) is attached artificially to sheep erythrocytes coated with γ-globulin.

rose water *Materials.* water that is tinctured with rose oil; used as a perfume.

Rosidae *Botany.* a large subclass within the division Dicotyledonae, containing woody and herbaceous species organized into between 14 and 21 orders, depending upon the system of classification; generally characterized by bisexual flowers possessing multiple stamens and developing centripetally.

rosiérésite *Mineralogy.* a yellow to brown amorphous mineraloid, a hydrous phosphate of lead, copper, and aluminum having a specific gravity of 2.2 and an undetermined hardness, occurring in recent stalactitic masseson copper mine walls at Rosiérés, France.

rosin *Materials.* the combustible, hard, brittle, translucent, yellowish to amber residual fragments after oil of turpentine is distilled from crude pine tree resin; insoluble in water and very soluble in alcohol; used in adhesives, varnishes, soldering compounds, and polyesters.

rosin joint *Electricity.* a soldered connection that uses rosin to clean the connection while heat is being applied.

rosin oil *Materials.* a viscous, odorless, strong-tasting liquid that is derived from the fractional distillation of rosin; insoluble in water, slightly soluble in alcohol, and soluble in ether and fatty oils; used as a lubricant and as an ingredient in hot-melt adhesives, printing inks, and varnishes.

rosin tin *Mineralogy.* a red or yellow type of cassiterite. Also, RESIN TIN.

Rosiwal analysis *Petrology.* a method of estimating mineral volume percentages in a rock, based on the Rosiwal principle, by examining thin sections under a microscope fitted with a micrometer for measuring the linear intercepts of each mineral along a particular set of lines.

Rosiwal principle *Petrology.* a quantitative mineralogic estimate based on an assumption that the area of a mineral on an exposed surface is proportional to its volume in a rock mass.

rosolic acid see AURIN.

Ross, Sir James Clark 1800–1862, English rear admiral and polar explorer; located the north magnetic pole; discovered the Ross Sea.

Ross, Sir Ronald 1857–1932, British physician; awarded the Nobel Prize for proposing the Anopheles mosquito as the agent of malaria.

Ross Barrier *Oceanography.* the high, abrupt seaward edge of the Ross Ice Shelf in Antarctica.

Rossby, Carl-Gustaf 1898–1957, Swedish-born American meteorologist; developed the Rossby diagram, equations, and number; discovered Rossby waves.

Rossby diagram *Thermodynamics.* a diagram by which constant equivalent potential temperature is plotted on a graph whose horizontal axis is the mixing ratio and the vertical axis is the potential temperature.

Rossby number *Fluid Mechanics.* a dimensionless ratio of the inertial force to the Coriolis force for a rotating fluid flow; $R_0 = U/fL$, where U is the characteristic velocity of the flow, f the Coriolis parameter, and L the characteristic length.

Rossby parameter *Fluid Mechanics.* the northward variation of the Coriolis parameter f; arises from the sphericity of the earth. Also, **Rossby term.**

Rossby regime *Fluid Mechanics.* a flow pattern in a rotating fluid subjected to differential radial heating in which the major radial transport of shear and momentum is brought about by horizontal eddies of low wave number; the regime occurs for low Rossby number, on the order of 0.1.

Rossby wave *Meteorology.* a wave on a uniform current in a two-dimensional, nondivergent fluid system, rotating at varying angular speed about the local vertical; a special case of a barotropic disturbance, conserving absolute vorticity.

Ross Ice Shelf *Geography.* a broad area of glacial ice that fills the southern part of the Ross Sea.

rossite *Mineralogy.* $CaV_2^{+5}O_6 \cdot 4H_2O$, a clear, glassy, yellow triclinic mineral, very rare and weakly radioactive, having a specific gravity of 2.43 to 2.45 and a hardness of 2 to 3 on the Mohs scale; found as small grains embedded in metarossite in carnotite-bearing sandstone.

rösslerite *Mineralogy.* $MgHAsO_4 \cdot 7H_2O$, a colorless or white, transparent, monoclinic mineral occurring as fibrous or fine-grained crusts, having a specific gravity of 1.93 to 1.94 and a hardness of 2 to 3 on the Mohs scale; found as a secondary mineral in arsenic-bearing ore deposits. Also, ROESSLERITE.

Ross objective *Optics.* a lens having a wide field of view; added to cameras used in astronomy.

Ross River virus *Virology.* a virus of the genus *Alphavirus*, of the family Togaviridae, that is found in birds and mosquitoes and infects a wide range of vertebrates, causing fever and polyarthritis in humans.

Ross Sea *Geography.* a broad arm of the South Pacific off the coast of Antarctica south of New Zealand.

rostellum *Biology.* a small beak or beak-shaped structure on a plant or animal, such as the abortive third stamen of an orchid flower. *Invertebrate Zoology.* **1.** in some tapeworms, a hooked projection on the head. **2.** in some insects, tubular mouthparts, generally adapted for sucking.

rosthornite *Mineralogy.* a brown to garnet-red, resinous variety of retinite that forms lenticular masses in coal.

Rostratulidae *Vertebrate Zoology.* the painted snipes, a family of birds of the order Charadriiformes, characterized by dark patterned plumage with a white belly and a deep resonant voice; inhabits tropical marshes and swamps.

Rostroconchia *Paleontology.* a small class of mollusks found in deposits worldwide between the early Cambrian and the end of the Permian; especially diverse in the early Ordovician; possibly ancestral to the bivalves.

rostrum *Biology.* the beak of a bird or a structure resembling a beak on certain other animals, in shape or position on the head. *Invertebrate Zoology.* an anterior pointed projection, such as the front of the carapace of a decapod crustacean carapace, or the beak or proboscis of many insects and worms.

rot *Biology.* any natural process of decay or decomposition, as by the action of bacteria or fungi. *Plant Pathology.* any of various plant diseases usually caused by bacteria or fungi; characterized by the deterioration and decay of plant tissue. *Molecular Biology.* a measurement used in RNA-driven hybridization, determined from the initial concentration of single-stranded RNA and the time required for reassociation.

Rotaliacea *Invertebrate Zoology.* a superfamily of foraminiferans in the suborder Rotaliina, having a perforated spiral test.

rotameter *Engineering.* a rate-of-flow meter consisting of a float that moves vertically in a tapered tube as a function of the flow velocity.

rotary *Mechanics.* relating to or involving a process of rotation; i.e., a process of motion about an internal axis of symmetry. *Mechanical Engineering.* **1.** a machine having a major rotating element, such as a rotary drill or rotary press. **2.** the mechanical assembly that supports and rotates a well-drilling machine.

rotary actuator *Mechanical Engineering.* an electrically powered instrument that transforms electric energy into controlled rotary force.

rotary air heater *Mechanical Engineering.* an air-heating device in which heat-transferring members are moved through gas and air streams in an alternate pattern.

rotary annular extractor *Mechanical Engineering.* a device used in liquid-liquid extraction processes, composed of two vertical cylinders, in which the inner cylinder rotates and two liquid streams move countercurrently through the space between the cylinders.

rotary atomizer *Mechanical Engineering.* a combined pump and nozzle used to spray fluids.

rotary beam *Electromagnetism.* a short-wave antenna system that can be made to rotate in any direction desired, by means of a rotary antenna.

rotary blower *Mechanical Engineering.* a machine in which the centrifugal force of rotating vanes creates air movement at a low pressure.

rotary boring *Mechanical Engineering.* a process in which a hollow cutting tool is mechanically rotated to bore a hole in rock.

rotary bottom *Agriculture.* a plow bottom having bent rotating blades to mix crop residue with the soil.

rotary compressor *Mechanical Engineering.* a machine having a rotating member that directly compresses fluid in an enclosed housing; the fluid pressure rises as the volume of the closed space decreases.

rotary crane *Mechanical Engineering.* a machine having a boom mounted on a fixed or movable structure that moves in a complete circle to handle heavy material.

rotary crusher *Mechanical Engineering.* a pulverizing machine, having an inverted cone that rotates at high speeds on a vertical shaft and presses the material against a surrounding shell; both the shell and the cone may be equipped with teeth, graduated in size from the top to the bottom of the assembly.

rotary current *Oceanography.* a continuously flowing tidal current whose direction of flow rotates through all points of the compass during a tide cycle.

rotary cutter *Mechanical Engineering.* a machine that slices fibrous material using one set of blades mounted on a rotating holder and another fixed on the casing enclosure.

rotary-disk contactor *Chemical Engineering.* a mechanically agitated, mass-contacting device that consists of a vertical shell separated into compartments with agitation provided in each compartment by a centrally located, rotating horizontal disk.

rotary dispersion *Optics.* a change in the angle through which a substance rotates the plane of polarization in a polarized light when the wavelength of the light changes. Also, ROTATORY DISPERSION.

rotary drill *Mechanical Engineering.* a drilling machine having a rigid, rotating, tubular string of rods that is connected to a rock-cutting bit.

rotary drilling *Mechanical Engineering.* the process of using a rotary drill in drilling for oil.

rotary dryer *Mechanical Engineering.* a cylindrically shaped furnace that tilts slightly horizontally and is rotated on bearings, using rising hot gases up to 800°F to extract moisture from material; used for cement production or calcining.

rotary engine *Mechanical Engineering.* an engine having a thermodynamic cycle mechanism that is powered and designed to move on a circular path; it is used for steam engines and in some internal-combustion (automotive) engines.

rotary fault *Geology.* a fault in which displacement is exhibited by a rotational movement about an axis perpendicular to the fault plane. Also, ROTATIONAL FAULT, PIVOTAL FAULT.

rotary feeder *Mechanical Engineering.* a machine having a revolving element or vane that emits a regulated flow of granular or pulverized material.

rotary furnace *Mechanical Engineering.* a circular furnace through which the material to be heated moves along a revolving path.

rotary gap see ROTARY SPARK GAP.

rotary kiln *Engineering.* a long, cylinder-shaped kiln, usually inclined and slowly rotated about its axis, that is fired by a burner set axially at its lower end; used in the manufacture of cement, the dead-burning of magnesite, and the calcination of fire clay.

rotary-percussive drill *Mechanical Engineering.* a drilling machine in which the bit is turned by continuous striking with a heavy tool.

rotary phase converter *Electricity.* a rotating machine that converts AC voltage of one or more phases to AC voltage of a different number of phases.

rotary plow *Agriculture.* a type of plow that consists of a rotating horizontal shaft fitted with a series of tines that pulverize the soil and create a seedbed. Also, ROTOTILLER.

rotary press *Graphic Arts.* a printing press on which the paper is printed as it passes between a curved plate and an impression roller; used in letterpress and offset lithography.

rotary pump *Mechanical Engineering.* a pump in which fluid is positively displaced at a constant rate by two gear- or screw-type elements that rotate in contact.

rotary rig *Petroleum Engineering.* a derrick equipped with the equipment used in rotary drilling, including prime movers or engines, hoisting and rotating machinery, drill pipe, drill collars and bits, and mud pumps.

rotary shear *Mechanical Engineering.* a machine that cuts sheet metal along a curved line, using a pair of rotating cutting wheels or disks that are mounted on parallel shafts and moved simultaneously.

rotary solenoid *Electromagnetism.* a solenoid having a rotatable armature, used to translate rotational motion into linear motion.

rotary spark gap *Electricity.* a spark gap in which sparks between electrodes project outward from the circumference of a motor-driven metal disk. Also, ROTARY GAP.

rotary stepping switch see STEPPING RELAY.

rotary swager *Mechanical Engineering.* a device that hammers the surface of a pipe or bar supported on a cylindrical axle to decrease its diameter or thickness.

rotary switch *Electricity.* a switch having movable contacts that can be set to two or more positions. Also, selector switch.

rotary system *Telecommunications.* a receiver-sender-based switching system in which dial pulses are received by a register that translate the registered information to control the group selector and final selector switches; the rotary finder and rotary selector switches are driven by a constantly rotating shaft; busy signals, ringing, and shutting of the talking path are performed by a motor-driven sequence switch. Used widely in Europe in the first half of the 20th century, but now largely replaced with newer systems.

rotary transformer *Electricity.* a motor-generator set designed to transform DC power from one voltage to another.

rotary vacuum filter see DRUM FILTER.

rotary valve *Mechanical Engineering.* a combined inlet and exhaust valve in an engine cylinder that rotates continuously or through an arc.

rotary-vane attenuator *Electromagnetism.* a waveguide device that introduces a variable amount of impedance by varying the angular position of a piece of resistive material inserted into the guide.

rotary-vane meter *Engineering.* a rate-of-fluid-flow meter using the flow to rotate a drum as an indicator of the rate.

rotary-wing aircraft *Aviation.* an aircraft, such as a helicopter or autogiro, having wings that provide lift and sometimes thrust by rotating about a vertical or near-vertical axis. Also, ROTORCRAFT.

rotate *Mechanics.* to carry on a process of rotation; move about an axis. *Botany.* of or relating to plant parts that are shaped like a wheel; used especially to describe flowers that have corolla lobes in a circular or actinomorphic arrangement.

rotating anode *Radiology.* an anode in the form of a disk with the target material annealed to its rim; as the anode is rotated, the electron stream strikes only a small part of the target at one time, thus allowing heat to dissipate.

rotating-anode tube *Electronics.* a tube in which a stream of high-velocity electrons bombard a revolving anode to produce large amounts of X-ray energy, without destroying the anode.

rotating band *Ordnance.* a band of soft metal near the base of a projectile; it is larger in diameter than the land diameter of the bore so that it will engage the rifling, thus rotating the projectile.

rotating-beam ceilometer *Engineering.* a ceilometer that uses the triangulation method in measuring cloud height.

rotating coordinate system *Mechanics.* a coordinate system that is rotating with respect to a fixed inertial frame of reference.

rotating crystal method *Solid-State Physics.* a method of diffraction analysis by which a narrow beam of X-rays or neutrons is directed at a single crystal, which is made to rotate about a crystallographic axis perpendicular to the beam; the reflections are recorded on a photographic plate.

rotating-cylinder method *Fluid Mechanics.* a method for measuring the viscosity of a fluid that is placed between two concentric cylinders in which the outer cylinder is rotated at a constant speed, and the torque on the inner stationary cylinder is measured.

rotating disk contactor *Biotechnology.* an agitating contactor used for liquid-liquid extraction that consists of a vertical shell fitted with horizontal plates and a central opening.

rotating joint *Electromagnetism.* a waveguide joint that permits one section of a transmission line to rotate continuously with respect to another while the waveguide is in operation.

rotating machinery *Electricity.* motors and generators.

rotating magnetic amplifier *Electricity.* a DC generator whose output can be controlled by small field input power.

rotating meter see VELOCITY-TYPE FLOWMETER.

rotating model *Oceanography.* a hydrodynamic laboratory model that can be rotated to simulate the earth's rotation.

rotating polarization see OPTICAL ACTIVITY.

rotating Reynolds number *Fluid Mechanics.* a Reynolds number used in rotating viscous flow, and also in problems in which the fluid is agitated vigorously by an impeller; $Re_r = d^2W/u$, where d is the diameter or characteristic dimension of the impeller, W its angular velocity, and u the kinematic viscosity of the fluid.

rotating wedge *Optics.* a wedge having a small refracting angle, which rotates and obstructs the path of various light rays in order to slightly divert the line of sight.

rotation *Mechanics.* **1.** the motion of a body about an internal axis of symmetry; this axis of rotation may be fixed or moving, and it may be actual or theoretical. *Rotation* is the movement of a body in relation to an internal point, and *revolution* is its movement in reference to an external point. Thus the earth *revolves* around the sun and *rotates* on its own axis. **2.** a description of such rotating motion, expressed as the change in a body's angular orientation over time, with respect to some relevant (usually inertial) frame of reference. Also, ROTATIONAL MOTION. *Astronomy.* one complete turn of a celestial body about its axis with reference to the stars. *Biochemistry.* the ability of a substance in solution to rotate the plane of polarized light of at a given wavelength. Also, OPTICAL ROTATION. *Computer Programming.* **1.** the turning of a computer-modeled object in computer graphics, a rotating that is relative to an origin point on a coordinate system. **2.** the degree of rotation of an object relative to a coordinate system. *Mathematics.* an isometry of Euclidean R^n that has, in R^2, only one fixed point and that does not reverse orientation. *Agriculture.* see CROP ROTATION.

rotational *Mechanics.* relating to or involving a process of rotation; i.e., a process of motion about an internal axis of symmetry.

rotational bomb *Volcanology.* a volcanic bomb having a spherical, tear, or spindle shape due to spiral motion or rotation during flight.

rotational casting *Engineering.* a casting that produces hollow plastic products derived from lattices and plastisols. The mold is rotated in one or two planes. Also, **rotational molding.**

rotational constant *Physical Chemistry.* the quantity, inversely proportional to the point of inertia in a linear molecule, that is used to calculate quantums in microwave spectroscopy.

rotational delay *Computer Technology.* the amount of time taken, expressed in milliseconds, for the recording device to rotate to the proper position under the read-write head.

rotational energy *Mechanics.* the kinetic energy of a body as measured in a nonrotating frame of reference in which its center of mass is stationary. *Physical Chemistry.* **1.** the energy generated by the rotation of a molecule along three axes. **2.** the difference between the energy in a molecule with more than one atom, and the energy in a theoretical molecule constructed by stopping the rotation of the nuclei without hindering their vibration, or hindering the movement of electrons.

rotational equilibrium *Mechanics.* the condition of a system in which there is either no rotation, or rotation at a constant angular velocity; e.g., a wheel is in rotational equilibrium if it is either at rest or is rotating about a fixed axis at a fixed speed.

rotational fault see ROTARY FAULT.

rotational flow *Fluid Mechanics.* a fluid flow in which the curl of the velocity vector is nonzero.

rotational impedance *Mechanics.* the ratio of the complex torque applied to a body to its complex angular velocity at a given frequency. Also, MECHANICAL ROTATIONAL IMPEDANCE.

rotational landslide *Geology.* a landslide in which shearing takes place on a curved shear surface, producing a backward rotation in the displaced material. Also, **rotational slump.**

rotational level *Physical Chemistry.* the energy level of a molecule with more than one atom (diatomic or polyatomic) that reflects both the value of the rotation of the nuclei and their angular momentum.

rotational motion *Mechanics.* see ROTATION, def. 2.

rotational movement *Geology.* an apparent displacement in which fault blocks have rotated relative to one another, disturbing the alignment of once parallel features.

rotational quantum number *Physical Chemistry.* the quantity (quantum number J) that represents the angular momentum of a molecule's nuclei.

rotational reactance *Mechanics.* the imaginary part of the rotational impedance, representing time lag in the system response. Also, MECHANICAL ROTATIONAL REACTANCE.

rotational resistance *Mechanics.* the real part of the rotational impedance, representing dissipation of energy. Also, MECHANICAL ROTATIONAL RESISTANCE.

rotational spectrum *Spectroscopy.* a molecular spectrum produced by rotational-level transitions; observed in the far-infrared region between 50 and 200 μm and in the microwave region.

rotational stability *Mechanical Engineering.* the property of the rotating part of a machine, by which the radial forces supplied by the bearings are sufficient to overcome the radial forces exerted by the axis, and to keep the resulting axis displacements below some acceptable threshold.

rotational strain *Mechanics.* a state of strain in which the orientation of the axes of strain changes.

rotational sum rule *Spectroscopy.* for a molecule that behaves like a symmetric top, a rule stating that the sum of the line strengths associated with transitions to or from a particular rotational level is proportional to the statistical weight of that level.

rotational transform *Physics.* the displacement of magnetic field lines in a plasma device such that the lines do not close on themselves after making a rotation about the circuit.

rotational transition *Physical Chemistry.* a transition between two molecular energy levels that differ only in the energy associated with molecular rotation.

rotational viscometer see COUETTE VISCOMETER.

rotational wave see SHEAR WAVE.

rotation anemometer *Engineering.* an anemometer that measures wind speed by the amount of rotation of an element either vertically or horizontally.

rotation axis *Crystallography.* see AXIS OF SYMMETRY.

rotation camera *Solid-State Physics.* a device used to record diffracted beams of X-rays or neutrons onto a cylindrical film whose axis coincides with the rotation axis of the rotating crystal method.

rotation coefficient *Mechanics.* a factor employed in computing the effects on range and deflection of a projectile due to the rotation of the earth; applicable only for firing at very long ranges.

rotation firing *Engineering.* the crushing of a small piece of rock with a first explosion, and timing subsequent blast holes to throw their burdens toward the space made by that and other preceding explosions. Also, ROW SHOOTING.

rotation function *Crystallography.* a function describing the rotation of a known group of vectors; e.g., those corresponding to a known portion of a structure, for comparison with a Patterson map. The orientation of a known part of a structure can be found through such comparison.

rotation group *Mathematics.* a subgroup of the group of isometries of Euclidean R^n that fixes the origin and does not reverse orientation. The rotation group is isomorphic to the special orthogonal group.

rotation-inversion axis *Crystallography.* an axis for which the symmetry operation is a rotation by $360°/n$ combined with inversion through a point lying on the axis. The special point is a center of symmetry only if n is odd. Also, ROTATORY-INVERSION AXIS, ROTOINVERSION AXIS.

rotation photograph *Crystallography.* a photograph of the diffraction pattern obtained by rotating a crystal continuously about a fixed axis normal to some set of reciprocal lattice planes.

rotation rate see ANGULAR SPEED.

rotation-reflection axis *Crystallography.* an axis about which some crystals can be rotated a specific amount and then reflected across the plane perpendicular to the axis, so that there is no physical distinction between the initial crystal and the result of this symmetry operation. Also, **rotoreflection axis.**

rotation spectrum *Physical Chemistry.* the amount of radiation absorbed if only the rotation, not the vibration, of a molecule is energized by electromagnetic radiation.

rotation therapy *Radiology.* the circular movement of a patient or radiation source and beam around a fixed anatomical axis during radiotherapy treatment.

rotation twin *Crystallography.* a twin crystal whose components would be the same if one is rotated a specific amount.

rotation-vibration spectrum *Physical Chemistry.* the amount of radiation absorbed when both the rotation and the vibration of a molecule are energized by electromagnetic radiation.

rotator *Mechanics.* something that rotates; any body that carries on a rotating motion. *Anatomy.* **1.** a muscle that rotates a part of the body. **2.** see ROTATOR CUFF. *Electromagnetism.* a device, such as a ferrite rotator or a twist in a rectangular waveguide, that rotates the plane of polarization. *Quantum Mechanics.* a quantum-mechanical system whose motion is the analogue of classical rotation. Also, TOP.

rotator cuff *Anatomy.* a bandlike muscle-tendon structure about the capsule of the shoulder joint, formed by the inserting fibers of four muscles, and providing mobility and strength to the shoulder.

Rotatoria *Invertebrate Zoology.* a former term for the Rotifera.

rotatory *Mechanics.* relating to or carying on a process of rotation. *Optics.* able to rotate the plane of polarization of polarized electromagnetic radiation.

rotatory dispersion see ROTARY DISPERSION.

rotatory-inversion axis see ROTATION-INVERSION AXIS.

rotatory power *Physical Chemistry.* a quantity that represents the specific rotation of an element times its atomic or molecular weight. *Optics.* a measure of the ability of a substance to rotate the plane of polarization of polarized electromagnetic radiation.

Rotavirus *Virology.* a genus of viruses of the family Reoviridae, having a double-layered, wheel-like capsid structure, that cause severe diarrhea in children; found in the feces of birds and mammals.

rotaxane *Organic Chemistry.* a compound consisting of two or more independent portions linked by a linear portion threaded through a ring and maintained in this position by large end groups.

rote learning *Psychology.* the process of learning material merely through repetition, at the cost of comprehension and meaning. *Artificial Intelligence.* learning of particular facts or of the classification of particular data sets.

rotenone *Organic Chemistry.* $C_{23}H_{22}O_6$, toxic, white needles; insoluble in water and soluble in alcohol and ether; melts at 165–166°C; used as an insecticide and in mothproofing agents. Also, TUBATOXIN.

rotenturm wind *Meteorology.* a warm, southerly wind that blows through the Rotenturm Pass in the Transylvanian Alps.

Rothia *Bacteriology.* a genus of bacteria of the family Actinomycetacea, occurring in the human oral cavity as Gram-positive, asporogenic cells in various filament forms.

rotifer *Invertebrate Zoology.* **1.** an organism belonging to the Rotifera. **2.** of or relating to this class of organisms.

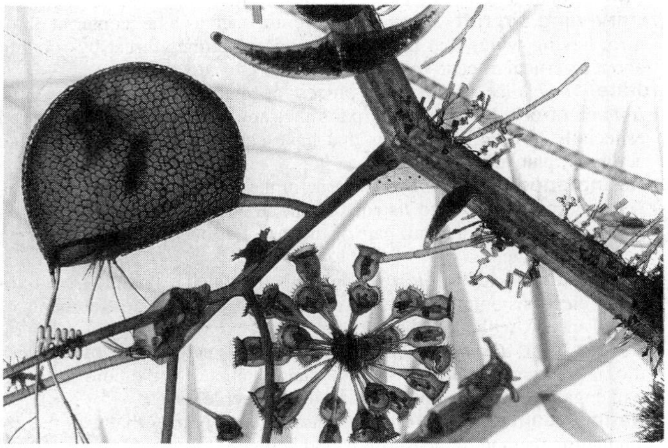

rotifers

Rotifera *Invertebrate Zoology.* the wheel animalcules, a class of freshwater wormlike or spherical animals in the phylum Aschelminthes; they are minute, often microscopic, but many-celled; the corona of cilia by which they move and feed gives the appearance of a revolving wheel. (From the Latin word for wheel.)

Rotliegende *Geology.* a European geologic series of the Lower and Middle Permian periods.

rotoflector *Electromagnetism.* a rotatable radar reflector that reflects a vertically directed beam into a horizontal position, so that the beam sweeps out a horizontal plane.

rotogravure [rō´tə grə vyůr´] *Graphic Arts.* **1.** a large-scale form of gravure printing using a web-fed press. **2.** a print made by this process.

rotoinversion axis see ROTATION-INVERSION AXIS.

roton *Physics.* a quantum of rotational motion in a liquid, analogous to the quantum of linear motion represented by a phonon.

rotor *Mechanical Engineering.* the rotating component of an electrical machine, turbine, compressor, blower, or contactor. *Engineering.* any circular object that undergoes rotational movement; it may be disklike or have blades, projections, or wire windings.

rotor assembly *Aviation.* the rotating airfoil assembly of a helicopter, consisting of long, narrow airfoils (rotor blades) set horizontally and attached centrally to the rotor hub; the turning of this assembly provides an autogiro with lift and a helicopter with both lift and thrust.

rotor blade *Aviation.* any one of the airfoils in the rotor of a rotary-wing aircraft.

rotor cloud *Meteorology.* an altocumulus cloud formation characterized by turbulence, and found in the lee of large mountain barriers, particulary in the Sierra Nevadas near Bishop, California.

rotorcraft see ROTARY-WING AIRCRAFT.

rotor disc *Aviation.* the space that is occupied by the rotating blades of a rotor.

rotor fermenter *Biotechnology.* a highly productive continuous pressure dialysis fermenter in which the usual simple membrane is replaced by a rapidly rotating membrane cylinder; the result is unusually high filtration rates.

rotor hub or **rotor head** *Aviation.* a device or apparatus serving as the central component of a rotor assembly, attached to the blades and mounted on the drive shaft.

rotorod *Biotechnology.* a device used for sampling airborne microflora.

rotor plate *Electricity.* a movable plate in a variable capacitor.

rotor spinning *Materials Science.* in textile processing, a method of forming twist in ply yarn by interrupting the flow of fibers at a point just ahead of spin twisting, making it possible to rotate the downstream end of the strand just below the point of interruption. Also, BREAK SPINNING, OPEN-ENDED SPINNING, SELF-TWIST SPINNING.

rototiller see ROTARY PLOW.

rotten ice *Hydrology.* soft, spongy ice in an advanced state of disintegration or melting, characterized by a honeycomb structure consisting of pockets filled with meltwater or seawater.

rotten spot *Hydrology.* see POTHOLE, def. 1.

rottenstone *Materials.* friable, siliceous stone in powder form; used as an abrasive on metals and woods and as a filler in molding compounds.

rotunda *Architecture.* a round hall or building, usually domed.

Rouché's theorem [roo shā′] *Mathematics.* let D be a closed simply connected domain of the complex plane. Then suppose that the functions f and g are meromorphic in some region containing D, have neither zeros nor poles on the boundary of D, and $|f(z) - g(z)| < |g(z)|$ for z on the boundary of D. Then $Z_f - P_f = Z_g - P_g$, where Z_f and Z_g are the numbers of zeros of f and g, and P_f and P_g are the numbers of poles of f and g, respectively, inside D and counting multiplicities. If f and g are both analytic in D, then they have the same number of zeros in D.

rouge [roozh′] *Materials.* a reddish powder composed mainly of ferric oxide and used as a polish for metals, glass, precious stones, and other materials.

rougemontite *Geology.* a coarse-grained igneous rock composed mainly of anorthite and titanaugite.

roughage *Nutrition.* the coarser parts or kinds of foods, such as bran and fruit skins, that stimulate the movement of food and waste through the intestines.

rough air *Aviation.* a term for flight turbulence.

rough-axed brick see AXED BRICK.

rough bark *Plant Pathology.* **1.** any of various virus diseases common to woody plants; characterized by abnormal roughness and splintering of the tree bark. **2.** a disease of apple trees caused by the fungus *Phomopsis mali*; characterized by scaly cankers on the bark.

rough burning *Space Technology.* an uneven burning of fuel and oxidizer in a rocket combustion chamber.

rough colony *Microbiology.* a bacterial colony that exhibits a rough, uneven surface, generally because the bacteria lack cell-wall-enveloping capsules. Also, R COLONY.

rough cut *Engineering.* the initial removal of overburden material before the final cuts are made.

rough endoplasmic reticulum *Cell Biology.* a system of intracellular membranes studded with numerous ribosomes that plays an essential role in protein synthesis, glycosylation, and secretion in eucaryotic cells.

rougher cell *Mining Engineering.* a flotation cell into which ore is introduced to remove the bulk of the gangue.

rough grinding *Mechanical Engineering.* a preliminary process of abrading the surface of a material without regard to finish, usually followed by another operation.

rough hardware *Engineering.* the unfinished nails, tacks, and the like, used for concealed construction work rather than for finishing work.

rough ice *Hydrology.* an expanse of ice having an uneven surface resulting from the formation of pressure ice, or the freezing in place of small fragments of floating glacier or sea ice.

roughing *Engineering.* **1.** the first step in the pumping down of a high vacuum device. **2.** preliminary machining without regard to finish.

roughing mill *Metallurgy.* a rolling mill in which steel ingots are converted into blooms, billets, or slabs.

roughing stand *Metallurgy.* in a hot-rolling mill, the first stand through which a billet is worked.

roughing tool *Engineering.* a lathe or planer tool for general removal before more detail work is performed on a piece.

rough machining *Mechanical Engineering.* the process of making heavy cuts on an object to achieve an approximate dimension, usually to remove excess metal material.

roughness *Fluid Mechanics.* a term for the height of the protrusions or irregularities on the nonsmooth interior of a pipe wall.

roughness elements *Oceanography.* structures that project from the bottom and sides of laboratory models, which are designed to affect the water flow in the same way as natural obstructions on the sea floor.

roughness factor *Fluid Mechanics.* a parameter used to determine flow resistance due to roughness of the surface over which the fluid flows.

roughness length see DYNAMIC ROUGHNESS.

roughness-width cutoff *Mechanical Engineering.* the maximum width in inches of surface irregularities that can be included in measurement of the roughness height.

rough threading *Engineering.* **1.** the removal of bulk material prior to threading metalwork. **2.** roughening a surface to aid adhesion prior to hot metal spraying.

rough turning *Mechanical Engineering.* a process of quickly removing excess material from a workpiece.

rouleau *Pathology.* erythrocytes that appear under the microscope as if stacked like coins.

roulette *Mathematics.* given two plane curves γ_1 and γ_2, the locus of points traced by a fixed point P on γ_1 as γ_1 rolls along γ_2 without slipping. (From its resemblance to a roulette wheel.)

round *Ordnance.* a single unit of ammunition, including all the components necessary to fire a single shot or detonate the munition once. *Military Science.* a number of shots intended to be fired simultaneously or with delay periods between shots. *Navigation.* a series of observations.

round cells *Immunology.* cells that have a single nucleus, particularly small lymphocytes.

round-face bit *Mechanical Devices.* any drill bit with a rounded cutting face.

round file *Mechanical Devices.* a long, slender, round, parallel, or tapered file; used for enlarging holes to finish concave surfaces, or for gulleting the teeth of large circular or pit saws. Also, CIRCULAR FILE.

round-head bolt *Mechanical Devices.* a short, solid bolt with male threads and a rounded head; used to seal the female end of fittings or valves.

round-head buttress dam *Civil Engineering.* a mass-concrete dam similar to the multiple-arch dam in appearance but of heavier construction; its walls are thickened at the water's end until they come together in a parallel buttress style.

round-head screw *Mechanical Devices.* a standard wood screw with threads along 3/4 of its shaft and typically a straight-slot, rounded or button-shaped head; used to secure items constructed of thin wood. Also, BUTTON-HEADED SCREW.

rounding *Mathematics.* the alteration of a radix representation of a real number, so that the true value of the number is represented to some desired accuracy. Also, **rounding off.** Any number whose radix representation is obtained in this manner is called a **radix approximation.** The following procedure is customarily followed: Let b be the base or radix used in the representation, and suppose the rightmost digit to be retained is followed by the digit x. If $x \le (b - 1)/2$, then the digits to the left of x are left unchanged, and x and the digits to its right are replaced with zeros. This is known as **rounding down,** since the result is less than or equal to the original number. If $x > (b - 1)/2$, then the rightmost digit to be retained is increased by 1 (carrying to the next place if necessary), and x and the digits to its right are replaced with zeros. This is known as **rounding up,** since the result is greater than or equal to the original number. Trailing zeros to the right of x are not written in the rounded number.

rounding error *Mathematics.* the difference between a number and some rounding of that number. Rounding errors can accumulate in a computation, and their resultant is collectively called the rounding error or **round-off error.**

round joint *Robotics.* a joint that articulates with four degrees of freedom.

round ligament *Anatomy.* **1.** the paired cordlike fibromuscular bands that extend from the front of the body to the uterus along the broad ligament and through the inguinal canal to the labia majora. **2.** the fibrous cords attaching the head of the femur to the acetabulum. **3.** the cordlike remnant of the umbilical vein that extends from the umbilicus to the liver.

roundness *Geology.* a term for the degree of abrasion of a sedimentary particle as indicated by the sharpness of its edges and corners.

round-nose chisel *Mechanical Devices.* **1.** a short, thick, hexagonal steel bar with a rounded point, used with a hammer to strike, cut, and chip curved grooves in such metals as brass, copper, aluminum, and unhardened steel. **2.** a chisel with a rounded cutting edge, used in wood and metal turning for finished curved and hollowed parts, and in cutting oil grooves on brass objects. Also, **round-nose tool.**

round of beam *Naval Architecture.* the rise or crown of a deck amidships, with respect to the sides of the deck, in transverse section. Also, CAMBER.

round of bearings *Navigation.* a series of bearings made in close succession.

round of sights *Navigation.* a series of celestial observations made in close succession.

round-robin scheduling *Computer Programming.* a time-sharing queue structure that puts arriving jobs in a single queue and allots each job a fixed amount of time on a first-come, first-served basis; jobs requiring more time are returned to the end of the queue for another turn.

rounds complete *Ordnance.* a term used in artillery and naval gunfire support to indicate that the requested number of rounds in fire for effect have been fired.

roundstone *Geology.* **1.** any naturally rounded fragment of rock with a diameter greater than 2 mm. **2.** a naturally rounded stone suitable for use in construction.

round-strand rope *Mechanical Devices.* a rope that consists of six twisted strands of hemp or wire around a central strand of individual wires.

round-the-world echo *Telecommunications.* a signal occurring every 1/7 second as a radio wave continuously travels around the earth at a speed of 186,000 miles per second.

round trip *Transportation Engineering.* a trip to a given destination and back to the original destination.

round-trip echo *Electromagnetism.* in radar, an echo signal that is strong enough to echo back to the target (after reflecting off the target and the radar station) and then back to the radar receiver to be detected; round-trip echoes appear as multiple integral values of the true range.

round wind *Meteorology.* a wind that gradually moves its direction through almost 180° during the day.

round window *Anatomy.* a small opening in the medial wall of the middle ear that leads to the cochlea and is covered by the secondary tympanic membrane.

roundworm *Invertebrate Zoology.* the common name for members of the phylum Nematoda.

Rous, Francis Peyton 1879–1970, American medical researcher; awarded the Nobel Prize for the discovery of oncogenic viruses.

Rouse-Bueche theory *Materials Science.* a theory postulating that a polymer chain can be considered to be a succession of equal submolecules, each one long enough to be considered a random coil in its own right; a precursor to the DeGennes reptation theory of molecular motion in polymers.

Rous sarcoma *Veterinary Medicine.* a transplantable malignant tumor occurring in fowl that is caused by a specific carcinogenic retrovirus; originally used to demonstrate that a virus can cause some cancers.

Rous sarcoma virus *Virology.* a leukovirus belonging to the Type C oncovirus group of the family Retroviridae that causes sarcoma in fowl, especially chickens, and contains the viral *src* oncogene.

route *Transportation Engineering.* the course traveled by a vehicle or passenger from the beginning to the end of a trip. *Navigation.* a generally agreed-upon lane for the passage of craft between certain points. Routes are established for traffic separation or general safety. *Aviation.* **1.** a defined and charted air path for flight vehicles. **2.** the cities or other points connected by an air carrier.

route assignment *Transportation Engineering.* the selection of one among alternate routes between a given origin and destination.

route component *Meteorology.* the average forecast wind component that is parallel to the flight path at flight level for an entire route; rated as positive if helping (tailwind) and negative if retarding (headwind).

route-control signaling *Transportation Engineering.* a railway signaling system in which all signals and points along a route are set simultaneously.

route forecast *Meteorology.* an aviation weather forecast for one or more air routes.

route locking *Civil Engineering.* the locking of switches, derails, and the like along a section of track to prevent conflicting train movements or incorrect switch positioning.

route mile *Transportation Engineering.* a mile of air route; the airport to airport distance of a route. The route miles of a given route differ from the actual flown miles due to airport circling patterns, evasion of turbulence, and other detours encountered in actual flight.

router *Mechanical Devices.* a canister-shaped tool with two handles and a bit that revolves at high speed beneath the center of the tool; used for fast and accurate cutting of wood and plastic with a smooth-cut surface along the edge of the material.

router patch *Mechanical Devices.* a plywood panel with parallel sides and rounded ends.

router plane *Mechanical Devices.* a plane with a flat metal plate in the bottom and a small blade in its center; used to form interior angles and recesses and to cut and smooth grooves in wood surfaces.

route split *Transportation Engineering.* the assignment or distribution of traffic among different routes between a given origin and destination.

routes to chaos *Chaotic Dynamics.* a particular sequence of bifurcations or of stable solutions leading to a chaotic solution as a parameter is changed.

route survey *Cartography.* a survey conducted specifically for the construction of linear transportation or communication systems, such as railroads, highways, pipelines, or power transmission lines.

Routh-Hurwitz criterion *Mathematics.* the condition that all the roots of the real polynomial

$$f(z) = \sum_{k=0}^{n} a_k z^k$$

have negative real parts if and only if the following sequence of determinants D_k ($k = 1, 2, \ldots, n$) are all positive:

$$D_1 = a_1;\; D_2 = \begin{vmatrix} a_1 & 0 \\ a_3 & a_2 \end{vmatrix};\; D_3 = \begin{vmatrix} a_1 & a_0 & 0 \\ a_3 & a_2 & a_1 \\ a_5 & a_4 & a_3 \end{vmatrix};\; \ldots;\; D_n = \begin{vmatrix} a_1 & a_0 & 0 & 0 & \ldots & 0 \\ a_3 & a_2 & a_1 & 0 & \ldots & 0 \\ a_5 & a_4 & a_3 & a_2 & \ldots & 0 \\ \ldots & \ldots & \ldots & \ldots & \ldots & \ldots \\ a_{2n-1} & \ldots & \ldots & \ldots & \ldots & a_n \end{vmatrix},$$

where $a_0 > 0$ and $a_m = 0$ if $m > n$.

Routh's rule of inertia *Mechanics.* the statement that the moment of inertia of a body about any axis of symmetry is equal to the product of its mass and the sum of the squares of its semiaxes normal to this axis, divided by: 3 if the body is rectangular (having 1, 2, or 3 dimensions); 4 if the body is elliptical (having 2 dimensions); and 5 if the body is ellipsoidal (having 3 dimensions).

routine *Computer Programming.* a sequence of instructions directing a computer to carry out a set of operations that form part of a program; a program, subroutine, or function.

routine library see SUBROUTINE LIBRARY.

routing *Transportation Engineering.* the process of assigning a route to a vehicle or train. *Telecommunications.* the process of assigning a path in a network for the purpose of sending information between stations. *Graphic Arts.* the process of cutting away nonimage areas of a printing plate to prevent accidental printing.

routing indicator *Telecommunications.* a coded indication showing or controlling the routing of a specified transmission.

routing message *Telecommunications.* the process at a central message processor of choosing the route, either direct or alternate, by which a message will continue to its next location.

routivarite *Geology.* a fine-grained, garnet-bearing igneous rock composed of orthoclase, plagioclase, and quartz.

rouvillite *Geology.* a light-colored theralite rock composed mainly of labradorite or bytownite and nepheline.

Roux, (Pierre Paul) Émile [roo] 1853–1933, French physician; with Behring and Kitasato, developed the first antitoxin (for diphtheria).

Roux, Wilhelm 1850–1924, German biologist; a pioneer in experimental embryology.

Roux flask *Biotechnology.* a flat-sided rectangular, glass, or plastic vessel that opens at one end via a short wide tube; commonly placed on its side when used for the growth of tissue cultures. Also, **Roux bottle.**

Rovac cycle see AIR CYCLE.

roving *Textiles.* **1.** in textile manufacturing, a soft strand of fiber that has been twisted, drawn, and cleaned before converting it into yarn. **2.** the stage in the manufacturing process during which this occurs.

roving artillery *Ordnance.* artillery that is moved and fired from varying positions to deceive the enemy as to strength and location. Similarly, **roving gun.**

row *Computer Programming.* **1.** a set of entries across an electronic spreadsheet. **2.** on a character display, a horizontal row of character positions. *Mathematics.* in an $m \times n$ matrix with entries a_{ij}, the n values of a_{ij} for fixed i are called, collectively, the ith row.

row address *Computer Programming.* an index array entry field containing the main storage address of a data block.

rowan *Botany.* **1.** the European mountain ash, *Sorbus aucuparia*, characterized by pinnate leaves and clusters of red, berrylike fruit. **2.** the American mountain ash, *Sorbus americana* or *S. sambucifolia*. **3.** the berry of these trees.

row crop *Agronomy.* a crop that is grown in widely spaced rows, such as corn or soybeans.

roweite *Mineralogy.* $Ca_2Mn_2^{+2}B_4O_7(OH)_6$, a light-brown, transparent orthorhombic mineral forming a series with fedorovskite, occurring as lathlike crystals, having a specific gravity of 2.935 and a hardness of about 5 on the Mohs scale; the only known specimen was found at Franklin, New Jersey.

rowen *Agriculture.* another name for the aftermath, a second growth of grass growing up after the first has been cut or plowed under.

Rowland, Henry Augustus 1848–1901, American physicist; determined the value of the ohm and the mechanical equivalent of heat; invented the Rowland diffraction grating for spectroscopy.

rowland *Spectroscopy.* a former unit of wavelength measurement, equal to 999.81/999.94 Å.

Rowland circle *Spectroscopy.* for a given concave diffraction grating in a spectrograph, the circle along which the entrance slit, grating, and focal points of various wavelengths lie.

Rowland ghost *Spectroscopy.* in a spectroscopic instrument, a false spectral line that results from periodic errors in groove positions on a diffraction grating.

Rowland grating see CONCAVE GRATING.

Rowland mounting *Spectroscopy.* in a spectrograph, an arrangement in which the concave grating and camera or the photographic plate lie at opposite ends of a diameter of the Rowland circle, and the slit lies anywhere on the semicircle between them; used in X-ray photoelectron spectroscopy.

Rowland ring *Electromagnetism.* a ring-shaped sample of magnetic material that is a test for its magnetic properties in a transformer core.

rowlock *Naval Architecture.* a fitting on a vessel's gunwale or rowing frame that holds an oar in place, while allowing it to pivot freely for rowing.

row-major order *Computer Science.* a method of storing a two-dimensional array so that elements of a row of the array are adjacent in memory.

row marker *Ordnance.* in land mine warfare, a natural, artificial, or specially installed marker located at the start and finish of a mine row.

row nucleation structure *Materials Science.* a structure found in natural rubber molecules, in which crystals nucleate around a central backbone and grow in a direction perpendicular to the main chain.

row operations *Mathematics.* rules for manipulating the rows of a matrix representing a linear transformation so that the image space of the linear transformation is unchanged.

row pitch *Computer Technology.* the distance between rows of holes on paper tape.

row planter *Agriculture.* a planter that sows seeds in a single row or a straight line, rather than scattering them.

row rank *Mathematics.* the dimension of the row space of a matrix; equivalently, the number of linearly independent rows of the matrix. The row rank of a matrix equals the column rank; thus the term rank is often used.

row shooting *Mining Engineering.* a method of blasting in which the row of holes nearest the face is fired first, followed by other rows behind it in succession.

row space *Mathematics.* given an $m \times n$ matrix $M = (a_{ij})$ over some field F, the row space of M is the subspace of F^n (the n-dimensional vector space over F) spanned by the m rows of M regarded as vectors of F^n. The row space is the same (up to a change of basis) as the image space of the linear transformation corresponding to M.

row vector *Mathematics.* a matrix with only one row.

royal jelly *Entomology.* a predigested food substance containing a high concentration of proteins that is secreted from the pharyngeal glands of worker honeybees; female honeybee larvae that are fed the substance by workers develop into queen bees. *Materials.* this viscous protein substance, rich in vitamin B, used in face creams.

royal paulowina *Materials.* the light, soft wood of the east Asian tree, *Paulowina tomentosa*; used for ceiling boards, decorative trim, musical instruments, and carvings.

RPC *Computer Science.* remote procedure call.

RPE rotating platinum electrode.

RPG report program generator.

R. Ph. Registered Pharmacist.

RPL robot programming language.

R plasmid see R FACTOR.

rpl gene *Genetics.* any of a number of genes that encode r-proteins of the large ribosomal subunit.

rpm or **RPM** revolutions per minute.

rpo gene *Genetics.* any of a number of genes that encode DNA-dependent RNA polymerase subunits.

r-process *Nuclear Physics.* the process by which various elements and nuclides merge in supernovas.

RPROM reprogrammable programmable read-only memory.

rps or **RPS** revolutions per second.

rps gene *Genetics.* any of a number of genes that encode r-proteins of the small ribosomal subunit.

RQ or **R.Q.** respiratory quotient.

RRDE rotating ring-disk electrode.

-rrhage *Medicine.* a suffix meaning "an abnormal or excessive flow." Also, **-rrhagia.**

-rrhaphy *Medicine.* a suffix meaning "suture" or "operative repair."

-rrhea *Medicine.* a suffix meaning "flow or discharge." Also, **-rrhoea.**

-rrhexis *Medicine.* a suffix meaning "rupture."

RR Lyrae star *Astronomy.* any of a group of pulsating giant variable stars often found in globular clusters and having periods of less than a day; sometimes called cluster Cepheids.

rRNA ribosomal RNA.

rRNA oligonucleotide cataloguing *Microbiology.* a method used to classify bacterial species phylogenetically, by compiling a catalogue of rRNA oligonucleotide sequences for a given species.

RRT rail rapid transit.

r.s. right side.

RS-232C *Telecommunications.* a standard used in serial connections for computers. The standard consists of four parts: interface mechanical characteristics, electrical signal characteristics, functional description of the signals, and a list of standard interface types. The official name of the standard is Interface between Data Terminal Equipment employing serial binary interface. The "C" denotes a revision.

R scope *Electronics.* a radarscope that produces an R display.

r selection *Evolution.* the selection of traits that stress rapid population growth; appropriate for species, like insects, that experience large population fluctuations or inhabit an area for only a short time.

R.S.M. Royal Society of Medicine.

RSNA or **R.S.N.A.** Radiological Society of North America.

R star *Astronomy.* an outmoded name for a carbon star, a cool stellar object characterized by a high abundance of carbon relative to oxygen and a surface temperature of approximately 2200 kelvins.

rt. right.

rt. or **rte.** route.

R tectonite *Petrology.* a tectonite characterized by a fabric that is thought to have resulted from rotation as opposed to slip.

RTF resistance transfer factor (R factor).

RTG radioisotope thermoelectric generator.

RTL resistor-transistor logic.

Ru the chemical symbol for ruthenium.

rubber *Materials.* **1.** any of various natural or synthetic high polymers characterized by their elasticity; natural rubber is *cis*-polyisoprene; biological protein rubbers include resilin, abductin, and elastin. **2.** describing materials in which rubber is a primary ingredient. Thus, **rubber fiber.**

rubber accelerator *Organic Chemistry.* any material that increases the speed of curing rubber.

rubber-base paint *Materials.* a paint using latex or a similar material as the binder. Also, **rubber-based paint.**

rubber belt *Mechanical Devices.* a conveyor belt made of a rubber-covered cotton, nylon, or other synthetic fiber reinforced with metal wire.

rubber blanket *Engineering.* a sheet of rubber used as a die in rubber formation.

rubber cement *Materials.* a viscous liquid adhesive consisting of unvulcanized rubber in an organic solvent such as benzene or gasoline.

rubber file *Mechanical Devices.* a heavy, coarse file for smoothing extremely rough surfaces.

rubber foam see FOAM RUBBER.

rubber hose *Mechanical Devices.* a hose pipe consisting of a strong, durable, vulcanized india rubber.

rubber hydrochloride *Organic Chemistry.* a nonflammable, nontoxic, white powdery or clear film derivative of rubber that is soluble in aromatic hydrocarbons and softens when heated to 110°C; used in food packaging, rainwear, shower curtains, and protective coverings.

rubber ice *Oceanography.* elastic young sea ice that is too weak to support the weight of a human standing still.

rubberize *Engineering.* to coat or impregnate with a rubber compound.

rubber mirror see ACTIVE MIRROR.

rubber-modified plastics *Materials.* a plastic with outstanding toughness; commercially prepared by mixing with or polymerizing a plastic in the presence of small rubber particles.

rubber-pad forming *Materials Science.* in sheet metal working, the use of a rubber pad in place of one of the dies of a conventional male and female die pair.

rubber plant *Botany.* **1.** a plant, *Ficus elastica,* of the family Moraceae, having oblong shiny leaves; native to southern Asia, where it serves as a source of rubber; cultivated elsewhere as an ornamental houseplant. **2.** any of various plants that resemble *Ficus elastica* or serve as a source of India rubber or caoutchouc.

rubber plate *Graphic Arts.* a printing plate made from molded hard rubber; used in long press runs where high image quality is not essential.

rubber plating *Engineering.* the application of a rubber coating to metalwork.

rubber sponge see FOAM RUBBER.

rubber-toughened plastic *Materials.* a material formed by combining rubber and plastic in an interpenetrating polymer network; properties include toughness and a built-in resistance to flow at high temperatures. Also, **rubberized plastic.**

rubber tree *Botany.* any of various trees that yield latex used in the production of rubber, especially *Hevea brasiliensis* of the family Euphorbiaceae, a South American tree that is the most important commercial source of rubber-producing latex.

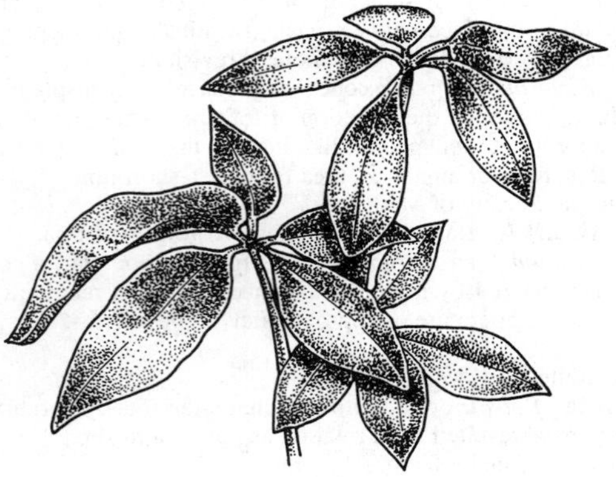

rubber tree

rubber wheel *Mechanical Devices.* a grinding wheel with rubber as the primary bonding material.

rubbery plateau *Materials Science.* a stage in glass-rubber transition in which the elastomer exhibits long-range rubber elasticity; it can be stretched several hundred percent and retract.

rubbery range *Materials Science.* in polymers, the temperature range in which large elastic elongation is evident; in this range, extensive deformation is possible, with rapid spring back to the original shape when stress is removed.

Rubbia, Carlo born 1934, Italian physicist; with van der Meer, shared the Nobel Prize for discovery of *W*(–), *W*(+), and *Z* particles, confirming the unity of electromagnetism and the weak force.

rubbing alcohol *Pharmacology.* a preparation of acetone, methyl isobutyl ketone, and about 70% ethyl alcohol; used in massaging and as a rubefacient.

rubble *Geology.* **1.** any accumulation or layer of loose, unconsolidated, irregular fragments of rock. **2.** the loose, fragmentary material that covers outcropping rock or forms the upper portion of a rock stratum in a quarry. *Building Engineering.* any old stone or masonry used as filler.

rubble drift *Geology.* **1.** a rubbly deposit formed by the slow downslope flow of saturated soil and other surficial material under periglacial conditions. **2.** a mass consisting of angular debris and large blocks set in an earthy matrix of glacial origin.

rubble ice *Oceanography.* fragments of floating or grounded sea ice, resulting from the breakup of larger formations. Also, **rubble.**

rubble-mound structure *Civil Engineering.* a structure, such as a breakwater, formed by rock or rubble of various sizes and shapes, and laid with gentle slopes to minimize shifting.

rubble tract *Geology.* the part of a reef flat that is covered over with pebbles and other coarse reef fragments.

rube- a combining form meaning "red."

rubeanic acid see DITHIOOXAMIDE.

rubefacient *Medicine.* reddening the skin. *Pharmacology.* an agent that reddens the skin by producing active or passive hyperemia.

rubefaction the process of making red; specific uses include: *Medicine.* the process or condition of reddening the skin. *Agronomy.* the development of red coloring in soils.

rubella *Pathology.* a contagious disease with an incubation period of two to three weeks, caused by a virus and characterized by a rash; it may produce congenital defects in infants born of mothers who contract the disease in the first trimester of pregnancy. Also, GERMAN MEASLES.

Rubella virus *Virology.* a virus belonging to the genus *Rubivirus,* of the family Togaviridae, that causes rubella (German measles) in humans. Also, GERMAN MEASLES VIRUS.

rubellite *Mineralogy.* a transparent, dark-pink to red-violet variety of elbaite; used as a gemstone.

rubene see TETRACENE.

rubeola *Medicine.* another name for the disease measles. See MEASLES.

Rubeola virus *Virology.* a virus belonging to the genus *Morbillivirus,* of the family Paramyxoviridae, that causes rubeola (measles) in humans. Also, MEASLES VIRUS.

Rubiaceae *Botany.* the madders, a large dicotyledonous family of the order Rubiales consisting of around 600 genera and 10,000 species of trees, shrubs, lianes, and herbs; tropical species are woody while temperate species are herbaceous; characterized by stipules and generally actinomorphic, hermaphrodite flowers.

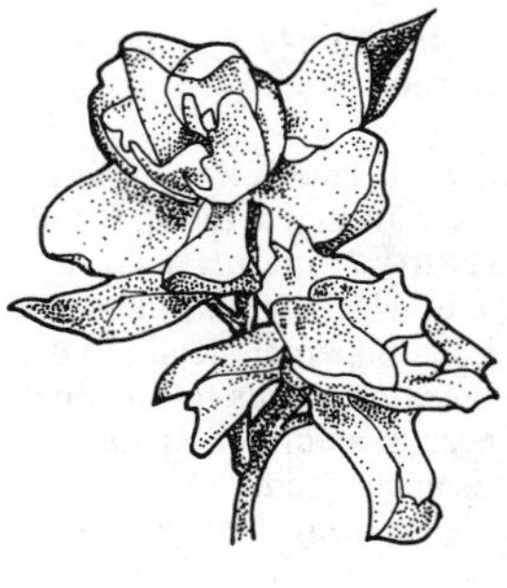

Rubiaceae (gardenia)

Rubiales *Botany.* an order of angiosperms of the subclass Asteridae, including the families Rubiaceae and Theligonaceae.

rubicelle *Mineralogy.* a yellow or orange-red transparent, gem variety of spinel.

rubidium [roo bid´ē əm] *Chemistry.* an alkali metal element having the symbol Rb, the atomic number 37, an atomic weight of 85.47, a melting point of 38°C, and a boiling point of 690°C; a silvery-white, very reactive metal that is obtained from lepidolite, leucite, and mineral waters; used in photocells and as a catalyst. (From the Latin word for the color red; because of the red lines in its spectrum.)

rubidium bromide *Inorganic Chemistry.* RbBr, colorless to white cubes or powder; soluble in water and insoluble in alcohol; melts at 693°C and boils at 1340°C; used in medicine.

rubidium chloride *Inorganic Chemistry.* RbCl, a white crystalline powder; soluble in water; melts at 718°C and boils at 1390°C; used in analysis and as a source of rubidium.

rubidium fluoride *Inorganic Chemistry.* RbF, white crystals that are soluble in water and insoluble in alcohol; melts at 795°C and boils at 1410°C; a strong irritant to tissue.

rubidium halide *Inorganic Chemistry.* any compound consisting of the element rubidium with an element from the halogen group (periodic table Group VIIa), such as rubidium bromide, RbBr.

rubidium halometallate *Inorganic Chemistry.* any compound of rubidium containing a halide and a metal, such as rubidium hexafluorogermanate, Rb_2GeF_6.

rubidium hexafluorogermanate *Inorganic Chemistry.* Rb_2GeF_6, a white crystalline solid that is slightly soluble in cold water and very soluble in hot water; melts at 696°C.

rubidium hydroxide *Inorganic Chemistry.* RbOH, a grayish-white mass that absorbs moisture and carbon dioxide from the air; soluble in water and alcohol; melts at about 300°C; a strong irritant to tissue. Also, **rubidium hydrate.**

rubidium magnetometer *Engineering.* a highly sensitive magnetometer capable of determining magnetic field strength to the submicrooersted by use of the spin precession principle. Also, **rubidium-vapor magnetometer.**

Dynamic Earth

Aurora borealis. Jack Finch/SPL/Photo Researchers.

Lightning strike. Keith Kent/SPL/Photo Researchers.

Volcanic eruption and lava flow. S. Summerhays/Photo Researchers.

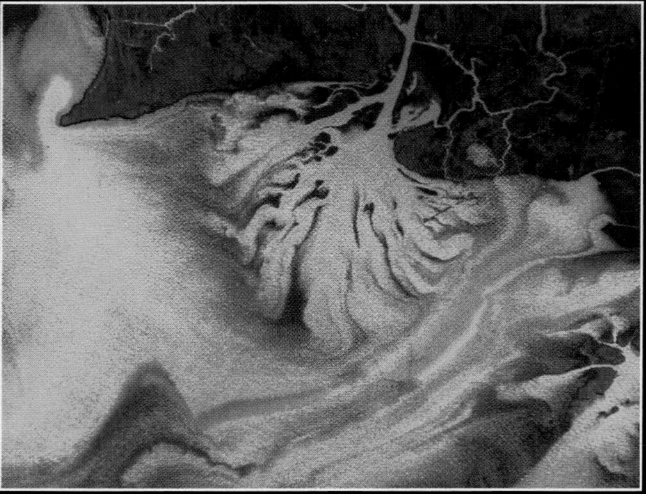

Calibrated airborne multispectral scanner image of a developing delta on the Louisiana coast. Courtesy of NASA.

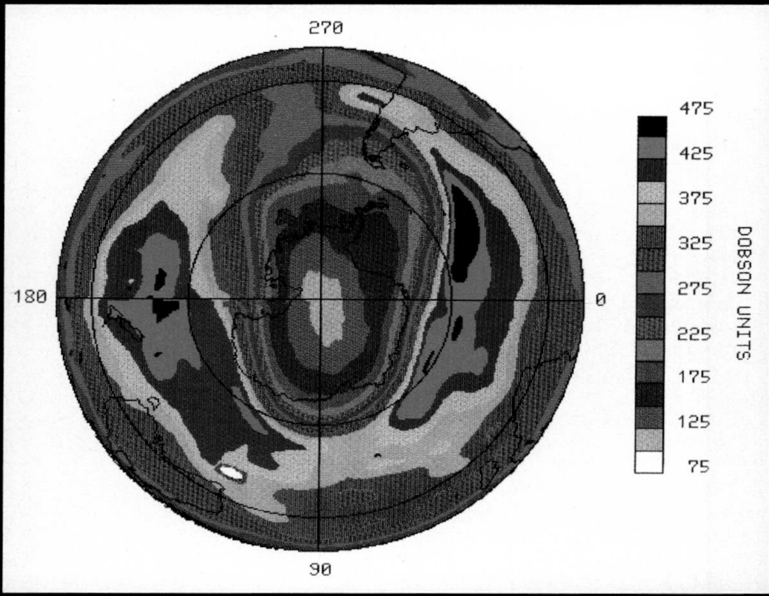

Image of the hole in the ozone layer of Earth's atmosphere above the South Pole. Taken on October 5, 1991, it is the deepest hole recorded to date. Courtesy of NASA, Goddard Space Flight Center Laboratory for Atmospheres, Ozone Processing Team.

Trees damaged by acid rain. Supplied by Carolina Biological Supply Company.

rubidium-strontium dating *Geology.* a method for determining the age, in years, of a rock or mineral based on the ratio of radioactive rubidium-87 to strontium-87 present in a sample.
rubidium sulfate *Inorganic Chemistry.* Rb_2SO_4, colorless rhombic crystals, very soluble in water; melts at 1060°C and boils at about 1700°C; used in medicine.
rubidium-vapor frequency standard *Physics.* a standard frequency defined by the atomic spectral line from an envelope containing rubidium vapor and a buffer gas: 6834 MHz.
Rubivirus *Virology.* a genus of viruses of the family Togaviridae, having as its single member the human rubella virus.
rub-proof *Graphic Arts.* of ink, thoroughly dried and able to resist abrasion.
rubratoxin *Biochemistry.* a mycotoxin produced along with fungal growth in certain strains of *Penicillium rubrim;* causes liver damage, brain lesions, and hemorrhages in the intestine when ingested by animals.
rubredoxin *Biochemistry.* an iron-sulfur protein believed to function in electron transport.
rubrene *Organic Chemistry.* $C_{42}H_{28}$, a red solid that is soluble in benzene; melts at 331°C; used in organic synthesis. Also, 9,10,11,12-TETRAPHENYLNAPHTHACENE.
rubricyte or **rubriblast** see PRONORMOBLAST.
ruby *Mineralogy.* the transparent, red gem variety of corundum, containing trace amounts of chromium as a coloring agent; very rare and one of the most valuable gemstones. *Materials Science.* a synthetically produced imitation of this mineral; used for watch movements and instrumentation.
ruby copper ore see CUPRITE.
ruby glass *Materials.* any reddish-colored glass, especially one that is colored by adding gold, copper, or selenium to the batch.
ruby laser *Optics.* a laser that uses a synthetic ruby to produce a beam of coherent red light.
ruby maser *Physics.* a type of maser in which the active medium is a ruby crystal.
ruby mica *Mineralogy.* a red mica mineral constituting the finest grade of Indian mica; it is used in electrical capacitors.
ruby silver **1.** see PROUSTITE. **2.** see PYRARGYRITE.
ruby spinel *Mineralogy.* a transparent, ruby-red variety of spinel, containing small amounts of chromium; it is not a variety of ruby. Also, SPINEL RUBY.
ruby zinc see ZINCITE.
rudaceous *Petrology.* describing sedimentary rock that is characterized by a significant quantity of fragments coarser than sand grains and at least 2 mm in diameter.
Rudbeckia *Botany.* a genus of North American herbs having composite flowers; *R. lacinata* is used as a tonic and diuretic.
rudder *Naval Architecture.* a flat piece of wood or metal that is attached vertically to the stern of a ship and operates to control the ship's direction. *Aviation.* a movable auxiliary airfoil at the rear of an airplane that controls its horizontal direction of flight.
rudder angle *Mechanical Engineering.* the acute angle formed by the rudder with reference to the fore-and-aft axis of a plane or ship.
rudder gudgeon *Naval Architecture.* any of the ringlike hinge fittings on a rudderpost from which a ship's rudder is hung; pintles on the rudder fit through the gudgeons to hold the rudder in place while allowing it to swing.
rudder lug *Naval Architecture.* any of the fittings on the forward edge of the rudderstock to which the rudder pintles are attached.
rudderpost *Naval Architecture.* the upright section of a ship's stern on which the rudder is hung; it may be the sternpost or abaft of the sternpost.
rudderstock *Naval Architecture.* the vertical member on which a nonbalanced rudder pivots; its bottom part is hinged to the rudderpost by gudgeons and pintles; its top part extends vertically above the main body of the rudder and connects to the steering mechanism.
rudder stop *Naval Architecture.* a fitting which limits the swing of a ship's rudder head, preventing the rudder from being put over to an excessive angle.
ruddy turnstone *Vertebrate Zoology.* a small shorebird, *Arenaria interpres,* of the family Scolopacidae, that turns stones over in search of food.
ruderal *Ecology.* living or thriving on disturbed sites. *Botany.* specifically, any plant that flourishes on a disturbed site, as on rubbish or in a wasteland area, often colonizing it. Also, ALETOPHYTE.
rudiment *Developmental Biology.* **1.** the first indication of a structure in the course of its development. **2.** a structure that has remained undeveloped, or one having little function at present but which was functionally developed earlier in the individual or in its phylogentic ancestors.
rudimentary *Developmental Biology.* **1.** not yet fully developed. **2.** remaining undeveloped; vestigial.
Rudimicrosporea *Invertebrate Zoology.* a class of minute, intracellular protozoans that are hyperparasites of gregarines, which are in turn parasites of marine annelids.
rudistids *Paleontology.* an unusual group of cemented bivalves that briefly challenged the status of corals in the Cretaceous as the dominant reef-building animal; mostly small but some were as high as 50 centimeters; extant in the Uppermost Cretaceous.
rudite *Geology.* any consolidated sedimentary rock, such as a conglomerate or breccia, composed of fragments that are coarser than sand.
Rudzki anomaly *Geophysics.* a gravity anomaly derived by positing a mirror image of the natural topography within the geoid.
rue *Botany.* an odiferous herb, *Ruta graveolens,* whose volatile oil is an irritant poison.
Ruffini, Paolo 1765–1822, Italian mathematician; contributed to the theory of solvability through work in higher-degree equations.
Ruffini corpuscles *Anatomy.* ovoid, encapsulated sensory nerve endings found in subcutaneous tissue. Also, **Ruffini's cylinders.** (Named for Angelo *Ruffini,* 1874–1929, Italian anatomist.)
ruffle *Textiles.* a strip of fabric that is gathered along one edge and used as trimming. *Biology.* a structure resembling this, as on the neck of a bird or mammal. Also, **ruff.** *Geology.* a ripple mark formed by an eddy. *Cell Biology.* any of numerous flattened protrusions extending from the leading edge of a migrating cell that curl back over the top of the cell if they do not adhere to the substratum, resulting in a rufflelike appearance. Thus, **ruffling.**
ruffled groove cast *Geology.* a groove cast that exhibits a feather pattern, and that consists of lateral wrinkles joining the main cast at an acute angle in the down current direction.
rufous *Biology.* reddish-brown.
ruga *plural,* **rugae.** *Anatomy.* a ridge, fold, or wrinkle, as of a mucous membrane.
ruggedized *Electronics.* describing electronic components that are intended to resist destructive elements, such as severe shock, extreme temperature fluctuations, or prolonged vibrations. *Computer Technology.* specifically, describing a computer, built to be operated in a hazardous environment, such as on board an aircraft, in space, or in a missile. Thus, **ruggedization.**
ruggedness number *Geology.* a number that is equal to the maximum basin relief multiplied by the drainage density within a drainage basin; used to express the geometric characteristics of a given drainage system.
Rugosa *Paleontology.* perhaps the most important coelenterate of the Paleozoic; one of the four main orders of coral, the Rugosa are stratigraphically important from their appearance in the Middle Ordovician to their extinction at the end of the Permian, when they and the Tabulata were replaced by the Octocorallia and Scleractinia.
rugose *Biology.* having a wrinkled surface. Also, **rugous.**
rugose mosaic *Plant Pathology.* an often fatal virus disease of potatoes characterized by stunted, roughened, and spotted leaves and other plant parts.
RUH *Aviation.* the airport code for Riyadh, Saudi Arabia.
ruin agate *Mineralogy.* a brown variety of agate whose polished surfaces or cut faces show markings that resemble the ruins of buildings.
ruin marble *Petrology.* a brecciated variety of limestone featuring a mosaic appearance, resembling ruins or ruined buildings, when cut and polished.
rule *Science.* **1.** a statement of conditions that are commonly observed in a given situation. **2.** a statement of a prescribed course of action designed to obtain a given result. *Artificial Intelligence.* a formalized means of expressing information in a cause-effect format; an "if . . . , then . . . " statement. *Mechanical Devices.* a flexible metal, tape, wood, or steel device marked in units and used to measure and/or mark off lengths. *Graphic Arts.* **1.** a type-high decorative or straight line. **2.** a piece of metal type used to print such a line.
rule application *Artificial Intelligence.* **1.** see FIRE. **2.** the application of a rule to particular data by instantiating it for that data.
rule-based shell *Artificial Intelligence.* a software engineering tool for building rule-based expert systems that may contain a database generator, inference engine, I/O functions, audit and recovery functions, etc. Also, PRODUCTION RULE SHELL.

rule-based system *Artificial Intelligence.* an expert system in which knowledge is stored in a production rule representation.

rule-based tool *Artificial Intelligence.* any of a variety of productivity tools to be used during any phase of building a rule-based system, such as interview aids, knowledge representation shells, and shells for database generation.

ruled surface *Mathematics.* any surface generated by the motion of a straight line; for example, a cone is a ruled surface.

rule induction *Artificial Intelligence.* a reasoning process in which rules are generated from specific examples whose outcomes are known; rules are then extended to other cases.

rule interpreter *Artificial Intelligence.* see INFERENCE ENGINE.

rule joint *Mechanical Devices.* a knuckle joint with projecting shoulders that connect between fully opened pieces thus allowing folding in one direction.

rule level *Mechanical Devices.* a parallel-sided strip of heavy metal, beveled on one edge for cutting or scribing; used to test the flatness of a surface and to cut a straight line.

rule of approximation *Mining Engineering.* a rule that applies to placer mining locations and entries upon surveyed lands, applied on the basis of ten-acre legal subdivisions.

rule of mixtures *Materials Science.* when applied to composites, a method of predicting the properties by considering the volume-weighted averages of the component properties; only accurate in certain simple situations.

rule of sixty *Navigation.* a method of roughly estimating the magnitude of the probable error in a bearing. It may be stated as, "the offset of the plotted bearing line from the observer's actual position is 1/60th of the distance to the object observed for each degree of error."

rule of the road *Navigation.* any regulation regarding the safe movement of ships in relation to one another.

rule of V's *Geology.* a method for determining the direction in which a formation lies beneath a stream, based on the direction of the acute angle formed when the outcrop of the formation crosses a valley.

ruler *Engineering.* a straight edge with a distance scale printed along its side; used in drawing and measuring straight lines.

rule-role orientation *Behavior.* the theory that behavior is shaped or influenced by the effect of societal codes and roles.

rules of engagement *Military Science.* military directives that specify the circumstances and limitations under which forces will initiate or continue combat engagement with other forces encountered, especially under conditions other than a declared state of war.

ruling engine *Spectroscopy.* a machine used to score equally spaced lines on a diffraction grating.

rum *Food Technology.* an alcoholic beverage distilled from molasses or other fermented sugarcane products.

rumble *Mechanical Engineering.* **1.** in an internal-combustion engine or liquid-propellant rocket engine, a form of combustion instability characterized by a low-pitched, low-frequency rumbling noise. **2.** the noise produced by this uneven combustion.

rumen *Vertebrate Zoology.* the first of the four compartments of the stomach of the ruminant mammals. Also, PAUNCH.

rumenitis *Veterinary Medicine.* inflammation of the rumen.

Rumford, Count see THOMPSON, BENJAMIN.

ruminant *Vertebrate Zoology.* an animal that chews its cud; one of the Ruminantia having even-toed hooves and a four-chambered stomach, such as the cow, buffalo, goat, deer, or llama.

Ruminantia *Vertebrate Zoology.* the mammalian suborder containing the ruminants; this includes the deer, giraffes, cattle, antelopes, goats, sheep, and chevrotains.

ruminate *Biology.* **1.** appearing to have been chewed. **2.** to chew the cud.

rumination *Vertebrate Zoology.* the process of digestion in ruminants, whereby the animal swallows food quickly, and then regurgitates and chews it more thoroughly at a later time until digestion is completed. *Medicine.* in human infants, the regurgitation of food after almost every meal, part of it being vomited and the rest swallowed. *Psychology.* a preoccupation with distressing thoughts for an excessive period of time.

Ruminococcus *Bacteriology.* a genus of Gram-positive, anaerobic bacteria, occurring in the rumen as cocci in pairs or chains possessing a fermentative metabolism.

rump *Vertebrate Zoology.* the hind quarters of an animal. *Physiology.* the buttock or gluteal region.

rumposome *Mycology.* a specialized organelle in a zoospore whose function is unknown.

run any of numerous actions or features that are thought of as analogous to the physical action of running, as by being swift or uninterrupted, by following a certain path, and so on; specific uses include: *Engineering.* any processing cycle or batch-treatment operation. *Mechanical Engineering.* a section of pipe that lies oriented in the same direction as the flow in the preceding section. *Chemical Engineering.* the amount of feed used in an industrial process during a specified time. *Transportation Engineering.* **1.** a single trip, especially of a commercial vehicle. **2.** the distance traveled by a vehicle or vessel in a given period of time. *Computer Programming.* **1.** the execution of a routine, program, or suite of programs to their completion. **2.** to execute a routine, program, or group of programs. *Navigation.* to sail with the wind nearly behind the vessel. Also, **run before the wind.** *Ordnance.* a single passage of a moving target across the target range. *Naval Architecture.* the tapering portion of a ship's underwater hull between the midbody and the stern. *Military Science.* **1.** in horizontal bombing, a steady, level flight over the target. **2.** in naval mine warfare, the transit of a sweeper-sweep combination or a minehunter operating its equipment through a single lap. *Cartography.* in aerial photography, the path followed by a plane making a strip of photographs. *Geology.* **1.** a ribbonlike, irregular, flat body of ore that follows the stratification of the host rock. **2.** a branching extension of the conduit through which magma passes to an igneous intrusion. *Mining Engineering.* a heading in a mine that cuts diagonally between the dip and the strike of a coal seam. Also, SLANT. *Hydrology.* a brook, small creek, or other small, swiftly flowing watercourse. *Statistics.* in nonparametric statistics, a sequence of identical occurrences or symbols that is preceded by and followed by different occurrences or symbols, such as a number of consecutive "heads" in the repeated toss of a coin.

runaround *Graphic Arts.* copy that is typeset so that it moves past or around an illustration. *Mining Engineering.* a passageway in the shaft pillar to allow safe passage from one side to the other within the shaft.

runaround crosstalk *Telecommunications.* the crosstalk that results from the coupling of the high-level end of one repeater and the low-level end of another.

runaway effect *Electronics.* a condition brought on by a rise in temperature that increases a transistor's current and eventually causes the device to malfunction.

runaway electron *Electronics.* an electron, in an ionized gas with an electric field, that absorbs energy from the field faster than it loses energy interacting with other particles in the gas.

runaway gun *Ordnance.* a defective automatic or semiautomatic gun that continues firing after the trigger is released.

runaway star *Astronomy.* an O-type or B-type star with a rapid space motion, thought to have been ejected from a binary star system when its companion exploded as a supernova.

runaway tape *Computer Technology.* a magnetic tape reel that has run off its spool, usually due to a drive malfunction.

runback *Petroleum Engineering.* in a refinery, piping that returns the overheads to the still instead of allowing it to be drawn off as product.

run book *Computer Programming.* documentation providing the necessary information for running a job, such as the problem statement, flow charts, coding, and operating instructions. Also, PROBLEM FILE, PROBLEM FOLDER.

run chart *Computer Programming.* a flow chart of one or more computer runs structured in terms of their input and output. Also, **run diagram.**

runcinate *Botany.* describing or referring to plants having leaves that are coarsely and sharply pinnatifid, with the lobes directed backward very abruptly.

Runcinoidea *Invertebrate Zoology.* an order of very small marine gastropod mullusks, which lack an operculum and either lack a shell or possess an extremely reduced external one.

rundown line *Chemical Engineering.* a pipe connecting the distillation column to the rundown tank in a petroleum refinery.

rundown tank *Chemical Engineering.* a tank that receives distillation products in a petroleum refinery.

rune *Archaeology.* a character in a runic alphabet. See RUNIC.

Runella *Bacteriology.* a genus of bacteria of the family Spirosomaceae that occur in fresh water as Gram-negative, pigmented, straight to curved, rod-shaped cells.

Runge, Carl David Tolme [rung´gə] 1856–1927, German mathematician; with W. Kutta, formulated Runge-Kutta methods.

Runge, Friedlieb 1794–1867, German chemist; discovered carbolic acid (phenol).

Runge-Kutta method *Mathematics.* any technique for numerically integrating the first-order ordinary differential equation $y' = f(x,y)$ that exhibits the property that no (partial or ordinary) derivatives of f are evaluated during the integration steps. Instead, f is evaluated at multiple points which themselves are determined by an iterative scheme.

Runge vector *Mechanics.* a vector describing certain unaltering features of a nonrelativistic two-body interaction obeying an inverse-square law, either in classical or quantum mechanics; the constancy reflects the symmetry inherent in the inverse-square interaction.

runic *Archaeology.* of or relating to a type of alphabet developed in various forms in Scandinavian and Germanic areas about 300 AD; associated with ceremonial artifacts and structures, but also seen as graffiti.

R unit see GERMAN R UNIT.

runite see GRAPHIC GRANITE.

run-length encoding *Computer Programming.* **1.** a method of data compression by suppressing repeated characters. **2.** a graphics encoding method in which the coding scheme describes the characteristics of each horizontal line of the graphics image.

runnel *Hydrology.* a rivulet, streamlet, or little brook. *Geology.* **1.** the channel eroded by such a rivulet or streamlet. **2.** a troughlike hollow formed by the action of waves or tides on a tidal sand beach. **3.** see SWALE.

runner *Botany.* a creeping stem formed by many rosette plants that may root or develop leaves upon contact with the soil surface or substrate as a mechanism for asexual reproduction. *Engineering.* **1.** a flat piece of timber placed above bars to connect them. **2.** a steel-shod poling board that is driven into unbroken but loose ground as excavation progresses. **3.** a portion of a gate assembly that connects the downgate, sprue, or riser with the casting. *Mining Engineering.* **1.** a vertical timber sheet pile for preventing excavation collapse. **2.** a worker who transports loaded mine cars by means of gravity from working places to gangways and who assists in delivering empty cars. *Metallurgy.* **1.** an open channel used to transfer molten metal from one location to another. **2.** in casting, the part of a mold that connects the sprue to the gate.

runner box *Metallurgy.* in casting, a manifold that distributes molten metal as it enters the mold.

running *Naval Architecture.* describing any object or assembly that may be moved in ordinary use. Thus, **running bowsprit, running gaff, running rigging.** *Cartography.* in surveying, a series of differences in elevations, measured in one direction along a section of a line of levels, that gives the difference in elevation between the points at the ends of the section.

running bird *Vertebrate Zoology.* any of the large flightless birds, usually classified as ratites, whose main method of locomotion is running.

running block see TRAVELING BLOCK.

running bond *Building Engineering.* see STRETCHER.

running fit *Design Engineering.* the dimensional difference of mating mechanical parts intentionally fabricated so as to permit movement relative to each other.

running fix *Navigation.* a position determined by crossing lines of position that have been determined at different times and advanced or retarded to a common time.

running gate *Metallurgy.* in casting, a gate through which the molten metal enters the mold.

running gear *Mechanical Engineering.* the equipment that supports a truck and its load, creating a rolling-friction contact as the truck moves along a surface.

running ground *Mining Engineering.* a term for insecure and plastic or semiplastic ground that deforms easily under pressure.

running head *Graphic Arts.* a headline placed at the top of every page or every other page in a book, usually the title of the book or chapter. Also, **running title.**

running-in *Engineering.* a quality control technique that entails operating products for a specific amount of time before shipment to ensure correct operation, and to provide an initial wearing and smoothing of moving parts.

running sand see QUICKSAND.

running water *Hydrology.* water that is moving in a stream or that is not stagnant or brackish.

run number *Materials Science.* the average number of like mer sequences occurring in a copolymer per 100 mers; used to determine the amount of order in a copolymer structure.

runoff *Hydrology.* **1.** the part of precipitation that flows over the land and eventually reaches surface streams. **2.** the amount of water that runs off.

runoff coefficient *Hydrology.* the percentage of precipitation appearing as runoff, expressed as a constant between zero and one.

runoff cycle *Hydrology.* the part of the hydrologic cycle that involves the movement of water from the time it reaches the land until it is discharged through stream channels or is lost through evapotranspiration.

runoff intensity *Hydrology.* the volume of water that is derived from a land surface per hour. Also, **runoff rate.**

runoff pit *Mining Engineering.* an area to which spillage can gravitate should it be necessary to dump contents of classifiers, thickeners, slurry pumps, or other mill machines. Also, SPILL PIT.

run-of-mill *Mining Engineering.* ore that is accepted for treatment subsequent to the rejection of waste and dense media. Also, MILL-HEAD ORE.

runout *Hydrology.* **1.** the area where an avalanche slows down or stops, depositing its debris. **2.** see WATER YIELD. *Metallurgy.* **1.** the undesirable, and often dangerous, flow of molten metal from a furnace, ladle, or mold. **2.** in a casting, a defect caused by such runout.

runout table *Metallurgy.* in metal fabrication, the table on which the product exiting the stand is received.

runout time *Industrial Engineering.* the time required for a machine tool to run back to a safe clearance position following an actual work process.

runover *Graphic Arts.* **1.** the second line of a heading or headline. **2.** indented lines in a flush-and-hang paragraph. **3.** see TURNOVER.

runt *Zoology.* a popular term for an animal that is abnormally small or weak.

run-time *Computer Science.* of or referring to something that happens during execution of a program. Also, **runtime.**

run-time data *Mechanical Engineering.* any data that aids in setting the running time for an apparatus or piece of equipment, especially when the time limit is based on economic, durability, or safety considerations. *Robotics.* sensorial data obtained during a robot's normal operations which may then be used to improve its performance.

run-time error *Computer Programming.* a logical or semantic program error that occurs when the program is executed.

run-time library *Computer Programming.* a set of standard input/output and other utility routines that provide a uniform run-time environment.

runting disease or **runt disease** *Veterinary Medicine.* A clinical disease found in laboratory animals which is characterized by immunodeficiency and failure to grow normally; associated with the injection of allogenic spleen cells into newborns. Also, WASTING DISEASE.

run-up *Aviation.* a preflight acceleration to test an airplane's engines.

runway *Aviation.* **1.** a clearly defined rectangular area for aircraft to use for landing and taking off. **2.** relating to or used in such an area. *Civil Engineering.* the pavement applied over a concrete slab, often in removable panels. *Geology.* a stream channel.

runway approach *Aviation.* the final stage of an aircraft's approach from a navigational facility to a specific runway.

runway lights *Aviation.* lights that mark the edges of a runway; they show through a limited angle.

runway observation *Meteorology.* a weather analysis of meteorological elements observed at a specified point on or near an airport runway, including temperature, wind speed and direction, ceiling, visibility, and icing conditions.

runway temperature *Meteorology.* the temperature of the air above the runway at an airport; used in the determination of density altitude.

runway visibility *Meteorology.* the visibility along a runway, determined from a specified point on the runway with the observer facing in the same direction as a pilot using the runway.

runway visual range *Meteorology.* the maximum distance along the runway at which the runway lights are visible to a pilot after touchdown.

Runyon groups *Microbiology.* a method of classification of *Mycobacteria* species, based on growth rate and pigment formation.

Rupelian *Geology.* a European geologic stage of the Early Oligocene epoch, occurring after the Priabonian and before the Chattian. Also, STAMPIAN.

Rupe rearrangement *Organic Chemistry.* a reaction that produces α,β-unsaturated ketones by treating ethynylcarbinols with strong formic acid.

rupicolous *Ecology.* of or relating to an organism that lives among or grows on rocks. Thus, **rupicole.**

Ruping process *Engineering.* a method of creosoting lumber for preservation by applying a treatment to the lumber under atmospheric pressure, and then applying a vacuum.

Ruppiaceae *Botany.* a monospecific family of glabrous, submerged, aquatic herbs of the order Naladales, chiefly found in alkaline and brackish water; commonly known as ditch grass.

rupture *Medicine.* any forcible tearing or disruption of tissue, such as a hernia. *Materials Science.* a mode of failure whereby a part or parts breaks open or bursts; for example, a water tank will rupture if it is pressurized beyond its limits.

rupture zone *Geology.* the region immediately adjacent to the edge of an explosion crater.

rural *Geography.* of or relating to agriculture or to the countryside.

rural community *Geography.* a dispersed settlement, especially one where the population is predominantly engaged in agriculture.

rural radio service *Telecommunications.* a radio service that provides a public message communication service between a central office and areas in which wire lines cannot be installed.

Rush, Benjamin 1745–1813, American physician, educator, and statesman; wrote a pioneering study of mental illness.

Rushton-Oldshue column *Chemical Engineering.* a mixing unit for two-phase contacting in continuous pipeline blending, containing separation plates, baffles, and mixing impellers.

Ruska, Ernst August Friedrich 1906–1988, German electrical engineer; shared the Nobel Prize for development of the electron microscope.

Bertrand Russell

Russell, Bertrand 1872–1970, British mathematician, philosopher, and social activist; the cofounder (with Alfred North Whitehead) of mathematical logic.

Russell, Henry Norris 1877–1957, American astronomer; worked in the theory of stellar evolution; developed the Hertzsprung-Russell diagram.

Russell bodies *Pathology.* mucoproteinous, globular blood cell inclusions, harboring gamma globulin, which are thought to generate from the thickening of intracellular secretions. Also, CANCER BODIES, FUCHSIN BODIES. (Named for William *Russell,* 1852–1940, Scottish physician.)

Russell diagram see HERTZSPRUNG-RUSSELL DIAGRAM.

Russell flask *Petroleum Engineering.* a vessel used to calculate the volume of sand grains alone from a volume of grains plus voids.

russellite *Mineralogy.* Bi_2WO_6, a yellow to green tetragonal mineral occurring in fine-grained, compact masses, having a specific gravity of 7.35 and a hardness of 3.5 on the Mohs scale.

Russell mixture *Astrophysics.* a mixture of elements in the same relative proportions as those found in the sun, the main constituent being hydrogen (90%).

Russell movable-wall oven *Chemical Engineering.* a coal carbonization test oven that has movable walls and is used when the correct process pressure is not known, so as to avoid damage to permanent process ovens.

Russell-Saunders coupling *Physics.* a process whereby the spin angular momentum vectors in a system are combined to form a resultant spin angular momentum vector, and likewise for the orbital angular momentum; then the two resultant vectors are combined to form a total angular momentum for the system. Also, LS-COUPLING.

Russell's paradox *Mathematics.* the paradox arising from the following: For any set *X*, either *X* is an element of itself or *X* is not an element of itself. Let *M* be the class of all sets *X* such that *X* is not an element of itself. (*M* is not empty. For example, the set of all books is not itself a book.) *M* is a proper class, i.e., not a set. For if *M* were a set, then either *M* is an element of itself or it is not an element of itself. But then the definition of *M* means that either alternative leads to a contradiction. Thus *M* is not a set. Russell's paradox forces a distinction between classes and sets. Also, **Russell's antinomy.**

Russell's viper *Vertebrate Zoology.* the common name for *Vipera russellii*, a highly venomous snake of India, having a light brown body with three rows of large, black-edged brown spots. Also, TIC POLONGA.

Russell-Vogt theorem see VOGT-RUSSELL THEOREM.

russet *Botany.* any of several varieties of fruits and vegetables, especially apples and potatoes, having a rough brownish skin.

Russian spring-summer encephalitis *Pathology.* an epidemic form of encephalitis acquired from tick bites, or by eating the flesh of infected mammals or drinking the milk of infected goats; characterized by deteriorative changes in organs other than those of the central nervous system, and ranging in severity from mild to fatal.

Russula *Mycology.* a genus of fungi of the family Russulaceae; they are often brightly colored, have fruiting bodies shaped like mushrooms with gills, and live on the roots of higher plants without causing damage.

Russulaceae *Mycology.* a family of fungi of the order Agaricales, characterized by brittle basidiocarps and composed of families of gill-forming mushrooms.

Russulales *Mycology.* in some classifications, an order of fungi of the class Hymenomycete, having sphaerocysts in their fruiting bodies.

rust *Metallurgy.* **1.** the oxidation product of any iron-base material. **2.** to undergo such a process. *Plant Pathology.* any of numerous plant diseases caused by the rust fungi of the order Uredinales and characterized by reddish-brown to black lesions on leaves or other plant surfaces.

rusticyanin *Biochemistry.* a copper-containing protein found in the gram-negative bacteria, *Thiobacillus ferrooxidans.*

rusting *Metallurgy.* the oxidation of any iron-base material; the formation of rust on iron-bearing metals and alloys.

rust joint *Engineering.* a joint used to achieve a rigid connection by packing an intervening space tightly with a stiff paste that oxidizes the iron, so the whole rusts together and hardens into a solid mass; it generally cannot be separated without destroying some of the pieces.

rusty blotch *Plant Pathology.* a barley disease caused by the fungus *Helminthosporium californicum* and characterized by irregular brown spots on the foliage.

rusty gold *Metallurgy.* native gold coated with iron oxide or silica.

rusty mottle *Plant Pathology.* a virus disease of cherry trees characterized by stunted growth in the spring, spotting and shot-holing on the foliage, and defoliation.

rut *Vertebrate Zoology.* **1.** a period of sexual excitability in many male mammals that occurs once or more each year and usually leads to mating; it corresponds to the estrus cycle of the female. **2.** see ESTRUS.

rutabaga *Botany.* **1.** a plant, *Brassica napus,* of the family Brassicaeae, with an edible tuber similar to a turnip. **2.** the tuber of this plant, used as a food.

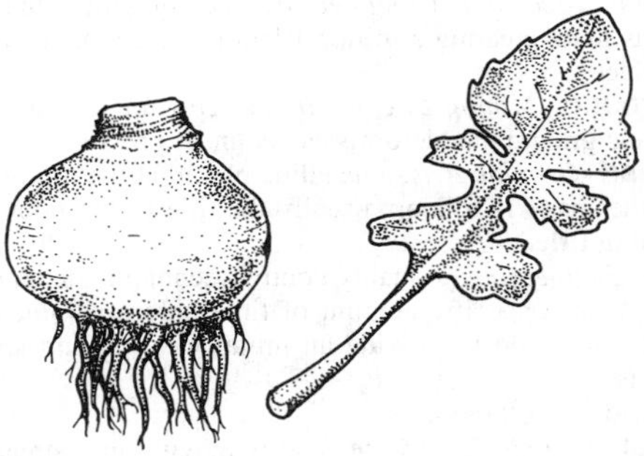

rutabaga

Rutaceae *Botany.* a dicotyledonous family of the order Sapindales, consisting of around 160 genera and 1700 species of primarily tropical, aromatic trees, shrubs, and herbs; includes the genus *Citrus.*

rutaceous *Botany.* relating to, belonging to, or resembling the family Rutaceae.

Rutenbergiaceae *Botany.* a monogeneric family of dull mosses of the order Isobryales that form loose mats on tree bark; characterized by bordered leaves, a reduced peristome, and hairy calyptrae; native to Madagascar and the surrounding islands.

ruthenium *Chemistry.* a rare metallic element of the platinum group having the symbol Ru, the atomic number 44, and an atomic weight of 101.07, with 7 stable isotopes, a melting point of 2310°C, and a boiling point of about 4000°C; it is a grayish-white brittle metal used as a hardener for platinum and palladium, as a catalyst, and in alloys. (From the region of *Ruthenia,* in western Ukraine, where it was first identified.)

ruthenium chloride *Inorganic Chemistry.* $RuCl_3$, a black deliquescent solid that is insoluble in cold water and decomposes in hot water; decomposes above 500°C; used in laboratory analysis. Also, **ruthenic chloride, ruthenium trichloride, ruthenium sesquichloride.**

ruthenium halide *Inorganic Chemistry.* any compound consisting of the element ruthenium with an element from the halogen group (periodic table Group VIIa), such as ruthenium chloride, $RuCl_3$.

ruthenium red *Inorganic Chemistry.* $Ru_2(OH)_2Cl_4 \cdot 7NH_3 \cdot 3H_2O$, a brownish-red powder that is soluble in water; used as a chemical stain and reagent. Also, **(ammoniated) ruthenium oxychloride.**

ruthenium tetroxide *Chemistry.* RuO_4, yellow crystals or needles; soluble in carbon tetrachloride; melts at 25.5°C and decomposes at 108°C; a fire hazard in contact with organic substances; used as an oxidizer.

Rutherford, Ernest 1871–1937, British physicist; discovered alpha, beta, and gamma rays; awarded the Nobel Prize for the theory of atomic transmutation.

Ernest Rutherford

rutherford *Nucleonics.* a unit of nuclear decay rate equivalent to 10^6 disintegrations per second.

Rutherford backscattering *Materials Science.* a technique for calculating the depth to which ions penetrate a target by measuring the energy lost when the ions enter and exit the target.

Rutherford backscattering spectrometry *Spectroscopy.* a spectroscopic technique in which the energy of ions backscattered by a solid sample is measured to determine the concentration of an element as a function of depth beneath the surface.

rutherfordine *Mineralogy.* $UO_2(CO_3)$, a rare, strongly radioactive, white to brownish or yellow orthorhombic mineral occurring as lathlike crystals and aggregates of minute fibers, having a specific gravity of 5.7 and an undetermined hardness; a secondary mineral found in uranium deposits and pegmatites.

rutherfordium *Chemistry.* a post-actinide element having the symbol Rf and the atomic number 104; its isotopes are unstable, with short half-lives. (From Ernest *Rutherford.*) Also, KURCHATOVIUM.

Rutherford nuclear atom *Atomic Physics.* a model of atomic structure in which an atom's mass is contained almost entirely in its nucleus, and its electrons take up most of the atom's volume.

Rutherford scattering *Atomic Physics.* the dispersal of charged particles caused by an electrostatic force (Coulomb field) due to the nucleus.

rutilated quartz *Mineralogy.* a sagenitic quartz that is penetrated by needlelike crystals of rutile.

rutile *Mineralogy.* TiO_2, a red to reddish-brown tetragonal mineral of the rutile group, trimorphous with anatase and brookite, occurring in commonly twinned prismatic crystals and granular masses, having a specific gravity of 4.23 and a hardness of 6 to 6.5 on the Mohs scale; found in metamorphic and acid igneous rocks and as residual grains in beach sands; an ore of titanium.

rutin *Organic Chemistry.* $C_{27}H_{32}O_{16}$, yellow needles that melt at 214–215°C; a plant product used medicinally to treat vascular disorders.

rutterite *Petrology.* a dark-pink, equigranular, medium-grained igneous rock consisting primarily of microperthite, microcline, and albite, with traces of nepheline, biotite, graphite, amphibole, and magnetite.

Ruzicka, Leopold 1887–1976, Swiss chemist; awarded the Nobel Prize for work with polymethylenes.

RV residual volume; recreational vehicle.

R-value see DISCREPANCY INDEX.

R_f value *Analytical Chemistry.* the ratio of the distance traveled by a substance to that traveled by the solvent front, both measured from the point of application in chromatography.

RV Tauri star *Astronomy.* a yellow supergiant star with an irregular variability and a period roughly between 100 and 150 days; often found in globular clusters.

RW radiological warfare.

RW Aurigae stars [ô rī´ gē] *Astronomy.* a class of stars that exhibit erratic variation in luminosity; they occur in groups and are always associated with regions full of interstellar nebulosity and ongoing star formation.

Rx or ℞. *Pharmacology.* **1.** a symbol meaning "take;" the first instruction in a prescription. **2.** broadly, a prescription. (An abbreviation for the Latin imperative form *recipe,* "take.") Also, $\mathbf{R_x}$.

ry rydberg.

rydberg *Atomic Physics.* a unit of energy equal to the square of the charge of the electron divided by twice the radius of the ground state orbit of the hydrogen atom in the Bohr theory (Bohr radius); equivalent to approximately 13.606 electron volts. (Named for Johannes *Rydberg,* 1854–1919, Swedish physicist.)

Rydberg atom *Atomic Physics.* an atom that has a very highly charged outer electron orbiting far from its nucleus.

Rydberg constant *Atomic Physics.* a constant used in atomic spectral analysis defined by the quantity $2\pi^2 me^4/ch^3$, where m is the rest mass of an electron, e is the charge of an electron, c is the speed of light, and h is Planck's constant; the value of the constant is approximately $1.0974 \times 10^5\ cm^{-1}$.

Rydberg correction *Atomic Physics.* a variable, added to a formula, that represents the energy of a single electron in the outermost shell; used to account for the inner electron shells' inability to completely shield the nuclear charge.

Rydberg series formula *Spectroscopy.* an empirical formula for determining the wave numbers of various series of spectral lines, such as for neutral hydrogen and alkali metals.

rye *Botany.* **1.** a cereal grass, *Secale cerale,* of the family Poaceae, characterized by single-nerved glumes and double-flowered or triple-flowered spikelets. **2.** the seed of this plant. *Food Technology.* made from or with rye grain. Thus, **rye bread, rye whiskey.**

rye grass *Botany.* any of various grasses of the genus *Lolium.*

Ryle, Sir Martin 1918–1984, English astronomer; awarded the Nobel Prize for work in radiotelescopy.

Rynchopidae *Vertebrate Zoology.* the skimmers, a monogeneric family of gull-like birds of the order Charadriiformes, specialized for catching fish by skimming the water with a long beak; found in tropics worldwide.

Rytionidae *Vertebrate Zoology.* a subfamily of serenians, mammals in the family Dugongidae.

Ryukyu Islands *Geography.* an archipelago that stretches between Kyushu, Japan, and Taiwan.

Rzeppa joint *Mechanical Engineering.* a Bendix-Weiss universal joint, consisting of four large balls that serve as transmitting elements and a center ball that serves as a spacer; used to transmit uninterrupted angular velocity through a single universal joint.

S *Behavior.* stimulus.
S the chemical symbol for sulfur.
S label. (From Latin *signa.*)
S or **S.** short; subject; Svedberg unit; siemens; steradian.
S or **s** sign. Also **S., s.**
S or **s.** south; southern.
s strange quark; a symbol for standard deviation.
s or **s.** an abbreviation for section; second; small; symmetrical.
S- *Aviation.* the U.S. military designation for antisubmarine aircraft, as in *S-2.*
s_1, s_2, s_3, etc. *Genetics.* a designation for successive generations of self-fertilized plants; s_1 refers to the offspring of the selfed parent; the s_2 generation results from the selfed s_1 parents, etc.
S-2 *Aviation.* a twin-engine antisubmarine aircraft designed to operate from an aircraft carrier and to detect, locate, and destroy enemy submarines.
^{35}S *Genetics.* a radioactive isotope of sulfur with a half-life of 81 days; used to label proteins in laboratory experiments.
SA *Ordnance.* the official designation for a Soviet missile series ranging from SA-1 to SA-13; see the code names such as GUILD and GUIDELINE.
SA without date (from Latin *sine anno,* "without year"); sinoatrial.
SA or **S.A.** South America, South Africa, South Australia.
s.a. according to standard practice. (From Latin *secundum artem,* "according to art.") Also, **S.A.**
Saalic orogeny *Geology.* a brief, Early Permian crustal deformation occurring between the Autunian and Saxonian stages.
Saanen *Agriculture.* a breed of milk goat that is white or creamy-white in color, short-haired, and hornless. (From the *Saanen* Valley in Switzerland.)
Saarinen, Eero [sär´ə nen] 1910–1961, American architect, born in Finland; designed Dulles Airport.

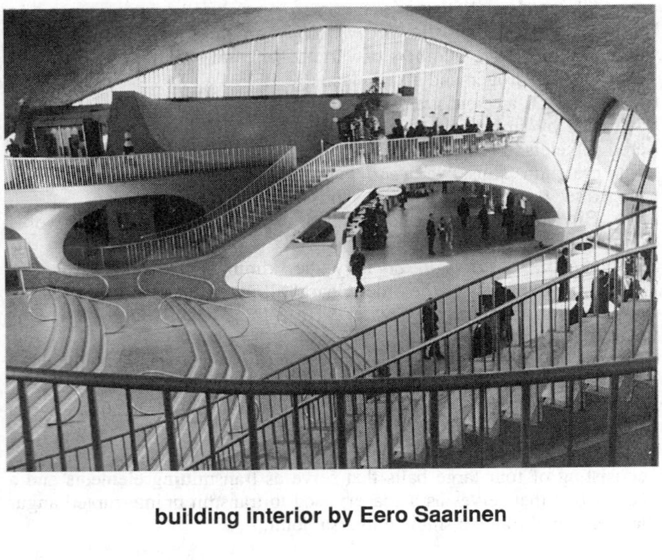

building interior by Eero Saarinen

Saarinen, Eliel 1873–1950, American architect, born in Finland, the father of Eero Saarinen; designed the Helsinki railroad station.
saba *Botany.* a liana of the banana family, *Saba comorensis,* characterized by fragrant flowers, fibers that are used for textiles, and a bananalike fruit that is edible when cooked; it is native to the Phillipines.
sabadilla [sä´bə dē´yə; sä´bə dil´ə] *Botany.* a grasslike plant, *Schoenocaulon officinale,* having long, grasslike leaves and bitter seeds; found in Mexico. *Materials Science.* the bitter seeds of this plant, used as an insecticide on cattle.
Sabathé's cycle [sä bə tā´] *Mechanical Engineering.* an internal combustion cycle in which the combustion process is partially explosive and partially at constant pressure.
Sabatier, Paul [sä bä tyā´] 1854–1941, French chemist; awarded the Nobel Prize for a method of hydrogenating oils.
Sabattier effect *Graphic Arts.* a part-negative and part-positive effect that is obtained by briefly exposing a photographic emulsion to white light in the middle of the development process. Also, PSEUDOSOLARIZATION. (Developed by Armand *Sabatier,* 1834–1910, French physician and scientist.)

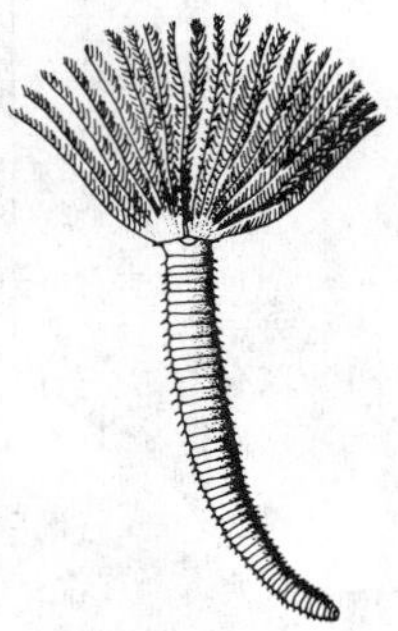

Sabellidae

Sabellidae *Invertebrate Zoology.* a family of fanworms, marine, tube-dwelling, sessile polychaete annelids with bright-colored, feathery, retractile arms on the head.
Sabellinae *Invertebrate Zoology.* an extensive subfamily of large fanworms in the family Sabellidae.
Saber *Ordnance.* a Soviet mobile, intermediate-range, ballistic-missile system delivering three 150-kiloton charges at a range of 3000 miles; officially designated **SS-20.**
Saberjet see SABREJET.
saber saw *Mechanical Engineering.* a hand-held electric jigsaw having a straight blade and a base plate.
saber-toothed tiger *Paleontology.* any of various species of prehistoric mammals of the cat family Felidae, characterized by very long, curved upper canine teeth thought of as resembling a saber; found in Europe, Asia, and the Americas; a noted North American species was *Smilodon californicus*; extant in the Oligocene to Pleistocene. Also, **saber-tooth, saber-toothed cat.**

Sabiaceae *Botany.* a family of tanniferous trees, shrubs, and woody vines usually assigned to the order Ranunculales; characterized by alternate, simple, or pinnately compound leaves without stipules and an embryo with two folded or uniquely coiled cotyledons; they are native to southeastern Asia and tropical America.

Albert Sabin

Sabin, Albert Bruce born 1906, American virologist, born in Poland; developed the oral polio vaccine and vaccines against dengue and encephalitis.

sabin *Acoustics.* a unit of sound absorption based on the Sabine equation, expressed as the amount of absorption that takes place over a surface area of 1 square foot of a perfectly absorbing material.

Sabinas *Geology.* a North American Gulf Coast provincial series of the Upper Jurassic period, occurring before the Coahuilan.

Sabine, Edward 1788–1883, English astronomer; studied the effects of the Gulf Stream; led a world magnetic survey; determined the length of a second's pendulum.

Sabine equation *Acoustics.* an equation that computes a value for the amount of sound absorption in a room, of the form

$$T = 60v/[1.086S_i(\alpha_{Sab})],$$

in which v represents the speed of sound in the medium, S_i represents the total area, and α_{Sab} represents the degree of decay. (Named after Wallace Clement *Sabine,* 1868–1919, American physicist.)

Sabinian *Geology.* a North American Gulf Coast stage of the Eocene epoch, occurring after the Midwayan and before the Claibornian.

sabinism *Toxicology.* poisoning due to ingestion of or exposure to savin (sabin) oil; the symptoms may include tissue damage.

Sabin vaccine *Immunology.* an oral vaccine, composed of attenuated live polio virus, that serves to provide protection against the disease poliomyelitis. (Named for Albert *Sabin.*)

sabkha or **sabka** see SEBKHA.

sable *Vertebrate Zoology.* **1.** a mammal, *Mustela zibellina,* of the family Mustelidae, an Old World, weasel-like carnivore that is prized and raised for its soft, glossy, dark brown fur, which is one of the most costly furs; found in cold regions of Eurasia and the North Pacific Islands. **2.** a marten, especially *Mustela americana,* the North American sable. **3.** the fur of any of these animals.

sable

sable antelope *Vertebrate Zoology.* a large antelope, *Hippotragus niger,* of the family Bovidae; characterized by large, sickle-shaped horns and, in the male, a black coat with white underparts; an endangered species found in Africa.

sabot [sab´ō; sə bō´] *Ordnance.* a lightweight case for a subcaliber projectile that allows the projectile to be fired from a larger caliber weapon; it fills the bore of the weapon and is usually discarded a short distance from the muzzle.

sabotage *Military Science.* any action that attempts to injure, interfere with, or obstruct the national defense of a country by injuring or destroying material, facilities, or utilities that are vital to such defense, including human or natural resources. (From an earlier meaning of the word *sabot,* a wooden shoe traditionally worn by peasants and workers in Europe; said to be from the practice by French workers in the early 1800s of throwing their shoes into machinery to cause a breakdown.)

Sabouraud's agar *Microbiology.* a dextrose-peptone culture medium that is used to grow certain fungi, sometimes containing antibiotics but often pathogenic to humans and other animals. Also, **Sabouraud's medium, Sabouraud's dextrose agar, Sabouraud's dextrose sugar.** (Named for Raymond *Sabouraud,* 1864–1938, French dermatologist.)

Sabrejet *Aviation.* a popular name for the F-86 fighter jet, which was widely used by the U.S. and its allies in and after the Korean War. Also, **Sabre.**

sabulous see ARENACEOUS.

sac *Biology.* a bag-shaped, soft-walled pouch, cavity, or indentation within a plant or animal body.

SAC Strategic Air Command.

sacbrood *Entomology.* a viral disease affecting honeybee larvae, which causes them to shrivel and manifest a scalelike appearance.

Saccardiaceae *Mycology.* a family of fungi of the order Asterinales, occurring primarily in tropical regions on plant leaves.

Saccardinulaceae *Mycology.* a family of fungi of the order Myriangiales, composed of nonparasitic species that occur on the leaves or branches of higher plants in tropical to temperate regions.

saccate *Botany.* of or related to a plant or plant part that is shaped like a pouch or sack.

sacchar- or **saccharo-** a combining form meaning "sugar."

saccharase *Biochemistry.* an enzyme that catalyzes the hydrolysis of the disaccharide sucrose to the monosaccharides glucose and fructose.

saccharic acid *Organic Chemistry.* $C_6H_{10}O_8$, needlelike, white crystalline solid or syrup; soluble in water; a dibasic acid, occurring in three optically different forms; usually derived from the oxidation of cane sugar, glucose, or starch.

saccharide [sak´ə rīd´] *Chemistry.* **1.** an organic compound consisting of one or more simple sugars; one of a series of carbohydrates, classified according to the number of simple sugar groups composing them. **2.** a simple sugar; a monosaccharide. **3.** an ester of sucrose.

saccharification *Biochemistry.* the process of converting starch or dextrin into sugar.

saccharimeter *Engineering.* an instrument that determines the sugar content of a solution from changes in its polarization.

saccharin [sak´ə rin] *Organic Chemistry.* a popular name for the anhydride of sulfobenzoic acid imide, occurring as a potentially carcinogenic, white crystalline powder; slightly soluble in water and ether, and soluble in alcohol; melts at 226–230°C; in dilute solution, it is 500 times as sweet as sugar; used as a noncaloric, nonnutritive sweetener.

saccharoidal *Petrology.* of or relating to a crystalline or granular texture resembling that of loaf sugar. Also, SUCROSIC, SUGARY.

saccharometer *Engineering.* an instrument that determines the sugar content of a solution from changes in the specific gravity of the gases produced during fermentation.

Saccharomonospora *Bacteriology.* a genus of soil bacteria of the order Actinomycetales that form substrate and aerial mycelia, in which spore formation is usually accompanied by development of a green pigmentation.

Saccharomyces *Mycology.* a genus of yeast fungi in the family Saccharomycetaceae, found in foods such as fruit juices, fruit, and alcoholic beverages, and in soil and human skin; the species *S. cerevisiae* is used in food processing (as "brewer's yeast") and in genetic research and development.

Saccharomycetaceae *Mycology.* a family of yeast fungi belonging to the order Endomycetales, found throughout nature, for example, in soil, water, flowers, and fruits, and in plant debris; some members live off nonliving organic matter, and others are used in food processing and in enzyme production.

Saccharomycetales *Mycology.* a former order of yeast fungi of the subdivision Hemiascomycetidae, now reclassified within the order Endomycetales.

Saccharomycetoideae *Mycology.* a subfamily of fungi of the family Saccharomycetaceae, characterized by spores of various shapes and composed of the single species, *S. ludwigii,* which occurs in wine, grape juice, and the slime flux of trees.

Saccharomycopsis *Mycology.* a genus of yeast fungi of the order Endomycetales, found in the tunnels of wood-boring beetles, in soil, in foods, and on fruit.

saccharopine *Chemistry.* an intermediate in the metabolism of lysine.

saccharopinuria *Medicine.* a congenital disorder of amino acid metabolism, thought to be a cause of mental retardation and characterized by an abnormally high level of saccharopine in the urine.

Saccharopolyspora *Bacteriology.* a genus of bacteria belonging in the order Actinomycetales, occurring as fragmenting substrate mycelia and spore-bearing aerial mycelia.

saccharose see SUCROSE.

Saccifoliaceae *Botany.* a monospecific family of pulvinate dicotyledonous subshrubs of the order Gentianales, native only to isolated mountains in the Guayana Highlands of southern Venezuela.

Saccoglossa *Invertebrate Zoology.* an order of gastropod mollusks in the subclass Opisthobranchia that have a single row of teeth on the radula and feed by sucking algae.

Saccopharyngiformes *Vertebrate Zoology.* a former order of bony fishes containing the deep-sea gulpers, now in the order Anguilliformes.

Saccopharyngoidei *Vertebrate Zoology.* a suborder of bony fishes of the order Anguilliformes, containing the deep-sea gulpers, swallowers, and monognathids.

saccular aneurysm *Medicine.* an aneurysm in which only a small area of the vessel, rather than the whole circumference, is distended, forming a sac or protrusion.

sacculate *Biology.* formed by a series of sacs or saclike expansions.

sacculus *Anatomy.* **1.** a small bag or sac. **2.** the smaller of two membranous divisions in the vestibule of the inner ear; it contributes to the sense of equilibrium. Also, **saccule.**

sac fungus *Mycology.* a common term for fungi belonging to the subdivision Ascomycotina.

Sachs, Julius von 1832–1897, German botanist; founder of plant physiology; fundamental contributions to the study of photosynthesis.

sackcloth or **sack cloth** *Textiles.* a coarse, heavy, unbleached cloth of cotton, linen, or animal hair. Also, **sacking.**

sack paper *Materials.* a material made from processed wood fiber and used to make bags.

sackungen *Geology.* a deep-seated, gradual settlement of a large slab of rock into an adjacent valley that results in a ridge-top trench.

Sackur-Tetrode equation *Physics.* an equation that gives the translational entropy of an ideal gas composed of free fermions; under certain approximations, it is

$$S_{tr} = R\{\ln [(2mkT)^{3/2}/h^3N]V + 5/2\},$$

where S_{tr} is the molar translational entropy, R is the gas constant, m is the molecular mass, k is the Boltzmann constant, T is the absolute temperature, h is the Planck constant, N is Avagadro's number, and V is the molecular volume.

sacral *Anatomy.* of or relating to the sacrum.

sacral block *Medicine.* a regional anesthesia administered by injection into the sacral canal.

sacral nerve *Anatomy.* any of five pairs of mixed spinal nerves that extend from the ventral and dorsal sacral foramina.

sacral vertebrae *Anatomy.* the five fused segments of the vertebral column that form the sacrum.

sacrificial anode *Physical Chemistry.* a layer of a lesser metal that is more reactive than an adjacent preferred metal, and will thus act as an anode in the process of sacrificial protection.

sacrificial metal *Physical Chemistry.* a metal that is used as a sacrificial anode.

sacrificial protection *Physical Chemistry.* the purposeful corrosion of a less desirable metal so that a preferred metal can be protected from corrosion; e.g., if a layer of zinc is placed in contact with steel in seawater, the surrounding water will serve as an electrolytic solution and current will flow between the two metals, making the more reactive zinc the anode and the less reactive steel the cathode.

sacro- a combining form meaning "sacrum."

sacrococcygeus *Anatomy.* the poorly developed and sometimes absent muscles on the surface of the coccyx and the sacrum.

sacrodynia *Medicine.* any pain in the sacrum, the triangular bone below the lumbar vertebrae.

sacroiliac [sak′rə il′ē ak] *Anatomy.* **1.** relating to the sacrum and the ilium. **2.** the joint where the sacrum and ilium meet.

sacrospinal *Anatomy.* relating to the sacrum and the vertebral column.

sacrum *Anatomy.* the triangular bone located just below the lumbar vertebrae, and formed usually by the five fused sacral vertebrae that are wedged dorsally between the two hip bones.

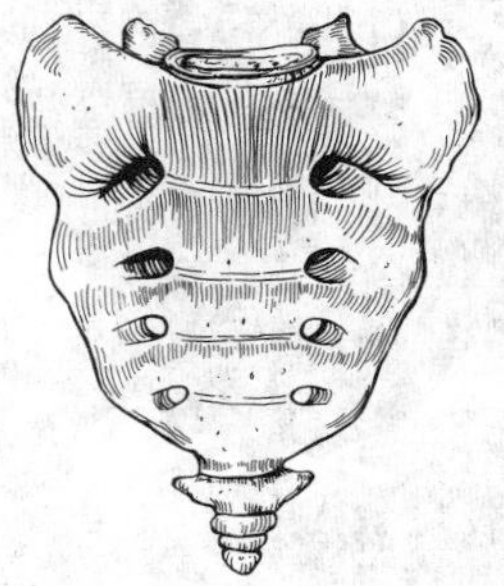

sacrum

saddle *Agriculture.* a seat used in riding a horse. *Geology.* **1.** a broad gap with gently sloping sides, forming a mountain pass and resembling a saddle in shape. **2.** a col or low point along the crest line of a ridge. **3.** a more or less flat ridge that joins the peaks or summits of two higher elevations. **4.** a depression or low point along the surface axis of an anticline. **5.** see COL, def. 2. *Mechanical Engineering.* a support shaped to fit a workpiece being held. *Mechanical Devices.* a portion of a lathe supporting the sliding bed between the headstock and the tailstock. *Ordnance.* the support for the trunnion on some gun carriages.

saddleback *Geology.* a hill or ridge having a concave outline along its crest. *Meteorology.* the cloudless air between towers of two cumulus congestus or cumulonimbus clouds and above a lower cloud deck.

saddleblock anesthesia *Medicine.* anesthesia produced by the introduction of anesthetic low in the dural sac to affect a region that roughly corresponds with the areas of the buttocks, perineum, and inner aspects of the thighs, which would impinge on a saddle in riding.

saddle flange *Mechanical Devices.* a curved, hollowed flange to fit a boiler or pipe.

saddle fold *Geology.* a fold having an additional flexure lying perpendicular to the main flexure at or near its crest.

saddle leather *Materials.* vegetable-tanned animal hide that is used to make horse saddles; similar leather is used to make other goods such as jackets or handbags.

saddle point *Mathematics.* a point in the domain of a function at which first partial derivatives vanish, but which is neither a (local) maximum nor minimum of the function; i.e., a critical point that is not an extremum. *Meteorology.* see COL.

saddle-point azeotrope *Physical Chemistry.* a specific mixture of liquids whose vapor and liquid phases have the same composition, in which the mixture's boiling point lies midway between the highest and lowest boiling points of its individual components.

Saddler *Ordnance.* a Soviet two-stage, intercontinental ballistic missile powered by liquid propellant motors and equipped with inertial guidance, delivering a 20-megaton thermonuclear warhead at a range of 6500 miles; it was replaced by submarine-launched ballistic missiles under SALT I; officially designated **SS-7.**

saddle reef *Geology.* a saddle-shaped mineral deposit generally lying in vertical succession along the bedding planes of an anticlinal fold. Also, **saddle vein.**

saddle stitching *Graphic Arts.* a method of binding a printed signature along its central fold with thread, silk, or wire.

saddle stone see APEX STONE.

saddle type turret lathe *Mechanical Engineering.* a lathe, with no ram, that has a multisided indexing tool holder mounted directly on a support saddle.

saddling *Metallurgy.* the process of fabricating a seamless ring by forging.

sadism [sā´diz əm; sad´iz əm] *Psychology.* **1.** a sexual deviation in which a person derives pleasure from inflicting pain or suffering on other people. **2.** any tendency or desire to direct pain or humiliation against others. (From the Marquis de *Sade,* 1740–1814, a French novelist who described this condition.)

sadist [sā´dist; sad´ist] *Psychology.* a person affected by sadism; one who derives sexual pleasure from inflicting pain or suffering on others. Thus, **sadistic.**

sad mutants *Genetics.* mutants that are defective in ribosomal subunit assembly. (An acronym for subunit asssembly defective.)

sadomasochism [sā´dō mas´i kiz əm] *Psychology.* **1.** a sexual deviation in which one partner derives pleasure from inflicting pain or suffering on the other, who also derives pleasure from the experience. **2.** any tendency to derive pleasure from inflicting or receiving pain or humiliation. Thus, **sadomasochist.**

SAE Society of Automotive Engineers. Also, **S.A.E.**

Saefftigen's pouch *Invertebrate Zoology.* an elongated pouch inside the genital sheath of many acanthocephalans.

SAE number *Engineering.* **1.** a standard designation of viscosity for crankcase and lubrication oils, set by the Society of Automotive Engineers. **2.** a screw-thread standard set by the SAE for use by the automotive industries. **3.** a system of numbering for classes of steel.

safe *Ordnance.* of a munition, designed and set in such a manner that it will not function accidentally.

safe area *Military Science.* a designated area in hostile territory that offers a friendly evader or escapee a reasonable chance of avoiding capture and of surviving until he can be evacuated.

safe burst height *Ordnance.* the height of burst at or above which the level of fallout or damage to ground installations is at a predetermined level acceptable to the military commander.

safe current *Military Science.* in naval mine warfare, the maximum current that can be supplied to a sweep that does not produce a danger area with respect to the mines being swept for.

safe depth *Military Science.* in naval mine warfare, the shallowest depth of water in which a ship will not actuate a bottom mine of a particular type. Similarly, **safe speed.**

safe distance *Military Science.* in naval mine warfare, the horizontal range from the edge of the explosion damage area to the center of the sweeper.

Safeguard *Ordnance.* a ballistic missile defense system.

safe level of protein intake *Nutrition.* the human daily requirement of 70–80 grams of protein; a distribution of amino acids in dietary proteins obtained by taking both animal and plant proteins at a ratio of 1:3–4, as determined by the Food and Agricultural Organization.

safelight *Graphic Arts.* one of a system of electric lights used in a photographic darkroom; each light is equipped with a screen to filter out particular light rays that would damage certain emulsions or other photosensitive materials.

safe load *Engineering.* the structural load that a soil or foundation can safely support.

safe period *Medicine.* the interval in the menstrual cycle when conception is considered least likely to occur, usually a few days before and after the onset of menstruation.

safe separation distance *Ordnance.* the minimum distance between a weapon and its delivery system beyond which the hazards associated with detonation are acceptable.

safety *Engineering.* the prevention of accidents by the use of special devices or equipment, by educational means (lectures, etc.), by training, and by improvement in working conditions, including machinery. *Ordnance.* a device on a weapon or munition that prevents it from functioning accidentally. Also, **safety device.** Similarly, **safety lock.**

safety belt *Engineering.* **1.** a belt that attaches to some fixed object to protect the wearer from falling or collision. **2.** a protective belt or harness with remote anchorage, which allows a drop of about six feet; used for working on sheer faces at a height.

safety block *Ordnance.* a block that prevents a fuse from functioning accidentally by limiting the motion of the firing pin.

safety bolt *Civil Engineering.* a gate or door lock that can be unlocked from only one side, allowing only one-way access.

safety button *Nucleonics.* a device worn on the clothing of personnel who work in the presence of radioactive materials, used to warn the individual of excessive radioactivity.

safety cable *Mining Engineering.* a mining machine cable that reduces the chance of shock hazard and arcing by cutting off power when the positive conductor insulation is damaged.

safety cage *Mining Engineering.* a cage, box, or platform that raises and lowers workers and equipment into and out of mines, fitted with a safety clutch to prevent a fall in the event of a break of the supporting cable.

safety can *Engineering.* a container for storing, handling, and transporting flammable liquids on a short-term basis.

safety capacity *Industrial Engineering.* the portion of a shop's work capacity that is ordinarily not used for production, enabling the shop to respond to contingencies such as increased demand.

safety catch *Ordnance.* an external device that is moved in order to activate or deactivate the safety mechanism on some weapons. *Mechanical Engineering.* see SAFETY STOP.

safety chuck *Mechanical Devices.* a drill chuck designed so that the heads of the drill set do not extend beyond the chuck.

safety engineering *Industrial Engineering.* the engineering design of a machine or facility to ensure safe operation and minimize accidents.

safety explosive *Materials Science.* an explosive that requires an extremely powerful detonating force, and thus is safe to handle under ordinary conditions.

safety factor *Ordnance.* **1.** the adjustment in range or elevation that must be made on a gun firing over friendly troops to ensure their safety. **2.** an overload factor incorporated in the design of a weapon or munition to ensure safety. *Electricity.* a rating of a component that is 10% to 20% lower than the maximum value it can handle, ensuring that it will operate for the longest possible time without needing service or replacement. *Mechanics.* see FACTOR OF SAFETY.

safety film *Graphic Arts.* modern photographic film on a plastic base that will not disintegrate or burn easily.

safety flange *Mechanical Devices.* a flange with tapered sides to keep a wheel intact in the event of breakage.

safety fork *Ordnance.* a metal clip that fits over the collar of a landmine fuse in order to prevent accidental detonation.

safety fuse *Ordnance.* a pyrotechnic that is used to transmit a flame to a detonator; it is encased in a flexible, weather-proof sheath and burns at a constant, timed rate.

safety glass *Materials Science.* **1.** a pane of glass made by laying transparent plastic or artificial resin between two panes of normal glass; the middle layer retains the glass fragments so that the pane will shatter rather than shear upon impact. **2.** glass that has been tempered by heating to the point of softness, then cooled rapidly, increasing resistance to impact. **3.** glass that is reinforced with wire mesh.

safety groove *Ordnance.* a groove incorporated into the design of a munition to ensure that any potential failure will occur in a selected, less critical area.

safety harness *Engineering.* see HARNESS.

safety hoist *Mechanical Engineering.* **1.** a hoist that uses differential pulleys. **2.** a hoist that has a catch or grip to prevent the rope from running if the tension is accidently released.

safety hook *Mechanical Devices.* a hook equipped with a spring-loaded latch which prevents loads from sliding off.

safety lamp *Mining Engineering.* in coal mining, a lamp deemed relatively safe for use even in atmospheres that may contain flammable gas.

safety lanes *Navigation.* sea lanes that are designated for use in transit by submarine and surface ships to prevent attack by friendly forces.

safety level of supply *Industrial Engineering.* the quantity of materials, in addition to the operating level of supply, that is required to be on hand in order to continue operations in the event of minor interruptions of normal replenishment or unpredictable fluctuations in demand. Also, **safety stock.** *Aviation.* specifically, a level of fuel supply that is sufficient to allow continued operations during a minor interruption in the normal flow of supply or to accommodate a sudden rise in demand.

safety lever *Ordnance.* **1.** a lever that sets the safety mechanism on some automatic weapons. **2.** a metal piece in a grenade fuse that prevents initiation after the safety pin is removed; it is discarded when the grenade is released, thus allowing the firing pin to initiate the detonation action.

safety line *Military Science.* in land-mine warfare, a demarcation line for trip-wire or wire-activated mines that is used to protect the laying personnel; after laying, it is neither marked on the ground nor plotted on the minefield record.

safety match *Engineering.* a match that lights only when struck on a special surface on the box.

safety paper *Graphic Arts.* paper designed to prevent the alteration of any printing or writing on its surface; used for bank checks, securities, contracts, and the like.

safety pinion *Horology.* a pinion on the center arbor of a timepiece, designed to unscrew itself from the arbor if the mainspring breaks, thus reducing the pressure and the potential for damage to the works.

safety plug *Engineering.* **1.** a fusible plug in high-pressure devices such as boilers that melts at a determined temperature. **2.** a plug in a system of water pipes for fire protection in buildings.

safety rod *Nucleonics.* a control rod that is magnetically suspended above the core of a reactor; used to rapidly shut down the system if the power level surpasses a predetermined level. Also, SCRAM ROD.

safety service *Telecommunications.* a radio communications service that is used for the purpose of maintaining the safety of human life and property.

safety shoe *Engineering.* **1.** a well-built shoe of leather or rubber provided with a sheet-steel or other strong, stiff toe. **2.** a special shoe without nails in the sole used by persons working near explosives.

safety stake *Ordnance.* a ground stake used to define the safe lateral limits of a weapon's fire.

safety stop *Mechanical Engineering.* a check on a hoisting apparatus that prevents a cage or lift from falling. Also, SAFETY CATCH.

safety valve *Engineering.* an automatic valve that releases steam from a boiler when the pressure rises above a preset pressure level.

safety zone *Military Science.* an area on land, sea, or air that is reserved for noncombat operations of friendly aircraft, surface ships, submarines, or ground forces.

safe yield *Civil Engineering.* the rate at which water can be taken from an aquifer without causing a long-term decline in the water table or piezometric surface.

safflorite *Mineralogy.* $CoAs_2$, a tin-white, metallic, orthorhombic mineral of the löllingite group occurring as fibrous masses and prismatic crystals, having a specific gravity of 7.2 and a hardness of 4.5 to 5 on the Mohs scale; found with other cobalt-nickel minerals in mesothermal veins.

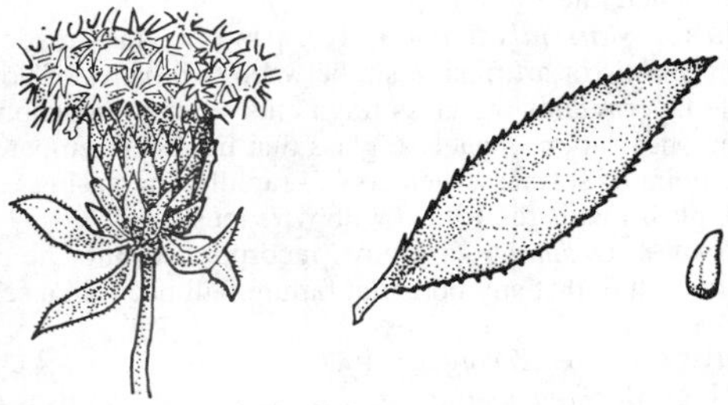

safflower

safflower *Botany.* a thistlelike herb, *Carthamus tinctorius,* family Compositae, having finely toothed, edible leaves, large orange-red flowers that yield a dye, and seeds from which an oil is extracted.

safflower oil *Food Technology.* an oil derived from safflower seeds; used in cooking, in salad dressings, and as an ingredient in medicines, paints, and varnishes.

saffron *Botany.* a crocus of the Iris family, *Crocus sativus,* characterized by showy purple flowers; its dried stigmas yield an orange condiment that is used to flavor and color foods and medicine.

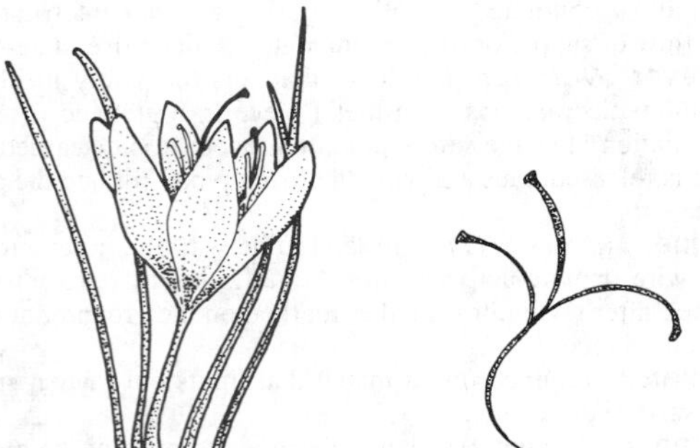

saffron

safing *Ordnance.* the process of changing a weapon or munition from a state of readiness for initiation to a safe condition.

safranine *Organic Chemistry.* any of several phenazine-based dyes that are sometimes used as a biological stain.

safrole *Organic Chemistry.* $C_{10}H_{10}O_2$, a toxic, colorless or pale-yellow liquid; insoluble in water and soluble in alcohol; melts at 11°C and boils at 234°C; used in soaps, perfumes, insecticides, and pharmaceuticals.

sag *Geology.* **1.** a saddle-shaped pass or gap in a ridge or mountain range. **2.** a shallow depression along a generally flat or gently sloping land area. **3.** a regional downwarp or broad, shallow basin having gently sloping sides. **4.** a depression formed near a fault by the downwarping of beds in the direction opposite that of the frictional drag. *Electricity.* the distance at the center of a span of wire between the actual wire and the position if the wire were straight, assuming a catenary or parabolic curve. *Metallurgy.* in casting, an undesirable dimensional variation caused by sagging sand.

Sagan, Carl born 1934, American astronomer and educator; author of noted popular works on science.

Sagartildae *Invertebrate Zoology.* a family of zoantharians in the order Actiniaria.

sag bolt *Mining Engineering.* a bolt installed at intersections in a mine roof, used to detect and measure sag.

sage *Botany.* **1.** any plant or shrub of the genus *Salvia* in the mint family, especially *Salvia officinalis,* whose grayish-green leaves are used in medicine, cooking, and perfumery. **2.** the leaves of this plant.

SAGE *Ordnance.* an air defense system that detects, identifies, and tracks air-breathing weapons threatening the U.S. and Canada and processes the information to produce weapon assignments and guidance orders. (An acronym for Semiautomatic Ground Environment System.)

sagebrush *Botany.* any of various bushy undershrubs of the genus *Artemisia,* family Compositae, especially *Artemisia tridentata*; characterized by silvery, wedge-shaped leaves with three teeth at the tip; found on the alkine plains of the western United States.

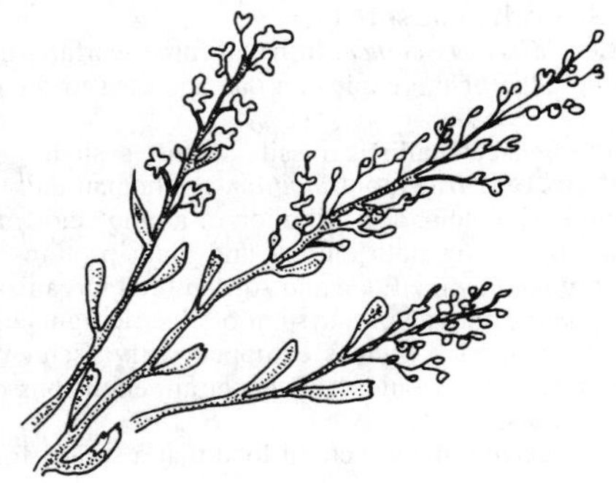

sagebrush

sage grouse *Vertebrate Zoology.* the largest American grouse, *Centrocercus urophasianus,* characterized by gray, buff, and black plumage; common on the plains of western North America.

sagenite *Mineralogy.* a variety of rutile forming complexly reticulated, twinned groups of acicular crystals, commonly enclosed in quartz or other minerals.

sagenitic *Geology.* occurring as needles or plates intersecting in a gridlike manner.

Sagenocrinida *Paleontology.* an order of inadunate crinoids in the subclass Flexibilia, characterized by an ovoid crown and a conical cup half the height of the crown; extant from the Silurian to Permian.

Sagenopteris *Paleontology.* a form genus of leaves of the Mesozoic plants known as Caytoniales, a group that resembles seed ferns and angiosperms; extant from the Triassic to Cretaceous.

sage oil *Materials.* a yellow-to-green essential oil occurring in two varieties: Clary and Dalmatian; used in perfumes and flavoring. Also, CLARY SAGE OIL, OIL OF SASSAFRAS.

sage sparrow *Vertebrate Zoology.* a small, grayish-brown finch, *Amphispiza belli*; found in dry, brushy areas of the southwestern U.S. and nearby Mexico.

sage thrasher *Vertebrate Zoology.* a grayish-brown and white thrasher, *Oreoscoptes montanus,* that is similar to the mockingbird; it is found in arid regions of the western U.S.

saggar *Materials.* **1.** a refractory case in which fine pottery and porcelain is fired. **2.** a hard, unlayered clay found beneath coal beds and used to make such a case. Also, **saggar clay.**

sagging *Naval Architecture.* a condition in which the midsection of a vessel droops with respect to the ends; the opposite of hogging.

Saghatherium *Paleontology.* an extinct genus of large hyracoids in the family Procaviidae; extant in the Lower Oligocene.

Sagitta [sə jit′ ə] *Astronomy.* the Arrow, a small constellation in the Milky Way between Aquila and Vulpecula.

sagitta *Vertebrate Zoology.* the larger of the two otoliths found in the ears of most bony fishes. *Mathematics.* the distance from the midpoint of an arc to the midpoint of its chord.

sagittal *Zoology.* of or relating to the longitudinal vertical plane dividing an animal's bilaterally symmetrical body into left and right halves. *Anatomy.* situated in the plane of the sagittal suture or parallel to it.

sagittal crest *Anthropology.* a ridge of bone along the midline length of the skull to which heavy musculature is attached; a characteristic of robust australopithecines and other heavy-jawed primates.

sagittal focus see SECONDARY FOCUS.

sagittal suture *Anatomy.* the line of junction between the two parietal bones.

Sagittariidae *Vertebrate Zoology.* the secretary bird, a monotypic family of large snake-eating raptors of the order Falconiformes, characterized by long legs and a crest of long feathers; found on African plains.

Sagittarius [saj′i târ′ē əs] *Astronomy.* the Archer, a bright zodiacal constellation of the northern hemisphere summer; it lies partly in the Milky Way and contains star clouds, globular clusters, and gaseous nebulae.

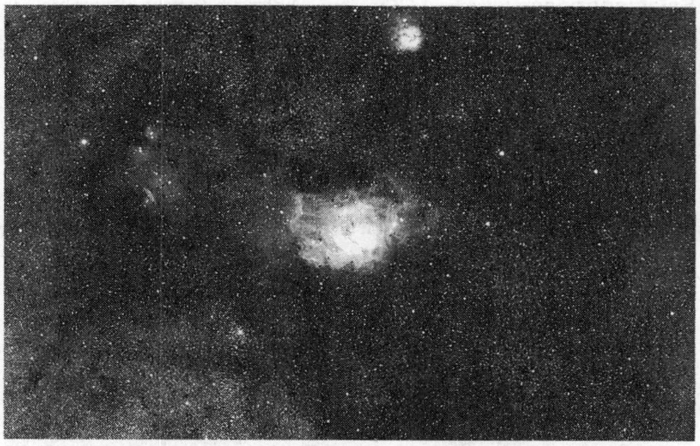

Sagittarius

Sagittarius A *Astronomy.* the radio source thought to mark the center of the Milky Way Galaxy.

Sagittarius arm *Astronomy.* one of the spiral arms of the Milky Way Galaxy that lies closer to the galactic center than earth does.

Sagittarius B2 *Astronomy.* a dense and massive complex of ionized hydrogen regions and molecular clouds, the strongest molecular radio source in our galaxy.

Sagittarius star cloud *Astronomy.* a major star cloud lying in the direction of Sagittarius; it stretches between 1500 and 6000 light-years from the sun.

sagittate *Botany.* arrow-shaped; used especially to describe a leaf with the basal lobes free and projected downward while the apex is pointed. Also, **sagittiform.**

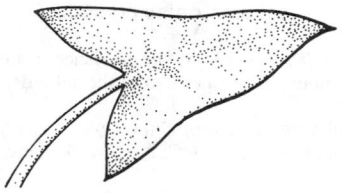

sagittate leaf

sagittocyst *Invertebrate Zoology.* in free-living flatworms of class Turbellaria, an epidermal sac containing a single spindle-shaped needle that can be shot out.

sago *Materials.* a starchy substance derived from the soft interior of the trunk of various palms and cycands; used to make a type of flour (**sago flour**) that is used in puddings and soups.

sago palm *Botany.* any of various Old World palms, especially of the genera *Metroxylon* and *Caryota*, from whose pith the substance sago is made; found in tropical regions.

sag pond *Geology.* a small body of water filling an enclosed depression along an active or recent fault in which differential displacement has confined drainage. Also, RIFT LAKE.

saguaro [sə gwär′ō; sə wär′ō] *Botany.* a very large branching cactus plant, *Carnegiea gigantea* (or *Cereus giganteus*), that grows 25 to 50 feet high and bears edible fruit; it is found in Arizona and neighboring areas. Also, **sahuaro.**

saguaro

Saha, Meghnad 1894–1956, Indian physicist; formulated thermal ionization equation, which he used to analyze stellar spectra.

Sahara [sə hâr′ə] *Geography.* the world's largest desert, stretching across northern Africa from the Atlantic Ocean to the Red Sea and covering an area of about 3.5 million square miles. (From Arabic; literally "desert" or "the desert.")

Saha's equation *Physics.* an equation that gives the amount of thermal ionization of a monatomic gas as a function of its pressure, temperature, ionization potential, and other such factors.

Sahel [sə hel′; sə hāl′] *Geography.* the southern Sahara desert area, an extremely arid region stretching across northwestern Africa from Senegal to Chad. *Meteorology.* also, **sahel.** a strong, dusty desert wind of the Sahara.

sahlinite *Mineralogy.* $Pb_{14}(AsO_4)_2O_9Cl_4$, a sulfur-yellow monoclinic mineral occurring in aggregates of thin scales, having a specific gravity of 8.0 and a hardness of 2 to 3 on the Mohs scale; found at Långban, Sweden.

sahlite see SALITE.

SAIDS simian AIDS.

saiga [sī′gə] *Vertebrate Zoology.* a goatlike antelope, *Saiga tatarica*, characterized by a peculiarly expanded muzzle that serves to filter dust and to warm the air breathed in; found in western Asia and eastern Russia.

sail *Naval Architecture.* a large piece of fabric or similar flexible material that is used to catch or deflect the wind, providing propulsive power for a vessel to move forward. *Navigation.* **1.** to handle a sail-driven vessel. **2.** to get underway or put to sea in any type of vessel, however propelled.

sailboat *Naval Architecture.* a small, sail-fitted vessel, historically a passenger vessel, cargo ship, or warship; today usually a pleasure craft. Many sailboats are also equipped with auxiliary motors.

sailcloth *Textiles.* any of various heavy, durable, canvas-type fabrics used for sails, such as cotton or Dacron.

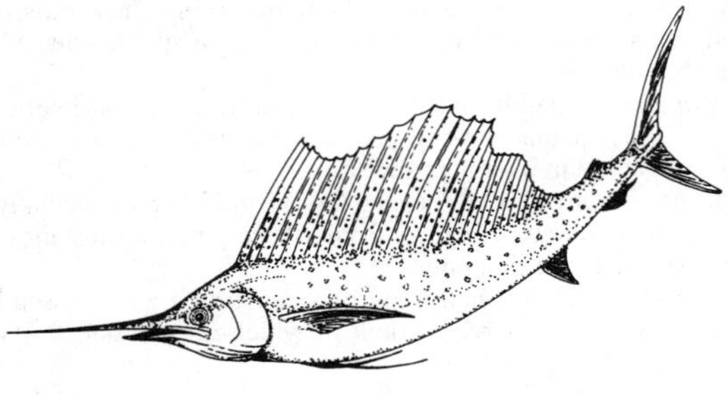

sailfish

sailfish *Vertebrate Zoology.* a large pelagic fish belonging to the genus *Istiophorus* in the family Istiophoridae, possessing an elongate bill used to stun prey, long pelvic fins, a very long, high dorsal fin, and a double keel on each side of the tail; found in tropical and subtropical waters.

sailing *Navigation.* **1.** the process, skill, or sport of guiding a sail-powered boat over water. **2.** any of various methods used to determine the course to be followed by a vessel in order to arrive at a particular point; it is typically calculated on paper, often using trigonometry to plot the course on a chart; collectively the methods are termed **the sailings**.

sailing chart *Navigation.* a small-scale nautical chart designed for offshore navigation.

sail plan *Naval Architecture.* a drawing, normally in side elevation, showing the positions and arrangement of a vessel's sails.

Saint For compound entries beginning with the word *Saint,* see also **St.** plus the appropriate word or words; e.g., the entry for the St. Lawrence River appears among the "stl" words.

Saint Anthony's fire see ERGOTISM.

Saint Augustine grass *Botany.* a branched, perennial, creeping grass, either *Stenotaphrum secundatum* or *Manisuris rugosa*, widely used as sod grass or as a sand binder; native to the southern United States. (Associated with the city of *St. Augustine,* Florida.)

Saint Elmo's fire *Electricity.* a glow accompanying discharges of atmospheric lightning, historically seen at the tops of church steeples or the tall masts of sailing ships, now also seen on the wing tips of aircraft. (From *Saint Elmo,* another form of the name of Saint Erasmus, the patron saint of early Mediterranean sailors.)

Saint-Hilaire method *Navigation.* a method of establishing a line of position by comparing the calculated zenith distance of a celestial body, based on an assumed position, with the true zenith distance, based on observation; the difference, called the *intercept*, is then plotted in relation to the body's azimuth, thus establishing a point through which the line of position is drawn. Also, INTERCEPT METHOD, MARCQ ST.-HILAIRE METHOD.

Saint Joseph retort process *Metallurgy.* an electrochemical process used to extract zinc.

Saint Louis encephalitis *Medicine.* an arborvirus infection of the central nervous system, spread by mosquito bite, and occurring in the central and western U.S. and in Florida.

Saint Peter's sandstone *Geology.* an Early Ordovician formation in the midwestern United States.

Saint Vitus' dance *Medicine.* **1.** a disease of the nervous system, often hereditary, more frequently found in the young and especially in girls, that causes convulsive, irregular, jerking movements, speech disturbances, depression, irritability, and mental impairment. It may occur epidemically and is frequently seen with or following rheumatic fever. Also, SYDENHAM'S CHOREA. **2.** more generally, any form of chorea. (From *Saint Vitus,* the patron saint of those afflicted with chorea.)

Sakharov, Andrei [säk´ə räf; säk´ə räv] 1921–1989, Russian physicist and political activist; studied thermonuclear chain reactions; developed USSR's hydrogen bomb.

Sakmann, Bert born 1942, German physiologist; with Erwin Neher, shared the Nobel Prize for discovering how ion channels regulate the passage of ions into and out of cells.

Sakmarian *Geology.* a European geologic stage of the lowermost Permian period, occurring after the Stephanian of the Carboniferous period and before the Artinskian.

Sakseneaceae *Mycology.* a family of fungi of the order Mucorales occurring primarily in soil in temperate regions.

sal *Science.* the Latin word for salt; used in chemistry and in pharmacy. *Petrology.* see SIAL.

salable coal *Mining Engineering.* the net output of a coal mine, equivalent to the total output less the quantity rejected or consumed during preparation for market.

Salado formation *Geology.* a redbed of rock salt and potash salts in southwest New Mexico, formed during the Permian period.

salal *Botany.* an evergreen shrub, *Gaultheria shallon*, of the heath family, characterized by leathery, oblong leaves, clusters of pink or white flowers, and purplish-black fruit; native to the western coast of North America.

Salam, Abdus born 1926, Pakistani physicist; awarded the Nobel Prize for independently establishing the unity of electromagnetism and "weak force."

salamander *Vertebrate Zoology.* a general term used to designate generally small, tailed amphibians of the order Caudata, characterized by a soft, moist, scaleless skin and usually an aquatic larval stage. *Metallurgy.* a mass of iron that collects at the bottom of a blast furnace, resulting from the escape of molten metal through the hearth.

salamander

Salamandridae *Vertebrate Zoology.* a holarctic family of amphibians of the suborder Salamandroidea containing newts and salamanders, characterized by complete metamorphosis and wholly or partly aquatic adults.

Salamandroidea *Vertebrate Zoology.* the suborder of the amphibian order Caudata containing newts, salamanders, mudpuppies, olms, and congo eels.

salami attack *Computer Programming.* a technique for computer crime, in which a small amount of money from many accounts, e.g., the fraction of a cent remaining when a bill is rounded to even cents, is added to the perpetrator's account.

sal ammoniac *Mineralogy.* NH_4Cl, a colorless or white, transparent cubic mineral occurring as dendritic to fibrous masses and trapezohedral crystals, having a specific gravity of 1.519 to 1.532 and a hardness of 1 to 2 on the Mohs scale; found as an encrustation around volcanic fumaroles, including those on Mt. Vesuvius, Italy.

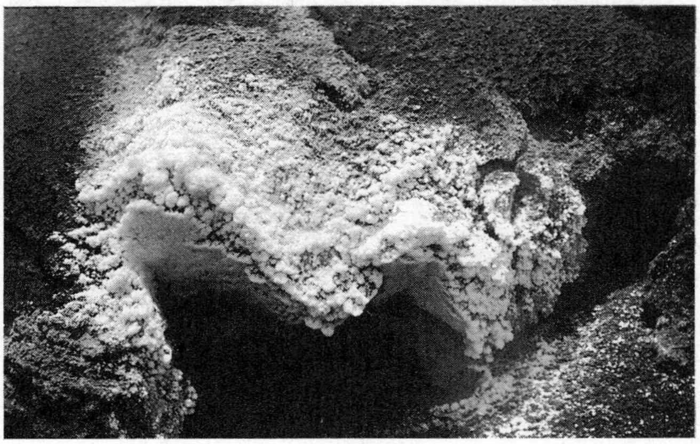

sal ammoniac crystals

Salangidae *Vertebrate Zoology.* the icefishes, a family of elongate, scaleless, anadromous, freshwater fish of the suborder Salmonoidei; native to the Orient.

salazosulfadimidine *Organic Chemistry.* $C_{19}H_{17}N_5O_5S$, a brown crystalline solid that melts at 207°C; used medicinally to treat ulcerative colitis.

salcrete *Geology.* a thin, hard crust of salt-cemented sand grains, formed on a marine beach that is intermittently saturated by salt water.

Saldidae *Invertebrate Zoology.* the shore bugs, a family of small, oval, predaceous hemipteran insects in the superfamily Saldoidea, living in marshes and grassy shores.

Saldoidea *Invertebrate Zoology.* a superfamily of true bugs in the order Hemiptera, including the shore bugs.

saléeite *Mineralogy.* $Mg(UO_2)_2(PO_4)_2 \cdot 10H_2O$, a lemon-yellow, monoclinic, pseudotetragonal mineral of the autunite group, occurring as rectangular platy crystals, and having a specific gravity of 3.27 and a hardness of 2 to 3 on the Mohs scale; found associated with autunite and carnotite.

Saleniidae *Invertebrate Zoology.* a family of sea urchins, echinoid echinoderms in the order Salenioida, having imperforate tubercles.

Salenioida *Invertebrate Zoology.* an order of sea urchins, echinoid echinoderms with the anus located off center due to a large periproct plate.

salesite *Mineralogy.* $Cu^{+2}(IO_3)(OH)$, a bluish-green, orthorhombic mineral occurring as small, thick prismatic crystals, having a specific gravity of 4.77 and a hardness of 3 on the Mohs scale.

sales register see CASH REGISTER.

salfemic rock *Geology.* a class of igneous rocks in which the ratio of salic minerals to femic minerals is less than 5:3 but greater than 3:5.

salic *Geology.* a mineral soil horizon enriched with soluble salts. *Mineralogy.* relating to igneous rocks having normative minerals that are rich in silicon, aluminum, or magnesium, or rocks having such minerals as constituents. (A combination of the terms *silica* and *aluminum.*)

Salicaceae *Botany.* the willow family, a dicotyledenous family of dioecious trees and shrubs belonging to the order Salicales, having fruit that is a two-valved to four-valved capsule; composed of two genera and around 335 species, including willows, aspens, and poplars.

Salicales *Botany.* an order of angiosperms with a single family, the Salicaceae.

salicin *Organic Chemistry.* $C_{13}H_{18}O_7$, colorless crystals or white powder; soluble in water and alcohol, and insoluble in ether; melts at 199–202°C; used in medicine and as a reagent.

salicyl alcohol *Organic Chemistry.* $C_7H_8O_2$, combustible white crystals; soluble in hot water and very soluble in alcohol and ether; melts at 86–87°C; obtained from the bark of the American aspen; used in medicine, chiefly as an antipyretic and an analgesic.

salicylaldehyde *Organic Chemistry.* C_6H_4OHCHO, a combustible, colorless liquid or dark-red oil; slightly soluble in water and soluble in alcohol and ether; melts at –7°C and boils at 196°C; used in perfumes and flavorings, and as a reagent.

salicylamide *Organic Chemistry.* $C_6H_4(OH)CONH_2$, white or slightly pink crystals; slightly soluble in cold water and soluble in hot water, alcohol, and ether; melts at 139–142°C and decomposes at 270°C; used in medicine. Also, *O*-HYDROXYBENZAMIDE.

salicylanilide *Pharmacology.* $C_{13}H_{11}NO_2$, a drug that acts against fungus; formerly used topically to treat ringworm of the scalp.

salicylate *Organic Chemistry.* a salt or an ester of salicylic acid.

salicylic acid [sal′i sil′ik] *Organic Chemistry.* HOC_6H_4COOH, a combustible, toxic, white powder; slightly soluble in water and soluble in alcohol and ether; melts at 159°C and boils at 211°C (20 torr); used in the manufacture of aspirin and resins, and as a reagent and fungicide. Also, *O*-HYDROBENZOIC ACID.

salicylism *Medicine.* a toxic overdose of salicylic acid or of its salts, usually marked by tinnitus, nausea, and vomiting.

salient *Geology.* **1.** a geological structure that juts outward or upward from the surrounding adjacent land. **2.** a region of a fold or series of folds in which the axial traces are convex in the direction of the outer margin of the folded belt.

Salientia *Vertebrate Zoology.* an alternate name for the order Anura, the tailless amphibia including frogs and toads.

salient pole *Electromagnetism.* a magnetic pole structure, such as a pole piece, that projects from the rest of a magnetic structure upon which a coil is mounted.

saliferous stratum *Geology.* a stratum of sedimentary or igneous rock that produces, contains, or is impregnated with salt.

salimeter see SALINOMETER.

salina *Geology.* **1.** a region, such as a salt pan or salt-encrusted playa, where deposits of crystalline salt are formed or found. **2.** an inland marsh subject to intermittent overflow by salt water. *Hydrology.* any body of water, such as a salt pond, having a high concentration of dissolved salts. *Engineering.* see SALTERN.

salinastone *Geology.* a sedimentary rock consisting primarily of saline minerals.

saline [sa′lēn] *Science.* composed of or like salt; salty. *Medicine.* a sterile solution of sodium chloride used to dilute medications or for intravenous therapy.

saline-alkali soil *Geology.* a soil unusable for agricultural purposes because it contains more than 15% exchangeable sodium, has a high content of soluble salts, and a pH less than 9.5. Also, **salina-alkali soil.**

salinelle *Geology.* a mud volcano that ejects saline mud.

saline-water reclamation *Chemical Engineering.* the demineralization of saline or brackish water.

salinity [sə lin′i tē] *Oceanography.* the weight ratio between dissolved salts and water in seawater. *Chemistry.* the amount of dissolved salts in any solution.

salinity current *Oceanography.* a density current in the ocean whose flow is caused by its relatively higher salinity, and therefore its greater density, in comparison to the surrounding water.

salinity-temperature-depth recorder *Engineering.* a device with sensors and a recorder, used to record salinity, temperature, and depth of ocean water. Also, CTD RECORDER, STD RECORDER.

salinization *Geology.* an accumulation of soluble salts in the soil of an arid, poorly drained region, as a result of the evaporation of the waters that carried them to the soil zone.

salinometer *Engineering.* an instrument that measures the percentage of salt in a solution, such as brine.

salite *Mineralogy* a grayish-green, deep-green, or black ferroan variety of diopside. Also, SAHLITE.

salitrite *Petrology.* a lamprophyre consisting primarily of titanite and acmite-bearing diopside, with accessory apatite, microcline, and sometimes anorthoclase and baddeleyite.

saliva *Physiology.* the clear, alkaline, digestive fluid of the mouth, secreted by the parotid, submaxillary, sublingual, and mucous glands, that softens, moistens, and begins the digestion of food.

salivaria *Invertebrate Zoology.* small pockets in an insect's mouth, holding the opening of the salivary ducts.

salivary *Anatomy.* relating to saliva.

salivary calculus see SIALOLITH.

salivary gland *Physiology.* one of the glands that produce saliva; the primary glands are near the masseter muscle, under the tongue and near the lower jawbone, and numerous smaller glands are found in the submucosa of the tongue, cheeks, lips, and palate.

salivary gland chromosomes *Cell Biology.* large polytene chromosomes resulting from repeated DNA replication without subsequent separation of the daughter strands, occurring in the salivary gland cells of *Dipteran* larva.

salivary stone see SIALOLITH.

salivation *Physiology.* **1.** the response of the salivary glands that produces a flow of saliva into the mouth. **2.** stimulation of the salivary glands. *Medicine.* the excretion of excessive amounts of saliva.

Salk, Jonas born 1914, American microbiologist; developed the first effective polio vaccine; founding director of the Salk Institute.

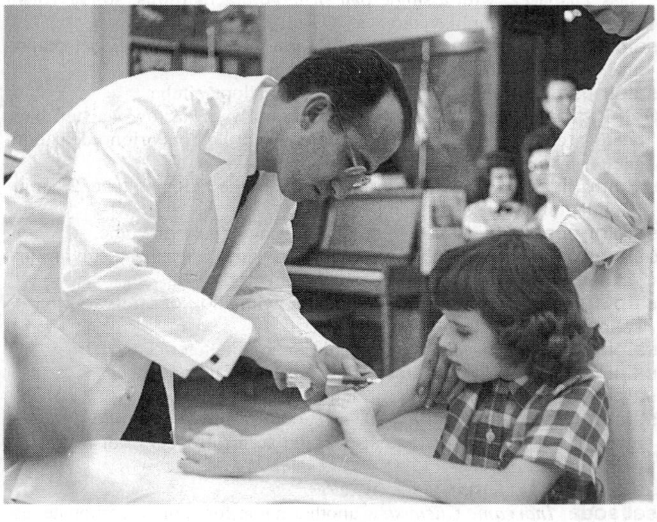

Jonas Salk

Salk vaccine *Immunology.* a vaccine, composed of formalin-killed polio virus, that serves to provide protection against the disease poliomyelitis.
S allele see SELF-STERILITY GENE.
sally see SORTIE.
sally port *Ordnance.* a large gate or passage in a fortified structure.

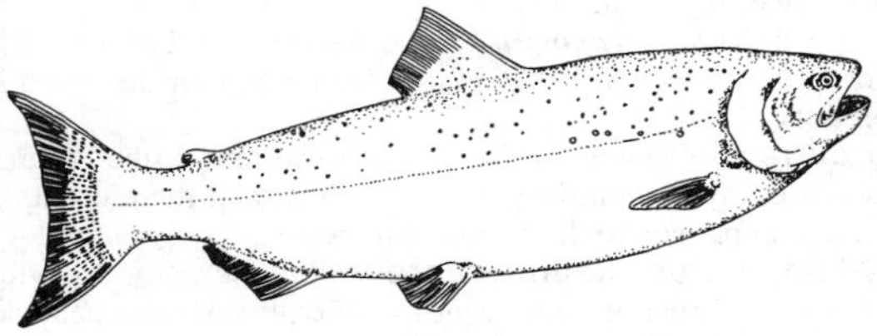

salmon

salmon *Vertebrate Zoology.* any of several medium to large anadromous fish of the order Salmonidae, native to the North Atlantic and North Pacific Oceans and spawning in adjacent streams of Europe and North America; an important food fish of North America and Europe, noted for the characteristic reddish-orange color of its flesh. Well-known North American species include the **chinook** (**king**) **salmon,** *Oncorhynchus tshawytscha*; the **coho** (**silver**) **salmon,** *O. kisutch*; the **sockeye** (**red**) **salmon,** *O. nerka*; and the **Atlantic salmon,** *Salmo salar*.
Salmonella *Bacteriology.* a genus of Gram-negative, usually motile, rod-shaped bacteria found in the intestines of humans and animals; can cause abdominal pain and violent diarrhea.
Salmonelleae *Bacteriology.* in former classifications, a tribe of Gram-negative, rod-shaped, anaerobic bacteria of the family Enterobacteriaceae, including the pathogenic genera *Salmonella* and *Shigella*.
salmonellosis *Medicine.* any infection caused by salmonella bacteria, usually manifested as food poisoning, and causing severe diarrhea and cramping.
Salmonidae *Vertebrate Zoology.* a family of anadromous fish of the suborder Salmonoidei, containing the salmon, trouts, chars, graylings, and whitefish.
Salmoniformes *Vertebrate Zoology.* an order of primitive, generalized, soft-rayed fishes, including pikes, salmon, trouts, and smelts.
Salmonoidei *Vertebrate Zoology.* the suborder of fish of the order Salmoniformes, containing the salmonids, smelts, ayus, galaxiids, and icefishes.
salmonsite *Mineralogy.* a mixture of jahnsite and hureaulite.
Salmopercae *Vertebrate Zoology.* an equivalent name for the order Percopsiformes, the pirate perch and cavefishes of North America.
salol see PHENYL SALICYLATE.
Salpida *Invertebrate Zoology.* an order of transparent tunicates in the class Thaliacea, with a single pair of large brachial gills; solitary or in long, floating, chainlike colonies.
salping- or **salpingo-** a combining form referring to a trumpet-shaped tube, especially the Fallopian tube.
salpingectomy *Surgery.* excision of the Fallopian tube.
Salpingidae *Invertebrate Zoology.* a family of narrow-waisted bark beetles, coleopteran insects in the superfamily Tenebrionoidea; both adults and larvae are predaceous.
salpingitis *Medicine.* **1.** an inflammation of the fallopian tube. **2.** an inflammation of the eustachian tube.
salpingocele *Medicine.* a hernial protrusion of a uterine tube.
salpingocyesis *Medicine.* a condition in which a fertilized egg has implanted in a uterine tube. Also, ECTOPIC PREGNANCY.
salpingo-oophoritis *Medicine.* an inflammation of an ovary and its Fallopian tube.
salpingopexy *Surgery.* the surgical fixation of a uterine tube.
salpingoplasty *Surgery.* the plastic repair of a uterine tube.
salpingorrhaphy *Surgery.* the suture of a uterine tube.
salpingostomy *Surgery.* **1.** the creation of an opening into a uterine tube for the purpose of drainage. **2.** the surgical opening of a blocked uterine tube.
salpinx see FALLOPIAN TUBE, OVARY.
sal soda *Inorganic Chemistry.* another name for sodium carbonate, especially sodium carbonate decahydrate, $Na_2CO_3 \cdot 10H_2O$. See SODIUM CARBONATE DECAHYDRATE.
salt *Materials.* NaCl, a crystalline compound, occurring as a mineral and as a constituent of sea water; used extensively since ancient times for seasoning food and as a preservative. *Chemistry.* the compound formed when the hydrogen atom of an acid is replaced by a metal atom or a positive radical, such as NH_4^+, characteristically having an ionic lattice and disassociating completely in a water solution. *Engineering.* to mix a retardant or accelerator into a cement. *Mining Engineering.* **1.** to introduce, with or without intention, extra amounts of either a valuable or waste mineral into a sample to be assayed. **2.** to enrich a mine, artificially and usually fraudulently, by covertly introducing valuable materials into some of the working places.

salt crystals (rock salt)

salt-affected soil *Geology.* any soil that is not suitable for crop cultivation or other agricultural use because it contains excessive amounts of salt, exchangeable sodium, or both.
salt-and-pepper sand *Geology.* a sand made up of a mixture of dark- and light-colored particles.
salt anticline *Geology.* a dome or anticlinal fold having a mobile, linear salt core that forces its way upward, rupturing the brittle overlying rocks. Also, SALT WALL.
sal tartar see SODIUM TARTRATE.
saltation *Neurology.* **1.** sudden, jerky, or leaping movement as seen in chorea. **2.** the conduction of nerve impulses in myelinated nerves in which the impulse jumps rapidly from one node of Ranvier to the next. *Evolution.* a sudden mutation producing a heritable variation in the morphology of a species that occurs more abruptly than other instances of comparable change and is difficult to explain through natural selection. Also, MACROGENESIS. *Geology.* a type of sediment transport in which particles are moved forward by water or air currents through a series of intermittent, short leaps and bounces off a bottom surface.
saltational speciation *Genetics.* see QUANTUM SPECIATION.
saltation load *Geology.* that part of a stream's load that is transported by saltation, or whose movement is affected by the impact of bouncing particles.
saltatorial *Zoology.* adapted or designed for leaping.
saltatory *Neurology.* of or relating to saltation. Also, **saltatoric.**
saltatory motion *Cell Biology.* the movement of certain organelles, such as mitochondria, within a eucaryotic cell; apparently occurring in association with microtubules.
saltatory replication *Molecular Biology.* a sudden amplification of a chromosome segment, yielding a vast number of copies of specific DNA nucleotide sequences.
salt bath *Metallurgy.* in the thermal treatment of steel, a molten salt bath used to heat parts.
salt bottom *Geology.* a flat or level area of low-lying land that is encrusted with salt.
saltbox *Architecture.* a wood frame house with two stories in front and one behind, and a gabled roof whose rear slope is much longer than its front slope. Also, **saltbox house.**

salt bridge *Physical Chemistry*. a tube saturated with a solution of potassium chloride and connected between the two half cells of an electrochemical cell; it minimizes the voltage that appears at the liquid junction between the half cells and also prevents the solutions from mixing.

salt burst *Geology*. a disintegration of rock as a result of the crystallization of soluble salts that enter the rock pores.

salt cake *Inorganic Chemistry*. a term for impure sodium sulfate, Na_2SO_4, especially when the compound is obtained as a by-product of the manufacture of hydrochloric acid from common salt and sulfuric acid; used in soaps, paper pulp, glass, and dyes.

salt crust *Hydrology*. a deposit of salt formed on a surface of ice as crystal growth forces salt out of young sea ice and pushes it upward.

salt dome *Geology*. a dome or anticlinal fold having a mobile, equidimensional salt core that forces its way upward due to density contrast, rupturing the brittle overlying rocks. Also, ACROMORPH.

salt-dome breccia *Geology*. a breccia appearing as a dome-shaped mass surrounding a salt plug in deep shale sequences.

salted weapon *Ordnance*. a nuclear weapon that captures neutrons at the time of the explosion and produces radioactive products over and above the usual radioactive weapon debris.

saltern *Engineering*. a building in which salt is made. Also, SALTWORKS, SALINA.

salt error *Analytical Chemistry*. an error in an analytical determination of a saline liquid, caused by the effect of the neutral ions in the solution on the color of the pH indicator, and thus upon the apparent pH.

salt field *Geology*. an area beneath which a workable salt deposit of economic value is present.

salt fingers *Chemistry*. an array of fluctuating columns of fluids that form in a liquid when two solutes that have different diffusion rates are separated by an interface.

salt flat *Geology*. a level or flat, salt-encrusted area at the bottom of a lake or pond that has dried up. Also, ALKALI FLAT.

salt-fog test *Metallurgy*. an accelerated test for corrosion in a marine environment, in which the specimen is sprayed with a sodium chloride or other corrosive solution. Also, **salt-spray test.**

salt garden *Engineering*. a saltern in which seawater is evaporated in large shallow basins by solar heat.

salt glacier *Geology*. a flow of salt down the slopes of a salt plug along the course of the pre-existing structure under the influence of gravity.

salt gland *Vertebrate Zoology*. in certain marine turtles, snakes, and birds, a tubular gland near the eyes and nostrils that secretes a watery fluid rich in salt.

salt glaze *Engineering*. a glaze produced on stoneware by firing it in a kiln that contains salt.

salt grainer *Chemical Engineering*. a type of surface crystallizer used to produce coarse salt from brine; salt crystals are collected from a heated brine solution and then allowed to drain on an inclined surface.

salt haze *Meteorology*. an atmospheric haze created by the existence of finely divided particles of sea salt in the air.

salt hill *Geology*. a hill of salt having sinkholes and pinnacles at its summit.

Salticidae *Invertebrate Zoology*. a family of jumping spiders, arachnids in the order Araneae; active diurnal hunters that stalk and then jump on prey.

saltierra *Geology*. a salt deposit that is left behind after a shallow inland lake evaporates.

salting *Food Technology*. the injection or introjection of salt to preserve a food product or improve its taste. *Chemical Engineering*. see SALTING-OUT EFFECT.

salting out *Crystallography*. the precipitating, coagulating, or separating of a substance from a solution by the addition of salt.

salting-out effect *Chemical Engineering*. the growth of crystals on heating-surface walls of a material that has increasing solubility with increasing temperature. Also, SALTING.

salt lake *Hydrology*. an inland body of water with no outlet to the sea, containing a high concentration of dissolved salts. Also, BRINE LAKE.

salt marsh or **saltmarsh** *Ecology*. **1.** a habitat, characterized by high salinity, found at the edge of aquatic ecosystems, most commonly in tropical or subtropical regions. **2.** a type of salt-tolerant vegetation that grows on the mud banks formed at a river's mouth.

saltmarsh plain *Ecology*. a salt marsh that has been raised above the water level and has become dry land.

salt mine *Mining Engineering*. a mine established to work rock-salt deposits.

salt of sorrel see POTASSIUM BITARTRATE.

salt of tartar see POTASSIUM CARBONATE.

salt pan *Geology*. a small, shallow, undrained, natural depression containing a salt deposit produced by the accumulation and subsequent evaporation of water; or a shallow lake of somewhat salty water occupying such a depression. *Chemistry*. a large pan used to collect salt by evaporation of a salt solution such as seawater.

saltpeter *Inorganic Chemistry*. a popular name for potassium nitrate, KNO_3. (Literally, "salt of rock," because it is often found as a saltlike crust on rocks.)

saltpeter cave *Geology*. a cave containing ore deposits of calcium nitrate or other nitrate minerals, usually in quantities sufficient to be mined commercially.

saltpeter earth *Geology*. a cave deposit containing calcium nitrate in quantities sufficient to be mined for conversion to saltpeter.

salt pillow *Geology*. an early stage in the formation of a salt dome, during which the dome is rising from the depths of its source bed.

salt pit *Geology*. a hole in the ground in which sea water accumulates and evaporates, and from which salt is obtained; or the body of water occupying such a hole.

salt plasma *Hematology*. blood plasma to which a neutral salt has been added to prevent clotting.

salt plug *Geology*. the nearly equidimensional salt core that has risen from a mother salt bed through the surrounding sediments in a salt dome.

salt polygon *Geology*. a surface of salt on a playa, having as many as eight ridged sides formed by the expansive forces of crystallizing salt.

salt-spray climax *Ecology*. a climax community occurring along exposed Atlantic and Gulf seacoasts that is inhabited by plants able to withstand the effects of salt transported from the sea by winds.

salt spring *Geology*. a mineral spring containing a large amount of common salt; a spring of salt water. Also, BRINE SPRING.

salt tectonics *Geology*. the study of the structure and processes involved in the formation of salt domes. Also, HALOKINESIS.

salt velocity meter *Engineering*. **1.** a rate-of-flow meter that determines the transit time between two points of a small sample of salt or a radioisotope by measuring electrical conductivity at those points. **2.** a special class of accelerometer.

salt wall see SALT ANTICLINE.

salt water or **saltwater** *Hydrology*. water containing a large amount of salt; seawater. *Biology*. also, **salt-water.** living or found in seawater, as opposed to fresh water.

salt-water front *Oceanography*. the interface between salt water and fresh water in an aquifer or estuary.

salt-water intrusion *Hydrology*. the displacement of fresh water by salt water as a result of salt water having the greater density.

salt-water underrun *Oceanography*. a density current of seawater moving under fresh water in an estuary.

salt-water wedge *Oceanography*. an intrusion of seawater into a tidal estuary, characterized by a marked increase in salinity from top to bottom, so that the bottom layers penetrate farther upstream than the upper layers.

salt-water well *Petroleum Engineering*. an oil or gas well from which salt water flows after the petroleum has been exhausted.

salt weathering *Geology*. the effects of saline solutions or salt-crystal growth in the granular disintegration or fragmentation of rock material.

salt well *Engineering*. a bored or driven well from which brine is obtained.

saltworks see SALTERN.

saluretic *Medicine*. a substance that promotes renal excretion of sodium and chloride ions.

Salvadoraceae *Botany*. a family of small dicotyledonous trees and shrubs of the order Celastrales, often having leathery leaves, mustard oils, and interxylary phloem; native to Africa, Madagascar, the Middle East, and southeastern Asia.

salvage *Navigation*. **1.** the process of saving or recovering a ship or its cargo from distress or danger at sea. **2.** the goods that are so recovered, or compensation paid for this recovery. *Engineering*. a similar process of rescue or recovery on land. *Archaeology*. specifically, an emergency survey and excavation in a site that is threatened by immediate destruction due to human development or natural phenomena.

salvage ethnography *Anthropology*. the rapid study and assessment of a culture that is being quickly assimilated into a larger, more dominant culture, as when an isolated hunting-and-gathering society comes in contact with a modern technological society; e.g., the aboriginal Australians.

salvage group *Military Science.* in an amphibious operation, a naval task organization that rescues personnel and salvages materiel.
salvage pathway *Biochemistry.* a pathway employing compounds, produced in catabolism for biosynthetic purposes, that are not true intermediates of a regular biosynthetic pathway.
salvage procedure *Navigation.* a procedure for recovering and cataloging abandoned military hardware and vessels at sea for repair or scrap. *Engineering.* the listing of obsolete machinery to determine its disposition.
salvage value *Navigation.* the net worth of material reclaimed from a salvage operation. *Engineering.* **1.** the value of reusable material remaining after some industrial operation. **2.** the value assigned to a structure, machine, or vehicle that is used to calculate annual maintenance cost or present worth.
salvage vessel *Naval Architecture.* a vessel fitted for use in ship salvage and operations. Salvage vessels are commonly tug types fitted with special pumps, lifting gear, and diving gear.
salvarsan *Organic Chemistry.* $C_{12}H_{12}N_2As_2O_2$, a powder that deliquesces at 185°C; once widely used in the treatment of syphilis. Also, ARSPHENAMINE.
Salviniaceae *Botany.* a monogeneric family of free-floating water ferns of the order Salviniales that exhibit heterospory; characterized by three whorled leaves, two that float and one that is dissected and dangles in place of roots below, and a rudimentary vascular system; widespread in warm regions.
Salviniales *Botany.* an order of water ferns that includes two families: Salviniaceae and Azollaceae.
salvo *Military Science.* **1.** in naval gunfire support, a method of fire in which a number of weapons are fired at the same target simultaneously. **2.** in close air support or air interdiction, a method of delivery in which all ordnance of a specific type are released or fired simultaneously. **3.** the shots fired by these methods.
sal volatile see SMELLING SALTS.
Salyut *Space Technology.* a Soviet series of manned earth-orbiting space stations initiated in 1971.
SAM or **Sam** *Ordnance.* a shorter term for a surface-to-air missile.
SAM *S*-adenosylmethionine.
samara *Botany.* a one- or two-winged dry, indehiscent fruit, usually containing a single seed; produced by such trees as the maple, ash, and elm.
samaria see SAMARIUM OXIDE.
samarium *Chemistry.* a rare-earth (lanthanide) metal having the symbol Sm, the atomic number 62, an atomic weight of 150.36, a melting point of 1080°C, and a boiling point of 1800°C; a rare-earth element that is fifth in the lanthanide series in the periodic table; a brittle, gray metal used in the magnet alloy, $SmCo_5$, as a neutron absorber, and as a dopant for laser crystals. (Named for *samarskite,* in which it was discovered.)
samarium chloride *Inorganic Chemistry.* $SmCl_3 \cdot 6H_2O$, greenish-yellow hygroscopic crystals; soluble in water; loses $5H_2O$ at 110°C.
samarium-cobalt magnet *Electromagnetism.* a type of permanent magnet manufactured with the rare-earth element samarium; the magnet typically has low leakage and high resistance to demagnetization.
samarium oxide *Inorganic Chemistry.* Sm_2O_3, a cream-colored powder that absorbs moisture and carbon dioxide from the air; insoluble in water and soluble in acids; used as a catalyst in ethanol production. Also, **samarium sesquioxide.**
samarskite *Mineralogy.* $(Y,Ce,U,Fe^{+3})_3(Nb,Ta,Ti)_5O_{16}$, a black to brown monoclinic mineral with a vitreous to resinous luster, moderately to strongly radioactive, massive in habit or as prismatic crystals, having a specific gravity of 5.15 to 5.69 and a hardness of 5 to 6 on the Mohs scale; found in granite pegmatites. (Named after the Russian mining engineer V.E. *Samarskii*-Bykhovets, 1803–1870.)
Sambonidae *Invertebrate Zoology.* a family of tongue worms of the phylum Linguatulida that are endoparasites of mammals, having larvae that infest the viscera and adults that infest the nasal cavity.
Sam-D *Ordnance.* surface-to-air missile defense; a term for the U.S. Patriot and other such missile systems intended to intercept enemy missiles in midair.
same-sense mutation *Genetics.* a mutation, usually involving only one nucleotide switch in a codon, that produces a redundant amino acid message and does not change the protein being produced.
Samlet *Ordnance.* the code name for the Soviet SSC-2B missile.
SAMOS program *Space Technology.* a series of U.S. Air Force reconnaissance satellites. (An acronym for satellite and missile observation system.)
sample *Science.* a representative segment of a larger whole that is studied in order to gain information about the characteristics of the whole. *Statistics.* a subset of a population that is observed in order to infer properties of the whole population.
sample-and-hold *Electronics.* **1.** a circuit that periodically measures the input signal in an analog-to-digital converter, then produces an output signal corresponding to the input signal's measurement. **2.** in an analog-to-digital converter, a circuit that suspends an oscillating voltage before it is converted into another form. *Computer Programming.* a procedure in which a sample of an analog computer's input signal is placed in a capacitor for processing.
sample function see EXPERIMENT.
sampleite *Mineralogy.* $NaCaCu_5^{+2}(PO_4)_4Cl \cdot 5H_2O$, a blue to blue-green orthorhombic mineral occurring as crusts of minute lathlike crystals, having a specific gravity of 3.20 and a hardness of about 4 on the Mohs scale; found in highly oxidized ore at Chuquicamata, Chile.
sample oxidizer *Biotechnology.* an automated analyzer used with particulate samples containing radioactive carbon or hydrogen for scintillation counting; the sample is ignited and fully oxidized.
sample pages *Graphic Arts.* examples of pages printed by a compositor to demonstrate the solutions to typographic problems (such as tables, captions, or symbols) in a printed piece.
sample path *Mathematics.* the function consisting of the range of values of a stochastic process $\{X_t(\omega)\}$ for fixed ω, as t varies over some real interval T. Also, **sample function.**
sampler *Engineering.* **1.** a device used to collect small samples of materials for analysis. **2.** specifically, a device for measuring air pollution in various locales and at specified intervals to provide frequent monitoring.
sample space *Statistics.* the set of all possible outcomes of an experiment. *Mathematics.* the measure space Ω of points ω, together with a sigma algebra F and a measure μ, that constitute a probability space. The points ω are called **sample points** and the sets of F (subsets of Ω) are called **events.**
sampling *Science.* the process of testing or analyzing samples. *Statistics.* the process of selecting a subset of a population. *Industrial Engineering.* the process of inspecting random items in a lot according to a set plan for the purpose of quality control. *Robotics.* the process of monitoring a function to obtain a set of values for particular moments in time. *Engineering.* **1.** a technique for checking air pollution by monitoring at various time intervals and in different locations. **2.** the collecting of small amounts of soil for laboratory analysis to determine the feasibility of testing the soil in situ for seismic exploration. **3.** a process in which a series of instantaneous wave values are obtained.
sampling bottle *Engineering.* a container that traps a water sample and transports it to the surface without contaminating it.
sampling circuit *Electrical Engineering.* a circuit whose output is a sampling of the values of the input at a series of points in time.
sampling distribution of the mean *Statistics.* the distribution of the means of all possible samples of a given size from the population.
sampling error see RANDOM ERROR.
sampling fraction *Statistics.* the portion of a given population represented by a selected sample.
sampling frame *Statistics.* a list of all the elements in the population considered in a survey. Also, FRAME.
sampling gate *Electronics.* a circuit that retrieves information from a signal only after another signal identifies it.
sampling plan *Industrial Engineering.* a procedure for sampling and testing for quality control, including the standards set out for acceptance or rejection of lots based on sampling.
sampling probe *Engineering.* **1.** a small tube containing a sensing element that can be lowered into a borehole to obtain samples. **2.** a probe that absorbs trace gases from an area under pressure and passes it to a leak detector at a reduced pressure.
sampling risk *Industrial Engineering.* the statistical chance that quality-control sampling will give an incorrect signal, leading to rejection of acceptable lots or acceptance of defective lots.
sampling spark chamber *Nucleonics.* a narrow-gap spark chamber that gives the coordinates of a single point along the track of an ionizing particle through the gap.
sampling switch see COMMUTATOR SWITCH.
sampling theorem *Telecommunications.* a theorem stating that a signal that varies continuously with time can be defined by its values at an infinite sequence of equally spaced times if the frequency of these times is more than twice the greatest frequency component of the signal. Also, SHANNON'S SAMPLING THEOREM.

sampling theory *Statistics.* a theory of inference in which probabilistic statements are based only on the randomness in the sampling mechanism.

sampling voltmeter *Engineering.* a voltmeter that uses various sampling techniques to measure high-frequency signals or signals in noise.

samsonite *Mineralogy.* $Ag_4MnSb_2S_6$, a black metallic monoclinic mineral occurring as prismatic crystals that may appear red in transmitted light, having a specific gravity of 5.51 and a hardness of 2.5 on the Mohs scale.

Samuelsson, Bengt born 1934, Swedish biochemist; with Sune K. Bergström, Swedish biochemist, and John R. Vane, British biochemist, shared the Nobel Prize for research in prostoglandins.

Samythinae *Invertebrate Zoology.* a subfamily of sessile polychaete annelid worms, in the family Ampharetidae, having a conspicuous dorsal membrane and retractile oral tentacles.

SAN *Organic Chemistry.* an acronym of styrene-acrylonitrile resin, a solid thermoplastic copolymer.

SAN *Aviation.* the airport code for Lindbergh Field, San Diego.

sanakite *Petrology.* a glassy variety of andesite containing bronzite, augite, magnetite, and a few large crystals of plagioclase and garnet.

sanatron circuit *Electronics.* a circuit consisting of two pentodes and two diodes with variable time delays.

sanbornite *Mineralogy.* $BaSi_2O_5$, a white to colorless orthorhombic mineral occurring as crude tabular crystals, having a specific gravity of 3.77 and a hardness of 5 on the Mohs scale.

sanction *Anthropology.* any institution that controls behavior in a society, either positive or negative; may be a formal sanction such as a law or an informal one such as eating etiquette.

sand *Geology.* **1.** a small, somewhat rounded fragment or particle of rock ranging from 0.05 to 2 mm in diameter, and commonly composed of quartz. **2.** a loose aggregate or more or less unconsolidated deposit, consisting essentially of sand-sized rock particles or medium-grained clastics. **3. sands.** a tract or region composed primarily of sand.

Sandal *Ordnance.* a Soviet medium-range ballistic missile powered by a liquid propellant motor and probably equipped with radio-inertial guidance; it delivers a nuclear or conventional warhead at a speed of 4300 mph and a range of 1100 miles; officially designated **SS-4.**

Sandalidae see RHIPICERIDAE.

sandalwood *Botany.* **1.** any of several Asian trees of the genus *Santalum,* family Santalaceae, noted for their aromatic wood; especially *S. album,* an Indian evergreen characterized by ovate leaves and flowers that change from yellow to red. **2.** the wood of such trees. **3.** any of various other trees noted for their aromatic wood, or the wood of such trees.

sand apron *Geology.* a sand deposit occurring along the shore of a reef lagoon.

sandarac *Botany.* a coniferous tree, *Tetraclinis articulata* (*Callitrus quadrivalvis*), that yields a resin and a hard, dark-colored wood used in building; found in northwest Africa.

sandarac gum *Materials.* a brittle, yellowish, aromatic resin derived from the bark of the sandarac tree; used in making varnish and as incense.

sand auger see DUST WHIRL.

sand avalanche *Geology.* the movement of large sand masses down the face of a dune.

sandbag *Engineering.* a bag filled with sand or small debris, used for building pack walls, for filling cavities behind timber, steel, or concrete roadway linings, or for flood control work near weakened dikes, etc. *Geology.* an accumulation of glacial debris occurring in the roof of a coal seam that was deposited after the coal formed.

sandbank *Geology.* **1.** a sandbar or a submerged ridge of sand in a body of water, usually exposed at low tide. **2.** a large accumulation of sand, especially in a shallow region near the shore.

sandbar *Geology.* a bar or low ridge of sand along the shore of a lake or sea that is built up to the surface by water currents or wave action. Also, SAND REEF.

sandblasting *Engineering.* a method of abrasive cleaning by spraying sand entrained in a high velocity air or stream upon the surface. *Geology.* a type of abrasion resulting from the impact of sand-sized grains of hard minerals driven against an exposed stationary surface by wind. Also, **sandblast action.**

sand boil *Hydrology.* a spring that bubbles through a river levee, ejecting sand and water as the river water is forced through permeable layers of sands and silts below the levee during flood stage. Also, BLOWOUT.

sand-cast *Metallurgy.* the attribute of a casting produced in a sand mold.

sand casting *Materials Science.* a process in which shapes are formed by pouring molten metal into a cavity made of sand.

sand cay see SANDKEY.

sand cone *Geology.* **1.** a cone-shaped sand deposit. **2.** a low, cone-shaped mound of snow or ice on a glacier, having a protective veneer of sand.

sand control *Metallurgy.* in sand casting, the quality control of the molding sand.

sand count *Petroleum Engineering.* a calculation of the total thickness of the permeable zone of a petroleum reservoir, excluding impermeable zones such as shale streaks; may be derived from electrical logs.

sand crack *Veterinary Medicine.* a crack or fissure starting at the bearing of a horse's hoof; caused by dryness of horn, it rarely causes lameness.

sand crystal *Geology.* a large crystal, containing as much as 60% inclusions of detrital sand, that develops in an incompletely cemented sandstone, usually during sedimentation.

sand dab or **sanddab** *Vertebrate Zoology.* any Pacific Ocean flatfish of the genus *Citharichthys* of the family Bothidae, favored as a food fish.

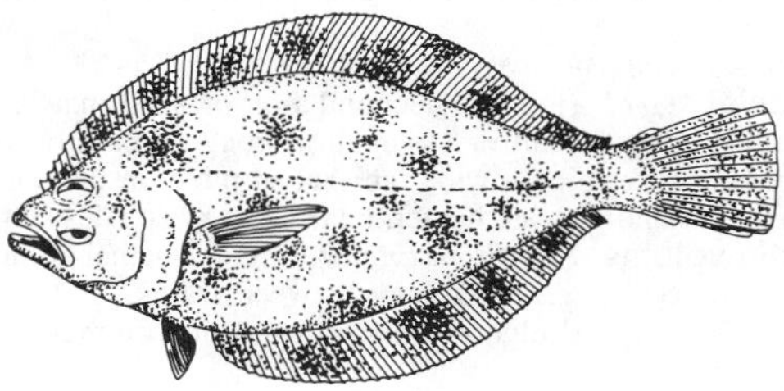

sand dab

sand devil see DUST WHIRL.

sand dike *Geology.* a sedimentary dike composed of sand that has been forced or injected upward into a fissure.

sand dollar *Invertebrate Zoology.* any of various irregular echinoids in the order Clypeasteroidea, having round, flat tests covered with fine spines.

sand drain *Civil Engineering.* a vertical hole filled with sand to facilitate vertical drainage of stratified soils and to relieve pore water pressures in compressible soils.

sand drift *Geology.* **1.** any surface movement of windblown sand, occurring in deserts or along coastlines. **2.** an accumulation of sand on the leeward side of a fixed obstruction.

sand drip *Geology.* a spherical or crescent-shaped feature on beach sand, formed by the quick absorption of overwash.

sand dune *Geology.* a mound or ridge of windblown sand frequently occurring along low-lying seashores above the level of high tide.

sand dunes

sand eel *Vertebrate Zoology.* any slender fish of the family Ammodytidea that burrows into submarine sand. Also, SAND LANCE.

Sand equation *Physical Chemistry.* an equation governing the current-time-concentration relations in constant current electrolysis in an unstirred solution.

sander *Mechanical Engineering.* a portable, motor-driven apparatus that passes a sheet of sandpaper or other abrasive over a surface to produce a smooth finish.

sandfall see SLIP FACE, def. 1.

sand filter *Civil Engineering.* a bed of sand arranged in layers of increasing grain size from top to bottom; used widely in purification of public water supplies and in lime soda water softeners.

sand finish *Engineering.* a buffing process using rottenstone and pumice to produce high polishes on such materials as German silver and white metal.

sand flat *Geology.* a sandy, barren tidal flat.

sand flood *Geology.* a large mass of sand that moves or is carried along a desert floor.

sandfly *Invertebrate Zoology.* **1.** a small, very hairy biting fly in the subfamily Phlebotominae, especially in the genus *Phlebotomus,* that breeds in damp sand; a carrier of phlebotomus fever. **2.** any of various other two-winged, blood-sucking flies in the families Heleidae and Simuliidae.

sand gall see SAND PIPE.

sand glacier *Geology.* **1.** an accumulation of sand that is moved by wind up the side of a hill or mountain, through a pass or saddle, and then spreads out forming a wide fan-shaped plain on the opposite side. **2.** a horizontal sand plateau that terminates at a steep talus slope.

sand-grain volume *Petroleum Engineering.* the actual volume of sand grains in a reservoir, excluding voids between grains.

sand hill *Geology.* a ridge of windblown sand commonly formed in desert regions.

Sandhoff disease *Genetics.* a hereditary disease resulting from a recessive allele on chromosome 5; homozygous individuals suffer from a deficiency of the enzyme hexosaminidase A, resulting in severe deterioration of the nervous system with symptoms similar to those of Tay-Sachs disease.

sandhog *Engineering.* a person who works underground or on underwater construction projects, usually in a medium of compressed air, as in a tunnel.

sand hole *Geology.* a small depression on a beach, having a raised border that is formed by waves expelling air from a previously waterlogged sand mass.

sand hopper *Invertebrate Zoology.* any of various small intertidal amphipod crustaceans that burrow and jump. Also, BEACH FLEAS, BEACH HOPPERS.

sand horn *Geology.* a triangular sand deposit extending from the shore into shallow water.

sanding *Engineering.* **1.** the process of smoothing or polishing with sand, sandpaper, or other abrasive. **2.** the process of covering or mixing with sand.

sanding block *Mechanical Devices.* a block of wood upon which an abrasive is bound, used for smoothing rough wooden surfaces.

sanding disk *Mechanical Devices.* a disk or drum lined with an abrasive for a power sanding application. Also, **sanding drum.**

sandkey *Geology.* a small sandy island lying parallel to the shore. Also, SAND CAY.

sand lance see SAND EEL.

sand levee see WHALEBACK DUNE.

sand-lime brick *Materials.* a hard brick manufactured by high-pressure molding and baking of silica sand with about 6% hydrated lime and water.

sand line *Mining Engineering.* a wire line used to raise and lower a bailer or sand pump to remove cuttings from a borehole.

sand load *Electromagnetism.* a transmission line attenuator or terminator used to dissipate power where the space between the conductors is filled with a sand-carbon mixture.

sand lobe *Geology.* a rounded deposit of sand covering an area that extends from the shore into shallow water.

Sandmeyer reaction *Organic Chemistry.* a reaction used to replace a diazonium group in a diazonium salt with a halide or cyonide group, by treatment with heat and cuprous chloride or cuprous bromide.

sand mill *Mechanical Engineering.* a mill in which grains of sand are used to grind aggregate down in size.

sand mist see BAI.

sandpaper *Materials.* **1.** heavy paper having sand or a similar abrasive substance glued to one side, used for smoothing or polishing. **2.** to use such paper for smoothing or polishing.

sand pavement *Geology.* a surface cover of sand, derived from coarse-grained sand ripples, that develops on the lower, windward side of a dune or rolling sand area during a period of intermittent light, variable winds.

sand pile *Civil Engineering.* a foundation consisting of sand that is rammed into a hole left by the removal of a solid pile, mostly used in soft soils.

sand pipe *Geology.* a roughly cylindrical, vertical structure filled with sand and some gravel in sedimentary rocks. Also, SAND GALL.

sandpiper *Vertebrate Zoology.* any bird of the family Scolopacidae, being a small shorebird with long legs and a long slender beak that breeds mostly in arctic regions but migrates into temperate latitudes.

sandpiper

sandpit *Civil Engineering.* an excavation made to obtain sand for use in paving, construction, or other purposes.

sand plain *Geology.* **1.** a plain covered with sand. **2.** a small, flat or gently sloping surface of sand deposited by the flow of meltwater streams from a glacier.

sand pump *Mechanical Engineering.* a centrifugal-type pump that removes sand- and gravel-laden liquids from a bore-hole without clogging. Also, SLUDGE PUMP.

sand reef see SANDBAR.

sand reel *Mechanical Engineering.* a drum hoist used in drilling operations to raise and lower a sand pump.

sand return *Petroleum Engineering.* the generally problematic return of injected sand to the wellbore subsequent to fracturing operations in the formation.

sand ridge *Geology.* **1.** any low, elongated ridge of sand occurring at some distance from the shore, either under or rising above the water. **2.** one of a series of long, very low, broad, parallel elevations of sand, thought to represent the eroded remains of former sand dunes. **3.** a crescent-shaped landform on a sandy beach. **4.** see SAND WAVE.

sand river *Geology.* a river that deposits a large portion of its load of sand along the middle of its course, to be removed later by the wind.

sandrock *Geology.* **1.** a sandstone that is not firmly cemented. **2.** see SANDSTONE.

sand roll see PSEUDONODULE.

sand run *Geology.* **1.** a movement of dry sand that resembles fluid motion. **2.** a body of dry sand in motion.

sand sea *Geology.* **1.** an extensive arrangement of various types of sand dunes in a sandy area, characterized by the absence of directional indicators and a wavelike configuration consisting of dunes separated by troughs. **2.** the relatively level, rain-smoothed plain of volcanic material on the floor of a caldera.

sand shadow *Geology.* an accumulation of sand formed in the shelter of and directly behind a fixed obstruction on the leeward side.

sandshale *Geology.* a sedimentary deposit characterized by thin, alternating beds of sandstone and shale.

sand-shale ratio *Geology.* the ratio in a stratigraphic section of the thickness or percentage of sandstone to shale, not including the nonclastic material.

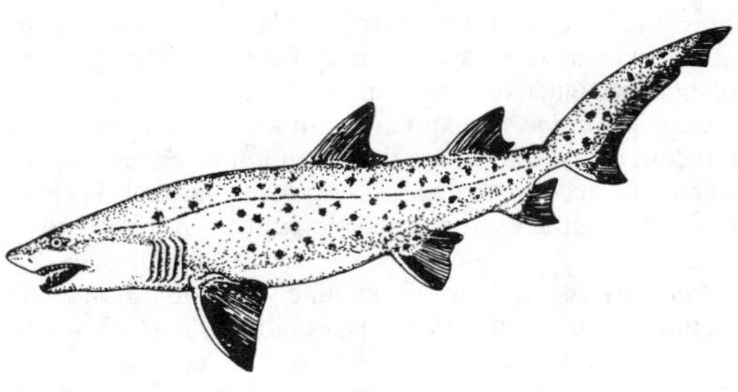

sand shark

sand shark *Vertebrate Zoology.* any of several sharks of the family Carcharinidae, including tiger sharks, requiem sharks, and blue sharks; characterized by shallow-water predation. Also, SAND TIGER. (So named because they are often seen close to the shore.)

sand sheet *Geology.* a thin layer of coarse sand or fine gravel having an extremely flat surface.

sand slinger *Mechanical Engineering.* an apparatus using centrifugal force to project sand at high velocities into molds.

sand snow *Hydrology.* dry, loosely packed snow that has fallen at a very low temperature (usually below –25°C) and has a surface consistency resembling dry sand.

sandspit *Geology.* a spit composed mainly of sand.

sand splay see FLOODPLAIN SPLAY.

sand stargazer see DACTYLOSCOPIDAE.

sandstone *Petrology.* a cemented or otherwise compacted detrital sedimentary rock consisting primarily of quartz and featuring sand-size grains between 0.06 and 2 mm in diameter; the consolidated equivalent of sand.

sandstone

sandstone dike *Geology.* a sedimentary dike consisting of sandstone or lithified sand and ranging in thickness from a fraction of an inch to several inches and in length up to several miles.

sandstone sill *Geology.* a tabular body of sandstone that has been forcibly injected parallel to the pre-existing structure or bedding.

sandstorm *Meteorology.* a strong wind that carries sand through the air, the diameter of most of the particles ranging from 0.08 to 1 mm; the sand particles are mostly confined to the lowest ten feet of this storm; caused by strong winds enhanced by surface heating, which tend to form during the day.

sand streak *Geology.* the linear, low ridges having a symmetric cross section that form at the interface of sand and air or water, and that are parallel to the direction of flow.

sand stream *Geology.* a small delta of sand deposited at the mouth of a gully, or an accumulation of sand along the bed of a small creek, produced by torrential rain.

sand strip *Geology.* a narrow, elongated sand ridge at each horn of a sand dune, extending a considerable length downwind.

sand tiger see SAND SHARK.

sand trap *Engineering.* a device in a conduit that traps sand or particles of soil transported by water.

sand tube see FULGURITE.

sand tumor see PSAMMOMA.

sandur see OUTWASH PLAIN.

sand verbena *Botany.* any of several low, mostly trailing annual plants of the genus *Abronia,* in the family Nyctaginaceae, that resemble (but are not related to) verbena and have fragrant pink, white, or yellow flowers.

sandwash *Geology.* a sandy or gravelly stream bed, lacking vegetation and filled with water only during sudden, heavy rainstorms.

sand wasp *Invertebrate Zoology.* any of various wasps of the subfamily Bembicinae that burrow in coastal sand.

sand wave *Geology.* **1.** a large, primary wavelike structure on the upper surface of a sedimentary bed, produced by high-velocity air or water currents. **2.** any wavelike bedform in sand.

sand wedge *Geology.* a vertical, wedge-shaped body of sand whose apex points downward.

sand wheel *Mechanical Engineering.* a wheel equipped with steel buckets about its circumference that is used for lifting sand and sludge out of a sump.

sandwich *Materials.* **1.** a layer of a material between two layers of another material, arranged in the manner of meat or other food between two pieces of bread in a sandwich. **2.** a composite material constructed of a lightweight low-density material surrounded by dense solid layers; it combines overall light weight with excellent stiffness.

sandwich construction *Design Engineering.* a composite construction of alloys, plastics, or wood with a foam layer formed into honeycomb patterns that is laminated and glued between two outer sheets. Also, **sandwich laminate.**

sandwich heating *Materials Science.* a process of heating both sides of a thermoplastic sheet simultaneously before shaping it.

sandwich molding *Materials Science.* in plastics, a process by which a skin core structure, typically of two different polymers, is produced in a twin injection process.

sandwich rolling *Metallurgy.* a process of rolling more than one metallic strip of sheet simultaneously.

sandwich technique *Immunology.* a technique used to determine the presence of antibodies or antibody-producing cells, in which the antibodies are attached to a fluorescent dye in tissue sections or smear samples.

Sandy *Military Science.* a term for a search-and-rescue operation.

sandy bentonite see ARKOSIC BENTONITE.

sandy chert *Petrology.* chert occurring in sandy beds and featuring oolithlike structures; produced when silica replaces cement or fills pore spaces.

sane *Psychology.* judged by legal authority to have normal mental health and abilities and thus capable of handling one's own affairs and legally responsible for one's actions.

Sanfilippo's syndrome *Medicine.* any of four forms of mucopolysaccharidosis characterized by the excretion of heparan sulfate in the urine; the onset is from two to six years of age, and the syndrome is marked by rapid mental deterioration and relatively mild somatic symptoms.

Sangamon *Geology.* the third interglacial stage of the Pleistocene epoch in North America, occurring between the Illinoian and Wisconsin glacial stages. Also, **Sangamonian.**

Sanger, Frederick born 1918, English biochemist; Nobel Prize for establishing amino acid sequence of a protein (insulin) and determining chemical structure of pieces of DNA.

Sanger method *Molecular Biology.* a method in which DNA is sequenced by radioactively labeling strands in enzymatic reactions terminating specifically at positions corresponding to each given base.

sangui- a combining form meaning "blood."

sanguimotor *Hematology.* of or relating to the circulation of blood.

sanguine [sang´gwin; sang´gwīn´] *Hematology.* abounding in blood. *Psychology.* **1.** one of the four basic personality types defined in ancient and medieval times, described as being characteristically cheerful and optimistic; said to result from the predominance of blood over the other bodily humors. Also, **sanguine type. 2.** of or relating to this personality type.

sanguineous *Hematology.* relating to or abounding in blood. Also, **sanguinous.**

sanguivorous *Zoology.* feeding mainly or exclusively on blood.

sanidaster *Invertebrate Zoology.* a rod-shaped sponge spicule having spines at intervals along its length.

sanidine *Mineralogy.* $(K,Na)(Si,Al)_4O_8$, a colorless or whitish monoclinic mineral of the feldspar group, with disordered Al-Si arrangement, occurring in glassy, prismatic crystals, having a specific gravity of 2.56 to 2.62 and a hardness of 6 on the Mohs scale; of common occurrence in acidic and alkaline volcanic and plutonic igneous rocks.

sanidinite *Petrology.* an igneous rock consisting primarily of sanidine or other alkali feldspar.

sanitarium *Medicine.* **1.** an institution for the promotion of health. **2.** an older term for a mental-health facility. Also, **sanitorium.**

sanitary *Medicine.* relating to, promoting, or conducive to health; used especially to denote cleanliness.

sanitary engineering *Civil Engineering.* the part of civil engineering related to public health and the environment, such as water supply, sewage, and industrial waste.

sanitary landfill *Civil Engineering.* a facility used for burying solid wastes, to seal off odors and stave off vermin, rats, and other disease carriers.

sanitary sewer *Civil Engineering.* a conduit or pipe that carries sanitary wastes excluding storm and ground waters.

sanitation *Civil Engineering.* the control of all those factors in the physical environment that exercise or may exercise deleterious effects on the environment; the establishment or maintenance of healthy environmental conditions.

sanitize *Medicine.* to clean and sterilize. Thus, **sanitized, sanitizer.**

sanity *Psychology.* a legal term for a condition in which an individual is judged to be capable of handling his own affairs and can be held responsible for his actions.

sannaite *Petrology.* an extrusive, porphyritic, igneous rock consisting of phenocrysts of barkevikite, clinopyroxene, and biotite in a fine-grained groundmass of alkali feldspar, acmite, chlorite, calcite, and mica.

Sannoisian see TONGRIAN.

sansar *Meteorology.* a northwest wind of Iran.

sansicl *Geology.* a loose, unconsolidated sediment made up of sand, silt, and clay, in which no constituent forms 50% or more of the whole mixture.

sans serif *Graphic Arts.* of a typeface, having no serifs; e.g., Helvetica.

Santa Ana *Meteorology.* a hot, dry, northeasterly desert wind that occurs west of the Sierra Nevada mountains. (Noted especially in the pass and river valley of *Santa Ana,* California.)

Santa Gertrudis *Agriculture.* a breed of beef cattle that is deep red in color and is valued for its tolerance of hot climates. (From the *Santa Gertrudis* section of the King Ranch in Kingsville, Texas, where the breed was developed.).

santal or **santalum** see SANDALWOOD.

Santalaceae *Botany.* the sandalwood family, a dicotyledonous family of the order Santales, consisting of 36 genera and 500 species of primarily tropical or dry hemiparasitic trees, shrubs, and herbs; the fruit is a nut or drupe.

Santalales *Botany.* an order of angiosperm plants consisting of ten families: Medusandraceae, Dipentodontaceae, Olacaceae, Opiliaceae, Santalaceae, Misodendraceae, Loranthaceae, Viscaceae, Eremolepidaceae, and Balanophoraceae.

santalol *Organic Chemistry.* $C_{15}H_{24}O$, a colorless sesquiterpene alcohol; insoluble in water and soluble in alcohol; boils at 165–170°C (15 torr); derived from sandalwood oil, and used in perfumes. Also, ARHEOL.

Santonian *Geology.* a European geologic stage of the Upper Cretaceous period, occurring after the Coniacian and before the Campanian.

α-santonin *Organic Chemistry.* $C_{15}H_{18}O_3$, a toxic, colorless, tricyclic, sesquiterpene keto-lactone; very slightly soluble in water and soluble in alcohol; melts at 174°C and sublimes at boiling point; used in medicine. Also, **santonic acid.**

santorinite *Petrology.* **1.** a light-colored, extrusive igneous rock containing 60–65% silica, with calcic plagioclase as the only feldspar. **2.** a hypersthene andesite consisting of plagioclase crystals with labradorite cores and sodic rims in a groundmass containing microlites of sodic oligoclase.

Santorio, Santorio 1561–1636, Italian physician; developed a clinical thermometer; the first to experiment with metabolism.

sanukite *Petrology.* an andesite featuring orthopyroxene and garnet as the mafic minerals and andesine as the plagioclase, all in a glassy groundmass.

SAO *Aviation.* the airport code for Sao Paolo, Brazil.

Sao Francisco *Geography.* a river in eastern Brazil, flowing 1987 miles northeast into the Atlantic.

sap *Botany.* the vascular fluid of woody plants, containing mineral salts and sugar dissolved in water; often applied specifically to visibly secreted plant fluids, such as pine resin or maple sugar.

SAP *Acoustical Engineering.* a feature in a television receiver or videocassette recorder that allows for the reception of either of two different audio signals to accompany the same video material, such as English and Spanish soundtracks for the same film. (An acronym for second audio program.)

sapele *Materials.* the dark, mahoganylike wood of various African trees of the genus *Entandrophragma*, especially, *E. cylindricum*; used in making furniture. Also, ABOUDIKRO.

saphena *Anatomy.* either of two large superficial veins of the leg.

saphenous nerve *Anatomy.* a branch of the femoral nerve that supplies the leg, ankle, and foot.

Sapindaceae *Botany.* a family of dicotyledonous plants of the order Sapindales consisting of 145 genera and about 1325 species of trees, shrubs, vines, and lianes found in tropical and warm climates; the fruit is fleshy or dry, and the seeds have arils.

Sapindales *Botany.* an order of angiosperms consisting of 15 families: Staphyleaceae, Melianthaceae, Bretschneideraceae, Akaniaceae, Sapindaceae, Hippocastanaceae, Aceraceae, Burseraceae, Anacardiaceae, Julianiaceae, Simaroubaceae, Cneoraceae, Meliaceae, Rutaceae, and Zygophyllaceae.

Sapir, Edward 1884–1939, American anthropologist; studied American Indian languages; major contributions to descriptive linguistics.

Sapir-Whorf hypothesis see WHORF HYPOTHESIS.

sapling *Botany.* a young tree.

saponiferous *Botany.* containing or bearing the glycoside saponin.

saponification *Chemistry.* a chemical reaction in which esters are hydrolyzed under basic conditions to give an alcohol and a salt of the acid; this process is usually carried out on fats, and the sodium salt formed is called a soap.

saponification equivalent *Chemistry.* the weight of an ester that consumes one gram-equivalent weight of base in saponification.

saponification number *Analytical Chemistry.* the number of milligrams of potassium hydroxide required to hydrolyze the fat, oil, or wax in a one-gram sample.

saponify *Chemistry.* to subject to a process of saponification.

saponin *Organic Chemistry.* any of a group of plant glycosides that foam in water; used in detergents, as an emulsifier, and as a foaming agent in beverages.

saponite *Mineralogy.* $(Ca/2,Na)_{0.3}(Mg,Fe^{+2})_3(Si,Al)_4O_{10}(OH)_2 \cdot 4H_2O$, a soft, soapy, monoclinic mineral of the smectite group, occurring in fine-grained masses of various colors; having a specific gravity of 2.24 to 2.30 and a hardness of 1 to 2 on the Mohs scale; found in serpentine and basaltic rocks. Also, GRIFFITHITE, PIOTINE, SOAPSTONE.

Sapotaceae *Botany.* a dicotyledonous family belonging to the order Ebenales, consisting of 116 genera and about 1100 species of mainly tropical trees and shrubs, typically producing a latex and a fleshy, indehiscent fruit.

sapote *Botany.* **1.** a tree of Mexico and Central America, *Pouteria sapota,* of the family Sapotaceae. **2.** the sweet edible fruit of this tree.

sapotoxin *Toxicology.* any of a group of toxic saponins found in such plants as *Quillaja saponaria* (of the family Rosaceae) and *Saponaria officinalis* (Caryophyllaceae).

sappare see KYANITE.

sapphire *Mineralogy.* any gem variety of corundum, Al_2O_3, that is not ruby red; most commonly applied to the transparent blue variety, whose color is due to traces of cobalt, chromium, and titanium; found, together with ruby, predominantly in Asia.

sapphire quartz *Mineralogy.* a rare, opaque variety of quartz, colored indigo blue by included fibers of silicified crocidolite; used as a semiprecious gemstone.

sapphirine *Mineralogy.* $(Mg,Al)_8(Al,Si)_6O_{20}$, a rare, green or pale-blue monoclinic mineral occurring most commonly in granular form and occasionally as tabular crystals, having a specific gravity of 3.4 to 3.5 and a hardness of 7.5 on the Mohs scale; a high-temperature metamorphic mineral.

sapping *Geology.* a natural erosion process occurring along the base of a cliff in which the wearing away of the softer, underlying layers weakens the upper mass, causing large blocks to break away and fall from the face of the cliff. Also, CLIFF EROSION, UNDERMINING.

Saprist *Geology.* a suborder of the soil order Histosol, characterized by a high content of almost completely decomposed plant matter and water saturation.

sapro- a combining form meaning "rotten" or "putrid;" often used to denote decay or a decaying material.

saprobe *Botany.* an organism, especially a fungus, that derives nourishment from decaying or decomposing organic matter; a fungal saprophyte.

Saprodinium *Invertebrate Zoology.* a genus of ciliate protozoans found in deposits in polluted water; very important in pollution ecology.

saprogen *Biology.* an organism that lives on nonliving organic matter.

saprogenic *Biology.* formed by or causing decay. Also, **saprogenous.**

saprogenic ooze *Geology.* an ooze derived from putrefying organic matter. Also, **saprogenous ooze.**

Saprolegnia *Mycology.* a genus of fungi of the order Saprolegniales, found in fresh water and in soil; it includes pathogens of aquatic vertebrates such as fish.

Saprolegniaceae *Mycology.* the water molds, a family of fungi of the order Saprolegniales, most species of which live in soil or fresh water, feeding on nonliving plant and animal remains; some are parasitic.

Saprolegniales *Mycology.* an order of fungi of the class Oomycetes, occurring in marine water, fresh water, or soil; some species are parasites of algae and fish.

saprolite *Geology.* a soft, reddish or brownish, earthy, clay-rich decomposed rock formed in place by chemical weathering, and generally retaining some evidence of the original structures that were present in the parent rock. Also, **saprolith.**

sapropel *Geology.* a slimy, unconsolidated organic ooze or sludge, derived mainly from plant remains, that develops under anerobic conditions on the shallow bottoms of lakes or seas.

sapropel-clay *Geology.* a sedimentary deposit having a higher percentage of clay than sapropel.

sapropelic coal *Geology.* any coal derived from the putrefaction of organic residues under anerobic conditions in bodies of stagnant or standing water. Also, **sapropelite.**

sapropel-peat see PEAT-SAPROPEL.

saprophage *Biology.* an organism that lives on decaying organic matter. Thus, **saprophagous.**

saprophile *Biology.* an organism that thrives in an environment of decaying organic matter. Thus, **saprophilous, saprophily.**

saprophyte *Botany.* a plant that derives nourishment primarily from decaying organic matter, such as certain fungi and bacteria; the term *saprobe* is often used for a fungal saprophyte. Thus, **saprophytic**.

Saprospira *Bacteriology.* a genus of filamentous bacteria of the family Cytophagacea that exhibit gliding motility, reproduce by fragmentation, and occur in aquatic habitats.

saprotroph *Biology.* a heterotrophic saprophyte or saprozoite that feeds on dead organic matter.

saprovore *Zoology.* animals that feed mainly or exclusively on dead or decaying matter.

saprozoite *Zoology.* an animal that feeds on dead or decaying organic matter. Thus, **saprozoic.**

sapsucker *Vertebrate Zoology.* any of several species of North American woodpeckers of the family Picidae, noted for drilling rows of holes in trees and feeding on oozing sap and insects.

sapsucker

sapwood *Botany.* a collective term for tissue of the secondary xylem system, distributed by the vascular cambium in an outer ring around the dead heartwood; active in the transport and conduction of sap. Although often referred to as living tissue, only a small percentage of sapwood cells are alive.

Sapwood *Ordnance.* an early Soviet intercontinental ballistic missile powered by a liquid-propulsion sustainer with strap-on boosters; officially designated **SS-6.**

Sapygidae *Invertebrate Zoology.* a family of rare wasps, hymenopteran insects belonging to the superfamily Scolioidea; the larvae parasitize leafcutter bees.

sarah or **SARAH** *Navigation.* a radio homing device used in recovering spacecraft at sea; it was originally designed for personnel rescue. (An acronym for search and rescue and homing.)

saran *Matrials Science.* the commercial name for a vinylidene chloride polymer that forms a tough coherent sheet with the properties of high tensile strength, chemical inertness, and resistance to abrasion when spread on a smooth surface.

sarc- a combining form meaning "flesh."

sarca *Meteorology.* a strong north wind that blows over Lake Garda in Italy.

Sarcina *Bacteriology.* a genus of Gram-positive, spheroid, anaerobic bacteria of the family Micrococcaceae, characterized by the manner in which cells divide into three planes and form clusters of at least eight cells; occur in grains and soil.

Sarcinochrysidales *Botany.* a small order of marine algae of the class Chrysophyceae in which the swarmer structure resembles that of the brown algae Phaeophyceae.

Sarcinosporon *Mycology.* a genus of imperfect fungi belonging to the class Oomycetes; its only species, *S. inkin,* is found in human skin wounds.

sarco- a combining form meaning "flesh."

sarcocarcinoma see CARCINOSARCOMA.

sarcocarp *Botany.* a fleshy fruit or the fleshy mesocarp of certain fruits, such as the plum or peach.

sarcochore *Botany.* a plant that disperses fleshy disseminules.

Sarcocystis *Invertebrate Zoology.* a genus of parasitic protozoans belonging to the order Sarcosporidia that encyst in the muscle tissues of mammals.

Sarcodiaceae *Botany.* a small, poorly characterized family of red algae of the order Gigartinales, having branched thalli of multiaxial construction.

Sarcodina *Invertebrate Zoology.* a superclass of aquatic protozoans with no pellicle, some having a secreted test, that use temporary pseudopodia to move and to gather food; free-living and parasitic forms; includes Amoebae, Foraminifera, and Radiolaria.

sarcoglia *Cell Biology.* a granular substance located at the junction of a nerve filament and a muscle fiber. *Neurology.* an accumulation of neurilemma cells at the motor endplate.

sarcoid *Anatomy.* fleshlike. *Medicine.* see SARCOIDOSIS. *Oncology.* a sarcomalike tumor.

sarcoidosis *Medicine.* a chronic, progressive, infectious disease of unknown cause in which granulomatous lesions form in the skin, lymph, eyes, lungs, bones, and other organs.

Sarcolaenaceae *Botany.* a family of usually evergreen dicotyledonous trees and shrubs of the order Theales, often having stellate hairs and having mucilage cells in the parenchymatous tissue; limited to Madagascar.

sarcolemma *Histology.* the thin layer of connective tissue surrounding a muscle fiber.

sarcoleukemia see LEUKOSARCOMA.

sarcoma *Oncology.* a malignant neoplasm arising in soft tissue and composed of closely packed cells embedded in a fibrillar or homogeneous substance.

sarcoma botryoides *Oncology.* a grape-shaped tumor found most often in young children or infants in the upper vagina, cervix of the uterus, or neck of the urinary bladder.

sarcomagenic *Oncology.* causing a sarcoma.

Sarcomastigophora *Invertebrate Zoology.* a subphylum of protozoans having pseudopodia or flagella or both.

sarcomatosis *Medicine.* a condition characterized by the formation of sarcomas.

sarcomere *Histology.* the basic contractile unit of muscle (skeletal and cardiac) that structurally is the portion of a myofibril between two adjacent Z lines, the unit repeated along the entire length of the myofibril.

Sarcophagidae *Invertebrate Zoology.* the blowflies and scavenger flies, a family of dipteran insects belonging to the subsection Calyptratae; its eggs are laid in decaying animal matter, which is eaten by the larvae; some species, especially those of the genus *Sarcophaga,* cause disease in humans and domestic animals. (From a Greek phrase meaning literally "flesh-eating.")

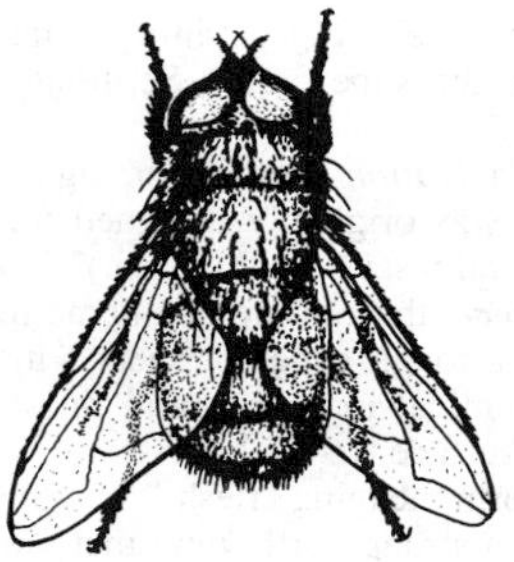

Sarcophagidae

sarcoplasm *Histology.* the cytoplasmic matrix of muscle fibers.
sarcoplasmic *Histology.* relating to, composed of, or containing sarcoplasm.
sarcoplasmic reticulum *Cell Biology.* a membranous organelle system of muscle cells, composed of vesicular and tubular components, that stores calcium ions involved in muscle contraction.
sarcoplast *Cell Biology.* an interstitial cell of a muscle that is itself capable of being transformed into muscle.
sarcopside *Mineralogy.* $(Fe^{+2},Mn^{+2},Mg)_3(PO_4)_2$, a colorless to brownish monoclinic mineral having a specific gravity of 3.79 and a hardness of 4 on the Mohs scale, which is found as fibrous masses in granite pegmatites.
Sarcopterygii *Vertebrate Zoology.* a subclass of the bony fish class Osteichthyes, containing lobefinned fishes including the lungfishes, the coelacanth, and the bichirs.
sarcoptic itch *Entomology.* a skin inflammation brought about by the itch mite, *Sarcoptes scabiei*; it causes secondary skin lesions, noted most commonly on the webs between the fingers.
Sarcoptiformes *Invertebrate Zoology.* a suborder of cheese, itch, and mange mites, tiny spherical arachnids in the order Acarina that lack a tracheal system and exchange gases through the body wall. See ASTIGMATA.
Sarcoscyphaceae *Mycology.* a family of fungi of the order Pezizales, occurring on wood and soil and including some plant parasites.
sarcosine *Organic Chemistry.* CH_3NHCH_2COOH, a crystalline sweet-tasting solid; deliquesces at 210°C; soluble in water and alcohol; used in making toothpaste. Also, *N*-METHYLGLYCINE.
sarcosoma *Invertebrate Zoology.* the fleshy part of an anthozoan. Also, **sarcosome.**
Sarcosomataceae *Mycology.* a family of fungi of the order Pezizales, occurring on wood and soil and including some plant pathogens.
sarcosome *Cell Biology.* a former name for one of the mitochondria of a muscle cell.
Sarcosporida *Invertebrate Zoology.* a subclass of protozoans in the class Sporozoa, which reproduce by spores; endoparasites in the muscles of vertebrates, including humans.
sarcosporidiosis *Veterinary Medicine.* a potentially fatal disease usually of grazing warm-blooded animals, presumably caused by infection with the protozoan parasite *Sarcocystis* and characterized by fever, anemia, paralysis, and cysts in the muscles.
sarcostyle see DACTYLOZOOID.
sarcotic *Physiology.* promoting the growth of flesh.
sarcotoxin *Biochemistry.* an antibacterial protein formed in larvae of the parasitic flesh fly, *Sarcophaga peregrinia,* after an injury or exposure to bacteria.
sarcotubule *Cell Biology.* one of the tubular components of the sarcoplasmic reticulum of a muscle cell.
sarcous *Anatomy.* relating to flesh or muscular tissue.
sard *Mineralogy.* a translucent brown, brownish-red, or deep orange-red variety of chalcedony, similar to but darker than carnelian; used as a gemstone. Also, **sardine, sardius.**
Sardic orogeny *Geology.* a brief, Late Cambrian crustal deformation.
sardine *Vertebrate Zoology.* a small marine fish of the family Clupeidae that is commercially canned; specifically, *Sardina pilchardus,* the young of the herringlike pilchard of the same family. (From its association with the island of *Sardinia.*)

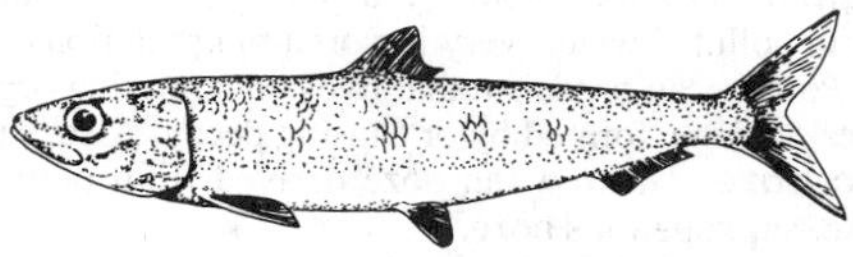

sardine

Sardinia *Geography.* a large island in the western Mediterranean (area: 9302 sq. mi.), south of Corsica and west of the Italian peninsula.
sardonyx *Mineralogy.* a variety of chalcedony containing parallel layers of deep orange-red or brownish sard alternating with white or black layers.
Sard's theorem *Mathematics.* let $f: U \to R^n$ be a smooth map defined on an open subset U of R^m, and let A be the set of all points x of U such that the rank of the derivative (Jacobian matrix) of f at x is less than n. Then the image of A in R^n has Lebesgue measure zero.
Sargassaceae *Botany.* a family of brown algae of the order Fucales, characterized by a branching thallus with lateral outgrowths that include leafy segments, air bladders, and spore-bearing structures.
Sargasso Sea *Oceanography.* an extensive area in the North Atlantic Ocean, northeast of the West Indies, characterized by unusually calm waters and by the presence of extensive areas of free-floating marine plants. (From the Portuguese name for a type of seaweed.)
Sargassum *Botany.* the largest genus of algae in the family Sargassaceae, commonly found throughout the tropics and in temperate waters; it includes *S. bacciferum,* the common gulfweed.

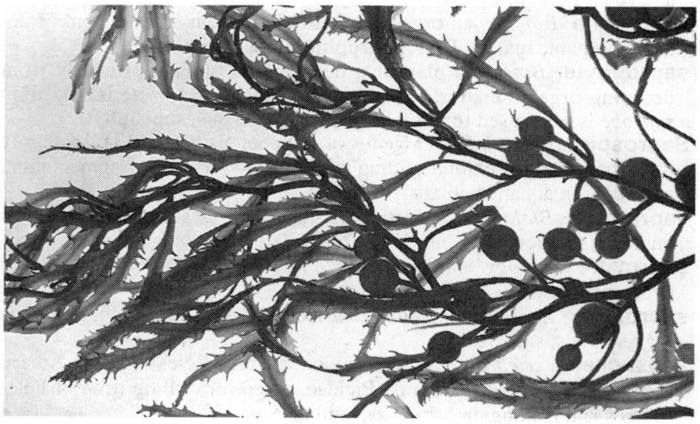

Sargassum

Sargent curve *Nuclear Physics.* a graph that shows the decay constraints of radioactive isotopes, which emit beta particles, against the maximum energy the particles produce, measured on a logarithmic scale.
Sargent cycle *Thermodynamics.* an ideal thermodynamic cycle with four reversible processes: a compression with no heat transfer; a constant-volume heating; an expansion with no heat transfer; and a constant-pressure cooling.
Sargentodoxaceae *Botany.* a monospecific family of twining, woody, dioecious vines belonging to the order Ranunculales, characterized by primitive, spirally arranged pistils but a more advanced solitary ovule and a large, fleshy fruiting receptacle; it is native to China, Laos, and Vietnam.
sarin *Organic Chemistry.* $[(CH_3)_2CHO](CH_3)FPO$, a nerve gas; boils at 147°C; toxic by inhalation and skin absorption; a cholinesterase inhibitor.
Sark *Ordnance.* the code name for the Soviet SS-N-4 missile.
sarking *Building Engineering.* a layer of boards positioned beneath tiles or other roofing materials so as to add a layer of thermal insulation and prevent the entrainment or movement of water into the roof system.

sarkinite *Mineralogy.* $Mn_2^{+2}(AsO_4)(OH)$, a yellow or red monoclinic mineral occurring as thick tabular crystals, having a specific gravity of 4.08 to 4.18 and a hardness of 4 to 5 on the Mohs scale.

sarkomycin *Bacteriology.* an antibiotic that is produced by certain actinomycete bacteria and used in the treatment of cancer.

Sarkosyl *Virology.* a trade name for *N*-lauroyl-sarcosine, an anionic detergent that disrupts virus particles or cells.

Sarmatian *Geology.* a European geologic stage of the Upper Miocene epoch, occurring after the Tortonian and before the Pontian.

sarmentocymarin *Biochemistry.* $C_{30}H_{46}O_8$, a steroid glycoside that yields sarmentogenin and sarmentose on hydrolysis.

sarmentogenin *Biochemistry.* $C_{23}H_{34}O_5$, the steroid aglycon of sarmentocymarin that contains a hydroxyl group at carbon number 11; a possible source of cortisone.

sarmentose *Botany.* having runners; developing a prostrate growth form. *Biochemistry.* a methyl ether of a 2-desoxyhexomethyl sugar from sarmentocymarin.

sarmentum *Botany.* a runner or other prostrate growth form.

sarmientite *Mineralogy.* $Fe_2^{+3}(AsO_4)(SO_4)(OH)\cdot 5H_2O$, a yellow to yellow-orange monoclinic mineral occurring as nodular masses of microscopic prismatic crystals, having a specific gravity of 2.58 and an undetermined hardness.

sarnaite *Mineralogy.* a feldspathoid-bearing syenite containing cancrinite and clinopyroxene

saros *Astronomy.* a cycle of lunar or solar eclipses, recurring at intervals of 6585.3 days or approximately 18 tropical years; used in ancient times to predict eclipses.

sarospatakite *Mineralogy.* a micaceous clay mineral having mixed layers of illite and montmorillonite. Also, **sarospatite.**

SARO thickness *Food Technology.* the measurement in mils (0.001 inches) of batter in a food portion.

Sarothriidae see JACOBSONIIDAE.

sarothrum *Entomology.* the brushlike structure on a bee's hind leg that collects pollen.

Sarraceniaceae *Botany.* the pitcher plant, a New World family of dicotyledonous plants in the order Nepenthales, consisting of insectivorous perennial herbs that live in nitrogen-poor soil regions, such as wetlands and bogs, and trap insects to compensate for nitrogen deficiency.

Sarraceniales *Botany.* a former order containing the family Sarraceniaceae; now generally recognized as Nepenthales.

Sarrett reagent *Organic Chemistry.* an alcohol oxidizing reagent consisting of chromium trioxide and pyridine.

sarsaparilla *Botany.* **1.** any of various climbing or trailing plants of the lily family, genus *Smilax,* native to tropical America and characterized by alternate leaves, umbels of flowers, and a root that is used in medicinal preparations. **2.** the root of this plant. *Food Technology.* a soft drink flavored with this root.

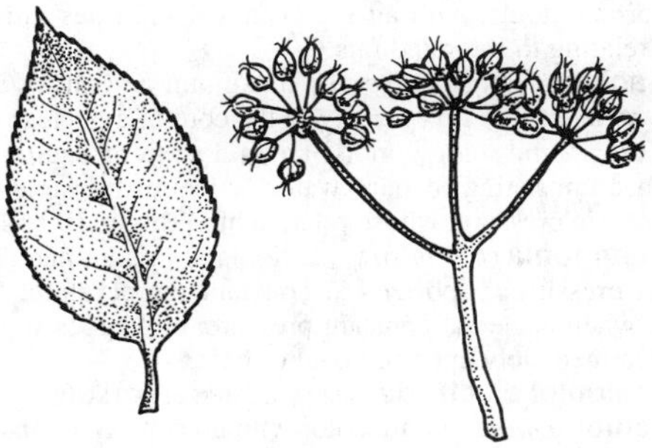

sarsaparilla

sartorite *Mineralogy.* $PbAs_2S_4$, a dark-gray, opaque, metallic monoclinic mineral occurring as prismatic crystals, having a specific gravity of 5.10 and a hardness of 3 on the Mohs scale; found in crystalline dolomite in Binnental, Switzerland.

sartorius *Anatomy.* a long muscle extending from the anterior superior iliac spine to the medial surface of the tibia; it flexes the thigh and leg.

Sartorya *Mycology.* a genus of asexual fungi of the order Eurotiales, termed *Aspergillus fumigatus* in its sexual phase.

SAS stability augmentation system.

SAS or **S.A.S.** Society for Applied Spectroscopy.

sash *Building Engineering.* a window frame in which glass is set.

sash bar *Building Engineering.* any of one or more wood strips that separate the narrow panes of glass in a window composed of several panes. Also, MUNTIN.

sash brush *Mechanical Devices.* a paintbrush having a small, often angled brush head; used for painting window sashes, sash bars, and other surfaces requiring fine lines.

sash cord *Building Engineering.* a cord or chain that connects a vertically sliding window sash with a counterweight. Also, **sash line.**

sash cramp *Mechanical Devices.* a device used to hold door frame parts during construction, consisting of a steel bar with two sliding brackets adjusted by a screw.

sash door *Building Engineering.* a door whose upper end is glazed.

sash fastener *Building Engineering.* a fastening device mounted to the meeting rails of sashes so as to hold both the top and bottom sash in a closed position. Also, **sash lock.**

sash plane *Mechanical Devices.* a plane having a notched cutter, used for shaving door frames and window sashes.

sash pocket chisel *Mechanical Devices.* a strong-bladed chisel having a narrow edge, used to cut a pocket in the pulley stile of a sash and frame.

sash saw *Mechanical Devices.* a steel strip having one toothed edge, used for water-power mill sawing operations.

Sasin *Ordnance.* a Soviet two-stage intercontinental ballistic missile powered by liquid propellant motors, delivering a 5-megaton thermonuclear warhead at a range of 6500 miles; officially designated **SS-8.**

Sa spiral *Astronomy.* a class of spiral galaxy characterized by tightly wound arms and large nuclei.

sassafras *Botany.* **1.** a tree of the laurel family, *Sassafras albidum,* characterized by egg-shaped leaves and long clusters of greenish-yellow flowers; its roots are covered with an aromatic bark that yields a pungent oil used for tea, flavorings, and medicinal purposes. **2.** the root bark of this tree.

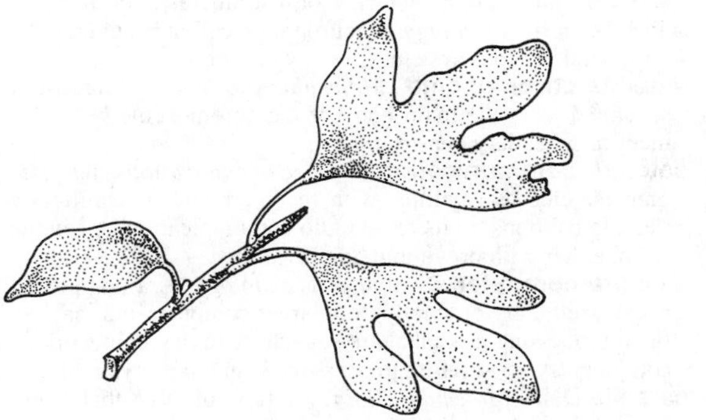

sassafras

sassolite *Mineralogy.* H_3BO_3, a white or gray triclinic mineral, native boric acid, occurring as tabular crystals or as small pearly scales, having a specific gravity of 1.46 to 1.52 and a hardness of 1 on the Mohs scale; found in natural waters and as a sublimation product. Also, **sassolin, sassoline.**

sastruga *Hydrology.* one of a series of low, parallel ridges formed on a relatively level snow surface by wind erosion, or cut into snow dunes previously deposited by the wind.

SAT *Aviation.* the airport code for San Antonio.

sat. saturated; saturate.

satan see DUST WHIRL.

Satan *Ordnance.* a Soviet two-stage intercontinental ballistic missile that is liquid-rocket powered, delivering one 20-megaton warhead or up to ten 500-kiloton MIRVs at a range of 6500 miles; officially designated **SS-18.**

satchel charge *Ordnance.* a charge composed of blocks of explosive taped to a board to which a rope or wire is attached for carrying and/or affixing the charge to the target.

satd. saturated.

sateen *Textiles.* a smooth, lustrous cotton fabric made in a close satin weave.

satellite *Astronomy.* any celestial body, natural or artificial, that revolves in an orbit about a planet. The moon is a natural satellite of the earth. *Space Technology.* an artificial device designed to orbit about the earth or some other celestial body. *Anatomy.* a vein that closely accompanies an artery, such as the brachial. *Medicine.* a minor lesion situated near a larger one. *Cell Biology.* **1.** a globoid mass of chromatin attached at the secondary constriction to the ends of the short arms of acrocentric autosomes. **2.** the posterior of a pair of gregarine protozoans undergoing syzygy. (From a Greek term meaning "follower" or "attendant.")

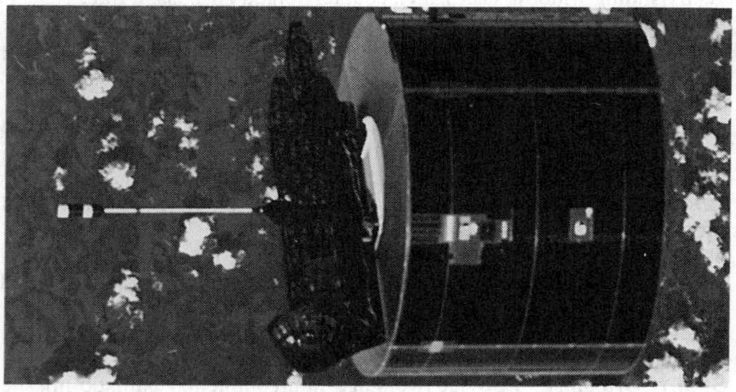

communications satellite

satellite and missile observation system see SAMOS PROGRAM.

satellite and missile surveillance *Military Science.* the tracking through space of satellites and inflight missiles of other nations.

satellite antenna *Electromagnetism.* an antenna that is mounted on a satellite and used to transmit or receive signals to or from an earth station.

satellite astronomy *Astronomy.* the study of celestial objects from above the atmosphere by means of instruments in orbit.

satellite cell *Neurology.* a neurilemma cell surrounding neurons in the peripheral nervous system.

satellite chromosome *Molecular Biology.* a chromosomal segment separated from the main body of the chromosome by a chromatic filament or secondary constriction.

satellite communications *Telecommunications.* the use of orbiting objects, such as satellites with receivers and transmitters or radio reflecting balloons, to increase radio communication by sending signals to earth, either with or without amplification.

satellite computer *Computer Technology.* **1.** a computer that is connected, remotely or locally, to a larger computer and that carries out defined processing tasks that are subsidiary to the main work of the central computer. Also, **satellite processor. 2.** an auxiliary, off-line computer.

satellite DNA *Molecular Biology.* a form of DNA that differs in density from that of the main DNA component and consists of repeating sequences of nucleotide pairs.

satellite galaxy *Astronomy.* a galaxy, usually small and irregular, that orbits a larger one; e.g., the Magellanic Clouds are satellite galaxies of the Milky Way.

satellite geodesy *Geodesy.* the technique of using artificial satellites to gather geodetic information about the earth.

satellite infrared spectrometer *Spectroscopy.* a spectrometer carried aboard Nimbus series satellites, used to measure radiation from atmospheric carbon dioxide at various wavelengths.

satellite male *Behavior.* a sexually mature male that does not have a dominant position within a group but occupies a location at the edge of or near a dominant male's territory. Also, PERIPHERAL MALE.

satellite meteorology *Meteorology.* a branch of meteorology that uses remote sensing satellites for instantaneous determination of the atmospheric state.

satellite tracking *Space Technology.* a system used to identify the positions and speeds of satellites through radio and optical means.

satellite triangulation *Cartography.* in surveying, the technique of using the simultaneous observations of a satellite from two ground stations to determine the angular relationship, and thereby the distance, between the stations.

satellite vehicle *Space Technology.* a rocket vehicle used to launch a satellite.

satellite virus *Virology.* a virus that is inherently dependent upon the presence of a helper virus in order to replicate in a host cell; believed to be defective in coding for capsid formation.

satellitic crater see SECONDARY CRATER.

satellitism *Bacteriology.* a phenomenon in which certain bacterial species grow more vigorously in the immediate vicinity of colonies of another unrelated species that produces a metabolite essential to the former.

satellitosis *Neurology.* a condition characterized by the accumulation of neuroglial cells around an injured neuron.

satelloid *Space Technology.* a satellite whose orbit is within the planetary atmosphere, thus requiring sustaining thrust to balance drag.

satiation *Behavior.* the tendency for an act or stimulus to become less attractive to the subject upon repetition.

satin *Textiles.* a silk or rayon fabric in a satin weave with a smooth, glossy face and a dull back. (Derived from the name of a city in China, which exported this type of fabric during the Middle Ages.)

satin finish *Metallurgy.* a metal finish that is very fine, but not reflecting.

satin ice see ACICULAR ICE.

satin spar *Mineralogy.* a translucent, white, fibrous variety of gypsum having a silky luster Also, **satin stone.**

satin weave *Textiles.* one of the three basic weaves (the others being plain and twill weave), in which the face of the fabric is almost completely formed by either warp threads that pass over several filling threads before weaving under one, or filling threads that pass over several warp threads before weaving under one. This type of weave produces a very smooth, solid surface.

satinwood see AYAN.

satisfiable *Artificial Intelligence.* see CONSISTENT.

satisficing *Artificial Intelligence.* the process of searching for a solution that is satisfactory, though not necessarily optimal.

satisfy *Mathematics.* to fulfill the requirements or conditions of an equation, theorem, and so on. *Artificial Intelligence.* to provide a set of data values that makes a logical formula or the antecedent of a rule true.

saturable able to undergo saturation; capable of being saturated.

saturable-core magnetometer *Engineering.* a magnetometer that measures magnetic field strength from variations in the permeability of a ferromagnetic core.

saturable reactor *Electromagnetism.* an iron-core reactor having an additional coil that carries direct current to adjust the magnetic saturation point of the core and thus change the alternating current winding reactance; also called a saturable core reactor or transductor.

saturable transformer *Electromagnetism.* a saturable core reactor having an additional winding that acts as a secondary winding in a transformer for voltage transformation.

saturate to cause to become saturated or undergo saturation.

saturated *Physical Chemistry.* describing a solution that is not able to dissolve additional solute; i.e., a solution that contains the maximum possible amount of a substance under the given conditions. *Organic Chemistry.* decribing an organic compound that has only single bonds between carbon atoms, with all the chemical affinities satisfied. *Nutrition.* of or relating to saturated fats.

saturated activity *Nucleonics.* the maximum amount of radioactivity obtainable by a definite flux density in the core of a nuclear reactor.

saturated air *Meteorology.* moist air in a state of equilibrium with a plane surface consisting of pure water or ice at the same temperature and atmospheric pressure, whose relative humidity is 100 percent.

saturated ammonia *Chemistry.* **1.** liquid ammonia that, when heated at constant pressure, vaporizes at constant temperature. **2.** ammonia vapor that, when heated at constant pressure, undergoes superheating; it condenses immediately upon removal of heat.

saturated calomel electrode see CALOMEL ELECTRODE.

saturated color *Optics.* a pure color with no admixture of white.

saturated compound *Organic Chemistry.* an organic compound that does not contain any double or triple bonds.

saturated diode *Electronics.* a diode operating under conditions such that its current cannot be increased by raising the voltage.

saturated fat *Nutrition.* any animal or vegetable fat that contains a saturated fatty acid; diets high in saturated fats have been associated with elevated serum cholesterol levels and may contribute to heart disease.

saturated fatty acid *Nutrition.* an organic compound of carbon, hydrogen, and oxygen that combines with glycerol to form fat; it has no double bonds in the carbon chain, is solid at room temperature, and is a component of common animal fats, such as butter and lard.

saturated hydrocarbon *Organic Chemistry.* an organic compound consisting of only carbon and hydrogen linked by single bonds.

saturated liquid *Chemistry.* a liquid solution that contains so much dissolved substance that no more will dissolve at a given temperature and pressure.

saturated mineral *Mineralogy.* any mineral capable of crystallization from rock magma in the presence of free silica.

saturated permafrost *Geology.* any permafrost in which all the pore spaces are filled with ice, so that it contains no more ice than the ground could hold if the water were in a liquid state.

saturated rock *Petrology.* an igneous rock consisting primarily of saturated minerals in the absence of excess silica.

saturated solution *Physical Chemistry.* a solution that contains a sufficient amount of a substance so that no more will dissolve under the given conditions; i.e., the concentration of dissolved solute is or would be in equilibrium with any excess undissolved solute; the undissolved solute need not actually be present for the description to apply.

saturated vapor *Thermodynamics.* a vapor that is sufficiently concentrated to be able to exist in equilibrium with the liquid form of the same substance.

saturated zone see ZONE OF SATURATION.

saturating signal *Electronics.* a radar signal whose amplitude is greater than the receiver's dynamic range.

saturation the process of becoming or causing to become saturated, or the state of being saturated; specific uses include: *Physical Chemistry.* the condition of a solution in which it already has taken in the maximum possible amount of a solute under the given conditions. *Ecology.* a condition in which a given habitat is filled to capacity by individuals of a particular population in terms of the amount of resources needed to support the population. *Physics.* a condition of maximum effect in which any increase in a particular external influence will not result in an increase in the response. *Electromagnetism.* specifically, the condition in which a body of magnetic material subject to an increasing magnetic field does not undergo an increase in magnetization. *Electronics.* **1.** a condition in which an input signal no longer generates a change in the output signal of a circuit. **2.** the condition in which a transistor is driven so hard that its collector reverses itself and becomes positive with respect to the base instead of negative. *Nucleonics.* the condition of a nuclear reactor in which the production rate and the decay rate of a particular radionuclide are equal. *Optics.* a subjective perception of the percentage of white in a given color; low saturation indicates a large percentage of white, while high saturation indicates little or no white. Also, COLOR SATURATION.

saturation-adiabatic process *Meteorology.* an adiabatic process in which the air is maintained at saturation by either the evaporation or condensation of water substance, with the latent heat being supplied by or to the air, respectively.

saturation bombing *Military Science.* an intensive bombing technique in which the target area is completely destroyed; it may employ many smaller bombs or a single massive bomb.

saturation current *Electronics.* **1.** the greatest amount of current that can be obtained under specific conditions, as when the maximum voltage has been applied. **2.** the current in the anode when it has received all of the electrons emitted by the cathode. **3.** the current, traveling between the base and a collector in a transistor, that is no longer affected by an increase in voltage. *Nucleonics.* the current produced in an ionization tube when the voltage is high enough to collect all ions produced in an ionizing event but not high enough to produce ionization by collision. *Radiology.* the amount of current in an X-ray tube when the voltage is strong enough to drive all the electrons produced from the cathode filament to the anode as fast as they are produced.

saturation curve *Geology.* a graph showing the weight of solids per unit volume of a saturated mass of soil in relation to its water content.

saturation deficit *Meteorology.* **1.** the difference between the actual vapor pressure and the saturation vapor pressure at the existing temperature. **2.** the additional amount of water vapor that is needed to reach saturation at the current temperature and pressure, expressed in grams per cubic meter.

saturation dispersal *Ecology.* a condition in which a large number of individuals emigrate from a population that is at or near its capacity; usually these individuals are the less fit members of the group.

saturation hybridization *Molecular Biology.* an in vitro (isolated) reaction in which an excess of one polynucleotide component exists, causing all complementary segments in another polynucleotide component to subsist into a duplex form.

saturation induction *Electromagnetism.* the maximum possible magnetic flux density obtainable from a body of magnetic material. Also, **saturation flux density.**

saturation limiting *Electronics.* a method of suppressing the output voltage in an electron tube by saturating its anode with current.

saturation line *Petrology.* the line, on a variation diagram of an igneous rock series, representing saturation with respect to silica; rocks to the right of the line are oversaturated, while those to the left are undersaturated.

saturation magnetization *Electromagnetism.* a condition whereby a sample of magnetic material exhibits negligible changes in magnetization intensity with an increase of magnetizing force.

saturation mixing ratio *Meteorology.* the value of the mixing ratio of saturated air at a given temperature and pressure.

saturation of forces *Physics.* a principle applicable to forces acting between particles in which a particle can interact strongly with only a limited number of other particles in a system of particles.

saturation point *Physical Chemistry.* the point at which a solution will receive no more of a certain solute under the given conditions. *Ecology.* the maximum number of a species or other category of organisms that can be supported by a particular range under optimum conditions.

saturation-presaturation hypothesis *Ecology.* the concept that individuals emigrating from a thriving population tend to be more fit and thus more apt to survive than those emigrating from a declining or oversaturated one.

saturation ratio *Meteorology.* the ratio of the actual specific humidity to the specific humidity of saturated air at a given temperature.

saturation scale *Optics.* a stepped array of the same color to which progressively more white has been added.

saturation signal *Electromagnetism.* a signal whose amplitude is greater than the dynamic range of the system that receives it.

saturation specific humidity *Thermodynamics.* the value of the specific humidity of a sample of saturated air when measured at a given temperature and pressure.

saturation spectroscopy *Spectroscopy.* a spectroscopic technique in which a laser beam is used to change the energy-level populations of a resonant medium over a narrow range of particle velocities in order to produce Doppler-free spectral lines.

saturation vapor pressure *Thermodynamics.* the vapor pressure of a thermodynamic system, measured at a specified temperature, in which the vapor phase and the liquid or solid phase exist in equilibrium.

saturator *Chemical Engineering.* a device that saturates one material with another, such as an inert gas with the vapors of a volatile liquid.

Saturn *Astronomy.* the sixth major planet from the sun, having a yellowish color as observed by the naked eye; it orbits at 9.5 astronomical units, has a period of revolution of 29.46 years, a mass of 95 earths, and an equatorial diameter of 120,500 kilometers. *Space Technology.* any of a series of high-thrust, multistage booster vehicles; used to launch the landing vehicles used in the NASA *Apollo* program. (From the name of an ancient Roman god of agriculture and fertility.)

Saturn

saturnigraphic coordinates *Astronomy.* the position of a point on the surface of Saturn, as determined by latitude from the planet's equator and longitude from a reference meridian.

Saturniidae *Invertebrate Zoology.* a family of bright-colored lepidopteran insects, including giant silkworm moths, in the superfamily Saturnioidea, some having a wingspan of six inches; larvae eat the leaves of trees, and the cocoons of some species are used for commercial silk.

Saturnioidea *Invertebrate Zoology.* a superfamily of moths, lepidopteran insects in the suborder Heteroneura, having reduced mouthparts, a reduced or absent frenulum, and comblike antennae.

saturnism see LEAD POISONING.

Saturn Nebula *Astronomy.* a double-ring planetary nebula located in the constellation Aquarius; designated **NGC 7009.**

Saturn's moons *Astronomy.* the 17 major satellites of Saturn whose existence, locations, and periods have been definitively confirmed: Atlas, Calypso, Dione, Enceladus, Epimetheus, Helene, Hyperion, Iapetus, Janus, Mimas, Pandora, Phoebe, Prometheus, Rhea, Telesto, Tethys, and Titan. Also, **Saturn's satellites.**

Saturn's rings see RINGS OF SATURN.

Satyrinae *Invertebrate Zoology.* a family of gray or brown butterflies, lepidopteran insects in the family Nymphalidae, including wood nymphs, satyrs, and arctics.

saucer crater *Astronomy.* a very shallow lunar crater, as is found on the floor of the crater Ptolemaeus.

saucerization *Surgery.* the surgical excision of tissue to make a shallow depression, usually performed on infected bone in order to promote drainage.

Saucesian *Geology.* a North American geologic stage of the Oligocene and Miocene epochs, after the Zemorrian and before the Relizian.

sauconite *Mineralogy.* $Na_{0.3}Zn_3(Si,Al)_4O_{10}(OH)_2 \cdot 4H_2O$, a reddish brown monoclinic clay mineral of the smectite group, very fine-grained and massive in habit, having a hardness of 1 to 2 on the Mohs scale and an undetermined specific gravity.

saucyite *Petrology.* a rhyolitic igneous rock containing large phenocrysts of sanidine in a gray groundmass of orthoclase microlites, and minute crystals of biotite, augite, titanite, zircon, and magnetite.

Sauer, Carl 1889–1975, American geographer; a pioneer in comparative human geography.

sault [soo] *Hydrology.* a rapids or waterfall within a stream. (From French; going back to a Latin word meaning "leap.")

sauna [sô´nə] *Building Engineering.* a closed building or room in which high levels of dry heat are produced for therapeutic purposes.

Sauria *Vertebrate Zoology.* the lizards, a suborder of reptiles of the order Squamata, usually having limbs, scaly bodies, eyelids, and external ear openings.

saurian *Vertebrate Zoology.* **1.** any member of the suborder Sauria. **2.** of or relating to the suborder Sauria or any of its members.

Saurichthyidae *Paleontology.* a family of pikelike chondrostean fishes in the extinct order Saurichthyiformes; about 3 feet long and characterized by symmetrical dorsal and anal fins near the tail; extant in the Triassic and Lower Jurassic.

Saurischia *Paleontology.* one of the two large orders of dinosaurs, characterized by a slightly more primitive arrangement of pelvic bones than the Ornithischia; the saurischians probably arose from a thecodont ancestor in the Middle or Early Triassic and flourished until the mass extinctions toward the end of the Cretaceous; it includes two main subgroups, the carnivorous theropods and the herbivorous sauropods.

saurischian *Paleontology.* **1.** any member of the order Saurischia. **2.** of or relating to the order Saurischia or any of its members.

sauroid *Zoology.* resembling a reptile.

sauropod *Paleontology.* **1.** any member of the infraorder Sauropoda. **2.** of or relating to the infraorder Sauropoda or any of its members.

Sauropoda *Paleontology.* an infraorder of quadrupedal saurischian dinosaurs; the sauropods were semiaquatic and herbivorous, and included the largest of the dinosaurs, such as Diplodocus and Apatosaurus; extant in the Jurassic and Cretaceous.

Sauropodomorpha *Paleontology.* a suborder of herbivorous saurischian dinosaurs, cousins of the theropods, that includes the infraorders Plateosauria and Sauropoda; extant from the Triassic to the Cretaceous.

Sauropterygia *Paleontology.* a superorder of aquatic diapsid reptiles, composed of the nothosaurs and plesiosaurs; the **sauropterygians** arose near the Permian-Jurassic boundary and became extinct in the Cretaceous.

Saururaceae *Botany.* a family of dicotyledonous plants of the order Piperales consisting of perennial herbs and shrubs of eastern Asia and North America; fruit is an apically dehiscent capsule.

sausage *Food Technology.* minced meat that is combined with seasonings and other ingredients and usually stuffed into a casing of intestine or other edible material.

sausage instability see KINK INSTABILITY.

Saussure, Ferdinand de [sô sür´] 1857–1913, Swiss linguist; the founder of structural linguistics; distinguished *langue* (the general state of a language at a given time) from *parole* (the speech of a given individual user of that language).

Saussure, Horace de 1740–1799, Swiss scientist; wrote extensively on Alpine geology, plants, and weather; invented hair hygrometer.

saussurite *Mineralogy.* a tough, compact mineral aggregate of white, greenish, or grayish color, consisting of a mixture of calcite, albite, sericite, prehnite, epidote, zoisite, and other silicates of calcium and aluminum; formed by alteration of calcic plagioclase.

saussuritization *Geology.* a metamorphic process in which plagioclase in basalts and gabbros is altered and replaced by a fine-grained aggregate of zoisite, albite, or epidote, together with varying amounts of calcite, sericite, and zeolites.

sauterelle *Engineering.* an implement used by a mason to trace and form angles.

Sauterne [sô tern´] *Food Technology.* **1.** a sweet white Bordeaux wine, made with mostly Sémillon grapes fermented in the "noble rot" technique. **2.** also, **sauterne.** a similar wine made elsewhere.

Sauvignon Blanc [sô´vin yōn´ blängk´] *Botany.* a green grape grown in Bordeaux, France, and elsewhere; often blended with Sémillon and other grapes to make wine. *Food Technology.* a dry California white wine made mostly from this grape. Also, **sauvignon blanc.**

SAV *Aviation.* the airport code for Savannah, Georgia.

Savage *Ordnance.* a Soviet three-stage intercontinental ballistic missile powered by solid-propellant motors, delivering a nuclear or thermonuclear warhead at a range of 5000 miles; officially designated **SS-13.**

savagery *Anthropology.* in the 19th-century evolutionary view of culture, the lowest stage of social development, preceding barbarism and civilization; nonliterate societies of hunter-gatherers were thus classified by early anthropologists.

savanna *Ecology.* a type of plain dominated by grasses, interspersed with tall shrubs and trees in an open formation; it may occur in a variety of tropical and subtropical habitats, but is most common in arid regions or ones with a dry season; e.g., East Africa. Also, **savannah.**

savanna climate see TROPICAL SAVANNA CLIMATE.

savanna woodland see TROPICAL WOODLAND.

savant [sə vänt´] *Psychology.* an individual who is mentally handicapped yet has a remarkable talent for some specific mental feat, such as rapid mathematical calculation or large-scale memorization, as of statistics or dates. Also, IDIOT SAVANT.

Savart, Félix [sä vär´] 1791–1841, French physicist; with Jean Baptiste Biot, formulated the Biot-Savart Law on intensity of electromagnetic fields.

savart *Acoustics.* a unit of pitch interval formerly used instead of the cent; 1 savart was defined as 1/300 of an octave, resulting in 25 savarts for a tempered semitone and 50 savarts for a whole tone.

Savart plate *Optics.* either of a matched pair of calcite plates in a Savart polariscope that detects polarized light.

Savart polariscope *Optics.* an instrument used to detect polarized light in which two Savart plates are mounted perpendicularly with a tourmaline crystal analyzer.

Savic orogeny *Geology.* a brief, Late Oligocene crustal deformation involving Phanerozoic rock, occurring between the Chattian and Aquitanian stages.

Savonius rotor *Mechanical Engineering.* an impeller consisting of two offset semicylindrical elements that rotate about a vertical axis.

Savonius windmill *Mechanical Engineering.* a windmill having two offset semicylindrical cups that rotate about a vertical axis.

savory *Botany.* any of various aromatic herbs of the mint family, genus *Satureja,* especially *S. hortensis* **(summer savory)** or *S. montana* **(winter savory),** having narrow leaves that are used as a condiment, as a tea, and for flavoring.

saw *Mechanical Devices.* any of a variety of tools used for cutting wood or metal, consisting of a flat metal blade with sharp teeth mounted in a handle or frame.

SAW surface acoustic wave.

saw bit *Mechanical Devices.* a drill bit with serrated teeth.

saw-cut *Geology.* a large canyon that cuts sharply across a terrace, so that it can be seen only from sites near its edge.
sawdust *Materials.* **1.** wood particles produced by sawing. **2.** of or relating to a material in which sawdust is a principal ingredient. Thus, **sawdust concrete.**
sawfish *Vertebrate Zoology.* a large sharklike cartilaginous fish of the genus *Pristis* in the family Pristidae; characterized by a long, flattened snout and lateral teeth, and found worldwide in tropical and subtropical waters.

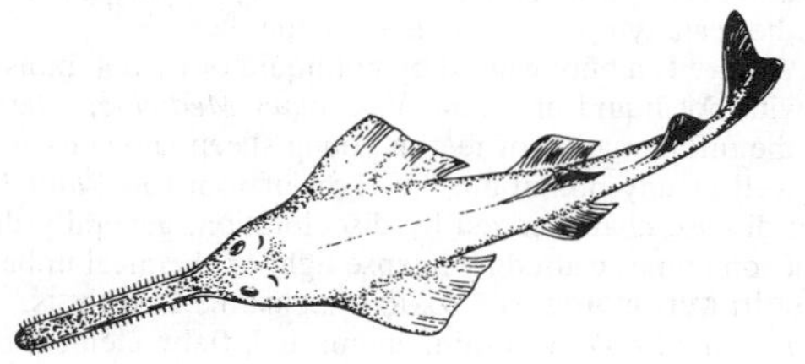

sawfish

Sawfly *Ordnance.* the code name for the Soviet SS-N-6 missile.
saw gumming *Mechanical Engineering.* a process in which burrs and machine marks are ground out of a saw blade.
sawhorse *Engineering.* a rack that holds wood while it is being sawed.
sawing *Mechanical Engineering.* the cutting of a piece or part with a band, blade, or circular disk having teeth.
sawmill *Mechanical Engineering.* **1.** a large machine that uses a saw or a series of saws to strip and cut logs. **2.** a building that houses such machines.

sawmill

saw set *Mechanical Devices.* a tool used for setting saw teeth, usually designed to increase the saws's kerf by putting a slight bend in each alternate tooth.
sawtooth *Mechanical Devices.* any of the teeth of a saw. *Architecture.* any of the constituent structures of a sawtooth roof. *Science.* having a profile resembling the teeth of a saw; serrate. Also, **saw-toothed.** *Electronics.* of a waveform, repeatedly rising and falling, such that its representation resembles the teeth of a saw.
sawtooth blasting *Mining Engineering.* a method of blasting a face in which a series of oblique, horizontal holes is blasted to form slabs that, in plan, resemble saw teeth.
sawtooth crusher *Mechanical Engineering.* a crushing apparatus that passes the material between two sawtooth-studded shafts and rotates it at different speeds; the product size is controlled by the peripheral speeds and the saw spacings.
sawtooth floor channeling *Mining Engineering.* the channeling of inclined marble beds by successive removal of right-angle blocks from the various beds, resulting in a floor with a sawtooth appearance.
sawtooth generator *Electronics.* a device that produces alternating voltage in a sawtooth waveform.
sawtooth modulated jamming *Electronics.* a radar countermeasure in which high-powered signals are transmitted into an enemy's receiver in order to obliterate signals received from a target.
sawtooth pulse *Electronics.* a signal characterized by a gradual rise in amplitude followed by an abrupt drop.
sawtooth roof *Architecture.* a roof system of parallel triangular roof structures, each having one steeper side glazed for southern exposure; its profile resembles sawteeth.
sawtooth stoping *Mining Engineering.* a method of overhand stoping used in the United States in which the line of advance is up the dip and the benches are advanced in a line parallel to the drift.
sawtooth wave *Electronics.* a periodic wave whose amplitude varies linearly between two values.
sax *Mechanical Devices.* a tool used for trimming the edges of roof slates, with a sharp chopping edge on one side and a spike for making nail holes on the other.
saxatile *Ecology.* living or growing among rocks or stones.
saxicole *Botany.* any plant that grows or thrives among rocks. Thus, **saxicolous.**
Saxifragaceae *Botany.* a family of dicotyledonous perennial herbs of the order Rosales, consisting of 37 genera and 475 species found primarily in north temperate and colder regions of the Northern Hemisphere; the fruit type is a septicidal capsule.
saxifrage [saks´i frəj] *Botany.* any plant of the family Saxifragaceae.
saxifrage family *Botany.* a popular name for the family Saxifragaceae.
saxitoxin *Toxicology.* a neurotoxin found in clams and mussels that have fed on members of the genus *Gonyaulax* and related dinoflagellates whose local population explosions are associated with a red tide. Also, SHELLFISH POISON, MYTILOTOXIN.
Saxonian *Geology.* a European geologic stage of the Middle Permian period, occurring after the Autunian and before the Thuringian.
saxonite *Petrology.* a peridotite consisting primarily of olivine and orthopyroxene, sometimes containing biotite, hornblende, chromite, and picotite.
saxophone *Electromagnetism.* a linear array antenna that produces a cosecant-squared radiation pattern.
Saybolt chromometer *Optics.* a device used to measure the natural color of hydrocarbon fluids, such as kerosene, gasoline, or naphthalene.
Saybolt color *Engineering.* a color standard determined by the Saybolt colorimeter.
Saybolt colorimeter *Engineering.* an apparatus that is used in the United States for determining the color of light oils.
Saybolt Furol viscosimeter *Engineering.* a Saybolt viscosimeter for determining the viscosity of heavy fuel and road oils.
Saybolt Furol viscosity *Fluid Mechanics.* the time in seconds it takes for 60 cm^3 of fluid to flow through a capillary tube in a Saybolt Furol viscosimeter at a specified temperature between 70 and 210°F (21 and 99°C).
Saybolt seconds universal *Fluid Mechanics.* a measurement unit in time dimension for Saybolt universal viscosity.
Saybolt universal viscosimeter *Engineering.* an apparatus that measures the viscosity of lubricating oils under varying temperatures.
Saybolt universal viscosity *Fluid Mechanics.* the time in seconds it takes for 60 cm^3 of fluid to flow through a capillary tube in a Saybolt universal viscosimeter at a certain temperature.
Saytzev (Zaitsev) rule *Organic Chemistry.* a rule predicting that during dehydration of alcohols, the order of removal of hydrogen atoms, when alternative paths exist, is tertiary in preference to secondary, and secondary in preference to primary.
SB or **S.B.** Bachelor of Science. (From Latin *scientiae baccalaureus.*)
Sb the chemical symbol for antimony. (From Latin *stibium.*)
sb stilb.
SBA *Aviation.* the airport code for Santa Barbara, California.
S band *Telecommunications.* a band of radio frequencies extending from 1550 to 5200 megahertz and corresponding to wavelengths of 19.37 to 5.77 centimeters.
SBN *Aviation.* the airport code for South Bend, Indiana.
Sb spiral galaxy *Astronomy.* a type of spiral galaxy characterized by extended arms and a small nucleus.
SC semilunar closure; secretory component.
Sc the chemical symbol for scandium.
Sc or **Sc.** stratocumulus.
sc or **sc.** science; scale; subcutaneous.
sc or **s.c.** small capitals.
sc. namely. (From Latin *scilicet,* "able to know.")

scab *Medicine.* **1.** the blood-formed crust of a superficial cut or scrape of the skin. **2.** to form such a crust. *Plant Pathology.* any of various diseases caused by a wide range of fungi and a few bacteria and characterized by roughened, crustlike areas on plant surfaces. *Metallurgy.* in a sand casting, a flaw consisting of an extra piece of metal partially attached to the surface.

SCAB *Oncology.* an acronym for an anticancer chemotherapy regimen of streptozocin, CCNU (lomustine), Adriamycin (doxorubicin), and bleomycin.

scabbard *Ordnance.* **1.** originally, a carrying case for swords or knives. **2.** in modern usage, a full-length, open-ended carrying case for a rifle, carbine, shotgun, or submachine gun; it is usually made of leather or canvas.

scabbing *Ordnance.* the chipping of fragments off the inside of a metal wall, caused by the impact of a projectile on the outside of the wall.

scabicide *Toxicology.* an agent that is destructive to *Sacroptes scabiei*; used in the treament of scabies.

scabies [skā´bēz] *Medicine.* a contagious disease caused by the itch mite, *Sacroptes scabiei,* which bores under the skin, resulting in itching of the skin and occasionally in secondary bacterial infection.

scabietic *Medicine.* relating to or afflicted with the disease scabies. Also, **scabetic.**

scabland [skab´land´] *Geology.* a relatively flat, elevated land area covered with basalt, a thin layer of soil, sparse vegetation, and usually deep, dry channels.

scabrock [skab´räk´] *Geology.* **1.** a scabland or section of a scabland that projects outward from the surrounding land. **2.** the weathered material on the surface of a scabland.

scabrous *Biology.* having a rough surface covered with short stiff hairs or scales.

scacchite *Mineralogy.* $MnCl_2$, a soft, colorless to rose-red trigonal mineral occurring in fumarolic areas as deliquescent crystalline crusts, having a specific gravity of 2.977 and an undetermined hardness.

scaffold *Mechanical Devices.* a temporary, movable platform structure having a wood or metal framework; used to support workers and materials above the ground during construction operations.

scaffolding *Mechanical Devices.* **1.** a scaffold or system of scaffolds. **2.** the component materials of a scaffold, often wood planks and a metal framework.

scaffolding protein *Biochemistry.* a protein or unit in a complex of proteins that gives temporary structural support during development, such as a bacteriophage head.

scaglia [skal´yə; skäl´yə] *Geology.* a dark, very fine-grained, generally calcareous shale, typically formed during the Upper Cretaceous and Lower Tertiary periods in the northern Apennines.

scala *Anatomy.* any of various stairlike structures, especially in the cochlea. (A Latin word meaning "staircase.")

scalability *Computer Science.* the ability of a technique or algorithm to work when applied to larger problems or data sets.

scala media *Anatomy.* the spiral membraneous cochlear duct of the inner ear, containing the organ of Corti on its floor. Also, COCHLEAR DUCT.

scalar [skā´lär] *Science.* relating to or represented on a scale. *Physics.* **1.** a quantity that has only a magnitude; i.e., it has no direction associated with it. Also, **scalar invariant.** **2.** of or relating to such a quantity. *Mathematics.* an element of the field (usually the real or complex numbers) by which elements of a vector space over the field are multiplied; such multiplication is called **scalar multiplication.** The field is also called the **scalar field** or **field of scalars**.

scalar data type *Computer Science.* a data type consisting of a single item of data (e.g., an integer or character), as distinguished from a record or array type.

scalar field *Physics.* any field of space and time in which the value at each point in space has a scalar value.

scalar function *Physics.* any function whose value at any point is a scalar quantity. *Mathematics.* a function from a vector space (module) to its field of scalars; a scalar-valued function.

scalariform *Biology.* shaped like a ladder; having transverse parts or markings like a ladder.

scalar meson *Particle Physics.* a meson having positive parity and zero spin; such a particle may be described by a scalar field.

scalar potential *Physics.* a scalar function in which a negative gradient describes a vector field.

scalar processor *Computer Science.* a type of central processing unit in which only one operation on data is executed at a time.

scalar product see INNER PRODUCT.

scalar triple product *Mathematics.* given vectors a, b, and c in three-dimensional Euclidean space, the scalar triple product $(a \times b) \cdot c$ is the real number (scalar) whose absolute value equals the volume of the parallelepiped spanned by a, b, and c. If a, b, and c form a right-handed system, the scalar triple product is positive, and it is negative otherwise; a, b, and c are linearly independent if and only if $(a \times b) \cdot c \neq 0$.

scala tympani *Anatomy.* the portion of the spiral canal of the cochlea that lies below the spiral lamina, is filled with perilymph, and is continuous with the scala vestibuli at the helicotrema.

scala vestibuli *Anatomy.* the portion of the spiral canal of the cochlea that lies above the spiral lamina, is filled with perilymph, and is continuous with the scala tympani at the helicotrema.

scald *Medicine.* **1.** a burn caused by hot liquid or by hot, moist vapor. **2.** to burn with hot liquid or steam. *Veterinary Medicine.* a term used to describe the inflammation of feet in young sheep and consequent lameness, as well as any nonprogresive form of foot rot. *Plant Pathology.* any plant disease characterized by discoloration, generally due to heat injury but sometimes caused by intense light or chemical imbalances.

scalded skin syndrome see TOXIC EPIDERMAL NECROLYSIS.

scale *Vertebrate Zoology.* a thin, compacted, flaky element of the protective covering of the body of fish, reptiles, the legs of birds, and the tails of a few mammals. *Botany.* a bract such as that on a catkin or cone. *Medicine.* **1.** a thin fragment of tartar or other concretion on the surface of a tooth. **2.** to remove such deposits from the teeth using an instrument. *Engineering.* to remove rust or salt from a surface. *Mining Engineering.* to remove loose rocks and debris from a tunnel surface after blasting. *Metallurgy.* the oxidation product occurring on a metallic surface during heating in a nonprotective atmosphere.

scale *Engineering.* a balance, or other such device or instrument, used for weighing. *Astronomy.* see LIBRA.

scale *Physics.* **1.** the order of magnitude of some quantity. **2.** a one-to-one correspondence between a physical quantity and some numbering or indexing system, or between two numbering systems, as in the centigrade and kelvin temperature scales. *Cartography.* a guide showing the relationship between a distance on a map or drawing and the actual distance. *Mechanical Engineering.* **1.** a graduated rule or instrument used in mechanical drawing to measure linear dimensions. **2.** to measure distance in this manner. *Acoustics.* a series of notes that are used for the composition of music; the most common types are the Pythagorean or tempered scale, and the diatonic or justly tempered scale. *Graphic Arts.* **1.** a device on a camera, designed to facilitate manual focusing of images. **2.** in lithography and photography, any of various devices having graduated markings or tonal variations. **3.** a percentage indicating the degree to which an image will be enlarged or reduced for printing.

scaleboard *Materials.* a very thin board, used as a veneer or as backing in picture frames. *Graphic Arts.* a thin strip of wood used in justifying lines of linotype.

Scaleboard *Ordnance.* a Soviet surface-to-surface tactical ballistic missile powered by a liquid-propellant motor and probably equipped with inertial guidance, delivering a thermonuclear warhead at a range of 450 miles; officially designated **SS-12.**

scale drawing *Engineering.* a drawing in which each element is shown in the same proportion to an existing or proposed object or structure.

scale effect *Fluid Mechanics.* the effect that is negligible for the prototype, but significant for the model, or vice versa in wind tunnel, hydraulic basin, or other experiments; arises from changing the scale of a body in the fluid flow. *Aviation.* a correction made to account for this effect, as in a wind-tunnel experiment or test.

scale factor *Engineering.* a factor used to multiply the reading of an instrument to give the true final value. *Crystallography.* see ABSOLUTE TEMPERATURE SCALE.

scale height *Geophysics.* a difference between height h and a given height h_0, used to determine the atmospheric pressure at h from the given pressure at h_0.

scale insect *Invertebrate Zoology.* the popular name for various homopteran insects in the family Coccidae; the female adults lose their limbs and attach to plants, secreting a waxy cover that looks like scales; some are serious plant pests, while others produce useful substances such as cochineal and shellac.

scale invariance *Chaotic Dynamics.* an essential feature of fractals, manifested by similarity of details on all scales of magnification; i.e., the small details resemble larger sections of the entire object.

scale leaf *Botany.* a usually thin leaf flattened against other leaves or against the stem, and serving in photosynthesis, in protecting buds, or in holding reserves.

scale model *Graphic Arts.* a three-dimensional model in which each element is shown in the same proportion to an existing or proposed object or structure.

scalene triangle [skā´lēn] *Mathematics.* a triangle having no two sides of equal length; equivalently, a triangle having no two angles equal.

scalenohedron *Crystallography.* a closed crystal form whose faces are scalene triangles.

scalenus *Anatomy.* a group of anterior thoracic muscles that extend from the cervical vertebrae to the upper ribs.

scale of impact see LIFE-CHANGE SCALE.

scale property *Mathematics.* the property that $\|\lambda x\| = |\lambda|\,\|x\|$, where $\|x\|$ indicates the norm of a vector x and λ is a scalar; also called homogeneity (of degree one).

scaler *Medicine.* an instrument used to remove deposits from tooth surfaces. *Electronics.* **1.** a device that generates an output pulse after a specified number of input pulses are received; used to count pulses in digital computers. Also, SCALING CIRCUIT. **2.** a device that lengthens the frequency range of another device. *Radiology.* an electronic instrument that counts radiation-induced pulses from a detector, amplifier, electric window, or other radiation detector.

scale scar *Botany.* a mark on a stem where a scale was once attached.

scale-up *Design Engineering.* a design process in which a model-sized prototype is used to design a large, usually commercial-sized unit.

Scalibregmidae *Invertebrate Zoology.* a family of deepsea mud-ingesting polychaete worms in the subclass Sedentaria.

scalid *Invertebrate Zoology.* a sense organ that contains ciliary receptors and sensory bristles.

scaling *Mechanics.* the expression of each variable and differential element in an equation in nondimensional form by dividing it by a characteristic value of the variable for the problem at hand. By setting these ratios to unity, the relative typical magnitudes of the terms may be directly compared, and those of small magnitude may be dropped for computational convenience but should be validated afterward. Also, NONDIMENSIONALIZING. *Mathematics.* the process of converting a quantity from one notation to another. *Electronics.* **1.** the process of counting pulses with a scaler; used when pulses arrive faster than conventional counting methods can handle. **2.** a method for converting the coefficient of a circuit so that it corresponds to its input-signal terminals. *Computer Science.* a process by which problem variables are related to machine variables in a computer. *Graphic Arts.* the process of calculating the printed size of a reproduction as a percentage of the original copy by any of several methods. *Geology.* see EXFOLIATION.

scaling circuit see SCALER.

scaling factor *Electronics.* the ratio of input pulses to output pulse needed by a scaler. Also, **scaling ratio.** *Engineering.* a factor used in heat-exchange measurements that allows for loss of heat conductivity of a material due to surface scales. *Physics.* a proportionality constant in some linear scaling relationships.

scaling hammer *Mechanical Devices.* a hammer having a keen edge and a ball or straight and cross panes, used to chip scale from steam boilers. Also, BOILERMAKER'S HAMMER.

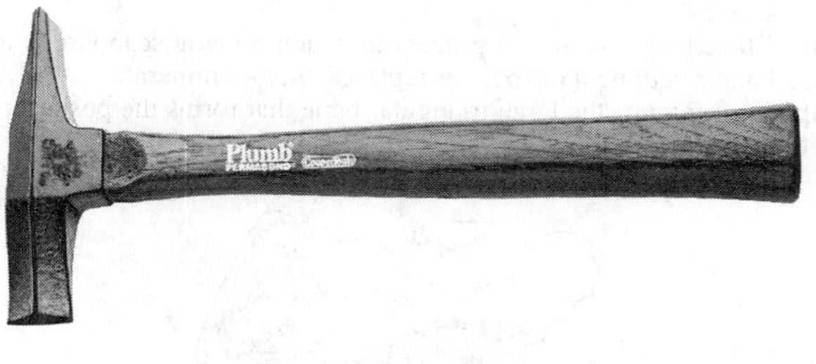

scaling hammer

scaling law *Physics.* a relationship of direct proportionality or linearity between two quantities.

scaling ratio *Engineering.* the ratio of a certain property of a laboratory model to the same property in the natural prototype.

scaling symmetry *Mathematics.* a symmetry of a mathematical object consisting of multiplying existing variables by fixed functions of a single scalar. For example, the partial differential equation $\partial^2 u/\partial x^2 = \partial u/\partial t$ remains invariant when x is replaced by αx and t is replaced by $\alpha^2 t$.

scallion [skal´yən] *Botany.* **1.** a young (green) onion that has not formed a large bulb; used in salads. **2.** see SHALLOT. **3.** see LEEK.

scallop [skal´əp; skäl´əp] *Invertebrate Zoology.* the popular name for various marine bivalve mollusks of the family Pectinidae that are valued as food; their shell has a distinctive arrangement of raised ribs radiating in a fanlike pattern; they swim by flapping the two valves of their shells together to produce a jet of water. Common North American species are the **bay scallop,** *Argopecten irradians,* and the **sea (deep-sea) scallop,** *Placopecten magellanicus. Geology.* **1.** see FLUTE. **2.** see SCALLOPING.

scalloped upland *Geology.* the area of land near or at the divide of an upland, into which glacial cirques have intruded from opposite sides.

scalloping *Geology.* a sedimentary structure having the appearance of an oscillation ripple mark whose concave side is always oriented toward the top of the bed. *Psychology.* the wavy, scalloped response curve caused by fixed-interval reinforcement schedules, where the subject's response rate lessens notably after reinforcement, then increases sharply just before the next reinforcement is scheduled to occur. *Navigation.* in certain navigational systems, an irregularity in the field pattern of the ground station caused by reflection from obstructions or terrain features; it creates cyclical variations in bearing error during flight.

scalp *Anatomy.* the part of the skin of the head, other than the face or ears, that normally is covered with hair. *Agronomy.* to create a furrow slice in an area of existing vegetation, usually so that a new plant species can be introduced. *Metallurgy.* to remove a surface layer from a billet or ingot. Thus, **scalped billet, scalped extrusion ingot.** *Mining Engineering.* **1.** to remove oversize lumps from a stream of bulk material. **2.** to remove large pieces of mine waste from run-of-mine coal on its way to a preparation plant.

scalpel *Surgery.* a small surgical knife having a straight handle and usually a blade with a convex edge. *Mechanical Devices.* a knife having a sharp, thin, usually detachable blade.

Scalpel *Ordnance.* a Soviet intercontinental ballistic missile propelled by a solid-rocket motor and carrying a 10-megaton warhead; officially designated **SS-24.**

Scalpellidae *Invertebrate Zoology.* a family of stalked barnacles in the class Cirripedia having peduncles covered with calcareous plates or scales.

scalpellum *Invertebrate Zoology.* the piercing mouthpart of true bugs and blood-sucking dipteran flies.

scaly anteater see PANGOLIN.

scaly leg *Veterinary Medicine.* a condition of poultry and pet birds caused by burrowing mites and characterized by the formation of rough, hard, irregular scales on the feet and legs.

Scamp *Ordnance.* a Soviet mobile intermediate-range ballistic missile system firing the Scapegoat missile.

scan *Electronics.* to sweep over a specified range, field, or dimension. *Telecommunications.* **1.** the process by which a FAX or television system analyzes or reproduces an image in a signal-line element. **2.** to determine the availability of a communication or data channel. **3.** a control in a radio that searches for each audible frequency in sequence, remaining tuned to each one for a few seconds before proceeding to the next. *Computer Technology.* **1.** to examine sequentially each item in a list, each record in a file, each point of a display, or each input or output channel of a communication link. **2.** a procedure for listing each node in a tree data structure. **3.** the process of reproducing an image from a document onto a display screen. *Medicine.* **1.** to use sensing equipment to gather information about or map a body region. **2.** a compilation of the data gathered by this process. **3.** a visual representation of these data; e.g., a CAT scan.

scan converter *Electronics.* **1.** a device that receives and converts radar images sent over telephone lines or narrow bandwidth radio circuits at an extremely fast rate, generally between three and ten kilohertz. **2.** a tube that can store radar, television, and data displays for long periods of time.

scandent *Botany.* of or relating to the development of a climbing growth habit; climbing.

scandia see SCANDIUM OXIDE.

Scandinavia *Geography.* a peninsula in northwestern Europe occupied by Norway and Sweden. Also, **Scandinavian Peninsula.**

scandium *Chemistry.* a metallic element having the symbol Sc, the atomic number 21, an atomic weight of 44.96, a melting point of 1540°C, and a boiling point of about 2780°C; a rare-earth-like silvery-white metal with a pink tinge that is obtained from thortveite and used in the semiconductor field. (Named for *Scandia* or *Scandinavia,* the source of the ore in which it was first identified.)

scandium chloride *Inorganic Chemistry.* $ScCl_3$, colorless crystals that are very soluble in water; melts at 939°C and sublimes at 800–850°C.

scandium fluoride *Inorganic Chemistry.* ScF_3, a compound derived from the reaction of scandium oxide with ammonium bifluoride, NH_4HF_2; used to prepare scandium metal.

scandium halide *Inorganic Chemistry.* a compound of scandium and a halogen, such as scandium chloride, $ScCl_3$, or scandium fluoride, ScF_3.

scandium oxide *Inorganic Chemistry.* Sc_2O_3, a white amorphous powder that is basic; insoluble in water and soluble in hot acids; used to prepare scandium fluoride. Also, SCANDIA.

scandium sulfate *Inorganic Chemistry.* $Sc_2(SO_4)_3$, colorless crystals; very soluble in hot water; decomposes on heating.

scandium sulfide *Inorganic Chemistry.* Sc_3S_3, a yellowish powder that decomposes in boiling water to give off hydrogen sulfide, H_2S.

scanistor *Electronics.* an optical scanning device that converts images into electrical signals.

scanner a device or system that scans; specific uses include: *Computer Technology.* an input device that surveys a spatial pattern, recognizes characters or other features, and generates signals corresponding to the pattern; used, for example, to input text data directly, rather than having it entered by keyboarding, or to create a computerized version of an existing photo or illustration. *Computer Programming.* a routine, as in a compiler, that scans a sequence of characters and collects them into meaningful units such as words and numbers. *Telecommunications.* **1.** the moving part of a radar antenna that causes the beam to scan. **2.** in FAX transmission, the section of the transmitter that systematically translates the densities of the elemental areas of the subject copy into signals. **3.** in switching systems, a high-speed multiplexed monitor designed to detect changes of state in lines, trunks, or service circuits. *Electronics.* see SCANNING RADIO. *Graphic Arts.* a photoelectric device that scans color artwork or copy and automatically produces color-separation negatives. *Medicine.* any of various devices or systems used to produce a visual representation of a body part, internal organ, and so on, as to examine the brain, locate a tumor, evaluate various body functions, and so on.

scanning the action of a person or device that scans; specific uses include: *Medicine.* the process of examining visually in detail; for example, a small area or different isolated locations. *Telecommunications.* **1.** in television, FAX, or picture transmission, the process of analyzing or synthesizing successively the light values or equivalent characteristics of elements constituting a picture area. **2.** in radar, the process of directing a beam of radio-frequency energy successively over the elements of a given region, or the corresponding process in reception. **3.** in switching systems, the process of sequentially monitoring a plurality of lines, trunks, or service circuits for changes in state.

scanning acoustic microscope *Acoustical Engineering.* a microscope that uses ultrasound, typically by means of a thin piezoelectric transducer to generate the ultrasound, a sapphire pellet to focus the sound, and a raster-scan cathode-ray tube to display an image of a small object with a high level of magnification. Thus, **scanning acoustic microscopy.**

scanning circuit see SWEEP CIRCUIT.

scanning electron microscope *Electronics.* an instrument that enlarges the image of a specimen by simultaneously and systematically sweeping an electron beam over the specimen and a cathode-ray tube display of the electric signal created when the beam first struck the specimen. Thus, **scanning electron microscopy.**

scanning head *Electronics.* a device consisting of a light source and a phototube; used to scan material in a photoelectric control system.

scanning HEED *Physics.* a technique used in electron diffraction analysis similar to that of reflection HEED, except that the diffraction pattern is read by scanning across the field of diffracted electrons with sensitive detectors, rather than recording the pattern on an emulsion. (An acronym for high-energy electron diffraction.)

scanning line *Telecommunications.* **1.** a path traced by a scanning or recording spot in a single sweep across the subject copy. **2.** in television transmission, a single thin strip produced by the scanning process.

scanning linearity *Electronics.* the consistency of the scanning speed in a television picture tube.

scanning radio *Electronics.* a device that automatically sweeps across radio bands and stops at the first preselected channel that is currently on the air. Also, SCAN, RADIO SCAN.

scanning radiometer *Engineering.* a radiometer with mirrors rotating at a 45° angle to the optical axis that image the entire path around the instrument.

scanning sequence *Telecommunications.* the order in which points are scanned.

scanning sonar *Engineering.* an echo-ranging system in which the outgoing ping is transmitted simultaneously throughout the entire search area and a rapidly rotating narrow beam scans for the returning echoes.

scanning switch see COMMUTATOR SWITCH.

scanning transmission electron microscope *Electronics.* an instrument that enlarges the image of a specimen by penetrating it with an extremely narrow electron beam, and then employs a detection device to produce an image by the electrons transmitted through the sample. Thus, **scanning transmission electron microscopy.**

scansorial *Biology.* describing a plant or animal that is characterized by climbing or adapted for climbing.

scantling *Building Engineering.* **1.** a small piece of timber 2–4 inches thick, used primarily for studding. **2.** a building stone of a length greater than 6 feet. *Naval Architecture.* **1.** a wooden or metal framing member. **2.** the thickness of plating or planking, and cross dimensions of frames, beams, and other structural members.

Scapaniaceae *Botany.* a family of liverworts of the order Jungermanniales, characterized by predominantly lateral-intercalary branching, stems having a cortex of thick-walled cells, scattered rhizoids, and sharply spreading complicate bilobed leaves.

Scapanorhychidae *Vertebrate Zoology.* an alternate name for the family Lamnidae, the mackerel sharks.

scape *Botany.* a leafless stem that rises from a rosette of leaves and bears a flower, several flowers, or an inflorescence, as on a dandelion. *Entomology.* **1.** the basal segment of the antenna in insects. **2.** the stalk of the haltere in dipteran insects.

scapegoat *Psychology.* a person or group toward whom displaced aggression is directed.

Scapegoat *Ordnance.* a Soviet two-stage missile powered by solid-propellant motors and delivering a nuclear or thermonuclear warhead at a range of 2200 miles; officially designated **SS-14.**

scapewheel see ESCAPE WHEEL.

scapha *Anatomy.* the long, curved depression on the external ear that separates the helix from the antihelix. Also, FOSSA HELICIS.

Scaphidiidae *Invertebrate Zoology.* a small family of shining fungus beetles, coleopteran insects in the superfamily Staphylinoidea that live among fungi, dead wood, rotting leaves, and bark.

Scaphites *Paleontology.* a heteromorphic genus of loosely coiled ammonoids; characterized by a straight segment between the last-formed coils and the first-formed coils; one of the later ammonoids, known only from the Cretaceous.

scapho- or **scaph-** a combining form meaning "boat."

scaphocephaly *Medicine.* a condition in which the head is abnormally long and narrow and shaped like a boat, as a result of premature closure of the sagittal suture; it usually results in mental retardation.

scaphoid *Anatomy.* **1.** shaped like a boat. **2.** the largest and most lateral of the proximal row of carpal bones.

Scaphopoda *Invertebrate Zoology.* the tusk or tooth shells, a class of burrowing mollusks, having a foot and tentacles projecting from a curved conical shell that is open at both ends.

scapolite *Mineralogy.* **1.** a group name for the tetragonal silicate series marialite, $Na_4Al_3Si_9O_{24}Cl$ to $Ca_4Al_6Si_6O_{24}(CO_3,SO_4)$. **2.** any member of this series.

scapolitization *Geology.* a process in which mineral scapolite is introduced into a sediment or rock, or replaces another mineral.

scapula *Anatomy.* the large triangular bone that forms the posterior part of the shoulder; the shoulder blade.

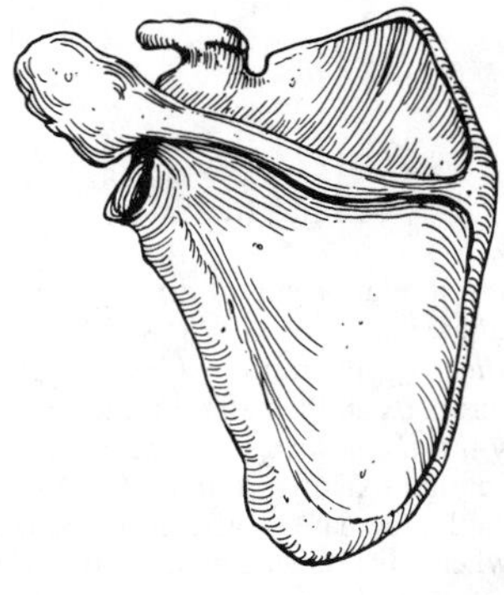

scapula

scapulet *Invertebrate Zoology.* in scyphozoan jellyfish, any of a number of projections on the outer surface of each arm.

scapulothoracic *Anatomy.* relating to the shoulder blade and the chest.

scapulus *Invertebrate Zoology.* in some sea anemones, a modification between the capitulum (upper, thin-walled) and the scapus (lower, thick-walled) body regions.

scapus *Botany.* having a leafless flowering stalk arising from the ground.

scar *Medicine.* a noticeable mark remaining after the healing of a wound. *Mycology.* in fungi, a mark on the organism where cell separation occurred as a result of fission or budding. *Geology.* **1.** a steep, rocky eminence or slope in which bare rock is prominently exposed. **2.** see SHORE PLATFORM.

SCARA *Robotics.* a Japanese robot that is designed to move on a horizontal plane while performing assembly tasks. (An acronym for Selective-Compliance Assembly Robot Arm.)

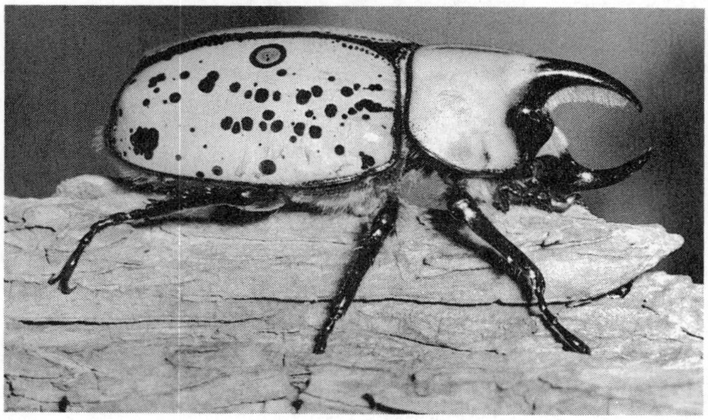

scarab (rhinoceros beetle)

scarab *Invertebrate Zoology.* **1.** any of various beetles belonging to the superfamily Scarabaeoidea, such as the rhinoceros beetle. **2.** specifically, *Scarabaeus sacer,* a large black Mediterranean dung beetle, the sacred scarab of ancient Egypt. Also, **scarab beetle.** *Archaeology.* an image or representation of a beetle, very common in ancient Egypt. *Mineralogy.* any cut gem whose shape suggests that of a beetle.

Scarab *Ordnance.* a Soviet short-range ballistic missile with a range of 75 miles; officially designated **SS-21.**

scarabaeid *Invertebrate Zoology.* **1.** relating or belonging to the family Scarabaeidae. **2.** any member of this family.

Scarabaeidae *Invertebrate Zoology.* the scarab beetles, a family of coleopteran insects in the superfamily Scarabaeoidea, often bright-colored, including dung, skin, earth-boring, flower, June, rhinocerus, Hercules, and elephant beetles and cockchafers; larvae of many species are plant pests; adults of many species feed on dung, which they roll into balls and bury.

Scarabaeoidea *Invertebrate Zoology.* a superfamily of beetles, coleopteran insects in the suborder Polyphaga, sometimes considered coextensive with the family Scarabaeidae, or with the superfamily Lamellicornia.

scarabiasis *Medicine.* an invasion of the intestinal tract by the dung beetle, usually found in children, and marked by intestinal disturbances, anorexia, and emaciation.

scarce mRNA *Molecular Biology.* a population of RNA molecules that consists of separate mRNA species. Also, COMPLEX MRNA.

scarf *Engineering.* the end of any piece in a scarf joint.

scarf cloud *Meteorology.* see PILEUS.

scarfing *Metallurgy.* the process of removing local surface defects from an ingot or billet.

scarf joint *Engineering.* **1.** an oblique joint formed by notched and lapped timbers placed end-to-end which are then bolted, riveted, or glued together. **2.** a tapered joint formed by applying a bevel to each part to be joined. *Materials Science.* an acute-angled butt joint, used in adhering composites.

Scaridae *Vertebrate Zoology.* the parrotfishes, a family of the suborder Labroidei in the order Perciformes; characterized by teeth and jaws that are fused into a parrotlike beak, which is used to scrape algae and coral from coral reefs.

scarifier *Engineering.* a machine with downward projecting teeth for breaking hard soil at quarries and opencast pits and on asphalt roads prior to rebuilding or resurfacing.

scarify *Medicine.* to make a series of small incisions or punctures in the skin, as for the introduction of smallpox vaccine. *Engineering.* to break up a surface, especially using a scarifier. Thus, **scarification.**

scarious *Botany.* of or relating to a plant or plant part that is thin and membranous, more or less transluscent.

scarlatina see SCARLET FEVER.

scarlet see SCARLET RED.

scarlet fever *Medicine.* an acute, contagious disease of childhood, caused by a strain of streptococcus, and giving rise to chills, sore throat, vomiting, rapid pulse, rash, and fever. (From the bright red color of the skin rash.) Also, SCARLATINA.

scarlet fever streptococcus antitoxin *Immunology.* a substance that is composed of a sterile aqueous solution of antitoxins derived from animals that have been immunized against group A beta hemolytic streptococci toxin; formerly used to treat and produce an immunity against the disease scarlet fever.

scarlet fever streptococcus toxin *Immunology.* a substance that causes the scarlet fever rash, and that consists of a toxic filtrate of cultures of *Streptococcus pyogenes.*

scarlet pimpernel see PIMPERNEL.

scarlet red *Organic Chemistry.* $C_{24}H_{20}N_4O$, a dark-brown powder that melts at 181–188°C, and is soluble in chloroform and petroleum ether; used as a dye and in making medicinal ointments. Also, SCARLET.

scarp *Geology.* **1.** a line of cliffs or a steep, straight, clifflike slope of any height rising above the surrounding land that is produced by faulting or by differential erosion. **2.** a steeply inclined surface on the undisturbed ground along the margins of a landslide.

scarp

Scarp *Ordnance.* **1.** a Soviet two-stage intercontinental ballistic missile that is powered by liquid-propellant motors and equipped with inertial guidance, delivering one 20-megaton thermonuclear warhead or three smaller charges at a range of 5500 miles; officially designated **SS-9. 2.** An alternate version designed to deliver a thermonuclear warhead into earth orbit from which it can be controlled and dropped onto its target; officially designated **FOBS**, an acronym for Fractional Orbital Bombardment System.

Scarpa's fascia *Anatomy.* the deep layer of subcutaneous abdominal fascia that borders the inguinal area. (Named for Anthony *Scarpa,* 1747–1832, Italian anatomist and surgeon.)

Scarpa's fluid see ENDOLYMPH.

scarped *Geology.* cut by or into steep slopes.

scarped plain *Geology.* a land surface characterized by a succession of faintly inclined or gently folded strata.
scarp-foot spring *Hydrology.* a spring flowing onto the surface of the land at or near the foot of an escarpment.
scarpland *Geology.* an area characterized by series of parallel or nearly parallel scarped ridges separated by ravines or vales.
scarplet see PIEDMONT SCARP.
scarp slope *Geology.* the steeper face of a cuesta, oriented in the direction opposite that of the dip of the strata. Also, INFACE, FRONT SLOPE.
scarp stream *Hydrology.* a stream flowing down a scarp in a direction opposite to that of the dip of the local strata.
scar tissue *Medicine.* the new tissue formed in the healing of a wound.
SCAT supersonic commercial air transport; sheep cell agglutination test.
scatchard plot *Molecular Biology.* a graph used to study molecular interaction in a mixed population of molecules in which different types of molecules are bound reversibly to one another. Also, **Scatchard plot.**
Scatopsidae *Invertebrate Zoology.* a family of minute, black scavenger flies, dipteran insects in the series Nematocera.
scatoscopy *Medicine.* the examination of feces for diagnostic purposes.
scatter *Radiology.* the deflection of some incidental radiation by a target material through which the rays pass.
scatterable mine *Ordnance.* a mine laid without regard to a pattern.
scatter band *Telecommunications.* in pulse interrogation systems, the total bandwidth inhabited by the frequency disseminated by various interrogations operating on the identical nominal radio frequency.
scatter diagram *Statistics.* a graph that displays the relationship between two variables by plotting pairs of observations. *Petrology.* see POINT DIAGRAM.
scattered *Meteorology.* of or relating to a sky cover of 0.1 to 0.5, used only when clouds or other obscuring phenomena aloft are present.
scattered rays *Radiology.* secondary rays whose direction has been changed by interaction with matter during passage through a subject.
scatter gun *Ordnance.* a term for a shotgun, especially a shotgun with a shortened barrel to widen the dispersion of the shot.
scattering *Physics.* a collisional process by which a particle (such as an electron, photon, or neutron) of known incident energy or direction is scattered off of a target specimen, resulting in a different energy and direction. *Electromagnetism.* the random diffusion of electromagnetic radiation from air masses, which aids the long-range propagation of radio signals over geographic obstacles such as mountains. *Acoustics.* the spreading of sound in many directions, due to irregularities in the transmission medium or due to an object placed in the sound field.
scattering amplitude *Quantum Mechanics.* a quantity specifying the far-field intensity of scattered radiation as a function of scattering direction.
scattering angle *Physics.* the angle between the incident and scattered directions in a scattering process. Also, **scatter angle.**
scattering centers *Materials Science.* any imperfection in a crystal lattice that causes scattering of the conduction electrons, resulting in increased electrical resistivity.
scattering coefficient *Physics.* the fractional decrease in intensity per unit length of penetration of a beam of particles or radiation due to scattering. *Electromagnetism.* a transmission or reflection coefficient used in a scattering matrix associated with a waveguide junction.
scattering cross section *Physics.* an area that measures the probability of scattering. *Electromagnetism.* the ratio of scattered electromagnetic power from an antenna to the intensity incident on the antenna.
scattering factor *Crystallography.* see ATOMIC SCATTERING FACTOR.
scattering function *Electromagnetism.* for a given direction, the intensity of electromagnetic radiation scattered from an object per lumen of flux incident on the object.
scattering layer *Oceanography.* a midwater layer of organisms that scatters sound and returns sonar echoes.
scattering length *Nuclear Physics.* a parameter used to evaluate the distribution of nuclear particles at low energies. Also, **scattering power.**
scattering loss *Electromagnetism.* a loss in electromagnetic power associated with scattering from within a medium or diffusion from the rough surface of a medium.
scattering matrix *Electromagnetism.* a square matrix whose elements are usually complex quantities representing reflection and transmission coefficients associated with a waveguide junction. *Quantum Mechanics.* a matrix that, when multiplied by the state vector of a two-particle system before scattering, gives the state vector after scattering.
scattering theory *Physics.* a theory concerned with the determination of the scattering, distribution, and amplitudes from the knowledge of the structure of the target, equations of motion, and the potential field.
scatter loading *Computer Programming.* a method of loading a program into main memory so that sequential segments of the program are not loaded into adjacent memory locations.
scatterometer *Engineering.* a radar with a wide sweep that maps a terrain.
scatter propagation *Electromagnetism.* the transmission of electromagnetic waves by scattering from the troposphere or ionosphere.
scatter read *Computer Programming.* the process of loading data into noncontiguous areas of storage from a single input record, or block. The reverse of this process is called **gather write.**
scatter reflections *Electromagnetism.* electromagnetic waves that are reflected from the troposphere and ionosphere at different virtual heights and thus interfere with each other, thereby causing rapid decay of the signal.
scaup *Vertebrate Zoology.* either of two wild diving ducks of the genus *Aythya*, characterized by a bluish-gray broad bill and a glossy head and neck; the **greater scaup,** *A. marila,* or the **lesser scaup,** *A. affinis.*

lesser scaup

scavenge *Metallurgy.* to clean molten metal by introducing a substance that combines chemically with unwanted gases.
scavenger *Ecology.* **1.** any organism that feeds on dead matter or other organic refuse. **2.** a carnivorous animal, such as a hyena or vulture, that feeds mainly on animals which are already dead rather than pursuing live prey. *Chemistry.* any chemically reactive substance added to a system to reduce trace impurities by competitive reaction with the impure species. *Metallurgy.* specifically, a material added to scavenge a molten metal or alloy.
scavenger system *Ordnance.* a component of a firearm that clears the smoke and gases from the chamber of the bore after firing.
scavenger well *Hydrology.* a well that is situated between a source of potential contamination and a well or wells that yield usable water, used to prevent the contaminated water from reaching the useful well or wells.
scavenging *Archaeology.* the nonscientific removal of archaeological materials from a site by later inhabitants of the same site. *Metallurgy.* the process of removing unwanted gases from a molten metal or alloy by adding a scavenger. *Ordnance.* the process of using a blast of air to clear the smoke and gases produced by the firing of a gun. *Mechanical Engineering.* the process of discharging exhaust gases from the cylinder during an internal combustion engine cycle and replacing it with combustible gases or air.
scavenging stroke see EXHAUST STROKE.
SCD sudden cardiac death.
ScD or **Sc.D.** Doctor of Science. (From Latin *scientiae doctor.)*
ScDP or **Sc.D.P.** scapulodextra anterior.
Scelionidae *Invertebrate Zoology.* a family of tiny, dark, shining wasps, hymenopteran insects in the superfamily Proctotrupoidea, with 11- or 12-segmented, elbowed antennae; some are parasites of insect plant pests.
scend *Navigation.* **1.** of a ship, to heave upward. **2.** a sudden upward heave of a vessel on the swell of a wave.
scene analysis *Robotics.* the use of image understanding to develop machine-vision systems for robots.
Scenedesmaceae *Botany.* a family of freshwater and terrestrial green algae of the order Chlorococcales, characterized by asexual reproduction by autospores that cohere to form variously shaped coenobia.

scenedesmus *Botany.* the most common green algae genus of the family Scenedesmaceae, distinguished by sexual reproduction in an otherwise asexual family.

scent gland *Vertebrate Zoology.* a specialized skin gland producing an odiferous substance in animals, such as the beaver or civet cat, that functions as a social or sexual signal or as a defensive weapon.

scent marking *Behavior.* the action of identifying a territory for other members of the species with an odorous substance, such as urine, feces, saliva, or glandular secretions.

scfh *Fluid Mechanics.* the cubic feet per hour of gas flow at specified standard temperature and pressure.

Schaeffer acid *Organic Chemistry.* $C_{10}H_6OHSO_3H$, white leaflets; soluble in water and alcohol; melts at 167°C; used as an azo dye intermediate. Also, β-NAPHTHOLSULFONIC ACID.

Schaeffer's salt *Organic Chemistry.* $C_{10}H_6OHSO_3Na$, a yellow to pink powder that is soluble in water; used in organic synthesis.

schafarzikite *Mineralogy.* $Fe^{+2}Sb_2^{+3}O_4$, a metallic, red to reddish-brown tetragonal mineral occurring as prismatic crystals, having a specific gravity of 4.3 and a hardness of 3.5 on the Mohs scale.

schairerite *Mineralogy.* $Na_{21}(SO_4)_7F_6Cl$, a colorless, transparent trigonal mineral occurring as minute crystals, having a specific gravity of 2.61 to 2.63 and a hardness of 3.5 on the Mohs scale; found at Searles Lake, California.

Schaller, George B. born 1933, American zoologist; a popular writer on large mammals and endangered species.

Schally, Andrew V. born 1926, Polish-born American physiologist; shared the Nobel Prize for research in pituitary hormone secretion.

schalstein *Petrology.* a slaty rock consisting of an altered basic or calcareous tuff with shear structures.

Schardinger enzyme see XANTHINE OXIDASE.

scharnitzer *Meteorology.* a cold northerly wind that blows for long periods over Tyrol, Austria.

Schatzki crankshaft motion *Materials Science.* a main chain motion found in hydrocarbon-based polymers in which CH groups rotate in the manner of an old-time crank shaft.

Schauder's theorem *Mathematics.* **1.** let *S* be a compact, convex subset of a Banach space; then any continuous function from *S* to itself has at least one fixed point. Also, **Schauder's fixed-point theorem. 2.** a linear operator between Banach spaces is compact if and only if its dual is compact.

Schawlow, Arthur L. born 1921, American physicist; shared the Nobel Prize for his contributions to laser technology and research.

Scheat *Astronomy.* Beta (β) Pegasi, an irregular variable red giant star in the constellation Pegasus.

schedule *Science.* any plan of procedure indicating the date and time at which certain events will or should take place. *Transportation Engineering.* specifically, a plan that designates the time and sequence of each trip on a line, route, or system; a timetable.

scheduled aircraft mile *Transportation Engineering.* a route mile in an airline's flight schedule, not including "second section" aircraft carrying overflow passengers on a given scheduled flight.

scheduled down time *Computer Technology.* the time when a computer device is inactive due to normal servicing.

scheduled fire *Military Science.* prearranged fire executed at a predetermined time.

scheduled target *Military Science.* **1.** in artillery and naval gunfire support, a planned target on which fire is to be delivered at a specific time. **2.** (**nuclear**) a planned target on which a nuclear weapon is to be delivered at a specific time during the operation of the supported force; the time may be specified in military time or in terms of the accomplishment of a predetermined movement or task.

schedule of fire *Military Science.* groups or series of fires delivered in a definite sequence according to a definite program; often referred to by a code name or other designation.

schedule of reinforcement *Behavior.* **1.** the relationship between the occurrence of a response and the provision of a reinforcing event. **2.** in operant conditioning, the prescription or rule for distributing reinforcements contingent upon behavior. Also, REINFORCEMENT SCHEDULE.

schedule of targets *Military Science.* in artillery and naval gunfire support, individual targets, groups, or series of targets to be fired on in a definite sequence according to a definite program.

scheduler *Science.* any person or device that plans or carries out a schedule. *Computer Programming.* specifically, a computer program, usually part of the operating system, that controls the scheduling, initiation, and termination of jobs.

scheduling algorithm *Computer Programming.* in a multiprogramming system, a set of rules within the scheduling routine of the operating system that will determine which of several processes will use nonsharable system resources.

Scheele, Karl 1742–1786, Swedish chemist; independently discovered oxygen and nitrogen; discovered chlorine, glycerin, and many acids.

Scheele's green see CUPRIC ARSENITE.

scheelite *Mineralogy.* $CaWO_4$, a yellowish-white to brown tetragonal mineral forming a series with powellite, occurring in octahedral or tabular crystals and in massive form, and having a specific gravity of 6.10 and a hardness of 4.5 to 5 on the Mohs scale; found in contact metamorphic deposits, veins, and pegmatites, and mined as an ore of tungsten.

Scheffel engine *Mechanical Engineering.* a multirotor engine having nine impellers that all rotate in the same direction.

schefferite *Mineralogy.* $(Na,Ca)(Fe^{+3},Mn^{+3})Si_2O_6$, a brown to black manganoan variety of aegirine, a monoclinic mineral of the pyrozene group.

schefflera *Botany.* **1.** any of various trees or shrubs of the genus *Schlefera* of the ginseng family, characterized by showy, glossy, palmately compound leaves; found in the tropics and cultivated widely as a house plant. **2.** a related plant, *Brassaia actinophylla.*

Scheibel extractor *Chemical Engineering.* a continuous liquid-liquid contacting extraction tower that uses mesh packed sections and mechanical agitation to disperse the liquids into droplets. Also, **Scheibel column, Scheibel-York extractor.**

Scheie's syndrome *Medicine.* a hereditary condition characterized by the excretion of excess chondroitin sulfate B in the urine, cloudiness of the cornea, unusual facial appearance, and aortic valve disorder. Also MUCOPOLYSACCHARIDOSIS III.

schema [skē´mə] *plural,* **schemata.** *Science.* a general structure for recording data or events. *Psychology.* **1.** any outline, structure, or plan that creates a framework for cognitive functioning. **2.** in Jean Piaget's theory, the earliest form of a cognitive concept, occurring before age two and sensorimotor in nature. *Computer Programming.* a description of the logical organization, structure, and content of a database. (From the Greek word for "form.")

schematic [skē mat´ik] *Science.* relating to or having the form of a scheme or schema. *Graphic Arts.* see SCHEMATIC DRAWING.

schematic diagram *Electricity.* a graphical representation of a circuit, using lines to represent wires and various symbols to represent components. Also, CIRCUIT DIAGRAM. *Graphic Arts.* see SCHEMATIC DRAWING.

schematic drawing *Graphic Arts.* a drawing showing the structural relationships among elements of an object or system.

scheme *Science.* a diagram, map, or other such graphic representation of information.

Scheme *Artificial Intelligence.* a small, elegant dialect of Lisp that is sometimes used for teaching Lisp and for applications.

scheme of maneuver *Military Science.* a tactical plan to be executed by a force in order to seize assigned objectives.

schemochrome *Zoology.* colored, as a feather, not through any inherent pigmentation, but due to the refraction of light.

Schering bridge *Electricity.* a bridge designed to measure capacitance and dielectric loss; used also for the study of insulation at high voltages.

Scheuchzeriaceae *Botany.* a monospecific family of rhizomatous and monocotyledonous bog herbs of the order Najadales, characterized by vessels confined to the roots and unique in the order in having a bract beneath each flower.

Schiaparelli, Giovanni 1835–1910, Italian astronomer; discovered the "canals" of Mars; showed that meteors traverse space in cometary orbits.

Schick, Béla 1877–1967, Hungarian-born U.S. physician; developed the Schick test.

Schick test *Immunology.* a skin test used to determine an individual's resistance to diptheria toxin.

Schiff base *Organic Chemistry.* any one of a class of compounds having the formula *RR'C=NR''*, some occurring as colorless crystalline solids hydrolyzed by water and strong acids to form carbonyl compounds and amines; used in liquid crystals in electronic display systems, dyes, and rubber accelerators.

Schiff's reagent *Analytical Chemistry.* an aqueous solution of sulfurous acid and rosaniline hydrochloride (fuchsin); used in the Schiff test for aldehydes, and also as a nuclear stain.

Schiff test *Analytical Chemistry.* a test using Schiff's reagent to distinguish aldehydes from ketones; a purplish color indicates a positive test for aldehydes.

Schilbidae *Vertebrate Zoology.* a family of slender, elongate catfishes of the order Siluriformes, characterized by a very long anal fin; found in the freshwaters of Africa, southern India, and southeastern Asia.

Schilder's disease *Medicine.* a group of subacute or chronic neurological diseases beginning in childhood, characterized by demyelination of the white matter of the brain, blindness, deafness, bilateral spasticity, and progressive mental deterioration. Also, PERIAXIAL ENCEPHALITIS.

schiller *Optics.* the iridescent play of color seen in certain crystals under suitable illumination that arise from the diffraction, reflection, and scattering of light in surface layers of the crystal by spherical particles.

Schiller *Astronomy.* an elliptical walled plain about 112 miles long and 60 miles wide, located in the third quadrant of the face of the moon.

schillerization *Optics.* the development of schiller in a crystal.

Schindleriidae *Vertebrate Zoology.* a family of small neotenic fish composing the suborder Schindlerioidei that are transparent and have many larval functions; found in the tropical Pacific Ocean.

Schindlerioidei *Vertebrate Zoology.* a suborder of fish of the order Perciformes, containing the single family Schindleriidae.

schindylesis *Anatomy.* a bony articulation in which the thin portion of one bone fits into a crevice in another bone, as in the articulation of the perpendicular plate of the ethmoid bone with the vomer.

schirmerite *Mineralogy.* a gray-black, opaque, metallic, orthorhombic mineral ranging in composition from $Ag_3Pb_3Bi_9S_{18}$ to $Ag_3Pb_6Bi_7S_{18}$, occurring in finely granular and massive form, and having a specific gravity of 6.737 and a hardness of about 2 on the Mohs scale; found disseminated in quartz in veins with galena.

schis- or **schiso-** a combining form meaning "split."

Schisandraceae *Botany.* a family of tropical and subtropical evergreen or deciduous, aromatic woody vines of the order Illiciales, characterized by arillate seeds and small flowers in spikes or racemes with overlapping tubular bracts subtending each flower.

schist [shist] *Geology.* a strongly foliated, crystalline, coarse-grained metamorphic rock with lamellar mineral constituents that allow it to split easily into thin flakes or slabs.

schist- or **schisto-** a prefix meaning "split."

schist-arenite *Petrology.* a light-colored sandstone with more than 20% fragments derived from rocks in an area of regional metamorphism.

Schistochilaceae *Botany.* a family of dioecious liverworts of the order Jungermanniales, often pendulous and possessing brown to purplish pigments.

schistose *Geology.* characteristic of, resembling, relating to, or having the nature of schist.

schistosis *Medicine.* fibrosis of the lungs that is caused by inhalation of dust from slate.

schistosity *Geology.* the tendency of a foliated rock to split into thin flakes as a result of the parallelism of its constituent mineral grains.

schistosome see BILHARZIA.

schistosome dermatitis *Medicine.* an itching dermatitis caused by larvae of the schistosoma parasitic fluke burrowing into the skin.

schistosomiasis [shis´tə sō mī´ə sis] *Medicine.* an infection with parasitic flukes of the genus *Schistosoma* that may affect the skin, intestines, liver, vascular system, or other organs. The infection, most prevalent in Asia and the tropics, is transmitted to humans through feces-contaminated water. Also, BILHARZIASIS.

Schistostegaceae *Botany.* a monotypic moss family of the order Bryales composed of the rare but widespread *Schistostega pennata*; characterized by small, delicate plants forming loose, golden, shiny tufts in caves and cavities and growing from a persistent protonema that forms chains and plates of vesicular cells under low light conditions. Popularly called **luminous moss** or **goblin gold.**

schiz- or schizo- a combining form meaning "split."

Schizaeaceae *Botany.* a diverse family of terrestrial ferns of the order Filicales, characterized by sporangia with apical annules and by a hairy or scaly short-creeping rhizome.

schizaxon *Neurology.* the axon of a neuron that is divided into two equal or nearly equal branches.

Schizoblastosporion *Mycology.* a genus of yeast fungi belonging to the class Hyphomycetes that contains a single species found in soil.

schizocarp *Botany.* a dry compound fruit that is segregated into single-seeded segments and splits apart upon maturity.

Schizocoela *Invertebrate Zoology.* a group of phyla of invertebrates with a type of coelom (body cavity) formed by the splitting of the embryonic mesoderm into two layers.

schizogamy *Biology.* reproduction in which a sexual form is produced by fission or budding from an asexual part, as in certain polychaetes.

schizogenesis *Biology.* reproduction by fission.

schizognathous *Vertebrate Zoology.* describing birds that have an open palate with a small vomer and ununited maxillopalatine bones.

Schizogoniales *Botany.* in some systems of classification, a small order of the green algae division Chlorophyta, containing submicroscopic and small macroscopic organisms that attach to rocks by rhizoids.

schizogony *Invertebrate Zoology.* asexual reproduction by multiple fission, as in the life cycles of some protozoans; in sporozoans, a sporozoite encysts in a host cell, becoming a schizont which fissions asexually to produce merozoites; each merozoite enters a new host cell, becoming a schizont; the cycle repeats up to three times, after which the schizont becomes either male or female and gamogony ensues.

schizoid [skit´soid] *Psychology.* **1.** relating to or having a schizoid personality disorder. **2.** having some symptoms of schizophrenia but not actually diagnosed as schizophrenic; other members of the family of a true schizophrenic are often so identified.

schizoid personality (disorder) *Psychology.* a personality disorder characterized by emotional coldness, indifference to the feelings of others, and general withdrawal; not generally considered a form of true schizophrenia. Also, **schizoid character.**

schizolite *Mineralogy.* a light-red to brown manganoan variety of pectolite.

Schizomeridaceae *Botany.* in some systems of classification, a family of green algae in the order Ulvales.

Schizomida see UROPYGI.

Schizomycetes *Bacteriology.* a former taxonomic class containing all bacteria, defined as unicellular microorganisms that reproduce by cell division; the ten orders include Actinomycetales, Beggiatoales, Caryophanales, Chlamydobacteriales, Eubacteriales, Hyphomicrobiales, Mycoplasmatales, Myxobacterales, Pseudomonadales, and Spirochaetales.

Schizomycophyta *Botany.* in former systems of classification treating bacteria as plants, the subdivision containing the bacteria.

schizont *Invertebrate Zoology.* in some parasitic sporozoans, a multinucleate trophozoite that divides into merozoites within a host cell.

Schizopathidae *Invertebrate Zoology.* a family of corals in the order Antipatharia, having an axial skeleton composed of a black, horny material bearing thorns.

schizopelmous *Vertebrate Zoology.* of certain birds in which the two flexor tendons of the toes are separate and the hallux flexor goes to the first toe only.

Schizophora *Invertebrate Zoology.* a division of dipteran flies in the superfamily Muscoidea.

schizophrenia [skit´sə fren´ē ə; skit´sə frē´nē ə] *Psychology.* a general term for a number of severe mental disorders involving disturbed thought processes, withdrawal from reality, and various emotional and behavioral symptoms; the four basic categories of schizophrenia are catatonic, hebephrenic, paranoid, and simple.

schizophrenic [skit´sə fren´ik] *Psychology.* **1.** a person affected with schizophrenia. **2.** of or relating to schizophrenia. (From a Greek phrase meaning literally "split mind.")

schizophrenogenic *Psychology.* causing schizophrenia.

Schizophyceae *Bacteriology.* in former classifications, the blue-green algae, which are now classified as the blue-green bacteria of the family Cyanophyceae and commonly termed Cyanobacteria, occurring in both salt and fresh water, and often causing pollution of drinking water.

Schizophyllaceae *Mycology.* a family of fungi belonging to the order Agaricales that occur on soil, dung, wood, and plants, or live in a symbiotic relationship with plants; used widely in genetic research.

Schizophyllum *Mycology.* the single genus of fungi belonging to the family Schizophyllaceae, which are characterized by fruiting bodies that are in the shapes of fans or brackets; they cause wood rot and human diseases.

Schizophyta *Botany.* in former systems of classification, the division containing the prokaryotes, including bacteria and blue-green algae.

schizopod *Invertebrate Zoology.* of or belonging to the former order Schizopoda, now divided into the orders Mysidacea, containing the opossum shrimps, and Euphausiacea, containing krill; soft-shelled shrimplike crustaceans with branched limbs.

schizopod larva *Invertebrate Zoology.* in crustaceans such as lobsters and crabs, a shrimplike larval stage.

Schizopteridae *Invertebrate Zoology.* a family of tiny hemipteran ground bugs that live in leaf mold and jump when disturbed.

Schizopyrenida *Invertebrate Zoology.* an order of naked, lobose amoebae that occur in soil and fresh water, feeding on bacteria; one parasitic form, *Naegleria fowleri*, causes fatal amoebic meningoencephalitis in humans.

schizorhinal *Vertebrate Zoology.* having a deep cleft on the posterior border of the bony nostrils, as in pigeons, shorebirds, and certain other birds.

Schizosaccharomyces *Mycology.* a genus of fungi belonging to the family Saccharomycetaceae that contains the fission yeasts; species are characterized by the division of each cell into two daughter cells of the same size and are found in fermented beverages and fruit juices.

Schizosaccharomyces pombe *Mycology.* a species of fungi of the genus *Schizosaccharomyces*; used widely in cell cycle research.

schizothecal *Vertebrate Zoology.* having the horny covering of the tarsus divided into scalelike plates, as in most birds.

schizothoracic *Invertebrate Zoology.* having a large prothorax that is loosely articulated with the thorax.

Schizothyriaceae *Mycology.* a family of fungi belonging to the order Asterinales that occur on leaves, stems, or twigs in temperate to tropical regions.

schizotrypanum *Invertebrate Zoology.* in some systems of classification, a genus of flagellate protozoans that cause Chagas' disease.

schizotypal personality (disorder) *Psychology.* a personality disorder characterized by eccentric thought, speech, and behavior and social withdrawal; similar to but not as severe as schizophrenia.

Schläfli's integral formulas *Mathematics.* representations of the classical orthogonal polynomials as contour integrals; obtained by inserting Rodrigues' formula for the polynomials into Cauchy's integral formula. The Schläfli integral formula for the *n*th Legendre polynomial is

$$P_n(z) = (1/2\pi i)\int_C [(t^2 - 1)^n]/[2^n(t - z)^{n+1}]dt.$$

Schleiden, Matthias Jakob [shlī´dən] 1804–1881, German botanist; discovered cytoblasts, thus laying the foundation for Schwann's cell theory.

Schlemm's canal *Anatomy.* a venous sinus in the sclera of the eye that encircles the cornea and through which the anterior aqueous humor leaves the eye.

schlepper *Immunology.* a macromolecular substance that combines with a weak antigen to increase its potency. Also, **Schlepper.** (From a Yiddish word meaning "to pull or drag.")

Schlernwind *Meteorology.* an east wind that blows down from the Schlern near Bozen in Tyrol, Austria.

schlicht function see UNIVALENT FUNCTION.

Schlick vibration formula *Naval Architecture.* a formula used to calculate hull vibration characteristics.

Schliemann, Heinrich [shlē´män] 1822–1890, German archaeologist; discovered and excavated Troy, Tiryns, Mycenae, and Ithaca.

schlieren *Petrology.* irregular streaks, several centimeters to tens of meters in length, occurring in igneous rocks as the result of differences in mineral ratios despite the same general mineralogic composition as the rock; boundaries with the rock tend to be transitional in an elongated flow. *Optics.* regions in a transparent substance in which a change in density is revealed by a change in refractive index, typically causing a light or dark area. (From a German word meaning "streak.")

schlieren arch *Geology.* an intrusive body of igneous rock having flow layers along its borders, but in whose interior such layers are poorly developed or lacking.

schlieren dome *Geology.* an intrusive body of igneous rock that is almost completely outlined by flow layers that terminate in one central area.

schlieren optics *Optics.* an optical system that detects inhomogeneities within a medium by recording the energy refracted by inhomogenous regions on a shadowgram.

schlieren photography *Optics.* a technique for photographing the flow of air or other gas by recording variations in the refractive index of density gradients under special illumination.

Schloenbachia *Paleontology.* a strongly ribbed genus of Mesozoic ammonoids, characterized by compressed form and tubercles on shoulders; extant in the Cretaceous.

Schlumberger dip meter *Engineering.* a dip meter that measures both the amount and direction of dip by readings taken in a borehole.

Schlumberger photoclinometer *Engineering.* an instrument that simultaneously measures both the amount and direction of the deviation of a borehole.

Schmeisser submachine gun *Ordnance.* one of several German Parabellum submachine guns designed by Hugo Schmeisser; the first version, designated **MP18-I**, was used in large quantities by German troops at the end of World War I.

Schmidt, Bernhard 1879–1935, German optician; invented the Schmidt reflector telescope, used for astronomical photography.

Schmidt-Cassegrain telescope *Optics.* a telescope using Schmidt optics in a Cassegrain telescope arrangement.

Schmidt corrector plate *Optics.* a transparent plate with an aspheric figure used in a Schmidt telescope or camera to distort incoming light in such a way that the spherical primary mirror can focus the light. Also, **Schmidt lens.**

Schmidt field balance *Engineering.* a magnetic instrument for prospecting on land, consisting of a magnet pivoted near its center of mass, so that the magnetic field of the earth creates a torque around the pivot opposed by the torque of the gravitational pull on the center; the angle at which equilibrium is reached depends on the strength of the field.

Schmidt net *Geology.* a coordinate system used in structural geology to plot a Schmidt projection.

Schmidt number *Fluid Mechanics.* a dimensionless quantity that is similar to the Prandtl number but is used in mass diffusivity rather than in thermal diffusivity.

Schmidt number 2 see SEMENOV NUMBER 1.

Schmidt number 3 *Physical Chemistry.* a dimensionless number that represents the product of the dielectric properties and the viscosity of a liquid divided by the product of its density, electrical conductivity, and the square of its length. Also, **Sc_3.**

Schmidt optical system *Optics.* a photographic or telescopic system characterized by a spherical primary mirror and an aspheric corrector plate; Schmidt systems commonly have low *f* ratios and a curved focal plane. Also, **Schmidt camera, Schmidt optics.**

Schmidt projection *Geology.* a Lambert azimuthal equal-area projection of a sphere's lower hemisphere onto the plane of a meridian; used in structural geology.

Schmidt reaction *Organic Chemistry.* a reaction that forms a primary amine from a carboxylic acid by treatment with sodium azide in sulfuric acid.

Schmidt reflector *Optics.* any telescope or camera using a Schmidt optical system.

Schmitt trigger *Electronics.* a bistable device in which the output changes when the input crosses a threshold level, and the threshold for a rising input signal is higher than the threshold for a falling signal. Also, **Schmitt circuit, Schmitt limiter.**

Schmorl's bacillus see NECROBACILLOSIS.

Schneider *Ordnance.* any of a variety of weapons manufactured by the Schneider Company of France and used by many countries; the company's founder, F. E. Schneider, developed the first successful centerfire cartridge case in 1861.

Schneiderian membrane *Anatomy.* the mucous membrane that lines the nasal cavity.

Schneider recoil system *Mechanical Engineering.* an artillery recoil mechanism, based on the hydropneumatic principle, that uses the action of both water and gas.

Schneider's index *Medicine.* a means of measuring the efficiency of circulation by comparing pulse rates and systolic blood pressure and observing the time for the pulse rate to change after changing levels of physical activity.

Schoenbiinae *Invertebrate Zoology.* a subfamily of moths, lepidopteran insects in the family Pyralididae, including the genus *Acentropus*, the most completely aquatic lepidopteran.

Schoenherr-Hessberger process *Chemical Engineering.* a nitrogen fixation process that reacts nitrogen and oxygen by using a long arc of electricity to provide an elevated temperature.

schoepite *Mineralogy.* $UO_3 \cdot 2H_2O$, a rare, strongly radioactive, yellow, orthorhombic mineral, occurring as tabular or prismatic crystals, having a specific gravity of 4.8 and a hardness of about 2.5 on the Mohs scale; found as an alteration product of uraninite.

schönfelsite *Petrology.* a dark-colored picritic basalt consisting of olivine and augite phenocrysts in a dense, fine-grained groundmass of apatite, bytownite, glass, magnetite, augite, and bronzite.

Schönflies crystal symbols *Crystallography.* the system of symbols used to designate the 32 classes of crystal symmetry, based on rotations that define these point groups but that do not indicate lattice type or symmetry.

school psychology *Psychology.* a branch of psychology concerned with counseling and advising elementary and secondary level children and improving learning conditions in school through research, testing, and guidance.

schooner *Naval Architecture.* a fore and aft rigged vessel with at least two masts; the second mast from the bow is normally the mainmast.

Schoop process *Engineering.* a zinc-coating process in which volatized zinc is sprayed at high pressure onto a metallic surface to produce a layer of any desired thickness.

schorl *Mineralogy.* $NaFe_3^{+2}Al_6(BO_3)_3Si_6O_{18}(OH)_4$, a black, translucent to opaque trigonal mineral of the tourmaline group, forming a series with dravite, occurring as prismatic crystals and in massive to fibrous form, and having a specific gravity of 3.10 to 3.24 and a hardness of 7 on the Mohs scale; found in granite pegmatites, pneumatolytic veins, and some metamorphic rocks. Also, **schorlite.**

schorlomite *Mineralogy.* $Ca_3Ti_2^{+4}(Fe_2^{+3}Si)O_{12}$, a black cubic mineral of the garnet group, forming a series with andradite, occurring as dodecahedral or trapezohedral crystals and in granular masses, and having a specific gravity of 3.76 and a hardness of 6.5 to 7.5 on the Mohs scale.

schott see SHOTT.

Schotten-Baumann reaction *Organic Chemistry.* a reaction used to acylate the hydroxyl and amino groups of organic compounds by treatment with an acid chloride in dilute alkaline solution; used in solution polymerization, in which phosgene is reacted with bisphenol A in the presence of an appropriate hydrogen chloride acceptor.

Schottky, Walter Hans 1886–1976, German physicist; conducted important research in the theory of electrons and ions; discovered the Schottky effect.

Schottky anomaly *Solid-State Physics.* an excess contribution to the heat capacity of a solid, particularly at low temperatures, due to the thermal population of discrete energy levels; this redistribution requires excess energy, which is manifested by a peak in the heat capacity.

Schottky barrier *Electronics.* a junction between a layer of semiconductor material and a layer of metal, characterized by hot carriers.

Schottky barrier diode *Electronics.* a device having a junction between the semiconductor and a metal contact instead of between materials with different polarities, as with most PN diodes. Also, HOT-CARRIER DIODE.

Schottky defect *Solid-State Physics.* a vacancy in a crystal produced by removing an atom from the bulk of the crystal and moving it to the surface. Also, VACANCY. *Materials Science.* see SCHOTTKY DISORDER.

Schottky disorder *Materials Science.* a structural imperfection in an ionic crystal, characterized by the presence of positive and negative vacant sites in the crystal lattice. Also, SCHOTTKY IMPERFECTION, SCHOTTKY DEFECT.

Schottky effect *Solid-State Physics.* an effect observed in thermionic emitters in which the work function is reduced when the emitter is subjected to an electric field.

Schottky imperfection see SCHOTTKY DISORDER.

Schottky line *Solid-State Physics.* a plot of the logarithm of the saturation current from a thermionic emitter as a function of the square root of the applied electric field in the Schottky effect; according to theory, such a plot should yield a straight line with slope $(e^{3/2}/kT)$, where e is the electronic charge, k is the Boltzmann constant, and T is the absolute temperature.

Schottky noise see SHOT EFFECT.

Schottky theory *Solid-State Physics.* a theory that explains the rectification properties at a semiconductor-conductor junction as the result of the formation of a barrier layer (depletion layer) at the surface of the junction.

Schottky transistor-transistor logic *Electronics.* a type of logic circuit in which a Schottky barrier diode is placed across a transistor's base-collector junction in order to enhance the circuit's speed. Also, **Schottky TTL.**

schreibersite *Mineralogy.* $(Fe,Ni)_3P$, a strongly magnetic, silver to tin-white, metallic, tetragonal mineral that tarnishes to brass-yellow or brown, occurring as tablets, needles, or rods, and having a specific gravity of 7.0 to 7.8 and a hardness of 6.5 to 7 on the Mohs scale; found in all iron meteorites. Also, RHABDITE.

Schreier's theorem *Mathematics.* any two subnormal (normal) series of a group G have subnormal (resp. normal) refinements that are equivalent. That is, there exist refinements S and T such that the nontrivial factors of S and the nontrivial factors of T can be put into one-to-one correspondence and corresponding factors are isomorphic. Similar results hold for modules.

Schrieffer, John Robert born 1931, American physicist; with John Bardeen and Leon B. Cooper, shared the Nobel Prize for the theory of superconductivity.

schriesheimite *Petrology.* an amphibole peridotite containing diopside, with hornblende enclosing the olivine.

schröckingerite *Mineralogy.* $NaCa_3(UO_2)(CO_3)_3(SO_4)F \cdot 10H_2O$, a greenish-yellow tricilinic secondary mineral with vitreous luster occurring as platy crystals and scaly aggregates, having a specific gravity of 2.55 and a hardness of 2.5 on the Mohs scale. Also, **schroeckingerite.**

Schröder, Ernst 1841–1902, German mathematician; introduced the concept of logical duality; wrote a comprehensive study of symbolic logic.

Schrödinger, Erwin 1887–1961, Austrian physicist; awarded the Nobel Prize for the Schrödinger wave equation and other work in wave mechanics.

Schrödinger equation *Quantum Mechanics.* the fundamental equation governing the behavior of quantum mechanical wavefunctions; the general form is written

$$E\{\hbar\}(d\psi/dt)=\{H\}\psi,$$

where $\{H\}$ is the system Hamiltonian and $\{\hbar\}$ is Planck's constant divided by 2π.

Schrödinger-Pauli equation *Quantum Mechanics.* a modified version of Schrödinger's equation that describes spin 1/2 particles.

Schrödinger picture *Quantum Mechanics.* a means of representing quantum mechanics in which operators are treated as stationary, and state vectors evolve with time. Also, **Schrödinger representation.**

Schrödinger's cat *Quantum Mechanics.* a hypothetical cat proposed by Erwin Schrödinger that is imagined to be in a box isolated from any outside observation; according to quantum theory it would be neither alive nor dead, but would exist in a superposition of these two states, until the box is opened.

Schrödinger's wave mechanics *Quantum Mechanics.* a formulation of quantum mechanics in which the state of a system is described by a wave function, and physical quantities or their expectation values are obtained by operation on the wave function by the appropriate operators; the equation describing the energy of a state as the expectation value of the Hamiltonian operator.

Schrödinger wave function *Quantum Mechanics.* the function of space and time, satisfying Schrödinger's equation, that completely describes the state of a quantum mechanical system.

Schroeder-Bernstein theorem *Mathematics.* **1.** let X and Y be sets. If X is equivalent to a subset of Y, and Y is equivalent to a subset of X, then X is equivalent to Y. **2.** in terms of cardinal numbers, if u and v are cardinal numbers such that $u \leq v$ and $v \leq u$, then $u = v$.

Schröter, Johann 1745–1816, German astronomer; founded the Lilienthal Observatory; studied features of the moon and planets.

Schröter effect *Astronomy.* a discrepancy between the theoretical and observed dichotomy of the appearance of Venus in its waning and waxing phases as seen from the earth.

schrötterite *Mineralogy.* an opaline, aluminum-rich variety of the mineral allophane, a hydrous aluminum silicate.

schrund line *Geology.* the line separating the steeper slope of the cirque wall from the gentler slope below at the base of the bergschrund.

Schubertellidae *Paleontology.* a family of foraminiferids in the superfamily Fusulinacea; extant in the Carboniferous and the Permian.

Schuermann series *Chemistry.* a series of metals ordered in a way in which the sulfide of a metal is precipitated at the expense of a metal lower in the series.

Schuler pendulum *Mechanics.* an apparatus that executes pendulum-like motion under gravity, having such a large moment of inertia that it is equivalent to a pendulum of a length equal to the earth's radius. Such a pendulum will, if initially at rest, always point toward the earth's center regardless of what motions its pivot is subjected to, making it a useful apparatus for navigation.

Schuler tuning *Engineering.* the setting of a gyroscopic device so that its periods of oscillation are 84.4 minutes.

schultenite *Mineralogy.* $PbHAsO_4$, a colorless, transparent monoclinic mineral occurring in flattened crystals resembling gypsum, having a specific gravity of 5.94 to 6.07 and a hardness of 2.5 on the Mohs scale; found with other secondary minerals at Tsumeb, Namibia.

Schultz-Dale reaction *Immunology.* a contraction of a smooth muscle-tissue sample that occurs when a small amount of antigen is reinjected into a previously sensitized sample.

Schultz-Hardy rule *Chemistry.* a rule stating that ions having the opposite charge of a colloid particle are the most effective in coagulating the particle, and further, that increased ionic charge results in increased coagulating ability.

Schultz syndrome see AGRANULOCYTOSIS.

Schulze's reagent *Analytical Chemistry.* an oxidizing mixture made up of a saturated aqueous solution of $KClO_3$ and HNO_3.

Schumag machines *Materials Science.* a machine used in the drawing of rod and wire.

Schumann region *Optics.* the part of the spectrum farthest toward the ultraviolet that is still capable of exposing a photographic emulsion.

schungite *Geology.* a hard, black, amorphous, carbon-rich material found interbedded among Precambrian schists. Also, **shungite.**

schuppenstructure see IMBRICATE STRUCTURE.

Schur's lemma *Mathematics.* let A be a simple R-module and B any R-module. Then: (a) every nonzero R-module homomorphism $f: A \rightarrow B$ is a monomorphism; (b) every nonzero R-homomorphism $g: B \rightarrow A$ is an epimorphism; and (c) the endomorphism ring $\mathrm{Hom}_R\ (A, A)$, i.e., the ring of all endomorphisms from A to A, is a division ring.

Schuster method *Spectroscopy.* a technique used to focus a prism spectroscope without employing a Gauss eyepiece or using a distant object.

Schwagerinidae *Paleontology.* a family of foraminiferids in the superfamily Fusulinacea; characterized by a two-layered wall in which the mural pores are enlarged to form alveoli; extant in the Upper Carboniferous and Permian.

Schwann, Theodor 1810–1882, German biologist; formulated the cell theory (based largely on Schleiden's work with plant cells); discovered Schwann cells, microbes, and pepsin.

Schwann cell *Neurology.* a cell that wraps around the axons of peripheral neurons forming the neurilemma.

Schwanniomycetes *Mycology.* a genus of yeast fungi belonging to the order Endomycetales which occurs in soils.

schwannitis *Neurology.* an inflammation and proliferation of the Schwann cells of peripheral nerves.

schwannoma see NEUROFIBROMA.

schwartzembergite *Mineralogy.* $Pb_6(IO_3)_2Cl_4O_2(OH)_2$, a yellow, orthorhombic, pseudotetragonal mineral occurring as flat, pyramidal crystals, having a specific gravity of 7.39 and a hardness of 2 to 2.5 on the Mohs scale; found as secondary crusts in oxidized lead ore.

Schwarz, Hermann Amandus 1843–1921, German mathematician; worked on the theory of functions and theory of minimal surfaces; devised Schwarz's lemma.

Schwarz-Christoffel transformation *Mathematics.* a transformation that conformally maps the interior of an arbitrary polygon in the complex plane into the upper half-plane bounded by the real axis. In particular, suppose that the polygon has n vertices w_k with corresponding interior angles α_k, and that the points w_k are to be mapped onto the points x_k of the real axis. Then the transformation is given by the integral

$$w = A \int (z - x_1)\beta_1(z - x_2)\beta_2 \cdots (z - x_n)\beta_n\, dz + B,$$

where the exponents are $\beta_k = \alpha_k/\pi - 1$, and where A and B are complex constants. Infinite open polygons can be considered as limiting cases of closed polygons.

Schwarz function *Mathematics.* an analytic function of the complex variable z that takes on the value $\bar{z}$ along a specified curve in the complex plane, where $\bar{z}$ indicates complex conjugate.

Schwarzian derivative *Mathematics.* in the theory of conformal mapping, the expression

$$[w'''/(w')^2] - [(/2)(w'')^2/(w')^3],$$

which is a third-order differential invariant for the group of all linear fractional transformations. The quotient of any two solutions of an arbitrary second-order linear ordinary differential equation has a Schwarzian derivative of zero.

Schwarz reflection principle *Mathematics.* let $f(z)$ be analytic in a region R of the complex plane, where the boundary of R contains a segment of the real axis on which $f(z)$ assumes real values. Then the analytic continuation of $f(z)$ into the region obtained by reflecting R across the real axis is $[f(\bar{z})]^*$, where $\bar{z}$ and * indicate complex conjugate.

Schwarzschild, Karl 1873–1916, German astronomer; discovered Schwarzschild singularities, now known as black holes.

Schwarzschild radius *Physics.* in general relativity, the distance given by $2MG/c^2$ where M is the mass of a black hole, G is the gravitational constant, and c is the speed of light.

Schwarzschild solution *Physics.* a solution in the theory of general relativity for a nonrotating black hole in empty space.

Schwarz's lemma *Mathematics.* suppose that f is analytic on the interior of the unit disk D of the complex plane with $|f(z)| \le 1$ for $z \in D$ and $f(0) = 0$. Then $|f(0)| \le 1$ and $|f(z)| \le z$ for all $z \in D$. That is, an analytic function of the unit disk to itself that fixes the origin is a contraction. If, in addition, $|f(0)| = 1$ or if $|f(z)| = |z|$ for some nonzero z, then there is a constant c such that $|c| = 1$ and $f(z) = cz$ for all $z \in D$.

Schwassman-Wachmann 1 *Astronomy.* a comet with a nearly circular orbit between Jupiter and Saturn, which has occasional outbursts in brightness.

Schweitzer's reagent *Organic Chemistry.* $Cu(NH_3)_4(OH)_2$, an aqueous solution of cupric ammonium hydroxide; used as a solvent for cellulose.

Schweitzer's reagent *Chemistry.* a dark-blue solution of cupric hydroxide in concentrated ammonia that dissolves cotton, silk, and linen; used as a test for wool and in the manufacture of rayon.

Schwinger, Julian S. born 1918, American physicist; with Richard P. Feynman and Julian S. Tomonaga, shared the Nobel Prize for their theory of quantum electrodynamics.

Schwinger's variational principle *Physics.* a principle commonly applied to electromagnetic theory in the calculation of an approximate value of a quadratic functional.

sci. science; scientific.

sciaenoid *Vertebrate Zoology.* **1.** relating or belonging to the family Sciaenidae. **2.** any member of this family.

Sciaenidae *Vertebrate Zoology.* the drums or croakers, a family of fish of the order Perciformes, which are coastal marine and freshwater fish of the Atlantic, Pacific, and Indian Oceans that produce sound using the swim bladder.

Sciara *Invertebrate Zoology.* a genus of tiny black fungus gnats, dipteran insects in the series Nematocera; larvae move in large groups, sometimes destroying mushroom beds and seedlings.

sciatic [sī at′ik] *Neurology.* relating to or involving the sciatic nerve or the associated areas of the body.

sciatica [sī at′i kə] *Neurology.* a syndrome characterized by pain along the sciatic nerve which extends from the sacral plexus down the leg; it may also include tenderness and burning or other abnormal sensations of the thigh and leg, and wasting away of the calf muscles.

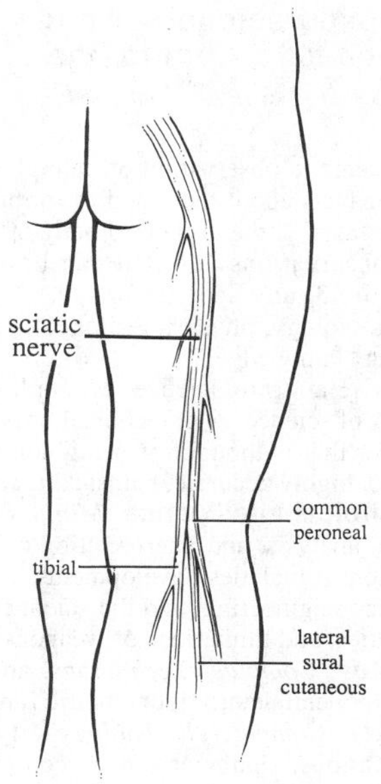

sciatic nerve

sciatic nerve *Neurology.* a large, long nerve arising from the sacral plexus and extending into the thigh.

SCID severe combined immunodeficiency.

Science

Science connotes both the knowledge contained in such disciplines as astronomy, physics, chemistry, biology, and geology and the activities involved in obtaining it. The latter sense yields a more secure definition than would an attempt to demarcate the fields to which the word applies. "We very much need a name to describe a cultivator of science in general," wrote William Whewell in 1840. "I should incline to call him a scientist." The term took.

Science, by common consent, has come to be understood as what scientists do. Its restriction to investigation of the material universe is largely, though not entirely, a modern development, and is in some measure peculiar to the English language. In the 17th and 18th centuries, what is now understood by science was called natural philosophy or natural history. In 18th-century dictionaries—Ephraim Chambers, Samuel Johnson, the *Encyclopédie*—the first meaning is philosophical: science is certain knowledge grounded on self-evident principles or on demonstration. It is set over against doubt, with opinion in between. The attribute of certainty carries over from scholastic philosophy, wherein *Scientia* is contrasted to belief.

Old dictionaries then give instances of further meaning with reference to bodies of knowledge that qualify as "Sciences." It would be a mistake to suppose, as is often done, that technicality has not conferred a measure of authority throughout the history of the word. The seven liberal arts of the medieval university were often called sciences, the lower trivium consisting of grammar, rhetoric, and logic, and the higher quadrivium of arithmetic, music, geometry, and astronomy. In Chambers's *Cyclopaedia* (1741) knowledge has two branches, "Natural and Scientifical" and "Artificial and Technical." Stemming from the first, nine of the fourteen specialties are what would now be called scientific, several with the same names, while ten of the sixteen on the second branch would be applied science or engineering. Clearly, Whewell's need for a word for the people doing such things professionally was recognition that by the 19th century the scientific enterprise had become a defining feature of modern civilization.

Current classifications normally distribute the learned disciplines into science (from which it would be pedantic to exclude mathematics), social science, and the humanities. In public opinion generally, and more particularly in economic and political circles, science is clearly thought to be a more effective form of knowledge than the others. Its prestige is compounded of small residues of appreciation for its supposed certainty, with large doses of evidence for its indispensability in technology of all sorts.

Earlier in the 20th century, positivistic philosophy thought to analyze the basis of its certainty, while sociology sought to emulate its methods. In recent years, leading investigators in both disciplines have taken an opposite tack. Certain philosophers dispute the view that statements of science correspond to natural reality and thereby partake of a validity unattainable in forms of discourse limited to human and social matters. In a similar way, a school of sociologists argues that the findings of science are determined, not by how well theories fit the world, but by the interactions of scientists, as are the outcomes of social undertakings in general.

Both critiques deny science the role of knowledge privileged by its bearing on nature. The thrust may be blunted by the consideration that an analysis, whether philosophical or sociological, denying the capacity of science to make true statements about the nature of the world reflexively vitiates the credibility of its own claim to make true statements about the nature of science. It is for less esoteric reasons that such skepticism makes little dent on the scientist's confidence that he or she is investigating the forces and structure of nature and on acceptance of that self-assessment by the public.

Charles C. Gillispie
Professor of History of Science Emeritus
Princeton University

science **1.** the systematic observation of natural events and conditions in order to discover facts about them and to formulate laws and principles based on these facts. **2.** the organized body of knowledge that is derived from such observations and that can be verified or tested by further investigation. **3.** any specific branch of this general body of knowledge, such as biology, physics, geology, or astronomy. (From the Latin word meaning "knowledge.")

scientific **1.** of or relating to science. **2.** employing or based on the methods or theories of science. **3.** of technical or practical activities, carried out in a manner that is thought of being comparable to science, as by being systematic, highly accurate, painstaking, and so on.

scientific and technical intelligence *Military Science.* data derived from the collection, analysis, and interpretation of foreign scientific and technical information; it includes developments in basic and applied research and in applied engineering, and the scientific and technical characteristics, capabilities, and limitations of weapons and materiel.

scientific calculator *Computer Technology.* an electronic calculator that can perform exponential and trigonometric functions.

scientific computer *Computer Technology.* a type of computer used in scientific applications, characterized by complex computations involving floating-point arithmetic.

scientific creationism the belief that the theory of creationism (i.e., that the universe was created in essentially its present form by a supernatural being) is supported by scientific evidence as well as by Biblical commentary; e.g., the argument that the fossil record does not provide evidence of evolution, or that modern cosmological theories such as the Big Bang theory do not explain the ultimate origin of matter.

scientific management *Industrial Engineering.* an approach to management that uses standards and procedures based upon systematic observation of individual situations.

scientific method *Science.* an organized approach to problem-solving that includes collecting data, formulating a hypothesis and testing it objectively, interpreting results, and stating conclusions that can later be evaluated independently by others.

scientific notation *Science.* a method of expressing a given quantity as a certain number multiplied by 10 to an appropriate power; e.g., the scientific notation for 4257 is 4.257×10^3. Scientific notation is often used to express the degree of accuracy required in a given situation; in such cases, 4257 might be expressed as 4.26×10^3.

scientific system *Computer Technology.* any computer system that is devoted primarily to computations.

scientism *Science.* **1.** the techniques, principles, and activities that are typical of science. **2.** the belief that scientific methods of evaluation and scientific terminology can (and should) be applied in other disciplines, such as the social sciences or the humanities.

scientist *Science.* a person who is trained in some field of science or who is engaged in the profession of science.

scil. namely. (From Latin *scilicet,* "able to know.")

scillaren *Pharmacology.* a cardiotonic drug formed from a mixture of scillaren A ($C_{36}H_{52}O_{13}$) and scillaren B (glucosides remaining after extraction of scillaren A); originally derived from the fresh squill plant.

Scincidae *Vertebrate Zoology.* the skinks, a cosmopolitan family of shiny-scaled lizards of the suborder Sauria, characterized by small or absent limbs.

Scinidae *Invertebrate Zoology.* a family of marine amphipod crustaceans in the suborder Hyperiidea; some are commensal on gelatinous invertebrates.

scintillation [sin´tə lā´shən] *Optics.* a light flash formed when rapidly traveling particles pass through matter. *Nucleonics.* the production of light by a phosphor that is struck and ionized by an energetic particle or photon. *Astronomy.* a rapid twinkling of stars caused by constant small changes in the atmosphere's density. *Electromagnetism.* the fluctuations of a radar target about its mean position.

scintillation camera *Nucleonics.* a camera that records radionuclide distribution over a particular region of a body in a single exposure, thus obviating the need for line-scanning techniques.

scintillation counter *Nucleonics.* a device that counts ionizing events by means of a photomultiplier tube, which receives scintillating light and converts it to an electrical signal. *Crystallography.* a similar device used for measuring the intensity of an X-ray beam. It makes use of the fact that X-rays cause certain substances to emit visible light by fluorescence. The intensity of this visible light is proportional to the intensity of the incident X-ray beam and is measured by a photomultiplier. Also, **scintillation detector, scintillometer.**

scintillation counter crystal *Nucleonics.* the active element that emits light in a scintillation counter device; the crystal is a fluor such as thallium-activated sodium iodide.

scintillation counting technique *Radiology.* a means of determining radioactive intensity with a scintillation crystal and accompanying electronic accessories.

scintillation spectrometer *Nucleonics.* a scintillation counter that is also equipped to sort the scintillations according to the energy.

scintillator *Nucleonics.* a material that emits optical photons when struck by ionizing radiation.

scintiscan *Radiology.* a graphic record of the gamma rays emitted by a radioisotope, revealing its relative concentration in various tissues. Also, **scintigram.**

scintiscanner *Radiology.* the equipment that is used to produce a scintiscan.

Sciomyzidae *Invertebrate Zoology.* the marsh flies, a family of dipteran insects in the subsection Acalyptratae; the larvae prey on aquatic snails.

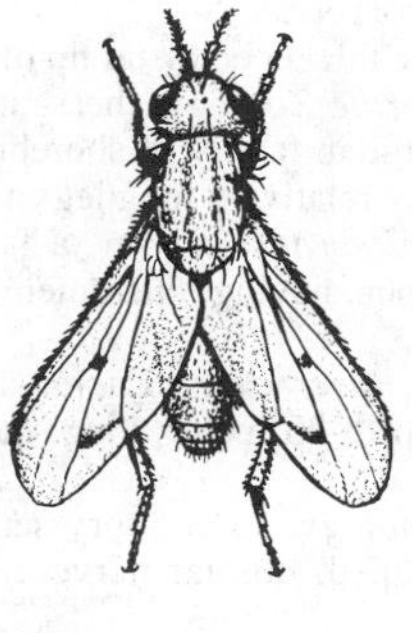

Sciomyzidae (marsh fly)

scion [sī´ən] *Botany.* a shoot prepared for grafting to a root.

sciophilous *Ecology.* thriving in a shaded environment. Thus, **sciophile, sciophily.**

sciophyte *Botany.* a plant that lives or thrives in a shaded environment.

scirocco see SIROCCO.

scirrhous *Oncology.* relating to or characteristic of a hard (scirrhous) carcinoma.

scirrhous carcinoma *Oncology.* a carcinoma having a hard structure due to the formation of dense connective tissue in the stroma. Also, **scirrhoma, scirrhus.**

scission *Chemistry.* the splitting of a molecule into two or more simpler molecules; cleavage. *Materials Science.* any degradation of polymers, such as interruptions in the chains.

scissoring see CLIPPING.

scissor jack *Mechanical Engineering.* a lifting jack driven by a horizontal screw that lengthens or shortens the links of the jack.

scissors *Mechanical Devices.* any of various cutting tools consisting of two beveled cutting blades attached to looped handles and joined in the middle by a pivot.

scissors crossover *Civil Engineering.* an intersection between two parallel tracks, shaped like a pair of scissors, that allows trains to cross in either direction from either track.

scissors fault *Geology.* a fault on which the offset or displacement along the strike increases in one direction and decreases in the opposite direction from an initial point of no displacement. Also, DIFFERENTIAL FAULT.

scissors jack *Mechanical Devices.* a portable automobile jack consisting of a hinged, diamond-shaped frame that is raised or lowered by a horizontal screw.

scissors truss *Building Engineering.* a truss used in pitched roof applications; it consists of two principal rafters that are braced by two other members, each joining the foot of a common rafter at one end and an intermediate point in a rafter at the other end; used primarily in the installation of roofs over halls and churches.

Scitek *Graphic Arts.* the trade name for a device that permits the digital alteration of existing graphic material; used, for example, in altering a photograph before publication to remove objectionable material.

Sciuridae *Vertebrate Zoology.* the squirrels, a nearly cosmopolitan rodent family of the suborder Sciuromorpha, being diurnal, arboreal, and terrestrial or semifossorial.

Sciuromorpha *Vertebrate Zoology.* the suborder of rodents that includes the squirrels, mountain beavers, pocket gophers, kangaroo rats, beavers, and springhares; considered the most primitve suborder of Rodentia.

SCL *Aviation.* the airport code for Santiago, Chile.

ScLA or **Sc.L.A.** scapulolaeva anterior.

scler- a combining form meaning "hard," as in *sclerema.*

sclera *plural,* **sclerae.** *Anatomy.* the tough, fibrous outer coating of the eyeball that forms the "white" of the eye, covering all of the eye except for the area covered by the cornea.

Scleractinia *Invertebrate Zoology.* the stony corals, an order of anthozoans in the subclass Zoantharia; they are colonial or solitary polyps that secrete a heavy calcareous skeleton attached to a solid substrate. Also, MADREPORARIA.

scleractinian *Invertebrate Zoology.* **1.** any of the corals of the order Scleractinia; this still-abundant order first appeared in the Triassic and was the first to replace the tabulate and rugose corals, which had disappeared at the end of the Permian. **2.** of or relating to this order.

scleradenitis *Medicine.* the inflammation and hardening of a gland.

scleral *Anatomy.* of or relating to the sclera.

Scleraxonia *Invertebrate Zoology.* a suborder of horny corals, anthozoans in the subclass Alcyonaria, having feather- or fan-shaped skeletons.

scleredema *Medicine.* a diffuse, symmetrical, woodenlike, nonpitting induration of the skin of unknown origin; typically it begins on the face, head, or neck and spreads progressively. The condition often follows a staphylococcal infection and usually occurs in association with diabetes mellitus.

sclerema *Medicine.* any of various hardened conditions of tissue, especially sclerema neonatorum.

sclerema neonatorum *Medicine.* a severe, often fatal disorder of adipose tissue occurring chiefly in premature, sick, or debilitated infants suffering from a serious underlying illness. The skin becomes cold, yellowish white, mottled, boardlike, and inflexible.

sclerencephaly *Neurology.* sclerosis and shrinkage of the brain. Also, **sclerencephalia.**

sclerenchyma *Botany.* a type of tissue composed of cells with thickened lignified secondary cell walls that function primarily in plant support.

sclerid *Botany.* a highly lignified, thickened cell with a secondary cell wall that may or may not be living; such cells give pears a grainy texture. Also, **sclereid.**

sclerite *Invertebrate Zoology.* any hard calcareous or chitinous plate or spicule, especially the rigid portions of the arthropod exoskeleton bounded by sutures of softer cuticle.

scleritis *Medicine.* an inflammation of the sclera, occurring either alone or with keratitis or uveitis. Also, **scleratitis.**

sclero- a combining form meaning "hard," as in *scleroderma.*

scleroblast *Invertebrate Zoology.* in sponges, the specialized cells that secrete spicules. Also, **sclerocyte.**

scleroblastema *Developmental Biology.* an embryonic tissue that takes part in the formation of bone structures. Thus, **scleroblastemic.**

sclerocaulous *Botany.* of a plant, having a hard, dry stem due to overdevelopment of sclerenchyma.

Sclerodactylidae *Invertebrate Zoology*. the sea cucumbers, a family of holothurian echinoderms in the order Dendrochirotida having a complex calcareous ring and branched tentacles.

scleroderm *Invertebrate Zoology*. the skeleton of stony corals.

scleroderma *Medicine*. a relatively rare disease of the autoimmune system, characterized by fibrous degeneration of the connective tissue of the skin, lungs, and internal organs, most common in middle-aged women. Also, **scleriasis.**

scleroderma neonatorum *Medicine*. see SCLEREMA NEONATORUM.

Sclerodermataceae *Mycology*. a family of fungi of the order Sclerodermatales, containing some species that exist in symbiosis with conifer trees and some that are common to subtropical regions.

Sclerodermatales *Mycology*. an order of nonparasitic fungi belonging to the class Gasteromycetes that occur in rotting wood, in soil, or in roots.

sclerodermatous *Medicine*. relating to or characterized by scleroderma. *Vertebrate Zoology*. having a hard outer covering, as in turtles and armadillos. *Invertebrate Zoology*. having an exoskeleton composed of hard plates or chitinous scales or, especially, of scleroderm, as in stony corals.

Sclerogibbidae *Invertebrate Zoology*. a family of rare wasps, hymenopteran insects in the superfamily Bethyloidea.

scleroma *Medicine*. a hardenened patch or induration, especially of the nasal or laryngeal tissues.

sclerometer *Engineering*. an instrument that tests hardness by scratching a line with a diamond point under a known pressure and measuring the width of the line.

sclerophyllous *Botany*. describing leaves that are stiff or thickened, possessing a multilayered cuticle and often associated with arid or dry climates.

scleroprotein *Biochemistry*. any of the class of albuminoid proteins (e.g., collagen, elastin, or keratins) that have a supporting or protective function in tendons, bones, cartilages, ligaments, and other tough parts of the body, and are soluble only in concentrated solutions of strong acids and alkalies.

sclerosarcoma *Oncology*. a firm, fibrotic sarcoma.

scleroscope *Engineering*. an instrument that measures hardness in terms of elasticity by the rebound height of a diamond-tipped hammer bouncing off a given material.

sclerose *Pathology*. to become sclerotic; harden.

sclerosed *Pathology*. affected with sclerosis.

scleroseptum *Invertebrate Zoology*. a calcareous radial portion in certain corals.

sclerosing adenomatosis *Medicine*. a form of dysplasia of the breast with encasement of the ducts in fibrous tissue, such that the appearance is that of cancerous formation. Also, FIBROSING ADENOSIS, ADENOMATOSIS.

sclerosing hemangioma *Medicine*. a type of benign hemangioma, occurring as a solidly cellular lesion thought to develop from a hemangioma by proliferation of endothelial and connective tissue cells.

sclerosis [sklə rō´sis] *Pathology*. a hardening or induration, especially the hardening of a body part from inflammation or in diseases of the interstitial substance.

Sclerospongiae *Invertebrate Zoology*. a class of sponges that secrete a basal skeleton of crystalline calcium carbonate.

sclerostomy *Surgery*. the procedure for creating a fistulalike passage through the sclera in order to alleviate pressure on the eye in glaucoma.

sclerotan *Biochemistry*. a glucan found in species of *Sclerotinia*.

sclerotic *Pathology*. affected with sclerosis. *Anatomy*. **1.** hardened or made firm. **2.** of or relating to the sclera of the eye.

Sclerotinia *Mycology*. a genus of fungi of the order Helotiales that cause such plant diseases as brown rot and clover rot.

Sclerotiniaceae *Mycology*. a family of fungi of the order Helotiales that live on decaying or living plant material and are sometimes pathogenic, causing such diseases as rot, mold, and blight.

sclerotinite *Geology*. a coal maceral of the internite group, composed of the sclerotia of fungi.

sclerotium *Bacteriology*. **1.** in fungi, a hardened structure that develops as a defense against unfavorable environmental conditions such as lack of moisture or nourishment; the plasmodium becomes a hard mass that may stay dormant until favorable conditions return. **2.** in certain protozoa, a multinucleate hard cyst into which the plasmodium divides in response to adverse conditions. Also, **sclerotis.**

Sclerotium *Mycology*. a genus of fungi of the class Myxomycetes that cause such plant diseases as onion rot.

sclerotization *Entomology*. the hardening and darkening of the chitin to form rigid plates with a horny consistency in the exoskeleton following molting; caused by cuticle secretions.

sclerotome *Developmental Biology*. the ventromedial portion of the somite, which eventually forms vertebrae and ribs after surrounding the notochord and spinal cord. *Anatomy*. the portion of a given bone that is supplied by a spinal segment. *Surgery*. an instrument used for incision of the sclera.

sclerotomy *Medicine*. a surgical incision into the sclera of the eye.

sclerous *Medicine*. hard; indurated.

ScLP or **Sc.L.P.** scapulolaeva posterior.

Sc.M. Master of Science.

scobiform *Botany*. resembling sawdust.

Sco-Cen association *Astronomy*. an association of very young stars in Scorpius and Centaurus, lying at a distance of 750 light-years in the Gould Belt; the brightest member is Antares.

scolecite *Mineralogy*. $CaAl_2Si_3O_{10}{\cdot}3H_2O$, a colorless or white, vitreous to silky monoclinic mineral of the zeolite group, occurring in radiating groups of fibrous or acicular crystals, having a specific gravity of 2.27 and a hardness of 5 to 5.5 on the Mohs scale; found in basalt cavities.

scolecodont *Paleontology*. describing the jaws of marine annelid polychaete worms; these fossils are as large as a few millimeters and are common in Ordovician to Triassic strata; they virtually died out at the end of the Cretaceous, but a few rare fossils have also been found in Tertiary strata.

scolex *plural*, **scoleces.** *Invertebrate Zoology*. the head of a tapeworm, bearing various forms of hooks and suckers that adhere to the host tissues.

Scoliidae *Invertebrate Zoology*. the hairy black wasps, a family of brightly spotted or banded hymenopteran insects in the superfamily Scolioidea that frequent flowers; the larvae are ectoparasites of beetle larvae.

scolio- a combining form meaning "twisted" or "crooked."

Scolioidea *Invertebrate Zoology*. a superfamily of ants and parasitic wasps, hymenopteran insects in the suborder Apocrita.

scoliosis *Medicine*. an abnormal lateral curvature of the spine, a common abnormality of childhood.

scolite *Geology*. any of various narrow, vertical, tubular trace-fossil structures in rock, especially sandstone, thought to be the fossilized burrows of marine worms. Also, **scolithus.**

s colony see SMOOTH COLONY.

scolop *Entomology*. the thickened distal tip of a scolopophore.

Scolopacidea *Vertebrate Zoology*. the sandpipers, snipes, and turnstones, a large cosmopolitan family of shorebirds of the order Charadriiformes characterized by relatively long legs and bill and pointed wings.

Scolopendridae *Invertebrate Zoology*. a family of centipedes in the order Scolopendromorpha, having long antennae and four ocelli in each eye.

Scolopendromorpha *Invertebrate Zoology*. an order of centipedes in the class Chilopoda, up to 20 inches long, including some of the most venomous species.

scolopophore *Entomology*. an auditory sense organ in insects composed of a spindle-shaped, bipolar nerve attached to the body wall. Also, **scolophore.**

Scolosaurus *Paleontology*. a genus of ornithischian dinosaurs in the family Ankylosauridae, covered with spines of various sizes from head to tail, and heavily armored; Scolosaurus, which is known only from North American deposits of the Upper Cretaceous, grew to about 18 feet long and weighed 2 tons.

Scolytidae *Invertebrate Zoology*. a family of small cylindrical coleopteran insects in the superfamily Curculionoidea, including bark or engraver beetles and ambrosia beetles; the larvae and adults both bore into bushes and trees and are serious pests of conifers.

Scombridae *Vertebrate Zoology*. the mackerels and tunas, a family of fast-swimming marine fish of the suborder Scombroidei, characterized by a spindle-shaped body with a pointed snout and a tapering tail; found in all tropical and temperate seas.

Scombroidei *Vertebrate Zoology*. a suborder of marine fish of the order Perciformes, including the tunas, mackerels, and billfishes; characterized by a nonprotrusible upper jaw and a widely forked caudal fin.

scombrotoxin *Toxicology*. a histaminelike toxin produced by bacteria in decaying fish belonging to the family Scombridae.

scoop *Mechanical Devices*. any large-sized shovel with a scoop-shaped blade, used for moving liquid or loose material. *Surgery*. a spoonlike instrument used to excavate cavities or remove foreign objects.

scopa *Entomology*. in some hymenopteran insects, a brushlike structure under the abdomen that collects pollen.
scope *Optics*. a popular term for any viewing instrument, such as a microscope, radarscope, or telescope.
-scope a combining form meaning "viewing instrument."
scope of a variable *Computer Programming*. the block of instructions within which a variable is defined and can be accessed.
Scopeumatidae *Invertebrate Zoology*. the dung flies, a family of dipteran insects in the subsection Acalyptratae.
Scopidae *Vertebrate Zoology*. the hammerheads, a monotypic family of wading birds of the order Ciconiiformes, characterized by relatively short legs and neck, a large head, and a long straight bill; native to tropical Africa and Madagascar.
scopolamine *Pharmacology*. $C_{17}H_{21}NO_4$, an alkaloid derived from belladonna and other plants; used to relieve parkinsonism, dilate the pupils, counteract toxic agents, reduce secretions such as sweat, paralyze the ciliary muscle in the eye, and prevent and alleviate motion sickness. Also, HYOSCINE.
scopoline *Organic Chemistry*. $C_8H_{13}NO_2$, a hygroscopic crystalline alkaloid; soluble in water and alcohol; melts at 108–109°C and boils at 248°C; used in medicine. Also, OSCIN.
scopometer *Optics*. a device used to measure the turbidity of a liquid by comparison of a sample target viewed through the liquid and a standard-brightness field adjacent.
scopophilia *Psychology*. see VOYEURISM.
scopula *Entomology*. a dense, brushlike tuft or row of hairs found on the feet of certain insects. *Cell Biology*. in certain ciliated protozoans, an adoral organelle consisting of a number of kinetosomes, each associated with a cilium.
scopulite *Geology*. a rod-shaped crystallite having terminal brushes or plumes.
scorching *Engineering*. the burning of an exposed surface in order to change certain physical characteristics without destroying the surface. *Materials*. the premature curing of rubber during a vulcanization process. *Plant Pathology*. a browning of plant tissue due to heat, parasites, infection, or lack or excess of some nutrient.
scorch time *Materials*. the time during which a rubber compound can be worked at a given temperature before curing begins..
score *Metrology*. a group of 20, such as 20 years. *Engineering*. to carry out a process of scoring.
scoria *Volcanology*. a cindery, vesicular crust formed on the surface of basaltic or andesitic lava as a result of the escape and expansion of gases before solidification. *Materials*. the refuse left after melting or smelting metal.
scoria cone *Volcanology*. a volcanic cone composed of scoria.
scoria mound *Volcanology*. a round volcanic hill composed of scoria.
scoria tuff *Geology*. an accumulation of fragmented scoria in a fine-grained matrix of tuff.
scorification *Metallurgy*. in precious metal recovery, the oxidation of molten lead containing such metals.
scoring *Engineering*. **1.** the marring or scratching of any formed part by metal pickup on the punch or die. **2.** the process of reducing the thickness of a material along a line to weaken it purposely along that line. *Geology*. **1.** an abrasion process in which rock fragments carried by a moving glacier create parallel scratches, lines, or grooves on the surface of the underlying bedrock. **2.** any scratch, line, or groove formed by such scoring action. Also, **score.** *Graphic Arts*. an impression made to facilitate the folding of a printed page or signature.
scorodite *Mineralogy*. $Fe^{+3}AsO_4 \cdot 2H_2O$, a pale green to brown, blue, or colorless, orthorhombic mineral of the variscite group, forming a series with mansfieldite, occurring as crusts and aggregates of pyramidal crystals, having a specific gravity of 3.278 and a hardness of 3.5 to 4 on the Mohs scale; found widespread as an oxidation product of arsenic-bearing minerals.
Scorpaenidae *Vertebrate Zoology*. the scorpionfishes, rockfishes, and turkeyfishes, a family of marine bony fish of the order Scorpaeniformes, noted for the presence of enlarged venomous spines; found in all tropical and temperate seas.
Scorpaeniformes *Vertebrate Zoology*. an order of bony fishes containing many venomous forms, including the scorpionfishes, stonefishes (the most venomous fish known), sea robins, velvetfishes, and prowfishes.
scorpamine *Toxicology*. a potent neurotoxin produced by scorpions, especially those belonging to the family Buthidae.
Scorpio *Astronomy*. see SCORPIUS.
scorpioid cyme *Botany*. a determinate, coiled inflorescence, named for its resemblance to the abdomen curl of a scorpion.
scorpion *Invertebrate Zoology*. the popular name for arachnids in the order Scorpionida having large clawlike pedipalps and a segmented abdomen that ends in a sting. The sting of some species is fatal to humans.

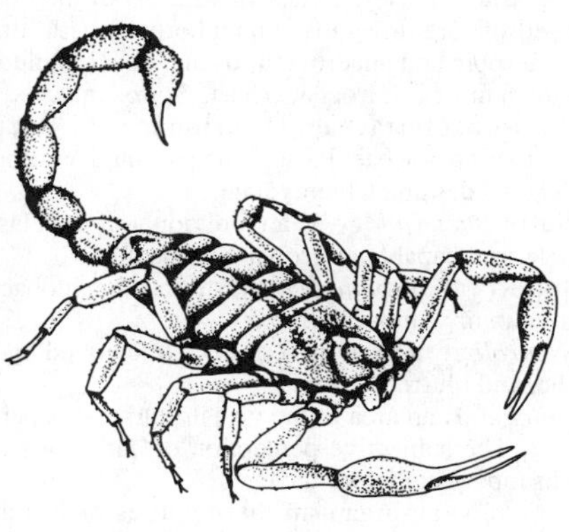

scorpion

Scorpion *Astronomy*. see SCORPIUS. *Aviation*. a popular name for the F-89 jet fighter plane.
scorpionfish *Vertebrate Zoology*. any fish of the family Scorpaenidae, especially those of the genus *Scorpaena*.
Scorpionida *Invertebrate Zoology*. the scorpions, an order of nocturnal viviparous arachnids with an elongated body, large clawlike pedipalps, book lungs, and a sting, which in some species is fatal to humans; found in most warm and tropical regions.
scorpionism *Toxicology*. poisoning due to a scorpion sting; symptoms may include muscle weakness, paralysis, and sometimes death.
Scorpius *Astronomy*. the Scorpion, a magnificent zodiacal constellation of the Southern Hemisphere; it contains rich star fields and the 1st-magnitude supergiant star Antares.
Scorpius X-1 *Astrophysics*. a compact source of soft X-rays lying at a distance of 1100 light-years; it is marked by day-to-day variations of brightness and has a period of 0.787 days with quasiperiodic oscillations of 165 seconds in its light curve.
scorzalite *Mineralogy*. $(Fe^{+2},Mg)Al_2(PO_4)_2(OH)_2$, a deep blue, monoclinic mineral of the lazulite group, having a specific gravity of 3.38 and a hardness of 5.5 to 6 on the Mohs scale; found as granular masses in granite pegmatites, associated with quartz.
scotch *Engineering*. **1.** a block or wedge shoved under an object to keep it from rolling. **2.** to wedge or block an object to prevent it from slipping or rolling.
scotch boiler *Mechanical Engineering*. a fire-tube boiler having at least one internal cylindrical furnace in which heat is transferred to water partly in the furnace and partly by the passage of hot gases through the tubes. Also, DRY-BACK BOILER.
Scotch bond see AMERICAN BOND.
Scotch derrick see STIFFLEG DERRICK.
Scotch mist *Meteorology*. a mixture of thick fog or mist and heavy drizzle that occurs frequently in Scotland and in parts of England.
Scotch pine *Botany*. a tree of the pine family, *Pinus sylvestris*, that is native to Eurasia; characterized by a reddish trunk and twisted, blue-green needles.

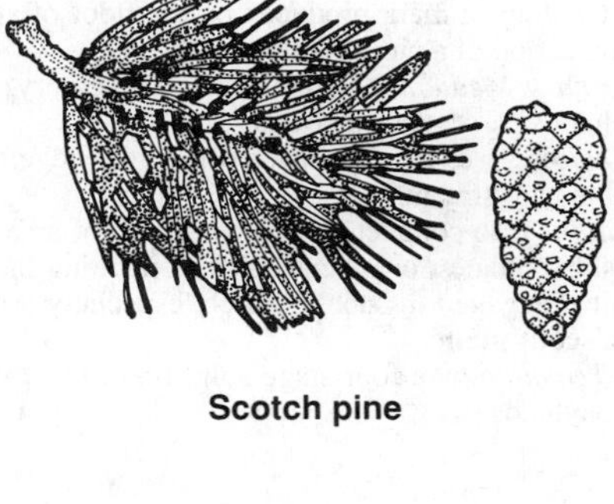

Scotch pine

Scotch-type volcano *Volcanology.* a volcanic form typically consisting of concentric cuestas and formed by the sinking of a cauldron.

Scotch yoke *Mechanical Engineering.* an apparatus with a four-bar linkage arrangement that converts rotary motion into simple harmonic motion.

scoto- a combining form meaning "darkness."

Scotobacteria *Bacteriology.* a class of bacteria of the division Gracilicutes composed of Gram-negative, nonphotosynthetic organisms. The class contains aerobic and anaerobic rods and cocci, including the medically important families Spirochaetaceae, Spirosomaceae, Pseudomonadaceae, Enterobacteriaceae, Vibrionaceae, Bacteroidaceae, Neisseriaceae, Legionellaceae, Pasteurellaceae, and Veillonellaceae, and the orders Rickettsiales and Chlamydiales.

scotobacterium *Bacteriology.* **1.** an individual of the class Scotobacteria. **2.** any bacterium capable of growing in the dark.

scotochromogen *Microbiology.* any strain of mycobacteria that can form pigment with or without light.

scotodinia *Neurology.* a form of vertigo characterized by sudden dizziness, headache, and blurred vision.

scotoma *Medicine.* **1.** an area in the visual field that is partially or completely blind. **2.** The subjective perception of dark obscure patches appearing and disappearing before the eyes.

scotophile *Biology.* an organism that requires or has an affinity for darkness.

scotophor *Materials.* material, usually potassium chloride, that darkens under electron bombardment and is used for screens of cathode-ray tubes.

scotopia *Physiology.* the adaptation of the eye to darkness; loss of color perception, with the ability to discern only shades of black and white. Also, DARK ADAPTATION, NIGHT VISION.

scotopic vision *Physiology.* of or relating to scotopia. Thus, **scotopic vision.**

scotoscope *Electronics.* a type of telescope that uses image intensifiers to improve night viewing.

Scott connection *Electronics.* in a power transmission, two single-phase transformers connected to operate in a three-phase system.

scour *Science.* to carry out a process of scouring. See SCOURING.

scour and fill *Geology.* a process in which the action of a stream or tide alternately excavates and refills a channel, particularly during a period of flooding.

scour cast see FLUTE CAST.

scour channel *Geology.* a large, grooved erosional feature produced in sediments by the scouring action of flowing water or ice.

scour depression *Geology.* a crescent-shaped cavity, in a stream bed, near the outside of a bend, produced by intense scouring below the grade of the stream.

scouring *Geology.* an erosion process involving the action of flowing air, ice, or water. *Mechanical Engineering.* the wet or dry cleaning or finishing of a hard surface by the mechanical application of an abrasive. *Agriculture.* the action of soil rubbing against and cleaning the blade of a plow as it passes through the ground, thereby reducing friction. *Textiles.* **1.** the process of cleaning dirt and natural grease from wool, using soaps and chemicals. **2.** the process of cleaning fabric before dyeing. **3.** the process of removing sizing and tint used on warp yarns in weaving.

scouring basin *Civil Engineering.* a reservoir that holds water to maintain a desired depth in an entrance channel by releasing quantities of water during low water conditions. Also, SLUICING POND.

scouring rush *Botany.* see HORSETAIL.

scouring velocity *Geology.* the speed of water required to forcibly remove stranded solids from a stream bed.

scour lineation *Geology.* a narrow, smooth, low ridge with symmetric ends, on a sedimentary surface, thought to result from the intense scouring action of a current of water.

scour mark *Geology.* a mark produced on the floor of a body of water by the scouring action of a current of water.

scours *Veterinary Medicine.* **1.** diarrhea or dysentery, particulary in newborn ruminants. **2.** see CALF SCOURS.

scourway *Geology.* a channel excavated near the edge of an ice sheet by a powerful glacier stream.

scout *Navigation.* **1.** to conduct an orderly search of an area, usually by a number of ships, planes, or other vehicles operating in a specific pattern. **2.** a vehicle engaged in such a search, especially a reconnaissance aircraft. Thus, **scout plane.**

Scout *Space Technology.* a four-stage solid-fuel rocket vehicle used to launch small payloads.

scout car *Ordnance.* a reconnaissance vehicle used for high-speed missions or cross-country operations; it is lightly armed and armored and may be wheeled or half-tracked.

scout hole *Mining Engineering.* **1.** a borehole, drilled through only the uppermost part of an ore body, used to analyze the surface configuration. **2.** a shallow borehole drilled to determine the presence of ore or to make a preliminary exploration of an area.

SCP single-cell protein.

SCR silicon-controlled rectifier.

Scrag *Ordnance.* a Soviet three-stage intercontinental ballistic missile developed with the SS-9 but never placed into service; officially designated **SS-10.**

scram *Nucleonics.* a term for the rapid shutdown of a nuclear reactor system by lowering safety rods into the core when a dangerous level of neutron flux density is reached.

scramble *Telecommunications.* to modify communications signals for the purpose of maintaining secrecy or protecting broadcast rights; depending upon the application, such modification may be according to a predetermined scheme, quasi-random, or random. *Aviation.* to take off as quickly as possible, usually receiving course and altitude instructions while in flight; used especially of military aircraft taking off from a base or aircraft carrier in a combat situation.

scramble competition see DIRECT COMPETITION.

scrambler *Electronics.* a circuit that separates voice frequencies into various ranges, then inverts and displaces them to produce unintelligible sounds; used to transmit secret communications. *Telecommunications.* a device or signal used to scramble television signals so that they cannot be decoded through unauthorized circuits.

scramjet *Aviation.* a ramjet engine in which combustion occurs at supersonic air velocities through the engine. (An acronym for supersonic combustion ramjet.)

scram rod see SAFETY ROD.

scram system *Nucleonics.* the electromechanical system used to shut down a nuclear reactor when a certain level of activity is reached.

scrap *Metallurgy.* metallic material that is off specification, generated as a waste stream in a processing operation, or discarded after service. *Engineering.* any recyclable refuse from a manufacturing process.

scraped-surface exchanger *Chemical Engineering.* a double-pipe heat exchanger in which a rotating element scrapes the inside surface of the exchanger to remove deposits; used for fluids that are subject to crystallization and severe fouling and also for solvent extraction and high viscosity fluids.

scraper *Mechanical Devices.* any of various instruments having a thin, flat steel blade and a raised edge, used to remove extraneous materials from surfaces, to shape materials, or to create grooves. *Civil Engineering.* an earth-collection device used for leveling operations. *Archaeology.* a flake tool that has been sharpened on one edge and left blunt on other edges to allow grasping; called a **side scraper** or **end scraper** depending on the sharpened edge; it was probably used to scrape animal hides.

scraper conveyor *Mechanical Engineering.* a conveyor consisting of chain-drawn scrapers or flights running in a trough through which the material is conveyed. Also, DRAG-LINK CONVEYOR, FLIGHT CONVEYOR.

scraper hoist *Mechanical Engineering.* a power-driven hoist that moves the scraper in a scraper-loader system.

scraper loader *Mechanical Engineering.* a combined scraper and transporting machine that loads coal or rock by pulling a scoop back and forth between the face and the loading point, then discharging the load from the scoop onto a car or conveyor.

scraper ring *Mechanical Engineering.* a piston-mounted ring that scrapes oil to prevent it from burning and coking on the inner surface of a cylinder.

scraper ripper *Mining Engineering.* a strip-mining machine with teeth on the lip for breaking or ripping the coal and a flight conveyor for carrying the broken material away from the lip.

scrapie *Veterinary Medicine.* a usually fatal disease of sheep causing degeneration of the central nervous system, tremors, intense itching, a tendency to scrape against objects for relief, and resulting wool loss.

Scraptidae see MELANDRYIDAE.

scratch *Computer Programming.* **1.** to erase or overwrite data on a storage device or memory. **2.** to delete the name of a file from a directory, allowing the reuse of the file's storage area. **3.** of or relating to storage that may be used as a temporary work space.

scratch awl *Mechanical Devices.* any of a number of sharp, pointed tools used to incise guide marks in materials.

scratchboard *Graphic Arts.* a piece of white paperboard that is coated black, then incised with a scratch knife to create a white-on-black line image.

scratch coat *Engineering.* an initial layer of plaster that is grooved to enable the second coat to bond more easily.

scratch file *Computer Programming.* a temporary file used as a work space during the execution of a program.

scratch filter *Acoustical Engineering.* a filter designed to reduce needle-scratch noises in a phonograph by reducing high frequency.

scratch hardness *Materials Science.* the resistance of a material, such as stone or metal, to scratching; the known hardnesses are standardized in a scale, such as the Mohs scale. Thus, **scratch-hardness test.**

scratch knife *Graphic Arts.* a tool used on scratchboard to create white line art; also used in retouching copy.

scratchpad *Computer Technology.* a small, fast-access semiconductor memory used for temporary storage of intermediate results or other information required during a computation. Also, **scratchpad memory.**

scratch tape *Computer Technology.* a reel of magnetic tape used temporarily by a job, but whose information is not saved after the job terminates. Also, WORK TAPE.

screamer *Vertebrate Zoology.* a swan-sized bird of the family Anhimidae that has a harsh, trumpeting call; characterized by long unwebbed toes, sharp spurs on the wings, and a layer of air-filled cells under the skin; found in South America.

screaming *Space Technology.* a form of moderately high-frequency combustion instability distinguished by a high-pitched noise; occurs most often in liquid-fuel rocket engines. Also, **screeching.**

scree *Geology.* **1.** weathered and broken rock fragments. **2.** a steep slope composed of such fragments. **3.** an accumulation of such fragments at the base of a mountain slope. **4.** a sheet of such fragments covering the slopes of a mountain or hillside. Also, TALUS.

screech owl *Vertebrate Zoology.* any of various small owls of the genus *Otus*, distinguished by a long, quavery cry and characterized by a usually gray or reddish-brown coat and hornlike tufts of feathers, as *O. asio*, of eastern North America.

screed *Mechanical Devices.* **1.** an arrangement of straight-edged wood or metal slats forming a template, placed on a wall as a guide for plastering to ensure uniform thickness of the plaster. **2.** a cylindrically shaped bar mounted on the wheels of a rolling machine; used for pavement smoothing. *Civil Engineering.* a rotary apparatus made of steel, used to strike off and smooth a surface.

screen *Engineering.* **1.** a coarse mesh of wire used to sift sand, gravel, or lime. **2.** a cover that protects against the elements, including light, heat, and debris. **3.** a filter that separates fluids from solids. *Electronics.* **1.** the surface upon which images from devices such as a television or oscilloscope become visible or are projected. **2.** a partition that shields a device from external electric or magnetic fields. *Computer Technology.* **1.** the surface of a video display device on which information can be displayed. **2.** in information retrieval, to make an initial selection from a set of data according to specified conditions. *Electromagnetism.* a metallic enclosure that is used to protect a device from electromagnetic fields. *Military Science.* **1.** an arrangement of ships, aircraft, or submarines to protect a main body or convoy. **2.** any natural or artificial material that is opaque to surveillance sensors and interposed between the sensors and an object for the purpose of camouflage or concealment. *Graphic Arts.* see HALFTONE SCREEN.

screen analysis *Engineering.* a method that determines the particle-size distribution of loose materials by measuring the percentage of particles passing through a screen with holes of different sizes.

screen angle *Electromagnetism.* an angle subtended in a vertical plane by a line extending from an antenna to the horizon and a horizontal line extending from the antenna, assuming a 4/3 earth's radius; the taller the antenna, the greater the angle. *Graphic Arts.* in color reproduction, the angle at which two or more halftone screens are juxtaposed in order to avoid undesirable moiré patterns.

screen deck *Mechanical Devices.* a surface with specified apertures used for screening operations.

screen dissipation *Electronics.* the thermal power released from a screen grid that has been struck by bombarding electrons.

screened cable *Electricity.* a flexible cable having a protective screen of conducting material that encloses each power core of the cable separately.

screen editor *Computer Programming.* an editor for files or data in which a screenful or page of the data is seen by the user and changes can be made to any part of the displayed data.

screened pan *Meteorology.* an evaporation pan with a top covered by wire-mesh screening, which reduces air circulation and insolation and results in a pan coefficient nearer to unity than that for unscreened pans.

screened photo print *Graphic Arts.* a photographic print shot through a halftone screen; halftone dots appear on such a print, but do not appear when it is rephotographed.

screened trailing cable *Electricity.* a flexible cable with a protective screen of conducting material that encloses each power core of the cable separately.

screen format *Computer Technology.* the appearance of data as displayed on the screen of a video display device, as contrasted with the appearance of data as printed in hard copy.

screen grid *Electronics.* the element placed between the grid and the anode in an electron tube to prevent electrons from bunching in the space between it and the cathode.

screen image buffer *Computer Technology.* a memory area containing information related to the image displayed on the screen; in order to change the display, the program must first write information into the screen image buffer.

screening *Engineering.* the process of separating various-sized particles using screens with different-sized openings by rotating, shaking, vibrating, or otherwise agitating the screen. *Industrial Engineering.* the process of examining job lots for defective parts. *Atomic Physics.* the reduction of the electric field surrounding the nucleus, caused by the net electric charge (space charge) of the surrounding electrons. *Electromagnetism.* the shielding of electromagnetic fields by a metallic enclosure around a device.

screening agent *Analytical Chemistry.* a nonchelating dye used to sharpen the colorimetric end point of a complexometric titration; two complementary colors indicate the metal and nonmetal forms of the end-point indicator.

screening constant *Atomic Physics.* a value that, when subtracted from an atom's atomic number, gives its effective atomic number; used to apply a hydrogenlike formula to X-ray spectra.

screening factor *Nuclear Physics.* the actual rate of a nuclear reaction in a dense plasma, divided by the rate that theoretically would apply if there were no free electrons screening repulsion between the nuclei.

screening smoke *Ordnance.* smoke used to conceal friendly troops or prevent observation by the enemy.

screening test *Science.* a test to determine the presence of a given substance or organism in a large number of samples.

screenless precession photography *Crystallography.* a method of measuring the diffraction pattern of a macromolecular crystal by the precession method without the use of a screen to filter out certain reflections. The indexing is done by computer after some preliminary photographs have been taken to determine the crystal orientation with high accuracy; used for data collection for crystalline macromolecules. Small precession angles (1° to 2°) are used so that many sets of photographs with different settings of the dial axis are required.

screen memory *Psychology.* in psychoanalytic theory, a form of resistance in which a nonthreatening memory, usually from early childhood, is used unconsciously to repress or screen out the memory of a more threatening event.

screen mesh *Engineering.* a network constructed of woven-wire cloth, metal bars, or perforated plates that separates materials.

screen pipe *Engineering.* a perforated pipe, wrapped in wire coils, that filters sand from well fluids.

screen printing *Graphic Arts.* a printing process in which the substrate is covered with a stencil and a porous screen made of fabric or mesh; using a squeegee, thick ink is applied to the screen and pressed through to the image areas not covered by the stencil. *Textiles.* a printing method in which color paste is forced through a fine mesh screen onto fabric, with each color in the pattern printed through a separate screen.

screen ruling *Graphic Arts.* the number of lines per inch on a halftone screen.

screen separation see DIRECT SEPARATION.

screen size *Engineering.* a standard for sizing particles by passing them through a series of screens with holes of descending size.

screw *Mechanical Devices.* **1.** a device consisting of an inclined plane incised into and spiraling around a tapered cylinder forming a helical groove; one of the simple machines, often used to fasten materials together or to transmit motion and apply pressure. **2.** a metal fastener of this type, having a tapered body with a spiraling thread and a slotted head, to be inserted into wood or other material with a screwdriver. Also, SCREW FASTENER.

screw anchor *Mechanical Devices.* an expandable metal shell placed into a screw hole that cuts into material, such as brick or stucco, and prevents the screw from coming loose. *Naval Architecture.* see MOORING SCREW.

screw auger *Mechanical Devices.* an auger that has a spiral shank suggesting a screw thread.

screw axis *Crystallography.* for a screw axis designated n_r, the symmetry operation is a rotation about the axis by $(360°/n)$ coupled with a translation parallel to the axis by r/n of the unit-cell length in that direction.

screw bean *Botany.* a tree, *Prosopis pubescens*, of the legume family, characterized by twisted pods that are used as fodder; found in the southwestern U.S. Also, TORNILLO.

screw bolt *Mechanical Devices.* a bolt with a mounted screw shank.

screw clamp *Mechanical Devices.* a carpentry clamp in which large screws loosen and tighten pieces of wood around the material being held. Also, MOORING SCREW, HAND SCREW, HAND CLAMP.

screw compressor *Mechanical Engineering.* a gas compressor in which the compression is achieved by two counter-rotating, intermeshing screws in an enclosed chamber.

screw conveyor *Mechanical Engineering.* a conveyor in which a spiral blade presses material forward as it rotates in a trough or housing.

screw-cutting lathe *Mechanical Devices.* a thread-cutting lathe with a slide rest and leadscrew.

screw dislocation *Crystallography.* a dislocation in a crystal form caused by slicing partway into the crystal and shearing it parallel to the slice by one atomic spacing.

screw displacement *Mechanics.* the rotation of a rigid body about an axis with a translation of the body along that axis.

screwdriver *Mechanical Devices.* a narrow, steel rod or bar used to turn screws in wood or other material, consisting of a handle attached to a wedge-shaped, flat head that fits into a slot on a screw head.

screw elevator *Mechanical Engineering.* a vertical screw conveyor for handling pulverized or granulated materials.

screw eye *Mechanical Devices.* a screw with a ring-shaped head.

screw fastener see SCREW, def. 2.

screwfeed *Mechanical Engineering.* a gear-and-ratchet mechanism in the swivel head of a diamond drill that controls the rate at which the bit penetrates a rock formation.

screw feeder *Mechanical Engineering.* an auger-type screw that transfers material from one piece of equipment to another.

screw head *Mechanical Devices.* the top part of a screw, having a slot that is engaged by a screwdriver for turning the screw.

screw hook *Mechanical Devices.* a small hook with a threaded shank used for hanging objects in wall paneling or woodwork.

screw ice *Hydrology.* **1.** heaps or ridges composed of small fragments of ice that result from the crushing together of ice cakes. **2.** a small formation consisting of pressure ice.

screw jack see JACKSCREW.

screw joint *Mechanical Devices.* a joint formed by screwing together the male and female mating screws.

screw machine *Mechanical Engineering.* a lathe used to turn and thread small screws from rods or bars that are fed through a hollow spindle.

screw nut *Mechanical Devices.* a nut threaded so as to receive a screw.

screw pile *Civil Engineering.* a pile with a wide projecting helix at the end, so that it bores its way into the earth as it is revolved; useful in alluvial ground.

screw plasticating injection molding *Materials Science.* a process of converting plastic from pellets to a viscous melt by an extruder screw of the molding machine.

screw press *Mechanical Engineering.* a press used for punching holes and stamping thin metal works by means of a square, threaded screw with a heavy cross piece for increased pressure. Also, FLY PRESS.

screw propeller *Mechanical Engineering.* a rotor with blades angled so that, as the rotor spins, the blades advance along the shaft.

screw pump *Mechanical Engineering.* a water pump consisting of rotating helical-screw impellers in a casing.

screw rivet *Mechanical Devices.* **1.** a threaded bolt first positioned into its functional location and then riveted at its end. **2.** a short rod threaded throughout the length of a shaft without access to the point.

screw spike *Mechanical Devices.* a railroad spike with a helical screw thread at its top end, allowing it to be screwed into a railroad tie.

screwstock *Mechanical Engineering.* a rod, bar, or wire that is to be die-cut. Also, DIESTOCK.

screw thread *Design Engineering.* a helical ridge formed on a cylindrical core, such as on screw fasteners and pipes; it may be right-hand threaded or left-hand threaded, either of which directs the screw's movement into a housing.

screw-thread gauge *Mechanical Devices.* any of several instruments used to determine the various dimensions of a screw thread, such as pitch, straightness, and major or minor diameters.

screw-thread micrometer *Mechanical Devices.* a device with an anvil and spindle shape, used to measure the pitch diameter of a screw.

screwworm fly *Entomology.* an insect pest that lays its eggs in open wounds or the nostrils of livestock and more rarely humans; successful eradications have involved the release of males sexually sterilized through radiation; indigenous to large parts of the Western Hemisphere.

screw wrench *Mechanical Devices.* a wrench with an adjustable jaw that is moved by a screw. Also, **screw spanner.**

scribe *Engineering.* to mark a material with a scriber.

scribe projection *Telecommunications.* a system of automatic information presentation in which information placed on a metallic-coated glass slide by a movable fine-pointed scribe is projected on a screen.

scriber *Mechanical Devices.* a pointed tool used for marking wood or other material, especially as a guide for cutting. *Graphic Arts.* a small, sharp hand tool used mainly to incise lines on the emulsion of a photographic negative.

scribing *Electronics.* a process in which a semiconductor wafer is etched with deep groves, so that it can be broken into smaller pieces. *Graphic Arts.* the use of a scriber to remove film emulsion, often in order to draw a line or rule on an image.

script *Graphic Arts.* a class of typefaces having connected letters resembling handwriting. *Psychology.* a stored scenario of events that are characteristic of a particular situation or social setting and the behavior appropriate to the situation, such as eating in a restaurant; necessary for an individual to understand and deal with any such similar circumstances. *Artificial Intelligence.* a framelike structure that describes a sequence of actions, e.g., the sequence of steps involved in ordering a meal and dining at a restaurant.

scRNA *Genetics.* small cytoplasmic RNA; occur as ribonucleoprotein particles in the cytoplasm of eukaryotic cells; their function is unknown.

scrobiculate *Biology.* furrowed or pitted.

scrod *Vertebrate Zoology.* a popular term used to describe any of several immature marine fish used for food, such as haddock or cod.

scrofula [skräf´yə lə] *Medicine.* a former term for tuberculosis of the lymphatic glands, sometimes including and affecting the bones and joints, accompanied by abscesses, usually of the cervical lymph nodes, now known as **tuberculous cervical lymphadenitis.**

scrofulous [skräf´yə ləs] *Medicine.* relating to, resembling, or affected by scrofula.

scroll *Computer Programming.* on a visual display unit, to move data vertically or horizontally within a display area, with new data appearing at one edge of the area as previous data disappears at the opposite edge. This is analogous to rolling and unrolling a paper scroll to change the visible part of the scroll. *Architecture.* any ornament, such as an Ionic volute, resembling a scroll of paper. *Geology.* one of a series of crescent-shaped accumulations of sediment formed by a stream on the inner bank of a shifting channel.

scroll gear *Mechanical Devices.* a flat, variable gear with teeth on only one face.

scrolling *Computer Programming.* **1.** the process of moving a display image vertically or horizontally so that new data appears at one edge as previous data disappears at the opposite edge. **2.** specifically, the process of using this scroll function in word processing to read consecutively through a document, as one would read consecutive pages of printed text.

scroll meander *Hydrology.* a meander in which the buildup of scrolls on the inner bank results in the lateral erosion of the outer bank, maintaining the width of the stream channel.

scroll saw *Engineering.* a thin-bladed saw used for cutting ornamental designs in wood.

Scrooge *Ordnance.* a Soviet mobile, surface-to-surface missile propelled by a solid rocket motor and having a range of 3500 miles; officially designated **SS-15.**

Scrophulariaceae *Botany.* the snapdragon family, a dicotyledonous family of the order Scropulariales, consisting of about 220 genera and 4450 species of trees, shrubs, and herbs native to temperate regions and tropical mountains; characterized by fruit that is usually a septicidal capsule.

Scrophulariales *Botany.* an order of the subdivision Asteridae, consisting of 12 families: Buddlejaceae, Oleaceae, Scrophulariaceae, Globulariaceae, Myoporaceae, Orobanchaceae, Gesneriaceae, Acanthaceae, Pedaliaceae, Bignoniaceae, Mendonciaceae, and Lentibulariaceae.

scrotum *Anatomy.* the sac that contains the testes and spermatic cord in most male mammals.

scrub *Ecology.* a general term for any type of vegetation, such as evergreen shrubs or dwarfed trees, growing in areas with poor soil or low rainfall. *Agriculture.* a domestic animal that is considered inferior because its parentage is mixed or unknown. *Space Technology.* to cancel a planned flight or military mission, either before or during countdown. *Computer Programming.* to examine a large amount of data, avoiding duplicate or unwanted information.

scrubber *Engineering.* **1.** a device that removes air-polluting particles of 1 to 5 micrometers, such as water-soluble gases; used in air-pollution control. **2.** a device that removes dust particles or liquid droplets entrained in fluid. **3.** a device that eliminates an extrinsic gas component from process gas streams. *Mining Engineering.* a device in which coarse and sticky material, such as ore or clay, is washed free of adherents or is mildly disintegrated.

Scrubber *Ordnance.* the code name for the Soviet SS-N-1 missile.

scrubbird *Vertebrate Zoology.* either of two passerine birds of the genus *Atrichornis* of the family Menurae that resemble the lyrebird, characterized by a loud voice and minimal powers of flight; found in Australia.

scrub fowl see MEGAPODE.

scrub jay *Vertebrate Zoology.* a type of crestless jay, *Aphelocoma coerulescens*, characterized by blue and gray plumage; found in the western and southern U.S. and Mexico.

scrub oak *Botany.* any of several small, scrubby oaks, such as *Quercus ilicifolia* and *Q. prinoides*, usually found in dry, rocky soil.

scrub pine *Botany.* any of several low pines, such as the jack pine, usually small and straggly but reaching a height of 100 feet in some parts of its range; found in dry, sandy soil.

scrub plane *Mechanical Devices.* a carpenter's plane with a rounded cutting edge, used to remove large amounts of stock.

scrub typhus *Pathology.* a highly infectious, sometimes fatal typhus-like disease of Asia and the southern and western Pacific that is spread by mites and characterized by fever, painful lymphatic swelling, scabbing, and large, dark-red lesions. Also, ISLAND DISEASE, TSUTSUGAMUSHI DISEASE, TROPICAL TYPHUS.

scruff *Metallurgy.* an oxide dross floating on a molten metallic bath.

scruple *Metrology.* a unit of the old apothecaries' system of weights, equal to 20 grains or 1.296 grams.

SCS silicon-controlled switch.

SCSI [skuz´ē; skiz´ē] *Computer Technology.* **1.** a standardized electrical and logical specification for peripheral device interfaces. **2.** describing a device that meets the SCSI standard; e.g., a **SCSI disk.** (An acronym for Small Computer Systems Interface.)

Sc spiral galaxy *Astronomy.* a type of spiral galaxy characterized by large, loosely coiled arms and a very small, concentrated nucleus.

SC star *Astronomy.* a star intermediate in type between S stars and carbon stars.

scuba [skoo´bə] *Mechanical Devices.* a portable breathing device, consisting of a mouthpiece joined by hoses to one or two tanks of compressed air strapped on the back; used by underwater swimmers. (An acronym for self-contained underwater breathing apparatus.)

scuba diving *Engineering.* the activity of diving or exploring underwater using a scuba that allows the diver to move freely underwater.

scud *Meteorology.* a mass of fractured, ragged clouds, usually stratus fractus, and typically moving rapidly beneath a layer of nimbostratus.

Scud *Ordnance.* a Soviet-made surface-to-surface missile powered by a liquid-propellant rocket motor and equipped with an inertial guidance system; it can deliver a conventional, chemical, or nuclear warhead at a speed of Mach 5 and a range of 80 miles; officially designated **SS-1.** The later versions **Scud-A** and **Scud-B** have greater range, of 185 and 280 miles, respectively. (An acronym for Subsonic Cruise Unarmed Decoy.)

Scud al Hussein *Ordnance.* an Iraqi version of the Soviet Scud missile modified for a longer range of 375 miles; used in the Persian Gulf War. (Named for the Iraqi leader Saddam *Hussein.*)

scuffing *Engineering.* a worn or rough spot on the surface of glass, usually resulting from abrasion.

scuffle hoe *Mechanical Devices.* a garden hoe whose blade is sharpened front and back to allow cutting in either direction.

sculpin *Vertebrate Zoology.* any one of several species of the fish family Cottidae, characterized by a large, spiny head and a broad mouth, and lacking scales; includes both marine and freshwater forms found in the Northern Hemisphere and near New Zealand, eastern Australia, and Argentina.

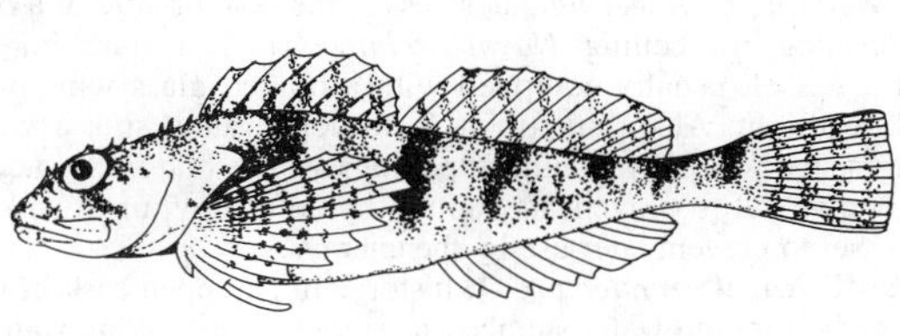

sculpin

Sculptor *Astronomy.* Sculptoris, a constellation of the Southern Hemisphere which contains no stars brighter than 4th magnitude; the South Galactic Pole lies in the constellation.

Sculptor Group *Astronomy.* a group of galaxies that lies at a distance of 10 to 13 million light-years.

Sculptor's Tool *Astronomy.* the constellation Caelum.

Sculptor system *Astronomy.* a dwarf elliptical galaxy in the Local Group, approximately 280,000 light-years distant.

sculpture *Geology.* to change the contour of (the land surface) by erosion.

scum *Materials.* **1.** a general term for foul or unwanted material that forms a film or layer on the surface of a liquid; it may be organic or inorganic. **2.** a similar excess layer of material on a solid, such as metal or cement. *Biology.* also, **pond scum.** a layer of free-floating algae that forms on the surface of still or stagnant water, appearing as a greenish film.

scum chamber *Biotechnology.* a container in an Imhoff tank that facilitates sewage digestion; it functions as an area for gasses to rise to the surface of sludge.

scumming *Graphic Arts.* during a press run, the buildup of ink on the nonimage areas of a plate.

scupper *Naval Architecture.* **1.** a drain in a deck or bulwark, leading overboard; used primarily to drain seawater or rainwater from the deck. **2.** to deliberately sink a vessel by making a hole in the underwater hull or allowing seawater into an existing hole.

scuppernong *Botany.* **1.** a large yellowish muscadine grape grown in the southern U.S. **2.** the vine bearing this fruit. (Named for the *Scuppernong* River in North Carolina.)

scurf *Medicine.* another term for dandruff; a flaky substance originating from the epidermis.

S-curve see REVERSE CURVE.

scurvy *Medicine.* a condition caused by vitamin C deficiency, resulting in weakness and anemia, the tendency of the mucous membranes to hemorrhage, and sponginess of the gums; historically, it was common among sailors during a long sea voyage because of their inadequate diets, until a practice of issuing rations of citrus juice was initiated.

scutch *Engineering.* a tool similar to a small pick with flat cutting edges for trimming bricks.

scutching *Materials Science.* in the processing of natural fibers, a mechanical process for separating fibers from the stem tissue by scraping.

scutate *Botany.* formed like a round shield. *Zoology.* having shieldlike parts or large scales.

scute *Invertebrate Zoology.* a scale or shield, such as the dorsal plate of an arthropod exoskeleton. Also, SCUTUM.

Scutechiniscidae *Invertebrate Zoology.* a family in the water bear phylum, Tardigrada, tiny (0.3 to 0.5 mm) animals that inhabit the water films surrounding the leaves of mosses and lichens.

Scutelleridae *Invertebrate Zoology.* a family of shield bugs, hemipteran insects in the superfamily Pentatomoidea, having some that are bright-colored; some species are plant-eating pests.

Scutellinia *Mycology.* a genus of fungi belonging to the order Pezizales, which is characterized by a border of expicular hair and occurs in soil or dung. Also EYELASH FUNGUS.

scutellum *Botany.* a thickened, food-storing cotyledon of grasses, such as corn. *Entomology.* a shield-shaped plate formed from the hindmost part of the notum of an insect or any similar dorsal covering. *Vertebrate Zoology.* one of the transverse scales or plates on the tarsi and toes of birds.

scutica *Cell Biology*. a hooklike structure that is found in certain stomate ciliates, composed of nonciliated kinetosomes and formed during stomatogenesis.

Scutigeromorpha *Invertebrate Zoology*. an order of centipedes in the class Chilopoda, including the common house centipede *Scutigera coleoptrata*.

scuttle *Building Engineering*. an access to the roof or attic of a building by an opening in a ceiling. *Naval Architecture*. **1.** a round hole in the side of a vessel, usually equipped with a hinged glass window and a hinged deadlight. Also, PORTHOLE. **2.** to deliberately sink a vessel by opening seacocks or by making openings in the bottom; e.g., to sink one's own vessel to prevent its capture by an enemy, or to sink a captured vessel to prevent later use by the enemy.

scuttlebutt *Naval Architecture*. **1.** historically, an open cask in which a ship's ready drinking water supply was stored. **2.** a drinking water fountain aboard ship. **3.** in nautical slang, a rumor, especially one that circulates aboard ship.

scutulum *Volcanology*. a very flat shield volcano.

scutum see SCUTE.

Scutum *Astronomy*. the Shield, a small constellation that lies to the north of Sagittarius and east of Aquila; it contains a bright star cloud of the Milky Way.

Scydmaenidae *Invertebrate Zoology*. a family of antlike stone beetles, shiny brownish to reddish-brown coleopteran insects in the superfamily Staphylinoidea.

scyelite *Petrology*. a coarse-grained, ultramafic hornblende peridotite with a poikilitic texture due to inclusion of olivine crystals in crystals of other minerals, especially amphiboles.

Scyliorhnidae *Vertebrate Zoology*. the cat sharks, the largest family of sharks in the order Carcharhiniformes, containing mostly bottom-dwelling harmless species, of tropical, temperate, and boreal seas.

Scyllaridae *Invertebrate Zoology*. the Spanish or shovel-nosed lobsters, a family of decapod crustaceans in the superfamily Scyllaridea, section Macrura, with a heavily armored exoskeleton and no rostrum or chelae.

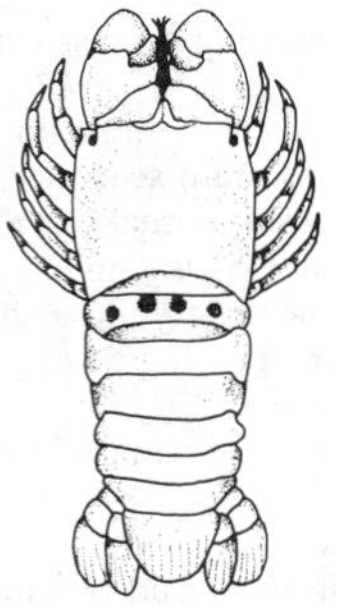

Scyllaridae (shovel-nosed lobster)

scyphistoma *Invertebrate Zoology*. a sessile polyp stage in the life cycle of scyphozoan jellyfish that buds asexually to form either new free-swimming jellyfish or more sessile scyphistomae. Also, **scyphopolyp.**

scyphomedusa *Invertebrate Zoology*. the jellyfish stage of scyphozoan coelenterates.

Scyphozoa *Invertebrate Zoology*. the true jellyfish, a class of the phylum Coelenterata; the umbrella is shaped like a bell or saucer, with many marginal tentacles and the mouth on a stalk dangling from the center. Also, **Scyphomedusae.**

scyphozoan *Invertebrate Zoology*. **1.** an organism belonging to the class Scyphozoa. **2.** of or relating to this class.

scyphus *Botany*. a cup-shaped part, especially such an enlargement of the podetium in lichens.

scythe [sīth} *Mechanical Devices*. a hand tool traditionally used for cutting grasses and grains, consisting of a long, curved blade sharpened on one edge that is attached to an irregularly curved handle with two knobs for grasping.

Scythian stage *Geology*. a European geologic stage of the Lower Triassic period, occurring after the Tatarian of the Permian period and before the Anisian. Also, WERFENIAN STAGE.

Scytonema *Bacteriology*. a genus of filamentous cyanobacteria, whose trichomes are composed of disc-shaped vegetative cells.

Scytopetalaceae *Botany*. a family of trees, shrubs, and woody vines of the order Theales; characterized by narrow-rayed phloem stratified into hard and soft layers, and by a dicotyledonous embryo embedded in abundant endosperm; found in tropical West Africa.

Scytosiphonales *Botany*. a monofamilial order of brown algae of the class Phaeophyceae, composed of the family Schytosiphonaceae; characterized by alternating phases of a parenchymatous macrothallus and a smaller pseudoparenchymatous microthallus.

SD skin dose; streptodornase.

SD or **S.D.** Doctor of Science.

sd. sound

s.d. standard deviation. Also, S.D.

SDA specific dynamic action.

SDA or **S.D.A.** sacrodextra anterior.

SDDC sodium dimethyldithiocarbamate.

SDF *Aviation*. the airport code for Louisville.

SDI Strategic Defense Initiative.

SDP or **S.D.P.** sacrodextra posterior.

SDS-PAGE *Biochemistry*. an electrophoretic method used to establish molecular weight, as in a protein or polypeptide chain, or for separating protein mixtures. (An acronym for sodium dodecyl sulfate-polyacrylamide gel electrophoresis.)

SDS-polyacrylamide gel *Biochemistry*. polyacrylamide gel in the presence of the denaturant, sodium dodecyl sulfate (SDS); used in electrophoresis.

SDT or **S.D.T.** sacrodextra transversa.

SE or **S.E.** southeast; southeastern.

S.E. standard error.

Se the chemical symbol for selenium.

se or **s.e.** southeast; southeastern.

sea *Geography*. **1.** all or any part of the great body of salt water covering most of the earth's surface. **2.** one of the subdivisions of this body of salt water that is smaller than an ocean, such as the Caribbean Sea, North Sea, or Mediterranean Sea. **3.** a very large saltwater lake, such as the Caspian Sea, Dead Sea, or Salton Sea. *Oceanography*. a collective term for waves still within their fetch and still under the influence of their generating wind.

SEA *Aviation*. the airport code for Seattle/Tacoma International.

sea-air-land team see SEAL.

sea-air temperature difference correction *Navigation*. an adjustment that is made to a sextant altitude reading in order to compensate for refraction caused by the difference in temperature between the water and the air.

sea anchor *Naval Architecture*. a drag-producing drogue, sometimes jury-rigged of spars and canvas, placed in the water and secured to a vessel; commonly used to keep a vessel's bow or stern facing into the wind and prevent excessive leeway in heavy weather.

sea anemone *Invertebrate Zoology*. the common name for various anthozoan coelenterates in the order Actiniaria that attach to surfaces by a foot; characterized by a cylindrical body that lacks a skeleton and numerous tentacles surrounding the mouth; they are often bright-colored and flowerlike.

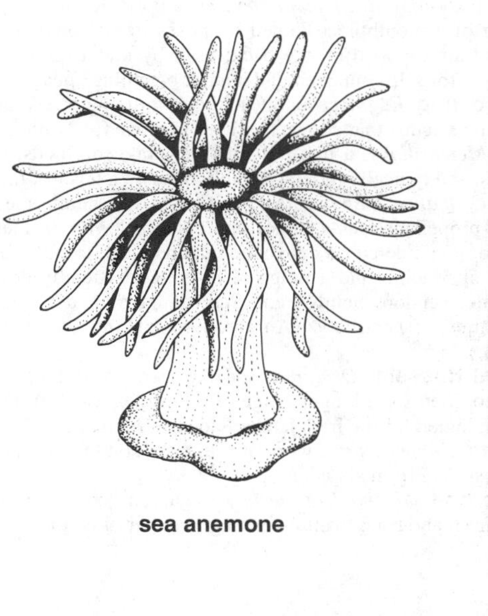

sea anemone

sea arch

sea arch *Geology.* a passageway through a headland consisting of a bridge of rock over water that is produced either by wave erosion or by solution. Also, **sea bridge.**

sea ball *Oceanography.* a mass of living or dead vegetation that has been compacted into a spherical shape by wave movement in shallow water.

sea bass or **seabass** *Vertebrate Zoology.* any of numerous marine food and game fishes of the family Serranidae, having a peculiar tail fin; found off the coast of the U.S.

Sea Bat *Aviation.* a helicopter armed with active-passive sonar and acoustic homing torpedoes; designed for antisubmarine warfare.

sea bed or **sea bottom** see SEA FLOOR.

Seaborg, Glenn born 1912, American chemist; shared the Nobel Prize with Edwin M. McMillan for the synthesis of plutonium and other transuranium elements.

sea breeze *Meteorology.* a coastal local wind that blows from the sea to the land, usually during calm, sunny, summer days, alternating with the weaker land breeze that blows in the opposite direction at night; caused by the temperature difference when the sea surface is colder than the adjacent land.

sea buoy *Navigation.* the buoy that marks the seaward end of a channel or harbor. Also, **farewell buoy.**

sea butterfly *Invertebrate Zoology.* the common name for a gastropod mollusk of the suborder Pteropoda.

sea cabbage *Invertebrate Zoology.* a brown alga, *Hedophyllum sessile*, characterized by a compact mass of fronds resembling a cabbage.

sea-captured stream *Hydrology.* a stream that flows parallel to the seashore and that has been separated into two parts as a result of marine erosion.

Seacat *Ordnance.* a British ship-based, surface-to-air and surface-to-surface missile designed for close-range combat; it is powered by a two-stage solid-propellant motor, equipped with radio command guidance, and delivers a high-explosive warhead at a range of about 2.2 miles.

sea cave *Geology.* a crack or small opening that is formed at the base of a sea cliff by the action of waves on easily weathered rock. Also, **sea chasm.**

sea channel *Geology.* a long, narrow, U-shaped or V-shaped shallow trough on the sea floor, typically found on a gently sloping fan or plain.

sea cliff or **seacliff** *Geology.* a cliff or slope formed by wave erosion at the seaward margin of a coast or at the landward side of a wave-cut platform, and representing the inner limit of beach erosion.

sea clutter *Electromagnetism.* the clutter of radar signals received on an airborne system due to reflections from the sea. Also, **sea return.**

sea-control operations *Military Science.* the employment of naval forces, supported by land and air forces as appropriate, to achieve military objectives in vital sea areas.

sea cow *Vertebrate Zoology.* **1.** any of several large, fishlike, plant-eating mammals of the family Dugongidae, including the manatee and the dugong. **2.** a former name for the hippopotamus.

sea cucumber *Invertebrate Zoology.* the common name for echinoderms in the class Holothuroidea, characterized by a long, flexible, leathery body, somewhat resembling a cucumber, with tentacles around the anterior end.

Sea Dart *Ordnance.* a British ship-based surface-to-air and surface-to-surface missile designed to intercept aircraft and air-launched missiles as well as ship-launched missiles; it is powered by a ramjet sustainer with solid propellant rocket booster, equipped with semiactive radar homing guidance, and delivers a high-explosive warhead at a range of at least 19 miles.

Sea Eagle *Ordnance.* an advanced British air-to-surface missile with sea-skimming, fire-and-forget capability; it is powered by a turbojet engine, equipped with active radar homing, and delivers an antishipping conventional warhead at approximately Mach 0.9 and a range that is probably greater than 30 miles.

sea echelon *Military Science.* in an amphibious landing, a part of the assault shipping force that does not participate in the landing and operates in a designated seaward area in an on-call or unscheduled status.

sea fan *Invertebrate Zoology.* the common name for corals of the order Gorgonacea. *Geology.* see SUBMARINE FAN.

sea feather see SEA PEN.

sea floor *Geology.* the area of the earth's crust underlying the oceans. Also, SEA BED, SEA BOTTOM.

sea-floor spreading *Geology.* in plate tectonics, a theory proposing that the oceanic crust is expanding as a result of convective upwelling of magma along oceanic ridges and the movement of this new material away from the ridges at a rate of 1–10 centimeters per year.

sea-floor trench *Geology.* see TRENCH, def. 3.

sea fog *Meteorology.* an advection fog formed when air that has been lying over a warm water surface is transported over a colder water surface, resulting in a cooling effect of the lower layer of air below its dew point. Also, SEA MIST.

sea gate *Civil Engineering.* a barrier, beach, or channel that provides protection from the sea, usually at a harbor or tidal basin. *Geography.* a navigable channel giving access to the sea; a river mouth.

sea glow *Oceanography.* the luminous cobalt-blue appearance of clear water in midocean, resulting from absorption of much of the red spectrum in light reflected upward from below the surface.

sea goat see CAPRICORNUS.

seagoing *Naval Architecture.* describing a vessel that is designed for open ocean service. Thus, **seagoing tug.**

sea grape *Botany.* **1.** a small tree, *Coccoloba uvifera*, of the buckwheat family, characterized by white flowers and attractive fall foliage and bearing grapelike clusters of edible purple berries. **2.** the berries of this tree that are made into jelly and a beverage.

seagrass *Botany.* any of several monocotyledonous grasslike marine plants of the genus *Zostera,* family Zosteraceae; not a true grass.

sea gully see SLOPE GULLY.

sea horse *Vertebrate Zoology.* any fish of the genus *Hippocampus* of the pipefish family, characterized by an elongated snout, a head bent at a right angle to the body, and a prehensile tail; found in warm waters.

sea horse

sea ice

sea ice *Oceanography.* **1.** ice formed by the freezing of seawater, as distinguished from glacier ice or other land ice. **2.** generally, any ice floating in the sea.

sea-ice shelf *Oceanography.* sea ice floating near where it was formed, separated from fast ice only by a tide crack or a system of tide cracks.

Sea Island cotton *Botany.* a fine, long-fibered cotton, *Gossypium barbadense,* that was originally grown in the Sea Islands off the coast of the southeastern U.S. and now grown chiefly in the West Indies. Also, **sea-island cotton.**

sea kale *Botany.* a plant, *Crambe maritima,* of the mustard family, characterized by fleshy, blue, cabbagelike leaves; found in maritime regions of Europe, where the stems are blanched and eaten as a vegetable; also used as a pot plant.

seakeeping *Naval Architecture.* the capability to carry out prolonged operations at sea in heavy weather; a key characteristic of many types of naval ships.

Sea Killer *Ordnance.* an Italian ship-based, surface-to-surface missile; the MK-2 version is powered by a solid-propellant, booster/sustainer rocket motor, equipped with beam-riding, radio-command guidance, and delivers a high-explosive, armor-penetrating warhead at a maximum range of 11.5 miles.

sea kindly *Naval Architecture.* describing a ship that sails easily in a rough sea. Thus, **sea kindliness.**

seal *Engineering.* **1.** a device that closes or shuts completely to make an object air- or liquid-tight. **2.** a nonpermeable coating that is applied to wood before painting. Seals may include gaskets, threaded fittings, or liquids.

seal *Vertebrate Zoology.* **1.** a marine, carnivorous mammal of the family Phocidae (the eared or fur seals) or Otariidae (the earless or hair seals) that lives chiefly in cool seacoasts but crawls ashore to breed and bear young; characterized by limbs modified as webbed flippers. **2.** generally, any pinniped other than a walrus.

SEAL or **Seal** *Military Science.* a member of a U.S. Navy group that is specially trained and equipped to conduct unconventional and paramilitary operations and to train allied personnel in such operations, including surveillance and reconnaissance in and from restricted waters, rivers, and coastal areas. (An acronym for sea-air-land team.)

Sealab *Science.* any one of several underwater vessels of the U.S. Navy used as underwater habitats for aquanauts.

sea lamprey *Vertebrate Zoology.* a parasitic lamprey, *Petromyzon marinus,* that spawns in fresh water along the Atlantic coast and in the Great Lakes, where it is highly destructive to such economically valuable fish as lake trout.

sea-lane *Navigation.* see SEAWAY.

sea-launched ballistic missile *Ordnance.* a ballistic missile that is launched from a surface ship or submarine. Similarly, **sea-launched cruise missile.**

sea lavender *Botany.* any of a group of mostly perennial herbs, such as the Old World *Limonium volgare* or *L. carolinianum,* of the eastern coast of North America; characterized by one-sided spikes of tiny lavender, yellow, or multicolored flowers that retain their color after drying.

sealed cabin *Aviation.* the occupied space of an aircraft or spacecraft containing its own means for maintaining the inside atmosphere; the walls are designed to prevent gaseous exchange between the inner atmosphere and the surrounding atmosphere.

sealed tube *Electronics.* a hermetically sealed electron tube.

sealer *Materials.* **1.** a substance that is applied to a porous surface in order to seal it, usually in preparation for painting, varnishing, or similar coating. **2.** a material used in a post-anodizing treatment.

Seale rope *Mechanical Devices.* a ropelike wire cable made up of six or eight strands; each strand consists of a large wire core around which nine small wires are wrapped in two layers.

sea level *Geology.* the average elevation of the sea surface over a 19-year period. Also, MEAN SEA LEVEL.

sea-level datum *Engineering.* the mean level of the oceans, taken as the mean level between high and low tides, and used as a standard for measuring heights and depths.

sea-level pressure *Meteorology.* the atmospheric pressure at mean sea level, either directly measured or empirically determined from the observed station pressure.

sea-level pressure chart see SURFACE CHART.

sea lily *Invertebrate Zoology.* the common name for echinoderms in the class Crinoidea.

sealing *Metallurgy.* the process of containing the surface porosity present in an anodic coating or in a casting.

sealing voltage *Electrical Engineering.* the voltage required to complete the movement of a magnetic circuit-closing device from the point at which the contacts first touch each other.

sealing-wax structure *Geology.* in an otherwise normal sedimentary succession, a primary sedimentary flow structure that is formed by slumping and occupies a zone of highly fluid contortion.

sea lion *Vertebrate Zoology.* a member of the subfamily Otariinae in the family Otarridae, such as *Eumetopias jubatus,* of the northern Pacific, and *Zalophus californicus,* of the Pacific coast of North America; eared marine mammals that are related to the fur seals, having well-developed guard hairs but no underfur.

sea lion

seal off *Engineering.* to shut off a tube or borehole with a sealant or cement. *Petroleum Engineering.* to prevent a formation from producing by injecting a drilling fluid into it.

sea louse *Invertebrate Zoology.* the common name for small, parasitic, marine isopod crustaceans.

SEAL team see SEAL.

seal weld *Metallurgy.* any of several types of welds that form a leak-proof seal.

seam *Geology.* **1.** a vein, bed, or thin layer of coal, ore, or other material. **2.** the line of separation between two distinct rock layers. **3.** a rock stratum that separates two distinct layers of different compositions. *Engineering.* a line formed by joining two separate pieces, as in the resistance seam-welding process. *Metallurgy.* **1.** a mechanical joint of sheet metal. **2.** in metal working, an undesirable fold caused by a defect in the stock.

seaman *Military Science.* in the U.S. Navy, an enlisted person ranking below a petty officer.

seamanite *Mineralogy.* $Mn_3^{+2}(PO_4)B(OH)_6$, a transparent, yellow orthorhombic mineral occurring in acicular crystals, having a specific gravity of 3.128 and a hardness of 4 on the Mohs scale.

seamanship *Navigation.* the ability to handle a ship at sea and conduct it from one point to another.

sea mark or **seamark** *Navigation.* **1.** originally, a navigational marker on land, such as a lighthouse. **2.** in modern usage, a floating marker, such as a buoy, that is designed to aid navigation; as distinguished from a landmark, such as a lighthouse.

sea marker *Engineering.* **1.** a line that marks the limit of the tide. **2.** a dye on the ocean surface that attracts rescue planes.

sea marsh *Ecology.* a salt marsh that is periodically submerged by the sea. Also, **sea meadow.**

seam blast *Mining Engineering.* a blast produced by introducing explosives along and within a crack between a solid wall and the rock or coal to be removed.

sea mile *Navigation.* an approximate value of the nautical mile equal to 6080 feet; it is the length of 1' of arc along the meridian at latitude 48°.

seaming *Metallurgy.* the process of joining two sheets or strips longitudinally by mechanical means.

sea mist see STEAM FOG.

seamless smoothly continuous in quality; having no seams. *Computer Science.* not requiring any special action by a user or program in order to cross boundaries; e.g., a seamless file system would allow files on a local disk and a network file server to be accessed in the same way.

seamless tube *Materials Science.* a hollow cylinder of material other than that made by bending over a flat strip and welding the edges; produced by piercing a hole through a billet and then rolling over a mandrel to form a tube of the required dimensions or, in the case of some nonferrous metals, by hot extrusion. Thus, **seamless tubing.**

sea moat *Geology.* see MOAT, def. 1.

sea moth *Vertebrate Zoology.* a dragonfish of the family Pegasidae, characterized by armor of bony rings and large, horizontal, fanlike pectoral fins; found in tropical Indo-Pacific waters.

seamount *Geology.* an upward projection of the sea floor with an elevation of 1000 meters or more and having either a flat or peaked top; found in all the major ocean basins; generally formed in the earth's crust and in the center of oceanic plates.

seamount chain *Geology.* a linear arrangement of several seamounts whose bases are separated by a relatively flat sea floor.

seamount group *Geology.* a linear or nonlinear arrangement of several seamounts spaced closely together.

seamount range *Geology.* an arrangement of several seamounts with connecting bases situated along a ridge or other elevation.

sea mouse *Invertebrate Zoology.* any of several large, marine annelids of the genus *Aphrodite* and related genera, characterized by a covering of long, fine, hairlike setae that give it a mouselike appearance.

seam roller *Mechanical Devices.* a hand tool, consisting of a roller in a handle, used for flattening the seams between panels of wallpaper.

sea mud *Geology.* a rich, slimy deposit occurring in a salt marsh or along a seashore. Also, **sea ooze.**

seam weld *Metallurgy.* a weld that joins two sheets longitudinally; used, for example, to make welded tubing.

sea oats *Botany.* a tall grass, *Uniola paniculata*, with oatlike panicles; found in coastal areas of southeastern North America; used to control sand erosion.

sea otter

sea otter *Vertebrate Zoology.* a rare marine mammal, *Enhydra lutria*, of the family Mustelidae, characterized by short legs, large, webbed hind feet, and a blunt cylindrical tail and having a very valuable fur; found in the north Pacific from the Kuril Islands to central California.

sea peak *Geology.* a seamount having a pointed summit.

sea pen *Invertebrate Zoology.* the common name for anthozoans in the order Pennatulacea. Also, SEA FEATHER.

sea plain see PLAIN OF MARINE EROSION.

seaplane

seaplane or **sea plane** *Aviation.* an aircraft capable of taking off and landing on water; supported on water by pontoons, or flats, or by a specially designed fuselage. Also, AIRBOAT.

seaport *Civil Engineering.* a port, harbor, or town accessible to seagoing ships.

seaquake *Geophysics.* an earth tremor with an epicenter beneath the ocean. Also, SUBMARINE EARTHQUAKE.

sear *Ordnance.* the lock or catch in a firearm that holds the hammer or firing pin in a cocked position until released by the trigger.

sea rainbow see MARINE RAINBOW.

sea raven *Vertebrate Zoology.* a large marine fish belonging to the genus *Hemitripterus,* such as *H. americanus*, common on the north Atlantic coast of America; characterized by a long, spiny dorsal fin.

search to go or look carefully to find something missing; specific uses include: *Navigation.* **1.** a systematic reconnaissance of a defined area at sea, employing one or more ships, craft, or other vehicles following specific course lines so that all parts of the area pass within visibility. **2.** an operation to locate an enemy force known or believed to be at sea. *Industrial Engineering.* in micromotion studies, the elemental motion of using the eyes to look for, or the hands to grope for, an object. *Computer Programming.* **1.** to scan a set of data items for those having a specific property. **2.** a method of locating a specific record or data item. *Artificial Intelligence.* any of a variety of methods applied by inference engines to obtain information that will lead to the desired goal. *Behavior.* of an animal, to selectively concentrate interest on a specific class of objects, such as a form of prey or other food source.

search and rescue *Engineering.* the use of ocean or land vehicles, radar, sonar, and aircraft to search for and rescue persons in distress.

search antenna *Electromagnetism.* an antenna in a radar system capable of directing a radiation beam throughout a specific region of interest.

search attack unit *Military Science.* one or more ships that are separately organized or detached from a formation as a tactical unit to search for and destroy submarines.

search field *Computer Programming.* the field in a record that contains the value being sought.

search function *Robotics.* a robotic function that uses sensory and decision-making capabilities to move from one position to another.

search gate *Electronics.* a signal that scans back and forth over a given range looking for specific conditions.

search image *Behavior.* an acquired image assumed to be held by a predator of the type of prey most profitable to it, based on the observed fact that many predator species will differentially select certain prey from among many types available.

searching behavior see APPETITIVE BEHAVIOR.

searching control *Engineering.* a mechanism that makes changes automatically in the azimuth and elevation of a searchlight, so that the beam constantly sweeps back and forth within certain limits.

searching fire *Military Science.* fire distributed in depth by successive changes in gun elevation.

search key *Computer Technology.* a particular value, such as a name, that is compared to each search field in order to obtain the desired item.

searchlight *Optics.* a device that projects a beam of light with high intensity and low divergency, often using a paraboloidal reflector to generate parallel rays from a light source at the focus.

search mission *Aviation.* an air reconnaissance by one or more aircraft dispatched to locate an object or objects known or suspected to be in a specific area.

search radar *Engineering.* a radar system that covers large areas of space and displays targets on screen immediately upon entry; used for early warning and air-traffic control.

search radius *Military Science.* in search and rescue missions, a radius centered on a datum point and having a length equal to the total probable error plus an additional safety length to ensure a greater than 50% probability that the target is in the search area.

search time *Computer Technology.* the average time required to find an item of data satisfying a specified condition.

search tree *Artificial Intelligence.* a tree structure, either explicit or implicit, of the states or goals considered in a search process. The initial state is the root of the tree; branches from a state correspond to operator applications or subgoals.

sea return see SEA CLUTTER.

sea rim *Astronomy.* the horizon as observed at sea.

searlesite *Mineralogy.* $NaBSi_2O_5(OH)_2$, a soft white monoclinic mineral occurring as prismatic crystals or fibrous spherulites, having a specific gravity of 2.460 and an undetermined hardness; found in clay at Searles Lake, California.

searle translocation *Molecular Biology.* an X-autosome chromosomal aberration or change in heterozygotes that exhibits X-chromosome inactivation; that is, the curbing of the two X chromosomes in the somatic cells of the female heterozygote.

sea robin *Vertebrate Zoology.* a bottom-dwelling marine fish of the family Triglidae, having armored plates and spiny fins used in crawling across the ocean bottom; found in tropical and temperate waters of the Americas. (Said to be named for the birdlike sounds made by its swim bladder and muscles.)

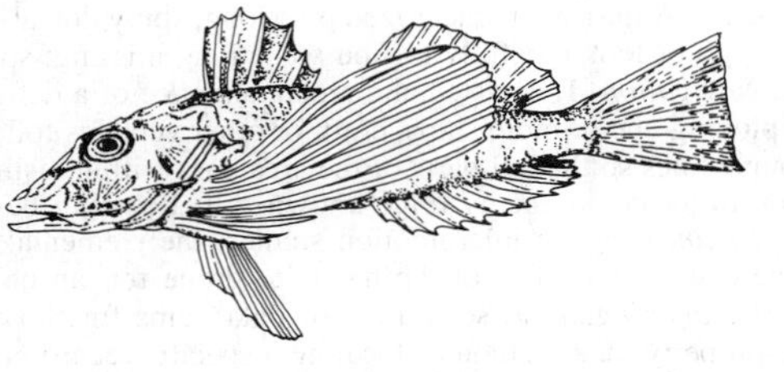

sea robin

sea room *Navigation.* the space necessary for safe and efficient movement at sea, as between ships or between a ship and the shore or shoals.

Sears, Paul Bigelow born 1891, American botanist; studied pollens; wrote extensively on applied ecology and the history of climate.

sea-run *Vertebrate Zoology.* having a habit of returning to a river from the sea for breeding, as salmon do.

sea salt *Oceanography.* the solid salt that is left behind when seawater is evaporated.

sea-salt nucleus *Oceanography.* a hygroscopic condensation nucleus produced by complete or partial desiccation of seawater droplets produced by wave spray or breaking air bubbles.

seascape *Oceanography.* the appearance of the surrounding sea from an observer's viewpoint.

seascarp *Geology.* a relatively long, high, linear submarine cliff or wall.

seashell *Invertebrate Zoology.* the common name for the shells of marine invertebrates, usually members of the phylum Mollusca.

seashore *Geology.* **1.** a narrow strip of land at the edge of an ocean or sea. Also, **seaside. 2.** the foreshore or ground between normal tide levels. Also, **seastrand.**

seasickness *Medicine.* a form of motion sickness caused by the movement of a boat or ship.

Sea Skua *Ordnance.* a British air-to-surface missile that provides ship-based helicopters with a relatively long-range weapon against small, mobile surface ships; powered by boost/sustainer motors, equipped with semiactive radar homing, and delivering a 44-pound (20 kg) blast/fragmentation warhead at high subsonic speed and a range of over 9.3 miles.

sea slope *Geology.* a gradual inclination of the land toward the sea.

sea slug *Invertebrate Zoology.* the common name for shell-less gastropod mollusks in the suborder Nudibranchia.

Seaslug *Ordnance.* a British ship-based, surface-to-air and surface-to-surface missile designed as a long-range antiaircraft weapon; it is powered by a solid-propellant rocket motor with four jettisonable boosters, equipped with a beam-riding guidance system, and delivers a high-explosive warhead.

sea smoke see STEAM FOG.

sea snail *Invertebrate Zoology.* the common name for any creeping gastropod mollusk with a spiral shell.

season *Meteorology.* a division of the year characterized by regularly occurring phenomena or weather conditions specific to a given area, e.g., the rainy season and the dry season in the tropics.

seasonal affective disorder *Psychology.* a mild, recurrent form of depression occurring usually in winter, characterized by irritability and loss of energy and sexual drive, often relieved by a change in season.

seasonal balancing *Chemical Engineering.* a seasonal variation of the percent volume of butane in gasoline released from the stabilizing tower to reduce the vapor pressure and volatility of gasoline for the summer season, so as to avoid vapor lock in motors.

seasonal current *Oceanography.* an ocean current that undergoes significant changes in speed or direction at certain times of the year because of seasonal winds.

seasonally frozen ground *Geology.* ground that is frozen only during the winter when seasonal temperatures are low. Also, FROST ZONE.

seasonal stream *Hydrology.* a stream that flows only during a certain climactic season.

seasonal thermocline *Oceanography.* a shallow thermocline that develops in summer because of warmer air temperatures and the effects of warm summer winds.

seasonal variation *Geophysics.* an alteration of a physical element of the upper atmosphere that occurs with a change of season and results in a variation in transmission of radio signals over large distances. *Statistics.* in time-series analysis, a cycle that is completed within one calendar year and repeated each year.

season crack *Metallurgy.* the fracture of certain copper-base alloys that are affected by internal or external stresses. *Ordnance.* specifically, a fine cracking that formerly occurred in brass cartridge cases long after production due to residual stress inherent in the manufacturing process; it has been essentially eliminated by modern manufacturing methods. Thus, **season cracking.**

sea sparrow see MURRELET.

Sea Sparrow *Ordnance.* an air-to-air, solid propellant missile with a nonnuclear warhead and electronic-controlled homing; officially designated **AIM-7**; the ship-launched surface-to-air version is designated **Sea Sparrow (RIM-7).**

Sea Sparrow missile

sea spider *Invertebrate Zoology.* the common name for small, marine, spiderlike arthropods of the class Pycnogonida.

sea squirt *Invertebrate Zoology.* the common name for sessile tunicates of the class Ascidiacea that, when disturbed, shoot streams of water from mouth and cloacal openings.

sea star *Invertebrate Zoology.* any echinoderm of the class Asteroidea, characterized by a body that is arranged radially in the shape of a star, with five or more rays or arms radiating from a central disk, a mouth under the disk, and rows of tubular walking feet; some are carnivorous, causing great damage to oyster beds. Popularly known as **starfish**.

sea superiority *Military Science.* in a sea battle, a degree of dominance of one force over another that permits the conduct of operations by the dominant force and its related land, sea, and air forces at a given time and place without prohibitive interference by the opposing force.

sea supremacy *Military Science.* a degree of sea superiority in which the opposing force is incapable of effective interference.

sea-surface slope *Oceanography.* a gradual change over some distance in the height of the sea surface, caused by the Coriolis effect or by wind.

sea swell *Meteorlogy.* a long, rectangular, surface-wave motion created by winds in a lengthy fetch.

seat a location in which something is placed, in the manner of a person sitting in a chair; specific uses include: *Engineering.* **1.** a support or holder for a mechanism. **2.** to fit correctly into a holder or prepared position. *Ordnance.* to position a weapon component properly; e.g., to seat a cartridge in the chamber of a gun. *Mechanical Engineering.* the part of a pressure valve that makes contact with the moving part of the valve in order to close it.

seat clay see UNDERCLAY.

sea terrace see MARINE TERRACE.

seating *Ordnance.* the distance that a projectile is rammed into the bore of a cannon; it is usually measured between the base of the projectile and the rear face of the breech.

seat load factor *Transportation Engineering.* in a public passenger vehicle, the percentage of seats occupied during a given period.

sea turn *Meteorology.* a wind blowing from the sea, characterized by mist; this term is limited mainly to New England.

sea turtle *Vertebrate Zoology.* any member of the superfamily Chelonioidea, tropical marine turtles whose forelimbs are modified into flippers; widely distributed in warm seas. Also, **sea tortoise.**

sea urchin *Invertebrate Zoology.* the common name for echinoderms in the class Echinoidea, with spherical bodies covered by movable spines.

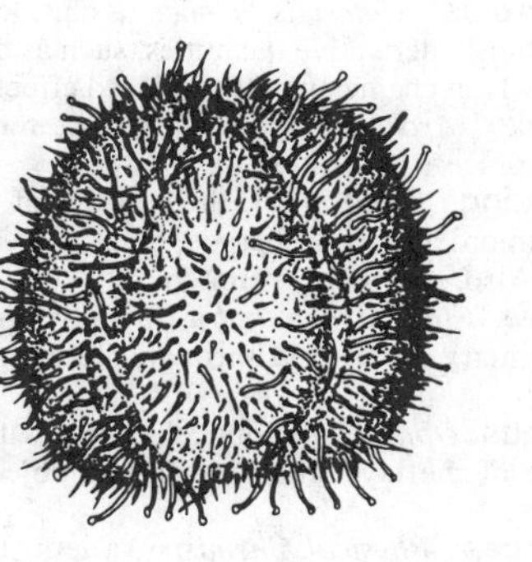

sea urchin

sea valley *Geology.* a shallow, wide depression sloping gently into the sea floor.

seawall or **sea wall** *Civil Engineering.* a wall or embankment that is erected to protect the shore from erosion or to act as a breakwater. Also, **seabank.** *Geology.* an extensive, steep-faced embankment that is built up by strong storm waves along a seacoast at the high-water mark.

seaward *Navigation.* the direction toward the sea and away from the land.

seawater *Oceanography.* ocean water, characterized by high salinity.

seaway *Navigation.* **1.** a route of travel over the open sea. Also, SEA-LANE. **2.** the forward motion of a vessel over the open sea. **3.** the condition of wave action at sea, used especially in the expression "heavy seaway" for high wave conditions. *Geography.* a canal large enough to accommodate ocean-going vessels, such as the St. Lawrence Seaway.

seaweed *Botany.* a general term applied to a plant or plants growing in a saltwater environment, especially marine algae.

Seawolf *Ordnance.* an advanced British ship-based, surface-to-air and surface-to-surface missile designed for close-range defense against supersonic aircraft and missiles, as well as surface ships and hovercraft.

seaworthiness *Naval Architecture.* the ability and readiness of a vessel to survive in varied and potentially hazardous sea conditions.

sebaceous [sə bā´shəs] *Medicine.* relating to or secreting sebum.

sebaceous gland *Anatomy.* a holocrine gland in the skin that exudes sebum and usually opens into a hair follicle. Also, HAIR GLAND, OIL GLAND.

sebacic acid *Organic Chemistry.* $HOOC(CH_2)_8COOH$, combustible, white leaflets; slightly soluble in water and soluble in alcohol and ether; melts at 134°C and boils at 295°C (100 torr); used in perfumes, paints, and hydraulic fluids, and to stabilize synthetic resins. Also, DECANEDIOIC ACID.

sebastianite *Petrology.* an igneous rock containing euhedral anorthite, biotite, augite, and apatite, but lacking quartz and feldspathoids.

Sebdeniaceae *Botany.* a family of red algae of the order Gigartinales, characterized by multiaxial leaflike thalli and restricted to the Mediterranean Sea.

Sebekidae *Invertebrate Zoology.* a family of tongue worms in the suborder Porocephaloidea.

sebkha or **sebka** *Geology.* **1.** an extensive, smooth, flat plain found in arid regions of North Africa, having a high salt content and sometimes becoming a temporary salt marsh or shallow lake after a rain. **2.** a salt flat or salt-encrusted plain in a coastal area, especially on the Arabian peninsula. Also, SABKHA, SABKA.

seborrhea [seb´ə rē´ə] *Medicine.* **1.** an excessive secretion of sebum. **2.** seborrheic dermatitis.

seborrheic [seb´ə rē´ik] *Medicine.* **1.** affected with seborrhea. **2.** referring to those areas of the body in which sebaceous glands are abundant, such as the scalp, face, chest, back, and groin.

seborrheic dermatitis *Medicine.* an inflammatory disease of the skin involving areas of the scalp, face, and ears, in which there is yellowish and oily scaling and itching of the affected areas. Also **seborrhea, seborrheic eczema, seborrhea sicca.**

sebum [sē´bəm] *Physiology.* a fatty exudate of the sebaceous glands, varying in consistency at different areas of the body.

sec secant.

SEC secondary electron conduction.

sec. section; second; secondary.

seca *Meteorology.* a dry wind or drought in Brazil.

secalose *Biochemistry.* a polysaccharide found in rye and oats; when dried, it forms a white, hygroscopic powder that is convertible by inversion into levulose.

secant [sē´kant; sē´kənt] *Mathematics.* **1.** the reciprocal of the cosine; denoted sec. **2.** a straight line that intersects a circle in two points; the segment of the line on the circle and its interior is called a chord. **3.** more generally, a straight line that cuts a curve in two or more points; segments of the secant between points of intersection are called chords. *Cartography.* in surveying, a great circle that intersects a parallel of latitude at the first and fifth mile stations.

secant coefficient see EULERIAN NUMBERS.

secant method *Cartography.* in surveying, a technique for determining the base line of a survey by measuring offsets from a great circle that intersects the base line at the first and fifth mile stations of a township boundary.

secbumeton *Organic Chemistry.* $C_{10}H_{19}N_5O$, a white solid that melts at 86°C; used as an herbicide.

Secchi, Pietro Angelo 1818–1878, Italian astronomer; developed the first spectral classification; spectrum analysis of the sun and 4000 stars.

Secchi classification *Astronomy.* a now-superseded system of stellar spectral classification with four classes.

Secchi disk *Engineering.* a white disk used to measure the clarity of seawater by lowering it into water until it can no longer be observed.

Secernentea *Invertebrate Zoology.* a class of nematode roundworms with phasmids; most are terrestrial and parasitic in plants and animals. Also, PHASMIDA.

sech *Mathematics.* a hyperbolic secant.

sechard *Meteorology.* a dry, warm wind that blows over Lake Geneva in Switzerland.

seclusion *Meteorology.* an unusual occurrence in the occlusion process, in which the point at which the cold front first overtakes the warm front is at some distance from the apex of the wave cyclone.

secobarbital sodium *Pharmacology.* $C_{12}H_{17}N_2NaO_3$, a white, slightly bitter, odorless powder; a short-acting barbiturate used as an oral sedative and to induce sleep.

secodont *Vertebrate Zoology.* relating to or having teeth modified for cutting.

second *Metrology.* the fundamental unit for the measurement of time, equal to one-sixtieth of a minute; now defined as the duration of 9,192,631,770 cycles of vibration of the radiation emitted at a specific wavelength by a cesium-133 atom. *Mathematics.* a unit of angular measurement equal to 1/3600 degree, or π/648,000 radian; symbolized ″. Sixty seconds equal one minute of arc.

secondary *Electromagnetism.* of or relating to the output characteristics of a transformer. *Geology.* **1.** of a mineral deposit or enrichment, formed by descending solutions. **2.** of or relating to the Mesozoic era.

secondary absorption see OBSTRUCTIVE ATELECTASIS.

secondary accent *Linguistics.* **1.** in a word with more than one accented syllable, a relatively weak accent, indicated in pronunciation by a lighter mark than a primary accent. In the word "secondary" itself, the first syllable has the primary accent and the third syllable has a secondary accent: [sek´ən der´ē]. **2.** the symbol indicating such an accent.

secondary air *Mechanical Engineering.* the air in a combustion chamber that meets with primary air to consume the fuel.

secondary alcohol *Organic Chemistry.* *RR′*CHOH, an organic alcohol in which *R* and *R′* are either identical or different groups.

secondary allocation *Computer Technology.* a storage area for a disk file when the major space allocation is too small to hold all the data in the file.

secondary amine *Organic Chemistry.* *RR′*NH, an organic compound in which *R* and *R′* designate either identical or different groups.

secondary amyloidosis *Medicine.* amyloidosis that follows a chronic inflammatory disease in which there has been suppuration; amyloid becomes deposited in the connective tissue, especially the spleen, liver, kidneys, adrenals, and arterioles.

secondary area *Aviation.* the area within an air-operations segment in which required obstacle clearance (ROC) decreases as distance from the prescribed course increases.

secondary axis *Optics.* a line formed by the principal ray of an oblique bundle of rays.

secondary battery see STORAGE BATTERY.

secondary blast injury *Military Science.* any injury that occurs due to the general effect of a blast but is not directly due to the blast effect; e.g., an injury caused by falling bricks or projected debris.

secondary calcium phosphate see DIBASIC CALCIUM PHOSPHATE.

secondary cambium *Botany.* one of the tissue layers that develops after the initial cambial layer in the roots of certain plants, such as the beet, producing a ring of tissue.

secondary carbon atom *Organic Chemistry.* a carbon atom bonded to two other carbon atoms by single bonds.

secondary cell *Electricity.* a chemical cell that requires recharging, such as the cells in an automobile battery. *Biotechnology.* a cell that is obtained by the repeated in vitro passage of primary cells.

secondary cell culture *Biotechnology.* a cell culture that is obtained by the repeated in vitro passage of primary cells; most of these cultures die after a limited numer of cell divisions, but some survive and form a continuous cell line that can be made indefinitely in vitro.

secondary circuit *Metallurgy.* in welding, the circuit that conducts the secondary current. *Electricity.* a circuit that is associated with a secondary winding of a transformer.

secondary clay *Geology.* a clay that has been moved from its site of origin and redeposited in another location.

secondary coast *Geology.* a mature coast or shoreline whose features are shaped by present-day marine processes such as wave erosion.

secondary cold front *Meteorology.* a secondary weather front that forms behind the primary frontal cyclone and within a cold air mass, characterized by a substantial horizontal temperature gradient.

secondary color *Graphic Arts.* a color produced by mixing or superimposing two primary colors; in painting, the secondaries are violet, orange, and green.

secondary consequent stream *Hydrology.* **1.** a tributary of a subsequent stream that developed after the subsequent stream, but which flows parallel to or down the same slope as the original consequent stream. **2.** a stream flowing down the side of an anticline or syncline.

secondary consolidation *Geology.* the compaction of sedimentary material at essentially constant pressure as a result of internal processes such as recrystallization.

secondary consumer *Ecology.* an organism that feeds on other organisms which consume plant products, thus representing the second stage of consumption in a food chain; e.g., rabbits or field mice (primary consumers) eat plants and they are in turn eaten by hawks, owls, snakes, and so on (secondary consumers).

secondary cosmic rays *Geophysics.* the radiation created by a collision of primary cosmic rays with molecules or atoms in the earth's atmosphere.

secondary crater *Geology.* a type of impact crater formed by the low-velocity impact of fragments ejected from a larger primary crater. Also, SATELLITIC CRATER.

secondary creep *Mechanics.* the second stage in a creep process characterized by a nearly constant strain rate.

secondary crusher *Mechanical Engineering.* a machine that further reduces the particle size of shale or rock after it has passed through the primary crusher.

secondary cyclone *Meteorology.* a cyclone that forms in association with a primary cyclone. Also, SECONDARY LOW.

secondary electrode see BIPOLAR ELECTRODE.

secondary electron *Electronics.* **1.** an electron produced by secondary emission. **2.** the electron that has the least amount of energy following a collision between two electrons. *Materials Science.* an electron emitted when primary electrons enter a specimen that are imaged in a scanning electron microscope; they are sensitive to depth changes and are therefore used for morphological investigations.

secondary-electron conduction *Electronics.* the movement of secondary electrons through holes in a porous material, guided by an externally applied electric field.

secondary emission *Electronics.* an action in which a beam of high-velocity electrons dislodges electrons still bound to atoms.

secondary enlargement *Mineralogy.* the deposition, about a classic mineral grain, of additional material of identical composition so that optical and crystallographic continuity are maintained; good crystal faces characteristic of the original mineral are often developed in this way.

secondary enrichment *Geology.* a process by which a mineral deposit is naturally enriched or concentrated by the weathering of material that originated later in time and was carried downward and redeposited by percolating water. Also, SUPERGENE ENRICHMENT.

secondary extinction *Crystallography.* see EXTINCTION.

secondary fermentation *Food Technology.* the process of continuing the fermentation of wine through the addition of sugar, either to increase the yield of alcohol or to produce natural carbonation. Use of this process in still-wine production is forbidden by law or covenant in many wine-growing regions.

secondary ferroics *Materials Science.* a class of materials in which orientations differ in derivative quantities, such as dielectric or magnetic susceptibility, which characterize the induced effects.

secondary flake *Archaeology.* a stone flake removed from a larger flake, as in the process of refining for a new use.

secondary flaking *Archaeology.* the removal of flakes from an existing stone tool in order to thin, sharpen, blunt, or otherwise modify it for a specific use. Also, **secondary retouch.**

secondary flow *Fluid Mechanics.* a flow configuration that is superposed on a primary fluid field through the action of friction at the boundary layer.

secondary focus *Optics.* in an optical system with astigmatism, a line at which some of the rays from an off-axis object are imaged. Also, SAGITTAL FOCUS.

secondary forces *Physical Chemistry.* a term for various relatively weak forces that produce the interaction of molecules, such as dispersion force or induction force, as distinguished from stronger forces such as hydrogen bonding.

secondary fragments *Ordnance.* fragments that result from the impact of projectile fragments or the breakup of the target under the force of impact.

secondary front *Meteorology.* a front that forms within a baroclinic cold air mass, which itself is separated from a warm air mass by a primary frontal system; the most common is the secondary cold front.

secondary geosyncline *Geology.* **1.** a geosyncline that occurs at the culmination of or subsequent to geosynclinal orogeny. **2.** a subordinate geosyncline that is separated from the primary geosyncline by a later development of geanticlines. Also, SEQUENT GEOSYNCLINE.

secondary glacier *Hydrology.* one of the small valley glaciers that flows into a larger, central glacier to form a trunk glacier.

secondary grid emission *Electronics.* an action in which electrons are ejected from atoms on an electron tube's grid when struck by high-velocity electrons or other charged particles.

secondary grinding *Mechanical Engineering.* the grinding of particulate matter after it has been reduced to approximately sand-grain size in rod or ball mills.

secondary hardening *Metallurgy.* the process of hardening certain steels by thermal treating at relatively low temperatures.

secondary hydrogen atom *Organic Chemistry.* a hydrogen atom that is bonded to a secondary carbon atom.

secondary immune response *Immunology.* the total immunological reaction that occurs as a response to a second or subsequent injection of a given antigen; it is characteristically high in antibody formation.

secondary inclusions *Materials Science.* inclusions in a frozen alloy that result from microsegregation; the size is controlled by metal composition and cooling rate; rapid quenching from the melt can make inclusions so small that they may actually improve mechanical properties of the alloy.

secondary index *Computer Programming.* an index used to access a physical or logical database by a path different from that provided by the database definition.

secondary institution *Anthropology.* any aspect of culture beyond the primary institutions of subsistence efforts and group organization, such as religion, the arts, or folklore.

secondary interstice *Geology.* an interstice that develops in a rock subsequent to the formation of the rock.

secondary limestone *Petrology.* limestone deposited from solution in cavities and crevices of other rocks.

secondary low see SECONDARY CYCLONE.

secondary lysosome *Cell Biology.* an intracellular organelle that is produced by the fusion of a lysosome with an endosome containing endocytosed materials to be digested.

secondary memory see LONG-TERM MEMORY.

secondary meristem *Botany.* a meristem that develops from cells which were at one time differentiated, functioning as part of a mature tissue system, and which then became meristematic again.

secondary metal *Metallurgy.* metal that is not derived from a natural source; e.g., scrap metal or the slags and drosses from processing.

secondary mineral *Mineralogy.* **1.** a mineral produced due to alteration of a primary mineral. **2.** a mineral that forms later than the rock in which it is enclosed, due to metamorphic, solution, or weathering activity; generally formed at the expense of an earlier primary mineral.

secondary motivation *Psychology.* a motivation that does not have an immediate biological source, such as competition or bonding.

secondary nondisjunction *Molecular Biology.* the nondisjunction of the sex chromosomes of an XXY individual, resulting in gametes having an XX, XY, X, or Y complement of sex chromosomes.

secondary oocyte see OOCYTE.

secondary optic axis *Crystallography.* either of two optic axes in a crystal along which all light rays travel with equal velocity.

secondary periderm *Botany.* any periderm layer other than the first, outermost layer.

secondary photocurrent *Electronics.* a photocurrent arising from ohmic contacts that is able to replenish charge carriers passing out of the opposite contact and whose maximum gain greatly exceeds unity.

secondary phloem *Botany.* phloem tissue formed from the vascular cambium rather than the apical meristem.

secondary polycythemia *Hematology.* any absolute increase in the total red cell mass other than polycythemia vera. It may be appropriate, adjusting for general tissue hypoxia, such as that associated with pulmonary disease, alveolar hypoventilation, cardiovascular disease, and prolonged exposure to high altitude, or occurring as a result of defective hemoglobin or drugs. It may also be inappropriate, reflecting excessive erythropoietin production due to renal or extrarenal disorders. Also, ERYTHROCYTOSIS.

secondary porosity *Geology.* the porosity that develops after the formation of a rock from processes such as dissolution or acidization.

secondary process *Psychology.* in psychoanalytic theory, conscious and rational mental activity that is characteristic of the ego and based on the reality principle.

secondary production *Ecology.* the energy acquired by primary consumers through the consumption of plant matter.

secondary radar *Electronics.* a radar system in which the target's reflection is beamed at a transponder that transmits its own signal back to the radar receiver.

secondary radiation *Physics.* radiation or particles produced by the interaction between matter and primary radiation (radiation coming directly from its source).

secondary rainbow *Optics.* a fainter rainbow that occasionally forms outside a primary rainbow, with its colors in reverse order from those of the primary.

secondary recovery *Hydrology.* the replenishment of groundwater during and after the wet season, accompanied by a rise in the level of the water table. *Petroleum Engineering.* the use of methods beyond normal flowing or pumping to recover additional oil from a depleted or nearly depleted reservoir. Also, **secondary oil recovery.**

secondary refuse *Archaeology.* unwanted objects or materials that are removed from the site at which they were used and are disposed of at a different location.

secondary reinforcer see CONDITIONED REINFORCER.

secondary root *Botany.* a branch of a primary root; a root that arises from a point other than the normal point of origin.

secondary sewage sludge *Civil Engineering.* a biological sludge, such as activated sludge or trickling filter humus.

secondary sexual characteristics *Anatomy.* the sexually distinguishing anatomical features not directly involved in the production of an egg or sperm, such as the facial hair of a male or the enlarged breasts of a female.

secondary shaft *Mining Engineering.* a mine shaft extending downward from but not in line with the primary shaft.

secondary sodium phosphate see DIBASIC SODIUM PHOSPHATE.

secondary speciation *Evolution.* the creation of new evolutionary lines through the hybridization of two separate lines.

secondary splits *Mining Engineering.* in mining ventilation, air splits that occur following the main air splits.

secondary standard *Science.* a copy of a primary standard that is used for standardizing laboratory measurements. *Physics.* a unit that is directly derived from the definition of a primary standard, as a kilometer or millimeter is defined in terms of a multiple or submultiple of the primary standard of the meter.

secondary station *Ordnance.* in gunnery, an observation point at the end of the baseline farthest from the gun or directing point. *Telecommunications.* in a data communication network, any station other than the net control station.

secondary storage see AUXILIARY STORAGE.

secondary stratification *Geology.* stratification that develops when sediments that were already deposited become suspended and are redeposited. Also, INDIRECT STRATIFICATION.

secondary stress *Mechanics.* a self-limiting stress that is caused by the constraints of a structure; it causes minor deformations that play no role in the failure of the structure.

secondary structure *Geology.* any rock structure, such as a fault or fold, that was formed after the original deposition or emplacement of the rock in which it is found. *Paleontology.* a duplicate body part, such as the secondary palate that led to partial separation of the breathing passage from the alimentary canal in mammals. *Biochemistry.* in protein structure, the folding, twisting, coiled, often springlike chain resulting when hydrogen bonds form between the adjacent parts of a molecule, as in an alpha helix.

secondary surveillance radar *Navigation.* the radar that operates in conjunction with the air traffic control radar beacon system.

secondary substances *Botany.* a term for plant chemicals that are not involved in known biosynthetic functions, but are detected in large amounts in leaves and other organs; they are thought to function in plant defense.

secondary succession *Ecology.* an ecological succession that follows the partial or total destruction of an earlier community.

secondary suture *Surgery.* **1.** a surgical wound closure whose execution is delayed because of contamination or expected infection. **2.** a resuture of a wound following disruption.

secondary syphilis *Medicine.* the second stage of syphilis, characterized by anemia, skin disease, inflammation of mucous membranes, and swelling of lymph glands.

secondary target *Ordnance.* a target that is attacked after the primary target has been destroyed or after it has become impossible to continue fire against the primary target.

secondary tectonite *Geology.* a type of tectonite characterized by a fabric that clearly reflects the historical deformation of the rock.

secondary tillage or **secondary till** *Agronomy.* additional tilling after the initial plowing of the soil, as to cut up crop residues, destroy weeds, or provide greater soil pulverization.

secondary tissue *Botany.* tissue that originates from the cambium or the cork cambium.

secondary trait *Psychology.* a specific, narrow personality trait, having limited influence on the overall personality of an individual. Also, **secondary disposition.**

secondary tympanic membrane *Anatomy.* the membrane that covers the round window of the middle ear.

secondary use *Archaeology.* an artifact or feature showing an alternate or later use that differs from its original function; for example, an abandoned well that becomes a refuse pit.

secondary voltage *Electromagnetism.* the voltage appearing across the secondary windings of a transformer; the output voltage of the transformer.

secondary wall *Botany.* a part of a plant cell wall that develops inside of and after the primary wall; usually constitutes most of the cell wall, consisting of several layers and having prominent rings, spirals, bars, or reticulations.

secondary wave *Optics.* a wave that propagates from a point on an already moving wavefront. *Geophysics.* see S WAVE.

secondary weapon *Ordnance.* an auxiliary weapon of a unit, aircraft, ship, or vehicle; it is usually smaller in caliber than the primary weapon and its purpose is to protect or supplement the main weapon's fire.

secondary winding *Electromagnetism.* the winding (coil) that is inductively coupled to the primary winding of a transformer.

secondary xylem *Botany.* xylem produced by the vascular cambium during secondary growth.

second audio program see SAP.

second boiling point *Physical Chemistry.* a temperature at which some substances change from a liquid to a gas while cooling.

second bottom *Geology.* the first terrace rising above the normal flood plain of a river.

second breakdown *Electronics.* a condition in which a surge in current, brought on by structural imperfections, causes destruction of the transistor.

second category *Mathematics.* a subset of a topological space is said to be of the second category if it is not of the first category; that is, it cannot be expressed as the union of a countable collection of nowhere dense sets. The Baire category theorem states that a complete metric space is of the second category.

second-class ore *Mining Engineering.* ore requiring preliminary treatment before a marketable grade or a grade suitable for further treatment is obtained. Also, MILL ORE.

Second Coming type *Graphic Arts.* an informal name for the largest and boldest headline type used by a particular newspaper; reserved for epochal news stories. (Said to be reserved for the headline of a story reporting the Second Coming, the return of Jesus Christ to earth.)

second condition of equilibrium *Mechanics.* the requirement that for a system to be in equilibrium, the net torque acting on the system must be equal to zero.

second countable *Mathematics.* a topological space is second countable if it satisfies the second axiom of countability; i.e., it has a countable base.

second-degree burn *Medicine.* in the classification of burns, moderate damage, characterized as a burn of the skin that is sufficiently severe to cause blistering.

second derivative *Mathematics.* the derivative of the derivative of a function.

second-derivative map *Geophysics.* a contour map of the second vertical derivative of a potential field.

second detector *Electronics.* in a superheterodyne receiver, the circuit that demodulates the IF signal.

second filial generation see F_2.

second-foot *Hydrology.* in the United States, a unit of stream discharge equal to one cubic foot of water per second.

second-foot day *Hydrology.* a unit of runoff volume or reservoir capacity equal to a flow of one cubic foot per second for 24 hours, or a total of 86,400 cubic feet of water.

second fundamental form *Mathematics.* **1.** a vector-valued bilinear form on the tangent space of a manifold that describes locally how the manifold is embedded in the overlying space. For example, for a hypersurface M (codimension 1) embedded in a Riemannian manifold N, let n be a unit vector field normal to M. Then, according to the Weingarten formulas, the (vector-valued) second fundamental form $s(X,Y)$ is given implicitly by

$$-< v, s(X,Y) > = < -\nabla'_X v, Y >,$$

where X and Y are tangent vectors at a particular point of M. **2.** sometimes a scalar second fundamental form II is defined for a hypersurface from the vector form by defining $\mathrm{II}(X,Y)n = s(X,Y)$.

second-generation *Industrial Engineering.* relating to or being a significantly improved or altered version of an original product or form of technology.

second-generation anaerobic digester *Biotechnology.* an advanced anaerobic bacterial fermentation device with high rates of gas production during biological decomposition of organic material.

second-generation computer *Computer Technology.* a computer developed in the late 1950s characterized by solid-state transistor circuitry in place of vacuum tubes, magnetic core main memory, and disks and drums as well as tape for auxiliary storage.

second growth *Forestry.* **1.** the smaller trees in a forest that are left after a major cutting or destruction. **2.** the trees remaining in a forest area after a cutting that are available for another logging.

second hand *Horology.* the hand that indicates seconds on a timepiece; it may be carried on the center arbor and powered by the dial train, or have a small dial of its own powered by the fourth wheel of the going train.

second harmonic generation *Spectroscopy.* the generation of spectral lines having twice the frequency of the lines of the fundamental spectrum. *Crystallography.* when a beam of red laser light is passed through a crystal, the emerging light may have two components: one red and the second ultraviolet with exactly twice the frequency of the incident light. This frequency doubling is described as the production of a second harmonic. Even higher order effects, third-order harmonics, for example, are found to occur with certain materials; substances showing these properties are referred to as **nonlinear optical materials** because their optical properties do not depend in a linear way on the intensity of the incident light.

second isomorphism theorem *Mathematics.* let I and J be ideals in a ring R. Then $I/(I$ « $J)$ is isomorphic to $(I + J)/J$. Similar results hold for groups and modules.

second law of motion see NEWTON'S SECOND LAW.

second law of thermodynamics *Thermodynamics.* the general statement that any natural process will tend to proceed in a preferred direction; this is expressed in various forms, such as: **1.** no device can operate in a cycle and permit a heat transfer from a colder body to a hotter body, unless some other effect takes place (Clausius statement). **2.** no device can operate in a constant-temperature cycle and convert the heat transfer it receives into work, unless some other effect takes place (Kelvin-Planck statement). **3.** all spontaneous processes tend to increase the entropy of an isolated system toward a maximum.

second-level address see INDIRECT ADDRESS.

second-level addressing see MULTILEVEL ADDRESSING.

second-level controller *Control Systems.* in a large-scale control system, a controller that controls the actions of first-level controllers so as to compensate for subsystem interactions in order to satisfy the overall objectives and constraints of the system.

second-level inference *Psychology.* a description of behavior that makes inferences from the observed actions to generalize about the subject's behavior in other similar situations.

second lieutenant *Military Science.* in the U.S. Army, Air Force, or Marine Corps, the lowest ranking noncommissioned officer.

second messenger *Cell Biology.* a molecule, such as cyclic AMP, that is generated as the result of the attachment of a hormone to its receptor, transmitting the stimulus to the cytoplasm and nucleus.

second-order *Cartography.* of or relating to any surveying work that has a next-to-the-highest level of accuracy.

second-order control *Cartography.* a two-level classification of control networks meeting the following criteria for accuracy and precision: for horizontal coordinates, the maximum values of the ratio of standard deviation of distances between survey points to the distances themselves; Class I—1:50,000, Class II—1:20,000; for vertical coordinates, the maximum values of the ratio of standard deviations of elevation differences in millimeters between survey points to the square root of the horizontal distance in kilometers between those points; Class I—1.0, for Class II—1.3.

second-order equation *Mathematics.* any differential equation that includes the second derivative of the unknown function but no derivatives of higher order.

second-order leveling *Cartography.* in surveying, a technique of leveling that does not meet the degree of accuracy of first-order leveling, but does meet the following criteria: the lines measured are between bench marks established by first-order techniques, and the error of closure does not exceed 8.4 mm times the square root of the length in meters of the line or section of the line.

second-order phase transformation *Materials Science.* a phase transition characterized by the absence of latent heat or hysteresis, such as the ferromagnetic or antiferromagnetic transition, the order-disorder transition in alloys, the superconducting transition, and the ferroelectric transition.

second-order reaction *Physical Chemistry.* a chemical reaction in which the rate of change of the reaction is proportional to the product of the concentrations of the two reactants involved, or to the square of the concentration if only one reactant is involved.

second-order transition *Thermodynamics.* a change of state in which the free energy of a substance and its first derivatives are continuous functions throughout the transition with respect to temperature or pressure.

second quantization *Quantum Mechanics.* the description of a field as an ensemble of particles and the dependent variables of the field as operators having specified commutation rules; the procedure facilitates the description of many-particle systems.

second radiation constant *Physics.* a constant appearing in the Planck radiation law, given by the product of the speed of light and Planck's constant divided by the Boltzmann constant: approximately 1.439 degree-centimeters.

second-site mutation see INTRAGENIC SUPPRESSION.

second sound *Acoustics.* a temperature wave that propagates through the superfluid phase of superfluid helium in which the density and pressure remain unchanged.

seconds pendulum *Horology.* in the escapement mechanism of a clock, a pendulum that swings in an arc with a duration of one second.

second-stage graphitization *Materials Science.* the second step in the heat treatment of malleable irons that are to have a ferritic matrix; the iron is cooled slowly from the first-stage temperature so that the austenite transforms into ferrite and graphite rather than pearlite.

second strike *Military Science.* the first counterstrike of a war; it is generally associated with nuclear operations.

second-strike capability *Military Science.* the ability to survive a first strike with sufficient resources to deliver an effective counterstrike; it is generally associated with nuclear weapons.

second-year ice *Oceanography.* sea ice that has not melted during its first summer. Also, TWO-YEAR ICE.

seco-steroid *Endocrinology.* a modified steroid, such as vitamin D_3, in which the B ring of the steroid nucleus is opened.

Secotiaceae *Mycology.* a family of fungi belonging to the order Cribbiaceae; well known for its species *Secotium*; found in the deserts of North America and Australia.

secotioid fungi *Mycology.* fungi whose fruiting bodies remain closed and have basidia that do not actively discharge spores; they include species of the family Secotiaceae.

secretary bird *Vertebrate Zoology.* a large, long-legged raptorial bird, *Sagittarius serpentarius*, that feeds on reptiles and other vertebrates; found in Africa. (So called because its crest suggests a group of pens stuck behind the ear, a habit once associated with secretaries.)

secretary bird

secrete [si krēt´] *Physiology.* to undergo or cause to undergo a process of secretion; to discharge or release some substance such as saliva.

secretin *Endocrinology.* a polypeptide hormone that is secreted by the S-cells of the duodenal and jejunal mucosa and that is released in response to increased acid in the intestinal lumen; it stimulates the secretion of water and bicarbonate by the exocrine pancreas. Also, INCRETIN, OXYKRININ.

secretion [si krē´shən] *Physiology.* **1.** the normal process by which various glands of the body release their substances. **2.** such a substance produced by a gland; for example, sweat, digestive juices, bile, mucus, or saliva. *Geology.* a secondary structure formed from dissolved substances that are redeposited within an empty rock cavity, usually on or parallel to the cavity walls.

secretor gene *Genetics.* the dominant autosomal gene in humans that permits the secretion of A and B type water-soluble blood group antigens into saliva and other body fluids.

secretory [sē´krə tôr´ē] *Physiology.* relating to or taking part in a process of secretion.

secretory component *Biochemistry.* a large glycoprotein, formed as a membrane protein in hepatocytes and in epithelial cells of certain mucous surfaces, that is synthesized by secretory gland epithelial cells and thought to restrain breakdown of IgA enzymes in the gut.

secretory granules *Cell Biology.* the specialized compartments in a cell involved in releasing substances from a cell by the process of secretion.

secretory IgA *Immunology.* a class of immunoglobulins containing alpha heavy chains, occurring in body secretions (such as saliva), and composed of an IgA dimer and a protein, known as a secretory piece. Also, **sIgA.**

secretory lobe *Invertebrate Zoology.* the lobe of any organ that secretes hormones or other fluid products.

secretory protein *Biochemistry.* a protein secreted from a cell to the exterior.

secretory structure *Botany.* plant cells or tissue that produce any of a variety of secretions.

secretory vesicle *Cell Biology.* a membrane-bound compartment that is involved in releasing substances from a cell by the process of secretion. Also, **secretory granule.**

secs. or **Secs.** sections; seconds.

sectile *Materials Science.* describing a mineral that can be sectioned; i.e., that can be cut with a knife without disintegrating or breaking into pieces.

section a distinct part of a larger whole, especially one that is cut off or separated from the whole; specific uses include: *Civil Engineering.* a topographical measure of land, equal to one square mile or 640 acres, one thirty-sixth of a township. *Cartography.* in leveling, any portion of a line of levels that is measured, and has its value recorded and abstracted, as a unit. *Surgery.* **1.** the process of cutting into a tissue or organ. **2.** a cut surface of a tissue or organ. **3.** a segment or subdivision of a tissue or organ. *Geology.* **1.** a natural or artificial cut through part of the earth's surface. **2.** a graphic representation or description of a series or succession of rock units or the geologic structures along a given vertical plane. **3.** see THIN SECTION. *Architecture.* a representation of something (such as a building) as if cut by a vertical plane. *Telecommunications.* an individual transmission span within a radio relay system. *Systematics.* a rank in the botanical taxonomic hierarchy, below a genus and above a series. *Military Science.* **1.** in the U.S. Army and Marine Corps, a tactical unit that is smaller than a platoon and larger than a squad; in some instances, it is the basic tactical unit. **2.** in the U.S. Navy, a tactical subdivision normally consisting of one-half of a division of ships or two aircraft. *Mathematics.* **1.** let E be a subset of a (Cartesian) product space $X \times Y$, where X and Y are spaces. If x is any point of X, then the set $E_x = \{y : (x, y) \in E\}$ is the section of E that is determined by x. The section of E that is determined by a point y in Y is similarly defined as the set $E_y = \{x : (x, y) \in E\}$. It is not always necessary to specify which particular point determines a section; in this case, we refer to an X-section (or Y-section), and understand that some unspecified point of X (or Y) determines the section. A section of a set in a product space is a subset of one of the component spaces. **2.** let f be a function defined on a subset E of a product space $X \times Y$. If x is any point of X, the function f_x, defined on the section E_x by $f_x(y) = f(x, y)$ is the section of f determined by x, or an X-section of f. A Y-section of f is similarly defined. **3.** a section of a bundle (E, B, π) is a mapping f from the base space B to the bundle E such that $\pi \circ f$ (the composition of the projection π and f) is the identity map on B.

sectional center *Telecommunications.* in the United States hierarchical long-distance network, the second-highest rank of long-distance telephone office, below a regional center.

sectional conveyor *Mechanical Engineering.* a conveyor belt system that can be shortened or lengthened by means of interchangeable sections.

sectional density *Ordnance.* of a bullet, its weight in pounds divided by the square of its diameter in inches.

sectionalized vertical antenna *Electromagnetism.* a vertical antenna that has isolated discontinuities (bridged by reactances or driving voltages) along its length, giving more desired radiation pattern in the vertical plane.

sectional radiography *Electronics.* the process of using X-rays or gamma rays to produce a detailed picture from a portion of an object. Also, PLANIGRAPHY.

section line *Civil Engineering.* a boundary line for a parcel of land, usually a section.

section modulus *Mechanics.* the moment of inertia of a cross section under flexural stress, divided by the greatest distance of an element on the section from the neutral axis of the section.

sector a definable part of a larger area; specific uses include: *Military Science.* a designated area within which a unit operates and for which it is responsible. *Civil Engineering.* a portion of military airspace clearly defined and for a particular purpose. *Computer Technology.* **1.** the smallest part of a track or band on a magnetic disk or magnetic drum that can be addressed by number. **2.** a pie-shaped area on the surface of a magnetic disk. **3.** a fixed-length block formatted on a magnetic tape. **4.** a logical division of main or cache memory. *Electromagnetism.* the azimuthal range of a radar. *Mathematics.* a region bounded by two radii of a circle and one of the arcs joining them. *Meteorology.* a representation on synoptic weather charts of a sector of a circle as considered in a warm sector that lies between both warm and cold weather fronts within a cyclonic disturbance.

sectoral horn *Electromagnetism.* an electromagnetic horn antenna having two opposite sides that are parallel and two sides that diverge.

sector boundary *Astrophysics.* the change in the dominant polarity of the sun's magnetic field (when drawn out by the solar wind) as solar rotation carries the field past the earth.

sector disk *Physics.* a disk that is opaque to radiation but has a section or hole cut in it in order to effectively reduce the intensity of a beam of radiation, as the beam is chopped while the disk is rapidly rotating.

sector display *Electronics.* a radar display that shows only a portion of the area under surveillance.

sectored light *Navigation.* a light in which sections of the same color are separated by darker sections.

sector gate *Civil Engineering.* a type of roller gate in which the roller is not cylindrical but a sector of a circle; used for regulating flow at dam spillways. Also, SLIDING GATES, ROLLER GATE.

sector gear *Mechanical Engineering.* a gear having teeth only on a portion of its circumference.

sectorial *Paleontology.* describing the single-cusped pointed tooth of nonhuman primates; a sectorial tooth often has a cutting edge and may be capable of a scissors-like cutting action in occlusion.

sector mark *Computer Technology.* a mark that can be sensed by hardware and delineates the boundary of a storage sector.

sector method *Materials Science.* a technique for studying photo polymerization in which light is flashed on the reaction vessel by a rotating sector placed in the path of the light source; used to gather information about polymerization kinetics.

sector of fire *Military Science.* an area that an individual, weapon, or unit is required to cover by fire.

sector scan *Electronics.* the rotation of a radar antenna through a limited angle, as distinguished from a complete rotation.

sector search *Navigation.* a sailing plan composed of three legs, with the turning points equidistant from a fixed or moving point; it may also apply to a flight plan.

sector structure *Astrophysics.* the observed interplanetary magnetic field polarity pattern during a solar rotation.

sector wind *Meteorology.* the average wind speed and direction at flight level for a given sector of an air route.

secular *Science.* continuing or taking place progressively throughout an age or ages, without an observed recurrence in a cycle.

secular acceleration *Astronomy.* an increase in the mean angular velocity of the moon such that its period shortens in response to a change in the earth's rotation speed.

secular equation *Quantum Mechanics.* an equation, in obtaining eigenvalues of an operator from its matrix representation, that satisfies the condition that the determinant of the coefficients vanishes for solution of the wave equation.

secular equilibrium *Nucleonics.* a condition in which the parent species in a radioactive series has a much longer half-life than its succeeding species, such that there is no significant change in its concentration during the time interval over which the shorter-lived species attain their equilibria, whereupon all species appear to decay at the same rate.

secular instability *Physics.* a nonperiodic instability in the properties of a system.

secular parallax *Astronomy.* a change in the apparent direction of a nearby celestial object due to the motion of the sun through space.

secular perturbation *Astronomy.* slow, continuous change in position of an orbit of a celestial body.

secular variable *Astronomy.* a star whose brightness changes over centuries.

secular variation *Geophysics.* **1.** any changes in the earth's magnetic field occurring over hundreds of years and caused by internal changes in the earth. **2.** a variation of any field or parameter that occurs over hundreds of years. Also, GEOMAGNETIC SECULAR VARIATION.

secund *Biology.* unilateral; arranged or occurring on one side.

secundine *Botany.* the inner or second coat or integument of an ovule.

secundine dike *Geology.* a dike that has been intruded into hot country rock.

secundines see AFTERBIRTH.

secure *Military Science.* to gain possession of a position or terrain feature and to take measures to prevent its destruction or loss.

secure visual *Telecommunications.* an encrypted digital signal that is made up of animated visual and audio data. Similarly, **secure voice.**

securinine *Pharmacology.* $C_{13}H_{15}NO_2$, an alkaloid prepared from the leaves and roots of *Securinega suffruticosa;* its nitrate salt is used to treat neurasthenia, heart failure, and impotence.

security *Military Science.* **1.** a collective term for the measures taken by a military unit, activity, or installation to protect itself against all acts that may impair its effectiveness. **2.** the condition that results from the implementation and maintenance of such measures. *Industrial Engineering.* the condition that prevents unauthorized individuals from having access to classified information. *Computer Programming.* **1.** the protection of data against accidental or intentional unauthorized disclosure, transfer, modification, theft, or destruction. **2.** measures to prevent unauthorized access to a computer.

security classification *Military Science.* a category to which national security information and material is assigned to denote the degree of damage that unauthorized disclosure would cause to the national defense or foreign relations and to denote the degree of protection required; there are three such categories: **top secret, secret, confidential.**

security clearance *Military Science.* an administrative determination by competent authority that an individual is eligible for access to classified information.

security glass *Materials.* a complex laminated glass that can withstand the impact of bullets or explosions.

security intelligence *Military Science.* intelligence regarding hostile organizations or individuals who are or may be engaged in espionage, sabotage, subversion, or terrorism.

security kernel *Computer Programming.* a part of an operating system that is protected from unauthorized access and implements basic system security.

sed or **sed.** sedimentation; sediment.

SED skin erythema dose.

sedation *Medicine.* the act or process of calming, often by the administration of sedative drugs.

sedative *Pharmacology.* **1.** allaying activity and excitement. **2.** any agent that calms and quiets nervous excitement in any part of the body, including the brain.

sedative poison *Toxicology.* any toxin that produces depression of the central nervous system.

Sedentaria *Invertebrate Zoology.* a grouping of sessile polychaete annelid worms with reduced sensory organs, living in tubes or burrows.

sedentary *Physiology.* **1.** relating to or characterized by a sitting posture. **2.** of inactive habits; sitting habitually. *Geology.* of a sediment or soil, formed in place and overlying its rock of origin.

sedge *Botany.* **1.** any grassy or rushlike plant of the genus *Carex,* family Cyperaceae, occurring in wet environments. **2.** broadly, any plant belonging to the Cyperaceae family, commonly called the **sedge family.**

sedifluction *Geology.* in the primary stages of rock formation, a process of subaqueous or subaerial movement of particles in unconsolidated sediments.

sediment *Chemistry.* solid particulate material that is deposited from suspension in a liquid. *Geology.* fragments of organic or inorganic material that are transported by, suspended in, or deposited by wind, water, or ice, and that then are accumulated in unconsolidated layers on the surface of the earth. (Going back to a Latin word meaning "to sit" or "to settle.")

sedimentary *Chemistry.* relating to or characteristic of sediment. *Geology.* formed by the deposition of sediment.

sedimentary breccia *Petrology.* a sedimentary rock consisting of angular fragments larger than 2 mm in diameter, usually featuring imperfect mechanical sorting and either the predominance of fragments from one local source or a great variety of randomly mixed materials. Also, SHARPSTONE CONGLOMERATE.

sedimentary cycle see CYCLE OF SEDIMENTATION.

sedimentary differentiation *Geology.* the gradual and progressive separation of a well-defined rock mass into distinct physical and chemical products that are resorted and deposited as sediments over relatively separate regions.

sedimentary dike *Geology.* a tabular mass of sedimentary material that cuts transversely through the structure or bedding of preexisting rock; formed by the filling in with sediments of a crack or fissure through forcible injection, by intrusion under abnormal pressure, or by simple infilling from above.

sedimentary facies *Geology.* for an areally restricted region of a given stratigraphic unit, a facies that differs significantly, both lithologically and paleontologically, from other regions of the unit.

sedimentary insertion *Geology.* the processes by which sedimentary material is emplaced among deposits or rocks already formed, such as infilling, injection, or intrusion.

sedimentary laccolith *Geology.* an intrusive mass of plastic sedimentary material that has been forced up under high pressure, penetrating the invaded formation in a direction parallel or nearly parallel to its bedding planes.

sedimentary lag *Geology.* the interval of time between the formation of a potential sediment by weathering and the removal and deposition of the sediment.

sedimentary peat *Geology.* a dark brown peat that is formed under water, mainly from partially decomposed algae and related forms. Also, DREDGE PEAT, LAKE PEAT, PULPY PEAT.

sedimentary petrography *Petrology.* the science concerned with description and classification of sedimentary rocks. Also, SEDIMENTOGRAPHY.

sedimentary petrology *Petrology.* the science concerned with the composition, characteristics, and origins of sediments and sedimentary rocks.

sedimentary quartzite see ORTHOQUARTZITE.

sedimentary ripple *Geology.* see RIPPLE.

sedimentary rock *Petrology.* a rock produced by the consolidation of layers of sediment. Also, DERIVATIVE ROCK, STRATIFIED ROCK.

sedimentary rock

sedimentary structure *Geology.* a structure that is formed within sedimentary rock at or near the time of deposition (primary sedimentary structure) or after deposition (secondary sedimentary structure).

sedimentary tectonics *Geology.* in a geosynclinal basin, folding and deformation that are caused by the sinking and buckling of rock strata of the geosyncline.

sedimentary trap *Geology.* an area, often between a high-energy and a low-energy environment, where sedimentary material accumulates and is not moved farther.

sedimentary tuff *Geology.* a tuff that contains a small amount of nonvolcanic detrital sediment.

sedimentary volcanism *Volcanology.* **1.** the processes whereby, under pressure, a mixture of sediment, water, and gas is expelled, extruded, or breaks through overlying formations. **2.** the processes whereby a sand or mud volcano is formed.

sedimentation *Geology.* **1.** any process by which particulate material accumulates in layers to form sedimentary deposits. **2.** the process by which sediment is mechanically deposited from suspension in a liquid. *Engineering.* see SETTLING. *Metallurgy.* the separation of metallic powder according to its rate of settling from a fluid suspension.

sedimentation analysis *Physical Chemistry.* the examination of the rate at which solid particles in a suspension settle out, either by gravity or, more often, through the use of a strong applied force, as with a centrifuge; this permits the determination of both the average molecular weight of the solution and the molecular weight of its individual components.

sedimentation balance *Analytical Chemistry.* an instrument designed to measure the weight of solid particles that settle out of a liquid and the rate at which settling occurs; used to determine molecular particle sizes.

sedimentation basin *Geology.* a wide, flat-bottom depression in the ocean floor where sediment accumulates. *Civil Engineering.* see SETTLING BASIN.

sedimentation coefficient *Physical Chemistry.* a factor that is characteristic of a given solute and solvent; used in sedimentation analysis, it represents the velocity of the boundary between the settled molecules below and the pure solvent above, divided by the square of the angular velocity of the system times the distance from the center of rotation.

sedimentation curve *Geology.* a curve that represents cumulatively the amount of sediment deposited or removed from an originally uniform suspension in successive units of time.

sedimentation diameter *Geology.* an expression of the size of a sedimentary particle, equal to the diameter of a hypothetical sphere having the same specific gravity and settling velocity as a given sedimentary particle in the same fluid.

sedimentation equilibrium *Analytical Chemistry.* a steady-state condition that occurs during centrifugation, indicated by no movement of the liquid-sediment boundary of the sample; used to determine molecular weight.

sedimentation radius *Geology.* one half of the sedimentation diameter of a given sedimentary particle.

sedimentation rate *Physical Chemistry.* the time within which a sediment is placed or deposited in a given volume of solution, particularly after being centrifuged. *Pathology.* specifically, the rate at which erythrocytes precipitate from a well-mixed specimen of venous blood; used in testing for anemia and various inflammatory diseases. Also, SED RATE. *Geology.* see RATE OF SEDIMENTATION.

sedimentation tank see SETTLING TANK.

sedimentation trend *Geology.* the direction in which sediments were deposited.

sedimentation trough *Geology.* a depression in the ocean floor having a narrow U-shaped or V-shaped bottom where sediment accumulates.

sedimentation unit *Geology.* a layer or deposit of sediment that was deposited during one distinct episode of sedimentation.

sedimentation velocity *Analytical Chemistry.* the rate at which solid particles suspended in a liquid settle out; used for molecular weight determinations.

sediment bulb *Engineering.* a bulb that contains sediment settled from liquid in a tank.

sediment charge *Hydrology.* the ratio of the weight or volume of sediment in a stream to the weight or volume of water passing a given cross section of the stream per unit time.

sediment concentration *Hydrology.* in a water-sediment mixture obtained from any body of water, the ratio of the dry weight of the sediment to the total weight of the mixture.

sediment corer *Engineering.* an apparatus used to cut sediment from the ocean floor in cylindrical sections.
sediment-delivery ratio *Geology.* for a given drainage basin, the ratio of sediment yield to the entire amount of sediment transported by sheet and channel erosion, expressed as a percentage.
sediment discharge *Hydrology.* the rate at which sediment passes a given cross section of a stream, measured by the dry weight or volume of sediment moved in a given time.
sediment-discharge rating *Hydrology.* the ratio between the discharge of sediment and the total discharge of the stream. Also, **sediment-transport rate.**
sediment load *Hydrology.* the solid material carried by a stream, expressed as the dry weight of all sediment that passes a given point over a given interval of time.
sedimento-eustasy *Oceanography.* a hypothetical state of equilibrium in the worldwide capacity of the ocean basins between the volume of sediment deposited and a possible increase in capacity due to local sinking of the sea floor caused by the weight of the sediment in some river deltas; in fact, the overall sedimentation rate is positive, but its effect on world sea level is still minor in comparison to the effects of glacio-eustatic and tectono-eustatic changes.
sedimentography see SEDIMENTARY PETROGRAPHY.
sedimentology *Geology.* the study of sediments and sedimentary rocks, including their formation, characteristics, classification, and origin.
sediment production rate *Geology.* for a given drainage basin, the sediment yield per unit of area, equal to the annual sediment yield divided by the area of the drainage basin.
sediment station *Hydrology.* a vertical cross section of a stream where load samples are collected systematically for the purpose of determining concentration and other characteristics.
sediment trap *Engineering.* a device to measure the accumulation rate of sediment on the bottom of a body of water, such as a pond or lake.
sediment vein *Geology.* a sedimentary dike formed when a fissure is filled in from above with sedimentary material.
sediment yield *Geology.* the quantity of material eroded from the surface of the land by runoff and transported to a stream system.
sedoheptulose *Biochemistry.* a seven-carbon ketose sugar, the phosphate of which is an intermediate in the pentose phosphate pathway and in the Calvin cycle of photosynthesis.
sed rate *Pathology.* see SEDIMENTATION RATE.
Seebeck coefficient *Electronics.* in a circuit exhibiting the Seebeck effect, the ratio of the open-circuit voltage to the temperature difference between the hot and cold junctions.
Seebeck effect *Electronics.* the generation of an electromotive force by the temperature differences between junctions in a circuit containing two different metals, alloys, or bodies; the phenomenon upon which thermocouples are based.
seed *Botany.* an embryonic sporophyte formed from a fertilized ovule and enclosed within a protective seed coat, which forms a new plant upon germination. *Agriculture.* to sow or scatter plant seed. *Meteorology.* to sow clouds with particles of silver iodide or other compounds in order to induce precipitation. *Chemistry.* a small amount of material that serves as a nucleus for initiating a desired reaction. *Crystallography.* a small crystal used to start the growth process of a large crystal.
seed beater *Archaeology.* an instrument usually made of wood or reeds that is formed into a racketlike shape; used to strike seeds from bushes.
seedbed *Agriculture.* an area of soil in which seeds are planted and take root.
seed charge *Chemistry.* a material that initiates precipitation when added to a supersaturated solution.
seed coat *Botany.* the outer covering of a seed that develops from the integuments.
seed core *Nucleonics.* a reactor core consisting of a small amount of highly active uranium embedded in a larger amount of natural uranium or thorium; the enriched uranium provides neutrons for the natural material, which in turn provides much of the power for the reactor.
seed corn *Agriculture.* corn (ears or kernels) that is set aside from a crop to be used as seed.
seed cotton *Agriculture.* unginned cotton that contains the seed with attached lint as picked from the boll.
seed drill *Agriculture.* see DRILL, def. 1.
seeder *Agriculture.* a machine or device used to plant or spread seeds.
seed fern *Paleontology.* any member of the order Pteridospermaphyta.
seed fibers *Textiles.* the fibers and hairs produced by plant seeds or seed pods, such as cotton and vegetable silk. Also, **seed-hair fibers.**
seedling *Botany.* **1.** any young plant grown from a seed. **2.** a young plant grown in a nursery for transplanting.
seedling blight *Plant Pathology.* a fungus disease that occurs in the early growth stages of grasses and legumes, in which the young plants wither and die without rotting.
seed nuclei *Astrophysics.* nuclei from which other nuclei are synthesized.
seed-snipe *Vertebrate Zoology.* any of various South American shorebirds of the family Thinocoridae, which bear some resemblance to quail. Also, **seedsnipe.**
seed stalk *Botany.* the stalk of a flowering plant that bears fruit.
seedstock *Agriculture.* **1.** seed, roots, or tubers that are set aside from a harvest and kept for planting. **2.** livestock or other animals that are selected and kept for breeding. Also, **seed stock.**
seed tree cutting *Forestry.* a method of harvesting timber in which a few scattered trees are left in the harvested area in order to provide a natural source of seeds.
seeing *Astronomy.* a term for the degree to which the image of a star as seen in a telescope is distorted by atmospheric disturbances.
seek *Military Science.* to move toward a target or similar object guided by phenomena emanating from the target or object, such as heat or light. *Computer Programming.* to position a read-write head over a specified track on a direct-access storage device. *Telecommunications.* a control in a radio that searches for the next audible frequency in sequence and tunes the radio to that signal.
seek area see CYLINDER.
seeker *Ordnance.* **1.** a weapon or other moving object, especially a missile, that is guided toward its target by phenomena emanating from the target. **2.** the device within a weapon that performs the seeking function.
seek time *Computer Programming.* the time taken by a direct-access storage device to position the read-write head over the specified track.
Seelandian *Geology.* a European geologic stage of the lowermost Upper Paleocene epoch.
seep *Geology.* a relatively small area in which water or oil percolates slowly to the land surface. *Petroleum Engineering.* an oil spring that yields a few drops to a few barrels daily; normally located at a low elevation where there is an accumulation of water.
seepage *Fluid Mechanics.* **1.** a slow movement of a fluid through a porous medium. **2.** the amount of fluid that so moves. Thus, **seepage loss.**
seepage face *Geology.* an area where water emerges from a slope at atmospheric pressure and flows downward, usually to a stream.
seepage lake *Hydrology.* **1.** a lake that has no surface outlet and that loses water mainly by seepage through the floor and walls of the lake's basin. **2.** a lake receiving its water mainly from seepage.
seepage spring *Hydrology.* a spring having only a small yield. Also, WEEPING SPRING.
seersucker *Textiles.* a lightweight, plain-weave fabric, usually of cotton, with permanently woven crinkle stripes made by alternating the tension of warp threads during weaving.
seg. segment.
Segas process *Chemical Engineering.* a petroleum-refinery oil gasification process that produces gases having caloric values similar to that of coal gas.
Seger cone *Materials Science.* a small cone of clay and oxide mixtures, calibrated to soften and bend within defined temperature ranges; used to indicate the temperature within a furnace.
segment a division, section, or part of a larger whole; specific uses include: *Mathematics.* **1.** the portion of a line between two distinct points on the line, including neither, one, or both of the points. Also, LINE SEGMENT. **2.** more generally, the portion of a curve between two distinct points on the curve, including neither, one, or both of the bounding points. **3.** the region of a circle between a chord and one of the arcs subtended by the chord. *Zoology.* any portion of a larger body or structure that is set off by natural or arbitrarily established boundaries, such as the parts of an arthropod body. *Computer Programming.* **1.** to divide a program into smaller units, storing some in internal storage and others in auxiliary storage, with the necessary instructions to go from one to another. **2.** the smallest executable unit of an overlay program. **3.** a large portion of the storage of a program, such as the code segment or data segment. *Transportation Engineering.* **1.** a single functional portion, fixed in location, of an instrument flight rules approach. **2.** a portion of a scheduled flight, which may or may not include intermediate stops.

segmental *Science.* **1.** relating to or forming a segment or a product of division, especially into a series of similar parts. **2.** undergoing segmentation.

segmental arch *Architecture.* an arch whose intrados is less than a semicircle.

segmental reflex *Physiology.* **1.** any neural reaction that uses the same pathway for both afferent and efferent impulses. **2.** a reflex controlled by a single segment or region of the spinal cord.

segmentary lineage system *Anthropology.* a hierarchy of lineage groups that is based on genealogical relation; usually functioning to create allies in crisis situations, such as warfare or food shortages.

segmentation a division into segments; specific uses include: *Biology.* **1.** a physical division into more or less similar parts, such as somites or metameres. **2.** see CLEAVAGE. *Computer Programming.* **1**. the division of a program into segments to allow the storage of only certain segments in main memory at a given time. **2.** the process of breaking up logical address space into several blocks, not necessarily of a particular size or mapped to a physical address. *Artificial Intelligence.* the process of dividing a visual image into regions that are visually coherent.

segmentation cavity see BLASTOCOEL.

segmentation nucleus see CLEAVAGE NUCLEUS.

segmentation sphere see BLASTOMERE.

segmented genome *Genetics.* a viral genome that is divided into two or more nucleic acid molecules.

segmented mirror telescope *Optics.* a telescope, such as the 10-meter Keck instrument on Mauna Kea, whose primary mirror is made up of interlocking separate mirrors.

segmented polymer *Materials Science.* any of a class of thermoplastic elastomers derived from aromatic dicarboxylic acids, polyalkylene-ether glycols, and short-chain diols; used for injection-molded parts for footwear and automobiles and for wire coating, pressure-sensitive adhesives, belts, and hoses.

segment mark *Computer Programming.* any character or mark used to delineate the boundaries of a segment of main or auxiliary memory.

segment saw *Mechanical Engineering.* a saw consisting of a flange with steel segments attached around its edge; used to cut veneer.

Sego *Ordnance.* a Soviet two-stage intercontinental ballistic missile powered by liquid propellant motors and equipped with inertial guidance; probably capable of carrying a 25-megaton thermonuclear warhead or a three-charge MRV with a range of 6500 miles; officially designated **SS-11.**

sego lily *Botany.* a lily, *Calochortus nuttallii,* having showy, bell-shaped flowers; native to the southwestern U.S., it is the state flower of Utah. Also, **sego.**

Segrè, Emilio born 1905, Italian-born American physicist; with Chamberlain, shared the Nobel Prize for discovery of the antiproton.

Segrè chart *Nuclear Physics.* a graph that displays all radioactive nuclides and their characteristics in ascending order, starting with those having the lowest atomic number.

segregated ice *Hydrology.* a film or layer of pure ice formed in frozen ground or permafrost by the drawing in of water as the ground becomes frozen. Also, **segregation ice.**

segregated vein *Geology.* a fissure filled with mineral matter that is derived from country rock by the action of percolating water. Also, EXUDATION VEIN.

segregating mixers *Materials Science.* in the mixing of particulate solids, a mixer in which the predominant mechanism is convection; single or multiple rotating mixing devices are used to transport groups of particles within the powder mass; examples are ribbon mixers and vertical screw mixers.

segregation a process of separating or isolating; specific uses include: *Genetics.* **1.** the separation of a pair of homologous chromosomes at meiosis so that only one member of each pair is present in any one gamete. **2.** any of various similar separation processes that occur during cell division, such as that of homologous chromatids or of nonhomologous plasmids. *Geology.* the formation of a secondary sedimentary feature by the chemical rearrangement of minor constituents within a sediment after its deposition. *Materials Science.* the presence of nonequilibrium composition differences in a material, often caused by insufficient time for diffusion during solidification. *Metallurgy.* a phenomenon during the solidification of steel ingots in which various components having the lowest freezing point concentrate in the last parts of the ingot to solidify, causing an uneven distribution of components. *Industrial Engineering.* the practice of separating different process streams.

segregational petite *Genetics.* a type of petite mutation occurring in a nuclear gene in yeast; it produces defective mitochondria, and has the ability to produce progeny consisting of both petite and wild-type cells.

segregation banding *Petrology.* the occurrence of compositional bands in gneisses due to the segregation of material originally from a homogeneous rock.

segregation distortion *Genetics.* the occurrence of abnormal segregation ratios due to lethality or disfunction of gametes carrying particular alleles; in a heterozygote, the normal segregation ration would be 1:1.

segregation lag *Molecular Biology.* a delay in phenotypic expression of a newly acquired gene mutation in bacterial cells containing two or more nucleoids.

segs. segments; segmented neutrophils.

seiche *Hydrology.* the pendulum-like movement of a body of water that continues after cessation of the originating force, which is usually wind but may be other atmospheric phenomena or seismic disturbances; a tide is a special case of a seiche.

Seidel, Philipp von 1821–1896, German mathematician and astronomer; discovered uniform convergence of series (independently of Stokes); developed theory of aberrations.

Seidel aberrations *Optics.* the aberrations in monochromatic light as described by the Seidel theory: coma, spherical aberration, distortion, astigmatism, and field curvature.

Seidel method see GAUSS-SEIDEL METHOD.

Seidel theory *Optics.* a theory that describes aberrations in an optical system using the first two terms of a Taylor series expansion to approximate the sine of the angle between a ray and the optical axis.

seif dune *Geology.* a sharp-crested longitudinal sand dune or chain of sand dunes whose profile consists of a succession of peaks and cols. Also, SWORD DUNE.

seine net *Engineering.* a large fishing net with floats along the top edge and weights on the bottom edge to keep the net in a perpendicular position.

seismic [sīz´mik] *Geophysics.* relating to a naturally or artificially induced earthquake or earth vibration.

seismic anisotropy *Geophysics.* the relation between direction of propagation and seismic velocity, if not constant.

seismic area see EARTHQUAKE ZONE.

seismic belt *Geophysics.* an elongated zone of earthquake activity on the earth.

seismic constant *Civil Engineering.* a specified amount of steady ground acceleration; used in designing structures that can withstand an earthquake.

seismic detector *Engineering.* a device that detects sound waves during seismic explorations.

seismic discontinuity *Geophysics.* **1.** a seismic boundary in which seismic waves abruptly change velocity. **2.** a boundary between seismic layers. Also, INTERFACE, VELOCITY DISCONTINUITY.

seismic efficiency *Geophysics.* the percentage of seismic energy released by an earthquake as compared to the total seismic strain.

seismic event *Geophysics.* **1.** an earthquake. **2.** a return generated by an explosion in seismic surveying.

seismic exploration *Engineering.* a process of determining the composition and density of a soil by producing an elastic sound wave in the earth and recording the time of its arrival at a detector.

seismic gap *Geophysics.* an area of a fault zone in which earthquake activity has not occurred at the expected rate or intensity.

seismic gradient *Geophysics.* a change in velocity of a seismic wave over distance in a specified direction.

seismic hazard *Geophysics.* the possibility of a large earthquake occurring in a specified area in some near future, usually quantified as up to twenty years.

seismic intensity *Geophysics.* the average rate at which seismic-wave energy flows through a unit cross section at right angles to the direction of wave propagation.

seismicity *Geophysics.* the frequency, intensity, and distribution of earthquake activity in a given area. Also, **seismic activity.**

seismic map *Geophysics.* a map presenting the space and time distribution of seismicity in a given area.

seismic-piezoelectric effect *Geophysics.* a variation in resistivity caused by elastic deformation of rock. Also, **seismic-electric effect.**

seismic profile *Geophysics.* an array of geophones, usually arranged along a line, used to make seismic surveys.

seismic prospecting *Mining Engineering.* the use of artificially generated seismic waves in prospecting for oil, gas, or mineral sources.

seismic risk *Geophysics.* **1.** a predicted range of earthquake effects, such as ground shaking, landslides, and casualties. **2.** the probability of the extent of social or economic consequences of earthquakes within a given region over a specified duration.

seismic sea wave see TSUNAMI.

seismic shooting *Mining Engineering.* the initiation of shock waves in rocks by firing an explosive charge at a known point.

seismic survey *Mining Engineering.* an exploration technique utilizing seismic shooting to delineate subsurface geologic structures of economic importance.

seismic trajectory *Geophysics.* the path along which seismic waves travel.

seismic velocity *Geophysics.* the rate, measured in kilometers per second, at which seismic waves propagate.

seismic vertical *Geophysics.* **1.** the point on the earth's surface that lies directly above the underground point where an earthquake impulse originates. **2.** the vertical line connecting this surface point to the point of origin.

seismic wave *Geophysics.* the spreading of a deformation started by the energy released when rocks within the earth fracture or slip abruptly along fault planes during an earthquake.

seismism [sīz´miz əm] *Geophysics.* any natural phenomenon associated with an earthquake.

seismo- a combining form meaning "earthquake."

seismocardiography *Cardiology.* the recording of cardiac vibrations.

seismogram [sīz´mə gram´] *Engineering.* a recording by a galvanometer of the actions of a seismograph.

seismograph [sīz´mə graf´] *Engineering.* an instrument that records vibrations in the ground and determines the location and strength of an earthquake.

seismology [sīz mäl´ə jē´] *Geophysics.* the study of earthquakes, including their origin, propagation, energy manifestations, and possible methods of prediction.

seismometer [sīz mäm´ə tər´] *Engineering.* an instrument that detects vibrations in the ground.

seismoscope *Engineering.* an instrument that records the time, but usually not the magnitude, of an earthquake. Also, **seismochronograph.**

Seisonidea *Invertebrate Zoology.* an order of marine rotifers, containing the single genus *Seison,* with an elongated leechlike body, commensal on some crustaceans. Also, **Seisonacea.**

seistan *Meteorology.* a strong north to northwesterly wind of monsoon origin that blows for a period of about 120 days in the Seistan district of eastern Iran and Afghanistan, beginning in late May or early June.

Seitz filter *Microbiology.* a bacterial filter containing an asbestos pad, used to filter microorganisms from solutions without utilizing heat.

seizure *Medicine.* **1.** the sudden attack or recurrence of a disease. **2.** specifically, an attack of epilepsy.

sejunction water *Hydrology.* capillary water completely surrounded by meniscuses and in static equilibrium within the soil above the capillary fringe.

sekaninaite *Mineralogy.* $(Fe^{+2},Mg)_2Al_4Si_5O_{18}$, a blue orthorhombic mineral, occurring as imperfect twinned crystals, having a specific gravity of 2.77 and a hardness of 7 to 7.5 on the Mohs scale; it forms a series with cordierite.

SEL *Aviation.* the airport code for Seoul, Republic of Korea.

Selachii *Vertebrate Zoology.* an alternate name for the subclass Elasmobranchii, which includes sharks, skates, and rays.

Selaginella *Paleontology.* one of the five living genera of club moss; one of the lycopsids, *Selaginella* is a smaller cousin of the extinct giant club moss of the Mesozoic; extant from the Silurian to the present.

Selaginellales *Botany.* an order of small club mosses of the subclass Lycopodineae, characterized by ligulate leaves, sporophylls in four-sided stobiles, and heterosporous reproduction.

selagite *Petrology.* a mica trachyte containing abundant tabular biotite crystals set in a holocrystalline groundmass of orthoclase, diopside, and sometimes olivine and secondary quartz.

selatan *Meteorology.* a strong, dry southerly wind that blows during the southeastern monsoon season over the Celebes Islands.

Selbornian see ALBIAN.

select *Computer Programming.* **1.** to choose one subroutine from a file of subroutines. **2.** to activate the control and data channels to and from an input/output unit.

select bit *Computer Programming.* in an input/output instruction word, a bit that selects the function of a specified device, e.g., odd or even parity selection.

selected mine *Ordnance.* an underwater mine that is connected to an onshore control station and can be tested, fired, or disarmed independently of the other mines in its group.

selected time *Industrial Engineering.* an observed time for a work element or process, chosen as a representative time for measurement or planning.

selecting circuit *Electricity.* a circuit that has the ability to separate a desired signal from those at other frequencies. Also, SELECTIVE CIRCUIT.

selection a process of choosing or of being chosen; specific uses include: *Evolution.* any natural or artificial limitation that favors the survival and reproduction of individuals in a population who possess a specific hereditary character over those that do not possess it. *Telecommunications.* the process of addressing a call to a certain station in a selective calling system. *Industrial Engineering.* in micromotion studies, the elemental motion of choosing a particular part from a collection of similar parts.

selection bias *Statistics.* in the selection of a sample from a population, a systematic pattern leading to nonrandom samples.

selection check *Computer Programming.* a check to verify that the correct device, such as a register, is selected in the interpretation of an instruction.

selection coefficient *Genetics.* a mathematical expression of the effect of natural selection against a certain genotype; usually expressed as the proportional reduction of genotypes left by that genotype in the succeeding generation relative to an alternative genotype that is not being selected against.

selection cutting *Forestry.* a method of harvesting timber in which small patches of mature trees are removed in order to make room for younger trees and new growth, thus leaving only small openings in a forest.

selection differential *Genetics.* in artificial selection, the difference in mean phenotypic value between the individuals selected as parents of the following generation and the whole population.

selection gain *Genetics.* in artificial selection, the difference in mean phenotypic value between the progeny of the selected parents and the parental generation.

selection pressure *Evolution.* any of the environmental or genetic factors that influence natural selection and consequently the direction of evolutionary change. *Genetics.* a measure of the intensity of natural selection on a population, as shown by the degree of alteration in the genetic makeup of that population over a given period of time.

selection restriction *Artificial Intelligence.* a semantic characteristic of a word that restricts the choice of other words to accompany it; e.g., the subject of the verb *walk* must be animate.

selection rules *Physics.* the rules that collectively dictate the possible changes in the quantum numbers of a quantum system during a transition from one state to another.

selection sort *Computer Science.* a sorting algorithm in which a record at the front of the file is exchanged with the record in the remainder of the file that has the minimum key value; this step is repeated with succeeding records until the end of the file is reached.

selection value see ADAPTIVE VALUE.

selective *Acoustical Engineering.* having good selectivity. *Pharmacology.* characterized by selectivity.

selective absorption *Electromagnetism.* the absorption of frequencies of a certain range while others are excluded.

selective acidizing *Petroleum Engineering.* the injection of acid into specific zones of an oil reservoir.

selective adsorbent *Chemical Engineering.* a chemical compound that is used to adsorb a selected compound from a mixture.

selective advantage *Genetics.* any favorable structural, functional, or behavioral characteristic that increases an individual's opportunity and likelihood of survival and reproduction over those of other individuals in the same population who lack this characteristic.

selective attention *Psychology.* the ability to concentrate on one significant stimulus in the present environment while excluding other more incidental stimuli.

selective avoidance see SELECTIVE INATTENTION.

selective breeding *Agriculture.* the breeding of animals or plants having desirable characters.

selective calling *Telecommunications.* any of several radio communications methods for calling only one of several receiving stations guarding the same frequency by modulating the transmitted signal according to predetermined codes.

selective circuit see SELECTING CIRCUIT.

selective cracking *Chemical Engineering.* a petroleum-refinery process that re-cracks gas oil from the fractionation column separately instead of mixing it with fresh feed in order to improve gas yield.
selective death *Ecology.* the death of an individual caused by a genetic trait which is harmful to that individual in that environment; e.g., a lighter-colored moth that is more visible to predators in an environment darkened by air pollution.
selective dissolution *Materials Science.* corrosion that attacks metals along a microstructural path, usually anodic to the rest of the material, or one metal selectively in an alloy; the result is increased weakness or failure of the material.
selective dump *Computer Programming.* a printout of a specified area of internal storage.
selective fading *Telecommunications.* fading that varies at different frequencies within the same modulation envelope.
selective filling *Mining Engineering.* a method of filling by hand such that only clean material is loaded, dirt and stone being rejected.
selective flotation see DIFFERENTIAL FLOTATION.
selective fracturing *Petroleum Engineering.* a means of creating multiple fractures in specific reservoir zones by plugging casing perforations or isolating the zones.
selective fusion *Geology.* the fusion of only a part of a mixture, such as a rock.
selective identification feature *Electronics.* a device that receives and transmits a limited range of identification signals from an aircraft, such as friend-or-foe or ground-to-air communications.
selective inattention *Psychology.* **1.** the tendency of an individual to restrict or avoid awareness of anxiety-producing experiences, so that they are ignored, misinterpreted, or forgotten. **2.** the tendency of an individual to avoid or ignore information that is incompatible with his own point of view.
selective interference *Telecommunications.* a type of interference whose energy is consolidated into a narrow band of frequencies.
selective jamming *Military Science.* a procedure that impairs only one channel of an enemy's communication system by bombarding it with electromagnetic radiation.
selective medium *Microbiology.* a bacterial culture medium that is used to grow a particular type of organism. Also, **selective culture.**
selective mining *Mining Engineering.* a mining technique in which the best ore is mined, leaving only low-grade ore that cannot be extracted profitably.
selective neutrality *Genetics.* for a given gene, a situation in which a mutant allele and the wild-type allele confer equal fitness.
selective species *Ecology.* any plant species that is rarely found outside a particular community.
selective vagotomy *Surgery.* division of the vagal nerve fibers to the stomach with preservation of the hepatic and celiac branches.
selective value see ADAPTIVE VALUE.
selectivity *Acoustical Engineering.* a measure of the ability of a radio tuner to reject nearby channels. *Pharmacology.* the ability of a given drug to produce desired effects in relation to its adverse effects. *Toxicology.* the relative differential toxicity of a poison.
selector *Computer Science.* in object-oriented programming, the name of a message action. The class contains the association between the selector and the corresponding method that performs that action for objects in the class. The selector can be considered to be a generic procedure name.
selector channel *Computer Technology.* a type of input/output channel that is electrically connected to only a single device, such as a disk or tape drive, at a time, and performs high-speed data transfers between that device and the central processing unit.
selector switch see ROTARY SWITCH.
selenate *Chemistry.* a salt or ester of selenic acid.
selene *Anatomy.* moon-shaped. (From *Selene,* the Greek goddess of the moon.)
selenic acid *Chemistry.* H_2SeO_4, a toxic white solid or crystals; soluble in water; melts at 58°C and decomposes at 260°C; a strong corrosive, resembling sulfuric acid.
selenide *Chemistry.* any compound of selenium with another element or radical, especially of bivalent selenium with a positive element or group.
selenious *Chemistry.* containing bivalent or tetravalent selenium.
selenious acid *Chemistry.* H_2SeO_3, toxic, transparent, colorless crystals, deliquescent, soluble in water and alcohol, that decompose at a melting point of 70°C; used as an analytic reagent. Also, SELENOUS ACID.
selenite broth *Microbiology.* a liquid medium for the enrichment of the bacterium *Salmonella* in fecal samples; used because of the growth inhibition of many other enteric bacteria.
selenium *Chemistry.* a nonmetallic element having the symbol Se, the atomic number 34, and the atomic weight 78.96; an essential mineral resembling sulfur, being a constituent of the enzyme glutathione peroxidase and believed to be closely associated with vitamin E and its functions. (From the Greek name for the moon; it was associated with tellurium, which is named for the earth.)
selenium dioxide *Chemistry.* SeO_2, toxic, white or yellowish-white to slightly reddish crystalline powder or needles, soluble in water and alcohol, that sublime at 315°C and melt at 340°C; used as an oxidizing agent, antioxidant, and catalyst.
selenium poisoning *Toxicology.* a poisoning of livestock feeding on plants that have absorbed high levels of selenium from the soil; found especially in the north central Great Plains of the U.S., and characterized by cirrhosis of the liver, anemia, hair loss, bone erosion, and emaciation.
selenium tRNAs *Biochemistry.* tRNAs containing pyrimidine bases that have been modified by the covalent attachment of a selenium atom.
seleno- a combining form meaning "moon."
selenocysteine *Biochemistry.* $C_3H_7NO_2$, the selenium analogue of cysteine, present in certain proteins.
selenodesy *Astronomy.* the study of the moon's precise shape and gravity field.
selenodont *Vertebrate Zoology.* having crescent-shaped molar cusps, as in ruminants.
selenographic coordinates *Astronomy.* a coordinate system that locates features on the moon's surface by latitude and longtitude.
selenographic latitude *Astronomy.* angular distance north or south of the lunar equator.
selenographic longitude *Astronomy.* angular distance measured westward from the prime lunar meridian.
selenography *Astronomy.* the study of the moon's surface and description of its topography.
selenomethionine *Biochemistry.* $C_5H_{11}NO_2Se$, an analogue of methionine in which selenium takes the place of sulfur; when added to proteins, it can substitute for methionine in cell-free systems.
Selenomonas *Bacteriology.* a genus of Gram-negative, anaerobic bacteria that exhibit flagellar motility, possess a fermentative mode of metabolism, and occur in rumen and the human oral cavity.
selenone *Organic Chemistry.* any of several organic selenium compounds having the general formula R_2SeO_2.
selenonic acid *Organic Chemistry.* any organic acid containing the group $-SeO_3H$; analogous to a sulfonic acid.
Selenophoma *Mycology.* a genus of fungi belonging to the class Coelomycetes.
selenoproteins *Biochemistry.* a group of proteins containing selenocysteine or selenomethionine residues; e.g., glutathione peroxidase.
selenosis see SELENIUM POISONING.
selenotrope *Engineering.* a geodetic surveying instrument that reflects the rays of the moon to distant points.
selenous acid see SELENIOUS ACID.
selenoxide *Organic Chemistry.* any of several organic selenium compounds having the general formula R_2SeO.
self *Psychology.* the totality of an individual's personal experience and expression. *Biology.* having a uniform quality or characteristic throughout, such as being one color. *Immunology.* relating to an individual's own tissue constituents.
self-absorption *Nucleonics.* the absorption of ionizing radiation by a material from which the radiation is emitted. *Spectroscopy.* in an emission line, a reduction in intensity near the center that results from selective absorption by the cooler parts of the radiation source. Also, **self-reduction, self-reversal.**
self-actualization *Psychology.* the full development and use of an individual's abilities and talents, satisfying his or her physical, social, intellectual, and emotional needs.
self-adaptation *Biology.* the innate ability of an organism to adapt in response to new conditions.
self-adjoint *Mathematics.* **1.** an operator on a Hilbert space or linear transformation on an inner product space is said to be self-adjoint if it is equal to its adjoint. A self-adjoint operator S can be moved from one side of an inner product to the other: $(Su, v) = (u, Sv)$. Also, HERMITIAN. **2.** a subalgebra of an involutive algebra is said to be self-adjoint if it contains the adjoint of each of its elements.

self-analysis *Psychology*. an attempt to examine one's own behavior, usually from a psychoanalytic perspective.
self antigen or **self-antigen** see AUTOANTIGEN.
self-assembly *Molecular Biology*. a spontaneous aggregation of multimers that forms weak chemical bonds between complementary surfaces.
self-bias *Electronics*. a bias voltage for a transistor or vacuum tube terminal produced by direct current flowing in an external resistor. Also, C BIAS, AUTOMATIC GRID BIAS.
self-biasing bias *Acoustical Engineering*. the addition of sound to a recording, especially a high-frequency signal, to produce full frequency coverage and prevent distortion due to the lack of a signal. Also, BIAS COMPENSATION.
self-bonded magnet wire *Materials*. a rigid, solid electrical conductor with a top-coat polymer that provides a self-bonding action for cementing turns of a coil.
self-capacitance see DISTRIBUTED CAPACITANCE.
self-centering chuck *Mechanical Engineering*. a lathe chuck that, when closed, maintains concentric positioning by means of scroll or radial screws. Also, UNIVERSAL CHUCK.
self-chambering *Ordnance*. describing a weapon that automatically loads cartridges into the chamber.
self-charge *Quantum Mechanics*. the contribution to a particle's observed charge from the interaction of its intrinsic charge with the vacuum polarization; this concept is used in charge renormalization.
self-checking code see ERROR-DETECTION CODE.
self-checking number *Computer Programming*. a number that contains a suffix figure, such as a check bit or modulo-N figure, that verifies a successful data transfer.
self-cleaning *Mechanical Engineering*. describing any mechanism designed to clean itself.
self-cloning experiment *Genetics*. a cloning experiment in which the same strain or species is both the source of the cloned DNA and the site of replication for the recombinant DNA.
self-complementing code *Computer Programming*. a binary code in which the bit combination for the complement of a decimal digit is the complement of the bit combination that represents the digit itself.
self-concept *Psychology*. an individual's most complete description of himself or herself, including both internal and external attributes.
self-conjugate particle *Particle Physics*. a particle whose charge, lepton number, baryon number, and hypercharge are all zero, thus indicating that it is identical with its antiparticle.
self-contained *Design Engineering*. of a machine, complete in itself; operating independently of other machines, systems, or power sources.
self-contained breathing apparatus *Engineering*. a portable breathing unit that permits freedom of movement for the user.
self-contained night attack *Military Science*. the capability of an aircraft to perform a solo night mission, including navigation to and from the target area, target location, and accurate attack.
self-control *Psychology*. the ability to be in command of one's behavior, especially the ability to repress or inhibit impulsive behavior.
self-correcting *Design Engineering*. of a machine or system, automatically correcting or adjusting to errors.
self-correcting code see ERROR-CORRECTING CODE.
self-corrosion *Materials Science*. corrosion of a metal in an underground environment, resulting from anodic and cathodic regions on the metal surface; divided into three types: corrosion in disturbed soil, corrosion in undisturbed soil, and microbial corrosion.
self-damping *Electrical Engineering*. denoting a decrease in the excursion of the variable of interest from the initial or steady-state value following a perturbation as a result of the internal dissipation of energy.
self-deception *Psychology*. an individual's failure to have insight into his or her own limitations.
self-defeating behavior *Behavior*. behavior that interferes with or counteracts the individual's own objectives and desires; e.g., trying so hard to please others that this ultimately alienates them.
self-destroying *Ordnance*. of or relating to a fuse or tracer that is designed to burst a projectile, missile, or rocket in the air if the weapon overshoots its target. Thus, **self-destroying fuse.**
self-destruct charge *Ordnance*. the explosive element of a missile destruct system.
self-destruction *Ordnance*. the destruction of a projectile, missile, or rocket initiated by a self-destroying fuse or tracer.
self-differentiation *Physiology*. cell differentiation stemming from an internal control that allows development only along a predetermined path.
self-diffusion *Solid-State Physics*. the spontaneous relocation of atoms in a crystal that move to occupy a new lattice site.
self-efficacy *Psychology*. an individual's comprehensive sense of his or her own capabilities; those with adequate self-efficacy can cope with the demands of a wide variety of situations.
self-energizing brake *Mechanical Engineering*. a braking mechanism that is designed to reinforce its power by applying friction between its lining and the drum.
self-energy *Physics*. **1.** in a quantum mechanical system, the energy associated with the virtual emission and absorption of particles. **2.** in a classical system, the energy associated with the interaction among parts within the system.
self-esteem *Psychology*. an attitude of acceptance and approval of oneself; one of the fundamental ingredients of mental health.
self-excitated vibration see SELF-INDUCED VIBRATION.
self-excited generator *Electricity*. a generator whose armature supplies the current for the field coils.
self-excited oscillator *Electronics*. a device that generates an alternating current entirely from the direct voltages applied to its electrodes and uses its own resonant circuits, rather than a crystal, to control frequency.
self-fertilization *Biology*. **1.** the union of female and male gametes produced by a single organism. **2.** breeding between closely related individuals; in-breeding.
self-field *Electromagnetism*. an electromagnetic field produced by an intense beam of charged particles; such fields tend to limit the intensity of the beam.
self-fulfilling *Psychology*. of or relating to a psychological condition that can be brought about or intensified by the belief that the condition is present.
self-fulfilling prophecy *Psychology*. a belief or expectation that appears to be responsible for its own realization; more likely, the individual has behaved, perhaps unconsciously, in such a way as to bring about the outcome.
self-handicapping (strategy) *Psychology*. an individual's deliberate reduction of his chances of succeeding at a task for which there is a fear of failure; e.g., not studying before a difficult exam, so that the failure can be attributed to circumstance rather than personal inability.
self-incompatibility see SELF-STERILITY.
self-induced transparency *Optics*. a phenomenon in which a pulse of coherent light, at a specified frequency, amplitude, and duration, passes through an ordinarily opaque medium.
self-induced vibration *Mechanics*. oscillations of a system due to the conversion of nonoscillatory excitation to oscillatory excitation within the system. Also, SELF-EXCITATED VIBRATION.
self-inductance *Electromagnetism*. the inductance associated with an isolated circuit that is characteristic of the circuit's physical design.
self-induction *Electromagnetism*. the generation of a voltage in a circuit due to self-inductance, the polarity of which tends to oppose the changing current in the circuit.
selfing *Genetics*. a process of breeding by self-fertilization or self-pollination.
self-instruction *Psychology*. a behavioral modification technique in which the individual is trained to control or modify the unacceptable behavior on his own.
self-interstitial *Materials Science*. of or relating to a point defect in which an atom occupies a nonlattice site.
selfish DNA *Molecular Biology*. a type of DNA that is found in higher eukaryotes and does not constitute genes; its function is still a matter of debate.
self-loading *Ordnance*. describing a firearm or gun that automatically extracts and ejects the empty cartridge case and loads the next charge into the chamber; the self-loading mechanism may be powered by recoil or the explosive gases; a self-loading weapon may be semiautomatic or fully automatic.
self-locking nut *Mechanical Devices*. a nut designed with a locking feature to resist loosening by vibration.
self-monitoring *Psychology*. the process of observing responses to one's behavior in a given situation and then modifying the behavior accordingly.
self-noise *Acoustics*. **1.** for a sound system, noise that is introduced to the recording or reproduction of the sound by the system itself. **2.** the underwater noise that originates on a vessel and is detected on it.
self-organization *Chaotic Dynamics*. the spontaneous formation of patterns or localized order in a system which begins in a more homogeneous state. Thus, **self-organizing.**

self-organizing function *Control Systems.* a function that automatically alters the structure of a large-scale control system in response to contingencies or changes in system objectives.

self-pollination *Botany.* the process of transferring pollen grains from an anther to a stigma of the same flower, of another flower on the same plant, or of the flower of a genetically identical plant.

self-propelled *Mechanical Engineering.* of or relating to the ability to achieve motion or power without the aid of external help. *Ordnance.* **1.** of or relating to a weapon that is mounted on a self-propelled vehicle and usually fired from the vehicle. Thus, **self-propelled antitank gun, self-propelled artillery, self-propelled howitzer. 2.** of or relating to a military unit equipped with such weapons. **3.** of or relating to a missile, such as a rocket, that is powered by self-contained fuel.

self-propelled gun *Ordnance.* a gun that is mounted on a self-propelled vehicle and fired from the vehicle.

self-psychology *Psychology.* any approach to psychology that focuses on the self as the central concept and interprets all behavior in reference to the self.

self-pulsing *Electronics.* of a device, employing a special circuit to automatically start and stop oscillation at the pulsing rate.

self-purification *Ecology.* the process of organic decay.

self-quenched detector *Electronics.* a circuit that amplifies a radio signal until it exceeds an audio frequency, then cuts it off; used to prevent a squealing condition in a receiver.

self-quenching *Materials Science.* a process of using the mass of metal underlying a heat-treated surface to quench the heated layer.

self-quenching oscillator *Electronics.* a circuit in which the generation of alternating current is intermittently switched off by the buildup of direct currents.

self-regulation *Ecology.* any internal response in a population that inhibits growth above or below a certain high level of density. *Behavior.* see BIOFEEDBACK. *Psychology.* see SELF-MONITORING.

self-regulation hypothesis *Ecology.* the theory stating that an unlimited population increase is not possible even in the absence of restricting external factors, because after a certain point internal changes in the group will occur that inhibit further growth.

self-reinforcement *Psychology.* the process of rewarding or punishing oneself psychologically according to one's appraisal of one's own actions.

self-repair *Computer Technology.* a form of automatic recovery in which a machine can detect a fault during its operation, switch over to a spare component, and initiate recovery software without human intervention.

self-replication *Molecular Biology.* the ability of most cells and some molecules, such as DNA and in some cases RNA, to make copies or duplicates of themselves.

self-rescuer *Mining Engineering.* a small filtering device carried by miners underground for immediate protection against carbon monoxide and smoke.

self-reset *Electricity.* of a relay or circuit breaker, automatically returning to its original position.

self-resetting loop *Computer Programming.* a loop that contains instructions that restore the initial conditions before exiting the loop.

self-reversal *Geophysics.* the acquisition by a rock of a remanent magnetism opposite in polarity to the ambient magnetic field at the time of formation.

self-rising ground *Geology.* in certain playas, a puffy, irregular surface or near-surface zone of loose granular sediment that is overlain by a thin clay rust; results from the capillary rise of groundwater.

self-saturation *Electronics.* the process of connecting half-wave rectifiers to a magnetic amplifier in order to obtain a higher voltage and a quicker response time.

self-scanned image sensor *Electronics.* a device that produces a television signal from an image without using an electron beam.

self-schema *Psychology.* an organized and well-defined component of identity that places the self in a specific role, such as an individual's clear perception of himself or herself as a teacher.

self-screening range *Electromagnetism.* the range of a radar system at which a target could be detected (with a specified probability) while transmitting a jamming signal.

self-sealing *Engineering.* a container with a substance that seals automatically, such as a gas-tank hole or tire puncture.

self-selection bias *Statistics.* an experimental bias that is created by allowing the subjects of an experiment to participate on a voluntary basis.

self-serving bias *Psychology.* an individual's tendency to take credit for his or her successes, while blaming failure on external factors.

self-shielding *Nucleonics.* exhibiting or characterized by self-absorption.

self-similar flow *Fluid Mechanics.* a fluid flow whose shape does not vary with time, such as a spherical expansion.

self-similarity *Chaotic Dynamics.* a characteristic of fractals that, when magnified, may resemble their unmagnified image. Structures and patterns observed at one magnification are found again at each magnification. *Mathematics.* a geometrical figure G is self-similar (or exhibits self-similarity) if there exists a 1-parameter (or n-parameter) group of conformal transformations of the overlying space that leaves G unchanged. For example, the plane spiral $r = e^q$ (polar coordinates) is unchanged when r is replaced by re^a and q is replaced by $q + a$, where a is a parameter.

self-splicing RNA *Molecular Biology.* in a split gene, RNA that can excise one or more class I intervening sequences, or introns.

self-starter *Mechanical Engineering.* a device, usually an electric motor driven by a battery, that starts an internal-combustion engine automatically.

self-starting synchronous motor *Electricity.* a synchronous motor that has bars embedded in the rotor like a squirrel-cage induction motor, so that it can start rotating without assistance when voltage is applied. Also, LINE-START SYNCHRONOUS MOTOR.

self-steering microwave array *Electromagnetism.* an antenna array that is capable of detecting the phases of incoming signals and then steering the array beam in the direction of the incoming signals.

self-sterility *Botany.* in certain hermaphroditic plants, a condition in which the plant is unable to fertilize itself or is sterile to its own pollen. Also, SELF-INCOMPATIBILITY.

self-sterility genes *Genetics.* genes found in monoecious plants that inhibit the growth of pollen tubes bearing the same alleles; this system favors the growth of pollen tubes from another plant, thereby minimizing the success of inbreeding and promoting cross-pollination. Also, S ALLELES.

self-suggestion see AUTOSUGGESTION.

self-synchronous repeater see SYNCHRO.

self-talk *Psychology.* a behavioral modification technique in which the individual carries on an internal dialogue in order to evaluate his or her behavior and select the desired course of action.

self-tapping screw *Mechanical Devices.* a specially hardened screw used in wood and soft metals that self-cuts its own threads into the material being worked on. Also, TAPPING SCREW, SHEET METAL SCREW.

self-theory *Psychology.* any personality theory that considers the self to be a central factor in personality organization and function.

self-thinning *Ecology.* the increase in mortality rate that occurs in a population due to its high density.

self-thinning rule *Ecology.* the characteristic relationship between plant density and the biomass of individual plants.

self-timer *Engineering.* a device that delays the closing of a camera shutter so that the photographer can be in the picture.

self tolerance *Immunology.* an inability to produce an immune response to autoantigens; typically acquired during the fetal stage of life.

self-transmissible plasmids *Molecular Biology.* plasmids that encode all the necessary functions of intercellular transmission by the union of two single-celled organisms. Also, CONJUGATIVE PLASMIDS.

self-treatment *Psychology.* any of various therapeutic techniques in which the subject is trained or encouraged to conduct independent treatment, rather than following the explicit directions of a therapist.

self-triggering program *Computer Programming.* a program that begins execution as soon as it is loaded into main memory.

self-tuning regulator *Control Systems.* an adaptive control system having two loops, one of which controls the process with an ordinary linear feedback regulator, while the other adjusts the parameters of the regulator with a recursive parameter estimator and a design calculation.

self-twist spinning see ROTOR SPINNING.

self-unloading ship *Naval Architecture.* a ship equipped to unload its cargo without any special harbor facilities, by use of hoisting rigs, endless belts, or other on-board gear.

Seliberia *Bacteriology.* a genus of Gram-negative bacteria that occur in soil as rod-shaped cells, accumulate iron, and propagate by budding.

seligmannite *Mineralogy.* $PbCuAsS_3$, a lead-gray to black, opaque, metallic orthorhombic mineral, occurring as imperfect twinned crystals, having a specific gravity of 5.38 to 5.44 and a hardness of 3 on the Mohs scale; forms a series with bournonite.

Seliwanoff's test *Analytical Chemistry.* a colorimetric method of analysis used to identify ketosis by the development of a red color when the test sample is mixed with resorcinol in dilute hydrochloric acid.

sellaite *Mineralogy.* MgF_2, a colorless to white, vitreous tetragonal mineral occurring as prismatic crystals and fibrous aggregates, having a specific gravity of 3.148 and a hardness of 5 on the Mohs scale; found associated with anhydrite and native sulfur.

sella turcica *Anatomy.* a shallow depression, located in the superior surface of the body of the sphenoid bone, that contains the pituitary gland.

sellite *Inorganic Chemistry.* Na_2SO_3, a solution of sodium sulfite that is used in the manufacture of the explosive TNT (trinitrotoluene).

Sellmeier's equation *Electromagnetism.* an equation used to determine the index of refraction as a function of wavelength in a medium whose molecules have several different oscillator frequencies.

selva *Ecology.* another name for a neotropical rain forest.

selvage *Petrology.* in an igneous rock mass, a marginal zone distinguished by fabric or composition, such as a chilled border of fine-grained or glassy texture in a dike rock. Also, SALBAND. *Textiles.* the lengthwise or warp edge of woven fabrics, often made with stronger yarns more tightly woven than the body to prevent raveling. Also, **selvedge.** *Geology.* see FAULT GOUGE.

Selwood engine *Mechanical Engineering.* an engine whose block is rotated by the action of two curved pistons opposite each other at 180° running in toroidal tracks.

Selye, Hans [zel´yə; sel´yā] 1907–1982, Austrian-born Canadian physician and physiologist; originated the specific analysis and treatment of stress.

SEM scanning electron microscopy.

sem or **sem.** semi-; semen.

Semaeostomeae *Invertebrate Zoology.* a suborder of common jellyfish, scyphozoans in the order Discomedusae, having frilled lobes and tentacles on the margin of the umbrella.

semantene *Telecommunications.* a language element that puts forth a definite idea or image, e.g., a data element. Also, LEXEME.

semantic *Linguistics.* **1.** of, relating to, or arising from the meanings of words or other symbols. **2.** of or relating to the field of semantics.

semantic analysis *Computer Programming.* **1.** in linguistics, the study of the relationships between words and symbols in a language and their interpretation. **2.** in a compiler, analysis of the meaning of a programming language statement, following syntactic analysis.

semantic code *Psychology.* a mental concept of an idea, object or event, based on its function or properties, as distinguished from its visual image.

semantic component *Linguistics.* any individual part of a word that distinguishes its meaning from other words, such as the suffix *-ed* in "walk" vs "walked."

semantic conditioning *Psychology.* the conditioning of a word to some object that the word represents.

semantic count *Linguistics.* an analysis of the relative frequency of use of the various individual meanings of a word.

semantic error *Computer Programming.* an error in the logic or an inconsistency of the intent of a program.

semantic extension *Computer Programming.* a feature of an extensible language that permits the definition of new kinds of objects, such as additional data types or operations.

semantic field *Linguistics.* a region of human perception, such as color, that is dependent upon and categorized by the vocabulary of a language.

semantic generalization *Psychology.* in learning, the tendency for an individual to offer a previously learned response to a second stimuli if it is similar in meaning or association to the first.

semantic grammar *Artificial Intelligence.* a grammar in which the phrases have meaning in the domain of application, e.g., part-name, as distinguished from being general syntactic units, e.g., noun-phrase.

semantic markers *Artificial Intelligence.* markers associated with a word, especially a noun, that describe semantic categories to which the word belongs, e.g., animate, human, countable.

semantic memory *Psychology.* general memory for anything significant or having meaning, including words, facts, and ideas.

semantic network *Artificial Intelligence.* a knowledge representation scheme in which objects are depicted as nodes in the network and the relationships between objects are represented by connecting arcs; the arcs are labeled by type of relationship such as "is," "belongs to," and so on.

semantic network model *Psychology.* a theory of the configuration of long-term memory, proposing that material is stored as independent units, such as words or concepts, which are then connected to one another.

semantic primitive *Artificial Intelligence.* see PRIMITIVE.

semantics *Linguistics.* the study of the systematic ways in which languages structure and present meaning, particularly in words and sentences.

semaphore *Transportation Engineering.* an apparatus that gives visual signals, usually by the position of a movable arm during daylight hours and by the color of lights at night. *Computer Science.* a synchronization mechanism used to coordinate the execution of two or more programs that share resources and to provide interprocess communications.

sematic *Biology.* functioning as a sign of danger, as warning odors and colors do.

Sematophyllaceae *Botany.* a large, tropical family of shiny mosses of the order Hypnobryales that form soft mats on soil, humus, tree trunks, and rocks; they are characterized by well-developed alar cells, leaves that are ecostate, complanate, or falcate-secund, and a hypnobryalean peristome.

semeiography *Medicine.* a description of the symptoms or signs of a disease.

semelparity *Ecology.* the condition of producing all offspring in a single reproductive event over one relatively short period.

sememe *Linguistics.* **1.** the meaning of a morpheme. **2.** any basic unit of meaning.

semen *Physiology.* a thick, whitish fluid that contains the sperm cells and is produced by the testes, the prostate, the seminal vesicles, and the bulbourethral glands.

Semenov, Nikolai 1896–1986, Soviet chemist; awarded the Nobel Prize for his research and theoretical work on the kinetics of chemical chain reactions.

Semenov number 1 *Physical Chemistry.* a dimensionless quantity that equals a mass transfer constant divided by a reaction rate constant. Also, S_M, SCHMIDT NUMBER 2.

Semenov number 2 *Physics.* a dimensionless quantity given by the diffusion coefficient divided by the thermal diffusivity; equal to the reciprocal of the Lewis number.

semi- a prefix meaning: **1.** half. **2.** partially.

semiactive homing guidance *Ordnance.* a system of missile guidance in which an outside source, such as ground-based or ship-borne radar, illuminates a target; the direction-finding receiver in the missile then homes in on the radiations reflected from the target. Also, **semiactive guidance.** Similarly, **semiactive tracking system.**

semialgorithm *Computer Programming.* a problem-solving method that will halt if a solution can be found in a finite number of steps, but will not necessarily halt if the problem has no solution.

semianechoic room *Acoustics.* a small room with a live-end-dead-end that absorbs most of the sound, preventing excessive reverberations, but does allow controlled echoes in certain areas.

semianthracite *Geology.* a coal having a fixed-carbon content of 86–92%, ranking between bituminous coal and anthracite.

semiaquatic *Biology.* growing or living near water or partially in water.

semiarid climate see STEPPE CLIMATE.

semiautomatic partly automatic. *Ordnance.* of or relating to a gun that is self-loading but requires a separate pressure on the trigger for each shot. Thus, **semiautomatic fire, semiautomatic pistol, semiautomatic rifle, semiautomatic weapon.**

semiautomatic flight inspection *Navigation.* an airborne system designed to record flight data in order to test the accuracy of the ground-based navigational system. Also, **SAFI.**

Semiautomatic Ground Environment System see SAGE.

semiautomatic supply *Industrial Engineering.* a supply system in which certain items are shipped automatically on the basis of periodic reports, while other items are shipped only on the basis of specific requisitions.

semiautomatic telephone system *Telecommunications.* a system that limits automatic dialing to only those subscribers who have the same exchange as the calling subscriber.

semiautomatic transmission *Mechanical Engineering.* an automobile transmission system that assists an operator in shifting from one gear to another.

semiautomatic welding *Metallurgy.* arc welding performed by an operator with the aid of a filler metal feeder.

semibituminous coal *Geology.* a coal that ranks between bituminous coal and semianthracite; it is harder and more brittle than bituminous coal, has a high fuel ratio, and burns without smoke. Also, SMOKELESS COAL.

semibolson *Geology.* a broad desert basin or valley that has a poorly developed or absent central playa and is drained by an intermittent stream flowing through canyons at each end to a surface outlet.

semibright coal *Geology.* a variety of banded coal containing 61–80% bright constituents, such as vitrain, clarain, and fusain, with clarodurain and durain composing the remainder.

semicarbazide *Organic Chemistry.* $H_2NCONHNH_2$, a crystalline solid; soluble in water and alcohol; melts at 96°C; used as a reagent. Also, AMINOUREA.

semicarbazide hydrochloride *Organic Chemistry.* $H_2NCONHNH_2 \cdot HCl$, toxic, white crystals; soluble in water and insoluble in absolute alcohol and ether; decompose at a melting point of 175°C; used as an analytical reagent. Also, CARBAMYLHYDRAZINE HYDROCHLORIDE.

semicarbazide-3-thio see THIOSEMICARBAZIDE.

semicarbazone *Organic Chemistry.* a condensation product of an aldehyde or ketone with semicarbazide, having the general formula $R_2C{=}NNHCONH_2$.

semicentrifugal casting *Materials Science.* a method of pressure-casting metal, in which a mold assembly is rapidly rotated, increasing the gravitational force for pulling molten metal into the mold; the centrifugal force distributes the metal uniformly, resulting in a dense casting.

semichemical pulp *Materials.* pulp produced from raw material by chemical and mechanical means.

semichemical pulping *Chemical Engineering.* a paper industry pulping process that uses a mild chemical treatment to partially remove lignin and soften the wood, then completes the fiber separation mechanically.

semicircle *Mathematics.* **1.** half of a circle; the arc from one end of a diameter to the other. Also, **semicircumference. 2.** anything that has or is arranged in the form of a half circle.

semicircular *Mathematics.* having the form of a half circle.

semicircular canal *Anatomy.* one of three fluid-filled canals forming part of the inner ear, which are active in detecting stimuli relating to equilibrium and movement.

semicircular deviation *Navigation.* a magnetic compass error in which the changes of deviation between east and west occur approximately 180° apart; it is a combination of a transient error caused by the vertical iron and a constant error caused by the ship's subpermanent magnetism.

semiclosed-cycle gas turbine *Mechanical Engineering.* a heat engine in which only a portion of the expanded gas is recirculated.

semicoma *Neurology.* a mild state of stupor in which reflexes can be roused through painful stimuli.

semiconducting compound *Solid-State Physics.* a semiconductor substance that is composed of two or more elements that are chemically combined.

semiconducting crystal *Solid-State Physics.* a crystalline sample of semiconductor material such as germanium or silicon.

semiconductive loading tube *Engineering.* a tube that prevents the premature detonation of static electric charges, used to load blasthole explosives.

semiconductor *Materials Science.* a crystalline material having intermediate values of electrical resistivity (in the approximate range of 10^{-2} to 10^9 ohm-cm, between the values for metals and insulators); the resistivity usually strongly decreases with increasing temperature. Semiconductors are the basic material of various electronic devices used in telecommunications, computer technology, control systems, and other applications.

semiconductor detector *Nucleonics.* a device capable of detecting ionization within the depletion layer of a reverse-biased p-n junction in a semiconductor.

semiconductor device *Electronics.* any of a wide variety of devices that employ the electrical properties of semiconductor materials, such as silicon, to control the flow of electrons; such devices include diodes, photocells, and transistors.

semiconductor diode *Electronics.* a two-terminal device incorporating a semiconductor.

semiconductor-diode parametric amplifier *Electronics.* a parametric amplifier that employs a semiconductor diode as the parametric (varying reactance) device.

semiconductor disk *Computer Technology.* a large semiconductor memory that emulates a disk drive in which action occurs more quickly, thereby upgrading performance. Also, NONROTATING DISK.

semiconductor grade silicon *Materials.* a very pure form of silicon used in a photovoltaic cell of a solar energy system to convert light energy to electrical energy.

semiconductor heterostructure *Electronics.* a design that joins two different semiconductors to achieve electrical characteristics not found in the individual conductors; used primarily in lasers and solar cells.

semiconductor intrinsic properties *Solid-State Physics.* the properties related to an ideal crystal of semiconductor material, free of impurities and defects.

semiconductor junction *Electronics.* the transitional region in a semiconductor that lies between materials with different electrical properties.

semiconductor laser *Optics.* a laser in which a semiconductor lasing element, such as gallium arsenide, emits a coherent beam of light. Also, DIODE LASER, LASER DIODE.

semiconductor memory *Computer Technology.* a type of memory that uses solid-state electronic components on integrated circuit chips as memory cells. Also, **semiconductor storage.**

semiconductor thermocouple *Electronics.* a thermocouple made up of a semiconductor device that can operate at extremely high temperatures.

semiconscious *Medicine.* only partially aware of one's surroundings.

semiconservative replication *Genetics.* replication that occurs when the strands of the parental DNA duplex are separated; each then acts as a template for synthesis of a complementary strand. *Molecular Biology.* a reproduction of a double-strand nucleic acid such that each daughter duplex contains one parental strand and one new strand.

semicontinuous function *Mathematics.* if for any arbitrary positive number ε a real-valued function f satisfies the relation $f(x) < f(x_0)+\varepsilon$ for all x in some neighborhood of x_0, then f is **upper semicontinuous** at x_0; if $f(x) > f(x_0)-\varepsilon$ for all x in some neighborhood of x_0, then f is **lower semicontinuous.**

semicontrolled mosaic *Cartography.* a photomosaic in which the photographs are all roughly the same scale, but have been oriented to a coordinate system other than horizontal ground control.

semiconvection *Fluid Mechanics.* a partial convective mixing that causes a region to become convectively stable before complete mixing is achieved.

semicratonic see QUASICRATONIC.

semicrystalline *Science.* partly or imperfectly crystalline.

semicrystalline polymer *Materials Science.* a solid polymer containing both crystalline and amorphous components.

semicubical parabola *Mathematics.* the graph in the plane of the equation $y^2 = x^3$.

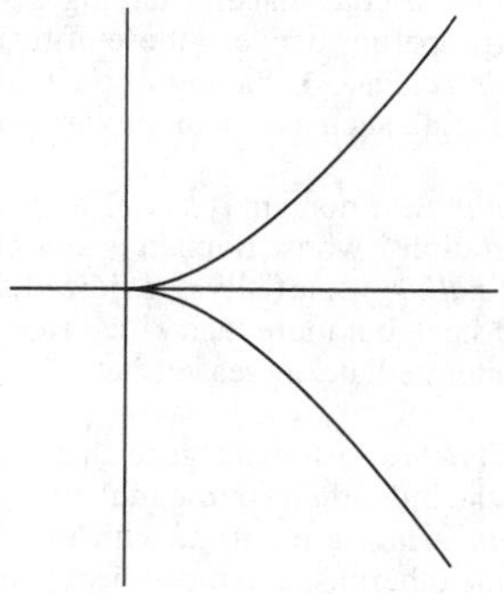

semicubical parabola

semidense list *Computer Programming.* a list that can be divided into two sections, one completely full and the other completely empty.

semidesert *Ecology.* a region that lies between a grassland or woodland and a true desert, composed primarily of shallow-rooted plants that can exploit surface precipitation before it evaporates.

semidetached binary *Astronomy.* a binary star system, such as Algol, whose secondary fills its Roche lobe but whose primary does not.

semidiameter *Mathematics.* one-half of a diameter; a radius. *Astronomy.* one-half the angle subtended by the disk of a celestial object.

semidiameter correction *Navigation.* a correction applied in using a sextant to calculate the altitude of the sun or moon; the edge of the celestial body is normally brought to the horizon, and the semidiameter must be added in order to obtain the true altitude, measured from the center of the body.

semidiesel engine *Mechanical Engineering.* **1.** a diesel engine that does not use compressed air to inject fuel. **2.** an internal-combustion engine whose fuel is ignited by spraying it onto a heated surface or by igniting a portion of the charge in an uncooled portion of the combustion chamber.

semidiscontinuous replication *Molecular Biology.* a type of DNA replication in which one new strand is synthesized continuously, and the other, discontinuously.

semidiurnal *Astronomy.* **1.** a period consisting of one-half of a day or 12 hours. **2.** relating to or occurring in such a time period.

semidiurnal current *Oceanography.* a tidal current that has two flood periods and two ebb periods during a tidal day.

semidiurnal tide *Oceanography.* a tide that consists of two low waters and two high waters during a tidal day.

semidomesticated *Agriculture.* living in association with humans but in a half-wild state; e.g., ducks or other water birds on a farm pond.

semidominance see PARTIAL DOMINANCE.

semidormancy *Botany.* a decrease in plant growth rate that may be seasonal or associated with unfavorable environmental conditions.

semidouble *Botany.* of a flower that has more than the normal number of petals but that retains some pollen-bearing stamens or perfect disk florets.

semidull coal *Geology.* a variety of banded coal composed primarily of carudurain and durain, with 21–40% bright constituents such as vitrain, clarain, and fusain.

semielliptical arch see BASKET-HANDLE ARCH.

semievergreen *Botany.* describing a plant that retains green, unwithered leaves during part of a winter or through a mild winter.

semifinishing *Metallurgy.* in metal working, any operation performed prior to finishing.

semifixed ammunition *Ordnance.* ammunition in which the cartridge case is not permanently attached to the projectile; the charge can be changed to obtain different ranges, but the round is loaded as a single unit.

semiflexion *Anatomy.* a posture in which a limb or joint is halfway between extension and complete flexion.

semifloating axle *Mechanical Engineering.* an automobile axle that carries wheel loads and torques at its outer end.

semiforbidden line *Spectroscopy.* a spectral line that corresponds to a semiforbidden transition.

semiforbidden transition *Atomic Physics.* an atomic transition for which the probability is reduced because of selection rules by a factor of the order of 10^6.

semifusinite *Geology.* a coal maceral having a well-defined woody structure and optical properties between those of fusinite and vitrinite.

semigloss *Materials Science.* **1.** having a moderately satiny luster. **2.** of or relating to a material, such as a paint or varnish, that produces such a luster.

semigroup *Mathematics.* a nonempty set G that is closed under a binary, associative operation * whose domain is all of G.

semikilled steel *Metallurgy.* partially deoxidized steel, containing less oxygen than rimmed steel, but more than killed steel.

semilate *Botany.* intermediate in season between midseason and late forms.

semilethal gene *Genetics.* a mutant gene that causes the death of at least half but not all the individuals of the mutant genotype.

semilive skid *Engineering.* a platform with two fixed wheels at one end and two legs at the other; used to move heavy materials.

Sémillon [se mē yōn´] *Botany.* a green grape of Bordeaux used in making dry white Bordeaux as well as Sauterne. *Food Technology.* usually, **Sémillon Blanc.** a white wine made from this grape.

semilogarithmic *Mathematics.* having one scale logarithmic and the other arithmetic or of uniform gradation.

semilogarithmic coordinate paper *Mathematics.* paper ruled with two sets of parallel lines: one set spaced according to logarithms of consecutive numbers, and the other set evenly spaced and perpendicular to the first.

semilunar *Science.* resembling a half-moon.

semilunar cartilage *Anatomy.* one of the half-moon-shaped cartilages found in the knee joint.

semilunar ganglion see GASSERIAN GANGLION.

semilunar valve *Anatomy.* one of three half-moon-shaped flaps of tissue that together form the valve preventing backward flow of blood between the right ventricle and the pulmonary trunk; a similar valve separates the left ventricle and the aorta.

semimagnetic controller *Electricity.* an electric controller having only a subset of its basic functions performed by electromagnetic devices.

semimajor axis *Astronomy.* half the major axis of the elliptical orbit that one celestial body describes around another.

semimat *Materials Science.* describing a finish that is between matte and glossy.

semimember *Civil Engineering.* a zero force member in a truss or frame, when the stresses in the member reverse.

semimembranosus *Anatomy.* one of the hamstring muscles on the medial posterior thigh that extends the thigh and flexes the leg.

semimetric *Mathematics.* given a set X, a symmetric, nonnegative real-valued function d defined on $X \times X$ that satisfies the triangle inequality. If, in addition, $d(x, y) = 0$ if and only if $x = y$, then d is a **metric**. Also, **semi-metric.**

semimicroanalysis *Chemistry.* an analysis in which the sample weighs between 10 and 100 milligrams.

semiminor axis *Astronomy.* half the minor axis of the elliptical orbit that one celestial body describes around another.

semimobile artillery *Ordnance.* artillery weapons that are mounted for mobility but must be partially disassembled before firing.

semimonocoque *Aviation.* a reinforced monocoque; that is, a three-dimensional form, such as a fuselage, in which longitudinal and circumferential members support the skin and help to carry the burden of stress; found on most modern aircraft. Also, STIFFENED-SHELL FUSELAGE.

seminal *Physiology.* of or relating to semen or the secretion of the male reproductive organs. *Botany.* relating to a seed.

seminal bursa *Invertebrate Zoology.* in free-living flatworms (turbellarians), a sperm storage center connected to the ovaries by an oviduct.

seminal fluid *Physiology.* the albuminous fluid containing spermatozoa that is secreted by the male reproductive organs.

seminal groove *Zoology.* either of a pair of grooves on the ventral surface of an earthworm that transport sperm from the male apertures to the clitellum region.

seminal vesicle *Anatomy.* one of two sacculated, glandular structures in the male, connected to the ductus deferens, that hold and secrete seminal fluid.

seminiferous *Anatomy.* conveying or containing semen. *Botany.* bearing or producing seed.

seminiferous tubule *Anatomy.* one of many convoluted tubules in the testis that are the sites of sperm production.

seminivorous *Zoology.* feeding mainly or exclusively on seeds.

seminoma *Oncology.* a radiosensitive, malignant neoplasm of the testis believed to be derived from primordial germ cells of the sexually undifferentiated embryonic gonad. Three variants are recognized: classical, anaplastic, and spermatocytic.

seminorm *Mathematics.* a nonnegative real-valued function (denoted by $\| \ \|$) defined by a vector (linear) space X that satisfies the triangle inequality and the scale property. If, in addition, $\|x\| = 0$ if and only if x is the zero vector, then $\| \ \|$ is a norm. Also, **semi-norm.**

semiochemical *Physiology.* describing a class of secretory substances, notably pheromones in insects, that regulate behavior in members of the same species.

Semionotiformes *Vertebrate Zoology.* the gars, a monofamilial freshwater fish order of North and Central America and Cuba, characterized by an elongate head and body and ganoid scales.

semiotics *Linguistics.* the study of communication through signs and signification.

semioviparous *Zoology.* bearing young in an incomplete state of development, as some marsupials.

semioxamazide *Organic Chemistry.* $H_2NCOCONHNH_2$, a crystalline solid; soluble in hot water, acids, and alkalies; decomposes at 220°C; used as a reagent.

semipalmate *Vertebrate Zoology.* having partially webbed anterior toes, as in some shorebirds.

semiparasite *Ecology.* a parasite that is independent of its host to some degree, as by being free living for part of its life cycle or deriving part of its nourishment from other sources. Also, PARTIAL PARASITE.

semipermanent mold *Metallurgy.* a permanent mold that contains a sand core.

semipermeable *Science.* permitting the passage of only certain molecules. Thus, **semipermeable membrane.**

semipermissive cells *Virology.* a cluster of cells of which only a portion allow lytic infection by a given virus.

semipersistent transmission *Virology.* a state of plant virus transmission that falls between nonpersistent and persistent transmission, in which there is no latent period and the vector is able to transmit the virus for a period of several days, as distinguished from minutes or hours in nonpersistent transmission.

semiplastic *Science.* in a state between rigidity and plasticity.

semiplastic explosive *Materials.* an explosive in which the mixture of liquid products is not compressible.

semipolar bond *Materials Science.* a bond in which both shared electrons come from one atom, giving the bond properties between those of ionic and covalent bonds. Also, COORDINATE BOND.

semipositive mold *Engineering.* a plastics mold that lets a small amount of excess plastic escape when it is shut.

semiprocessed steel *Materials.* a highly permeable, economically beneficial steel that contains primarily manganese and smaller amounts of silicon and aluminum.

semiquinone *Chemistry.* any of the class of free radicals formed as intermediates in the oxidation of a hydroquinone to a quinone.

semiquinone

semiregular variable star *Astronomy.* a variable star that has roughly periodic light variations but is inconsistent in its period and the shape of the light curve from one cycle to the next.

semiring *Mathematics.* a nonempty class P of sets that is closed under intersection and that has the following property: if E and F are sets in P such that E is contained in F, then there exists a finite sequence $\{C_0, C_1, \ldots, C_n\}$ of nested sets in P such that

$$E = C_0 \subset C_1 \subset \cdots \subset C_n = F$$

and $C_i - C_{i-1}$ is also in P. For example, the class P of all real intervals of the form $[a,b]$ (i.e., left-closed and right-open) forms a semiring.

semischist *Petrology.* a partially metamorphosed, somewhat foliated, sedimentary rock.

semiselective ringing *Telecommunications.* in telephony, a party-line service in which the bells of two of as many as eight stations are rung at the same time.

semisilica brick *Materials.* a firebrick made of silica, alumina, titania, iron oxide, magnesia, lime, and alkalies; can be used in temperatures up to 2700°F and has a high-load and carrying capacity.

semisimple *Mathematics.* **1.** a ring R is said to be semisimple if its Jacobson radical (the intersection of all maximal ideals of R) is zero. Also, **Jacobson semisimple**. **2.** a nonzero module A over a ring R is said to be semisimple if it satisfies one of the following equivalent conditions: (a) A is the sum of a family of simple submodules; (b) A is the (internal) direct sum of a family of simple submodules; or (c) for every nonzero element $a \in A$, $Ra \neq 0$, and every submodule B of A is a direct summand of A. A semisimple module is sometimes called completely reducible.

semisoft cheese *Food Technology.* any cheese containing between 40 and 50% moisture by volume, such as Limburger or Roquefort.

semispecies *Systematics.* any of the component species of a superspecies.

semispinalis *Anatomy.* a group of muscles that extend from the transverse processes of certain vertebrae to the spines of other vertebrae, helping to maintain upright posture and to twist the spine from side to side.

semisplint coal *Geology.* a variety of banded coal whose composition and character fall between those of bright-banded coal and splint coal, with 20–30% opaque attritus and more than 5% anthraxylon.

semisteel *Metallurgy.* steel made in a cupola furnace by adding steel scrap to pig iron.

semisubmersible rig *Petroleum Engineering.* a self-propelled barge mounted on partially submerged legs supported on underwater pontoons that serves as a work base in deep offshore drilling.

semitendinosus *Anatomy.* one of the hamstring muscles on the medial posterior thigh that extends the thigh and flexes the leg.

semiterrestrial *Biology.* not growing or living entirely on land, such as plants that grow on boggy ground.

semitone *Acoustics.* a half-step separation between two subsequent tones in a Pythagorean (tempered) musical scale, with a frequency interval ratio of 256/243 rather than the 9/8 ratio of a whole step. Also, HALF TONE, HALF STEP.

semitrailer *Engineering.* a hauling vehicle with two axles at the rear.

semitransparent photocathode *Electronics.* an electrode that generates electricity on one side when its opposite side is struck by light.

Semliki Forest virus *Virology.* an arbovirus of the genus *Alphavirus* of the family Togaviridae that is found in African mosquitoes and may be pathogenic in humans.

Semmelweis, Ignaz 1818–1865, Hungarian physician; discovered the cause of puerperal fever, anticipating Lister's work in antiseptic surgery.

semolina *Food Technology.* a granular, starchy product obtained from the endosperm of durum wheat; used to make pasta.

Semper's cells *Entomology.* four glycogen-containing cells that form the crystalline cone of each unit of a compound eye. Also, CONE CELLS.

Semper's larva *Invertebrate Zoology.* a cylindrical larva in the life history of certain zoanthid corals, characterized by a hole at each end and an annular or longitudinal band of long cilia.

sempervium *Botany.* any crassulaceous plant of the genus *Sempervivum* of the stonecrop family, characterized by leaves in dense basal rosettes and a compact, flat-topped cluster of flowers; includes the houseleek, *S. tectorum*, and other species.

semple plunger *Ordnance.* a centrifugal plunger designed to keep a fuse in a safe position until centrifugal force unlocks the firing pin.

semple vaccine *Immunology.* a vaccine consisting of phenol-inactivated rabies virus that has passed through a rabbit brain; used to treat and produce immunity against rabies.

sems *Mechanical Devices.* a fastener assembly consisting of a screw and an attached washer.

semseyite *Mineralogy.* $Pb_9Sb_8S_{21}$, a gray to black monoclinic mineral with a metallic luster occurring as aggregates of tabular or prismatic crystals, having a specific gravity of 6.03 to 6.08 and a hardness of 2.5 on the Mohs scale; found in hydrothermal veins with pyrite, stibnite, and other sulfides.

semustine *Oncology.* an antineoplastic agent of the nitrosourea group that is used in the treatment of brain tumors, Hodgkin's disease, colon and gastric cancers, and melanoma.

sen *Materials.* the light, soft wood of the east Asian trees *Acanthopanax ricinifolius* or *Kalopanax pictus;* used for interior trim, cabinets, doors, musical instruments, furniture, sporting goods, and veneers.

senaite *Mineralogy.* $Pb(Ti,Fe,Mn)_{21}O_{38}$, a black trigonal mineral of the crichtonite group occurring as rough crystals and rounded fragments, having a specific gravity of 5.301 and a hardness of 6 on the Mohs scale; found in diamond-bearing placer deposits near Diamantina, Brazil.

senarmontite *Mineralogy.* Sb_2O_3, a colorless, white, or gray cubic mineral dimorphous with valentinite occurring in massive form and as as octahedral crystals, having a specific gravity of 5.50 and a hardness of 2 to 2.5 on the Mohs scale; used as a paint pigment and in flame-proofing.

Sendai virus *Virology.* a virus of the genus *Paramyxovirus* of the family Paramyxoviridae that causes natural infection in mice; commonly used to induce cell fusion.

sender *Telecommunications.* the equipment in a signaling terminal that receives pulses from a dial and then controls all the other operations that are needed to establish a proper telephone connection.

sending-end impedance *Electricity.* the input impedance of a transmission line.

S_1 endonuclease see ENDONUCLEASE S_1.

Sendust *Materials.* a magnetic alloy composed primarily of iron with some silicon and aluminum.

Seneca, Lucius c. 3 BC–65 AD, Roman author and statesman; wrote *Natural Questions*, on geography, astronomy, and other sciences.

Senecan *Geology.* a North American provincial series of the lower Upper Devonian period, occurring after the Erian and before the Chatauquan.

seneciosis *Toxicology.* poisoning due to ingestion of plants of the genus *Senecio,* characterized by liver damage. Also, **senecio disease.**

senescence *Biology.* **1.** the aging process. **2.** the study of the biological changes related to aging. *Geology.* in the erosion cycle or developmental sequence of a landform, the stage at which old age begins.

senescent *Biology.* growing old; aging.

senescent lake *Hydrology.* a lake that is nearing extinction.

senesland *Geology.* a land surface at a stage of development between a matureland and a peneplain.

sengierite *Mineralogy.* $Cu_2(UO_2)_2V_2O_8 \cdot 6H_2O$, a very rare, strongly radioactive, yellow-green monoclinic mineral, having a specific gravity of 4.05 and a hardness of 2.5 on the Mohs scale; found as platy crystals with other secondary copper, iron, cobalt, and nickel minerals.

sengkuang *Materials.* the gray-yellow to gray-brown nondurable wood of the trees of the *Dracontomelum* species of the Far East; used for furniture, plywood, veneers, and cabinets. Also, DAO, LAMIO.

senile [sē´nil] *Medicine.* **1.** of or relating to old age. **2.** manifesting the physical and mental deterioration commonly associated with old age, especially memory loss and other intellectual impairment. *Geology.* describing a stage in the cycle of erosion in which erosion of a land surface has reached a minimum and base level has been approached. (From a Latin word meaning "old man.")

senile dementia *Psychology.* a psychological disorder resulting from progressive, irreversible degeneration of the brain, usually starting after age 65, and characterized by impairment of intellectual ability, memory, and judgment, and radical changes in personality and behavior, such as apathy, withdrawal, and irritability.

senile eczema *Medicine.* eczema caused by old age.

senile emphysema *Medicine.* emphysema caused by old age.

senile gangrene *Medicine.* gangrene caused by old age.

senile paraplegia see SPASTIC PARAPLEGIA.

senile psychosis *Psychology.* any of a group of chronic disorders of the aged, resulting primarily from degeneration of the brain and characterized by gradual, continual deterioration of mental functions until death.

senile stage *Geology.* a stage in the cycle of erosion in which the erosion of a land surface has reached a minimum and base level has been approached.

senile vaginitis *Medicine.* vaginitis caused by old age.

senility [sə nil´i tē] *Medicine.* the feebleness of body and mind commonly attributed to the effects of old age; not a specific condition but a general term describing a variety of conditions, such as Alzheimer's disease. *Geology.* in the cycle of erosion of a land surface, the stage in which erosion has reached a minimum and base level has been approached.

senna *Pharmacology.* the dried leaflets of certain senna plants, used to promote evacuation of the bowels.

Senonian *Geology.* a European geologic stage of the Upper Cretaceous period, occurring after the Turonian and before the Danian; it includes the Coniacian, Santonian, Campanian, and Maastrichtian.

sensate *Science.* perceived by the senses.

sensate focus *Psychology.* **1.** a focus on particular sensual experiences. **2.** a means of treating sexual dysfunction by developing the individual's ability to experience sensual pleasure through touching.

sensation *Physiology.* an awareness of a stimulus as a result of its perception by sensory receptors.

sensation-seeking *Psychology.* the degree to which individuals pursue and enjoy participating in activities that produce high levels of sensation.

sensation unit *Acoustics.* a unit based on a signal level expressed in decibels, computed by the equation $dB = 20 \log I/I_0$, where I is the measured sound intensity, and I_0 is the reference sound; used for units such as the phon.

sense *Biology.* **1.** any of the body's mechanisms that receive and recognize stimuli, such as sight, hearing, touch, taste, and smell. **2.** to detect a stimulus by means of a sense. *Navigation.* **1.** the direction from which a radio signal arrives. **2.** to detect such a signal. *Computer Technology.* **1.** to test the state of some part of hardware either manually or during the operation of the device. **2.** to study data relative to a set of criteria. **3.** to read physical data. *Engineering.* to determine the layout of a device, the value of a quantity, or the location of a problem in an apparatus or machine.

sense amplifier *Electronics.* **1.** a circuit that detects low-level voltage, which it then increases to the level required by the system. **2.** a circuit that detects and adjusts for changes in a radio signal's phase or voltage.

sense antenna *Electromagnetism.* an auxiliary antenna used in conjunction with a directional receiving antenna to resolve the 180° ambiguity in the receiving direction. Also, **sensing antenna.**

sense codon *Genetics.* any one of the 61 mRNA codons that designate an amino acid.

sense datum *Psychology.* a stimulus or object of sensation.

sense finder *Navigation.* in a direction finder, the component that determines which of two possible directions is the source of a radio signal; this determination, called **sensing the bearing,** is necessary because a simple loop antenna cannot distinguish between two directions 180° apart.

sense light *Computer Technology.* a light on a computer console that may be turned on or off and that may be interrogated by a program. Similarly, **sense switch.**

sense organ *Physiology.* any organ receiving stimuli that give rise to sensations, translating certain forms of energy into nerve impulses which are perceived as special sensations; includes the visual, olfactory, vestibulocochlear, and olfactory organs, each characterized by highly specialized neuroreceptors. Also, **sensory organ.**

sense strand see CODING STRAND.

sensibility *Psychology.* a responsiveness or susceptibility to sensory stimuli. *Physics.* **1.** the realignment of a magnetic compass pointer along a magnetic field line after the pointer has been deflected. **2.** in weighing an object on a balance beam, a measure of the ability to reproduce the same reading on the pointer scale.

sensible atmosphere *Meteorology.* the part of the atmosphere that offers resistance to a body passing through it.

sensible heat *Thermodynamics.* **1.** the thermal energy that is absorbed by a substance during a change in temperature that occurs without a change of state. **2.** see ENTHALPY.

sensible-heat factor *Thermodynamics.* a quantity that is used in calculations for air conditioning, given by the ratio of the space's sensible thermal energy to the space's total thermal energy.

sensible-heat flow *Meteorology.* the poleward transport of sensible heat in the atmosphere across a given latitude belt by fluid flow.

sensible horizon *Astronomy.* the great circle of the celestial sphere whose poles are the nadir and zenith. Also, **astronomical horizon.**

sensible temperature *Meteorology.* the temperature at which average indoor air of moderate humidity would cause the same sensation of comfort in a lightly clothed person as that caused by the actual environment.

sensillum *plural,* **sensillae.** *Zoology.* a small, simple sense organ or sensory receptor. *Invertebrate Zoology.* in insects, a sensory cell or group of cells with chitinous supporting structures, often on the antennae; may be in the form of a hair, plate, or pit.

sensing array *Robotics.* a row of detectors set on a common base.

sensing element see SENSOR.

sensing matrix camera *Robotics.* a camera based on a densely packed, two-dimensional matrix of visual sensors.

sensing threshold *Robotics.* the lowest limit of force and moment that a sensor is designed to detect.

sensistor *Electronics.* a device whose resistance to current is determined by temperature, power, and time.

sensitive *Neurology.* **1.** responding to a stimulus. **2.** possessing acute responsiveness to stimuli, especially keen or exaggerated perception. *Physiology.* having a low tolerance for a specific physical or psychic stimulus, as in certain allergic responses. *Military Science.* of or relating to an installation, individual, document, or activity that requires special protection from unauthorized disclosure.

sensitive altimeter *Engineering.* an altimeter that responds to pressure changes with a high degree of sensitivity.

sensitive clay *Geology.* a clay whose shear strength is lessened to a very small fraction of its former value when it is remolded at constant moisture content.

sensitive fern *Botany.* a common fern, *Onoclea sensibilis,* characterized by spore-containing beadlike spikes and large, triangular leaves; the segments of the fronds tend to droop and fold together when detached. (Named for the sensitivity of the foliage to frost.)

sensitive period or **sensitive phase** *Behavior.* a phase of development during which an animal or an individual is especially receptive to experiences that affect learning. Also, CRITICAL PERIOD.

sensitive time *Nucleonics.* the time interval sufficient for the formation of a track produced by an energetic particle in a cloud chamber.

sensitive volume *Nucleonics.* the region in a radiation counter tube in which a specific ionizing event would be registered by the device.

sensitivity *Science.* **1.** the ability of a device or organism to respond to external stimuli. **2.** the degree to which a device or organism is affected by a particular stimulus. *Neurology.* the quality of being responsive to a stimulus, often referring to abnormally keen perception. *Electronics.* the ability of a device to respond to a low-level stimulus. *Engineering.* **1.** the ability of the output of a device or system to respond to an input stimulus as measured by speed or degree of responsiveness. **2.** a quality that expresses the ease at which a given material explodes. *Acoustical Engineering.* a measure of the ability of a radio tuner to pick up weak signals. *Acoustics.* a rating for a transducer, such as a microphone or loudspeaker, which relates output voltage to input sound, or vice versa, usually in terms of volts and micropascals. *Geology.* the effects of remolding on the consistency of a clay or a cohesive soil, regardless of the physical nature of the causes of such effects.

sensitivity agar *Microbiology.* any solid medium that is used to perform an antibiotic-sensitivity test with a given microorganism.

sensitivity analysis *Electrical Engineering.* an analysis to determine the variations of a given function when changes about a selected reference value are made in its parameters.

sensitivity time control *Electronics.* a circuit that increases a radar receiver's sensitivity after a pulse has been transmitted, so that reflections from distant objects are received and those of nearby objects are obscured.

sensitivity training *Behavior.* a technique in which members of a group are taught to observe and understand both their personal interactions and the group process as a whole.

sensitization *Behavior.* **1.** a procedure for eliminating a problem behavior by pairing a negative stimulus with the stimulus that elicits the behavior. **2.** the process by which a response to a given stimulus increases due to repeated exposure to the stimulus that evoked the response. *Metallurgy.* precipitation of carbides in an austenitic stainless steel, causing localized corrosion in certain liquid environments.

sensitized cell *Immunology.* **1.** a cell that produces an immune response in reaction to a specific antigen. **2.** a cell that has been coated with an antibody.

sensitometer *Engineering.* an instrument that measures the response of photographic material exposed to radiant energy, such as to light.

sensor *Engineering.* the component of an instrument that converts an input signal into a quantity that is measured by another part of the instrument and changed into a useful signal for an information-gathering system. Also, SENSING ELEMENT.

sensorial *Neurology.* of or relating to the sensorium.

sensorimotor *Psychology.* of or relating to motor activity that is caused by sensory stimuli.

sensorimotor stage see SENSORY MOTOR STAGE.

sensorineural *Medicine.* relating to or affecting a sensory nerve.

sensorineural deafness *Medicine.* deafness in which sound is conducted normally through the outer and middle ear, but is distorted by a defect of the inner ear.

sensorium *Neurology.* **1.** broadly, any sensory nerve center. **2.** specifically, the center of sensations in the brain, often referring to the part of the brain that receives and combines impressions transmitted to the individual sensory centers. Also, PERCEPTORIUM.

sensory *Physiology.* of or relating to sensation or the perception of a stimulus.

sensory adaptation *Behavior.* a decrease in sensitivity to stimuli due to prolonged stimulation. Also, NEGATIVE ADAPTATION.

sensory aphasia *Medicine.* the inability to understand the meaning of communication symbols due to dysfunction in the auditory and visual word centers of the nervous system.

sensory area *Physiology.* a region in the cerebral cortex concerned with receiving and interpreting sensations.

sensory cell *Physiology.* any nerve cell whose dendrite connects with a sensory receptor.

sensory control *Robotics.* the use of sensor readings to control or alter a robot's actions.

sensory-controlled robot *Robotics.* a robot that can use sensor data to modify its programmed instructions.

sensory deafness *Medicine.* a condition in which the ear functions normally to detect sound, and the subject hears the sound, but fails to understand the meaning of words due to a lesion in the brain's auditory center. Also, MENTAL DEAFNESS, PSYCHIC DEAFNESS, SOUL DEAFNESS.

sensory deprivation *Behavior.* a reduction or alteration in the level of stimulation so that it no longer conforms to an organism's normal range or kind of stimulation.

sensory hierarchy *Robotics.* a relationship between sensors in which the readings of those on the lower level are used as input for those on the higher level.

sensory memory *Psychology.* the part of the human memory system that stores a representation of the sensory image by means of which information first enters the system, lasting just a few seconds after removal of the stimulus. Also, **sensory register.**

sensory motor stage or **period** *Psychology.* the first of Jean Piaget's four stages of cognitive development, from birth to about age two, when the child first develops a sense of self and understands objects and their manipulation, but cannot think abstractly.

sensory nerve *Physiology.* an outer axon or group of axons carrying stimuli from the sensoria to the central nervous system.

sensory nerve cells *Physiology.* cells that convey nerve impulses from the periphery to the central nervous system.

sensory pits *Invertebrate Zoology.* in insects, a depression in the cuticle containing sensory cells, probably olfactory.

sensory preconditioning *Behavior.* a procedure in which two conditioned stimuli are paired during preconditioning sessions; one conditioned stimulus is then paired with the unconditioned stimulus during the conditioning stage of the experiment, after which the other conditioned stimulus is tested for conditioning.

sensory setae *Entomology.* tactile, auditory, or chemical receptor hairs with nerve cells leading from their bases, arising from an insect's epidermis and cuticle.

sentence *Linguistics.* a word or group of words that is grammatically complete and that forms a statement, command, question, request, etc. *Computer Programming.* **1.** in COBOL, a sequence of statements that are terminated by a period and a space. **2.** a complete statement in a language, such as a programming language. **3.** a string generated by a formal grammar.

sentence symbol *Computer Science.* a distinguished nonterminal symbol in a formal grammar that represents a complete statement (sentence) in the language.

sentential connectives *Mathematics.* in the theory of statement calculus, the logic operators NOT, AND, OR, if...then, and if and only if.

sentinel *Computer Programming.* see FLAG, def. 1.

sepal *Botany.* one of the modified leaves that forms the calyx of a flower.

sepaloid *Botany.* of or relating to a sepal.

separable extension *Mathematics.* an extension field F of a field K is a separable extension if every element of F is algebraic over K.

separable graph *Mathematics.* a graph that is not connected or has at least one cut vertex; equivalently, a graph of connectivity at most 1. A connected graph with no cut vertices is a **nonseparable graph,** or a **block.**

separable polynomial *Mathematics.* a polynomial f of degree n is said to be separable if it has n distinct roots. In particular, suppose f is a polynomial over a field K but is irreducible in K. Then f is separable if, in some splitting field of f over K, every root of f is a simple root.

separable space *Mathematics.* a space that has a countable dense subset.

separated ammunition *Ordnance.* a type of ammunition that is used when the ammunition is too large to handle in fixed form; a closed cartridge case containing the primer and propelling charge is loaded with the projectile in a single operation.

separated-function synchrotron *Nucleonics.* a proton synchrotron having separate magnet groups that serve to focus the beam into a circular path.

separated sets *Mathematics.* two subsets of a topological space with the property that the intersection of either with the closure of the other is empty.

separate-loading ammunition *Ordnance.* ammunition in which the projectile and charge are loaded separately; such ammunition has no cartridge case.

separately excited generator *Electricity.* a generator whose field current is supplied from outside the generator.

separating calorimeter *Physics.* a device that measures the amount of moisture in a sample of steam.

separating family *Mathematics.* a family of seminorms P on a vector space X is said to be separating if for each nonzero vector x in X, there exists at least one seminorm in P that is nonzero at x.

separating power *Chemical Engineering.* the ability of a process or system to achieve a certain separation.

separating set *Mathematics.* a vertex cut.

separation *Engineering*. the segregation of solid particles by screening or phases, such as gas-liquid. *Chemical Engineering*. an operation that separates the feed into its components. *Mining Engineering*. any treatment to separate and remove gangue from valued material. *Acoustical Engineering*. the separation, expressed in decibels, between channels in a multichannel system, especially in a stereo tuner. *Geology*. the distance between any two sections of a plane that has been divided by a fault; the apparent relative displacement of the two surfaces of a fault, measured in any direction. Also, APPARENT RELATIVE MOVEMENT. *Graphic Arts*. see COLOR SEPARATION. *Aviation*. **1.** a designated lateral, longitudinal, or vertical distance between aircraft under positive control. **2.** a collapse of the attached fluid flow around a body into gross turbulence that occurs at a specific time (as in a stall) or place (a **separation point**). Equilibrium can be sustained with attached (laminar or turbulent) flow upstream and complete separation downstream. *Space Technology*. **1.** the movement of a rocket stage or other fallaway section away from the primary body of a vehicle. **2.** the moment when this occurs. *Transportation Engineering*. **1.** (highways) a physical separation of traffic streams, either by a median divider or (for intersecting routes) a bridge or other grade separation. **2.** (aviation) the separation of aircraft by horizontal position and altitude.

separation anxiety *Psychology*. excessive concern, generally among young children, when separated from a parent, usually the mother, or from the home environment.

separation axioms *Mathematics*. five conditions on a topological space X. (a) T_0: Given distinct points x and y, there exists an open set that contains exactly one of x and y. (b) T_1: Given distinct points x and y, there exists an open set that contains y but not x. (c) T_2: Given distinct points x and y, there exist disjoint open sets O_1 and O_2 such that $x \in O_1$ and $y \in O_2$. (d) T_3: In addition to T_1, given a closed set F and a point x not in F, there exist disjoint open sets O_1 and O_2 such that $x \in O_1$ and F is contained in O_2. (e) T_4: In addition to T_1, given disjoint closed sets F_1 and F_2, there exist disjoint open sets O_1 and O_2 such that F_1 is contained in O_1 and F_2 is contained in O_2. All five separation axioms are satisfied in a metric space. A topological space satisfies T_1 if and only if every set consisting of a single point is closed.

separation disc *Botany*. a layer of gelatinous material between two adjacent negative cells in some blue-green algae, associated with hormogonium formation.

separation energy *Nuclear Physics*. the amount of energy required to dislodge a particle, such as a proton or a neutron, from a nucleus.

separation factor *Nucleonics*. a quantity that gives a measure of the effectiveness of an isotope separation process, given by the quotient of the abundance ratio after processing divided by the abundance ratio before processing.

separation filter *Electronics*. a filter circuit with two outputs that separates one band of frequencies from another; for example, to transmit carrier and voice frequencies over different communication paths.

separation layer *Botany*. a distinct layer of specialized cells containing abundant starch and dense cytoplasm, located at the base of a leaf, flower, fruit, or other part of a plant; the cells in such a layer are smaller and different in shape from those above and below it; when they disintegrate, they cause that part of the plant to separate.

separation negative *Graphic Arts*. any of a set of three or four negatives produced in color separation, including one each for yellow, magenta, cyan, and usually black.

separation of variables *Mathematics*. **1.** a technique for solving partial differential equations by selecting (a) a coordinate system, and (b) a fixed functional form for the solution comprising unknown functions of each of the coordinate functions separately (usually a product or sum). The partial differential equation is re-expressed in the new coordinates and the presumed functional form inserted into the equation. If the partial differential equation then reduces to a set of ordinary differential equations, the equation is said to be solvable by separation of variables. **2.** a technique for solving an ordinary differential equation by initially using algebra to convert the equation into the form $f(y)dy = g(x)dx$ (treating the parts dx and dy of the derivative as algebraic objects) and then solving for y by integrating both sides of the equation and performing more algebra.

separation syndrome or **trauma** *Psychology*. see DEPRIVATION.

separation theorem *Control Systems*. in optimal control theory, a theorem stating that the solution to the linear quadratic Gaussian problem separates into an optimal controller for the corresponding problem without noise, in which the state used is obtained as the output of an optimal state estimator.

separative work unit *Nuclear Physics*. a measure of the work required to divide a quantity of isotopic mixture into two parts, with one having a greater fraction of the desired isotope.

separator *Computer Programming*. a special character, such as a semicolon, that marks the division between two logical units of data, such as fields or records. *Electricity*. a layer of material in a battery or capacitor that is used to prevent the battery's positive or negative poles, or capacitor plates, from coming into direct physical contact. *Electronics*. a circuit that removes one type of signal from another type. *Engineering*. **1.** a machine that divides materials with different specific gravities. **2.** a device that intercepts moisture and condensation before steam reaches the turbines in a power plant. *Mechanical Engineering*. **1.** a support device in a ball-bearing race that keeps an equal separation distance between the balls or rollers. Also, CAGE. **2.** any process vessel used to separate two or more phases of an operation. *Petroleum Engineering*. a pressure vessel used to separate the gaseous and liquid components of reservoir fluids into gas, oil, and water. Also, GAS SEPARATOR, GAS-OIL SEPARATOR, OIL-FIELD SEPARATOR, OIL-GAS SEPARATOR, OIL SEPARATOR.

separator-filter *Engineering*. a device that filters entrained liquids and solid particles from a fluid or gas stream by means of a baffle with a filter.

separator page *Computer Technology*. the page before or after a hardcopy output, providing information such as the time, date, and job.

separatory funnel *Chemistry*. a glass funnel container with a stopcock that allows venting and controlled draining of liquids; used in an extraction process to separate two immiscible liquids.

sepatrix *Control Systems*. in the phase plane of a control system, a curve that represents the solution to the equations of motion of the system that would cause the system to become unstable.

sepetir *Materials*. the brown oily wood of the trees of the *Sindora* species of the Far East; used for furniture, veneers, and cabinets. Also, MAKATA, GO.

sepia [sē´pē yə; sēp´yə] *Materials*. a brown pigment obtained from the secretion of cuttlefish; used as ink.

Sepioidea *Invertebrate Zoology*. the cuttlefish, an order of cephalopod mollusks with a squidlike body, well-developed eyes and brain, separated posterior fins, an interior shell, and ten tentacles around the head, the fourth pair is extra long and retractile; when disturbed, the fish jets backward and releases clouds of ink, from which the pigment sepia was once prepared.

sepiolite *Mineralogy*. $Mg_4Si_6O_{15}(OH)_2 \cdot 6H_2O$, a white to light-gray or light-yellow orthorhombic claylike mineral, having a specific gravity of about 2 and a hardness of 2 to 2.5 on the Mohs scale; found in fibrous or earthy masses as an alteration product of serpentine or magnesite; used in tobacco-pipe bowls and ornamental carvings. Also, MEERSCHAUM.

seps- or **sepso-** a combining form denoting infection.

Sepsidae *Invertebrate Zoology*. black scavenger flies and spiny-legged flies, a family of dipteran insects in the subsection Calyptratae that frequent manure piles.

sepsis *Medicine*. the presence in the blood or in other tissues of pathogenic microorganisms or their toxins; infection or contamination.

sept *Anthropology*. a clan believing themselves descended from a common ancestor.

septal *Anatomy*. relating to a septum.

septal filament *Invertebrate Zoology*. in the gut of anthozoan sea anemones, the free edges of the mesentery containing gland cells and nematocysts (stinging cells).

septal ostium *Invertebrate Zoology*. any opening in the septa of anthozoans.

septarian *Geology*. of or relating to a septarium, to the irregular polygonal pattern of cracks produced in a septarium, or to the mineral deposits that fill these cracks. Thus, **septarian boulder.**

septarium *plural*, **septaria.** *Geology*. a large, generally spheroidal concretion, usually made up of argillaceous carbonate, having irregular polyhedral blocks that are internally cemented together by crystalline minerals. Also, BEETLE STONE, SEPTARIAN NODULE, TURTLE STONE.

septate *Biology*. being divided or partitioned by crosswalls; having a septum.

septate coaxial cavity *Electromagnetism*. a coaxial cavity having a vane or septum placed between the inner and outer conductors so that it behaves like a rectangular cross-section cavity with a transverse bend.

septate junction *Cell Biology*. an invertebrate cell junction composed of a regular arrangement of proteins that bind the cells together.

septate waveguide *Electromagnetism*. a waveguide that has metal vanes (septa) that act to control microwave power transmission.

septectomy *Surgery.* the removal of all or part of a septum, especially the nasal septum.
septemia see SEPTICEMIA.
Septibranchia *Invertebrate Zoology.* an order of bivalve mollusks with equal valves, and gills reduced to muscular partitions of the gill chamber.
septic *Medicine.* produced by or due to decomposition by microorganisms; putrefactive.
septic abortion *Medicine.* termination of pregnancy in which there is infection of the uterus.
septic embolus *Medicine.* an embolus formed by bacteria.
septicemia [sep´ti sē´mē ə] *Medicine.* a toxic condition caused by both bacterial infection and the products of bacteria in the body; symptoms include chills, fever, and extreme exhaustion. Also, **septic infection.**
septicemic plague *Pathology.* a virulent form of plague in which the infecting organisms invade the bloodstream.
septicidal *Botany.* of or related to a fruit that opens or divides along a septum.
septic tank *Civil Engineering.* a tank designed and constructed to separate solid waste from liquids, digesting organic matter through a period of detention with aerobic or anaerobic action.
septillion *Mathematics.* **1.** in American usage, the number 10^{24}. **2.** in British and German usage, the number 10^{42}.
septinary number *Mathematics.* a number expressed in base 7.
septineuritis *Neurology.* neuritis due to the existence of pathogenic microorganisms or their toxins in blood or other tissues.
Septobasidiaceae *Mycology.* the only family of fungi belonging to the order Septobasidiales, occurring on the stems and leaves of woody plants and composed of some species that form fungal enclosures around scale insects.
Septobasidiales *Mycology.* an order of fungi belonging to the subclass Heterobasidiomycetidae, composed of some members that are parasitic to scale insects.
Septoria *Mycology.* a genus of fungi belonging to the order Sphaeropsidales, which causes diseases to such plants as celery, tomatoes, wheat, and azaleas.
septoria *Plant Pathology.* a plant disease caused by fungi of the family Sphaeropsidaceae, characterized by fibrous growths on the plant leaves.
septostomy *Surgery.* a surgical incision that creates an opening in a septum.
septulum *Anatomy.* a small partition, septum, or separating wall.
septum *plural,* **septa.** *Anatomy.* a dividing wall or partition. *Electromagnetism.* a metal vane inserted into a waveguide and designed to give inductive, capacitive, or resitive characteristics.
septum pellucidum *Anatomy.* a thin double membrane that stretches from the corpus callosum to the fornix in the brain, thereby separating the anterior horns of the lateral ventricles.
septum primum *Developmental Biology.* a partition in the embryonic heart that divides the primitive atrium into its left and right chambers.
septum secundum *Developmental Biology.* a septum that lies to the right of, and eventually fuses with, the septum primum in the embryonic heart.
seq or **seq.** **1.** the following (one or ones). (From Latin *sequens, sequentia.*) **2.** that which follows. (From Latin *sequela.*)
seq luc or **seq. luc.** the following day. (From Latin *sequenti luce.*)
Sequanian *Geology.* a British substage of the Upper Jurassic period, occurring after the Rauracian substage and before the Kimmeridgian.
sequela *Medicine.* a lesion or other effect following or caused by an episode of a disease.
sequenator *Biotechnology.* an automatic device used to determine the amino-acid sequences of proteins or peptide fragments.
sequence *Engineering.* a set of quantities, elements, or events arranged in a fixed and repeatable order. *Geology.* **1.** a series of geological events, processes, or strata, arranged in chronological order. **2.** an informal lithostratigraphic unit or geographically discrete series of units that can be traced over large areas of a continent, or that were deposited under related environmental conditions. Also, STRATIGRAPHIC SEQUENCE. *Computer Programming.* a group of items in a specified order, according to identifiable keys. *Mathematics.* **1.** a mapping s of the natural numbers into a given set X. The image under s of a natural number n is usually denoted x_n, where $x_n \in X$. **2.** a set of objects $\{x_n\}$ indexed over the natural numbers (or some subset of the natural numbers); also denoted (x_n) when the (index) ordering is being emphasized. If the index set of the sequence has a finite number of elements, then the sequence is said to be finite; otherwise, it is said to be infinite.
sequence analysis *Genetics.* any study that documents the sequence of codons or nucleotides in DNA or RNA, or the sequence of amino acids in a certain protein.
sequence check *Computer Programming.* the process of checking to verify that data items in a file are in ascending or descending order.
sequence-checking routine *Computer Programming.* a routine that checks every executed instruction and prints out specific data, such as register contents or transfer instructions.
sequence dating *Archaeology.* the dating of a group of similar objects according to their archaeological sequence.
sequence monitor *Computer Technology.* the monitoring by a computer of the step-by-step actions required of an operator during the start-up or shutdown of a power unit.
sequence of current *Oceanography.* in regard to semidiurnal currents, the order in which the four tidal current strengths occur during a tidal day, with special reference to whether the greater ebb immediately follows or precedes the greater flood.
sequence of tide *Oceanography.* in regard to semidiurnal tides, the order of occurrence of the four tides of a tidal day, with special reference to whether the lower low water immediately follows or precedes the higher high water.
sequencer *Computer Technology.* **1.** a machine that orders items according to certain requirements. **2.** a circuit that retrieves information from the control store memory according to external events or conditions. *Engineering.* a mechanical or electronic device that starts a sequence. *Biochemistry.* a device that is able to sequence nucleic acids or protein.
sequence register see PROGRAM COUNTER.
sequence robot *Robotics.* a preprogrammed robot that must follow the programmed instructions.
sequence timer *Metallurgy.* in resistance welding, a timer that controls the welding cycle parameters, except for weld time and heat time.
sequencing *Industrial Engineering.* the arrangement of jobs in the order in which they will be run by a production facility. *Transportation Engineering.* the ordering of aircraft in a carrier's schedule. *Genetics.* the specification of the order of the nucleotides in an RNA or DNA molecule.
sequencing equipment *Telecommunications.* a device that allows messages received from various teletypewriter circuits to be selected and then retransmitted over a more limited number of trunks.
sequencing gel *Molecular Biology.* a gel that has the ability to separate single-stranded DNA or RNA fragments that differ in length by only one nucleotide.
sequent geosyncline see SECONDARY GEOSYNCLINE.
sequential access *Computer Programming.* **1.** the process of writing or reading data serially. **2.** of or relating to files that must be searched serially from the beginning to find a particular record.
sequential analysis *Statistics.* any of a variety of methods for inference and decision in situations in which data are collected one at a time, and a decision is made after each observation whether to continue or stop.
sequential circuit *Electricity.* a switching circuit whose function depends on past and present inputs.
sequential color television *Telecommunications.* in color television, a system by which the primary color components of a picture are both transmitted and reproduced in sequence. Also, SEQUENTIAL SYSTEM.
sequential control *Computer Programming.* a mode of computer operation in which instructions are executed in consecutive order, which permits branching to an instruction other than the next one in the program sequence.
sequential landform *Geology.* any of a series of smaller landforms produced by the erosion, weathering, and mass wasting of a larger initial landform. Also, **sequential form.**
sequential logic element *Electronics.* a logic device in which the output at any instant is determined by the input at that instant and the sequence of inputs that preceded it.
sequentially compact *Mathematics.* a topological space X is said to be sequentially compact if every sequence in X has a subsequence that converges to a point in X.
sequential machine *Computer Technology.* a mathematical model of a sequential switching circuit with an input, an output, and a current internal state.
sequential network *Computer Technology.* a model of a switching circuit that reflects only the logical properties of the circuit and not the electronic properties.

sequential operation *Computer Programming.* the consecutive execution of two or more operations.
sequential organization see SERIAL FILE.
sequential processing *Computer Programming.* the processing of records in a file in alphabetical or numerical order.
sequential sampling *Industrial Engineering.* a quality-control sampling procedure in which sampling is repeated until a decision can be made about acceptance or rejection of the lot.
sequential scanning see PROGRESSIVE SCANNING.
sequential scheduling system *Computer Technology.* an operating system in which the job scheduler reads one input stream and executes one job step at a time from that input stream.
sequential search *Computer Programming.* a search method in which the key value of each record in an ordered file is compared in turn to a desired key, until a match is located or the end of the file is reached.
sequential selection *Telecommunications.* the process of choosing successive elements of a given message from a set of all possible elements.
sequential system see SEQUENTIAL COLOR TELEVISION.
sequential test *Statistics.* a procedure used for testing a hypothesis based on data collected sequentially, usually consisting of a stopping rule and a decision rule for acceptance or rejection after stopping.
sequential trials *Statistics.* a technique of sequential analysis in which the results of one stage of analysis are known prior to advancing to the next stage.
sequestering agent *Biochemistry.* any substance that is able to remove a metal ion from a solution system by forming a new complex ion, which does not have the same chemical reactions of the original metal ion. *Chemistry.* specifically, a phosphate compound that functions as a builder of complex ions, thus reducing the number of free ions in solution.
sequestrectomy *Surgery.* the removal of dead spicules or portions, especially of bone.
sequestrum *plural,* **sequestra.** *Medicine.* a portion of dead bone that is wholly or partially separated from the main bone mass.
Sequoia [sə kwoi´ə] *Botany.* a genus of very large coniferous trees of the family Taxodiaceae, characterized by both linear and awl-shaped leaves, grooves in the trunk, and winter buds with overlapping scales. (Named for the Cherokee Indian leader *Sequoya.*)
sequoia *Botany.* a tree belonging to the genus *Sequoia* or to the related genus *Sequoiadendron,* such as the redwood, *Sequoia sempervirens,* or the **giant sequoia,** *Sequoiadendron giganteum*; they are among the oldest and largest of all living things.

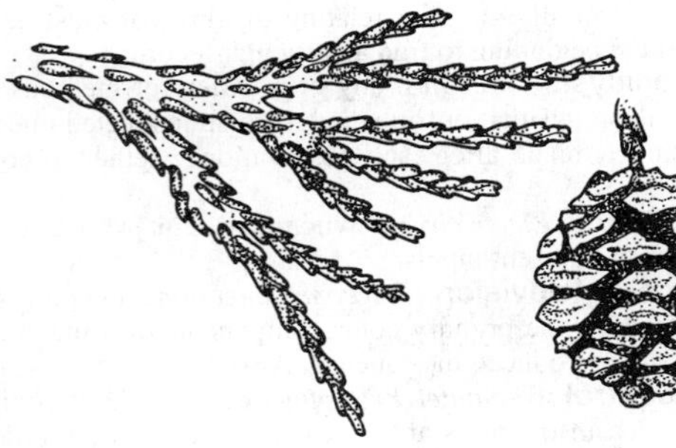

sequoia

Sequoya [sə kwoi´ə] 1760?–1843, Cherokee Indian leader and scholar, noted for inventing a system of writing for the Cherokee language.
Ser serine.
sera *Hematology.* the plural of serum. See SERUM.
serac *Hydrology.* a ridge, pinnacle, or irregular block of ice occurring on the surface of a glacier, usually at the intersection of crevasses, and formed by the breaking of the glacier as it passes over a steep slope.
seral *Ecology.* of or relating to a sere.
seralbumin *Hematology.* serum albumin.
serandite *Mineralogy.* $Na(Mn^{+2},Ca)_2Si_3O_8(OH)$, a pink triclinic mineral forming a series with pectolite, occurring as thick tabular or prismatic crystals, and having a specific gravity of 3.41 and a hardness of 4 to 5.5 on the Mohs scale; found in nepheline syenites.
Serb *Ordnance.* the code name for the Soviet SS-N-5 missile.
serclimax *Ecology.* a stable plant community that persists at a stage before the subclimax of an ecological succession.
sere *Ecology.* one of a series of plant communities that appear during the different developmental stages of plant succession in an evolving ecosystem.
serein *Meteorology.* a phenomenon characterized by very fine rain falling from an apparently clear sky, with clouds, if any, that are too thin to be visible.
Serengeti *Geography.* a plain in northern Tanzania noted for its extensive and varied wildlife population.
Sereny test *Microbiology.* a diagnostic test to determine the invasive capacities of a bacterial pathogen by infecting a guinea pig conjunctival sac with the pathogen and subsequently examining the degree of corneal ulceration.
serfdom *Anthropology.* a social and political system prevalent in Europe in the Middle Ages and extending into the 19th centurty in some areas of Eastern Europe, in which farm workers (**serfs**) were generally attached to a certain tract of land and required to serve the owner of that land. (From the Latin word for "slave.")
serge *Textiles.* **1.** a widely used twill-weave fabric with flat, diagonal ribbing, made of wool, worsted, cotton blends, or other fabrics. **2.** specifically, a finished worsted suiting fabric that is dyed navy blue.
sergeant *Military Science.* in general, a noncommissioned officer of the highest rank; there are five grades in the U.S. Army and U.S. Marine Corps and six grades in the U.S. Air Force; specifically, the term is applied to the lowest sergeant grade in each of the services.
Sergeant *Ordnance.* a mobile, inertially guided, surface-to-surface missile; it has nuclear warhead capability and a range of 75 nautical miles; officially designated **MGM-29A.**
Sergestidae *Invertebrate Zoology.* a family of pelagic shrimps in the infraorder Caridea.
SERI Solar Energy Research Institute.
serial *Computer Programming.* **1.** of or relating to the sequential handling of data items or instructions, as opposed to parallel handling. **2.** of or relating to the sequential performance of two or more activities in a single device or channel.
serial access see SERIAL MEMORY.
serial addition *Computer Programming.* the process of adding two numbers one digit at a time, moving from right to left.
serial bit *Computer Programming.* a binary digit transmitted as part of a sequential data transfer.
serial computer *Computer Technology.* **1**. originally, a computer that processed each digit or word bit sequentially, one at a time. **2**. a computer that processes a single instruction or a single word at a time.
serial correlation see AUTOCORRELATION.
serial data *Computer Programming.* data that is transmitted sequentially, one bit at a time.
serial dot character printer *Computer Technology.* a dot matrix printer with a movable print head that prints one dot or column of dots at a time.
serial file *Computer Programming.* a simple file organization in which items are arranged, searched, and entered sequentially. Also, SEQUENTIAL ORGANIZATION.
serial homology *Zoology.* a basic anatomical conformity by a series of similar structures within the body of a single organism, such as the vertebrae or ribs.
serial input/output *Computer Programming.* a method of data transfer between a computer and a peripheral unit in which data is transmitted, bit by bit, over a single circuit. Also, SERIAL TRANSFER.
serial interface *Computer Programming.* an interface in which serial transfer of data occurs over a single circuit, one bit at a time.
serial learning *Behavior.* a type of learning in which the subject is presented with a series of words or other items and is required to recall the items in the exact order in which they were originally presented. Also, **serial-order learning.**
serially reusable *Computer Programming.* of or relating to a subroutine that is reusable by another task only after the current task has finished with it.
serial marriage *Anthropology.* a form of marriage that is characterized by several short successive marriages during a lifetime. Also, SERIAL MONOGAMY.
serial memory *Computer Technology.* a type of memory in which data is essentially accessible only in the same sequence as it was stored. Also, SERIAL ACCESS, SERIAL STORAGE.
serial monogamy see SERIAL MARRIAGE.

serial observation *Oceanography.* a sequence of temperature readings and water samples taken at several levels between the surface and the bottom.

serial operation *Computer Programming.* the sequential performance of an arithmetic operation or data transmission one digit or character at a time.

serial-parallel *Computer Programming.* **1.** of or relating to a combination of serial and parallel actions; for example, the serial handling of characters and the parallel handling of the bits that compose the characters. **2.** of or relating to a device that transforms a serial input into a parallel output.

serial port *Computer Technology.* an input/output port through which data is received and transmitted sequentially, one bit at a time.

serial printer *Graphic Arts.* a printing element, for example, on a word processor or typewriter, that moves back and forth along the line of type producing one character at a time.

serial processor *Computer Technology.* a computer that processes data sequentially in individual units. Thus, **serial processing.**

serial programming *Computer Programming.* programming in which only one arithmetic or logical operation can be executed at one time.

serial sampling *Statistics.* **1.** a method that assumes positive correlation for sequential batches in production and, through a stochastic process, obtains a sentencing rule for a batch. **2.** a method of taking samples, as at fixed time or space intervals, that will ensure randomness.

serial storage see SERIAL MEMORY.

serial transfer see SERIAL INPUT/OUTPUT.

serial transmission *Telecommunications.* the process of sending the elements of a signal one after another, as distinguished from parallel transmission.

seriate *Geology.* of a granular igneous rock, textured with grains or crystals that vary gradually in size.

seriation *Archaeology.* the organization of artifacts or features into a sequence according to changes over time in their attributes or frequency of appearance; used as a relative dating method.

sericeous *Botany.* of or relating to a plant or plant part that is closely covered with fine silky hairs.

sericin *Biochemistry.* a protein that binds the two strands of a raw silk fiber.

sericite *Mineralogy.* a white fine-grained member of the mica group, usually muscovite, but may also be paragonite or illite; occurring in minute flakes with a silky luster, and formed as an alteration product of various aluminum silicates in metamorphic rocks and the wall rocks of ore deposits.

sericitic sandstone *Petrology.* a sandstone in which the voids between quartz grains are filled with sericite, derived from decomposed feldspar, and finely divided quartz.

sericitization *Geology.* a metamorphic or hydrothermal process in which sericite is introduced into a sediment or rock, or replaces another mineral.

sericulture *Agriculture.* the growing of silkworms, especially for the commercial production of silk.

seriema *Vertebrate Zoology.* either of two birds of the family Cariamidae, *Cariama cristata,* a large, broad-billed, long-legged, crested screamer, found in Brazil, or *Chunga burmeisteri,* a somewhat smaller related bird, found in Argentina; both have limited flying ability.

series a group of related items that progress in some sequence; specific uses include: *Electricity.* **1.** the arrangement of connecting components in a circuit to form a single path for current. **2.** of, relating to, or operating with such an arrangement. Thus, **series feed, series coil, series modulation, series peaking,** and so on. *Geology.* **1.** a group of rocks, minerals, or fossils that can be arranged in a natural sequence on the basis of characteristics such as composition, occurrence, or growth patterns. **2.** a temporal-stratigraphic unit between a system and a stage. *Electricity.* describing a single path connecting components in a circuit. *Systematics.* **1.** a rank in the botanical taxonomic hierarchy, below a section and above a species. **2.** a collection of specimens of a particular species or other taxon. *Mathematics.* **1.** an expression of the form

$$x_0 + x_1 + x_2 + \cdots \text{ or } \cdots + x_{-2} + x_{-1} + x_0 + x_1 + x_2 + \cdots;$$

the terms x_i may be real or complex numbers, functions, and so on, denoted, respectively, by $\sum_{i=0}^{\infty} x_i$ or $\sum_{i=-\infty}^{\infty} x_i$. Also denoted by $\sum_i x_i$, if the limits of summation are understood. If all but a finite number of the terms are zero, the series is said to be a **finite series**; otherwise it is an **infinite series**. **2.** a former term for a sequence. See SEQUENCE, def. 1.

series circuit *Electricity.* a circuit in which each component is joined end-to-end successively with the next, so that the same current flows through each component. Also, **series connection.**

series compensation see CASCADE COMPENSATION.

series disintegration see DECAY SERIES.

series excitation *Electricity.* the field excitation achieved in a motor or generator when the armature current flows through the field winding.

series-fed vertical antenna *Electromagnetism.* a vertical antenna that is isolated from ground and fed at its base. Also, END-FEED VERTICAL ANTENNA.

series feed *Electronics.* a method for powering an electron tube by applying direct current and alternating current in series.

series firing *Engineering.* the process of passing the total supply current through each detonator in a round of shots.

series loading *Electronics.* the placement of reactive elements in series with conductors along a transmission line in order to improve performance.

series motor *Electricity.* a motor in which the field winding and armature winding are connected in series. Also, **series generator.**

series multiple *Electricity.* a switchboard jack arrangement in which a single line circuit appears before two or more operators, all appearances being connected in series.

series of lines *Spectroscopy.* a set of spectral lines or line groups whose separation and intensity decreases regularly toward shorter wavelengths.

series parallel or **series-parallel circuit** see PARALLEL SERIES.

series parallel firing *Engineering.* the process of firing a round of detonators by dividing the total supply current into branches and feeding it to a number of detonators wired in series.

series-parallel switch *Electricity.* a switch designed to toggle the connections of devices from series to parallel, or vice versa.

series peaking *Electronics.* the placement of a small inductor coil in series with a resistor in a video amplifier to produce a high-frequency response that is capable of compensating for a power loss.

series radio tap *Telecommunications.* a type of telephone tapping in which a small radio transmitter is placed in series with one wire of the target pair, so that the power for the transmitter is procured from the central battery of the telephone.

series reactor *Electricity.* a reactor designed for protection in AC power systems against very large currents during short-circuit conditions.

series regulator *Electricity.* a device, such as a Zener diode, that is used to regulate the output voltage in a power supply.

series reliability *Engineering.* the condition of a system that consists of elements arranged so that failure of one element causes failure of the entire system.

series resonance *Electricity.* the resonance in a series connection of inductors, capacitors, and sometimes resistors.

series resonant circuit *Electricity.* any circuit exhibiting series resonance.

series-shunt network see LADDER NETWORK.

series termination error *Crystallography.* an effect that results from a limitation in the number of terms in a Fourier series. Ideally, an infinite amount of data is required in a Fourier series; in practice, only a finite number of data are measured in a diffraction pattern. This causes a truncation of the Fourier series, and peaks in the resulting Fourier syntheses may be surrounded by series of ripples, which are especially noticeable around a heavy atom. The use of difference syntheses obviates most of the effects of series termination errors.

series transistor regulator *Electronics.* in a DC power supply, a transistor placed in series with the output voltage independent of the current.

series-tuned circuit *Electricity.* an inductance-capacitance circuit in which the impedance is almost zero at the resonant frequency, but rises as the frequency deviates from resonance.

series ventilation *Mining Engineering.* a system that uses a single air current to ventilate a number of faces.

series welding *Metallurgy.* the process of simultaneously performing more than one resistance welding operation.

series winding *Electricity.* **1.** the part of an autotransformer winding that is connected in series between the input and output circuits. **2.** the field winding of a series motor or generator.

series-wound *Electricity.* describing a commutator motor in which the field circuit and armature circuit are connected in series. Thus, **motor-wound motor, motor-wound generator.**

serif *Graphic Arts.* a small finishing stroke appended from the main stroke of the characters in certain typefaces. According to whether or not they contain such appendages, typefaces are classified as **serif** or **sans serif**; for example, Times Roman is a serif face, and Helvetica is a sans serif face. Compare the four typefaces below:

Times Roman (serif)	Helvetica (sans serif)
A B C D E F G H I J	A B C D E F G H I J
Garamond (serif)	Optima (sans serif)
A B C D E F G H I J	A B C D E F G H I J

serigraph *Graphic Arts.* a poster or print created by the silkscreen printing process.
serigraphy *Graphic Arts.* the art or process of using silkscreen printing to make posters or prints.
serin *Vertebrate Zoology.* a small finch, *Serinus serinus*, closely related to the canary; found in Europe and northern Africa.
serine *Biochemistry.* $C_3H_7NO_3$, one of the twenty amino acids that are incorporated into proteins; it may be synthesized from glycine; used as a dietary supplement, in biological tests, and in culture media.
serine carboxypeptidase *Enzymology.* one of the enzymes that hydrolyze single amino-acid residues from the C-terminus of peptide chains and have a serine residue at the active site.
serine proteases *Enzymology.* a group of enzymes that hydrolyze proteins and have a serine and histidine involved at the active site of the catalysis; includes enzymes active in digestion, blood coagulation, immune reactions, and fertilization of the ovum. Also, **serine proteinase.**
serioscopic *Nucleonics.* relating to or obtained by serioscopy.
serioscopy *Nucleonics.* a method of three-dimensional radiography in which two components of the three-component system (subject, film, and tube) are moved; this results in blurring all images not contained in the particular plane.
seriscission *Surgery.* a surgical procedure to divide soft tissues by pulling tightly an encircling silk ligature.
seritinous *Ecology.* coming late; occurring in the latter part of a day, growing season, and so on.
seroalbuminous *Hematology.* **1.** containing both serum and albumin. **2.** containing serum albumin.
seroconversion *Immunology.* a change from positive to negative in a serology test, which occurs when antibodies are formed in response to infection or immunization.
serodiagnosis *Medicine.* diagnosis by the use of sera. Also, SERUM DIAGNOSIS.
serodiagnostic *Medicine.* relating to or based on serodiagnosis.
seroepidemiology *Medicine.* the study and identification of antibodies to specific antigens in populations of individuals.
serofibrinous *Physiology.* made up of both serous and fibrinous substances.
seroglobulin *Hematology.* serum globulin; the globulin of the blood serum.
Serolidae *Invertebrate Zoology.* a family of isopod crustaceans with greatly flattened bodies that burrow in sandy intertidal areas.
serological *Immunology.* relating to any test in which an antibody-antigen reaction takes place under artificial conditions, rather than in a living organism.
serology *Biology.* the study of immunological phenomena; in plants, the branch of science analyzing the relationships among proteins in different taxa; in animals, the branch of science dealing with the properties and reactions of blood sera.
seronegative *Pathology.* serologically negative; that is, displaying a negative diagnosis to the basic antigen-antibody reaction in vitro.
seropositive *Pathology.* serologically positive; that is, displaying a positive diagnosis to the basic antigen-antibody reaction in vitro.
seropurulent *Medicine.* both serous and purulent at once.
seroresistance *Pathology.* a post-treatment failure of the serological titer, or volume ratio, to decrease sufficiently.
serosa *Anatomy.* **1.** a serous membrane. **2.** the outermost layer of tissue covering the stomach and intestines. *Developmental Biology.* a serous membrane, such as the chorion or the outermost covering of some embryos.
serose *Hematology.* an albumose obtained from serum albumin.
serotherapy *Medicine.* the treatment of disease by injection of blood serum from immune subjects directly into the blood stream of the affected subject. Also, SERUM THERAPY.
serotinous *Botany.* **1.** of or related to a plant that grows or blooms late in the season. **2.** related to pinecones that open late to release seeds.
serotonin *Biochemistry.* $C_{10}H_{12}N_2O$, a hormone that is a vasoconstrictor and neurotransmitter, present in the brain, intestinal tissues, and blood platelets. *Endocrinology.* 5-hydroxytryptamine, a biogenic amine that is produced by the pineal gland and is also released by platelets; its role in the regulation of vascular tone is unclear.
serotype *Microbiology.* the process of grouping microorganisms into a type based on serological criteria, namely the antigens that they contain. Also, **serovar.**
serous [sēr´əs] *Medicine.* **1.** resembling serum. **2.** containing or secreting serum.
serous cystadenoma *Medicine.* a benign cystic tumor of the ovary, made up of cylindrical cells resembling those of the uterine tube. Also, SEROUS CYSTOMA, ENDOSALPINGIOMA.
serous cystoma see SEROUS CYSTADENOMA.
serous gland *Physiology.* an organ that secretes a watery fluid containing proteins and enzymes.
serous membrane *Histology.* a thin, secretory membrane lining certain internal closed cavities in the body.
Serozem see SIEROZEM.
Serpens *Bacteriology.* a genus of Gram-negative bacteria occurring in freshwater pond sediments and composed of rod-shaped cells or filaments having polar and lateral flagellation and a respiratory metabolism.
Serpentes *Vertebrate Zoology.* a nearly cosmopolitan suborder of the reptilian order Squamata, containing the snakes; characterized by limbless bodies, an extensible forked tongue, and a carnivorous diet.
serpentine [ser´pin tīn´] *Vertebrate Zoology.* relating to or characteristic of snakes. *Mineralogy.* **1.** a subgroup of monoclinic and orthorhombic minerals, part of the kaolinite-serpentine group, with the general formula $A_3Si_2O_5(OH)_4$, where $A = Mg, Fe^{+2}, Ni$; commonly variegated green in color with a greasy or silky luster; formed by the alteration of magnesium-rich silicates, and occurrring as compact to fibrous masses in igneous and metamorphic rocks; massive translucent varieties are used as an ornamental stone, fibrous varieties as asbestos. **2.** any mineral of this subgroup. *Mathematics.* see SERPENTINE CURVE.

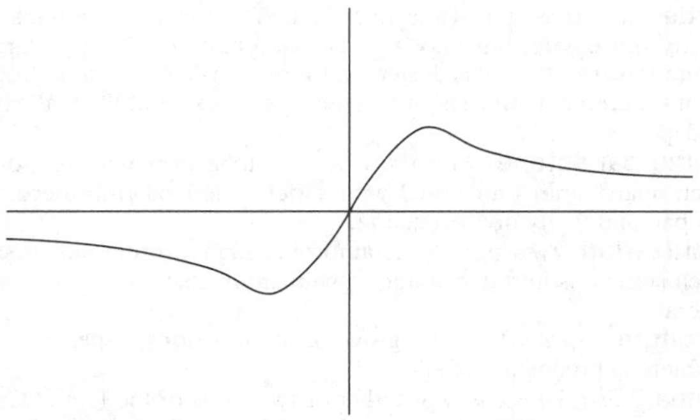

serpentine curve

serpentine curve *Mathematics.* the graph in the plane of the equation $y = abx/(x^2 + a^2)$; in polar coordinates, $x = a \cot q$ and $y = b \sin q \cos q$. The curve passes through the origin, is symmetric about the line $y = -x$, and is asymptotic to the x-axis in both directions.
serpentine jade *Mineralogy.* a variety of serpentine resembling jade, specifically bowenite; used as an ornamental stone and gemstone.
serpentine locomotion *Vertebrate Zoology.* the gliding type of locomotion characteristic of snakes and limbless lizards, by which the animal moves forward by pushing against irregularities in the substrate with the ventral part of its body.
serpentinite *Petrology.* a rock consisting almost entirely of serpentine minerals altered from previously existing olivine and pyroxene. Also, **serpentine rock.**
serpentinization *Geology.* a hydrothermal process by which magnesium-rich silicate minerals are converted into or replaced by serpentine minerals.
serpierite *Mineralogy.* $Ca(Cu^{+2},Zn)_4(SO_4)_2(OH)_6 \cdot 3H_2O$, a sky-blue monoclinic mineral occurring as tufts of minute tabular crystals, having a specific gravity of 3.07 and an undetermined hardness; dimorphous with orthoserpterite.

Serpula *Mycology.* a genus of fungi belonging to the order Agaricales that cause various rots, such as brown dry rot of lumber.

Serpulidae *Invertebrate Zoology.* a family of fanworms, polychaete annelids in the group Sedentaria, that live in secreted calcareous tubes.

Serranidae *Vertebrate Zoology.* the sea basses, a family of mostly marine, predatory fish of the order Perciformes, characterized by a heavy to elongate body, a large protractile mouth, and hermaphroditism.

serrate *Biology.* having a notched or saw-toothed margin with sharp teeth pointing forward or toward the apex. *Geology.* of a topographic feature, having a notched or toothed edge or a saw-edged profile. Also, SAW-TOOTHED.

serrated pulse *Electronics.* a signal that is divided into a series of short pulses in order to synchronize the vertical and horizontal elements of a television picture.

serrate ridge see ARETE.

Serratia *Bacteriology.* a genus of Gram-negative, motile, facultatively anaerobic bacteria of the family Enterobacteriaceae that form red, pink, or white pigments; found in water and soils and on plants; species include opportunistic pathogens that cause infections in such locations as wounds, blood, and the urinary tract. Also, **Serratieae.**

serrefine *Surgery.* a small spring forceps for compressing blood vessels.

Serret-Frenet formulas see FRENET-SERRET FORMULAS.

Serridentinus *Paleontology.* an extinct genus of bunodont mastodons related to the gomphotheres; extant in the Miocene.

Serritermitidae *Invertebrate Zoology.* a family of termites, insects in the order Isoptera, with a single genus, *Serritermes.*

serrodyne *Electronics.* a device that changes the frequency of a signal by means of a phase shift or a time delay.

Serropalpidae see MELANDRYIDAE.

Sertoli cell *Histology.* a cell in the seminiferous tubules that supports and nourishes developing sperm.

Sertoli-Leydig cell tumor see ARRHENOBLASTOMA.

serum [sēr´əm] *Zoology.* the thin, waterlike part of any animal fluid. *Immunology.* **1.** the liquid portion of blood remaining after the removal of clotting proteins. **2.** a preparation of this substance that contains antibodies against a specific antigen; obtained from an immunized donor, either a human or an animal such as a horse. (From the Latin word for *whey,* the watery portion of milk that separates from the curds after coagulation.)

serum accident *Immunology.* a serious hypersensitivity reaction that occurs immediately after a foreign serum is introduced into a hypersensitive individual.

serum albumin *Biochemistry.* the principal protein found in blood plasma (approximately 60% of the total), which is responsible for much of the plasma colloidal osmotic pressure and serves as a transport protein carrying large organic anions, such as fatty acids, bilirubin, and many drugs, and also carrying certain hormones, such as cortisol and thyroxine, when their specific binding globulins are saturated. Low serum levels occur in protein malnutrition, active inflammation, and serious hepatic and renal disease.

serum cholinesterase see CHOLINESTERASE.

serum diagnosis see SERODIAGNOSIS.

serum-fast *Immunology.* relating to the resistance to destructive effects of serum.

serum globulin *Biochemistry.* the globulin fraction of blood serum, from which it can be separated by precipitation or by electrophoresis.

serum hepatitis see HEPATITIS B.

serum hepatitis virus see HEPATITIS B VIRUS.

serum killing *Microbiology.* the killing of certain Gram-negative bacterial strains by serum, usually the result of initiation of the alternate complement pathway.

serum proteins *Hematology.* proteins of blood serum, i.e., all plasma proteins except fibrinogen.

serum shock *Medicine.* anaphylactic shock resulting from administration of foreign serum to a sensitized individual.

serum sickness *Medicine.* a hypersensitive reaction to foreign serum or serum proteins caused by the formation of circulating antigen-antibody complexes deposited in tissues, which trigger tissue injury mediated by complement and polymorphonuclear leukocytes. It is marked by fever, urticaria, arthralgia, edema, and lymphadenopathy. Because most animal-derived antisera have been replaced with human immune globulins, serum sickness is now rare. Also, **serum disease.**

serum therapy see SEROTHERAPY.

serv. service.

serv. keep, preserve. (From Latin *serva.*) Also, **Serv.**

server *Computer Science.* any device that is connected to a network and provides a service, such as printing or file storage and retrieval, in response to requests from computers connected to the network.

Servetus, Michael 1511–1553, Spanish physiologist and theologian; discovered and described pulmonary circulation.

service *Engineering.* the performance of maintenance, supply, and installation of instruments, systems, and vehicles.

serviceability *Engineering.* the degree to which a machine or product is designed for ease of service and inspection.

service ammunition *Ordnance.* ammunition intended for combat, rather than for training purposes.

service area *Telecommunications.* the geographic area that is effectively served by a telecommunications provider.

service band *Telecommunications.* a band of frequencies assigned to a specific class of radio service.

service bit *Telecommunications.* an overhead bit, such as one used for a numbering sequence, that is used to monitor data transmission rather than to convey information.

service ceiling *Aviation.* the height above sea level at which an aircraft is unable to climb faster than a designated rate (in the U.S., 100 feet per minute) under standard atmospheric conditions; a basic performance parameter usually applied to military aircraft.

service engineering *Engineering.* the process of determining the integrity of material and services in order to measure and maintain operational reliability, approving modifications of design, and following through by conforming with specifications and standards.

service factor *Engineering.* a measurement of the continuity of an operation in a chemical or petroleum processing system, obtained by dividing the actual running time with the total elapsed time.

service life *Engineering.* the length of time a machine, system, or tool can be used before breakdown or obsolescence.

service mine *Ordnance.* a mine capable of a destructive explosion, as distinguished from a dummy mine.

service pipe *Civil Engineering.* a branch pipe between a utility main and the primary meter.

service program *Computer Programming.* a computer program that performs a supplemental function for the control system, such as a utility routine or a language translator. Also, **service routine, service utility program.**

service road *Civil Engineering.* a supplementary carriageway parallel to a main road; used primarily by local traffic.

service routine see UTILITY ROUTINE.

service shaft *Mining Engineering.* a shaft used only to transport workers and materials between the surface and underground.

service test *Military Science.* a test of an item, technique, or system of materiel that is conducted under simulated or actual operational conditions to determine whether the specified military requirements or characteristics are satisfied.

service troops *Military Science.* units that provide services required by air and ground combat units to effectively carry out their combat mission; such services include supply, maintenance, transportation, evacuation, and hospitalization.

service velocity *Ordnance.* the muzzle velocity used in computing range tables; based on a projectile of standard weight fired under standard temperature conditions from a new gun.

service volume *Navigation.* the cylindrical airspace surrounding certain aviation navigation systems, within which a clear, usable signal exists.

service wires *Electricity.* a wire system that distributes electricity to homes and commercial buildings for lights, heating and air conditioning, and other appliances.

servicing time *Computer Technology.* the time during which a machine is out of service due to testing, repairs, or preventive maintenance.

serving *Electricity.* a wrapping around the core of a cable before it is leaded.

servo see SERVOMOTOR.

servo amplifier *Electronics.* the component that increases the amount of energy stored in a servomoter before it is used to power a mechanical device, known as a servomechanism.

servoarm attachment *Mechanical Engineering.* a device that augments the maximum distance the manipulator of a robot can travel.

servo brake *Mechanical Engineering.* a braking mechanism whose braking power is enhanced by a power-driven mechanism or by an increase of pressure on one of the shoes by the motion of the vehicle.

servocontrol *Control Systems.* control by means of a servomechanism.

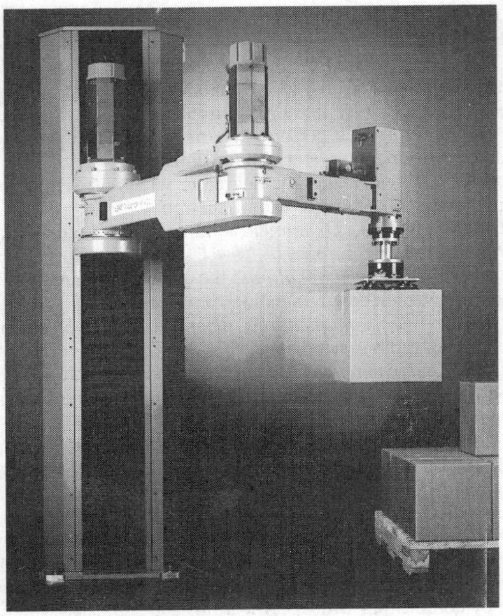

servo-controlled robot

servo-controlled robot *Robotics.* a robot that is driven by servomechanisms; usually a continuous-path or point-to-point robot.
servolink *Control Systems.* a mechanical power amplifier that uses low-power signals to control devices requiring relatively large amounts of power.
servo loop see SINGLE-LOOP SERVOMECHANISM.
servomechanism *Control Systems.* a motor-driven, automatic-control amplifier that generates feedback control signals and is used to control mechanical motion.
servomotor *Control Systems.* an electric, hydraulic, or other type of motor that drives a servomechanism. Also, SERVO.
servonoise *Engineering.* the hunting action of a radar servosystem produced by backlash in the gears and shafts of the mount.
servovalve *Mechanical Engineering.* a valve used to switch large quantities of fluids from one part of a hydraulic system to another with a comparatively small amount of initial force.
sesame [ses´ə mē] *Botany.* a tropical herb, *Sesamum indicum*, native to Asia and now cultivated in the southern United States; grown for its seeds, which are widely used to flavor bread, rolls, cakes, candy, and other foods, and which are the source of sesame oil. (From the Greek word for "seed.")
sesame oil *Materials.* a yellow oil expressed from the seeds of the sesame plant; used in cooking, medicines, and cosmetics.
sesamoid *Anatomy.* shaped like a sesame seed; i.e., small and flat.
sesamoid bone *Anatomy.* any one of various small, flat bones embedded within a tendon that moves over a bone; the largest is the patella.
sesqui- a prefix meaning "one and one-half."
sesquilinear *Mathematics.* a complex function f of two variables is said to be sesquilinear if it is linear in one variable and conjugate linear in the other; e.g., $f(\alpha x, y) = \alpha f(x,y)$ and $f(x, \alpha y) = \overline{\alpha} f(x,y)$, where $\overline{\alpha}$ indicates the complex conjugate of α. The usual example is a Hermitian inner product on a complex vector space.
sesquioxide *Chemistry.* a compound of oxygen and a metal in the proportion 3:2, as in Fe_2O_3, or vice versa.
sesquisalt *Chemistry.* a compound of an acid and base in the proportion 3:2, as in $Fe_2(SO_4)_3$.
sesquiterpene *Organic Chemistry.* any terpene with the formula $C_{15}H_{24}$; that is, 1.5 times the general terpene formula.
sessile [ses´əl; ses´īl] *Botany.* attached directly onto a base without a stalk, petiole, peduncle, or other support. *Zoology.* not free-moving; permanently fixed to a substrata. Also, SEDENTARY. *Invertebrate Zoology.* of eyes, not stalked.
sessile dislocation *Crystallography.* in a metallic crystal, a dislocation that cannot migrate easily.
sessile drop method *Fluid Mechanics.* a method to determine the surface tension of a drop resting on a surface that it does not wet; the depth and mass of the drop are measured and used to determine its shape and thus its surface tension.
Sessilina *Invertebrate Zoology.* a suborder of ciliate protozoans in the order Peritrichida, including sessile and sedentary forms having a stalk at the tapered end of the body.
sessoblast *Invertebrate Zoology.* in freshwater bryozoans, a bud that remains fixed to the adult.
seston *Oceanography.* a collective term for the suspended mineral matter, the living and dead animals, and the other organic debris found in the ocean.
set *Engineering.* **1.** any arrangement of connected units or parts that perform operational functions. **2.** to adjust a variable device, such as a clock, timer, meter, gauge, and so on, at a specific level, quantity, etc. *Electronics.* a piece of equipment that serves as an integral part of an electronic system, such as a television or radio set. *Chemistry.* **1.** to solidify or harden. **2.** the solidification or hardening of a substance. *Materials Science.* **1.** to stabilize a material or to become stabilized. **2.** the conversion of a liquid plastic material into a solid by a process such as gelling. **3.** a permanent deformation of a material when stressed beyond the elastic limit. *Mechanics.* see PERMANENT SET. *Mathematics.* a collection of distinct objects (called *elements* or *members*), such that for any object, it can be determined whether the object belongs to the collection. In Gödel-Bernays set theory, a class A is a set if and only if there is a class B such that $A \in B$. *Computer Programming.* **1.** a collection of data items having a feature in common. **2.** to put a circuit, flag, or variable into a specified state, usually nonzero. *Graphic Arts.* **1.** in typesetting, to compose or arrange characters. **2.** in a given font, the measure in points of the width of a capital M or of an entire lowercase alphabet. *Navigation.* **1.** to establish a course. **2.** to hoist and sheet a sail with the wind behind it. *Astronomy.* the passage of a celestial object below the western horizon. *Oceanography.* the direction of flow of an ocean current. *Naval Architecture.* the displacement of a ship. *Geology.* a basic group of essentially conformable strata or cross-strata consisting of two or more beds of the same lithology that are distinguished from adjacent sedimentary units by surfaces of erosion or nondeposition, or by a sudden change in character.
seta [sē´tə] *plural,* **setae** [sē´tē]. *Biology.* a slender, usually rigid, bristle-like hair. *Invertebrate Zoology.* any of various types of hair, bristles, or spines projecting from an arthropod exoskeleton or the cuticle of a worm. Also, CHAETA.
setaceous *Biology.* appearing like bristles or having bristles. *Invertebrate Zoology.* bearing setae; bristly or bristle-like. Also, **setigerous.**
setal *Biology.* relating to or in the form of setae.
setback *Mechanics.* relative rearward motions of component parts in a projectile undergoing forward acceleration during launching. These motions and the force producing them are used to promote events that participate in the arming and eventual functioning of the fuse. *Architecture.* a technique in which the upper part of a building is recessed from the ground-level building line or from the part immediately below it; often a building has a series of such recessions, forming a steplike pattern.
setback force *Mechanics.* the rearward force of inertia opposing the forward acceleration of a projectile during launching; it is proportional to the acceleration and the mass of the parts being accelerated.
set bit *Mechanical Devices.* a drill bit into which cutting materials such as diamonds are set.
set casing *Civil Engineering.* the process of cementing between the casing and wall of a hole that seals off intermediate formations and prevents fluids from entering the hole.
set composite *Electricity.* a signaling circuit in which two signaling legs may be superimposed on a two-wire interoffice trunk by using a balanced pair of high-impedance coils hooked up to each side of the line with a capacitor network.
set condition *Electronics.* the condition in which a flip-flop circuit's internal state is 1.
set copper *Metallurgy.* a crude copper containing substantial amounts of oxygen.
setf [set´ef] *Artificial Intelligence.* in Lisp, a macro that takes as arguments code to retrieve a component of a data structure and a value, and produces code to store the value into the specified component of the data structure.
set-forward *Mechanics.* relative forward motions of component parts of a projectile in flight when impact occurs. These motions are due to inertia and are opposite in direction to setback.

set-forward force *Mechanics.* the forward force of inertia that is produced by the deceleration of a projectile when impact occurs; it is proportional to the deceleration and the mass of the parts being decelerated. Also, IMPACT FORCE.

set-forward point *Mechanics.* the point on the course of the target at which the target is expected to arrive at the end of the time of flight.

set function *Mathematics.* a function whose domain of definition is a class of sets.

set hammer *Mechanical Devices.* **1.** any hammer into which the handle is simply driven, rather than wedged. **2.** a blacksmith's hammer used to straighten and shape metal bars. **3.** a hollow-faced hammer used to flatten rivet heads.

setigerous *Invertebrate Zoology.* possessing or producing setae or bristles. Also, **setiferous.**

setoff *Graphic Arts.* the inadvertent transfer of ink from a printed sheet onto the back of the sheet piled above it.

set of support *Artificial Intelligence.* in theorem proving, a restriction that every proof step must involve a goal clause or a clause derived from a goal clause.

seton *Surgery.* a thread of silk, linen, or other finely drawn material used to penetrate a sinus, fistula, or epithelial tract, often to facilitate the introduction of dilating instruments.

set point *Control Systems.* a predetermined position or value to be implemented by an automatic controller, as in a thermostat. *Physiology.* an internal regulatory system that maintains stability (e.g., of body temperature) through changing external conditions.

set pressure *Mechanical Engineering.* the inlet pressure at which a relief valve is required to open by the standard applicable to the particular system.

set pulse *Electronics.* a pulse signal that will set a flip-flop to the set condition, independent of the presence of a clock signal.

setq [set´kyoo´] *Artificial Intelligence.* in Lisp, the function that assigns a value to a variable.

setscrew *Mechanical Devices.* **1.** a headless machine screw of various threads and points, used to tightly jam one part on another so as to prevent movement between the parts. **2.** a screw used to adjust a valve setting, meter setting, or the like.

setscrew wrench *Mechanical Devices.* a wrench that is designed to turn square fasteners and setscrews.

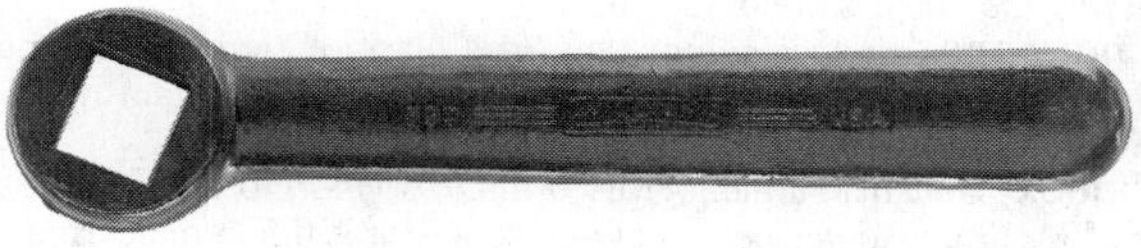

setscrew wrench

set solid *Graphic Arts.* describing type that is set in lines without any spacing between lines other than that provided within the typeface itself; for example, 10/10 type (10-point type set on a 10-point line).

set theory *Mathematics.* the study of sets, especially their logical foundations; includes the study of axioms, construction of cardinal and ordinal numbers, order, the axiom of choice and its equivalents, etc.

setting angle *Mechanical Engineering.* an angle, usually of 90°, between the axis of the shank of a working tool, such as a drill bit, and the straight section of the working piece.

setting circle *Engineering.* a coordinate scale on a telescope or transit.

setting coat *Building Engineering.* in plastering, the finishing coat; it is approximately 1/8 in. thick. Also, SKIMMING COAT, SKIM COAT.

setting gauge *Engineering.* a standard gauge for testing a limit gauge or an adjustment. Also, **setting gage.**

setting punch see RIVET SET.

setting rounds *Ordnance.* rounds fired at various elevation angles in order to push the gun mount firmly into the ground.

setting temperature *Materials Science.* the temperature at which a liquid resin sets, hardens, cures, or gels.

setting time *Materials Science.* the time required for a solid to reach a certain average percent consolidation; i.e., the time it takes a material to harden, gel, or cure.

settleable solids test *Civil Engineering.* a determination of the ability of suspended solids to precipitate, made by measuring the volume of solids settled out of a measured volume of sample in a specific interval of time. Also, IMHOFF CONE TEST.

settled *Meteorology.* describing a state of weather in which no storms occur for several days or longer.

settled ground *Mining Engineering.* ground that has fully subsided over the waste area of a mine.

settled snow *Hydrology.* old snow that has been greatly compacted and metamorphosed.

settlement *Geography.* **1.** a sedentary group of people living in a particular place. **2.** the group of houses, huts, or other structures in which they live. Also, COMMUNITY. *Geology.* the sinking of surface material as a result of compaction. *Mining Engineering.* the gradual lowering of overlying strata within a mine as coal or stratified mineral is removed.

settlement pattern *Archaeology.* the study of ancient human occupation and activity patterns within a specified area.

settlement system *Archaeology.* the entire set of settlements used by a community; for example, all the camps used by a band of hunter-gatherers.

settler *Engineering.* a container for the partial separation of a mixture made by density difference.

settling *Engineering.* the separation of suspended solids from fluid by gravitational force. Also, SEDIMENTATION. *Geology.* **1.** the sag in outcrops of rock layers as a result of rock creep. Also, OUTCROP CURVATURE. **2.** the deposition of sediment.

settling basin *Civil Engineering.* an enlargement or basin within a water conduit that provides for the settling of suspended matter and is usually equipped with some means of removing the accumulation. Also, SEDIMENTATION BASIN, SETTLING RESERVOIR, SAND TRAP.

settling chamber *Engineering.* a container in which heavy liquids or solid particles separate from the fluid as a result of gravity during processing or storing.

settling pond *Mining Engineering.* a natural or manmade pond used to recover the solids from washery effluent.

settling tank *Engineering.* a compartment that provides for particles in a fluid stream to move downward solely by the force of gravity, and thus reduces the downward speed of the particles. Also, GRAVITY SETTLING CHAMBER, SEDIMENTATION TANK.

settling time *Robotics.* the time it takes for a robot or end effector to become steady after moving from one predetermined point to another.

settling velocity *Fluid Mechanics.* **1.** the velocity at which suspended solid particles in a fluid precipitate or deposit. **2.** the velocity at which a gas bubble, liquid droplet, or solid particle rises or falls in a fluid medium under the counteracting forces of gravity and drag.

setup an organization or arrangement; specific uses include: *Industrial Engineering.* **1.** an arrangement of work stations and machines to suit the requirement of a particular production job. **2.** the act of preparing work stations and machines to perform a particular production job. *Electronics.* in a television picture, the ratio between the degree of blackness and the degree of whiteness, generally expressed as a percentage. *Telecommunications.* a camera position or point of view. *Medicine.* the arrangement of teeth on a denture base.

setup time *Industrial Engineering.* the time required to prepare personnel and equipment for operation, either before a job or between jobs. *Robotics.* the time required to prepare a robot to carry out a task, including the time expended to obtain tools, end effectors, and workpieces.

Seuratiaceae *Mycology.* a family of fungi belonging to the order Myriangiales which is characterized by rounded or pulvinate colonies and occurs on the leaves or branches of higher plants.

707 *Aviation.* a pioneer medium-to-long-range jet airliner first flown in 1954; powered by four turbofan engines and designed to carry about 195 passengers.

727 *Aviation.* a jet airliner powered by three rear-mounted turbofan engines. Designed by Boeing, it was the most popular of all Western jetliners during its years of production (1963–1984).

737 *Aviation.* a short-to-medium-haul commercial transport airplane powered by two wing-mounted turbofan engines. A later version, the **737-300,** is longer and wider.

747 *Aviation.* a widebody jumbo jet; the world's largest airliner, designed to carry about 490 passengers.

757 *Aviation.* a slim-body jetliner designed to carry about 200 passengers for short and medium hauls.

767 *Aviation.* a medium-haul wide-body jetliner designed to carry about 255 passengers.

seven-eighths rule *Navigation.* a method of approximating the distance to an object broad on the beam of a ship; the distance to the object is approximately 7/8 the distance traveled between the relative bearing of 30° and 60° or between 120° and 150°.

seventeen-year locust or **cicada** *Invertebrate Zoology.* a North American cicada, *Magicicada septendecim,* so named because its nymphs spend up to seventeen years living underground and feeding on roots before emerging to cast their skins and lay eggs as short-lived winged adults. Also, PERIODIC LOCUST.

seven-tenths rule *Navigation.* a method of approximating the distance to an object broad on the beam of a ship; the distance to the object is approximately 7/10 the distance traveled between the relative bearing of 22-1/2° and 45° or between 135° and 157-1/2°.

seven-thirds rule *Navigation.* a method of approximating the distance to an object broad on the beam of a ship; the distance to the object is approximately 7/3 the distance traveled between the relative bearing of 22-1/2° and 26-1/2°, between 67-1/2° and 90°, between 90° and 112-1/2°, or between 153-1/2° and 157-1/2°.

seventy-five-degree line *Military Science.* a theoretical line between the final bomb release line and the point on the ground where antiaircraft guns can deliver effective fire on the bomb release line at an angle of 75°.

severe mental retardation *Psychology.* a level of mental retardation characterized by an IQ ranging from 25 to 39; those affected have limited motor coordination, cannot achieve any level of academic learning, and require constant supervision. However, they are able to talk and can be taught the rudiments of self-care.

severe storm *Meteorology.* any destructive storm, especially local storms such as thunderstorms, hail storms, and tornadoes. Thus, **severe-storm observation.**

severity factor *Chemical Engineering.* a measure of the intensity of the reaction conditions in a chemical process, such as the temperature or pressure in a catalytic cracker.

Sevier orogeny *Geology.* a series of crustal deformations along the eastern edge of the Great Basin in Utah, culminating early in the Late Cretaceous period.

sewage *Civil Engineering.* any liquid-born waste that contains animal or plant matter in suspension or solution, chemicals in solution, or soils and storm water.

sewage disposal plant *Civil Engineering.* a system of structures and appurtenances that receive raw sewage and reduce the organic and bacterial content of the waste, thus rendering it less dangerous and odorous. Also, **sewage treatment plant.**

sewage farm *Agriculture.* a farm that is fertilized with sewage treatment residue.

sewage gas *Materials Science.* a combustible gas self-generated from the digesting of sewage sludge and having a slow rate of flame propagation.

sewage sludge *Chemical Engineering.* a slime that is produced by the precipitation of solid matter from liquid sewage in sedimentation tanks.

sewage system *Civil Engineering.* a multitude of conduits, culverts, channels, and drainage receptacles for carrying sewage and surface runoff to a sewage disposal plant.

sewage treatment *Biotechnology.* any process to which sewage is subjected in order to remove or alter its dangerous and objectional constituents by reduction in the organic and bacterial content.

sewellel [sə wel′əl] *Vertebrate Zoology.* a primitive fossorial rodent, *Aplodontia rufa,* of the family Aplodontidae, having a rudimentary tail, small eyes and ears, and a superficial resemblance to the ground squirrel; it lives in communal burrows near water in northwestern United States mountains. (From an American Indian name for the skin of this animal.) Also, MOUNTAIN BEAVER.

sewellel

sewer *Civil Engineering.* an open channel or underground conduit to convey refuse matter to a place of disposal.

sewerage *Civil Engineering.* the entire works required to collect, treat, and dispose of sewage, including the sewer system, pumping stations, and treatment plants.

sewing machine *Mechanical Engineering.* a mechanical device that uses a double-pointed or an eye-pointed needle to stitch together cloth, paper, leather, or other materials.

sewing press *Graphic Arts.* a machine used to bind a book's signatures with cotton or nylon thread.

sex *Biology.* **1.** the fundamental distinction among members of a species, found in most species of animals and plants and based on the types of gametes produced by the individual or on the category into which the individual fits on the basis of that criterion. Ova, or macrogametes, are produced by the female, and spermatozoa, or microgametes, are produced by the male. The union of these distinctive germ cells is the natural prerequisite for the production of a new individual in sexual reproduction. **2.** the sum of characteristics by which the male and female of a given species are differentiated. **3.** to determine the sex of a given individual. *Physiology.* sexual intercourse or other sexual activity.

sex- a combining form meaning "six."

sexadecimal *Mathematics.* of or relating to a base-16 number system.

sexagesimal *Mathematics.* relating to base 60. For example, sexagesimal measure of angles refers to the division of each degree into sixty minutes and each minute into sixty seconds.

sex cell see GAMETE.

sex change see SEX REVERSAL.

sex chromatin see BARR BODY.

sex chromosomes *Genetics.* chromosomes that determine the sex of an individual or carry genes that determine sex; designated X and Y in humans, with normal females having two X chromosomes and normal males having an X and a Y chromosome; such chromosomes may also carry genes that determine other characteristics.

sex-conditioned character *Genetics.* a phenotypic character whose expression is conditioned by the sex of an individual. Also, **sex-influenced inheritance.**

sex cord *Developmental Biology.* see MEDULLARY CORD, def. 2.

sex determination *Genetics.* the process in a given species that determines the sex of the species.

sexduction *Genetics.* the transfer of genes from one bacterium to another by the process of conjugation.

sex hormone *Endocrinology.* any of a class of steroid hormones required for the sexual development and function of an organism, such as estradiol or testosterone.

sex index *Genetics.* a number used in *Drosophila* to indicate the number of X chromosomes per set of autosomes; it is 0.5 in males and 1.0 in females.

sex-limited character *Genetics.* a phenotype, due to either a sex-linked or an autosomal gene, that is expressed in only one sex.

sex-linkage *Genetics.* the type of linkage that occurs when a gene that produces a certain phenotypic trait is located on a sex chromosome. Also, **sex-linked inheritance.**

sex organs *Anatomy.* the reproductive organs, including the ovary, testis, and external genitals.

s-expression *Artificial Intelligence.* a symbolic expression in Lisp, consisting of a symbol or a list structure whose components are s-expressions.

sex ratio *Biology.* the relative proportion of males and females in a population.

sex reversal *Biology.* a change from one sex to the other by natural, pathological, or artificial means. Also, SEX CHANGE.

sex role *Behavior.* an attitude or pattern of behavior that is expected because it is believed to be typical or appropriate for members of that sex. Also, PSYCHOSEXUAL ROLE.

Sextans *Astronomy.* the Sextant, a faint, small, and barren constellation located at the southern border of Leo; its brightest star is only 4th magnitude.

sextant *Navigation.* an double-reflecting optical instrument designed to determine latitude and longitude at sea by measuring angular distance; it is primarily used to measure the altitudes of the sun, moon, and stars. *Mathematics.* **1.** a unit of angular measure equal to 60°, or $\pi/3$ radian. **2.** a division of a circle into six equal pieces analogous to quadrant, octant, and so on.

sextant altitude *Navigation.* the altitude of a celestial body as measured by a sextant prior to applying various adjustments.

sextant error *Navigation.* an error in sextant altitude that is caused by phenomena inherent in the sextant rather than by external physical phenomena or operator error. Also, **octant error.**

sextic *Mathematics.* a polynomial of degree six (not necessarily homogeneous).

sextile aspect *Astronomy.* two celestial bodies positioned 60° apart from each other.

sextillion *Mathematics.* **1.** in American usage, the number 10^{21}. **2.** in British and German usage, the number 10^{36}.

sex-typing *Psychology.* the cultural designations of certain behaviors as feminine or masculine and the social influencing of children to conform to such roles.

sexual *Biology.* **1.** relating to sex. **2.** involving or between the sexes. **3.** having sex organs. *Psychology.* describing a person in terms of sexual behavior and relations.

sexual access *Anthropology.* rules or codes of behavior defining permissible sexual conduct, especially in the form of taboos against relations between family members.

sexual behavior *Behavior.* all thoughts, feelings, and actions that are related to sexual organs and other erotogenic zones.

sexual cycle *Physiology.* the periodic changes taking place in the sex organs of females when pregnancy has not occurred. *Biology.* the interval of sexual proliferation in an organism that also propagates asexually.

sexual dimorphism *Biology.* diagnostic morphological differences between the sexes.

sexual generation *Biology.* the sexual phase in some animals and plants that exhibit the alternation of generations; a gametophyte.

sexual imprinting *Psychology.* an individual's experiences in early life that determine or influence later sexual responses.

sexual isolation *Ecology.* a failure of members of related species or semispecies to bear hybrid offspring because of noncorresponding behaviors that prevent successful mating.

sexuality *Psychology.* **1.** the totality of an individual's sexual attributes, behavior, and tendencies. **2.** the quality of being sexual.

sexually transmitted disease *Medicine.* any disease that is primarily or typically transmitted by sexual contact, such as syphilis.

sexual reproduction *Biology.* reproduction involving the union of gametes from two parental cells.

sexual script *Psychology.* an individual's expectations about and attitudes toward sexual behavior, developed through and reflecting the influence of socialization.

sexual selection *Evolution.* the component in the process of natural selection that tends to perpetuate the hereditary characteristics that attract one sex to the other. *Genetics.* the selection of characteristics in males that enhance their ability to compete with other males for reproduction with females, even if those characteristics confer no other survival advantage to the possessor.

sexual spore *Biology.* a spore resulting from conjugation of gametes of opposite sex.

Seyfert galaxy *Astronomy.* a type of spiral galaxy with a compact, bright nucleus and broad emission lines in its spectrum; they are often bluish in color and emit radio energy.

Seyfert's Sextet *Astronomy.* a compact group of galaxies that surround NGC 6027, with most interacting physically.

Seymouriamorpha *Paleontology.* an extinct suborder of labyrinthodont amphibians in the order Anthracosauria; mostly terrestrial and so far known definitely only from North American deposits; extant in the Permian.

Sézary syndrome *Medicine.* a form of cutaneous T-cell lymphoma, characterized by scaling dermatitis and by abnormal hyperchromatic mononuclear cells in the skin, lymph nodes, and peripheral blood. (Named for Albert *Sézary,* 1880–1956, French dermatologist.)

S factor *Ordnance.* the deflection adjustment, measured in mils, that is necessary to keep the burst on the observer-target line when the range is adjusted 100 yards along the line. *Psychology.* see SPECIFIC FACTOR.

SFC specific fuel consumption.

SFC/FT-IR *Spectroscopy.* an analysis wherein supercritical fluid chromatography is combined with on-line Fourier transform-infrared instrumentation and detection.

sferics [sfer´iks; sfēr´iks] *Meteorology.* the study of atmospherics, especially from a meteorological perspective; involves techniques of locating and tracking atmospherics sources and evaluating received signals in terms of source.

sferics observation *Meteorology.* an analysis, from sferics receivers, of the location of weather conditions with which lightning is associated.

sferics receiver *Meteorology.* an instrument that measures electronically the direction of arrival, intensity, and rate of occurrence of atmospherics; consists basically of two orthogonally crossed antennas.

sfm surface feet per minute.

SFO *Aviation.* the airport code for San Francisco.

SG Surgeon General.

S-G Sachs-Georgi (test).

s.g. specific gravity.

s-g subgenus.

S0 galaxies *Astronomy.* in Hubble's classification, galaxies that lie intermediate between spirals and ellipticals; their disks have no structure.

s-gg subgenera.

S-glass *Materials.* a magnesia-alumina-silicate glass used for making high-strength fibers.

SGO Surgeon General's Office.

SGOT serum glutamic oxaloacetic transaminase.

Sgt or **Sgt.** Sergeant.

Sgt Maj or **Sgt. Maj.** Sergeant Major.

SH serum hepatitis; social history.

SHA *Aviation.* the airport code for Shanghai, China.

SHA sidereal hour angle.

shackle *Mechanical Devices.* an open (U-shaped) or closed metal connecting link that can accommodate a pin or bolt through the holes of its flattened ends or legs.

shackle bolt *Mechanical Devices.* a pin or stud having a flat end that fits through the loops of a shackle and attaches to a chain or certain parts of tackle in sailing vessels.

Shackleton, Sir Ernest Henry 1874–1922, British explorer; led an early Antarctic expedition; discovered the south magnetic pole.

shad *Vertebrate Zoology.* an anadromous marine fish of the genus *Alosa* and related genera of the family Clupeidae, especially *A. sapidissima,* characterized by a deeper body than most herrings and used as food fishes in Europe and North America.

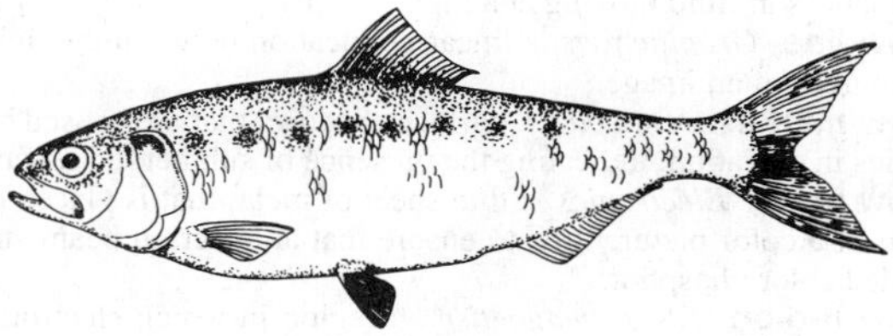

shad

Shaddock *Ordnance.* the code name for the Soviet SSC-1 missile.

shade *Optics.* the color of a dye or pigment in relation to the admixture of black dye or pigment that it contains.

shaded-pole motor *Electricity.* an AC inductance motor that is started by a pole with a single-turn copper ring placed about its end.

shaded relief *Cartography.* a technique of portraying three-dimensional relief on a map or chart by adding the shadows that would be cast by a light shining from the northwest over the same terrain shown on the map at the same scale.

shade error *Navigation.* an error in sextant altitude caused by refraction in the shade glasses.

shade glass *Optics.* a tinted piece of glass used in a sextant to reduce the sun's brightness for solar observations.

shade-tolerant *Botany.* of a plant, able to grow with relatively little direct sunlight.

shading *Graphic Arts.* the representation of different values of light and dark in a photograph, drawing, etc. *Electronics.* **1.** a process that compensates for unwanted signals produced by a camera tube. **2.** a change in a television picture's brightness that is caused by the camera tube.

shading ring *Acoustical Engineering.* a heavy copper coil that is placed around the pole of a dynamic loudspeaker in order to suppress hum. *Electromagnetism.* a single-turn copper ring that is placed about the end of a core of an alternating-current coil and used to delay the change in the flux for purposes of engine ignition, chatter reduction, and the like. Also, **shading coil.**

shading signal *Electronics.* a signal produced by a television camera that increases the power in the amplifier when the electron beam focuses on a dark portion of the screen.

shadow *Optics.* the dark region lying behind an illuminated opaque object. *Physics.* a region that is blocked by an object so that radiation is unable to reach the region directly. *Graphic Arts.* the darker area or part of a subject, negative, or reproduction. *Medicine.* an attenuated image of an actual object, such as a faded or colorless erythrocyte. *Radiology.* a figure or image that is captured on film or other recording media when light or other rays are interrupted by radiopaque structures. *Psychology.* in Carl Jung's theory, an individual's primitive instincts, often the unconscious opposite of the individual's conscious awareness. *Artificial Intelligence.* in a frame system, to override, by means of a specified slot value in an instance frame, a normal or default value that is specified in the class to which the instance belongs; e.g., if the color of a particular elephant is specified as pink, it shadows the normal elephant color of gray. Also, OVERRIDE.

shadow attenuation *Electromagnetism.* the attenuation of electromagnetic energy caused by obstacles intervening in the path of propagation.

shadow bands *Astronomy.* an atmospheric phenomenon that occurs just before totality during a solar eclipse, in which faint shadows caused by differences in optical density of the atmosphere ripple quickly across the ground.

shadow effect *Telecommunications.* diminution of power of an ultrahigh-frequency signal caused by a massive object between the points of transmission and reception.

shadow factor *Optics.* a factor used in analyzing the height of an object in a photograph; it combines solar declination, time of day, and latitude. *Electromagnetism.* the ratio of the electric field intensity of an electromagnetic wave that propagates over the surface of a sphere to that of a wave that propagates over a plane, with all other factors held constant.

shadowgram *Optics.* a graphic representation of a shadow as projected onto a two-dimensional surface.

shadowgraph *Optics.* a technique in which light moving through a fluid is refracted by density gradients in the liquid, causing light and dark areas to appear on a screen placed behind the fluid; used to detect disturbances in fluid moving at a high velocity.

shadow line *Graphic Arts.* a linear delineation between the lighter and darker areas of an image.

shadow marks *Archaeology.* surface shadows that are caused by irregularities in elevation, indicating the presence of submerged features.

shadow mask *Electronics.* a thin sheet of metal that is placed over the screen of a color picture tube to ensure that an electron beam strikes its intended color phosphor.

shadow region *Electromagnetism.* a region in which electromagnetic radiation is blocked due to the placement of an obstacle, nontransparent to the radiation, in the path of propagation.

shadow zone *Acoustics.* **1.** a depth range, created by a positive temperature gradient above a negative temperature gradient that, according to Snell's law, causes the refraction of sound away from high-velocity regions and creates a space that is centered at the point of maximum velocity between the two gradients, into which sound cannot penetrate in appreciable amounts. **2.** a space in which sound is diminished by several decibels, due to excessive attenuation around the space, such as a barrier, temperature gradient, wind, or the like. *Geophysics.* a region, 103–144° from the epicenter of an earthquake in which there is no direct penetration of seismic waves, due to the refractive and absorptive qualities of the low-velocity zone inside the earth's core boundary.

Shafrir *Ordnance.* an Israeli air-to-air missile that is based on early U.S. Sidewinder missiles; it is powered by a solid-propulsion motor, equipped with an optical system and infrared seeker, and delivers a conventional 11-kg warhead at supersonic speed and a range of 5 km.

shaft *Geology.* a vertical or nearly vertical passage in a cave. *Mining Engineering.* an excavation of small area in comparison to depth, used in mining operations for such purposes as ventilation, finding or mining material, and raising or lowering water, material, equipment, or workers. The term is often applied specifically to vertical or approximately vertical areas of this type to distinguish them from inclines or inclined areas. *Mechanical Engineering.* a cylindrical piece of metal that rotates or provides an axis of revolution, and upon which rotating machine parts are mounted to transmit power or motion. *Architecture.* the main part of a column or pilaster, between the base and capital. *Anatomy.* a long, slender part, such as a diaphysis.

shaft alley *Naval Architecture.* the watertight tunnel through which a ship's propeller shaft runs from the engine room to the point near the stern at which the shaft projects from the hull.

shaft allowance *Mining Engineering.* the space between the excavation diameter and the finished diameter, into which will fit the permanent shaft lining.

shaft balancing *Design Engineering.* the process by which a mass is redistributed and attached to a rotating body to reduce centrifugal force induced by vibrations.

shaft column *Mining Engineering.* a series of pipes within a mine shaft used for pumping, for compressed air, or for hydraulic stowing.

shaft coupling *Mechanical Engineering.* see COUPLING.

shaft encoder *Robotics.* a transducer that converts position data on shaft angles into digital form for use by a controller.

shaft furnace *Mechanical Engineering.* a blast furnace having a tapered structure through which hot gas rises.

shaft hopper *Mechanical Engineering.* a device that feeds rods and shafts to threaders, grinders, screw machines, or tube benders.

shaft horsepower *Mechanical Engineering.* **1.** the power supplied to a compressor or pump. **2.** the power produced by the shaft of a motor or engine.

shafting *Mechanical Engineering.* the hardware used to transmit the rotary motion of an engine or motor to a driven element.

shaft kiln *Engineering.* a vertical kiln charged at the top and discharged at the bottom; used for the calcination of flint, dolomite, and fire clay.

shaft lining *Mining Engineering.* a structure composed of concrete, steel, brick, or timber, fixed about the perimeter of a shaft to support the walls.

shaft pillar *Mining Engineering.* a large portion of a seam of coal or ore left unworked near the shaft bottom as protection against subsidence.

shaft plumbing *Mining Engineering.* **1.** a survey operation by which two plumb bobs are oriented both at surface and at depth so as to transfer the bearing underground. **2.** a means of ensuring that a shaft lies in a true vertical line by transferring one or more points at the surface to plumb-line positions at the bottom.

shaft-position encoder *Electronics.* a device that converts a mechanical rotation into digital form.

shaft siding *Mining Engineering.* the station at the bottom of the winding shaft, arranged for full and empty buckets or tubs.

shaft spillway *Civil Engineering.* a funnel-shaped, vertical shaft that provides an outlet for overflows in a reservoir.

shaft station *Mining Engineering.* an enlargement of a level close to a shaft, from which mined material may be hoisted and supplies unloaded.

shaft strut *Naval Architecture.* a strut supporting the protruding portion of a vessel's propeller shaft.

shagreen *Materials.* a rough, untanned leather, especially one prepared from the hide of a shark.

shake *Materials Science.* a crude shingle or siding formed by handcutting a log into tapered radial sections. *Nucleonics.* a time interval defined as 0.01 microsecond (10^{-8} second); originally applied to the average time between the release of a neutron and its subsequent capture by a fissionable nucleus in a chain reaction.

shake culture *Microbiology.* **1.** a method used to isolate anaerobic bacteria by shaking the liquid culture of an agar before the medium solidifies. **2.** a culture in a liquid medium that is aerated by constant shaking.

shakedown *Engineering.* the process of determining the performance or operational characteristics of a new or repaired vehicle, ship, or system. Thus, **shakedown cruise, shakedown flight, shakedown test.**

shakeout *Metallurgy.* the process of removing a sand casting from a mold.

shaker *Electromagnetism.* an electromagnetic device that is capable of applying controlled vibratory motion to an object. Also, **shake table.**

shake-table test *Engineering.* a procedure for testing a given object by subjecting it to controlled mechanical vibrations.

shake wave see S WAVE.

shaking-out *Petroleum Engineering.* a petroleum-refinery testing process in which oil is centrifuged to determine its sediment and water content.

shaking screen *Mechanical Engineering.* a mesh screen with holes of varying dimensions that is moved with a back and forth or rotary motion to separate material by size.

shaking table *Mining Engineering.* a flat rectangular table that can be tilted along the longitudinal axis, in which a vibrator provides rapid reciprocating motion as unclassified sands are washed down the table against horizontal baffles; heavier minerals work across the table while lighter minerals gravitate downward to various discharge zones. Also, WIFFLEY TABLE.

shale *Petrology.* a fine-grained, laminated or fissile, red, brown, black, or gray, sedimentary rock; produced by the consolidation of clay, silt, or mud, and composed roughly of one-third quartz, one-third clay materials, and one-third other miscellaneous materials such as carbonates, iron oxides, feldspars, and organic matter.

shale ball *Geology.* a meteorite that has been partly or completely converted to iron oxides as a result of weathering. Also, OXIDITE.

shale break *Geology.* a thin layer or band of shale occurring between harder strata or within a bed of limestone or sandstone.

shale crescent *Geology.* a current crescent formed by the filling in of a ripple-mark trough with shale.

shale ice *Hydrology.* a collection of thin, brittle plates of ice on a lake or river, formed by the breaking up of skim ice into small pieces.

shale oil *Petroleum Engineering.* petroleum distilled from oil shale; characterized by a large percentage of unsaturated hydrocarbons, alkenes, and dialkenes.

shale shaker *Petroleum Engineering.* a vibrating screen used to separate rock cuttings from drilling fluid; fluid returning from the downhole is passed through the screen, trapping the cuttings on the mesh.

shallot [shal´ət] *Botany.* any of several perennial herbs of the onion family, especially *Allium ascalonium* or *A. cepa aggregatum,* having a divided bulb and often used in cooking.

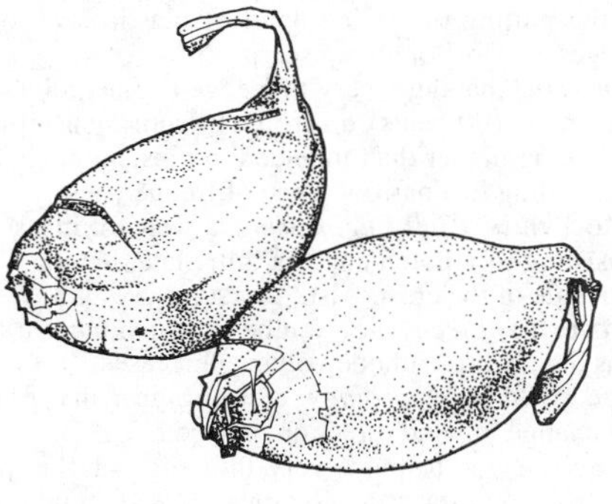

shallot

shallow *Hydrology.* not deep; of little depth. *Aviation.* making an angle of only a few degrees, usually with the horizontal; used to form compounds such as **shallow climb** or **shallow glide.** A turn with the wings only slightly banked is often called a **shallow turn.**

shallow-focus earthquake *Geophysics.* an earthquake whose focus occurs within the first 70 km of the earth's crust, the most common type of earthquake.

shallow fog *Meteorology.* a low-lying fog that does not obstruct horizontal visibility at a level six feet or more above the surface of the earth; almost always a form of radiation fog.

shallow fording *Ordnance.* the ability of a self-propelled gun or ground vehicle equipped with built-in waterproofing, with its wheels or tracks in contact with the ground, to negotiate a water obstacle without the use of a special waterproofing kit. Also, **shallow-fording capability.**

shallow inland sea *Geology.* an epicontinental sea that periodically covers a cratonic area as a result of continental subsidence or eustatic rises in sea level.

shallow knowledge *Artificial Intelligence.* causal information or relationships, often gained by experience or by trial and error, that lacks a deep or fundamental justification.

shallow marginal sea *Geology.* an epicontinental sea along the margin of a craton.

shallow model *Artificial Intelligence.* an expert system model that contains only experiential knowledge as distinguished from a deep model that is based on fundamental knowledge. Also, SURFACE MODEL.

shallows *Hydrology.* an expanse of shallow water or a relatively shallow place in a lake, stream, or other body of water.

shallow water *Hydrology.* water of such limited depth that the bottom topography influences the formation of surface waves.

shallow-water wave *Hydrology.* a surface wave whose length is at least 25 times greater than the water depth, and whose shape and velocity are affected by the depth of the water. Also, LONG WAVE.

shallow well *Hydrology.* **1.** a water well that draws upon the shallowest aquifer in the area. **2.** a well having a low enough water level to permit the use of a suction pump.

shaluk *Meteorology.* a hot desert wind other than the simoom.

shaly *Geology.* relating to, composed of, or containing shale.

shaly bedding *Geology.* a bedding consisting of 2–10 mm laminae.

shamal *Meteorology.* a sudden northwest wind that blows into the lower valley of the Tigris and Euphrates and the Persian Gulf, lasting 1–5 days during the daylight hours, except for June and early July, when it blows continuously for as many as 40 days.

shaman [shä´mən; shā´mən] *Anthropology.* in certain tribal societies, a priestly figure who is believed to have supernatural or magical powers for healing disease and for invoking or communicating with spirits.

shamanism *Anthropology.* **1.** a belief in the power or influence of shamans. **2.** a religious system based on this belief.

shamanistic *Anthropology.* relating to or believing in shamans.

shandite *Mineralogy.* $Pb_2Ni_3S_2$, a brass yellow, metallic, opaque, trigonal, pseudocubic mineral, occurring as minute grains embedded in serpentine, having a specific gravity of 8.72 and a hardness of 4 on the Mohs scale.

Shane-Wirtanen catalogue *Astronomy.* a photographic survey made at Lick Observatory that examined a million galaxies brighter than 19th magnitude.

shank *Zoology.* a leg or a leglike part. *Mechanical Devices.* **1.** the projecting end of any tool or device such as a knob, eye, loop of a button, which is secured to a handle or other part. **2.** see BIT BLANK.

shank cutter *Mechanical Devices.* a trimming tool for the outside of the shank of a shoe.

shank-type cutter *Mechanical Devices.* a cutter that has a projecting end which can be attached to a handle, brace, or machine spindle.

Shannon, Claude E. born 1916, American mathematician; a founder of information theory.

shannon *Telecommunications.* a unit of information quantity equal to the information contained in a message represented by one of two equally probable, exclusive or exhaustive states of anything used to store or convey information.

Shannon formula *Telecommunications.* a formula stating that the maximum number of binary digits per second that can be transmitted with an arbitrarily small occurrence of error is equal to the product of the bandwidth and $\log_2(1 + R)$, where R is the signal-to-noise ratio.

Shannon limit *Telecommunications.* in Shannon's formula, the greatest signal-to-noise ratio improvement that can be reached by the best modulation technique.

Shannon's sampling theorem see SAMPLING THEOREM.

shantung [shan tung´; shan tŭng´] *Geology.* a monadnock that is in the process of being buried by huangho deposits. *Textiles.* a plain weave silk fabric with a rough texture produced by the use of uneven yarns and yarns with knots, lumps, and other imperfections. (Named for the province of *Shantung* (Shandong), in eastern China.)

Shantung soil see NONCALCIC BROWN SOIL.

shaped-chamber manometer *Engineering.* a manometer that measures the difference between two liquid pressures.

shaped charge *Ordnance.* a charge that is shaped so as to concentrate its explosive force in a particular direction; it usually has a hollow area toward the front and produces a long thin jet of hot gases that is effective in penetrating armor. Also, CAVITY CHARGE.

shaped crystal growth *Crystallography.* any method for producing single crystals of specified shapes.

shape factor *Optics.* a mathematical factor used to describe a lens; it is calculated by the formula

$$(r_1 + r_2)/(r_2 - r_1),$$

where r_1 and r_2 are the radii of the lens's front and back surfaces, respectively. Also, CODDINGTON SHAPE FACTOR. *Fluid Mechanics.* **1.** the ratio of pressure force to viscous force. **2.** the area of a sphere having the same volume as that of a certain solid particle, divided by the actual surface area of that particle. *Electricity.* see FORM FACTOR. *Electronics.* **1.** the ratio between the highest and lowest attenuation in a filter's frequency range. **2.** the variable that accounts for a coil's shape when computing impedance.

shape isomer *Nuclear Physics.* a state in which an isomer maintains a high level of energy over an unusually long period of time, due to it having a shape that differs radically from the lower energy state into which it is allowed to decay.

shape memory *Materials Science.* a property possessed by certain alloys that return to their original shape upon heating after having been deformed.

shape-memory alloy *Metallurgy.* any alloy having a high degree of shape memory, such as certain copper-zinc-aluminum and certain nickel-titanium alloys.

shaper *Mechanical Engineering.* **1.** a planing machine tool whose cutting edge travels over the surface of a stationary flat workpiece. **2.** a machine with one or two vertically mounted cutters; used in woodworking for forming moldings and other irregular shapes. *Metallurgy.* a machine that forms patterns in sheet metal by striking or stamping.

shape recovery see IMAGE UNDERSTANDING.

shaping a process of giving form to some object or entity; specific uses include: *Mechanical Engineering.* **1.** the process of producing a flat surface on a workpiece using a single-point tool. **2.** the machining of an edge profile or edge pattern on the side or periphery of a piece of wood, plastic, or other material. *Behavior.* a method for teaching a new behavior by rewarding any action tending in the direction of the desired behavior; the actions become progressively closer to the desired behavior until the behavior is mastered. Also, SUCCESSIVE APPROXIMATION, APPROXIMATION CONDITIONING. *Electromagnetism.* the adjustment of a radar screen pattern (plan-position indicator pattern) that is set up by a rotating magnetic field.

shaping dies *Mechanical Engineering.* a set of dies that shape work pieces into desired forms by bending, pressing, extruding, etc.

shaping network or **shaping circuit** see CORRECTIVE NETWORK.

Shapley, Harlow 1885–1972, American astronomer; explained Cepheid variables; calculated the size and shape of the Milky Way galaxy.

Harlow Shapley

shapometer *Engineering.* an instrument that measures the shape of sedimentary particles.

shard *Archaeology.* a fragment of broken pottery; a potsherd. *Geology.* a fragment of glassy pyroclastic material having a characteristically curved surface of fracture.

share *Agriculture.* see PLOWSHARE.

shared control unit *Computer Technology.* a peripheral-device control mechanism that serves more than one device.

shared file *Computer Technology.* a direct-access storage device that may be used by two systems simultaneously, sometimes serving to link the two systems.

shared logic *Computer Programming.* the simultaneous use of a computer system, and usually of common software, by more than one user.

shared memory *Computer Technology.* **1.** the use of the same section of memory by two distinct processors. **2.** the section of memory shared in this manner. Also, MULTIPORT MEMORY.

shared resource *Computer Technology.* any computer resource, such as memory or peripherals, that can be shared by several users or programs.

shareware see FREEWARE.

shark *Vertebrate Zoology.* any elasmobranch fish of the superorder Squalomorphii, being mostly marine and carnivorous, and characterized by a fusiform body, heterocercal tail, and a tough skin with tubercules; of worldwide distribution.

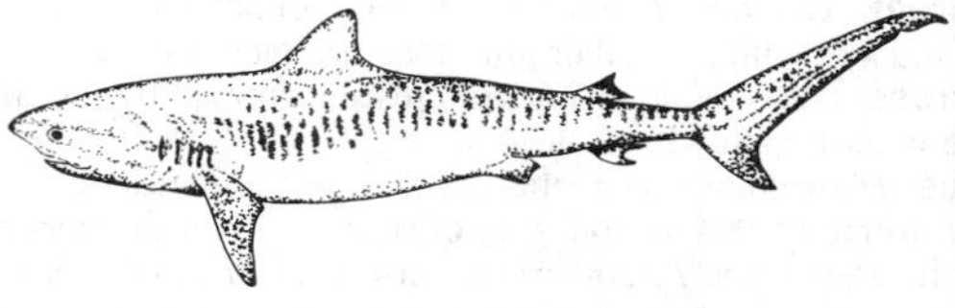

tiger shark

sharkskin *Textiles.* **1.** a textured worsted or woolen fabric in a twill weave using two colors of yarn to create a finished effect resembling the pebbled skin of a shark; used extensively for men's suits. **2.** a lighter-weight, basket weave acetate or rayon fabric with a crisp texture.

sharkskin pahoehoe *Geology.* a type of pahoehoe whose surface is covered by numerous tiny spines or spicules, which are produced by escaping gas bubbles.

shark-tooth projection *Geology.* a narrow, sharp-pointed structure formed by the pulling or tearing apart of plastic lava, often occurring along the edge of a flow or slump scarp.

sharp *Engineering.* having a very thin edge or fine point. *Acoustics.* **1.** a musical half step (100 cents) up in pitch from a specified note. **2.** the condition of being higher than intended in musical pitch. *Electricity.* relating or responding to a narrow range of frequencies.

sharp-crested weir *Civil Engineering.* a weir for measurement of discharge, consisting of a notch cut in a thin plate, having a sharp edge on the upstream side of the crest. Also, MEASURING WEIR.

sharp-cutoff tube *Electronics.* a tube in which current flow decreases uniformly as the grid-to-cathode voltage decreases.

sharp-edged gust *Meteorology.* a strong gust that represents an instantaneous change in wind direction or speed.

sharpen *Engineering.* to put a very thin edge or fine point on a tool used for cutting or piercing. *Graphic Arts.* to make halftone dots smaller, often by adding an etching chemical to separated or screened negatives; to decrease the strength of an image; the opposite of *thicken*.

sharpened Patterson function *Crystallography.* a Patterson map computed with values of $|F|^2$ modified by an exponential or similar function that enhances those reflections with high values of sin θ/λ. The resulting interatomic vectors appear as sharper peaks and the Patterson map may therefore be simpler to interpret.

sharpening stone *Engineering.* a whetstone used for sharpening by hand.

sharp freezing *Food Technology.* the earliest and least expensive form of freezing technology, in which the product is placed in an insulated room where the temperature is kept at a range of –23° to –30°C.

sharp iron *Engineering.* a tool that slits seams for caulking.

sharpite *Mineralogy.* $Ca(UO_2)_6(CO_3)_5(OH)_4 \cdot 6H_2O$, a very rare, strongly radioactive, greenish-yellow orthorhombic mineral, occurring as fibrous crusts and rosettes, having a specific gravity of 4.45 and a hardness of about 2.5 on the Mohs scale.

sharpness *Graphic Arts.* the degree of detail defined in a photographic print or negative.

sharpness of resonance *Electricity.* a means of comparing different resonant circuits by specifying the frequency band in which the circuit responds; the narrower the band, the sharper the resonance.

sharps *Food Technology.* the bran residue from wheat milling. Also, MIDDLINGS.

sharp sand *Geology.* an angular-grained sand that is wholly or partly free of clay, loam, and other foreign particles.

sharp series *Spectroscopy.* for alkali elements, a set of spectral lines that represent a change in total orbital angular momentum accompanying a transition from the *s* state to the *p* state.

sharpstone *Geology.* any rock fragment that is larger than sand (more than 2 mm in diameter) and has angular edges and corners.

sharpstone conglomerate see SEDIMENTARY BRECCIA.

sharp tuning *Electricity.* the precise adjustments of the resonant frequency of an inductance-capacitance circuit using individually variable inductors or capacitors.

sharp V thread *Design Engineering.* a screw thread whose included angle is approximately 60°, thus producing a sharp crest and root.

Shatt-al-Arab *Geography.* a tidal river formed by the confluence of the Tigris and Euphrates and flowing 120 miles to the Persian Gulf.

shatter breccia *Petrology.* a tectonic breccia consisting of angular fragments that exhibit little rotation. Also, CRACKLE BRECCIA.

shatter cone *Geology.* a distinctively striated conical rock fragment along which fracturing has occurred; most commonly found in limestone, dolomite, sandstone, shale, and granite. Also, SHEAR CONE, PRESSURE CONE.

shatter cone

shattering *Materials Science.* the breaking up of a very hard material into pieces of irregular angular blocks, under very high stresses. *Fluid Mechanics.* the breakup of a liquid into extremely fine drops, caused by the action of outside repeated and/or strong forces.

shatter zone *Geology.* a belt of rock characterized by random fissures or cracks forming a network of veins that may be filled by mineral deposits.

shattuckite *Mineralogy.* $Cu_5^{+2}(SiO_3)_4(OH)_2$, a translucent blue, orthorhombic mineral occurring as aggregates of thin prismatic crystals, and in massive form having a specific gravity of 4.11 and a hardness of 3.5 on the Mohs scales; found as an alteration product of other secondary copper minerals.

Shaula *Astronomy.* Lambda (λ) Scorpii, the second-magnitude white star at the tip of the Scorpion's stinger.

shave hook *Mechanical Devices.* a cutting tool used by plumbers or metal workers consisting of a sharp, circular, steel plate with a shank sunk into a handle, used for cutting pipe and lead materials.

shaving *Mechanical Engineering.* the process of removing a surface layer from a workpiece, usually to clean it or to reduce its size by a small amount.

sheaf *plural,* **sheaves.** *Agriculture.* a bundle, especially of cereal grain such as wheat or rye. *Ordnance.* in artillery and naval gunfire support, planned planes or lines of fire that produce a desired pattern of bursts with rounds fired by two or more weapons. *Mathematics.* a presheaf F over a topological space X is called a sheaf if the following conditions hold for every collection $\{U_i\}$ of subsets of X whose union is an open set U: (a) If

$$s,t \in F(U) \text{ and } r_{U_i}^{U}(s) = r_{U_i}^{U}(t)$$

for all i, then $s = t$; and (b) if $s_i \in F(U_i)$ and if for $U_i \cap U_j \neq \emptyset$, one has

$$r_{U_i \cap U_j}^{U_i}(s_i) = r_{U_i \cap U_j}^{U_j}(s_j)$$

for all i, then there exists an s in $F(U)$ such that $r_{U_i}^{U}(s) = s_i$ for all i. The first condition for a sheaf means that data defined on large open sets U can be determined by examining it locally, and the second condition means that local data of a given presheaf can be pieced together to give global data of the same kind in the presheaf. Sheaves are useful in the study of differentiable manifolds, since such manifolds are locally Euclidean; they are also used in information theory.

sheaf of planes *Mathematics.* a collection of planes having a point or line in common; analogous to a pencil of lines.

sheaf structure *Geology.* a structure in which the bundled crystals are arranged in bundles, characteristic of certain fibrous minerals such as stibnite.

shear *Agriculture.* to cut the wool from a sheep's coat. *Mechanical Devices.* see SHEARS. *Mechanics.* **1.** the movement of parallel surfaces of a solid body, in such a way that they remain parallel to each other; i.e., the surfaces slide against each other. The shape of the body is thus changed, but the volume remains unchanged. **2.** to move in such a manner. **3.** the deformation in shape that results from a shear force. **4.** the measurement of this force; shear stress.

shear angle *Mechanical Engineering.* the angle formed by the shear plane and the work surface.

shear center see CENTER OF TWIST.

shear cleavage see SLIP CLEAVAGE.

shear cone see SHATTER CONE.

shear deformation *Mechanics.* see SHEAR, def. 2.

shear diagram *Mechanics.* a plot of the shear force at every point along a structural member against the position of the point on the member.

shear drag see SHEAR RESISTANCE.

shear fold *Geology.* a minor fold that results from differential displacement along closely spaced cleavage or fracture planes. Also, GLIDE FOLD, SLIP FOLD.

shear fracture *Mechanics.* the failure of a structural member due to excessive shear; a common cause of failure of a ductile material under tensile stresses.

shear-gravity wave *Geophysics.* a wave disturbance that forms at the interface between atmospheric layers of different densities, causing the wave to move at different velocities.

shearing *Agriculture.* the process of removing wool from a sheep with a cutting device. *Textiles.* a finishing operation used on woolens, worsteds, and other fabrics in which the nap on a cloth is leveled, a pattern is cut in the pile, or uneven threads are removed from a fabric surface by a machine with rotating spiral blades. *Mechanical Engineering.* the cutting or breaking up of materials by applying shear stress. Thus, **shearing machine, shearing punch.** *Mining Engineering.* a method of attacking a face of coal by means of vertical side cutting, used in combination with holing or horizontal undercutting.

shearing die *Mechanical Engineering.* a die equipped with a punch so that the workpiece may be removed from the stock by shearing.

shearing field *Physics.* a magnetic field configuration that is used in a rotational transform whereby the transform angle varies with distance from the magnetic axis, so as to provide hydrodynamic stability.

shearing instability see HELMHOLTZ INSTABILITY.

shearing interferometer *Optics.* an interferometer that produces interference between rays that are sheared in the sample object by a small lateral distance.

shearing stress see SHEAR STRESS.

shearing tool *Mechanical Devices.* a cutting tool designed to operate with a large angle between its face and the edge of a workpiece.

shear joint *Geology.* a joint that is formed by a shearing action and constitutes a potential plane of shear. Also, SLIP JOINT.

shear line *Meteorology.* a line characterized by maximum horizontal wind shear, a narrow zone across which there is an abrupt change in the horizontal wind component parallel to the zone.

shear lip *Metallurgy.* a narrow ridge along the edge of a fracture surface.

shear modulus *Mechanics.* the strength factor for a material under shear stress, expressed by the relationship of the shear force applied to it to the change in position produced by this force. Also, MODULUS OF RIGIDITY.

shear moraine *Geology.* a debris-laden surface or zone found along the margin of any ice sheet or ice cap, dipping in toward the center of the ice sheet but becoming parallel to the bed at the base; thought to form through the refreezing of meltwater at the base of a glacier.

shear pin *Mechanical Devices.* any pin used to connect or hold parts together in place, such as fuses to electrical wiring or a power train winch assembly; designed to break apart when shear stresses or impact forces become too great for the specified load. Also, BREAK PIN.

shear plane see SHEAR SURFACE.

shear rate *Fluid Mechanics.* stresses caused by a velocity gradient in a viscous flow; the off-diagonal components of the rate of deformation tensor.

shear resistance *Fluid Mechanics.* the tangential stress caused by fluid viscosity that takes place along a flow boundary and acts in the tangential direction to the local motion. Also, SHEAR DRAG.

shears *Mechanical Devices.* **1.** any of a number of large cutting tools having a pair of handled cutting blades joined in the center by a pivot. **2.** a machine tool that is used for cutting sheet metal. **3.** an apparatus used for lifting heavy objects vertically, consisting of a pair of uprights crossed and secured at the top, held in position by guys, and equipped with tackle.

shear slide *Geology.* a landslide, especially one resulting from shear failure along a plane of weakness such as a bedding or cleavage plane.

shear sorting *Geology.* a sorting of sediments in which the smaller grains tend to move toward the zone of greatest shear strain, and the larger grains tend toward the zone of least shear strain.

shear spinning *Mechanical Engineering.* a method of forming sheet metal parts with rotational symmetry over a mandrel, by using a roller so that the diameter of the original blank does not change but its thickness decreases in accordance with the mandrel angle.

shear strain *Mechanics.* the movement of parallel surfaces of a solid body, as determined by the change in angle between points on the two surfaces that were originally perpendicular to each other and both perpendicular to the plane of the movement. Also, DETRUSION.

shear strength *Mechanics.* **1.** the maximum shear stress that a material can withstand without rupture. **2.** the ability of a material to support shear stresses.

shear stress *Mechanics.* the force per unit area that tends to cause the shear strain. Also, SHEARING STRESS, TANGENTIAL STRESS.

shear structure *Geology.* a rock structure resulting from the shearing of rock.

shear surface *Mechanics.* a surface along which relative displacement has taken place parallel to the surface. Also, SHEAR PLANE.

shear test *Civil Engineering.* a measure of the shear or rupture displacement of a mass of soil beneath a footing.

shear thickening *Fluid Mechanics.* the phenomenon in which non-Newtonian fluids subjected to shear stress will undergo a viscosity increase; e.g., a dilatant material such as a starch suspension in water.

shear thinning *Fluid Mechanics.* the phenomenon in which non-Newtonian fluids subjected to shear stress will undergo a viscosity reduction; pseudoplastic materials are examples, such as certain polymeric materials, or suspensions of pigments.

shear transformation see MARTENSITIC TRANSFORMATION.

shear-viscosity function *Fluid Mechanics.* an expression for viscometric flow of a viscous, non-Newtonian fluid in terms of its shear stress and velocity gradient.

shearwater *Vertebrate Zoology.* any of several marine, pelagic birds of the family Procellariidae, having long wings and tubular nostrils; found in all oceans.

shear wave *Mechanics.* a wave whose propagated disturbance is a shear strain in an elastic medium. Also, ROTATIONAL WAVE. *Geophysics.* see S WAVE.

shear yielding *Materials Science.* the ability of glassy polymers to deform without breaking under stress.

shear zone *Geology.* a tabular region of rock showing evidence of shear stress in the form of crushing and brecciation by many parallel fractures.

sheath *Biology.* a tubular structure enclosing or surrounding another part or organ. *Mechanical Devices.* a cover or protective holder, as for a blade. *Electricity.* a protective covering applied to a cable. *Electromagnetism.* in a waveguide, the external conducting surface of the waveguide wall. *Electronics.* the charge produced by the cluster of ions that appear near an electrode in a gas tube.

sheathed bacteria *Microbiology.* any bacteria that are covered with a hyaline envelope or sheath.

sheathing paper *Materials Science.* a flexible waterproof bitumenous material reinforced with fiber and faced with thick kraft paper; used as a lining in construction.

sheath-reshaping converter *Electromagnetism.* a waveguide structure in which, by use of conductive metal sheets and gradual changes in the contour of the waveguide sheath, particular desired modes can be achieved.

sheave *Mechanical Devices.* **1.** a pulley or wheel having a grooved rim around which a rope or chain rotates. **2.** the disk of an eccentric part keyed directly on the shaft of a steam engine.

shed *Building Engineering.* a small, usually roughly built structure used for shelter or storage.

shedding *Biology.* the natural sloughing or casting off of hair, feathers, leaves, and so on. *Textiles.* an operation of a loom in which some warp threads are raised by the heddles and others are left down, creating a path for the shuttle to pass through to insert filling threads.

Sheehan's syndrome *Medicine.* postpartum necrosis of the pituitary gland, and the resultant reduced functioning of the pituitary, thyroid, adrenal cortex, and gonads, caused by uterine hemorrhaging.

sheen *Optics.* the shine or luster given off by a surface.

sheep *Vertebrate Zoology.* any of several wild or domesticated species of large ruminants in the genus *Ovis*, characterized by a stocky build and heavy wool coat. The common domestic sheep, *O. aries*, is bred in a number of different varieties.

sheep (Cotswold ram)

sheepback rock or **sheepback** see ROCHE MOUTONNÉE.

sheepsfoot roller *Mechanical Devices.* a rolling disk or drum having studs or spikes set in it; used to perforate or compact earth or pavement.

sheepshead clock *Horology.* a lantern clock having a very large dial.

sheepskin *Materials.* the skin of a sheep, usually dressed with the wool on and used in making blankets and garments.

sheepskin wheel *Mechanical Devices.* a polishing wheel whose surface is covered with sewn or quilted sheepskin.

sheer *Geology.* very steep, as a precipice or a steep face of a cliff. *Materials.* unmixed with other substances. *Textiles.* thin and nearly transparent.

sheer *Navigation.* **1.** to deviate a ship from its course; swerve. **2.** to move or position a ship at anchor so that it keeps clear of the anchor. *Naval Architecture.* an upward curve at either end of a vessel's sheer line.

sheering batten *Naval Architecture.* one of a series of long wooden strips that serve as a guides for applying the shell plating in proper relationship to the sheer line.

sheer line *Naval Architecture.* the fore and aft line formed by the upper edge of a vessel's side or bulwarks. A vessel whose sheer line curves up toward bow or stern is said to have sheer.

sheer pole *Naval Architecture.* a crossbar joining and stabilizing a vessel's shrouds, normally fitted just above the deadeyes. Also, **sheer batten.**

sheer strake *Naval Architecture.* the uppermost strake of a vessel's side planking or plating, forming the sheer line of the hull.

sheet *Textiles.* a large rectangular piece of cotton, linen, or other material used as a bed covering. *Materials.* a similar broad, thin piece of some other material, such as paper, glass, or metal. *Cartography.* any single printed map, whether self-contained or part of a set or series. *Geology.* **1.** a thin coating of calcite deposited in a cave by floating water. **2.** a tabular igneous intrusion. *Hydrology.* see SHEET ICE. *Naval Architecture.* a line leading aft from the lower edge of a sail, used to secure it against the force of the wind. *Mathematics.* **1.** in topology, a connected subset of a covering space (such as a Riemann surface) whose projection covers the entire base space. **2.** a component of a disconnected (usually two-dimensional) manifold, such as a parabolic hyperboloid.

sheet anchor *Naval Architecture.* **1.** in very large ships, the heaviest anchor carried for use when the bower anchors cannot hold the vessel. **2.** an informal term for ship-board security.

sheet cavitation *Fluid Mechanics*. the type of cavitation in which cavities of gaseous or vapor fluid form on a solid boundary and remain attached, unless the condition that led to the formation is altered. Also, STEADY-STATE CAVITATION.

sheet crack *Geology*. a planar crack thought to be produced by shrinkage of sediment due to dewatering.

sheet deposit *Geology*. a horizontal stratiform mineral deposit that is extensive in area relative to its thickness.

sheet drift *Geology*. a flat, evenly spread deposit of glacial drift that did not significantly alter the form of the underlying rock surface.

sheeted fissure *Geology*. any of a group or series of closely spaced fissures.

sheeted vein *Geology*. any of a series of closely spaced parallel veins.

sheeted zone *Geology*. an area characterized by mineral deposits occurring in sheeted veins.

sheet erosion *Geology*. an erosion process in which thin, relatively uniform layers of surface material are gradually removed by the action of continuous sheets of water flowing over an extensive region of land. Also, **sheetform erosion, sheetwash.**

sheet explosive *Ordnance*. plastic explosive provided in a sheet form.

sheet-fed press *Graphic Arts*. a printing press designed to accept one sheet of paper at a time, as distinguished from presses that print on a continuous roll.

sheet film see CUT FILM.

sheet flood or **sheetflood** *Hydrology*. a short-duration flood that spreads over a large area as a thin, continuous film, rather than being concentrated in well-defined channels.

sheet flow or **sheetflow** *Hydrology*. the downslope movement of surface runoff over relatively smooth land surfaces in the form of a thin, continuous film that is not concentrated in channels.

sheet frost *Hydrology*. a thick covering of rime that forms on hard surfaces, such as windows.

sheet glass *Materials*. glass produced by drawing a continuous filament from a molten bath and, after cooling, cutting it into sheets.

sheet grating *Electromagnetism*. a three-dimensional grating that is constructed by conductive metal sheets placed longitudinally along a waveguide and having lengths of approximately one wavelength; used to filter modes of propagation.

sheet ice *Hydrology*. a relatively smooth, thin layer of ice that is formed as the surface of a body of water freezes rapidly.

sheeting *Textiles*. a plain-weave fabric made from carded or combed yarns in medium and heavy weights, usually in cotton but also in nylon, silk, and other fibers; used for bed sheets, upholstery, window shades, floor cloths, etc. *Materials*. a film of plastic over 250 μm in thickness; usually made by extrusion, casting, and calendering.

sheeting plane *Geology*. the primary cleavage plane or parting in an igneous rock.

sheeting structure *Geology*. a type of rock fracture or jointing that is formed by pressure-release jointing or exfoliation. Also, SHEET JOINT, EXFOLIATION JOINT, EXPANSION JOINT.

sheet joint see SHEETING STRUCTURE.

sheet lightning *Geophysics*. a common cloud or cloud-to-cloud discharge whose illumination has been obscured by the surrounding cloud, giving the discharge a solid, diffuse appearance. Also, LUMINOUS CLOUD.

sheet line see NEATLINE.

sheet metal *Metallurgy*. a flat rolled product that is thinner than a plate, thicker than a foil, and wider than a strip.

sheet-metal forming *Metallurgy*. the mechanical deformation of a metal sheet to obtain parts of desired shapes; some shapes may impart rigidity to parts or structural components.

sheet-metal screw see SELF-TAPPING SCREW.

sheet pile *Civil Engineering*. any of a line of timber, steel, or reinforced concrete piles used in sheet piling.

sheet piling *Civil Engineering*. a line of sheet piles forming a wall that serves to resist lateral pressures of moving earth or water.

sheet polarizer *Optics*. a sheet of material that linearly polarizes light passing through it.

Sheetrock *Building Engineering*. the trade name for a plasterboard commonly used in housing construction, consisting of a sheet of pressed gypsum covered on each face with paper.

sheet sandstone *Geology*. a thin, regionally extensive, blanketlike deposit of sandstone.

sheet separation *Metallurgy*. in spot, seam, or projection welding, the separation between the joined parts after welding.

sheet spar *Geology*. a sheet crack filled with spar.

sheet structure *Materials Science*. a characteristic microstructure in which there is strong bonding in two dimensions, producing platelike layers, or sheets; allows a material to be split into very thin sheets, as in mica, or permits mobility between sheets, giving graphite its lubricating and semimetallic properties.

sheet texture *Materials Science*. a preferred orientation obtained in rolling or extrusion processes; most of the grains are aligned so that a crystallographic direction is parallel to the rolling direction and a preferred crystallographic plane is parallel to the sheet surface.

sheet train *Mechanical Engineering*. the entire machine assembly used for forming plastic sheets.

sheetwash or **sheet wash** *Geology*. **1.** the detritus transported and deposited by a sheetflood. **2.** see SHEET EROSION. *Hydrology*. see SHEET FLOOD.

sheetwise *Graphic Arts*. a printing method in which different forms are used to print the two sides of a sheet of paper. Also, WORK-AND-BACK.

sheldgoose *Vertebrate Zoology*. a water bird of the family Anatidae, not a true goose; includes several tropical and subtropical species worldwide having short, small bills, long legs, and white and black feathers.

shelduck *Vertebrate Zoology*. any of various particolored ducks, especially of the genus *Tadorna*. The male is called a **sheldrake.**

shelf *Geology*. **1.** the solid rock underlying an alluvial deposit. **2.** a flat-surfaced layer or projecting ledge of flat rock. **3.** that part of a stable cratonic region of sedimentation that is bordered by a geosyncline or other more mobile area of sedimentation.

shelf angle *Civil Engineering*. a section of angle iron or steel that is welded or otherwise secured to an I beam or channel section to provide support for the formwork, the hollow tiles of a concrete slab, or precast concrete.

shelfback *Graphic Arts*. see SPINE.

shelf break *Geology*. an obvious, abrupt increase in gradient that marks the boundary between the continental shelf and the continental slope.

shelf channel *Geology*. a shallow valley along a continental shelf.

shelf edge *Geology*. an increase in gradient (less steep than a shelf break) that marks the boundary between a continental shelf and a continental slope.

shelf facies *Geology*. sedimentary facies produced in the neritic environments of marginal shelf seas that border a low, stable land surface; characterized by carbonate rocks and fossil shells. Also, FORELAND FACIES, PLATFORM FACIES.

shelf ice *Hydrology*. floating freshwater ice that has formed in an ice shelf or broken away from an ice shelf. Also, BARRIER ICE.

shelf life *Engineering*. the time it takes for such items as food, batteries, or chemicals to age or deteriorate beyond use.

shelf sea see EPICONTINENTAL MARGINAL SEA.

shelfstone *Geology*. a horizontally projecting ledge of mineral material that is formed at the edge of a cave pool.

shell *Botany*. the hard outer covering of a nut, seed, or fruit. *Zoology*. **1.** the calcareous, horny, or chitinous covering or encasement of certain animals such as mollusks. **2.** the hard, protective calcium-based covering of an egg. *Design Engineering*. any object resembling an animal or egg shell, usually having a hemispherically shaped hollow structure. *Mechanical Engineering*. the outer skin, casing, or external covering of any machine or device such as a boiler. *Chemistry*. the orbits of the electrons as they revolve around the nucleus of an atom; depending on the atomic number, an atom of a particular element has from one to seven shells, each of which represents a successively higher energy level. *Architecture*. the curved structure forming a dome or vaulted roof. *Building Engineering*. **1.** the unfinished framework of a building. **2.** a building or structure without partitions. *Geology*. **1.** the crust of the earth, or any of the continuous underlying layers making up the earth's interior. **2.** a thin, generally hard layer of rock. *Ordnance*. **1.** a hollow projectile containing or designed to contain a high-explosive or other charge. **2.** a shotgun cartridge; it is sometimes applied to any cartridge for artillery or firearms. **3.** to fire upon with shells. **4.** of or relating to shells. Thus, **shell burst, shell case, shellfire, shell hole, shell holder, shellproof, shell room.** *Computer Programming*. **1.** a program that connects and interprets commands written by users, calling programs from memory and executing them sequentially. **2.** a reuseable software tool for building knowledge-based and expert systems that provides inference engines, user interfaces, explanation facilities, report generators, etc. Also, SKELETAL PROGRAM. **3.** a function of certain application programs by which the user can temporarily exit the program and return to the operating system without closing the files in the application program.

shellac *Materials.* **1.** a purified lac, generally formed into thin flakes or sheets. **2.** a thin, relatively clear varnish made by dissolving this material, usually in alcohol.

shellac wheel *Mechanical Devices.* a grinding or polishing wheel whose abrasive surface contains shellac.

shell-and-tube exchanger *Engineering.* a device that transfers heat from a hot fluid to a cooler fluid.

shell bit *Mechanical Devices.* a gougelike boring tool fitted with a brace used for making screw holes. Also, GOUGE BIT.

shell broach *Design Engineering.* a broach whose body consists of roughing and intermediate sections.

shell capacity *Engineering.* the amount of liquid that a tank car or truck shell can hold.

shell clearance *Design Engineering.* in a reaming shell, the difference between the outside diameter of a bit or core barrel and the outside set or gage diameter.

shell-destroying tracer *Ordnance.* a tracer that is designed to burst a projectile, missile, or rocket in the air if the weapon overshoots its target.

shell drill *Design Engineering.* a drill similar to a twist drill but having three or four flutes and tapered bores.

shell filler *Ordnance.* the charge or filler of a projectile; it may be high-explosive, chemical, atomic, nuclear, inert, or other material.

shellfish *Invertebrate Zoology.* the common name for various aquatic invertebrates having a shell (mollusks) or a hard exoskeleton (crustaceans); used especially to refer to such animals that are edible by humans; e.g., oysters, crabs, and lobsters.

shell gland *Invertebrate Zoology.* **1.** in shelled mollusks, an organ that secretes the materials for the shell's formation. **2.** in many animals, a gland in the oviduct that secretes an egg's covering.

shell ice *Hydrology.* ice that formed originally on a body of water and which remains unbroken after the water level has dropped, leaving a cavity between the ice and the surface of the water. Also, CAT ICE.

shell innage *Transportation Engineering.* the depth of the liquid in a tank car or truck shell.

shell knocker *Engineering.* a device that strikes the outside surface of a horizontally rotating vessel, such as a kiln or dryer, and loosens solid materials that have built up on the inner walls of the shell.

shell marl *Geology.* a light-colored calcareous deposit that forms on the bottoms of freshwater lakes, composed of uncemented mollusk shells, precipitated calcium carbonate, and hard parts of minute organisms.

shell membrane *Cell Biology.* a covering that encloses an egg directly beneath the shell.

shell model *Nuclear Physics.* a model of an atom in which nucleons of a given type are postulated to reside in groups of roughly the same level of energy in a succession of orbits, or shells, around the nucleus.

shell molding *Metallurgy.* forming a mold with sand containing a thermosetting resin and heat curing prior to pouring molten metal.

shell mound or **shell midden** *Archaeology.* a refuse mound of discarded shells that gives evidence of early human use of certain mollusks.

shell outage *Transportation Engineering.* the unfilled portion of a tank car or truck shell.

shell pump *Mechanical Engineering.* a simple sand or sludge pump, consisting of a hollow cylinder with a ball or clack valve at the bottom and used with a flush of water.

shell reamer *Mechanical Devices.* an economically designed, hollow reamer mounted on a mandrel or a removable arbor.

shell roof *Architecture.* see SHELL.

shell sand *Geology.* any loose aggregate containing shell fragments, especially a marine sand that contains up to 5% shell fragments.

shell shock *Psychology.* a term used during World War I for a disorder caused by prolonged exposure to the heavy artillery shelling of battle; now described as post-traumatic stress disorder. Thus, **shell-shocked.**

shell star *Astronomy.* hot stars, such as novae and supernovae, that are surrounded by extensive atmospheres or expanding gas envelopes.

shell still *Chemical Engineering.* a large horizontal cylinder mounted over a furnace, used to distill crude oil in a petroleum refinery.

shell structure *Nuclear Physics.* in the shell model, the configuration of the nucleus that places nucleons at roughly the same energy level in the same quantum state.

shell tap *Design Engineering.* a shankless, high-speed steel tap that is bored along its entire length and mounted on a driving device.

shell-type transformer *Electromagnetism.* a transformer constructed such that the magnetic circuit completely surrounds the coils.

shelly *Geology.* **1.** of a sediment or sedimentary rock, containing the shells of animals. **2.** of land, abounding in or covered with shells.

shelly facies *Geology.* a nongeosynclinal sedimentary facies characterized by an abundance of calcareous fossil shells, carbonate rocks, and mature orthoquartzitic sandstones and by a paucity of shales.

shelly pahoehoe *Volcanology.* a type of pahoehoe whose surface is characterized by open tubes and blisters.

shelterbelt *Ecology.* a barrier of trees or shrubs that prevents erosion and offers shelter from wind.

shelter cave *Geology.* a cave that extends only a short distance under the ground and whose roof of overlying rock usually extends beyond its sides. Also, ROCK CAVE.

shelter deck *Naval Architecture.* a full-length superstructure deck above the main deck; similar to the spar deck of a sailing vessel.

shelter porosity *Geology.* a type of porosity that occurs when relatively large sedimentary particles prevent the infilling of underlying pore spaces by finer clastic particles.

shelterwood cutting *Forestry.* a method of harvesting timber in which trees that require shade to develop are removed in several stages over a period of 10 to 20 years.

shepherd satellite *Astronomy.* a satellite that confines ("herds") a ring or ring arc by means of its gravity.

sherd see POTSHERD.

shergottite *Geology.* an achondritic stony meteorite that is composed primarily of pigeonite and maskelynite.

sheridanite *Mineralogy.* a colorless to pale-green, talclike variety of clinochlore.

Sheridan tank *Ordnance.* a light tank that can be transported by air and delivered by parachute; it is armed with a 152-mm gun launcher that can fire conventional rounds or Shillelagh missiles.

Sherman tank *Ordnance.* a medium-sized (34-ton) tank that is armed with a 75-mm gun and carries a crew of four; the main U.S. fighting tank of World War II.

Sherrington, Sir Charles 1857–1952, English physiologist; pioneered the functional analyses of neurons (Nobel Prize), synapses, and reflexes.

Sherrington phenomenon *Neurology.* a slow muscle contraction in response to stimulation of the sciatic nerve after nerve degeneration.

sherry *Food Technology.* a slightly sweet fortified wine, originally produced in the region of Jerez de la Frontera, Spain. The best-known types of sherry are the pale dry **Fino sherry** and the sweet, often nutty **Amontillado sherry.** (An Anglicized version of the name *Jerez.*)

sherry topaz *Mineralogy.* a variety of the mineral topaz, characterized by an amber color similar to sherry wine.

Sherwood effect *Quantum Mechanics.* the property by which a particle's wave function collapses in the presence of perpendicularly directed electric and magnetic fields.

Sherwood number see NUSSELT NUMBER.

Shetland Islands *Geography.* an island group northeast of Great Britain, north of the Orkney Islands.

Shetland pony *Agriculture.* a breed of small, sturdy ponies originally raised in the Shetland Islands.

Shetland sheep *Agriculture.* a breed of sheep raised in the Shetland Islands, distinguished by its fine wool.

Shewanella *Bacteriology.* a proposed genus of bacteria that would include species of *Alteromonas* and species isolated from deep-sea habitats.

SHF super high frequency; sensible heat factor.

shield a protective structure or device; specific uses include: *Building Engineering.* a metal barrier placed around certain parts of equipment to protect the operator. *Electromagnetism.* a metallic enclosure that houses an electromagnetic device and prevents it from interacting electromagnetically with other devices. *Nucleonics.* an absorbing material that surrounds a nuclear reactor to aid in reducing radiation leakage from the system. *Metallurgy.* **1.** any barrier that shields from radiant heat. **2.** in electroplating, a nonconductive material placed at selected locations to modify the distribution of the electric current. *Ordnance.* an armored plate mounted on a gun carriage to protect the gun and crew. *Mining Engineering.* a steel cylinder, used in excavating a tunnel in soft material, with a diameter equal to that of the tunnel. *Zoology.* a protective plate, scale, or other such structure on the body of an animal. *Geology.* **1.** a broad, stable area of the earth's crust consisting mostly of Precambrian rocks surrounded by sediment-covered platforms that have been generally unaffected by later orogenic episodes. **2.** see CONTINENTAL SHELF. **3.** see PALETTE.

shield basalt *Geology.* basaltic lava that flowed from a group of small, closely spaced shield-volcano vents and coalesced to form a single unit.
shield cone *Geology.* a dome-shaped volcano built up by successive outpourings of lava.
shielded arc welding *Metallurgy.* arc welding in which the weld is protected by a suitable atmosphere or flux.
shielded cable *Electricity.* a cable with conductors that are shielded, either in groups or as a whole.
shielded-conductor cable *Electricity.* a cable made up of strands of shielded wire and covered with a tough protective material such as polyethylene or rubber.
shielded joint *Electricity.* a shielded junction or connection of two components.
shielded line *Electromagnetism.* a transmission line in which the elements confine all radiation to a finite space in the vicinity of the line.
shielded pair *Electricity.* a two-wire transmission line that is surrounded by a metallic sheath.
shielded wire *Electricity.* a conductor that has electromagnetic shielding to prevent radiation.
shield factor *Telecommunications.* in a telephone circuit, the ratio of noise, voltage, or current when the point of origin of shielding is present to the corresponding amount when the shielding is absent.
shield grid *Electronics.* a structure in a glass tube that protects the anode and cathode from heat generated by the control electrode.
shield-grid thyratron *Electronics.* a tube in which both the cathode and the shield grid emit electrons to trigger current flow.
shielding *Electromagnetism.* see SHIELD. *Metallurgy.* **1.** any barrier that shields from radiant heat. **2.** in electroplating, placing a nonconductive material at selected locations to modify the distribution of the electric current. *Nucleonics.* see SHIELD.
shielding factor *Geophysics.* the ratio of the actual strength of a magnetic field to the recorded strength of the field on a directional compass after the materials of that compass have dissipated the strength of the field.
shielding gas *Metallurgy.* any gas that protects an operation such as arc welding from oxidation.
shielding layer *Meteorology.* the layer of air nearest the earth that acts as a shield from circulation patterns and other weather activity in the free atmosphere above, or vice versa, such as the relatively unstable layer beneath a subsidence inversion.
shielding ratio *Electromagnetism.* a ratio given by the electromagnetic field strength of a region near a shielded device divided by the strength of the field when the shield is not present.
shield volcano *Geology.* a broad, gently sloping volcano resembling a flattened dome, generally formed by overlapping and interfingering basaltic lava flows. Also, BASALTIC DOME, LAVA DOME.
shift to change in position, rate, status, and so on; specific uses include: *Mechanical Engineering.* **1.** to change the gear ratio in an engine to obtain a desired rotational speed. **2.** see GEAR SHIFT. *Computer Technology.* **1.** to change the interpretation of characters, such as to change from lowercase to uppercase letters. **2.** the movement of a bit pattern in a bit string to the left or right. *Metallurgy.* in casting, a flaw caused by the lack of registry of mold assembly components. *Microbiology.* any alteration in the environment of a microorganism that causes an increase or a decrease in the rate of growth. A change that increases the growth rate is a shift up; one that slows growth is a shift down. *Genetics.* a structural aberration that results in the transposition of a chromosomal segment to another area of the same chromosome. *Spectroscopy.* for a spectral line, a small change in wavelength or frequency associated with a change in pressure, density, or some other cause. Also, **shift of spectral line.** *Geology.* the relative displacement of faulted rock masses lying outside the fault zone itself. *Building Engineering.* in masonry, a type of brick and building stone layup wherein vertical joints are noncontinuous. Also, BREAKING JOINT. *Industrial Engineering.* a set period during which a group of workers is on duty; around-the-clock production typically consists of three shifts: a **day shift,** an **evening shift** (**swing shift**), and a **night shift** (**graveyard shift**).
shift down *Microbiology.* see SHIFT.
shift factor *Materials Science.* a factor used in calculations of glass-rubber transition behavior; relates the time of transition and melt viscosity to temperature.
shifting *Geology.* **1.** a change in the position of a coastline due to changes, fluctuations, or oscillations of sea level. **2.** see MIGRATION. *Hydrology.* the movement of a stream divide away from a more actively eroding stream toward a weaker stream. Also, MIGRATION.
shifting cultivation or **shifting agriculture** *Agronomy.* a simple farming method in which parcels of land are cultivated extensively for short periods, left fallow for several years, and used in later years. Also, EXTENSIVE CULTIVATION.
shifting fire *Military Science.* fire delivered at constant range at varying deflections; it is used to cover the width of a target that is too great to be covered by an open sheaf.
shifting theorem *Mathematics.* **1.** if $F(x)$ is the Fourier transform of $f(t)$, then $\exp(iax)F(x)$ is the Fourier transform of $f(t - a)$. **2.** if $L(y)$ is the Laplace transform of $f(x)$, then $\exp(-ay)L(y)$ is the Laplace transform of $f(x - a)$.
shift instruction *Computer Technology.* a computer instruction that causes the word stored in a register to be displaced a specified number of bit positions to the left or right.
shift joint *Building Engineering.* a vertical joint that is positioned on either a heavy-duty or a strong structural member of a course directly beneath it.
shift key *Computer Technology.* a key that shifts the type to uppercase letters and also controls other functions depending on the program.
shift of butts *Naval Architecture.* an arrangement of planking in which the butts, or ends of the planks, are staggered horizontally instead of in line vertically.
shift-reduce parser *Computer Science.* a parser that operates by alternately shifting input elements onto the top of a stack or reducing a group of elements from the top of the stack to a larger element representing a phrase.
shift register *Computer Technology.* a special-purpose register capable of moving a sequence of stored bits to the right or left a specified number of positions.
shift-register generator *Computer Technology.* a random-number generator that uses a shift register to perform linear transformations over a vector space to produce a series of pseudorandom integers.
shift up *Microbiology.* see SHIFT.
Shigella *Bacteriology.* a genus of Gram-negative, facultatively anaerobic, nonmotile, coliform bacteria of the family Enterobacteriaceae; the causative agent of human and animal dysenteries.
shii-take [shē´i tä´kē; shē´ē tä´kā] *Mycology.* a fungal mushroom, *Lentinula edodes,* that is edible and is cultivated in Japan and China. (From a Japanese term meaning "oak mushroom.")
shikimic acid *Biochemistry.* $C_7H_{10}O_5$, a hydroxylated, unsaturated cyclohexane derivative that serves as an intermediate in the biochemical pathway between phosphoenolpyruvic acid and tyrosine.
Shillelagh [shə lā´lē] *Ordnance.* a missile system mounted on the main battle tank and assault reconnaissance vehicle for employment against enemy armor, troops, and field fortification; it includes a gun launcher with a fire control system and operates at a range of around 2.5 miles; officially designated **MGM-51.** (From the *shillelagh,* an Irish walking stick also used as a weapon.)
shim *Mechanical Devices.* **1.** thin, graded, metal sheets, usually of metal, used to fill, fit, align, or adjust the spacing between bearings on a shaft or other parts. **2.** a sharp, horizontal attachment to a cultivator, used for weeding and surface scraping between rows of crops. **3.** thin tapered metal or other material used to fill in cracks, crevices, or spaces between stones or railway ties to account for wear or leveling. *Engineering.* a narrow strip glued to a panel of wood or into the lumber core during the manufacture of plywood.
shim joint *Materials Science.* a joint used in joining composites, in which thin metallic interleaved layers reinforce the composite material at the joint.
shimmer *Meteorology.* to appear to be wavering or shimmering as a result of atmospheric refraction variations along one's line of sight; an effect usually associated with heat waves in the atmosphere. Also, TERRESTRIAL SCINTILLATION.
shimmy *Mechanics.* excessive vibrations in the front wheels of a motor vehicle, reinforced by resonance at critical speeds.
shim rod *Nucleonics.* a control rod used to make coarse adjustments to the reactivity in a reactor. Also, **shim control.**
shin *Anatomy.* the anterior aspect of the leg from the knee to the ankle.
shinbone *Anatomy.* a popular name for the tibia. See TIBIA.
shin-bone fever see TRENCH FEVER.
Shine-Dalgarno sequence *Genetics.* the specific sequence of nucleotides (AGGAGG) in an mRNA molecule that is precisely complementary to a sequence at the 3' end of an rRNA molecule and helps to align a ribosome with the AUG initiation codon. Also, **Shine-Dalgarno region.**

shingle *Building Engineering.* **1.** a thin piece of wood, slate, or other material, usually rectangular and 8–10 inches across; laid in overlapping rows to cover the roof or walls of a building. **2.** to cover a building surface with shingles.

shingle *Geology.* **1.** a small, well-rounded waterworn pebble, stone, or other coarse detrital material typically occurring on the higher parts of a beach. **2.** see SHINGLE BEACH.

shingle beach *Geology.* a narrow, steeply sloping beach composed of shingle. Also, COBBLE BEACH, SHINGLE.

shingle-block structure see IMBRICATE STRUCTURE.

shingle nail *Building Engineering.* a nail whose thickness varies between a half to a full gauge greater than that of a common nail.

shingle oak *Botany.* an oak, *Quercus imbricaria,* whose wood is used for shingles, clapboard, and the like.

shingle rampart *Geology.* a narrow, wall-like ridge of shingle built up by wave action along the seaward edge of a reef.

shingle ridge *Geology.* a steeply sloping bank of shingle piled up parallel to the shore.

shingles *Medicine.* a popular name for herpes zoster, a condition of painful skin eruptions caused by a viral infection. See HERPES ZOSTER.

shingle structure see IMBRICATE STRUCTURE.

shingling see IMBRICATION.

shingling hatchet *Mechanical Devices.* **1.** a hatchet with a perpendicular notch on one side of its cutting blade for extracting nails, and a hammer head on the opposite side. **2.** a hand boring tool with a single cutting edge with a spiral body, with or without a screw, and without a spur on its outer cutting tip.

shinleaf *Botany.* any of several low perennial herbs of the genus *Pyrola,* especially *P. elliptica,* characterized by creeping underground stems, evergreen leaves, and racemes of white or purplish flowers; found in North America. (So called because its leaves were formerly used for shinplasters.)

shin splints *Medicine.* a painful condition of the shin, caused by strain of the flexor digitorum longus muscle, commonly occurring after strenuous athletic activity such as running.

ship *Naval Architecture.* **1.** any vessel that moves on water. **2.** specifically, a vessel, normally seagoing, that is larger than a boat and capable of independent operation. There is no exact distinction between small ships and large boats. **3.** a sailing vessel with a full ship rig of at least three square rigged masts. **4.** to take anything aboard a vessel, deliberately or otherwise, from cargo to seawater, as in the phrase "shipping water" used to describe a leaking vessel taking on water. **5.** to set a fitting in place, as to "ship the rudder."

ship-based missile (Crossbow missile)

ship-based missile *Ordnance.* a missile designed to be launched at a target from a seagoing vessel.

shipbuilding *Civil Engineering.* the conception, design, and construction of ships.

ship counter *Ordnance.* a device in a naval mine that prevents the mine from detonating until the counter registers a preset number of actuations.

ship drift *Oceanography.* a method of measuring an ocean current by using a ship's navigational instruments to trace its movements while the ship is drifting in a current.

shipfitter *Civil Engineering.* a person who lays out the location of rivet holes and weld paths, the bevelings upon the plates and steel structures, and the fittings of a ship under construction.

ship heading marker *Navigation.* a marker on a compass that indicates the direction in which the ship is heading.

ship influence *Military Science.* in naval mine warfare, the magnetic, acoustic, and pressure effects of a ship, or a minesweep simulating a ship, that are detectable by a mine or other sensing devices.

ship motion *Navigation.* any form of transitional and rotational motion of a ship in a wave system, such as heave, surge, sway, roll, pitch, and yaw.

shipping *Naval Architecture.* the term for an aggregate supply of ships and other significant vessels belonging to a nation or port, or available for a given purpose.

shipping designator *Telecommunications.* in a nautical environment, a code word, usually consisting of four letters and a number, that is used as an address on shipments to a particular overseas location.

shipping fever *Veterinary Medicine.* **1.** pneumonic pasteurellosis; a usually fatal viral disease affecting cattle and sheep, often occurring after stress associated with prolonged transport, intense change in weather, castration, or change in diet. **2.** an acute, highly fatal form of hemorrhagic septicemia.

shipping lane *Navigation.* a sea route that is regularly used, especially by commercial vessels. Also, SEA LANE.

shipping time *Engineering.* the time between sending and receiving a cargo by ship.

ship report *Meteorology.* a report of a marine weather observation that is encoded, usually under World Meteorological Organization or National Oceanic and Atmospheric Administration weather standards, and then transmitted to a weather-reporting facility on land.

ship's bell clock *Horology.* a clock that strikes every half hour, progressing from one to eight bells over a four-hour period, in a manner similar to the system used to mark the progression of watches aboard ships. True ship's bell clocks vary the pattern to include irregular watches.

ship's clock *Horology.* a timepiece using a lever escapement and fitted into a round brass case; designed to be used aboard ship.

ship's field error *Navigation.* an error in a ship's radio direction finder caused by the location of the direction-finding antenna in an area where the ship's primary electromagnetic field is not parallel to the center line of the vessel.

ship's head *Navigation.* the direction that a ship is pointing at a particular instant, measured in degrees. Also, HEADING.

ship synoptic code *Meteorology.* a modification of the international synoptic code used for marine weather observation.

ship-tended acoustic relay *Navigation.* a navigational system in which an acoustic transducer under the hull of a ship broadcasts a signal to a transponder located on the bottom of the ocean, which then rebroadcasts the signal to the ship; by measuring the time between broadcast and reception, the system then determines the distance to the transponder. Also, STAR.

ship-to-shore *Engineering.* operating between a ship at sea and a land installation, as a radio or other form of communication.

ship-to-shore movement *Military Science.* the deployment of the landing force in an amphibious operation from the assault ship to designated landing areas.

shipway *Civil Engineering.* a launching course and repair support for ships.

ship will adjust *Military Science.* in naval gunfire support, a control method in which the ship can see the target and will adjust with the concurrence of the spotter.

shipworm *Invertebrate Zoology.* the common name for various wormlike mollusks in the family Teredinidae that bore into and destroy marine timbers.

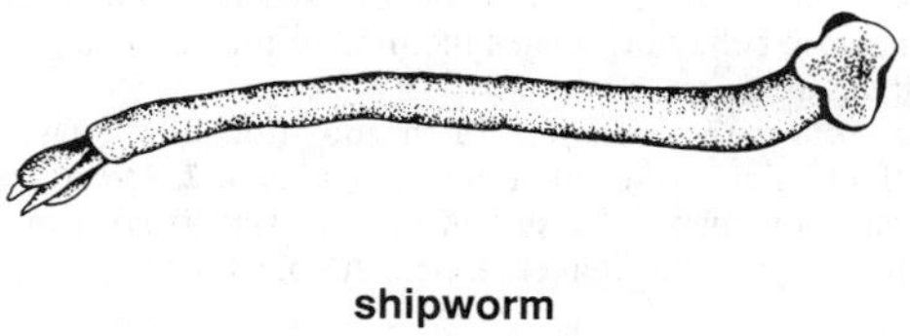

shipworm

shipwright *Civil Engineering.* a person skilled in building and repairing vessels, working chiefly in steel and wooden structures.

shipyard *Civil Engineering.* a place in which ships are built, repaired, or put out of commission.

shiran *Electronics.* an instrument that measures the distance of radio signals with extreme accuracy.

Shire *Agriculture.* the largest breed of draft horse, characterized by a brown or bay coat with white markings and long hairs on the sides and backs of the legs. (From an Old English word meaning "district." Many domestic animals are named for the English *shire,* or county, where they originated.)

shish-kebab morphology see FIBRILLAR CRYSTALLIZATION.

shish-kebab structure *Materials Science.* a specific configuration of polymers in which the random coil chain is interlaced with aggregated crystalline segments; the result is a chain structure that resembles the arrangement of the way food is arranged in a shish-kebab.

SHM simple harmonic motion.

shmoo cell *Cell Biology.* a yeast cell with a protrusion at one end, giving it a pearlike shape, usually resulting from exposure of the cell to a mating pheromone. (From its resemblance to the shape of a *shmoo,* an imaginary animal appearing in the comic strip *Li'l Abner,* by the American cartoonist Al Capp, 1909–1979.)

shoal *Geology.* a submerged accumulation of sediments in a shallow body of water. *Hydrology.* a shallow place in an ocean, lake, stream, or other body of water.

shoal breccia *Petrology.* a submarine breccia, commonly of limestone, produced by wave and tide action on a shoal and resulting from diastrophism or aggradation.

shoaling *Oceanography.* the effect of the bottom on deep-water waves as they approach the shore, including an initial decrease in wave height followed by a rapid increase before the wave breaks.

shoal patches *Oceanography.* irregularly scattered elevations of the bottom, at depths of 10 fathoms (18 meters) or less, composed of materials other than coral or rock.

shoal reef *Geology.* a reef that develops in irregular masses amid submerged shoals of calcareous reef detritus.

shoal water *Oceanography.* shallow water; water in which the bottom is 10 fathoms deep or less.

shoat *Agriculture.* a young hog that has been weaned, especially one that is less than a year old. Also, **shote.**

shock *Medicine.* **1.** a sudden and violent disturbance of physical, emotional, or mental equilibrium. **2.** a condition of profound disturbance of the body's vital processes, characterized by failure of the circulatory system to maintain adequate perfusion of vital organs; it may result from inadequate blood volume (**hypovolemic shock**), inadequate cardiac function (**cardiogenic shock**), or inadequate vasomotor tone (**neurogenic shock, septic shock**). The symptoms of shock include lowered blood pressure, weak pulse, rapid and shallow breathing, restlessness, anxiety or mental dullness, nausea with or without vomiting, and reduced body temperature. *Physiology.* the effect on the body of an electric current passing through it. *Mechanics.* any pulse or abrupt transient motion or sudden force that can initiate a mechanical response in a system. *Geology.* the physical effect of an earthquake.

shock absorber *Mechanical Engineering.* a device, such as a spring or hydraulic damper, that is designed to absorb the energy associated with impact so as to reduce the amount of shock to the frame or support. Informally called **shocks.**

shock action *Military Science.* a mobile attack strategy in which the first impact of the attack is sudden, violent, and massive, thus producing the main offensive effect; it is employed especially by tank units.

shock breccia *Petrology.* a fragmental rock produced by shock waves, such as suevite formed by the impact of a meteorite.

shock bump *Mining Engineering.* **1.** a rock bump due to the sudden collapse of a strong deposit, such as thick sandstone. **2.** a bump felt in a mine when overlying rocks break, causing the effect of a hammer blow.

shock diamonds *Physics.* a term for shock waves that are present in the exhaust stream of a rocket engine, forming a diamond configuration when viewed from the side.

shock excitation *Electricity.* the excitation generated by a pulse; used to begin oscillation in an oscillator's resonant circuit.

shock front *Physics.* a pressure front; the boundary that separates the zero gauge pressure from the peak gauge pressure as a shock wave propagates through a fluid medium.

shock heating *Physics.* the heating of a fluid by the passage of a shock wave through it.

shock isolation *Mechanical Engineering.* the use of shock absorbers or other devices to relieve the shock to a specific element.

Shockley, William 1910–1989, American physicist; shared the Nobel Prize for the development of the transistor; he also proposed a controversial theory of heredity arguing that intelligence varies by race.

Shockley diode *Electronics.* a device that converts alternating current into direct current by employing the electrical properties of silicon; commonly used to trigger electrical circuits.

Shockley partial dislocation *Solid-State Physics.* a partial dislocation in a crystal in which the translation vector of dislocation (Burger's vector) lies parallel to the fault plane, such that gliding is possible.

shock lithification *Geology.* a process by which loose fragmental materials are converted into coherent aggregates by the action of shock waves.

shock loading *Geophysics.* the process of exposing geological materials to artificially created shock waves generated by explosions, or the exposure of these materials to the impact of a meteorite.

shock melting *Geophysics.* the fusion of geological materials at high temperatures produced by high-energy shock waves.

shock metamorphism *Petrology.* the sum of the permanent changes in a rock, including physical, chemical, mineralogic, and morphologic, that are caused by high-pressure shock waves acting over a period from a few microseconds to a fraction of a minute.

shock mount *Mechanical Engineering.* a structure upon which a shock-absorbing device is mounted to prevent the transmission of shock motion to sensitive equipment.

shock organ *Immunology.* an organ of the body that displays a hypersensitive response to an antibody-antigen reaction, such as the lungs in allergic asthma.

shock resistance *Engineering.* the ability of a metal to resist load impact.

shock stall *Aviation.* a stall brought on by a separation of flow aft of a shock wave.

shock strut *Aviation.* a shock-absorbing strut on the landing gear of a flight vehicle.

shock test *Ordnance.* a test to determine the effect of impact of high explosives on an armor sample.

shock treatment *Psychology.* a treatment of severe mental disorders, usually depression, by using electric current or drugs to induce convulsions or unconsciousness. Also, **shock therapy.**

shock troops *Military Science.* a body of troops that take part in a shock action.

shock tube *Fluid Mechanics.* a long tube divided into two zones by a diaphragm with a high-pressure compression chamber on one side and an expansion chamber on the other; when the diaphragm is ruptured, a shock wave is formed across the diaphragm and travels down the tube; the choices of the drive (compression) gas and the driven gas greatly affect the strength and shape of the shock wave.

shock tunnel *Engineering.* a wind tunnel with hypervelocity that generates shocks with Mach numbers of 6 to 25.

shock wave *Physics.* a pressure wave passing through a fluid medium in which the pressure, density, and particle velocity undergo drastic changes. The source of the shock wave moves at a speed greater than the speed of sound in the fluid.

shock-wave lip *Physics.* a shock wave produced by the lip of a jet nozzle, due to an inability to match the stream pressure to the exhaust pressure.

shock zone *Geology.* a volume of rock in or near an impact or explosion crater that displays distinct evidence of shock metamorphism.

shoe any of various devices or contrivances thought to resemble a covering for the foot; specific uses include: *Engineering.* **1.** a glassmaking crucible in a furnace for heating the blowing irons. **2.** a metal block used for support in bending operations. **3.** a piece used to break rock in crushing machines. *Mechanical Engineering.* **1.** a part of a brake that applies pressure to the drum rim. Also, BRAKE SHOE. **2.** a socket or cap that acts as a receptacle for beams and trusses to bear the thrust of the loads. *Mining Engineering.* **1.** a piece of steel fitted over a guide of a mine cage to guide the cage as it moves. **2.** the wedge-shaped piece attached to the bottom of tubing for sinking through quicksand.

shoebill *Vertebrate Zoology.* a large, grayish wading bird, *Balaeniceps rex,* of the family Balaenicipitidae, related to the herons and storks and characterized by a broad bill shaped somewhat like a shoe; found in central Africa, especially along the White Nile.

shoe brake *Mechanical Engineering.* a brake that operates on friction between an element (shoe) and the surface of a rotating drum.

shoestring *Geology.* a long, relatively straight and narrow sedimentary body having a width/thickness ratio of less than 5:1.
shoestring rill *Geology.* one of several long, narrow, closely spaced, generally parallel, uniform channels produced by overland flow and barely scratching the surface of a relatively steep slope of bare soil or weak, clay-rich bedrock.
shoestring sand *Geology.* a shoestring of sand or sandstone usually found in deposits of mud or shale. Also, **shoestring sandstone.**
shonkinite *Petrology.* a dark-colored syenite consisting primarily of augite and orthoclase, with smaller amounts of olivine, hornblende, biotite, and nepheline.
shoot *Ordnance.* **1.** to discharge a firearm, gun, or similar weapon, or to project a missile. **2.** to hit a target with a bullet, projectile, or missile. *Engineering.* to detonate an explosive that breaks coal loose in a seam, or to blast open a borehole. *Geophysics.* the geologic exploration of an area by detonating high explosives that generate seismic waves, which are analyzed to determine the makeup of the materials through which they have passed. *Botany.* **1.** the aerial portion of a young plant, composing the stem and leaves; new or young growth that arises from some portion of a plant. **2.** to grow from the ground; to put forth buds or shoots. *Hydrology.* **1.** a place where a stream plunges downward or flows rapidly. **2.** a natural or artificial passageway through which water moves to a lower level. **3.** a rapids or a rush of water down a steep slope. *Geology.* see ORE SHOOT.
shooting board *Mechanical Devices.* a device with a swivel and groove or a runway used with a shooting plane for accurately holding a workpiece while making an angular cut between 0° and 90°. Also, MITER SHOOTING BOARD. *Engineering.* **1.** a table and plane upon which printing plates are trimmed. **2.** a fixture used for planing boards.
shooting flow see RAPID FLOW.
shooting plane *Mechanical Devices.* the side plane of a shooting board, which secures a workpiece during cutting.
shooting star *Astronomy.* a popular name for a meteor, which in earlier times was thought to be a star moving across the sky. *Botany.* a perennial of the genus *Dodecatheon*, especially *D. meadia*, characterized by a cluster of nodding, rose, purple, or white flowers with reflexed petals and stamens forming a pointed beak; found in North America.
Shooting Star *Aviation.* a popular name for the F-80 jet fighter.
shooting stick *Mechanical Devices.* **1.** a spiked stick with a beveled top edge for mounting a hunting rifle. **2.** a short wooden or metal bar with a notch used to loosen or tighten a printing quoin.
shop *Industrial Engineering.* a general term for any relatively small building or facility where a specific type of work is done, especially manual work.
shop drawing *Graphic Arts.* a scale drawing used as a guide in the manufacture of a product.
Shope papilloma *Veterinary Medicine.* a transmissible, viral-induced, naturally occurring tumor on the skin of a rabbit.
shop fabrication *Engineering.* the process of making parts and assemblies in a shop rather than at a work site.
shop standards *Engineering.* the written procedures for installation.
shop weld *Engineering.* a weld made in a workshop before a piece is delivered; other welds may be necessary during assembly in the field.
shoran *Navigation.* a precise short-range electronic navigation system utilizing pulse-type transmissions from two or more fixed stations to measure slant-range distance from the stations. (An acronym for short-range navigation.)
shore *Geology.* **1.** a narrow strip of land bordering a body of water, especially a sea or large lake. **2.** the region between the low-water mark and the landward limit of effective wave action. *Engineering.* **1.** a timber that serves as a temporary prop in excavations and buildings. **2.** to support as by a stout timber, usually by a prop.
shorebird *Vertebrate Zoology.* any bird that frequents the shoreline between the ocean or large lakes and the land, especially a bird of the suborder Charadrii, such as a sandpiper, plover, or snipe.
shore bombardment line *Military Science.* a ground line established to delimit bombardment by friendly surface ships.
shore bug see SALDIDAE.
shore current *Hydrology.* a water current near a shore.
shore drift see LITTORAL DRIFT.
shore effect *Electromagnetism.* an effect in which radio waves are bent toward a shoreline when passing over a water-land boundary, presumably due to differences in wave velocities over land and water.
shoreface *Geology.* the narrow, steeply sloping, permanently submerged zone lying seaward (or lakeward) from the low-water shoreline.
shoreface terrace *Geology.* in the shoreface zone of a body of water, a wave-built terrace composed of gravel and coarse sand that was swept from the wave-cut bench into deeper water.
shore fire-control party *Military Science.* a unit that controls naval gunfire in support of shore troops; it consists of a spotting team to adjust fire and a naval gunfire liaison team to work with the supported battalion commander.
shore ice *Oceanography.* pieces of ice that have been beached by wind, tides, ice pressure, or currents.
shore lead *Oceanography.* a lead between pack ice and the shore, or between pack ice and a narrow fringe of fast ice; wind, tides, or currents may narrow it to a tide crack.
shoreline *Geology.* **1.** the intersection of a body of water with the shore. Also, STRANDLINE, WATERLINE. **2.** the outline of the shore.
shoreline cycle *Geology.* the series of progressive changes or stages through which coastal features pass during the normal development of a shoreline, beginning with the establishment of a water level up to the time when the water can no longer make changes in the land. Also, MARINE CYCLE.
shoreline-development ratio *Geology.* a ratio indicating the degree of irregularity of a lake shoreline, expressed as the ratio of the length of the shoreline to the circumference of a circle whose area is equal to that of the lake.
shoreline of depression *Geology.* a shoreline resulting from the absolute subsidence or emergence of a landmass.
shoreline of elevation *Geology.* a shoreline resulting from an absolute rise or emergence of the floor of an ocean or lake.
shoreline of emergence *Geology.* a straight or gently curving shoreline that results from the dominant relative emergence of the floor of an ocean or a lake. Also, EMERGED SHORELINE, NEGATIVE SHORELINE.
shoreline of submergence *Geology.* an irregular shoreline, characterized by bays, promontories, and other minor features, that results from the dominant relative submergence of a landmass. Also, POSITIVE SHORELINE, SUBMERGED SHORELINE.
shore party *Military Science.* an organization of the landing force that facilitates the landing and movement off the beaches of troops and materiel, evacuates casualties and prisoners of war, and facilitates the beaching, retraction, and salvaging of landing ships and craft. Also, BEACH PARTY.
shore platform *Geology.* the flat or gently rolling surface developed along a shore by wave erosion, especially a wave-cut beach. Also, SCAR.
shore polynya *Oceanography.* a stable open patch of water, usually oblong, that is bounded on one side by the shore. Also, **shore polyn'ya.**
shore protection *Civil Engineering.* the protection of shores from scour by breakwaters, graded filters, groynes, and every sort of revetment.
shore reef see FRINGING REEF.
shore tank *Petroleum Engineering.* a shoreside tank in which liquid petroleum products released by tankers are stored.
shore terrace *Geology.* **1.** a terrace produced along the shore of a sea or lake by the action of waves and currents. **2.** see MARINE TERRACE.
shore-to-shore movement *Military Science.* the assault movement of personnel and materiel directly from a shore staging area to the objective.
shoring *Engineering.* the process of providing temporary support for a foundation during an underpinning operation.
short *Electricity.* see SHORT CIRCUIT. *Engineering.* the failure to fill a mold completely in the injection molding of plastics. Also, SHORT SHOT. *Ordnance.* in artillery and naval gunfire support, a spotting or observation to indicate that a burst occurred short of the target in relation to the spotting line; it may also describe the shot or projectile so spotted or observed.
short antenna *Electromagnetism.* an antenna whose length is less than one-tenth of a wavelength and whose current is assumed to be roughly constant over its length so it may be treated as an elementary dipole.
short-baseline system *Aviation.* a system for measuring a trajectory that uses a baseline that is short in comparison to the distance of the actual object being tracked.
short-chain branching *Materials Science.* a polymer structural variable characterized by branches that are usually only four carbons long; affects a polymer chain's degree of crystallinity and density.
short circuit *Electricity.* **1.** a path of low resistance through which a current will flow rather than through the load circuit **2. short-circuit.** to bring about such a current path, usually inadvertently and often causing damage and excessive current flow.

short-circuit impedance *Electricity.* the input impedance of a circuit when the output terminals are connected together (short-circuited).

short column *Civil Engineering.* a column whose diameter is so large that bending under load may be neglected, and whose failure would occur in compression; commonly assumed as a column of height less than 20 diameters.

short-contact switch *Electricity.* a selector switch with make-before-break contacts; used to avoid noise during the process of switching.

short-crested wave *Oceanography.* **1.** a wave whose crest length is finite; that is, a wave that actually occurs in nature. **2.** a wave whose crest length is of the same order of magnitude as its wavelength.

short-day plant *Botany.* a plant that requires relatively short daily exposure to light in order to bloom or change in some other way.

short-delay blasting *Engineering.* a technique in which a series of charges are detonated over brief intervals.

short-distance navigation *Navigation.* **1.** marine navigation at distances of less than three miles. **2.** aviation navigation using navigational aids located less than 200 miles from the aircraft.

shortening *Food Technology.* a usually semisolid emulsion made of animal or vegetable fats; used in baking to produce a crispy, flaky effect by dispersing air and oil throughout the batter.

short exact sequence *Mathematics.* an exact sequence of the form

$$0 \to A \to B \to C \to 0;$$

that is, an exact sequence with only three nonzero modules.

short flashing light *Navigation.* a light in which each flash is less than two seconds in duration.

short fuse *Engineering.* **1.** an electrical fuse designed to protect a system against shorts. **2.** a primer fuse that is not long enough to stretch from a charge to the borehole, but is dropped into the hole with the fuse attached.

short-gate gain *Electronics.* a short-range gate's video gain.

short-haul convoy *Military Science.* a convoy between two countries that travels primarily through coastal waters.

Shorthorn *Agriculture.* a breed of beef cattle that originated in England, having red, white, or roan coloring and characteristic short horns; the **Polled Shorthorn** is raised for beef; the **Milking Shorthorn** produces milk as well as beef. Also, DURHAM.

short hundredweight see HUNDREDWEIGHT, def. 1.

short ink *Graphic Arts.* thin or buttery ink, so called because it will not stretch into long strands when compressed and released.

shortite *Mineralogy.* $Na_2Ca_2(CO_3)_3$, a colorless or yellow orthorhombic mineral occurring as wedge-shaped crystals, having a specific gravity of 2.60 to 2.63 and a hardness of 3 on the Mohs scale; found with calcite and pyrite in Eocene clay in Wyoming.

short leg *Engineering.* either of two wires on a blasting cap that is shortened to prevent splices from coming together and causing a short circuit.

short-line seeking *Computer Technology.* a printer optimization technique in which the print element moves only as far as necessary, left or right, to begin printing the next line.

short-long flashing light *Navigation.* a light in which a flash of approximately 0.4 second alternates with a flash of approximately 1.6 seconds, so that the second flash is four times the length of the first flash; this short-long combination is repeated between six and eight times each minute.

shortness *Metallurgy.* a loss of ductility; described as **hot-shortness** if occurring at elevated temperatures and as **cold-shortness** if occurring at ambient temperature.

short noise see SHOT EFFECT.

short oil *Materials.* a term used to describe varnishes and similar substances that contain less than 40% oil.

short-period comet *Astronomy.* a comet whose orbital period is less than 200 years.

short-period interspersion *Genetics.* a genomic pattern in which repetitive DNA sequences, about 300 base pairs in length, alternate with longer, nonrepetitive ones.

short-pulse laser *Optics.* a laser that emits a pulse of light lasting a few nanoseconds or less.

short-range attack missile *Ordnance.* an air-to-surface missile, armed with a nuclear warhead and designed to be launched from a B-52 or F-111 aircraft; the missile's range, speed, and accuracy allow the aircraft to launch it from outside enemy defenses; officially designated **AGM-69.**

short-range ballistic missile *Ordnance.* a ballistic missile with a range up to approximately 600 nautical miles.

short-range force *Physics.* any force between two particles that is considered to be ineffective when the distance between the particles exceeds a certain distance, typically 10^{-13} cm for interactions between nucleons.

short-range forecast *Meteorology.* a weather forecast made for a time period usually not greater than eighteen hours in advance, and often a daily forecast.

short-range order *Materials Science.* a structure in which the arrangement of atoms, typically in a liquid, is only regular and predictable over a short distance, usually comparable to the average distance between atoms.

short-range radar *Engineering.* a radar with a line-of-sight range of 50 to 150 miles.

short residuum *Chemical Engineering.* a type of residual oil that cannot be volatized by vacuum distillation and is refined to make very viscous lubricants.

short round *Ordnance.* **1.** the unintentional or inadvertent delivery of munitions on friendly troops, installations, or civilians by a friendly weapon system, especially by a projectile or missile that fails to travel the expected distance. **2.** a defective cartridge in which the bullet or projectile has been seated too deeply.

short run *Metallurgy.* in casting, the partial filling of a mold.

shorts *Agriculture.* livestock feed that consists of by-products of the milling process, including fine particles of wheat bran, germ, and flour. *Food Technology.* wheat husk fractions that are finer than bran. *Engineering.* the oversized particles left on a screen after the fine particles have been sifted through a screen.

short-scale contrast see HIGH CONTRAST.

short-scope buoy *Navigation.* a navigational buoy that remains nearly vertical over its sinker.

short shot *Engineering.* see SHORT.

short-slot coupler see THREE-DECIBEL COUPLER.

short spin process *Materials Science.* a process for producing staple synthetic fiber using equipment with reduced vertical dimension; uses a shorter air-quench zone and free-fall length and operates at a much lower spinning speed than conventional spinning equipment, but maintains productivity by using spinnerets with a large number of holes.

short stop *Chemical Engineering.* a substance added to terminate the reaction during a polymerization process.

short takeoff and landing see STOL.

short-term memory *Behavior.* the part of the human memory system that stores information shortly after the material is presented; characterized by rapid decay and a limited capacity. Also, PRIMARY MEMORY, SHORT-TERM STORE.

short-term repeatability *Robotics.* the ability of a robot to make the same movements over precisely the same path for a short time period.

short-term store see SHORT-TERM MEMORY.

short-time rating *Electricity.* a rating that defines the load that a machine can carry for a given short duration.

short ton see TON, def. 1.

Shortt pendulum clock *Horology.* a precision timepiece that uses a free pendulum system in which the escapement function is performed by a slave clock, thus allowing the master pendulum to swing without resistance, keeping nearly perfect time. (Named for its inventor, William H. *Shortt*.)

short-tube vertical evaporator *Chemical Engineering.* a process evaporator that consists of a vertical cylinder containing vertical tubes in horizontal tube sheets; used mainly to evaporate cane sugar juice. Also, CALANDRIC, ROBERTS, or STANDARD EVAPORATOR.

shortwall *Mining Engineering.* **1.** a mining method in which relatively small areas are worked individually. **2.** a coal face intermediate in length between a stall and a longwall face, usually 5–30 yards long; used generally in pillar methods of working.

shortwall coal cutter *Mining Engineering.* a coal-cutting machine used to undercut coal by cutting more or less continuously right to left across a heading.

short wave or **shortwave** *Electricity.* a radio wave shorter than that used in AM broadcasting, corresponding to frequencies of over 1600 kilohertz; used for long-distance reception or transmission. *Physics.* a wave of electromagnetic radiation that is no longer than the wavelength of visible light.

shortwave or **short-wave** *Telecommunications.* relating to or used in shortwave broadcasting. Thus, **shortwave radio.**

shortwave broadcasting *Telecommunications.* a radio frequency located in the band between 1600 and 30,000 kilohertz.

shortwave converter *Electronics.* a device that is connected between a radio's antenna and receiver, allowing the receiver to accept signals at a higher-than-normal frequency.

short-wavelength limit *Materials Science.* the shortest wavelength or highest energy radiation emitted or reflected from a material, particularly when X-rays are used, so that there are no heat losses.

shortwave propagation *Telecommunications.* the propagation of radio waves located in the band between 1600 and 30,000 kilohertz.

shortwave radiation *Electromagnetism.* the electromagnetic radiation in the approximate wavelength range of 0.4 to 1.0 micrometer (visible to near infrared).

short word *Computer Programming.* in a system capable of handling words of two different lengths, the fixed word of lesser length.

shoshonite *Petrology.* an alkali basalt, intermediate in composition between absarokite and banakite, containing phenocrysts of olivine and augite in a groundmass that includes labradorite with orthoclase rims, olivine, augite, a small amount of leucite, and some dark-colored glass.

shot *Ordnance.* **1.** a single instance of firing a gun or other weapon; the projectile that is fired. **2.** a solid projectile for a cannon, without a bursting charge. **3.** small lead balls or pellets used as the charge in a shotgun. **4.** a gunnery report that indicates a gun or guns have been fired. *Space Technology.* an act or instance of firing a probe or other unguided ballistic rocket, usually from the earth's surface. *Engineering.* **1.** a charge of any kind of explosive. **2.** the yield from one complete molding cycle. **3.** the small steel balls that form the cutting agent of a shot drill. **4.** any tiny spherical-shaped pieces of steel. *Mining Engineering.* **1.** an explosive charge placed in a hole drilled in coal, ore, or stone to break apart the material. **2.** coal that has been broken down by blasting or other means.

shot bit *Mechanical Devices.* a short length of heavy-walled steel tubing with diagonal slits cut into its bottom edge.

shot blasting *Metallurgy.* the process of impinging hard shots on a metallic surface, usually for descaling.

shot boring *Engineering.* the process of producing a borehole with a shot bit.

shot break *Engineering.* the electrical pulse that records the exact moment that an explosion occurs, used in seismic prospecting.

shot capacity *Engineering.* in plastic forming, the maximum weight of melted resin that an accumulator can push during one forward stroke of the ram.

shot copper *Mineralogy.* small, rounded particles of naturally occurring copper that resemble shot in size and shape.

shotcreting *Engineering.* a process in which mortar or concrete is sprayed from a hose at high velocity onto a surface in order to strengthen the bond between the material and the surface.

shot depth *Engineering.* the distance from the surface to where a charge explodes.

shot drill see CALYX DRILL.

shot effect *Electronics.* excess voltage generated in a tube by unstable electrons, producing popping sounds in a radio broadcast or snow in a television picture. Also, SCHOTTKY NOISE, SHOT NOISE.

shot elevation *Engineering.* the height a dynamite charge reaches in a shot hole.

shot feed *Mechanical Engineering.* a mechanism for delivering chilled steel shot to the drill bit of a calyx drill (shot drill).

shot firing *Engineering.* the process of detonating any type of explosive; blasting.

shot-firing cable *Electricity.* a current-carrying medium in which individual charge carriers cause noise impulses as they move from atom to atom; the noise is known as the shot effect.

shot-firing circuit *Electricity.* in certain radio receivers, circuits in the first amplifying stage that produce a shot effect.

shotgun *Ordnance.* a smoothbore, shoulder-supported firearm that is designed to fire a charge of small lead balls or pellets; it may be single-barreled or double-barreled.

shotgun cloning *Molecular Biology.* an experiment in which the DNA of an organism is divided into fragments and then cloned into a single vector. Also, **shotgun experiment.**

shotgun shell *Ordnance.* a fixed round of ammunition designed to be fired from a shotgun; it usually consists of a plastic or stiff paper case with a thin brass head at one end, a folded crimp at the other, and primer, powder charge, shot charge, and various wads inside. Also, **shotgun cartridge.**

shothole *Engineering.* a borehole in which explosives are inserted for blasting. *Plant Pathology.* a symptom of various diseases that are caused by bacteria, virus, fungi, or spray injury; characterized by small spots on leaves of stone fruits and related plants that drop out, leaving holes that make the leaves look as if they are riddled with shot. Also, **shot hole, shot-hole.**

shothole casing *Engineering.* a lightweight pipe threaded on both ends to prevent the shothole from caving.

shothole drill see BLASTHOLE DRILL.

shot noise see SHOT EFFECT.

shot pattern *Ordnance.* the design created on the surface of a target or other object by the shot scattered from a single shotgun shell or by a series of shots fired under similar conditions. Also, **shot group.**

shot peening *Materials Science.* the process of introducing compressive residual stresses into the surface of a material by bombarding it with shot; the residual stresses may improve the overall performance of the material.

shot point *Engineering.* the point at which an explosion begins generating vibrations in the ground.

shot rock *Engineering.* any rock that has been blasted.

shott or **schott** *Geography.* a shallow brackish lake or saline marsh, usually dry by late summer; scattered throughout the Algerian plateau and south of the Atlas mountains.

shotting *Metallurgy.* the process of producing shots (small steel balls), usually by water quenching molten metal droplets.

shot tongs *Ordnance.* a tool for lifting and moving heavy projectiles in a horizontal position.

shoulder *Anatomy.* the part on each side of the body at which there is a junction of the arm and the trunk. *Robotics.* the point at which the arm of an industrial robot is attached to the base or pedestal. *Design Engineering.* a portion of a shaft or of a stepped or flanged object that has an increasing diameter. *Civil Engineering.* the edge of a road or highway. *Geology.* **1.** a short, rounded spur that projects laterally from the slope of a mountain or hillside **2.** the sloping region that lies below the summit of a mountain or hill. **3.** a bench on the side of a glaciated valley, located where the steep sides of the inner glaciated valley meet the more gradual slope above the level of glaciation. **4.** a structure on the face of a joint formed by the intersection of plume-structure ridges with fringe joints. *Graphic Arts.* the nonprinting area around a type character, image, or image area.

shoulder blade or **shoulderblade** *Anatomy.* a popular name for the scapula, the large triangular bone that forms the posterior part of the shoulder.

shoulder bolt *Mechanical Devices.* a bolt with a hardened, ground shank used in fitting a bearing or bush. Also, **stripper bolt.**

shoulder guard *Ordnance.* a shield over the firing mechanism of a weapon that is designed to protect the gunner when operating in small spaces, such as the inside of a tank.

shoulder-hand syndrome *Medicine.* a symptomatic syndrome characterized by shoulder pain and stiffness, with swelling and pain of the hand on the same side; the condition occurs most often following a myocardial infarction, but may also be produced by other causes.

shoulder harness *Engineering.* a safety apparatus in a vehicle that fastens over the shoulders and chest to prevent the wearer from being thrown forward in a collision or sudden stop.

shoulder joint *Robotics.* the joint that links the arm to the body of a robot.

shoulder nipple *Mechanical Devices.* a nipple that is used in plumbing; slightly longer than a closed nipple with approximately one-quarter inch of unthreaded space between the threads on both ends.

shoulder screw *Mechanical Devices.* a screw with an unthreaded edge beneath its head which acts as a fulcrum for a lever placed on it.

shoulder stock *Ordnance.* see STOCK.

shoulder weapon *Ordnance.* any weapon that is designed to be supported on the shoulder when firing; e.g., a rifle, carbine, or shotgun. Also, **shoulder arm.**

shoulder-wing monoplane *Aviation.* a monoplane on which the wing halves are mounted near the top of the fuselage.

shovel *Mechanical Devices.* any of a number of hand tools having a broad, flat scoop and a hollowed-out blade attached to a handle, used to scoop and lift loose material such as earth, asphalt, or coal. *Mechanical Engineering.* a machine for excavation in which large amounts of earth are moved by scooping action.

shovel loader *Mechanical Engineering.* a machine capable of scooping loose material, elevating it, and ejecting it to the rear of the machine.

shovel-shaped incisor *Anthropology.* upper incisor teeth that have reinforced enamel on the inner edges, forming a cuplike interior; a frequently occurring feature of certain indigenous groups from Asia and North America.

shovel test *Archaeology.* a shovel-sized sample taken at various intervals across a site.

Showalter stability index *Meteorology.* a numerical representation of the determined local static stability of the atmosphere, expressed as an index.

shower *Meteorology.* brief, sudden precipitation from a convective cloud; characterized by the suddenness with which it begins and ends, by the rapid changes in intensity, and by the associated rapid changes in the sky's appearance. *Nuclear Physics.* see COSMIC-RAY SHOWER.

shower meteor *Astronomy.* a meteor that is part of a group traveling in the same orbit.

shower unit *Nucleonics.* the mean free path distance of travel that a relativistic particle traces in passing through a medium before its energy is reduced by a factor of 0.5; one shower unit is related to the radiation length times 0.693.

show-through *Graphic Arts.* the degree to which printing on one side of a sheet can be seen from the other side under normal reading and lighting conditions.

showy orchis *Botany.* a wild orchid, *Orchis spectabilis*, characterized by a spike of showy flowers with pink-purple or white sepals and petals united into a galea and a white lip; found in eastern North America.

shrapnel *Ordnance.* **1.** small lead or steel balls encased in an artillery projectile and scattered by a powder charge actuated by a time fuse. **2.** in general, any small scattered ammunition fragments, as from an artillery shell. (Named after Henry *Shrapnel*, 1761–1842, a British army officer who invented an early projectile of this type.)

shrew

shrew *Vertebrate Zoology.* a small, mouselike, chiefly nocturnal, insectivorous mammal of the genus *Sorex* and related genera of the family Soricideae, having a long pointed snout, small eyes, and velvety fur.

shrike *Vertebrate Zoology.* any of numerous grayish to brown oscine birds of the family Lanidae, having sharp bills and claws and known for impaling their prey on thorns.

Shrike *Ordnance.* an air-to-surface antiradiation missile designed to home on and destroy radar emitters; officially designated **AGM-45.**

shrimp *Invertebrate Zoology.* the common name for many free-swimming decapod crustaceans of the suborder Natantia, having long legs and a compressed abdomen; most are small, with larger species sometimes called **prawns**. Certain species are used for food.

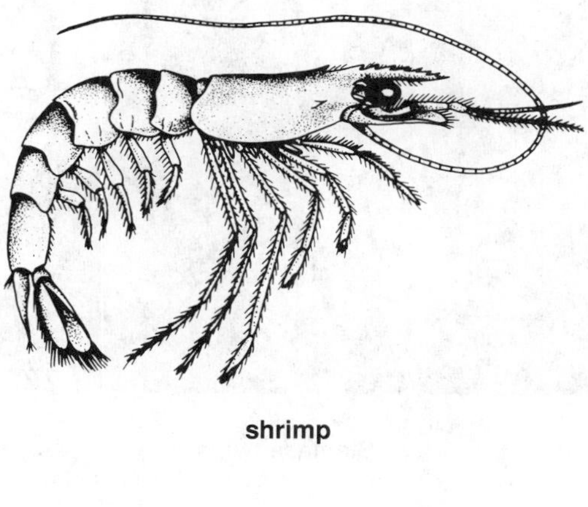

shrimp

shrimpfish *Vertebrate Zoology.* a small, compressed, East Indian marine fish of the family Centriscidae, having a flat, bladelike body with a transparent platelike covering and a sharp spine down the belly (thus called **razorfish** in Australia); rests in a head-up or head-down position.

shrinkage *Textiles.* a reduction in width, length, or overall size, caused by washing, steaming, or other treatments. *Engineering.* the contraction of a molded material when cooling. *Geology.* the decrease in the volume of a soil or sediment as a result of mechanical compaction, pressure from superimposed loads, natural consolidation, or drying.

shrinkage crack *Geology.* a small crack formed in fine-grained sediment or rock by the loss of moisture during drying or dehydration. *Metallurgy.* a fracture associated with solidification shrinkage.

shrinkage limit *Geology.* the amount of moisture in a soil for which a decrease in the moisture content will not result in a decrease in soil volume, while an increase in the moisture content will result in an increase in volume.

shrinkage pore *Geology.* an irregular pore formed in muddy sediment by shrinkage. Also, FENESTRA.

shrinkage ratio *Geology.* the ratio of a change in soil volume to the change in moisture content above the shrinkage limit.

shrinkage stoping *Mining Engineering.* **1.** a stoping method in which ore is mined in successive slices, working upward from the level. **2.** a modified method of overhead stoping in which part of the ore is used for support and as a working platform. Also, BACK STOPING.

shrinkage stress *Materials Science.* in a fibrous composite, interfacial pressure that occurs during curing or solidification of the matrix around the fibers.

shrink fit *Design Engineering.* a tight interference fit of mating parts made by heating the outer member to expand the bore for ease of assembly, then cooling it with dry ice so the outer member contracts.

shrink forming *Design Engineering.* the forming of metal in which a piece undergoes shrinkage during a cooling period immediately following the application of heat, pressure, or cold upset.

shrink ring *Mechanical Devices.* a ring used to fasten or fix assembled parts, such as the commutator bars of a dynamo, by placing it when hot and allowing it to cool and shrink on an assembly for tight adhesion.

shrink template *Design Engineering.* a contour template allowing for shrinkage; used in making plaster models that must be larger than the master layout to account for shrinkage when the dies are cast.

shrink wrap or **shrinkwrap** *Materials Science.* **1.** a type of plastic film used in packaging, which develops tension when subjected to a hot-air blast in a chamber or continuous-motion tunnel, causing the film to try to return to its original dimensions. Also, **shrink film, shrink wrapping. 2. shrink-wrap.** to wrap a product in shrink wrap.

Shropshire [shräp′shər] *Agriculture.* a breed of mutton and medium-wool sheep that are hornless, brown-faced, and completely covered with wool from the tip of the nose to the feet. (From the former county of *Shropshire*, in England.)

shroud *Aviation.* on a parachute, any of the lines that attach the harness or load to the canopy. Also, **shroud line.** *Engineering.* a housing or jacket, especially housing around gear wheels. *Horology.* the disk on either end of the lantern pinion into which the pins of the pinion are fitted. *Naval Architecture.* a line extending from a masthead to either side of a ship, serving as a lateral support for the mast.

shroud ring *Aviation.* a circular casing or enclosure, especially a band used to enclose and strengthen the blades of a turbine wheel.

shrub *Botany.* a general term for a woody plant, smaller than a tree, usually with several permanent stems branching from or near the ground rather than a main trunk.

shrub-coppice dune *Geology.* a small, streamlined dune formed on the leeward side of bush-and-clump vegetation on a smooth surface of very shallow sand.

shrund see BERGSCHRUND.

Shubnikov-de Haas effect *Solid-State Physics.* an effect due to the quantized energy levels of electrons in a metal or semiconductor, manifested by oscillations of the resistance values or Hall coefficients when the magnetic field acting on the material is increased.

Shubnikov groups *Solid-State Physics.* point groups and space groups of crystals having magnetic moments.

shufflon *Molecular Biology.* within a plasmid, a group of DNA segments that are able to invert independently or in groups, effecting rearrangements of the DNA.

shuga *Oceanography.* spongy, white lumps of ice a few centimeters thick; an early stage of sea ice in rough water, shuga may grade into pancake ice if the water becomes calmer. Also, BALL ICE.

Shultz-Charlton reaction *Immunology.* a skin reaction that indicates the presence of the disease scarlet fever; when a scarlet fever antitoxin or convalescent serum is injected into an area of the skin showing a bright red rash, blanching occurs at the site of the injection.

Shumardia *Paleontology.* a genus of small (less than 5 mm long), blind ptychopariid trilobites in the superfamily Conocoryphacea; extant in the Uppermost Cambrian to Ordovician.

Shumway, Norman born 1923, American surgeon; a pioneer in heart-transplant operations.

shungite *Geology.* a hard, black, amorphous, coal-like material containing more than 98% carbon. Also, **schungite**.

shunt to turn to one side; divert or bypass; specific uses include: *Surgery.* **1.** a channel or passage, natural or artificial, by which fluid passes between two natural channels, as in a bypass between two arteries to divert the blood flow from one part of the body to another. **2.** the operation of forming a shunt. *Electricity.* **1.** a parallel connection. **2.** to create a parallel connection. *Electromagnetism.* a piece of iron in a magnetic circuit providing a parallel path for magnetic flux about an air gap.

shunt-excited *Electromagnetism.* a condition in which field windings are connected across armature terminals, such as in a DC motor.

shunt-excited antenna *Electromagnetism.* a tower antenna whose feeder is connected at a point about one-fifth of the way up the antenna.

shunt generator *Electricity.* a self-excited generator in which the armature and field windings are connected in parallel.

shunting *Electricity.* a parallel connecting of one component across another component or group of components.

shunt loading *Electricity.* a reactance connected across the output of a device, generally for the purpose of impedance matching.

shunt motor *Electricity.* a self-excited motor in which the armature and field windings are connected in parallel.

shunt neutralization see INDUCTIVE NEUTRALIZATION.

shunt peaking *Electronics.* a method of compensating for frequency loss by placing a coil between two circuits whose load capacities differ.

shunt reactor *Electricity.* a reactor that has a fairly high inductance and is connected across the line input; used to neutralize the charging current of the line to which it is connected.

shunt regulator *Electricity.* a voltage-regulating device connected in parallel with the output of a power supply.

shunt resistor *Electricity.* a small-value resistor that is connected in parallel with an ammeter to increase its range.

shunt transition *Electricity.* the process of switching the connection of motors from series to parallel.

shunt valve *Engineering.* a valve that provides fluid under pressure an easy escape route.

shunt-wound *Electricity.* of a device, having its generator field coils in parallel with the main armature circuit.

shut-in well *Petroleum Engineering.* a well whose wellhead valves have been closed, ceasing production.

shutoff head *Mechanical Engineering.* the pressure developed in a system when the flow in a centrifugal or axial flow pump is terminated.

shutter *Optics.* a mechanical device that controls the time during which light is admitted to an optical system. *Nucleonics.* a plate of radiation- or neutron-absorbing material that can be moved into place to cover a hole where a beam of radiation or neutron flux may emerge. *Ordnance.* an obstruction or interrupter that opens or closes like a shutter; used to stop a detonation wave and to maintain fuse safety.

shutter dam *Civil Engineering.* a dam in which flowthrough is allowed by the lowering or revolving of a series of gates.

shuttered fuse *Ordnance.* a fuse in which inadvertent initiation of the detonator will not initiate the booster or the burst charge.

shutterridge *Geology.* a ridge formed by displacement of a fault across a ridge-and-valley topography, such that the displaced part of the ridge encloses an adjacent ravine or canyon.

shuttle *Textiles.* a canoe-shaped device on a loom, used for weaving the filling yarn through the space between raised and depressed warp threads. *Transportation Engineering.* a vehicle that regularly travels back and forth along the same (often short) route. *Mechanical Engineering.* the back and forth motion of a machine that is facing in one direction. *Nucleonics.* a small container of material that is intended to be exposed to the radiation or neutron flux in a nuclear reactor by being driven through a tube that passes through the core.

shuttle bombing *Military Science.* a method of bombing in which a bomber formation takes off from its home base, bombs its target, flies on to a second base, reloads, and returns to its home base, bombing a target on its return flight if required.

shuttle box *Behavior.* an experimental apparatus having two compartments, one in which a subject animal receives electric shocks, and another into which the subject learns to escape in order to avoid shocks.

shuttle conveyor *Mechanical Engineering.* a conveyor in a self-contained structure in which the motion is parallel to the flow of material along a defined path.

shuttleless loom *Textiles.* a loom that forces the threads through the warp threads by means of air or water jets rather than by a shuttle.

shuttle vector *Molecular Biology.* a vector molecule that can replicate in more than one organism, and can therefore be implemented to carry or "shuttle" genes from one organism to another.

Shwartzman reaction *Immunology.* an immunological reaction in which symptoms occur hours after the administration of a second dose of a two-dose series of endotoxin. Also, **Shwartzman-Sanarelli phenomenon.**

SI soluble insulin; sacroiliac; stimulation index.

SI or **S.I.** *Metrology.* the International System of Units. (From French *Système International.*)

Si the chemical symbol for silicon.

Siacci method *Mechanics.* an analytic method for calculating the trajectories of high-velocity missiles with low quadrant angles of departure, assuming that atmospheric density anywhere on the trajectories is constant, and that the angle of departure is less than about 20°.

sial *Petrology.* the upper layer of the earth's crust, consisting of silica- and alumina-rich rocks. Also, GRANITIC LAYER, SAL.

sial- or **sialo-** a prefix meaning: saliva; salivary glands.

sialadenitis *Medicine.* an inflammation of a salivary gland. Also, **sialoadenitis.**

sialagogue *Pharmacology.* any agent that promotes the flow of saliva. Also, PTYALAGOGUE.

sialic acid *Biochemistry.* any of a family of acylated neuraminic acids that are components of lipids, polysaccharides, and mucoproteins.

sialidase *Microbiology.* an enzyme located in the surface coat of myxoviruses; eradicates the neuraminic acid of cell surfaces during attachment, thus inhibiting hemagglutination. Also, NEURAMINIDASE.

sialography *Medicine.* the process of creating an X-ray image of the salivary ducts by injection of substances opaque to X-rays into the salivary tract. Thus, **sialogram.**

sialolith *Pathology.* a calcareous concretion or calculus formed within the salivary duct or gland. Also, SALIVARY CALCULUS, SALIVARY STONE.

sialolithiasis *Medicine.* a condition characterized by the formation of salivary calculi.

Siamese blow *Materials Science.* a method of plastics blow molding, in which two or more parts are produced in a single blow and then separated. (Thought of as analogous to surgically separating Siamese twins.)

Siamese twins *Medicine.* twins who are physically, congenitally joined. Also, CONJOINED TWINS. (From the conjoined twins Eng and Chang, 1811–1874, who were *Siamese,* or Thai, and who were widely known through their appearance in exhibitions and photographs.)

Siamese twins

SIAR or **S.I.A.R.** Society for Industrial and Applied Mathematics.
sib *Anthropology*. see CLAN.
SIB screen image buffer.
Siberian high *Meteorology*. an area of high pressure that forms over Siberia in the wintertime and is particularly apparent on mean charts of sea-level pressure; this anticyclone is magnified by the surrounding mountains, which prevent the cold air from easily flowing away. Also, **Siberian anticyclone.**
Siberian husky *Vertebrate Zoology*. any of a Siberian breed of strong, medium-sized sled-dogs, characterized by a soft, thick black, gray, or tan coat with white markings and a brush tail.
Siberian subregion *Ecology*. a large zoogeographical region that includes the area of Siberia and other areas west of the Urals, north of the Himalayas, and west of Alaska.
Siberian tick typhus *Medicine*. a mild, acute fever found in north, central, and east Asia, caused by *Rickettsia siberica* transmitted by ticks, and characterized by a diffuse rash, headache, and conjunctival inflammation.
Siberian wallflower *Botany*. a plant, *Erysimum asperum*, of the mustard family, characterized by orange-yellow flowers; found in North America.
siberite *Mineralogy*. a violet-red or purple variety of elbaite found in Siberia.
sibicide or **siblicide** see FRATRICIDE.
sibling *Biology*. one of two or more offspring of the same parents, but not necessarily of the same birth. *Computer Science*. in a tree, a node having the same parent as a given node.
sibling rivalry *Psychology*. **1.** jealousy between siblings based on competition for parental affection and attention. **2.** any form of competition among siblings.
sibling species *Biology*. any of two or more species that are almost morphologically identical but that do not interbreed. *Systematics*. see CRYPTIC SPECIES.
Sibson's notch *Cardiology*. in acute pericardial effusion, an inward bending at the left height limit of precordial dullness.
Sicilian *Geology*. a European geologic stage of the Upper Pleistocene period, occurring after the Emilian.
Sicily *Geography*. a large island (area 9926 square miles) in the central Mediterranean, off the southwestern tip of the Italian Peninsula.
sickle *Mechanical Devices*. a metal, hook-shaped hand tool consisting of a sharp, curved blade fitted into a handle; used for cutting plants.
Sickle *Astronomy*. an asterism in the constellation of Leo consisting of Alpha, Eta, Gamma, Zeta, Mu, and Epsilon Leonis. *Ordnance*. a Soviet intercontinental ballistic missile propelled by a solid rocket motor and carrying one nuclear warhead; officially designated **SS-25.**
sickle bar *Mechanical Devices*. the cutting mechanism of a grain harvesting machine, made up of a sickle, guard and ledger plates, and a bar.
sickle cell *Medicine*. an abnormal red blood cell that has a crescent shape like that of a sickle. Also, **sickled cell.**

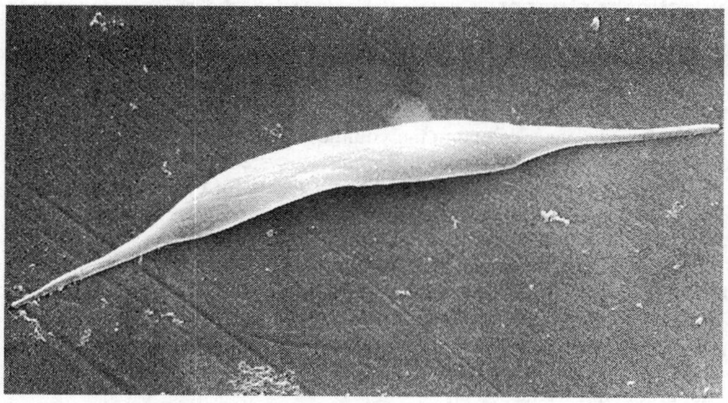

sickled red blood cell

sickle-cell anemia *Medicine*. a hereditary disease of the blood in which the red blood cells take on a sicklelike appearance, characterized by the homozygous presence of S hemoglobin in the red blood cells, and occurring almost exclusively in black people. Symptoms include arthralgia, acute attacks of abdominal pain, and ulcerations of the lower extremities. Also, **sickle-cell disease.**
sickle-cell hemoglobin *Pathology*. abnormal hemoglobin that occurs in a crescent-shaped or sickle-shaped cell and causes reduced solubility at low oxygen tension.
sickle-cell trait *Genetics*. a condition in which an individual is heterozygous for the codominant sickle-cell gene; red blood cells of such individuals contain both normal hemoglobin and abnormal (sickle-cell) hemoglobin; these individuals suffer only mild debilitation and apparently have a greater resistance to malaria, thus ensuring the continuation of the sickle-cell gene in the population.
sicklerite *Mineralogy*. $Li(Mn^{+2},Fe^{+3})(PO_4)$, a yellowish to dark-brown orthorhombic mineral occurring in cleavable masses, having a specific gravity of 3.36 to 3.45 and a hardness of 4 on the Mohs scale; found in the weathering zones of granite pegmatites.
s.i.d. *Medicine*. once a day. (From Latin *semel in die.*)
sidearm or **side arm** *Ordnance*. a weapon that is worn at the side or in a belt when not in use; e.g., a pistol.
sideband *Electromagnetism*. the frequency bands found on either side of a carrier frequency including all the sum and difference frequency components between the carrier signal and the audio signal.
side bar *Engineering*. **1.** a part of a pair of elastic wooden supports that are positioned lengthwise, from which the bodies of buggies or other light vehicles are suspended. **2.** a bar that functions as a support for molding pins.
side canyon *Geology*. a ravine or other valley smaller than a canyon, through which a tributary flows into the main stream.
side chain *Mechanical Devices*. in an automobile, one of two chains that pass over the pinions of a countershaft and drive-wheel gears in a final chain drive. *Organic Chemistry*. a grouping of similar atoms that branches off from a straight-chain or cyclic molecule.
side-chain motion *Materials Science*. specific chain movement and relaxation assigned to the polymer side chains during glassy-state transitions.
side-channel spillway *Civil Engineering*. a spillway used on dam sites located in narrow valleys or canyons, in which the water passes over the spillway crest into a channel and then flows out parallel to the crest, so that the initial flow and final flow are approximately at right angles to each other.
side chisel *Mechanical Devices*. a wood turning and cutting chisel with an oblique cutting edge, used to attain a clean-cut surface on large-diameter workpieces.
side circuit *Electricity*. either of two pairs of wires used to set up a phantom circuit; one side circuit connects the phantom with the "go" channel and the other connects it with the return channel.
side direction *Mechanics*. in stress analysis, a term for any direction that is normal to the plane of symmetry of an object.
side dressing *Agronomy*. the process of applying fertilizer between rows of growing plants.
side echo *Electromagnetism*. any echo that is produced by the side lobe of an antenna pattern.
side effect *Medicine*. **1.** any effect of medication or therapy other than the intended effect. **2.** specifically, a negative effect of a drug that accompanies the desired effect, such as nausea, drowsiness, dizziness, and so on. *Computer Programming*. a result of a procedure that is peripheral or in addition to the computed results; often a change in variables not local to the procedure or an external activity such as printing.
side-facing tool *Engineering*. a finishing tool having a cutting point at an angle of less than 60°.
side frequency *Telecommunications*. the frequency of one of the sidebands that originates with the modulation of a carrier wave.
sidehill bit *Mechanical Devices*. a drill bit set off-center to allow the cutting of a hole with a diameter wider than the bit itself.
sidehill plow *Agriculture*. a plow that turns all furrows toward the downhill side of a slope.
side keelson *Naval Architecture*. a longitudinal stiffening girder running just inside a vessel's bilge plating on either side of the keelson.
side lobe *Acoustics*. any of several minor beams formed around the main beam of passive sonar and steered to the sound source of interest. *Electronics*. in an antenna, a direction other than the intended main direction in which there is appreciable radiation, though less than in the main direction.
side-lobe blanking *Electronics*. a technique that contrasts the signal strength between a stationary antenna and one that rotates.
side-lobe cancellation *Military Science*. an electronic counter-countermeasure designed to negate or reduce the effect of jamming signals aimed at the sides or rear of a receiving system.

side-looking radar *Engineering.* a type of airborne radar in which antennas face the boundaries of the flight path; it produces high-resolution detailed maps and is used primarily to map enemy territory or to identify the presence of submarines amid sea clutter.

side milling *Mechanical Engineering.* the use of a side-milling cutter to mill a vertical surface over the edge of a workpiece or to cut grooves or slots.

side-milling cutter *Mechanical Devices.* a narrow milling tool with cutting blades and teeth lined around its circumference and on each side.

side moraine see LATERAL MORAINE.

side plates *Mining Engineering.* in timbering, the cap and the sill in situations where both are used and the posts act as spreaders.

sider- see SIDERO-.

sideraerolite see STONY-IRON METEORITE.

side rake *Mechanical Engineering.* an angle between a reference plane and the tool face of a single-point turning tool.

side reaction *Chemistry.* a reaction that takes place in addition to a primary chemical reaction, usually reducing the yield of the desired product.

sidereal [si dēr′ē əl] *Astronomy.* of or relating to the stars. *Horology.* derived from the stars; commonly used in defining units of time.

sidereal clock *Horology.* a clock that measures time based on the successive passages of a star across the local meridian.

sidereal day *Astronomy.* the period between two successive meridian transits of the vernal equinox; it amounts to 23 hours, 56 minutes, and 4.091 seconds.

sidereal hour angle *Astronomy.* the angle measured westward from the hour circle of the vernal equinox to that of a celestial body.

sidereal month *Astronomy.* one lunar revolution around earth with reference to the stars: 27.32166 days.

sidereal noon *Astronomy.* the instant the vernal equinox crosses the meridian.

sidereal period *Astronomy.* the time it takes a planet or moon to make one journey around its primary body, reckoned with respect to the stars.

sidereal table *Ordnance.* a servo-driven table designed to compensate for the effect of the earth's rotation on artillery gunfire settings.

sidereal time *Astronomy.* the hour angle of the vernal equinox. *Horology.* a system of time measurement based on the successive passages of a star across the local meridian; the mean sidereal day is 23 hours, 56 minutes, 4.091 seconds of mean solar time.

sidereal year *Astronomy.* the time it takes earth to revolve once around the sun with respect to the stars: approximately 365.25636 mean solar days.

side relief angle *Design Engineering.* the angle that the flank of a cutting tool makes with a plane normal to a base beneath its cutting edge.

siderite *Mineralogy.* $Fe^{+2}CO_3$, a brownish, gray, greenish, or yellowish trigonal mineral of the calcite group, occurring as rhombohedral crystals or in massive to granular form, having a specific gravity of 3.96 and a hardness of 4 on the Mohs scale; found widespread in sedimentary beds, hydrothermal ore veins, basic volcanics, and some metamorphic rocks. Also, CHALYBITE, SPARRY IRON, IRON SPAR.

sidero- a combining form meaning "iron." Also, SIDER-. (From the Greek word for iron.)

sidero- a combining form meaning "star." Also, SIDER-. (From a Latin word meaning "star group.")

Siderocapsaceae *Microbiology.* a family of Gram-negative bacteria consisting of organisms that are able to obtain energy from inorganic sources and are known for depositing manganese or iron compounds on or around their cells.

Siderococcus *Bacteriology.* a genus of Gram-negative soil bacteria of the family Sidercapsaceae, occurring as small, unencapsulated coccoidal cells in colonies that may be covered with ferric hydroxide deposits.

siderocyte *Cell Biology.* an erythrocyte containing free iron not bound in hemoglobin.

sideroferrite *Geology.* a type of native iron that occurs as grains in petrified wood.

siderofibrosis *Medicine.* fibrosis of the spleen with deposits containing iron compounds.

siderograph *Engineering.* a combination clock and navigation device that keeps the time of the Greenwich longitude.

siderolite see STONY-IRON METEORITE.

sideromelane see TACHYLYTE.

sideromycin *Microbiology.* an iron-chelating antibiotic produced by *Actinomycetes subtropicus* with the cellular uptake of sideramines. Also, ALBOMYCIN.

sideronatrite *Mineralogy.* $Na_2Fe^{+3}(SO_4)_2(OH)\cdot 3H_2O$, a yellow to orange, fibrous, orthorhombic mineral, having a specific gravity of 2.28 and a hardness of 1.5 to 2.5 on the Mohs scale; found with other secondary sulfate minerals.

sideronitic texture *Geology.* a texture in mineral deposits consisting of a mesh of silicate minerals that have been so shattered and pressed that solutions and other volatiles have been forced out.

sideropenic *Nutrition.* relating to or characterized by iron deficiency.

siderophile *Biochemistry.* a siderophilous cell, tissue, or structure. Also, **siderophil.**

siderophile element *Geochemistry.* a term for iron or for an element having a chemical affinity for iron; the classification is based on the distribution of elements in meteorites.

siderophilin *Biochemistry.* any of a group of iron-chelating glycoproteins occurring in vertebrates; includes transferrin.

siderophilous *Biochemistry.* having a tendency to absorb iron.

siderophyllite *Mineralogy.* $KFe_2^{+2}Al(Al_2Si_2)O_{10}(F,OH)_2$, a blue-green, monoclinic mineral of the mica group, having a specific gravity of 3.0 to 3.27 and a hardness of 2.5 to 3 on the Mohs scale; found with other secondary sulfate minerals.

siderophyre *Geology.* a stony-iron meteorite containing bronzite and tridymite crystals in a network of nickel-iron. Also, **siderophyry.**

siderosilicosis *Medicine.* pneumoconiosis caused by inhalation of dust containing iron-ore and silica.

siderosis *Medicine.* **1.** pneumoconiosis due to inhalation of iron dust, or the particle dust of other metals. **2.** excess iron in the blood. **3.** the deposit of iron in tissues or in parts of the eye.

siderostat *Optics.* a movable mirror that feeds light from a celestial object into a fixed telescope.

siderostat telescope *Optics.* a rotating mirror telescope in which a plane mirror is rotated once in 24 hours about a solar axis, set so that light from a celestial object is reflected along the polar axis.

siderotic *Medicine.* relating to or characterized by siderosis.

siderotil *Mineralogy.* $Fe^{+2}SO_4\cdot 5H_2O$, a white to yellowish or pale-green triclinic mineral of the chalcanthite group, occurring as fibrous crusts and groups of divergent, acicular crystals, having a specific gravity of 2.1 to 2.2 and an undetermined hardness; formed as a dehydration product of melanterite.

sideslip *Aviation.* **1.** a movement of an aircraft in which a relative flow of air moves along the lateral axis, resulting in a sideways movement from a projected flight path, especially a downward slip toward the inside of a banked turn. **2.** a lateral movement of a descending parachute.

side slope *Mechanical Engineering.* in vehicle testing, a test course set up to determine factors such as the vehicle's lateral stability, steering, and carburation.

side spray *Ordnance.* fragments of a bursting projectile that are thrown to either side of the line of flight, as opposed to nose spray or base spray, which is thrown to the front or back respectively.

sidestream *Chemical Engineering.* a liquid stream taken from a contacting tower at any intermediate point. *Hydrology.* see TRIBUTARY.

sidestream stripper *Chemical Engineering.* the process equipment used to further distill an intermediate stream from a contacting tower.

sidetone *Telecommunications.* the unwanted sound of a caller's own voice as heard in the caller's telephone receiver.

sidetone level *Telecommunications.* the ratio, expressed in decibels, of the volume of the sidetone to the volume of the caller's voice.

sidetone ranging *Telecommunications.* a system of introducing a time marker in a radio signal broadcast by a satellite for the purpose of navigation. Various audio tones of different frequencies are transmitted, and the phases of the tones sent from the satellite are contrasted with the tone phases that are received from the ground station.

sidetracking *Engineering.* the movement of railroad cars from a main track to a short side track or spur in order to provide clearance of the main line or to temporarily store the cars. *Petroleum Engineering.* the drilling of a second oil well to bypass an obstruction in an adjacent, nonproducing well.

sidewalk *Civil Engineering.* any pedestrian footway or roadway, usually paved.

sidewall *Engineering.* the side of a pneumatic tire, between the rim of the wheel and the top edge of the tread.

sidewash *Aviation.* an airflow that is deflected laterally by an airfoil or other body.

sidewash angle *Aviation.* the acute angle between the sidewash from a body and the undisturbed airstream; this angle is positive when the deflection is to port.

sidewinder *Vertebrate Zoology.* a small, horned rattlesnake, *Crotalus cerastes,* of the family Viperidae, native to the southwestern U.S. and Mexican deserts; characterized by its light gray color and sideways locomotion across the sand in a series of S-shaped loops.

sidewinder

Sidewinder *Ordnance.* an air-to-air solid-propellant missile with a nonnuclear warhead and infrared, heat-seeking homer; designated **AIM-9;** the ground-to-air version is the Chaparral; designated **MIM-72.**

siding *Building Engineering.* any of various facings used to protect wood frame buildings from the elements. *Transportation Engineering.* a short length of side line onto which one train from the main line may be shunted to allow the passage of another train on the main line.

SIDS sudden infant death syndrome (crib death).

Siegbahn, Kai born 1918, Swedish physicist; awarded the Nobel Prize for the development of high-resolution electron spectroscopy.

Siegbahn, Karl 1886–1978, Swedish physicist; awarded the Nobel Prize for work in X-ray spectroscopy; the father of Kai Siegbahn.

siegbahn *Spectroscopy.* a former unit of length equal to $(1.00202 \pm 0.00003) \times 10^{-13}$ meter, used to express the wavelengths of X-rays. Also, X-RAY UNIT.

siege-howitzer *Ordnance.* a short, heavy, large-caliber howitzer designed to destroy fortifications.

Siegenian *Geology.* a European geologic stage of the Lower Devonian period, occurring after the Gedinnian and before the Emsian.

siegenite *Mineralogy.* $(Ni,Co)_3S_4$, a gray, opaque, metallic, cubic mineral of the linnaeite group, occurring in massive form or as octahedral crystals, having a specific gravity of 4.5 to 5.8 and a hardness of 4.5 to 5.5 on the Mohs scale; found in hydrothermal vein deposits.

Siemens, Ernst Werner [sē´mənz] 1816–1892, German engineer and inventor; invented an electroplating process and the dial telegraph.

Siemens, Karl Wilhem 1823–1883, German electrical engineer; active in Great Britain as Sir Charles William Siemens; with his brother Ernst, established many early electrical-engineering projects and manufacturing facilities.

siemens [sē´mənz] *Electricity.* the SI electromagnetic unit of conductance. One siemen is the conductance at which a potential of one volt forces a current of one ampere. (Named for Karl Wilhem *Siemens.*)

Siemens' electrodynamometer *Electromagnetism.* an electromagnetic instrument in which current flows through all the coils in series.

sienna *Materials.* any of various iron-containing earthy substances used as pigments; **raw sienna** is yellow brown; **burnt sienna,** roasted in a furnace, is reddish-brown. (From *Siena,* a city in Tuscany, Italy, noted as an early source of this substance.)

Sierozem *Geology.* a group of zoned soils that develop in cool to temperate regions, characterized by a brownish-gray surface horizon overlying a lighter accumulation of carbonate or a layer of hardpan. Also, GRAY DESERT SOIL, GRAY EARTH, SEROZEM.

Sierpinski gasket *Mathematics.* a fractal that is a two-dimensional version of the Cantor set, constructed recursively as follows: Inscribe a triangle T_1 in an equilateral triangle T by connecting the midpoints of the sides of the triangle. Remove the inscribed triangle T_1, leaving three congruent equilateral triangles. Apply the same process to each of the remaining three triangles, removing triangles T_2, T_3, T_4 and leaving nine smaller congruent equilateral triangles. Continue ad infinitum, leaving 3^n congruent triangles at the nth step. The Sierpinski gasket is the set of points

$$T - \bigcup_{k=1}^{\infty} T_k;$$

its fractal dimension is log 3/log 2.

sierra *Geography.* in Spain and its former colonies, a range of hills or mountains, especially one rising in peaks resembling saw teeth. (From a Spanish word meaning "saw.")

Sierra Madre *Geography.* the mountain system of Mexico, containing three ranges: **Sierra Madre del Sur** along the southern Pacific coast, **Sierra Madre Occidental** along the northern Pacific coast, and **Sierra Madre Oriental** in the northeast, inland from the Gulf of Mexico; highest peak: Orizaba (Citlaltépetl) (18,700 feet).

Sierra Nevada *Geography.* **1.** a mountain range in eastern California; highest peak: Mt. Whitney (14,495 feet). **2.** a mountain range in southern Spain; highest peak: Mulhacén (11,411 feet).

sieve *Mechanical Devices.* **1.** a screen through which unconsolidated dry material is refined, liquid is strained, and soft solids are pulverized. **2.** a screen with openings of uniform size, used to sort unconsolidated solids according to particle size.

sieve analysis *Mechanical Engineering.* a mechanical analysis of unconsolidated material, whereby material is put through sieves of various sizes to determine the size distribution of the component particles. Also, **sieve classification.**

sieve area *Botany.* a region of sieve element wall containing clusters of pores through which the protoplasts of adjacent sieve elements are interconnected.

sieve cell *Botany.* in lower vascular plants and gymnosperms, an elongated phloem cell whose walls include sieve areas that function in the translocation of sugars and other nutrients.

sieve deposition *Geology.* a process in which coarse-grained lobate masses are formed on an alluvial fan that is composed of coarse, permeable material permitting complete infiltration of water. Similarly, **sieve lobe.**

sieve diameter *Engineering.* the measurement of a single opening in a sieve, through which a specific-sized particle can barely pass.

sieve fraction *Engineering.* the quantity of unconsolidated material that passes through a standard sieve and is retained by a finer sieve.

sieve mesh *Design Engineering.* in a sieve or screen, the standard opening defined by four boundary wires (warp and woof); the size is measured in the number of parallel wires per linear inch.

Sieve of Eratosthenes *Mathematics.* a systematic method of obtaining all primes up to any given number x, devised by the ancient Greek astronomer and geographer Eratosthenes and proceeding as follows: List the integers from 2 to x in their natural order. Cross out all multiples of 2 except 2 itself. The next number following 2 that has not been crossed out is 3; keep 3 and cross out all other multiples of 3 not already crossed. Continue in this manner; at each stage keep the next number not yet crossed and cross out all its multiples. The integers being kept are the prime numbers less than or equal to x.

sieve plate *Chemical Engineering.* a plate used in a contacting tower in which the liquid flows directly across the tray while the gas flows upward through holes in the tray. Also, **sieve tray.** *Botany.* a wall or part of a wall between adjoining sieve cells that has one or more sieve areas.

sievert *Radiology.* the SI unit of measure defined as the quantity of absorbed radiation that induces the same biologic effect in a specified tissue as 1 gray of high-energy X-rays; equal to approximately 8.38 roentgens; 1 sievert = 100 rem. Also, **sievert unit.** (Named for Rolf Maximilian *Sievert,* 1896–1966, Swedish radiologist.)

sieve shaker *Chemical Engineering.* a machine used for sieve testing that holds several test sieves vertically while it imparts both a circular and a tapping motion on the sieves to check the accuracy of their sifting.

sieve texture see POIKILOBLASTIC.

sieve tube *Botany.* a vertical series of sieve tube elements, joined end-to-end at sieve plates, whose function is to translocate sugars and other organic compounds.

sieve tube element *Botany.* in angiosperms, a vascular phloem cell, more complex than a sieve cell, composed of a sieve plate, a nonlignified secondary cell wall, and a living enucleate protoplast; a number of such cells are joined at their ends to form a sieve tube.

sieve tube member *Botany.* one of the component cells of a sieve tube; found primarily in flowering plants and typically associated with a companion cell.

sieving see SIEVE ANALYSIS.

siffanto *Meteorology.* an often violent, southwesterly wind that occurs over the Adriatic Sea.

sift *Food Technology.* to pass a material such as flour through one or more screens in order to filter out the coarser fractions of the material. *Computer Programming.* to extract desired items of information from a large volume of data.

Sig or **Sig.** *Medicine*. let it be written or labeled; a direction to be written on a prescription label for the patient's information. (From Latin *signetur*.)
sig. signal; signature.
sIgA *Immunology*. see SECRETORY IGA.
Siganidae *Vertebrate Zoology*. the rabbitfishes, a family of marine, herbivorous bony fishes of the order Perciformes, found in the Indo-Pacific and Mediterranean seas. In some classification systems, an equivalent term for Teuthidae.
Sigatoka *Plant Pathology*. a serious leaf-spot disease of bananas caused by the sooty mold fungus *Mycosphaerella musicola,* and occurring primarily in the American tropics.
sight *Physiology*. the ability to see, or the process of seeing; vision. *Navigation*. the measurement of the altitude of a celestial body, usually made with a sextant at Greenwich Mean Time. Also, CELESTIAL OBSERVATION. *Ordnance*. **1.** a mechanical or optical device for aiming a firearm or laying a gun or launcher. Similarly, **sighting instrument, sighting system. 2.** to aim or lay a weapon using such a device, or to look through such a device to determine the angular direction of a point.
sight aperture *Ordnance*. an adjustable opening on a rear sight through which the front of the gun is viewed in aiming at a target or other object.
sight base *Ordnance*. **1.** a mount for a gunsight. **2.** of an optical ring sight, the distance between the rings and the gunner's eye.
sight bracket *Ordnance*. a clamp that holds a detachable sight in position on a gun.
sight distance *Optics*. the distance over which an observer can keep an object at eye level in view.
sight extension *Ordnance*. any device that raises a gunsight or extends it to the side.
sight glass *Engineering*. a transparent section in a pipe or tank wall through which the liquid level in a container may be viewed or the flow of the liquid observed.
sighting *Optics*. **1.** the process of aiming a device by means of a sight unit. **2.** the act of visually observing something.
sighting angle *Ordnance*. in bombing, the angle between the line-of-sight to the aiming point and the vertical.
sighting shot *Ordnance*. a trial shot to determine if the sights are adjusted correctly.
sight leaf *Ordnance*. the hinged component of a rear sight; it may be raised and set to a given range or folded flat.
sight reduction *Navigation*. the process of determining a line of position from the raw observational data obtained with a sextant; the term may also be applied to the data itself. Also, REDUCING AN OBSERVATION.
sight rule see ALIDADE.
sigill. seal. (From Latin *sigillum*.)
Sigillaria *Paleontology*. a lycopsid tree of the Devonian and Carboniferous whose height averaged around 20 meters, with some specimens as high as 34 meters; it had a crown of large pods on a slender, branchless stalk.
sigma the 18th letter of the Greek alphabet, written Σ, σ.
sigma algebra *Mathematics*. sigma algebra of sets is a nonempty class of sets closed with respect to the formation of complements and countable unions; equivalently, a sigma ring closed under complementation. It is common to speak of a sigma algebra on a given set X, consisting of a collection of subsets of X that form a sigma algebra. The empty set is a member of every sigma algebra. Also, **σ-algebra, sigma-algebra, sigma field.**
sigma bond *Physical Chemistry*. a covalent bond that forms at the point where the electron orbitals of two adjoining atoms overlap. Also, **σ-bond.**
sigma factor *Enzymology*. a subunit of bacterial RNA polymerase that does not take an active part in the catalytic activity of the enzyme, but is necessary for the recognition of and binding to specific sites during the initiation of RNA transcription. Also, **σ-factor.**
sigma finite *Mathematics*. a measure m defined on a measure space (X, A, m) is said to be sigma finite if X is expressible as a countable union of sets $A_i \in A$, such that $m(A_i) < \infty$. Also, **σ-finite, sigma-finite.**
sigma function *Mathematics*. an odd, entire, doubly quasi-periodic function $\sigma(z)$ that satisfies the ordinary differential equation $d(\log \sigma)/dz = \zeta(z)$, where $\zeta(z)$ is the (Weierstrassian) zeta function. Sometimes four sigma functions are defined in analogy with four theta functions. *Thermodynamics*. a function that is given by the difference between the enthalpy of an air-water vapor mixture and the product of the specific humidity and the enthalpy of liquid water measured at the wet-bulb temperature.
sigma hyperon *Particle Physics*. an unstable elementary particle belonging to the baryon group; exists in positive, negative, and neutral charge states with (respectively) 2328, 2343, and 2333 times the mass of an electron. Also, **sigma particle.**
sigma-minus hyperonic atom *Atomic Physics*. an atom that is characterized by a negatively charged semistable baryon (sigma hyperon) orbiting about an ordinary nucleus.
sigma phase *Metallurgy*. a hard and brittle intermetallic compound that impairs the ductility of several alloys.
sigma pile *Nucleonics*. a system of moderating material and a neutron source used in research to study the properties of neutron interaction with properties of the material.
sigma ring *Mathematics*. a ring R of sets is called a sigma ring if R is closed under countable unions of subsets; that is,

$$\text{if } A_i \in R \ (i = 1, 2, 3, \ldots), \text{ then } \bigcup_{i=1}^{\infty} A_i \in R.$$

Also, **σ-ring, sigma-ring.**
sigmaspire *Invertebrate Zoology*. a spiral sponge spicule in the form of a C or an S.
sigma-t *Oceanography*. a shortened notation for the density of a seawater sample, using the relationship sigma-t = (density − 1) × 10^3, so that normal seawater, whose density is 1.02680, has a sigma-t of 26.8.
sigmate *Science*. shaped like a Roman letter S or a Greek Σ.
sigmatism *Medicine*. an incorrect, difficult, or excessive use of the *s* sound. Also, **sigmasism.**
sigmatron *Nucleonics*. a combination of a betatron and a cyclotron operating together to produce X-rays on the order of 10^9 volts.
sigma two formula *Crystallography*. a formula used in the direct method of structure determination. It relates the phases of three strong reflections to one another. Also, **Σ_2 formula.**
Sigma virus or **Sigmavirus** *Virology*. a possible member of the family Rhabdoviridae that infects adult fruit flies and is transmitted by female flies to their offspring; usually nonpathogenic except upon exposure of the infected fly to pure carbon dioxide, which results in paralysis and death. Also, CARBON DIOXIDE SENSITIVITY VIRUS.
sigmoid *Science*. S-shaped; doubly curved. *Biology*. relating to the sigmoid colon. Also, **sigmoidal.**
sigmoidal dune *Geology*. an S-shaped, steep-side, crested sand dune formed by the action of alternating and opposing winds having approximately equal velocities.
sigmoidal fold *Geology*. a recumbent fold that resembles the letter S, due to the curvature of its axial surface.
sigmoid colon *Anatomy*. the lower portion of the descending colon that is shaped like the letter S. Also, **sigmoid flexure.**
sigmoid curve *Statistics*. an S-shaped curve that monotonically increases between two horizontal asymptotes; this often occurs in growth studies.
sigmoiditis *Medicine*. an inflammation of the sigmoid colon.
sigmoidoscope *Medicine*. a speculum instrument designed for examination of the sigmoid flexure. Also, **sigmoscope.**
sigmoidoscopy *Medicine*. inspection of the sigmoid flexure through a sigmoidoscope. Also, **sigmoscopy.**
sign anything that conveys meaning or indicates or stands for something; specific uses include: *Medicine*. a definite indication of the existence of some condition; i.e., objective evidence that is perceptible to the examining physician, as opposed to the subjective sensations (symptoms) of the patient. *Mathematics*. **1.** the sign of a nonzero real number x is + (read "positive" or "plus") if $x > 0$ and is − (read "negative" or "minus") if $x < 0$. **2.** the sign of a permutation π, denoted sgn π, is +1 if the permutation is even and is −1 if the permutation is odd. In both definitions, the sign is a group homomorphism onto the integers modulo 2.
signage *Graphic Arts*. a graphic representation designed for exterior display, such as a large outdoor advertisement, the name and logo of a store, or a directional sign.
signal anything that serves to direct, guide, warn, and so on; specific uses include: *Behavior*. a message that passes through a channel of communication from a sending organism to a receiving organism; the message may be optical, acoustic, chemical, tactile, or electrical. *Electronics*. any transmitted electrical impulse. *Transportation Engineering*. an acoustic or visual apparatus that attracts attention by giving off an appropriate indication, such as a foghorn or signal light. *Telecommunications*. **1.** data that are transferred over a given communications system by visual or aural means. **2.** any coded message or text that is conveyed via electrical, acoustical, or electronic means.

signal aspect *Transportation Engineering.* the operative view of a cab or wayside signal.
signal conditioning *Telecommunications.* the processing of the mode of a signal in order to make it compatible with a certain device, such as a data transmission line.
signal control *Transportation Engineering.* the control of traffic by signal lights.
signal correction *Engineering.* a figure used in seismic analysis to adapt for time differences between reflection times, caused by changes in an outgoing signal from shot to shot.
signal-detection theory *Psychology.* a mathematical theory of the detection of stimuli, based on the assumption that an individual's sensitivity to a signal is not merely a result of its intensity but is also influenced by the degree of environmental distraction and the motivation of the individual.
signal distance see HAMMING DISTANCE.
signal distortion generator *Electronics.* an instrument that generates distorted signals; used to test or adjust communications equipment.
signal effect *Engineering.* an occurrence in seismology, in which variations in an outgoing signal cause the arrival times of reflections taken with the same filter settings to vary.
signal flare *Navigation.* a pyrotechnic device of distinctive character and color that is used for signaling. It is extensively used as a distress signal.
signal generator *Engineering.* an electrically powered instrument that tests the performance of radio-receiving equipment by producing known voltages over a range of frequencies, generally modulated in amplitude, frequency, or pulse. Also, TEST OSCILLATOR.
signal hypothesis *Biochemistry.* a hypothesis for the way secretory proteins are chosen by the rough endoplasmic reticulum and sent through it.
signaling rate *Telecommunications.* the speed at which signals are sent.
signal intensity *Telecommunications.* the electric-field power of the electromagnetic wave that is sending a signal.
signalization *Transportation Engineering.* the process of equipping an intersection or series of intersections with traffic signals.
signal level *Acoustics.* the decibel magnitude of a sound, referenced to a zero decibel level for a reference sound pressure such as one micropascal. *Telecommunications.* the difference between the level of a signal at a certain location in a transmission system and the level of a preselected reference signal.
signal light *Engineering.* any form of illumination that is used as a sign, such as a flare, a navigation light, or a traffic signal.
signal panel *Aviation.* a strip of colored cloth or other material that is laid out on the ground to signal airborne flight vehicles.
signal peptidase *Biochemistry.* an endopeptidase that excises the signal peptide during translocation, its catalytic site occurring at the membrane exterior.
signal peptide *Biochemistry.* a sequence of amino acid residues present in newly synthesized membrane proteins that guides the protein toward, or through, the proper membrane, and is usually excised during translocation. Also, SIGNAL SEQUENCE, LEADER PEPTIDE.
signal phase *Transportation Engineering.* the state of a traffic signal for a given traffic stream, such as green, red, yellow, left turn only, and so on.
signal processing *Computer Programming.* a computer manipulation of data derived from sensory receptions (such as a microphone or TV camera) to modify the data, interpret it, or to derive information from it.
signal-processing system *Computer Technology.* a computerized system for analyzing acoustic or other analog signal data to identify the source or acquire other information.
signal-recognition particle *Biochemistry.* a multisubunit protein of eukaryotes, located in the cytosol, that assists in the translocation of proteins across the endoplasmic reticulum. Also, **signal-recognition protein.**
signal regeneration *Telecommunications.* the restoring of a waveform that represents a signal to its original shape and amplitude; sometimes referred to as signal reshaping. Also, **signal reshaping.**
signal rocket *Ordnance.* a rocket that emits a color or other visual phenomenon that can be interpreted as a message according to a preestablished code.
signal security *Military Science.* a generic term that includes both communications security and electronic security.
signal sequence see LEADER SEQUENCE.
signal-shaping network *Electronics.* a circuit that is added to a telegraph, normally at the receiving end, to improve reception.
signals intelligence *Military Science.* a category of intelligence information that includes all communications intelligence, electronics intelligence, and telemetry intelligence.
signal speed *Telecommunications.* in a given communications system, the rate at which code elements are transmitted.
signal standardization *Telecommunications.* the utilization of one signal to promulgate another that meets preset standards for amplitude, shape, and timing; also, **signal normalization.**
signal station *Telecommunications.* a shore location from which signals are sent out to sea.
signal strength *Electromagnetism.* the amplitude of a signal received by an antenna, typically measured in volts and sometimes in volts per meter of effective height of the receiving antenna.
signal-strength meter *Electronics.* a device that calibrates the strength (level) of an incoming signal in a communication receiver, generally expressed in decibels or arbitrary *S* units.
signal-to-interference ratio *Electronics.* the ratio of the amplitude of a desired signal to that of a signal interfering with its reception.
signal-to-noise ratio *Electronics.* the ratio of the power of any desired signal to the cumulative power of any unwanted signal. *Acoustics.* specifically, the ratio (in decibels) of an acoustic signal from a specific source to the ambient noise or system noise; used as a measure of the fidelity of a tuning system.
signal tower *Transportation Engineering.* a tower where railroad signals are switched, displayed, and controlled.
signal tracer *Electronics.* an instrument that monitors a signal passing through a circuit in order to detect faulty wiring or components; commonly used in radio receivers and audio amplifiers.
signal voltage *Electricity.* the amplitude of a signal expressed in terms of its voltage, measured as peak, peak-to-peak, and root-mean-square values.
signal wave *Telecommunications.* a wave that conveys some intelligence or message; often referred to simply as signal.
signal winding *Electricity.* the control winding in a magnetic amplifier.
sign-and-magnitude code *Computer Programming.* a method of representing negative numbers by attaching a separate sign bit to a string of digits that represents the magnitude of the number.
signature *Electromagnetism.* the unique characteristics of a target, such as a ship or aircraft, that make it identifiable by a detection or classification system, such as radar. *Pharmacology.* the part of a prescription that gives the patient directions for use. *Graphic Arts.* a sheet that is printed and folded so that when trimmed it will display a number of pages (usually 32) in their proper order. *Mathematics.* the signature of a quadratic form (i.e., symmetric bilinear form or Hermitian matrix) is the number of positive eigenvalues minus the number of negative eigenvalues. By Sylvester's theorem, the signature is invariant under change of basis.
sign bit *Computer Programming.* **1.** a single bit that denotes the algebraic sign of a number, usually 0 for positive and 1 for negative. **2.** the highest-order bit in the output of the arithmetic logic unit.
sign convention *Optics.* a decision made in optical system computations when the designer chooses which quantities (such as radii of curvature) will be positive or negative.
sign digit *Computer Programming.* a digit, composed of one or more bits, that is in the sign position indicating the algebraic sign of a value.
signed *Computer Programming.* describing a number or group of numbers that contain a digit indicating the algebraic sign. Thus, **signed binary, signed decimal, signed field, signed integer.**
signed field *Computer Programming.* a field containing a digit used to indicate the algebraic sign of a number.
signed measure *Mathematics.* an extended real-valued, countably additive set function μ on the class of all measurable sets of a measurable space (X, S), such that: (a) $\mu(\varnothing) = 0$, and (b) μ assumes at most one of the values $+\infty$ and $-\infty$. By the Hahn decomposition theorem, any signed measure can be written as the difference of two (positive) measures.
signed minor *Mathematics.* see COFACTOR.
signed number *Mathematics.* a number having an algebraic sign; i.e., a positive or negative number.
signet-ring cell *Histology.* a cell that resembles a signet ring due to the presence of a large, central, fat-filled vacuole, which squeezes the nucleus against the periphery; an adipocyte.
sign flag *Computer Programming.* a condition code in the central processing unit status register that is set to 1 if the result of an arithmetic operation is negative and is cleared to 0 if the result is positive.

significance arithmetic *Computer Programming.* a method of monitoring the accuracy of arithmetic operations on approximated operands by estimating the number and position of the resulting significant digits.
significance level *Statistics.* in hypothesis testing, the probability of rejecting a hypothesis that is correct, that is, the probability of a Type I error.
significant *Statistics.* probably resulting from something other than chance; probably having a systematic cause.
significant digit *Mathematics.* in an *n*-ary representation (such as binary or decimal) of a number that represents an uncertain or approximate quantity, a digit is significant if its value would not change were the quantity to vary between its limits of uncertainty. For example, the number 3100, representing rounding to the nearest 100, has two significant digits.
significant wave *Oceanography.* a statistical term referring to the largest one-third of all waves in a particular series of waves.
signing *Linguistics.* the process of communicating by or with the deaf or hearing-impaired, through the use of a sign language.

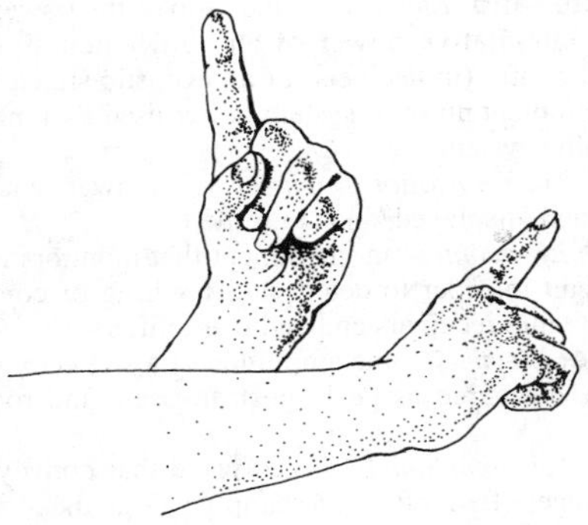

sign language

sign language *Linguistics.* any of several systems of communication in which hand gestures are used to represent letters or words; this may be a system used by the deaf or hearing-impaired, such as American Sign Language, or it may be a system used by people who do not speak the same language, as was the case among the Plains Indians.
sign of the zodiac *Astronomy.* **1.** any of the 12 irregularly sized constellations that lie along the ecliptic. **2.** any of 12 ecliptic divisions spanning a uniform 30° each.
sign position *Computer Programming.* the position of the bit or bits indicating the sign of a number, usually located at one end of the number.
sign stimulus *Behavior.* an external stimulus that elicits some species-specific behavior that is not a reflex response.
sign test *Statistics.* any nonparametric test that uses the signs of observed differences, rather than magnitudes, to test whether two populations differ.
signum *Mathematics.* the complex function sgn(z) defined by sgn$(0) = 0$ and sgn$(z) = z/|z|$ for nonzero z. If z is a nonzero real number, sgn(z) is $+1$ or -1 according to whether z is positive or negative.
sigua *Meteorology.* a horizontally moving monsoon gale of the Philippines.
sikimitoxin *Toxicology.* a poisonous substance found in the sikimi plant, *Illicium religiosum.*
sikussak *Oceanography.* an Inuit word for very old sea ice trapped in fjords and resembling glacier ice because snowfall accumulates on its surface and contributes to its make-up.
silage *Agriculture.* any material that is harvested while still green and placed in a silo for fermentation. Also, ENSILAGE.
silage cutter *Agriculture.* a farm machine that cuts green crops into short lengths and places them in a silo. Also, SILO FILLER.
silane *Inorganic Chemistry.* **1.** SiH_4, a colorless gas that has a repulsive odor and is a dangerous fire hazard; freezes at −185°C and boils at −111.8°C; used as a doping agent for solid-state devices. Also, SILICANE, SILICON TETRAHYDRIDE. **2.** any of various other compounds of silicon and hydrogen, Si_nH_{2+n}, analogous to alkanes or saturated hydrocarbons; e.g., disilane, Si_2H_6, trisilane, Si_3H_3, or tetrasilane, Si_4H_{10}. Also, SILICON HYDRIDE.
silcrete *Geology.* a conglomerate of surficial sand and gravel cemented together by silica.
silencer *Ordnance.* a device for reducing or eliminating the sound made by the firing of a gun, by slowing the escape of gasses from the muzzle.
silent *Pathology.* describing a pathological condition that produces no detectable signs or symptoms. Thus, **silent infection, silent tumor.**
silent allele *Genetics.* an allele whose effects are not detectable, or that does not produce a detectable protein.
silent mutation *Genetics.* a gene mutation that does not affect or change the wild-type phenotype.
silent period *Telecommunications.* in telephony, time intervals between bursts of ringing current from a central office to a called subscriber. *Navigation.* the preset, hourly time that ship and shore radio stations must stay silent in order to listen for distress calls.
silent speed *Engineering.* the rate at which silent motion pictures are put through a projector, equal to 16 frames per second, as opposed to the speed for sound films, which equals 24 frames per second.
silent stock support *Mechanical Engineering.* a flexible tube that encases a stock tube for an automatic screw machine; the support acts to deaden the sound and reduce vibration.
silex *Materials Science.* **1.** a heat-resistant and shock-resistant silica brick used to line grinding mills in order to avoid contamination by abraided steel. **2.** a finely ground form of quartz that is used as a filler.
silexite *Geology.* a chert that occurs in calcareous beds. *Petrology.* an igneous rock, or rock group, containing 60–100% primary quartz.
silhouette [sil′oo wet′; sil′ə wet′] *Graphic Arts.* **1.** an outline of an object, especially a person's head in profile, seen in black or another solid color against a light background. **2.** any dark image outlined against a light background. (Named for Etienne de *Silhouette,* 1709–1767, French minister of finance; supposedly as a mockery of his economic policies.)
silhouette halftone *Graphic Arts.* a halftone illustration from which all or part of the background has been cut away.
silica *Mineralogy.* silicon dioxide, SiO_2, occurring naturally in crystalline form as low and high quartz, cristobalite, and tridymite, and as coesite and stishovite; in microcrystalline form as chalcedony and chert; and in amorphous form as lechatelierite and opal; used to make glass, ceramics, and concrete, in pharmaceuticals, cosmetics, and insecticides, and for many other purposes.
silica acid soil *Agronomy.* soil having a pH value of less than 7.0 due to the presence of active hydrogen and/or aluminum ions; most soils having a pH of less than 6.6 are silica acid soils.
silica brick *Materials Science.* a refractory that is composed mainly of silica and also contains lime, alumina, and ferric oxide; it is resistant to spalling and to attack by acid and slags.
silica flour *Metallurgy.* in casting, a fine silica powder added to molding sand.
silica gel *Inorganic Chemistry.* a colloidal, highly absorbent form of silicon dioxide; widely used as a dehumidifying and dehydrating agent and in many chemical processes.
silica glass *Materials Science.* glass made of pure silica, characterized by a very low coefficient of thermal expansion.
silicane see SILANE, def. 1.
silica sand *Geology.* a sand containing a very high percentage of silicon dioxide.
silica stone *Petrology.* any sedimentary rock consisting of siliceous minerals.
silicate *Inorganic Chemistry.* **1.** any of a wide variety of compounds containing silicon, oxygen, and one or more metals with or without hydrogen. Most rocks and many minerals are silicates. **2.** any salt of silica or of one of the silicic acids. *Mineralogy.* any mineral having a crystal lattice containing SiO_4 tetrahedra, either isolated or joined through one or more of the oxygen atoms; rock-forming silicates include micas, pyroxenes, and feldspars.
silicate cement *Materials Science.* a bonding material used in dentistry; made from aluminosilicate powder mixed with phosphoric acid.
silicate grinding wheel *Mechanical Devices.* a grinding wheel whose abrasive surface bonds contain sodium silicate and other fillers.
silicate network *Materials Science.* in the structure of silicates, a three-dimensional structure in which tetrahedra having oxygen at the corners and silicon ions in the center are the principal component.
silicate of soda see SODIUM SILICATE.
silica tile *Materials Science.* a material made from amorphous, high-porosity, silica fibers, which are bonded with a slurry, sintered into blocks, and coated with borosilicate glass for waterproofing; used in the thermal protection systems of space vehicles.

silication *Geology.* the process by which mineral constituents in a rock or sediment are converted to or replaced by silicates.
silicatosis see SILICOSIS.
siliceous [si lish´əs] *Mineralogy.* consisting of, containing, or resembling silica. *Petrology.* of rock, rich in silica, especially free silica. *Botany.* growing in soil that is rich in silica. Also, **silicious.**
siliceous dust *Mining Engineering.* dust arising from dry workings of sandstone, sand, trap, granite, and other igneous rocks; it is insoluble in body fluids, and its inhalation may lead to a form of pneumoconiosis known as silicosis.
siliceous earth *Geology.* a loose, friable, porous, lightweight, fine-grained, and silica-rich sediment having a soft, dry, earthy feel; usually derived from the remains of organisms.
siliceous limestone *Petrology.* **1.** a dark, dense, usually thin-bedded variety of limestone consisting of a closely associated, simultaneous admixture of calcium carbonate and chemically precipitated silica. **2.** a silicified limestone with calcite replaced by silica.
siliceous ooze *Geology.* a silica-rich ooze composed of the skeletal remains of organisms.
siliceous sediment *Geology.* a sediment that is composed of fragmental, concretionary, or precipitated silica-rich organic or inorganic materials.
siliceous shale *Petrology.* a hard, fine-grained, silica-rich rock with a shaly texture, composed of as much as 85% silica and formed either by silicification of normal shale or by accumulation of organic material at the time the clay was deposited.
siliceous sinter *Mineralogy.* a white, lightweight, porous, opaline variety of silica occurring as a precipitate at the mouth of a geyser or hot spring. Also, FIORITE, GEYSERITE.
silicic *Petrology.* of igneous rock or magma, having a silica content generally exceeding 65%. Also, OVERSATURATED, PERSILICIC. *Chemistry.* **1.** of or relating to silica or its acids. **2.** of or relating to silicon.
silicic acid *Inorganic Chemistry.* $SiO_2 \cdot nH_2O$, any of various jellylike, amorphous precipitates obtained when a silicate is acidified; e.g., orthosilicic acid, $SiO_2 \cdot 2H_2O$. The amount of H_2O varies with the method of preparation and decreases with heat gradually until relatively pure SiO_2 remains. Silicic acids are used to bleach fats, waxes, and oils and as a reinforcing agent in rubber.
silicide *Chemistry.* any binary compound containing silicon and a metal.
silicide resistor *Electronics.* a device that employs the residual silicide of molybdenum or chromium from a direct current in an integrated circuit to protect the circuit from radiation.
siliciferous *Mineralogy.* producing, containing, or combined with silica.
silicification *Geology.* the process by which silica is introduced into the pores of a rock or sediment, or replaces existing minerals, generally in the form of quartz, chalcedony, or opal. Also, SILIFICATION.
silicified wood see PETRIFIED WOOD.
silicify *Geology.* to undergo or cause to undergo silicification.
silicinate *Geology.* **1.** the silica cement of a sedimentary rock. **2.** of or relating to such cement.
siliclastic *Petrology.* describing clastic noncarbonate rocks that are almost exclusively silicon-bearing, either as quartzes or silicates.
silicle *Botany.* a short silique, nearly equal in length and width.
silicoblast *Invertebrate Zoology.* in siliceous sponges, a specialized cell that secretes spicules.
Silicoflagellata *Botany.* a class of marine flagellates of the order Chrysomonadina, characterized by a flagella and an internal siliceous skeleton.
silicoflagellate *Botany.* any of a group of unicellular, planktonic marine algae of uncertain affiliation, characterized by a flagella and an internal siliceous skeleton; represented by a single present-day genus, *Dictyocha.*
Silicoflagellida *Invertebrate Zoology.* an order of plantlike, flagellate marine protozoa belonging to the class Phytomastigophorea (subphylum Mastigophora) having one flagellum and a star-shaped siliceous skeleton that envelops the body.
silicol *Pharmacology.* an organic silica compound, silicic oxide casein metaphosphate, used in the treatment of tuberculosis.
silicomagnesiofluorite *Mineralogy.* $Ca_4Mg_3Si_2O_5(OH)_2F_{10}$, a mineral consisting of a basic silicate and fluoride of calcium and magnesium.
silicomanganese *Metallurgy.* a master alloy containing about 20% silicon and 70% manganese; extensively used in steelmaking.
silicon [sil´i kän; sil´i kən] *Chemistry.* a nonmetallic element having the symbol Si, the atomic number 14, an atomic weight of 28.09, a melting point of 1412°C, and a boiling point of about 2480°C in the crystalline state. The most important semiconducting element, silicon occurs primarily as an oxide or silicates; it forms many complex compounds on the earth's surface, and it is used as a semiconductor in solid-state devices and in alloys to impart hardness. Also, **silicium.** (From *silica;* going back to a Latin word meaning "stone" or "boulder.")
silicon bromide *Inorganic Chemistry.* $SiBr_4$, a fuming, colorless liquid with a suffocating odor, strongly irritating to tissue; it decomposes in water with the introduction of heat; melts at 5.4°C and boils at 154°C. Also, SILICON TETRABROMIDE, TETRABROMOSILANE.
silicon bronze *Metallurgy.* a copper-base alloy containing 2.0–2.6% silicon (UNS C65100) or 2.8–3.8% silicon (UNS C65500).
silicon burning *Nuclear Physics.* a process by which stars synthesize certain elements, primarily those from the iron group, through a series of nuclear reactions that include bombarding silicon-28 with gamma rays.
silicon capacitor *Electronics.* a device that acts as a capacitor utilizing the properties of a reverse biased silicon P-N junction.
silicon carbide *Inorganic Chemistry.* SiC, bluish-black crystals with excellent thermal conductivity; insoluble in water and alcohol; sublimes with decomposition at about 2700°C; used as an abrasive, in light-emitting diodes to produce green or yellow light, and for various other purposes.
silicon chip *Electronics.* see CHIP.
silicon chloride *Inorganic Chemistry.* $SiCl_4$, a fuming, toxic, highly mobile liquid that is corrosive to most materials in the presence of water; decomposed by water and alcohol; melts at –70°C and boils at 57.6°C; used in smoke screens and in various industrial processes. Also, SILICON TETRACHLORIDE, TETRACHLOROSILANE.
silicon-controlled rectifier *Electronics.* a device that employs the electrical properties of silicon and an externally applied voltage to control the conversion of alternating current to direct current.
silicon-controlled switch *Electronics.* a device that employs the electrical properties of silicon in order to control the flow of current in a circuit in a variety of ways, such as by amplifying it, converting it, or blocking it.
silicon copper *Metallurgy.* a copper-base master alloy containing about 25% silicon.
silicon diode *Electronics.* any diode utilizing the properties of a silicon P-N junction.
silicon dioxide *Inorganic Chemistry.* SiO_2, colorless crystals or a white powder; insoluble in water; melts at about 1700°C and boils at about 2230°C; toxic on inhalation; it occurs widely in nature in various forms known collectively as silica. See SILICA.
silicone [sil´i kōn´] *Materials Science.* any of a large number of polymers consisting of alternate silicon and oxygen atoms combined with various organic groups; available in the form of oil, grease, or plastic, and characterized by resistance to water, heat, and the passage of electricity. It is a widely used industrial material for lubricants, waterproof and protective coatings, sealants, electrical insulators, paints, adhesives, coolants, and many other purposes. It is also used in medicine for artificial body parts; its use for surgical breast implants has become a subject of debate.
silicone elastomer *Materials.* any of various elastic silicone polymers, having excellent temperature, chemical, and oil resistance.
silicone rubber *Materials.* a synthetic rubber made from silicone elastomers and characterized by resistance to chemicals, weak acids, and bases; used for gaskets and electrical insulation.
silicon fluoride *Inorganic Chemistry.* SiF_4, a colorless, fuming gas with a suffocating odor; decomposed by water and soluble in absolute alcohol; freezes at 90.2°C and boils at 86°C; used in chemical analysis and to seal water out of oil wells during drilling. Also, SILICON TETRAFLUORIDE, TETRAFLUOROSILANE.
silicon halide *Inorganic Chemistry.* a compound consisting of silicon and an element from the halogen group (periodic table Group VIIa), such as fluorine or chlorine.
silicon hydride see SILANE, def. 2.
silicon image sensor *Electronics.* a type of television camera in which images are projected onto light-sensitive elements bonded to a silicon substrate. Also, **silicon imaging device.**
silicon iodide *Inorganic Chemistry.* SiI_4, colorless to white crystals that decompose in water; melts at 120.5°C and boils at 287.5°C; toxic and an irritant. Also, SILICON TETRAIODIDE.

siliconizing *Metallurgy.* the process of diffusing silicon into the outer layer of a steel.

silicon-manganese deoxidation *Materials Science.* in the primary processing of steels, a process in which molten steel is deoxidized to specified levels by the addition of manganese and silicon to produce semikilled steel.

silicon micromachining *Chemical Engineering.* the use of various chemical and plasma techniques to fabricate miniaturized devices and components through the use of a single-crystal silicon material.

silicon monoxide *Inorganic Chemistry.* SiO, an amorphous solid that is hard and abrasive; insoluble in water; melts above 1700°C and boils at 1880°C; used to form a thin, protective surface on optical parts, dielectrics, insulators, and mirrors.

silicon nitride *Inorganic Chemistry.* Si_3N_4, an amorphous gray powder; insoluble in water; melts under pressure at 1900°C; used in solid-state devices, high-strength fibers, and corrosion-resistant coatings and for various other industrial purposes.

silicon-on-sapphire *Electronics.* a process in which metaloxide semiconductor components are mounted onto a thin single-crystal silicon film growing atop a synthetic sapphire substrate.

silicon-oxygen tetrahedron *Physical Chemistry.* the basic unit of the microstructure of silicates, composed of a silicon ion in the center surrounded by four equidistant oxygen ions, each with a valence of –2; the oxygen ion may be shared between adjacent silica tetrahedra, producing chains plates, or a three-dimensional network of silica tetrahedra.

silicon P-N junction *Nucleonics.* a semiconductor device in which ionizing radiation is detected in the depletion layer of a reverse-biased junction; manufactured by diffusing phosphorus about 2 micrometers into the surface of a P-type silicon base.

silicon rectifier *Electronics.* a device that employs the electrical properties of silicon to convert AC current to DC at extremely high voltages.

silicon resistor *Electronics.* a resistor manufactured as part of an integrated circuit.

silicon solar cell *Electronics.* a device that employs the electrical properties of silicon to convert light into electricity.

silicon steel *Metallurgy.* any of several steels containing up to 5% silicon, mainly used to manufacture transformer cores.

silicon surface barrier detector *Nucleonics.* a semiconductor detector that detects ionizing radiation; made by etching and treating the surface of an N-type silicon crystal to form a P-type layer, which is then contacted with a gold lead.

silicon-symmetrical switch *Electronics.* a device that permits current to flow in both directions; used to control motor speed.

silicon tetrabromide see SILICON BROMIDE.

silicon tetrachloride see SILICON CHLORIDE.

silicon tetrafluoride see SILICON FLUORIDE.

silicon tetrahydride see SILANE, def. 1.

silicon tetraiodide see SILICON IODIDE.

silicon transistor *Electronics.* a device that employs the electrical properties of silicon to amplify an electrical signal.

silicon wafer *Electronics.* see CHIP.

silicosiderosis *Medicine.* pneumoconiosis from the inhaled dust of silica and iron.

silicosis [sil´i kō´sis] *Medicine.* a lung disease, a form of pneumoconiosis, acquired from inhaled silica dust, found in conditions where stone, flint, or sand is ground into dust; the symptoms may include destruction of macrophages and damage to lung tissue. Also, SILICATOSIS.

silicospiegel *Metallurgy.* a master alloy consisting of pig iron with about 20% manganese and 10% silicon.

silification see SILICIFICATION.

silique *Botany.* a long, narrow, many-seeded capsule with two valves that split from the bottom, characteristic of the Brassicaceae family.

silk *Textiles.* a fine, glossy fiber, noted for its strength, elasticity, resiliency, and luster; produced by the larva of a cultivated silkworm as it is forming a cocoon. A single filament of 300 to 1600 yards is unreeled from each silkworm cocoon after the cocoon is softened in warm water. *Mineralogy.* microscopic needle-shaped inclusions of rutile crystals in a natural gem that reflect a whitish sheen resembling that of a silk fabric.

silk cotton *Materials.* another name for kapok, the silky, elastic down or fiber covering the seeds of certain tropical trees of the bombax family, used for stuffing pillows and cushions.

silk cotton tree *Botany.* any of several tropical trees belonging to the genus *Ceiba* of the bombax family, particularly *C. pentandra*, the kapok tree, which is characterized by palmately compound leaves and large pods from which silk cotton is obtained.

silk gland *Invertebrate Zoology.* a gland in various insects and arachnids that secretes silk; an abdominal gland in spiders and a salivary gland in insects.

silk oak *Botany.* any of several trees of the genus *Grevillea*, especially *G. robusta*, characterized by feathery leaves and orange or yellow flowers; native to Australia and grown as an ornamental street tree in Florida and California. Also, **silky oak.**

silk paper *Materials.* paper containing silk fibers; used in making postage and revenue stamps.

silkscreen *Graphic Arts.* **1.** an image created by silkscreen printing. **2.** to print by this process.

silkscreen printing *Graphic Arts.* screen printing in which color is forced through a stenciled silk fabric stretched over a heavy wooden frame. Also, SERIGRAPHY.

silk tree *Botany.* an Asian tree, *Albizia julibrissin*, of the legume family, having broad, spreading branches and plumelike pink flowers; widely cultivated as an ornamental tree.

silkworm *Invertebrate Zoology.* the common name for the larvae of several species of moths, especially *Bombyx mori*, that secrete a large amount of fine, strong silk for building their cocoons, harvested in the silk industry.

silky flycatcher *Vertebrate Zoology.* a New World waxwing bird of the family Bombycillidae, characterized by a long tail, a crested head, and red, yellow, and white patches on the wings and tail.

silky fracture *Metallurgy.* a type of fracture that appears fine grained and denotes reasonable toughness.

sill *Building Engineering.* a horizontal timber or block that serves as a foundation of a wall, house, or other structure. *Civil Engineering.* **1.** a horizontal member forming the bottom of a window or exterior door frame. **2.** the horizontal overflow line of a measuring weir. Also, LOCK SILL. *Mining Engineering.* the floor of a mine gallery or passageway. *Geology.* a submarine ridge at relatively shallow depth that separates a partly closed basin from another basin or from an adjacent sea. Also, THRESHOLD. *Petrology.* a tabular igneous intrusion concordant with the planar structure of the surrounding rock. *Robotics.* robotic articulation with three degrees of freedom.

sill anchor *Building Engineering.* a projection from a foundation wall or slab that secures the sill to the foundation.

sillar *Volcanology.* tuff that has been hardened by recrystallization rather than by welding.

sill depth *Oceanography.* the depth of the lowest point on a ridge or rise between two ocean basins, as at Gibraltar, or between a fjord and the adjacent sea floor.

silled basin see RESTRICTED BASIN.

sillénite *Mineralogy.* $Bi_{12}SiO_{20}$, a soft greenish, waxy, translucent cubic mineral occurring as earthy masses, having a specific gravity of 9.3 and an undetermined hardness; found as a secondary mineral at Durango, Mexico.

Silliman, Benjamin 1779–1864, American chemist and geologist; the founding editor of the *American Journal of Science*.

sillimanite *Mineralogy.* Al_2SiO_5, a white, gray, brown, or pale-green orthorhombic mineral with vitreous luster occurring as slender prismatic crystals and in fibrous masses, having a specific gravity of 3.23 to 3.27 and a hardness of 6.5 to 7.5 on the Mohs scale; trimorphous with andalusite and kyanite and found in granite, schist, and gneiss.

silo *Agriculture.* an above-ground, air-tight structure that is used to store green crops and convert them into silage for livestock feed. *Space Technology.* an underground missile housing, usually closed by a hardened lid and provided with shock-isolating missile supports; equipped with facilities for raising the missile to a launch position or for launching directly from the housing.

silo filler see SILAGE CUTTER.

siloxane *Organic Chemistry.* any of a class of silica-based polymers having the formula R_2SiO, in which R is an alkyl group; occurs in the form of oils, greases, rubbers, resins, or plastics.

Silphenylene *Materials.* a trade name for a dimethylsiloxane oligomer that is used as a material to replace extraoral portions of the flexible tissues of the face.

Silphidea *Invertebrate Zoology.* carrion beetles, a family of large, brightly colored coleopteran insects in the superfamily Staphylinoidea, which feed and breed in decaying plant and animal matter.

Silsbee effect *Physics.* for a superconductor that carries an electric current, an effect in which the superconductivity is quenched when the magnetic field produced by the current reaches the critical field strength.

silt *Geology.* **1.** a very small rock fragment or mineral particle, smaller than a very fine grain of sand and larger than coarse clay; usually described as having a diameter from 0.002 millimeter to 0.06 millimeter; the smallest soil material that can be seen with the naked eye. **2.** sedimentary material suspended in running or standing water, especially sediment composed of silt-sized particles. **3.** see SILT SOIL.

silting *Civil Engineering.* **1.** any process in which silt is deposited in a reservoir, river, seabed, lake, or overflow area. **2.** specifically, the process of filling old mine workings hydraulically with fine waste material.

silting index *Engineering.* a measurement of the tendency of a gel or fluid to produce fine sedimentary deposits in a structure such as a pipe or valve.

siltite see SILTSTONE.

silt loam *Geology.* a soil that contains 50–88% silt, 0–27% clay, and 0–50% sand.

silt shale *Petrology.* a consolidated shale containing 10% or less sand and at least twice as much silt as clay.

silt soil *Geology.* a soil composed of at least 80% silt, but not more than 12% clay and 20% sand.

siltstone *Geology.* an indurated silt having the texture and composition of shale, but lacking fine lamination or fissility. Also, SILTITE.

silttil *Geology.* a friable, brownish to buff, open-textured silt, derived from chemically decomposed eluviated till.

Silurian *Geology.* **1.** a period of the Paleozoic era, occurring after the Ordovician and before the Devonian, dating from about 438 to 408 million years ago. **2.** the rock strata formed during this period.

Siluridae *Vertebrate Zoology.* the Eurasian catfishes, a primitive family of the order Siluriformes, characterized by an elongate body, a small or absent dorsal fin, and no adipose fin; found throughout Eurasian freshwaters.

Siluriformes *Vertebrate Zoology.* the catfishes, a cosmopolitan order of mostly freshwater fish characterized by strong spines in one or more fins and the production of a pheromone.

Siluroidei *Vertebrate Zoology.* in some classification systems, a suborder of the order Cypriniformes, containing the catfishes.

Silvanidae *Invertebrate Zoology.* a family of flat grain beetles in the order Coleoptera that includes many grain-infesting pests.

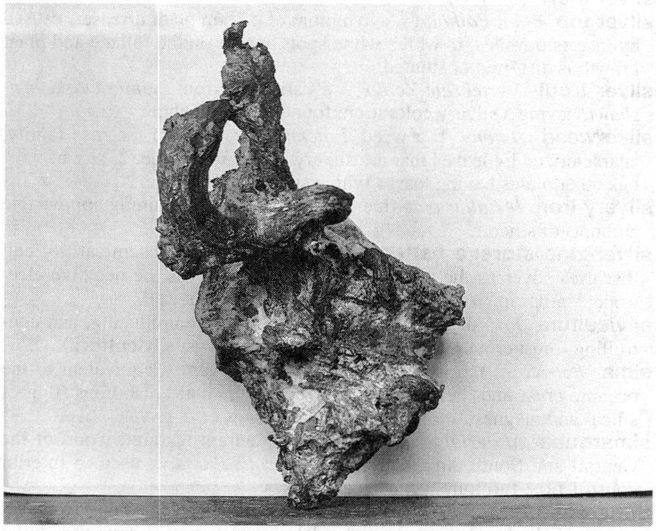

silver

silver *Chemistry.* a metallic element having the symbol Ag (from the Latin *argentum*), the atomic number 47, an atomic weight of 107.87, a melting point of 961°C, and a boiling point of 2210°C; a ductile, white lustrous metal with the highest electrical and thermal conductivity of all metals. It is used in the production of silver compounds for photography and catalysis, electrical components, jewelry, water distillation, and scientific and medical equipment; it is also used for plating. (From the Old English word for this metal.)

silver acetate *Organic Chemistry.* CH_3COOAg, toxic, deliquescent, white crystals or powder; soluble in water and dilute nitric acid; used as a reagent and an oxidizing agent.

silver acetylide *Inorganic Chemistry.* Ag_2C_2, an unstable, toxic, white powder, a salt of acetylene; insoluble in water and soluble in acids; highly explosive and used in detonators.

silver arsenite *Inorganic Chemistry.* Ag_3AsO_3, a yellow powder; insoluble in water; decomposes at 150°C; used in medicine.

silver arsphenamine *Pharmacology.* an arsenic-containing drug formerly used in the treatment of nervous complications of syphilis. Also, SILVER SALVARSAN.

silver battery *Electricity.* a battery having cells similar to silver-oxide and silver-zinc cells, but using various electrolytes.

silver bell *Botany.* a shrub or small tree of the genus *Halesia* of the Storax family, characterized by white, bell-shaped flowers and toothed leaves; found in North America. Also, **silver-bell tree.**

silverberry *Botany.* a shrub, *Elaeagnus commutata* or *E. argentea*, characterized by silvery leaves, flowers, and drupelike fruit; found in north-central North America.

silver bichromate see SILVER DICHROMATE.

silver blight see SILVER LEAF.

silver brazing *Metallurgy.* a process of brazing with one of several fillers containing silver.

silver-brazing alloy see SILVER SOLDER.

silver bromate *Inorganic Chemistry.* $AgBrO_3$, a poisonous, colorless to white light-sensitive powder; decomposed by heat; used as a reagent.

silver bromide *Inorganic Chemistry.* AgBr, pale yellow crystals that darken on exposure to light, finally turning black; insoluble in water; melts at 432°C and decomposes above 1300°C; used in photographic films or plates.

silver-cadmium storage battery *Electricity.* a storage battery that combines the desired space and weight attributes of silver-zinc batteries with the long shelf-life of nickel-cadmium batteries.

silver carbonate *Inorganic Chemistry.* Ag_2CO_3, a yellow light-sensitive powder; insoluble in water and alcohol; decomposes above 218°C; used as a laboratory reagent.

silver chlorate *Inorganic Chemistry.* $AgClO_3$, white crystals, slightly soluble in water and sensitive to light; melts at 230°C and decomposes at 270°C; used in organic synthesis. Also, ARGENTOUS CHLORATE.

silver chloride *Inorganic Chemistry.* AgCl, a white granular powder that darkens on exposure to light, finally turning black; very slightly soluble in water; melts at 455°C and boils at 1550°C; used in optics, photochromic glass, photography, and silver plating and as an antiseptic.

silver chromate *Inorganic Chemistry.* Ag_2CrO_4, a reddish-brown powder; insoluble in water and soluble in acids and some solutions of alkali chromates; used as an analytical reagent.

silver coating see SILVER PLATING.

silver cyanide *Inorganic Chemistry.* AgCN, an odorless, tasteless, toxic white powder that darkens on exposure to light; insoluble in water and decomposes at 320°C; used in medicines and in silver plating.

silver dichromate *Inorganic Chemistry.* $Ag_2Cr_2O_7$, a dark red powder; soluble in acids and decomposes in hot water and when heated; a possible carcinogen.

silver-disk pyrheliometer *Engineering.* a device that measures solar radiation; it is composed of a silver disk at the bottom of a diaphragm tube that receives radiation for an attached calorimeter; the temperature variations of the calorimeter indicate the intensity of the radiation.

silvered mica capacitor *Electronics.* a device that employs the electrical properties of silver and mica, rather than a conducting metal foil, to store an electrical charge.

silver fir *Botany.* a coniferous European tree, *Abies alba*, characterized by leaves that are dark green on top and have white bands below.

silverfish *Invertebrate Zoology.* any of various members of the primitive insect order Thysanura, especially *Lepisma saccharina*, a wingless, silvery pest that inhabits houses and feeds on starchy substances, often damaging books and clothing.

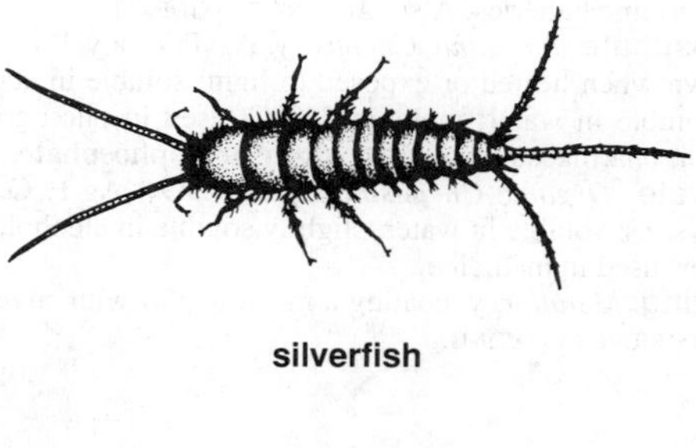

silverfish

silver fluoride *Inorganic Chemistry.* $AgF{\cdot}H_2O$, a very hygroscopic yellow or brown solid that is light-sensitive and a strong irritant; soluble in water; melts at 435°C and boils at about 1160°C; used in medicine.

silver frost *Meteorology.* a frosty glaze that builds up on trees, shrubs, and other exposed areas during freezing precipitation such as an ice storm. Also, SILVER THAW.

silver hake *Vertebrate Zoology.* a common hake, *Merluccius bilinearis,* a food fish inhabiting waters off the Atlantic coast of North America; whiting.

silver halide *Inorganic Chemistry.* a compound consisting of silver and an element from the halogen group (periodic table Group VIIa), such as fluorine or chlorine.

silver iodate *Inorganic Chemistry.* $AgIO_3$, a white powder; slightly soluble in water; melts above 200°C; used in medicine.

silver iodide *Inorganic Chemistry.* AgI, a pale yellow, odorless, tasteless powder that darkens on exposure to light; melts at 558°C and boils at 1506°C; used in medicine, photography, and cloud seeding.

silver lactate *Organic Chemistry.* $AgC_3H_5O_3$, a white or gray crystalline or powdery solid; soluble in water and alcohol; used in medicine.

silver leaf *Plant Pathology.* a symptom of certain plant diseases, particularly a disease of trees and shrubs caused by the basidiomycete *Stereum purpurteum,* which is characterized by abnormal silver-colored leaves. Also, SILVER BLIGHT.

silverline system *Invertebrate Zoology.* a series of pellicle and subpellicle structures in ciliate protozoans that show lines of stain in silver impregnation methods.

Silverman needle *Surgery.* an instrument for taking tissue specimens, consisting of an outer cannula, an obturator, and an inner split needle with longitudinal grooves in which the tissue is retained when the needle and cannula are withdrawn. (Named for Irving *Silverman,* born 1904, American surgeon.)

silver maple *Botany.* **1.** a maple, *Acer saccharinum,* characterized by leaves that are light green on top and silvery white below. **2.** the hard, close-grained wood of this tree, used in making furniture.

silver migration *Electricity.* a process that reduces insulation resistance and causes dielectric failure by ionically transporting silver from one area to another.

silver nitrate *Inorganic Chemistry.* $AgNO_3$, colorless, transparent, light-sensitive crystals; soluble in cold water and very soluble in hot water; melts at 212°C and decomposes at 444°C; used in photographic film, hair dyes, silver plating, and ink manufacture, as a reagent, and to cauterize wounds.

silver nitride *Inorganic Chemistry.* Ag_3N, colorless crystals or powder; insoluble in water; highly explosive when disturbed in any way.

silver nitrite *Inorganic Chemistry.* $AgNO_2$, yellow or grayish needles that darken on exposure to light; soluble in hot water and insoluble in alcohol; decomposes at 140°C; used as a chemical reagent.

silver oxide *Inorganic Chemistry.* Ag_2O, a dark-brown or black powder; insoluble in water and alcohol; decomposes above 300°C; used as a catalyst, as a reagent, and in purifying drinking water. Also, ARGENTOUS OXIDE.

silver(II) oxide see SILVER SUBOXIDE.

silver oxide cell *Electricity.* a dry cell that can be made with tiny dimensions and used in cameras, watches, miniaturized calculators, and similar devices.

silver perchlorate *Inorganic Chemistry.* $AgCiO_4$, colorless deliquescent crystals; soluble in water; decomposes at 486°C; highly explosive and used in manufacturing explosive materials.

silver permanganate *Inorganic Chemistry.* $AgMnO_4$, dark violet crystals that are light-sensitive and decomposed by alcohol or heat; a dangerous explosion risk when shocked or heated; used in antiseptics and in gas masks.

silver peroxide *Inorganic Chemistry.* Ag_2O_2, an unstable grayish-black powder; insoluble in water; decomposes above 100°C; a strong oxidizing agent and a serious fire and explosion risk; used in manufacturing silver-zinc batteries. Also, ARGENTIC OXIDE.

silver phosphate *Inorganic Chemistry.* Ag_3PO_4, a yellow powder that turns brown when heated or exposed to light; soluble in acids and very slightly soluble in water; melts at 849°C; used in photographic emulsions and in pharmaceuticals. Also, **silver orthophosphate.**

silver picrate *Organic Chemistry.* $C_6H_2O(NO_2)_3Ag{\cdot}H_2O$, very toxic, yellow crystals; soluble in water, slightly soluble in alcohol, and insoluble in ether; used in medicine.

silver plating *Metallurgy.* coating a metallic part with silver or a silver alloy. Also, SILVER COATING.

silver poisoning see ARGYRIA.

silver potassium cyanide *Organic Chemistry.* $KAg(CN)_2$, very toxic, white crystals; soluble in water and alcohol; used in silver plating and as a bactericide and an antiseptic.

silverprint or **silver print** *Graphic Arts.* **1.** a photographic print made on paper that has been sensitized with silver halide or silver chloride salts. **2.** a photocopy used as a proof in photocomposition.

silver protein *Organic Chemistry.* a brown hygroscopic powder containing 7.5–8.5% silver; made by reaction of a silver compound with gelatin in an alkaline solution; used as an antibacterial.

silver salvarsan see SILVER ARSPHENAMINE.

silver scurf *Plant Pathology.* **1.** a disease of potato plants caused by the fungus *Spohdylocladium atrovirens,* in which silvery spots occur on the skin. **2.** a disease of citrus fruits caused by an insect known as the thrip, causing the fruit to become silvery in appearance. Also, **silver scab.**

silver selenide *Inorganic Chemistry.* Ag_2Se, a gray powder; insoluble in water but soluble in ammonium hydroxide; melts at 880°C and decomposes at its boiling point.

silversides *Vertebrate Zoology.* any of several small fishes of the family Atherinidae, such as the grunion, having a silvery stripe along each side; found along the Atlantic and southern Pacific coasts of the U.S. Also, **silverside.**

silver solder *Metallurgy.* any of several filler metals used for silver brazing. Also, SILVER-BRAZING ALLOY.

silverstat regulator *Electricity.* a multitapped resistor whose taps are connected to single-leaf silver contacts.

silver storm see ICE STORM.

silver suboxide *Inorganic Chemistry.* a common name for silver(II) oxide, AgO, a gray powder used in manufacturing silver oxide-zinc alkali batteries.

silver sulfate *Inorganic Chemistry.* Ag_2SO_4, lustrous white crystals or powder; turns gray on exposure to light; soluble in hot water and acids and insoluble in alcohol; melts at 652°C and decomposes at about 1085°C; used as a laboratory reagent.

silver sulfide *Inorganic Chemistry.* Ag_2S, a grayish-black powder; insoluble in water and soluble in concentrated sulfuric acid; melts at 825°C and decomposes at its boiling point; used in ceramics and inlaying in metal work.

silver thaw see SILVER FROST.

silver top *Plant Pathology.* a symptom of certain plant diseases, caused by insects or mites, in which white spots appear on the foliage and plant growth is distorted or stunted.

silver trout *Vertebrate Zoology.* a variety of trout, *Salmo clarki henshawi,* having a silvery coloration; found in Lake Tahoe.

silverweed *Botany.* **1.** a weed, *Potentilla anserina* of the rose family, characterized by leaves that are silvery on the underside. **2.** any of various other plants having leaves with a silvery appearance.

silvery iron *Metallurgy.* a fine-grained cast iron containing appreciable amounts of silicon.

silver-zinc storage battery *Electricity.* an electrochemical dry cell that uses silver as the positive electrode and zinc as the negative electrode; similar in size and capacity to the silver oxide cell.

silviculture *Forestry.* the science of cultivating, reproducing, and controlling forest crops and the growth of forests. Thus, **silvicultural.**

sima *Petrology.* the lower layer of the earth's crust, equivalent to the oceanic crust and the lower part of the continental crust, consisting of silica- and magnesia-rich rocks. Also, INTERMEDIATE LAYER.

simarouba *Materials.* the nondurable, medium-textured wood of the Central and South American tree, *Simarouba amara*; used in interior construction, furniture, veneers, and boxes.

Simaroubaceae *Botany.* a family of mostly tropical dicotyledonous trees and shrubs of the order Sapindales; many species have very bitter wood, bark, and seeds used for various medicinal purposes.

SIMD [sim´dē´] *Computer Science.* a parallel computer architecture in which the same instruction is simultaneously executed on multiple data. (An acronym for single instruction, multiple data.)

simian [sim´ē ən] *Vertebrate Zoology.* **1.** an ape or monkey. **2.** relating to or resembling an ape or monkey.

simian AIDS *Veterinary Medicine.* a condition of unknown origin in rhesus monkeys, characterized by weight loss, diarrhea, fever, anemia, lymphopenia, and hepatitis; similar to AIDS in humans. Also, SAIDS.

simian virus 40 *Virology.* a virus of the genus *Polyomavirus,* family Papovaviridae, that silently infects monkeys and that was originally isolated from kidney cells of the rhesus monkey; it causes tumors in experimental rodent hosts.

similar *Mathematics.* **1.** two figures are similar if one can be obtained by a conformal transformation of the other; for example, polygons having the same number of sides are said to be similar if their corresponding angles are equal and corresponding sides are in proportion; i.e., one polygon may be obtained from the other by a similarity transformation. **2.** two $(n \times n)$ matrices A and B for which there exists an invertible matrix P such that $B = PAP^{-1}$ are said to be similar.

similar fold *Geology.* one of a series of successive congruent folds whose limbs are thinner than their axes.

similarity-based learning *Artificial Intelligence.* learning based upon the similarities and differences of training instances.

similarity coefficient *Systematics.* a mathematical representation of the presence or frequency of specified characters in organisms, measuring the degree of association or similarity between those organisms.

similarity transformation *Mathematics.* **1.** let $L(V)$ be the space of linear transformations on a vector space V. Then a similarity transformation is any automorphism of $L(V)$ of the form $f(A) = PAP^{-1}$, where A is an element of $L(V)$ and P is an invertible element of $L(V)$. **2.** if V is a vector space of dimension n and a basis is chosen, then $L(V)$ is represented as the space of $(n \times n)$ matrices. A similarity transformation of a matrix A has the form

$$\phi(A) = PAP^{-1},$$

where P is an invertible matrix. If A represents a linear transformation on V, and P is a matrix representing a change of basis, then $\phi(A)$ is the matrix representing the linear transformation in the new basis. **3.** a conformal transformation; i.e., any geometric transformation of Euclidean space obtained from a finite sequence of translations, rotations, stretchings, and shrinkings.

similitude a resemblance or similarity; specific uses include: *Engineering.* a commercial process that has been developed from a smaller-scale laboratory process. *Physics.* the application of scaling principles to infer the behavior of systems that are either very large or very small compared to systems whose behavior is already known.

Simmental or **Simmenthal** *Agriculture.* a breed of beef cattle, characterized by a coat varying from red to fawn with white markings. (From the *Simme* Valley in Switzerland.)

Simmonds' disease *Medicine.* a disease caused by total atrophy of the pituitary body, found mostly in women, characterized by premature aging, emaciation, premature senility, hair loss, loss of sexual desire, and depressed basal metabolic rate. (Named for Morris *Simmonds,* German physician, 1855–1925.)

Simmondsiaceae *Botany.* the jojoba family, a monospecific family of xeromorphic, evergreen, dioecious shrubs of the order Euphorbiales, distinguished by an unusual liquid wax contained in the seeds; native to the southwest United States and adjacent Mexico.

Simmons-Smith reaction *Organic Chemistry.* a reaction that delivers methylene stereospecifically to a carbon-carbon double bond using a carbenoid organozinc compound.

simo chart *Industrial Engineering.* a detailed time-and-motion chart that displays the elements which a given worker will perform with each hand throughout a work cycle. (An acronym for simultaneous motion chart.)

Simonsiella *Bacteriology.* a genus of gliding bacteria of the family Simonsiellaceae, occurring in the human oral cavity as flat, filamentous cells that reproduce by fragmentation.

Simonsiellaceae *Bacteriology.* a family of gliding bacteria that belong to the order Cytophagales, characterized by cells forming flat filaments.

Simon's theory *Engineering.* a drilling theory stating that the rate at which a chisel-shaped bit penetrates rock is determined by the equation

$$R = NAf_v/\pi D,$$

in which R equals the rate of penetration, N equals the number of wings on the bit, A equals the cross-sectional area to be drilled, f_v equals the number of impacts per unit time, and D equals the diameter of the drill bit.

simoom *Meteorology.* a strong, dry, dusty desert wind that blows in the Sahara, Israel, Syria, and the Arabian desert, increasing the air temperature to over 130°F with a relative humidity often below 10%.

simple *Biology.* **1.** composed of not more than one anatomically or morphologically identical unit. **2.** not branched. **3.** consisting of identical units, as a simple tissue. *Chemistry.* consisting of only one substance or element; not mixed.

simple alternative *Statistics.* in hypothesis testing, the alternative hypothesis in a case in which the null hypothesis is a simple hypothesis, i.e., completely defines the distribution of the random variable.

simple balance *Engineering.* a balance consisting of a beam rotating on a support point and having pans suspended from both its ends, used to determine the mass of an object; the object to be measured is placed in one pan, known weights are placed in the other pan, and a small horizontal weight slides across the beam until it balances horizontally.

simple branched tubular gland *Anatomy.* an exocrine gland having tubular secretory portions arranged in a single order of branching.

simple buffering *Computer Programming.* an input/output buffering method in which a single buffer is assigned to one file for the duration of the process.

simple carbohydrate *Nutrition.* a compound of carbon, hydrogen, and oxygen that is an important and immediate source of energy for the body; found in sucrose.

simple closed curve *Mathematics.* a closed curve that does not intersect itself.

simple conic map projection *Cartography.* a type of map projection in which the surface of the sphere being mapped is projected on the surface of a cone tangent to the sphere; the cone is then unrolled to form a plane. Thus, **simple conic chart.**

simple continuous reaction *Biotechnology.* an improvement on the standard batch-process system by the use of pasteurized feed pumped into an agitated vessel containing highly productive cell cultures.

simple crater *Geology.* a meteorite impact crater of relatively small diameter, a uniformly concave-upward shape, a maximum depth in the center, and a lack of central uplift.

simple cross-bedding *Geology.* cross-bedding formed solely by deposition, having nonerosional lower bounding surfaces.

simple cubic lattice see CUBIC UNIT CELL.

simple dike *Petrology.* an igneous dike occurring as the result of a single intrusion of magma.

simple engine *Mechanical Engineering.* an engine whose cycle includes expansion in a single phase, followed by the exhaust of the working fluid.

simple event see ELEMENTARY EVENT.

simple extension *Mathematics.* a field F is said to be a simple extension of a field K if F is generated by K and a single element u which is in F but not in K.

simple fracture *Medicine.* a fracture that does not create an open wound in the skin.

simple fruit *Botany.* a fruit derived from a single pistil of a single flower.

simple function *Mathematics.* a real-valued or complex-valued function f defined on a measure space (X, A, m) is said to be simple or finitely valued if f is zero except on a finite number n of disjoint sets $A_i \in A$, where each A_i has finite measure $m(A_i)$ and f equals a finite constant k_i on each A_i. The product and finite linear combinations of simple functions are also simple functions. Also, STEP FUNCTION.

simple gland *Anatomy.* an exocrine gland whose secretory portions are unbranched.

simple goiter *Medicine.* simple hyperplasia, i.e., enlargement, of the thyroid; found in areas where there is low iodine intake.

simple graph *Mathematics.* a graph with no loops or multiple edges.

simple group *Mathematics.* a group that has no proper normal subgroups.

simple harmonic current *Electricity.* a current with a sinusoidal waveform.

simple harmonic electromotive force *Electricity.* an alternating electromotive force that is equivalent to the product of a constant and the cosine or sine of an angle which varies linearly with time.

simple harmonic motion see HARMONIC MOTION.

simple harmonic oscillator *Mechanics.* see HARMONIC OSCILLATOR.

simple harmonic voltage *Electricity.* a voltage with a sinusoidal waveform.

simple hypothesis *Statistics.* a hypothesis that completely specifies the distribution of a random variable.

simple leaf *Botany.* a leaf with one undivided blade; or a lobed leaf that is not divided to the midrib.

simple lens *Optics.* any lens consisting of only a single element.

simple machine *Mechanics.* one of a set of elementary mechanisms that are regarded as the basis for all other forms of machines; the standard group of six simple machines includes the lever, pulley, inclined plane, wheel and axle, wedge, and screw.

simple mastectomy *Surgery.* the removal of a breast, including only breast tissue, nipple, and part of the skin, to treat breast cancer.
simple microscope *Optics.* a single lens that produces large images from small objects; for example, a magnifying glass.
simple order see LINEAR ORDER.
simple ore *Geology.* an ore composed of a single metal.
simple oscillator *Mechanics.* see HARMONIC OSCILLATOR.
simple pendulum *Mechanics.* an idealized device consisting of a small massive body suspended by a rigid massless support from a fixed horizontal axis, about which it can rotate freely.
simple pistil *Botany.* a pistil composed of a single carpel.
simple pit *Botany.* a thin, depressed area in a cell wall that does not have the overarching margin characteristic of a bordered pit.
simple protein *Biochemistry.* one of a group of proteins, including globulins, glutelins, histones, prolamines, and protamines, that yield only amino acids upon hydrolysis.
simple root *Mathematics.* a root of multiplicity 1. In particular, if f is a polynomial over a ring R, an element r of R is a simple root of f if $(x-r)$ is a factor of f but $(x-r)^2$ is not.
simple salt *Chemistry.* a salt that contains no replaceable H^+ or OH ion.
simple schizophrenia *Psychology.* a form of schizophrenia that is characterized mainly by withdrawal from personal relationships and external interests, often accompanied by extreme dissatisfaction and indifference.
simple sequence DNA see CENTROMERIC DNA.
simple shear *Geophysics.* a type of plane-strain deformation caused by differential movement on a single set of parallel planes.
simple sound source *Acoustics.* a sound source whose dimensions are small in comparison to the wavelength of the emitted sound, so that at great enough distances, the geometry of the source is irrelevant.
simple stomach *Anatomy.* a saclike, distensible portion of the digestive tube that is not subdivided into specialized chambers.
simple tone *Acoustics.* a tone having only a single sinusoidal frequency, as distinguished from complex sound.
simple tubular gland *Anatomy.* an exocrine gland whose secretory portion is tubular
simple urethritis see NONGONOCOCCAL URETHRITIS.
simple valley *Geology.* a valley that remains in constant relation to the general structure of the underlying strata.
simplex [sim´pleks] *plural,* **simplexes** or **simplices** [sim´plə sēz´]. having or consisting of only one element; simple; specific uses include: *Electricity.* communication in which transmission can go in only one direction at a time. Thus, **simplex channel, simplex transmission.** *Mathematics.* an ***n*-simplex** σ^n or **simplex of dimension *n*** is the convex hull of $(n+1)$-independent points $X_1, \ldots, X_{n+1}$ in some Euclidean space R^m (where $m \geq n$); i.e., the intersection of all closed convex sets that contain these points. The points $X_1, \ldots, X_{n+1}$ are said to span s^n. Equivalently, s^n is the set of all points

$$X = \lambda_1 X_1 + \cdots + \lambda_{n+1} X_{n+1},$$

where the λ_i (the barycentric coordinates of the point) are nonnegative scalars whose sum is 1. For example, a 1-simplex is a line segment, a 2-simplex is a triangle, and a 3-simplex is a tetrahedron. By theorem, any two n-simplexes (or n-simplices) are homeomorphic. A face of an n-simplex is any $(n-1)$-simplex spanned by n points of the n-simplex. *Medicine.* see HERPES SIMPLEX.
simplex circuit *Electrical Engineering.* the circuit formed using two wires in parallel with ground return.
simplex concrete pile *Civil Engineering.* a structural pile constructed by driving a steel pipe into the ground, placing reinforcement into the pipe, and introducing concrete as the pipe is vibrated out of the ground. This prevents water ingress.
simplex method *Mathematics.* in linear programming, a classical iterative technique for optimizing a linear objective function.
simplex pump *Mechanical Engineering.* a pump that has only one water cylinder and one steam cylinder.
simplex structure *Computer Technology.* a computer system configured with the minimum number of components needed to accomplish its function.
simplex uterus *Anatomy.* a uterus consisting of a single cavity, not divided into horns or extensions.
Simplexvirus *Virology.* a genus of viruses of the family Herpesviridae, subfamily Alphaherpesvirinae, of which herpes simplex 1 is the type member.
simplicial complex *Mathematics.* a topological space K that is a union of a finite number of subspaces, each of which is homeomorphic to some simplex $(\sigma_i)^n i$ (that is, the ith simplex, having dimension n_i) and subject to the following conditions: (a) if $(\sigma_i)^n i$ is a simplex of K with face $(\sigma_i)^n i^{-1}$, then $(\sigma_i)^n i^{-1}$ is also a simplex of K; (b) if $(\sigma_i)^n i$ and $(\sigma_k)^n k$ are simplexes of K, then either they are disjoint or their intersection $(\sigma_i)^n i \cap (\sigma_k)^n k$ is also a simplex of K; and (c) a topology on K is given by defining a subset C of K to be closed if and only if $C \cap (\sigma_i)^n i$ is closed in $(\sigma_i)^n i$ for each i. The dimension of a complex K is the maximum value of the dimensions n_i of the component simplexes. For example, a simplicial complex that is the union of 2-simplexes is called a **triangulation** and has dimension 2. By theorem, any simplicial complex is homeomorphic to a subset of some Euclidean space R^m.
simplicial homology see SINGULAR HOMOLOGY.
simplicial subdivision *Mathematics.* a decomposition of one or more of the simplices in a simplicial complex so that the simplices involved are the union of simplices of smaller dimension.
simplification the fact or result of simplifying; specific uses include: *Artificial Intelligence.* the replacement of a formula or expression by one that is equivalent but simpler.
simply connected *Mathematics.* **1.** a subset A of a topological space X is said to be simply connected if its boundary is connected. A simply connected set A has no holes; that is, any closed curve contained in A can be continuously deformed to a point without passing through a point in the complement $X - A$ of A. **2.** a topological space X is said to be simply connected if it is both connected and locally connected. Equivalently, X is simply connected if its fundamental group has exactly one element.
simpsonite *Mineralogy.* $Al_4(Ta,Nb)_3(O,OH,F)_{14}$, a colorless or light-yellow, transparent, trigonal mineral occurring as tabular or short, prismatic crystals, having a specific gravity of 6.70 and a hardness of 7 to 7.5 on the Mohs scale; found in granite pegmatites.
Simpson's rule *Mathematics.* the Newton-Cotes formula in the case where $n = 2$. In particular, if $f(x)$ is a real-valued integrable function, then the integral $\int_a^b f(x)dx$ is approximately equal to

$$(b-a)[f(a)+4f((a+b)/2)+f(b)]/6.$$

Petroleum Engineering. the application of this formula in determining a reservoir's gas-bearing or oil-bearing, net-pay volume after analysis of contour lines on the reservoir's subsurface geological map.
SIMS secondary ion mass spectrometer.
SIMSCRIPT *Computer Programming.* a programming language used for discrete event simulation.
SIMULA *Computer Science.* a programming language used for specifying discrete-event simulation models; originated object-oriented programming.
simulate *Engineering.* to produce a likeness of all or part of a process or system; applied particularly to computers or physical models, but also involves physical scale models of complex plants and refineries.
simulated annealing *Artificial Intelligence.* a technique, related to hill climbing, that attempts to find optimal assignments of values to multiple parameters by doing a hill-climbing search starting with a large step size ("temperature"), then reducing the step size and doing another search, starting from the previously determined optimum.
simulation *Transportation Engineering.* a model, such as a computer program, that simulates a traffic flow.
simulation language *Computer Programming.* a high-level, problem-oriented programming language specifically designed for discrete-event or continuous simulation applications.
simulation model *Industrial Engineering.* a mathematical representation of variables and their relationships in specific organizational situations.
simulation program *Transportation Engineering.* a computer program used for traffic flow simulation.
simulator *Engineering.* a device that reproduces a system or process to demonstrate or test its performance under various conditions. It can also be used to train operators. *Computer Technology.* **1.** a computer system, device, or program that imitates the behavior of a physical or abstract system. **2.** a routine that allows a certain computer to behave like a different computer.
Simuliidae *Invertebrate Zoology.* the black flies and buffalo gnats, a family of dipteran insects in the series Nematocera that breed near running water; females bite viciously, often in swarms, and some transmit disease-causing parasites.

simultaneity [sī´mul tə nā´i tē] the fact of being simultaneous; the occurrence of two events at the same time; specific uses include: *Mechanics.* a condition in which two separate events are perceived to occur at exactly the same time by a given observer; this same perception may not be shared by a second observer in a different frame of reference, especially when extremely high speeds are involved (i.e., close to the speed of light). Thus simultaneity is a relative condition rather than an absolute one. *Computer Technology.* the ability of a computer system to perform several operations concurrently.

simultaneous altitudes *Astronomy.* the instantaneous observation of two or more celestial bodies from a given location.

simultaneous color television *Electronics.* a television system in which three primary colors are activated simultaneously, rather than sequentially.

simultaneous computer *Computer Technology.* a computer composed of separate units that concurrently process different parts of a computation and are interconnected in a manner determined by the computation.

simultaneous conditioning *Behavior.* a classical conditioning procedure in which the conditioned and unconditioned stimuli are presented and terminated at the same time.

simultaneous equations *Mathematics.* two or more given equations that are required to have a common (or simultaneous) solution.

simultaneous lobing see MONOPULSE RADAR.

simultaneous motion chart see SIMO CHART.

sin sine.

SIN *Aviation.* the airport code for Singapore.

Sinanthropus pekinensis *Anthropology.* a former scientific name for Peking man.

sincosite *Mineralogy.* $CaV_2^{+4}(PO_4)_2(OH)_4 \cdot 3H_2O$, a soft, leek-green tetragonal mineral, occurring as tabular crystals, scales, or plates, having a specific gravity of 2.970 and an undetermined hardness; found in siliceous gold ores and carbonaceous shales.

Sindbis virus *Virology.* a virus of the genus *Alphavirus* of the family Togaviridae, occurring in Africa, Asia, Australia, and Europe; it infects vertebrates, primarily birds, and is transmitted by mosquitoes.

sindle poison *Toxicology.* a chemical agent that disrupts cell division by preventing the assembly of microtubules within the cell.

sine [sīn] *Mathematics.* **1.** an odd, periodic, real-valued analytic function of a real angle (rotation) with period 2π. The sine of q, written $\sin q$, is the ordinate of the point on the unit circle obtained by moving $q \geq 0$ units counterclockwise along the circle from the point (1,0). (For $q < 0$, move $|q|$ units in a clockwise direction.) For real q, the sine function can be shown to be the imaginary part of the function e^{iq}. **2.** in elementary trigonometry, the sine of an acute angle q, written $\sin q$, is defined to be the ratio of the side opposite to q to the hypotenuse of a right triangle having vertex angle q. When $0 < q < \pi/2$, these two definitions are equivalent.

sine bar *Mechanical Devices.* a highly accurate device consisting of a steel straight edge mounted between two cylindrical rollers, whose centers are either 5 or 10 inches apart and equidistant from the straightedge; used for measuring angles and precision layout work on angle plates.

sine condition see ABBÉ SINE CONDITION.

sine-cosine encoder *Electronics.* a device that translates the sine of a rotating shaft angle into sequential digital pulses.

sine-cosine generator see RESOLVER.

sine curve *Mathematics.* the graph that is in the plane of the equation $y = \sin x$.

sine galvanometer *Engineering.* an instrument used to measure the intensity of a magnetic field, consisting of a small magnet hung between two Helmholtz coils; when known currents pass through the coils, the movement of the magnet is measured to determine magnetic intensity.

Sinemurian *Geology.* a European geologic stage of the Lower Jurassic period, occurring after the Hettangian and before the Pliensbachian.

sine potentiometer *Electronics.* a device designed to move a shaft between two resistors in order to provide an adjustable resistance that is proportional to the sine of the angle through which the shaft has rotated.

sinew *Anatomy.* another term for the tendon of a muscle, especially in older use.

sine wave *Physics.* a wave whose variation is purely sinusoidal; that is, having a single frequency of oscillation.

sine-wave modulated jamming *Electronics.* a method of disabling a radar system by flooding it with radio signals that are modulated with waves of pure alternating current.

sine-wave oscillator see SINUSOIDAL OSCILLATOR.

sine-wave response see FREQUENCY RESPONSE.

sing or **sing.** of each. (From Latin *singulorum.*)

singeing [sinj´ing] *Textiles.* a finishing process in which fabric or thread is passed over a gas flame or heated copper plates to singe off protruding fibers or fuzz and yield a smooth surface.

single-bevel groove weld *Metallurgy.* a groove weld that is not symmetrical, because one of the parts is beveled.

singing *Control Systems.* a term for a continuous, unwanted oscillation in a system or component due to excessive positive feedback.

singing fish *Vertebrate Zoology.* see MIDSHIPMAN.

singing margin *Control Systems.* the difference, in decibels, between the singing point and the normal gain of a system or component.

singing point *Control Systems.* the minimum amount of gain in a system or component that will cause singing.

singing sand see SOUNDING SAND.

singing-stovepipe effect *Electricity.* the reception and reproduction of radio signals by metal pieces in contact with each other, such as stovepipe sections.

single-acting *Mechanical Engineering.* describing the operation of a piston or plunger that is driven by only one side, as in most early engines.

single action *Ordnance.* a revolver or other weapon using single-action firing.

single-action *Ordnance.* describing a firing system in which the hammer must be manually cocked for each shot in a separate motion from that applied to the trigger, as in older revolvers.

single-action press *Mechanical Engineering.* a press having only one slide.

single-address instruction *Computer Programming.* an instruction that references only one operand location.

single-axis gyroscope *Engineering.* a gyroscope that is supported by a single gimbal having bearings that form the gyroscope's output axis.

single-beam spectrometer *Spectroscopy.* a spectrographic instrument that employs a single beam of electromagnetic radiation to induce the absorption spectrum that is recorded.

single blind *Science.* an experiment or clinical trial in which the researchers know whether or not a particular subject is receiving the indicated drug or treatment, but the subjects themselves do not know this; distinguished from a double blind, in which neither the researchers nor the subjects know who is receiving treatment and who is not. Thus, **single-blind experiment, single-blind test.**

single-block brake *Mechanical Engineering.* a friction brake operated by a lever on a fulcrum that applies a block to the contour of a braking surface.

single-board computer *Computer Technology.* a complete computer on one printed circuit board; includes ROM, RAM, CPU, and I/O interface.

single bond *Physical Chemistry.* a conventional bond in which there is one shared electron pair, as opposed to a multiple bond. Also, SINGLE BONDING, SINGLE COVALENT BOND.

single-bond energy see BOND DISSOCIATION ENERGY.

single-burst experiment *Virology.* a process in which a cell population is infected with viruses and manipulated in order to study the burst size of a single lytic cycle.

single bus *Electricity.* a communication path that handles only one signal.

single-button carbon microphone *Acoustical Engineering.* a microphone having a small buttonlike container filled with carbon granules on one side of a diaphragm that is vibrated by sound waves, thus producing an electroacoustic signal by compressing and releasing the button.

single-carrier theory *Solid-State Physics.* a theory that deals with cases where only one carrier type is responsible for the conduction across the rectifying barrier of a metal-semiconductor junction.

single-cell protein *Biotechnology.* the biomass component of a fermentation process that is produced, usually by commercial culturing, for food or animal feed; cultivated from single cell cultures of various microorganisms such as bacteria, yeast, mold, or microalgae; large-scale production is extremely energy intensive.

single-channel multiplier *Electronics.* a device that generates large amounts of current by beaming light through a cylinder to dislodge electrons from the material coating its sides.

single-channel queueing *Transportation Engineering.* a queueing system having one queue and one service (exit) point.

single-channel simplex *Telecommunications.* a simplex procedure that allows for nonsimultaneous radio communications between stations that are on the same frequency channel.
single completion *Petroleum Engineering.* a well with one tubing string, used to recover gas or oil from a single reservoir zone or level.
single-contact sensor *Robotics.* a sensor that only allows one-dimensional measurement and can only transmit two different signals.
single-copy DNA *Genetics.* in a genome, a DNA sequence that occurs only once, or a very few times.
single-copy gene *Genetics.* in a haploid genome, a unique (nonrepeated) gene.
single-copy plasmid *Genetics.* a plasmid that is maintained in bacteria at a ratio of one plasmid for every host cell.
single covalent bond see SINGLE BOND.
single cross *Genetics.* a cross between two inbred strains.
single-current transmission *Telecommunications.* in telegraphy, a transmission in which current moves in a single direction during marking intervals, and no current moves during spacing intervals.
single-cut *Mechanical Devices.* having a single set of parallel cutting edges diagonally oriented on its face. Thus, **single-cut file.**
single-cycle mountain *Geology.* a mountain that has been folded and later destroyed without reelevation of its important parts.
single-degree-of-freedom gyro *Mechanics.* a gyroscope whose axis of spin can only rotate about one of the orthogonal axes.
single-density *Computer Technology.* describing the storage capacity of the first generation of diameter floppy disks.
single diffusion test *Immunology.* a method used to analyze an antibody-antigen mixture, in which an antigen solution is placed above a column of antibody incorporated in a gel, and the components are observed as they diffuse toward each other.
single-drift flight *Aviation.* a flight in which one drift correction is used to compensate for all drift factors between the departure point and the destination.
single-effect evaporation *Chemical Engineering.* an evaporation process completed using a single heating unit or entirely in one vessel.
single-end amplifier *Electronics.* an amplifier in which the voltage of the output signal is referenced to ground; the opposite of a push-pull amplifier.
single-ended *Electricity.* describing a circuit or transmission line that is grounded on one side.
single-ended ferry *Naval Architecture.* a ferry having different hull forms at bow and stern; designed to proceed normally in one direction only, in contrast to a double-ended ferry having an interchangeable bow and stern.
single-ended output see UNBALANCED OUTPUT.
single-ended spread *Engineering.* a series of sensing instruments that detect vibrations in areas of rock, soil, or ice, having a shot point at one end of the series.
single-gun color tube *Electronics.* a television picture tube in which a single electron beam moves sequentially across the primary-color phosphor dots to produce a color element.
single-hand drilling *Engineering.* a rock-drilling process in which a hand-held drill steel is hit with a hammer and rotated between the blows; used in areas where movement is restricted.
single-hit kinetics see ONE-HIT KINETICS.
single-hit theory (of hemolysis) *Hematology.* the theory that hemolysis results from a single complement-induced lesion of the erythrocyte surface, rather than that lesions at several sites are necessary. Similarly, **single-hit theory of carcinogenesis.**
single-hop transmission *Telecommunications.* a radio link that is reflected down from the ionosphere only once between the transmitter and the receiver.
single-hung window *Building Engineering.* a window having two sashes, only one of which is movable.
single-impulse welding *Metallurgy.* a process of spot, projection, or upset welding that is performed with a single current impulse.
single in-line package *Electronics.* any integrated circuit that has a single row of terminals or lead wires along one edge of the package.
single-knit fabric *Textiles.* a textile made on a circular knitting machine with only one set of needles and with a simple interlooping process. Single-knits can be made much faster than double-knits.
single knock-on *Solid-State Physics.* an atom in a crystal lattice that is struck by an energetic particle passing through the solid, and subsequently recoils but does not gain sufficient energy from the collision to become free and to strike another atom.
single-length *Computer Programming.* of or relating to the representation of binary numbers so that each number can be contained in a single computer word.
single lens see SIMPLE LENS.
single-level address see DIRECT ADDRESS.
single-line stream *Cartography.* a term for a stream that is too narrow to be shown at a given scale with separate lines for each bank.
single-loop feedback *Control Systems.* a type of feedback that may only occur along one electrical path.
single-loop servomechanism *Control Systems.* a servomechanism that has only one feedback loop. Also, SERVO LOOP.
single-origin hypothesis *Anthropology.* the theory that all humans are descended from a single set of ancestors.
single packing *Mining Engineering.* conventional strip packing on a longwall face, with the widest pack located along the roadside.
single-pass weld *Metallurgy.* a weld completed in one pass.
single-path system *Navigation.* a navigational system in which a circular line of position is established by measuring the absolute time between transmission and reception of a radio signal.
single-phase *Electricity.* of a circuit or device, powered by one phase of AC voltage.
single-phase flow *Chemical Engineering.* a flow in which only one phase is present.
single-phase material *Materials Science.* a material composed of only one phase, which may be a solid solution.
single-phase meter *Engineering.* a power-factor meter that moves to show an angle between current and voltage; composed of a stationary coil containing a load current and crossed coils that are connected to the load voltage.
single-phase motor *Electricity.* an AC inductance motor that is designed to operate from a single-phase voltage source.
single-phase rectifier *Electronics.* a device that converts alternating current into direct current from a single voltage rather than several voltages.
single-photon emission computed tomography *Medicine.* a process that measures the emission of single photons of a given energy from radioactive tracers in the human body.
single-piece milling *Mechanical Engineering.* a machining technique in which a workpiece is milled in a single machine cycle.
single-plane service *Transportation Engineering.* a service in which one aircraft flies the entire length of a route, so that passengers continuing to the end of the route do not have to change planes. Also, THROUGH PLANE SERVICE.
single-plate lapping *Design Engineering.* the lapping in machines that have a single, revolving cast-iron or alloy-bonded abrasive lap for grinding wheel applications.
single-point grounding *Electricity.* a grounding system that attempts to restrict all return currents to a network; the ground point functions as the circuit reference.
single-point tool *Engineering.* a cutting tool that has a single face and one uninterrupted cutting edge.
single-polarity pulse *Electricity.* a pulse that is only positive or only negative.
single-pole double-throw *Electricity.* a switch that makes one connection to either of two circuits.
single-pole single-throw *Electricity.* a switch that makes one connection to one circuit.
single-precision *Computer Programming.* referring to or describing the use of one computer word to represent a number, as distinguished from double-precision or multiprecision. Thus, **single-precision arithmetic, single-precision number.**
single refraction *Optics.* the refraction that occurs in an isotropic crystal.
single sampling *Industrial Engineering.* a quality-assurance sampling procedure in which a single sample is taken from each lot and tested as the basis for acceptance or rejection of the lot, as opposed to sampling schemes in which one outcome of a sample test can be used to test a further sample.
single-screw extruder *Materials Science.* a screw pump used for continuously shaping or mixing polymers as well as incorporating additives such as color concentrates, fillers, reinforcing fibers, stabilizers, lubricants, antioxidants, foaming agents, and cross-linking agents.
single-sheet *Computer Technology.* describing a printer that accepts individual sheets of paper, as opposed to a roll or fan-folded continuous forms. Thus, **single-sheet feed.**

single-shot *Ordnance.* **1.** describing a firearm that is loaded manually for each shot. Thus, **single-shot weapon, single loader. 2.** of or relating to an automatic weapon in semiautomatic operation, so that the trigger must be squeezed separately for each shot.
single-shot blocking oscillator *Electronics.* an oscillator that produces a single cycle of alternating current, rather than a continuous sinusoidal signal.
single-shot camera *Optics.* a camera, often 35 mm in format and used with a microscope, designed to take a single frame before being reloaded.
single-shot exploder *Engineering.* an explosion-generating device that is operated by twisting the firing key one half turn.
single-shot probability *Ordnance.* the likelihood that a single projectile fired under certain conditions will hit its target.
single-shot survey *Petroleum Engineering.* a directional record or log used in an oil well to obtain a single reading from within the drill pipe or a nonmagnetic drill collar.
single-shot trigger circuit *Electronics.* a circuit that delivers a single electric pulse to initiate one complete cycle. Also, SINGLE-TRIP TRIGGER CIRCUIT.
single-sideband communication *Telecommunications.* a communications system in which one of the two sidebands that are employed in amplitude modulation is suppressed. Thus, **single-sideband modulation, single-sideband transmission.**
single-sided amplifier see SINGLE-END AMPLIFIER.
single-sided disk *Computer Technology.* a disk on which data can be recorded on only one side.
single-signal receiver *Electronics.* a device that accepts a narrow range of ultrahigh radio signals from a densely crowded band of signals; used for code reception.
single-stage compressor *Mechanical Engineering.* a machine designed to compress gases without using any multiple elements or stages in a sequence.
single-stage pump *Mechanical Engineering.* a pump whose head is developed by one impeller in one housing.
single-stage rocket *Space Technology.* a rocket or rocket missile consisting of or propelled by a single rocket motor having one or more chambers.
single-stand mill *Metallurgy.* a mill in which the stock is not simultaneously worked by more than two rolls.
single-step operation *Computer Programming.* a mode of operation, often used during debugging, in which only one computer instruction executes at a given time in response to an external signal such as a keypress.
single-strand binding protein *Biochemistry.* any of a class of four-part proteins that bind specifically and cooperatively to single-stranded DNA at the Y-fork, thereby protecting the DNA (by preventing transcription) during replication; includes *E. coli* DNA binding protein I. Also, HELIX DESTABILIZING PROTEIN, UNWINDING PROTEIN.
single-stranded DNA *Biochemistry.* a linear polymer of deoxyribonucleotides in which the β-D-deoxyribofuranose residues are connected by 5', 3' phosphate linkages to form the backbone of a DNA molecule and the purine bases are attached as side chains, one to each deoxyribose residue. Also, SSDNA.
single-stub transformer *Electromagnetism.* a section of coaxial line that is short-circuited to a main section of coaxial line at a point where a discontinuity occurs to provide impedance matching.
single-stub tuner *Electromagnetism.* a portion of transmission line that is terminated by a short-circuiting plunger attached to a main transmission line to provide match impedance.
single thread *Design Engineering.* a screw thread having a single helix in which the lead and pitch are equal.
single-throw switch *Electricity.* a switch that makes one connection in one or more circuits.
singleton *Mathematics.* a set consisting of exactly one element. Also, **singleton set.**
single-tone keying *Telecommunications.* a type of keying in which the modulating function is responsible for bringing about the state of the carrier being modulated with a single tone for one condition (either marking or spacing) and unmodulated for the other condition.
singlet oxygen *Chemistry.* an excited state of oxygen in which all electrons are paired.
single-trip trigger circuit see SINGLE-SHOT TRIGGER CIRCUIT.
singlet state *Quantum Mechanics.* the electronic state in which the total spin angular momentum is zero.
single-tuned amplifier *Electronics.* a device that amplifies signals only at a given frequency.
single-tuned circuit *Electricity.* a circuit that can be represented by one inductance and one impedance, along with associated resistances.
single-tuned interstage *Electronics.* a circuit that is positioned between two other circuits and is resonant at one frequency only.
single-unit semiconductor device *Electronics.* a device that passes current along a single path.
single-wire line *Electricity.* a line feeding an antenna, with the return provided by earth ground.
single yarn *Textiles.* a simple strand of yarn as it comes from a spinning machine, silk reel, or spinneret, and before it has been twisted with any other yarn.
singly linked ring *Computer Programming.* a circular data structure in which the elements have forward pointers, thus permitting searches in one direction, but not in both directions.
singular *Mathematics.* **1.** an operator, map, transformation, and so on that is not one-to-one, i.e., that is not injective, is said to be singular. **2.** a (square) matrix of determinant zero is called a **singular matrix;** equivalently, a noninvertible matrix is singular. Any matrix representing a singular transformation is singular. **3.** the **singular values** of a square (complex) matrix M with adjoint matrix M^* are the positive square roots of the eigenvalues of M^*M. **4.** two measures μ and ν defined on a measure space X are said to be **mutually singular** (denoted $\mu \perp \nu$) if there exist disjoint sets A and B such that

$$X = A \cup B \text{ and } \mu(A) = \nu(B) = 0.$$

It can be said that μ is singular with respect to ν, or that ν is singular with respect to μ. The notion of absolute continuity is antithetical to that of singularity.
singular arc *Control Systems.* in an optimal control problem, the portion of the optimal trajectory in which the Hamiltonian function is not an explicit function of the control inputs, such that higher-order necessary conditions must be applied in the solution process.
singular cohomology *Mathematics.* the cohomology groups that are computed for a manifold using the graded modules S^p consisting of functions (singular p-cochains) that assign a real number to each singular p-simplex. For $f \in S^p$, the differential (coboundary operator) is defined by setting $df(s) = f(\partial\sigma)$, where σ is a $(p + 1)$-simplex. There is a natural isomorphism between the singular cohomology groups of a manifold and the dual spaces of the singular homology groups.
singular corresponding point *Meteorology.* a reappearing characteristic that appears on successive charts of constant pressure as a center of elevation or depression, and on charts of constant height as a center or high or low pressure.
singular homology *Mathematics.* the homology groups that are computed for a manifold using (differentiable or continuous) p-simplices as the graded modules and the boundary operator ∂ as the differential. ∂_p takes each p-simplex to a $(p-1)$-simplex, and the pth homology group is equal to the kernel of ∂_p (the p-cycles) modulo the image of ∂_{p-1} (the p-boundaries). A nonzero pth homology group indicates the presence of p-dimensional "holes" in the manifold.
singular integral *Mathematics.* an integral with infinite limits of integration or an integrand with singularities. A **singular integral equation** is an equation involving a singular integral.
singularity *Physics.* a point in space-time where the space-time curvature becomes infinite. *Meteorology.* a meteorological condition that tends to occur on or near a specific calendar date more frequently than pure chance would seem to allow. *Mathematics.* **1.** a point at which a given function of a complex (real) variable is not analytic (resp., differentiable). If a point in the domain of a function is not a singularity, it is said to be regular. Also, SINGULAR POINT. **2.** a point on a curve at which there is no unique tangent; e.g., a cusp, isolated point, or multiple point. *Robotics.* a function of the kinetic equations that describe the condition of the axes of a robot's arm as they pertain to alignment while moving.
singular point *Mathematics.* **1.** see SINGULARITY, def. 1. **2.** a point that is a singularity for one or more coefficient functions appearing in a differential equation.
singular solution *Mathematics.* an ordinary differential equation of degree n without boundary conditions possesses n arbitrary constants in its general solution. A solution having fewer than n arbitrary constants, if it exists, is called a singular solution. Clairaut's equation is a well-known example.
singultus *Medicine.* a technical name for a hiccup.

sinhalite *Mineralogy.* $MgAlBO_4$, a colorless, yellow, or brown, transparent orthorhombic mineral, having a specific gravity of 3.475 to 3.50 and a hardness of about 6.5 to 7 on the Mohs scale; found in limestone contact zones and alluvial deposits; sometimes used as a gemstone.

sinistral *Science.* of or relating to the left side. *Neurology.* **1.** favoring the left side of the body. **2.** a left-handed person. *Invertebrate Zoology.* of a structure or opening, especially a gastropod shell aperture, inclined to the left side.

sinistral fault see LEFT-LATERAL FAULT.

sinistral fold *Geology.* an asymmetrical fold whose long limb appears to be displaced to the left when viewed along its dip.

sinistrality *Neurology.* a favoring of the left side of the body, rather than the right, in voluntary motor acts or in guiding bimanual movements; especially refers to use of the left hand.

sinistro- or **sinistr-** a combining form meaning "left."

sinistromanual *Neurology.* a favoring of the left hand, rather than the right, in voluntary motor acts.

sinistropedal *Neurology.* a favoring of the left foot, rather than the right, in voluntary motor acts.

sinistrorse *Biology.* a counterclockwise helical twining growth pattern that is characteristic of some stems.

sink a structure whose form suggests that of a kitchen sink; to move downward; specific uses include: *Geology.* **1.** a land surface depression characterized by a central playa or a saline lake having no outlet. **2.** see SINKHOLE. *Volcanology.* a circular or ellipsoidal depression formed by a collapse on the side of or near to a volcano. Also, PIT CRATER, VOLCANIC SINK. *Engineering.* to dig, drill or otherwise drive a shaft, slope, or hole. *Mining Engineering.* **1.** a preliminary pit or excavation to be worked and enlarged to the point of becoming a full-sized shaft. **2.** to excavate strata straight downward for purposes of winning and working valuable materials. *Physics.* a device that is used to readily absorb an extensive quantity, such as a heat sink that absorbs excessive heat. *Telecommunications.* **1.** the part of a system in which transmissions are considered to be received. **2.** a recording or memory device in which data can be kept for future use. *Electromagnetism.* the region in a Rieke diagram in which the rate of change in frequency with respect to the phase of the reflection coefficient is maximized; operation in this region may produce undesirable results. *Mathematics.* **1.** in a directed graph, a vertex with outdegree 0. **2.** in a flow graph, a vertex that may have a net flow into the vertex. In this case, the outdegree of the vertex may exceed 0.

sinkage *Engineering.* the process of sinking, or the extent to which something sinks. *Graphic Arts.* blank space left at the top of a printed page, especially on the first page of a chapter.

sink flow *Fluid Mechanics.* **1.** a point into which fluid from all directions flows uniformly in three-dimensional flow. **2.** a straight line into which fluid from all directions perpendicular to the line flows uniformly in two-dimensional flow.

sink habitat *Ecology.* a type of habitat in which a population would decline and eventually go extinct without recruitment from outside sources, due to low reproductivity or high mortality.

sinkhole *Geology.* in a region of karst topography, a funnel or saucer-shaped surface depression produced by the solution of surface limestone or by the collapse of underground caverns. Also, SINK.

sinkhole

sinkhole plain *Geology.* a regionally extensive plain or plateau characterized by well-developed karst features.

sinking *Oceanography.* the downward movement of surface water, caused by an increase in the density of a water mass or by converging currents. *Optics.* a mirage effect that causes an object that lies above the geometrical horizon to disappear from sight as if below the horizon.

sinking pump *Mining Engineering.* a long, narrow, electrical pump used to keep a shaft dry during shaft sinking.

sinking speed *Aviation.* the speed at which an aircraft loses altitude, especially in a glide in still air under given conditions of equilibrium.

sinking stream *Hydrology.* a surface stream that disappears underground. Also, LOST STREAM, DISAPPEARING STREAM.

sink mark *Engineering.* in injection molding, a small indentation on a plastic object that forms when the molding surface collapses, following internal shrinkage after the gate of the mold is sealed.

sino- a combining form meaning "sinus."

Sino- a combining form meaning "Chinese."

sinoatrial *Anatomy.* relating to the sinus venosus and the atrium of the heart. Also, **sinoauricular.**

sinoatrial node *Anatomy.* a mass of specialized cardiac tissue that lies at the junction of the superior vena cava with the right atrium; it acts as the "pacemaker" of the heart in initiating each cardiac contraction.

Sinografin *Radiology.* a trademark for preparations of diatrizoate meglumine; used as a contract medium during X-ray.

sinoite *Mineralogy.* Si_2N_2O, a colorless to light gray orthorhombic mineral, occurring as lathlike crystals and irregular grains, having a specific gravity of 2.84 and an undetermined hardness; found only in one chondritic meteorite.

Sinope *Astronomy.* the outermost moon of Jupiter, having a retrograde orbital period of 758 days and a diameter of 17 miles.

sinople *Mineralogy.* a variety of quartz with inclusions of red hematite.

Sino-Tibetan family *Linguistics.* in some classification schemes, a language family that encompasses languages spoken in Burma, Tibet, and parts of China and Thailand.

sinter *Materials Science.* to carry out a sintering process. *Petrology.* a chemical sedimentary rock occurring as a hard incrustation deposited on rocks or on the ground as a result of precipitation from hot or cold mineral waters; may contain silica (siliceous sinter) or calcium carbonate (travertine).

sintering *Materials Science.* the process in which fine particles of a material become chemically bonded at a temperature that is sufficient for atomic diffusion.

sintering map *Materials Science.* a diagram showing sintering mechanisms over ranges of temperature and extent of sintering; areas on such diagrams correspond to regimes in which particular mechanisms dominate, and the boundary lines between two areas represent conditions under which the two mechanisms occur.

sinuate *Botany.* of a leaf, having a wavy, indented margin.

sinus a cavity or channel; specific uses include: *Anatomy.* **1.** a cavity, recess, or depression in an organ, tissue, or other part of a human or animal body. **2.** specifically, one of the cavities in the skull connecting to the nasal cavities. *Pathology.* an abnormal channel or fistula permitting the escape of pus. *Botany.* the indentation or space between two lobes or divisions, such as in a lobed leaf.

sinusal *Anatomy.* of or relating to a sinus.

sinus arrest *Cardiology.* a brief failure of the sinoatrial node to emit an electrical impulse, resulting in a pause in the heartbeat.

sinus arrhythmia *Cardiology.* a normal irregularity of the heartbeat caused by the speeding up of the heartbeat during inhalation and its slowing down during exhalation; more noticeable in the young.

sinus gland *Invertebrate Zoology.* the center for hormone release in most higher crustaceans, located between the two basal optic nerves in the eyestalk.

sinus hair see VIBRISSA.

sinusitis *Medicine.* an inflammation of a sinus.

sinus of Morgagni *Anatomy.* one of several grooves or pockets in the columnar region of the anal canal. Also, CRYPTS OF MORGAGNI.

sinusoid *Anatomy.* **1.** resembling a sinus. **2.** a blood channel lined with reticuloendothelium found in various organs, such as the bone marrow, liver, spleen, and adrenal gland. *Electricity.* a waveform that can be represented by $K \sin(2\pi ft + \theta)$, where K is the magnitude, f is the frequency in Hz, t is the time in seconds, and θ is the phase angle in relationship to a reference signal. *Mathematics.* the graph in the plane of any function of the form $y = a \sin(bx + c)$, where a, b, and c are constant.

sinusoidal *Physics.* having a single frequency of oscillation.

sinusoidal map projection *Cartography.* a type of equal-area map projection that uses the equator as the standard parallel, and shows all other geographic parallels as accurately spaced straight lines along which exact scale is preserved.

sinusoidal oscillator *Electronics.* an oscillator whose output is a continuous sinusoidal signal. Also, HARMONIC OSCILLATOR, SINE-WAVE OSCILLATOR.

sinusoidal spiral *Mathematics.* the graph in the plane of the equation in polar coordinates $r^b = a^b \cos bq$, where b is rational and a is any nonzero real number.

sinusoidal wave see SINE WAVE.

sinus rhythm *Cardiology.* the normal rate and rhythm of the nerve signals from the sinus node, which cause the heart muscle's contractions.

sinus venosus *Developmental Biology.* the venous receptacle attached to the posterior wall of the primitive atrium that receives the allantoic, common, and vitelline veins. *Vertebrate Zoology.* the membranous chamber attached to the heart that receives venous blood in fish, amphibians, and reptiles, and transmits it to the right atrium.

SIP single inline package.

sipapu *Archaeology.* in a kiva (ceremonial room) of the U.S. Southwest, a hole in the center of the floor that symbolizes the emergence of the spirit into another world.

Siphinodontallidae *Invertebrate Zoology.* a family of tusk or tooth shells, burrowing marine mollusks in the class Scaphopoda.

sipho- or **siphono-** a prefix meaning "tube."

siphon [sī´fən] *Engineering.* **1.** a bent tube or pipe used to transfer liquid from a receptacle, in which atmospheric pressure forces the liquid up the shorter leg of the vessel while the weight of the excess liquid in the longer leg causes a continuous downward flow. **2.** to transfer liquid by means of such a device. *Botany.* a tubular element in certain algae, including *Polysiphonia. Invertebrate Zoology.* a tubelike structure such as the water intake and output tubes of bivalves, a sucking proboscis in arthropods, or the locomotive jet of cephalopods. *Geology.* a part of a cave passage that connects with a water trap.

siphonaceous *Botany.* of or relating to a plant having large, tubular, multinucleate cells that lack cross walls. Also, **siphoneous.**

siphonage [sī´fə nij] *Medicine.* the use of a siphon, as in gastric lavage or in draining the bladder.

Siphonales *Botany.* a large order of mostly marine green algae, characterized by coenocytic, nonseptate filaments.

Siphonaptera *Invertebrate Zoology.* fleas, a large order of small, sclerotized, wingless insects with bloodsucking mouthparts and strong jumping legs; more than 800 species have been described; ectoparasites on birds and mammals.

siphon barograph *Engineering.* a siphon barometer that also records atmospheric pressure.

siphon barometer *Engineering.* a J-shaped instrument containing mercury that has its short leg open to the atmosphere and its longer leg closed; used to measure atmospheric pressure.

Siphonia *Paleontology.* an extinct genus of pear-shaped demosponges in the order Lithistida; extant in the Cretaceous.

Siphonocladaceae *Botany.* a family of green algae in the order Siphonocladales.

Siphonocladales *Botany.* an order of green algae, characterized by multicellular thalli attached to the substrate by rhizomes.

siphonogamous *Botany.* of or relating to a plant fertilized by a pollen tube that transfers male nuclei to the embryo sac.

siphonoglyph *Invertebrate Zoology.* in anthozoan sea anemones, a ciliated groove at one or both ends of the slit-shaped mouth, thought to help circulate water in the gullet.

Siphonolaimidae *Invertebrate Zoology.* a family of nematode roundworms in the superfamily Monhysteroidea, having a mouth that is modified into a long, narrow protrusible structure.

Siphonolaimoidea *Invertebrate Zoology.* a superfamily of nematode roundworms.

Siphonophora *Invertebrate Zoology.* an order of marine hydrozoan coelentrates, both medusoid and polypoid individuals joining to form free-swimming colonies, often large and brightly colored, with highly toxic stings; includes the Portuguese man-of-war.

siphonophore *Invertebrate Zoology.* belonging to the Siphonophora.

siphonosome *Invertebrate Zoology.* the lower part of a siphonophore medusoid colony to which the nutritive and reproductive individuals attach.

siphonostele *Botany.* a hollow, central cylinder of xylem and phloem tissues that usually encloses a pith.

Siphonostomatoida *Invertebrate Zoology.* an order of copepod crustaceans with a mouth cone for sucking blood and tissue juices; parasitic on fishes and whales.

Siphonotretacea *Paleontology.* a superfamily of inarticulate brachiopods in the extinct suborder Acrotretidina; characterized by an elongate, tear-shaped shell; extant from the Upper Cambrian to Middle Ordovician.

siphonozooid *Invertebrate Zoology.* a zooid in colonial corals of subclass Alcyonaria that lacks tentacles and gonads; its function is to propel water through the colony.

siphon recorder *Engineering.* a delicate automatic receiving instrument used in submarine telegraphy, consisting of a small siphon that discharges ink to make a record that corresponds with the dots and dashes of the Morse code.

siphon spillway *Civil Engineering.* a spillway in which water must rise to the crest of the siphon to prime it; as a result the water begins to flow out and does not stop until it falls below the siphon inlet.

Siphoviridae *Virology.* a family of virulent or temperate phages, having long, noncontractile tails and particles that contain a linear dsDNA genome.

siphuncle *Invertebrate Zoology.* in some cephalopods, a cord of body tissue extending from the body mass through the divisions (septa) of the shell, thought to regulate gas secretion and hence control buoyancy.

Siphunculata see ANOPLURA.

Siporex *Materials.* a trade name for a light-density building material with excellent insulation properties; composed of sand, cement, and a catalyst and manufactured under high-pressure steam conditions.

Sipuncula *Invertebrate Zoology.* a phylum of peanut worms having an elongated body with a retractible anterior section (introvert); they inhabit marine substrates, burrowing in sand, mud, or tropical reef limestone.

sipunculid *Invertebrate Zoology.* belonging to the Sipunculida.

Sipunculida *Invertebrate Zoology.* the peanut worms, a phylum or annelid class of marine burrowing worms, having a tentacled mouth in an eversible proboscis. Also, **Sipunculoidea.**

sire *Agriculture.* a horse that is the father of a foal.

siren *Acoustical Engineering.* a device that produces a loud, penetrating howl or whistle, often used as a warning signal, by forcing compressed air against a rotating perforated disk or cylinder.

siren *Vertebrate Zoology.* any of the three species of aquatic salamanders of the family Sirenidae, having tiny forearms, no hindlegs or pelvis, and permanent external feathery gills; able to survive up to two months of dry conditions buried in mud; found in the coastal plain and lower Mississippi Valley. (From the *Sirens,* sea nymphs in Greek mythology whose haunting singing lured sailors to their death.)

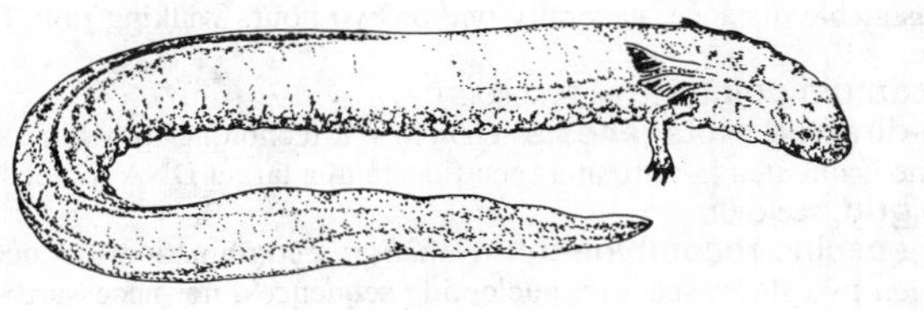

siren

Sirenia *Vertebrate Zoology.* an order of fully aquatic, herbivorous mammals containing the manatee, sea cow, and dugong; characterized by a large head and a massive fusiform body that tapers into the tail; found in large rivers and adjacent coastal waters of the tropical regions of the Atlantic, Pacific, and Indian Oceans.

Siricidae *Invertebrate Zoology.* horntails, a family of large, wasplike, metallic hymenopteran insects in the suborder Chalastogastra; the females deposit eggs in living wood using a stout, hornlike ovipositor.

Siricoidea *Invertebrate Zoology.* a superfamily of horntails and wood wasps, hymenopteran insects in the suborder Chalastogastra.

siriometer *Astronomy.* an older unit of length equal to a million astronomical units.

Sirius [sēr´ē əs] *Astronomy.* Alpha (α) Canis Majoris, the brightest star in the night sky; it has a magnitude of –1.47 and lies approximately 8.7 light-years from earth. Also, DOG STAR.

Sirobasidiaceae *Mycology.* a family of fungi belonging to the order Auriculariaceae that occur on decaying plant material in tropical regions.

sirocco [sə räk´ō] *Meteorology.* a warm, south or southeast wind that occurs ahead of an eastward advancing depression across the southern Mediterranean Sea or North Africa; in crossing the Mediterranean, this wind picks up moisture because of its high temperature and reaches Malta, Sicily, and southern Italy as a very hot, humid wind. Also, SCIROCCO.

Sirolpidiaceae *Mycology.* a family of fungi belonging to the order Lagenidales that is composed primarily of marine parasites of mollusks and algae.

SIRS satellite infrared spectrometer.

sis *Genetics.* an oncogene originally identified as the transforming determinant in simian sarcoma.

sisal *Botany.* an agave plant native to the Yucatan Peninsula, *Agave sisalana*, whose leaves yield a strong, coarse, durable fiber used in making ropes and brushes. *Materials.* the stiff yellow fibers of the leaves of this plant; used to make ropes, cords, sacks, brush bristles, and twine.

siserskite see IRIDOSMINE.

SISI short-increment sensitivity index.

sister chromatid exchange *Cell Biology.* an exchange of parts of DNA molecules between sister (genetically identical) chromatids during meiosis or mitosis.

sister chromatids *Cell Biology.* two identical DNA molecules, resulting from DNA replication plus the proteins that package them. Sister chromatids separate from one another at the anaphase stage of mitosis.

sister groups *Systematics.* the coordinate branches in a cladogram or evolutionary tree. Also, **sister clade.**

sister hook *Mechanical Devices.* **1.** one of a pair of matching hooks mounted on a ring, which form a ring when the open sides face each other. **2.** a set of such hooks.

SIT self-induced transparency; static induction transistor.

site a certain place in which something is located; specific uses include: *Geography.* the precise location of a geographic feature; often distinguished from its spatial relationship to other features (or situation). *Archaeology.* any defined location that demonstrates past human activity, as evidenced by the presence of artifacts, features, or other material remains. *Engineering.* **1.** the location of an object or structure. **2.** the location where a structure will eventually be built or where a structure once stood. *Computer Technology.* the location of a system, such as a room, building, or complex, usually interconnected by local privately owned communications facilities. *Genetics.* in a gene, the smallest subunit having the capacity for independent mutation. *Molecular Biology.* the place occupied by a mutation.

site catchment or **site territory** *Archaeology.* the area surrounding a settlement or camp that is habitually used by the inhabitants as a source of materials for food, toolmaking, and the like. Also, CATCHMENT AREA.

site catchment analysis *Archaeology.* a method of reconstructing the economy of a site by studying the resources that are available within a reasonable distance, generally one or two hours walking time from the site.

site datum point see DATUM POINT.

site-directed mutagenesis *Genetics.* a technique by which a cloned gene is mutated in vitro at a specific site in a target DNA molecule.

site grid see GRID.

site-specific recombination *Genetics.* recombination that occurs between two short, specific, nucleotide sequences, not necessarily on homologous chromosomes; examples are phage integration/excision or resolution of cointegrate structures during transposition.

Sitka spruce *Botany.* **1.** a tall spruce, *Picea sitchensis*, characterized by long, silvery-white needles; found in the Pacific Coast region from Alaska to California. **2.** the soft, light-colored wood of this tree; used in furniture and construction.

sito- a combining form meaning "food" or "grain."

sitology *Medicine.* a branch of medicine dealing with nutrition and dietetics.

sitophilous *Microbiology.* of or relating to any microorganism growing on food.

sitotoxin *Toxicology.* a toxin found in food, produced by microorganisms in decaying grain.

sitotoxism see FOOD POISONING.

situation *Geography.* the location of a geographic feature in relation to its surroundings, including other features; thus a feature's relative location as distinguished from its absolute location (or site).

situational attribution *Psychology.* the assigning of external or circumstantial causes, rather than inherent motives, to one's own or someone else's behavior.

situational factors *Psychology.* those influences on a person's behavior in a given situation that are external factors of the situation, rather than internal traits of character or personality.

situation-display tube *Electronics.* a tube that displays critical data required for an air-defense mission, such as speed and direction.

situationism *Behavior.* a theory defining behavior as the interaction between an organism and its environment as it occurs at a certain moment.

situation-specific behavior *Behavior.* behavior that is determined by or limited to a particular situation.

situs inversus *Medicine.* a lateral transposition of the viscera of the thorax and abdomen.

SI unit *Metrology.* one of the basic measurements of the International System of Units; these include the meter, kilogram, second, ampere, kelvin, candela, and mole.

Sivapithecus *Anthropology.* a fossil primate found in Turkey and Pakistan in the 1980s. *Paleontology.* a genus of hominoid primates in the family Pongidae; this group of the Middle to Upper Miocene is now considered to include Ramapithecus, Bramapithecus, and possibly Kenyapithecus; now generally considered closer to the line leading to the orangutan than to the hominid line.

six-axis system *Mechanical Engineering.* a robot with three rectangular and three rotational degrees of freedom.

six-j-symbol *Quantum Mechanics.* a coefficient that is used in the recoupling of angular momenta; it differs by at most a sign from the Racah coefficient.

six-phase circuit *Electricity.* an AC circuit having two three-phase waves at the same frequency, with one leading the other by a phase angle of 60°.

six-phase rectifier *Electronics.* a rectifier that operates from six voltage sources, whose phases differ by one-sixth of a cycle, or 60°.

Six's thermometer *Engineering.* a type of maximum and minimum thermometer composed of a U-shaped tube with a bulb at each end; one bulb is filled with creosote in contact with a column of mercury that moves according to the expansion and contraction of the creosote, pushing a small steel index in front of it at either end; the indexes remain at their extreme positions, providing a maximum and minimum temperature reading.

Sixteen Personality Factor Test *Psychology.* Raymond Cattell's complex inventory for assessing normal and abnormal personality traits, yielding 16 scores determined by factor analysis on various traits, such as submissive vs dominant or emotional vs calm. Also, **16-PF.**

sixth-power law *Fluid Mechanics.* a law stating that the size of particles that can be carried by a stream is proportional to the sixth power of its velocity.

size *Science.* the dimensions of an object or system. *Materials.* **1.** any of various pasty or gelatinous substances, such as glues, resins, or flour mixtures, that are used to coat, fill, or stiffen porous materials such as paper, cloth, leather, and plaster; such substances are also used as a base for the adhesive in fixing gold leaf to books. **2.** an organic substance used to coat glass fibers to improve bonding and moisture resistance in fiberglass. **3.** to apply any of these substances. *Metallurgy.* to press to close tolerances.

size analysis *Geology.* see PARTICLE-SIZE ANALYSIS.

size classification *Engineering.* see SIZING, def. 2.

size constancy *Psychology.* the ability to perceive that an object's size is constant and not based on its distance from the observer; used as a measure of childhood cognitive development.

size control *Electronics.* a mechanism in a television receiver that varies the size of the picture, either horizontally or vertically.

size dimension *Design Engineering.* the specific dimension, e.g., diameter, width, length, height, that directly defines the size of an object.

size effect *Metallurgy.* any property variations caused by size.

size enlargement *Chemical Engineering.* any process that forms larger particles from smaller ones, while the original particles can still be identified, such as bricks.

size-exclusion chromatography *Materials Science.* a separation method for polymers in which a chromatographic column is fitted with a rigid porous gel having pores the same size as the dimensions of the polymer molecule; used as a method for estimating molecular-weight distribution.

size-frequency analysis *Geology.* see PARTICLE-SIZE ANALYSIS.

size reduction *Mechanical Engineering.* the use of crushing and grinding equipment to reduce bulk material to a more marketable or manageable size.

sizing *Science.* the process of establishing the dimensions of an object or system. *Engineering.* **1.** the process of treating a surface with glutinous material to fill up poreholes and prepare the surface for the application of a coating or adhesive. Also, **sizing treatment. 2.** the process of separating a mixture according to its various sized particles, using a series of screens. Also, SIZE CLASSIFICATION. *Graphic Arts.* **1.** a viscous substance added to paper to make it moisture resistant and to provide a smoother writing surface. **2.** the process of fitting illustrations into a page layout. *Textiles.* starch, glue, wax, paraffin, or other such substances applied to fabric or yarns to add stiffness, smoothness, resistance, and weight. *Mechanical Engineering.* the final finishing of a workpiece in order to make it conform to certain specifications and tolerances.

sizing chisel *Mechanical Devices.* a wood-turning tool with an attached gauge to determine the wood size.

sizing screen *Mechanical Devices.* a wire screen of standard mesh used to separate particles of different sizes.

SJC *Aviation.* the airport code for San Jose, California.

SJO *Aviation.* the airport code for San José, Costa Rica.

sjögrenite *Mineralogy.* $Mg_6Fe_2^{+3}(CO_3)(OH)_{16}\cdot 4H_2O$, a white or yellowish transparent hexagonal mineral of the manasseite group, dimorphous with pyroaurite, occurring as thin hexagonal plates, and having a specific gravity of 2.11 and a hardness of 2.5 on the Mohs scale; found with pyroaurite as a low-temperature hydrothermal mineral.

Sjögren's syndrome [shō´grənz] *Medicine.* a symptom complex of unknown cause, usually affecting middle-aged or older women, characterized by inflammation of the cornea and conjunctiva and connective tissue disease. (Named for Henrik Samuel Conrad *Sjögren,* Swedish physician.)

SJU *Aviation.* the airport code for San Juan, Puerto Rico.

SK *Aviation.* the airline code for SAS (Scandinavian Airlines System).

skarn *Geology.* an assembly of lime-bearing silicates, derived from limestone and dolomite with the introduction of large amounts of silicon, aluminun, iron, and magnesium.

skate *Vertebrate Zoology.* any of many species of cartilaginous rays belonging to the genus *Raja* of the family Rajidae, having a dorsoventrally flattened body and usually a pointed snout, as *R. binoculata,* growing as long as 8 feet and inhabiting waters along the Pacific Coast of the United States.

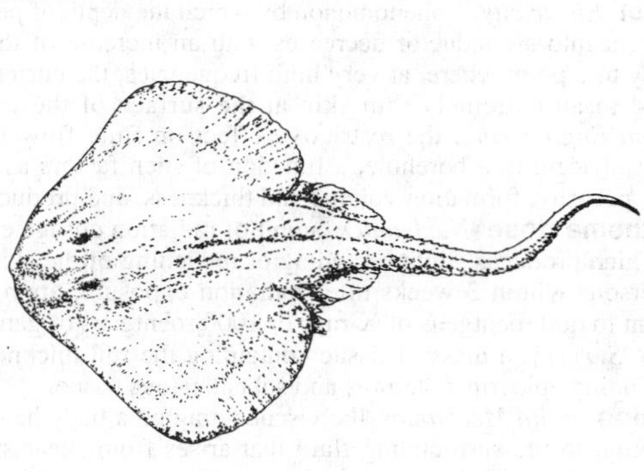

skate

skatole *Organic Chemistry.* C_9H_9N, a white crystalline substance; soluble in hot water and alcohol; melts at 95°C and boils at 265–266°C; derived from the African civet cat, and used as a fixative in perfumes. Also, 3-METHYLINDOLE.

skauk *Hydrology.* a vast field of crevasses in a glacier.

Skean *Ordnance.* a Soviet intermediate-range ballistic missile powered by a liquid-propellant motor, delivering a nuclear or thermonuclear warhead at a range of 2000 miles; officially designated **SS-5**.

skeet gun *Ordnance.* a shotgun used for skeet shooting; it may be .410, 12-, 20-, or 28-gauge and is usually equipped with two 26- or 28-inch over-and-under barrels with a choke between that of a true cylinder and an improved cylinder.

skeg *Naval Architecture.* **1.** a stiffening knee connecting a ship's keel and sternpost. **2.** a finlike projection under a vessel's hull, especially at the stern; used to improve directional handling and often serving as a protective cover for the rudder or the propeller shaft.

skein [skān] *Textiles.* a standardized quantity of a continuous strand of yarn, silk, wool, and the like, wound onto a reel, then taken off and loosely twisted or looped.

skeletal coding *Computer Programming.* the framework of a routine in which some parts, such as addresses, must be specified in detail each time the routine is used.

skeletal muscle *Anatomy.* the muscles under voluntary control that are attached to and move the skeleton.

skeletal program *Computer Programming.* see SHELL.

skeletal soil see LITHOSOL.

skeletal system *Anatomy.* the bones and attached cartilaginous structures that compose the bony framework of the body.

skeleton *Biology.* the bones of a human or animal body, fitted together in their natural places, that together form the framework of the body, supporting the muscles and organs and protecting the soft inner organs. *Invertebrate Zoology.* a similar hard supporting or covering framework in an invertebrate, such as the shell of a crustacean. *Engineering.* the framework of a building or other structure. *Mathematics.* see SPANNING FOREST.

skeleton framing *Building Engineering.* in skyscraper construction, a steel framing that bears all the gravity loading of the structure.

skeleton grain *Geology.* a relatively stable grain of soil material that is not readily displaced but can be concentrated or reorganized by soil-forming processes.

skeleton layer *Oceanography.* of or relating to the structure of the underside of freezing sea ice, which consists of vertical platelets of ice separated by thin layers of brine.

skeleton of the heart *Cardiology.* the dense, mainly fibrous structure that supports the musculature of the heart.

skeleton steps *Building Engineering.* in stairway or staircase construction without risers, the steps of a stair with treads fixed at suitable positions atop one another between side supports.

skeleton texture *Petrology.* a texture of limestone containing an in situ accumulation of skeletal crystals of hard-shelled organisms.

skelp *Metallurgy.* the stock in strip form that is used to make welded pipes or tubes.

skerry *Geology.* a low, small, rugged island, reef, or rocky islet rising above the surface of the water. (From a Scottish dialectal word meaning "rock in the sea.")

sketchmaster *Photogrammetry.* an instrument used to superimpose an image of a photograph on a map or map manuscript for cartographic purposes; the sketchmaster may or may not rectify the image.

Skevas-Zerfus disease see SPONGE-DIVER'S DISEASE.

skew *Science.* **1.** to move aside; take an oblique course. **2.** to distort, as figures or statistics. *Mechanical Engineering.* of or relating to gearing that is out of alignment, neither parallel nor intersecting. *Cartography.* to produce a map projection of which the neatlines do not align with a general north-south orientation. *Electronics.* **1.** a signal received by a facsimile that results when the scanner and recorder are not synchronized. **2.** the degree of nonsynchronous elements in a magnetic tape. *Computer Programming.* **1.** the percentage of bit errors that are caused by a computer's reading of ones and zeros. **2.** in computer graphics or optical scanning, the situation in which the characters or lines are neither parallel nor at right angles to the leading edge. **3.** the tendency for parallel bits to reach an interface at differing times. **4.** the tendency for a clock signal to reach different parts of the central processing unit at different times.

skew back *Civil Engineering.* the courses of stones from which an arch springs (on the top of a pair) and in which the upper and lower beds are diagonal to each other.

skew bridge *Civil Engineering.* a structural bridge that is diagonal to the valley or area to which it allows crossing and is, therefore, longer than the crossing.

skew chisel *Mechanical Devices.* a wood-turning chisel with a straight cutting edge angled to the shank.

skewed *Science.* **1.** slanting; oblique. **2.** distorted or misaligned. **3.** inaccurate or biased.

skewed density function *Statistics.* a nonsymmetrical probability density function.

skewed distribution *Statistics.* a frequency distribution that is nonsymmetrical.

skew failure *Computer Technology.* a failure of an operation that results from the misalignment of a machine-readable document in the reading unit.

skew field *Mathematics.* a division ring; equivalently, a field in which multiplication is noncommutative.

skew Hermitian matrix *Mathematics.* a square matrix that equals the negative of its adjoint.

skew level gear *Mechanical Devices.* a level gear that has spaced axes.
skew lines *Mathematics.* two lines that are not coplanar.
skewness the fact of being skewed; specific uses include: *Statistics.* a lack of symmetry of a distribution about a central measure; e.g., right skewness corresponds to a right tail declining more slowly than the left tail.
skew surface *Mathematics.* a ruled surface having Gaussian curvature equal to zero at each point.
skew-symmetric *Mathematics.* **1.** a (square) matrix M that equals the negative of its own transpose; i.e., $M = -M^t$ is said to be skew-symmetric. **2.** a tensor with the property that interchanging any two indices changes only the sign of the tensor is said to be skew-symmetric.
skialith *Petrology.* a shadowy relic of country rock occurring in granite, obscured but not yet fully replaced by granitization.
skiaphilic *Biology.* requiring or having an affinity for shade; living or thriving in shade. Thus, **skiaphily.**
skiascope *Optics.* a device that measures refractive properties within the eye.
skiascopy *Optics.* the process of using a skiascope to measure refractive properties within the eye.
skiatron see DARK-TRACE TUBE.
skid *Mechanical Engineering.* **1.** the slipping of a wheel on a rail, due to insufficient friction. **2.** a brake for a power machine. **3.** a slightly raised wooden pallet for storing or carrying loads. Also, **skid platform.** *Engineering.* **1.** a length of timber or steel placed under a heavy object when it is being moved, to facilitate sliding and prevent it from sinking into the ground. **2.** an object, usually made of iron, that is attached to a chain and positioned under a wheel to prevent its movement when descending a steep incline. *Aviation.* **1.** a skyward movement of an aircraft in flight, especially a sideways movement away from the center of a turn due to insufficient banking; the opposite of a sideslip. **2.** a metal slide or runner on an aircraft landing gear. *Mining Engineering.* an arrangement upon which certain coal-cutting machines travel along a working face.
skid boulder *Geology.* on the floor of a playa, an isolated boulder derived from an outcrop near the playa margin that is found in conjunction with a trail or mark indicating that the boulder has recently slid across the mud surface.
Skiddavian see ARENIGIAN.
skidder a device or object that skids; specific uses include: *Mechanical Engineering.* a tractor used to haul logs and other loads over rough terrain.
skidding *Forestry.* **1.** a process in which loads are hauled by sliding from stump to roadside, skidway, or landing. **2.** the process of laying skids across the surface of a road.
skid-mounted *Engineering.* describing an object or system that is set upon a movable platform.
skidway *Forestry.* an inclined landing on which skids are laid; used for piling logs that are to be sawed or loaded.
skill *Psychology.* an acquired or learned ability to perform some task or comprehend some concept.
skill acquisition *Artificial Intelligence.* the ability of a program or system to learn or refine its skills within a specific domain and to improve its performance over time.
skill score *Meteorology.* a measure of the degree of skill of a set of forecasts, expressed in terms of some standard, such as forecasts based on persistence or climatology.
skim coat *Building Engineering.* see SETTING COAT.
skim gate *Metallurgy.* in casting, a gate designed to retain slag when molten metal flows into the mold.
skim ice *Hydrology.* a thin layer of ice that forms on the surface of water as it begins to freeze. Also, SKIN.
skimmer *Vertebrate Zoology.* a gull-like bird of the family Rhynchopidae, distinguished by an ability to catch fish by skimming and characterized by black and white plumage, a long red beak, pointed wings, and a short pointed tail; found worldwide pantropically.
skim milk or **skimmed milk** see NONFAT MILK.
skimming *Hydrology.* **1.** the redirection of water from a stream or conduit by means of shallow overflow. **2.** the removal of a thin body of fresh groundwater floating on salt water.
skimming plant *Chemical Engineering.* a petroleum refinery that recovers only the light products of crude oil.
skim sweeping *Military Science.* in naval mine warfare, a technique of wire sweeping to a fixed depth over deep-laid moored mines, in order to cut any mines that are shallow enough to endanger surface shipping.

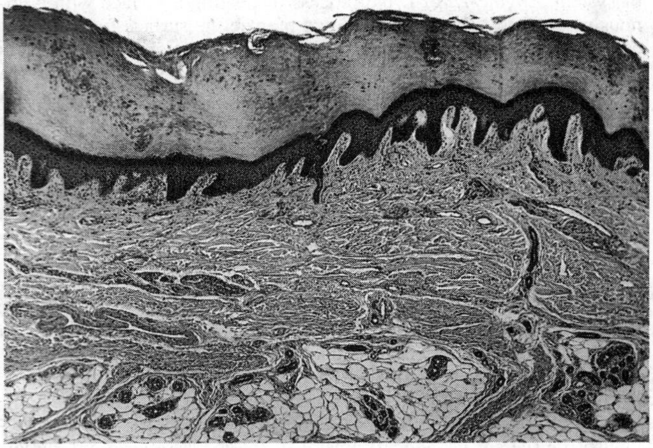

skin tissue

skin *Anatomy.* the outer covering of the body, consisting of a superficial epidermis and a deeper dermis. Also, INTEGUMENT, INTEGUMENTARY SYSTEM. *Engineering.* any outer covering of an object that is thought of as comparable to this. *Aviation.* **1.** the surface or covering of a flight vehicle or of a member such as a fuselage or wing. **2.** the body shell of a monocoque or semimonocoque fuselage. *Building Engineering.* the outside of a building. *Metallurgy.* the superficial layer of a metallic product, if different from the underlying material. *Hydrology.* see SKIM ICE.
skin antenna *Electromagnetism.* an aircraft antenna that is flush mounted to the body and isolated from the rest of the aircraft by an appropriate insulating material.
skin depth *Electromagnetism.* the distance to which an electromagnetic wave travels below the surface of a conductor, at which the current density is reduced by one neper below that at the surface of the conductor; it is, in general, dependent upon the frequency of the incident radiation.
skin effect *Electricity.* a phenomenon by which the depth of penetration of a current into a conductor decreases with an increase of the current frequency to a point where, at very high frequencies, the current flow is restricted to an extremely thin skin at the surface of the conductor. *Petroleum Engineering.* the restrictive effect on fluid flow through a reservoir adjacent to a borehole, a function of such factors as reservoir pressure, porosity, formation volume and thickness, and product rate.
skin erythema dose *Nucleonics.* a unit of radiation dosage equivalent to that which produces slight browning or reddening of the skin of 80% of all persons within 3 weeks after radiation exposure; approximately equivalent to 600 roentgens of X-rays or 1000 roentgens of gamma rays.
skin flap *Surgery.* a mass of tissue containing the full thickness of the skin, including epidermis, dermis, and subcutaneous tissue.
skin friction *Fluid Mechanics.* the viscous drag on a body having a relative motion to the surrounding fluid that arises from shear stresses in the boundary layer. *Aviation.* specifically, the drag caused by the friction of air against the outside of a moving flight vehicle.
skin-friction coefficient *Meteorology.* a mathematically dimensionless drag coefficient that expresses the proportionality between the frictional force per unit area and the square of the surface wind speed.
skin graft *Surgery.* skin tissue, either full-thickness or split-thickness, transplanted to replace a lost portion of skin.
skink *Vertebrate Zoology.* a diurnal, terrestrial, and insectivorous member of the cosmopolitan lizard family Scincidae, characterized by shiny scales and small limbs with reduced or absent digits.

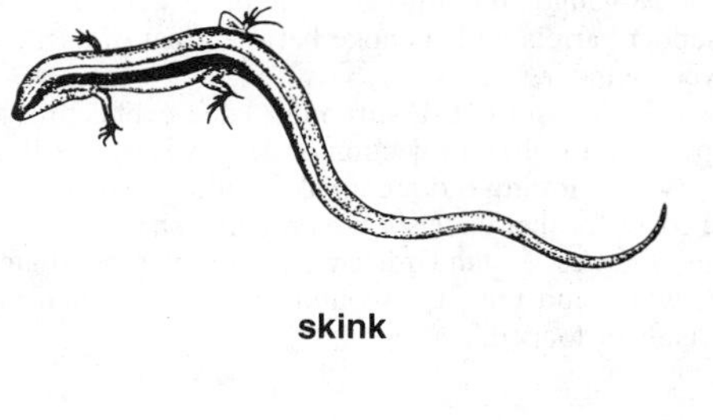

skink

skin lamination *Metallurgy.* in flat rolling, a surface flaw caused by the separation of subsurface laminations.

Skinner, B(urrus) F(rederic) 1904–1990, American psychologist and writer; best known for his research into the learning process and for his writings on ideal planned societies.

B. F. Skinner

Skinner box *Behavior.* the popular term for a type of test chamber associated with the experimentation of B. F. Skinner, typically consisting of a box or container in which a subject animal is confined, a bar or other device to which the animal can respond, and a dispenser for delivering a reinforcement such as food or water.

Skinnerian [ski nâr′ē ən] *Psychology.* **1.** of or relating to B. F. Skinner or his theories, especially the principle that all human behavior is a learned response to positive or negative reinforcement. **2.** a psychologist who follows or supports the behavioral theories of B. F. Skinner.

skin resistance *Electricity.* for an alternating current having a given frequency, the direct-current resistance of a layer at the surface of a conductor whose thickness is equivalent to the skin depth. *Naval Architecture.* the frictional resistance that is produced by water passing a vessel's underwater hull; this is the primary source of resistance for relatively slow speed ships.

skin test *Immunology.* a test used to determine the cellular immunity of an individual to a given substance, in which the substance is applied or injected into the skin.

skintle *Civil Engineering.* a brickwork laid so that the resulting face is irregular.

skin tracking *Electromagnetism.* the process of tracking the motion of an object with radar, without using a beacon or other signal device on board the object being tracked.

skiograph *Electronics.* an instrument that measures the intensity of X-rays.

skip *Computer Programming.* **1.** to ignore one or more computer instructions in a sequence of instructions. **2.** an instruction that will either execute the following instruction or ignore it, depending on a tested condition. **3.** to bypass one or more positions on a data medium. **4.** in word processing, the feature of a machine that allows bypassing of recorded text. *Mechanical Engineering.* see SKIP HOIST.

skip bombing *Military Science.* a bombing method in which a bomb is released from such a low altitude that it slides or glances along the surface of the water or ground and strikes the target at or above water or ground level.

skip cast *Geology.* the cast of a skip mark.

skip distance *Electromagnetism.* the minimum distance over the surface of the earth in which a signal is transmitted from an antenna, reflected from the ionosphere, and returned to the earth, measured at a particular time and frequency.

skip effect *Telecommunications.* the presence of a circular area surrounding a radio transmitter within which no radio signals can be received.

skip fading *Electromagnetism.* the fading of electromagnetic radio waves that reflect from the ionosphere due to variations in the shape or height of the reflecting layer.

skip flag *Computer Programming.* a 1-bit in a specific position that suppresses the transfer of data into main memory until the count equals 0.

skip hoist *Mechanical Engineering.* a bucket or car that operates up and down a defined path to receive, elevate, and discharge bulk materials. Also, SKIP.

skipjack *Vertebrate Zoology.* any of various fishes that swim at and sometimes leap above the surface of the water, such as the tuna *Euthynnus pelamis* or the bonito or the bluefish. *Naval Architecture.* an American one-masted sailing vessel.

skip-keying *Electronics.* the process of reducing the frequency at which a radar system normally operates, in order to reduce interference between it and other radar systems.

skiplane *Aviation.* an airplane, modified with skis in place of wheels and landing gear, designed for taking off and landing on ice or snow. Also, **skip plane**.

skip logging *Acoustical Engineering.* a phenomenon in which a cycle is skipped during sonic logging due to low acoustical energy. Also, CYCLE SKIP.

skip mark *Geology.* one of a sequence of regularly spaced, crescent-shaped marks produced by an object as it skips along the bottom of a stream.

skip-stop service *Transportation Engineering.* a service in which alternate vehicles call at alternate stops along the same route.

skip trajectory *Mechanics.* a trajectory composed of ballistic phases alternating with skipping phases; one of the basic trajectories for the unpowered portion of the flight of spacecraft reentering the earth's atmosphere.

skip vehicle *Space Technology.* a space vehicle designed to climb after reentering the atmosphere in order to cool its body and increase its range.

skip zone *Acoustics.* a region between convergence zones, in which there is a drop of 10 to 20 decibels more than there would be in a free field. *Telecommunications.* the ring-shaped region within a transmission range in which signals from a transmitter cannot be received; it is the area between the outermost limit of reception of radio high-frequency ground waves and the innermost limit of reception of refracted sky waves. Also, ZONE OF SILENCE.

Skiron *Meteorology.* a northwest wind in Greece that is cold in winter and hot in summer.

skirting block *Building Engineering.* **1.** a hidden block that holds a baseboard in place. **2.** a corner block to which a base strip and vertical frame meet. Also, BASE BLOCK, PLINTH BLOCK.

skirting board *Building Engineering.* a board covering wall plaster at the floor level. Also, BASEBOARD, MOPBOARD, WASHBOARD.

skirting plate *Ordnance.* a thin plate in front of the main armored plate that offers passive resistance to the jet of shaped-charge ammunition, thus lessening its ability to penetrate the main plate.

skirt roof *Building Engineering.* the illusion of roofing overhanging or projecting horizontally outward between stories of a multistory building.

skiver *Materials.* a thin, soft leather prepared from sheepskin; used for hat linings and book bindings.

skiving *Mechanical Engineering.* a machining operation in which the face of the cutting tool is at an angle to the direction in which the workpiece is moving. *Metallurgy.* the process of removing a layer of a metallic material with a cutting tool.

skleropelite *Petrology.* an argillaceous or allied rock that has been subject to low-grade metamorphism; denser and more massive than shale, and differentiated from slate by the absence of cleavage.

Skoda *Ordnance.* a delayed-blowback, water-cooled machine gun developed in Czechoslovakia in 1888 and adopted by the Austrian Army in 1893; it was further developed in the early 20th century and produced in calibers ranging from 6.5 to 11 mm. (Developed by Emil von *Skoda*, 1839–1900, Czech engineer.)

Skolem constant *Artificial Intelligence.* in mathematical logic, a constant that is introduced as a substitute for an existentially quantified variable.

Skolem function *Artificial Intelligence.* in mathematical logic, a function introduced as a substitute for an existentially quantified variable that depends on universally quantified variables.

Skolemization *Artificial Intelligence.* the process of eliminating existential quantifiers by replacing them with Skolem constants and Skolem functions.

skolite *Mineralogy.* a dark-green, scaly glauconite mineral typically rich in aluminum and calcium, but deficient in ferric iron.

skomerite *Petrology.* a compact, fine-grained, extrusive igneous rock consisting of microscopic grains and crystals of augite and olivine, phenocrysts of decomposed plagioclase, and a plagioclase groundmass more calcic than the phenocrysts; an altered andesite.

skot *Optics.* a unit used to measure luminance at low light levels; equal to one-thousandth of an apostilb.

Skraup synthesis *Organic Chemistry.* a commercial method for the preparation of synthetic quinoline by heating aniline and glycerol in the presence of sulfuric acid and an oxidizing agent.

skua *Vertebrate Zoology.* any of several cosmopolitan predatory seabirds of the genus *Catharacta* in the family Stercorariidae, characterized by dark plumage with a light belly and powerful, swift flight; the only bird to breed at both the North and South Poles.

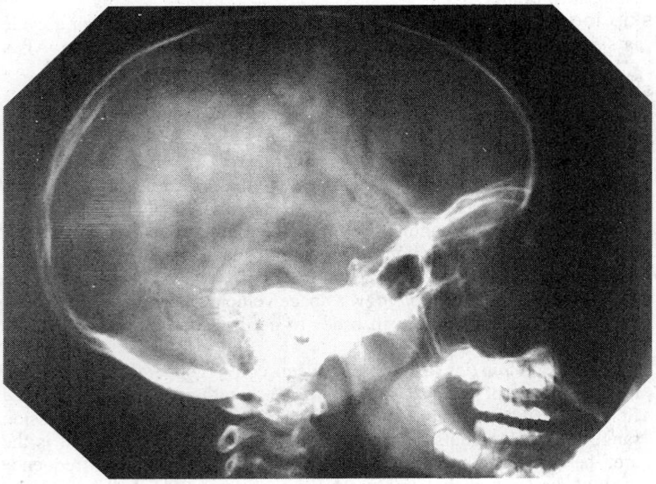

skull X-ray

skull *Anatomy.* the bony framework of the head, including the cranial and facial regions.

skull-cracker *Engineering.* an informal name for a wrecking ball used to demolish old buildings.

skull crucible *Metallurgy.* a metallic, water-cooled crucible used to melt reactive metals and alloys, such as titanium. During the operation, a part of the charge freezes on the crucible wall, forming a skull. Thus, **skull melting.**

skunk *Vertebrate Zoology.* an omnivorous New World mammal, *Mephitis mephitis,* of the family Mustelidae; noted for the malodorous product of its scent glands and its warning coloration of black with white markings, usually in a stripe down the back.

striped skunk

skunk cabbage *Botany.* the popular name for a common low-growing plant of North America, *Symplocarpus foetidas,* having large, broad leaves resembling cabbage. (So called because of the foul skunklike odor it emits.)

skutterudite *Mineralogy.* $CoAs_{2-3}$, a tin-white to silver-gray, metallic, cubic mineral, occurring as cubes, octahedra, and in massive form, having a specific gravity of 6.1 to 6.9 and a hardness of 5.5 to 6 on the Mohs scale; forms a series with nickel-skutterudite; found in medium- to high-temperature hydrothermal veins with other cobalt-nickel minerals.

sky *Astronomy.* the apparent flattened dome seen as extending above the earth, appearing as blue in sunlight; the upper atmosphere of the earth.

Skybolt *Ordnance.* a two-stage, air-to-surface ballistic missile with a range of 1000 nautical miles and hypersonic velocity; it is no longer produced.

sky compass *Navigation.* an astrocompass that determines direction based on the polarization of sunlight during the arctic twilight; in order for the reading to be accurate, the sun must be between 6.5° and 10° above the horizon. Also, TWILIGHT COMPASS.

sky cover *Meteorology.* the amount of sky that is covered but not necessarily concealed by clouds, the amount of sky concealed by obscuring phenomena that reach the ground, or the amount of sky covered or concealed by a combination of such factors; classified as clear, scattered, broken, overcast, partial obscuration, and obscuration.

sky diagram *Astronomy.* a chart indicating the apparent positions of celestial objects using horizon coordinates.

Sky Flash *Ordnance.* a British air-to-air missile powered by a solid-propulsion motor and equipped with a monopulse seeker; it delivers a 66-pound continuous-rod conventional warhead at a speed of Mach 4 and a range of 31 miles.

Skyhawk *Aviation.* a carrier-based single-engine U.S. turbojet attack aircraft capable of delivering nuclear and nonnuclear weapons, providing troop support, and carrying out reconnaissance missions.

skyhook *Mechanical Engineering.* any of various devices designed to lift heavy loads from high above, as from a helicopter or towering crane. *Mining Engineering.* to drive bolts into overlying rock strata as reinforcement for a mine roof.

Skyhook *Aviation.* a large, plastic constant-level free balloon developed by the U.S. Navy for rocket launchings and for meteorological observations at very high altitudes.

Skylab *Space Technology.* the first U.S. space station; in three manned missions (1973–1974), it orbited the earth for 28, 59, and 84 days.

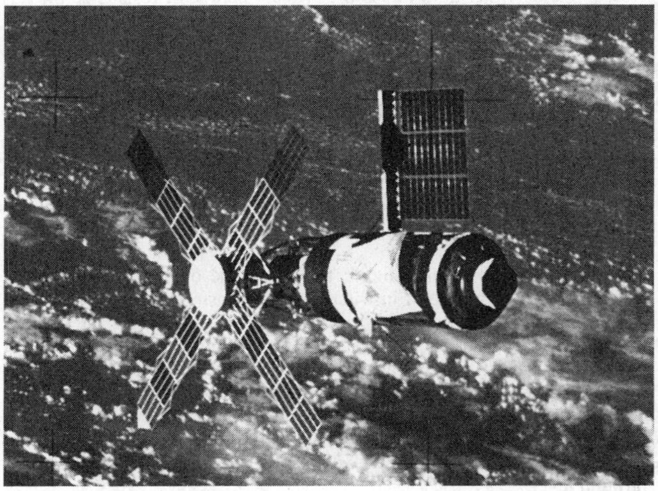

Skylab

skylight *Building Engineering.* an opening in a roof that is fitted with translucent or transparent glass or plastic in order to admit sunlight. *Astronomy.* diffuse light that reaches the earth, scattered by molecules in the atmosphere, as opposed to direct radiation from the sun.

skyline *Architecture.* the general outline of the tall buildings of a city against the sky. *Forestry.* a heavy cable that is tightly stretched between two spar trees and used as a track along which logs are hauled during a logging operation.

skyline logging or **skyline hauling** *Forestry.* a method of power-cable logging in which logs, with one end in the air, are hauled in by a heavy cable that is tightly stretched between two spar trees. Also, CARRIER-CABLE LOGGING, HIGH-LINE LOGGING.

sky map *Meteorology.* the pattern of variable brightness observed on the underside of a cloud layer due to differences in the reflectivity of material on the land surface that is directly below; used mainly in the polar regions.

sky noise *Electromagnetism.* the electromagnetic radiative noise presumed to originate from galactic sources.

Skyraider *Aviation.* a carrier-based, general-purpose U.S. attack aircraft having a single reciprocating engine and capable of delivering relatively long-range, low-level nuclear and nonnuclear weapons and torpedos, laying mines, and carrying out reconnaissance and troop support; designated **A-1.**

Skyray *Aviation.* a single-engine, single-pilot, supersonic limited all-weather U.S. jet fighter that is designed for operation from aircraft carriers and used for interception and destruction of enemy aircraft; can be armed with the Sidewinder missile; designated **F-6.**

skyrmion *Particle Physics.* a particle with spatial extension, which can be either a boson or a fermion and yet is entirely made up of boson fields.

skyscraper *Architecture.* a tall, multistoried building, usually a commercial office building; made possible in modern times by the development of elevators and steel-frame construction.

Skytrain *Aviation.* a popular name for the C-47 troop transport.

Skytrooper *Aviation.* a popular name for the C-53 troop transport.

Skywarrior *Aviation.* a twin-engine, all-weather U.S. attack turbojet aircraft capable of delivering nuclear and nonnuclear weapons, carrying out reconnaissance missions, or laying mines; able to be refueled in midair; manned by a crew of four; designed to operate from aircraft carriers; designated **A-3.**

sky wave *Electromagnetism.* an electromagnetic wave that is transmitted from earth and into the ionospheric layer, where it suffers multiple reflections within the layer before reaching its destination back on earth. Also, IONOSPHERIC WAVE.

sky-wave correction *Electronics.* a technique for correcting positional errors created when radio waves, generated by the ionosphere, interfere with a plane's navigational equipment.

sky-wave synchronized loran *Navigation.* a loran that uses signals reflected from the ionosphere to synchronize the transmitting stations; this provides greater range and improved night operation. Also, SS LORAN.

sky-wave transmission delay *Electromagnetism.* a difference in the time of transmission between waves propagated as sky waves and waves propagated as ground waves from a transmitter antenna to a receiver antenna.

SL or **S.L.** sea level; south latitude.

sl. slow; slightly.

s.l. without place (of publication). (From Latin *sine loco.*)

SLA sacrolaeva anterior.

slab a broad, generally flat piece of material; specific uses include: *Civil Engineering.* a reinforced concrete floor that is supported at intervals on beams or columns, or laid directly on the ground surface. *Metallurgy.* a flat metal or alloy that is thicker than a plate but thinner than an ingot. *Geology.* a tabular plate or fragment of hard, evenly stratified stone or finely parallel jointed rock. Also, **slabstone.** *Hydrology.* a rigid layer of snow forming a large, solid mass. *Forestry.* an irregular outside section of a log that is cut to square the log or as a preparatory step before sawing the log into boards. *Electronics.* a thick chunk of quartz crystal from which blanks are cut. *Mining Engineering.* a skip or slice of material taken from the rib of a mine room or entry.

slab avalanche *Hydrology.* one of the basic types of snow avalanche, in which a large, solid mass of snow breaks loose and splits into smaller pieces as it slides.

slabbing *Mining Engineering.* a method of mining pillars after rooms have been finished; successive slabs are cut from one side or rib of the pillar until no more material can safely be recovered.

slabbing cutter *Mechanical Engineering.* a milling tool used for face-cutting large, deep cuts in a steel slab.

slabbing mill *Metallurgy.* a mill used to produce slabs from ingots.

slab jointing *Geology.* a structure produced in rock by the formation of many closely spaced parallel joints that divide the rock into thin slabs.

slab pahoehoe *Geology.* a type of pahoehoe whose surface is broken up into an irregular arrangement of flow-crust slabs.

slack *Mechanical Engineering.* the looseness of interconnecting parts in a mechanism, such as gear parts, chains, and belts. *Geology.* a depression between lines of shore dunes or in a sandbank or mudbank on a shore. *Oceanography.* see SLACK WATER. *Industrial Engineering.* see SLACK TIME, def. 1.

slack barrel *Petroleum Engineering.* a vessel used to store and ship petroleum paraffin; it is lighter than the standard oil barrel but similarly shaped, having a capacity of 235–245 lb net.

slack ice *Hydrology.* open pack ice that is floating on still or slow-moving water.

slackline *Forestry.* a heavy cable that is suspended loosely between two spar trees; used especially to haul logs downhill or across steep gullies and canyons. Thus, **slackline logging.**

slackline cableway *Mechanical Engineering.* a cable excavator with a track cable that is loosened to lower the bucket and tightened to raise it.

slack quenching *Materials Science.* an incomplete hardening of a steel by quenching slower than the rate to completely transform it to martensite, resulting in an additional transformation product.

slack time *Industrial Engineering.* **1.** the amount of time between the anticipated date of completion of an order and the order due date, used as an indication of the urgency of the order. Also, SLACK. **2.** a pretested time measurement for a process in a critical-path-method network or PERT, which represents the difference between the earliest possible completion time and the latest time at which the activity must be completed to keep on schedule with the entire project.

slack water *Oceanography.* the period when a tidal current's speed is near zero, so that there is little or no horizontal movement of water; usually occurring between ebb and flood. *Hydrology.* a quiet or still part of any normally active body of water, such as the inside of a bend in a river, where the current is slight.

slade *Building Engineering.* an inclined footpath.

slaframine *Mycology.* a toxin produced by the fungus *Rhizoctonia leguminicola,* causing diseases such as slobber syndrome in livestock.

slag *Metallurgy.* the layer covering molten metal during smelting and refining; in refining, it is composed of oxidized impurities; in smelting, it consists of the flux and gangue minerals. *Mining Engineering.* waste that remains after re-sorting coal. *Volcanology.* see SCORIA.

slag cement *Materials.* cement that is composed of approximately 80% granulated slag and 20% hydrated lime.

slagging *Metallurgy.* **1.** the process of removing slag from a metallic bath. **2.** the process of forming a slag over a metal bath, as by adding a flux or scavenger.

slag inclusions *Metallurgy.* particles trapped in the solid metal after melting under a slag.

slagsitall *Materials.* a glass ceramic derived from steelmaking slags to which sand or clay is added; it is processed into rolled sheets and used for interior construction, such as flooring. Also, **slagceram.**

slake *Geology.* to undergo or cause to undergo a process of slaking.

slaked lime *Organic Chemistry.* a commercial form of calcium hydroxide, $Ca(OH)_2$; so called because it is obtained by the slaking action of water on lime (calcium oxide).

slaking *Geology.* **1.** the crumbling and disintegration of earth materials when exposed to air or moisture. **2.** the disintegration of indurated soil or dried clay when saturated with or immersed in water. **3.** the disintegration of tunnel walls in swelling clay.

SLAM or **Slam** *Ordnance.* a modification of the Harpoon antiship missile, deployed from carrrier-based aircraft and powered by an air-breathing turbojet engine; it delivers a 500-pound high-explosive penetration warhead at a range of 50 miles. (An acronym for standoff land attack missile.)

Sláma, Karel born 1934, Czech entomologist; conducted important research in insect pheromones; a pioneer in nonpoisonous pest control.

slamming *Naval Architecture.* a term for a condition of severe pitching in which a vessel's bow rises entirely out of the water before descending and impacting on the water.

slamming stile *Building Engineering.* an upright member or vertical strip of a door case against which a door abuts when closed and into which the bolt of a rim lock engages.

SLAN without a specified place, year, or name. (From Latin *sine loco, anno, nomine.*)

slant *Design Engineering.* an inclined line, surface, or direction of movement. *Graphic Arts.* a variation to a basic typeform, tilting it to the right, as in italics. *Mining Engineering.* **1.** a short, inclined crosscut driven to connect the entry with its air course so as to facilitate haulage. **2.** see RUN. *Microbiology.* a sloping surface of agar in a test tube.

slant culture *Microbiology.* a method of growing bacteria cultures by placing solid medium in a tube, which is then slanted and streaked with the inoculum. Also, SLOPE CULTURE.

slant depth *Design Engineering.* the distance between the crest and root of spur thread when measured along the angle formed by a screw thread's flank.

slant plane *Ordnance.* in antiaircraft artillery, the plane defined by the target course line and pintle center of the gun.

slant range *Navigation.* **1.** the line of sight distance between two points located at different elevations. Also, **slant distance. 2.** the line of sight distance at which a radar or radio will operate.

slant tube fermenter *Microbiology.* a highly efficient tower fermenter in which the sugar solution enters the base of a slanted tube and fermentation takes place progressively up the tube.

slant visibility *Navigation.* the degree of air clarity in a direction other than horizontal.

slash *Forestry.* an open area in a forest strewn with the residue of trees (such as logs, uprooted stumps, twigs, bark, and chips) after felling or because of fire, storm, wind, or poisoning.

slash-and-burn *Agronomy.* a simple farming method in which existing vegetation is cut away and a field is then cleared by burning; the land is farmed for a few years and then abandoned or left fallow to regenerate. Thus, **slash-and-burn agriculture, slash-and-burn method, slash-and-burn technique.**

slash-sawing see BACKSAWING.

slat *Design Engineering.* a long, thin strip of wood, metal, or other material. *Aviation.* **1.** any of certain movable vanes or auxiliary airfoils, usually set along the leading edge of a wing but able to be lifted away at certain angles of attack. **2.** the forward position of a slotted airfoil with a fore-located slot.

slat conveyor *Mechanical Engineering.* a conveyor to which one or more endless chains with nonoverlapping, spaced slats are attached.

slate *Petrology.* a compact, fine-grained metamorphic rock derived from mudstone, siltstone, and other clayey sediment; characterized by perfect fissility or slaty cleavage that allows it to split into slabs and thin plates.

Slater determinant *Quantum Mechanics.* the determinant of an $N \times N$ matrix of one-particle wavefunctions, used as a wavefunction for N particles, for example, in second quantization.

slate ribbon *Geology.* any of a series of varicolored stripes across the cleavage surface of slate, usually a trace of bedding.

Slater's rule *Electronics.* the rule that the ratio of a magnetron's cathode radius to its anode radius is approximately $(N - 4)/(N + 4)$, where N is the number of resonators.

slaty cleavage see FLOW CLEAVAGE.

slave *Anthropology.* someone who is considered to be the personal property of another person and completely subject to that person's control. *Engineering.* any unit, such as a machine or computer device, that is under the control of another similar unit (the master). *Invertebrate Zoology.* see SLAVE ANT.

slave ant *Invertebrate Zoology.* an ant that is raised as a slave by a slave-making ant.

slave antenna *Electromagnetism.* a directional antenna whose direction is controlled by a servosystem that receives information from a tracking and positioning system.

slave arm *Engineering.* a component that copies the motions of another component (the master arm), sometimes varying the scale or the force.

slave clock *Horology.* **1.** a clock that releases electrical impulses to the master clock of a Shortt pendulum clock or other free-swinging pendulum system. **2.** a dial driven by a master clock; more accurately called an impulse dial.

slaved gyro magnetic compass *Navigation.* a gyroscopic compass in which a flux valve maintains the orientation to magnetic north.

slave-making ant *Invertebrate Zoology.* any of various ant species that take the eggs and pupae of other ant species to hatch and raise as slaves.

slave robot *Robotics.* a robot that duplicates the movements of a master robot, but not always with the same scale of displacement or force.

slavery *Anthropology.* a social system, existing widely from ancient times through the mid-1800s and still found in some societies, in which certain persons are regarded as being owned by others and are forced to work without pay. *Invertebrate Zoology.* a practice of certain ant species that invade the nests of other ant species, taking pupae to raise as slaves or indenturing the invaded nest; seen especially in the blood-red ant *Formica sanguinea* and slave-making ant *Polyergus rufescens.* (Related to the word *Slav*; in medieval times many Slavic people were held as slaves in Europe.)

slave station *Cartography.* in surveying, any station for which the control is determined or based on another station, rather than on direct observation. *Navigation.* in a hyperbolic navigation system, such as loran, a station whose transmissions are controlled by a master station. Also, SECONDARY STATION.

slavikite *Mineralogy.* $NaMg_2Fe_5^{+3}(SO_4)_7(OH)_6 \cdot 33H_2O$, a greenish-yellow, trigonal mineral, occurring as crusts of small tabular crystals, having a specific gravity of 1.89 to 1.99 and an undetermined hardness; found as a secondary mineral formed by the alteration of pyrite.

S layer *Cell Biology.* a regular array of protein subunits that form the outermost layer surrounding certain bacterial cells, such as archaebacteria and eubacteria.

SLBM submarine-launched ballistic missile.

SLC *Aviation.* the airport code for Salt Lake City, Utah.

SLE systemic lupus erythematosus.

sled *Mechanical Engineering.* a vehicle that moves on runners to transport people or goods over ice and snow. *Mining Engineering.* a vehicle used to convey coal along the road to the chute or to the area where it is loaded into cars. Also, SLEDGE, SLYPE.

sledge *Mechanical Devices.* see SLEDGEHAMMER. *Mechanical Engineering.* see SLED.

sledgehammer *Mechanical Devices.* a heavy, long-handled hammer, typically weighing 5–14 pounds; wielded with both hands and used for driving, breaking, and pounding materials such as rock boulders or stone.

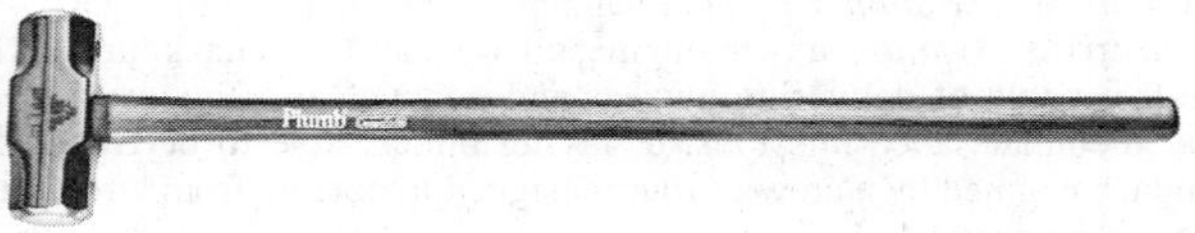

sledgehammer

sled kiln *Materials Science.* a fast-firing kiln used in firing advanced ceramics; the sled loaded with the unfired products serves as the kiln base, and is continuously pushed through the kiln during firing; especially suitable for sintering fragile, large, and complicated shapes.

sleep *Neurology.* a natural state of rest during which the body's physiological powers are restored; characterized by minimal physical movement, partial or complete suspension of volition and consciousness, and a diminished, but readily reversible, awareness of the surrounding environment and reactivity to external stimuli. *Computer Technology.* **1.** a state in which a computer or program is caught in an endless loop and appears to be doing nothing. **2.** a state in which a program will not be executed for a certain period of time.

sleep drunkenness see SOMNOLENTIA.

sleeper *Vertebrate Zoology.* any of several tropical marine and freshwater fish species of the family Eleotridae, very similar to gobies, but lacking united pelvic fins; named for their habit of lying quietly on the bottom. *Building Engineering.* a long piece of wood, stone, or metal that is laid horizontally and used as a footing. *Transportation Engineering.* see SLEEPING CAR.

sleeper effect *Psychology.* the principle that a persuasive message does not have its greatest impact immediately upon transmittal, but rather after some time has elapsed; often applied to the effects of propaganda or mass media communications.

sleeping car *Transportation Engineering.* a railroad car for passengers to sleep in; usually fitted with berths or compartments.

sleeping sickness *Medicine.* an East African infectious disease caused by *Trypanosoma rhodesiense* and transmitted by the tsetse fly, characterized by the presence of parasites in the circulating blood, inflammation of the lymph nodes, accumulation of fluid in the tissues, and inflammation of the heart muscle. Also, RHODESIAN TRYPANOSOMIASIS.

sleep paralysis *Neurology.* temporary paralysis that occurs when the subject is falling asleep or awakening.

sleep therapy see NARCOSIS THERAPY.

sleet *Meteorology.* **1.** transparent or translucent ice pellets that are 5 mm or less in diameter. **2.** precipitation in the form of such pellets. **3.** precipitation in the form of a mixture of rain and snow.

sleeve *Mechanical Devices.* a cylindrically shaped tubular piece that fits over a rod, joint, or shaft in the manner of a sleeve of a garment fitting over an arm. Also, **sleeving.** *Electricity.* the grounded outer part of a male phone connector.

sleeve antenna *Electromagnetism.* a vertical half-wave antenna in which the center of a coaxial line is connected to the quarter wavelength portion, while the lower half, connected to the outer conductor of the feed line, is a metal sleeve.

sleeve bearing *Mechanical Engineering.* an unjointed machine bearing race in which the shaft rotates and is lubricated inside a sleeve.

sleeveboard *Mechanical Devices.* a small ironing board used to press sleeves.

sleeve coupling *Mechanical Devices.* a hollow cylindrical tube with an internal thread at either end, used to connect lengths of pipe.

sleeve dipole antenna *Electromagnetism.* a dipole antenna whose central portion is enclosed by a coaxial sleeve.

sleeve joint *Mechanical Devices.* a device formed by forcing the ends of two wires or cables into both ends of a hollow sleeve; used for joining such cables or wires.

sleeve nut *Mechanical Devices.* a right or left nut.

sleeve valve *Mechanical Engineering.* a thin steel hollow sleeve that fits inside the cylinder of an internal-combustion engine and has a slot in its side that, when properly aligned with a port in the cylinder, admits and exhausts gases during certain stages of the engine cycle.

sleigh [slā] *Mechanical Engineering.* a light vehicle that is drawn on runners, often by horses; a large sled. *Ordnance.* the part of a gun carriage that slides with the recoil; it supports the gun barrel and the recoil mechanism.

slenderness ratio *Civil Engineering.* the ratio of effective length or height of a wall, column, beam, or pier to the radius of gyration; used as a means of assessing stability. *Aviation.* the aspect ratio of an aircraft landing ski or other elongated surface; calculated by dividing the square of the length of the surface by the area of the surface. *Space Technology.* a dimensionless number expressing the ratio of a missile's length to its diameter.

slewing *Engineering.* the action of rapidly moving a radio antennae or sonar transducer in the horizontal or vertical plane.

slewing mechanism *Engineering.* a device that provides for a quick change in the vertical or horizontal position of an instrument.

slewing motor *Electricity.* a motor designed to drive a high-speed radar antenna for slewing to monitor a target.

slew rate *Electronics.* the highest rate of change of signal level that an amplifier can produce under given conditions; the response rate of a signal. *Robotics.* **1.** the greatest speed at which a robot can follow a command. **2.** the greatest velocity with which a Cartesian-coordinate robot can move a tool tip or a manipulator can move a joint.

sley *Textiles.* the number of warp threads per inch on a fabric.

slice *Geology.* an arbitrary division of a stratigraphic unit based on some uniform standard, such as thickness; used for purposes of analytic study. *Engineering.* **1.** a thin, broad piece of material that is cut off, such as a portion of coal cut from a pillar. **2.** to extract ore by cutting off successive slices.

slice bar *Engineering.* a steel bar having a broad, flat blade used for scraping or chipping operations.

slice drift *Mining Engineering.* the crosscuts driven between every other slice 18–36 feet apart during sublevel caving.

slice method *Meteorology.* a method of evaluation for static stability of a limited area at a reference level in the atmosphere, with consideration of both upward and downward air movements.

slicer see CLIPPER-LIMITER.

slicer amplifier see AMPLITUDE GATE.

slicing *Electronics.* the transmission of the portion of a waveform that falls in a given amplitude range.

slicing method *Mining Engineering.* a mining method in which horizontal layers are removed from a massive ore body.

slick *Hydrology.* a smooth, glassy streak or patch on the water's surface; usually caused by an almost undetectable film of organic material on the surface; temporary slicks may also be caused by objects in the water, such as a turning ship.

slickens *Geology.* a layer of extremely fine-grained material, such as silt deposited by a stream during a flood. *Mining Engineering.* **1.** the light soil extracted by means of sluicing during hydraulic mining. **2.** the tailings resulting from the operation of a hydraulic mine or stamp mill.

slickenside *Geology.* the polished, grooved, smoothly striated surface produced on a rock as a result of friction along a fault plane. Also, POLISHED SURFACE, SLIP-SCRATCH.

slickolite *Geology.* a vertically discontinuous surface that develops on sharply dipping limestone bedding as a result of slippage and shearing, forming the molding on the wall of a solution cavity.

slide *Geology.* **1.** a mass movement of earth, snow, or rock that results from failure under shear stress. **2.** the mass of material carried or laid down by such a slide; the track left by such a slide. **3.** a fault produced in association with folding, being conformable with the fold limb or axial surface. *Engineering.* **1.** a sloping trough with a flat bottom end; used to transport goods from a high level to a lower level. **2.** a piece of a mechanism that moves linearly over a surface between guides. *Mechanical Engineering.* the main reciprocating member of a mechanical press to which the die is fastened. *Ordnance.* **1.** in certain weapons, the part of the breech mechanism that is moved back and forth to clear and reload the weapon. **2.** in some automatic weapons, the sliding part of the receiver. **3.** in semiautomatic pistols, a metal sleeve over the barrel or the breech mechanism, which is driven back by recoil and returned to firing position by a spring.

slide-action see PUMP-ACTION.

slide-back voltmeter *Electronics.* a tube that determines the power of an unknown voltage indirectly by adjusting a calibrated voltage source until it equals the unknown voltage.

slide caliper *Mechanical Devices.* a caliper with one or two moveable jaws set at right angles to a graduated beam. Also, CALIPER SQUARE.

slide conveyor *Engineering.* an inclined chute through which small objects, liquids, or mixtures move in a downward direction.

slide gate *Civil Engineering.* a type of spillway crest gate with a vertical lift, usually restricted to small spillways. Also, ROLLER GATE.

slide projector see OPTICAL LANTERN.

slider *Electricity.* the lever on a slide potentiometer that sets the variable resistor by moving up and down, or back and forth.

slider coupling *Mechanical Engineering.* a mechanism that connects two rotating shafts that are not longitudinally aligned. Also, OLDHAM'S COUPLING, DOUBLE-SLIDER COUPLING.

slide rest *Mechanical Engineering.* an adjustable device upon which a cutting tool is mounted.

slide rule *Mathematics.* a mechanical device based on the concept of nomography; used for quick numerical computations such as multiplication, extracting roots, and so on. The typical logarithmic slide rule is a ruler marked on both edges with a logarithmic scale, and one or more tongues, also marked with a logarithmic scale, that can be shifted with respect to the other scales. The electronic calculator has now almost completely supplanted the slide rule because of its greater ease of use and wider variety of functions.

slide-rule dial *Engineering.* an indicating part of an instrument, having a pointer that moves over a calibrated straight line in a manner similar to the scales of a slide rule.

slide valve *Mechanical Engineering.* a valve mechanism that covers or uncovers ports by sliding a member over the hole; used to admit or discharge fluid.

slide-wire bridge *Electricity.* a type of Wheatstone bridge used extensively for measuring electrical resistances.

slide-wire potentiometer *Electricity.* a potentiometer similar to a variable resistor, used for low-current applications in DC circuits.

sliding-chain conveyor *Mechanical Engineering.* a conveyor system that transports pipes, rods, or other materials over a plane of parallel chains.

sliding fit *Design Engineering.* a fit between two machined parts that slide together.

sliding friction see KINETIC FRICTION.

sliding gear *Mechanical Devices.* a change gear that allows gears to slip in and out of mesh by sliding them along their axes, thus changing the speed of the driven shaft.

sliding growth *Botany.* the sliding of one cell wall past that of another cell during fiber growth.

sliding pair *Mechanical Engineering.* a two-part mechanism in which one part moves in a specified path with respect to the other, which is constrained to move in accordance with the design of the pair, as in a piston and cylinder combination.

sliding pin *Robotics.* a device that gives a robot two degrees of freedom.

sliding ram *Design Engineering.* in a shaping machine, the arm or ram that drives the cutting tool forward and then retracts it, usually horizontally, with the work carried by a flat bed.

sliding sash *Building Engineering.* in a horizontally sliding window, either of two sashes that can slide to an open or closed position; in heavy load applications, the sash is often provided with nylon rollers to aid in operation, or it can be suspended from rollers operating on overhead tracks.

sliding scale *Science.* any standard or system that varies according to existing conditions, such as a wage scale that varies according to the cost of living, or a fee scale that varies according to the ability to pay. *Meteorology.* specifically, a set of combinations of ceiling and visibility data, constituting the operational weather limits at an airport.

sliding-vane compressor *Chemical Engineering.* a rotary compressor that has moving vanes to compress the fluid in stages.

sliding vector *Mechanics.* a vector with a specified line of application but an arbitrary point of application.

sliding way *Civil Engineering.* the heavy timbers or steel beams forming the lower part of the cradle that slides down the standing ways when a ship is launched.

slime *Geology.* another term for mud, especially soft, watery mud. *Biology.* any organic matter resembling a thin layer of mud, such as the secretion of slugs or certain fishes. *Engineering.* a similar moist, sticky layer of industrial material. *Metallurgy.* specifically, finely particulated material that is formed during the electrorefining of certain metals, such as copper; it is often reprocessed to recover precious metals or other valuable substances.

slime bacteria *Microbiology.* a common term for gliding bacteria of the order Myxobacterales that leave slime deposits on surfaces.

slime disease *Plant Pathology.* any of several bacterial plant diseases characterized by a slimy rot on plant parts.

slime flux *Plant Pathology.* a symptom of various diseases or injuries of deciduous trees, characterized by a viscous bleeding of the bark or wood. Also, WETWOOD.

slime gland *Zoology.* any gland that extrudes slime or mucous.

slime mold *Mycology.* a common term for fungi belonging to the order Myxomycetes, including both cellular slime molds and acellular slime molds, which undergo both moldlike and ameboid stages during their life cycle. Also, **slime fungus.**

slim hole *Engineering.* a drill hole of minimum size, created for structure tests, to be used as a seismic shothole, or occasionally for stratigraphic tests.

slimicide *Microbiology.* an antimicrobial substance that inhibits certain slime-producing microorganisms.

sliming *Oceanography.* the formation of a film of bacteria or diatoms or both on a submerged object.

sling *Anthropology.* an early weapon used to propel a stone or other missile, consisting of a leather strap with a long string at each end; it is whirled around the head and one end of the string is then released to hurl the stone forward. *Medicine.* a strap or bandage used to suspend a part of the body, especially a wide bandage worn around the neck and supporting an injured arm or hand. *Engineering.* a chain, strong rope, or wire used to attach an object to a crane hook.

sling psychrometer *Engineering.* an instrument used to determine the relative humidity of the atmosphere; it consists of a wet and dry bulb thermometer that can be whirled in the air so that wet and dry bulb temperature measurements can be taken at the same time.

sling thermometer *Engineering.* a thermometer that is mounted on a frame so that it may be whirled by hand.

slip a slight or gradual movement; specific uses include: *Civil Engineering.* a small landslide. *Geology.* the actual relative displacement along the fault plane between two formerly adjacent points on either side of a fault. Also, ACTUAL RELATIVE MOVEMENT, TOTAL DISPLACEMENT. *Naval Architecture.* **1.** a space between two piers where a vessel is tied when not in use; associated primarily with small craft. **2.** see CABLE STOPPER. *Crystallography.* **1.** the movement of one part of a crystal in relation to another. **2.** a plastic deformation of a metallic crystal. *Fluid Mechanics.* the relative velocity between a solid surface and a fluid particle at a point just outside the surface. *Mechanics.* yielding behavior in which the yield occurs in discrete surfaces called yield surfaces, with little or no strain between the surfaces; the material deforms in thin slabs that slip across one another. *Materials Science.* a clay suspension containing enough water so that it can be poured; used in slip casting. *Electricity.* a method in which wires are connected between switching units so that they are used sequentially.

slip an error or shortcoming; specific uses include: *Electromagnetism.* the percentage by which the rotor speed of an induction motor falls below the rotation of the magnetic field in which it operates. *Telecommunications.* a distortion that appears in a FAX image when the mechanical drive system misfunctions.

slipband see SLIP LINE.

slip bedding *Geology.* the contortion of stratification planes caused by gliding.

slip block *Geology.* an isolated rock mass that has slid away from its original position and come to rest down the slope without undergoing significant deformation.

slip casting *Materials Science.* a ceramic shape-forming process in which a suspension of ceramic particles and water is poured into a porous mold; some of the water from the cast material then diffuses into the mold, creating a solid shape.

slip cleavage *Geology.* a type of cleavage superimposed on slaty cleavage or schistosity, in which thin tabular bodies of rock occur between finite-spaced cleavage planes. Also, CLOSE-JOINTS CLEAVAGE, CRENULATION CLEAVAGE, SHEAR CLEAVAGE.

slip coupling *Mechanical Devices.* **1.** a coupling that is used on slip carriages. **2.** a load-reducing coupling for a driving unit under a heavy load.

slip direction *Crystallography.* the direction of movement of one part of a crystal with respect to another in a dislocation.

slip face or **slipface** *Geology.* **1.** the steeply sloping leeward surface of a dune. Also, SANDFALL. **2.** the leeward surface of a sand wave. Also, SLIP SLOPE.

slip flow *Fluid Mechanics.* a rarefied gas flow regime in which the mean free path of the gas molecules is on the order of the characteristic length of the conduit or of the body in the fluid stream.

slip fold see SHEAR FOLD.

slip form *Civil Engineering.* a mobile formwork for forming concrete; used to increase the rate at which concrete is placed in highway paving and in the construction of canals and tunnels.

slip forming *Engineering.* a stretch-forming operation used in plastic sheet manufacturing, in which a part of the plastic sheet slips through mechanically operated clamping rings.

slip friction clutch *Mechanical Engineering.* a clutch that is designed to slip when too much power is applied to it, in order to protect the drive mechanism.

Slipher, Vesto 1875–1965, American astronomer; calculated the radial velocities of nebulae.

slip joint *Civil Engineering.* a joint between two sections of wall, which allows one section to move relative to the other and can sometimes be kept waterproof. *Engineering.* **1.** a process used in flexible bag molding, in which the edges of plastic sheet veneers are cut to partially or totally overlap the scarfed area. **2.** a type of coupling between objects that permits only limited endwise movement, as between pipes and ducts. *Geology.* see SHEAR JOINT.

slip-joint pliers *Mechanical Devices.* pliers whose arms are joined on an adjustable pivot so as to make the opening of the jaws wide or narrow.

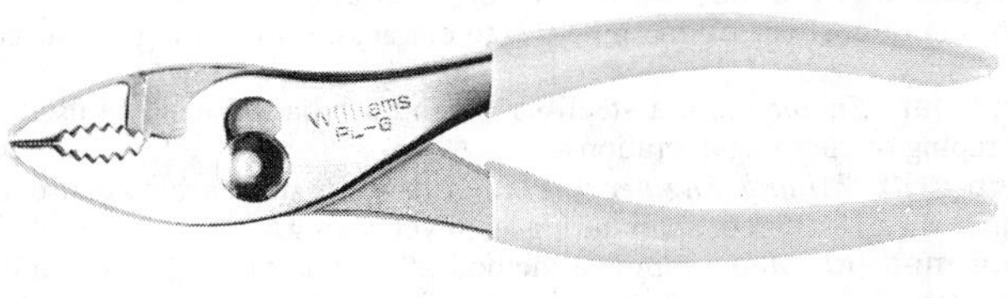

slip-joint pliers

slip line *Crystallography.* a line that appears in a crystal as a result of the parts moving with respect to one another during slip. *Metallurgy.* specifically, one of a series of parallel lines that are visible on a previously polished surface of a piece of metal, due to the step created by plastic deformation along slip planes; such lines disappear when the surface is repolished. Also, SLIPBAND.

slipmouth *Vertebrate Zoology.* a small, slimy-bodied fish of the Leiognathidae family, characterized by a laterally compressed body with protrusible mouth, a luminescent abdomen, and spine-sheathed fins; found primarily in the Indo-West Pacific. Also, PONYFISH.

slip-off slope *Geology.* the long, low, gently sloping face on a meander.

slip of the tongue see LAPSUS LINGUAE.

slippage the process or result of slipping; specific uses include: *Engineering.* the amount of fluid that leaks between the plunger and the bore of a pump piston. Also, SLIPPAGE LOSS. *Petroleum Engineering.* the movement of gas through or around a liquid-phase reservoir front, occurring in lieu of driving the liquid forward; may be observed in gas-lift oil-well bores or gas-drive reservoirs.

slippage loss *Engineering.* **1.** an unintentional shift in the position of the faces of two adjacent solid objects. **2.** see SLIPPAGE.

slipper brake *Mechanical Engineering.* a plate that is placed against the face of a moving part to slow or stop its motion.

slippery elm *Botany.* **1.** an elm tree, *Ulmus rubra,* having a hard wood and fragrant inner bark that becomes mucilaginous when moistened and in early spring. **2.** the inner bark of this tree, used in medicine as a demulcent.

slipping cut *Mining Engineering.* a drill-hole pattern that is well suited for use in wide, rectangular tunnel faces, in which successive vertical lines of shots are fired, each time breaking to the face made by the previous round, so that the entire face is broken in successive slabs from one side to the other. Also, SLABBING CUT, SWING CUT.

slipping sensor *Robotics.* a sensor that detects movement, within the gripper, of an object being gripped.

slip plane *Crystallography.* in a crystalline material, the closest packed crystal plane on which dislocations glide. *Materials Science.* a flat surface that can be seen in transparent material only under reflected light; it results from improper welding and a decrease in the size of the material during the cooling process. *Geology.* a planar slip surface.

slip ratio *Mechanical Engineering.* a measure of the efficiency of a screw propeller, which relates its actual advance to a theoretical advance based on spin and pitch.

slip ring *Electricity.* a conductive rotating ring that, in conjunction with a stationary brush, gives a continuous electrical connection between rotating and stationary conductors.

slip-scratch see SLICKENSIDE.

slip sheet *Geology.* a gravity-collapse structure consisting of a stratum or rock unit on the limb of an anticline that has slid down and away from the anticline.

slip slope see SLIP FACE, def. 2.

slipstream *Aviation.* the turbulent flow of air driven backward by a propeller or downward by a rotor.

slip surface *Geology.* the surface along which the movement of a landslide occurs.

slip system *Materials Science.* the direction and plane on which slip occurs.

slip tongue *Engineering.* an elongated part of a horse-drawn vehicle that is fastened by slipping it between two plates joined to the forecarriage.

slip velocity *Fluid Mechanics.* the difference in velocities between liquids and solids, or between gases and liquids, in the vertical flow of two-phase mixtures through a pipe, tower, or contactor due to the slip between the two phases.

slipway *Naval Architecture.* **1.** an above-water shipyard work area fitted with a sloping floor or track that leads into the water. Most large vessels are built in a slipway, then launched by sliding the hull down into the water. **2.** a ramp in the stern of a whaling ship up which whales are hoisted for processing. **3.** see MARINE RAILWAY.

slit *Design Engineering.* a long, narrow opening that allows radiation or particle streams to move through certain instruments.

slitless spectrograph *Optics.* a spectrograph that makes monochromatic images of the light source itself.

slit sampler *Biotechnology.* an apparatus that samples airborne microflora; air is drawn in through a narrow slit and microflora are deposited on an adhesive collecting surface.

slit-scan *Computer Technology.* an optical character reader consisting of a narrow column of photoelectric cells, used to obtain the horizontal and vertical components of the input character.

slit spectrograph *Optics.* a spectrograph that displays lines of absorption or emission in the spectrum of a light source, which are monochromatic images of the slit through which light enters the instrument.

slitter *Mechanical Engineering.* a rotary cutter with a synchronizing mechanism for the precise cutting of materials into strips.

slitting *Mechanical Engineering.* the cutting of sheet materials, such as plastic, rubber, paper, or sheet metal, with a slitter.

sliver a long, very thin piece of material; specific uses include: *Metallurgy.* a flaw consisting of a thin piece of metal attached to the base at one end. *Textiles.* a long, loose, untwisted textile fiber, after the fiber has been carded or combed but before drawing, roving, or spinning.

slop *Chemical Engineering.* an informal term for a petroleum product that is of inferior quality and must be rerun.

slop culture *Botany.* a method of growing plants in which a nutrient solution is poured over sand or other medium in which the plants are growing, and the surplus is allowed to run through the medium.

slope *Geology.* **1.** the surface of any inclined part of the earth's crust, especially a relatively slight incline. **2.** a broad area of a continent that descends toward an ocean. *Mathematics.* **1.** given a straight line in a plane, the (constant) value $m = (y_2-y_1)/(x_2-x_1)$, where (x_1,y_1) and (x_2,y_2) are any two points on the line; sometimes referred to as **rise over run.** The angle of inclination made by the line, measured with respect to the *x*-axis, equals Arc tan *m* and is called the **slope angle.** A vertical line is said to have **infinite slope,** or **no slope**; a horizontal line has **zero slope.** **2.** the slope of a curve at a point *P* on the curve is the slope of the tangent to the curve at *P*. *Aviation.* the projection of a flight path in the vertical plane. *Electronics.* the gradual rise in attenuation over a range of frequencies. Also, ROLL-OFF.

slope angle *Aviation.* see GLIDE SLOPE.

slope control *Metallurgy.* in welding, the gradual variation of the current, when it is controlled according to a prescribed schedule.

slope conveyor *Mechanical Engineering.* a conveyor system, that may be a screw, ladder, bucket, closed-flight, tray, or other type of conveyor; used to transport materials up or down steep grades.

slope correction see GRADE CORRECTION.

slope course *Engineering.* a raised area having roads of variable angles down its sides; used to test the slope performance of vehicles, military and otherwise.

slope culture see SLANT CULTURE.

slope deviation *Aviation.* in a flight path, the difference between the planned slope and the actual slope; usually expressed in angular terms.

slope failure *Geology.* the gradual or rapid downward and outward movement of soil or rock beneath a natural slope or other inclined surface.

slope gully *Geology.* a small, discontinuous underwater valley, usually formed by slumping along a fault scarp or along the slope of a river delta. Also, SEA GULLY.

slope of fall *Mechanics.* the ratio of the drop of a projectile to its horizontal movement.

slope stability *Geology.* the resistance of a slope or other inclined surface to failure by sliding or slopewash.

slope stake *Mining Engineering.* **1.** a stake marking the point at which the finished side slope of an excavation or embankment meets with the surface. **2.** a stake marking the line at which a cut or fill meets with the original grade.

slope wash *Geology.* **1.** the process of mass-wasting, by which rock and soil material are transported down a slope as a result of sheet erosion. **2.** the material that is or has been transported by such a process.

slosh test *Engineering.* a test used to measure the degree to which a control system of a liquid-propelled missile is affected by the movement of liquid in its fuel tanks.

slot a long, narrow opening; specific uses include: *Design Engineering.* a narrow opening into which some object is inserted, such as a coin, to activate or operate a device or machine. *Aviation.* **1.** a long, narrow, spanwise gap in a wing, usually near the leading edge, designed to improve air flow at high angles of attack. **2.** any long, narrow opening designed to improve air flow. *Computer Technology.* a socket in the backplane of a computer into which a plug-in circuit board can be installed. *Artificial Intelligence.* a named entry in a frame representation in which data about a particular aspect of the object represented by the frame is stored; e.g., a person frame might have an age slot. A slot may have multiple facets, including a value facet, if-needed or if-added methods, data type, documentation, and so on.

slot antenna *Electromagnetism.* an antenna that is fed by a coaxial line and consists of a metal surface having one or more narrow slots.

slot-bound *Computer Technology.* of or relating to a computer with no empty slots, thus with no capacity for further expansion.

slot coupling *Electromagnetism.* a method of injecting electromagnetic waves into a waveguide by means of cutting a slot in the sheath of a coaxial cable and another in the sheath of a waveguide; the slots are positioned in coincidence.

slot distributor *Engineering.* a long, narrow passage in a pipe through which sheet material, such as plastic, is discharged.

slot dozing *Engineering.* a process in which a bulldozer repeatedly uses the same path so that excess material from the sides of the blade builds up on both sides, forming a slot; additional loads are then pushed through the slot, which prevents the spreading of the material and allows the bulldozer to move larger loads.

slot extrusion *Materials Science.* a process in which a plastic-film sheet is formed by forcing a molten, thermoplastic mixture through a straight, narrow opening.

sloth [slōth; släth] *Vertebrate Zoology.* an herbivorous mammal of the family Bradypodidae, almost completely arboreal and noted for its extremely slow movement; characterized by a cuboid skull, long and slender limbs, and two or three exposed digits with strong, recurved claws; found in neotropical forests of the Americas. (From an earlier meaning of the word *sloth,* "laziness" or "sluggishness," with reference to the animal's lack of movement.)

sloth

slothbear *Vertebrate Zoology.* the insectivorous bear species *Melursus ursinus* of the family Ursidae, characterized by a dark, shaggy coat, a yellow snout, a V-shaped mark on the chest, reduced teeth, and huge claws used to climb trees and dig into insect nests; found in forests of India and Sri Lanka. Also, HONEY BEAR.

slot-mask picture tube *Electronics.* a color television picture tube in which the screen is painted with vertical phosphor stripes, and a mask with short vertical slots is used to direct the electron beam.

slot radiator *Electromagnetism.* the primary radiating element in the form of a narrow slot cut in a conductive surface.

slotted flap *Aviation.* a flap that, when depressed, exposes a slot and increases air flow between itself and the rear edge of the wing.

slotted-head screw *Mechanical Devices.* a screw with a groove cut across its head diameter.

slotted nut *Mechanical Devices.* a hexagonal nut with grooves cut into its head for the insertion of a cotter pin to hold it in place.

slotted section *Electromagnetism.* a section of waveguide that has a nonradiating slot cut into its surface for measuring and coupling. Also, **slotted line, slotted waveguide.**

slotted templet *Photogrammetry.* a graphical representation of a photograph, in which the radials are represented by slots cut in a sheet of cardboard, metal, or plastic.

slotted-templet triangulation *Photogrammetry.* a graphical method of radial triangulation in which the radials are obtained from slotted templates.

slotter *Mechanical Engineering.* a vertically reciprocating planing machine used to make slots and mortises in a workpiece or to shape the sides of apertures. Also, **slotting machine.**

slotting *Mechanical Engineering.* the process of using a slotting machine to cut mortises or slots in a workpiece.

slot washer *Mechanical Devices.* **1.** a lock washer with a cut in its rim to allow a screw or nail to be driven through it once the bolt or nut has been secured. **2.** a washer with a cut through it allowing it to be inserted or removed without removing a secure nut or bolt.

slot wedge *Electricity.* the wedge that contains the windings in a slot in a rotor or in the stator core of an electrical machine.

slot weld *Metallurgy.* a weld made in an elongated hole.

slough [sluf] *Medicine.* **1.** necrotic tissue in the process of separating from viable portions of the body. **2.** to shed or cast off. *Zoology.* a general term for a part that is shed or cast off, such as the skin of a snake. *Engineering.* earth material that crumbles off the sides of a mine working or drill hole. Also, SLUFF.

slough [slou; sloo] *Hydrology.* a general term for a marsh, swamp, or bayou, or for a sluggish body of water in such an area. *Geology.* any soft, muddy, or waterlogged ground or mudhole.

slough ice *Hydrology.* slushy snow or ice.

slow-acting relay *Electromagnetism.* a relay device that is designed to allow some specified delay time before energizing the coil and pulling up the armature.

slow axis *Optics.* in the vibration of a light wave passing through a birefringent material, the vector that has a high refractive index and is situated perpendicular to the fast axis.

slow-blow fuse *Electricity.* a protective device designed to open only with a continued overload, as occurs with a short circuit.

slow death *Electronics.* a term for a decline in a transistor's effectiveness caused by the gradual buildup of ions on its surface.

slow dipping *Materials Science.* a coating process in which an object is immersed in the coating fluid and then slowly withdrawn; used primarily for decorative coating of small objects, using fast-drying, high-solid-content, viscous paints.

slowdown *Nucleonics.* a process by which an energetic particle loses kinetic energy during a collision with a nucleus.

slowdown density *Nucleonics.* the rate at which neutrons fall below a specified energy level due to collision in a reactor.

slowed-down video *Electronics.* a technique by which radar information is transmitted over a narrow frequency range.

slow fire *Military Science.* **1.** in small-arms fire, a rate used in instruction or target practice in which there is no time limitation between shots. **2.** artillery or naval gunfire in which the guns of a battery fire together at regular intervals.

slow igniter cord *Engineering.* a plastic-coated cord that passes a flame along its length to ignite safety fuses in a series; it is composed of a copper wire covered by a plastic incendiary material strengthened with iron wire.

slowing of clocks *Physics.* the time-dilation effect; a phenomenon of special relativity in which time is perceived as elapsing more slowly when a clock is moving with respect to the observer than when an identical clock is at rest with respect to the observer; this leads to the conclusion that measurements of time are not absolute, but are dependent on the frame of reference of the observer.

slow ion see LARGE ION.

slow match *Engineering.* a match or fuse designed to burn slowly and evenly at a known rate; used for firing explosives.

slow memory see SLOW STORAGE.

slow-motion videodisk recorder *Electronics.* a device that stores video images, so that they may be replayed at normal speed, slow motion, or stop action.

slow neutron *Nuclear Physics.* **1.** a neutron that has a low level of kinetic energy, no more than 100 eV. **2.** see THERMAL NEUTRON.

slow-neutron spectroscopy *Physics.* the analysis of scattered neutrons having energies up to about 100 eV, commonly used in determining magnetic structures at the atomic level.

slow nova *Astronomy.* a nova whose brightness grows over days or weeks rather than hours, as is typical for a classical nova.

slow ray *Optics.* the element of light that has the least velocity and the highest index of refraction.

slow reactor *Nucleonics.* a nuclear reactor in which fission results from slow (thermal) neutrons having energies up to about 100 eV.

slow-release relay *Electromagnetism.* a relay in which there is a time delay between the de-energizing of the coil and release of the armature.

slow roll *Aviation.* a roll executed principally by movement of the ailerons, with the rudder and elevators used only for trimming and the flight path remaining relatively straight. Also, AILERON ROLL.

slow sand filter *Civil Engineering.* a shallow basin partly filled with sand and provided with an under-drainage system; used for the purification of surface waters intended for domestic use.

slow-scan television *Telecommunications.* a horizontal-scanning system through which television pictures are transmitted over regular telephone circuits at a rate less than is required for the perception of continuous motion; used for transmitting printed matter, photographs, and illustrations.

slow storage *Computer Programming.* the portions of memory whose access rate is relatively slow. Also, SLOW MEMORY.

slow time scale *Computer Technology.* the time scale of a data-processing simulation in which the modeled system is simulated at a speed slower than real time. Also, EXTENDED TIME SCALE.

slow-vibration direction *Optics.* the direction in which the electric field vector in a ray of light will travel when it passes through an anisotropic crystal at its lowest velocity.

slow virus *Virology.* a virus that causes a slow, progressive disease in humans or animals.

slow virus infection *Virology.* an infection in humans or animals, having a long incubation period followed by a slow progression of symptoms, that may be fatal. Also, **slow disease.**

slow wave *Electromagnetism.* an electromagnetic wave whose phase velocity is less than that of the speed of light in a vacuum.
slow-wave sleep see SYNCHRONIZED SLEEP.
sloyd knife *Mechanical Devices.* a single-bladed carving knife used for carving and trimming.
SLP sacrolaeva posterior.
SLR single-lens reflex; straight leg raising.
SLSI super-large-scale integration.
slub *Textiles.* a tuft, lump, or uneven twist in a yarn, usually considered an imperfection.
slud *Geology.* **1.** the muddy material that is transported downslope as a result of solifluction. **2.** any ground that behaves as a viscous fluid.
sludge *Biotechnology.* a combination of solids and liquids resulting from sewage-treatment processes without thickening or physical or chemical pretreatment. *Chemical Engineering.* any undesirable solids settled out from a treatment process. *Engineering.* **1.** a mixture of cuttings and water formed at the bottom of a borehole after drilling. **2.** a muddy deposit that accumulates in a steam boiler. **3.** a deposit that settles out from oils. *Hydrology.* a thin surface layer of ice crystals separated from each other or only slightly frozen together, more solid than lolly ice but not yet a coherent ice rind. Also, SLUSH, GREASE ICE, CREAM ICE. *Geology.* a soft, soupy or muddy stream-bottom deposit, especially a black ooze formed on the bottom of a lake.
sludge assay *Mining Engineering.* the use of chemicals to assay drill cuttings for a specific metal or metal group.
sludge cake *Oceanography.* a mass of sludge ice that has hardened enough to bear the weight of a person.
sludge coking *Chemical Engineering.* a process used to recover sulfuric acid from sludge acid, in which the acid is mixed with water and heated until the acid tar and the acid oil layers separate from the acid.
sludged blood *Medicine.* blood in which the red blood cells have collected together into masses, slowing the flow of blood where it occurs in the smaller vessels. Also, VASCULAR AGGLUTINATION.
sludge ice *Oceanography.* a soupy mass of newly formed sea ice, consisting of floating frazil that forms a thin layer on the sea surface and gives it a grayish or leaden color.
sludge lump *Oceanography.* an irregular mass of sludge that a strong wind has driven together.
sludge pit or **sludge pond** see SLUSH PIT.
sludge pump *Mechanical Engineering.* **1.** a short iron pipe or tube with a valve at its lower end, with which sludge is extracted from a borehole. **2.** see SAND PUMP.
sludgeworm *Invertebrate Zoology.* a small worm, *Tubifex tubifex*, that often inhabits sewage sludge and the muddy bottoms of freshwater lakes, rivers, and pools.
sluff *Engineering.* see SLOUGH.
slug *Invertebrate Zoology.* the common name for any of various terrestrial gastropod mollusks of the family Limacidae, having a rudimentary shell within the mantle and a long, fleshy body; some are plant-eating pests, and some are carnivorous; they are found in damp environments worldwide. *Graphic Arts.* **1.** a line of type produced in one piece by a linecasting machine. **2.** six-point or twelve-point type leads used as line spacing. **3.** a single line of type used as a heading to identify a piece of copy, as on a galley proof. *Electromagnetism.* a heavy copper ring placed about the core of a relay-to-delay operation. *Metallurgy.* in forging or extruding, a component inserted in the die. *Metrology.* a unit of mass, equal to the mass that will accelerate at one foot per second per second when acted on by a force of one pound, or about 32.17 pounds. *Nucleonics.* a short fuel rod of uranium. *Mining Engineering.* to introduce a cement, slurry, or liquid containing shredded materials into a borehole in order to seal off openings in the rocks of the borehole wall to restore circulation.

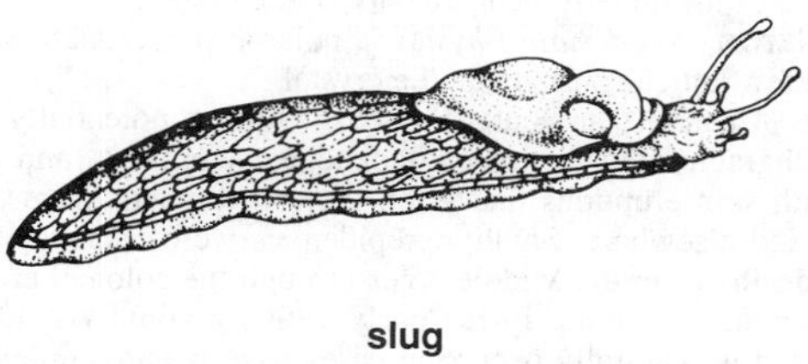

slug

slugging *Metallurgy.* in welding, the undesirable practice of adding a piece of solid material to the weld without allowing it to fuse.
slug tuner *Electromagnetism.* a waveguide tuning device that consists of one or more longitudinally movable pieces of conductive or dielectric material; used for impedance matching.
slug tuning *Electromagnetism.* the varying of the resonant frequency of a resonant circuit by introducing slug material into regions of electromagnetic fields produced by the current.
sluice [sloos] *Civil Engineering.* **1.** a channel for the passage of water fitted with a vertically sliding gate for flow control. **2.** a channel for passing water at a high velocity for discharge of minerals. **3.** a body of water held back or controlled by a sluice gate.
sluice box *Mining Engineering.* a long, inclined trough or launder, through which a current of water is run in placer mining, having riffles in the bottom to trap heavy minerals in ore concentration.
sluice gate *Civil Engineering.* a vertically sliding gate that regulates a flow of water.
sluicing *Mining Engineering.* **1.** the process of washing gold-bearing soil through sluices. **2.** the process of separating minerals from a running stream. **3.** the process of moving rock or mineral materials, such as sand and gravel, by means of flowing water.
slump *Geology.* **1.** a landslide characterized by the downward slipping of one or more masses of rock or unconsolidated debris along curved slip surfaces, accompanied by a backward tilting about a horizontal axis that is parallel to the slope from which the mass descends. **2.** a mass of submarine material that moves by slumping.

slump

slump ball *Geology.* a relatively flattened mass of thinly laminated sandstone, having internal contortions and a smooth or lumpy external form, that is formed by subaqueous slumping.
slump basin *Geology.* a shallow basin, often containing a short-lived lake, that is formed near the base of a canyon wall or on a shale hill or ridge as a result of small, irregular slumps.
slump bedding *Geology.* any bedding that has been disturbed or deformed by the slumping or lateral movement of newly deposited sediment. Also, **slurry bedding.**
slump fault see NORMAL FAULT.
slump fold *Geology.* a minor fold produced by the slumping of soft sediments in a sedimentary layer.
slumping *Geology.* the downsliding of a mass of sediment shortly after its deposition on a submarine slope.
slump overfold *Geology.* a fold that consists of hook-shaped, sandstone masses produced during slumping.
slump scarp *Volcanology.* a low cliff or rim of thin solidified lava that forms along the margins of a lava flow and against the valley walls or around isolated protrusions of bedrock, as a result of the collapse of the central part of the lava crust.
slump sheet *Geology.* a well-defined, horizontally extensive bed of limited thickness in which slump structures have developed.
slump structure *Geology.* any sedimentary structure produced by subaqueous slumping.
slump test *Engineering.* a process used to approximate the consistency of freshly mixed concrete, in which the concrete is poured into a conical mold that is turned upside down and removed; the distance that the concrete sinks is the slump.

slurry *Materials.* **1.** a thin paste produced by mixing an insoluble substance, such as cement or clay, with enough water or other liquid to allow the mixture to flow viscously. **2.** to prepare such a paste. *Petroleum Engineering.* **1.** a suspension of pulverized solid material in water or oil, pumped or poured into pipelines. **2.** a thin, free-flowing mixture of water and cement pumped into an oil well through the hollow drill pipe to support the casing and seal the wellbore.
slurrying *Engineering.* **1.** a process by which mud, slime, or a suspension is formed from insoluble material and liquid. **2.** the process of filling in joint holes with slurry.
slurry mining *Mining Engineering.* the use of drill-hole equipment in hydraulically breaking down a subsurface ore matrix, resulting in a slurry that is then educed to the surface for processing.
slurry preforming *Engineering.* a wet-processing operation used to prepare reinforced plastic preforms.
slurry slump *Geology.* a slump in which the incoherent mass of sliding material is mixed with water and disintegrates to form a highly mobile, quasiliquid suspension of finely divided insoluble matter (slurry).
slurry truck *Engineering.* a vehicle that carries dry blasting ingredients and mixes them, so that they may be input as an explosive slurry into blastholes.
slush *Hydrology.* **1.** snow or ice that has been mixed with or saturated by rain or meltwater, or that has become watery as a result of warm temperatures or treatment with chemicals. **2.** see SLUDGE.
slush avalanche *Geology.* a far-reaching avalanche composed of rock debris released by snow that is supersaturated and mixed with meltwater; characteristic of a catastrophic opening of ice- and snow-dammed brooks to the spring flood. Also, WET-SNOW AVALANCHE.
slush ball *Hydrology.* an extremely compacted buildup of snow, ice particles, and frazil produced by wind, wave action, or turbulent river flow.
slush casting *Metallurgy.* casting a hollow by filling a round mold and pouring out the liquid before solidification is completed.
slush field *Hydrology.* a soupy area of water-drenched snow. Also, SNOW SWAMP.
slushflow *Hydrology.* **1.** a sudden discharge of water-drenched snow along a stream course as a result of more meltwater being produced than can drain through the snow. **2.** a flow of clear slush on a glacier.
slush icing *Meteorology.* the ice and water that accumulates on exposed aircraft surfaces when the aircraft is flown through wet snow or snow and liquid drops at temperatures near 0°C.
slushing *Engineering.* a process in which filler is sprayed over castings, such as sinks, to prepare them for enamel coating; the excess filler is removed by shaking the castings.
slushing compound *Materials.* a rust-inhibiting liquid composed of mineral oil and anticorrosive additives.
slush molding *Engineering.* a process used to form a molded plastic object, in which a thin plastic paste is poured into a hollow heated mold that is rotated until a viscous skin forms in a desired shape, excess slush is drained off, and finally the molded object is extracted.
slush pit *Engineering.* a diked area or an area that is hollowed so that it may hold the sludge, mud, or other debris that originates from an oil well. Also, MUD PIT, SLUDGE PIT, SLUDGE POND.
slush pond *Hydrology.* a pool or lake of slush found on the ablation surface of a glacier.
slush pump *Petroleum Engineering.* a large, reciprocating pump used to circulate drilling mud during rotary drilling.
slype *Mining Engineering.* see SLED.
S-M or **S/M** sadomasochism, sadomasochist. Also, **SM.**
SM or **S.M.** Master of Science. (From Latin *Scientiae Magister.*)
Sm the chemical symbol for samarium.
Smale's horseshoe *Chaotic Dynamics.* a topological transform resulting from the repeated stretching of a space in one direction, shrinking it in another, then folding the stretched-compressed space in a C-shape, not unlike a mechanical taffy-maker, until it is intricately self-embedded, as a visual analogue of sensitive dependence on initial conditions.
small *Ordnance.* the thinnest part of the stock of a small-arms weapon; it is usually designed to be held by the right hand and is located behind the receiver and trigger assembly.
small-amplitude approximation *Mechanics.* an approximation used in analyzing many physical systems that exhibit small-amplitude motion, in which the restoring force is approximated by a linear restoring force. This enables equations of motion to be solved analytically; the approximation is valid as long as the amplitude of vibration is sufficiently small.
small-angle grain boundary *Crystallography.* in a crystal, an array of dislocations causing a small misorientation between adjacent portions.
small-angle scattering *Physics.* the scattering of particles or radiation through small angles due to interaction with cavities or other particles. *Crystallography.* the study of matter by analysis of the diffraction of X-rays with diffraction angles smaller than a few degrees; that is, θ less than 1° for copper radiation. This scattering occurs when the sample is composed of particles with dimensions of the order of several hundred to several thousand Å. Measurement of the intensity distribution gives information on the low-resolution structure of the diffracting material; for example, it will give the radius of gyration of the particle.
small arms *Ordnance.* **1.** in the U.S. military, all arms of a caliber up to and including 20 mm (0.787 inch); nonmilitary definitions may range up to a caliber of one inch; small arms include pistols, revolvers, rifles, carbines, shotguns, machine guns, and other automatic weapons. **2.** of or relating to such weapons. Thus, **small-arms ammunition.**
small-bore *Ordnance.* of or relating to 22-caliber firearms.
small capitals or **small caps** *Graphic Arts.* capital letters that are typeset in a size that is smaller than the normal capital letters for a given font, equal to or slightly higher than the x-height (the height of a lowercase letter x). Note the examples below for Times Roman:

lowercase letters REGULAR CAPITALS SMALL CAPITALS

small circle *Geodesy.* a circle on the surface of the earth, such as any parallel of latitude other than the equator, whose plane does not pass through the earth's center; any parallel of latitude except the equator is a small circle.
small computer systems interface see SCSI.
small-craft warning *Meteorology.* a weather statement that warns marine interests of impending wind conditions, having wind speeds of at least 32 miles per hour. Also, **small-craft advisory.**
small-diameter blasthole *Engineering.* a hole made in low-face quarries that holds a heavy charge of explosives and measures 3.8 to 7.6 cm.
small diurnal range *Oceanography.* the average difference in height between mean lower high water and mean higher low water.
small grain *Agronomy.* a term used to classify grain crops such as wheat, oats, barley, and rye.
small hail see ICE PELLETS.
small ice floe *Oceanography.* an ice floe from 10 to 200 meters across.
small intestine *Anatomy.* the region of the digestive tube lying between the stomach and the colon.
small ion *Meteorology.* a strongly mobile atmospheric ion that collectively is the major agent of atmospheric conduction. Also, FAST ION, LIGHT ION.
small-ion combination *Meteorology.* the disappearance of small ions by one of two processes: by an Aitken nucleus that forms a union with a small ion, thus creating a larger ion, or by neutralization of a large ion by a small ion.
Small Magellanic Cloud see MAGELLANIC CLOUDS.
small-mouth bass *Vertebrate Zoology.* a freshwater game fish, *Micropterus dolomieu*, having a body that is yellowish-green above and lighter below and characterized by a lower jaw that extends to the eye; found in North America. Also, **small-mouthed bass.**
small nuclear RNA *Genetics.* an abundant species of RNA found in the nuclei of eukaryotic cells and consisting of approximately 100 to 300 nucleotides, generally occurring in small ribonucleoprotein particles (snRNPs or snurps) and believed to play an important role in mediating and regulating RNA processing after transcription, as in the splicing of introns from nuclear pre-mRNA.
small perturbation *Physics.* a disturbance imposed on a steady-state system that is small enough so that its effect can be calculated from knowledge of the unperturbed behavior of the system.
small polaron *Solid-State Physics.* a polaron defect that extends over at most a few lattice constants in the crystal.
smallpox *Medicine.* an acute, highly infectious, potentially fatal viral disease characterized by lower backaches, vomiting, and chills and fever, with skin eruptions that often leave permanent scars (pocks) on the face and elsewhere. Smallpox epidemics were responsible for millions of deaths from the Middle Ages through the colonial era; vaccination programs beginning in the early 1800s gradually eradicated the disease, and no naturally occurring cases have been recorded since the late 1970s. (So named in contrast with the *great pox*, syphilis.) Also, VARIOLA.

smallpox vaccine *Immunology.* an antigen preparation, derived from the vesicular eruptions of vaccine on the skin of a calf or sheep, that is administered to produce immunity against the smallpox virus; smallpox vaccine was originally developed in 1796 by the English physician Edward Jenner.

smallpox virus *Virology.* a virus of the genus *Orthopoxvirus* of the family Poxviridae, which causes smallpox in humans. Also, VARIOLA VIRUS.

small-scale integration *Electronics.* an integration technology that produces integrated circuits of considerably less complexity than those produced by medium-scale integration, usually having less than the equivalent of 12 gates.

small-scale map *Cartography.* any map or chart with a scale of 1:600,000 or smaller.

small-scale yielding *Materials Science.* in nonlinear fracture mechanics, a condition that occurs in the region surrounding the crack tip when the elastic singularity fields afford a good approximation to the actual fields in an annular region surrounding the tip.

small-signal *Electrical Engineering.* **1.** a general term describing a low-amplitude signal. **2.** a signal that is sufficiently small so that the system coefficients can be treated as constants and the system equations are linear.

small-signal method *Electronics.* a method of analyzing the operation of a nonlinear device, assuming that the signal represents small variations of current or voltage about the quiescent point; the response to the signal is then calculated by a linear approximation.

small-signal parameter *Electronics.* a parameter characterizing electronic equipment functionality for small input values.

SMALLTALK *Computer Programming.* a programming language based on object-oriented programming and a user interface designed for ease of use, which allows the user to select processes by pointing at icons on the screen and pressing a mouse button.

small tropic range *Oceanography.* the vertical difference between tropic lower high water and tropic higher low water.

smalt *Materials.* a coloring agent made of blue glass in which silica has been fused with potassium carbonate and cobalt oxide; used in powdered form to color vitreous materials such as glass and ceramics.

smaltite *Mineralogy.* an arsenic-deficient variety of skutterudite.

smaragd *Mineralogy.* another name for an emerald, especially in older use. See EMERALD.

smaragdine *Mineralogy.* of or relating to emeralds.

smaragdite *Mineralogy.* a light-green, thin-foliated to fibrous variety of actinolite basic igneous containing some alumina; often pseudomorphous after pyroxene in basic igneous rocks.

smart *Computer Technology.* describing a device or system that has the ability to process data; intelligent.

smart bomb *Ordnance.* a popular term for a bomb guided to the target by its own on-board laser or video system.

smart card *Computer Technology.* a plastic credit card containing a microprocessor.

smart terminal see INTELLIGENT TERMINAL.

smart tool *Robotics.* a fixed tool or end effector on a robot that uses sensors to determine the relative position of the tool and an actuator to adjust the tool's position.

smashing *Graphic Arts.* in binding, the process of compressing sewn book blocks to a uniform thickness in order to counter the swelling effect of sewing signatures.

smaze *Ecology.* a mixture in the atmosphere of smoke and haze.

sm. c. small capitals.

SMC X-1 *Astronomy.* the most luminous X-ray pulsar detected; it is located in the Small Magellanic Cloud and consists of a neutron star and a supergiant in a binary star system.

SMC X-2 *Astronomy.* a hot O-type star in the Small Magellanic Cloud that is a highly variable source of X-rays; it was the second X-ray emitter found in the Cloud.

SMC X-3 *Astronomy.* a hot O-type star in the Small Magellanic Cloud that is a highly variable source of X-rays; it was the third X-ray emitter found in the Cloud.

SMD senile macular degeneration.

SME or **S.M.E.** Society of Mining Engineers.

smear *Biology.* a preparation for microscopic examination made by spreading a drop of fluid, such as blood, across a slide and using the edge of another slide to leave a uniform film. *Electronics.* television picture distortion in which the images on the screen appear to be blurred or stretched beyond their normal size and shape.

Smeaton, John 1724–1792, British engineer; considered the founder of modern civil engineering in Britain; built noted bridges and canals.

smectic *Physical Chemistry.* of liquid crystals, having a structure of separate layers or planes that exhibit little viscosity. Thus, **smectic liquid crystal, smectic phase.**

smectic-A *Physical Chemistry.* a subclass of smectic liquid crystal in which the free-flowing molecules move across the molecule layer in a direction perpendicular to it.

smectic-B *Physical Chemistry.* a subclass of smectic liquid crystals in which the free-flowing molecules form a tightly packed lattice within the molecule layer and move in a direction perpendicular to the layer.

smectic-C *Physical Chemistry.* a subclass of smectic liquid crystals in which the free-flowing molecules within the molecule layer move with their axes oriented toward the layer.

smectite *Mineralogy.* **1.** a group name for monoclinic silicate clay minerals of the general formula $X_{0.3}Y_{2-3}Z_4O_{10}(OH)_2 \cdot nH_2O$, where X (exchangeable ions)=Ca/2,Li,Na; Y=Al,Cr^{+3},Cu^{+2},Fe^{+2},Fe^{+3},Li,Mg, Ni,Zn; and Z=Al,Si; characterized by swelling properties and high capacities for cation exchange. **2.** any mineral of this group.

smectogenic *Physical Chemistry.* describing a solid material that when heated forms a smectic phase. Thus, **smectogenic solid.**

smegma *Physiology.* a thick white exudate of a sebaceous gland, especially under the foreskin of the penis. (From a Greek word for "soap.")

smell *Physiology.* **1.** the sensation recognized by the olfactory receptors in the nasal mucus membranes and a specialized area of the cortex. **2.** the characteristics of substances that stimulate this smell sensation. **3.** to recognize a substance by its smell.

smelling salts *Pharmacology.* the popular name for a preparation of ammonium carbonate with sweet-smelling substances such as lavender or nutmeg added, to be sniffed as a general stimulant and restorative. Also, SAL VOLATILE.

smell prism *Psychology.* a prism-shaped representation of the six primary odors (putrid, burned, spicy, resinous, ethereal, and fragrant) and the relations among them.

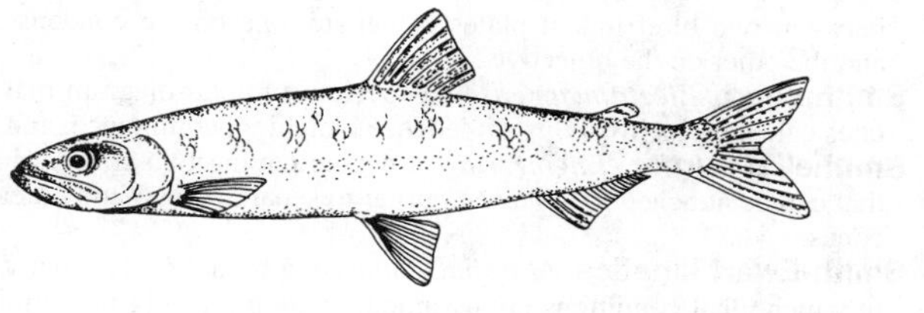

smelt

smelt *Vertebrate Zoology.* any of various small salmonid fishes of the family Osmeridae, resembling the trout; characterized by silvery skin, a large mouth and powerful jaws, and delicate oily flesh; found in cold ocean waters and spawning in fresh water; North American species include the **American** or **rainbow smelt**, *Osmerus mordax*, and the **pond smelt**, *Hyposmesus olidus*.

smelt *Metallurgy.* to carry out a process of smelting.

smelter *Metallurgy.* **1.** any of various installations used for smelting ores or concentrates. **2.** a person or device involved in a process of smelting.

smelting *Metallurgy.* any of the various metallurgical methods by which ores or concentrates are heat-processed to yield a crude metal, which is then reduced or refined.

S-meter see SIGNAL-STRENGTH METER.

SMF *Aviation.* the airport code for Sacramento Metropolitan Field, Sacramento, California.

SMF sodium motive force.

Smilacaceae *Botany.* a cosmopolitan family of monocotyledonous climbing, herbaceous, or slender woody vines or branching shrubs of the order Liliales, characterized by starchy rhizomes and commonly having vessels in all vegetative organs.

smilagenin *Pharmacology.* $C_{27}H_{44}O_3$, a steroid precursor derived from any of several species of *Smilax* and used in the manufacture of compounds of the pregnane series.

Smilax *Botany.* a genus of climbing plants of the family Smilacaceae, species of which produce the flavoring agent sarsaparilla and the steroid precursor smilagenin.

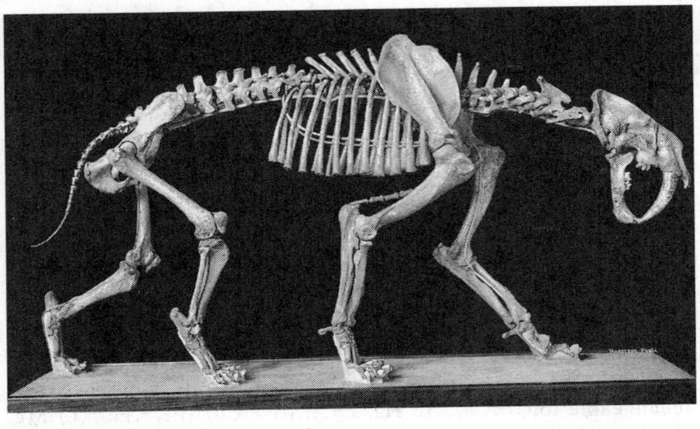

Smilodon

Smilodon *Paleontology.* a genus of large saber-tooth tigers of the Pleistocene having canine teeth more than 6 inches long; they were found worldwide except in Australia and Antarctica.

Sminthuridae *Invertebrate Zoology.* a family of springtails, small primitive jumping insects in the order Collembola, with a short rounded body; they are terrestrial or semiaquatic.

Smith, Hamilton O. born 1931, American geneticist; shared Nobel Prize for research in restriction enzymes.

Smith, Theobald 1859–1934, American pathologist; identified agents of cattle fever; differentiated bovine and human tuberculosis.

Smith, William 1769–1839, British geologist; founded stratigraphy and geological mapping; demonstrated relations of fossils to strata.

Smith and Wesson *Ordnance.* any of numerous firearms manufactured by Smith and Wesson, Inc., founded by Horace Smith and Daniel Wesson in the mid-1800s; Smith & Wesson .38-caliber service pistols have been widely used by law-enforcement agencies.

Smith-Baker microscope *Optics.* a microscope that places the object between two birefringent plates of calcite, one on the condenser lens, and the other on the objective.

Smith chart *Electromagnetism.* a specialized polar diagram that facilitates calculations involving impedance matching, admittance, and so on.

Smithell's burner *Engineering.* a device having two concentric tubes that can be attached to a bunsen burner to separate inner and outer flame cones.

Smith-Ewart kinetics *Materials Science.* a model of polymer kinetics in which ideal conditions are assumed, including negligible termination of free radicals and adequate diffusion currents.

smith forging *Metallurgy.* a process of forging by hand, as by a blacksmith.

Smith-Helmholz law *Optics.* the law stating that, in a spherical lens, the product of the refractive index, off-axis distance, and off-axis angle at the image point is equal to the corresponding product computed at the image plane.

smithite *Mineralogy.* $AgAsS_2$, a red monoclinic mineral occurring as equant or tabular crystals, having a specific gravity of 4.88 and a hardness of 1.5 to 2 on the Mohs scale; found in crystalline dolomite at the Legenbach Quarry, Switzerland.

Smith-Petersen nail *Medicine.* a nail with a flange used for holding the head of the femur in place in the procedure to repair a fracture of the femur's neck.

Smithson, James 1765–1829, British chemist and mineralogist; gave funds to the United States to found the Smithsonian Institution.

smithsonite *Mineralogy.* $ZnCO_3$, a variously colored, trigonal mineral of the calcite group, occurring as rhombohedral crystals and in botryoidal and massive form, having a specific gravity of 4.2 to 4.45 and a hardness of 4 to 4.5 on the Mohs scale; found in the oxidized zone of zinc-bearing ore deposits. Also, ZINC CARBONATE, CALAMINE, SZASKAITE.

smog *Meteorology.* **1.** originally, a mixture of smoke and fog, as produced in large urban or industrial centers (e.g., London) from the open burning of coal and other fuel. **2.** in current use, a fog contaminated by manmade industrial pollutants such as carbon monoxide as a by-product from automobile exhaust; especially prevalent in large urban areas surrounded by higher elevations (e.g., Los Angeles, Denver, Mexico City). **3.** in general, any form of urban air pollution, regardless of the source or composition. (A blend of the words smoke and fog.)

smoke *Physical Chemistry.* **1.** the visible mixture given off by a burning substance, consisting of a suspension of tiny particles of carbon in hydrocarbon gases or in air. **2.** any visible cloud of small particles suspended in a gaseous medium. *Food Technology.* to carry out the process of preserving or flavoring food (e.g., ham or fish) by exposing it to smoke, as from the burning of a certain type of wood such as hickory, maple, or ash. *Medicine.* to inhale into the mouth and exhale the smoke from cigarettes or other forms of tobacco; especially, to do this regularly or as a habit. *Ordnance.* of or relating to a weapon or other device that produces a cloud of smoke; the smoke may be used to mask friendly or enemy installations or maneuvers, to mark a target, or as an antipersonnel irritant. Thus, **smoke bomb, smoke candle, smoke canister, smoke generator, smoke grenade, smoke pot, smoke projectile, smoke rocket, smoke signal,** and so on.

smokebox *Mechanical Engineering.* an external vessel that traps the unburned products of a combustion boiler.

smoke chamber *Engineering.* a section of a fireplace extending from the top of the throat to the base of the flue.

smoke detector *Engineering.* a device that sets off an alarm when smoke exceeds a predetermined density; used as a warning signal in homes, offices, hotel rooms, and so on. Also, **smoke alarm.**

smoke horizon *Meteorology.* the top of a smoke layer that is confined by a low-level temperture inversion so as to give the appearance of the horizon when seen from above against the sky; the true horizon in such cases is usually obscured by the smoke layer.

smokehouse *Food Technology.* a building where fish or specialty meat products are smoked.

smoke jumper *Forestry.* a firefighter who parachutes from an airplane or helicopter to a fire area that is difficult to reach by land.

smokeless *Ordnance.* describing ammunition that produces little or no smoke when used in the weapon for which it is designed. Thus, **smokeless powder, smokeless propellant.**

smokeless coal see SEMIBITUMINOUS COAL.

smokeless powder *Materials.* any of various substances used as a substitute for gunpowder because of their relatively minimal production of smoke.

smoke point *Engineering.* in kerosene testing, a measurement that represents the maximum flame height, in millimeters, at which kerosene can burn without smoking.

smoke printing see ELECTROSTATIC PRINTING.

smoker *Food Technology.* an enclosed box or compartment used to smoke foods. *Medicine.* a person who regularly or habitually smokes cigarettes or other forms of tobacco.

smokes *Meteorology.* a weather condition signaling the approach of the harmattan, characterized by dense white haze with dust clouds during the dry season on the Guinea coast of Africa.

smokescreen *Military Science.* a layer of dense smoke that is intentionally created to obscure the vision of the enemy, so as to conceal the position of ships or vehicles, hide the movement of troops, and so on.

smokestack *Engineering.* a funnel or chimney through which the gaseous products of combustion are discharged, such as the chimney of a locomotive.

smoke technique *Fluid Mechanics.* a method of calculating very low speed air velocity wherein smoke particles are used to enable observation of fluid motion; the particles are timed over a measured distance along an airway of constant cross-sectional area to determine the flow velocity.

smoke test *Engineering.* **1.** a test that determines the maximum flame height at which a material can burn without producing smoke. **2.** a test to determine areas of pipe leakage, in which smoke is forced into a pipe system, and leakage can be determined by smoke oozing through the pipe or by an odor penetrating into the room.

smoke tree *Botany.* **1.** an American tree, *Cotinus obovatus,* of the family Anacardiaceae, having ovoid leaves and large clusters of small white flowers. **2.** a Eurasian shrub, *Cotinus coggygria,* having elliptical leaves and clusters of purple flowers.

smoke washer *Engineering.* an instrument that forces smoke upward into a downward spray of water so that solid particles are removed from the smoke.

smoking *Food Technology.* a process used to preserve some foods and improve their flavor by exposing them to wood smoke for extended periods at low temperatures. Hardwoods such as oak, hickory, or ash are used most often because their smoke contains aldehydes and phenols, which aid in preservation. *Medicine.* the practice or habit of smoking cigarettes or other forms of tobacco.

smoky quartz *Mineralogy.* a smoky-yellow, smoky-brown, or brownish-gray, often transparent, crystalline variety of quartz with inclusions of carbon dioxide colored by exposure to a radioactive source; used as a semiprecious stone. Also, **smokestone.**

S Monel *Metallurgy.* a nickel-base alloy nominally containing 30% copper and 4% silicon.

smooth having a flat, even surface; specific uses include: *Oceanography.* describing an area of relatively calm water between heavy seas. *Statistics.* to modify observed data for the purpose of creating a smooth curve connecting the plotted values. *Mathematics.* **1.** infinitely differentiable, such as an infinitely differentiable map or infinitely differentiable manifold; denoted by C^∞. **2.** r-continuously differentiable; that is, continuously differentiable at least r times, where r is large enough to satisfy the requirements of the given context; denoted by C^r.

smooth blasting *Engineering.* a blasting process that produces uniformly even rock surfaces without creating cracks in the rock.

smooth-bore *Ordnance.* **1.** a weapon having an unrifled bore; such weapons include shotguns, most mortars, and older guns such as the musket or blunderbuss. **2.** of or relating to such weapons.

smooth chert *Geology.* a hard, dense, homogeneous chert having a smooth, conchoidal-to-even fracture surface, and lacking crystallinity, granularity, or other distinctive structure.

smooth colony *Microbiology.* a bacterial colony exhibiting a smooth, even surface, generally because the bacteria have cell-wall-enveloping capsules. Also, S COLONY.

smooth drilling *Mining Engineering.* a type of fast drilling process that results in a high recovery of core with vibration-free rotation of the drill stem.

smooth endoplasmic reticulum *Cell Biology.* a system of intracellular membranes that lack bound ribosomes, important in synthesis of steroid hormones.

smoothing *Engineering.* the process of making a uniformly even surface. *Mathematics* **1.** the process of approximating a given function by one of a greater degree of differentiability. **2.** the process of finding a curve that passes between discrete data points so as to minimize an arbitrarily chosen deviation. Also, **smoothing of measurement results.**

smoothing choke *Electronics.* an iron-core choke coil designed to remove ripple in the output current of a vacuum-tube rectifier or direct-current generator.

smoothing mill *Mechanical Engineering.* a polishing wheel used for smoothing and beveling glass or stone.

smoothing trowel *Mechanical Devices.* a large masonry trowel used to finish cemented or plastered surfaces.

smooth muscle *Anatomy.* a general term for the elongated, spindle-shaped muscles not under voluntary control, such as those of the digestive tract, the blood vessels, the urogenital system, glands, the hair follicles, and so on, but not those of the heart.

smooth muscle fiber *Histology.* any of the elongate, nucleated cells making up the smooth (involuntary) muscles.

smooth phase *Geology.* a means of sediment transport in a stream, in which a mass of sedimentary material travels as a sheet whose density gradually increases from the surface downward.

smooth-rough variation *Cell Biology.* a reversible alteration in the cell surface composition of many bacterial strains, involving a spontaneous change from a capsulated, or smooth, form to a noncapsulated, or rough, form.

smooth sea *Oceanography.* a sea whose waves are no greater than ripples or small wavelets.

smothered bottom *Geology.* a sedimentary surface on which complete, well-preserved, often fragile and delicate fossils were preserved as a result of instantaneous burial by an influx of mud.

smother kiln *Engineering.* a heated enclosure in which pottery is blackened by introducing smoke during the firing process.

SMR standard mortality (or morbidity) ratio.

SMSA Standard Metropolitan Statistical Area.

smudge *Agriculture.* a smoky fire built to repel insects or to protect fruit trees from frost.

smudge pot *Agriculture.* a heavy cast-iron pot used for burning oil or other fuels to produce protective smudge.

smudging *Agriculture.* the process by which an orchard or garden is protected from frost by the introduction of heavy smoke or heat.

smut *Plant Pathology.* any of various destructive plant diseases caused by smut fungi and characterized by large sooty masses of dark spores on plant organs. *Metallurgy.* a reaction product left on a metallic part after chemical attack or after electroplating.

smut fungus *Mycology.* any fungus causing smut, especially one of the family Ustilaginaceae.

smut grass *Botany.* a tufted, wiry grass, *Sporobolus poiretii,* with narrow, fungus-infected panicles; native to the West Indies but common in the southern United States. Also, CARPET GRASS.

Smyth sewing [smīth] *Graphic Arts.* the most common method of sewing signatures for case binding; the signatures are fed in order into a sewing press, where they are saddle stitched and attached at the spine edge to adjacent signatures. (Developed by the American inventor David *Smyth*, 1833–1907.)

Smyth-sewn *Graphic Arts.* describing a book that has signatures bound together by the Smyth sewing method.

Sn the chemical symbol for tin. (From Latin *stannum.*)

SN or **S.N.** according to nature. (From Latin *secundum naturam.*)

SN *Aviation.* the airline code for Sabena World Airlines.

S/N signal-to-noise (ratio).

SNA *Aviation.* the airport code for John Wayne International Airport, Orange County, California.

snag *Textiles.* a pulled yarn creating a defect in a knit fabric.

snagging *Mechanical Engineering.* the process of rough grinding with a wheel to remove large surface defects or burrs from a workpiece.

snagline mine *Ordnance.* a contact mine with a buoyant line that may be caught up and pulled by the hull or propellers of a ship.

snail *Invertebrate Zoology.* any of numerous creeping terrestrial and aquatic gastropod mollusks having a spiral shell; some are plant-eating pests, and some are carnivorous; they are found worldwide in a variety of environments. *Horology.* a cam that is shaped something like the profile of a snail's shell; used to regulate the number of times a clock strikes or chimes. A snail used to regulate striking is called an **hour snail**, while one used to regulate chiming is called a **quarter snail**.

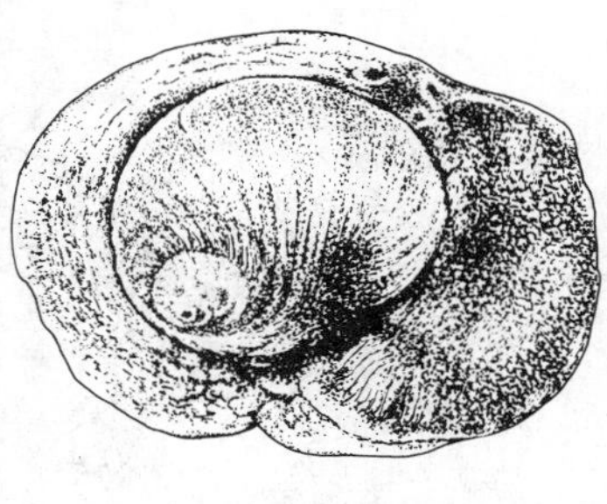

great moon snail

snail fever see SCHISTOSOMIASIS.

snake *Vertebrate Zoology.* any of the numerous scaly, limbless reptiles of the suborder Serpentes, some of which are poisonous; they are found worldwide in a variety of environments, typically in tropical or temperate climates. *Mechanical Devices.* any of various devices having the thin, elongated shape of a snake, such as a flexible metal tube with various attachments used to remove clogs in pipes, or a long wire with an attached hook used to pull electrical wire through small openings. *Metallurgy.* **1.** in hot rolling, the product resulting from twisting a rod prior to reentering the mill stand. **2.** in metal bending, a flexible mandrel used to support the walls of the workpiece.

snakehead *Vertebrate Zoology.* any of various elongated fishes of the family Channidae, noted for their ability to breath air.

snake hole *Engineering.* **1.** a hole drilled beneath a boulder in preparation for blasting. **2.** any type of drill hole made for quarrying or bench blasting.

snake venom *Toxicology.* the venom of any poisonous snake.

snaking *Engineering.* the process of moving a large load by dragging it with a long cable.

snaking stream see MEANDERING STREAM.

snap *Mechanical Devices.* a kind of clasp or hasp consisting of a ball and socket with each attached to a separate, unique part. Also, **snap fastener.** *Meteorology.* a brief period of extreme weather, usually cold weather, that sets in suddenly.

SNAP *Nucleonics.* a small-scale nuclear power plant that provides power for telemetry, spacecraft instrumentation, etc. (An acronym for systems for nuclear auxiliary power.)

snap-action switch *Electricity.* a single-pole, single-throw toggle switch; it snaps to "on" and "off" to reduce arcing.

snapback experiment *Molecular Biology.* a procedure in which a DNA strand is recovered from an RNA-DNA hybrid and is shown to contain repetitive DNA by self-annealing in reassociation kinetic experiments.

snapback forming *Engineering.* a method of molding a plastic sheet in which the sheet is stretched out, heated, and finally allowed to contract over a form to create the desired shape.

snapdragon *Botany.* the popular name for various plants belonging to the genus *Antirrhinum* of the family Scrophulariaceae, having large, variously colored flowers that are thought to look like a dragon's head, especially the **common snapdragon,** *A. majus*, which is widely cultivated in North America.

snap flask *Metallurgy.* in casting, a flask that is removable prior to pouring.

snap gauge *Mechanical Devices.* a preset gauge used to measure and regulate the tolerance for correct internal and external sizes in production work.

snap-off diode *Electronics.* a silicon diode that is processed so a charge is stored near the junction when the diode is conducting; when reverse voltage is applied, the stored charge causes the diode to **snap off** (switch into its blocking state).

snap-on ammeter *Electricity.* a portable AC ammeter having a magnetic core in the form of hinged jaws that can be snapped around the current-carrying wire.

snapper *Vertebrate Zoology.* any of several large marine food fishes, especially those of the family Lutjanidae, such as the red snapper. *Engineering.* a bucket used to collect deep-sea samples, snapping shut to prevent loss of the samples as the bucket moves to the surface.

snapping turtle *Vertebrate Zoology.* **1.** either of two large freshwater turtles of the family Chelydridae, of North America and Central America, having a large head and powerful jaws. **2.** see ALLIGATOR SNAPPING TURTLE.

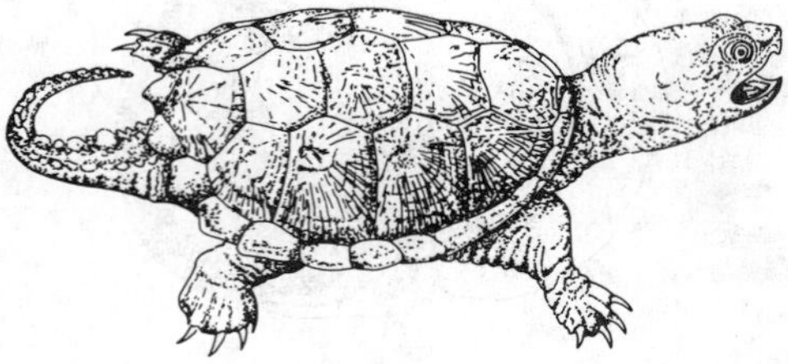

snapping turtle

snap ring *Mechanical Devices.* a deformed, elastic, broken metal ring that fits into the groove of a shaft in a loaded or stressed position and is then allowed to snap to its retaining or stop position.

snap roll *Aviation.* a quick roll executed by bringing an aircraft sharply nose-up, then sharply moving the rudder in the direction of the roll.

snapshot dump *Computer Programming.* a memory dump performed at selected times during the execution of a program.

snare *Surgery.* a surgical device having a wire loop that can be closed to remove a polyp or tumor at its base.

Snark *Ordnance.* the first U.S. intercontinental missile; it was equipped with all-inertial guidance and delivered a nuclear warhead at a maximum range of 5500 nautical miles. (Named for the *Snark,* an imaginary animal in a poem by the English humorist Lewis Carroll.)

snarling iron *Mechanical Devices.* a long-beaked tool used to raise metal on a workpiece by repercussion.

snatch block *Mechanical Devices.* a wood or metal block, one side of which is attached to a hinge, allowing it to receive the bight of a rope, thus fastening it to a scaffold.

snatch plate *Mining Engineering.* a thick steel plate with a hole in the middle, designed to slip over a drill rod while one edge is fastened to a securely anchored chain; used as a protective locking mechanism during drilling processes in which high-pressure water may be encountered.

S-N curve *Materials Science.* a graph representing the results of fatigue testing of a given material; relates the applied stress, S, to the number of cycles to failure, N.

S-N diagram *Engineering.* a graph showing the parameters of cycles and the number of stresses needed to produce fatigue failure during fatigue testing of materials.

sneakernet *Computer Technology.* a slang term referring to the transfer of data from one system or computer to another by physically carrying it on a removable medium such as a tape or disk.

sneeze *Physiology.* **1.** to push air out of the nose and mouth forcefully, usually as a result of irritation of the nose. **2.** the process of forcing out air in this manner.

Snell, George born 1903, American immunologist; awarded the Nobel Prize for the discovery (in mice) of major histocompatibility complex.

Snell, Willebrord 1591–1696, Dutch astronomer and mathematician; discovered Snell's Law and the triangulation method of geodesy.

Snellen chart *Medicine.* any chart that is used in the Snellen test of visual perception; the familiar letter chart with a large "E" at the top is a Snellen chart.

Snellen test *Medicine.* **1.** a test of visual acuity, using a chart on which letters, numbers, or symbols are arranged in decreasing size from top to bottom. The subject stands at a given distance from the chart and reads as many of the symbols as possible; the results of this test are expressed as a ratio comparing the subject's performance with a statistical norm. A person who has 20/20 vision can read what the average person could read from a distance of 20 feet. **2.** a test for pretended blindness in one eye, in which the subject is requested to look at alternate red and green letters; the admittedly sound eye is covered with a red glass and if the green letters are read, evidence of fraud is present. (From George *Snellen,* 1834–1908, Dutch ophthalmologist.)

Snell's law (of refraction) *Optics.* a law stating that, when a light ray enters a medium, the entering ray and the refracted ray lie in the same plane as the normal to the surface, and are on opposite sides of the normal, and their angles of incidence and refraction have sines that are in a constant ratio to each other. *Acoustics.* a law stating that sound refraction occurs in such a way that the ratio between sound velocity and the cosine of the angle of incidence as a sound wave travels into a space with a different sound velocity is constant:

$$C_v = v_1/\cos\theta_1 = v_2/\cos\theta_2.$$

snezhura see SNOW SLUSH.

SNG synthetic natural gas.

snifter valve *Engineering.* a regulating device on a cylinder or pump that permits air to escape or enter, and allows for the release of accumulated water. Also, **snifting valve.**

snipe *Vertebrate Zoology.* **1.** any of several long-billed game birds of the genera *Gallinaago* (*Capella*) and *Limnocryptes*, characterized by barred and striped white, brown, and black plumage; inhabiting marshy areas in Eurasia and North America. **2.** any of various other similar shore birds, as some sandpipers or the jacksnipe.

snipe

snipe eel *Vertebrate Zoology.* a marine fish of the family Nemichthyidae, characterized by a long (up to 3 feet), eel-like body with a speckled back and blackish belly and anal fin.

snipefish *Vertebrate Zoology.* any of several fishes of the family Macrorhamphosidae, having long, tubular snouts that resemble a snipe's beak; found in tropical and temperate seas.

sniperscope *Ordnance.* a telescoping sight designed to mount on a rifle or carbine; it allows the gunner to see and aim at a target in total darkness.

snips *Mechanical Devices.* short, heavy shears, used especially to cut sheet metal.

snivet *Electronics.* a straight or twisted vertical line found near the right-hand edge of a television screen; caused by a discontinuity in the plate current; characteristic of a horizontal amplifier tube under zero-bias conditions.

SNM or **S.N.M.** Society of Nuclear Medicine.

SNN *Aviation.* the airport code for Shannon, Republic of Ireland.

SNOBOL *Computer Programming.* a high-level language designed to process character strings and to match patterns.

snook *Vertebrate Zoology.* a large tropical marine fish of the family Centropomidae, resembling an elongated bass and having a lateral line to the tail fin; a popular food and game fish. Also, ROBALO.

snooperscope *Electronics.* a night-viewing device composed of an infrared source, an infrared image converter, and a battery-operated, high-voltage, DC source; allows the user to see objects obscured by darkness.

snorkel *Naval Architecture.* originally, a set of tubes allowing a submarine to remain underwater for extended periods, providing for the intake of air for the diesel engines and for breathing by the crew, and also for the exhaust of stale air. Modern nuclear submarines do not employ such an air-supply system. *Engineering.* in current use, a tube that allows a swimmer to breathe while swimming face down just below the surface of the water. (From the German name for this device, literally meaning "to snort;" it was originally developed by the German navy.)

snout *Vertebrate Zoology.* the part of an animal's head that projects forward, containing the nose and jaws; muzzle. *Entomology.* an anterior prolongation of the head, bearing the mouth parts, as in snout beetles; rostrum. *Geography.* a protrusion of a feature, especially the lower extremity of a glacier.

snow *Meteorology.* precipitation composed of white or translucent ice crystals in a flake or starlike hexagonal form, often coated with rime and matted with ice needles. *Electronics.* a speckled background produced on a television or radar screen; caused by random noise, usually indicating a weak signal.

Snow, John 1813–1858, English physician; pioneered the study and administration of chloroform and ether; studied the transmission of cholera.

snow accumulation *Meteorology.* the actual measured depth of snow on the ground at a given point during or after a snowstorm or series of storms.

snow avalanche *Hydrology.* a large mass of relatively pure snow that falls or slides very rapidly under the force of gravity. Also, SNOWSLIDE.

snowbank *Hydrology.* a large mass of snow, as from the compacting of a snowdrift or the plowing of snow from a roadway.

snowbank glacier see NIVATION GLACIER.

snow banner *Meteorology.* snow that is being blown from the crest of a mountain. Also, SNOW PLUME.

snow barchan *Hydrology.* a crescent-shaped or horseshoe-shaped dune of wind-transported snow whose ends point downwind.

snow bin *Engineering.* a compartment designed to measure the amount of snowfall.

snowbird see SNOW BUNTING.

snow blight *Plant Pathology.* a disease of conifer trees caused by the fungus *Phacidium infestans*; the needles become infected under snow, turn brown, and become covered with a white fungal growth.

snow blindness *Medicine.* the usually temporary loss of sight due to prolonged exposure to the brightness of sunlight on snow.

snow blink *Meteorology.* a bright, white glare that appears on the underside of midlevel clouds as a result of light reflected from a snow-covered surface; used primarily in the polar regions. Also, SNOW SKY.

snow blower or **snowblower** *Mechanical Engineering.* a motor-driven machine used to remove snow from road surfaces by drawing it in with a screwlike blade and ejecting it out to the side. Also, SNOW THROWER.

snowbreak *Civil Engineering.* a barrier to hold drifting snow, thus safeguarding highways, railways, and other such projects.

snowbridge or **snow bridge** *Hydrology.* a connecting layer or fragment of ice or snow across a glacial crevasse.

snow bunting *Vertebrate Zoology.* a small finch, *Plectrophenax nivalis*, having white plumage with brownish and black markings; found in northern parts of the Northern Hemisphere. Also, SNOWBIRD, SNOWFLAKE.

snow cap or **snowcap** *Hydrology.* **1.** snow that covers only the peak or ridge of a mountain on which there is no snow at lower elevations. **2.** a blanket of snow on the frozen surface of a lake.

snow climate see POLAR CLIMATE.

snow cloud *Meteorology.* any cloud from which snow falls.

snow concrete *Hydrology.* snow that has been highly compacted at a low temperature by a heavy object, making the snow strong and tough. Also, **snowcrete**.

snow course *Hydrology.* a line or series of observation stations along a snow-covered terrain at which samples of snow are taken for measuring its depth, density, water equivalent, and other characteristics.

snow cover *Hydrology.* **1.** all the snow, including snowfall, snowdrift, and avalanches, that has accumulated on the ground in a given location. Also, SNOW MANTLE. **2.** in such a location, the percentage of total ground area that is covered with snow, or the average depth of the deposited snow.

snow-cover chart *Meteorology.* a synoptic chart highlighting snow-covered areas by contour lines of varying snow depth.

snow crab *Vertebrate Zoology.* an edible spider crab, *Chionoecetes opilio*, found in the North Pacific and important as a frozen seafood product.

snowcreep *Hydrology.* the slow, internal change in form of a snow cover as a result of stress under its own weight and the metamorphism of snow crystals.

snow crust *Hydrology.* a layer of hard snow or snow and ice overlying a layer of softer snow, formed by periodic thawing and refreezing.

snow crystal *Meteorology.* a single ice crystal found in snow, as distinguished from a snowflake, which is usually an aggregate of many snow crystals.

snow cushion *Hydrology.* a deep, soft, unstable accumulation of snow deposited on the leeward side of an overhanging ledge of ice or snow on a steep mountain slope.

snow density *Hydrology.* the mass of snow per unit volume.

snowdrift or **snow drift** *Hydrology.* an accumulation or mass of snow piled up by the wind on the leeward side of an obstruction or irregularity in the surface.

snowdrift glacier *Hydrology.* a small glacier in a mountain area nourished mainly by windblown snow. Also, DRIFT GLACIER.

snowdrift ice *Hydrology.* a mass of ice formed by the accumulation of drifted snow.

snowdrop *Botany.* any of various bulbous plants of the genus *Galanthus* of the amaryllis family, especially *G. nivalis*, which blooms early in the spring and has drooping white flowers with green markings; native to Eurasia.

snow dune *Hydrology.* an accumulation of windblown snow that resembles a sand dune.

snow dust *Meteorology.* fine snow crystals that become fragmented as they are driven by the wind.

snow eater *Meteorology.* **1.** a warm wind, usually a foehn, that blows over a snow surface. **2.** a fog whose moisture effects are such that a snow cover beneath it rapidly disappears in its presence.

snowfall *Meteorology.* **1.** a snowstorm. **2.** the rate at which snow falls, usually expressed as inches of snow depth per six-hour period.

snow fence *Civil Engineering.* **1.** a barrier for impounding snow in areas where melting in place contributes to soil moisture. **2.** a barrier erected on the windward side of a road or structure for protection against drifting snow.

snowfield *Hydrology.* **1.** a broad, continuous, relatively smooth and uniform expanse of snow covering the ground or ice in mountain areas or high latitudes throughout the year. **2.** any region of permanent snow cover, such as the accumulation area of a glacier. **3.** an accumulation of perennial ice and snow that is too small to be called a glacier.

snowflake *Meteorology.* an ice crystal or an aggregation of crystals that falls from a cloud; simple snowflakes or single crystals show exquisite variety of form, but broken single crystals or clusters of fragments are more typical of actual snowfalls. It is generally accepted that no two snowflakes can be exactly alike in shape, though obviously this cannot be empirically verified. *Vertebrate Zoology.* see SNOW BUNTING.

snowflake obsidian *Petrology.* a variety of obsidian containing white, gray, or reddish spherulites ranging in size from microscopic dimensions to greater than one meter in diameter.

snow flurry *Meteorology.* a very brief and light snow shower.

snowflush *Geology.* an accumulation of drifted snow, windblown soil, and wind-transported seeds on a lee slope, characteristically marked during the winter by a hard patch of soil.

snow-forest climate *Meteorology.* a major category of climatic classification having a coldest-month mean temperature of less than 26.6°F and a warmest-month mean temperature of greater than 50°F; characterized by a cold winter with at least a month of snow-covered ground; the coldest of the tree climates.

snow garland *Hydrology.* a looped or curved rope of snow held together by surface tension, found occasionally hanging from trees or other objects.

snow gauge *Hydrology.* a device used to measure the water equivalent in a given snow sample.

snow geyser *Meteorology.* fine, powdery snow that is blown aloft by a snow tremor.

snow glide *Hydrology.* the slow movement of a snow cover over an inclined surface under its own weight.

snow goose *Vertebrate Zoology.* a white wild goose, originally classified as *Chen hyperborea* but now considered the light color phase of the blue goose, *C. caerulescens*; found in North America.

snow grains *Meteorology.* a solid equivalent of drizzle characterized by small, white to opaque ice particles resembling snow pellets, but having a flattened and elongated aspect and generally a diameter of less than 1 millimeter.

snow ice *Hydrology.* ice formed by the freezing of a mixture of snow and water.

snow leopard *Vertebrate Zoology.* a long-haired, wild cat, *Panthera uncia*, characterized by a relatively small head and a thick, creamy-gray coat with rosette spots; an endangered species found in the mountains of central Asia.

snow line *Geography.* **1.** the line of perpetual snow, above which snow does not usually melt in summer. **2.** the line above which snow lies at any given time.

snow load *Civil Engineering.* the live load allowed by local building code, used to design roofs in areas subject to snowfall.

snow mantle see SNOW COVER.

snowmelt or **snow melt** *Hydrology.* water from melting snow. Also, SNOW WATER.

snow-melting system *Civil Engineering.* a system for melting snow from a roadway, usually consisting of heating pipes or electric cables imbedded below the roadway.

snowmobile *Mechanical Engineering.* a small motorized vehicle for one or two persons, used to travel over snow or ice; it has two steerable skis in front and an engine-powered revolving track at the rear.

snow mold *Mycology.* certain fungi, such as the species *Calonectria graminicola* and *Fusarium nivale*, that infect grass and winter crop plants such as wheat and rye. *Plant Pathology.* a fungus disease of grasses or cereals, characterized by large superficial patches of white fungal growths. Also, **snow rot.**

Snow Mountain agent *Virology.* a small, round, nonenveloped member of the Norwalk group of viruses that causes acute gastroenteritis in humans.

snow niche see NIVATION HOLLOW.

snowpack or **snow pack** *Hydrology.* **1.** any cover of snow. **2.** the amount of snow that accumulates annually at higher elevations, especially in the mountainous regions of the western United States.

snow-patch erosion see NIVATION.

snow pea *Botany.* a variety of small, green pea, *Pisum sativum macrocarpon*, characterized by flat, edible pods that are used in cooking. Also, SUGAR PEA.

snow pellets *Meteorology.* white, opaque ice particles with a round or conical aspect and a crisp, delicate form, of 2 to 5 millimeters in diameter, rebounding when they fall on a hard surface and then often breaking up.

snow penitente see NIEVE PENITENTE.

snow pillow *Engineering.* a fluid-filled device, resembling the form of a pillow, that is attached to a pressure transducer or pipe and float; it is used to record the variation in the weight of a snow cover.

snowplow *Mechanical Engineering.* a vehicle fitted with a large plow at the front, used to clear snow from streets, highways, airport runways, railroad tracks, and so on.

snow plume see SNOW BANNER.

snow point *Physical Chemistry.* the temperature at which the vapor pressure of a sublimable component in a gas mixture equals its partial pressure in the mixture; akin to the dew point.

snowquake see SNOW TREMOR.

snow resistograph *Engineering.* a device that produces a graphic representation of the hardness profile of an area of snow, by measuring the pressure required of a blade to move upward through the snow.

snow roller *Hydrology.* a cylindrical mass of wet, cohesive snow, accumulating as wind-blown snow rolls down a slope.

snow sampler *Hydrology.* a hollow tube used to extract a sample or core of deposited snow that extends the depth of the snow cover.

snowshed *Civil Engineering.* a covering structure designed to cover unprotected areas from snow accumulation in roadways or railways. *Hydrology.* a drainage basin that is supplied mainly by water from melting snow.

snowshoe hare *Vertebrate Zoology.* a large-footed hare, *Lepus americanus*, that is white in winter and dark brown in summer; found in North America.

snowshoe hare

snow sky see SNOW BLINK.

snowslide see SNOW AVALANCHE.

snow sludge *Oceanography.* sludge consisting primarily of snow that has fallen on freezing seawater.

snow slush *Hydrology.* **1.** see SLUSH. **2.** watery new snow that falls at above freezing temperatures. **3.** slush formed by snow falling into water at a temperature below that of the snow. Also, SNEZHURA.

snow stage *Meteorology.* the thermodynamic process of sublimation of water vapor into snow in an ideal adiabatic lifting of moist air.

snow stake *Engineering.* a wooden, calibrated scale used to measure the depth of deep snow. Also, **snow scale.**

snow static *Electromagnetism.* the static interference caused by the build up and discharge of electric charge due to falling snow.

snowstorm *Meteorology.* any storm in which snow falls.

snow survey *Hydrology.* the determination of snow density, depth, and water equivalent in a particular area by taking samples along a snow course.

snow swamp see SLUSH FIELD.

snow thrower see SNOW BLOWER.

snow tremor *Hydrology.* the sudden collapse or disturbance of one or more layers of snow in a snowfield. Also, SNOWQUAKE.

snow water see SNOWMELT.

SNR signal-to-noise ratio.

snRNA small nuclear RNA.

snRNP *Molecular Biology.* small nuclear ribonucleoprotein; consists of species of small nuclear RNA associated with a specific protein, and used in excising introns from mRNA.

SNRP small nuclear ribonucleic particle.

SNS somatic nervous system; sympathetic nervous system.

SNU solar neutrino unit.

snub *Mining Engineering.* **1.** to increase the height of an undercut by explosive or other means. **2.** to check a car's descent by the turn of a rope about a post, or to check the descent of any hand-lowered object.

snubber *Mechanical Engineering.* a device that is automatically activated by a predetermined displacement to limit the operation of a system.

S1 nuclease *Molecular Biology.* an endonuclease purified from the fungus *Aspergillus oryzae* that cleaves ssRNA and ssDNA to produce 5'-nucleoside monophosphates.

snuffles *Medicine.* a catarrhal discharge from an infant's nasal mucous membranes, usually from congenital syphilis.

snurp *Genetics.* any of various small ribonucleoprotein particles in which small nuclear RNA generally occurs. (An acronym for small nuclear ribonucleoprotein.)

Snyder sampler *Engineering.* a mechanical sampler consisting of a revolving cast-iron plate with an inclined spout passing through the flange; the spout acquires samples when it comes in line with a stream of fine-grained solids, such as ore.

So. or **so.** south; southern.

soak cleaning *Metallurgy.* a process of electrochemical cleaning carried out while the workpiece is immersed in a liquid.

soaked zone *Hydrology.* the area of a glacier where significant surface melting takes place in summer, so that meltwater soaks through the whole mass of the snow layer.

soaking *Metallurgy.* in thermal treatment, the process of holding at a specific temperature for a considerable time.

soaking drum *Chemical Engineering.* a chamber in a petroleum-refinery that holds hot oil during the time necessary for it to complete cracking. Also, CRACKING CHAMBER.

soap *Materials Science.* **1.** a water-soluble cleansing substance generally consisting of sodium or potassium salts derived by the action of an alkali (usually sodium hydroxide or potassium hydroxide) on a fat or fatty acid, with glycerol as a by-product. **2.** any metallic salt of an acid derived from a fat.

soap bubble test *Engineering.* a type of leakage test, in which a soap solution is applied to the exterior of a vessel being tested; a gas leak from the vessel causes the formation of soap bubbles.

soap builder *Materials.* any of various substances, such as rosin or sodium phosphate, that are mixed with soap to improve cleansing properties, modify alkali content, or impart water-softening qualities.

soapstone *Mineralogy.* **1.** a massive compact variety of the mineral talc, commonly dark gray or green, with a greasy, soaplike feel. Also, **soaprock. 2.** see SAPONITE.

soar *Aviation.* to fly without engine power, utilizing thermals or other upcurrents to maintain altitude. Thus, **soaring.**

SOB shortness of breath.

Sobemovirus group *Virology.* a group of plant viruses with icosahedral virions containing ssRNA; they have a narrow host range and are transmitted either mechanically or by seed or beetle. (From so̲uthern be̲an mo̲saic vi̲rus, the type member for this group.).

sobole *Botany.* an underground or near-ground shoot that becomes a creeping stem.

Sobolev inequality *Mathematics.* suppose that $1 \leq p < n$ and that q is defined by $1/q = 1/p - 1/n$. Then there is a positive constant C depending on n and p such that, for every infinitely differentiable function f with compact support in R^n,

$$\|f\|L^q \leq C \{ \sum_{k=1}^{n} \int |\partial f/\partial x_k| p dx \}^{1/p},$$

where $\|f\|L^q$ is the L^q norm of f.

Sobolev space *Mathematics.* given an open set U of R^n, the Sobelev space $H^{m,p}(U)$ is the space of functions in $L^p(U)$, all of whose partial derivatives up to and including order m are also in $L^p(U)$; usually $p = 2$ and one writes $H^m(U)$ for $H^{m,2}(U)$. Using duality, spaces $H^{-m,p}(U)$ can be defined.

social *Anthropology.* of or relating to human society; involving the interaction of people in groups. *Behavior.* **1.** relating to or designating a species whose members live in groups rather than in isolation from one another. **2.** of or relating to behavior involving the interaction of members of the same species. Often used in compounds, such as **social attraction, social communication, social drive, social dynamics, social signal.**

social accommodation *Psychology.* see ACCOMMODATION.

social adaptation *Behavior.* a modification in the behavior of an animal species that facilitates interactions among its members.

social animal *Zoology.* any animal that lives in groups with others of its kind for mutual protection, food gathering, and mating.

social anthropology *Anthropology.* a major branch of anthropology that examines social relationships within human groups, such as sex roles, family relationships, courtship and marriage practices, leadership functions, and forms of conflict.

social anxiety *Psychology.* an individual's feelings of stress concerning his or her social status and behavior.

social attribution see ATTRIBUTION.

social behavior *Ecology.* the pattern of interaction between unrelated or distantly related members of the same species; usually does not include reproductive or parenting behaviors. *Behavior.* **1.** any human or animal activity that involves some form of interaction with others. **2.** the behavior of a group, social unit, or social organization.

social comparison theory *Psychology.* the belief that people appraise their own behavior, attitudes, and values by comparing them with those of others.

social Darwinism *Anthropology.* **1.** a 19th-century belief that societies develop in a manner similar to biological evolution and that the rule of "survival of the fittest" applies to culture; i.e., a society that has more advanced technology and greater military and economic power is inherently superior. **2.** any belief that individuals or societies achieve success because of inherent superiority and a resulting competitive advantage.

social desirability *Psychology.* a tendency to respond to questionnaires or polling inquiries with answers that are socially acceptable, as by avoiding responses that could be construed as revealing prejudice, selfishness, and the like.

social dominance *Ecology.* physical domination of one individual over another within a population, initiated and sustained by aggression.

social environment *Behavior.* the totality of the interactions that individuals have with members of their own species.

social facilitation *Ecology.* augmentation of the behavior of an individual due to the presence of other individuals who share the same behavior.

social geography *Geography.* a branch of human geography dealing with the relationships between geography and human social organization including population distribution and settlements.

social grooming see ALLOGROOMING.

social group *Ecology.* an interdependent collection of individuals of the same species, banding together for their mutual benefit.

social hierarchy *Behavior.* the presence of dominant and subordinate individuals in a local population of vertebrates, based on sex, strength, aggressiveness, and size.

social-impact assessment *Building Engineering.* a study of the potential social effects a proposed construction project will have on a community in terms of population change, economic impact, and the like.

social influence *Psychology.* the ability of a person or group to affect or direct the behavior, attitude, and values of another person or group.

social inhibition *Behavior.* the inhibition of an action by one individual because of the presence of another or others of the same species.

social insect *Invertebrate Zoology.* describing insects that live in large communities, such as ants and bees.

social insect colony *Invertebrate Zoology.* see COLONY.

socialization *Behavior.* the process by which an animal acquires competence in social interaction through contact with other members of its species. *Psychology.* specifically, the process by which an individual, starting in infancy, learns and internalizes the behavior, traditions, and values appropriate to his or her particular culture.

social-learning theory *Psychology.* **1.** a behavioral approach to learning that stresses the importance of cognitive processes, the influence of other people as role models, and reinforcement in the development of behavior. **2.** the theory that most behavior develops from reinforcement and social influences.

social motive *Psychology.* any learned motive, as opposed to a physiological drive.

social norm *Psychology.* an established pattern of acceptable behavior, as set forth and approved by a particular group or society, and utilized by group members to interpret and sanction activities and surroundings.

social organization *Anthropology.* the patterns and institutions under which a society is structured.

social parasitism *Behavior.* the habit of some birds (especially cuckoos and cowbirds) to deposit their eggs in another species' nest and let the foster parents raise the young.

social pathology *Ecology.* a pattern of disturbances within a social group, both behavioral and physiological, caused by overcrowding.

social phobia *Psychology.* an anxiety disorder characterized by a fear of having to do something in public and thereby being subject to scrutiny by others, such as speaking before a group.

social psychiatry *Psychology.* a branch of psychiatry that studies the effects of social relationships on the treatment of mental illness and the ways in which changes in such relationships can alleviate certain mental disorders. Thus, **social psychiatrist.**

social psychology *Psychology.* a branch of psychology concerned with the effects of group membership upon individual behavior, the interactions of groups, and other social aspects of human behavior. Thus, **social psychologist.**

social releasor see RELEASER STIMULUS.

social role *Psychology.* the position and function of an individual in a group situation.

social-skills training *Psychology.* a therapeutic technique for shy or inhibited people that uses assertiveness training as well as behavior and cognitive rehearsal to train them to behave with confidence and ease.

social structure *Behavior.* the organization of a group or species in terms of rank, spatial distribution, interrelationships, and any other factors that differentiate it from others. Also, **social system.**

society *Anthropology.* a relatively large group of people regarded as forming a distinct group, sharing certain common characteristics such as language, ethnicity, geographical location, and cultural traditions. *Behavior.* a communal group of animals characterized by such social traits as parental care of young, division of labor, mutual protection, and regulation of activities. *Ecology.* a group of plants, generally of the same species, that form part of a community but are not the dominant form of vegetation in that community.

sociobiology *Biology.* an interdisciplinary branch of biology and social sciences that is concerned with the biological bases of social behavior in all organisms, including humans. Thus, **sociobiologist.**

sociofact *Archaeology.* an object whose primary function is to express social rank or establish social order, rather than to serve practical or ideological needs; e.g., an axe that is used as a symbol of chiefdom rather than as a weapon.

sociofunction *Archaeology.* the use of an object for social purposes; for example, the wearing of a certain garment to convey high social status.

sociofunctional *Archaeology.* of or relating to sociofunction; used to express social status or organization. Also, **sociotechnic.**

sociolinguistics *Linguistics.* the ethnographic study of cultural and subcultural patterns of speech and how they vary within different social contexts (such as social status, relationship, sex, age, and occasion). Thus, **sociolinguist.**

sociologist a person who is trained in sociology or whose profession is sociology.

sociology the systematic study of the development, organization, and activities of human societies.

sociopath *Psychology.* a person having a sociopathic personality. Also, PSYCHOPATH.

sociopathic personality *Psychology.* an older term for an antisocial personality, a mental disorder characterized by chronic antisocial behavior, rejection of authority and discipline, and lack of social responsibilities or close emotional ties. Also, PSYCHOPATHIC PERSONALITY.

socked in *Meteorology.* a term for weather conditions at an airport that are characterized by an extremely low ceiling and very limited horizontal visibility, so that operations are shut down; commonly used in the early days of aviation.

socket *Electricity.* a form of jack into which a device with one or more prongs is designed to fit.

socket chisel *Mechanical Devices.* a heavy carpenter's chisel whose tang is designed to fit a hollow, tapering handle.

socket-head screw *Mechanical Devices.* a screw whose head is recessed or socketed to fit a wrench, for turning and providing a secure fit.

socket plug *Mechanical Devices.* a plug having a recess or socket in its end face to receive a spanner; used to close off the ends of pipes.

socket punch *Mechanical Devices.* a hollow punch, such as a belt punch, with a cutting edge forming a closed curve.

socket washer *Mechanical Devices.* a countersunk washer with a bolt head or nut.

socket wrench *Mechanical Devices.* a wrench having a recessed or hollowed head into which a nut or bolt can be inserted.

soda *Inorganic Chemistry.* a popular term for various compounds of sodium, especially the compounds sodium bicarbonate (**baking soda**), sodium carbonate (**washing soda**), and sodium hydroxide (**caustic soda**). *Food Technology.* soda water, or a drink made with this. See SODA WATER. (From an Arabic word describing a plant from which soda was produced.)

soda-acid extinguisher *Engineering.* a type of fire extinguisher that sprays water at a high rate; it is generated by an interior mixing of sulfuric acid and sodium bicarbonate that produces carbon dioxide.

soda alum see SODIUM ALUMINUM SULFATE.

soda ash *Inorganic Chemistry.* the commercial form of anhydrous sodium carbonate, Na_2CO_3, a grayish-white powder or lumps, soluble in water and insoluble in alcohol; a widely produced chemical with many industrial uses, as in soaps, detergents, and cleansers, water treatment, paper and glass manufacture, petroleum refining, and aluminum production. Also, CALCINED SODA.

soda crystals see SODIUM CARBONATE MONOHYDRATE.

soda-granite *Petrology.* **1.** a soda-rich variety of granite, with soda more abundant than potash. **2.** a granite containing soda-plagioclase rather than the orthoclase found in normal granite.

soda lake *Hydrology.* a salt lake containing a high concentration of dissolved sodium salts, especially sodium carbonate together with sodium chloride and sodium sulfate. Also, NATRON LAKE.

soda lime *Materials.* a mixture of calcium oxide with sodium hydroxide or potassium hydroxide; occurs as toxic whitish or grayish-white granules; used as a drying agent, carbon dioxide absorbent, and reagent.

soda-lime glass *Materials.* a glass composed primarily of SiO_2, Na_2O, and CuO, with smaller additions of MgO and Al_2O_3; characterized by low thermal shock resistance and used for bottles, windows, and other glassware.

sodalite *Mineralogy.* $Na_8Al_6Si_6O_{24}Cl_2$, a generally blue, cubic mineral of the sodalite group, with vitreous luster, occurring as dodecahedral crystals and in massive form, having a specific gravity of 2.14 to 2.4 and a hardness of 5.5 to 6 on the Mohs scale; found in nepheline syenites and metasomatized calcareous rocks.

sodamide see SODIUM AMIDE.

soda monohydrate see SODIUM CARBONATE MONOHYDRATE.

soda niter see SODIUM NITRATE.

soda pulping process *Chemical Engineering.* a chemical pulping process that uses sodium hydroxide as the alkaline pulping agent to remove the lignin portion of wood.

soda water *Food Technology.* carbonated water used in soft drinks.

sod crop *Agronomy.* a crop, such as certain grasses and legumes, that forms a close, heavy growth over the entire soil surface rather than growing in separated rows.

sod cropping *Agronomy.* the practice of growing a sod crop to control erosion.

Soddy, Frederick 1877–1956, English chemist; formulated theories of atomic disintegration (with Ernest Rutherford) and isotopes; was awarded the Nobel Prize.

soddyite *Mineralogy.* $(UO_2)_2SiO_4{\cdot}2H_2O$, a rare, strongly radioactive, yellow, orthorhombic mineral, occurring in fine-grained aggregates of bipyramidal crystals and in massive form, having a specific gravity of 4.70 and a hardness of 3.5 on the Mohs scale; found as a secondary mineral in pegmatites.

Soddy's displacement law *Nuclear Physics.* a law stating how much charge a given nuclide will lose during decay, as in alpha decay, where its atomic number will decrease by two.

sodium *Chemistry.* a metallic element having the symbol Na, the atomic number 11, an atomic weight of 22.99, a melting point of 97°C, and a boiling point of 883°C; a tetragonal, crystalline, soft, silvery-white solid that does not occur in elemental form in nature due to high reactivity; it has excellent electrical conductivity and high heat-absorbing capacity; used as a supplier of electrons in laboratory reactions, as a conductor in cables, in alloys, as a polymerization catalyst, and in non-glare lighting. (From the word *soda*; so named by Sir Humphry Davy, who isolated it from caustic soda.)

sodium-24 *Nuclear Physics.* an isotope of sodium, produced by deuteron bombardment, with a half-life of 14.96 hours; it decays into magnesium with emission of beta rays.

sodium acetate *Organic Chemistry.* $NaC_2H_3O_2$, combustible, colorless crystals, soluble in water and ether, and slightly soluble in alcohol; melts with decomposition at 326°C; used as a food additive and reagent, in pharmaceuticals, soaps, and medicines, and in meat preservation.

sodium acetone bisulfite see ACETONE SODIUM BISULFITE.

sodium acid carbonate see SODIUM BICARBONATE.

sodium acid chromate see SODIUM DICHROMATE.

sodium acid fluoride see SODIUM BIFLUORIDE.

sodium acid phosphate see MONOBASIC SODIUM PHOSPHATE.

sodium acid pyrophosphate *Inorganic Chemistry.* $Na_2H_2P_2O_7{\cdot}6H_2O$, a white crystalline powder that is soluble in water; decomposes at 220°C; used in electroplating, metal cleaning, baking powders, food processing, and for various other purposes. Also, DISODIUM DIPHOSPHATE, DISODIUM (DIHYDROGEN) PYROPHOSPHATE.

sodium acid sulfate see SODIUM BISULFATE.

sodium acid sulfite see SODIUM BISULFITE.

sodium acid tartrate see SODIUM BITARTRATE.

sodium alginate *Organic Chemistry.* $C_6H_7O_6Na$, a combustible, colorless or yellowish granular or powdery solid; insoluble in alcohol and ether; forms a viscous colloid in water; extracted from brown seaweeds and used in food thickeners and stabilizers, medicine, experimental ocean-floor covering, and textile printing.

sodium aluminate *Inorganic Chemistry.* $NaAlO_2$, a white powder that is soluble in water and insoluble in alcohol; melts at 1800°C; strongly basic in solution and a strong irritant; used in water purification, in soaps and cleansers, and for various other purposes.

sodium aluminosilicate *Inorganic Chemistry.* $xNa_2O \cdot xAl_2O_3 \cdot xSiO_2$, any of a series of hydrated sodium aluminum silicates of variable composition; a white amorphous powder or beads; insoluble in water, alcohol, and organic solvents; used as an anticaking agent in foods. Also, SODIUM SILICOALUMINATE.

sodium aluminum phosphate *Inorganic Chemistry.* $Na_3Al_2H_{15}(PO_4)_8$ or $NaAl_3H_{14}(PO_4)_8 \cdot 4H_2O$, a white odorless powder; insoluble in water and soluble in hydrochloric acid; used as a food additive. Also, **(acidic) sodium aluminum phosphate.**

sodium aluminum sulfate *Inorganic Chemistry.* $NaAl(SO_4)_2 \cdot 12H_2O$, colorless crystals; soluble in water; melts at 61°C; used as a mordant and for waterproofing textiles, as a food additive, and for matches, ceramics, tanning, and water purification. Also, SODA ALUM.

sodium amalgam *Inorganic Chemistry.* Na_xHg_x, a silver-white crystalline mass that decomposes in water; a dangerous fire hazard; used to prepare hydrogen and as a reducing agent and reagent.

sodium amide *Inorganic Chemistry.* $NaNH_2$, a white crystalline powder having an ammoniacal odor; decomposed by water and hot alcohol; melts at 210°C and decomposes at 400°C; a dangerous fire hazard; used as a reagent and in organic synthesis. Also, SODAMIDE.

sodium ammonium (hydrogen) phosphate *Inorganic Chemistry.* $NaNH_4HPO_4 \cdot 4H_2O$, colorless crystals that give off ammonia and water on heating and that are soluble in water and insoluble in alcohol; decomposes at 79°C; used as a reagent. Also, MICROCOSMIC SALT, PHOSPHOROUS SALT.

sodium antimonate *Inorganic Chemistry.* $NaSbO_3$ or $NaSb(OH)_6$ or $Na_2H_2Sb_2O_7 \cdot H_2O$ (sodium pyroantimonate) or $2NaSbO_3 \cdot 7H_2O$ (sodium meta-antimonate); composition is both variable and uncertain, a toxic white granular powder or colorless crystals; insoluble to slightly soluble in water; used in enamels. Also, ANTIMONY SODIATE.

sodium arsanilate *Organic Chemistry.* $C_6H_4NH_2(As \cdot OH \cdot ONa)$, a highly toxic, white crystalline powder, soluble in water and slightly soluble in alcohol; used in organic synthesis and in medicine.

sodium arsenate *Inorganic Chemistry.* $Na_3AsO_4 \cdot 12H_2O$, toxic, clear, colorless crystals; soluble in water and slightly soluble in alcohol; melts at 86.3°C; used in dyeing and printing, in making arsenates, and as a germicide. Also, SODIUM ORTHOARSENATE.

sodium arsenite *Inorganic Chemistry.* $NaAsO_2$, a toxic grayish powder that absorbs carbon dioxide from the air; very soluble in water and slightly soluble in alcohol; used in insecticides, herbicides, antiseptics, and dyes. Also, SODIUM METAARSENITE.

sodium ascorbate *Organic Chemistry.* $C_6H_7O_6Na$, the white crystalline sodium salt of ascorbic acid; soluble in water; decomposes at 218°C; used as an antioxidant in food and in vitamin C deficiency therapy.

sodium aurichloride see SODIUM GOLD CHLORIDE.

sodium aurothiomalate or **sodium aurothiosulfate** see GOLD SODIUM THIOSULFATE.

sodium azide *Inorganic Chemistry.* NaN_3, highly toxic, colorless crystals that decompose on heating; soluble in water and slightly soluble in alcohol; used for air bag inflation and in preservatives and explosives.

sodium barbiturate *Organic Chemistry.* $C_4H_3N_2O_3Na$, a toxic, white to yellow powder; soluble in water and in dilute mineral acid; used as an intermediate and a catalyst.

sodium benzoate *Organic Chemistry.* C_6H_5COONa, a combustible, colorless crystalline solid or white granular powder; soluble in water and alcohol; used as a food preservative, as an intermediate, and in medicine, pharmaceuticals, and tobacco.

sodium benzoylacetone dihydrate *Organic Chemistry.* a metal chelate that melts at 115°C; soluble in acetone.

sodium bicarbonate *Inorganic Chemistry.* $NaHCO_3$, a white powder or crystalline lumps; loses carbon dioxide on heating to 270°C and decomposes slowly in moist air; soluble in water and insoluble in alcohol. It is widely used as an antacid, in baking powder, in effervescent beverages, in fire extinguishers, and for many other purposes. Also, BAKING SODA, BICARBONATE OF SODA, SODIUM ACID CARBONATE.

sodium bichromate see SODIUM DICHROMATE.

sodium bifluoride *Inorganic Chemistry.* $NaHF_2$, a colorless or white crystalline powder that decomposes on heating; soluble in water; highly toxic and a strong irritant; used as an antiseptic and preservative.

sodium biphosphate see MONOBASIC SODIUM PHOSPHATE.

sodium bismuthate *Inorganic Chemistry.* $NaBiO_3$, a yellow or brown amorphous powder, insoluble in cold water and decomposes in hot water and acids; melts at 92.5°C and boils at 200°C; used in pharmaceuticals and as a reagent. Also, SODIUM METABISMUTHATE.

sodium bisulfate *Inorganic Chemistry.* $NaHSO_4$ or $NaHSO_4 \cdot H_2O$, colorless crystals or white fused lumps; acidic in solution and a strong irritant; the anhydrous form melts above 315°C and the hydrated form melts at about 59°C; used in manufacturing various industrial products and in many chemical processes. Also, SODIUM ACID SULFATE, SODIUM HYDROGEN SULFATE, NITER CAKE.

sodium bisulfide *Inorganic Chemistry.* $NaSH \cdot 2H_2O$, colorless needles or lemon-yellow flakes that give off toxic hydrogen sulfide when combined with acids; soluble in water and alcohol; decomposed by heat; used in manufacturing artificial fabrics and as a reagent. Also, SODIUM HYDROSULFIDE, SODIUM HYDROGEN SULFIDE, SODIUM SULFHYDRATE.

sodium bisulfite *Inorganic Chemistry.* $NaHSO_3$, white crystals or a crystalline powder; decomposed by air or heat; very soluble in water; used in textiles, bleaches, and antiseptics, as a reducing agent or reagent, and for many other purposes. Also, SODIUM ACID SULFITE, SODIUM HYDROGEN SULFITE.

sodium bisulfite test *Analytical Chemistry.* the addition of sodium bisulfite to a sample for the purpose of detecting aldehydes; a positive test is indicated by the formation of a white crystalline precipitate.

sodium bitartrate *Organic Chemistry.* $NaHC_4H_5O_6 \cdot H_2O$, a combustible, white crystalline powder; soluble in water and slightly soluble in alcohol; loses water at 110°C and decomposes at 219°C; used in analysis and as a nutrient medium. Also, SODIUM ACID TARTRATE.

sodium borate *Inorganic Chemistry.* $Na_2B_4O_7 \cdot 10H_2O$, toxic white crystals that lose water when heated and fuse into a glassy mass (borax glass); very soluble in water and insoluble in alcohol; used in bleaches, detergents, cleansers, fertilizers, herbicides, and for many other purposes. Also, SODIUM TETRABORATE, SODIUM PYROBORATE.

sodium borate hydrate see BORAX.

sodium borohydride *Inorganic Chemistry.* $NaBH_4$, a white to gray crystalline powder; decomposes when heated to 400°C; decomposes in water with the liberation of hydrogen; a dangerous fire hazard; used in many chemical and industrial processes.

sodium bromate *Inorganic Chemistry.* $NaBrO_3$, toxic white crystals or powder; soluble in water and insoluble in alcohol; decomposes when heated to 381°C; an oxidant and dangerous fire hazard in contact with organic substances; used as an analytical reagent.

sodium bromide *Inorganic Chemistry.* NaBr or $NaBr \cdot 2H_2O$, a toxic white crystalline powder that hardens after absorbing water from the air; soluble in water and slightly soluble in alcohol; the anhydrous form melts at 747°C and boils at 1390°C; used in photography and medicine and to produce bromides.

sodium carbolate see SODIUM PHENATE.

sodium carbonate *Inorganic Chemistry.* **1.** Na_2CO_3, the disodium salt of carbonic acid, especially its commercial form, known as soda ash. See SODA ASH. **2.** any of various other carbonates of sodium. See SODIUM BICARBONATE, SODIUM CARBONATE MONOHYDRATE, SODIUM SESQUICARBONATE, and so on.

sodium carbonate decahydrate *Inorganic Chemistry.* $Na_2CO_3 \cdot 10H_2O$, white crystals that are very soluble in water and insoluble in alcohol; melts at 32.5–34.5°C and loses water at this temperature; a pure form of sodium carbonate that is widely used as a general cleanser, for washing and bleaching, and also in pharmaceuticals. Also, SAL SODA, WASHING SODA.

sodium carbonate monohydrate *Inorganic Chemistry.* $Na_2CO_3 \cdot H_2O$, white crystals or powder; loses water on heating to 100°C; soluble in water and insoluble in alcohol; used in photography, in glass manufacture, as a cleanser, and for various other purposes. Also, SODA MONOHYDRATE, CRYSTAL CARBONATE, SODA CRYSTALS.

sodium carbonate peroxide *Inorganic Chemistry.* $2Na_2CO_3 \cdot 3H_2O_2$, a white crystalline powder that decomposes on heating to give oxygen; soluble in water and decomposes rapidly at 100°C; an oxidant and a fire hazard in contact with organic substances; used as a source of hydrogen peroxide and in bleaches, cleansers, detergents, and dyes.

sodium caseinate *Organic Chemistry.* a coarse, white powder that is soluble in water; used as a food additive, emulsifier, and stablilizer.

sodium chlorate *Inorganic Chemistry.* $NaClO_3$, colorless, odorless crystals; decomposes on heating to the boiling point and melts at 248–261°C; a strong oxidant and dangerous fire hazard in contact with organic substances. It is used as an oxidizing agent and bleach, in matches, explosives, and fireworks, and for many other purposes.

sodium chloride *Inorganic Chemistry.* NaCl, common table salt; colorless transparent crystals or a white crystalline powder; soluble in water and absorbs some water from the air; melts at 801°C and boils at 1413°C. In addition to its ancient and universal use for seasoning and preserving food, it is used in producing other chemicals such as soda ash and chlorine, in ceramic glazes, in deicing and water softening, and for a variety of purposes in medicine.

sodium chloride solution see NORMAL SALINE.

sodium chloride structure *Materials Science.* a crystal structure having a face-centered cubic unit cell with a two-ion motif (one sodium and one chloride ion) at each lattice point.

sodium chlorite *Inorganic Chemistry.* $NaClO_2$, white crystals or a crystalline powder that absorbs some water from the air and decomposes on heating to 180–200°C; it is a dangerous fire hazard and explosive, and a strong irritant in solution; used as an oxidizing agent and reagent.

sodium chloroacetate *Organic Chemistry.* $ClCH_2COONa$, a nonhygroscopic, noncombustible, white powder; soluble in water; used in the manufacture of dyes, pharmaceuticals, and weed killers, and as a defoliant.

sodium chloroplatinate *Inorganic Chemistry.* Na_2PtCl_6, yellow crystals or powder; absorbs water from the air; soluble in water and alcohol; used in etching and photography and as a catalyst. Also, SODIUM PLATINICHLORIDE, SODIUM HEXACHLOROPLATINATE.

sodium chromate *Inorganic Chemistry.* $Na_2CrO_4\cdot 10H_2O$, toxic, translucent, yellow crystals; soluble in water and loses water to the air; melts at 19.9°C; used in inks, dyes, and leather tanning.

sodium citrate *Organic Chemistry.* $C_6H_5O_7Na_3\cdot 2H_2O$, combustible, white crystals or granular powder; soluble in water and insoluble in alcohol; loses water at 150°C and decomposes with strong heating; used in making cheeses, soft drinks, and medicines.

sodium cobaltinitrite *Inorganic Chemistry.* $Na_3Co(NO_2)_6$, a yellow to brownish crystalline powder; very soluble and decomposes in cold water; used as a reagent.

sodium-cooled reactor *Nucleonics.* a nuclear reactor that uses liquid metallic sodium as a coolant.

sodium cyanate *Inorganic Chemistry.* NaCNO, a white powder that is soluble in water and insoluble in alcohol; decomposes in a vacuum at 700°C; used in organic synthesis, in steel fabrication, and in various medicines.

sodium cyanide *Inorganic Chemistry.* NaCN, a highly toxic, white, crystalline powder; soluble in water and slightly soluble in alcohol; strongly basic in solution; decomposes on standing; melts at 563.7°C and boils at 1496°C; used in electroplating, in treating metals, in extracting gold and silver, and for various other industrial purposes.

sodium cyanoaurite see SODIUM GOLD CYANIDE.

sodium cyclamate *Organic Chemistry.* $C_6H_{11}NHSO_3Na$, white crystals that are freely soluble in water and practically insoluble in alcohol; it is thirty times sweeter than sucrose and was once used as a nonnutritive sweetener. It is a suspected carcinogen and is now prohibited for food use in the U.S.

sodium dehydroacetate *Organic Chemistry.* $C_8H_7O_4Na\cdot H_2O$, a white powder that is soluble in water and insoluble in most organic solvents; used in pharmaceuticals and as a fungicide, plasticizer, and food preservative.

sodium diacetate *Organic Chemistry.* $CH_3COONa\cdot x(CH_3COOH)$, combustible, white crystals; soluble in water and slightly soluble in alcohol; decomposes above 150°C; used as a buffer, varnish hardener, food preservative, mordant, and mold inhibitor.

sodium diatrizoate *Organic Chemistry.* $C_6I_3(COONa)(NHCOCH_3)_2$, white crystals that are soluble in water; used in medicine as a radiopaque medium.

sodium dichloroisocyanurate *Organic Chemistry.* $C_3N_3O_3Cl_2Na$, white crystals that are soluble in water; used in swimming pools as an algicide and bactericide.

sodium dichromate *Inorganic Chemistry.* $Na_2Cr_2O_7\cdot 2H_2O$, red-orange crystals that absorb water from the air; soluble in water and insoluble in alcohol; loses water at 100°C and decomposes at 400°C; toxic and an irritant; used in colorimetry and as a complexing agent and oxidation inhibitor. Also, SODIUM BICHROMATE, SODIUM ACID CHROMATE.

sodium diethyldithiocarbamate *Organic Chemistry.* $(C_2H_5)_2NCS_2Na$, a solid that is soluble in alcohol and water; its trihydrate is used to detect small amounts of copper and to separate copper from other metals.

sodium dihydrogen phosphate see MONOBASIC SODIUM PHOSPHATE.

sodium dimethyldithiocarbamate *Organic Chemistry.* $(CH_3)_2NCS_2Na$, an amber to light-green liquid; used as a fungicide, a rubber accelerator, and an intermediate.

sodium dinitro-*o*-cresylate *Organic Chemistry.* $CH_3C_6H_2(NO_2)_2ONa$, a toxic, orange-yellow dye; used as a fungicide and an herbicide.

sodium dithionite see SODIUM HYDROSULFITE.

sodium diuranate *Organic Chemistry.* $Na_2U_2O_7\cdot 6H_2O$, a yellow-orange solid that is insoluble in water and soluble in dilute acids; used in making fluorescent uranium glass and for glazing ceramics. Also, URANIUM YELLOW.

sodium dodecyl benzenesulfonate *Organic Chemistry.* $C_{12}H_{25}C_6H_4SO_3Na$, combustible, biodegradable, white to light-yellow flakes, granules, or powder; used as an anionic detergent.

sodium dodecyl sulfate-polyacrylamide gel electrophoresis see SDS-PAGE.

sodium ethoxide see SODIUM ETHYLATE.

sodium ethylate *Organic Chemistry.* C_2H_5ONa, a white powder that readily hydrolyzes to alcohol and sodium hydroxide; synthesized from ethanol; important in organic synthesis. Also, SODIUM ETHOXIDE, CAUSTIC ALCOHOL.

sodium 2-ethylhexyl sulfoacetate *Organic Chemistry.* $C_8H_{17}OOCCH_2SO_3Na$, combustible, light cream-colored flakes that are soluble in water and have good foaming properties; used as a solubilizing agent in soapless shampoos and electroplating detergents.

sodium ethyl xanthate *Organic Chemistry.* $C_2H_5OC(S)SNa$, a yellowish powder that is soluble in water and alcohol; used as an ore flotation material. Also, SODIUM XANTHOGENATE.

sodium ferricyanide *Inorganic Chemistry.* $Na_3Fe(CN)_6\cdot H_2O$, red crystals that absorb water from the air; soluble in water and insoluble in alcohol; used in pigments, dyeing, and printing. Also, RED PRUSSIATE OF SODA.

sodium ferrocyanide *Inorganic Chemistry.* $Na_4Fe(CN)_6\cdot 10H_2O$, yellow crystals that lose water to the air; used in the manufacture of sodium ferricyanide and in various industrial processes. Also, YELLOW PRUSSIATE OF SODA.

sodium fluorescein see URANINE.

sodium fluoride *Inorganic Chemistry.* NaF, toxic and strongly irritant clear crystals or white powder, sometimes dyed blue; soluble in water; melts at 993°C and boils at 1695°C; widely used in the fluoridation of water, in toothpastes and other dental preparations, as an insecticide and fungicide, and for many other purposes.

sodium fluoroacetate *Organic Chemistry.* FCH_2COONa, a toxic, nonvolatile, fine white powder; soluble in water and insoluble in most organic solvents; decomposes at 200°C; restricted use as a rodenticide and for predator elimination.

sodium fluoroborate *Inorganic Chemistry.* $NaBF_4$, a white powder that decomposes on heating; soluble in water and slightly soluble in alcohol; used in electrochemical processes and as a fluorinating agent. Also, **sodium fluoborate.**

sodium fluorosilicate *Inorganic Chemistry.* Na_2SiF_6, a colorless to white, free-flowing powder that is soluble in water and decomposed by heat; it is toxic and a strong irritant; it is used in fluoridation, in metallurgy, as a preservative, and for many other purposes. Also, **sodium fluosilicate.**

sodium folate *Organic Chemistry.* $C_{19}H_{18}N_7O_6Na$, an orange liquid; soluble in water; used medicinally in folic acid therapy. Also, FOLIC ACID SODIUM SALT.

sodium formaldehyde sulfoxylate *Organic Chemistry.* $NaHSO_2\cdot CH_2O\cdot 2H_2O$, a hygroscopic, white crystalline solid; soluble in water and alcohol; melts at 64°C; used as a bleaching agent for soap and molasses, and as a textile stripping agent.

sodium formate *Organic Chemistry.* $NaCHO_2$, a slightly hygroscopic, white crystalline powder; soluble in water, slightly soluble in alcohol, and insoluble in ether; melts at 253°C; used as a reducing agent, reagent, mordant, and buffering agent.

sodium glucoheptonate *Organic Chemistry.* $HOCH_2(CHOH)_5COONa$, a light tan crystalline powder used in mercerizing cotton cloth, cleaning metals, stripping paint, and etching aluminum.

sodium gluconate *Organic Chemistry.* $C_6H_{11}O_7Na$, a white to yellowish crystalline powder that is soluble in water and slightly soluble in alcohol; used as a metal cleaner and in pharmaceuticals and foods.

sodium glutamate see MONOSODIUM GLUTAMATE.

sodium gold chloride *Inorganic Chemistry.* $NaAuCl_4\cdot 2H_2O$, a yellow crystalline powder, soluble in water and alcohol; used to color glass and decorate porcelain. Also, SODIUM AURICHLORIDE.

sodium gold cyanide *Inorganic Chemistry.* $NaAu(CN)_2$, a toxic yellow powder; soluble in water; used for goldplating jewelry, electronics, and tableware. Also, SODIUM CYANOAURITE.

sodium-graphite reactor *Nucleonics.* a nuclear reactor that operates with liquid metallic sodium as a coolant and graphite as a moderator material.

sodium halide *Inorganic Chemistry.* any compound consisting of sodium with an element from the halogen group (periodic table Group VIIa), such as fluorine or chlorine.

sodium halometallate *Inorganic Chemistry.* any compound composed of sodium with a metal and an element from the halogen group (periodic table Group VIIa).

sodium hexachloroplatinate see SODIUM CHLOROPLATINATE.

sodium hexylene glycol monoborate *Organic Chemistry.* $C_6H_{12}O_3BNa$, an amorphous white solid that is soluble in nonpolar solvents and melts at 426°C; used as a flame retardant, corrosion inhibitor, and lubricating oil additive.

sodium hydride *Inorganic Chemistry.* NaH, a silvery-white powder that reacts violently with water to produce hydrogen; decomposes at 800°C; a dangerous fire hazard; used as a condensing and alkylating agent and for descaling metals.

sodium hydrophosphate or **sodium hydrogen phosphate** see MONOBASIC SODIUM PHOSPHATE.

sodium hydrosulfate or **sodium hydrogen sulfate** see SODIUM BISULFATE.

sodium hydrosulfide or **sodium hydrogen sulfide** see SODIUM BISULFIDE.

sodium hydrosulfite or **sodium hydrogen sulfite** see SODIUM BISULFITE.

sodium hydroxide *Inorganic Chemistry.* NaOH, a white solid that absorbs water and carbon dioxide from the air; soluble in water and alcohol; melts at 318.4°C and boils at 1390°C; corrosive, toxic, and a strong irritant. One of the more widely produced chemicals in the U.S., it is used in chemical manufacture, soaps and detergents, food processing, pharmaceuticals, and for various other purposes. Also, **sodium hydrate.**

sodium hypochlorite *Inorganic Chemistry.* $NaOCl \cdot 5H_2O$, pale greenish crystals having a strong odor; decomposes in air or hot water; melts at 18°C; a strong oxidant and dangerous fire hazard in contact with organic substances; used as a germicide, disinfectant, fungicide, water purifier, and for various other purposes.

sodium hypophosphite *Inorganic Chemistry.* $NaH_2PO_2 \cdot H_2O$, white granules that absorb water from the air and decompose on heating to give toxic, spontaneously flammable phosphine gas; explosive in contact with oxidizing materials; used in pharmaceuticals and nickel plating, and as a reagent and a meat preservative.

sodium iodate *Inorganic Chemistry.* $NaIO_3$, a white crystalline powder that is soluble in water and insoluble in alcohol; an oxidant and a fire hazard in contact with organic substances; used in antiseptics, disinfectants, and feed additives.

sodium iodide *Inorganic Chemistry.* NaI, colorless to white crystals or powder that turns brown when exposed to air and absorbs water from the atmosphere; soluble in water and alcohol; used in photography, as an expectorant, and as a feed additive and reagent.

sodium iodide scintillator *Nucleonics.* a crystalline substance used to study gamma-ray energy by scintillation; the crystal is composed of sodium iodide activated with thallium.

sodium isopropyl xanthate *Organic Chemistry.* $(CH_3)_2CHOC(S)SNa$, deliquescent light-yellow crystals; soluble in water; decomposes at 150°C; used as an herbicide and a flotation agent for ores.

sodium lactate *Organic Chemistry.* $CH_3CHOHCOONa$, a combustible, very hygroscopic, colorless or yellowish syrupy liquid; soluble in water; melts at 17°C and decomposes at 140°C; used as a hygroscopic agent, glycerol substitute, plasticizer, and corrosion inhibitor in antifreeze.

sodium lauryl sulfate *Organic Chemistry.* $C_{12}H_{25}SO_4Na$, small white or light-yellow crystals; soluble in water; used as a detergent, wetting agent, and food additive.

sodium lead thiosulfate or **sodium lead hyposulfate** see LEAD SODIUM THIOSULFATE.

sodium mercaptoacetate see SODIUM THIOGLYCOLATE.

sodium meta-antimonate see SODIUM ANTIMONATE.

sodium meta-arsenite see SODIUM ARSENITE.

sodium metabismuthate see SODIUM BISMUTHATE.

sodium metaborate *Inorganic Chemistry.* $NaBO_2$, white lumps; soluble in water; melts at 966°C and boils at 1434°C; used as an herbicide.

sodium metaphosphate *Inorganic Chemistry.* $(NaPO_3)_n$, a series of cyclic and polymeric compounds, such as **sodium trimetaphosphate,** $(NaPO_3)_3$, or **sodium hexametaphosphate,** $(NaPO_3)_6$, having various industrial uses, as in dentistry and in food and textile processing.

sodium metasilicate *Inorganic Chemistry.* Na_2SiO_3, colorless crystals that are soluble in water and melt at 1088°C; used in soaps, detergents, and industrial cleansers.

sodium metavanadate *Inorganic Chemistry.* $NaVO_3$, toxic, colorless crystals or a pale green crystalline powder; soluble in water; melts at 630°C; used in photography, in inks and dyes, and as a mordant and fixer.

sodium methiodal *Organic Chemistry.* ICH_2SO_3Na, a white crystalline powder; soluble in water, very soluble in methanol, and practically insoluble in ether; decomposes on exposure to light; used as a radiopaque contrast medium.

sodium methoxide *Organic Chemistry.* $CH_3ONa \cdot 2CH_3OH$, a white powder that is soluble in water and alcohol; used in organic synthesis. Also, **sodium methylate.**

sodium *N*-methyldithiocarbamate dihydrate *Organic Chemistry.* $CH_3NHC(S)SNa \cdot 2H_2O$, a toxic, white crystalline solid; soluble in water and alcohol, and decomposes in dilute aqueous solution; used as a fungicide, pesticide, herbicide, and soil fumigant.

sodium molybdate *Inorganic Chemistry.* **1.** Na_2MoO_4, often of variable composition; small opaque crystals, soluble in water; melts at 687°C; an irritant; used as a reagent and pigment and in a variety of chemical processes. **2.** $Na_2MoO_4 \cdot 2H_2O$, the dihydrate form, which is used commercially.

sodium 12-molybdophosphate *Inorganic Chemistry.* $Na_3PMo_{12}O_{40}$, white or yellow crystals; soluble in water and a strong oxidant in solution; used as a catalyst and pigment and in various industrial processes. Also, SODIUM PHOSPHO-12-MOLYBDATE.

sodium monoxide *Inorganic Chemistry.* Na_2O, a caustic white powder; sublimes when heated to 1275°C; reacts with water to give sodium hydroxide; strongly basic, caustic, and a strong irritant; used as a polymerizing and dehydrating agent. Also, SODIUM OXIDE.

sodium motive force *Biochemistry.* a source of energy similar in function to proton motive force; the energy connected with a transmembrane electrochemical gradient of sodium ions.

sodium naphthalenesulfonate *Organic Chemistry.* $NaC_{10}H_7SO_3$, combustible, toxic, yellowish crystalline plates or white scales; soluble in water and insoluble in alcohol; used as a liquefying agent and in organic synthesis.

sodium naphthionate *Organic Chemistry.* $NaC_{10}H_6(NH_2)SO_3$, a white crystalline or yellow granular solid; soluble in water, alcohol, and ether; melts at 232°C; used in chemical analyses for nitrous acid. Also, RIEGLER'S REAGENT.

sodium nitrate *Inorganic Chemistry.* $NaNO_3$, toxic, colorless, transparent crystals with a slightly bitter taste; soluble in water and alcohol; melts at 306.8°C, decomposes at 380°C, and explodes when heated to 537°C; a fire risk in contact with organic substances; used in matches, explosives, and fireworks, in glass manufacture, in fertilizers, and as a preservative and color fixative for cured meats, and formerly used in medicine. Also, SODA NITER.

sodium nitrite *Inorganic Chemistry.* $NaNO_2$, colorless to yellowish crystals or powder; oxidized by air and soluble in water; melts at 271°C, decomposes at 320°C, and explodes when heated to 537°C. It is used as a pain reliever and poison antidote, in photography, and as an analytical reagent; formerly widely used in cured meat and fish products, but now restricted because of carcinogenic results in laboratory animals.

sodium nitroferricyanide *Inorganic Chemistry.* $Na_2Fe(CN)_5NO \cdot 2H_2O$, red transparent crystals that decompose slowly in water; used as an analytical reagent. Also, **sodium nitroprussiate, sodium nitroprusside.**

sodium oleate *Organic Chemistry.* $C_{17}H_{33}COONa$, a combustible white powder; soluble in water with slight decomposition, and soluble in alcohol; used in ore flotation and waterproofing textiles, and as an emulsifier.

sodium orthoarsenate see SODIUM ARSENATE.

sodium orthophosphate see MONOBASIC SODIUM PHOSPHATE, DIBASIC SODIUM PHOSPHATE, TRIBASIC SODIUM PHOSPHATE.

sodium oxalate *Organic Chemistry.* $Na_2C_2O_4$, a toxic, white crystalline powder; soluble in water and insoluble in alcohol; melts (decomposes) at 250–270°C; used as a reagent and in tanning leather, textile finishing, and blue printing.

sodium oxide see SODIUM MONOXIDE.

sodium paraperiodate *Inorganic Chemistry.* $Na_3H_2IO_6$, a white crystalline solid that is slightly soluble in water; used as a selective oxidant and in processing paper and tobacco. Also, SODIUM TRIPARAPERIODATE.

sodium pentaborate *Inorganic Chemistry.* $Na_2B_{10}O_{16}\cdot 10H_2O$, free-flowing white crystals or powder; used in glassmaking, fireproofing, and as an herbicide. Also, **sodium pentaborate decahydrate.**

sodium pentachlorophenate *Organic Chemistry.* C_6Cl_5ONa, a toxic, white or tan powder; soluble in water, alcohol, and acetone; used as a fungicide and herbicide.

sodium perborate *Inorganic Chemistry.* $NaBO_3$, a toxic white powder or crystals; a strong oxidizing agent, and a fire hazard in contact with organic materials; used in denture cleaners and bleaching agents, and as a source of oxygen.

sodium perchlorate *Inorganic Chemistry.* $NaClO_4$, white crystals that absorb water from the air; soluble in water and alcohol; decomposes at 482°C; a dangerous fire and explosive hazard in contact with organic materials or sulfuric acid; used in explosives and jet fuel, and as an analytical reagent.

sodium permanganate *Inorganic Chemistry.* $NaMnO_4\cdot 3H_2O$, toxic purple crystals or powder; very soluble in water and decomposes on heating to 170°C; a strong oxidant and a dangerous fire hazard in contact with organic substances; used as a poison antidote, oxidizing agent, disinfectant, and in saccharin manufacture.

sodium peroxide *Inorganic Chemistry.* Na_2O_2, a white to yellowish powder that absorbs water and carbon dioxide from the air; soluble in cold water and decomposes in hot water; on heating it turns yellow and decomposes.It is an irritant, a strong oxidant, and a dangerous fire and explosive hazard in contact with water and many other substances. It is used in bleaches, pharmaceuticals, deodorants, and antiseptics, in medicinal soaps, and for various industrial purposes.

sodium persulfate *Inorganic Chemistry.* $Na_2S_2O_8$, a white crystalline powder that is soluble in water and decomposes in moist air; a strong irritant; used as a bleaching agent and in batteries and polymerization. Also, **sodium peroxydisulfate.**

sodium phenate *Organic Chemistry.* C_6H_5ONa, white crystals, deliquescent, that are soluble in water and alcohol, and decomposed by carbon dioxide in the air; used as an antiseptic and in organic synthesis. Also, SODIUM CARBOLATE.

sodium phenylacetate *Organic Chemistry.* $C_6H_5CH_2\cdot COONa$, a pale yellow solution, soluble in water and insoluble in alcohol; used in the production of penicillin G and as an intermediate for producing heavy metal salts.

sodium phenylphosphinate *Organic Chemistry.* $C_6H_5PH(O)(ONa)$, crystals that are soluble in water and melt at 355°C; used as an antioxidant and a heat and light stabilizer.

sodium phosphate *Inorganic Chemistry.* a general term applied to a wide variety of compounds of sodium and phosphate. See MONOBASIC SODIUM PHOSPHATE, DIBASIC SODIUM PHOSPHATE, TRIBASIC SODIUM PHOSPHATE, SODIUM METAPHOSPHATE, SODIUM PYROPHOSPHATE, and SODIUM TRIPOLYPHOSPHATE.

sodium phosphite *Inorganic Chemistry.* $Na_2HPO_3\cdot 5H_2O$, a white crystalline powder that absorbs water from the air and decomposes on heating to 200–250°C; very soluble in hot water and insoluble in alcohol; used as a poison antidote.

sodium phospho-12-molybdate see SODIUM 12-MOLYBDOPHOSPHATE.

sodium phosphotungstate see SODIUM TUNGSTOPHOSPHATE.

sodium phytate *Organic Chemistry.* $C_6H_9O_{24}P_6Na_9$, a hygroscopic powder; soluble in water; used as a chelating agent and in medicine. Also, INOSITOLHEXAPHOSPHORIC ESTER, SODIUM SALT.

sodium picramate *Organic Chemistry.* $NaOC_6H_2(NO_2)2NH_2$, a toxic, yellow crystalline solid; soluble in water; derived from the neutralization of picramic acid with caustic soda; used in dye manufacture and organic synthesis.

sodium platinichloride see SODIUM CHLOROPLATINATE.

sodium plumbite *Inorganic Chemistry.* Na_2PbO_2, a solution of lead oxide, PbO, in sodium hydroxide, NaOH; highly toxic and corrosive; used to improve the odor of petroleum products.

sodium polymannuronate see SODIUM ALGINATE.

sodium polysulfide *Inorganic Chemistry.* Na_2S_x, a yellow-brown, granular polymer; used in dyes, insecticides, synthetic rubber, petroleum additives, and electroplating.

sodium-potassium ATPase *Enzymology.* the adenosine triphosphatase found in cell membranes that functions in the active transport of sodium and potassium ions.

sodium-potassium pump *Molecular Biology.* a membrane protein channel through which sodium ions are moved from cells and replaced with potassium ions. Also, **sodium pump.**

sodium propionate *Organic Chemistry.* $C_3H_5O_2Na$, a white granular powder that is soluble in water and alcohol; used as a fungicide and mold inhibitor.

sodium pyroantimonate see SODIUM ANTIMONATE.

sodium pyroborate see SODIUM BORATE.

sodium pyrophosphate *Inorganic Chemistry.* **1.** $Na_4P_2O_7$, toxic white crystals, soluble in water; melts at 880°C. **2.** $Na_4P_2O_7\cdot 10H_2O$, the decahydrate form, toxic colorless crystals; soluble in water and insoluble in alcohol; loses water at 93.8°C; used in detergents, water softeners, rubber manufacture, and textile dyeing. Also, NORMAL SODIUM PYROPHOSPHATE. **3.** see SODIUM ACID PYROPHOSPHATE.

sodium rhodanate or **sodium rhodanide** see SODIUM THIOCYANATE.

sodium saccharin *Organic Chemistry.* $C_7H_4NNaO_3S\cdot 2H_2O$, a white crystalline or powdery solid; very soluble in water and slightly soluble in alcohol; it is 500 times sweeter than sugar and is used as a nonnutritive sweetener, but is a possible carcinogen.

sodium salicylate *Organic Chemistry.* HOC_6H_4COONa, combustible, lustrous, white crystalline scales or amorphous powder, soluble in water and alcohol; used as a food preservative and in pharmaceuticals.

sodium selenate *Inorganic Chemistry.* $Na_2SeO_4\cdot 10H_2O$, toxic white crystals that are soluble in water; used as a reagent and insecticide.

sodium selenite *Inorganic Chemistry.* $Na_2SeO_3\cdot 5H_2O$, toxic white crystals that are soluble in water and insoluble in alcohol; used in glass, as a reagent, to test seeds, and to decorate porcelain.

sodium sesquicarbonate *Inorganic Chemistry.* $Na_2CO_3\cdot NaHCO_3\cdot 2H_2O$, colorless to white crystals that are soluble in water and decompose on heating; an irritant; used in detergents, soaps, and cleansers.

sodium sesquisilicate *Inorganic Chemistry.* $Na_6Si_2O_7$, a toxic white granular powder that is soluble in water; used in industrial cleaning and in textile processing.

sodium silicate *Inorganic Chemistry.* **1.** any of various compounds of sodium, silicon, and oxygen having the general formula $xNa_2O\cdot ySiO_2$, often hydrated. Sodium silicates have various physical properties and are used in soaps, detergents, bleaches, and adhesives, in medicine, and for a wide range of industrial purposes. Also, SOLUBLE GLASS, WATER GLASS. **2.** see SODIUM METASILICATE. **3.** see SODIUM SESQUISILICATE.

sodium silicoaluminate see SODIUM ALUMINOSILICATE.

sodium silicofluoride see SODIUM FLUOROSILICATE.

sodium stannate *Inorganic Chemistry.* $Na_2SnO_3\cdot 3H_2O$ or $Na_2Sn(OH_6)$, toxic, colorless to whitish crystals; decomposes in air and loses water on heating to 140°C; used in glassmaking, ceramics, as a mordant, and for various other purposes.

sodium stearate *Organic Chemistry.* $C_{18}H_{35}O_2Na$, a white powder; soluble in hot water and hot alcohol, slowly soluble in cold water and cold alcohol, and insoluble in many organic solvents; used in toothpastes and cosmetics, and as a waterproofing agent and stablizer.

sodium subsulfite see SODIUM THIOSULFATE.

sodium succinate *Organic Chemistry.* $C_4H_4O_4Na_2\cdot 6H_2O$, a white granular or powdery solid that is soluble in water and alcohol; loses water at 120°C; used in pharmaceuticals.

sodium sulfate *Inorganic Chemistry.* Na_2SO_4, white crystals or powder; soluble in water and insoluble in alcohol; melts at 884°C; a widely produced chemical used to manufacture glass, paper, pharmaceuticals, and textiles, as an analytical reagent, and for various other purposes.

sodium sulfate decahydrate or **sodium sulfate crystals** see GLAUBER'S SALT.

sodium sulfhydrate see SODIUM BISULFIDE.

sodium sulfide *Inorganic Chemistry.* Na_2S, yellow to red lumps, flakes, or crystals; absorbs water from the air; soluble in water and slightly soluble in alcohol; a strong irritant and a dangerous fire and explosion hazard; reacts with acids to give toxic hydrogen sulfide gas. It is used in dyes, synthetic fabrics, leather and paper making, and for various other purposes. Also, SODIUM SULFURET.

sodium sulfite *Inorganic Chemistry.* Na_2SO_3, white crystals or powder; loses water and decomposes on heating; used in photography, paper making, water treatment, and for various other purposes. Also, **sodium sulfuret.**

sodium sulfite process *Chemical Engineering.* a chemical wood pulping process that uses a pulping liquor of sulfuric acid and a salt of sulfurous acid to remove the lignin portion of wood.

sodium sulfocyanate or **sodium sulfocyanide** see SODIUM THIOCYANATE.

sodium tartrate *Organic Chemistry.* $C_4H_4O_6Na_2{\cdot}2H_2O$, white crystals or granules; soluble in water and insoluble in alcohol; loses $2H_2O$ at 150°C; used as a reagent, food additive, and stabilizer. Also, SAL TARTAR.

sodium TCA see SODIUM TRICHLOROACETATE.

sodium tetraborate see SODIUM BORATE.

sodium tetrafluorescein. see EASIN.

sodium tetraphenylborate *Organic Chemistry.* $[(C_6H_5)_4B]Na$, white crystals that are soluble in water and acetone; used as a reagent for detemining alkali metal ion.

sodium tetrasulfide *Inorganic Chemistry.* Na_2S_4, yellow crystals that absorb water from the air; soluble in water and alcohol and decomposed by heat; an irritant and a fire risk in contact with flame; used in insecticides and fungicides and in various chemical processes.

sodium thiocyanate *Inorganic Chemistry.* NaSCN, light-sensitive, colorless crystals that absorb water from the air; soluble in water; melts at 287°C; used in dyeing and printing and as a reagent.

sodium thioglycolate *Organic Chemistry.* $HSCH_2COONa$, combustible hygroscopic crystals; soluble in water and slightly soluble in alcohol; used in bacteriology and as a depilatory and reagent. Also, SODIUM MERCAPTOACETATE.

sodium thiosulfate *Inorganic Chemistry.* $Na_2S_2O_3{\cdot}5H_2O$, a white translucent crystalline powder that absorbs water in moist air, loses water in warm, dry air, and decomposes on heating; soluble in water and insoluble in alcohol; used in photography and medicine, in extracting silver from its ores, and as a bleach, mordant, and reagent. Also, SODIUM SUBSULFITE.

sodium *p*-toluenesulfochloramine see CHLORAMINE-T.

sodium trichloroacetate *Organic Chemistry.* CCl_3COONa, a toxic substance used as a pesticide and herbicide.

sodium 2,4,5-trichlorophenate *Organic Chemistry.* $C_6H_2Cl_3ONa{\cdot}1.5HOH$, buff to light-brown flakes; soluble in water, acetone, and methanol; used as an industrial preservative.

sodium triparaperiodate see SODIUM PARAPERIODATE.

sodium tripolyphosphate *Inorganic Chemistry.* $Na_5P_3O_{10}$, a white granular powder or crystals; dissolves in water and decomposes on heating; used in water softening and in various chemical treatments.

sodium tungstate *Inorganic Chemistry.* $Na_2WO_4{\cdot}2H_2O$, colorless crystals that are soluble in water and insoluble in alcohol; loses water on heating to 100°C and melts at 698°C; used in the preparation of tungsten compounds, in fireproofing, and as a reagent and precipitant. Also, SODIUM WOLFRAMATE.

sodium tungstophosphate *Inorganic Chemistry.* $2Na_2O{\cdot}P_2O_5{\cdot}12WO_3{\cdot}18H_2O$, a yellowish-white granular powder; very soluble in water and alcohol; used as a reagent, in fur and leather treatment, and in making water-resistant films, cements, and adhesives. Also, SODIUM PHOSPHOTUNGSTATE.

sodium undecylenate *Organic Chemistry.* $CH_2{=}CH(CH_2)_8COONa$, a combustible white powder; soluble in water and decomposes above 200°C; used in pharmaceuticals and cosmetics.

sodium-vapor lamp *Electronics.* a gas-discharge lamp containing sodium vapor; used mainly for highway illumination.

sodium wolframate see SODIUM TUNGSTATE.

sodium xanthate or **sodium xanthogenate** see SODIUM ETHYL XANTHATE.

sofar *Acoustical Engineering.* the technique of using explosions for tracking sound through deep ocean layers and in sound channels. (An acronym for sound fixing and ranging.)

soffione *Geology.* a jet of steam, usually together with other vapors, that is emitted from vents or fissures in a volcanic area.

soffit *Architecture.* the underside of a structural member or other feature, such as a beam, stairway, or arch; especially the highest part of the underside of an arch.

soffosion knob see FROST MOUND.

soft not hard; specific uses include: *Materials.* **1.** easily yielding to pressure. **2.** easily penetrated or divided. *Physics.* of a beam of radiation or particles, having low energy and little penetrative power. *Chemistry.* of water, having low concentrations of calcium and of magnesium compounds. *Acoustics.* of a sound, low in volume. *Graphic Arts.* of an image, lacking in sharpness or contrast. *Metallurgy.* **1.** of a metal, easily magnetized and demagnetized. **2.** of an alloy or metal, fully annealed.

soft acid *Chemistry.* a Lewis acid or accepter of larger size, lower positive charge, and higher polarizability.

soft agglomerate *Materials.* a uniform, free-flowing powder produced by spray drying and used as die fill for automated pressing operations.

soft automation *Engineering.* automation achieved primarily through computer software with little reliance on hardware.

softback *Graphic Arts.* a book that has a paper cover; a paperback book.

soft base *Chemistry.* a Lewis base or donor of large size and high polarizability, such as one containing carbon, phosphorus, or iodine.

soft cataract *Medicine.* a cataract that has no hard nucleus.

soft chancre see CHANCROID.

soft cheese *Food Technology.* any cheese containing more than 50% moisture by volume, such as Camembert or cottage cheese.

soft coal see BITUMINOUS COAL.

soft copy *Computer Programming.* any output data in the form of a display image, an audio signal, or other impermanent form.

soft-copy terminal see DISPLAY CONSOLE.

soft coral *Invertebrate Zoology.* any coral of the order Alcyonacea, with polyps embedded in masses of gelatinous material and a skeleton of separate spicules.

softcover or **soft cover** *Graphic Arts.* of a book, having a cover made of coated paper; paperbound.

soft crash see GRACEFUL DEGRADATION.

soft dot *Graphic Arts.* a halftone dot having excessive halation around its perimeter.

soft drink *Food Technology.* a sweet, often carbonated beverage containing fruit or other flavorings and about 55–65% sugar syrup by weight.

softener *Chemical Engineering.* a substance that reduces the hardness of water by removing or trapping calcium and magnesium ions. *Materials.* **1.** a substance used to reduce friction and facilitate processing when dry powders are added to a polymeric material. **2.** a fat-liquoring substance used to soften leather, or an oil or fatty alcohol used in textile finishing. Also, **softening agent.**

softening *Metallurgy.* the second stage of the typical lead refining process, including the removal of lead-hardening impurities such as arsenic, antimony, and tin; it involves air oxidation in a reverberatory furnace or circulation through a caustic soda melt to which sodium nitrate is added.

softening point *Physics.* the temperature at which viscous flow becomes plastic flow; this term is used for materials that do not have definite melting points.

soft error *Computer Technology.* an error that occurs only once or at infrequent, unpredictable intervals; may be theoretically eliminated by repeating the operation. Also, TRANSIENT ERROR.

soft-fail system see FAIL-SOFT SYSTEM.

soft failure *Computer Technology.* a spontaneous change in the contents of computer memory, often caused by a cosmic ray or radioactive decay particles.

soft ferrite *Materials Science.* any of a class of ferrimagnetic ceramics characterized by high electrical resistivity, a wide range of magnetic properties, and ease of fabrication using standard ceramics techniques; used for high-frequency inductors and transformers, switching cores for memory applications, and magnetic elements in microwave components.

soft ground *Mining Engineering.* **1.** the portion of a mineral deposit that is minable without drilling or shooting hard rock. **2.** heavy ground; rock near underground openings that does not stand well, thus requiring heavy timbering.

soft hail see SNOW PELLETS.

soft hammer *Mechanical Devices.* a hammer having a head made of a soft substance such as plastic or rubber; designed to prevent damage while working on finished surfaces. *Archaeology.* any object of relatively soft material (such as bone, antler, or wood) that was used as a percussion tool.

soft-hammer percussion *Archaeology.* the use of a soft hammer to chip or shape a stone. Also, **soft-hammer technique.**

soft hyphen *Computer Technology.* a hyphen that is printed only if it is the last character on the printed line; used only to break a word between syllables; e.g., the hyphen in "high-frequency" is a hard hyphen, meaning it is part of the spelling of the word; this hyphen in the word "because" is a soft hyphen that appears only to divide the word at the end of the line.

soft-iron ammeter *Engineering.* an instrument used to measure alternating currents, composed of a charged coil that magnetizes two pieces of magnetic material within it so that they repel one another; one magnet is fixed and the other is connected to a pointer that moves to indicate the alternating current measurement.

soft landing *Space Technology.* a gentle landing on the surface of the moon or other body in space at such slow speed as to avoid damage to the landing vehicle or its payload.
soft limit *Computer Technology.* an operating limit that is imposed by software and can be relaxed by changing the controlling instructions.
soft limiting *Electronics.* a type of limiting in which there is a considerable increase in output for increases in input signal strength up to the range in which limiting action occurs.
soft magnet *Electromagnetism.* magnetic material that does not retain its magnetization and is easily demagnetized.
soft missile base *Military Science.* a missile base, usually above ground, that is relatively vulnerable to enemy attack.
soft palate *Anatomy.* the posterior, muscular portion of the palate that forms an incomplete septum between the mouth and the oropharynx and between the oropharynx and the nasopharynx.
soft patch *Computer Programming.* a temporary revision of machine code in memory, made to correct an error in the program. *Mechanical Engineering.* a patch designed to cover a crack in a container such as a steam boiler; composed of soft material that is covered by a metal plate bolted to the container.
soft point *Ordnance.* a soft lead bullet with a hard metal case that is open at the nose to create greater expansion upon impact; it is designed for big game hunting and there are prohibitions against its use for combat and law-enforcement weapons. Also, **soft-point bullet.**
soft-point side-slit bullet *Ordnance.* a soft-point bullet with slits down the sides to create even greater expansion.
soft radiation *Physics.* radiation of low energy that does not readily penetrate most materials.
soft ray see SOFT X-RAY.
soft return *Computer Programming.* a conditional line feed and carriage return command combination issued automatically by a word processing program when the current word does not fit on the line; part of the word-wrap feature.
soft rime *Hydrology.* an opaque coating of rime deposited mainly on vertical surfaces.
soft rock *Petrology.* **1.** any sedimentary rock. **2.** any rock that offers little resistance to erosion.
soft roller *Graphic Arts.* in the inking mechanism of a press, a rubber or glycerine roller used in combination with a rider to break down the ink.
soft rot *Materials Science.* wood deterioration caused by various bacteria and fungi; it occurs under environmental conditions in which growth of decay fungi would not be expected, such as in preservative-treated wood or very wet or water-submerged wood. *Plant Pathology.* a rapid, slimy, mushy, odorous disintegration of plant parts, caused by either fungi or bacteria.
soft science *Science.* a term for any science, such as psychology or anthropology, whose criteria for investigating the world are less readily measurable and verifiable than those of the so-called hard sciences, such as biology or chemistry.
soft sector *Computer Technology.* a section on a disk that is marked by a magnetic pattern written on the disk, and used by the disk controller to locate specific areas; as opposed to hard sectors of fixed size.
soft-shell(ed) crab *Invertebrate Zoology.* a crab that has just molted and thus has a relatively soft, edible shell; blue crabs in this condition are harvested commercially.
soft-shell(ed) turtle *Vertebrate Zoology.* any aquatic turtle of the family Trionychidae, whose shell is covered with leathery skin rather than horny plates.
soft shower *Physics.* a cosmic-ray shower made up primarily of electrons and positrons too weak to penetrate deeply into lead.
soft solder *Metallurgy.* any solder, usually an alloy of lead and tin, that is fusible at temperatures below 700°F.
soft tube *Electronics.* **1.** an X-ray tube having a vacuum of about 0.000002 atmosphere. **2.** a tube into which a small amount of gas has been placed in order to obtain desired characteristics.
software *Computer Programming.* any computer instructions written to be executed on hardware, including operating systems, utility programs, and application programs.
software compatibility *Computer Technology.* the capability of a computer to run software written for use in other computers.
software engineering *Computer Programming.* the application of the techniques and disciplines of engineering to the development of high-quality, large-scale software systems; includes controlled design, structured program methodology, productivity aids, and project management tools.
software flexibility *Computer Programming.* the ability of software to be changed easily to meet the needs of the user and the system.
software floating point *Computer Programming.* the capability through software routines of performing floating-point arithmetic operations on hardware designed for integer or fixed-point operations.
software interface *Computer Programming.* the communication of system software and users programs.
software maintenance *Computer Programming.* the updating of existing software to correct errors, to enhance performance or add capabilities, to adapt to new hardware, etc.
software monitor *Computer Programming.* **1.** a software program that monitors computer performance. **2.** the operating system.
software multiplexing see MULTIPROGRAMMING.
software package *Computer Programming.* a collection of prewritten computer programs for a particular application.
software path length *Computer Programming.* the specific number of machine-language instructions needed to perform a particular routine. Also, PATH LENGTH.
software piracy *Computer Technology.* the act of copying proprietary software without permission.
software protection *Computer Technology.* any means of guarding against unauthorized software copying.
software reusability *Computer Technology.* **1.** the suitability of a software package for reuse. **2.** the study of software reuse and features of software and compiler technology that foster reuse.
software reuse *Computer Programming.* the use of a program or abstract algorithm for an application different from the one for which it was originally written.
soft water *Chemistry.* water in which the concentration of calcium and magnesium compounds is extremely low.
soft-wired numerical control see COMPUTER NUMERICAL CONTROL.
softwood or **soft wood** *Materials.* the wood from coniferous or needle-leaved trees, such as pines, firs, and redwoods, used for construction purposes and as the pulpwood to produce cellophane and paper products. The softwood/hardwood distinction is based on the type of tree and does not necessarily indicate the relative hardness of the wood. Some softwood trees have harder wood than certain hardwoods.
soft X-ray *Electromagnetism.* an X-ray of long wavelength, low energy, and little penetrative power.
soft X-ray absorption spectroscopy *Spectroscopy.* a spectroscopic method of obtaining information about unoccupied states above the Fermi level in metals, or about empty conduction bands in inoculators. Soft X-rays are of low photon energy (~1000 eV).
soft X-ray appearance potential spectroscopy *Spectroscopy.* an electron spectroscopic technique in which small, abrupt changes in the intensity of X-ray emissions are detected as the energy with which a solid surface is bombarded by monochromatic electrons is varied.
Sohm Abyssal Plain *Geography.* a deep basin in the North Atlantic Ocean, situated between Newfoundland and the mid-Atlantic ridge.
Sohncke's law *Physics.* a law stating that the stress applied normally to a crystallographic plane, required to produce a fracture in the crystal, is a constant characteristic of the crystalline substance.
soil *Geology.* **1.** all loose, unconsolidated, weathered, or otherwise altered rock material above bedrock. **2.** a natural accumulation of organic matter and inorganic rock material that is capable of supporting the growth of vegetation.
soil amendment *Agronomy.* the process of adding substances to the soil to improve plant growth.
soil association *Geology.* a mapping unit used in detailed soil surveys, consisting of two or more distinguishable soils in a given geographic area that are grouped together on the basis of their areal distribution. Also, **soil complex.**
soil atmosphere *Geology.* the air occupying the pore spaces in soil. Also, **soil air.**
soil blister see FROST MOUND.
soil-borne wheat mosaic virus group see FUROVIRUS GROUP.
soil cement *Materials.* a mixture of soil and cement that is used as a base or surface for roads and pavement.
soil chemistry *Geochemistry.* the study of the composition and chemical properties of soils, including investigation of the chemical processes that occur as a result of the action of hydrological, geological, and biological agents on soil.
soil colloid *Geology.* the inorganic and organic matter in soils characterized by small particles having relatively large surface areas per unit of mass.

soil compactor

soil compactor *Mechanical Engineering.* a tractor having a heavy roller in place of front wheels; used to compact soil for road construction and the like.
soil conservation *Agronomy.* any of various methods of land management that seek to protect the soil from erosion and chemical decay, so as to maintain its fertility.
soil creep *Geology.* the gradual, almost imperceptible, steady downhill movement of soil and loose rock debris on a slope. Also, SURFICIAL CREEP.
soil depletion *Agronomy.* see DEPLETION.
soil ecology *Ecology.* the study of the relationship between the activities of soil organisms and the overall soil environment.
soil erosion *Geology.* the detachment and transport of topsoil by the action of wind and running water.
soil fall see DEBRIS FALL.
soil flow or **soil fluction** see SOLIFLUCTION.
soil genesis *Geology.* the mode, processes, and factors responsible for the origin and development of a true soil from unconsolidated parent material. Also, PEDOGENESIS.
soil geography *Geography.* the branch of physical geography dealing with the geographical distribution of various soil types.
soil horizon *Geology.* a layer of soil differentiated from adjacent layers by specific physical characteristics..
soil ice *Hydrology.* ice that forms from the freezing of moisture already present in the ground rather than from precipitation.
soil management *Agronomy.* the preparation and mangement of the soil for efficient crop production and conservation of soil and water.
soil map *Agronomy.* a map showing the distribution of soil types in relation to the prominent physical and cultural features of the earth's surface.
soil mark *Archaeology.* any visible irregularity in the appearance of the soil surface; a darker area may indicate human wastes, or a lighter area a former road or trail.
soil mechanics *Geology.* the mechanical reaction of a given soil to natural phenomena or to human activities such as compacting, shearing, or the laying and use of pipe. *Civil Engineering.* the scientific study of such reactions in various soils.
soil microbiology *Microbiology.* the branch of microbiology that deals with microorganisms found in the soil, including their functions and their effects on the growth and maintenance of plant life.
soil moisture see SOIL WATER.
soil physics *Geophysics.* the study of the character and chemical composition of the earth's soil.
soil pipe *Civil Engineering.* a pipe that carries the discharge from water closets or similar fixtures to the sanitary sewer system.
soil-pipe cutter *Mechanical Devices.* a specially adapted cutter for replacing sections of deteriorated or plugged soil pipes.
soil profile *Geology.* a vertical section of a soil that displays the succession of horizons from the surface to the parent material.
soil rot *Plant Pathology.* any plant rot caused by microorganisms in the soil.
soil science *Geology.* the study of the formation, properties, distribution, and classification of soil as a natural resource. Also, PEDOLOGY.
soil separate *Geology.* a group of rock and mineral particles separated from a sample of soil in which the particles have equivalent diameters of less than 2 mm and range between specified size limits.
soil series *Geology.* a group of soils that developed from a particular kind of parent material and which have similar profile characteristics and arrangements, except for the texture of the surface horizon.
soil shear strength *Geology.* the maximum resistance of a given soil to shearing stresses.
soil stack *Building Engineering.* a vertical soil pipe that serves as a main to carry wastewater to all fixtures in a structure.
soil sterilization *Agronomy.* the introduction into the soil of a material that will make the soil poisonous to plants.
soil stripes *Geology.* a patterned ground consisting of alternating parallel bands of fine rock debris and coarse rock material that are considerably finer in texture than the bands in stone stripes formed by frost action or slopes.
soil structure *Geology.* the combination of primary soil particles into aggregates or clusters, classified on the basis of the characteristics of the constituent particles.
soil survey *Agronomy.* the systematic examination, description, classification, and mapping of soils in a given area.
soil thermograph *Engineering.* an instrument that creates a photographic representation of the heat of a designated underground area; it consists of a sensing element that transmits data to a remote recording device.
soil thermometer *Engineering.* an instrument, commonly a mercury-in-glass thermometer, that is used to measure the temperature of soils. Also, EARTH THERMOMETER.
soil water *Hydrology.* water in the zone of aeration immediately below the ground surface. Also, RHIZIC WATER, VADOSE WATER.
soil-water belt see BELT OF SOIL WATER.
sol *Chemistry.* a liquid colloid or mixture in which solid particles, small enough to pass through filter membranes, are dispersed in a liquid phase.
sol. solution; soluble. Also, **Sol.**
-sol *Geology.* a combining form meaning "soil."
solaire *Meteorology.* any of several easterly winds in central and southern France.
sol-air temperature *Meteorology.* the temperature that under conditions of no direct solar radiation and no air motion would cause the same heat transfer into a house as that caused by the combination of all existing atmospheric conditions.
Solanaceae *Botany.* a large family of shrubs, trees, lianes, and herbs of the order Solanales, characterized by alternate leaves and often showy flowers with five stamens and a two-celled ovary with many ovules in each cell; it includes many economically important plants such as the tomato, potato, pepper, and tobacco.
Solanales *Botany.* an order of usually dicotyledonous plants of the subclass Asteridae, often producing alkaloids and characterized by stems with internal phloem, perfect flowers, and an ovary surrounded by an annular nectary disk. Also, POLEMONIALES.
solano *Meteorology.* a southeasterly or easterly wind on the southeastern coast of Spain in the summertime; associated weather conditions can be hot and humid with rain, or dusty and dry.
solar *Astronomy.* relating to, proceeding from, or influenced by the sun. *Engineering.* describing a device that utilizes solar energy, especially to provide heat. *Horology.* of time, determined by or referring to the sun. (From *sol,* the Latin word for sun.)
solar absorption index *Geophysics.* the relation of the position of the sun at various latitudes to its absorption in the ionosphere.
solar activity *Astronomy.* any variable and short-lived disturbance occurring on the disk of the sun, such as sunspots, prominences, and solar flares.
solar air mass *Meteorology.* for any given position of the sun in the sky, the optical air mass penetrated by light from the sun.
solar antapex *Astronomy.* the apparent direction (in the constellation Columba) away from which the sun and solar system are moving in their orbit around the galaxy's center.
solar apex *Astronomy.* the point on the celestial sphere toward which the sun and solar system are moving; currently this point is approximately 10° southwest of the star Vega.
solar array *Electrical Engineering.* a panel containing solar cells that are electrically interconnected in such a way that they provide a direct output voltage with a current capacity proportional to the number of elements.

solar atmospheric tide *Geophysics.* the regular expansion or contraction of the atmosphere due to the gravitational or thermal effects of the sun.
solar attachment *Engineering.* a device that is attached to a surveyor's compass or transit to determine the true meridian directly from the sun.
solar battery *Electronics.* a battery that is composed of an array of solar cells.
solar bridge *Astronomy.* a division of a large sunspot's umbra into two or more parts showing a bright but narrow shape with streaks through it.
solar burst *Astrophysics.* a sudden enhancement of nonthermal radio emission from the solar corona immediately following the eruption of a solar flare.
solar calendar *Astronomy.* a calendar that conforms to the seasons on the earth and that is derived from the tropical year, which lasts 365.24220 days.
solar cavity see HELIOSPHERE.
solar cell *Electrical Engineering.* a photosensitive cell or combination of cells designed to generate voltage by direct conversion of light to electricity when exposed to a source of light. Also, SOLAR GENERATOR.
solar climate *Meteorology.* the theoretical climate that would prevail on a uniform solid earth with no atmosphere; a climate of temperature alone and determined only by the amount of received solar radiation.
solar collector *Engineering.* any of various devices that absorb solar radiation and store it as usable energy.
solar constant *Meteorology.* the rate at which radiant energy from the sun is received outside the earth's atmosphere on a surface normal to the incident radiation, and at the earth's mean distance from the sun.
solar corona see CORONA.
solar cycle *Astronomy.* the periodic cycle, lasting about 11 years, in which solar activity waxes and wanes; reckoning by the magnetic polarity reversal in sunspots, the cycle is 22 years long.
solar day *Astronomy.* the 24-hour interval between two successive transits of the sun across a given meridian on earth; this is approximately 3 minutes and 56 seconds longer than a sidereal day.
solar dermatitis see SUNBURN.
solar distillation *Chemical Engineering.* a distillation process that uses solar radiation to evaporate seawater in order to obtain salt. Also, SOLAR EVAPORATION.
solar eclipse *Astronomy.* the passage of the moon between the sun and the earth; it is total where the umbra of the moon's shadow reaches the ground and partial outside the line drawn by the umbra. When the moon lies too far from the earth for the shadow to reach the ground, the eclipse is termed annular, because an annulus, or ring, of uneclipsed sun surrounds the moon.
solar energy *Astrophysics.* energy received by the earth from the sun in the form of radiation.
solar energy unit *Electrical Engineering.* any device or apparatus designed to convert solar energy into electrical energy, typically consisting of a connected series of solar cells.

solar energy unit

solar engine *Electrical Engineering.* any engine that converts thermal energy from the sun into mechanical or electrical energy.
solar evaporation *Hydrology.* the evaporation of water by the heat of the sun. *Chemical Engineering.* see SOLAR DISTILLATION.
solar-excited laser see SUN-PUMPED LASER.
solar faculae see FACULA.
solar flare *Astrophysics.* a region of exceptionally high temperature and brightness that suddenly develops in the solar chromosphere near a sunspot; it is often associated with complex magnetic fields.
solar flux unit *Astrophysics.* a unit of solar radio emission; 1 solar flux unit = 10^{-22} watts per square meter per hertz.
solar furnace *Engineering.* a heating unit that uses the rays of the sun, concentrated by a concave mirror, to produce high temperatures.
solar generator see SOLAR CELL.
solar heating *Engineering.* the process of converting solar energy into heat for domestic or industrial uses.
solar heat storage *Engineering.* a process in which heat produced from the conversion of solar radiation is stored, generally through the heating of water.
solarimeter *Engineering.* **1.** an instrument used to measure radiation from the sky; composed of a Moll thermopile that is protected by a bell glass. **2.** see PYRANOMETER.
solarization *Physics.* the loss of color or transparency in glass due to extensive exposure to sunlight or ultraviolet radiation. *Graphic Arts.* a process in which an image is deliberately or accidently overexposed, causing a total or partial reversal of normal light gradations. *Engineering.* to convert or adapt a building or the like to the use of solar energy.
solar kiln *Materials Science.* a system for drying wood in which solar energy is used; some solar kilns have a simple greenhouse design combined with a thermal storage system, while others use forced convection.
solar lake *Hydrology.* a lake that has no connection to the sea and whose water temperature and salinity increase with depth.
solar magnetic field *Astrophysics.* the dipolar field indigenous to the ionized gases of the sun.
solar magnetograph *Engineering.* an instrument that measures the intensity and polarity of magnetic fields at the sun's surface by using the Zeeman effect; it is composed of a telescope, a spectrograph, a differential analyzer, and a device for visual representation of differencing and recording.
solar modulation *Astrophysics.* an 11-year periodic change in the cosmic-ray intensity detected at earth due to the solar cycle: when the sun is active, cosmic rays are less intense.
solar month *Astronomy.* any of 12 time intervals making up a solar year.
solar motion *Astronomy.* **1.** the movement of the sun with respect to other observed stars. **2.** the motion of the sun considered alone. **3.** the velocity of the sun as it rotates around the center of the Milky Way Galaxy, approximately 19 kilometers per second.
solar nebula *Astronomy.* the cloud of interstellar gas and dust out of which the sun and planets formed, roughly 4.5 billion years ago.
solar neutrino *Astrophysics.* a neutrino generated by nuclear reactions in the sun.
solar neutrino unit *Astrophysics.* a unit of measure of neutrinos captured per atom of target material; 1 solar neutrino unit = 10^{-36} solar neutrinos.
solar noise *Electromagnetism.* the radio-frequency noise produced by the sun, which increases with the occurrence of sun spots and solar flares.
solar nutation *Astronomy.* a perturbation in the motion of the earth's rotation axis caused by the changing position of the sun relative to the earth's equatorial bulge.
solar orbit *Astronomy.* any orbit that has the sun at one focus.
solar panel *Astronomy.* a solar collector in the form of a blackened metal plate, typically placed on the roof of a house.
solar parallax *Astronomy.* the angle subtended by the earth's equatorial radius at the center of the sun; it amounts to 8.79 seconds of arc.
solar phase angle *Astronomy.* the angular separation between the earth and the sun at a given planet.
solar physics *Astrophysics.* the study of all physical activities associated with the sun.
solar plexus *Anatomy.* a network of nerves lying on the front and sides of the aorta at the origins of the celiac trunk and superior mesenteric and renal arteries. Branches of the solar plexus extend along all the adjacent arteries.
solar power *Engineering.* any power obtained by converting radiation emitted by the sun into useful power.

solar probe *Space Technology.* a space probe designed to gather data about the sun.
solar prominence *Astronomy.* a large eruption in the sun's chromosphere that often appears as flames when observed at the sun's limb.
solar propagation *Botany.* a method of rooting cuttings in which a modified hotbed is heated by radiation of stored heat from the sun by bricks or stones in the bottom of the hotbed frame.
solar propulsion *Space Technology.* the use of solar energy to power a spacecraft, either directly by photon pressure on solar sails or indirectly from solar cells used to generate electrical energy.
solar pumping *Optics.* a technique in which sunlight is focused onto a laser rod to initiate the lasing action.
solar radiation *Astrophysics.* all the constituents making up the sun's emission, including electrons, protons, neutrinos, and rare and heavy atomic nuclei.
solar-radiation observation *Geophysics.* the observation and evaluation of solar radiation received at a specified point.
solar radio emission *Astrophysics.* the sum total of all radio-frequency electromagnetic emission from the sun.
solar radio noise see SOLAR NOISE.
solar sail *Space Technology.* on a spacecraft, a vane designed to collect sunlight, either to generate power or to measure the flow of solar particles. Also, PHOTON SAIL.
solar satellite *Space Technology.* a space vehicle designed to orbit about the sun.
solar sector *Astronomy.* a region in the solar wind that is characterized by one magnetic polarity.
solar sensor *Electronics.* a diode that transmits a signal to a spacecraft when it senses the sun.
solar spectrum *Astrophysics.* a continuum spectrum that is marked with dark absorption lines extending over the entire electromagnetic spectrum.
solar spicule see SPICULE.
solar still *Chemical Engineering.* a still used to produce drinking water from salty or brackish sources, using the sun's radiation to evaporate the water and leaving the salt behind.
solar system *Astronomy.* **1.** the sun and all the planets and other bodies that travel around it. **2.** the region within which these bodies move.
solar system abundances *Astrophysics.* the relative proportion of elements in the solar system, usually measured by reference to solar element abundances.
solar telescope *Optics.* a telescope that minimizes optical distortions caused by solar heating; it typically has a relatively modest aperture and a long focal length, and is often designed to image the sun in monochromatic light.
solar-terrestrial phenomena *Geophysics.* any of the various phenomena observable on earth that are caused by the influence of the sun, such as atmospheric tides or the aurora borealis.
solar tide *Oceanography.* the harmonic tidal constituent attributable to the tide-producing force of the sun.
solar time *Horology.* **1.** local time as measured by an instrument such as a sundial. **2.** a system of measurement based on the position of the earth in relation to the sun.
solar-topographic theory *Meteorology.* a theory stating that the changes of climate through geologic time have been due to changes of land and sea distribution and orography, combined with fluctuations of solar radiation.
solar tower *Astronomy.* a solar telescope that is mounted on a tall tower in order to place the optics above the image-distorting layer of sun-heated air near the ground.
solar-type stars *Astronomy.* stars with a G-type spectrum and having the same luminosity class.
solar ultraviolet radiation *Astrophysics.* the portion of the solar electromagnetic radiation spectrum that has wavelengths ranging from 4 to 400 nanometers.
solar velocity *Astronomy.* the velocity of the sun relative to those of the nearby stars.
solar wind *Geophysics.* the movement of ionized particles, mostly helium and hydrogen, from the sun through the solar system with velocities of 300–1000 kilometers per second.
solar year see TROPICAL YEAR.
Solasteridae *Invertebrate Zoology.* the sun stars, a family of echinoid echinoderms belonging to the order Spinulosida; long-spined, many-armed starfish that are typically brightly colored; some species prey on other starfish.
solation *Physical Chemistry.* the process by which a substance changes from a gel into a sol.
solder *Metallurgy.* **1.** to join metal objects without melting them by fusing a metal alloy that has been applied to the joint between them. **2.** any of several alloys used in this process.
solder bridge *Electricity.* an undesirable electrical connection formed when solder inadvertently connects two wires, pins, or similar devices.
soldering embrittlement *Metallurgy.* ductility impairment caused by intergranular penetration of solder.
soldering flux *Metallurgy.* a flux used in soldering.
soldering gun *Engineering.* a gun-shaped tool used to apply heat in the process of soldering.
soldering iron *Engineering.* an elongate copper tool having a grip at one end and a wedge or point at the other end; used to apply heat in the process of soldering.
soldering lug *Electricity.* a small metal eyelet attached to a component for connecting another component or wire using solder.
soldering pencil *Engineering.* a small soldering iron, similar in size and weight to a standard pencil, used to join and take apart the joints on printed wiring boards.
solderless contact see CRIMP CONTACT.
solder track *Electronics.* a conducting track on a printed circuit board.
soldier *Military Science.* **1.** any person who serves in the military, especially the army. **2.** an enlisted man or woman, as distinguished from an officer. *Invertebrate Zoology.* a member of a sterile caste of termites or ants, having an enlarged head and jaws and serving as a fighter to protect the colony. (From *solidus,* a gold coin used in ancient Rome to pay members of the army.)
soldier ant *Invertebrate Zoology.* a member of a caste of ants having a large head and jaws, specialized for defending the colony; some species can eject formic acid up to six inches.
soldierfish *Vertebrate Zoology.* any of several squirrelfishes of the family Holocentridae having rough scales and sharp spines.
soldier's heart see NEUROCIRCULATORY ASTHENIA.
soldier termite see NASUTE.
sole *Anatomy.* the bottom of the foot. *Geology.* **1.** the relatively flat undersurface or bottom of a sedimentary stratum or another body of rock. **2.** the more gently inclined middle and lower portions of the shear surface of a landslide. **3.** the major fault plane underlying a thrust nappe. Also, **sole plane.** *Hydrology.* the lower or underlying ice of a glacier, often containing fragments of rock that give it a dirty appearance. *Building Engineering.* see SOLE PLATE. *Electronics.* in a magnetron or backward-wave oscillator, an electrode transmitting a current that generates a magnetic field in the desired direction.
sole *Vertebrate Zoology.* a popular name for any edible flatfish; first applied to the cosmopolitan family Soleidae, especially the **Dover** or **channel sole,** *Solea solea,* a popular food fish of Europe, a brown flatfish with eyes on the right side of its head. (From *sole* in the sense of the bottom of a shoe or the foot; because of its flat shape.)

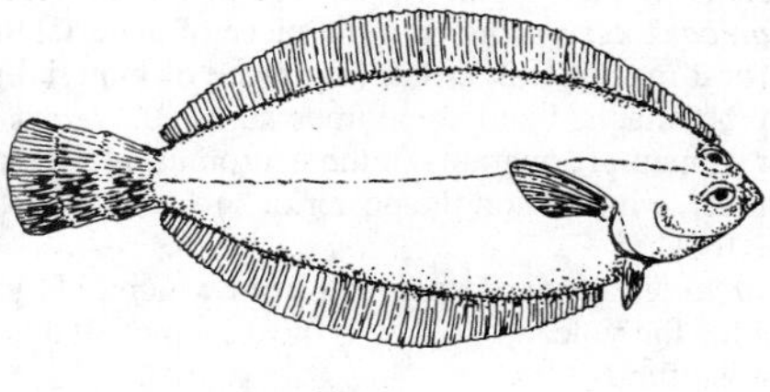

Dover sole

soleil compensator *Optics.* a modification to the Babinet compensator, which has two quartz prisms mounted so as to slide over each other in a way that maintains phase changes across the entire field of view.
sole injection *Geology.* an intrusion of igneous rock that was put in place along a thrust-fault plane.
sole mark *Geology.* a small, wavelike irregularity or penetration on the original undersurface of a sandstone or siltstone bed overlying and in contact with a softer, finer-grained layer, such as shale.
Solemyidae *Invertebrate Zoology.* a family of burrowing marine clams having oblong or oval valves and a large foot for digging into sandy or muddy bottoms.

Solenichthyes *Vertebrate Zoology.* an alternate name for the suborder Syngnathoidei, which contains the pipefishes, sea horses, snipefishes, shrimpfishes, trumpetfishes, and flutemouths.

solenium *Invertebrate Zoology.* **1.** see STOLON, def. 2. **2.** a tube connecting adjacent polyps in some colonial anthozoans.

solenocyte *Invertebrate Zoology.* a type of elongated flame bulb having a single long flagellum.

solenodon *Vertebrate Zoology.* any member of the family Solenodontidae, primitive ratlike insectivorous mammals that are found in Cuba and Hispaniola.

Solenodontidae *Vertebrate Zoology.* a family of very rare, large ratlike insectivores of the mammalian group Lipotyphia having a long snout, short ears, a long scaly tail, hard fur, and forty teeth.

Solenogastres see APLACOPHORA.

solenoid [sōl′i noid; säl′i noid] *Electromagnetism.* an electromagnetic coil wound in the shape of a hollow cylinder or spool, often containing a movable iron core that is pulled into the coil when current flows through the wire turns, thus allowing it to move other devices such as relays and circuit breakers. *Mechanical Engineering.* a switch or other device that is activated by such a coil, as in an automobile starting system. *Meteorology.* a tube formed in space by the intersection of surfaces of equal pressure and density.

solenoidal *Electromagnetism.* relating to or powered by a solenoid. *Mathematics.* a vector field F defined in a simply connected domain is said to be solenoidal if its divergence vanishes at every point of the domain. Also, SOURCE-FREE.

solenoidal index *Meteorology.* a mathematical expression of the difference between the mean virtual temperature from the ground to a known altitude aloft at 55° latitude and the mean virtual temperature for the corresponding layer averaged at 35° latitude.

solenoid brake *Mechanical Engineering.* an electromechanical braking device in which the brake toggle is operated by the plunger of a solenoid.

solenoid model *Genetics.* a model proposed to explain the supercoiled nature of chromatin, postulating that DNA is compacted and coiled around histone molecules to form nucleosomes of 30-nm diameter; common to most chromosomes.

solenoid valve *Mechanical Engineering.* a valve actuated by the magnetic field produced in a solenoid to control the flow of gas or fluid in a pipe.

Solenopora *Paleontology.* a genus of calcite-secreting algae in the extinct family Solenoporaceae; extant from the Cambrian to Cretaceous.

Solenoporaceae *Paleontology.* a family of calcareous red algae classified in the phylum or division Rhodophyta; they formed nodular masses of calcite consisting of tightly packed vertical tubes; extant in the Cambrian to Miocene.

solepiece *Civil Engineering.* **1.** any horizontal member used to distribute the loads from one or more uprights. **2.** a member that supports the foot of a raking shore. *Naval Architecture.* an attachment to the foot of a rudder that aligns with a false keel.

sole plate or **soleplate** *Building Engineering.* the lower surface of the body of a plane or plate upon which studding is erected. Also, SHOE, SOLE. *Mechanical Engineering.* **1.** a flat piece of material that serves as a foundation for a machine. **2.** a flat, thin piece of material upon which a bearing may be attached and sometimes adjusted. *Neurology.* an obsolete term for subneural apparatus of the neuromuscular junction.

soleus *Anatomy.* a muscle on the posterior surface of the tibia that plantar flexes the foot.

solfatara *Volcanology.* a volcanic vent from which only gases are emitted. (Named for the volcano *Solfatara*, near Naples, Italy; from the Italian word for "sulfur.")

solfataric stage *Volcanology.* the final stage of a volcanic eruption, during which only gases are emitted from the vent.

sol-gel coating *Materials Science.* a coating produced by the sol-gel process of glassmaking, in which glass is formed at low temperatures from suitable compounds by chemical polymerization in a liquid phase; a gel is formed from which glass may be derived by the successive elimination of interstitial liquid and the collapse of the resulting solid residue by sintering.

sol-gel process *Materials Science.* a processing technique in which a fibrous gel is drawn from a solution at near room temperature and converted into glass or ceramic fibers at several hundred degrees Celsius.

soliciting or **solicitation** *Zoology.* the movements and postures of a female animal that attract a male to her for copulation. Also, PROCEPTIVE BEHAVIOR.

solid *Physics.* one of the three fundamental states of matter, along with liquids and gases. Of these three forms, a solid has the greatest tendency to resist forces that would alter its shape; thus its shape and volume are fixed and are not affected by the space available to it. In comparison with liquids and gases, solids have closely packed molecules; their normal condition is a crystalline structure. *Mathematics.* a closed and bounded subset of three-dimensional space having positive volume.

solid angle *Mathematics.* a measure on the space of rays emanating from a point in Euclidean 3-space; equal to the area of the intersection of the set of rays with the surface of the unit sphere centered at the point. The set of all rays emanating from a point has solid angle equal to 4π steradians.

solid coupling *Mechanical Engineering.* a nonflexible connection between two shafts that forms a permanent joint designed to bear a full load of rotation or transmission.

solid cutter *Mechanical Devices.* the cutting part of a machine tool, made from a single piece of material.

solid die *Mechanical Devices.* an internally threaded, screw-cutting tool, constructed of a single piece of material.

solid drilling *Engineering.* a process used in diamond drilling, in which the entire face of an area is ground, and no core is extracted for sampling.

solid electrolytes *Materials Science.* materials that conduct electricity by ionic diffusion, including crystalline, vitreous, polymeric, or electrolyte-colloidal-particle composites; used as thin-membrane separators of two reactants, as in batteries.

solid electrolytic capacitor *Electricity.* a capacitor that uses a solid electrolyte for one plate.

solid explosive *Materials.* an explosive in the form of a powder, a granulated mass, or solid sticks.

solid geometry *Mathematics.* the geometric study of space figures such as polyhedra, cylinders, cones, and spheres, including the notions of similarity, congruence, and computation of area and volume.

solid helium *Physics.* a solid phase of helium that is only obtained with the application of about 25 atmospheres of external pressure while at a temperature near absolute zero.

solidification *Physics.* the transition of a liquid or a gas to the solid phase; the process of becoming solid.

solidification inclusion *Materials Science.* a defect in a metal casting resulting from the inclusions of generally nonmetallic materials, such as slag, that can affect the mechanical properites by acting as stress raisers.

solidification shrinkage *Metallurgy.* in casting, the shrinkage occurring during solidification.

solidify *Physics.* to undergo or cause to undergo solidification.

solid injection system *Mechanical Engineering.* a diesel-engine injection system in which a pump forces the fuel through a line and an atomizing nozzle into the combustion chamber.

solid insulator *Electricity.* any dielectric material with high mechanical strength that is used to separate conductors without allowing electric current to flow.

solid laser *Optics.* see SOLID-STATE LASER.

solid-liquid equilibrium *Physical Chemistry.* **1.** the thermodynamic relationship between a solid and its melt when vapor pressure remains constant. **2.** the thermodynamic relationship between the concentration of a solid and a solvent, other than the melt of that solid. Also, LIQUID-SOLID EQUILIBRIUM.

solid logic technology *Electronics.* a computer design technology that incorporates miniaturized modules, resulting in faster circuitry due to the reduced distances that electric current must travel.

solid lubricant *Materials.* a thin film of solid material interposed between two surfaces to reduce friction and wear under severe operating or environmental conditions; includes solid inorganic compounds such as graphite, solid organic compounds such as soaps and waxes, and metal surface coatings such as chemically deposited oxide films, soft metal films, and bonded coatings.

solid moment of inertia *Physics.* a quantity applicable to a solid having a definite volume; used to describe the rotational inertia of the solid about some specified axis.

solid-phase sequenter *Biotechnology.* a device used to determine the amino acid sequence in a protein; the sample is covalently attached to a solid-phase glass or styrene bead and packed in a micro-column prior to degradation.

solid-phase welding *Metallurgy.* any of several welding processes in which joining occurs by solid-state diffusion and a filler metal is not used. Also, SOLID-STATE WELDING.

solid-piled *Materials Science.* of or relating to plywood that is piled onto a flat base directly after being released from clamps or a hot press and weighted down until it reaches its normal conditions of temperature and moisture content.

solid propellant *Materials.* a rocket propellant in solid form that usually contains a combination of fuel and oxidizer. Also, **solid rocket fuel, solid rocket propellant.**

solid propellant binder *Materials Science.* the single ingredient of a propellant that acts to hold all the other ingredients together.

solid-propellant rocket engine *Space Technology.* a rocket engine consisting of a combustion chamber fueled by a solid propellant, a nozzle for the exhaust jet, and other components such as grines or liners. (In strict usage, such a device is called a **solid-propellant rocket motor,** with the term "rocket engine" reserved for liquid-propellent engines.)

solid rocket *Space Technology.* a rocket powered by a solid propellant.

solid rocket booster *Space Technology.* a solid-propellant rocket that is strapped on a missile or launch vehicle to accelerate it during lift-off.

solid Schmidt telescope *Optics.* a Schmidt camera whose body is made from a solid piece of glass; affords increased photographic speed, but due to problems of chromatic aberration, solid Schmidts can be used only with monochromatic light, such as that from a spectrograph.

solid shafting *Mechanical Engineering.* a solid bar that supports rollers or wheels.

solid shank tool *Engineering.* a cutting instrument in which the central steel bit and attached cutting parts are made from a single piece.

solid solution *Physics.* a homogeneous solid mixture whose constituent particle concentrations may vary over specified ranges. *Crystallography.* a solid mixture of two components; if crystalline, the second molecule must replace the first in the crystal structure or else fit between the molecules.

solid solution hardening *Materials Science.* the restriction of plastic deformation in alloys by forming solid solutions; the foreign atoms in the solid solution serve as obstacles to dislocation motion.

solid solution series *Mineralogy.* two or more crystalline compounds (minerals) whose physical properties vary along a smooth curve over a definite compositional range. Also, SERIES, ISOMORPHOUS SERIES.

solid stage *Volcanology.* in the cooling of a magma, the stage during which the magma solidifies completely.

solid state *Physics.* **1.** the solid phase of a substance. **2.** the physical properties of a solid.

solid-state *Physics.* of or relating to solid-state physics. *Electronics.* of or relating to a device, component, or system, such as a semiconductor, transistor, crystal diode, or ferrite core, that can control electric or magnetic phenomena through solids without heated filaments, moving parts, or vacuum gaps.

solid-state battery *Electricity.* a battery consisting of radioactive material and a photovoltaic cell; the radiation causes the cell to generate electricity.

solid-state circuit *Electronics.* a complete circuit made from a single block of semiconductor material.

solid-state circuit breaker *Electronics.* a circuit breaker in which a solid-state device is used to determine when load terminal voltage goes beyond a safe value; such a device can trip an electromechanical circuit breaker that disconnects the load from the power line.

solid-state component *Electronics.* a component whose operation is dependent on the control of electric or magnetic phenomena in solids, such as a transistor, crystal diode, or ferrite device.

solid-state computer *Computer Technology.* a computer assembled primarily from solid-state electronic circuit elements.

solid-state counter *Nucleonics.* a radiation counter that has a crystalline solid for its active material, such as a sodium iodide scintillator.

solid-state device *Electronics.* a device whose operation is dependent on the movement of electrons within a solid piece of a semiconductor material, as distinguished from a vacuum or gas device.

solid-state image sensor see CHARGE-COUPLED IMAGE SENSOR.

solid-state joining *Materials Science.* a process for joining material that does not involve a liquid; the bond is achieved by diffusion, generally at elevated temperature and under pressure, to bring the two surfaces into intimate contact.

solid-state laser *Optics.* a laser in which a semiconductor material is used to generate a coherent beam of light.

solid-state maser *Physics.* a maser that utilizes a semiconductor material to produce a coherent microwave output.

solid-state memory *Computer Technology.* a storage device composed of integrated circuits.

Solid-State Physics

Solid-state physics, a term that did not come into general use until after World War II, denotes the part of physics concerned with understanding the properties of condensed matter on an atomic scale. A variety of experimental techniques ranging from crystal growth for materials preparation to scanning tunneling microscopy, cyclotron resonance, photon absorption, and scattering by X rays and neutrons is used for measuring materials properties. The theoretical techniques needed to understand these properties derive from quantum and statistical mechanics.

The electrical and magnetic properties of metals were extensively studied during the years immediately following the development of quantum mechanics. The rapid early progress is documented in books and review articles by A. Sommerfeld and H. A. Bethe (1933), H. Fröhlich (1936), and N. F. Mott and H. Jones (1936). F. Seitz's *Quantum Theory of Solids* (1940) was the first to present an overview of the entire field as then defined. C. Kittel's *Introduction to Solid State Physics* (1953) played a defining role by attracting students who subsequently developed expertise in areas ranging from electronics to metallurgy. By 1955 the field had expanded to a point that an expanded version of Seitz's book was impractical. Instead, *Solid State Physics,* a series of volumes describing *Advances in Research and Applications,* edited by F. Seitz and D. Turnbull (1955), was founded to fill the need for timely updates.

Solid-state physics has contributed substantially to fundamental understanding. For example, the BCS theory of superconductivity, the Kondo Effect, and the fractional quantum Hall effect have deepened perceptions of the many-body problem. To a larger extent its growth is related to its technological applications. With the invention of the transistor, solid-state lasers, magnetic tape, optical fibers, liquid crystals, and many other materials and devices leading to microelectronics, and innovations ranging from information processing to medicine, the number of researchers in this field has grown enormously. For many interested in practical matters, *solid-state physics* is now *materials science*; for those involved in understanding physically novel phenomena like high temperature super-conductivity and phase transitions in liquid crystals it is *condensed matter physics.* Use of the term *solid-state physics* is declining, but the basic underlying physical theory, as already explicated in the early literature, remains intact to a remarkable degree.

Henry Ehrenreich
Clowes Professor of Science
Harvard University

solid-state physics *Physics.* a branch of physics that is primarily devoted to the study of matter in its solid phase, especially at the atomic level.

solid-state polymerization *Materials Science.* a chain-growth polymerization reaction that is initiated by exposing the monomer or mixture of monomers to ionizing or electromagnetic (microwave) radiation.

solid-state relay *Electronics.* a relay composed exclusively of solid-state semiconductor devices.

solid-state sintering *Materials Science.* the bonding of particles without the formation of a liquid phase; occurs when powder particles are compacted and heated in an atmosphere in which they are stable to a suitably high temperature below the melting point.

solid-state switch *Electronics.* a microwave switch whose switching element is composed of solid-state material.
solid-state synthesis *Biotechnology.* a commonly used method for the chemical synthesis of polypeptides or polynucleotides.
solid-state thyratron *Electronics.* a solid-state device that has the switching speed and power-handling abilities of a gaseous thyratron tube.
solid-state welding see SOLID-PHASE WELDING.
solid stowing *Mining Engineering.* the packing of stone and earth to fill completely the waste area behind a longwall face. Also, **solid packing.**
solid substrate fermentation *Biotechnology.* fermentation in which microorganisms are grown on solid material rather than in a liquid medium; growth can occur either on the surface of the substrate or in heaps of substrate, which are churned by hand.
solid tantalum capacitor *Electricity.* an electrolytic capacitor that uses tantalum oxide as a dielectric.
solidungulate *Zoology.* describing an animal, such as a horse, that has a single, undivided hoof on each foot. Also, SOLIPED.
solidus *Physical Chemistry.* the points on a phase diagram that indicate the temperature below which a given component will freeze while being cooled, or above which it will melt while being heated.
solidus curve *Physical Chemistry.* **1.** the line on a phase diagram that represents the composition of a substance's liquid phase when it is in equilibrium with its solid phase. **2.** a curve that shows the temperature at which a series of alloys are completely solid.
solid-web girder *Civil Engineering.* a structural steel girder or light beam having a solid web; usually cold formed from sheet or another rolled section, but not a lattice.
Solieriaceae *Botany.* a small family of red algae of the order Gigartinales, having erect branched thalli of multiaxial construction and spermatangia produced in superficial clusters; some genera are used as food or as a source of agar.
solifluction *Geology.* the slow, viscous, downslope movement of waterlogged soil and other water-saturated surface materials in periglacial areas at high elevations. Also, **soil fluction, solifluxion.**
solifluction lobe *Geology.* an isolated, tongue-shaped land feature having a steep front and a smooth upper surface that was formed by more rapid solifluction on certain sections of a slope showing variations in gradient. Also, **solifluction tongue.**
solifluction mantle *Geology.* the locally derived, unsorted, water-saturated surficial material transported downslope by solifluction. Also, FLOW EARTH.
solifluction sheet *Geology.* a broad, extensive deposit of a solifluction mantle.
solifluction stream *Geology.* a narrow, laterally confined deposit of a solifluction mantle that resembles a stream.
Solifugae see SOLPUGIDA.
solion *Electricity.* a circuit element that uses ions in solution in place of electrons as charge carriers; the basic solion is a diode composed of platinum electrodes in an aqueous solution of iodine and potassium iodide.
soliped see SOLIDUNGULATE.
solipsism [sōl´ip siz´əm; säl´ip siz´əm] *Psychology.* the belief that the only thing an individual can truly know is his or her own personal experiences and, by extension, the belief that nothing else exists outside of these experiences. Thus, **solipsistic, solipsist.**
soliquid *Physical Chemistry.* a liquid in which solid particles are dispersed.
solitaire *Vertebrate Zoology.* a flightless bird, *Pezophaps solitaria* of the family Rapidae, closely related to the dodo but with a smaller bill and more coordination; it inhabited the Mascarene Islands in the Indian Ocean and became extinct in the 1700s.
solitary *Behavior.* relating to or designating a species in which the individuals live in isolation as adults rather than forming lasting pairs or groups, such as the tiger.
solitary sandpiper *Vertebrate Zoology.* a sandpiper, *Tringa solitaria,* characterized by dark wings, a brownish-gray, white-spotted back, and a white breast; found in inland wetlands of North America.
solitary vireo *Vertebrate Zoology.* a vireo, *Vireo solitarius,* characterized by a bluish-gray head; found in North and Central America. Also, BLUE-HEADED VIREO.
solitary wave *Physics.* a wave that, unlike a shock wave, does not produce unusually large or drastic changes in the pressure, density, or particle velocity as it passes through a fluid medium.
soliton *Physics.* a wavelike solution to a nonlinear differential equation that maintains a constant shape as it propagates.
soln. solution.
solo *Aviation.* **1.** a solo flight. **2.** to make a solo flight.
solo flight *Aviation.* a flight in which the pilot is alone in the aircraft.
Solo man *Paleontology.* the name given to the fossil remains of several hominids found between 1931 and 1933 at Ngandong, Java, about 100 km from Trinil, the site of the "Java man" find; because of a cranial capacity range of 1035 to 1255 cc, Solo man may represent a transitional stage between *Homo erectus* and *H. sapiens*; extant in Late Pleistocene.
Solomon R unit *Nucleonics.* a unit of rate of radiation dosage, equivalent to 2100 roentgens per hour of X-ray radiation.
Solonchak soil *Geology.* a group of light-colored intrazonal halomorphic soils that develop in semiarid or desert regions and have a high content of soluble salts.
Solonetz soil *Geology.* a group of intrazonal black alkali soils derived from a Solonchak soil by the leaching of salts that exhibit a columnar structure.
solore *Meteorology.* a cold evening wind of the mountains along the course of the Drome River in southeastern France.
Soloth soil *Geology.* a group of intrazonal halomorphic soils derived from the saline material of a Solonetz soil and characterized by a soft, friable surface horizon overlying a lighter-colored, leached horizon above a darker horizon. Also, **Solod soil.**
Solpugida *Invertebrate Zoology.* an order of arachnids, sun spiders, or windscorpions, generally large, pale, and active; nonvenomous but with large chelicerae for seizing and crushing prey. Also, SOLIFUGAE.
solstice [sōl´stis; säl´stis] *Astronomy.* the moment that the sun reaches its greatest extent in declination, north or south.
solstitial colure *Astronomy.* the hour angle that passes through the solstices on the celestial sphere.
solstitial points *Astronomy.* the points on the ecliptic at which the sun reaches its maximum north or south declination.
solstitial tidal currents *Oceanography.* the tidal currents associated with solstitial tides, when the tropic diurnal inequality increases.
solstitial tides *Oceanography.* the tides at or near the solstices, when the sun reaches its greatest declination north or south and the tropic range reaches its maximum.
solubility [säl´yə bil´i tē] *Physical Chemistry.* the ability or tendency of one substance to dissolve into another at a given temperature and pressure; generally expressed in terms of the amount of solute that will dissolve in a given amount of solvent to produce a saturated solution.
solubility coefficient *Physical Chemistry.* an expression of the quantity of solvent needed to dissolve a quantity of gas or solid at a given pressure and temperature.
solubility curve *Physical Chemistry.* a line depicting the concentration of a substance in a saturated solution at a given temperature.
solubility parameter *Materials Science.* the square root of the cohesive energy density; a measurement used to predict miscibility of solutions by their enthalpies of vaporization.
solubility product *Physical Chemistry.* the equilibrium constant for a solid in equilibrium with the concentration of its ions in a saturated solution. Also, **solubility product constant.**
solubility test *Analytical Chemistry.* **1.** any test for the ability of one substance to blend uniformly with another. **2.** a test for the degree of solubility of asphalts and other bituminous material in solvents.
soluble [säl´yə bəl] *Chemistry.* capable of being dissolved in a particular solvent; this quality is often dependent upon temperature.
soluble antigen *Virology.* a virus-specific antigen that is not the virion itself.
soluble barbital see BARBITAL SODIUM.
soluble cutting oil *Materials.* a petroleum oil that mixes easily with water; used as a coolant for metal-cutting power tools.
soluble gas process *Materials Science.* a process for producing powders from molten metal, in which hydrogen gas is charged into a melt, and the hydrogenated alloy is then exposed to a vacuum where, as the temperature falls and hydrogen solubility decreases, the hydrogen is expelled, resulting in self-atomization.
soluble glass see SODIUM SILICATE.
soluble ligature *Surgery.* a ligature, consisting of animal membrane, that will be absorbed by the body.
soluble mold casting *Materials Science.* a shape-forming process that uses a wax mold to fabricate complex shapes.
soluble oil *Materials.* an oil that easily forms a stable emulsion in water. Also, EMULSIFYING OIL.

soluble RNA *Genetics.* an RNA molecule of low molecular weight, consisting principally of transfer RNA; all RNA is actually water-soluble, regardless of molecular weight. Also, TRANSFER RNA.

soluble starch *Materials.* a product of the hydrolysis of starch that is obtained by treating starch with dilute acids, by boiling with glycerin, or by the action of enzymes; used as an emulsifying agent and paper coating.

solum *Geology.* the surface soil and subsoil composing the upper part of a soil profile in which soil-forming processes take place. Also, TRUE SOIL.

solute *Chemistry.* a substance that is dissolved in a solvent; the part of a solution that is uniformly dispersed in another substance.

solution *Chemistry.* a mixture of two or more substances uniformly dispersed throughout a single phase, so that the mixture is homogeneous at the molecular or ionic level. The phases may be: a solid dissolved in a liquid, such as sodium chloride (salt) in water; a liquid dissolved in another liquid, such as water and alcohol; a gas dissolved in a liquid, such as carbon dioxide in water in carbonated beverages; a gas dissolved in another gas, such as nitrogen and oxygen in air; a gas dissolved in a solid, such as hydrogen in platinum; or a solid mixed with another solid, such as an alloy of carbon and iron. *Mathematics.* any value of a variable (or variables) that makes an equation true. The set consisting of all solutions to a given equation is called the **solution set.** Also, ROOT.

solution ceramic *Electricity.* a ceramic insulating coating that can be applied to wires at relatively low temperatures.

solution-dyeing *Textiles.* a process of coloring synthetic yarn in which a pigment or dye is mixed into a spinning solution before it is extruded through a spinneret and formed into yarn, filament, or staple. Also, SPIN-DYEING.

solution gas *Petroleum Engineering.* natural gas dissolved in crude oil as a result of reservoir pressures. Also, DISSOLVED GAS.

solution-gas drive *Petroleum Engineering.* a primary method of oil recovery in which the reservoir gas originally dissolved in the oil is expanded, thereby displacing the oil and forcing it into the wellbores. Also, DISSOLVED-GAS DRIVE, GAS DEPLETION DRIVE, INTERNAL GAS DRIVE.

solution-gas reservoir *Petroleum Engineering.* an oil reservoir whose production is driven principally by the expansion of the natural gas dissolved in the oil; the original pressure of such a reservoir is equal to or greater than the bubble-point pressure of the gas-oil mixture.

solution groove *Geology.* one of a sequence of continuous, more or less parallel furrows that develop on inclined or vertical surfaces of soluble, homogeneous rocks, such as limestone, as a result of the slow corrosive action of trickling water.

solution heat treatment *Metallurgy.* a thermal treatment that causes solid-state dissolution of one or more phases; usually followed by rapid cooling to avoid reprecipitation of such phases.

solution mining *Mining Engineering.* a means of recovering soluble minerals by injecting fluids into subsurface strata and extracting the mineral-laden solutions.

solution poison *Nucleonics.* a substance that is added to the coolant of a nuclear reactor to control reactivity by providing extra absorption of neutron flux density; usually used during shutdown periods and chemically removed prior to start up.

solution polymerization *Materials Science.* a process in which the mixture of monomers and the initiator are dissolved in a nonmonomeric liquid solvent; this is used to create polymers such as lacquers and adhesives.

solution pool *Geology.* a pool that forms in a rock by dissolution in ocean water.

solution porosity *Petroleum Engineering.* a general term for porosity induced by solution action on a reservoir formation.

solution potholes *Geology.* potholes produced by dissolution of carbonate rocks.

solution pressure *Physical Chemistry.* **1.** a measure of the tendency for atoms or molecules to cross a phase boundary and enter a solution. **2.** a measure of the tendency of hydrogen, metals, and certain nonmetals to enter a solution as ions.

solution process *Chemical Engineering.* a petroleum refinery process that involves the washing of gasoline with caustic solution to remove mercaptans.

solution softening *Metallurgy.* softening that occurs in certain alloys when heat-treated to dissolve precipitated constituents.

solution transfer *Geology.* a process in which the pressure solution of detrital mineral grains is followed by the chemical redeposition of the dissolved material on the less strained parts of the grain surfaces.

solution treatment *Materials Science.* the first step in the age-hardening heat treatment, in which the alloy is heated above the solvus temperature to dissolve any second phase and to produce a homogeneous, single-phase structure.

solutizer *Chemistry.* any admixture to a substance that promotes or increases its solubility or the solubility of any of its components.

solutizer-air regenerative process *Chemical Engineering.* a petroleum-refinery sweetening process that dissolves mercaptans from light distillates by dissolving them with a solubility promoter, such as isobutyric acid, along with caustic soda, then regenerates the solution with blowing air. The **solutizer-steam regenerative process** is similar except that steam is the regenerating agent.

solutizer-tannin process *Chemical Engineering.* an early form of the solutizer-air regenerative process that regenerates the solution by a tannin-catalyzed oxidation.

Solutrean *Archaeology.* a culture of the Upper Paleolithic period in Europe, from about 20,000 to about 15,000 years ago; characterized by the use of projectile points, especially the laurel-leaf blade. (From *Solutré,* a site in central France.) Also, **Solutrian.**

solutrope *Chemistry.* a partially miscible mixture of three components, of which two are liquid phases; one component is distributed in one or both of the liquid phases, depending upon its concentration.

solv. solvent.

solv. dissolve. (From Latin *solve.*)

solvable group *Mathematics.* a group G is said to be solvable if it possesses a subnormal series of the form

$$G = G_0 > G_1 > \cdots > G_n = <e>,$$

such that each factor G_i/G_{i+1} is Abelian, where $<e>$ is the trivial subgroup consisting only of the identity element. Also, **solvable series.**

solvable Lie algebra *Mathematics.* a Lie algebra is said to be solvable if its derived series of ideals $L = L^{(0)} \geq L^{(1)} \geq L^{(2)} \geq \cdots$ has $L^{(n)} = 0$ for some n, where $L^{(k+1)} = [L^{(k)}, L^{(k)}]$, i.e., the commutator algebra of $L^{(k)}$.

Solvan *Geology.* a European geologic stage of the Middle Cambrian period, occurring after the Caerfaian and before the Menevian.

solvate *Chemistry.* the compound formed by the interaction of a solvent and a solute.

solvation *Chemistry.* the binding to an ion or polar molecule of one or more molecules of solvent.

Solvay process *Chemical Engineering.* a commercial process that is used to produce soda ash (sodium carbonate) Na_2CO_3, in which sodium bicarbonate separates as a solid, is calcined, and then is changed to soda ash. (Devised in 1861 by the brothers Ernest and Edward *Solvay,* Belgian chemists.)

solvent *Chemistry.* the continuous phase of a solution, in which the solute is dissolved; water is the most common liquid solvent; others include alchol, acetone, and ether.

solvent deasphalting *Chemical Engineering.* a petroleum-refinery process that uses a solvent to remove asphalt by dissolving the oil in the solvent so that the asphalt settles out by gravity. Also, **solvent deresining.**

solvent dewaxing *Chemical Engineering.* the use of a solvent to dissolve waxes from an oil solution, in which the wax solution is chilled and removed by filtration.

solvent extraction *Chemical Engineering.* a process for separating components in solution by transferring mass from one liquid phase into a second immiscible liquid phase.

solvent-flattening see DENSITY MODIFICATION.

solvent molding *Engineering.* a method used to make thermoplastic objects, in which a mold is immersed in a resin solution and the solvent eventually evaporates, leaving a plastic film stuck to the mold.

solvent recovery *Chemical Engineering.* the reclamation of an extraction solvent from a stream, usually by distillation.

solvent-refined *Chemical Engineering.* describing a product that has been purified by the use of chemical solvents.

solvent-refined oil *Materials.* a high-quality, lubricating oil that has been solvent treated during refining.

solvent refining *Chemical Engineering.* the treatment of oil or other material with chemical solvents to remove undesirable compounds.

solvmanifold *Mathematics.* a homogeneous space G/H, where G is a solvable group and H is a closed subgroup of G.

solvolysis *Chemistry.* a chemical reaction in which a solute and a solvent combine to form a new compound or compounds; involves decomposition of the solvent.

solvus *Physical Chemistry.* the points on a phase diagram that indicate the temperature at which a given solid will freeze or melt.

som- or **soma-** a prefix meaning "body." Also, **somat-, somato-.**

soma *Biology.* the whole vegetative body of an organism, excluding the reproductive cells. *Physiology.* the body as distinguished from the mind. *Cell Biology.* a cell body.

Somali Current *Geography.* a current of the Indian Ocean that flows northward along the coast of Somalia in summer and southwestward during the rest of the year.

Somasteroidea *Invertebrate Zoology.* a subclass of starfish, echinoderms in the subphylum Asterozoa.

somasthesia *Physiology.* bodily awareness or consciousness. Thus, **somasthetic.**

somatic *Biology.* **1.** of or relating to the body part of a plant or animal. **2.** made up of or relating to somatic cells, as distinguished from the germ cells.

somatic aneuploidy *Cell Biology.* an abnormal number of chromosomes, typically too many, in a somatic cell.

somatic capillary *Physiology.* any capillary other than those of the lungs.

somatic cell *Cell Biology.* any cell of a multicellular organism that does not participate in the production of gametes; thus any cell except a germ line cell. Also, VEGETATIVE CELL.

somatic cell hybrid *Cell Biology.* a cell formed by the fusion of two somatic cells.

somatic copulation *Mycology.* a form of reproduction in fungi belonging to the subdivisions Ascomycotina and Basidiomycotina, characterized by certain cells joining sexually.

somatic crossing-over *Cell Biology.* a rare event involving a double break, exchange, and rejoining of two strands of homologous chromosomal DNA during mitosis in somatic cells.

somatic death *Biology.* the cessation of characteristic life functions.

somatic hybridization *Cell Biology.* a technique involving the fusion of two different somatic cells to form a somatic cell hybrid.

somatic mesoderm *Developmental Biology.* the outer of the two layers into which the lateral plate mesoderm splits.

somatic mutation *Cell Biology.* a mutation that occurs in a somatic cell rather than a germ line cell, often with no phenotypic effect since the gene in which the mutation occurs may not be expressed in that particular cell.

somatic mutation

somatic mutation theory *Genetics.* a theory stating that the process of aging is the result of a build-up of mutations in somatic tissue.

somatic nervous system *Physiology.* the portion of the neural structure that provides nerve connections to the skin, skeleton, and musculature of the body, excluding the viscera, blood vessels, and glands.

somatic pairing *Cell Biology.* the association of homologous chromosomes in somatic cells, most frequently occurring in polytene chromosomes.

somatic reflex system *Physiology.* an involuntary reflex system involving skeletal muscles.

somatization *Psychology.* the organic expression of a mental disorder.

somatoblast *Invertebrate Zoology.* **1.** a developmental cell that gives rise to body cells, especially in annelid worms. **2.** a cleavage cell descended from a primary somatoblast.

somatochrome *Cell Biology.* a nerve cell that exhibits intense staining properties upon treatment with basic aniline dyes.

somatocyst *Invertebrate Zoology.* in siphonophores, an air cavity in the float.

somatoderm *Invertebrate Zoology.* in Mesozoa, the layer of tissue surrounding the reproductive cell.

somatogenic *Medicine.* originating in the cells of the body. Also, **somatogenetic.**

somatoliberin see GROWTH HORMONE-RELEASING HORMONE.

somatology *Anthropology.* a branch of anthropology that deals with human physical characteristics.

somatomammotropin see CHORIONIC SOMATOMAMMOTROPIN.

somatomedin *Endocrinology.* one of a family of polypeptide growth factors that are produced by the liver in response to growth hormone and are trophic for many cell types, including hepatocytes, adipocytes, lymphocytes, and chondrocytes. Also, INSULINLIKE GROWTH FACTOR.

somatomegaly *Medicine.* abnormally large size of the body; gigantism.

somatometry *Anatomy.* **1.** measurement of the human body. **2.** the classification of individuals based on measurements of their bodies.

somatoplasm *Cell Biology.* the cytoplasm of a somatic cell, as distinguished from germ plasm.

somatopleure *Developmental Biology.* the outer body wall formed by the ectoderm and the somatic mesoderm.

somatopsychic *Medicine.* of both body and mind, used to describe a physical disorder that produces mental symptoms.

somatosensory *Biology.* of or relating to bodily sensations not associated with the eyes, ears, tongue, and other primary sense organs.

somatostatin *Endocrinology.* $C_{76}H_{104}N_{18}O_{19}S_2$, a polypeptide hormone that is produced by the hypothalamus and inhibits the release of growth hormone by the anterior pituitary; it is also produced in the delta cells of the endocrine pancreas, where it inhibits the secretion of insulin and glucagon, and decreases intestinal motility and absorption. Also, SOMATOTROPIN RELEASE-INHIBITING FACTOR.

somatostatinoma *Oncology.* a rare islet-cell tumor of the pancreas that secretes somatostatin; associated with diabetes mellitus or abnormal glucose tolerance.

somatotropic *Biology.* stimulating or promoting overall growth and nutrition. Also, **somatotrophic.**

somatotropin see GROWTH HORMONE.

somatotropin release-inhibiting factor see SOMATOSTATIN.

somatotropin-releasing factor see GROWTH HORMONE-RELEASING HORMONE.

somatotype *Psychology.* an individual's body type according to the theory proposed by W. H. Sheldon that relates physique to personality; the three basic types are endomorph (round), ectomorph (thin), and mesomorph (muscular). Thus, **somatotypology.** Also, **somatype.**

somatyping *Anthropology.* a former practice of classifying humans on the basis of standard body types.

Sombrero galaxy *Astronomy.* an Sa/Sb-type spiral galaxy located in the constellation Virgo; it lies about 60 million light-years distant. Also, **M.104.** (Thought of as having the shape of a Mexican *sombrero.*)

somesthesis *Physiology.* the sensory awareness of one's body.

somma *Geology.* **1.** a circular or crescent-shaped ridge that represents the rim of an ancient volcano. Also, **somma ring. 2.** relating to a volcanic crater that has a central cone surrounded by such a ridge.

Sommelet process *Organic Chemistry.* the production of thiophene aldehydes by treating thiophene with hexamethylenetetramine.

Sommerfeld, Arnold 1868–1951, German physicist; modified Bohr's atomic model; formulated the theory of spectral lines.

Sommerfeld formula *Electromagnetism.* a formula that determines the strength of an electromagnetic field produced by an antenna for relatively short distances. Also, **Sommerfeld equation.**

Sommerfeld law for doublets *Atomic Physics.* a law that is derived from the Bohr-Sommerfeld theory and claims that frequency splitting of relativistic doublets occurs according to

$$a^2R(Z-s)^4/n^3(l+1),$$

where a is the fine structure constant, R is the Rydberg constant, Z is the atomic number, s is the screening constant, and l is the orbital angular momentum.

Sommerfeld model or **Sommerfeld theory** see FREE-ELECTRON THEORY OF METALS.

Sommerfield-Kossel displacement law see SPECTROSCOPIC DISPLACEMENT LAW.

somn- or **somno-** a prefix meaning "sleep." Also, **somni-.** (From *Somnus,* the ancient Roman god of sleep.)

somnambulism [säm nam´byə liz´əm] *Physiology.* the practice of sleepwalking; i.e., leaving one's bed while asleep and moving about without fully awakening, especially when this occurs without dreaming and without later memory of the event.

somnambulist [säm nam´byə list] *Physiology.* a person who sleepwalks; someone who is affected by somnabulism.

somnolence [säm´nə ləns] *Neurology.* **1.** drowsiness or sleepiness. **2.** a level of semiconsciousness approaching coma.

somnolent [säm´nə lənt] *Neurology.* **1.** sleepy or drowsy. **2.** tending to cause sleep.

somnolentia [säm´nə len´shə] *Neurology.* **1.** a condition of sleepiness. **2.** specifically, a condition of being partially awake and unsteady on one's feet, staggering about as if drunk; characterized by excited or violent behavior and a loss of orientation. Also, SLEEP DRUNKENNESS.

somnus *Medicine.* another term for sleep.

son *Computer Science.* see CHILD.

son- or **sono-** a prefix meaning "sound."

sonar [sō´när] *Engineering.* **1.** any system that uses underwater sound waves to determine the location of objects or for communication. **2.** see SONAR SET. *Acoustics.* specifically, a system that uses transmitted acoustic signals and echo returns, as well as acoustic signals originating from other sources, for navigating and determining position and bearing. (An acronym for <u>so</u>und <u>n</u>avigation <u>a</u>nd <u>r</u>anging.)

sonar array *Electronics.* a configuration of sonar transducers or sonar projectors.

sonar attack plotter *Ordnance.* a tactical system that employs sonar data and information from a ship's gyrocompass to graphically present a plan of attack against enemy submarines.

sonar beacon *Acoustical Engineering.* an independent sound-transmitting device, used for underwater homing, tracking, or navigation.

sonar boomer transducer *Acoustical Engineering.* a part of a sonar system used for locating oil under the ocean floor, consisting of an underwater transducer that transmits low-frequency sound (comparable to a sonic boom) into the ocean bottom and receives the echo returns of these transmissions.

sonar capsule *Acoustical Engineering.* a sonar enclosure on a naval vessel, housing a transducer element or array of elements, often physically separated from the main hull structure in order to minimize pickup of the ship's noise.

sonar countermeasures *Military Science.* the measures used to neutralize, deny, or destroy an enemy's effective use of sonar.

sonar dome *Engineering.* a protective compartment for sonar equipment that interferes only minimally with sound reception and transmission.

sonar navigation see SONIC NAVIGATION.

sonarography *Radiology.* a form of ultrasonic scanning that produces a two-dimensional image corresponding to sections of acoustic interfaces in tissues.

sonar projector *Acoustical Engineering.* a transducer, usually a crystal or magnetostriction transducer, that is used underwater to convert electromagnetic energy to acoustic energy with active sonar.

sonar receiver *Electronics.* a device used to receive and display sound signals from an underwater source. Also, **sonar detector.**

sonar resolver *Electronics.* a resolver used in conjunction with echo-ranging and depth-determining sonar in order to determine the horizontal range of a sonar target, as required for depth-bombing.

sonar self-noise *Electronics.* any undesired sonar signals that are generated by the sonar receiving equipment itself.

sonar set *Engineering.* an equipment system that utilizes underwater sound waves to receive or transmit information underwater, or to locate or track objects.

sonar target *Acoustical Engineering.* a vessel at which active or passive sonar is deployed to determine range, course, and other vessel parameters.

sonar transducer *Acoustical Engineering.* a hydrophone or projector used underwater to convert acoustic energy into electromagnetic energy, or electromagnetic energy into acoustic energy.

sonar transmission *Acoustical Engineering.* the emission of underwater sound produced by an active sonar system.

sonar transmitter *Electronics.* a device that generates and transmits electrical signals to a sonar transducer or sonar projector, which in turn generates intelligence-carrying sound waves of the same frequency in water.

sonar window *Acoustical Engineering.* **1.** an acoustically transparent cover over an underwater sonar transducer, hydrophone, or projector that allows the transmission of sound waves with little attenuation. **2.** the domain of effectiveness of a sonar system, such as range, depth, and the like.

sondage see CONTROL PIT.

sonde *Engineering.* a device that measures and transmits physical and meteorological information from high altitudes to a ground station; often attached to a satellite, rocket, or balloon.

Sondheimer effect *Materials Science.* oscillations of the magnetoresistance of a metal plate in a magnetic field; this effect provides information on electron paths as the magnetic field changes the shape and sizes of electron trajectories.

sone *Acoustics.* a unit of loudness defined as the loudness of a 1000-Hz tone at a 40-decibel sound pressure level as heard by both ears, in which the number of sones doubles with every 10-decibel increase above 40 decibels and halves with every 10-decibel decrease below 40 decibels.

song *Behavior.* a series of sounds made by a bird, or by a whale, frog, or other animal, distinguished from the animal's ordinary calls by length and complexity.

song sparrow *Vertebrate Zoology.* a small sparrow, *Melospiza melodia*, having a brown body and a streaked breast with a dark central spot; found widely in North America.

song sparrow

sonic *Acoustics.* of or relating to sound. *Mechanics.* of or relating to a speed equal to that of sound. *Aviation.* specifically, designed to perform at or above the speed of sound.

sonic altimeter *Engineering.* a device that measures the altitude of an aircraft by measuring the time it takes for sound waves to reach the earth and then return to the aircraft.

sonic anemometer *Engineering.* an instrument that measures wind speed, using the principle that the speed of electromagnetic wave motion of a sound wave within a moving medium is equivalent to the speed of sound in relation to the medium plus the speed of the medium.

sonication *Biochemistry.* the process of subjecting a substance, cell, virus, etc., to high-frequency sound-wave energy, usually for purposes of dispersal or separation.

sonic barrier *Aviation.* a sharp, sudden increase in aerodynamic drag on an aircraft approaching the speed of sound. Also, SOUND BARRIER, TRANSONIC BARRIER.

sonic boom *Acoustics.* a noise heard on the ground as a loud boom, resulting from strong pressure waves that are created by high-speed aircraft in supersonic flight.

sonic chemical analyzer *Engineering.* an instrument that analyzes a material's molecular composition by observing how sound waves are altered when they pass through the material.

sonic cleaning *Engineering.* a process of purifying contaminated material, in which the material is immersed in a liquid that is acted upon by intense sound.

sonic delay line see ACOUSTIC DELAY LINE.

sonic depth finder *Engineering.* a device that determines the depth of a body of water or the location of an underwater object by measuring the time interval between the emission of a sound and the return of the sound after reflection. Also, ECHO SOUNDER. *Acoustics.* see FATHOMETER.

sonic drilling *Mechanical Engineering.* a process of cutting materials with an abrasive slurry that is driven by a reciprocating tool vibrating at sonic frequency.

sonic flaw detection *Engineering.* the process of identifying areas of imperfection in a solid substance by observing the way in which the substance reflects sound waves or the way sound waves are altered as they travel through the substance.

sonic navigation *Navigation.* navigation that utilizes audible or nonaudible sound waves. Also, ACOUSTIC NAVIGATION, SONAR NAVIGATION. Similarly, **sonic bearing, sonic fix, sonic line of position.**

sonic nucleation *Chemical Engineering.* the use of sonic energy to cause coagulation of submicrometer droplets.

sonic pump *Petroleum Engineering.* a lifting pump used to extract crude from shallow wells by means of surface-generated harmonic vibrations, which cause several hundred strokes per minute, each of which moves the fluid a fraction of an inch.

sonic radiation *Acoustics.* waves that are created by acoustic energy, described by the equation $c = \lambda\nu$, where c is the speed of sound in a medium, λ is the wavelength of the sound, and ν is the frequency of the sound.

sonics *Acoustics.* the technology of sound processing and analysis, including all noncommunicative processes.

sonic sifter *Mechanical Engineering.* a high-speed device that separates particles by size using vibrating motion at sonic frequencies.

sonic sounding *Engineering.* a process in which ocean depth is measured by observing the time interval between sound transmission from a ship and the return of the echo from the ocean bottom.

sonic spark chamber *Nucleonics.* a spark chamber equipped with two microphones placed so that they can determine the position at which a spark occurs, by measuring the time required for the sound of the spark to reach the microphones.

sonic thermometer *Engineering.* an instrument that measures temperature by calculating the change in the speed of a sound wave as it passes through the material to be measured.

sonic well logging *Engineering.* a process that determines the size or holding capacity of a well, using a pulse-echo system that measures the distance between a sound-originating instrument and a sound-reflecting surface.

Sonnar lens *Optics.* a modified triplet photographic lens.

Sonne dysentery *Medicine.* a form of dysentery occurring in temperate regions, caused by the Sonne strain of the Shigella dysenteria bacteria, *Shigella sonnei.*

Sonnenschein's reagent *Analytical Chemistry.* a solution of phosphomolybdic acid that is used to test for alkaloid sulfates; a yellow precipitate indicates a positive test.

Sonneratiaceae *Botany.* a family of tanniferous, dicotyledonous trees of the order Myrtales, characterized by internal phloem and large, regular flowers pollinated by bats; native to the Old World tropics.

Sonn-Muller reaction *Organic Chemistry.* a reaction that involves condensation of an acid chloride with aniline and conversion of the anilide with phosphorus pentachloride into an imino chloride.

sonobuoy *Acoustical Engineering.* a buoy containing a hydrophone and radio transmitter, attached to a parachute and dropped into a body of water by an aircraft; it picks up underwater sounds, generally from submarines, and transmits them back to the aircraft. Also, RADIO SONOBUOY.

sonobuoy reference system *Navigation.* an automatic electronic system that locates sonobuoys. Also, SRS.

sonocatalysis *Chemistry.* the process of catalyzing a reaction by ultrasonic radiation.

sonochemistry *Chemistry.* chemistry in which reactions occur under ultrasound.

sonogram *Radiology.* a record or display obtained by ultrasonic scanning.

sonograph *Engineering.* an instrument that records sounds or seismic vibrations and translates them into phonetic symbols.

sonography see ULTRASONOGRAPHY.

sonolucency *Radiology.* the property of being sonolucent.

sonolucent *Radiology.* of or relating to a substance that permits the passage of ultrasound waves without creating echoes.

sonoluminescence *Physics.* a luminescent phenomenon that is produced in certain materials by high-frequency sound waves or phonons.

sonometer *Engineering.* an instrument used to determine rock stress by measuring the change in pitch of a piano wire that is stretched between two bolts in the rock and then destressed.

sonora *Meteorology.* a thunderstorm that occurs in the mountains and deserts of southern California and Baja, Mexico, in the summertime.

sonoscan *Engineering.* a microscope that analyzes sound waves by observing, with the aid of a laser beam, how an unfocused acoustic beam affects a liquid-solid interface.

soot *Materials.* a powdery black substance, composed chiefly of carbon, and formed by the incomplete combustion of wood, coal, oil, or other material.

soot blower *Engineering.* a blasting system of air or steam jets that removes ash and slag deposits from the tubes of a steam generator.

soot luminosity *Optics.* the amount of light in a flame that comes from incandescent solid particles of soot.

sooty bark *Plant Pathology.* a disease of the sycamore tree caused by *Cryptostroma corticale.*

sooty blotch *Plant Pathology.* **1.** a fruit disease, especially of apples and pears, that is caused by the fungus *Gloeodes pomigena*; characterized by sootlike spots on the fruit. **2.** a clover disease caused by the fungus *Cymadothea trifoli*; characterized by crusty black blotches on the underside of the leaves.

sooty mold *Mycology.* fungi belonging to the subdivision Ascomycotina that occur most frequently on dung, fruit, twigs, needles, stems, and leaves, on which they form sooty-looking patches without being parasites. *Plant Pathology.* a plant disease common to citrus trees, caused by fungi of the family Capnodiaceae and characterized by a thick, velvety coating on foliage and fruit.

SOP or **S.O.P.** standard operating procedure.

sophisticated *Engineering.* describing a computer, electronic device, or the like that is regarded as highly advanced and as capable of performing complex or inticate operations.

sophisticated robot *Robotics.* a robot that uses the latest technology and is under microprocessor control.

sophisticated vocabulary *Computer Programming.* an advanced and elaborate set of instructions that enables a computer to go beyond the basic tasks of addition, multiplication, and subtraction to more complex tasks, such as line organization and square root extraction.

sophoretin see QUERCETIN.

sopor *Neurology.* an unnaturally deep sleep; stupor.

soporous *Neurology.* relating to or causing sopor; comatose; stuporous.

s.op.s. or **S.op.s.** if needed. (From Latin *si opus sit.*) Also, **S.O.S., s.o.s.**

SOR stimulus-organism-response.

Sorapillaceae *Botany.* a monogeneric family of sparsely branched mosses of the order Dicranales; characterized by dictichous and complanate leaves with bases that clasp the stem, short immersed capsules, and lateral sporophytes.

sorb *Physical Chemistry.* to undergo a process of sorption.

sorbent *Materials.* any material, generally a mineral such as a clay or a silicate, that absorbs or adsorbs; sorbent clays are used to decolorize and clarify various kinds of oil, sugar cane, beer, and wine, to absorb industrial chemicals, and as pet litter.

sorbic acid *Organic Chemistry.* $CH_3CH{=}CHCH{=}CHCOOH$, a combustible, white crystalline solid; slightly soluble in water and many organic solvents; melts at 134°C and boils (decomposes) at 228°C; used as a fungicide, food preservative, and intermediate. Also, 2,4-HEXADIENOIC ACID.

sorbide *Chemistry.* a collective term for dianhydrosorbitols, having the general formula $C_6H_8O_2(OH)_2$ and derived from sorbitals.

sorbitol *Organic Chemistry.* $HOCH_2(CHOH)_4CH_2OH$, a hygroscopic, white crystalline powder; soluble in water and almost insoluble in most organic solvents; the anhydrous form melts at 110–112°C; used in toothpastes, cosmetics, pharmaceuticals, resins, and as a food additive. Also, HEXAHYDRIC ALCOHOL.

sorbose *Biochemistry.* a carbohydrate prepared by fermentation used in the production of vitamin C. Also, **sorbin.**

sorcerer *Anthropology.* a person who practices sorcery; a woman who does this is usually called a **sorceress.**

sorcery *Anthropology.* an action undertaken to bring harm to another by means of supernatural powers; often the personal belongings of the victim or other materials are used in the ritual.

Sordaria *Mycology.* a genus of fungi belonging to the order Sordariales that are found on soil and dung and produce black or brown spores called **ascospores** in dark spore-bearing bodies.

Sordariaceae *Mycology.* the only family of fungi belonging to the order Sordariales; these fungi grow rapidly on plant debris, dung, and seeds, and are used extensively in biochemical and genetic research.

Sordaria fimicola *Mycology.* a species of fungi belonging to the genus *Sordaria* that are homothallic or self-fertilizing and are used widely for biological research.

Sordariales *Mycology.* an order of fungi belonging to the class Pyrenomycetes that are found on dung, soil, and rotting wood; some members of this order are used in biochemical and genetic research.

sordawalite see TACHYLITE.

soredium *plural,* **soredia.** *Botany.* a vegetative propagation structure of some lichens that occurs on the surface of the thallus and is composed of algal cells enclosed within fungal hyphae; upon separation from the original thallus, it grows into a new thallus.

Sörensen, Sören Peter 1868–1939, Danish biochemist; developed the pH system for measuring the concentration of hydrogen ions in a solution.

Sörensen acid indicator *Analytical Chemistry.* a substance that indicates by its color the hydrogen-ion concentration of a solution.

Sörensen titration *Analytical Chemistry.* the use of a Sörensen acid indicator in a titration.

sore shin *Plant Pathology.* a disease of cotton, tobacco, and other plants caused by a fungus, especially *Rhizoctonia solani,* and characterized by a fungal growth that encircles the base of the stem.

Soret coefficient *Physics.* a quantity frequently used in the study of binary thermal diffusion given by the coefficient of thermal diffusion divided by the coefficient of ordinary diffusion, and equal to the product of the thermal diffusion constant and the mole fractions of the two molecular weight components.

Soret effect *Physics.* the phenomenon of thermal diffusion.

sorghum *Botany.* **1.** one of two cereal grasses, *Sorghum vulgare* or *S. bicolor,* with broad, cornlike leaves, a tall pithy stem, and grain borne in a dense terminal cluster; cultivated mainly for stock feed and syrup. **2.** the syrup obtained from sorghum.

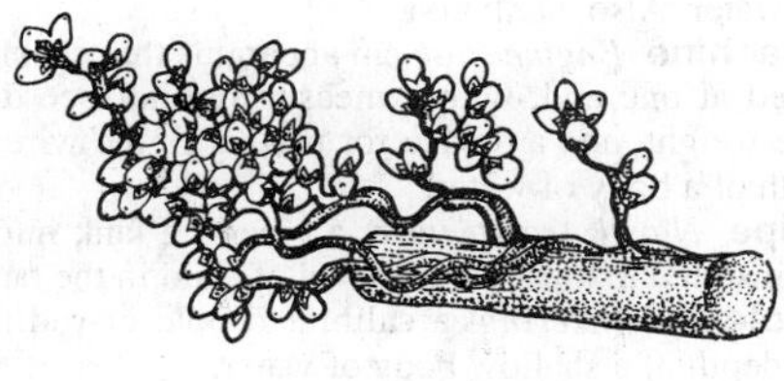

sorghum

Soricidae *Paleontology.* the shrews, a large family of insectivores in the suborder Erinaceomorpha (Lipotyphla); first appeared in the Late Eocene. *Vertebrate Zoology.* the shrews, a family of mouselike insectivores belonging to the suborder Lipotyphla and having a worldwide distribution.

soroban *Mathematics.* a Japanese form of abacus in which the upper section has one marker per wire, and the lower section has four markers per wire.

sorocarp *Mycology.* a kind of fungal fruiting body found in certain slime molds belonging to the family Acrasieae, consisting of a stalk carrying spores.

Sorocarpaceae *Botany.* a family of brown algae of the order Ectocarpales, characterized by a prostrate system of filaments giving rise to simple, erect filaments ending in phaeophycean hair.

sorocyst *Mycology.* a kind of fungal fruiting body similar to a sorocarp but without a stalk.

sorogen *Mycology.* the mass of fungal cells that forms a sorocarp.

sororal polygyny *Anthropology.* the practice in which one man is married to two or more sisters simultaneously, often with one joint household.

sororate *Anthropology.* the practice in some societies in which a woman must marry the husband of her deceased sister.

sorosilicate *Mineralogy.* any of a group of silicate minerals, such as epidote, whose structures are characterized by linked pairs of SiO_4 tetrahedra sharing an oxygen atom, forming Si_2O_7 groups.

sorosis *Botany.* a fleshy multiple fruit composed of the ovaries, receptacles, and associated parts of an entire cluster of flowers, as in the pineapple and mulberry.

sorption *Physical Chemistry.* a general term for the various processes by which one substance binds to another, especially the processes of absorption (taking in the other substance completely) or adsorption (holding the other substance on the surface).

sorption isotherm *Materials Science.* the expression of the hygroscopicity of wood as a curve that relates moisture content with a given relative-vapor pressure.

sorption pumping *Engineering.* a process used to lower atmospheric gas pressure, in which gas is collected and held on a granular sorbent material, the material is immersed in liquid nitrogen, and the gas is sorbed.

sorption water see PELLICULAR WATER.

sorrel *Botany.* **1.** any of various perennial herbs belonging to the genus *Rumex* of the buckwheat family; characterized by acid edible leaves that are used in salads and sauces. **2.** any of various similar plants.

sorrel tree *Botany.* a tree, *Oxydendrum arboreum,* of the heath family, characterized by drooping clusters of white flowers and acid-tasting leaves. Also, SOURWOOD.

sort *Computer Programming.* **1.** to organize a group of records in a specified order, such as chronological or alphabetical order, to increase the efficiency of locating specific records. **2.** a utility routine that sorts records stored on tape or disk. **3.** a feature of certain word-processing programs that permits items in a document to be arranged in alphabetical order according to the first word of each paragraph. **4.** a verb in COBOL that is used to invoke sorting.

sort algorithm *Computer Programming.* a method used to arrange data items into a particular sequence according to precise rules.

sorted *Geology.* **1.** relating to patterned-ground features displaying a border of stones, usually alternating with or surrounded by smaller particles. **2.** of or relating to an unconsolidated sediment or a cemented detrital rock that consists of particles of essentially uniform size or lying within limits of a single grade. **3.** see GRADED, def. 2.

sorted polygon *Geology.* a patterned ground characterized by a polygonal mesh, whose sorted appearance results from a border of stones surrounding finer material. Also, STONE POLYGON, STONE MESH.

sort field *Computer Programming.* a field in a record that dictates the sequence of the sort. Also, **sort key.**

sort generator *Computer Programming.* a routine that produces a sort program for use in a production run.

sortie *Aviation.* a flight or mission, especially by a single military aircraft into enemy territory. *Military Science.* **1.** a sudden attack made from a defensive position. Also, SALLY. **2.** to depart from a port or anchorage, especially for an operation or maneuver.

sortie number *Engineering.* a number that represents the images detected by all the sensors during a single reconnaissance flight of an aircraft.

sortiite *Geology.* a type of meteorite similar to the pallasites, with troilite substituting for olivine.

sorting *Geology.* the process by which particles of a particular sediment are selected and separated according to size, shape, or specific gravity from associated but dissimilar particles by natural agents of transportation, especially running water. *Computer Programming.* see SORT.

sorting coefficient *Geology.* an expression of the geometric spread of the central half of the particle-size distribution of a sediment, indicating the range of conditions of the transporting medium and to some extent the distances over which the particles have been transported.

sorting index *Geology.* a measure of the degree of sorting or particle-size uniformity in a sediment.

sorting out *Cell Biology.* the phenomenon by which mixed aggregates of different embryonic cell types eventually differentiate so that similar cells come together, often with one cell type forming a central mass surrounded by the other cell type.

sorting table *Engineering.* a conveyor belt along which workers standing on either side sort out materials.

sort/merge *Computer Programming.* a program that sorts a set of records and then merges them with a master file.

sort/merge program *Computer Programming.* a generalized program designed to produce many different sorting or merging programs according to control information specified by the user.

sort order *Computer Programming.* the desired sequence of the file, either ascending or descending.

sort pass *Computer Programming.* a single pass through the items of a list during the execution of a sort program.

sortworker *Computer Programming.* a temporary file created to hold the intermediate results of a sort when the amount of data is greater than the available storage space.

sorus *Botany.* **1.** in vascular plants, a cluster of sporangia. **2.** in lichens, a mass of soredia. **3.** in algae, a cluster of antheridia. *Mycology.* in fungi, a cluster of spores.

SOS *Telecommunications.* the international radiotelegraph distress signal; in Morse code it consists of the letters SOS, represented by the sequence of three dots, three dashes, and three dots.

SOS *Electronics.* silicon-on-sapphire.

SOS gene *Molecular Biology.* any of a class of specialized repair genes that are normally repressed by the lexA gene but become activated when DNA is damaged or inhibited.

sosoloid *Physical Chemistry.* a type of solid in which there are dispersed particles from another solid.

SOS repair *Molecular Biology.* a process by which damaged DNA is acknowledged and mended; it involves a number of enzymes and is therefore subject to error.

SOS system *Molecular Biology.* a system by which the expression of certain genes is inititated or enhanced in response to an inhibition of DNA replication or damage to DNA.

Sothic cycle *Astronomy.* a 1460-year calendric cycle used in ancient Egypt. Also, **Sothic period.** (From *Sothis,* the Greek name for *Sirius,* the dog star, whose heliacal rising determined the fixed year of the ancient Egyptians.)

souma *Veterinary Medicine.* an unscientific name for a highly fatal disease of horses and other herbivores, due to infection by trypanosomes transmitted by tsetse flies and characterized by anemia, intermittent fever, and gradual weight loss.

sound *Acoustics.* **1.** a pressure disturbance that propagates through a medium due to stress or displacement of the medium from its equilibrium state. **2.** the auditory perception that is induced by such a disturbance; something heard by the ears. *Geography.* a broad sea passage, especially between an island and the mainland. *Surgery.* a probelike surgical instrument that may be inserted into a body to detect foreign matter or to dilate a stricture, as of a urethra. *Computer Science.* describing a theorem-proving technique or method of reasoning that is guaranteed to derive only valid conclusions.

sound absorption *Acoustics.* the decay of sound energy as its distance from the sound source increases, due to conversion of the acoustic energy to heat generated in the medium.

sound analyzer *Engineering.* a device used to measure sound energy in bands of various frequencies; composed of an arrangement of stable electrical filters or one tunable electrical filter, amplifiers, and an indicating meter.

sound-and-flash ranging *Ordnance.* a combined method of locating the position of enemy fire or adjusting the range of friendly fire by utilizing the techniques of sound ranging and flash ranging.

sound attenuation *Acoustics.* the decrease of sound energy due to absorption, scattering, and spreading in a medium as the sound moves farther from its source; the intensity can be described by the equation $I = I_0e^{-\alpha r}$, where I_0 represents the original sound intensity, α represents the attentuation coefficient of the medium, and r is the distance.

sound band pressure level *Acoustics.* the sound pressure level for the selected frequency band of a frequency analyzer.

sound barrier see SONIC BARRIER.

sound buoy *Navigation.* a buoy that emits a recognizable, audible sound; designed to be used as a navigational aid.

sound carrier *Telecommunications.* the television carrier that is frequency modulated by the aural part of a television broadcast.

sound channel *Electronics.* in a television receiver, a transmission channel through which sound is received. *Acoustics.* a depth range in an ocean or other large body of water in which sound is trapped between two upper and lower boundaries of maximum sound velocity, with a depth of minimum sound velocity at the center of the channel so that sound refracts away from the boundaries toward the central axis.

sound detection *Acoustics.* the discrimination of a specific sound of interest separated from background noise.

sound effects *Acoustical Engineering.* artificially created sounds used to simulate real sounds during an audio or video recording or in a radio broadcast.

sound energy *Acoustics.* the energy associated with sound waves, expressed quantitatively by the difference between the total energy of a sound field and the energy that would exist in the same region in the absence of sound. Also, ACOUSTIC ENERGY.

sound energy density *Acoustics.* for an enclosed volume, a quantity expressed by the amount of sound energy within the volume divided by the volume.

sound energy flux *Acoustics.* the average amount of sound energy transmitted in a specific direction perpendicularly across a unit area at the point considered.

sounder *Telecommunications.* an instrument that emits a series of short electromagnetic pulses toward the surface of a planet or satellite; the return signal indicates electrical conductivity below the surface, providing information on the subsurface structure and topography.

sound film *Acoustical Engineering.* motion-picture film that has an accompanying audio track near the edge of one side of the film.

sound filmstrip *Acoustical Engineering.* a filmstrip with an audio accompaniment that is separate from the film and that must be synchronized with the visual images.

sound gate *Acoustical Engineering.* a gate through which the audio track of a sound film passes in a movie projector, for the pickup and amplification of the audio portion of the movie.

sound head *Acoustical Engineering.* the part of a movie projector that picks up and converts the electroacoustic signals to audible signals.

sound image *Acoustics.* the amplitude pattern of acoustic energy around a sound source due its radiated sound.

sounding *Engineering.* **1.** a technique for measuring the depth of a body of water by using an echo-sounder or sounding line. **2.** a term for any procedure that involves penetrating the natural environment to make observations. *Meteorology.* an observation of weather conditions aloft, above the range of the normal surface weather observation; used in the analysis of upper-air charts. Also, UPPER-AIR OBSERVATION, UPPER-AIR SOUNDING. *Mining Engineering.* a technique for measuring the depth of bedrock by driving a steel rod into the soil.

sounding balloon *Engineering.* a small, free balloon sent into the upper atmosphere containing transducers to measure and self-registering instruments to record and transmit meteorological reports to a ground station.

sounding lead *Engineering.* a heavy lead weight attached to a calibrated line; used aboard ships to determine depth measurements.

sounding line *Engineering.* a strong cord, calibrated at intervals and weighted at one end with lead, that is used to take depth measurements in bodies of water. Also, LEAD LINE.

sounding machine *Engineering.* an apparatus that consists of a reel of wire weighted at one end, with a measuring and recording device attached to the weight, and a device for pulling in the wire; used to measure the depth of a body of water.

sounding pipe *Naval Architecture.* a pipe in a tank into which a dipstick can be inserted to measure the level of fluid in the tank.

sounding pole *Engineering.* a calibrated pole or rod that is used to measure the depth of a shallow body of water.

sounding rocket *Space Technology.* a rocket designed to carry equipment used for studying the upper atmosphere, usually having an almost perfectly vertical trajectory.

sounding sand *Geology.* a dry, usually clean sand that emits peculiar musical, humming, or crunching sounds when disturbed. Also, SINGING SAND.

sounding velocity *Acoustics.* the velocity of sound in water, used to compute water depth when taking a sounding; for most electronic fathometers, it is calibrated at approximately 4800 feet per second.

sounding wire *Engineering.* strong wire used in a sounding machine.

sound intensity *Acoustics.* the loudness of a given sound, described as the rate of sound energy transmission in a specified direction at a given point, expressed by the equation $dB = 2\log(I/I_0)$, in which I_0 represents the reference level expressed in phons.

sound irradiator *Acoustics.* a device that is designed to emit and focus sound into a point of high intensity at its focus.

sound lag *Acoustics.* the difference in time between the emission and reception of a sound wave.

sound level *Acoustics.* **1.** the relative loudness of a sound being recorded or broadcast. **2.** the root-mean-square energy equivalent of a sound as measured with a meter calibrated to a scale, as in A-weighting. **3.** in underwater sound, the pressure level of a specific sound as measured by a sonar system.

sound-level meter *Engineering.* an apparatus that measures sound level intensities; consists of a microphone, an amplifier, frequency-weighting networks, and a meter calibrated in decibels or volume units.

sound locator *Acoustical Engineering.* an electroacoustic device with high directivity that is commonly used to find a sound source.

sound masking *Acoustics.* a phenomenon in which one sound blocks out another sound, in which the sound that is heard is not necessarily the loudest, but the one that has the greatest psychological effect on the listener; particularly effective at masking when close in frequency to the original frequency.

sound navigation and ranging *Acoustics.* see SONAR.

sound power *Acoustics.* the power due to sound energy that is emitted from a sound source, normally measured in watts and calculated with one of several equations, depending on the medium and its theoretical model. Also, ACOUSTIC POWER.

sound-powered telephone *Acoustical Engineering.* a telephone that uses highly inductive and durable transducers, such as single-button carbon microphones, for communication, and that derives its power from the acoustic energy of the voice.

sound pressure *Physics.* the difference between the pressure at a point in a medium through which a sound wave is passing and the static pressure of the medium at that point. *Acoustics.* see EFFECTIVE SOUND PRESSURE.

sound pressure level *Acoustics.* the decibel measurement of sound force per unit area, equal to 20 times the logarithm (base ten) of a nonreference pressure divided by a reference pressure, as described by the equation $dB = 20 \log(P/P_0)$.

sound production *Acoustical Engineering.* the conversion of one form of energy, usually mechanical energy, into acoustic, or audible, energy.

sound ranging *Acoustical Engineering.* the use of sound, as employed naturally by bats, to determine the range and bearing of an object by measuring the time between the transmission and reception of a sound reflection from the target.

sound ray diagram *Acoustics.* a diagram of the predicted behavior of various sound wavefronts over a given region, based on measurements of temperatures or velocities over this region.

sound reception *Acoustical Engineering.* the detection and recognition of an audio signal by a sound sensor, as by human hearing or passive sonar.

sound recognition *Artificial Intelligence.* the process of reasoning about acoustic signals, such as alarms or spoken words, and producing readable output.

sound recording *Acoustical Engineering.* the preservation of sound, either by mechanical means such as a disk recording, electromagnetically by means of flux patterns on electromagnetic tape, or by optical means, as on a compact disk.

sound reduction factor *Acoustics.* an attenuation coefficient; a measure of the rate of reduction of sound intensity as a function of frequency for a specific medium or for specific environmental factors such as humidity.

sound-reinforcement system see PUBLIC-ADDRESS SYSTEM.

sound-reproducing system *Acoustical Engineering.* an electroacoustical system, such as a tape player, phonograph, compact disk player, or public address system, that converts electroacoustical representations of sound into a facsimile of the original sound. Also, AUDIO SYSTEM, SOUND SYSTEM.

sound reproduction *Acoustical Engineering.* the re-creation of electromagnetically, mechanically, or optically recorded sound with a sound system such as a tape player, record player, or compact disk player.

sound signal *Telecommunications.* any sound that conveys intelligence, as distinguished from noise.

sound spectrogram *Acoustical Engineering.* an instrument that displays the intensity-level distribution of frequency and time for a short speech sample.

sound spectrum *Acoustics.* **1.** the specific frequency coverage of a frequency analyzer and similar frequency-dependent sound-measurement equipment. **2.** generally, a broad theoretical range, such as the human audible spectrum of 20 Hz to 20 kHz.

sound speed *Engineering.* the rate at which a sound motion picture film moves through a projector; 24 frames per second. The speed of sound, however, is about 1100 feet per second at sea level.

sound system see SOUND-REPRODUCING SYSTEM.

soundtrack or **sound track** *Acoustical Engineering.* **1.** a narrow band along the edge of a motion-picture film on which the dialogue and other sound accompanying the film is recorded. **2.** this accompanying sound itself; the audio portion of a motion picture. **3.** a portion of a tape on which the electroacoustic signal from one channel of a sound system is recorded.

sound transmission *Acoustics.* the movement of acoustic energy through a medium due to a mechanical disturbance that propagates through the medium.

sound transmission coefficient *Acoustics.* the ratio of incident sound energy to that of transmission in the event that a sound wave strikes a boundary of different material; the coefficient is dependent upon the material and the angle of incidence.

sound type system *Computer Science.* a type system of a programming language in which it is guaranteed that the value of a variable at runtime can only be of the type that was determined for that variable at compile time; i.e., there can be no runtime type errors.

sound velocity *Acoustics.* the speed of sound in a specific medium at a given temperature and pressure, according to the formula $c = \lambda\nu$, where c is the sound velocity, λ is a specific wavelength, and ν is a corresponding frequency of the sound at this wavelength.

sound volume velocity *Acoustics.* the flow rate of a substance resulting from the passage of sound waves through the medium.

sound wave *Physics.* any of the longitudinal progressive vibrations in an elastic medium, by which sounds are transmitted.

sour *Physiology.* describing a food or other substance that has a sharp, acidic taste like that of a lemon or of vinegar. *Chemistry.* a substance containing large amounts of sulfur compounds such as hydrogen sulfide or mercaptans; used to neutralize alkalinity or decompose residual bleach in laundry and in textile operations.

source the point or object from which something emanates; specific uses include: *Thermodynamics.* any body, device, or system that provides energy or material. *Physics.* a point, line, or area in space from which lines of force in a vector field originate. *Acoustics.* a region of space in contact with a fluid medium where new acoustic energy is being generated and radiated outward uniformly in all directions as sound waves. *Electricity.* any active component, battery, or generator that supplies energy. *Electronics.* **1.** a circuit or device that transmits signal power or electric energy to a transducer or load circuit. **2.** in a field effect transistor, the terminal that receives the flow of charge carriers. *Spectroscopy.* an arc, spark, lamp, laser, or other device used to supply radiant energy for a spectroscope. *Nucleonics.* a sample of radioactive material that is prepared for experimental or industrial use. *Mathematics.* **1.** in a directed graph, a vertex with indegree 0. **2.** in a flow graph, a vertex which may have a net flow out of the vertex. In this case, the indegree of the vertex may exceed 0. **3.** in a network, any of a set of vertices designated as sources.

source address *Computer Programming.* the address of the device or memory location from which data is being transferred.

source bed *Geology.* the original stratigraphic horizon of a sedimentary environment from which secondary sulfide minerals were derived.

source code *Computer Programming.* any symbolic computer instructions that are written in a high-level language and cannot be directly executed without first being compiled into object code.

source data automation *Computer Technology.* a technique for automatically collecting input or source data for a computerized process that does not require manual transcription of the data into machine-readable form.

source data capture *Computer Programming.* the collection and transformation of source data prior to computer processing.

source data entry *Computer Programming.* the automatic entry of source data through techniques such as optical scanning and reading of human-readable documents without transcription.

source document *Computer Programming.* an original document from which data is extracted.

source field *Acoustics.* the sound field radiated by a source; if the flow outward is $Se^{-i\omega t}$, the pressure radiated by the source can be described by the following equation:

$$p(r) = (ik\rho c/4\pi r)Se^{ik(r-ct)},$$

where $k = \omega/c$, ω = the radian frequency, ρ = the density of the medium, c = the sound velocity in the medium, and r = the distance from the source to the field point.

source flow *Fluid Mechanics.* **1.** a point from which a fluid flows uniformly to all directions in three-dimensional flow. **2.** a straight line from which a fluid flows uniformly along all directions perpendicular to the line in two-dimensional flow.

source-focused illumination *Optics.* illumination that images light onto the surface of an object.

source-follower amplifier see COMMON-DRAIN AMPLIFIER.

source function *Astronomy.* the ratio of emissivity to opacity for stellar or other radiating material.

source habitat *Ecology.* a type of habitat in which surplus individuals are produced and from which these individuals disperse to other habitats.

source impedance *Electricity.* the output impedance of a source in a circuit, such as that of a generator or transducer.

source language *Computer Programming.* the original language in which a program is prepared before compilation.
source level *Acoustics.* the sound pressure level, measured in decibels, of a detected target for a passive sonar system in the following equation: $DT = SL - TL + DI$, with DT representing the detection threshold, TL representing one-way transmission loss, and DI representing the directivity index of the sonar system; for an active sonar system, the source level is expressed in the equation $SL - 2TL + TS - RL + AG = DT$, with $2TL$ representing the two-way transmission loss, RL representing the reverberation loss, and AG representing the array gain.
source library *Computer Programming.* a collection of source programs available to several users.
source listing *Computer Programming.* a printed copy of a source program.
source material *Nucleonics.* a substance from which fissionable material can be obtained.
source module *Computer Programming.* a unit of source code capable of being compiled as a single unit.
source program *Computer Programming.* a program coded so that it must be translated into machine language prior to use.
source program optimizer *Computer Programming.* a compiler program designed to restructure the source program so as to optimize computer performance.
source region *Meteorology.* an extensive area of a surface characterized by uniform weather conditions as a result of stagnant air taking on the temperature and water vapor properties intrinsic to the surface, such as an ocean, and imparting these properties to an overlying air mass.
source rock *Geology.* **1.** a sedimentary rock whose organic material was transformed over time into liquid or gaseous hydrocarbons under pressure and heat. **2.** the rock from which a sediment is derived. Also, PARENT ROCK.
source time *Computer Technology.* see ACCESS TIME.
source trait *Psychology.* a basic, underlying personality trait that determines part of an individual's behavior pattern.
source transition loss *Electronics.* the transmission loss occurring at the junction between an energy source and a transducer.
sourcing *Electronics.* a redesign or restructuring of existing equipment in order to remove radio-frequency interference; when sourcing is not feasible, engineers resort to suppression, filtering, or shielding.
sour corrosion *Petroleum Engineering.* corrosion in gas or oil wells in which there is an iron-sulfide corrosion product and foul-smelling hydrogen sulfide is detected in the recovered fluid.
sour cream *Food Technology.* a milk product made of pasteurized cream that has been mildly fermented with *Lactobacilli* and *Leuconostoc* to give it a sharp taste.
sour dirt *Petroleum Engineering.* earth containing sulfate or smelling of sulfur dioxide or hydrogen sulfide; usually indicates oil in the region. Also, COPPER DIRT.
sour gum *Botany.* a tree, *Nyssa sylvatica*, having elliptic leaves, dark-blue fruit; found in eastern North America. *Materials.* the wood of this tree.
sour rot *Plant Pathology.* a disease of citrus trees caused by the fungus *Oospora citriaurantii* and characterized by a soft, slimy decay of the fruit.
sourwood *Botany.* a tree, *Oxydendrum arboreum,* of the family Ericaceae, characterized by long, pointed, finely toothed leaves with an acid flavor, one-sided racemes of white flowers clustered in drooping panicles, and fruit that is a five-angled gray capsule.
south *Geodesy.* a reference direction relative to the earth, indicated by the negative direction of a line lying in a plane through the earth's axis of rotation; the opposite of north.
South African jade see TRANSVAAL JADE.
South African subregion *Ecology.* a distinct zoogeographical region that includes the Republic of South Africa and adjacent areas of southern Africa.
South African tick-bite fever *Medicine.* a form of typhus, spotted fever occurring in South Africa, transmitted by the ticks, *Haemaphysalis leachi* and *Amblyomma hebraeaum.*
South America *Geography.* the southern continent of the Western Hemisphere.
South American blastomycosis *Medicine.* a chronic infectious disease of the skin, occurring in South America, that begins its attack on the cheek tissue, spreads to adjacent skin, and then to the tonsils, liver, spleen, and gastrointestinal lymphatic system. Also, PARACOCCIDIOIDAL GRANULOMA, ALMEIDA'S DISEASE, LUTZ-SPLENDORE-ALMEIDA DISEASE.
South Atlantic Current *Oceanography.* the eastward-flowing southern segment of the South Atlantic gyre, continuous with the northern edge of the Antarctic Circumpolar Current between 40°W and 5°W at about the latitude of Buenos Aires. Also, **Southern Ocean Current.**
South Australian faunal region *Ecology.* the marine creatures living in the geographical region that runs along the southwestern coast of Australia.
south celestial pole *Astronomy.* the point on the celestial sphere that the earth's southern rotation pole would intersect if extended.
South China Sea *Geography.* an arm of the western Pacific between southeastern China, Taiwan, Indochina, the Philippines, Borneo, and Malaya.
Southdown *Agriculture.* a breed of small, medium-wool and mutton sheep; one of the oldest breeds. (From the *South Downs*, a region of low hills in southern England, where the breed was developed.)
Southeast Drift Current *Oceanography.* the eastern segment of the North Atlantic gyre, flowing generally southward from about 45°N along the western side of the Iberian Peninsula and as far as the Canary Islands, where it becomes the Canary Current.
southeaster *Meteorology.* a southeasterly wind, particularly a strong gale, such as the winter southeast storms of the San Francisco Bay.
South Equatorial Current *Oceanography.* the generic name given to the westward-flowing northern segment of the oceanic gyre in the South Pacific, Indian, and Atlantic Oceans; in the Indian and Atlantic it lies well below the equator, but in the mid-Pacific it extends north of the equator to around 3°N.
souther *Meteorology.* a south wind often reaching gale force conditions.
southerly burster *Meteorology.* a cold, southerly wind of Australia.
southern blight *Plant Pathology.* a disease caused by a fungus, *Sclerotium rolfsii*, that affects the roots of peanut, tomato, and other plants, resulting in rapid wilting.
Southern blot *Molecular Biology.* a process used to move DNA segments from an agrarose gel to a nitrocellulose filter paper sheet through capillary action. The segments are then probed with a radioactive, complementary nucleic acid to help determine their positions.
Southern Cross *Astronomy.* a southern constellation between Centaurus and Musca. Also, CROSS.
Southern Crown *Astronomy.* the constellation Corona Australis.
Southern Fish *Astronomy.* the constellation Piscis Austrinus.
Southern hybridization *Molecular Biology.* a technique used to find a specific sequence or sequences in DNA that has been cleaved by restriction endonuclease and subjected to Southern blotting.
southern lights *Astronomy.* a popular name for the aurora australis. See AURORA AUSTRALIS.
Southern Triangle *Astronomy.* the constellation Triangulum Australe.
south foehn *Meteorology.* a foehn condition consisting of a strong south to north air flow across a transverse mountain barrier; e.g., the south foehn of the Alps.
south frigid zone *Geography.* the area between the Antarctic Circle and the South Pole. Also, ANTARCTIC ZONE.
south geographic pole *Geography.* see SOUTH POLE.
south geomagnetic pole *Geophysics.* see SOUTH POLE.
South Indian Current *Oceanography.* the eastward-flowing southern segment of the South Indian gyre, continuous with the northern edge of the Antarctic Circumpolar Current between the Agulhas and West Australian Currents.
South Island *Geography.* the southern island of the two main islands of New Zealand; area: 58,092 square miles.
south magnetic pole *Geophysics.* see SOUTH POLE.
South Pacific Current *Oceanography.* the eastward-flowing southern segment of the South Pacific gyre, continuous with the northern edge of the Antarctic Circumpolar Current between New Zealand and Chile.
south point *Astronomy.* the southern of the two points at which the meridian intersects the horizon.
south polar distance *Astronomy.* an archaic term referring to the angular distance between a celestial object and the south celestial pole, as measured along the shortest meridian between them.
south pole or **South Pole** *Geography.* the southernmost point on the earth, near the middle of Antarctica. Also, SOUTH GEOGRAPHIC POLE. *Geophysics.* the point in the Southern Hemisphere at which the geomagnetic axis intersects the earth's surface; located at 90° south latitude, of the earth's rotational axis. Also, SOUTH MAGNETIC POLE, SOUTH GEOMAGNETIC POLE. *Electromagnetism.* the pole of a dipole or magnet at which the magnetic field lines are defined to converge.

South Seas *Geography.* the parts of the oceans south of the equator, especially the South Pacific.
south temperate zone *Geography.* the area between the Tropic of Capricorn and the Antarctic Circle.
south tropical disturbance *Astronomy.* an intermittent feature in the South Tropical Zone of Jupiter, which consists of an elongated dark band near the latitude of the Great Red Spot.
southwester *Meteorology.* a southwest wind often reaching gale force. Also, **sou'wester.**
souzalite *Mineralogy.* $(Mg,Fe^{+2})_3(Al,Fe^{+3})_4(PO_4)_4(OH)_6 \cdot 2H_2O$, a dark green, fibrous, triclinic mineral, having a specific gravity of 3.087 and a hardness of 5.5 to 6 on the Mohs scale; forms a series with gormanite.
sow [sō] *Agriculture.* to scatter or plant seeds.
sow [sou] *Agriculture.* an adult female swine, especially one that has produced a litter. *Vertebrate Zoology.* the adult female of certain other large animals, such as the bear. *Metallurgy.* a large pig.
sow block *Metallurgy.* in forging, an equipment component inserted in the anvil to avoid excessive wear of the anvil.
sower [sō´ər] *Agriculture.* a machine that is used for broadcast seeding.
SO_x sulfur oxide.
Soxhlet extractor *Chemistry.* a laboratory apparatus used in the continuous solvent extraction of solids, consisting of a glass distillation flask, a condenser, and a porous cup for collecting extract.
soybean *Botany.* **1.** an erect annual dicotyledonous plant of the legume family, *Glycine max*, native to Asia but widely cultivated in other parts of the world. **2.** the edible seed of this plant, a leading source of protein for the human diet and for animal feed.

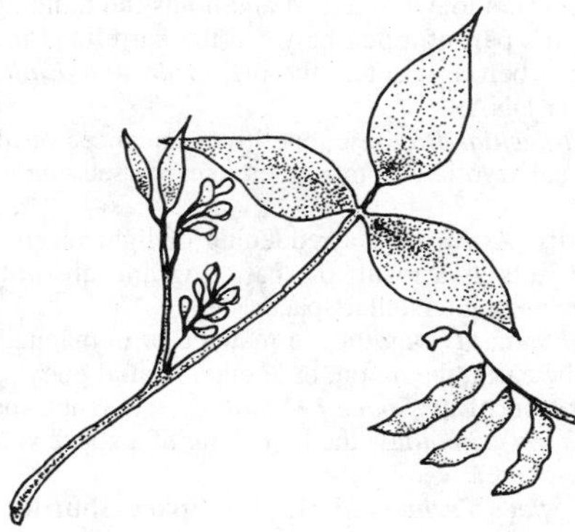

soybean

soybean oil *Materials.* a pale yellow oil that is derived from soybeans and used as a food, and in soap, candles, paints, plasticizers, and varnishes. Also, **soya bean oil, soy oil.**
Soyuz *Space Technology.* a Soviet program of manned space flights initiated in 1967 and used chiefly to ferry cosmonauts between earth and Salyut space stations.
SP or **s.p.** without issue. (From Latin *sine prole.*)
SP self-propelled; specialist; systolic pressure; single pole.
sp. or **sp** species; special; specific; specimen.
spa *Hydrology.* **1.** a mineral spring regarded as having medicinal qualities. **2.** a place having such a spring. (From *Spa*, a noted resort of this type in Belgium.)
space *Astronomy.* **1.** the three-dimensional expanse in which all matter is located and all events take place, extending in all directions and variously described as extending indefinitely or as finite but immeasurably large. **2.** also, **outer space.** the portion of this expanse lying beyond the earth's atmosphere; this is the sense of the term *space* as it is used in compounds such as *space technology, space shuttle, spacecraft,* and so on. *Graphic Arts.* **1.** a blank area created by spacing. **2.** in hot-metal typesetting, a piece of metal type used to create such an area; a quad. *Mathematics.* a set on which a topology or some other structure is defined. *Telecommunications.* in telegraphic communication, the open-circuit condition of modulation, occurring when there is a presence of negative voltage, line to ground, or current or tone is shut off; as distinguished from mark, the closed-circuit condition.
Space Age *Space Technology.* the era in modern history including the period of space exploration, usually considered to begin with the launching of the Soviet *Sputnik* satellite on October 4, 1957.
space biology *Biology.* the various biological sciences and disciplines that are concerned with the study of living things in the space environment.
space booster *Space Technology.* a booster, such as the Titan, designed to carry a payload from the earth's surface to a designated location or orbit in space.
space capsule *Space Technology.* an environmentally controlled container, usually unmanned, designed to safely transport an organism or equipment in space.
space centrode see CENTRODE.
space character see BLANK CHARACTER.
space charge *Geophysics.* an excess of either negatively or positively charged ions in a layer of the atmosphere, giving that layer either a negative or positive space charge. *Electricity.* an electric charge resident in a cloud of electrons lying between a cathode and plate within an electron tube. *Materials Science.* a region of electrical deficiency in the interior of a material that balances the concentration of electrons at the surface; often the cause of a surface charge in semiconductors.
space-charge balanced flow *Electronics.* a technique of directing an electron beam in the interaction region of a traveling-wave tube.
space-charge debunching *Electronics.* a phenomenon in which the bunched electrons in a microwave tube are spread out, due to the mutual interactions between electrons in the stream.
space-charge effect *Electronics.* a phenomenon in which the electrons emitted from the cathode of a thermionic electron tube are repulsed by electrons accumulated in the space charge near the cathode, causing a reduction in anode current.
space-charge grid *Electronics.* a grid driven at a low positive potential and positioned between the cathode and the control grid of an electron tube in order to minimize the limiting effect of space charge on the current through the tube.
space-charge layer see DEPLETION LAYER.
space-charge limitation *Electronics.* the maximum value of a given current passing through a vacuum between a cathode and an anode.
space-charge polarization *Electricity.* the polarization of a dielectric that occurs when charge carriers are present and migrate an appreciable distance through the dielectric; under such a condition the charge carriers become trapped and are unable to discharge at an anode. Also, INTERFACIAL POLARIZATION.
space-charge region *Electronics.* a region of a semiconductor device in which the net charge density is nonzero.
space communications *Telecommunications.* communication between the earth and a satellite in outer space or between stations outside the earth's atmosphere.
space cone *Mechanics.* a cone, fixed in space, swept by the precessing angular-velocity vector of a rotating free body executing Poinsot motion. Also, HERPOLHODE CONE.
space coordinates *Photogrammetry.* in aerial photography, any system of three-dimensional coordinates used to specify the position of a point in object space, rather than the image of that point on the photograph.
spacecraft *Space Technology.* any vehicle or other device that travels into or through space, especially one that carries a payload.
spacecraft ground instrumentation *Space Technology.* any devices located on the earth that locate, track, monitor, or communicate with manned space vehicles. Also, GROUND INSTRUMENTATION.
spacecraft launch *Space Technology.* the launching of a space vehicle with sufficient power to escape the earth's gravitational forces.
spacecraft propulsion *Space Technology.* the powering of space vehicles by means of rocket engines.
spacecraft tracking *Space Technology.* the process of determining the location and movement patterns of space vehicles with the use of radio and optical instruments.
space current *Electronics.* in an electron tube, the total amount of current traveling between the cathode and all other electrodes.
spaced antenna *Electromagnetism.* a system of receiving antennas spaced significantly apart to minimize fading effects at shortwave receiving stations.
space defense *Military Science.* a collective term for all defensive measures designed to destroy attacking enemy vehicles while in space, or to nullify or reduce the effectiveness of such an attack.
space detection and tracking system see SPADATS.
space diversity reception *Electromagnetism.* the placing of several receiving antennas in various locations to reduce the effects of radio signal fading.

space division switching *Telecommunications.* a technique of interconnecting switching-system terminals by activating spatially segregated connecting elements, linking the terminals to be connected, as in a crossbar switch.

space environment *Astronomy.* the flux of particles, fields, and radiation that permeates space.

space factor *Electromagnetism.* **1.** a ratio of the volume of a conductive material in the windings of a coil to the total volume of the windings. **2.** the ratio of the volume of iron in a magnetic core to the total volume of the core.

space-filling curve *Mathematics.* a continuous curve in the plane that has positive content (i.e., passes through every point of some subset of the plane having positive content). Also, PEANO CURVE.

space fixed reference *Navigation.* an oriented reference system located in space that establishes position independent of the earth.

space flight or spaceflight *Space Technology.* the flight of a manned or unmanned spacecraft into or through space, either in orbital flight around the earth or (usually) outside the earth's atmosphere.

space food *Food Technology.* the types of food designed for space travel; usually concentrated, dehydrated, and packaged in quick-serve, lightweight containers.

space frame *Building Engineering.* a three-dimensional frame that is naturally stable or resistant to wind loads and does not have to be propped against another structure.

space group *Crystallography.* a set of transformations that leave a triply periodic, discrete set of labeled points (atoms) unchanged, excluding those transformations that are purely integral lattice translations. There are 230 such groups that can be identified from systematic absences in the diffraction pattern. The space groups can be derived by the addition of translational symmetry operations to the 32 point groups appropriate for structures arranged on lattices; the operations include simple translations, screw axes, and glide planes. Thus a space group may be considered the group of operations that converts one molecule or asymmetric unit into an infinitely extending three-dimensional pattern.

space guidance *Navigation.* a guidance operation in outer space, including ascent from the earth to an orbit or trajectory in space, travel through space, and descent to the earth, moon, or other body.

space headway *Transportation Engineering.* the spatial separation between successive vehicles along a given route; e.g., cars in a traffic lane or aircraft along an airway.

space heater *Engineering.* a small, self-contained heater, often having a fan, used to heat the room or space in which it is placed.

space-hold *Telecommunications.* the process of sending a consistent space signal over a transmission line that is not carrying any traffic.

Spacelab *Space Technology.* a self-contained, manned laboratory in space, developed by the European Space Agency, that is carried aboard an orbiting space shuttle.

space lattice *Building Engineering.* a space frame built with lattice girders. *Crystallography.* one of seven types of unit cells existing in crystals that uniquely describes the lattice points.

spacelike surface *Physics.* a surface in three dimensions of four-dimensional space-time having the property that no point (event) on the surface lies within the absolute future or absolute past of any other point on the surface.

spacelike vector *Physics.* any vector in Minkowski space having the property that its spatial components collectively have a greater magnitude than that of the product of the time component and the speed of light.

space medicine *Medicine.* the branch of medicine concerned with the physical conditions of life in space and its effects, such as those resulting from little or no gravity or rapid acceleration, from special conditions of diet, exercise, or unusual exposure to radiation, and from exposure to unusual sensory experience.

space mission *Space Technology.* a manned or unmanned space flight outside the earth's atmosphere.

space modulation *Telecommunications.* the process of mixing signals in space to form one signal of predetermined characteristics.

space motion *Astronomy.* the velocity and direction of motion of a star or celestial object with respect to the sun. Also, PECULIAR MOTION.

space perception *Physiology.* the consciousness of the mass and three-dimensional aspects of an object or environment; human binocular vision produces two distinct images which, when combined, give a subjective visual impression of the object.

space permeability *Electromagnetism.* the factor given by the ratio of magnetic induction in a vacuum to the magnetizing force.

space physiology *Physiology.* the science and study of the effects of space travel and attendant phenomena (such as weightlessness and high-velocity projection) on the cells and tissues of human beings and sometimes other animals.

space platform see SPACE STATION.

spaceport *Space Technology.* a land-based facility designed to test, launch, and service space vehicles.

space power system *Space Technology.* an on-board system that enables a spacecraft or space station to generate and distribute electrical energy.

space probe *Space Technology.* an unmanned instrumented spacecraft that is rocketed into space, often around or near a spatial body, in order to obtain scientific data.

space processing *Space Technology.* a general term for experiments performed aboard a spacecraft that exploit its low-gravity, high-vacuum environment.

space propulsion *Space Technology.* the intermittent or sustained propulsion of a spacecraft through space outside the earth's sensible atmosphere.

space quadrature *Physics.* the distance over which two waves of the same frequency are out of phase by 90°; that distance defined by one-quarter of a wavelength.

space quantization *Quantum Mechanics.* the quantization of one component of a system's angular momentum.

spacer *Mechanical Devices.* **1.** any component or device that holds two parts at a distance from each other. Also, **spacer block. 2.** a wooden piece that fits between charges to lengthen an explosive column. *Engineering.* **1.** in an explosive apparatus, a piece of metal wire that secures the explosive in a shothole at one end and holds the tamping in place at its other end. **2.** the part of a pug between the barrel and the die, where clay is compacted before it enters the die. *Molecular Biology.* a single segment of spacer DNA.

spacer DNA *Molecular Biology.* repetitive sequences of untranscribed DNA, found in eukaryotic chromosomes interspersed among functional cistrons.

space reddening *Astronomy.* a reddening of light received from distant celestial objects as a result of scattering and absorption by dust clouds and particles in interstellar space.

space satellite *Space Technology.* a manned or unmanned vehicle designed to orbit the earth, the moon, or another spatial body.

spaceship or **space ship** *Space Technology.* a manned spacecraft.

space shot *Space Technology.* the launching of a space vehicle beyond the earth's atmosphere.

space shuttle *Space Technology.* **1.** also, **Space Shuttle.** any of a series of U.S. space vehicles that are launched into space by solid rocket boosters and return to earth under their own power. **2.** any space vehicle designed to transport people and equipment between earth and a space station.

space shuttle

space simulator *Space Technology.* a sealed vacuum chamber used to test space systems or components by recreating the space environment on earth.

space station *Space Technology.* a permanent space facility used to carry out scientific and technological studies, earth-oriented applications, and astronomical observations and to service other vehicles and their crews in space. Also, SPACE PLATFORM.

spacesuit *Space Technology.* a suit having life-support provisions that make it possible for the wearer to leave the area of a pressurized cabin while either in space or in an area on earth having low ambient atmospheric pressure.

space suppression *Computer Technology.* the act of eliminating unnecessary spaces from data that is transmitted, printed, or displayed, so as to conserve memory.

space technology *Aviation.* the use of scientific methods and research, engineering, and technology in the study, exploration, and utilization of space.

Space Technology

Space technology is the application of imagination, mathematics, science, and engineering to create and apply the means to go beyond the earth's atmosphere to explore, to increase knowledge, and to develop resources for humankind. The Space Age began with Sputnik, the first artificial earth satellite, on October 4, 1957. Space technology had its birth and continues to flower because of the visions of many people. Examples are Konstantin Tsiolkovsky of Russia (the theoretical basis for high-performance rockets), Hermann Oberth of Germany (*The Rocket into Interplanetary Space*), and Robert Goddard of the United States (the design and test of rockets). Writers such as Jules Verne (*From the Earth to the Moon*), H. G. Wells (*War of the Worlds*), and Arthur C. Clarke (who forecasted the development of the communications satellite) fostered an evolution in aspirations for space technology and for the exploration of space by humans. Ballistic-missile technology subsequently was applied in the first launch of a U.S. earth satellite. A team led by the visions of President John F. Kennedy and Wernher von Braun and motivated by international competition developed the space technology evident in the Saturn V launch vehicle and the Apollo Spacecraft, which placed Neil Armstrong and Edwin "Buzz" Aldrin on the lunar surface at the Sea of Tranquility on July 20, 1969.

The fruits of space technology applications are innumerable. The earth is monitored from space for weather, ocean conditions, and environment. We are served constantly by communication satellites in earth orbit. Space technology has revolutionized astronomy by providing views unfettered by the earth's atmosphere. Many things that we use on earth originated as space technology. Our perspective of earth is forever altered by space technology.

The future of space technology includes advances in launch and space transportation systems; improved space power systems; computer technologies, artificial intelligence, robotics, telepresence, and robot satellites; reliable long-term life support systems with closed-loop ecologies; tailored materials for applications in space and on extraterrestrial surfaces; and capabilities of the moon and Mars. In the 21st century, an outpost on the moon will include advanced telescopes for astronomy, laboratories for geological investigations, mining and manufacturing installations, facilities for human habitation, and perhaps hotel facilities for those wishing to visit our neighbor in the heavens, either on short trips from earth or as a stop on the way to Mars.

Stewart W. Johnson
Principal Engineer
Advanced Basing Systems
BDM International, Inc.

space-time *Physics.* a four-dimensional space consisting of three dimensions corresponding to the ordinary spatial coordinates of length, width, and height, and a fourth dimension corresponding to time; frequently used in the theory of relativity, which specifies that the three spatial dimensions cannot be accurately described except with reference to the time of the description. Also, **space-time continuum.**

space-to-mark transition *Telecommunications.* in telegraphic communications, the changeover from the space condition to the mark condition.

Spacetrack *Military Science.* a global system of radar, optical, and radiometric sensors linked to a computation and analysis center that is designed to detect, track, and catalogue all manmade objects in earth orbit; constitutes the Air Force portion of the North American Air Defense Command Space Detection and Tracking system; the Navy portion of the system is called **Spasur.**

Space Tracking and Data Acquisition Network *Engineering.* an arrangement of ground stations, controlled by the National Aeronautics and Space Administration, that receive information about the location and movement of unmanned satellites with self-registering devices.

space vehicle *Space Technology.* a spacecraft, especially one designed for sustained space flight.

space velocity *Chemical Engineering.* the volumetric feed flow rate per unit of reactor volume in chemical reactor design.

space walk or **spacewalk** *Space Technology.* the travel of an astronaut, protected by a pressurized spacesuit, outside a spacecraft during space exploration.

space wave *Electromagnetism.* the part of a ground wave that travels directly through space to its destination, either from the transmitting antenna to the receiving antenna or by experiencing one reflection from the surface of the earth.

space weapon *Ordnance.* any weapon that travels beyond the earth's atmosphere; it may return through the atmosphere to attack a ground, sea, or air target, or it may attack a target in space.

spacing *Graphic Arts.* in typesetting, the insertion of blank areas to separate words, sentences, and columns, to indent paragraphs, and to justify lines.

spacistor *Electronics.* a semiconductor device, comparable to a transistor, that is capable of generating very high frequencies of up to about 10,000 megahertz by injecting electrons into a space-charge layer, which rapidly forces these carriers to a collecting electrode.

spackling *Engineering.* the process of using a paste (**spackle**), made from a dry powder and water mixture, to fill the cracks of a surface.

SPADATS *Space Technology.* a system that determines the location and movement of space vehicles and transmits the information to a central control station. (An acronym for space detection and tracking system.)

spade *Mechanical Devices.* a flat-bladed tool attached to a long or short handle that is driven into the ground with the hand or foot; used for digging and overturning earth.

spade bolt *Mechanical Devices.* a bolt having a broad, flat head with a transverse hole cut through it; used to secure electronic devices such as coils or capacitors to a chassis.

spade drill *Mechanical Devices.* a drill used to cut holes 1 inch in diameter or larger, consisting of a cutting bit, a shank, and a fastener to secure the bit to the shank.

spadefish *Vertebrate Zoology.* a laterally compressed, deep-bodied, spiny-finned, small-mouthed food fish of the family Ephippidae; found in warmer parts of the western Atlantic. (So called with reference to its shape.)

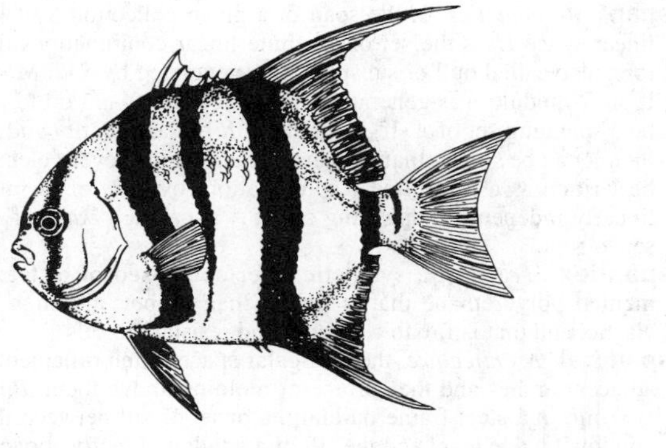

spadefish

spadefoot toad *Vertebrate Zoology.* a burrowing toad of the family Pelobatidae, characterized by a sharp-edged horny sheath in the inner bone of the tarsus, used for digging; native to Europe, Asia, northwest Africa, and North America; the **plains spadefoot,** *Scaphiopus bombifrons,* the **Western spadefoot,** *S. hammondi,* and the **Eastern spadefoot,** *S. holbrooki,* are found in North America.

spadefoot toad

spade grip *Ordnance.* a D-shaped handle used in aiming certain flexible automatic weapons; it is usually located behind the receiver.

spade lug *Mechanical Devices.* **1.** a pronged electrical connector designed to be slipped under a screw or nut. **2.** a broad, flat steel attachment to the rim of a tractor drive that prevents slipping.

spadix *Botany.* an inflorescence consisting of sessile flowers borne on a fleshy spike or head, usually enveloped by a spathel; mainly applied to the family Araceae.

spaghetti *Electricity.* a slender tubing that is made of a varnished cloth or plastic and used as a sleeve over insulation for wires and brush wires; thought to resemble the common food.

spall *Engineering.* **1.** to break off thin fragments in layers parallel to the surface of a hard material, such as ore. **2.** to reduce stone blocks to a desired size by chipping. **3.** the fragments that are a result of these processes. *Geology.* a fragment that has been removed from a rock surface by weathering, especially a relatively thin, sharp-edged fragment produced by exfoliation. *Ordnance.* a fragment that has broken away from the surface of an armor plate, from either the force of impact or the force of explosion.

spallation *Materials Science.* the cracking and breaking away of surface layers in ceramics, resulting from exposure to heat or cold, sometimes combined with a chemical action. *Nuclear Physics.* a type of nuclear reaction in which the high-energy level of incident particles causes the nucleus to eject more than three particles, thus changing both its mass number and its atomic number. Also, NUCLEAR SPALLATION.

spallation reaction *Nuclear Physics.* a nuclear reaction in which a high-energy level results in the production of a large number of nucleons.

Spallazani, Lazzaro 1729–1799, Italian biologist; disproved spontaneous generation; pioneered research in digestion and fertilization.

span *Engineering.* the distance between the supporting ends of a structure, such as a bridge or arch. *Aviation.* **1.** the linear distance between lateral extremities of an aircraft or other winged body. **2.** the operative radial distance from root to tip of a rotating airfoil.

span *Mathematics.* **1.** the span of a given collection *S* of vectors (in a linear space *L*) is the set of all finite linear combinations of those vectors; also called hull or subspace of *L* generated by *S*. **2.** More generally, if an *R*-module *A* is generated by a set *Y*, then *Y* is said to span *A*, or to be a spanning set of *A*. If, in addition, *R* has an identity and *A* is unitary, then it can be shown that *Y* spans *A* if and only if every element of *A* can be written as a finite linear combination (over *R*) of members of *Y*. A linearly independent spanning set of *A* is called a *basis* of *A*. *Statistics.* see RANGE.

spandex *Textiles.* a synthetic fiber composed of at least 35% segmented polyurethane that is lighter than rubber, although not quite as elastic, and that is flexible, strong, and resistant to oils.

spandrel *Architecture.* the triangular space, often ornamented, between adjacent arches and the cornice or molding above them. *Building Engineering.* in a steel frame building, a panel or sill between the head of a window on one level and the sill of a window directly above it.

spandrel wall *Building Engineering.* a wall carried on the pillars or mantlelike projection between two adjoining arches.

spangle *Metallurgy.* a term used to describe the form of coarse crystals visible on the surface of steel that has been hot-dipped in zinc.

spangolite *Mineralogy.* $Cu_6^{+2}Al(SO_4)(OH)_{12}Cl \cdot 3H_2O$, a dark-green, vitreous, trigonal mineral occurring as prismatic or tabular rhombohedral crystals, having a specific gravity of 3.14 and a hardness of about 3 on the Mohs scale; found as a secondary mineral in the oxidized zones of copper deposits.

Spanish fly *Invertebrate Zoology.* the popular name for a bright-green blister beetle, *Cantharis* (*Lytta*) *vesicatoria*, whose dried and powdered body yields a preparation used medicinally as a counterirritant and diuretic, and also regarded as a potent aphrodisiac; found in Europe. Also, **Spanishfly.**

Spanish moss *Botany.* a mosslike, epiphytic plant, *Tillandsia usneoides*, that grows in long streamers draping the branches of trees; characterized by narrow, grayish leaves; found widely in the southern United States.

Spanish windlass *Surgery.* an improvised tourniquet consisting of a band of fabric or handkerchief tied around a part of the body and tightened by twisting a stick passed under it.

spanker *Naval Architecture.* a fore-and-aft sail set on the mizzenmast.

Spanker *Ordnance.* a Soviet two-stage, intercontinental ballistic missile that is liquid-rocket powered, delivering four 750-kiloton MIRVs at a range of 6200 miles; officially designated **SS-17.**

spanned record *Computer Programming.* a format in which a large, variable-length record can be stored as a number of smaller, fixed-length blocks.

spanner *Engineering.* **1.** a cross beam or brace that lies horizontally. **2.** an attachment for a sextant that provides an artificial horizon for use when the sea horizon is obscured. **3.** a tool for removing circular-shaped parts. *Mechanical Devices.* **1.** a flat wrench with a crescent-shaped head equipped with fittings for various screws, bolts, or couplings. Also, **spanner wrench. 2.** in British use, any wrench.

spanning forest *Mathematics.* the union of spanning trees of each of the components of a graph. Also, SKELETON OF THE GRAPH.

spanning subgraph *Mathematics.* given a graph *G*, a subgraph *H* of a graph *G* that contains all the vertices of *G*.

spanning tree *Mathematics.* a spanning subgraph *T* of a connected graph *G* that is also a tree. The cospanning tree T^* of a spanning tree *T* in *G* is the (spanning) subgraph of *G* having all the vertices of *G* and exactly those edges of *G* not in *T*. The edges of *T* are called **branches** and those of T^* are called **links** or **chords.**

span of detonation *Ordnance.* the duration between the earliest possible detonation of an atomic demolition munition and the latest possible detonation, resulting from a timer error; the beginning and end of the duration are termed the **early time** and **late time,** respectively; the time the munition will detonate should the timer function precisely without error is termed the **fire time.**

spar *Naval Architecture.* a beam or bar forming part of a vessel's rigging. Spars include masts, yards, booms, and gaffs. *Aviation.* a primary longitudinal structural member in an airfoil, especially in a wing; may be an assembly or a single piece. *Mineralogy.* any readily cleavable, transparent or translucent, light-colored crystalline mineral with vitreous luster, such as fluorspar and feldspar. *Mining Engineering.* a small vein of clay within a coal seam.

sparagmite *Geology.* the Late Precambrian fragmental rocks of Scandinavia, especially the feldspathic sandstones of the Swedish Jotnian.

Sparassis *Mycology.* a genus of fungi belonging to the family Sparassidaceae, including the edible species *Sparassis crispa*, which is found around the trunks of pine trees.

spar buoy *Navigation.* a fixed, floating upright pole used to mark a specific location, such as the edge of a channel, as an aid to navigation; it is painted a bright color that indicates its significance.

SPARC *Computer Technology.* one of the reduced instruction set computer (RISC) processors available in modern workstations. (An acronym for scalable processor architecture.)

spar cap *Aviation.* a flat strip on the upper or lower edge of certain spars to which the airfoil covering is attached.

spardeck *Naval Architecture.* a light, unarmed complete deck above a sailing ship's main deck, usually intended as a working area for crewmembers tending the rigging. Also, **spar deck.**

Sparganiaceae *Botany.* a monocotyledonous family of the order Pandales, characterized by an inflorescence of globose heads, a vestigial perianth, and fruits that are sessile or nearly sessile achenes.

sparganosis *Veterinary Medicine.* a disease caused by infestation of the body with plerocercoids (second larval stage) of tapeworms of the genus *Spirometra*; characterized by inflammation and fibrosis of subcutaneous tissues.

sparganum *Invertebrate Zoology.* a large, wormlike larval stage in tapeworms. Also, PLEROCERCOID.

sparging *Chemical Engineering.* the process of injecting vapors into a liquid-solid suspension in order to mix the materials.

Sparidae *Vertebrate Zoology.* the porgies, a family of mostly tropical and subtropical marine perciform fish, characterized by an oblong, deep, compressed body, a large head, and well-developed teeth.

spark *Electricity.* a short-lived electric discharge produced by a sudden breakdown of air or some other dielectric separating two terminals, accompanied by a momentary flash of light. Also, **spark discharge, sparkover.**

Spark *Ordnance.* a program to develop an advanced antitank missile that can penetrate armor at a speed so high that it does not need a warhead. (An acronym for solid-propellant, advanced ramjet, kinetic energy.)

spark arrester *Engineering.* a component that prevents the escape of sparks from a smokestack, such as a wire framework. *Electricity.* a resistor, capacitor, diode, or a combination of such units that is employed to minimize or eliminate sparking between make-and-break contacts, such as those of a relay. Also, **spark killer, spark suppressor.**

spark capacitor *Electricity.* a capacitor that is connected across a pair of contact points or across the inductance that causes the spark, in order to diminish sparking at these points.

spark chamber *Nucleonics.* an enclosed chamber equipped with a device that generates a series of sparks in an array of spark gaps; used to observe the trajectory of an energetic particle passing through the chamber.

spark coil *Electromagnetism.* a high-voltage induction coil used to produce a spark for ignition systems.

spark counter *Nucleonics.* a particle detector that counts events of collected charge liberated by ionizing radiation passing through an argon-methane gas mixture.

spark excitation *Spectroscopy.* an emission technique in which a high-voltage electric spark is used to excite the atoms of a sample.

spark-explosion method *Analytical Chemistry.* a method of mixing a test sample with an oxidant, exploding it, and analyzing the combustion products; used to analyze hydrogen content.

spark gap *Electricity.* a device consisting of two metal points, tips, or balls separated by a small air gap; a high voltage applied to the electrodes causes a spark (or, with an AC voltage, a train of sparks) to jump across the gap.

spark-gap arrester *Electricity.* a lightning arrester that uses an air gap to reduce voltage surges; when the voltage across the gap becomes excessive, an arc forms to create a low-resistance path across the gap.

spark-gap generator *Electricity.* a high-frequency generator in which a capacitor is repeatedly charged to a high voltage and allowed to discharge through a spark gap into an oscillatory circuit, thus generating successive trains of damped, high-frequency oscillations.

spark-ignition combustion cycle see OTTO CYCLE.

spark-ignition engine *Mechanical Engineering.* an internal combustion engine in which an electrical discharge ignites a fuel-air mixture within a cylinder.

spark knock *Mechanical Engineering.* a knock that occurs before the piston reaches the top dead-center position in an internal-combustion engine; often due to faulty timing or poor fuel characteristics.

spark lead *Mechanical Engineering.* the amount by which a spark occurs before a piston reaches its compression dead-center position in the cylinder of a combustion-engine cycle.

sparkling wine *Food Technology.* an effervescent table wine, especially one carbonated by secondary fermentation; e.g., champagne.

spark machining *Metallurgy.* machining by repeated electrical sparking between an electrode and the part to be machined.

sparkover-initiated discharge machining *Mechanical Engineering.* an electromachining process in which a potential is impressed between a tool and workpiece that are separated by a dielectric material, so that when the potential is sufficient to cause the dielectric to rupture, a discharge current flows through the ionized path.

sparkover voltage see FLASHOVER VOLTAGE.

spark plate *Electricity.* a metal plate insulated from the chassis of an auto radio by a thin sheet of mica and connected to the battery lid, to bypass noise signals picked up by battery wiring in the engine compartment.

sparkplug or **spark plug** *Electricity.* a device that is fitted into the cylinder of an internal-combustion engine to provide a pair of electrodes, between which an electrical discharge is passed to ignite an explosive mixture such as gasoline.

sparkproof *Engineering.* **1.** describing material that has been treated so that it will not be damaged or set aflame by sparks. **2.** describing equipment that does not generate sparks.

spark range *Military Science.* a range at which a free-flying missile can be photographed under the illumination produced by a spark that is actuated by the passage of the missile.

spark recorder *Engineering.* a type of recording instrument composed of a paper that moves between a metal plate and a moving metal pointer; sparks from an induction coil pass through the paper periodically and burn holes in it to form the recording.

spark sintering *Materials Science.* a method of hot pressing in powder metallurgy in which an electric current is passed through the powder charge in a die via the top and bottom punches; this causes localized arcing and heating at contact points, and the heated powder is then pressed.

spark spectrum *Spectroscopy.* the emission spectrum produced by an ion as a result of vaporization by a high-voltage discharge between metallic electrodes. Also, IONIC SPECTRUM.

spark transmitter *Electronics.* a radio transmitter whose source of radio-frequency power is the oscillatory discharge of a capacitor through an inductor and a spark gap.

spark voltage *Electricity.* the voltage required to create an arc across the gap of a spark plug. Also, **sparking voltage.**

Sparnacean *Geology.* a European geologic stage of the Upper Paleocene epoch, occurring after the Thanetian and before the Ypresian of the Eocene epoch.

sparrow *Vertebrate Zoology.* a popular name for any small seed-eating songbird with a conical bill, especially those of the families Emberizidae and Passeridae; the common **house** or **English sparrow,** *Passer domesticus,* is found in almost all populated areas of North America.

sparrow

Sparrow see SEA SPARROW.

sparrowhawk *Vertebrate Zoology.* any of various small hawks or falcons noted for preying on sparrows and other small birds, especially the Old World *Accipternisus* of the family Accipitridae, and the New World *Falco sparverius* of the family Falconidae.

sparry calcite *Mineralogy.* calcite occurring as clean, coarse-grained crystals.

sparry cement *Geology.* a relatively coarse-grained calcite occurring in the interstices of sedimentary rock.

sparry iron see SIDERITE.

sparse *Mathematics.* **1.** a matrix whose entries are mostly zeros is said to be a **sparse matrix**. **2.** a sparse programming problem involves many variables, but each objective or constraint function involves relatively few of the variables. Such a problem is also said to exhibit **sparseness.**

Spartan *Ordnance.* a nuclear surface-to-surface guided missile formerly deployed as part of the Safeguard ballistic missile defense weapon system; it is designed to intercept strategic ballistic reentry vehicles in the exoatmosphere.

sparteine *Organic Chemistry.* $C_{15}H_{26}N_2$, an oily alkaloid; soluble in water, alcohol, and ether; boils at 173°C (8 torr); used in pharmaceuticals. Also, LUPINIDINE.

spar tree *Forestry.* a tall, stout, well-reinforced tree that is trimmed and used for supporting the heavy cables in skyline logging.

spar varnish *Materials.* **1.** an oleoresinous, weather-resistant varnish applied to unpainted wooden areas, especially on ships. **2.** a varnish made from sulfur, rosin, and linseed oil.

spasm [spaz´əm] *Neurology.* a sudden, involuntary muscle contraction, such as in a hiccup, a stutter, or a tic, causing pain or discomfort and impairing normal motor functions. Also, **spasmus.** *Medicine.* a sudden, transient constriction of a passage, canal, or orifice.

spasmodic [spaz mäd´ik] *Neurology.* of or relating to a spasm.

spasmodic strabismus see SPASTIC STRABISMUS.

spasmodic turbidity current *Geophysics.* a single, rapidly developed turbidity current.

spasmogenic *Neurology.* capable of producing a spasm.

spasmolysis *Neurology.* the relief or arrest of a spasm or convulsion.

spasmoneme *Invertebrate Zoology.* the contractile fibril in the stalk of some ciliates.

spasmophilia *Medicine.* a condition of muscular sensitivity, in which the motor nerves respond very quickly and strongly to stimuli, overreact, and tend to spasm, with convulsions and tetany.

spasmophilic *Neurology.* being affected by or having a tendency to produce spasm.

spastic [spas´tik] *Neurology.* relating to or affected by sudden, involuntary contractions of the muscles; having or involving spasms.

spastic colitis see MUCOUS COLITIS.

spastic colon see IRRITABLE BOWEL SYNDROME.

spastic diplegia see LITTLE'S DISEASE.

spastic ileus *Medicine.* an obstruction of the bowel resulting from constant contraction of the intestinal musculature.

spasticity *Neurology.* a state of increased muscular tone characterized by heightened resistance to the stretching of muscles and exaggerated reflexes.

spastic paralysis *Medicine.* a condition of paralysis in which there is rigidity of the musculature and heightened reflex sensitivity.

spastic paraplegia *Medicine.* a form of paraplegia which is characterized by tonic spasm of the paralyzed muscles, usually caused by transverse lesions of the spinal cord or by anterolateral sclerosis. Also, SENILE PARAPLEGIA, TETANOID PARAPLEGIA.

spastic strabismus *Medicine.* strabismus due to spasm of the eye muscles. Also, SPASMODIC STRABISMUS.

spat *Aviation.* any of a set of streamlined fairings set around the wheel of a landing gear. Also, WHEEL SPATS.

Spatangoida *Invertebrate Zoology.* an order or suborder of more or less flattened, heart-shaped sea urchins; some are sand-burrowing. Also, **Spatangina, Spatangida, Spatangoidea.**

spate see FRESHET, def. 1.

spathe *Botany.* a bract or pair of bracts, often large and colorful, that subtends or envelops an inflorescence.

Spathulosporales *Mycology.* in some classifications, an order of fungi belonging to the subdivision Ascomycotina; its only genus, *Spathulospora,* is a parasite of marine algae.

spatial [spā´shəl] *Physics.* **1.** of or relating to space; existing or occurring in some space. **2.** specifically, involving relations or measurements that have to do with space, but that do not involve a description of time.

spatial analysis *Archaeology.* the statistical study of concentrations of human activity in a defined space.

spatial coherence *Physics.* the coherence of two or more points in a wave train; the coherence length gives the distance over which spatial coherence exists.

spatial data management *Computer Technology.* a user interface technique that allows access to information by pointing to representative pictures on a display screen.

spatial dendrite *Meteorology.* a spherically shaped, complex ice crystal with branches extending in many directions from a central nucleus. Also, **spatial dendritic crystal.**

spatial difference *Archaeology.* a method of defining variation in artifacts by their location in an activity area.

spatial dimension *Archaeology.* a characteristic of an artifact based on its location in an activity area.

spatial filter *Optics.* a filter that has an extremely small aperture.

spatial linkage *Robotics.* linkage that allows motion in three dimensions.

spatial resolution *Robotics.* a measure of a system's ability to control a robot's tool tip.

spatio- *Space Technology.* a combining form meaning "spatial."

spatiotemporal *Physics.* relating to both space and time simultaneously.

spatter *Metallurgy.* a term for the metal particles ejected during a process of welding.

spatter cone *Volcanology.* a low, steep-sided mound or cone of small pyroclastic fragments built up on a fissure or around a volcanic vent. Also, AGGLUTINATE CONE, VOLCANELLO.

spatter dash *Civil Engineering.* a rich mixture of portland cement and coarse sand, thrown onto a background with a trowel, scoop, or other appliance so as to form a thin, coarse-textured, continuous coating.

spatterdock *Botany.* any of various water lilies of the genus *Nuphar,* characterized by large, erect leaves and rounded yellow flowers; common in stagnant waters of the eastern U.S.

spatter rampart *Geology.* a low, circular ridge of pyroclastic material built up along the margins of a small volcano.

spatulate *Biology.* shaped like a spoon or spatula. *Botany.* having a wide, rounded end and a narrow, attenuate base, as a **spatulate leaf.**

spavin *Veterinary Medicine.* **1.** a disease of horses, characterized by enlargement of the hock joint caused by collected fluids (**bog spavin**), bony growth (**bone spavin**), or distention of the veins (**blood spavin**). **2.** the enlargement so formed.

spawn *Zoology.* **1.** the mass of eggs that is deposited by fishes, amphibians, mollusks, crustaceans, and the like. **2.** to deposit eggs or sperm directly into the water. *Mycology.* **1.** the mycelium of common edible mushrooms, usually combined with dried manure and other organic matter, used to start new mushroom beds. **2.** to plant with mushroom mycelium.

spay *Veterinary Medicine.* to remove surgically the ovaries and uterus of a female animal, usually a pet cat or dog, to prevent pregnancy.

SPCA serum prothrombin conversion accelerator.

SPCA or **S.P.C.A.** Society for the Prevention of Cruelty to Animals.

SPE specific dynamic effect.

SPE or **S.P.E.** Society of Petroleum Engineers.

speaker *Acoustical Engineering.* a shorter term for a loudspeaker. See LOUDSPEAKER.

speaker efficiency *Acoustical Engineering.* the ratio of acoustic power transmitted by a loudspeaker to the electrical power fed to the speaker.

Spearman's correlation coefficient *Statistics.* a measure of dependence between two variables based on the correlation between the rank of the observations.

Spearman's correlation test *Statistics.* a significance test that uses ranks to test for correlation between two measurable characteristics that exist either in each individual of the sample group or at the same time in different individuals.

spearmint *Botany.* **1.** an herbaceous plant, *Mentha spicata,* of the family Labiatae, characterized by lance-shaped leaves that yield an aromatic oil used for tea and as a flavoring. **2.** the leaves of this plant.

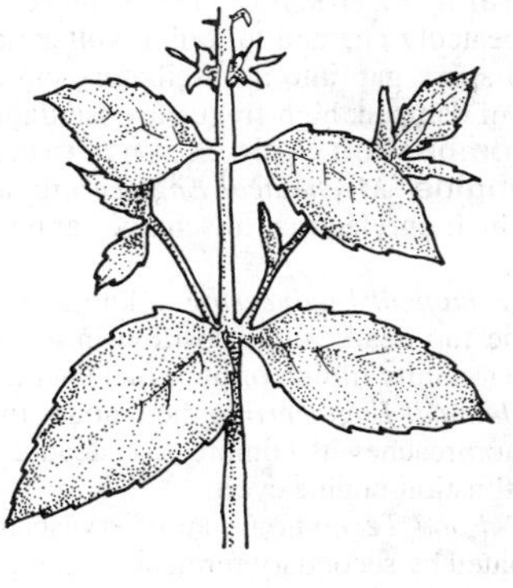

spearmint

spearthrower or **spear-thrower** *Anthropology.* a device used to hurl a spear or missile with greater force and distance than the arm alone; consisting of a short rod to which the spear is hooked at one end.

spearwort *Botany.* any of several buttercups having lance-shaped leaves and small flowers, such as *Ranunculus ambigens,* of the eastern U.S.

spec or **spec.** specification; special.

special character *Computer Programming.* **1.** a character in a character set other than a letter, digit, or space. **2.** in COBOL, a character that is neither numeric nor alphabetic.

special creation a theory proposing that biological species did not evolve, but were created individually in basically their present form.
special factor see SPECIFIC FACTOR.
special functions *Mathematics.* a loosely defined class of functions that recur in different mathematical applications often enough to have achieved recognition. These include solutions of standard differential equations; e.g., Bessel functions, spherical harmonics, Laguerre polynomials, Mathieu functions, etc. In modern mathematics, most are recognizable as matrix entries in the irreducible representations of the classical Lie groups that are symmetries of the problem being solved.
specialization *Evolution.* the evolutionary adaptation of an organism to a specific habitat at the expense of flexibility in surviving in different habitats.
specialized *Biology.* adapted by structure or function to a certain environment or habitat through the process of evolution.
specialized farm *Agriculture.* a farm on which a single type of crop or livestock is grown. Thus, **specialized farming.**
specialized transduction *Molecular Biology.* a selective transfer of DNA from a host (donor) cell to a recipient cell using a bacteriophage as a vector.
special linear group *Mathematics.* the group of linear transformations on a vector space V that have determinant equal to $+1$; the special linear group is a subgroup of the group of all nonsingular linear transformations on V.
special meander corner see MEANDER CORNER.
special nuclear material *Nucleonics.* any of certain materials that are directly controlled by the Nuclear Regulatory Commission, including isotopes uranium-233 and uranium-235 and plutonium.
special observation *Meteorology.* an aviation weather observation that is taken to report important changes in one or more weather elements since the previous observation was taken.
special operations *Military Science.* operations conducted by specially trained, equipped, and organized forces against strategic or tactical targets in pursuit of military, political, or economic objectives; they may be independent or in support of conventional operations.
special orthogonal group *Mathematics.* the group of orthogonal transformations on a vector space V that have determinant equal to $+1$; the special orthogonal group is a subgroup of the Lie group of orthogonal transformations on V.
special-purpose buoy *Navigation.* a buoy used to mark a particular area being used for a special, often temporary, purpose; for example, a quarantine area or dredging operation.
special-purpose computer *Computer Technology.* a computer that is designed to solve a specific class or narrow range of problems.
special-purpose item *Engineering.* an object that has been designed for a specific purpose and is of limited usage, such as a tool that is designed solely for use on a designated model of a piece of machinery. Similarly, **special-purpose vehicle.**
special-purpose language *Computer Programming.* a programming language designed for a single objective such as simulation or string processing.
special-purpose robot *Robotics.* a robot designed for difficult or unusual tasks.
special-purpose vehicle *Engineering.* a vehicle specifically designed to fulfill a specific task, such as a tractor.
special relativity *Physics.* a field of study that deals with the laws of physics as formulated by inertial observers in relative motion. See RELATIVITY THEORY.
special sheaf *Ordnance.* in artillery and naval gunfire support, any sheaf other than parallel, converged, or open.
special-use airspace *Transportation Engineering.* a zone of airspace set aside for a specific use, from which other aircraft are barred.
special weapons *Ordnance.* weapons grouped for special procedures, for security, or for other reasons.
special weather report *Meteorology.* the encoded data from a special observation that is transmitted between weather reporting facilities.
speciation *Evolution.* the process in which new species develop as a result of evolutionary forces and processes.
species [spē´shēz] *plural,* **species.** *Systematics.* **1.** a group of organisms of common ancestry that are able to reproduce only among themselves and that are usually geographically distinct. **2.** a fundamental rank in the taxonomic hierarchy, indicating the limit of organisms able to interbreed. In botany, the species falls below a series; in other kingdoms, the species falls directly below a genus. *Chemistry.* a specific kind of molecule, atom, or ion. *Nuclear Physics.* see NUCLIDE.
species area curve *Ecology.* a common pattern in which the number of species on islands increases as island area increases.
species diversity *Ecology.* an index combining the number and relative frequency of different species in an area or community.
species equilibrium *Ecology.* a condition in which the number of new species immigrating into an area is balanced by the rate of extinction of species already existing there.
species nova *Systematics.* in taxonomic literature, a notation used to indicate a new species.
species population *Ecology.* a group of organisms living in a particular area at a given time that have similar characteristics.
species recognition *Behavior.* the capacity of an organism to distinguish between members of its own species and organisms belonging to other species.
species richness *Ecology.* the number of different species in an area or community.
species-specific *Biology.* limited to or affecting only one species; e.g., a disease that affects only humans.
species-specific behavior *Behavior.* **1.** patterns of behavior that are shown by most members of the same species of the same sex when they are in the same or a similar situation. **2.** complex, stereotyped behavior appearing in most members of a species of the same sex with no evidence of prior opportunity to learn it, thus assumed to be innate.
species-specific defense reaction *Behavior.* an innate behavior that occurs when a new or sudden stimulus is presented, sometimes preventing escape or avoidance learning.
species-specific response see FIXED-ACTION PATTERN.
specif. specifically.
specific *Biology.* peculiar to or characteristic of a species. *Physics.* **1.** an adjectival term indicating that a quantity is expressed as an amount per unit length, area, volume, mass, charge, and so forth. **2.** a term indicating that a quantity is being compared to a standard quantity, such as specific gravity, which is given by the volume density of a substance divided by the volume density of water at 4°C.
specific acoustic impedance *Acoustics.* the complex ratio of the pressure to velocity of a sound phasor, that includes both acoustic resistance and acoustic reactance.
specific acoustic reactance *Acoustics.* the reactive portion of specific acoustic impedance that is complex.
specific acoustic resistance *Acoustics.* a real quantity ρc, where ρ is the equilibrium density and c is the sound velocity; the quantity represents an analogy to resistance in electrical circuits.
specific action potential *Behavior.* see TENDENCY.
specific activity *Enzymology.* the number of enzyme units per milligram of protein. *Nucleonics.* the intensity of radioactivity per unit mass of radionuclide, either by mass of isotope among several isotopes in an element, by mass of pure radionuclide, or by mass of a compound material.
specific adsorption *Physical Chemistry.* a process of adsorption on a surface caused by specific interactions between the adsorbate and the adsorbing surface, rather than by simple electrostatic forces; e.g., the adsorption of chloride ions on mercury.
specifications *Industrial Engineering.* the designed or required characteristics of an item or machine, including dimensions, materials of fabrication, maximum power input or output, and the like. *Engineering.* a written list of specific information required for construction materials and operations, including dimensions, quality, and manner in which work is to be conducted.
specific burnup *Nucleonics.* the fission energy released per unit mass of fuel expressed as megawatt days per ton or per kilogram (MWd/t or MWd/kg).
specific charge see CHARGE-MASS RATIO.
specific code see ABSOLUTE PROGRAMMING.
specific cryptosystem *Telecommunications.* a general cryptosystem combined with a key or keys that are used to control a cryptographic operation.
specific cutting energy *Materials Science.* the energy per unit volume in metal cutting, expressed as

$$u = F_P v / vbt = F_P / bt,$$

where vbt is the volume rate of material removal.
specific dynamic effect *Nutrition.* the stimulating effect upon the metabolism by the ingestion and subsequent digestion of food, especially proteins, causing the metabolic rate to rise above basal levels.

specific energy *Thermodynamics*. a quantity given by the internal energy of a substance per unit mass. *Hydrology*. the energy of water at any cross section of a stream, equal to the sum of the water depth and the velocity head at that cross section.

specific evolution *Anthropology*. cultural change within individual societies in their particular environments.

specific factor *Psychology*. a particular ability or condition that influences an individual's performance on one test or task.

specific fuel consumption *Mechanical Engineering*. the required rate at which fuel is consumed to produce a unit of power or thrust.

specific gravity *Mechanics*. the ratio of the density of a material at a given temperature and pressure to the density of some standard material. For solids and liquids, the standard is usually pure water at a temperature of 3.98°C and standard atmospheric pressure; for gases, the standard is often hydrogen, oxygen, or air at a specified temperature and pressure. Also, RELATIVE DENSITY.

specific gravity bottle *Engineering*. a bottle-shaped instrument containing a thermometer; used to measure and compare the densities of liquids and solids. Also, **specific gravity flask.**

specific gravity hydrometer *Engineering*. a floating instrument that measures the specific gravity of a liquid as it corresponds to water at a designated temperature.

specific growth rate *Microbiology*. the growth of a given microbial population as a function of time.

specific growth-rate constant *Microbiology*. a proportionality constant for the growth of unicellular microorganisms undergoing binary fission.

specific heat *Thermodynamics*. a partial derivative with respect to temperature; under certain conditions it can be expressed as a ratio of the heat transfer to increase the temperature of a unit mass of a substance by 1° to the amount of heat transfer to increase the temperature of a unit mass of water by the same temperature increment, while subject to constant conditions of volume or pressure. Constant-volume specific heat and constant-pressure specific heat are macroscopic properties of a substance.

specific humidity *Meteorology*. in any system of moist air, the ratio of the mass of water vapor to the total mass of the system.

specific immunity *Immunology*. a resistance to infection or disease that results from a specific prior recognition of an antigen.

specific impulse *Space Technology*. a performance standard for rocket propellants, equal to the effective exhaust velocity divided by gravity and to the thrust in pounds divided by the weight flow rate in pounds per second. Also, SPECIFIC THRUST.

specific infectivity *Virology*. a measure of the degree of infectivity of a virus relative to the unit weight or number of virus particles.

specific insulation resistance see VOLUME RESISTIVITY, def. 1.

specific ionization *Nucleonics*. the number of ion pairs that an energetic particle passing through a gas produces per unit length of track. Also, TOTAL SPECIFIC IONIZATION, IONIZATION COEFFICIENT.

specific locus test *Genetics*. a test designed to detect recessive, induced mutations in diploid organisms by crossing a strain carrying known recessive, homozygous mutants with a nonmutant strain treated to induce mutations in its germ cells; induced recessive mutations allelic with those of the test strain will be expressed in the progeny.

specific modulus *Materials Science*. the modulus of elasticity divided by the density.

specific power *Nucleonics*. the amount of power per unit mass of nuclear fuel in a reactor.

specific productivity index *Petroleum Engineering*. barrels of oil produced daily per drop in bottom-hole pressure per foot of effective reservoir formation thickness.

specific propellant consumption *Space Technology*. with a given propellant in a given rocket, the flow needed to produce one pound of thrust; the reciprocal of specific impulse, calculated by dividing weight flow rate by thrust.

specific repetition rate *Electronics*. **1.** the pulse repetition rate of a pair of transmitting stations of a navigation system. **2.** any of a set of closely spaced repetition rates that are derived from a basic rate and associated with a specific set of synchronized loran stations.

specific retention volume *Analytical Chemistry*. the relationship between retention volume, void volume, and adsorbent weight; used to standardize gas chromatography adsorbents.

specific rotation *Optics*. the rotation to its plane of polarization that a beam of light of a given wavelength undergoes as it passes through a solution of a given density, path length, concentration, and temperature.

specific routine *Computer Programming*. a routine designed to solve a particular data-handling problem, in which each address refers to specified registers and locations.

specific speed *Mechanical Engineering*. a factor by which the performance of a centrifugal or axial pump or a hydraulic turbine can be computed, given as the speed in revolutions per minute. *Mechanical Engineering*. a quantity N_s that gives a measure of the performance of a centrifugal or axial pump or a hydraulic turbine; for the pumps, the specific speed is given by $N_s = NQ^{1/2}/H^{3/4}$, where N is the number of revolutions per minute, Q is the volume flow rate, P is the shaft horsepower, and H is the head measured in feet; for the turbine, $N_s = NP^{1/2}/H^{5/4}$.

specific stiffness *Materials Science*. a property of materials, represented as the ratio of the modulus of elasticity to the weight density, that is important in gauging the effectiveness of fibers and fiber composites; for example, the strength of structural steel in fiber form has a tensile strength eight times that of the same material in bulk form.

specific strength *Mechanics*. the strength per unit volume or per unit mass of a substance. *Materials Science*. the strength of a material divided by the density, expressed as specific strength = s/r, where s is the yield strength and r is the density.

specific surface *Chemical Engineering*. an average particle size measurement used in size reduction and enlargement processes; the exterior of the surface of a unit of weight or volume of a particle is based on size distribution data.

specific susceptibility see MAGNETIC SUSCEPTIBILITY.

specific thrust see SPECIFIC IMPULSE.

specific viscosity *Fluid Mechanics*. the ratio of fluid viscosity to that of water at 20°C.

specific volume *Mechanics*. the volume per unit mass; the reciprocal of density.

specific-volume anomaly see ANOMALY OF SPECIFIC VOLUME.

specific weight *Mechanics*. weight per unit volume.

specific yield *Hydrology*. a measure of water available to wells, equal to the percentage by volume of gravity water that will drain from a given mass of saturated rock or soil.

specimen *Science*. an item or part that typifies a larger group to which it belongs; a sample that is taken for testing or study. *Systematics*. specifically, a living or dead sample of an organism collected and used to study and describe the organism for taxonomic purposes.

speck *Plant Pathology*. a bacterial or fungal disease of rice characterized by shriveled, dotted grains.

speckle *Optics*. the instantaneous image of a point light source, such as a star, that is distorted by atmospheric turbulence.

speckle interferometry *Optics*. a process that removes the effects of scintillation on a star's image; for a few sufficiently large nearby stars, the process can measure the diameter of the star or even produce an image of the star's actual photosphere.

specs *Engineering*. a shorter term for specifications.

spectacle *Zoology*. **1.** a transparent scale that covers the eyes of snakes. **2.** colored rings surrounding the eyes of an animal, such as the raccoon.

spectacle frame *Naval Architecture*. a structure that projects from the after-end of the hull and serves as a support for the shafts of a twin-screw vessel.

spectator ion *Chemistry*. an ion that is present in a solution in which a reaction is taking place but is not directly involved in the net reaction.

spectinomycin *Pharmacology*. $C_{14}H_{24}N_2O_7$, an antibiotic drug obtained from the actinomycete *Streptomyces spectabilis* or produced synthetically; used especially in treating strains of gonorrhea resistant to penicillin.

spectra [spek´trə] *Science*. a plural of spectrum. See SPECTRUM.

spectral [spek´trəl] *Science*. relating to or produced by a spectrum.

spectral bandwidth *Spectroscopy*. the minimum range of radiant-energy frequencies over which a spectroscopic instrument is accurate; the width of the transmitted band measured at half maximum transmittance value. Also, EFFECTIVE BANDWIDTH.

spectral centroid *Optics*. the average wavelength for a given filter's spectral energy transmission.

spectral characteristic *Optics*. a quantity, such as brightness or sensitivity, measured as a function of wavelength.

spectral classification *Astronomy*. a classification of stellar spectra using the Henry Draper system, which is based on the star's spectral characteristics.

spectral color *Optics*. the name given to a color at a specific wavelength, or to a color that lies on the chromaticity diagram between the no-hue point and the rim.

spectral density *Electromagnetism.* a distribution of electromagnetic energy over a range of frequencies, expressed as a function of wavelength or frequency. Also, **spectral energy distribution.**

spectral density function *Mathematics.* let $x(t)$ be a stochastic process defined on a probability space (Ω, F, P); let $X(\omega)$ denote the Fourier transform of $x(t)$; and let $X_T(\omega) = X(\omega)$ for $-T < t < T$ and $X_T(\omega) = 0$ otherwise. Then the spectral density function of $x(t)$ is

$$S_{xx}(\omega) = \lim_{T \to \infty} \{|X_T(\omega)|^2 / 2\pi T\}.$$

This is used in predicting statistical characteristics of random processes. Also, **power spectrum, autospectral density function.**

spectral directional reflectance factor *Analytical Chemistry.* the ratio of energy diffused by an object under analysis to the energy diffused by an absolute standard; used in colorimetry.

spectral emissivity *Thermodynamics.* the ratio of the energy radiated in a very narrow frequency band by a body to the energy of the same frequency band radiated by a blackbody at the same temperature; this ratio is a function of the wavelength.

spectral energy distribution *Electromagnetism.* a distribution of energy by electromagnetic radiation expressed as a function of wavelength (or frequency).

spectral extinction *Optics.* the absorption of light in a medium, such as water, as a function of wavelength.

spectral factorization *Mathematics.* the expression of a rational function $f(z)$ of a complex variable as a product $f(z) = f_L(z)f_R(z)$, where $f_L(z)$ and $f_R(z)$ are each functions having all their zeros and poles in the left and right (complex) half-planes, respectively.

spectral hygrometer *Engineering.* an instrument that measures the amount of precipitable moisture in the atmosphere; it is composed of a collimated energy source and a sensing element that measures the decrease in radiant energy caused by the absorption bands of water vapor.

spectral irradiance *Optics.* the flux given off by a luminous object as a function of wavelength.

spectral line *Spectroscopy.* in a spectrum, a discrete value associated with a particular quantity, such as wavelength or frequency. Also, SPECTRUM LINE.

spectral locus see SPECTRUM LOCUS.

spectral luminous efficacy *Optics.* **1.** the sensitivity of the average human retina to light of various colors. **2.** the ratio of luminous flux produced by a monochromatic light in lumens to its radiant flux in watts.

spectral measure *Mathematics.* a resolution of the identity; i.e., an operator-valued measure based on the spectrum of a given (usually normal) operator.

spectral photography *Optics.* a technique that uses a special narrowband pass film to intensify small color effects caused by mineralization and alteration; used in airborne surveys of possible mineral sites.

spectral pyrometer see NARROW-BAND PYROMETER.

spectral radiance factor *Analytical Chemistry.* the directions for analysis of energy reflected from an object are all essentially the same in spectrophotometric colorimetry.

spectral radius *Mathematics.* given the spectrum $s(T)$ of an operator T, the least upper bound of the set $\{|\lambda| : \lambda \in \sigma(T)\}$.

spectral reflectance *Analytical Chemistry.* the diffusion or reflectance of energy from an object in all directions.

spectral regions *Spectroscopy.* the arbitrary ranges of frequencies or wavelengths into which the electromagnetic spectrum is divided; e.g., infrared or ultraviolet.

spectral sensitivity *Physics.* the relative magnitude of the response of a device that detects monochromatic light considered as a function of the wavelength or frequency. Also, **spectral response.** *Graphic Arts.* the relative reaction of a photographic emulsion to each color of the spectrum. *Electronics.* the relation between the radiant sensitivity in a camera tube and the wavelength of incident radiation under given conditions, usually measured with a collimated beam.

spectral series *Spectroscopy.* any set of spectral lines or line groups occurring in sequence.

spectral temperature *Optics.* the temperature of an object, in degrees kelvin, that is obtained by assuming that it radiates like a blackbody.

spectral theorems *Mathematics.* every bounded normal operator T on a Hilbert space induces a resolution E of the identity on the Borel subsets of its spectrum $s(T)$. In addition, T can be reconstructed from E by an integral of a particular type.

spectral transmission *Optics.* the varying percentage of light transmitted by a substance as a function of its wavelength.

spectral type *Astronomy.* one of the seven classes (based on spectral characteristics) into which nearly all stars can be fitted; in order of decreasing temperature, the types are O, B, A, F, G, K, and M.

spectrin *Hematology.* a contractile protein attached to glycophorin at the cytoplasmic surface of the cell membrane of erythrocytes, considered to be important in the determination of red-cell shape.

spectrobolometer *Spectroscopy.* a device consisting of a spectrometer and a bolometer that is used to measure radiation from stars within a narrow range of electromagnetic frequencies.

spectrofluorometer *Spectroscopy.* a spectroscopic instrument that uses a monochromator to increase the selectivity of a fluorometer in the recording of a fluorescence emission spectrum. Also, **spectrofluorimeter.**

spectrogram [spek´trə gram´] *Spectroscopy.* a photographic record of a spectrum produced by a spectrograph.

spectrograph [spek´trə graf´] *Spectroscopy.* a spectroscopic instrument in which a camera or other device is used to record a spectrum on photographic film.

spectrographic [spek´trə graf´ik] *Spectroscopy.* relating to or obtained by the use of a spectrogram.

spectrography [spek träg´rə fē] *Spectroscopy.* the use of a camera or other device to make a photographic record of a spectrum.

spectroheliocinematograph *Optics.* a telescope that produces monochromatic films of the solar photosphere or inner corona.

spectroheliogram *Astronomy.* a photograph of the sun taken in the light of one particular wavelength, or color.

spectroheliograph *Optics.* a device that records monochromatic images of the sun.

spectrohelioscope *Optics.* a device that provides a view of the sun at a single wavelength of light.

spectrometer [spek´trə mēt´ər] *Spectroscopy.* any device that produces a spectrum by dispersion and is calibrated to measure transmitted energy with respect to wavelengths of radiation, refractive indices of prism materials, or radiant intensities at various wavelengths.

spectrometry [spek träm´ə trē] *Spectroscopy.* the use of spectroscopic techniques to measure wavelengths and intensities of spectral lines or bands and to identify the atoms, molecules, or ions that produce them.

spectrophone *Analytical Chemistry.* a microphone-equipped cell that has windows through which a laser beam enters; used in the optoacoustic detection method.

spectrophotometer *Spectroscopy.* a spectroscopic device used to photometrically measure a quantity of light absorbed by a sample or the relative intensities of spectral lines or bands produced by a sample with respect to wavelength.

spectrophotometric analysis see ABSORPIOMETRIC ANALYSIS.

spectrophotometric titration *Analytical Chemistry.* a titration in which the radiant-energy absorption is measured spectrophotometrically after each increment of titrant is added.

spectrophotometry *Spectroscopy.* the photometric measurement of either the electromagnetic wavelengths absorbed or emitted by a sample or the relative intensities of spectral lines or bands produced by a sample. Also, **spectrophotometric analysis.**

spectropolarimeter *Optics.* a device that measures the amount of polarization in a light beam as a function of its wavelength.

spectropyrheliometer *Spectroscopy.* in astronomical spectroscopy, a device used to measure the distribution of radiant energy from the sun in the ultraviolet and visible regions of the electromagnetic spectrum.

spectroscope [spek´trə skōp´] *Spectroscopy.* a device for analyzing dispersed light producing an observable spectrum; it consists of a slit, prism, collimator lens, object lens, and grating.

spectroscopic [spek´trə skäp´ik] *Spectroscopy.* relating to or obtained by the use of a spectroscope.

spectroscopic binary star *Astronomy.* a binary star that cannot be resolved by a telescope, but whose spectral lines periodically double, thus revealing (by Doppler shift) the existence of a binary star system.

spectroscopic displacement law *Spectroscopy.* a rule stating that the spectrum of a neutral (un-ionized) atom is similar to that of a singly ionized atom of the element whose atomic number is one greater than the neutral atoms, and to that of a doubly ionized atom of the element whose atomic number is two greater, and so on. Also, SOMMERFIELD-KOSSEL DISPLACEMENT LAW.

spectroscopic parallax *Astronomy.* a measurement of stellar distance that uses a star's spectrum to infer its intrinsic luminosity, which is then compared with its apparent brightness to derive the distance.

spectroscopic variable see SPECTRUM VARIABLE.

Spectroscopy

Spectroscopy encompasses the various activities involved in utilizing and interpreting all the aspects of the phenomena known as "spectra" through the use of a spectroscope. The term "spectra" was first applied in 1672 by Newton to the remarkable discovery that the sun's image could be broken into all the visible colors when passed through a prism. He derived this term from the Latin *spectrum*, an appearance or image.

The word "spectroscope" is, in turn, derived from the Latin root for "spectrum" and the Greek *skopein*, to view. A spectroscope is composed of three elements, viz., a source of electromagnetic energy, a means of dispersion of the energy into its many components, and a detector of the dispersed energy.

The first spectroscope was developed in 1814 by the German optician Fraunhofer, who utilized a combination of a telescope, a prism, and a device for measuring the angle at which the various components of light from stars were deflected. The initial application of spectroscopy was by Bunsen and Kirchoff to phenomena in flames. Such studies have continued to the present day, and have permitted a greater understanding of the processes involved in the combustion of diverse materials over a broad range of pressures—from less than 0.01 atmosphere in some flames to the >100 atmospheres characteristic of the combustion of the complex mixtures making up solid and liquid propellants for use in rocket motors. At the same time, the range of the electromagnetic spectrum over which spectroscopy has been utilized has spread enormously from the visible region observed by Newton. The development of tools to disperse electromagnetic energy and to record it have led to the extension of the wavelengths being studied from hundredths of Angstroms to hundreds of meters; fifteen orders of magnitude overall! In addition, dispersion is now being achieved by matter on the atomic level, rather than by a prism.

The development of improved optics and of data acquisition and processing techniques has also led to the ability to study processes characteristic of high-speed engines through the technique known as CARS (Coherent Anti-Stokes Raman Spectroscopy) and of analyzing the surfaces of materials to depths at the atomic level through the incorporation of magnetic fields to yield NMR (Nuclear Magnetic Resonance) spectra and X-ray radiation to yield Auger electron spectra.

R. H. Woodward Waesche
Principal Scientist
Virginia Propulsion Division, Atlantic Research Corp.

spectroscopy [spek träs´kə pē] *Physics.* the science that is concerned with the measurement of the emission and absorption spectra of light and other forms of electromagnetic radiation.

spectrum [spek´trəm] *plural,* **spectra** or **spectrums.** *Science.* the complete range or array of entities in which some phenomenon or system exists. *Optics.* also, **visible spectrum.** the visible portion of all electromagnetic radiation; i.e., the band of colors that is produced by white light passing through a prism, ranging from about 750 to about 400 nanometers and generally perceived as appearing in a sequence of red, orange, yellow, green, blue, indigo, and violet. *Physics.* the range of wavelengths or frequencies produced when any other form of electromagnetic radiation is dispersed, such as gamma rays, X-rays, or radio waves; these lie beyond the visible spectrum. *Electronics.* a continuous, usually wide range of frequencies having a given common characteristic; the range of frequencies in a given system. *Mathematics.* **1.** let T be a continuous operator on a Banach space, and I the identity operator. The spectrum $\sigma(T)$ of T is the union of the point, continuous, and residual spectra of T. By theorem, $\sigma(T)$ is nonempty and is contained in the closed disc of the complex plane of radius $\|T\|$. **2.** The set of all prime ideals in a ring R is called the (prime) spectrum of R. (A term coined by Sir Isaac Newton; a special use of a Latin word meaning "appearance" or "image.")

spectrum analysis *Optics.* the study of the range of light from a given source; this is useful in determining the composition and energy state of the source. *Physics.* in general, for any emission of complex waves, the measurement of all the frequencies and their corresponding amplitudes making up the waveform. *Acoustics.* the measurement of energy or power in a sound source as a function of frequency; that is, the quantitative measurements of sound plotted vertically against the sound frequency on the horizontal axis.

spectrum analyzer *Spectroscopy.* an instrument designed primarily to indicate the distribution of energy in the frequencies produced by a pulse magnetron; it also measures the magnetron's cold impedance and the Q of resonant cavities and lines.

spectrum level *Telecommunications.* the level of a given signal component at a given frequency, contained within a frequency band 1 hertz wide and centered at the specified frequency.

spectrum line see SPECTRAL LINE.

spectrum locus *Optics.* the locus of points that represent pure spectral colors in a chromaticity diagram. Also, SPECTRAL LOCUS.

spectrum of turbulence *Astrophysics.* the relationship between the size of violent eddies in the sun's atmosphere and their velocity.

spectrum-selectivity characteristic *Electronics.* the increase in the minimum input-signal power over the minimum detectable signal required to produce an indication on a radar indicator.

spectrum signature *Electronics.* the spectral characteristics of the transmitter, receiver, and antennae of a communications system.

spectrum signature analysis *Electronics.* a study of electromagnetic interference from transmitting and receiving devices, made in order to determine the operational functionality and environmental compatability of the device.

spectrum variable *Astronomy.* stars that exhibit periodic changes in their spectra and brightness over periods of 1 to 25 days. Also, SPECTROSCOPIC VARIABLE.

specular *Optics.* **1.** in or relating to the direction in which a mirror reflects light. **2.** relating to or involving the use of a mirror.

specular hematite *Mineralogy.* Fe_2O_3, a variety of hematite featuring a blue-gray to gray color and a splendent metallic luster.

specularite *Mineralogy.* α-Fe_2O_3, a black or gray variety of hematite with a brilliant metallic luster, occurring as tabular crystals, or in foliated or micaceous masses. Also, **specular iron.**

specular reflection *Optics.* reflection from a smooth surface. Also, MIRROR REFLECTION, REGULAR REFLECTION. *Physics.* any reflection in which the angle of incidence is equal to the angle of reflection.

specular reflection factor *Optics.* the proportion of light falling on a surface that is reflected specularly.

specular reflector *Optics.* a surface capable of reflecting a direct image of a source; a mirror.

specular transmittance *Electromagnetism.* the ratio of energy transmitted by a body to that incident on it, not including the scattered emergent energy.

speculum *Optics.* a mirror whose reflective surface is either polished metal or a metal coating applied to some other substance such as glass. *Medicine.* an instrument for exposing a cavity or channel in the body by enlarging the opening to allow viewing. *Vertebrate Zoology.* an iridescent patch of color on the wings of certain ducks and other birds. (From a Latin word for *mirror.*)

speculum examination *Medicine.* **1.** the viewing of a canal of the body, using a speculum. **2.** specifically, a viewing of the vagina and cervix with a vaginal speculum.

speech *Linguistics.* **1.** the system of vocal sounds by which humans produce understandable words to communicate their thoughts with one another. **2.** the process of communicating in this way; the act of speaking. **3.** the power of communicating in this way; the ability to speak.

speech amplifier *Acoustical Engineering.* an electronic device designed to increase the amplitude, to clip the peaks, or to otherwise enhance electromagnetic wave input derived from sound in the range of the human voice, about 1000 to 3000 hertz.

speech bandwidth *Telecommunications.* the range of speech frequencies that are sent by a carrier telephone system.

speech clipper *Acoustical Engineering.* a speech amplifier designed to limit excessively high amplitude of input, thereby preventing distortion.

speech clipping *Acoustics.* the limiting of the amplitudes of certain signals in speech processing devices; used in tests for analyzing the intelligibility of speech.

speech compression *Telecommunications.* a type of modulation that uses certain properties of the speech signal in order to send adequate tranmissions over a frequency band that is narrower than would otherwise be necessary.

speech dysfunction see SPEECH IMPEDIMENT.

speech frequency see VOICE FREQUENCY.

speech impediment *Medicine.* any ongoing defect or impairment that causes a person's speaking pattern in the native language to be abnormally different from other native speakers of the language, often so as to interfere with communication; e.g., aphasia, stuttering, lisping, slurred speech, inappropriate voice level, and so on; the causes include a variety of physical conditions and also emotional or psychological factors. Also, SPEECH DYSFUNCTION.

speech-interference level *Acoustics.* a decibel measurement of the amount of noise interference with speech, calculated as the arithmetic average of the sound pressure level over a frequency band of 600 to 4800 hertz.

speech interpolation *Telecommunications.* a system of acquiring more than one voice channel for each voice circuit by providing each subscriber a speech path in the proper direction only when his or her speech requires it.

speech pathology *Medicine.* **1.** the study of the nature and causes of speech abnormalities and of malfunction of the organs of speech. **2.** see SPEECH THERAPY. Thus, **speech pathologist.**

speech recognition *Linguistics.* the ability to recognize the communicative meaning of vocally produced sounds. *Artificial Intelligence.* a process by which a computer can recognize but not necessarily interpret a spoken word, phrase, or sentence.

speech therapist *Medicine.* a person who is trained in the profession of speech therapy.

speech therapy *Medicine.* the treatment of speech impediments; i.e., of disorders affecting the ability to carry on normal oral communication.

speech understanding *Artificial Intelligence.* a process by which a computer cannot only recognize a spoken word, phrase, or sentence, but also ascertain its meaning.

speed *Physics.* in general, the time-rate at which any physical process takes place. *Mechanics.* the time-rate of change of position without regard to the direction of motion; i.e., the ratio of the distance traveled to the time elapsed. *Optics.* see FOCAL POWER. *Graphic Arts.* a measurement of the light sensitivity of a photographic emulsion.

speed circle *Navigation.* a circle drawn around a given center with the radius equal to a given speed.

speed cone *Mechanical Engineering.* a cone-shaped pulley or a series of pulleys on a common axle having increasing diameters to form a stepped cone. Also, SPEED PULLEY.

speed-density control *Aviation.* a system that meters the fuel supply to an engine in response to variations in rpm, air density, intake manifold pressure and temperature, and exhaust back pressure. Also, **speed-density metering.**

speed-density-volume relationships *Transportation Engineering.* a pattern of relationships observed among traffic speed, density, and total traffic volume. At low densities, traffic volume is nearly proportional to density while speed remains relatively constant. As roadway capacity is approached, further increases in density lead to reduced speeds rather than increased traffic volume.

speed distribution *Transportation Engineering.* the range of speeds observed in vehicles passing a given point or using a stretch of road.

speed error *Navigation.* an acceleration error caused by a craft's change in speed.

speed lathe *Mechanical Engineering.* a hand-controlled lathe with no carriage or back gears.

speed-length ratio *Naval Architecture.* a vessel's speed divided by the square root of its length; used to compare the performance of hull forms of different sized ships.

speed line *Navigation.* a line of position drawn perpendicular to the course and used to calculate the speed made good.

speed made good *Navigation.* the actual average speed, measured in knots, that a vessel maintained along its course from one point to another. Also, SPEED OVER THE GROUND.

speed matching buffer *Computer Technology.* see BUFFER.

speed of advance *Navigation.* **1.** in naval usage, the speed expected to be made good over the ground. **2.** the average speed that a vessel must maintain in order to arrive at its destination at a particular time.

speed of light *Electromagnetism.* the speed at which electromagnetic radiation propagates through a vacuum, which is measured at approximately 299,792.458 kilometers per second (or about 186,282 miles per second).

speed of response *Physics.* the reaction time of a system stimulated by some external influence.

speed of sound *Acoustics.* **1.** the scalar rate at which sound travels through a given medium, usually measured in feet per second or meters per second; this rate varies with temperature and according to the nature of the medium. **2.** specifically, the speed at which sound moves through dry air at standard atmospheric pressure and a temperature of 0°C; this is approximately 331 meters per second (about 740 miles per hour).

speed of travel *Metallurgy.* in welding, the speed of the operation in the longitudinal direction.

speedometer *Engineering.* a device in an automobile or other vehicle that indicates the rate at which the vehicle travels, generally in miles per hour, kilometers per hour, or knots.

speed over the ground *Aviation.* the speed of an aircraft relative to the earth's surface, as opposed to airspeed. Also, GROUND SPEED. *Navigation.* see SPEED MADE GOOD.

speed-payload trade-off *Mechanical Engineering.* the difference between the maximum weight of a workpiece and the maximum speed at which a machine can move it.

speed-power product *Electronics.* the product of the gatespeed of an electronic circuit and its power dissipation.

speed pulley see SPEED CONE.

speed reducer *Mechanical Engineering.* a train of gears between a motor and the machinery it drives to reduce the speed of power transmission.

speed regulator *Electricity.* a device that maintains the speed of a motor or other device at a predetermined value, or varies it in accordance with a predetermined plan.

speed-reliability trade-off *Industrial Engineering.* a term for the balance between the speed of an operation and the reliability or quality of the operation or product so produced.

speed triangle *Navigation.* a vector diagram in which one vector represents the actual course and speed of the reference craft and the second vector represents the relative course and speed of a craft being tracked, as determined, for example, by radar plotting. The third vector then represents the actual course and speed of the craft being tracked.

speiss [spīs] *Metallurgy.* an intermediate product obtained during the smelting of certain ores; it consists of arsenides and antimonides.

Spelaeogriphacea *Invertebrate Zoology.* an order of tiny primitive crustaceans in the superorder Peracarida, having a cylindrical, elongated body; the single known member is *Spelaeogriphus lepidops*, a South African cave dweller.

spelean *Geology.* of or relating to a feature in a cave.

speleologist *Geology.* a scientist who studies and explores caves.

speleology *Geology.* the scientific study and exploration of caves.

speleothem *Geology.* a secondary mineral deposit formed in a cave by the action of water.

spell-checker *Computer Programming.* a feature of many word-processing programs and some electronic typewriters that allows a user to analyze a document in progress to identify words that may be misspelled; the program has the ability to identify any word in the document that does not appear in its internal dictionary and usually also to suggest alternative spellings for such words. Also, **spell-check, spelling checker.**

spelter *Metallurgy.* crude zinc obtained by smelting zinc ore.

spelter's fever *Toxicology.* poisoning caused by exposure to zinc fumes during the smelting process; characterized by malaise, fever, and muscular pain; a form of metal chill fever. Also, SPELTER CHILLS, ZINC CHILL, ZINC FUME FEVER.

spelter solder *Metallurgy.* a brazing filler that contains copper and zinc.

spelunker *Geology.* a person who explores caves as a hobby.

Spemann, Hans 1869–1941, German zoologist; awarded the Nobel Prize for discovery of the organizer function in embryonic development.

spencerite *Mineralogy.* $Zn_4(PO_4)_2(OH)_2 \cdot 3H_2O$, a white, monoclinic mineral occurring in platy masses and as radiating aggregates of tabular crystals, having a specific gravity of 3.145 and a hardness of 3 on the Mohs scale; found in oxidized zinc ore.

spending beach *Geology.* the beach in a wave basin on which the entering waves spend themselves.

Spenocleaceae (bellflower)

Spenocleaceae *Botany.* a family of flowering plants characterized by bisexual, bell-shaped flowers with usually united petals. Also, CAMPANULACEAE.

spent fuel *Nucleonics.* a term for fuel that can no longer sustain a chain reaction in a nuclear reactor because of the excessive amount of accumulated poisons and the consumption of fissionable isotopes.

spergenite *Geology.* a biocalcarenite that contains ooliths and fossil debris with a maximum quartz content of 10%. Also, BEDFORD LIMESTONE, INDIANA LIMESTONE.

sperm *Biology.* **1.** the fluid of a male that fertilizes the eggs of the female; semen. **2.** one of the sperm cells in the semen; spermatozoon.

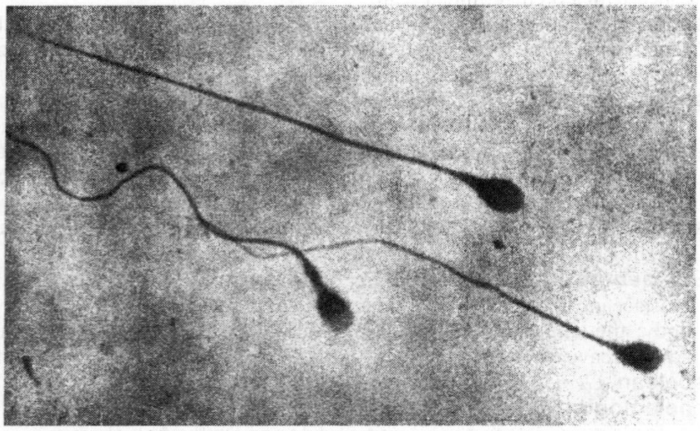

sperm

spermaceti [spur´mə set´ē] *Materials.* a white, translucent solid that is obtained from the oil in the head of the sperm whale; used in cosmetics and candles. Also, CETACEUM, SPERMACETI WAX.

spermagonium *Mycology.* a structure found in certain fungi that forms male sex cells; it is termed **pycnium** in rust fungi. Also, **spermogonium, spermatogonium, spermagone.**

spermatheca *Zoology.* a bulbous sac found mostly in female insects that stores sperm for fertilization.

spermatic cord *Anatomy.* the cordlike structure passing through the inguinal canal and connecting the testis to the abdomen; consists of the ductus deferens, arteries and veins, including the pampiniform plexus, lymphatics, and sheaths of connective tissue.

spermatid *Histology.* an immature sperm cell whose development is nearing completion.

spermatin *Biochemistry.* an albuminoid material that is found in semen.

spermatoblast *Developmental Biology.* a sperm cell that arises by the process of mitotic division from the germ cell, eventually becoming a primary spermatocyte.

spermatocele *Medicine.* a cyst in the scrotum that contains spermatozoa.

Spermatochnaceae *Botany.* a family of slime-producing brown algae of the order Chordariales, characterized by sporophytes with terete, irregularly branched thalli, apical growth, and primary assimilatory filaments that cover the cortex.

spermatocidal *Medicine.* destructive to sperm. Also, **spermicidal.**

spermatocyte *Histology.* an immature sperm cell that is still undergoing the special cell divisions that reduce its chromosome number by one-half.

spermatogenesis *Cell Biology.* the process by which undifferentiated male germ cells form into mature spermatozoa; in the first stage, termed **spermatocytogenesis,** the germ cells divide into spermatocytes, which in turn give rise to spermatids; in the second stage, termed **spermiogenesis,** the spermatids develop into spermatozoa.

spermatogonium *Histology.* the first developmental stage of cells in the seminiferous tubules that will give rise to sperm.

spermatoid *Medicine.* resembling or involving sperm.

spermatolytic *Medicine.* relating to the destruction or solution of sperm.

spermatopathy *Medicine.* any disease of the sperm.

spermatophore *Zoology.* a capsulized mass of sperm transferred by male to female, found mainly in invertebrates such as annelids and arthropods.

spermatophyte *Botany.* a plant of the division Spermatophyta, which includes the gymnosperms and the angiosperms; a seed-producing plant.

spermatorrhea *Medicine.* too frequent, involuntary discharge of semen without stimulation.

spermatozoid *Botany.* a motile male gamete produced in an antheridium, by which the female organs are fertilized.

spermatozoon [spur´ma tə zō´ən] *plural,* **spermatozoa.** *Histology.* a mature sperm cell.

spermaturia *Medicine.* the condition of having semen present in the urine.

spermidine *Biochemistry.* $C_7H_{19}N_3$, a triamine found bound to DNA in semen and other animal tissues.

spermine *Biochemistry.* $C_{10}H_{26}N_4$, a tetramine found bound to DNA in semen and other body tissues.

sperm nucleus *Botany.* one of the two male nuclei that arise from the generative nucleus of a pollen grain and travel through the pollen tube to the female gametophyte.

sperm oil *Materials.* a thin, yellow water-insoluble liquid that is obtained from the sperm whale; used mainly as a lubricant in light machinery such as watches and clocks.

Spermphthoraceae *Mycology.* a family of fungi belonging to the order Endomycetales, characterized by needle-shaped or spindle-shaped ascospores and occurring in soil, water, flowers, fruits, and insects; some of its species are plant pathogens.

sperm whale *Vertebrate Zoology.* a large, square-snouted, toothed whale, *Physeter catodon*, of the family Physeteridae, characterized by a large closed cavity in the head holding a fluid mixture of spermaceti and oil, a blubber that yields superior oil, and the production of ambergris in the intestines; found worldwide, although reduced in numbers and now rare in some areas. Also, BLACK WHALE.

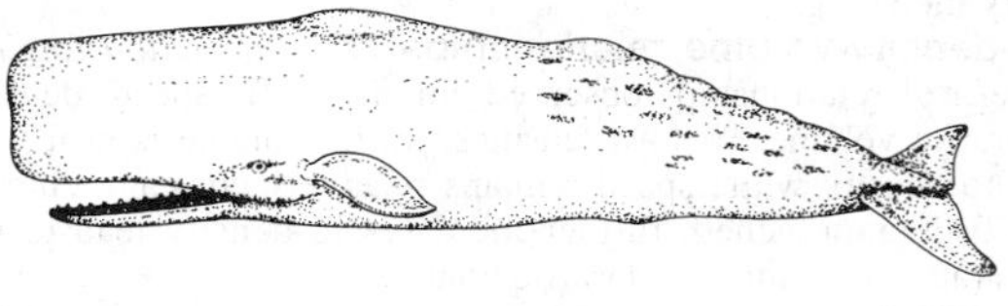

sperm whale

Sperner's lemma *Mathematics.* form any simplicial triangulation of a given triangle *ABC*. Vertices of the triangulation are points chosen in an arbitrary manner on the sides of the triangle or in its interior. Label the points on the side *AB* either *A* or *B*, the points on the side *BC* either *B* or *C*, the points on the side *CA* either *C* or *A*, and the interior points *A*, *B*, or *C* in an arbitrary way. Then the number of small triangles labeled *ABC* is odd.

speromagnetic state *Solid-State Physics.* a state of an amorphous substance in which the atomic magnetic moments are arranged in a more or less random order and are fixed in these positions while the substance remains below the magnetic ordering temperature.

Sperry, Elmer Ambrose 1860–1930, American inventor; developed a gyroscope (**Sperry gyroscope**) that became a standard instrument in navigation.

Sperry, Roger Wolcott born 1913, American neurophysiologist; awarded the Nobel Prize for his discovery of the "split brain."

sperrylite *Mineralogy.* $PtAs_2$, a very rare, tin-white cubic mineral of the pyrite group, with metallic luster, occurring as cubes and cubo-octahedrons, and having a specific gravity of 10.46 to 10.6 and a hardness of 6 to 7 on the Mohs scale; the most widespread platinum mineral, found in all types of deposits.

Sperry process *Chemical Engineering.* an electrolytic process for producing white lead, in which lead anodes are electrolytically corroded in a sodium acetate electrolyte that contains Na_2CO_3.

spessartine *Mineralogy.* $Mn_3^{+2}Al_2(SiO_4)_3$, a red to brown cubic mineral of the garnet group occurring as dodehedral or trapezohedral crystals and in massive or granular form, having a specific gravity of 3.8 to 4.25 and a hardness of 7 to 7.5 on the Mohs scale; forms a series with almandine; found in granite pegmatite, schist, gneiss, and quartzite. Also, **spessartite.**

spessartite *Petrology.* a dioritic or gabbroic lamprophyre consisting of phenocrysts of green hornblende or clinopyroxene in a groundmass of sodic plagioclase, with traces of olivine, biotite, apatite, and opaque oxides.

SPF or **spf** see SUN-PROTECTION FACTOR.

sp. gr. specific gravity.

sph. spherical.

Sphacelariaceae *Botany.* a cosmopolitan family of marine and freshwater, filamentous, tufted brown algae in the order Sphacelariales; characterized by erect axes topped by a large apical cell, which is often dark brown due to tannin accumulating around the central nucleus.

Sphacelariales *Botany.* an order of filamentous, tufted brown algae of the class Phaeophyceae, characterized by multibranched axes arising perpendicular to a prostrate base or stolon and often having a tiered appearance due to regular patterns of segmentation.

Sphacelia *Mycology.* a genus of fungi of the class Hyphomycetes whose sexual counterpart or teleomorph comprises the genus *Claviceps.*

Sphaceloma *Mycology.* a genus of fungi of the class Coelomycetes composed of species that cause such plant diseases as rose leaf spot.

Sphaeractinoidea *Paleontology.* a group of marine hydrozoans similar to the Stromatoporoidea and the Spongiomorphida; extant in the Late Paleozoic.

Sphaeriales *Mycology.* in some classifications, an order of fungi of the subdivision Ascomycotina occurring on wood and bark and including both plant parasites and nonparasites that obtain nutrients from dead organic matter.

Sphaeridiotrichaceae *Botany.* a family of the order Phaeothamniales, composed of simple or ramified filamentous forms with mucilaginous walls and reproduction by vegetative cell division.

Sphaeriidae *Invertebrate Zoology.* the pea shells, a cosmopolitan family of minute freshwater bivalve mollusks in the order Heterodonta.

Sphaerioidaceae see SPHAEROPSIDACEAE.

Sphaeriparaceae *Botany.* a family of nonphotosynthetic flagellates of the order Syndiniales; parasites of appendicularians and one radiolarian.

Sphaerirhynchia *Paleontology.* an extinct genus of articulate brachiopods in the order Rhynchonellida; strongly biconvex, almost spherical, around 2 cm across; extant in the Silurian and Devonian.

sphaerite *Mineralogy.* a light-gray or blue mineral consisting of a hydrous phosphate of aluminum, occurring in global concretions.

sphaero- a variant of *sphero-,* a combining form meaning "round" or denoting a relationship to a sphere.

Sphaerobolaceae *Mycology.* a family of fungi belonging to the order Tulostomatales, commonly known as the gun or cannonball fungus and occurring primarily in tropical regions.

Sphaerobolus *Mycology.* a genus of fungi belonging to the order Sclerodermatales, occurring on dung and wood and typically ejecting its mature spores.

Sphaerocarpaceae *Botany.* a family of small winter-annual liverworts of the order Sphaerocarpales, characterized by numerous hyaline rhizoids and numerous sex organs that thickly cover the dorsal thallus surface; includes the genus *Sphaerocarpos,* the study of which led to the discovery of sex-correlated chromosomes in plants.

Sphaerocarpales *Botany.* a small order of liverworts, characterized by a large involucre surrounding each archegonium.

Sphaeroceridae *Invertebrate Zoology.* a family of small black dung flies, dipteran insects in the subsection Acalyptratae.

sphaerocobaltite *Mineralogy.* $CoCO_3$, a rose-red trigonal mineral of the calcite group, occurring as small spherical masses and crusts, having a specific gravity of 4.13 and a hardness of 4 on the Mohs scale. Also, COBALTOCALCITE.

Sphaerococcaceae *Botany.* a monogeneric family of red algae of the order Gigartinales, characterized by laterally branched, erect, flattened thalli and by spermatangia that develop in small depressions.

sphaerocyst *Cell Biology.* a round or ovoid cell found in certain genera of fungi.

Sphaerocytophaga *Bacteriology.* a genus of Gram-negative bacteria that exhibit a gliding motility, occur in the human oral cavity, and may be identical to species of *Capnocytophaga.*

Sphaerodoridae *Invertebrate Zoology.* a family of small, bottom-dwelling polychaete annelid worms in the Errantia.

Sphaerolaimidae *Invertebrate Zoology.* a family of mostly marine nematode roundworms in the superfamily Monhysteroidea, having a smooth or transversely ridged body and a large mouth.

sphaerolitic see SPHERULITIC.

sphaeromastigote *Cell Biology.* a cell form occurring in cells of *Trypanosoma* at certain stages of its life cycle.

Sphaeromatidae *Invertebrate Zoology.* a family of marine isopod crustaceans in the suborder Flabellifera, having flat bodies and some fused abdominal segments.

Sphaerophorus *Bacteriology.* in former classifications, a genus of Gram-negative anaerobic bacteria that are now included in the genera *Bacteroides* and *Fusobacterium.*

Sphaeropleaceae *Botany.* the single, monogeneric family of green algae in the order Sphaeropleales.

Sphaeropleales *Botany.* an order of green algae of the class Chlorophyceae, characterized by an unbranched filamentous thallus that floats in fresh water, thin lateral cell walls, and the lack of a gelatinous sheath.

Sphaeropleineae *Botany.* a suborder of green algae in the order Ulotrichales, characterized by long coenocytic cells with numerous chloroplasts and by heterogametes produced in undifferentiated vegetative cells.

Sphaeropsidaceae *Mycology.* a family of rust fungi belonging to the order Sphaeropsidales, characterized by dark fruiting bodies shaped like cones, lenses, or flasks. Also, SPHAERIODACEAE, PHOMACEAE.

Sphaeropsidales *Mycology.* an order of fungi belonging to the class Coelomycetes, characterized by spores or conidia that form on the surface within the host or on the matter upon which they rest. Also, PHOMALES, PHYSARALES, PHYLLOSTICTALES.

Sphaerosepalaceae *Botany.* a family of dicotyledonous, tanniferous, deciduous trees and shrubs of the order Theales, characterized by mucilage cells in the parenchymatous tissues and stratified phloem with wedge-shaped rays; confined to Madagascar.

Sphaerotilus *Bacteriology.* a genus of Gram-negative, aerobic, rod-shaped bacteria that are found in organically polluted freshwater habitats and sewage water as filamentous cells coated with iron oxide.

Sphagnaceae *Botany.* the single, monogeneric family of mosses in the order Sphagnales, occurring chiefly in bogs.

Sphagnales *Botany.* an order of mosses constituting the subclass Sphagnobrya and containing the single family Sphagnaceae.

Sphagnobrya *Botany.* a subclass of mosses in the division Bryopsida, characterized by grayish-green plants with numerous spirally arranged branches; common in bogs and other wet habitats, it includes the single order Sphagnales.

Sphagnopsida *Botany.* a class of bryophytes characterized by a thallus protonema that produces a single gametophore, rhizoids on the thallus only, equally forked stems, and elevated capsules.

sphagnum [sfag´nəm] *Botany.* any moss of the genus *Sphagnum,* in the family Sphagnaceae; used as a packing material; e.g., in potting plants.

sphagnum bog *Ecology.* an area of moist soil, rich in organic debris from acid-tolerant mosses.

sphalerite *Mineralogy.* (Zn,Fe)S, a colorless to yellow, black, or brown, resinous, cubic mineral of the spalerite group, occurring as tetrahedral crystals and commonly in coarse to fine granular masses, having a specific gravity of 3.9 to 4.1 and a hardness of 3.5 to 4 on the Mohs scale; found with galena in carbonate rocks, hydrothermal veins, and contact metasomatic deposits; the most abundant ore of zinc. Also, BLENDE, ZINC BLENDE.

S phase *Genetics.* the restricted part of the eukaryotic cell cycle during which replication of DNA occurs.

Sphecidae *Invertebrate Zoology.* the mud daubers or digger wasps, solitary hymenopteran insects in the superfamily Sphecoidea that build mud nests.

Sphecoidea *Invertebrate Zoology.* a superfamily of solitary predaceous wasps, hymenopteran insects in the suborder Clistogastra, including sand wasps, digger wasps, and mud daubers. Also, **Spheciformia.**

Sphenacodontidae *Paleontology.* a family of large, predatory, advanced synapsids in the extinct order Pelycosauria; they were characterized by high, long-snouted skulls and were probably ancestral to the therapsids; extant in the Upper Carboniferous and Permian.

sphene see TITANITE.

sphenethmoid *Anatomy.* the curved plate of bone in front of the lesser wing of the sphenoid bone. *Vertebrate Zoology.* in many amphibians, single or paired bones that surround the anterior portion of the brain. Also, **sphenoethmoid.**

sphenic *Science.* wedge-shaped.

Spheniscidae *Vertebrate Zoology.* the penguins, a family of flightless swimming birds composing the order Sphenisciformes, characterized by wings that are modified into flat flippers and by distinctive black-and-white plumage.

Sphenisciformes *Vertebrate Zoology.* a monofamilial order of flightless aquatic birds containing the penguins and belonging to the superorder Impennes.

spheno- or **sphen-** a combining form meaning "wedge" or denoting a relationship to the sphenoid bone.

sphenochasm *Geology.* a triangular gap in the oceanic crust between two continental blocks that converges to a point.

Sphenodontidae *Vertebrate Zoology.* the tuatura, a family of lizard-like reptiles of the order Rhynchocephalia; found on New Zealand and adjacent islands.

sphenoid *Science.* wedge-shaped. *Anatomy.* of or relating to the sphenoid bone. Also, **sphenoidal.** *Crystallography.* an open wedge-shaped crystalline form obtained by repetition of a face around a twofold axis.

sphenoid bone *Anatomy.* a large skull bone that forms the central floor of the cranial cavity, bordered by the frontal, zygomatic, ethmoid, parietal, temporal, and occipital bones, and the vomer.

sphenoid sinus *Anatomy.* either of two openings in the sphenoid bone, connecting with the nasal cavity.

sphenolith *Geology.* a wedgelike igneous intrusion that is partly concordant and partly discordant with the country rock.

Sphenomonadales *Botany.* an order of euglenoids with one or two emergent flagellum, colorless cells that are osmotrophic or phagotrophic, and a rigid cell with grooves or keels.

sphenopalatine *Anatomy.* relating to the sphenoid and palatine bones.

sphenopalatine foramen *Anatomy.* the foramen formed between the lower surface of the sphenoid bone and the sphenopalatine notch of the palatine bone that transmits the sphenopalatine artery and nasopalatine nerve.

Sphenophyllales *Paleontology.* an extinct order of vascular plants that have been roughly placed up to now in the genus *Sphenophyllum*; sometimes referred to as "articulates;" extant in the Carboniferous and the Permian.

Sphenopsida *Botany.* the horsetails, a class of the Pteridophyta characterized by a sporophyte with roots, small whorled leaves at distinct stem nodes, and sporangia borne on sporangiophores.

spher- a combining form meaning "round" or denoting a relationship to a sphere.

spherator *Physics.* a plasma device used for studying fusion processes.

sphere [sfēr] *Science.* a round body or figure. *Mathematics.* the mathematical description of such a figure; the set of points in a metric space that are a constant distance, called the radius, from a fixed point. For example, in Euclidean n-space, the unit $(n-1)$ sphere centered at the origin is the set of points $(x_1, \ldots, x_n)$ that satisfy the equation

$$x_1^2 + \cdots + x_n^2 = 1.$$

The interior of the sphere, i.e., the set of points satisfying

$$x_1^2 + \cdots + x_n^2 < 1,$$

is called the **open ball.** *Physics.* any globular or rounded object approximating the form of a geometric sphere. *Astronomy.* **1.** any planet or star. **2.** see CELESTIAL SPHERE.

sphere gap *Electricity.* a spark gap in which the spark passes between two polished metal balls.

sphere of attraction *Physical Chemistry.* the area between two molecules within which the energy generated by their mutual attraction is significant enough to be distinguished from the average energy of other molecules in the system.

sphere of reflection *Crystallography.* a construction for considering conditions for diffraction in terms of the reciprocal rather than the real lattice. It is a sphere, of radius $1/\lambda$, with the incident beam along a diameter. The origin of the reciprocal lattice is positioned at the point where the incident beam emerges from the sphere. Whenever a reciprocal lattice point touches the surface of the sphere, the conditions for a diffracted (or reflected) beam are satisfied. Thus, for any orientation of the crystal relative to the incident beam, it is possible to predict which reciprocal lattice points, and thus which planes in the crystal, will be in a "reflecting position" (in the sense of the term used by Bragg). Also, EWALD SPHERE.

spheres of Eudoxus [yoo däks´ əs] *Astronomy.* a cosmological theory dating back to 400 BC, in which the planets, the sun, and the moon were described as being carried on a series of concentric spheres rotating inside one another on various axes.

spherical [sfēr´i kəl; sfer´i kəl] *Science.* having the form of a sphere; shaped like a ball or globe.

spherical aberration *Optics.* a defect of all spherically figured optical systems; rays focused by the edge of a lens or mirror come to a focus in front of rays focused by the center of the lens, making it impossible to form a sharp image.

spherical antenna *Electromagnetism.* an antenna whose shape is that of a sphere; primarily used in theoretical studies.

spherical capacitor *Electricity.* a capacitor made of two concentric metal spheres, with a dielectric filling the area between the spheres.

spherical-coordinate robot *Robotics.* a robot whose working envelope is defined by the coordinates of a sphere. Also, SPHERICAL ROBOT.

spherical coordinates *Mathematics.* a curvilinear coordinate system in three-dimensional Euclidean space; particularly useful in centrally symmetric problems. A point is given by the coordinates (r, ϕ, θ) (i.e., radius, colatitude, longitude) whose relation to ordinary Cartesian coordinates (x, y, z) is given by

$$x = r \sin \phi \cos \theta,\ y = r \sin \phi \sin \theta, \text{ and } z = r \cos \phi.$$

Sometimes the coordinates (radius, longitude, latitude) are used. Also, SPHERICAL POLAR COORDINATES.

spherical cyclic curve see CYCLIC CURVE.

spherical degree *Mathematics.* a unit of solid angular measure equal to 1/90 of a spherical right angle; i.e., 1 spherical degree = $\pi/180$ steradian.

spherical-earth attenuation *Electromagnetism.* the attenuation of a signal propagating over an imperfectly conducting earth considered as a sphere as compared to that over a perfectly conducting plane.

spherical-earth factor *Electromagnetism.* the ratio of the electric field strength in a wave that propagates over the surface of an imperfectly conductive spherical earth to that of a wave propagating over a perfectly conductive plane.

spherical excess *Mathematics.* the spherical excess of a spherical triangle with angles α, β, and γ is

$$\varepsilon = \alpha + \beta + \gamma - \pi,$$

i.e., the amount by which the sum of the interior angles of a spherical triangle exceeds 180°. L'Huillier's formula gives the spherical excess of a spherical triangle in terms of its sides.

spherical grade *Mathematics.* a unit of solid angular measure equal to 1/100 of a spherical right angle; i.e., 1 spherical grade = $\pi/200$ steradian. Also, **spherical grad.**

spherical harmonics *Mathematics.* **1.** solutions of Laplace's equation, expressed in spherical coordinates. **2.** more generally, a spherical harmonic of order $r = 0, 1, 2, \ldots$ is any homogeneous harmonic polynomial of degree r defined on the unit 1-sphere in R^n.

spherical indicatrix *Mathematics.* the set of points on the surface of the unit sphere corresponding to radii that are parallel in the positive direction to some geometric object, such as the binormals to a given space curve or a moving envelope of tangents.

spherical mirror *Optics.* a mirror whose reflecting surface is a segment of a sphere.

spherical pendulum *Mechanics.* a simple pendulum whose shaft can rotate about its support in any direction; thus the movement of the bob describes a spherical surface.
spherical polar coordinates see SPHERICAL COORDINATES.
spherical robot see SPHERICAL-COORDINATE ROBOT.
spherical sailing *Navigation.* any of the various sailings that compensate for the spherical shape of the earth.
spherical sector *Mathematics.* the solid of revolution obtained by rotating a sector of a circle about a diameter of the circle passing through the sector.
spherical segment *Mathematics.* **1.** the solid of revolution that is obtained by rotating a segment of a circle about a diameter of the circle passing through the segment. **2.** in general, the (solid) region that is bounded by a sphere and two parallel planes intersecting or tangent to the sphere.
spherical separator *Petroleum Engineering.* a spherical tank used to separate gas and oil.
spherical stress *Mechanics.* the portion of total stress corresponding to an isotropic hydrostatic pressure; equal to the mean of the three principal stresses at a point.
spherical triangle *Mathematics.* the region on a sphere bounded by three intersecting great circular arcs.
spherical trigonometry *Mathematics.* the study of geometry on the sphere; in particular, the study of spherical triangles.
spherical wave *Physics.* a wave that has spherically concentric wave fronts; commonly generated by a uniform spherical source.
spherical weathering see SPHEROIDAL WEATHERING.
spherical wedge *Mathematics.* the (solid) region determined by the intersection of a sphere and a solid angle determined by a lune and the center of the sphere.
spherical working envelope *Robotics.* a working envelope whose coordinates are defined by the major portion of a sphere.
sphericity *Science.* **1.** the condition of being spherical. **2.** the degree to which a shape is spherical.
sphero- a combining form meaning "round" or denoting a relationship to a sphere.
spherochromatism *Optics.* a change in the amount of spherical aberration visible in a lens that occurs with a change in the color of light used to test the lens.
spherocylindrical lens *Optics.* a lens that has one spherical surface and one cylindrical surface.
spherocyte *Hematology.* a tiny, spherical, fully hemoglobinated blood cell that is missing the normal lighter center; usually seen in congenital spherocytosis but frequently present in acquired hemolytic anemia.
spherocytic anemia *Medicine.* a type of anemia, often hereditary in origin, characterized by the presence of spherocytes in the blood; symptoms include jaundice, splenomegaly, and fragility of the red blood cells.
spherocytosis *Medicine.* the presence of spherocytes in the blood.
spherodite *Materials Science.* a mixture of reduced particles of cementite, Fe_3C, in an alpha ferrite matrix.
spheroid [sfēr´oid] *Science.* **1.** having a shape that is approximately spherical. Also, **spheroidal. 2.** any globular body, or one resembling a sphere. *Mathematics.* an ellipsoid of revolution; i.e., the surface obtained by rotating an ellipse about one of its axes, especially one whose axes are very nearly equal. *Geodesy.* a mathematically constructed imaginary figure closely resembling the geoid in size and shape, and used as a surface of reference during geodetic surveys. *Virology.* an occlusion body produced late in the replication cycle of *Entomopoxvirus* infections, consisting of mature virions within a crystalline matrix of protein and spindles.
spheroidal *Mathematics.* **1.** relating to or shaped like a spheroid. **2.** relating to ellipsoidal coordinates or trigonometric properties of an ellipsoid. For example, a spheroidal triangle is the region on a spheroid bounded by three geodesic lines; the spheroidal excess of a spheroidal triangle with angles α, β, and γ is

$$\varepsilon = \alpha + \beta + \gamma - \pi;$$

i.e., the amount by which the sum of the interior angles of a spheroidal triangle exceeds 180°; and spheroidal harmonics are solutions of Laplace's equation expressed in ellipsoidal coordinates.
spheroidal galaxy see ELLIPTICAL GALAXY.
spheroidal recovery *Geophysics.* the theoretical recovery of the earth to a spherical shape after it has become distorted.
spheroidal weathering *Geology.* chemical weathering in which concentric or spherical shells of decayed rock are successively loosened and removed from a fine-grained, well-jointed block of rock by the action of water. Also, SPHERICAL WEATHERING.
spheroidicity *Science.* **1.** the condition of being spheroidal. **2.** the degree to which a shape is spheroidal.
spheroidin *Virology.* the matrix protein of *Entomopoxvirus* occlusion bodies.
spheroidized carbides *Metallurgy.* in a metallic microstructure, carbide particles that are mainly equiaxed.
spheroidized steel *Metallurgy.* a steel that has been heat treated to form globular carbides, usually improving toughness.
spheroidizing *Metallurgy.* a thermal treatment designed to change the shape of carbide particles.
spheroid of reference see REFERENCE SPHEROID.
spherometer *Engineering.* an instrument used to measure the curvature of a surface.
spherop *Geodesy.* any equipotential surface in the normal gravity field of the earth, differing from the geoid in that a spherop may be determined for any elevation; the geoid is a spherop at mean sea level. Also, **spheropotential surface**.
spheroplast *Cell Biology.* an osmotically sensitive cell that forms a sphere after its cell wall has been partially or completely removed.
spherotoric lens *Optics.* a lens in which one portion of the surface is a sphere and the other is a torus.
spherulin *Immunology.* an antigen used in skin tests to identify individuals who are coccidiodin-positive; obtained from the spherule-endosphore phase of *Coccidioides immitis* organisms.
spherulite *Geology.* a spherical body or coarsely crystalline aggregate having a radial internal structure arranged around one or more centers.

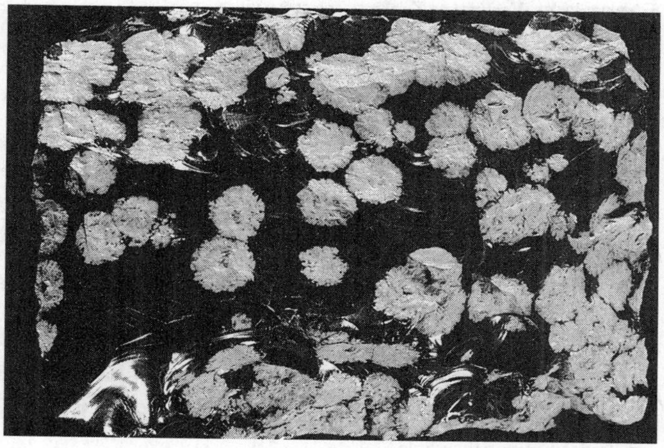

spherulites

spherulitic *Petrology.* describing a rock texture having abundant spherulites. Also, GLOBULAR, SPHAEROLITIC.
spherulitic crystallization *Materials Science.* in the solidification of polymer melts, the formation of polycrystalline aggregates consisting of chain-folded lamellae, e.g., radiating outward from nucleating centers; spherulites are important structural units that fix the scale at which there is coordination in local orientation and cooperative plasticity during deformation.
spherulitic graphite cast iron see NODULAR CAST IRON.
sphincter [sfingk´tər] *Anatomy.* a circular band of muscle fibers that surrounds an opening and closes it by contracting.
sphincteral [sfingk ter´əl] *Anatomy.* relating to or affecting a sphincter muscle.
sphincterectomy *Surgery.* the surgical cutting or excision of any sphincter, especially the sphincter of the iris.
sphincter of Oddi *Anatomy.* the sheath of muscle fibers investing the associated bile and pancreatic ducts as they traverse the wall of the duodenum.
sphincteroplasty *Surgery.* surgery to repair a defective sphincter.
sphincterotome *Surgery.* a surgical device designed specifically to cut a sphincter.

Sphinctozoa *Paleontology.* an order of calcareous sponges, characterized by segmented, beadlike, and possibly exterior skeletons; possibly present in the Cambrian but at least by the Ordovician; long thought to be extinct by the Late Cretaceous, but a single surviving species was discovered in 1977.

Sphindidae *Invertebrate Zoology.* a family of small beetles, coleopteran insects belonging to the superfamily Cucujoidea, that live in dry tree fungi.

Sphingidae *Invertebrate Zoology.* the hawk moths and hummingbird moths, the only family in the lepidopteran superfamily Sphingoidea, having a large heavy body, pointed antennae, a long proboscis, and narrow wings with a wing span up to eight inches, on which they hover while feeding from flowers.

Sphingobacterium *Bacteriology.* a proposed genus of bacteria composed of Gram-negative, aerobic strains of *Flavobacterium.*

Sphingoidea *Invertebrate Zoology.* a superfamily of large moths, lepidopteran insects in the suborder Heteroneura, containing a single family, Sphingidae.

sphingolipid *Biochemistry.* any lipid, such as sphingomyelin, that yields sphingosine or one of its derivatives on hydrolysis; a major component in cellular membranes.

sphingolipidosis *Medicine.* any of various lysosomal storage diseases characterized by abnormal storage of sphingolipids, such as Niemann-Pick disease or Gaucher's disease. Also, **sphingolipodystrophy.**

sphingomyelin *Biochemistry.* a phospholipid made up of choline, sphingosine, phosphoric acid, and a fatty acid; an important component in the cell membranes of animals.

sphingomyelin lipidosis see NIEMANN-PICK DISEASE.

sphingosine *Biochemistry.* $C_{18}H_{37}NO_2$, a compound that produces a variety of the long-chain amino alcohols present in many sphingolipids, such as the sphinogosine and dihydrosphingosine found in mammals.

Sphrynidae *Vertebrate Zoology.* see HAMMERHEAD.

sp. ht. specific heat.

sphygmic *Cardiology.* of or relating to the pulse.

sphygmo- or **sphygm-** a combining form meaning "pulse."

sphygmobolometer *Cardiology.* a device for measuring and recording the power of the pulse, and therefore of the systole.

sphygmocardiogram *Cardiology.* a graphical representation of heartbeat and pulse waveforms made by a sphygmocardiograph.

sphygmocardiograph *Cardiology.* an instrument for measuring and recording heartbeat and pulse waveforms at once.

sphygmogram *Cardiology.* the graphic recording of the arterial pulse curve, showing movement, form, and force of a pulse.

sphygmograph *Cardiology.* an instrument used for recording the arterial pulse curve.

sphygmography *Cardiology.* the process by which the arterial pulse is graphically recorded with the use of a sphygmograph.

sphygmomanometer *Cardiology.* any of several instruments that measure blood pressure in the arteries.

sphygmophone *Cardiology.* a device that amplifies the vibrations of the pulse to make it audible.

Sphyraenidae *Vertebrate Zoology.* the barracudas, a monogeneric family of marine perciform fish having slender bodies, a pointed snout, and a large mouth with fanglike teeth; found in all tropical and subtropical seas.

Sphyriidae *Invertebrate Zoology.* the fish maggots, a family of crustacean ectoparasites that bury their heads in the gill tissue of fish.

spica *Medicine.* a bandage having the form of a figure 8, with turns that cross one another, usually at the shoulder or hip.

Spica *Astronomy.* Alpha (α) Virginis, a blue-white eclipsing binary star of magnitude 1.0, and a distance of 270 light-years.

spice *Food Technology.* an aromatic seasoning usually made from the root, stem, or seeds of certain plants such as nutmeg, allspice, cinnamon, and black pepper.

spicule *Science.* a minute, sharp, needlelike body or part. *Invertebrate Zoology.* a tiny skeletal part of sponges, radiolarians, sea cucumbers, and others, composed of calcium carbonate or silica; often a simple needle shape, sometimes radial, irregular, or joined in a network. *Astronomy.* any of numerous vertical, spiked jets that emanate from the sun's chromosphere and protrude into the corona.

spiculite *Petrology.* a spindle-shaped belonite believed to have resulted from the coalescence and agglomeration of globulites.

spiculum *Invertebrate Zoology.* any needlelike structure, such as the dart of a snail, an echinoderm's spines, or a nematode's copulatory bristle.

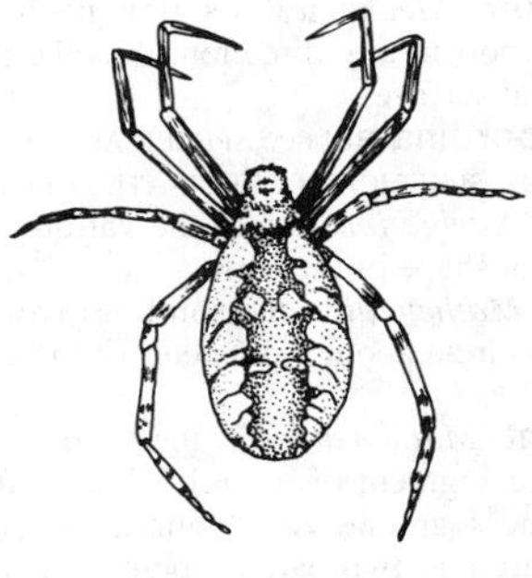

orb-web spider

spider *Invertebrate Zoology.* any predaceous arachnid in the order Araneae, having eight legs, a two-part body, and spinnerets; most spiders spin webs to trap prey, and many have poison glands; some have a bite that is harmful to humans, e.g., the black widow. They exist in an estimated 50,000 species and are found in virtually all habitats.

spider any of various objects or devices thought of as comparable to a spider, as in shape or movement; specific uses include: *Design Engineering.* any machine or part whose shape suggests that of a spider, usually having radiating arms or spokes. *Acoustical Engineering.* a flexible, washer-shaped component that centers the voice coil of a speaker on its pole piece; designed to allow the coil to travel freely in and out. *Electricity.* a structure that supports the rotor poles on the shaft of an electric rotating machine; composed of a hub, spokes, and a rim. *Petroleum Engineering.* a hinged latching device made of steel and attached to an elevator, through which hoisting pipe, casing, and tubing are moved into and out of a well.

Spider *Ordnance.* a Soviet short-range ballistic missile that delivers one nuclear warhead at a range of 300 miles; officially designated **SS-23.**

spider angioma see NEVUS ARACHNOIDEUS.

spiderling *Invertebrate Zoology.* the young of a spider.

spider mine *Ordnance.* an air-delivered, anti-personnel mine with eight nylon threads extending approximately eight yards from the mine; touching any of the threads actuates the mine.

spider nevus see NEVUS ARACHNOIDEUS.

spider templet *Photogrammetry.* a reusable mechanical template, constructed by attaching slotted steel arms, representing the radials, to a central hub.

spider venom *Toxicology.* a poison produced by a spider such as *Lactrodectus, Atrax, Ctenus,* or *Lycosa.*

spiderweb antenna *Electromagnetism.* a receiving antenna that has several different-length dipole elements for receiving a broader band of frequencies.

spiderweb clot *Hematology.* the fine fibrin clot that forms when a sample of fluid from a subject with tuberculous meningitis is allowed to stand, especially when it is warmed to 37°C for a few hours.

spiegeleisen *Metallurgy.* a pig iron that contains 15–30% manganese.

Spiegler's test *Pathology.* a method of identifying the presence of albumin using a water-glycerin solution of mercuric chloride, sodium chloride, and tartaric acid filtered with acetic acid; a white ring will appear at the point where the liquids meet if albumin is present.

spigot *Mechanical Devices.* an object or device that controls or stops the flow of liquid from a pipe or from a container. *Ordnance.* a tube that is attached to the warhead of a spigot mortar and inserted into the mortar.

spigot mortar *Ordnance.* a mortar that fires a warhead of larger bore than that of the mortar, using a spigot tube to hold the warhead in place.

spike *Mechanical Devices.* a large, square-sectioned or cylindrically shaped nail, three inches or longer, used to join parts or fix them in place. *Botany.* an inflorescence in which sessile or nearly sessile flowers or spikelets are borne on an elongated unbranched axis. *Virology.* a protein or glycoprotein projection that extends from the surface of a virus particle and helps bind the particle to the cell surface. *Physics.* a transient of very short duration and relatively large peak value. *Solid-State Physics.* a method of ejecting target atoms from a material in which the number of atomic collisions within the material is allowed to become significant.

spikelet *Botany.* a basic unit of the inflorescence in grasses and sedges composed of one to several florets borne on a thin axis and subtended by two glumes.

spike microphone *Acoustical Engineering.* a microphone used for secret surveillance purposes, consisting of a spikelike shaft that is driven into the wall of a room being monitored, with a diaphragm transducer on the other side of the wall for pickup and amplification.

spike potential see ACTION POTENTIAL.

spike-tooth harrow *Agriculture.* a harrow that has spiked teeth mounted vertically on horizontal bars; used to smooth the ground immediately after plowing.

spile *Mechanical Devices.* a wooden peg or plug, such as one used for a spigot. *Forestry.* a spout used to tap the sap from a maple tree. *Building Engineering.* a large timber put into the ground as a foundation; a pile. *Mining Engineering.* a plank driven ahead of a timber face, used to support the roof.

spilite *Petrology.* an altered basaltic rock, typically amygdaloidal or vesicular, containing albitic feldspar along with chlorite, calcite, epidote, or other low-temperature hydrous crystallization products; frequently occurs as submarine lava flows featuring pillow structures.

spill *Engineering.* an unwanted occurrence in which material becomes released from its container. *Nucleonics.* specifically, a nuclear accident in which radioactive materials are released into the environment.

spilling *Oceanography.* the process in which the white water at the crest of a gradually breaking wave slowly moves ahead of the lower levels of the wave and rolls over without the sudden crash of a plunging wave.

spillover *Telecommunications.* the reception of a radio signal that is of a different frequency from the one to which the receiver is tuned.

spill pit see RUNOFF PIT.

spill stream see OVERFLOW STREAM, def. 1.

spillway *Civil Engineering.* a structure that passes flood water through, over, or around a dam.

spillway apron *Civil Engineering.* an antierosion structure built at the bottom of a spillway; usually made of concrete, rocks, or timbers.

spillway channel *Civil Engineering.* a course running from the outlet of a spillway.

spillway dam see OVERFLOW DAM.

spillway gate *Civil Engineering.* a gate located on the spillway crest of a dam to regulate the flood discharge and thus maintain or lower the water level. Also, ROLLER GATE, SLIDING GATE.

spin *Mechanics.* the rotation of a body about an axis through the body. *Quantum Mechanics.* **1.** the angular momentum possessed intrinsically by elementary particles and nuclei even while at rest. **2.** a quantum number that describes the rotation of an elementary particle on its axis or of a system of such particles in orbital motion that is responsible for angular momentum and magnetic moment. Spin is always described as a half or a whole quantum: . . . , –1, –1/2, 0, 1/2, 1, 3/2, and so on. *Aviation.* a maneuver or action in which an airplane descends at an angle of attack greater than maximum lift, usually with its nose pointed downward. *Space Technology.* to intentionally cause a rocket or guided missile to go into a roll. *Textiles.* to make fibers into yarn, usually by drawing out, twisting, and winding them.

spina *Cell Biology.* a hollow, rigid, apparently nonprosthecate appendage projecting from some prokaryotic cells.

spina bifida [spī´nə bif´i də] *Medicine.* a developmental anomaly of the spinal column in which spinal membranes and sometimes the spinal cord protrude through the vertebral column. Also, **spinal dysrhaphis.**

spinaceous *Botany.* relating to or resembling spinach or other plants of the family Chenopodiaceae.

spinach *Botany.* an annual herb, *Spinacea oleracea,* of the family Chenopodiaceae in the order Caryophyllales; cultivated extensively for its edible, crinkly or flat leaves.

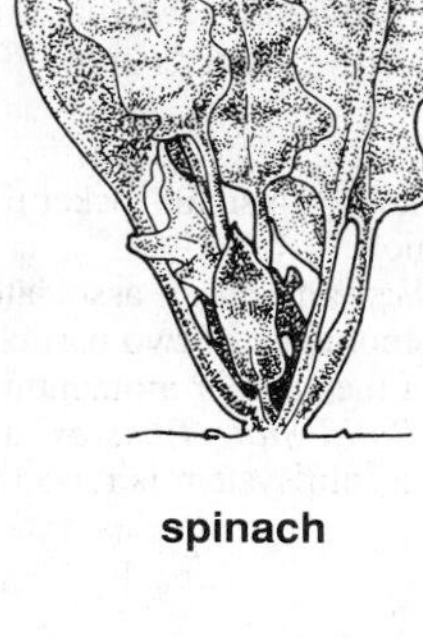

spinach

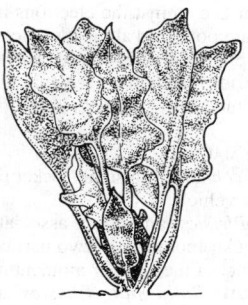

spinal *Anatomy.* relating to the spine or vertebral column. *Medicine.* involving or affecting the spine or spinal nerves. Thus, **spinal anesthesia.**

spinal canal see NEURAL CANAL.

spinal cavity *Anatomy.* a bony cavity formed by the vertebrae of the backbone, containing the spinal cord and the beginnings of the spinal nerves.

spinal column *Anatomy.* see SPINE, def. 1.

spinal cord *Anatomy.* the column of nerve tissue that extends from the medulla oblongata through the canal created by the vertebral foramina of the vertebral column.

spinal embolism *Neurology.* the blocking of an artery of the spinal cord by a clot or foreign material.

spinal foramen *Anatomy.* the large cavity in a vertebra between the arch and body. Also, VERTEBRAL FORAMEN.

spinal fusion see SPONDYLOSYNDESIS.

spinal ganglion *Anatomy.* the mass of sensory neuron cell bodies located on the dorsal or posterior root of each spinal nerve.

spinalgia *Neurology.* pain that is related to or that emanates from the spinal cord.

spinal meningitis *Medicine.* an inflammation of the meninges of the spinal cord.

spinal nerve *Anatomy.* any of the 31 pairs of nerves arising from the spinal cord and extending between the vertebrae.

spinal puncture or **spinal tap** *Medicine.* the insertion of a needle into the lumbar region of the spine for the purpose of removing spinal fluid or introducing substances. Also, LUMBAR PUNCTURE.

spinal reflex *Physiology.* any reflex whose arc is connected with a center in the spinal cord.

spinal shock *Neurology.* the usually temporary loss or impairment of spinal reflexes affecting the muscles below the level of the lesion after injury to the spinal cord.

spin angular momentum *Quantum Mechanics.* the intrinsic angular momentum of a particle due to its spin.

spin axis *Physics.* any axis of rotation for a rigid body.

spin-decelerating moment *Mechanics.* a moment about the axis of a projectile to decrease its rate of spin.

spin-density wave *Solid-State Physics.* for a metal in its ground state, a sinusoidal variation of the spin density of the conduction electrons; the periodicity of such a wave is unrelated to the lattice constant but is instead determined by the nature of the Fermi surface.

spin-dependent force *Physics.* a force that acts between two particles and is dependent on the relative angle between the spin orientations of the particles or sometimes on the angles between the spin orientations and the line connecting the particles.

spindle *Textiles.* a long, thin rod with tapered ends that is held vertically on the side of a spinning frame and used for twisting and winding threads or fibers as they are spun. *Mechanical Devices.* **1.** any large, pointed rod, bar, or pin upon which an object spins, turns, or is fixed. **2.** the bar that actuates the latch in a lock. *Cell Biology.* a network of fibrous microtubules formed during mitosis that connects the poles (centrosomes) with the kinetochores. It mediates both the movement of the sister chromatids to the poles and the further movement apart of the poles themselves Also, NUCLEOSPINDLE, MITOTIC SPINDLE. *Virology.* a proteinaceous inclusion body that is produced during infection by an *Entomopoxvirus*; it contains no virus particles.

spindle fiber *Cell Biology.* one of the many microtubule filaments constituting the mitotic spindle.

spindle poison *Toxicology.* a chemical agent that disrupts cell division by preventing the assembly of microtubules within the cell.

spindle pole body *Cell Biology.* in a dividing yeast cell, a structure located within the nuclear membrane and attached to the microtubules of the mitotic spindle.

spindle tuber *Plant Pathology.* a virus disease of potatoes characterized by elongated, pointed tubers and dwarfed, spindly foliage.

spindle virosis *Virology.* an *Entomopoxvirus* infection characterized by the production of spindle-shaped inclusion bodies.

spine *Anatomy.* **1.** the row of bones supporting the back; the vertebral column. Also, SPINAL COLUMN. **2.** a sharp process on a bone. *Vertebrate Zoology.* any stiff, pointed, projection or external process of an animal body, as on a porcupine or in some fishes' fins. *Invertebrate Zoology.* a stiff, narrow, pointed process or structure, such as the articulated projections of a sea urchin used for locomotion and defense. *Botany.* a stiff, sharply pointed outgrowth on a plant, such as that on a cactus. *Graphic Arts.* the part of a book where the front and back covers are connected and where the pages and the binding are stitched together.

spin-echo technique *Nuclear Physics.* a type of nuclear magnetic resonance in which the radio-frequency field is applied in two pulses.

spinel *Mineralogy.* **1.** $MgAl_2O_4$, a variously colored cubic mineral of the spinel group, occurring as octahedral crystals and in compact to granular masses, having a specific gravity of 3.58 and a hardness of 7.5 to 8 on the Mohs scale; found in metamorphic and basic igneous rocks, and used as a gemstone. **2.** a group name for cubic oxide minerals of the general formula AB_2O_4, where A=Co,Cu,Fe^{+2},Ge,Mg,Mn^{+2},Ni,Ti,Zn and B=Al,Cr^{+3},Fe^{+2},Fe^{+3},Mg,Mn^{+3},Ti,V^{+3}. Also, **spinelle, spinell.**

spineless dogfish see DALATIIDAE.

spinellid *Mineralogy.* any mineral of the spinel group.

spinel ruby see RUBY SPINEL.

spin filter *Electronics.* in a Lamb-shift polarized ion source, a filter that is designed to cause those atoms having an undesired nuclear spin orientation to decay from their metastable state to the ground state, while those with the desired spin orientation are allowed to continue without decay.

spin-flip laser *Optics.* a laser in which a fixed-frequency laser pumps a semiconductor crystal kept at a low temperature and in a strong magnetic field; adjusting the magnetic field controls the frequency of the laser by altering the spin states of electrons scattered off the crystal by photons from the pumping laser.

spin-flip scattering *Quantum Mechanics.* the scattering of a spin 1/2 particle that reverses the spin direction.

spin glass *Solid-State Physics.* a magnetic material with a frozen-in, disordered arrangement of interacting electron spins; such an alloy will not retain magnetization if it is magnetized above the transition temperature and cooled through the spin glass temperature.

spin isomer *Nuclear Physics.* a state in which an isomer maintains a high level of energy over an unusually long period of time, due to its having a spin that differs radically from the spin of the state into which it is allowed to decay.

spin label *Physical Chemistry.* a molecule in which an atom or group of atoms exhibit an electron spin that causes the atoms to bond to another molecule, and that can be detected by spectroscopy.

spin-lattice relaxation *Solid-State Physics.* the magnetic relaxation of electron spins in a crystal wherein the excess energy of the spins is transferred to the lattice.

spin magnetism *Solid-State Physics.* ferromagnetic or paramagnetic behavior exhibited by a material due to the magnetic polarization of the electron spins in the material.

spinnaker *Naval Architecture.* a large, unsupported triangular sail used by a yacht sailing with the wind; it is set in front and to the opposite side of the mainsail; when the wind dies, the spinnaker collapses.

spinnbarkeit *Medicine.* **1.** the formation of a thread by cervical mucus when spread on a glass slide. **2.** the stringy, elastic quality of cervical mucus during the ovulatory period that allows it to be formed into a long thread.

spinnbarkeit relaxation *Fluid Mechanics.* a rheological effect that develops when an object that has been immersed in a viscoelastic fluid is pulled out, and liquid threads are also pulled away.

spinner *Engineering.* **1.** a radar antenna that rotates automatically in coordination with adjacent radar system equipment. **2.** a device on a mechanical scanner that rotates about an axis, usually at high speeds. *Aviation.* a streamlined faring that fits over a propeller boss on an aircraft and rotates with the propeller.

spinneret *Invertebrate Zoology.* a body mechanism through which spiders and certain insect larvae produce silky filaments used variously in webs and cocoons. *Materials Science.* an extrusion die containing many small openings through which the hot or molten polymer is forced to produce filaments; rotation of the spinnerette twists the filaments into a yarn. Also, **spinnerette**.

spinner rocket *Space Technology.* a spin-stabilized rocket.

spinnery *Textiles.* a spinning mill.

spinney *Ecology.* a small grove or thicket.

spinning *Textiles.* the process of making thread or yarn from fiber by any of several methods, most commonly by drawing out and twisting fibers together, or by extruding a synthetic spinning solution through a spinneret. *Invertebrate Zoology.* the process of secreting and positioning filaments, as in the construction of a cocoon or web. *Metallurgy.* the process of shaping sheet metal into round hollow-ware by pressing the metal sheets to molds or forms revolving on a special lathe.

spinning-band column *Analytical Chemistry.* a glass distillation column that contains a series of driven spinning bands; used to increase the amount of liquid-vapor contact.

spinning-cup sequenter *Biotechnology.* an apparatus for the liquid phase sequencing of peptides or proteins; a protein sample is added to a spinning cup, from which volatile liquids evaporate, and amino acids are removed from the remaining thin layer by solvent extraction.

spinning frame *Textiles.* a machine designed to draw, twist, and wind yarn.

spinning jenny *Textiles.* the first spinning machine having more than one wheel; patented by James Hargreaves in 1764.

spinning machine *Textiles.* **1.** any machine designed to spin thread or yarn. **2.** specifically, an earlier machine of this type used during the Industrial Revolution. *Metallurgy.* a machine that shapes sheet metal. *Mechanical Engineering.* a machine that winds insulation on electric wire.

Arkwright's water-frame spinning machine (1769)

spinning wheel *Textiles.* an early spinning machine, dating from the 1200s and consisting of a large wheel and a spindle that were connected by a band or belt and powered by a foot pedal.

spinoblast *Invertebrate Zoology.* a free-floating statoblast having external barbs or hooks, as in Pectinatella.

spinodal structure *Metallurgy.* one of several two-phase structures characterized by outstanding homogeneity of each phase.

spinode *Mathematics.* see CUSP.

spinor *Physics.* a vector having two complex components, used to represent the spins of fermions.

spin-orbit coupling *Quantum Mechanics.* the interaction between a particle's spin and its orbital angular momenta.

spin-orbit multiplet *Physics.* a collection of atomic or nuclear states whose individual energy levels may vary within the system because of spin-orbit coupling.

spinous *Biology.* covered with spines or thorns; spiny. Also, **spinose.**

spinous process *Anatomy.* the prominent dorsal projection from each vertebra.

spin paramagnetism *Solid-State Physics.* paramagnetism caused by the polarization of electron spins in a solid.

spin-polarized low-energy electron diffraction *Solid-State Physics.* a low-energy electron diffraction technique used to study magnetic properties of surface atoms; the electrons in the beam are oriented such that their spins are polarized along a particular direction.

spinproof *Aviation.* of an airplane, highly resistant to spinning.

spin quantum number *Quantum Mechanics.* an integer or half-integer giving the magnitude of a particle's spin, in units of Planck's constant divided by 2π.

spin resonance see MAGNETIC RESONANCE.

spin rocket *Space Technology.* a small rocket that transmits a spin to a missile or other flight vehicle.

spin-spin energy *Physics.* an energy associated with the interaction between the spin angular momenta of two particles, found to be proportional to the dot product of the angular momentum vectors.

spin-spin relaxation *Solid-State Physics.* a relaxation process in which the temperature of a spin system is raised upon the application of weak magnetic fields.

spin-stabilized *Space Technology.* of a projectile, steadied in flight by a rotating motion about its longitudinal axis, as imparted by the rifling of a gun barrel or by the slanting exhaust nozzles of a **spin-stabilized rocket.**

spin state *Quantum Mechanics.* a particle state that is an eigenstate of the two spin operators, corresponding to the magnitude squared of the spin, and to one spin component.

spin temperature *Solid-State Physics.* a temperature assigned to a population of spin energy levels in an electron spin system of a lattice such that the population is given by the Boltzmann factor exp $(-E_i/kT)$, where E_i is the energy level of the spin of the ith electron, k is the Boltzmann constant, and T is the absolute temperature.

spinthariscope *Electronics.* an instrument used along with a microscope to view scintillations of alpha particles on a luminescent screen.

Spintheridae *Invertebrate Zoology.* a family of small polychaete annelid worms in the Errantia.

spintherometer *Radiology.* an apparatus for measuring changes in the vacuum of an X-ray tube and the associated power of ray penetration. Also, **spintometer.**

spin-trapping *Physical Chemistry.* the detection of an unstable free radical, usually by electron spin resonance spectroscopy, by reacting it with a compound that forms a more stable radical.

Spinulosida *Invertebrate Zoology.* an order of starfish, echinoderms in the class Asteroidea, having small plates bordering the arms, gills on both surfaces, few pedicellariae, and groups of spines on the upper surface. Also, **Spinulosa.**

spin-up *Astronomy.* a constantly increasing frequency in a pulsar.

spin wave see MAGNON.

spin welding *Engineering.* a process of joining two objects by forcing them together while one of them is spinning, thus creating a heated friction at their point of interface, and then stopping the spinning and holding the objects together until they are welded.

spiny anteater see ECHIDNA.

spiny dogfish *Vertebrate Zoology.* any of several dogfish sharks of the genus *Squalus*, in the family Squalidae, characterized by a spine in front of each dorsal fin.

spiny-headed worm *Invertebrate Zoology.* the common name for any member of the phylum Acanthocephala.

spiny lobster *Invertebrate Zoology.* any of several marine crustaceans of the family Palinuridae, resembling true lobsters but lacking large pincers and having a spiny shell.

spiny-rayed fish *Vertebrate Zoology.* any member of the subclass Actinopterygii, characterized by fins supported by stiff, narrow unbranched rays.

Spionidae *Invertebrate Zoology.* a family of bristleworms, small burrowing and tube-dwelling polychaetes in the Sedentaria, having gills in the form of dorsal cirri and no tentacles; numerous in sandy tidal areas.

spioniform worm *Invertebrate Zoology.* any of a group of families of polychaete annelids having a pair of grooved palpi near the mouth.

spir- a combining form meaning: **1.** coil; spiral. **2.** breath; breathing.

spiracle *Vertebrate Zoology.* **1.** a respiratory opening in cartilaginous and ganoid fishes for passage of water. **2.** the excurrent gill opening in amphibian tadpoles. **3.** the nasal opening in whales, porpoises, and dolphins; the blowhole. *Invertebrate Zoology.* in insects and other terrestrial arthropods, any of several tracheal openings in the exoskeleton used in respiration. *Volcanology.* a vent formed in a lava flow as a result of a gaseous explosion.

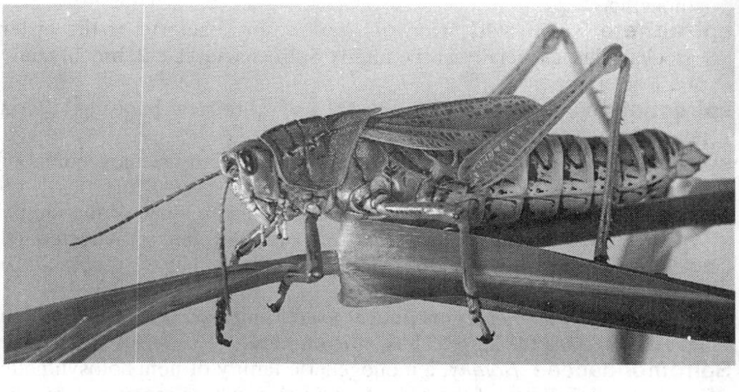

abdominal spiracles on a grasshopper

spiradenoma *Oncology.* an adenoma of the sweat glands. Also, ADENOMA SUDORIPARUM, HIDRADENOMA.

spiral *Mathematics.* a continuous plane curve that winds outward from a point or inward to a point. *Science.* any object or movement having or suggesting such a form or direction.

spiral arms *Astronomy.* the dust, gas, and stars that extend outward in a disk from the nucleus of a galaxy.

spiral band *Meteorology.* radar echoes with a spiral-shaped aspect that are received from precipitation zones within intense tropical cyclones, curving cyclonically inward toward the storm center with a merging effect, to form a wall around the eye of the storm. Also, HURRICANE BAND.

spiral bevel gear *Mechanical Devices.* a bevel gear with teeth in a curved line that converge with another gear on the axis of rotation. Also, **spiral gear.**

spiral binding *Graphic Arts.* a form of binding consisting of a continuous spiral wire that is threaded through holes punched along the margin of pages; used primarily for looseleaf notebooks. Thus, **spiral-bound.**

spiral chute *Mechanical Devices.* a gravity chute consisting of multiple curves around a pillar or column; used to convey material from an upper to a lower level.

spiral cleavage *Developmental Biology.* a cleavage pattern in which the planes of cleavage are at an oblique angle to the major axis of the egg.

spiral coupling *Mechanical Devices.* a jaw coupling that is engaged only when rotating in one direction.

spiral cutterhead *Mining Engineering.* a rotary digging apparatus used to dislodge alluvial sand or gravel and feed it to a suction dredge.

spiral delay line *Electromagnetism.* a transmission line that has a helical inner conductor.

spiral flow tank *Civil Engineering.* an aeration chamber used in the activated sludge process in which air is introduced to the sludge through a spiral helical movement by diffusers.

spiral flow test *Materials Science.* a process whereby the flow behavior of a thermoplastic resin is determined by measuring the weight and flow of the resin as it passes through a spiral hole.

spiral four cable *Electricity.* a quad cable in which the four conductors are twisted around a common axis; the two sets of opposite conductors are used as pairs.

spiral galaxy *Astronomy.* a system of roughly 100 million stars arranged in a lenslike nucleus of reddish, older stars and a disk with spiral arms consisting mainly of dust and gas clouds and blue young stars.

spiral galaxy

spiral ganglion *Anatomy.* a long mass of bipolar nerve cells on the vestibular cochlear nerve that conduct impulses from the organ of Corti to the brain.

spiral gear *Mechanical Engineering.* a stepped, helical, or worm gear which transmits power from one shaft to another that is not parallel to the first.

spiraling *Invertebrate Zoology.* in certain ciliates, a swimming pattern that approximates a spiral as the organism moves through the water.

spiral-jaw clutch *Mechanical Engineering.* a clutch having a helical-toothed section that allows gradual meshing between the mating faces.

spiral layer see EKMAN LAYER.

spiral ligament *Anatomy.* the thickened outer periosteal lining of the cochlear duct, to which the basal membrane is attached.

spirally wound module *Biotechnology.* a membrane configuration used for reverse osmosis and ultrafiltration; a sandwich of membrane and spacers is wound into a spiral.

spiral-mold cooling *Engineering.* a process of cooling down a molding device by running a liquid through a spiral opening inside the mold.

spiral nebula *Astronomy.* a former name for a spiral galaxy.

spiral of Archimedes see ARCHIMEDEAN SPIRAL.

spiral organ see ORGAN OF CORTI.

spiral pipe *Mechanical Devices.* a steel pipe having a continuous coiled seam along its length.

spiral plate exchanger *Chemical Engineering.* a heat-transfer device of compact design that consists of a pair of plates that are rolled to provide two passages for the fluids to flow countercurrently.

spiral pressure gauge *Engineering.* a device that measures pressure, composed of a hollow-tube spiral that unwinds in direct relation to its inside pressure. Also, **spiral gauge.**

spiral ramp system *Mining Engineering.* a system of mine haulageways driven at moderate inclines between the surface and ore horizons underground.

spiral ratchet screwdriver *Mechanical Devices.* a mechanical screwdriver, with a removable head for fitting screws, that operates by applying downward pressure in either direction of its reciprocating handle motion.

spiral scanning *Engineering.* a type of radar scanning in which the area of maximum radiation is in a single direction and in the form of a section of a spiral.

spiral spring *Mechanical Devices.* a wire spring composed of an elastic material such as steel, brass, or iron, whose coils are of equal diameter to allow for compression and expansion in its flat wound state about a central axis; designed to provide power to a shaft with ends fastened to a link.

spiral thermometer *Engineering.* a thermometer consisting of a pair of metal strips, each of a different material, forming spirals that expand and contract at different rates and change in relation to the temperature by tightening or untightening.

spiral thickening *Botany.* in tracheary elements, a growth process in which the secondary cell wall is laid down in a helical pattern. Also, HELICAL THICKENING.

spiral-tube heat exchanger *Engineering.* a concurrent heat-exchange instrument consisting of a set of spiral coils that are joined by manifolds and arranged concentrically; generally used in air-separation plants for very low temperature exchanges.

spiral valve *Vertebrate Zoology.* **1.** a spiral structure in the intestines of sharks, rays, and lungfish that slows the passage of food and increases absorption. **2.** a twisted, valvelike structure in the truncus arteriosus of certain amphibians.

spiral-welded pipe *Mechanical Devices.* a steel pipe consisting of welded steel plates that form helical seams.

spiral wire column *Analytical Chemistry.* a distillation column having a spiral wire running down its interior, which serves as a surface for liquid-vapor contact.

spiramycin *Microbiology.* an antibiotic that is produced by the bacterium *Streptomyces ambofaciens;* it is similar to erythromycin in both structure and use.

spiran *Organic Chemistry.* a polycyclic compound containing two rings having one carbon atom in common.

spiraster *Invertebrate Zoology.* a spiral sponge spicule.

spire *Architecture.* a slender, steeply tapering pyramidal roof surmounting a tower; a feature of many European medieval cathedrals or of contemporary churches built in this style. *Botany.* a slender, tapering leaf or stem, such as a blade of grass.

spiricle *Botany.* a thin, coiled, threadlike structure of some seed coats that uncoils when moistened.

Spiridentaceae *Botany.* a family of large mosses of the order Bryales that form horizontal to hanging tufts on trees in the Old World tropics; characterized by irregularly branching stems with a strong central shoot and unusual spreading leaves.

Spirifer *Paleontology.* a genus of large, biconvex, articulate brachiopods in the suborder Spiriferidina; characterized by strophic hinges and laterally directed spires; extant in the Carboniferous.

Spiriferida *Paleontology.* an order of articulate brachiopods that appeared in the Ordovician, became very widespread in the Devonian, and persisted into the Jurassic; characterized by biconvex shells and spiral brachidia (arm-bearers); included both punctate and impunctate forms.

Spiriferidina *Paleontology.* a suborder of articulate brachiopods in the order Spiriferida, characterized by strophic hinges and laterally directed spires; extant in the Lower Silurian to Lower Jurassic.

Spirillaceae *Bacteriology.* in former classifications, a family of spiral-shaped, motile bacteria including the genera *Spirillum* and *Campylobacter.*

Spirillinacea *Invertebrate Zoology.* a superfamily of foraminiferan protozoans in the suborder Rotaliina, having a spiral or conical test.

Spirillospora *Bacteriology.* a genus of bacteria of the family Actinoplanaceae, occurring as branching substrate and aerial mycelia, with the mycelium developing into spherical sporangia; found in soil and leaf litter.

spirillum *Bacteriology.* **1.** a member of the genus *Spirillum.* **2.** any of several microorganisms that resemble *Spirillum.*

Spirillum *Bacteriology.* a genus of Gram-negative, anaerobic or microaerophilic bacteria of the family Spirillaceae, occurring as spiral-shaped cells with polar flagella; typically found in aquatic habitats.

spirit *Chemistry.* **1.** any volatile or distilled liquid. **2.** a solution of a volatile material in alcohol.

spirit compass *Navigation.* a liquid compass in which the card floats in a mixture of 45% alcohol and 55% distilled water.

spirit duplicating *Graphic Arts.* an older form of duplication in which an alcohol-moistened paper is brought into contact with a dyed master (**spirit master**) so that some of the dye from the master will be transferred to the paper.

spirit level *Mechanical Devices.* a device consisting of a glass tube set in a straightedge with centering marks and filled with a liquid and an air bubble; it is used to determine the relative position of a surface with respect to a true horizontal or vertical plane when the bubble appears between the centering marks. Also, LEVEL.

spirit leveling *Cartography.* in surveying, the technique of determining elevations of points with respect to each other or to a common datum by use of spirit levels to establish horizontal lines of sight.

spirit of hartshorn see AROMATIC SPIRIT OF AMMONIA.

spirits of turpentine see TURPENTINE, def. 3.

spirit-soluble dye *Organic Chemistry.* any dye that is soluble only in an organic solvent.

spirit thermometer *Engineering.* a thermometer consisting of a closed capillary tube with a liquid-filled bulb at one end; the liquid in the bulb travels up the capillary according to the temperature of the bulb.

spiro- a combining form meaning: **1.** coil; spiral. **2.** breath; breathing.

Spirobolida *Invertebrate Zoology.* an order of chiefly tropical millipedes with bodies composed of 35 to 60 ringlike segments.

Spirobrachiidae *Invertebrate Zoology.* a family of deep-sea, tube-dwelling beard worms in the order Thecanephria.

Spirochaeta *Bacteriology.* a genus of Gram-negative, anaerobic or facultatively anaerobic bacteria of the family Spirochaetaceae, occurring as large helical or rod-shaped cells in mud and water containing hydrogen sulfide.

Spirochaetaceae *Bacteriology.* a family of anaerobic, facultatively anaerobic, or microaerophilic bacteria of the order Spirochaetales, occurring in spirals and characterized by undulating motion.

Spirochaetales *Bacteriology.* an order of Gram-negative, spiral-shaped, motile bacteria found in soil and water; several species are pathogenic.

spirochete [spī´rə kēt] *Microbiology.* **1.** any bacteria of the order Spirochaetales, especially of the family Spirochaetaceae. **2.** broadly, any spiral bacterium.

spirochetemia *Medicine.* the presence of spirochete parasites (blood flukes) in the blood.

spirograph *Medicine.* an instrument used to record the movements of respiration. Thus, **spirogram.**

Spirogyra *Botany.* a large genus of scumlike green algae of the family Zygnemataceae that grow in freshwater ponds and tanks; characterized by cells with one or more ribbonlike chloroplasts that wind spirally to the right.

spirometer *Medicine.* an instrument used to measure the air taken into and exhaled from the lungs. Thus, **spirometry.**

Spiromondaceae *Botany.* a monogeneric family of nonphotosynthetic freshwater flagellates of the order Syndiniales, whose members are parasite-specific on the cryptomonad *Chilomonas.*

spironolactone *Pharmacology.* $C_{24}H_{32}O_4S$, a synthetic steroid that increases excretion of water and sodium and decreases secretion of potassium; used to reduce edema due to heart failure and liver and kidney disease, to treat low levels of potassium in the blood, and, in combination with other drugs, to treat high blood pressure.

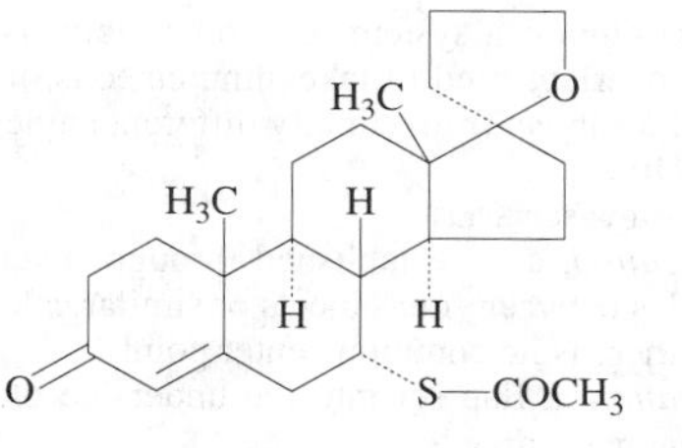

spironolactone

Spiroplasma *Bacteriology.* a genus of facultatively anaerobic bacteria of the family Spiroplasmataceae that occur as helical filaments or coccoid cells bounded by a membrane but lacking a true cell wall; parasitic or pathogenic for a variety of invertebrates and plants.

Spiroplasmataceae *Bacteriology.* a family of bacteria of the order Mycoplasmatales, occurring as helical filaments that lack cell walls and require sterol for growth.

c3 Spiroplasmavirus group *Virology.* a proposed genus of viruses, isolated from *Spiroplasma,* that structurally resemble members of the family Podoviridae.

spiro ring system *Organic Chemistry.* a structural formula consisting of two rings having one carbon atom in common.

Spirosoma *Bacteriology.* a genus of Gram-negative, nonmotile, rod-shaped bacteria of the family Spirosomaceae, containing a yellow pigment and occurring as helices or long coiled filaments; found in soil and fresh water.

Spirosomaceae *Bacteriology.* a family of Gram-negative, aerobic, pigmented, rod-shaped bacteria found in soil and aquatic habitats.

Spirosteptida *Invertebrate Zoology.* an order of tropical millipedes that contains the largest species known, reaching 28 cm in length.

Spirostomum *Invertebrate Zoology.* a genus of free-swimming ciliate protozoans in the order Spirotricha having a large, elongated body.

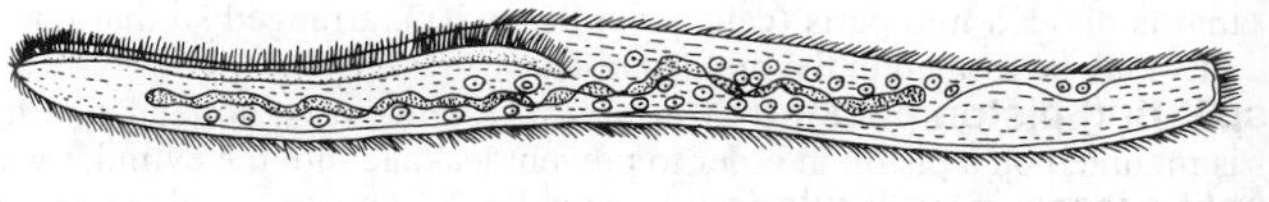

Spirostomum

Spirotricha *Invertebrate Zoology.* an order of ciliate protozoans having membranelles around the mouth and few cilia elsewhere on the body.

Spirulidae *Invertebrate Zoology.* a family of small, deep-sea cephalopod mollusks that includes the species *Spirula spirula,* which has a chambered, coiled shell.

Spirulina *Bacteriology.* a genus of phototrophic cyanobacteria that exhibit gliding motility and form extensive mats of filamentous cells in warm, saline environments.

Spiruria *Invertebrate Zoology.* a subclass of nematode roundworms in the class Secernentea.

Spirurida *Invertebrate Zoology.* an order of nematodes in the class Phasmidea having a two-part esophagus; occur as endoparasites in vertebrates, with insects and crustaceans as intermediate hosts; includes the Guinea worm and filarial worm. Also, **Spiruroidea.**

spit *Geography.* a narrow point of land extending out into the water.

spite *Behavior.* a term for a behavior pattern that decreases the fitness of the actor as well as the recipient.

Spitfire *Aviation.* a single-engine fighter plane that was used by the British R.A.F. throughout World War II.

Spitsbergen *Geography.* an island group in the Arctic Ocean north of Norway.

Spitsbergen Current *Oceanography.* a branch of the Norway Current that first flows northward toward Spitsbergen but then turns westward and joins the southward-flowing East Greenland Current.

spitting rock *Engineering.* a body of rock under pressure from which fragments break and shoot off at relatively high speeds.

Spitzenkorper *Mycology.* in fungi, a small body, dark when stained, occurring near the cytoplasmic membrane in the tip of a growing hyphal filament.

Splachnales *Botany.* an order of large mosses belonging to the division Bryophyta that form spectacular loose to dense tufts and mats on soil or decaying material; characterized by erect stems bearing terminal sporophytes, and by soft and flaccid leaves.

Splachnidiaceae *Botany.* a monotypic family of brown algae of the order Chordariales, characterized by thalli consisting of cylindrical, little-branched axes and by unilocular sporangia within conceptacles; distribution limited to South Africa.

splanchnic *Anatomy.* of or relating to the viscera.

splanchnic mesoderm *Developmental Biology.* the inner of the two layers into which the lateral plate mesoderm splits.

splanchnic nerve *Anatomy.* any of several sympathetic nerves that arise from thoracic and lumbar ganglia and supply the visceral organs.

splanchno- or **splanch-** a combining form denoting a relationship to a viscus or to a splanchnic nerve.

splanchnocranium *Anatomy.* the part of the skull that arises embryonically from the branchial arches, consisting mainly of the facial bones.

splanchnopleure *Developmental Biology.* the layer formed by the joining of the splanchnic mesoderm with the endoderm, from which muscles and connective tissue of the digestive tube develop. Also, **splanchnoderm.**

splash *Ordnance.* **1.** in artillery and naval gunfire support, a word transmitted to the observer or spotter five seconds before the estimated time of impact of a salvo or round. **2.** in air interception, the destruction of a target that has been verified by radar or visual observation.

splashdown *Space Technology.* **1.** the termination of a space flight on water, especially on the ocean. **2.** the moment or location where such termination occurs.

splash erosion *Geology.* a process in which soil particles are dislodged and moved by the impact of falling raindrops.

splash lubrication *Mechanical Engineering.* a system of engine lubrication in which the connecting-rod bearings dip into oil troughs and splash the oil onto the cylinder and piston rods.

splash zone *Ecology.* see SUPRALITTORAL ZONE.

spleen *Anatomy.* the largest lymphatic organ in the body; it contains reticuloendothelial tissue and lies in the upper left part of the abdominal cavity, between the stomach and diaphragm; it serves as a reservoir of blood, disintegrates red blood cells, produces lymphocytes and plasma cells, and has other functions that are not fully understood. Also, **splen.**

spleen colony-forming cell *Cell Biology.* a bone marrow cell that is injected into an irradiated mouse, proliferates, and replaces the dead, irradiated, original bone marrow cells.

splenalgia *Medicine.* any pain in the spleen.

splenectomy *Surgery.* the surgical removal of the spleen.

splenic *Anatomy.* of or relating to the spleen. Also, **splenetic.**

splenic fever see ANTHRAX.

splenic flexure *Anatomy.* the left bend in the colon that demarcates the transverse colon from the descending colon.

splenitis *Medicine.* an inflammation of the spleen, usually produced by pyemia, characterized by enlargement of the spleen with pus, and marked by severe local pain.

splenium *Medicine.* a bandage or compress. *Anatomy.* the rounded posterior end of the corpus callosum. Also, **splenium corporis callosi.**

spleno- or **splen-** a combining form meaning "spleen."

splenolymphatic *Anatomy.* relating to the spleen and the lymph nodes.

splenoma *Oncology.* a tumor of the spleen. Also, **splenoncus.**

splenomegaly *Medicine.* an enlargement of the spleen.

splenopathy *Medicine.* any disease of the spleen.

splenopexy *Surgery.* surgical fixation of a mobile spleen.

splenoptosis *Medicine.* a downward displacement of the spleen.

splenorrhagia *Medicine.* a hemorrhage from the spleen.

splenorrhaphy *Surgery.* the surgical repair of the spleen.

splent coal see SPLINT COAL.

splice to join or unite; a joint or union; specific uses include: *Molecular Biology.* **1.** to connect a DNA segment of an organism to that of another. **2.** to insert an altered DNA segment or gene into an organism with the intention of changing the organism's genetic makeup. **3.** to make such a connection or insertion. *Engineering.* to weave together two components, such as two pieces of rope, in order to form an uninterrupted length. *Electricity.* specifically, to join two wires by tightly winding and often soldering the ends together for a short and applying electrical tape or other material for insulation.

splice acceptor junction *Molecular Biology.* a segment of DNA at the 3' end of an intron that facilitates excision and splicing reactions. Also, **splice acceptor site.**

splice donor junction *Molecular Biology.* a segment of DNA at the 5' end of an intron that facilitates excision and splicing reactions. Also, **splice donor site, splice junction.**

spliced vein *Geology.* a vein that pinches out and is overlapped at that point by another parallel vein.

spliceosome *Molecular Biology.* nucleoprotein particle that aids in the splicing of pre-mRNAs in higher eukaryotes.

splice plate *Civil Engineering.* a plate located over a joint and fastened to the pieces being joined, to provide stiffness.

splicing *Molecular Biology.* **1.** see DNA SPLICING. **2.** see RNA SPLICING.

spline *Engineering.* a thin, elongate piece of wood, metal, or plastic; some splines are flexible and can be bent into curved shapes. *Mechanical Devices.* a long, rectangularly shaped keyway or key with a shaft and gear or pulley hub that provides a means of sliding when a key motion is applied. Also, FEATHER KEY. *Graphic Arts.* a tool made of flexible wood or metal that is used in drawing curved lines.

spline broach *Mechanical Engineering.* a tool used for cutting straight splines or keyways.

splined shaft *Design Engineering.* a shaft having longitudinal gearlike ridges along its interior or exterior surface.

spline function *Mathematics.* given a set K_n of points $\{x_0, x_1, \ldots, x_n\}$ where $a = x_0 < x_1 < \cdots < x_n = b$, then a function $S_k(x)$ defined on the interval $[a, b]$ is called a spline function of degree $k \geq 0$ over K_n if: (a) $S_k(x) \in C^{k-1}([a, b])$; i.e., $S_k(x)$ is continuously differentiable $k - 1$ times on $[a, b]$; and (b) for $x \in [x_{i-1}, x_i]$, $S_k(x)$ is a polynomial of degree at most k. The set of spline functions over K_n is often denoted by $S_n(K_n)$. Spline functions are used to approximate a given function on an interval. Also, INTERPOLATING SPLINE FUNCTIONS.

spline joint *Mechanical Devices.* a joint consisting of wood, metal, or plastic pieces inserted into grooves cut into the sides of adjacent pieces for added strength.

splint *Medicine.* **1.** an apparatus used to immobilize a displaced or movable part or to keep in place and protect a part. **2.** the act of placing such an apparatus on the body.

splint coal *Geology.* a hard, dull, blocky, grayish-black, banded bituminous coal that is characterized by a rough, uneven fracture and granular texture and that burns with intense heat.

split to divide into parts; a division into parts; specific uses include: *Chemistry.* to divide by cleavage. *Systematics.* to raise the rank of one or more taxa and remove them from some superior taxon. *Computer Programming.* **1.** to divide a file or other data into two or more parts. **2.** the creation of two ordered files from one regular file. *Geology.* a coal seam that cannot be mined as a single unit because it is separated from the main stream by a parting of other sedimentary rock. *Mathematics.* **1.** let F be a field and $f \in F[x]$ a polynomial of degree $n \geq 1$. Then f is said to split over F or to be split in $F[x]$ if f can be written in the form

$$f = u_0(x - u_1)(x - u_2) \cdots (x - u_n),$$

where $u_i \in F$, $0 \leq i \leq n$. For example, the polynomial $x^2 - 2$ splits over the field of real numbers but not over the field of rational numbers. **2.** let R be a ring. The short exact sequence of R-module homomorphisms

$$0 \to A \xrightarrow{j} B \xrightarrow{\pi} C \to 0$$

is said to split or to be **split exact** if one of the following equivalent conditions holds: (a) there exists an R-module homomorphism $h: C \to B$ with $\pi h = 1_C$, the identity homomorphism on C; (b) there exists an R-module homomorphism $k: B \to A$ with $kj = 1_A$, the identity homomorphism on A; or (c) B is isomorphic to the direct sum of A and C, that is, $B \cong A \oplus C$. (The assertion that the three conditions are equivalent is a theorem.)

split-anode magnetron *Electronics.* a magnetron having an anode that is divided into two equal segments, usually by parallel slots.

split barrel *Mechanical Devices.* a core barrel that is divided lengthwise into two pieces, which allows for sample removal. Thus, **split-barrel sampler.**

split-base concept *Military Science.* a tactic in which the resources of a deployed combat unit are divided between two bases.

split beam *Electronics.* in a cathode-ray tube, an electron beam that emanates from a single gun but is divided so that two traces appear on the screen.

split bearing *Mechanical Devices.* a bolted, two-piece shaft bearing. Also, DIVIDED BEARING, SPLIT BOX.

split-brain *Psychology.* relating to theories or knowledge of the differing functions of the hemispheres of the brain. See LEFT-BRAIN, RIGHT-BRAIN. *Medicine.* denoting an injury to or severance of the corpus callosum.

split cameras *Optics.* a system of two or more fixed cameras whose fields of view overlap; used to take simultaneous or closely sequenced photographs of a subject from slightly different angles. Thus, **split-camera photography.**

split coal *Geology.* see SPLIT.

split fix *Navigation.* a fix established through sextant readings of two horizontal angles between four objects or similar calculations on a chart; in either case, there is no common center point.

split flap *Aviation.* a flap set into the underside of a wing; depressed downward to increase drag.

split fountain *Graphic Arts.* a method of printing in multiple colors by dividing the ink fountain into sections and filling each with a different color of ink.

split gene *Genetics.* a gene that contains coding regions (exons) that alternate with noncoding regions (introns); characteristic of most eukaryotic genes.

split-lens interference *Optics.* interference produced on a screen by light rays passing through two parts of a divided lens.

split link *Design Engineering.* a metal link that consists of a two-turn helix pressed together.

splitnut *Design Engineering.* a type of screw nut that is laterally cut into two equal sections so that it may be opened and adjusted easily.

split personality *Psychology.* a popular term for a multiple personality syndrome. See MULTIPLE PERSONALITY.

split-phase motor *Electricity.* a single-phase induction motor that has an auxiliary winding connected in parallel with a main winding; it is magnetically positioned away from the main winding so as to produce the required rotating magnetic field for starting; the auxiliary circuit is generally opened when the motor has attained a predetermined speed.

split pin *Mechanical Devices.* a pin or cotter with a split in one end bent to meet the opposing end to prevent backslipping.

split-ring (core) lifter *Mechanical Devices.* a hardened steel ring with a serrated inside edge and an outside taper designed to allow a core to travel downward through it during drilling, but when lifted up, it contracts to tightly grip the core barrel.

split-ring mold *Engineering.* a plastics mold having a small hollow that is divided into parts (called a **split cavity**), arranged so that undercuts may be formed in the molded object.

split-ring piston packing *Mechanical Engineering.* a metal ring that is mounted on a piston in order to prevent leakage into the cylinder wall.

split screen or **split window** *Computer Technology.* a video display on which a portion of a file is shown above, below, or beside another portion of the same file, or a different file, so that the user can refer to other data or work on two or more files simultaneously.

split shovel *Mechanical Devices.* a tool consisting of separate troughs of equal size and spacing, which can be pushed into the ground to extract soil or ore.

split-stator variable capacitor *Electronics.* a variable capacitor whose rotor section serves two separate stator sections.

split stream *Hydrology.* a stream that divides into two channels as it passes around an island or other such obstruction.

splitter vanes *Engineering.* in a gas pipe, an arrangement of curved, parallel vanes in a sharp bend that facilitates the movement of gas around the bend.

split-thickness graft *Surgery.* a skin graft that uses epidermis and a portion of the dermis only.

splitting a process of dividing; a division; specific uses include: *Quantum Mechanics.* a phenomenon in which states whose energies would otherwise be exactly the same assume slightly different energies due to some additional outside influence, such as the nuclear magnetic moment in the atom or an external field; the removal of degeneracy by the application of an external perturbation. *Mathematics.* see FACTORING. *Artificial Intelligence.* in natural deduction theorem proving, a method of proof by cases, i.e., dividing a theorem to be proved into multiple cases and proving each case separately. *Systematics.* the process of creating subdivisions in a species based on minor morphological or behavioral differences.

splitting enzyme *Enzymology.* any enzyme that breaks a bond in a molecule or breaks a molecule into two portions.

splitting field *Mathematics.* **1.** let K be a field and $f \in K[x]$ be a polynomial of degree $n \geq 1$. Then F is said to be a splitting field over K of the polynomial f if: (a) F is an extension field of K; (b) f splits in $F[x]$; and (c) $F = K(u_1, \ldots, u_n)$, where $u_1, \ldots, u_n$ are the roots of f in F; i.e., F is generated over K by the roots of f. **2.** let K be a field and S be a set of polynomials of positive degree in $K[x]$. Then an extension field F of K is said to be a splitting field over K of the set S if every polynomial in S splits in $F[x]$ and F is generated over K by all the roots of all the polynomials in S.

split transducer *Engineering.* an instrument powered by electricity from one system that supplies acoustic power to a second system; characteristically the components are arranged to maintain electrical separation of each division.

split-word operation *Computer Programming.* an operation that can be performed with parts of computer words, such as half-words.

sp. nov. species nova. Also, **sp. n.**

Spock, Benjamin M. born 1903, American physician; the author of a widely used handbook on infant and child care.

spod- or **spodo-** a combining form meaning "waste materials."

spodic horizon *Geology.* a soil horizon characterized by illuvial accumulation of amorphous materials having a high cation-exchange capacity.

spodosol *Geology.* one of the 10 major soil classifications; characterized by a surface horizon that contains an accumulation of organic matter and aluminum oxide; coarse textured and not naturally fertile.

spodumene *Mineralogy.* $LiAlSi_2O_6$, a translucent or transparent, vitreous to dull, monoclinic mineral of the pyroxene group occurring as prismatic crystals of various colors, sometimes quite large, having a specific gravity of 3.0 to 3.2 and a hardness of 6.5 to 7.5 on the Mohs scale; found in granite pegmatites. *Materials.* $Li_2O \cdot Al_2O_3 \cdot Si_4O_2$, a form of this mineral found in South Dakota and North and South Carolina; used as a flux in ceramics and chinaware and as gems.

spoil *Mining Engineering.* **1.** a term for the overburden or debris removed from a mine. **2.** a stratum consisting of a mixture of coal and dirt.

spoilage *Food Technology.* any chemical or physical alteration of food that makes it unfit or unsafe to eat.

spoil bank *Mining Engineering.* **1.** an accumulation of overburden in surface mining. **2.** the location at which spoil is collected. Also, **spoil heap.**

spoil dam *Mining Engineering.* an earthen dam that forms a depression for collecting and retaining returns from a borehole.

spoiler *Aviation.* a movable, long, narrow plate along the upper surface of an airplane wing; used to reduce lift and increase drag by breaking or spoiling the smoothness of the airstream flow. A **flap spoiler** is hinged along one edge and is even with the wing when inoperational; a **retractable spoiler** recedes edgewise into the wing. *Electromagnetism.* a grating constructed of rods and used with a parabolic reflector to change the pencil beam pattern to a cosecant-squared pattern.

spoiling attack *Military Science.* a tactical maneuver employed to seriously impair a hostile attack while the enemy is in the process of forming or assembling for an attack.

spoke *Design Engineering.* a bar or rod that projects from the center of a wheel.

spokeshave *Engineering.* a small tool that is used to smooth curved surfaces.

spondyl- or **spondylo-** a combining form meaning "spine," "vertebrae."

spondylalgia *Medicine.* any pain in a vertebra.

spondylarthritis *Medicine.* arthritis of the spine.

spondylitis *Medicine.* **1.** any inflammation of the spine. **2.** see POTT'S DISEASE.

spondyloarthropathy *Medicine.* any disease affecting the joints of the spine.

spondylolisthesis *Medicine.* the forward displacement of a spinal vertebra relative to the vertebra below, usually the fifth lumbar vertebra over the body of the sacrum, or of the fourth lumbar over the fifth.

spondylomalacia *Medicine.* the softening of vertebrae.

Spondylomoraceae *Botany.* a family of flagellates of the order Volvocales, characterized by a coenobia with cells arranged in alternating tiers, no gelatinous matrix, isogamous reproduction, and a quadriflagellate zygote that eventually loses its flagella, becomes spherical, and develops a thick wall.

spondylosyndesis *Medicine.* the surgical immobilization or ankylosis of the spine. Also, SPINAL FUSION.

sponge *Invertebrate Zoology.* the popular name for any of numerous aquatic animals of the phylum Porifera, characterized by a porous structure and a tough, elastic skeleton of interlaced fibers; they are free swimming as larvae, but later attach in colonies to stones or plants, usually at the bottom of the ocean. *Materials.* **1.** the soft, porous, fibrous framework of certain animals or colonies of this group, especially of the genera *Spongia* and *Hippospongia*, from which the living matter has been removed; it readily absorbs water and remains soft when wet; used in the bathing and cleansing of surfaces. **2.** a similar absorbent made of artificial material. *Chemical Engineering.* a catalyst made of wood shavings coated with iron oxide; used in processes removing hydrogen sulfide from industrial gases.

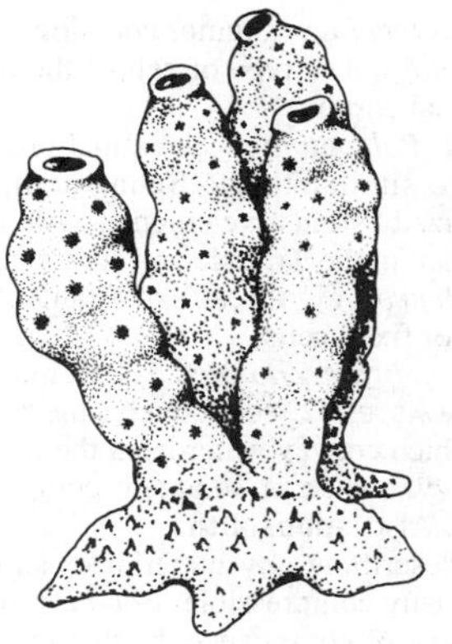

sponge

sponge-diver's disease *Toxicology.* poisoning due to the stinging tentacles of sea anemones of the genera *Sagartia* and *Actinia*; symptoms may include burning, itching, ulceration, and necrosis. Also, SKEVAS-ZERFUS DISEASE.

sponge hook *Mechanical Devices.* a curved hook with three prongs equally spaced at the end of a long handle, used for removing sponge growths in seawater.

sponge iron *Metallurgy.* iron made from low-grade, sandy iron ores containing little carbon; used to replace scrap steel in steel products and pig iron in steel furnaces.

sponge metal *Metallurgy.* any metal or alloy that is very porous.

sponge rubber *Materials.* a light rubber, similar to foam rubber, that is made from dried natural rubber or latex by adding bubbling carbon dioxide; used for padding, insulation, cushions, and gaskets.

spongework *Geology.* an entangled net of small irregular interconnecting cavities on walls of limestone caves.

Spongiidae *Invertebrate Zoology.* a family of large irregular sponges, including the familiar bath sponges, with a spongin skeleton.

Spongilla *Invertebrate Zoology.* a genus of common freshwater sponges in the family Spongillidae, usually green; found encrusting submerged objects.

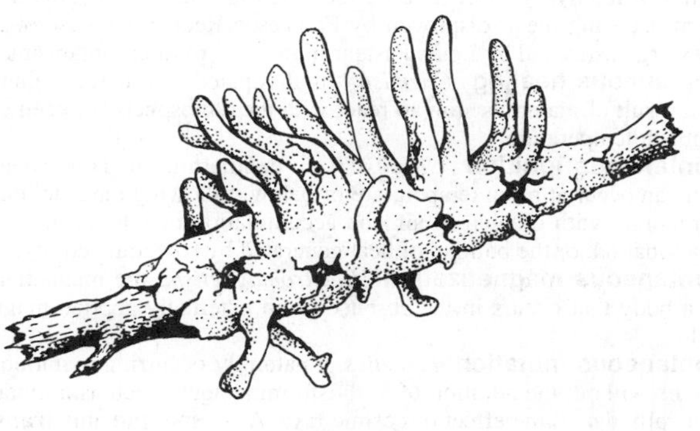

Spongilla

Spongillidae *Invertebrate Zoology.* a family of siliceous freshwater sponges in the class Demospongiae, having encrusting and branching forms; may be gray, brown, white, or green in color.

spongin *Invertebrate Zoology.* a fibrous, sulfur-containing protein, similar to collagen, that composes the interconnecting fibers of the skeleton of some sponges.

sponging *Textiles.* a shrinking process applied to woolens and worsteds before they are cut, in which the cloth is dampened with a wet sponge or with steam.

spongioblast *Invertebrate Zoology.* in some sponges, a specialized cell that secretes spongin to form the skeleton. Also, **spongoblast.** *Developmental Biology.* any or all of the embryonic epithelial cells from the neural tube that are transformed into either neuroglial or ependymal cells.

spongioblastoma *Oncology.* a tumor containing spongioblasts.

spongiocyte *Histology.* **1.** a type of cell of the neuroglia. **2.** a sponge-like cell of the adrenal cortex.

Spongiomorphida *Paleontology.* an extinct order of reef-building hydrozoans that were similar to the Sphaeractinoidea and Stromatoporoidea; characterized by massive colonies with radial pillars joined by horizontal bars; extant in the Triassic and Jurassic.

spongioplasm *Neurology.* threadlike filaments of protoplasm that form the reticulum of fixed cells.

spongocoel *Invertebrate Zoology.* **1.** the main central cavity of a sponge. Also, PARAGASTER. **2.** collectively, the canals and cavities of a complex sponge, which convey water from the flagellated chambers.

spongolite *Geology.* a rock or sediment composed principally of the remains of sponges. Also, **spongolith.**

spongy *Science.* describing any material or substance that is elastic, soft-textured, and easily compressible. *Robotics.* of or relating to a robot that has a high degree of compliance, so that the application of a small force results in a large motion.

spongy parenchyma *Botany.* a layer of irregularly shaped cells with relatively few chloroplasts and large intercellular air spaces, lying below the palisade layer of a leaf. Also, **spongy mesophyll.**

sponson *Ordnance.* a structure projecting from the side or main deck of a vessel to support a gun. *Aviation.* a short, winglike protuberance on each side of a seaplane; designed to increase lateral stability.

sponson mount *Ordnance.* a gun mount positioned on the sponson of a tank or combat vehicle.

spontaneous *Physics.* relating to processes that occur without stimulation from some external influence, but are caused by inherent properties.

spontaneous abortion *Medicine.* a spontaneously occurring termination of pregnancy before the 20th week, with expulsion of the fetus.

spontaneous amputation *Medicine.* the loss of a part that occurs without surgical intervention, as in leprosy or diabetes mellitus.

spontaneous behavior *Behavior.* behavior that is not caused by any observable external stimulus.

spontaneous combustion *Mechanical Engineering.* the ignition of a substance produced by the chemical action of its own constituents without heat from an external source.

spontaneous emission *Electromagnetism.* the emission of electromagnetic radiation from an atom or molecule that is stimulated by an internal agency.

spontaneous fission *Nuclear Physics.* a type of decomposition that does not involve particles from outside the nucleus.

spontaneous generation *Biology.* the former theory that living matter can come to life spontaneously from a nonliving organic source within a relatively short time period, such as flies or maggots arising from decaying meat; disproven by Francesco Redi in the case of complex organisms and by Louis Pasteur in the case of microorganisms.

spontaneous heating *Chemistry.* undissipated heat that accumulates as a result of materials slow to react with the atmospheric oxygen at the ambient temperature.

spontaneous ignition *Chemistry.* an exothermic oxidation reaction that can occur at room temperature or below, following the reaction of a compound with oxygen in air, the accumulation of internal heat from autoxidation, or the buildup of heat generated by bacterial activity.

spontaneous magnetization *Electromagnetism.* the magnetization of a body that occurs in the absence of an externally applied magnetic field.

spontaneous mutation *Genetics.* a naturally occurring mutation that occurs without the addition of a transforming agent such as a mutagen; generally due to the effect of cosmic rays. Also, **spontaneous transformation.**

spontaneous nucleation *Meteorology.* a purely nucleating phase change of a substance devoid of any seeding nuclei within or in contact with it, such as pure vapor condensing to pure liquid, pure liquid freezing to pure solid, and a pure solution crystallizing to pure solute crystals.

spontaneous pneumothorax *Medicine.* the accumulation of air or gas in the pleural space, occurring spontaneously and not as a result of trauma or a pathological process.

spontaneous polarization *Electricity.* the electric polarization that a substance possesses in the absence of an external electric field.

spontaneous process *Thermodynamics.* a thermodynamic process in a system that occurs due to properties that are completely inherent in the system; no external agency is required to execute the process.

spontaneous recovery *Behavior.* the reappearance of an extinguished conditioned response following rest.

spontaneous remission *Psychology.* the disappearance of symptoms and general improvement of an individual without treatment.

spontaneous symmetry breaking *Physics.* a condition in which a solution to a set of apparently symmetric physical equations exhibits asymmetry, as in the case of a magnet that exhibits a sense of direction whereas the physical equations describing the metal are generally isotropic.

spoofing *Electronics.* any electronic operation designed to deceive or mislead an enemy, such as resuming transmission on a frequency after it has been jammed, using decoy radar transmitters to provoke a futile jamming effort, or transmitting radio messages bearing false information for intentional interception.

spool *Textiles.* **1.** a small cylindrical piece of wood or other material on which yarn or thread is wound for spinning or sewing. **2.** to unwind yarn or thread from a spool. *Mechanical Engineering.* a drum or cylinder onto which a wire, tape, chain, rope, or other such item is wound. *Graphic Arts.* **1.** in a camera, a cylinder designed to hold a roll of film. **2.** the wound cylinder of paper used to feed a web press. *Computer Programming.* a program or device that provides a spooling capability. Also, **spooler.**

spooling *Computer Programming.* **1.** the use of auxiliary storage as a buffer to reduce processing delays during the transfer of data between peripheral equipment and the central computer; e.g., **print spooling** allows the computer to be freed for another task before a printing operation has been completed. **2.** the reading of input data streams and the writing of output data streams on auxiliary storage devices during a job execution in a format that allows later processing or output operations. (An acronym for simultaneous peripheral operations on line, taken from the common verb sense of *spooling* thread or other material.)

spoon *Mechanical Devices.* any of a number of multipurpose tools with a small, shallow cup or basin attached at right angles to a handle, as in the common eating implement; used for such purposes as scraping, extracting, or excavating.

spoonbill *Vertebrate Zoology.* **1.** a spatulate-billed wading bird with long legs and a bare head of the family Threskiornithidae and the subfamily Plataleinae, related to the ibises; found in the tropical regions of the world. **2.** any of various birds having a similar bill, as the shoveler duck.

spoonbill

spoonleaf *Plant Pathology.* a disease of the red currant, caused by a strain of the raspberry ringspot virus; characterized by leaves that become roundish and lack teeth in their margins.

spor- or **sporo-** a combining form meaning "seed." Also, **spori-.**

sporadic E layer *Geophysics.* a layer of intense ionization, usually of limited horizontal extent, occurring intermittently in the E region of the ionosphere, whose variables include time of occurrence, height, geolographical distribution, and ionization density.

sporadic meteor *Astronomy.* a random meteor not part of any recognized stream or shower.

sporadic reflections *Electromagnetism.* sharply defined reflections from the sporadic E layer of the ionosphere that exhibit substantial intensity.

sporadin *Invertebrate Zoology.* a developmental stage of gregarine sporozoan parasites in the class Apicomplexa.

sporangiolum *Mycology.* a fungal structure containing one or only a few spores and resembling a sporangium. Also **sporangiole.**

sporangiophore *Botany.* a structure of plants or fungi that bears a sporangium or spores.

sporangiospore *Botany.* a spore produced within a sporangium.

sporangium *Botany.* **1.** a specialized case, sac, capsule, or other structure that produces spores. **2.** a hollow unicellular or multicellular sac, case, or capsule in which spores are produced.

spore *Biology.* a one-celled or multicellular, asexual, reproductive or resting body that is resistant to unfavorable environmental conditions and that is capable of developing into an adult without fusion with another cell when the environment is favorable. *Invertebrate Zoology.* see SPOROCYST.

spore mother cell *Botany.* a diploid cell of a sporophyte that divides meiotically to give rise to four haploid spores. Also, SPOROCYTE.

spore print *Mycology.* a fungal spore deposit used to identify mushrooms that is obtained from glass, paper, or other flat surfaces upon which a mature mushroom cap has been placed, gills down.

Spörer minimum *Astronomy.* a period of low sunspot activity lasting from approximately 1420 to 1570 AD.

Spörer's law *Astronomy.* a law stating that the first-appearing sunspots of a new cycle emerge first at high solar latitudes and, over the cycle, migrate toward the solar equator.

Sporichthya *Bacteriology.* a genus of bacteria of the family Streptomycetaceae, consisting of soil organisms that form aerial mycelia and produce spores that become motile in water.

Sporidiales *Mycology.* in some classifications, an order of fungi belonging to the class Ustilaginomycetes; most are not parasitic.

Sporidiobolus *Mycology.* in some classifications, a genus of fungi belonging to the order Sporidiales, which is characterized by the formation of budding yeast cells.

sporidium *Mycology.* a small fungal spore found in rust or smut fungi.

sporinite *Geology.* a maceral of coal within the exinite group composed of spore exines that have been compressed parallel to the stratification.

sporistasis *Mycology.* a condition in which germination and outgrowth of a viable spore are prevented by chemical or other factors.

sporoactinomycetes *Mycology.* certain fungi included in the order Actinomycetales, which includes spore-forming organisms that are more complex than the nocardioform actinomycetes.

sporoblast *Invertebrate Zoology.* in the life cycle of many Sporozoa, one of the bodies formed by asexual fission of a sporont; each secretes a membrane to become a sporocyst and then fissions again to produce several sporozoites.

sporobola *Mycology.* the horizontal shoot of a spore.

Sporobolomyces *Mycology.* a genus of fungi belonging to the class Hyphomycetes, which occurs on the leaves of cereals and produces budding yeast cells that are forcibly discharged.

Sporobolomycetaceae *Mycology.* a family of fungi belonging to the order Sporobolomycetales, which is characterized by budding yeast cells and ballistospores or spores forcibly ejected at maturity.

Sporobolomycetales *Mycology.* in some classifications, an order of fungi belonging to the class Blastomycetes, which includes so-called mirror or shadow yeasts that possess characteristics of yeasts and molds, such as reproduction by budding.

sporocarp *Botany.* a multicellular structure of fungi, lichens, ferns, or other plants in or on which spores are formed.

Sporochnales *Botany.* a monofamilial order of brown algae of the class Phaeophyceae, composed of the family Sporochnaceae; characterized by unique tufts of unbranched filaments at the end of each branch; most common in warm waters of the Southern Hemisphere.

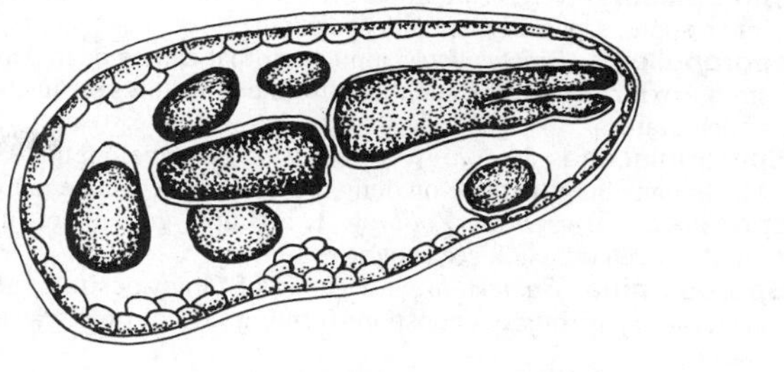

sporocyst

sporocyst *Botany.* a unicellular resting body from which asexual spores arise, usually formed as a resistant phase of the life cycle by certain algae and fungi. *Invertebrate Zoology.* **1.** in the life cycle of Sporozoa, a secreted membrane containing a sporoblast or, after fission, several sporozoites. Also, SPORE. **2.** a saclike larval stage of many trematode worms, usually embedded in the tissue of an intermediate host snail.

sporocyte see SPORE MOTHER CELL.

Sporocytophaga *Bacteriology.* a genus of gliding bacteria of the family Cytophagaceae that occur as pigmented, rod-shaped cells characterized by the ability to produce a type of resting cell called a **microcyst.**

sporodochium *Mycology.* a fungal spore-bearing body shaped like a cushion or pad, upon which the hyphal spore masses rest.

sporogenesis *Biology.* **1.** reproduction by means of spores. **2.** the process of making spores.

sporogony *Biology.* **1.** multiple fission of a cell to produce many dormant spores. **2.** reproduction by means of spores. *Invertebrate Zoology.* part of the life cycle of Sporozoa in which a sexually produced zygote encysts to form an oocyst, which leaves the host and becomes a sporont; the sporont fissions asexually to produce several sporoblasts, each of which encases itself to form a sporocyst and then fissions asexually to produce sporozoites, which are eaten by a new host; schizogony ensues.

Sporolactobacillus *Bacteriology.* a genus of Gram-positive, microaerophilic, rod-shaped bacteria of the family Bacillaceae that form endospores and do not contain heme compounds; found in soils.

Sporomusa *Bacteriology.* a genus of Gram-negative, anaerobic bacteria composed of curved, rod-shaped, endospore-producing cells with a flagellar motility; found in mud or soil.

sporont *Invertebrate Zoology.* **1.** a zygote formed during the protozoan life cycle, especially in sporozoans. **2.** in certain sporozoans, a sexually produced life stage that fissions asexually to produce sporoblasts.

Sporopachydermia *Mycology.* a genus of yeast fungi belonging to the family Saccharomycetaceae, which reproduces by multilateral budding and carries out aerobic metabolism.

sporophore *Mycology.* in fungi, a spore-bearing structure.

sporophyll *Botany.* a modified leaf that bears or subtends one or more sporangia.

sporophyte [spôr´ə fīt] *Botany.* in plants with alternation of generations, the diploid plant that produces haploid spores.

sporophytes

sporoplasm *Invertebrate Zoology.* a mass of protoplasm that develops into a spore, especially in Cnidosporidia.

sporopollenin *Biochemistry.* a polymer that can be degraded only with strong oxidizing agents such as chromic acid; present in the exine of pollen grains.

Sporormiaceae *Mycology.* a family of fungi belonging to the order Melanommatales that live on dung or decaying vegetable matter.

sporosac *Invertebrate Zoology.* **1.** a simple gonophore in some hydroids. **2.** see REDIA. **3.** see SPOROCYST.

Sporosarcina *Bacteriology.* a genus of Gram-positive, aerobic soil bacteria, occurring as endospore-forming coccoid cells in tetrads and packets.

Sporospirillum *Bacteriology.* a genus of motile, helical, endospore-forming bacteria that are found in the intestinal contents of tadpoles.

Sporothrix *Mycology.* a genus of fungi belonging to the class Plectomycetes, which includes some species that are nonparasitic, obtaining nutrients from organic dead matter, such as fallen wood and soil, and others that are skin parasites.

sporotrichosis *Medicine.* a chronic fungal infection caused by the fungus *Sporotthrix schenchii,* giving rise to lesions of the skin that spread along the lymph channels, sometimes also affecting internal organs and bones. The fungus is found in soil and decaying vegetation and is acquired through a break in the skin or by inhalation.

Sporotrichum *Mycology.* a term formerly used to designate a genus of fungi belonging to the class Hyphomycetes; the genus is now reclassified as Sporothrix.

Sporozoa *Invertebrate Zoology.* a class or subphylum of spore-forming parasitic protozoans with no cilia, flagella, or pseudopodia; cell structure is simple, but the life cycle is complex with both sexual and asexual phases, generally in more than one host.

sporozoite *Invertebrate Zoology.* a sporozoan that has been liberated from the oocyst and is ready to penetrate a new host cell.

sporting gun *Ordnance.* **1.** in general, any gun designed for sporting purposes, such as hunting or target shooting, as distinguished from defense, police, or military weapons. **2.** specifically, a class of shotguns that includes guns with 30-inch full-choke barrels.

sports medicine *Medicine.* a field of medicine specifically concerned with injuries sustained in athletic activities, especially injuries that are typically associated with such activities, such as shin splints, tennis elbow, tendinitis, and so on.

sporulation *Biology.* the process of producing or liberating spores.

sporule *Biology.* a spore, especially a small one.

spot *Ordnance.* to observe fire, visually or electronically, in order to determine deviations of ordnance from the target for the purpose of supplying necessary information for the adjustment of fire. Thus, **spotting.** *Graphic Arts.* any decorative element used in typography. Also, **spot illustration.** *Electronics.* in a cathode-ray tube, the luminous area that is instantaneously affected by the impact of an electron beam.

SPOT satellite positioning and tracking.

spot blotch *Plant Pathology.* a barley disease caused by the fungus *Helminthosporium sativum*; characterized by long, dark markings on the foliage.

spot check *Industrial Engineering.* a random check made of some part of a production process or operation to ensure that work is being carried out properly.

spot color see FLAT COLOR.

spot desmosome *Cell Biology.* one of many intercellular junctions that occur as discrete points of contact, binding two epithelial cells together.

spot drilling *Mechanical Engineering.* the process of drilling a small indentation in a work surface as a guide for subsequent machining operations.

spot elevation *Cartography.* a spot on a map or chart, such as the peak of a mountain, for which the elevation above or below a specified reference datum, such as a contour, is noted by a graphic symbol accompanied by the elevation value. Also, **spot height**.

spot facing *Mechanical Engineering.* **1.** the process of machining a flat seat for a bolt head or nut at the end of and at right angles to the axis of a previously made hole. **2.** the process of truing a relatively small surface or spot, as forming a true seat for the underside of a screw head.

spot film *Radiology.* a radiograph of a limited anatomic area obtained by rapid exposure during fluoroscopy to provide a record of a transiently observed abnormality, or by limitation of penetrable radiation passing through an area to give improved definition and detail of the image produced.

spot gluing *Engineering.* after gluing a structure, the process of applying dielectric heating to the areas that are most likely to be disturbed by movement, in order to facilitate the setting of the glue.

spot group *Astronomy.* a collection of sunspots having a complex form with several umbrae and a common penumbra.

spot-hover *Aviation.* to stay in place over one point on the ground while airborne, as in a training maneuver in which a helicopter remains in stationary, low-level flight over one point while executing four successive turns at headings 90° apart.

spot jammer *Electronics.* a jammer that attempts to interfere with reception on a specific channel or frequency.

spotlight *Electricity.* a flat light, usually equipped with a lens and reflectors to produce a narrow, strongly focused beam.

spot-noise factor *Electronics.* in a transducer at a given frequency, the ratio of the total output noise power to the portion attributable to thermal noise in the input termination. Also, **spot-noise figure.**

spot-size error *Electronics.* an error in interpreting a radarscope presentation, occurring when the spot on the screen is so big that two or more objects appear as one.

Spotted *Agriculture.* a breed of swine that has a spotted black and white coat. Also, **Spotted Swine.**

spotted aphid *Invertebrate Zoology.* a pale yellowish aphid, *Therioaphis maculata,* characterized by black spots and fine spines on the back; very destructive of alfalfa and a pest of some other legumes, such as clover; found in the southern U.S., especially west of the Mississippi River. Also, **spotted alfalfa aphid.**

spotted cavy see PACA.

spotted cowbane *Botany.* a water hemlock, *Cicuta maculata* of the parsley family, characterized by a mottled stem, white flowers, and tuberlike roots that are deadly poisonous; found in North America. Also, **spotted hemlock.**

spotted fever *Pathology.* any of various fevers characterized by the appearance of spots on the skin, especially cerebrospinal meningitis, typhus, or Rocky Mountain spotted fever.

spotted guitarfish see GUITARFISH, def. 1.

spotted skunk *Vertebrate Zoology.* a small, nocturnal skunk of the genus *Spilogale*, characterized by a white forehead patch, a white-tipped tail, and a luxuriant coat of broken stripes and spots; includes *S. putorius* of temperate North America and *S. pygmaea* of Mexico.

spotted wilt *Plant Pathology.* a viral disease of a wide range of crops and wild plants, characterized by a bronze discoloration and a downward curl of the foliage.

spotter *Ordnance.* a person who observes and reports the results of gunfire to the firing group, and who may also be employed in designating targets. *Optics.* the person in an optical fabrication plant who locates and marks the center of a mirror or lens element.

spotter-tracer *Ordnance.* a bullet or projectile that is equipped with a tracer and contains a filler that may be used for spotting.

spot test *Analytical Chemistry.* a simple method of qualitative analysis that involves the addition of a drop of test substance to a drop or two of a reagent; a color change or the formation of a precipitate indicates a positive test.

spottiness *Electronics.* a condition in which bright spots are scattered irregularly over the reproduced image in a television receiver, usually due to static interference entering the system.

spotting *Engineering.* **1.** the process of marking a die part with an oil color so that it will fit properly with a mated part. **2.** the process of laying stabilizer bags down at regular intervals on ground that is to be stabilized. *Graphic Arts.* the use of a pencil or brush and dye to retouch details or blemishes on a photographic negative or print. *Ordnance.* the activity of an artillery spotter in reporting on fire and selecting targets.

spotting board *Ordnance.* a device that converts information provided by spotters into firing data.

spotting line *Ordnance.* a line used in observing artillery or naval gunfire; it may be the gun-target line, the observer-target line, or a reference line used by the spotter in making spot corrections.

spotting pistol *Ordnance.* an automatic or semiautomatic firearm mounted coaxially with a larger-caliber gun; it fires a spotter-tracer projectile that is ballistically matched with the larger projectile in order to conserve the larger ammunition and improve the first-round accuracy of the larger gun. Similarly, **spotting rifle.**

spotting scope *Optics.* a telescope that provides an erect image; it is portable and typically has a magnification under 100×.

spotty ore *Mining Engineering.* ore in which the valuable material is distributed in irregular concentrations of small particles.

spot welding *Metallurgy.* welding that is confined to small circular areas.

spot wind *Meteorology.* wind direction and speed that are observed or forecast at a particular altitude over a fixed location; used in air navigation.

spout hole *Vertebrate Zoology.* **1.** a nostril of walruses and seals. **2.** the blowhole of whales, porpoises, and dolphins.

spouting *Engineering.* the process of inserting or ejecting fine- or coarse-grained solid material with the use of inclined or vertical discharge spouts.

spouting horn *Geology.* a sea cave with a rearward or upward opening through which water spurts after waves enter the cave.

spp. or **spp** species (plural).

sprag *Engineering.* a short piece of wood placed between the spokes of a wheel or dug into the ground to prevent the movement of a vehicle. *Mining Engineering.* a wooden prop positioned on a slant to support a mine roof or prevent coal from flying during blasting.

sprag road *Mining Engineering.* a mine road so steep that mine cars require sprags between their wheel spokes to slow their descent.

sprain *Medicine.* **1.** an injury to the supporting tissues of a joint, characterized by swelling, heat about the joint, discoloration, and pain; treatment includes support, rest, and alternating cold and heat. **2.** to incur such an injury to a body part.

sprain fracture *Medicine.* the separation of a ligament or tendon from its insertion point, where the displaced portion carries with it a segment of the bone to which it is connected.

spraing *Plant Pathology.* a disease of potato plants caused by the tobacco rattle virus and characterized by the appearance of reddish-brown arcs on the flesh of the tubers.

spray *Mechanics.* **1.** a stream of water or other liquid broken up into minute particles and propelled through air or another gas. **2.** to move a liquid in such a manner. *Engineering.* a mechanism or device used to discharge a liquid in this way, as in applying paint, dispensing medicine, spreading insecticide, and so on. *Astronomy.* an explosive release of gas during a solar flare that extends from the sun's chromosphere in all directions uniformly, with speeds up to 1500 km per second.

spray angle *Fluid Mechanics.* the angle formed by the cone of liquid droplets discharged from an atomizer.

spray chamber *Mechanical Engineering.* **1.** any vessel or chamber in which a spray of liquid is used to effect contacting, humidification, or the collection of solid particles and dusts. **2.** the chamber in an air-conditioning unit in which humidification takes place.

spray coatings *Materials Science.* any dispersion of liquid in a gas used in coating applications, most commonly in spray painting; spray coating systems typically consist of an air compressor, a fluid supply, a spray gun, connector hoses, and regulators, and allow rapid deposition, saving time and labor.

spray dome *Ordnance.* the mound of water spray thrown into the air when the shock wave from an underwater detonation of a nuclear weapon reaches the surface.

spray dryer *Mechanical Engineering.* a machine that dries an atomized mist by direct contact with hot gases. *Food Technology.* specifically, an apparatus for making dried food products from liquids such as milk. The liquid is sprayed as fine mist into a hot-air chamber, where it dehydrates; its solids fall to the bottom of the chamber as dry powder.

spray-drying *Materials Science.* a ceramic processing used to create a uniform die fill for automated pressing; the slurry is fed into the spray drier, where circulating hot air evaporates the water; the resulting powder is collected for further processing.

sprayed metal mold *Engineering.* a type of plastics mold generally used to form plastic sheets; produced by spraying molten metal onto a form to create a shell of a desired thickness, removing the shell after it cools, and then backing the shell with a strong material such as plaster.

sprayer plate *Engineering.* a rotating metal disk designed to improve the process of atomization in an oil burner.

spray freezer *Food Technology.* a machine used in the quick freezing of fruits, vegetables, and fish; it sprays a very cold liquid such as nitrogen, brine, or sugar solution over food moving on a conveyor past a spray station.

spray gun *Mechanical Devices.* a device that delivers an atomized stream of liquid through a control orifice.

spray nozzle *Mechanical Devices.* a device that ruptures a continuous stream of liquid and disperses it in a mist of small atomized droplets.

spray painting *Engineering.* the process of applying a fine coat of paint using a spraying mechanism.

spray point *Electricity.* any of a series of sharp points that are aligned and charged to a high direct-current potential; used to charge and discharge the conveyor belt in a Van de Graaff generator.

spray pond *Engineering.* a reservoir in which water is cooled by the action of nozzles in the reservoir that spray water into the air to produce an evaporative cooling effect.

spray probe *Engineering.* an instrument used to locate a gas leak in a vacuum system; the leak is in the form of a jet spray.

spray quenching *Metallurgy.* the process of quenching in which a sprayed liquid is used as the cooling medium.

spray region see FRINGE REGION.

spray torch *Engineering.* a thermal spraying mechanism designed for the application of self-fluxing alloys.

spray tower *Chemical Engineering.* an industrial gas absorber for the mass transfer of highly soluble gases, consisting of a large open chamber into which liquid is sprayed downward and a gas that flows upward or passes horizontally by the liquid.

spray transfer *Metallurgy.* in consumable arc welding, the transfer of fine drops of filler metal.

spray-up *Engineering.* any process in which a spray gun is used as a processing tool.

spread *Engineering.* **1.** the area covered at a given thickness by a given quantity of material. **2.** see SENSITIVITY. *Statistics.* see RANGE.

spreader a device that spreads; specific uses include: *Civil Engineering.* **1.** any appliance that distributes water uniformly in or from a channel. **2.** a machine fitted with wide plates for spreading soil, subsoil, or rock excavated from a pond, drainage ditch, or other cut. *Mining Engineering.* **1.** a timber placed horizontally below the cap of a set to stiffen the legs and support the brattice when two air courses are in the same gangway. **2.** a piece of timber set across a shaft for temporary wall support. *Electricity.* an insulator used to separate the wires of an air-spaced transmission line. *Mechanical Engineering.* a tool used in sharpening machine drill bits.

spreader beam *Engineering.* a stiff beam suspended from a crane hook, having ropes or chains hanging from it at intervals; used to lift long concrete piles or fragile objects.

spreader stoker *Mechanical Engineering.* a system of mechanical feeders that loads and evenly distributes a thin layer of fuel over the grate in a coal-burning furnace.

spread footing *Civil Engineering.* a rectangular prism of reinforced concrete larger in lateral dimensions than the column or wall it supports; designed to distribute the load of the column or wall to the subgrade soil.

spreading activation *Artificial Intelligence.* the principle that knowledge, in humans or machines, may be organized as a network of related concepts, and that activation of nodes that are explicitly mentioned can spread to activate other nodes; thus, "red" and "siren" may suggest "fire engine." *Psychology.* in cognitive psychology, the principle that a particular remembrance will spread to similarly associated memories, which are themselves often activated as well.

spreading coefficient *Thermodynamics.* an expression specifying the amount of work involved in spreading a liquid over a unit area of the surface of another liquid; it is equal to the difference between the work of adhesion between the two liquids and the work of cohesion of the spreading liquid.

spreading center *Geology.* a linear zone on the floor of the sea from which adjacent crustal plates are moving apart and along which magma rises.

spreading fire *Ordnance.* in naval gunfire support, a notification by the fire control agency, either the ship or the spotter, to indicate that fire is about to be distributed over an area.

spread plate *Microbiology.* a petri dish on which a liquid inoculum is spread over the entire surface, yielding a microbial lawn upon incubation.

spread reflection *Electromagnetism.* the reflection of electromagnetic radiation from a rough surface with large irregularities.

spread spectrum *Electronics.* a communications method in which many signal waveforms are transmitted; used to spread normally narrow-band signals over a wide band of frequencies, thereby allowing multiple signal access and increasing immunity to noise and interference.

spread-spectrum transmission *Electronics.* a communications method in which many signal waveforms are transmitted in a spread spectrum; used for security and privacy, to prevent jamming, and to utilize signals buried in noise.

spring *Mechanics.* the quality of a material that will cause it to return quickly to its original position after being moved or to its original shape after being compressed. *Mechanical Devices.* a simple machine consisting of an elastic helical coil that when stressed or bent will return to its original form; used for storing potential energy, absorbing shocks, isolation from vibration, etc. *Engineering.* **1.** a stored energy device or system that absorbs and releases energy to provide a level ride. After release it returns to its original shape (**elastic spring**) or original position (**pneumatic spring**). **2.** to enlarge the base of a borehole by applying small charges of a high explosive, so that space is created for an eventual full charge.

spring *Hydrology.* any place where a concentrated, natural discharge of groundwater issues forth as a definite flow onto the surface of the land or into a body of surface water.

spring *Astronomy.* a season between the vernal equinox and the summer solstice, in which days become progressively longer in northern latitudes and progressively shorter in southern latitudes.

springback *Metallurgy.* the recovery of elastic deformation occurring after the force that has caused plastic deformation has been removed.

spring balance *Engineering.* an instrument that measures the intensity of the force applied to stretch or compress a coiled spring.

spring bearings *Naval Architecture.* the bearings that support a vessel's propeller shaft through the interior of the hull.

spring binder *Mechanical Devices.* a loose-leaf binder used to secure or clamp papers with a long, clamplike spring that forms its spine. Also, **spring clamp.**

springbok *Vertebrate Zoology.* a gazelle, *Antidorcas marsupialis,* that springs almost straight up into the air when excited or disturbed; found in southern Africa. Also, **springbuck.**

springbok

spring bolt *Mechanical Devices.* a lock bolt that shoots into place through spring action when it is released.

spring box *Mechanical Devices.* a hollow box with a helical spring mounted sideways to actuate a lever or pawl when a device is brought in line with the box.

spring box mold *Engineering.* a type of mold that shapes material through heat and compression; it contains a spacing fork that is removed after the material has been partially compressed.

spring brass *Metallurgy.* a copper-base alloy containing 30% zinc; it is cold worked to impart very high hardness.

spring caliper *Engineering.* a caliper that has two legs joined together with a spring and a pivot.

spring clip *Mechanical Devices.* **1.** any small electrical connector coupled with a spring. **2.** a U-shaped clip that attaches a leaf spring or axle to a vehicle.

spring clock *Horology.* any type of clock in which a wound spring provides the power to the going train.

spring collet *Mechanical Devices.* a bushing slotted at the front and tapered, so that when the collet is drawn back on a screwing machine, the bushing closes around the workpiece preventing slippage; the same spring action is used to release the grip. Also, **spring chuck.**

spring constant *Mechanics.* see FORCE CONSTANT.

spring contact *Electricity.* **1.** an electric contact that is actuated by a spring. **2.** a relay or switch contact mounted on a flat metal spring.

spring coupling *Mechanical Engineering.* a coupling that has elastic parts for flexibility and resilience.

spring cramp *Mechanical Devices.* a spring shaped like a broken ring and made of tensile steel.

spring crust *Hydrology.* a snow crust usually formed in late winter and spring when loose firn becomes recemented as the temperature decreases.

spring die *Mechanical Devices.* a die for cutting screw threads, consisting of a cut-away hollow cylinder with internal pronglike cutting teeth and a taper-threaded external surface.

spring equinox see VERNAL EQUINOX.

spring faucet *Engineering.* a faucet that is kept closed by the action of a spring; force is applied against the spring to open the faucet, which closes when the force is removed.

Springfield rifle *Ordnance.* a U.S. .30-caliber, breech-loading, bolt-operated, magazine-fed, 5-shot repeating rifle manufactured by the Spingfield Armory; it was adopted as the standard U.S. military rifle in the early 20th century and various versions were used up to World War II; officially designated **U.S. .30 caliber rifle Model 1903.**

spring gravimeter *Engineering.* an instrument that measures relative gravity, using the formula $g_2 - g_1 = k(s_2 - s_1)$, with g representing gravity, and s representing the elongation of the spring of the instrument.

spring hammer *Mechanical Engineering.* **1.** a steam hammer with a spring-suspended head that acts as a shock absorber. **2.** a power hammer driven by a compressed spring or compressed air.

springhare *Vertebrate Zoology.* a small, nocturnal rodent, *Pedetes capensis* of the famiy Pedetidae, resembling a small kangaroo with large broad feet, shortened forelimbs, and a bushy tail; native to southern Africa. Also, **springhaas.**

spring hinge *Mechanical Devices.* a door hinge fitted with a spring so that the opened door will close automatically.

spring hook *Mechanical Devices.* a hook whose end is equipped with a spring-action snap to prevent slipping. Also, SNAP HOOK.

spring-joint caliper *Mechanical Devices.* a caliper with a heavy-duty spring at its leg joint, which is activated by a knurled nut.

spring line *Hydrology.* a line of springs marking the point at which the water table intersects with the land surface. *Naval Architecture.* a mooring line leading to a pier at an acute angle and used to control forward or aft movement of a moored vessel; it runs forward from the after section of the vessel or aft from the forward section.

spring-loaded meter *Engineering.* an instrument that measures the variable flow rate of a liquid through a tube, by measuring the difference in pressure created by the liquid flowing past an obstruction in the tube and the pressure created in a spring attached to the obstruction.

spring-loaded regulator *Mechanical Engineering.* a relief pressure valve that is controlled by a calibrated spring; when a certain value is exceeded, the pressure acts against the spring tension to activate the valve.

spring modulus see FORCE CONSTANT.

spring needle *Textiles.* a knitting-machine needle with a long, flexible hook on one end and a butt on the other. Also, BEARDED NEEDLE.

spring overturn *Ecology.* the mixing of the waters in a lake due to the melting of ice and snow, resulting in uniform water temperatures at every level of the lake.

spring peeper *Vertebrate Zoology.* a small, brown tree frog, *Hyla crucifer,* characterized by an X-shaped mark on the back and a shrill call that is heard early in the spring; found near ponds and swamps of eastern North America.

spring pin *Mechanical Engineering.* **1.** an iron rod that maintains constant pressure on the axle of a locomotive. **2.** a hollow dowel pin manufactured from spring steel.

spring range *Oceanography.* the mean semidiurnal tide range at the time of spring tides. Also, MEAN SPRING RANGE.

spring rise see MEAN SPRING RISE.

spring scale *Engineering.* a scale in which weight measurements of a load are determined by the change in shape or dimensions of the spring component of the scale.

spring seepage see SEEPAGE SPRING.

spring shackle *Engineering.* a U-shaped fastening device on the end of a spring that allows the spring to change in length as it bends.

spring snow *Hydrology.* a wet, coarse, granular snow formed during the spring as a result of alternate thawing and freezing. Also, CORN SNOW.

spring switch *Civil Engineering.* a railroad switch designed to be activated by the last wheels of a railroad car passing over it.

springtail *Vertebrate Zoology.* any of various minute, wingless insects of the order Collembola; characterized by forked, tail-like appendages at the end of the abdomen which act as a spring, giving them a nearly perpetual springing pattern.

spring temper *Metallurgy.* the attribute of a metal or alloy that has a hardness intermediate between full-hard and extra-spring.

spring tidal currents *Oceanography.* the currents of increased velocity that are associated with spring tides.

spring tide *Oceanography.* a tide of increased range that occurs twice monthly at the new and full phases of the moon.

spring-tooth harrow *Agriculture.* a variation of the spike-tooth harrow, consisting of long, curved teeth mounted on horizontal bars; used to level rocky ground after plowing.

springwood *Botany.* the less dense, larger-celled xylem that composes the first wood of a growing season. Also, EARLY WOOD.

sprinkle *Meteorology.* of rain, to fall lightly in fine drops. *Engineering.* to distribute water or another liquid in a similar light shower or spray.

sprinkler *Engineering.* a device that dispenses water or another liquid in the form of a light shower or spray. *Agronomy.* specifically, such a device used to dispense water in an irrigation system.

sprinkler irrigation *Agronomy.* any method of irrigation in which water is distributed by means of an aerial mist or spray rather than applied directly on the soil surface.

sprinkler system *Engineering.* a system of pipes that conveys water or other fire-extinguishing fluid to sprinkler heads that open automatically when a predetermined level of heat is detected.

Sprint *Ordnance.* a high-acceleration, nuclear, surface-to-air guided missile that was formerly deployed as part of the Safeguard ballistic missile defense weapon system; it is designed to intercept strategic ballistic reentry vehicles in the endoatmosphere.

sprite *Computer Technology.* a small, high-resolution object that can be moved about the screen independently of any text or images; used in computer animation.

s-process *Nuclear Physics.* a process by which various elements, mainly from the iron group, merge over long periods of time; involves the capture of slow neutrons produced by the interaction of alpha particles with carbon-15 and neon-21.

sprocket *Mechanical Engineering.* **1.** a toothed wheel that engages a cable or power chain. Also, **sprocket wheel. 2.** one tooth of such a wheel. *Ordnance.* in naval mine warfare, an antisweep device that allows a sweep wire to pass through a mine mooring without parting the mine from its sinker.

sprocket chain *Mechanical Engineering.* a continuous chain whose links are meshed with the teeth of a sprocket so that power can be transmitted from one sprocket to another. Also, PITCH CHAIN.

sprocket hole *Engineering.* any of a series of perforations on the border of a motion-picture film or paper roll, designed to fit into the projections of a sprocket, which moves the material through a device such as a film projector.

S protein *Immunology.* a complement system regulatory protein.

sprout *Botany.* **1.** new or young growth from a plant; a shoot or bud. **2.** to shoot off such growth. Thus, **sprouting. 3. sprouts.** the young shoots of alfalfa, soybeans, and the like, eaten as a raw vegetable.

sprouting

sprouting prevention *Food Technology.* the treatment of potatoes in storage to inhibit sprouting, either with low-level irradiation or with ambient chemical sprays such as vaporized nonyl alcohol.

spruce *Botany.* any evergreen, coniferous tree of the genus *Picea* of the family Pinaceae, characterized by short, angular needles attached singly around twigs and hanging cones with persistent scales; found in North America and Europe. *Materials.* the wood of this tree, having a straight, white grain; used for lumber, boxes, and papermaking.

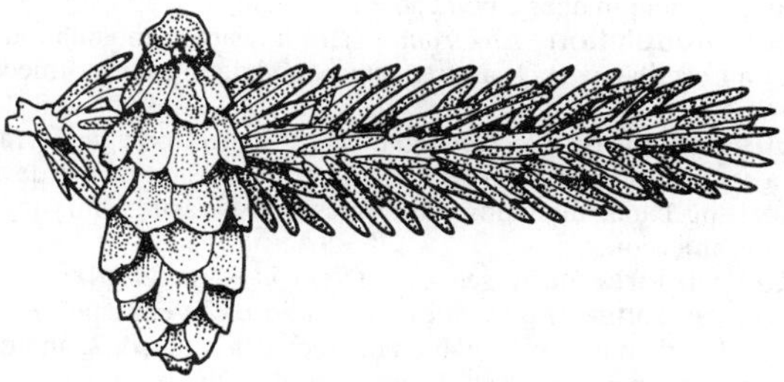

spruce

spruce pine *Botany.* **1.** a tall coniferous tree, *Pinus glabra*, having a smooth, gray bark and needles in bundles of two; found in the southeastern U.S. **2.** any of various American pines, such as *P. eliottii* or *P. virginiana*, having light, soft wood.

sprue *Medicine.* **1.** a chronic form of malabsorption syndrome occurring in both tropical and nontropical forms, manifesting a wide range of symptoms including diarrhea, weakness, poor appetite, weight loss, and anemia. **2.** in a dental casting, an opening in the investment through which the molten alloy or metal can reach the mold after the wax has been eliminated. *Engineering.* **1.** the hole through which molten metal or plastic is poured to pass through the mold gate and then into the mold. **2.** the waste piece that solidifies in the mold gate.

sprue bushing *Engineering.* a piece of steel in an injection mold that contains the sprue and supports an injection-cylinder nozzle.

sprue gate *Engineering.* the channel through which molten metal or plastic passes from a nozzle to a mold cavity.

sprung axle *Mechanical Engineering.* an axle, supported by a spring-suspension system, that carries the rear wheels of an automobile.

sprung weight *Mechanical Engineering.* the weight of a vehicle that is supported by a spring suspension.

spud *Mechanical Devices.* **1.** any of a number of digging or gouging tools with a short, blunt, often curved blade. **2.** a small shovel with a crowbar at one end used to dig out stumps. Also, STUMP SPUD. **3.** a short length of pipe that connects a meter to a supply line. **4.** a surgical instrument used to remove foreign objects such as dust from the eye. *Engineering.* a nail, similar to a horseshoe nail, marking a surveying station. *Naval Architecture.* a piling used to secure a scow or dredge in place.

Spumavirinae *Virology.* a subfamily of viruses of the family Retroviridae that infect mammals, usually resulting in persistent infection but no known disease. Also, FOAMY VIRUSES.

spunbonding *Materials Science.* a process for producing nonwoven polymer fabrics in which polymer melts are transformed into fabric in one continuous operation; continuous filaments are extruded in web form onto a collection belt and then consolidated. Polyolefin, polyester, and polyamides are spunbonded for commercial applications such as geotextiles, packaging materials, and medical fabrics.

spun silk *Textiles.* a silk yarn made of short lengths of silk wastes, rather than from continuous filaments.

spun yarn *Textiles.* a yarn made by spinning short natural or synthetic fibers into a continuous strand.

spur *Mechanical Devices.* a U-shaped device with projecting points, worn over the heel of a boot by the rider of a horse and used to goad the horse forward. *Civil Engineering.* see SPUR TRACK. *Botany.* **1.** a slender, tubular projection from a part of a flower, usually functioning in the production or storage of nectar. **2.** a short, projecting root of a tree. **3.** a short, fruit-producing tree branch. *Geology.* a subordinate ridge or rise projecting sharply from a larger elevation feature. *Hydrology.* see RAM. *Physics.* an aggregate of ionized molecules formed by a primary ionization process (due to an energetic charged particle passing through the cluster) and subsequent secondary ionization (due to the electrons which are released in the primary ionization). *Mathematics.* see TRACE, def. 1.

spur blight *Plant Pathology*. a common disease of raspberries and blackberries caused by the fungus *Oidymella applanata*, which is fatal to fruit spurs and produces dark markings on the cane.
spurge *Botany*. any of numerous plants of the genus *Euphorbia*, characterized by flowers with no petals or sepals and an acrid, milky juice that has purgative or medicinal properties. (From French *espurgier*, meaning "to purge.")
spur gear *Mechanical Devices*. a wheel or cog with teeth projecting radially and a face parallel with its axis.
spurious disk *Optics*. the apparent disk produced by diffraction when an optical system images a point source of light such as a star.
spurious modulation *Electronics*. any unwanted modulation occurring in an oscillator, such as frequency modulation due to mechanical vibration.
spurious radiation *Electromagnetism*. the electromagnetic radiation from a transmitting antenna with frequencies outside of the designated broadcasting band, including harmonics and parasitic emission. Also, **spurious emission.**
spurious reinforcement see ACCIDENTAL REINFORCEMENT.
spurious response *Electronics*. **1.** a radio receiver response to a frequency other than the one to which the receiver is tuned. **2.** an undesired response from an electric transducer or similar device.
spur line *Cartography*. in surveying, a secondary line of levels run from the main line in order to determine the elevations of stations not easily included in the main line or to connect the main line with previously established bench marks in order to check accuracy. *Civil Engineering*. see SPUR TRACK.
spurrite *Mineralogy*. $Ca_5(SiO_4)_2(CO_3)$, a light-gray monoclinic mineral occurring in granular masses or as anhedral crystals, having a specific gravity of 3.0 and a hardness of 5 on the Mohs scale; dimorphous with paraspurrite.
Spurr's medium *Microbiology*. a medium that contains vinylcyclohexene dioxide and is used to embed specimens for electron microscopic examination.
spur track *Civil Engineering*. a short secondary track that extends off the main track of a railroad. Also, SPUR, SPUR TRACK.
spur traverse *Cartography*. in surveying, any short traverse that extends from the established traverse for the purpose of reaching an additional point or position. Also, STUB TRAVERSE.
Sputnik [sput´nik] *Space Technology*. **1.** a Soviet unmanned artificial satellite that was launched on October 4, 1957; the first artificial satellite ever to achieve and maintain an earth orbit. **2.** any of a series of similar unmanned Soviet satellites of this same name launched from November 1957 to March 1961. (From a Russian word meaning literally "fellow traveler.")
sputter-coating *Microbiology*. a technique used to cover a specimen uniformly with a film of metal prior to examination by scanning electron microscopy.
sputtering *Materials Science*. a phenomenon that occurs when energetic ionized particles impinge on the surface of a solid or liquid target, causing the emission of particles and erosion of the surface of a solid; the sputtered target particles from the target can appear as charged or neutral atoms or molecules, atom clusters, or chunks; the controlled deposition of sputtered particles to form thin films and coatings has industrial applications in electronics, optics, and corrosion and wear-resistant coatings. *Electronics*. this phenomenon occurring in a cathode-ray tube, causing a dislocation of atoms or groups of atoms from the surface of the cathode.
sputum *Physiology*. the product of coughing and clearing the respiratory tract; may contain mucus, blood, pus, cellular debris, and microorganisms.
spyglass *Optics*. a popular name for a small, hand-held telescope.
SQ *Aviation*. the airline code for Singapore Airlines.
SQ subcutaneous.
sq. the following. (From Latin *sequens*.)
sq. or **sq** square.
sq. ft. or **sq ft** square foot; square feet.
sq. in. or **sq in** square inch; square inches.
sq. m. or **sq m** square meter; square meters.
sq. mi. or **sq mi** square mile; square miles.
sqq. the following (ones). (From Latin *sequentia*.)
sq. r. or **sq r** square rod; square rods.
squad *Military Science*. a military unit usually made up of 10 privates, a staff sergeant, and a corporal; the smallest military unit. There are usually four squads in a platoon.
squadron *Military Science*. **1.** the basic administrative aviation unit of the U.S. Air Force, Navy, Marine Corps, and Army. **2.** an organization consisting of two or more divisions of ships, or two or more Navy divisions or flights of aircraft, usually of the same type.
squalene *Biochemistry*. $C_{30}H_{50}$, an unsaturated hydrocarbon that is an intermediate in cholesterol biosynthesis by way of lanosterol; found in the liver oil of sharks and certain other fishes, as well as in other animals including humans.
Squalidae *Vertebrate Zoology*. the spiny dogfishes, a diverse worldwide family of small to moderate fish of the order Squaliformes, having small dorsal fins, and a body densely covered with small denticles.
Squaliformes *Vertebrate Zoology*. an order of mostly benthic marine fishes characterized by a subcylindrical body, a depressed to conical head, labial folds, and a groove at the mouth corners.
squall *Meteorologyy*. a sudden, violent wind that lasts only a few minutes and then has a sudden decrease in speed; in the U.S., a squall is reported only if a wind speed of over 16 knots is sustained for at least two minutes.
squall cloud *Meteorology*. a small cloud with an eddy circulation that may be formed between the updraft and downdraft currents beneath the leading edge of a thunderstorm.
squall line *Meteorology*. a nonfrontal line or narrow band of active thunderstorms, with or without squalls.
squama *Science*. a scale or scalelike part, as of epidermis or bone.
Squamariaceae see PEYSSONNELIACEAE.
Squamata *Vertebrate Zoology*. the order of reptiles containing lizards, snakes, and amphisbaenids, noted for extraordinary adaptive radiation in a wide range of habitats.
squames *Cell Biology*. stacks of dead, flattened cells that are filled with keratin filaments and occupy the outermost layers of the mammalian epidermis.
squamosal bone *Vertebrate Zoology*. a membrane bone found in the posterolateral region of the skull of most vertebrates, often contributes to jaw suspension. *Anatomy*. specifically in humans, the lateral flat portion of the temporal bone.
squamous *Biology*. covered with or composed of small scales; scaly.
squamous-cell carcinoma *Medicine*. a slow-growing carcinoma that grows from the squamous portion of the epithelium, forming a red, horny, painless nodule; it is often found in the lungs and skin, but also occurs in the larynx, nose, cervix, anus, bladder, and other sites. Also, **squamous carcinoma.**
squamous epithelium *Histology*. epithelial tissue consisting of one or more layers of thin, flattened cells.
squamulose *Biology*. a lichen growth form with numerous, small, loosely attached lobes.
square *Mathematics*. **1.** a quadrilateral having all four sides and all four angles equal; equivalently, a rectangle with equal sides or a rhombus with a right angle. **2.** the square of a quantity x is $x \cdot x = x^2$. For example, the square of 3 is 9 and the square of $(ax + b)$ is

$$(ax + b)^2 = a^2x^2 + 2abx + b^2.$$

Metrology. relating to or denoting a system of area measurement in which each of the sides is of the stated size; a square meter is the area covered by a square whose sides are each one meter long. *Mechanical Devices*. an L-shaped instrument used to draw sets of perpendicular lines, or to test surfaces for perpendicularity.
square bacteria *Microbiology*. any flat, frequently square-shaped bacterial cells that are typically found in hypersaline aquatic environments.
square centimeter *Metrology*. a unit of area equal to 0.0001 square meter, or 0.155 square inch.
square decimeter *Metrology*. a unit of area equal to 100 square centimeters, or 15.5 square inches.
square degree *Mathematics*. a unit of solid angular measure equal to the square of one (circular) degree; i.e., 1 square degree = $(\pi/180)^2$ steradian.
square dekameter *Metrology*. a unit of area equal to 100 square meters, or 119.6 square inches.
square dome see COVED VAULT.
squared rectangle *Mathematics*. a rectangle with integer length sides partitioned into a finite number $n \geq 2$ of squares, also of integer length sides. If the squares are all of different sizes, the rectangle is called perfect. Electrical circuit theory is used to find a partition into squares of a given rectangle with integer side lengths, and only certain such rectangles have partitions of this kind.

square-edged orifice *Engineering.* an orifice plate having sharp edges perpendicular to the orifice; used for measuring fluid flow from the differential pressure on the two sides of the orifice.

square engine *Mechanical Engineering.* an engine in which the stroke of the piston is equal to the cylinder bore.

square foot *Metrology.* a unit of area equal to 144 square inches, or 0.093 square meter.

square grade *Mathematics.* a unit of solid angular measure equal to the square of one (circular) grade; i.e., 1 square grade = $(\pi/200)^2$ steradian. Also, **square grad.**

square hectometer *Metrology.* a unit of area equal to 10,000 square meters, or 2.471 acres.

square inch *Metrology.* a unit of area equal to a square measuring one inch on each side, equivalent to 6.452 square centimeters.

square-jaw clutch *Mechanical Engineering.* a positive clutch having square- or perpendicular-sided teeth that mesh together when aligned.

square key *Mechanical Devices.* a machine key whose cross section is square.

square kilometer *Metrology.* a unit of area equal to 100,000 square meters, or 0.3861 square mile.

square-law detector *Electronics.* a detector whose output voltage is proportional to the square of the radio-frequency input voltage; operation of this type of circuit is dependent on nonlinearity of the detector characteristic, rather than on rectification. Also, **square-law demodulator.**

square league *Metrology.* a historic measure of distance, equal to 3 square miles.

square-loop ferrite *Electromagnetism.* a ferrite material whose hysteresis loop is approximately rectangular.

square-loop magnet *Materials Science.* a magnetic property of a material whose remnant induction and saturation induction are equal, resulting in a square-loop hysteresis diagram; such materials are important where it is necessary to magnetize easily to a certain value and retain most of the flux density when the power is off, so that it can be used as "memory" to activate the controls when called on, as in computers.

square matrix *Mathematics.* a matrix with the same number n of rows and columns. Also, $\boldsymbol{n \times n}$ **matrix.**

square mesh *Mechanical Devices.* a wire-cloth mesh of equal squares vertically and horizontally; used in textile work.

square meter *Metrology.* a unit of area equal to 10,000 square centimeters, or 1550 square inches.

square mile *Metrology.* a unit of area equal to 640 acres or 2.59 square kilometers.

square millimeter *Metrology.* a unit of area equal to 0.01 square centimeter, or 0.002 square inch.

squareness ratio *Electromagnetism.* for a magnetic material in a symmetrically cyclically magnetized state, the ratio of the flux density at zero magnetizing force to the maximum flux density.

square rod *Metrology.* a unit of area equal to 30.25 square yards, or 25.29 square meters.

square root *Mathematics.* given a quantity x, a quantity r, if it exists, such that $r^2 = x$.

square root law *Statistics.* a law stating that the standard deviation of the ratio of the number of successes to the number of trials is inversely proportional to the square root of the number of trials.

square search *Navigation.* a search conducted over a pattern of courses that cover an expanding square; it may be centered on a fixed or moving point.

square-serif *Graphic Arts.* a typeface or character having serifs that are as heavy as or heavier than the main strokes.

square set *Mining Engineering.* a set of timbers consisting of a cap, girt, and post that meet to form a 90° angle; they are framed at their intersection to form a compression joint with three other sets.

square thread *Design Engineering.* a screw thread having a square cross section and a width that is equal to the pitch or distance between threads.

square wave *Electricity.* a periodic wave that alternately assumes two different fixed values for equal lengths of time; the transition time is negligible in comparison with the duration of each fixed value.

square-wave amplifier *Electronics.* a resistance-coupled amplifier (in effect, a wideband video amplifier) that is used to amplify a square wave with a minimum amount of distortion.

square-wave generator *Electronics.* a signal generator that generates a square-wave or rectangular-wave output voltage.

square wheel *Mechanical Devices.* a wheel with a flat spot on its rim.

square yard *Metrology.* a unit of area equal to 9 square feet, or 0.836 square meter.

squaring circuit *Electronics.* **1.** a circuit that changes a sine or other wave into a square wave. **2.** a circuit that contains nonlinear elements and generates an output voltage proportional to the square of the input voltage.

squaring shear *Mechanical Engineering.* a machine tool used to cut sheet metal or plate, consisting of a fixed cutting knife and another cutting knife mounted on the front of the reciprocally moving crosshead.

squaring the circle *Mathematics.* the problem of constructing, with compass and straightedge alone, a square having an area exactly equal to that of a given circle. Also, **quadrature of the circle.** It is one of a collection of impossible constructions attempted by the ancient Greeks; others include duplicating the cube and trisecting the angle.

s quark see STRANGE QUARK.

squarrose *Botany.* of or related to a plant that is rough or ragged, with projecting scales or rigid leaves; describing especially a plant having stiff, spreading bracts.

squarrulose *Botany.* somewhat squarrose or tending to become squarrose.

squash *Botany.* **1.** any of several vinelike, tendril-bearing, annual plants of the genus *Curcurbita* of the gourd family, such as the **butternut squash,** *C. moschata,* or the **acorn squash,** *C. pepo.* **2.** the fruit of such a plant, eaten as a vegetable or made into a pie.

squash bug *Invertebrate Zoology.* a large, foul-smelling, dark-brown bug, *Anasa tristis,* that sucks the sap from leaves of squash, pumpkin, and other plants of the gourd family; found in North America.

Squatinidae *Vertebrate Zoology.* a monogeneric family of sharklike, cartilaginous, benthic fish of tropical and warm temperate seas, composing the order Squatiniformes and the superorder Squatinomorphii.

squatting *Naval Architecture.* the lowering of the stern of a vessel moving at high speed, due to the vessel's tendency to draw more water astern when in motion than when stationary.

squawfish *Vertebrate Zoology.* any of the edible species of the genus *Ptychocheilus* in the family Cyprinidae that are related to carps and minnows; voracious carnivores and important sportfish found in western U.S. and Canada. The **Colorado squawfish,** *P. lucius,* is an endangered species.

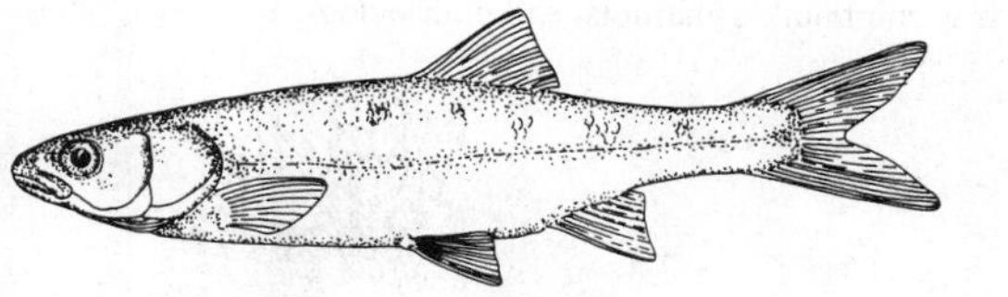

squawfish

squawk *Transportation Engineering.* the response signal emitted by an aircraft transponder in response to an air traffic control radar impulse.

squealing *Electronics.* in a radio receiver, a high-pitched sound that occurs along with the desired signal; usually caused by interference between stations or oscillations in some receiver circuit.

squeegee [skwē´jē´] *Mechanical Devices.* a T-shaped implement with a long rubber or leather piece across its head, used for pressing or wiping liquids across a surface such as the windshield of an automobile.

squeezable waveguide *Electromagnetism.* a waveguide whose dimensions can be altered periodically.

squeeze *Engineering.* **1.** to force grout into a bore hole under high pressure. **2.** the expansion of soft rock in the walls of a borehole or mine working that reduces the diameter of the opening. *Mining Engineering.* **1.** the settling of a mine roof, without breakage, over a considerable working area. **2.** the gradual upheaval of a mine floor resulting from the weight of overlying strata. **3.** see CREEP.

squeeze-bore see TAPERED-BORE.

squeeze casting *Materials Science.* a permanent-mold process that combines the features of casting and forging in one operation, in which solidification is accomplished under high pressure.

squeeze roll *Mechanical Engineering.* a machine that shapes material by passing it between two rollers.

squeeze section *Electromagnetism.* a section of waveguide whose critical dimensions can be altered.

squeeze time *Metallurgy.* in resistance welding, the time between application of pressure and application of current.
squegging *Electronics.* a self-blocking condition in an electron-tube-oscillator circuit.
squelch *Electronics.* to quiet a receiver by reducing its gain in direct response to a specified input characteristic.
squelch circuit see NOISE SUPPRESSOR.
squib *Ordnance.* a small, low-explosive pyrotechnic device; it is usually electrically initiated and may be used to fire a rocket or for some similar purpose.
squibload *Ordnance.* **1.** a defective cartridge or shell that does not reach a normal level of pressure. **2.** a delay in the firing of a cartridge or shell.
squid *Invertebrate Zoology.* the popular name for various carnivorous marine cephalopod mollusks belonging to the family Loliginidae of the order Teuthoidea, having a torpedo-shaped body, ten suckered arms around the mouth, a vestigial internal shell, an ink sac, and a pair of posterior fins; the common **Atlantic long-finned squid,** *Loligo pealeii,* is sometimes used as food; the giant squid may attain a length of over 50 feet and is often cited as the origin of legends about sea serpents.
Squid *Ordnance.* a British ship-based, medium-range, antisubmarine mortar system that fires three charges in a three-dimensional pattern ahead of the target.
SQUID superconducting quantum interference device.
Squillidae *Invertebrate Zoology.* a family of predatory, bottom-dwelling, burrowing mantis shrimps, malacostracan crustaceans in the order Stomatopoda, up to 20 cm in length.
squint *Electromagnetism.* **1.** the angle between the full-right and full-left positions in a conical-scan radar beam. **2.** the angle between two major lobes in a lobe-switching antenna. **3.** the angle between the lobe axis and a selected characteristic axis such as a reflector axis. *Medicine.* another term for strabismus, an abnormal condition of the eye in which there is noncoincidence of the optic axes, so that both eyes do not focus on the same object at the same time.
squirrel *Vertebrate Zoology.* **1.** any of numerous small, agile, aboreal, nocturnal rodents of the families Sciuridae and Anomaluridae, characterized by a bushy tail and long strong hindlimbs; found nearly worldwide. Among those common in North America are the **Eastern gray squirrel** and **Western gray squirrel,** *Sciurus carolinensis* and *S. griseus*, and the **Red squirrel**, *Tamiasciurus hudsonicus*. **2.** any of various other members of the family Sciuridae other than the true squirrels, such as chipmunks, marmots, and prairie dogs.

gray squirrel

squirrel cage *Electricity.* a rotor constructed of bars embedded around the periphery of a laminated, iron-core rotor and joined at each end by continuous rings; the most common type of induction-motor rotor due to its inherent low cost. Also, **squirrel-cage rotor.**
squirrel-cage motor *Electricity.* an induction motor with a squirrel-cage winding. Also, **squirrel-cage induction motor.**
squirrel-cage winding *Electricity.* a permanently short-circuited winding usually uninsulated around the periphery of the rotor and joined by continuous end rings.
squirrelfish *Vertebrate Zoology.* any of several bright-colored, nocturnal fishes of the family Holocentridae, characterized by large eyes and sharp spines and scales; found in shallow waters of tropical reefs, especially *Holocentrus ascensionis* of the West Indies. (Possibly named for the sound similar to a squirrel's call that it makes out of the water.)
squirrel monkey *Vertebrate Zoology.* either of two small, long-tailed monkeys, the endangered *Saimiri oerstedii* of Central America and *S. sciureus* of South America, having a bushy, nonprehensile tail, a small white face with a black muzzle, and brown, gold, or greenish fur.
squirt can *Engineering.* a flexible can from which a fluid, such as oil, may be forced through the spout by pressing the bottom of the can.
squirt gun *Engineering.* a hand device with a nozzle, from which liquid is squirted by squeezing a bulb or pressing a piston.
squitter *Electronics.* in radar, the random firing of the transponder transmitter in the absence of interrogation.
sq. yd. or **sq yd** square yard; square yards.
SR *Aviation.* the airline code for Swissair.
SR sedimentation rate; stimulation ratio; sigma reaction.
S-R or **SR** *Behavior.* stimulus-response.
Sr the chemical symbol for strontium.
sr steradian.
SRAM short-range attack missile.
s rays see GOLDSTEIN'S RAYS.
SRBC sheep red blood cell.
SRBM short-range ballistic missile.
SRC *Oncology.* a family of proto-oncogenes (v-src, c-src, e-erb) that produce a protein product that stimulates protein-tyrosine-kinases located at the cell surface or in the cytoplasm.
SRC stored response chain.
src gene *Genetics.* an oncogene of the Rous retrovirus, which causes sarcomalike tumors in chickens.
SRF skin reactive factor; somatotropin-releasing factor.
SRF or **S.R.F.** Smithsonian Research Foundation.
SRH somatotropin-releasing hormone.
SRIF somatostatin.
Sri Lanka see CEYLON.
SRMS structure resonance modulation spectroscopy.
sRNA soluble RNA.
4S RNA *Molecular Biology.* an RNA molecule that transfers an amino acid to a growing polypeptide chain during translation; among the smallest known biologically active nucleic acids.
5S RNA *Molecular Biology.* an RNA molecule composed of 120 nucleotides; a component of a large ribosomal subunit.
5.8S RNA *Molecular Biology.* an RNA molecule composed of 120 nucleotides that functions as a component of the nucleolus organizer.
SRO *Microbiology.* any of a number of bacterial strains of *Spiroplasma* that infect species of the fly *Drosophila,* causing infected female flies to produce only female progeny. (An abbreviation for sex-ratio organism.)
SRP signal-recognition particle or signal-recognition protein.
SRQ *Aviation.* the airport code for Sarasota/Bradenton, Florida.
S-R theory *Behavior.* a theory that views behavior as a series of responses to specific kinds of stimuli.
SS *Ordnance.* the official designation for a Soviet missile series ranging from SS-1 to SS-25; see the code names such as SABER and STILETTO.
SS steamship; stainless steel.
ss. or **ss** one-half. (From Latin *semis.*)
SS.11 *Ordnance.* a French surface-to-surface tactical missile powered by a two-stage, solid-propellant rocket motor; equipped with wire guidance and having a cruising speed of 360 mph and a range of 1650 to 10,000 feet; it can be equipped with an armor-piercing warhead, a high-explosive, semi-armor-piercing warhead, or a high-fragmentation, antipersonnel warhead.
SS.12 *Ordnance.* a French surface-to-surface tactical missile that is a larger version of the SS.11 and can carry a similar range of warheads; it has a speed of 425 mph and a range of 20,000 feet.
SS-16 *Ordnance.* a Soviet three-stage intercontinental ballistic missile powered by a solid-rocket motor and capable of delivering a single one-megaton warhead at a range of 5600 miles.
SS-22 *Ordnance.* a Soviet short-range ballistic missile that delivers one 500-kiloton nuclear warhead at a range of 600 miles.
SS 433 *Astronomy.* a highly reddened stellar object lying amidst a supernova remnant, which is identified with a variable nonthermal radio source, an infrared source, and an X-ray source known as A 1909+04; it is a binary star system that has undergone extensive evolution. Also, **Stephenson-Sanduleak 433.**

SSA or **S.S.A.** Seismological Society of America.
SS agar *Microbiology.* a medium that is used to select for and differentiate between the two bacteria *Salmonella* and *Shigella.*
SSB single sideband.
SSB protein single-strand binding protein.
ssc *Molecular Biology.* sister-strand crossover exchange of segments between sister chromatids.
SSC-1 *Ordnance.* a Soviet ground-launched cruise missile designed as a coastal defense weapon, powered by a turbojet engine and two solid-propellant boosters, and equipped with infrared or active radar homing; it carries a one-kiloton nuclear or 1000-pound conventional warhead at a speed of Mach 1.5 and a range of 280 miles.
SSC-2 *Ordnance.* a Soviet subsonic ground-launched cruise missile powered by a turbojet cruise motor, equipped with radar homing, and designed to deliver a conventional warhead.
SSC-2A *Ordnance.* a version of the SSC-2 missile used as a tactical weapon by the Russian Army.
SSC-2B *Ordnance.* a version of the SSC-2 missile used as a coastal defense weapon.
SSC-4 *Ordnance.* a Soviet ground-launched cruise missile with a range of 3000 km.
ss(c)DNA a single-stranded circular DNA.
SSD source-skin distance; steady-state distribution.
ssDNA single-stranded DNA.
SSE or **S.S.E.** south-southeast. Also, **s.s.e.**
SSI small-scale integration.
SSIE or **S.S.I.E.** Smithsonian Scientific Information Exchange.
SSM surface-to-surface missile.
SS-N-1 *Ordnance.* a Soviet ship-based cruise missile that delivers a conventional warhead at a range of 150 miles.
SS-N-2 *Ordnance.* a Soviet surface-to-surface cruise missile powered by a sustainer and a solid-propellant booster that can be jettisoned; believed to be equipped with radar homing, delivering a conventional warhead at an estimated range of 30 miles.
SS-N-3 *Ordnance.* a ship-based version of the Soviet SSC-1 missile.
SS-N-4 *Ordnance.* a Soviet two-stage missile for launching from submarines or surface ships against surface targets; powered by solid-propellant rocket motors and delivering a nuclear warhead at a range of 380 miles.
SS-N-5 *Ordnance.* a Soviet two-stage ballistic missile for launching from submarines or surface ships against surface targets; powered by solid-propellant rocket motors and delivering a one-megaton nuclear warhead at a range of 750 miles.
SS-N-6 *Ordnance.* a Soviet two-stage ballistic missile for launching from submarines or surface ships against surface targets; powered by solid-propellant rocket motors and delivering a nuclear or thermonuclear warhead at a range of 1750 miles.
SS-N-7 *Ordnance.* a Soviet underwater-launched cruise missile equipped with autopilot and terminal homing, having a cruising speed of Mach 1.5 and a range of 35 miles.
SS-N-8 *Ordnance.* a Soviet submarine-launched ballistic missile powered by a liquid-propellant rocket motor and equipped with inertial guidance, delivering a one-megaton or 800-kiloton nuclear warhead at a range of 4800 miles.
SS-N-9 *Ordnance.* a Soviet sea-launched medium-range cruise missile equipped with autopilot and radio-command guidance, delivering a conventional warhead at a speed of Mach 1.4 and a range of 175 miles.
SS-N-10 *Ordnance.* a Soviet sea-launched, short-range cruise missile equipped with radar guidance and antiradiation passive homing, having a speed of Mach 1.2 and a range of 30 miles.
SS-N-11 *Ordnance.* a Soviet sea-launched, short-range cruise missile powered by a solid rocket booster and equipped with active radar homing, having a speed of Mach 0.9 and a range of 30 miles.
SS-N-12 *Ordnance.* a Soviet submarine-launched cruise missile powered by a turbojet engine, having a speed of Mach 2.5 and a range of 1550 miles.
SS-N-13 *Ordnance.* a Soviet submarine-launched, antiship ballistic missile powered by a solid rocket engine and having a range of 625 miles.
SS-N-14 *Ordnance.* a Soviet sea-launched cruise missile.
SS-N-15 *Ordnance.* a Soviet sea-launched missile.
SS-N-17 *Ordnance.* a Soviet submarine-launched ballistic missile that delivers a one-megaton nuclear warhead with a relatively short range.
SS-N-18 *Ordnance.* a Soviet submarine-launched ballistic missile that carries three to seven nuclear warheads.
SS-N-19 *Ordnance.* a Soviet sea-launched missile.
SS-N-20 *Ordnance.* a Soviet submarine-launched ballistic missile that carries six to nine nuclear warheads at a range of 5000 miles.
SSP static spontaneous potential.
ssp or **ssp.** subspecies.
SSPE subacute sclerosing panencephalitis.
SSR solid-state relay.
ssRNA single-stranded RNA.
SSS specific soluble substance.
s.s.s. layer upon layer. (From Latin *stratum super stratum.*)
SSSA or **S.S.S.A.** Soil Sciences Society of America.
SST supersonic transport.
S star *Astronomy.* a red giant star with irregular long periods of variation; exhibits strong bands of heavy-metal molecules such as zirconium.
s-state *Quantum Mechanics.* a state of a single particle whose orbital angular momentum quantum number is zero.
SSV under a poison label. (From Latin *sub signo veneni.*) Also, **S.S.V., s.s.v.**
SSW or **S.S.W.** south-southwest. Also, **s.s.w.**
ST or **S.T.** standard time.
S.T. or **s.t.** short ton.
St. or **St** street; strait. Also, **st.**
st. stone (weight); stratum.
STA serum thrombotic accelerator.
sta. station; stationary.
stab culture *Microbiology.* a culture in which a tube of solid medium is stabbed with a wire or needle covered with the bacterial inoculum. Also, NEEDLE CULTURE.
stabilate *Microbiology.* a frozen or freeze-dried preparation of viable microorganisms.
stabilator *Aviation.* a movable horizontal tail that combines the actions of a stabilizer and elevator, increasing longitudinal stability while creating a pitching moment.
stability the fact of being stable; steadiness, balance, consistency, and so on; specific uses include: *Engineering.* **1.** the ability of a structure to maintain its intended position and orientation. **2.** the tendency of a vehicle such as a ship or aircraft to maintain its normal orientation, or to return to that orientation when disturbed. *Mechanics.* a condition in which a dynamical system slightly displaced from its equilibrium configuration always tends to return to this configuration. *Fluid Mechanics.* the ability to restore to the original state in the presence of either a pulsed or periodic perturbation. *Chemistry.* the tendency of a species to remain in a particular state or condition without changing spontaneously. *Geology.* **1.** the extent to which a structure, slope, or clay embankment resists sliding, collapsing, or other conditions of stress. **2.** the extent to which a structure resists weathering or chemical alteration. **3.** the extent to which the earth's crust resists deformation or mountain-building forces. **4.** the ability of a mineral to resist chemical alteration or destruction. *Electronics.* in a power-generating system in which two or more synchronous machines are connected through an electrical network, the condition in which the relative angular positions of the machines' rotors either remain constant or (following a disturbance) become constant. *Control Systems.* the property of a system that results in a bounded output signal for any bounded input signal. *Robotics.* an absence of oscillation in the end effector of a robot.
stability augmentation system *Aviation.* in a manual control system, an auxiliary subsystem that enables a pilot to input responses to the control devices, thus providing a preselected vehicle response through the selection of variable gains in a standard feedback loop on control-surface output; most commonly found on modern fighter aircraft. Also, **stability augmentors.**
stability chart *Meteorology.* a synoptic weather chart that highlights the stability index distribution.
stability constant *Chemistry.* an equilibrium constant for the formation of complex ions, used to calculate equilibrium concentrations of the complex and other reaction species.
stability criterion *Control Systems.* any necessary and sufficient condition for a system to be stable.
stability exchange principle *Control Systems.* a principle stating that the complex frequency in a linear system varies with the parameter so that the real and imaginary parts pass through zero simultaneously.
stability factor *Electronics.* at any given point in an electrical system, the ratio of the stability limit to the nominal power flow, expressed as a specified quantity such as maximum load.
stability index *Meteorology.* the indication of local static stability of a layer of air.

stability limit *Electronics.* the maximum power that can flow through a given point in an electrical system when the system is stable.
stability matrix see STIFFNESS MATRIX.
stabilivolt *Electronics.* a device consisting of a gas-filled tube containing a number of concentric, coated iron electrodes; designed to provide constant voltage for an electrical apparatus operating at low current.
stabilization the process of making or becoming stable; specific uses include: *Chemical Engineering.* a petroleum-refinery process that removes dissolved gases from liquid hydrocarbons by fractional distillation, thereby reducing the vapor pressure. *Materials Science.* the addition of titanium or niobium to stainless steels; the titanium and niobium combine preferentially with carbon, thus preventing the precipitation of chromium carbides and the resulting loss of chromium in solid solution which leads to intergranular corrosion. *Electromagnetism.* the treatment of a magnetic material in order to increase its magnetic retentiveness. *Electronics.* feedback that is introduced into electron-tube or transistor-amplifier stages in order to reduce distortion, by making the amplification substantially independent of electrode voltages and tube constants. *Control Systems.* see COMPENSATION.
stabilize to make stable; specific uses include: *Aviation.* **1.** to provide an aircraft or other vehicle with stability. **2.** to fit out a piece of equipment, such as a gyroscope or radar antenna, so that it maintains a desired attitude regardless of the orientation of the vehicle in which it is located.
stabilized winding *Electricity.* an auxiliary winding used in star-connected transformers to stabilize the neutral point of the fundamental frequency voltage or to protect the transformer and the system from excessive third-harmonic voltage. Also, TERTIARY WINDING.
stabilized zirconia *Materials.* zirconia with materials such as lime added to stabilize the zirconia at all temperatures, which causes good electrical insulation, resistance to acids and bases, and a low coefficient of expansion.
stabilizer a substance, structure, or device that makes something stable; specific uses include: *Materials.* any substance added to another substance, compound, or emulsion to prevent deterioration, decomposition, or loss of specific properties. *Chemical Engineering.* a petroleum-refinery chemical added to oil to neutralize undesirable effects such as oxidation and discolorization or the fractionation column used to remove gases for stabilization. Also, INHIBITOR, PROTECTIVE AGENT. *Aviation.* a usually fixed airfoil, or combination of airfoils, used to provide stability for an aircraft; the term usually denotes a horizontal stabilizer unless a vertical stabilizer is specified. *Naval Architecture.* any device, active or passive, designed to reduce rolling or pitching, such as antiroll fins or tanks. *Engineering.* a hardened, splined bushing that extends slightly beyond the outer diameter of a core barrel back head mounted below it.
stabilizing factor *Ecology.* an environmental factor that enhances the stability of a system.
stabilizing magnetic field *Physics.* a magnetic field that is provided for a plasma device so as to add stability to the contained plasma.
stabilizing selection *Evolution.* a process of natural selection in which genetic variation is selected against, resulting in a population from which ill-adapted or peripheral variants are eliminated. Also, CENTRIPETAL SELECTION. *Genetics.* any natural selection that reduces the genetic variability of a population by removing alleles that cause deviation from the mean phenotype.
stabilizing treatment *Metallurgy.* any of several thermal treatments that stabilize a structure, thereby assuring dimensional stability and optimum properties in service.
stabistor *Electronics.* a diode component incorporating closely controlled conductance, controlled storage charge, and low leakage, as required for clippers, clamping circuits, bias regulators, and other logic circuits that demand tight voltage-level tolerances.
stable not likely to fall, give way, change, fail, and so on; steady or consistent; specific uses include: *Physics.* **1.** a condition in which a system is not subjected to any change due to external influences. **2.** describing particles that are incapable of spontaneous radioactive decay. *Mechanics.* relating to or displaying stability. *Physiology.* describing a condition in which vital processes such as respiration, pulse rate, temperature, and blood pressure are within an acceptable range and are not undergoing significant change. *Medicine.* a patient status in which the vital signs are normal and there is no significant change in the patient's condition. *Computer Science.* of a sorting algorithm, preserving the relative order of records with equal key values.
stable *Agriculture.* a farm building where animals are housed and fed, especially one having stalls or compartments.
stable age distribution *Ecology.* the age distribution attained by a population in a constant, stable environment, with no disproportionate number of younger or older individuals.
stable element *Engineering.* a component, typically a gyroscope, of a rotating or revolving assembly such as a gun or radar mounted on a vessel to maintain a constant orientation for reference. *Aviation.* specifically, any such device used to stabilize equipment mounted in a flight vehicle.
stable equilibrium *Mechanics.* the condition of a system in equilibrium, such that any net force or torque that the system experiences will eventually restore the system to its original equilibrium state.
stable homeomorphism conjecture *Mathematics.* let $h: R^n \to R^n$ be any orientation-preserving homeomorphism from R^n to itself. The stable homeomorphism conjecture asserts that each such h can be expressed as the composition of homeomorphisms on R^n, each of which is the identity homeomorphism on some nonempty open subset of R^n.
stable isobar *Nuclear Physics.* one of two or more stable nuclides having the same mass number but differing in atomic number.
stable isotope *Nuclear Physics.* an isotope that does not undergo radioactive decay.
stable isotope analysis *Ecology.* a technique using a stable isotope as a marker to quantitatively or quasi-quantitatively measure the movement of certain materials in biological processes. Isotopes are atoms of the same element having the same chemical properties but differing in mass and mass-dependent physical properties.
stable nucleus *Nuclear Physics.* a nucleus that does not undergo radioactive decay.
stable orbit see EQUILIBRIUM ORBIT.
stable oscillation *Electrical Engineering.* a time-varying parameter that is bounded and repeats itself indefinitely.
stable platform *Aviation.* a gyroscopic mechanism used to maintain a desired orientation in space, independent of craft motion.
stable population *Ecology.* a population in which the birth rate and death rate are approximately equal.
stable strobe *Electronics.* a sequence of strobes that behave as if caused by a single jammer.
stable vertical *Engineering.* a stable element in a tilting assembly that maintains a constant vertical orientation.
stachybotryotoxicosis *Pathology.* a fungus poisoning observed in cattle and horses, caused by eating fodder overgrown by the fungus *Stachybotoys artra;* humans may be affected by inhaling the toxin from hay or absorbing it into the skin.
Stachybotrys *Mycology.* a genus of fungi belonging to the class Hyphomycetes, characterized by dark spore-bearing structures.
Stachyuraceae *Botany.* a monogeneric family of dicotyledonous, tanniferous shrubs and small trees of the order Violales, containing proanthocyanins and ellagic acid; native to Japan.
stack a large, generally vertical structure, device, or feature; specific uses include: *Engineering.* **1.** a chimney, pipe, or other such vertical exhaust channel, especially a large one. **2.** a collection of items placed on a single column in a more or less orderly pile. **3.** to place a group of items in this fashion. *Building Engineering.* the part of a chimney that extends above the roof. *Metallurgy.* the upper portion of a blast furnace where the charge is preheated and prereduced. *Geology.* **1.** an isolated, rocky island or columnar mass of bedrock projecting above the ocean surface near a headland or cliffy shore from which it has been separated by wave erosion. Also, SEA STACK, MARINE STACK. **2.** a steep-sided rocky mass on a slope or hill that rises above its surroundings on all sides. Also, ROCK STACK. *Aviation.* see AIR STACK. *Computer Programming.* an area of storage reserved for the temporary storage of data, usually a linear list in which all insertions and deletions take place at one end, such that the last item added is the first item removed (access on a last-in, first-out basis).
stack burn *Food Technology.* in food storage, discoloration or other damage that occurs when newly processed cans are stacked too close together without sufficient space for ventilation and cooling.
stacked array *Electromagnetism.* an array in which the antennas are arranged one above the other and connected in phase to increase gain. Also, **stacked antennas.**
stacked-beam radar *Engineering.* a radar system in which separate beams are sent out at the same azimuth but at different elevation angles, thus providing altitude readings of targets.
stacked-job processing *Computer Programming.* a processing method in which multiple independent jobs are placed end-to-end and input into a system as a continuous job stream.

stacked loops *Electromagnetism.* two or more loop antennas arranged one above another on a vertical supporting structure and connected in phase to increase gain.

stacked plate reactor *Biotechnology.* a reactor designed to promote large-scale growth of mammalian cells; circular glass or stainless steel plates are attached vertically to a rotating central shaft that is immersed in a growth medium.

stack effect *Mechanical Engineering.* a pressure difference created by the impulse of a heated gas to rise in a vertical passage, as in a chimney or a small enclosure.

stacker *Mechanical Engineering.* a machine that may be hand, electrically, or hydraulically operated to lift objects on a platform or fork and arrange them in layers.

stacker-reclaimer *Mechanical Engineering.* the machinery that transports and forms stockpiles of industrial materials, then recovers and transports the materials to processing plants.

stack frame see ACTIVATION RECORD.

stack gas *Engineering.* the exhaust of gases emitted through a stack or chimney.

Stackhousiaceae *Botany.* a family of dicotyledonous, tanniferous, and rhizomatous herbs of the order Celastrales, characterized by small, fleshy or leathery leaves, a cupular hypanthium lined by a thin nectary disk, and seeds having an oily endosperm; native to Australia and New Zealand.

stacking any process in which items are arranged vertically one above the other; specific uses include: *Electromagnetism.* the positioning of antennas or array elements one above another and connected in phase so as to increase the gain of the system. *Molecular Biology.* the lining up of neighboring planar purine and pyrimidine bases in a DNA double helix.

stacking fault *Materials Science.* a surface defect in face-centered, cubic and hexagonal close-packed metals caused by the improper stacking sequence of close-packed planes; can result from the passing of a partial dislocation.

stacking-fault strengthening *Materials Science.* a mechanism in the precipitation strengthening of an alloy that involves forces arising from differences between the stacking-fault energies of the precipitate and matrix phases.

stacking sequence *Crystallography.* the sequence in which close-packed planes are stacked; if the sequence is ABABAB, i.e., the atoms in the third plane are above those in the first plane, a hexagonal close-packed crystal structure is produced; if the sequence is ABCABCABC, a face-centered cubic structure is produced.

stack operation *Computer Programming.* **1.** the insertion or deletion of items in a stack. **2.** a type of central processing unit organization in which the top two locations of the stack are processor registers and the rest of the stack is stored in memory.

stack pointer *Computer Programming.* a special central processing unit register containing the address of the top item in a stack.

stack pollutants *Engineering.* a term for various polluting gases, particulates, or other materials that are emitted from a stack or chimney.

stack processsing *Computer Programming.* a processing method that uses a pushdown stack implemented in hardware as memory for programs and data; permits addressless instructions to be used and makes programs more compact.

stack welding *Metallurgy.* the process of simultaneously resistance spot welding a stack of sheets.

stactometer see STALAGMOMETER.

STADAN *Space Technology.* a system of ground stations, controlled by the National Aeronautics and Space Administration, that receive information about the location and movement of unmanned satellites with self-registering devices. (An acronym for Space Tracking and Data Acquisition Network.)

stade *Geology.* a substage of a glacial stage represented by a readvance of glaciers. Also, STADIA.

Stader splint *Veterinary Medicine.* a splint used for external-internal fixation of fractures, made of metal bars with pins fixed at right angles to the bars for insertion into the bone fragments of a fracture to maintain their position. Also, **Stader's splint.**

stadia *Cartography.* **1.** in surveying, a graduated rod used in trigonometric methods of determining distances, by forming a side of known length in a right triangle between the observer and the rod. Also, **stadia rod. 2.** of or relating to any surveying procedure making use of a stadia rod. Thus, **stadia survey, stadia distance, stadia method.** *Geology.* see STADE.

stadia constant *Cartography.* **1.** in surveying, a numerical constant multiplied against the stadia interval to give the distance in meters from the stadia rod to the focal point of the observing instrument. **2.** a numerical constant multiplied against the sum of the stadia intervals for all sights in a running to give the length in kilometers of the running.

stadia hairs *Engineering.* two horizontal reference lines in the focal plane of a stadia telescope, set at such a distance apart that they subtend a particular known angle at the eye of the observer. Also, **stadia wires.**

stadia interval *Cartography.* in surveying, the vertical length of the stadia rod visible between the top and bottom crosswires of the viewing instrument.

stadial moraine see RECESSIONAL MORAINE.

stadia tables *Engineering.* tables by which the apparent dimensions observed through a transit scope are converted into distance values.

stadimeter *Engineering.* a survey instrument by which the distance of an object may be read if its height is known.

stadiometry *Robotics.* a method of finding the distance to an object by measuring the size of the image.

Stadler, Lewis 1896–1954, American geneticist; a pioneer in the study of mutations.

staff *Military Science.* a group of officers who do not themselves have command authority, but assist a commanding officer.

Staffellidae *Paleontology.* a family of large foraminiferids in the superfamily Fusulinacea; appeared in the Lower Carboniferous and persisted into the Permian.

staff gauge *Engineering.* a river gauge, consisting of a vertical staff bearing water-level markings.

Staffordian *Geology.* a European geologic stage of the Middle Upper Carboniferous period, occurring after the Yorkian and before the Radstockian.

staff sergeant *Military Science.* in the U.S. Army, Air Force, and Marine Corps, a noncommissioned officer ranking immediately above a sergeant.

stag *Vertebrate Zoology.* an adult male deer, or the male of various similar animals.

stage a single step or interval in some ongoing process; specific uses include: *Transportation Engineering.* a portion of a trip made in a single aircraft; distinguished from a leg or segment by the need for passengers to change planes. *Geology.* **1.** a phase in a cycle of erosion marked by distinct developments of landscape features. **2.** an interval of time in the historical development of a geologic feature. **3.** a primary subdivision of a glacial epoch. **4.** a time-stratigraphic division of a series that represents the rocks formed during a geologic age. *Hydrology.* the elevation or height of a water surface above a given plane. Also, GAUGE HEIGHT. *Aviation.* a step or process through which a liquid fuel or other fluid passes during compression or expansion. *Space Technology.* **1.** a propulsion unit of a rocket, especially one of the units of a multistage rocket, containing its own fuel and tanks and capable of being jettisoned upon burnout. **2.** in a rocket's flight, the period during which such a unit is operative. **3.** of such a rocket propulsion unit, to disengage from the remaining body in order to move along its own flightpath. *Electronics.* **1.** a circuit containing a single section of an electron tube or equivalent device, or two or more similar sections connected in parallel, push-pull, or push-push. **2.** one step of an amplifier, especially if part of a several-step process. *Mining Engineering.* **1.** a landing or platform. **2.** a thin, narrow dike, particularly one composed of soft material. *Optics.* the platformlike part of a microscope on which an object is examined.

stage-and-a-half *Space Technology.* a liquid propellant rocket that jettisons only part of its propulsion unit during flight, as when booster rockets fall away but leave the sustainer engine to consume the remaining fuel.

stage-discharge curve *Hydrology.* a graph illustrating the relationship between river stage and volume of discharge. Also, RATING CURVE, DISCHARGE-RATING CURVE.

stage gain *Electronics.* the ratio of the output power of an amplifier stage to the input power, usually expressed in decibels.

stage loader see FEEDER CONVEYOR.

stage screw *Mechanical Devices.* a tapered screw resembling a corkscrew, used to fasten stage braces, framing, or posts to a floor.

stage theory *Physiology.* a theory of color vision that proposes the existence of three types of cone receptors interacting at some stage between the retina and visual cortex of the cerebrum, such that strong signals from one of the receptors inhibits signals from the others.

staggered cut *Molecular Biology.* a product that results from breaking the two strands of duplex DNA at different locations.

staggered ends *Genetics.* sticky ends of DNA that self-anneal. Also, COHESIVE ENDS.

staggered tuning *Electronics.* a method of producing a wide bandwidth by aligning successive tuned circuits to slightly different frequencies.

staggers *Veterinary Medicine.* a popular name for any of various diseases associated with damage to the central nervous system, marked by lack of coordination, such as gid and loco disease.

stagger-tooth cutter *Mechanical Engineering.* a milling cutting tool having teeth with alternating helical angles; used for side milling.

stagger-tuned amplifier *Electronics.* a filter whose different sections are tuned to different center frequencies, in order to achieve an overall wider passband.

stagger-tuned filter *Electronics.* a filter consisting of a cascade of amplifier stages with tuned coupling networks whose resonant frequencies and bandwidths may be easily adjusted to achieve a specified transmission function.

staghead *Plant Pathology.* a condition of a tree in which it dies from the top downward, as in oak wilt and other diseases.

staging an occurrence or process involving separate stages; specific uses include: *Oncology.* the classification of the extent and severity of a malignant disease. *Space Technology.* **1.** the disengaging of a rocket propulsion unit. **2.** the moment when this occurs. *Computer Programming.* the moving of blocks of data from an off-line or low-priority device back to an on-line or higher-priority device on the demand of the system or at the request of a user. *Graphic Arts.* the process of applying a coating to particular areas on a plate or negative in order to prevent any subsequent chemical treatment from altering the coated areas.

staging area *Military Science.* **1.** a general area between the mounting area and the objective of an amphibious or airborne expedition; it may be used for refueling, for regrouping of ships, or for exercise, inspection, and redistribution of troops. **2.** a general area for the concentration of troop units and transient personnel between movements.

staging base *Military Science.* **1.** an advanced naval base for anchoring, fueling, and refitting transports and cargo ships, and for replenishing mobile service squadrons. **2.** a landing and takeoff area with minimum servicing, supply, and shelter provided for the temporary occupancy of military aircraft.

stagnant *Meteorology.* **1.** describing water or air that is without any flow or motion. **2.** made stale or foul by such lack of motion.

stagnant basin *Hydrology.* an isolated or barren basin containing motionless water, capable of supporting only anaerobic life forms.

stagnant glacier *Hydrology.* a glacier or part of a glacier that has stopped flowing and is wasting away in place, no longer having an accumulation area. Also, DEAD GLACIER.

stagnant ice *Hydrology.* **1.** glacier ice that no longer flows forward or receives material from an accumulation area. **2.** blocks of ice left behind by glacial recession. Also, DEAD ICE.

stagnant water *Hydrology.* a body of water that is not flowing or otherwise moving.

stagnation *Hydrology.* **1.** the condition of a body of water that is not flowing in a stream or is otherwise motionless, being unstirred by currents or waves. **2.** the condition of a glacier that has ceased to flow.

stagnation point *Fluid Mechanics.* the point in a flow field about a body at which the fluid has zero velocity relative to the body.

stagnation region *Space Technology.* on a rocket vehicle in flight, the area around the front of the nose cone where the air has little relative velocity.

stagnum *Hydrology.* a small lake or pool having no outlet.

Stahl, Georg Ernst 1660–1734, German chemist; proposed the phlogiston theory of combustion; distinguished soda from potash.

stain *Materials.* **1.** a solution including a dye or pigment that is used to color wood, textiles, or other materials. **2.** a dye used to color microscopic specimens for laboratory study.

stained glass *Materials.* glass manufactured using colored, translucent material and assembled into a mosaic.

stain figure *Physics.* markings or some other indication appearing on the surface of a body which has undergone a deformation due to some applied stress.

stainierite see HETEROGENITE.

staining *Microbiology.* any procedure that facilitates the microscopic identification or detection of a specific organism or portion of that organism, such as treatment with a colored or fluorescent dye.

stainless-clad steel *Metallurgy.* any steel, usually carbon alloy or low alloy, that is cladded with stainless steel.

stainless steel *Metallurgy.* any of a group of ferrous alloys that contain at least 12% chromium, providing good corrosion resistance in the presence of oxygen and absence of chloride ions.

stair *Civil Engineering.* **1.** a single step. **2.** a series of steps or flights of steps connecting two or more levels. Also, **stairway.**

stake *Mechanical Devices.* **1.** a pointed, cylindrical piece of wood or metal that is driven into the ground to mark boundaries or act as an anchor or tether for a tie, rope, string, or cable. **2.** any of a number of types of small T-shaped metal anvils used by sheetmakers for bending and forming metals. *Engineering.* **1.** the valves in a water barrel that allow water to pour back into the sump. **2.** to hold open with a chain. *Electricity.* an iron peg used as a power electrode to transfer current into the ground in electrical prospecting.

stalactite [stə lak´tīt] *Geology.* a mineral deposit that hangs down from the roof of a cave, having a shape similar to an icicle and formed by the dripping of water containing calcite (calcium carbonate, $CaCO_3$) or other minerals through cracks in the roof of the cave.

stalacto-stalagmite *Geology.* a pillar in a cave, formed by a stalactite extending downward to meet with a complementary stalagmite extending upward.

stalagmite [stə lag´mīt] *Geology.* a mineral deposit that extends upward from the floor of a cave, having the shape of a cone or post and formed by the dripping of water containing calcite (calcium carbonate, $CaCO_3$) or other minerals onto the floor of the cave.

stalagmites

stalagmometer *Engineering.* an instrument used to measure the size of liquid drops suspended from a capillary tube. Also, STACTOMETER.

stalk *Invertebrate Zoology.* a slender, upright, supporting or connecting part of an animal or one of its body parts, such as the peduncle of a crustacean eye. *Botany.* **1.** the main axis or stem of a plant, usually rising from the root and supporting leaves, flowers, or fruits. **2.** any slender structure that supports or connects plant parts. *Mycology.* the structure in a fungus that supports the organs growing above ground. Also STIPE. *Microbiology.* **1.** a prostheca or holdfast that is an extension of a cell and used for attachment of the cell to an object. **2.** a nonprosthecate appendage of a bacterial cell, such as those protruding from species of *Gallionella.* **3.** the supporting part of the fruiting body formed by some members of Myxobacterales.

stalked barnacle *Invertebrate Zoology.* a barnacle with a fleshy stalk used to attach to floating objects; includes goose barnacles.

stalk rot *Plant Pathology.* a disease of corn caused by diplodia fungi and characterized by reddish-purple or dark brown blotches on leaf stalks and sheaths.

stall *Mechanical Engineering.* **1.** of an engine that is in operation, to abruptly stop working, so that the vehicle or device it operates has no power. **2.** the occurrence of such a stoppage and loss of power. *Aviation.* **1.** the gross loss of lift when the angle of attack increases to the point at which the flow of air breaks away from a wing or other airfoil, causing it to drop. **2.** a maneuver initiated by the steep raising of an aircraft's nose, resulting in a loss of velocity and an abrupt drop.

stall flutter *Aviation.* a vibrating and oscillating motion (flutter) due to the periodic stalling and unstalling of an airfoil.

stalling speed *Aviation.* the airspeed at which a given aircraft under certain conditions will stall.

stallion *Vertebrate Zoology.* **1.** an adult male horse, especially one used for breeding. **2.** the adult male of other animals kept for stud, such as sheep or goats.

stall torque *Mechanical Engineering.* the limiting torque at which an engine or other rotating device will stall.

stall warning *Aviation.* any condition, such as a change in control force, indicating that a stall is likely or about to occur.

stall warning indicator *Aviation.* a mechanism that alerts the pilot to an impending stall by determining the critical angle of attack for a given aircraft under prevailing conditions.

STALO *Electronics.* a highly stable, local radio-frequency oscillator used for heterodyning signals to generate an intermediate frequency in radar moving-target indicators; only echoes that have changed slightly in frequency due to reflection from a moving target actually produce an output signal. (An acronym for stable local oscillator.)

stamen [stā´mən] *Botany.* the male reproductive structure of a flowering plant, a pollen-producing organ typically consisting of an anther at the tip of a filament; however, in primitive flowering plants, such as *Degeneria*, the stamen is relatively broad and flat without a distinct anther or filament.

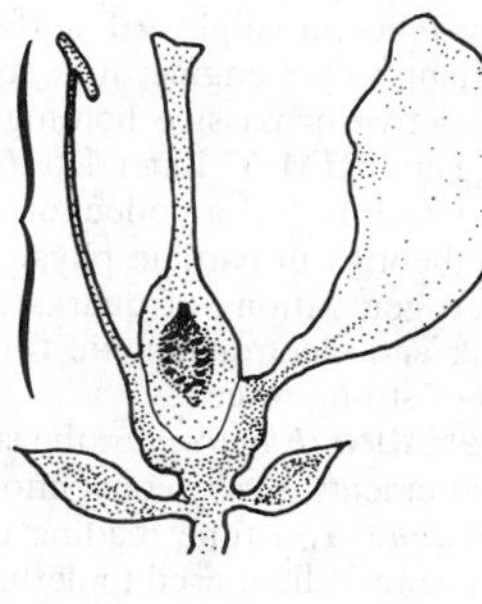

stamen

stamen blight *Plant Pathology.* a disease of blackberries caused by the fungus *Hapalosphaeria deformans*; characterized by a powdery, gray mass of spores covering the anthers.

Stamey test *Medicine.* a test of differential urinary excretion to detect vascular disease in one kidney.

staminate flower *Botany.* a flower that has stamens but not pistils.

staminode *Botany.* a sterile or nonproductive stamen.

stammering *Neurology.* a speech disorder characterized by hesitation and repetition of words, or mispronunciation and transposition of certain consonants, especially *l, r,* and *s.*

stamp *Mechanical Engineering.* to exert downward pressure on a material, as to compress, shape, or mark it. *Graphic Arts.* **1.** to impress or imprint a surface in relief. **2.** a device used in this process. **3.** the design or mark produced by such a process.

stamp battery *Mining Engineering.* a machine used to crush very hard rocks or ores by dropping a crushing member on a die, crushing the material in water between the member and the die. Also, GRAVITY STAMP.

stamp copper *Mining Engineering.* copper recovered from copper-bearing rock that is stamped and washed before smelting.

stamper a device that stamps; specific uses include: *Acoustical Engineering.* one of two assemblies used in record manufacturing, consisting of an electroplated mother reinforced with a solid metal-plate backing.

stamp head *Mining Engineering.* a heavy, nearly cylindrical, cast-iron head attached to the bottom of a stamp rod, shank, or lifter to add weight during ore stamping.

Stampian see RUPELIAN.

stamping *Mechanical Engineering.* a general term for a pressing operation, such as shearing, blanking, drawing, bending, hot or cold forming, or coining. *Mining Engineering.* a method of reducing valuable material to a desired fineness in a stamp mill; the resulting grains are generally not as fine as those produced by grinding in pans. *Electronics.* a transformer lamination cut out of a sheet of metal by a punch press.

stamping die *Graphic Arts.* a printing plate fitted with a steel die, used to emboss an image on cover material. Also, BINDER'S DIE.

stamp mill or **stamping mill** *Mining Engineering.* **1.** a machine powered by water or steam; used to crush ore or rock by means of descending pestles (stamps). **2.** the building that houses such a machine. Also, CRUSHING MILL.

stamukha *Oceanography.* a Russian term for an individual piece of ice stranded on a shoal.

stanchion [stan´shən; stan´chən] *Naval Architecture.* an upright support, especially for a vessel's railing or deck.

stand *Systematics.* a classification of different plant species commonly found growing together in a number of different comparable communities. *Ecology.* any group of plants of the same type growing together in the same area. *Forestry.* **1.** specifically, the growing trees in a given area, particularly those of the same species, grade, or size, which form a silvicultural unit and which can be treated in a consistent manner. **2.** the amount of timber standing on an area, usually measured in volume units per unit area. *Oceanography.* the period of time during which slack water occurs, when there is no discernible rise or fall of the tide.

stand-alone machine *Computer Technology.* a computer that is self-contained and runs independently of other systems.

standard *Science.* **1.** any set of conditions that describe the normal, desired, or ideal state of something, and that can be used as a means of describing or evaluating actual examples of this thing. **2.** conforming to such an established set of conditions. *Metrology.* a reference quantity or sample established as an accepted unit to which physical quantities can be compared; e.g., the carbon-12 isotope is the standard to which the atomic weights of all other elements are referred, in a scale in which it is assigned 12.00. *Botany.* **1.** the large upper petal of certain flowers. Also, VEXILLUM. **2.** one of the three upper petal-like segments of an iris.

Standard *Ordnance.* **1.** see STANDARD ARM. **2.** see STANDARD MISSILE.

standard advanced base units *Military Science.* personnel and materiel organized to function as advanced base units, including the functional components that are employed in the establishment of naval advanced bases; such units may establish repair bases, supply bases and depots, airfields and air bases, or other naval shore establishments at overseas locations.

standard air munitions package *Ordnance.* the quantity of nonnuclear air-to-ground munitions required to support a particular type of aircraft for thirty days.

standard antenna *Electromagnetism.* an antenna that consists simply of an open single-wire, including the lead-in wire, with an effective height of four meters.

Standard Arm *Ordnance.* an air-launched antiradiation missile able to home on and destroy radar emitters; officially designated **AGM-78.**

standard artillery atmosphere *Ordnance.* a set of parameters used to describe atmospheric conditions on which ballistic computations are based: i.e., no wind, a surface temperature of 15°C, a surface pressure of 1000 millibars, a surface relative humidity of 78%, and a lapse rate that yeilds a prescribed density-altitude relation.

standard artillery zone *Meteorology.* a subdivision of the standard artillery atmosphere; a layer of air of prescribed thickness and altitude.

standard atmosphere *Mechanics.* see STANDARD ATMOSPHERIC PRESSURE. *Meteorology.* **1.** a hypothetical, vertical distribution of pressure, temperature, and density that is taken to be representative of the actual conditions of the earth's atmosphere; used for such purposes as pressure altimeter calibrations, aircraft performance calculations, aircraft and missile design, and the generation of ballistic tables. **2.** as recommended for meteorological use, a model atmosphere based on climatological averages figured under several numerical constants, including a surface temperature of 15°C, a surface (sea-level) pressure of 760 millimeters of mercury, a lapse rate of 6 degrees per kilometer of altitude, and an isothermal stratospheric lapse rate as high as 24 kilometers.

standard atmospheric pressure *Mechanics.* a standard unit of atmospheric pressure that is taken to be the earth's normal air pressure at an altitude of sea level, equal to the pressure of a column of mercury 760 millimeters high (about 29.92 inches) and established as 101.325 kilopascals (1.01325×10^5 newtons per square meter), or about 14.7 pounds per square inch. Also, STANDARD ATMOSPHERE, STANDARD PRESSURE.

standard ballistic conditions *Ordnance.* a set of ballistic conditions chosen to be the standard for the computation of the firing table.

standard batch process reactor *Biotechnology.* a bioreactor that was once commonly used to produce ethanol; now largely replaced by more advanced and efficient stirred tank reactors.

standard branching *Robotics.* the programmed alteration of the path of a robot at specified points.

standard broadcast band see BROADCAST BAND.
standard calomel electrode see CALOMEL ELECTRODE.
standard capacitor *Electricity.* a capacitor whose capacitance value is not likely to vary; used mainly in capacitance bridges.
standard cell *Electricity.* a highly refined primary cell that serves as a standard of voltage; used to supply an accurate DC voltage for electronic measurement, having a voltage of 1.018636 volts at 20°C. Also, WESTON STANDARD CELL.
standard compass *Navigation.* a compass used for primary course readings and the adjustment of other compasses on a vessel; it is precisely calibrated and placed as far as possible from magnetic influences.
standard conditions *Physics.* for a gas, the temperature 0°C and the pressure 1.00 atmosphere. Also, STANDARD TEMPERATURE AND PRESSURE. *Solid-State Physics.* the conditions of room temperature (25°C) and atmospheric pressure that determine the properties of the most commonly occurring allotropic form of a substance.
standard coordinates *Astronomy.* a coordinate system used in photographic astrometry to compute the angular separation between two stars from the linear separation of their images on a photographic plate.
standard corner *Cartography.* in surveying, any corner mark situated on a standard parallel or a baseline.
standard deviation *Statistics.* a measure of variability representing an average distance of the data from the mean; its square is the variance. Also, STANDARD ERROR, ROOT-MEAN-SQUARE ERROR.
standard drill *Transportation Engineering.* the standard procedures employed in various phases of a flight.
standard electrode potential *Physical Chemistry.* **1.** the electrode potential of the standard hydrogen electrode, which is arbitrarily assigned as zero in order to describe the potential of other electrodes. **2.** the electrode potential of an actual electrode in its standard state, expressed in relation to this. Also, **standard electrode voltage, standard potential, standard reduction potential.**
standard elemental time *Industrial Engineering.* the standard time, based on prior research or estimate, that an elemental work motion is expected or scheduled to take.
standard error see STANDARD DEVIATION.
standard error of estimate *Statistics.* in bivariate regression analysis, the scatter of observed values, for a fixed value of the independent variable, around the computed value on the regression line.
standard error of the mean *Statistics.* the standard deviation of the sampling distribution of the mean; a measure of sampling error.
standard fit *Design Engineering.* a fit of machine parts with a standardized allowance and tolerance.
standard form *Computer Programming.* a floating-point number representation in which the mantissa lies within a range of specified values; a normalized floating-point number.
standard free energy increase *Thermodynamics.* the amount of increase in Gibbs free energy for a chemical reaction, when the reactants and the products of the reaction are measured in their standard states.
standard gauge *Mechanical Devices.* any device or template used to measure the dimensions of standardized parts. *Design Engineering.* a gauge used as a reference for calibrating other gauges. *Civil Engineering.* a generally accepted uniform width for the inner faces of railway of 1.435 meters, reduced to about 1.432 meters on sharp curves. Also, **standard gage.**
standard gold *Metallurgy.* a gold alloy accepted for coinage.
standard gravity *Mechanics.* the standard value of gravitational acceleration at sea level on earth, taken to be 9.80665 meters per second per second (or about 32 feet per second per second).
standard heat of formation *Thermodynamics.* the total amount of heat transfer required to form one mole of a compound in its standard state from its elements in their standard states.
standard hole *Design Engineering.* a hole with zero allowance plus a specific tolerance; fit allowance is provided for by the shaft in the hole.
standard illuminant *Optics.* one of three standardized sources of light, designated A, B, and C, that are used for accurate color measurement; A is an incandescent filament at a color temperature of 2854 K, B represents noon sunlight at 4810 K, and C represents overcast sky light at a color temperature of 6770 K.
standard inductor *Electromagnetism.* an air-core or iron-core inductor with a high degree of stability in inductance value with respect to current, frequency, and temperature.
standard interface *Computer Technology.* **1.** a specification of the physical means of interconnecting diverse peripheral units to a central processor. **2.** an interface that meets the specification.
standardization *Engineering.* the establishment of standard dimensions or characteristics for fittings or components, permitting interchangeability. *Design Engineering.* the use of these generally accepted industry standards in the design of a product or facility.
standardize *Engineering.* to make according to established industry standards. *Computer Programming.* to normalize a floating-point number.
standardized product *Design Engineering.* a product that conforms to specifications and standards in accordance with technical requirements dictated by industry standards.
standardized test statistic *Statistics.* a test statistic that has been transformed into standard normal variables.
standardized units see STANDARD NORMAL VARIABLE.
standard leak *Engineering.* the deliberate release of a measured quantity of gas into a leak detector in order to calibrate it.
standard lens *Optics.* a camera lens whose focal length is approximately equal to the diagonal of the film frame. Also, NORMAL LENS.
standard line *Cartography.* in surveying, a baseline, standard parallel, principal meridian, or guide meridian.
standard load *Design Engineering.* a preplanned dimensioned load classified by weight and balance and designated by a number.
standard meridian *Cartography.* a meridian on a map or chart along which the scale does not deviate from the stated scale.
standard mineral *Mineralogy.* a mineral whose presence is theoretically possible in a given rock, based on chemical analysis of that rock. Also, NORMATIVE MINERAL.
Standard Missile *Ordnance.* a shipboard, surface-to-surface/air missile with a solid propellant rocket engine; it is equipped with a nonnuclear warhead and semiactive or passive homing; officially designated **RIM-66** Medium Range and **RIM-67** Extended Range.
standard model *Particle Physics.* a collection of established experimental knowledge and theories in particle physics that summarizes the field; includes the three generations of quarks and leptons, the electroweak theory of weak and electromagnetic forces, and the quantum chromodynamic theory of strong forces.
standard noise temperature *Electronics.* the standard reference temperature for noise measurements, set by convention at 290 K.
standard noon *Astronomy.* **1.** a time reading of 12 o'clock noon as measured by the reference meridian; used to define standard time. **2.** the moment at which the sun appears highest over the standard meridian.
standard normal distribution *Statistics.* a normal distribution having a mean of 0 and a standard deviation of 1. The density is given by the equation $c \exp\{-x^2\}$, where c is the normalizing constant square root $(2p)$.
standard normal variable *Stastics.* the variable Z that is produced by standardizing the normally distributed variable X with the equation $Z = (X - \mu)/s$, where μ is the mean and s is the standard deviation of X.
standard operating procedure *Engineering.* a set of fixed instructions for carrying out usually routine operations. *Aviation.* specifically, a predetermined set of instructions for responding effectively to a given situation commonly encountered in flight; used unless circumstances dictate otherwise, but can be overridden if necessary. Also, STANDING OPERATING PROCEDURE.
standard output *Industrial Engineering.* the output of a worker performing at standard or 100% productivity, expressed as units per time period.
standard oxidation-reduction potential see REDOX POTENTIAL.
standard parallel *Cartography.* **1.** a parallel of latitude on a map or chart along which the scale does not deviate from the stated scale. **2.** a parallel of latitude used as baseline or control line when establishing a map projection. **3.** in surveying, a line on the earth's surface corresponding to an astronomical parallel, usually established at intervals of 24 miles north and south of the baseline, and along which township corners are positioned.
standard pattern *Ordnance.* in land mine warfare, the pattern in which mines are normally laid.
standard performance *Industrial Engineering.* the level of worker performance, based on measurable output, established as a standard for work scheduling and evaluation.
standard pitch *Acoustics.* **1.** an international standard frequency for a musical note; most commonly, 440 hertz for the note A on the treble clef. **2.** a standard tone, such as 1000 hertz, used as a reference for determination of an acoustic unit, such as the sone.
standard plane *Crystallography.* a plane whose Miller indices are (111).

standard pressure *Mechanics.* see STANDARD ATMOSPHERIC PRESSURE. *Meteorology.* a hypothetical atmospheric pressure of 1000 millibars, to which adiabatic processes are compared to allow for the determination of potential temperature or equivalent potential temperature. Also, NORMAL PRESSURE.

standard project flood *Hydrology.* in a particular geographic region, the high-water stage that is expected to result from the severest combination of hydrologic and meteorologic conditions characteristic of the region.

standard propagation *Electromagnetism.* the propagation of radio waves over the earth, which is considered as smooth and spherical, having a constant conductivity and dielectric constant, while subject to standard refraction conditions.

standard rate turn *Aviation.* a turn in which the heading of a flight vehicle changes 3° per second, or 360° in two minutes.

standard refraction *Electromagnetism.* refraction that occurs in an idealized atmosphere in which the index of refraction decreases at a uniform rate of 39×10^{-6} per kilometer.

standard route *Navigation.* in naval control of shipping, a preplanned single track connecting positions within the main shipping lanes; it is assigned a code name.

standard score *Statistics.* the deviation of a value of a random variable from the mean, measured in units of the standard deviation.

standard shaft *Design Engineering.* a shaft with zero allowance minus a specified tolerance.

standard solution *Physical Chemistry.* a solution of known concentration, usually expressed in terms of the amount of solute in a given amount of solvent or solution.

Standard SSM (ARM) *Ordnance.* a surface-to-surface antiradiation missile equipped with a conventional warhead and designed for antiship missions; officially designated **RGM-66D.**

standard star *Astronomy.* a star used as a reference for observing the positions of other celestial objects, because of its precisely known position, motion, and constant brightness.

standard state *Physics.* the state in which a substance in its pure and stable form exists at standard pressure and temperature.

standard subroutine *Computer Programming.* a subroutine whose function has applications in more than one program.

standard target *Electromagnetism.* a radar target that will produce an echo of known power under various conditions.

standard temperature and pressure *Physics.* see STANDARD CONDITIONS.

standard terminal arrival route *Aviation.* a preplanned and coded air traffic control arrival routing; it is printed for pilot use in text or text and graphic form. Also, **STAR.**

standard test frequencies *Electronics.* a group of carrier frequencies chosen to test transmission in an AM broadcast band, usually including 600, 1000, and 1400 kHz and sometimes also 540, 800, and 1200 kHz.

standard test tone *Electronics.* a signal used to test signal transmission, usually a 1-mW 1000-Hz signal applied to the 600-ohm audio portion of a circuit at a zero transmission level reference point.

standard test-tone power *Electronics.* one milliwatt (0 decibels above one milliwatt) at 1000 hertz.

standard time *Astronomy.* the standardized local civic time within a band of longitude roughly 15° wide. *Industrial Engineering.* the normal or scheduled time that a work cycle is expected to take, including normal allowances.

standard ton see TON.

standard trajectory *Mechanics.* the calculated trajectory that a projectile is expected to follow under given conditions of weather, position, and materials used. Firing tables are based on standard trajectories.

standard virus *Virology.* a nondefective virus that must be mutated to form a defective virus; its presence is necessary for replication of a defective virus because it is the wild-type infectious form.

standard visibility see METEOROLOGICAL RANGE.

standard volume *Physics.* the volume occupied by one mole of an ideal gas at 0°C and 1.00 atmosphere of pressure.

standard waveguide *Electromagnetism.* any one of several rectangular waveguides whose dimensions have been specified by various agencies and are in general use.

standard wire rope *Mechanical Devices.* a wire rope with six strands of wire wrapped around a sisal core. Also, HEMP-CORE CABLE.

standby battery *Electricity.* a storage battery that is held in reserve for an emergency power source in the event of failure of regular power supplies at a hospital, radio station, or other such facility.

standby block *Computer Programming.* a storage location reserved for communication with buffers.

standby computer *Computer Technology.* a duplicate computer used in a dual or duplex system that can take over the real-time processing load in case of equipment malfunction or overload.

standby power source *Electricity.* a power source that is not dependent on current supply from an outside commercial source; e.g., a battery system within a computer that provides power to run the clock while the computer is unplugged.

standby register *Computer Programming.* a register that stores accepted or verified data for rerun in the event of a program error or machine malfunction.

standby time *Industrial Engineering.* the time during which a worker or a machine is available but not working due to uncontrollable circumstances such as inclement weather, equipment failure, inadequate supplies, or the like. *Computer Programming.* **1.** the time that elapses between entering an inquiry and receiving a reply. **2.** the time during which two or more computers are connected and available for use. **3.** the time between system setup and actual use.

stand fire *Forestry.* a forest fire that ignites tree trunks.

standing cloud *Meteorology.* a cloud that maintains its position over a mountain peak or ridge.

standing crop *Ecology.* a term for the total weight of all individuals of one species or all organisms living in an ecosystem at a given moment.

standing friction see STATIC FRICTION.

standing ground *Mining Engineering.* a term for ground that stands firm and does not require timbering.

standing-on-nines carry *Computer Programming.* a procedure used to speed up parallel addition in which a carry out of a digit (low-order) position is generated by a carry into the digit position, bypassing the normal adding circuit; thus, if the digit position is nine, the nine is changed to zero.

standing operating procedure *Aviation.* see STANDARD OPERATING PROCEDURE.

standing rigging *Naval Architecture.* those lines in a ship's rigging that are fixed in place and only occasionally require adjusting or tightening, as distinguished from running rigging.

standing valve *Petroleum Engineering.* a stationary (as distinguished from traveling) discharge valve on a sucker-rod pump used in an oil well.

standing water *Hydrology.* water that has has no motion or current; stagnant water.

standing wave *Physics.* a nonprogressive wave in which the motion of any point along the waveform may be represented by a function of space multiplied by a sinusoidal function of time.

standing-wave detector *Electromagnetism.* an instrument that is capable of detecting the presence of a standing wave in a waveguide or transmission line. Also, **standing-wave indicator, standing-wave meter.**

standing-wave loss factor *Electromagnetism.* a ratio of the loss in a waveguide with unmatched impedance to the loss in a waveguide that has matched impedance.

standing-wave method *Electromagnetism.* a method of determining the wavelength or frequency of an electromagnetic wave by making measurements of distances between nodes or antinodes in a standing wave.

standing-wave producer *Electromagnetism.* a probe that is inserted into a slotted waveguide so as to support a standing-wave pattern within the guide.

standing-wave ratio *Physics.* the ratio of the maximum amplitude to the minimum amplitude of a specified point along a standing wave.

stand method *Forestry.* the practice of maintaining reproduction from self-sown seed brought on by successive cuttings of different-age trees in a stand; leads to a stand of trees of the same age.

standoff insulator *Electricity.* an insulator, usually of the post type, employed to hold a wire or other component away from a chassis or base.

standoff jammer *Electronics.* **1.** an airborne jamming device. **2.** an aircraft that beams such a device into a target air space and performs high-power jamming of both flight vehicles and acquisition or tracking devices.

standoff target acquisition system *Ordnance.* an airborne radar system designed to accurately locate moving targets in enemy territory while the aircraft maintains a relatively safe distance from the target area. Also, **SOTAS.**

standoff weapon *Ordnance.* a weapon that may be launched from a distance that allows the operating personnel to evade fire from the target area.

stand on *Navigation.* to maintain the same course.

standpipe *Engineering.* a vertical storage pipe or tank, used to even out surges and maintain constant pressure in a water flow or pumping system.

standstill *Astronomy.* a period in a variable star's light curve when its brightness temporarily remains virtually constant.

stand-still feature *Robotics.* a feature designed to prevent a robot's controller from being affected by false signals, such as fluctuations in the power supply, which might cause the robot to move haphazardly.

stanfieldite *Mineralogy.* $Ca_4(Mg,Fe^{+2},Mn^{+2})_5(PO_4)_6$, a red-amber monoclinic mineral found only in meteorites as irregular grains and veinlets, having a specific gravity of 3.15 and a hardness of 4 to 5 on the Mohs scale.

Stanford Achievement Test *Psychology.* an educational achievement test designed to measure primary and secondary school learning levels in reading, mathematics, languages, science, social studies, and listening skills.

Stanford-Binet (intelligence) test *Psychology.* a widely used intelligence test for children and adults that gives an individual's mental age; this is then divided by the individual's chronological age, the result being an IQ score. (A revision of the *Binet*-Simon test done at *Stanford* University.)

Stanley, Wendell Meredith 1904–1971, American biochemist; discovered that viruses are nucleoproteins; awarded the Nobel Prize for preparing enzymes and virus proteins in pure form.

stannane see TIN HYDRIDE.

stannic *Chemistry.* **1.** of or relating to tin, which is also called stannum. **2.** describing various compounds of tin, especially those in which the element has a valence of 4.

stannic acid or **stannic anhydride** see STANNIC OXIDE.

stannic bromide *Inorganic Chemistry.* $SnBr_4$, an irritant; white crystals that produce fumes in air; soluble in water and alcohol; melts at 31°C and boils at 202°C; used in mineral separations. Also, TIN TETRABROMIDE.

stannic chloride *Inorganic Chemistry.* $SnCl_4$, a colorless, fuming, and corrosive liquid that becomes a crystalline solid, $SnCl_4 \cdot 5H_2O$, on addition of water; soluble in cold water and decomposed by hot water; freezes at –33°C and boils at 114.1°C; used as a mordant, in electroconductive coatings, in soaps and ceramics, and for many other purposes. Also, TIN CHLORIDE.

stannic chromate *Inorganic Chemistry.* $Sn(CrO_4)_2$, a toxic, brownish-yellow crystalline powder; slightly soluble in water and decomposed by heat; used to color porcelain and china. Also, TIN CHROMATE.

stannic iodide *Inorganic Chemistry.* SnI_4, yellow to reddish crystals that sublime on heating and are decomposed by hot water; melts at 144.5°C and boils at 364.5°C. Also, TIN IODIDE.

stannic oxide *Inorganic Chemistry.* SnO_2 or its various hydrated forms, $SnO_2 \cdot nH_2O$; a white powder that is insoluble in water; sublimes at 1800–1900°C; used in tin salts, as a catalyst, in ceramics, perfumes, and textiles, and for various other purposes. Also, TIN PEROXIDE, TIN DIOXIDE, FLOWERS OF TIN.

stannic sulfide *Inorganic Chemistry.* SnS_2, a yellow-brown powder; insoluble in water; decomposes on heating to 600°C; used in imitation gilding and as a pigment. Also, ARTIFICIAL GOLD, MOSAIC GOLD, TIN BRONZE, TIN DISULFIDE.

stannite *Mineralogy.* Cu_2FeSnS_4, a steel-gray or iron-black, metallic, tetragonal mineral of the stannite group, having a specific gravity of 4.3 to 4.5 and a hardness of 4 on the Mohs scale; found in hydrothermal vein deposits. Also, TIN PYRITES, BELL-METAL ORE.

stannous *Chemistry.* **1.** of or relating to tin, which is also called stannum. **2.** describing various compounds of tin, especially those in which the element has a valence of 2.

stannous bromide *Inorganic Chemistry.* $SnBr_2$, a yellow powder that turns brown in air; soluble in water, alcohol, and ether; melts at 215.5°C and boils at 620°C; a skin irritant. Also, TIN DIBROMIDE.

stannous chloride *Inorganic Chemistry.* $SnCl_2$ or $SnCl_2 \cdot 2H_2O$, white crystals; soluble in water and alcohol; the anhydrous form melts at 246°C and boils at 652°C, and the hydrated form melts at 37.7°C and decomposes on heating; absorbs oxygen from the air to give the oxychloride. It is used as a reducing agent, chemical intermediate, catalyst, antisludging agent, and food preservative, and for various other purposes. Also, TIN DICHLORIDE, TIN CRYSTALS, TIN SALT.

stannous chromate *Inorganic Chemistry.* $SnCrO_4$, a toxic brown powder, essentially insoluble in water; used to decorate porcelain. Also, TIN CHROMATE.

stannous 2-ethylhexoate *Organic Chemistry.* $Sn(C_8H_{15}O_2)_2$ a toxic, light-yellow liquid; insoluble in water and soluble in petroleum ether; used as a lubricant, catalyst, and stabilizer.

stannous fluoride *Inorganic Chemistry.* SnF_2, a white, lustrous, crystalline powder; slightly soluble in water; toxic and a strong irritant; widely used as a fluoride source in toothpastes and other dental products. Also, TIN FLUORIDE.

stannous oxalate *Organic Chemistry.* SnC_2O_4, a white crystalline powder that is insoluble in water and soluble in acids; melts (decomposes) at 280°C; used in textile manufacture and as a catalyst. Also, TIN OXALATE.

stannous oxide *Inorganic Chemistry.* SnO, a brownish-black powder; unstable in air and insoluble in water; decomposed by heating to 1080°C; used as a reducing agent, to make stannous salts, in pharmaceuticals, and as a mild abrasive. Also, TIN PROTOXIDE.

stannous sulfate *Inorganic Chemistry.* $SnSO_4$, toxic white to yellowish crystals that rapidly decompose in aqueous solution and lose SO_2 when heated above 360°C; used in dyeing and tin plating. Also, TIN SULFATE.

stannous sulfide *Inorganic Chemistry.* SnS, a toxic dark gray to black crystalline powder; insoluble in water; melts at 882°C and boils at 1230°C; used in bearing materials, as a polymerization catalyst, and as a reagent. Also, TIN SULFIDE.

stannum *Chemistry.* the Latin name for tin; the basis for the chemical symbol Sn for tin.

Stanton diagram *Fluid Mechanics.* a log-log plot of friction factor against the Reynolds number; the same sort of diagram is also credited to Moody.

Stanton number *Thermodynamics.* a dimensionless quantity, N_{St}, used in forced convection calculations, given by the heat-transfer coefficient divided by the product of the specific heat at constant pressure, the fluid velocity, and the fluid density.

stapedectomy *Medicine.* the surgical removal of the stapes bone (the stirrup-shaped bone of the middle ear) and replacement with a surrogate, to restore hearing in the treatment of otosclerosis.

stapedius muscle *Anatomy.* a small muscle that dampens the movement of the stapes.

stapes *Anatomy.* the innermost of the three ossicles of the middle ear; its base fits into the oval window.

staph [staf] *Medicine.* an informal term for staphylococci bacteria or for conditions caused by these bacteria. Thus, **staph infection.**

Staph. *Staphylococcus.*

staphyl- or **staphylo-** a combining form meaning: **1.** cluster. **2.** uvula.

Staphyleaceae *Botany.* a family of dicotyledonous, mucilaginous, tanniferous shrubs and trees of the order Sapindales, characterized by pinnate or trifoliate leaves, small flowers borne in drooping panicles or racemes, and seeds with a copious, oily endosperm; found in the Americas and Eurasia.

staphylectomy *Surgery.* excision of the uvula. Also, UVULECTOMY.

Staphylinidae *Invertebrate Zoology.* the rove beetles, a large family of coleopteran insects in the superfamily Staphylinoidea, having an elongated body, short forewings, and long legs; larvae and adults are predators or scavengers.

Staphylinoidea *Invertebrate Zoology.* a superfamily of coleopteran insects in the suborder Polyphaga, including the rove beetles, carrion beetles, and allies.

staphylitis *Medicine.* an inflammation of the uvula.

staphylococcal pneumonia *Medicine.* pneumonia caused by staphylococcal bacteria, many strains of which are antibiotic-resistant; this infection has a strong tendency to spread.

staphylococcemia *Medicine.* the presence of staphylococci in the blood.

staphylococcin *Bacteriology.* a bacteriocin produced by certain strains of *Staphylococcus aureus.*

Staphylococcus [staf´ə lō käk´əs] *Bacteriology.* a genus of Gram-positive, facultatively anaerobic bacteria of the family Micrococcaceae, order Eubacteriales, consisting of usually unencapsulated cocci; often pathogenic, causing local lesions and serious infection. (Taken from a Greek term meaning "a bunch of grapes;" with reference to the appearance of a group of these bacteria under a microscope.)

staphylococcus *plural,* **staphylococci.** an organism of the genus *Staphylococcus.*

Extinctions & Continuities

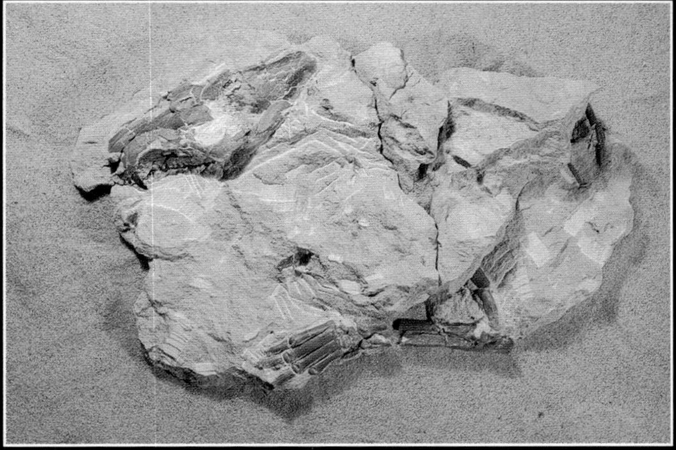

Animal fossil. San Diego Natural History Museum.

Eocene fish fossil. Richard H. Gross/Biological Photography.

Plant fossil. San Diego Natural History Museum.

California condor. Zoological Society of San Diego.

Staphylococcus aureus *Bacteriology.* a species of bacteria that is a major human pathogen, composed of nonmotile cocci, and distinguished by the yellow pigment of the colonies.

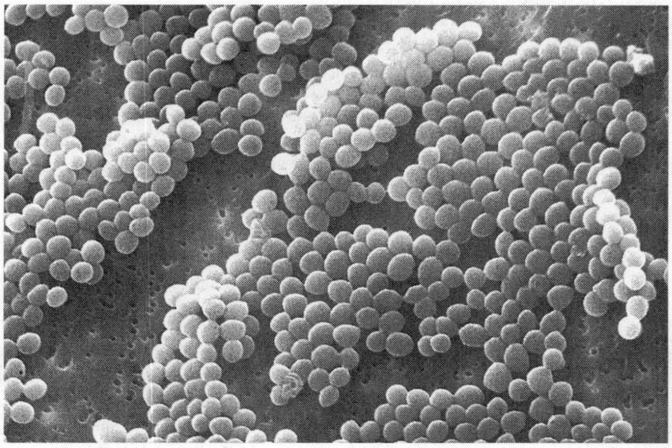

Staphylococcus aureus

Staphylococcus epidermis *Bacteriology.* a coagulase-negative bacterial species that ferments glucose and produces colonies that are usually white; commonly found on human skin.
staphyloderma *Medicine.* a skin infection that is caused by staphylococci; e.g., carbuncles.
staphylokinase *Biochemistry.* an enzyme found in filtrates of *Staphylococcus* cultures; helps dissolve human blood clots by catalyzing the activation of plasminogen (profibrinolysin) to plasmin (fibrinolysis).
staphylomycin *Microbiology.* an antibiotic that is produced by a strain of *Actinomyces* bacteria and is active against certain acid-fast bacilli and Gram-positive organisms.
staphyloslide *Microbiology.* a microscope slide upon which strains of *Staphylococcus aureus* can be examined for the presence of a coagulating substance, based on their ability to clump sheep erythrocytes coated with fibrinogen.
Staphylothermus *Bacteriology.* a genus of archaebacteria that grow optimally at high temperatures under anaerobic conditions; its type species was isolated from a deep-sea hydrothermal vent.
staphylotoxin *Biochemistry.* any of several exotoxins produced by *Staphylococcus aureus,* or a mixture of these. Also, **staphylococcal toxin.**
staphyloxanthin *Microbiology.* the main carotenoid pigment produced by the bacterium *Staphylococcus aureus.*
staple *Mechanical Devices.* **1.** a U-shaped metal or wire fastener, pointed at both ends, used to hold material such as a joint or sheets of paper together or attached to another device. **2.** to fasten something with such a device. *Textiles.* a standard length of textile fibers, representing the average length taken collectively.
staple fibers *Textiles.* **1.** any short textile fiber that when spun and twisted forms a yarn rather than a filament. **2.** specifically, synthetic continuous filaments that have been deliberately cut to the same length as various natural fibers. Staple fibers may be spun alone or blended with a natural fiber to form a yarn, according to the spinning system used for that fiber.
staple punch *Mechanical Devices.* a punch with two hole-making points to allow a staple to be driven into a material or object such as cloth or sheets of paper.
stapler *Mechanical Devices.* **1.** a device used to drive staples into paper and other similar materials. **2.** a wire-stitching machine, particularly one used in bookbinding. **3.** a hand-powered tool used to drive staples into wood and other materials. Also, **stapler gun.**
star *Astronomy.* a large, gaseous, self-luminous celestial body such as the sun, held together by self-gravity and powered by thermonuclear reactions that derive mainly from the fusing of hydrogen nuclei into helium nuclei. *Nucleonics.* a group of ionization tracks originating from a common point in a bubble chamber, cloud chamber, or nuclear emulsion. *Electricity.* **1.** a polyphase circuit, in which each phase connects to a common or neutral point. A **wye** is a three-phase star. **2.** a network with three or more branches that connect to a common node.
star atlas *Astronomy.* a chart of the heavens, usually printed in a book or portfolio format, whose pages or plates each depict a portion of the sky and the objects contained therein.
starboard [stär′bərd; stär′bôrd′] *Navigation.* on or to the right side of a vessel or aircraft, when facing forward. (Originally, "steer board;" not related to the word *star.*)
starburst galaxy *Astronomy.* a blue compact dwarf galaxy that is undergoing strong bursts of star formation.
star catalog *Astronomy.* a listing of stars usually by right ascension with observational data elements such as coordinates, brightness, motions, distance, and so on.
starch *Biochemistry.* any of a group of polysaccharides that store energy in plants, having the general formula $(C_6H_{10}O_5)_n$, occurring in the form of minute granules in seeds, tubers, and other parts of plants; starch forms an important constituent of rice, corn, wheat, beans, potatoes, and other vegetable foods. *Textiles.* a commercial preparation of this substance used to stiffen fabrics in laundering.

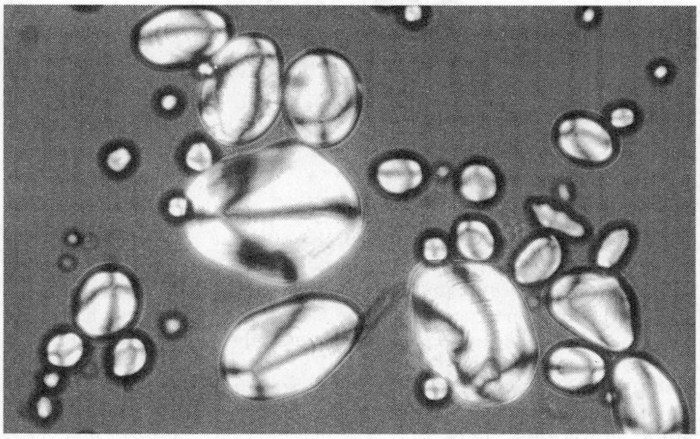

starch

star chain *Navigation.* a group of navigational transmitting stations including a master station with three or more slave stations located symmetrically around it.
star chart *Astronomy.* a chart or map showing the relative apparent positions of the stars as viewed from the earth. Also, STAR MAP.
starch cellulose see AMYLOPECTIN.
starch gel electrophoresis *Biochemistry.* a form of electrophoresis used to detect sickle-cell anemia.
starch gum *Materials.* see DEXTRIN.
starch lesion *Virology.* a local accumulation of starch in a virus-infected leaf.
star cloud *Astronomy.* a region of the Milky Way with a higher-than-average star density; often appears hazy to the naked eye.
star cluster *Astronomy.* a gravitationally bound group of stars having anywhere from a few dozen to hundreds of thousands of members; there are globular clusters and galactic (or open) clusters.
star color *Astronomy.* a measure of a star's temperature, with the hotter stars being bluer.
star-connected circuit *Electricity.* a polyphase circuit in which all the current paths within the region that delimits the circuit extend from each of the entry points of the phase conductors to a common conductor.
star connection *Electricity.* see STAR NETWORK.
star count *Astronomy.* the number of stars that appear in a given region of sky; usually counted on a photographic plate or CCD exposure.
star day *Astronomy.* the time from one merdian passage of a star to the next.
star density *Astronomy.* the number of stars contained per unit of volume of space.
star drift *Astronomy.* the relative motion of two groups of stars in the Milky Way moving in opposite directions. Also, STAR STREAMING.
star drill *Mechanical Devices.* a manually operated or machine drill with a cross or four-pointed tip, used for drilling rock by applying a hammer while the drill is rotated.
star dune *Geology.* a large isolated hill of sand rising above a surrounding plain, having sharp-crested ridges in the shape of a star that converge to form a central peak.

star finder *Astronomy.* a device such as a star map or a celestial globe used by beginners to locate star positions in the night sky. Also, **star identifier.**

starfish *Invertebrate Zoology.* the popular name for predatory marine echinoderms of the class Asteroidea, characterized by a body consisting of five or more rays radiating from a central disk, a mouth under the disk, and rows of tubular walking feet. Also, SEA STAR.

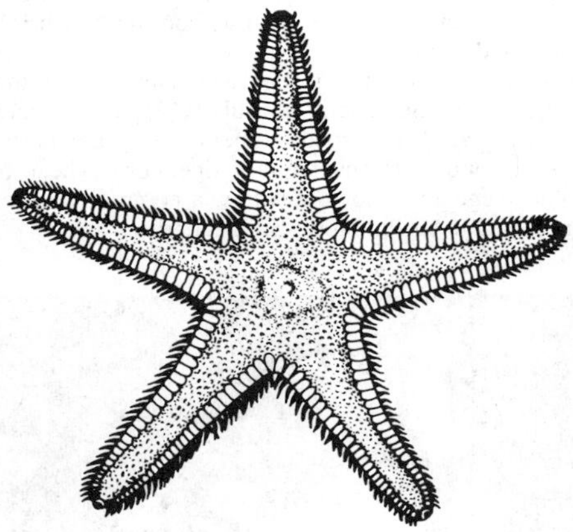

starfish

stargazer *Vertebrate Zoology.* any fish of the two families Dactyloscopidae and Uranoscopidae, having electrical organs on the head, and mouth and eyes positioned upward; bottom-dwellers found in tropical and temperate waters worldwide.

star group *Astronomy.* a collection of stars, numbering from 10 to a million, that share the same direction of motion.

Stark, Johannes 1874–1957, German physicist; awarded the Nobel Prize for his discovery of the Stark effect.

Stark effect *Spectroscopy.* the splitting of spectral lines, arising from rotational absorption, in an intense electric field.

Stark-Lunelund effect *Electromagnetism.* an effect that exhibits the polarization of light emitted by a beam of atoms or molecules in a region free of any electromagnetic fields.

star lamp *Electricity.* a high-pressure xenon arc that produces a small, intense point of light, which is focused through thousands of lenses and pinholes and projected to a planetarium's dome.

Starlifter *Aviation.* a popular name for the C-141 cargo transport.

starlike region *Mathematics.* **1.** a region R is said to be starlike or star-shaped with respect to a point P in the region if every ray emanating from P cuts the boundary of R in exactly one point. **2.** a region R is said to be starlike or star-shaped if there exists a point $P \in R$ such that R is starlike with respect to P. **3.** a region is said to be starlike or star-shaped with respect to a ball B contained in the region if the region is starlike with respect to every point of B.

starling *Vertebrate Zoology.* a medium-sized, gregarious, European songbird, *Sturnus vulgaris,* of the family Sturnidae, characterized by dark iridescent plumage, a pointed bill, and a short, square tail; naturalized in the United States, southern Canada, Australia, and New Zealand.

starling

Starling, Ernest 1866–1927, English physiologist; isolated secretin (with William Bayliss); developed the hormone concept.

Starling's law of the heart *Physiology.* a general principle stating that the force of the muscle fibers contracting the heart is proportional to their length at the beginning of systole.

star map see STAR CHART.

star motion *Astronomy.* any movement of a star, although the reference point is usually specified, for example, with respect to local stars or in a circular orbit around the Milky Way Galaxy's center.

star names *Astronomy.* the highly varied nomenclature used to identify stars, including proper names, Greek letters, numbers, catalogue identifiers, and the like; bright stars often have several designations.

star network *Electricity.* a network with many branches connected at a point. Also, STAR CONNECTION. *Computer Science.* a form of computer network in which there is a central hub computer to which every other computer in the network is directly connected.

star place *Astronomy.* the position of a star on the celestial sphere in which the star coordinates are referred to an equator and equinox at a specified epoch.

star polymers *Materials Science.* polymers that exhibit a characteristic starlike macromolecular construction, in which branches radiate from a central atom or group of atoms, such as the star di-block copolymer.

starquake *Astrophysics.* in a pulsar or neutron star, an abrupt change in the star's quasisolid surface that makes a minute but detectable change (a glitch) in the pulse period; in an ordinary star, the term refers to the acoustic vibrations studied by helioseimology.

Starria *Bacteriology.* a genus of cyanobacteria, or blue-green algae, that grow as filamentous cells, in which a cross section of the trichomes shows them to be triradiate.

star ruby *Mineralogy.* an asteriated variety of ruby mineral, usually having six chatoyant rays.

star sapphire *Mineralogy.* an asteriated variety of sapphire characterized by six starlike rays.

star shell *Ordnance.* a shell that bursts in the air to illuminate enemy positions.

starspot *Astronomy.* an area, hypothesized to exist on the surfaces of some stars, that is cooler and darker than the surrounding stellar photosphere; similar to a sunspot.

star streaming see STAR DRIFT.

start bit *Computer Programming.* the first of a series of bits that signifies the beginning of a character group; used to maintain synchronization between transmitting and receiving terminals.

start codon see INITIATOR CODON.

star telescope *Optics.* a telescope mounted on a sextant, having a large objective for greater field of view and illumination in taking star sightings.

starter something that begins or initiates a process; specific uses include: *Electricity.* **1.** an igniter electrode in an ignitron. **2.** a device used to start an electric motor and accelerate it to its operating speed. *Electronics.* an auxiliary electrode used in a gas tube to establish sufficient ionization to reduce the anode breakdown voltage. *Engineering.* a relatively large-diameter drill used to begin a hole, which is then continued by a smaller-diameter drill called the follower. *Food Technology.* a culture of microorganisms used to initiate a fermentation process; obtained either as pure culture from a commercial laboratory or from a portion of previously cultured product. Also, **starter culture.**

starter fertilizer *Agronomy.* a chemical fertilizer used especially with new plantings.

starter punch *Mechanical Devices.* a bluntly pointed metal bar that is driven into wood to start a hole for a screw.

star test *Optics.* a test that uses a point source of light, either a real or an artificial star, to evaluate the aberrations in a lens.

starting block *Mechanical Devices.* a wood or metal frame secured to the ground with adjustable, triangularly shaped blocks mounted in parallel; used by runners to provide a maximum takeoff speed in a race.

starting box *Electricity.* a special rheostat that is used to start a motor gradually, in steps; designed with an electromagnet for holding the arm in the maximum speed position and releasing it when power is interrupted.

starting reactor *Electricity.* an iron core conductor connected in series with a machine stator winding; used to decrease or limit the starting current of electric motors.

starting resistance *Mechanical Engineering.* the force required to produce a thin layer of oil in the shaft of the journal bearings of a locomotive starting from rest.

starting sheet *Metallurgy.* in electrorefining, a sheet of the material being refined, onto which the product accretes during operation.
starting taper *Design Engineering.* a short-end taper on a reamer used to begin a borehole.
startle display *Behavior.* see DEIMATIC BEHAVIOR.
startle response *Physiology.* a series of responses manifested when an individual experiences a sudden frightening stimulus; this includes a tightening of muscles, changes in body stance, cardiovascular changes directed by the sympathetic nervous system, brain activity stimulation, and emotional excitement.
start-to-leak pressure *Mechanical Engineering.* the amount of inlet pressure at which a leak becomes apparent for a relief valve.
start-up curve *Industrial Engineering.* the curve of rising efficiency and reduced delays or costs seen as experience is gained in a new production process or operation.
starvation *Computer Science.* a term for a situation in which a process never receives service due to an unfair scheduling algorithm.
starved basin *Geology.* a term for a depression in the land surface in which the rocks dip toward a common center, characterized by a rate of subsidence that exceeds the rate of sedimentation.
starved joint *Engineering.* a joint to which insufficient adhesive has been applied. Also, HUNGRY JOINT.
Star Wars an informal name for the Strategic Defense Initiative, a proposed U.S. weapons system to intercept and destroy incoming enemy missiles, using various forms of ordnance and technology including missiles and lasers based in outer space. (Because its technology is thought to resemble that of the popular science-fiction film *Star Wars.*)
stasigenesis *Evolution.* a state of little evolutionary change in a form or a group over an extended period of time, implying a particularly well-adapted form in an exceptionally stable habitat or a very generalized form.
stasimorphic *Evolution.* **1.** of or relating to an ancestral, hereditary character retained by a more recent form of an organism. **2.** of or relating to a deviant form resulting from arrested development.
stasipatric speciation *Genetics.* speciation resulting from the beneficial effects and increased fitness conferred by chromosomal rearrangements that establish homozygotes which are adaptively superior in one part of the geographical range of a species.
stasis [stas´is; stā´sis] *Medicine.* **1.** a state of balance or equilibrium among opposing forces. **2.** the stoppage or slowing of a flow of material in a body channel, as of blood through the vascular system or digested matter in the intestine.
stasis dermatitis *Medicine.* an often chronic inflammation of the skin of the legs, with edema, eczema, and sometimes ulceration, due to stoppage or slowing of the venous blood flow.
stat *Nucleonics.* a unit of rate of disintegration, equivalent to the rate produced by an amount of radon that gives rise to one statcoulomb of charge in one second while in air under standard conditions. *Graphic Arts.* a shorter term for a photostat. See PHOTOSTAT.
stat or **STAT** *Medicine.* quickly; immediately. (From Latin *statim.*)
stat- *Electricity.* a prefix that indicates an electrical unit in the electrostatic centimeter-gram-second system of units.
-stat a combining form used in the names of devices that make constant what is specified in the initial element. Thus, **thermostat.** (From Greek meaning "to make stand.")
statampere *Electricity.* the centimeter-gram-second electrostatic unit of current; equal to 3.3356×10^{-10} ampere. Also, **statA.**
statcoulomb *Electricity.* the unit of charge in the electrostatic centimeter-gram-second system of units; equal to the charge that exerts a force of 1 dyne on an equal charge at a distance of 1 centimeter under vacuum; equal to 3.3356×10^{-10} coulomb. Also, **statC.**
state *Physics.* **1.** the phase of matter: solid, liquid, or gas. **2.** the conditions of a system as specified as completely as possible by some given scheme, such as the specification of the quantum numbers of an electron in a particular orbital. *Quantum Mechanics.* specifically, the condition of a system as described by a wavefunction satisfying Schrödinger's equation. *Control Systems.* the minimum set of numbers required to express a system's history and compute its future behavior.
state-dependent learning *Psychology.* the principle that learning and memory occur in association with a particular biological or psychological state and cannot be readily retrieved in states unrelated to the original.
state description *Artificial Intelligence.* a data structure or set of values containing the information that specifies a particular state in a state-space search.
state equations *Physical Chemistry.* see EQUATIONS OF STATE. *Control Systems.* a set of equations that describe the state of a system and its output as a single valued function of the system's input at that time plus the initial state of the system.
state estimator see OBSERVER.
state feedback *Control Systems.* a feedback control law stating that the control inputs at a given time are determined solely by the values of the state variables at that time, and not by their history.
state graph *Computer Technology.* a directed graph of a sequential machine in which nodes represent internal states and labeled arrows represent transitions to other states.
statement *Linguistics.* a single sentence that gives some fact or point of information. *Computer Programming.* a computer program instruction that specifies one or more actions to be performed during the execution of a program or provides specifications concerning the program.
statement calculus *Mathematics.* the analysis of those logical relations among sentences that depend solely on their composition from constituent sentences using sentential (or logical) connectives. Given an initial set of sentences (or prime sentences), the following assumptions are made: (a) each prime sentence is a statement; i.e., a truth value may be assigned to each prime sentence; and (b) each sentence under consideration is composed from prime sentences using sentential connectives, and, for a given assignment of truth values of these prime sentences, receives a truth value in accordance with definitions of the sentential connectives. Also, SENTENTIAL CALCULUS, PROPOSITIONAL CALCULUS.
state observer see OBSERVER.
state of matter *Physics.* the specification of a sample of matter in terms of its existence as a solid, liquid, or gas.
state of readiness *Ordnance.* the state of a demolition target: **state 1, safe** has the demolition charges in place, the firing or initiating circuits installed but not connected to the charge, and the detonators or initiators neither connected nor installed; **state 2, armed** is ready for immediate firing, with the demolition charges in place, and the firing and priming circuits installed and complete.
state of strain *Mechanics.* a complete description of the strain tensor with components e_{ij} (where $i,j = 1, 2, 3$) that a deformed body experiences.
state of stress *Mechanics.* a complete description of the stress tensor with components s_{ij} (where $i,j = 1, 2, 3$) that a deformed body experiences.
state of the sea *Oceanography.* a standardized numerical or written description of ocean surface roughness. Also, **sea state.**
state of the sky *Meteorology.* the current condition of cloud cover as observed from the ground, taking into account the amounts, types, movement, and heights of all clouds.
state society *Anthropology.* a form of political and social organization in which there is an ongoing government or centralized power base that regulates the society.
state space *Control Systems.* all the possible values for the state vector of a system. *Artificial Intelligence.* the set of all attainable states of a problem, often represented by a directed graph with nodes as states and arcs as operators, that transform one state into another.
state-space search *Artificial Intelligence.* a problem-solving technique in which all possible alternative courses of action are tried so that a successful or best action is identified.
state table *Computer Technology.* a tabular representation of a sequential machine in which the rows correspond to the internal states, the columns correspond to the input, and the table entries represent the next internal state.
state transition equation *Control Systems.* an equation that describes a state transition matrix.
state transition matrix *Control Systems.* a matrix in which the product of the matrix and the state vector at an initial time will give the state vector at a later time.
state variable *Thermodynamics.* a quantity that describes and defines the thermodynamic state of a system, and that is used in an equation of state to specify the condition of the system; a change in such a variable is independent of the path the system follows in changing from one state to another but is, instead, dependent on the initial and final conditions. Also, THERMODYNAMIC FUNCTION OF STATE. *Control Systems.* any variable that belongs to the minimum set of variables required to describe a system's history and compute its future behavior. *Artificial Intelligence.* a variable that is added to predicates, in a predicate calculus representation used for planning, to account for the possibility that facts may be true in some states but not in others.

state vector *Computer Programming.* in a multitasking system, the state of the processor that is saved when a task is interrupted and execution is switched to another task. *Control Systems.* a column vector, the components of which are the state variables of a system. *Quantum Mechanics.* a vector in a Hilbert space that specifies the state of a system.

statfarad *Electricity.* the centimeter-gram-second electrostatic unit of capacitance; equal to 1.112650×10^{-12} farad. Also, **statF.**

stathenry *Electromagnetism.* the unit of inductance in the centimeter-gram-second system of units; approximately equivalent to 8.987554×10^{11} henries. Also, **statH.**

stathmokinesis *Cell Biology.* an inhibition or interruption of the cellular process of mitosis.

static *Mechanics.* not experiencing motion; at rest; stationary. *Telecommunications.* radio or television broadcast interference that is produced by static or atmospheric electricity having a wavelength within that of the broadcast frequencies, usually manifested as a crackling or hissing noise.

static accuracy *Control Systems.* in the steady state of an automatic control system, the degree to which actual position response reflects a desired position, or the difference between the former and the latter.

static algorithm *Computer Programming.* mathematical software whose operation is known in advance, such as arithmetic functions.

static analysis *Computer Science.* an analysis of a program by examining it but not running it.

static balance *Mechanics.* the condition of a rotating body whose center of mass lies on its axis of rotation.

static bed *Chemical Engineering.* in a process vessel, an arrangement of granular solids among which there is no relative motion.

static characteristic *Electronics.* in an electron tube or other device, the relationship between a pair of variables, such as electrode voltage and electrode current, with all other operating voltages maintained constant; usually presented as a graph.

static charge *Electricity.* the accumulated electric charge that is present on an object.

static check *Computer Technology.* the process of testing equipment at setup time.

static contact and wear *Materials Science.* the contact between two surfaces that are not in motion relative to each other; static contact results in shear stress, causing plastic deformation and therefore wear, even under light loading.

static deflection *Robotics.* load deflection that applies only to loads that are motionless.

static discharger *Electricity.* a device that is attached to the trailing edges of the surface of an aircraft to discharge static electricity in flight.

static dump *Computer Programming.* a dump executed when a program reaches an end-of-run condition or some other distinguishable stage within the run.

statice [stat´ə sē; stat´is] *Botany.* any of various herblike plants of the genus *Limonium* of the leadwort family, characterized by rosettes of narrow evergreen leaves and globular heads of pink, purple, or white flowers that retain their color when dried.

static electricity *Electricity.* energy in the form of a stationary electric charge such as that stored in capacitors and thunderclouds, or produced by friction or induction.

static eliminator *Electronics.* any of various devices used to reduce the effects of atmospheric static interference in a radio receiver. *Graphic Arts.* see STATIC NEUTRALIZER.

static equilibrium *Mechanics.* the condition of a system at rest with respect to its inertial frame, so that the net force on the system is equal to zero.

static error *Statistics.* an error that is independent of time-related variations in observation of the variable.

static evaluation function *Artificial Intelligence.* a function that computes an approximate value of a state (e.g., a position in a board game) without performing a search.

static fatigue *Materials Science.* the growth of subcritical cracks analogous to fatigue at a stress lower than that required for instantaneous fracture. Also, STRESS RUPTURE.

static firing *Space Technology.* a test firing of a rocket in a hold-down position on a test stand or launch pad.

static fluid column *Fluid Mechanics.* an unchanging height of fluid in a vertical tube or container.

static friction *Mechanics.* the force of friction, existing between two surfaces that are in surface contact and at rest, in reponse to an external force tending to slide one over the other. Also, STANDING FRICTION.

static gel buildup *Fluid Mechanics.* a method for determining fluid thixotropy by viscometric measurement of its gel strength.

static granitization *Petrology.* the formation of a granitic rock by metasomatism in the absence of compressive forces.

static grizzly *Mining Engineering.* a grizzly, consisting of a stationary bar screen, set up to screen ore or rock by sorting out unwanted small material or rejecting oversize material.

static head *Fluid Mechanics.* the pressure on a fluid due to the head of the static fluid above it; thus it is also the height to which the fluid will rise in a piezometer.

static induction transistor *Electronics.* a transistor that can function at high current and voltage levels, having current-voltage characteristics that do not saturate and are similar in form to those of a vacuum triode.

static inverter *Electricity.* an electronic device that converts a DC voltage to an AC voltage; used in motor controllers and for uninterruptible power. Also, INVERTER.

staticize *Computer Programming.* **1.** to capture time-dependent data in non-time-variant form, such as a serial bit stream captured as a parallel bit word. **2.** to fetch an instruction from memory and store it in the central processing unit.

static jet thrust *Aviation.* see STATIC THRUST.

static level *Fluid Mechanics.* the elevation of a static liquid surface. *Hydrology.* **1.** the height or level to which water will rise in a well under its full pressure head. Also, HYDROSTATIC LEVEL. **2.** the water level of a well that is not affected by the removal of groundwater.

static line *Aviation.* a cord or webbing that is attached to a parachute pack and hooked inside an aircraft in order to deploy the parachute automatically when its load is dropped from the aircraft.

static load *Mechanics.* the constant load that is exerted by a mass due to its weight. *Aviation.* a load imposed on a flight vehicle under static conditions, such as the weight of the wings; often contrasted with the dynamic load imposed on the same craft by acceleration.

static machine see ELECTROSTATIC GENERATOR.

static map *Cartography.* any map or chart that shows information limited to a specific date or time; most maps are static maps.

static metamorphism *Geology.* a process of regional metamorphism induced by heat and solvents at high lithostatic pressures, rather than by orogenic deformation.

static moment *Mechanics.* see MOMENT.

static neutralizer *Graphic Arts.* an attachment on a printing press that is designed to carry away any static electricity which may build up on the surface of paper as sheets move through the press. Also, STATIC ELIMINATOR.

static oceanography *Oceanography.* the study of the nondynamic aspects of oceanography; e.g., the structure of the ocean floor and the physical and chemical composition of seawater.

static pressure *Acoustics.* a steady or constant pressure that would exist in a medium in the absence of sound. *Fluid Mechanics.* **1.** the normal compressive stress at a point when a fluid is static. **2.** the average of any three mutually perpendicular components of a normal compressive stress at a point when a fluid is in motion.

static-pressure tube see STATIC TUBE.

static random-access memory *Computer Programming.* random-access memory in which data does not need to be refreshed and is retained as long as the computer is on.

static reaction *Mechanics.* a force or moment exerted on a body by other bodies to keep it in equilibrium.

static recovery *Materials Science.* in microstructures of metals at elevated temperatures, a process that occurs between intervals of straining or after metal flow has ceased, leading to a loss of dislocations and therefore to a decrease in yield strength or flow stress.

static recrystallization *Materials Science.* in the microstructure of metals at elevated temperatures, the recrystallization that takes place between intervals of straining or after metal flow has ceased.

static reflex *Physiology.* any response that maintains the body posture and that is produced by a stationary stimulus, such as lateral pressure on the body.

static regulator *Electronics.* a transmission regulator whose adjusting mechanism is in self-equilibrium at any setting; control power must be applied to change the setting.

statics *Mechanics.* the branch of mechanics that deals with bodies or particles at rest.

static sensitivity *Electronics.* the quotient of the direct anode current divided by the incident radiant flux of constant value in phototubes.

static split tracking see MONOPULSE RADAR.

static stability *Physics.* a condition in which a system is in equilibrium and tends to remain in that state unless an external influence is applied, whereupon the system would react so as to return to equilibrium. *Meteorology.* a condition in which an atmosphere maintains hydrostatic equilibrium with respect to vertical displacements, usually considered by the parcel method; the criterion for stability is that the displaced parcel be subject to a buoyant force opposite to its displacement, e.g., that a parcel displaced upward be colder than its new environment. Also, HYDROSTATIC STABILITY, VERTICAL STABILITY.

static storage *Computer Programming.* a memory storage in which each location has a fixed space, and data in each location is available at any time.

static subroutine *Computer Programming.* a subroutine that does not require parameters other than operand addresses.

static switching *Electricity.* a circuit that is switched by means of magnetic amplifiers, semiconductors, and other devices that have no moving parts.

static test *Engineering.* **1.** a test of a structure or system carried out under unchanging environmental conditions. **2.** a test of a rocket or jet engine that is fixed in place during the test.

static thrust *Aviation.* the thrust produced by a jet engine, rocket motor, or the like, when fired in a hold-down position.

static tube *Engineering.* a tube attached to a fluid flow line and used to measure the static, rather than dynamic or kinetic, pressure in the line. Also, STATIC-PRESSURE TUBE.

static-type checking *Computer Science.* a checking or determination of the types of variables in a language at compile time, eliminating the need for dynamic-type checking, but requiring that a variable have only a single type.

static universe *Astronomy.* a closed universe of finite volume with a constant radius of curvature.

static variable *Computer Programming.* a local variable that retains its value between invocations of the routine that defines it.

static vehicle envelope *Transportation Engineering.* the space occupied by a stationary vehicle.

static weapon *Ordnance.* a weapon that is employed in a particular position and does not move during the period of its employment.

static work *Industrial Engineering.* work that is performed with the hands or arms, involving little or no movement on the part of the worker, such as holding.

station *Science.* a place where a phenomenon can be observed and studied in its normal or natural environment. *Transportation Engineering.* a loading and unloading facility on a transportation line, especially a rail or bus line. *Space Technology.* **1.** a facility or place in space or on the ground where some activity is carried out. **2.** in a stationary orbit, the point on the earth's surface over which a satellite's position is maintained. *Electronics.* **1.** a location at which radio, television, radar, or other electric equipment is installed. **2.** an input or output in a communications system. *Engineering.* **1.** in an industrial facility, the location or position of a given machine. **2.** a location on a survey line or street center line, usually at 100-foot increments. *Military Science.* **1.** any military or naval activity at a fixed land location. **2.** a section of the ocean patrolled by the U.S. Navy. **3.** any place of duty, post, or position in the field to which an individual, group, or unit may be assigned. *Mining Engineering.* **1.** an excavation adjoining the shaft at any level, at which material and workers are delivered or removed. **2.** an enlargement of a shaft or gallery, providing room for landing and receiving loaded mine cars. *Cartography.* in surveying, any distinct point on the surface of the earth, which may or may not be indicated by a mark or monument, and of which the exact location has been determined by surveying methods. *Neurology.* the position taken while standing. *Industrial Engineering.* see WORK STATION. *Computer Technology.* see DATA STATION.

stationary *Mechanics.* having a fixed, unchanging position; not moving; motionless.

stationary distribution *Physics.* any distribution of a scalar quantity that is independent of time.

stationary engine *Mechanical Engineering.* an engine that is fixed in place for prolonged use.

stationary front see QUASI-STATIONARY FRONT.

stationary grid *Radiology.* a radiographic grid placed in apposition to a roentgenographic film, producing a well-defined image with grid lines.

stationary noise *Electronics.* a random noise for which the probability that the noise voltage lies within any given interval remains constant with time.

stationary orbit *Space Technology.* an orbit of a satellite about a primary body at the angular rate at which the primary body rotates on its axis; thus, from the primary body, the satellite appears to be stationary over a particular point. With respect to the earth, a stationary orbit can be referred to as a "24-hour orbit."

stationary phase *Mechanics.* the condition of a body or system at rest. *Microbiology.* in a bacterial culture, the period when the cell population is relatively stable because the bacteria are not growing or the number of bacteria dying equals approximately the number dividing. *Biotechnology.* the solid packing or support material used in chromatography; used in the form of gels, sheets, or powders.

stationary point *Astronomy.* the point in the movement of a planet at which it appears to stand still against the stars; this is an optical illusion resulting from the relative motion between the planet and earth.

stationary population *Ecology.* a population with stable age distribution among its various groups.

stationary satellite see FIXED SATELLITE.

stationary state *Physics.* an energy state that an atom may occupy without emitting electromagentic radiation.

stationary stochastic process *Mathematics.* a continuous parameter or countable stochastic process $\{X(t)\}$ having the property that each joint probability distribution of the process is invariant with respect to t.

stationary time series *Statistics.* a time series whose probability distribution is not affected by incremental changes in the time period defining it.

stationary wave see STANDING WAVE.

station buoy *Navigation.* a secondary buoy moored near a lighted buoy or lightship to mark the station if the primary navigational aid breaks its mooring. Also, MARKER BUOY, WATCH BUOY.

station continuity chart *Meteorology.* a continuity chart on which one coordinate is time, and the other is one or more of the meteorological elements observed at that station.

station control *Control Systems.* a module that controls a work station and is controlled by a cell-control module.

station density *Transportation Engineering.* the relative number of stations in an area, measured as a ratio of stations to land area, population, line length, or other criteria.

station elevation *Meteorology.* the vertical elevation of a station above mean sea level, used as the reference datum level for current measurements of atmospheric pressure at the station.

station keeping *Space Technology.* a program or series of corrective actions that must be taken periodically in order to maintain a satellite in its predetermined, programmed, or desired orbit or station.

station mark *Cartography.* in surveying, some physical indication of the exact location of a survey station; may be permanent or temporary, and artificial or natural.

station model *Meteorology.* the pattern used to record on a weather map the meteorological symbols representing the weather at a particular observing station.

station pole *Civil Engineering.* one of a number of rods or poles used in surveying to measure elevations, to sight points, or to mark stations.

station pressure *Meteorology.* atmospheric pressure as computed with respect to station elevation.

station stock level *Military Science.* the maximum quantity of supplies allowed at any particular installation, expressed in days of supply.

statismospore *Mycology.* a fungal spore that is not forcibly discharged.

statistic *Statistics.* **1.** any function or summary of a sample, such as an estimator or a test. **2.** broadly, any item of information in numerical or mathematical form.

statistical *Statistics.* **1.** expressed in the form of statistics; consisting of data, especially data in numerical form. **2.** of or relating to the science of statistics.

statistical analysis *Statistics.* an analysis of data derived from a sample in order to predict the characteristics of the population under study; many analytical procedures and mathematical models may be applied.

statistical copolymers *Materials Science.* a polymer that is specific in that its chain consists of statistically distributed monomeric units, as compared with an alternating copolymer, in which the molecules consist of two species of monomers distributed in alternating sequence.

statistical estimation *Statistics.* a method of estimating population parameters from sample data with varying degrees of precision.

statistical forecast *Meteorology.* a weather forecast that is based on a statistical analysis of recent weather patterns, rather than on purely thermodynamic and hydrodynamic considerations.

statistical hypothesis *Statistics.* see HYPOTHESIS.

statistical independence *Statistics.* the case in which the occurrence of one event does not affect, nor is affected by, the occurrence of another event.

statistical inference *Statistics.* the process of using mathematical methods to gain greater understanding of a population's characteristics by technical manipulation of sample data, either by estimation or hypothesis testing.

statistically significant *Statistics.* describing evidence in which the discrepancy between the sample evidence and the null hypothesis is too large or improbable to be attributed to chance. Thus, **statistical significance.**

statistical mechanics *Physics.* a branch of physics that is concerned with the macroscopic behavior of a system based on the theory and treatment of microscopic interactions between the constituent parts of the system, in which the number of constituent parts is usually very large.

statistical monitor *Computer Programming.* a utility program that collects system performance data.

statistical multiplexer *Electronics.* a multiplexer utilizing the statistical properties of the various channels to vary the way they are combined into a single channel, rather than using a fixed combining scheme, resulting in increased capacity of the multiplexed channel.

statistical outlook *Meteorology.* a long-term mathematical projection of temperature or precipitation. Also, **statistical projection.**

statistical parallax *Astronomy.* the mean parallax for a star group (in which individual stars are assumed to lie at the same distance) as determined from each radial velocity and proper motion and averaged over the group.

statistical quality control *Industrial Engineering.* a quality-control procedure that employs statistical extrapolation from a test sample to determine the quality of entire lots.

statistical universe *Statistics.* see UNIVERSE.

statistical weight *Statistics.* a number assigned to a value or interval that represents the number of individual observations of a random variable which assume that value or fall within that interval; a measure of frequency. *Physics.* the number of microstates corresponding to a particular macrostate.

statistician *Statistics.* **1.** a person who compiles statistics or is an expert on statistics. **2.** a professional in the feld of statistics.

statistics *Science.* **1.** in general, any collection of data, especially data expressed or presented in numerical form. **2.** the body of techniques used to facilitate the collection, organization, analysis, and interpretation of data. **3.** the branch of science that deals with data in this manner, using mathematical probability theories.

statmho see STATSIEMEN.

statoblast *Invertebrate Zoology.* an asexually produced, encased bud of a freshwater bryozoan, which germinates in spring after the disintegration of the parent colony in fall.

statocone *Invertebrate Zoology.* a microscopic cone-shaped statolith in a statocyst of some mollusks and other invertebrates.

statoconium see OTOLITH.

statocyst *Botany.* a cellular cyst containing one or more statoliths in a fluid medium. *Invertebrate Zoology.* a small organ of balance in many invertebrates, consisting of a fluid-filled sac containing one or more statoliths that helps indicate position when the animal moves. Also, LITHOCYST, OTOCYST.

statohm *Electricity.* the unit of resistance in the centimeter-gram-second electrostatic system of units; one statohm is approximately equal to 8.98766×10^{11} ohms.

statokinetic *Physiology.* concerned with providing balance and stability of the body during motion.

statolith *Botany.* a starch grain or other solid body contained within a plant cell that, by a change in position within the cell, causes an organ or part of the plant to change position. *Invertebrate Zoology.* any of the granules of lime or sand, suspended in fluid, that are contained in a statocyst.

stator [stā´tər] *Mechanical Engineering.* the stationary part of a machine around which a rotor turns. *Electricity.* **1.** the stationary part of a motor or generator. **2.** the stationary plate of a rotating capacitor. Also, STATOR PLATE. **3.** the conducting surfaces of a switch. *Aviation.* the system of stationary airfoils in the compressor of a jet engine.

stator armature *Electricity.* the winding on the stator.

stator blade *Aviation.* a blade or vane that remains in a fixed position relative to a rotating blade, such as a fixed vane in a gas turbine.

statoreceptor *Physiology.* any of numerous specialized cells receiving stimuli and reacting to the body position in space, such as the utricle and sacculus of the human inner ear.

stator plate *Electricity.* see STATOR, def. 2.

statoscope *Engineering.* an instrument used to measure small variations in gas pressure or small changes in altitude.

statospore *Botany.* a resting spore such as the thick-walled resting spore formed within frustules of diatoms.

statsiemen *Electricity.* the unit of conductance, admittance, and susceptance in the centimeter-gram-second electrostatic system of units; one statsiemen is approximately equal to 1.1126×10^{-12} siemen. Also, STATMHO.

stattesla *Electromagnetism.* the unit of magnetic flux density in the centimeter-gram-second system of units; one stattesla is approximately equal to 2.9979×10^{6} tesla.

Statistics

Statistics is the science and art of dealing with uncertainty in all its forms. It borders on each of the other sciences to the extent that statements in them are uncertain, and on philosophy in that much of modern epistemology concerns inference under uncertainty.

There are several important approaches to uncertainty being pursued in statistics. The oldest, called Bayesian (after Reverend Thomas Bayes, 1702–1761), has been revived through the work of Ramsey (1903–1930), of deFinetti (1906-1985), and of Savage (1917–1971). In its modern form, the Bayesian approach generally takes probability models to be subjective statements about the world. Probability distributions are conditioned on information as it becomes available, transforming *prior* distributions to *posterior* distributions.

The classical, or sampling theory, school rests on the work of Neyman (1894–1981) and Pearson (1895–1980), of Fisher (1890–1962), and of Gossett (1876–1937). The probabilistics basis of sampling theory is the limiting frequency of an imagined infinite sequence of independent, identical trials, and is intended to be "objective." Typical methods include *hypothesis testing* and *confidence intervals.* While data are conceived of as having distributions, parameters are to be taken as unknown constants not having distributions. By contrast, in the Bayesian view there is no essential distinction between data and parameters, only between what is known and what is not.

Modern statistics uses computing heavily, and often performs data analyses that do not square easily with either of the approaches mentioned above. This pragmatic challenge to solve practical problems of inference is often in tension with the desire to unify the theoretical basis of statistics.

Some of the important journals in statistics include the *Journal of the American Statistical Association, Biometrika,* the *Annals of Statistics,* the *Journal of the Royal Statistical Society,* and the *Canadian Journal of Statistics.*

Joseph B. Kadane
Leonard J. Savage Professor,
Statistics and Social Sciences
Carnegie Mellon University

statuary bronze *Metallurgy.* any of several copper-tin alloys suitable for casting art works.

status [stat´əs; stāt´əs] *Anthropology.* a position in society in which one person is compared to others and that may be achieved through accomplishments, granted by birth, or otherwise ascribed by the society. *Behavior.* the relative position of an animal within a group.

status asthmaticus *Medicine.* a particularly severe episode of asthma, usually requiring hospitalization, that does not respond adequately to ordinary therapeutic measures.

status byte see STATUS REGISTER.

status epilepticus *Medicine.* a series of epileptic seizures with no periods of consciousness between episodes.

status line *Computer Programming.* **1.** a location on a display screen reserved for the display of processing status information. **2.** a connector on an input/output bus that transmits the status of the device to the central processing unit.

status marmoratus *Neurology.* a condition marked by excessive myelinization of the nerve fibers of the corpus striatum; characteristic of Vogt's syndrome. Also, MARBLE STATE.

status register *Computer Programming.* a register designed to store information about the condition of hardware components and computational results. Also, STATUS BYTE.

status spongiosis see SUBACUTE SPONGIFORM ENCEPHALOPATHIES.

status word see PROCESSOR STATUS WORD.

statute league see LEAGUE, def. 1.

statute mile see MILE, def. 1.

statvolt *Electricity.* the unit of voltage in the centimeter-gram-second electrostatic system of units; one statvolt is approximately equal to 2.997925×10^2 volts.

statweber *Electromagnetism.* the unit of magnetic flux in the centimeter-gram-second electrostatic system of units; one statweber is approximately equal to 2.9979×10^2 webers.

Staudinger, Hermann 1881–1965, German chemist; awarded the Nobel Prize for his research in polymers.

Staudinger's hypothesis *Materials Science.* a theory stating that polymers consist of linear chains that may or may not be crosslinked.

Staurastrum *Botany.* a genus of unicellular desmids (green algae) of the family Desmidiaceae.

staurolite *Mineralogy.* $(Fe^{+2},Mg,Zn)_2Al_9(Si,Al)_4O_{22}(OH)_2$, a reddish-brown to black, monoclinic, psuedo-orthorhombic mineral, occurring as commonly twinned prismatic crystals, and having a specific gravity of 3.65 to 3.83 and a hardness of 7 to 7.5 on the Mohs scale; found most commonly in metamorphic rocks. Also, GRENATITE.

Stauromedusae *Invertebrate Zoology.* an order of sessile scyphozoan coelenterates that attach to seaweeds or stones by a stalk with a sucker.

Staurosporae *Mycology.* a fork-shaped or star-shaped spore found in certain imperfect fungi.

stave *Mechanical Devices.* **1.** one of a series of narrow strips of wood or metal that, when placed together, form the sides, lining, or covering of a structure such as a barrel. **2.** a projection that cradles a crankshaft during forging. *Design Engineering.* a rung of a ladder.

stay *Engineering.* a tensile structural member, typically a wire or cable. *Naval Architecture.* **1.** a rope or wire that supports a mast in a fore-and-aft direction. **2.** a structural reinforcing member of a rudder. **3.** the position of an anchor chain leading away from a vessel; a nearly vertical chain is at **short stay**; a gently curving chain is at **long stay**.

stayed-cable bridge *Civil Engineering.* a modified cantilever bridge supported by sloped cables connected to a tower or number of towers, usually consisting of girders and trusses cantilevered both ways from the central tower.

stbd. starboard.

STD recorder see SALINITY-TEMPERATURE-DEPTH RECORDER.

std. standard.

steadite *Metallurgy.* a constituent of cast iron consisting of ferrite and iron phosphide.

steady flow *Fluid Mechanics.* a flow in which all properties (such as fluid density) and values (such as fluid velocity at a point) are constant with respect to time.

steady state *Physics.* a state of a system in which the conditions do not change in time. *Control Systems.* a condition in which all transient responses have died out, and all values either remain virtually constant or recur in a cyclical pattern. *Electricity.* **1.** a condition in direct-current circuits in which all currents and voltages are essentially constant. **2.** a condition in alternating-current circuits in which the phase, magnitude, and waveform of all currents and voltages are essentially constant.

steady-state conduction *Thermodynamics.* the heat conduction through a medium that is constant when the temperature of each point in the medium does not change with time.

steady-state current *Electricity.* an electric current whose periodic waveform does not change with time (for alternating current), or whose value does not change with time (for direct current).

steady-state diffusion *Materials Science.* atomic diffusion that takes place under a concentration gradient that does not change with time, such as the diffusion of a gas through a thin metal membrane where a high gas pressure is maintained on one side and a low pressure on the other side.

steady-state distribution *Analytical Chemistry.* the equilibrium condition between phases in each step of a countercurrent liquid-liquid extraction procedure.

steady-state error *Control Systems.* an error that persists in a control system after the elimination of transient conditions.

steady-state flow *Chemical Engineering.* a fluid flow with no change in composition or phase equilibria relationships over time.

steady-state infection *Virology.* the simultaneous occurrence of virus replication and cell multiplication in cell culture.

steady-state model *Astronomy.* see STEADY-STATE THEORY. *Petroleum Engineering.* an electric or electrolytic model representing steady-state fluid flow through porous reservoir media; examples are gel, blotter, liquid, and potentiometric models.

steady-state reactor *Nucleonics.* a nuclear reactor that operates under conditions that do not significantly change in time.

steady-state theory *Astronomy.* a theory originally propounded by Herman Bondi, Thomas Gold, and Fred Hoyle, according to which the universe has no beginning and no end but has average properties, constant in space and time; it assumes that new matter is constantly and spontaneously created to maintain these average values in the face of the expansion of the universe. This theory contrasts with, and has generally been supplanted by, the Big Bang theory, which postulates an original episode of extremely rapid expansion and continuously changing properties. Also, STEADY-STATE MODEL.

steady-state vibration *Mechanics.* vibration in which the velocity of each particle in a system varies continuously and periodically, with constant amplitude.

steady-state wave motion *Physics.* a type of simple wave motion in which all quantities that undergo the wave action do so in a periodic manner.

Stealth bomber *Aviation.* a flying-wing aircraft, about 17 feet high and 69 feet long, with a wingspan of about 172 feet, that carries a crew of two; it has the capability to deliver both nuclear and conventional weapons against fixed and mobile targets, and has a very good range, employing low observables, or **Stealth,** technology.

Stealth bomber (B-2)

steam *Physical Chemistry.* **1.** water vapor, especially such vapor having a temperature above the boiling point of water. **2.** in general, the vapor of any liquid. *Mechanical Engineering.* water in this vapor form used for the purpose of producing mechanical energy.

steam accumulator *Mechanical Engineering.* a pressure vessel on a boiler in which water is heated by steam during low demand periods and regenerated as steam when required.

steam attemperation *Mechanical Engineering.* the control of the maximum temperature of superheated steam by water injection or submerged cooling.

steamboat *Naval Architecture.* a small vessel, especially a vessel used on inland waters, that is powered by steam.

steamboat

steam boiler *Mechanical Engineering.* a vessel or plant for generating steam in which water is vaporized by transferring heat from a source of higher temperature. Also, STEAM GENERATOR.

steam bronze *Metallurgy.* a former designation of a copper-base alloy nominally containing 6% tin, 4.5% zinc, and 1% lead; used for shaped castings.

steam cock *Engineering.* a valve for opening or closing off a steam line.

steam condenser *Mechanical Engineering.* a device that condenses exhaust steam from an engine or turbine into water.

steam cracking *Chemical Engineering.* a petroleum-refinery process that uses steam to provide the heat for cracking oil.

steam cure *Engineering.* the use of steam to cure concrete or mortar.

steam cycle see RANKINE CYCLE.

steam distillation *Chemical Engineering.* a petroleum-refinery distillation that uses steam to minimize cracking, lowering the boiling point of the oils being distilled by adding the vapor pressure of the steam. Also, STEAM STRIPPING.

steam drive *Mechanical Engineering.* any device in which the pressure of expanding steam generates the power to move a machine or machine part.

steam dryer *Mechanical Engineering.* a device that removes moisture droplets from the steam-vapor phase.

steam emulsion test *Engineering.* a test in which steam is used to demulsify oil and water; the emulsion is measured at five-minute intervals, ending when the emulsion is reduced below 3 milliliters.

steam engine *Mechanical Engineering.* a reciprocating engine, worked by the force of steam on the piston, in which the steam expands from the initial pressure to the exhaust pressure in a single stage.

steam engine indicator *Engineering.* an instrument that plots steam pressure and piston displacement in a reciprocating steam engine.

steamer *Engineering.* a steamship or other vessel or device powered by steam. *Microbiology.* an apparatus in which a material can be exposed to steam at atmospheric pressure.

steam fog *Meteorology.* a fog that forms when water vapor mixes with air that is much colder than the vapor's source, as when very cold air drifts across warm water. Also, SEA SMOKE, ARCTIC SMOKE, STEAM MIST.

steam gauge *Engineering.* a gauge that determines steam pressure.

steam generator see STEAM BOILER.

steam hammer *Mechanical Engineering.* a vertical mechanical hammer whose ram is raised by a steam cylinder and, added to gravitational action, may be driven down by the cylinder for extra force; used extensively for forging operations.

steam-heated evaporator *Mechanical Engineering.* a device used for evaporating liquid on one side of a tube wall by condensing steam on the other side; the heat of condensation is transferred to the liquid side and utilized in the vaporization process.

steam heating *Mechanical Engineering.* any system, especially a home heating system, that uses steam as the primary medium for providing heat.

steaming *Textiles.* **1.** a finishing process used to shrink and condition fabrics by passing steam through the cloth by means of a perforated steam box. **2.** the process of passing steam over a dyed or printed fabric to set the colors.

steam jacket *Mechanical Engineering.* a jacket formed around a steam-engine cylinder that is supplied with live steam to prevent excessive condensation of the working steam in the cylinder.

steam jet *Engineering.* a jet of high-pressure steam. *Mining Engineering.* **1.** a mine-ventilating system using several jets of steam that blow constantly from pipes at the bottom of the shaft. **2.** a jet of steam that moistens the intake current of a mine, keeping the coal dust wet.

steam-jet cycle *Mechanical Engineering.* a thermodynamic cycle used in refrigeration in which a high-velocity steam jet produces a vacuum in the evaporator, which in turn lowers the boiling point of water.

steam-jet ejector *Mechanical Engineering.* a system for accelerating and ejecting fluid entrainment by a steam jet.

steam locomotive *Mechanical Engineering.* a self-propelled railway steam engine and boiler that is mounted on a frame and fitted with wheels driven by the engine.

steam loop *Engineering.* a simple steam condensation system consisting of condensation pipes forming an inverted U shape that returns to a boiler without relying on an injector or pump.

steam mist see STEAM FOG.

steam molding *Engineering.* the use of steam heat to mold plastic onto a form.

steam nozzle *Mechanical Engineering.* a specially shaped passage for expanding steam, creating kinetic energy of flow with minimum loss.

steam plate *Mechanical Devices.* a cored mounting plate used for molds to allow circulation of steam.

steam point *Thermodynamics.* the true boiling point of water; the temperature at which a mixture of water and steam is in equilibrium at standard atmospheric pressure; i.e., 100°C (212°F).

steam pump *Mechanical Engineering.* a pump that is driven by steam acting on a piston in a cylinder.

steam reheater *Mechanical Engineering.* a component of a steam boiler that adds heat to intermediate-pressure steam which has lost some of its energy through the high-pressure turbine.

steam roller *Mechanical Engineering.* a road roller that is powered by steam.

steam separator *Mechanical Engineering.* a device that removes the liquid phase from a steam volume, leaving only the vapor phase. Also, **steam purifier.**

steamship *Naval Architecture.* an oceangoing vessel that is powered by steam.

steamshovel or **steam shovel** *Mechanical Engineering.* a power shovel that is driven by a steam engine.

steam still *Chemical Engineering.* a petroleum-refining still in which steam provides most of the heat, so that distillation can be achieved at a lower temperature than in standard equipment.

steam stripping see STEAM DISTILLATION.

steam superheater *Mechanical Engineering.* a steam generator component that adds additional heat to the steam after it has been vaporized.

steam tracing *Engineering.* the use of steam to keep a fluid line hot, typically by jacketing the line in a coiled steam line to prevent liquids from solidifying.

steam trap *Mechanical Engineering.* an automatic device that allows water or air to pass but prevents the passage of steam; used to drain the water of condensation that accumulates in steam pipes.

steam-tube dryer *Mechanical Engineering.* a rotary cylindrical dryer that uses the heat of steam contained in tubes for drying; the tubes run the entire length of the cylinder and rotate with it.

steam turbine *Mechanical Engineering.* a machine in which high-pressure steam is made to do work by acting on and rotating blades in a cylinder.

steapsin *Biochemistry.* an enzyme in pancreatic juice that catalyzes the hydrolysis of fats.
stear- or **stearo-** a prefix meaning "fat." Also, **steat-, steato-.**
stearamide *Organic Chemistry.* $CH_3(CH_2)_{16}CONH_2$, colorless leaflets; insoluble in water and slightly soluble in alcohol and ether; melts at 109°C and boils at 251°C (12 torr); used as a corrosion inhibitor. Also, OCTADECANAMIDE.
stearane *Organic Chemistry.* a cycloalkane derived from a sterol.
stearate *Organic Chemistry.* a salt or ester of stearic acid having the formula $CH_3(CH_2)_{16}COOX$, where X is a monovalent radical.
stearic acid *Organic Chemistry.* $CH_3(CH_2)_{16}COOH$, a combustible, waxy solid that is insoluble in water and soluble in alcohol and ether; it melts at 69.6°C and boils at 213°C (5 torr); it is the most common fatty acid occurring in natural animal and vegetable fats; it is used in soaps, pharmaceuticals, cosmetics, and food packaging, and as a lubricant. Also, *n*-OCTADECANOIC ACID.
stearin *Organic Chemistry.* $C_3H_5(C_{18}H_{35}O_2)_3$, colorless crystals or powder; insoluble in water and soluble in alcohol; melts at 72°C; used in soaps, candles, candies, metal polishes, and water-proofing paper. Also, GLYCERYL TRISTEARATE.
stearolic acid *Organic Chemistry.* $CH_3(CH_2)_7C{\equiv}C(CH_2)_7COOH$, a crystalline solid; soluble in ether; melts at 48°C and boils at 189–190°C (1.8 torr). Also, 9-OCTADECYNOIC ACID.
stearone *Organic Chemistry.* $(C_{17}H_{35})_2CO$, a combustible, leaflike solid that is insoluble in water and melts at 88°C; used as an antiblocking agent. Also, 18-PENTATRIACONTANONE.
stearyl alcohol *Organic Chemistry.* $CH_3(CH_2)_{16}OH$, combustile white flakes or granules; insoluble in water and soluble in alcohol and ether; melts at 59°C and boils at 210°C (15 torr); used in cosmetics, perfumes, lubricants, and resins. Also, OCTADECANOL.
steatite *Petrology.* a compact, massive, fine-grained rock composed predominantly of talc, but with a significant amount of other materials; essentially an impure, talc-rich rock.
steatization *Geology.* a process in which talc, or steatite, is introduced into a sediment or rock, or replaces other material, especially the hydrothermal alteration of ultrabasic rock into talcose rock.
steatocystoma *Medicine.* an epithelial cyst.
steatoma *Oncology.* **1.** a fatty mass within a sebaceous gland. **2.** see LIPOMA.
steatomatosis *Oncology.* the simultaneous occurrence of numerous steatomas of the sebaceous glands.
steatopygia *Anthropology.* a condition in certain populations in which individuals have broad masses of fat protruding from the buttocks and thighs with an additional slant to the sacrum; often noted among women of the Bushmen and Hottentot groups of Africa.
Steatornithidae *Vertebrate Zoology.* the oilbird, a monotypic family of nocturnal, fruit-eating birds of the order Caprimulgiformes, characterized by a large head with a strong bill and by young that get extremely fat, sometimes reaching twice their adult weight; native to forests of northern South America.
steatorrhea *Medicine.* excessive amounts of fat in the stool, as in malabsorption syndromes.
Steckel rolling *Metallurgy.* a process of rolling in a two-high strip mill in which the rolls are not mechanically driven, but in which the product is pulled from either end.
S tectonite *Petrology.* a tectonite characterized by a fabric that is dominated by planar elements caused by deformation, such as slate.
Stedman disintegrator see CAGE MILL.
steel *Metallurgy.* any of various alloys of the elements iron and carbon containing less than 2.5% carbon, usually also with lesser amounts of other elements, having substantial qualities of strength, hardness, and malleability.
steel bronze *Metallurgy.* a former designation for a copper-tin alloy suitable for making guns.
steel-clad rope *Mechanical Devices.* a hoisting rope with a ribbon of flat steel wrapped around each of its six strands, to reduce wear and add flexibility.
steel converter *Metallurgy.* a furnace, such as a Bessemer furnace, used for steelmaking.
steel-die engraving *Graphic Arts.* a form of intaglio printing in which an image is engraved in reverse on the surface of a steel die or plate; used to produce sharply defined characters on stationery, invitations, and the like.
steel emery *Metallurgy.* abrasive material consisting of quenched iron particles.
Steelflex coupling *Mechanical Engineering.* a flexible coupling consisting of two grooved steel hubs, each fitting into its respective shaft, joined together by a tempered steel member.
steel foil *Metallurgy.* a very thin sheet or strip of any steel.
steel-grinder's disease see SIDEROSIS.
steel jack *Mineralogy.* a popular name for sphalerite (Zn,Fe)S, a black or brown, resinous, cubic mineral consisting of zinc sulfide, occurring commonly in coarse to fine granular masses, and having a specific gravity of 3.9 to 4.2 and a hardness of 3.5 to 4. on the Mohs scale.
steel-jacket(ed) *Ordnance.* of or relating to small-arms ammunition with a steel outer shell. Similarly, **steel-case(d).**
steelmaking *Metallurgy.* the art and science of producing steel from iron, scrap, and alloying elements.
steel sets *Mining Engineering.* steel structures with I-beams for caps and H-beams for posts or wall plates; used in shafts in metal mines and in main entries in coal mines.
steel wool *Metallurgy.* ferrous filaments, irregularly bundled; used for scouring, polishing, and the like.
steelyard *Engineering.* a weighing device that uses a counterbalance arm to measure the weight of a suspended load.
steen *Civil Engineering.* the process of lining a well or soakaway with stones or bricks laid usually dry without mortar.
steenbok see STEINBOK.
steepest descent method *Mathematics.* a numerical technique for approximating functions of the form

$$g(\alpha) = \int_C e^{\alpha f(z)} dz$$

near saddle points, where α is assumed to be large, positive, and real, and C is a path in the complex plane such that the ends of the path do not contribute significantly to the value of the integral. The integrand is approximated by use of a Taylor series approximation for $f(z)$ near its maximum or other saddle point.
steeple *Architecture.* a church tower and its spire.
steer *Navigation.* to direct the movement of some vehicle or craft.
steer *Agriculture.* a male bovine animal, especially one that has been castrated.
steerable antenna *Electromagnetism.* an antenna whose major lobe can be easily positioned over a range of directions.
steering *Meteorology.* any effect that an outside force has on the direction of movement of an atmospheric disturbance, such as the influence of an isotherm on a surface pressure system.
steering arm *Mechanical Engineering.* an arm in an automotive vehicle that transmits the rotation movement of the steering wheel to the drag link.
steering brake *Mechanical Engineering.* a brake that slows or stops a tracked vehicle from one side.
steering column *Mechanical Engineering.* a column that connects the steering wheel of an automotive vehicle to the steering gear assembly.
steering compass *Navigation.* the compass used by the helmsman to keep a ship on the set course.
steering engine *Naval Architecture.* an engine used to turn a ship's rudder in response to turns of the wheel.
steering flow *Meteorology.* a basic fluid flow that strongly influences the direction of movement of a disturbance embedded within it. Also, **steering current.**
steering gear *Mechanical Engineering.* **1.** the directional guidance system of a vehicle, including the drop arm, drag link, steering arm, and track rod. **2.** the slewing gear for guiding a ship.
steering level *Meteorology.* a hypothetical level at which the velocity of the basic flow is directly proportional to the velocity of an atmospheric wave-disturbance motion embedded in it; used in forecasting the motion of surface pressure systems.
steering repeater *Navigation.* a remote device that repeats the reading of the gyrocompass for use by the helmsman.
steering wheel *Mechanical Engineering.* a hand-operated wheel for controlling the direction of a vehicle or vessel.
Stefan, Josef 1835–1893, Austrian physicist; made investigations in fluid mechanics; stated the Stefan-Boltzmann law.
Stefan-Boltzmann constant *Physics.* the constant of proportionality in the Stefan-Boltzmann law, symbolized by σ; approximately equal to 5.672×10^{-5} erg/cm^2-kelvin^4-sec. Also, **Stefan's law of radiation.**
Stefan-Boltzmann law *Physics.* a law giving the relation between the energy flux E radiated by a blackbody having the absolute temperature $T : E = \sigma T^4$, where σ is the Stefan-Boltzmann constant.

Stefan number *Thermodynamics.* a dimensionless number used in calculations involving radiant heat transfer, given by the product of the Stefan-Boltzmann constant, the absolute temperature raised to the third power, and the thickness of a layer divided by the thermal conductivity of the layer.

Stefan's formula *Oceanography.* an equation set up by Josef Stefan in 1891 for the growth of thickness *h* of a sea ice cover at various freezing temperatures, in the form $h = (2K/Ld)E$, where *K* is the coefficient of thermal conductivity, *L* is the latent heat of fusion, *d* is the density of the ice, and *E* is the cold sum (days of freezing exposure since first ice formation); the solution disregards the snow cover, any radiation imbalance, and heat transport through the water.

Steganopodes *Vertebrate Zoology.* a former term for an order including swimming birds with totally webbed feet.

Stegodontidae *Paleontology.* a family of proboscideans in the extinct suborder Mammutoidea; characterized by low-crowned teeth and the absence of lower-jaw tusks; extant in the Miocene and Pliocene.

Stegosauria *Paleontology.* a suborder of herbivorous dinosaurs in the order Ornithischia; characterized by a spinal row of plates or spines; extant almost worldwide from the Lower Jurassic to the Lower Cretaceous, when they may have been replaced by ankylosaurs and other smaller dinosaurs.

Stegosaurus *Paleontology.* a genus of herbivorous ornithischian dinosaurs in the suborder Stegosauria; known only from the Upper Jurassic of North America.

steigerite *Mineralogy.* $AlV^{+5}O_4 \cdot 3H_2O$, an inadequately described, rare, canary-yellow monoclinic mineral occurring as cryptocrystalline coating on sandstone of the Colorado-Utah uranium district.

Stein, William Howard 1911–1980, American biochemist; shared the Nobel Prize for the method of determining amino-acid sequences in proteins.

steinbok *Vertebrate Zoology.* a small antelope, *Raphicerus campestris,* characterized by a reddish-brown coat and short vertical horns in the male; found in southern and eastern Africa. Also, STEENBOK.

Steiner, Jakob 1796–1863, Swiss-born German mathematician; developed projective geometry; proved the Poncelet-Steiner theorem.

Steiner's theorem see PARALLEL-AXIS THEOREM.

Steiner tunnel test *Materials Science.* a test of the flammability of elastomers; used in carpet and rug underlay.

Steinheil lens *Optics.* an aplanatic magnifying lens that sandwiches a biconvex element of crown glass between two flint-glass elements.

Steinheim man *Paleontology.* an early form of *Homo sapiens*, known from a fairly well-preserved skull without the lower jaw found at Steinheim, Germany; the brow ridge is prominent, the upper jaw is strongly built, and the cranial capacity is between 1100 and 1200 cc, above the lower limits of skull variation in *H. sapiens*; the age has been estimated at about 400,000 years Before Present.

steinkern *Geology.* **1.** rock material that is composed of consolidated mud or sedimentary deposits that filled a fossil shell or other hollow organic structure. **2.** the fossil that results when such a mold is dissolved. Also, ENDOCAST, INTERNAL CAST.

Stein-Leventhal syndrome *Medicine.* a complex of symptoms found in female development, in which there are enlarged ovaries with many cysts, cessation of the menses or abnormal uterine bleeding, or both, and in many cases abnormal hairiness; sometimes also found with retarded breast development and obesity. (Named for the American gynecologists Irving *Stein* and Michael *Leventhal.*) Also, POLYCYSTIC OVARY DISEASE, POLYCYSTIC OVARY SYNDROME.

Steinmann pin *Medicine.* a metal rod for the fixation of the ends of fractured bones, such as the femur or tibia.

Steinmetz, Charles Proteus 1865–1923, German-born American electrical engineer; formulated the law of hysteresis; devised a method of calculating alternating current.

Steinmetz coefficient *Electromagnetism.* the proportionality constant found in Steinmetz's law. *Physics.* see HYSTERESIS COEFFICIENT.

Steinmetz's law *Electromagnetism.* a law stating that during a cyclic magnetization change, the amount of heat generated per unit volume per cycle is proportional to the maximum magnetic induction raised to the power of 1.6 and the constant of proportionality is dependent only on the material.

stela *plural,* **stelae.** *Archaeology.* an upright stone monument with carved inscriptions or relief figures; often found in Mesoamerican sites.

stele [stē´lē; stēl] *Botany.* a central cylinder of vascular tissue, pith, and pith rays in the stem and root of vascular plants. *Architecture.* a prepared surface on the face of a building or a rock bearing an inscription.

Stelenchopidae *Invertebrate Zoology.* a family of polychaete annelid worms in the group Myzostomaria, having a broad body and limited or no segmentation; parasites of echinoderms.

Stella *Bacteriology.* a genus of Gram-negative bacteria occurring as nonmotile, prosthecate, chemoorganotrophic cells; found in fresh water, sewage, and soil.

stellar *Astronomy.* of, relating to, or consisting of stars.

stellar association *Astronomy.* a sparsely populated and loosely bound group of young stars; they have a common origin and space velocity, but the group is soon pulled apart by the gravity of nearby stars and clouds of gas and dust.

stellar atmosphere *Astronomy.* the outer envelope of tenuous gas and plasma that surrounds a star; characterized by pressure, temperature, density, and opacity at varying altitudes and usually having a chemical composition of 90% hydrogen and 9% helium.

stellarator *Physics.* an experimental high-temperature plasma device designed in a closed figure-eight or racetrack configuration; so-called because the reactions produced resemble those in stars.

stellar camera *Photogrammetry.* a camera specifically designed to photograph stars; may be calibrated to and used in conjunction with one or more mapping cameras in an aircraft, enabling the absolute altitude of photographic plane to be accurately determined. If used to photograph the sun, a stellar camera is referred to as a **solar camera.**

stellar cannibalism *Astronomy.* a process by which one star in a binary system pulls in material from its companion as the latter evolves.

stellar core *Astronomy.* the central region of a star in which energy is produced by thermonuclear fusion.

stellar distribution *Astrophysics.* the number of stars, often of a specified type, in a given volume of space.

stellar evolution *Astrophysics.* the gradual changes in spectrum, luminosity, physical state, and chemical composition that take place during the life cycle of an individual star.

stellar flare *Astronomy.* an explosion of material from the surface of a star, lasting from between a few minutes to over an hour.

stellar guidance *Navigation.* a system in which a guided missile may follow a predetermined course based on the relative position of the missile and certain preselected celestial bodies. Also, CELESTIAL GUIDANCE.

stellar inertial guidance *Navigation.* **1.** the guidance of a flight-borne vehicle, using a combination of stellar guidance and inertial guidance. **2.** the guidance equipment itself.

stellar interferometer *Optics.* a device on a telescope that measures the angular diameter of a star by the spacing of its interference fringes.

stellar interior *Astronomy.* the part of a star beneath its photosphere.

stellar light *Astronomy.* the illumination of night sky from the light of countless stars.

stellar lightning *Geophysics.* a series of lightning discharges that seem to originate from a single point.

stellar luminosity *Astronomy.* the absolute measure of brightness of a star in some unit such as ergs per second or solar luminosities.

stellar magnetic field *Astrophysics.* the field, presumably similar to the sun's, that many stars are known to possess.

stellar mass *Astrophysics.* the total material constituting a star, often using the sun as a unit of measure.

stellar model *Astrophysics.* a theoretical and mathematical solution to the equations that describe the physical conditions, such as temperature, density, and pressure, at various depths inside a star.

stellar parallax *Astronomy.* **1.** the angle formed by the radius of the earth's orbit observed at the location of a given star. **2.** more generally, the star's distance from earth as indicated by this angle.

stellar photometry *Astronomy.* the precise measurement of a star's brightness.

stellar population *Astronomy.* see POPULATION I and POPULATION II.

stellar pulsation *Astrophysics.* the swelling and shrinking of a star as it evolves from the main sequence to the red-giant stage, with periodic variations in surface temperature and brightness.

stellar rotation *Astronomy.* the spinning of a star on its axis; it can range from several kilometers to 500 kilometers per second as measured at the equator.

stellar spectroscopy *Astronomy.* the study of stars' compositions and conditions from their spectra.

stellar spectrum *Astronomy.* the intensity or flux of electromagnetic radiation plotted by wavelength.

stellar structure *Astrophysics.* the study of the two main zones of a star held together by its own gravitational attraction and powered by thermonuclear reactions.

stellar system *Astronomy.* a galaxy or other similarly large collection of stars held together by gravity.

stellar temperature *Astrophysics.* the temperature (measured in kelvins) at various points inside a star; measurements of temperature form valuable clues to processes operating inside stars.

stellar wind *Astrophysics.* the outflow of particles and plasma, similar to the solar wind, that a star emits at varying rates throughout its lifetime.

stellate ganglion *Anatomy.* a ganglion of the sympathetic trunk lying at the level of the seventh cervical vertebra, behind the subclavian artery.

stellate reticulum *Histology.* the middle portion of the epithelial dental organ, composed of stellate cells having elongate processes.

Stellatosporea *Invertebrate Zoology.* a class of parasitic spore-forming protozoans that have electron-dense particles in their cytoplasm.

Stelleroidea *Invertebrate Zoology.* in some classifications: starfish and brittle stars, a class of echinoderms coextensive with the subphylum Asterozoa.

Steller's jay *Vertebrate Zoology.* a large, crested jay, *Cyanocitta stelleri*, characterized by a blackish head and foreparts and blue wings, tail, and belly; found widely in western North America. (Named in honor of the German naturalist George *Steller,* 1709–1746.)

stellite *Materials.* a cobalt alloy containing chromium, tungsten, molybdenum, and silicon; used for jet engine parts and for cutting. *Metallurgy.* a former designation of any cobalt-chromium alloy that has a starlike brilliance when polished.

St. Elmo's fire *Electricity.* see SAINT ELMO'S FIRE.

stem *Botany.* the main ascending axis of a vascular plant, usually located above the ground and bearing leaves, buds, and flowers. *Engineering.* **1.** the central tube within a vacuum tube, containing the leads and filament. **2.** a thick rod that links the bit and the balance on the string tool of a churn rod. **3.** to ram material into a shothole. *Naval Architecture.* the upright portion of a vessel's bow. *Navigation.* to travel forward in the water against a resisting force such as a current.

STEM scanning transmission electron microscopy.

stem-and-loop structure *Molecular Biology.* in a nucleic-acid molecule, a secondary structure in which complementary sequences in a strand base-pair form a stem, and nucleotides between these sequences form a single-stranded loop. Also, **stem-loop structure.**

stem blight *Plant Pathology.* any of various fungus blight diseases that directly attack the plant stem.

stem cabbage see KOHLRABI.

stem canker *Plant Pathology.* any of various canker diseases that directly attack the plant stem.

stem cell *Developmental Biology.* a cell, capable of both indefinite proliferation and differentiation into specialized cells, that serves as a continuous source of new cells for such tissues as blood and testes.

stem cell leukemia *Medicine.* a form of leukemia in which the predominating cell is so immature and primitive that its classification becomes extremely difficult. Also, UNDIFFERENTIATED CELL LEUKEMIA, HEMOCYTOBLASTIC LEUKEMIA, HEMOBLASTIC LEUKEMIA.

stem correction *Thermodynamics.* a correction imposed on a thermometer-measurement reading to account for the portion of the thermometric fluid in the column of the thermometer not in the region at which the measurement is being made.

stem-end rot *Plant Pathology.* a disease of fruits caused by any of several fungi of the genera *Diplodia* and *Phomopsis*; characterized by discoloration, shriveling, and decay of the stem and adjacent parts.

stemflow *Hydrology.* water that reaches the ground by flowing down from tree trunks or plant stems.

Stemonaceae *Botany.* a family of monocotyledonous erect herbs, herbaceous vines, or low shrubs of the order Liliales, commonly producing unique lactone alkaloids; native to Asia, Malaysia, northern Australia, and the southeastern United States.

Stemonitaceae *Mycology.* the only family of fungi within the order Stemonitales.

Stemonitales *Mycology.* an order of fungi belonging to the class Myxomycetes.

Stemonitis *Mycology.* a genus of fungi belonging to the class Myxomycetes which composes slime molds found on humus and rotting trees, having stalks that bear dark fruiting bodies.

Stemonitomycetidae *Mycology.* a subclass of fungi belonging to the class Myxomycetes, characterized by a unique sporophore development and growth pattern; composed of slime molds that occur on melting snowbanks, decaying wood, and tree bark.

stem rot *Plant Pathology.* any of various bacterial and fungal rot diseases that directly attack the plant stem.

stem rust *Plant Pathology.* any of various fungal diseases affecting a wide range of plants but especially grasses; characterized by black or reddish-brown lesions on the stem.

stem succulent *Botany.* a plant with a succulent, photosynthetic stem and small leaves or spines, such as cacti.

stem-winding *Mechanical Engineering.* of or relating to a timepiece that is wound by an internal mechanism turned by an external knob.

stencil *Graphic Arts.* **1.** a sheet of plastic, paper board, or other material that is cut with an image, through which ink or paint is pressed onto the image area of a substrate; commonly used in screen printing. **2.** to mark a surface by means of this technique.

stencil printing *Textiles.* a method of printing using a stencil through which color is applied onto fabric by brushing, spraying, or sponging.

Stenetrioidea *Invertebrate Zoology.* a group of marine isopod crustaceans in the suborder Asellota, with a fused first pair of pleopods.

stengel gneiss see PENCIL GNEISS.

Sten gun see STEN SUBMACHINE GUN.

Steno, Nicolaus 1631–1686, Danish-born Italian anatomist and geologist; performed important early research on fossils, strata, and geological history.

steno- *Biology.* a combining form meaning "narrow" or "small."

stenobathic *Ecology.* describing marine or freshwater life that can exist in only a narrow range of depth.

Stenocephalidae *Invertebrate Zoology.* a family of true bugs, hemipteran insects in the group Pentatomorpha.

stenocephaly *Medicine.* an abnormal narrowness of the head.

stenode circuit *Electronics.* a highly selective superheterodyne receiving circuit in which a piezoelectric unit is used in the intermediate-frequency amplifier to remove all signals except those having a specified intermediate-frequency value.

Stenoglossa *Invertebrate Zoology.* a group of marine snails, gastropod mollusks with one comblike gill and a gill-like osphradium, including cone shells, olive shells, and whelks.

stenohaline *Ecology.* describing an organism that is able to tolerate only a narrow range of saline levels.

stenohydric *Ecology.* describing an organism that is able to tolerate only a narrow range of moisture levels.

stenohygric *Ecology.* describing an organism that is able to tolerate only a narrow range of humidity levels.

Stenolaemata *Invertebrate Zoology.* a class of marine colonial bryozoans, each with a circular lophophore and a calcified cylindrical exoskeleton fused to those of adjoining zooids.

stenometer *Engineering.* an instrument that measures distance using a stereo telescope.

Stenonasteridae *Paleontology.* an extinct family of irregular oval and heart-shaped euechinoids in the order Holasteroida; Cretaceous.

stenopetalous *Botany.* having narrow petals.

stenophagous *Ecology.* describing an organism that feeds on a narrow range of species. Also, **stenophagic.**

stenophyllous *Botany.* having narrow leaves.

stenoplastic *Biology.* exhibiting only a limited ability to vary phenotypically under different environmental conditions.

Stenopodidea *Invertebrate Zoology.* the banded or cleaner shrimps, small decapod crustaceans in the suborder Natantia, having one pair of long, strong legs and two pairs of weaker legs; usually found in tropical tidepools.

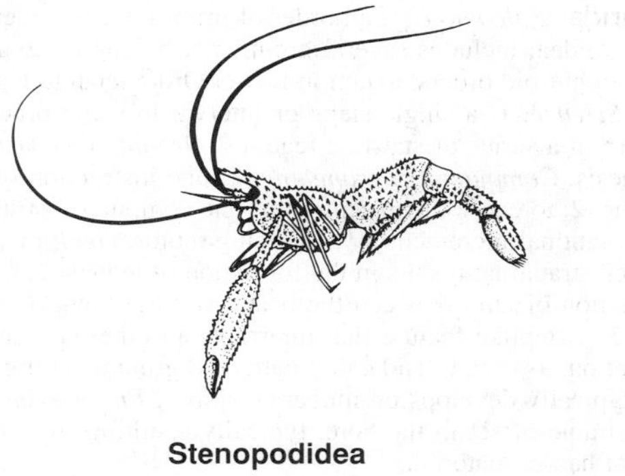

Stenopodidea

stenosis *Medicine.* a narrowing or stricture of a duct or canal, such as the aorta, the pyloric orifice, or the larynx.

Stenostomata see STENOLAEMATA.

stenotele see PENETRANT.

stenotherm *Biology.* an organism that can live only within a narrow range of temperatures. Thus, **stenothermic.**

Stenothoidae *Invertebrate Zoology.* a family of amphipod crustaceans in the suborder Gammaridea, including parasitic and commensal forms.

stenotopic *Ecology.* describing an organism that is able to exist only in a narrow range of habitats.

stenotropic *Ecology.* describing an organism that has only a limited ability to adapt to changes in its environment.

stenoxenous *Microbiology.* having a very limited range of hosts.

Stensioellidae *Paleontology.* a monogeneric family of placoderms usually classified in the order Petalichthyida, but now placed in the monogeneric order Stensioellida; possibly the most primitive of all placoderms; characterized by an absence of large, bony plates, and by a superficial covering of tesserae; has pelvic fins and large pectoral fins; extant in the Lower Devonian.

Sten submachine gun *Ordnance.* a British 9-mm, Parabellum blowback-operated submachine gun with selective full automatic/semiautomatic fire at a rate of 550–600 rounds per minute. (An acronym derived from the names of its designers, R.V. Shepherd and H. J. Turpin, and from Enfield, a borough of London, England, where it was produced.)

stent *Surgery.* a molded device used to maintain a skin graft in position or to provide support for tubular structures for anastomosis. (Named for Charles *Stent,* a 19th-century English dentist.)

Stentor *Invertebrate Zoology.* a genus of large, sessile, trumpet-shaped ciliate protozoans of the class Heterotrichida, often bright-colored; found in stagnant water; one species, *S. coeruleus,* contains a blue pigment; other species may be yellow, red, or brown. (Named for *Stentor,* a Greek hero in the Trojan War, whose voice was described as being as loud as that of 50 ordinary men; because of the association of a trumpet with a loud voice.)

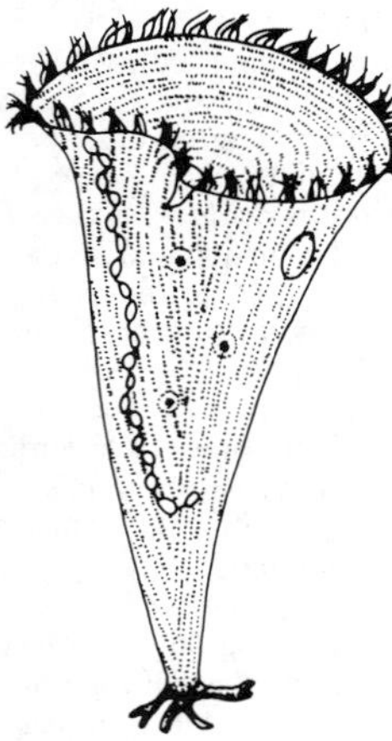

Stentor coeruleus

stentorin *Biochemistry.* the suspected photoreceptor in *Stentor coeruleus*; found in subpellicullar membranous vesicles or granules.

Stenurida *Paleontology.* an order of primitive echinoderms in the class Ophiuroidea; includes several groups not belonging to any of the three basic ophiuroid orders; extant in Lower Ordovician to Upper Devonian.

step *Science.* **1.** a single stage or interval in some process. **2.** a single feature in a series of stairlike features. *Organic Chemistry.* a stage in a synthesis. *Computer Programming.* **1.** one instruction or operation in a routine. **2.** to cause the execution of one computer instruction. *Robotics.* one rotational increment by a stepping motor. *Geology.* **1.** a dislocation of rock strata that results in the formation of ledges. **2.** a nearly horizontal section of land between the beach and the low-tide shoreline. Also, TOE. **3.** a steplike feature that interrupts an otherwise smoothly sloping formation, as on a hillside. **4.** a patterned ground having a steplike form that typically develops on moderate slopes. *Engineering.* in drilling, an undesirable offset in the bore, typically resulting from a slanting interface of harder material.

step aeration *Biotechnology.* a procedure in which increments of sewage are added along the line of flow in the aeration tanks of an activated sludge plant.

step-and-repeat *Graphic Arts.* a process in which an image is projected from roll film directly onto a printing plate; used especially for multiple impositions of the same image.

step-and-repeat camera *Optics.* a camera mounted so as to take a motor-driven series of exposures in a predetermined pattern.

step angle *Electricity.* the angle between two stepping positions of a stepping motor.

Stepanov method *Materials Science.* a method for shaped crystal growth using a wettable shaping die for growing single crystal ribbons from a liquid melt in a crucible; the liquid melt climbs up the die, which controls the final crystal form, and spreads out over its surface, forming a crystal as the die is raised from the melt.

step attenuator *Electronics.* an attenuator in which the attenuation can be adjusted in precisely known steps by the use of switches.

step bearing *Mechanical Engineering.* a bearing that supports the bottom of a shaft while allowing it to rotate about its axis. Also, FOOTSTEP BEARING, PIVOT BEARING.

step brazing *Metallurgy.* a process of brazing several joints at different, decreasing temperatures.

step-by-step switch *Electricity.* a bank-and-wiper switch in which the wipers are moved by individual electromagnetic ratchet mechanisms.

step-by-step system *Control Systems.* a control system that moves the drive motor in discrete steps even when the input varies continuously.

step change *Electronics.* the instantaneous change of a variable from one value to another. Also, **step function.**

step counter *Computer Programming.* a register used to count the steps in shift instructions and in arithmetic instructions, such as division and multiplication, that involve a repetition of operations.

step-down transformer *Electricity.* a transformer in which the voltage on the secondary winding is less than the voltage on the primary winding.

step fault *Geology.* one of a series of closely associated, parallel faults having the same direction of displacement. Also, MULTIPLE FAULT, DISTRIBUTIVE FAULT.

step function *Mathematics.* **1.** another term for a simple function. **2.** a function f on an interval I in the real numbers such that there are disjoint subintervals $I_1, I_2, \ldots$ of I whose union is I, such that f is constant on I_j for each j. The interval I may be finite or infinite in length.

step-function generator *Electronics.* a function generator whose output waveform suddenly increases and decreases in steps.

step gauge *Mechanical Devices.* **1.** a plug or male gauge made up of several cylindrical gauges with graduated diameters mounted on a rod. **2.** a gauge with a perpendicularly sliding blade mounted to a handle; used to measure shoulders or steps.

step generator *Electronics.* a device used to test the linearity of an amplifier by applying a step wave to the input and observing the waveform at the output with an oscilloscope.

step-growth polymerization *Materials Science.* polymerization in which a simple compound such as water is eliminated from two monomer molecules when they are combined.

Stephanian *Geology.* a European geologic stage of the Upper Carboniferous period, occurring after the Westphalian and before the Sakmarian of the Permian period.

Stephanidae *Invertebrate Zoology.* a widely distributed family of slender ichneumon flies, hymenopteran insects in the superfamily Ichneumonoidea, having filamentous, many-segmented antennae and a long ovipositor; the larvae are parasitic on wood-boring insects.

stephanite *Mineralogy.* Ag_5SbS_4, an iron-black orthorhombic mineral with metallic luster occurring as short prismatic crystals and in massive form, having a specific gravity of 6.25 to 6.26 and a hardness of 2 to 2.5 on the Mohs scale; found in silver vein deposits. Also, BRITTLE SILVER ORE.

Stephanoascus *Mycology.* a genus of yeast fungi belonging to the family Saccharomycetaceae that includes only one species, *S. ciferrii,* which is found in soil and in clinical specimens of humans and other animals.

stephanokont *Cell Biology.* a cell that is encircled at one pole by a ring of flagella of equal length.

Stephanothecaceae *Mycology.* a family of fungi belonging to the order Asterinales that occur primarily in tropical regions, on the surfaces of plant leaves or stems.

stephanotis *Botany.* any tropical twining vine of the genus *Stephanotis* of the milkwood family, characterized by fragrant, waxy, white or cream-colored flowers and leathery leaves.

Stephan's Quintet *Astronomy.* a cluster of five peculiar galaxies at a distance of 400 million light-years in the constellation of Pegasus; includes NGCs 7317, 7318A, 7318B, 7319, and 7320.

Stephen reaction *Organic Chemistry.* a reaction that reduces an aromatic nitrile to an aldehyde by treating the nitrile with stannous chloride and hydrogen chloride.

Stephenson, George 1781–1848, English inventor and engineer; designed the *Rocket* and other early steam locomotives.

Stephenson, Robert 1803–1859, English engineer and architect; the son of George Stephenson; designed noted railways, bridges, and viaducts; e.g., the first railway into London.

stepladder *Mechanical Devices.* a self-supported set of portable steps with flat treads and a hinged platform that can be raised and lowered.

step lake see PATERNOSTER LAKE.

step multiplier *Electricity.* a multiplier in which a feedback network is controlled in a stepwise manner to vary the gain of the amplifier proportional to one variable; the other variable is fed into the amplifier and the output is the required product.

step-out time *Geophysics.* the time it takes a given point on a reflected or refracted wave generated during seismic prospecting to reach various detection devices placed at different locations on the surface.

step-out well *Petroleum Engineering.* a well drilled near a proven well but in an unproven area, so as to determine the boundaries of a partially developed formation. Also, OUTPOST WELL, DELAYED-DEVELOPMENT WELL.

steppe or **steppes** *Ecology.* a treeless, midlatitude, level grassland such as that covering vast areas of Central Asia; similar to a prairie but with a short-grass cover associated with semiarid climates. (From a Russian or Ukranian word for this terrain; not related to the English word *step*.)

steppe climate *Meteorology.* the characteristic climate of steppe regions, in which precipitation is very slight but sufficient to support the growth of short, sparse grasses. Also, SEMIARID CLIMATE.

stepped footing *Civil Engineering.* **1.** a wall footing with horizontal steps to accommodate sloping grade or bearing stratum. **2.** a column or wall footing composed of two or more steps on top of one another to distribute load.

stepped gear wheel *Mechanical Devices.* a gear wheel having two or more sets of parallel, adjacent teeth on the same rim, arranged so that each set forms a step.

stepped leader *Geophysics.* a lightning streamer that advances intermittently, giving it a stepped appearance.

stepped screw *Mechanical Devices.* an interrupted screw with some thread divisions removed so that the remaining threads form steps.

stepper motor *Electricity.* a motor that typically rotates in fixed angular steps rather than rotating continuously. Also, MAGENTIC STEPPING MOTOR, STEP-SERVO MOTOR. *Robotics.* specifically, a robotic motor that moves one discrete, rotational increment when a pulse of electricity is applied. Also, **stepping motor.**

stepping reflex *Physiology.* a walking motion produced when an infant is held upright and tilted forward, with the feet pressed against a flat surface; the response is present at birth and evident until about six weeks of age.

stepping relay *Electricity.* a multiposition relay whose moving-wiper contacts mate with successive positions of fixed contacts in a series of steps, as a pulse of current is received and an electromechanical ratchet mechanism advances to the next contact position. Also, ROTARY STEPPING SWITCH, STEPPING SWITCH.

stepping-stone states *Artificial Intelligence.* in a state-space search, states that are known to be necessary intermediate states between the start state and a goal.

stepping switch see STEPPING RELAY.

step pulley *Mechanical Engineering.* a series of pulley wheels of various diameters all arranged on a concentric support; used to vary the velocity ratio of shafts. Also, CONE PULLEY.

step-recovery diode *Electronics.* a varactor in which forward voltage is used to inject carriers across the junction; before the carriers can combine, voltage reverses, and carriers return to their origin in a group; the result is an abrupt cessation of reverse current and a harmonic-rich waveform.

step response *Control Systems.* the behavior of a system in response to a step input.

step rocket see MULTISTAGE ROCKET.

step scale *Graphic Arts.* in the marginal trim area of a printed sheet, a visual scale showing the gradations of gray tones; used as a guide to color density to appear in halftones.

step-servo motor see STEPPER MOTOR.

step soldering *Metallurgy.* soldering several joints at different, decreasing temperatures.

step strobe marker *Electronics.* a strobe marker in which the discontinuity is in the form of a step in the time base.

step tap *Mechanical Devices.* a tap with sections of tapering diameters to provide a cutting surface for heavy, internal threads.

step test *Graphic Arts.* a trial-and-error method used to find an appropriate camera exposure by gradually increasing exposure time.

steptoe *Geology.* an isolated protrusion or island of bedrock occurring within a lava flow. Also, DAGALA, KIPUKA.

step trench *Archaeology.* an excavation technique in which a series of stairlike cuts are made; often used to uncover sections of a deep deposit, such as a mound.

step trench

step-up *Graphic Arts.* a technique used to enlarge a small photograph by making a series of reproductions that gradually increase in size.

step-up transformer *Electricity.* a transformer in which the voltage on the secondary winding is greater than the voltage on the primary winding.

step voltage regulator *Electricity.* a regulating transformer in which the voltage of the regulated circuit is controlled by means of steps and without interruption of the load.

step wedge *Radiology.* a block of absorbing material, commonly aluminum, machined in steps of varying thickness; used to measure the penetrability of X-rays.

steradian *Mathematics.* a unit of solid angular measurement; abbreviated **sr.** The solid angular measure of the steregon, i.e., the solid angle subtended by a unit sphere centered at a point, is equal to 4π steradians. A solid angle of $\pi/2$ steradian is called a **spherical right angle**. Also, **sterad.**

steradiance see RADIANCE.

Sterba curtain *Electromagnetism.* a stacked dipole antenna array having one or more phased half-wave sections with a quarter-wave section at each end; the array is highly directional and can be either center or end fed.

stercobilin *Biochemistry.* a form of urobilin that is formed by intestinal bacteria from bilirubin; a component of brown fecal pigment.

stercobilinogen *Biochemistry.* a colorless reduction product of stercobilin that is found in feces.

Stercorariidae *Vertebrate Zoology.* the skuas and jaegers, a monogeneric, worldwide family of marine gull-like birds belonging to the order Charadriiformes, characterized by light and dark color phases, powerful flight, and the ability to swim.

stercorite *Mineralogy.* $H(NH_4)Na(PO_4)\cdot 4H_2O$, a white to yellowish or brownish, water-soluble, triclinic mineral, occurring as crystalline masses, and having a specific gravity of 1.55 to 1.57 and a hardness of 2 on the Mohs scale; found in guano deposits on islands off the coast of Peru.

Sterculiaceae *Botany.* a large family of mainly tropical dicotyledonous trees, shrubs, and herbs, characterized by overlapping or contorted petals, ten or more stamens, and two-celled anthers.

steregon *Mathematics.* the solid angle subtended by a sphere; the set of all rays emanating from a single point. It has solid angular measure equal to 4π steradians.

stere- or **stereo-** a combining form meaning: **1.** solid. **2.** having three dimensions. **3.** firmly established.

stereo *Acoustical Engineering.* a shorter term for stereophonic sound or a stereophonic sound system. See STEREOPHONIC.

stereo amplifier *Acoustical Engineering.* an electroacoustical device designed to amplify the electroacoustical signals on two separate stereophonic channels that pass through it.

stereoanesthesia *Neurology.* an inability to recognize the shape, size , weight, and texture of objects due to tactile impairment.

stereoblastula *Developmental Biology.* a solid blastula, lacking a fluid blastocoele.

stereo camera see STEREOSCOPIC CAMERA.

stereochemistry *Physical Chemistry.* the branch of chemistry that studies how atoms or molecules are affected by their three-dimensional spatial arrangement; e.g., the study of stereoisomers.

stereocilia *Cell Biology.* a cluster of highly elongated immobile microvilli that are found on the surface of epithelial hair cells of the inner ear and are involved in motion detection; also found in the epididymis of the male reproductive tract.

stereocomparagraph *Optics.* a device that superimposes two aerial photos to produce a three-dimensional effect.

stereo comparator *Optics.* a device that allows two photographs of the sky taken at different times to be optically superimposed so that changes in star brightness or moving objects stand out clearly.

stereocompilation *Cartography.* the act of compilation using stereoscopic plotting instruments.

stereo effect *Acoustics.* the creation of a spatially correct image in three dimensions of multiple sound sources, such as musical instruments, that appear to be coming from different directions.

stereofluoroscopy *Electronics.* a fluoroscopic technique that yields three-dimensional images.

stereogastrula *Developmental Biology.* a solid structure that follows the blastula and consists of the ectoderm and mesentoderm.

stereognosis *Psychology.* the knowledge or recognition of objects by touch.

stereogram *Graphic Arts.* a pair of images produced and mounted to be viewed through a stereoscope.

stereographic *Graphic Arts.* of or relating to stereography.

stereographic chart *Cartography.* any chart based on a stereographic map projection.

stereographic map projection *Cartography.* a conformal map projection in which the points on the sphere are projected onto a plane tangent to the sphere, having the point of projection at the opposite end of the diameter of the sphere that passes through the point of tangency with the plane. The **stereographic horizon map projection** has its center on some selected parallel of latitude other than the equator; the **stereographic meridional map projection** has its center on the equator.

stereographic projection *Mathematics.* the conformal mapping of a sphere onto a plane tangent to the sphere at a point P; accomplished by associating to each point z on the sphere the point on the tangent plane where the line through z and the antipode of P intersects the tangent plane. In this way, only the antipode itself fails to have a corresponding point on the tangent plane. In complex analysis, the sphere is usually taken to be tangent to the origin, and the mapping is holomorphic. *Cartography.* see STEREOGRAPHIC MAP PROJECTION. *Crystallography.* a crystal surrounded by an imaginary sphere in which the coordinate centers of the crystal and the surrounding sphere are the same; lines perpendicular to each crystal face are drawn outward, touching the surface of the sphere, and the results are transferred to a two-dimensional plot with the points above (north of) the equatorial plane joined to the lower (south) pole; the points at which these lines intersect with the equatorial plane are marked with filled circles for the northern hemisphere and open circles for the southern hemisphere; used to demonstrate the symmetry of a crystal, without the need for a three-dimensional model.

stereography *Graphic Arts.* the process or art of depicting solid objects on a plane surface.

stereoisomers *Organic Chemistry.* any compounds that have the same kinds and numbers of atoms, but differ in the spatial arrangement of their constituent atoms.

stereome *Botany.* a general term for any mechanical or strengthening tissue of a plant.

stereometer *Photogrammetry.* a measuring device in a stereoscopic viewing system that uses a micrometer to measure differences in the separation of index marks on pairs of stereoscopic photographs in order to determine the difference in parallax. Also, PARALLAX BAR.

stereometric camera *Cartography.* an assembly of two identical cameras, with parallel optical axes and synchronized shutters, mounted on a rigid connecting rod; used to make stereoscopic pairs of photographs.

stereometric map *Cartography.* a relief map prepared by viewing aerial photographs through a stereoscope to achieve a three-dimensional effect, then recording the elevations on the map by some graphic technique.

stereomicrography *Optics.* a technique in which a subject is photographed through a microscope from two slightly different angles to provide three-dimensional viewing.

stereomicrometer *Engineering.* an optical micrometer fitted into a telescope or similar optical system.

stereophonic [ster´ē ə fän´ik] *Acoustical Engineering.* of or relating to the reproduction of sound by means of two or more separately processed channels of electroacoustic information (typically, one right and one left channel), in order to give spatial qualities and a sense of realism to the reproduced sound. Thus, **stereophonic sound system.** (A term formed from two words of Greek origin, meaning literally "three-dimensional sound.")

stereophonics *Acoustical Engineering.* the use or study of spatial qualities of sound in sound system engineering, due to binaural effect of two auditory sensors (the ears) in animals.

stereo pickup *Acoustical Engineering.* a two-channel (left and right side) magnetic or stylus pickup for recording or playback of sounds.

stereoplanigraph *Engineering.* a device used to draw topographic maps using images provided by a stereo comparator.

stereo preamplifier *Acoustical Engineering.* the portion of a stereo amplifier designed to boost input signals from right and left stereo channels prior to further amplification.

stereopsis *Robotics.* a method of determining distance through the use of binocular vision in machine applications.

stereo record *Acoustical Engineering.* a record in which the groove was created by input right and left stereo channels, so that the stylus movement toward the right or left produces a signal on the appropriate channel in the stereo sound-reproduction system.

stereoregularity *Materials Science.* a regular steric configuration of monomer units in a polymer chain.

stereoregular polymer see STEREOSPECIFIC POLYMER.

stereorubber *Organic Chemistry.* a stereospecific polymer that forms a synthetic rubber.

stereoscope *Optics.* an instrument that gives the viewer a sense of depth when viewing a photograph taken with a stereoscopic camera.

stereoscopic camera [ster´ē ə skäp´ik] *Optics.* a camera that records a scene in stereo by taking two photographs separated by a short distance. Also, STEREO CAMERA.

stereoscopic pair *Graphic Arts.* a pair of images having sufficient overlap to be viewed through a stereoscope.

stereoscopic parallax see ABSOLUTE STEREOSCOPIC PARALLAX.

stereoscopic photography *Optics.* the act of taking pictures with a stereoscopic camera.

stereoscopic plotting instrument *Cartography.* an instrument that creates three-dimensional models from stereoscopic pairs of photographs, which are then viewed through the instrument and plotted on a map sheet.

stereoscopic power *Optics.* the product of the magnifying power in a stereo system, such as binoculars, and the ratio of the distance from the objective axes to the eyepiece axis.

stereoscopic radius *Physiology.* the greatest distance at which three-dimensional depth of view is possible due to a slight difference in the visual images formed by the two eyes.

stereoscopic rangefinder *Optics.* an instrument that determines the distance to an object by the amount of adjustment needed to make the two images of the object formed by separate lenses coincide optically. Also, **stereo rangefinder, stereoscopic height finder.**

stereoscopy [ster´ē äs´kə pē] *Physiology.* the observation of an object or body part when it is X-rayed with a stereoscope, a machine that produces two roentgenograms. Seen together, these images provide a three-dimensional view of the object.

stereoselective reaction *Organic Chemistry.* a chemical reaction in which one stereoisomer is produced or decomposed more rapidly than the other.

stereo sound system see STEREOPHONIC.

stereospecificity *Organic Chemistry.* the condition of a polymer whose monomers are positioned so as to confer crystalline properties on the polymer.

stereospecific polymer *Organic Chemistry.* a polymer in which the specific order of the spatial arrangement of monomers permits close packing of molecules and leads to a high degree of polymer crystallinity. Also, STEREOREGULAR POLYMER.

stereospecific synthesis *Organic Chemistry.* a synthetic reaction of a single stereoisomeric monomer that produces a specific stereoisomeric polymer.

Stereospondyli *Paleontology.* a former term used to designate a group of late temnospondyl reptiles; no longer accepted as a separate grouping.

stereo tape *Acoustical Engineering.* a two-track or four-track tape with at least one track used for recording a right stereo channel and the other track, recorded in the same direction, used for the left stereo channel.

stereo tape recorder *Acoustical Engineering.* a tape-recorder system consisting of at least one set of two-channel (stereo) tape heads, a stereo amplifier, and a separate speaker housing for both stereo channels.

stereotaxis *Biology.* a directed orientation response of a motile organism to stimulation by contact with a solid surface.

stereotemplet *Photogrammetry.* a slotted template, constructed from the templates of a stereoscopic pair of photographs and adjustable in scale; used to do aerotriangulation with a stereoscopic plotter not designed for stereotriangulation.

stereotriangulation *Cartography.* a technique of triangulation in which pairs of stereographic photographs are arranged by use of a stereoscopic plotting instrument into continuous strips, allowing for the extension of horizontal or vertical control.

stereotriplet *Photogrammetry.* a series of three photographs in which the center photograph has a common field of view with the two adjacent ones, allowing complete stereoscopic viewing of the center photograph.

stereotropism *Biology.* the growth or orientation response of a nonmotile organism to stimulation by contact with a solid surface.

stereo tuner *Acoustical Engineering.* an FM tuner capable of receiving both channels of a stereo broadcast.

stereotype *Behavior.* a fixed pattern of behavior that is characteristic of a species or group. Also, **stereotypy, stereotyped behavior.** *Psychology.* a category into which people are placed on the basis of their group identification rather than any evaluation of individual behavior. *Graphic Arts.* a metal printing plate cast from the papier-mâché mold of a page of type.

Stereum *Mycology.* a genus of fungi belonging to the order Agaricales, which occurs on nonliving organic matter and is characterized by leathery fruiting bodies.

sterhydraulic *Mechanical Engineering.* of or relating to a hydraulic system in which pressure is created by introducing a solid body into a liquid-filled cylinder.

steric anomaly see ANOMALY OF SPECIFIC VOLUME.

steric effect *Physical Chemistry.* the influence on a chemical reaction of the spatial arrangement of a reacting substance. Also, **steric factor.**

steric hindrance *Chemistry.* the inteference to a reaction that is caused by the spatial arrangements of the atoms within the reacting molecules.

stericooling *Food Technology.* a type of cooling treatment in which spinning cans are subjected to a coolant for a period of time to effectively reduce the heat.

steriflamme process *Food Technology.* in canning, a heat treatment in which spinning cans of food are exposed to gas flames, heated to 120–130°C, held at this temperature for a specified time (depending on the size of the can), and then cooled.

sterigma *Botany.* a small stalk that bears a basidiospore, a sporangium, or a conidium.

sterigmatocystin *Biochemistry.* a carcinogenic, hepatotoxic poison produced by various species of *Aspergillus.*

Sterigmatomyces *Mycology.* a genus of yeast fungi belonging to the class Hyphomycetes whose cells carry out aerobic metabolism; it occurs in the Atlantic and Indian Oceans.

sterilant *Microbiology.* any compound that can sterilize an object or material under the appropriate conditions.

sterile [ster´əl] *Biology.* not able to reproduce; not producing offspring; barren. *Microbiology.* lacking any living microorganisms. *Surgery.* of instruments or materials, free from microorganisms; aseptic.

sterile distribution *Ecology.* an area in which a species is able to exist and to mate but not able to produce viable offspring.

sterile layer *Archaeology.* an excavation layer in which there are no cultural materials or evidence of human occupation.

sterility [stə ril´i tē] *Biology.* the fact of being unable to produce offspring; the failure to reproduce. *Microbiology.* the fact of being devoid of living microorganisms.

sterilization *Microbiology.* the process of destroying all living microorganisms in or on a material. *Physiology.* the process of making a person or animal unable to produce children, as by removing the sex organs.

sterilize *Microbiology.* to destroy all living organisms in or on a material, such as food or surgical instruments, by using chemicals, heat, radiation, filtration, or by other means. *Physiology.* to render a person or animal unable to produce young, by removing the sex organs or inhibiting their functions. *Agriculture.* to make land barren or unproductive. *Ordnance.* **1.** in naval mine warfare, to permanently render a mine incapable of firing by means of a device within the mine. **2.** in covert and clandestine operations, to remove marks or devices that can identify material as emanating from the sponsoring nation or organization.

sterilizer *Engineering.* a device using heat, radiation, or other means to sterilize objects or materials.

Sterkfontein Valley *Anthropology.* the location of several important sites near Johannesburg, South Africa, where many australopithecines were found in caves in the 1940s.

sterling silver *Materials.* a silver-base alloy containing 92.5% silver and 7.5% copper; used for plating utensils, jewelry, and electrical contacts.

Sterling submachine gun *Ordnance.* a British light, shoulder-fired, 9-mm Parabellum submachine gun with a firing rate of 555 rounds per minute and a muzzle velocity of 1280 feet per second; it was adopted as a standard British military weapon in 1953.

stern *Naval Architecture.* the rear end of a vessel. *Aviation.* the tail or aft end of an aircraft, especially of an airship; used to form compounds such as **stern attack** and **sternpost.**

Stern, Otto 1888–1969, German-born American physicist; awarded the Nobel Prize for detecting and measuring the magnetic momentum of protons.

sternal *Anatomy.* relating to or involving the breastbone (sternum).

sternalgia *Medicine.* **1.** any pain in the sternum. **2.** angina pectoris.

sternbergite *Mineralogy.* $AgFe_2S_3$, a golden-brown, metallic orthorhombic mineral occurring as aggregates of platy crystals having a specific gravity of 4.1 to 4.25 and a hardness of 1 to 1.5 on the Mohs scale; found in silver mines; dimorphous with argentopyrite.

sternebra *Vertebrate Zoology.* individual segments of the sternum of a vertebrate.

Sterneedle *Medicine.* a trade name for a multiple-point, controlled-depth apparatus used in the diagnosis of tuberculosis.

Sterneedle test *Medicine.* a test for tuberculosis, in which the Sterneedle points are dipped into tuberculin P.P.D. and used to prick the skin; a positive reaction appears as a palpable, coalescing induration within three to seven days, indicating infection. Also, HEAF TEST.

stern end bulb *Naval Architecture.* a bulb-shaped stern fitting designed to reduce resistance that would create waves and eddies.

Stern-Gerlach effect *Atomic Physics.* a phenomenon in which a beam of atoms or molecules is split as it passes through an intromageneous magnetic field.

Stern-Gerlach experiment *Quantum Mechanics.* a series of experiments conducted in the early 1920s by the German physicists Otto Stern and Walther Gerlach, that demonstrated the spatial quantization of electron spin, in which electrons were passed through a region of intromageneous magnetic field and observed to be deflected with only two angles, corresponding to the two possible z components of spin.

sternite *Invertebrate Zoology.* a ventral plate of an arthropod segment; there may be several per segment, forming a sternum.

Stern layer *Physical Chemistry.* in an electrical double layer in an electrolytic solution, the ion layer directly at the electrode surface.

sternoclavicular *Anatomy.* relating to the breastbone and collarbone.

sternocleidomastoid *Anatomy.* the large muscle, extending from the clavicle and the manubrium of the sternum to the mastoid process of the temporal bone, that flexes the neck and rotates the head.

sternocostal *Anatomy.* relating to the sternum and the ribs.

sternohyoid *Anatomy.* a muscle, extending from the manubrium of the sternum to the hyoid bone, that depresses the hyoid bone and larynx.

sternopericardial *Anatomy.* relating to the sternum and the pericardium.

Sternoptychidae see HATCHETFISH.

Sternorrhyncha *Invertebrate Zoology.* a suborder of plant pests in the order Homoptera, with long antennae and a beak appearing to project from the thorax between the forelegs; includes aphids and scale insects.

sternoschisis *Medicine.* a congenital fissure of the sternum.

sternothyroid *Anatomy.* a muscle, extending from the manubrium of the sternum to the thyroid cartilage of the larynx, that depresses the larynx.

sternotomy *Surgery.* an incision through the sternum.

sternovertebral *Anatomy.* relating to the sternum and the vertebrae.

Sternoxia *Invertebrate Zoology.* a group of beetles characterized by the extended prothorax that forms a posterior projection; includes click beetles, false click beetles, and metallic wood-borers.

stern post *Naval Architecture.* the aftermost upright structural member of a vessel's stern.

stern tube *Naval Architecture.* **1.** the bearing through which a propeller shaft passes through the hull's shell plating. **2.** an aft-mounted, aft-firing torpedo tube.

sternum *Anatomy.* the long flat breastbone, consisting of three portions: the manubrium, body, and xiphoid process; articulates with the clavicles and with costal cartilages of the upper seven ribs. *Invertebrate Zoology.* **1.** the ventral plate or plates on each segment of an arthropod thorax. **2.** the ventral plates of the cuticle in members of the phylum Kinorhyncha.

sternutatory *Toxicology.* an agent that causes sneezing.

sternway *Navigation.* the backward movement of a vessel in the water, whether intended or not.

stern wheel *Naval Architecture.* a paddle wheel mounted at a vessel's stern, a fitting particularly associated with riverboats.

Stern-Zartman experiment *Physics.* an experiment for measuring the velocity distribution of silver atoms in a heated oven to test the validity of the Maxwell-Boltzmann distribution law; it involves a molecular beam exiting the oven and encountering a rotating drum with a slit opening through which the atoms may pass and be deposited on the opposite drum face.

steroid [ster´oid; stēr´oid] *Biochemistry.* **1.** any of a class of lipid proteins, such as sterols, bile acids, sex hormones, or adrenocortical hormones, containing a cyclopentanoperhydrophenanthrene nucleus; most have specific physiological action. **2.** specifically, a shorter term for anabolic steroids. See ANABOLIC STEROIDS.

steroid bioconversion *Biochemistry.* any of a variety of commercially produced steroid hormones and related compounds, such as cortisone, that are formed by combining chemical and microbial reactions.

steroid hormone *Endocrinology.* one of a group of biologically active organic compounds that are secreted by the adrenal cortex, testis, ovary, and placenta and are characterized by the presence of a cyclopentanoperhydrophenanthrene ring, including the estrogens, androgens, mineralocorticoids, progestogens, and glucocorticoids.

steroid hormone binding globulin *Endocrinology.* one of a group of serum proteins that bind steroids in blood and serve as carrier molecules for the transport of steroid hormones to their target tissues.

steroid hydroxylase *Enzymology.* an enzyme that catalyzes the formation of a hydroxyl group on a substrate by incorporation of oxygen from O_2. Also, **steroid 11β-monooxygenase.**

steroidogenesis *Endocrinology.* the biosynthesis or production of steroids through enzymatic pathways.

steroid receptors *Endocrinology.* specific intracellular molecules that are found in the nucleus and cytoplasm and are activated by the binding of specific steroid hormones, mediating the biological effects of the steroid hormones.

sterol [ster´ôl; stēr´ôl] *Biochemistry.* any of a group of solid, chiefly unsaturated steroid alcohols, such as cholesterol or ergosterol, present in animal and plant tissues.

sterol carrier protein *Biochemistry.* any of the cytoplasmic proteins that bind to certain sterols and transfer them between endoplasmic reticulum-bound enzymes during metabolism.

stertorous *Physiology.* describing heavy breathing that is characterized by a snoring sound.

steth- or **stetho-** a combining form meaning "chest."

stethometer see THORACOMETER.

stethoparalysis *Neurology.* paralysis of the respiratory muscles.

stethoscope [steth´ə skōp´] *Medicine.* an instrument that is used in medical observation to transmit low-volume internal bodily sounds, such as the heartbeat, or intestinal, venous, or fetal sounds, to the ear of the observer; it consists of two earpieces connected by means of flexible tubing to a diaphragm, which is placed against the patient's skin.

stethospasm *Neurology.* a spasm of the chest muscles.

Stevens, John Cox 1749–1838, American engineer; invented the first seagoing steamboat.

Stevens, Robert Livingston 1787–1856, American engineer; the son of John Cox Stevens; a pioneer in steamship design and construction.

Stevens rearrangement *Organic Chemistry.* an intramolecular rearrangement that occurs in a quaternary ammonium salt so that the benzyl group migrates smoothly from nitrogen to carbon.

Stevin, Simon 1548–1620, Dutch mathematician; founder of hydrostatics; formulated the theory of static equilibrium; advanced the use of decimals; demonstrated the equal rate of falling bodies.

stewartite *Geology.* a steel-gray, ash-rich, fibrous variety of bort, characterized by iron content and magnetic properties. *Mineralogy.* $Mn^{+2}Fe_2^{+3}(PO_4)_2(OH)_2 \cdot 8H_2O$, a colorless or yellow to brownish-yellow triclinic mineral consisting of a hydrous phosphate of manganese, occurring as minute triclinic crystals or fibrous tufts, having a specific gravity of 2.94 and an undetermined hardness; found in pegmatites.

sthene [sthēn] *Mechanics.* a unit of force equal to 10^3 newtons.

Sthenurus *Paleontology.* an extinct genus of large, short-faced kangaroos in the family Macropodidae; *Sthenurus* grew as large as 3 meters tall; they inhabited Australia from the Pliocene to Pleistocene.

stibarsen *Mineralogy.* SbAs, a tin-white to gray, opaque, metallic trigonal mineral of the arsenic group, having a specific gravity of 5.8 to 6.2 and a hardness of 3 to 4 on the Mohs scale; commonly found intergrown with native arsenic or antimony.

Stibellaceae *Mycology.* the sole family of fungi belonging to the order Stilbellales, which occurs on decaying organic matter and is composed of some species that are parasitic on flowering plants and insects.

stibialism *Toxicology.* poisoning due to exposure to antimony and its compounds; symptoms may include severe gastrointestinal disturbances and hemolysis of red blood cells.

stibiconite *Mineralogy.* $Sb^{+3}Sb_2^{+5}O_6(OH)$, a usually white to pale yellow cubic mineral occurring as fine-grained crusts and masses, having a specific gravity of 3.3 to 5.5 and a hardness of 3 to 7 on the Mohs scale; found as an alteration product of stibnite.

stibine see ANTIMONY HYDRIDE.

stibiocolumbite *Mineralogy.* $SbNbO_4$, a brown orthorhombic mineral occurring as brittle, prismatic crystals, having a specific gravity of 5.98 and a hardness of 5.5 on the Mohs scale; forms a series with stibiotantalite; found in pegmatites.

stibium *Chemistry.* the Latin name for antimony; the basis for the chemical symbol Sb for antimony.

stibnite *Mineralogy.* Sb_2S_3, a lead-gray, opaque, metallic orthorhombic mineral occurring as thin prismatic crystals and in bladed masses, having a specific gravity of 4.63 to 4.66 and a hardness of 2 on the Mohs scale; found mainly in lower-temperature hydrothermal veins and replacement deposits; the prinicipal ore of antimony.

stibnite

stibophen *Pharmacology.* $C_{12}H_{18}Na_5O_{23}S_4Sb$, an antimony compound active against parasites; used especially in treatment of schistosomiasis and granuloma inguinale. Also, NEOANTIMOSAN.

Stichaeidae *Vertebrate Zoology.* the pricklebacks, a family of marine fishes of the order Perciformes, characterized by a long, slender, compressed body, small or no scales, and a spiny dorsal fin; found in intertidal zones in the North Pacific and North Atlantic Oceans.

stichobasidial *Mycology.* describing a fungal spore-bearing structure in which the nuclear spindles of the basidia lie parallel to the longitudinal axis.

stichochrome *Neurology.* a former term for the arrangement of a nerve cell's chromophil, or stainable substance, in somewhat parallel rows.

Stichococcus *Botany.* a genus of unbranched filamentous green algae of the family Klebsormidiaceae that are a common contaminant of marine cultures.

Stichocotylidae *Invertebrate Zoology.* a family of trematode endoparasites in the subclass Aspidogastrea; adults are slender and elongated.

Stichopodidae *Invertebrate Zoology.* a family of sea cucumbers, holothurian echinoderms in the order Aspidochirotida, having respiratory trees, modified dorsal tube feet, and a warty appearance.

stichtite *Mineralogy.* $Mg_6Cr_2(CO_3)(OH)_{16}\cdot 4H_2O$, a translucent, lilac to pink, trigonal mineral of the hydrotalcite group, dimorphous with barbertonite, having a specific gravity of 2.16 and a hardness of 1.5 to 2 on the Mohs scale; found as foliated to fibrous masses in serpentine rock.

stick *Ordnance.* a series of bombs or missiles released successively from a single aircraft.

stick gauge *Engineering.* a vertical staff with level marks, used to measure the depth of liquid in a vessel or receptacle.

sticking coefficient *Physical Chemistry.* the ratio between the number of particles impacting a surface and the number actually adsorbed there.

Sticklan reaction *Biochemistry.* an amino acid fermentation reaction involving the coupled decomposition of two or more amino acid substrates, one of which is oxidized and the other reduced.

stickleback *Vertebrate Zoology.* a member of the freshwater fish family Gasterosteidae, characterized by two or more free spines anterior to the dorsal fin and a male who builds and guards the nest; native to the northern hemisphere. Also, PRICKLEBACK.

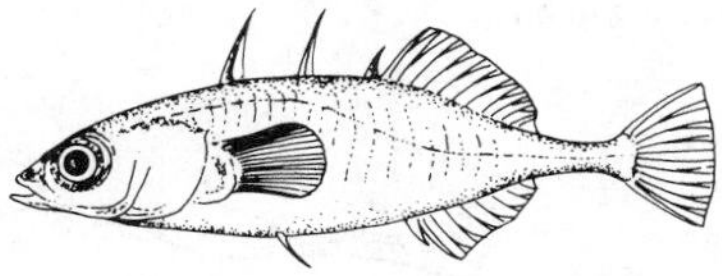

stickleback

sticky ends see COHESIVE ENDS.

sticky grenade *Ordnance.* a small explosive charge covered with adhesive that holds it in place until detonation by a time fuse. Also, **sticky charge.**

Sticta *Botany.* a genus of tropical lichens of the family Stictaceae, characterized by foliaceous lobed thalli that are usually coriaceous in texture.

Stictidaceae *Mycology.* the only family of fungi belonging to the order Ostropales, which is composed of species that live on decaying wood, as lichens, or as plant parasites.

stiction *Mechanics.* the friction that prevents relative motion between two movable parts at their null position.

Stictodiscaceae *Botany.* a family of marine diatoms of the order Centrales, characterized by circular valves with radiating internal ribs and labiate structures at the valve center.

stieda body *Cell Biology.* a structure that is associated with the sporocyst in certain protozoan species of the phylum Apicomplexa.

Stieltjes integral see LEBESGUE-STIELTJES INTEGRAL.

Stifel, Michael 1487–1567, German mathematician; wrote *Arithmetica integra,* a comprehensive survey of arithmetic and algebra.

stiffener *Civil Engineering.* a bar angle or channel shape attached to a steel slender beam or column web to increase its resistance to buckling. *Textiles.* any substance, such as buckram or starch, that serves to stiffen fabrics.

stiffleg derrick *Mechanical Engineering.* a derrick supported in a vertical position by a fixed tripod of sturdy materials. Also, DERRICK CRANE, SCOTCH DERRICK.

stiffness *Mechanics.* in a mechanical vibrating system, the restoring force per unit displacement. *Acoustics.* specifically, a constant value, k, for an oscillator, representing a measure of the difficulty of displacing the oscillator from its equilibrium position; along with the mass of the oscillator, determines the frequency of the oscillator's vibration.

stiffness coefficient *Mechanics.* the ratio of a reactive force to the resulting displacement from equilibrium for a linear mechanical system, such as a linear spring.

stiffness constant see COMPLIANCE CONSTANT.

stiffness matrix *Mechanics.* an $n \times n$ matrix K corresponding to a system having n generalized coordinates, which relates the potential energy V of a system to a set of small displacements from equilibrium by the relation $V = (1/2)q^TKq$, where q is the generalized coordinate vector, and q^T is its transpose. Also, STABILITY MATRIX.

stigeoclonium *Botany.* a genus of filamentous green algae of the family Chaetophoraceae, distinguished by branches that are approximately of equal diameter.

stigma *plural,* **stigmata.** *Botany.* an expanded apex of a pistil, supported by the style; the part of the pistil that receives pollen. *Invertebrate Zoology.* **1.** in butterflies, a distinctive marking or spot. **2.** in arthropods, a respiratory aperture. **3.** in tunicates, a gill slit. **4.** in protozoa, an eye spot. *Medicine.* the mark or manifestation of some disease or condition. *Psychology.* see STIGMATIZATION. *Metrology.* a unit of length equal to 10^{-12} meters. Also, BICRON.

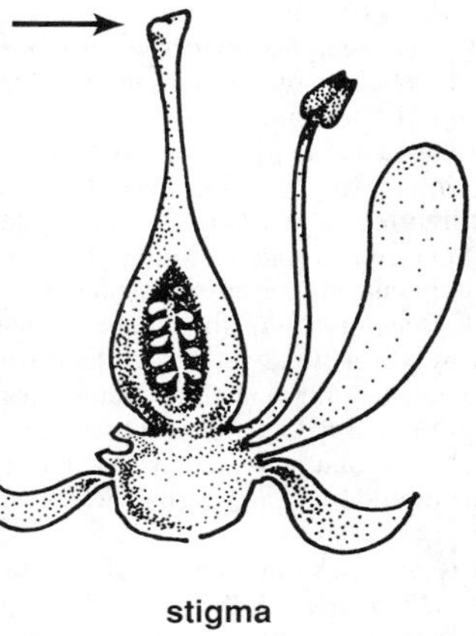

stigma

Stigmaria *Paleontology.* a genus of lycopsid underground axes, part of the same plant whose trunk is called Lepidodendron; extant in the Carboniferous.

stigmatic *Science.* relating to or being a stigma or identifying mark. *Optics.* a lens whose focal length is the same for all axial light rays passing through it.

stigmatic concave grating *Optics.* a concave surface ruled with many closely spaced concentric lines that can simultaneously form an image and disperse light into a spectrum; used in space optics and in food and metal analysis.

stigmatism *Pathology.* possessing stigmata, the marks or manifestations of certain diseases or conditions. *Physiology.* the accurate focusing of light rays upon the retina.

stigmatization *Psychology.* the appearance, without apparent physical cause, on an individual's body of marks corresponding to Jesus Christ's crucifixion wounds; the marks are known as **stigmata.**

stigmatomycosis *Plant Pathology.* a cotton boll rot caused by the fungus *Eremothecium cymbalariae* or by any fungus belonging to the genus *Nematospora.*

stilb *Optics.* a unit of luminance equal to one candela per square centimeter.

Stilbellales *Mycology.* an order of fungi belonging to the class Hyphomycetes, characterized by its fertile hyphae that are intertwined to form synemmata.

stilbene *Organic Chemistry.* $C_6H_5CH{=}CHC_6H_5$, combustible, colorless or slightly yellow crystals; insoluble in water, slightly soluble in alcohol, and soluble in ether; in the transform melts at 124–125°C and boils at 306–307°C; used in the manufacture of optical bleaches and dyes. Also, TOLUELENE.

stilbesterol see DIETHYLSTILBESTEROL.

stilbite *Mineralogy.* $NaCa_2Al_5Si_{13}O_{36}\cdot 14H_2O$, a white, brown, red, or yellow monoclinic and triclinic mineral of the zeolite group occurring in sheaflike aggregates of tabular crystals with pearly luster, having a specific gravity of 2.09 to 2.20 and a hardness of 3.5 to 4 on the Mohs scale; found in amygdules in basaltic rocks. Also, DESMINE.

stile *Building Engineering.* a vertical piece of a window or door in framing or paneling.

Stiletto *Electronics.* an electronic subsystem used in U.S. strike aircraft to detect, identify, and locate ground-based radars. *Ordnance.* a Soviet two-stage, intercontinental ballistic missile that is capable of delivering one 10-megaton warhead or six 550-kiloton MIRVs at a range of 6200 miles; officially designated **SS-19.**

still *Chemical Engineering.* **1.** a distilling apparatus used to separate liquids by heating, then cooling to condense the vapor. **2.** specifically, in popular use, such an apparatus used in distilling alcoholic beverages.

stillage *Food Technology.* grain residue left in a still after removal of the distilled alcohol; used in the production of animal feed.

stillat. *Medicine.* by drops; in small quantities. (From Latin *stillatim.*)

stillbirth *Medicine.* **1.** the birth of a dead child. **2.** a fetus dead at birth. Also, **stilldeath.**

stillborn *Medicine.* born dead. Also, **stilldead.**

stilling basin *Engineering.* a portion of a channel that is wider or deeper than the main channel, reducing the flow speed through that portion. Also, **stilling box.**

Still's disease *Medicine.* a form of polyarthritis affecting children, characterized by lymph node enlargement and irregular fever. Also, JUVENILE RHEUMATOID ARTHRITIS.

Stillson wrench *Mechanical Devices.* a pipe wrench with sliding, adjustable L-shaped, serrated jaws on a pivot that firmly grip a workpiece.

stillstand *Geology.* **1.** the condition of a land mass that remains stationary or stable in relation to the inner earth or to sea level. **2.** a time during which this condition exists. Also, STANDSTILL.

still water or **stillwater** *Hydrology.* **1.** a nearly level reach of a stream having no visible current or other motion. **2.** a stream whose water appears sluggish, showing little or no turbulence.

still-water level *Oceanography.* the level the ocean's surface would be if there were no wave or wind action: in deep water, approximately at the midpoint of wave height; in shallow water, nearer to the trough.

still well *Meteorology.* a device consisting of a brass cylinder mounted over a hole in a triangular galvanized-iron base; used in evaporation measurements to provide an undisturbed water surface and support an evaporation pan.

still wine *Food Technology.* any nonsparkling table wine.

stilpnomelane *Mineralogy.* $K(Fe^{+2},Mg,\ Fe^{+3})_8(Si,Al)_{12}(O,OH)_{27}$, a black, brown, or greenish-black monoclinic and triclinic mineral, having a specific gravity of 2.59 to 2.96 and a hardness of 3 on the Mohs scale; found as foliated plates and in fibrous form in iron-ore deposits and schists.

stilt *Vertebrate Zoology.* a cosmopolitan wading bird of the family Recurvirostridae, especially *Cladorhynchus leucocephalus* and *Himantopus himantopus,* noted for having the longest legs compared to body size of any bird; characterized by strongly contrasting black and white plumage, bright pink legs, and a long, slender, straight bill; feeds on insects, seeds, and crustaceans by probing the mud.

stilt root *Botany.* a branch root given off before entering the soil, thus forming an additional support; a prop root or buttress root.

Stimson anchor *Naval Architecture.* a cast-iron anchor having an airfoil-like shape that is designed to bury itself in soft bottom material when subjected to a dragging force.

stimulant *Physiology.* any agent that causes an increase in the functional activity of the body or some part of the body. *Pharmacology.* specifically, a drug that has this effect, such as caffeine, nicotine, or amphetamines.

stimulate *Physiology.* to temporarily excite the functional activity of the body or some part of the body, especially a nerve or gland. *Behavior.* to bring about a behavioral response.

stimulated emission *Atomic Physics.* a process in which an incident photon causes an atom to drop to a lower energy level, thus allowing for the emergence of two photons of the same frequency; such phenomena are critical in laser technology. *Electromagnetism.* the emission of electromagnetic radiation from an atom or molecule as the result of being stimulated by an external radiation field of the same frequency.

stimulated-emission device *Electronics.* any device, such as a laser, that uses the principle of amplification of electromagnetic waves by stimulated emission.

stimulation *Biology.* the act of stimulating or the fact of being stimulated; any reaction to a stimulus, either a physiological reaction of a body part or a behavioral reaction of the whole organism.

stimulation deafness *Medicine.* deafness that is brought on by noise, which causes changes in the chemical interchange between the canals of the cochlea as well as nerve destruction.

stimulation treatment *Petroleum Engineering.* any technique, such as acidizing or hydraulic fracturing, used to increase gas or oil production from a reservoir formation.

stimulator *Medicine.* any agent that excites functional activity.

stimulus *Physiology.* any type of force able to produce a response in a sensorium or any responsive tissue; an alarm that triggers a response. *Behavior.* a condition or change in the environment that produces a behavioral response. *Control Systems.* a signal that is acted upon by the controlled variable in a control system.

stimulus control *Behavior.* the degree to which a stimulus can determine the occurrence of a conditioned response.

stimulus filtering *Behavior.* the selection of only the stimuli that elicit or regulate behavior from the great number of stimuli that are present in a given situation.

stimulus generalization *Behavior.* see GENERALIZATION.

stimulus overload *Psychology.* see OVERLOAD.

stimulus-response theory see S-R THEORY.

stimulus summation see HETEROGENEOUS SUMMATION.

sting *Medicine.* a sharp, painful wound that is inflicted by penetration of the skin by a specialized organ, often venom-bearing, of a bee, wasp, ant, mosquito, scorpion, jellyfish, or other such organism. *Zoology.* see STINGER. *Botany.* a stiff, sharp glandular hair on certain plants, such as nettles, that emits an irritating fluid when touched.

stinger *Zoology.* **1.** a puncturing organ, as in a bee or wasp, usually capable of delivering venom. **2.** an organism that is capable of delivering a sting.

Stinger *Ordnance.* a portable, shoulder-fired, air-defense artillery missile weapon for low-altitude air defense of forward area combat troops.

stinging cell or **stinging capsule** *Zoology.* another term for a nematocyst. See NEMATOCYST.

stingray *Vertebrate Zoology.* any of various rays, especially those of the family Dasyatidae and related families, having a long, whiplike tail that bears one to several sharp, barbed, often poisonous spines capable of inflicting severe wounds.

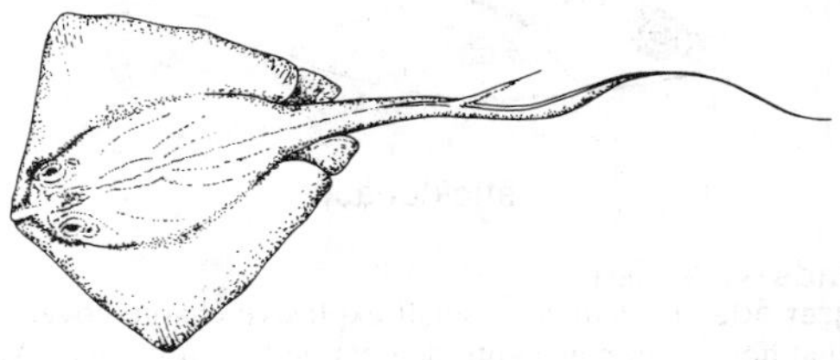

stingray

stinkhorn *Mycology.* any of various unpleasant-smelling, brown-capped fungi of the genus *Phallus,* especially *P. impudicus*; often found growing in backyards and under open stairways.

stinkstone *Geology.* a term for a stone that emits an unpleasant odor when struck, rubbed, or broken, because it contains decomposing organic matter. *Petrology.* see SWINESTONE.

stipe *Botany.* a stalk or support structure of a mushroom, plant, or plant organ, such as the petiole of a fern, the stalk of a seaweed, or the caudex of a palm. *Zoology.* a stemlike part or organ, as a footstalk.

stipel *Botany.* a secondary stipule situated at the base of a leaflet or compound leaf.

stipes *plural,* **stipites.** *Invertebrate Zoology.* the second joint in a maxilla of crustaceans and insects.

Stipitococcaceae *Botany.* a family of freshwater yellow-green algae of the order Rhizochloridales, distinguished by cells that produce a delicate, hyaline, vase-shaped lorica on a threadlike stipe; includes species that are epiphytic on filamentous algae.

stippler *Mechanical Devices.* an artist's paint brush used to gently touch a paintboard, making dots or points.

stipule *Botany.* one of a pair of leaflike or membranous structures at the base of a leaf petiole.

Stirling, James 1692–1770, Scottish mathematician; formulated the Stirling approximation and the Stirling series of numbers.

Stirling, Robert 1790–1878, Scottish clergyman and inventor; developed the Stirling engine (1816).

Stirling cycle *Thermodynamics.* a thermodynamic cycle that consists of four processes: a constant-temperature compression; a constant-volume heat transfer to the working fluid; a constant-temperature expansion; and a constant-volume heat transfer out of the system. This is an ideal cycle whose efficiency could equal that of a Carnot cycle; it is used as a model for actual gas turbines.

Stirling engine *Mechanical Engineering.* an actual example of the Stirling cycle; an external-combustion reciprocating engine in which work is performed as air in the cylinder is heated and expands, driving the working piston.

Stirling numbers *Mathematics.* **1. Stirling numbers of the first kind** are a triangular array of integers $s(r,n)$ such that: (a) $r,n \geq 0$; (b) the initial conditions are $s(0,0) = 1$ and $s(r,0) = 0$ for $r \neq 0$; and (c) the recursion relation is

$$s(r + 1,n) = s(r,n - 1) - rs(r,n).$$

2. Stirling numbers of the second kind are a triangular array of positive integers $S(r,n)$ such that: (a) $r,n \geq 1$; (b) the initial conditions are $S(r,n) = 0$ for $r < n$ and $S(r,n) = 1$ for $n = 1$ or $r = n$; and (c) the recursion relation is

$$S(r + 1, n) = nS(r, n) + S(r, n - 1).$$

The Stirling number $S(r,n)$ of the second kind is the number of ways that r distinct objects can be distributed among n nondistinct cells with no cell left empty. Changes of basis between factorial polynomials in x and ordinary powers of x are given by the matrices of Stirling numbers of the first and second kinds.

Stirling's approximation or **Stirling's formula** *Mathematics.* an approximation for the factorial of a positive integer n; i.e.,

$$n! \approx (n/e)^n(2\pi n)^{1/2}.$$

That is, it can be shown that

$$\lim_{n\to\infty} n!/(n/e)^n(2\pi n)^{1/2} = 1.$$

Stirodonta *Invertebrate Zoology.* an order of echinoids with gills, a rigid test, and keeled teeth.

stirred tank fermenter *Biotechnology.* a widely used fermentation vessel or tank in which the vessel is fitted with a mechanically driven impeller and ports for input and output.

stirring *Physics.* a mechanical process by which fluids are mechanically mixed by introducing an increased amount of turbulence.

stirring effect *Electromagnetism.* the circulation of molten conductive substances giving rise to an effective electric current; the circulation is incurred by a combination of pinch and motor effects.

stirrup *Mechanical Devices.* a loop or ring suspended from the saddle of a horse and used as a support for the rider's foot. *Anatomy.* another name for the stapes, the innermost of the three small bones of the middle ear; it has a shape somewhat like the stirrup on a saddle. *Civil Engineering.* a lateral reinforcement formed of individual units, open or closed, or of continuously wound reinforcement; usually used in reinforced concrete to resist shear. *Engineering.* any of various devices resembling the stirrup on a horse's saddle, used to hold, support, or clamp something. *Naval Architecture.* a short rope with an eye at the end that is hung from a yard to support a footrope.

stishovite *Mineralogy.* SiO_2, a colorless, transparent tetragonal mineral of the rutile group, occurring with coesite in granular aggregates of submicrometer size, having a specific gravity of 4.28 and an undetermined hardness; polymorphous with quartz, tridymite, cristobalite, and coesite and the only polymorph in which silicon is in sixfold coordination.

stitch *Textiles.* **1.** an in-and-out movement through fabric made by a threaded needle. **2.** a single loop of yarn formed by a knitting needle or crochet hook. **3.** the various methods or styles of stitching, including backstitch, basting, cross stitch, hemming stitch, running stitch, and slipstitch. *Surgery.* **1.** to loop a thin material (such as thread, catgut, or wire) by means of a needle through tissues to join or support them. **2.** a loop so made.

stitch abscess *Surgery.* a localization of pus that forms at or adjacent to a suture.

stitch bonding *Engineering.* a method of placing wire on an integrated circuit and making connections in which the wire is extruded through the center of the welding electrode.

stitching *Engineering.* a technique in which two materials are welded together by electrodes heated with radio waves in a machine that resembles a standard sewing machine; used to bind rubberized fabric and thermoplastic resins.

stitch rivet *Engineering.* one of a line or series of rivets used to join parallel members such as double-plated girders.

stitch welding *Metallurgy.* spot resistance welding performed in many, spaced locations.

STL *Aviation.* the airport code for St. Louis International.

St. Lawrence *Geography.* a river flowing 1900 miles northeast from Lake Ontario to the Atlantic via the Gulf of St. Lawrence in southeastern Canada.

STO *Aviation.* the airport code for Stockholm, Sweden.

stoa *Architecture.* in Greek architecture, a portico.

Stobbe reaction *Organic Chemistry.* a reaction of succinic esters with aldehydes and ketones in the presence of alkoxides to form monoesters of α-alkylidenesuccinic acids.

stochastic *Mathematics.* random; of or relating to random variables.

stochastic automaton see PROBABILISTIC AUTOMATON.

stochastic control theory *Control Systems.* a branch of control theory that attempts to predict and minimize the magnitudes and limits of random deviations in a control system by optimizing the design of the controller.

stochastic error see RANDOM ERROR

stochastic matrix *Mathematics.* a square matrix with nonnegative real entries such that the sum of the entries of each row equals one. If, in addition, the sum of the entries in each column equals one, the matrix is said to be **doubly stochastic.**

stochastic model *Statistics.* a mathematical model that contains uncertainties or random variables.

stochastic process *Statistics.* a sequence of random variables, often describing the evolution of a phenomenon over time. *Mathematics.* see PROCESS.

stochastic sequential machine see PROBABILISTIC SEQUENTIAL MACHINE.

stochastic variable *Statistics.* see RANDOM VARIABLE.

stock *Industrial Engineering.* **1.** supplies on hand, of either materials used in work or of the finished product. **2.** lumber or metal in standard cut sizes or shapes. *Graphic Arts.* any paper used in printing, or the paper used in a particular job. *Textiles.* the raw materials or fibers involved in factory processing, including raw fibers, waste materials, mixtures, and staples. *Agriculture.* farm animals as a group; livestock. *Invertebrate Zoology.* a colony of connected individuals, such as anthozoans. *Botany.* **1.** a stem or root into which a scion or bud is inserted for grafting. Also, ROOTSTOCK, UNDERSTOCK. **2.** a plant strain that is maintained for propagation. **3.** the trunk or main stem of a plant, especially a tree, as opposed to the leaves and smaller branches. *Ordnance.* the component of a shoulder weapon to which the barrel is attached and that allows the weapon to be supported on the shoulder; it is usually made of hardwood, but may be made of plastic, nylon, or other material. Also, SHOULDER STOCK. *Geology.* **1.** a circular or elliptical intrusive body of igneous rock, having a relatively small surface exposure (usually less than 40 square miles) and no determinable floor. **2.** see PIPE.

stock control see INVENTORY CONTROL.

stock cover see COVER STOCK.

stock dyeing *Textiles.* the process of dyeing fibers before they are spun into yarn; used especially with wool.

stocker *Agriculture.* a stock animal, especially a steer.

stocking cutter *Mechanical Engineering.* a cutting tool having a side rake or curved edges for cutting the rough tooth spaces on a gear blank.

stockpile-to-target sequence *Ordnance.* the order of events involved in removing a nuclear weapon from storage and assembling, testing, transporting, and delivering it on the target.

stockwork *Geology.* a mineral deposit forming a pattern of small interlacing veins, which diffuse into the country rock.

stoichiometric [stoi´kē ə met´rik] *Physical Chemistry.* **1.** of or relating to stoichiometry. **2.** describing the quantitative relationship between chemical substances, in which proportions of reactants and products are characteristic for each reaction and are determined from such information as molecular weight, chemical formulas, and equations.

stoichiometric compound *Materials Science.* a compound with integer atom or ion ratios.

stoichiometry [stoi´kē äm´ə trē] *Physical Chemistry.* **1.** the branch of chemistry that applies the laws of definite proportions and conservation of matter and energy to chemical processes. **2.** the science or study of the proportional relationships of two or more substances during a chemical reaction.

stoke *Mechanical Engineering.* to tend and supply fuel to a fire, especially of a boiler or furnace.

stoke *Fluid Mechanics.* a unit of measurement for kinematic viscosity in the centimeter-gram-second system, equal to the poise divided by the density in grams per cubic centimeter. (Named for Sir George *Stokes*.)

stokehold or **stokehole** see FIREROOM.

stoker *Mechanical Engineering.* a continuous mechanical apparatus for feeding fuel into a boiler or furnace, handling the by-products, and controlling the air ventilation to the fuel.

Stokes, George Gabriel 1819–1903, British mathematical physicist; formulated Stokes' law; described fluorescence; contributed to wave theory of light.

Stokes-Adams syndrome *Medicine.* see ADAMS-STOKES DISEASE.

Stokes drift *Fluid Mechanics.* the drift of particles in a gravity wave, occurring because the particles have periodic velocities with a nonzero mean.

Stokes flow *Fluid Mechanics.* a low Reynolds number flow where inertial forces are negligible compared with pressure and viscous forces; the nonlinear terms in the Navier-Stokes equation can thus be ignored.

Stokes frequency *Optics.* in Raman scattering, any frequency of light that is lower than that of the stimulating beam.

Stokes' integral theorem *Mathematics.* a theorem stating that $\int_C d\omega = \int_{\partial C} \omega$; i.e., the operators d and ∂ are adjoints of each other.

stokesite *Mineralogy.* $CaSnSi_3O_9 \cdot 2H_2O$, a colorless orthorhombic mineral occurring in acute pyramidal crystals, having a specific gravity of 3.185 and a hardness of 6 on the Mohs scale.

Stokes' law *Fluid Mechanics.* a law stating that the drag on a spherical body moving with a constant relative velocity is equal to $3pmUD$, where p is π (3.1416), m is the viscosity, U is the relative velocity, and D is the diameter of the body; valid in very low Reynolds number flow. *Spectroscopy.* a law stating that the wavelength of luminescent radiation is always longer than that of the original exciting radiation.

Stokes' lemma *Mathematics.* in an odd-dimensional space, the integrals of a 1-form ω, along any two curves encircling the same tube of characteristic lines of ω, are equal.

Stokes lines *Spectroscopy.* in Raman spectroscopy, spectral lines produced by scattered light having longer wavelengths (lower frequencies) than those of the original incident light.

Stokes number 1 *Fluid Mechanics.* a dimensionless number used in examining the dynamics of a particle in a fluid that is equal to the product of viscosity and the particle's vibration time, divided by the product of fluid density and the square of a characteristic length.

Stokes number 2 *Fluid Mechanics.* a dimensionless number used in calibrating rotameters, equal to

$$1.042\, m_f g\rho(1 - \rho/\rho_f)R^3/\mu^2,$$

where m_f and ρ_f are the mass and density of the float, ρ and μ are the density and dynamic viscosity of the fluid, and R is the ratio of the radius of the tube to the radius of the float.

Stokes parameters *Optics.* the four qualities that fully describe a beam of polarized light: I is the intensity of the beam, M is the part of the beam polarized in the horizontal plane, C is the part polarized in the +45° direction, and S is the part that is circularly polarized.

Stokes shift *Spectroscopy.* in luminescent spectroscopy, a displacement of spectral lines or bands toward longer wavelengths (lower frequencies) than those of the corresponding absorption lines or bands.

Stokes stream function *Fluid Mechanics.* a stream function for an axially symmetric flow or a flow about a body of revolution.

Stokes stretcher *Medicine.* a stretcher constructed of tubular steel and wire mesh forming a stiff basket that acts as a splint for the entire body.

Stokes vector *Optics.* the matrix whose elements indicate the polarization state and relative intensity of a beam of light.

STOL [stōl; stäl] *Aviation.* **1.** the ability of an aircraft to clear a 50-foot obstacle within 1500 feet of beginning a takeoff, or to stop within 1500 feet after clearing a 50-foot obstacle during landing. **2.** see STOL AIRCRAFT. (An acronym for short takeoff and landing.)

STOL aircraft *Aviation.* **1.** an aircraft that can take off and land within the STOL parameters. **2.** broadly, any aircraft that can take off and land within a relatively short distance.

stolon *Botany.* a creeping stem or runner that lays on or just below the surface of the ground and roots at the nodes or the tip to produce new plants. *Mycology.* the creeping hyphae of certain fungi that give rise to new individuals. *Invertebrate Zoology.* **1.** in some colonial bryozoans and hydrozoans, a stemlike, cylindrical, horizontal structure from which individuals grow. **2.** an outgrowth of a colonial zooid from which new buds develop. Also, SOLENIUM.

stolonate *Biology.* **1.** having stolons. **2.** developing from a stolon.

Stolonifera *Invertebrate Zoology.* an order of anthozoans in the subclass Alcyonaria, with polyps growing upright from a stolon base or thin mats; small colonies adhere to rocks in shallow or tidal waters; includes the organ pipe coral.

stoloniferous *Biology.* having or developing stolons.

stolonization *Biology.* the production of stolons.

Stolonoidea *Paleontology.* a minor order of graptolites; includes sessile and encrusting forms, usually characterized by irregular stolons; extant in the Ordovician.

stolzite *Mineralogy.* $PbWO_4$, a yellow to reddish-brown tetragonal mineral, dimorphous with raspite occurring in dipyramidal or thick tabular crystals, having a specific gravity of 7.9 to 8.34 and a hardness of 2.5 to 3 on the Mohs scale; found in the oxidized zones of tungsten deposits.

stoma *Biology.* any small opening or pore in a surface. *Botany.* any of numerous tiny intercellular openings in the epidermis of a leaf or stem epidermis, usually concentrated on the undersurface of leaves; this provides the main route for both photosynthetic gas exchange and for the water vapor lost in transpiration. *Zoology.* a usually small, simple mouth or other ingestive opening. *Surgery.* the opening established in the abdominal wall by surgical procedure such as colostomy or ileostomy, or the opening between two portions of the intestine in an anastomosis.

stomach *Anatomy.* the saclike, expandable digestive organ located between the esophagus and small intestine. *Zoology.* any similar or analogous organ in lower organisms.

stomacher *Biotechnology.* a device for separating bacteria from food samples, fabrics, or swabs.

stomachic *Anatomy.* of or relating to the stomach. Also, **stomachal.** *Pharmacology.* describing any medicine that promotes the functional activity of the stomach.

stomate *Anatomy.* relating to or having a mouth or stoma. Also, **stomatal, stomatiferous, stomatose, stomatous.**

stomatitis *Medicine.* any inflammation of the mouth.

stomato- or **stomat-** a combining form meaning "mouth." Also, **stomo-, stom-.**

Stomatochroon *Botany.* a genus of branched green algae of the order Trentepohliales, characterized by a thallus reduced to a few cells that emerge from stomatal chambers of various tropical and subtropical tree leaves.

stomatodynia *Medicine.* a pain in the mouth.

stomatogenesis *Cell Biology.* the development of the oral structures of a stomate ciliate during binary fission.

stomatology *Medicine.* the branch of medical science concerned with the study of the mouth and its diseases.

stomatomycosis *Medicine.* any mouth disease caused by a fungus.

stomatoplasty *Surgery.* plastic surgery to repair or reconstruct the mouth.

Stomatopoda *Invertebrate Zoology.* the mantis shrimps, an order of malacostracan crustaceans in the superorder Hoplocarida, having a flattened carapace, stalked eyes, and five pairs of thoracic appendages, the second pair of which are enlarged for catching prey; they are active predators that grow up to 12 inches long.

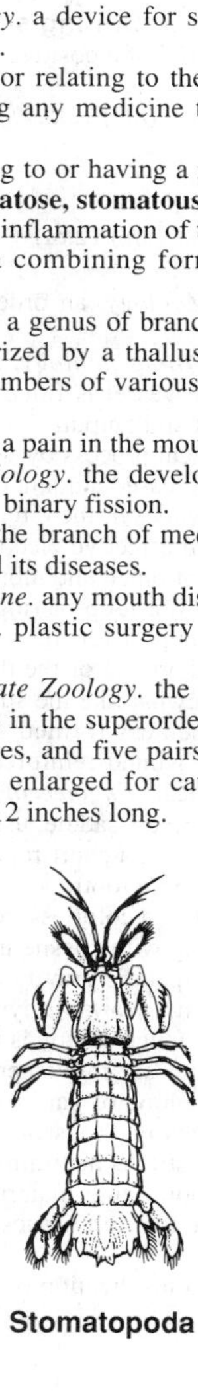

Stomatopoda

stomatorrhagia *Medicine.* a hemorrhage from the mouth.
Stomiatoidei *Vertebrate Zoology.* a suborder of fishes of the order Salmoniformes that includes bristlemouths, lightfishes, hatchetfishes, viperfishes, and dragonfishes; characterized by the presence of luminescent organs, and usually found in deep marine waters. Also, **Stomioidei.**
stomium *Botany.* **1.** group of thin-walled cells in the wall of a fern sporangium where rupture occurs at maturity. **2.** an opening in an anther where it splits apart, usually between lip cells.
stomodeum *Invertebrate Zoology.* the anterior part of the gut in many invertebrates, the part of the digestive system that includes the mouth, pharynx, and esophagus which may be enlarged to form a crop. Also, FOREGUT. *Developmental Biology.* an invagination of the ectoderm that develops into the mouth cavity. Also, **stomodaeum.**
stone *Geology.* **1.** any small fragment of rock or mineral matter. **2.** a large fragment within a variable matrix of sedimentary rock. **3.** see STONY METEORITE, STONY-IRON METEORITE. *Botany.* the hard endocarp of a drupe, as of a cherry or peach. *Pathology.* see CALCULUS. *Metrology.* a traditional British measure of weight, equal to 14 pounds.
Stone Age *Archaeology.* a time period, before 3000 BC, designated in the Three-Age System as preceding the Bronze Age and the Iron Age; defined by the concept that stone tools were the basic human technology during this period.
stoneboat *Mining Engineering.* a flat, runnerless sled or drag used to transport heavy materials.
stone-bordered strips see STONE STRIPES.
stone bubble see LITHOPHYSA.
stone canal *Invertebrate Zoology.* in echinoderms, a short canal connecting a madreporite to the ring canal around the mouth.
stone cell see BRACHYSCLEREID.
stone coal see ANTHRACITE.
stone crab *Invertebrate Zoology.* a crab, *Menippe mercenaria,* of the Gulf of Mexico, widely harvested for its edible meat.
stone curlew *Vertebrate Zoology.* any of several tropical wading birds of the family Burhinidae having a thickened joint between the femoral and tibiotarsal bones. Also, THICK-KNEE.
stone dust *Mining Engineering.* any inert dust used in coal mine roadways to absorb heat and thereby reduce the danger of coal-dust explosions.
stonefish *Vertebrate Zoology.* a tropical scorpionfish, *Synanceja verrucosa,* that secretes a deadly poison from spines on its dorsal fin.
stone fruit see DRUPE.
Stonehenge *Archaeology.* an ancient monument on Salisbury Plain (Wiltshire, England), the remains of four massive trilithons surrounded by concentric circles of megaliths, probably constructed about 1500 BC and used for astronomical observation and prediction, possibly in connection with some form of sun worship.

Stonehenge

stone ice see GROUND ICE.
stone lattice *Geology.* a distinctive cell-like or honeycomb pattern produced on a desert rock face by the effects of windblown sand. Also, **stone lace.**
stone pine see UMBRELLA PINE.
stone polygon or **stone mesh** see SORTED POLYGON.
stone ring *Geology.* **1.** a round or polygonal border of coarse rock debris that encloses an area of finer material. **2.** see SORTED POLYGON.
stone river see ROCK STREAM.
Stone's theorem *Mathematics.* a theorem stating that every Boolean ring is isomorphic to a ring of subsets of some set.
stone stripes *Geology.* a patterned ground consisting of downslope parallel bands of coarse rock debris alternating with wider bands of finer material. Also, ROCK STRIPES.
stoneware *Materials.* **1.** a hard, opaque, glassy ceramic ware that is fired at high temperatures and composed primarily of clay, silica, and feldspar. **2.** any of various highly vitrified ceramics with low tensile strengths made of selected clays, feldspar, and silica; resistant to all acids, except hydrofluoric acid; used especially in industrial chemical equipment.
Stone-Weierstrass theorem *Mathematics.* let A be a collection of continuous real-valued functions on a compact space K such that: (a) A contains the constant function $e(x) = 1$ for $x \in K$; and (b) for any pair of distinct points x and y of K, there exists a function f in A such that $f(x) \neq f(y)$. Then any continuous real-valued function on K can be uniformly approximated by polynomials (with real coefficients) of functions in A.
stonework *Building Engineering.* **1.** masonry construction using stone. **2.** the preparation or setting of stone for building or paving.
stonewort *Botany.* any plantlike green algae of the class Charophyceae, especially those of the genus *Chara,* characterized by a jointed filamentous or thalloid body, often encrusted with calcium carbonate.
Stoney, G. Johnstone 1826–1911, Irish physicist; analyzed and named the electron.
stony coral *Invertebrate Zoology.* any reef-building colonial anthozoan of the order Sclerectinia; a true coral.
stony-iron meteorite *Geology.* a relatively rare variety of meteorite, composed of large and roughly equal amounts of nickel-iron and heavy basic silicates such as olivine and pyroxene; one of the three major types of meteorites. Also known by various other names, such as IRON-STONY METEORITE, LITHOSIDERITE, and SIDEROLITE.
stony meteorite *Geology.* a meteorite composed exclusively or primarily of silicate minerals, chiefly olivine, pyroxene, and plagioclase; one of the three major types of meteorites. Also known by various other names, such as METEORIC STONE, AEROLITE, and ASIDERITE.

stony meteorite

stooping *Meteorology.* an atmospheric optical phenomenon in which the image of a distant object is foreshortened in the vertical; caused by a decrease with elevation in the curvature of light rays due to atmospheric refraction; the opposite of towering.
stop a device that halts or obstructs; specific uses include: *Optics.* **1.** a difference in the amount of exposure given to an emulsion in a camera, with each successively higher stop providing half the exposure of the preceding stop; usually controlled by an adjustable diaphragm. **2.** any of the aperture settings marked on a camera lens. **3.** a piece of thin metal with a fixed-size hole used in place of an adjustable diaphragm to control the amount of light passing through a camera lens. *Linguistics.* a sound produced by obstructing the flow of air through the mouth as in the sounds of [k], [t], or [p] or as in a glottal stop. Also, PLOSIVE. *Mechanical Engineering.* a projecting piece at one end of the top of a height-adjustable bench; used to steady a workpiece in the process of being planed. *Building Engineering.* a wood or metal piece that is attached to the frame or base of a door or window to prevent motion beyond a given point. *Robotics.* the final position of a robot's trajectory.

stop bath *Graphic Arts.* a chemical bath used to halt the development process of photographic film by neutralizing the developer.
stop bead *Building Engineering.* a molding strip on the pulley stile along the inside edge of a window frame, which holds a sliding sash.
stop bit *Computer Programming.* the last element of a character that serves as a signal to prepare a receiving device for the reception of a subsequent character or block.
stopcock *Engineering.* a valve body having a fitted plug.
stop code *Computer Programming.* a code that stops a reader and suspends machine operations. In an interactive system, it is usually used as an index for a **stop list** for operator changes or correction.
stop codon see NONSENSE CODON.
stopdown *Optics.* the act of reducing the amount of light passing through a lens by closing the lens diaphragm to a smaller aperture.
stope *Mining Engineering.* **1.** an excavation for removing ore in a series of steps, usually from vertical or highly inclined veins. **2.** to excavate ore by driving a series of horizontal workings into a vein, each immediately above or below the previous. **3.** any underground ore extraction.
stope board *Mining Engineering.* a timber stage established on the floor of a stope; used for setting a rock drill and tilted to allow the bottom holes to be drilled in the same inclined direction.
stope hoist *Mining Engineering.* a portable hoist powered by compressed air; used in operating a scraper-loader or in positioning heavy timbers, often in narrow stopes.
stope pillar *Mining Engineering.* a column of ore left standing as support for the stope.
stoping *Mining Engineering.* the process of excavating a mine; working a stope. *Geology.* the process by which magma moves upward in the earth, by detaching and engulfing blocks of country rock.
stoping drill *Mining Engineering.* a small, air- or electric-powered drill, generally mounted on an extensible column, used to work stopes, raises, and narrow workings. Also, **stoper.**
stoping ground *Mining Engineering.* a portion of an ore body that has been opened by drifts and raises and is ready for working.
stop instruction see PROGRAM STOP.
stop log *Civil Engineering.* a balk, plank, precast concrete beam, or steel joist that fits between vertical grooves in walls or piers to close up a spillway or other water channel.
stop loop *Computer Programming.* a small nonterminating loop used for operator convenience to halt program execution in the event of an error or other unusual condition.
stop needle *Surgery.* a needle with a shoulder to prevent insertion past a certain distance.
stop number see F-NUMBER.
stop nut *Mechanical Devices.* **1.** a nut on an adjusting apparatus, such as a screw for a pliers tool, that limits the amount the adjusting screw can travel. **2.** a nut equipped with a nonmetallic fiber or plastic insert that acts to stabilize a screw or bolt, eliminating the need for a washer.
stoppage *Ordnance.* a condition in which a weapon cannot be fired due to a faulty mechanism or faulty ammunition; the term is applied especially to automatic weapons.
stopped-flow method *Chemistry.* a technique that allows the rates of fast chemical reactions to be monitored, accomplished by suddenly blocking the flow of reaction and measuring, e.g., the optical density of a portion of solution by a rapid response method such as photoelectric photometry.
stopper *Mechanical Devices.* any cork, plug, or other object used to close a bottle, drain, tube, and so on.
stopper rod *Metallurgy.* a rod that controls the flow of molten metal from a bottom-pouring ladle.
stopping *Mining Engineering.* a masonry or brick wall, or less commonly a brattice, constructed in a mine across an old heading, chute, or airway; used to confine an air current to certain passages, to lock up the gas in old workings, or to smother a fire. *Nucleonics.* the reduction of kinetic energy of a particle passing through matter due to ionizations and other interactions.
stopping down *Graphic Arts.* the process of reducing the size of a camera lens aperture.
stopping-off *Metallurgy.* **1.** a coating intentionally applied to a portion of a metallic part to prevent chemical attack or coating. **2.** in casting, the process of filling a part of a mold cavity to prevent molten metal from entering it.
stopping-out *Graphic Arts.* **1.** the application of opaquing material to film negatives. **2.** the application of protective coatings to printing plates so that those areas will not attract ink.
stopping potential *Electronics.* the voltage required to halt the outward movement of electrons generated by a given photoelectric or thermionic action.
stopping power *Metallurgy.* in ion-implantation metallurgy, the slowing down of ions in a solid, represented as the energy loss per unit distance traveled in the solid. *Nucleonics.* the kinetic energy lost per unit path length along the track of an energetic particle passing through a sample of matter. Also, ATOMIC STOPPING POWER. *Ordnance.* the relative ability of a hunting weapon to bring down a charging animal with one shot, or by extension, the relative ability of a military weapon to bring down an advancing enemy.
stopping range *Nuclear Physics.* the distance to which an energetic particle can penetrate a medium before coming to rest, which is directly proportional to the medium's equivalent thickness; measured in milligrams per square centimeter, and inversely proportion to its density.
stopping rule *Statistics.* in sequential sampling, a rule that determines when a study is to be terminated.
stopping-up *Graphic Arts.* the process of increasing the size of a camera lens aperture.
stop valve *Mechanical Engineering.* any valve that can be set to completely block off flow.
stopwatch *Horology.* a watch having a sweep second hand and sometimes a minute counter and employing stop, start, and flyback controls; used to measure the duration of events rather than to tell the time of day.
stopwater *Naval Architecture.* a canvas watertight fitting that is placed in a gap in a metal hull frame.
stopway *Aviation.* a cleared area beyond the end of a runway, provided as a safety measure for aircraft that must stop after an aborted takeoff.
stopwork *Horology.* in a timepiece, a device composed of a wheel with irregular teeth and a pointer from the barrel arbor, designed to limit both the winding tension that can be applied to the mainspring and how far the mainspring can unwind.
storable *Space Technology.* of a rocket fuel, able to be kept safely in a fuel tank without the need of special measures for temperature or pressure control. Thus, **storable propellant.**
storage the fact of putting something aside for future use; specific uses include: *Computer Technology.* a device or medium capable of receiving and retaining information for an indefinite period of time and allowing access to that information upon command. *Psychology.* a term used for the capacity or processes of memory, based on the concept that human memory is analogous to information storage in a computer.
storage address register see ADDRESS REGISTER.
storage alignment *Computer Science.* **1.** a requirement of many central processing units that certain data have addresses which fall at even memory word boundaries, so that the data will be contained in a single memory word. **2.** in a compiler or assembler, the adjustment of memory addresses so that data will be properly aligned.
storage allocation *Computer Programming.* the process of assigning specific areas of storage to programs or blocks of data.
storage and retrieval *Computer Science.* a term used to describe the ability of a computer to retain data that has been input into the system (storage) and then later to access this data so that it may be consulted, manipulated, altered, and so on (retrieval).
storage and retrieval system *Computer Science.* an organized information-management system that allows data to be retrieved from memory; i.e., a computer.
storage area *Computer Programming.* a location in a storage medium designated for storing specified programs or data.
storage battery *Electricity.* a voltaic battery constructed of storage cells. Also, ACCUMULATOR BATTERY, RECHARGEABLE BATTERY.
storage block *Computer Programming.* a contiguous section of storage.
storage buffer register see MEMORY BUFFER REGISTER.
storage camera see ICONOSCOPE.
storage capacity *Computer Technology.* the amount of data that a storage device can hold, measured in binary digits, bytes, characters, words, or other units of data.
storage cell *Electricity.* a cell that can be recharged with electricity; e.g., lead-acid and nickel-cadmium cells. Also, SECONDARY CELL. *Computer Technology.* the smallest physical component of information storage. Also, **storage element.**
storage compacting *Computer Programming.* see COMPACTION.
storage cycle *Computer Programming.* the recurrent sequence of storing, sensing, and regeneration of data that occurs during the transfer of data to or from the storage device of a computer.

storage density *Computer Technology.* the number of units of data that can be stored per unit length or area of a storage medium, such as the number of characters per inch on a magnetic tape.

storage device *Computer Technology.* any system component used to permanently or temporarily store data and programs.

storage disease *Medicine.* a metabolic disorder, formerly called thesaurismosis, in which some substance accumulates or is stored in certain cells in unusually large amounts; the stored substances may be lipids, proteins, carbohydrates, or other substances.

storage dump see MEMORY DUMP.

storage equation *Hydrology.* an equation applied to the routing of floods through a reservoir or stream reach, stating that the rate of inflow minus the rate of outflow to a particular area is equal to the change in storage.

storage fill *Computer Programming.* the storing of a pattern of characters in storage areas not to be used during a particular machine run; causes an error condition if inadvertently accessed.

storage fragmentation see FRAGMENTATION.

storage hierarchy *Computer Technology.* a hierarchy of storage technologies used to implement data storage, with smaller amounts of faster and more expensive memory at the top of the hierarchy and larger amounts of slower, less expensive memory at lower levels.

storage integrator *Computer Technology.* in an analog computer, an integrator used to store a voltage in the hold condition for future use, while the computer simultaneously assumes another control state.

storage key *Computer Programming.* a set of bits designed to be related to every word or character in a block of storage, allowing tasks with a matching set of protection key bits to use that block of storage.

storage location *Computer Programming.* **1.** a memory cell that contains one machine word and usually has a specific address. **2.** a character field within a word in a machine that allows character addressing. Also, STORED WORD.

storage mark *Computer Programming.* an indicator that defines the character space immediately to the left of the most significant character in accumulator storage. In the following example, the letter "a" would be the storage mark: a5245257.

storage matrix *Computer Programming.* storage in which a specific location or circuit element is addressed by the use of two or more coordinates.

storage medium *Computer Technology.* the physical material onto which information can be recorded and stored for later retrieval; examples are magnetic tape and optical and magnetic disks.

storage organ *Botany.* any swollen plant part in which food material is stored, such as a root.

storage oscilloscope *Electronics.* an oscilloscope in which the cathode-ray tube stores an image for a predetermined duration, or until it is deliberately erased to make space for a new image.

storage pit *Archaeology.* a deep pit in the ground in which food or other precious items were stored or hidden.

storage pool *Computer Technology.* **1.** a list of available storage areas in main memory. **2.** a group of similar storage devices, such as a set of tape or disk drives.

storage print see MEMORY DUMP.

storage protection *Computer Programming.* **1.** the limitation of access to a storage device to prevent the overwriting of existing data or programs. **2.** the limitation of access to main memory so that only the program to which a memory area is assigned can access it.

storage register see MEMORY BUFFER REGISTER.

storage reservoir see IMPOUNDING RESERVOIR.

storage ring *Nuclear Physics.* a device for storing a beam of accelerated particles fed from an accelerator, consisting of a set of magnets set in a circular track that are adjusted to keep the particles circulating until they are used.

storage time *Electronics.* **1.** the time needed for excess minority carriers stored in a forward-biased P-N junction to be removed after the junction is switched to reverse bias; therefore, the time interval between the application of reverse bias and the cessation of forward current. **2.** the time required for extra charge carriers in the collector region of a saturated transistor to be removed when the base signal is changed from a maximum to a zero level, and hence for the collector current to cease.

storage tube *Electronics.* **1.** an electron tube that uses cathode-ray-beam scanning and charge storage for the introduction, storage, and removal of information. **2.** a cathode-ray tube that stores images on a separate storage screen behind the viewing screen in the tube; images remain on the viewing screen until the storage screen is erased.

Störber crystal growth method *Materials Science.* a large-scale production in which the whole melt, rather than a saturated solution, is solidified, producing crystals of high optical quality with volumes greater than 10^3 cm^3.

store *Computer Programming.* **1.** to place data in a storage device. **2.** to copy data from a processor register to internal storage.

stored-energy welding *Metallurgy.* in welding, a process in which the electrical energy is slowly accumulated and suddenly discharged.

stored program *Computer Programming.* a set of instructions held in main memory awaiting execution. Also, **stored routine.**

stored-program computer *Computer Programming.* a computer that operates according to a set of instructions in the computer memory specifying the operations to be performed and the location of the data on which the operations are to be performed.

stored-program logic *Computer Programming.* a program stored in the memory that contains commands to perform identical processes on all problems.

stored-program numerical control see COMPUTER NUMERICAL CONTROL.

stored word see STORAGE LOCATION.

storethrough see WRITE-THROUGH.

store transmission bridge *Electricity.* a telecommunication signal bridge that is composed of four identical impedance coils separated by two capacitors and that electrostatically joins the outgoing and incoming telephone lines in an AC current; used to transmit voice-frequency currents while maintaining a separation between the two lines for the transmission of direct current for speaking purposes.

stork *Vertebrate Zoology.* a large, long-legged, wading bird of the family Ciconiidae, characterized by loose plumage, long legs, and a long neck and bill; found worldwide, mostly in tropical and temperate zones. The only stork found in North America is the **wood stork,** *Mycteria americana*; the stork of Europe that is associated in folklore with childbirth is the **white stork,** *Ciconia ciconia.*

stork

storm *Meteorology.* **1.** any atmospheric disturbance, especially one affecting the earth's surface, generally involving destructive or otherwise unpleasant weather. **2.** in synoptic meteorology, a complete individual disturbance, such as a tornado or cyclone, identified as a complex of pressure, precipitation, wind, clouds, etc. **3.** the transient local conditions produced by such a complex, such as a rainstorm or snowstorm. **4.** a wind whose speed is 56–63 knots (64–72 miles per hour).

storm beach *Geology.* **1.** a low ridge of coarse gravel, cobbles, and boulders built up at or behind the inner edge of a beach by storm waves. Also, STORM TERRACE. **2.** a beach that has recently undergone removal or deposition of material following a powerful storm.

storm center *Meteorology.* within a cyclone, the area or position having the lowest atmospheric pressure.

storm choke *Petroleum Engineering.* a safety valve installed below the surface in the tubing string of an oil well to shut in the well when the flow reaches a specified rate and to act as an automatic shutoff if the control valve is damaged. Also, TUBING SAFETY VALVE.

storm delta see WASHOVER.

storm door *Building Engineering.* an extra outside door that adds protection during harsh weather conditions; used especially in winter to minimize heat escape at the entrance of a building.

storm drain *Civil Engineering.* a drain used to convey rainwater, subsurface water, condensate, or similar discharge, but not sewage or industrial waste. Also, **storm sewer.**

storm ice foot *Oceanography.* an ice foot formed higher up the shore than the high-water level, due to a storm or heavy sea.

storm microseism *Geophysics.* any microseism that is caused by the action of ocean waves and lasts 25 seconds or longer.

storm model *Meteorology.* a three-dimensional physical representation of the movement of air and water vapor in a storm; used especially in hydrometeorology to compute the effective precipitable water from the surface dew point.

storm petrel *Vertebrate Zoology.* **1.** any of several small seabirds of the family Hydrobatidae, characterized by a tube nose and usually having black or sooty-brown plumage with a white rump. **2.** the British petrel, *Hydrobates pelagicus,* of the eastern Atlantic Ocean. Also, **stormy petrel.**

storm petrel

storm seepage see INTERFLOW.

storm sewage *Civil Engineering.* the material flowing in combined sewers or storm sewers as a result of rainfall.

storm surge *Oceanography.* an exceptionally high water caused by high winds or the convergence of wind-driven currents; may also include the effect of a drop in atmospheric pressure caused by hurricane winds. Also, SURGE WAVE, STORM WAVE.

storm terrace see STORM BEACH, def. 1.

storm tide *Oceanography.* the increase in high-water level above the level predicted in tide tables, caused by a storm surge.

storm track *Meteorology.* the path followed by a storm center.

storm transposition *Meteorology.* the process of transferring precipitation patterns or values from an actual to a potential position; used in hydrometeorology to augment the storm history of an area.

storm warning *Meteorology.* a weather forecast designed to alert the public to impending dangers, especially to potentially dangerous wind conditions at sea; classified in four types: small-craft warning, gale warning, whole-gale warning, and hurricane warning. *Navigation.* a message broadcast in plain language to navy and merchant ships, giving information on the position, movement, and intensity of a storm center.

storm-warning signal *Meteorology.* a pattern of flags (or, at night, of lanterns) on a coastal storm-warning tower that gives information about an impending storm.

storm-warning tower *Meteorology.* a sturdy metal tower used to display coastal storm-warning signals.

storm wind *Meteorology.* a wind having a speed of 56–63 knots (64–72 miles per hour).

storm window *Building Engineering.* an extra plane of glass mounted on the exterior of the main window or sash as protection from severe weather conditions.

stormy clot *Bacteriology.* a turbid coagulation produced in milk medium by lactose fermentation by the bacterium *Clostridium perfringens*; the clot is subsequently disrupted and broken apart due to the formation of a large volume of gas from this same fermentation process.

story *Building Engineering.* **1.** a complete horizontal section of a building; the space between two floors. **2.** the set of rooms on the same floor or level of a building. *Architecture.* each of a number of tiers or rows of columns, windows, or other building parts placed horizontally one above the other. Also, **storey.**

storyboard *Graphic Arts.* a board or panel containing a series of small drawings or sketches that roughly depict the sequence of action for a script to be filmed, as for a motion picture, television commercial, music video, or the like.

stoss *Geology.* describing the side of a hill or prominent rock that faces the upstream side of a glacier or the direction from which a glacier or ice sheet has moved. (From a German word meaning "to push.")

stout *Food Technology.* an alcoholic brew made with roasted malt, usually very dark and bitter.

stove *Engineering.* a closed chamber that can be heated with a fuel-air mixture to a high temperature for cooking, curing, or other applications.

stove bolt *Mechanical Devices.* a fastener used in stoves having a notched, countersunk, or hemispheric head and an attached square nut with coarse threads.

stovepipe *Engineering.* a sheet-metal pipe of relatively large diameter. *Space Technology.* the outside skin of a rocket or missile.

stoving *Engineering.* see BAKING.

stow *Naval Architecture.* to put cargo, provisions, or equipment in the places intended for them. *Aviation.* **1.** to retract an aircraft landing gear or other component and lock it in place. **2.** to pack or store cargo in the cargo area of an aircraft.

stowage *Naval Architecture.* **1.** the act of stowing. **2.** a place in which something is or may be stowed.

stowage diagram *Naval Architecture.* a scale drawing showing the exact cargo, clearances, hatches, and lifting gear for each cargo deck or platform.

stowage factor *Naval Architecture.* the cubic feet occupied by one long ton of a given cargo.

stowage plan *Naval Architecture.* a plan showing a ship's exact cargo load, indicating the destination port of each cargo item.

stowboard *Mining Engineering.* a mine heading used to collect waste.

STP standard temperature and pressure.

STR self-tuning regulator.

Str. *Streptococcus.*

str. strait.

strabismus [strə bis′məs] *Medicine.* the medical name for the condition known popularly as cross-eye, an involuntary misdirection of the eye and its visual axis, resulting from unequal ocular muscle tone, weak eyesight, or oculomotor nerve lesion. (Derived from a Greek word meaning "to squint.") Also, HETEROTROPIA, MANIFEST DEVIATION.

Strabo 63? BC–23? AD, Greek geographer and historian; wrote the 17-volume *Geographia,* describing much of Europe, Asia, and Africa.

Strachey, James 1671–1743, English geologist; identified and described strata.

straddle milling *Mechanical Engineering.* the simultaneous face milling on both sides of a workpiece, by using two cutters spaced as needed.

strafe *Ordnance.* to deliver automatic-weapons fire by aircraft on ground targets. (From a German word meaning "to punish.")

straggling *Physics.* random variations of some characteristic property of a substance, which can be accounted for by the passage of ions through the matter.

straight beam *Engineering.* in ultrasonic materials testing, a longitudinal wave produced by the ultrasonic search unit traveling perpendicularly to the test surface.

straightedge *Mechanical Devices.* a long, flat piece of wood, metal, or plastic with one or more straight edges used to test the straightness of edges or lines. *Mathematics.* specifically, such a ruler used in Euclidean constructions for constructing the line determined by two points, not for making measurements.

straight-flow turbine *Mechanical Engineering.* a low-head hydraulic turbine in which the reservoirs are connected by a straight tube into which the runners are integrated, and with the generator placed directly on the edge of the runners.

straight-flute drill *Mechanical Devices.* a rigid drill with a cone-shaped tip, backed-off cutting edges, and longitudinal shank flutes; used for cutting soft metals.

straight-in landing *Transportation Engineering.* a landing made as nearly directly as possible (within 30° from the aircraft's direction of arrival in an airport area).

straight-line code *Computer Science.* a sequence of computer instructions that does not contain any branches and is executed in sequence.

straight-line mechanism *Mechanical Engineering.* any mechanism that is designed to produce motion in a straight or nearly straight line over at least a portion of its cycle.

straight-line motion *Robotics.* robotic motion in which an end effector moves in a straight line until it reaches a way point, then stops for a moment before moving in another direction.

straight-pane hammer *Mechanical Devices.* a hammer with a flat face at one end and a blunt, chisel-like face at the other end in line with the shaft.

straight-peen hammer *Mechanical Devices.* a hammer having a long, narrow, rounded striking end parallel with its handle.

straight polarity *Metallurgy.* arc welding in which the electrode is negative.

straight-run *Chemical Engineering.* gasoline in a petroleum refinery that has been produced from direct distillation of crude oil in a still and is very low in unsaturated hydrocarbons. Also, **straight-run gasoline.**

straight-run distillation *Chemical Engineering.* a petroleum-refinery process that distills crude oil to produce straight-run gasoline.

straight-shank drill *Mechanical Devices.* a small drill with a round, parallel shank used to self-center chucks.

straight sinus *Anatomy.* a venus sinus of the dura mater of the brain, located at the junction of the cerebral falx and the cerebellar tentorium.

straight strap clamp *Mechanical Devices.* a flat clamp with an elongated slot, secured with a T-bolt and nut to allow positioning in various operations.

straight-tube boiler *Mechanical Engineering.* a boiler in which all the water tubes are straight and connected in parallel, so that they require appropriate connectors from the headers for circulation through the system. Also, HEADER-TYPE BOILER.

straight turning *Mechanical Engineering.* the process of turning a workpiece on a lathe to produce a shaft with a constant diameter over its entire length.

straight vertical antenna *Electromagnetism.* an antenna consisting of a straight vertical wire.

straightway pump *Mechanical Engineering.* a pump designed to provide direct fluid flow through suction and discharge valves along a straight path.

straight whiskey *Food Technology.* unblended, unadulterated single-grain whiskey, usually 100 proof.

straight wire *Biotechnology.* a device used to inoculate a subculture from an isolated colony or to prepare a stable culture; a rod-shaped metal handle holds a small straight piece of platinum or nickel-steel wire.

strain *Mechanics.* a relative change in the dimensions of a body in response to an applied force, expressed as the ratio of the distortion of a body dimension to some undistorted dimension (not necessarily the same one); for example, a change in length in a given direction, per unit original length of the same or some other direction. *Materials Science.* the manifestation of this change in an actual body; the deformation of a material under a stress; it may be **elastic strain** (deformation disappears when stress is removed) or **plastic strain** (deformation is permanent). (Going back to a Latin word meaning "to tie" or "to draw tight.")

strain *Genetics.* **1.** a group of organisms of the same species possessing distinctive hereditary characters that distinguish them from other such groups. **2.** specifically, an artificial group of this kind, maintained by humans through inbreeding or genetic controls to retain or emphasize specific characteristics, as in a domestic animal or agricultural plant. *Microbiology.* a population of microorganisms that are distinguished in some way, such as biochemically or genetically, from other organisms of the same taxonomical classification. *Virology.* a virus that has major properties in common with other viruses within its category or type but that differs in minor properties such as vector specificity or serological or genetic properties. (From an Old English word meaning "tribe" or "stock.")

strain aging *Metallurgy.* aging promoted by cold work.

strain-anneal method *Materials Science.* a process, used to make large single crystals of pure metals, solid solutions, and intermetallic phases, in which a fine-grained polycrystal is strained uniformly in small amounts and heated gradually, so that one recrystallizing nucleus consumes the entire specimen, producing a crystal that can be as large as 100 cm^3.

strain bursts *Mining Engineering.* rock bursts that manifest spitting, flaking, and sudden fracturing at the face, signaling increased pressure.

strained-layer superlattice *Electronics.* a structure consisting of alternating layers of two different semiconducting materials, each several nanometers in thickness, in which a mismatch between the lattice spacings of the two materials is accommodated by elastic strains in the thin layers without the occurrence of mismatch defects.

strain ellipsoid *Mechanics.* the locus of points in a homogeneously strained elastic solid that, prior to strain, lay on a unit sphere. The axes of the ellipsoid are the principal axes of strain. Also, DEFORMATION ELLIPSOID.

strain energy *Mechanics.* the energy stored in a body due to an elastic deformation, equal to the work done to produce this deformation. Also, ELASTIC ENERGY, POTENTIAL ENERGY OF DEFORMATION.

strainer *Engineering.* a device that uses a mesh to remove particles above a given size from a fluid flow.

strain figure *Materials Science.* markings or other indication appearing on the surface of a body that has undergone a deformation due to some applied stress.

strain gauge *Engineering.* a transducer/sensor that determines pressure by measuring electrical resistance variations in a stressed wire.

strain-gauge rosette *Robotics.* the combined measurement of a number of strain gauges cemented to a surface in a geometric pattern.

strain hardening *Materials Science.* an increase in resistance to hardness produced by plastic deformation. Also, WORK HARDENING.

strain hardening coefficient *Materials Science.* an indication of the effect of straining on the resulting strength of a material.

strain-induced crystallization *Materials Science.* the crystallization in polymers that results from realignment of crystalline or polymer units.

strain-induced transformation *Materials Science.* a solid-state transformation in alloys caused by plastic deformation.

strain insulator *Electrical Engineering.* an insulator that is inserted into the span wire of an overhead-wire system; designed to transmit the entire pull of the conductor to a transmission tower or other support mechanism while insulating it.

strain localization *Materials Science.* the confinement of strain under certain stress conditions to one region of a component at which deformation or early fatigue cracks can occur preferentially.

strain point *Materials Science.* the temperature at which glass becomes rigid; corresponds to a viscosity of $10^{14.5}$ poise.

strain rate *Materials Science.* the rate at which a material is deformed.

strain-relief method *Mining Engineering.* a technique for determining absolute stress and strain in rock in situ, by inserting a gage for measuring diametral deformation into a smooth hole bored into the rock.

strain-restoration method *Mining Engineering.* a technique for determining absolute stress and strain in rock in situ, in which strain gauges are installed on the rock surface and a slot is cut into the rock between the gauges, allowing the surface rock to expand; hydraulic pressure is applied until the rock returns to its original state of strain, and the pressure required for this is taken as the original rock stress.

strain rosette *Mechanics.* a set of intersecting lines on a surface along which strains are measured to determine the stress at the intersecting point; usually, strains on three intersecting lines are measured.

strain shadow see PRESSURE SHADOW.

strain-slip *Geology.* a break in rock structure that produces a minor displacement.

strain tensor *Mechanics.* a tensor of rank two that has the Cartesian components e_{ij} arising from the difference between a deformed line element squared (ds^2) and its original (ds_0^2), expressed in terms of the Cartesian components of the line element vector either after (dx_i) or before ($dx_i^{(0)}$) the deformation; it is a symmetric tensor, $e_{ij} = e_{ji}$, and has six independent components.

strain theory *Organic Chemistry.* a theory stating that the ease with which a carbon ring forms depends on the amount the bond must deviate from its normal tetrahedral angle of 109°28'; the amount of deviation is designated as the strain in the ring.

strait or **straits** *Geography.* a narrow sea passage connecting two larger sea or ocean areas.

Straits of Florida *Oceanography.* the water passage between Florida and Cuba, through which the Florida Current flows before it joins the Gulf Stream.

strake *Mining Engineering.* **1.** a wide trough or sluice set on a slope and covered with a blanket to catch coarse gold or other valuable material. **2.** a trough for washing ore, gravel, or other raw material; a launder. *Naval Architecture.* a row of planks or plates running fore and aft along a ship's hull.

strand *Engineering.* **1.** a single element of a wire or cable. **2.** a wire having a uniform and symmetrical cross section. *Geology.* **1.** a shore or beach, especially one that borders the ocean. **2.** an area of land that is covered by water at high tide but exposed at low tide. *Navigation.* to hit bottom, especially on a beach or shoal due to bad weather; the term implies a severe grounding in which the vessel is damaged or cannot readily be gotten underway.

strand casting *Metallurgy.* any of several continuous casting processes that yields elongated shapes, often as multiple strands.

strand displacement *Molecular Biology.* in certain viruses, a process in which one DNA strand is removed from a duplex template as a new strand is synthesized.

stranded ice *Oceanography.* floating ice that has become fast by being grounded, usually describing relatively large pieces.

stranded ice foot *Oceanography.* an ice foot formed or increased by ice floes or small icebergs that are grounded and frozen fast to the shore. Also, **stranded floe ice foot.**

stranded wire *Electricity.* a conductor made of a collection of wires or a combination of groups of wires, usually in a twisted or braided arrangement. Also, **stranded conductor.**

stranding machine see CLOSING MACHINE.

strandline *Geology.* **1.** a beach that is elevated from the current sea level. **2.** the elevation a body of water reaches to become level with the land.

strandline see SHORELINE, def. 1.

strange *Particle Physics.* a term used in describing or classifying quarks. See STRANGE QUARK.

strange attractor *Chaotic Dynamics.* the phase space attractor for a chaotic dynamical solution. A bounded region of phase space to which nearby trajectories are asymptotically attracted. Yet in some directions nearby trajectories diverge exponentially, leading to sensitive dependence on the initial conditions.

strangeness conservation *Particle Physics.* a principle in which the sum of the strangeness numbers in a system of hadrons is conserved; the principle is violated under weak interactions.

strangeness number *Particle Physics.* a quantum number assigned to elementary particles that are conserved by the strong nuclear force; given by the difference between the hypercharge and the baryon number. (Originally described as *strangeness* because such particles have a much longer lifetime than that of previously known particles.)

strange particle *Particle Physics.* a particle thought to contain just one strange quark; the remaining quarks in strange particles are either up or down quarks.

strange quark *Particle Physics.* the third identified type, or flavor, of quark. (So-called in contrast with the previously known *up* and *down* particles, the two types found in ordinary matter.) Also, S QUARK.

strangles *Veterinary Medicine.* an acute infectious disease of horses, especially those under five years of age; characterized by nasal discharge, loss of appetite, fever, and enlargement and suppuration of the lymph nodes; caused by *Streptococcus equie.*

strangulated hernia *Medicine.* a hernia that is constricted and therefore necrotic or likely to become necrotic.

strangulation *Medicine.* **1.** a choking or throttling arrest of respiration, due to obstruction or closing off of the air passage. **2.** the arrest of the circulation in a part, due to compression.

strap *Mechanical Devices.* **1.** a band or strip of flexible material, used especially to hold or attach parts to each other. **2.** to fasten or secure with one or more straps. **3.** a metal fitting that surrounds and holds together other parts of a mechanical assembly.

strap bolt see LUG BOLT.

strap hammer *Mechanical Engineering.* a belt-driven hammer with the head slung from a leather strap.

strap hinge *Mechanical Devices.* a door or gate hinge having long flaps that attach to the door or gate and the molding.

strap joint *Mechanical Devices.* a joint consisting of two butted metal or wood pieces each overlaid with a riveted metal strip.

strapped-down inertial navigation equipment *Navigation.* inertial navigational equipment in which the gyros and accelerometers are attached directly to the carrier rather than to a stable platform and gimbal system; a computer uses data from the gyros to stabilize the carrier.

strapped magnetron *Electronics.* a multicavity magnetron in which resonator segments that have the same polarity are hooked together by small conducting strips in order to minimize any undesired oscillation modes.

strapped wall see BATTENED WALL.

strapping *Mechanical Engineering.* **1.** any collection or arrangement of straps. **2.** the material from which straps are made. *Electronics.* in a multicavity magnetron, the process of connecting together resonator segments having the same polarity in order to suppress any undesired modes of oscillation. *Surgery.* the application of overlapping strips of adhesive tape or plaster to protect and compress an injured or affected area. *Petroleum Engineering.* the process of measuring the circumference of a petroleum tank with steel measuring tapes (straps) at various heights to determine its volume at each height.

strapping option *Computer Programming.* a reconnection of the jumpers on a printed circuit board to make a hardware feature operative or inoperative.

strapping table *Petroleum Engineering.* a record of tank volume versus height; by measuring the liquid depth in a tank, one can read or interpolate the liquid's volume from such a table. Also, TANK TABLE.

Strassmann, Fritz 1902–1980, German chemist; with Otto Hahn, split the uranium atom and produced barium; with Hahn and Lise Meitner, explained nuclear fission.

strata [strat´ə; strāt´ə] the plural of *stratum.*

strategic *Military Science.* of or relating to strategy; having to do with with actions, ideas, and weapons that are intended to enhance a nation's war-making ability or impair the war-making ability of an enemy or potential enemy. Thus, **strategic (air) warfare, strategic intelligence, strategic bombing, strategic mission, strategic (nuclear) weapon, strategic target, strategic material, strategic transport.**

strategic aircraft *Aviation.* any aircraft designed or used to carry out strategic operations.

strategic airlift *Aviation.* in support of remote military units, a planned movement of airborne units, personnel, and materiel, either between area commands or between the home state and overseas.

strategic concentration *Military Science.* the assembly of designated forces in areas from which they are best disposed to initiate the plan of campaign.

Strategic Defense Initiative see STAR WARS.

strategic mobility *Military Science.* in general, the capability to deploy and sustain military forces worldwide in support of national strategy; specifically, the capability of a force to move readily in advance of engagement with a hostile force.

strategic reserve *Military Science.* **1.** material that is placed in a particular location due to strategic considerations or in anticipation of major interruptions in the supply distribution system. **2.** a reinforcing force that is not committed in advance to a specific major command but may be deployed to any region; a force held in reserve as a strategic measure. *Aviation.* in aviation supply, a reserve level of supply over and above the stockage objective, held in a certain geographic location in anticipation of a major interruption in the flow of supplies.

strategic warning *Military Science.* a notification that enemy-initiated hostilities may be imminent; the time between the receipt of warning and the beginning of hostilities is called the **strategic warning lead time.**

strategy *Military Science.* **1.** the development and use of military, political, economic, and psychological forces during peace and war, to support national policies and increase the probability and favorable consequences of victory. **2.** actions, ideas, and weapons designed to prevent an enemy or potential enemy from carrying out hostile actions, as by destroying selected materials, facilities, personnel, and the like that are essential to the enemy's war-making ability. *Statistics.* a plan of action in which a single act or decision is assigned to each possible outcome of an experiment or series of experiments. *Mathematics.* in game theory, a specified collection of moves that covers all possible situations in the complete play of a game.

strath *Geology.* **1.** an extensive, elongated depression on the continental shelf, having steep sides. **2.** a broad, flat valley floor that results from degradation.

strath terrace *Geology.* an extensive remnant of a flat valley floor belonging to a former erosion cycle that has been dissected by a stream following uplift.

stratification *Geology.* **1.** a process in which sedimentary material forms, accumulates, or is deposited in layers. **2.** a process in which sedimentary rocks are arranged in layered units or beds. **3.** the larger structure or formation that results from such layering processes. *Hydrology.* **1.** the arrangement of water masses in a lake or other body of water into two or more horizontal layers having different characteristics. **2.** the formation of layers in snow, ice, or firn as a result of snow sedimentation or other processes.

stratification index *Geology.* a measurement for a given stratigraphic unit that represents the number of beds in a 100-foot section of the unit. Also, BEDDEDNESS INDEX.
stratification plane see BEDDING PLANE.
stratified *Geology.* formed, arranged, or laid down in layers or strata. *Statistics.* of a population, divided into parts on the basis of features such as age or geographical location.
stratified clot see LAMINATED CLOT.
stratified drift *Geology.* any sorted and layered material produced as a result of glacier activity, either deposited by a meltwater stream or settled from suspension in a body of water adjacent to the glacier.
stratified flow *Fluid Mechanics.* a two-phase fluid flow in a horizontal pipe in which the gas phase flows above the liquid phase.
stratified fluid *Fluid Mechanics.* a fluid with varying densities along the axis of gravity.
stratified rock *Petrology.* any rock characterized by internal stratification. All stratified rock is sedimentary, but not all sedimentary rock is stratified.

stratified rock

stratified sampling *Statistics.* a sampling technique in which the population is first divided into distinct subpopulations (or strata), each of which is then sampled randomly. Also, **stratified random sampling.**
stratified society *Anthropology.* a society having hierarchically ranked groups to which people belong, often by birth, and which create unequal access to basic life resources; members also have differing access to power and prestige.
stratified squamous epithelium *Histology.* epithelial tissue composed of a basal layer of cuboidal cells and several to many layers of thin, flattened, superficial cells.
stratified volcano see COMPOSITE CONE.
stratiform *Geology.* **1.** describing an igneous or sedimentary mineral deposit that occurs in layers. **2.** describing any structure consisting of parallel bands, layers, or sheets. *Meteorology.* of a cloud, having a predominantly horizontal development; the opposite of *cumuliform.*
stratiformis *Meteorology.* a cloud species characterized by an extensive horizontal layer or series of layers, either continuous or noncontinuous; the most common form for the altocumulus and stratocumulus genera, and occasionally found in cirrocumulus.
stratify *Geology.* to form layers or strata.
stratigrapher *Geology.* a geologist who specializes in the study of stratigraphy.
stratigraphic *Geology.* of or relating to stratigraphy. Also, **stratigraphical.**
stratigraphic column see GEOLOGIC COLUMN.
stratigraphic cutoff *Geology.* see CUTOFF.
stratigraphic geology see STRATIGRAPHY.
stratigraphic map *Geology.* a map that displays the areal distribution, shape, or aspect of a stratigraphic unit or surface.
stratigraphic oil fields *Geology.* an area in which hydrocarbon reserves occupy sedimentary traps.
stratigraphic section see GEOLOGIC SECTION.
stratigraphic sequence *Geology.* see SEQUENCE.
stratigraphic throw *Geology.* the thickness of the strata formerly separating two faulted beds, equal to the distance between the two parts of the disrupted strata, measured perpendicular to their plane. Also, **stratigraphic separation.**
stratigraphic trap *Geology.* a sealed oil or gas reservoir that results from changes in the physical properties of a formation, rather than from structural deformation.
stratigraphic unit *Geology.* any rock stratum or body of strata that can be recognized or classified as a unit on the basis of its characteristics, properties, or attributes.
stratigraphy *Geology.* a branch of geology that is concerned with the systemized study, description, and classification of stratified rocks, including their origins, composition, characteristics, distribution, and correlation with one another.
Stratiomyidae *Invertebrate Zoology.* a family of brightly colored or iridescent soldier flies, larvae usually found in decaying vegetation, dung, or mud puddles. Also, **Stratiomyiidae.**
stratisphere *Geology.* the part of the earth's crust that contains stratified rocks.
strato- *Meteorology.* a combining form representing *stratus* (as in *stratocumulus*) or *stratosphere* (as in *stratopause*).
stratocumulus *Meteorology.* a principal cloud classification, characterized by gray or white, usually stratiform layers that nearly always have dark patches and a nonfibrous aspect; usually arranged in orderly groups, lines, or waves, and composed of small water droplets, sometimes accompanied by larger droplets, soft hail, and (rarely) snowflakes.
Stratofortress *Aviation.* a popular name for the B-52 strategic bomber.
Stratofreighter *Aviation.* a popular name for the C-97 military transport or the KC-97 strategic tanker-freighter.
Stratojet *Aviation.* a popular name for the B-47 bomber.
stratopause *Meteorology.* the top of the stratosphere; a boundary in the upper atmosphere marking a temperature increase with altitude and dividing the stratosphere from the mesosphere.
stratoscope *Optics.* a telescope that operates by remote control and is lifted by balloon to high altitudes so as to reduce the effect of atmospheric turbulence on the observations.
stratosphere [strat´ə sfēr´] *Meteorology.* **1.** the region of the upper atmosphere lying above the troposphere and below the mesosphere; characterized by a slight increase in temperature with height and by stable, dry, and cloudless conditions. **2.** in former use, all of the atmosphere above the troposphere.
stratospheric [strat´əs fēr´ik] *Meteorology.* relating to or occurring in the stratosphere.
stratospheric coupling *Meteorology.* the interaction of weather phenomena, especially disturbances, between the stratosphere and the troposphere.
stratospheric radiation *Geophysics.* the infrared radiation generated by infrared exchange constantly occurring within the stratosphere.
stratospheric steering *Meteorology.* the steering of disturbances along the contour lines at the tropopause, thought to follow the direction of the wind at that level.
Stratotanker *Aviation.* a popular name for the KC-135 strategic tanker-freighter.
stratotype *Geology.* the sequence of strata that was originally described for a given location, and that serves as a standard against which other parts of the stratigraphic unit are compared. Also, TYPE SECTION.
stratovolcano see COMPOSITE CONE.
stratum [strat´əm; strāt´əm] *plural,* **strata** [strat´ə; strāt´ə]. *Geology.* a distinct homogeneous layer of rock or unconsolidated sedimentary material that is visibly separable from layers above and below it. *Biology.* a layer; i.e., a sheetlike mass of substance of nearly uniform thickness, especially when the layer is one of several associated layers, as in tissue. *Statistics.* see SUBPOPULATION.
stratum basale *Histology.* the deepest layer of cells in a stratified epithelium that contact the basement membrane.
stratum corneum *Histology.* the outermost, keratinized layer of dead cells in the epidermis.
stratum disjunctum *Histology.* the exfoliating layer of cells of the stratum corneum that is constantly sloughed off.
stratum germinativum *Histology.* the basal layer of the epidermis, which provides new cells through cell division.
stratum granulosum *Histology.* a deep layer of epidermal cells that accumulates keratohyalin granules.
stratum lucidum *Histology.* a layer of transparent epidermal cells beneath the stratum corneum in the palms and soles.

stratus [strat´əs; strāt´əs] *Meteorology.* a principal cloud genus characterized by a gray layer having a relatively uniform base; often occurring in the form of ragged patches or fragments (**stratus fractus**) and usually composed of fairly widely dispersed water droplets; similar to stratocumulus, but lower and lacking the latter's uniform relief.

Strauss reaction *Immunology.* a swelling of the scrotum of male hamsters and guinea pigs that occurs following an injection of *Pseudomonas mallei,* a causative agent of the bacterial disease glanders.

straw *Botany.* a single stalk or stem of certain species of grain, including oats, wheat, rye, and barley. *Agriculture.* the dried stems and leaves of grain crops that remain after threshing. *Materials.* a material made from plant stalks that have been dried and often pressed; commonly used to weave hats, baskets, and mats.

strawberry *Botany.* **1.** any stemless plant of the genus *Fragaria,* in the family Rosaceae, noted for its juicy red fruit consisting of an enlarged fleshy receptacle bearing achenes on its outside surface. **2.** the fruit of such plants.

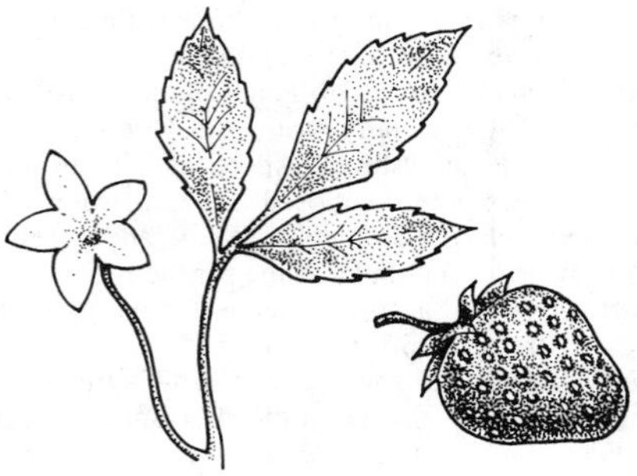

strawberry

strawberry foot-rot *Veterinary Medicine.* the colloquial name for an epidermal infection of sheep similar to lumpy wool, but which affects the lower portions of the legs and is caused by the fungus *Dermatophilus pedis.* Also, DERMATOPHILOSIS.

strawwalker *Agriculture.* a device that pushes the straw to the back of a combine or thresher.

stray *Geology.* an incidentally occurring lenticular formation that is distinctive in lithology or hardness from an adjacent formation, and that is discovered when drilling an oil well or a gas well.

stray capacitance *Electronics.* **1.** any undesired capacitance between circuit wires, between wires and the chassis, or between components and the chassis of electronic equipment. **2.** the capacitance that is introduced into a circuit by the leads and wires connecting the circuit components.

stray current *Electricity.* **1.** a portion of the total current that flows over unintended paths. In a battery, it can cause electrochemical corrosion of metals in contact with electrolytes. **2.** an undesirable current generated by the discharge of static electricity; often occurs in the loading and unloading of petroleum fuels and other volatile chemicals.

stray emission *Physics.* radiation that cannot be utilized.

stray field *Electromagnetism.* magnetic flux leakage that spreads out from a coil and does no useful work.

stray line *Electrical Engineering.* the section of line closest to the end of a current pole, having no distance marks; used to allow the pole to stabilize before measurement is begun.

stray losses *Electrical Engineering.* losses that are not associated with primary function of the circuit or transmission line; for example, core losses in a transformer or heating losses in a transmission line associated with the series resistance or the conductance of the surrounding dielectric.

stray sand *Geology.* a stray that is composed of sandstone.

streak *Mineralogy.* the color of the powder of a mineral, as determined by scratching the surface directly or by rubbing the mineral on an unglazed porcelain surface, called a **streak plate**; this color may differ from that of the mineral sample itself. *Plant Pathology.* any plant disease that is caused by a virus and characterized by irregular linear markings on the plant, such as tomato streak.

streaking *Microbiology.* the process of inoculating a petri plate containing solid medium with a given microorganism by drawing a bacteriological loop across the surface of the plate.

streak lightning *Geophysics.* a lightning discharge exhibiting a single, usually straight, distinct channel.

streak line *Fluid Mechanics.* an instantaneous line traced by those particles that have passed through a specified fixed point in the fluid at a previous instant.

streak plate *Microbiology.* a culture plate of solid medium that has been inoculated by streaking, often in order to obtain discrete colonies of bacteria. *Mineralogy.* see STREAK.

stream *Hydrology.* **1.** any body of water, together with its dissolved or suspended materials, that moves within a well-defined channel to progressively lower levels under the influence of gravity. **2.** the water that flows in such a manner. **3.** in popular use, a moving body of fresh water that is smaller than a river. *Fluid Mechanics.* any flow of liquid or fluid. *Computer Programming.* **1.** a data structure in the form of a list, e.g., a sequence of characters constituting a file, in which the elements are processed sequentially. **2.** see JOB STREAM.

stream anchor *Naval Architecture.* a relatively small anchor dropped from the stern to limit a vessel's swinging at anchor.

streambed *Hydrology.* a popular term for an active or former stream channel; i.e., a depression in which a stream flows or formerly flowed.

stream-built terrace see ALLUVIAL TERRACE.

stream capacity *Hydrology.* a measurement of a certain stream's ability to transport unconsolidated sediment, expressed in terms of the amount of material transported past a given point per unit of time.

stream channel *Hydrology.* the extensive, narrow depression or bed through which a natural stream flows or may flow. Also, STREAMWAY.

stream-channel form ratio *Hydrology.* the mathematical relationship between the volume and the perimeter of a stream channel.

stream current *Hydrology.* **1.** a steady current within a river or stream. **2.** a more or less narrow, deep, fast-moving ocean current.

stream day *Chemical Engineering.* a 24-hour operating period of a flow-processing unit in an industrial setting.

streamer *Geophysics.* the initial stroke of a lightning discharge that, by generating an ion-dense area immediately ahead of itself, finds and establishes the path for subsequent strokes of lightning. *Electricity.* any electrical discharge resembling a lightning streamer, i.e., emanating in a narrow path from a point of high potential on a charged body. *Astronomy.* any of numerous long extensions of the solar corona.

stream erosion *Hydrology.* the gradual and progressive process by which a stream removes and carries exposed material from the surface of its channel.

stream feeder *Graphic Arts.* a device on a press that maintains a steady flow of slightly overlapping sheets running toward the grippers.

streamflood *Hydrology.* a flood of water in an arid region in which the water flow is mainly confined to a definite channel that is normally dry.

streamflow or **stream flow** *Hydrology.* the movement of surface runoff within a stream channel.

stream frequency *Hydrology.* a measurement of topographic texture, equal to the ratio between the number of streams in a given drainage basin and the area of the basin. Also, CHANNEL FREQUENCY.

stream function see LAGRANGE STREAM FUNCTION.

stream gauge see RIVER GAUGE.

stream gradient *Hydrology.* the shape of a stream bed, equal to the angle between a horizontal plane and the water surface or channel floor of the stream, measured in the direction of flow. Also, STREAM SLOPE.

stream gradient ratio *Hydrology.* the ratio of the stream gradient of a stream channel of one order to the gradient of the next higher order channel within the same drainage basin. Also, CHANNEL GRADIENT RATIO.

streaming *Fluid Mechanics.* any process or instance of flowing. *Computer Programming.* a condition of a terminal or modem that has locked into a constant carrier signal, thus preventing the normal flow of data.

streaming current *Electricity.* the resulting electric current from liquid force fed through a diaphragm, capillary, or porous solid.

streaming flow *Hydrology.* **1.** glacial flow in which the ice moves in a relatively smooth and unbroken stream rather than breaking or cracking into blocks. **2.** see TRANQUIL FLOW.

streaming potential *Electricity.* the electropotential gradient that results from a unit velocity of liquid fed through a capillary or diaphagm structure or fed past an interface.

streaming tape *Computer Programming.* a continuous magnetic tape drive that is used primarily for backing up hard disk drives.

stream-length ratio *Hydrology.* within a particular drainage basin, the ratio of the average stream length of a given order to the average stream length of the next lowest order stream.

streamline *Fluid Mechanics.* **1.** a curve whose tangent, at any point, is in the direction of the fluid velocity at that point. **2.** a line on which all fluid particles have a velocity vector that is tangent to the line at a given instant. *Design Engineering.* to reduce resistance to motion through a fluid; make streamlined. *Industrial Engineering.* to change a process or operation to simplify it or make it more efficient.

streamlined *Design Engineering.* of a component or assembly, contoured so as to reduce its resistance to motion through a fluid; e.g., the long, needlelike nose of the Concorde, a supersonic jet aircraft.

streamline flow *Fluid Mechanics.* a flow characteristic of a viscous fluid in which fluid moves by sliding against adjacent layers; there is no turbulent motion, and particles move in definite and observable paths in parallel lines. Also, LAMINAR FLOW, VISCOUS FLOW.

stream load *Geology.* **1.** the total material other than water that is actually carried by a stream. **2.** the total material that is transported past a given point in a stream at a given time, or over a given period of time.

stream measurement *Transportation Engineering.* any direct measurement made of the flow of traffic along a stretch of road.

stream morphology see RIVER MORPHOLOGY.

stream order *Hydrology.* the use of a series of increasing consecutive integers to indicate the relative position of the segments of a stream within a drainage-basin network, where the smallest, unbranched tributary that terminates at an outer point is designated order #1. Also, CHANNEL ORDER.

stream profile *Hydrology.* a line representing the upper edge of a vertical section of a stream that follows the winding of the valley along the length of the stream from its source to its mouth. Also, LONGITUDINAL PROFILE, RIVER PROFILE.

stream segment *Hydrology.* the portion of a stream that extends between the junctions of specified tributaries. Also, CHANNEL SEGMENT.

stream slope see STREAM GRADIENT.

stream takeoff *Aviation.* a column of aircraft directed to take off in close succession.

stream terrace *Geology.* any of a series of long, narow, flat surfaces bordering a stream valley relatively parallel to the stream channel and above the level of the water; formed during a former stage of erosion or deposition.

stream tin *Geology.* pebbles that occur along valleys or are deposited by streams, and that are composed of the mineral cassiterite.

stream transport *Hydrology.* any of various processes by which weathered and eroded rock material is moved in and by a stream.

stream tube *Fluid Mechanics.* an imaginary tube made by all streamlines passing through a closed curve.

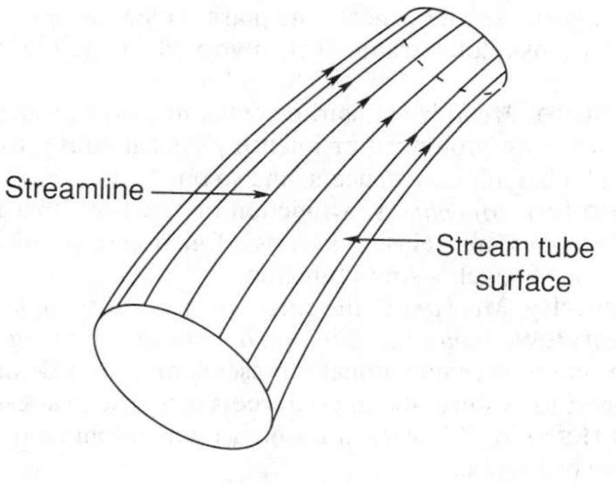

stream tube

stream underflow *Hydrology.* percolating water in the bed of a stream, flowing parallel to the stream.

streamway see STREAM CHANNEL.

Streblidae *Invertebrate Zoology.* the bat flies, a small but widely distributed family of dipteran insects in the section Pupipara that are ectoparasites of bats.

Strecker amino acid synthesis *Organic Chemistry.* the synthesis of an α-amino acid through the reaction of an aldehyde or ketone with a mixture of ammonium chloride and sodium cyanide followed by acid hydrolysis of the amino nitrile formed.

Strecker degradation *Organic Chemistry.* a general reaction in which any compound containing an aldehyde or a keto group in conjugation with another carbonyl group or nitro group reacts with an α-amino acid to give an aldehyde and a carbon dioxide.

streetcar *Transportation Engineering.* any public rail car, such as a trolley car, that runs along city streets.

street elbow *Mechanical Devices.* a pipe elbow having a female thread on one end and a male thread on the other.

street virus *Virology.* a wild-type strain of rabies virus found in an animal that is naturally infected, as opposed to a laboratory-adapted strain.

Strelitziaceae *Botany.* a family of giant to medium-size monocot herbs and unbranched trees of the order Zingiberales, characterized by perfect flowers with five functional stamens and without an evident hypanthium; in some systems of classification, plants of the Strelitziaceae are considered part of the banana family, Musaceae.

stremmatograph *Engineering.* an instrument that measures the longitudinal stress in rails as a train passes over them.

strengite *Mineralogy.* $Fe^{+3}PO_4 \cdot 2H_2O$, a pale-red to violet, orthorhombic mineral of the variscite group, dimorphous with phosphosiderite,occurring as crystals and in small fibrous aggregates, having a specific gravity of 2.87 and a hardness of 3.5 to 4.5 on the Mohs scale; an alteration product of iron-rich phosphate minerals.

strength the quality or quantity of force, power, concentration, resistance, etc.; specific uses include: *Mechanics.* the ability of a material to resist stress without yielding or fracture, especially as described by the stress level at which it fractures or fails. *Acoustics.* the maximum amplitude of a signal, giving a measure of the volume displacement produced by a sound source. *Chemistry.* the concentration of a solution. *Mathematics.* given a graph G, the strength of G is

$$\min_{S \subseteq E(G)} |S|/\omega(G-S) - \omega(G),$$

where, for any graph H, $\omega(H)$ is the number of components of H, $|S|$ is the number of elements in set S, and the minimum is taken over all subsets S of $E(G)$ for which $\omega(G-S) > \omega(G)$. Graphs whose strength and fractional arboricity are equal are useful in the design of communication networks that are to be survivable against attack.

strength of current *Oceanography.* the phase of a tidal current when its speed is at a maximum, usually given with the time at which it occurs; also, the speed of such a current.

strength of ebb *Oceanography.* the ebb tidal current at the time of its greatest speed; also, the speed of such a current.

strength of ebb interval *Oceanography.* the period of time between the transit (upper or lower) of the moon over the local or Greenwich meridian and the strength of ebb at a given place.

strength of enemy forces *Military Science.* a term used to describe an enemy unit of force in terms of amount of personnel, weapons, and equipment.

strength of figure *Cartography.* a determination of the precision of the lengths of a triangulation net, indicated by the relative strength of the triangles as a function of their shapes, and calculated using the formula for probable error of a triangle, which is independent of the accuracy of the observations.

strength of flood *Oceanography.* the flood tidal current at the time of its greatest speed; also, the speed of such a current.

strength of flood interval *Oceanography.* the period of time between the transit (upper or lower) of the moon over the local or Greenwich meridian and the strength of flood at a given place.

Strep. or **Strep** *Streptococcus.* Also, **strep.**

strephosymbolia *Neurology.* **1.** a visual disorder in which objects appear reversed, as if seen in a mirror. **2.** an inability to distinguish written or printed letters that are similar but extend in opposite directions, such as p and d, or n and u.

strepogenin *Biochemistry.* a peptide that is obtained by partial hydrolysis of various proteins and stimulates the growth of certain microbes, principally lactic acid bacteria.

Strepsiptera *Invertebrate Zoology.* an order of tiny beetles, coleopteran insects with larvae that are parasitic on other insects; the female adults remain wingless endoparasites in the host, while the males develop hindwings and fly to a new host to mate.

Strepsirhini *Paleontology.* a proposed suborder of primates that would include the lemuroids and their extinct relatives; the suborders Strepsirhini and Haplorhini would correspond roughly to the generally accepted suborders Prosimii and Anthropoidea, respectively.

strept- or **strepto-** a combining form meaning "twisted."

streptaster *Invertebrate Zoology.* a sponge spicule with rays radiating from a common axis.

streptavidin *Biochemistry.* a protein that is derived from *Streptomyces avidinii* and can be used in place of avidin.

strep throat *Medicine.* an informal term for an infection of the pharynx and tonsils caused by a species of *Streptococcus,* and characterized by sore throat, chills, fever, swollen lymph nodes, and occasionally nausea and vomiting.

Streptoalloteichus *Bacteriology.* a genus of bacteria of the order Actinomycetales that form substrate and aerial mycelia which give rise to spores.

Streptobacillus *Bacteriology.* a genus of Gram-negative, anaerobic bacteria of uncertain affiliation, composed of nonmotile, rod-shaped cells or filaments, and typically found in the mouth and throat of rats.

streptobacillus *plural,* **streptobacilli.** *Bacteriology.* an organism belonging to the genus *Streptobacillus.*

streptobiosamine *Biochemistry.* $C_{13}H_{23}NO_9$, a nitrogen-containing disaccharide obtained when streptomycin undergoes acid hydrolysis.

Streptococcaceae *Bacteriology.* a family of Gram-positive, spheroid, usually nonmotile, facultatively anaerobic bacteria that metabolize nutrients through fermentation; formerly classified as a tribe of bacteria belonging to the family Lactobacillaceae.

streptococcemia *Medicine.* the presence of streptococci in the blood.

Streptococcus [strep´tō käk´əs] *Bacteriology.* a genus of nonmotile, nonspore-forming cocci occurring in pairs or chains, assigned to the family Streptococcaceae, including many serious human and animal pathogens that are responsible for conditions such as bacterial pneumonia, rheumatic fever, scarlet fever, tonsillitis, and strep throat.

streptococcus *plural,* **streptococci.** *Bacteriology.* an organism belonging to the genus *Streptococcus.*

streptodornase *Enzymology.* a DNase produced by hemolytic streptococci; used in combination with and having effects similar to streptokinase.

streptokinase *Enzymology.* an enzyme that is secreted by streptococci and changes plasminogen into plasmin; used in medicine to dissolve blood clots, pus, and other waste matter associated with infection.

streptolysin *Bacteriology.* an exotoxin that is released by certain streptococci and lyses red blood cells.

Streptomyces *Bacteriology.* a genus of funguslike aerobic bacteria of the family Streptomycetaceae; most of its several hundred species are soil forms, but some are parasitic on plants and animals. More than half of the antibiotics used in medicine are produced from species of *Streptomyces*; one species, **Streptomyces vinaceus,** produces vitamin B_{12}.

Streptomycetaceae *Bacteriology.* a family of bacteria of the order Actinomycetales; found in soils.

streptomycin [strep´tə mī´sən] *Microbiology.* a water-soluble antibiotic produced by the bacterium *Streptomyces griseus*; effective against a broad range of bacteria, including those that cause plague and tuberculosis.

Streptosporangium *Bacteriology.* a genus of Gram-positive, spore-producing bacteria of the family Actinoplanaceae that tend to grow as branching substrate and aerial mycelia in soil and water.

streptothricin *Microbiology.* an antibiotic produced by the bacterium *Streptomyces lavendulae,* active against a variety of Gram-negative and Gram-positive organisms. The first antibiotic isolated, it was found too toxic for systemic use.

Streptoverticillium *Bacteriology.* a genus of aerobic bacteria of the order Actinomycetales, growing as substrate and aerial mycelia and distinguished by the formation of spores in verticils.

streptozocin *Oncology.* $C_8H_{15}N_3O_7$, an antineoplastic antibiotic derived from *Streptomyces achromogenes* or produced synthetically; used against islet-cell tumors of the pancreas and other endocrine tumors. Also, **streptozotocin.**

stress *Mechanics.* **1.** an external force that acts on a material and tends to change the dimensions of the material, by compressing it, stretching it, or causing it to shear. **2.** a description of this effect, expressed as the force per unit area acting on a material surface element (at an interior point or bounding of a continuum bulk) that has normal vector n, with the Cartesian force components $f_i = s_{ik}n_k$ $(i,k = 1,2,3)$, where s_{ik} depend on position and time, but not on n; s_{ik} constitute the Cartesian components of the stress tensor. *Materials Science.* the amount of this effect acting on a given material, measured in terms of the force exerted per unit area and usually expressed in pascals (Pa). *Physics.* any factor acting to disturb the equilibrium of a system. *Behavior.* **1.** an unusual environmental condition that causes physiological, emotional, behavioral, or cognitive changes in an individual. **2.** the changes or effects resulting from such environmental demands. *Psychology.* **1.** specifically, a physical or emotional reaction to a situation perceived as unfamiliar, threatening, harmful, and so on. **2.** the negative situation itself.

STRESS *Computer Programming.* a programming language used for solving engineering problems. (An acronym for structural engineering system solver.)

stress amplitude *Mechanical Engineering.* one-half the algebraic difference between the maximum and the minimum stress imposed on a test object in one fatigue test cycle.

stress analysis *Physics.* the study of stresses and strains imposed on a body that is subjected to external forces.

stress-assisted transformation *Materials Science.* a solid-state transformation in alloys in which nucleation occurs on the same initial nucleation sites that trigger the spontaneous transformation usually on cooling, but is assisted thermodynamically by applied stress.

stress birefringence *Optics.* birefringence induced by the application of mechanical stress. Also, MECHANICAL BIREFRINGENCE.

stress concentration *Mechanics.* high localized stress intensity at a geometric discontinuity of a body, as in the vicinity of a notch or hole or at a change in the diameter of a shaft. The maximum stress can be many times higher than the stress value when there is no geometric discontinuity.

stress-concentration factor *Mechanics.* the ratio of the greatest stress at the site of a stress concentration to the average stress over a cross-sectional area; depends on the geometry of the flaw and of the stress component, but not on the properties of the material.

stress corrosion *Metallurgy.* corrosion occurring preferentially in stressed locations.

stress-corrosion cracking *Metallurgy.* failure caused by stress corrosion.

stress crack *Mechanics.* an internal or external crack developed in a body due to tensile, compressive, or shearing loads.

stress difference *Mechanics.* the difference between the largest and the smallest of the three principal stresses.

stressed skin construction *Civil Engineering.* a type of construction in which surface material is structural and not, like slate, a mere weather-resisting cladding.

stress ellipsoid *Mechanics.* a mathematical representation of the state of stress at a point when all the principal stresses are positive. The directions of the semiaxes of the ellipsoid represent the principal directions of stress, and the distance from the center of the ellipsoid to the surface along any direction is inversely proportional to the square root of the normal stress at the point along that direction.

stress exponent *Materials Science.* a constant used in determining the plastic deformation of ceramics under stress; it varies with the specific ceramic material.

stress fibers *Cell Biology.* bundles of microfilaments composed of actin and myosin that lie beneath the plasma membrane of certain cells, forming a cytoskeletal framework involved in maintenance of cell shape.

stress fracture *Medicine.* a hairline crack in a bone, usually of the leg or foot, caused by prolonged or intense physical stress; found often in runners and other athletes, dancers, and so on.

stress function *Mechanics.* a function or functions that uniquely describe the stresses in an elastic body as a function of position; for example, Airy's or Maxwell's stress function.

stress intensity *Mechanics.* the extent of stress at a single given point.

stress interview *Industrial Engineering.* an interviewing technique in which the interviewer intentionally presents hostile or demanding questions in order to evaluate the interviewee's response to stress.

stress marks *Graphic Arts.* lines on a photographic emulsion caused by pressure or friction.

stress mineral *Mineralogy.* a former term for a mineral whose formation in metamorphosed rock is supposedly favored by shearing stress; for example, amphiboles, chlorite, and talc.

stressor *Behavior.* any environmental factor that produces stress.

stress raiser *Metallurgy.* any discontinuity that serves to intensify local stresses.

stress range *Mechanics.* the algebraic difference between the maximum and minimum stress in a periodically varying stress field.

stress ratio *Mechanics.* the ratio of the minimum stress to the maximum in fatigue testing.

stress-related *Medicine.* describing a disease or condition that is identified as being caused or aggravated by stress; e.g., stomach ulcers, high blood pressure, asthma.

stress relaxation *Materials Science.* a reduction of the stress acting on a material over a period of time at constant strain due to viscoelastic deformation.

stress relief *Materials Science.* a reduction or elimination of stress during metal annealing.

stress relief anneal *Materials Science.* a recovery-stage process during annealing, in which residual stresses are relieved without substantially reducing the mechanical properties of the material.

stress relief cracking *Metallurgy.* in a welded structure, intergranular cracking occurring upon a thermal treatment designed to relieve internal stresses.

stress relieving *Metallurgy.* heat treating to decrease internal stresses.

stress-rupture *Materials Science.* a time-dependent fracture of a material under a constant load, usually at elevated temperatures.

stress rupture test *Metallurgy.* a high-temperature mechanical testing method that determines the time elapsed between the application of a constant tensile lead or stress and the rupture of the specimen.

stress sensor *Robotics.* a contact sensor that sends a feedback signal when it touches something.

stress-strain curve *Materials Science.* a graph, usually the result of a tensile test of a material, indicating the relationship between the applied stress and the resulting strain.

stress-strain relation see DEFORMATION CURVE.

stress tensor *Mechanics.* a second rank tensor with its Cartesian components designated as s_{ik} $(i,k = 1,2,3)$, whose three diagonal components $(i = k)$ represent normal stresses, and whose off-diagonal elements $(i \neq k)$ represent shear stresses.

stress test *Medicine.* a test of heart function using an electrocardiograph to measure heart performance before, during, and after measured, increasingly strenuous physical exercise such as running on a treadmill or peddling a stationary exercise bicycle. *Computer Technology.* a test of hardware or software in which the system is subjected to extremely heavy loading.

stress to break *Materials Science.* a mechanical property of an elastomer describing the amount of stress needed to fracture a specimen.

stretch *Hydrology.* a term for a continuous, uninterrupted expanse of water. *Petroleum Engineering.* the extra length of oil-well casing or tubing when it is suspended freely in a fluid.

stretch blow molding *Materials Science.* a thermoplastic forming process in which an extruded or injection molded parison or molded preform is stretched axially at a temperature close to the softening one and blown, either simultaneously or in two stages, to produce a final product of desired dimensions.

stretched pebble *Geology.* one of a number of small pebbles in sedimentary rock that have changed from a round to an elongated shape, as a result of deformation.

stretcher *Medicine.* a light framework typically consisting of a strong fabric such as canvas supported by two long poles, used in emergency medicine to carry an injured or sick person by hand by two or more attendants. *Building Engineering.* a masonry unit laid with its length parallel with the face of the wall liner. Thus, **stretcher bond.** *Mining Engineering.* a bar wedged against or pocketed into the sides of mine roadways for roof support.

stretcher bar *Mining Engineering.* a single screw column used in small drifts to hold one machine drill.

stretcher leveling *Metallurgy.* the process of increasing the flatness of a sheet by mild stretching.

stretch fabric *Textiles.* a cloth that can be elongated under tension and recover its original shape when released; a condition imparted to cloth by the use of stretch nylon, polyester, rubber, or spandex yarns, or by chemical and mechanical treatments.

stretch former *Mechanical Engineering.* a machine used for stretch-forming operations.

stretch forming *Mechanical Engineering.* a material-forming process in which a sheet is heated until it exhibits plastic flow and then wrapped about a die that defines its shape as it cools. Also, WRAP FORMING.

stretch marks *Medicine.* light, streaky lines appearing on the skin, especially of the abdomen and thighs, as a result of the stretching of the skin tissue over a short period of time, as from pregnancy, rapid weight gain, muscle expansion from weightlifting, and so on.

stretch reflex *Physiology.* a response that contracts a muscle as it is stretched. Also, MYOTATIC REFLEX.

stretch thrust *Geology.* a reversed fault that is produced by shearing through the middle limb of an overturned fold. Also, **stretch fault.**

stretch wrapping *Materials Science.* the use of a stretch "frozen" film, such as polyethylene, as a wrapping material, which upon heating gives better surface coverage, stronger clinging properties, and more security than strapping.

stria [strī´ə] *plural,* **striae** [strī´ē]. *Biology.* narrow bands, lines, grooves, streaks, or channels that are arranged more or less parallel. *Geology.* one of the furrows or grooves in a pattern of striation.

striae

Striariaceae *Botany.* a family of brown algae of the order Dictyosiphonales, characterized by cylindrical axes that are sparingly to abundantly branched and by striated cortical cells.

striated ground see STRIPED GROUND.

striated muscle *Histology.* muscle tissue whose fibers possess microscopically visible cross striations.

striation *Histology.* any of the alternating light and dark crossbands that are visible in certain muscle fibers and are produced by the distribution of contractile proteins. *Geology.* **1.** one of a series of shallow, generally parallel furrows, scratches, or grooves produced on a rock surface by geologic agents, such as the abrasive action of a glacier carrying rock debris or the movement of a fault. **2.** a small, narrow mark cut below a sedimentary surface. *Electronics.* in low-pressure gas discharge, a phenomenon by which alternately luminous and dark striae are formed in the line between electrodes; sometimes observed in the positive column of a glow-discharge tube near the anode.

striation technique *Acoustics.* a method of producing a visible image of a sound wave pattern by taking advantage of the ability of a sound wave to refract light.

stricture *Medicine.* the abnormal narrowing of a channel from the deposit or formation of abnormal tissue.

striding compass *Engineering.* a compass mounted on a survey theodolite to orient it.

stridor *Medicine.* a harsh, high-pitched respiratory sound as of blowing wind, often heard in acute restriction of the breathing passages.

stridulation *Invertebrate Zoology.* in some insects such as grasshoppers and crickets, the production of sound through friction caused by rubbing together modified body parts.

Strigaeidae *Invertebrate Zoology.* a family of trematode flukes with a flattened or concave anterior section and cylindrical posterior, endoparasites of birds, mammals, and reptiles.

Strigidae *Vertebrate Zoology.* the typical owls, a family of nocturnal birds of the order Strigiformes, characterized by soft brown to black plumage, a large head with a round facial disk, and a short, strong, hooked bill; found worldwide.

Strigiformes *Vertebrate Zoology.* the order of birds containing the owl, a nocturnal raptor characterized by a large head, heavy hooked bill, forward-directed eyes, and excellent night vision and directional hearing.

strigil *Entomology.* **1.** a comblike structure on certain insect legs often used for cleaning the antennae. Also, **strigilis. 2.** the abdominal appendages in Corixidae.

strigose *Biology.* covered with straight, stiff, pointed hairlike scales or bristles.

strigovite *Mineralogy.* $Fe_3^{+2}(Al,Fe^{+3})_3Si_3O_{11}(OH)_7$, a dark-green chlorite mineral consisting of a basic silicate of iron and aluminum and occurring as incrustations.

Strigulaceae *Botany.* a family of Ascolichenes, characterized by the formation of extensive crusts on or under the cuticle of leaves.

strike *Military Science.* an attack designed to inflict damage on, seize, or destroy an objective. Similarly, **strike force.** *Geology.* the direction of a line formed by the intersection of a rock stratum or other geologic structure with a horizontal plane. Also, LINE OF STRIKE. *Metallurgy.* **1.** a thin electrodeposit applied prior to the final coating. **2.** a plating solution suitable for applying such thin deposit.

strike board *Mining Engineering.* in shaft sinking, a board at the shaft top from which buckets are tipped. Also, STRIKE TREE.

strike fault *Geology.* a fault whose strike is parallel to the strike of the affected strata.

strike joint *Geology.* a joint whose strike is parallel to the bedding or cleavage planes of the affected rock.

strike-off board *Engineering.* a straight-edged board used to scrape off excess plaster or cement.

strike-on typesetting see IMPACT TYPESETTING.

strike plate *Mechanical Devices.* a metal plate screwed into a door jamb that receives the door lock bolt as it travels to its stop hole as the door is closed. Also, **striking plate.**

strike plating *Metallurgy.* the process of applying a thin electrodeposit prior to final coating.

striker *Ordnance.* a firing pin or rod on the hammer of a firearm that, when released by the trigger, strikes the primer and initiates the explosive train.

striker plate *Ordnance.* a plate in the breech of a weapon that supports the base of the cartridge; the striker hits the primer through a hole in the plate.

strike separation *Geology.* the distance of the apparent displacement between two formerly adjacent beds on either side of a fault surface, as measured parallel to the strike of the fault.

strike-separation fault see LATERAL FAULT.

strike slip *Geology.* the component of the slip or movement in a fault that lies parallel to the strike of the fault. Also, HORIZONTAL DISPLACEMENT, HORIZONTAL SEPARATION.

strike-slip fault *Geology.* a fault in which the actual displacement along the fault plane is horizontal, being parallel to the strike of the fault. Also, **strike-shift fault.**

strike stream see SUBSEQUENT STREAM.

strike tree see STRIKE BOARD.

striking *Horology.* **1.** in a timepiece, the act of ringing a bell or sounding an alarm on the hour only, rather than chiming or ringing on the quarter hour. **2.** of or relating to the gear train in a timepiece that controls the striking process.

striking hammer *Engineering.* a hammer used to drive a rock drill.

striking platform *Archaeology.* the area on a stone core that receives the force when a flake is detached in tool-making. Also, PLATFORM.

striking potential *Electronics.* **1.** the voltage necessary to start an electric arc. **2.** the lowest grid-to-cathode potential value at which plate current begins flowing in a gas-filled triode.

striking train *Horology.* a set of gears and pinions, usually synchronized to the going train, that causes chimes or bells in a timepiece to be sounded on the hour.

string any of various features thought of as resembling a piece of string, as by having a long, thin, continuous form; specific uses include: *Computer Programming.* a contiguous set of characters that are treated as a unit. *Mechanics.* a term for a solid body whose cross-sectional dimension is very much smaller than its longitudinal dimension, and which has no stiffness. *Particle Physics.* a specific effective form for the gluonic force between two colored particles. *Geology.* see STRINGER. *Engineering.* in drilling, a section of pipe, casing, or other fitting or fittings lowered into a bore hole. *Building Engineering.* a wooden joist with a slope that supports steps in wooden stairs.

string bean *Botany.* **1.** any of various plants of the pea family that bear long, green or yellow pods, such as the snap bean. **2.** the unripe pods of this plant containing smooth, somewhat flat seeds and eaten as a food, such as the green bean, usually after stripping off a fibrous thread along the side.

string break *Computer Programming.* the point in sorting at which no more records with sufficiently high control keys can be found to fit on the current output string.

string character *Computer Programming.* a string constant consisting of a single character.

string constant *Computer Programming.* any sequence of letters, numbers, and symbols that are enclosed in quotation marks; normally representing a constant string to be used in a program or a piece of written text.

string course *Building Engineering.* a brick or stone course that projects horizontally from a wall for asthetic reasons; typically narrower than other courses and extending across the facade of a structure or encircling pillars. Also, BELT COURSE.

string electrometer *Engineering.* an electrometer composed of a conducting fiber stretched between two oppositely charged plates whose electrostatic forces move the fiber laterally in proportion to the voltage between the plates.

stringent control *Biochemistry.* control of the rate of RNA synthesis that occurs when cells are deprived of an essential amino acid and show a quick decline in the synthesis of stable RNA and protein. *Genetics.* a mechanism by which certain bacterial cells are able to limit the number of any one type of plasmid, usually to less than ten per cell.

stringent mutant *Genetics.* a mutant bacterium that abruptly ceases the synthesis of RNA when its growth medium becomes depleted of certain nutrients.

stringent response *Molecular Biology.* a termination of tRNA and ribosome synthesis by bacteria in response to nitrogen starvation.

stringent washing *Molecular Biology.* a laboratory procedure performed after hybridization of probe DNA; all probe DNA that is not bound by pairing to a substantial length of DNA is removed by washing with a series of progressively diluted saline solutions.

stringer *Civil Engineering.* a long horizontal member that ties together trestle heads in a bridge or provides support under a rail and parallel to it in a rail bridge. *Aviation.* a lightweight longitudinal structural member of a fuselage, nacelle, or similar body. *Geology.* a very small, threadlike deposit of mineral material within a host rock, usually occurring as one of a series extending from a larger vein. Also, STRING. *Metallurgy.* in a microstructure, foreign material aligned in the direction of working. *Materials Science.* any inclusion in a material that is deformed or aligned with the direction of deformation.

stringer lode *Geology.* a shattered area of host rock consisting of a network of stringers.

stringer plate *Naval Architecture.* a plate forming part of a vessel's fore and aft stiffening structure.

string galvanometer *Engineering.* a galvanometer that measures oscillating current using a silver-plated quartz fiber placed under tension in a magnetic field. Also, EINTHOVEN GALVANOMETER.

stringhalt *Veterinary Medicine.* a nerve disorder of horses that causes an exaggerated flexing or jerking movement of the hind legs.

stringing *Petroleum Engineering.* the connecting of lengths of tubing or casing so that the resulting string is long enough to reach a desired depth in a borehole.

string milling *Mechanical Engineering.* a time-saving milling technique in which work pieces are mounted in a row and milled in sequence by a single cutter or series of cutters.

string-processing language *Computer Programming.* a programming language designed for manipulating strings using such operations as concatenation, substring extraction, transformation, and pattern matching. Also, **string-manipulation language.**

string shot *Petroleum Engineering.* a method of stimulating an oil well by firing a string of explosive opposite the producing zone down a wellbore, removing deposits such as gypsum, mud, and paraffin from the formation face.

stringy floppy *Computer Technology.* a fast, magnetic-tape storage device consisting of thin tape enclosed in a cartridge.

striocerebellar *Neurology.* of or relating to both the corpus striatum and the cerebellum.

strip *Mining Engineering.* to extract coal, rock, earth, or other material from a working that is at or near the surface. *Engineering.* to remove the insulator from an electrical wire. *Metallurgy.* any flat metallic material narrower than a sheet. *Ordnance.* to disassemble a weapon for cleaning, repair, or transport. *Aviation.* a long, narrow surface for the takeoff and landing of aircraft; used in compounds such as **airstrip, landing strip.** *Graphic Arts.* to position film for printing in a process of stripping.

strip-borer drill *Mechanical Engineering.* a skid- or caterpillar-mounted drill that is operated by an electric motor or diesel engine and used to bore blasting holes having diameters of about 5 inches and depths up to 100 feet.

strip-chart recorder *Engineering.* any device, such as a seismograph, that records data output graphically on a moving chart medium.

strip coordinates *Cartography.* in aerial photography, the coordinates of any point in a flight path, either on the ground or in the object space, that are referred to the coordinate system used in the first overlap in the strip.

strip cropping

strip cropping *Agronomy.* the practice of growing different crops in alternate strips along contours to control erosion. Also, **strip farming, strip planting.**

striped bass *Vertebrate Zoology.* a important gamefish, *Morone saxatilis*, characterized by blackish stripes along the sides; found in coastal waters of North America.

striped ground or **striped soil** *Geology.* a patterned ground consisting of an arrangement of alternating bands of fine and coarse rock or soil material, produced by frost action on an inclined surface. Also, STRIATED GROUND.

stripe smut *Plant Pathology.* a fungus disease of grasses, caused by several smut fungi of the genera *Urocystis* and *Ustilago*, in which individual leaf blades curl and show black stripes that run parallel with the veins and have black powdery spores. Also, FLAG SMUT.

strip irrigation see BORDER IRRIGATION.

stripline *Electromagnetism.* a strip transmission line that has a conductive center strip separated from two outer conductive sheets by dielectric strips.

stripline circuit *Electromagnetism.* a circuit that utilizes the characteristics of strip transmission lines in microwave practice; the strip transmission line can act as a filter or another circuit component.

strip marker *Ordnance.* a marker located at the start and finish of a land mine strip; it may be natural, artificial, or specially installed.

strip method *Forestry.* the practice of conserving lumber by clearing timber in narrow strips throughout a forest; reproduction on the cleared strips is brought about by seed sown from the adjoining woods.

strip mine *Mining Engineering.* an opencut mine in which coal beds are worked by removing overburden before extracting coal.

strip mining *Mining Engineering.* the surface mining of coal rather than metalliferous ores.

strip mosaic *Cartography.* in aerial photography, a mosaic made from the photographs taken on a single flight.

stripped atom *Atomic Physics.* an atom that has many fewer electrons than it has protons, caused by ionization.

stripped bedding plane *Geology.* the exposed top surface of resistant rock that forms a stripped structural surface when extended over a relatively large area. Also, **stripped plane** or **plain.**

stripped plain *Geology.* the exposed top surface of a hard rock layer that forms a stripped structural surface when covering a relatively large area. Also, **stripped bedding plain.**

stripped structural surface *Geology.* in a region consisting of horizontal or gently inclined strata of differing resistances, the exposed relatively smooth surface of underlying resistant rock produced by the erosion of the overlying softer strata. Also, **stripped surface.**

stripper *Mechanical Devices.* a plierlike tool that removes the insulation from wires. Also, WIRE STRIPPER. *Agriculture.* a device in a harvester that removes grain from the stalks of grasses and hay. Also, HARVEST STRIPPER. *Chemical Engineering.* a process vessel that uses a gas stream to remove gaseous compounds from a liquid stream. *Petroleum Engineering.* a nearly depleted oil well, one that yields fewer than ten barrels a day. Also, **stripper well.** *Graphic Arts.* a person who carries on a process of stripping film into position for printing.

stripper rubber *Petroleum Engineering.* a pressure-activated seal used in a low-pressure well to control gas pressure in the casing-tubing annulus during insertion or withdrawal of tubing.

stripping *Graphic Arts.* **1.** the process of assembling photographic positives or negatives on a lithographic flat. **2.** the process of peeling film emulsion from its base for transfer to another negative; used in platemaking to combine or repair images. Also, IMAGE ASSEMBLY. *Textiles.* a process of removing dye from a fabric by chemical means. *Metallurgy.* removing a coating from a metallic material. *Chemical Engineering.* the removal of gaseous compounds from a liquid stream by flash or steam-induced vaporation. Also, DESORPTION. *Ordnance.* **1.** a condition in which the jacket of a bullet leaves metal residue in the bore of a gun because the bullet does not properly engage the rifling. **2.** see STRIP.

stripping analysis *Analytical Chemistry.* determination of metals by the electrodeposition and then stripping away from an electrode, usually a hanging mercury drop, by measurement of stripping current or by weighing the metal deposited.

stripping a shaft *Mining Engineering.* **1.** the process of extracting the timber from an abandoned shaft. **2.** the process of squaring or otherwise trimming the sides of a shaft.

stripping excavation *Archaeology.* an excavation technique that reveals the horizontal details of a site.

stripping film *Graphic Arts.* photographic film having emulsion that can be peeled away; used especially in composite photographs.

stripping ratio *Mining Engineering.* the unit amount of waste that must be removed to access a similar unit amount of ore or mineral material.

stripping reaction *Nuclear Physics.* a reaction in which a high-energy deuteron grazes the edge of a target nucleus so that the deuteron's proton is stripped off, to be captured in the target nucleus or deflected by a magnetic field, as in a cyclotron, while the neutron continues in a straight path with most of its initial momentum; useful in determining the angular momenta of nuclear states.

strip pit *Mining Engineering.* **1.** a coal or other type of mine that is worked by stripping methods. **2.** an open-pit mine.

strip planting *Forestry.* a pattern of planting in which trees are set out in two or more parallel lines in a long narrow strip of cleared land.

strip printer *Engineering.* a device that prints a data output on a long narrow strip; similar to a ticker-tape machine.

strip radial plot *Cartography.* a type of direct radial triangulation in which aerial photographs are first plotted in flight strips without reference to ground control, then adjusted to one another and to ground control. Also, **strip radial triangulation.**

STRIPS *Artificial Intelligence.* an early planning system that represented the capabilities of a robot as a set of STRIPS operators and used search to find a sequence of operators that could achieve a goal. (An acronym for Stanford Research Institute Problem Solver.)

STRIPS operator *Artificial Intelligence.* a representation of a possible action of a robot in terms of a precondition, an add list, and a delete list.

strip survey *Forestry.* a survey of one or more sample strips of forest land, chosen as average samples from which to estimate the value of a large area.

strip transmission line *Electromagnetism.* a microwave transmission line that is constructed by printed circuit techniques, consisting of a narrow, thin metal strip mounted on a dielectric material, with wide ground planes on the other sides of the dielectric.

strobe *Electronics.* **1.** a shorter term for a strobe light or a stroboscope. See STROBE LIGHT, STROBOSCOPE. **2.** any of various devices having pulsing action similar to a stroboscope. **3.** a signaling pulse of very short duration. *Computer Technology.* a control signal used to synchronize data transfer between independent units, such as a central processing unit with an input/output device.

strobe circuit *Electronics.* a circuit that generates an output pulse only at certain times or under certain conditions; for example, a gating or coincidence circuit.

strobe light *Electronics.* **1.** a lamp that is capable of producing an extremely brief and intense flash of light; used in conjunction with a stroboscope, as in high-speed photography of rapidly moving objects. **2.** a similar bright, flashing light used for visual effects, as for a theatrical performance, rock concert, and so on.

strobe marker *Electronics.* a small bright spot, a short gap, or some other discontinuity produced on the line trace of a radar display to indicate the part of the time base that is receiving attention.

strobe pulse *Electronics.* a pulse whose duration is less than the time period of a recurrent phenomenon; used for making a close investigation of that phenomenon; the frequency of the strobe pulse bears a simple relation to that of the phenomenon, and the relative timing is usually adjustable.

strobe speed *Electronics.* the rate at which a strobe light emits its brilliant flashes of light. Also, **strobe frequency.**

strobila *Invertebrate Zoology.* **1.** the body of a tapeworm, excluding the scolex, consisting of linearly arranged individual segments called proglottids. **2.** a stage in Scyphozoa resembling a pile of saucers, each of which becomes an individual free-swimming immature medusa.

strobilation *Invertebrate Zoology.* in tapeworms and jellyfish, asexual reproduction by which identical body segments formed by budding separate to produce new individuals.

strobilocercus *Invertebrate Zoology.* a larval tapeworm that everts from its bladder after strobilation while remaining in the intermediate host.

strobilus *Botany.* a cluster of overlapping, scalelike sporophylls arranged in a cone-shaped structure, such as that of the horsetail, club moss, and gymnosperm.

strobing *Computer Technology.* a technique used in asynchronous data transfer in which a time-synchronized gate is opened at the optimum time for data readout.

stroboscope *Electronics.* **1.** a device that uses a strobe light to measure the speed of a rotating shaft or machine; when the strobe speed is equal to the rotation speed, the rotation appears to stop. **2.** a device that brings a moving object into view by alternately illuminating it with an intense light and blocking the light with a shutter; capable of making high-speed vibrations visible.

stroboscopic *Electronics.* relating to or shown by means of a stroboscope.

stroboscopic direction finder *Navigation.* a radio direction-finding system in which a rotating directional-finding element, such as an antenna or goniometer, is connected to a stroboscope that indicates the bearing of the received signal.

stroboscopic disk *Engineering.* a disk having lines or dots, used with a strobe to test the speed of a phonograph turntable.

stroboscopic lamp see FLASH LAMP.

strobotron *Electronics.* a cold-cathode, gas-filled arc-discharge tube containing one or more internal or external grids; used in a stroboscope to start current flow and produce intensely bright flashes of light. Also, **stroboscopic tube.**

stroke *Medicine.* **1.** a cerebrovascular accident; a sudden loss of consciousness, often with resulting paralysis, caused by hemorrhage into the brain, by the blockage of blood flow to the brain by an embolus or thrombus, or by the rupture of an artery exterior to yet supplying the brain, causing loss of blood supply to the brain. Also, APOPLEXY, CEREBROVASCULAR ACCIDENT. **2.** any sudden and severe attack; used in compound terms, such as **heat stroke, paralytic stroke,** or **sun stroke.** *Computer Programming.* a line segment, point, or other mark that is used in the formation of characters in written form. *Electronics.* a pen-like motion of an electron beam in cathode-ray-tube displays. *Mechanical Engineering.* the linear distance traveled in either direction by a piston or rod in an engine.

stroke analysis *Computer Programming.* the dissection of a character into certain specified elements for purposes of identification.

stroke-bore ratio *Mechanical Engineering.* the ratio of the distance traveled by a piston in a cylinder to the diameter of the cylinder.

stroke center line *Computer Programming.* the midline between two average-edge lines that describes the stroke's direction of travel.

stroke density *Geophysics.* the number and density of lightning discharges in a given area at a given time.

stroke edge *Computer Programming.* a continuous line that traces the outermost part of a stroke.

stroke output *Cardiology.* the volume of blood pumped by a ventricle during a heartbeat. Also, **stroke volume.**

stroke width *Computer Programming.* the distance between the stroke edges and a line drawn perpendicular to the center line of the stroke.

stroma *Histology.* the connective tissue portion of an organ or structure that supports and nourishes the functional tissues or parenchyma. *Mycology.* in fungi, certain masses of hyphae, on or within which the spores are formed.

Stromateoidei *Vertebrate Zoology.* a suborder of marine perciform fish including the medusafishes, driftfishes, squaretails, and butterfishes, distinguished by an expanded lachrymal bone that covers most of the maxillary; found in the tropical and temperate regions of the Atlantic, Pacific, and Indian Oceans.

stromatite *Geology.* a type of composite rock consisting of two or more dissimilar textural elements arranged in generally parallel layers. Also, STROMATOLITH.

stromatolite *Geology.* a calcareous organosedimentary body, having a laminated concentric structure and varying in shape from horizontal to spherical; formed by sediment trapping or precipitation as a result of the growth and activity of microogranisms, especially blue-green algae.

stromatolith *Geology.* **1.** a tabular igneous intrusion having a complex structure that is interfingered with strata of sedimentary rock. **2.** see STROMATITE.

Stromatoporoidea *Paleontology.* a group of primitive reef-building organisms of uncertain affinities; some authorities have considered them blue-green algae, foraminiferids, or bryozoans, but they are now generally considered coelenterates, either hydrozoans, sponges, or a separate class; they arose in the Precambrian and persisted through the Mesozoic, dying out at the end of the Cretaceous.

Stromatopteridaceae *Botany.* a monotypic family of small terrestrial ferns of the order Filicales; characterized by very little differentiation among rhizomes, roots, and fronds.

Strombacea *Invertebrate Zoology.* a superfamily of snails, gastropod mollusks in the order Mesogastropoda, that cement objects to the shell as camouflage; includes pelican's foot shell and queen conch.

Strömberg's asymmetrical star streams *Astronomy.* stars with large radial velocities tending to fall into either of two streams that have opposite motions perpendicular to a line toward the center of the galaxy.

Strombolian or **strombolian** *Volcanology.* describing a volcanic eruption characterized by the ejection of fire fountains of fluid lava from a central crater, and consisting of a continuous series of low- to moderate-intensity explosions that emit scoria along with a white vapor cloud. Thus, **Strombolian-type eruption.** (From *Stromboli,* a noted volcano on an island off the northeastern coast of Sicily.)

stromeyerite *Mineralogy.* AgCuS, an orthorhombic mineral occurring as metallic-gray compact masses with a blue tarnish, having a specific gravity of 6.2 to 6.3 and a hardness of 2.5 to 3 on the Mohs scale; found in silver ore deposits.

Strömgren, Svante [strōm´grən] 1870–1947, Swedish astronomer; studied the orbit of comets.

Strömgren radius *Astronomy.* the radius of an emission nebula within which hydrogen is nearly completely ionized; its size depends on the temperature of the central star and the density of the nebular gas.

Strömgren sphere *Astronomy.* a roughly spherical region of ionized hydrogen surrounding a hot star; some regions may be more ragged than spherical due to the effects of interstellar dust, gas, motions, and nonuniformities in mass.

strong acid *Chemistry.* an acid that dissociates completely into ions in a given solvent.

strongbark *Botany.* any of several shrubs or small trees of the genus *Bourreria* of the borage family, especially *B. ovata* of southern Florida and the West Indies, characterized by elliptic leaves, fragrant, white flowers, and a strong, hard, brown wood streaked with orange. Also, **strongback.**

strong base *Chemistry.* a base that dissociates completely into ions in a given solvent.

strong breeze *Meteorology.* a wind having a speed of 22–27 knots (25–31 miles per hour).

strong electrolyte *Physical Chemistry.* a compound that is completely or largely dissociated in an aqueous solution; most soluble mineral salts, such as sodium chloride or potassium nitrate, are strong electrolytes.

stronger topology *Mathematics.* a topology S on a space X is said to be stronger than a topology T on the same space X if every open set of T is also an open set of S. T is said to be weaker than S.

strong fix *Navigation.* a fix determined from well-placed horizontal sextant angles.

strong flour *Food Technology.* any wheat flour having a high gluten content and tight chemical structure; generally grown in dry, cool climates.

strong force *Particle Physics.* the short-range force and interaction between quarks that is carried by the gluon; the strong force also dominates the behavior of interacting mesons and baryons, and accounts for the strong binding among nucleons. Also, **strong interaction.**

strong gale *Meteorology.* a wind having a speed of 41–47 knots (47–54 miles per hour).

strongly typed language *Control Systems.* a high-level programming language in which each variable must be declared according to its type at the beginning of the program, and the language enforces rules concerning the manipulation of such variables.

strong topology *Mathematics.* the norm topology on a normed space; that is, the metric topology induced by the norm.

strong typing *Computer Science.* a system of static type checking in which the types of all variables must be declared and correct use of types is enforced by the compiler.

strongyle *Invertebrate Zoology.* **1.** any nematode worm belonging to the family Strongylidae; parasitic on and causing disease in mammals, especially serious in horses. **2.** a type of large sponge spicule with rounded ends.

Strongylina *Invertebrate Zoology.* a suborder of nematodes equivalent to the superfamily Strongyloidea.

Strongyloidea *Invertebrate Zoology.* the hookworms and lungworms, a superfamily of nematodes belonging to the order Rhabditida, including endoparasites of humans and domestic animals. Also, **Strongylida.**

strongyloidiasis *Medicine.* an infestation of the intestines with the parasitic nematode roundworm *Strongylus stercoralis,* found in tropical and subtropical countries; the larvae penetrate the skin from contact with infected soil, are carried in the bloodstream to the lungs, causing hemorrhage, and eventually reach the small intestine, causing diarrhea and ulceration. Also, **strongyloidosis.**

strongylote *Invertebrate Zoology.* a type of sponge spicule with one rounded end.

strontia see STRONTIUM OXIDE.

strontianite *Mineralogy.* $SrCO_3$, a white, yellow, gray, or pale-green orthorhombic mineral of the aragonite group occurring as prismatic to acicular crystals and in massive form, having a specific gravity of 3.76 to 3.785 and a hardness of 3.5 on the Mohs scale; found as veins and geodes in limestone.

strontium [strän´shē əm; strän´tē əm; strän´shəm] *Chemistry.* an alkaline-earth metal and element having the symbol Sr, the atomic number 38, an atomic weight of 87.62, a melting point of 770°C, and a boiling point of 1380°C, with two stable isotopes and two radioactive isotopes; silvery crystals that are obtained from strontianite and celestite ores; used in alloys and as a scavenger in electron tubes. (From *Strontian,* the name of a mining area in Scotland where this element was first identified.)

strontium-90 *Nuclear Physics.* a heavy radioactive isotope of strontium with a half-life of 29.1 years, formed as a fission product and present in the fallout from nulear explosions; it poses a significant hazard in humans and animals because, like calcium, it can be assimilated and deposited in the bones.

strontium acetate *Organic Chemistry.* $Sr(C_2H_3O_2)_2 \cdot 1/2H_2O$, a white crystalline solid; soluble in water and loses water of hydration at 150°C; used as an intermediate.

strontium bromide *Inorganic Chemistry.* $SrBr_2 \cdot 6H_2O$, toxic white crystals or powder; absorbs water from the air; loses $4H_2O$ at 88.6°C, and $6H_2O$ above 180°C; used in medicine and as a reagent.

strontium carbonate *Inorganic Chemistry.* $SrCO_3$, a white powder; soluble in acids and slightly soluble in water; loses carbon dioxide when heated to 1340°C; used as a catalyst, in television tube glass, to make ceramic ferrites, and in pyrotechnics.

strontium chlorate *Inorganic Chemistry.* $Sr(ClO_3)_2$ or $Sr(ClO_3)_2 \cdot 8H_2O$, a white crystalline powder that decomposes on heating; soluble in water and slightly soluble in alcohol; a strong oxidant and a dangerous explosive hazard with organic materials, shock, heat, or friction; used in pyrotechnics and tracer bullets.

strontium chloride *Inorganic Chemistry.* $SrCl_2$ or $SrCl_2 \cdot 6H_2O$, white crystals with a sharp, bitter taste; soluble in water and alcohol; the anhydrous form melts at 875°C and boils at 1250°C, and the hexahydrate loses its water on heating to 100°C; used in pyrotechnics, electron tubes, and in making other strontium salts.

strontium chromate *Inorganic Chemistry.* $SrCrO_4$, a light yellow corrosion-resistant powder, toxic and a carcinogen; used in coatings, colorants, electroplating, and pyrotechnics.

strontium dioxide see STRONTIUM PEROXIDE.

strontium fluoride *Inorganic Chemistry.* SrF_2, a toxic white powder; insoluble in water; melts at 1473°C and boils at 2489°C; used in electronics, optics, lasers, and lubricants.

strontium hydroxide *Inorganic Chemistry.* $Sr(OH)_2$ or $Sr(OH)_2 \cdot 8H_2O$, colorless crystals that absorb water and carbon dioxide from the air; slightly soluble in cold water and soluble in hot water; the anhydrous form melts at 375°C and decomposes at 710°C, and the octahydrate form loses its water at 100°C; used in lubricant soaps and greases and as a stabilizer for plastics and glass. Also, **strontium hydrate.**

strontium iodide *Inorganic Chemistry.* SrI_2 or $SrI_2 \cdot 6H_2O$, toxic white crystals that yellow in air and are decomposed by moist air and by heat; used in medicine and as a chemical intermediate.

strontium nitrate *Inorganic Chemistry.* $Sr(NO_3)_2$, a white powder; soluble in water; melts at 570°C; a strong oxidant and dangerous fire hazard in contact with organic substances, explosive when shocked or heated; used in fireworks, signal flares, and matches.

strontium oxalate *Inorganic Chemistry.* $SrC_2O_4 \cdot H_2O$, a white odorless powder; insoluble in cold water; loses its water at 150°C; toxic on ingestion; used in fireworks and in tanning.

strontium oxide *Inorganic Chemistry.* SrO, a gray-white powder or lumps; reacts with water to give the hydroxide; melts at 2430°C and boils at about 3000°C; used in the manufacture of strontium salts and in pigments, greases, and soaps. Also, STRONTIA.

strontium peroxide *Inorganic Chemistry.* SrO_2 or $SrO_2 \cdot 8H_2O$, a white powder; slightly soluble in cold water and decomposed by hot water; the anhydrous form decomposes at 215°C and the octahydrate loses its water at 100°C; it is a dangerous fire and explosive hazard in contact with organic substances and is sensitive to heat, shock, catalysts, and reducing agents; used as a bleach and antiseptic and in pyrotechnics. Also, STRONTIUM DIOXIDE.

strontium salicylate *Organic Chemistry.* $Sr(C_7H_5O_3)_2 \cdot 2H_2O$, white crystals or powder, soluble in water and alcohol, that decomposes when heated; used in pharmaceuticals and in chemical manufacture.

strontium sulfate *Inorganic Chemistry.* $SrSO_4$, white crystals; insoluble in alcohol and almost insoluble in water; melts at 1605°C; used in pyrotechnics, ceramics, and the manufacture of glass and paper.

strontium sulfide *Inorganic Chemistry.* SrS, a gray powder; slightly soluble in water, with a disagreeable odor in moist air; melts above 2000°C. It is an irritant and a fire and explosive hazard; used in depilatories and luminous paints, and in making strontium compounds. Also, **strontium monosulfide.**

strontium titanate *Inorganic Chemistry.* $SrTiO_3$, a powder; insoluble in most solvents; melts at 2060°C; used in electronics and electrical insulation.

strontium unit *Nucleonics.* a concentration unit of strontium-90 relative to calcium, equivalent to that which produces 10^{-12} curie per gram of calcium.

strontiuresis *Toxicology.* the elimination of strontium from the body in the urine.

strophanthin *Pharmacology.* a drug derived from *Strophanthus kombé,* occurring as a white or yellow powder and formerly injected as a heart tonic of rapid onset and short duration.

strophiole *Botany.* a small outgrowth arising from a seed coat, such as a caruncle.

strophoid *Mathematics.* the graph in the plane of the equation

$$y^2 = x^2(a - x)/(a + x),$$

where a is constant. In polar coordinates, this equation has the form $r = a \cos 2\theta \sec \theta$. The curve may be viewed as the locus of a point P on a ray l revolving about its endpoint $A = (a,0)$. At each position of l, P is that point on l that is the same distance from A as the y-intercept of l is from the origin. The strophoid is symmetric about the x-axis and asymptotic in both directions to the line $x = -a$. It lies between that line and the y-axis except for a loop on the other side of the y-axis, having a double point at the origin and passing through the point $(a,0)$.

Strophomena *Paleontology.* an early articulate brachiopod, the type genus of the extinct order Strophomenida, having one valve concave and the other convex; extant in the Ordovician.

Strophomenida *Paleontology.* an order of articulate brachiopods that arose in the Early Ordovician, became widespread in the Triassic, and became extinct in the Early Jurassic; includes more genera than any other order of brachiopods; generally characterized by one convex shell and one flat or concave.

Strophomenidina *Paleontology.* a suborder of articulate brachiopods in the extinct order Strophomenida; extant from the Ordovician to the Triassic.

strophulus *Pathology.* a usually harmless, pimply eruption of the skin, occurring especially in infants in several forms.

Strouhal number *Mechanics.* a dimensionless number used in the study of a vibrating body in a fluid flow; equal to a characteristic dimension of the body times the vibrating frequency, divided by the relative fluid velocity. Also, REDUCED FREQUENCY.

struck *Veterinary Medicine.* an enterotoxemia caused by *Clostridicum perfringens* and affecting adult sheep in Britain.

struck joint *Civil Engineering.* a masonry joint in which excess mortar is removed by the stroke of a trowel.

structural of or relating to structure; specific uses include: *Chemistry.* of or relating to the arrangement of atoms in a molecule. *Biology.* relating to organic structure. *Geology.* relating to geologic structure.
structural adhesive *Materials.* a strong, high-performance modified epoxy, neoprene-phenolic, or vinyl formal-phenolic adhesive that is used to bond metal to wood, plastics, or metal.
structural analysis *Engineering.* an analysis of the stresses and strains to which a structure will or might be subjected. *Petrology.* see STRUCTURAL PETROLOGY.
structural anthropology *Anthropology.* see STRUCTURALISM.
structural approach *Psychology.* in psychoanalytic theory, the belief that individual personality is divided into three structures, the id, ego, and superego. Also, **structural model.**
structural bench *Geology.* a bench representing the erosional remnant of the edge of a structural terrace. Also, ROCK BENCH.
structural ceramics *Materials Science.* ceramics, such as silicon nitride, Si_3N_4, and silicon carbide, SiC, that have high strength at elevated temperatures offset by their brittleness; used as structural materials in applications such as ball bearings, turbine blades, and piston caps in combustion engines.
structural change *Genetics.* a chromosomal aberration or mutation that involves a change in the form or shape of the chromosome.
structural colors *Entomology.* bright iridescent colors produced by the reflection of light off thin transparent plates rather than by pigments; found on the wings of some butterflies and beetles, and on the bodies of dragonflies.
structural connection *Civil Engineering.* a connection that involves the individual members of a structure to make it a complete stable structural unit.
structural contour map see STRUCTURE-CONTOUR MAP.
structural datum see DATUM HORIZON.
structural deflection *Mechanics.* any deformation or movement of a structural member from its original position.
structural drawing *Graphic Arts.* any of a set of schematic or scale drawings that depict the design elements of a building or other structure.
structural drill *Mechanical Engineering.* a diamond or rotary drilling apparatus mounted on a truck for mobility; used to drill holes to obtain information about the composition of subsurface strata. Thus, **structural drilling.**
structural engineering *Civil Engineering.* the application of scientific methods and engineering principles to civil engineering problems, including selection of the appropriate construction material to be used for buildings, bridges, and other structures and determination of the most economical load-bearing members.
structural formula *Chemistry.* a molecular formula that shows the order of atoms in the molecule; a plane representation of the atomic arrangement of the molecule.
structural-functionalism *Anthropology.* a theory stating that while culture is the product of the needs of the members, their behaviors maintain the overall social structure rather than satisfying their individual needs.
structural gene *Genetics.* a gene that codes for a product such as an enzyme, protein, or RNA, rather than serving as a regulator.
structural geology *Geology.* a branch of geology that is concerned with the architecture of the earth's crust, including its form, movements, and internal structures and the arrangements of its rocks.
structural high *Geology.* a geological feature having positive relief, or the crust or uppermost part of an anticlinal structure.
structural integrity *Aviation.* the ability of an airframe to carry its designated loads.
structuralism *Anthropology.* the study of cultural systems as an expression of the underlying structure of the mind, focusing on how the behavior and beliefs of individuals are expressed in the institutions of a society; associated with Claude Levi-Strauss. Also, STRUCTURAL ANTHROPOLOGY. *Psychology.* **1.** an approach to psychology, active in Germany and the United States between 1890 and 1920, that evaluated human experience from the point of view of the individual, through introspective analysis of the structural components of mental processes. Also, **structural psychology. 2.** any psychological theory that focuses on the structure of mental functions and of the mind. *Linguistics.* see STRUCTURAL LINGUISTICS.
structural isomers *Organic Chemistry.* two or more compounds having the same number and kinds of atoms and the same molecular weight, but differing in the order in which the atoms are attached to one another.
structuralist *Science.* a person who follows or supports a theory of structuralism, as in linguistics, psychology, or anthropology.
structural linguistics *Linguistics.* **1.** in general, descriptive linguistics; i.e., the formal description of language based on the systematic observation of actual usage. **2.** specifically, a movement in language study originated by Ferdinand deSaussure in the early 1900s and later associated in the U.S. with Leonard Bloomfield, Edward Sapir, Charles C. Fries, and others, based on the concept that each existing language has its own unique set of patterns (structures) by which sounds are used to form sentences, and maintaining that the patterns of a given language should be described in absolute terms rather than relative to any other language or to a supposed ideal form of the given language.
structural loading *Design Engineering.* a process in which a load is placed on a structural beam in order to maximize its use and strength.
structural low *Geology.* a geological feature having negative relief, or the saddle or lowermost part of a synclinal structure.
structural nose *Geology.* see NOSE, def. 1.
structural petrology *Petrology.* the study of rock fabric on a thin-section or micro scale. Also, STRUCTURAL ANALYSIS.
structural protein *Biochemistry.* **1.** a protein that serves mainly as a framework component of cells and tissues. **2.** see SCLEROPROTEIN.
structural riveting *Engineering.* the process of joining structural members with rivets through punched holes.
structural shape *Metallurgy.* an iron or steel product suitable for structural applications that is designed according to accepted standards.
structural terrace *Geology.* a steplike or shelflike formation occurring where the underlying, otherwise uniformly inclined rock strata flatten locally.
structural trap *Geology.* an area in a reservoir bed in which oil or gas is trapped as a result of folding, faulting, or other deformation.
structural unconformity see ANGULAR UNCONFORMITY.
structural valley *Geology.* a valley that is produced and shaped directly by the movement and processes of the underlying geologic structure.
structural weight *Aviation.* the weight of an aircraft's airframe. *Space Technology.* the weight of a rocket or rocket vehicle not including fuel, cargo, or crew.
structure *Science.* the arrangement of parts in an object or organism. *Chemistry.* **1.** the components and manner of arrangement of atoms in a molecule. **2.** a diagram or model of a molecular arrangement. *Engineering.* something that is constructed; a combination of elements that are fabricated and interconnected in accordance with a design and intended to support vertical and horizontal loads; e.g., a building, bridge, or radio tower. *Geology.* **1.** the overall attitude, disposition, or relationship of the rock masses in a designated area; e.g., bedding, cleavage, or brecciation. **2.** an assemblage of rocks that has been or is being affected by agents of erosion. *Mineralogy.* the form in which a mineral occurs; e.g., fibrous, massive, or tabular. *Aviation.* the construction or makeup of a flight vehicle, including its fuselage, wings, empennage, nacelles, and landing gear but not its power plant, furnishings, or equipment.
structure cell *Crystallography.* see UNIT CELL.
structure contour *Geology.* a line connecting points of equal elevation that represents the upper or lower surface of a particular stratum, indicating the presence of a structural formation, such as a fault. Also, SUBSURFACE CONTOUR.
structure-contour map *Geology.* a map that shows subsurface configurations by means of projecting structure contour lines onto a horizontal plane. Also, **structure map.**
structured data tape *Computer Programming.* a collection of individual data items of the same or different data types; examples are arrays (same data types) and records (different data types).
structured interview *Psychology.* an interview conducted according to a predetermined set and order of questions or topics.
structured programming see MODULAR PROGRAMMING.
structured walkthrough *Computer Programming.* a technical review of a program by the programmer for colleagues to detect errors.
structure factor *Crystallography.* a factor that determines the intensity of a reflected beam in crystal diffraction analysis; given by the summation of the atomic scattering factors weighted by appropriate phase factors when summed over the atoms in the unit cell. The magnitude of the structure factor $|F|$ is the ratio of the amplitude of X-rays scattered in a particular direction by the contents of one unit cell of a crystal to that scattered by a classical point electron at the origin of the unit cell under the same conditions.
structure ground see PATTERNED GROUND.

structure invariant *Crystallography.* a Bragg reflection or group of Bragg reflections whose phase angles do not change when the origin of the unit cell of a crystal is changed.

structure modulation spectroscopy *Spectroscopy.* an infrared spectroscopic technique in which scattered visible light near a structure resonance is modulated to determine the absorption spectrum of an aerosol particle.

structure number *Design Engineering.* a number from 0 to 15 used to identify the relationship of spacing of abrasive grains in a grinding wheel relative to their grit size.

structure resonance *Spectroscopy.* for a small aerosol particle, a very narrow resonance occurring at one of the natural electromagnetic frequencies at which the dielectric sphere oscillates.

structure section *Geology.* an exposed surface or graphic representation that shows a vertical segment of a geologic structure as it appears or would appear if cut by a vertical plane through the earth's crust.

struma *Medicine.* see GOITER.

struma lymphomatosa *Medicine.* thyroid enlargement of unknown cause, marked by epithelial regression and lymphoid hyperplasia. Also, HASHIMOTO'S STRUMA, LYMPHADENOID GOITER.

struma ovarii *Medicine.* a rare ovarian teratoma composed mainly or entirely of thyroid tissue.

strumectomy *Surgery.* the surgical removal of a goiter.

strut *Engineering.* any of a wide variety of usually external structural support members used to sustain compression and provide support. *Aviation.* specifically, on an aircraft, a rigid structural member designed to hold two or more components together and support a load, such as a bar used to brace a wing or landing gear of an aircraft. *Naval Architecture.* a support for a propeller shaft, hydrofoil, etc., that protrudes from a vessel's underwater hull.

Struthionidae *Vertebrate Zoology.* a monospecific family of birds containing the ostrich.

Struthioniformes *Vertebrate Zoology.* a monofamilial order of ratite birds of the suborder Palaeognathae, containing the ostrich.

struthious *Vertebrate Zoology.* relating to or resembling the ostrich. Also, **struthioid.**

Struve, Otto 1897–1963, Russian-born American astronomer; researched spectroscopic binaries, stellar rotation, and stellar evolution.

struvite *Mineralogy.* $(NH_4)MgPO_4 \cdot 6H_2O$, a colorless to yellow, orthorhombic mineral, occurring as crystals of varied habit, having a specific gravity of 1.71 and a hardness of 2 on the Mohs scale; found in bat guano in Australia, South Africa, and California.

struvite stones *Pathology.* concretions made up of ammonia magnesium phosphate, having hard crystals similar to the mineral struvite.

strychnine [strik´nīn; strik´nin] *Organic Chemistry.* $C_{21}H_{22}N_2O_2$, a toxic, hard, white crystalline alkaloid with a bitter taste; slightly soluble in water, alcohol, and ether; it melts at 285°C and boils at 270°C (5 torr); derived from *Nux vomica* and related plants; used as a pesticide and medicinally as a tonic and stimulant for the central nervous system. (From the Greek name for one of the plants from which it is derived.)

strychninism *Toxicology.* poisoning from chronic exposure to strychnine; symptoms include increased sensory acuity, followed by tonic convulsions and vomiting, and, in severe cases, respiratory paralysis and death.

strychninization *Medicine.* the application of strychnine to the skin to increase nervous excitability in order to facilitate the study of neural connections.

STS serologic test for syphilis.

STS or **S.T.S.** Society of Thoracic Surgeons.

STU skin test unit.

Stuart factor *Biochemistry.* a procoagulant, similar to prothrombin, that is found in normal plasma. Also, **Stuart power factor.**

Stuart's transport medium *Microbiology.* a colorless medium used for the transport or temporary storage of delicate anaerobic bacteria such as *Neisseria gonorrhoeae.*

Stuart windmill see FALES-STUART WINDMILL.

stub *Computer Technology.* a dummy component used temporarily in top-down program development so that coding may continue. *Electromagnetism.* an impedance matching device consisting of a short section of transmission line, either open or shorted at its far end, and connected in parallel with a transmission line whose impedance is to be matched to an antenna or a transmitter.

stub angle *Electromagnetism.* a 90° coaxial elbow for a radio-frequency transmission line whose inner conductor is supported by a quarter wavelength stub.

stub axle *Mechanical Engineering.* **1.** an axle designed to carry only one wheel. **2.** a dead axle that carries the wheels and the swivel pins of a vehicle.

stubble *Agronomy.* the short stalk of a plant remaining after the top part has been cut off in harvesting or cropped by a grazing animal.

stubble cropping *Agronomy.* **1.** a method of growing where seed is sown on grain stubble after harvest for plowing under as green manure the following spring. **2.** a process of seeding on top of the previous year's crop residues without cultivation.

stubble mulch *Agronomy.* crop residues left on farm land as a ground cover before and during land preparation for the purpose of erosion control.

stubble plow *Agriculture.* a type of plow that has a short, sharply curved moldboard.

stubborn disease *Plant Pathology.* a virus disease of citrus trees characterized by bushlike growth of twigs, acorn-shaped fruit, and chlorotic leaves.

stub cable *Electricity.* a short branch off of a principal cable whose end is usually sealed until used.

stub entry *Mining Engineering.* a short, narrow entry that is driven into solid coal, turned from another entry but not connected to other workings.

stub matching *Electromagnetism.* the use of stub to match the impedance between a transmission line and an antenna or transmitter.

stub mortice *Building Engineering.* a mortice that extends only partway through a wooden member. Also, **stub mortise.**

stub-supported coaxial *Electromagnetism.* a coaxial line whose inner conductor is positioned and supported by quarter wavelength, short-circuited stubs.

stub-supported line *Electromagnetism.* a transmission line that is supported by quarter wavelength (infinite reactance) short-circuited stubs.

stub switch *Engineering.* a short, free-moving rail switch, used in mining and some narrow-gauge industrial railway applications.

stub tenon *Building Engineering.* a tenon placed in a stub mortice.

stub traverse see SPUR TRAVERSE.

stub tube *Mechanical Engineering.* a short tube attached to a pressurized vessel for connecting additional piping or equipment, if necessary.

stub tuner *Electromagnetism.* an impedance matching stub that is terminated by a movable short-circuiting means.

stucco *Materials.* **1.** a rough plaster generally made of cement, sand, and lime that is used as a coating for exterior walls. **2.** a fine plaster that is used as a finished coating for interior walls, moldings, or decorative purposes. **3.** to apply either of these coatings.

stud *Building Engineering.* an upright structural member composed of wood, steel, or other materials that forms the frame of a wall or partition in a framed building, to which wallboards, lathing, or paneling is fastened or nailed and is then covered over with plaster or siding. *Mechanical Devices.* **1.** a short rod that projects from another member or device. **2.** a bolt that is threaded on both ends. **3.** a nail with a blunt head. *Agriculture.* **1.** a male animal kept for breeding. **2.** a farm where stallions are kept for breeding, usually for racing or hunting.

stud bolt *Mechanical Devices.* a headless bolt having threads at each end; used to screw one end into a fixed part, such as a wall, and to receive a nut on the opposite end.

stud driver *Mechanical Devices.* a tool used for driving or firing in a stud.

Student's t distribution *Statistics.* a type of sampling distribution for a random variable. A normal distribution divided by the square root of a chi-square distribution divided by its degrees of freedom has a t distribution with that degrees of freedom. There are noncentral, vector, and matric-t distributions. Its square has an F distribution with one and υ degrees of freedom. (Named for *Student,* the pseudonym of William Sealy Gosset, 1876–1937, English statistician.)

Student's t statistic *Statistics.* a standardized sample mean that is used as a test statistic when population variances are not known. A typical example is in the test for equality of means across populations.

Student's t test *Statistics.* a test based on a Student's t statistic.

stud finder *Mechanical Devices.* a tool used to locate studs in walls inside a wall or other structure.

stud link chain *Naval Architecture.* a chain whose links are fitted with studs to prevent kinking.

stud wall *Building Engineering.* a timbered wall.

stud welding *Metallurgy.* a welding process in which the arc is struck between a metal stud and the part.

stuffing *Mechanical Engineering.* grease or other lubricating material forced into a stuffing box and sealed. Also, PACKING.

stuffing box *Mechanical Engineering.* a box containing lubricating materials, positioned around a piston rod or valve. Also, PACKING BOX.

stuffing nut *Mechanical Engineering.* a nut used to seal and adjust the lubricating pressure in a stuffing box.

stull *Mining Engineering.* **1.** a timber prop stretched between the walls in a stope. **2.** a timber platform used to support workers or to carry waste or ore.

stump *Forestry.* the standing lower part of a tree trunk after the upper part and the branches have been cut off. *Surgery.* the distal end of a body limb that is left after amputation. *Mining Engineering.* any small pillar of coal, especially as may be left between the gangway or airway and the breasts to protect these passages.

stun *Neurology.* to render unconscious or immobile by a cerebral trauma, such as by a sharp blow to the head.

stunt *Plant Pathology.* any of various diseases producing an abnormal reduction in size and vigor in a plant.

stupe *Medicine.* a cloth or sponge that is soaked in hot water, wrung until nearly dry, and then made irritant or otherwise medicated.

stupefacient *Medicine.* inducing stupor. *Pharmacology.* an agent that induces stupor.

stupor *Neurology.* a state of impaired consciousness accompanied by diminished responsiveness to external stimuli. *Psychology.* any disorder marked by reduced responsiveness.

stupose *Biology.* having a tuft or mat of long hairs; woolly. Also, **stupeous.**

stupp *Mining Engineering.* a black residue produced in the distillation of mercury ores, consisting of mercury metal, mercuric oxide, hydrocarbons, soot, dust, and particles of ore.

sturgeon *Vertebrate Zoology.* a primitive, medium to large, mostly freshwater fish of the family Acipenseridae; characterized by a spindle-shaped body, bony scutes on the body, and four barbels around the mouth; found in the Northern Hemisphere.

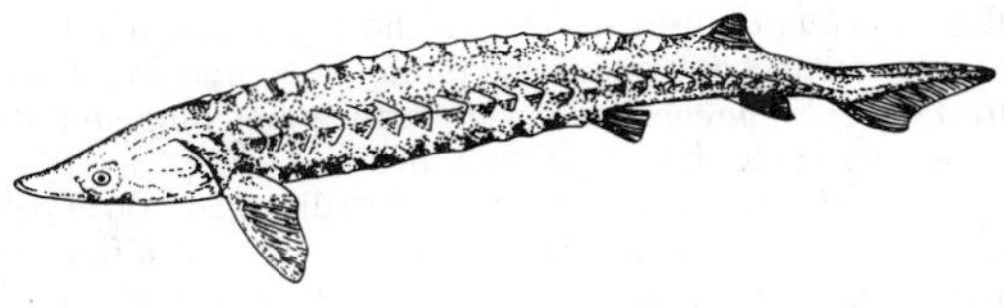

sturgeon

Sturm, Jacques 1803–1855, Swiss-born French mathematician; formulated Sturm's theorems on the real roots of an equation.

Sturm-Liouville operator *Mathematics.* a partial differential operator on R^n of the form

$$L = -\sum_{i=1}^{n} \partial/\partial x_i\,(p\partial/\partial x_i) + q,$$

where p and q are arbitrary functions; this can be shown to be elliptic and self-adjoint.

Sturm-Liouville problem *Mathematics.* the problem of solving a linear differential equation of even order $2n$ with $2n$ boundary conditions. The given differential equation together with its boundary conditions is called a **Sturm-Liouville system.**

Sturm's theorem *Mathematics.* an algorithm for determining the number of real roots of a polynomial $p(x)$. Let $h(x)$ be the greatest common factor of $p(x)$ and $p'(x)$, which can be found using the Euclidean algorithm; then the polynomial $g(x) = p'(x)/h(x)$ has the same roots as $p(x)$, except with multiplicity 1. Again using the Euclidean algorithm, the following sequence of polynomials $g_k(x)$, called the **Sturm chain** or the **Sturm sequence,** is constructed:

$$g(x) = q_1(x)g'(x) - g_1(x)$$
$$g'(x) = q_2(x)g_1(x) - g_2(x)$$
$$g_1(x) = q_3(x)g_2(x) - g_3(x)$$
$$\cdots\cdots\cdots\cdots\cdots\cdots\cdots$$
$$g_{m-2}(x) = q_m(x)g_{m-1}(x) - g_m(x)$$
$$g_{m-1}(x) = q_{m+1}(x)g_m(x)$$

The last nonzero remainder $g_m(x)$ is the least common divisor of $g(x)$ and $g'(x)$ and must be a nonzero constant, since $g(x)$ has only simple roots. Let $w(\xi)$ denote the number of sign changes in the Sturm chain evaluated at a real number ξ, where zero values are omitted. Then Sturm's theorem states: If a and b are not roots of $g(x)$, then the number of real roots of $g(x)$ in the closed interval $[a, b]$ equals $w(a) - w(b)$. Newton's rule is used to find a closed interval containing all real roots of the polynomial $p(x)$, and hence of $g(x)$.

Sturnidae *Vertebrate Zoology.* the starlings, a passerine family of birds of the suborder Oscines, often having metallic plumage and building nests around large buildings and other structures; native to Europe but introduced into the United States, Australia, and New Zealand.

Sturtevant, Alfred 1891–1970, American biologist; discovered technique for mapping genes; the coauthor of *The Mechanism of Mendelian Heredity,* a founding work of modern genetics.

sturtite *Mineralogy.* a black, amorphous mineral of doubtful status that is apparently a hydrous manganese.

sturzstrom see ROCKFALL.

stutter *Neurology.* **1.** the characteristic speech pattern of stuttering. **2.** to speak in such a manner.

stuttering *Neurology.* a speech disorder characterized by spasmodic enunciation of words, including excessive hesitations, repetition of the same syllables, stumbling, and prolongation of sounds. Stuttering may result from disease, defect, or injury, but in most cases the cause is emotional or psychological; it is often distinguished from stammering, which is characterized by blocking or involuntary pauses in speech. Also, PSELLISM.

Stüve diagram or **chart** *Meteorology.* a thermodynamic diagram used for plotting upper-air soundings; its abscissa is temperature, and its ordinate is pressure to the power 0.286, increasing downward. Also, PSEUDOADIABATIC CHART.

St. Venant's compatibility equations see COMPATIBILITY CONDITIONS.

St. Venant's principle *Mechanics.* the principle that the effects of a system of forces which vectorially add to zero, when applied to a region of a solid, will only be felt in the vicinity of the region, decaying to insignificance with sufficient distance.

S-twist *Textiles.* a left-handed yarn twist in which the spiral formed by twisting slants in the same direction as the center part of the letter S, as opposed to a Z-twist. Also, LEFT TWIST.

stye *Medicine.* a localized staphylococcal inflammation of the connective tissue of the eyelids near a hair follicle. Also, HORDEOLUM.

Styginae *Invertebrate Zoology.* a subfamily of bright-colored butterflies, lepidopteran insects in the family Lycaenidae; males have nonfunctional prothoracic legs.

Stygocaridacea *Invertebrate Zoology.* an order of crustaceans in the superorder Syncarida, containing four little-known species of blind and colorless animals that live between freshwater sand grains.

Stylasterina *Invertebrate Zoology.* a suborder of colonial hydrozoan coelenterates in the order Hydrocorallina, with an upright, branching, calcareous exoskeleton, often pink or purple; a component of coral reefs.

style *Botany.* the slender, elongated part of a pistil, supporting the stigma. *Zoology.* any of various similar small, pointed parts or processes. *Medicine.* see STYLET. *Horology.* see GNOMON.

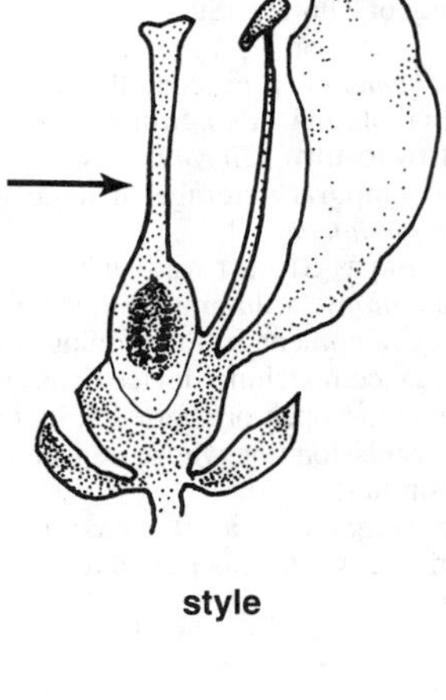

style

-style *Architecture.* a suffix meaning "a column" or "columned," as in *peristyle.*

stylet *Invertebrate Zoology.* **1.** a sharp, slender, hollow process found on the mouthparts of certain insects usually for sucking, stinging, or piercing. **2.** a short unjointed abdominal appendage of some insects. **3.** in nematode and nemertean worms, a small, pointed needlelike process usually near or in the mouth, used for feeding or defense. *Graphic Arts.* a thin, pointed tool used to engrave plates and other lithographic materials. *Medicine.* **1.** a wire inserted into a catheter or cannula to stiffen it or to remove debris. **2.** a slender probe. Also, STYLE, STYLUS.

style zone *Archaeology.* an area in which the artifacts of various communities show the same stylistic attributes.

Stylidiaceae *Botany.* a family of dicotyledonous herbs and undershrubs of the order Canpanulales, usually scapose with a basal rosette of grasslike leaves; native to Australia, southeastern Asia, and southern South America.

stylistic attribute *Archaeology.* a characteristic of an artifact that relates to its surface appearance, such as color, decoration, and texture.

stylistic seriation *Archaeology.* the organization of artifacts by sequence according to changes over time in their stylistic attributes.

stylo- a combining form meaning: **1.** style or styloid, as in *stylohyoid.* **2.** column or pillar, as in *stylolite.*

Stylococcaceae *Botany.* a family of rhizopodal protoplasts of the order Chrysamoebidales, characterized by species in which the cell is surrounded by a lorica.

styloglossus *Anatomy.* a muscle, extending from the styloid process of the temporal bone to the margin of tongue, that retracts and raises the tongue.

stylohyoid *Anatomy.* a muscle, extending from the styloid process of the temporal bone to the hyoid bone, that elevates the hyoid bone.

styloid *Biology.* long, slender, and pointed; resembling a style.

stylolite *Geology.* a thin seam or surface occurring in certain carbonate rocks marked on both sides by irregular interlocking columns, pits, or toothlike projections; caused by dissolution. Also, **stylolite seam.**

stylomastoid *Anatomy.* relating to the styloid and mastoid processes of the temporal bone.

Stylommatophora *Invertebrate Zoology.* the land snails and slugs, a suborder of gastropod mollusks in the subclass Pulmonata having eyes on retractile tentacles; includes common garden snails and slugs.

stylopodium *Botany.* a disclike or conical enlargement at the base of the style, as in plants of the Umbelliferae family.

stylostixis see ACUPUNCTURE.

Styloviridae *Virology.* a proposed name for the family of phages that are now called Siphoviridae.

stylus *Graphic Arts.* a tool similar to a pen that does not emit ink, used to make marks on a reproducible master. *Medicine.* **1.** see STYLET. **2.** a pencil-shaped medicinal preparation, such as a styptic pencil or stick of caustic. *Acoustical Engineering.* see NEEDLE.

stylus printing see MATRIX PRINTING.

styolite *Geology.* a lengthwise striated extension of a rock body that interlocks into a corresponding socketed extension.

S-type asteroid *Astronomy.* an asteroid with a reddish surface color, a moderately high reflectivity, and chemical elements of pyroxene and olivine silicates in a mixture of metallic iron.

S-type magma *Geology.* a magma originating from sedimentary source material.

styphnic acid *Organic Chemistry.* $C_6H(OH)_2(NO_2)_3$, yellow crystals, slightly soluble in water and soluble in alcohol and ether, that melt at 179°C; used as a priming agent in explosive devices. Also, 2,4,6-TRINITRORESORCINOL.

Stypocapitellidae *Invertebrate Zoology.* a family of polychaete annelid worms in the grouping Sedentaria, containing a single genus found in Germany.

Stypocaulaceae *Botany.* a family of coarsely filamentous brown algae of the order Sphacelariales, characterized by alternate, distichous branching; usually appear very harsh and wiry or shaggy and woolly.

styptic *Medicine.* **1.** astringent; serving to arrest bleeding by mechanical or chemical processes. **2.** a hemostatic and astringent compound.

styptic pencil *Medicine.* a stylus made of a paste containing a styptic agent, usually alum, used to stanch bleeding in minor cuts, as from shaving, or relieve the itch of insect bites.

Styracaceae *Botany.* a family of dicotyledonous tanniferous trees and shrubs of the order Ebenales, characterized by a resinous bark and a usually scaly indumentum; native to warm areas in the Americas, the Mediterranean, Southeast Asia, and Malaysia.

styramate *Pharmacology.* $C_9H_{11}NO_3$, a drug that relaxes the skeletal muscles for a relatively long period.

styrene *Organic Chemistry.* $C_6H_5CH{=}CH_2$, a fragrant liquid; soluble in alcohol and ether; boils at 145–146°C; used in making polystyrene plastics and rubbers. Also, **styrol, styrolene, styrene monomer.**

styrene-acrylonitrile resin *Organic Chemistry.* a solid thermoplastic copolymer of styrene and acrylonitrile that resists scratching, chemicals, and stress.

styrene oxide *Organic Chemistry.* $C_6H_5C_2H_3O$, a combustible, toxic, colorless to pale straw-colored liquid that is miscible with ether and boils at 194–195°C; used as an organic intermediate.

styrene plastic *Organic Chemistry.* a plastic compound of polymerized styrene or copolymers containing styrene.

Styrofoam [stī′rə fōm′] *Materials Science.* the trade name for a widely used form of polystyrene foam having various industrial applications, as in insulating materials, ice chests, drinking cups, and flotation devices.

Styx *Ordnance.* the code name for the Soviet SS-N-2 missile.

SU *Aviation.* the airline code for Aeroflot.

SU strontium unit.

sub *Naval Architecture.* a shorter name for a submarine.

sub- a combining form meaning "under" or "below."

sub- *Mathematics.* a mathematical object A is said to be a **subobject** of some given object U if each element of A is also an element of U and if A is the same kind of object as U, with the same algebraic operations. For example, A is a **subgroup** (**subring, subfield, submanifold, subcover, subset,** and so on) of a group (resp. ring, field, manifold, cover, set, etc.) U if A is contained in U and A is also a group (resp. ring, field, manifold, cover, set, and so on). If, in addition, there is some element of U that is not an element of A, then A is a **proper subgroup** (subring, subfield, submanifold, subcover, subset, and so on) of U.

subacute sclerosing panencephalitis *Pathology.* an uncommon and devastating type of leukoencephalitis mostly afflicting children and adolescents, characterized by lesions of the intracytoplasm and demyelination, with invasive inclusion bodies to the nerve cells and oligodendroglia, implying a viral etiology. Aggressively virulent, it generates cerebral dysfunction within a matter of weeks, and death usually follows within one year.

subacute spongiform encephalopathies *Pathology.* a degenerative disease of the brain characterized by extensive vacuolization, or the forming of cavities, in the cerebral cortex. Also, STATUS SPONGIOSIS.

subadditive function *Mathematics.* **1.** a function f with the property that $f(x + y) \leq f(x) + f(y)$ for all x and y in the domain of definition of f. **2.** an extended real-valued set function μ that is defined on a class E of sets with the property that $\mu(E \cup F) \leq \mu(E) + \mu(F)$, whenever E, F, and $E \cup F$ are in E. More generally, μ is said to be **finitely subadditive** if, for every finite class $\{E_1, \ldots, E_n\}$ of sets in E whose union is also in E,

$$\mu(\bigcup_{i=1}^{n} E_i) \leq \sum_{i=1}^{n} \mu(E_i).$$

Also, μ is said to be **countably subadditive** if, for every (infinite) sequence $\{E_i\}$ of sets in E whose union is also in E,

$$\mu(\bigcup_{i=1}^{\infty} E_i) \leq \sum_{i=1}^{\infty} \mu(E_i).$$

subaerial *Geology.* describing conditions, processes, or deposits that are formed, situated, or occur in contact with the open air, directly adjacent to or on the land surface.

subage *Geology.* a subdivision of a geologic age corresponding to the time-rock unit substage.

subalkaline *Geochemistry.* **1.** describing an igneous rock containing no alkali minerals other than feldspars and having a lower alkali to silica ratio than alkaline rocks. **2.** describing a soil having a pH level between 8.0 and 8.5.

subalphabet *Computer Programming.* a subset of a character set.

subalpine *Botany.* describing plant growth on mountains below the limit of tree growth and above the foothill zone, usually between 4000 and 5500 feet in most parts of the Temperate Zone.

subalternate *Botany.* alternate, but tending to become grouped oppositely.

Subantarctic Intermediate Water *Oceanography.* a layer of water in the South Atlantic, and to some extent in the South Indian and Pacific Oceans, just to the north of the Antarctic Convergence at a depth of 700–850 meters above the deep-water layer.

subapical *Anatomy.* located below the apex, or highest point.

subaqueous *Science.* describing conditions, processes, features, or deposits occurring in or under water. Thus, **subaqueous mining.**
subaqueous dune *Geology.* a sand dune formed in water or beneath the water's surface as a result of water flow.
subarachnoid *Anatomy.* of, relating to, or situated below the arachnoid membrane.
subarachnoid block see INTRASPINAL BLOCK.
subarachnoid hemorrhage *Medicine.* a hemorrhage between the pia mater and the arachnoid of the brain, caused by trauma, the rupture of an aneurysm, or other bleeding from a blood vessel, and sometimes by hypertension. Also, **SAH.**
subarachnoid space *Anatomy.* the space located between the arachnoid and the pia mater, containing the cerebrospinal fluid.
subarctic *Geography.* those regions of the Northern Hemisphere associated with taiga or subarctic climates, having temperatures below freezing for over half the year but often over 80°F during the summer.
subarctic climate see TAIGA CLIMATE.
subarid *Meteorology.* moderately arid.
subarkose *Geology.* a type of sandstone, intermediate in composition between arkose and pure quartz sandstone, containing less feldspar than an arkose. Also, FELDSPATHIC QUARTZITE.
subartesian water *Hydrology.* confined ground water that is under sufficient pressure to rise above the water table but not to reach the ground surface. Thus, **subartesian well.**
subassembly *Engineering.* a portion of a structure or assembly, as of electronic or machine parts, that can be installed, removed, or generally treated as a unit, but that also has a structure or assembly.
subatmospheric heating system *Mechanical Engineering.* a steam-heating system with a differential pressure controller and vacuum pump that maintains a vacuum differential between the supply and return of the system.
subatomic particle *Physics.* an elementary particle, which may be a constituent particle in an atom.
subaural *Anatomy.* located below the ear.
subbase *Mathematics.* a family S of subsets of a set X is called a subbase for a topology if every $x \in X$ belongs to at least one set in the family S. Then it can be shown that the collection of all finite intersections of sets in S forms a base for a topology on X.
subbituminous coal *Geology.* a black coal that ranks between lignite and bituminous coal, having a higher carbon content and lower moisture content than lignite.
subboreal *Ecology.* of or relating to a climate or region with near-freezing conditions.
subbottom depth recorder *Engineering.* an instrument that uses a low-frequency underwater spark discharge to measure the characteristics of the strata below a lake or sea bottom.
subbottom reflection *Geophysics.* the reflection of sound waves back from below the sea bottom.
subboundary structure *Metallurgy.* a series of interconnecting low-angle boundaries that are usually less than one degree, found within the main crystals of a metallographic structure.
subcaliber *Ordnance.* of or relating to firing ammunition of a caliber smaller than that which is standard for the gun in which it is used, usually for practice purposes; an adapter may be fitted into the gun or onto the ammunition. Thus, **subcaliber ammunition, subcaliber equipment, subcaliber firing, subcaliber tube, subcaliber rocket.**
subcaliber gun *Ordnance.* a smaller gun mounted above the barrel of a larger gun; used to fire subcaliber ammunition in aiming practice with the larger gun. Similarly, **subcaliber mount.**
subcaliber range *Ordnance.* a target range designed for practice firing with subcaliber ammunition; it may be equipped with miniature landscape targets or other devices to simulate the use of regular ammunition.
subcapillary interstice *Geology.* an interstice, smaller than a capillary interstice, in which the molecular attraction of its walls spans the entire opening, so that water inside is held immovable under the normal pressure affecting submerged water.
subcardinal vein *Vertebrate Zoology.* either of a pair of veins in mammalian embryos that partly replace the postcardinal veins in the abdomen during development.
subcarrier oscillator *Electronics.* **1.** in a color television receiver, the crystal oscillator that operates at the chrominance subcarrier or burst frequency of 3.58 MHz. **2.** in a telemetering system, an oscillator used to translate variations in an electrical quantity into variations of an FM signal at a given subcarrier frequency.
subchondral *Anatomy.* located beneath a cartilage.
subclass *Systematics.* a taxonomic rank immediately below a class. *Artificial Intelligence.* in a frame system or object-oriented system, a category that is a subset of a given class and inherits properties from it; e.g., dog is a subclass of mammal.
subclavian *Anatomy.* located beneath the clavicle.
subclavian artery *Anatomy.* the large artery, branching from the brachiocephalic trunk on the right and the aortic arch on the left, that passes under the clavicle and produces many major branches including the vertebral artery, the thyrocervical trunk, the suprascapular artery, and the internal thoracic artery; it becomes the axillary artery.
subclavian vein *Anatomy.* the large vein that is the direct continuation of the axillary vein and which passes under the clavical and joins the infernal jugular vein to form the brachiocephalic.
subclavius *Anatomy.* a small muscle that extends from the first rib to the clavicle and either depresses the clavicle or raises the first rib.
subclimax *Ecology.* the last stage of succession in an evolving community before development stops.
subclinical infection *Medicine.* a latent or dormant infection. Also, COVERT INFECTION, INAPPARENT INFECTION, SILENT INFECTION.
subclutter visibility *Electronics.* **1.** a measure of the effectiveness of moving-target indication radar, equivalent to the ratio of the signal from a fixed target that can be canceled to the signal from a just-visible moving target. **2.** broadly, the ability of a radar to see through clutter.
subcomponent *Design Engineering.* a part or element of a component that has characteristics of that component.
subcompound *Chemistry.* a compound in which an element exhibits a lower valency than is found in its other, more common compounds.
subconchoidal *Geology.* describing a break in a rock structure with an undulating form resembling that of the inside surface of a bivalve shell.
subconscious *Psychology.* **1.** the material in the mind that is not conscious but is capable of being brought to consciousness. **2.** of or relating to such material.
subconsciousness *Neurology.* a state of partial consciousness in which mental processes occur without the mind's perceiving its own activity.
subconsequent stream see SECONDARY CONSEQUENT STREAM.
subcontinent *Geography.* **1.** a very large, relatively distinct subdivision of a continent, such as South Africa or India. **2.** usually, **the Subcontinent.** another name for the Indian Peninsula. **3.** a large land mass such as Greenland that is smaller than the generally recognized continents. Australia and Antarctica are sometimes, but now rarely, deemed subcontinents.
subcontinental *Geography.* **1.** of or relating to a subcontinent. **2.** below or south of a continent.
subcontract *Engineering.* a contract signed by the prime contractor on a project rather than by the final customer, calling for another contractor to perform part or all of the work.
subcontractor *Engineering.* a contractor who is contracted with and responsible to a general contractor, rather than to the final customer, for performance of part or all of a job.
subcooled liquid *Thermodynamics.* a liquid that exists in a state at a temperature lower than the saturation temperature corresponding to its pressure or at a pressure greater than the saturation pressure corresponding to its temperature. Also, COMPRESSED LIQUID.
subcooled water see SUPERCOOLED WATER.
subcosta *Invertebrate Zoology.* a longitudinal vein in the anterior portion of the wing of an insect.
subcostal *Anatomy.* situated or occurring below a rib or ribs.
subcostal incision *Surgery.* an incision on one side of the anterior abdominal wall, parallel to and just below the costal margin.
subcritical *Nucleonics.* having or using less than the amount of fissionable material needed to sustain a chain reaction.
subcritical crack growth *Materials Science.* a growth of cracks that begin at flaws in a material and slowly propagate until fracture occurs; caused by stress lower than that required for instantaneous fracture.
subcritical flow see SUBSONIC FLOW.
subcritical mass *Nucleonics.* the mass of a sample of fissionable material that is less than the critical mass; such a quantity is stable and does not support a chain reaction within the sample.
subcritical reactor *Nucleonics.* a nuclear reactor operating with a multiplication factor that is less than unity, such that self-sustaining fission is not possible. Also, **subcritical assembly.**
subcrop *Geology.* an occurrence of subsurface strata within a stratigraphic unit that are in contact with the undersurface of a younger, unconformable layer marked by a conspicuous overstep.

subculture *Anthropology.* the shared customs and beliefs of a smaller group within a larger, dominant culture. *Microbiology.* **1.** a microbial culture that is prepared from another culture, usually by inoculating a fresh medium with an aliquot from the existing culture. **2.** to prepare such a culture.

subcutaneous *Anatomy.* located beneath the skin.

subcutaneous connective tissue *Histology.* the layer of loose connective tissue that lies beneath the skin.

subcutaneous emphysema *Medicine.* the abnormal accumulation of air or a gas in the subcutaneous tissue, sometimes seen in the region of the neck in divers.

subcutaneous injection *Medicine.* the injection by hypodermic needle of a small amount of medication below the skin into the subcutaneous tissue, usually on the upper arm, thigh, or abdomen.

subcutis *Anatomy.* the deeper layer of the dermis, containing mostly fat and connective tissue.

subcycle generator *Electronics.* a frequency-reducing device employed in telephone equipment to furnish ringing power at a submultiple of the power-supply frequency.

subdelirium *Neurology.* mild or intermittent delirium.

subdiagonal of a matrix *Mathematics.* given an $n \times m$ matrix with entries a_{ij}, the entries of the form $a_{i+1,i}$, $i = 1, 2, \ldots, \min(n-1, m)$; i.e., the entries just below the main diagonal.

subdirectory *Computer Science.* a directory and set of files that are contained within another directory. The set of directories and subdirectories forms a tree structure.

subdivided capacitor *Electricity.* a capacitor in which several capacitors, called sections, are mounted so that they can be used individually or in series.

subdivision survey *Cartography.* a survey of an area to be developed in which all necessary corners and boundary lines are marked or monumented in order to determine legal boundaries of lots, streets, rights-of-way, and so on.

subdominance *Behavior.* a situation in which an animal is submissive to the reigning male in a group but dominant toward others.

subdominant *Ecology.* **1.** a species that appears to dominate a region, but which in fact does not, as in a savannah where grasses dominate, yet trees and shrubs seem more common. **2.** any species that flourishes yet is not the dominant species in its habitat. *Behavior.* relating to or showing subdominance.

subdrainage *Civil Engineering.* open-jointed or perforated pipes laid in a trench or at the bottom of an excavation, to drain the ground.

subdrilling *Engineering.* a term for any borehole drilled from a broken base one or more feet below a quarry floor.

subduction *Geology.* a process in which one lithospheric plate descends beneath another, often as a result of folding or faulting.

subduction zone *Geology.* the region along which a lithospheric plate descends relative to another plate into the asthenosphere. Also, ZONE OF SUBDUCTION.

subdural *Anatomy.* situated or occurring between the dura mater and the arachnoid.

subdural hematoma *Medicine.* the accumulation of blood between the dura mater and the arachnoid in the brain, usually due to trauma to the head and often resulting in depression of consciousness, seizures, and focal neurologic defects.

subdural hemorrhage *Medicine.* bleeding between the dura mater and the pia mater; usually results from injury to the meningeal vessels.

subdural space *Anatomy.* the space between the dura mater and the arachnoid.

subdurite *Geology.* a type of basic basalt that is made up of augite, hyposthene, bytownite, and magnetite; it typically has a pillow structure.

subdwarf star *Astronomy.* a star that is smaller and less luminous than a dwarf (main sequence) star of the same spectral type; the subdwarf is usually older with fewer metals in its composition.

subdwarf symbiotic star *Astronomy.* a binary star composed of a cool, red giant and a small, hot subdwarf star that is evolving into a white dwarf star. Also, PLANETARY NEBULA SYMBIOTIC STAR.

subendothelial layer *Histology.* a layer of collagenous and elastic fibers in the tunica intima of larger arteries and veins.

subependymoma *Oncology.* an ependymoma with diffuse, proliferated, fibrous astrocytes among the ependymal tumor cells.

suberic acid *Organic Chemistry.* $HOOC(CH_2)_6COOH$, combustible, colorless crystals; partially soluble in water and ether and soluble in alcohol; melts at 143°C and boils at 230°C (15 torr); used in organic synthesis. Also, OCTANEDIOIC ACID.

suberin *Biochemistry.* a fatty substance present in the cell walls of cork and other plant tissues.

suberinite *Geology.* a variety of provitrinite consisting of corky tissue.

suberization *Botany.* the impregnation of plant cell walls with suberin, resulting in corklike tissue.

suberose *Botany.* having the form or nature of cork; corklike.

suberosis *Medicine.* a form of allergic alveolitis caused by inhalation of and tissue reaction to moldy cork dust containing a species of *Penicillium.*

subfamily *Systematics.* a taxonomic rank immediately below a family.

subfeldspathic *Petrology.* describing a mature lithic wacke or arenite, predominantly composed of quartz grains and containing less than 10% feldspar grains.

subfloor *Building Engineering.* a rough-textured, secondary-floor surface beneath a finished floor. Also, BLIND FLOOR.

subform *Systematics.* the lowest rank in the botanical taxonomic hierarchy.

subg. subgenus. Also, **subgen.**

subgelisol *Geology.* a zone of unfrozen ground underlying permanently frozen ground.

subgenomic mRNA *Genetics.* a messenger RNA that is shorter than normal, produced in a eukaryotic cell infected by an RNA virus.

subgenus *Systematics.* a taxonomic rank immediately below a genus.

subgeostrophic wind *Meteorology.* a wind whose speed is less than that of the geostrophic wind for a given pressure gradient; not necessarily a subgradient wind.

subgiant star *Astronomy.* a star with a lower absolute magnitude than a normal giant star of its spectral type; spectral classes G and K are the most common.

subgingival *Medicine.* located or occurring under the gums.

subglacial *Hydrology.* referring to or describing the area immediately beneath a glacier or ice sheet. Also, INFRAGLACIAL. *Geology.* describing a deposit or feature occurring in or formed by the bottom parts of a glacier or ice sheet. Thus, **subglacial eruption, subglacial drainage.**

subglacial moraine see GROUND MORAINE.

subglossal see SUBLINGUAL.

subgrade *Civil Engineering.* **1.** the soil prepared and compacted to support a structure or pavement system. **2.** the elevation of the bottom of a sewer or pipe trench.

subgradient wind *Meteorology.* a wind whose speed is less than that required by a given pressure gradient and centrifugal force.

subgrain *Metallurgy.* a part of a crystal or grain with a slightly different orientation than a neighboring portion of the same crystal, which is separated by low-angle boundaries.

subgraywacke *Petrology.* an argillaceous sandstone similar to graywacke, but containing less feldspar and more abundant, better-rounded quartz grains.

subharmonic *Physics.* a sinusoidal component in a complex wave that has a frequency equal to an integral submultiple of the fundamental frequency. *Acoustics.* specifically, a sound wave whose frequency is the inverse of a whole number of the frequency of a specified fundamental; thus, the subharmonics of a frequency of 4000 hertz are 2000, 1000, 500, 250, and so on.

subharmonic function *Mathematics.* a function f on an open subset D of R^n is said to be subharmonic if $-f$ is superharmonic. Equivalently, f is upper semicontinuous, locally integrable, and bounded above on D, and the average of f at any point x of D (computed as an integral over an open ball contained in D and centered at x) is greater than or equal to $f(x)$. Examples include functions satisfying $\nabla^2 f \leq 0$. It can be shown that if f is harmonic in D, then $|f|$ is subharmonic.

subharmonic triggering *Electronics.* a method of frequency division that uses a triggered multivibrator having a period of one cycle, allowing triggering by a pulse that is an exact integral number of input pulses from the last effective trigger only.

subhead *Graphic Arts.* a heading within a chapter or article that is subordinate in prominence to the main title or headline; e.g., a paragraph heading within a chapter.

subhedral *Mineralogy.* of or relating to a mineral that is bounded in part by its own crystal faces and in part by surfaces formed against preexisting minerals; applied to minerals of igneous rocks and describing a condition intermediate between euhedral and anhedral.

subhumid climate *Meteorology.* a humidity province having a moisture index value between –20 and +20; characterized by grassland or prairie vegetation and very susceptible to drought conditions. Also, GRASSLAND CLIMATE, PRAIRIE CLIMATE.

subhymenium *Mycology.* a fungal tissue located under the hymenium or tissue forming the outer layer of the fruiting body.

subida *Geology.* a rock-floored belt produced by wind scour, extending to the base of a mountain range.

subidiomorphic see HYPOCRYSTALLINE.

subimago *Entomology.* the stage unique in mayflies before the imago (adult) in which a thin, dull skin covers the body and wings.

subinvolution *Medicine.* the inability of a part to return to its normal state after enlargement, as the uterus following childbirth.

subirrigation *Agronomy.* a method of irrigation in which the distribution of water takes place below the surface, as by a series of underground pipes. Also, **subsurface irrigation.**

subj. subject.

subjacent *Geology.* **1.** describing a rock stratum that lies directly under or beneath a specific higher stratum or below an unconformity. **2.** describing a formation that is oriented lower than, but not necessarily lying directly beneath, another formation.

subjacent igneous body *Geology.* an igneous rock solidified beneath the surface that has no evidence of a floor; in theory it increases in size with depth.

subjamming visibility *Military Science.* a term used in describing the capability of a particular radar antijamming technique to penetrate jamming signals.

subject contrast *Radiology.* a contrast that results from differences in the absorption of radiation by various parts of the subject.

subject copy *Graphic Arts.* any text material that has been or is to be typeset, transmitted, entered in a computer, and so on.

subjective organization *Psychology.* **1.** the group classification of words, pictures, or events that an individual perceives to belong together. **2.** the tendency to recall associated items in the same order in which they are presented.

subjective probability *Statistics.* a measure of an individual's degree of belief in the occurrence of a certain event. Also, PERSONALISTIC PROBABILITY.

subjective sound see AUDITORY SYNESTHESIA.

subkiloton weapon *Ordnance.* a nuclear weapon producing a yield below one kiloton.

sublacustrine *Geology.* describing a structure that exists or is formed under the water or on the bottom of a lake.

sublacustrine channel *Geology.* a channel that was eroded in a lake bed by a surface stream before the lake was formed, or by a powerful current within the lake.

sublethal gene *Genetics.* an allele that causes death in less than 50% of the individuals who possess it.

subleukemic or **subleukaemic** *Medicine.* a less than leukemic state of the blood; the term is usually used to refer to temporary suppression of leukemic symptoms.

sublevel *Mining Engineering.* **1.** a secondary level used in working ore, formed by top slicing and sublevel caving. **2.** an intermediate level driven at a short depth below a main level or below the top of an ore body, preliminary to caving the ore immediately above it. *Atomic Physics.* see SUBSHELL.

sublevel drive *Mining Engineering.* a drive made in an underground section, particularly in gently inclined deposits, to divide a deposit into narrower panels and zones.

sublimate *Psychology.* to divert the energy of an undesirable impulse from its immediate goal to a goal more socially, morally, or aesthetically acceptable. *Thermodynamics.* to carry on a process of sublimation; sublime. *Chemistry.* the product obtained from sublimation.

sublimation *Psychology.* **1.** a defense mechanism in which an individual substitutes approved goals for unapproved ones, thus redirecting repressed impulses into more socially acceptable channels. **2.** any redirection of energy or attention from the socially unacceptable to the acceptable. *Thermodynamics.* a phase-transition phenomenon in which a solid is transformed into a gas while bypassing the intermediate liquid phase; the term applies to the reverse process as well. *Meteorology.* the transition of water vapor to ice crystals, or vice versa, without apparent liquefaction.

sublimation cooling *Thermodynamics.* the heat transfer from a gas resulting in sublimation.

sublimation curve *Thermodynamics.* a graph of the vapor pressure of a solid plotted against the temperature.

sublimation energy *Thermodynamics.* the amount of heat transfer required to convert one mole or unit mass of a solid directly into a gas under conditions of constant temperature and pressure.

sublimation heat *Physics.* the heat that must be added to a substance in the process of sublimation from the solid phase to the vapor phase while held at constant pressure and temperature.

sublimation nucleus *Meteorology.* a particle upon which an ice crystal may grow as a result of sublimation.

sublimation point *Thermodynamics.* a temperature at which the total pressure of a gaseous vapor is equal to the vapor pressure of the solid phase of the same substance.

sublimation pressure *Thermodynamics.* the vapor pressure associated with a solid at which sublimation occurs.

sublimation vein *Geology.* a deposit of foreign minerals within a rock fracture or joint that is formed by direct condensation from a vapor.

sublimatography *Analytical Chemistry.* fractional sublimation in which a solid mixture is separated into bands along a condensing tube with a temperature gradient.

sublimator *Chemistry.* an apparatus consisting of a evacuation chamber, a vacuum, and a cooling surface for crystal formation; used in the sublimation process in which a solid is purified by direct conversion to the vapor phase and then converted back to solid form.

sublime *Thermodynamics.* to carry on the action of sublimation; to pass directly from a solid phase to a gaseous phase or vice versa, without passing through an intermediate liquid phase.

sublimed sulfur *Pharmacology.* a fine yellow powdered form of sulfur prepared by condensing sulfur vapor; applied topically to treat scabies and other parasitic skin conditions. Also, FLOWERS OF SULFUR.

subliminal [sub lim´i nəl] *Behavior.* relating to or caused by stimuli that are so faint as to be below the level of consciousness, but which can still influence later behavior. Thus, **subliminal perception, subliminal stimulation.**

sublingual *Anatomy.* situated under the tongue. Also, SUBGLOSSAL.

sublingual glands *Anatomy.* paired salivary glands; the smallest of the salivary glands, located in the floor of the mouth under the tongue.

sublitharenite *Petrology.* a sandstone that is intermediate in composition between litharenite and pure quartz sandstone, consisting of 65–95% quartz, quartzite, and chert, 5–25% fine-grained rock fragments, and less than 10% feldspar.

sublittoral *Ecology.* relating to or describing an organism living immediately below low-tide level.

sublittoral zone *Ecology.* the deeper part of a lake below the area in which rooted plants grow. *Oceanography.* the benthic environment from the extreme low-water level to around 200 meters deep; from the lowest tide area to the edge of the continental shelf.

subluminous star *Astronomy.* any star fainter than a main sequence star, such as white dwarfs, subdwarfs, and the central stars in planetary nebulae.

sublunar point *Astronomy.* the point on earth where the moon lies at the zenith.

submachine gun *Ordnance.* a lightweight, short-barreled, automatic or semiautomatic firearm; it uses pistol-type ammunition and is designed to be fired from the shoulder or hip. Also, **submachine weapon.**

submandibular duct *Anatomy.* the duct that transports saliva from the submandibular salivary gland to the mouth, opening near the frenulum of the tongue.

submandibular gland *Anatomy.* one of two salivary glands located between the digastric muscle and the angle of the mandible. Also, SUBMAXILLARY GLAND.

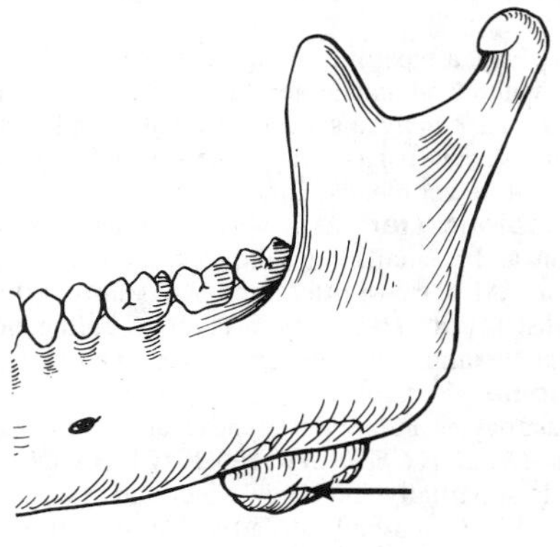

submandibular gland

submarginal *Biology.* near, but not on, the edge or margin of an organism.

submarine *Science.* located or occurring beneath the surface of the water, especially the ocean water. *Naval Architecture.* any vessel that is designed to operate primarily underwater. Most are naval combat ships, although some small submarines are used in underwater research and in rescue and salvage operations. Modern combat submarines have a self-contained nuclear power plant and do not need to come to the surface periodically for air as earlier diesel-powered submarines did.

submarine base *Naval Architecture.* an operating base with special facilities for handling submarines.

submarine bell *Navigation.* a bell that emits a signal that is transmitted through the water for use as a navigational aid.

submarine blast *Engineering.* a blast set off in boreholes drilled below a lake, river, or sea bottom to dislodge hazardous projections or to increase channel depth.

submarine cable *Electricity.* a long-distance, coaxial, lead-covered cable that is laid along the sea floor with submerged repeaters at various intervals and used to amplify signals; in shallow water, the cable may be armored or buried in the seabed to mitigate dangers from trawler anchors; in deeper water, lightweight cables without armoring are used with a central core of high steel.

submarine canyon *Geology.* a steep-sided, winding, V-shaped trench or valley on the deep-sea floor, which cuts into the continental shelf or continental slope. Also, **submarine valley.**

submarine delta see SUBMARINE FAN.

submarine earthquake see SEAQUAKE.

submarine false target *Ordnance.* a pyrotechnic decoy fired by a submarine to confuse underwater echo-ranging equipment and create a deceptive surface wake.

submarine fan *Geology.* a cone-shaped or fan-shaped mass of sediment, occurring seaward of a large river or at the lower end of a submarine canyon. Also, SEA FAN, SUBSEA APRON.

submarine gate *Engineering.* an edge gate with an opening from the runner located below the printing line or mold surface.

submarine geology see GEOLOGICAL OCEANOGRAPHY.

submarine isthmus *Geology.* an elevated feature of the ocean floor that connects two land areas and divides two basins or depressions.

submarine mine *Mining Engineering.* a mine working for removing minerals or ores from the sea.

submarine navigation *Navigation.* **1.** navigation of a submarine on or below the surface of the water. **2.** see UNDERWATER NAVIGATION.

submarine oscillator *Acoustical Engineering.* a powerful diaphragm horn used to produce intense sound signals underwater.

submarine peninsula *Geology.* an elevated area of the ocean floor having a form similar to that of a peninsula.

submarine pipeline *Engineering.* a pipeline laid along a sea or lake bottom, commonly used to transport petroleum or natural gas products.

submarine pit *Geology.* a small hole on the sea bottom. Also, **submarine well.**

submarine relief *Geology.* the relative elevations of the ocean floor, or a graphic representation of those elevations.

submarine rescue ship *Naval Architecture.* a ship, designed for underwater rescue and retrieval operations, that is fitted with hoists and flotation devices and is capable of supplying air and materials to the crew of a submerged vessel.

submarine rescue ship

submarine ridge *Geology.* a long, steep elevation of the deep-sea floor.

submarine rocket *Ordnance.* an underwater, submarine-launched surface-to-surface rocket with a nuclear depth charge or homing torpedo payload; it is designed primarily as an antisubmarine weapon; officially designated **UUM-44A.**

submarine sentry *Engineering.* a kitelike device towed underwater to test seabed clearances; it rises to the surface if it strikes an obstruction.

submarine signal *Navigation.* a navigational aid for use in poor visibility; an electric or electronic oscillator attached to the underside of a buoy or light vessel emits an underwater signal that is picked up by receivers located on either side of the bow of a ship; as the ship approaches the buoy, it adjusts back and forth depending upon which side of the bow receives the stronger signal.

submarine spring *Hydrology.* a freshwater spring that emerges off the seashore.

submarine striking force *Military Science.* a force of missile-carrying submarines joined for a specific offensive mission, especially a strategic nuclear mission.

submarine topography *Geology.* the arrangement and form of an underwater surface, such as the sea bottom or any surface existing within the water mass.

submarine wave recorder *Engineering.* a sonar-type device mounted on the upper side of a submarine and used when submerged to obtain a surface echo in order to measure the changing water wave level above the submarine.

submarine weathering *Geology.* a slow process by which chemical, thermal, and biological activity causes alteration of the shape, texture, and constituents of the sea floor. *Geochemistry.* see HALMYROLOSIS.

submaxillary *Anatomy.* situated beneath the maxilla.

submaxillary gland see SUBMANDIBULAR GLAND.

submeander *Hydrology.* a meandering spring contained within the banks of the main channel.

submerged arc-welding *Metallurgy.* an arc-welding process in which fusion is achieved by heating with an electric arc located between a bare metal electrode and the work.

submerged breakwater *Oceanography.* a breakwater whose top is below the still-water level; it reflects part of a wave's energy seaward and largely dissipates the rest of the wave's energy in a breaker, a multiple crest system, or a simple wave system.

submerged coastal plain *Geology.* a section of the continental shelf that represents the seaward continuation of a coastal plain on the land.

submerged-combustion *Engineering.* describing a device in which a gas burner submerged within a liquid serves as a heat source. Thus, **submerged-combustion evaporator, submerged-combustion heater.**

submerged culture *Microbiology.* a liquid culture medium of aerobic bacteria in which the bacteria grow away from the surface of the liquid.

submerged fermentation *Microbiology.* an industrial method of producing such substances as antibiotics and enzymes by growing submerged cultures.

submerged-foil hydrofoil *Naval Architecture.* a hydrofoil vessel with a foil that is entirely submerged below the surface during hydrofoil operation, and maintained stable by an autopilot system.

submerged land *Geology.* **1.** an area that is underwater during any stage of the tide, especially as distinguished from land that is submerged at high tide and exposed at low tide. **2.** an area that is at the bottom of a lake or that is covered by water when the lake is at its mean high level.

submerged screw log *Naval Architecture.* an instrument for measuring a vessel's speed or distance traveled, consisting of a small underwater propeller (screw) which is turned by the force of water rushing past due to the vessel's movement.

submerged shoreline *Geology.* **1.** a former shoreline that is now underwater, often identifiable at shallow depths. **2.** see SHORELINE OF SUBMERGENCE.

submerged weir *Civil Engineering.* a weir in which the tail water level is higher than the crest water level.

submergence *Geology.* a change in the levels of water in relation to the land, so that formerly dry land becomes inundated due to sinking of the land or a rise in water level.

submersible *Mechanical Engineering.* able to be submerged; able to operate effectively while underwater.

submersible pump *Mechanical Engineering.* a centrifugal pump unit, usually driven by electricity and made of corrosion-resistant bronze and stainless steel with sealed motor stator windings, that can be operated underwater.

submersion *Mathematics.* **1.** a mapping $f: X \to Y$, where X and Y are topological spaces, such that for every $x \in X$ there exists a neighborhood N of x on which f is a homeomorphism. **2.** in general, if X^q is a differentiable manifold, and $f: X^q \to Y^n$ is a differentiable mapping of rank n for every point $x \in X^q$ (so that $q \geq n$), then f is called a submersion. (If f is of rank q for every point $x \in X^q$, then f is called an **immersion.** In this case, it follows that $q \leq n$.)

submetacentric *Genetics.* describing a chromosome whose centromere is located near, but not directly on, one of the ends.

submetallic *Optics.* a luster that is less than metallic yet not nonmetallic.

submillimeter wave *Electromagnetism.* an electromagnetic wave having wavelengths less than one millimeter (frequencies greater than 300 gigahertz).

subminiature tube *Electronics.* any small electron tube generally used in miniaturized equipment, typically about 4 centimeters long and 1 centimeter in diameter.

submission see APPEASEMENT BEHAVIOR.

submissive behavior see APPEASEMENT BEHAVIOR.

submodular *Mathematics.* a function f on subsets of a set S is submodular if

$$f(A \cup B) + f(A \cap B) \leq f(A) + f(B)$$

for all subsets A and B of S. The rank of a matroid is a submodular function.

submucosa *Histology.* the layer of connective tissue lying between the mucosa and muscularis layers in the intestinal tract and other organs.

submucosal plexus or **submucous plexus** *Anatomy.* a plexus of autonomic nerve fibers located in the submucosa of the intestine. Also, MEISSNER'S PLEXUS.

submultiple *Mathematics.* a number or quantity that divides another number or quantity without remainder.

submultiple resonance *Physics.* a resonance condition in which the resonant frequency is an integral submultiple of the driving frequency.

submunition *Ordnance.* any munition that separates from a parent munition in order to perform its task; there are usually several submunitions carried and released by the parent munition. Similarly, **submissile.**

subnarcotic *Pharmacology.* describing or referring to a drug or agent that is mildly narcotic.

subneural apparatus *Neurology.* the area of the muscle fiber beneath the axon terminal, marked by a series of folds of the sarcolemma lining the primary synaptic cleft.

subnormal *Science.* below normal.

subnormal series *Mathematics.* a subnormal series of a group G is a chain of subgroups $G = G_0 > G_1 > \cdots > G_n$ such that G_{i+1} is normal in G_i for $0 \leq i < n$. Sometimes called a **normal tower;** the quotient groups G_i/G_{i+1} are called the factors of the series; and the number of nonidentity factors (i.e., the number of strict inclusions) is called the length of the series. If, in addition, G_i is normal in G for all i, the series is said to be normal. A subnormal series is not necessarily a normal series.

subnuclear particle *Physics.* any elementary particle, including those that appear only as the result of a high-energy collision of other particles or nuclei.

subnucleon *Physics.* a hypothetical constituent of an elementary particle.

subnucleus *Neurology.* a smaller or secondary nucleus formed by the division of a larger nerve nucleus.

suboccluding ligature *Surgery.* a ligature that obstructs primary blood supply but does not damage portions that are capable of establishing capillary blood flow.

suboptimization *Engineering.* the process of fulfilling a particular objective that constitutes a portion of a larger project.

suborbital *Space Technology.* describing a spacecraft that does not achieve an altitude and velocity necessary for a full orbit. *Anatomy.* situated below the orbit of the eye.

suborder *Systematics.* a taxonomic rank immediately below an order.

subordinate *Behavior.* **1.** being in a position of lesser status in a group; not dominant. **2.** an animal having such a status.

subordinate station *Oceanography.* a place for which tide or tidal current predictions are made by applying a correction to the predictions for a reference station.

subordination *Behavior.* a situation in which an animal consistently defers or gives way to another of higher status in a group in terms of access to food, space, or mates. Also, **subordinacy.**

Suboscines *Vertebrate Zoology.* in former classification systems, a name used to designate four suborders of passeriform birds: the Eurylaimi, Furnarii, Tyranii, and Menurae.

subpellicular tubules *Invertebrate Zoology.* in some protozoans, tiny tubular structures just beneath the pellicle.

subphrenic *Anatomy.* situated under the diaphragm.

subphylum *Systematics.* a zoological taxonomic rank immediately below phylum.

subpolar anticyclone see SUBPOLAR HIGH.

subpolar glacier *Hydrology.* a glacier having a temperature below freezing throughout most of its mass, but on which some surface melting occurs in the summer.

subpolar high *Meteorology.* a high-pressure system that forms over a cold continental land mass in a subpolar latitude. Also, POLAR ANTICYCLONE, POLAR HIGH, SUBPOLAR ANTICYCLONE.

subpolar low-pressure belt *Meteorology.* a low-pressure region that extends, in the mean, between 50° and 70° of latitude.

subpopulation *Statistics.* a subgroup of a population, usually identified on the basis of a common feature. Also, STRATUM.

subpotency *Genetics.* a reduced capacity to transmit inherited characteristics.

subpress see DIE SET.

subproblem *Artificial Intelligence.* a smaller problem that needs to be solved as part of a larger problem.

subprogram *Computer Programming.* a routine that performs a specific function and is part of a larger program.

subpubic *Anatomy.* situated below the pubic arch.

subpulmonary *Anatomy.* situated below the lung.

subpulse *Astronomy.* the weaker component in the pulse of a pulsar, measured relative to the whole of its emission cycle.

SubQ subcutaneous.

subrefraction *Electromagnetism.* refraction in which radar waves are refracted upward by the atmosphere due to humidity gradients or temperature gradients.

SUBROC *Ordnance.* a rocket containing a nuclear depth charge; it can be launched underwater from a submarine torpedo tube. (An acronym for submarine rocket.)

subroutine *Computer Programming.* a subprogram that is called by name from another program or subroutine and may have arguments passed to it in the call.

subroutine library *Computer Programming.* a collection of subroutines, written for general application, that can be incorporated into different programs when needed. Also, ROUTINE LIBRARY.

subsatellite *Space Technology.* a satellite carried into orbit by a spacecraft, then ejected.

subscapular *Anatomy.* located beneath or on the deep surface of the scapula, as a muscle or artery.

subscapularis *Anatomy.* a large muscle that extends from the subscapular fossa of the scapula to the humerus and rotates the humerus.

subschema *Computer Programming.* a description of a subset of a database that includes only those elements required by a particular program or user.

subscriber line *Electricity.* the telecommunications link that connects a subscriber's telephone station or branch exchange to a central office. Also, **subscriber loop.**

subscript *Graphic Arts.* a number or letter that is written slightly below, and usually to the immediate right or left of, another character, such as the 2 in H_2O.

subsea apron see SUBMARINE FAN.

subsection *Systematics.* a botanical taxonomic rank immediately below section.

subsequence *Mathematics.* let (b_n) be a given sequence and $(n(k))$ a strictly increasing sequence of natural numbers. Then $(b_n(k))$ is called a subsequence of (b_n). Alternatively, a subsequence of a sequence (b_n) is an ordered subset of the (ordered) set $\{b_n\}$, the elements of which appear in the same order as in $\{b_n\}$.

subsequent *Geology.* describing any geologic or topographic feature that developed after a consequent feature of which it is a part.

subsequent drainage *Hydrology.* a drainage system that developed after the system of which it is a part.

subsequent stream *Hydrology.* a tributary formed after the original stream into which it flows, having a valley that developed along a belt of underlying weak rock. Also, LONGITUDINAL STREAM, STRIKE STREAM.

subsequent valley *Geology.* a valley that has been worn away by a subsequent stream, or that contains such a stream.

subsere *Ecology*. **1.** a temporary plant community that springs up in a denuded area. **2.** a stage of plant colonization halted by soil conditions or interference from other organisms.

subseries *Systematics*. a botanical taxonomic rank immediately below a series.

subset *Mathematics*. if every element of a set X is also an element of a set Y, then X is said to be a subset of Y, or X is said to be contained in Y; denoted by $X \subseteq Y$ or $Y \supseteq X$. If, in addition, there exists at least one element of Y that is not in X, then X is said to be a **proper subset** of Y; denoted by $X \subset Y$ or $Y \supset X$. In particular, $X = Y$ means that $X \subseteq Y$ and $Y \subseteq X$.

subshell *Atomic Physics*. the electrons that occupy the same energy level or shell in an atom and have the same azimuthal quantum number. Also, SUBLEVEL.

subshrub *Botany*. a small shrub with only partially woody stems.

subsidence *Geology*. **1.** a local mass movement in which a portion of the earth's surface gradually settles or is displaced vertically downward, with little or no horizontal movement. **2.** a local or regional sinking of the ground with respect to its surroundings. *Mining Engineering*. specifically, the sinking of strata, including the surface, as a result of underground excavations. *Meteorology*. a descent of air in the atmosphere, usually over a wide area.

subsidence break *Mining Engineering*. a rock fracture overlying a coal seam or mineral deposit, due to mining operations.

subsidence inversion *Meteorology*. a temperature inversion caused by the adiabatic warming of subsiding air.

subsidiary conduit *Civil Engineering*. an underground conduit that runs from a building to a manhole and is the ending run.

subsidiary station *Cartography*. in surveying, a station established only as a means to avoid some natural impediment to the progress of the survey, rather than to determine the position of that station itself.

sub-sieve analysis *Metallurgy*. in powder metallurgy, the size distribution of particles that will pass through a 44-μm standard sieve.

sub-sieve fraction *Metallurgy*. in powder metallurgy, the particles that will pass through a 44-μm standard sieve.

subsistence farming *Agriculture*. the production of crops and livestock primarily for one's own use, rather than for sale.

subsistence stages *Anthropology*. an earlier theory of cultural evolution maintaining that culture progresses through three distinct phases of subsistence, from a hunting-and-gathering base to an agricultural base to an industrial base.

subsoil *Agronomy*. the part of the soil that lies below the topsoil. Also, UNDERSOIL.

subsoiler *Agriculture*. a device for breaking up the subsoil, consisting of a long pointed shank that is attached to a tractor.

subsoil plow *Agriculture*. a plow with chisel-like tines that penetrate and break up hard soils beneath the surface of the ground.

subsolar point *Astronomy*. the point on the earth or a planet at which the sun appears in the zenith.

subsolvus *Physical Chemistry*. a substance that forms more than one solid phase from an original homogeneous phase.

subsonic *Physics*. relating to aerodynamic velocities that are less than the speed of sound. *Aviation*. relating to flights at speeds below the speed of sound or flight vehicles that operate at such speeds. *Acoustics*. relating to sound waves having frequencies below the normal range of human hearing (about 20 hertz).

subsonic flight *Aviation*. the movement of a flight vehicle through the atmosphere at subsonic speed.

subsonic flow *Fluid Mechanics*. a flow in which the particle velocity in the fluid never exceeds that of sound in the same fluid, i.e., $M < 1$. Also, SUBCRITICAL FLOW.

subsonic inlet *Engineering*. an inlet designed to admit fluid flowing at less than the speed of sound in the fluid.

subsonic speed *Fluid Mechanics*. a speed that is less than the speed of sound in the same fluid medium under the same conditions.

subspace *Mathematics*. a subset S of a vector space V such that S is a vector space with the same addition, division ring of scalars, and scalar multiplication as V.

subspecies *Systematics*. a taxonomic rank immediately below species, indicating a group of organisms that is geographically isolated from and may display some morphological differences from other populations of a species, but is nevertheless able to interbreed with other such groups within the species where their ranges overlap.

substance *Physics*. any physical entity; matter, commonly homogeneous, that occurs in macroscopic amounts.

substance abuse *Pathology*. **1.** a pathological, long-term misuse of alcohol or drugs, characterized by daily intoxication and an inability to reduce intake, usually resulting in diminished functioning. **2.** more generally, any form of alcohol or drug addiction or misuse.

substance P *Biochemistry*. $C_{63}H_{98}N_{18}O_{13}S$, a neurotransmitter composed of amino acids; a member of the tachykinin family of peptide hormones that is present in nerve cells and in certain endocrine cells and that induces contraction of intestinal smooth muscle and vasodilation; in the central nervous system, it acts as a neurotransmitter in the pain pathway.

substandard propagation *Electromagnetism*. the propagation of radio energy under conditions of substandard refraction in the atmosphere.

substantia *Anatomy*. a nucleus of neurons in the ventral portion of the midbrain, thought to contribute to regulating muscular movements.

substantial runoff see BASE RUNOFF.

substantive dye *Organic Chemistry*. a direct dye that is applied by immersing the fiber or cloth in a hot solution of the dye in water.

substellar point *Astronomy*. the point on the earth at which a given star stands in the zenith.

substernal *Anatomy*. situated beneath the sternum.

substituent *Organic Chemistry*. any atom or radical that replaces another in a molecule as the result of a reaction.

substitute *Chemistry*. to replace one or more elements or groups in a compound with other elements or groups. *Behavior*. see SURROGATE.

substitute behavior see DISPLACEMENT BEHAVIOR.

substitute center *Cartography*. in aerial photography, a point selected for use as a radial center because it is easier to identify on overlapping photographs than the principal point.

substitute mode *Computer Programming*. a method of exchange in which an element of information is replaced by some other element of information.

substitute object *Behavior*. any object to which a behavior pattern is directed in the absence of the usual focus of the behavior.

substitution *Mathematics*. **1.** the process of replacing a variable with a fixed value, new variable, or expression known or presumed to be equivalent to the original variable. **2.** a former term for permutation. *Chemistry*. a chemical reaction in which an element, atom, or group of atoms is replaced in a compound by an equivalent element, atom or group. *Artificial Intelligence*. **1.** the creation of an instance of a formula or pattern in which particular values are substituted for variables in the pattern. **2.** a set of variables and the corresponding values to be substituted for them.

substitutional defect *Materials Science*. a point defect produced when a solute atom is substituted in a site regularly occupied by a solvent one of regular lattice.

substitutional impurity *Solid-State Physics*. an impurity atom in a semiconductor that occupies the site of a host atom in the lattice.

substitution mutation *Genetics*. a mutation that involves the replacement of either a single nucleotide or codon in a DNA sequence, or a single amino acid in a protein with a different unit of the same class.

substitution reaction *Chemistry*. a reaction in which one or more atoms replace another atom or group of atoms in a molecule.

substitution solid solution *Metallurgy*. a solid alloy, having random distribution, in which the solute atoms are found at some of the lattice points of the solvent.

substitution transfusion see EXCHANGE TRANSFUSION.

substitution vein see REPLACEMENT VEIN.

substitutive nomenclature *Organic Chemistry*. a system of naming organic compounds that uses the substituent as a prefix or suffix added to the parent compound; for example, in 1-bromobutane, a bromine has replaced a hydrogen atom on the terminal carbon of the butane chain.

substrate *Engineering*. the structural surface beneath paint or other coverings. *Graphic Arts*. the paper or other surface upon which something is printed. *Electronics*. the support material on which an integrated circuit is constructed or to which it is attached. *Organic Chemistry*. a compound that reacts with a reagent. *Biochemistry*. the reactant in any enzyme-catalyzed reaction.

substrate-accelerated death *Microbiology*. a phenomenon in which the death rate of a population of starving bacteria is increased upon addition of the last substrate encountered prior to their starvation.

substrate-level phosphorylation *Biochemistry*. the conversion of ADP to ATP, or the conversion of other nucleoside diphosphates, by a cytoplasmic process involving the concurrent hydrolysis of another high energy phosphate compound.

substrate mycelium *Mycology.* hyphal masses or mycelium that grow on the surface to which they are attached. Also, VEGETATIVE MYCELIUM.
substratosphere *Meteorology.* the atmospheric region that is of indefinite extent beneath the stratosphere.
substratum *Geology.* any solid layer occurring beneath the true soil or superficial deposits. *Agriculture.* another term for the subsoil. *Biology.* the base on which a nonmotile organism lives.
substring *Computer Programming.* a contiguous portion of a string.
substructure *Civil Engineering.* **1.** the part of any structure that is below ground. **2.** specifically, the foundation or piers of a bridge.
subsumed clause *Artificial Intelligence.* in predicate calculus, a clause that can be deleted because another stronger clause is present; e.g., $P(x)$ subsumes $P(x) \vee Q(x)$ because whenever $P(x)$ is true, then $P(x) \vee Q(x)$ must also be true.
subsurface contour see STRUCTURE CONTOUR.
subsurface current *Oceanography.* an ocean current flowing beneath the surface, usually below the main thermocline and often in a different direction from that of the surface currents.
subsurface flow see INTERFLOW.
subsurface geology *Geology.* a branch of geology concerned with the study and correlation of geologic features and processes occurring beneath the earth's surface. Also, UNDERGROUND GEOLOGY.
subsurface hydrology *Hydrology.* the branch of hydrology dealing with the study of underground water, as opposed to surface water.
subsurface ice see GROUND ICE.
subsurface tillage *Agriculture.* a form of plowing in which underground blades stir the soil, but leave surface vegetation in place. Also, **subsurface till.**
subsurface waste disposal *Engineering.* a method of disposing of wastes by pumping them into porous rock and certain other formations.
subsurface wave *Electromagnetism.* an electromagnetic wave that propagates through land or water rather than on the earth's surface; attenuation is significant for such waves above about 35 kilohertz.
subsynchronous *Electricity.* having a frequency that is a submultiple of the driving frequency.
subsystem *Science.* a portion of a system that can be treated as a single element in the main system, but that can also be considered a distinct system itself.
subtend *Botany.* of a plant or plant part, to grow beneath and close to another plant or part, often embracing or enclosing it. *Mathematics.* to extend under, stretch across, or be opposite to, as a chord and an arc.
subtense bar *Engineering.* in surveying, a horizontal bar used for distance measurement by determining the arc it subtends as viewed from the measuring point.
subtense technique *Civil Engineering.* a technique used in surveying to measure distances, in which a transit angle subtended by the subtense bar permits the calculation of the distance between the transit and bar.
subterranean [sub´tə rā´nē ən] *Geology.* relating to or located in the region beneath the surface of the earth, especially the land surface; underground.
subterranean ice see GROUND ICE.
subterranean stream *Hydrology.* a stream flowing beneath the surface through a cave or group of connected caves.
subterranean water see GROUNDWATER.
subtilin *Microbiology.* an antibiotic that is produced by the soil bacterium *Bacillus subtilis*; effective against organisms such as Gram-positive bacteria and the tubercle bacillus.
subtilisin *Biochemistry.* an extracellular protease produced by bacteria of the genus *Bacillus* (such as *B. amyloliquefaciens* or *B. subtilis*) that digest protein molecules; used as an active agent in detergents and in protein-structure research. Also, NAGAROSE.
subtract *Mathematics.* to carry on a process of subtraction; remove one number or quantity from another.
subtracter *Computer Technology.* a circuit that performs arithmetic subtraction operations on binary numbers.
subtraction *Mathematics.* the process by which, for two given numbers or quantities, one is removed from the other to give a third; the operation that is inverse to addition. It is a noncommutative binary operation denoted by the minus sign "–"; $a - b = a + (-b)$, where $-b$ is the additive inverse of b. "+" usually refers to the operation of an Abelian group or the group operation in a ring, module, or vector space over which the other operations distribute.
subtractive primaries *Optics.* the colors that are complementary to red, blue, and green in subtractive printing processes; respectively, cyan, yellow, and magenta.
subtractive process *Optics.* the process that reproduces colors by means of transparent inks or dyes in their complementary colors; widely used in photography and printing.
subtractor *Electronics.* an operational amplifier circuit whose output is determined by the difference between the input signals.
subtrahend *Mathematics.* a quantity that is to be subtracted from another quantity (the minuend).
subtribe *Systematics.* a taxonomic rank immediately below a tribe.
subtropical *Meteorology.* **1.** of or relating to the subtropics. **2.** bordering on the tropics; nearly tropical.
subtropical anticyclone see SUBTROPICAL HIGH.
Subtropical Convergence *Oceanography.* the boundary zones between warm subtropical and cold subpolar waters where some ocean currents tend to converge; the term generally refers to the well-defined southern convergence zone around 40°S, but it may also refer to the ill-defined northern zone between 20°N and 28°N.
subtropical cyclone *Meteorology.* a low-level atmospheric graphical representation of a cutoff low on a surface chart.
subtropical easterlies see TROPICAL EASTERLIES.
subtropical easterlies index *Meteorology.* a scale used to measure the strength of easterly winds in the northern subtropics; calculated from the average sea-level pressure difference between 20°N and 35°N, and expressed in meters per second as the east-west component of the corresponding geostrophic wind.
subtropical high *Meteorology.* any of the semipermanent high-pressure systems that lie over the oceans of the subtropical high-pressure belt and are best developed during the summer season. Also, SUBTROPICAL ANTICYCLONE, OCEANIC ANTICYCLONE, OCEANIC HIGH.
subtropical high-pressure belt *Meteorology.* either of two belts of high atmospheric pressure that are formed by subtropical highs and centered, in the mean, near 30° latitude N and S. Also, **subtropical ridge.**
subtropical westerlies see WESTERLIES.
subtropics *Meteorology.* either of two latitudinal belts between the tropics and the temperate regions, roughly between 35° and 40° in latitude but varying in width according to continental influences.
subulate *Botany.* of or related to a leaf that is slender, cylindrical, and tapering from base to a fine-pointed apex.
Subulitacea *Paleontology.* a large superfamily of Paleozoic gastropods belonging to the subclass Prosobranchia of the order Mesogastropoda; generally characterized by a spindle or awl shape, usually 3–4 centimeters high; extant from the Ordovician to the Permian.
Subuluridea see HETERAKIDAE.
subumbrella *Invertebrate Zoology.* the concave undersurface of a coelenterate medusa.
suburb *Civil Engineering.* a smaller community, usually mainly residential, that lies adjacent to or at the edge of a major city or town.
subvariety *Systematics.* a botanical taxonomic rank immediately below species or subspecies.
subversion *Military Science.* action designed to weaken the military, economic, psychological, or political strength of a nation, especially by undermining the morale, loyalty, or reliability of its people.
subvert *Military Science.* to carry on an operation of subversion.
subviral agent *Microbiology.* any viroid or other acellular infectious entity that lacks at least one essential feature of a virus.
subviral particle *Virology.* an intermediate form of virus particle, lacking in certain components, that is formed during the replication cycle.
subvolution *Surgery.* the turning up of a flap of mucous membrane so that the cutaneous surface comes in contact with the raw surface of dissection to prevent adhesions, especially for a pterygium.
subwaking *Neurology.* a state of awareness between sleep and complete consciousness.
subway *Civil Engineering.* **1.** an underground rail transport system. **2.** a train that is part of such a system. **3.** especially in British use, an undergound passage, as beneath an urban street.
subway-type transformer *Electricity.* a transformer that is constructed so that it can be installed in an underground vault.
Sucaryl *Organic Chemistry.* a trade name for sodium cyclamate, an artificial sweetener formerly in wide use.
succession *Ecology.* **1.** the progressive replacement, on a single site, of one type of community by another. **2.** the process by which change takes place in a community, which can be induced by environmental factors or by the species' own intrinsic characteristics.
succession cropping *Agronomy.* the growing of two or more crops one after the other on the same land in one growing season.
successive approximation *Behavior.* see SHAPING.

successive-approximation converter *Computer Programming.* an analog-to-digital converter that successively considers each bit position, setting the bit equal to 0 or 1 on the basis of the output of a comparator.

successive-fracture treatment *Petroleum Engineering.* the use of a series of fracturing operations in a new zone of an oil reservoir.

successor *Mathematics.* **1.** given a set x, the set x^+ obtained by adjoining x to the elements of x; i.e., $x^+ = x \cup \{x\}$. This is used in constructing the natural numbers, as follows: let $0 = \varnothing$ (the empty set); $1 = 0^+ = \{0\}$; $2 = 1^+ = \{0, 1\}$; $3 = 2^+ = \{0, 1, 2\}$; and so on. **2.** the successor of a vertex v in a directed graph is any vertex u such that there exists an arc from v to u. If u is a successor of v, then v is called the **predecessor** of u. **3.** let a be an element of a linearly ordered set A with ordering $\leq$. The immediate successor of a, if it does exist, is the least element in the set $\{x \in A : a \leq x \text{ and } a \neq x\}$. If A is well-ordered by $\leq$, then at most one element of A has no immediate successor.

successor job *Computer Programming.* a task that requires the results of a predecessor job as input before it can begin execution.

successor set *Mathematics.* a set A such that: (a) $\varnothing$ (the empty set) is an element of A, and (b) if $x \in A$, then the successor (x^+) of $x \in A$. By the axiom of infinity, such a set A exists. The intersection of every nonempty family of successor sets is also a successor set, and thus it makes sense to consider the intersection of all successor sets of the set A. It can be shown that this set, denoted ω, can be shown to be the only successor set that is included in every other successor set; that is, ω is the minimal successor set. The elements of ω are called natural numbers.

succinamide *Biochemistry.* $C_4H_8N_2O_2$, an amide of succinic acid that is used primarily in the synthesis of organics.

succinate *Organic Chemistry.* a salt or ester of succinic acid.

succinic *Organic Chemistry.* relating to or derived from amber.

succinic acid *Organic Chemistry.* $HOOCCH_2CH_2COOH$, combustible, colorless crystals that are slightly soluble in water and soluble in alcohol and ether; melts at 187–189°C and boils at 235°C with partial conversion into the anhydride; used in organic synthesis and the manufacture of pharmaceuticals, dyes, and perfumes. Also, BUTANEDIOIC ACID.

succinic acid dehydrogenase *Enzymology.* an enzyme that catalyzes the dehydrogenation of succinic acid to fumaric acid.

succinic acid 2,2-dimethylhydrazide *Organic Chemistry.* $(CH_3)_2NNHCOCH_2CH_2COOH$, white crystals that are soluble in water and insoluble in simple hydrocarbons; melts at 154°C; used as a growth regulator for many crops. Also, DIAMINOZIDE.

succinic anhydride *Organic Chemistry.* $(CH_2CO)_2O$, colorless or light-colored needles or flakes; insoluble in water and soluble in alcohol; melts at 120°C, boils at 261°C, and sublimes at 115°C (5 torr); used in the manufacture of chemicals and pharmaceuticals. Also, BUTANEDIOIC ANHYDRIDE.

succinimide *Organic Chemistry.* $(CH_2CO)_2NH$, colorless crystals or thin, light-tan flakes; soluble in water, slightly soluble in alcohol, and insoluble in ether; melts at 126°C and boils at 287–289°C; used in organic synthesis and plant growth stimulators.

Succinimonas *Bacteriology.* a genus of Gram-negative, anaerobic, motile bacteria of the family Bacteroidaceae occurring as rod-shaped cells with rounded ends and polar flagella; found in the rumen of cattle.

succinite *Mineralogy.* **1.** a former name for amber, especially that from the Baltic Sea area. **2.** a pale yellow, amber-colored variety of the mineral grossular, of the garnet group.

Succinivibrio *Bacteriology.* a genus of Gram-negative, anaerobic, motile bacteria belonging to the family Bacteroidaceae, occurring as curved, rod-shaped cells with polar flagella, characterized by the fermentative production of succinic acid; found in the rumen of cattle and sheep.

succinoxidase *Enzymology.* an enzyme system, involving succinic dehydrogenase and cytochromes, that catalyzes the conversion of succinate to fumarate.

succinylcholine chloride *Biochemistry.* $C_{14}H_{30}Cl_2N_2O_4$, a crystalline solid that is soluble in water and melts at 162°C; a basic compound with curarelike actions that is used intravenously as a powerful muscle relaxant, e.g., during procedures such as endotracheal intubation or as an adjunct to surgical anesthesia.

succinyl CoA *Enzymology.* a high-energy intermediate, formed in the tricarboxylic acid cycle by the oxidation of α-ketoglutaric acid, that undergoes deacylation to form succinic acid.

succinyl CoA synthetase *Enzymology.* an enzyme that catalyzes the formation of succinyl CoA.

succinylsulfathiazole *Pharmacology.* $C_{13}H_{13}N_3O_5S_2$, an antibacterial drug administered orally to treat gastrointestinal infections and given prior to gastrointestinal surgery. Also, SULFASUXIDINE.

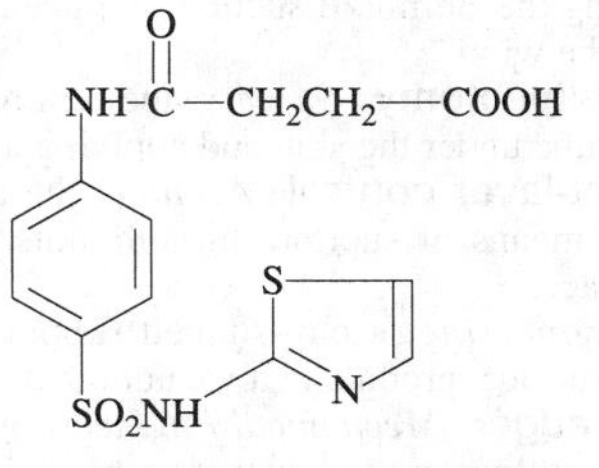

succinylsulfathiazole

succubous *Botany.* describing leaves that overlap, with the base of each leaf covering part of the leaf under it.

succulent [suk′yə lənt] *Botany.* **1.** a plant, such as a cactus, with soft, thick tissues that store water; such tissues have a juicy, fleshy appearance. **2.** of, relating to, or resembling such a plant.

succus entericus *Physiology.* intestinal fluid; the alkaline product of numerous glands in the lining of the small intestine.

suchovei see SUKHOVEI.

sucker *Botany.* **1.** a shoot that develops from a root or a subterranean stem to form an independent plant. **2.** see HAUSTORIUM, def. 1. *Zoology.* any of various organs or parts that carry on a sucking action or that adhere to a surface by suction. *Vertebrate Zoology.* any of several freshwater food fishes of the Catostomidae family that suck in food or have thick, fleshy lips that suggest sucking; found in North America.

suckerfish see REMORA.

sucker rod *Petroleum Engineering.* one of a string of steel rods that connect a downhole oil-well pump to a pumping jack on the surface.

sucker-rod pump *Petroleum Engineering.* a pump with a cylinder-piston arrangement, used to drive oil into the tubing of a well and subsequently to the surface.

sucking louse *Invertebrate Zoology.* the common name for the Anoplura, a group of parasitic insects characterized by piercing mouthparts adapted to suck the fluids of the host, a flattened body, and legs with strong claws for clinging to the host.

sucrase see INVERTASE.

sucrose [soo′krōs] *Organic Chemistry.* $C_{12}H_{22}O_{11}$, table sugar, a substance in the form of combustible, hard, dry, white crystals; soluble in water and slightly soluble in alcohol; decomposes at 160–186°C; produced in most plants and harvested from sugarcane and sugar beets; used as a sweetener in foods and beverages, and as an intermediate and emulsifying agent. Also, SACCHAROSE. *Enzymology.* an enzyme that catalyzes the hydrolysis of the disaccharide sucrose to the monosaccharides glucose and fructose. (From the French word for sugar.)

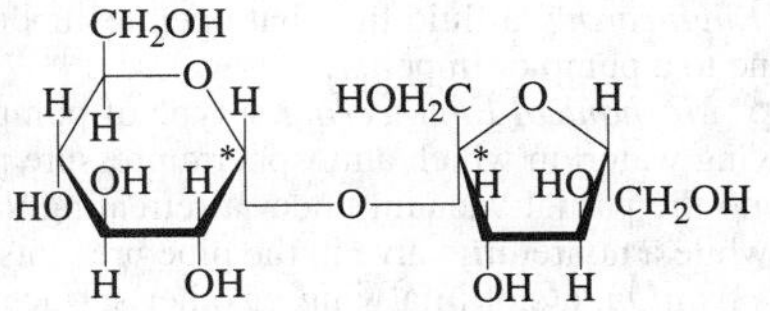

sucrose

sucrose gradient centrifugation *Biotechnology.* a technique that separates biological components from cellular homogenates by centrifugation with a density gradient of sucrose.

sucrose octoacetate *Organic Chemistry.* $C_{12}H_{14}O_3(OOCCH_3)_8$, a combustible, hygroscopic, intensely bitter, white crystalline solid; slightly soluble in water; melts at 86°C, boils at 260°C (1 torr), and decomposes at 285°C; used as a plasticizer, adhesive, insecticide, and denaturant.

sucrosic see SACCHAROIDAL.

suction *Fluid Mechanics.* **1.** a force that produces a condition of reduced pressure in a given region, so that a fluid in this region will flow to an adjacent region of greater pressure, or a rigid body will be held in place on a surface. **2.** the movement of a fluid or adhering of a body to a surface produced by such a condition. *Surgery.* specifically, the extraction of a gas or liquid from a wound or cavity by mechanical aspiration.

suction and curettage *Surgery.* a process in which a fetus is extracted through a suction tube; used to perform abortions during the early stages of pregnancy.

suction anemometer *Engineering.* an instrument that measures wind speed by measuring the Bernoulli suction produced on a closed tube set at right angles to the wind.

suction-assisted lipectomy *Surgery.* the removal of fatty tissue by placing a narrow tube under the skin and applying a vacuum.

suction boundary-layer control *Aviation.* the control of boundary-layer air flow by means of suction through slots or perforations in a wing or other surface.

suction cup *Engineering.* a cup-shaped rubber device that can be pressed against a surface, producing a vacuum that will hold it in place.

suction-cutter dredger *Mechanical Engineering.* a dredger in which rotary blades cut and loosen the material to be drawn up by suction for excavation.

suction dredge *Naval Architecture.* a dredge that uses a centrifugal pump to draw up loose seabed material.

suction hose *Engineering.* a flexible hose that sucks up pieces of material to deposit in another location.

suction hose

suction lift *Mechanical Engineering.* the inlet head required for a pump to raise the liquid from a supply well to the level of the pump. Also, **suction head.**

suction line *Engineering.* a fluid line that operates under suction, such as the input line to a pump or impeller.

suction pump *Mechanical Engineering.* a type of pump often used for raising or moving water, in which atmospheric pressure pushes the fluid to be raised into the partial vacuum under a retreating valved piston on the upstroke, while a nonreturn valve in the pipe prevents reflux.

suction slot *Aviation.* a slot in a wing or other surface through which boundary-layer air is removed by suction.

suction stroke *Mechanical Engineering.* a piston stroke that draws a fresh charge into the cylinder of an engine or the receiver tank of a compressor. Also, INDUCTION STROKE, INTAKE STROKE.

Suctoria *Invertebrate Zoology.* a class of highly evolved aquatic protozoans, composed of ciliated free-living immature stages and sessile adults, having long, piercing and sucking tentacles with which they catch and feed on prey.

suctorial *Zoology.* **1.** relating to or used for sucking. **2.** having a sucking organ or part.

suctorian *Invertebrate Zoology.* **1.** an organism that is a member of the class Suctoria. **2.** of or relating to this class.

sucupina *Materials.* the durable, coarse-textured wood of the Central and South American trees of the *Bowdichia* species; used for heavy construction and railroad crossties.

sudamen *plural,* **sudamina.** *Medicine.* **1.** a whitish vesicle caused by the retention of sweat in the sweat ducts or the layers of the skin. **2.** an eruption of these vesicles.

Sudan black B *Microbiology.* a blue-black biological dye that dissolves in and therefore stains cellular lipids.

sudburite *Geology.* a type of basalt containing augite, hypersthene, bytownite, and magnetite; it typically has a pillow structure.

sudden commencement *Geophysics.* a series of magnetic storms that begin rapidly and simultaneously all over the earth.

sudden infant death syndrome see CRIB DEATH.

sudden ionospheric disturbance *Geophysics.* a disturbance of the ionosphere that begins within a few minutes after the appearance of a solar flare, causing a disruption in long-distance, short-wave communications.

sudomotor *Physiology.* **1.** stimulating the sweat glands. **2.** of or relating to the nervous response of sweating.

sudoriferous *Physiology.* producing perspiration or sweat.

suede [swād] *Materials.* leather that is finished with a soft nap on the flesh side; used for shoes, gloves, and other such items. (From a French term meaning "of Sweden;" originally referring to gloves made in Sweden of this material.)

suede cloth *Textiles.* a woven or knitted fabric of cotton, wool, or rayon with a nap finished to resemble leather suede.

Suess, Eduard [zyoos] 1831–1914, Austrian geologist; an expert on structural geology; wrote a definitive treatise on the evolution of the earth's surface features.

suestada *Meteorology.* a strong, southeast wind that blows along the southeastern coast of South America in winter, generating heavy seas, fog, and rain.

suevite *Geology.* a yellowish or grayish fragmental rock composed of angular shock-metamorphosed fragments and glassy inclusions; it is generally associated with meteorite impact craters.

Suez Canal *Geography.* a canal in northeastern Egypt, connecting the Red Sea with the Mediterranean.

suffix stripping see MORPHOLOGICAL ANALYSIS.

Suffolk *Agriculture.* **1.** a breed of chestnut-colored draft horse. **2.** a breed of large mutton sheep, characterized by a black polled head and ears and black legs. (From *Suffolk,* a county in England.)

suffosion *Hydrology.* a process by which underground water from partially melted ground ice bursts through the upper surface, carrying with it a deposit of mud, sand, rock, or other such material.

suffrutescent *Botany.* of or relating to a plant having a base that is partially or slightly woody and that does not die down each year.

suffruticose *Botany.* of or relating to a plant that is woody at the base and herbaceous above.

sugar *Food Technology.* **1.** any of various carbohydrates of animal or vegetable origin that are aldehyde or ketone derivatives of polyhydric alcohols, and that are widely used to sweeten candy, ice cream and other desserts, soft drinks, and many other types of food. The two main groups of sugars are the disaccharides ($C_{12}H_{22}O_{11}$), such as sucrose or maltose, and the monosaccharides ($C_6H_{12}O_6$), such as dextrose and fructose; all sugars are white crystallizable solids that are soluble in water and dilute alcohol. **2.** any of various artificial substances used as substitutes for natural sugars.

sugar alcohol *Organic Chemistry.* any of several acyclic linear polyhydric alcohols, classified according to the number of hydroxyl groups in the molecule.

sugar beet *Botany.* a beet, *Beta vulgaris,* that has a white root and is cultivated for the sugar it produces.

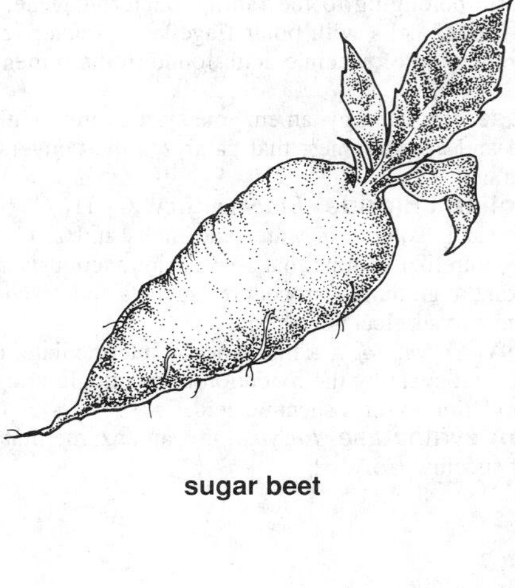

sugar beet

sugar berg *Oceanography.* an iceberg composed of glacier ice formed at very low temperatures; it is relatively porous and disintegrates rapidly.
sugar bloom *Food Technology.* a discoloration produced on chocolate surfaces by the crystallization of sugar in the product; caused by fluctutations in temperature and humidity during storage.
sugarcane *Botany.* a tall grassy plant, *Saccharum officinarum*, characterized by a stout, jointed stalk and cultivated for its sugar content; native to tropical and subtropical regions.
sugar iceberg *Hydrology.* an iceberg consisting of porous glacier ice formed at very low temperatures.
sugarloaf bullet see PICKET BULLET.
sugarloaf sea *Oceanography.* a type of wave that produces cross swell when it is superimposed on others. Also, PYRAMIDAL SEA, INTERSECTING WAVES.
sugar maple *Botany.* any of several maple species that have sweet sap, especially *Acer saccharum*, characterized by a short trunk with gray furrowed bark, long curving branches, and sharp-pointed scaly winter buds.

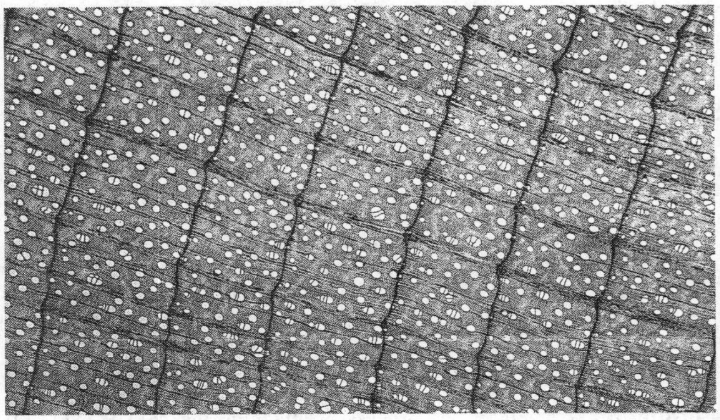

sugar maple

sugar of lead see LEAD ACETATE.
sugar pea see SNOW PEA.
sugar refining *Food Technology.* any process by which raw sugar is broken down into a usable form.
sugar snow see DEPTH HOAR.
suggestion *Psychology.* the process of influencing another individual to accept an idea without using direct persuasion and without the individual's evaluating the validity of the idea.
sugi *Materials.* the light, soft wood of the *Cryptomeria japonica* tree of Asia, used for construction, paneling, ceiling boards, and furniture.
Suhl amplifier *Solid-State Physics.* a parametric amplifier that operates on the instabilities of certain spin waves in a ferromagnetic substance subjected to an intense microwave field.
Suhl effect *Electronics.* a process by which the holes injected into an N-type semiconducting filament are deflected to the surface when a strong transverse magnetic field is applied to the filament; the holes may then recombine rapidly with electrons or be withdrawn by a probe.
suicide gene *Genetics.* any gene whose expression in a cell is lethal for that cell.
suicide inhibitor *Biochemistry.* a substrate that changes to such a degree that it uncharacteristically binds to an enzyme by generating a reactive group, which binds it covalently to a neighboring enzyme group.
suicide substrate *Cell Biology.* a substrate that is lethal for an enzyme, because its processing forms a covalent complex that inactivates the enzyme.
Suidae *Vertebrate Zoology.* a nonruminant family of the order Artiodactyla including swine, boars, pigs, and hogs; native to the Eastern Hemisphere but now found in domesticated, wild, or semiwild form in most temperate and tropical locations.
Suida process *Organic Chemistry.* a process that uses tar oil to extract acetic acid from the vapor produced by the destructive distillation of wood.
Suiformes *Vertebrate Zoology.* the swine, peccaries, and hippopotamuses, a nonruminant suborder of the order Artiodactyla.
Suipoxvirus *Virology.* a virus genus of the subfamily Chordopoxvirinae, family Poxviridae, that includes the swinepox virus.
suite *Computer Technology.* **1.** a group of several related programs run in sequence as an operational job. **2.** a set of connected hardware components that make up a complete system.
sukhovei *Meteorology.* a hot, dry, dusty wind that blows across the steppes of southern Russia, usually from the east, often causing crop damage and drought. (A Russian word meaning "dry wind.") Also, SUCHOVEI.
Sula *Vertebrate Zoology.* the boobies, a genus of tropical seabirds of the family Sulidae having webbed feet and a sharp, often brightly colored bill.

Sula (booby)

Sulawesi see CELEBES.
sulcal *Biology.* relating to or a part of a groove or depression.
sulcate *Biology.* **1.** grooved or channeled, as plant stems. **2.** furrowed or cleft, as horses' hooves.
sulculus *Anatomy.* a small sulcus.
sulcus *Anatomy.* a groove or furrow found in the body, especially between two convolutions of the brain. *Biology.* any groove or furrow.
sulf- *Chemistry.* a prefix used in naming chemical compounds, indicating the presence of divalent sulfur or of a sulfo group.
sulfa *Pharmacology.* a shorter term for sulfonamide or sulfanilamide.
sulfadiazine *Pharmacology.* $C_{10}H_{10}N_4O_2S$, a sulfonamide used to treat a variety of infections, including meningitis, chancroid, urinary tract infections, and the early stages of a certain type of chlamydia and of a type of malaria caused by organisms resistant to chloroquine.
sulfa drug see SULFONAMIDE.
sulfafurazole see SULFISOXAZOLE.
sulfaguanidine *Pharmacology.* $C_7H_{10}N_4O_2S$, a sulfonamide formerly used to treat gastrointestinal infections, especially bacillary dysentery.
sulfallate *Organic Chemistry.* $C_8H_{14}NS_2Cl$, an oily liquid; boils at 128–130°C; used as an herbicide.
sulfamate *Chemistry.* a salt of sulfamic acid.
sulfamerazine *Pharmacology.* $C_{11}H_{12}N_4O_2S$, a readily absorbed sulfonamide usually used in combination with other sulfonamide drugs to reduce renal toxicity.
sulfamic acid *Inorganic Chemistry.* NH_2SO_2OH, toxic, colorless, rhombic crystals; soluble in water and alcohol; melts at 200°C; decomposes at the boiling point; used in cleaning, dyeing, and bleaching operations and as a catalyst.
sulfanilamide [sul′fə nil′ə mīd] *Pharmacology.* $C_6H_8N_2O_2S$, a powerful sulfonamide that was formerly used to treat various bacterial infections; the first sulfonamide discovered and the basis for most other sulfonamides, it has been replaced by derivatives and antibiotics that are more effective and less toxic. Also, SULPHANILAMIDE.

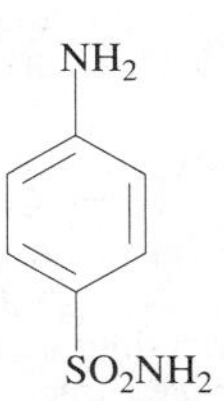

sulfanilamide

sulfanilic acid *Organic Chemistry.* $H_2NC_6H_4SO_3H{\cdot}H_2O$, combustible, grayish-white crystals; slightly soluble in water and very slightly soluble in alcohol and ether; melts (chars) at 280–300°C; used in medicine, organic syntheses, and making dyes, and as a reagent.

sulfapyridine *Pharmacology.* $C_{11}H_{11}N_3O_2S$, a sulfonamide administered orally to suppress dermatitis herpetiformis, a skin disease; it was formerly used to treat pneumonia and streptococcal infections.

Sulfasuxidine *Pharmacology.* the trade name for preparations of succinylsulfathiazole; used as a gastrointestinal antibacterial agent.

sulfatase *Enzymology.* an enzyme that catalyzes the hydrolysis of an organic sulfate to an alcohol and inorganic sulfate.

sulfate *Chemistry.* a salt of sulfuric acid; a compound containing the sulfate group, $(SO_4)^{-2}$, such as magnesium sulfate, $MgSO_4$, or calcium sulfate, $CaSO_4$. Also, SULPHATE.

sulfate mineral *Mineralogy.* any of a group of nonsilicate minerals whose structures are characterized by the presence of the anionic sulfate group $(SO_4)^{-2}$; generally colorless or white, soft, and earthy or massive; usually tabular when crystalline.

sulfate process *Materials Science.* a process for making wood pulp, in which wood chips are digested in an alkaline liquor consisting primarily of caustic soda and sodium sulfate; one of the chief methods of processing wood for paper pulp. Thus, **sulfate pulp, sulfate paper.**

sulfate pulping *Chemical Engineering.* a chemical procedure to carry out the sulfate process; i.e., to solubilize and remove the lignin portion of wood for paper pulp. Also, KRAFT PROCESS, KRAFT PULPING.

sulfathiazole *Pharmacology.* $C_9H_9N_3O_2S_2$, a sulfonamide once used widely as an antibacterial drug but now replaced by less toxic sulfonamides and antibiotics.

sulfatide *Biochemistry.* a sulfate ester of galactocerebroside that is composed of fatty acids with 22 to 26 carbon atoms; found in brain tissue and especially in myelin; it may accumulate in the white matter of the brain in metachromatic leukodystrophy. Also, SULPHATIDE.

sulfatide lipidosis *Pathology.* a disruption in the white matter of the brain characterized by an abundance or stockpiling of a sphingolipid (sulphatide) in neural and nonneural tissue with a subsequent loss of myelin in the central nervous system and resulting in blindness, mental deterioration, and eventual acute dementia. Also, METACHROMATIC LEUKODYSTROPHY.

sulfating *Electricity.* the accumulation of lead sulfate particles on the plates of a lead-storage battery, reducing the energy-storing ability of the battery and ultimately leading to its premature failure.

sulfation *Chemistry.* **1.** the formation of lead sulfate, $PbSO_4$, in a lead storage battery. **2.** the formation of sulfuric acid esters from alcohols or olefins.

sulfenic acid *Organic Chemistry.* an oxide acid of sulfur with the general formula *R*SOH.

sulfenyl chloride *Organic Chemistry.* a derivative of a sulfenic acid with the general formula *R*SCl.

sulfhemoglobin *Biochemistry.* an abnormal greenish form of hemoglobin containing sulfur that is bound to heme.

sulfhydrate *Chemistry.* **1.** the HS^- anion. **2.** a salt containing this anion.

sulfhydryl compound *Chemistry.* a compound containing the –SH group. Also, THIO COMPOUND.

sulfhydryl group *Chemistry.* the –SH group. Also, MERCAPTO GROUP.

sulfide *Chemistry.* any compound of sulfur and another element, usually made by direct reaction of the elements; heavy metal sulfides are generally insoluble.

sulfide dye *Organic Chemistry.* any of a group of water-insoluble dyes, produced by heating various organic compounds with sulfur, that are soluble in a 0.25–0.50% sodium disulfide solution; used to dye cotton. Also, SULFUR DYE.

sulfide mineral *Mineralogy.* any of a group of generally opaque minerals, including the majority of ore minerals, such as chalcopyrite or galena, characterized by the presence of sulfur in combination with one or more metals or semimetals.

sulfinate *Organic Chemistry.* **1.** a compound containing the R_2SX_2 grouping, in which *X* is a halide. **2.** a salt of sulfinic acid having the general formula *R*S=O(OH).

sulfinic acid *Organic Chemistry.* any of a class of monobasic organic acids of sulfur having the general formula *R*S=O(OH).

sulfinyl bromide see THIONYL BROMIDE.

sulfisoxazole *Pharmacology.* $C_{11}H_{13}N_3O_3S$, a sulfonamide used as an orally administered antibacterial drug used against a wide variety of infections, especially of the urinary tract. Also, GANTRISIN, SULFAFURAZOLE.

sulfite *Inorganic Chemistry.* any salt of sulfurous acid, H_2SO_3, containing the ion SO_3^-.

sulfite process *Materials Science.* a process for making wood pulp, in which wood chips are digested in an acid liquor consisting primarily of sulfurous acid and a salt; one of the chief methods of processing wood for paper pulp. Thus, **sulfite pulp, sulfite paper, sulfite pulping.**

sulfo- *Chemistry.* a prefix used in naming chemical compounds, indicating the presence of divalent sulfur or of a sulfo group.

sulfoborite *Mineralogy.* $Mg_3B_2(SO_4)(OH)_8(OH,F)_2$, a colorless, transparent, orthorhombic mineral, occurring as short to long prismatic crystals, having a specific gravity of 2.44 and a hardness of 4 to 4.5 on the Mohs scale; found in salt deposits.

sulfobromophthalein sodium *Pharmacology.* $C_{20}H_8Br_4Na_2O_{10}S_2$, a disodium salt occurring as a white, odorless, crystalline powder; soluble in water; used to test liver function.

sulfocarbanilide see THIOCARBANILIDE.

sulfocyanic acid see THIOCYANIC ACID.

sulfofication *Geochemistry.* in soils, the formation of sulfates by bacterial action resulting in the oxidation of sulfur and its compounds.

sulfo group *Chemistry.* the univalent group $-SO_3H$ (or $-SO_2OH$).

sulfolane *Organic Chemistry.* $(CH_2CH_2)_2SO_2$, a toxic, white or creamy-white crystalline powder; very slightly soluble in water and freely soluble in alcohol; melts at 27.4–27.8°C and boils at 285°C; used as a curing agent and in medicine and in liquid-vapor extractions.

Sulfolobus *Bacteriology.* a genus of Gram-negative, coccoid, thermophilic bacteria of uncertain affiliation that inhabit sulfur-rich hot springs and oxidize sulfur for energy production.

sulfonamide *Organic Chemistry.* any of a class of organic sulfur compounds having the general formula RSO_2NH_2. *Pharmacology.* any preparation of such a compound, active against a wide variety of bacteria; formerly widely used in the treatment of wounds, burns, and infections, especially in the World War II era, but now usually replaced by antibiotics, which are more effective and less toxic. Also, SULFA DRUG.

sulfonate *Chemistry.* **1.** to treat an aromatic hydrocarbon with fuming sulfuric acid. **2.** a salt or ester of sulfonic acid, containing the SO_2–O structural group.

sulfonation *Chemistry.* the process of attaching the $-SO_3H$ (SO_2–O) group to a carbon or nitrogen atom of an organic compound.

sulfone *Organic Chemistry.* any of a class of compounds having the general formula RSO_2R; used in the treatment of leprosy.

sulfonic *Chemistry.* containing a sulfo group.

sulfonic acid *Organic Chemistry.* any of a large group of strong acids having the structure RSO_2OH; used in synthesis of dyes and phenols.

sulfonic acid anhydride *Organic Chemistry.* any of a class of compounds having the formula $(RSO_2)_2O$.

sulfonium *Chemistry.* the group H_3S^+.

sulfonium salt *Organic Chemistry.* a class of compounds having the formula R_3SX or $R_3S^+X^-$.

sulfonyl see SULFURYL.

4,4'-sulfonyldianiline *Pharmacology.* see DAPSONE.

sulfophile element *Geochemistry.* an element that tends to concentrate in sulfide minerals, or sulfide-rich phases. Also, THIOPHILE ELEMENT, CHALCOPHILE ELEMENT.

5-sulfosalicylic acid *Organic Chemistry.* $C_6H_3OH(SO_2OH)COOH{\cdot}2H_2O$, colorless crystals or colored pink with traces of iron; very soluble in water; melts at 120°C when dry and decomposes at higher temperatures; used as a reagent and an intermediate.

sulfotransferase *Enzymology.* an enzyme that catalyzes the transfer of sulfate from a donor molecule to an acceptor. Also, SULFUTRANSFERASE.

sulfoxide *Organic Chemistry.* **1.** $C_{18}H_{28}O_3S$, a brown pyrethroid liquid, insoluble in water and miscible with most organic solvents; used as an insecticide synergist. **2.** any of a class of compounds having the general formula R_2SO or the group =SO.

sulfoxism *Toxicology.* poisoning caused by contact with sulfuric acid.

sulfur *Chemistry.* a nonmetallic element having the symbol S, the atomic number 16, and an atomic weight of 32.06; pure sulfur exists in two stable crystalline forms and in at least two liquid forms. *Mineralogy.* a yellow orthorhombic mineral that is the native form of this element; occurring as thick tabular or bipyramidal crystals and as granular to powdery masses, having a specific gravity of 2.07 and a hardness of 1.5 to 2.5 on the Mohs scale; found in volcanic or hot springs deposits, in sedimentary beds, and in salt domes; used in sulfuric acid production, in rubber vulcanization, and paper manufacture, and in gunpowder, fertilizers, and pharmaceuticals. (From the Latin name for this substance; its existence has been known since ancient times.) Also, SULPHUR.

sulfur-35 *Nuclear Physics.* a radioactive isotope of sulfur produced synthetically in a nuclear reactor, with a half-life of 87.5 days; widely used as a tracer in both medicine and industry.

sulfurate *Chemistry.* to combine or charge with sulfur. Thus, **sulfurated, sulfuration.**

sulfurated potash *Pharmacology.* a mixture of potassium polysulfides and potassium thiosulfate, containing 12.8% sulfur in combination as sulfide; used in pharmaceutical preparations as a source of sulfur, and in veterinary medicine as a bath for mange.

sulfur bacteria *Microbiology.* any bacteria that oxidize inorganic sulfur compounds.

sulfur ball *Geology.* **1.** a thin covering of sulfurous mud enclosing a bubble of hot volcanic gas that hardens on contact with the air. **2.** a pyritic impurity in coal.

sulfur bromide *Inorganic Chemistry.* S_2Br_2, a red liquid; decomposes in water; melts at –40°C; boils at 54°C; a strong irritant to tissues. Also, SULFUR MONOBROMIDE.

sulfur cement *Materials.* a cement made of sulfur and pitch mixed in equal parts and used to fix iron work.

sulfur chloride *Inorganic Chemistry.* S_2Cl_2, a yellow to red liquid; soluble in alcohol and decomposes in water; melts at –80°C and boils at 135.6°C; toxic on inhalation and ingestion; used in insecticides, pharmaceuticals, and rubber vulcanizing, and as a reagent. Also, SULFUR MONOCHLORIDE.

sulfur cycle *Chemistry.* the complex sequence of reactions brought about by bacteria in water and soil, in which sulfur is changed from organic sulfur compounds in plants and animals to elemental sulfur and sulfates, and then returned to organic sulfur.

sulfur dichloride *Inorganic Chemistry.* SCl_2, a toxic, dark red liquid; decomposes in water and alcohol; melts at –78°C; decomposes at 59°C; used in metallurgy and the manufacture of organic chemicals and insecticides. Also, **sulfur bichloride.**

sulfur dioxide *Inorganic Chemistry.* SO_2, a toxic, colorless gas; soluble in water and alcohol; freezes at –72.7°C; boils at –10°C; an oxidizing and reducing agent; a dangerous constituent of smog; used in metal refining, pharmaceuticals, and bleaching, and as a food additive. It is formed naturally by volcanic activity and organic decay.

sulfur dome *Metallurgy.* an inverted domelike container that is mounted over a melting pot to prevent burning of molten magnesium during magnesium alloy casting.

sulfur dye see SULFIDE DYE.

sulfureted hydrogen see HYDROGEN SULFIDE.

sulfuretum *Biochemistry.* a section of a pond or marine sediment where a number of complementary organisms metabolize sulfur and related compounds through cycles of oxidation and reductions.

sulfur hexafluoride *Inorganic Chemistry.* SF_6, a colorless gas; slightly soluble in water and soluble in alcohol; freezes at –50.5°C and sublimes at –63.8°C; used as a dielectric.

sulfur hexameter *Engineering.* an instrument that monitors the quantity of sulfur hexafluoride in devices, such as waveguides where the compound is used as a dielectric.

sulfuric [sul fyůr´ik] *Chemistry.* **1.** of or relating to sulfur. **2.** describing various compounds of sulfur, especially those in which the element has a valence of 6. **3.** relating to or derived from sulfuric acid. Also, SULPHURIC.

sulfuric acid *Inorganic Chemistry.* H_2SO_4, a colorless liquid; miscible with water and decomposes in alcohol; freezes at 10.36°C; its addition to water generates heat and explosive spattering. It is the most widely used industrial chemical, and is also used in fertilizers, dyes, electroplating, and industrial explosives, and as a reagent. Also, HYDROGEN SULFATE, BATTERY ACID, ELECTROLYTE ACID.

sulfuric acid alkylation *Chemical Engineering.* a petroleum-refinery process that combines butene and isobutene in the presence of a sulfuric acid catalyst to produce isooctane, a very high-grade gasoline.

sulfuric anhydride see SULFUR TRIOXIDE.

sulfuric chloride see SULFURYL CHLORIDE.

sulfuric chlorohydrin see CHLOROSULFONIC ACID.

sulfuric ether see ETHYL ETHER.

sulfur iodide *Inorganic Chemistry.* S_2I_2, dark gray, lustrous, brittle crystals, with an odor of iodine; insoluble in water; used in medicine. Also, **sulfur iodine.**

sulfur monobromide see SULFUR BROMIDE.

sulfur monochloride see SULFUR CHLORIDE.

sulfur monoxide *Inorganic Chemistry.* SO, a colorless gas; decomposes in cold water; stable only at reduced pressures.

sulfur number *Analytical Chemistry.* the number of milligrams of sulfur in each 100 milliliters of sample, determined by electrometric titration; used for petroleum products.

sulfurous [sul fyůr´əs] *Chemistry.* **1.** of or relating to sulfur. **2.** describing various compounds of sulfur, especially those in which the element has a valence of 3. Also, SULPHUROUS.

sulfurous acid *Inorganic Chemistry.* H_2SO_3, a toxic colorless solution of sulfur dioxide in water; the solutions are easily oxidized; used in the manufacture of textiles, paper, and wines, and as an antiseptic, a reagent, and a preservative.

sulfurous acid anhydride see SULFUR DIOXIDE.

sulfur oxide *Inorganic Chemistry.* any compound made up of only sulfur and oxygen.

sulfur oxychloride or **sulfurous oxychloride** see THIONYL CHLORIDE.

sulfur respiration *Biochemistry.* energy-producing metabolism in which sulfur serves as the terminal electron acceptor.

sulfur stream *Hydrology.* a spring whose water contains sulfur compounds, especially enough hydrogen sulfide to smell and taste.

sulfur subchloride see SULFUR CHLORIDE.

sulfur test *Petroleum Engineering.* any of various analytical tests designed to measure the sulfur content of petroleum products; e.g., by combustion in a bomb or by controlled lamp combustion.

sulfur trioxide *Inorganic Chemistry.* SO_3, silky, fibrous needles; decomposes in water; melts at 16.83°C and boils at 44.8°C; very toxic and a strong irritant; used in the sulfonation of organic compounds, and in detergents and solar energy collectors. Also, SULFURIC ANHYDRIDE.

sulfuryl *Chemistry.* containing the sulfone group, $-SO_2^-$. Also, SULFONYL.

sulfuryl chloride *Inorganic Chemistry.* SO_2Cl_2, a colorless liquid; decomposes in water; soluble in acid; melts at –54.1°C; boils at 69.1°C; a strong irritant; used in pharmaceuticals and dyes, and as a catalyst. Also, **sulfonyl chloride, sulfuric chloride.**

sulfuryl fluoride *Inorganic Chemistry.* SO_2F_2, a toxic, colorless gas; slightly soluble in water; soluble in alcohol; melts at –136.7°C; boils at –55.4°C; used as an insecticide and fumigant.

sulfutransferase see SULFOTRANSFERASE.

Sulidae *Vertebrate Zoology.* the gannets and boobies, a worldwide family of medium to large marine birds of the order Pelecaniformes, having long pointed wings, a strong pointed beak, a small gular pouch, and excellent flight and soaring abilities.

sulky plow *Agriculture.* an early form of plow, having a seat for the plowman to ride on.

Sullivan, Harry Stack 1892–1949, American psychiatrist; formulated interpersonal theory of personality; studied schizophrenia.

Sullivan, Louis 1856–1924, American architect; inspired the Chicago School of urban architecture; associated with the dictum that "form follows function;" i.e., the shape and appearance of a building should be an expression of its purpose and of its physical and cultural environment.

Louis Sullivan's Carson Pirie Scott Building (Chicago)

Sullivan angle compressor *Mechanical Engineering.* a two-stage compact compressor in which the low-pressure cylinder is horizontal and the high-pressure cylinder is vertical; it can be belt-driven or connected directly to an electric motor or diesel engine.

Sullivan reaction *Organic Chemistry.* the reaction of cysteine with 1,2-naphthoquinone-4-sodium sulfate in a highly alkaline-reducing medium, producing a red-brown color change.

sulph- see SULF-.

sulphanilamide see SULFANILAMIDE.

sulphate see SULFATE.

sulphatide see SULFATIDE.

sulpho- see SULFO-.

sulphohalite *Mineralogy.* $Na_6(SO_4)_2FCl$, a colorless, gray, pale greenish or yellow cubic mineral, occurring as dodecahedral and octahedral crystals, having a specific gravity of 2.505 and a hardness of 3.5 on the Mohs scale; found in salt deposits. Also, **sulfohalite.**

sulphur *Chemistry.* another spelling of sulfur, especially in British or older U.S. use. See SULFUR.

sulphuric see SULFURIC.

sulphurous see SULFUROUS.

sultriness *Meteorology.* a sultry state of the weather, often enhanced by calm air and cloudiness.

sultry *Meteorology.* oppressively hot and humid.

sulvanite *Mineralogy.* Cu_3VS_4, a bronze-gold, metallic, cubic mineral, occurring as cubic crystals or in massive form, having a specific gravity of 3.86 to 4.00 and a hardness of 3.5 on the Mohs scale; found in hydrothermal copper deposits containing vanadium.

Sulzberger-Chase phenomenon *Immunology.* a condition in which there is a lack of an immune response to a chemical sensitizing agent, which can cause a delayed hypersensitivity reaction.

Sulzer two-cycle engine *Mechanical Engineering.* a two-cycle diesel engine in which effective scavenging and charging takes place by means of a system developed by the Sulzer company.

sum *Mathematics.* **1.** a quantity obtained by the addition of two or more other quantities or summands. The process of determining a sum is called summation. **2.** the sum of two functions defined on a common domain is the function whose value at each point of the domain is the sum of the values of the two functions at that point. **3.** the sum of an infinite series is the limit, if it exists, of the sequence of partial sums of the series. If this limit exists, the series is said to be **summable** or to converge; otherwise the series is said to diverge. Various notions of summation may also be defined for divergent series; e.g., Abel, Cesàro, and Hölder summation.

sum. *Medicine.* take; let someone take; to be taken. (From Latin *sume, sumat, sumendum.*)

SUM surface-to-underwater missile.

sumac *Botany.* any of various plants of the genus *Rhus,* especially the nonpoisonous species as distinguished from poison sumac.

sumac wax see JAPAN WAX.

Sumatra *Geography.* a large island (area: 166,789 square miles) in western Indonesia, south of the Malay peninsula.

sumatra *Meteorology.* a squall that blows through the Malacca Strait between Malay and Sumatra during the southwest monsoon season (April-November), usually at night, with winds sometimes exceeding 30 miles per hour.

sumi ink see INDIA INK.

summand *Mathematics.* any of a group of numbers to be added together.

summary recorder *Computer Programming.* any output equipment that records a summary of the information handled.

summation *Mathematics.* the process of determining a sum. *Physiology.* the cumulative effects of a number of stimuli applied to a muscle, nerve, or reflex arc.

summation check *Computer Programming.* the verification of data integrity accomplished by the comparison of check sums computed on the same data at different times.

summation convention *Mathematics.* a shorthand notation employed in tensor (Ricci) calculus. In particular, if the same index appears as both a superscript and subscript of a term, the term is understood to be summed over the range of the index. For example:

$$a_i b^i = \sum_{i=1}^{n} a_i b^i; \text{ and } a_i^i = \sum_{i=1}^{n} a_i^i; \text{ and } a_i^j b_j^{ik} = \sum_{i=1}^{n} \sum_{j=1}^{n} a_i^j b_j^{ik}.$$

In the expression $\partial x^i/\partial x_k$, i and k are regarded as upper and lower indices, respectively. Also, **Einstein's summation convention.**

summation principle *Meteorology.* a rule stating that the sky cover at any given level is equal to the sum of the sky cover at all lower levels plus the additional sky cover provided at the layer in question.

summation tone *Acoustics.* a tone whose frequency results from the addition of two overlapping frequencies; used extensively in underwater acoustic systems.

summation wave *Physiology.* the added effect of various stimuli when combined into an oscillation flowing through an organ or body part.

summative fractionation *Materials Science.* a fractionation method in which the point of fractionating is varied through a series of polymer solutions of different concentrations; it provides data about molecular weights with a minimum of preliminary calibration and equipment expense.

summer *Astronomy.* the period between the summer solstice (which occurs about June 22) and the autumnal equinox (which occurs about September 23); it includes the months of June through August in the Northern Hemisphere, and December through February in the Southern Hemisphere. *Meteorology.* the characteristic climate associated with this season in a given area, which tends to vary with latitude. Near the tropics, the summer is nearly always hot but may be markedly wet or dry; moving poleward, the intensity of storm systems becomes a crucial characteristic; and near the poles, the greater duration of sunlight is the most important factor.

summer *Building Engineering.* a main beam or girder, such as one spanning girts to support joists. Also, **summertree.**

summer balance *Hydrology.* the change in mass of a glacier from its maximum value of a given year to the minimum value of that year.

summer fallow *Agronomy.* the plowing of a field prior to or during the summer and cultivating it enough to prevent weeds in preparation for another crop.

summer solstice *Astronomy.* the moment when the sun reaches its farthest northern declination; this occurs about June 22 in the Northern Hemisphere (when the sun is overhead on the Tropic of Cancer) and about December 22 in the Southern Hemisphere (when the sun is overhead on the Tropic of Capricorn).

summer spore *Mycology.* in such fungi as rust fungi, a spore that grows and germinates rapidly. Also, UREDINOSPORE, UREDIOSPORE.

summertime *Meteorology.* the summer season.

Summer Triangle *Astronomy.* a trianglar asterism formed by the stars Vega in Lyra, Deneb in Cygnus, and Altair in Aquila which lies high overhead on summer and early autumn evenings.

summerwood *Botany.* wood formed late in the growing season; i.e., in summer; latewood.

summing amplifier *Electronics.* an amplifier that generates an output voltage proportional to the sum of two or more input voltages.

summing network *Electricity.* a network with a voltage output proportional to the sum of two or more input voltages.

summit *Science.* a peak or highest point, as of a mountain or anatomical structure. *Ordnance.* the highest altitude above mean sea level that a projectile reaches in its flight from the gun to a target; it is the algebraic sum of the maximum ordinate and the altitude of the gun.

summit plain see PEAK PLAIN.

Sumner, James Batcheller 1887–1955, American biochemist; awarded the Nobel Prize for the first crystallization of an enzyme (jackbean urease).

Sumner line *Navigation.* a line of position established by joining two or more positions computed at different assumed latitudes or longitudes; the line is actually an arc of a very large circle with the geographical position of the observed body at its center and the true zenith of the body as its radius. Similarly, **Sumner method.**

sum of states or **sum over states** see PARTITION FUNCTION.

sump *Mechanical Engineering.* the lowest portion of a machine or structure, serving as a collecting basin for wastes or fluids, such as lubricating oil, that are to be recirculated. *Mining Engineering.* a basin or pit, as at the bottom of a shaft, where water is allowed to collect. *Hydrology.* any pit, pool, or other depression in which water collects, especially waste water.

sump fuse *Engineering.* a blasting fuse used for underwater work.

sump pump *Mechanical Engineering.* a simple, single-stage vertical pump used to remove sludge from a sump or to drain shallow pits.

sum rule *Quantum Mechanics.* a formula relating one quantity of a system to the sum, over all the possible states of the system, of another quantity whose magnitude may be different for each possible state; useful primarily in studying the excitation or decay modes of a system.

sum. tal. *Medicine.* take one like this. (From Latin *sumat talem.)*

sun or **Sun** *Astronomy.* the star that is the central celestial body in the solar system; it has a diameter of about 1.4 million kilometers (about 109 times that of earth), a mass about 330,000 times that of earth (99.8% of the total mass of the solar system), a volume about 1.3 million times that of earth, and a surface temperature of 6000 kelvins. It consists of about 75% hydrogen, almost 25% helium, and traces of at least 70 other elements.

sunbeam snake *Vertebrate Zoology.* an elusive burrowing snake, *Xenopeltis unicolor,* having a dark, iridescent upper body and yellow belly; the only member of the family Xenopeltidae of Asia.

sunbear *Vertebrate Zoology.* a bear of Southeast Asia, *Helarctos malayanus,* of the family Ursidae; the smallest of bears, standing about 4 feet tall, it has a powerful body and paws, and dark glossy fur with a white chest patch.

sunbird *Vertebrate Zoology.* any of numerous small, brilliantly colored songbirds of the family Nectariniidae, native to southern Asia and Australia and characterized by a long tail and a long, pointed bill; it is similar to a hummingbird, but does not hover to feed on nectar.

sun bittern *Vertebrate Zoology.* a wading bird of South America, *Eurypyga helias,* of the family Eurypygidae.

sunblock *Pharmacology.* a highly effective sunscreen, usually having a sun protection factor of 15 or above so that it blocks tanning rays as well as burning rays.

sunburn *Medicine.* an injury to the skin caused by exposure to ultraviolet rays, usually of the sun; effects vary depending on the severity of the burn, ranging from mild reddening to erythema, tenderness, and blistering. Also, DERMATITIS SOLARIS, SOLAR DERMATITIS.

sunburst light see FANLIGHT.

sun compass *Navigation.* a device similar to a sundial that utilizes the shadow cast by the sun to determine direction.

sun crack see MUD CRACK.

sun cross *Meteorology.* an optical phenomenon in which bands of white light intersect at right angles above the sun; usually caused by the superposition of a perhelic circle and a sun pillar.

sun crust *Hydrology.* a thin, smooth-surfaced snow crust formed by the refreezing of melted surface snow.

Sundas *Geography.* the western islands of the Malay Archipelago, including the **Greater Sundas** (Sumatra, Borneo, Celebes, and Java) and the **Lesser Sundas** (from Bali to Timor).

sundew *Botany.* any of several small insectivorous bog plants of the genus *Drosera,* characterized by leaves with long sticky hairs that trap insects.

sundial

sundial *Horology.* an early timekeeping device, consisting of a gnomon or style that indicates apparent solar time by casting a shadow across a graduated disk or cylindrical surface.

sun dog see PARHELION.

sun drawing water *Meteorology.* a phenomenon in which bright bands appear when the sun shines through scattered openings in a layer of clouds and into a layer of turbid air; similar to crepuscular rays.

sun drying *Food Technology.* the use of direct sunlight in the preservation of foods such as fruits and nuts. It is limited by climatic conditions.

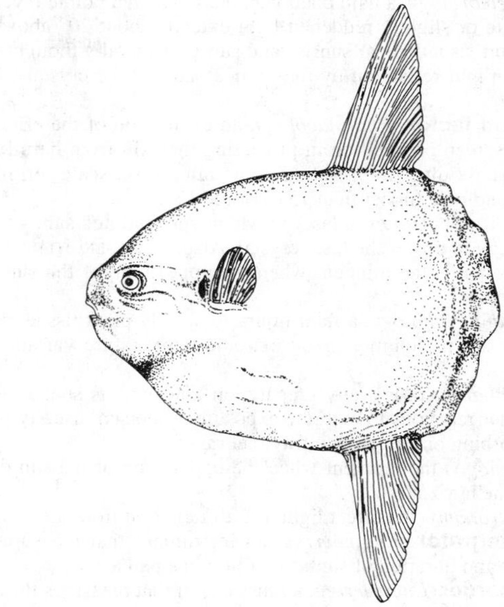

sunfish

sunfish *Vertebrate Zoology.* a fish of the North American freshwater perciform family Centrarchidae, characterized by a deep compressed body and usually brilliant metallic coloration.

sunflower *Botany.* an annual plant, *Helianthus annuus,* characterized by a long stem, large, composite flowers with yellow petals, and broad ovate leaves.

sunflower

sun follower *Electronics.* a photoelectric pickup and an associated servomechanism used to maintain a sun-facing orientation, as for a space vehicle. Also, **sunseeker.**

sun gear *Mechanical Engineering.* the central gear in a planetary set.

sun-grazing comet *Astronomy.* a comet that comes within about one diameter of the sun's surface at perihelion; no sun-grazer survives many perihelion passages.

sunk *Mining Engineering.* of a shaft, drilled or excavated downward.

sunken rock *Navigation.* an underwater rock, located below the defined level of a rock awash, that presents a danger to shipping.

sunken stream see SINKING STREAM.

sunlamp *Electricity.* a special tungsten-filament lamp that utilizes a mercury-vapor gas-discharge tube in order to provide a high degree of ultraviolet output; used for therapeutic or cosmetic tanning purposes, though this is now in dispute because of possible skin damage.

sunlight *Astronomy.* the direct light from the sun.

sun line *Navigation.* a line of position determined by observation of the sun with a sextant. Similarly, **sun sights, shooting the sun.**

sunlit aurora *Geophysics.* an aurora occurring in a part of the atmosphere above the earth's shadow and exposed to sunlight.

sun opal see FIRE OPAL.

sun pillar *Meteorology.* a halo phenomenon in which a pointed, vertical streak of white or slightly reddened light extends about 20° above and beneath the sun, usually near sunrise and sunset; generally thought to be a result of sunlight reflection by the vertical axes of ice crystals. Also, LIGHT PILLAR.

sun-protection factor *Pharmacology.* an expression of the effectiveness of a sunscreen preparation in protecting the skin from harmful effects of the sun's ultraviolet radiation, as rated on a scale, originally from 2 to 15 and now higher than 15. Also, SPF.

sun-pumped laser *Optics.* a laser in which concentrated sunlight provides the energy to power the laser crystal. Also, SOLAR-EXCITED LASER.

sunrise *Astronomy.* the moment when the upper limb of the sun rises above the horizon.

sunscald *Plant Pathology.* a fatal injury to woody plant tissue caused by too much sun in summer or by extreme temperature variations in winter.

sunscreen *Pharmacology.* any preparation that inhibits sunburn, skin cancer, or other reactions to extreme exposure to the sun, usually by selectively absorbing or reflecting ultraviolet rays.

sunset *Astronomy.* the moment when the upper limb of the sun disappears below the horizon.

sunshine *Astronomy.* the direct light and all radiation from the sun.

sunshine integrator *Engineering.* an instrument that measures the total intensity and duration of sunlight falling in an area.

sunshine recorder *Engineering.* an instrument that measures the duration of daylight, without regard to the intensity of sunlight.

sun spike see NIEVE PENITENTE.

sunspot *Astronomy.* a relatively cool and dark area on the sun's photosphere, characterized by a strong magnetic field; sunspots are cyclically variable in number and magnetic polarities. It has been widely speculated that variations in sunspot activity may affect weather and climate on earth.

sunspot cycle *Astronomy.* the roughly 11-year cycle during which the frequency of sunspots varies from a maximum to a minimum level and back again; reckoning by magnetic polarities, however, the full cycle is two ordinary sunspot cycles, or 22 years.

sunspot maximum *Astronomy.* the period in the solar sunspot cycle in which the maximum number of sunspots appears and solar activity is at its peak.

sunstone *Mineralogy.* a brilliant, translucent, aventurine feldspar, usually oligoclase, with minute included hematite flakes. Also, HELIOLITE.

sun strobe *Electronics.* the signal display that is seen on the screen of a radar plan-position indicator when the radar antenna is directed at the sun; this pattern is similar to that produced by continuous-wave interference, and is due to radio-frequency energy radiated by the sun.

sunstroke *Medicine.* a form of heat stroke occurring from overexposure to the sun, resulting in exhaustion, coma, convulsions, and elevation of body temperature. Also, INSOLATION, THERMIC FEVER.

sun's way *Astronomy.* the direction of motion that the solar system, taken as a whole, maintains through space.

SUn symmetry see UNITARY SYMMETRY.

sun-synchronous orbit *Space Technology.* the orbit of a synchronous satellite.

suntan *Physiology.* a darkening of the skin resulting from exposure to natural sunlight or to artificial light from a sunlamp.

Suoidea *Vertebrate Zoology.* a superfamily of the suborder Paleodonta that contains pigs and peccaries.

sup. superior; supra.

super- or **supero-** a combining form meaning: **1.** above. **2.** excess.

Super 530 *Ordnance.* an advanced French all-weather, air-to-air missile designed to provide the launch aircraft with the capability to destroy a target aircraft flying at a significantly different altitude.

superacid *Chemistry.* a solution of a strong acid in a very acidic solvent; a highly effective protanating agent.

superadiabatic lapse rate *Meteorology.* an environmental lapse rate that is greater than the dry-adiabatic lapse rate.

superaerodynamics *Fluid Mechanics.* the dynamics of rarefied gases whose densities are so low that the molecular mean free path is not negligible and the gases no longer behave as a continuous fluid.

superalloy *Materials Science.* a nickel-, iron-, and cobalt-base alloy that has good heat, creep, and oxidation resistance.

supercalendered finish *Materials Science.* a glossy finish applied to paper by passing it through a supercalender.

supercalendering *Engineering.* the process of placing a high gloss on paper using steam and high pressure.

supercapillary interstice *Geology.* an interstice, larger than a capillary interstice, that is too large to hold water above a free water surface by surface tension.

supercavitating propeller *Naval Architecture.* a marine propeller designed so that at high speed the entire back side of each blade is inside a smooth cavitation bubble.

supercentrifuge *Mechanical Engineering.* a centrifuge designed to rotate at speeds faster than those of a normal centrifuge in order to achieve difficult or exotic separations.

supercharge *Engineering.* to force air into an internal-combustion engine at greater than atmospheric pressure.

supercharge method *Engineering.* a method of determining the maximum knock-limited power of an aviation fuel by burning it under supercharged conditions.

supercharger *Mechanical Engineering.* a blower that increases the intake pressure of an engine, so as to make fuel burn more quickly and increase engine power.

supercharging *Mechanical Engineering.* a method of introducing air into the cylinder of an internal-combustion engine at a pressure above atmospheric pressure.

supercilium *plural,* **supercilia.** *Anatomy.* **1.** the eyebrow. **2.** an individual hair of the eyebrow.

supercirculation *Aviation.* the use of an air compressor or other device to augment air flow over an airfoil and thereby increase lift.

superclass *Systematics.* a taxonomic rank immediately above a class.

supercluster *Astronomy.* a cluster of clusters of galaxies.

supercobalt drill *Mechanical Devices.* a drill used to work especially hard grades of stainless steel, composed of 8% cobalt high-speed steel.

supercoil *Molecular Biology.* a covalently closed, circular double-stranded DNA molecule coiled so that the helix axis crosses itself one or more times.

supercoiling of DNA *Molecular Biology.* a DNA configuration in which the double helix is organized into large whirls, or "coils."

supercompressibility factor see COMPRESSIBILITY FACTOR.

supercomputer *Computer Technology.* **1.** any of a category of extremely powerful, large-capacity mainframe computers that are capable of manipulating massive amounts of data in an extremely short time. **2.** any computer that is one of the largest, fastest, and most powerful available at a given time.

superconducting *Physics.* capable of superconductivity; superconductive.

superconducting accelerator *Nucleonics.* a particle accelerator that uses superconducting magnets for increased magnetic field strength.

superconducting alloy *Metallurgy.* an alloy, as of lead and bismuth, that exhibits superconductivity.

superconducting circuit *Physics.* an electric circuit that has some or all of its elements in a superconductive state.

superconducting computer *Computer Technology.* a computer whose circuits are made of superconducting materials that increase speed and performance.

superconducting cyclotron *Nucleonics.* a cyclotron whose only superconducting element is its main coil, which effectively increases the magnetic field strength by a factor of about 3 over that of a nonsuperconductive coil; this allows for the reduction in size of the device.

superconducting gyroscope see CRYOGENIC GYROSCOPE.

superconducting magnet *Physics.* an electromagnet that uses coils of a type II superconducting material capable of producing a very high magnetic field at a high transition temperature.

superconducting magnetic energy storage *Electricity.* the storing of electrical energy as a circulating current in a massive superconducting coil or magnet, for use by a power utility during a peak load period.

superconducting memory *Electronics.* a computer memory consisting of thin-film devices that operate only under cryogenic conditions and can be made to change from a low-temperature superconducting state to a normal resistive state by the application of a magnetic field. The memory dissipates power only during the read/write operation.

superconducting quantum interference device *Electronics.* a superconducting ring that couples with one or two Josephson junctions; used in high-sensitivity magnetometers, near-magnetic-field antennas, and small current or voltage measurement.

superconducting thin film *Physics.* a thin film of some superconductive element such as indium or tin; used in cryogenic devices for switching or storing.

superconductive *Physics.* relating to or having the capacity for superconductivity.

superconductivity *Physics.* a phenomenon shown by certain metals, alloys, and other compounds of having negligible resistance to the flow of electric current at temperatures approaching absolute zero. Each material has a critical temperature T_c above which it is a normal conductor, operating as a superconductor only under cryogenic conditions; recent experiments, however, have revealed materials that are superconductive at temperatures hundreds of degrees above absolute zero.

superconductor *Physics.* any material that can exhibit superconductivity. Also, **superconducting material.**

supercool *Thermodynamics.* **1.** to bring a substance to a temperature below which a change of state would normally occur, without this change taking place. **2.** also, **supercooled.** describing a substance in this condition.

supercooled cloud *Meteorology.* an unstable cloud composed of supercooled water droplets; the principal source of aircraft icing.

supercooled water *Meteorology.* water that remains in the liquid state below 0°C. Also, SUBCOOLED WATER.

supercooling *Thermodynamics.* a condition in which a pure substance is cooled below its freezing point, condensation point, or sublimation point, without experiencing the corresponding change of state that this cooling would normally bring about.

supercritical *Thermodynamics.* a condition in which a substance experiences a change in its pressure, temperature, or volume that would normally carry the system through its critical point although no phase change is observed. *Nucleonics.* of, relating to, having, or operating with a multiplication factor greater than one.

supercritical field *Physics.* a field (electric, gravitational, and so on) that is strong enough to break down a perfect vacuum (completely void of real particles) to a medium in which real particles are present.

supercritical flow *Fluid Mechanics.* see SUPERSONIC FLOW. *Hydrology.* see RAPID FLOW.

supercritical fluid *Thermodynamics.* the mobile phase of a substance intermediate between a liquid and a vapor, maintained at a temperature greater than its critical point.

supercritical fluid chromatography *Spectroscopy.* a technique in chromatographic separation using a supercritical fluid, rather than a liquid or a gas, and having highly desirable transport properties as well as low decomposition rates.

supercritical fluid extraction *Chemical Engineering.* a solvent-extraction process that uses a supercritical fluid as the selective solvent.

supercritical mass *Nucleonics.* the mass of a sample of fissionable material that is greater than the critical mass; such a quantity is unstable and supports a chain reaction within the sample.

supercritical reactor *Nucleonics.* a nuclear reactor operating with a multiplication factor that is greater than unity, such that self-sustaining fission is possible.

supercritical wing *Aviation.* a wing that allows subsonic aircraft to attain maximum efficient cruise speed, usually very close to the speed of sound, by using a supercritical airfoil section; it is distinguished by its bluff leading edge, flattish top, bulged underside, and downcurved trailing edge which develops a weak shock wave at transonic speeds as distinguished from the very strong upper surface shock that normally occurs with the more conventional airfoil.

supercurrent *Solid-State Physics.* an electric current that flows without resistance due to the motion of the superconductive electron pairs in a superconducting material.

superdeformed nuclear state *Nuclear Physics.* a nucleus in a highly excited state, having a shape corresponding to an ellipsoid with an axis ratio near 2:1.

superdiagonal of a matrix *Mathematics.* given an $n \times m$ matrix with entries a_{ij}, the entries of the form $a_{i,i+1}$, $i = 1, 2, \ldots, \min(n, m-1)$; i.e., the entries just above the main diagonal.

superdominance see OVERDOMINANCE.

superego *Psychology.* in psychoanalytic theory, the part of the mind that internalizes parental and cultural values, restrictions, and inhibitions, and that is responsible for self-imposed standards of behavior and for restraining the id from gratifying itself in unacceptable ways.

superelastic collision *Physics.* a collisional process between an electron and a target atom that is in an excited state, by which the electron absorbs the energy released by the atom as it decays to a lower energy state and thus leaves the collision site with greater kinetic energy than before the collision.

superelevation *Ordnance.* **1.** in antiaircraft gunnery, an angle of elevation added to compensate for the effect of gravity on a projectile. **2.** the degree of angle that a gun must be elevated above the gun-target line.

superexchange *Solid-State Physics.* a strong antiferromagnetic coupling between two positive ions whose interaction is mediated by the electrons in their common neighborhood.

superfamily *Systematics.* a taxonomic rank immediately above a family.

superfecundation *Physiology.* the insemination of two or more ova, resulting from coitus with different partners and giving rise to littermates or siblings with different fathers.

superfemale *Genetics.* **1.** an abnormal female mammal that possesses more than two X chromosomes. **2.** in *Drosophila,* a female phenotype of reduced viability, characterized by the presence of more than one X chromosome for each set of autosomes. Also, METAFEMALE.

superficial *Science.* situated on or near the surface. *Medicine.* specifically, describing an injury or condition affecting the skin surface.

superficial cleavage *Developmental Biology.* a form of cleavage in which only the surface part of the egg participates.

superficial deposit see SURFICIAL DEPOSIT.

superficial expansion see COEFFICIENT OF SUPERFICIAL EXPANSION.

superficial palmar arch *Anatomy.* an arch of arteries, created by connections between the ulnar and radial arteries, that produce the arterial supply to the palm and fingers.

superfines *Metallurgy.* in powder metallurgy, the portion of a powder sample that is less than 10 micrometers.

superfinishing *Metallurgy.* a form of honing in which the abrasive stones are spring supported.

superfluid *Physics.* a collection of particles that exhibit zero viscosity and zero entropy; the particles are known to obey the Bose-Einstein statistics.

superfluidity *Physics.* a property possessed by liquid helium in which a portion of the liquid exhibits no viscosity when its temperature is lowered below the λ-point (approximately 2.19 K).

Superfortress *Aviation.* a popular name for the B-29 bomber.

Superfortress

supergalaxy *Astronomy.* a cluster of galaxies with an ellipsoidal shape but a nonuniform distribution of members.

supergene *Genetics.* a sequence of genes preserved by selection and prevented from crossing over by inversions or translocations; it is thus transmitted without transformation to succeeding generations.

supergene *Mineralogy.* **1.** of or relating to minerals and mineral deposits or enrichments formed near the surface, generally by descending solutions. **2.** those descending solutions and that general environment. Also, HYPERGENE.

supergene enrichment see SECONDARY ENRICHMENT.

supergeostrophic wind *Meteorology.* wind having a speed greater than the geostrophic wind required by a given pressure gradient; not necessarily a supergradient wind.

supergiant star *Astronomy.* the largest stars known; found among all spectral classes, they have enormously extended and tenuous atmospheres and very high luminosities and evolve rapidly on scales of a few millions of years.

superglacial *Hydrology.* describing the upper surface of an ice sheet or a glacier.

supergradient wind *Meteorology.* a wind having a speed greater than the gradient wind required by a given pressure gradient and centrifugal force.
supergranulation cells *Astronomy.* cells of convective activity in the solar photosphere, typically about 30,000 kilometers in diameter; uniformly distributed over the solar sphere, they last about 20 hours.
supergravity *Physics.* a supersymmetry that unifies quantum theory and general relativity.
superharmonic function *Mathematics.* let f be a lower semicontinuous, locally integrable function on an open subset D of R^n that is bounded below. If the average of f at any point x of D (computed as an integral over an open ball contained in D and centered at x) is less than or equal to $f(x)$, then f is said to be superharmonic. Examples include functions that satisfy $\nabla^2 f \geq 0$. It can be shown that if f is harmonic in D, then $-|f|$ is superharmonic.
superheat *Thermodynamics.* **1.** to bring a substance to a temperature above that at which a change of state would normally occur, without this change taking place; e.g., to heat a liquid above its normal boiling point without producing vaporization. **2.** the temperature of a substance in this state; i.e., a temperature greater than the normal saturation temperature for the existing pressure. **3.** see SUPERHEATED STEAM.
superheated *Thermodynamics.* at a temperature above which a change of state would normally occur (e.g., the change of water from a liquid to a vapor), without this change having taken place.
superheated steam *Thermodynamics.* water vapor that has been heated, at constant pressure and out of contact with the water from which it was formed, to a temperature significantly above its saturation point, so that it does not recondense to water as readily as ordinary steam; used to provide increased efficiency in steam engines.
superheated vapor *Thermodynamics.* a vapor that is heated to a temperature that exceeds the boiling point of the liquid phase.
superheater *Mechanical Engineering.* a coil or other device through which steam from a boiler passes and is heated above its saturation temperature.
superheating *Thermodynamics.* a condition in which a pure substance is heated above its melting point, boiling point, or sublimation point, without experiencing the phase change that this heating would normally produce.
superheavy *Chemistry.* of or relating to a superheavy element.
superheavy boson *Particle Physics.* a conjectured particle that would be responsible for interactions between quarks and leptons in the early universe. Also, X BOSON.
superheavy element *Chemistry.* a proposed element with an atomic number greater than 106; such elements have never been produced experimentally. Also, SUPER TRANSURANIC.
superhelix *Biochemistry.* a helix formed by a strand that is itself a helix, as in DNA.
superheterodyne detector see HETERODYNE DETECTOR.
superheterodyne receiver *Electronics.* a receiver in which all received modulated radio-frequency carrier signals are translated to a common intermediate-frequency carrier value for additional amplification and selectivity before demodulation.
superhighway *Civil Engineering.* a large roadway system for high-speed traffic, such as a freeway, turnpike, or expressway.
superimposed *Geology.* describing rocks that are layered or stratified.
superimposed back pressure *Mechanical Engineering.* the static pressure at the outlet of a pressure relief valve that results from pressure in the discharge system.
superimposed drainage *Hydrology.* a drainage system consisting of superimposed streams.
superimposed fan *Geology.* a fan-shaped body of material deposited by a stream that has a steeper gradient than the older fan on which it has been deposited.
superimposed metamorphism see POLYMETAMORPHISM.
superimposed stream *Hydrology.* a stream that developed on a new surface and maintained its course despite the existence of differing geologic structures and lithologies as it eroded downward into the underlying rock strata. Also, **superinduced stream.**
superimposed valley *Geology.* a valley that contains a superimposed stream or has been worn away by such a stream.
superincumbent *Geology.* describing a superjacent rock layer, especially one that exerts pressure on the layer beneath it.
superinfection *Virology.* the infection of an already virus-infected host with a second virus, usually a different strain from the first infecting virus.
superinfection exclusion *Virology.* the ability of a virus-infected cell to resist or restrict the growth of a second virus by means of DNA breakdown or phage-induced changes in the host cell membrane that prevent the second virus from penetrating or replicating. Also, **superinfection immunity.**
superinsulation *Chemical Engineering.* a heavy-duty insulation used for low-temperature applications.
superinvolution *Medicine.* the return of a part to a size much smaller than normal after enlargement, as occurs in the uterus of a nursing mother. Also, HYPERINVOLUTION.
superionic conduction *Solid-State Physics.* a solid displaying an ionic conductivity that is unusually high, typically on the order of 100 times that normally observed.
superionic conductor *Materials Science.* a solid in which the ionic conductivity approaches a value that is typical of a molten salt or of proton conduction in a strong acid.
superior *Science.* situated above or directed upward. *Botany.* **1.** of or relating to a plant part or organ that is situated above some other part or organ. **2.** of a calyx, apparently arising from the ovary. **3.** of an ovary, free from the calyx. *Anatomy.* **1.** describing an organ or part that is in a higher place or position. **2.** toward the head. *Materials.* of higher grade or quality. *Graphic Arts.* describing a letter, number, or other character that is above the general level of other characters on the same line, and usually also smaller, such as the "n" in R^n or the "2" in $E = mc^2$.
superior air *Meteorology.* a very dry air mass produced by subsidence, usually aloft (often above tropical maritime air) but sometimes reaching the earth's surface during extreme subsidence.
superior alveolar canals *Anatomy.* canals in the maxilla for passage of the superior alveolar vessels and nerves that enter the root canals of the maxillary teeth.
superior cervical ganglion *Anatomy.* the largest and uppermost ganglion of the sympathetic trunk, located near the base of the skull between the internal carotid artery and the internal jugular vein.
superior colliculus *Anatomy.* one of a pair of rounded eminences in the mesencephalon, primarily concerned with visual reflexes.
superior conjunction *Astronomy.* the point at which a planet lies directly on the other side of the sun.
Superior, Lake *Geography.* the largest (area: 31,820 square miles) and westernmost of the Great Lakes.
superior mesenteric artery *Anatomy.* a large unpaired artery that arises from the abdominal aorta and delivers blood to the small intestine, cecum, ascending colon, and part of the transverse colon.
superior mirage *Optics.* an image of an object produced by refraction that appears above the object's actual position; the opposite of an inferior mirage.
superior planet *Astronomy.* any planet whose orbit is farther from the sun than that of the earth; i.e., Mars, Jupiter, Saturn, Uranus, Neptune, or Pluto.
superior tide *Oceanography.* a tide in the hemisphere in which the moon is above the horizon.
superior vena cava *Anatomy.* the large vein that drains blood primarily from organs above the diaphragm and empties into the right atrium of the heart.
superjacent *Geology.* describing a rock stratum that lies directly upon or over a specific lower stratum or above an unconformity.
superjacent waters *Oceanography* the ocean overlying a continental shelf.
super-large-scale integration *Electronics.* the production of very complex integrated circuits that are highly dense, consisting of transistors and other components on the order of 10^6 total components.
superlattice *Solid-State Physics.* a multicomponent solid system in which there exists a regular array of atoms of one species interspersed among the lattice structure of another species; the two species need not form a compound. Also, **superstructure.** *Electronics.* a structure that consists of alternating layers of two different semiconductor materials, each several nanometers in thickness.
superline *Computer Programming.* a unit of text that is longer than an ordinary line.
super-long play *Mechanical Engineering.* a videotape recording speed that is slower than standard speed, to allow six hours of recording time for a T-120 tape, as distinguished from the standard two-hour time.
superluminal radio source *Astronomy.* an energetic knot of gas in a galactic jet, which has a high velocity and is moving in a direction such that perspective effects give it the appearance of faster-than-light motion.

superlunary *Astronomy.* situated above or beyond the moon.

supermale *Genetics.* in *Drosophila*, a male phenotype of reduced viability, characterized by cells containing three sets of autosomes, but only one X chromosome. Also, METAMALE.

supermassive star *Astronomy.* a star whose mass exceeds 50 times that of the sun.

supermature *Geology.* describing a clastic sediment at its final stage of textural maturity, characterized by well-sorted grains that have become rounded.

supermaxilla *Anatomy.* the technical name for the upper jaw.

super-metal-rich stars *Astronomy.* type K stars characterized by low luminosities and strong CN bands.

supermicro *Computer Technology.* a high-performance microcomputer.

supermini *Computer Technology.* a minicomputer that has been enhanced to the size and performance of a mainframe computer.

supermode laser *Optics.* a laser in which the output beam is divided in two and recombined with one channel modulated to give it the same frequency as the other, but with its phase rotated 180°; the result is a greatly strengthened signal.

supermolecule *Physical Chemistry.* a theoretical construct consisting of an aggregate that forms momentarily when reacting molecules collide and then disintegrates after the collision.

supermultiplet *Quantum Mechanics.* a multiplet that is split into more than three levels.

supernatant *Biotechnology.* the overlying fluid layer that remains after precipitation of a solid component through centrifugation.

supernatant liquor *Engineering.* the liquid that forms above a settled solid precipitate or sludge.

supernormal stimulus *Behavior.* a stimulus that is unusually effective in producing a specific response.

supernova *Astronomy.* the explosion of a massive star in which the core undergoes gravitational collapse and the upper layers of its atmosphere are blown off; the energy released can temporarily rival that of an entire galaxy.

supernova remnant *Astronomy.* the luminous shell of gas and heavy elements that are thrown into space by a supernova; they give off strong radio and X-ray emission as the gas slams into the interstellar medium and heats up.

supernumerary chromosome *Genetics.* any of one or more extra chromosomes found inconsistently in wild populations of certain species of animals; such chromosomes are not homologous to members of the regular set of chromosomes and apparently exert little influence on the phenotypic effect. Also, ACCESSORY CHROMOSOME, B CHROMOSOME. *Cell Biology.* any chromosome that is additional to the normal chromosomal complement of a cell.

supernumerary rainbow *Optics.* an arc of faint colors occasionally seen inside a primary rainbow; typically, several such arcs merge inward and blend.

superorder *Systematics.* a taxonomic rank immediately above an order.

superorganicism *Anthropology.* the concept that culture is a separate force or entity that acts independently from individuals and to which individuals are subordinate.

superovulate *Biology.* to produce more than the normal number of ova at one time, as through hormonal treatment.

superoxide *Chemistry.* a chemical compound with the O_2^- ion, such as NaO_2 or KO_2; the odd number of electrons, 13 in the O_2^- ion, makes such compounds paramagnetic. *Biochemistry.* the species of dioxygen formed by reduction.

superoxide anion *Biochemistry.* a radical formed when an oxygen molecule gains an electron.

superoxide dismutase *Enzymology.* an enzyme that catalyzes the reaction of two molecules of the potentially harmful compound superoxide to form oxygen and hydrogen peroxide.

superoxol test *Microbiology.* an assay for the enzyme catalase in a given microorganism, involving the addition of a high percentage (usually 20–30%) of hydrogen peroxide and the observation of bubbles in catalase-positive cultures.

superparamagnetism *Materials Science.* the tendency of fine particles to behave independently of one another in a manner similar to paramagnets.

superparasitism *Ecology.* the infestation of a host with several individuals of one or many species of parasites.

superphosphate *Materials Science.* a mixture of calcium sulfate and dihydrogen calcium phosphate made by treating bone ash or basic slag with sulfuric acid; used as an agricultural fertilizer.

superplastic forming *Materials Science.* an advanced metal manufacturing process that permits large deformations without necking under controlled temperature and strain rates.

superplasticity *Materials Science.* the ability of a material to deform uniformly by an exceptionally large amount, usually at elevated temperatures.

superpolymer *Materials Science.* an extensive setting up of crosslinks between molecular chains, resulting in an infusible polymer, such as in a thermosetting resin.

superposed *Botany.* of or relating to a plant part or organ that is growing or situated above or upon another part or organ.

superposition *Physics.* a principle that may be applied to systems in which individual influences act linearly: the resultant effect on the system is equivalent to the summation of the effects of the individual influences that are acting on the system. *Geology.* **1.** the natural order of an undisturbed rock succession, in which sedimentary layers accumulate one above the other, so that the higher beds or layers are relatively younger than those below. **2.** the process by which such a layering takes place, so that younger layers are deposited on older, lower layers. *Archaeology.* the principle that artifacts found at a lower level of a site predate those at a higher level.

superposition eye *Invertebrate Zoology.* an insect eye in which all light rays except those entering the central facet of a facet group are intercepted.

superposition integral *Control Systems.* an integral that sums the response of the system with a given input in terms of the impulse or the step response of the system.

superposition principle *Mathematics.* the statement that any sum of solutions of a linear homogeneous equation is again a solution; e.g., two waves that are solutions of the homogeneous wave equation can pass through each other (because their sum is a solution) without distortion (because each wave is unchanged by the presence of the other wave). Further, new solutions may be obtained by summation or integration of members of a family of solutions. For example, suppose that $u(x_1, \ldots, x_n;k)$ is a family of solutions depending on a parameter k. Then

$$v(x_1, \ldots, x_n) = \sum_k u(x_1, \ldots, x_n;k) \text{ or}$$

$$v(x_1, \ldots, x_n) = \int u(x_1, \ldots, x_n;k)dk$$

is also a solution, where k is a discrete or continuous parameter, respectively, and where the sum or integral may be either finite or infinite.

superposition principle or **theorem** *Physics.* see PRINCIPLE OF SUPERPOSITION.

superprint *Geology.* see OVERPRINT.

superradiant scattering *Physics.* the scattering of radiation that exhibits collective phase coherence and therefore has an intensity that is proportional to the square of the number of participating photons.

superregeneration *Electronics.* a regeneration process in which the oscillation is broken up at a frequency slightly above the upper audibility limit of the human ear by a separate oscillator circuit that is connected between the grid and anode of the amplifier tube, to prevent regeneration from going beyond the greatest useful amount.

superresolution *Oceanography.* the analysis of a tide into components of different frequencies, without measuring the full extent of the longest-period component.

Super Sabre *Aviation.* a popular name for the F-100 fighter plane.

supersaturate *Physical Chemistry.* to saturate abnormally; to cause to undergo a process of supersatuation.

supersaturated solid solution *Materials Science.* a solid solution formed when a material is cooled, usually by quenching, from a high-temperature, single-phase region to a low-temperature, two-phase region without the second phase precipitating.

supersaturated solution *Physical Chemistry.* a solution that holds more of a dissolved solute than is required to produce equilibrium with its undissolved solute. In nature, bees' honey is an example of a supersaturated solution of sugars (fructose and glucose) in water, formed by the evaporation of some of the water.

supersaturation *Physical Chemistry.* a condition in which a solution contains more solute than is normally necessary to achieve saturation under the same conditions. Also, SUPERSOLUBILITY. *Meteorology.* a local atmospheric condition in which the relative humidity is greater than 100%, containing more water vapor than needed to produce saturation with respect to a plane surface of pure water or ice; this occurs in the absence of condensation nuclei or other wettable surfaces.

superscalar *Computer Science.* a type of central-processing unit design in which certain operations that are adjacent in a single instruction stream can be executed concurrently.

super-Schmidt camera *Optics.* a camera that uses two spherical shell lenses to achieve a very low focal ratio (f/0.67); originally designed to photograph faint meteors.

superscript *Graphic Arts.* a superior character; i.e., a number or letter that is written slightly above, and usually to the immediate right or left of, another character, such as the "2" in πr^2.

supersensitive relay *Electricity.* a relay that operates on an extremely low current, usually below 250 microamperes.

supersensitivity *Neurology.* abnormally increased sensitivity.

superset *Mathematics.* a set Y is a superset of a set X if X is a subset of Y, and Y is said to contain X. This is denoted by $Y \supseteq X$ or $X \subseteq Y$. If, in addition, there is at least one element of Y that is not in X, then Y is a proper superset of X. This is denoted by $Y \supset X$ or $X \subset Y$. *Computer Programming.* a programming language that includes extensions to its original command repertoire.

supersolubility *Physical Chemistry.* see SUPERSATURATION.

supersonic *Physics.* relating to aerodynamic velocities that are greater than the speed of sound. *Aviation.* relating to flights at speeds above the speed of sound or flight vehicles that can operate at such speeds; used to form compounds such as **supersonic aircraft, supersonic transport, supersonic airfoil,** and so on. *Acoustics.* relating to sound waves having frequencies above the normal range of human hearing (about 20,000 hertz).

supersonic aerodynamics *Fluid Mechanics.* a branch of aerodynamics that deals with flows at supersonic speeds.

supersonic compressor *Mechanical Engineering.* a compressor in which a high-pressure rise is obtained by passing fluids over a rotor blade or a stator blade at supersonic velocities which produce oblique shock waves.

supersonic diffuser *Mechanical Engineering.* a diffuser that increases pressure and reduces the velocity of fluid moving at supersonic velocities.

supersonic flow *Fluid Mechanics.* a flow in which the particle velocity in the fluid exceeds that of sound in the same fluid, i.e., $M > 1$. Also, SUPERCRITICAL FLOW.

supersonic nozzle *Aviation.* a converging-diverging nozzle designed to accelerate a fluid to supersonic speed.

superspecies *Systematics.* a taxon consisting of a group of closely related allopatric species. A superspecies name is given in brackets between the genus and the species name.

superstandard propagation *Electromagnetism.* the propagation of radio waves under conditions of superstandard refraction in the atmosphere.

superstition *Psychology.* a false belief that certain negative events can be averted, or positive events brought about, by various charms, signs, ritual actions, and the like; resulting from accidental reinforcement or from cultural traditions. (Derived from a Latin phrase meaning "to stand over;" from the ancient idea of the gods watching over human events.)

superstitious *Psychology.* relating to or caused by superstition.

superstitious behavior *Psychology.* repetitive behavior that is intended to bring about or ward off an event, motivated by superstition.

superstring theory *Particle Physics.* an elementary-particle theory using supersymmetry, in which the particles are one-dimensional closed curves with zero thickness.

superstructure *Solid-State Physics.* a periodic arrangement of atoms of one type superimposed in the lattice of atoms of another type while the two species are not identified as a compound. Also, SUPERLATTICE. *Civil Engineering.* that part of a building or other structure above the foundation. *Naval Architecture.* all partial decks, deckhouses, bridges, or other structures other than masts that project above a vessel's upper deck.

supersuppressor *Molecular Biology.* a dominant external suppressor, either partial or total. Also, MULTISITE SUPPRESSOR.

supersymmetry *Particle Physics.* a proposed theory of elementary particles in which the spin property of the particle is used; supersymmetry relates bosons, which have integral spin, with fermions, which have half-integral spin.

supertanker *Naval Architecture.* the common, informal name for a tanker of the largest type, having a tonnage exceeding about 100,000 tons. Such vessels are formally classed as VLCC (Very Large Crude Carrier) or ULCC (Ultra Large Crude Carrier).

super transuranic see SUPERHEAVY ELEMENT.

supertrees *Forestry.* a term for superior trees that are straighter and faster growing than most of their species and that possess healthy, high-quality wood.

superturbulent flow *Fluid Mechanics.* a fluid flow whose frictional energy loss is so great that the Reynolds transition criterion from laminar-to-turbulent flow does not apply.

supervised learning *Artificial Intelligence.* learning in which a human instructor provides training instances and their correct classifications.

supervisor *Computer Programming.* the part of an operating system that coordinates the use of system resources and maintains processing unit operations. Also, **supervisory program.**

supervisor call *Computer Programming.* an instruction that interrupts normal processing and calls upon the supervisor to perform a task that only the supervisor is able, or permitted, to do.

supervisor interrupt *Computer Programming.* an interruption caused by an instruction issued by the program to turn control over to the supervisor.

supervisor mode *Computer Programming.* a state of operation allowing certain instructions, such as memory-protection modification instructions or input/output operations, that are prohibited to the user program. Also, MASTER MODE.

supervisory computer *Computer Technology.* a minicomputer that accepts results and transfers new programs to satellite minicomputers, and may communicate with a larger computer.

supervisory control *Industrial Engineering.* a control panel or station that provides readouts of critical operating conditions throughout a plant or facility and allows control of the operation.

supervisory-controlled manipulation *Industrial Engineering.* a computer-aided manufacturing technique in which an operator performs a task, his or her motions being coded and stored for machine-operated repetition of the motions.

supervisory-controlled robot *Robotics.* a robotic system in which a human operator can use a computer to tell the manipulator the patterns of motion it will later use to accomplish some task.

supervisory routine *Computer Programming.* a coordinating routine that controls the operations of other routines or programs.

supervisory signal *Electricity.* a signal used to indicate to an attendant various states of circuitry in a switching apparatus or those for a particular connection.

supervisory system *Electricity.* all devices operating between the stations of an electric power distribution system that constitute supervisory control, indications, and telemeter selection, and all of the complementary devices located in remote stations that utilize a common interconnecting channel to transmit indicator signals between stations.

supervital gene *Genetics.* an allele that raises the viability of the individual possessing it above that of the wild type.

supervitaminosis see HYPERVITAMINOSIS.

supervoltage *Electricity.* a voltage that has a value of at least 500 kilovolts but that may be as high as 2000 kilovolts; used for X-ray tubes.

supination *Physiology.* **1.** the act of lying flat on one's back, with the face upward. **2.** the act of turning the palm forward or upward.

supine *Physiology.* lying with the face upward.

supine position see DORSAL POSITION.

supplementary angle *Mathematics.* given an angle $\theta < 180°$, the angle $\phi = 180° - \theta$. The angles θ and ϕ are said to be **supplementary**.

supplementary contour *Cartography.* a contour line added to a relief map, usually between intermediate contour lines in areas of exceptionally low relief, to make what topographic variation there is clear.

supplementary control *Cartography.* in surveying, control established by determining by ground surveying methods the position of a point or points visible in aerial photographs, in order to relate the photograph to the geodetic control system used. The points used to establish supplementary control are **supplementary control points.**

supplementation *Nutrition.* the addition of adequate levels of essential amino acids to plant food products, so as to produce products of increased nutritional value, such as the addition of lysine to wheat flour.

supplied-air respirator *Engineering.* a respirator in which air is provided from a tank or outside source; used in oxygen-deprived or toxic atmospheres.

supply current *Geophysics.* an electric current in the atmosphere that, by transporting negative charges downward or positive charges upward, balances the observed air-earth current in fair-weather regions of the earth.

supply voltage *Electricity.* the voltage that is supplied from a power source to operate a circuit or electronic device.

support *Military Science.* **1.** the action of a force that aids, protects, complements, or sustains another force. **2.** specifically, a unit that aids another unit in combat; e.g., artillery, air, or naval gunfire may be a support for infantry. *Medicine.* **1.** a device or appliance that helps to maintain a part or structure. **2.** the foundation upon which a denture rests. *Mathematics.* **1.** the support of a function f is the smallest closed set outside which f vanishes identically, i.e., outside which f is zero; denoted by supp f. In particular, supp f is the closure of the set $\{x: f(x) \neq 0\}$. **2.** in a measure space (X, A, μ), the support of the measure μ is the unique closed set E_μ such that: (a) if $U \in A$ (i.e., an open set) and $E_\mu \cap U \neq \Delta$, then $\mu(E_\mu \cap U) > 0$; and (b) $\mu(E_\mu') = 0$, where E_μ' is the complement of E_μ in X. By theorem, such a set always exists; it is sometimes called the **carrier** or **spectrum** of μ.

support gripper *Robotics.* an end effector, such as a hook, that can hold an object.

supporting *Military Science.* of or relating to actions, forces, weapons, and materiel that are used to support other forces in combat. Thus, **supporting aircraft, supporting arms, supporting artillery, supporting attack, supporting fire, supporting operations,** and so on.

support item *Ordnance.* any item that is needed to operate, service, repair, or overhaul an end item; e.g., parts, tools, and test equipment.

support list *Artificial Intelligence.* in a truth maintenance system, a list of concepts related to a given concept that must be either in or out for the given concept to be true (in).

suppository *Pharmacology.* a medicated mass, solid at room temperature and melting at body temperature, shaped to facilitate insertion in rectal, vaginal, or urethral body openings.

suppressant *Pharmacology.* an agent that halts some process such as secretion, excretion, or discharge.

suppressed-zero instrument *Engineering.* an instrument having a readout in which the zero value is below the scale markings.

suppression a reduction, elimination, or inhibition; specific uses include: *Psychology.* **1.** the conscious inhibition of an unacceptable idea or activity. **2.** any conscious elimination of certain behavior. *Physiology.* the sudden stoppage of a secretion, excretion, or normal discharge. *Genetics.* the silencing of an aberrant gene activity, resulting in the restoration of a lost or mutationally altered genetic function. Also, GENE SUPPRESSION. *Military Science.* the temporary degradation of the performance of a weapons system to a level below that necessary to fulfill its mission objectives; it is accomplished by an opposing force to allow the force to fulfill its own mission objectives without danger from the weapons system. Thus, **suppression mission, suppressive fire.** *Electronics.* **1.** the elimination of any component of a transmission, such as a particular frequency or group of frequencies in an audio frequency of a radio-frequency signal. **2.** the reduction of noise pulses generated by a motor. *Computer Programming.* a function that prevents the printing of certain selected characters under specified conditions.

suppression amblyopia *Medicine.* a condition of amblyopia resulting from suppression of vision in one eye to avoid diplopia.

suppression shield *Nucleonics.* any material that is used to totally or partially block radiation for a particular region.

suppressor *Electricity.* **1.** a component such as a capacitor or resistor, or both, that is used in an electric circuit to dampen high-frequency oscillation, when a current is broken at a contact point. **2.** a resistor that is used in conjunction with a spark plug or distributor in an internal combustion engine to suppress spark noise from the ignition system, which would interfere with radio reception. *Electronics.* a device used to reduce or eliminate noise or other signals that interfere with the operation of a communication system.

suppressor cells *Cell Biology.* differentiated T lymphocytes that inhibit the response of other lymphocytes, either B or T cells, to an antigen. Also, **suppressor T cells.**

suppressor grid *Electronics.* a grid placed between two positive electrodes in order to reduce the flow of secondary electrons from one electrode to the other; usually between the screen grid and the anode.

suppressor mutation *Genetics.* **1.** an intergenic mutation that codes for an abnormal tRNA whose anticodon interprets the mutated codon either normally or as an acceptable substitute for the original meaning. **2.** an intragenic mutation that restores the orginal reading frames after a frameshift mutation. Also, **suppressor gene.**

suppressor tRNA *Molecular Biology.* any mutant transfer RNA species that recognizes and utilizes any of the natural nonsense codons.

suppressor T set *Immunology.* a subpopulation of T lymphocytes that weakens the immune responses of other T or B cells.

suppurate *Medicine.* to form or discharge pus.

suppuration *Medicine.* **1.** the formation or discharge of pus. **2.** the matter so formed or discharged.

supra- a prefix meaning: **1.** above or over, as in *supraneural.* **2.** beyond the limits of or outside, as in *supranormal.*

supracardinal veins *Vertebrate Zoology.* either of two transient veins in mammalian embryos and various lower vertebrate adults; located in the dorsolateral thoracic and abdominal regions, and giving rise to the azygous and hemiazygous veins and part of the inferior vena cava.

supracostal *Anatomy.* situated above or upon a rib or ribs.

supracrustal *Geology.* relating to or located at the surface of the earth.

supracrustal rocks *Geology.* rocks that lie directly above the oldest rocks (or basement) of a unit.

suprahepatic *Anatomy.* situated above the liver.

suprahyoid muscle *Anatomy.* any of several muscles that extend upward from the hyoid bone to structures in the head; including the digastric, mylohyoid, stylohyoid, and geniohyoid muscles.

supralateral tangent arc *Meteorology.* either of two oblique luminescent arcs that are produced by refraction in hexagonal columnar ice crystals and appear concave to the sun and tangent to the halo of 46° at points above the sun's altitude.

supraliminal *Physiology.* above the threshold of consciousness.

supralittoral see SUPRATIDAL.

suprameatal triangle see MASTOID FOSSA.

supranasal *Anatomy.* situated above the nose.

supranormal *Medicine.* greater than normal; occurring in excess of normal amounts or values.

supranuclear *Anatomy.* describing those regions of the nervous system situated rostral to the motor neurons of the cranial and spinal nerves.

supraoccipital *Anatomy.* situated above the occipital bone or occipital region of the skull.

supraoptic *Anatomy.* situated above the eye.

supraoptic nucleus *Physiology.* either of a pair of nuclei of nerve cells located in the hypothalamus on each side of the third ventricle, near the optic tract.

supraorbital *Anatomy.* situated above the orbit. Thus, **supraorbital ridge.**

suprapelvic *Anatomy.* situated above the pelvis.

suprapermafrost layer *Hydrology.* the layer of ground or soil lying above the permafrost, including the seasonally frozen layer as well as any other frozen or unfrozen ground. Also, **supragelisol.**

suprapubic *Anatomy.* situated above the pubic arch.

suprapubic cystotomy *Surgery.* cutting into the bladder with an incision above the pubic symphisis. Also, LAPAROCYSTOTOMY.

suprarenal *Anatomy.* situated above a kidney.

suprarenal gland see ADRENAL GLAND.

suprascapular *Anatomy.* situated on the upper part of the scapula.

suprascleral *Anatomy.* situated on the outer surface of the sclera.

suprasegmental reflex *Physiology.* a complex reflex involving numerous neurons that fix body positions and integrate and coordinate complex muscular movements.

supraspinal *Anatomy.* situated above or upon a spine.

supraspinous *Anatomy.* situated above a spine or a spinous process.

suprasternal *Anatomy.* situated above the sternum.

suprasternal notch *Anatomy.* the large notch in the superior border of the manubrium of the sternum.

suprasternal space *Anatomy.* the small space above the manubrium of the sternum through which the anterior jugular veins pass.

supratemporal *Anatomy.* situated above the temporal bone, temporal fossa, or temporal region.

suprathoracic *Anatomy.* situated above the thorax.

supratidal *Ecology.* of or relating to the shore area adjoining and just above the high-tide level. Also, SUPRALITTORAL.

supratidal sediment *Geology.* unconsolidated sedimentary material that is deposited directly above the high-tide level. Also, **supralittoral sediment.**

supratidal zone *Ecology.* the region of the shore that is just above the highest water level, within reach of waves or spray. Also, **supralittoral zone.**

supratonsillar *Anatomy.* situated above a tonsil.

supravaginal *Anatomy.* situated above or outside of a sheath, especially above the vagina.

supravital *Biology.* describing the staining of living cells after removal from a living animal or of still living cells from a recently killed animal.

supremum *Mathematics.* another term for least upper bound; written lub or sup. See LEAST UPPER BOUND.

supremum principle *Mathematics.* the principle that every nonempty set of real numbers that has an upper bound also has a supremum (least upper bound.).
sur *Meteorology.* a cold wind in Brazil.
sur- a prefix meaning: **1.** over; above. **2.** in addition.
suracon *Meteorology.* a very cold, rainy wind in Bolivia.
surbase *Architecture.* the crown molding of a pedestal or baseboard.
surcharge *Civil Engineering.* **1.** any load above the ground surface. **2.** any load above the top of a retaining wall.
surcharged wall *Civil Engineering.* a retaining wall carrying a surcharge, such as an embankment, usually above the top of the wall.
surf *Oceanography.* ocean waves breaking on the shore or within the area from the shore to the farthest oceanward breaker; wave activity in the surf zone.
surface *Science.* the outer part or an external aspect of an object; e.g., the earth or a body of water. *Physics.* any boundary of a body or system that is in physical contact with something external to it. *Mathematics.* a two-dimensional submanifold of three-dimensional space, in particular, of R^3. *Aviation.* an airfoil that provides an outer contour to perform a function, such as a control surface or lifting surface.
surface acoustic wave *Acoustics.* a sound wave that travels close to the surface boundary of a medium, due to surface-boundary reflections; e.g., a shock wave produced by an earthquake or a sound wave trapped near the ocean surface due to a positive sound velocity gradient near the surface.
surface acoustic-wave device *Electronics.* any device that employs surface acoustic waves with frequencies in the range of 10^7–10^9 Hz to process electronic signals.
surface-active agent *Materials Science.* any substance that when dissolved reduces the surface tension or the interfacial tension between a solid and liquid.
surface-active glass *Materials.* any of various glasses developed to form a bond with living tissues by means of controlled chemical reaction at the surface of an implant; used as coatings for high-strength ceramics and surgical alloys for implants in which strength is required, such as bone and tooth replacements. Also, BIOGLASSES.
surface-air leakage *Mining Engineering.* the volume of surface air that enters a mine fan by passing through the casing at the top of the upcast shaft, at air-lock doors, and at fan-drift walls.
surface analysis *Computer Technology.* a method of testing the surface of a disk to detect damaged areas.
surface area *Mathematics.* the two-dimensional surface measure of an object.
surface barrier *Electronics.* on the surface of a semiconductor junction, a barrier sometimes produced by the diffusion of charge carriers.
surface-barrier transistor *Electronics.* a transistor triode in which surface barriers are formed by training two jets of electrolyte against opposite surfaces of N-type semiconductor material to etch and then electroplate the surfaces.
surface biopsy *Pathology.* the removal and diagnostic examination of cells most commonly scraped from the surface of lesions suspected of cervical cancer.
surface boundary layer *Meteorology.* a thin layer of air, extending from near the earth's surface up to the base of the Ekman layer (less than 300 feet), within which shearing stresses are nearly constant and wind distribution is determined largely by the vertical temperature gradient and the nature and contours of the underlying surface. Also, GROUND LAYER, SURFACE LAYER, FRICTION LAYER.
surface carburetor *Mechanical Engineering.* a carburetor that operates by passing air over the surface of a gasoline charge to absorb vapor.
surface-charge transistor *Electronics.* an integrated-circuit transistor element which is used to control the transfer of stored electric charges along the surface of a semiconductor.
surface chart *Meteorology.* a synoptic weather chart showing the distribution of sea-level pressure and the location, intensity, and movement of fronts and air masses; the type of chart most commonly used as a "weather map." Also, SURFACE MAP, SEA-LEVEL PRESSURE CHART.
surface checking *Materials Science.* a defect that appears as surface fissures during the drying of lumber, as a result of stresses created when the surface layer loses moisture and begins to shrink before the inner core.
surface chemistry *Physical Chemistry.* the branch of chemistry that measures and analyzes the factors and forces that act at the surface of solids, liquids, and gases, or at the interfaces between two phases; e.g., the study of surface tension in liquids.
surface-coated mirror *Optics.* a mirror whose reflective coating lies on its surface. Also, FIRST-SURFACE MIRROR.
surface combustion *Engineering.* a nonluminous combustion method in which burning takes place just above a porous surface through which the combustible gas is passed.
surface condenser *Mechanical Engineering.* a heat-transfer device in which exhaust is condensed by contact with metal surfaces cooled by a flow of cold water on the sides opposite the condensing surface.
surface contamination *Nucleonics.* radioactive contamination that exists only at the surface layers of a body.
surface-controlled avalanche transistor *Electronics.* a transistor in which avalanche breakdown voltage is controlled by an external field applied through surface-insulating layers; permits operation at frequencies in the 10-gigahertz range.
surface creep *Geology.* a slow advance of sand grains downcurrent, whereby the grains move each other along the surface.
surface current *Oceanography.* **1.** a current whose core of maximum velocity is near the surface. **2.** a current that extends no more than 3 meters deep in nearshore areas and no more than 10 meters deep in deep water.
surface density *Physics.* the amount of a quantity distributed over a surface area divided by the unit area, such as in a surface-charge density which may be expressed with the units of coulombs/m^2.
surface deposit see SURFICIAL DEPOSIT.
surface detention *Hydrology.* water from precipitation that exists temporarily on the surface, producing overland flow. Also, DETENTION, DETENTION STORAGE.
surface drainage *Hydrology.* the removal of excess water from the ground surface by any natural or artificial means, such as grading, terracing, and ditch digging.
surface drilling *Mining Engineering.* a borehole or series of boreholes collared at the earth's surface rather than underground or underwater.
surface duct *Geophysics.* an atmospheric duct whose lower boundary is the surface of the earth.
surface-effect ship *Naval Architecture.* a vessel that operates on water only and is supported by low-pressure, low-velocity air.
surface energy *Fluid Mechanics.* the work per unit area required to bring fluid molecules to the interface between two immiscible liquids or between a liquid and a gas. *Materials Science.* see INTERFACIAL FREE ENERGY.
surface exclusion *Molecular Biology.* a condition in which a conjugative plasmid in a cell limits the cell's ability to acquire a similar plasmid by conjugation from another cell.
surface fermentation *Biotechnology.* a fermentation process in which the microorganism grows on the surface of the static fermentation liquor.
surface finish *Engineering.* the degree of smoothness of a surface.
surface fire *Forestry.* a forest fire that burns only surface litter, forest floor debris, and undergrowth.
surface flow see OVERLAND FLOW.
surface force *Mechanics.* an external force exerted only on the surface of a body. Also, SURFACE TRACTION.
surface friction *Geophysics.* the drag exerted on the atmosphere by the motion of the earth, usually manifested in the shearing action of the wind.
surface front *Meteorology.* see FRONT.
surface gauge *Mechanical Devices.* **1.** an adjustable steel scriber mounted on a metal block or surface plate, used by machinists for testing the accuracy of planed surfaces or to mark off castings. **2.** a gauge used to determine the distance of points on a plane from a reference plane.
surface geology *Geology.* a branch of geology that is concerned with the study and correlation of the geologic features at the surface of the earth.
surface grafting *Materials Science.* a chemical modification of a polymer surface by grafting a thin layer of a second polymer.
surface grinder *Mechanical Engineering.* a machine that uses a high-speed abrasive wheel to produce a plane surface.
surface hardening *Metallurgy.* a process that increases hardness at the metal surface, such as casehardening, cold drawing, or tumbling.
surface hoar *Hydrology.* a hoarfrost, consisting of leaf-shaped or plate-shaped ice crystals, formed directly on a surface of snow or on some smooth, plane surface.
surface hydrology *Hydrology.* the branch of hydrology that deals with the study of water on the land surface, as opposed to underground water.

surface ignition *Engineering.* gas ignition triggered by contact with a hot surface, such as a heating coil.
surface integral *Mathematics.* an integral of a 2-form over a two-dimensional surface.
surface inversion *Meteorology.* a temperature inversion that begins at the earth's surface, usually occurring over land prior to sunrise and in winter over a high-latitude continental interior. Also, GROUND INVERSION.
surface irrigation *Agronomy.* any method of irrigation in which the water is distributed over the soil by flooding or in furrows.
surface layer see SURFACE BOUNDARY LAYER.
surface leakage *Electricity.* a current that passes over, rather than through, an insulator.
surface lift *Mining Engineering.* the heaving and uplift of the surface around a shaft resulting from the freezing method of shaft sinking.
surface luster *Optics.* see GLOSS, def. 2.
surface magnetic wave *Electromagnetism.* a magnetostatic wave that propagates along the surface of a magnetic material.
surface map see SURFACE CHART.
surface mining *Mining Engineering.* any mining performed at or near the earth's surface.
surface model see SHALLOW MODEL.
surface mount assembly *Electronics.* **1.** a method of mounting electronic circuit components and their electrical connections on the surface of a printed board, instead of through holes. **2.** a circuit board produced by this method.

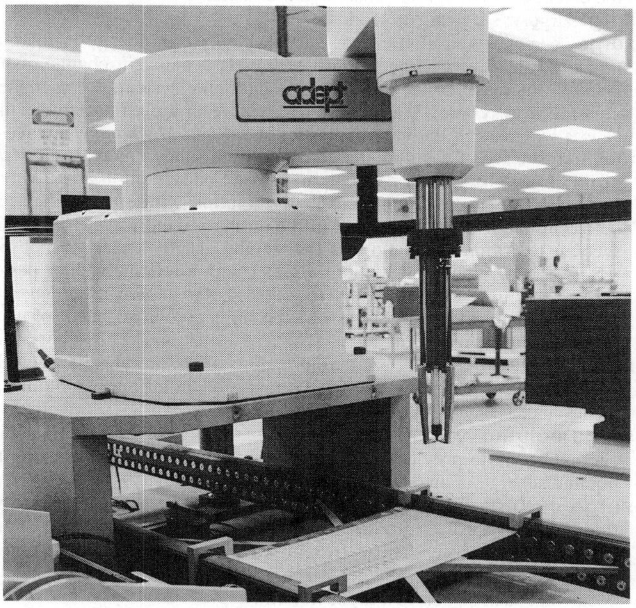

surface mount assembly

surface navigation *Navigation.* any navigation on the surface of the earth, especially on the surface of the water.
surface noise *Electronics.* the noise component in the electric output of a record-player pickup caused by irregularities in the contact surface of the groove; i.e., from physical wear on the record rather than improper functioning of the equipment.
surface observation *Meteorology.* a weather observation made from the earth's surface, as distinguished from an upper-air observation; used in preparing surface synoptic charts. Also, SURFACE WEATHER OBSERVATION.
surface of discontinuity *Meteorology.* an interface that occurs in the atmosphere.
surface of revolution *Robotics.* a surface created by the rotation of an object around an axis at a specific angle. *Mathematics.* any surface produced by rotating a plane curve about a line (called the **axis of revolution**) in its plane. Sections of such a surface perpendicular to the axis of revolution are circles.
surface of section see POINCARÉ SURFACE OF SECTION.
surface orientation *Physical Chemistry.* an arrangement of molecules on the surface of a liquid with one specific end of the molecules toward the liquid, as in a monofilm.
surface passivation *Electronics.* a method of surface coating of a P-type wafer for a diffused junction transistor with an oxide compound, to prevent penetration of the impurity in any undesired areas.
surface phase *Geochemistry.* a thin rock layer whose properties differ from those homogeneously, physically distinct layers on either side. Also, VOLUME PHASE.
surface physics *Solid-State Physics.* the study of physical phenomena occurring at or near the surface layers of atomic planes in a crystal.
surface-piercing hydrofoil *Naval Architecture.* a hydrofoil vessel whose foils skim along the surface during normal hydrofoil operation, maintained stable by hydrodynamic balance of the submerged and non-submerged portions of the foils.
surface pipe *Petroleum Engineering.* the first casing set into a well, cemented into place as a foundation for subsequent operations and as a means of shutting off shallow water formations from contamination by deeper, saline waters.
surface plasmon *Solid-State Physics.* a quantum of longitudinal excitation in the electron gas of a metal that is located near the boundary and therefore subject to boundary conditions.
surface plate *Mechanical Devices.* a precision steel instrument with one or two flat surfaces at right angles, used as a standard for flatness to locate testing fixtures. *Graphic Arts.* the most common type of lithographic printing plate, consisting of a water-receptive metal sheet coated with a photosensitive, ink-receptive substance that hardens when exposed to light; after exposure, the soft coating is removed from nonimage areas, but the hard coating remains on the image area.
surface pressure *Physics.* a film pressure; the force per unit length that is analogous to a gas pressure (force per unit area) except that the fluid molecules are restricted to two-dimensional motion, such as a monomolecular layer of oil on water. *Meteorology.* **1.** the atmospheric pressure at any given point on the earth's surface. **2.** see STATION PRESSURE. **3.** see SEA-LEVEL PRESSURE.
surface printing see PLANOGRAPHY.
surface reaction *Chemistry.* a chemical reaction that occurs on an adsorbent or solid catalyst.
surface recombination velocity *Solid-State Physics.* a quantity given by the electron or hole current density component normal to the surface of a semiconductor, divided by the excess electron or hole volume charge density near the surface; the quantity provides a measure of the rate of recombination occurring at the semiconductor's surface.
surface resistivity *Electricity.* the resistance on the surface, between two opposite edges of a square of film.
surface road *Transportation Engineering.* **1.** any road that is at ground level rather than elevated or submerged. **2.** specifically, a local road of this type, as opposed to a limited- access expressway.
surface runoff *Hydrology.* the part of precipitation that travels over the ground surface into the nearest surface streams.
surface-set bit *Mechanical Devices.* a drill bit set with a single layer of diamonds which protrude from the steel surface of the bit. Also, **surface-layer bit.**
surface slope *Hydrology.* for flowing water, the difference between the angle of the water surface and a theoretical horizontal surface.
surface soil *Geology.* the soil of the earth's surface, usually defined as extending to a depth of 12 to 20 centimeters.
surface state *Solid-State Physics.* in a semiconductor, an electron state that is restricted to a surface layer.
surface steps *Materials Science.* atomic-level defects on the surface of a crystalline solid that take the form of steps.
surface storage *Hydrology.* the portion of precipitation which is stored temporarily at or above the ground surface during or after a rainfall. Also, **surface retention.**
surface structure *Linguistics.* in Noam Chomsky's theory of transformational grammar, the actual (written or spoken) syntactic form of a sentence, as opposed to deep structure (the underlying rules employed intuitively by users of a language in forming sentences).
surface survey see PEDESTRIAN SURVEY.
surface temperature *Meteorology.* the air temperature at or near the earth's surface.
surface tension *Fluid Mechanics.* the stretching force required to form a liquid film; equal to the surface energy of the liquid per unit length of the film at equilibrium; the force tends to minimize the area of a surface. Also, **surface tensity.**

surface tension number *Fluid Mechanics.* a dimensionless number used in the study of mass transfer in packed columns; $T_s = m^2L/Asr$, where m is the liquid viscosity, L the perimeter of a packing element, A the surface area of the packing element, s the surface tension of the liquid, and r the density of the liquid.

surface thermometer *Engineering.* a thermometer used to measure the temperature of the surface layer of water; typically it is placed in a bucket and lowered to the surface.

surface-to-air (guided) missile *Ordnance.* a surface-launched missile designed to operate against a target in the atmosphere.

surface-to-air missile envelope *Ordnance.* the air space that is within the kill capabilities of a specific surface-to-air missile system.

surface-to-surface (guided) missile *Ordnance.* a surface-launched missile designed to operate against a target on the surface; the term "surface" may refer to either the ground or the surface of a ship.

surface-to-surface missile (U.S. Scout missile)

surface traction see SURFACE FORCE.

surface trait *Psychology.* a specific, superficial personality trait that derives from the more fundamental source traits.

surface treating *Materials Science.* a general term for a variety of treatments designed to protect a material, such as waterproofing, or to make the material more receptive to other materials, such as paint.

surface vibrator *Mechanical Engineering.* a vibrating machine that is used on pavement to compact and solidify the concrete.

surface visibility *Meteorology.* visibility as determined from a point on the ground, as distinguished from that obtained from an elevated control tower or from an airborne aircraft.

surface wash see SHEET EROSION.

surface water *Hydrology.* all the bodies of fresh water, salt water, ice, and snow on the earth's surface. *Oceanography.* see MIXED LAYER.

surface wave *Fluid Mechanics.* a wave that distorts the free surface separating two fluid phases. *Mechanics.* a wave that propagates on the surface of a solid; first studied by Rayleigh in connection with earthquake vibrations propagating in the earth's outer shell. Also, RAYLEIGH WAVE. *Electromagnetism.* an electromagnetic wave that propagates along an interface between two different mediums without radiation. *Oceanography.* a progressive gravity wave in which the particle movement within the fluid mass is confined to the upper layers and is greatest at the surface. *Telecommunications.* see GROUND WAVE.

surface-wave transmission line *Electromagnetism.* a low-loss, single-conductor transmission line with a dielectric sheath that is typically three or more times greater in diameter than the conductor and with a conical horn at each end.

surface weather observation see SURFACE OBSERVATION.

surface wind *Meteorology.* the wind measured at a surface observation station, measured at a standard height (usually 10 meters) above the ground in order to minimize distortions from local obstacles and terrain.

surface zero see GROUND ZERO.

surfacing *Materials Science.* a welding technique by which a hard or corrosion-resistant material is deposited on the surface of a second material, producing a laminar composite.

surfactant *Materials Science.* any surface-active agent or substance that modifies the nature of surfaces, often reducing the surface tension of water; surfactants are used as wetting agents, detergents, penetrants, and emulsifiers.

surfactin *Biochemistry.* the extracellular product of *Bacillus subtilis* that dissolves red blood cells and certain protoplasts.

surf beat *Oceanography.* a series of irregular oscillations of the nearshore water level, with periods of the order of several minutes, caused by the irregular arrival of groups of relatively large waves.

surficial *Geology.* relating to or occurring on or near a surface, usually a land surface.

surficial creep see SOIL CREEP.

surficial deposit *Geology.* loose accumulations of unconsolidated material that overlie bedrock or occur on or near the earth's surface of the earth, derived from processes of streams, glaciers, or deformation. Also, SUPERFICIAL DEPOSIT, SURFACE DEPOSIT.

surficial geology *Geology.* a branch of geology dealing with surficial deposits, including soils and also bedrock at or near the surface.

surfperch *Vertebrate Zoology.* any of several marine fishes of the family Embiotocidae found in shallow waters along the Pacific coast of North America. Also, **surffish.**

surf ripple *Geology.* an undulating surface feature on a sandy beach, formed by wave-generated currents.

surf zone *Oceanography.* the area between the outermost breakers and the extent of wave uprush. Also, BREAKER ZONE.

surg. surgery; surgical; surgeon.

surge *Electricity.* a momentary, sudden increase or change in the current or voltage of a circuit. *Engineering.* **1.** a transient change, especially an increase, in the pressure of a hydraulic system. **2.** the highest pressure in a system. **3.** a buildup of pressure in a plastic extruder that causes waviness in the hollow plastic tube. *Fluid Mechanics.* a wave at the free surface of a liquid characterized by a sudden increase in depth of the flow across the wavefront, and severe eddy motion at the wavefront. *Oceanography.* **1.** an ocean wave motion with a period halfway between that of the tide and that of the ordinary wind wave (roughly 30 seconds to 60 minutes), and of a low height, usually under 10 centimeters. **2.** the horizontal oscillation of water with a relatively short period accompanying a seiche. **3.** see STORM SURGE. *Astronomy.* a sudden, violent event that sends a jet of gas into the sun's corona at speeds of up to 200 kilometers per second.

surge admittance *Electricity.* the reciprocal of the ratio between the voltage and current of a wave, which travels on an infinitely long line of the same characteristics as the relevant line.

surge column *Petroleum Engineering.* a large-diameter pipe with enough height to provide a static head capable of absorbing the surging liquid discharge of an attached process tank.

surge current *Electricity.* **1.** a high-amperage electric current that flows initially into a capacitor when a charging voltage is applied. **2.** a short-duration, large electric-current wave that may sweep through a power transition network when electrical activity in the atmosphere is occurring, such as during a thunderstorm.

surge electrode current see FAULT ELECTRODE CURRENT.

surge generator see IMPULSE GENERATOR.

surge impedance *Electricity.* the impedance of a transmission line seen by a pulse; it is approximately equal to $\sqrt{L/C}$, where L and C are the inductance and capacitance per unit length. *Telecommunications.* see CHARACTERISTIC IMPEDANCE.

surge line *Meteorology.* a line marking a discontinuity in wind speed, sometimes accompanied by a change in direction; the wind speed is usually stronger upstream from the surge line.

surgeon *Surgery.* a licensed practitioner of surgery; a physician who treats disease, injury, dysfunction, or deformities by manual or operative means.

surgeonfish *Vertebrate Zoology.* a coral-reef fish of the family Acanthuridae, characterized by one or more long, sharp spines near the base of the tail fin; found in tropical seas. (So-called because of the resemblance of its spines to a surgeon's instruments.)

surgeon's knot *Surgery.* a knot made by passing suture material through an initial loop twice to prevent slippage.

surge protection *Electrical Engineering.* any form of protection in which electrical and electronic equipment are protected from damage caused by sudden voltage excursions. Thus, **surge protector.**

Surgery

Surgery is that branch of medicine concerned with the treatment of injuries and other disorders or diseases by means of manual and instrumental operations. Historically, surgery is one of the oldest forms of therapy. The earliest known surgical procedures were performed in prehistoric times with crude, sharpened stones as emergency operations, including trephining, bloodletting, incision of abscesses, ligation of blood vessels, and circumcision. The objective of surgeons at that time was to extirpate the pathologic process.

In the Stone Age, amputations were performed with saws made of stones or bones. The ancient Hindus, using tongs, hooks, forceps, sharp scalpels, needles and threads, rectal speculums, and magnets, performed such major operations as lithotomy, amputations, ophthalmic surgical procedures, hemorrhoidectomy, alloplasty, and rhinoplasty some thousand years before the Greeks and Romans, who also had skilled surgeons. With the decline of their civilizations, unfortunately, their surgical techniques were lost.

During the Middle Ages, surgical knowledge regressed. A rift that developed between physicians and surgeons resulted in the unskilled and uneducated practicing surgery. These practitioners were called barber-surgeons because they were both barbers and surgeons. The organization of guilds of barber-surgeons in England brought some respect to them, but surgery did not attain true professional status until the eighteenth century. Notable surgeons of the Renaissance (fourteenth through sixteenth centuries) included Girolamo Fracastoro, Ambroise Paré, Guido Lanfranchi, Andreas Vesalius, and Paracelsus.

Modern elective surgery began in the nineteenth century with the introduction of antisepsis and asepsis, anesthesia, and the roentgen-ray. Among those associated with the doctrine of asepsis were Ignacz Semmelweis; Louis Pasteur, who determined that microbes caused infection and disease; Sir Joseph Lister, who in 1860 introduced carbolic acid as a disinfectant; and Robert Koch, who is known as the founder of bacteriology. Asepsis eventually replaced antisepsis. Crawford Long and William Thomas Green Morton discovered the use of ether to induce anesthesia. Another important development was steam sterilization under pressure for instruments, which Ernst von Bergmann introduced. Other knowledge that advanced surgery included discovery of the causes of shock, blood-group typing, blood transfusion techniques, blood-clotting, anticoagulants, and antibiotics. Roentgenography and fluoroscopy were invaluable in determining or verifying the diagnosis and in helping the surgeon to plan surgical procedures.

Thus, surgery's initial objective of extirpation of the diseased tissue or organ was replaced with preservation or restoration of function and cure of the underlying disease. Throughout history, intelligent, imaginative, and observant surgeons have contributed through research to the profound body of medical knowledge that has been amassed by man, but the striking surgical advances that occurred during the twentieth century with such rapidity contributed enormously to the current status of surgery as a prestigious scientific discipline with its various subspecialties (urology, obstetrics and gynecology, neurosurgery, orthopedic surgery, ophthalmology, otorhinolarygology, and plastic surgery).

During the last half of the twentieth century, aggressive cardiovascular research made possible the successful surgical treatment of previously fatal coronary artery disease, carotid artery disease, and aneurysmal and occlusive arterial disease by resection and graft replacement. Organ transplantation restored the function of vital organs to patients whose kidneys, heart, lungs, liver, pancreas, or bones were deteriorating rapidly.

Michael E. DeBakey, M.D.
Chancellor and Chairman
Department of Surgery
Baylor College of Medicine

surgery *Medicine.* **1.** the branch of medicine that treats injuries, deformities, and diseases by operative or manual methods, often with incision of the body for internal manipulation. **2.** the place in a hospital or doctor's or dentist's office where such medical treatment is performed. **3.** the operative or manual procedures performed in this way. (Going back to a Greek term meaning, literally, "hand work.")

surge stress *Mechanics.* stress on process equipment or systems that arises from a surge in the fluid flow rate or pressure.

surge suppressor *Electronics.* a component that responds to the rate of change of a current or voltage in order to prevent a quick rise above a predetermined value.

surge tank *Engineering.* a tank placed in a hydraulic system to relieve pressure variations or to handle the highest surge height, similar in function to a standpipe. *Mining Engineering.* a relatively small tank used to maintain a constant feed of ore pulp into a pump.

surgical *Surgery.* of, relating to, or correctable by surgery.

surgical current *Surgery.* an electric current applied by a surgical instrument to dissect or fulgerate tissue.

surgical isolator *Surgery.* a large, clear, plastic bag surrounding and enclosing a medical team during a surgical procedure so that infective contamination of the patient is prevented.

surgical needle *Medicine.* a sewing needle used for suturing in surgical operations.

surgical triangle *Surgery.* a triangular anatomical region, in which certain nerves, vessels, and organs are located, established for reference in surgical operation.

surging *Navigation.* the back-and-forth movement of a ship or vessel, especially when moored.

surging glacier *Hydrology.* a glacier that alternates between short periods (1 to 4 years) of very rapid flow, or surges, and longer periods (10 to 100 years) of stagnation or near stagnation.

Surianaceae *Botany.* a family of dicotyledonous tanniferous shrubs and trees of the order Rosales, characterized by alternate and simple leaves, flowers without nectary disks, and indehiscent fruits; native to Australia and other warm tropical areas.

suricate *Vertebrate Zoology.* a small, burrowing, carnivorous mammal, *Suricata suricatta*, of the civet family, characterized by a grayish color with dark bands across the back and a social behavior similar to the prairie dog. Also, **suricat.**

Surirellaceae *Botany.* a family of marine and freshwater benthic diatoms of the order Pennales, characterized by a dual-slitted raphe that extends around the valve and is separated by nodules.

surjection *Mathematics.* a mapping or function that is onto.

surmullet see GOATFISH.

suroet *Meteorology.* a rainy southwest wind that blows persistently along the west coast of France.

surprint *Graphic Arts.* a photographic print made by superimposing a print from one negative onto an image created from a previously printed negative.

surprinting *Graphic Arts.* **1.** the process of creating a surprint. **2.** the process of making a second imposition to print line art over halftone art or text over a tinted area.

surra *Veterinary Medicine.* a usually fatal blood disease affecting a variety of mammals, such as camels, horses, dogs, and cattle, that is caused by a blood-infecting protozoan parasite, *Trypanosoma evansi*, and transmitted by bloodsucking flies and fleas; characterized by fever, anemia, emaciation, and paralysis.

surrogate [sur′ə gət; sur′ə gāt′] *Behavior.* an unrelated animal that functions or is regarded as a parent. *Psychology.* a person who serves as a conscious or unconscious substitute for someone, usually a parent.

surrogate genetics *Genetics.* experiments that utilize the oocytes of nonhuman vertebrate subjects, especially *Xenopus*, to study the mechanics of gene activity and regulation.

surround-sound *Acoustical Engineering.* a term for a sound reproduction system in which four or more loudspeakers are variously placed to provide a listener with the sense of a natural sound environment.

sursassite *Mineralogy.* $Mn_2^{+2}Al_3(SiO_4)(Si_2O_7)(OH)_3$, a reddish-brown to copper-red monoclinic mineral, massive or fibrous in habit, and having a specific gravity of 3.256 and an undetermined hardness; found in iron-manganese ores.

surveillance [sur vā′ləns] *Aviation.* the use of photography, radar, infrared, or visual methods to study an area from an overhead aircraft or satellite, for military intelligence and other purposes.

surveillance approach *Aviation.* an instrument approach conducted in accordance with directions issued by a controller using a surveillance radar display.

surveillance radar *Aviation.* a ground-radar system that assists air-traffic controllers in converting random signals to regular landings and in positioning aircraft for low approaches.

survey *Science.* **1.** a detailed analysis of an area or situation, obtained through a careful and comprehensive examination or by the collection of data from individuals. **2.** the results of such an analysis. **3.** to carry out such an analysis. *Engineering.* **1.** to accurately measure and delineate the features of a land area, for mapping or as a preliminary to a construction project. **2.** the process of such measurement and delineation, or the information obtained by this process.

survey foot *Metrology.* a unit of measure equal to about 1.000002 feet.

surveying *Cartography.* the theories and techniques of determining the relative positions of points on, under, or above the surface of the earth (or another celestial body), for the purpose of establishing legal boundaries or constructing accurate charts and maps. *Geodesy.* the theories and techniques of determining the size and shape of the earth and of its gravity field.

survey instrument *Nucleonics.* a portable instrument used to make measurements of radioactivity in an area surrounding a source of radiation. Also, **survey meter.**

survey net *Cartography.* a system of closed loops or circuits extending over an area formed from observations taken from survey station to survey station. A horizontal control survey net consists of arcs of triangulation, sometimes with lines of traverse, connected together; a vertical control survey net consists of lines of levels connected together.

Surveyor *Space Technology.* a series of lunar probes, launched by the U.S. between 1966 and 1968, that brought back thousands of photographs and soil samples from the moon's surface.

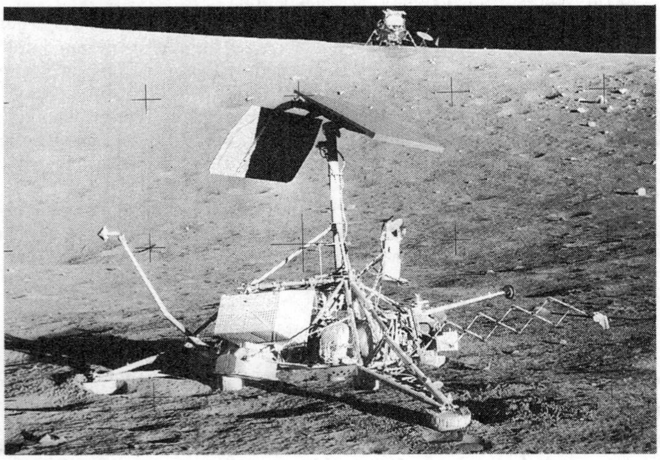

Surveyor III

surveyor *Engineering.* a person whose occupation or profession is surveying.

surveyor's compass *Engineering.* an instrument used to determine horizontal angles and bearings in survey work.

surveyor's cross *Engineering.* an instrument consisting of two bars crossing at right angles, each with sights, used in laying out rectilinear lines in survey work.

surveyor's level *Engineering.* a telescope fitted with a spirit level and adjusting screws for use in survey work.

surveyor's measure *Engineering.* a system of length and distance units used in survey, based on the surveyor's chain of 66 feet with 100 links of 7.92 inches.

survivability *Military Science.* the capability, especially of a weapons system, to survive an attack or the residual effects of an attack while maintaining the ability to perform the designated mission.

survival *Biology.* the fact of remaining alive, especially under adverse conditions. *Ecology.* see SURVIVORSHIP.

survival curve *Statistics.* a curve plotted on a graph indicating the survival rates of a specific population over time. *Nucleonics.* specifically, a plot of the number of organisms that survive a given amount of radiation over time.

survival function *Statistics.* the probability of surviving until time t as a function of t. The complement of the cumulative failure distribution.

survival of the fittest *Evolution.* the elimination, by natural selection, of members of a population that are least adapted to the environment, with the fittest (or best adapted) individuals surviving to contribute to the genetic makeup of succeeding generations.

survival probability *Military Science.* the likelihood that a target will survive a specific attack or operation.

survival ratio *Biology.* a comparison of the number of organisms surviving ionizing radiation to the number of organisms exposed.

survivorship *Ecology.* the percentage of newborn individuals that can be expected to survive to a given age. Also, SURVIVAL.

survivorship curve *Ecology.* a graph used to show the probability of survival for a given individual based on age.

Surwell clinograph *Engineering.* an instrument lowered by line into a well, consisting of a level, compass, thermometer, and clock, with a camera to record their readings; used to measure inclination from the vertical and other characteristics.

susannite *Mineralogy.* $Pb_4(SO_4)(CO_3)_2(OH)_2$, a colorless to yellowish or greenish trigonal mineral occurring as rhombohedral crystals, having a specific gravity of 6.55 and a hardness of 2.5 to 3 on the Mohs scale; trimorphous with leadhillite and macphersonite; found in the oxide zones of lead deposits.

suscept *Plant Pathology.* relating to a plant that is especially susceptible to infection by a disease-producing agent.

susceptance *Electricity.* the imaginary part of admittance; measured in siemens.

susceptibility [sə sep′tə bil′i tē] *Military Science.* **1.** the vulnerability of a weapons system or other device to attack due to an inherent weakness. **2.** the vulnerability of a target audience to specific forms of psychological operations approach. *Virology.* a condition in a host that allows the replication of a virus or other pathogen within it. *Electricity.* the ratio of the polarization of a body to the electric field imposed on the body. *Electromagnetism.* a ratio of the magnetization of a body to the magnetizing force imposed on the body.

susceptometer *Engineering.* an instrument used to measure the paramagnetic, diamagnetic, and ferromagnetic susceptibility of a material.

sus mutant *Genetics.* a suppressor-sensitive mutant.

suspect *Ecology.* a plant or animal thought to be susceptible to or hosting a disease organism.

suspended acoustical ceiling *Building Engineering.* a ceiling structure with acoustical characteristics in its materials, suspended from inside the roof structure or from a ceiling fixed into the building above it.

suspended ceiling *Building Engineering.* in flat ceiling surface fabrication, the use of cross-furring members attached to main structural runners composed of 3/4- to 1.5-inch cold-rolled steel channels, which lie perpendicular to the furring and are suspended from the floor system by heavy galvanized wires.

suspended load *Geology.* the part of the deposit carried by a stream that is in suspension for a relatively extensive period of time. Also, SUSPENSION LOAD.

suspended span *Civil Engineering.* a span supported by two cantilevers or a cantilever and a column pier; used mainly in bridges and some roofs.

suspended transformation *Thermodynamics.* a change in a thermodynamic system that fails to materialize after a change in the system conditions, such as the metastable equilibria resulting from supercooling or superheating.

suspension *Chemistry.* a system in which very small particles of solid, semisolid, or liquid matrial are more or less evenly dispersed in a liquid or gas phase; e.g., fog is a suspension of liquid (water droplets) in a gas (air); milk is a suspension of solid fat particles in a liquid. *Engineering.* **1.** a small spring mechanism that stabilizes the needle or moving element of a meter so that it can be read more easily and accurately. **2.** a system of springs, shock absorbers, or similar devices connecting the axles to the chassis of an automobile, railroad car, or other vehicle; designed to reduce unwanted motion transmitted from the road, railroad track, or other riding surface. *Mining Engineering.* a method used to bolt rock fragments and sections that might otherwise loosen and fall.

suspension bridge *Civil Engineering.* a bridge having a road or deck hung from a pair of steel cables, each carried by two towers, one at each bank; e.g., the Golden Gate Bridge over San Francisco Bay.

suspension cable *Engineering.* a freely hanging cable that may carry an evenly distributed load.

suspension culture *Cell Biology.* a cell culture that is maintained in a liquid medium, growing in suspension rather than attached to the surface of the culture container.

suspension feeder *Zoology.* an organism that feeds on small particles, living or dead, that are suspended in the water.

suspension insulator *Electricity.* a freely hanging insulator composed of units connected in series, by which an overhead line is suspended from the arm of a transmission-line tower; it withstands only the force of tension.

suspension load see SUSPENDED LOAD.

suspension method *Radiology.* a method of administering radiation therapy to the bladder wall by instilling a radioactive solution or suspension into the vesical lumen using a catheter.

suspension polymerization *Materials Science.* a polymerization process in which monomers are dispersed by mechanical agitation in a liquid phase, and monomer droplets are polymerized in the shapes of drops, beads, or irregular granules.

suspension roast see FLASH ROAST.

suspension roof *Building Engineering.* the use of cabling to suspend a roof; it is often made of a thick gauge steel in roof applications.

suspension system *Mechanical Engineering.* a system that supports the frame and load of an automotive vehicle, by means of springs, shock absorbers, and other hardware.

suspension test *Microbiology.* any qualitative or quantitative assay of the effectiveness of a given disinfectant upon a liquid culture of microorganisms.

suspensor *Botany.* in seed plants and club mosses, a group of cells attached to the embryo that pushes the embryo deeper into the embryonic sac and into contact with nutritive tissue.

suspensory *Surgery.* **1.** serving to hold up a part or structure. **2.** a bone, ligament, muscle, sling, or bandage so acting.

sussexite *Mineralogy.* $Mn^{+2}BO_2(OH)$, a white monoclinic or orthorhombic mineral occurring in fibrous veins, having a specific gravity of 3.30 and a hardness of 3 to 3.5 on the Mohs scale; it forms a series with szaibelyite.

sustainable yield. *Ecology.* the crop or yield that can be harvested repeatedly from a population without diminishing the stock.

sustained oscillation *Control Systems.* oscillation that continues over a period of time. *Physics.* oscillation that is driven and continued by an external force.

sustained rate of fire *Ordnance.* the rate of fire that a weapon can deliver for an indefinite length of time without overheating.

sustained yield *Biology.* the replacement of a harvest yield by growth or reproduction before another harvest occurrs. *Forestry.* **1.** the balance that is maintained between the growth of wood and the annual harvest to ensure a continuous timber supply. **2.** the yield that a forest can continuously produce.

sustainer engine *Space Technology.* an engine designed to maintain or increase the speed of a rocket or missile that first derives its programmed cruising speed from a booster or other engine.

sustentacular cell *Histology.* an epithelial cell that provides support and nourishment to adjacent functional cells, such as the Sertoli cells of seminiferous tubules.

sustention *Aviation.* the process of sustaining altitude in an aircraft.

SU_3 symmetry see UNITARY SYMMETRY.

Sutherland, Earl W., Jr. 1915–1974, American physiologist; awarded the Nobel Prize for research on hormonal control of human metabolism.

Sutherland's formula *Physics.* a relation for the absolute viscosity of a gas at a particular temperature *T*,

$$\eta = \eta_0(T/273.1)^{3/2}[(C + 273.1)/(C + T)],$$

where η_0 is the viscosity of the gas at 0°C and *C* is a constant related to the gas.

Sutro weir *Civil Engineering.* a measuring weir in which the head above the crest is directly proportional to the discharge.

Sutton-Rendu-Osler-Weber syndrome see HEREDITARY HEMORRHAGIC TELANGIECTASIA.

suture [sooch´ər] *Medicine.* **1.** a fibrous joint in which the opposed surfaces are closely united, as in the skull. **2.** a threadlike material, often silk or catgut, used to stitch a wound. **3.** the stitch or series of stitches made to close a surgical or accidental wound, or the process of stitching a wound. *Biology.* the line resulting from the fusion of contiguous parts; the seam at which dehiscence or splitting occurs, as in a fruit.

sutured *Petrology.* of or relating to a rock texture in which mineral grains or irregularly shaped crystals interfere with one another, resulting in interlocking, suturelike contacts without interstitial spaces.

SV simian virus.

S.V. or **s.v.** sailing vessel.

SV40 simian virus 40.

Sv sievert.

svabite *Mineralogy.* $Ca_5(AsO_4)_3F$, a colorless to yellow or grayish-green, hexagonal mineral of the apatite group, occurring as short prismatic crystals and in massive form having a specific gravity of 4 to 5 and a hardness of 4.5 on the Mohs scale.

svanbergite *Mineralogy.* $SrAl_3(PO_4)(SO_4)(OH)_6$, a colorless to yellow, rose, or reddish-brown trigonal mineral of the beudantite group, occurring as rhombohedral crystals with a vitreous to adamantine luster, and having a specific gravity of 3.22 and a hardness of 5 on the Mohs scale.

Svedberg, Theodor 1884–1971, Swedish chemist; awarded the Nobel Prize for developing the ultracentrifuge and using it to determine the molecular weights of proteins and colloid particles.

svedberg *Physical Chemistry.* a unit of time that is equal to 10^{-13} second, used to measure sedimentation rate in a colloidal solution and thus determine molecular weight. Also, **svedberg unit.** (Named for Theodor *Svedberg.*)

Svedberg equation *Physics.* an equation stating that the mean square displacement of a particle undergoing Brownian motion is proportional to the time the motion has proceeded.

sverdrup *Fluid Mechanics.* a unit of measurement for the flow of ocean current; 1 sverdrup = 10^6 m^3/sec.

sv40 virus *Virology.* a virus of the genus *Polymavirus* of the family Papovaviridae, originally isolated from kidney cells of the rhesus monkey, whose antibodies are often found in humans in contact with monkeys; important in recombinant DNA and cancer research.

SW or **S.W.** southwest; southwestern. Also, **s.w.**

SW switch.

swab *Medicine.* a mass of absorbent material wrapped around one end of a supporting stick, wire, or flexible fiber; used for various purposes dealing with a small or localized area; e.g., cleansing a wound, applying medication, collecting specimens for microscopic examination, and so on. *Mining Engineering.* any of various instruments or devices used to clean or treat a small area, such as the inside of a borehole or a well casing.

swage [swāj] *Mechanical Devices.* **1.** a tool used to shape metal, set saw teeth, straighten pipes, and other similar tasks. **2.** to use this tool for such a purpose.

swage block *Mechanical Devices.* a heavy metal block with grooves and perforations, used in forging and shaping metal bolts or bars.

swage bolt *Mechanical Devices.* a masonry bolt with indentations for gripping.

swaged needle *Surgery.* a needle permanently attached to suture material.

swaging *Metallurgy.* the process of forming a taper-shaped reduction on metal products by forging, squeezing, or hammering.

swale *Geology.* **1.** a shallow, sometimes swampy, depression in generally flat or level land. **2.** a shallow depression in a gently rolling ground moraine, caused by uneven glacial deposition. **3.** a long, narrow, generally shallow, troughlike depression between two beach ridges, oriented generally parallel to the coastline. Also, RUNNEL.

swallow *Vertebrate Zoology.* **1.** any of numerous small, weak-footed passerine birds of the family Hirundinidae, characterized by a long, pointed, deeply forked tail and distinguished by their swift, graceful flight and the size, distance, and regularity of their migrations; distributed worldwide. Among the common species of the Americas are the **barn swallow,** *Hirunda rustica*, the **cliff swallow,** *Petrochelidon pyrrhonota*, and the **tree swallow,** *Iridoprocne bicolor*. **2.** any of several swifts that resemble the swallow, such as the chimney swift. (From an Old English word for such a bird; not derived from the common sense of swallowing food or drink.)

swallow float *Engineering.* a cylindrical buoy having neutral buoyancy, tracked by a ship, and used to measure current velocity. Also, **swallow buoy.**

swallow hole *Geology.* **1.** a hole, cavity, or depression through which a stream disappears underground. **2.** an underwater lake outlet.

swallowing *Physiology.* the process of passing a substance from the mouth through the esophagus into the stomach.

swallowtail *Invertebrate Zoology.* any of several large butterflies of the genus *Papilio*, characterized by tail-like extensions of the rear wings that resemble the tail of a swallow.

swallowtailed kite *Vertebrate Zoology.* a graceful black and white hawk, *Elanoides forficatus*, having a long, black, forked tail; found in the southern U.S. and south to Argentina.

Swammerdam, Jan 1637–1680, Dutch biologist; the first to observe red blood cells; studied animal respiration; a founder of entomology.

Swammerdam's glands *Biology.* in amphibians, glands that secrete calcareous nodules on each side of the vertebral column.

swamp *Ecology.* a general term for an area that is waterlogged and covered with abundant vegetation, especially shrubs and trees. *Navigation.* to fill with water and sink, as a small boat.

swamp buggy *Mechanical Engineering.* a vehicle with large, low-pressure tires that is designed to operate in shallow water, mud, and swamp areas.

swamper *Mining Engineering.* the person who operates the rear brakes in a metal mine. *Forestry.* a person who trims felled trees into logs.

swamp fish *Vertebrate Zoology.* a small striped fish, *Chologaster cornuta*, that is related to the cavefish, characterized by small but functional eyes and almost transparent skin; found inhabiting swamps and streams of the Atlantic coastal plain.

swamping resistor *Electronics.* in transistor circuits, a resistor placed in the emitter lead to suppress the effects of temperature on the emitter-base junction resistance.

swan *Vertebrate Zoology.* a large, graceful aquatic bird of the genus *Cygnus* in the family Anatidae, characterized by a heavy body, a long, slender, curving neck, and a male that is usually pure white; species of North America are the **trumpeter swan,** *C. buccinator*, the **mute swan,** *C. olor*, and the **tundra swan,** *C. columbianus*.

swan

Swan bands *Astrophysics.* spectral bands that arise from the carbon radical C_2, which is common in objects such as carbon stars.

Swan-Ganz catheter *Surgery.* a type of balloon catheter that is used for measuring pulmonary arterial pressures. It is introduced into the basilic, internal jugular, or subclavian vein and is guided by blood flow into the superior vena cava, the right atrium and ventricle, and the pulmonary artery.

Swan Nebula see OMEGA NEBULA.

swan-neck *Building Engineering.* a downspout connector used between a gutter and a downpipe, beneath the eaves.

swan-neck flask *Science.* a spherical glass flask with a long narrow neck bent in an S curve to prevent access of airborne microorganisms to the sterile contents of the flask; used by Louis Pasteur to refute the doctrine of abiogenesis.

Swanscombe man *Paleontology.* an early form of *Homo sapiens*, known from three skull fragments found at Swanscombe, England, associated with artifacts of the Acheulian hand-axe culture; Swanscombe man seems to be anatomically similar to modern humans; cranial capacity is around 1300 cc; about 450,000 years Before Present.

swap *Computer Programming.* **1.** to temporarily unload a program from main memory to auxiliary storage and load a different program into memory; used in time sharing and in systems with no memory-protection hardware. **2.** to interchange the contents of an area of main storage with the contents of an area in auxiliary storage. **3.** to bring a new page of data from auxiliary storage into main memory in a virtual memory system, writing out the old page first if it has been modified.

swap in *Computer Programming.* to load a program or page into memory.

swap out *Computer Programming.* to write a program or page to auxiliary storage.

swap space *Computer Science.* an area of secondary memory, such as a disk, that is reserved for swapping memory pages with main memory in a virtual memory system.

swarf *Engineering.* chips, shavings, and other fine material produced during a grinding operation, ground from either the work piece or the grinding wheel.

swarm *Invertebrate Zoology.* a large mass of insects that fly or move about together. *Microbiology.* a colony of bacterial cells that are able to spread or migrate across the surface of moist agar medium. *Geophysics.* a series of minor earthquakes, none of which may be identified as the main shock, that occur in a particular area over a limited time.

swarm cell *Microbiology.* a daughter cell having a long filament termed a flagellum with which it breaks away from the mother cell and which is used for locomotion; generally occurs in fungi and algae. Also, **swarm spore.**

swarming *Invertebrate Zoology.* a synchronous behavioral phenomenon involving the congregation of large numbers of individuals for such purposes as mating (polychaete worms), migration in search of food (locusts), or establishing a new colony (honeybees).

Swartkrans *Anthropology.* an important human fossil site in South Africa's Sterkfontein Valley.

Swarts reaction *Organic Chemistry.* a reaction between metallic fluorides and chlorinated hydrocarbons that produces chlorofluorohydrocarbons.

swartzite *Mineralogy.* $CaMg(UO_2)(CO_3)_3 \cdot 12H_2O$, a very rare, strongly radioactive, green monoclinic mineral occurring as crusts of small prismatic crystals, having a specific gravity of 2.3 and an undetermined hardness.

swash *Geology.* **1.** a narrow sound or channel of water within a sandbank or separating a sandbank from the shore. **2.** a sandbar that is oriented in such a way that the waves of the sea wash over it. Also, **swash bar.** *Oceanography.* the uprush of water onto the beach from a wave. Also, RUN-UP.

swash mark *Geology.* a thin line or very small ridge of fine sand and bits of debris on a beach, indicating the farthest limit of wave movement onto the beach. Also, DEBRIS LINE.

swash plate or **swash bulkhead** *Naval Architecture.* a perforated plate or partial bulkhead inside a tank, designed to minimize the sloshing effect of liquids in the tank. Also, BAFFLE PLATE.

swash-plate pump *Mechanical Engineering.* a rotary pump having cylinders on an incline, so that piston reciprocation occurs when the angle between the incline and the cylinder system is varied.

swath [swäth] *Agriculture.* the path or row of material that is left behind a mower or combine as it moves through a field.

SWATH ship *Naval Architecture.* a double-hulled vessel with an underwater hull connected to an above-water hull by surface-piercing struts; the underwater hull provides buoyancy while the above-water hull provides functional space. (An acronym for small-waterplane-area twin-hulled ship.)

S wave *Geophysics.* a seismic body wave that propagates with a shearing motion which oscillates particles at right angles to the direction of propagation, capable of moving only through solids with a speed of 3.0–4.0 kilometers per second in the earth's crust and 4.4–4.6 kilometers per second in the earth's mantle. Also, SECONDARY WAVE, SHEAR WAVE, TRANSVERSE WAVE.

sway *Naval Architecture.* **1.** the movement of a vessel's hull to one side or the other due to sea action; a sideways displacement of the entire hull with respect to its longitudinal axis. **2.** of a vessel, to move in this manner. **3.** to pull on a line in order to hoist an object. Also, **sway away.**

swayback *Medicine.* **1.** an abnormal downward curvature of the back, found especially in horses. **2.** an abnormally increased curvature of the lumbar and cervical spine as viewed from the side. **3.** a congenital locomotor ataxia of lambs, thought to be associated with copper deficiency.

sway brace *Civil Engineering.* a diagonal brace used to resist wind, earthquake, or other lateral forces. Also, CROSS-BRACING.

sway frame *Civil Engineering.* a frame designed to resist lateral forces, usually consisting of sway bracing.

sweat *Physiology.* **1.** the liquid produced by the sweat glands; perspiration; it consists of water, sodium chloride, and traces of albumin, urea, phosphate, ammonia, and other waste products. **2.** to produce liquid in this manner; perspire. *Meteorology.* a collection of drops of moisture or dampness on a surface that occur because of condensation from the air. *Metallurgy.* specifically, small beads or globules made up of a single ingredient that has a low melting point, appearing on the surface of a solid mass or alloy mixture during heat treatment. *Chemical Engineering.* the entrapped oil and low-melting waxes that drain off from the filter cake during a dewaxing process.

sweat gland *Physiology.* one of the many organs in the skin that produce and excrete sweat.

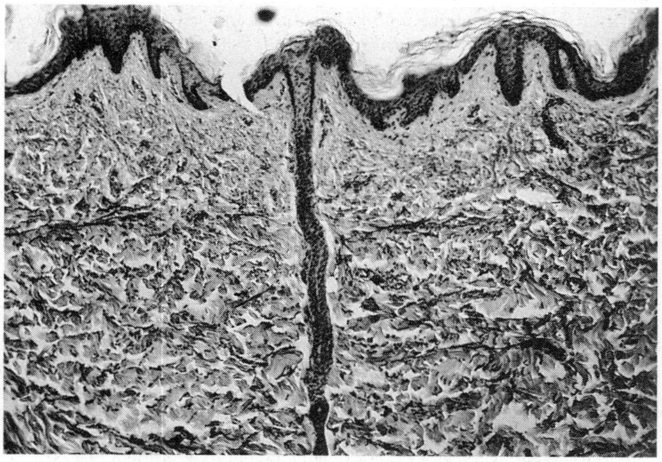

sweat gland

sweating *Chemical Engineering.* a petroleum-refinery term used to describe the separation of paraffin oil from paraffin wax by fractional fusion. Also, EXUDATION.

swedenborgite *Mineralogy.* $NaBe_4SbO_7$, a colorless to yellow hexagonal mineral occurring as short, prismatic crystals, having a specific gravity of 4.285 and a hardness of about 8 on the Mohs scale; found in skarn at Långban, Sweden.

sweep *Military Science.* **1.** to cover a wide area with gunfire. **2.** to detect, explode, or remove mines in a body of water. **3.** to employ technical methods to uncover surveillance devices. *Electronics.* **1.** the steady transition of an electron beam across the screen of a cathode-ray tube, thereby producing a steady bright line when no signal is present; the line is straight for a linear sweep and circular for a circular sweep. **2.** the crossing of a range of values of a quantity in order to delineate, sample, or control another quantity. *Agriculture.* a triangular blade used for subsurface tillage. *Metallurgy.* a template used for shaping sand molds or cores by hand. *Aviation.* the slant of a wing or other airfoil relative to a plane perpendicular to the longitudinal axis of the aircraft.

sweep amplifier *Electronics.* an amplifier stage used to increase the amplitude of the sweep voltage.

sweepback *Aviation.* **1.** a backward inclination of an airfoil from root to tip in a way that causes the leading edge and often the trailing edge to meet relative wind obliquely. **2.** the degree of inclination on such an airfoil. Also, BACKSWEEP.

sweep circuit *Electronics.* the sweep oscillator, sweep amplifier, and any other stages that generate the deflection voltage for a cathode-ray tube. Also, SCANNING CIRCUIT.

sweep-frequency reflectometer *Electromagnetism.* a reflectometer device that measures standing-wave ratios and insertion losses over a range of frequencies.

sweep generator *Electronics.* an instrument that produces a voltage or current as a prescribed function of time; the resulting waveform is then used as a time base to be applied to the deflection system of an electron-beam device, such as a cathode-ray tube. Also, **sweep oscillator.**

sweep hand *Horology.* a second hand that is mounted with the minute and hour hands at the center of a timepiece; it sweeps over the entire dial once every minute. Also, **sweep-second hand.**

sweeping *Navigation.* the process of locating the presence of underwater obstacles or clearing an area of underwater mines.

sweeping receivers *Electronics.* a set of receivers whose tuning is automatically and continuously swept across a frequency band in order to detect certain transmitted signals.

sweep jamming *Electronics.* the act of jamming an enemy radarscope by sweeping the radar beam coverage area with electromagnegic waves that have the same frequency as those received by the radarscope.

sweep rake *Agriculture.* a machine having long teeth that extend forward, used for picking up windrowed hay. Also, BUCK RAKE, BULL RAKE.

sweep rate *Electronics.* the frequency with which a radar radiation pattern rotates, usually expressed as the duration of one complete rotation.

sweepstakes route *Ecology.* a term used to account for animal migration to isolated areas by means that would not normally be used, such as crossing a sea on natural rafts and thus colonizing oceanic islands. Similarly, **sweepstakes dispersal, sweepstakes bridge.**

sweep test *Electronics.* **1.** a test performed by displaying the variation of a quantity as a function of an independent variable (usually time or frequency) varied by a sweep voltage. **2.** the use of an oscilloscope to check attenuation on coaxial cable.

sweep-through jammer *Electronics.* a jamming transmitter that passes through a radio-frequency band in short steps in order to jam each frequency for a short time.

sweep voltage *Electronics.* the periodically varying voltage used to deflect an electron beam.

sweet almond *Botany.* the variety of the almond tree, *Prunus amygdalus dulcis,* whose fruit is edible.

sweet alyssum *Botany.* a garden plant, *Lobularia maritima*, of the mustard family, characterized by clusters of small, fragrant, white or purple flowers and narrow leaves.

sweet bay *Botany.* **1.** another name for the European laurel, *Laurus nobilis.* **2.** an American magnolia, *Magnolia virginiana*, having round, fragrant white flowers; found in swamps along the Atlantic coast.

sweet birch *Botany.* a tree, *Betula lenta*, having a hard, dark-colored, closed-grain wood used for furniture and an aromatic bark and twigs that are a source of methyl salicylate; found in eastern North America.

sweet chocolate a confection made of chocolate liquor, cocoa butter, sugar, and usually other ingredients such as vanilla.

sweet corrosion *Petroleum Engineering.* a term for corrosion in gas or oil wells in which there is no iron-sulfide corrosion product and no hydrogen sulfide can be detected in the recovered fluid.

sweetening *Chemical Engineering.* a petroleum-refinery process that improves the odor of oils by removing sulfur compounds from the oils with the use of an alkaline oxidizing agent.

sweetgum *Botany.* a tall aromatic tree of the witch hazel family, *Liquidambar styraciflua*, characterized by star-shaped leaves and fruits in rounded burlike clusters. *Materials.* **1.** the reddish-brown wood of this tree, used to make furniture. **2.** the amber balsam exuded by this tree, used in perfumes and medicines.

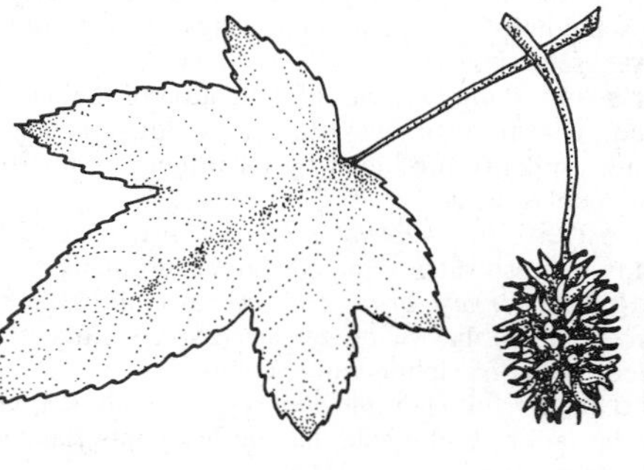

sweetgum

sweet potato *Botany.* **1.** a trailing, twining perennial vine, *Ipomoea batatas,* of the morning glory family, characterized by purplish flowers and thick, starchy, tuberous roots. **2.** the roots of this plant, grown as a vegetable.

sweet water *Hydrology.* another term for fresh water, which is thought of as sweet in contrast to salt water.

swell *Oceanography.* ocean waves that have traveled a considerable distance from their fetch; characterized by a relatively long and regular period and by flatter crests than waves within their fetch. *Geology.* **1.** an increase in the volume of a soil, as a result of its removal from a compacted bed or absorption of water. **2.** an enlargement of a section of an orebody. **3.** see RISE. *Mining Engineering.* **1.** sandstone or shale found within a channel in a coal seam cut by flowing water. **2.** an enlarged or thickened portion of a vein or ore deposit.

swell-and-swale topography *Geography.* undulating landscape associated with plains.

swelled ground *Geology.* a soil or rock that increases in size when it mixes with water.

swell forecast *Oceanography.* a prediction of the size and frequency of a swell at some remote location from its characteristics at its place of origin.

swell head *Veterinary Medicine.* a sheep and goat poisoning caused by ingestion of the leaves of the plant *Agave lechuguilla*, resulting in lesions of the kidney and liver and edema of the face and ears. Also, GOAT FEVER, LECHUGUILLA FEVER, LECHUGUILLA POISONING.

swelling *Medicine.* a transient enlargement or protuberance, such as that resulting from edema, that is not caused by the proliferation of cells. *Materials Science.* an increase in the volume of a gel or solid associated with the uptake of a liquid or gas.

swelling clay *Geology.* any clay that can absorb large quantities of water and thus increase greatly in volume, such as bentonite.

swelling coefficient *Materials Science.* a numerical value representing nondimensional fiber volume change when a fiber-reinforced composite is subjected to hygrothermal stress.

swelling ground *Geology.* a clay-rich soil or rock that greatly increases in volume as it absorbs water. Also, **swelled ground.**

swelling pressure *Geology.* the pressure exerted by a swelling clay when it absorbs water in a confined space.

swep *Organic Chemistry.* $C_8H_7Cl_2NO_2$, the generic name for methyl-3,4-dichlorocarbanilate; toxic, white crystals that are insoluble in water; melts at 112–114°C; used as an herbicide.

sweptback wing *Aviation.* an aircraft wing having sweepback on both its leading and trailing edges.

swept-frequency reflectometer *Electromagnetism.* a reflectometer that is capable of measuring standing-wave ratios and insertion losses by making measurements of amplitude and phase relationships over a wide range of frequencies.

swept path *Military Science.* in naval mine warfare, the width of the lane swept by the mechanical sweep at all depths less than the sweep depth.

sweptwing *Aviation.* a sweptback wing; used to form compounds such as **sweptwing aircraft** and **sweptwing fighter.**

swidden agriculture see SLASH-AND-BURN.

swift *Vertebrate Zoology.* a small bird of the family Apodidae, characterized by long, strong wings and rapidity of flight; related to the hummingbird and resembling the swallow. The **chimney swift,** *Chaeutura pelagica,* is widely found in the eastern half of North America; it is noted for its habit of nesting in chimneys and other hollow man-made structures.

swim bladder *Vertebrate Zoology.* an elongate, gas-filled sac dorsal to the digestive tract in most bony fish, used primarily as a hydrostatic device, but sometimes functioning in hearing or sound production.

swimmeret see PLEOPOD.

swimmer's ear *Pathology.* an inflammation of the outer ear caused by prolonged swimming or by failure to adequately dry the ears after swimming; typically produced by an infectious microorganism in a swimming pool or lake.

swimmer's itch *Invertebrate Zoology.* an itching rash produced on human skin by freshwater trematode larvae, normally parasitic on birds.

swimming bell *Invertebrate Zoology.* any umbrella-shaped or bell-shaped coelenterate that swims by successive contractions, especially a specialized zooid of a siphonophore colony.

swimming crab *Invertebrate Zoology.* the common name for marine crabs of the family Portunidae, having leg joints flattened and fringed for use as swimming fins.

swimming pool *Building Engineering.* a tank or artificial basin, usually of concrete, metal, or plastic, used for swimming and diving.

swine *plural,* **swine.** *Vertebrate Zoology.* a member of the family Suidae, including pigs, hogs, and boars; characterized by a stout body, short legs, a thick skin, and a large, somewhat mobile snout; native to the Eastern Hemisphere but now found in temperate and tropical regions worldwide. Also, HOG, PIG.

swine erysipelas *Veterinary Medicine.* a contagious disease affecting mainly pigs, characterized by high fever, reddish or purplish spots on the skin, and hemorrhaging of certain internal organs in acute cases; marked by lameness, difficulty in breathing, and endocarditis in chronic cases.

swine fever see HOG CHOLERA.

swine flu *Veterinary Medicine.* an acute, highly contagious but not usually fatal viral infection of pigs, caused by an orthomyxovirus; characterized by fever, coughing, sneezing, labored breathing, and sometimes delirium. Also, **swine influenza.**

swine plague *Veterinary Medicine.* **1.** hemorrhagic septicemia of swine, caused by *Pasteurella suiseptica* and characterized by pleuropneumonia. **2.** the pneumonic form of swine fever (hog cholera).

swine pox *Veterinary Medicine.* a mild viral disease of mostly young pigs, probably spread by pig lice; characterized by the formation of pustules on the belly, groin, and inner surfaces of the legs.

swinestone *Petrology.* a variety of limestone that contains black bituminous matter and gives off a foul odor when broken or rubbed. Also, STINKSTONE.

swing *Electricity.* the maximum variation of an electrical quantity, e.g., voltage swing.

swing-around trajectory *Space Technology.* a trajectory or orbit by which a spacecraft uses the gravitational field of the destination body to propel about the body and, if so programmed, back to earth.

swing bridge *Civil Engineering.* a movable bridge that can turn about a pivot, to allow the passage of a vessel.

Swingfire *Ordnance.* a British antitank missile powered by a solid-propellant rocket motor and equipped with wire guidance; it delivers an extremely effective, hollow-charge, high-explosive warhead at a range up to 13,000 feet.

swing-frame grinder *Mechanical Engineering.* a grinding machine suspended by a chain at the center point, so that it can be turned in any direction for surface grinding.

swing-hammer crusher *Mining Engineering.* a rock-crushing machine in which hammers, mounted loosely on a rapidly revolving shaft, break up ore by impelling it against breaker plates.

swinging buoy *Navigation.* a buoy used in swinging a ship; the bow of the ship is attached to the buoy and the ship is moved around it by an auxiliary vessel, such as a tug, or by attachment to additional buoys.

swinging choke *Electricity.* an iron-core, choke coil inductor with an air gap in its magnetic circuit, designed so that its effective inductance decreases as the current passing through it increases. Also, **swinging reactor.**

swinging compass *Navigation.* a precise, portable magnetic compass used for aircraft compass calibration.

swinging ship *Navigation.* the process of moving a ship through each or every other compass point to check for deviations of the magnetic compass and prepare a deviation table or deviation card. Also, COMPASS CALIBRATION.

swinging the arc *Navigation.* the process of tilting a sextant up and down through the image of a celestial object to determine the lowest point of the circle through the body. Also, ROCKING THE SEXTANT.

swinging traverse *Ordnance.* a method of machine gun fire in which the traverse clamp is loosened, allowing the gunner to move the weapon rapidly from side to side; used against dense troops or moving targets.

swing joint *Mechanical Devices.* a pipe joint that allows an attachment to move or rotate relative to another part or about its own axis.

swing pipe *Engineering.* the intake portion of a discharge pipe, swiveled so that it can be positioned at higher or lower levels in a tank.

swing saw *Mechanical Devices.* a circular saw mounted overhead on a hinged frame.

swingwing *Aviation.* **1.** a wing whose horizontal angle to the fuselage centerline can be adjusted in flight to vary aircraft motion at differing speeds. **2.** an aircraft having such wings.

swirl defects *Materials Science.* the impurity segregation and precipitation defects in crystals that appear as a characteristic swirl pattern, caused by interaction of the convective mixing with the rotation of the crystal.

swirl error *Navigation.* a magnetic compass error caused by friction in the compass liquid while the vessel is turning.

swirl flowmeter see VORTEX PRECESSION FLOWMETER.

Swiss cheese *Food Technology.* an elastic hard cheese made of raw milk and ripened by *Propionibacterium shermanii,* which produces its characteristic holes and sweet nutlike flavor; originally made in Switzerland and now produced in other countries as well.

Swiss pattern file *Mechanical Devices.* a file used for very fine work such as watch parts and dies.

switch *Electricity.* a device that is used to open or close an electric circuit; it can be actuated manually or automatically. *Civil Engineering.* a device for diverting moving trains or rolling stock from one track to another. *Robotics.* **1.** a mechanical device that indicates which of several alternatives has been chosen. **2.** a conditional branch in a robotic control program. *Computer Programming.* **1.** a point in a program at which the program takes one of several courses of action, depending on a condition established in the program. **2.** a statement in a programming language that changes among alternative courses of action depending on a condition. **3.** a built-in device that can be queried in order to select a course of action. **4.** a hardware or software device capable of altering program flow or changing the state of an instruction or device.

switch angle *Civil Engineering.* the angle formed by the stock and switch rails of a railway, measured at the point of intersection between the gage lines.

switchback *Transportation Engineering.* a zigzag arrangement of rail tracks or roadway to facilitate travel up or down a steep grade, as on a mountainside.

switchblade knife *Mechanical Devices.* a pocketknife with a spring-loaded blade that can be quickly released by pressing a button.

switchboard *Electricity.* a large panel of assembled switches, circuit breakers, meters, fuses, and terminals that are primary to the operation of electronic equipment. *Telecommunication.* an assembly of switch panels that is manually operated at a telephone exchange, where all subscriber circuits and other exchanges terminate so that operators establish communication links between two subscribers onto the same or different exchanges.

switchboard model *Materials Science.* a theory explaining the structure of a crystalline polymer in terms of random or irregular folding of the chains and reentering lamellae.

switch control *Robotics.* the use of mechanical switches to control a machine or operation.

switch function *Electronics.* a logic function in which the two logic levels represent an open or closed switch.

switch gear *Electricity.* a general term for switching, metering, interrupting, and regulating devices; any electrical equipment whose primary function is to open and close electric circuits, for the transmission and distribution of electric power.

switch hook *Electricity.* a switch on a telephone set that is located on the structure supporting the receiver or handset, operated by the removal or replacement of the receiver or handset on the support.

switch horn *Ordnance.* in a naval mine, a switch operated by a projecting spike.

switching *Electricity.* the process of making, breaking, or altering connections in an electric circuit.

switching circuit *Electricity.* **1.** a circuit that performs a switching function. **2.** combinations of switching circuits used to perform a logical operation.

switching device *Engineering.* a device used to switch the flow of power, fluid, or raw material to any of several machines, circuits, or other destinations. It can also bring another device or circuit into operating or nonoperating mode. Also, **switching mechanism.**

switching diode *Electronics.* a crystal diode that performs the same function as a switch; below a specified voltage it has high resistance corresponding to an open switch, while above that voltage it changes to the low resistance of a closed switch.

switching gate *Electronics.* an electronic circuit in which an output having constant amplitude is registered if a particular combination of input signals exists.

switching key *Electricity.* see KEY.

switching pad *Electronics.* a transmission loss pad that is automatically inserted into or removed from a toll circuit for different desired operating conditions.

switching power supply *Electricity.* a regulated power supply that works by switching current into a high-frequency transformer or inductor at frequencies much higher than the input voltage.

switching reactor *Electromagnetism.* a saturable-core reactor having multiple input control windings and at least one output winding; the device serves the purpose of a relay.

switching regulated power supply *Electricity.* a regulated power supply with a rectified line voltage that is converted to a pulse-width-modulated high-frequency voltage, such as 20,000 hertz, stepped up or down to the target voltage by a transformer, then rectified and filtered.

switching substation *Electricity.* a power substation whose equipment is used for circuit connections, interchanges, and interconnections; does not include transformers.

switching surface *Robotics.* the place in bang-bang control systems at which the robot stops.

switching theory *Electronics.* the theory of circuits composed of digital devices.

switching time *Electronics.* in a circuit or device with two stable states, the delay between the application of an input that will cause the device to change state and the time the new state is achieved.

switching transistor *Electronics.* a transistor used for on/off switching.

switching tube *Electronics.* a gas tube used for switching high-power radio-frequency energy in the antenna circuits of radar and other pulsed radio-frequency systems.

switch plate or **switchplate** *Electricity.* a plate, usually of metal or plastic, that covers a switch so that the operating toggle or knob protrudes. *Mining Engineering.* an iron plate mounted on mining tramroads to change direction of movement.

switch pretravel *Electricity.* the movement in a switch-operating lever made before the actuation of the switch to open or close its contacts.

switch register *Computer Technology.* a toggle switch on the operator console used for the manual entry of addresses and data into computer memory and for manual intervention in program execution.

switch train *Electricity.* a series of switches in a tandem configuration, through which connection must be made when a circuit is established between a calling telephone and a called telephone.

swivel *Design Engineering.* **1.** a part that rotates freely on a headed pin or bolt. **2.** to move with a rotating motion by means of such a device. Thus, **swivel chair.** *Robotics.* articulation involving three degrees of freedom in a spherical-coordinate robot.

swivel block *Mechanical Devices.* a block and shackle whose hook is attached to a pivot which allows a free rotation.

swivel coupling *Mechanical Engineering.* a coupling in which one link can be rotated independently of the other links.

swivel gun *Ordnance.* a gun mounted so that it can be moved freely from side to side or up and down.

swivel head *Mechanical Engineering.* a mechanical or hydraulic assembly for a diamond drill machine that controls the rotation and advance of the drill.

swivel hook *Mechanical Devices.* a hook attached to a pivot on its base or eye, which allows a free rotation.

swivel joint *Mechanical Devices.* any joint with a swivel mechanism to allow one joint section to move relative to another.

swivel pin see KINGPIN.

swivel vise *Mechanical Devices.* a vise mounted on a rotating swivel plate to allow optimum positioning for a workpiece.

swoon *Neurology.* a fainting spell, characterized by mild temporary loss of consciousness. Also, SYNCOPE.

sword bayonet *Ordnance.* a short sword with a narrow blade that can be attached to the muzzle of a gun and used as a bayonet in close combat.

sword dune see SEIF DUNE.

swordfish *Vertebrate Zoology.* a very large saltwater food fish, *Xiphias gladius,* of the family Xiphiidae, having a long, swordlike bone protruding from its upper jaw.

Swordfish *Astronomy.* the constellation Dorado.

swordtail *Vertebrate Zoology.* any of several small, bright-colored, freshwater fish of the genus *Xiphophorus,* having the lower part of the caudal fin elongated into a swordlike structure; native to Central America and often kept in aquariums.

SWU *Physics.* separative work unit; a unit of work used in discussing what is required to separate a given amount of an element into its isotopes.

SXAPS soft X-ray appearance potential spectroscopy.

syanamorph *Mycology.* any of two or more anamorphs or fungi in their asexual stage that correspond to fungi in their sexual or teleomorphic stage.

sycamore *Botany.* **1.** any of several North American deciduous trees of the genus *Platanus*, especially *P. occidentalis.* **2.** a Near Eastern tree, *Ficus sycomorus,* noted for its edible fruit.

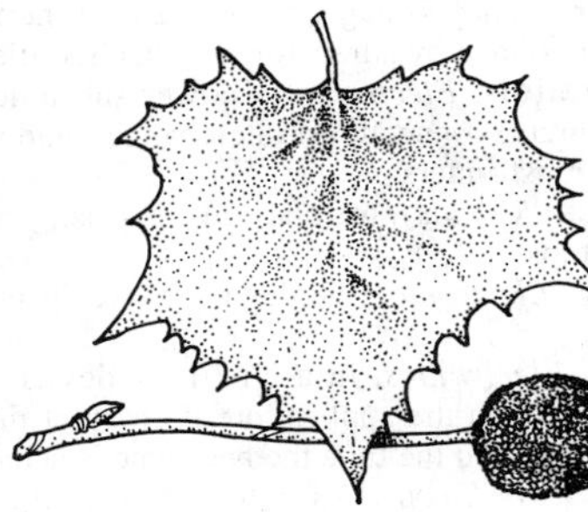

sycamore

sycamore maple *Botany.* a Eurasian maple tree, *Acer psuedoplatanus*, widely planted as a shade tree.

Sycettida *Invertebrate Zoology.* an order of sponges in the subclass Calcaronea, having a calcareous skeleton and flagellated chambers.

Sycettidae *Invertebrate Zoology.* a widely distributed family of bristly calcareous marine sponges in the order Sycettida, found worldwide from tidal zones to 250 meters deep.

Sycidiales *Paleontology.* an extinct order of charophytic algae having gyrogonites in vertical units; characterized by a wall pierced by thin pores, uncommon in charophytes; Silurian to Lower Carboniferous.

sycon *Invertebrate Zoology.* **1.** a type of canal system found in complex sponges, with small flagellated chambers opening into the central cavity and two sets of radial canals. **2.** any sponge of the order Sycettida (formerly Syconosa).

syconium *Botany.* any multiple fleshy fruit developed from an enlarged succulent receptacle, such as a fig.

sycosis *Medicine.* a chronic inflammatory disease of the hair follicles, especially of the facial hair, manifesting as crusting pustules, papules, and tubercles which are perforated by the hairs.

SYD *Aviation.* the airport code for Sydney, Australia.

Sydenham, Thomas 1624–1689, English physician; wrote descriptions of measles, influenza, fevers, gout, and venereal diseases.

Sydenham's chorea see SAINT VITUS' DANCE.

syenite *Petrology.* a crystalline plutonic rock, or rock group, characterized by granular texture and consisting principally of alkali feldspar, with small amounts of plagioclase and mafic minerals, and possibly traces of quartz; it is the intrusive equivalent of trachyte.

syenodiorite *Petrology.* a plutonic rock, or rock group, intermediate in composition between syenite and diorite, containing sodic plagioclase, orthoclase, and ferromagnesian minerals; may be equivalent to monzonite.

syenogabbro *Petrology.* an intrusive plutonic rock, or rock group, similar to gabbro but containing orthoclase in addition to plagioclase and mafic minerals; an orthoclase gabbro.

syllabic [si lab′ik] *Linguistics.* relating to, being, or consisting of a syllable or syllables.

syllabic compandor *Electronics.* a compandor in which the effective gain variations are performed at speeds allowing response to the syllables of speech but not to individual cycles of the signal wave.

syllabic writing *Linguistics.* a form of writing in which a symbol is used to represent a syllable, represented by some forms of Egyptian hieroglyphics and Japanese writing systems.

syllable [sil′ə bəl] *Linguistics.* a basic unit of speech consisting of one or more sounds, including a nucleus that is usually a vowel, and generally perceived as having no interrupting pause within it.

Syllidae *Invertebrate Zoology.* a family of small, free-swimming marine polychaete annelid worms in the group Errantia, having a long thin translucent body and reproducing by asexual budding.

Syllinae *Invertebrate Zoology.* a subfamily of free-swimming polychaete annelids in the family Syllidae.

syllogism [sil′ə jiz′əm] *Science.* a model of deductive reasoning consisting of a major premise, one or more minor premises, and a conclusion; e.g., all dogs have four legs (major premise); all beagles are dogs (minor premise); therefore, all beagles have four legs (conclusion).

syllogistic [sil′ə jis′tik] *Science.* relating to or in the form of a syllogism; using deductive reasoning.

Sylonidae *Invertebrate Zoology.* a family of parasitic decapod crustaceans, in the order Rhizocephala, having no appendages; rootlike structures grow into the tissues of the host.

Sylopidae *Invertebrate Zoology.* a family of beetles, coleopteran insects in the superfamily Meloidea, having large, fan-shaped hindwings and forewings reduced to small leathery flaps.

Sylow subgroup *Mathematics.* a subgroup P of a group G is said to be a **Sylow *p*-subgroup** (where p is prime) if P is a maximal p-subgroup of G; that is, a maximal subgroup such that every element of P has order p. Sylow p-subgroups always exist (e.g., the subgroup consisting of the identity element is a Sylow p-subgroup for every p), and every p-subgroup is contained in a Sylow p-subgroup.

Sylow theorems *Mathematics.* **1. first Sylow theorem:** let p be prime, $n \geq 1$, and m be a positive integer not a multiple of p. Then (a) any group G of order $p^n m$ contains a subgroup of order p^i for each positive integer $i \leq n$, and (b) every subgroup of G of order $p^i (i < n)$ is normal in some subgroup of G of order p^{i+1}. **2. second Sylow theorem:** let H be a p-subgroup of a finite group G, and P be any Sylow p-subgroup of G. Then there exists an element x of G such that H is normal in xPx^{-1}. In particular, any two Sylow p-subgroups of G are conjugate. **3. third Sylow theorem:** let p be a prime and G a finite group. Then the number of Sylow p-subgroups of G divides the order $|G|$ of G and is of the form $kp + 1$, for some nonnegative integer k.

sylvanite *Mineralogy.* $(Au,Ag)_2Te_4$, a gray, white, or yellow, monoclinic mineral with a brilliant metallic luster, occurring as thick tabular or short prismatic crystals, having a specific gravity of 8.16 and a hardness of 1.5 to 2 on the Mohs scale; found in vein deposits with gold and other tellurides.

sylvatic plague *Veterinary Medicine.* a form of plague affecting wild rodents, transmitted by the rodents and their fleas, and occurring in many rural and forested parts of the world; it rarely affects humans.

Sylvester's theorem *Mathematics.* the theorem that if V is a vector space over an ordered field K with a nondegenerate quadratic form g over K, there exists a nonnegative integer r such that, for any orthogonal basis $\{v_1, \ldots, v_n\}$ of V, exactly r of the n elements $g(v_1), \ldots, g(v_n)$ of K are greater than 0 and $n - r$ are less than 0. In particular, the signs of eigenvalues (the inertia) of a quadratic form are unchanged under a change of basis. Also, **Sylvester's law of inertia.**

Sylvian sulcus *Anatomy.* a shallow furrow on the surface of the human cerebrum, separating adjoining convolutions; it is a deeper groove in most primate brains and is a deep cleft in the most primitive monkeys, such as the marmoset Callitrix. (Named for Jacobus *Sylvius,* the Latin name of Jacques Dubois, 1478–1555, French anatomist.)

Sylviidae *Vertebrate Zoology.* the warblers, a diverse family of small passerine birds of the suborder Oscines, related to the thrushes and found mostly in the Old World, with a few in North and Central America.

sylvite *Mineralogy.* KCl, a colorless to white, gray, or reddish, transparent, cubic mineral, occurring in sedimentary deposits with other evaporite minerals and as a sublimation product near volcanoes, having a specific gravity of 1.993 and a hardness of 2 on the Mohs scale; the chief ore of potassium compounds used for fertilizers. Also, **sylvine, sylvin.**

sym. symbol; symmetrical.

sym- a combining meaning "together." Also, **sy-, syl-, sys-.**

sym- *Organic Chemistry.* a prefix describing a compound whose constituents are symmetrical with respect to a functional group or to the carbon skeleton.

symballophone *Medicine.* a double stethoscope, one element feeding into each ear, that employs the user's auditory acuity to precisely locate or determine the characteristics of an internal sound.

Symba process *Biotechnology.* a fermentation process that uses starch wastes fermented by two type of yeasts, *Saccharomyces fibuligera* and *Candida utilis,* to produce a protein.

symbiogenesis *Microbiology.* the theoretical process by which mitochondria and chloroplasts developed from intracellular symbiotic organisms.

symbiont *Ecology.* an organism that forms a close association with another organism. *Computer Science.* see DEMON.

symbiosis [sim′bē ō′sis] *Ecology.* a relationship in which two dissimilar organisms live in close association with each other. *Psychology.* a mutually reinforcing relationship between two persons who are dependent on each other.

symbiotic [sim´bē ät´ik] *Ecology.* associated in symbiosis; living together.

symbiotic star *Astronomy.* a close binary star system in which a large, cool, highly evolved star is transferring mass to a smaller, hot companion in a prenova state. Each element in such a system may be called a **symbiotic object.**

symblepharon *Medicine.* an adhesion between the eyeball and the eyelid.

symbol a letter, word, or object that represents or refers to something else; specific uses include: *Chemistry.* the standard one-letter or two-letter abbreviation for the name of a chemical element; e.g., C for carbon or Hg for mercury. *Linguistics.* an arbitrary representation of an object, event, or entity. *Computer Programming.* a set of contiguous characters, uninterrupted by a space or special character, that form a symbolic name. *Artificial Intelligence.* **1.** a data structure that corresponds to an object in the real world, to a word, or to a concept. A symbol typically has a print name, a set of properties, and relationships with other symbols. **2.** an instance of the symbol data type in the Lisp language.

symbolic *Science.* relating to, serving as, or expressed by a symbol.

symbolic address *Computer Programming.* a label that is chosen by a programmer to identify a data item or an instruction. Also, SYMBOLIC NUMBER.

symbolic algebraic manipulation *Artificial Intelligence.* the manipulation of mathematical formulas, expressed as symbolic structures within the computer, to produce new formulas or derived information; e.g., symbolic integration, solving equations simultaneously, or producing a graph of an equation.

symbolic algebraic manipulation language *Artificial Intelligence.* a language that is specialized for symbolic algebraic manipulation and provides facilities for manipulation of formulas. Also, **symbolic mathematical computation.**

symbolic anthropology *Anthropology.* the branch of anthropology in which societies are examined as to how they interpret their culture through cultural traits and institutions such as customs, dress, arts, religion, and recreation.

symbolic assembly language listing *Computer Programming.* a printout that may be produced during the compiling of an assembly language program.

symbolic assembly system see ASSEMBLER.

symbolic coding *Computer Programming.* coding using symbolic notation for operators and operands.

symbolic control *Control Systems.* **1.** the use of alphanumeric or pictorial symbols in the control of equipment. **2.** a control device that employs such symbols.

symbolic debugger *Computer Programming.* a program that facilitates debugging at the machine level by accepting commands and printing results in a more easily understood symbolic form.

symbolic inference *Artificial Intelligence.* a derivation of new facts or conclusions by manipulation of facts and rules that are represented by structures of symbols; e.g., from "all men are mortal" and "Socrates is a man," derive "Socrates is mortal."

symbolic information *Artificial Intelligence.* a representation in terms of symbols and their properties and relationships, as opposed to purely numeric data.

symbolic integration *Artificial Intelligence.* the mathematical integration of a given formula, with the result expressed as another formula.

symbolic language *Computer Programming.* a language that expresses instructions in symbols familiar to humans rather than in machine language.

symbolic logic *Mathematics.* the formal study of symbolic representations in mathematical logic.

symbolic manipulation *Artificial Intelligence.* **1.** the processing of symbolic information to produce derived conclusions or new symbolic structures. **2.** symbolic algebraic manipulation.

symbolic name *Computer Programming.* **1.** a name designated by a programmer to represent a location in a program. **2.** a name given to a variable in a high-level language program.

symbolic number see SYMBOLIC ADDRESS.

symbol input *Computer Programming.* all of the contextual symbols that may appear in a source text.

symbol table *Computer Programming.* a table created by a compiler, assembler, or program that contains symbolic names and their properties; examples are variables, statement identifiers, constants, subprogram names, and often symbolic addresses related to their absolute addresses, relevant to the translation process.

symmetrical [si me´tri kəl] *Science.* relating to or exhibiting symmetry; having a shape or configuration in which parts that lie on opposite sides of an actual or imaginary central line are identical in size, shape, and structure. *Chemistry.* of a compound, having atoms or groups at equal intervals in the molecule. Also, **symmetric.**

symmetrical achromat lens *Optics.* a lens composed of two identical achromats placed back to back around the aperture diaphragm.

symmetrical alternating quantity *Physics.* any quantity that varies periodically in such a way that each half cycle variation is equal and opposite in sign to its preceding half cycle; examples are the sine and cosine functions.

symmetrical attenuator *Electronics.* an attenuator whose circuit would be the same if the input and output terminals were interchanged.

symmetrical avalanche rectifier *Electronics.* an avalanche rectifier that can be triggered in either direction, after which it has a low impedance in the triggered direction.

symmetrical band-pass filter *Electronics.* a band-pass filter whose attenuation is symmetrical with respect to a frequency at the center of the pass band.

symmetrical band-reject filter *Electronics.* a band-rejection filter whose attenuation as a function of frequency is symmetrical with respect to a frequency at the center of the rejection band.

symmetrical clipper *Electronics.* a clipper in which the maximum and minimum limits on the amplitude of the output signal are positive and negative values of equal magnitude.

symmetrical deflection *Electronics.* a technique of electrostatic deflection in which voltages that are equal in magnitude and opposite in sign are applied to the two deflector plates.

symmetrical fold *Geology.* a fold in a rock structure in which both limbs dip at about the same angle in relation to the axial surface. Also, NORMAL FOLD.

symmetrical H-attenuator *Electronics.* an H-attenuator in which the impedance close to the input terminals is equivalent to the corresponding impedance near the output terminals. Similarly, **symmetrical O-attenuator, symmetrical π-attenuator, symmetrical t-attenuator.**

symmetrical inductive diaphragm *Electromagnetism.* an inductive diaphragm for a waveguide, consisting of two plates having a small space at their centers.

symmetrical lens *Optics.* a lens having identical sets of elements on either side of its center.

symmetrical transducer *Electronics.* a transducer that is symmetrical relative to a specified pair of terminations, the interchange of which will not affect the transmission in any way.

symmetric architecture *Computer Technology.* a form of computer design that allows any type of data to be used with any type of instruction.

symmetric connection *Mathematics.* a connection (with associated covariant differentiation ∇) whose torsion tensor T vanishes identically; i.e., for vector fields X and Y,

$$T(X,Y) = \nabla_X Y - \nabla_Y X - [X,Y].$$

Equivalently, in a coordinate basis, the connection coefficients Γ^i_{jk} are symmetric in the two lower indices j and k.

symmetric difference *Mathematics.* the symmetric difference of two sets A and B is $(A - B) \cup (B - A)$; denoted $A \Delta B$ or $A + B$.

symmetric function *Mathematics.* a function of two or more variables that is invariant after any permutation of its variables. A function may also be **partially symmetric**, i.e., symmetric with respect to some, but not all, of its variables, or with respect to particular pairs of variables.

symmetric group *Mathematics.* the permutation group on k objects is called the symmetric group S_k of degree k. S_k has $k!$ elements.

symmetric laminates *Materials Science.* laminated fibrous composites with symmetrical stacking sequences; for each ply above the midplane there is a ply identical in composition and orientation at an equal distance below the midplane.

symmetric list *Computer Programming.* a linked list in which each item contains links to both its predecessor and its successor. Also, DOUBLY LINKED LIST.

symmetric matrix *Mathematics.* a square matrix A equal to its own transpose; that is, $A = A^t$.

symmetric operator *Mathematics.* a linear operator T such that $(Tx, y) = (x, Ty)$ for all x and y in the domain of definition of T. A theorem of functional analysis states that any symmetric linear operator on a Hilbert space is continuous.

symmetric relation *Mathematics.* given a set A, a relation R on $A \times A$ is said to be symmetric if $(x, y) \in R$ implies $(y, x) \in R$ for all x and y in A. R is an equivalence relation if it is reflexive, symmetric, and transitive.

symmetric ripple mark *Geology.* a ripple mark that is similarly shaped on both sides of its crest, thus having a symmetric cross-section profile.

symmetric top molecule *Physical Chemistry.* a molecule that rotates around one axis of threefold or greater symmetry.

Symmetrodonta *Paleontology.* an order of therian mammals belonging to the extinct infraclass Trituberculata, including the well-known Kuehneotherium; characterized by nearly symmetrical upper and lower molar teeth; may be ancestral to mammals; extant in the Upper Triassic to Upper Cretaceous.

symmetrogenic *Cell Biology.* of or relating to a process of longitudinal fission practiced by flagellates in which the two resulting daughter cells are mirror images of each other.

symmetry [sim´ə trē] *Science.* an even or regular correspondence of parts, such as a shape or configuration in which parts that lie on opposite sides of a central line are identical in size, shape, and structure. *Biology.* the correspondence of organs and other constituent parts of the body with respect to imaginary axes. *Physics.* a property indicating invariance under certain transformations. *Mathematics.* an element of a symmetry group. For example, the equilateral triangle exhibits symmetry with respect to rotations of 120° and reflections across an axis; these are the elements of the dihedral group, acting as a group of isometries of the plane.

5-3-2 symmetry see ICOSAHEDRAL SYMMETRY.

symmetry axis see AXIS OF SYMMETRY.

symmetry breaking *Physics.* a deviation from exact symmetry exhibited in many physical systems; typically classified as either **explicit symmetry breaking** or **spontaneous symmetry breaking.** *Chaotic Dynamics.* specifically, the creation of solutions that are less symmetrical than the equations, variables, or coordinates of the model.

symmetry element *Crystallography.* **1.** a geometrical entity, such as a point, line, or plane, with respect to which a particular symmetry operation is performed. **2.** the set of symmetry operations associated with this entity.

symmetry group *Mathematics.* given a mathematical object (such as a differential equation), a group action (such as coordinate transformations of the independent and dependent variables), and a property to be preserved (such as the solution space of an equation or the equation's functional form), a symmetry group is the subgroup (of the acting group) that preserves the given property.

symmetry number *Physical Chemistry.* the quantity of identical movements that a molecule exhibits when rotated around symmetry axes.

symmetry operation *Crystallography.* an object is symmetrical if it can be converted into itself by a symmetry operation or a series of symmetry operations. In crystal structures (assumed infinite in extent), the possible symmetry operations include axes of rotation and rotatory inversion, screw axes, and glide planes, as well as lattice translation. Proper operations are translation and rotation, while improper operations are reflection and inversion.

symmetry operation of the second kind see IMPROPER SYMMETRY OPERATION.

symmetry principle *Physics.* a principle stating that the physical quantities within a physical law remain unchanged during certain transformations, such as reflections or time reversals.

symmict *Geology.* describing a sedimentation unit that lacks structure and is composed of various-sized particles that are mixed to a greater extent in the lower section. Also, SYMMINCT.

symmictite *Petrology.* a homogenized eruptive breccia consisting of a mixture of country rock and intrusive rock.

symmicton see DIAMICTON.

symminct see SYMMICT.

Symon's cone crusher *Mining Engineering.* a modified gyratory rock crusher used in secondary ore crushing; a conical crushing head gyrates within a downward-flaring bowl, the main shaft being gyrated by means of a long eccentric driven by bevel gears.

Symon's disk crusher *Mining Engineering.* a mill with two cup-shaped plates revolving at the same speed in the same direction on shafts set at a small angle to each other; ore is fed from the center and is gradually crushed as it approaches the edge before being thrown out by centrifugal force.

sympathectomy *Surgery.* the transection, resection, or other interruption of some portion of the sympathetic nervous pathways.

sympathetic *Neurology.* of or relating to the sympathetic nervous system.

sympathetic detonation *Engineering.* an explosion triggered by the detonation wave of another explosion.

sympathetic ink see INVISIBLE INK.

sympathetic nervous system *Anatomy.* the portion of the autonomic nervous system consisting of nerves that originate in the thoracic and lumbar regions, which functions especially in processes associated with response to alarm, such as acceleration of the heart rate, constriction of the blood vessels, and elevation of the blood pressure.

sympathetic ophthalmia *Medicine.* an inflammation in an uninjured eye in reaction to an injury to the other eye.

sympathetic vibration *Physics.* a resonant oscillation in an acoustical or mechanical system that is driven by an adjacent system operating at the same frequency.

sympathicoblast *Neurology.* an embryonic sympathetic nerve cell that later develops into a sympathetic ganglion cell. Also, **sympathoblast.**

sympathicoblastoma *Oncology.* a malignant tumor containing sympathicoblasts.

sympathicotropic cell *Histology.* a cell that exerts an effect upon the sympathetic nervous system.

sympathin *Biochemistry.* a neurohormone of the sympathetic nerve endings that serves various organs as a chemical mediator; e.g., epinephrine and norepinephrine.

sympathoblast *Neurology.* an embryonic sympathetic nerve cell that later develops into a sympathetic ganglion cell.

sympathochromaffin cell *Histology.* a cell in the adrenal medulla that develops into a sympathetic or medullary cell.

sympatholytic *Pharmacology.* see ANTIADRENERGIC.

sympathomimetic *Pharmacology.* any agent that produces effects similar to those of impulses conveyed by adrenergic postganglionic fibers of the sympathetic nervous system.

sympatric *Ecology.* describing different species or populations that live in the same geographical area. Thus, **sympatry.**

sympatric speciation *Evolution.* the evolutionary change of an ancestral population inhabiting a single geographic region into two distinct gene pools as a result of genetic mutation and variation.

sympetalous see GAMOPETALOUS.

symphile *Entomology.* any insect that lives protected and fed within the nest of its ant or termite hosts, often providing secretions favored by the hosts, including rove beetles, claviger beetles, and certain mirid bugs.

symphily *Ecology.* a harmonious relationship between an organism and a colony of social insects with whom it lives. Thus, **symphilic.**

symphogenous development *Mycology.* the process by which fungal structures, such as pycnidia, develop from the intertwining of distinct hyphae.

Symphyla *Invertebrate Zoology.* a small class of tiny, whitish, cylindrical terrestrial arthropods of the Myriapoda, having 12 pairs of legs and 15–22 body segments, that typically hide in damp, dark places; includes the garden centipede, *Scutigerella immaculata.*

symphyseal *Anatomy.* of or relating to a symphysis. Also, **symphysial.**

symphysic *Medicine.* characterized by abnormal fusion of adjacent parts.

symphysis *Anatomy.* **1.** a joint between two bones in which the joint cavity is filled by fibrocartilage; e.g., the pubic symphysis or junction between the paired pubic bones. Also, FIBROCARTILAGINOUS JOINT. **2.** any fusion point between two structures.

symphysodactyly *Medicine.* abnormal fusion of the fingers or toes.

Symphyta see CHALASTOGASTRA.

symplasma *Bacteriology.* an aggregate formed by certain bacterial cells that may be enclosed within a translucent sheath.

symplast *Botany.* of protoplasts, the continuum that is linked from cell to cell by plasmodesmata and bounded by the plasmalemma.

symplastic growth *Botany.* a type of tissue growth in which touching walls grow equally, thus maintaining cell contact. Also, COORDINATED GROWTH.

symplectic *Mathematics.* **1.** an exterior differential closed 2-form F of even rank $2n$ defined on a smooth manifold X of dimension $2n$ is called a **symplectic form**. Darboux's theorem gives an expression for F in terms of differentials of canonical coordinates about each point of X. **2.** A diffeomorphism f on a smooth manifold X of even dimension $2n$ is said to be symplectic or a **symplectomorphism** if it leaves the symplectic form F invariant; i.e., $f^*F = F$ on X.

symplectic group *Mathematics.* a (Lie) group of linear transformations S on a $2n$-dimensional vector space V that will satisfy the relation $S^tKS = K$, where K is a fixed linear transformation having n eigenvalues equal to 1 and the remaining n eigenvalues equal to −1. Classically, K is represented as the $2n \times 2n$ matrix having the $n \times n$ identity matrix in the upper right and lower left quadrants, and zero in the upper left and lower right quadrants; the S and S^t are represented as $2n \times 2n$ matrices.

symplektite *Mineralogy.* an intimate intergrowth of two different minerals, commonly quartz and feldspar, producing textures found in certain igneous and metamorphic rocks; sometimes restricted to secondary minerals. Also, **symplectite.**

symplesite *Mineralogy.* $Fe_3^{+2}(AsO_4)_2 \cdot 8H_2O$, a blue to green or greenish-black, triclinic mineral, occurring as small imperfect crystals and as fibrous spherical aggregates, having a specific gravity of 3.01 and a hardness of 2.5 on the Mohs scale.

sympodia *Biology.* fusion of the lower extremities.

sympodial *Biology.* relating to, resembling, or characterized by sympodia.

sympodial growth *Botany.* plant growth characterized by the formation of sympodia, as on a grapevine. *Invertebrate Zoology.* in hydroid polyps, a type of growth where the hydrocaulus is formed of the bases of successive hydranths.

sympodium *plural,* **sympodia.** *Botany.* a stem bundle and associated leaf traces that resemble a single stem.

symport *Biochemistry.* the electrogenic or nonelectrogenic, coupled transmembrane movement of two substances in the same direction.

symptom [simp′təm] *Medicine.* any subjective evidence of a disease or of a patient's condition, such as that perceived by the patient; e.g., pain, numbness, dizziness, nausea, and so on.

symptomatic [simp′tə mat′ik] *Medicine.* relating to or being a symptom; indicative of a certain condition.

symptomatology *Medicine.* the branch of medicine concerned with the study of symptoms of disease.

symptom substitution *Psychology.* the theory that an individual will replace one neurotic symptom with another if therapy does not resolve the deeply rooted unconscious conflicts responsible for the symptoms.

syn- a prefix meaning "together."

syn- *Organic Chemistry.* a prefix designating cisoid geometrical isomerism in compounds containing N=N and C=N double bonds.

synadelphite *Mineralogy.* a colorless to red or brown triclinic, pseudo-orthorhombic mineral with an approximate formula of $(Mn^{+2},Mg,Ca,Pb)_9(As^{+3}O_3)(As^{+5}O_4)_2(OH)_9 \cdot 2H_2O$, occurring as prismatic crystals and in massive form, having a specific gravity of 3.57 to 3.79 and a hardness of 4.5 on the Mohs scale. Also, HEMAFIBRITE.

synalgia see REFERRED PAIN.

Synallactidae *Invertebrate Zoology.* a family of deep-sea sea cucumbers, holothurian echinoderms in the order Aspidochirotida, having respiratory trees and modified dorsal tube feet.

synandrous *Botany.* of or relating to a plant with several united stamens.

synangium *Botany.* a mass of fused sporangia, as in some ferns. *Vertebrate Zoology.* a peripheral arterial trunk from which branches arise in lower vertebrates.

synantectic *Mineralogy.* describing a primary mineral whose formation is due to the reaction of two other minerals. Also, **synantetic.**

synantexis *Geology.* an alteration of igneous rock that occurs during the later stages of its formation, resulting directly from the consolidation of a magma. Also, DEUTERIC ALTERATION.

Synanthales *Botany.* an order of monocotyledonous herbs or somewhat woody plants, characterized by leaves with a forked, expanded blade.

synapse [sin′aps] *Neurology.* a junctional site between two nerve cells, or between a nerve cell and a muscle or gland cell; the site where the electrical nerve impulse is converted to a chemical signal for transmission to the adjacent cell.

synapsid *Paleontology.* of or relating to the mammal-like reptiles of the subclass Synapsida.

Synapsida *Paleontology.* the mammal-like reptiles, identified by a single aperture in the skull below the postorbital squamosal bones; the extinct subclass Synapsida includes the orders Pelycosauria (Permian) and Therapsida (chiefly Triassic, descendants of the pelycosaurs and ancestors of mammals); extant from the Carboniferous to Jurassic.

synapsis *Cell Biology.* the gene-by-gene pairing of homologous chromosomes that occurs during the first prophase of meiosis. Also, SYNDESIS.

synaptic *Neurology.* of, relating to, or communicated by a synapse. *Cell Biology.* of or relating to synapsis.

synaptic transmission *Neurology.* the circulation of electrical stimuli or chemical material among the neurons or from neurons to muscle through a special link known as a synapse.

synapticulum *Invertebrate Zoology.* in tunicates and others, any of numerous bars of cartilage or calcareous material connecting the gill-bars.

synaptic vesicle *Neurology.* small, round, membrane-bound organelles within the presynaptic neuron near the surface of the synapse; believed to contain a certain protein-bound substance that acts as a neurotransmitter for carrying nerve impulses across the synaptic junction.

Synaptidae *Invertebrate Zoology.* a family of wormlike burrowing sea cucumbers, holothurian echinoderms in the order Apodida, lacking tube feet and respiratory trees, and usually sticky to the touch.

synaptology *Neurology.* the branch of neurology that deals with the study of synapses and synaptic connections in the nervous system.

synaptonemal complex *Cell Biology.* a specialized structure appearing during meiosis to hold the homologous chromosomes of a bivalent together, presumably facilitating the process of crossing over.

synaptosome *Neurology.* a membrane-bound sac containing synaptic vessels that break away from a synaptic terminal after brain tissue has been homogenized in sugar solution.

synarthrosis *Anatomy.* a skeletal articulation in which the joint cavity is filled by sufficient fibrous or cartilaginous material to prevent movement between the bones.

Synbranchidae *Vertebrate Zoology.* the swamp eels, a pantropical family of eel-like fishes constituting the order Synbranchiformes, characterized by accessory breathing capabilities that enable them to survive out of the water for long periods of time.

Synbranchiformes *Vertebrate Zoology.* the single-family, pantropical order of freshwater fish that includes the swamp eels. Also, **Synbranchii.**

sync. or **synch.** synchronization; synchronizing.

Syncarida *Invertebrate Zoology.* a superorder of small wormlike freshwater crustaceans belonging to the subclass Malacostraca, with biramous nonchelate thoracic appendages and no carapace or brood pouch.

syncarp *Botany.* a fleshy aggregate fruit having united carpels.

syncarpous *Botany.* **1.** of, relating to, or resembling a syncarp. **2.** having carpels united in a compound ovary.

syncelom *Anatomy.* the perivisceral cavities of the body considered as one structure, including the pleural, cardiac, and peritoneal cavities and the tunica vaginalis.

Syncephalastraceae *Mycology.* a family of fungi of the order Mucorales, characterized by abundant, branched coenocytic mycelia and occurring in the Northern Hemisphere on soil or dung.

synchisite see SYNCHYSITE-(CE).

synchondrosis *Anatomy.* a temporary or permanent skeletal articulation in which the joint cavity between the bony elements is filled by hyaline cartilage.

synchorology *Ecology.* the science or study of the distribution and classification of plant communities into regional units, such as a province, a sector, or a district.

synchro [sing′krō] *Electricity.* a small motorlike device that, when electrically connected, provides a means of remote control. The rotor of the remote syncro will line up with the rotor of the controlling syncro. Also, SELF-SYNCHRONOUS REPEATER. *Robotics.* an encoder that transmits and receives angular shaft positions over wires.

synchro- a prefix meaning "at the same time."

synchro control transmitter *Electricity.* a precision synchro transmitter that has high-impedance windings.

synchrocyclotron *Nucleonics.* a cyclotron in which the frequency of the accelerating voltage is modulated so as to maintain synchronism with the particles, which spiral outward to relativistic velocities.

synchro differential generator *Electricity.* a synchro unit that receives an order from a synchro generator at its primary terminals, changes this order by mechanical means by any desired amount, and transmits the new order from its secondary terminals to other synchro units. Also, **syncro differential transmitter.**

synchro differential motor *Electricity.* a motor that is electrically similar to a synchro differential generator, except that a damping device is added to prevent oscillation; its rotor and stator are both connected to synchro generators, and its purpose is to indicate the difference between the signals transmitted by the two generators.

synchro generator *Electricity.* a synchro that has an electrical output proportional to the angular position of its rotor.

synchromesh *Mechanical Engineering.* a subsystem of the transmission system of an automotive vehicle that minimizes clashing by equalizing rotational speeds between the gears before engaging the clutch.
synchro motor *Electricity.* a synchro receiver that outputs the difference in a mechanical angle after two electric angles are compared with one being subtracted from the other. Also, **synchro differential receiver.**
synchronal *Science.* occurring at the same time or rate; synchronous.
synchronic linguistics *Linguistics.* an approch to language study that focuses on the state of a language at the time of study, without reference to its historic state.
synchronism the fact or quality of occurring at the same time; a simultaneous occurrence; specific uses include: *Physics.* a condition in which two or more oscillating quantities have the same frequency and a constant phase difference. *Electricity.* **1.** the condition between two signals of the same waveform and at the same frequency when the phase angle between them is equal to zero. **2.** a condition in which pulsed signals are in step with each other; the phase relationship between two or more signals whose phase difference is zero.
synchronization *Horology.* the adjustment of two timepieces or machines to show the same time or operate on the same time cycle. *Engineering.* **1.** the adjustment of beams in a television camera and display tube so that their scan patterns are identical. **2.** the maintenance of one operation in step with another.
synchronization indicator *Engineering.* an instrument that provides a visual display of two varying quantities, or of the motion of two vehicles or moving components.
synchronized blocking oscillator *Electronics.* a blocking oscillator that is synchronized with pulses occurring at a rate faster than its own frequency.
synchronized shifting *Mechanical Engineering.* the shifting of gears in an automobile transmission system by equalizing the gear speeds before making the shift.
synchronized sleep see NREM SLEEP.
synchronizer *Computer Technology.* a storage device that serves as a buffer to maintain synchronization between devices transmitting at different rates. *Electronics.* the component of a radar set that produces the timing voltage for the complete set.
synchronizing generator *Electronics.* a generator that provides synchronizing pulses to television studio and transmitter equipment. Also, **sync generator, sync-signal generator.**
synchronizing reactor *Electricity.* a current-limiting reactor used for a momentary connection across open contacts of a circuit-interrupting device, to allow synchronization to occur.
synchronizing relay *Electricity.* a programming relay that is used to initiate the closing of a circuit breaker between two alternating current sources, in order to synchronize with voltages between the two sources.
synchronous occurring at the same time or rate; specific uses include: *Engineering.* describing two or more devices, circuits, or operations that occur at the same time or have identical periods; e.g., an earth satellite with a one-day orbital period is synchronous with the earth. *Electricity.* having the same frequency and zero phase difference. *Geology.* describing geologic features or formations that occur, exist, or are deposited at the same time.
synchronous bombing *Military Science.* a method of bombing in which the telescope is focused on the target and synchronized with the movement of the airplane while the plane's course is determined by adjustment of the bombsight; the scope and sight determine the dropping angle and time of release. Similarly, **synchronous radar bombing.**
synchronous booster convertor *Electricity.* a synchronous converter that has a mechanically connected, alternating-current reversible booster or generator connected in series with the alternating-current supply circuit in order to adjust output voltage at the commutator of the converter.
synchronous booster inverter *Electricity.* an inverter that has a mechanically connected, reversible synchronous booster connected in series to adjust the output voltage.
synchronous capacitor *Electricity.* a synchronous device, operating without a mechanical load, that supplies or absorbs reactive power to or from a power system; acts as a capacitor, thereby improving the power factor and voltage regulation of an alternating-current power system.
synchronous communication *Computer Programming.* the transfer of serial binary data between computer systems or between a computer system and a peripheral device in which transmitter and receiver are synchronized and the binary data is transmitted at a fixed rate.
synchronous computer *Computer Technology.* a computer in which each event or the execution of each operation is controlled by equally spaced signals from a clock.
synchronous converter *Electricity.* a converter with motor and generator windings combined on one armature and excited by one magnetic field; used to change alternating current to direct current.
synchronous culture *Microbiology.* a population of cells that are all simultaneously in the same stage of the cell cycle.
synchronous detector *Electronics.* a detector that inserts a missing carrier signal which is in exact synchronization with the original carrier at the transmitter. Also, **synchronous demodulator.**
synchronous flashing *Entomology.* waves of light that appear to run from the top of a tree down, or from one tree or shrub to another; caused by a species of firefly that congregates and synchronizes their production of light.
synchronous gate *Electronics.* a time gate in which the output intervals are synchronized with respect to an incoming signal.
synchronous generator *Electricity.* a device that generates an alternating voltage when its armature or field is rotated by a motor; the output frequency is proportional to the speed of the generator.
synchronous growth *Microbiology.* a phenomenon in which all cells in a bacterial population divide at the same time; the characteristic growth pattern of a synchronous culture.
synchronous machine *Electricity.* an AC machine that has an average speed exactly proportional to the frequency of the system to which it is connected; if the machine is two pole, it rotates at a speed determined by supply frequency; if it is four pole, it rotates at half that speed.
synchronous motor *Electricity.* an alternating-current electric motor that is designed to run at a synchronous speed; its stator windings can have the same arrangement as an induction motor.
synchronous motor clock *Horology.* a type of clock in which a synchronous electric motor turns a simple set of reduction gears to move the hands; a very accurate type of clock as long as the frequency of the current to the motor remains constant. Also, **synchronous clock, synchronous electric clock.**
synchronous network control *Transportation Engineering.* an electronic headway control system using slots that move along guideways at regular, predetermined speeds, thus controlling scheduled departures.
synchronous operation *Electronics.* a function that occurs regularly with respect to the occurrence of a particular event in another process.
synchronous pacemaker *Cardiology.* an implanted pacemaker whose induced ventricular rhythm is synchronized to meet the needs of the atrium.
synchronous phase modifier *Electricity.* a large synchronous machine which is used solely for altering power factor at the receiving end of a transmission line so as to maintain the constant voltage under all load conditions.
synchronous pluton *Geology.* a pluton, or body of intrusive igneous rock, that was emplaced simultaneously with a major mountain-building event.
synchronous rectifier *Electronics.* a circuit that rectifies an alternating current by passing the positive and negative parts of the current through different paths by means of switches whose operation is synchronized with the AC frequency.
synchronous rotation *Astronomy.* the rotation of a satellite whose rotation period matches its period of revolution; like earth's moon, it thus keeps one side turned perpetually toward its primary.
synchronous satellite see FIXED SATELLITE.
synchronous speed *Electromagnetism.* the angular velocity, commonly expressed in revolutions per minute, of rotating magnetic field in a synchronous machine given by the AC frequency in Hertz times 120, divided by the number of poles in the device.
synchronous switch *Electronics.* a thyratron circuit used to control the process of ignitrons in applications like resistance welding.
synchronous timing *Metallurgy.* a method of regulating a welding-transformer current so that the first half-cycle begins at the proper time in relation to the voltage for a balanced current wave, the following half-cycles are identical to the first, and the last half-cycle is of opposite polarity to the first.
synchronous vibrator *Electromagnetism.* an electromagnetic vibrator that converts low DC voltage to low AC voltage and then applies it to a power transformer, where the output is a high-voltage AC signal to be rectified.
synchronous working *Computer Technology.* the clock-controlled operation of a synchronous computer.

synchro receiver *Electricity.* a transformer that is electrically similar to a synchro transmitter, but develops a torque that increases with the difference in angular alignment of the transmitter and receiver rotors in a direction to reduce the difference between the secondary windings of two devices interconnected toward zero. Also, RECEIVER SYNCHRO, SELSYN MOTOR, SELSYN RECEIVER.

synchroscope *Electronics.* a cathode-ray oscilloscope used to display a short-duration pulse by using a fast sweep that is synchronized with the pulse signal to be observed. *Engineering.* any of various other devices using lamps to determine synchronization of two systems.

synchro-shutter *Engineering.* a camera shutter that is set to open at the instant that a light flash is triggered.

synchro system *Electricity.* an electric system used for transmitting angular position or motion through the use of one or more synchro transmitters, synchro receivers, or synchro control transformers; it may also include differential synchro machines.

synchro transmitter *Electricity.* a transmitter with relatively rotatable primary and secondary windings such that the output of the secondary winding consists of two or more voltages that vary with, and completely identify, the relative angular position of the primary and secondary windings. Also, SELSYN GENERATOR, SELSYN TRANSMITTER, TRANSMITTER SYNCHRO.

synchrotron *Nucleonics.* a machine that accelerates charged particles in circular orbits by varying the frequency of the accelerating voltage and also magnetic field so as to maintain a circular orbit; combining features of the cyclotron and betatron, it can produce up to 70 million volts.

synchrotron radiation *Electromagnetism.* the radiation produced by relativistic charged particles orbiting in a magnetic field. It consists of pulses (on a nanosecond scale) with a very high intensity and a high degree of polarization. The radiation is produced in the direction of travel of the electrons (tangential to the curve of the ring) and is characterized by a continuous spectral distribution. Also, **synchrotron emission.**

synchysite-(Ce) *Mineralogy.* $Ca(Ce,La)(CO_3)_2F$, a weakly radioactive, yellow to brown, translucent, orthorhombic, pseudohexagonal mineral occurring as small, thin to thick tabular crystals, having a specific gravity of 3.902 to 4.15 and a hardness of 4.5 on the Mohs scale. Two additional minerals have been identified: in **synchysite-(No),** cerium is replaced by neodymium; in **synchysite-(Y),** lanthanum is replaed by yttrium. Also, SYNCHISITE.

Synchytriaceae *Mycology.* a family of fungi of the order Chytridiales occurring in land and water; they live off algae, higher plants, or flowering plants and are pathogenic to some crops.

synclinal *Geology.* of or relating to a syncline. *Anatomy.* of two parts, bent or inclined together.

synclinal axis see TROUGH SURFACE.

synclinal valley *Geology.* a valley that is produced by or that follows the site of a syncline.

syncline *Geology.* a downward fold of stratified rocks in which the sides slope downward toward each other.

syncline

synclinorium *Geology.* a composite fold of great extent consisting of a series of minor folds arranged in the general form of a syncline.

synclitic *Hematology.* relating to or marked by synclitism.

synclitism *Hematology.* normal, synchronous maturation of the nucleus and cytoplasm of blood cells. Also, **syncliticism.**

synclonus *Neurology.* **1.** a tremor, or clonic spasm, of several muscles simultaneously. **2.** any disease or disorder characterized by muscular tremors or clonic contractions of a group of muscles.

Syncom *Space Technology.* any of a series of commercial radiotelevision communications satellites placed in synchronous orbit around the earth's equator.

Synconosa *Invertebrate Zoology.* an order of simple sponges in the class Calcarea, having thickened, folded internal walls and tissues; the flagellated chambers are small pouches off the main cavity.

syncope [sing´kə pē] *Medicine.* a faint or swoon; the temporary suspension of consciousness due to cerebral ischemia, marked by the sudden onset of pallor, coldness of the skin, and lightheadedness.

syncytial *Cell Biology.* of, relating to, or producing a syncytium.

syncytial ciliate theory *Evolution.* a theory of metazoan origin that proposes that metazoans evolved from a multinucleate protozoan stock that cleaved to become cellular after it had become specialized.

syncytioma malignum *Oncology.* a tumor that originates at the placental site or the puerperium and is composed of large cells from the outermost layer of the placenta and smaller cells from the epithelium of the chorionic villi.

syncytiotrophoblast *Cell Biology.* a syncytial layer that forms the outermost fetal component of a mammalian placenta, acting as the barrier between fetal and maternal tissue. Also, **syncytial trophoblast.**

syncytium *Cell Biology.* a multinucleate mass of protoplasm, such as a giant cell, that is produced by the merging of cells.

syndactyly *Anatomy.* a congenital, developmental anomaly of the hand or foot marked by fusion of two or more fingers or toes.

syndesis *Cell Biology.* see SYNAPSIS. *Surgery.* see ARTIFICIAL ANKYLOSIS.

syndesmectomy *Surgery.* the excision of a ligament or a segment of a ligament.

syndesmoma *Oncology.* a tumor composed of connective tissue.

syndesmorraphy *Surgery.* the suture or repair of a ligament.

syndesmosis *Anatomy.* a skeletal articulation in which the connection between the bony elements is maintained by fibrous connective tissue in the form of an interosseous membrane or ligament; for example, the connection between the distal ends of the tibia and fibula.

Syndiniaceae *Botany.* a family of nonphotosynthetic flagellates of the order Syndiniales, which are internal parasites in marine copepods, appendicularians, crabs, and sardines.

Syndiniales *Botany.* an order of marine dinoflagellates composing the subclass Syndiniophycidae.

Syndiniophycidae *Botany.* a subclass of nonphotosynthetic, marine dinoflagellates of the class Dinophyceae that are parasitic on various vertebrates, nonvertebrates, and other dinoflagellates; characterized by a multinucleate parasitic stage.

syndiotactic *Materials Science.* of or relating to polymers in which nonsymmetrical groups in the monomer alternate from one side of the chain to the other. Thus, **syndiotactic polymer.**

syndnone *Organic Chemistry.* a class of mesoionic compounds having a 1,2,3-oxadiazole ring structure.

syndrome *Medicine.* a group of symptoms or signs that appear together and that tend to indicate, with some consistency, the presence of a certain disease or other condition.

syndynamics *Ecology.* the scientific study of the causes and patterns of successive changes in a plant community.

synechia *plural,* **synechiae.** *Medicine.* any adhesion of parts of the body.

Synechococcus *Bacteriology.* a genus of photosynthetic cyanobacteria occurring as rod-shaped cells dividing by binary fission and found in a number of habitats, including marine, hypersaline, and hot-spring environments.

Synechocystis *Bacteriology.* a genus of cyanobacteria that grow as coccoid cells and reproduce by binary fission in more than one plane, resulting in cell clusters.

synecology *Ecology.* the scientific study of the relationships among communities of organisms and their environment.

synecthry *Ecology.* a relationship between two organisms in which both openly display hostility toward each other, yet one allows the other to feed off it.

syneresis *Chemistry.* the contraction of a gel and the resulting exudation of liquid from the gel.
synergic curve *Space Technology.* a trajectory designed to minimize air resistance, thus providing optimum fuel economy with optimum speed during the ascent of a rocket vehicle or missile; initially moving vertically, it bends toward the horizontal between 20 and 60 miles altitude to minimize the thrust required for vertical ascent.
synergid *Botany.* one of the two sterile cells without cell walls that lie beside the ovum in the embryo sac of angiosperms; along with the ovum, the two compose the egg apparatus.
synergism [sin´ər jiz´əm] *Science.* an interaction of elements such that their combined effect is greater than the sum of their individual effects. *Ecology.* specifically, the cooperative action of two or more organisms so that the effect of their collective effort is greater than would be the sum of any individual actions.
synergistic [sin´ər jis´tik] relating to or exhibiting synergism.
synergy [sin´ər jē´] a combined action or functioning; synergism; specific uses include: *Pharmacology.* the cooperation or correlated action of two or more drugs, such that their combined biological effect is greater than the sum of their individual effects. *Physiology.* the cooperative action of two or more muscles, nerves, or the like.
synesthesia *Neurology.* a disorder of sensory perception characterized by a secondary and subjective perception that accompanies an actual perception of a different sensory mode, such as a subjective perception of color or sound induced by an actual sensation of taste.
synfuel see SYNTHETIC FUEL.
Syngamidae *Invertebrate Zoology.* a family of hookworms, nematodes in the order Strongyloidea, including tapeworms and gullet worms; endoparasites of birds and mammals.
syngamy *Biology.* sexual reproduction involving the union of two gametes (egg and sperm); fertilization.
syngas see SYNTHESIS GAS.
Synge, Richard Laurence Millington [sing] born 1914, English biochemist; with Archer J. P. Martin, shared the Nobel Prize for developing partition chromatography.
syngenesious *Botany.* of or related to a flower with anthers united about the style in the form of a cylinder, as in the family Compositae.
syngenesis *Biology.* formation of the germ by fusion of the male and female elements, so that the substance of the embryo is derived from both parents; sexual reproduction. *Geology.* a process by which loose, unconsolidated sedimentary deposits are produced or accumulated in place.
syngenetic *Geology.* **1.** describing an ore or mineral deposit formed at the same time and by the same processes as the host rock. Also, IDIOGENOUS. **2.** describing a primary sedimentary structure formed at the same time as the deposition of the sediment.
syngenic *Genetics.* describing identical twins or other organisms that are genetically identical, and hence share identical body chemistry.
syngenite *Mineralogy.* $K_2Ca(SO_4)_2 \cdot H_2O$, a colorless or white monoclinic mineral occurring in tabular or prismatic crystals, having a specific gravity of 2.603 and a hardness of 2.5 on the Mohs scale; found in salt deposits and as a product of vulcanism.
Syngnathidae *Vertebrate Zoology.* the seahorses and pipefishes, a family of tropical and warm-temperate marine fish of the suborder Syngnathoidei, characterized by an elongate body encased in bony armor, a tubular snout with a toothless mouth, and a specialized structure in males for storing eggs and caring for young.
syngraft see ISOGRAFT.
synkaryon *Cell Biology.* a diploid cell nucleus formed by the fusion of two nuclei. *Genetics.* this nucleus that occurs in the formation of a zygote and in the fusion of nuclei of somatic cells of different genotypes during genetic experiments.
synkinematic see SYNTECTONIC.
synkinesia *Physiology.* an uncontrolled motion or reflex that occurs at the same time as a voluntary motion.
synkinesis *Neurology.* an involuntary movement that occurs simultaneously with a similar voluntary movement in another part of the body, such as swinging the arms while walking. Thus, **synkinetic.**
synnema *Mycology.* an erect bundle of spore-bearing filaments or hyphae in certain fungi, particularly of the order Stilbellales.
synnematospores *Mycology.* spores formed on a synnema.
synodic *Astronomy.* of or relating to a conjunction between two celestial objects.
synodic curve *Space Technology.* a theoretical earth satellite having the same orbiting period as the moon.
synodic month *Astronomy.* the month defined by the time from one new moon to the next: approximately 29.3 days.
synodic period *Astronomy.* the time elapsed between two successive conjunctions or oppositions of a planet with the sun as perceived by an observer on earth.
synonym *Linguistics.* **1.** a word having the same or nearly the same meaning as another word in the same language. **2.** a word or expression accepted as another name for something. *Systematics.* specifically, any of two or more taxonomic names assigned to the same taxon.
synonymous relating to or being a synonym.
synonymous codon *Molecular Biology.* a codon that can be used in place of another to encode the same amino acid.
synonymy *Linguistics.* **1.** the quality of being equivalent in meaning. **2.** the study of synonyms. *Systematics.* the group of synonyms by which a taxon is known, in addition to the proper name; synonymy usually indicates a difference in taxonomic classification schemes among taxonomists.
synophthalmia see CYCLOPIA.
synopsis *Systematics.* a concise description or abstract of the significant or distinguishing features of some topic or entity.
synoptic *Meteorology.* **1.** obtained simultaneously over a wide area in order to afford a simultaneous overall view. Thus, **synoptic data, synoptic forecasting, synoptic observation, synoptic report,** and so on. **2.** relating to, constituting, or providing an overall view.
synoptic chart *Meteorology.* any meteorological chart or map describing the state of the atmosphere over a large area at a particular moment and presenting the results of weather data and analyses by the graphical representation of weather data elements.
synoptic climatology *Meteorology.* the study and analysis of climate of a given locality in a particular synoptic condition, as distinguished from the study and analysis of climatic parameters extending over many or all synoptic conditions.
synoptic code *Meteorology.* any code used to communicate synoptic weather observations, such as the U.S. airways code.
synoptic meteorology *Meteorology.* a subscience of meteorology in which analyses are made simultaneously of weather observations at several points in the atmosphere.
synoptic model *Meteorology.* a graphical representation of the spatial distribution of meteorological elements such as clouds, precipitation, humidity, temperature, pressure, and wind.
synoptic oceanography *Oceanography.* the study of ocean phenomena from a comprehensive point of view, using primarily data recorded at a particular time at widely separated places; in recent years, increasingly relying on satellite photography.
synoptic scale see CYCLONIC SCALE.
synoptic terrain photography *Photogrammetry.* a photographic representation of large areas of the earth's surface taken from an orbiting spacecraft.

synoptic terrain photography (from Gemini 5)

synoptic wave chart *Oceanography.* a chart of an area that shows synchronous wave reports from vessels and other sources, as well as predicted wave heights for areas not reported on; may also have wave-height isolines drawn in, along with meteorological data.

synoptic weather observation *Meteorology.* a periodic surface weather observation of sky cover, state of the sky, cloud height, sea-level atmospheric pressure, temperature, dew point, wind speed and direction, precipitation, and any special phenomena that are present at the time of the observation or that have been observed since the previous observation; as specified by the World Meteorological Organization, such observations usually occur at regular three- and six-hour intervals.

synorchidism *Medicine.* a fusion of the testes. Also, **synorchism.**

synorogenic *Geology.* describing a geologic process or event that took place at the same time as a mountain-building event, or a rock or feature formed during that time.

synostosis *Anatomy.* **1.** the osseous union of adjacent bones or adjacent parts of a single bone that were separated by fibrous or cartilaginous material before they were joined. **2.** the osseous union of bones that are normally distinct. Also, **synosteosis.**

synovial fluid *Physiology.* the transparent liquid produced by the synovial membranes of a joint, and acting as a lubricant. Also, **synovia.**

synovial joint see DIARTHROSIS.

synovial membrane *Histology.* a layer of connective tissue that lines joint cavities, bursae, and tendon sheaths. The membrane produces synovial fluid that has a lubricating function.

synovioma *Medicine.* a tumor originating in the synovial membrane.

synoviosarcoma *Oncology.* a malignant neoplasm in the synovial membrane contained in joint cavities, bursae, and tendon sheaths.

synovitis *Medicine.* an inflammation of a synovial membrane, usually as the result of an aseptic wound or an injury, characterized by swelling due to the accumulation of fluid around the capsule; the joint is tender and painful, and motion is restricted.

synpelmous *Vertebrate Zoology.* having the two main flexor tendons of the toes united beyond the branches to each digit, as in certain birds.

synphylogeny *Ecology.* the science or study of the historical and evolutionary trends and changes that take place within plant communities.

synphysiology *Ecology.* the science or study of the evolutionary trends and changes that take place within plant communities.

synsepalous see GAMOSEPALOUS.

syntactic [sin tak´tik] *Linguistics.* having to do with syntax. Also, **syntactical.**

syntactic ambiguity *Artificial Intelligence.* the possibility that a natural-language sentence can be parsed in more than one way; e.g., "I saw John's dog driving to work today."

syntactic analysis *Computer Science.* **1.** a compilation step that includes lexical analysis, statement parsing, and intermediate code generation. **2.** the parsing of a language to generate a parse tree. Also, SYNTAX ANALYSIS.

syntactic extension *Computer Programming.* a feature of an extensible language that permits the definition of new notations for existing or user-defined mechanisms.

syntactics *Linguistics.* the branch of semiotics that deals with the formal characteristics of symbols and signs.

syntax [sin´taks] *Linguistics.* **1.** the set of natural rules or patterns that govern how message-bearing units (i.e., words and word parts such as prefixes) are combined in a language to form meaningful sentences. **2.** the study of these rules or patterns. *Artificial Intelligence.* a formal statement of the grammatically acceptable forms of a natural language; used in language analysis. *Computer Programming.* a set of rules specifying the structure of a programming language.

syntax analysis see SYNTACTIC ANALYSIS.

syntax checker *Computer Programming.* a program that tests source statements in a programming language for violations of its syntax. Also, SYNTAX SCANNER.

syntax diagram *Computer Programming.* a diagram representing the structural rules of a programming language.

syntax-directed compiler *Computer Programming.* a compiler whose activities are based on syntactical relations found in parsing the input program. Also, **syntax-oriented compiler.**

syntax-directed translation *Computer Science.* the translation of a statement, for example, in a programming language, that is guided by the syntactic form of the statement.

syntax-directed translator *Computer Science.* a program, such as a compiler, that performs translation from a language into another form in a syntax-directed manner.

syntax error *Computer Programming.* a violation of the grammatical rules of a programming language statement, such as misspelling the name of a command. Also, **syntactic error.**

syntaxial overgrowth *Mineralogy.* an optically oriented crystal overgrowth on a detrital mineral grain of identical composition, formed during diagenesis.

syntax scanner see SYNTAX CHECKER.

syntechny *Ecology.* a resemblance between unrelated organisms due to their adaptation to similar environments. Thus, **syntechnic.**

syntectic *Geology.* describing a magma formed by the melting of two or more rock types and the assimilation of country rock.

syntectonic *Geology.* describing a process, event, or geological feature that occurred or was formed during any kind of tectonic activity. Also, SYNKINEMATIC.

Synteliidae *Invertebrate Zoology.* the sap-flow beetles, a superfamily of coleopteran insects in the superfamily Histeroidea.

syntenic genes *Genetics.* genes that are thought to be located on the same chromosome because they are lost concurrently with a specific marker gene that is known to be located on that chromosome.

synteny *Genetics.* the condition of genetic loci that lie on the same chromosome.

Syntexidae *Invertebrate Zoology.* a family of wasps, hymenopteran insects in the superfamily Siricoidea, having a single species *Syntexis libocedrii*; larvae are found in North American incense cedars.

syntexis *Geology.* a process by which magma is generated by the melting of two or more rock types and the assimilation of country rock.

synthase see LYASE.

synthesis [sin´thə sis] *Science.* **1.** the combining of constituent elements into a unified entity. **2.** a unified whole formed by combining. *Chemistry.* specifically, the process of building chemical compounds from more elementary substances by means of one or more chemical reactions, or by nuclear change. *Control Systems.* the use of available components to plan and construct a system that will perform in a specified manner. Also, SYSTEM DESIGN.

synthesis gas *Chemical Engineering.* a mixture of hydrogen and carbon monoxide obtained by the reforming of methane. Also, SYNGAS.

synthesis of continuity *Surgery.* the union of the edges of a wound or the ends of a fractured bone.

synthesize [sin´thə sīz´] *Chemistry.* to carry on a process of synthesis; form a single or unified entity from constituent elements.

synthesized attribute *Computer Science.* an attribute of a structure, e.g., a phrase in a programming language statement, that is derived from the attributes of its components. For example, the sum of two floating-point quantities will also be floating point.

synthesized translation *Computer Science.* a method of translating statements, e.g., in a programming language, such that the translation of a phrase is built up from the translations of its components.

synthesizer *Electronics.* **1.** any of various instruments that integrate simple elements or functions to produce more complex entities. **2.** specifically, an electronic device that simulates the sound of various musical instruments.

synthetase see LIGASE.

synthetic *Science.* relating to, produced by, or involving synthesis. *Chemistry.* relating to compounds formed artificially by chemical synthesis. *Engineering.* in general, describing any product or item that is the result of human technology rather than something that exists in nature.

synthetic aperture *Engineering.* a technique by which the phase relationship of two receiving systems, such as radar receivers or telescopes, is used to emulate the resolving power of an instrument having an aperture equal to their separation.

synthetic aperture radar *Engineering.* a radar technique in which a timed pulse signal is sent out from an aircraft flying a very steady course; the phase of returned pulses is analyzed to emulate the characteristics of a radar with an antenna size of up to a mile.

synthetic basic motion times *Industrial Engineering.* standard time values that are obtained from predetermined systems or that are determined by measuring similar task elements, rather than by directly measuring the work at hand, because they apply to basic motions that are common to many tasks. Also, BASIC MOTION TIMES.

synthetic detergent *Materials.* any synthetic substance other than soap that is an effective cleanser and functions equally well as a surface-active agent in hard or soft water.

synthetic diet *Biochemistry.* a diet that consists only of known chemical ingredients.

synthetic division *Mathematics.* a rapid method of evaluating a polynomial

$$p(x) = a_n x^n + a_{n-1}x^{n-1} + \cdots + a_1 x + a_0$$

at a fixed argument c and of factoring out $x - c$ from $p(x)$ if $p(c) = 0$, accomplished as follows: (a) three rows of "detached coefficients" will be required. The first consists of the $n + 1$ coefficients $a_n a_{n-1} \cdots a_1 a_0$ of the polynomial, with zeros inserted for missing powers of x. (b) the space in the second row below the leading coefficient of the polynomial is left blank, and the leading coefficient a_n is copied below itself to begin the third row. (c) the process continues inductively, multiplying the latest entry of the third row by c, writing the product in the succeeding column in the second row, adding that product to the corresponding entry of the first row, and writing the sum in the third row of that column. (d) the process ends when the last column is used up. The third row of detached coefficients has $n + 1$ entries. The first n are the coefficients of the polynomial of degree $n - 1$ that is the quotient of $p(x)$ with $(x - c)$; the rightmost entry is both the remainder term and $p(c)$. The example below illustrates the method:

$p(x) = 2x^3 - 4x + 1$; $c = 3$.

2	0	–4	1
	6	18	42
2	6	14	43

$(2x^3 - 4x + 1)/(x - 3) = 2x^2 + 6x + 14 + 43/(x - 3)$; $p(3) = 43$.

synthetic fiber *Textiles.* a generic term for textile fiber made by chemical synthesis, such as nylon, polyester, acrylic, and vinyl. The term does not specifically include cellulose fibers such as rayon and acetate, although it is often applied to them.

synthetic fuel *Materials.* a fuel derived from coal, oil shale, oil sands, or biomass, used as a substitute for oil or natural gas. Also, SYNFUEL.

synthetic language *Computer Programming.* a fabricated language.

synthetic medium see DEFINED MEDIUM.

synthetic oil of bitter almond *Organic Chemistry.* a popular name for benzaldehyde, C_6H_5CHO, an almond flavoring agent. See BENZALDEHYDE.

synthetic peptide *Biochemistry.* a peptide that originates through artificial chemical processes rather than naturally.

synthetic polymer *Materials.* a polymer that is manmade, such as plastics, elastomers, and adhesives.

synthetic resin *Organic Chemistry.* an amorphous, organic, semisolid or solid material produced by polymerization of unsaturated monomers such as ethylene.

synthetic rubber *Materials.* a rubberlike compound produced by the polymerization of unsaturated hydrocarbons or by the copolymerization of such hydrocarbons with styrene or butadiene.

synthol process *Chemical Engineering.* a reaction of carbon monoxide and hydrogen with an iron and sodium carbonate catalyst that produces a mixture usable as a synthetic gasoline.

syntony *Electricity.* a condition occurring when two oscillating circuits have the same resonant frequency.

syntopic *Ecology.* of or relating to different populations or species that occupy the same general habitat.

Syntrophiidina *Paleontology.* a suborder of articulate brachiopods in the extinct order Pentamerida; characterized by an impunctate and subcircular shell with a dorsal fold; extant in the Cambrian and Ordovician.

Syntrophobacter *Bacteriology.* a genus of Gram-negative, anaerobic, nonmotile bacteria, characterized by a cross-feeding association with a second, hydrogen-scavenging bacterial strain.

syntrophoblast *Developmental Biology.* the outer syncytial layer of the trophoblast that covers each chorionic villus of the placenta. Also, **syncytiotrophoblast.**

Syntrophomonas *Bacteriology.* a genus of Gram-negative, anaerobic, rod-shaped bacteria that use protons as electron acceptors during oxidative metabolism and are only able to grow in the presence of hydrogen-scavenging systems.

Syntrophus *Bacteriology.* a proposed genus of Gram-negative, anaerobic, rod-shaped bacteria whose growth is inhibited by the presence of hydrogen, and thus requires the presence of hydrogen-utilizing bacteria for growth.

syntrophy *Ecology.* a relationship in which two organisms feed on each other, with each deriving one or more essential nutrients from the other. Thus, **syntrophic, syntrophism.**

syntype *Systematics.* any member of a type series for which no holotype is designated.

synulosis *Surgery.* a process of complete scar formation.

synulotic *Surgery.* **1.** acting to promote the formation of a scar. **2.** an agent that promotes the formation of a scar.

synura *Botany.* a genus of biflagellate free-swimming flagellates of the family Synuraceae, characterized by sequentially developed scales and a basal flagellar swelling; often produce odors and fishy flavors in water supplies.

Synuraceae *Botany.* a family of unicellular or colonial flagellates of the order Ochromonadales, characterized by a complex, specialized cell envelope composed of an armor of silica scales or needles, and having a well-developed flimmer flagellum; common in freshwater phytoplankton worldwide.

synusia *Ecology.* a community made up of plants that share a similar life form, occupy the same habitat, and have similar ecological requirements.

Synxiphosurina *Paleontology.* an extinct suborder of chelicerate arthropods in the order Xiphosurida; Paleozoic cousins of the modern horseshoe crab, *Limulus*; characterized by free abdominal segments; exact affinities uncertain because of the incomplete fossil record; extant from the Ordovician to the Devonian. Also, **Synziphosurina.**

syphilis [sif′i ləs] *Medicine.* a chronic infectious disease, usually transmitted by sexual contact, but also acquired in utero or by direct contact with infected tissues or blood, caused by infection with the spirochete *Treponema pallidum.* Untreated, it progresses through distinct stages over a period of years and may involve many organs and tissues, leading to mental and physical disability and premature death.

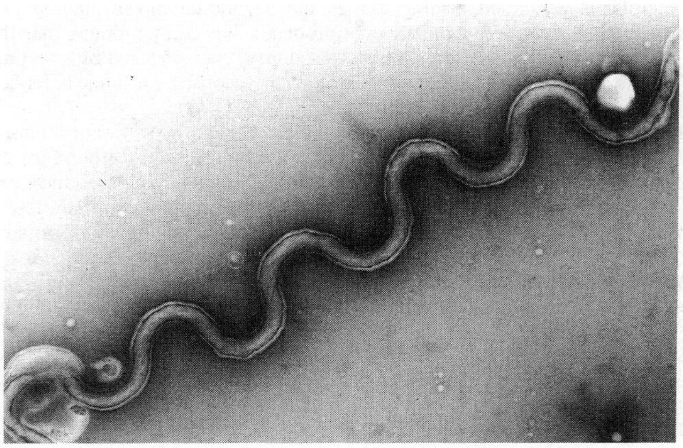

syphilis

syphilitic [sif′i lit′ik] *Medicine.* relating to, produced by, or affected by syphilis.

syphilitic aneurysm *Cardiology.* an abnormal localized saclike swelling or dilation of the thoracic aorta related to or caused by cardiovascular syphilis.

syphilitic aortic insufficiency *Cardiology.* a condition related to or caused by syphilis and marked by failure of a cardiac valve to close normally, permitting regurgitation of blood past the closed valve.

syphilitic aortitis *Cardiology.* inflammation of the thoracic aorta caused by syphilis; a manifestation of tertiary syphilis; characterized by the replacement of elastic tissue by vascular scars commonly resulting in dilation and the formation of aneurysms.

syphilophobia *Psychology.* an irrational fear of syphilis.

SYR *Aviation.* the airport code for Syracuse, New York.

Syracosphaeraceae *Botany.* a family of flagellate marine algae of the order Coccosphaerales, characterized by coccoliths of the caneolith type.

Syrah *Botany.* a pungent black grape of the northern Rhône Valley, used as a blending grape in wines such as Côte Rôtie and Hermitage.

syringe [si rinj′; sēr′inj] *Medicine.* an instrument for injecting liquids or withdrawing them.

syringeal *Medicine.* of, relating to, or connected with the syrinx.

syringectomy *Surgery.* the excision of the walls of a fistula.

syringobulbia *Medicine.* the presence of cavities within the medulla oblongata in the midbrain.

syringoma *Medicine.* adenoma of the sweat glands. Also, **syringocystadenoma, syringocystoma.**

syringomycin *Biochemistry.* a peptide-inclusive phytotoxin, produced by bacterial *Pseudomonas* species, that breaks down plant and fungal cells.

syringomyelia *Medicine.* an abnormal syndrome of the spinal cord characterized by the presence within the cord of liquid-filled cavities; generally limited to the cervical region but sometimes extending to the medulla oblongata or the thoratic region, resulting in muscular weakness, atrophy, or sensory impairment.

Syringophyllidae *Paleontology.* a family of corals in the extinct subclass Tabulata and order Sarcinulida; extant in the Ordovician.

syringopontia *Neurology.* a disorder in which a congenital cavitation of the medulla oblongata extends into the pons.

Syringopora *Paleontology.* a genus of colonial tabulate corals in the extinct suborder Syringoporina, having horizontal or funnel-shaped tabulae; extant in the Silurian and Carboniferous. Also, ORGAN-PIPE CORAL.

Syringothyris *Paleontology.* a genus of articulate brachiopods in the superfamily Spiriferacea; characterized by an impunctate shell, a pyramidal shape, and an interior syrinx; sometimes semi-infaunal, around 5 centimeters across; extant in the Devonian and Carboniferous.

syringotome *Surgery.* a surgical instrument used to cut the walls of a fistula.

syringotoxin *Biochemistry.* a phytotoxin produced by citrus-infecting bacterial *Pseudomonas* species.

syrinx *Vertebrate Zoology.* the sound-producing organ of birds, found at the bifurcation of the trachea. *Paleontology.* a tube that surrounds the pedicle of some extinct brachiopods; the tube represents an evolutionary modification of the deltidial plates of earlier brachiopods.

Syrphidae *Invertebrate Zoology.* the hover flies, flower flies, and drone flies, bright-colored dipteran insects in the series Aschiza; their legless larvae eat aphids, and the adults feed on flower nectar.

syserstkite see IRIDOSMINE.

sysgen or **SYSGEN** system generation.

SYSIN *Computer Programming.* **1.** a system input stream. **2.** the data definition name of a data set in the input stream. Similarly, **SYSOUT**.

syssiderite see STONY-IRON METEORITE.

Systellomatophora *Invertebrate Zoology.* an order of slugs, gastropod mollusks belonging to the subclass Pulmonata, having nonretractile eyestalks.

system *Science.* an assemblage of parts that form a unified whole. *Physics.* a collection of particles or interacting components considered to form a distinct physical entity for the purpose of study or identification. *Engineering.* a combination of components, elements, subsystems, and operating procedures, functioning together to achieve some objective; for example, an antiaircraft missile system may include radars, computers, launchers, and training simulators, as well as missiles. *Geology.* **1.** a time-stratigraphic unit of worldwide significance, higher than a series and representing the rocks formed during a geologic period. **2.** any group of natural structures, features, or forces that are related.

system analysis *Control Systems.* an analytic method that uses mathematics to describe how a set of interconnected components will react to a given input or set of inputs.

systematic analog network testing approach *Electronics.* an online computer-based system with an integrated database and optimal human intervention that provides computer printouts used in automatic testing of networks. Also, **SANTA.**

systematic desensitization *Behavior.* a behavior therapy technique that is used to reduce or eliminate anxiety, in which deep muscle relaxation is paired with imagined scenes of anxiety-evoking situations that increase in intensity.

systematic distortion *Electricity.* a periodic or constant distortion such as bias or characteristic distortion.

systematic error *Engineering.* an error that is characteristic of a device or system, arising from a physical law or resulting from some bias in the measuring process. *Statistics.* a sampling error that causes the resulting measurement to be incorrect in a systematic way.

systematic geography *Geography.* the study of the areal distribution or differentiation of individual features or elements of the earth's geography, as distinguished from regional geography, which studies numerous features and elements in a limited area.

systematic joints *Geology.* a continuous, patterned arrangement of joints created at right angles to the boundaries of the constituent rock unit.

systematics *Biology.* the classification of living organisms in regard to their natural relationships, including the description, naming, and classifying of these organisms.

Systematics

Systematics is the science that deals with the naming and classification of all kinds of organisms, past and present. By thus codifying the branching patterns of relationship among species within a genus, or genera within a family, and so on, systematics underpins all studies of the evolution and diversity of life on earth.

The subject goes back at least to Aristotle, who outlined a scheme for classifying plants and animals in a hierarchical "chain of being" (by contrast, modern evolutionary biologists see present species as the co-equal twig-tips of a tree whose branches and roots lie in the past). In a sense, many earlier myths and legends prefigure a wish to see orderly relations in the living world. But the foundation of modern systematics dates only from the work of Linneus in Uppsala, who in 1758 gave us the still-used scientific classification by Latin binomials (*Genus species,* as in *Homo sapiens*). Linneus recorded some 9,000 species of plants and animals.

It is remarkable that systematics as a science originated a full century after Newton. After all, systematically classifying the creatures we share the world with is more straightforward than understanding the gravitational laws governing planetary motion. This fact holds little-explored implications for the history of ideas, and of how we see ourselves in relation to the natural world.

Even today, the total number of plant and animal species that have been named and recorded is uncertain, ranging from 1.5 million to 1.8 million or more; there is, amazingly, no central catalogue. The global total is much more uncertain, with reasonable estimates ranging from as low as 3 million to as high as 30 million or more. I would estimate around 5–10 million.

In earlier times classifications and inferred evolutionary relations were based on comparative studies of the shape, anatomy, physiology, and ecology of organisms. Recent advances have made available many other kinds of comparisons. In particular, it is now possible to quantify similarities and differences among amino-acid sequences in particular molecules (such as the hemoglobins that supply oxygen to the blood, or, at the most fundamental level, the DNA molecules that code for the self-assembly of living things).

At the same time, earlier and more intuitive ways of perceiving patterns of relationship are being replaced by computer-based statistical techniques for constructing the most parsimonious patterns. All this has brought new excitement of the subject, suggesting new answers to old questions (for example, are humans more closely related to chimpanzees or gorillas?).

As tropical deforestation accelerates, and other human impacts emperil or extinguish species at a rate probably without precedent in the geological record, systematics is increasingly relevant as the science undergirding efforts to conserve biological diversity. We are probably extinguishing species today faster than we are recording them. In this sense, systematics is unique among the fields covered in this Dictionary in that it is a discipline with a time limit.

Robert M. May
Royal Society Research Professor
Department of Zoology
University of Oxford

systematic sampling *Mining Engineering.* the extraction of samples from a unit of coal in fixed quantities or at evenly spaced time intervals.
systematic support *Mining Engineering.* the placement of steel or timber supports at fixed intervals regardless of the condition of the roof and sides.
system bandwidth *Control Systems.* the difference between frequencies at which the gain of a system is 0.707 times its peak value.
system calendar *Computer Programming.* a register maintained by the operating system that provides the date and year on request by a program.
system capacity *Transportation Engineering.* **1.** the maximum number of vehicles that can operate in a system at a given time. **2.** the number of passengers that these vehicles can carry.
system chart *Computer Programming.* a flow chart emphasizing system component operations, including those performed either by machines or by people. Also, RUN DIAGRAM, SYSTEM FLOWCHART.
system command *Computer Programming.* a statement that directs the operating system to perform a system operation, such as logging the user in or out, saving a file to disk, or running a program.
system design *Computer Programming.* the specifications of the working relations among all the parts of a system in terms of their characteristic actions.
system documentation *Computer Programming.* the technical description of an overall system, including preliminary and final design descriptions, testing procedures and results, and user manuals.
system effectiveness *Engineering.* the overall ability of a system to perform its intended function, elements of which include reliability, efficiency, and safety.
Systeme International d'Unités *Metrology.* the French version of the term International System of Units.
system flowchart see SYSTEM CHART.
system generation *Computer Programming.* **1.** the process of tailoring a generalized operating system for a specific user's computer and peripherals, compiling the appropriate programs, and making an operating system version. **2.** in general, the process of making a complete version of a large software system from its parts.
systemic [sis tem′ik] *Biology.* affecting or relating to an organism as a whole. *Medicine.* relating to or affecting the whole body. *Plant Pathology.* relating to a type of plant pesticide that is harmless to the plant itself, but is absorbed into the sap and enables the whole plant to resist the attack of pests such as aphids or mites.
systemic circulation *Physiology.* the pathway taken by blood from the heart's left ventricle through the aorta and its branching arteries to the capillaries within the tissues and its arrival back at the right atrium, after flowing through the veins and the venae cavae.
systemic context *Archaeology.* the status of an artifact or feature that is still in active use by humans.
systemic infection *Virology.* an infection caused by the spread of a virus from the original site of infection to disparate cells of the organism.
systemic lupus erythematosus *Medicine.* a chronic inflammatory disease of unknown cause that affects many systems in the body, but is characterized by involvement of the skin, joints, kidneys, nervous system, and mucous membranes; the disease occurs mainly among middle-aged women.
system improvement time *Computer Technology.* the machine down time needed for installing and testing new components, modifying existing components, and running test programs on a modified machine.
system integration *Computer Technology.* the process of linking all hardware components and separately developed software modules together for final testing.
system-level timer *Computer Technology.* a device that can be set to interrupt a system after a specified interval or at a specified time.
system library *Computer Programming.* a collection of programs maintained in auxiliary storage and retrievable by the operating system as needed.
system life cycle *Engineering.* the entire series of stages a system goes through from conception to abandonment, including development, initial deployment, operational use, upgrading, and so forth.
system loader *Computer Programming.* an operating system program that locates programs and copies them into main memory.
system of linear equations *Mathematics.* a linear system.
systemoid *Oncology.* a tumor that is made up of a variety of tissues.
system operation *Computer Technology.* the complete administration and operation of a data-processing system.
system reliability *Engineering.* the probability that a system will operate as required when called upon to perform its function.
system safety *Engineering.* the optimum, overall safety level of a system in operation, including safety of operating personnel, surroundings, and so forth.
systems analysis *Engineering.* the techniques of analyzing a system, breaking its requirements down into component elements and functions, and reintegrating these into the overall system. Thus, **systems analyst.**
systems approach *Psychology.* the theory that an item of knowledge or behavior can be studied only within the context of how it fits into other larger and smaller systems.
systems definition *Computer Technology.* a document that provides a general description of a computer-based system for processing data. Also, **systems specification.**
systems ecology *Ecology.* a method of study of ecosystems that utilizes a combination of approaches to analyze the entire system and its subsystems.
systems engineering *Engineering.* the branch of engineering that deals with the analysis, development, and operation of complete systems.
systems implementation test *Engineering.* a test that exercises a complete system under operational conditions.
system software *Computer Programming.* any of a variety of programs that facilitate and monitor computer use, such as operating system programs, utilities, compilers, and database management systems; usually provided by either the hardware vendor or a software supplier. Also, **systems progam.**
systems programmer *Computer Science.* a programmer who writes or maintains systems software, such as compilers or operating systems.
systems programming *Computer Programming.* the writing and maintenance of systems software.
system supervisor see SUPERVISOR.
system test *Computer Programming.* **1.** the running of an entire system using test data. **2.** a simulation of an actual running system. **3.** a test of an interconnected set of components to check proper functioning and interconnection.
systle *Architecture.* an intercolumniation of two diameters.
systole [sis′tə lē] *Physiology.* the portion of the cardiac sequence when the heart muscle is contracting; it takes place between the first and second heart sounds as the blood flows through the aorta and pulmonary artery.
systolic [sis täl′ik] *Cardiology.* pertaining to systole, a contraction of the heart, particulary the ventricles.
systolic array *Computer Technology.* an array of processing elements of cells connected to a memory that pulses data through the array, allowing each data item to be used effectively at each cell it passes along the array.
syzygy [siz′i jē] *Astronomy.* the moment of the New or Full Moon, or the time of opposition or solar conjunction for a planet. *Invertebrate Zoology.* **1.** in protozoans, especially sporozoans, the pairing of gamonts or gametocytes prior to sexual union of the gametes. **2.** end-to-end attachment of the sporonts of some gregarine protozoans. **3.** close union or fusion of organs or skeletal elements, as in some crinoids.
szaibelyite *Mineralogy.* $MgBO_2(OH)$, a white to yellow monoclinic mineral occurring as nodules and fibrous veinlets or masses, having a specific gravity of 2.60 to 2.62 and a hardness of 3 to 3.5 on the Mohs scale; found as coatings on serpentine.
szaskaite see SMITHSONITE.
Szechtman cell *Chemical Engineering.* an electrolytic cell used in the manufacture of chlorine.
Szent-Györgi, Albert von [sent jôr′ jē] 1893–1986, Hungarian biochemist, living in the United States; awarded the Nobel Prize for his research in biological oxidation.
Szilard, Leo [sil′ärd] 1898–1964, Hungarian-born American physicist; with Enrico Fermi, developed the self-sustaining nuclear reactor.
Szilard-Chalmers effect *Nucleonics.* the separation of radioactive isotopes by chemical means.
szmikite *Mineralogy.* $Mn^{+2}(SO_4){\cdot}H_2O$, a gray, red, or brownish-white monoclinic mineral of the kieserite group, having a specific gravity of 2.845 to 3.21 and a hardness of 1.5 on the Mohs scale.
szomolnokite *Mineralogy.* $Fe^{+2}SO_4{\cdot}H_2O$, a colorless, yellow, brown, or bluish monoclinic mineral of the kieserite group, occurring as bipyramidal crystals, and in globular or stalactitic form, having a specific gravity of 3.03 to 3.07 and a hardness of 2.5 on the Mohs scale; found with other secondary iron sulfates.

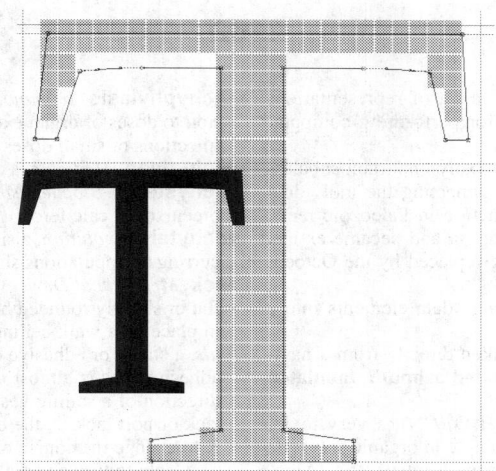

T *Space Technology.* time; i.e., the moment at the end of a countdown when a rocket is due to be launched; used in expressions such as "T minus 40."

T the chemical symbol for tritium.

T or **T.** tesla; thymine; thoracic vertebrae; intraocular tension.

$\mathbf{T_3}$ triiodothyronine.

$\mathbf{T_4}$ thyroxine.

$\mathbf{t_{1/2}}$ *Nucleonics.* the symbol for half-life.

t 1/2 or **T 1/2** half-time.

t or **t.** target; technical; temperature; tension; tertiary; ton; translocation.

t. in the time of. (From Latin *tempore*.)

T&A tonsilladenoidectomy.

T.A. or **T.A.T.** toxin-antitoxin.

Ta the chemical symbol for tantalum.

tab *Computer Programming.* any of various markers used in controlling the movement of a print or display mechanism. The simplest tabs, also called **tab stops,** are used to set a keyboard's column indents.

tabacosis *Toxicology.* poisoning by tobacco, especially by the inhalation of tobacco dust; symptoms may include lung damage. Also, **tabacism, tabagism.**

Tabanidae *Invertebrate Zoology.* the horseflies and deerflies, a family of dipteran insects in the series Brachycera; the larvae are aquatic; male adults feed on plant juices, females on the blood of mammals, sometimes transmitting disease.

tab-card cutter *Mechanical Devices.* a cutting die for paper stock that cuts tabulating-card-sized pieces.

Taber ice see SEGREGATED ICE.

tabes [tā´bēz] *Neurology.* **1.** any decay or wasting away of the body or any part of the body; progressive atrophy. **2.** see TABES DORSALIS.

tabes dorsalis *Neurology.* a slow, progressive deterioration of the posterior columns and posterior roots and ganglia of the spinal cord, occurring 15 to 20 years after the onset of syphilis and characterized by sharp, stabbing pains, ataxia, urinary incontinence, impaired ankle and knee reflexes, and Romberg's sign. Also, LOCOMOTOR ATAXIA, TABETIC NEUROSYPHILIS.

tabetisol see TALIK.

table *Mechanical Engineering.* the part of a machine on which an object is placed for operation. *Mining Engineering.* **1.** a platform or other flat surface used to clean and sort coal or ore. **2.** a wide, shallow sluice box used in placer mining to extract gold or other valuable minerals from screened gravel. *Geography.* see TABLELAND. *Anatomy.* the inner or outer hard layer or any of the flat bones of the skull. *Mathematics.* a list or array of function values at corresponding points in the domain. The entries in a table often are numerical or computed values (for example, a table of logarithms) rather than exact values. *Computer Programming.* a collection of records or other data values that are arranged or labeled for ready reference and access.

tableau [ta blō´] *Mathematics.* in the simplex method for solving a linear program with n variables and m linear constraints of the form

$$a_{i_1}x_1 + a_{i_2}x_2 + \cdots + a_{in}x_n \le b_i,$$

the $(m + 1) \times (n + m + 2)$ augmented matrix whose first m rows come from the linear constraints with slack variables and whose last row (or sometimes the first row) comes from the objective function.

table-driven *Computer Programming.* of a program or component, using table look-up as a primary means of computation or control. Thus, **table-driven compiler, table-driven program.**

table flotation *Mining Engineering.* the flotation of a slurry of ore on a shaking table, in which floatable particles roll across the table with the aid of low-pressure jets of air before being discharged nearly opposite the feed end.

table iceberg see TABULAR ICEBERG.

tableland *Geography.* an elevated plain; a plateau, especially one rising sharply from adjacent lowlands. Also, TABLE.

table look-up *Computer Programming.* a method of retrieving information in which a data item is found by looking up its known identifying label within a table.

table look-up device *Computer Technology.* a digital device or circuit that performs complex calculations or logic operations with the help of function values previously stored at specific addresses in memory.

table mountain *Geography.* a mountain with a relatively flat summit.

table of connections *Artificial Intelligence.* in general-purpose problem solving, a table connecting a difference between a given state and a goal state with operators that may be relevant in reducing the difference.

table of contents *Computer Programming.* see DIRECTORY, def. 1.

table reef *Geology.* a small, isolated, flat-topped organic reef that does not surround a lagoon.

tablespoon *Metrology.* a unit of measure equal to 3 teaspoons or one-half fluid ounce. Also, **tablespoonful.**

table sugar *Food Technology.* ordinary sugar; sucrose. See SUCROSE.

tableting *Engineering.* a punch-and-die technique used for the preparation of pills, pellets, and similar compacted, formed items.

table tripod *Engineering.* a low mount or stand designed to hold a camera. Also, HIGH HAT.

table wine *Food Technology.* **1.** any wine containing less than 14% alcohol by volume, usually meant to be served with food. **2.** loosely, a nonvintage, relatively inexpensive wine.

taboo *Anthropology.* a social prohibition of objects, persons, words, or actions, usually stemming from early cultural conceptions of the item as magical or sacred and therefore associated with power or danger, and by extension, the unclean. Also, **tabu.**

taboparesis *Neurology.* general paresis occurring simultaneously with tabes dorsalis. Also, **taboparalysis.**

tabtoxin *Toxicology.* a phytotoxin that is produced by bacteria of the *Pseudomonas* species, which contributes to wildfire disease in tobacco.

tabula *plural,* **tabulae.** *Invertebrate Zoology.* in some corals and hydroids, one of the flat horizontal partitions of the vertical canals. *Paleontology.* an internal, usually horizontal plate or bar between the walls of some marine colonial organisms, such as the archaeocyathids and tabulate corals. (From the Latin word for "slate" or "tablet.")

tabular *Geology.* describing the slablike shape of a sedimentary particle or body, having a large surface area in relation to its thickness.

tabular crystal *Crystallography.* a crystal that has two large parallel faces, giving it a broad and flattened appearance.

tabular iceberg *Oceanography.* a relatively flat iceberg, typical of Antarctic icebergs, sometimes hundreds of kilometers across. Also, TABLE ICEBERG.

tabular interpolation *Mathematics.* any interpolation based on function values obtained from a table.

tabular language *Computer Programming.* a set of representations and rules composed of decision tables used for performing computations and for communicating with computers.

Tabulata *Paleontology.* a large and stratigraphically important subclass of colonial corals; characterized by tabulae connecting the individual corallites; often served as the framework structure in Paleozoic reefs; they arose in the Cambrian just before the Rugosa and became extinct (along with the Rugosa) in the Permian, being replaced by the Octocorallia and Scleractinia.

tabulate *Computer Programming.* **1.** to arrange data elements into a table format. **2.** to print totals.

tabulated altitude *Navigation.* an altitude taken directly from a navigational sight reduction table. Similarly, **tabulated azimuth, tabulated azimuth angle.**

tabun *Organic Chemistry.* $(CH_3)_2NP(O)(C_2H_5O)(CN)$, a very toxic, combustible liquid; miscible with water and soluble in organic solvents; boils at 240°C and freezes at −50°C; used as a nerve gas. Also, DIMETHYLPHOSPHORAMIDOCYANIDIC ACID, ETHYL ESTER.

TAB vaccine *Immunology.* an antigen preparation that is administered to induce an immune response against *Salmonella typhi* and *Salmonella paratyphi* A and B; it consists of killed and preserved cells of both *Salmonella* organisms. Also, TYPHOID VACCINE.

TAC *Immunology.* a transferrin receptor on an activated lymphocyte, which is detected by a monoclonal antibody.

TAC Tactical Air Command.

tacan or **Tacan** *Navigation.* a short-range, ultrahigh-frequency, electronic air-navigation system that provides a continuous indication of bearing and distance to a station, using common components for bearing and distance determination. (An acronym for tactical air navigation.)

Taccaceae *Botany.* a family of monocotyledonous, perennial, tuberous herbs of the order Liliales, characterized by scattered raphide cells and scalariform vessels in the roots only, and lacking secondary growth; best developed in southeastern Asia and Polynesia.

tach- or **tacho-** a combining form meaning "speed."

tacheion *Ecology.* a division of aquatic organisms that includes both crawling and free-swimming forms.

tache noire *Medicine.* a black crusted ulcer on the skin arising from the bite of a tick carrying scrub typhus, Boutonneuse fever, or South American tick-bite fever.

Tachinidae *Invertebrate Zoology.* the tachina flies, a large family of bristly black or gray dipteran insects, having larvae endoparasitic in caterpillars and insects; used in biological control of forest pests.

tachistoscope *Mechanical Engineering.* a machine that displays a series of rapid visual images in order to measure attention, perception, and learning.

tachometer [ta käm´i tər] *Mechanical Engineering.* an instrument used to measure the rotational speed of a shaft or machine, as of an engine. *Electricity.* a precision DC generator that produces a voltage proportional to velocity; used to measure the velocity of various forms of movement or flow, as of a machine, a river, or the blood.

tachy- a combining form meaning "rapid."

tachyarrhythmia *Cardiology.* an irregular heartbeat.

tachycardia *Cardiology.* rapid beating of the heart, usually applied to rates over 100 per minute. Also, **tachyrhythmia.**

tachy-electromagnetic pulse *Electromagnetism.* a pulse of electromagnetic radiation generated within a microsecond of a high-altitude nuclear explosion by high-energy electrons.

Tachyglossidae *Vertebrate Zoology.* a family of egg-laying mammals of the order Prototheria that includes the echnida and spiny anteater; native to Tasmania, Australia, New Guinea, and adjacent islands.

tachyhydrite *Mineralogy.* $CaMg_2Cl_6 \cdot 12H_2O$, a colorless or yellow trigonal mineral occurring as rounded masses, having a specific gravity of 1.667 and a hardness of 2 on the Mohs scale; found in potash-rich saline deposits.

tachylyte or **tachylite** *Mineralogy.* a black to brown or green volcanic glass of basaltic composition, readily fusible and of a high luster; common as chilled margins of dikes, sills, or flows. Also, SIDEROMELANE, BASALT GLASS, HYALOBASALT, JASPOID, WICHTISITE.

tachymeter *Cartography.* any of several surveying instruments used to determine distance, direction, and difference of elevation.

Tachyniscidae *Invertebrate Zoology.* a family of dipteran insects in the subsection Acalyptratae.

tachyon *Particle Physics.* a hypothetical particle whose speed is greater than that of light, $c = 3 \times 10^5$ km/sec.

tachyphagia *Medicine.* a habit or condition of abnormally rapid eating.

tachyphylasis *Immunology.* a condition in which an individual's reaction to doses of organ extract or serum decreases rapidly due to previous injections of small doses of the same preparation.

tachypnea *Medicine.* a condition of excessively rapid respiration.

tachysterol *Biochemistry.* $C_{28}H_{44}O$, a product of vitamin D that is the precursor of calciferol.

tachytely *Evolution.* a significantly fast rate of evolutionary change, occurring as populations shift from one major adaptive zone to another.

tack *Mechanical Devices.* a small, sharply pointed metal fastener, with a flat or slightly rounded head, used to hold light materials, such as paper, in place on a wall, ceiling, or other vertical surface or structure. *Materials.* a sticky or adhesive quality; used especially to describe a coating or adhesive that is almost dry. Also, **tacky, tackiness.** *Navigation.* **1.** the direction of a sailing vessel relative to the wind; for example, starboard tack or port tack. **2.** the distance a vessel has sailed on a particular tack. **3.** to change the course of a vessel back and forth between starboard and port tack; to turn a vessel into the wind in order to change the tack. Similarly, **tacking.**

tack claw *Mechanical Devices.* a small hand tool with a handle and claw at one end; used to remove tacks.

tack coat *Civil Engineering.* a thin coat of road tar, bitumen or emulsion that enhances the bonding of two layers of asphalt.

tack hammer *Mechanical Devices.* a light hammer used to push tacks into a mounting surface.

tacking *Metallurgy.* the process of making tack welds.

tackle *Mechanical Engineering.* an assembly of pulleys, ropes, and blocks used to lift or lower an object. *Naval Architecture.* such an arrangement of lines and blocks set up for a particular purpose, such as anchor tackle or cargo-handling tackle.

tack range *Materials.* the period of time during which an adhesive remains semidry but sticky.

tack welds *Metallurgy.* small, scattered welds that are made to hold a weldment in the correct alignment while the final welds are being made.

Taconic orogeny *Geology.* a mountain-building event of the latter part of the Ordovician period that produced the northern Appalachians. Also, **Taconian orogeny.** (From the *Taconic* mountains of eastern New York State, which are part of this range.)

taconite *Geology.* an iron-containing chert or unleached iron formation of the Lake Superior district that contains magnetite, hematite, siderite, and hydrous iron silicates.

tactical *Military Science.* **1.** of or relating to the employment of units in combat. Thus, **tactical troops, tactical unit, tactical vehicle.** **2.** of or relating to the arrangement and maneuver of units in relation to one another and to the enemy, in order to utilize their full potentials. Thus, **tactical intelligence, tactical map, tactical mobility, tactical locality, tactical reserve.** **3.** of or relating to specific combat objectives, usually of a limited scope and confined to a limited area, and to weapons and methods designed to accomplish those objectives. Thus, **tactical mine field, tactical nuclear weapon, tactical target.**

tactical air-control system *Aviation.* the equipment and organization necessary to plan and execute tactical air operations and to coordinate air services with other facilities. Thus, **tactical air-control center.**

tactical aircraft *Aviation.* aircraft designed for or used in operations against a hostile force in the air or especially on the surface.

tactical air operations *Aviation.* a military operation integrating air power with ground or naval forces.

tactical air reconnaissance *Aviation.* the use of flight vehicles to secure information relating to terrain, weather, and the composition, movement, and installations of enemy forces.

tactical air support *Aviation.* any air activity that is coordinated with surface forces to provide assistance in a land or sea conflict.

tactical call sign *Telecommunications.* a call sign that identifies a tactical communications facility.

tactical command ship *Naval Architecture.* a ship fitted with specialized gear and facilities in order to serve as command and control vessel of an operation.

tactical communications system *Telecommunications.* a military communications system of fixed-size, self-contained assemblages that provides internal communications within tactical facilities.

tactical control radar *Engineering.* a radar that directs the operation of an antiaircraft weapon system; distinct from target-acquisition radar, though the equipment may be of the same type.

tactical diameter *Navigation.* the minimum distance a vessel travels to the right or left of its original course in the process of making a 180° turn at full speed.

tactical electronic warfare *Military Science.* the military doctrine of the use of electronic systems to achieve several tactical goals: enhance the operational capability of friendly forces through the use of electronic systems, such as radar, sonar, and communications systems; impair the operational capability of enemy systems through the use of countermeasures systems; and negate the effectiveness of enemy countermeasures equipment through the use of counter-countermeasures systems.

tactical fire control *Military Science.* the control of fire including target selection, opening, suspending, and ceasing fire, and establishing classes of fire.

tactical frequency *Telecommunications.* a radio frequency that is assigned to a military unit for the purpose of carrying out a tactical mission or missions.

tactical missile *Military Science.* a missile that is directed toward a specific and limited target. Also, **tactical ballistic missile.**

tactical range *Military Science.* a range in which realistic targets are in use and a certain freedom of maneuver is allowed.

tactical range recorder *Ordnance.* a sonar recording device used aboard a surface antisubmarine warfare vessel to track the time and range coordinates of a target submarine.

tactical warning *Military Science.* a warning of a threatening or hostile act that is received after initiation of the act; contrasted with strategic warning, which occurs prior to initiation.

tacticity *Materials Science.* the quality or degree of orderliness in the succession of repeating configurational units in the main chain of a polymer molecule.

tactic polymer *Organic Chemistry.* a polymer whose molecules are regularly and symmetrically arranged, as in a stereospecific polymer.

tactics *Military Science.* **1.** the arrangement or movement of troops in combat. **2.** the ideas and methods employed to accomplish a specific combat objective. **3.** also, **tactic.** any specific plan or procedure dealing with a limited and immediate objective, as distinguished from large-scale or long-range planning.

tactile [tak´təl; tak´tīl] *Physiology.* **1.** relating to or involving the sense of touch. **2.** detectable by the sense of touch.

tactile amnesia see ASTEROGNOSIS.

tactile cells *Invertebrate Zoology.* sensory cells that are specialized to detect touch stimuli.

tactile corpuscle *Physiology.* any of numerous oval organs made of flattened cells and encapsulated nerve endings, occurring in sensitive, hairless parts of the skin such as the tips of the fingers and toes, and functioning as a touch receptor. Also, **tactile bud.**

tactile sensor *Robotics.* a transducer in the end effector of a robot that is sensitive to touch.

tactite *Petrology.* a coarse-grained rock featuring a complex mineralogical composition, formed by contact metamorphism and metasomatism of carbonate rocks, and composed of garnet, iron-rich pyroxene, epidote, wollastonite, and scapolite.

tactoid *Biochemistry.* a spindle-shaped particle found under a polarizing microscope in mosaic virus, fibrin, and myosin, characteristic of the sickle-cell hemoglobin in sickle-cell anemia.

tactor *Neurology.* a sensitive mechanoreceptor or tactile sense organ.

tactoreceptor *Physiology.* a specialized nerve cell that responds to touch sensation.

TAD *Oncology.* an acronym for a cancer chemotherapy regimen that includes 6-thioguanine, Ara-C (cytarabine), and daunomycin.

tadpole *Vertebrate Zoology.* the aquatic, fishlike larva of a frog or toad, characterized by a rounded body, a fin-covered tail, and external gills that soon become covered, after which limbs and lungs are developed and the tail and gills are lost.

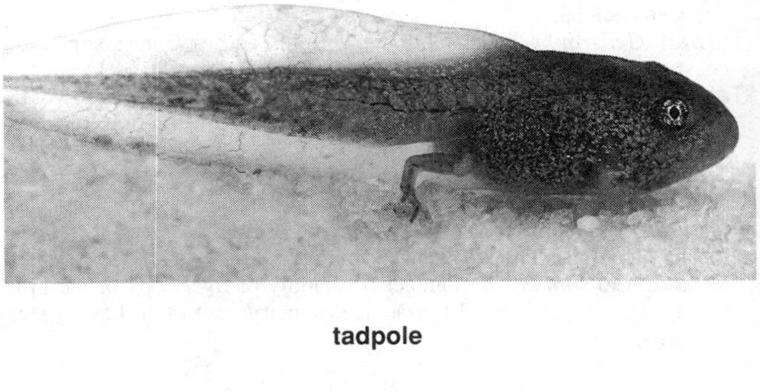

tadpole

tadpole shrimp *Invertebrate Zoology.* the common name for tiny crustaceans of the order Notostraca, found in seasonal pools; eggs can lie dormant for years until suitable hatching conditions occur.

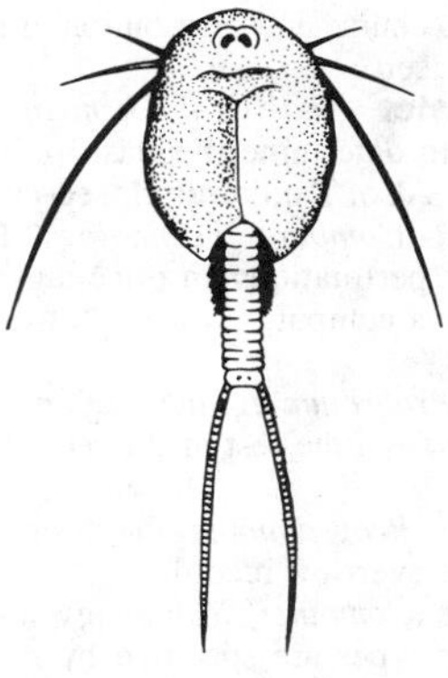

tadpole shrimp

taenia *Anatomy.* **1.** any flat, elongate band of tissue resembling a tapeworm. **2.** a tapeworm.

taenia coli *Anatomy.* flat, elongate bands of smooth muscle in the outer wall of the colon.

taeniafuge *Medicine.* an agent or medicine that expels tapeworms from the body. Also, TENIAFUGE.

taeniasis *Pathology.* infestation with tapeworms. Also, TENIASIS.

taenie coli see TENIAE COLI.

Taeniodonta *Paleontology.* an extinct order of placental mammals of western North America; the taeniodonts were highly specialized but also retained many primitive eutherian characteristics; the Paleocene genus *Onychodectes* was similar to the opossum, and the advanced taeniodont Stylinodon (Eocene) was bear-sized; extant in the Paleocene and Eocene.

Taenioidea or **Taenoidea** *Invertebrate Zoology.* the largest order of tapeworms in the class Cestoda, with both suckers and hooks; includes species that infect humans from undercooked beef and pork, and common cat and dog parasites. Also, CYCLOPHYLLIDAE.

Taeniolabidoidea *Paleontology.* a suborder of rodentlike multituberculate mammals; the type genus *Taeniolabis* (Paleocene) was similar to a beaver; known mainly from Asian deposits from the Cretaceous to the Eocene.

taeniolite *Mineralogy.* $KLiMg_2Si_4O_{10}F_2$, a brownish or colorless monoclinic mineral of the mica group occurring as tabular crystals and scales, having a specific gravity of 2.82 to 2.90 with a hardness of 2.5 to 3.5 on the Mohs scale; found in nepheline-syenite pegmatites.

taenite *Mineralogy.* V–(Ni,Fe), a silver-white to grayish, metallic, magnetic, face-centered cubic mineral that is found only in meteorites, occurring intergrown with kamacite and in massive form, and having a specific gravity of 7.8 to 8.22 and a hardness of 5 to 5.5 on the Mohs scale; a natural alloy of nickel and iron, with 27–74% nickel and a little cobalt.

taenoglossate radula *Invertebrate Zoology.* in certain mollusks, a long, flexible tonguelike organ with small horny or chitinous teeth in rows of seven, used for grasping food.

Tafel equation *Physical Chemistry.* an empirical equation that shows the type of logarithmic dependence of current density on overpotential for an electrochemical process with activation polarization.

Tafel plot *Physical Chemistry.* a graph that shows the relation between overpotential and the logarithm of current density; the slope of this plot gives the transfer coefficient.

Tafel slope *Electricity.* a graphical representation of a slope of a curve above a potential or electrolytic polarization in volts as plotted against the logarithm of current density.

taffeta *Textiles.* a crisp, smooth, plain-weave group of fabrics, usually having a lustrous face and a very fine cross rib; originally produced of silk, cotton, and wool, but synthetic fibers are now also used.

taffrail log *Naval Architecture.* **1.** the upper part of the stern of a ship. **2.** a railing running along the stern of a ship. *Engineering.* a type of log, traditionally cast over the taffrail or stern rail of a ship, consisting essentially of a rotator, a braided log line, and a distance-recording device. Also, PARENT LOG.

tag *Computer Programming.* **1.** the portion of a machine instruction that identifies the index register to be used in modifying the address in the instruction. **2.** see FLAG. **3.** a code attached to a data word or instruction that identifies the type of the data. *Nucleonics.* an isotope of a radioactive substance that is incorporated into another substance for the purpose of following the course and distribution of the sample substance or carrier; an isotopic tracer.

Tag closed-cup tester *Analytical Chemistry.* a device used by the petroleum industry to determine the flash point of petroleum liquids flashing below 79°C. Also, **Tagliabue closed-cup tester.**

tag-converting unit *Computer Technology.* **1.** a device designed to read data encoded as perforations on price tags and automatically convert and transfer it to a computer system. **2.** the tag part of a data word or instruction.

tag field *Computer Programming.* the section of a record that contains information different from the rest of the record, so that it may be used as a marker or label.

tag format *Computer Programming.* the design of records so that tags are created to retrieve overflow records.

tagged architecture *Computer Technology.* a kind of computer architecture in which data types are specified by tag fields associated with the data, and in which the processing of data by the central processing unit is partly determined by these tags.

tagged molecule *Chemistry.* a molecule containing a radioactive atom whose behavior can be traced in chemical and biological reactions.

Taghanican *Geology.* a North American geologic stage of the uppermost Middle Devonian period, occurring after the Tioughniogan and before the Fingerlakesian.

tagma *plural,* **tagmata.** *Invertebrate Zoology.* one of the major regions of an arthropod's body, consisting of two or more similar segments, such as the thorax or abdomen.

tagmatization *Invertebrate Zoology.* the evolutionary development of specialized body regions in arthropods.

tag polymerase *Enzymology.* a polymerase that is linked to a radioactive or otherwise labeled group for identification.

Tag-Robinson colorimeter *Engineering.* an instrument used to determine the color of an oil, such as a lubricating oil, by varying the thickness of an oil column until its color matches a color standard.

tag sort *Computer Programming.* the process of arranging or ordering data so that only the record labels or addresses are moved during the comparison process; indexing, as distinguished from actual record sorting.

tagua palm *Botany.* a tropical palm, *Phytelpas macrocarpa,* grown commercially for its seeds, which are used in place of ivory.

tahoma *Geology.* a high snowy or glacier-clad mountain, especially in the Pacific Northwest. (Based on an American Indian name for Mount Rainier.)

taiga [tī´gə] *Ecology.* a region of boreal forest, the largely evergreen forest vegetation of northern areas of the Northern Hemisphere, below the arctic and subarctic tundra regions.

taiga climate *Meteorology.* the climate characterized by taiga vegetation; it is milder than a tundra climate and moist enough to support considerable vegetation, but it is too cold for abundant tree growth. Also, SUBARCTIC CLIMATE.

tail *Vertebrate Zoology.* **1.** a posterior extension from the body of a vertebrate containing the caudal vertebrae. **2.** the caudal fin of any aquatic vertebrate. **3.** the feathered uropygium of a bird. **4.** the body portion behind the vent of a limbless reptile.

tail any of various parts, structures, or formations thought of as resembling the tail of an animal; specific uses include: *Aviation.* the rear part of an aircraft, rocket, or other body. *Electronics.* **1.** the slowly decaying trailing edge of a pulse waveform. **2.** a spurious, lower-amplitude pulse waveform following the main part of the pulse. *Graphic Arts.* the lower portion of a type character, as of g or Q.

tail *Mathematics.* **1.** given a sequence or series, the object obtained by deleting the first *n* terms for some *n*. For example, a tail of the series

$$\sum_{k=0}^{\infty} a_k \text{ is } \sum_{k=n}^{\infty} a_k;$$

and a tail of a stochastic process $\{X_t(\omega)\}$ is $\{X_t(\omega) : t \geq t_0\}$. **2.** the vertex *u* from which a directed edge is directed; the edge is directed toward the head *v* and is denoted (u,v).

tail assembly see EMPENNAGE.

tailbone *Anatomy.* a popular name for the coccyx, the beaklike bone at the lower end of the spinal column.

tail cone *Aviation.* **1.** an exhaust cone. **2.** a stream of burning exhaust gases appearing as a cone of fire from the tail of an aircraft. **3.** any cone-shaped section at the rear of something, such as the tapering rear section of an aircraft.

tail drive *Space Technology.* the most common form of rocket propulsion, in which the exhaust is ejected from the tail of the rocket.

tailed frog *Vertebrate Zoology.* a frog, *Ascaphus truei,* in which the cloaca of the male is modified into a tail-like copulatory organ; found in the northwestern United States.

tailed phage *Virology.* any of a group of several hundred DNA-containing bacterial viruses, characterized by a cubic head and helical tail; the group includes the families Myoviridae, Siphoviridae, and Podoviridae.

tail fin or **tailfin** *Aviation.* a fin, such as a vertical stabilizer, set at the end of a rocket or other body to provide guidance or stability. *Design Engineering.* a similar decorative element on the rear of a motor vehicle, as on American automobiles of the 1950s. *Invertebrate Zoology.* see CAUDAL FIN.

tailgate *Civil Engineering.* the gates at the low-level end of a lock or canal. *Engineering.* a gate at the rear of a station wagon, truck, or gondola car that is swung down to open.

tail house *Chemical Engineering.* a refinery installation that contains facilities for sampling, a look box, and controls to divert the products to other locations in the refinery for further processing or to storage tanks.

tailing *Molecular Biology.* the process of adding nucleotides onto the ends of a restriction fragment.

tailings *Engineering.* **1.** relatively light particles that pass across a sieve. **2.** refuse material from processed ore. *Mining Engineering.* the decomposed outcrop of a bed or vein of valuable material.

tailpiece *Graphic Arts.* a small illustration or typographic ornament printed at the end of copy, especially of a book chapter.

tail pipe or **tailpipe** *Engineering.* an exhaust at the rear of a motor vehicle or aircraft powered by an internal-combustion engine.

tail pulley *Mechanical Engineering.* a pulley or roller in the tail section of a belt conveyor, around which the belt runs. Also, FOOT-SECTION PULLEY.

tailrace *Engineering.* the outflow duct or channel from a water wheel or water turbine.

tail rotor *Aviation.* a rotor mounted on the tail of a helicopter and used to decrease torque.

tailshaft *Naval Architecture.* the rear portion of a propeller shaft which protrudes from the stern tube, and to which the propeller is fitted.

tail spin or **tailspin** *Aviation.* a spin, especially a controlled spin.

tailstock *Mechanical Engineering.* a part of a lathe that holds only the end of a workpiece so it can be rotated freely.

tail surface *Aviation.* the portion of the tail of an aircraft or missile designed to provide stability and control.

tail unit see EMPENNAGE.

tailwind *Meteorology.* a wind that blows in the same direction as a moving object, such as an aircraft in flight, rendering the object's ground speed greater than its indicated airspeed; the opposite of a headwind. Also, FOLLOWING WIND.

taino *Meteorology.* a West Indian term for hurricane, used especially in the Greater Antilles. (Originally the name of an extinct Arawakan Indian tribe.)

Tainter gate *Civil Engineering.* a gate located on the spillway crest of a dam with a curved upstream face, which is horizontally pivoted; the curved surface is made concentric to the pivot or pin, so that the resultant water pressure passes through the pivot.

Tait's conjecture *Mathematics.* a conjecture that a plane cubic 3-connected graph is Hamiltonian. This was shown to fail by W. T. Tutte; Barnette's conjecture is a weakening of Tait's conjecture which has not yet been decided.

Taiwan *Geography.* an island in the western Pacific off the southeastern coast of China; area: 13,885 square miles.

Taiwan red pine *Materials.* the easily worked, hard, heavy wood of the Asian *Pinus taiwanensis* tree; used for building construction, furniture, musical instruments, and pulp.

Taka-diastase or **takadiastase** *Enzymology.* the trademark for an amylolytic enzyme formed by the action of the spores of the fungus *Aspergillus oryzae* on the bran of wheat and used as a digestant. (Named for the Japanese chemist Jokichi *Takamine.*)

Takakiaceae *Botany.* a monogeneric family of liverworts of the order Calobryales, characterized by radial symmetry, and often having leafy branches.

Takayanagi model *Materials Science.* a representation of the two-dimensional (shear, elongation) behavior of a plastic, using a sequence of elastic and viscous units joined in series or parallel, to predict and explain the mechanical behavior of multiphase polymers, polyblends, blocks, grafts, copolymers, or interpenetrating networks.

take *Graphic Arts.* one of several parts of a manuscript that are divided among different compositors, or among different operators at the same compositor, for typesetting.

take-all *Plant Pathology.* a disease of cereal grasses caused by the fungus *Ophiobolus graminis* that is characterized by foot-rot, a loss of color in the stalks, heads, and leaves, and empty or only partly filled plant heads.

take down or **takedown** *Computer Technology.* to deliberately stop computer operations; e.g., to reboot to correct a problem. *Computer Programming.* the physical process of completing one job or cycle and preparing for the next; for example, dismounting tapes or disks, removing printouts, or entering console commands.

takedown time *Computer Programming.* the period during which equipment is prepared for the next operating cycle or job.

take off *Aviation.* to carry out the process of a takeoff.

takeoff *Aviation.* the complete process of getting an aircraft off the ground and into the air, particularly at the moment it breaks contact with the ground. *Space Technology.* a rocket launch.

takeoff assist *Aviation.* **1.** added thrust given to an aircraft during takeoff. **2.** a rocket motor or other device used to provide such added thrust.

takeoff distance *Aviation.* the distance required for a given aircraft to take off under given conditions; measured from the craft's point of rest to the point at which it breaks contact with the ground.

takeoff mass *Space Technology.* the cumulative mass of a rocket vehicle and its payload at takeoff.

takeoff weight *Aviation.* **1.** the actual weight of a flight vehicle at takeoff, including its structure, equipment, fuel, and payload. **2.** the maximum weight a flight vehicle is designed or permitted to carry at takeoff.

take the ground *Navigation.* to become grounded, especially when due to a receding tide. Also, **take the bottom.**

takeup *Mechanical Devices.* a tensioning device in a belt-conveyor system that takes up slack. *Textiles.* the contraction of fabric that results from the wet operations in the finishing process, especially fulling.

takeup pulley *Mechanical Engineering.* an adjustable pulley mounted in a belt-conveyor system that accommodates changes in the length of the belt in order to maintain proper belt tension.

takeup reel *Engineering.* the reel onto which a recording tape or film is wound as it is played.

taku *Meteorology.* a strong east-northeasterly wind that blows in southern Alaska from October to March; characterized by wind gusts, some of which attain hurricane force. (Occurring in the region of the *Taku* Glacier and River.)

Tal. such a one. (From Latin *talis.*)

TALAR *Navigation.* a modular, air navigational microwave radar system with step-scan capabilities. (An acronym for tactical landing approach radar.)

Talaromyces *Mycology.* a genus of fungi belonging to the order Eurotiales that also occurs in its asexual or anamorphic stage in the genus *Penicillium.*

Talbot, William Henry Fox 1800–1877, English inventor; invented calotype (**talbotype**), the first negative–positive photographic process.

talbot *Optics.* a unit of luminous energy flux equal to 1 lumen per second.

Talbot's bands *Optics.* the dark bands in a continuous spectrum appearing when one places a thin glass plate over the jaw of the spectroscope that lies on the blue side of the spectrum.

Talbot's law *Optics.* a law stating that an object that flashes 10 times a second or faster has an apparent brightness that equals its actual brightness times the ratio of the "on" duration to the total duration.

talbutal *Pharmacology.* $C_{11}H_{16}N_2O_3$, an orally administered barbiturate used as a hypnotic and sedative.

talc *Mineralogy.* $Mg_3Si_4O_{10}(OH)_2$, a white, green, or gray, very soft monoclinic and triclinic mineral with a pearly luster and greasy feel, occurring in compact masses and as thin tabular crystals, and having a specific gravity of 2.58 to 2.83 and a hardness of 1 on the Mohs scale; widespread as a hydrothermal alteration product of ultrabasic rocks and in metamorphosed dolomite. *Materials.* a commerical form of this substance, a soft, friable white material composed of natural hydrous magnesium silicate; used in cosmetics, as a filler for paints and plastics, and for paper coatings.

talcose [tal′kōs] *Geology.* relating to, resembling, or containing talc.

talcose rock *Petrology.* any rock characterized by a soft, soapy feel similar to that of talc. Thus, **talcose slate.**

talcose slate

talcosis *Medicine.* a disease caused by talc dust inhalation or by the absorption or implantation of talc dust in the skin; characterized by lung damage.

talc schist *Petrology.* a schistose rock consisting primarily of talc, with associated quartz and mica.

tale see FOLK TALE.

talfan disease see TESCHEN DISEASE.

talik *Geology.* a permanent or temporary layer of unfrozen ground occurring above, within, or below permanently frozen ground. Also, TABETISOL. (Originally a Russian term.)

talipes [tal′i pēz] *Medicine.* another term for clubfoot, a congenital deformity in which the foot is twisted out of shape or position.

talipes cavus *Medicine.* a form of talipes characterized by exaggeration of the planar arch.

Talitridae *Invertebrate Zoology.* the sandhoppers, a family of amphipod crustaceans in the suborder Gammaridea, modified for jumping; most found along seashores, although some are terrestrial.

talking *Mining Engineering.* a term for a series of cracking noises or small bumps emanating from mine walls, indicating that the rock is beginning to yield to stress.

talk-listen switch *Acoustical Engineering.* a buttonlike switch on a sound-powered phone that is depressed to allow the user to speak into the mouthpiece and raised to allow the user to listen.

tall grass or **tallgrass** *Botany.* a high-growing grass characteristic of the prairie region of central North America; e.g., big bluestem, yellow Indian grass.

tallgrass prairie *Ecology.* **1.** another term for the historic prairie region of central North America, noted for its abundant growth of grasses. **2.** specifically, the eastern portion of this region, where relatively high annual rainfall produced a higher growth of grasses than in the western part.

tall oil *Materials.* a resinous secondary product from the manufacture of chemical wood pulp; used in making soap, paint, and the like. (From Swedish *tallolja* meaning "tall-pine oil.")

tallow *Materials.* the solid, relatively hard fat of cattle, sheep, or other animals that has been rendered from the surrounding tissue; used to make candles, soap, and similar products.

talon *Vertebrate Zoology.* a well-developed hooked claw, found especially on birds of prey; used to capture food. *Mechanical Devices.* the projection on the bolt of a lock against which the key engages in sliding the bolt.

Talos *Ordnance.* a shipborne, surface-to-air missile with a solid-propellant rocket/ramjet engine; it is equipped with a nuclear or nonnuclear warhead, and command, beam-rider homing guidance; officially designated **RIM-8.**

Talpidae *Vertebrate Zoology.* the moles and desmans, a family of insectivorous, blind, burrowing mammals of the suborder Lipotyphla.

talus *Anatomy.* the highest of the tarsal bones that articulates with the tibia and fibula to form the ankle joint. Also, ASTRAGALUS, ASTRAGALOID BONE, ANKLE, ANKLE BONE. *Geology.* **1.** see SCREE. **2.** see TALUS SLOPE.

talus creep *Geology.* the generally slow movement of talus downslope.
talus fan see ALLUVIAL FAN.
talus glacier see ROCK GLACIER.
talus slope *Geology.* a steep, concave, downward sloping formation, formed by the accumulation of coarse, angular rock debris at the base of a cliff or steep slope. Also, DEBRIS SLOPE.
talweg see THALWEG.
tamarack *Botany.* a North American larch tree of the family Pinaceae, *Larix laricina,* grown commercially for its wood, which is used as a building material.

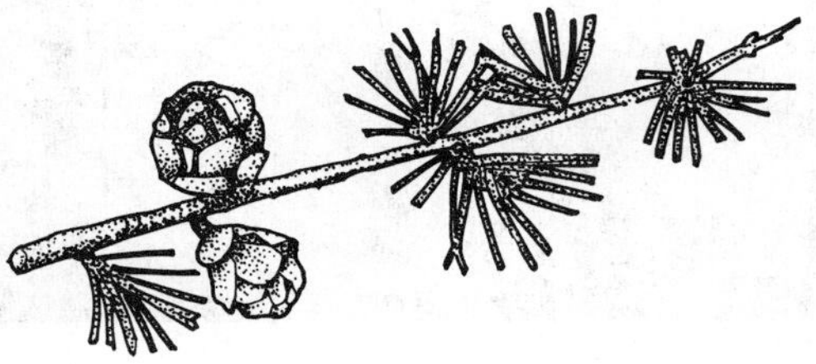

tamarack

Tamaricaceae *Botany.* a family of mostly halophytic or xerophytic dicotyledonous shrubs and small trees of the order Violales, characterized by slender branches and small, scalelike leaves; native to Eurasia and Africa.
tamarin *Vertebrate Zoology.* a marmoset of the genera *Saguinus* or *Leontopithecus* of the family Callithricidae, characterized by silking fur and a nonprehensile tail; found in South America, some species being threatened or endangered.
tamarind *Botany.* a tropical evergreen tree of the family Leguminosae, *Tamarindus indica,* native to Asia and now grown in the southern United States; its seed pods produce a juicy acid brown pulp that is used in drinks and foods.
tamarugite *Mineralogy.* $NaAl(SO_4)_2 \cdot 6H_2O$, a colorless, water-soluble, monoclinic mineral with vitreous luster, occurring as thin tabular crystals and in massive form, having a specific gravity of 2.06 and a hardness of about 3 on the Mohs scale; found as an alteration product of mendozite.
Tamm, Igor 1895–1971, Soviet physicist; shared the Nobel Prize for the theoretical explanation of Cherenkov radiation.
Tamm-Dancoff method *Quantum Mechanics.* a method of approximating the wavefunction of a system of interacting particles by an algebraic sum of some possible wavefunctions, the number of such functions determined according the accuracy desired.
Tamm-Horsfall protein *Biochemistry.* a high-molecular-weight mucoprotein contained in the mucus found on human uroepithelial cells.
tamoxifen citrate *Oncology.* $C_{26}H_{29}NO \cdot C_6H_8O_7$, an antiestrogen administered orally in metastatic breast cancer therapy; also used to stimulate ovulation in infertility.
tamp *Engineering.* **1.** to pound earth in order to settle or compress it. **2.** to place materials over a blasting charge to contain and intensify the explosion.
tamper *Mechanical Devices.* any of various hand-operated or power-driven machines used to tamp materials. *Ordnance.* any substance in a weapon or munition that offers resistance for a split second, so that the active materials behind the substance can build up greater pressure. *Nucleonics.* a reflecting material such as water, graphite, or beryllium that surrounds the core of a nuclear reactor so as to reduce the loss of neutrons from the core.
tamping bar *Engineering.* a bar used to tamp down a blasting charge. Also, **tamping rod.**
tamping pick *Engineering.* a wide, flat-headed pick used to drive ballast under railroad ties.
tamping plug *Engineering.* a plug placed over a bore hole to tamp a blasting charge.
tampion *Mechanical Devices.* a cone-shaped wooden tool used to remove or open the end of a lead pipe. Also, **tampin.**
tampon *Medicine.* **1.** a plug or cylinder of absorbent material inserted into a wound or natural body opening to absorb secretions or stop the flow of blood. **2.** specifically, an object of this type used to absorb menstrual flow.
tamponade *Surgery.* **1.** the surgical application of a tampon or compressive device to arrest bleeding. **2.** pathologic compression of a part, such as the compression of the heart by pericardial fluid.
Tamworth *Agriculture.* a breed of hog that is golden-red in color and that has a long, narrow body and face and large, erect ears. (From *Tamworth,* a borough in Staffordshire, England.)
tan *Engineering.* to carry out a tanning process; convert a hide into leather. *Medicine.* **1.** to darken the skin or become darker by exposure to the sun or to ultraviolet light. **2.** the brownish color of the skin resulting from such exposure.
tan. or **tan** tangent.
tanager *Vertebrate Zoology.* the common name for the bird family Thraupidae, colorful passerines characterized by a short tail and a conical bill, living mostly in forests of the New World, and feeding on insects, fruits, and flowers. Two common species of North America are the **scarlet tanager,** *Piranga olivacea,* and the **summer tanager,** *P. rubra,* both of which have bright red coloring in the male; the male of the **western tanager,** *P. ludoviciana,* has a red head and a yellow body.

tanager

Tanaidacea *Invertebrate Zoology.* an order of tiny malacostracan crustaceans in the superorder Peracarida, having a small carapace; mostly found on sea floors, usually in burrows or tubes.
Tanaostigmatidae *Invertebrate Zoology.* a family of wasps, hymenopteran insects in the superfamily Chalcidoidea, parasites of insects.
Tanapox virus *Virology.* a virus of the family Poxviridae that causes a febrile illness with localized skin lesions in humans and also infects monkeys.
tandem *Aviation.* of controls, wings, landing gears, and such, having two or more like elements or units arranged one behind the other; used to form compounds such as **tandem boosters** and **tandem-powered.** *Electricity.* describing or relating to two-terminal pair networks in which the output terminals of one network are connected to the input terminals of the other. Also, CASCADE.
tandem accelerator *Nucleonics.* an accelerator device in which a negative hydrogen ion is accelerated from ground to a high-voltage potential; then the two electrons of the ion are removed by passing through a gas cell or thin foil, thus leaving the remaining proton to be further accelerated to ground potential.
tandem airplane *Aviation.* an airplane having two or more seats and/or sets of controls set one behind the other, usually in a tandem cockpit.
tandem central office see TANDEM OFFICE.
tandem compensation see CASCADE COMPENSATION.
tandem duplication *Cell Biology.* the arrangement of one or more identical copies of a segment of DNA, lying one behind the other along a chromosome.
tandem genes *Molecular Biology.* two genes that adjoin one another on a chromosomal segment.
tandem harrow see DOUBLE-DISK HARROW.
tandem hoisting *Mining Engineering.* a method of hoisting used in deep shafts in which two skips are employed in one shaft, the lower being suspended from the tail rope of the upper.
tandem launch *Space Technology.* the launching of two or more satellites using a single launch vehicle.
tandem mill *Metallurgy.* a rolling mill having two or more stands that are arranged to keep metal being processed in a straight line from stand to stand.
tandem office *Telecommunications.* a telephone facility that makes connections between local offices in a region where there are too many such offices to economically establish a full set of direct connections. Also, TANDEM CENTRAL OFFICE.

tandem propellers *Naval Architecture.* an arrangement of two propellers on a single shaft.
tandem roller *Mechanical Engineering.* a road roller with two rolls of similar diameter running on the same track.
tandem switching *Telecommunications.* a type of telephone call routing in which calls travel through one or more intermediate switches to reach their destination.
tandem system *Computer Technology.* a system composed of two computers both of which are executing the same problem at the same time; may be either a redundant system or a master/slave configuration.
tang *Mechanical Devices.* the shank or protruding end of a file or other tool, to which a handle may be attached. *Archaeology.* a projection or stem that protrudes from the end of a projectile point as a barb or grip.
Tanganyika, Lake *Geography.* a long lake in central Africa, between Zaire and Tanzania (area: 32,893 square miles).
tangelo *Botany.* **1.** a tree hybrid that is a cross between a tangerine or mandarin orange and a shaddock or grapefruit. **2.** the edible citrus fruit of this hybrid.
tangent *Mathematics.* **1.** the trigonometric function of an angle θ, given by tangent $\theta = \sin\theta/\cos\theta$; abbreviated tan θ. **2.** in a right triangle with hypotenuse length *hyp*, the side opposite angle θ of length *opp*, and the side adjacent to angle θ of length *adj*, $\tan\theta = opp/adj$. **3.** the **tangent line** to a curve at a point P of the curve is the limiting position (if it exists) of the secant passing through P and a point Q on the curve as Q approaches P. An orientation is chosen so that the positive direction of the tangent points in the positive direction of the curve. If the curve is given by a differentiable function $y = f(x)$, then the tangent line at $P = (x_0, y_0)$ has slope $f'(x_0)$. **4.** the **tangent plane** to a surface at a point P of the surface is the plane (if it exists) passing through P and containing every line tangent to some curve on the surface at the point P.
tangent arc *Meteorology.* any of various halo arcs that form as loci tangent to other halos.
tangent bending *Metallurgy.* the process of wiping metal around a radius so that it forms identical bands with parallel axes in a single part.
tangent bifurcation *Chaotic Dynamics.* as a region of positive curvature (second derivative), a map changes from two intersections with $x_{n+1} = x_n$, to a single point tangency, to a region of near tangency.
tangent bundle *Mathematics.* let X be a differentiable manifold; a tangent bundle is the space $T(X^n)$ of all pairs (x, v_x), where x is a point of the differentiable manifold X^n and v_x is a tangent vector in the tangent space $T_x(X^n)$. The tangent bundle then has fiber bundle structure. If in addition, X^n is a differentiable manifold of class C^k, then $T(X^n)$ is a differentiable manifold of class C^{k-1} and is called a **differentiable** (C^{k-1}) **tangent bundle.**
tangent formula *Mathematics.* for a triangle with sides a, b, and c and angles α, β, and γ opposite those sides, the formula

$$\tan\gamma = c\sin\alpha/(b - c\cos\alpha) = c\sin\beta/(a - c\cos\beta).$$

tangent galvanometer *Engineering.* a galvanometer consisting of a compass placed horizontally in a vertical coil; the deflection of the compass from its normal bearing indicates the strength of the current.
tangential [tan jen´shəl] *Mathematics.* **1.** relating to or characteristic of a tangent. **2.** being or moving in the direction of a tangent.
tangential acceleration *Mechanics.* **1.** the component of acceleration perpendicular to the radial and axial directions in a cylindrical or polar coordinate system; it is equal to the angular acceleration multiplied by the radius. **2.** the component of a particle's acceleration vector that is parallel to the velocity.
tangential component *Mathematics.* at a point on a hypersurface in Euclidean n-space, any n-vector can be expressed as the sum of a vector tangent to the hypersurface and a vector normal to the hypersurface, called the **tangential component** and **normal component** of the vector, respectively.
tangential focus see PRIMARY FOCUS.
tangential stress see SHEAR STRESS.
tangential velocity *Mechanics.* **1.** the component of velocity that is perpendicular to the radial and axial directions in a cylindrical or polar coordinate system; it is equal to the angular velocity multiplied by the radius. **2.** the velocity of a particle on a curved path; it is tangent to the path.
tangential wave see S WAVE.
tangential wave path *Electromagnetism.* the path of a radio wave that propagates parallel to the surface of the earth; its path is curved due to atmospheric refraction.
tangent latitude error *Navigation.* in certain gyro compasses, the angle between the local meridian and the spin axis or setting position.
tangent offset *Engineering.* a method of laying out survey traverse lines, in which a constant is multiplied by the tangent of a given base angle.
tangent point see POINT OF TANGENCY.
tangent screw *Engineering.* a fine-adjustment screw that acts along an arc, such as the screw for fine adjustment of a sextant.
tangent space *Mathematics.* the vector space $T_x(X)$ of tangent vectors to a differentiable manifold X at a point x of the manifold. Vector addition and scalar multiplication are defined by

$$(\alpha u_x + \beta v_x)(f) = \alpha u_x(f) + \beta v_x(f),$$

where u_x and v_x are any tangent vectors to X at x, and a and b are any real numbers; also known as the **tangent vector space** to X at x.
tangent vector *Mathematics.* **1.** at a point P on a curve, a vector pointing in the positive direction of the tangent at P. **2.** a tangent vector to a differentiable manifold X at a point x of the manifold is a real-valued linear functional v_x on the space of functions that are differentiable at some neighborhood of x, such that, for all functions f and g on X differentiable at x and all real α, β: (a)

$$v_x(\alpha f + \beta g) = \alpha v_x(f) + \beta v_x(g)$$

(bilinearity); and (b)

$$v_x(fg) = f(x)v_x(g) + g(x)v_x(f)$$

(Leibnitz' rule). Tangent vectors are elements of the tangent vector space $T_x(X)$, and can be viewed as equivalence classes of curves tangent to X at x.
tangerine *Botany.* **1.** any of several varieties of the citrus tree *Citrus reticulata,* in the mandarin group of the family Rutaceae. **2.** the fruit of the tree, grown as a food crop.
tangerine oil see MANDARIN OIL.
tangling *Solid-State Physics.* a method of reducing the motion of dislocations by increasing the number of dislocations so that they interfere with one another.
tangoreceptor *Physiology.* a sense organ that is responsive to tactile stimuli.
tanh *Mathematics.* a hyperbolic tangent.
tank *Engineering.* a general term for any large vessel, closed or open, used for holding a fluid such as water, compressed air, gasoline, or other fuel, and so on. *Ordnance.* a heavily armored combat vehicle providing mobile firepower and crew protection for offensive combat, typically armed with a cannon and machine guns and moving on caterpillar treads. *Electronics.* **1.** see TANK CIRCUIT. **2.** an acoustical delay-line storage unit consisting of multiple channels.
tankage *Engineering.* the fluid contained in a storage tank.
tank balloon *Engineering.* an expansible bladder that holds vapor released from a gasoline tank; on cooling, the condensing vapor returns from the bladder into the tank.
tank barge *Naval Architecture.* a barge fitted with tanks for oil or other fluids.
tank battery *Petroleum Engineering.* a system of two or more storage tanks connected together to receive the oil produced from one or more wells on a lease.
tank bottom *Chemical Engineering.* a mixture of stored liquid, rust and other sediments below the level of the outlet pipe in a tank.
tank car *Transportation Engineering.* a railroad car having a cylindrical body, used to carry liquids.
tank circuit *Electronics.* a circuit that contains capacitance and inductance and is capable of storing electrical energy by passing it alternately between the two circuit elements; tank circuits operate at a relatively narrow band of frequencies centered about a single resonant frequency.
tankdozer *Ordnance.* a standard tank that is equipped with a detachable bulldozer blade.
tanker *Naval Architecture.* a vessel designed to carry a liquid cargo. Crude oil tankers are the most common type, and are the largest vessels in existence. Also, **tankship.**
tank farm *Botany.* a plant-growing area that employs hydroponics. Thus, **tank farming.** *Petroleum Engineering.* an area of up to several square miles on which a series of large storage tanks hold crude oil or other petroleum products.

tank gauge *Engineering.* a vertical staff in a tank, used for measuring liquid level.
tank periscope *Optics.* a device in which a series of lenses and/or mirrors allows the crew of a tank to look outside without opening the hatch.
tank reactor *Nucleonics.* a nuclear reactor whose core is suspended within a closed tank.
tank scale *Engineering.* a scale supporting a tank, used for weighing bulk fluids.
tank switch *Petroleum Engineering.* an automatic system controlling oil-field lease tanks and their associated lines.
tank truck *Mechanical Engineering.* a truck carrying or towing a cylindrical vessel mounted on a frame and wheels, used for carrying liquids.
tannage *Engineering.* the process or product of leather tanning.
tannase *Enzymology.* an esterase that catalyzes the hydrolysis of tannic acid to gallic acid; found in various tannin-bearing plants or produced in cultures by *Asperigillus niger* and *Penicillium glaucum.*
tannic acid *Organic Chemistry.* $C_{76}H_{52}O_{46}$, combustible, toxic, lustrous, faintly yellowish amorphous powder, glistening scales, or spongy mass; soluble in water and alcohol, and almost soluble in ether; decomposes at 210–215°C; a natural substance derived by extraction of powdered nutgalls with water and alcohol; used in tanning, textiles, organic synthesis, and as an alcohol denaturant. Also, GALLOTANNIC ACID.
tanniferous *Botany.* containing or yielding tannin.
tannin *Chemistry.* the popular name for tannic acid or similar compounds.
tanning *Engineering.* the curing process used to convert animal hide into leather. *Medicine.* the process in which the skin becomes darker as a result of exposure to sunlight. *Entomology.* see SCLEROTIZATION.
tan rot *Plant Pathology.* a disease of strawberries caused by the fungus *Pezizella lythri* and characterized by light brown sunken spots on the fruit.
tantalite [tan´tə līt] *Mineralogy.* a member of the series ferrotantalite ($Fe^{+2}Ta_2O_6$)−manganotantalite ($Mn^{+2}Ta_2O_6$); brownish-black to black orthorhombic minerals occurring as aggregates of tabular or prismatic crystals and in compact masses, having a specific gravity of 8.0 to 8.2 and a hardness of 6 to 6.5 on the Mohs scale; found in granite pegmatites.
tantalum [tan´tə ləm] *Chemistry.* a chemical element that is a rare metal of the vanadium family, having the symbol Ta, the atomic number 73, an atomic weight of 180.95, a melting point of 2996°C, and a boiling point of 5560°C; a malleable gray solid that is resistant to corrosion, considered a refractory metal because of its high melting point; used for electric light filaments, in dental and surgical instruments, and as a catalyst. (Named for King *Tantalus,* a character in Greek mythology who was punished (tantalized) by being forced to stand with food and drink barely out of his reach; the difficulty of extracting this metal from its compounds made its existence seem tantalizing.)
tantalum capacitor *Electricity.* an electrolytic capacitor with a tantalum or sintered-slug anode, such as solid tantalum, tantalum-foil electrolytic, and tantalum-slug capacitors; the weight and volume are less than those for comparable aluminum electrolytic capacitors of the same rating.
tantalum carbide *Inorganic Chemistry.* TaC, black cubic crystals; insoluble in water; melts at 3880°C and boils at 5500°C; highly resistant to chemical reaction; used in cutting tools and dies.
tantalum chloride *Inorganic Chemistry.* $TaCl_5$, a light-yellow vitreous crystalline powder; soluble in alcohol and decomposes in cold water; melts at 216°C and boils at 242°C; used as a chlorination agent and chemical intermediate and in production of the pure metal. Also, **tantalic chloride, tantalum pentachloride.**
tantalum-foil electrolytic capacitor *Electricity.* an electrolytic capacitor whose anode and cathode are composed of tantalum foils; the anode foil is oxide-coated.
tantalum nitride *Inorganic Chemistry.* TaN, brown to black hexagonal crystals, insoluble in water; melts at approximately 3360°C.
tantalum-nitride resistor *Electronics.* a resistor made by depositing a thin layer of tantalum nitride on an insulating substrate.
tantalum oxide *Inorganic Chemistry.* Ta_2O_5, colorless, rhombic crystals; insoluble in water and acid; melts at approximately 1872°C; used in electronics, lasers, and optical glass, and in the production of tantalum metal. Also, **tantalic acid anhydride, tantalum pentoxide.**
tantalum-slug electrolytic capacitor *Electricity.* an electrolytic capacitor whose anode is a sintered tantalum slug.
T antenna *Electromagnetism.* an antenna having one or more horizontal wires that are fed by a lead-in wire connected at their centers.
tanteuxenite-(Y) *Mineralogy.* $(Y,Ce,Ca)(Ta,Nb,Ti)_2(O,OH)_6$, a brown or black, orthorhombic, resinous mineral, having a specific gravity of 5.4 to 5.9 and a hardness of 5 to 6 on the Mohs scale; found in tin placer deposits of western Australia.
T antigen *Immunology.* **1.** a protein antigen that occurs in the cell walls of streptococci. **2.** an antigen that occurs in tumor cells. Also, TUMOR ANTIGEN. **3.** an antigen that determines an individual's immune reactions to grafts. Also, TRANSPLANTATION ANTIGEN. **4.** an antigen that occurs in human red blood cells and is detected during enzyme treatment.
tantiron *Metallurgy.* a cast iron that is high in silicon content (approximately 14%) and highly resistant to corrosion.
tanycyte *Neurology.* an ependymal cell located near the third ventricle with processes that extend to the capillary plexus of the portal circulation.
tap to draw off air, water, or another fluid; a device used in doing this; specific uses include: *Surgery.* to drain off fluid from a body cavity by means of a hollow, pointed rod or needle. *Mining Engineering.* to bore or cut into old workings or water-bearing strata so that gas or water may be removed or proved. *Mechanical Devices.* **1.** a faucet or stopper used to close an opening in a cask, barrel, or other container. **2.** to draw liquid, e.g., beer, by opening such a faucet or stopper. **3.** a hardened steel tool having male threads grooved to form cutting edges; used for cutting female screw threads, as in a pipe. *Electricity.* **1.** a connection that extends from a winding at a connector point. **2.** a branch in a conductor, such as a battery tap. *Telecommunications.* to make a secret connection to a telephone or other communications line for the purpose of eavesdropping or surveillance.
tap bolt *Mechanical Devices.* a bolt that has a head on one end and screw threads on the other, used without the need for a nut. Also, TAP SCREW.
tap changer *Electricity.* a device used to alter the ratio of input and output voltages of a transformer to any of a defined number of steps.
tap changing *Electricity.* a technique used to vary the voltage ratio of a transformer by tapping windings so that the turn ratio is altered.
tap crystal *Electronics.* a type of semiconductor that is capable of storing electrical energy when stimulated by incident light, and that releases energy in the form of light when mechanically stimulated, such as by tapping.
tap density *Metallurgy.* in powder metallurgy, the density of a powder formed when the receptacle is vibrated during loading.
tap drill *Mechanical Devices.* a drill used to make a precisely sized tap hole.
tape *Materials.* a long thin strip of flexible material that is coated with adhesive and used for binding, sealing, or attaching objects together. *Computer Technology.* a strip of plastic or other flexible material that is coated with a magnetic film and used for data input, storage, and output. *Acoustical Engineering.* **1.** a magnetic tape that is used to record sound or to play back prerecorded sound. **2.** to record on magnetic tape or videotape. *Mechanical Devices.* a shorter term for a tapemeasure. See TAPEMEASURE.
tape alternation see PING-PONG.
tape bootstrap routine *Computer Programming.* a set of instructions that are stored in the first data block of a tape and serve to load the operating system and other resident software.
tape cartridge *Acoustical Engineering.* a closed-loop reel and housing, developed primarily for eight-track recordings, which operate at 4-3/4 inches per second; they are no longer manufactured due to their relatively poor quality of recording and reproduction and lack of durability, in comparison with cassette recordings.
tape cassette *Acoustical Engineering.* see CASSETTE.
tape-controlled carriage *Computer Technology.* an automatic paper feed on a line printer that is controlled by a punched paper tape.
tape crease *Computer Technology.* on a magnetic tape, a local fold or bend that may cause an error in the reading or writing of data.
tape deck *Acoustical Engineering.* a tape recorder and player without a dedicated power amplifier and speaker, usually integrated into a sound system with other components such as a turntable, CD player, and tuner.
tape drive *Acoustical Engineering.* a motor-driven mechanism for moving tape past the recording or playback heads of a tape recorder. *Computer Technology.* a device designed to control the movement of magnetic tape past the heads, which enables the tape to be read, recorded (written upon), or rewound. Also, TAPE STATION, TAPE TRANSPORT, TAPE UNIT. *Mechanical Engineering.* a device that uses pulleys and flexible tapes to transmit power from an actuator to a remote mechanism.

tape editor *Computer Programming.* a program that edits and updates magnetic tapes.
tape-float gauge *Engineering.* a gauge for measuring liquid levels in a tank or basin, employing a float attached to a tape and winding-unwinding device; the tape provides the level reading.
tape gauge *Engineering.* a tide gauge composed of a float attached to a tape and winding-unwinding device, placed in a perforated box or cylinder that releases short-term wave effects while registering the longer-term rise and fall of the tide.
tape grass *Botany.* a flowering aquatic plant, *Vallisneria spiralis,* of the family Hydrocharitaceae, characterized by its long, thin leaves.
tape joint *Mechanical Devices.* a paper or gauze tape used to cover plaster or wall joints prior to coating the surface.
tape label *Computer Programming.* **1.** a special record that appears at the beginning of a reel of magnetic tape and serves to identify the tape and its contents. **2.** a printed identifying label that is attached to a tape reel.
tape library *Computer Technology.* **1.** a collection of magnetic tapes that are arranged in a logical order. **2.** the room in which magnetic tapes are stored under secure and environmentally controlled conditions.
tape loop *Acoustical Engineering.* a length of magnetic tape having ends joined together to create an endless loop; can be used for a repeating message or some other function for which it is unnecessary to rewind the tape.
tape mark *Computer Programming.* a code that is recorded on magnetic tape to indicate the end of a file.
tapemeasure or **tape measure** *Mechanical Devices.* a narrow strip of strong, flexible material that is marked with linear dimensions and used for measuring short distances.
tape player *Acoustical Engineering.* a simple tape recorder used for playback, often not designed to record.
tape pool *Computer Technology.* a group or set of tape drives.
taper *Aviation.* on an airfoil, blade, or the like, a gradual decrease in chord length or thickness, usually from root to tip. *Electricity.* a continuous or gradual change of an electric property.
taper bit *Mechanical Devices.* a cone-shaped, noncoring drill bit used in drilling blastholes or in wedging and reaming operations.
tape reading *Computer Programming.* the sensing of data that occurs when an encoded magnetic tape passes across the heads, thus converting the information to electrical signals.
tape-record *Acoustical Engineering.* to carry out a tape recording process; to record sound on magnetic tape.
tape recorder *Acoustical Engineering.* an electroacoustical device designed for recording electroacoustic signals on a magnetic tape and, usually, for playback of such tapes as well, including a set of recording, erasing, and playback heads, an amplifier, and one or more speakers. Also, **tape machine.**
tape recording *Acoustical Engineering.* the preservation of electromagnetic flux patterns, representing sound signals, onto a tape coated with magnetic particles by means of a tape recorder; the sound can later be reproduced by a tape player or the playback head on a tape recorder.
tapered-bore *Ordnance.* describing a weapon whose bore is tapered, either throughout the length of the bore or near the muzzle; the tapering is usually designed to project a smaller projectile at higher velocity. Also, SQUEEZE-BORE.
tapered core bit *Mechanical Devices.* a core bit that tapers upward from the cutting face to the shank and that has cutting diamonds set into its surface.
tapered file *Mechanical Devices.* a file whose edges converge at one end.
tapered joint *Mechanical Devices.* a joint between pipes formed by screw threads that have a tapering diameter, making for a tight, waterproof connection.
tapered pipeline *Petroleum Engineering.* a pipeline in which the pressure grade is altered by varying material or wall thickness in certain sections to compensate for a drop in working pressure.
tapered-thrust bearing *Mechanical Devices.* a wedge-shaped bearing design that is utilized in high load construction or other mechanical thrust bearing machinery.
tapered waveguide *Electromagnetism.* a waveguide or transmission line whose electrical or physical characteristics continuously change with distance along the axis of the waveguide. Thus, **tapered transmission line.**
tapered wheel *Mechanical Devices.* a flat-faced grinding wheel whose hub is thicker than its face.
tape-relay station *Telecommunications.* in a communications center, a component that receives and forwards messages via the tape-relay mode of operation.
tapering gutter *Mechanical Devices.* a parapet gutter whose inner surface gradually widens in the direction of flow.
taper-in-thickness ratio *Aviation.* a gradual change in the thickness ratio along a wing span with no change in chord.
taper key *Mechanical Devices.* a removable machine key, rectangular in shape and tapering toward one end, that is inserted between a shaft and a hub along its depth in order to prevent movement of the shaft and hub relative to each other while transmitting torque.
taper pin *Mechanical Devices.* a small rod of metal or wood, tapering toward one end, used to join parts together or to hold parts in place.
taper ratio *Aviation.* on an airfoil, the ratio of the tip chord length to the root chord length.
taper reamer *Mechanical Devices.* a reamer tool having a tapering, fluted surface; used to enlarge or clean tapered holes.
taper-rolling bearing *Mechanical Engineering.* a roller bearing with tapered rollers and coned races, designed to withstand end thrust.
taper shank *Mechanical Devices.* a cone-shaped protrusion on a tool, which fits into the jaws, chuck, spindle or other tapered holding device on a machine or mechanism.
taper tap *Mechanical Devices.* a tapered tool formed from hardened tool steel, used in conjunction with other tools for the cutting of female screw threads.
tape skip *Computer Programming.* a tape drive instruction that is executed when a defect is detected on a magnetic tape, automatically skipping the damaged area.
tape speed *Acoustical Engineering.* the constant speed at which an audio tape travels past the tape heads for recording or playback; standard speeds are normally 15 inches per second (ips), 7-1/2 ips, 3-3/4 ips, and 1-7/8 ips.
tape squeal *Acoustical Engineering.* an unwanted high-pitched sound produced during playback as a tape moves past dirty or glazed pressure pads.
tape station *Computer Technology.* see TAPE DRIVE.
tapestry *Textiles.* a heavy, ornamental woven fabric with pictorial designs created in the weaving process rather than embroidered or dyed into the fabric. Originally made by hand, tapestries can now be woven by machine.
tape-to-tape conversion *Computer Programming.* a method of changing data from one type of tape format or density to another.
tape transmitter *Telecommunications.* **1.** a code-transmitting device that is activated by prepunched paper tape; used in high-speed transmissions because the tape can be fed into the machine at a greater speed than when it was originally punched. **2.** in FAX communications, a transmitter that sends subject copy printed on narrow tape.
tape transport *Acoustical Engineering.* the structure that houses all mechanical components of a tape drive that move an audio tape past the recording and playback heads. Also, MOTORBOARD. *Computer Technology.* see TAPE DRIVE.
tapetum *Anatomy.* **1.** a membranous covering or layer of cells. **2.** a layer of fibers in the brain extending from the corpus callosum to the roof and lateral walls of the inferior and posterior horns of the lateral ventricles. *Botany.* a layer of cells surrounding the spore mother cells in vascular plants, providing nutrition to the developing spores. (A Latin word meaning "small carpet.")
tape unit *Computer Technology.* see TAPE DRIVE.
tapeworm *Invertebrate Zoology.* the popular name for various segmented parasitic flatworms of the class Cestoda. Those infecting humans are mainly of the genera *Taenia, Diphyllobothrium, Hymenolepis, Echinococcus,* and *Dipylidium.*

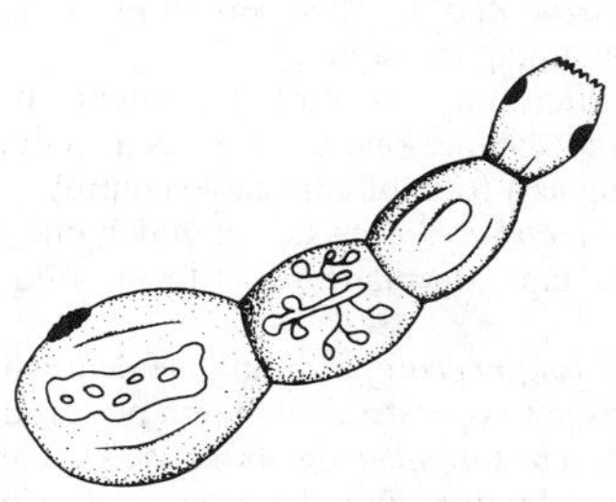

tapeworm

tape-wound core *Electromagnetism.* a length of ferromagnetic material in tape form, wound in such a way that each turn falls directly over the previous turn.

tap hole *Metallurgy.* a hole in a furnace, ladle, or pot that is used in the tapping process of molten metal.

taphonomy *Paleontology.* **1.** the process of formation of a fossil assemblage. **2.** the reconstruction of the history of an organism between the time of its death and the time when it is studied as a fossil.

Taphrina *Mycology.* a genus of fungi belonging to the order Taphrinales; it includes parasites found on trees such as oak, peach, plum, cherry, and birch.

Taphrinaceae *Mycology.* the only family of fungi in the order Taphrinales; includes species that are parasitic and pathogenic to higher plants and ferns.

Taphrinales *Mycology.* an order of fungi belonging to the class Hemiascomycetes; includes plant parasites that cause serious diseases to higher plants, such as witches' brooms and galls.

taphrogeny *Geology.* a process in which essentially vertical movements in the earth's crust produce major high-angle or block faulting and associated subsidence resulting in the formation of rift or trench phenomena. Also, **taphrogenesis.**

taphrogeosyncline *Geology.* a downward flexure in the earth's crust occurring as a rift or trough between parallel faults.

taping *Engineering.* **1.** any process of recording electronic data, including sound or video tracks, on a magnetic tape. **2.** the process of measuring distances with a tape, as in survey work.

tapioca *Food Technology.* a starchy food product made from the cassava root; used as a thickener, especially in tapioca pudding. *Botany.* another name for the cassava plant.

tapioca snow see SNOW PELLETS.

tapiolite *Mineralogy.* any black, weakly radioactive, tetragonal mineral of the series ferrotapiolite, $(Fe^{+2},Mn^{+2})(Ta,Nb)_2O_6$, to manganotapiolite, $(Mn^{+2},Fe^{+2})(Ta,Nb)_2O_6$; having a specific gravity of 7.82 and a hardness of 6 to 6.5 on the Mohs scale; found as prismatic crystals in granite pegmatites or derived detrital deposits.

tapir [tā´pər] *Vertebrate Zoology.* any of three species of large, nocturnal ungulates in the genus *Tapirus* of the family Tapiridae; characterized by a heavy, sparsely haired body; a long, mobile snout; and an uneven number of toes, three on the hind feet and four on the front; native to Central and South America, Malaya, and Sumatra.

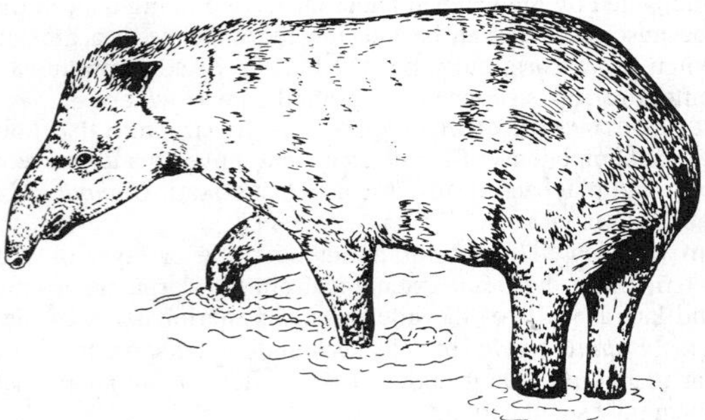

tapir

Tapiridae *Vertebrate Zoology.* the tapirs, a family of the mammalian ungulate order Perissodactyla.

Tapiroidea *Vertebrate Zoology.* a superfamily of the mammalian order Perissodactyla, containing the tapirs.

tapped control *Electronics.* a circuit element, usually a resistor or transformer winding, having one or more connection points along the element and serving as a form of adjustable control.

tapped resistor *Electricity.* a resistor in which one or more intermediate connections or taps are made so as to provide certain resistance steps.

tappet *Mechanical Engineering.* a sliding element whose motion is controlled by a cam; used to operate a valve or a push rod.

tappet rod *Mechanical Engineering.* a rod used to carry a tappet.

tapping *Mechanical Engineering.* the process of using a tap to form an internal screw thread.

taproot *Botany.* the main axis of a root system, often penetrating quite deeply into the ground; used in some plants, such as the carrot, for food storage.

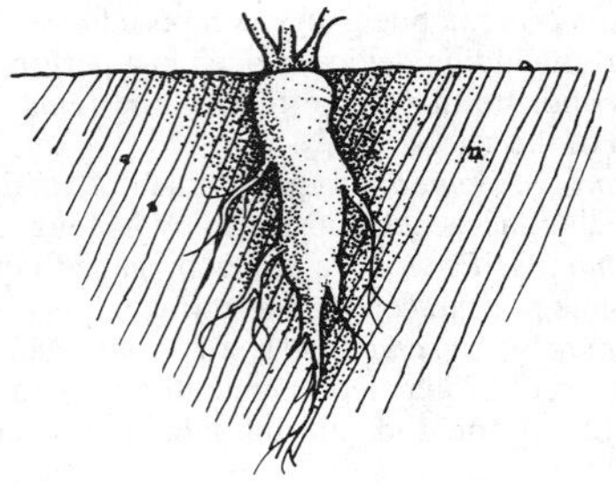

taproot

tap screw see TAP BOLT.

tap switch *Electricity.* a multiposition switch that is used to connect an external circuit to numerous taps on a component or device.

tap wrench *Engineering.* a T-shaped tool used to hold and turn a tap.

tar *Materials.* **1.** a dark, viscous, usually pungent substance derived from the destructive distillation of various organic compounds such as wood, coal, or shale; used for various industrial purposes. **2.** derived from or containing tar. Thus, **tar distillate.**

tar acid *Organic Chemistry.* any of numerous very weakly acidic substances, such as phenol, cresols, and xylenols, soluble in sodium hydroxide, that are separated during the industrial fractionation of coal tar.

taranakite *Mineralogy.* $(K,NH_4)Al_3(PO_4)_3(OH)\cdot 9H_2O$, a very soft, white, yellowish-white, or gray, trigonal mineral occurring in massive form and as tiny lathlike crystals, having a specific gravity of 2.06 to 2.09 and an undetermined harness; found within clays adjacent to guano deposits.

tarantata *Meteorology.* a strong northwest breeze of the Mediterranean region.

tarantism *Medicine.* a disorder that occurred in southern Italy in the 1400s to 1600s, characterized by hysteria and frenzied, uncontrolled dancing; associated with the bite of a tarantula, but probably psychosomatic in origin.

tarantula *Invertebrate Zoology.* the popular name for any of various large, hairy, ground-dwelling, predaceous spiders of the suborder Orthognata whose bite is painful but not generally dangerous to humans. (Originally associated with the ancient Greek city of *Taranto,* in southern Italy.)

tarapacáite *Mineralogy.* K_2CrO_4, a bright-yellow, transparent, water-soluble orthorhombic mineral occurring in thick tabular crystals, having a specific gravity of 2.74 and an undetermined hardness; found in nitrate deposits in Chile.

tar base *Organic Chemistry.* any of various weakly alkaline substances, such as pyridine, picoline, lutidine, and quinoline, soluble in dilute sulfuric acid, that are separated during the industrial fractionation of coal tar.

tarbuttite *Mineralogy.* $Zn_2(PO_4)(OH)$, a colorless to pale-yellow, red, or green triclinic mineral occurring in aggregates of short prismatic or equant crystals, having a specific gravity of 4.14 to 4.19 and a hardness of about 4 on the Mohs scale; found as a secondary mineral in zinc mines.

tar camphor see NAPHTHALENE.

Tardigrada *Invertebrate Zoology.* the water bears, a class or subphylum of minute arthropods having sucking mouthparts and four pairs of stubby clawed legs; they are able to survive dehydration and are distributed from sea floors to ditches and gutters.

Tardigrada

tardive dyskinesia *Medicine.* an abnormal condition characterized by twitching of the face and tongue, and involuntary movement of the body and limbs, occurring primarily as a side effect of the prolonged use of phenothiazine drugs for parkinsonism or various antipsychotic drugs.

tare *Engineering.* the weight of a container or vehicle when empty, subtracted from the gross weight to obtain the net weight of the material inside the container. *Fluid Mechanics.* the drag produced by interference in wind-tunnel testing.

tare effect *Fluid Mechanics.* the air flow affected by the support assembly (called interference), which also produces a drag (or tare) in wind tunnel testing.

target a point, object, or goal at which something else is directed; specific uses include: *Military Science.* **1.** a specific individual, object, location, or group of objects at which fire is directed. **2.** a geographical area, complex, or installation planned for capture or destruction by military forces. **3.** a country, area, installation, agency, or individual against which intelligence activities are directed. *Atomic Physics.* **1.** any material or substance bombarded by particles in order to induce a nuclear reaction, usually in an accelerator. **2.** an atom that has been irradiated. *Electronics.* **1.** an object detected by a radar set at a signal strength sufficient for it to be detected on the radar screen. Also, RETURN. **2.** the surface in a camera tube that produces an output signal when scanned by an electron beam. *Radiology.* the plate of an X-ray tube that focuses the electrons and from which the X-rays are released. Also, ANTICATHODE. *Medicine.* a cell or organ that is selectively affected by a particular agent, such as a hormone or drug. *Computer Programming.* **1.** an end goal, such as a program under development, or a language into which a progam is to be translated. **2.** of or relating to such a goal or language, or an end user.

target acquisition *Electronics.* the receipt of a radar target of sufficient signal quality that it can be tracked and otherwise monitored and analyzed. *Military Science.* the detection, identification, and location of a target in sufficient detail to permit effective employment of weapons.

target-acquisition radar *Military Science.* the radar in an antiaircraft weapons system that is responsible for detecting and identifying possible targets; distinct from the tactical control radar, though the equipment may be of the same type.

target analysis *Military Science.* an examination of potential targets to determine military importance, attack priority, and weapons required to obtain the desired level of damage or casualties.

target angle *Navigation.* the bearing to a target, measured right or left of a (true or relative) reference line. *Ordnance.* the angle at the target subtended by the observing baseline.

target-approach point *Aviation.* a navigational checkpoint marking the final turn-in to a landing zone or drop zone.

target area *Space Technology.* a designated impact or landing zone for a missile, reentry vehicle, or the like.

target array *Military Science.* a graphic representation of enemy forces, personnel, and facilities in a specific situation, accompanied by a target analysis.

target baseline *Military Science.* a line connecting prime targets along the periphery of a geographic location.

target bearing *Military Science.* **1. true target bearing.** the true compass bearing of a target from a firing ship. **2. relative target bearing.** the bearing of a target measured in the horizontal from the bow of one's own ship clockwise from 0° to 360°, or from the nose of one's own aircraft in hours of the clock.

target computer *Computer Technology.* a computer for which a program is written, although the program may be compiled and debugged on a larger and more capable system.

target concentration *Military Science.* a grouping of geographically proximate targets; a geographically integrated series of such groupings is a target complex.

target configuration *Computer Technology.* the complete computer system, including all peripherals and auxiliary memory units, for which a program is written.

target-designating system *Electronics.* a system used to assign a target that has been located at one operator station to another operator station for tracking, fire-control computation, or other appropriate processing.

target deviation *Ordnance.* the distance from the point of impact or burst to the target.

target discrimination *Electronics.* the ability of a radar system to distinguish between two or more targets that are close together in range or azimuth.

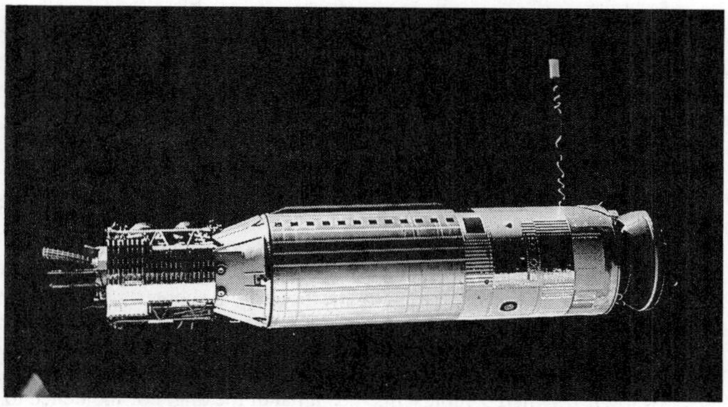

Agena target docking vehicle

target docking vehicle *Space Technology.* an orbiting artificial satellite, such as a space station, that serves as a landing site for another spacecraft.

target dossier *Military Science.* a file of assembled target intelligence about a specific geographic area; an assembly of intelligence on a specific target is a target folder.

target drone *Aviation.* a pilotless aircraft operated by a remote control device and used as a target for antiaircraft weapons.

target echo *Electromagnetism.* the radar signal returned after reflecting off a target object.

targeted mutagenesis *Molecular Biology.* the use of a specific mutagen to cause a mutation at a specific site on a DNA molecule.

target identification *Military Science.* the process, usually involving electronic techniques, of determining the nature of a target and whether it is friendly or hostile.

target-indicating system *Ordnance.* a system that communicates with the tracker of an antiaircraft automatic weapon, to indicate the approach direction of a suitable target.

target information center *Military Science.* an onshore or shipborne intelligence center that processes target information for air, artillery, and naval gunfire support.

target intelligence *Military Science.* information regarding the nature, location, vulnerability, and relative importance of a target or target complex.

target language *Linguistics.* **1.** in foreign-language learning, the new language that the learner seeks to acquire, as opposed to his or her native language. **2.** a second language into which a text is translated. *Computer Programming.* a computer language into which another language is converted.

target noise *Electromagnetism.* the noise component of a received radar signal due to reflections from a target that has several reflecting elements oriented randomly in space.

target of opportunity *Military Science.* **1.** a target that is observed or detected during combat, that is within range of available weapons, and against which fire had not previously been directed or requested. **2.** any previously unselected target observed after an operation begins.

target organ *Medicine.* the organ or structure toward which something is primarily directed, such as a therapeutic dose of radiation, the effects of a drug or hormone, or the greatest concentration of a radioactive tracer.

target pack *Computer Technology.* a disk pack used to store systems software; often used to test proposed modifications in a system control program before the master program is changed.

target phase *Computer Programming.* the occasion when a target program is run, usually after compiling has occurred.

target priority *Military Science.* a grouping of targets with the indicated sequence of attack.

target program *Computer Programming.* a computer program in a target language.

target range *Military Science.* an area designated for instruction and practice in shooting rifles and other small arms.

target response *Military Science.* the effect on personnel and materiel of blast, heat, light, and radiation resulting from the explosion of a nuclear weapon.

target routine see OBJECT PROGRAM.

target seeker see SEEKER.

target selector *Ordnance.* an off-carriage observation instrument that is included in both a target-designating system and a target-indicating system; in the former, it electrically moves the gun mount into approximate firing position for the selected target; in the latter, it indicates to the tracker the approach direction of the selected target.
target signature *Electronics.* those individual attributes a radar or sonar target displays that can be useful in classifying the target in terms of its military characteristics.
target spot *Plant Pathology.* any plant disease that causes the production of lesions in the form of concentric circles on plant parts.
target spotter *Ordnance.* a small black disk attached to a practice target after a hit to indicate the bullet's location to the shooter.
target strength *Acoustical Engineering.* a measure of the amplification of the active transmission from an active sonar system or torpedo; used in the calculation of the echo signal level and computed by the equation $T = 10 \log I_T/I$, where I represents the signal at the angle of incidence to the target and I_T represents the reflected signal one yard from the target.
target system *Military Science.* **1.** all the functionally related targets in a particular geographic area. **2.** a group of related targets whose combined destruction will produce a particular effect desired by the attacker.
target theory *Cell Biology.* a theory that attempts to explain the mechanism by which ionizing radiation produces mutations in DNA; it postulates that one or more ionizing events, called hits, are necessary within a particular sensitive region of the cell in order to produce a mutation. Also, HIT THEORY.
target timing *Navigation.* the timing and plotting of successive target positions to determine the target's track and ground speed.
target-type flowmeter *Engineering.* an instrument for measuring flow through a pipe or conduit, consisting of a small circular component placed across the flow and attached to a counterbalance arm; the degree of displacement corresponds to the level of flow.
target vulnerability *Military Science.* a measure of a potential target's vulnerability, indicated by a number on a standard scale.
Targhee *Agriculture.* a breed of sheep developed in the United States by cross-breeding from the Lincoln and Rambouillet breeds.
Targioniaceae *Botany.* a widespread family of typical liverworts of the order Marchantiales, characterized by purplish or brownish pigments, dictomous or ventral branches, simple pores in the epidermis, and a solitary sporophyte with a dehiscent capsule.
tarichatoxin *Toxicology.* a potent neurotoxin, identical to tetrodotoxin, that is produced in the epidermal glands of the California newt, *Taricha torosa.*
tariff *Industrial Engineering.* **1.** a government-imposed charge on imported goods. **2.** a list or table of charges of this kind. **3.** a document filed by a common carrier with a regulatory body or commission that defines the services which are offered, establishes a pay rate, and defines the general obligations of the customer and the common carrier. **4.** the actual rate charged by a common carrier for a service.
tarmac or **Tarmac** *Materials.* a bituminous substance used in paving. *Aviation.* an airport surface paved with this substance, especially a runway or an apron fronting a hangar. (A trade name for tar macadam.)
tarn see CIRQUE LAKE.
tarnish *Metallurgy.* **1.** a film of corrosion, usually made up of sulfides, that discolors the surface of a metal. **2.** of metal, to become discolored by such a film.
tarpon *Vertebrate Zoology.* the largest of the herringlike marine fishes of the family Megalopidae, up to 8 feet in length; found in the Atlantic and Indo-Pacific Oceans; the **Atlantic tarpon,** *Megalops atlantica,* is a popular U.S. game fish.
tarragon *Botany.* a bushy perennial plant, *Artemisia dracunculus,* of the family Compositae. *Food Technology.* the dried leaves of this plant, used to flavor vinegar and foods.
tarring *Engineering.* the process of coating an object or surface with tar.
tarry stools see MELENA.
tarsal *Anatomy.* **1.** relating to the tarsus of an eyelid or to the instep of the foot. **2.** any of the bones of the tarsus of the foot.
tarsal gland *Anatomy.* one of the sebaceous glands between the tarsi and the conjunctiva of the eyelid that empty onto the margin of the eyelid.
tar sand *Geology.* a bituminous oil sand or sandstone whose pores are filled with dense, viscous asphaltic residue, remaining after the lighter fractions of crude oil have escaped.
tar seep *Geology.* natural tar near the ground surface that flows slowly through cracks in the earth or from between rocks, frequently forming pits or pools of tar.
tarsier *Vertebrate Zoology.* any large-eyed, arboreal mammal of the genus *Tarsius* in the primate family Tarsiidae, being about the size of a small squirrel and having soft grayish fur, a tufted tail, and adhesive pads on the fingers and toes; native to the East Indies.

tarsier

Tarsiidae *Vertebrate Zoology.* a family of large-eyed arboreal primates of the suborder Haplorhini, characterized by adhesive pads on expanded ends of the fingers and toes. *Paleontology.* a proposed monogeneric family of primates in the order Tarsiiformes, based on the tentative placement of Afrotarsius with the Tarsiiformes; extant in the Oligocene.
Tarsonemidae *Invertebrate Zoology.* a family of small, soft-bodied, usually pale mites in the suborder Tromiformes, including plant and grain eaters and insect predators.
tarsoptosis *Medicine.* a falling of the tarsus; a fallen arch.
tarsus *plural,* **tarsi.** *Anatomy.* **1.** the region of articulation between the leg and foot; it is composed of seven bones. **2.** any of the plates of connective tissue forming the framework of an eyelid. *Invertebrate Zoology.* in many arthropods, the part of a leg farthest from the body, consisting of one or more segments.
Tartaglia, Nicolo [tär täl´yə] 1499?–1559?, Italian mathematician; developed methods for solving cubic and quartic equations; studied trajectories.
tartar *Medicine.* the popular name for dental calculus, a hard, rough deposit of food residue and various minerals collecting on the teeth; it is the calcified form of dental plaque. *Food Technology.* the lees, or sediment, in a wine cask, consisting of crude potassium bitartrate (cream of tartar).
Tartar *Ordnance.* a shipborne, surface-to-air missile with a solid-propellant rocket engine and nonnuclear warhead; designated **RIM-24.**
tartar emetic *Organic Chemistry.* $K(SbO)C_4H_4O_6 \cdot 1/2H_2O$, a crystalline solid that is soluble in water; used to attract and kill moths. Also, ANTIMONY POTASSIUM TARTRATE.
Tartarian see TATARIAN.
tartaric acid *Organic Chemistry.* $HOOC(CHOH)_2COOH$, a transparent, colorless crystalline solid or white crystalline powder; soluble in water, alcohol, and ether; melts at 170°C; occurs naturally in wine lees and also made synthetically; used in chemicals, baking powder, ceramics, photographic chemicals, tanning, mirror silvering, and metal coloring. Also, DIHYDROXYSUCCINIC ACID, THREOIC ACID.
tartary buckwheat *Botany.* one of the three major species of buckwheat, *Fagopyrum tataricum;* widely grown as a food crop.
tartrate *Organic Chemistry.* a salt or ester of tartaric acid.
tartrated *Organic Chemistry.* containing tartar or tartaric acid.
tartrazine *Organic Chemistry.* $C_{16}H_9N_4O_9S_2Na_3$, an allergenic, bright orange-yellow powder; soluble in water; used as a dye in foods, drugs, and cosmetics.
task *Computer Programming.* **1.** a unit of work for a central processing unit; e.g., loading a program or buffering output data. **2.** a program in a multiprogramming (multitasking) system.

task descriptor *Computer Programming.* in a multitask system, a description of a task at each point in its execution, including the task address space, an instruction pointer designating the next instruction to be executed, a protection code, and the current machine condition.

task fleet *Military Science.* a mobile command consisting of ships and aircraft necessary to accomplish a specific major task or tasks, either of a temporary or continuing nature.

task force *Military Science.* **1.** a temporary grouping of units under one commander to carry out a specific operation or mission. **2.** a semipermanent organization of units under one commander to carry out a continuing specific task. **3.** a component of a fleet organized to accomplish a specific task or tasks; further subdivisions are, in descending order of size: **task group, task unit, task element**; any of these, including a task force, may be termed a **task component.**

task management *Computer Programming.* the portion of an operating system that regulates the use of system resources (other than I/O devices) among current tasks.

task-oriented language *Robotics.* a computer language that is not tied to any particular processing system.

task program *Robotics.* a computer application program that directs a robot in carrying out a specific task.

task programmer *Computer Programming.* a person who designs, writes, and tests low-level routines.

Tasmania *Geography.* a large island (area: 26,215 square miles) off the southeastern coast of Australia.

Tasmanian devil *Vertebrate Zoology.* a powerful, medium-sized burrowing marsupial, *Sarcophilus ursinus*, characterized by a black coat with a white chest; native to Tasmania.

Tasmanian wolf *Vertebrate Zoology.* a rare, carnivorous, doglike marsupial, *Thylacinus cynocephalus*, of the family Thylacinidae; once common in Australia but now limited to Tasmania. Also, **Tasmanian tiger.**

tasmanite *Geology.* an impure coal that represents a transitional stage between cannel coal and oil shale. Also known by various other names, such as COMBUSTIBLE SHALE, YELLOW COAL, and WHITE COAL.

tassel *Botany.* a drooping, male inflorescence found especially in corn.

T association *Astronomy.* a type of stellar association that contains cool dwarf T Tauri stars embedded in dark nebulosity or bright nebulae; the prototype star, T Tauri, is irregularly variable with hydrogen emission lines in its spectrum.

tastant *Neurology.* any substance that stimulates the sense of taste or elicits gustatory excitation.

taste *Physiology.* **1.** the quality by which food, drink, and other substances are described when taken into the mouth; technically, the impression produced by stimulating the gustatory receptors in the tongue and oropharynx. **2.** the sense of perceiving food and other substances in this way. In Western cultures, the taste of something is generally described as being bitter, sweet, salty, or sour.

taste-aversion study *Behavior.* a type of experiment in which the subject learns to dislike a particular taste that is paired with a noxious stimulus.

taste blindness *Genetics.* the inability to taste certain substances used in genetic studies; e.g., individuals who are **taste blind** to phenylthiocarbamide (PTC) are homozygous for an autosomal recessive allele.

tastebud or **taste bud** *Anatomy.* any of numerous oval concentrations of sensory nerve endings; concentrated mostly on the tongue, but also located on the surfaces of the palate and pharynx.

taster *Food Technology.* a person who is skilled at or who is employed in tasting various substances for quality; e.g., wine, coffee, tea, cheese. *Genetics.* an individual who is capable of tasting a particular substance, such as phenylthiocarbamide (PTC); used in genetic studies.

TAT thematic apperception test; toxin-antitoxin; tyrosine aminotransferase.

TATA box *Molecular Biology.* a sequence that occurs upstream from the region of a eukaryotic promoter to which RNA polymerase II binds. Also, HOGNESS BOX, GOLDBERG-HOGNESS BOX.

Tatarian *Geology.* a European geologic stage of the Upper Permian period, occurring after the Kazanian and before the Scythian of the Triassic period. Also, CHIDERUAN, TARTARIAN.

T attenuator *Electricity.* a composite of three resistors with one end of each connected together forming a T-network, the free ends of two resistors are connected to an input and output terminal, while the free end of the third resistor is connected to the common input and output terminals.

tatting *Textiles.* a delicate knotted lace made by hand using a small shuttle.

Tatum, Edward Lawrie 1909–1975, American biochemist; awarded the Nobel Prize for showing how genes control chemical reactions (with George Beadle) and discovering transduction (with Joshua Lederberg).

Tatumella *Bacteriology.* a genus of Gram-negative, facultatively anaerobic, rod-shaped bacteria of the family Enterobacteriaceae; usually found in the human respiratory tract.

tau [tou] the nineteenth letter of the Greek alphabet, written T, τ.

Taube, Henry born 1915, American chemist; awarded the Nobel Prize for research on electron transfer in chemical reactions.

Tauber's theorem *Mathematics.* **1. Tauber's first theorem.** the statement that if $\sum_{n=1}^{\infty} a_n$ is Abel summable to S and if

$$\lim_{n\to\infty} a_n/n = 0$$

(the **Tauberian condition**), then $\sum_{n=1}^{\infty} a_n$ converges to S in the conventional sense. **2. Tauber's second theorem.** the statement that if $\sum_{n=1}^{\infty} a_n$ is Abel summable to S, then $\sum_{n=1}^{\infty} a_n$ is summable to S in the conventional sense if

$$\lim_{n\to\infty} \left(\sum_{k=1}^{\infty} na_n\right)/n = 0.$$

Tauber test *Analytical Chemistry.* a test for detecting pentose sugars by heating the sample with a solution of benzidine in glacial acetic acid; a cherry red color indicates a positive test.

Tauber trap *Biotechnology.* a cylindrical vessel with a specialized lid containing a central hole that is used to trap airborne microflora.

tau lepton *Particle Physics.* an elementary particle in the lepton group with a mass 3500 times that of an electron but with similar properties. Also, **tau, tauon.**

Taung child *Anthropology.* the first australopithecine to be recognized as an ancient descendent of humans; found in the 1920s by Raymond Dart at a quarry in Taung, South Africa.

tau protein *Biochemistry.* a protein associated with the assembly, stability, or enhancement of the polymerization of microtubules.

Taurids *Astronomy.* an annual meteor shower that reaches its peak activity in the first week of November and whose members appear to radiate from the constellation of Taurus.

taurine *Organic Chemistry.* $NH_2CH_2CH_2SO_3H$, a crystalline aminosulfonic acid; soluble in water and insoluble in alcohol; decomposes at about 300°C; found in bile combined with cholic acid, it is isolated from ox bile and used in biochemical research, pharmaceuticals, and wetting agents. Also, 2-AMINOETHANESULFONIC ACID.

tauro- a combining form denoting a relationship to a bull or to taurine.

taurocholic acid *Biochemistry.* $C_{26}H_{45}NO_7S$, a common bile acid that is the amide formed between cholic acid and taurine.

taurodont *Anatomy.* an abnormal molar tooth that has an elongate body and a shortened root.

taurodontism *Anthropology.* the tendency for the molar teeth to have large pulp cavities; a characteristic of Classic Neanderthal populations.

Taurus [tôr´əs] *Astronomy.* the Bull, a zodiacal constellation of northern autumn and winter that contains the bright star Aldebaran and the famous Pleidaes and Hyades star clusters.

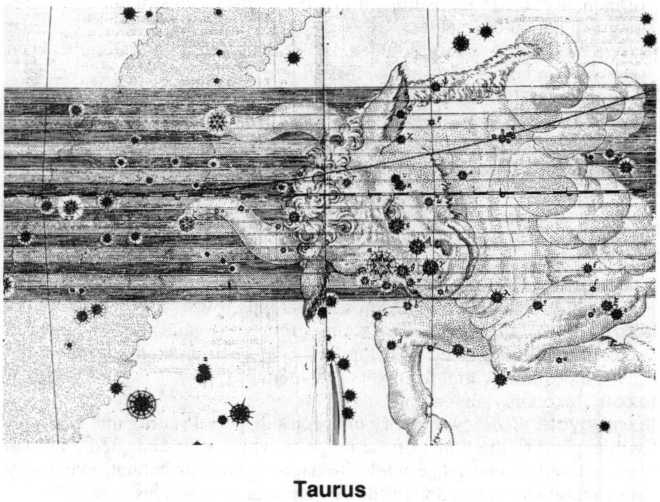

Taurus

Taurus dark cloud *Astronomy.* an aggregate of dust and gas involved in star formation that lies at a distance of approximately 500 light-years, with many T Tauri stars and Herbig-Haro objects.

Taussig, Helen Brooke 1898–1986, American physician; discovered cause of blue baby syndrome and helped develop corrective surgery.

taut-line cableway *Mechanical Engineering.* a cableway with one carrier that operates on a wire cable stretched between two towers not more than 3000 feet apart; the traction can be reeved at the carrier so that loads can be raised and lowered.

tauto- a combining form meaning "the same."

tautological [tô tə läj′i kəl] *Science.* **1.** relating to or being a tautology. **2.** needlessly repeating the same idea or argument.

tautology [tô täl′i jē] *Science.* **1.** a logical formula that is always true; i.e., a logic formula whose value is true for all possible assignments of truth value to its prime components. Also, VALID FORMULA. **2.** an argument that is true only because it restates the original premise with different wording; e.g., "Species survive because they are fit. Fitness is the ability to survive."

tautomer *Chemistry.* any chemical compound that is capable of exhibiting tautomerism.

tautomerism *Chemistry.* an isomerism in which the migration of a hydrogen atom gives rise to two or more structures that spontaneously convert from one form into the other with such ease that they usually exist together in equilibrium. Thus, **tautomeric.**

tautomerization *Chemistry.* the process of undergoing tautomerism. *Molecular Biology.* in DNA, the movement of a hydrogen atom in a purine or pyrimidine base from one position to another, which alters the base-pairing specificity of the affected base and can lead to mutations during replication.

tau-value *Meteorology.* the time rate of change of *D*-values at a fixed point, as defined by the relation $\tau = \Delta_t D/\Delta t$, where Δt is the change in time, and $\Delta_t D$ is the change in *D*-value during this time. Also, **τ-value.**

taxa the plural of *taxon.*

Taxaceae *Botany.* a family of primarily North American nonresinous evergreen shrubs and trees of the order Pinatae, characterized by spirally arranged needlelike or linear leaves, scaly staminate cones that occur solitarily or in spikes, and single or paired ovules.

Taxales *Botany.* an order of evergreen shrubs and trees, characterized by solitary ovules surrounded by an aril and borne on short terminal shoots.

taxi *Transportation Engineering.* also, **taxicab.** an automobile that hires out to carry passengers, usually at a metered rate per distance and time. *Aviation.* **1.** to drive an aircraft on the ground or other surface under its own power, except during takeoff and landing runs. **2.** the movement occurring in such an operation.

taxi run *Aviation.* a high-speed run of an aircraft on the ground, and possibly a very short distance off the ground, in order to test its performance.

taxis *Physiology.* the development or motion produced by a plant or animal in response to a designated physical stimulus, such as temperature, light, nearness to a surface, etc. The response can be positive, as when the organism grows or moves toward the source of the stimulus, or negative, as manifested in withdrawal from the stimulus. *Surgery.* the use of manual force to reposition an injured or displaced structure, as in reducing a hernia or fracture.

taxiway *Aviation.* a path or other surface designated for taxiing between a runway and other parts of an air field.

Taxocrinida *Paleontology.* an order of echinoderms in the extinct subclass Flexibilia; characterized by a conical cup (1 centimeter across) and elongate crown; extant from the Ordovician to the Carboniferous.

Taxodiaceae *Botany.* a family of tall, deciduous or evergreen trees of the order Pinatae, characterized by spirally arranged leaves, clustered or solitary staminate cones with spirally arranged scales, and solitary pistillate cones of spirally arranged scales; contains some of the oldest and tallest conifers.

Taxodonta *Invertebrate Zoology.* an order or subclass of bivalve mollusks with shell teeth of uniform size in a long row.

taxon *plural,* **taxa.** *Systematics.* any group of organisms that has been scientifically designated as belonging to a specific taxon, and has been given a position within the taxonomic hierarchy.

taxon. taxonomy; taxonomic.

taxon cycle *Ecology.* a theory of species dispersal stating that a species will expand to the limits of the range to which it has adapted and then be confined to that range while it adapts to another habitat, eventually speciating; an explanation of the evolution of new species.

taxonomic [tak′sə näm′ik; tak′sə nōm′ik] *Systematics.* of or relating to taxa or taxonomy. Also, **taxonomical.**

taxonomic rank *Systematics.* the position of a taxon in a hierarchy of categories. Also, **taxonomic position.**

taxonomist [tak sän′ə mist] *Systematics.* a specialist in taxonomy.

taxonomy [tak sän′ə mē] *Systematics.* the theories and techniques of describing, naming, and classifying organisms. Also, **taxology.** *Artificial Intelligence.* in a frame system or object-oriented system, the tree structure formed by the class-superclass relations.

Taxopodida *Invertebrate Zoology.* an order of bilaterally symmetrical, small biflagellate, planktonic marine protozoa of the class Heliozoa, having siliceous spines and swimming by the rowing action of axopodia arranged in parallel longitudinal rows.

taxospecies *Systematics.* a species determined solely by the application of numerical taxonomic methods.

Tayassuidae *Vertebrate Zoology.* the peccaries, a family of nonruminant mammals of the order Artiodactyla.

Taylor, Brook 1685–1731, English mathematician; formulated calculus of finite differences, Taylor's series, and Taylor's theorem.

Taylor, Frederick Winslow 1856–1915, American engineer; called the father of scientific management; developed time and motion study.

Taylor, Sir Geoffrey 1886–1975, English mathematician; worked in theoretical hydrodynamics; designed giant anchors for D-Day.

Tayloran *Geology.* a North American Gulf Coast stage of the Upper Cretaceous period, occurring after the Austinian and before the Navarroan.

Taylor effect *Fluid Mechanics.* the phenomenon that the relative motion of a homogeneous rotating liquid tends to be the same in all planes perpendicular to the axis of rotation; a consequence of the Taylor-Proudman theorem.

Taylor instability *Fluid Mechanics.* the formation of rolls generated in a column of fluid bounded by differentially rotating walls; governed by the Taylor number.

Taylor number *Fluid Mechanics.* a dimensionless number used in problems associated with a rotating-cylinder viscometer and a viscous fluid: $Ta = r_i w^2 h^3/v^2$, where r_i is the radius of the inner cylinder, w is the rate of rotation, h is the depth of the fluid, and v is the kinematic viscosity.

Taylor-Orowan dislocation see EDGE DISLOCATION.

Taylor process *Metallurgy.* a method of manufacturing fine wire in which wire is inserted through a glass tube, and both tube and wire are elongated under high temperature.

Taylor-Proudman theorem *Fluid Mechanics.* in an incompressible rotating fluid, vorticity is conserved on cylinders; an obstruction generates a vertical cylindrical circulation pattern called a Taylor column.

Taylor's series *Naval Architecture.* a set of hull resistance tables based on model tests in which individual dimensions have been modified, feature by feature, to provide relative performance values for a series of related hull forms. *Mathematics.* given a function f that is analytic in a neighborhood of a (real) point x_0, a Taylor series for f is the infinite series representation of f at points x in some (possibly smaller) neighborhood of x_0, given by:

$$f(x) = \sum_{k=0}^{\infty} f^{(k)}(x_0)\,(x - x_0)^k/k!.$$

Also called the **Taylor expansion** about x_0. In the case $x_0 = 0$, it is called the **Maclaurin series** representation of f. For example, the function $f(x) = (1 + x^2)^{-1}$ is analytic everywhere on the real line, but the Taylor (Maclaurin) series expansion about 0 is $1 - x^2 + x^4 - x^6 + x^8 - \ldots$, which converges only for $|x| < 1$.

Taylor's theorem *Mathematics.* suppose the function f is continuous in a neighborhood N of a point x_0, and that the first $n + 1$ derivatives of f exist in N. Then for any $x \in N$, there is a **Taylor expansion** for $f(x)$; that is,

$$f(x) = \sum_{k=0}^{n} f^{(k)}(x_0)\,(x - x_0)^k/k! + R_n(x).$$

Lagrange's form of the remainder is

$$R_n(x) = f^{(n+1)}(x_0 + h(x - x_0))\,(x - x_0)^{n+1}/(n+1)!,$$

for a number $h \in (0,1)$. If f is infinitely differentiable and $\lim_{n\to\infty} R_n(x) = 0$, then f can be represented by its Taylor series. The case $n = 0$ is the Mean Value Theorem. When $x_0 = 0$, the Taylor expansion is called **Maclaurin's formula.**

Taylor theory *Materials Science.* a theory of strain hardening in polycrystalline microstructures, predicting that the flow stress is inversely proportional to the average dislocation spacing, with the assumption that dislocation sources are distributed uniformly over the entire grain-boundary area.

Tay-Sachs disease *Medicine.* a genetic disorder of lipid metabolism that affects the brain and nerves, occurring almost exclusively among people of eastern European Jewish ancestry; symptoms appear by 6 months of age and result in progressive mental and physical retardation, convulsions, blindness, and finally death between 2 and 5 years of age. (Named for Warren *Tay,* 1843–1927, English physician, and Bernard *Sachs,* 1858–1944, American neurologist.)

tb tablespoon; tablespoonful.

Tb the chemical symbol for terbium.

TB tuberculosis; tubercle bacillus; trial balance.

T-beam *Civil Engineering.* a composite beam of steel rebar and concrete, usually used in slab floors where the horizontal parts of the T act as part of the floor.

TBG thyroxine-binding globulin.

Tbilisi phage *Virology.* a phage of the family Podoviridae that replicates in various strains of *Brucella abortus.*

T bolt *Mechanical Devices.* a clamping bolt whose head is T-shaped in cross section, mounted into a similarly shaped slot of a machine tool table or a drill head.

TBP tributyl phosphate.

tbs tablespoon; tablespoonful. Also, **tbsp.**

TBW total body weight.

tc tierce.

Tc the chemical symbol for technetium.

TCA trichloroacetic acid.

TCBM transcontinental ballistic missile.

T cell see T LYMPHOCYTE.

T-cell replacing factor *Immunology.* a cell derived from the thymus that is needed by B lymphocytes to produce antibodies, when T lymphocytes (T cells) are damaged or lacking.

tchitola *Materials.* the decorative, reddish-brown wood of the African *Oxystigma oxyphyllum* tree; used for interior decoration. Also, TOLA.

Tchuprow-Neymann allocation *Statistics.* a method of stratified sampling in which the size of the sample from each stratum is proportional to the stratum population times the stratum standard deviation; used to achieve unbiased estimates of parent mean values while maintaining minimum variance.

T circulator *Electromagnetism.* a T-shaped waveguide structure whose branches are identical; power enters the junction from one of the three branches, then exits through only one of the other two, due to the use of a ferrite post or wedge at the center.

TCMI T-cell-mediated immunity.

TCP tool center point; tricresyl phosphate.

Td tetanus and diphtheria toxoids, adult type.

t distribution see STUDENT'S *t* DISTRIBUTION.

TDN total digestible nutrients.

T-DNA *Genetics.* a group of genes that are always found in crown gall disease cells of plants, and that become integrated into the nuclear DNA of the host plant when it is infected with the Ti plasmid during tumor induction.

t.d.s. to be taken three times a day. (From Latin *ter die sumendum.*)

TDY temporary duty.

Te tetanus.

Te the chemical symbol for tellurium.

tea *Botany.* an evergreen tree, *Thea sinensis,* of the family Theaceae, widely cultivated in Asia and elsewhere for its leaves. *Food Technology.* **1.** a popular beverage obtained by brewing the dried leaves of this plant. **2.** a similar beverage obtained from various other plants.

TEA tetraethylammonium; triethnolamine.

TEAC tetraethylammonium chloride. Also, **TEA chloride.**

teach *Robotics.* to program a robot by guiding it through a set of motions that are then recorded and stored for future use.

teach-by-doing *Robotics.* a method of programming in which an operator physically guides the manipulator through its intended motions. Also, **teach-through.**

teach-by-driving *Robotics.* the process of using a teach pendant to teach a robot how to carry out a task.

teaching *Behavior.* any behavior that tends to alter the behavior pattern of another individual of the same species toward some standard of performance. *Robotics.* the process of programming a robot.

teaching interface *Robotics.* devices and hardware, such as the teach pendant, that are used to instruct a robot and take it through the proper movements to perform some task.

teach mode *Robotics.* a mode of operation in which a robot is led through its motions with a teach pendant.

teach pendant *Robotics.* a handheld device used to teach a robot by guiding it manually through a series of points or patterns. Also, **teach box, teach gun.**

teallite *Mineralogy.* $PbSnS_2$, a grayish-black orthorhombic mineral occurring in masses of thin, flexible folia, having a specific gravity of 6.36 and a hardness of 1.5 on the Mohs scale; found in tin veins.

tear [tēr] *Physiology.* also, **tears.** a watery, slightly alkaline and saline secretion of the lacrimal glands that serves to moisten the conjunctiva. *Materials.* a small, naturally formed droplike mass of a gum or resin.

tear [târ] *Engineering.* to pull apart or to pieces by force. *Medicine.* **1.** to wound or injure, especially by ripping apart; lacerate. **2.** a laceration.

tear-down *Engineering.* the process of dismantling a machine setup or drilling rig for relocation or conversion to another job. Thus, **tear-down time.**

teardrop balloon *Aviation.* a sounding balloon that resembles an inverted teardrop when operationally inflated, a shape designed to improve stability in ascension.

tear fault *Geology.* a steeply inclined to vertical fault in the hanging wall of a low-angle overthrust fault whose strike is perpendicular to that of the overthrust fault.

tear gland see LACRIMAL GLAND.

tear strength *Mechanics.* the force necessary to initiate or continue the tearing of a sheet or fabric.

tear test *Materials Science.* any destructive evaluation of a sheet material, especially paper; a measure of the energy expended in the controlled, relatively slow propagation of a failure line.

teaspoon *Metrology.* a unit of measure equal to one-third tablespoon. Also, **teaspoonful.**

teat *Agriculture.* the small protuberance on the udder of a cow or other milk-giving animal, through which the milk passes. *Biotechnology.* a rubber bulb that is compressed by hand to create suction and draw liquids into a pipet.

TeBG testosterone-estradiol-binding globulin.

tech. technical; technology; technician.

technetium *Chemistry.* an element having the symbol Tc, the atomic number 43, an atomic weight of its isotope with longest half-life (^{98}Tc) of 97.90; it does not occur naturally on earth but has been detected in stars and obtained by deuteron bombardment of molybdenum and in fission products of uranium and plutonium; used as a metallurgical tracer and in cryochemistry and nuclear medicine. (Related to the word *technical;* because it is artificially obtained.)

technic **1.** see TECHNIQUE. **2.** see TECHNICS.

technical *Science.* of or relating to an art, science (especially an applied science), trade, profession, method, or technique. *Engineering.* relating to the engineering or operational aspects of a device or system, as opposed to its physical makeup or design. Thus, **technical characteristics, technical evaluation, technical information, technical inspection, technical maintenance, technical specifications,** and so on.

technical atmosphere *Mechanics.* a unit of pressure equal to 1 kilogram-force per square centimeter.

technical cohesive strength *Metallurgy.* the fracture stress in a notch tensile test.

technical escort *Military Science.* individuals who are technically qualified and properly equipped to accompany designated material requiring a high degree of safety and security during transport.

technical intelligence see SCIENTIFIC AND TECHNICAL INTELLIGENCE.

technical sergeant *Military Science.* in the U.S. Air Force, a noncommissioned officer ranking above a staff sergeant and below a master sergeant.

technical survey *Military Science.* a complete electronic and physical inspection to ascertain that a location where classified information is discussed is free of monitoring systems.

technician *Engineering.* a person trained in and expert in a given applied field, or in the performance of one or more mechanical processes.

Technicolor *Graphic Arts.* the trade name for a process of making color films by superimposing each of the primary colors onto a separate black-and-white negative; widely used in American commercial films from the 1930s to the 1950s. *Particle Physics.* also, **technicolor.** a proposed theory that claims the existence of a new force to explain the breakdown of symmetry in the unified electroweak interaction.

technics *Engineering.* the science or study of the mechanical and industrial arts.
technihadron *Particle Physics.* any of a class of hypothetical particles bound together in the same manner that the color force binds quarks, antiquarks, and gluons into hadrons.
technique *Science.* **1.** the method or procedure in any mechanical process, surgical operation, and so on. **2.** the body of procedures, methods, and operations used in a given field.
techno- a combining form meaning "technique," "technical," or "technology."
technoeconomic base *Anthropology.* the concept that all human societies have a particular strategy for using tools and obtaining food in their particular environment; a means of classifying societies without the idea of progress inherent in earlier evolutionary definitions.
technofact *Archaeology.* an object that is used for a practical function, such as providing food, shelter, or defense, rather than to express social standing or ideological beliefs.
technofunction *Archaeology.* the use of an object for practical purposes; for example, the wearing of a certain garment simply to provide warmth or protection.
technofunctional *Archaeology.* of or relating to technofunction; used in a practical way. Also, **technomic.**
technol. technology; technological.
technological *Science.* relating to or involving the use of technology. *Anthropology.* designating a society that practices modern industrial techniques.
technological attribute *Archaeology.* a characteristic of an artifact that is directly the result of how it was made or the raw material which formed it.
technologist *Science.* a person who applies scientific information for practical uses, for example, in industry.
technology *Science.* the application of scientific knowledge for practical purposes; the employment of tools, machines, materials, and processes to do work, produce goods, perform services, or carry out other useful activities. *Anthropology.* more generally, any use of objects by humans to do work or otherwise manipulate their environment.
technosphere *Ecology.* a collective term for all those aspects of the physical environment that have been created or altered by humans.
tecnetron *Electronics.* a type of field-effect transistor having high-power applications.
Tectibacter *Bacteriology.* a genus of Gram-negative, flagellated, rod-shaped bacteria of uncertain affiliation that occur as endosymbionts in strains of *Paramecium aurelia.*
Tectibranchia *Invertebrate Zoology.* a suborder of gastropod mollusks in the order Opisthobranchia, having a large head and a thin shell embedded in the mantle, or no shell; includes the bubble shells, sea slugs, and sea hares.
tectite see TEKTITE.
Tectiviridae *Virology.* a family of lipid-containing bacteriophages having large isometric particles that contain a linear dsDNA genome.
Tectivirus *Virology.* the single genus of phages of the family Tectiviridae.
tectofacies *Geology.* a facies distinguished from laterally equivalent strata on the basis of its tectonic aspects.
tectogene *Geology.* **1.** a long, deep, narrow, folded or faulted downwarp of the earth's crust, believed to be related to mountain-building processes. Also, GEOTECTOGENE. **2.** the portion of an orogene that is downfolded. Also, DOWNBUCKLE.
tectogenesis see OROGENY.
tectonic *Geology.* relating to or resulting from tectonics.
tectonic axis see FABRIC AXIS.
tectonic breccia *Petrology.* a breccia formed from brittle rocks by crustal tectonic movement and pressure planes. Also, PRESSURE BRECCIA.
tectonic cycle see OROGENIC CYCLE.
tectonic fabric see DEFORMATION FABRIC.
tectonic framework *Geology.* **1.** the interplay in space and time of waning, stable, and increasing deformational movements of the earth's crust in a sedimentary source area. **2.** the structural elements of a region.
tectonic gap see LAG FAULT.
tectonic land *Geology.* any land feature, particularly linear fold ridges and volcanic islands, that existed temporarily in the interior of an orogenic belt during the early stages of mountain-building activity.
tectonic lens *Geology.* a long, cylindrical body of rock produced by the distortion of a continuous incompetent layer encased between competent layers.
tectonic map *Geology.* a map that displays the positions, attitudes, and configurations of the upper part of the earth's crust, including some indication of its composition and history.
tectonic moraine *Geology.* an accumulation of boulders that has been incorporated into the base of an overthrust mass.
tectonic patterns *Geology.* the configuration of the primary structural units of the earth's crust, such as mountain systems, basins, and arches.
tectonic plate *Geology.* see PLATE.
tectonic profile *Geology.* see PROFILE.
tectonic rotation *Geology.* a fine-scale internal rotation occurring in a tectonite in the direction of transport.
tectonics *Geology.* **1.** a branch of geology that is concerned with the study of the features, deformational movements, and processes of the earth's crust, including their mutual relationships, origins, and historical development. **2.** see PLATE TECTONICS. (From a Greek word meaning "building" or "construction.") Also, GEOTECTONICS.
tectonite *Petrology.* any rock whose deformation history is revealed in its fabric.
tectonoeustasy *Oceanography.* tectonic movements of the earth's crust that result in changes in the shape and capacity of ocean basins, causing changes in sea level. Also, **tectonoeustatism.**
tectonomagnetism *Geophysics.* the study of magnetic anomalies caused by the earthquakes.
tectonometer *Engineering.* a surface probe containing a microammeter, used to measure the characteristics of underlying rock layers.
tectonophysics *Geophysics.* the study of the forms, movements, and other related phenomena dealing with the earth's crust.
tectoquinone *Organic Chemistry.* $C_{15}H_{10}O_2$, yellow needles; soluble in alcohol, benzene, and acetic acid; melts at 177–179°C; used as an insecticide. Also, 2-METHYLANTHRAQUINONE.
tectorial membrane *Anatomy.* **1.** a ligamentous connection between the occipital bone and the bodies of the second and third cervical vertebrae. **2.** a gelatinous membrane overlying the organ of corti in the cochlea.
tectosilicate *Mineralogy.* any of a group of silicate minerals, such as quartz or feldspar, having a structure in which the SiO_4 tetrahedra share all four oxygen atoms with neighboring tetrahedra to form a three-dimensional network with a Si:O ratio of 1:2.
tectosome *Geology.* a body of strata that represents an area characterized by a uniform tectonic environment.
tectosphere *Geology.* the outer shell or layer of the earth's crust, consisting of lithospheric plates, in which crustal or tectonic movements originate.
tectum *Anatomy.* **1.** the corpora quadrigemina of the midbrain. Also, CORPORA QUADRIGEMINA. **2.** any rooflike or covering structure.
tedder *Agriculture.* a machine that stirs and loosens a swath of hay in order to speed up the drying process.
tee *Engineering.* a plumbing fitting having a T-shaped configuration, with a branch extending at right angles to a line or pipe.
tee junction see T JUNCTION.
teeming *Materials Science.* the process of shaping a material, such as glass or steel, by pouring it into a mold, table, or roll.

teeming steel

teepleite *Mineralogy.* $Na_2B(OH)_4Cl$, a white, transparent tetragonal mineral occurring as tabular crystals, having a specific gravity of 2.076 and a hardness of 3 to 3.5 on the Mohs scale; found in tin veins.

teeth the plural of tooth.

Teflon *Materials.* a trade name for a polymer of polytetrafluoroethylene, characterized by extreme chemical inertness, withstanding the attack of all reagents except molten alkali metals; a tough, heat-resistant fluorocarbon resin used in packing, bearings, filters, electrical insulation, cooking utensils, and plumbing sealants.

TEG triethylene glycol.

tegafur *Oncology.* a cancer chemotherapy agent now under investigation; it is similar to 5-fluorouracil in structure and potential uses.

tegmen *Botany.* a thin coating that surrounds the embryo in a seed. *Invertebrate Zoology.* **1.** the leathery forewing of orthopteran and some homopteran insects such as grasshoppers. **2.** the leathery upper surface around the mouth of a crinoid echinoderm.

tegmentum *Botany.* the outer scaly coat of a leaf bud, or one such scale. *Anatomy.* **1.** the portion of the midbrain ventral to the cerebral aqueduct and covering the substantia nigra. **2.** any rooflike or covering structure. *Invertebrate Zoology.* the upper calcareous layer of the plate-shells of crustaceans in the class Amphineura.

tegula *Invertebrate Zoology.* either of a pair of small thoracic sclerites that cover the bases of the wings in Lepidoptera and Hymenoptera.

teichoic acid *Biochemistry.* a polymer formed from ribitol or glycerol phosphate and other compounds, such as glucose; found in the cell walls of certain bacteria.

Teiidae *Vertebrate Zoology.* a large family of active, diurnal lizards of the suborder Scincomorpha, characterized by a flat, elongate scaly tongue ending in two points; found primarily in the American tropics.

teineite *Mineralogy.* $Cu^{+2}Te^{+4}O_3 \cdot 2H_2O$, a deep-blue orthorhombic mineral occurring as aggregates of prismatic crystals, having a specific gravity of 3.80 and a hardness of 2.5 on the Mohs scale; found in quartz-chalcedony and barite veins at the Teine Gold Deposit, Japan.

TEIRESIAS *Artificial Intelligence.* a knowledge-acquisition aid developed to help create and revise rules and to provide an expanded explanation feature for a specific expert system while using the EMYCIN shell. (Named for *Teiresias* (Tiresias), a blind prophet in Greek mythology and drama.)

Teisserenc de Bort, Léon Philippe 1855–1913, French meteorologist; identified and named the troposphere and stratosphere.

tektite *Geology.* a body of nonvolcanic silicate glass believed to be of extraterrestrial origin, occurring in various widely separated regions of the earth's surface and bearing no relation to the surrounding geologic features; characterized by a rounded form and jet-black to greenish or yellowish coloration. Also, TECTITE, OBSIDIANITE.

telain see PROVITRAIN.

telalgia see REFERRED PAIN.

telamon *Invertebrate Zoology.* in various male nematode worms, part of the copulatory apparatus.

telangiectasis *Medicine.* the permanent dilation of superficial capillaries, arterioles, and venules, creating small focal red lesions, usually in the skin or mucous membranes; causes include actinic damage, hereditary vascular anomaly, elevated estrogen levels, and vascular disease.

telangiectatic fibroma see ANGIOFIBROMA.

telautograph *Telecommunications.* in telegraphy, a system by which handwriting can be transmitted over wires; a hand-held pen at the transmitting position varies the current in two circuits in such a manner as to promulgate corresponding movements of an automatically controlled pen at a receiving unit.

tele- or **tel-** a combining form meaning "distant," often denoting transmission over a long distance.

telecardiography *Medicine.* the recording of an electrocardiogram by transmission of impulses to a site at a distance from the patient. Thus, **telecardiogram.**

telecast *Telecommunications.* a television broadcast.

telecentric system *Optics.* a telescope system in which the stop is located at one of the foci of the objective lens.

teleceptor *Physiology.* a sense organ receptive to objects distant from the body; it may be stimulated by sound waves, light waves, molecules evoking the smell response, and so on.

telechir *Robotics.* a remote manipulator that works like a human hand.

telechirics *Robotics.* the use of remote manipulators or teleoperators.

telecine camera *Electronics.* a television camera that converts visible images from a motion-picture projector or slide projector into video signals for broadcasting by a television transmitter.

telecommunications *Electronics.* **1.** any process or group of processes that allows a correspondent to relay printed or written matter, moving or fixed pictures, or other visible or audible signals via an electromagnetic system. **2.** the study or body of knowledge relating to such processes. **3.** broadly, any communications over long distances.

Telecommunications

Modern telecommunications began in 1876 with the invention of the telephone by Alexander Graham Bell. Crude electromagnetic devices converted voice to and from electrical signals that could be sent over copper wires from place to place. Initially, human operators enabled manual selection of destinations, a switching function that became the heart of telephone networks. Electromagnetic relay switches of increasing sophistication were developed, but since 1947 when Bell Telephone Laboratories invented the transistor, which in turn spawned the integrated circuit, switches have become electronic with steadily increasing speed.

In the underlying technologies, analog representations of information have been rapidly giving way to digital, offering higher quality and performance. In transmission, the development of the optical fiber provides a high-speed replacement for copper wire. And software has become all-pervasive for operating the switches, the networks, and the operations support systems. A revolution is also occurring in the customer premises equipment—from the simple telephone to multi-function terminals, personal computers, and work stations.

Increasingly, because of advances in computers and computer networking, the software intelligence for the management of the communications not only resides in the network but is distributed among the terminals as well.

The trends to higher-speed capabilities are also leading to the possibility of ubiquitous image and video networking services. Another recent, rapidly growing trend is the use of portable radio with cellular networks for mobile use, automobile or pedestrian, offering great convenience to the user.

These technologies are increasingly available to anyone so that different networks are proliferating, both private and public. The growing number of network interfaces poses severe challenges to the technologists seeking to facilitate universal networking services; the current direction is toward "intelligent" networks offering great flexibility, even enabling users to design their own services.

A major paradigm shift is clearly under way in telecommunications, from voice to information, from Universal (Voice) Service to Information Networking, where people and their machines will be able to communicate with each other, securely and easily, anywhere, any time, in any medium—voice, data, image, or video—and at acceptable cost.

Overall, the societal impact of telecommunications is incalculable. It is a major technological element pacing the development of global society.

Allan G. Chynoweth
Vice President, Applied Research
Bellcore (Bell Communications Research)

teleconference *Telecommunications.* a conference, often a business meeting, between persons in two or more different locations who are connected by a telecommunications system. Thus, **teleconferencing.**
telecurietherapy *Radiology.* treatment with a radioactive source, such as radium, located at a distance from the body.
Telegeusidae *Invertebrate Zoology.* the long-lipped beetles, a family of coleopteran insects in the superfamily Cantharoidea.
telegony *Genetics.* a former belief that the genes from a male that previously mated with a female could influence the offspring resulting from subsequent matings with other males.
telegram *Telecommunications.* **1.** a message transmitted by telegraph. **2.** to send a message by telegraph.
telegraph *Telecommunications.* **1.** a device designed to transmit information over a distance by means of electricity, using Morse code or another code. In its most basic form, it consists of a key to open and close the circuit and a battery-powered sounder to form the code. The telegraph was the first device to allow immediate communication beyond the range of human sight and hearing, and was the dominant form of such technology in the second half of the 19th century. Modern forms of communication such as the teletypewriter and the FAX (facsimile) machine are adaptations of the telegraph system. **2.** to send a message by means of a telegraph device. (From a Greek term meaning "distant writing.")
telegraph bandwidth *Telecommunications.* in telegraphy, the difference between the minimum and maximum limiting frequencies of a channel utilized to send signals.
telegraph buoy *Engineering.* a buoy marking the location of an underwater telegraph cable.
telegraph cable *Electricity.* a uniformly conductive circuit that is used to carry telegraph signals; it consists of twisted pairs of insulated wires, coaxially shielded wires, or combinations of each.
telegraph carrier *Telecommunications.* in telegraphy, a frequency that is modulated for sending signals.
telegraph circuit *Telecommunications.* in a telegraph system, the complete radio or wire circuit over which signal currents move between transmitting and receiving devices.
telegraph code *Telecommunications.* in telegraphy, a system of symbols used for transmitting messages; each character is represented by a group of long and short electrical pulses, pulses of opposing polarities, or by time gaps of equal length in which a signal is either present or absent. Also, **telegraph alphabet.**
telegraph concentrator *Electricity.* a switching configuration that operates by means of a number of branch or subscriber lines or station sets that are connected to a smaller number of trunk lines, operating positions, or instruments through an automatic or manual switching device system to obtain more efficient use of facilities.
telegraph distributor *Electricity.* a device that effectively associates one direct current or carrier telegraph channel in rapid succession with electronic components defined by one or more signal transmitting or receiving devices.
telegraph emission *Telecommunications.* a signal sent by a telegraph system.
telegraph equation *Mathematics.* a partial differential equation having the form

$$(\partial^2 u/\partial t^2) = c^2(\partial^2 u/\partial x^2) - r(\partial u/\partial t) - ku + q(x,t),$$

where the constants satisfy $c > 0$, $r \geq 0$, $k \geq 0$ (but r and k are not both 0). The equation may be interpreted as governing the small vibrations of a string under the action of a damping force $r(\partial u/\partial t)$, a restoring force $-kt$, and an applied force $q(x,t)$.
telegraph grade *Telcommunications.* the class of communication circuits that are capable of sending only telegraphic signals; consists of the lowest level of circuits with regard to accuracy, cost, and speed.
telegraphic *Telecommunications.* relating to or sent by a telegraph.
telegraph interference *Telecommunications.* any unwanted electrical energy that hinders the reception of telegraph signals.
telegraph receiver *Electricity.* the conversion of telegraph signals into a pattern of holes on a tape, printed letters, or other data elements when used in conjunction with a reperforator, teletype printer, or similar equipment.
telegraph repeater *Electricity.* an arrangement of apparatus and circuitry that receives telegraph signals from one line and retransmits corresponding signals to another line; the signals are amplified with a repeater inserted in intervals on long telegraph lines.
telegraph signal distortion *Telecommunications.* in telegraphy, the time displacement of transitions between marking and spacing conditions, as measured against their correct relative positions in perfectly timed signals.
telegraph transmitter *Electricity.* a device used to control a source of power so as to allow the formation of telegraph signals.
telegraphy [tə leg′rə fē] *Telecommunications.* **1.** the use of a telegraph system; communication over a distance by means of code signals that are composed of electrical or electromagnetic pulses and that are sent over wires or by radio. **2.** the technology associated with this form of communication.
telekinesis see PSYCHOKINESIS.
telemagmatic *Geology.* describing a hydrothermal mineral deposit that is located at a considerable distance from its magmatic source.
telemeteorograph *Engineering.* a meteorological instrument or station that provides its data readout to a location remote from the sensing instruments.
telemeteorography *Engineering.* the design and operation of telemeteorographs.
telemeteorometry *Meteorology.* the process and methods of making remote meteorological measurements.
telemeter *Engineering.* **1.** an instrument that provides its data readout to a location remote from the sensing device. Also, **telemetry system, telemetering system.** **2.** to transmit data to a remote point.
telemetering *Engineering.* the operation of instruments that provide data readout to a location remote from their sensing devices.
telemetering antenna *Electromagnetism.* a directional antenna controlled by a servosystem; used for tracking and receiving telemetering signals from guided missiles or spacecraft.
telemetering receiver *Electronics.* a radio receiver and associated equipment used as part of a telemetry system.
telemetering transmitter *Electronics.* a radio transmitter and associated equipment used as part of a telemetry system.
telemetering wave buoy *Engineering.* a buoy that senses wave motions and transmits that data to a remote recording or observation location.
telemetry [tə lem′ə trē] *Engineering.* the use of telemeters for measuring and transmitting data.
telemotor *Engineering.* a hydraulic, mechanical, or electric system of remote control, especially such a system actuating a ship's rudder.
telencephalization *Neurology.* during the evolutionary process, the transfer to the telencephalon of the control of the more complex nerve reactions and motor functions.
telencephalon *Developmental Biology.* the anterior of the two vesicles formed from the prosencephalon that gives rise to the cerebral cortex, olfactory lobes, and corpora striata.
teleneurite *Neurology.* the slender end expansion of a nerve fiber.
teleneuron *Neurology.* **1.** in general, any nerve ending. **2.** specifically, the end neuron or neuronal process in a reflex arc or pathway that involves several neurons.
teleo- a combining form meaning: **1.** perfect. **2.** complete.
teleology [tē′lē äl′ə jē] *Science.* the study of natural phenomena in terms of an overall purpose or design.
teleomorph *Mycology.* the sexual or so-called perfect growth stage or phase in fungi.
teleoperator *Robotics.* a device that is controlled by a human operator through sensors and actuators to function in remote and hazardous environments.
Teleosauridae *Paleontology.* a family of aquatic crocodilians in the suborder Mesosuchia; characterized by forelimbs half as long as the hind limbs, but otherwise similar to their terrestrial relatives; extant in the Jurassic and Lower Cretaceous.
Teleostei *Vertebrate Zoology.* in some classification systems, an infraclass of the subclass Actinopterygii, evolutionarily advanced ray-finned fishes that possess cycloid scales, a homocercal caudal fin, and a swim bladder; it includes most species of bony fishes.
telepathy [tə lep′ə thē] *Psychology.* the supposed ability to have direct communication from one person's mind to another, without the use of speech or other ordinary forms of communication.
telephone *Telecommunications.* a device that converts human speech into variations of electric current for the purpose of transmission by line, fiber, or radio to a remote or local point where the signals are reconverted into sound waves; patented by Alexander Graham Bell in 1876 and now the leading form of long-distance communication. (From a Greek term meaning "distant sound.")

telephone-answering system *Telecommunications.* a private-branch exchange or automatic call-distribution system used by a facility that answers telephone calls for subscribers.
telephone carrier current *Electricity.* a carrier current that is used for telephone communication in order to obtain numerous channels through a single pair of wires.
telephone channel *Telecommunications.* a one- or two-way path that is suitable for sending audio signals between two stations.
telephone circuit *Electricity.* a complete circuit consisting of insulated conductors between two telephone subscribers; includes audio and signaling currents that travel between subscribers as they communicate with each other through a telephone system.
telephone data set see DATA SET COUPLER.
telephone dial *Engineering.* a hand-rotated dial used to make a telephone connection, with the amount of rotation corresponding to each digit in a telephone number; now generally supplanted by push-button devices.
telephone induction coil *Electricity.* a small telephone two-line, impedance-matching transformer used in telephone systems so that the line impedance matches the telephone transmitter or receiver.
telephone influence factor *Telecommunications.* **1.** a function that provides the relative interfering effect of sound promulgated in a telephone circuit by current or voltage in a power circuit. **2.** a measurement of the harmonic content of a power signal.
telephone line *Electricity.* a line that carries telephone signals; any one of many conductors that extend between telephone subscriber stations and the central offices used to route telecommunications.
telephone modem *Telecommunications.* a device that converts computer-system digital signals into telephone-system-compatible signals (and vice versa) for communication between computer equipment over telephone lines.
telephone pickup *Electricity.* an induction-coil device that is used to monitor telephone conversations for recording or listening purposes; it usually fits over a telephone receiver, or the entire telephone may rest upon it.
telephone pole *Electricity.* a utility pole that supports telephone wires.
telephone receiver *Acoustical Engineering.* the portion of a telephone headpiece placed on or near the ear that converts the electrical impulses into sound.
telephone repeater *Electronics.* an electronic apparatus used in a telephone system to compensate for line losses by amplifying and redistributing the signals at one or more points along the line.
telephone repeating coil *Electricity.* a coil that is used to offset a direct telephone circuit connection by inductively coupling two sections of a line.
telephone ringer *Electromagnetism.* an electromagnetic device that actuates a clapper in a telephone bell.
telephone set *Acoustical Engineering.* the device used in a telephone system, consisting of a headset or headpiece with a transmitter and one or two receivers.
telephone transmitter *Acoustical Engineering.* the portion of a telephone set used for converting voice sounds into electromagnetic signals; also capable of functioning as a receiver.
telephonic [tel´ə fän´ik] *Telecommunications.* relating to or sent by a telephone system.
telephony [tə lef´ə nē] *Telecommunications.* **1.** the making and operation of telephone systems. **2.** the use of telephonic equipment in the transmission of sound between points, with or without wires.
telephotographic *Telecommunications.* a photograph of distant objects using a telephoto lens. Also, **telephoto.**
telephotography *Telecommunications.* the photography of distant objects using a telephoto lens.
telephoto lens *Optics.* a camera lens constructed so that it is physically shorter than its nominal focal length, in order to produce a relatively large image; used to photograph small or distant objects. Also, **telephographic lens.**
telephotometer *Engineering.* an instrument that measures the light received from a distant light source.
telephotometry *Optics.* the measurement of a source's brightness by means of a photometer attached to a telescope.
telepresence *Computer Science.* the ability to operate a device by remote control, including perceptual data and sensory feedback transmitted to the operator and motor commands transmitted from the operator, such that it appears to the operator as if the operator were present at the site of the remote device and operating it directly.
teleprinter *Telecommunications.* in telegraphy, an instrument that possesses a signal-actuated mechanism that automatically prints messages; used for transmitting, receiving, and printing data.
teleprinter code *Telecommunications.* a code used with typewriters or teleprinters, in which each letter is represented by a number of time gaps of equal length, during which current is either absent or present.
teleprocessing *Computer Programming.* the transmission of data between remote locations and a central processor or between two computer systems, especially over telephone lines.
teleprocessing monitor *Computer Programming.* in teleprocessing, a software program designed to facilitate interfacing between remote users and a central processor.
TelePrompTer *Telecommunications.* the trade name for a device used in television broadcasting to prompt actors or other on-camera performers by displaying a magnified script out of camera range.
telepsychrometer *Engineering.* an atmospheric-humidity detecting instrument that transmits its data readout to a remote location.
teleradium treatment see RADIUM-BEAM THERAPY.
teleran [tel´ə ran´] *Navigation.* a navigational system that employs ground-based search radar equipment to locate aircraft and then utilizes television to transmit information concerning these aircraft to pilots, including information regarding landing approach. (An acronym for tele-vision and radar navigation system.)
telerecording bathythermometer *Engineering.* an instrument that detects the temperature and water depth of deep underwater layers, and then transmits the data readout to a remote location, typically on the surface.
telering [tel´ə ring´] *Electronics.* a device that generates a telephone-ringing signal.
telescope *Optics.* an instrument that produces a distinct, generally magnified image of a distant object (such as the moon or another celestial object) by modifying its light rays with a system of lenses or mirrors. *Science.* to become shortened or condensed. (From a Greek term originally meaning "distant watcher.")
Telescope *Astronomy.* another name for the constellation Telescopium.
telescope structure *Geology.* a composite alluvial fan structure made up of younger fans with flatter gradients, occurring among older fans with steeper gradients.
telescopic [tel´ə skäp´ik] *Optics.* **1.** relating to or involving the use of a telescope. **2.** obtained or seen by means of a telescope. *Engineering.* decribing a device in which the parts fit together in the manner of a jointed telescope, so that it can be extended or shortened.
telescopic alidade *Engineering.* a telescopic sight mounted on a plane table with fittings and accessories, used in survey work to measure angles and distances.
telescopic comet *Astronomy.* a term for a comet too faint to be seen without the aid of a telescope.
telescopic derrick *Engineering.* a derrick whose upper support base can extend from or retract into the lower section.
telescopic loading trough *Mining Engineering.* a shaker conveyor trough with two sections, one nested within the other; used near the face to advance the trough line without adding a length of pan following each cut.
telescopic sight *Engineering.* a sight that magnifies distant objects. *Ordnance.* specifically, a gunsight in which a refracting telescope serves as the primary sighting device. Also, **telescope sight.**
telescopic slide *Mechanical Devices.* two or more hollow tubes fitted into each other to provide a long, extended support.
telescopic tripod *Engineering.* a tripod support for a camera or other instrument fitted with legs that can be extended or retracted as required.
telescoping gauge *Mechanical Devices.* a gauge coupled with a collapsible ring, used in conjunction with a micrometer to measure the inside diameter of boreholes.
telescoping valve *Mechanical Engineering.* a valve that uses telescoping members to regulate water flow in a pipe with minimal streamline disturbance.
Telescopium [tel´ ə skō´ pē əm] *Astronomy.* the Telescope, a dim and inconspicuous southern constellation lying between the constellations of Ara and Corona Austrinus.
teleseism *Geophysics.* the motion generated by a distant earthquake.
teleseismology *Geophysics.* the study of seismic data recorded at a distance far from the site of the disturbance.
Telestacea *Invertebrate Zoology.* an order of anthozoans in the subclass Alcyonaria, having erect, branching colonies of long axial polyps attached by basal stolons and bearing lateral polyps.

telestereoscope *Optics.* an instrument for viewing distant objects in three dimensions; an optical range finder.

Telesto [te les´tō] *Astronomy.* Saturn XIII, a moon of Saturn discovered in 1980 and positioned in the orbit of Tethys, but 60° of orbital longitude ahead of it; Telesto has an irregular shape and measures about 25 km in diameter.

telesynd *Electronics.* a mechanical-motion repeater device that is used in telemetry and remote-control applications to reproduce mechanical motions.

teletext *Telecommunications.* a system that allows a limited number of pages of text to be sent by a television broadcasting station along with the associated program emissions, using the unoccupied lines in a regular video signal.

teletherapy *Medicine.* treatment in which the source of the therapeutic agent is at a distance from the body, as in radiation therapy provided by a machine positioned at some distance from the patient.

telethermal *Geology.* describing a hydrothermal mineral deposit formed some distance from the source at low depth and relatively mild temperatures, and displaying little or no alteration of the wall rock.

telethermometer *Engineering.* a thermometer that transmits its data readout to a remote location.

telethermometry *Meteorology.* the process and methods of making remote temperature measurements.

telethermoscope *Engineering.* a thermometer on the exposed exterior of a weather station that provides its data readout on the inside.

Teletype *Telecommunications.* the trade name for an early type of teletypewriter.

teletypesetter *Telecommunications.* an early form of electronic typesetting, employing punched paper tape to automatically operate a Linotype machine or other such typesetting device.

teletypewriter *Telecommunications.* an electric typewriter that emits coded electric signals corresponding to manually typed characters and automatically types messages upon reception of coded signals from another machine. Also, TWX MACHINE.

teletypewriter code *Telecommunications.* character codes used by teletypewriters, in which (originally) each code group consists of five elements of equal length, called marking or spacing impulses; the five-element, start-stop code is made up of five signal impulses that are introduced by a start impulse and followed by a stop impulse (baudot code or three-row code); subsequently 7 characters (ASCII or four-row code) were added to implement a richer set of characters.

teletypewriter exchange service *Telecommunications.* a national or international telephone service in which subscribers can communicate by means of a teletypewriter. Also, TWX SERVICE.

televise *Telecommunications.* to transmit images by means of television broadcasting.

television *Telecommunications.* **1.** an electrical system that converts visual images into signals and transmits the signals by means of wires or carrier waves to remote receivers, at which the signals are converted to the original images. **2.** a device for receiving such transmissions; a television set. (From a Greek term meaning "distant viewing.")

television antenna *Electromagnetism.* the receiving antenna for a television that picks up the signals transmitted by television stations.

television bandwidth *Telecommunications.* the difference between the upper and lower limiting frequencies of a television channel.

television broadcast band *Telecommunications.* several groups of channels, each of which contain a cluster of 6-megahertz channels, that can be utilized for television broadcasting.

television broadcasting *Telecommunications.* the transmission of television programs by means of electromagnetic waves.

television camera *Telecommunications.* a nonphotographic camera that converts a scene being viewed into video signals, which are then transmitted to a television receiver.

television channel *Telecommunications.* a band of frequencies, measuring 6 megahertz in width and located in the television broadcast band, that can be utilized for television broadcasting.

television emission see TELEVISION SIGNAL.

television film scanner *Telecommunications.* a device that permits film to be viewed on television, by adapting the 24 frame-per-second speed of standard film to the 30 scan-per-second rate of television.

television monitor *Telecommunications.* a special-purpose television receiver that accepts a direct video signal (rather than modulated radio-frequency signals), such as closed-circuit observation of security areas, remote handling of hazardous materials, or microsurgery; also used in television stations to monitor visual and sound output.

television network *Telecommunications.* a cluster of communication channels that connect groups of television broadcasting stations or closed-circuit television users in various locations for simultaneous program viewing.

television pickup station *Telecommunications.* a land mobile station that transmits television programs from the actual scene of an event to a television broadcast station.

television receiver see TELEVISION SET.

television reconnaissance *Telecommunications.* reconnaissance using television to transmit a scene from the point of origin to another position, either on the surface or in the air.

television recording *Telecommunications.* any permanent record of a video signal, recorded by electronic or photographic means.

television repeater *Telecommunications.* in a television distribution system, an apparatus used to compensate for line losses by amplifying and redistributing the signals at one or more points along the line. Also, **television relay system.**

television scanning *Telecommunications.* the process of examining the brightness of each element of detail or pixel that exists in a television image.

television screen *Telecommunications.* **1.** the viewing surface of a television picture tube. **2.** a viewing surface onto which a television image is projected.

television set *Telecommunications.* a type of receiver that accepts modulated radio-frequency signals containing video intelligence and converts these into images on the screen of a television picture tube; most television sets also receive audio signals and produce associated sound.

television signal *Telecommunications.* the visual and aural signals that are broadcast together in order to produce the picture and sound components of a television program. Also, TELEVISION EMISSION.

television station *Telecommunications.* the location where television transmissions are sent or received.

television studio *Telecommunications.* a room or complex of rooms designed for the origination and broadcasting of live or recorded television programs.

television transmitter *Electronics.* electronic equipment used to transmit radio-frequency television signals, which are modulated by composite video, synchronizing, and audio information.

television tuner *Electronics.* a device used to select the channel frequency at which a television set is to operate: a manual tuner typically consists of a rotating cylindrical device on which resonant circuit elements are mounted; a remote-control tuner consists of a battery-powered device that sends an electronic signal to the receiver. When either device is activated, the tuner receives the signal and switches in the correct circuit element for the selected channel frequency.

telewriter *Telecommunications.* a system in which writing movement at the transmission end promulgates a simultaneous and corresponding movement of a writing instrument at the receiving end.

telex *Telecommunications.* **1. Telex.** an international teleprinter exhange system that provides direct send and receive teleprinter links between subscribers. **2.** any similar system. **3.** a machine used in such a system. **4.** to send a message by means of such a system. (Short for teleprinter exchange.)

Telford, Thomas 1757–1834, Scottish civil engineer; an innovative designer and builder of roads, bridges, waterways, and harbors; e.g., the Caledonian Canal.

telford pavement *Civil Engineering.* a smooth, surface pavement built on a foundation of large rocks. (Named for Thomas *Telford.*)

telinite *Geology.* a variety of coal within the vitrinite group that contains plant cell-wall structures.

Teliomycetes *Mycology.* a class of fungi in the subdivision Basidiomycotina, which is composed of the orders Ustilaginales and Uredenales, otherwise known as the smut and rust fungi.

teliospore *Mycology.* a thick-walled spore found in the fungal orders Uredenales and Ustilaginales, and commonly termed a resting spore or a winter spore, which forms a spore-bearing structure called a basidium. Also, TELEUTOSPORE.

telium *Mycology.* a sorus that produces teliospores in certain types of rust fungi.

tell *Archaeology.* a large mound that is the result of continuous habitation; used mainly to describe such features in ancient settlement sites in the Middle East.

Teller, Edward born 1908, Hungarian-born American physicist; instrumental in the development of the first hydrogen bomb.

Tellerette *Chemical Engineering.* a type of inert packing used in gas-absorption operations to enlarge the surface area to increase contact between falling liquid and rising vapor.

Teller-Redlich rule *Physical Chemistry.* a rule stating that for two isotopic molecules, the geometrical structure of the molecule and the masses of the atoms, and not the potential constants, determine the product of the frequency ratio values of all vibrations of a given symmetry type.

telltale *Engineering.* an external indicator of the internal liquid level in a tank; typically a vertical glass tube connected to the tank at top and bottom.

telltale compass *Navigation.* a magnetic compass in the master's cabin of a merchant ship; it is usually an inverted type suspended face downward from the deck above. Also, HANGING COMPASS, OVERHEAD COMPASS.

telltale float *Civil Engineering.* any device designed to indicate the movement of a medium, such as a foundation pile or reservoir.

tellurate *Chemistry.* a salt of a telluric acid.

telluric *Science.* of, relating to, or proceeding from the earth or soil. *Chemistry.* relating to or containing tellurium.

telluric acid *Inorganic Chemistry.* H_6TeO_6, toxic, white cubic crystals; soluble in water and insoluble in alcohol; melts at 136°C; used as a chemical reagent. Also, HYDROGEN TELLURATE.

telluric line *Spectroscopy.* in solar or stellar spectra, any line or band produced by the absorption of light from a star by the earth's atmosphere.

tellurinic acid *Organic Chemistry.* an acidic compound of tellurium having the general formula R_2TeOOH.

tellurion *Science.* a model of the moving earth and the sun, showing how this rotation and revolution produces the changes in the seasons and the alteration of day and night.

tellurite *Mineralogy.* TeO_2, a white or yellowish orthorhombic mineral occurring as acicular crystals and spherical masses, having a specific gravity of 5.90 and a hardness of 2 on the Mohs scale; dimorphous with paratellurite; found as an oxidation product of tellurium minerals.

tellurium *Chemistry.* an element having the symbol Te, the atomic number 52, an atomic weight of 127.6, a melting point of 451°C, and a boiling point of about 1200°C; silvery-white, brittle semimetal, with poisonous compounds; used in alloys. (From a Latin term meaning "of the earth;" so named in 1798 by its discoverer, German chemist Martin Klaproth, in contrast with uranium, "of the heavens," which he had discovered nine years earlier.)

tellurium dibromide *Inorganic Chemistry.* $TeBr_2$, toxic brown to gray-green needles; slightly soluble in alcohol and decomposes in cold water; melts at 210°C and boils at 339°C; Also, **tellurium bromide, tellurous bromide.**

tellurium dichloride *Inorganic Chemistry.* $TeCl_2$, toxic black crystals or amorphous mass; soluble in acid and decomposes in water; melts at approximately 209°C and boils at 327°C. Also, **tellurium chloride, tellurous chloride.**

tellurium dioxide *Inorganic Chemistry.* TeO_2, toxic white tetragonal or rhombic crystals; insoluble in water; melts at 733°C and boils at 1245°C. Also, **tellurous acid anhydride.**

tellurium disulfide *Inorganic Chemistry.* TeS_2, a toxic red to black powder or amorphous mass; insoluble in cold water and acid. Also, **tellurium sulfide.**

tellurium hexafluoride *Inorganic Chemistry.* TeF_6 a toxic colorless gas with an offensive odor that is slowly hydrolyzed by water to give telluric acid.

tellurium monoxide *Inorganic Chemistry.* TeO, toxic black crystals or amorphous mass; insoluble in water and decomposes at 3709°C in CO_2; decomposes at boiling point.

tellurobismuthite *Mineralogy.* Bi_2Te_3, a pale lead-gray, metallic, trigonal mineral of the tetradymite group, occurring in foliated masses or irregular plates, and having a specific gravity of 7.815 and a hardness of 1.5 to 2 on the Mohs scale; forms a series with tellurantimony.

tellurometer *Engineering.* a microwave instrument that determines distances by measuring the time it takes for a radio wave to travel from one point to another and then return; it functions very much like radar. Thus, **tellurometric.**

tellurous acid *Inorganic Chemistry.* H_2TeO_3, toxic white flakes that are slightly soluble in water, soluble in acid, and insoluble in alcohol; decomposes at 40°C.

telmatology *Ecology.* the science or study of wetland and swamp habitats.

telmicolous *Ecology.* living or thriving in freshwater marshes. Thus, **telmicole.**

telo- a combining form meaning "final."

teloblast *Invertebrate Zoology.* in most annelid embryos, a special large cell that divides rapidly and produces lines of many small cells.

telobranchial body see ULTIMOBRANCHIAL BODY.

telocentric *Cell Biology.* of or relating to a chromosome containing a terminally located centromere. Thus, **telocentric chromosome.**

telocoel *Developmental Biology.* the cavity of the telencephalon.

telodendron *Anatomy.* the arborization at the end of an axon. Also, **telodendrion.**

telodynamic *Mechanics.* of or relating to the transmission of mechanical power over significant distances.

telogen *Physiology.* the quiescent or resting phase of the hair cycle, following catagen, in which the hair grows no further.

teloglia *Neurology.* Schwann cells that encapsulate the motor nerve endings at a neuromuscular junction, sometimes projecting into the synaptic cleft.

telolecithal *Cell Biology.* of or relating to an egg with a large yolk accumulated at one pole, designated the vegetal pole, as in the large-yolked eggs of reptiles and birds.

telolemma *Neurology.* a composition of sarcolemma and an extension of Henle's sheath, forming a covering of a motor end plate.

telomere *Cell Biology.* the end section of a eukaryotic chromosome, composed of several hundred base pairs.

telomers *Organic Chemistry.* polymers that are shorter than ordinary polymers, usually containing only from 10 to 50 monomers.

telopeptide *Materials.* on the N-terminal and C-terminal ends of chains of a type I collagen, the areas that are sites of lysine residue and are susceptible to proteolytic attack.

telophase *Cell Biology.* the final stage of mitosis or meiosis, during which a nuclear membrane forms around each set of daughter chromosomes.

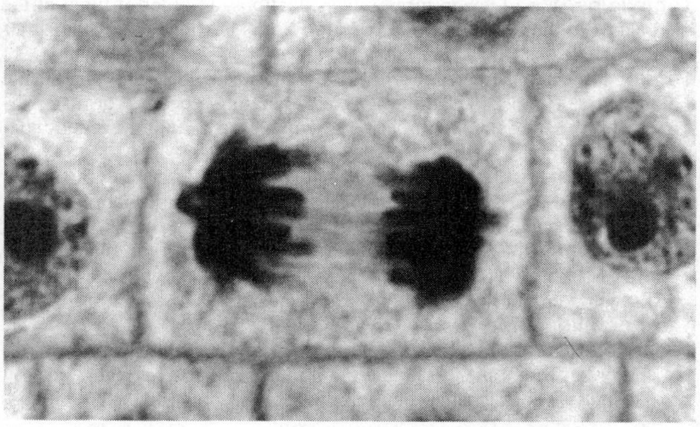

telophase

telopodite *Invertebrate Zoology.* an arthropod appendage excluding the segment attached to the body.

Telosporidia *Invertebrate Zoology.* a subclass of protozoans in the class Sporozoa, having elongated sporozoites; disease-causing in humans and animals.

telotaxis *Biology.* movement of an organism away from an external stimulus.

telotroch *Invertebrate Zoology.* **1.** a ring or tuft of cilia surrounding the anus of some aquatic larvae. **2.** in some protozoans and annelids, a stage with anterior and posterior rings of cilia.

telpher *Mechanical Engineering.* **1.** a hoist that is operated from an overhead, raised cab. **2.** a car, cab, or other system running from electric cables.

Telsmith breaker *Mechanical Engineering.* a gyratory crusher with a fixed spindle mounted in a long eccentric sleeve whose rotation gives a gyratory motion to the crushing head, but a parallel stroke; used for primary crushing.

telson *Invertebrate Zoology.* in certain arthropods, the last segment of the abdomen or terminal structure of the posterior section, such as the sting of a scorpion or part of a crayfish tail fan.

Telstar

Telstar *Space Technology.* any of a series of commercial communications satellites financed by private industry but launched by NASA beginning in 1962.

Telvar *Organic Chemistry.* $ClC_4H_6NHCON(CH_3)_2$, a trade name for 3-(*para*-chlorophenyl)-1,1-dimethylurea; used as an herbicide and soil sterilizer.

TEM triethylenemelamine.

Temellales *Mycology.* in former classifications, an order of fungi of the subclass Phragmobasidiomycetidae, now reclassified within the order Eutremellales.

Temin, Howard Martin born 1934, American virologist; shared the Nobel Prize (with David Baltimore and Renato Delbecco) for discoveries regarding the interaction between tumor viruses and the genetic material of host cells.

Temkin isotherm *Physical Chemistry.* a relation describing the extent of adsorption of a species on an electrode surface as a function of concentration and electrode potential.

Temnocephalida *Invertebrate Zoology.* a suborder of turbellarian flatworms in the order Neorhabdocoela, having adhesive tentacles and disks; ectocommensal on freshwater crustaceans, snails, and turtles.

Temnochilidae *Invertebrate Zoology.* see OSTOMIDAE.

Temnopleuridae *Invertebrate Zoology.* a family of small sea urchins of the order Temnopleuroida, having no arms, a subspherical rigid test, and usually short, lightly banded spines and conspicuous spineless areas.

Temnopleuroida *Invertebrate Zoology.* an order of echinoid echinoderms with the mouth bridged by bars, and many gill slits or depressions along the peristomial margin of the test.

Temnospondyli *Paleontology.* the largest order of labyrinthodont amphibians; widespread in aquatic and semiaquatic environments (e.g., the plagiosaurs); at least one group, Trematosauroidea, was chiefly marine; extant from the Cretaceous to the Uppermost Triassic, and possibly the Lower Jurassic.

temp. temperature; temporary.

temper *Metallurgy.* **1.** to soften and thus toughen hardened steel by reheating it to some temperature below the eutectoid temperature. **2.** the process of treating a metal in this manner. **3.** the degree of hardness and toughness of a metal, especially as imparted by tempering. *Engineering.* **1.** to heat and then cool glass in order to increase its strength and make it less brittle. **2.** to moisten and mix clay, plaster, or mortar to an appropriate consistency. **3.** to modify the color of paint by the addition of a binder or medium. *Materials Science.* the nonplastic material that is mixed into clay to strengthen it and lessen shrinkage during firing. *Acoustical Engineering.* to tune a musical instrument, such as a piano or harpsichord, by adjusting the notes to approximate a diatonic scale.

tempera *Materials.* a paint whose binder or medium consists of an emulsion of egg yolk and water or oil; used in art, especially in the Middle Ages. Also, **egg tempera.**

temperament *Psychology.* **1.** an individual's predisposition toward certain patterns of emotional reactions, present from an early age and considered by some to have a basis in heredity. **2.** in ancient and medieval thought, one of four personality types, choleric, melancholic, sanguine and phlegmatic, characterized in terms of bodily fluids, or humors. *Acoustical Engineering.* the process of tempering a piano, harpsichord, or other musical instrument.

temperate *Meteorology.* moderate; not subject to prolonged extremes of hot or cold. *Virology.* of a virus, not virulent; rarely causing lysis in host cells.

temperate and cold savannah *Ecology.* a habitat commonly found in North America and in the higher altitudes of Eurasia; characterized by an extensive carpet of lichens interspersed with clumps of trees, primarily evergreens, and mosses.

temperate and cold scrub *Ecology.* a habitat with extremely moist soil, arising from either a permanent mist, seasonal rains, or snow melt, and characterized by evergreen shrubs, large-leaved herbs, and an absence of trees. Also, BOSQUE, FOURRÉ.

temperate belt *Meteorology.* either of two broad belts around the earth, between the hot belt and the cold caps, within which the annual mean temperature is less than 20°C and the mean temperature of the warmest month is above 10°C.

temperate climate *Meteorology.* the climate found in the middle latitudes of the northern and southern hemispheres, characterized by relatively moderate temperatures year-round.

temperate forest see TEMPERATE WOODLAND.

temperate glacier *Hydrology.* a temperate-zone glacier composed of ice and firn near the melting point at the end of the ablation season.

temperate grassland *Ecology.* any of various flat, largely treeless areas known locally as prairies, steppes, pampas, veldt, etc.; characterized by rainfall intermediate between a forest and a desert and dominated by grasses and large grazing mammals.

temperate mixed forest *Ecology.* a woodland area containing a mixture of spruces and other northern conifers with broadleaf species such as maples and beeches; commonly found in the north temperate zone, as in wilderness areas of the northeastern United States.

temperate phage see BACTERIOPHAGE LAMBDA.

temperate rain forest *Ecology.* a classification of vegetation that grows in temperate areas with an abundant and evenly distributed rainfall,as is found in northwest Washington State, in which evergreens with small leaves and large tree ferns flourish. Also, MOSS FOREST.

temperate rainy climate *Meteorology.* a rainy climate in which the coldest-month mean temperature is greater than 26.6°F and below 64.4°F, while the warmest-month mean temperature is greater than 50°F.

temperate virus see CRYPTIC VIRUS.

temperate-westerlies index *Meteorology.* a measure of the strength of a temperate westerly, calculated from the average sea-level pressure difference between 35°N and 55°N, and expressed in meters per second as the west-to-east component of the geostrophic wind.

temperate westerly *Meteorology.* a west wind in the southern two-thirds of the westerly belt, that is, between 35°N and 55°N.

temperate woodland *Ecology.* any of various types of forest area found in temperate regions; may be entirely broadleaf, entirely evergreen, or mixed.

Temperate Zone *Meteorology.* either of two latitudinal zones lying between 23°27' and 66°32' N and S; that is, between the Tropic of Cancer and the Arctic Circle in the Northern Hemisphere (**North Temperate Zone**) and between the Tropic of Capricorn and the Antarctic Circle in the Southern Hemisphere (**South Temperate Zone**); characterized by a warm climate in summer and cold in winter, with relatively moderate conditions in the spring and fall. Also, VARIABLE ZONE.

temperature *Thermodynamics.* a quantity that is a measure of the average kinetic energy due to the thermal agitation of particles in a system, expressed in terms of units that are designated by a value on a scale. The temperature of a system determines the direction of heat flow with respect to the system; heat will flow from a region of higher temperature to an adjacent region of lower temperature. *Meteorology.* the level of sensible heat energy of the atmosphere as measured by a thermometer and expressed on a given temperature scale, usually Celsius or Fahrenheit. *Physiology.* a measure of the heat energy present in a living body; in humans it is normally maintained at a constant level of 98.6°F (37°C). *Medicine.* an informal term for a higher-than-normal level of body heat; a fever. (From a Latin term meaning "to temper;" "regulate;" "keep at a certain level.")

temperature bath *Thermodynamics.* a relatively large body of matter that is maintained at a constant temperature so that an object placed in thermal contact with this reservoir remains at the same constant temperature.

temperature belt see TEMPERATURE ZONE.

temperature coefficient *Physics.* **1.** the ratio of the change of electrical resistance in a wire when the temperature of the wire is increased by 1°C to the resistance of the wire at 0°C. **2.** in evacuated filament tubes, the ratio of the filament voltage drop to the temperature gradient between the filament and the ambient air temperature or the temperature of the glass envelope. *Microbiology.* an indication of the increase expected in the death rate of a population of microorganisms that is exposed to a disinfectant upon raising the ambient temperature by a given amount.

temperature coefficient of reactivity *Nucleonics.* the change in reactivity per degree of temperature in a nuclear reactor brought on by a change in the operating temperature of the system.

temperature color scale *Thermodynamics.* a scale that gives the temperature of an incandescent material as a function of the color or wavelength of the light emitted by the material.

temperature-compensated *Electronics.* having features or components that reduce the effects of variations or extremes in ambient temperature.

temperature-compensating capacitor *Electricity.* a capacitor with predictively varying capacitance due to temperature; used extensively in oscillator circuits as a compensating device for changes in values of other components with temperature.

temperature compensation *Electronics.* the process of compensating for the effects of variations or extremes in the ambient temperature of electronic equipment; often accomplished by including circuit components possessing temperature characteristics opposite to those of the critical circuit components.

temperature control *Engineering.* a device that detects the temperature of a space or system and maintains it within desired limits by actuating heating or cooling units as required.

temperature correction *Cartography.* in surveying, a correction applied to correct the error possibly introduced into measurements by the difference between the temperature at which an instument was standardized and that at which it was used in the field.

temperature efficiency *Meteorology.* see THERMAL EFFICIENCY.

temperature factor *Crystallography.* an exponential expression by which the scattering of an atom is reduced as a consequence of vibration (or a simulated vibration resulting from static disorder). For isotropic motion the exponential factor is $\exp(-B_{iso}\sin^2\theta/\lambda^2)$, with B_{iso} called the **isotropic temperature factor.** It equals $8\pi^2\langle u^2\rangle$, where $\langle u^2\rangle$ is the mean square displacement of the atom from its equilibrium position. For anisotropic motion the exponential expression contains six parameters, the anisotropic vibration or displacement parameters, which describe ellipsoidal rather than isotropic (spherically symmetrical) motion or average static displacements.

temperature gradient *Thermodynamics.* **1.** a vector that is perpendicular to a constant-temperature surface and whose magnitude is equal to the rate of change in the temperature with distance away from the surface, measured at the point. **2.** a difference in temperature per unit distance within the same body of matter, thus giving rise to the heat transfer from the hotter region to the colder region. *Meteorology.* the rate of temperature change over any horizontal or vertical distance.

temperature-gradient zone melting *Materials Science.* a purification technique of moving a small molten zone of higher impurity content through a relatively pure solid to achieve unusual impurity distribution. In nature this process occurs when droplets of brine trapped in sea ice migrate toward the warmer surface of the ice.

temperature-humidity index *Meteorology.* a measure of the combined effects of temperature and moisture on humans during a warm season; computed by multiplying the sum of the dry-bulb and wet-bulb Fahrenheit temperature readings by 0.4 and adding 15 to that figure. Also, COMFORT INDEX, DISCOMFORT INDEX.

temperature inversion *Meteorology.* see INVERSION. *Oceanography.* any anomalous condition in a body of water that causes the temperature of a water layer to increase with depth.

temperature log *Petroleum Engineering.* a well log used to record temperature versus depth in a borehole. Also, THERMAL LOG.

temperature profile recorder *Engineering.* a portable instrument that is lowered into a lake to determine the water temperature at various depths.

temperature province *Meteorology.* a climatic classification highlighting a major division of temperature as determined by the temperature-efficiency index or the potential evapotranspiration over a given area; one of C. W. Thornethwaite's schemes of climatic classification.

temperature resistance coefficient *Electricity.* the change of electrical resistance in a wire per degree Centigrade change of temperature.

temperature-salinity diagram *Oceanography.* a plot of temperature as ordinate and salinity as abscissa on which a curve joins the points observed at a single serial station; the diagram may identify water masses, sigma-*t* values, and the stability of the water column. Also, T-S CURVE, T-S RELATION.

temperature saturation *Electronics.* a condition in which the cathode of a thermionic electron tube can no longer increase electron emission for an increase in cathode temperature; in this condition anode current remains constant as filament voltage is increased while other electrode voltages are held constant. Also, FILAMENT SATURATION.

temperature scale *Meteorology.* a metered scale used to measure and display on a thermometer the temperature of ambient air or water, especially one using either the Celsius or Fahrenheit scale. *Thermodynamics.* a one-to-one relation between the temperature and a range of numerical values as measured by instrumentation; the relation may be empirical, in which case the number assignment is based on a convenient property, such as the freezing and boiling point of water, or it may be an absolute scale of temperature.

temperature-sensitive gene *Biotechnology.* a conditional gene that is not expressed when it is exposed to a temperature that falls outside certain limits.

temperature-sensitive mutant *Genetics.* a mutation that demonstrates phenotypic expression only at certain temperatures.

temperature transducer *Engineering.* a device that converts a temperature value into some other form, such as an electric current value or a mercury level in a thermometer, which then activates a temperature-control mechanism.

temperature wave *Physics.* a temperature disturbance that propagates through a medium. Also, THERMAL WAVE.

temperature zone *Meteorology.* any region of the earth sharing common temperature characteristics, such as a hot belt, polar cap, or a temperate rainy climate. Also, TEMPERATURE BELT.

temper brittleness *Metallurgy.* brittleness that occurs when some steels are held within a certain range of temperature below the transformation range.

temper carbon *Materials Science.* clumplike nodules of graphite that are formed during the heat-treatment stage in the production of malleable iron.

tempered martensite *Materials Science.* the process of heating martensite to change the tetragonal to a cubic structure and in later stages to cause carbide formation.

tempered scale see PYTHAGOREAN SCALE.

tempering air *Engineering.* cooler air that is directed into a hot air stream in order to moderate its temperature.

temper rolling *Metallurgy.* the process of cold-rolling sheet metal, generally steel, to obtain a small thickness reduction as a means of preventing stretcher strains when the metal is later formed by stretching or bending and to obtain a desired texture and mechanical properties.

tempers *Metallurgy.* a notation using letters and numbers to describe the processing of an alloy. H tempers refer to cold-worked alloys; T tempers refer to age-hardening treatments.

template a pattern or mold; specific uses include: *Engineering.* **1.** a plate of a particular shape that serves as a guide for drafting, sawing, or cutting. **2.** a flat plate that provides a variety of curves, used in drafting. *Photogrammetry.* a sheet of transparent material (usually plastic) used to record the directions from the radial center of a photograph to the images of objects on the photograph, or on adjacent photographs, by cutting a slot in the sheet; the direction may then be transferred accurately to a map or chart. *Computer Programming.* **1.** a frequently used pattern specifying the format or contents of a spreadsheet or database program. **2.** the structure and specifications of a data type. **3.** a plastic card with cutouts, used by programmers to draw flowcharting symbols. **4.** a pattern of data that may be used in matching input or in generating output. *Molecular Biology.* a macromolecular pattern or mold for the synthesis of another macromolecule. *Genetics.* a structure composed of a sequence of nucleotides, of either DNA or RNA, that serves to specify the nucleotide sequence of another structure, which may then serve as an antitemplate in the formation of a new structure having the same sequence as the original structure. Also, **templet.**

template following *Robotics.* a method of machining in which a template is used to guarantee greater accuracy.
template matching *Artificial Intelligence.* the process of comparing the structure of unknown data with an established data structure in order to identify the unknown data type. *Robotics.* the identification or inspection of a workpiece by matching it against a picture or pattern.
temple *Architecture.* a building of worship for various religious groups.
temple *Anatomy.* the lateral region on either side of the upper part of the head. *Invertebrate Zoology.* an analogous region on the upper or posterior part of the sides of an insect's head.
temporal *Science.* of or relating to time. *Anatomy.* of, relating to, or near the temporal region or temporal bone.
temporal bone *Anatomy.* a bone on either lateral side of the skull that also forms the upper part of the cheek and houses the middle and inner ear.
temporal coherence *Physics.* a temporary constructive interference pattern resulting from the summation of several waves of different frequencies and phases and having a finite coherence length and time.
temporal decomposition *Control Systems.* the process of dividing control of a large-scale control system into a number of subproblems based on the time scales required to carry out each subprocess.
temporale *Meteorology.* a rainy wind that blows from southwest to west along the Pacific coast of Central America during July and August.
temporal lobe *Anatomy.* the lower lateral lobe of each cerebral hemisphere.
temporal-lobe epilepsy see PSYCHOMOTOR EPILEPSY.
temporal logic *Computer Science.* a kind of mathematical logic that allows quantification over time and allows temporal reasoning; e.g., one event must follow another, or a certain event will eventually occur.
temporary base level *Geology.* any base level, other than sea level, below which a land surface cannot be worn away by ordinary erosion processes. Also, LOCAL BASE LEVEL.
temporary bench mark *Cartography.* a bench mark established to temporarily hold the end of a completed section of a line of levels and to serve as a starting point from which the next section is run.
temporary file *Computer Programming.* a file that is created by the operating system or a program and is later destroyed.
temporary hardness *Chemistry.* hardness in water due to the presence of bicarbonate ion; upon heating, the bicarbonate precipitates as calcium or magnesium carbonate.
temporary storage *Computer Technology.* those portions of internal memory that are reserved for intermediate data and partial results of computations. Also, WORKING STORAGE.
temporary stream see INTERMITTENT STREAM, def. 1.
temporomandibular *Anatomy.* relating to the temporal bone and the mandible.
temporomandibular joint syndrome see TMJ SYNDROME.
ten- a combining form meaning "tendon," as in *tenalgia.*
tenacity *Behavior.* see FIDELITY. *Textiles.* see TENSILE STRENGTH.
tenaculum *Surgery.* **1.** a hooked surgical instrument used to seize and secure tissue. **2.** any fibrous band for maintaining structures in place. *Entomology.* a pair of partially fused catchlike appendages on the third abdominal segment of a springtail, which holds the furcula in place.
tenalgia *Medicine.* any pain in a tendon.
tendency *Behavior.* the probability that a specific behavior will occur. Also, SPECIFIC ACTION POTENTIAL. *Meteorology.* **1.** the rate of change in air pressure at a given location over a given time interval, usually three or six hours. Also, PRESSURE TENDENCY. **2.** the rate of change of any vector or a scalar quantity at a given location over a given time interval.
tendency chart *Meteorology.* a meteorology chart depicting the amount and direction of a tendency.
tendency equation *Meteorology.* a mathematical representation of a pressure tendency; derived by combining the equation of continuity with an integrated form of the hydrostatic equation.
tendency interval *Meteorology.* a unit of time that becomes the definition by which a tendency is measured, such as the standard three-interval between atmospheric pressure readings.
tender *Medicine.* exhibiting abnormal sensitivity to touch or pressure. *Food Technology.* of meat, having a texture that is soft and easily chewed. Thus, **tenderness.**
tender *Naval Architecture.* **1.** a service boat assigned to a ship. **2.** a naval ship that serves as a floating base for other vessels, most commonly submarines or destroyers, or for seaplanes. *Transportation Engineering.* a railcar that carries fuel and water, usually attached to the rear of a steam locomotive.
tenderizer *Food Technology.* any substance used in tenderizing.
tenderizing *Food Technology.* any process that improves the tenderness of meat, usually by using an enzymatic tenderizer such as papain to catalyze the hydrolysis of one or more proteins.
tenderometer *Food Technology.* in canning, an instrument that determines the ripeness of peas by measuring the force needed to shear them.
tendinitis *Medicine.* inflammation of a tendon due to irritation or strain. Also **tendonitis.**
tendinous *Anatomy.* relating to, characteristic of, or resembling a tendon.
tendinous chords see CHORDAE TENDINEAE.
Tendipedidae *Invertebrate Zoology.* the midges, a family of tiny, delicate, nonbiting mosquitolike flies in the series Nematocera; adults swarm in the evenings.
tendon *Anatomy.* a fibrous cord or sheath of connective tissue that attaches a muscle to a bone or other body part.
tendon sheath *Anatomy.* the membranes around some tendons that form a synovial cavity at points where tendons cross over bones, cushioning the tendon.
tendoplasty *Surgery.* plastic surgery of a tendon. Also, **tendinoplasty.**
tendotomy see TENOTOMY.
tendril *Botany.* a threadlike, leafless organ of climbing plants, often growing in spiral form, that twines itself around another body in order to support the plant.

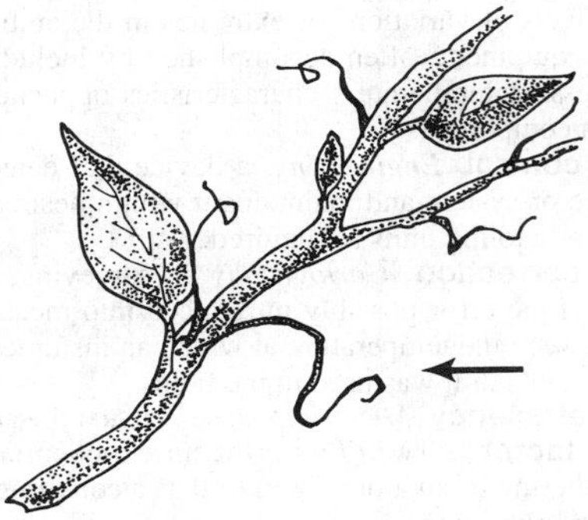

tendril

tenebrescence *Physics.* the absorption of light into a material so as to darken it (using primary X-rays or cathode rays) or bleach it (using heat or photons of an appropriate wavelength).
Tenebrionidae *Invertebrate Zoology.* the darkling beetles, a family of slow-moving, wingless, usually black or brownish-blue nocturnal coleopteran insects that raise the abdomen when alarmed and have a disagreeable odor; plant scavengers and pests in foods, including mealworms and flour beetles.
Tenebrionoidea *Invertebrate Zoology.* a superfamily of beetles, coleopteran insects in the suborder Polyphaga.
tenggara *Meteorology.* a strong, dry, east or southeasterly wind that blows during the monsoon in the Spermunde Archipelago, characterized by a hazy atmospheric condition.
teniae coli *Histology.* a group of three longitudinal bands of smooth muscle tissue formed by the longitudinal fibers in the muscular tunic of the large intestine and extending from the root of the vermiform appendix to the rectum. Also, TAENIE COLI.
teniposide *Oncology.* an antineoplastic agent derived from podophyllotoxin.
Tennant, Smithson 1761–1815, English chemist; demonstrated the identity of diamonds and carbon; isolated osmium and iridium.
tennantite *Mineralogy.* $(Cu,Fe)_{12}As_4S_{13}$, a lead-gray to iron-black, metallic, cubic mineral of the tetrahedrite group, occurring as tetrahedral crystals and in massive form, having a specific gravity of 4.59 to 4.75 and a hardness of 3 to 4.5 on the Mohs scale; forms a series with tetrahedrite; found in hydrothermal veins and contact metamorphic deposits; Also, GRAY COPPER ORE.
teno- a combining form meaning "tendon," as in *tenoplasty.* Also, **tenoto-, tenot-.**
tenodynia *Medicine.* pain in a tendon. Also, **tenontodynia.**

tenomyoplasty *Surgery*. plastic surgery involving a tendon and muscle. Also, **tenontomyoplasty.**

tenon *Engineering*. a wooden peg or projection that fits into a mortise.

tenon saw see DOVETAIL SAW.

tenontitis or **tenonitis** see TENDINITIS.

tenoplasty *Surgery*. plastic surgery of the tendons; operative repair of a defect in a tendon. Also, **tenontoplasty.**

tenorite *Mineralogy*. $Cu^{+2}O$, a monoclinic, metalic mineral occurring as black compact to earthy masses or shining black or steel-gray scales, having a specific gravity of 6.45 and a hardness of 3.5 on the Mohs scale; found in the oxidized zones of copper deposits.

tenorrhaphy *Surgery*. the union of a divided tendon by a suture. Also, **tenosuture.**

tenosynovitis *Medicine*. the inflammation of a tendon sheath; may be caused by strain or trauma, rheumatoid arthritis, gout, calcium deposits, high levels of blood cholesterol, or gonorrhea.

tenotomy *Surgery*. the cutting of an extraocular tendon to correct a strabismum. Also, **tenontotomy.**

tenpounder *Vertebrate Zoology*. either of two game fish of the genus *Elops* in the family Elopoidei, *E. saurus* or *E. affinis,* inhabiting waters of warm seas and resembling a small tarpon, to which they are closely related.

tenrec *Vertebrate Zoology*. any of several shrewlike, usually terrestrial and tailless insectivorous mammals of the family Tenricidae of Madagascar.

Tenrecidae *Vertebrate Zoology*. a family of insectivorous mammals in the suborder Lipotyphla having a long, pointed snout.

TENS *Medicine*. a portable electronic device designed to relieve chronic pain by sending electrical impulses through electrodes covering the afflicted area on the body. (An acronym for transcutaneous (or transdermal) electrical nerve stimulator.)

ten's complement *Mathematics*. given a number x expressed in base 10 with n digits to the left of the decimal point, the (base 10) number $10^n - x$.

tense *Physiology*. exhibiting tension; drawn tight; rigid.

tensile [ten´səl] *Mechanics*. of or relating to tension. *Materials*. capable of being stretched or drawn out.

tensile bar *Materials Science*. a bar-shaped test specimen of a material, used in tensile-strength tests. Also, **tensile specimen.**

tensile modulus *Mechanics*. the strength factor for a material under tensile stress, expressed by the ratio of the stress applied to it, to the change in length produced by this force.

tensile strain *Mechanics*. an increase in the length of a body that occurs when it is subjected to an external force.

tensile strength *Mechanics*. the maximum tensile stress (stretching) that a material can withstand without failure. *Materials Science*. the maximum engineering stress encountered during a tensile test of a given material. *Textiles*. the greatest stress that a fabric or thread can bear before it breaks or pulls apart, measured in force per unit of a cross-sectional area of the original sample. Also, TENACITY.

tensile stress *Mechanics*. **1.** an external force that acts on a body at both ends, parallel to the length, and that if unopposed will tend to elongate it; i.e., the material will stretch. **2.** a measure of this force per unit area of the material.

tensile test *Materials Science*. any test of the strength of a material under stress by extension, often using a **tensile test machine** designed to stretch a tensile bar to the breaking point.

tensimeter *Engineering*. a device that measures the differences in vapor pressure between two liquids that are placed in separate sealed, evacuated bulbs joined by a differential manometer.

tensiometer *Engineering*. **1.** an instrument for measuring longitudinal stress, as in a wire or structural element. **2.** an instrument that measures the surface tension of a liquid. Also, **tensometer.**

tensiometer method *Fluid Mechanics*. a method of determining the surface tension of a liquid by measuring the force necessary to lift a ring of known radius from the liquid surface.

tension *Mechanics*. **1.** the force exerted at the end points of a stretched object. **2.** the condition of an object being stretched between two points. *Fluid Mechanics*. the partial pressure of a gas in a fluid. *Electronics*. another term for voltage, as in the term "high-*tension* wire." (Going back to a Latin word meaning "to stretch.")

tension crack *Geology*. **1.** a break or separation caused by a stretching of the earth's crust (crustal tension). Similarly, **tension fault. 2.** a break or separation along the plane of least compression. Also, **tension fracture.**

tension joint *Geology*. a joint that develops perpendicular to the direction of greatest deformational stress.

tension pulley *Mechanical Engineering*. a pulley around which an endless rope passes; the pulley is mounted on a trolley or other movable bearing so that the slack of the rope can be taken up by the pull of the weights.

tension wood see REACTION WOOD.

tensor *Anatomy*. any muscular tissue that tightens or elongates the part to which it is connected. Also, TENSOR MUSCLE. *Mathematics*. an element of a vector space (more generally, R-module) created by taking the **tensor product** of vector spaces (R-modules) and their duals. In practice, a tensor may be viewed as a multilinear function of a fixed finite sequence of copies of a vector space V and its dual V^*. The **components of a tensor** are obtained by fixing a basis $\{e_i\}$ (and its dual basis $\{{*e^i}\}$) and then evaluating the tensor at the basis elements. The **degree of a tensor** is the number of arguments involved. For example, the degree 3 tensor $T: V \times V \times V^* \to C$ has the components $T_{ij}{}^k = T(e_i, e_j, {*e^k})$ and T may be written as the sum

$$T = \Sigma_{ijk} T_{ij}{}^k {*e^i} \otimes {*e^j} \otimes_k .$$

The basis tensor ${*e^i} \otimes {*e^j} \otimes e_k$ is the tensor (multilinear functional) that equals 1 when applied to the argument $(e_i, e_j, {*e^k})$ (with i, j, k fixed) and that equals zero when applied to any other ordered set of basis vectors. One can also think of the tensor T applied to only the first (one or) two of its arguments. In that case,

$$T(u\,, v\,, -) = \Sigma_{ijk} T_{ij}{}^k {*e^i}(u)\, {*e^j}(v) e_k$$

becomes a linear combination of basis vectors from V. Thus, the same tensor T can be thought of as a multilinear function $T: V \times V \to V$ or as $T: V \to V^* \otimes V$.

tensor analysis *Mathematics*. any of several methods for performing (differential and integral) calculus of tensor fields, especially the Ricci calculus. Also, **tensor calculus.**

tensor component *Mathematics*. the value of a tensor evaluated on vectors and covectors from a basis and dual basis, respectively. For example, suppose the vector space V has basis $\{e_i\}$, and the space of covectors V^* has dual basis $\{\mathrm{a}^i\}$. Then the components of a 5-tensor T with respect to those bases might take the form $T^{ij}{}_{kl}{}^m = T(\alpha^i, \alpha^j, e_k, e_l, \alpha^m)$. T can be reconstructed from its components via summation:

$$T(\omega^1, \omega^2, X_1, X_2, \omega^3) = \Sigma_{ijklm} T^{ij}{}_{kl}{}^m\, e_i(\omega^1)\, e_j(\omega^2)\, \alpha^k(X_1)\, \alpha^1(X_2) e_m(\omega^3).$$

In the Ricci calculus, the components of tensor fields are often calculated with respect to the coordinate basis $\{\partial/\partial x^i\}$ and dual basis $\{dx^i\}$.

tensor contraction *Mathematics*. a linear operation on tensors which reduces the contravariant and covariant degrees (number of arguments) of a tensor each by one. For example, from a tensor of covariant degree two and contravariant degree one, such as $T: V \times V \times V^* \to C$, one can form a new tensor $N: V \to C$ by contraction. To perform the contraction, one chooses a basis $\{e_i\}$ for V and the dual basis $\{{*e^i}\}$ for V^*. N is then given by

$$N(u) = \Sigma_i T(u, e_i, {*e^i}).$$

The resulting tensor N can be shown to be independent of the basis chosen for V.

tensor differentiation *Mathematics*. covariant differentiation; i.e., differentiation of a tensor field defined in such a way that the resulting quantity is also a tensor field.

tensor field *Mathematics*. a tensor defined on a neighborhood of a point on a differentiable manifold; i.e., a multilinear function on a fixed finite sequence of copies of the tangent space T_xM and the cotangent space T_x^*M. Tensor fields are required to be linear not only with respect to the base field of the tangent vector space but also with respect to functions on the manifold. For example, the degree 3 tensor field

$$T: (T_xM) \times (T_xM) \times (T_x^*M) \to R$$

is required to satisfy $T(fu, gv, h\omega) = fghT(u, v, \omega)$. Since tensor fields are often defined using differential operators, this remarkable linearity property depends on a precise cancellation of differentiated terms, and the resulting tensor fields carry a great deal of information about the manifold.

tensor force *Nuclear Physics.* a spin-dependent, noncentral force generated by the spin between nucleons that gives rise to the deuteron magnetic moment.

tensor muscle *Anatomy.* see TENSOR.

tensor product *Mathematics.* **1.** the tensor obtained by multiplying two or more given tensors; the components of the tensor product are the products of the corresponding components of the given tensors. **2.** in general, a multiplicative operation between two R-modules (usually vector spaces) A and B, obtained as follows: let F be the free Abelian group on the set $A \times B$, and K the subgroup of F generated by elements of the forms

$$(a + a', b) - (a, b) - (a', b),\ (a, b + b') - (a, b) - (a, b'),\ (ar, b) - (a, rb),$$

where $a, a' \in A$; $b, b' \in B$; and $r \in R$. Then the quotient group F/K is called the tensor product of the R-modules A and B; denoted $A \otimes_R B$. If R is understood, the notation $A \otimes B$ is also used. The coset $(a, b) + K$ is denoted $a \otimes b$; the coset $(0, 0) + K$ is denoted 0. This definition can be generalized to tensor products of algebras.

TENS unit *Medicine.* see TENS.

tent *Engineering.* a portable shelter consisting of fabric or other material that is stretched over a supporting framework or draped from a central pole; usually held in place by ropes that are attached to pegs driven into the ground. *Surgery.* **1.** a similar covering of material used to enclose a hospital space, as one placed over a patient's bed to administer oxygen. **2.** a conical and expansible plug of soft material, used to dilate an orifice or to keep a wound open in order to prevent healing except at the bottom. Also, TURUNDA.

tentacle *Zoology.* a long, slender whiplike appendage that may function in prehension and feeding or as a sense organ.

Tentaculata *Invertebrate Zoology.* a class of comb jellies having long tentacles, including most Ctenophora.

tentaculocyst *Invertebrate Zoology.* in certain jellyfish, any of usually eight small, club-shaped sense organs along the margin of the umbrella. Also, RHOPALIUM.

tentaculozooid *Invertebrate Zoology.* in some hydrozoans, a modified polyp type that serves as a tentacle.

tentative *Science.* of a hypothesis, design, or the like, experimental and subject to change.

tent caterpillar see EGGER.

tented ice *Oceanography.* a form of pressure ice in which two floes are forced upward, leaving a cavity between a more or less flat-sided arch and the open water below; if the process continues, the pack may become very rough and thick.

tenter *Textiles.* a framework used for stretching and drying fabric in a tentering process.

tenterhook *Textiles.* a hook or bent nail used to secure fabric on a tenter.

tentering *Textiles.* a finishing process in which a machine stretches a damp fabric to its desired width and then dries it.

Tenthredinidae *Invertebrate Zoology.* a family of sawflies, often bright-colored hymenopteran insects in the superfamily Tenthredinoidea; females use ovipositors to saw open leaves and fruits for egg deposits; caterpillars eat vegetation, harming forests and agriculture.

Tenthredinoidea *Invertebrate Zoology.* the sawflies, a superfamily of hymenopteran insects in the suborder Chalastrogastra, including harmful plant pests.

tentillum *Invertebrate Zoology.* a branch of a tentacle.

tenting *Oceanography.* the formation of tented ice.

tentorium *Anatomy.* **1. tentorium of cerebellum.** the process of dura mater that supports the occipital lobes and covers the cerebellum. **2.** any of various other parts resembling a tent or a covering. *Entomology.* the internal skeleton of an insect's head.

tentoxin *Biochemistry.* a cyclic tetrapeptide toxin, produced by spore *Alternaria* species, that contributes to plant chlorosis.

Tenuipalpidae *Invertebrate Zoology.* a small family of mites in the suborder Trombiformes, having a respiratory opening near the base of the chelicerae.

Tenuivirus group *Virology.* a genus of plant viruses having filamentous particles that contain linear ssRNA.

TEP triethyl phosphate.

tepary bean *Botany.* a twining or bushy plant, *Phaseolus acutifolius* var. *latifolius,* of the legume family, characterized by white or violet-colored flowers and edible seeds; native to Mexico and Arizona, and cultivated widely as a food crop.

tepee *Anthropology.* a tent that was commonly used by American Indian tribes, especially those living on the Great Plains, made by stretching buffalo or other animal skins over a conical framework of poles; the framework extends above the covering, allowing for ventilation. Also, TIPI. (From a Dakota Sioux word meaning "dwelling place.")

tepee

tepee structure *Geology.* an inverted cone-shaped fold within a sedimentary rock structure.

tephigram *Meteorology.* a thermodynamic diagram in which the coordinates are temperature and logarithm of potential temperature, and isobars are gently curved lines.

tephra *Volcanology.* any fragmental pyroclastic material that is ejected from a volcano.

tephrite *Petrology.* a basaltic igneous rock, or rock group, consisting primarily of calcic plagioclase, augite, and nepheline or leucite, with traces of sodic sanidine.

tephrochronology *Geology.* a scientific process whereby the age of a layer of volcanic ash is determined as a means of establishing the sequence of geologic events.

tephroite *Mineralogy.* $Mn_2^{+2}SiO_4$, a gray, green or reddish-brown orthorhombic mineral of the olivine group, forming a series with fayalite, occurring as short prismatic crystals and in massive or granular form, and having a specific gravity of 3.9 to 4.11 and a hardness of 6 on the Mohs scale; found in iron-manganese ore deposits.

tephromyelitis *Neurology.* inflammation of the gray matter of the spinal cord.

Tepuianthaceae *Botany.* a monogeneric family of dicotyledonous resinous trees and shrubs of the order Sapindales, native to the Guyana Highlands of northern South America.

ter- a combining form meaning "three."

tera- *Mathematics.* a prefix indicating one trillion, or 10^{12}; symbolized by T. (From a Greek term meaning "marvel" or "monster," with the sense that this is a huge and marvelous quantity.)

terahertz *Physics.* 10^{12} cycles per second; 10^{12} Hz.

teraohmmeter *Engineering.* an instrument that provides resistance readings in the teraohm range; used for testing the resistance of high-strength insulators.

terap *Materials.* the yellowish-brown, easily worked wood of various Asian trees of the genus *Artocarpus*; used for plywood, boxes, and joinery.

T-E ratio temperature-efficiency ratio.

terato- a combining form meaning "malformed" or "monstrous."

teratoblastoma *Oncology.* a neoplasm containing embryonic elements and differing from a teratoma in that its tissue does not represent all the germinal layers.

teratocarcinogenesis *Oncology.* a malignant neoplasm consisting of elements of teratoma combined with elements of embryonal carcinoma or choriocarcinoma; generally found in the testis. Also, **teratocarcinoma.**

teratogen *Medicine.* any agent or process that interferes with the normal development of the embryo or fetus, causing physical abnormalities.

teratogenesis *Medicine.* the development of an abnormal mass of cells composed of diverse, differentiated, and undifferentiated cell types during fetal development, causing physical defects in the fetus.

teratogenic *Cell Biology.* of, relating to, or tending to teratogenesis.

teratology *Medicine.* the study of the causes and effects of abnormal development and congenital defects in embryonic and prenatal development.

teratoma *Oncology.* a neoplasm composed of different types of tissue, none of which is native to the area where the neoplasm occurs; generally found in the ovary or testis.

teratophobia *Psychology.* a fear of deformed people or of any deformation of the body.

Teratornithidae *Paleontology.* a family of neognathan, condorlike vultures in the order Ciconiiformes; the type genus *Teratornia*, known from Southern California deposits of the Pleistocene, attained a wing span of more than 12 feet; there are two other genera in the family, one North American and the other South American; extant in the Miocene and Pleistocene.

terawatt *Physics.* 10^{12} joules per second.

terbacil *Organic Chemistry.* $C_9H_{13}ClN_2O_2$, colorless crystals that melt at 175°C; used as an herbicide.

terbia *Inorganic Chemistry.* Tb_2O_3, a white powder; insoluble in water and soluble in dilute acids; derived by ignition of hydroxides or salts of oxy-acids. Also, TERBIUM OXIDE.

terbium *Chemistry.* an element having the symbol Tb, the atomic number 65, an atomic weight of 158.9, a melting point of 1360°C, and a boiling point of about 2960°C; a rare-earth (lanthanide) element; extremely reactive and must be handled in an inert atmosphere; used as a phosphor activator and as a dope for solid-state devices. (From Ytterby, the Swedish town where it was originally found.)

terbium chloride *Inorganic Chemistry.* $TbCl_3 \cdot 6H_2O$, colorless, prismatic crystals; deliquescent; very soluble in cold water; melts at 588°C. Also, **terbium chloride hexahydrate.**

terbium nitrate *Inorganic Chemistry.* $Tb(NO_3)_3 \cdot 6H_2O$, colorless monoclinic crystals that are soluble in cold water; melts at 893°C; a strong oxidant and dangerous fire hazard in contact with organic materials.

terbium oxide see TERBIA.

terbutol *Organic Chemistry.* the comon name for 2,6-D-*tert*-butyl-*p*-tolylmethylcarbamate; used to kill crabgrass in turf lawns.

terbutryn *Organic Chemistry.* $C_{13}H_{19}N_5S$, a colorless powder that melts at 104–105°C; used as an herbicide.

terbutylhylazine *Organic Chemistry.* $C_9H_{16}N_5Cl$, a white solid that melts at 177–179°C; used as an herbicide.

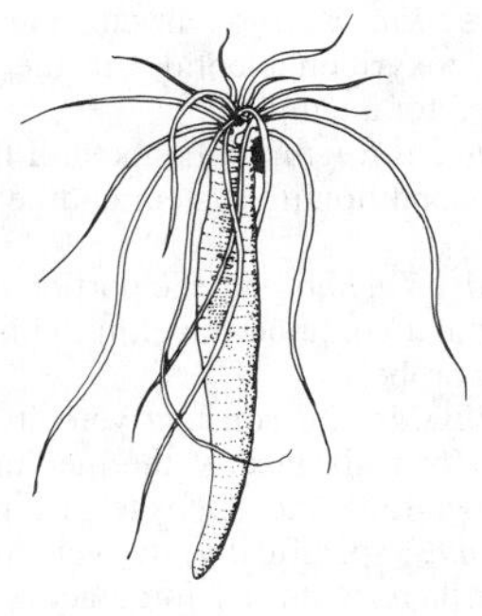

Terebellidae

Terebellidae *Invertebrate Zoology.* a large family of long, thick-bodied, marine polychaete annelid worms with prominent filamentous anterior gills, which move about freely from their burrows or tubes.

terebene *Pharmacology.* a colorless liquid consisting of a mixture of terpenes; used in medicine as an expectorant.

terebic acid *Chemistry.* $C_7H_{10}O_4$, an acid formed by the oxidation of certain terpenes; important in the discovery of the structures of many terpenes. Also, **terebinic acid.**

Terebratellidina *Paleontology.* a suborder of articulate brachiopods in the order Terebratellida; extant since the Triassic and characterized by a long loop and a median septum; two typical genera are the type genus *Terebratella* and the well-known modern *Magellania*.

terebration *Neurology.* **1.** the act of piercing or boring. **2.** a piercing pain.

Terebratula *Paleontology.* an extinct genus of large articulate brachiopods in the order Terebratulida and the suborder Terebratulidina, characterized by a short loop and an elongate oval shell; some genera were as long as 11 centimeters; extant in the Miocene and Pliocene.

Terebratulida *Invertebrate Zoology.* an order of brachiopods in the class Articulata, usually with smooth, pear-shaped, convex shells and a stalk; the arms are supported by a calcareous loop.

Teredinidae *Invertebrate Zoology.* shipworms, bivalve mollusks in the order Eulamellibranchia, which bore into wood with a shell reduced to a small but effective tool, destroying marine timbers.

tereno see TERRENHO.

terephthalic acid *Organic Chemistry.* $C_6H_4(COOH)_2$, combustible, white crystals or powder; insoluble in water and ether and slightly soluble in alcohol; sublimes above 300°C; used as a reagent and poultry-feed additive and in making polyester resins for fibers and films. Also, *p*-PHTHALIC ACID.

terephthaloyl chloride *Organic Chemistry.* $C_6H_4(COCl)_2$, combustible, colorless needles; soluble in ether and decompose in water and alcohol; melts at 83–84°C and boils at 266°C; used in the manufacture of resins, dyes, pharmaceuticals, and synthetic fibers.

Terfeziaceae *Mycology.* a family of fungi belonging to the order Tuberales, including species commonly called truffles, which occur in forest soil and have an aromatic odor.

tergeminate *Botany.* of or relating to a compound leaf that has a basal leaflet pair and then forks, with a leaf pair on each branch.

tergite *Invertebrate Zoology.* the dorsal plate of each body segment of an arthropod.

tergum *Invertebrate Zoology.* **1.** in arthropods, the dorsal surface of a body segment. **2.** in barnacles, a calcareous dorsal plate of the carapace.

terlinguaite *Mineralogy.* Hg_2ClO, a yellow to greenish-yellow monoclinic mineral occurring as small, prismatic crystals and massive aggregates, having a specific gravity of 8.7 to 9.27 and a hardness of 2.5 on the Mohs scale; found in mercury deposits of the *Terlingua* District, Texas.

term *Mathematics.* **1.** a summand; one of several addends. **2.** a member of a sequence. **3.** in an axiomatic theory, a symbol such as a letter or word that represents the basic objects discussed by the theory. *Artificial Intelligence.* in predicate calculus, an argument of a predicate that represents an object in the domain over which formulas are interpreted. *Spectroscopy.* a set of atomic states having a definite configuration and spin and orbital angular momentum quantum numbers.

terminal *Transportation Engineering.* a building where passengers embark or disembark, as on a railroad, bus line, or airline. *Electricity.* **1.** a point of connection, such as a screw, lug, or other point, for two or more conductors in an electrical circuit. **2.** a device that is attached to a conductor so as to facilitate connection with another conductor. **3.** equipment situated at the end of a microwave relay system or other communication link. *Computer Technology.* an input/output device having a keyboard for communicating with a computer, and often a display and/or a printer. *Architecture.* the terminating and often ornamental detail on a building, member, or object.

computer terminal

terminal ballistics *Ordnance.* the science concerned with the termination of a projectile's flight, including the striking and penetration of a target.
terminal bar *Cell Biology.* a junctional complex formed between adjacent epithelial cells, consisting of several types of specialized cell junctions, such as adherent junction, occluding junction, and desmosomes.
terminal block *Telecommunications.* in telephony, a group of five captive screw terminals in which a telephone pair terminates. Three terminals are provided for a talking pair (tip and ring) and ground, and two spares for adding auxiliary equipment.
terminal board *Electricity.* an insulating base or slab with associated terminals used to connect wires. Also, TERMINAL STRIP.
terminal box *Electricity.* an enclosure in which cable pairs are brought out to terminal points for connections; may include a cover and such accessories as mounting hardware, brackets, locks, and conduit fittings; used to protect one or more terminals or terminal boards.
terminal bud *Botany.* a bud that develops at the end of a shoot. Also, APICAL BUD.

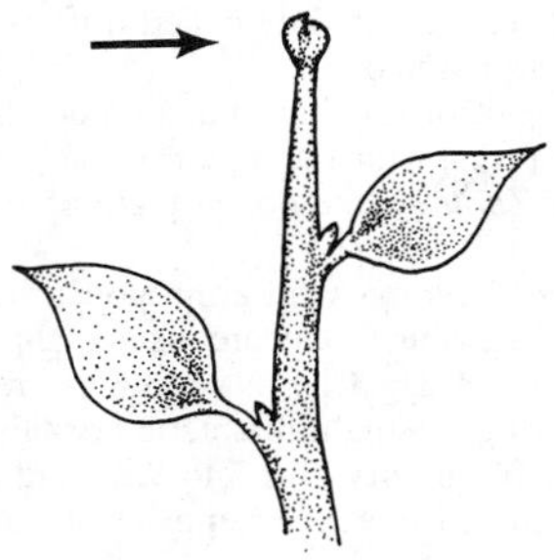

terminal bud

terminal cell *Cell Biology.* a cell located at the top of an embryo.
terminal chiasmata *Molecular Biology.* the end-to-end relationship of homologous chromosome arms that occurs as a result of terminalization.
terminal clearance capacity *Transportation Engineering.* the maximum amount of freight or the number of passengers that can be handled by a transportation terminal per day.
terminal control *Space Technology.* control over a space probe or one of its systems during the period when its main payload is programmed to function.
terminal control area *Transportation Engineering.* a zone of airspace surrounding a major airport, from which aircraft not under ground control are restricted or barred.
terminal curvature *Mining Engineering.* the bending of strata, indicating the direction of bed displacement with reference to faults.
terminal cutout pairs *Electricity.* an associated pair of accessible terminals that are numbered and designated to be brought out from a cable at a specific location.
terminal deletion *Molecular Biology.* the loss of the end region of a chromosome.
terminal equipment *Telecommunications.* **1.** communications equipment that is needed to transmit or receive a signal on a channel or circuit. **2.** equipment at a user or subscriber's terminal.
terminal extinction *Evolution.* the loss of an ancestral species that is not replaced by a newer species in the same phyletic lineage, resulting in an overall decrease in the number of species.
terminal guidance *Space Technology.* guidance of a spacecraft in the final stage of its main mission; that is, as it travels from some designated point to the encounter for which it was designed.
terminal hair *Anatomy.* adult hair, as opposed to the downy hair found on a newborn infant.
terminal ileitis see REGIONAL ENTERITIS.
terminal immigrant *Ecology.* an organism that can survive outside its natural habitat, but cannot reproduce there.
terminalization *Molecular Biology.* a successive shift between diplotene and metaphase of the first meiotic division in the apportionment of chromosomal sites. Also, CHIASMA TERMINALIZATION.
terminal leg see TERMINAL STUB.
terminal ligature *Surgery.* a ligature that ties off the end of a severed vessel.
terminal moraine *Geology.* an end moraine that spreads across a glacial valley in the form of an arc or crescent-shaped ridge, marking the farthest point to which the glacier has advanced.
terminal nerve *Anatomy.* the delicate nerve fibers that pass from a region near the olfactory bulb through the cribriform plate of the ethmoid bone into the mucosal lining of the nasal cavity; their precise function is not known.
terminal network *Computer Technology.* the communication network that connects a group of remote terminals to a central system.
terminal object see COUNIVERSAL OBJECT.
terminal operations *Engineering.* the loading, unloading, temporary storage, dispatch, and related operations of a transportation terminal.
terminal pair *Electricity.* an associated pair of accessible terminals, such as the input and output terminals of a device or network.
terminal point *Transportation Engineering.* the point at which an airline route originates or terminates, as distinguished from intermediate stops.
terminal pressure *Engineering.* the pressure drop observed across a device, such as a filter press, that is operating at maximum allowed pressure.
terminal radar control *Transportation Engineering.* a radar-control service provided in the vicinity of an airport.
terminal radar service area *Transportation Engineering.* the geographical area served by a terminal radar control system.
terminal redundancy *Molecular Biology.* the existence of similar base-pair sequences at the two ends of a linear DNA molecule.
terminal repeater *Telecommunications.* the repeater found at the terminus of an amplified trunk.
terminal room *Telecommunications.* in telephony, the room that contains the main distribution frames, relays, and similar equipment for lines and trunks within a central office building.
terminal sinus *Developmental Biology.* a vein that encircles the vascular portion of the blastoderm.
terminal station *Telecommunications.* **1.** the receiving equipment and associated multiplex equipment employed at the ends of a radio-relay system. **2.** in a submarine cable system, the physical plant located near the landing point of a sea-cable system, containing the terminal equipment.
terminal strip see TERMINAL BOARD.
terminal stub *Electricity.* an accessory cable that comes with a cable terminal and is used for splicing into the main terminal as necessary. Also, TERMINAL LEG.
terminal symbol *Computer Science.* a symbol in a phrase-structure grammar that is a part of the language described by the grammar, such as a word or character of the language.
terminal transferase *Enzymology.* an enzyme that catalyzes the sequential addition of deoxyribonucleotides to the 3'-OH end of a DNA strand without the need for a template.
terminal unit *Mechanical Engineering.* the unit that lies at the end of a branch duct of an air-conditioning system and delivers air to the conditioned space.
terminal vehicle *Space Technology.* the portion of a rocket vehicle that is the last to separate and continue traveling with a payload such as an earth satellite or lunar probe.
terminal velocity *Physics.* the constant velocity of a falling body attained when the force of resistance of the medium through which it is falling is equal in magnitude and opposite in direction to the force of gravity. *Fluid Mechanics.* specifically, the velocity of an object falling through a fluid when there is no net force acting on it, i.e., where the gravity and drag forces are equal. *Ordnance.* the remaining speed of a projectile at the point in its downward path at which it is level with the muzzle of the weapon. *Aviation.* the maximum speed that a particular aircraft or other body can achieve under given conditions.
terminal vertex *Mathematics.* in graph theory, a vertex of degree one. Also, ENDPOINT, LEAF.
terminal very-high-frequency omnirange *Navigation.* a low-power VOR-type air navigation system designed for use as an approach aid at or near an airport. Also, **terminal VOR, TVOR.**
terminal web *Cell Biology.* a specialized area found at the apex of an intestinal epithelial cell near the base of the microvilli, containing a network of spectrin proteins and actin filaments and thought to permit the movement of the microvilli.
terminated line *Electricity.* the transmission line that terminates in an impedance equal to the prevailing impedance of the line, in order to negate reflection or standing waves.

terminating the fact of bringing to an end; specific uses include: *Electricity.* a circuit closure at either end of a transmission line or transducer by the connection of some type of device.

termination the place or part where anything ends; specific uses include: *Materials Science.* the steps in a chain reaction in which reactive reaction intermediates are destroyed or rendered inactive by terminators, thus ending the chain. *Computer Science.* **1.** the end of the execution of a program, as indicated by a halt instruction or returning of control to the operating system. **2.** a property of a program that it will eventually terminate. *Electromagnetism.* the use of a load at the end of a transmission line or other device whose impedance, if matched to that of the line, will create no reflections.

termination codon see NONSENSE CODON.

termination factor *Molecular Biology.* **1.** a protein that recognizes termination codons and stimulates the release of polypeptides during genetic translation. **2.** a protein that causes transcription by *E. coli* RNA polymerase to terminate at specific sites in the DNA template.

termination sequence *Molecular Biology.* a DNA sequence that gives the signal to terminate transcription by RNA polymerase.

terminator *Astronomy.* the line of sunrise or sunset on the surface of a moon or planet. *Molecular Biology.* a DNA sequence where the synthesis of mRNA is terminated.

terminator codon *Genetics.* a nonsense codon that signals the end of a growing polypeptide chain.

termitarium *Invertebrate Zoology.* a nest, mound, or colony inhabited by termites; some massive earthen structures constructed by termites may be as high as 25 feet.

termite *Invertebrate Zoology.* any of numerous social insects of the order Isoptera that live in highly organized colonies in the manner of ants; many species feed on wood and are highly destructive to trees, to the wooden framework of buildings, and to other objects of wood or of paper. Termites are popularly known as white ants, though they are not a type of ant.

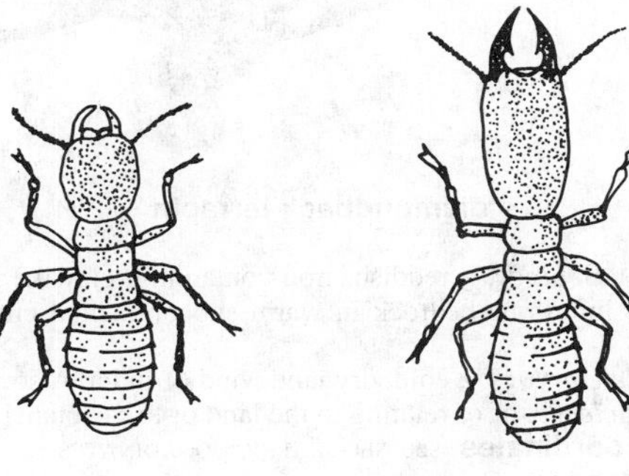

termite

termite shield *Building Engineering.* a protective shield of noncorrosive metal or galvanized iron positioned between the top of foundation walls and wood plates, pipes, or sills; it extends two inches over the wall with a 45° bend-down angle, thus allowing termites to pass without getting into the structural foundation of a building.

termitic *Invertebrate Zoology.* **1.** of or relating to termites. **2.** caused by termites.

termiticole *Ecology.* an organism that inhabits a termite's nest.

Termitidae *Invertebrate Zoology.* a large family of tropical and subtropical termites, including most of the order Isoptera, elaborate mound or nest builders.

Termitomycetes *Mycology.* a genus of fungi belonging to the family Amanitaceae, which occurs only in termite nests.

termitophile *Ecology.* **1.** an organism that spends part of its life among termites, often preying on them or stealing their food. **2.** an organism that lives among termites and forms a mutually beneficial relationship with them. Thus, **termitophilic, termitophilous.**

Termopsidae *Invertebrate Zoology.* a family of termites in the order Isoptera that favor damp wood.

term splitting *Quantum Mechanics.* the separating of the energy levels of the different states all corresponding to the same orbital angular momentum in the atom.

tern *Vertebrate Zoology.* any of numerous cosmopolitan seabirds of the subfamily Sterninae of the family Laridae, related to gulls and characterized by white and black plumage, a long, deeply forked tail, and pointed wings, and having a graceful flight, especially those of the genus *Sterna*; they are found in Eurasia and the Americas, often in large colonies.

tern

ternary *Science.* consisting of three parts; based upon the number three. *Mathematics.* **1.** relating to base three. For example, **ternary expansion** is the process of representing a real number in base three; the notation used is **ternary notation. 2.** relating to division into three parts. For example, the Cantor set is sometimes called Cantor's ternary set because it is constructed by repeated division of an interval into thirds. The function that maps the Cantor set onto the unit interval is called the Cantor ternary function. **3.** relating to three objects. For example, a **ternary relation** is a set of ordered 3-tuples. *Systematics.* see TRINOMEN.

ternary alloy *Metallurgy.* an alloy containing three principal elements.

ternary code *Telecommunications.* a code in which each code event or element may be any one of three specific levels or values.

ternary compound *Chemistry.* a chemical compound composed of three elements.

ternary diagram *Petrology.* a triangular diagram depicting the composition of a three-component mixture.

ternary incremental representation *Computer Programming.* a type of incremental representation in which only three changes (or increments) in the values of the variables, +1, 0, or –1, are used.

ternary system *Chemistry.* any system having three components.

ternate *Botany.* a structure having three parts, especially a compound leaf with three leaflets.

terne *Metallurgy.* steel coated with an alloy that is four parts lead to one part tin. Also, **terne metal.**

Terne *Ordnance.* a Norwegian rocket-propelled, antisubmarine depth charge that carries a 110-pound, high-explosive warhead; target data are preset before firing.

terneplate *Metallurgy.* a thin sheet of iron or steel coated with terne.

Ternifine man *Paleontology.* a set of remains (three mandibles, teeth, and a parietal bone) thought to belong to the species *Homo erectus*, found at Ternifine, Algeria, in 1954–55; very similar to the remains found at Zhoukoudian, China, which are usually referred to as Peking man; possibly also related to the remains known as Heidelberg man; dated at about 700,000 years Before Present.

terpene *Organic Chemistry.* **1.** any one of a class of hydrocarbons having the formula $(C_5H_8)_n$, found especially in plant oils, resins, and balsams. **2.** $C_{10}H_{16}$, an unsaturated liquid recovered from essential oils in plants and used in synthesizing menthol, camphor, and terpineol.

terpene alcohol *Organic Chemistry.* an alcohol compound derived from or related to a terpene hydrocarbon.

terpene hydrochloride *Organic Chemistry.* $C_{10}H_{16}\cdot HCl$, a solid that melts at 125°C; used as an antiseptic. Also, ARTIFICIAL CAMPHOR, TURPENTINE CAMPHOR.

terpenism *Toxicology.* poisoning due to ingestion of or skin contact with a terpene, characterized by vomiting, convulsions, and heart and lung problems.

terpenoid *Organic Chemistry.* a compound having a structure similar to that of a terpene hydrocarbon.

terphenyl *Organic Chemistry.* 1,2-$(C_6H_5)_2C_6H_4$, a crystalline solid; soluble in acetone, benzene, chloroform, and methanol; melts at 58°C. Also, **1,2-diphenylbenzene.**

***o*-terphenyl** *Organic Chemistry.* 1,2-$(C_6H_5)_2C_6H_4$, a crystalline solid; soluble in acetone, benzene, chloroform, and methanol; melts at 58°C and boils at 337°C. Also, **1,2-diphenylbenzene.**

***p*-terphenyl** *Organic Chemistry.* $(C_6H_5)_2C_6H_4$, a combustible, toxic liquid; melts at 213°C and boils at 389°C or 250°C (45 torr); polymerized with styrene to make plastic phosphors, and single crystals used as scintillation counters. Also, **1,4-diphenylbenzene.**

terpineol *Organic Chemistry.* $C_{10}H_{17}OH$, a colorless liquid; soluble in alcohol and water; boils at 214°C and is derived from pine oil; used in making soaps, perfumes, and disinfectants and as a solvent, food flavoring, and antioxidant.

α-terpineol *Organic Chemistry.* $C_{10}H_{17}OH$, combustible, colorless liquid terpene or transparent crystals; slightly soluble in water and glycerol; boils between 214–224°C and solidifies at 31–36°C; derived from pine oil and used in making soaps, perfumes, and disinfectants and as a solvent, food flavoring, and antioxidant.

terpin hydrate *Organic Chemistry.* $CH_3(OH)C_6H_9C(CH_3)_2OH{\cdot}H_2O$, combustible, lustrous, rhombic, colorless crystalline prisms or white powder; slightly soluble in water and soluble in alcohol and ether; melts at 115–117°C and boils at 258°C; used in medicine and pharmaceuticals and as a source of terpineol. Also, DIPENTENE GLYCOL.

terpinolene *Organic Chemistry.* $C_{10}H_{16}$, a flammable, water-white to pale-amber liquid; insoluble in water and soluble in alcohol and ether; boils at 183–185°C; used as a solvent and in the manufacture of synthetic resins and flavors.

terpinyl acetate *Organic Chemistry.* $C_{10}H_{17}OOCCH_3$, a combustible, colorless liquid; slightly soluble in water and soluble in alcohol; boils at 220°C and freezes at –50°C; used in food flavors and perfumes.

terpolymer *Organic Chemistry.* a polymer composed of three different monomers.

terra *Science.* the Latin word for "land" or "earth." *Astronomy.* a term for the highland (nonmare) areas on the moon; brighter than the maria, they are very heavily cratered.

terrace *Architecture.* **1.** a flat roof or open platform adjoining a building, usually used for dining or leisure; a deck or balcony. **2.** a level embankment, often paved or planted and balustraded. *Building Engineering.* a tiered building effect in masonry or turf where levels rise one above another via vertical or sloping sides of brick. *Geology.* **1.** a generally flat, long and narrow surface bordered on one side by a steep ascending slope and on the other side by a steep descending slope. **2.** a horizontal or gently sloping ridge of earth built along a hillside to decrease erosion or conserve moisture. **3.** a narrow section of coastal land that dips gently toward the water. **4.** a benchlike formation on the ocean floor. *Agronomy.* a long, low ridge of earth constructed across a slope with a flat or graded channel to stop soil erosion and control the runoff of water.

terracette *Geology.* a minor ledge or steplike formation produced on the surface of a slumped soil mass along a grassy slope.

terracing *Agronomy.* the practice of making a ditch with a ridge below, across a slope at various vertical intervals to control runoff.

terra cotta *Materials.* an unglazed, lightweight clay material that is typically red in color and used as floor tile and roofing, for sculpture, or for various decorative applications. (An Italian term meaning literally "baked earth.")

terrain *Geography.* a tract of land having or considered in terms of certain natural features and characteristics.

terrain-avoidance system *Navigation.* a radar system that provides the pilot or navigator of an aircraft with a display of the ground or obstacles that project above either a horizontal plane through the aircraft or a parallel plane. Also, **terrain-avoidance radar.**

terrain-clearance system *Navigation.* a system that provides the pilot or autopilot of an aircraft with climb or dive signals, so that the craft will maintain a selected height over flat ground and clear the peaks of undulating ground; it differs from the terrain following system in that the craft will not descend into a valley to follow the ground contour.

terrain error *Navigation.* a navigational error due to using a wave that has been distorted by the terrain over which it has traveled.

terrain exercise *Military Science.* a hypothetical exercise in which a stated military situation is solved on the ground using imaginary troops; the solution is usually written.

terrain flight *Navigation.* flight close to the earth's surface during which airspeed, height, or altitude is adapted to the contours and cover of the ground in order to avoid enemy fire.

terrain-following system *Navigation.* a radar system that provides the pilot or autopilot of an aircraft with climb or dive signals that will enable the aircraft to maintain a selected height above the contour of the ground. Also, **terrain-following radar.**

terrain intelligence *Military Science.* processed data on the military significance of natural and manmade characteristics of an area.

terrain profile recorder see AIRBORNE PROFILE RECORDER.

terrain sensing *Engineering.* the use of instruments such as radar, photography, infrared sensing, and the like, to survey a land area from an overhead satellite or aircraft.

terral *Meteorology.* a light, evening land breeze alternating with a strong daytime sea breeze (virazon) in western South America.

terral levante *Meteorology.* an occasionally squall-like foehn out of the northwest, which most often blows as a land breeze over Spain and Brazil.

terrane *Geology.* a region in which a particular rock or group of rocks occurs.

terrapin *Vertebrate Zoology.* any of several edible semiaquatic turtles of the family Emydidae; characterized by a low shell and webbed feet; usually inhabiting fresh or brackish waters, especially the **diamondback terrapin** of the Atlantic and Gulf coasts; found in North America. Some species are endangered.

diamondback terrapin

terra rossa *Geology.* a reddish, iron-containing, claylike soil that develops over limestone bedrock in warm subcontinental climates. Also, **terra rosa.**

terrenho *Meteorology.* a cold, dry land wind of India. Also, **tereno.**

terrestrial *Science.* of or relating to the land or to the planet earth.

terrestrial coordinates see GEOGRAPHIC COORDINATES.

terrestrial deposit *Geology.* **1.** an accumulation of sedimentary material occurring on land, beyond the border to which the tides extend. Also, TERRESTRIAL SEDIMENT. **2.** any accumulation of sedimentary material on land that does not result from the action of water.

terrestrial dynamical time *Astronomy.* a system of time that is tied to International Atomic Time; used in geocentric ephemerides.

terrestrial electricity *Geophysics.* all natural electrical phenomena of the earth and its atmosphere, including atmospheric electricity.

terrestrial frozen water *Hydrology.* water on the surface of the earth that is either seasonally or perennially frozen.

terrestrial globe *Cartography.* a sphere on which selected geographical features or political divisions of the earth's surface are represented to scale and in their relative positions.

terrestrial gravitation *Geophysics.* the gravitation and gravitational phenomena generated by the earth.

terrestrial meridian see ASTRONOMICAL MERIDIAN.

terrestrial photogrammetry *Photogrammetry.* the process of photogrammeteric measurement based on photographs taken from land-based stations, rather than from remote stations (aircraft or satellites).

terrestrial planet *Astronomy.* all planets nearest the sun and characterized by solid silicate surfaces and high mean densities: Mercury, Venus, Mars, and Earth.

terrestrial radiation *Geophysics.* the infrared energy emitted by the earth and its atmosphere, measured at wavelengths that are determined by their temperature.

terrestrial reference guidance *Navigation.* a guidance technique in which a missile is provided with characteristics of the surface over which it is flown, such as magnetic fields or atmospheric pressure, thereby achieving flight along a predetermined path. Similarly, **terrestrial reference flight.** Also, **terrestrial guidance.**

terrestrial refraction *Optics.* the bending of light rays toward the zenith, caused by density variations in the atmosphere.

terrestrial scintillation *Optics.* flashes of light seen in a light source within the earth's atmosphere. Also, ATMOSPHERIC SHIMMER, OPTICAL HAZE. *Meteorology.* see SHIMMER.

terrestrial sediment see TERRESTRIAL DEPOSIT, def. 1.

terrestrial telescope *Optics.* a generally portable telescope having inverting lenses or an eyepiece that produces an upright image.

terrestrial triangle *Navigation.* a navigational triangle on the surface of the earth.

terricolous *Biology.* living in or on the ground.

Terrier *Ordnance.* a surface-to-air missile with solid-fuel rocket motor, radar-beam rider or homing guidance, and a nuclear or nonnuclear warhead; officially designated **RIM-2.**

Terrier land weapon system *Ordnance.* a surface-to-air missile system developed specifically for amphibious operations; it utilizes the Terrier RIM-2B or RIM-2C missile with ground-launching and guidance equipment.

terrigenous deposit *Geology.* an accumulation of sedimentary material derived from land erosion and deposited in shallow ocean areas. Also, **terrigenous sediment.**

territorial *Behavior.* of or relating to animal territory. Often used in compounds such as **territorial display, territorial fighting, territorial marking, territorial song.**

territorial behavior *Behavior.* any behavior that is used by an animal to advertise or defend its territory, including marking, threat behaviors, and fighting.

territoriality *Behavior.* **1.** the fact or concept of an animal's having an established territory. **2.** the behaviors related to the demarcation and protection of such an area by an individual or group of individuals of the same species, for the purposes of feeding, mating, and caring for the young.

territory *Behavior.* an area that is defined or established in some way, occupied by an animal or animals, and defended against the intrusion of others of the same species.

terrycloth or **terry cloth** *Textiles.* a highly absorbent cotton or cotton-blend fabric with uncut loops on one or both sides.

tert- *Organic Chemistry.* the abbreviation for tertiary, indicating replacement of a third degree, particularly replacement of three of the hydrogens in a methyl radical by three other carbon atoms, i.e., $R_1R_2R_3C$–.

tertian *Pathology.* describing a fever, such as malaria, characterized by paroxysms that recur every other day.

tertiary [tur´shē âr´ē] *Science.* **1.** third in a sequence or order. **2.** third in importance; neither primary nor secondary.

Tertiary *Geology.* **1.** the geologic subera or period of the Cenozoic era extending from the end of the Mesozoic era to the beginning of the Quaternary period, covering a time span from about 63 million to 1.6 million years ago. **2.** describing the rocks formed during that time.

tertiary air *Mechanical Engineering.* the preheated air that is added to primary and secondary air in a combustion process.

tertiary alcohol *Organic Chemistry.* $RR'R''COH$, a trisubstituted organic alcohol in which R, R', and R'' are either identical or different groups, as in *tert*-butyl alcohol.

tertiary amine *Organic Chemistry.* $RR'R''N$, a trisubstituted amine in which R, R', and R'' are either identical or different groups.

tertiary amyl alcohol see AMYLENE HYDRATE.

tertiary calcium phosphate see TRIBASIC CALCIUM PHOSPHATE.

tertiary carbon atom *Organic Chemistry.* a carbon atom attached to three other carbon atoms with single bonds.

tertiary circulation *Meteorology.* a usually small, localized atmospheric circulation such as a thunderstorm or local wind.

tertiary consumer *Ecology.* an organism that feeds on other organisms, which are themselves secondary consumers; e.g., field mice (primary consumers) eat plants and are in turn eaten by snakes (secondary consumers), which then may be eaten in turn by hawks (tertiary consumers). Many predators function as both secondary and tertiary consumers, depending on the prey.

tertiary crushing *Mining Engineering.* **1.** the preliminary crushing of run-of-mine ore and, less often, coal. **2.** the third stage of crushing, subsequent to primary and secondary crushing.

tertiary grinding *Mining Engineering.* the ore grinding performed by the third in a series of ball mills; used for very fine grinding.

tertiary hydrogen atom *Organic Chemistry.* a hydrogen atom that is attached to a tertiary carbon atom.

tertiary pyroelectricity *Solid-State Physics.* polarization in pyroelectric and nonpyroelectric crystals that have no polar direction, resulting from temperature gradients and mechanical stresses induced by a nonuniform heating of the crystal.

tertiary recovery *Petroleum Engineering.* the third major stage of oil production, employing techniques designed to change crude oil properties; one such technique is micellar flooding, used to reduce the surface tension of crude and free it from the porous formations to which it adheres.

tertiary sodium phosphate see TRIBASIC SODIUM PHOSPHATE.

tertiary storage see ARCHIVAL STORAGE.

tertiary structure *Biochemistry.* the three-dimensional folding of polypeptide chains in a protein molecule.

tertiary triuranium octoxide *Inorganic Chemistry.* U_3O_8, radioactive, poisonous, greenish black crystals; insoluble in water and decomposes on heating; used in nuclear technology and preparation of uranium compounds. Also, URANOUS-URANIC ACID, URANYL URANATE.

tertiary winding see STABILIZED WINDING.

terylene *Organic Chemistry.* $(–COC_6H_4COOCH_2CHOH–)_n$, a polymer used in producing a textile commercially called Dacron.

teschemacherite *Mineralogy.* $(NH_4)HCO_3$, a colorless, white, or yellow orthorhombic mineral occurring as compact, crystalline masses, having a specific gravity of 1.58 and a hardness of 1.5 on the Mohs scale; found in guano deposits in Peru and South Africa.

teschen disease *Veterinary Medicine.* an acute viral infection of swine that is characterized by nervous symptoms, ataxia, anorexia, convulsions, and paralysis; resembles poliomyelitis in humans. Also, INFECTIOUS PIG PARALYSIS, INFECTIOUS PORCINE ENCEPHALOMYELITIS, TALFAN DISEASE.

teschenite *Petrology.* a granular, hypabyssal igneous rock consisting primarily of calcic plagioclase, augite, and sometimes hornblende, with smaller amounts of brotite and analcime.

Tesla, Nikola 1856–1943, American electrical engineer, born in Croatia (Austria-Hungary); responsible for many inventions including the Tesla coil and induction motor.

tesla *Electromagnetism.* the meter-kilogram-second unit of magnetic induction equivalent to one weber per square meter. (Named for Nikola *Tesla.*)

Tesla coil *Electromagnetism.* a high-frequency, high-voltage air-core transformer used in conjunction with a capacitor and spark gap.

Tessaratomidae *Invertebrate Zoology.* a family of large tropical true bugs, hemipteran insects in the group Pentatomorpha.

Tessar lens *Optics.* a triplet-based camera lens with a rear element composed of cemented positive and negative elements.

Tessarolax *Paleontology.* a genus of gastropods in the order Mesogastropoda; characterized by long, thin, curving spinelike processes of the lips that sometimes grew to 7 cm across; extant in the Jurassic and Cretaceous.

tessellate *Design Engineering.* having markings arranged in a pattern of small squares, like a checkerboard, as in mosaic tile. *Botany.* having markings arranged in a checkerboard pattern. Also, **tesselated.**

tessellation *Mathematics.* a tiling of the plane in which the tiles are congruent polygons. Also, **tesselatation.**

tessera *plural,* **tesserae.** *Architecture.* a small square of marble, glass, stone, or tile used in making mosaics. (A Greek word meaning "four.")

test *Science.* **1.** an examination or trial, especially one by which the presence, quality, or composition of something is determined. **2.** to carry out such an examination or trial. *Chemistry.* **1.** a significant chemical reaction, especially one that by its nature indicates the presence or composition of a substance under study. **2.** a reagent used to initiate such a reaction. **3.** an indication, obtained by such a reaction, of the presence or composition of a given substance.

test *Invertebrate Zoology.* an external protective or skeletal covering, usually calcareous, siliceous, chitinous, fibrous or membranous, secreted, or built. *Botany.* see TESTA.

testa *Botany.* the hard outer seed coat that is several layers thick and derived from the integuments of the ovule. Also, TEST.

testability *Engineering.* the degree to which the requirements of a given piece of hardware or software can be tested.

Testaceafilosida *Invertebrate Zoology.* an order of filose amoebas having a test or shell.

Testacealobosa *Invertebrate Zoology.* a subclass of amoebae with lobose pseudopodia and tests. Also, **Testacealobosia.**
testaceous *Biology.* **1.** covered by or made of a protective shell-like material; of the nature of a test. **2.** being dull reddish-brown in color.
Testacida or **Testacea** see ARCELLINIDA.
testalgia *Medicine.* pain in the testis.
test-and-set *Computer Science.* a CPU instruction that both tests the value of a lock variable and sets it to a locked state in a single memory cycle; used to protect a critical region.
testate *Biology.* having or covered by a test.
test bed *Aviation.* **1.** a flight vehicle used to test other objects in flight. Also, FLYING TEST BED. **2.** a stand, frame, or base designed for testing an engine or other equipment. *Computer Programming.* a program that is used to test the parameters or functioning of a new program.
testboard *Electricity.* a switchboard equipped with testing apparatus arranged so that connections are made from it to telephone lines or to central-office equipment for testing purposes.
test chamber *Aviation.* the section of a wind tunnel where flight vehicles are tested and observed. Also, TEST SECTION. *Space Technology.* an environmentally controlled sealed chamber, simulating the stratosphere or hard vacuum with space solar radiation, in which a person or object is subjected to an experiment.
test clip *Electricity.* a spring clip device that is fastened to the end of an insulated wire so as to allow for a quick, temporary connection when circuits or devices are in a testing mode.
test cross *Genetics.* a cross in which an individual of unknown genotype is crossed to a homozygous recessive organism, in order to determine the unknown genotype.
test data *Computer Programming.* sample data used to test a program or module, such that the expected results are known.
test driver *Computer Programming.* a program that invokes and executes another program or routine, and provides to it a set of test data.
testectomy see ORCHIECTOMY.
tester strain *Genetics.* a strain that is homozygous for one or more recessive alleles and is used for test crosses.
testes the plural of testis.
test failure *Space Technology.* in the development phase of a rocket vehicle or missile, a relative failure to achieve instrumentation data or other desired information as a result of an abort, equipment breakdown, or miscalculation.
test file *Computer Programming.* a sample data file used to test a program under development, ideally covering the full range of information that the new program is designed to process.
test fire *Forestry.* a controlled fire that is started to evaluate such factors as fire behavior, detection efficiency, and suppression measures.
test firing *Space Technology.* a live or a static firing of a rocket engine, performed in order to make controlled observations of the engine or of one of its components.
test flight *Aviation.* a flight that takes place under controlled conditions, made to observe the operation or performance of a flight or its components or equipment, especially when the object of the test has recently been developed or modified.
test function *Mathematics.* In the theory of distributions or generalized functions, test functions are infinitely differentiable (C^∞) functions with compact support. A distribution f is then interpreted as a linear functional that assigns the value $\int_\Omega f \cdot \phi$ to every test function ϕ.
test glass *Optics.* an optical element, having a curvature equal but opposite to the desired curvature of a lens, used to check the accuracy of lenses while they are being ground and polished; the test glass is placed in contact with the lens, the two are viewed under monochromatic light, and any imperfection in the lens shape results in visible interference in the light waves (Newton's rings).
test hole *Mining Engineering.* a hole drilled or a shallow excavation dug for purposes of testing an ore body or extracting rock samples for analysis.
testicle see TESTIS.
testicular *Anatomy.* relating to a testis.
testiculoma *Oncology.* a tumor of the testicle and testicular tissue.
testing *Science.* the process of carrying out a test or series of tests.
testing level *Electricity.* a reference power level value represented by 0.001 watt in a 600-ohm application.
testis *Anatomy.* either of the pair of primary male reproductive organs that produce both sperm cells and testosterone. Also, TESTICLE.
test jack *Electricity.* a jack with contacts oriented in series with a circuit in order to locate faults through the use of a testing device.
test lead *Electricity.* a flexible insulated wire attached to a test prod; used primarily for connecting meters and test instruments to monitor a circuit under test.
test of hypothesis see HYPOTHESIS TESTING.
test of significance *Statistics.* a measure of the amount of support lent by an experiment to a statistical hypothesis, independent of any alternative hypothesis.
testolactone *Oncology.* $C_{19}H_{24}O_3$, an antineoplastic agent derived from testosterone or progesterone; used as an adjunct to breast cancer therapy in postmenopausal women.
testolone acetate *Oncology.* $C_{21}H_{30}O_3$, an antineoplastic agent and androgenic steroid.
test oscillator see SIGNAL GENERATOR.
testosterone [tes täs´tə rōn´] *Endocrinology.* the major androgenic steroid hormone, produced by the Leydig cells of the male testes and in smaller quantities by the adrenal glands of both sexes, and necessary for the development and maintenance of normal sexual function in the male. *Pharmacology.* $C_{29}H_{28}O_2$, this substance produced synthetically from cholesterol or isolated from bull testes; used in treating male hypogonadism and symptoms of male climax, and sometimes administered for its anabolic properties.

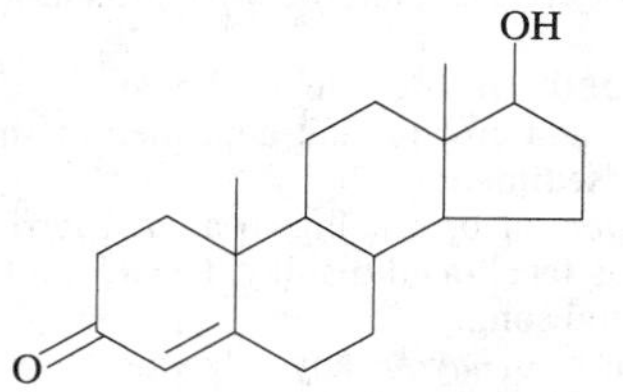

testosterone

test pattern *Telecommunications.* in television, a chart that consists of combinations of squares, circles, and lines and that serves to check or adjust receiving equipment. Also, RESOLUTION CHART.
test pile *Civil Engineering.* a pile to which a load is applied to determine the load-settlement characteristics of the pile and the surrounding ground for further design.
test pilot *Aviation.* the pilot of a test flight.
test pit *Civil Engineering.* an excavation for subsurface soil investigation to determine the type of soil, or to prospect for minerals. *Archaeology.* see CONTROL PIT.
test plant *Virology.* any plant species that is used for diagnosing a plant virus or evaluating its degree of infectivity.
test point *Electricity.* a terminal or plug-in connector provided in a circuit and used for the repair, troubleshooting, or debugging of a circuit; often labeled by the letters TP followed by a numeral, such as TP1 or TP2.
test prod *Electricity.* a sharp metal probe with a flexible insulated lead terminating in a plug or lug for attachment to an instrument; used to make a touch connection circuit to a terminal and has the means for electrically connecting a point to a test lead.
test program *Computer Programming.* a sample program file used to test computer hardware under development, ideally covering the full range of applications that the new hardware is designed to process.
test reactor *Nucleonics.* a nuclear reactor designed for conducting research on the properties of materials to be subjected to temperatures and radiation found in an operating reactor.
test record *Computer Programming.* any record in a test file.
test run *Computer Programming.* a run, using test data, that is performed to determine whether a program is operating correctly.
test section see TEST CHAMBER.
test set *Electronics.* a set of one or more units of equipment used for the testing and maintenance of electronics equipment; usually intended to have some degree of specialization in application, such as testing a particular type or model of equipment or performing a particular series of tests.
test specimen *Materials Science.* see TENSILE BAR.
test square *Archaeology.* an excavation unit, typically a 2-meter square, excavated as a test or sample.
test stand *Aviation.* **1.** a stationary stand or platform on which a reciprocating aircraft engine is tested; may include associated testing equipment. **2.** a stand or platform on which a jet or rocket engine is mounted for testing and measuring thrust. Also, PROVING STAND, THRUST STAND.

test statistic *Statistics.* the function of the observed sample on the basis of which the tested hypothesis is accepted or rejected.

test-tube baby *Developmental Biology.* a popular term for a human embryo that is produced through insemination outside the body of the mother.

testudinal *Vertebrate Zoology.* relating to or resembling a tortoise or a tortoise shell.

Testudinellidae *Invertebrate Zoology.* a family of free-swimming rotifers in the suborder Flosculariacea having two rings of cilia on the corona.

Testudines *Vertebrate Zoology.* an order of reptiles that includes all living turtles, tortoises, and terrapins; characterized by a shell consisting of a dorsal carapace and a ventral plastron. Also, **Testudinata.**

Testudinidae *Vertebrate Zoology.* the tortoises, a family of terrestrial reptiles of the infraorder Cryptodira in the order Testudines; characterized by elephantine feet with reduced digits and usually having domed shells; found in grasslands and deserts almost worldwide.

test-under-mask *Computer Programming.* a computer instruction to compare the part of the contents of a memory location or register enabled by a specific bit pattern or mask with a specific value.

test well *Petroleum Engineering.* an exploratory well used to determine the presence and quantity of oil and the feasibility of its production.

tetanic *Medicine.* of or relating to tetanus or tetany.

tetanic contraction *Medicine.* see TETANUS.

tetaniform *Medicine.* resembling tetanus or tetany.

tetanigeneous *Medicine.* producing tetanus or tetany.

tetanode *Medicine.* the unstimulated period between the tetanic contractions in tetanus.

tetanoid paraplegia see SPASTIC PARAPLEGIA.

tetanolysin *Biochemistry.* the hemolytic exotoxin produced by *Clostridium tetani,* the etiologic agent of tetanus.

tetanophobia *Psychology.* an irrational fear of contracting tetanus.

tetanospasmin *Biochemistry.* the neurotoxic component of tetanus toxin; one of the more powerful poisons known, it is a highly potent protein that binds to gangliosides and blocks the synaptic terminals of the central nervous system, causing the muscle spasms characteristic of tetanus.

tetanus [tet´nəs; tet´ə nəs] *Pathology.* an acute, potentially fatal infection caused by the endoneurotoxin of the bacteria *Clostridium tetani,* which most often enters the body through a puncture wound; characterized by headache, fever, muscular rigidity, paralysis, and painful muscle spasms resulting in lockjaw. Also, LOCKJAW. *Neurology.* a state of sustained muscular contraction, without periods of relaxation, caused by repetitive stimulation of the motor nerve trunk at frequencies so high that individual muscle twitches are fused and cannot be distinguished from one another. Also, PHYSIOLOGICAL TETANUS, TONIC CONTRACTION, TETANIC CONTRACTION, TETANY.

tetanus antitoxin *Immunology.* a serum used to treat and provide immunity against the infectious disease tetanus, consisting of antibodies that neutralize the tetanus toxin.

tetanus toxin *Toxicology.* a potent exotoxin produced by the bacterium *Clostridium tetani,* consisting of two components, the neurotoxin tetanospasmin and the hemolysin tetanolysin.

tetanus toxoid *Immunology.* an antigen preparation, administered to provide immunity against the infectious disease tetanus, that consists of detoxified tetanus toxin.

tetany *Neurology.* see TETANUS. *Pathology.* a condition characterized by painful spastic muscular contractions, convulsions, twitching of the muscles, and sharp flexion of the wrist and ankle joints; caused by an abnormality of calcium metabolism and associated with vitamin D deficiency, parathyroid hypofunction, or alkalosis.

tetany of the newborn *Pathology.* tetany in the newborn infant, a condition that is caused by insufficient calcium assimilation during gestation.

Tethinidae *Invertebrate Zoology.* a family of dipteran insects in the subsection Acalyptratae.

Tethys *Astronomy.* Saturn III, a moon of Saturn discovered in 1684; it orbits at a distance of 295,000 kilometers from Saturn and measures about 1060 kilometers in diameter.

Teton Range *Geography.* a group of peaks in the Rocky Mountains of northwestern Wyoming and southeastern Idaho (highest peak: Grand Teton, 13,770 ft).

tetra *Vertebrate Zoology.* any of several tropical freshwater fishes of the genus *Hyphessobrycon* in the family Characidae, notably the **red tetra** or **flame tetra,** *H. flammeus,* a popular aquarium fish.

tetra- or **tetr-** a combining form meaning "four."

tetra-allelic *Genetics.* describing a polyploid individual who possesses four different alleles for a certain gene.

tetraamylbenzene *Organic Chemistry.* $(C_5H_{11})_4C_6H_2$, a liquid that boils at 320–350°C; used as a solvent.

tetrabasic *Chemistry.* containing four atoms of replaceable hydrogen.

Tetrabranchia *Invertebrate Zoology.* a subclass of cephalopod mollusks, having two pairs of gills and a coiled, chambered external shell; *Nautilus* is the only living genus.

tetrabromobisphenol A *Organic Chemistry.* $(C_6H_2Br_2OH)_2C(CH_3)_2$, a white solid, soluble in ether, that melts at 180–184°C; used as a flame retardant for paper, plastics, and textiles.

***sym*-tetrabromoethane** see ACETYLENE TETRABROMIDE.

tetrabromofluorescein *Organic Chemistry.* $C_{20}H_8Br_4O_5$, a red crystalline powder that is insoluble in water and soluble in alcohol; melts at 295°C; used in cosmetics, ink, and dyes; its sodium salt, eosin (bromoeosin), is an important biological stain.

tetrabromophthalic anhydride *Organic Chemistry.* $C_6Br_4C_2O_3$, a pale-yellow crystalline solid; melts at 279°C; used as a flame retardant for paper, plastics, and textiles.

tetrabromosilane see SILICON BROMIDE.

tetrabutylthiuram monosulfide *Organic Chemistry.* $[(C_4H_9)_2NCS]_2S$, a combustible, brown liquid; insoluble in water and soluble in acetone, benzene, gasoline, and ethylene dichloride; used as a rubber accelerator.

tetrabutyltin *Organic Chemistry.* $(C_4H_9)_4Sn$, a combustible, colorless or slightly yellow liquid; insoluble in water and soluble in most organic solvents; boils at 145°C, freezes at –97°C, and decomposes at 265°C; used as a lubricant, fuel additive, and polymerization catalyst.

tetrabutyl titanate *Organic Chemistry.* $Ti(OC_4H_9)_4$, a combustible, colorless to light-yellow liquid; decomposes in water and soluble in most organic solvents; boils at 312°C; used to improve adhesion of paints, rubber, and plastics to metal surfaces, and as a catalyst. Also, titanium butylate.

tetrabutyl urea *Organic Chemistry.* $(C_4H_9)_4NCON(C_4H_9)_2$, a combustible liquid that is insoluble in water; boils at 305°C and freezes below –60°C; used as a plasticizer.

tetracaine hydrochloride *Organic Chemistry.* $C_{15}H_{24}O_2N_2 \cdot HCl$, a crystalline solid that is soluble in water; melts at 148°C; used as a local anesthetic.

tetracene *Organic Chemistry.* $C_{18}H_{12}$, an orange solid with a slight green fluorescence in daylight; not easily soluble; melts at 341°C; used in organic synthesis. Also, NAPHTHACENE, RUBENE.

Tetracentraceae *Botany.* a dicotyledonous tree belonging to the order Trochodendrales, characterized by a superior ovary, a perianth with an indumentum that is often stellate, and palmately veined leaves.

Tetraceratops *Paleontology.* a genus of advanced pelycosaurs or early therapsids; recent reexamination of *T. insignis* indicates that *Tetraceratops* was midway between the reptilian pelycosaurs and the more advanced therapsids, which were the precursors of mammals; extant in the Lower Permian.

1,2,3,4-tetrachlorobenzene *Organic Chemistry.* $Cl_4C_6H_2$, combustible, white crystals; insoluble in water; melts at 46.6°C and boils at 254°C; used in synthesis and as a component of dielectric fluids.

***sym*-tetrachlorodifluoroethane** *Organic Chemistry.* $FCCl_2CFCl_2$, a nonflammable, toxic, white solid or colorless liquid; insoluble in water and soluble in alcohol; melts at 26°C and boils at 93°C; used as a degreasing solvent.

***sym*-tetrachloroethane** *Organic Chemistry.* $CHCl_2CHCl_2$, a liquid that is soluble in water, alcohol, and ether; boils at 147°C; used as an organic solvent, metal cleaner, paint remover, and weed killer.

2,3,4,6-tetrachlorophenol *Organic Chemistry.* Cl_4C_6HOH, toxic, nonflammable, brown flakes; soluble in alcohol and ether; melts at 69–70°C and boils at 150°C (15 torr); used as a fungicide.

tetrachlorophthalic acid *Organic Chemistry.* $C_6Cl_4(COOH)_2$, colorless crystalline plates; soluble in hot water and slightly soluble in cold water; melts to form the anhydride at 250°C; used in dyes and intermediates.

tetrachlorophthalic anhydride *Organic Chemistry.* $C_6Cl_4(CO)_2O$, a nonhygroscopic, white powder that is slightly soluble in water; melts at 254–255°C and boils at 371°C; used in dyes, pharmaceuticals, and plasticizers, and as a flame retardant.

tetrachlorosilane see SILICON CHLORIDE.

Tetracoralla *Paleontology.* a term formerly used to refer to the rugose corals. Also, **Tetracorallia.**

tetracosane *Organic Chemistry.* $CH_3(CH_2)_{22}CH_3$, combustible crystals; insoluble in water and soluble in alcohol; melts at 51.5°C and boils at 391°C; used in organic synthesis.

Tetractinomorpha *Invertebrate Zoology.* a subclass of sponges in the class Demospongiae, having tetraxon spicules.

tetracyanoethylene *Organic Chemistry.* $(CN)_2C{=}C(CN)_2$, colorless crystals; melts at 298–200°C, boils at 223°C, and sublimes above 120°C; used in organic synthesis and dyes.

tetracycline [tet´rə sī´klēn; tet´rə sī´klin] *Microbiology.* any of a group of biosynthetic antibiotics of low toxicity isolated from certain *Streptomyces* bacteria such as *S. aurofaciens* or produced semisynthetically; they are effective against a broad range of bacteria and other microorganisms, including *Mycoplasma, Chlamydia, Brucella, Rickettsia,* and *Vibrio.*

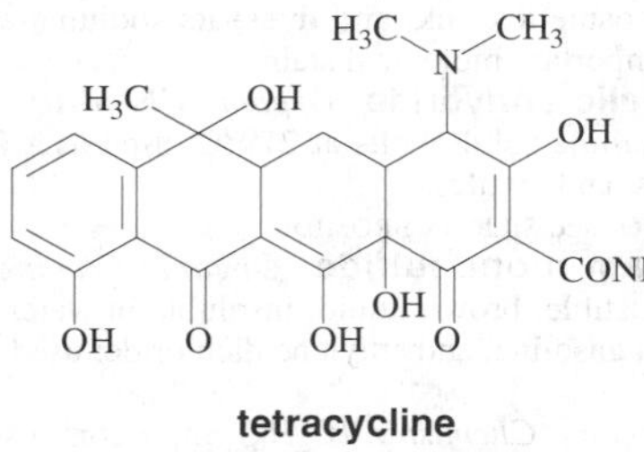

tetracycline

tetrad *Cell Biology.* a group of four homologous chromatids paired together during the first meiotic prophase. *Science.* any group of four items.

tetradactylous *Vertebrate Zoology.* having four fingers or toes.

tetradactyly *Vertebrate Zoology.* the condition of having four fingers or toes.

tetrad analysis *Genetics.* the study of the crossing-over and segregation of genes during meiosis by separately germinating and growing each haploid spore of certain fungi and observing the expression of a particular gene locus; fungi commonly used for such analyses include *Neurospora, Aspergillus,* and *Saccharomyces.*

***n*-tetradecane** *Organic Chemistry.* $CH_3(CH_2)_{12}CH_3$, a combustible, colorless liquid that is insoluble in water and soluble in alcohol; melts at 5.5°C and boils at 253°C; used as a solvent and in organic synthesis.

1-tetradecanol see MYRISTIC ALCOHOL.

1-tetradecene *Organic Chemistry.* $CH_2{=}CH(CH_2)_{11}CH_3$, a combustible, colorless liquid; insoluble in water but soluble in alcohol and ether; boils at 251°C and freezes at –12°C; used as a solvent in perfumes, flavors, and medicines.

tetradecylamine *Organic Chemistry.* $C_{14}H_{29}NH_2$, a combustible, white solid; insoluble in water and soluble in alcohol and ether; melts at 41–42°C and boils at 291°C; used as an intermediate.

tetradecyl thiol *Organic Chemistry.* $CH_3(CH_2)_{13}SH$, a toxic, combustible liquid that melts at 6.5°C and boils at 176–180°C (22 torr); used as an organic intermediate and in processing synthetic rubber. Also, MYRISTYL MERCAPTAN.

tetradentate ligand *Inorganic Chemistry.* a ligand attached at four points to the metal ion in a coordination compound.

tetradic *Mathematics.* an older term for a linear transformation between dyadics (contravariant 2-tensors).

tetrad of Fallot see TETRALOGY OF FALLOT.

tetradymite *Mineralogy.* Bi_2Te_2S, a pale steel-gray, metallic, trigonal mineral of the tetradymite group, occurring in granular or massive form, having a specific gravity of 7.1 to 7.5 and a hardness of 1.5 to 2 on the Mohs scale; found in gold-quartz veins and contact metamorphic deposits.

tetraethanolammonium hydroxide *Organic Chemistry.* $(HOCH_2CH_2)_4NOH$, a white crystalline solid; soluble in water; melts at 123°C; used as an alkaline catalyst and a solvent for dyes and metal-plating solutions.

tetraethylammonium chloride *Organic Chemistry.* $(C_2H_5)_4NCl$, anhydrous, hygroscopic, colorless crystals; soluble in water and alcohol and slightly soluble in ether; used in medicine. The tetrahydrate melts at 37.5°C.

tetra-(2-ethylbutyl) silicate *Organic Chemistry.* $[(C_2H_5)C_4H_8O]_4Si$, a colorless liquid; insoluble in water and miscible with most organic solvents; boils at 238°C (50 torr) and freezes at –100°C; used as a heat-transfer medium, lubricant, and hydraulic fluid.

tetraethylene glycol *Organic Chemistry.* $HO(C_2H_4O)_3C_2H_4OH$, a combustible, hygroscopic, colorless liquid; soluble in water; boils at 327.3°C and freezes at –4°C; used as a solvent for plasticizers and lacquers.

tetraethylene glycol dimethacrylate *Organic Chemistry.* a combustible, colorless to yellow liquid; insoluble in water; boils at 200°C (1 torr); used as a plasticizer.

tetraethylenepentamine *Organic Chemistry.* $NH_2(CH_2CH_2NH)_3CH_2CH_2NH_2$, a combustible, viscous, hygroscopic liquid; soluble in water and most organic solvents; boils at 333°C and freezes at –30°C; used as a solvent, as an intermediate for oil additives, and in the manufacture of synthetic rubber.

tetraethyl lead *Organic Chemistry.* $Pb(C_2H_5)_4$, a combustible, toxic, colorless liquid; insoluble in water and dilute acids; boils at 198–202°C and freezes at –136°C; decomposes slowly at room temperature and rapidly at 125–150°C. Tetraethyl lead was once widely used as an antiknock agent in gasoline motor fuel, but this use has significantly declined (and has been prohibited in some areas) because of the pollution it produces; other ingredients are now added to gasoline to produce the antiknock effect.

tetraethyl pyrophosphate *Organic Chemistry.* $(C_2H_5)_4P_2O_7$, a toxic, hygroscopic, water-white to amber liquid; miscible with water and most organic solvents; used as an insecticide.

tetrafluoroethylene *Organic Chemistry.* $F_2C{=}CF_2$, a colorless flammable gas; insoluble in water; much heavier than air; boils at –78°C and freezes at –142°C; used as a monomer; its polymer is known commercially as Teflon.

tetrafluorohydrazine *Inorganic Chemistry.* F_2NNF_2, an irritating colorless liquid or gas, explosive on contact with reducing agents or at high pressure; used in organic synthesis and rocket fuel.

tetrafluoromethane see CARBON TETRAFLUORIDE.

tetrafluorosilane see SILICON FLUORIDE.

tetragon *Mathematics.* a polygon having four sides or angles; a quadilateral or quadrangle. Thus, **tetragonal.**

tetragonal unit cell *Crystallography.* a unit cell in which there is a four-fold rotation axis parallel to one axis (arbitrarily chosen as c); as a result the lengths of a and b are identical and all interaxial angles are 90° ($a = b$, $\alpha = \beta = \gamma = 90°$). Also, **tetragonal lattice.**

tetragonal zirconia polycrystal *Materials Science.* a fine-grained, fully dense, single-phase ceramic material that is used to make such objects as cutting tools and golf club inserts.

Tetragonidiales *Botany.* a monofamilial order of freshwater cryptonomads of the class Cryptophyceae, having only a brief motile phase, cells that range in color from yellow to brown to green, and coccoid cells with a strong wall thought to be made of cellulose; composed of the family Tetragonidiaceae.

Tetragoniomyces *Mycology.* a genus of fungi belonging to the class Hymenomycetes that are found in temperate climates.

Tetragraptus *Paleontology.* a genus of Ordovician graptolites having four stipes and simple thecae; an important guide fossil because of the wide distribution of the many species and their restriction to the Lower Ordovician.

tetrahedrite-tennantite series *Mineralogy.* $(Cu,Fe)_{12}Sb_4S_{13}$, a grayish-black to black, metallic, cubic mineral of the tetrahedrite group, occurring as tetrahedral crystals and in massive or granular form, having a specific gravity of 4.6 to 5.1 and a hardness of 3 to 4.5 on the Mohs scale; forms two series with tennantite and with freibergite; found in lower temperature veins and in contact metamorphic deposits. Also, FAHLORE, GRAY COPPER ORE.

tetrahedron *Mathematics.* a polyhedron having four (triangular) sides. The regular tetrahedron is one of the five Platonic solids, and is a topological 3-simplex. *Crystallography.* an actual figure having this shape; a solid body having four equilateral triangular faces. *Aviation.* a landing direction indicator having this shape.

tetrahexahedron *Crystallography.* a 24-sided crystal form consisting of six four-sided pyramids whose square bases are connected to form a cube.

tetrahydrocannabinol *Organic Chemistry.* $C_{21}H_{30}O_2$, the active hallucinogenic agent in marijuana (*Cannabis sativa*); animal tests indicate that it can retard cancer growth and may also promote acceptance of organ transplants.

tetrahydrofolic acid *Biochemistry.* the coenzyme form of folic acid.

tetrahydrofuran *Organic Chemistry.* C_4H_8O, a flammable, toxic, water-white liquid; soluble in water and organic solvents; boils at 66°C and freezes at –65°C; used as a solvent, intermediate, and monomer.

tetrahydrofurfuryl acetate *Organic Chemistry.* $C_4H_7OCH_2OOCCH_3$, a combustible, colorless liquid; soluble in water, alcohol, and ether; boils at 194–195°C; used as a flavoring.

tetrahydrofurfuryl alcohol *Organic Chemistry.* $C_4H_7OCH_2OH$, a combustible, hygroscopic, colorless liquid; miscible with water; boils at 178°C; used as a solvent for resins, in making nylon, leather dyes, and in organic synthesis.

tetrahydrofurfuryl oleate *Organic Chemistry.* $C_{17}H_{33}COOCH_2C_4H_7O$, a combustible, colorless liquid; insoluble in water; boils at 240°C (2 torr) and freezes at –30°C; used as a plasticizer.

tetrahydrofurfuryl phthalate *Organic Chemistry.* $C_6H_4(COOCH_2C_4H_7O)_2$, a combustible, colorless liquid; insoluble in water; melts below 15°C; used as a plasticizer.

tetrahydrolinalool *Organic Chemistry.* $C_{10}H_{21}OH$, a combustible, colorless liquid; soluble in alcohol; boils at 196°C; used in perfumes and flavoring.

tetrahydronaphthalene *Organic Chemistry.* $C_{10}H_{12}$, a combustible, narcotic, colorless liquid; insoluble in water and miscible with most solvents; boils at 206°C and freezes at –25°C; used as a chemical intermediate, solvent, and substitute for turpentine.

Tetrahymena *Invertebrate Zoology.* a genus of ciliate protozoa in the suborder Tetrahymenina; used extensively in physiological and genetic studies, they have been shown to be capable of parasitic existence when injected into various hosts; *T. limacis* and *T. pyriformis* are representative species.

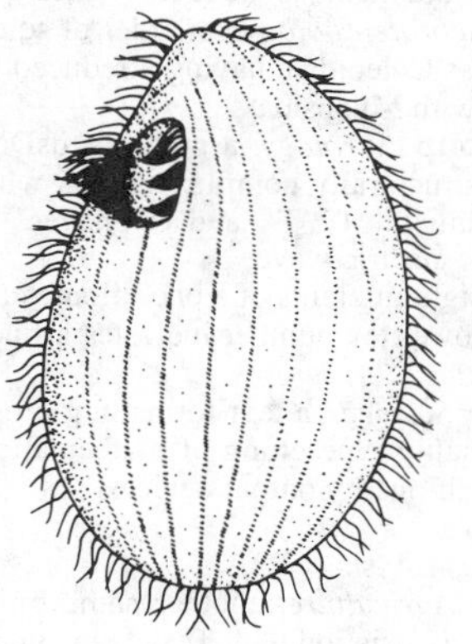

Tetrahymena

Tetrahymenina *Invertebrate Zoology.* a suborder of protozoans belonging to the order Hymenostomatida of the subclass Hymenostomatia, characterized by uniform body ciliation, three oral membranelles on the left, and an undulating or paroral membrane on the right; most are free-living in fresh water, but a few species are symbiotic, mainly in invertebrates.

tetraiodisilane see SILICON IODIDE.

tetraiodoethylene *Organic Chemistry.* $I_2C{=}CI_2$, a light-yellow crystalline solid that turns brown on exposure to light; insoluble in water and soluble in most organic solvents; melts at 187°C; used as a fungicide and in antiseptic ointments and surgical dusting powder.

tetraiodofluorescein *Organic Chemistry.* $C_{20}H_8O_5I_4$, a red powder; insoluble in water and slightly soluble in alcohol and ether; used in analytical chemistry. Also, IODEOSIN.

tetraisopropylthiuram disulfide *Organic Chemistry.* $([(CH_3)_2CH]_2NCS)_2S_2$, a tan powder, insoluble in water and soluble in benzene, chloroform, and gasoline; melts from 95 to 99°C; used as a rubber accelerator.

tetrakis(hydroxymethyl)phosphonium chloride *Organic Chemistry.* $(HOCH_2)_4PCl$, a crystalline compound made from phosphine, formaldehyde, and hydrochloric acid; used as a flame retardant for cotton fabrics.

tetralite see TETRYL.

tetralogy of Fallot [fa lō´] *Medicine.* a combination of four congenital heart defects usually found in adults: pulmonic stenosis, ventricular septal defect, dextroposition of the aorta such that it receives blood from both ventricles, and hypertrophy of the right ventricle. (Named for the French physician Etienne-Louis Arthur *Fallot,* 1850–1911.)

Tetralophodon *Paleontology.* a genus of proboscideans in the extinct family Gomphotheriidae that attained a height of about 8 feet; extant in the Miocene and Pliocene.

tetramer *Organic Chemistry.* a polymer that is formed from four identical monomers. *Science.* any entity that is composed of four parts.

Tetrameristaceae *Botany.* a family of dicotyledonous trees and shrubs belonging to the order Theales, characterized by scattered raphide cells in parenchymatous tissues and by large seeds with abundant endosperm; native to Malaysia and the Guayana Highlands of northern South America.

tetramerous *Botany.* of a flower, having four members or multiples of four in each whorl. *Science.* having, characterized by, or composed of four parts.

***sym*-tetramethylbenzene** see DURENE.

tetramethylbiarsine see CACODYL.

tetramethyldiaminobenzophenone *Organic Chemistry.* $[(CH_3)_2NC_6H_4]_2CO$, combustible, white to greenish crystalline leaflets; soluble in water, alcohol, and ether; melts at 172°C and boils (decomposes) at 360°C; used in the manufacture of dyes. Also, MICHLER'S KETONE.

tetramethylene see CYCLOBUTANE.

tetramethylethylenediamine *Organic Chemistry.* $(CH_3)_2NCH_2CH_2N(CH_3)_2$, a combustible, colorless liquid that is soluble in water and most organic solvents; boils at 121–122°C and freezes at –55.1°C; used in polyurethane formation, and as a corrosion inhibitor, a reaction solvent, and a textile finishing agent.

tetramethyl lead *Organic Chemistry.* $(CH_3)_4Pb$, a flammable, toxic, colorless liquid, insoluble in water and slightly soluble in alcohol and ether; boils at 110°C and freezes at 27.5°C; formerly used as an antiknock additive in gasoline.

2,6,10,14-tetramethylpentadecane see PRISTANE.

tetramethyl rhodamine see RHODAMINE B.

tetramethylsilane *Organic Chemistry.* $(CH_3)_4Si$, a flammable, colorless liquid; insoluble in water and soluble in most organic solvents; boils at 26.5°C; used as aviation fuel and as the internal standard for NMR analytical instruments.

tetramethylthiuram monosulfide *Organic Chemistry.* $[(CH_3)_2NCS]_2S$, a combustible, yellow powder; insoluble in water and gasoline; melts at 104–107°C; used as a fungicide, insecticide, and rubber ultra-accelerator. Also, BISDIMETHYLTHIOCARBAMYL SULFIDE.

tetramethylurea *Organic Chemistry.* $(CH_3)_2NCON(CH_3)_2$, a liquid that is soluble in water and organic solvents; boils at 176.5°C; used as a reagent and solvent.

Tetramitus *Invertebrate Zoology.* a genus of naked, colorless, aquatic amoebas having four flagella.

tetranitromethane *Organic Chemistry.* $C(NO_2)_4$, a toxic, colorless liquid; insoluble in water and miscible with alcohol and ether; melts at 13.8°C, boils at 125.7°C; used as a rocket fuel, an oxidant or monopropellant, diesel fuel booster, and in organic synthesis.

Tetranychidae *Invertebrate Zoology.* a family of spider mites, silk-spinning plant pests in the superfamily Trombidoidea. Also, RED SPIDERS.

Tetraodontiformes *Vertebrate Zoology.* an order of bony fishes characterized by a lack of parietals, nasals, infraorbitals, and lower ribs; includes the ocean sunfishes, trigger fishes, puffer fishes, porcupine fishes, and trunk fishes. Also, PLECTOGNATHI.

tetraodontoxin *Toxicology.* see TETRODOTOXIN.

Tetraonidae *Vertebrate Zoology.* the grouses, a family of chickenlike game birds of the order Galliformes, characterized by usually mottled plumage, inflatable brightly colored neck patches, and short legs with large feet; found in the Northern Hemisphere.

9,10,11,12-tetraphenylnaphthacene see RUBRENE.

tetraphenyltin *Organic Chemistry.* $(C_6H_5)_4Sn$, a white powder; insoluble in water and soluble in hot benzene, toluene, and xylene; melts at 225–228°C and boils at about 420°C; used as a mothproofing agent, stabilizer, and intermediate.

Tetraphidaceae *Botany.* the single plant family composing the order Tetraphidales.

Tetraphidae *Botany.* a subclass of mosses of the class Bryopsida, characterized by the presence of a thalloid protonema and a four-toothed peristome composed of multiple tiers of whole cells.

Tetraphidales *Botany.* a monofamilial order of mosses characterized by high, tufted erect stems and a capsular peristome with only four teeth.

tetraphosphorus hexasulfide see PHOSPHORUS TRISULFIDE.

tetraphosphorus trisulfide see PHOSPHORUS SESQUISULFIDE.

Tetraphyllidea *Invertebrate Zoology.* an order of tapeworms in the class Cestoda, having four suckers; the larvae are parasites in crustaceans, and the adults are parasites in sharks and rays. Also, PHYLLOBOTHRIOIDEA.

tetraphyllous *Botany.* having four leaves or leaflets.

tetraplegia *Medicine.* a condition of paralysis affecting all four limbs; quadriplegia.

tetraploid *Cell Biology.* **1.** a nucleus, cell, or organism having four sets of chromosomes. **2.** of or relating to tetraploidy.

tetraploidy *Cell Biology.* the state of having four sets of chromosomes.

tetrapod *Vertebrate Zoology.* an animal having four feet or four legs; used chiefly with reference to higher vertebrates.

Tetrapoda *Vertebrate Zoology.* a grouping, often recognized as a superclass of vertebrates, that includes all forms above fishes, i.e., amphibians, reptiles, birds, and mammals.

tetrapropylene see 1-DODECENE.

tetrapterous *Biology.* having four wings or winglike appendages.

Tetrarhynchia *Paleontology.* an extinct genus of articulate brachiopods in the order Rhynchonellida, characterized by a breadth greater than the length; found in Europe and North America; extant in the Jurassic.

tetrasodium pyrophosphate see SODIUM PYROPHOSPHATE.

tetrasomy *Cell Biology.* the state of having at least one tetraploid chromosome in an otherwise diploid nucleus. Thus, **tetrasomic.**

Tetraspora *Botany.* a genus of freshwater, pseudoflagellate green algae of the family Tetrasporaceae that form gelatinous masses up to 15 cm long.

Tetrasporaceae *Botany.* a family of freshwater green algae of the order Tetrasporales, having cells that form gelatinous colonies with flagella at the anterior end; colonies are often epibiotic or sometimes planktonic.

Tetrasporales *Botany.* the freshwater and marine algae of the Chlorophyta, characterized by nonmotive vegetative cells that may become temporarily motile; found in nonfilamentous colonies.

tetrasporangium *Botany.* a sporangium containing four asexual spores.

tetraspore *Botany.* one of the four asexual spores produced within a tetrasporangium

tetrastichous *Botany.* **1.** arranged in a spike of four vertical rows, as flowers. **2.** having four rows of flowers, such as a spike.

tetraterpene *Organic Chemistry.* a class of terpene compounds that are composed of light isoprene units; includes the carotenoid pigments of plants.

tetratohedral *Crystallography.* of or relating to a crystal that has one-quarter of the maximum number of sides designated by the crystal system to which it belongs.

tetravalent *Chemistry.* having a valence of four; quadrivalent.

Tetraviridae *Virology.* a family of viruses that infect insects and have unenveloped, isometric virions containing ssRNA. Also, NUDAURIN B VIRUS GROUP.

Tetravirus *Virology.* the single genus of viruses of the Tetraviridae family.

tetraxon *Invertebrate Zoology.* a type of sponge spicule having four separate rays.

tetrazene *Organic Chemistry.* $H_2NC(=NH)NHNHN=NC(=NH)NHNHNO$, a slightly hygroscopic, colorless or pale-yellow solid; almost insoluble in water, alcohol, and ether; used as an explosive initiator.

tetrazole *Organic Chemistry.* CH_2N_4, plates that are derived from alcohol as solvent; soluble in water and alcohol; melts at 156°C and sublimes at boiling point. *Pharmacology.* **1.** a crystalline acidic parent compound containing a ring that is composed of one carbon atom and four nitrogen atoms. **2.** any of the various derivatives of this parent compound.

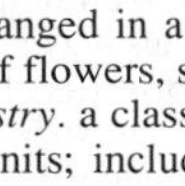

tetrazole

tetrazolium salt see TRIPHENYLTETRAZOLIUM CHLORIDE.

Tetrigidae *Invertebrate Zoology.* the grouse locusts or pygmy grasshoppers, orthopteran insects with reduced forewings and an elongate prothorax.

tetrode *Electronics.* an electron tube containing four electrodes: cathode, control grid, screen grid, and anode.

tetrode junction transistor see DOUBLE-BASE JUNCTION TRANSISTOR.

tetrode thyratron *Electronics.* a gas-filled, four-element thermionic electron tube containing a cathode, two grids, and an anode.

tetrode transistor *Electronics.* **1.** a bipolar transistor with four terminal leads: one for the emitter, one for the collector, and two base leads; the extra base connection reduces interterminal capacitance and lead inductance. **2.** a four-terminal, field-effect transistor containing one source, one drain, and two gates.

tetrodotoxin *Toxicology.* $C_{11}H_{17}N_3O_3$, a pure, crystalline, highly lethal neurotoxin found in many species of puffer fish and in newts of the genus *Taricha.* Also, FUGUTOXIN, TETRAODONTOXIN.

tetrodotoxism *Toxicology.* a pathological condition caused by the ingestion of puffer fish containing tetrodotoxin; the symptoms may include almost immediate malaise, dizziness, and tingling near the mouth region, which may be followed by ataxia, convulsions, respiratory paralysis, and death. Also, FUGUISM.

tetrose *Biochemistry.* any sugar that has four carbon atoms, such as erythrose.

tetryl *Organic Chemistry.* $(NO_2)_3C_6H_2N(NO_2)CH_3$, yellow crystals; insoluble in water and soluble in alcohol and ether; melts at 130–132°C and explodes at 187°C; used as a detonating agent. Also, TETRALITE.

Tettigoniidae *Invertebrate Zoology.* a family of delicate, greenish orthopteran insects, having long slender antennae and jumping hindlegs, including katydids, Mormon crickets, and longhorned grasshoppers.

Teuthidae *Vertebrate Zoology.* in some classification systems, a family of perciform fishes that includes the rabbit fishes. Also, SIGANIDAE.

Teuthoidea *Invertebrate Zoology.* an order of squids, cephalopod mollusks in the subclass Coleoidea, having a reduced internal shell and ten arms; coextensive with Myopsida.

T-even phage group *Virology.* a group of dsDNA-containing bacteriophages having structurally complex virions with characteristic contractile tails, including the T2, T4, and T6 phages.

TE wave transverse electric wave.

tex *Textiles.* the weight in grams of a one-kilometer length of yarn, fiber, or other strand; a lower tex number indicates a finer yarn. (A shortened form of *textile.*)

TEX [tek] *Computer Science.* a text-formatting program that is designed to produce high-quality typesetting of mathematical formulas for laser printers or other high-quality output devices.

Texales see TAXALES.

Texas fever see BABESIASIS.

Texas Longhorn *Agriculture.* another name for Longhorn cattle, because of their early association with Texas ranching. See LONGHORN.

Texas root rot see COTTON ROOT ROT.

Texas tower *Engineering.* an offshore radar station erected on a base that resembles an offshore oil platform.

text *Graphic Arts.* **1.** printed matter consisting of words, as opposed to photographs, illustrations, and other artwork. **2.** the main body of a book or other work, as distinguished from the supplementary sections at the front and back including the introduction, table of contents, index, and so on. **3.** also, **Text.** a typeface that is based on handwriting. *Computer Programming.* **1.** in word processing, the portion of a file that will be readable on a screen or printed page, sometimes but not always including incidentals such as headers. **2.** a file or portion of data that is to be interpreted as characters. **3.** in an ASCII file, the characters between a STX ("start text") control character and an ETX ("end text") control character.

text area or **text page** see TYPE AREA.

text-editing system *Computer Programming.* a system that allows the generation, addition, deletion, and reformatting of textual material, including source programs.

text file *Computer Programming.* **1.** see ACSII FILE. **2.** any file whose contents are to be interpreted as characters.

text generation *Artificial Intelligence.* the construction by computer of natural-language phrases, sentences, and paragraphs.

textile *Textiles.* **1.** any material that has been woven, knitted, bonded, or otherwise made into cloth from natural or synthetic fibers. **2.** the raw material that is used for this weaving, knitting, and so on. *Materials Science.* of, relating to, or used in the production of cloth, especially by weaving. Thus, **textile oil, textile softener.** (From a Latin word meaning "to weave" or "woven fabric.")

textile microbiology *Microbiology.* the scientific study of the effects of microorganisms on textile substances.

Textiles

If, conventionally, *textiles* connotes all cloth structures produced from fiber, there is a movement toward restricting *textiles*—as a term—to those woven on a loom. Where this is accepted, the umbrella term is *fabric*—to include not only those cloths, such as knits, made from a single yarn, but the simple or complex plaited mats and baskets I classified in *Interlacing: The Elemental Fabric,* as well as the meshed fibers of felt and paper and pliable films such as plastic or rubber sheeting. Interestingly enough, those who adhere to this more restrictive use of the term still house their fabric collections in textile museums or textile departments of larger institutions.

There is still a larger connotation for textiles: that of soft goods as a block, to include apparel, bed and bath linens, industrial fabrics, rugs and carpets. A glance at this dictionary's index would indicate that this is the case here. A further extension of textiles—not a small one—is the making and processing of fibers and yarns *before* cloth production. On the other side (also ancient, universal, and sizeable) is the dyeing and embellishment of textiles after production. These techniques include printing, embroidery, embossing, brushing, felting, et cetera.

The production of textiles is the oldest industry and, at least, one of the largest, employing vast numbers of workers on every continent. If from earliest times, the cloth trade was facilitated by ease of transport, the traditional exchange was from groups with higher technologies to those more formative, this trend is nowadays somewhat reversed: many resplendent cloths such as handwoven silks move *from* developing nations to those otherwise subsisting on mass production.

Beyond all that, textiles have always been symbols of rank, wealth, and religion. Along with other fabrics, textiles are today a major craft and an important art form.

Jack Lenor Larsen
Chairman and Design Director
Jack Lenor Larsen Incorporated

textiles the study of the manufacture of cloth, including the raw fibers used, both natural and synthetic, and of the various methods of production involved.

text paper *Graphic Arts.* high-quality paper used in printing books.

text-to-speech synthesizer *Acoustical Engineering.* a digital electroacoustical device designed to optically scan printed matter and convert it into electroacoustical signals representing the printed words; used for recording or broadcast by a speaker.

text type *Graphic Arts.* the type used to set the body of a printed piece, usually 14 points or smaller. Also, BODY TYPE.

Textulariina *Invertebrate Zoology.* a suborder of foraminiferan protozoans with a test built of particles glued together with secretions.

texture *Geology.* the physical characteristics of a soil that are determined according to the size of the constituent particles and the relative proportions of sand, clay, and silt it contains. *Petrology.* the general physical appearance or character of a rock, including the sum total of the smaller, megascopic, or microscopic surface features of a homogeneous rock or mineral aggregate, such as grain size, shape, and arrangement; as distinguished from larger features.

textured yarn *Textiles.* a continuous filament, synthetic fiber yarn that has been treated to give it new physical or surface properties, such as added stretchiness or spun yarn effects; sometimes designates any yarn that has been modified to give it a different texture.

texturing *Textiles.* the mechanical or chemical production of textured yarns. *Materials Science.* specifically, in textile production, any process that imparts to a polymer properties that are usually found in staple fiber yarns, such as breathability, permeability, comfort, and ease of care. The texturing processes involve deforming, heating to relax stresses, cooling, undeforming, and relaxation to allow redeformation. Methods of texturizing include gear crimping, the stuffing-box process, the air-jet process, edge crimping, and knitting-deknitting.

texturizer *Food Technology.* any substance added to foods to improve their consistency or texture, such as calcium chloride (added to canned fruits and vegetables).

TF task force; transfer factor.

TFE tetrafluoroethylene.

TG or **t.g.** type genus.

TGF transforming growth factor.

TGT thromboplastin generation test.

tgt. target.

TGZM temperature-gradient zone melting.

Th the chemical symbol for thorium.

Thaad *Ordnance.* a U.S. missile system that is intended to intercept and destroy short-range enemy missiles at high altitudes. (An acronym for theater high-altitude area defense.)

thalamencephalon *Anatomy.* the diencephalon.

thalamus *Anatomy.* the largest portion of the diencephalon of the brain, consisting of several nuclear masses of gray matter on either side of the third ventricle; a major relay center for both sensory and motor signals.

thalassemia *Medicine.* a group of hereditary anemias marked by microcytic, hypochromic, and short-lived red blood cells, caused by the deficient synthesis of hemoglobin; occurring most often among people of Mediterranean origin.

thalassic *Oceanography.* **1.** of or relating to inland seas or minor seas such as the Aegean. **2.** of or relating to the deposition of fine-grained sediments in relatively deep, still water far from land.

Thalassinidea *Invertebrate Zoology.* a group of mud shrimps, burrowing decapod crustaceans in the Macrura, having a thin carapace and asymmetrical claws.

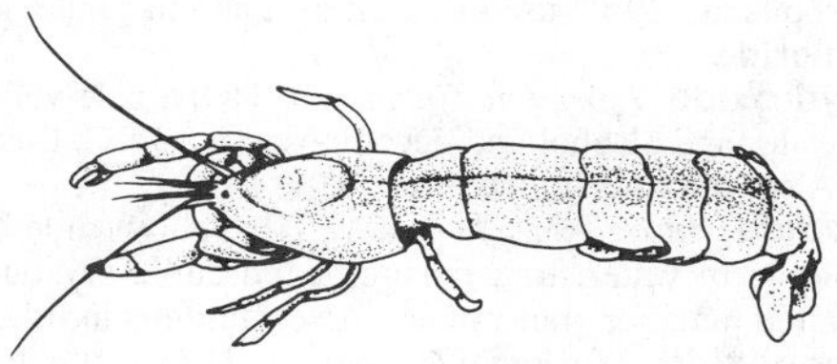

Thalassinidea

Thalassiosiraceae *Botany.* a family of planktonic diatoms of the order Centrales, characterized by circular valves with one or more labiate processes and multiple strutted processes in marginal rings; members are an important component of marine phytoplankton.

thalasso- a combining form meaning "sea."

thalassocratic *Geology.* **1.** of or relating to a thalassocraton. **2.** describing an interval of time in the geologic past during which the sea level was high.

thalassocraton *Geology.* a former continental creation that has been thinned and sunk to submarine depths.

thalassography *Oceanography.* the study of smaller bodies of water, such as bays, sounds, and gulfs.

Thalassomycetales *Botany.* an order of dinoflagellates of the class Dinophyceae that are nonphotosynthetic and ectoparasitic on arthropods and annelids; composed of the family Thalassomycetaceae.

thalassophile element *Geochemistry.* an element found specifically in higher concentrations in seawater than in continental waters.

thalassophobia *Psychology.* an irrational fear of the sea.

thalassostatic *Hydrology.* relating to a lower portion of a river subject to tidal influences.

Thalattosaurus *Paleontology.* a genus of primitive lepidosaurian reptiles of the extinct order Thalattosauria and the family Thalattosauridae; characterized by a slitlike upper fenestra; extant in the Triassic.

Thales (of Miletus) 625?–547? BC, Greek philosopher who accurately predicted a solar eclipse; introduced geometry to Greece and studied natural laws; considered the earliest known Western philosopher.

Thaliacea *Invertebrate Zoology.* a class of pelagic tunicates with a cylindrical body; long chains of individuals are produced by budding.

thalidomide [thə lid′ə mīd′] *Pharmacology.* $C_{13}H_{10}N_2O_4$, a crystalline, slightly water-soluble solid; a hypnotic, tranquilizer, and sedative that causes severe congenital fetal abnormalities, such as amelia and phocomelia, when taken during the early stages of pregnancy; it is therefore no longer in use.

thall- or **thallo-** a combining form meaning "branch," "shoot."

thalline *Organic Chemistry.* $C_9H_6N(OCH_3)H_4$, a colorless crystalline solid that is soluble in water and melts at 40°C.

thallitoxicosis *Toxicology.* poisoning due to ingestion of thallium or thallium-containing substances; symptoms may include hair loss and pain in the extremities. Also, THALLOTOXICOSIS.

thallium *Chemistry.* a metallic element having the symbol Tl, the atomic number 81, an atomic weight of 204.83, a melting point of 304°C, and a boiling point of 1460°C; a bluish-white metal that oxidizes in air at room temperature and forms toxic compounds when in contact with moisture; used in thallium salts, optical glass, low melting alloys, and rodenticides. (From a Greek term meaning "green stalk," because of the green line in its spectrum.)

thallium acetate *Organic Chemistry.* $TlOCOCH_3$, highly toxic deliquescent, white crystals; soluble in water, alcohol, and chloroform; melts at 131°C; used as an ore flotation solvent and in medicine.

thallium-beam clock *Horology.* an atomic clock in which a beam of thallium atoms is stimulated by an alternating magnetic field; when the frequency of the field equals the natural frequency of the atoms, it can be used for extremely precise timekeeping; thallium is less frequently used in atomic-beam clocks than cesium because it is more difficult to detect and deflect.

thallium bromide *Inorganic Chemistry.* TlBr, toxic, yellowish-white cubic crystals; slightly soluble in water and soluble in alcohol; melts at 480°C and boils at 815°C; used in infrared detectors and transmitters. Also, **thallous bromide.**

thallium carbonate *Inorganic Chemistry.* Tl_2CO_3, colorless, monoclinic crystals; soluble in water and insoluble in alcohol; melts at 273°C; used in analysis and artificial diamonds. Also, **thallous carbonate.**

thallium chloride *Inorganic Chemistry.* TlCl, white crystals; slightly soluble in water, insoluble in alcohol, and decomposes in acid; melts at 430°C and boils at 720°C; used as a catalyst and in suntan lamps. Also, **thallous chloride.**

thallium hydroxide *Inorganic Chemistry.* TlOH, pale yellow needles; soluble in water and alcohol and decomposes at 139°C; used for testing for ozone. Also, **thallous hydroxide.**

thallium iodide *Inorganic Chemistry.* TlI, yellow rhombic crystals; slightly soluble in water; transposing to red cubic crystals at 170°C; used in infrared radiation transmitters. Also, **thallous iodide.**

thallium monoxide *Inorganic Chemistry.* Tl_2O, deliquescent, black crystals; soluble in alcohol and acid and decomposes in water; melts at 300°C and boils at 1080°C; used in analysis for ozone, artificial gems, and optical glass. Also, **thallous oxide.**

thallium nitrate *Inorganic Chemistry.* $TlNO_3$, toxic colorless rhombic crystals; soluble in water and insoluble in alcohol; melts at 206°C and boils at 430°C; a fire and explosive hazard; used in analysis and pyrotechnics. Also, **thallous nitrate.**

thallium sulfate *Inorganic Chemistry.* Tl_2SO_4, toxic, colorless rhombic crystals; soluble in water; melts at 632°C and decomposes at the boiling point; used in analysis, measuring ozone levels, and pesticides. Also, **thallous sulfate.**

thallium sulfide *Inorganic Chemistry.* Tl_2S, toxic, lustrous, blue-black tetragonal crystals; slightly soluble in water and soluble in acid; used in infrared-sensitive photocells. Also, **thallous sulfide.**

Thallobacteria *Bacteriology.* a class of Gram-positive bacteria of the division Firmicutes, composed of branching, filamentous cells.

thallofide cell *Electricity.* a photoconductive cell with a thallus-sulfide, radiation-sensitive surface; it has a maximum response at the red end of the visible spectrum and in the near-infrared.

thalloid *Botany.* resembling or consisting of a thallus.

Thallophyta *Botany.* one of the two major divisions of the plant kingdom, whose members have no specialized organs, such as roots, stems, and leaves, and are characterized by thalli. Also, **Thallobionta.**

thallospore *Botany.* an asexual fungal spore formed on or in a thallus.

thallotoxicosis see THALLITOXICOSIS.

thallus *plural,* **thalli.** *Botany.* a simple plant body undifferentiated into true stems, roots, or leaves, ranging from a collection of filaments to a complex plantlike form; characteristic of the Thallophytes.

thalweg *Hydrology.* **1.** a line joining the lowest points along the entire length of a streambed or valley, whether or not the bed or valley lies underwater. Also, VALLEY LINE. **2.** a subsurface groundwater stream that percolates under the surface and in the same general direction as the surface stream. *Geology.* a line representing the profile of a land surface that crosses all contour lines at right angles. *Geography.* the midpoint in the navigable channel of a waterway serving as a boundary between states. Also, TALWEG. (From a German phrase meaning "valley way.")

THAM tromethamine.

Thamnidiaceae *Mycology.* a family of fungi belonging to the order Mucorales, which is found primarily on dung, although some live in soil and are parasitic to Mucorales.

thanat- or **thanato-** a combining form meaning "death."

thanatocoenosis *Paleontology.* a taphonomic association of organisms after their death; the fossils thus grouped did not necessarily belong to the same functional community while they were alive.

thanatology *Medicine.* the medical and legal study of death and conditions affecting dead bodies. Thus, **thanatological.**

thanatophidia *Toxicology.* a collective term for all venomous snakes. Also, TOXICOPHIDIA.

thanatophobia *Psychology.* an abnormal fear of death.

Thanatos *Psychology.* the instinct for death and destruction, one of Freud's two primary motives for human conduct.

thanatosis *Behavior.* the act of feigning death, usually in order to avoid or escape a predator.

Thanetian *Geology.* a European geologic stage of the uppermost Paleocene epoch, occurring after the Montian and before the Ypresian of the Eocene epoch.

Thaumaleidae *Invertebrate Zoology.* a family of dipteran insects in the series Nematocera, having long antennae.

Thaumastellidae *Invertebrate Zoology.* a family of Ethiopian true bugs, hemipteran insects in the group Pentatomorpha.

Thaumastocoridae *Invertebrate Zoology.* the single family in the superfamily Thaumastocoroidea.

Thaumastocoroidea *Invertebrate Zoology.* a superfamily of land bugs, hemipteran insects in the subdivision Geocorisae, having conspicuous antennae.

Thaumatomastigaceae *Botany.* a family of flagellates of the order Rahidomonadales that are saprophytic or phagotrophic and lack chlorophyll; most species occur in fresh water.

Thaumatoxenidae *Invertebrate Zoology.* a family of dipteran insects in the series Aschiza.

thaw *Meteorology.* a period of weather that is warm enough to melt ice and snow.

thawing *Science.* the process of bringing a frozen material, e.g., ice, to an unfrozen state by heat application, usually gradual.

thawing index *Meteorology.* the number of degree-days between the lowest and highest points on the cumulative degree-days time curve for one thawing season.

thawing season *Meteorology.* the time elapsed between the lowest and highest points on the time curve of cumulative degree-days above and below 32°F; the opposite of the freezing season.

THC tetrahydrocannabinol.

Theaceae *Botany.* the tea family, a family of dicotyledonous trees and shrubs of the order Theales, characterized by simple, usually alternate leathery leaves, mostly bisexual flowers with many stamens, and a superior ovary.

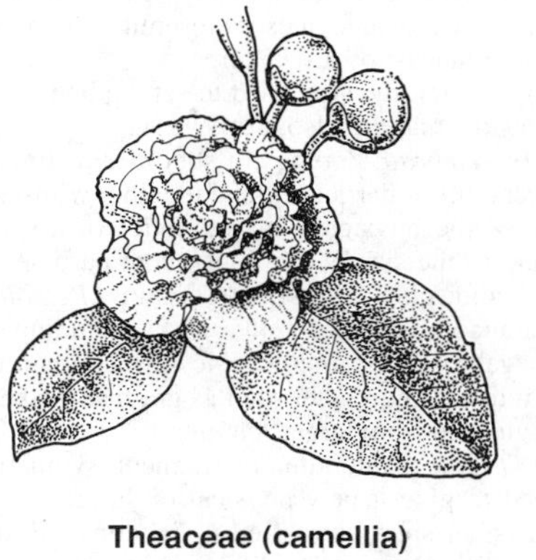

Theaceae (camellia)

theaism *Toxicology.* poisoning due to ingestion of excessive amounts of tea; symptoms may include nausea, headache, insomnia, vomiting, and muscular spasms. Also, THEINISM.

Theales *Botany.* an order of dicotyledonous trees, shrubs, and woody climbers of the subclass Dilleniidae that have essentially simple leaves and bisexual flowers.

theater *Military Science.* a large area in which military operations are carried out: the European *theater* during World War II. Thus, **theater of operations, theater of war.**

theater television *Electronics.* a television receiver system that projects the picture onto a large screen; designed for viewing by large audiences, as in a theater.

thebaine *Pharmacology.* $C_{19}H_{21}NO_3$, a poisonous, crystalline, anodyne alkaloid from opium that produces convulsions similar to those produced by strychnine.

Thebe [thē´bē] *Astronomy.* Jupiter XIV, a moon of Jupiter discovered in 1979, which orbits 222,000 kilometers from Jupiter and has a diameter of about 100 kilometers.

theca *Anatomy.* a membranous covering or sheath, as of an ovarian follicle or tendon. *Invertebrate Zoology.* a protective covering such as the test of some protozoans or the chitinous case of an insect pupa. *Botany.* a case or sac protecting plant parts, such as the pollen sac of an anther or a capsule in mosses.

Thecadiniaceae *Botany.* a family of marine, thecate, motile dinoflagellates of the order Peridiniales, characterized by a slightly displaced cingular groove situated near the anterior end of the motile cell, resulting in a small epitheca and a large hypotheca.

theca folliculi *Histology.* the capsule surrounding an ovarian follicle.

Thecanephria *Invertebrate Zoology.* an order of deep-sea animals in the phylum Brachiata, tube-dwellers with long bodies and tentacles.

thecate *Biology.* **1.** having a sac or saclike cell. **2.** having a case or sheathe such as the covering of an insect pupa. *Invertebrate Zoology.* possessing or resembling a theca.

Thecideidina *Paleontology.* a still-extant group of brachiopods; generally small and cemented, they appeared in the Triassic, possibly descended from strophomenids, but some consider them terebratulids.

thecium *Mycology.* a fungal tissue or hymenium that forms the outer layer of the fruiting body of fungi belonging to the subdivision Ascomycotina.

Thecodontia *Paleontology.* an order of primitive archosaurs that includes the Proterosuchia, Ornithosuchia, Phytosauria, and possibly the Aetosauria; probably ancestral to the dinosaurs, they arose in the upper Permian, flourishing and diversifying in the Triassic (when they may have become warm-blooded).

Thecosmilia *Paleontology.* a genus of colonial scleractinian corals in the suborder Faviina; an important reef-builder in the Jurassic; extant from the Middle Jurassic to the Cretaceous.

Thecosomata *Invertebrate Zoology.* a group of sea butterflies, gastropod mollusks usually with a calcareous shell; cosmopolitan, pelagic plankton-feeders.

Theiler, Max 1899–1972, South African physician and virologist; awarded the Nobel Prize for the development of yellow fever vaccine.

Theileria *Invertebrate Zoology.* a genus of parasitic sporozoan protozoans that parasitize mammals, transmitted by ticks; one species causes cattle fever in Africa.

theileriasis see EAST COAST FEVER.

theinism see THEAISM.

Thelastomidae *Invertebrate Zoology.* a family of parasitic nematode worms in the superfamily Oxyuroidea.

Thelebolaceae *Mycology.* a family of fungi belonging to the order Pezizales, which occurs primarily on dung.

Thelephoraceae *Mycology.* a family of fungi belonging to the order Agaricales, which occurs on soil, dung, wood, and plant material or in a symbiotic relationship with plants; some species are parasitic to vascular plants or other fungi.

theleplasty *Surgery.* plastic or reconstructive surgery of the nipple.

Theligonaceae *Botany.* a monogeneric family of tanniferous dicotyledonous herbs of the order Rubiales, characterized by a xylem cylinder in the stem and leaves that are basally opposite but spiral elsewhere by suppression of one of each pair; native to eastern Asia, the Mediterranean region, and the Canary Islands.

thelitis *Medicine.* inflammation of the nipple.

Thelodus *Paleontology.* a primitive jawless fish associated with Jamoytius; having a broad head and toothlike scales that lightly armored its body; extant in the Upper Silurian and Lower Devonian.

Thelohania *Invertebrate Zoology.* a genus of intracellular protozoan parasite found in shrimp.

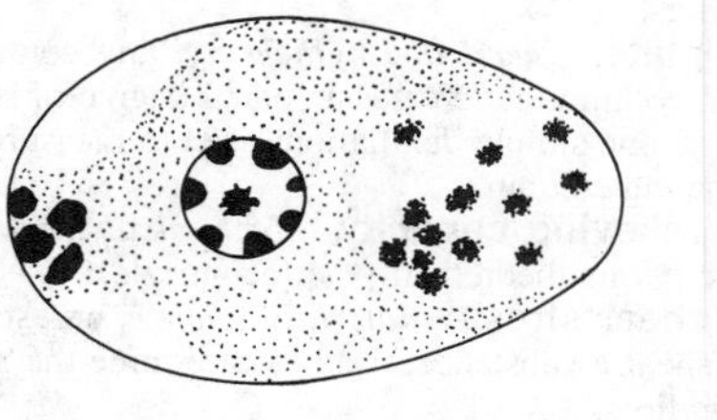

Thelohania

thelytoky *Zoology.* a type of parthenogenic reproduction in which only females are produced.

Thematic Apperception Test *Psychology.* a projective test for assessing elements of the personality by means of an individual's interpretation of a series of ambiguous pictures.

Themis [thē´məs] *Astronomy.* asteroid 24, discovered in 1853, and having a diameter of 220 kilometers; belonging to type C, it is the main body in the asteroid family that bears its name.

thenar *Anatomy.* **1.** relating to the palm of the hand. **2.** relating to the muscular bulge at the base of the thumb.

Thénard, Louis 1777–1857, French chemist; with Guy-Lussac, analyzed sugar, starch, and acids and showed that sulfur is an element.

thenardite *Mineralogy.* Na_2SO_4, a colorless to white or brownish orthorhombic mineral occurring as tabular or dipyramidal crystals and efflorescent crusts, having a specific gravity of 2.664 and a hardness of 2.5 to 3 on the Mohs scale; found in salt lake and playa deposits.

thenyl *Organic Chemistry.* $C_4H_3SCH_2^-$, an organic group derived from methylthiophene.

theobromine *Organic Chemistry.* $C_7H_8N_4O_2$, a crystalline solid; insoluble in water; melts at 351°C; a toxic alkaloid found in cocoa, chocolate products, cola nuts, and tea. Also, 3,7-DIMETHYLXANTHINE. *Pharmacology.* this powder used chiefly as a diuretic, myocardial stimulant, and vasodilator.

theodolite *Optics.* a telescope that allows accurate measurements of horizontal and vertical angles; used primarily in surveying and for tracking weather balloons and rockets.

theophobia *Psychology.* an irrational fear of God.

Theophrastaceae *Botany.* a tropical family of dicotyledonous woody plants of the order Primulales, characterized by alternate exstipulate leaves and bisexual or unisexual flowers found in corymbs, racemes, or panicles,

Theophrastus [thē´ō fras´təs] c. 372–287 BC, Greek philosopher who wrote leading classical works on botany as well as texts on physics and geology.

theophylline *Organic Chemistry.* $C_7H_8N_4O_2$, a white crystalline alkaloid that is soluble in water and alcohol; the monohydrate melts at 272°C; it is extracted from tea leaves and used in medicine. Also, 1,3-DIMETHYLXANTHINE.

theor. theory; theorem.

Theorell, Hugo 1903–1982, Swedish biochemist; awarded the Nobel Prize for his pioneering research on oxidation enzymes.

theorem [thēr´əm] *Mathematics.* the last statement of a formal proof; a mathematical assertion that can be proven.

theorem of corresponding states *Physics.* a theorem stating that two substances that have identical reduced pressures and reduced temperatures also have the same reduced volumes.

theorem proving *Artificial Intelligence.* the process of proving a mathematical theorem, or solving a problem stated by means of axioms in predicate calculus, by mechanically deriving new formulas using rules of inference.

theoretical *Science.* **1.** of or relating to theory or to a given theory. **2.** speculative rather than proven by experimentation and observation; hypothetical.

theoretical air *Engineering.* the quantity of air needed to provide total combustion of a given quantity of combustible material.

theoretical cutoff frequency *Electricity.* the frequency at which the attenuation constant of an electric system changes from zero to a positive value, or vice versa; it disregards the effects of dissipation. Also, **theoretical cutoff.**

theoretical draft *Fluid Mechanics.* a draft in a ducted space, neglecting flow losses due to fluid friction.
theoretical physics *Physics.* a division of physics devoted primarily to calculations and formulations that seek to codify or extend experimental findings.
theoretical plate *Chemical Engineering.* any contacting device in a fractionating column that creates an equal degree of separation of vapor from liquid as one simple distillation; used to measure the effectiveness of a fractionating column.
theoretical relieving capacity *Mechanical Engineering.* the volumetric capacity of a theoretically perfect nozzle.
theoretical shear stress *Materials Science.* an estimate of the stress required to shear a substance; used to determine the resistance of material to plastic flow.
theory *Science.* an explanation for some phenomenon that is based on observation, experimentation, and reasoning. *Mathematics.* the collection of definitions, principles, theorems, and rules of calculation associated with a class of mathematical objects. (From a Greek term meaning "a viewing;" "a way of considering.")
Theory, in popular use, is often assumed to imply conjecture or speculation, as in "Evolution is just a theory." In scientific use, however, to describe an explanation as a *theory* does not indicate it is uncertain. Progressing in degrees of uncertainty, a *law* is based on many observations of a natural phenomenon that demonstrate it to be true without exception; e.g., the second *law* of thermodynamics states that heat cannot flow spontaneously from a cooler body to a warmer one. A *theory* is developed from extensive observation and reasoning to explain a phenomenon; e.g., Einstein's *theory* of relativity postulated that the speed of light represents an upper limit of velocity for any particle having mass; this has since been experimentally confirmed. A *hypothesis* is a preliminary explanation for an observed phenomenon that may or may not be supported by more extensive observation; e.g., the current *hypothesis* that the earth's climate is gradually becoming warmer because of the effects of burning fossil fuels.
theory For entries dealing with a theory, see also the term itself; e.g., for **theory of relativity,** see RELATIVITY THEORY.
theory of elasticity see ELASTIC THEORY.
theory of equations *Mathematics.* the study of polynomials and their roots. The algebraic portion of the theory of equations (i.e., the study of symbolic manipulation of polynomials) includes Galois theory and parts of invariant theory. The analytic portion includes results concerning the distribution of roots and numerical methods for approximating them.
theory of games see GAME THEORY.
theory of recapitulation *Evolution.* the proposal that an organism's embryonic development recapitulates the major events in its evolutionary history.
theory of reentry *Cardiology.* in a premature heartbeat, the theory that the sinus impulse preceding the heartbeat initiates the beat, but the beat is delayed by the delaying of the activation wave in an area of diminished sensitivity.
theory of runs *Statistics.* a concept used to test the randomness of test samples that includes the sequence of occurrences as well as the frequency.
theralite *Petrology.* a granular, mafic igneous rock, or rock group, consisting primarily of pyroxene, andesine, and nepheline, with or without smaller amounts of hornblende, biotite, and olivine.
therapeutic [ther´ə pyoot´ik] *Medicine.* **1.** acting as a cure or a relief; curative. **2.** of or relating to therapeutics.
therapeutic abortion *Medicine.* an abortion induced to save the life or health (mental or physical) of a pregnant woman, such as after rape.
therapeutic index *Pharmacology.* a method of assessing the safety of a drug, originally defined as the ratio of the maximum tolerated dose of a drug to the minimum curative dose; now, to account for the variability of individual response, the ratio of the median lethal dose to the median effective dose.
therapeutics *Medicine.* **1.** the art and science of healing; the treatment and cure of disease. **2.** a scientific account of the treatment of a specific disease.
Theraphosidae *Invertebrate Zoology.* a family of very large hairy spiders of the suborder Orthognatha, found in temperate to tropical regions; only one species, *Sericopelma communis,* has a venom that is poisonous to humans, but others may inflict painful bites.
therapist *Medicine.* a person skilled in the treatment of disease; often used in combined terms, such as **speech therapist, physical therapist.**
Theraponidae see PUFFER.
Therapsida *Paleontology.* an order of mammal-like reptiles in the subclass Synapsida; ancestral to mammals; some genera were as long as four meters; the therapsids appeared in the Middle Permian, they underwent a great radiation in the Late Permian and the Triassic, and at least one genus is known from the Jurassic.
therapy *Medicine.* **1.** the treatment of diseases and disorders. **2.** specifically, the treatment of mental disorders; psychotherapy.
therblig *Industrial Engineering.* in time-motion studies, a single basic body motion by a worker. (An anagram of *Gilbreth,* after the inventors of the concept, Frank and Lillian Gilbreth.)
therblig chart *Industrial Engineering.* a chart showing basic human work motions or elements (e.g., *find, grasp*) and the times required to perform them.
therblig system *Industrial Engineering.* in time-motion studies, a system for identifying and classifying worker hand motions.
Therevidae *Invertebrate Zoology.* the stiletto flies, a family of hairy predaceous dipteran insects in the series Brachycera.
Theria *Vertebrate Zoology.* in some classifications, a subclass of mammals that includes all living mammals that give birth to living young and do not lay eggs. *Paleontology.* the ancient members of this subclass, including the marsupials and placentals; they first appeared in the Middle Jurassic, the eupantotheres being the main primitive group; the first marsupials and placentals appeared in the Cretaceous.
Theridiidae *Invertebrate Zoology.* a family of comb-footed spiders, arachnids in the suborder Dipneumonomorphae, having a small round body and slender legs, web-builders that trap their prey in sticky silk; includes the black widow and common house spiders.

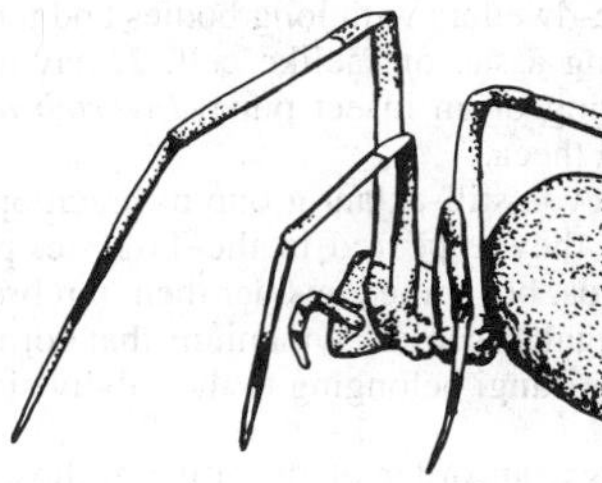

Theridiidae

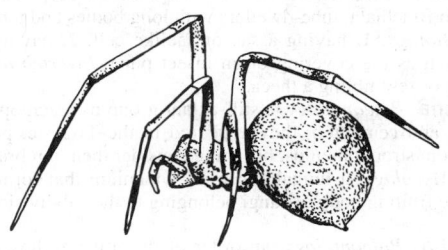

Theridomyidae *Paleontology.* an extinct family of rodents in the suborder Sciurognathi and the superfamily Theridomyoidea; extant in the Eocene and Miocene.
theriodont *Paleontology.* any member of the group of therapsids that were directly ancestral to the cynodonts and ultimately to the mammals; extant in the Permian and Triassic.
therm *Thermodynamics.* a unit of energy equivalent to 10^5 international table British thermal units, or approximately 1.055×10^8 joules.
therm. thermometer.
therm- a combining form meaning "heat."
thermal *Thermodynamics.* of or relating to heat or heat transfer. *Meteorology.* a localized air current that rises aloft when the lower atmosphere is heated by earth's surface enough to produce an absolute instability over a certain area; a common source of low-level clear-air turbulence.
thermal agitation *Solid-State Physics.* the random motion of free electrons in a conductor due to heat energy.
thermal analysis *Analytical Chemistry.* any test using heat for the analysis of thermodynamic or physical properties of materials; includes distillation, calorimetry, and boiling and freezing point determinations. *Metallurgy.* specifically, a test used to determine metal transformations based on the temperatures at which thermal arrests occur.
thermal arrest *Materials Science.* a discontinuity on the cooling curve during the solidification of a material, usually caused by the evolution of heat.
thermal-arrest calorimeter *Engineering.* an instrument that measures the heat of fusion of a material in a vacuum; the sample is frozen in a vacuum and then allowed to melt while maintained in the vacuum.
thermal barrier *Aviation.* **1.** the efficiency with which a flight vehicle engine converts the total heat energy of its fuel into propulsive energy. **2.** the limitation imposed by aerodynamic heating on the speed of a flight vehicle. Also, HEAT BARRIER.
thermal battery *Electricity.* a combination of thermal cells.

thermal belt *Ecology.* any of several belts of different vegetation types along the side of a mountain, resulting from vertical temperature variation.

thermal black *Chemistry.* a black pigment or dye that is derived from natural gas by a thermatomic process and has a largely carbon composition.

thermal blooming *Optics.* the defocusing of a collimated beam of light caused when it heats the medium, such as air, through which it is traveling.

thermal bond *Nucleonics.* thermally conductive material that is used in a nuclear reactor to provide maximum heat transfer between the reactor core and the protective shield.

thermal breeder reactor *Nucleonics.* a breeder reactor whose fission activity is maintained by thermal (slow) neutrons.

thermal bremsstrahlung *Physics.* the emission of electromagnetic radiation from high-temperature plasma; the radiation is produced as electrons are scattered by protons in a hot plasma.

thermal bulb *Engineering.* an instrument used to measure the temperature of a liquid, consisting of a bulb and a spiral Bourdon tube; as the liquid heats it is forced into the tube, which gradually unwinds; the degree of unwinding provides a temperature reading.

thermal burn *Medicine.* an injury to the skin caused by exposure to heat.

thermal capacitance *Thermodynamics.* a quantity given by the amount of entropy increase of a body divided by the amount of rise in temperature.

thermal capacity *Thermodynamics.* see HEAT CAPACITY.

thermal cell *Electricity.* **1.** a cell configured such that it is activated when heat is applied to melt a solidified electrolyte. **2.** a reserve cell.

thermal charge *Thermodynamics.* see ENTROPY.

thermal climate *Meteorology.* a climate as defined by temperature, divided regionally into temperature zones.

thermal column *Nucleonics.* a column of moderating material that extends into the reflector of a nuclear reactor so as to provide slow neutrons for experiments.

thermal compressor *Mechanical Engineering.* a device used to compress steam at greater than atmospheric pressure.

thermal conductance *Thermodynamics.* a quantity given by the heat transfer through a body divided by the temperature difference across the surfaces of the body.

thermal conductimetry *Thermodynamics.* the study and practice of measurement of thermal conductivities.

thermal conduction *Thermodynamics.* the flow of thermal energy through a material; this is affected by the amount of heat energy present, the nature of the heat carrier in the material, and the amount of dissipation.

thermal conductivity *Thermodynamics.* the material property that quantifies the rate of flow of thermal energy through a material in the presence of a temperature gradient, given by the amount of heat transfer across a unit area in a unit amount of time, divided by the negative of the space rate of change in temperature in the direction perpendicular to the unit area. Most materials with high electrical conductivities tend also to have high thermal conductivities; e.g., copper and silver. Also, HEAT CONDUCTIVITY.

thermal conductivity cell see KATHAROMETER.

thermal conductivity gauge *Engineering.* a vacuum pressure gauge that detects changes in the thermal conductivity of a heated wire passing through the gas; as the pressure changes, the thermal conductivity of the wire varies.

thermal conductor *Thermodynamics.* a material that permits a heat transfer by thermal conduction; many metals and some ceramics fit this description.

thermal convection *Thermodynamics.* the energy transfer by the physical movement of a fluid. *Meteorology.* the rising of solar-heated air from the earth's surface into the atmosphere; part of the process by which convection clouds are formed. Also, FREE CONVECTION, GRAVITATIONAL CONVECTION.

thermal converter see THERMOELECTRIC CONVERTER.

thermal coulomb *Thermodynamics.* a unit of entropy given by 1 joule per degree kelvin.

thermal cracking *Chemical Engineering.* a method of petroleum processing that applies heat without the use of catalysts to decompose, rearrange, or combine hydrogen molecules.

thermal cross section *Nucleonics.* the cross section associated with a thermal neutron.

thermal current *Meteorology.* see THERMAL.

thermal death point *Microbiology.* the lowest temperature that is lethal for a population of a given microorganism after a 10-minute exposure.

thermal death time *Microbiology.* the time required to kill a population of a given microorganism in aqueous solution at a given temperature.

thermal degradation *Chemistry.* a chemical process in which substances of high molecular weight are broken down by heating in the absence of air.

thermal denaturation *Molecular Biology.* a loss of the native configuration of a macromolecule that results from heat treatment.

thermal diffusion *Physical Chemistry.* a process in which temperature differentiation within a fluid mixture causes one constituent to flow differently than the mixture as a whole. *Chemical Engineering.* this phenomenon used as a method to separate isotopes: a hot wire is placed in the center of a long, vertical tube so that when a gas or liquid flows through the tube, the lighter molecules cluster around the wire, while the heavy molecules cluster near the tube surface.

thermal diffusivity *Materials Science.* the ratio of thermal conductivity to the heat capacity per unit volume for a material; an important criterion for ceramics that are used as thermal insulators or thermal conductors.

thermal drift *Electricity.* usually undesirable changes in voltage or current caused by temperature change.

thermal ecology *Ecology.* the study of the effects of thermal conditions on living organisms in a naturally heated environment.

thermal efficiency *Meteorology.* an expression of the effectiveness of temperature in determining the rate of plant growth, assuming sufficient moisture. Also, TEMPERATURE EFFICIENCY. *Thermodynamics.* the ratio of the amount of work performed by a heat engine in one cycle to the amount of thermal energy input required to operate the engine over one cycle. Also, **thermal efficiency ratio.**

thermal electromotive force *Electricity.* a voltage that develops across the junction of two unlike conductors or semiconductors when the junction is heated. Also, SEEBECK EFFECT.

thermal emission see THERMIONIC EMISSION.

thermal emissivity *Thermodynamics.* see EMISSIVITY.

thermal energy *Nucleonics.* the order of magnitude of the energy of thermal neutrons, about 0.025 electron volt at room temperature.

thermal equator see HEAT EQUATOR, def. 1.

thermal equilibrium *Thermodynamics.* a condition in which all portions of a thermodynamic system are at a uniform temperature consistent with its surroundings.

thermal etching *Materials Science.* **1.** in ceramics, surface cracks that occur between different phases or along grain boundaries in one face; one of the sources of microcracks and brittle fracture. **2.** a preferential evaporation.

thermal excitation *Physics.* a process in which collisions that occur between particles cause atoms or molecules to obtain additional kinetic energy.

thermal expansion *Physics.* an increase in the size of a substance when the temperature of the substance is increased.

thermal-expansion coefficient *Physics.* the proportionality constant that relates an increase in length, area, or volume to the original length, area, or volume as the result of an increase in temperature.

thermal exposure *Ordnance.* the total normal component of thermal radiation striking a given surface throughout the course of a detonation, expressed in calories per square centimeter and/or megajoules per square meter.

thermal farad *Thermodynamics.* the unit of thermal capacitance wherein a body experiences a temperature increase of 1 K when its entropy is increased by 1 joule/K.

thermal fatigue *Materials Science.* the result of rapid increases to and decreases from operating temperatures, which cause nonuniform dimensional changes that can lead to distortion or fracture.

thermal flasher *Electricity.* an electric device that automatically opens and closes a circuit at regularly occurring intervals, as a result of alternate heating and cooling of an associated bimetallic strip in series with a subject circuit.

thermalgesia *Neurology.* a disorder of sensation in which a heat stimulus causes pain.

thermalgia see CAUSALGIA.

thermal gradient *Geophysics.* the graded rate of temperature increase at depths below the surface of the earth.

thermal henry *Thermodynamics.* the unit of thermal inductance equivalent to the quantity of 1 K-sec divided by the rate of entropy flow of 1 joule/sec-K (1 watt per degree kelvin).

thermal high *Meteorology.* a high-pressure air parcel that occurs as a result of cooling of the earth's surface while the parcel remains virtually stationary over a given area.

thermal hysteresis *Thermodynamics.* the phenomenon of a property showing a certain temperature dependence upon heating, and showing a different dependence upon cooling over the same temperature range.

thermal imagery *Electronics.* the production of an image of an object or scene by mapping the differences detected in infrared radiation emitted from various objects and areas within the scene; used in many branches of science such as astronomy, medicine, and geology. Also, THERMOGRAPHY.

thermal inactivation point *Virology.* the temperature at which a virus loses its infectivity; ranges from 45°C to 95°C.

thermal inductance *Thermodynamics.* a quantity given by the temperature difference across the surfaces of a body times the time divided by the rate of entropy flow.

thermal instability *Fluid Mechanics.* an instability resulting in natural or free convection when a fluid is heated at a boundary.

thermal instrument *Engineering.* any instrument that employs the heating effect of an electrical current, such as a thermocouple.

thermal jet *Meteorology.* an atmospheric region characterized by strong thermal winds; represented on an upper-air chart by isotherms packed closely together.

thermal jet engine *Aviation.* a jet engine, such as a turbojet, that requires heat to expand the air or other gases used for rearward ejection.

thermal limit *Electricity.* the maximum current or power a system can deliver for an extended time without overheating. Also, CONTINUOUS DUTY RATING.

thermal-liquid system *Chemical Engineering.* a system used for process cooling and heating in which a special liquid functions in the system as a heat sink or heat source.

thermal low *Meteorology.* an area of low atmospheric pressure as a result of high temperatures caused by strong surface heating, common over continental land masses of the subtropics in the summertime. Also, HEAT LOW.

thermal magnon *Solid-State Physics.* a short-wavelength magnon, typically on the order of 10^{-8} meters.

thermal melting profile *Molecular Biology.* a curve that describes, for a given period, the amount of dissociation of the strands in a duplex that occurs as a function of temperature.

thermal metamorphism *Petrology.* a metamorphism in which heat is the principal agent of chemical reconstitution. Also, THERMOMETAMORPHISM.

thermal microphone *Acoustical Engineering.* a microphone that uses the heat generated by an electrically resistive element that is modulated by the variation of sound waves as the energy source for converting sound to electroacoustical signals.

thermal neutron *Nucleonics.* a neutron that is in thermal equilibrium at room temperature; the energy distribution of such neutrons is Maxwellian, and the average kinetic energy at room temperature is 0.025 electron volt. Also, SLOW NEUTRON.

thermal noise *Electronics.* electronic noise signals that are produced by the random thermal motion of charges in circuit elements such as resistors.

thermal noise generator *Electronics.* a unit of test equipment that produces a calibrated electronic noise signal derived from thermal noise generated within an electron tube.

thermal ohm *Thermodynamics.* a unit of thermal resistance equivalent to that of a body having a temperature difference across its surface of 1 K and experiencing an entropy flow rate of 1 watt per degree kelvin.

thermal overload relay *Electricity.* a heat-sensitive switch used as an overcurrent protective device; it opens the circuit of an electric motor or other device when current flowing through the relay and to the device exceeds a specified level.

thermal photography see THERMOGRAPHY.

thermal pollution *Ecology.* any increase in the temperature of a body of water that is caused by human activity (e.g., a discharge of heated water from a power plant or of heated industrial waste) and that has a harmful effect on water quality or on water-dwelling life.

thermal polymerization *Petroleum Engineering.* a thermal procedure used in petroleum refining for converting lighter hydrocarbon gases into liquid fuels.

thermal potential difference *Thermodynamics.* the difference between two temperatures measured at two different points.

thermal power plant *Electrical Engineering.* any electrical power plant, such as a fossil-fuel or nuclear plant, that uses heat energy to produce electricity.

thermal probe *Engineering.* **1.** a liquid-cooled tube that is inserted into a furnace in order to measure its heat absorption rate. **2.** an instrument that measures the amount of heat that flows outward from sediment at the bottom of an ocean.

thermal process *Chemical Engineering.* any process that uses heat without a catalyst to achieve chemical change.

thermal pulse method *Solid-State Physics.* a technique for measuring insulating and conducting properties in crystals wherein a heat pulse, initiated by a laser beam, is observed to propagate through the crystal.

thermal radiation *Thermodynamics.* see HEAT RADIATION. *Ordnance.* the heat and light produced by a nuclear explosion.

thermal reactor *Nucleonics.* a nuclear reactor that operates at a sufficiently low temperature that fission can be induced primarily by slow (thermal) neutrons. *Chemical Engineering.* any system, vessel, or device in which heat-induced chemical reactions take place.

thermal reforming *Chemical Engineering.* a procedure in petroleum refining in which heat is used to create a molecular readjustment of a low-octane naphtha to produce high-octane motor gasoline without a catalyst.

thermal relay *Electricity.* a relay that is actuated by heat.

thermal resistance *Thermodynamics.* a quantity expressing the ability of a substance to prevent a heat transfer; equal to the temperature difference across the surfaces of the body divided by the rate of heat transfer. *Electricity.* the ratio of the temperature rise to the rate at which heat is generated within an electric device under steady-state conditions.

thermal resistivity *Thermodynamics.* a quantity given by the reciprocal of the thermal conductivity of a substance.

thermal resistor *Electricity.* a device whose resistance varies with temperature, such as a thermistor or germanium diode.

thermal Rossby number *Fluid Mechanics.* a dimensionless ratio of the inertial force due to thermal wind to the Coriolis force in a fluid flow heated from below.

thermal runaway *Electronics.* a cumulative condition in a transistor in which localized heating at the base-collector junction causes collector current to increase, thus producing more localized heat, which in turn produces more current. This condition can continue until the transistor is destroyed.

thermal scattering *Solid-State Physics.* the scattering of X-rays, electrons, or neutrons when passing through a crystal due to thermal agitation of the constituent atoms.

thermal shield *Nucleonics.* a shield of thermally conductive material that is placed near the reflector of a nuclear reactor so as to absorb thermal neutrons and radiation.

thermal shock *Engineering.* any equipment or system failure caused by a sharp change in temperature. *Materials Science.* in ceramics, fracture caused by rapid temperature change. The ability of a material to withstand a given temperature change depends on a combination of thermal expansion, thermal conductivity, overall geometry, and the inherent brittleness of that material. Resistance to thermal shock is an important property of materials such as the glass-ceramics.

thermal spring *Hydrology.* a spring whose water temperature is greater than the average annual atmospheric temperature in the vicinity of the spring.

thermal steering *Meteorology.* a process in which an atmospheric disturbance is moved toward the thermal wind in its vicinity.

thermal stratification *Hydrology.* the division of a lake or other body of water into horizontal layers of differing densities, as a result of variations in temperature at different depths.

thermal stress *Materials Science.* stress developed in a restricted material as a result of temperature changes when the material is unable to expand or contract accordingly.

thermal-stress cracking *Materials Science.* the cracking of materials due to overexposure to sudden temperature changes or large temperature differentials.

thermal structure *Petrology.* a distinct structural pattern, such as an anticline or dome, formed by the arrangement of zones of increasing metamorphic grade produced by a localized heat source associated with orogenesis.

thermal switch *Electricity.* a switch that is temperature controlled. Also, THERMOSTAT, THERMOSWITCH.

thermal tide *Meteorology.* the changes in atmospheric pressure caused by the diurnal differential heating of the atmosphere by the sun.
thermal transducer *Engineering.* a device that converts other forms of energy, such as electrical or kinetic energy, into heat.
thermal tuning *Electricity.* the process of adjusting the frequency of a cavity resonator through the use of thermal expansion to vary its shape.
thermal utilization factor *Nucleonics.* a factor representing the probability of a thermal neutron being absorbed in a useful (fissionable) process.
thermal value *Thermodynamics.* the energy released in a combustion process, usually expressed in units of calories per gram or BTUs per pound.
thermal volt see KELVIN.
thermal vorticity *Meteorology.* the vector component equivalent to the curling characteristic or vorticity of a thermal wind.
thermal vorticity advection *Meteorology.* the transportation or advection of thermal vorticity by a thermal wind.
thermal wattmeter *Engineering.* an instrument that uses resistance heat, as detected by a thermocouple, to measure electrical current.
thermal wave *Solid-State Physics.* a short-wavelength sound wave passing through a solid. *Meteorology.* see TEMPERATURE WAVE.
thermal wind *Meteorology.* **1.** the vector difference between the geostrophic wind at a given level in the upper atmosphere and the wind at a lower level. **2.** the sheer or change in direction of the geostrophic wind with height.
thermal X-ray *Electromagnetism.* electromagnetic radiation having frequencies of the low-energy or soft X-ray region of the spectrum.
thermal zone *Meteorology.* **1.** see TEMPERATURE ZONE. **2.** see THERMAL BELT. **3.** see FROSTLESS ZONE.
thermanesthesia *Neurology.* an inability to recognize or differentiate between hot and cold temperatures.
thermic *Thermodynamics.* of or relating to heat; thermal.
thermic boring *Engineering.* a method of boring holes into concrete using heat, usually by driving burning steel wool into the concrete. Also, JET PIERCING.
thermic fever see SUNSTROKE.
thermion *Electronics.* an electron that has been emitted from a heated body such as the hot cathode of an electron tube.
thermionic *Electronics.* of or relating to a thermion; describing the emission of electrons from a heated body, especially from the hot cathode of an electron tube.
thermionic cathode see HOT CATHODE.
thermionic converter *Electronics.* a device that converts heat energy into electrical energy through thermionic emission from a heated cathode. Also, THERMIONIC GENERATOR, THERMOELECTRON ENGINE.
thermionic current *Electronics.* an electrical current consisting of the flow of electrons emitted from a heated surface, such as the plate current in a thermionic electron tube.
thermionic detector *Electronics.* a type of detector circuit in which a thermionic electron tube extracts the modulation signal from a radio-frequency carrier and discards the carrier; for example, the detector stage in a vacuum-tube radio receiver.
thermionic diode *Electronics.* a two-element electron tube device containing a heated cathode and an anode; it is characterized by the ability to pass an electric current in one direction only, from the cathode to the anode.
thermionic emission *Electronics.* the phenomenon by which electrons or ions are emitted from a heated object, such as the cathode of a thermionic electron tube. Also, THERMAL EMISSION.
thermionic fuel cell *Electronics.* a device that converts heat energy directly into electrical energy, using an ionizing gas to enhance efficiency.
thermionic generator see THERMIONIC CONVERTER.
thermionics *Electronics.* the study of the phenomenon by which electrons or ions are emitted from a heated object, such as the cathode of a thermionic electron tube.
thermionic tube *Electronics.* any electron tube whose cathode must be heated in order to emit electrons or ions for normal operation. Also, HOT-CATHODE TUBE.
thermionic work function *Electronics.* the energy required to emit electrons from a cathode.
thermistor *Electronics.* a temperature-sensitive semiconductor device possessing a negative temperature coefficient (i.e., resistance decreases as temperature increases); commonly used as a bolometer to measure temperatures and, indirectly, microwave energy levels. (An acronym for therm-ally sensitive res-istor.)
thermite *Materials.* a mixture of ferric oxide with powdered metallic aluminum that produces very high temperatures when ignited; used in thermite processing, welding, and incendiary bombs. Also, **thermit.**
thermite process *Metallurgy.* the action of aluminum on a metallic oxide when they are heated together, making a product that can be used in the production of metallic iron manganese, chromium, tungsten, and uranium, among others.
thermo- a combining form meaning "heat."
thermoacidophile *Biology.* **1.** growing well in a warm acid environment. **2.** easily stained with warm acid dyes.
thermoacoustic array *Acoustics.* a light beam, usually that of a laser, that is modulated by the frequency of an acoustic signal; the resulting sound pattern has a directivity that is perpendicular to the light beam's axis and has no side lobes.
Thermoactinomyces *Bacteriology.* a genus of Gram-positive, thermophilic bacteria of the family Micromonospora growing as both substrate and aerial mycelia, producing heat-resistant spores, and found in soil, compost, and water.
thermoalgesia see THERMALGESIA.
thermoammeter *Engineering.* an ammeter that employs a thermocouple to measure radio-frequency current.
thermoanesthesia see THERMANESTHESIA.
thermoanalysis see THERMAL ANALYSIS.
Thermobacteroides *Bacteriology.* a genus of Gram-negative, thermophilic bacteria occurring as rod-shaped cells with a strictly anaerobic respiration.
thermobalance *Analytical Chemistry.* a weighing scale modified to measure the weight changes of a material that is being heated.
thermocauterectomy *Surgery.* the excision of an organ or tissues by thermocautery.
thermocautery *Surgery.* the use of a heated wire or point to burn away tissue during a surgical procedure.
thermochemistry *Physical Chemistry.* the branch of chemistry that studies the heat changes which accompany chemical reactions and changes of state. Thus, **thermochemical, thermochemist.**
thermocline *Geophysics.* a layer in a body of water at which the rate of temperature decrease with depth is at a maximum.
thermocoagulation *Medicine.* a process of halting growths by causing coagulation, using high-frequency electrical current.
Thermococcus *Bacteriology.* a genus of coccoid bacteria of the order Thermoproteales, occurring in marine habitats and able to metabolize peptides for energy generation.
thermocompression *Mechanics.* any process involving both heat and pressure.
thermocompression bonding *Engineering.* the use of both heat and pressure to bond two objects together.
thermocompression evaporator *Mechanical Engineering.* a system designed to reduce the energy requirements for evaporation by compressing vapor from a single-effect evaporator so that it may be used as the heating medium in the next evaporation stage or in the same single-effect evaporator stage.
thermocouple *Electrical Engineering.* a device consisting of two different metallic conductors that are connected at both ends, producing a loop in which heat is converted into electrical current when there is a difference in temperature between their two junctions; used to measure the temperature of a third substance by connecting it to both junctions and measuring the voltage generated between them. Also, THERMOELECTRIC COUPLE.
thermocouple ammeter see THERMOAMMETER.
thermocouple converter *Electronics.* see THERMOELECTRIC CONVERTER.
thermocouple vacuum gauge *Engineering.* an instrument that uses the effect of thermal conductivity on a thermocouple to measure the density of rarified gas; typically used in the range of 10^{-1} to 10^{-3} millimeters of mercury.
thermocyclogenesis *Meteorology.* a cyclogenetic model in which a disturbance initiated in the stratosphere is reflected in the development of a disturbance in the lower troposphere.
Thermodiscus *Bacteriology.* a genus of disc-shaped bacteria of the order Thermoproteales.
thermoduric bacteria *Microbiology.* bacteria that are capable of surviving the high temperatures involved in the pasteurization process but do not grow in those temperatures.
thermodynamic *Physics.* **1.** of or relating to thermodynamics. **2.** supplying, using, or accepting energy.

thermodynamic cycle *Thermodynamics.* a series of processes that occur in a cyclic pattern acting on a substance in which one form of energy is converted to another form.

thermodynamic diagram *Meteorology.* any chart or graph on which values of pressure, density, temperature, and water vapor are represented so as to satisfy the equation of state, the Clapeyron-Clausius equation, and the first law of thermodynamics for adiabatic and saturation or pseudoadiabatic processes; used to display isobars, isotherms, vapor lines, dry adiabats, and pseudoadiabats.

thermodynamic equation of state *Thermodynamics.* used in conjunction with reversible processes, an equation that relates the pressure, temperature, or volume to these processes.

thermodynamic equilibrium *Thermodynamics.* the condition of a thermodynamic system that is in thermal, mechanical, and chemical equilibrium.

thermodynamic function of state *Thermodynamics.* see STATE VARIABLE.

thermodynamic laws see FIRST (SECOND, THIRD) LAW OF THERMODYNAMICS.

thermodynamic potential see FREE ENTHALPY.

thermodynamic potential *Thermodynamics.* an additive quantity whose value is independent of the history of a thermodynamic system and depends on the instantaneous state; under specified conditions, the potential is minimized when the system is in thermodynamic equilibrium.

thermodynamic principles *Thermodynamics.* a general term for the laws that are basic to the conversion of energy into various other forms.

thermodynamic probability *Thermodynamics.* the total number of microstates (different arrangements of particles), usually symbolized by W, that produce the same macrostate (the condition of a thermodynamic system as specified by its macroscopic properties); the thermodynamic probability is related to the entropy by the relation $S = k \ln W$, where k is the Boltzmann constant.

thermodynamic process *Thermodynamics.* a process resulting in the change in any macroscopic property of a system or body.

thermodynamic property *Thermodynamics.* see MACROSCOPIC PROPERTY.

thermodynamics **1.** the branch of science that is concerned in the study of energy and with the relationship of heat transfer and work to other forms of energy. Thermodynamics deals with the behavior of systems in which temperature is a significant factor. **2.** see FIRST (SECOND, THIRD) LAW OF THERMODYNAMICS.

thermodynamic system *Thermodynamics.* a quantity of matter (such as a collection of atoms, molecules, and particles) of fixed mass and identity on which attention is focused for thermodynamic study; each system is separated from its surroundings by the system boundaries, which may be fixed or movable.

thermodynamic temperature scale *Thermodynamics.* a temperature scale whose zero point coincides with absolute zero, such that the ratio of the temperatures of two heat transfer reservoirs used in a Carnot cycle is equal to the ratio of heat transfer from the high-temperature reservoir to that amount rejected to the low-temperature reservoir; examples of this are the Kelvin and Rankine temperature scales.

thermodynamic variable *Thermodynamics.* see STATE VARIABLE.

thermoelastic *Physics.* relating to the thermodynamic effects produced by deformation of an elastic substance.

thermoelastic inversion effect *Materials Science.* the tendency for a reduction in stress due to thermal expansion.

thermoelasticity *Physics.* a topic in thermodynamics that deals with the compressibility of substances.

thermoelectric *Electronics.* of or relating to thermoelectricity.

thermoelectric converter *Electronics.* a device to convert heat energy directly into electrical energy, typically consisting of a thermoelectric junction in contact with a heat source; the voltage developed across the junction is a function of the temperature of the source. Also, THERMOELECTRIC GENERATOR, THERMAL or THERMOCOUPLE CONVERTER.

thermoelectric cooler *Engineering.* an electronic heat pump used to produce cooling by the thermoelectric effect, consisting of a thermocouple with its "cold" end in the chamber to be cooled and its "hot" end in an outside heat sink. Thus, **thermelectric cooling.**

thermoelectric couple see THERMOCOUPLE.

thermoelectric effect *Physics.* the production of current or voltage (thermal electromotive force) generated by a thermocouple.

thermoelectric generator see THERMOELECTRIC CONVERTER.

thermoelectric heating *Engineering.* the use of the thermoelectric effect to produce heating by application of a current through a thermocouple. The current is reversed from that in thermoelectric cooling.

thermoelectricity *Electricity.* electricity produced by the direct action of heat or the direct conversion of heat into electricity, as in a thermocouple.

thermoelectric junction see THERMOJUNCTION.

thermoelectric laws *Engineering.* the laws governing the thermoelectric effect of direct conversion of heat into electric current and vice versa; the laws that govern the behavior of a thermocouple.

Thermodynamics

Thermodynamics originally dealt with problems of converting heat into maximal useful work in engines of the early 1800's. In the next hundred years the field was broadened into a discipline dealing with the macroscopic variables pertaining to equilibrium conditions (thermostatics) and to processes occurring not far from equilibrium (thermodynamics).

Over the last six decades, statistical thermodynamics has been extensively used to correlate microscopic (molecular) properties with macroscopic features of a system. In all of the above theories one must distinguish between quantities such as heat and work, whose magnitudes generally depend on the specific path by which a system is taken from an initial to a final state, and equilibrium properties of the system whose difference does not depend upon the path.

The four laws of thermodynamics are as follows: (0) Any two systems in thermal equilibrium with a third are in equilibrium with each other: this forms the basis for thermometry. (1) Every system in equilibrium is characterized by its total internal energy E, whose value can be changed only by transfer of heat, work, or matter across its boundaries. Heat flow occurs in response to a temperature difference across the system and surroundings; work is executed in response to unbalanced forces acting on the system. As a corollary, energy can only be transformed from one type to another or shifted from the system to the surroundings by execution of work and by transfer of heat and/or matter. E is thus conserved in an isolated system; this rules out the possibility of constructing perpetual-motion machines. (2) There exists another independent characteristic property, the entropy S, which for reversible processes responds to the ratio Q/T, where Q is the heat transferred across the boundary, and T is the absolute temperature. For a system at fixed energy, S also increases (but never decreases) in response to any spontaneous process which occurs solely within the system. As a corollary, when the system reaches equilibrium its entropy is at a maximum. This law leads to upper limits on the efficiency achievable with engines operating in cyclic processes. (3) As $T \to 0$, S approaches a least value (≥ 0) in such a manner that one cannot attain $T = 0$ in a finite process. Alternative formulations of the four laws are frequently cited but lack the generality of the above statements.

J. M. Honig
Professor of Chemistry
Purdue University

thermoelectric material *Electronics.* a material, such as cesium sulfide and lead telluride, that is capable of converting thermal energy to electric energy or of producing cooling when electric energy is applied to it.

thermoelectric nuclear battery *Nucleonics.* a low-voltage battery that converts heat generated by a radioactive isotope to electrical energy by means of thermocouples; the radioactive source is hermetically sealed in a capsule.

thermoelectric properties *Physics.* properties relating to the direct relations between heat and electricity as stated by the Thomson relations.

thermoelectric pyrometer *Engineering.* an instrument that uses thermocouples to measure high temperatures, ranging from 425° to 1315°C. Also, **thermocouple pyrometer.**

thermoelectric solar cell *Electronics.* an apparatus used in the indirect conversion of solar energy into electrical energy; solar energy is first transformed into heat energy, which is then converted to electrical energy in a cell made from a semiconductor material.

thermoelectric thermometer *Engineering.* an instrument that measures temperatures by inserting one thermocouple into a constant-temperature bath and employing another as a measuring junction.

thermoelectron *Electronics.* **1.** an electron emitted from a hot cathode. **2.** a negative thermion.

thermoelectron engine see THERMIONIC CONVERTER.

Thermofilum *Bacteriology.* a genus of rod-shaped or filamentous thermophilic bacteria of the order Thermoproteales that metabolize peptides for energy production; found in Icelandic solfataric environments.

thermoforming *Materials Science.* a category of processing methods that use vacuum, air pressure, or mechanical energy to force a heated thermoplastic sheet into the shape of a mold; after cooling, the plastic part is removed from the mold and trimmed; used for boats, and components ranging from blister or bubble packaging to fender liners on automobiles.

thermogalvanometer *Engineering.* an instrument that measures small, high-frequency currents by use of the thermoelectric effect; a DC galvanometer is connected to a thermocouple that is heated by a filament carrying the current to be measured.

thermogenesis *Science.* the production of heat, especially in an animal body by physiological processes.

thermogram *Engineering.* the temperature record produced by a thermograph or recording thermometer.

thermograph *Meteorology.* a self-recording thermometer that measures both air and soil temperature. *Engineering.* an instrument, operating at infrared wavelengths, that measures, senses, and records variations in atmospheric temperature. *Radiology.* the instrument used in radiological thermography.

thermograph correction card *Engineering.* a table used to correct the reading of a thermograph to make it consistent with the reading of a dry-bulb thermometer.

thermographics *Graphic Arts.* a reprographic process in which heat is applied to a heat-sensitive paper, which darkens in the image areas.

thermography *Engineering.* the process of temperature measurement, using radiation emitted by an object. *Radiology.* specifically, a technique in which an infrared camera is used to photographically portray the surface temperatures of the body, based on the self-emanating infrared radiation; sometimes used to diagnose underlying pathologic processes. *Electronics.* see THERMAL IMAGERY. *Graphic Arts.* a printing process that simulates engraved printing by using a slow-drying ink that is coated with a resinous powder, which is then heat-dried to produce a raised image.

thermogravimetric analysis *Analytical Chemistry.* the measurement of weight changes of a substance as a function of increased temperature; used in quantitative analysis and to determine decomposition.

thermogravitational column *Mechanical Engineering.* a device, used to achieve binary separations, in which a concentration gradient is formed as a result of thermal diffusion caused by countercurrent flow of hot and cold fluids.

thermohaline *Oceanography.* of or relating to the joint action of temperature and salinity in a water mass.

thermohaline convection *Oceanography.* vertical movement of a layer of water caused by changes in the temperature-salinity relationship between it and the adjoining layers, usually because one layer has become colder or more saline, or both, and thus become heavier.

thermohypesthesia *Neurology.* a disorder of sensation characterized by diminished sensitivity to high temperatures.

thermoinducible lysogen *Biotechnology.* a bacterial strain that contains a prophage whose repressor is temperature sensitive; an elevated temperature leads to further phage development.

thermointegrator *Engineering.* a device used to measure the total supply of heat present in a soil sample during a given time interval.

thermoisopleth *Meteorology.* an isopleth of temperature as recorded on a climatic diagram.

thermojet *Aviation.* **1.** a jet of heated gases ejected from a jet engine to provide thrust. **2.** a thermal jet engine.

thermojunction *Electronics.* the junction between the two dissimilar metal conductors in a thermocouple and the part of the thermocouple that comes in direct contact with the object being monitored. Also, THERMOELECTRIC JUNCTION.

thermojunction battery *Electricity.* a battery that converts heat directly into electrical energy.

thermokarst topography *Geology.* an area in a permafrost region characterized by an irregular land surface, developed by the melting of ground ice.

thermokinematics *Science.* the study of motion caused by heat.

thermokinetic analysis *Analytical Chemistry.* an analysis that involves the rapid and continuous addition of a titrant while measuring temperature; the endpoints are mathematically mapped and used for determining reaction rates.

thermolabile *Biochemistry.* liable to destruction or loss of characteristic properties by the action of moderate heat, as certain toxins.

thermoluminescence *Atomic Physics.* the glow or emission of light produced by the application of heat; used to monitor the radiation dose to which a substance has been exposed.

thermoluminescent dating *Archaeology.* a method of dating by measuring the rate of release of light energy from an object; often used to establish the date when a pottery artifact was last heated in antiquity. Also, TL DATING.

thermolysin *Biochemistry.* a neutral zinc-containing metalloprotease that remains active up to 80°C and is the most heat-stable protease available commercially; produced by a strain of the bacteria *Bacillus stearothermophilus* (or *B. thermoproteolyticus*). Also, THERMOPHILIC BACTERIAL PROTEASE.

thermolysis *Chemistry.* a process of chemical dissociation by heat. *Physiology.* the dissipation of body heat by processes such as radiation or evaporation.

thermomagnetic effect *Physics.* joule heating by currents that are induced in conductors by changing magnetic fields. *Thermodynamics.* see MAGNETOCALORIC EFFECT.

thermomastography *Radiology.* the use of thermography in the diagnosis of breast lesions.

thermomechanical treatment *Metallurgy.* the use of a combination of mechanical deformation and temperature-controlled transformation to provide improved mechanical properties; applied to steels, nickel-based superalloys, and aluminum alloys.

thermometamorphism see THERMAL METAMORPHISM.

thermometer *Engineering.* an instrument that measures and indicates temperature; usually consisting of a narrow tube filled with a liquid, such as mercury, that rises (expands) as the temperature rises and falls (contracts) as the temperature falls. The level of the liquid corresponding to a calibrated scale on the tube indicates the temperature.

thermometer anemometer *Engineering.* an anemometer that uses two thermometers to record the cooling rate promoted by an airstream; this value may then be converted to air speed velocity.

thermometric *Engineering.* using or obtained by a thermometer.

thermometric analysis *Physical Chemistry.* an examination of the physical changes that a substance undergoes as it is heated or cooled at a constant rate.

thermometric depth *Oceanography.* the depth at which paired, protected and unprotected thermometers attached to a Nansen bottle are reversed; the unprotected thermometer records higher temperature readings because of the effect of hydrostatic pressure on it, and the water's depth at the time of reversal may be calculated by comparing the two readings.

thermometric property *Thermodynamics.* a property related to the expansion of materials by a heat transfer.

thermometric titration *Analytical Chemistry.* a calorimetric titration that yields a plot of temperature versus the volume of titrant; used in precipitation reactions.

thermometry *Thermodynamics.* the study and application of temperature measurement.

Thermomicrobium *Bacteriology.* a genus of Gram-negative, aerobic, thermophilic bacteria of uncertain affiliation that grow as nonmotile, rod-shaped, pink-pigmented cells in hot-spring habitats.

thermomigration *Electronics.* a semiconductor manufacturing method in which exact quantities of impurities are added to a semiconductor material through the migratory action of the impurities when the material is heated.

Thermomonospora *Bacteriology.* a genus of thermophilic bacteria of the family Micromonosporaceae that form substrate and aerial mycelia and are characterized by metabolic versatility; found in soil and compost.

thermonatrite *Mineralogy.* $Na_2CO_3 \cdot H_2O$, a colorless, white, grayish, or yellowish orthorhombic mineral occurring as a crust or efflorescence, having a specific gravity of 2.255 and a hardness of 2.1 to 1.5 on the Mohs scale; found in saline lake deposits and fumaroles.

thermonuclear *Nucleonics.* of or relating to fusion processes between light nuclei involving high temperatures and great amounts of liberated energy.

thermonuclear bomb *Ordnance.* a bomb whose explosive power derives from a thermonuclear reaction; i.e., a hydrogen bomb.

thermonuclear device *Nucleonics.* a fusion bomb used for peaceful or experimental purposes.

thermonuclear reaction *Nuclear Physics.* a nuclear fusion reaction, occurring most readily between isotopes of hydrogen, in which nuclei are brought so close together by enormous temperatures and pressures (available in the sun, stars, and exploding fission bombs) that they are able to overcome their mutual electrostatic repulsion to form a more complex nucleus; the new rest mass is less than the sum of the rest masses of the original nuclei, since the rest mass difference is liberated as energy.

thermonuclear rocket *Nucleonics.* a hypothetical rocket whose thrust is generated by a controlled thermonuclear reactor (CTR); this form of propulsion would be very desirable, short of matter-antimatter propulsion, in view of the available energy.

thermonuclear weapon *Ordnance.* a nuclear weapon in which destruction is caused by a fusion explosion accompanied by a large amount of liberated heat.

thermoperiodicity *Botany.* the response of plants to normal changes in temperature within a given period.

thermophile *Microbiology.* a fungus, bacteria, or other microorganism that grows best at temperatures above 50°C. Thus, **thermophilic.**

thermophilic bacterial protease see THERMOLYSIN.

thermophobia *Psychology.* an irrational fear of heat.

thermophone *Acoustical Engineering.* a transducer that converts electrical energy (produced by driving an alternating current superimposed on a direct current) into heat by the dissipation of energy in a resistive element, thereby producing sound.

thermopile *Electricity.* an array of thermocouples connected in parallel, having greater sensitivity than a single thermocouple; used to convert radiant energy into electrical energy and for detecting and measuring radiant energy.

thermopile generator *Electricity.* an electricity source that is often connected to a thermopile to generate small amounts of electric current; it is powered by the heating of an electrical resistor connected to a thermopile.

thermoplacentography *Radiology.* the use of thermography to determine the site of placental attachment as revealed by a temperature rise at a site corresponding to the excessive blood flow.

Thermoplasma *Bacteriology.* a genus of aerobic, thermophilic, coccoid to filamentous bacteria of the order Mycoplasmatales that are found in hot springs or burning coal; generally considered to be among the archaeobacteria.

Thermoplasmales *Bacteriology.* in some systems of classification, an order of aerobic, thermophilic bacteria distinguished by the lack of a cell wall and classified among the archaeobacteria.

thermoplastic *Materials Science.* any of a group of polymers that can be easily, repeatedly processed and soften on heating, since no by-products are formed during their production. *Textiles.* of or relating to textile fibers and resins that have been softened at high temperatures; all synthetic fibers are thermoplastic.

thermoplastic recording *Electronics.* a form of recording in which the information is permanently impressed on plastic tape by the physical distortion of the tape, through thermal means, to produce irregularities along the edges.

thermoplastic rubber see ELASTOPLASTIC.

thermoplegia *Medicine.* heatstroke or sunstroke.

thermopower *Electricity.* in a conductor, the measurement of temperature-induced voltage.

Thermoproteales *Bacteriology.* an order of anaerobic, thermophilic, sulfur-dependent bacteria occurring as spheroid, rod-shaped, or irregularly shaped cells and found in hot-spring habitats; classified among the archaeobacteria.

Thermoproteus *Bacteriology.* a genus of thermophilic bacteria of the order Thermoproteales, occurring as rod-shaped or filamentous cells in Icelandic solfataric habitats.

thermoradiotherapy *Radiology.* the application of ionizing radiation to an anatomical area whose tissue temperature has been artificially elevated on the theory of increasing its radiosensitivity.

thermoreception *Physiology.* the response of certain differentiated body cells to a rise or fall in body temperature. *Entomology.* behavior that is due to changes in temperature; for example, worker honeybees responding to hotter than usual temperatures by beating their wings to cool the hive.

thermoreceptor *Physiology.* a sense organ that is stimulated by heat or cold.

thermoregulation *Physiology.* the various physiological processes by which the body regulates its internal temperature.

thermoregulator *Engineering.* a highly accurate and sensitive thermostat.

thermoregulatory center *Physiology.* the hypothalamic center regulating and controlling body heat.

thermoremanent magnetization *Geophysics.* the magnetic orientation, conforming to the ambient magnetic field at the time, that an igneous rock acquires at the time of its cooling from the Curie point to room temperature.

Thermosbaenacea *Invertebrate Zoology.* an order of tiny malacostracan crustaceans in the superorder Pancarida, found in salt and hot springs and in underground waters.

Thermosbaenidae *Invertebrate Zoology.* a family of tiny malacostracan crustaceans in the order Thermosbaenacea, found in hot springs.

thermoset *Materials Science.* describing any of a group of polymers that soften when initially heated, then harden and condense in bulk and retain a permanent shape; they cannot be softened or reprocessed by reheating.

thermosiphon *Mechanical Engineering.* **1.** a method of establishing circulation of a cooling liquid by utilizing the slight difference in density of the hot and cool portions of the liquid. **2.** a density-driven circulation system used to achieve high-heating fluxes in reboilers.

thermosiphon reboiler *Chemical Engineering.* a liquid reheater in which the natural circulation of a boiling liquid is achieved by sustaining an adequate liquid head; used for distillation-column bottoms.

thermosis *Ecology.* any change that takes place in an organism because of the effect of heat.

thermosphere *Meteorology.* the atmospheric layer, constituting essentially all of the atmosphere above the mesosphere, in which temperature increases with height; includes the exosphere and most or all of the ionosphere.

thermostable *Biochemistry.* capable of withstanding a moderate degree of heat without loss of characteristic properties, as certain toxins and enzymes.

thermostasis *Biology.* the maintenance of body temperature in warm-blooded animals.

thermostat *Engineering.* **1.** a device that regulates temperature by opening or closing a circuit when the temperature deviates from a set value. **2.** specifically, such a device used to control the temperature in a room, building, or other enclosure; hotter or cooler air is supplied as necessary to maintain the temperature at the same level as the setting on the thermostat. Also, **thermorelay.** *Electricity.* see THERMAL SWITCH.

thermostat theory *Neurology.* the theory that a decrease in body temperature activates the feeding center in the brain, and that an increase in body temperature activates the satiety center.

thermosteric anomaly *Oceanography.* the anomaly of specific volume of a water sample, calculated for the recorded temperature and salinity but at a standard pressure of 1 atmosphere.

thermoswitch see THERMAL SWITCH.

thermotaxis *Biology.* the movement of an organism in response to a source of heat. *Physiology.* the regulation of body temperature.

thermotherapy *Medicine.* therapy by application of heat. *Virology.* specifically, the use of heat to cure a virus infection in a host or cell line.

Thermothrix *Bacteriology.* a genus of facultatively anaerobic, rod-shaped or filamentous thermophilic bacteria of uncertain affiliation that utilize sulfur compounds in metabolism; found in hot springs.

thermotropic liquid crystal *Physical Chemistry.* a liquid crystal produced by heating a substance.

thermotropic model *Meteorology.* in numerical forecasting, a model atmosphere used to construct a surface prognostic chart, characterized by a constant-pressure surface height (typically 500 mb) and a single temperature (usually the mean temperature between 100 and 500 mb).

thermotropism *Biology.* the tendency to turn or bend in response to the sun or any other source of heat.

thermovac evaporator *Food Technology.* an evaporating machine that uses both heat and a vacuum to extract water from a food substance; often used repeatedly, with each stage further concentrating the food.

thermovoltmeter *Engineering.* a voltmeter in which a thermocouple attached to a heater wire generates a voltage that may then be measured.

Thermus *Bacteriology.* a genus of Gram-negative, aerobic, rod-shaped or filamentous thermophilic bacteria of uncertain affiliation, distinguished by the formation of yellow, orange, or red carotenoid pigments, and found in hot springs.

therophyte *Ecology.* a plant that accelerates its life cycle when conditions are favorable and weathers unfavorable conditions as a seed; typically found in deserts or on cultivated land. *Botany.* a plant that lives only one year or growing season.

Theropoda *Paleontology.* a suborder of carnivorous saurischian dinosaurs that spanned a great range of types and sizes; the theropods arose in the Triassic and quickly achieved a worldwide distribution; in the Cretaceous several families developed into the large, predatory forms typified by Allosaurus and Tyrannosaurus.

Theropsida *Paleontology.* a proposed ordinal term that, along with Sauropsida, would divide all the reptiles into two groups rather than the generally accepted division according to temporal openings in the skull: Anapsida, Synapsida, Diapsida, and Euryapsida.

therosaur *Paleontology.* of or relating to the therapsid carnivores that were ancestral to the more mammal-like cynodonts; extant from the Upper Permian to Lower Jurassic.

thesaurismosis see STORAGE DISEASE.

theta [thāt´ə] the eighth letter of the Greek alphabet, written Θ, θ.

theta antigen see THY-1 ANTIGEN.

theta functions *Mathematics.* a set of entire functions of a complex variable z that are periodic with a real period and quasiperiodic with a complex period. This is useful in the study of elliptic functions, since any elliptic function can be written as a quotient of theta functions. The complex quasiperiod can be regarded as a second variable τ, in which case the theta functions are modular functions in τ. The standard set of theta functions is usually taken to be the following, where $q = e^{i\pi\tau}$ and $\mathrm{Im}(\tau) > 0$:

$$\theta_1(z|\tau) = 2\sum_{n=0}^{\infty}(-1)^n q^{(n+1/2)^2}\sin(2n+1)z$$

$$\theta_2(z|\tau) = 2\sum_{n=0}^{\infty} q^{(n+1/2)^2}\cos(2n+1)z$$

$$\theta_3(z|\tau) = 1 + 2\sum_{n=1}^{\infty} q^{n^2}\cos(2nz)$$

$$\theta_4(z|\tau) = 1 + 2\sum_{n=1}^{\infty}(-1)n q^{n^2}\cos(2nz).$$

thetagram *Thermodynamics.* a plot of the temperature as a function of pressure for a thermodynamic system.

theta pinch *Physics.* a plasma fusion device that operates with a high degree of stability; it utilizes the magnetic field generated by its own currents (which are induced by pulsing) to confine the plasma for several microseconds; the device is configured so that the self-generated magnetic field runs parallel to a longitudinal axis.

theta polarization *Electromagnetism.* polarization in which the electric field vector is tangent to meridian lines of some specified spherical frame of reference.

theta rhythm *Physiology.* a pattern of brain waves having a regular frequency of 4 to 7 cycles per second as recorded by an electroencephalograph and occurring naturally in children up to about 12 years of age, but considered abnormal in adults.

theta temperature *Materials Science.* in polymer solutions, the temperature at which the second virial coefficient disappears and the coiled polymer molecules expand to their full contour length and become rod-shaped.

theta-theta *Navigation.* a general term for electronic navigation systems that determine position based on bearings from transmitters located at different, accurately known positions.

theta wave *Physiology.* any of the waves constituting a theta rhythm.

Thevenin's theorem *Electricity.* a theorem stating that a linear network with two output terminals can be represented as a single voltage source in series with a single impedance, in series with the terminals. The actual network can have any number of impedances and voltage and current sources, although the sources must be at the same frequency. The source and impedance, called an equivalent circuit, can be used to show the effect of the network on any device connected to the output terminals. Also, **Helmholtz's theorem.**

THF tetrahydrofuran; thymic humoral factor.

THFA tetrahydrofolic acid.

THI temperature-humidity index.

thiabendazole *Organic Chemistry.* $C_{10}H_7N_3S$, white to tan crystals; slightly soluble in water and alcohol; melts at 304°C; used as a fungicide and nematicide.

thiacetic acid see THIOACETIC ACID.

thiamine or **thiamin** [thī´ə min] *Biochemistry.* a water-soluble component of the B complex of vitamins; found in beans, green vegetables, sweet corn, egg yolk, liver, corn meal, and brown rice, and in blood plasma and cerebrospinal fluid in the free state; necessary for the breakdown of carbohydrates. Also, VITAMIN B_1.

thiamine hydrochloride *Organic Chemistry.* $C_{12}H_{17}ClN_4OS{\cdot}HCl$, a hygroscopic, white crystalline solid; soluble in water and decomposes at 247°C; it is usually administered as a vitamin B_1 supplement.

thiamine pyrophosphate *Biochemistry.* the coenzyme form of thiamine that participates in a range of enzymatic reactions in which aldehyde groups move from a donor to an acceptor molecule.

thianaphthene *Organic Chemistry.* C_8H_6S, a crystalline solid; insoluble in water and soluble in alcohol and ether; melts at 32°C and boils at 221°C; used in pharmaceuticals.

Thiaridae *Invertebrate Zoology.* a family of freshwater snails, gill-breathing gastropod mollusks in the order Pectinibranchia.

thiazine *Chemistry.* any of a class of compounds, each having a ring composed of four carbon atoms, one sulfur and one nitrogen atom; the parent substances of certain dyes.

thiazole *Organic Chemistry.* C_3H_3NS, a colorless or pale-yellow liquid; slightly soluble in water, soluble in alcohol and ether; boils at 116.8°C; used in the synthesis of fungicides, dyes, and rubber accelerators.

thiazole dye *Organic Chemistry.* any one of a class of dyes whose chromophore groups are –S–C= and =C=N–; used primarily for dyeing cotton.

thick-bedded *Geology.* describing a bed of sedimentary rock that ranges from 60 to 120 centimeters in thickness.

thicken to make or become thick; specific uses include: *Graphic Arts.* to make halftone dots larger and more dense, thereby increasing the strength of the image; the opposite of *sharpen*.

thickener any material or device that thickens; specific uses include: *Chemical Engineering.* a centrifugal or gravity device that removes liquid from a liquid-solid slurry.

thickening *Chemical Engineering.* the process of concentrating solid particles in a suspension so that a fraction having a higher degree of solid concentration than the original slurry may be recovered. *Mining Engineering.* the process of concentrating dilute slime pulp by rejecting substantially solid-free liquid.

thick-film capacitor *Electricity.* a capacitor in which two overlapping thick-film layers of conducting material are separated by a deposited dielectric film.

thick-film circuit *Electronics.* an electronic assembly in which some circuit elements such as resistances and capacitances are formed on an insulating substrate, and the active devices such as amplifiers and gates are added as separate packages.

thick-film processing *Materials Science.* in the processing of hybrid electronic circuits, a technology in which conductor, resistor, and dielectric patterns are printed onto ceramic substrates, augmented by discrete components such as chip capacitors.

thick-film resistor *Electricity.* a fixed resistor with a resistance element constituted by a film whose thickness exceeds 0.001 inch or 25 micrometers.

thickhead *Vertebrate Zoology.* any of several birds of the genus *Pachycephala*, related to the flycatcher and having a melodious call; found especially in various islands of the Pacific Ocean. Also, WHISTLER.

thick-knee see STONE CURLEW.

thickness *Meteorology.* the vertical depth (in geometric or geopotential units) of an atmospheric layer bounded by surfaces of two different values of the same physical quantity, usually constant-pressure surfaces.

thickness chart *Meteorology.* a synoptic weather chart representing the thickness of a particular atmospheric layer, such as the layer between two constant-pressure surfaces.

thickness line *Meteorology.* an isopleth of thickness; a line connecting all geographic points at which the thickness of given atmospheric layer is the same. Also, RELATIVE CONTOUR, RELATIVE ISOHYPSE.

thickness pattern *Meteorology.* the pattern of thickness lines on a thickness chart. Also, RELATIVE TOPOGRAPHY, RELATIVE HYPSOGRAPHY.

thickness ratio *Aviation.* on a given airfoil section, the ratio of maximum thickness to chord length.

thick-tailed bushbaby *Vertebrate Zoology.* the popular name for *Galago crassicaudatus*, a member of the primate family Lorisidae, noted for its long bushy tail. Also, GREAT GALAGO.

thick-thin chart see ISENTROPIC THICKNESS CHART.

thief *Petroleum Engineering.* a device used in petroleum sampling, featuring a metal or glass cylinder with a spring valve that can be tripped to capture a sample from any depth within a tank.

thief ant *Invertebrate Zoology.* a small red ant, *Solenopsis molesta*, that steals eggs and young larvae from, and nests in the walls of, the nest of a larger species; found in North America.

Thiele coordinates *Chemical Engineering.* a graphic method for determining the solvent-free content of two components undergoing separation by solvent extraction.

Thiele-Geddes method *Chemical Engineering.* a calculation procedure used for predicting the distribution of products from a distillation system that involves multiple components; the distribution of components between distillate and bottoms is predicted for a specified number of stages using tearing techniques.

Thiele melting-point apparatus *Analytical Chemistry.* a stirred, spaced test tube used in the determination of the melting point of a crystalline substance.

Thiele modulus *Chemical Engineering.* a characteristic ratio of the rate of reaction to diffusion for heterogeneous catalysis in a porous substrate.

Thiele reagent *Organic Chemistry.* an oxidizing reagent used in organic reactions, composed of chromium trioxide, acetic acid, and sulfuric acid.

Thiersch's graft see OLLIER-THIERSCH GRAFT.

Thiessen polygon method *Meteorology.* a method of extrapolating and comparing rainfall values for areas surrounding a number of measuring stations; polygons are formed by plotting the stations on a map and drawing lines between them, then dividing the lines with perpendicular bisectors and locating one station at the center of each polygon.

thigh *Anatomy.* the portion of the lower limb between the pelvic girdle and the knee.

thigh bone *Anatomy.* the popular name for the femur, the large, strong bone extending from the pelvis to the knee.

thigmotaxis *Biology.* movement of an organism in response to any object that provides a mechanical stimulus.

Thigmotrichida *Invertebrate Zoology.* an order of ciliated protozoans in the subclass Holotrichia, commensal in the mantle cavity of aquatic bivalves.

thigmotropism *Biology.* a tendency of an organism or part of an organism to bend or turn in response to a touch stimulus, as a vine coiling around a string support.

thill see UNDERCLAY.

thimble *Mechanical Devices.* **1.** a small cap of metal or plastic worn on a finger to protect it when pushing the needle while sewing. **2.** any of various similar small metal tubes, rings, sleeves, bushings, or machine fittings used for protection or reinforcement. *Computer Technology.* a printing element that is used for letter-quality printing; shaped like a thimble, with character slugs around its surface.

thimble chamber *Radiology.* a minute ionization cell, usually with thin walls of organic material, containing a central pin and a thimble-shaped cap as electrodes; used to gauge a radiation dose.

thimerosal *Organic Chemistry.* $C_9H_9HgNaO_2S$, a cream-colored, water-soluble, crystalline powder; an antimicrobial agent used as a topical anti-infective and in biological reagents as a preservative.

thin *Meteorology.* of a sky cover, predominantly transparent.

thin-bedded *Geology.* describing a bed of sedimentary rock that ranges from 5 to 60 centimeters in thickness.

thin film *Electronics.* any of several thin layers (generally less than 1 μm each) of insulating, conducting, or semiconductor material that are deposited successively on a supporting substrate in precise patterns to collectively form all or part of an integrated circuit; the deposition can be performed by mechanical, chemical, or high-vacuum evaporation methods. Thus, **thin-film circuit, thin-film integrated circuit, thin-film material, thin-film semiconductor.**

thin-film capacitor *Electricity.* a capacitor that is fabricated by electrodepositing a thin film of metal on each side of a grown layer of oxide in a thin-film integrated circuit.

thin-film cryotyron *Electronics.* a solid-state switching device that operates at very low temperatures to utilize the phenomenon of superconductivity, and in which the on/off state of a superconducting thin-film gate element is controlled by current through a lead control wire.

thin-film ferrite coil *Electromagnetism.* an inductor manufactured by depositing a thin film of conductive metal, such as gold, in a spiral about a ferrite substrate.

thin-film memory *Computer Technology.* a computer storage medium in which a magnetic thin film on a nonmagnetic substrate is magnetized to represent data.

thin-film resistor *Electricity.* a resistor that is fabricated by depositing an alloy, metal, carbon, or other thin film; used in thin-film integrated circuits.

thin-film solar cell *Electronics.* an apparatus for the conversion of solar energy into electrical energy, formed by the deposition of semiconductor material onto a thin metal or plastic substrate.

thin-film storage *Computer Technology.* a magnetic memory technology consisting of thin layers of magnetic material deposited on a nonmagnetic base; largely replaced by semiconductor technology.

thin-film transducer *Solid-State Physics.* a molecular film evaporated onto a crystal substrate that converts microwave radiation into hypersonic sound waves in the crystal.

thin-film transistor *Electronics.* a field-effect transistor formed by the deposition of one or more thin films of semiconductor and insulating material onto a substrate rather than onto a semiconductor chip.

things-to-do list *Psychology.* a therapeutic technique to build a sense of accomplishment and self-esteem in an individual by providing a list of finite, relatively simple tasks that must be accomplished daily.

thinking *Psychology.* cognitive behavior that uses symbols, concepts, and formulas for activities including reasoning, discrimination, abstraction, generalization, and imagination.

think time *Computer Technology.* in a time-sharing computer system, the time that a user spends at a terminal but is not engaged in actual use of the processor.

thin-layer chromatography *Analytical Chemistry.* a separation technique in which a single drop of the test solution is put on a glass plate coated with a thin layer of a stationary phase (usually silica) and the edge is dipped in a solvent; the solvent carries each of the various components of the test solution for a different distance up the plate, allowing for selective separation and identification.

thin-layer gel filtration *Biotechnology.* a highly sensitive method for separating and identifying compounds; a small sample is applied to a horizontal plate with a thin layer of xerogel, and the separated compounds are detected under ultraviolet light.

thin lens *Optics.* a lens whose thickness can be ignored in establishing the distance to the imaged object or the focal plane.

thin list see LOOSE LIST.

thinner *Materials.* any liquid, such as turpentine, that is used to dilute another liquid, such as a paint or varnish.

thinning *Agronomy.* the removal of plants to reduce overcrowding. *Forestry.* a felling made in an immature stand or crop, designed to improve the form and diameter growth of the remaining trees without permanently breaking the canopy.

Thinocoridae *Vertebrate Zoology.* the seed snipes, a family of South American shorebirds of the order Charadriiformes, characterized by a short strong bill, a large head on a short neck, and long narrow wings.

thin-out *Geology.* a process in which a stratum, vein, or other body of rock gradually decreases in thickness until it disappears. Also, PINCH-OUT, WEDGE-OUT.

thin section *Geology.* a slice of rock or mineral that is cut and mechanically ground to a thickness of about 0.03 millimeter, rendering it transparent or translucent so that its optical properties may be studied.

thin-skinned structure *Geology.* a large-scale feature of the earth's crust thought to have been formed by the movement of rocks in the upper layers and to lie above a detachment layer.
thin space *Graphic Arts.* **1.** on a linecaster, a 1/4-em space; a thin copper spacing blank. **2.** loosely, a hairspace.
thio *Chemistry.* **1.** containing sulfur. **2.** a substance containing sulfur.
thio- *Chemistry.* a prefix denoting the presence of sulfur; in chemical nomenclature, it indicates the replacement of oxygen by divalent sulfur.
thioacetamide *Organic Chemistry.* CH_3CSNH_2, combustible, toxic, colorless leaflets; soluble in water, alcohol, and ether; melts at 115°C; used as a source of H_2S in qualitative analysis.
thioacetic acid *Organic Chemistry.* CH_3COSH, a combustible, yellow liquid; soluble in water, alcohol, and ether; boils at 81.8°C and freezes at –17°C; used as an analytical reagent and as a lachrymator. Also, THIACETIC ACID.
thioaldehyde *Organic Chemistry.* any of a class of aldehydes that contain the –CHS group and have the suffix *-thial,* as in *propanethial.*
thioaminopropionic acid see CYSTEINE.
Thiobacillus *Bacteriology.* a genus of Gram-negative, aerobic, motile, rod-shaped bacteria of uncertain affiliation that oxidize sulfur compounds in soil, marine habitats, and hot springs.
Thiobacteriaceae *Bacteriology.* in former classifications (and now of undecided classification), a family of Gram-negative, rod-shaped anaerobic bacteria of the suborder Pseudomonadineae that oxidize inorganic sulfur compounds.
thiobarbiturate *Pharmacology.* a salt or derivative of thiobarbituric acid; many are similar to barbiturates in effect.
thiobarbituric acid *Organic Chemistry.* $C_4H_4N_2O_2S$, a platelike solid; soluble in water and alcohol; melts at 235°C; the parent compound of thiobarbiturates. Also, MALONYL THIOUREA.
thiobenzyl alcohol see BENZYL MERCAPTAN.
Thiocapsa *Bacteriology.* a genus of purple sulfur bacteria of the family Chromatiaceae that are photosynthetic, deposit sulfur intracellularly, and occur as nonmotile coccoid cells surrounded by a slime capsule.
thiocarbamide see THIOUREA.
thiocarbamizine *Pharmacology.* $C_{21}H_{17}AsN_2O_5S_2$, a white crystalline powder formerly used as an antiamebic. Also, **thiocarbamisin.**
thiocarbanilide *Organic Chemistry.* $CS(NHC_6H_5)_2$, a combustible, gray powder, insoluble in water and soluble in alcohol, and ether, that melts at 152–155°C; used in intermediates and dyes, and as an ore flotation agent and rubber accelerator.
thiocarbarsone *Pharmacology.* $C_{11}H_{13}AsN_2O_5S_2$, a white crystalline powder formerly used as an antiamebic.
thio compound see SULFHYDRYL COMPOUND.
thiocresol *Pharmacology.* C_7H_8S, an antiseptic and dermatologic agent.
thiocyanate *Inorganic Chemistry.* any salt of thiocyanic acid, HSCN, containing the SCN– ion. Also, **thiocyanide.**
thiocyanic acid *Inorganic Chemistry.* HSCN(HNCS), a colorless liquid or gas; very soluble in cold water and alcohol; melts above 110°C; becomes solid at –90°C; used in paper and insecticides. Also, RHODANIC ACID, SULFOCYANIC ACID.
thiocyanogen *Organic Chemistry.* NCS–SCN, an unstable liquid that freezes at –2°C and behaves like a halogen; decomposes in cold water; soluble in alcohol; usually used in solution to estimate the degree of unsaturation in oils.
thiodiglycol *Organic Chemistry.* $S(CH_2CH_2OH)_2$, a combustible, viscous, colorless liquid; soluble in water and alcohol, and slightly soluble in ether; boils at 283°C and freezes at –10°C; used as a solvent and an intermediate. Also, DIHYDROXYETHYL SULFIDE.
thiodiglycolic acid *Organic Chemistry.* $HOOCCH_2SCH_2COOH$, combustible, colorless crystals; soluble in water and alcohol; melts above 128°C; used as an analytical reagent.
thiodiphenylamine see PHENOTHIAZINE.
thiodipropionic acid *Organic Chemistry.* $HOOCCH_2CH_2{\cdot}S{\cdot}CH_2CH_2COOH$, leaflike crystals; soluble in water and alcohol; melts at 135°C; used as an antioxidant.
3,3'-thiodiproprionic acid *Organic Chemistry.* $(CH_2CH_2COOH)_2S$, a crystalline solid; melts at 134°C; soluble in water, acetone, and alcohol; used as an antioxidant for soap products and polymers of ethylene.
thiododecyl alcohol see LAURYL MERCAPTAN.
thioester *Biochemistry.* any compound having the general formula *R*–CO–S–*R*', formed by joining a carboxyl group of one molecule with a sulfhydryl group of another through the elimination of a molecule of water, as in acetyl coenzyme A.
thioether *Organic Chemistry.* any of a class of organic compounds having the general formula *RSR,* in which the *R* groups may be the same or different.
thioether hydrolase *Enzymology.* an enzyme that catalyzes the hydrolysis of a compound, such as methionine, in which a sulfur atom carries two alkyl groups.
thioethyl alcohol see ETHANETHIOL.
thioflavine T *Organic Chemistry.* $CH_3C_6H_3N(HCl)SCC_6H_4\,N(CH_3)_2$, a yellow basic dye that fluoresces yellow to yellowish-green when excited by ultraviolet light; used on textiles and in fluorescent paints.
thiofuran see THIOPHENE.
thioglycolic acid *Organic Chemistry.* $HSCH_2COOH$, a combustible, toxic, colorless liquid; miscible with water, alcohol, or ether; boils at 123°C (29 torr) and freezes at –16.5°C; used as a reagent and in the manufacture of pharmaceuticals. Also, MERCAPTOACETIC ACID.
thioguanine *Oncology.* $C_5H_5N_5S{\cdot}xH_2O$, an antineoplastic anti-metabolite agent related to mercaptopurine and used in leukemia therapy.
2-thiohydantoin *Organic Chemistry.* $C_3H_4N_2OS$, crystals or tan powder; slightly soluble in water and insoluble in alcohol and ether; melts at 229°C; used as an intermediate.
Thiokol see POLYSULFIDE ELASTOMER.
thiol *Organic Chemistry.* any of a group of organic compounds resembling alcohols, but with the oxygen of the hydroxyl group replaced by sulfur.
thiolactic acid *Organic Chemistry.* $CH_3CH(SH)COOH$, an oily liquid; soluble in water, alcohol, and acetone; boils at 102°C (16 torr) and freezes at 10–14°C; used in toiletries. Also, 2-MERCAPTOPROPIONIC ACID.
thiolase *Enzymology.* in beta oxidation, the catalytic enzyme that cleaves a carbon-carbon bond of a thiol compound.
thiolester hydrolase *Enzymology.* an enzyme that catalyzes hydrolysis of a compound containing the –CO–S– group.
thiol protease *Enzymology.* an enzyme that catalyzes the hydrolysis of proteins and has a cysteine residue at the active site.
thiolutin see AUREOTHRICIN.
thiomalic acid *Organic Chemistry.* $HOOCCH(SH)CH_2COOH$, combustible white crystals or powder; soluble in water, alcohol, and ether; melts at 149–150°C; used in biochemical research and as an intermediate, rust inhibitor, and tackifier for synthetic rubber. Also, MERCAPTOSUCCINIC ACID.
Thiomicrospira *Bacteriology.* a genus of Gram-negative bacteria of uncertain affiliation that occur as curved or spiral rod-shaped cells able to oxidize sulfur compounds; found in marine habitats including hydrothermal vents.
thionic acid *Organic Chemistry.* an organic acid that contains the –CSOH group. *Inorganic Chemistry.* $H_2S_nO_6$ *(n* = 2 or more), any of a series of acids of which only $H_2S_2O_6$ has been isolated; others are known by their salts.
thionyl bromide *Organic Chemistry.* $SOBr_2$, a red liquid that boils at 68°C (40 torr). Also, SULFINYL BROMIDE.
thionyl chloride *Organic Chemistry.* $SOCl_2$, a pale-yellow to red liquid; soluble in benzene and carbon tetrachloride; boils at 79°C and decomposes (fumes) at 140°C; used in pesticides, and as a chlorinating agent and catalyst. Also, SULFUR(OUS) OXYCHLORIDE.
thiopental sodium *Organic Chemistry.* $C_{11}H_{17}O_2N_2NaS$, a yellow crystalline solid; soluble in water; a rapidly acting barbiturate that is administered intravenously for general anesthesia and for hypnosis. Also, **thiopentone sodium.**
thiophanate *Organic Chemistry.* $C_{14}H_{18}N_4O_4S_2$, a brown solid; soluble in water; decomposes at 195°C; used as a fungicide.
thiophene or **thiophen** *Organic Chemistry.* C_4H_4S, a flammable, colorless liquid; insoluble in water and soluble in alcohol and ether; boils at 84°C and freezes at –38.5°C; used in organic synthesis and in solvent, dye, and pharmaceutical manufacture. Also, THIOFURAN.
thiophenol *Organic Chemistry.* C_6H_5SH, a combustible, water-white liquid; insoluble in water and soluble in alcohol and ether; boils at 168.3°C; used in pharmaceutical synthesis. Also, PHENYL MERCAPTAN.
thiophile element see SULFOPHILE ELEMENT.
Thioploca *Bacteriology.* a genus of Gram-negative gliding bacteria of the family Beggiatoaceae, occurring as bundles of filaments in aquatic habitats; they are able to oxidize hydrogen sulfide and deposit intracellular sulfur.
thioredoxin *Biochemistry.* a heat-stable, 108-amino-acid protein that functions as a hydrogen donor, in assimilatory sulfate reduction, in reducing ribonucleoside diphosphates to deoxyribonucleoside diphosphates, and in reducing thiol groups in proteins.

Thiorhodaceae *Bacteriology.* in former classifications, a family of anaerobic bacteria now reclassified as Chromatiaceae.

thiosalicylic acid *Organic Chemistry.* $2HOOCC_6H_4SH$, a combustible, yellow crystalline solid; slightly soluble in hot water and soluble in alcohol and ether; melts at 165.5°C; used in making dyes and as a reagent and intermediate. Also, 2-MERCAPTOBENZOIC ACID.

thiosemicarbazide *Organic Chemistry.* $NH_2CSNHNH_2$, a white crystalline powder; soluble in water and alcohol; melts at 182–184°C; used as an analytical reagent and in photographic chemicals and rodent poisons. Also, AMINOTHIOUREA, SEMICARBAZIDE-3-THIO.

thiosemicarbazone *Pharmacology.* any of a class of compounds that inhibit poxvirus replication by interfering with the synthesis of a structural protein and blocking the assembly of normal particles.

thiospinel *Mineralogy.* any mineral having a spinel structure and the general formula AR_2S_4.

Thiospira *Bacteriology.* a genus of Gram-negative, sulfur-oxidizing bacteria of uncertain affiliation that occur as spiral cells possessing polar flagella, and typically contain sulfur granules.

Thiospirillum *Bacteriology.* a genus of aquatic, purple sulfur bacteria of the family Chromatiaceae that occur as spiral-shaped, motile cells able to carry out photosynthesis and deposit sulfur granules intracellularly.

thiostreptone *Microbiology.* a sulfur-containing antibiotic produced by the bacterium *Streptomyces azureus*; effective against Gram-positive bacteria and other organisms, and used extensively in veterinary medicine.

thiosulfate *Inorganic Chemistry.* any salt of thiosulfuric acid, containing the $S_2O_3^{2-}$ ion.

thiosulfate sulfurtransferase see RHODANESE.

thiosulfonate *Organic Chemistry.* any of a class of compounds having the general formula $R(SO_2)SR$.

thiosulfonic acid *Organic Chemistry.* any of a class of organic acids having the general formula RS_2OOH.

thiosulfuric acid *Inorganic Chemistry.* $H_2S_2O_3$, in solution only; an unstable acid that easily decomposes to sulfur and sulfurous acid.

thiotepa *Pharmacology.* a cytotoxic alkylating agent of the ethylenimine group; used as an antineoplastic by intracavitary administration for malignant pleural or pericardial effusions.

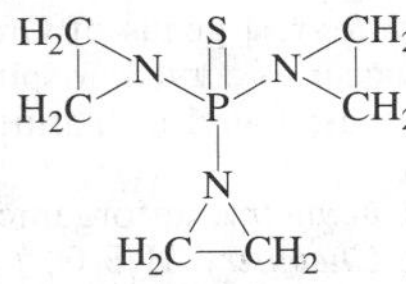

thiotepa

Thiothrix *Bacteriology.* a genus of sulfur-oxidizing gliding bacteria of the family Leucotrichaceae that occur as sheathed filaments in sulfide-rich aquatic habitats.

thiouracil *Pharmacology.* $C_4H_4N_2OS$, a thiourea derivative, formerly used to depress thyroid hormone synthesis in hyperthyroidism and to treat congestive heart failure and angina pectoris.

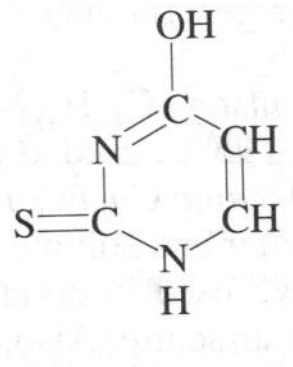

thiouracil

thiourea *Organic Chemistry.* $(NH_2)_2CS$, lustrous white crystals that are soluble in cold water and alcohol and nearly insoluble in ether; melts at 180–182°C and sublimes at boiling point; used as a rubber accelerator, and in medicine and the manufacture of photographic and photocopying chemicals. Also, THIOCARBAMIDE.

third contact *Astronomy.* the instant that totality ends in a total solar or total lunar eclipse.

third-degree burn *Medicine.* in the classification of burns, one involving the most severe damage, in which the epidermis and the dermis are destroyed and damage extends into the underlying tissue.

third-generation computer *Computer Technology.* a class of computer technology characterized by use of small-scale integrated circuits.

third harmonic *Physics.* three times the fundamental frequency of a resonating system.

third isomorphism theorem *Mathematics.* let I and J be ideals in a ring R such that I is contained in J. Then J/I is an ideal in R/I and $(R/I)/(J/I)$ is isomorphic to (R/J). Similar results hold for groups and modules.

third law of motion *Mechanics.* see NEWTON'S THIRD LAW.

third law of thermodynamics *Thermodynamics.* the statement that the entropy of a perfect crystalline substance in thermodynamic equilibrium will approach zero as its temperature approaches absolute zero.

third-level inference *Psychology.* a description of behavior that employs the observed actions to develop a theory about the underlying motivation for the subject's behavior.

third-order *Cartography.* describing survey work that has the lowest allowable level of accuracy and precision.

third-order control *Cartography.* a classification of control networks meeting the following criteria for accuracy and precision: for horizontal coordinates, the maximum value of the ratio of standard deviations of distances between survey points to the distances themselves; Class I—1:10,000, Class II—1:5000; for vertical coordinates, the maximum value of the ratio of standard deviations of elevation differences in millimeters between survey points to the square root of the horizontal distance in kilometers between those points; 2.0.

third-order leveling *Cartography.* in surveying, a technique of leveling that does not meet the degree of accuracy of second-order leveling, but does meet the following criteria: the lines of third-order leveling do not extend more than 30 miles from lines of first-order or second-order leveling; the error of closure does not exceed 12 millimeters times the square root of the length in meters of the line or section of line.

third-order reaction *Physical Chemistry.* a chemical reaction in which the rate of change of the reaction is proportional to the product of the concentrations of three separate reactants.

third-order traverse *Cartography.* in surveying, a traverse that forms a closed loop in itself, or that runs between first-level or second-level survey stations, and that has a closing error in position of one part in 5000 or better as related to its length.

third-order triangulation *Cartography.* a survey network that is based on third-order, and sometimes higher, control.

third rail *Civil Engineering.* a system of electrical traction for an electric locomotive where current is fed to the electric tractor from an insulated conductor rail running parallel with the track.

Thirion relaxation *Materials Science.* in an elastomeric network, the reversible relaxation of trapped entanglements; one of five categories of causes of stress relaxation and creep in polymers.

thirst *Physiology.* **1.** the craving for liquid, especially water. **2.** the stimulus arising from a deficiency of water.

thirty-two nucleus *Meteorology.* a freezing nucleus that becomes active when a supercooled cloud is cooled to about –32°C.

Thisbe [thiz´bē] *Astronomy.* asteroid 88, discovered in 1866, having a diameter of 232 kilometers; it belongs to type C.

thistle *Botany.* **1.** any of various prickly plants having showy, purple flower heads, especially of the genera *Cirsium, Carduus,* or *Onopordum.* The **Canada thistle,** *Cirsium arvense,* and the **bull** or **common thistle,** *C. vulgare,* are widely found as weeds throughout North America. **2.** any of various other prickly plants.

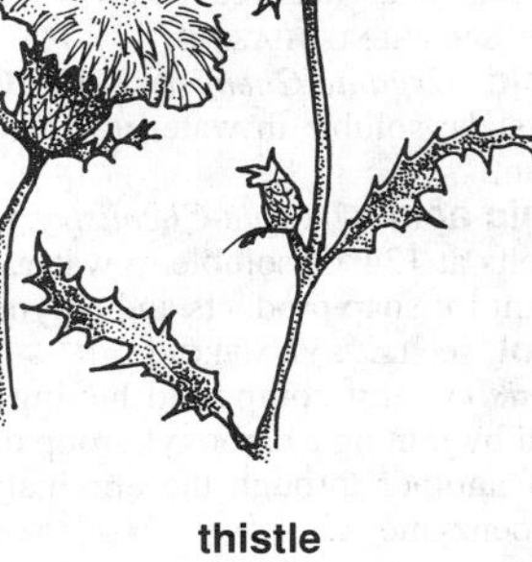

thistle

thiuram *Organic Chemistry.* any of a class of compounds containing the $-R_2NC(=S)-$ group.
thiuronium salt *Organic Chemistry.* any of a class of compounds having the formula $RSC(=NH)NH_2$.
thixocasting *Metallurgy.* a method of metal forming that permits metal ingots and components to be shaped in the semisolid state. Vigorous agitation in the early stages of solidification results in a semisolid slurry that flows homogeneously; after solidification, the slurry is reheated for forming. Also, **thixoforging.**
thixotropic clay *Geology.* a clay that weakens or changes from a semisolid state to a more liquid state when disturbed, and that strengthens or changes to a semisolid state when left standing.
thixotropy *Physical Chemistry.* a structural property that causes certain gels to liquefy when shaken and to solidify again when left standing; the breakdown to a liquid state is a function of time as well as shear rate, while the return to solidity occurs over time in the absence of shear.
Thlipsuridae *Paleontology.* a family of ostracods in the order Platycopida; extant in the Silurian and Devonian.
Thogoto virus *Virology.* a virus of the genus *Bunyavirus,* family Bunyaviridae, that infects ticks, cattle, camels, and humans, causing severe optic neuritis and meningoencephalitis in humans.
tholeiite *Petrology.* a silica-oversaturated basaltic rock, or rock group, containing phenocrysts of plagioclase, pyroxene, and iron oxide minerals in a glassy groundmass.
tholos *Architecture.* **1.** in Greek architecture, a circular building. **2.** a domed rotunda. **3.** a beehive-shaped tomb, as in Mycenae. Also, **tholus.**
Thomas cyclotron *Nucleonics.* a cyclotron that has variable azimuthal control over the magnetic field such that radial and axial focusing can both be maintained at cyclotron resonance.
Thomas-Fermi atom model *Atomic Physics.* a model that views the electrons in an atom as a gas and derives atomic structure in terms of the electrostatic potential and the electron density in the ground state.
Thomas-Fermi equation *Atomic Physics.* a differential equation used to calculate the electrostatic potential in the Thomas-Fermi atom model.
Thomas meter *Engineering.* a device used to determine the flow rate of a gas by measuring the rise in gas temperature produced by the addition of a known amount of heat.
Thomas precession *Physics.* the angular precession of a vector in an accelerated frame of reference when the vector appears to be constant to an observer in the system; in the Lorentz frame, which is instantaneously at rest with respect to accelerated frame (having velocity v and acceleration a), the angular precession rate is $\omega = (1/2c^2)(a \times v)$.
Thompson, Benjamin, Count Rumford 1753–1814, British scientist and administrator, born in America; studied ocean circulation; demonstrated the nature of heat; improved gunpowder.
Thompson, Sir Eric 1898–1975, English archaeologist; reconstructed temple at Chichen Itza; correlated the Mayan and Gregorian calendars.
Thompson submachine gun *Ordnance.* a .45-caliber, air-cooled submachine gun; the military model has selective automatic or semiautomatic fire and a firing rate of 600–700 rounds per minute. (Named for its coinventor, American army officer John T. *Thompson,* 1860–1940.)
Thomsen, Christian 1788–1865, Danish archaeologist; devised the three-age system of prehistory (Stone, Bronze, and Iron Ages).
thomsenolite *Mineralogy.* $NaCaAlF_6 \cdot H_2O$, a colorless to white monoclinic mineral occurring in small, prismatic or tabular crystals, having a specific gravity of 2.981 and a hardness of 2 on the Mohs scale; dimorphous with pachnolite; found associated with cryolite.
Thomson, Elihu 1853–1937, American electrical engineer; his inventions include electric welding, the AC generator, and the high-frequency transformer and generator.
Thomson, Sir George Paget 1892–1975, English physicist, son of Sir Joseph John Thomson; awarded the Nobel Prize for the independent discovery of electron diffraction by crystals.
Thomson, Sir Joseph John 1856–1940, English physicist; awarded the Nobel Prize for discovery of the electron; independently discovered photoelectricity; discovered the first isotopes (of neon).
Thomson, William see KELVIN, BARON WILLIAM THOMSON.
Thomson-Berthelot principle see BERTHELOT PRINCIPLE.
Thomson coefficient *Physics.* the ratio of the Thomson voltage to the temperature difference between the two points across which the Thomson voltage is measured.
Thomson cross section *Electromagnetism.* the scattering cross section used in Thomson scattering, given by the quantity $(8/3)p(q^2/mc^2)^2$, where q is the charge in electrostatic units, m is the mass of the scattering particle, and c is the speed of light.
Thomson effect *Physics.* a phenomenon in which a voltage develops across the junction of a bimetal contact when there is a temperature gradient across the junction.
Thomson formula *Electromagnetism.* a formula that gives the intensity of scattered radiation as a function of scattering angle in Thomson scattering: $I = k(1 + \cos^2 f)$, where k is a proportionality constant.
Thomson heat *Physics.* the reversible heat emitted or absorbed that arises in a current-carrying conductor along which there is a temperature gradient.
thomsonite *Mineralogy.* $NaCa_2Al_5Si_5O_{20} \cdot 6H_2O$, a colorless to white, yellowish, or green, orthorhombic mineral of the zeolite group, occurring as masses of radiating crystals with a vitreous luster, having a specific gravity of 2.25 to 2.4 and a hardness of 5 to 5.5 on the Mohs scale; found in amygdules in basaltic rocks.
Thomson parabolas *Electromagnetism.* parabolas that appear on a photographic plate and assemble to detect an ion beam (of a particular element) when the beam is passed through a region where there exists an electric field and magnetic field in the same direction, yet transverse to the beam direction; each parabola corresponds to a different charge-to-mass ratio.
Thomson scattering *Electromagnetism.* the classical, nonrelativistic scattering of electromagnetic radiation by loosely bound or free charged particles.
Thomson's gazelle *Vertebrate Zoology.* a common medium-sized gazelle, *Gazella thomsoni,* inhabiting the East African savannah. (Named for James *Thomson,* 1858–1895, British explorer.)
Thomson voltage *Physics.* the thermal electromotive force developed across a bimetal or semiconductor junction when there is a temperature gradient across the junction.
Thom's theorem *Mathematics.* in catastrophe theory, the theorem that only seven types of nontrivial singularities (the elementary catastrophes) can occur in the control space of a dynamical system.
thonzylamine hydrochloride *Pharmacology.* $C_{16}H_{23}ClN_4O$, an antihistaminic agent.
Thor *Ordnance.* a rocket-powered, surface-to-surface, intermediate-range ballistic missile that can carry nuclear warheads and be launched from a satellite; it is equipped with an all-inertial guidance system, uses liquid-propellant fuel, and has a range of 2000 miles. *Space Technology.* this missile adapted as a booster for launch vehicles such as the Delta, Agena, and Ablestar. (Named for *Thor,* the ruler of the sky in Norse mythology, who was noted for his great strength and fighting ability.)
thoracalgia *Anatomy.* pain in the chest wall.
thoracectomy *Surgery.* the incision of the chest wall and resection of a portion of a rib.
thoracentesis see PLEUROCENTESIS.
thoracic [thô ras´ik; thō ras´ik] *Anatomy.* of, relating to, or affecting the chest (thorax). Also, **thoracal.**
Thoracica *Invertebrate Zoology.* an order of barnacles belonging to the subclass Cirripedia, having calcareous plates and six pairs of thoracic appendages; the adults attach to floating or submerged objects.
thoracic aorta *Anatomy.* the proximal portion of the descending aorta.
thoracic duct *Anatomy.* the largest lymphatic duct of the body, beginning at the cisterna chyli of the abdomen and passing up to the left subclavian vein, where it empties lymphatic fluid collected from below the diaphragm and from the upper left quadrant of the body.
thoracic vertebra *Anatomy.* any of the twelve vertebrae that attach to the ribs and form the posterior medial limit of the thoracic cavity.

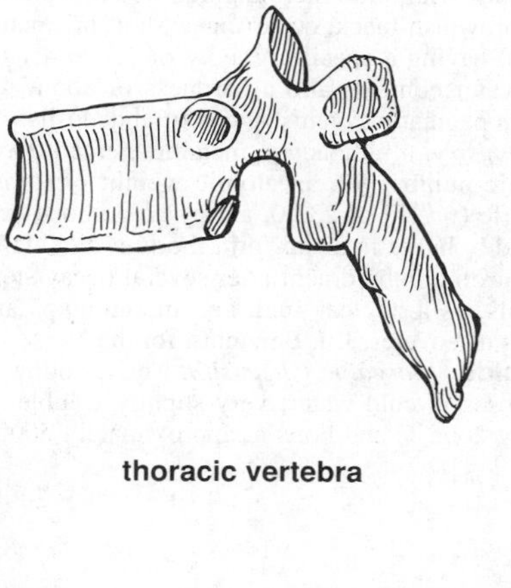

thoracic vertebra

thoraco- or **thorac-** a combining form meaning "chest" or "thorax."
thoracoabdominal *Zoology.* relating to the thorax and abdomen.
thoracoabdominal breathing *Vertebrate Zoology.* breathing movements in reptiles, birds, and mammals that use muscles of the thorax and abdomen to suck air into the lungs.
thoracocentesis see PLEUROCENTESIS.
thoracodynia *Medicine.* any pain in the chest.
thoracolaparotomy *Surgery.* the surgical incision of both the abdomen and the thorax to gain access to the diaphragmatic region.
thoracolumbar *Anatomy.* **1.** relating to thoracic and lumbar vertebrae. **2.** relating to the thoracic and lumbar regions of the trunk.
thoracolumbar fascia *Anatomy.* the tough sheath of fibrous connective tissue that covers the deep muscles of the back, extending from the lower ribs and thoracic vertebrae to the sacrum and iliac crests. Also, LUMBODORSAL FASCIA.
thoracolysis *Surgery.* the freeing of adhesions of the chest wall.
thoracometer *Medicine.* an instrument for measuring the circular dimension or expansion of the chest. Also, STETHOMETER.
thoracopathy *Pathology.* any disorder of the chest or thoracic organs.
thoracoplasty *Surgery.* the surgical removal of ribs, allowing the chest wall to move inward and collapse a diseased lung.
thoracopod *Invertebrate Zoology.* any of the thoracic appendages of crustaceans.
Thoracostomopsidae *Invertebrate Zoology.* a family of freeliving marine nematodes in the superfamily Enoploidea.
thoracostomy *Surgery.* **1.** the creation of an opening in the wall of the chest for the purpose of drainage. **2.** the opening so created.
thoracotomy *Surgery.* a surgical incision of the chest wall.
Thoraeus filter *Nucleonics.* a radiation filter composed of a primary layer of tin, a secondary layer of copper, and a third layer of aluminum; the copper filter absorbs radiation from the tin, and the aluminum filter absorbs radiation from the copper; used to harden X rays.
thorax [thôr´aks; thō´raks] *Anatomy.* the portion of the trunk between the base of the neck and the diaphragm. *Invertebrate Zoology.* the middle of three principal divisions of the body of certain arthropods; in insects, three leg-bearing segments; in crustaceans and arachnids, sometimes fused with the head to form a cephalothorax.
Thor-Delta *Space Technology.* another name for the Delta launch vehicle, which employs a Thor booster.
Thoreaceae *Botany.* a small family of freshwater red algae of the order Nemaliales, characterized by erect, branched thalli arising from a filamentous base.
thoreaulite *Mineralogy.* $Sn^{+2}Ta_2O_6$, a brown monoclinic mineral, forming a series with foordite, occurring as prismatic crystals and irregular masses, having a specific gravity of 7.6 to 7.9 and a hardness of 5.5 to 6 on the Mohs scale; found in grantic pegmatites.
thorianite *Mineralogy.* ThO_2, a strongly radioactive, black to reddish-brown or yellow cubic mineral with a resinous luster, occurring as cubic and cubo-octahedral crystals, having a specific gravity of 9.7 to 9.991 and a hardness of 6.5 to 7 on the Mohs scale; found as a primary mineral in pegmatites and in derived detrital deposits; it forms a series with uraninite.
thoriated tungsten filament *Electronics.* a tungsten filament that contains a small quantity of thorium oxide to enhance electron emission; used as a directly heated cathode in an electron tube. Also, **thoriated emitter.**
Thorictidae *Invertebrate Zoology.* a family of ant beetles, coleopteran insects in the superfamily Dermestoidea, that live in ant nests.
thorite *Mineralogy.* $(Th,U)SiO_4$, a strongly radioactive tetragonal mineral, dimorphous with huttonite, ranging in color from brownish-yellow to orange or brownish-black, occurring as short prismatic crystals and in massive form, having a specific gravity of 6.7 to 4.1 (decreasing with increase of water content) and a hardness of about 4.5 on the Mohs scale; found in pegmatites, veins, and detrital deposits.
thorium *Chemistry.* a radioactive metallic element having the symbol Th, the atomic number 90, an atomic weight of its isotope with the longest half-life (^{232}Th) of 232.0; a gray amorphous or crystalline, soft mass that readily burns in air to form thorium dioxide; forms uranium-233 upon neutron bombardment after several decay steps; used in photoelectric cells, as a nuclear fuel, and in sunlamps and X-ray tubes. (Named by its discoverer, J. J. Berzelius, for the Norse god *Thor.*)
thorium carbide *Inorganic Chemistry.* ThC_2, yellow tetragonal crystals; decomposes in cold water; very slightly soluble in acid; melts at approximately 2665°C and boils at approximately 5000°C; used as nuclear fuel.
thorium chloride *Inorganic Chemistry.* $ThCl_4$, colorless or white deliquescent rhombic crystals; very soluble in water and soluble in alcohol and acid; melts at approximately 770°C, sublimes at 820°C, and decomposes at 928°C; used in incandescent lighting. Also, **thorium tetrachloride.**
thorium dioxide *Inorganic Chemistry.* ThO_2, white cubic crystals; insoluble in water and acid; melts at approximately 3220°C and boils at 4400°C; used in ceramics, medicine, nuclear fuel, and as a catalyst. Also, **thoria, thorium anhydride, thorium oxide.**
thorium fluoride *Inorganic Chemistry.* ThF_4, a toxic white cubic powder; melts above 900°C; used in the production of thorium metal, alloys, and high-temperature ceramics.
thorium nitrate *Inorganic Chemistry.* $Th(NO_3)_4 \cdot 12H_2O$, radioactive colorless crystals; deliquescent; very soluble in cold water, alcohol, and acid; decomposes at its melting point; a dangerous fire and explosive hazard; used as a reagent and in thoriated tungsten filaments.
thorium oxalate *Organic Chemistry.* $Th(C_2O_4)_2 \cdot 2H_2O$, a white powder; insoluble in water and most acids; decomposes above 300°C; used in ceramics.
thorium reactor *Nucleonics.* a breeder reactor in which an enriched uranium core is surrounded by thorium.
thorium series *Nuclear Physics.* a naturally occurring transformation series of three radioactive isotopes that form a sequence of parent-daughter relationships through alpha and beta decay, beginning with thorium-232 and ending in the stable isotope of lead-208.
thorium sulfate *Inorganic Chemistry.* $Th(SO_4)_2 \cdot 8H_2O$, white needles or crystalline powder; soluble in water and insoluble in acids.

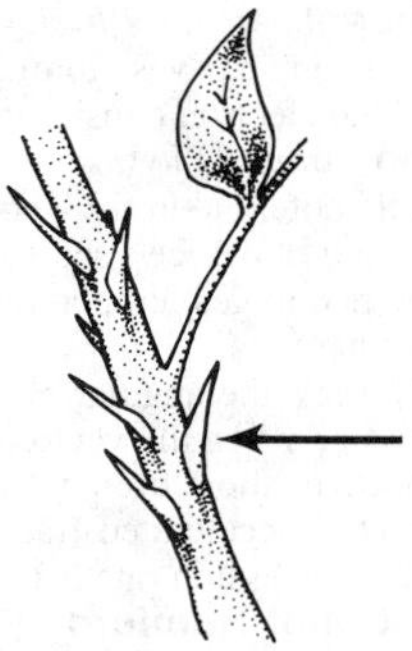

thorn

thorn *Botany.* **1.** a sharp, rigid, reduced, leafless branch connected to a plant's vascular system. **2.** a plant characterized by such structures, such as the hawthorn.
thornback *Vertebrate Zoology.* a European ray, *Raja clavata,* having spines on its dorsal surface.

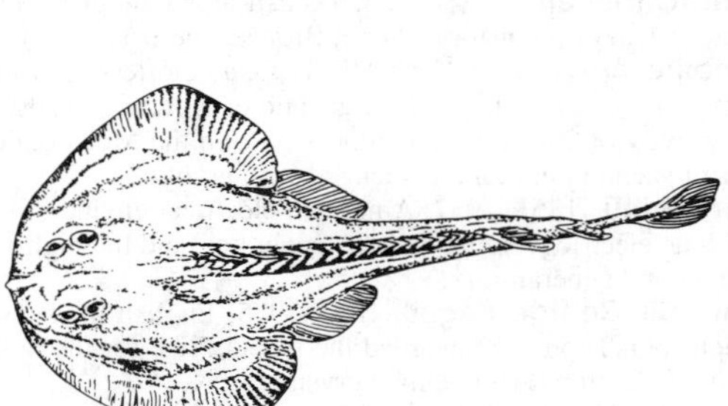

thornback

thornbush *Ecology.* a class of vegetation dominated by tall succulents and smooth-barked deciduous hardwoods, found in a variety of regions, including many in the Caribbean and Central Africa, which have a warm, arid climate interspersed with periods of short intense rainfall.
Thorndike, Edward Lee 1874–1949, American educational psychologist; made noted studies of animal intelligence and of mental development and learning in children.

thorn forest *Ecology.* a woodland area characterized by arid climate, and by vegetation composed of thorny shrubs, bushy trees, a few sparsely scattered tall trees, and little or no grass; commonly found in tropical and subtropical areas, or between desert and steppe areas.

thorny *Botany.* having or characterized by thorns; spiny.

thorogummite *Mineralogy.* $Th(SiO_4)_{1-x}(OH)_{4x}$, a yellowish, gray-green, or white tetragonal, secondary silicate mineral occurring in fine-grained aggregates, having a specific gravity of 3.2 to 5.4 and an undetermined hardness; found as an alteration product of thorite, thorianite, and yttrialite.

thoron *Nuclear Physics.* the descriptive name for a type of radioactive emanation that exists as the radon isotope in the thorium series with a half-life of 55 seconds.

Thorpe reaction *Organic Chemistry.* a reaction that produces cyclic ketones by condensing dinitriles to cyclic iminonitriles that can be processed to ketones.

thortveitite *Mineralogy.* $(Sc,Y)_2Si_2O_7$, a very rare, weakly radioactive monoclinic mineral occurring in greenish-gray, distorted, prismatic crystals, having a specific gravity of 3.58 and a hardness of 6 to 7 on the Mohs scale; found in granite pegmatites; a source of scandium.

thought *Psychology.* **1.** the result of thinking or an instance of thinking; an idea, belief, or image held in the mind. **2.** the process of thinking; mental activity.

thought broadcasting *Psychology.* an individual's feeling that his or her thoughts are being transmitted out loud and heard by others; a common symptom of schizophrenia.

thought disturbance *Psychology.* a general term for disorderly thought processes that negatively affect communication, language, or mental functioning.

thought experiment *Physics.* see GEDANKEN EXPERIMENT.

thought insertion *Psychology.* an individual's feeling that external thoughts have been inserted into his or her mind by others; a common symptom of schizophrenia.

thousandth mass unit *Physics.* 1/1000 of an atomic mass unit.

THPC tetrakis(hydroxymethyl)phosphonium chloride.

Thr threonine.

thrasher *Vertebrate Zoology.* any of various insectivorous American songbirds belonging to the genus *Toxostoma* of the family Mimidae, related to the mockingbird and resembling the thrush; characterized by a long body and tail, brown plumage, and a long, downcurved bill.

thrasher

thrashing *Computer Programming.* in a virtual memory system, a condition that occurs when swapping of pages between disk and main memory becomes very frequent, resulting in poor performance.

Thraupidae *Vertebrate Zoology.* the tanagers, a family of chiefly arboreal, tropical New World birds of the passerine suborder Oscines, characterized by colorful plumage.

Thraustochytriaceae *Mycology.* a family of fungi belonging to the order Saprolegniales, which lives off of decaying organic matter; found in marine and salt lake areas.

thread *Textiles.* a slender, continuous strand composed of two or more twisted filaments of cotton, silk, wool, or other fibers. *Design Engineering.* a continuous helical rib on a screw of pipe. *Geology.* an extremely small, thin vein, narrower than a stringer. *Hydrology.* **1.** a thin stream of water. **2.** the middle of a stream of water. *Mining Engineering.* a relatively straight line of stall faces without cuttings, loose ends, fast ends, or steps.

thread blight *Plant Pathology.* a disease of cocoa, tea, and other tropical and semitropical woody plants, caused by fungi of the species *Pellicularia* and *Marasmius*; characterized by filamentous fungal growths on twig and leaf surfaces.

thread chaser *Mechanical Devices.* a multipoint tool, one of a set of four in a die head used to cut screw threads.

thread contour *Design Engineering.* the shape of a thread design when observed via cross section along its major axis, such as a squared or rounded shape.

thread cutter *Mechanical Engineering.* a tool used to cut screw threads on a pipe or bolt.

threaded *Computer Programming.* of a program, consisting of calls to many separate subprograms.

threaded code *Computer Programming.* a style of writing application programs by means of calls to subprograms.

threaded sleeve *Mechanical Devices.* a metal, cylindrically shaped shell used to connect two pipe pieces or rods.

threader *Mechanical Devices.* **1.** a device that cuts screw threads. **2.** a device that pulls sewing thread through the eye of a needle. **3.** a device that draws a line through one or more narrow openings or channels.

threadfin *Vertebrate Zoology.* a bony marine or brackish water carnivorous fish in the family Polynemidae; found in all tropical and subtropical seas.

thread gauge *Mechanical Devices.* a device that measures the design characteristics of screw threads such as pitch, diameter, or thread angle. Also, **thread pitch gauge.**

threading die *Mechanical Engineering.* a die used to produce an external thread on a part.

threading machine *Mechanical Engineering.* a machine that cuts or forms threads on the inside or outside of a cylinder or cone.

thread-lace scoria *Geology.* a delicate network of threads formed in volcanic rock by the collapse of the vesicle walls. Also, RETICULITE.

thread plug *Engineering.* a mold part that places an internal thread into a molded object and is removed from the finished piece.

thread plug gauge *Mechanical Devices.* a thread gauge used to measure female screw threads.

thread protector *Engineering.* a threaded ring that may be screwed onto a pipe end to protect the pipe threads during shipping.

thread rating *Engineering.* the maximum internal pressure that a threaded pipe can withstand.

thread ring gauge *Mechanical Devices.* a thread gauge used to measure male screw threads.

thread rolling *Metallurgy.* a metal-forming process for producing threads or annular forms in cylindrical or tapered workpieces. An impression is made into the workpiece as it is rotated, causing ridging between the contact points, which flow together to produce a crest or maximum thread diameter larger than the original diameter of the workpiece.

threat behavior *Behavior.* behavior that repels but does not injure an opponent in a situation of conflict.

threatened species *Ecology.* a species of animal or plant that has been identified as facing a possible threat of extinction, due to a gradual reduction in its overall population or significant changes in its environment; not identified as facing as great a threat as an endangered species.

three-address code *Computer Programming.* a type of multiaddress code in which three storage addresses are specified in one instruction: two addresses from which data is taken, and one address where the result is entered.

three-address instruction *Computer Programming.* a type of machine instruction containing three parts: two for the operands and one for the results of the operation.

Three-Age System *Archaeology.* the division of human prehistory into three successive stages (the Stone Age, Bronze Age, and Iron Age), based on the main type of material used in tools of the period.

three-alpha process *Astrophysics.* a nuclear reaction common to red giant stars in which three helium-4 nuclei (alpha particles) and an excited carbon-12 nucleus fuse and release a gamma ray. Also, TRIPLE-ALPHA PROCESS.

three-arm protractor *Navigation.* an instrument used to determine a ship's position when two horizontal angles are given between three known points; it consists of a graduated metal circle with a hole in the center and one fixed and two movable arms; when the arms are placed on the points and set at the correct angles, the center hole indicates the ship's position. Also, STATION POINTER.

three-body problem see MANY-BODY PROBLEM.

three-chambered heart *Cardiology.* a heart with only three rather than the normal four chambers, due to the absence of either an atrial septum or a ventricle septum; a developmental abnormality in which there is only one ventricle or one atrium.

three-decibel coupler *Electromagnetism.* a waveguide junction in which an H-wall is shared and there is an H-type aperture coupling; the design allows one-half of the power to be passed. Also, SHORT-SLOT COUPLER.

three-decibel down point *Acoustical Engineering.* the point at which the sound level is one-half the maximum intensity, used extensively in transducer directivity or response curve diagrams. Also, **three-db down point.**

three-dimensional *Science.* having the dimensions of length, width, and depth. Also, **3-D.**

three-dimensional computing *Computer Technology.* computer hardware, software, or databases having sufficient capacity and designed to display and manipulate realistic surfaces and color-shaded, three-dimensional objects at a speed that approximates smooth motion. Also, **3-D computing.**

three-dimensional display system *Electronics.* a radar display that presents aircraft targets in three dimensions: distance (range), bearing (azimuth), and height (elevation).

three-dimensional flow *Fluid Mechanics.* a general flow in which fluid velocity is composed of three components in mutually perpendicular directions.

three factor cross *Genetics.* a cross in which alleles of three different X-chromosome genes are segregated.

three-finger gripper *Robotics.* an end effector that uses three opposing fingers to grasp an object.

three halves power equation see CHILD'S LAW.

three-input adder see FULL ADDER.

three-input subtractor see FULL SUBTRACTER.

three-jaw chuck *Mechanical Devices.* a device for securing a drill rod with three movable, serrated jaws.

three-j number *Quantum Mechanics.* a coefficient used in coupling two angular momenta; these numbers are related to the Clebsch-Gordan coefficients, but have greater symmetry. Also, **3-j symbol.**

three-junction transistor *Electronics.* a four-region (PNPN), three-junction transistor having three external connections: emitter, base, and collector; one of the four regions has no external connection and serves no operational purpose.

three-kiloparsec arm *Astronomy.* a strong hydrogen absorption feature detected in 21-centimeter radio surveys, which is seen projected against the Sagittarius A source; it lies approximately 3000 parsecs (10,000 light-years) on this side of the galactic center.

three-layer diode *Electronics.* a two-terminal, three-layer avalanche semiconductor diode characterized by its ability to conduct symmetrically with voltages of either polarity; used mainly in AC power-control applications. Also, DIAC, TRIGGER DIODE.

three-level laser *Optics.* a laser that operates with three energy levels, the ground state and two higher levels; most lasing occurs between the intermediate state and the ground level.

three-level maser *Physics.* a maser that operates on three energy states: the ground state E_1 is pumped to the short-lived state E_3, which then decays to the metastable state E_2; stimulated emission occurs in the transition from E_2 to E_1.

three-phase circuit *Electricity.* a circuit operated from three AC voltages that differ in phase by 1/3 cycle or 120 electrical degrees.

three-phase current *Electricity.* three currents in a three-phase circuit, each differing in phase from the other by 1/3 cycle or 120°.

three-phase four-wire system *Electricity.* an AC supply with three conductions carrying the three-phase supply and a fourth, neutral conductor, which is often connected to ground.

three-phase magnetic amplifier *Electronics.* a magnetic amplifier designed to accept a three-phase AC input from either a delta- or wye-source.

three-phase motor *Electricity.* an AC motor that is operated from a three-phase circuit.

three-phase rectifier *Electricity.* a rectifier for a three-phase voltage with one or two diodes for each phase.

three-phase seven-wire system *Electricity.* a system of alternating current supply from groups of three single-phase transformers connected in a Y; thus a three-phase, four-wire, grounded-neutral system, with a relatively higher voltage is attained; the neutral wire is common to both systems.

three-phase system *Physics.* a system that operates on three signals of the same frequency; each signal is symmetrically out of phase by 120° with the other two signals.

three-phase three-wire system *Electricity.* three conductors, each carrying one phase of a three-phase supply. Also, THREE-WIRE SYSTEM.

three-phase transformer *Electricity.* a transformer used in a three-phase circuit, containing three sets of primary and secondary windings on a single core.

three-piece set *Mining Engineering.* a timber set consisting of a cap and two supporting posts, which are often spread apart at the bottom to provide greater stability.

three-plus-one address *Computer Programming.* a machine instruction address section that contains the location of both operands, the result, and the next machine instruction.

three-point bending *Metallurgy.* a process of bending in which a metal is placed upon two supports and a force is applied between them.

three-point cross *Genetics.* a cross-over experiment involving three pairs of linked genes on the same chromosome.

three-point method *Geology.* a method for calculating the strike and dip of a structural surface by using three points of different elevation along the surface.

three-point problem *Engineering.* in surveying, the process of determining the location of the horizontal position of an observation post using two known angles and three known triangle lengths.

three-pulse cascaded canceler *Electronics.* a circuit configuration used in moving-target indicators of radar sets; a pair of moving-target-indicator cancelers are connected serially to permit comparison of three successive returns from the same target.

three-sided file *Mechanical Devices.* see TRIANGULAR FILE.

three-spine stickleback *Vertebrate Zoology.* a stickleback, *Gasterosteus aculeatus,* widely distributed in marine, brackish, or fresh waters through the Northern Hemisphere. Also, **three-spined stickleback.**

three-toed sloth *Vertebrate Zoology.* a variety of sloth of the genus *Bradypus,* characterized by very long forelimbs and three toes on each limb.

three-toed woodpecker *Vertebrate Zoology.* either of two woodpeckers of the genus *Picoides,* having a yellow crown and lacking the inner hind toe on each foot; found in the Northern Hemisphere.

three-way switch *Electricity.* a switch that can connect one conductor to either one of two other conductors.

three-wire generator *Electricity.* an electric generator that contains a balance coil, whose midpoint provides the potential of the neutral wire in a three-wire system.

three-wire system *Electricity.* **1.** a system of electric supply that is made up of three conductors; one is called a neutral wire and is maintained at a potential halfway between the other two (each called an outer conductor). Also, TWO-PHASE THREE-WIRE CIRCUIT. **2.** see THREE-PHASE THREE-WIRE SYSTEM.

thremmotology *Biology.* the science of breeding or propagating animals and plants under domestication.

threoic acid see TARTARIC ACID.

threonine *Biochemistry.* one of the twenty amino acids that make up proteins.

thresh *Agriculture.* to separate the grain or seed from the rest of the plant by beating.

thresher *Agriculture.* a farm machine that threshes crops. Also, **threshing machine.**

thresher shark *Vertebrate Zoology.* a dangerous, pelagic, predaceous shark of the genus *Alopias,* especially *A. vulpinus,* of the family Alopiidae, having very large eyes and a long, whiplike caudal fin with which it threshes the water to drive together small fish on which it feeds; found in temperate and tropical seas worldwide.

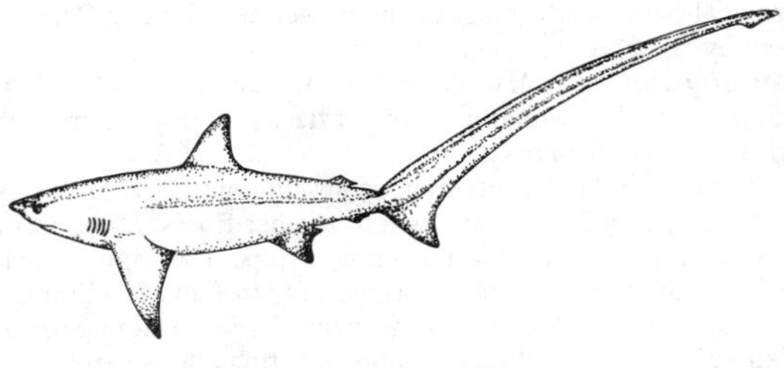

thresher shark

threshold *Building Engineering.* **1.** the sill of a doorway. **2.** the entry or exit area of a building or structure. *Physics.* a value or quantity of some parameter at which a reaction, breakdown, transition, or avalanche effect may occur. *Physiology.* **1.** the smallest amount of excitement needed to produce a certain reaction in a cell or test subject under the best test conditions. **2.** see THRESHOLD OF AUDIBILITY. *Engineering.* the least amount of input into a system required to produce a detectable output. *Electronics.* the lowest point or value at which an electronic effect is produced or an indication is observed; for example, a high-temperature alarm system adjusted to operate at a threshold of 150°F produces an alarm indication only when it senses a temperature above the 150° threshold. *Chemistry.* **1.** the point at which a chemical process or effect begins. **2.** the lowest detectable chemical value of a material. *Geology.* **1.** see RIEGEL. **2.** see SILL, def. 1. *Mathematics.* **1.** given a positive integer N, a logic operator that has the value True if at least N of a given set of statements are true, and the value False otherwise. N is sometimes referred to as the threshold. **2.** given a real number α, a logic operator that has the value True if a given stochastic process takes on a value greater than α. α is sometimes referred to as the threshold.

threshold closer see AUTOMATIC DOOR BOTTOM.

threshold contrast *Optics.* the minimum difference in brightness that a human eye can detect under given conditions.

threshold current *Electrical Engineering.* the minimum current required to initiate the desired response from a nonlinear circuit or device; for example, the current required to initiate laser action or reverse the polarization of a magnet.

threshold detector *Nuclear Physics.* an isotope that is used to ascertain the types of neutrons and their energy levels produced by nuclear reactions.

threshold dose *Radiology.* the minimum dose of ionizing radiation that will produce a detectable degree of any given effect.

threshold element *Computer Technology.* **1.** a circuit, such as a Schmitt trigger, that produces a "True" output when its analog input is above a given threshold value. **2.** see LOGIC CIRCUIT.

threshold frequency *Electronics.* in a photodetector device, the lowest frequency of photo energy that can be detected.

threshold illuminance *Optics.* the lowest level of brightness detectable by the human eye under given conditions.

threshold lights *Aviation.* a system of lights that indicate the longitudinal limits of a section of a runway or of another area that is suitable for landing.

threshold limit value *Medicine.* the maximum concentration of a chemical to which workers can be exposed for a specific period, such as an eight-hour working day, without physical damage.

threshold of audibility *Physiology.* the point at which a sound can just be heard; it serves as a reference point for gauging the loudness of sounds. Also, **threshold of detectability, threshold of hearing.**

threshold of feelings *Physiology.* the minimal amount of sound pressure of a specified signal that will stimulate the ear sufficiently to elicit a sensation of touch, tickling, discomfort, or pain, normally expressed in decibels relative to microbars.

threshold of reaction *Physics.* **1.** the state of a system that is in metastable equilibrium. **2.** the state of a system that is on the verge of reaction.

threshold signal *Electromagnetism.* the minimum intensity of a signal that can be detected and recognized.

threshhold speed *Engineering.* the minimum current speed for which a current meter will respond at its rated reliability.

threshold switch *Electronics.* a semiconductor switching system that responds to preset threshold values for any of several quantities such as temperature, pressure, or moisture. The switch is controlled by a voltage value that is an analog of the measured quantity; when the measured quantity increases to the threshold value, the control voltage is at a sufficient level to cause the switch to change abruptly from a high resistance to a very low resistance.

threshold treatment *Chemical Engineering.* the addition of minute quantities of dehydrated phosphates to water to prevent furring and corrosion in pipes and containers.

threshold value *Computer Programming.* a value at which the operating system makes a change in the operation being processed; for example, an error rate above which the system suspends processing on the assumption that there is some type of software or hardware failure. *Control Systems.* the minimum input that is required by an automatic control system to produce a corrective action. *Robotics.* the level of brightness that is required to produce a change in quantized image brightness.

threshold velocity *Geophysics.* the minimum velocity required by wind or water to move fragments of soil or sand under given circumstances in a given location.

threshold voltage *Electronics.* the lowest voltage at which an effect is produced or an indication is observed; e.g., the voltage at the base-emitter junction of a transistor must be in excess of a minimum threshold value in order for the junction to become forward biased and allow the transistor to turn on. *Nucleonics.* the minimum voltage across an ionization counter tube that will register an ionizing event within the tube.

Threskiornithidae *Vertebrate Zoology.* the ibises and spoonbills, a family of long-legged wading birds of the order Ciconiiformes; characterized by strong flight with an extended neck; found near water in tropic and temperate zones worldwide.

thrill *Medicine.* an abnormal vibration or tremor detected in a body part.

Thripidae *Invertebrate Zoology.* a large family of thrips, tiny insect plant pests in the order Thysanoptera, having fringed wings and sucking mouthparts.

thrips *Invertebrate Zoology.* the common name for any member of the order Thysanoptera, tiny plant pest insects that infest and feed on a wide variety of weeds and crop plants; some varieties spread fungus and virus.

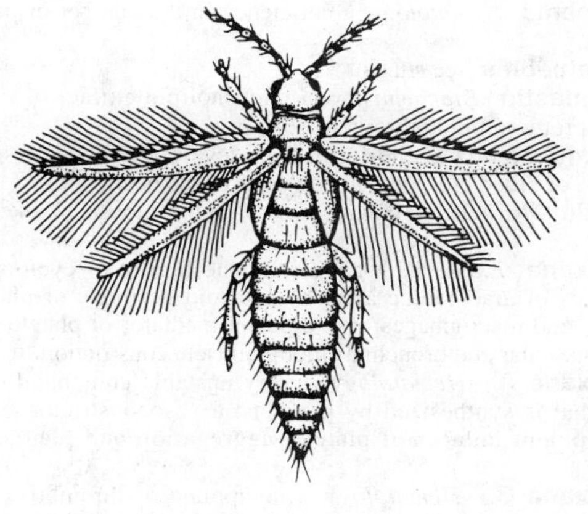

thrips

throat *Anatomy.* **1.** the front part of the neck. **2.** the region of the body from the mouth to the stomach or lungs, especially the pharynx. *Botany.* the upper opening of the gamophyllous corolla or calyx. *Design Engineering.* the narrowest portion of a constricted passage or area, such as of a nozzle. *Engineering.* **1.** the passageway in a fireplace that lies between the firebox and the smoke chamber, often closed by a damper. **2.** the inner edge of a railway flange. *Ordnance.* see LEADE.

throatable *Design Engineering.* describing a nozzle in which the size and shape of its throat can be altered to accommodate changes in the exhaust-gas concentration stream.

throat depth *Metallurgy.* the distance from the centerline of the electrodes in a resistance welding machine to the nearest point of interference for flatwork.

throat microphone see LARYNGOPHONE.

thromb- or **thrombo-** a combining form meaning "blood clot."

thrombase see THROMBIN.

thrombectomy *Surgery.* the surgical removal of a blood clot from a blood vessel.

thrombectomy of an artery *Surgery.* the surgical removal of a thrombus from an artery or vein.

thrombin *Enzymology.* an enzyme responsible for blood clotting by converting fibrinogen into fibrin. Also, FIBRINOGINASE, THROMBASE.

thromboangiitis obliterans *Medicine.* the development of blood clots in the blood vessels and the resulting restriction of circulation due to repeated attacks of inflammation of the vessels; commonly affecting the legs. Also, BUERGER'S DISEASE.

thrombocyte *Histology.* see PLATELET.

thrombocytopenia *Hematology.* an abnormally low blood-platelet level, which may be caused by various factors, including a too-active role of the platelet attacking antibodies, by the effects of radiation therapy and chemotherapy, or by leukemia; the condition results in delayed blood-clotting time.

thrombocytopenic purpura *Medicine.* a bleeding disease considered to manifest an autoimmune response, marked by multiple bruises and hemorrhages in the skin, mucous membranes, or serosal surfaces.

thrombocytosis *Hematology.* an unusually high number of thrombocytes (platelets) in the blood. Also, **thrombocythemia.**

thromboembolectomy *Surgery.* the surgical removal of a thromboembolism from a blood vessel.

thromboembolic *Medicine.* relating to an embolus that has resulted from detachment of a blood clot from its site of formation.

thromboembolism *Medicine.* the obstruction of a blood vessel by a thrombus that has dislodged from the site of its formation in another blood vessel.

thromboendaterectomy *Surgery.* a surgical procedure involving the removal of an occluding thrombus along with intima and atheromatus material from within an artery.

thrombogenesis *Medicine.* the formation of blood clots.

thrombokinase *Biochemistry.* the enzyme that changes prothrombin into thrombin. Also, AUTOPROTHROMBINE.

thrombolymphangitis *Medicine.* an inflammation of a lymph vessel due to a blood clot.

thrombolysis *Medicine.* the dissolution of a blood clot.

thrombolytic *Medicine.* **1.** dissolving or breaking up a blood clot. **2.** an agent that dissolves or breaks up a blood clot.

thrombopenia *Hematology.* a deficiency in the number of platelets in the blood.

thrombophlebitis see PHLEBITIS.

thromboplastin *Biochemistry.* a phospholipid extract of tissue that converts prothrombin to thrombin.

thrombosis *Medicine.* the formation, development, or presence of a thrombus, or blood clot, in a blood vessel or the heart.

thrombotic *Medicine.* **1.** having to do with thrombosis. **2.** like a thrombus.

thromboxane *Endocrinology.* one of the family of cyclooxygenase metabolites of arachidonic acid; an autacoid hormone of platelets, fibroblasts, and macrophages, it is a potent mediator of platelet aggregation and vascular and bronchial smooth muscle constriction.

thromboxane A_2 *Hematology.* a very unstable compound of thromboxane that is synthesized by platelets; a vasoconstrictor and an extremely potent inducer of platelet aggregation and platelet-release reaction.

thromboxane B_2 *Hematology.* a compound of thromboxane that is formed by nonenzymatic hydrolysis from thomboxane A_2; it is inactive with a half-life of 30 seconds.

thrombus *Medicine.* an aggregation of blood factors, including platelets and fibrin, clotting factors, and other cellular elements, attached to the interior wall of a blood vessel, sometimes causing vascular obstruction. (From a Greek term meaning "clot" or "lump.")

Throscidae *Invertebrate Zoology.* a family of wood-boring beetles, coleopteran insects in the superfamily Elateroidea.

throttle *Engineering.* a choke valve that regulates the flow of steam, gas, or other fuel to an engine. Also, **throttle valve.**

throttled flow *Fluid Mechanics.* a fluid flow forced to pass through a constricted area so that the velocity increases.

throttling *Thermodynamics.* an irreversible thermodynamic process in which a gas under pressure is allowed to expand by passing into a chamber of lower pressure. *Space Technology.* the process of changing the thrust of a rocket engine during powered flight, for example, by adjusting nozzle expansion or thrust chamber pressure. *Control Systems.* a control that has a number of intermediate steps between full on and full off. *Ordnance.* of or relating to a device or mechanism in the recoil system of a gun that controls the flow of recoil oil in order to regulate the length or resistance of the recoil motion. Thus, **throttling bar, throttling groove, throttling rod, throttling valve.**

throttling calorimeter *Engineering.* an instrument that measures the moisture content of steam, utilizing the principle of constant enthalpy expansion.

through arch *Civil Engineering.* an arch bridge with a roadway passing through the arch structure rather than over it; for example, the Hell Gate (New York) or Sydney Harbor bridges.

through glacier *Hydrology.* a glacier having two ends, formed by two valley glaciers that flow outward in opposite directions from the same depression.

through plane service *Transportation Engineering.* a service provided by aircraft flying the entire length of a route, so that passengers are not required to change planes en route. Also, SINGLE PLANE SERVICE.

throughput the movement of a material through a system; specific uses include: *Mining Engineering.* the quantity of mined material passing through a mill or a part of a mill in a given time or at a given rate. *Chemical Engineering.* a measure of the quantity of gas or vapor passing through a piece of equipment or section of a pipe or pump line during a specified time. *Computer Technology.* the general productivity of a system, especially the rate at which it handles work, often measured in jobs per day.

through repeater *Electronics.* in microwave-link communications, a repeater that retransmits only received signals and does not provide any services to local customers.

through street *Civil Engineering.* a street on which the movement of vehicle traffic is unimpeded or given priority over intersecting streets.

through transmission *Engineering.* a method of testing the structural characteristics of an object by transmitting ultrasonic frequencies through it and measuring at the far end.

through valley *Geology.* **1.** a flat-floored depression that was cut across a divide by glacier ice or meltwater streams. **2.** a valley formed by the action of a double-ended glacier.

through weld *Metallurgy.* a weld made by arc or gas welding through the surface of one member of a lap or tee joint, joining it to another member.

throw *Geology.* a measurement of the vertical displacement on a fault, or the vetical component of the net slip. *Agronomy.* **1.** the layer of soil that is pushed to one side by the plowing of a furrow slice. **2.** to move soil in this way. *Textiles.* to twist and combine filaments in producing thread or yarn. *Engineering.* to move a lever to activate, connect, disconnect, or turn off an apparatus or mechanism.

throwaway device *Electronics.* any electronic component or assembly for which repairs are impractical, due to cost considerations or other factors, and which is discarded when defective.

throwing *Textiles.* the process of combining and twisting one or more filaments into thread.

throwing power *Metallurgy.* in electroplating, the ability of a plating solution to deposit metal uniformly on an irregularly shaped cathode.

thrown silk *Textiles.* a raw silk that has been reeled and twisted into yarn.

throwout *Mechanical Engineering.* a mechanism in an automotive vehicle that separates the driven and driving plates of a clutch.

throw-out spiral see LEAD-OUT GROOVE.

thrush *Vertebrate Zoology.* **1.** any of numerous, medium-sized, migratory songbirds of the family Turdinae, characterized by an upper part that is a dull, solid brown and a spotted or colored breast; includes the wood thrush, robin, bluebird, and veery. **2.** any of various similar birds, such as the water thrush.

thrush *Medicine.* an outbreak of candidiasis in the mouth, characterized by the development of white crumbly or creamy plaques which leave a raw surface when stripped off; it is most common in sick, weak infants and the elderly in poor health. *Veterinary Medicine.* the popular term for a degenerative condition of the horn in the frog of the horse's foot and the accompanying thick black discharge; usually resulting from poor management and hygiene.

thrust *Geology.* **1.** the upward displacement or overriding movement of one crustal unit over another. **2.** see THRUST FAULT. *Mechanical Engineering.* the force exerted by a fluid jet or powered screw or impeller. *Aviation.* the pushing or pulling force exerted by a power plant such as an aircraft engine or rocket engine. *Building Engineering.* the horizontal or diagonal outward force or pressure of one member on another. *Mining Engineering.* a crushing of coal pillars resulting from excess stress due to the weight of overlying rocks.

thrust augmentation *Aviation.* a usually short-term increase in the thrust of an engine or power plant over the thrust normally developed; usually accomplished by liquid injection, afterburning, or introducing additional air into the induction system. Thus, **thrust augmenter.**

thrust-augmented *Space Technology.* of a launch vehicle, having strap-on motors to augment the thrust developed by its internal engines.

thrust axis *Aviation.* a line or axis through a flight vehicle or power plant along which the thrust acts. Also, THRUST LINE, AXIS OF THRUST, CENTER OF THRUST.

thrust bearing *Mechanical Engineering.* a bearing designed to carry axial loads on a shaft.

thrust block *Naval Architecture.* a fixture located in the stern of a vessel and designed to accept the thrust of the propeller and transmit it to the ship. *Geology.* see THRUST SHEET.

thrust coefficient see NOZZLE THRUST COEFFICIENT.

thruster *Space Technology.* an auxiliary reaction engine used to provide booster power or maneuverability on a spacecraft.

thrust fault *Geology.* a break in a rock structure having a dip of generally less than 45°, in which the hanging wall apparently moves upward in relation to the footwall. Also, THRUST SLIP FAULT, REVERSE SLIP FAULT.

thrust fault

thrust horsepower *Aviation.* **1.** the actual thrust, expressed in horsepower, of an engine-propeller combination; equal to the product of brake horsepower and propeller efficiency. **2.** the thrust of a jet engine or rocket expressed in horsepower; equal to the product of thrust pounds and airspeed (in miles per hour) divided by 375. *Naval Architecture.* an effective horsepower value, equal to the speed of advance of the propeller through the water in feet per second times the propeller thrust in pounds, divided by 550.

thrust line see THRUST AXIS.

thrust load *Mechanical Engineering.* a load parallel to the shaft of a vehicle.

thrust meter *Engineering.* an instrument that measures the static thrust of an engine.

thrust moraine see PUSH MORAINE.

thrust output *Aviation.* the net thrust (gross thrust minus intake drag) generated by a jet or rocket engine.

thrust-pound *Aviation.* a unit for measuring the thrust generated by a jet or rocket engine; equal to the product of fluid mass flow in pounds per second and relative jet velocity in feet per second divided by a gravity acceleration of 32.18 feet per second.

thrust power *Aviation.* the total driving force of a flight vehicle; equal to the product of thrust (or net thrust) and airspeed.

thrust reverser *Aviation.* a device designed to redirect the exhaust stream, and thereby the thrust, from a jet or rocket engine.

thrust section *Space Technology.* the part of a rocket vehicle containing the combustion chamber or chambers and nozzles. *Aviation.* loosely, the propulsion system of a flight vehicle.

thrust sheet *Geology.* a mass of rock composing the hanging wall of a large-scale thrust fault that has a level or gently dipping surface. Similarly, **thrust nappe, thrust plate, thrust slice.** Also, THRUST BLOCK.

thrust slip fault see THRUST FAULT.

thrust stand see TEST STAND.

thrust terminator *Space Technology.* a device designed to cut off the thrust of a rocket engine by stopping the flow of propellant (in a liquid-fuel rocket), or by diverting the flow of gases from the nozzle (in a solid-fuel rocket).

thrust-weight ratio *Aviation.* a standard used especially with combat aircraft to evaluate engine performance; equal to the thrust output of the engine divided by the engine weight without fuel.

thrust yoke *Mechanical Engineering.* a diamond drill part that connects the piston rods of the feed mechanism to the thrust block.

Thrydidae *Invertebrate Zoology.* a small family of moths belonging to the suborder Heteroneura.

thryptophyte *Ecology.* any plant whose relationship with its host is on the whole symbiotic, except that the host experiences a sublethal modification of its tissues.

Thuban *Astronomy.* Alpha (α) Draconis, a 4th-magnitude A-type star that lies in the constellation Draco; around 3000 BC, it was the pole star.

thucolite *Geology.* a brittle mixture of hydrocarbons, uraninite, and some sulfides found in ancient sedimentary rocks. Also, **thucholite.**

Thuidiaceae *Botany.* a family of large but delicate, matted dull mosses of the order Hypnobryales, characterized by usually prostrate and creeping stems, pinnate or bipinnate branching, and lateral sporophytes; occurring on rocks, soil, and tree bases.

Thule [thoo´lē; too´lē] *Geography.* in the ancient world, the region believed to be the northernmost point on earth, variously identified as Iceland, Jutland, the Shetland Islands, or other then-known northern areas. *Astronomy.* asteroid 279, discovered in 1888 and having a diameter of 135 km; belonging to type D, it orbits at the outer edge of the main belt of asteroids with a sidereal period 0.75% that of Jupiter.

thulite *Mineralogy.* a pink, rose-red, or purple-red, manganese-bearing variety of zoisite; used as an ornamental stone.

thulium *Chemistry.* an element having the symbol Tm, the atomic number 69, and an atomic weight of 168.93; a rare-earth (lanthanide) element with no stable isotopes; its solid has a metallic luster and its dust is flammable; used as an X-ray source. (Named after the region of *Thule.*)

thulium chloride *Inorganic Chemistry.* $TmCl_3 \cdot 7H_2O$, deliquescent, green crystals; very soluble in cold water and alcohol; melts at 824°C and boils at 1440°C.

thulium oxalate *Organic Chemistry.* $Tm_2(C_2O_4)_3 \cdot 6H_2O$, a greenish solid; soluble in dilute acids and alkaline oxalate solutions; melts at 50°C; used to separate thulium from common metals.

thulium oxide *Inorganic Chemistry.* Tm_2O_3, a greenish-white powder; slightly soluble in acid; used to make thulium metal. Also, **thulia.**

thumbnail *Graphic Arts.* a small, rough replica of a page layout, illustration, and so on.

thumb plane *Mechanical Devices.* a small woodworking plane with a one-inch-wide iron mounting; used with thumb pressure for very fine work in a limited space, as on wood models.

thumbscrew *Mechanical Devices.* a screw with a vertically flattened head that can be turned by the thumb and forefinger.

thumbtack *Mechanical Devices.* a tack with a large, round, flat head that can be pushed into a fairly soft surface by thumb pressure.

thump *Acoustical Engineering.* a term for a transient, low-frequency noise in a sound system or components.

Thunburg technique *Biochemistry.* a method of using methylene blue to indicate oxidation in dehydrogenation reactions.

thunder *Geophysics.* the sound created during an electrical discharge, caused by the rapid expansion of the air surrounding the channel of the discharge.

Thunderbird *Ordnance.* a British surface-to-air missile designed as the standard heavy antiaircraft weapon of the British Army; the **MK2** version is powered by a solid-propellant rocket motor with four jettisonable boosters, is equipped with semiactive radar homing, and delivers a high-explosive warhead.

thunderbolt *Geophysics.* a common term for an air-to-ground electrical discharge that is accompanied by the sound of thunder.

Thunderchief *Aviation.* a popular name for the F-105 supersonic tactical fighter.

thundercloud *Meteorology.* **1.** another name for a cumulonimbus cloud. **2.** the cloud mass associated with a thunderstorm.

thunderhead *Meteorology.* **1.** the incus of a cumulonimbus cloud. **2.** an entire cumulonimbus. **3.** the upper portion of a cumulus cloud.

thunderhead clouds

Thunderjet *Aviation.* a popular name for the F-84 jet fighter.

thundersquall *Meteorology.* a combined thunderstorm and squall, the latter being associated with the downrush phenomenon occurring in a well-developed thunderstorm.

thunderstorm *Meteorology.* a usually brief, localized storm in which thunder and lightning occur, often accompanied by strong gusts of wind and heavy rain; produced by a cumulonimbus cloud. Also, ELECTRICAL STORM.

thunderstorm cell *Meteorology.* the convection cell intrinsic to a cumulonimbus cloud mass.

thunderstorm charge separation *Geophysics.* the means by which particles within a thundercloud receive their charge and then accumulate in separate areas of the cloud.

thunderstorm cirrus cloud see FALSE CIRRUS CLOUD.

thunderstorm day *Meteorology.* an observation day during which thunder is heard at a weather station.

Thunderstreak *Aviation.* a popular name for the F-84 dual-purpose fighter/reconnaissance aircraft.

thunk *Computer Programming.* **1.** a compiler-generated, machine-language subroutine that evaluates an expression argument passed to a subroutine; used in the Algol language. **2.** the set of instructions in such a subroutine.

Thunnidae *Vertebrate Zoology.* an equivalent name for the family Scombridae, including the tunas and mackerels.

Thuringian *Geology.* a European geologic stage of the Upper Permian period, occurring after the Saxonian and before the Triassic period.

thuringiesin *Microbiology.* an exotoxin elaborated by the bacterium *Bacillus thuringiensis*; toxic to certain vertebrates and insects.

Thurniaceae *Botany.* a family of monocotyledonous plants of the order Juncales, distinguished by an inflorescence of one or more dense heads, vascular bundles of the leaf in vertical pairs, and silica bodies in the leaf epidermis.

thwart *Behavior.* to prevent a consummatory response.

Thy thymine.

Thy-1 antigen *Immunology.* a cell-surface isoantigen of mice that occurs on lymphocytes in the thymus, on peripheral lymphoid tissues, or in the central nervous system. Also, THETA ANTIGEN.

Thylacinidae *Vertebrate Zoology.* the Tasmanian wolves, a monotypic family of carnivorous, Australian, doglike marsupials belonging to the order Marsupicarnivora.

Thylacoleo *Paleontology.* the "marsupial lion," a genus of Australian marsupials in the family Phalangeridae; formerly thought to have been a carnivore, but now thought probably to have been herbivorous like other phalangeroids; *Thylacoleo* was catlike and large (estimated at 450 pounds); extant in the Pliocene and Pleistocene.

thylakoid *Cell Biology.* an extensive membrane system composed of stacks of flattened vesicles, occurring within the chloroplasts of plant cells and containing the components of photosynthesis.

thyme [tīm] *Botany.* any of various plants of the genus *Thymus,* of the mint family, including the garden herb *T. vulgaris,* which is cultivated commercially as a spice.

thyme camphor *Organic Chemistry.* a popular name for thymol.

thymectomy *Surgery.* the excision of the thymus gland.

Thymelaeceae *Botany.* a family of dicotyledonous trees, shrubs, and herbs of the order Myrtales; characterized by entire opposite or alternate leaves, bisexual and unisexual flowers found in racemes or umbels, and a superior ovary with a single ovule.

-thymia *Psychology.* a suffix denoting a condition of the mind, usually a mental disorder.

thymic *Anatomy.* relating to the thymus.

thymic acid see THYMOL.

thymic aplasia *Medicine.* the absence of the thymus gland.

thymic corpuscle *Histology.* a rounded, acidophilic mass of cells in the medulla of the thymus.

thymic humoral factor *Endocrinology.* a polypeptide factor extracted from bovine thymus that may be involved in the development of the cells of the thymus-dependent immune system. Also, THYMOLYMPHOPOIETIC FACTOR.

thymidine *Biochemistry.* $C_{10}H_{14}N_2O_5$, the 2'-deoxynucleoside produced by combining glycosylation of thymine with 2-deoxyribose; its 5'-phosphate is a component of DNA.

thymidine kinase *Enzymology.* an enzyme that catalyzes the phosphorylation of thymidine.

thymidylic acid *Biochemistry.* the phosphate ester of thymidine whose residues appear in DNA. Also, **thymidine 5'-phosphate.**

thymine [thī´mēn; thī´min] *Biochemistry.* the pyrimidine base in DNA that pairs with adenine.

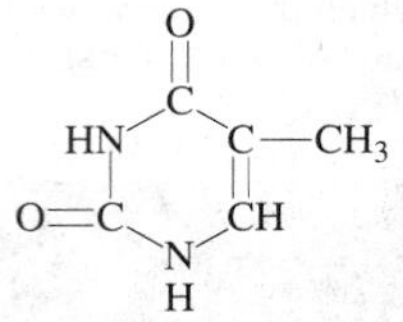

thymine

thymine dimer *Molecular Biology.* a condensation product that is formed by two adjacent thymine bases in a DNA strand and retards future DNA replication.

thymiosis see YAWS.

thymitis *Medicine.* an inflammation of the thymus.

thymocyte *Histology.* a lymphocyte that is produced in the thymus.

thymol *Organic Chemistry.* $(CH_3)_2CHC_6H_3(CH_3)OH$, combustible white crystals; slightly soluble in water and soluble in alcohol and ether; melts from 48 to 51°C and boils at 233°C; occurs naturally in the thyme plant; used as a fungicide, preservative, flavoring, and reagent, and in perfumes. Also, ISOPROPYL-*m*-CRESOL, THYMIC ACID, THYME CAMPHOR.

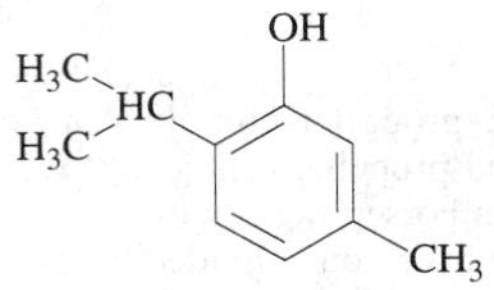

thymol

thymol blue *Organic Chemistry.* a popular name for thymolsulfonephthalein.

thymol iodide *Organic Chemistry.* $[C_6H_2(CH_3)(OI)(C_3H_7)]_2$, a combustible reddish powder; insoluble in water, slightly soluble in alcohol, and soluble in ether; used as feed additive and an antifungal agent.

thymolphthalein *Organic Chemistry.* $C_{28}H_{30}O_4$, a white powder that is insoluble in water and soluble in alcohol; melts at 253°C; used in medicine and as an acid-base indicator.

thymolsulfonephthalein *Organic Chemistry.* $C_{27}H_{30}O_5S$, brownish-green powder or crystals; insoluble in water and soluble in alcohol; melts (decomposes) at 223°C; used as an acid-base indicator.

thymolymphopoietic factor see THYMIC HUMORAL FACTOR.

thymoma *Medicine.* a usually benign tumor of the thymus gland that evolves from the epithelium or lymphoid tissue.

thymometastasis *Oncology.* the transfer of malignant cells from the thymus gland to another unconnected organ or part.

thymopathy *Medicine.* any disease of the thymus gland.

thymopharyngeal duct *Developmental Biology.* the third pharyngobranchial duct.

thymopoietin *Endocrinology.* a polypeptide hormone that is produced in the thymus gland and induces the differentiation of prothymocytes into T-lymphocytes .

thymosin *Endocrinology.* any of the family of polypeptide hormones or factors that are produced by the thymus gland and act as endogenous T-cell mitogens and differentiation factors.

thymus *Anatomy.* a lymphoid gland consisting of two pyramidal lobes, situated in the mediastinal cavity above the heart; the site of T-lymphocyte production, which is regulated by hormones produced by thymic epithelial cells. The thymus is most fully developed in children, reaching its maximum development around puberty and then undergoing gradual involution, resulting in a slow decline of immune function throughout adulthood.

thymusectomy *Surgery.* an excision of the thymus.

thyratron *Electronics.* a gas-filled, three-element thermionic electron tube containing a cathode, grid, and anode; a voltage pulse applied to the grid ionizes the gas and causes anode current to flow. Once initiated, anode current continues to flow and cannot be changed by grid voltage; the tube can be extinguished only by removing anode voltage or externally opening the anode current path.

thyratron gate *Electronics.* a logic AND gate circuit consisting of a gas-filled thermionic electron tube containing a cathode, anode, and one or more grids; coincident voltage pulses applied to all grids cause anode current to flow. Once initiated, anode current continues to flow, independent of the voltages on the grids; the tube can be extinguished only by externally removing anode voltage or anode current.

thyratron inverter *Electronics.* a circuit that uses thyratron electron tubes to convert DC power to AC power.

thyrector *Electronics.* a semiconductor switching diode used for surge protection in AC power applications; the diode presents a very high resistance until its breakdown voltage level is reached, at which time it abruptly changes to a low resistance device.

thyristor *Electronics.* a bistable semiconductor switch containing three or more junctions, used chiefly in power control applications; silicon-controlled rectifiers are the most common type of thyristor.

thyro- or **thyr-** a combining form denoting a relationship to the thyroid gland. Also, **thyreo-, thyre-.**

thyrocalcitonin see CALCITONIN.

thyrocardiac disease *Cardiology.* a disease of the heart associated with hyperthyroidism, characterized by atrial fibrillation, enlargement of the heart, and congestive heart failure.

thyroglobulin *Biochemistry.* an important protein of the thyroid gland that functions in the storage of iodine, and from which thyroxine and triiodothyronine are formed.

thyroglossal cyst *Pathology.* a cyst in the neck caused by persistence of portions of, or by lack of closure of, the primitive thyroglossal duct.

thyroglossal duct *Developmental Biology.* a passage extending between the thyroid primordium and the posterior part of the tongue.

thyrohyoid *Anatomy.* relating to the thyroid gland or cartilage and the hyoid bone.

thyroid [thī′roid] *Anatomy.* of or relating to the thyroid gland or to some condition of this gland, such as hyperthyroidism. *Pharmacology.* a compound including the cleaned, dried, and powdered thyroid gland, extracted from domesticated animals and formerly used in the treatment of hyperthyroidism in humans. (From a Greek word meaning "shield," because it was thought of as having the shape of a warrior's shield.)

thyroid cartilage *Anatomy.* either of two paired cartilages forming the major anterior portions of the larynx; popularly known as the Adam's apple.

thyroidectomy *Surgery.* the surgical removal of all or the major portion of the thyroid gland.

thyroid gland *Anatomy.* an endocrine gland located at the base of the neck whose secretions are important in regulating many aspects of metabolism and mineral balance.

thyroid hormone *Endocrinology.* **1.** either of the two hormones (thyroxine and triiodothyronine) that are secreted by the thyroid gland and that are required for the proper growth, development, and function of tissues such as those of the skeletal system, the central nervous system, the cardiovascular system, and the reproductive system, as well as the regulation of the basal metabolic rate of the body. **2.** any other hormone secreted by the thyroid gland and functioning in metabolism.

thyroiditis *Medicine.* an inflammation of the thyroid gland.

thyroliberin *Endocrinology.* a tripeptide hormone that is formed in the hypothalamus and that stimulates the release of thyrotropin from the anterior pituitary gland; it also possesses prolactin-releasing activity. Also, **thyrotropin-releasing hormone.**

thyroma *Oncology.* a tumor of the thyroid gland.

thyronine *Endocrinology.* the parent compound for the iodinated thyroid hormones (triiodothyronine and thyroxine), a compound that is not produced biologically but that would be the molecule that remained if thyroxine or triiodothyronine were completely deiodinated.

thyropathy *Pathology.* any disease of the thyroid gland.

Thyropteridae *Vertebrate Zoology.* the disk-winged bats, a family of Central and South American bats of the suborder Microchiroptera, characterized by a well-developed sucking disk on the soles of the feet.

thyrotoxic heart disease *Cardiology.* a disease of the heart due to the presence of excessively high concentrations of thyroid hormones.

thyrotoxic myopathy *Medicine.* a chronic disease causing muscle atrophy, associated with hyperthyroidism.

thyrotoxicosis *Medicine.* any condition resulting from the presentation to the tissues of excessive quantitites of thyroid hormones; it may be caused by overproduction by the thyroid gland, or by loss of storage function and leakage from the gland.

thyrotropic *Medicine.* having an influence on the thyroid gland. Also, **thyrotrophic.**

thyrotropin *Endocrinology.* a glycoprotein hormone that is produced by the thyrotrophs of the anterior pituitary gland and that stimulates the release of thyroxine and triiodothyronine from thyroglobulin in the thyroid gland; it acts as a general trophic factor for the thyroid gland itself. Also, **thyrotropic hormone, thyroid-stimulating hormone.**

thyroxine *Endocrinology.* the tetraiodothyronine hormone of the thyroid gland, produced by the coupling of two molecules of diiodotyrosine and stored in thyroglobulin within the thyroid gland; it is released in greater abundance and is longer-acting than triiodothyronine, but is less potent.

thyroxine binding globulin *Endocrinology.* a plasma carrier interglobulin that transports thyroxine in the blood.

thyrse *Botany.* a dense, panicle-like inflorescence whose main axis is racemose and whose latter axes are cymose; it has the appearance of a cluster of grapes.

Thysanoptera *Invertebrate Zoology.* the thrips, an order of tiny, slender insects having scraping and sucking mouthparts, short antennae, and narrow fringed wings; it includes many serious plant pests; parthenogenesis is common, and some species have no males. Also, PHYSOPODA.

Thysanura *Invertebrate Zoology.* an order of primitive wingless insects, including the silverfish and bristletails, having long antennae and cerci, a median tail filament, and a soft, scaly body.

THz terahertz.

Ti the chemical symbol for titanium.

TIA transient ischemic attack.

Tian Shan see TIEN SHAN.

tiba *Organic Chemistry.* $C_7H_3I_3O_2$, a colorless solid that melts at 226°C; used as a growth regulator for fruit.

tibia *Anatomy.* the inner and larger bone of the leg between the knee and ankle; the shinbone.

tibial *Anatomy.* relating to the tibia. Also, **tibialis.**

tibialgia *Medicine.* a condition marked by pain of the tibia, with lymphocytosis and eosinophilia.

tibiofemoral *Anatomy.* relating to the tibia and the femur.

tic *Neurology.* an involuntary, compulsive, repetitive, stereotyped movement, usually of the face and shoulders, that resembles a purposeful movement in that it is coordinated and involves muscles in their normal synergistic relationships.

tic douloureux see TRIGEMINAL NEURALGIA.

Tichocarpaceae *Botany.* a monospecific red algae family of the order Cryptonemiales, occurring in the northwest Pacific and having erect branched thalli with flattened axes.

tick *Telecommunications.* a pulse, broadcast at one-second intervals, that indicates the exact time.

tick *Invertebrate Zoology.* the common name for any arachnid in the superfamily Ixodoidea, bloodsucking ectoparasites that often transmit disease. Ticks are larger than their relatives, the mites. They belong to two families, Ixodidae (the **hard ticks**) and Argasidae (the **soft ticks**).

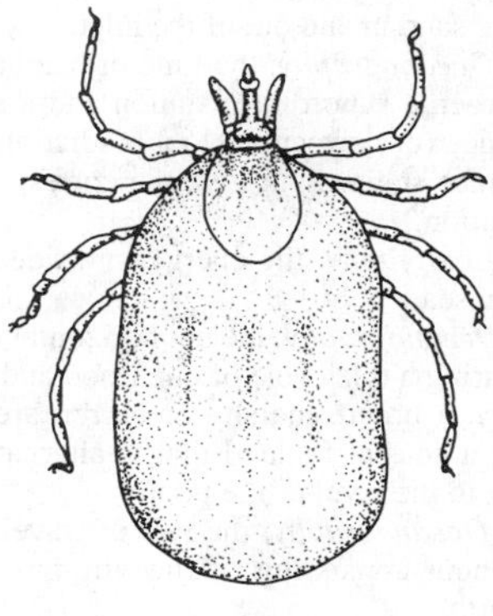

tick

tick-bite paralysis *Medicine.* a form of motor paralysis that affects many domestic species as well as humans, is caused by toxins present in the saliva of various species of ticks, and moves upward toward the brain. Also, **tick paralysis.**

tick-borne typhus fever (of Africa) *Medicine.* a form of typhus fever prevalent in Africa and transmitted by ticks.

tick fever or **tick typhus** see ROCKY MOUNTAIN SPOTTED FEVER.

ticking *Textiles.* a strong, twill-weave cotton cloth, generally characterized by alternating white and colored lengthwise stripes; used chiefly in upholstery and to cover pillows and mattresses.

tickle *Physiology.* **1.** to experience or cause to experience a light touch, usually found unpleasant, upon a sensitive area of the skin, producing laughter as a nervous response and a retreat from the stimulus. **2.** the feeling of being tickled.

tickler coil *Electronics.* a winding used for positive feedback between the input and output of a vacuum-tube stage; the winding is connected in series with the plate of the vacuum tube and is inductively coupled to a winding in the control-grid circuit.

tic polonga or **tic-polonga** see RUSSELL'S VIPER.

tidal *Oceanography.* of or relating to tides. *Geography.* describing low-lying land along a coast that is usually flooded at high tide, such as a **tidal flat** or **tidal marsh.** *Physiology.* of or relating to changes in body fluids.

tidal air *Physiology.* the air making up the tidal volume.

tidal basin *Oceanography.* a natural or artificial body of water that is subject to tidal action, or that is open to another body of water (e.g., a river) subject to tidal action.

tidal bench mark *Cartography.* in surveying, a bench mark close enough to a tidal station that the mark's elevation is referenced to mean sea level at that station, and the tidal gauge at that station is referenced to the mark. Also, PRIMARY BENCH MARK.

tidal bore *Oceanography.* see BORE.

tidal channel see TIDAL INLET.

tidal constants *Oceanography.* tidal relations that are essentially constant for a given location, including harmonic and nonharmonic constants.

tidal correction *Geophysics.* a correction made in gravity measurements to account for the effects of earth tides.

tidal creek *Hydrology.* a tidal inlet or estuary of a small river or brook.

tidal current *Oceanography.* the cyclical horizontal movement of water associated with the rise and fall of the tide due to the astronomical tide-producing forces.

tidal current chart *Oceanography.* a chart that shows, by numbers and arrows, the average direction and speed of currents at each phase of the current cycle, usually published together, one for each hour of the cycle.

tidal current tables see CURRENT TABLES.

tidal cycle see TIDE CYCLE.

tidal datum *Oceanography.* the permanently established level of the sea to which soundings or heights of tide are referenced, usually low water. *Cartography.* in surveying, any tide level that is used as a surface of reference for depth measurements in the sea and as a plane of reference for the determination of elevations above sea level. Also, CHART DATUM, DATUM PLANE, REFERENCE PLANE.

tidal day *Oceanography.* the interval between two successive upper transits of the moon over a local meridian, averaging 24.84 hours. Also, LUNAR DAY.

tidal delta *Geology.* a delta at the mouth of an inlet, formed by changing tides that move sand in and out of the inlet.

tidal difference *Oceanography.* the time or height difference of a high or low water between a subordinate station and a reference station; applying the difference (or correction) to a prediction given in the tide tables for the reference station will give the corresponding prediction for the subordinate station.

tidal energy *Oceanography.* the energy in a tide moving from a tidal basin into the open sea.

tidal excursion *Oceanography.* the net horizontal distance traversed by a water particle during a tide cycle of one flood and one ebb.

tidal flat *Geology.* a broad, marshy or barren area of unconsolidated sediment forming a coastal flatland that is alternately covered and uncovered according to the level of the tides.

tidal frequency *Oceanography.* the rate of travel, in degrees per day, of a tidal component consisting of the tide-producing forces of the moon, sun, and earth.

tidal friction *Oceanography.* the frictional drag of the tidal wave on the bottom of shallow seas; although this drag is too small to verify thus far, there is some evidence that it slows the earth's rotational velocity slightly, enough to be appreciable over millions of years.

tidal glacier see TIDEWATER GLACIER.

tidal harbor *Oceanography.* a harbor that is open to the sea and therefore affected by tides, as distinguished from a harbor whose water level is controlled artificially.

tidal hypothesis *Astronomy.* an archaic theory for the origin of the solar system which proposed that the planets formed from gas pulled out of the sun by the tidal effect of a passing star.

tidal inlet *Geology.* a waterway from open water into a lagoon through which water flows landward with the rising tide and seaward with the falling tide. Similarly, **tidal outlet.**

tidalite *Geology.* an accumulation of mineral or organic matter that is carried and deposited by tidal currents.

tidal light *Navigation.* a light placed at a harbor entrance to indicate tidal conditions within the harbor. Similarly, **tide signal.**

tidal platform ice foot see ICE FOOT.

tidal pool see TIDEPOOL.

tidal potential *Oceanography.* an expression of tidal forces as components of a vector field.

tidal prism *Oceanography.* the total volume of water that flows into a tidal basin with a flood tide and then out again with the ebb; as measured over considerable time, it represents the difference between the basin's mean high-water volume and its mean low-water volume.

tidal quay *Civil Engineering.* a depression in a harbor deep enough so that ships are not required to leave before low tide.

tidal radius *Astronomy.* the distance from the center of a planet to the point at which its gravitational pull equals that of the sun.

tidal scour *Geology.* the erosion of the ocean floor by powerful tidal currents, causing the removal of inshore sediments and the development of various-sized depressions in the sea floor.

tidal volume *Physiology.* the amount of air taken in during one breath in normal respiration.

tidal water see TIDEWATER.

tidal wave *Oceanography.* **1.** the wave motion of a tide. **2.** in popular use, an unusually high and destructive ocean wave; a tsunami or storm surge. Also, TIDE WAVE.

tidal wind *Meteorology.* a light breeze that blows onshore as a high tide rises and offshore as the tide ebbs.

tide *Oceanography.* the cyclical rise and fall of the water level in the oceans and other large bodies of water that is caused by the interaction of the rotating earth with various forces, principally the gravitational attraction of the moon and the sun. *Physiology.* a physiological variation or increase of a certain constituent in a body fluid, such as the increase of fat in the lymph following a meal. (From the Old English word for "time.")

tide amplitude *Oceanography.* one-half of the vertical difference between a high water and the following low water; half the tide range.

tide-bound *Navigation.* of or relating to a vessel that cannot move, especially out of a harbor, due to shallow water caused by a receding tide.

tide crack *Oceanography.* a narrow opening between shore ice and an ice shelf, caused by tidal water level fluctuations; may occur in groups and may widen to form a shore lead.

tide curve see MARIGRAM.

tide cycle *Oceanography.* a complete set of tidal conditions, such as a tidal day, lunar month, or Metonic cycle. Also, TIDAL CYCLE.

tide gate *Civil Engineering.* a swinging gate that excludes water at high tide and allows drainage at low tide; usually placed below the high tide level on the outside of the drainage conduit.

tide gauge *Engineering.* an instrument used to determine the level of the tide.

tidehead *Oceanography.* the inland limit of the effect of a tide.

tide hole *Oceanography.* a hole made in the sea ice for tidal observations.

tide indicator *Engineering.* the part of a tide gauge that indicates the height of the tide.

tideland *Geography.* **1.** land that is flooded at high tide and exposed at low tide. **2.** see TIDEWATER.

tideline *Ecology.* the area of a coastline that is intermittently submerged as a result of the regular rise and fall of the water level.

tide machine *Engineering.* an apparatus that computes future tide heights through a summation of the harmonic components of which the tide is composed.

tidemark *Oceanography.* **1.** the highest point reached by a high water (or, infrequently, the lowest point reached by a low water). **2.** a physical mark left by a high tide.

tide notes *Oceanography.* tidal information at key locations on tidal charts, often including the mean range or diurnal range, mean tide level, and extreme low water.

tide pole *Engineering.* a graduated measuring pole that is used to measure the rise and fall of the tide. Also, **tide staff.**

tidepool or **tide pool** *Oceanography.* **1.** a pool of seawater that remains in a depression on a beach or reef after the tide has receded. **2.** specifically, a pool that is regularly submerged at high tide and exposed at low tide; noted as the habitat of a wide variety of organisms. Also, TIDAL POOL.

tide prediction *Oceanography.* the predetermined time and height of high or low water at a certain place, used to compile tide tables; may be computed years in advance by computations based on the harmonic tidal constituents.

tide-producing force *Geophysics.* the centrifugal force that holds two astronomical bodies apart.

tide race *Oceanography.* **1.** a very rapid tidal current. **2.** the narrow channel or passage through which such a current flows.

tide range *Oceanography.* the general term for the difference in height between a high water and the following low water; specific tide ranges include the several types of apogean, diurnal, mean, perigean, and tropic ranges.

tide-rode *Navigation.* describing the position of a vessel that has been moved away from anchor by the force of the tide.

tide station *Oceanography.* a site at which measurements of tides are made. A **primary tide station** is one at which observations have been made for more than 19 years, a **secondary tide station** is one at which observations have been made for more than one but fewer than 19 years, and a **tertiary tide station** is one at which observations have been made for more than 30 days but less than one year.

tide table *Oceanography.* a table of daily tide predictions, usually published annually, giving times and heights of high and low water at a number of reference stations; usually supplemented by lists of tidal differences and constants to be used to obtain predictions for other locations.

tidewater *Geography.* **1.** coastal lowlands, often cut by rivers and inlets. Also, TIDELAND. **2. the Tidewater.** the coastal lowlands of eastern Virginia and North Carolina. *Oceanography.* **1.** water affected by the tides. **2.** water overlying tidelands.

tidewater glacier *Hydrology.* a glacier that terminates in the ocean, usually forming a vertical wall of ice from which icebergs break off. Also, TIDAL GLACIER.

tide wave see TIDAL WAVE.

tideway *Oceanography.* a channel in which a tidal current runs.

tie *Engineering.* a beam, post, or angle used to fasten objects together. *Electricity.* a bracket, clamp, clip, ring, or strip used to hold several wires together as a cable or bundle. *Civil Engineering.* **1.** a tension member such as that in a truss or frame. **2.** a loop of retaining bars around the longitudinal steel in reinforced concrete to add shear reinforcement. **3.** a transverse wooden or concrete beam on which the rails of a railroad track rest. Also, CROSSTIE. *Mining Engineering.* a roof support in a coal mine.

tieback *Mining Engineering.* a beam similar in function to a fend-off beam but fixed at the opposite side of the shaft or inclined road.

tie bar *Civil Engineering.* **1.** a bar used in reinforced concrete to add shear reinforcement and to hold longitudinal bars in place. **2.** a bar used to connect two switch rails on a railroad. *Geology.* see TOMBOLO.

tie beam *Mechanical Devices.* a wooden beam that ties together the lower ends of a roof truss or rafter to prevent spreading.

tie cable *Electricity.* **1.** an interconnecting cable between two distributing points in a telephone system. **2.** any cable that interconnects two circuits.

tied arch *Civil Engineering.* a structural arch with horizontal reactions supported by ties at the arch ends.

tied concrete column *Civil Engineering.* a concrete column that is reinforced with rectangular or circular tie bars placed around longitudinal bars to increase load-carrying capacity.

Tiedenmann's body *Invertebrate Zoology.* any of the small amebocyte-producing glands in the ring canals of many echinoderms.

tie-down diagram *Engineering.* a drawing that prescribes the method of securing an object to a particular vehicle.

tie-down point *Engineering.* a point of attachment used to secure an object to a vehicle.

tied rank *Statistics.* in the ordering of a set of objects, the relationship among objects having the same value.

tie flight see CONTROL STRIP.

tie line *Physical Chemistry.* a line on a phase diagram linking the liquid and vapor phase compositions that are in equilibrium. Also, CONODE. *Telecommunications.* **1.** a leased communication circuit. **2.** a trunk between PBXs.

tiemannite *Mineralogy.* HgSe, a dark-gray cubic mineral of the sphalerite group, occurring as tetrahedral crystals and compact to granular masses, having a specific gravity of 8.19 to 8.27 and a hardness of 2.5 on the Mohs scale; found in hydrothermal veins with calcite and other selenides.

tie molecule *Materials Science.* in a polymer microstructure, any of a number of molecules that begin to crystallize more or less simultaneously on two adjacent crystallites, or whose lamellae are partly trapped within one crystallite while building onto another; such molecules represent a departure from regular chain folding by providing a molecular tie between crystals.

Tien Shan *Geography.* a mountain range in northwestern China and western Turkestan; highest peak: Pobeda Peak (24,406 feet). Also, TIAN SHAN.

tie plate *Mechanical Engineering.* a plate used to connect the tie rods in a furnace. *Civil Engineering.* a steel plate that acts as reinforcement for a railroad tie and holds the rail in place.

tie point *Electricity.* **1.** a lug, screw, or other terminal resident at a junction to which wires are connected. **2.** an insulated distributing point. *Cartography.* **1.** in aerial photography, an image point that can be identified in the overlapping area of adjacent photographs, and that serves to connect the photographs into a single unit and a common network. **2.** in surveying, the point of closure, either of the survey on itself or on another survey.

tier [tēr] *Architecture.* a row, usually one of a layered series of rows, of architectural members or objects such as seats.

tier array *Electromagnetism.* a vertical array of antenna elements positioned one above another.

tie rod *Mechanical Engineering.* a rod used as a mechanical or structural support between parts of a machine. *Civil Engineering.* a purely tensile structural member. *Mining Engineering.* a vertical rod mounted on overlying horizontal timbers in a shaft.

Tierra del Fuego *Geography.* **1.** an archipelago off the southern tip of South America, south of the Strait of Magellan. **2.** the largest island in this group; area: 18,530 square miles. (Literally, "Land of Fire;" so named by Ferdinand Magellan, after he sighted the huge bonfires built there by the local Indians.)

tie strip see CONTROL STRIP.

tie trunk *Telecommunications.* a telephone line or channel that directly connects two private branch exchanges.

Tietze extension theorem *Mathematics.* let f be a real-valued continuous function on a closed subset A of a normal (T_4) space X. Then there is a continuous function F on X such that $F(x) = f(x)$ for all x in A. F is called an extension of f.

tie wire *Electricity.* a short section of wire used to tie an open-line wire to an insulator.

Tiffonian *Geology.* a North American land-mammal stage of the Middle Paleocene epoch, occurring after the Torrejonian and before the Clarkforkian. Also, LATTORFIAN, SANNOISIAN.

tiger

tiger *Vertebrate Zoology.* a large carnivorous cat of Asia, *Panthera tigris,* the largest of the cat family; characterized by a tawny coat transversely striped with black and a long, black-ringed tail.

Tiger *Aviation.* a popular name for the F-11 supersonic jet fighter.

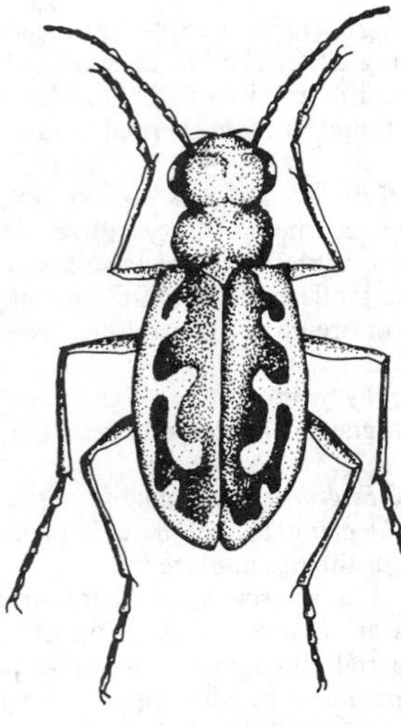

tiger beetle

tiger beetle *Invertebrate Zoology.* the popular name for any of the bright-colored beetles in the family Cicindelidae.

Tigercat *Aviation.* a popular name for the F7-F twin-engine fighter. *Ordnance.* a British surface-to-air missile system designed for close-range air defense; it utilizes Seacat missiles and includes a three-round launcher trailer and a director trailer.

tiger salamander *Vertebrate Zoology.* a moderate-sized North American salamander, *Ambystoma tigrinum,* of the family Ambystomatidae, characterized by brown or black skin with vertical yellow blotches.

tiger's-eye *Mineralogy.* a yellowish-brown, translucent, chatoyant variety of crystalline quartz, used as an ornamental stone and gemstone.

tiger shark see SAND SHARK.

tiger swallowtail *Invertebrate Zoology.* a yellow swallowtail butterfly, *Papilio glaucus,* distinguished by black stripes on the forewings.

tight *Mechanical Engineering.* **1.** describing a minimum, or sometimes inadequate, clearance between working parts of a machine. **2.** describing an absence of leaks in a pressure system. *Mining Engineering.* describing rock in which a hole will stand without crumbling. *Graphic Arts.* of a photographic negative, extremely overexposed.

tight-binding approximation *Solid-State Physics.* an approximation model used for calculating the band structure for electrons in a solid, in which it is assumed that overlap of electron wavefunctions between neighboring atoms is small.

tight coupling see CLOSE COUPLING.

tight fit *Mechanical Engineering.* a snug fit between male and female mating parts, so that a light to moderate force is required to accommodate the fit.

tight fold see CLOSED FOLD.

tight ion pair *Organic Chemistry.* an ion pair whose individual ions maintain their stereochemical positions without allowing solvent molecules to separate them.

tight junction *Cell Biology.* a bandlike intercellular junction commonly found on the surface of epithelial cells that seals together adjacent cells and prevents leakage from the epithelial sheet.

tightly coupled *Computer Technology.* of processors, linked together in such a way that they are dependent upon each other, usually communicating directly, sharing storage, and controlled by the same program.

tight rate *Industrial Engineering.* a standard time that is less than the time required by a qualified worker working at a normal pace.

tight sand *Geology.* an impermeable sand whose interstices are packed with finer grains or matrix material. Also, CLOSE SAND.

tigress *Vertebrate Zoology.* a female tiger.

Tigris *Geography.* a river in southwestern Asia flowing about 1150 miles from the Taurus Mountains in eastern Turkey to the Shatt-al-Arab River near the Persian Gulf.

tigroid *Neurology.* a condition in which a nerve cell's Nissl substance, when stained, may appear in a striped pattern similar to that of a tiger.

tilasite *Mineralogy.* $CaMg(AsO_4)F$, a green, gray, or gray-violet monoclinic mineral, having a specific gravity of 3.77 and a hardness of 5 on the Mohs scale; occurring as complex crystals found in veinlets with braunite in crystalline limestones.

tile *Materials.* **1.** a relatively thin, flat or curved unit of stone, concrete, or fired clay that is used for paving, roofing, flooring, or decorative purposes. **2.** a thin, square or rectangular unit of linoleum, rubber, or similar material that is used as a floor or wall covering. **3.** a drain made of earthenware pipe, semicircular stone, or fired clay tiles. **4.** a hollow unit of shale, gypsum, concrete, or fired clay used in construction. Also, HOLLOW TILE. **5.** to install any of these materials. *Mathematics.* a connected bounded region, together with its boundary, that may be translated, rotated, and possibly reflected as desired.

Tiliaceae *Botany.* a family of tanniferous dicotyledonous trees, shrubs, and rarely herbs of the order Malvales, having simple or stellate hairs or peltate scales and phloem stratified into hard and soft layers with wedged rays; native to tropical and subtropical regions and including the linden tree and a species producing jute fibers.

tiling the plane *Mathematics.* the mathematical problem of covering an arbitrarily large area of the plane with a given set of nonoverlapping tiles (i.e., tiles intersecting only at their boundaries). It is usually required that there be only finitely many noncongruent tiles. A tiling is periodic if there is a nontrivial translation that carries the tiling into itself.

till *Agriculture.* to work the soil with implements or machinery to make its condition more favorable for plant growth. *Geology.* unsorted and unstratified material deposited directly by glacier ice; it consists of a heterogeneous mixture of clay, silt, sand, gravel, and boulders. Also, BOULDER CLAY, GLACIAL TILL.

tillage *Agriculture.* the process of tilling the soil. Also, TILTH.

till billow *Geology.* an accumulation of glacial drift deposited by glacier ice in an irregular, undulating mass.

tiller *Agriculture.* **1.** an implement that uses a rotor to stir up the soil. **2.** a sprout that grows outward from the base of a plant. *Naval Architecture.* a horizontal shaft or handle used to turn a vessel's rudder; it may be powered mechanically or by hand.

Tilletiaceae *Mycology.* a family of smut fungi of the order Ustilaginales that form spores called basidiospores.

tilleyite *Mineralogy.* $Ca_5Si_2O_7(CO_3)_2$, a white monoclinic mineral occurring as grains or in massive form, having a specific gravity of 2.84 and an undetermined hardness; found in limestone contact zones.

tillite *Petrology.* a sedimentary rock produced by lithification of pre-Pleistocene glacial till.

Tillodontia *Paleontology.* an order of quadrupedal mammals, known only from incomplete fragments; the largest was about a meter long; characterized by rodentlike incisors; holarctic distribution; extant in the Paleocene and Eocene.

tilloid *Geology.* an assemblage of unsorted rocks of nonglacial origin, resembling till in appearance. *Petrology.* a rock that resembles tillite.

till plain *Geology.* a broad, generally flat or gently rolling surface that is underlain by till.

till sheet *Geology.* any layer or bed of till that at times may form a ground moraine.

Tilopteridaceae *Botany.* a family of tufted brown algae constituting the order Tilopteridales, occurring sublittorally throughout the North Atlantic and Arctic Oceans.

Tilopteridales *Botany.* a monofamilial order of loosely branched, filamentous, tufted brown algae of the class Phaeophyceae, characterized by pronounced primary axes with opposite branching that becomes unilateral above; generally 10 to15 centimeters tall.

tilt a condition or degree of inclining, sloping, or slanting; specific uses include: *Aviation.* the inclination of an aircraft or component from the horizontal or from its normal position. *Optics.* the angle that forms between a lens element and the optical axis. *Cartography.* in aerial photography, the dihedral angle between the plane of the photograph and a horizontal plane. *Meteorology.* in synoptic meteorology, the inclination of some significant feature of a circulation or pressure pattern, or of a temperature or moisture field.

tilt angle *Electromagnetism.* the angle between the vertical radiation axis and the horizontal plane; the reference axis.

tilt block *Geology.* a crustal unit bounded by faults that has become tilted, presumably by rotation on a hinge line.

tilt boundary *Materials.* a small angle grain boundary composed of an array of edge dislocations.

tilt displacement *Cartography.* in aerial photography, radial displacement from the isocenter of images on the photograph, resulting from the photograph's tilt.

tilted iceberg *Oceanography.* a tabular iceberg that has become unbalanced, so that its surface is inclined.

tilted interface *Geology.* an area where accumulations of oil and water meet; the water travels in a horizontal direction beneath the oil accumulation, causing the interface to tilt.

tilt error *Navigation.* in electronic navigation systems utilizing the ionosphere-reflecting layer, the component of the ionospheric error that is due to the tilt or varying height of the ionosphere.

tilth *Agriculture.* see TILLAGE. *Agronomy.* the physical state of soil that determines its ability to grow plants, taking into account texture, structure, consistency, and pore space.

tilt hammer *Mechanical Devices.* an old-fashioned blacksmith's hammer with a pivoted shaft and projecting teeth from a revolving cam or wheel which is pressed at its shaft for lifting, then falls of its own weight. Also, TRIP HAMMER, HELVE HAMMER.

tilting dozer *Mechanical Engineering.* a bulldozer whose blade can be pivoted on a horizontal center pin to cut low on either side.

tilting fillet see EAVES BOARD.

tilting idlers *Mechanical Engineering.* an arrangement of idler rollers that permits the carrier frame to lean slightly in the direction of belt travel and keep the belt moving on its correct course.

tilting level *Mechanical Devices.* a level having a telescoping tube that can be adjusted and leveled without the need to set the rotation axis to vertical.

tilting mixer *Mechanical Engineering.* a rotating drum used for mixing solids or liquid-solid mixtures that is tilted to discharge the contents.

tilting rotor *Aviation.* a rotor having blades that can be tilted upward.

tiltmeter *Engineering.* a device that determines changes in the inclination of the earth's surface, based upon seismologic and volcanic evidence.

tilt mold *Metallurgy.* a book-type mold whose bottom is nearly vertical at the start of pouring, but is gradually tilted back to a horizontal position during pouring in order to reduce agitation while metal is being poured into the mold.

tilt/rotate code *Engineering.* a code that tells a ball-type printing element the proper tilt and rotation angles for printing a given character.

tiltwing or **tilt wing** *Aviation.* **1.** an airplane wing designed to pivot on its longitudinal axis; used, for example, on convertiplanes to allow deactivation of the fixed wings when the rotor wings are in use. **2.** having or characterized by such wings. Thus, **tiltwing aircraft.**

timber *Forestry.* **1.** the wood of growing trees. **2.** the growing trees themselves. **3.** wood, whether growing or cut, especially when suitable for use in construction and carpentry. **4.** to fell trees. *Building Engineering.* a wooden member, usually consisting of a single piece of wood. *Naval Architecture.* any of a set of curved pieces of wood that extend upward and outward from a keel.

timber connector *Forestry.* a metallic fitting used for joining timbers, consisting of a metal plate with spikes and bolts on one or both sides.

timbered stope *Mining Engineering.* a stope in which square-set timbering or any variation thereof is used.

timber harvesting *Forestry.* the process of gathering trees that have been felled, especially in a clear-cutting operation.

timber harvesting

timbering *Forestry.* the process of felling trees. *Building Engineering.* structural work formed from timbers. Also, **timberwork.** *Mining Engineering.* **1.** the process of setting timber supports into mine shafts or underground workings. **2.** the timber structure supporting the faces of an excavation during construction.

timbering machine *Mining Engineering.* an electric machine used to saw support posts to desired lengths and then to raise and hold timber in place while the posts are being set.

timberline *Ecology.* **1.** the highest area of tree growth found in a mountain range. **2.** the northernmost extent of tree growth found in an arctic region. Also, TREE LINE.

timber mat *Mining Engineering.* the broken timber that constitutes the roof of an ore deposit being extracted by a caving method.

timber puller *Mining Engineering.* a machine for removing the supports or timbers in a mine.

timber trolley *Mining Engineering.* a strong, low-profile carriage, having U-shaped arms and a timber or steel base mounted on wheels; used to transport timber to underground workings from the surface stockyard.

timbre *Acoustics.* the musical quality of a tone resulting from the combination of all the harmonics associated with the fundamental frequency.

time *Physics.* **1.** a fundamental dimensional quantity by which physical events can be placed in an ordered sequence. **2.** a given instant within this sequence that may refer to a specific event. **3.** a continuous process made up of the complete range of such instants, regarded as constantly advancing and irreversible in nature. *Horology.* an expression of the quantity of time in terms of a specific hour, minute, and so on.

time-and-altitude azimuth *Navigation.* in celestial navigation, an azimuth obtained with an assumed or given meridian angle, polar distance or declination, and altitude.

time-and-motion study *Industrial Engineering.* a study of human work processes that seeks to determine the time and individual motions required in each stage of the work. The goal of such a study is to design tasks more efficiently, so as to take less time, be less fatiguing to workers, and so on.

time-assignment speech interpolation *Telecommunications.* a modulation technique based on the fact that human speech is a series of short signals or voice spurts rather than a continuous stream of information; the periods between the speech signals are used to transmit other voice spurts to conserve channels. TASI has been used in long-distance undersea telephone-cable systems.

time azimuth *Navigation.* in celestial navigation, an azimuth obtained with an assumed or given meridian angle, polar distance or declination, and latitude.

time base *Electronics.* **1.** the ramping voltage that is developed in a cathode-ray tube sweep circuit and that determines the time duration of the sweep and the rate at which the beam traverses the screen. **2.** the duration of the horizontal trace that appears on the screen of an oscilloscope; it can usually be varied by setting front panel controls.

time bias *Electrical Engineering.* an offset in the scheduled net interchange that is proportional to the time deviation for a control area.

time break *Engineering.* a mark on a seismogram that indicates the exact detonation time of a device.

time-change component *Mechanical Engineering.* a piece of equipment, such as a jet engine, that is designed to be replaced or serviced after a given length of time.

time-code generator *Electronics.* a precision timing pulse generator with digital outputs accurately representing the time of day and the date; used in applications, such as telemetry, where accurate timing of measurements and observations is required.

time compression *Acoustics.* a decrease in the apparent time required for sound waves striking a transducer to produce an electrical signal, as in electronic processing or tape recording, in order to derive original sound frequencies.

time constant *Physics.* for the exponential growth or decay in time of a quantity Q, the amount of time required for the quantity to grow from 0 to $(1 - 1/e)Q$ or to decay from Q to Q/e, where $e = 2.71828$ is the base of the natural logarithm.

time correlation *Geology.* the correspondence, or process of determining the correspondence, in age or age relations of two or more beds of rock in separated areas.

time-current characteristics *Electricity.* a time relationship between a fuse and the root-mean-square alternating or direct current.

time delay *Physics.* the amount of time required for a signal to travel from one point to another in an electrical circuit.

time-delay circuit *Electronics.* any circuit having the capability of delaying a signal or an event by some predetermined time period.
time-delay fuse *Electricity.* a fuse that opens some time after current exceeds the fuse rating.
time-delay relay see DELAY RELAY.
time-derived channels *Telecommunications.* channels that are the direct result of time-division multiplexing of a channel.
time-distance operation *Transportation Engineering.* a procedure for maintaining headways by setting minimum intervals between block entries.
time-division multiplexing *Telecommunications.* a method of multiplexing in which two or more signals are transmitted over a common path by allowing each signal to use the path exclusively for a short period.
time-division sound *Telecommunications.* a time-division multiplexing method of transmitting the audio and video signals of a television broadcast over the same channel.
time-division switching *Telecommunications.* a technique for transmitting information from several sources or channels over the same communication link by time-sharing access to the link; each channel is serviced sequentially and allotted a specific time interval for the transmission, after which the next channel is switched on.
time-domain reflectometer *Electronics.* an instrument used to test the operation of waveguides and antennas in a microwave system by comparing radio-frequency signals injected into the transmission system with the resulting reflections, as shown on an oscilloscope screen.
timed signal service *Telecommunications.* a radio-communication service for the transmission of time signals of stated high precision, intended for general reception.
timed turn *Aviation.* the process of turning a flight vehicle for a designated interval at a constant rate of change in heading.
time fire *Military Science.* fire using munitions that are equipped with fuses set to function between firing and impact.
time formula *Industrial Engineering.* a formula used for determining the standard time for a work task, based on input factors relating to the task.
time fuse *Engineering.* a fuse that contains a timing element in order to regulate the time interval in which the fuse is allowed to function.
time gate *Electronics.* a circuit that is enabled only during specified time periods.
time-height section *Electronics.* a graph showing cloud heights over a period of time, as computed by a cloud-detection radar.
time-interval radiosonde see PULSE-TIME-MODULATED RADIOSONDE.
time-invariant system *Control Systems.* a system in which behavior remains constant over time, so that the same input will produce the same behavior at different times.
time lag *Ordnance.* the interval between the actual time it takes a bomb to fall and the time of fall of a hypothetical ideal bomb released under the same conditions.
time-lapse photography *Graphic Arts.* a method of producing a motion picture by exposing individual frames in a predetermined timed sequence, so that action taking place over a long period of time can be viewed in the space of minutes or seconds.
timelike vector *Physics.* any vector in Minkowski space having the property that the magnitude of the product of the speed of light and the time component is greater than the magnitude of the spatial component.
time line or **timeline** *Geology.* **1.** a line in a geologic cross-section or correlation diagram that represents equal geologic age. **2.** a rock unit that is represented graphically by a time line.
time-mark generator *Electronics.* a circuit or piece of test equipment that generates pulses accurately spaced at specific time intervals; usually for calibration or measurement of time intervals on a cathode-ray tube screen.
time measurement *Horology.* the measurement of the duration between regularly repeating phenomena; the period between repetitions may be subdivided, such as dividing a day based on the rotation of the earth into hours, minutes, and seconds; alternatively, the repetitions may be added to form a basic unit of measurement, such as defining the atomic second as 9.192,631,770 oscillations of the cesium atom.
time meridian *Astronomy.* the standard longitude for a given time zone.
time modulation *Telecommunications.* a type of modulation in which the time of occurrence of a specific part of a waveform varies in accordance with a modulating signal.
time-of-day clock see DAY CLOCK.
time of delivery *Telecommunications.* the time at which an addressee receipts for a message.
time of flight *Mechanics.* the elapsed time between the firing of a projectile and the instant at which it reaches the target or explodes.
time-of-flight mass spectrometer *Spectroscopy.* a mass spectrometer in which charged particles are separated by mass on the basis of the differences in their velocities.
time of origin *Telecommunications.* the time at which a message is sent.
time of periapsis *Astronomy.* the moment when an orbiting object comes nearest the center of attraction.
time of perihelion *Astronomy.* the time when an object in solar orbit comes closest to the sun.
time of receipt *Telecommunications.* the time at which a receiving station completes the reception of a communication.
time on target *Military Science.* **1.** in artillery and naval gunfire support, a method of firing in which a number of units time their fire so that the initial rounds strike the target simultaneously. **2.** a measure of the ability of a gunner or fire-control system to maintain a weapon's aim on a target. **3.** see TIME OVER TARGET, def. 1.
timeout or **time out** *Behavior.* a term for the time, usually expressed in minutes, during which a particular behavior characteristically does not occur. *Robotics.* a method of debugging robotic software in which the robot is halted when there is a problem with the program and continues when it is corrected.
time-out technique *Behavior.* a behavior modification technique that entails the isolation or ignoring of an individual after the occurrence of an inappropriate or problem behavior.
time over target *Military Science.* **1.** the scheduled time or the actual time at which aircraft attack or photograph a target. **2.** the time at which a nuclear detonation is planned at a specified ground zero.
time-over-target conflict *Military Science.* the scheduling of two or more delivery vehicles such that their proximity violates established separation criteria for yield, time, and distance.
timepiece *Horology.* any mechanical, electric, or electronic watch or clock designed to measure and display the passage of time.
time projection chamber *Nucleonics.* a large-scale particle detector used in heavy-ion experiments in which several thousand gaps are arrayed to give position and drift time information from an ionizing event.
time pulse distributor *Electronics.* a circuit designed to distribute timing signals to the appropriate user circuit or device in a predetermined sequence.
timer *Electronics.* any device, circuit, or equipment designed to control the operation of another device, circuit, or equipment at predetermined times or intervals, such as the on-off control of home lights or lawn sprinklers, the start of transmission and synchronism of indicator scopes in a radar system, or the automatic sequencing of launch countdown and flight operations of satellites and rockets. *Mechanical Engineering.* an element that controls the timing of the ignition spark in an internal-combustion engine. *Computer Technology.* a clock device that is designed to change its value or generate an interrupt after a predetermined time interval.
timer clock *Computer Technology.* a computer device that measures the interval between events.
time reference scanning beam *Navigation.* **1.** a radio or radar beam that sweeps a sector of space such that the interval between reception of successive pulses by an aircraft indicates the craft's altitude or azimuth. **2.** a system that generates such beams at microwave frequencies.
time-resolved laser spectroscopy *Spectroscopy.* a spectroscopic technique in which a sample is exposed to extremely short-duration (subnanosecond or subpicosecond), intense pulses of laser light in order to study transient phenomena when light interacts with matter.
time-resolved spectrometry *Spectroscopy.* the study of spectra in which measurements are made in the time domains of nanoseconds and even picoseconds, thus allowing the observation of processes such as recombination, relaxation, and excitation.
time reversal *Physics.* a transformation operating on time in which time t is replaced by $-t$.
time-reversal invariance *Physics.* a principle stating that a quantity remains unchanged under a time-reversal transformation.
time-rock unit see TIME-STRATIGRAPHIC UNIT.
time scale *Computer Programming.* in an analog-computer simulation, the ratio of the duration of an event, as simulated by the computer, to the actual duration of the event in a natural or experimental system. Also, **time-scale factor.**

time-sensitive target *Military Science.* a target that requires immediate response because it poses a clear and present danger to friendly forces or because it affords only a limited time of response.
time series *Statistics.* a set of observations of values in chronological order that is used to determine the effect of time on the values observed; frequently used to predict future or past values outside the observable time span.
time-series analysis *Industrial Engineering.* a forecasting technique involving the study of trends and patterns in past consumer behavior. *Mathematics.* application of the methods of stochastic processes to data sampled at (usually regular) time intervals.
time-share *Computer Programming.* to share the use of a processor or device by alternating use of it.
time-shared amplifier *Electronics.* an amplifier used for amplification of signals from several sources or channels at different times by time-sharing access to the amplifier; a synchronous switch allows each channel to be serviced sequentially and allotted a specific time interval for amplification, after which the next channel is connected to the amplifier.
time-sharing *Computer Programming.* a technique of organizing use of a computer so that it can be accessed by several users for different purposes simultaneously.
time sight *Navigation.* in celestial navigation, a traditional sight taken in the morning or afternoon to determine longitude; with known or assumed altitude, latitude, and declination, the local hour angle is obtained and compared with the Greenwich hour angle.
time signal *Telecommunications.* in radio, an accurate time-of-day signal that is broadcast at regular intervals. *Transportation Engineering.* a signal that controls train speed by regulating the time spent in an approach block.
time slice *Computer Programming.* an interval of processing time allocated to a task so that each task will be served and no one task can monopolize the processor.
time slicing *Computer Programming.* an operating-system technique to allow multitasking by allocating time slices to user or application tasks.
time standard *Horology.* **1.** a natural phenomenon that occurs with such regularity that the period between the occurrences can be used as the fundamental value in a timekeeping system. **2.** a fundamental value, such as the atomic second, by which other timekeeping systems and devices are calibrated. *Industrial Engineering.* an established standard duration for the completion of a work task, used in setting production schedules and in some worker compensation schemes.
time-stratigraphy see CHRONOSTRATIGRAPHY.
time study see TIME-AND-MOTION STUDY.
time switch *Engineering.* any switch controlled by a timing device.
time-temperature-transformation diagram see TTT DIAGRAM.
time tick *Telecommunications.* **1.** a radio-broadcast time signal that consists of short, audible beats. **2.** the time mark output of a time-controlled system.
time-transgressive *Geology.* see DIACHRONOUS.
time-varying system *Control Systems.* a system in which behavior changes over time so that the same input will induce different behavior at different times.
time zone *Astronomy.* any of the 24 bands of longitude, each about 15° wide, into which the globe is divided and within which the same civil time is generally observed.
timing *Engineering.* the process of observing and recording the duration of an action or series of actions. *Mechanical Engineering.* the process of setting the valve-operating mechanism of an engine so that the valves open and close in correct relation to the crank during the cycle to produce maximum power. *Ordnance.* the adjustment of a firearm so that it will function according to a predetermined cycle.
timing belt *Mechanical Engineering.* a positive drive belt with transverse teeth on the inner side that prevents slippage and allows for accurate timing and synchronized operation of machine parts.
timing-belt pulley *Mechanical Engineering.* a pulley having grooves that accommodate the traverse teeth of a timing belt.
timing chain *Mechanical Engineering.* a chain that imparts motion from the crankshaft to the camshaft of an internal-combustion engine.
timing error *Computer Programming.* an error due to a mismatch between input/output speed and a program's ability to process the data.
timing gears *Mechanical Engineering.* a gear train that relates camshaft speed to crankshaft speed in a reciprocating engine.
timing motor *Electricity.* in timing and clock mechanisms, a motor that functions from an alternating-current power system synchronously with the alternating-current frequency. Also, CLOCK MOTOR.
timing relay *Electricity.* a motor-driven time-delay relay; used primarily to introduce a definite time delay in the performance of a function.
timing screw *Horology.* in a timepiece with a balance escapement, any of several screws in the rim of the balance wheel that function as weights and are used to adjust the center of balance of the wheel, causing it to oscillate evenly and keep accurate time.
timing signal *Electronics.* **1.** any signal intended to provide timing control or to synchronize the operation of an equipment or function. **2.** any signal that provides an indication of the time an event occurs. *Computer Technology.* see CLOCK PULSES.
Timken film strength *Mechanical Engineering.* a test that determines the amount of pressure a film of gear lubricant will withstand before rupturing.
Timken wear test *Mechanical Engineering.* a test that determines the abrasiveness of a gear lubricant.
TIMM *Artificial Intelligence.* a program used to aid in designing and constructing expert systems, both rule-based and infer-from-example types.
Timmiaceae *Botany.* a Northern Hemisphere monogeneric family of large, circumboreal, dull or shiny matted mosses belonging to the order Bryales, characterized by stiff erect stems with terminal sporophytes, an unusual endostome of 64 filaments, and lanceolate leaves arising from a reddish or hyaline sheathing base.
timothy *Botany.* the popular name for *Phleum pratense,* a hay grass of the order Cyperales, having cylindrical spikes, and cultivated as animal fodder.
tin *Chemistry.* a metallic element having the symbol Sn (from the Latin *stannum*), the atomic number 50, and an atomic weight of 118.7; a post-transition metal in the periodic table. Tin has two allotropic forms: the common beta form, a silvery-white ductile solid, which changes slowly to the alpha form, a brittle gray solid, at 180°C; used as solder and in alloys and in tin plate. So-called **tin cans** are thinly coated with this metal rather than composed entirely of it. (From an Old English word; tin was mined in areas of Cornwall, England, in ancient times.)
Tinamidae *Vertebrate Zoology.* the tinamous, a family of birds composing the order Tinamiformes, characterized by a gallinaceous appearance, short-distance flight, and a ratitelike palate; native to Mexico, Central America, and South America.
Tinamiformes *Vertebrate Zoology.* an order of Central and South American quail-like and chickenlike birds.
tinamou *Vertebrate Zoology.* any of various species of fowl-like, flightless birds of the family Tinamidae; characterized by strong running, long legs; short, weak wings; a plump brown body; and beautifully colored eggs.
Tinbergen, Nikolaas 1907–1988, Dutch-born British zoologist; shared the Nobel Prize for research on the social behavior of animals.
tin brass *Materials.* a copper-zinc alloy with tin added to improve corrosion resistance and color; used as an inexpensive material for making springs.
tin bromide or **tin bisulfide** **1.** see STANNIC BROMIDE. **2.** see STANNOUS BROMIDE.
tin bronze *Inorganic Chemistry.* see STANNIC SULFIDE. *Metallurgy.* any of various copper-based alloys containing 2–20% tin; commonly used as foundry alloys, often modified by the addition of nickel or phosphorus.
tinc. or **tinct.** tincture.
tincal see BORAX.
tincalconite *Mineralogy.* $Na_2B_4O_5(OH)_4{\cdot}3H_2O$, a white trigonal mineral occurring as a fine-grained powder having a specific gravity of 1.88 and an undetermined hardness; found as a hydration or dehydration product of other borates in alkaline lake deposits. Also, MOHAVITE OCTAHEDRAL BORAX.
tin chloride **1.** see STANNIC CHLORIDE. **2.** see STANNOUS CHLORIDE.
tin chromate see STANNIC CHROMATE.
tin crystals see STANNOUS CHLORIDE.
tinctorial *Materials.* of or relating to coloring or dye.
tinctorial strength *Materials Science.* the coloring power of a pigment, measured by the mass of the pigment required to tint a given mass of a particular white pigment to a standard shade; also related to particle size and refractive index.
tincture *Materials.* **1.** a substance that dyes or colors. **2.** a slight amount of a substance in a solution or mixture. **3.** to dye, color, or infuse with such a substance.
tincture of iodine *Pharmacology.* a mixture of iodine and sodium iodine in diluted alcohol, used to protect against possible infections of the skin.

tinderbox *Mechanical Devices.* a metal box used to hold kindling materials.
tin dibromide see STANNOUS BROMIDE.
tin dichloride see STANNOUS CHLORIDE.
tin difluoride see STANNOUS FLUORIDE.
tin dioxide see STANNIC OXIDE.
tin disulfide see STANNIC SULFIDE.
tinea *Medicine.* the medical name for ringworm, a group of common fungal skin diseases.
tinea capitis *Medicine.* ringworm of the scalp caused by various species of *Microsporum* and *Trichophyton*; characterized by circular, bald patches with slight erythema, scaling, and crusting.
tinea corporis *Medicine.* ringworm of the skin, characterized by the presence of one or more scaly macules and central healing, most common in hot, humid climates.
tinea favosa see FAVUS.
tinea pedis *Medicine.* the medical name for athlete's foot, a chronic superficial fungal infection of the skin of the foot, especially between the toes and on the soles.
tinea unguium see ONYCHOMYCOSIS.
Tineidae *Invertebrate Zoology.* a family of small moths, lepidopteran insects in the superfamily Tineoidea, with narrow, dull-colored, fringed wings, including clothes moths.
Tineoidea *Invertebrate Zoology.* a superfamily of moths having narrow, fringed wings; it includes the leaf miners, clothes moths, and carpet moths.
tin fluoride see STANNOUS FLUORIDE.
tinfoil *Materials.* a long, thin sheet of tin or tin-lead alloy, usually dispensed in the form of a roll and widely used as a wrapping for foods and other materials.
Ting, Samuel Chao Chung born 1936, American physicist; shared the Nobel Prize for the independent discovery of the J (or psi) particle.
Tingidae *Invertebrate Zoology.* the lace bugs, a family of hemipteran insects in the superfamily Tingoidea, having a raised, lacy pattern on dorsal surfaces; includes some plant pests.
tingling *Neurology.* a prickling sensation, usually repetitive, caused by cold or stimulus to a nerve, or as a result of various diseases of the nervous system.
Tingoidea *Invertebrate Zoology.* a superfamily of true bugs, hemipteran insects in the subdivision Geocorisae.
tin hydride *Inorganic Chemistry.* SnH_4, an unstable gas that decomposes at 145–150°C and boils at –52°C. Also, STANNANE.
tin monosulfide see STANNOUS SULFIDE.
tinned wire *Metallurgy.* a steel wire having a thin coating of tin, which is obtained by passing the wire through molten tin.
tinner's mallet *Mechanical Devices.* a hammer made of wood or rubber, used in sheet metal work for flattening or shaping.
tinner's rivet *Mechanical Devices.* a flat-headed metal fastener used in sheet metal work.
tinning *Metallurgy.* the process of coating metal with a thin layer of molten filler metal.
tinnitus *Medicine.* a continuous or repetitious noise in the ears, such as ringing, buzzing, roaring, or clicking.
tin oxalate see STANNOUS OXALATE.
tin oxide **1.** see STANNIC OXIDE. **2.** see STANNOUS OXIDE.
tin perchloride see STANNIC CHLORIDE.
tin peroxide see STANNIC OXIDE.
tin pest *Metallurgy.* the disintegration of tin to a gray powder at low temperatures. Also, **tin disease.**
tinplate or **tin plate** *Metallurgy.* sheet iron or steel that is cleaned by pickling in acid, then submerged in a molten tin coating.
tin protochloride see STANNOUS CHLORIDE.
tin protosulfide see STANNOUS SULFIDE.
tin protoxide see STANNOUS OXIDE.
tin pyrites see STANNITE.
tin salt see STANNOUS CHLORIDE.
tinsel cord *Electricity.* a metal film strip interwoven with fabric threads about a strong central cord; often used in headphones.
tin snips *Mechanical Devices.* short, heavy shears used to cut sheet metal.
tin sulfate see STANNOUS SULFATE.
tin sulfide see STANNOUS SULFIDE.
tin sweat *Metallurgy.* beads of a tin-rich, low-melting temperature alloy that appear on the surface of bronze castings that were poured from a melt containing hydrogen.
tint *Optics.* a color variation characterized by low saturation, produced by the addition of white to a pure color. *Graphic Arts.* **1.** to add color to a photographic positive or negative. **2.** a tint block.
tint block *Graphic Arts.* an area that is printed in a single color, usually without detail, to be surprinted in black; a colored background on a printed page. See page 2204 of this book for an example.
tin tetrabromide see STANNIC BROMIDE.
tin tetrachloride see STANNIC CHLORIDE.
tin tetraiodide see STANNIC IODIDE.
tinticite *Mineralogy.* $Fe_4^{+3}(PO_4)_3(OH)_3{\cdot}5H_2O$, a creamy-white monoclinic mineral, having a specific gravity of about 2.8 and a hardness of about 2.5 on the Mohs scale; found as earthy masses coating walls in a limestone cave at *Tintic*, Utah.
Tintinnida *Invertebrate Zoology.* an order of ciliated protozoans in the subclass Spirotrichia, having a cylindrical shell and a conical or trumpet-shaped cell.
tint of passage *Optics.* a color that appears in a transparent plate placed between crossed polarizing filters when the plate has the property of rotating the plane of polarization of the light passing through it.
tintometer *Optics.* an instrument that measures the strength of a colored fluid by comparing the color to a set of standardized transparencies or solutions.
tinzenite *Mineralogy.* $(Ca,Mn^{+2},Fe^{+2})_3Al_2BSi_4O_{15}(OH)$, a brown, blue, yellow, green, or gray triclinic mineral with a distinct cleavage and highly glassy luster, having a specific gravity of 3.29 and a hardness of 6.5 to 7 on the Mohs scale.
Tioughniogan *Geology.* a North American geologic stage of the Middle Devonian period, occurring after the Cazenovian and before the Taghanican.
tip an apex or extremity; specific uses include: *Anatomy.* a pointed extremity of a body part such as the nose or tongue. *Design Engineering.* a small piece of material at the point or extremity of an object, designed to be an end, cap, or cutting edge. *Electricity.* the end of a probe, plug, or jack forming the contacting point when it touches a designated port or receptacle. *Mechanical Engineering.* in welding, the part of the torch where the gas burns.
tipburn *Plant Pathology.* a disease common to cultivated plants such as lettuce and potatoes, caused by a rapid and excessive loss of water and characterized by browning of leaf margins.
tip clearance *Naval Architecture.* the distance between the rotating tips of a propeller and the nearest point on the hull shell.
Tiphiidae *Invertebrate Zoology.* a family of small, black hairy wasps, hymenopteran insects belonging to the superfamily Scolioidea; the larvae are ectoparasites of beetle larvae and some are used in biological control of pests.
tipi see TEPEE.
tip-in *Graphic Arts.* **1.** a piece of material that is printed separately from the larger work and then pasted or stapled into the work during binding, such as a subscriber card in a magazine or a four-color insert in a one-color book. **2.** to paste or otherwise attach a printed sheet within a printed work. Also, **tip in.**
ti plasmid *Molecular Biology.* a tumor-producing plasmid that is found in the soil bacterium *Agrobacterium tumefaciens* and that causes crown gall disease of broad-leaved plants.
tip-of-the-tongue phenomenon *Psychology.* the inability to recall a name or word that one senses one knows.
tipper *Mining Engineering.* **1.** a haulage vehicle with a body that can be raised at one end or side to discharge its load. Also, **tipcart. 2.** an apparatus used to turn mine cars upside down and then right side up, so that their loads may be discharged with a minimum of manual labor.
tipping-bucket rain gauge *Engineering.* a recording rain gauge that empties itself by tipping after filling to a set level and then rights itself; each tip is recorded and the total number of tips corresponds to the amount of precipitation.
tipple *Mining Engineering.* **1.** the location at which mine cars are tipped and their coal discharged. **2.** a cradle dump. **3.** see TIPPER, def. 2.
tip side *Electricity.* the conducting tip of a plug or the tip spring of a jack. Also, **tip wire.**
Tipulidae *Invertebrate Zoology.* the crane flies or daddy-longlegs, a family of dipteran insects in the series Nematocera that are slender and long-legged, resembling giant mosquitoes.
tire *Mechanical Engineering.* a continuous ring of rubber or other material, usually filled with air, that encircles the rim of a wheel of an automobile, truck, or other vehicle, and that serves to support the weight of the vehicle, absorb shock, provide traction, and so on.

tire aspect ratio or **tire profile** *Mechanical Engineering.* see ASPECT RATIO.

tire iron *Mechanical Devices.* a crowbarlike tool having ends with various shapes, used to remove a pneumatic tire from its rim.

Tiros *Space Technology.* any of a series of meteorological satellites first launched by the United States in 1960; designed to take television pictures of cloud cover and transmit them on ground command. (An acronym for television infrared observation satellite.)

tirril burner *Engineering.* a type of Bunsen burner that allows for variations in the air-gas mixture.

Tischenko reaction *Organic Chemistry.* a reaction that produces an ester by using an aluminum-alkoxide catalyst to cause two molecules of an aldehyde to undergo an oxidative-reduction condensation.

Tiselius, Anne 1902–1971, Swedish biochemist; Nobel Prize for research on serum proteins and isolation of mouse paralysis virus.

Tissierella *Bacteriology.* a proposed genus of bacteria that would include some species currently classified in *Bacteroides.*

tissue *Histology.* a group of similarly specialized cells that perform a common function; tissues compose the organs and other structures of living organisms. *Materials.* see TISSUE PAPER. *Textiles.* a light, gauzy woven fabric. (From a French word meaning "to weave;" related to the word *textile.*)

tissue culture *Cell Biology.* a technique for the in vitro maintenance, proliferation, and sometimes differentiation of plant or animal cells. *Agriculture.* the commercial use of this technique for plant propagation.

tissue dose *Nucleonics.* the dosage of X-ray or gamma-ray radiation absorbed by tissue in a particular region of interest; usually expressed in units of roentgens.

tissue equivalent material *Radiology.* a material whose radiation absorption and scattering properties simulate closely that of a specified human tissue.

tissue paper *Materials.* a very thin, gauzy, translucent paper used for wrapping, protecting, and tracing.

tissue plasminogen activator *Enzymology.* an enzyme that catalyzes the conversion of plasminogen into the active form plasmin, which is responsible for the destruction and removal of fibrin clots.

tissue typing *Immunology.* a procedure used to determine the presence of histocompatibility antigens, in which the cytotoxic effect of antisera or monoclonal antibodies on white blood cells is observed.

tit *Vertebrate Zoology.* **1.** see TITMOUSE. **2.** any of various other small birds.

Titan *Astronomy.* Saturn VI, the largest and brightest moon of Saturn, discovered in 1655; it has a diameter of 5150 kilometers and a dense atmosphere that is rich in hydrocarbons. *Ordnance.* a two-stage, rocket-powered intercontinental ballistic missile fueled by liquid propellant and equipped with an all-inertial guidance and control system; it carries a nuclear warhead and is designed for deployment in hardened and dispersed underground silos; officially designated **LGM-25C.** (From the *Titans,* the earliest group of gods in Greek mythology.)

titanate *Inorganic Chemistry.* any of a group of salts formed by heating titanium dioxide with metal oxides or hydroxides; these salts are divided into metatitanates and orthotitanates.

titanaugite *Mineralogy.* a titanium-bearing variety of the mineral augite, a member of the pyroxene group.

titanello or **titanellow** see TITANIUM TRIOXIDE.

titania or **titanic anhydride** see TITANIUM DIOXIDE.

Titania *Astronomy.* Uranus III, a moon of Uranus discovered in 1787; 1580 kilometers in diameter, it orbits 436,000 kilometers from Uranus.

titanic *Chemistry.* relating to or containing titanium, especially in its tetravalent state.

titanic acid or **titanic hydroxide** *Inorganic Chemistry.* H_2TiO_4, white crystals; decomposes in water and at melting point; used as a mordant in dyeing. Also, METATITANIC ACID.

titanic chloride see TITANIUM TETRACHLORIDE.

titanic iron ore see ILMENITE.

titanite *Mineralogy.* $CaTiSiO_5$, a sometimes weakly radioactive, monoclinic mineral occurring as wedge-shaped crystals and compact masses of various colors having a specific gravity of 3.45 to 3.55 and a hardness of 5 to 5.5 on the Mohs scale; found as an accessory mineral in igneous rocks, and in schists and gneisses. Also, SPHENE.

titanium [tī tā′nē əm] *Chemistry.* a metallic element of the first transition series having the symbol Ti, the atomic number 22, an atomic weight of 47.88, a melting point of 1660°C, and a boiling point of 3290°C; a silvery solid or dark gray powder; used in alloys, in powder metallurgy, and in production of pure hydrogen. (From the word *Titan.*)

titanium boride *Inorganic Chemistry.* TiB_2, hexagonal crystals; melts at 2900°C; highly resistant to oxidation; used in metallurgy and as an electrical conductor.

titanium butylide see TETRABUTYL TITANATE.

titanium carbide *Inorganic Chemistry.* TiC, gray metallic cubes that are insoluble in water; melts at approximately 3140°C and boils at 4820°C; used in the production of cutting tools and in coating dyes.

titanium dichloride *Inorganic Chemistry.* $TiCl_2$, light brown to black deliquescent hexagonal crystals; decomposes in cold water, sublimes at its melting point, and decomposes at 475°C; a dangerous fire hazard. Also, **titanium chloride.**

titanium dioxide *Inorganic Chemistry.* TiO_2, white rhombic crystals that are insoluble in water and acid; used as a pigment. Also, TITANIUM WHITE, TITANIA, TITANIC ANHYDRIDE.

titanium hydride *Inorganic Chemistry.* TiH_2, a gray powder; decomposes at 400°C; a dangerous fire hazard, the dust may explode when exposed to oxidants; used in metallurgy and as a hydrogenation agent.

titanium nitride *Inorganic Chemistry.* TiN, yellowish-bronze cubic crystals that are insoluble in water; melts at 2930°C; used in semiconductors, alloys, and rectifiers.

titanium oxalate *Organic Chemistry.* $Ti_2(C_2O_4)_3 \cdot 10H_2O$, yellow prisms that are soluble in water and insoluble in alcohol and ether; used in making titanic acid and titanium metal. Also, **titanous oxalate.**

titanium sulfate *Inorganic Chemistry.* $TiOSO_4$, a white or slightly yellow powder that decomposes in cold water; used in dyeing, pigment production, and laundry chemicals. Also, BASIC TITANIUM SULFATE, TITANYL SULFATE.

titanium tetrachloride *Inorganic Chemistry.* $TiCl_4$, a toxic, colorless liquid that is soluble in cold water and alcohol; decomposes in hot water; melts at –25°C and boils at 136.4°C; a strong irritant; used in smokescreens, in pigments, and as a catalyst. Also, TITANIC CHLORIDE.

titanium trichloride *Inorganic Chemistry.* $TiCl_3$, dark-violet deliquescent crystals that are soluble in water and very soluble in alcohol; decomposes at 440°C; used as a reducing agent and a cocatalyst. Also, **titanous chloride.**

titanium trioxide *Inorganic Chemistry.* TiO_3, a yellow powder that is soluble in acid, used in dentistry and ceramics. Also, TITANELLO, PERTITANIC ACID.

titanium white see TITANIUM DIOXIDE.

Titanoideidae *Paleontology.* a monogeneric family of mammals in the extinct order Pantodonta; the rhinoceros-size *Titanoides* was an early pantodont of North America; extant in the Paleocene.

titanous sulfate *Inorganic Chemistry.* $Ti_2(SO_4)_3$, a green crystalline powder; insoluble in water and alcohol; used in textile production.

titante ceramics *Materials.* a ceramic material having a high dielectric constant, composed of a mixture of titanium dioxide and the oxide of a metal.

titanyl sulfate see TITANIUM SULFATE.

titer *Chemistry.* **1.** a standard test for waxes and oils in which the solidifying point of the fatty acids resulting from saponification is determined. **2.** the reacting strength or concentration of a solution as determined by titration with a standard. *Virology.* **1.** the concentration of a particular virus in a sample, determined by bioassay. **2.** the concentration of an antibody relative to antigens in a sample, determined by endpoint titration. Also, TITRE.

Tithonian *Geology.* the southern European equivalent of the Portlandian stage: Upper Jurassic, after the Kimmeridgian and before the Berriasian of the Cretaceous.

Titius, Johann 1729–1796, German mathematician; formulated Bode's Law (Titus-Bode Law).

Titius-Bode law see BODE'S LAW.

title page *Graphic Arts.* the page at the beginning of a book that gives the title, the author or editor's name, and the name and location of the publisher.

titmouse *Vertebrate Zoology.* a small, very active, social songbird of the family Paridae and the genus *Parus,* having a rounded head and fluffy underbelly; found in North America. Also, TIT.

titrant *Analytical Chemistry.* a solution of precisely known concentration, used for titrations against a sample of unknown concentration.

titration *Analytical Chemistry.* any of a number of methods used to determine the concentration of a test substance in a solvent, performed by adding a standard solution to the test solution. The end point (e.g., indicated by a certain reaction, such as a color change) is used to calculate the composition of the sample.

titre see TITER.

titubation *Neurology.* an unsteadiness of gait, often accompanied by tremors of the trunk and head, commonly seen in cerebellar disease.

tityra *Vertebrate Zoology.* any of several songbirds of the genus *Tityra,* characterized by gray, white, and black plumage and a large swollen bill; found in the American tropics.

T junction *Electronics.* **1.** an electrical connection of three branches in a manner that either physically or schematically resembles the letter T. **2.** a waveguide junction in which one waveguide joins another at right angles in the form of the letter T. Also, TEE JUNCTION.

TKD tokodynamometer.

TKG tokodynagraph.

Tl the chemical symbol for thalium.

TLC total lung capacity; thin-layer chromatography.

TL dating see THERMOLUMINESCENT DATING.

TLH *Aviation.* the airport code for Tallahassee, Florida.

TLV *Aviation.* the airport code for Tel Aviv-Yaffa, Israel.

T lymphocyte *Immunology.* a cell derived from the thymus that plays a major factor in a variety of cell-mediated immune reactions. Also, T CELL.

TM trademark.

Tm *Chemistry.* the symbol for thulium. *Biochemistry.* a symbol used to denote DNA's melting temperature.

tm melting temperature.

TMA trimethylamine.

TMEDA tetramethylethylenediamine.

T method *Botany.* a budding method in which a T-shaped cut is made through the bark at the innernode of the stock, and the scion, separated and removed from the xylem, is forced into the incision.

TMJ syndrome *Medicine.* temperomandibular joint syndrome, a symptom complex consisting of partial deafness; a stuffy sensation in the ears; tinnitus, clicking, and snapping in the temperomandibular joint; dizziness; headache; and burning pain in the ear, nose, throat, and tongue. Numerous causes have been proposed, including mandibular overclosure and stress, but some researchers question the physiological evidence to justify this syndrome. Also, MYOFASCIAL PAIN DYSFUNCTION.

TMS *Artificial Intelligence.* truth maintenance system.

TMV tobacco mosaic virus.

tn. or **tn** ton.

T network *Electricity.* a network composed of three branches, in which one end of each is connected to a central junction point and the three remaining ends are connected to an input terminal, an output terminal, and a common terminal.

TNF tumor necrosis factor.

TNT *Materials.* a shorter name for the explosive trinitrotoluene. See TRINITROTOLUENE.

TNT equivalent *Nucleonics.* a measure of the explosive potential of a nuclear bomb, expressed in terms of the weight of TNT that would release the equivalent amount of energy in an explosion; based on the figure of 10^9 calories per ton of TNT.

TO tincture of opium.

toad *Vertebrate Zoology.* the common name for any of several froglike amphibians, especially those of the family Bufonidae, characterized by warty, glandular skin, relatively short hind legs used in hopping, and often swellings containing glands that secrete an irritating fluid in defense; terrestrial or semiterrestrial in habit.

Great Plains toad

toadfish *Vertebrate Zoology.* any of several fish of the family Batrachoididae, such as *Ospanus tau*; characterized by a thick head, wide mouth, and slimy skin without scales; found near the bottom of tropical and temperate oceans.

toadstool *Mycology.* **1.** in popular use, a poisonous or inedible mushroom, as opposed to an edible one. **2.** any mushroom, especially one that has an umbrellalike cap, such as those of the genus *Amanita.* (Probably from the humorous idea that toads seated themselves on such plants.)

Toarcian *Geology.* a European geologic stage of the Lower Jurassic period, occurring after the Pliensbachian and before the Bajocian.

tobacco *Botany.* any of various plants of the genus *Nicotiania* of the nightshade family, especially *N. tabacum,* grown commercially for its leaves, which are prepared for smoking, for chewing, or as snuff.

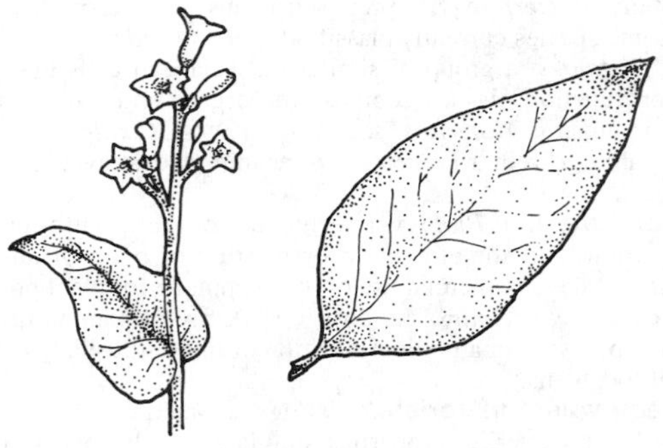

tobacco

tobacco mosaic *Plant Pathology.* any of several highly infectious virus diseases of tobacco and other plants; characterized by a dark green to yellow mottling of the foliage.

tobacco streak virus group see ILARVIRUS GROUP.

tobacco worm *Invertebrate Zoology.* the large, green larva of the hawk moth, *Manduca sexta,* having white markings and a hornlike process near the end of the body; it feeds on the leaves of tobacco and other plants of the nightshade family. Also, **tobacco hornworm.**

Tobamovirus group *Virology.* a genus of viruses that infect tobacco and other plants, having rigid rod-shaped particles that contain linear ssRNA; they are transmitted mechanically, by contact, and by seed. (From tobacco mosaic virus, the type member for this group.)

Tobravirus group *Virology.* a genus of plant viruses having rigid tubular particles that contain linear ssRNA; they have a wide host range and are transmitted primarily by nematodes. (From tobacco rattle virus, the type member for this group.)

toc- or **toco-** a combining form meaning: **1.** childbirth. **2.** labor.

tocography *Medicine.* the process of recording contractions of the uterus.

tocolysis *Medicine.* the inhibition of uterine contractions in premature labor.

tocopherol *Biochemistry.* any of the alcohols that make up dietary vitamin E; found in such foods as wheat germ oil, cottonseed oil, lettuce, spinach, and egg yolks.

α-tocopherol *Biochemistry.* $C_{29}H_{50}O_2$, 5,7,8-trimethyltocol, a tocopherol that functions as an antioxidant and is essential in mammalian food.

***d*-tocopherol** *Organic Chemistry.* $C_{29}H_{50}O_2$, a viscous, clear-yellow oil that is insoluble in water and soluble in alcohol and ether; boils at 200–210°C (0.1 torr); the most important of the vitamin E group, it is synthetically derived and used as a biological antioxidant and in meat curing and medicine.

Todd, Lord Alexander born 1907, British chemist; synthesized and analyzed vitamin B_1, nucleosides, nucleotides, and coenzymes; awarded the Nobel Prize for his work in biochemistry and genetics.

Todd-Hewitt broth *Microbiology.* a bacteriological nutrient medium containing beef-heart infusion, peptone, glucose, sodium chloride, and buffer; used with various species of *Streptococcus.*

T odd phage group *Virology.* a group of dsDNA-containing bacteriophages having characteristic noncontractile tails, including the T1, T3, T5, and T7 phages.

Todd unit *Immunology.* an eosinophilic structure found in the cytoplasm of red blood cells of certain amphibians.

Todidae *Vertebrate Zoology.* the todies, a monogeneric family of birds of the order Coraciiformes, having brilliantly colored red and green plumage; found in the West Indies.

tody *Vertebrate Zoology.* a tiny insectivirous bird belonging to the genus *Todus* of the family Todidae, native to the West Indies and related to the kingfisher.

toe *Anatomy.* any of the five digits of the human foot. *Zoology.* **1.** the analogous part of the foot of a vertebrate animal. **2.** the forepart of the foot or hoof.

toe any of various parts or structures thought of as comparable to the human toe; specific uses include: *Geology.* **1.** the leading margin of a thrust sheet. **2.** the lowermost portion of a slope or cliff, or the downslope end of an alluvial fan. **3.** the lowermost margin of landslide material on the undisturbed portion of a slope. **4.** see STEP, def. 2. *Civil Engineering.* the section projecting at the base of a dam or retaining wall. *Mining Engineering.* **1.** the base of the coal, ore, or overburden face in a quarry or open-pit mine. **2.** the burden of material situated between the borehole bottom and the free face. **3.** the bottom of a borehole. **4.** a small pillar of coal.

toe crack *Metallurgy.* a base-metal crack at the toe of a weld.

toecut *Engineering.* the cut obtained when utilizing a toe hole as the blast hole.

toe hole *Engineering.* a type of blasting hole that is horizontal or inclines slightly at the base.

toe-in *Mechanical Engineering.* a slight forward convergence of the front wheels of a motor vehicle that promotes vehicle stability and equalizes tire wear.

toenail *Engineering.* to drive a nail at an angle in order to connect two pieces of lumber.

toe-out *Mechanical Engineering.* the outward inclination of the front wheels of a motor vehicle that occurs during a turn.

Toepler-Holtz machine *Electricity.* a continuously charged electrical device that operates at a high voltage by electrostatic induction; now considered to be replaced by the Winhurst machine. Also, HOLTZ MACHINE.

toe-to-toe drilling *Mining Engineering.* the process of drilling large-diameter blasting holes in open pits or quarries.

tofan *Meteorology.* a violent spring storm that occurs in the mountains over Indonesia.

to-from indicator *Navigation.* a supplementary device used with an omnibearing selector to indicate whether an aircraft is flying toward or away from an omnidirection radio-range ground station. Also, SENSE INDICATOR.

tofu *Food Technology.* a soft, bland, white cheeselike food made from curdled soybean milk that is very high in protein and low in fat; originally used in Asian cookery, but now found in many Western dishes. Also, BEAN CURD.

Togaviridae *Virology.* a family of viruses that infect many kinds of vertebrates, having enveloped spherical particles that contain linear ssRNA; they can also infect arthropods, which then act as vectors.

Toggenburg *Agriculture.* a breed of milk goat, varying from light to dark brown in color, with white ears and white streaks on the face. (From *Toggenburg,* a district in Switzerland.)

toggle *Mechanical Devices.* a pin, rod, or bolt put through a loop in a rope or a link of a chain to keep it in place or to hold two ropes together. *Mechanical Engineering.* a jointed mechanism used to amplify force. *Electronics.* **1.** to change to the opposite state in a bistable device or circuit; for example, a flip-flop toggles to the opposite state when a pulse is applied to its clock input. **2.** to operate a toggle switch.

toggle bolt *Mechanical Devices.* an expansive bolt consisting of a nut with flanged wings which are pressed to the bolt and, after insertion in a thin or hollow wall, spread open through spring pressure, thus anchoring it to the wall.

toggle chain *Mechanical Devices.* **1.** a chain, consisting of a ring and toggle hook at one end and a ring at the other, that regulates the length of a binding chain for a load of logs. **2.** a chain with a small end hook that is inserted in a loose link.

toggle condition *Electronics.* either of two possible states of a bistable device or circuit.

toggle press *Mechanical Engineering.* a power press that uses a toggle joint to activate the slide.

toggle switch *Mechanical Devices.* a switch that opens or closes a circuit by moving a spring-loaded lever or rocker button from one position to another.

toile *Textiles.* a sheer linen and silk cloth.

toilet *Surgery.* the process of cleansing, as of a wound after an operation or of an obstetrical patient after childbirth. *Mechanical Devices.* a bathroom fixture used for defecation and urination.

tokamak *Physics.* a toroidal plasma device designed to maintain high plasma temperatures, densities, and confinement times.

Tokay or **Tokai** *Food Technology.* a sweet white wine of Hungarian origin made with a blend of grapes fermented by the "noble rot" technique.

token *Computer Programming.* **1.** a group of bits that can easily be distinguished from other groups within a sequence of characters. **2.** a symbol or icon representing an object in a program language. **3.** a coded electrical signal used to link computer equipment to a network. **4.** an abstract representation of an item, as in a Petri net or token ring.

token economy *Psychology.* a behavior-modification technique in which a desired behavior is immediately reinforced with a number of tokens that may later be exchanged for rewards.

token ring *Computer Science.* a form of computer network in which the computers are connected in a ring structure. A token is passed around the ring from a computer to its successor; possession of the token gives a computer the right to use the network to communicate with another computer. Also, RING ARCHITECTURE.

TOL *Aviation.* the airport code for Toledo, Ohio.

tola see TCHITOLA.

tolazoline hydrochloride *Organic Chemistry.* $C_{10}H_{12}N_2 \cdot HCl$, a white crystalline solid that is soluble in water and that melts at 173°C; used in medicine.

tolbutamide *Pharmacology.* $C_{12}H_{18}N_2O_3S$, an orally administered hypoglycemic agent used in the treatment of diabetes mellitus.

tolerance *Ecology.* **1.** the ability of an organism to survive extreme conditions. **2.** the range of an environmental condition within which an organism is able to survive. *Agronomy.* the ability of a certain crop to resist adverse conditions such as drought, extreme temperatures, or disease. *Pharmacology.* the capacity of an individual for enduring a drug or similar agent without exhibiting the expected effects. *Immunology.* see IMMUNOLOGIC TOLERANCE. *Engineering.* the allowable range of deviation from design specifications, expressed as a percentage of the nominal value. *Design Engineering.* specifically, the allowable variations in the dimensions of machine parts.

tolerance chart *Design Engineering.* a graphical representation of a sequence of operations to which a part must be manufactured with identifying specified dimensions indicating tolerance limits.

tolerance limits *Design Engineering.* the minimum and maximum values that are permitted by the tolerance of a production part, expressed in units of plus or minus dimensions according to angles, width, height, or thickness.

tolerance unit *Design Engineering.* the degree of tolerance allowed in fitting cylinders into cylindrical holes; expressed as a unit of length in micrometers.

tolerant *Agronomy.* of a plant, able to resist conditions that are harmful to growth. Often used in combinations such as **drought-tolerant** or **heat-tolerant.** *Medicine.* **1.** able to endure or resist the action of a drug, poison, or other harmful substance. **2.** manifesting low levels of immune response to a normally immunogenic substance.

tolerogen *Immunology.* a type of antigen that causes tolerance when injected.

toll *Telecommunications.* **1.** the charge for toll calls. **2.** any part of a telephone plant, circuits, or services for which toll charges are made.

toll call *Telecommunications.* any call to a region outside the area within which telephone calls are covered by a flat monthly rate or are charged on a message-unit basis.

toll center *Telecommunications.* in the United States, within the hierarchy of the long-distance network, a class-4 office in which trunks from end offices are joined to the long-distance system (toll-connecting trunks), and operators provide assistance in completing incoming calls. Also, **toll office.**

toll-connecting trunk *Telecommunications.* a trunk that connects one or more end offices to a toll office as the initial stage of concentration for intertoll traffic.

Tollen's aldehyde test *Analytical Chemistry.* a test to distinguish aldehydes from ketones by adding a solution of silver oxides in ammonia to the test sample. The silver plates out to form a mirror if an aldehyde is present; ketones either do not react or react slowly.

toll line *Telecommunications.* a telephone line that provides a link for different telephone exchanges.

toll-switching system see TRUNK EXCHANGE.

toll-switching trunk *Telecommunications.* a trunk that connects one or more end offices to a toll office as the initial stage of concentration for intertoll traffic.

toll terminal loss *Telecommunications.* on a toll connection, that part of the overall transmission loss that can be attributed to facilities from the toll center through the tributary office, to and including the subscriber's equipment.

toluelene see STILBENE.

toluene *Organic Chemistry.* $C_6H_5CH_3$, a flammable, toxic, colorless liquid; insoluble in water and soluble in alcohol and ether; boils at 110°C and freezes –94.5°C; used in explosives, high-octane gasoline, and organic synthesis. Also, METHYLBENZENE, PHENYLMETHANE.

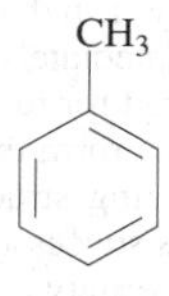

toluene

toluene 2,4-diisocyanate *Organic Chemistry.* $CH_3C_6H_3(NCO)_2$, a combustible, toxic, water-white to pale-yellow liquid that is soluble in ether, acetone, and other organic solvents; it melts at 19.4–21.5°C and boils at 251°C; it is used in making polyurethane foams and other elastomers.

***p*-toluenesulfonic acid** *Organic Chemistry.* 4-$CH_3C_6H_4SO_3H$, combustible, colorless leaflets; soluble in water, alcohol, and ether; melts at 107°C and boils at 140°C (26 torr); used as an organic catalyst, and in dyes and organic synthesis.

α-toluic acid see PHENYLACETIC ACID.

***m*-toluic acid** *Organic Chemistry.* 3-$CH_3C_6H_4COOH$, combustible, white to yellow crystals; slightly soluble in water and soluble in alcohol and ether; melts at 109°C and boils at 263°C; used as a base for insect repellants and in organic synthesis.

***o*-toluic acid** *Organic Chemistry.* 2-$CH_3C_6H_4COOH$, combustible white crystals; slightly soluble in water and soluble in alcohol and chloroform; melts at 103.5–104°C and boils at 258–259°C; used as a bacterial inhibitor.

***p*-toluic acid** *Organic Chemistry.* 4-$CH_3C_6H_4COOH$, combustible transparent crystals; slightly soluble in water and soluble in alcohol and ether; melts at 180°C and boils at 275°C; used in agricultural chemicals and as a supplement in animal feeds.

α-toluic aldehyde see PHENYLACETALDEHYDE.

***m*-toluidine** *Organic Chemistry.* $CH_3C_6H_4NH_2$, a combustible, toxic, colorless liquid; slightly soluble in water and soluble in alcohol and ether; boils at 203.3°C and freezes at –31.5°C; used in dyes and in the manufacture of organic chemicals.

***o*-toluidine** *Organic Chemistry.* $NH_2C_6H_4CH_3$, a combustible, toxic, light yellow liquid; very slightly soluble in water and soluble in alcohol and ether; boils at 200°C and freezes at –16°C; used in dyes and organic synthesis and as a vulcanization accelerator.

***p*-toluidine** *Organic Chemistry.* $CH_3C_6H_4NH_2$, combustible, toxic, white lustrous leaflets or plates; very slightly soluble in water and soluble in alcohol and ether; melts at 45°C and boils at 200°C; used as an analytical reagent and in dyes.

toluol *Chemistry.* the commercial form of toluene.

tolyl group *Organic Chemistry.* any one of three radicals that are derived from toluene, having the formula $CH_3C_6H_4^-$.

***p*-tolylsulfonylmethylnitrosamide** *Organic Chemistry.* $C_8H_{10}N_2O_3S$, a yellow crystalline solid; soluble in ether, benzene, and chloroform; melts at 62°C; used in making diazomethane and compounds for gas chromatography analysis.

tom- or **tomo-** a combining form meaning "to cut."

Tomahawk *Aviation.* a popular name for the P-40 pursuit (fighter) aircraft. *Ordnance.* a cruise missile that can be launched from land, air, ship, or submarine; as a land attack weapon, it carries a conventional or nuclear warhead; as a tactical antiship weapon, it carries a conventional warhead.

tomatine *Organic Chemistry.* $C_{50}H_{83}NO_{21}$, a white crystalline solid; soluble in alcohol and dioxane; melts at 270°C; it is a glycosidal alkaloid prepared from the leaves and stems of tomato plants; used as a fungicide and as a precipitating agent for cholesterol. Also, LYCOPERSICIN.

tomato *Botany.* **1.** any of various plants belonging to the genus *Lycopersicon* of the family Solanaceae, especially the widely cultivated species *L. esculentum*; native to Mexico and Central and South America, but now grown in many varieties in the United States, Europe, and elsewhere as a food crop. **2.** the slightly acid, pulpy, usually red fruit of the plant, which is eaten raw as a fruit or cooked as a vegetable. (From a native American word for this plant; it was originally cultivated by American Indian tribes, probably in the valleys of the Andes.)

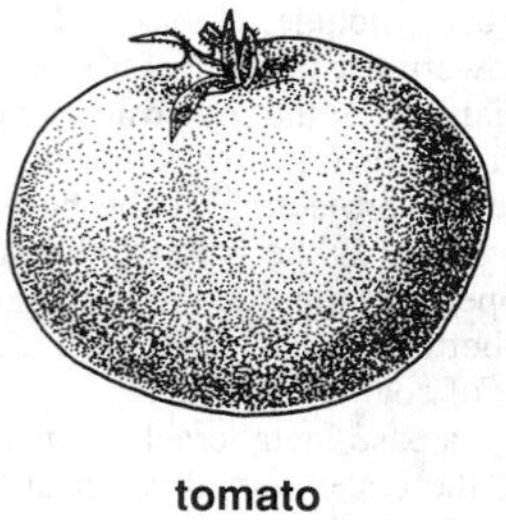

tomato

tomato paste *Food Technology.* highly concentrated tomato pulp containing about 25% to 32% tomato solids by volume.

tomato pulp *Food Technology.* skinned and seeded ground tomatoes, containing about 6% tomato solids by volume.

tomato purée *Food Technology.* concentrated tomato pulp containing about 11% to 16.5% tomato solids by volume.

Tomato spotted wilt virus group *Virology.* a genus of plant viruses having enveloped spherical virions that contain ssRNA; they are similar to bunyaviruses and are transmitted by thrips.

tomb *Archaeology.* a burial structure or chamber.

tombac *Metallurgy.* an alloy that consists of 70–92% copper with zinc and sometimes tin and other materials; used to imitate gold, usually in inexpensive jewelry. Also, **tomback, tombak.**

Tombaugh, Clyde born 1906, American astronomer; discovered the planet Pluto in 1930.

tombolo *Geology.* a sand or gravel bar that joins two islands, or connects an island to the mainland. Also, TIE BAR, CONNECTING BAR, TYING BAR.

Tombusvirus group *Virology.* a genus of plant viruses having isometric particles that contain linear ssRNA; they are found in the cytoplasm and nuclei of infected cells and are transmitted both mechanically and through soil. (From tomato bushy stunt virus, the type member for this genus.)

Tomcat *Aviation.* a popular name for the F-14 fighter interceptor.

-tome *Surgery.* a combining form referring to a surgical instrument used for cutting.

tomentose *Botany.* having a covering of dense, matted hairs.

tomentum *Anatomy.* a network of small blood vessels located between the cerebral cortex and pia mater. *Botany.* a soft down of longish, soft, entangled hairs pressed close to the surface.

tomium *Vertebrate Zoology.* a technical name for the sharp, cutting edge of a bird's beak.

Tommaselli's disease see CINCHONISM.

Tommy gun see THOMPSON SUBMACHINE GUN.

tomogram [tō´mə gram´] *Radiology.* an X-ray of a selected layer of the body made by tomography.

tomograph [tō´mə graf´] *Radiology.* an apparatus that moves an X-ray source in one direction as the film is moved in the opposite direction, so as to show a predetermined plane of tissue in detail while blurring detail in other planes.

tomographic [tō´mə graf´ik] *Radiology.* relating to or produced by means of a tomograph.

tomography [tə mäg´rə fē] *Radiology.* the recording of internal body images at a predetermined plane using a tomograph; a widely used technique for the discovery and identification of space-occupying lesions, as in the brain.

Tomonaga, Sin-itiro 1906–1979, Japanese physicist; awarded the Nobel Prize for empirical confirmation of the quantum electrodynamic theory.

Tomopteridae *Invertebrate Zoology.* the glass worms, a family of free-swimming polychaete annelids in the group Errantia, having long, branched parapodia.

-tomy *Surgery.* a combining form meaning "incision."

ton *Metrology.* **1.** a traditional unit of weight used in the United States for heavy items, equal to 2000 pounds; equivalent to 907.18 kilograms. Also, SHORT TON, NET TON. **2.** any of various similar large measures of weight, such as the long ton (2240 pounds) or the metric ton (2204.62 pounds). See LONG TON, METRIC TON. *Naval Architecture.* **1.** originally, a measure of a vessel's internal capacity based on the stowage space taken up by a large wine cask called a **tun**. **2.** in modern usage, a measure of a vessel's internal capacity, with a register ton being equal to 100 cubic feet of enclosed hull space. **3.** in the computation of a vessel's displacement tonnage, the volume of water weighing one shipping ton (2240 pounds), or 35 cubic feet. *Mechanical Engineering.* a unit of refrigerating capacity that is equivalent to the rate of latent heat extraction when one ton (2000 pounds) of ice is produced from water at 32°F in 24 hours, with the latent heat of fusion taken to be 144 BTUs per pound. Also, STANDARD TON. *Nucleonics.* a unit of energy released by one metric ton of high-explosive material, based on the figure of 1000 calories per gram or 10^9 calories per metric ton.

tonalite see QUARTZ DIORITE.

tonal language *Linguistics.* a language in which words that are otherwise phonetically identical are distinguished from each other by their tone, or pitch; e.g., Vietnamese. Also, TONE LANGUAGE.

tonA protein *Biochemistry.* an outer membrane protein in *Escherichia coli* that acts as a receptor; it is encoded by the *tonA* gene and participates in the ingestion of albomycin and ferrichrome. Also, FHUA PROTEIN.

tonB protein *Biochemistry.* a periplasmic protein in *Escherichia coli* that apparently acts as a transmitter of, for example, ferric-iron-chelate complexes, vitamin B_{12}, and metabolic energy.

tondal see METRIC TON.

tone *Acoustics.* a sound oscillation having a discernible pitch. *Graphic Arts.* **1.** also, **tone value.** the relative density or strength of grays between black and white. **2.** any distinguishable shade within this range.

tone-and-voice pager *Telecommunications.* in a radio-paging system, a receiver equipped with a speaker that broadcasts a short spoken message from a telephone caller.

tone burst *Acoustical Engineering.* a short wave train, 10 to 50 milliseconds in length, usually produced by a tone generator; used to test speaker response or other electroacoustical device characteristics.

tone code ranging *Navigation.* a method of transmitting a precise, high-resolution timing signal from a satellite, using an audio frequency to phase an audio oscillator in the user equipment.

tone control *Electronics.* a control, typically a potentiometer connected to one or more fixed capacitances, that is designed to change the frequency response of an audio amplifier to emphasize or deemphasize either the bass or treble tones, as desired.

tone deafness see AMUSIA.

tone dialing see PUSH-BUTTON DIALING.

Tonegawa, Susumu born 1939, Japanese immunologist; awarded the Nobel Prize for discovering how the body can produce specific antibodies against new agents.

tone generator *Electronics.* a piece of test equipment that provides an audio-frequency signal for testing electronic equipment; the output frequency and amplitude can usually be varied by setting front panel controls.

tone language see TONAL LANGUAGE.

Tonelli, Leonida 1885–1946, Italian mathematician; worked on the calculus of variations.

tone-modulated waves *Telecommunications.* waves obtained from continuous waves by amplitude-modulating at audio frequency on a regular basis.

tone modulation *Telecommunications.* a code-signal transmission that is achieved by allowing the radio-frequency carrier amplitude to vary at a set audio frequency.

tone-only pager *Telecommunications.* in a radio-paging system, a receiver that indicates a call from a specific telephone number, but does not provide any means of voice transmission.

toner *Graphic Arts.* **1.** in photography, a chemical bath used to change the color of a print. **2.** in printing, a concentrated pigment used to fortify inks. **3.** in reprographics, the electrostatic compound that acts as the "ink," forming the image on a plain-paper copy.

tone reversal *Telecommunications.* in FAX technology, a distortion of the recorder copy that causes the various shades of black and white to be out of proper order.

tongara *Meteorology.* a southeast wind that blows in the Macassar Strait, usually producing a hazy atmospheric aspect.

tong-hold *Metallurgy.* in drop-hammer and press-type forging, relating to the end portion of a forging billet that is gripped by the welder's tongs.

Tongrian *Geology.* a European geologic stage of the Lower Oligocene epoch, occurring after the Ludian of the Eocene epoch and before the Rupelian. Also, LATTORFIAN, SANNOISIAN.

tongs *Mechanical Devices.* any of various long-handled plierlike tools for gripping and holding materials or objects.

tongue *Anatomy.* the movable, muscular organ located in the mouth that is associated with taste, the manipulation of food, and the enunciation of vocalizations.

tongue any of various features or structures thought of as comparable to the human tongue, as by extending out from a larger area; specific uses include: *Geology.* **1.** a long, low-lying, narrow strip of land extending out from the mainland into a body of water. **2.** any branch or offshoot of a larger body, especially a lava flow extending from a larger flow. **3.** see LENTIL, def. 1. *Oceanography.* **1.** a projection of an ice edge, sometimes several kilometers long. **2.** a protrusion of a water mass into a different water mass, especially one of different temperature. *Hydrology.* see ICE TONGUE.

tongue and groove *Design Engineering.* a joint in which a projecting rib on the edge of one board fits into a groove in the edge of an adjacent board. Also, **tongue-and-groove joint.**

tongue-and-groove plane *Mechanical Devices.* a specially adapted plane used to cut edges of wood boards into tongues and grooves.

tongue-and-groove pliers *Mechanical Devices.* a common type of pliers having grooved jaws connected by a tongue-shaped arm.

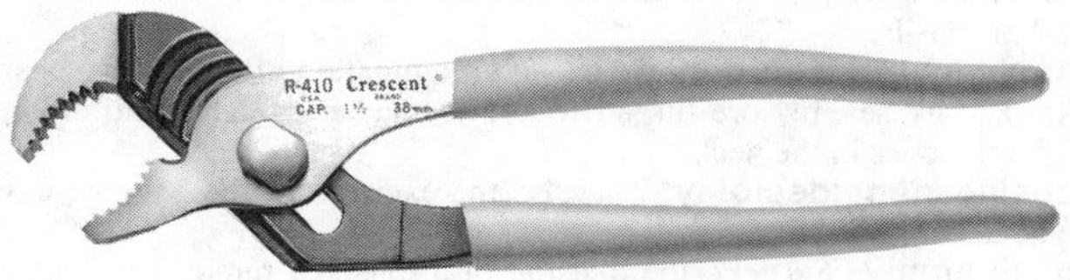

tongue-and-groove pliers

tongue-and-lip joint *Design Engineering.* a tongue-and-groove joint in which the tongue member provides a means of concealing the joint by its flush bead.

tongue depressor *Medicine.* a small, broad, thin piece of wood used to hold a patient's tongue during examination of the mouth and throat. Also, **tongue blade.**

tongue sole *Vertebrate Zoology.* any of several flatfishes of the family Cynoglossidae, having a tail that tapers to a point. Also, **tonguefish.**

tongue worm *Invertebrate Zoology.* any of a group of small tongue-shaped parasites of the phylum Pentastomida, having two pairs of hooks at the sides of the mouth; some reach adult stage in the nasal passages and sinuses of mammals.

-tonia a combining form meaning "stretching" or "tension."

tonic [tän´ik] *Neurology.* of or relating to a state or condition characterized by continuous or prologed muscle tension, without relaxation. *Medicine.* in popular use, a liquid preparation, usually nonprescription, that is regarded as having curative or enervating properties. *Acoustics.* of or relating to a tone or tones.

tonic contraction see TETANUS.

tonic epilepsy *Medicine.* an epileptic seizure characterized by generalized rigidity.

tonic labyrinthine reflexes *Physiology.* any of several responses resulting from the stimulation of stationary receptor cells in the labyrinth.

tonic neck reflexes *Physiology.* the response of receptors specialized to maintain basic neck posture.

tonic postural epilepsy *Medicine.* a form of epilepsy of uncertain cause, with seizure episodes that last from a few seconds to several minutes, characterized by continuous muscular contractions, resulting in a rigid pose, with extended arms and legs. Thus, **tonic convulsion.**

tonic spasm *Medicine.* a spasm in which there is sustained muscular contraction without relaxation.

tonic water or **tonic** see QUININE WATER.

ton-kilometer *Mining Engineering.* a unit of measurement of the work done in transporting mined material, equivalent to the product of material weight, in tons, multiplied by the number of kilometers driven.

ton-mile *Civil Engineering.* the movement of one ton of freight for a distance of one mile.

tonnage *Naval Architecture.* **1.** the traditional measure of a vessel's capacity. **2.** the internal capacity of a vessel in register tons; various formulas, with differing exemptions, are used to compute register tonnage; it is the normal measure for commercial vessels. **3.** the weight of a vessel or its equivalent in displaced seawater; it is the normal measure for naval vessels.

tonnage deck *Naval Architecture.* in older systems of tonnage measurement, the deck marking the upper side of the enclosed space considered in tonnage calculations. This was the upper deck of a two-decked ship, and the deck above the lowest deck of a ship with more than two decks.

tonne see METRIC TON.

tonneau see METRIC TON.

tonoclonic *Neurology.* characterized by muscular spasms that are intermittently clonic and tonic.

tonofibril *Cell Biology.* any of a number of fibers made up of intermediate filament proteins involved in cytoskeletal or supporting structures of a cell.

tonofilament *Cell Biology.* any of a number of cytoplasmic intermediate filaments that form an intracellular network in certain cells, especially epithelial cells; believed to be the principal precursor of keratin.

tonometer *Medicine.* an instrument that measures tension or pressure, especially intraocular pressure.

tonoplast *Botany.* a semipermeable membrane surrounding the vacuole of a plant cell.

tonsil *Anatomy.* a small, rounded mass of lymphatic tissue located in the pharynx and fauces. Also, **tonsilla.**

tonsillectomy [tän′si lek′tə mē] *Surgery.* the surgical removal of a tonsil or tonsils.

tonsillitis [tän′si li′tis] *Medicine.* an inflammation of the tonsils, characterized by severe sore throat, fever, headache, earache, and enlarged lymph nodes in the neck.

tonsilloadenoidectomy *Surgery.* the excision of the palative tonsils and adenoids.

tonsillotomy *Surgery.* the incision of a tonsil or removal of part of a tonsil.

tonstein *Geology.* an argillaceous rock predominantly composed of the clay mineral kaolinite and some detrital and carbonaceous material; it occurs as thin bands in coal seams.

tonus *Physiology.* the natural, slightly tense status of muscles, countering gravitational force and supporting the body skeleton, while being ready for motion.

tool *Mechanical Devices.* a portable and usually hand-held instrument, either unpowered or powered, that is used to increase the efficiency of a work effort; e.g., a saw, hammer, screwdriver, chisel, drill, and so on. *Archaeology.* in general, any existing physical object that is in some way fashioned or altered by humans and employed for a specific task or purpose.

tool bit *Engineering.* a piece of high-strength steel that is used as a cutting tool.

toolbox *Computer Programming.* a collection of routines designed to help a programmer in the development of software. Also, **toolkit**.

tool-center point *Robotics.* the point on the end effector or tool of a manipulator that represents the coordinates of the object being manipulated.

tool changer *Robotics.* the equipment on the wrist of a robot that can hold a number of end-of-arm tools.

tool-check system *Industrial Engineering.* a control procedure in which tools are checked out to workers as needed, and checked back in by them when done.

tool-coordinate system *Robotics.* a method of control based on the *x*, *y*, and *z* coordinates for a robot's end-of-arm tooling.

tool design *Design Engineering.* the division or department of mechanical engineering that focuses on the design of tools.

tool-dresser *Mechanical Engineering.* a toolstone-grade diamond set in a metal shank and used to trim or form the face of a grinding wheel.

tool extractor *Engineering.* a device that is used to withdraw broken drilling tools from a bore hole.

tool function *Robotics.* any programmed command that selects and controls the tools used by a robot in machining operations.

toolhead *Mechanical Engineering.* the tool-carrying part of a machine tool.

tooling *Mechanical Engineering.* the process of shaping metal with cutters, as opposed to shaping with grinders. *Robotics.* the tools and equipment that a robot uses in the process of doing its job.

tool joint *Engineering.* a device that is used as a drill pipe stem support and coupling element.

tool kit or **toolkit** *Engineering.* a box in which tools are kept. Also, **toolbox.** *Archaeology.* **1.** all the tools used by a given culture for its technology. **2.** a set of tools used together for a specific task, such as skinning and butchering an animal.

tool-length compensation *Control Systems.* the programming of machining operations so that any tools to be used are correctly positioned ahead of time.

toolmaker's vise see UNIVERSAL VISE.

tool mark *Geology.* a current mark produced on a muddy bottom by the impact or abrasive action of solid objects, such as pebbles, shell fragments, or fish bones, that are carried by the current. Also, **tool marking.**

tool offset *Mechanical Engineering.* the adjustment of a tool position in a machine to compensate for wear, finishing, or displacement.

tool path *Robotics.* the path or contour that is followed by a robotic tool in a machining operation.

tool post *Mechanical Engineering.* a clamp that holds a lathe or shaping machine tool in the slide rest or ram.

tool steel *Materials Science.* any of a group of high-carbon steels that provide combinations of high hardness, toughness, or resistance at elevated temperatures.

tool tip *Robotics.* the part of the end effector that comes in contact with the workpiece.

tool use *Behavior.* a behavior pattern in which an object is manipulated by an animal for some purpose, such as food acquisition; e.g., the use of a stone to break open shellfish by the North American sea otter.

tooth *plural,* **teeth.** *Anatomy.* any of the hard, conical structures that are embedded in the upper and lower jaw and that are used to chew food and, in the case of predatory animals, to seize and kill prey. *Invertebrate Zoology.* any of the various processes on any part of an invertebrate that function like or resemble vertebrate teeth. *Design Engineering.* **1.** any of a series of projections equally spaced on the edge or face of a gear wheel. **2.** an angled projection on a tool or other implement, such as a rake or saw. *Graphic Arts.* the texture of a paper that is especially receptive to ink or other printing agents.

toothache *Medicine.* any pain in a tooth, usually resulting from decay or disease in the tooth.

tooth decay *Medicine.* the localized destruction of a tooth surface by decalcification of the enamel, leading to the formation of a cavity which may penetrate the enamel and dentin and reach the pulp.

tooth point *Design Engineering.* the chamfered cutting edge of a work blade on a face mill.

tooth rest *Design Engineering.* a device that clamps onto a worktable and holds a tooth in position when sharpening a cutter.

tooth shell *Invertebrate Zoology.* the common name for any mollusk in the class Scaphopoda, especially those with a shell shaped like an elephant's tusk.

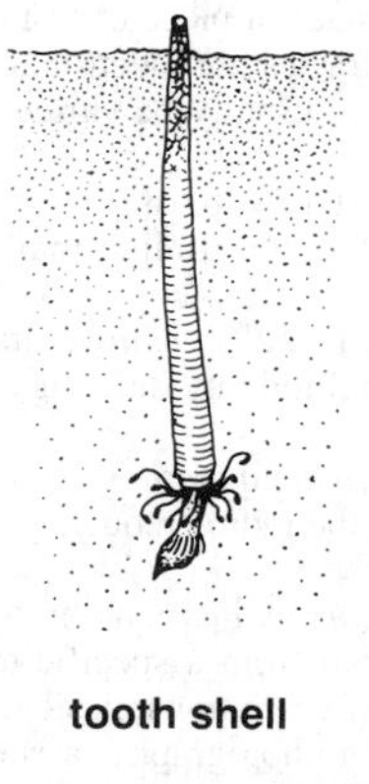

tooth shell

top *Geology.* **1.** the uppermost surface of a formation that is hit during drilling, identified by its distinctive features. **2.** see OVERBURDEN, def. 1. *Mechanics.* a body typified by the common toy top; i.e., it is solid and axisymmetric, it is fixed at some point along its axis, and it can rotate about the axis. *Particle Physics.* an as-yet unconfirmed property that is expected to characterize the sixth flavor of quark; topness is conserved in strong and electromagnetic interactions but not in weak interactions. Also, TRUTH. *Quantum Mechanics.* see ROTATOR.

top- a combining form meaning "place" or "position."
topagnosis *Neurology.* the inability to identify the site of stimulation to the body.
top-and-bottom process *Metallurgy.* a process in which copper and nickel sulfides are separated by adding sodium sulfate and carbon, causing the melt to separate into two layers with the bottom layer containing most of the nickel.
topaz *Mineralogy.* $Al_2SiO_4(F,OH)_2$, a colorless, white, pale-blue, pale-yellow, red, green, or brown orthorhombic mineral occurring as prismatic crystals and in massive form, having a specific gravity of 3.49 to 3.57 and a hardness of 8 on the Mohs scale; found in pegmatites, siliceous igneous rocks, and high-temperature quartz veins; widely used as a gemstone.
top benching *Mining Engineering.* a method by which a bench is extracted from above, as by means of a dragline.
top-blown converter *Materials Science.* a device used to convert hot metal to steel by oxidizing the carbon to carbon monoxide. Oxygen is blown vertically through a water-cooled lance down onto the surface of the melt and is consumed in the oxidation of impurities in the metal.
top dead center *Mechanical Engineering.* the dead-center position of an engine piston and its crankshaft arm when at the top or outer end of a stroke.
top-down analysis *Computer Programming.* a software development process that begins with general or high-level requirements and proceeds through progressively more detailed requirements and design.
top-down design *Industrial Engineering.* a design procedure that begins with the overall configuration or function of the finished product, and progressively divides this overall problem into subproblems for solution. *Computer Programming.* **1.** a program or system design based on top-down analysis. **2.** see DECOMPOSITION.
top-down parsing *Computer Science.* a predictive form of parsing, such as recursive descent, in which the parse tree of a statement is constructed starting at the root (sentence symbol).
top-down reasoning see BACKWARD REASONING.
top-dressing *Agronomy.* the process of applying fertilizer to the surface of a seedbed.
top grafting *Botany.* a propagation method in which the scion of one type of tree is inserted into the stock of another.
top hamper *Naval Architecture.* **1.** the upper sails of a ship, along with their gear and spars. **2.** broadly, any abovedecks rigging.
topholipoma *Oncology.* a lipoma that contains tophi.
tophus *plural,* **tophi.** *Medicine.* a chalky deposit of sodium urate occurring in gout, forming most often around joints in cartilage, bone, and bursae, and in the external ear. *Geology.* see TUFA.
topical *Medicine.* referring to a particular surface area; e.g., a topical anti-infective is applied to a certain area of the skin and affects only the area to which it is applied.
top-loaded vertical antenna *Electromagnetism.* a vertical antenna having a larger top portion so as to modify the current distribution and thus produce a more desirable radiation pattern.
topmark *Navigation.* a characteristic object located on the top of a buoy or beacon to indicate a channel or danger; for example, a cage, ball, or cone.
topminnow *Vertebrate Zoology.* **1.** a small fish of the family Cyprinodontidae that inhabits shallow waters. Also, KILLIFISH. **2.** a vivaparous freshwater and brackish water fish of the family Poeciliidae, several species of which are used in mosquito control; native to the New World.

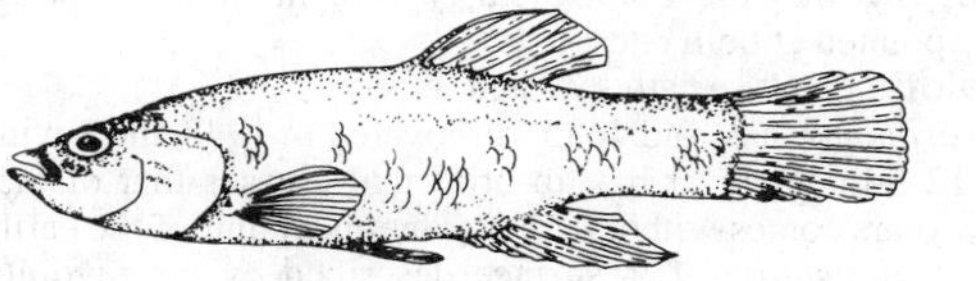

topminnow

Topo *Robotics.* the trade name of a personal or home robot.
topo- a combining form meaning "place" or "position."
topoanesthesia *Neurology.* the inability to identify the site of any tactile sensations.
topoangulator *Photogrammetry.* an instrument used to measure vertical angles in the principal plane of an oblique photograph.
topocentric *Cartography.* describing any set of measurements or coordinates that has the observer on the surface of the earth as its origin.
topocentric coordinate system *Cartography.* any set of coordinates having its origin at the observer's position on the surface of the earth; distinguished from geocentric coordinates, which have their origin at the center of the earth.
topochemical *Chemistry.* occurring in certain, usually limited, zones or fields of force.
topochemical control *Chemistry.* the inhibition of a reaction due to the structural arrangement of the atoms in at least one of the molecules.
topochemical polymerization *Materials Science.* polymerization that is activated either thermally or by ionizing radiation.
topochemical reaction *Chemistry.* a reaction that can occur only at certain sites on a molecule where reactive groups are available, as on the surfaces of a crystalline area.
topocline *Ecology.* a series of physiological or morphological characteristics of closely related organisms that correspond with a designated geographical zone.
topog. topography; topographical.
Topogon lens *Optics.* a lens with strongly curved lens elements that cover a wide field of view at a moderate photographic speed.
topographic *Geology.* relating to or affected by topography.
topographic anatomy *Anatomy.* the study of the morphological relationships among all of the structures in a specific body region, such as the thorax.
topographic climax *Ecology.* an ecological community whose stability is controlled by topographic features, such as hills, rivers, or undrained depressions.
topographic contrast *Materials Science.* a scanning electron microscope image of the morphology of a specimen produced from variations in the number of secondary electrons that emerge from the sample.
topographic deflection *Geodesy.* any portion of the deflection of the vertical that is caused by the gravitational pull of topographic masses.
topographic infancy *Geology.* see INFANCY.
topographic latitude see GEODETIC LATITUDE.
topographic map *Cartography.* a map that shows the vertical as well as the horizontal positions of surface features by use of some system of quantitative symbols (e.g., numbered contour lines); a map that shows relief as well as planimetric detail.
topographic maturity *Geology.* see MATURITY, def. 1.
topographic old age *Geology.* see OLD AGE, def. 1.
topographic relief *Geology.* see RELIEF.
topographic survey *Engineering.* a survey that determines the physical features of the surface of the earth.
topographic youth *Geology.* see YOUTH.
topography *Geology.* the configuration of a land surface, including its relief. *Cartography.* the representation of such a configuration on a contour map. Thus, **topographic.**
topoisomerase *Enzymology.* an enzyme that converts one topological isomer of DNA into another.
topological dimension *Mathematics.* any of several conventions for ascribing a dimension to a topological space; e.g., Hausdorff dimension, fractal dimension, or real dimension.
topological dual *Mathematics.* the space X' of all continuous complex-valued linear forms on a topological vector space X.
topological group *Mathematics.* a topological space X on which there is defined a binary operation $\circ$ such that the elements of X form a group under the operation $\circ$, and such that the map $\phi: X \times X \rightarrow X$ given by $\phi(x,y) = xy^{-1}$ is continuous.
topological isomer *Molecular Biology.* any of two or more covalently closed, circular DNA molecules that are of the same length but have different linking numbers.
topologically closed set see CLOSED SET.
topologically close-packed phase *Materials Science.* a microstructural phase consisting of layers of close-packed atoms separated by relatively large interatomic spaces. Such phases occur with platelike morphology and usually precipitate near grain boundaries where elements critical to their formation concentrate.
topological mapping or **topological transformation** *Mathematics.* see HOMEOMORPHISM.
topological product *Mathematics.* the Cartesian product of topological spaces. The result is a topological space, and the induced topology is the product topology.
topological property *Mathematics.* a property of a topological space X that is also true in any space homeomorphic to X.

topological space *Mathematics.* a set X on which a topology has been defined, allowing for definitions such as neighborhoods, continuity, compactness, connectedness, and so on; usually referred to as "the topological space X," or "the topological space (X, U)" if it is important to emphasize the topology U.

topological vector space *Mathematics.* a topological space X that is also a vector space, usually over the complex or real numbers, and in which the operations of vector addition and scalar multiplication are continuous.

topology *Computer Technology.* the physical pattern of connectivity within a computer network or between processors, memories, and peripherals. *Mathematics.* **1.** a collection U of subsets of a (nonempty) set X that contains both the empty set and the set X and is closed under unions and finite intersections; also called a topology on the set X. The sets in U are called open sets. The topology comprising only the empty set and X is called the **trivial topology,** and the topology comprising all subsets of X is called the **discrete topology.** If U_1 and U_2 are two topologies on X such that $U_1 \supset U_2$ (so that every open set in U_2 is also an open set in U_1), then U_2 is said to be coarser than U_1 and U_1 is finer than U_2. **2.** the study of those properties that are unaltered by homeomorphisms.

toponium *Particle Physics.* a hypothetical particle made up of a top quark bound to its antiquark.

toponym *Cartography.* a place name; the name of a natural or artificial feature of the surface of the earth. *Anatomy.* the name of a region of the body, as distinguished from the name of an organ.

topophobia see KENOPHOBIA.

topotype *Systematics.* a specimen that is collected from the same site or locality as the type series, but is not considered part of the series.

topping *Chemical Engineering.* the process of removing benzene or light distillate from heavier crude fractions.

topple *Mechanics.* a sudden change in the motion of a precessing body, occurring when the rotation rate declines below a minimum value necessary to sustain gyroscopic precession.

topple axis *Mechanics.* the axis around which topple occurs.

top quark *Particle Physics.* a hypothetical quark postulated as having a mass of approximately 30–50 GeV, baryon number 1/3, zero isotopic spin, strangeness, and charm. Also, T QUARK, TRUTH QUARK.

TOPS terminal operating system.

topset *Geology.* a nearly horizontal depositional layer of sedimentary material that overlies a growing delta. Also, **topset bed.**

top shell *Invertebrate Zoology.* the common name for any of the marine snails in the family Trochidae, having a pyramidal or cone-shaped, flat-based shell.

Topside Sounder *Space Technology.* a U.S.-Canadian meteorological satellite designed to obtain data on the ionosphere from above.

top slicing *Mining Engineering.* a stoping method for removing ore by excavating a series of horizontal or inclined timbered slices alongside each other, starting at the top of the ore body and progressing downward.

top slicing and cover caving *Mining Engineering.* a form of the top slicing mining method in which the overburden or cover is caved after a unit is mined.

topsoil *Agronomy.* the top portion of agricultural soil, usually richer in organic matter than the subsoil. *Geology.* the surface layer of any soil. Also, A HORIZON.

topspin *Mechanics.* a spinning motion that is imparted by an external source to a ball or other projectile, causing it to rotate forward. Also, OVERSPIN.

top steam *Chemical Engineering.* the steam added close to the top of a shell still to purge it, and to guard against the formation of a vacuum while the liquid contents are pumped out.

TOPV trivalent oral poliovirus vaccine.

tor *Geography.* a high rock, pile of rocks, or rocky peak. (A term of Celtic origin; one of the few indigenous, pre-Anglo-Saxon words still remaining in English.)

tor see TORR.

toral subalgebra *Mathematics.* a subalgebra of a Lie algebra consisting entirely of semisimple elements. By theorem, a toral subalgebra is Abelian.

TORAN *Navigation.* a digital radio-location receiver used with the Omega long-range navigation system.

torbanite *Geology.* a substance intermediate between oil shale and coal, having a high carbon content and considered to be a type of boghead coal. Also, KEROSENE SHALE, BITUMENITE.

torbernite *Mineralogy.* $Cu^{+2}(UO_2)_2(PO_4)_2 \cdot 8\text{–}12H_2O$, a green, strongly radioactive, tetragonal mineral of the autunite group occurring in tabular crystals and scaly aggregates, having a specific gravity of 3.22 and a hardness of 2 to 2.5 on the Mohs scale; found in association with other secondary uranium minerals. Also, CUPROURANITE.

torch *Engineering.* a lamplike gas burner producing a hot flame that is used to braze, cut, or weld.

toric surface *Mathematics.* a surface of revolution generated by rotating an arc of a circle about a coplanar line not containing the center of the circle.

tornado *Meteorology.* **1.** an intense, violently destructive windstorm consisting of a rotating column of air that descends from a cumulonimbus cloud and is almost always visible as a funnel-shaped cloud; the most destructive of all atmospheric phenomena, characterized by cyclonic winds reaching speeds of 300 miles per hour; most common in Australia and the midwestern U.S. **2.** a violent thundersquall, such as those that occur along the west coast of Africa during the summer.

tornado

Tornado *Aviation.* a supersonic, multipurpose military aircraft produced jointly by Germany, Britain, and Italy; a two-seater capable of flying in darkness and bad weather.

tornado belt *Meteorology.* a geographic region in which tornadoes are frequent, especially the lowland areas of the United States between the Rocky Mountains and the Appalachians.

tornado cloud see TUBA.

tornado echo *Meteorology.* a radar precipitation echo that frequently appears on a plan-position-indicator scope as a figure six, especially in the southwestern quadrant of a storm system; it suggests the existence of numerous tornadoes.

tornadotron *Electronics.* a device designed to produce microwave energy through interaction between an orbiting cloud of electrons and a radio-frequency field, when pulsed by a strong magnetic field.

tornaria *Invertebrate Zoology.* a free-swimming, ciliated larva of some acorn worms (Enteropneusta).

tornillo see SCREW BEAN.

tornote *Invertebrate Zoology.* a type of monoaxon sponge spicule, sharply pointed at both ends.

Tornwaldt's cyst see PHARYNGEAL BURSA.

Toro *Astronomy.* asteroid 1685, discovered in 1948 and having a diameter of 12 kilometers; it has an orbit that crosses that of the earth and every 8 years comes within 0.15 astronomical unit of the earth.

toroid *Mathematics.* **1.** a surface described by the revolution of any closed plane curve about an axis lying in its plane. **2.** the solid enclosed by such a surface.

toroidal *Science.* having the shape of a toroid; doughnut-shaped.

toroidal core *Electromagnetism.* a ring-shaped magnetic core in a toroidal magnetic circuit.

toroidal magnetic circuit *Electromagnetism.* a magnetic circuit in which a coil is wound about a toroidal core whose permeability is sufficiently high to confine virtually all the magnetic field lines within the core. Also, **toroidal coil.**

toromatic transmission *Mechanical Engineering*. a semiautomatic transmission that contains a planetary gear train and a torque converter.

torose *Invertebrate Zoology*. having knobby protrusions on the surface; bulging. *Botany*. cylindrical with alternate swellings and constrictions at intervals; knobbed. Also, **torous.**

torose load cast *Geology*. any of a group of elongated load casts that alternately contract and expand along their length and are likely to terminate down current in globular, tear-drop, or spiral formations.

Toroviridae *Virology*. a family of viruses, having enveloped tubular particles containing an ssRNA genome, that cause intestinal infection in humans, horses, and cattle.

Torpedinidae *Vertebrate Zoology*. the torpedoes or electric rays, a family of raylike sharks of the order Torpediniformes, found in the Atlantic and Pacific Oceans.

Torpediniformes *Vertebrate Zoology*. an order of batoid sharks including the torpedo or electric ray.

torpedo *Ordnance*. a self-propelled, explosive missile designed to attack targets in or under the water; it usually contains a homing device and may be launched from a ship, submarine, or aircraft. *Engineering*. an encased explosive used to blast clear a borehole. *Vertebrate Zoology*. another name for the electric rays.

torpedo

torpedo air flask *Ordnance*. a cylinder with separate compartments containing the compressed air, fuel, water, and chemicals that are combined to form the propelling charge of a torpedo.

torpedo boat *Naval Architecture*. **1.** a small, fast attack vessel armed primarily with torpedoes. **2.** specifically, a PT boat. See PT BOAT.

torpedo defense net *Ordnance*. a net that is used to close an inner harbor to torpedoes fired from seaward or to protect an individual ship, either at anchor or underway.

torpedo exercise head *Ordnance*. a device attached to a torpedo for a practice run, having instruments to record data on its performance.

torpedo exploder mechanism *Ordnance*. an electrical or mechanical device that actuates the explosive train of a torpedo warhead; it may be triggered by impact or by an influence signal.

torpedo tube *Ordnance*. a device on a surface ship or submarine through which a torpedo is launched, either by an explosive charge or by compressed air.

torpedo warhead extension *Ordnance*. a metal cylinder used to change a torpedo's center of gravity; it may be inert or contain an explosive.

torpid *Biology*. characterized by a dormant, inactive condition or state.

torpor *Neurology*. a state of stupor resulting from damage to or dysfunction of the central nervous system. *Physiology*. a stress reaction resulting in lethargy.

torque [tôrk] *Mechanics*. **1.** the tendency of a force applied to an object to cause the object to rotate about a given point. **2.** this tendency expressed by the equation $\tau = r \times F$, where τ is the torque, F is the vector of the force, and r is the position vector from the point of origin to the point of application of the force. *Transportation Engineering*. this expression of force acting in a rotational sense, used as a basic measure of the propulsive effect of a powered wheel.

torque amplifier *Computer Technology*. in an analog computer, a device designed to rotate the output shaft without creating any significant torque on it, so that its position matches that of the input shaft.

torque arm *Mechanical Engineering*. an arm that takes the torque of the rear axle in an automotive vehicle.

torque-coil magnetometer *Engineering*. a magnetometer that utilizes the torque developed by a current produced by turning a coil in the magnetic field to be measured.

torque constant *Electricity*. the ratio of the output torque of a motor to its input current. Also, TORQUE SENSITIVITY.

torque converter *Mechanical Engineering*. **1.** a hydraulic coupling that utilizes slipping to multiply torque. **2.** a device that acts as an infinitely variable gear.

torque-free precession *Mechanics*. the precession observed in a rigid, rotating body free from torques, undergoing Poinsot motion.

torque-load characteristic *Engineering*. the armature torque developed on an electric motor rotating at a constant speed, as plotted against the load.

torquemeter *Engineering*. an instrument that measures torque.

torque motor *Electromagnetism*. an electric motor that is designed to exert a torque while rotating slowly or stalled.

torque reaction *Mechanical Engineering*. the reaction between the bevel pinion and the bevel ring gear of a shaft-driven vehicle.

torque ripple see COGGING.

torque sensitivity see TORQUE CONSTANT.

torque sensor *Robotics*. a force sensor that measures torque.

torque-speed characteristic *Electricity*. the relationship between developed torque and armature speed of electric motors.

torque-tube flowmeter *Engineering*. a device that measures liquid flow characteristics by recording the bellows motion of a flexible tube, which is created by the differential pressure drop of a liquid flow.

torque-type viscometer *Engineering*. an instrument that converts measures of torque needed to rotate a paddle in a viscous fluid into viscosity.

torque winding diagram *Mechanical Engineering*. a diagram that shows the variation of winding load on a winch drum; used to determine the best method of balancing.

torque wrench *Mechanical Devices*. **1.** a wrench fitted with a special gauge, capable of both turning and tightening a nut or bolt and measuring the amount of torque used in the task. **2.** a wrench that may be preset to deliver a set amount of torque.

torr or **Torr** *Metrology*. a unit of pressure, equivalent to the amount of pressure that will support a column of mercury 1 millimeter high at 0°C and standard gravity; equal to 133.32 pascals. (Named for Evangelista *Torricelli*.)

Torrejovian *Geology*. a North American continental stage of the Lower Paleocene epoch, occurring after the Dragonian and before the Tiffonian.

Torrendiaceae *Mycology*. a family of fungi belonging to the order Melanogastrales, common to marine and terrestrial environments.

Torrert *Geology*. a suborder of the soil order Vestisol that develops in arid regions; characterized by deep, wide surface cracks that generally remain open throughout the year.

torreyite *Mineralogy*. $(Mg,Mn^{+2})_9Zn_4(SO_4)2(OH)_{22}\cdot 8H_2O$, a bluish-white, monoclinic mineral, massive to granular in habit, having a specific gravity of 2.665 and a hardness of 3 on the Mohs scale; found at Sterling Hill, New Jersey.

Torrey pine *Botany*. a rare pine tree, *Pinus torreyana*, characterized by a rounded crown of branches; found only in certain limited areas on the southern coast of California. (Named for John *Torrey*, 1796–1873, American botanist.)

Torricelli, Evangelista 1608–1647, Italian physicist and mathematician; invented the mercury barometer; the first to create a vacuum.

Torricellian barometer see MERCURY BAROMETER.

Torricellian vacuum *Fluid Mechanics*. the enclosed space above a mercury column in an inverted close-end tube whose open end is submerged in a bath of mercury; mercury vapor fills the space that is otherwise a vacuum.

Torricelli's law of efflux *Fluid Mechanics*. a law stating that the speed of efflux of a fluid particle from an orifice in a container is equal to the speed it would have when it falls from rest from the surface of a reservoir to the depth of the orifice. Also, **Torricelli's theorem.**

Torridincolidae *Invertebrate Zoology*. a family of beetles, coleopteran insects in the suborder Myxophaga.

Torrid Zone *Meteorology*. the portion of the earth's surface that lies between the Tropic of Cancer and the Tropic of Capricorn; one of the three divisions of the mathematical climate, along with the Temperate Zone and Frigid Zone.

Torrox *Geology.* a suborder of the soil order Oxisol that develops during a period of high rainfall; characterized by red coloration, low organic content, dryness, and good drainage.

torsiometer *Mechanical Engineering.* a device that measures the power transmission of a rotating shaft by determining the slight twist of the loaded shaft. Also, TORSIONMETER.

torsion *Mechanics.* a twisting or turning motion of a solid body about its axis of symmetry, produced by the application of opposing forces or torques at opposite ends of the body. *Invertebrate Zoology.* in gastropod mollusks, a shift during development of the most prominent portion of the mantle cavity from its originally posterior position to a location above the head, reversing the position of heart, gills, and anus.

torsional *Mechanics.* relating to or involving torsion.

torsional angle *Mechanics.* the relative rotation of the ends of a cylindrical bar subjected to torsion.

torsional modulus *Mechanics.* the ratio of the torsional rigidity of a cylindrical body to its length. Also, MODULUS OF TORSION.

torsional pendulum *Mechanics.* a device consisting of a disk or other body with a large moment of inertia, suspended by a thin rod or wire having a linear ratio of twist to torque for small amplitude vibrations. The disk is analogous to the bob of a standard pendulum, and the wire torsion provides the restoring torque, causing the device to oscillate in a manner similar to a standard pendulum. Also, **torsion pendulum.**

torsional rigidity *Mechanics.* the ratio of the applied torque for a cylindrical body to its torsional angle.

torsional vibration *Mechanics.* the motion of a bar in which it twists back and forth about its longitudinal axis.

torsional wave *Physics.* a wave in which the disturbance propagates by a twisting motion.

torsion angle *Crystallography.* the angle of rotation about the bond B-C in a series of bonded atoms A-B-C-D needed to make the projection of the line B-A coincide with the projection of the line C-D, when viewed along the B-C direction. The positive sense is clockwise. If the torsion angle is 180°, the four atoms lie in a planar zigzag (Z-shaped). If the torsion angle is 60°, one end atom is twisted 60° out of the plane of the other three. Enantiomers have torsion angles of equal absolute value but opposite sign. Also, CONFORMATIONAL ANGLE, ANGLE OF TWIST.

torsion balance *Engineering.* an instrument that measures weak gravitational, electrostatic, or magnetic forces by determining the amount of torsion they cause in a wire or filament. Also, **torsional balance.**

torsion bar *Mechanical Engineering.* a spring flexed by twisting about its axis.

torsion damper *Mechanical Engineering.* a device used to reduce torsional vibration of an automobile engine.

torsion element *Mathematics.* **1.** any group element of finite order. **2.** in an R-module M, any element m for which there exists a nonzero $r \in R$ such that $rm = 0$.

torsion fault see WRENCH FAULT.

torsion-free *Mathematics.* describing any group or module that has no elements of finite order except for the identity element.

torsion galvanometer *Engineering.* a device that measures the force between fixed and moving systems by measuring the angle between the moving system's present location and its zero location.

torsion group *Mathematics.* a group, all of whose elements have finite order. The subgroup G_t of all torsion elements of a given group G is called the **torsion subgroup** of G.

torsion hygrometer *Engineering.* a device in which the rotation of the hygrometric element is a function of humidity.

torsionmeter see TORSIOMETER.

torsion module *Mathematics.* a module A over a ring R is called a torsion module if for each $a \in A$, there exists a nonzero element r of R such that $ra = 0$; that is, each element of A is a torsion element. The submodule M_t of all torsion elements of a given R-module M is called the **torsion submodule** of M .

torsion of a curve *Mathematics.* the normalized rate at which a curve in R^3 departs from its oscillating plane. If the torsion of the curve is identically zero, then the curve lies in a plane.

torsion string galvanometer *Engineering.* a galvanometer in which the moving system is supported by two fibers that tend to wind around each other.

torso *Anatomy.* the trunk of the body, not including the head or limbs.

torso mountain see MONADNOCK.

torticollic *Medicine.* characterized by torticollis.

torticollis *Medicine.* a contracted state of the cervical muscles, causing twisting of the neck and unnatural position of the head.

tortoise *Vertebrate Zoology.* any one of several large terrestrial turtles, especially in the family Testudinidae, characterized by stumpy legs and a high, arched shell; the **gopher tortoise,** *Gopherus polyphemus,* and the **desert tortoise,** *G. agassizii,* are found in warm, dry areas of North America.

desert tortoise

tortoise core *Archaeology.* in stone toolmaking, a distinctive core having the shape of a tortoise shell.

Tortonian *Geology.* a European geologic stage of the Upper Miocene epoch, occurring after the Helvetian and before the Sarmatian.

Tortricidae *Invertebrate Zoology.* a family of moths, lepidopteran insects in the superfamily Tortricoidea, with fringed wings and threadlike antennae; their caterpillars include leaf rollers, a serious tree pest.

Tortricoidea *Invertebrate Zoology.* a superfamily of moths, lepidopteran insects in the suborder Heteroneura.

tortuosity *Hydrology.* the difference between the actual length of a meandering stream and a theoretical straight course.

torula *Food Technology.* a highly nutritious yeast grown in a medium made of waste from papermaking and fruit canning; used as a supplement in animal and poultry feeds and in pharmaceuticals. Also, **torula yeast.**

Torula *Mycology.* a genus of fungi of the class Hyphomycetes, occurring as a dark growth on dead wood; the species *T. ligniperda* causes stain in hardwoods.

Torulaspora *Mycology.* a genus of yeast fungi belonging to the family Saccharomycetaceae, found in soil, feces, wines, fermenting cucumber brines, and fruit juices.

Torulopsis *Mycology.* a genus of imperfect fungi of the family Cryptococcaceae, normally inhabiting the respiratory and gastrointestinal tracts and urogenital region in humans.

torulopsosis *Pathology.* an opportunistic disease of the tissue, not unlike histoplasmosis, that is caused by *Torulopsis glabrata* and is usually found in patients with another underlying disease.

torus *Anatomy.* a rounded bulge, projection, or swelling. *Botany.* **1.** the receptacle of a flower. **2.** the slight thickening of the pit membrane in the bordered pits of conifers and some gymnosperms. *Architecture.* a large convex molding, as on the base of an Ionic column. *Mathematics.* **1.** a doughnut-shaped surface of genus one; obtained by rotating a circle about an axis lying in the plane of the circle but not intersecting it. **2.** the topological space obtained by identifying opposite sides of a rectangle, i.e., by regarding opposite sides as being the same set of points. **3.** in general, the Cartesian product of a finite but arbitrary number n of circles; usually called an ***n*-torus.**

Torymidae *Invertebrate Zoölogy.* the seed flies, a family of wasps, hymenopteran insects in the superfamily Chalcidoidea; the larvae destroy the seeds of conifers.

TOS tape operating system.

tosca *Meteorology.* a southwest wind that blows over Lake Garda in northern Italy.

toscanite see RHYODACITE.

toss bombing *Military Science.* a method of bombing in which an aircraft flies toward the target, pulls up in a vertical plane, and releases the bomb at an angle that will compensate for the effect of gravity; it is similar to loft bombing but unrestricted in altitude.

tosylation *Organic Chemistry.* the reaction of *p*-toluenesulfonyl chloride (or tosyl chloride) with alcohols; an important reaction used in preparing the tosyl derivatives of carbohydrates.

TOT time on target.

total air *Engineering.* the amount of fuel supplied to a boiler for the purpose of fuel combustion.

total binding energy see BINDING ENERGY.

total biomass growth *Ecology.* the entire amount of organic matter produced by a population over a specific time and within a given area or volume, including both elimination and weight loss.
total bond energy see BOND ENERGY.
total carbon *Metallurgy.* the sum of the free and combined carbons in a ferrous alloy.
total composition *Nutrition.* the qualitative and quantitative makeup of natural and processed food substances.
total conductivity *Geophysics.* the aggregate electrical conductivity of positive and negative ions found in any given portion of the atmosphere.
total correctness *Computer Science.* the property of a program that is guaranteed to terminate and to produce the correct result.
total curvature see GAUSSIAN CURVATURE.
total curvature of a lens *Optics.* the arithmetic difference between the reciprocals of the radii of curvature of both surfaces of a lens.
total deadlock see DEADLOCK.
total differential *Mathematics.* for a function *f* having several variables $(x_1, \ldots, x_n)$, the quantity $df = \Sigma_{i=1}^{n}(\partial f/\partial x_i)dx_i$. Also, DIFFERENTIAL OF F.
total digestible nutrients *Agriculture.* a standard assessment of the energy value of a particular feed for livestock, expressed as a percentage, including all the digestible organic nutrients.
total displacement *Geology.* see SLIP.
total dosage attack *Military Science.* a chemical operation that does not involve a time limit within which to produce the required toxic level.
total drift *Navigation.* in a marine gyroscope, the sum of the apparent drift and the real drift.
total eclipse *Astronomy.* a condition occurring when one celestial body completely obscures or shades another.
total harmonic distortion *Electronics.* a measure of the quality of an audio amplifier system; the ratio (usually expressed as a percentage or in decibels) of the combined power of all the harmonics present to the power of the fundamental frequency, as measured at the output of the amplifier.
total heat see ENTHALPY.
total hysterectomy *Surgery.* the surgical removal of the entire uterus and cervix. Also, PANHYSTERECTOMY.
total impulse *Aviation.* the total thrust generated, as by an engine, over a given interval of time; the product of thrust and duration expressed in force-seconds. *Space Technology.* the total thrust generated by a rocket engine over the entire time that its fuel is burning.
total internal reflection *Optics.* the compete reflection that occurs within a substance when the angle of incidence of light striking the boundary surface is less than the critical angle.
totality *Astronomy.* the period of total obscuration of a celestial body such as the sun (during a solar eclipse), or the moon (during a lunar eclipse).
total lung capacity *Physiology.* thê entire amount of air in the lungs when they are fully inflated.
totally bounded *Mathematics.* a metric space *X* is said to be totally bounded if, for every real number $\varepsilon > 0$, *X* can be covered by a finite number of open balls of radius ε. Total boundedness is a uniform but not a topological property.
totally bounded set see PRECOMPACT SET.
totally disconnected *Mathematics.* **1.** a graph of at least two vertices whose edge set is empty is said to be totally disconnected. Also, NULL GRAPH. **2.** a topological space *X* of at least two points is said to be totally disconnected if, given any point *P*, the largest connected subset of *X* containing *P* contains no points other than *P*.
totally ordered set *Mathematics.* a set that is linearly ordered. Also, CHAIN.
total magnitude *Astronomy.* the integrated light from a star cluster, nebula, or galaxy.
total materiel requirement *Military Science.* the sum of the peacetime force materiel requirement and the war reserve materiel requirement.
total order see LINEAR ORDER.
total organic carbon *Biotechnology.* the total quantity of carbon contained in a measured sample; often used as an indicator of the amount of organic pollution in a water sample.
total oxygen demand *Biotechnology.* the amount of oxygen consumed in the total combustion of all the organic carbon in a sample, occurring at about 900°C.
total porosity *Geology.* the percentage of the total bulk volume of a given sediment or rock that consists of interstices.
total pressure *Fluid Mechanics.* the sum of static pressure, which is measured relative to the pressure at a reference height, and dynamic pressure, which arises from fluid motion. *Mechanics.* the gross applied load on a surface.
total range *Ecology.* the area over which an individual animal travels in its entire lifetime.
total runoff see RUNOFF.
total solidification time *Materials Science.* the time required for a casting to completely solidify after it has been poured.
total solids *Chemistry.* the entire content of all solids, dissolved and suspended, in water.
total specific ionization see SPECIFIC IONIZATION.
total transfer *Computer Programming.* the carry-forward of more than one class of totals (minor, intermediate, major) during the preparation of reports.
total variation *Mathematics.* **1.** the total variation of a real-valued function *f* on the interval $[a,b]$ is

$$T_a^b(f) = \sup \{ \sum_{k=1}^{n} |f(x_k) - f(x_{k-1})| \},$$

where $a = x_0 < x_1 < \cdots < x_n = b$ is a partition of the interval $[a,b]$. The supremum is taken over all partitions of $[a,b]$. Intuitively, the total variation $T_a^b(f)$ of a function *f* is the sum total of the rises and falls of *f* on $[a,b]$, all counted as positive. If $T_a^b(f) < \infty$, *f* is said to be of **bounded variation** on $[a,b]$. For example, $f(x) = \sin(1/x)$ is not of bounded variation on the interval $[0,1]$. **2.** given a measure space (X,A,μ), the total variation of the signed measure μ, denoted $|\mu|$, is the measure defined for every measurable set *E* by $|\mu|(E) = \mu^+(E) + \mu^-(E)$. μ^+ and μ^- are the upper and lower variations of μ, respectively. Note that, in general, $|\mu|(E) \neq |\mu(E)|$, unless μ is nonnegative everywhere.
total vorticity *Fluid Mechanics.* the magnitude of the vorticity vector.
totem *Anthropology.* **1.** a sacred animal, plant, or object from which a clan or society considers itself to be descended, or with which it is closely associated. **2.** a representation or symbol of such an object, serving as the emblem of the group.
totemism *Anthropology.* the belief that the members of a clan or society are related to the totem, from which the group is descended and named; it is usually forbidden to eat or make common use of the totem.
toti- a combining form meaning "whole," "entire."
totipalmate *Vertebrate Zoology.* having all four toes connected by webbing, as in many birds.
totipotency *Developmental Biology.* the capacity of a cell or nucleus to develop into any type of cell or nucleus.
Totiviridae *Virology.* a family of fungal viruses having isometric particles that contain dsRNA.
Totivirus group *Virology.* the single genus of the family Totiviridae.

toucan

toucan *Vertebrate Zoology.* a fruit-eating, mostly tropical American bird of the family Ramphastidae, having a very large, colorful beak.
Toucan *Astronomy.* another name for the constellation Tucana.

touch *Physiology.* **1.** the tactile sensation; the ability to perceive that one's finger, hand, or other body part is in contact with some external object or surface. **2.** the examination of an object employing this tactile sense. *Aviation.* to bring an aircraft's landing gear in contact with the ground, especially at the beginning of a landing run. Also, **touch down.** *Metallurgy.* an official mark put on precious metal to indicate its purity.

touch control *Electricity.* a control circuit that is activated when a person's finger or hand bridges two metal areas.

touchdown *Aviation.* an act or instance of touching down; used to form compounds such as **touchdown site** and **touchdown speed.**

touchdown zone *Aviation.* **1.** for fixed wing aircraft, the first 3000 feet of runway, beginning at the threshold. **2.** for rotary wings and vectored thrust aircraft, the portion of the helicopter landing area or runway used for landing.

touch feedback *Robotics.* feedback in which the force-generating mechanism, such as a claw with fingers, is controlled by resistance-sensing elements.

touch screen *Computer Technology.* a display screen on which the user can select commands by touching specified areas of the screen. Also, **touch-sensitive screen.**

touch sensor *Robotics.* a sensor that sends a feedback signal upon making contact with an object.

touch-tone dialing *Electronics.* a technique for placing a telephone call in which an individual, by pressing a button, transmits an audio signal that is read as a digit by the telephone circuit. Also, TONE DIALING, PUSH-BUTTON DIALING.

touch-tone telephone *Telecommunications.* a telephone that uses push buttons rather than a rotary dial to activate a system of distinct tones at a central circuit to call a number.

tough *Mathematics.* for a positive real number k, a graph G is k-tough if and only if the number of components of $G - S$ is at most $|S|/k$ for every vertex cut S of G. By theorem, if G is 2-tough, then G is Hamiltonian, and if G is 3/2-tough, then G has a 2-factor.

toughness *Materials Science.* the ability of a material to absorb energy from plastic deformation without fracturing; the converse of brittleness.

Toulouse *Agriculture.* a large, heavy breed of goose, characterized by a large head, gray plumage, and a short thick bill without a knob.

touraco *Vertebrate Zoology.* any of several brightly colored birds of the family Musophagidae, related to the cuckoo and having a helmetlike crest; found in Africa. Also, TURAKOO, TURACO.

tourbillion *Horology.* in a watch, an escapement mechanism mounted on a carriage that travels around the works in an epicycle and assumes all the vertical positions in one minute, thus neutralizing position errors.

Tourette's syndrome *Neurology.* a nervous disorder characterized by spasms of the facial muscles, shoulders, and extremities, sometimes accompanied by grunts, barks, or words, especially obscenities. Also, **Tourette's disease, Tourette syndrome.** (After Georges Gilles de la *Tourette*, 1857–1904, French neurologist, who described it.)

touriello *Meteorology.* a strong foehn wind that blows, usually for three or four days at a time, from the Pyrenees Mountains into the Ariege Valley of southern France.

tourmaline *Mineralogy.* any of a group of trigonal borosilicate minerals of the general formula $XY_3Z_6(BO_3)_3Si_6O_{18}(O,OH,F)_4$; where X = Ca, K, Na; Y = Al, Fe^{+2} ,Fe^{+3}, Li, Mg, Mn^{+2}; and Z = Al, Cr^{+3}, Fe^{+3}, V^{+3}. The clear pink, blue, and green varieties are used as gemstones.

tourmaline

Tournaisian *Geology.* a European geologic stage of the Lower Carboniferous period, occurring after the Famennian of the Devonian period and before the Viséian.

tournament *Mathematics.* a directed graph in which any two vertices are joined by exactly one arc. Such graphs can be used to record the results of games in which draws are not allowed. Each team is represented by a vertex, and the arc joining any two given teams is directed toward the loser; the number of wins for a particular team is the out-degree of the vertex representing the team. A tournament cannot contain more than one source or sink.

tourniquet [turn´i kit] *Medicine.* a device for controlling hemorrhage in an extremity by compression of a blood vessel, thereby preventing the flow of blood to or from a part of the body. (Derived from a French word meaning "to turn.")

Toussaint's formula *Meteorology.* an equation used to determine the linear decrease of temperature with height below 11,000 meters in the atmosphere, expressed as $t = 15 - 0.0065z$, where t is the temperature in degrees centigrade, 15 is the temperature at mean sea level (15°C), and z is the geometric height in meters above mean sea level.

Tovariaceae *Botany.* a monogeneric family of dicotyledonous coarse herbs and soft shrubs of the order Capparidales, characteristically producing mustard oil and having a compound plurilocular ovary; native to tropical America.

tow *Engineering.* **1.** to pull by a rope, chain, metal bar, or other device. **2.** an act or instance of this.

tow *Textiles.* **1.** a long, loose, untwisted rope of manufactured filaments. **2.** broken or short fibers used for yarn or stuffing. **3.** yarn or cloth made from such fibers.

TOW see TOW MISSILE.

tow bar or **towbar** *Engineering.* a bar that connects to a vehicle so that it may be towed or attached to a load that is to be towed.

towboat *Naval Architecture.* a boat designed to tow barges, or to push them from behind; found normally on rivers and inland waters.

towed artillery *Ordnance.* artillery that is designed to be towed by a vehicle or animal; it is not equipped for self-propulsion and must be adjusted before being placed in firing position. Similarly, **towed antitank gun, towed howitzer.**

tower *Architecture.* a structure that is much taller than it is wide; it may be freestanding or form part of a building. *Engineering.* a freestanding, open structure having this form and used to support an antenna, transmission line, missile, or the like. *Transportation Engineering.* see CONTROL TOWER. *Chemical Engineering.* a vertical metal cylinder from 6 inches to 20 feet in diameter situated between the boiler and the condenser in distillation units; it may be used for extraction and gas-liquid contacting such as adsorption.

tower bioreactor *Biotechnology.* any of a variety of highly efficient reactors consisting of a vertical cylinder with a conical base, topped by a large-diameter settling zone with baffles.

tower crane *Engineering.* a machine mounted in a tower for raising, shifting, and lowering heavy materials; commonly used in the construction of tall structures.

tower excavator *Engineering.* a cableway excavator having fixed or movable towers, originally designed for levee work but also used in surface mining to strip overburden, spoil, or waste.

tower fermenter *Biotechnology.* a reactor whose height is several times its diameter; the result is productivity much greater than that of simple batch fermenters.

towering *Meteorology.* an atmospheric refraction phenomenon in which the downward curvature of light rays increases with elevation such that the visual image of a distant object appears to be stretched vertically; the opposite of stooping.

towering cumulus see CUMULUS CONGESTUS.

tower launcher *Ordnance.* a tall, vertical missile launcher that provides directional stability to the missile.

tower loader *Mining Engineering.* a front-end loader having a bucket that is lifted along tracks on a vertical or nearly vertical tower.

tower loading *Electricity.* the environmental and load-induced effects on a transmission tower including its own weight, the weight of the wires with or without ice upon them, insulation material, the wind force, and pressure perpendicular to the line.

tower radiator *Electromagnetism.* a tall metal structure that acts as a transmitting antenna.

tower telescope *Astronomy.* a telescope (usually for observing the sun) that is mounted high on a tower to place its optics above the distorted air close to the sun-warmed ground.

TOW missile *Ordnance.* a 48-pound missile powered by a two-stage solid-propellant motor that is the munition component of a crew-portable guided missile system. (An acronym for tube-launched, optically tracked, wire-command link.)

town *Geography.* a settlement that is larger than a village but smaller than a city.

Townes, Charles Hard born 1915, American physicist; shared the Nobel Prize for major contributions to maser and laser technology.

townscape *Geography.* the general characteristics or aspect of an urban area, especially the form and configuration of its buildings and open space.

Townsend characteristic *Electronics.* the characteristic current versus voltage curve for a Townsend discharge phototube at a fixed value of illumination and at voltage levels insufficient to allow a self-sustained discharge (a discharge without photo excitation of the cathode) to occur.

Townsend coefficient *Electronics.* a value representing the requirements for an ionization breakdown in a Townsend discharge device, such as a radiation counter; the coefficient is expressed as the average number of ionizing collisions by an electron per centimeter of path length as it drifts in the direction of the applied electric field.

Townsend discharge *Electronics.* a discharge that occurs in a gas at voltage levels insufficient for it to be maintained by the electric field alone; the discharge must be started and sustained by ionization produced by stimuli such as photons, radioactivity, or cosmic rays.

township *Cartography.* in surveying, a unit of public land in the United States, usually a quadrangle measuring six miles on each side.

township line see RANGE LINE.

towrope *Engineering.* a strong line used to tow a vehicle or vessel.

towrope resistance *Naval Architecture.* the total resistance of a vessel to motion, equivalent to the resistance transmitted to a towrope when the vessel is towed at a given speed.

tow target *Ordnance.* a target that is towed behind an aircraft for anti-aircraft or aerial gunnery practice.

tox. toxicology.

tox- a combining form meaning "poison."

toxa *Invertebrate Zoology.* a curved sponge spicule.

toxalbumin *Toxicology.* any poisonous albumin.

Toxasteridae *Paleontology.* a family of irregular euechinoids in the order Spatangoida; extant from the Upper Jurassic to Upper Cretaceous.

toxemia *Toxicology.* poisoning due to the presence in the bloodstream of a toxin. *Medicine.* a condition resulting from metabolic disturbances, such as toxemia of pregnancy.

toxemia of pregnancy *Medicine.* a group of metabolic disturbances occurring in pregnant women, manifested by preeclampsia and fully developed eclampsia.

toxenzyme *Toxicology.* any poisonous enzyme.

toxic *Toxicology.* relating to or caused by a poison or toxin; poisonous. *Pathology.* manifesting the symptoms of severe infection. (Going back to a Greek word meaning "arrow poison.")

toxic amaurosis *Toxicology.* loss of vision that results from poisoning.

toxic anemia *Toxicology.* anemia caused by toxic effects. Also, **toxanemia.**

toxicant *Toxicology.* any poisonous agent; a poison.

toxic attack *Military Science.* an attack directed against humans, animals, or crops, using injurious agents of radiological, biological, or chemical origin.

toxic bullous epidermolysis see TOXIC EPIDERMAL NECROLYSIS.

toxic edema *Toxicology.* any swelling or retention of fluid caused by a poison.

toxic epidermal necrolysis *Pathology.* a rare skin disease, occurring as a severe exfoliative reaction to various etiologic factors, primarily drugs, but also including viral, bacterial, and fungal infections, neoplastic disease, graft-versus-host reaction, and chemical exposures; it is characterized by epidermal erythema, superficial necrosis, and skin erosions resulting in a scalded appearance. Also, NONSTAPHYLOCOCCAL SCALDED SKIN SYNDROME, TOXIC BULLOUS EPIDERMOLYSIS, LYELL'S DISEASE.

toxic equivalent *Toxicology.* the amount of a poison per kilogram of body weight.

toxic hepatitis *Toxicology.* an inflammation of the liver caused by the effects of hepatotoxins, such as carbon tetrachloride, yellow phosphorus, and a variety of drugs.

toxicide *Toxicology.* any antidote to a toxic or poisonous substance.

toxicity *Toxicology.* the quality or degree of being poisonous.

toxic neuritis *Toxicology.* neuritis caused by a poison.

toxico- or **toxic-** a combining form meaning "poison."

toxicodendrol *Toxicology.* a poisonous oily resin found in members of the genus *Toxicodendron* (*Rhus*), including poison ivy and poison oak, that causes severe dermatitis in sensitive individuals.

Toxicodendron see RHUS.

toxicodermatitis *Toxicology.* any skin disease caused by a poison.

toxicodynamics *Toxicology.* the relationship between the concentration of a toxin and its effects.

toxicogenic *Toxicology.* capable of producing a poison.

toxicoid *Toxicology.* **1.** similar to a poison. **2.** temporarily poisonous.

toxicokinetics *Toxicology.* the study of the absorption, metabolic distribution, and elimination of a toxic compound in a living organism.

toxicologist *Toxicology.* a person who studies the toxic effect of toxicants, or poisons.

toxicology *Toxicology.* the scientific study of poisons, their actions, their detection, and the treatment of conditions produced by them.

Toxicology

Toxicology is defined as the study of adverse effects of chemicals on biological systems. In the modern world, which depends heavily on chemical production and use, the importance of toxicology is obvious. The public is continuously bombarded with reports about chemical pollutants in water and air, pesticide residues in food and mother's milk, ecological disasters caused by oil and chemical spills, chemical dangers in the workplace, indoor air pollution, adverse reactions to prescription and over-the-counter drugs, and so on.

It is the scientific discipline of toxicology that combines the elements of chemistry and biology to help elucidate and explain the potential for harmful effects of chemicals on living organisms. It is the toxicologist who determines how much chemical it takes to cause harm and the types of toxic effects that are produced—such as cancer, birth defects, nervous system damage, or immunosuppression.

Once toxic effects have been identified for a particular chemical, it is important to understand, at the molecular and cellular level, the mechanisms by which these toxic effects are produced. Thus, toxicologists must utilize the latest developments in molecular and cellular biology to elucidate these mechanisms. An equally important aspect of toxicology is the utilization of information about chemicals to make reasonable predictions of whether or not harmful effects will occur in various exposure circumstances. This activity, called "risk assessment," assists in the establishment of rules and regulations aimed at protecting and preserving human health and the environment.

Toxicology has a long and interesting past and a bright future. It evolved with early man's use of poisons for hunting and waging war. Today, the emphasis is to understand why chemicals are toxic and to determine how they can be used safely. Toxicologists will continue to be involved in the acquisition of new knowledge, evaluation of the safety of new chemicals and products, teaching the principles of toxicology and research design, and development of regulations to govern the production, use, and disposal of hazardous and potentially hazardous materials.

I. Glenn Sipes
Professor and Head
Pharmacology and Toxicology
The University of Arizona

toxicomania *Toxicology.* an addiction to a drug, or a craving for any substance that produces a euphoric effect.
toxicopathic *Toxicology.* of or relating to any pathological condition produced by a toxin.
toxicopectic *Toxicology.* of or relating to any substance or process that neutralizes or inactivates a toxin in the body. Also, TOXOPEXIC.
toxicophidia see THANATOPHIDIA.
toxicophobia *Psychology.* an irrational fear of being poisoned. Also, TOXIPHOBIA.
toxicosis *Toxicology.* any disease condition caused by a poison or a toxic reaction. Also, TOXINOSIS, TOXIPATHY, TOXONOSIS.
toxic psychosis *Psychology.* psychosis caused by toxicity, especially by the toxic effect of a drug or chemical.
toxic shock syndrome *Pathology.* a severe, sometimes fatal illness, caused by infection with *Staphylococcus aureus,* which produces a unique toxin, enterotoxin F; the syndrome almost exclusively affects menstruating women using tampons, although it has been seen in newborn infants, children, and men. It is manifested as sudden high fever, vomiting, diarrhea, and myalgia, followed by hypotension and, in severe cases, shock. In the acute phase, a sunburnlike rash appears, especially on the palms and soles, accompanied by peeling skin. Also, TSS.
toxic tremor *Toxicology.* a tremor caused by a poison.
toxic unit see MINIMUM LETHAL DOSE.
toxicyst *Invertebrate Zoology.* a type of trichocyst, found in Protozoa, that may induce paralysis or lysis of prey.
toxiferine *Toxicology.* any of a group of potent curare poisons produced by the tropical tree *Strychnos toxifera.*
toxiferous *Toxicology.* conveying, containing, or producing a poison.
toxigenic see TOXICOGENIC.
toxigenicity *Microbiology.* the ability of a microorganism to cause disease, as determined by the toxin it produces, which partly determines its virulence.
toxignomic *Toxicology.* characteristic of the toxin action of a poison.
toximetry *Toxicology.* the branch of toxicology that deals with the quantitative measurement of toxicity.
toxin *Toxicology.* any poisonous agent, especially a poisonous substance produced by one living organism that is poisonous to other organisms; e.g., snake venom.
toxinemia *Toxicology.* the presence of a toxin in the blood or circulatory system; blood poisoning.
toxinology *Toxicology.* the branch of toxicology that studies toxins, especially those produced by pathogenic bacteria, certain higher plants, and animals.
toxinosis see TOXICOSIS.
toxin unit see MINIMUM LETHAL DOSE.
toxipathic hepatitis *Medicine.* inflammation of the liver caused by the effect of poison on the liver.
toxipathy see TOXICOSIS.
toxiphobia see TOXICOPHOBIA.
toxo- a combining form meaning "poison."
toxodont *Paleontology.* any member of the suborder Toxodontia.
Toxodontia *Paleontology.* a suborder of three-toed, rhinoceroslike South American ungulates belonging to the order Notoungulata; they ranged in length from 4.5 feet to 9 feet, and some were almost 9 feet high at the shoulder; characterized by nasal openings on the top of the skull; extant from the Paleocene to the Quaternary.
toxogen *Toxicology.* any organism that produces a poison.
toxoglobulin *Toxicology.* a toxic globulin.
Toxoglossa *Invertebrate Zoology.* a group of carnivorous marine gastropod mollusks in the suborder Pectinibranchia; the radula has two long, hollow teeth per row, which inject toxin into prey. Also, **Toxiglossa.**
toxoglossate radula *Invertebrate Zoology.* a radula in certain carnivorous gastropods, having long spearlike teeth with poison gland ducts.
toxoid *Toxicology.* a bacterial exotoxin that has been treated so it is no longer poisonous but is still able to elicit the production of an antitoxin.
toxoid-antitoxoid *Toxicology.* a toxoid mixed with an equivalent amount of antitoxoid serum, the precipitate being suspended in saline.
toxolecithin *Toxicology.* a lecithin combined with a poison, as found in the venom of certain snakes, such as cobras. Also, **toxolecithid.**
toxonosis see TOXICOSIS.
toxopexic see TOXICOPECTIC.
toxophilic *Toxicology.* characterized by a susceptibility to toxins.
toxophore *Toxicology.* the atomic group within the molecule of a toxin that actually produces the specific effect on an organism.
Toxoplasma *Invertebrate Zoology.* a genus of protozoans in the order Toxoplasmida, intracellular parasites of birds and mammals, including domestic cats and humans.

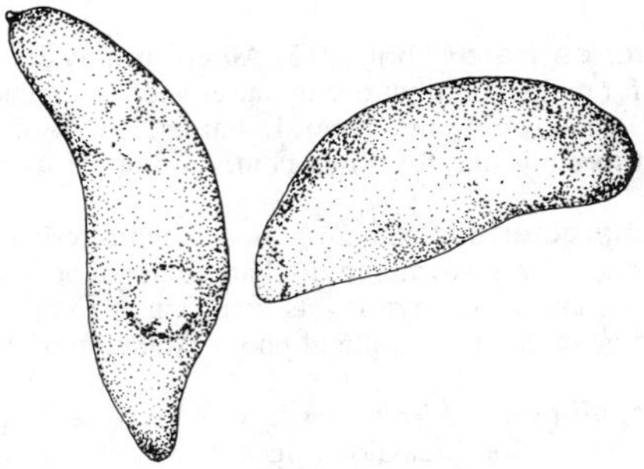

Toxoplasma

Toxoplasmea *Invertebrate Zoology.* a class of endoparasitic protozoans in the subphylum Sporozoa.
Toxoplasmida *Invertebrate Zoology.* an order of sporozoan protozoans, crescent-shaped endoparasites of vertebrates, including humans, that cause toxoplasmosis, a sometimes fatal disease of the nervous system and eyes.
toxoplasmin *Biochemistry.* an antigen prepared from mouse cells infected with *Toxoplasma gondi*; used in a skin test to show hypersensitivity to toxoplasmosis.
toxoplasmosis *Medicine.* a widespread disease of animals and humans caused by the protozoan *Toxoplasma gondii*; symptoms range from a mild disease resembling mononucleosis to an extensive fulminating disease that may cause damage to the brain, eyes, skeletal and cardiac muscles, liver, and lungs; the severe forms occur most often in fetuses infected by the mother and in those with impaired immune systems.
Toxopneustidae *Invertebrate Zoology.* a family of sea urchins, echinoid echinoderms in the order Temnopleuroida.
toxoprotein *Toxicology.* **1.** any poisonous protein. **2.** any mixture of a toxin and a protein.
Toxothrix *Bacteriology.* a genus of gliding bacteria of uncertain affiliation that occur as colorless cylindrical cells forming a filament, and are found in iron-rich cold-water environments.
Toxotidae *Vertebrate Zoology.* the archerfishes, a monogeneric family of small, perciform fishes living in fresh, brackish, and salt water and characterized by a highly protractile mouth used to spit water at insects to knock them into the water for food.
toyon *Botany.* a tree or shrub, *Photinia arbutifolia,* belonging to the rose family and having evergreen leaves and bright red berries resembling those of holly; found on the Pacific Coast of North America.
TP teleprocessing.
TPA *Aviation.* the airport code for Tampa International, Florida.
TPHA *Treponema pallidum* hemagglutination assay.
T phage *Virology.* any of a series of tailed DNA-containing phages (T1–T7) that lyse susceptible host cells. Notable members include the **T2 phage**, the type species of the T-even phage group, having particles with contractile tails that contain a linear dsDNA genome; the **T4 phage**, a complex member of the T-even phage group, family Myoviridae, having particles with a DNA-filled head; and the **T7 phage group**, a genus of phages of the family Podoviridae, having dsDNA-containing particles with an isometric head and a short tail; the virus infects female entero-bacteria.
tpm tons per minute.
TPN total parenteral nutrition; triphosphopyridine nucleotide.
TPNH reduced triphosphopyridine nucleotide.
TPP triphenyl phosphate.
TPTG oscillator see TUNED-GRID TUNED-ANODE OSCILLATOR.
t quark see TOP QUARK.
tr or **Tr** trace.
trabeated *Architecture.* denoting ancient Greek post-and-lintel architecture, in contrast to Roman arches.
trabecula *Anatomy.* **1.** a supporting cord of fibrous tissue that usually extends from the capsule surrounding an organ into its interior. **2.** any of numerous small interconnecting rods of bone making up a mass of spongy bone.

trace *Science.* **1.** a very small amount of a substance, often too small to be measured. **2.** the surviving evidence of the former occurrence or action of some event or agent. *Meteorology.* precipitation of less than 0.005 inch (0.127 mm). *Archaeology.* any physical characteristic of an artifact that can be described. *Geology.* a line representing the intersection of a geological surface with another surface, such as the trace of a bedding surface on a cleavage surface. *Engineering.* a record made by an instrument's recording element. *Computer Programming.* **1.** a record of the actions carried out by a computer or program. **2.** to create or initiate such a record. *Electronics.* the visible pattern observed on the face of a cathode-ray tube indicator. *Mathematics.* **1.** the sum of the (main) diagonal elements of a square matrix M; denoted tr M or Tr M. Also, SPUR. **2.** the trace of a linear transformation T on a finite-dimensional vector space is the trace of the matrix associated with T.

trace analysis *Analytical Chemistry.* the analysis of very low concentrations or small amounts of a substance in a mixture or solution, using such techniques as spectroscopy or polarography.

trace conditioning *Behavior.* a type of classical conditioning in which an unconditioned stimulus is presented following a conditioned stimulus that is no longer present, with a brief time interval between stimuli.

trace element *Geochemistry.* an element occurring in very small quantities in a mineral or rock and not essential to the formation of the material. Also, ACCESSORY ELEMENT, GUEST ELEMENT. *Biochemistry.* an element that is essential to nutritional or physiological processes, found in an organism in such minute quantities that analysis yields a presence of virtually zero amounts; e.g., copper, cobalt, manganese, zinc.

trace fossil *Geology.* a sedimentary structure consisting of a fossilized remnant or mark formed in soft sediment by the movement of an animal and pressed in sedimentary rock. Also known by various other names, such as ICHNOFOSSIL, VETIGIOFOSSIL, or BIOGLYPH.

trace interval *Electronics.* the time interval between the start and the end of the trace on the screen of a cathode-ray-tube indicator.

trace mineral *Nutrition.* a mineral that occurs in minute amounts but is still essential for growth, development, and health; a trace element.

trace program *Computer Programming.* a program designed to create trace records.

tracer *Mechanical Devices.* any device by which the outline of an object or the direction and extent of movement of a part may be graphically recorded. *Chemistry.* a radioactive isotope that is added to a material in order to facilitate monitoring the behavior of the material during a reaction. Also, **tracer element.** *Ordnance.* see TRACER BULLET.

tracer bullet *Ordnance.* a bullet that contains a pyrotechnic composition designed to burn during flight, thus making the trajectory visible in light or darkness. Similarly, **tracer mixture.**

tracer gas *Engineering.* a gas that is placed within a vacuum system so that leaks from the system can be detected.

tracer milling *Mechanical Engineering.* the duplication of a three-dimensional form using a mastic or hard plastic to direct the tracer-controlled cutter.

tracery *Architecture.* **1.** the ornamental and structural curved stone mullions in a Gothic window. **2.** similar work in other materials.

trace sensitivity *Electronics.* the degree of sensitivity of an oscilloscope; the greater the trace sensitivity, the lower the value of input voltage that must be applied to produce a visible trace on the oscilloscope display.

trace slip *Geology.* the component of the net slip on a fault that lies parallel to the trace of the bedding plane or other index plane along the plane of the fault.

trace-slip fault *Geology.* a fault in which the net slip is parallel to the trace of an index plane.

trace statement *Computer Programming.* a statement or list of traces, provided by some programming languages for debugging purposes.

trachea [trā´kē ə] *plural,* **tracheae.** *Anatomy.* the air tube supported by cartilaginous rings that stretches from the pharynx into the thorax, where it divides into the bronchial tubes. *Invertebrate Zoology.* any of the air-conveying tubules composing the complex branched respiratory system in insects, myriapods, and some arachnids. *Botany.* a vascular conducting tube located in the xylem.

tracheal *Biology.* of or relating to a trachea.

trachealgia *Medicine.* a pain in the trachea.

tracheal gills *Entomology.* specialized thin body-wall extensions that permit breathing in mayfly nymphs and other aquatic insects.

tracheary element *Botany.* a xylem cell in a vascular plant that acts in water conduction after its protoplasm dies and its cell walls become thickened with lignin.

tracheid *Botany.* a woody, cylindrical, and elongate conducting cell that is found in the secondary xylem and conducts water and nutrients throughout the plant.

Trachelomonas *Invertebrate Zoology.* a genus of green, yellow, or brownish freshwater protozoans having one flagellum and often a spiny shell.

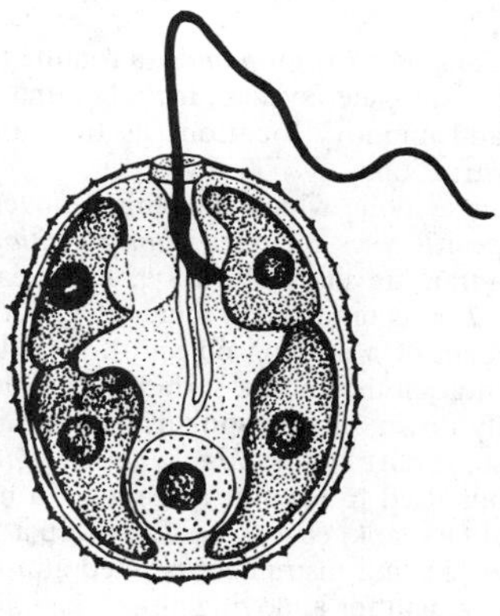

Trachelomonas

Trachelophyllum *Invertebrate Zoology.* a genus of voracious, carnivorous ciliate protozoans belonging to the order Holotrichia; the long, flexible cell has a narrow necklike portion.

tracheole *Entomology.* any of the smallest branches of an insect trachea.

tracheomalacia *Medicine.* a softening of the tracheal cartilages.

Tracheophyta *Botany.* in some systems of classification, a large division of plants characterized by the possession of a vascular system.

tracheoplasty *Surgery.* surgical repair of the trachea.

tracheorrhagia *Medicine.* a hemorrhage from the trachea.

tracheoscopy *Medicine.* an examination of the interior of the trachea.

tracheostomy *Surgery.* **1.** the surgical formation of a new opening into the windpipe from the neck. **2.** the opening so created.

tracheotome *Surgery.* an instrument used for incising the trachea.

tracheotomy [trā´kē ät´ə mē] *Surgery.* the process of surgically cutting into the windpipe, especially to provide an emergency opening in the airway below an obstruction that is inhibiting respiration.

trachodont *Paleontology.* relating to or being a hadrosaur of the genus *Anatosaurus,* the North American counterpart of Iguanodon; extant in the Upper Cretaceous.

trachoma *Medicine.* a chronic infectious disease affecting the conjunctiva and cornea of the eye, caused by a strain of the bacteria *Chlamydia trachomatis* and resulting in inflammation, oversensitivity to light, pain, tears, redness, granulation and hypertrophy, and scarring of the conjunctiva. Also, **trachomatous conjunctivitis.**

trachybasalt *Petrology.* an extrusive rock, intermediate in composition between trachyte and basalt, containing calcic plagioclase, sanidine, augite, olivine, and sometimes minor analcime or leucite.

Trachylina *Invertebrate Zoology.* an order of mostly marine hydrozoans; the medusa stage is generally large, the hydroid polyp generation reduced or absent.

Trachymedusae *Invertebrate Zoology.* a suborder of marine hydrozoans in the order Trachylina, with tentacles around the edge of the umbrella. Also, **Trachomedusae.**

Trachypodaceae *Botany.* a family of dull mosses of the order Isobryales that form loose mats on tree trunks and branches; characterized by large plicate leaves, creeping stems with scale leaves, and ascending to pendent secondary branches.

Trachypsammiacea *Invertebrate Zoology.* an order of colonial anthozoan coelenterates having a treelike skeleton.

Trachypteridae *Vertebrate Zoology.* the dealfish, a family of marine fishes having long, compressed bodies.

trachyte *Petrology.* a light-colored, aphanitic, extrusive igneous rock, or rock group, consisting primarily of alkali feldspar with minor mafic minerals; the extrusive equivalent of syenite.

trachytoid texture *Geology.* a texture of certain extrusive igneous rocks in which the mineral microlites of the groundmass have a subparallel or randomly divergent arrangement.

tracing the action of something that traces; specific uses include: *Engineering.* **1.** the process of drawing by following lines visible through a transparent paper. **2.** any graphic record produced by such a process. *Robotics.* a debugging technique that records information about each step of a programmed process during its execution.

tracing distortion *Acoustical Engineering.* the distortion in sound reproduced by a record player, due to an imbalance in one or more mechanical components, such as the stylus arm, turntable drive motor, or turntable.

tracing routine *Computer Programming.* a routine that records historical data regarding a computer system, including the status of all operational registers and memory locations as they are affected by the execution of an instruction.

track a path or course along which something moves; to follow such a path or course; specific uses include: *Transportation Engineering.* the rail along which a train travels. *Navigation.* **1.** the path of a ship on the surface of the sea. **2.** a recommended route on a nautical chart. **3.** to observe or plot the path of a moving object. **4.** to follow a desired path. *Aviation.* a projection on the earth's surface of the path along which an aircraft has actually flown; often distinguished from its projected course or flight path. Also, FLIGHT TRACK. *Mechanical Engineering.* either of a pair of roller chains used to support and propel a machine such as an earth-mover or military tank. *Ordnance.* **1.** to keep a gun properly aimed or to keep a target-locating instrument pointed at a moving target. **2.** a metal part forming a path for a moving object; e.g., the track around the inside of a vehicle for moving a machine gun. *Nucleonics.* the visible trail left behind a particle passing through a bubble chamber, cloud chamber, spark chamber, or nuclear emulsion. *Mining Engineering.* **1.** the slide or rack on which a diamond-drill swivel head can be moved to positions clear of the collar of a borehole. **2.** a groove cut into a rock. *Computer Technology.* a predefined path on which information is stored on a magnetic recording medium such as a disk or tape. *Computer Programming.* to record or follow the movement of a display screen cursor, mouse, or other user input device.

track address *Computer Technology.* a set of binary codes that are used to locate tracks on a disk.

track angle *Navigation.* a method of indicating a track by measuring from a reference direction, either clockwise or counterclockwise; it is expressed with the reference followed by the measurement; e.g., S15°E is 15 degrees east of south.

track ball *Computer Technology.* a freely rotating ball that is mounted in a box or casing, and used by an operator to position the cursor on a video display.

track cable *Engineering.* a rope that supports the wheels of a cableway carriage.

track circuit *Transportation Engineering.* a set of electrical equipment (including track rails) forming a closed system that detects trains on the rails and may also signal them.

track crawling *Navigation.* the process of following a track by making frequent minor adjustments in heading.

Tracker *Aviation.* a popular name for the S-2 antisubmarine aircraft.

track gauge *Transportation Engineering.* the separation between the inner top edges of the rails of a railroad track. Standard track gauge is 1.435 meters.

track homing *Navigation.* the process of following a line of position known to pass through a target or other objective.

track hopper *Engineering.* a hopper-shaped receptacle used in the unloading of railroad cars.

tracking the process of something that tracks; specific uses include: *Engineering.* the process of following the path of a moving object, either visually or by following a point of radiation. *Navigation.* **1.** the precise, continuous position-finding of targets by radar, optical, or other means. **2.** the process of navigating by closely following the movement of another craft, without reference to future positions. *Acoustical Engineering.* **1.** the ability of a tape recorder or record player to accurately record or reproduce a sound. **2.** a control in a videocassette recorder that makes slight electronic adjustments in the manner in which the tape control head picks up signals from the videotape, to provide a clearer playback. *Electronics.* **1.** a condition in which ganged or interrelated controls maintain the proper relationship over the entire adjustment range. **2.** a condition in which a component or several interrelated components maintain the correct alignment over the entire operating range. *Electricity.* the excessive leakage of current between two insulated points as a result of moisture or other environmental effects. *Robotics.* the continuous monitoring of a robot's movements by its controller.

tracking beam *Ordnance.* in antimissile operations, a beam that is directed at the target in order to obtain information that is then transmitted over the guidance beam to the counterattacking guided missile.

tracking cross *Computer Technology.* **1.** on a display screen, a cross-shaped cursor that can be positioned by a light pen or mouse. **2.** in computer graphics, a cross-shaped cursor used to designate points, lines, or arcs. Also, **tracking cursor.**

tracking error *Acoustical Engineering.* a nonlinear distortion caused by a deviation in the tracking by a stylus of a recorder player, usually due to misalignment of the stylus arm or cartridge.

tracking filter *Electronics.* a narrow bandwidth filter that is capable of passing a single desired input signal by automatically adjusting its own center frequency to track the changing frequency of the input signal.

tracking jitter *Engineering.* a fluctuation in the position of tracking radar.

tracking network *Engineering.* a group of synchronized tracking stations used to follow objects in outer space.

tracking radar *Engineering.* a type of radar used to monitor the movements of objects high in space, such as rockets and satellites.

tracking range *Robotics.* the outer limits of a robot's range of motion.

tracking satellite *Space Technology.* an artificial satellite designed to record and transmit data on the movements of other objects in space.

tracking satellite

tracking station *Engineering.* a fixed station equipped with tracking radar or some other means of precisely locating an airborne object.

tracking system *Engineering.* the apparatus that is used to precisely and continuously locate an object.

tracking telescope *Optics.* a telescope that gathers performance data while following the ascent of a rocket or weather balloon.

track in range *Electronics.* to set the range gate of a radar set so as to permit tracking of a target that is changing in range with respect to the radar set.

track-laying vehicle *Ordnance.* any vehicle that moves on two endless tracks; e.g., a tank.

trackless mine *Mining Engineering.* a mine without track haulage, in which rubber-tired vehicles are relied upon for haulage and transport.

trackless tunneling *Mining Engineering.* a method of tunneling that employs loaders mounted on caterpillars.

track made good *Aviation.* the representation, or actual path, of a flight vehicle as it moves over the surface of the earth.

track pitch *Computer Technology.* the distance between the tracks on which information is stored on a magnetic recording medium.

track-return power system *Electricity.* a power distribution system utilized by locomotives or other electric vehicles, using the track rails as an uninsulated return conductor.

track shifter *Transportation Engineering.* an apparatus that is used to shift a railway track laterally.

track telling *Telecommunications.* the communication of tactical data information and air surveillance between command and control systems and operations within the organization.

track-to-track access time *Computer Technology*. the time it takes a disk read-write head to move from one track position to another.

track via missile see TVM.

track-while-scan *Electronics*. a radar system equipped to scan for targets while simultaneously tracking one or more targets and computing their speed and future position.

tract *Anatomy*. **1.** a number of organs, arranged in a series, serving a common function; e.g., the digestive tract or the respiratory tract. **2.** a collection or bundle of nerve fibers having the same origin, function, and termination. *Geography*. any definable region or stretch of land or water. *Building Engineering*. a real-estate development containing tract houses.

tract house *Building Engineering*. any of a number of houses of similar design and appearance that are built and sold as part of a tract.

traction *Mechanics*. the force of adhesive friction exerted on a body moving over a surface. *Medicine*. a deliberate and prolonged exertion of a pulling force, as to stretch, separate, or position a structure; for example, **orthopedic traction**, a procedure in which a patient is maintained in a system of ropes, pulleys, and weights to correct and immobilize a fracture. *Surgery*. the process of pulling something through or out of a body cavity, as in a forceps delivery of an infant. *Geology*. a method of transport in which sedimentary particles are moved along and parallel to a bottom surface, as by rolling, sliding, dragging, or pushing. (From a Latin word meaning "to draw out.")

tractional force *Fluid Mechanics*. the force exerted by the current on particles that are below the surface of flowing water.

traction diverticulum *Medicine*. a localized distortion, angulation, or funnel-shaped bulging of the full thickness of the wall of the esophagus, caused by adhesions resulting from some external lesion.

traction load *Geology*. the part of the stream load that is carrried along the stream bed. Also, **tractional load.**

traction meter *Engineering*. a device that detects the force-generating capacity of a locomotive.

traction tube *Engineering*. an instrument that determines the minimum water velocity necessary to move a specified size of sand grain.

tractor *Mechanical Engineering*. **1.** an automotive vehicle on crawler tracks or large pneumatic tires that is used to pull agricultural or construction equipment. **2.** a piece of mechanical equipment that converts engine power into tractive power to move itself and other vehicles. *Aviation*. a propeller that is mounted at the front of an airplane. *Surgery*. an instrument used to apply surgical traction.

tractor

tractor drill *Mechanical Engineering*. a drill that has a crawler-mounted feed guide on an extendable arm.

tractor feed *Computer Technology*. a sprocketed mechanism that pulls continuous-feed paper through a printer by means of pins that engage perforations in the edges of the paper. Thus, **tractor-feed printer.**

tractor group *Military Science*. in an amphibious operation, a group of ships that carry the amphibious vehicles of the landing force.

tractor loader *Mechanical Engineering*. a tractor equipped with a tipping bucket that is used to dig, elevate material, and then dump it. Also, **tractor shovel.**

tractotomy *Surgery*. the process of severing or incising a nerve tract.

tractrix *Mathematics*. the locus of points P such that the segment of the tangent at P lying between P and the x-axis has constant length a; i.e., the evolute of the catenary. The equation of the tractrix in Cartesian coordinates is

$$\pm x = a \cosh^{-1}(a/y) - (a^2 - y^2)^{1/2},$$

where a is a positive real constant and $0 < y \leq a$. The curve is symmetric about the y-axis and has a cusp at $(0, a)$. Also, PURSUIT CURVE.

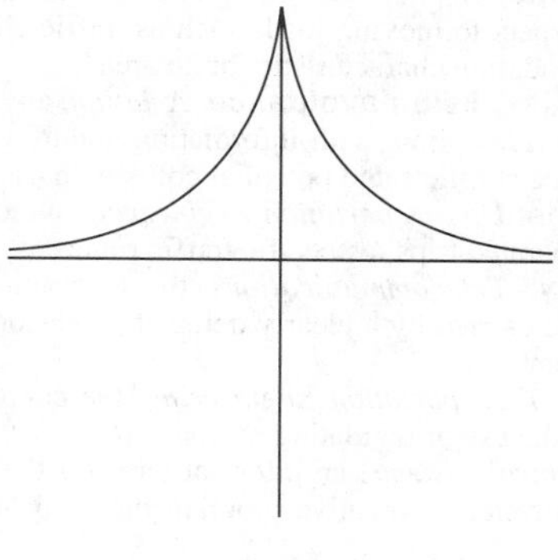

tractrix

trade *Meteorology*. **1.** see TRADE WIND. **2.** of or relating to a trade wind, or to an area where such winds occur.

trade air *Meteorology*. the air comprising a trade wind, characterized by the trade-wind inversion.

trade paperback *Graphic Arts*. a high-quality, often large-format paperbound book that is distributed only through bookstores, as distinguished from a mass-market paperback.

trades *Meteorology*. the wind system constituting the northeast and southeast trade winds. Also, **trade winds.**

trade wind *Meteorology*. **1.** any of the nearly constant easterly winds that occupy most of the tropics, blowing from the subtropical highs toward the equatorial trough, mainly from the northeast in the Northern Hemisphere and the southeast in the Southern Hemisphere; a major component of the general circulation of the earth's atmosphere. Also, TRADE. **2.** any of the prevailing summer westerlies in California. **3.** any of the northwest winds that blow in Oregon in summer. **4.** any wind that blows in a nearly continuous direction. (From an older use of the word *trade* to mean "a course or path;" reinforced by the idea that early trading ships sailed under the power of these winds.)

trade-wind belt *Meteorology*. the belt of latitude occupied by the trade winds, extending from about 30° N and S toward the equator.

trade-wind cumulus *Meteorology*. the type of cumulus cloud that is characteristic of the trade belt, generally based at about 2000–2500 feet altitude and rising 5000–7000 feet, ending abruptly at the lower stratum of the trade-wind inversion. Also, **trade cumulus.**

trade-wind desert *Meteorology*. **1.** a land area, such as the Sahara and Kalahari deserts, over which trade winds blow, bringing high temperatures and low rainfall. **2.** any of the dry western coastal regions of Africa and the Americas.

trade-wind inversion *Meteorology*. a temperature inversion that is characteristic of the trade-wind streams over the eastern portions of tropical oceans, especially in the Southern Hemisphere; formed when air from high altitudes in subtropical highs descends to meet low-level maritime air flowing in the same direction toward the equator. Below the inversion, the air is very moist and filled with trade-wind cumulus clouds; above it, the air is warm and very dry. Also, **trade inversion.**

tradition *Anthropology*. an acquired pattern of behavior or a body of information or beliefs, transmitted from one generation to another within a cultural group; e.g., ceremonies relating to birth, death, marriage, and passage to adulthood. *Archaeology*. a culture that exists for an extended period of time and usually over an extended area; used especially to designate specific New World cultures such as the **Arctic Small Tool Tradition, Big Game (Hunting) Tradition, Mississippi(an) Tradition, Woodland Tradition,** and **Desert Tradition.**

traditional *Anthropology*. relating to or being a cultural tradition. *Metrology*. of or relating to the traditional (English) system of measurement, as opposed to the metric system or another modern system.

traditional system or **traditional measure** *Metrology.* another name for the English system of measurement, because of its historic use in Great Britain and elsewhere. See ENGLISH SYSTEM.

traersu *Meteorology.* a violent east wind that blows over Lake Garda, in northern Italy.

traffic *Transportation Engineering.* the flow of vehicles, persons, or objects along a specified route. *Telecommunications.* **1.** the messages transmitted and received in a particular channel of communications. **2.** the intensity of messages transmitted, switched, or received in a network or network element.

trafficability *Civil Engineering.* the quality of bearing strength of the ground with respect to moving loads such as traffic along a road, functioning as a foundation characteristic for an area.

traffic alert and collision avoidance *Aviation.* an aircraft-to-aircraft system that provides pilots with information about nearby aircraft and warns of airspace conflicts and potential collision hazards.

traffic assignment *Transportation Engineering.* a modeling procedure that assigns generated trips to specific traffic routes.

traffic capacity *Telecommunications.* the maximum traffic per unit time that can be carried by a clearly delineated telecommunication system or subsystem.

traffic control *Transportation Engineering.* the control of the flow of traffic through the use of regulating mechanisms.

traffic cop *Control Systems.* an informal term for the portion of a programmable controller's executive program that is devoted to input and output.

traffic demand *Transportation Engineering.* the public's desire to make trips, which determines the capacity required of highways and other traffic systems. However, traffic demand itself responds to supply conditions, as when anticipated heavy congestion reduces demand.

traffic density *Transportation Engineering.* the average amount of cars occupying a certain designated length of roadway, expressed in vehicles per mile.

traffic diagram *Telecommunications.* a chart that illustrates the movement and control of traffic over a given communications system.

traffic distribution *Telecommunications.* **1.** the forwarding of communications traffic through a terminal to a dialing center or a switchboard. **2.** the traffic intensity as a function of time, geographic source, etc.

traffic engineering *Transportation Engineering.* a branch of transportation engineering concerned with the application of scientific methods and civil engineering principles to the analysis, design, construction, and operation of transportation systems.

traffic flow *Transportation Engineering.* the average number of vehicles that pass a given point in a given time, expressed in vehicles per hour.

traffic flow security *Telecommunications.* the transmission of a continuous flow of random text between two stations without indicating to an unauthorized receptor what transmission component contains an encrypted message and what component is random filler.

traffic forecast *Transportation Engineering.* the process or technique of estimating future traffic requirements, based on existing traffic patterns and anticipated new trip generating developments, planned additional highways, and related factors. *Telecommunications.* a prediction of communications traffic level, used as a basis for management decisions and engineering or designing plans.

traffic generation *Transportation Engineering.* an increase in traffic arising from population growth or new land development.

traffic intensity *Telecommunications.* the amount of message traffic divided by the time of observation; this equals the average number of resources that are busy during the same time period. Also, **traffic load.** *Computer Technology.* the ratio of data insertion rate to deletion rate.

traffic noise *Engineering.* a term for a background sonar noise that is produced by ships that are present in an area.

traffic pan see PRESSURE PAN.

traffic pattern *Aviation.* the prescribed or normal flight path of aircraft above a controlled or uncontrolled airdrome.

traffic signal *Civil Engineering.* an electric signal to direct traffic, operated by a preset time sequence or activated by the passage of traffic over or past some form of detector.

Tragulidae *Vertebrate Zoology.* the chevrotains or mouse deer, a family of very small ruminant mammals of the order Artiodactyla, characterized by the lack of horns or antlers; found in Africa and Asia in dense brush areas.

Traguloidea *Vertebrate Zoology.* a superfamily of ruminant mammals containing the chevrotains.

tragus *Anatomy.* the cartilaginous projection in front of the external auditory meatus.

trail a path or track; specific uses include: *Ecology.* a track made across rough country by humans or animals. *Astronomy.* the streak of a meteor on a photograph. *Aviation.* see CONDENSATION TRAIL. *Ordnance.* **1.** the manner in which a bomb follows behind the release aircraft, assuming the aircraft does not change its velocity after release; it may be defined by a line between the point of impact of the bomb and a point directly below the aircraft at the moment of impact. **2.** the rear part of a gun carriage that connects it to a towing vehicle and stabilizes it during firing. *Geology.* **1.** a linear deposit of rock fragments which were picked up by glacial ice at a localized outcropping and scattered along a generally defined tract as the glacier advanced or melted. **2.** the tracks of an organism. **3.** the crushed material along a fault giving evidence of motion. *Mathematics.* in a graph, a walk in which no edge is repeated.

trail angle *Aviation.* the angle between the vertical and the line of sight to an object over which an aircraft has passed; measured from the aircraft and expressed in mils.

trailer *Mechanical Engineering.* a towed carrier that rests on its own front and rear wheels. *Electronics.* **1.** a length of nonmagnetic tape attached to the end of a reel of magnetic recording tape to permit handling of the tape reel without damaging the recorded material. **2.** the visible effects of poor video channel response shown on the picture tube screen of a television set and manifesting itself a bright or dark streak positioned to the right of a light or dark area in the scene being viewed. *Molecular Biology.* a nontranslated segment at the right, or 3', end of messenger RNA following a termination codon.

trailer record *Computer Programming.* a record that follows a set of data and contains summary or supplementary information about the set. Also, **trailer label.**

trail formation *Transportation Engineering.* an arrangement in which vehicles or aircraft move in a column, one directly behind the other, and usually separated by a fixed distance.

trailing antenna *Electromagnetism.* an aircraft antenna with a weighted end that is allowed to hang free from the craft while in flight .

trailing arbutus *Botany.* a creeping plant, *Epigaea repens,* of the family Ericaceae, having leathery leaves and clusters of fragrant white or pink flowers.

trailing control surface *Aviation.* a control surface at the trailing edge of an airfoil within the empennage of an aircraft.

trailing edge *Aviation.* **1.** the rear edge of an airfoil; the edge of an airfoil over which the airflow normally passes last. **2.** the rear section of a multisection airfoil. *Electronics.* the portion of the waveform of a pulse that occurs last, such as the high-to-low excursion of a positive rectangular pulse.

trailing zero *Mathematics.* in the decimal representation of a real number, a digit of zero that is both right of the decimal point and also right of all nonzero digits.

trail rope see DRAG ROPE.

train *Transportation Engineering.* **1.** a connected, self-propelled group of rolling stock; a railroad train. **2.** a line of vehicles traveling together. *Mechanical Engineering.* see GEAR TRAIN. *Astronomy.* the luminous gas or ionized particles that linger behind the head of a meteor. *Ordnance.* a series of bombs dropped in short intervals or sequence. *Military Science.* a service force or group of service elements that provides logistic support; e.g., an organization of naval auxiliary ships or the vehicles and personnel who furnish supply, evacuation, and maintenance services to a land unit. *Electromagnetism.* to point a radar antenna toward the azimuth. *Mining Engineering.* to trace an alluvial mineral deposit to its place of origin. *Psychology.* to carry out a process of training.

train bombing *Military Science.* a bombing method in which two or more bombs are released at predetermined intervals through a single actuation of a bomb-release mechanism.

trainer *Engineering.* any piece of equipment used to instruct personnel in the operation or maintenance of complex systems (e.g., radar, sonar, or flight vehicles), usually through the use of equipment mock-ups or realistic computer-controlled equipment simulators.

training *Psychology.* **1.** any system or process of instruction; teaching. **2.** specifically, the process of teaching a person to carry out a specific physical task or operation or to perform in a certain job or activity. **3.** the sum of such instruction and practice experienced by an individual in a given area. **4.** relating to or utilized for instruction; used to form numerous compounds such as **training aid, training film, training session,** and so on. *Behavior.* the process of teaching a desired behavior to an animal or human.

training data *Robotics.* data entered into a robot's computer by an operator at the beginning of an operation.
training instance *Artificial Intelligence.* a fact represented in predicate calculus or an expert statement that the learning element uses to improve the knowledge base.
training time *Computer Technology.* the part of machine time that is used to train operators or conduct necessary demonstrations.
training wall *Civil Engineering.* any structure designed to influence the flow, scouring, or silting capacity of a river.
train printer see CHAIN PRINTER.
train protection *Transportation Engineering.* equipment or procedures that guard a train against collisions, derailment, or other hazards.
trait *Genetics.* any genetically determined characteristic. *Behavior.* an individual difference in behavior that is stable and consistent. *Pathology.* the heterozygous state of a recessive disorder; e.g., the sickle cell trait.
trait complex *Anthropology.* the interrelated qualities of function, form, use, and meaning centered in a particular object that express the behavioral patterns of a culture.
trait group. *Ecology.* a subdivision of the population within which selection for a trait occurs at both the level of the individual and the level of the group.
trajectory *Mechanics.* the path described by a body as it moves through space. *Ordnance.* specifically, the path of a missile, bullet, shell, or other such projectile between the firing point and the point of impact. *Geophysics.* the path along which seismic waves travel. *Surgery.* an instrument used to locate a bullet within a wound.
trajectory chart *Ordnance.* a chart providing side-view patterns and numerical measurements of the trajectories of a projectile fired at varying elevations under standard conditions; different charts are provided for different guns, projectiles, and fuses.
trajectory control or **trajectory planning** *Robotics.* a method of continuous-path control in which the controller computes a robot's path based on mathematical models of joint acceleration, arm loads, and actuation signals. Also, PATH CONTROL, PATH PLANNING.
trajectory-measuring system *Engineering.* a system that provides three-dimensional information pertaining to the path along which an object is traveling.
traJ gene *Genetics.* the gene that controls the transfer operon.
tram *Mechanical Engineering.* **1.** a vehicle or cage that is suspended from cables in a tramway system. **2.** a usually open-sided bus used to carry passengers at low speeds for short distances, as at a zoo or amusement park. *Mining Engineering.* a vehicle used to haul loads in a mine.
trama *Mycology.* in fungi belonging to the subdivision Basidiomycotina, fungal hyphal tissue that is interwoven between hymenia.
Trametes *Mycology.* a genus of fungi of the order Agaricales, occurring on wood and including the parasitic species *T. pini,* which attacks living trees.
trammel *Engineering.* a board having two intersecting perpendicular grooves in which the ends of a beam compass can slide, used to draw ellipses.
tramontane *Meteorology.* **1.** a violent polar wind that blows from the northwest through Languedoc and Roussillon, in southern France. **2.** any north wind that issues from a mountainous region. Also, **tramontana.**
tramp *Ecology.* any species that has an extensive geographical range and the ability to travel around it, but is usually absent from more mature or complete communities. Also, **tramp species.**
tramp metal *Mining Engineering.* unwanted metal that has become mixed with the run-of-mine coal or ore.
tramway *Mechanical Engineering.* a suspended cable system along which passengers or freight are transported. Also, AERIAL TRAMWAY, AERIAL CABLEWAY. *Mining Engineering.* **1.** an elevated track used for mine haulage. **2.** any roadway in a mine.
trance *Neurology.* **1.** a state of altered consciousness characterized by heightened focal awareness and decreased peripheral awareness. **2.** a sleeplike state of reduced consciousness and activity.
tranexamic acid *Pharmacology.* $C_8H_{15}NO_2$, a hemostatic drug, used experimentally in the treatment of severe hemorrhage associated with excessive fibrinolysis.
tranquil flow *Hydrology.* a term for water flowing at a rate that is lower than the velocity of a long surface wave in still water. Also, STREAMING FLOW, SUBCRITICAL FLOW.
tranquilizer *Pharmacology.* any drug or other substance that calms or soothes a person, such as an antidepressant.
trans *Organic Chemistry.* a form of a molecule in which the unsaturated bonds of a monomer are located on opposite sides of the molecule. *Genetics.* a configuration in which mutants of two nonallelic genes are present on separate homologous chromosomes.
trans- *Science.* a prefix meaning "through," "across," or "beyond." *Organic Chemistry.* a prefix denoting the presence of certain atoms or radicals on opposite sides of the molecule. *Genetics.* a prefix denoting the *trans* configuration.
trans. transfer; transverse; transition; transportation.
transabdominal *Physiology.* through the abdominal wall.
trans-acting factor *Molecular Biology.* any protein that acts on DNA molecules after diffusing through the cytoplasm into the nucleus; e.g., catabolite activator protein. Similarly, **trans-acting locus.**
transaction *Psychology.* an interaction of one person with the environment, especially with one or more other people. *Computer Programming.* any single message that is input to a computer system, especially an input designed to update a file or database.
transactional analysis *Psychology.* a form of psychotherapy that emphasizes the analysis of interpersonal relationships (transactions).
transactional theory (of perception) *Psychology.* the belief that fundamental perceptions are dependent on knowledge derived from interactions with the environment, and that this knowledge is then used to interpret new experiences.
transaction data *Computer Programming.* transient data relating to a particular incident in a data-processing application that will be combined with the master file. Also, WORK FILE.
transaction processing system *Computer Technology.* a computer system that processes multiple transactions in real time as they occur, such as an airline reservation system.
transaction record see CHANGE RECORD.
transactor *Computer Technology.* a system that transmits data from several sources in different locations to a processing center, where it is immediately processed by a computer.
transadmittance see TRANSCONDUCTANCE.
transalpine *Ecology.* found on the northern side of the Alps mountain range.
transaminase see AMINOTRANSFERASE.
transamination *Biochemistry.* the reversible transfer of an amino group from an amino acid to what was originally an α-keto acid, forming a new keto acid and a new amino acid.
transannular reaction *Organic Chemistry.* a reaction, normally involving adjoining atoms, that occurs in an alicyclic compound having eight or more members in the ring and actually takes place at sites separated by more than one ring atom.
transaxle *Mechanical Engineering.* an assembly containing the transmission and differential systems of a motor vehicle.

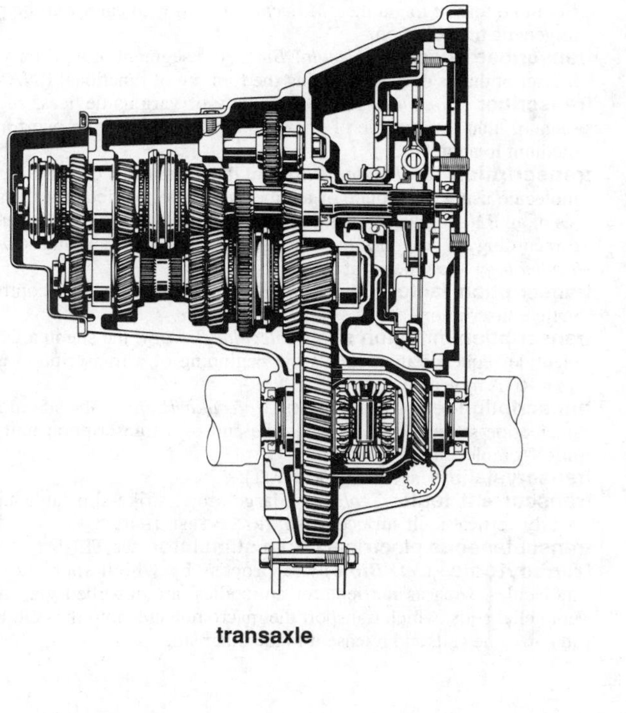

transaxle

transattack period *Military Science.* in nuclear warfare, the period from the initiation of the attack to its termination.

transcapsidation see PHENOTYPIC MIXING.

transceiver *Electronics.* a single unit comprising a radio transmitter and receiver and having some common components for both transmitting and receiving functions. *Computer Technology.* a terminal that can both transmit and receive data.

transcellular fluid *Physiology.* the portion of the extracellular fluid produced by active cellular secretion.

transcendental element *Mathematics.* let F be an extension field of a field K. An element u of F is said to be transcendental over K if it is not the root of any nonzero polynomial with coefficients in K; that is, u is not algebraic over K. For example, $\sqrt{2}$ is algebraic over the rational numbers, but π is not.

transcendental field extension *Mathematics.* an extension field F of a field K is called a transcendental extension if at least one element of F is transcendental over K. If F is not a transcendental extension, it is an algebraic extension of K. A **transcendence base** of F with respect to K is a minimal subset S of F that is algebraically independent over K; the number of elements of S is the **transcendence degree** of F with respect to K.

transcendental function *Mathematics.* a nonalgebraic function; i.e., a function that cannot be written as an algebraic expression of its argument(s). Examples include trigonometric, exponential, and hyperbolic functions and their inverses. An equation in which the arguments appear in a transcendental function is called a **transcendental equation,** and the graph of a transcendental function is called a **transcendental curve.**

transcendental number *Mathematics.* a real number that is not algebraic; i.e., one that is not the root of a polynomial with rational coefficients. Thus $\sqrt{2}$ is algebraic, while π is transcendental.

transcipient *Cell Biology.* describing a cell that has received DNA from another cell.

transconductance *Electronics.* a value expressing the performance of an amplifying device, defined as the ratio of the change in output current to the change in input voltage. Also, MUTUAL CONDUCTANCE, TRANSADMITTANCE.

transcontinental ballistic missile *Ordnance.* a ballistic missile that has a range of 12,500 miles or greater, thus allowing it to reach any point on the earth's surface from any other point.

transcortical *Neurology.* **1.** of or relating to connections between areas of the cerebral cortex. **2.** passing through the cortex of the cerebrum or cerebellum.

transcortin see CORTICOSTEROID-BINDING GLOBULIN.

transcribe *Computer Programming.* **1.** to record information, in any form, on any appropriate storage medium. **2.** to copy digital data from one storage medium to another or between a computer memory and its storage medium, in either direction; this may involve translating data from one format to another. *Molecular Biology.* to carry out the process of genetic transcription.

transcribed spacer *Molecular Biology.* a segment of a primary rRNA transcript that is discarded during the forming of functional RNAs.

transcriber *Computer Technology.* any of various devices, such as a scanner, that are capable of converting information from one format or medium to another.

transcription *Molecular Biology.* the formation of a nucleic acid molecule using a template of another molecule, particularly the synthesis of an RNA molecule using a DNA template. *Acoustical Engineering.* a recording of program material used in radio broadcasting. *Computer Technology.* the process of transcribing data.

transcription factor *Molecular Biology.* any factor that controls genetic transcription.

transcription initiation site *Molecular Biology.* the site in a DNA nucleotide sequence that signals the beginning of a transcription unit for an mRNA molecule.

transcription termination site *Molecular Biology.* the site in a DNA nucleotide sequence that signals the end of a transcription unit for an mRNA molecule.

transcrystalline see INTRACRYSTALLINE.

transcurrent fault *Geology.* a large-scale, strike-slip fault having a nearly vertical fault surface. Also, TRANSVERSE THRUST.

transcutaneous electrical nerve stimulator see TENS.

transcytosis *Cell Biology.* a process by which specific macromolecules, such as nutrients or antibodies, are absorbed via polarized epithelial cells, which transport the macromolecule into the cell, transfer it across the cell, and release it to the other side.

transdermal *Medicine.* entering through the skin. Also, **transcutaneous.**

transdermal patch *Pharmacology.* a small, bandagelike, adhesive patch containing medication and applied to the skin, allowing the medication to seep gradually into the skin; used, for example, to control motion sickness, inhibit overeating, end or reduce a smoking habit, and so on.

transdetermination *Cell Biology.* an alteration in a determined cell, causing the cell to follow a pathway of differentiation other than that originally determined for it.

transdifferentiation *Cell Biology.* a rare alteration in the state of differentiation of a cell.

transduced element *Cell Biology.* a fragment of DNA that has been transferred via a bacteriophage vector from one cell to another.

transducer *Engineering.* a device that converts a signal from one form of energy to another. *Acoustical Engineering.* specifically, a device that converts electrical energy into acoustic energy, such as a loudspeaker, or that converts acoustic energy into electrical energy, such as a microphone. *Robotics.* a device that converts one type of signal into another for the purpose of feedback.

transducer loss *Electronics.* an indication of the efficiency of a transducer defined as a ratio of the power into the transducer to the power out of the transducer.

transducing phage *Molecular Biology.* any bacteriophage that can facilitate genetic recombination between bacteria by transferring segments of chromosomes from a donor cell to a recipient.

transduction *Genetics.* the transfer of a bacterial gene from one bacterium to another by a bacteriophage. Thus, **transducing phage.**

transductional shortening *Molecular Biology.* in a transduction, a phenomenon in which a transduced plasmid becomes smaller than the donor plasmid due to the process of encapsidation.

transdural *Neurology.* passing through or across the dura mater of the brain and spinal cord.

transect *Science.* to cut across. *Ecology.* a line used to survey the distribution of organisms in a specific area.

transept *Architecture.* the major transverse part of the body of a church.

transesterification *Organic Chemistry.* the transformation of an organic acid ester into another ester of that same acid.

transf. or **trf.** transfer.

transfection *Genetics.* the process of inserting foreign DNA into a culture of eukaryotic cells by exposing them to naked DNA. *Virology.* the artificial infection of animal or bacterial cells after they have been inoculated with viral nucleic acid, resulting in the replicated nucleic acid.

transfer a movement from one place to another; a shift; specific uses include: *Transportation Engineering.* the process of changing planes by a passenger. At a hub airport (e.g., Dallas-Fort Worth, Chicago O'Hare), transfer traffic is a large proportion of the total internal traffic flow. *Robotics.* the conveyance of an object from one place to another by a robot. *Computer Programming.* **1.** to change control, or jump from one sequence of instructions to another sequence in another location. **2.** to move data from one storage location or medium to another. *Mining Engineering.* a vertical or inclined connection used as an ore pass between two or more levels in a mine. *Navigation.* the minimum distance a vessel moves perpendicular to its original course in making a turn of 90°; it is sometimes used in reference to a turn of less than 90°.

transfer admittance *Electronics.* a value expressing the performance of a transducer or network under dynamic conditions; defined as the value of the AC component of the current at one set of terminals divided by the AC component of the voltage applied to the other pair of terminals.

transfer area *Military Science.* in an amphibious operation, the water area in which troops and supplies are transferred from landing craft to amphibious vehicles.

transferase *Enzymology.* an enzyme that transfers a chemical group from a donor compound to an acceptor compound. Also, **transferring enzyme.**

transfer caliper *Mechanical Devices.* a caliper whose legs can be adjusted or removed as needed, so that the inside recesses of workpieces or other confined areas can be measured.

transfer cell *Cell Biology.* a specialized plant cell, containing an involuted plasma membrane, that facilitates the flow of solutes in the vascular system.

transfer chamber *Engineering.* a chamber in which thermosetting plastic is softened by heat and pressure before final curing.

transfer characteristic *Electronics.* **1.** the general relationship between the voltage and current at two different electrodes in an active device; usually shown in graph form. **2.** the function or characteristic of a circuit or device that describes its performance, usually expressed as a numeric value; for example, the voltage gain of an amplifier or the sensitivity of a television camera tube expressed as the relationship between output signal and input illumination.

transfer check *Computer Technology.* a check, usually automatically performed, on the accuracy of a data transfer.

transfer coefficient *Physical Chemistry.* **1.** a quantity indicating the rate at which mass, heat, or momentum is transferred in a physical or chemical process. **2.** in an electrochemical process, the fraction of the potential difference at the electrode surface that assists in charge transfer in one direction, but inhibits it in the opposite direction; related to the slope of a plot of log current versus potential.

transfer constant *Engineering.* a transducer rating, equal to one-half the natural logarithm of the complex ratio of the product of the effort to the effect variable. Also, TRANSFER FACTOR. *Electronics.* see IMAGE TRANSFER CONSTANT.

transfer devices *Robotics.* equipment used to move parts, components, and workpieces from one place to another.

transfer ellipse see TRANSFER TRAJECTORY.

transference *Medicine.* the passage or transfer of a symptom or disorder from one part of the body to another. *Psychology.* **1.** the tendency for an individual to transfer thoughts and feelings from earlier personal relationships, such as with a parent or other significant figure in childhood, to a present relationship with a therapist. **2.** broadly, any situation in which an individual transfers unresolved feelings from earlier significant relationships to a present interaction with another or others.

transference number *Materials Science.* the fraction of the total conductivity of a material that is contributed by charge carriers such as charged ions, electrons, or electron holes. *Physical Chemistry.* the fraction of the electrical current carried by a single ion in an electrolytic solution. Also, HITTORF NUMBER, TRANSPORT NUMBER.

transfer factor *Immunology.* a substance derived from the white blood cells of humans having delayed hypersensitivity; this hypersensitivity may be transferred by injecting the substance into normal individuals. *Engineering.* see TRANSFER CONSTANT.

transfer function *Control Systems.* the ratio of a linear system's output to its input in either the complex frequency or the frequency domain.

transfer grafting *Materials Science.* a method of polymerization in which the backbone monomer is reacted with an initiator in the presence of a second monomer; used to produce graft polymers.

transfer impedance *Electricity.* in a network, the ratio of the voltage applied to one pair of terminals and the current produced by it through another pair of terminals that are shorted; in general, expressed as a complex number.

transfer-in-channel command *Computer Technology.* a command used to transfer channel control to a location in main memory when the next location does not contain a channel command.

transfer instruction *Computer Programming.* an instruction, such as a jump or branch, that causes transfer of control to another sequence of instructions.

transfer machine *Mechanical Engineering.* a device used to move equipment within a factory. *Robotics.* a robotic device that holds and carries a workpiece through all stages of the manufacturing process.

transfer matrix *Control Systems.* **1.** a matrix in which the product of the vector representing input variables multiplied by the matrix produces the vector representing output variables. **2.** the multiple input/multiple output representation of a transfer function.

transfer molding *Materials Science.* a thermoset-molding process primarily used for thermosets but also used for thermoplastic composites in which the molding compound is first softened by heat in a transfer chamber, then forced under high pressure into one or more mold cavities for final curing.

transfer of control *Transportation Engineering.* the shift of responsibility for an aircraft from one ground control center to the next one along its route.

transfer of fire *Military Science.* the change of fire from one target to another, using the firing data adjustments for the first target on the firing data for the second target.

transfer of training *Behavior.* the theory that an acquired proficiency in one skill or branch of knowledge helps in the acquisition of a skill or knowledge in another field.

transfer operation *Computer Programming.* any operation used to move data from one location or medium to another.

transfer orbit see TRANSFER TRAJECTORY.

transfer printing *Textiles.* a printing technique in which a color or pattern is printed on paper, then transferred to fabric by means of heat and pressure.

transfer rate *Computer Technology.* the speed at which data can be transferred between an input/output device and a processing unit, usually expressed in bits per second.

transfer reaction *Nuclear Physics.* a nuclear reaction in which the target nucleus and the incident projectile exchange nucleons.

transferred electron *Electronics.* a free electron that has been transferred from one minimum to another within the same zone; the effective mass of the electron changes to that of the new minimum, but not its location.

transferred-electron amplifier *Electronics.* a microwave amplifier circuit in which the two main elements are a transferred-electron diode and a suitable resonator such as a tuned cavity.

transferred-electron device *Electronics.* a semiconductor device, usually a diode made from gallium arsenide or indium phosphide, whose operation depends on the negative resistance characteristics associated with transferred electrons; it is capable of operating in the microwave range of frequencies when combined with an appropriate cavity or a microstrip integrated circuit.

transferred-electron diode *Electronics.* a transferred-electron device that produces microwave energy directly from a DC input when combined with a cavity or microstrip integrated circuit.

transferred-electron effect *Solid-State Physics.* an effect that is observed in the variation of the effective carrier mobility of a semiconductor when a significant number of electrons are transferred between low- and high-mobility valleys in the conduction band.

transferrin *Hematology.* serum β-globulin that binds and transports iron. Several types (C, B, D, and many others) have been distinguished on the basis of electrophoretic mobility and related as the products of corresponding dominant somatic genes, Tf^C, Tf^B, and Tf^D.

transferring enzyme see TRANSFERASE.

transfer RNA *Genetics.* any of at least 20 species of RNA having a low molecular weight and a polynucleotide chain containing approximately 70 to 80 bases; transfers amino acids to the growing end of a polypeptide chain during translation.

transfer robot *Robotics.* a fixed-sequence robot with the single function of moving objects from one location to another.

transfer switch *Electricity.* a type of air switch arranged so that a conductor connection is transferable from one circuit to another without interruption to the current flow.

transfer trajectory *Space Technology.* the flight path taken by a spacecraft as it moves from an orbit or trajectory to another trajectory; tangent to both a departure orbit and a target orbit. Also, TRANSFER ORBIT, TRANSFER ELLIPSE.

transfer unit *Chemical Engineering.* the relationship in fixed-bed sorption operations between the overall rate coefficient, the fluid volumetric flow rate, and the column volume.

transfilter induction *Cell Biology.* a technique with which the ability of one cell type to induce differentiation in a second cell type in the absence of cell contact is examined by placing a pore-sized filter between the two growing cell layers in vitro. *Mathematics.* a technique of deductive reasoning equivalent to the axiom of choice, which is given in the following theorem: A subset S of a well-ordered set X is equal to X if whenever all the (strict) predecessors of an element x of X are in S, then x is also in X. Transfinite induction differs from mathematical induction in the following ways: (a) there is no assumption concerning a starting point; and (b) the inductive step involves passing to each element from the set of all its predecessors instead of from its immediate predecessor (this is important because elements of a well-ordered set may not have immediate predecessors).

transfinite *Science.* transcending the finite; going beyond what is finite.

transfinite number *Mathematics.* **1.** any ordinal number that is not a natural number; the least such number is the set ω of all natural numbers. **2.** any cardinal number that is not finite; the least such number is X_0, the cardinal number of the set of natural numbers.

transfixion *Surgery.* a surgical procedure in which tissue is cut from within outward, such as an amputation.

transfluxor *Electromagnetism.* a binary magnetic core that has two or more apertures and thus three or more legs of the magnetic circuits; used as a computer memory element or control element.

transform *Electricity.* to change an electric current to a higher or lower voltage, from alternating to direct current, or from direct to alternating current. *Computer Programming.* to change the format or composition of data without significantly altering its meaning or value. *Mathematics.* a generic term for a mapping that takes one space of functions into another space of functions; e.g., Laplace, Mellin, Fourier transforms (each of which happens to be linear), and so on.

transformation a process or instance of change; specific uses include: *Electricity.* **1.** the changing of a voltage or current from one magnitude or type to another. **2.** in a network, the interchanging of voltages and currents or of impedances and admittances, to develop an equivalent network. *Mathematics.* **1.** a map, function, or operator. **2.** a linear function on a vector space or module over a division ring. Also, LINEAR TRANSFORMATION. **3.** functions defined on a coordinate space that express one set of coordinates in terms of another. Also, COORDINATE TRANSFORMATION. *Robotics.* a method of making coordinate-system conversions that gives robots the capability of tracking objects or describing the position and orientation of two or more objects. *Linguistics.* in transformational grammar, the process that changes a deep-structure representation into a surface-language representation; e.g., "John hit the ball" and "The ball was hit by John" are related by a passive transformation. *Genetics.* **1.** in general, any alteration in the properties of a cell that is inherited by its progeny. **2.** the conversion of eukaryotic cells to a state of unrestrained growth in culture, resembling or identical with the characteristics of cancerous growth. *Molecular Biology.* a genetic change involving the incorporation into a cell of purified DNA from another cell. *Oncology.* the change of a cell from a normal to a malignant state.

transformational grammar *Linguistics.* a theory of language proposed by the American linguist Noam Chomsky, in which a distinction is made between the surface structure of a language (i.e., the sentences that are actually spoken and written by users of the language) and its deep structure (i.e., the underlying rules that these users employ to form such sentences). According to this theory, young children's ability to create an infinite variety of new sentences derives from an intuitive grasp of this consistent deep structure, rather than from imitation of the language of adults. Also, **transformational-generative grammar.**

transformation assay *Virology.* a test that determines the frequency with which virus-infected cells become tumorigenic.

transformation constant see DECAY CONSTANT.

transformation efficiency *Genetics.* efficiency of transforming a cell by DNA transfer.

transformation group *Mathematics.* any group whose elements can be interpreted as transformations on some space; the group operation is interpreted as a composition of functions.

transformation matrix *Electromagnetism.* a 2×2 matrix that is used to describe the relationship between the amplitudes of a traveling wave on either side of a waveguide junction.

transformation series *Evolution.* the sequence of changes in a single hereditary character over the course of its phylogenetic evolution, usually from plesiotypic to apotypic.

transformation temperature *Metallurgy.* the temperature at which a phase change occurs; this term is also used to show the limiting temperature of a transformation range. Also, **transformation point.**

transformation temperature ranges *Metallurgy.* the ranges in which austenite forms during heating and transforms during cooling.

transformation theory *Quantum Mechanics.* the study of coordinate and other transformations in quantum mechanics and how they affect properties of the system.

transformation toughening *Materials Science.* the expansion or increase in the volume of a material encased by a matrix that does not itself expand; this expansion results in tiny microcracks in the matrix that absorb stresses and greatly increase fracture toughness.

transformer *Electromagnetism.* an electromagnetic device that consists of an input coil (primary) and an output coil (secondary), both of which are wound on a common core; an alternating input signal is inductively coupled to the output and will generally have different voltage and current but will have the same frequency as the input.

transformer bridge *Electricity.* a network consisting of a transformer and two impedances; the input signal is applied to the transformer primary and the output signal is taken between the secondary center-tap and the junction of the impedances that connect to the outer leads of the secondary. Used as a crystal filter, a capacitor supplies one impedance to balance the static capacitance of the crystal (the other impedance), to avoid transmission except in the vicinity of crystal resonance.

transformer-coupled amplifier *Electronics.* an amplifier, usually an audio-frequency amplifier, that uses transformers as a means of coupling between stages.

transformer coupling *Electronics.* the use of a transformer as a means of coupling signals between stages in an amplifier or similar electronic device.

transformer input voltage see PRIMARY VOLTAGE.

transformer load loss *Electricity.* the losses in a transformer that occur when it carries a load; they include resistance loss in windings due to load current, stray loss due to stray fluxes in the windings, and core clamps.

transformer loss *Electricity.* the ratio of the signal power that an ideal transformer of the same impedance ratio would deliver to the load impedance, to the signal power delivered by the actual transformer; expressed in decibels.

transformer rectifier *Electricity.* a device that converts alternating current into direct current; a combination transformer and rectifier.

transformer substation *Electricity.* an electric power substation with numerous transformers.

transformer voltage ratio *Electricity.* under specific load conditions, the ratio of the primary voltage to the secondary voltage.

transform fault *Geology.* **1.** a strike-slip fault along which displacement terminates abruptly or changes direction, resulting in the ridge being offset; often associated with mid-oceanic ridges. **2.** a fault zone along which two plates slip past each other with no creation or destruction of the lithosphere taking place.

transforming growth factor *Biochemistry.* one of two polypeptides that generate the phenotype of transformed cells from the phenotype of certain normal cells, also furthering angiogenesis.

transforming principle *Microbiology.* DNA that is responsible for transforming bacterial cells.

transforming section *Electromagnetism.* a section of waveguide or transmission line that is used for impedance transformations.

transfusion *Medicine.* the infusion of whole blood or of blood components, such as plasma, platelets, or packed red cells, directly into the bloodstream; usually to replace blood lost from a wound or injury, during surgery, and so on.

transfusion cell *Botany.* a thin-walled plant cell in the vascular bundles of leaves that allows water distribution.

transfusion reaction *Medicine.* an adverse reaction to blood received in a transfusion, typically because of the incompatibility in some respect of the donor's blood with that of the recipient.

transgenesis *Genetics.* the introduction of new DNA sequences into the germ line, resulting in the production of transgenic animals.

transgenic animal *Molecular Biology.* any animal into which cloned genetic material has been transferred.

transgenic organism *Genetics.* an organism that has become transformed after the introduction of novel DNA into its genome.

transglycosylase see GLYCOSYLTRANSFERASE.

transgranular fracture *Materials Science.* fracture of a material through the grains rather than along the grain boundaries.

transgression *Geology.* a change that creates deep-water conditions in areas that formerly had shallow-water conditions. Also, INVASION. *Oceanography.* a more or less permanent landward advance of the sea, usually caused by a rise in sea level.

transgressive deposit *Geology.* a deposit of sedimentary material laid down in an onlap arrangement when the sea invades a large land area, or when an area of land sinks. Also, **transgressive sediment.**

transhumance *Anthropology.* the seasonal movement of animals, and the humans who herd or hunt them, from one ecological zone to another, as from alpine meadows to lowland valleys.

transhybrid loss *Electricity.* the transmission loss at a given frequency measured across a hybrid circuit.

transhydrogenase *Enzymology.* an enzyme that catalyzes the transfer of hydrogen between NADH and $NADP^+$.

transient *Physics.* **1.** a changing mechanical or electrical property that becomes progressively smaller in magnitude. **2.** any value that changes, decays, or disappears, such as a pulse or oscillation. *Acoustics.* a brief sound that does not occur regularly or predictably. *Military Science.* of or relating to personnel, ships, or craft that stop temporarily at a post, station, or port to which they are not assigned or attached, or to forces that pass through the area of responsibility of another command. Thus, **transient forces.** *Computer Programming.* describing a program that is stored in secondary storage, from which it is moved into and out of main storage as needed for execution.

transient distortion *Electronics.* the distortion of a waveform produced by the inability of a transducer or amplifier to reproduce quick changes in level.

transient error see SOFT ERROR.

transient lunar phenomena *Astronomy.* various mists, reddish glows, flashes of light, and temporary clouds on the moon that a few observers have claimed to see from time to time. Also, LUNAR TRANSIENT PHENOMENA.

transient motion *Physics.* the motion of an object that is suffering the effects of any transients as it approaches its steady-state motion.

transient overshoot *Physics.* the maximum value of a quantity in response to a sudden application of an external agency.

transient phenomena *Electricity.* rapidly changing events that occur in a circuit during a time interval between switch closure and steady-state conditions, or any other temporary events that occur after a change in the circuit or its constants.

transient polymorphism *Genetics.* the temporary presence of two or more allelic variants in a population while one variant is replacing another; an analysis of the gene pool shows only the variants that have survived.

transient problem see INITIAL-VALUE PROBLEM.

transient situational disturbance *Psychology.* an adjustment reaction of acute symptom formation in response to a crisis or overwhelming situation.

transistance *Electronics.* the basic property of active electronic devices by which they produce gain; transistance makes possible such active devices and circuits as amplifiers, oscillators, bistable circuits, and switching devices.

transistor *Electronics.* **1.** an active semiconductor device, usually made from germanium or silicon and possessing at least three terminals (typically, a base, emitter, and collector); characterized by its ability to amplify current and used in a wide variety of equipment such as amplifiers, oscillators, and switching circuits. **2.** relating to, using, or operated with transistors. Thus, **transistor radio, transistor amplifier,** and so on. (A shortened form of transfer resistor.)

transistor biasing *Electronics.* the process of applying a steady voltage to a transistor, normally between the base and emitter electrodes; the bias voltage determines the operating point of the transistor under quiescent conditions.

transistor characteristics *Electronics.* the parameters that describe the operation of a transistor, such as values of current gain or input resistance and impedance.

transistor chip *Electronics.* a shaped and processed piece of semiconductor material that forms a transistor.

transistor circuit *Electronics.* any electronic circuit employing one or more transistors as the active element(s); found, for example, in amplifiers, oscillators, logic gates, and bistable circuits.

transistor clipping circuit *Electronics.* a circuit that uses transistors to prevent a signal from reaching the peak amplitude it would otherwise attain; this results in a distorted waveform whose peaks appear clipped.

transistor gain *Electronics.* the amplification factor of a transistor amplifier stage; the ratio of signal output voltage or current to signal input voltage or current.

transistor input resistance *Electronics.* the resistance presented by a transistor to the input signal. Also, INPUT RESISTANCE.

transistor magnetic amplifier *Electronics.* a power control circuit consisting of a magnetic amplifier combined with a transistor amplifier; the transistor acts as a preamplifier for the control current to the magnetic amplifier.

transistor-transistor logic *Electronics.* a logic circuit that uses transistors to form the logic elements; circuits are configured as NAND gates, NOR gates, or logic-level inverters. Also, **TTL.**

transit *Transportation Engineering.* the conveyance or transportaion of persons or goods from one place to another, as in public transportation (**mass transit**). *Engineering.* **1.** to rotate a telescope on its horizontal axis 180°, reversing its direction. Also, PLUNGE. **2.** an instrument that is used to determine the passage of a celestial body across a meridian. **3.** an instrument fitted with a telescope that measures vertical and horizontal angles, used in surveying. Also, TRANSIT THEODOLITE. *Astronomy.* **1.** the passage of a celestial body across a meridian or through the field of view of a telescope. **2.** the passage of the planet Mercury or Venus across the sun's disk. **3.** the passage of a moon across the disk of a planet. *Navigation.* a worldwide radio navigation system in which fixes are determined by measuring the Doppler shift of continuous-wave, very ultrahigh-frequency waves transmitted from low-orbit satellites.

Transit *Space Technology.* any of a series of navigation satellites first launched by the U.S. Navy in 1960.

transit bearing *Navigation.* a bearing determined by noting the time at which two features on the earth's surface have the same relative bearing.

transit circle *Engineering.* an astronomical instrument used to measure the zenith distance or the declination of a celestial object. Also, MERIDIAN TRANSIT.

transit declinometer *Engineering.* a surveyor's transit that is used to sight on a mark and determine true direction.

transit instrument see TRANSIT TELESCOPE.

transition *Thermodynamics.* a phase change between the solid, liquid, or gas states. *Quantum Mechanics.* a change of a quantum-mechanical system from one energy state to another. *Telecommunications.* a sequential series of actions that takes place when a process changes from one state to another in response to an input; for example, the change from mark to space or from space to mark. *Molecular Biology.* a mutation in which one purine is substituted for another or one pyrimidine for another in DNA.

transitional adaptive zone *Evolution.* an ecological pathway between two major adaptive zones over which organisms pass during the evolution from a form adapted to one environment to one adapted to another.

transitional epithelium *Histology.* the specialized, distendable stratified epithelium that lines the urinary tracts.

transitional fit *Design Engineering.* a fit with varying clearances contingent upon specific tolerance requirements on a shaft or sleeve and hole.

transitional flow *Fluid Mechanics.* a flow in which fluid particles move from a well-behaved laminar pattern into adjacent layers in a random manner that grows in amplitude; the Reynolds stresses are of an order of magnitude equal to the viscous stresses.

transitional frequency *Quantum Mechanics.* the frequency of radiation emitted during a radiative transition.

transition altitude *Aviation.* the altitude at or below which the vertical position of an aircraft is controlled by true altitude. Also, **transition layer.**

transition bands *Materials Science.* in the microstructure of a deformed metal, the boundaries between different regions within a grain that achieve the same imposed strain with a different set of slip systems.

transition element *Chemistry.* in the periodic table, any of a group of elements representing a gradual shift from the strongly electropositive elements of groups IA and IIA to the more electronegative elements of groups IB and IIB; such elements have an atomic structure in which the inner orbitals are systematically filled, providing some unpaired electrons. Also, **transition metal.** *Electromagnetism.* an element or junction that couples one type of transmission system to another, as in the transition from a transmission line to a waveguide.

transition frequency *Acoustical Engineering.* the frequency during a disk recording at which the recording characteristic changes from constant amplitude to constant velocity. Also, CROSSOVER FREQUENCY.

transition function *Computer Technology.* in a sequential machine, the function that determines the next state from the present state and given input, while the output function determines the output.

transition lattice *Metallurgy.* a crystallographic, unstable lattice that forms during solid-state reactions such as eutectoid decomposition.

transition level *Aviation.* the lowest flight level available for use above the transition altitude.

transition loss *Electricity.* a loss at any point in a transmission system; the ratio of the available power from the part of the system ahead of the point under consideration to the power delivered to the part of the system beyond the point under consideration.

transition moment *Quantum Mechanics.* a multipole moment related to the probability of a radiative transition between two states.

transition operators *Artificial Intelligence.* a set of rules that determine all possible next states from a given state of the problem.

transition point *Thermodynamics.* the temperature at which a phase transition takes place. Also, TRANSITION TEMPERATURE. *Aviation.* the point or line in a flow, such as the boundary layer of an airfoil, where laminar flow becomes turbulent. *Electromagnetism.* a point in a circuit along which certain constants change such that there is a reflection of the wave propagating along the circuit.

transition probability *Quantum Mechanics.* the probability per unit time that a particular transition will occur. *Mathematics.* in Markov processes, the fixed (conditional) probability of passing from one state to another.

transition radiation detector *Particle Physics.* a device used to measure the radiation emitted by particles as they cross an interface between different materials; the intensity and other characteristics of these radiations serve as the basis of a velocity measurement.

transition region *Solid-State Physics.* the region between two homogeneous semiconductors where the impurity concentration undergoes a change. *Astronomy.* a layer of the solar atmosphere a few hundred kilometers thick that lies between the chromosphere and the corona, within which temperatures rise from about 10,000 to over one million kelvin.

transition temperature *Metallurgy.* in a notched bar impact test or a notched tensile test, the temperature at which the change from tough to brittle fracture occurs. *Thermodynamics.* see TRANSITION POINT.

transition time *Analytical Chemistry.* the time interval needed for a indicator electrode to become polarized in chronopotentiometry.

transition to chaos *Chaotic Dynamics.* the specification of parameters at which a normally behaved system becomes a chaotic system.

transition zone *Geology.* **1.** the region of the earth's upper mantle, equivalent to the C layer, in which density and seismic-wave velocities increase. **2.** the region within the earth's outer core, equivalent to the F layer, that is transitional to the inner core. *Fluid Mechanics.* the limited region in which flow changes from laminar to turbulent.

transitive relation *Mathematics.* given a set A, a relation R on $A \times A$ is said to be transitive if $(x,y) \in R$ and $(y,z) \in R$ imply that $(x,z) \in R$ for all x, y, and z in A. For example, $<$ ("less than") is a transitive relation. R is an equivalence relation if it is reflexive, symmetric, and transitive.

transitory target *Military Science.* a target that may exist or be able to be attacked for only a limited period of time.

transitron *Electronics.* an electron-tube circuit whose operation depends on the negative resistance characteristic of the suppressor grid of a pentode.

transitron oscillator *Electronics.* an oscillator whose operation depends on the negative resistance characteristics of a pentode vacuum tube in which the suppressor-grid voltage is more positive than the plate voltage and the suppressor grid is capacitively coupled to the screen grid.

transit survey *Engineering.* a method of ground surveying in which a transit is used to determine direction and distance to a fixed point.

transit telescope *Optics.* a telescope fixed in the meridian that is used to time the transit of a celestial object. Also, TRANSIT INSTRUMENT.

transit theodolite *Engineering.* see TRANSIT, def. 3.

transit time *Electronics.* the time it takes an electron to travel from the cathode to the plate in an electron tube, or for a charge carrier to travel from the emitter to the collector in a bipolar transistor.

transit-time microwave diode *Electronics.* a solid-state diode in which the transit time of charge carriers is sufficiently fast to permit operation at microwave frequencies.

transit-time mode *Electronics.* in a transferred-electron diode, any of several operating modes in which the frequency of operation of the device is limited by the dimensions of the drift region between the cathode and the anode.

transketolase *Enzymology.* an enzyme that catalyzes the transfer of a unit of glycolaldehyde from a ketose to an aldehyde acceptor.

translate *Computer Programming.* **1.** to convert from one computer language to another. **2.** generally, to convert information from one form to another without altering meaning or function.

translating roller *Ordnance.* a double-threaded screw on which the breechblock of a large-caliber gun moves back and forth.

translation *Computer Programming.* a process of converting from one computer language or form of data to another. *Mechanics.* **1.** the motion of a rigid body or system without any rotation or change in orientation, so that all points of the body move in the same direction. **2.** the act of moving a system to a new position without rotating it or changing its shape or structure. **3.** the motion of a particle in space, usually a linear or curvilinear motion. *Molecular Biology.* protein synthesis, occurring at the ribosome, in which the sequence of codons in mRNA is used to specify the sequence in which amino acids are added to a growing polypeptide chain. *Mathematics.* **1.** a function on a group given by operating on each group element by a fixed group element. **2.** a rigid motion (isometry) of the points of the Euclidean plane or space that carries any line segment AB to a parallel, similarly oriented, congruent line segment $A'B'$. In particular, if the group is the set of points in a coordinate space under the group operation of addition, a translation represents a coordinate shift parallel to one or more of the coordinate axes.

translational *Science.* of or relating to a movement or process of translation, especially the motion of a body without rotation.

translational amplification *Molecular Biology.* a series of translation processes in which a single mRNA directs the synthesis of many polypeptide chains.

translational energy *Mechanics.* the kinetic energy of a body or system, neglecting rotation, which is equal in the nonrelativistic limit to one-half the mass times the velocity of the center of mass squared.

translational equilibrium *Mechanics.* the condition of a system in which there is either no change in position or a change in position at a constant rate; e.g., a car is in translational equilibrium if it is either at rest or moving in a straight line at a constant speed.

translational fault *Geology.* a segment of a fault in which there has been uniform movement in one direction without rotation, so that the dip in the two walls does not change. Also, TRANSLATORY FAULT.

translation algorithm *Computer Programming.* a set of rules, steps, and procedures designed to translate a set of statements from one programming language to another.

translational lift *Aviation.* the lift exerted on helicopter blades by an increase in speed or a change from one type of flight to another, such as from hovering to horizontal flight.

translational motion *Mechanics.* the motion of a rigid body or system in space without any rotation. Also, PURE TRANSLATION.

translational movement *Geology.* the apparent movement of a fault-block uniformly and in one direction without rotation, so that features that were parallel prior to movement remain parallel afterward.

translation error *Robotics.* an error in the location of an assembled part.

translation glide *Materials Science.* a process in which slip occurs on crystal planes as a result of plastic deformation.

translation operation *Physics.* an operation that involves displacement without rotation.

translator *Computer Programming.* **1.** a program that can convert from one computer language to another. **2.** any program that can convert language statements to machine code, such as compiler, assembler, and interpreter programs. Also, TRANSLATOR ROUTINE. *Electronics.* **1.** any device or unit that converts the format of the intelligence contained in electronic signals to another format. **2.** in a television broadcast distribution system, the receiver-transmitter equipment used to rebroadcast received television signals on a different channel in order to optimize program coverage.

translatory fault see TRANSLATIONAL FAULT.

translatory motion see TRANSLATIONAL MOTION.

transliterate *Computer Programming.* **1.** to convert characters or words from one language to corresponding characters or words of another language. **2.** to convert data from one form to another so that the characters correspond to each other exactly.

translocase *Enzymology.* **1.** an enzyme that transports a substrate across a membrane or within a biological fluid. **2.** an enzyme that catalyzes the shift of the ribosome one codon along the mRNA molecule during protein synthesis.

translocation *Botany.* the movement of soluble nutrients from one part of a plant to another. *Cell Biology.* the movement of a segment of DNA from one chromosome to another, resulting in a change in the position of the segment.

translocational control *Biochemistry.* **1.** the control of protein synthesis at the level of translation. **2.** the control of a gene's expression based on the regulation of the rate of translation of the mRNA delineated by that gene.

translocational mapping *Molecular Biology.* a technique for mapping the positions of loci on a chromosome by using markers to locate the position of the breakage.

translocation breakpoint *Genetics.* a breakpoint in a chromosome leading to its movement to another position in the genome.

translucency *Materials Science.* the ability to transmit light; an important optical property of most porcelains, determined by the level of interaction between the incident electromagnetic radiation and the electrons within the material.

translucent *Optics.* having the properties of translucency or of a translucent medium.

translucent attritus *Geology.* a constituent of coal formed by humic degradation that is capable of transmitting light. Also, HUMODURITE.

translucent medium *Optics.* a medium that transmits light but does not transmit an image. Also, TRANSPARENT DIFFUSER.

translucidus *Meteorology.* a cloud variety that is sufficiently transparent or translucent so as not to obscure the sun, moon, or higher clouds; found in the genera altocumulus, altostratus, stratocumulus, and stratus.

translunar *Astronomy.* relating to or located in the region of space beyond the orbit of the moon.
transmethylase *Enzymology.* an enzyme of the transferase class that transfers a methyl group from one compound to another.
transmethylation *Biochemistry.* a reaction that causes methyl groups to move from one compound to another.
transmethylation factor see CHOLINE.
transmigration see DIAPEDESIS.
transmissibility *Mechanics.* the measure of a system's ability to amplify or suppress an input vibration; it is equal to the ratio of transmitted amplitude to the applied amplitude.
transmission *Mechanical Engineering.* the system of gears by which power is transmitted from the engine of an automobile or other motor vehicle to the driving axle (or axles). *Medicine.* the transfer of a disease or condition from one person to another. *Genetics.* the communication of inheritable qualities from parent to offspring. *Virology.* the process of transferring a virus from an infected organism to a noninfected organism. *Electronics.* the process of transferring electrical signal energy and information from one place to another by various media such as wires, radio waves, sound waves, and visible light (including laser). *Computer Programming.* the process of moving data from one place to another. *Electromagnetism.* the ratio of electromagnetic radiant power transmitted by a body to the total power incident on the body. Also, TRANSMITTANCE.
transmission band *Electromagnetism.* for a waveguide or transmission line, the frequency range whose lower limit is the cutoff frequency.
transmission coefficient *Physics.* the ratio of the amplitude of a quantity that is transmitted through a medium to the amplitude of the incident quantity. Also, TRANSMISSION FACTOR, TRANSMITTANCE. *Quantum Mechanics.* see PENETRATION PROBABILITY.
transmission control character *Telecommunications.* a functional character that controls or facilitates the transmission of data over communications networks. Also, DATA COMMUNICATION CONTROL CHARACTER.
transmission diffraction *Analytical Chemistry.* the passing of an electron beam through a fine powder for the purpose of electron-diffraction analysis.
transmission dynamometer *Engineering.* an instrument that measures torque and power between a power plant and the driving mechanism.
transmission electron microscope *Electronics.* an electron microscope in which an image is formed as a result of the scattering of electrons by the specimen; focusing is achieved by the use of magnetic lenses.
transmission electron radiography *Electronics.* a microradiography technique for obtaining images of very thin specimens; hardened X-rays are directed to the specimen, which is placed on the photographic plate and covered with a lead foil and a light-tight shield.
transmission factor *Crystallography.* a measure of the amount of light transmitted through a transparent object; it is the reciprocal of the linear absorption coefficient. *Physics.* see TRANSMISSION COEFFICIENT. *Materials Science.* see TRANSMITTANCE.
transmission function *Geophysics.* a mathematical model showing the relation between atmospheric transmission of infrared radiation, path length, and concentration of absorbing gases.
transmission gate *Electronics.* a gate circuit that normally blocks the transfer of signals, except when the gate is enabled by a control pulse and any signal at the input is allowed to pass to the output.
transmission genetics *Genetics.* the study of the mechanisms involved in the passage of a gene from one generation to the next.
transmission grating *Optics.* a diffraction grating that transmits light instead of reflecting it.
transmission interface converter *Computer Technology.* a device that converts data to or from a format suitable for transmission over a channel between a computer and a device or another computer.
transmission level *Telecommunications.* the ratio of the test-tone power at a given point in a transmission system to the test-tone power at a point in the system that has been chosen as the reference point. Also, RELATIVE TRANSMISSION LEVEL.
transmission line *Electricity.* a structure that forms a continuous path from one location to another; used to transmit electric power or signals between two points.
transmission-line admittance *Electricity.* the proportion of current flowing in a transmission line to the voltage across the line; both are expressed in phasor notation.
transmission-line cable *Electricity.* a cable that forms a transmission line and is specified by size and materials, such as a coaxial cable, a waveguide, or a microstrip.
transmission-line current *Electricity.* the amount of electrical energy or electrical charge that passes through a specified point in a transmission line at a specific time.
transmission-line efficiency *Electricity.* the ratio of electrical power at the source to that at the load end of a transmission line.
transmission-line impedance *Electricity.* the proportion of voltage across a transmission line to the current flowing within; both voltage and current are expressed in phasor notation.
transmission-line parameters *Electricity.* those quantities that specifiy impedance and admittance per unit length between various conductors of a transmission line. Also, LINEAR ELECTRICAL PARAMETERS.
transmission-line power *Electricity.* the amount of energy that passes a point in a transmission line per unit time; the power flowing in a transmission line.
transmission-line reflection coefficient *Electricity.* the ratio between the voltage that is reflected from the load at the end of a transmission line and the incident voltage.
transmission-line theory *Electricity.* the application of electrical and electromagnetic theory to the characteristics, perturbations, and phenomena of transmission lines.
transmission-line transducer loss *Electricity.* a proportion of power delivered by a transmission line to that produced at the generator, as expressed in decibels; it is equivalent to the sum of the line's attenuation and the mismatch loss.
transmission-line voltage *Electricity.* an instantaneous point at which a certain degree of force is required to transport a unit electrical charge between two designated conductors of a transmission line.
transmission loss *Telecommunications.* the diminution in power between any two points; usually expressed in decibels.
transmission modulation *Electronics.* modulation of the reading beam in a charge-storage tube; charge patterns on the storage-surface amplitude modulate the beam as it passes through holes in the storage surface.
transmission plane *Optics.* the plane of vibration that will pass through a polarizing filter in a polarized light.
transmission primaries *Telecommunications.* in a color television picture signal, the three signals, labeled L, Q, and Y, that may be combined to produce picture, chrominance, and luminance.
transmission regulator *Telecommunications.* a device used in communication systems for maintaining constant transmission levels.
transmission security *Telecommunications.* a component of communications security that results from all measures designed to protect transmissions from unauthorized interception, traffic analysis, and imitative deception.
transmission speed *Telecommunications.* the amount of informational events sent per unit time; this may be expressed as characters, bands, bits, word groups, or records per second.
transmission substation *Electricity.* an electric-power substation that outputs high voltages for transmission to other substations.
transmission time *Telecommunications.* the absolute time interval from transmission to reception of a signal.
transmissivity *Electromagnetism.* the ratio of transmitted electromagnetic radiation to radiation normally incident on the surface of the boundary between two media.
transmissometer *Engineering.* an instrument that measures the extraction coefficient of the atmosphere and determines visual range. Also, HAZEMETER.
transmissometry *Optics.* the study of the transmission properties of a medium; carried out by shining a light of known intensity and color into the medium and measuring the emerging electromagnetic radiation.
transmit *Physics.* **1.** to cause light, sound, heat, or another form of energy to pass through a medium. **2.** to convey force or motion from one part of a body or mechanism to another. **3.** to allow the passing through of a form of energy; to conduct. *Telecommunications.* to send out signals by radio, television, or wire. *Computer Programming.* to move data from one location to another via wire, radio, laser beams, optical fibers, or other media.
transmit-receive tube *Electronics.* in a radar system waveguide, a gas-filled tube that functions as an automatic switching device to connect the radar transmitter to the antenna when the transmitter is producing an output; at all other times it directs echos received from the antenna to the receiver. Also, DUPLEXER, TR TUBE, TR BOX.

transmittability *Telecommunications.* the capacity of standard electronic and mechanical elements or automatic communications equipment to maintain a code under several different signal-to-noise ratios.

transmittance *Materials Science.* the ratio of transmitted to incident radiant flux. Also, TRANSMISSION FACTOR. *Analytical Chemistry.* the amount of radiant energy transmitted by a sample being analyzed during absorption spectroscopy. *Electromagnetism.* see TRANSMISSION.

transmittancy *Electromagnetism.* the ratio of transmittance of a solution of a given thickness to that of the pure solvent of this thickness.

transmitted wave see REFRACTED WAVE.

transmitter *Telecommunications.* **1.** equipment that produces and modulates radio-frequency energy for the purpose of radio communication. **2.** a device that converts sound waves to electrical waves, such as the microphone of a telephone. **3.** in telegraphy, a device that reads perforated tape and then sends out teletypewriter signals.

transmitter-distributor *Electricity.* a motor-driven device used to translate teletypewriter code combinations from perforated tape to electrical impulses; these impulses are then transmitted to one or more receiving stations. Also, TD.

transmitter off *Telecommunications.* in data transmission, a control signal from receiver to transmitter to inhibit transmission.

transmitter on *Telecommunications.* in data transmission, a control signal from receiver to transmitter to allow transmission.

transmitting loop loss *Telecommunications.* the amount of loss to a signal detected and delivered by the station set, battery supply circuit, and subscriber line on the transmission end.

transmitting mode *Computer Technology.* **1.** any of several methods used to transmit data, such as **simplex** (one way), **simplex with back channel** (one way with limited handshaking), **half duplex** (two way, but one at a time), or **full duplex** (two way, simultaneously). **2.** the state of a device that is actively reading or writing data.

transmittivity *Electromagnetism.* the internal transmittance of a sample of nondiffusive matter of unit thickness.

transmutation *Physics.* any process in which a nuclide changes into a nuclide of a different value, either naturally, as by radioactive disintegration, or artificially, as by bombardment with neutrons. *Nucleonics.* the use of such a transformation process to neutralize nuclear waste. *Biology.* the transformation of one species into another.

transobuoy *Engineering.* a seagoing, automatic weather station that may be either free-floating or moored.

transolver *Electricity.* a synchro with a two-phase cylindrical rotor resident within a three-phase stator; used as a precision transmitter or a control transformer.

transom *Building Engineering.* **1.** a cross member that separates a door from a window or fanlight above it. **2.** a window that is built directly above and in line with the door beneath it. **3.** a window that is divided horizontally by a wood or stone crossbar. *Naval Architecture.* the flat end of a vessel above the waterline, squared off athwartship but often at an angle to the vertical. Transom sterns were common in sailing ships, and transom bows were found in traditional Chinese junks.

transonic *Physics.* relating to phenomena that occur at velocities close to the speed of sound, roughly 0.8 to 1.2 times the speed of sound.

transonic barrier see SONIC BARRIER.

transonic flight *Aviation.* the movement of flight vehicles in the transonic range, characterized by a marked increase in drag and sudden changes in moment.

transonic flow *Fluid Mechanics.* a flow around a body that is partly subsonic and partly supersonic.

transonic range *Fluid Mechanics.* for a body moving through a fluid, the range of speeds between the speed at which one point on the body reaches supersonic and that at which all points reach supersonic.

transonic speed *Fluid Mechanics.* the speed of a body relative to the surrounding fluid at which the flow past the body is transonic.

transonic wind tunnel *Engineering.* a wind tunnel designed to test airflows at speeds near the speed of sound.

transorbital lobotomy *Surgery.* an incision into a frontal brain lobe, which is performed by cutting through the bony orbit. Also, LOBOTOMY.

transosonde *Engineering.* **1.** a balloon equipped with radiosonde that transmits transoceanic meteorological information. **2.** the trajectory of a balloon whose level remains constant, established by following it with a radio-direction finding system.

transovarial transmission *Medicine.* the transmission of pathogens from the maternal organism, by invasion of the ovary and infection of eggs, to individuals of the next generation, as in infections of arthropods, especially mites and ticks.

transparency *Materials Science.* optical transmission through a material because of incident electromagnetic radiation being unable to move the electrons of the substance from their initial energy level to a different energy level; an important property of many ionic ceramics, such as glass. *Graphic Arts.* a positive color or black-and-white photographic image produced on a clear base and designed to be viewed by transmitted light. *Optics.* the extent to which a medium transmits electromagnetic radiation. *Astronomy.* the degree to which the atmosphere is free of dust, haze, water vapor, or other light-scattering and reflecting aerosols. *Artificial Intelligence.* the ability of an expert system to clarify for the user its reasoning, actions, and conclusions.

transparent *Physics.* having the property to allow radiation or particles to pass through with little interference. *Computer Programming.* **1.** of or relating to the technical complexities of or actions taken by a computer system that the user is not aware of, or does not need to know about. **2.** of or relating to an emulation or interface that looks exactly the same as the emulated device or interface.

transparent diffuser see TRANSLUCENT MEDIUM.

transparent medium *Optics.* a medium through which light or another electromagnetic radiation can pass and be detected on the other side.

transparent sky cover *Meteorology.* the portion of sky cover through which the sun, moon, blue sky, or higher clouds may be observed; the opposite of opaque sky cover.

transpassive region *Physical Chemistry.* the portion of an anodic polarization curve in which a metal anode becomes inert.

transpecific evolution see MACROEVOLUTION.

transpiration *Biochemistry.* the process by which water in plants passes into the air as water vapor, mostly through stomata. *Meteorology.* the amount of water so transferred, and thus present in the atmosphere.

transpiration cooling *Space Technology.* a method of cooling a space vehicle or other solid body by extruding a coolant fluid through pores in its surface. Also, SWEAT COOLING.

transplant *Agriculture.* **1.** to transfer a growing plant from one place to another. **2.** a seedling or plant that has been moved one or more times. *Medicine.* **1.** to transfer tissues from one part to another. **2.** an organ or tissue taken from the body for grafting into another area of the same body (e.g., skin or hair from one part of a person's body to another) or into another individual (e.g., a kidney from a healthy donor to a diseased person). **3.** an operation in whch such a transfer takes place.

transplantation *Biology.* the act of transferring an organ or part from one plant or animal to another.

transplantation antigen see T ANTIGEN.

transplantation disease *Medicine.* a disease that results from an immunological graft-versus-host reaction, occurring after transplantation of adult lymphoid cells to incompatible recipients who cannot reject them.

transplanter *Agriculture.* a machine that makes a furrow and then places growing plants in the soil, watering each plant as it is dropped.

transplutonium element *Inorganic Chemistry.* any element with an atomic number greater than 94, the atomic number of plutonium.

transpolarizer *Electricity.* a ferroelectric dielectric impedance that is controlled electrostatically.

transponder *Telecommunications.* **1.** a radio transmitter-receiver that automatically transmits identifiable signals upon receiving the proper interrogation. **2.** in a satellite, a device that accepts signals from the earth, amplifies them, changes their frequency band, and retransmits them back to earth. *Electronics.* an electronic device that emits a radio code in response to a detected radar signal; the response is detected by the radar and provides additional information to the radar operator.

transponder set *Electronics.* a piece of navigation or military identification equipment consisting of a transmitter and receiver that transmits the appropriate reply whenever it receives the proper coded interrogating signal from a distant station, such as a ship or aircraft.

transponder suppressed time delay *Electronics.* the inherent time delay between the receipt of an interrogation signal and transmission of a reply signal from a transponder.

transport *Engineering.* equipment used for transporting or conveying something from one place to another. *Naval Architecture.* a vessel used to carry troops, equipment, or supplies for military or naval operations. In addition to purpose-built military transports, converted merchant ships are also pressed into service as transports. *Computer Technology.* **1.** to move data physically as a whole from one device to another or over a network. **2.** see TAPE DRIVE. **3.** one of the levels in the seven-layer network model. *Mathematics.* the covariant derivative $\nabla_u v$ of a vector v in the direction of a vector u. Also, **parallel translation.**

transportable computer *Computer Technology.* a small computer that can easily be carried; a portable computer.
transport area *Military Science.* in amphibious operations, an area in which a transport organization unloads troops and equipment.
transportation *Science.* **1.** the act of being transported; any form of motion. **2.** specifically, the movement of people and goods by land vehicles and by air and water craft. *Geology.* a phase in the sedimentation cycle that involves the actual movement of sediments or any loose, broken material from one place on or near the earth's surface to another, by means of the action of natural agents such as gravity, air, water, or ice.
transportation emergency *Engineering.* a situation characterized by a shortage of load-carrying vehicles.
transportation engineering *Engineering.* the branch of engineering that is concerned with the movement of goods and people.

Transportation Engineering

Transportation engineering is the application of technology and scientific principles to the planning, design, operation, and management of modes of surface transportation—primarily streets, highways, and transit systems. The primary responsibility of transportation engineers is to provide safe, rapid, comfortable, convenient, economical, and environmentally compatible movement of people and goods. Traffic engineering is the phase of transportation engineering that deals with the planning, geometric design, and traffic operations of roads, streets, and highways, their networks, terminals abutting lands, and relationships with other modes of transportation.

Originally a subset of civil engineering involving primarily the design of roads and bridges, transportation engineering has evolved into a major discipline in its own right. Today's transportation engineers have backgrounds not only in civil engineering, but also in computers and electronics, planning, public administration, management, and the social sciences.

Transportation engineers are employed by federal, state, regional, and local government bodies, regional planning agencies, and public authorities responsible for transit, parking, and toll roads. They are also employed by private consulting firms, universities, and private industry.

Typical transportation engineering activities include planning the future mix of transportation facilities for a region, designing a new or improved highway, evaluating the potential of transit or ridesharing options, and operating a computerized freeway management system or computerized signal system. Advanced technologies are becoming an increasingly important tool of the transportation engineer, and will be used in the future to electronically link our vehicles and roadways.

Today's transportation engineer is a key player in many current and future issues of concern to the public. For example, whether it is retiming a traffic signal or planning a major new transit system, actions taken by transportation engineers impact our ability to achieve clean air and energy conservation objectives. Transportation engineering has therefore become one of the most human of all the engineering disciplines—a discipline that rewards not only technical expertise but also the ability to promote the preferred solutions in the public arena.

Mark R. Norman
Deputy Executive Director
Institute of Transportation Engineers

transportation lag see DISTANCE/VELOCITY LAG.
transportation priorities *Engineering.* a system of numbers assigned to all eligible traffic that establishes movement precedence, as on sea or in the air.
transportation problem *Industrial Engineering.* the problem of optimizing delivery from multiple sources to multiple customers, treated mathematically as a flow from multiple sources to multiple sinks.
transport capacity *Engineering.* the number of people or the amount of equipment a vehicle may safely hold under a given set of conditions.
transport lag see DISTANCE/VELOCITY LAG.
transport mean free path *Nucleonics.* **1.** a path length traveled by a neutron in a nuclear reactor, given as three times the diffusion coefficient of the neutron flux density; used when Fick's law is applicable. **2.** a modification imposed on the mean free path of an ionizing particle to account for anisotropic collisions.
transport medium *Microbiology.* any medium designed specifically to transport or temporarily store an organism or a material from which an organism will be isolated.
transport network *Transportation Engineering.* the system of routes available to the vehicles in an area.
transport number see TRANSFERENCE NUMBER.
transport properties *Physics.* properties associated with the irreversible flow of energy, momentum, or matter, such as viscosity, thermal conductivity, and electrical conductivity.
transport stream *Military Science.* a formation of transport vehicles proceeding in column formation, one behind the other.
transport systems *Physiology.* any of the channels through which the body's nutrients and wastes circulate; the cardiovascular and lymphatic systems collectively.
transport vehicle *Mechanical Engineering.* any vehicle used to move people or cargo. *Ordnance.* specifically, a motor vehicle designed and used, without modification to the chassis, to provide general transport service in the movement of military personnel and cargo.
transposable element *Molecular Biology.* a DNA sequence that can move from one site in a chromosome to another, or between different chromosomes. Also, TRANSPOSON.
transposase *Molecular Biology.* an enzyme that aids in catalyzing the movement of a transposable element.
transpose of a matrix *Mathematics.* for an $(n \times m)$ M matrix with entries (a_{ij}), the $(m \times n)$ matrix M^t with entries (a_{ji}); that is, the matrix obtained by interchanging the rows and columns of the given matrix.
transposition an act of interchanging; specific uses include: *Surgery.* **1.** displacement of a viscus to the opposite side. **2.** the surgical removal of a tissue flap without severing it from the original site until it is viable from nutrients obtained from the new site. *Molecular Biology.* a process in which a transposable element is excised from one site and inserted into a second site on the same or another DNA molecule. *Telecommunications.* **1.** an interchange of positions of conductors in open-wire or multi-pair cable transmission circuits at regular intervals; used to lessen inductive interference on communication circuits. **2.** a defect in transmission in which, during one character period, one or more signals are altered from one significant condition to the other, and a corresponding number of elements are altered in a reverse fashion. *Mathematics.* a cycle of length 2; i.e., a permutation that interchanges exactly two objects. Also, 2-CYCLE. *Behavior.* the fact of reacting to the relationships among stimuli rather than to the absolute properties of the stimuli.
transposition cipher *Telecommunications.* an encrypted message constructed by changing the order of the characters in the original message.
transposon *Genetics.* **1.** see TRANSPOSABLE ELEMENT. **2.** specifically, a transposable element that, in addition to those genes necessary for transposition, carries genes with other functions, such as resistance to antibiotics.
transradar *Telecommunications.* a bandwidth compression system that is used for the long-range, narrow-band transmission of radio emissions from a radar receiver to a local or remote region.
transrectification *Electronics.* the rectification that occurs in one circuit due to the application of an alternating voltage applied to another circuit.
transrectification characteristic *Electronics.* a family of curves showing the degree of transrectification that occurs in an electron tube for several values of applied alternating voltage; each curve plots the DC voltage-current curve for an electrode at different values of AC voltage applied to a different electrode, with voltages on all other electrodes held constant.

transrectifier *Electronics.* a device, typically an electron tube, in which rectification takes place in the circuit on one electrode when an alternating voltage is applied to another electrode.
transsexual *Psychology.* an individual who identifies with the opposite biological sex and who desires or undertakes medical and surgical procedures to become a member of the opposite sex.
trans structure *Materials Science.* a monomer microstructure in which the unsaturated bonds are located on opposite sides of the molecule.
transtemporal *Neurology.* **1.** passing through the temple. **2.** extending along the temporal lobe.
transthalamic *Neurology.* passing through or crossing the thalamus.
transthoracic *Medicine.* performed through the wall of the thorax, or through the thoracic cavity.
transuranic *Inorganic Chemistry.* describing any element with an atomic number greater than 92, the atomic number of uranium. Thus, **tranuranic (transuranium) element.**
transurethral *Anatomy.* performed through the urethra.
transurethral resection *Surgery.* the resection of the prostate gland by means of instruments passed through the urethra.
Transvaal jade *Mineralogy.* a light-green variety of material grossular garnet, from the Transvaal, Republic of South Africa; used as an ornamental and gem. Also, GARNET JADE, SOUTH AFRICAN JADE.
transvector *Toxicology.* an organism that transmits a poison that was produced outside its own body, such as a mollusk that transmits the poison produced by dinoflagellates in a red tide.
transversal *Mathematics.* **1.** a line that intersects each member of some family of lines; in particular, a line that intersects two parallel lines. **2.** let E be a nonempty finite set and $S = \{S_1, \ldots, S_m\}$ a family of nonempty (not necessarily distinct) subsets of E. A subset T of E that contains m distinct elements, one from each set S_i, is called a transversal of S. By theorem, S has a transversal if and only if the union of any k of the subsets S_i contains at least k elements ($1 \leq k \leq m$).
transverse *Anatomy.* placed crosswise; situated at right angles to the long axis of a part. *Cartography.* of or relating to a map projection in which a geodetic meridian is used as a fictitious equator.
transverse bar *Geology.* a partially submerged sand ridge that lies at approximately right angles to the shoreline.
transverse basin see EXOGEOSYNCLINE.
transverse chromatic aberration *Optics.* the component of chromatic aberration that lies at right angles to the optical axis.
transverse Doppler effect *Electromagnetism.* an optical Doppler effect that occurs when the source is moving in a direction perpendicular to the observer's line of sight; the observer will detect a frequency that is reduced by a factor of $[1 - (v/c)^2]^{1/2}$, where v is the source speed and c is the speed of light.
transverse dune *Geology.* an elongated, asymmetrical sand dune lying lengthwise at right angles to the prevailing winds, and having a gentle windward slope and steep leeward slope.

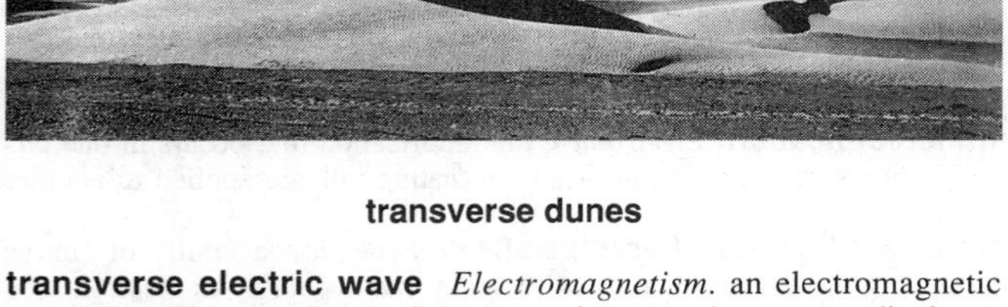

transverse dunes

transverse electric wave *Electromagnetism.* an electromagnetic wave in which the electric field vector is everywhere perpendicular to the direction of propagation.
transverse electromagnetic wave *Electromagnetism.* an electromagnetic wave in which both the electric and magnetic field vectors are everywhere perpendicular to the direction of propagation.
transverse equator *Cartography.* the plane that is perpendicular to the axis of a transverse map projection.
transverse fault *Geology.* a fault that strikes diagonally or at right angles to the general structure of the region.
transverse fracture *Medicine.* a fracture occurring at right angles to the axis of the bone.
transverse frame *Naval Architecture.* a structural stiffening frame that runs at right angles to a vessel's longitudinal axis. The frames of wooden vessels and steel cargo ships are primarily transverse frames.
transverse graticule *Cartography.* a type of fictitious graticule in which a geodetic meridian is used as the fictitious equator.
transverse interference *Electronics.* interference caused by coupling between adjacent wires or components.
transverse joint see CROSS JOINT.
transverse latitude *Cartography.* the angular distance measured from the transverse equator on a transverse map projection.
transverse longitude *Cartography.* the angular distance measured from the prime transverse meridian to any other transverse meridian on a transverse map projection.
transverse magnetic wave *Electromagnetism.* an electromagnetic wave in which the magnetic field vector is everywhere perpendicular to the direction of propagation.
transverse magnetization *Acoustical Engineering.* the magnetization of tape in a direction perpendicular to the normal direction of travel.
transverse magnetoresistance *Electromagnetism.* an effect in which an electric current perpendicular to a magnetic field gives rise to a potential gradient along the direction of the current.
transverse map projection *Cartography.* any map or chart that uses a fictitious graticule in which a geodetic meridian is selected for use as a fictitious equator.
transverse mass *Physics.* the quotient of the force acting on a relativistic particle perpendicular to the particle's velocity divided by the acceleration resulting from the force.
transverse Mercator chart *Cartography.* any chart based on a transverse Mercator map projection.
transverse Mercator projection *Cartography.* a conformal cylindrical map projection that is equivalent to the regular Mercator projection turned 90° in azimuth, so that its meridians are oriented to an east-west geodetic axis, rather than a north-south geodetic axis.
transverse meridian *Cartography.* any line representing a great circle that is perpendicular to a transverse equator in a fictitious graticule. The transverse meridian of reference is the **prime transverse meridian.**
transverse metacenter *Naval Architecture.* the axis of rotation for a vessel's roll. While a vessel also has a longitudinal metacenter (the axis of pitching motion), the transverse metacenter plays such an important role in stability calculations that it is usually referred to simply as "the metacenter."
transverse microtubules *Cell Biology.* a ribbon of microtubules, found in ciliated protozoans, that extends upward toward the pellicle, perpendicular to the kineties.
transverse parallel *Cartography.* a line or circle on a map or chart that connects all points of equal transverse latitude.
transverse pole *Cartography.* either of the two points 90° from the transverse equator in a fictitious graticule.
transverse polyconic map projection *Cartography.* a map projection in which a great circle perpendicular to the prime geographic meridian is substituted as a fictitious prime meridian and control axis.
transverse process *Anatomy.* a process that projects laterally from a vertebra.
transverse recording *Electronics.* a method of magnetic tape recording of television signals using rotating, quadruple-pickup recording heads that are set at right angles to the tape surface.
transverse rhumb line *Cartography.* a line or direction on the surface of the earth that follows a single compass bearing and makes the same angle with all fictitious meridians on a transverse Mercator projection, resulting in a diagonal straight line.
transverse ripple mark *Geology.* a ripple mark occurring at right angles to the flow of the current.
transverse stability *Engineering.* the ability of a ship or a plane to right itself after having rolled.
transverse thrust see TRANSCURRENT FAULT.
transverse valley *Geology.* **1.** a valley that cuts perpendicularly across a geological structure, such as a ridge or chain of mountains. Also, CROSS VALLEY. **2.** a valley whose direction is perpendicular to the general strike of the strata it overlies.

transverse vibration *Mechanics.* vibration of a rod in which the motion is perpendicular to its axis.

transverse wave *Physics.* a wave in which the motion of the disturbance is perpendicular to the direction of propagation of the wave; an S wave.

transversion *Molecular Biology.* a mutation caused by the substitution of a purine for a pyrimidine or a pyrimidine for a purine in DNA or RNA.

transvestism *Psychology.* the practice of dressing in clothes associated with the opposite sex, especially for the purpose of sexual excitement or gratification. Also, **transvestitism.**

transvestite *Psychology.* a person who practices transvestism; someone who makes a practice of dressing in the clothing of the opposite sex.

trap *Mechanical Devices.* a device used to capture wild animals, as for food, for fur, or to eliminate pests; typically consisting of a metal bar or set of jaws that springs shut. *Civil Engineering.* a plumbing fixture designed to prevent the flow of air and gasses; a bend in the pipe stays full of water creating the seal. *Space Technology.* in a solid-propellant rocket engine, a component that is used to keep the propellant grain in place and block its escape through the nozzle. *Geology.* see OIL TRAP. *Mechanical Engineering.* **1.** a device that reduces the effect of the vapor pressure of oil or mercury on the high-vacuum side of a diffusion pump. **2.** a vessel or pump used to collect certain discharges from industrial and process equipment, such as condensate from steam plants or heat exchangers. *Electronics.* **1.** a selective circuit designed to reject certain signals while accepting others; selectivity is usually obtained by the use of tuned circuits, resistance-capacitor networks, and the like. **2.** a physical irregularity in a semiconductor material that effectively holds charge carriers by preventing their migration through the semiconductor material. *Solid-State Physics.* a site in the crystal lattice of a semiconductor where a conduction electron or hole may be held for some length of time before it is released by thermal agitation. *Computer Programming.* an interruption in a computer's operation, especially one involving a transfer of control to a predetermined location that is executed by hardware or software when conditions such as errors are detected.

trap *Petrology.* any dark, finely crystalline, hypabyssal or extrusive, igneous rock. Also, TRAPP, TRAPPIDE, TRAP ROCK.

Trapaceae *Botany.* a monogeneric family of aquatic annual, tanniferous herbs of the order Myrtales, including the water chestnut species, that are often free-floating; characterized by submerged stems with elongated, green leaflike organs and an aerial stem with crowded rosulate leaves; native to tropical and subtropical Africa and Eurasia.

trap address *Computer Programming.* the location to which control is transferred when a trap condition is detected.

TRAPATT diode *Electronics.* a solid-state diode that operates as an oscillator in the microwave frequency spectrum; it depends for its operation on the synchronous interchange of energy between an external high-Q resonator and an electron-hole plasma created in the P or N region by the avalanche breakdown of the diode junction. (An acronym for trapped plasma avalanche transit time diode.)

trap crop *Agronomy.* a crop grown in order to attract harmful insects from another more valuable crop nearby. Thus, **trap-cropping.**

trapdoor or **trap door** *Building Engineering.* a lifting door on hinges or sliders, which covers an opening in a building structure along a roof, ceiling, or floor.

trapdoor fault *Geology.* a circular fault that is hinged at one edge.

trap-door spider *Botany.* any of various spiders belonging to the family Ctenizidae, noted for their habit of digging a burrow in the ground that is closed over with a lid fashioned of silk and mud.

trapeziform *Biology.* shaped like a trapezoid.

trapezium *Mathematics.* a quadrilateral having no two sides parallel. *Anatomy.* the most lateral bone of the distal row of carpal bones. Also, **trapezium bone.**

Trapezium *Astronomy.* a group of four, very hot, young stars in the center of the Orion Nebula.

trapezius *Anatomy.* a large flat triangular muscle of each side of the back that rotates the scapula to raise the shoulder in abduction of the arm.

trapezohedron *Crystallography.* a geometric solid in which all faces are trapezium faces, that is, they are plane figures with four sides, no two of which are parallel.

trapezoid [trapʹə zoid] *Mathematics.* a quadrilateral having exactly two parallel sides. The parallel sides are called bases. *Anatomy.* the bone in the distal row of carpal bones lying between the trapezium and capitate bones. Also, **trapezoid bone.**

trapezoidal *Mathematics.* relating to or having the form of a trapezoid.

trapezoidal excavator *Mechanical Engineering.* a digging machine that removes earth in trapezoidal cross sections.

trapezoidal generator *Electronics.* a circuit or piece of test equipment that generates a trapezoidal waveform for testing and alignment of electromagnetic sweep circuits in television receivers and similar circuits; the frequency, pulse-width, and amplitude are usually adjustable.

trapezoidal pulse *Electronics.* a pulse shape in which the plot of the voltage consists of an abrupt increase from a baseline level to some other level, followed by a linear, ramped increase to a higher level, and then an abrupt decrease back to the baseline level.

trapezoidal rule *Mathematics.* given a real-valued function f defined on a real interval $[a,b]$, a numerical integration method in which the interval of integration is divided into n equal subintervals and the integral $\int_a^b f(x)dx$ is approximated by

$$[\sum_{k=1}^{n-1} 2f(x_k) + f(a) + f(b)](b-a)/n,$$

where $a = x_0$ and $x_k = x_{k-1} + (b-a)/2n$. Also, **trapezoidal integration.**

trapezoidal wave *Electronics.* a recurring wave shape in which the plot of the voltage for one-half cycle consists of a trapezoidal pulse.

trap gun *Ordnance.* a shotgun designed for trapshooting; usually equipped with a 12-gauge, full-choke bore, with a barrel that is 30 inches or longer.

trap mine *Ordnance.* a land mine that is designed to explode when an enemy attempts to move an object.

trapp *Petrology.* see TRAP.

trapped *Robotics.* of a controller, no longer able to scan data.

trapped-air process *Engineering.* a procedure for blow-molding closed plastic objects.

trapped entanglements *Materials Science.* any physical cross-linking in a polymer apart from ordinary covalent cross-linking; causes permanent loops or entanglements in the network structure.

trapped fuel *Engineering.* the fuel in an engine system that is not in the fuel tank.

trapped radiation *Geophysics.* any radiation from space that has become trapped in the earth's magnetic field.

trapped radicals *Materials Science.* the occlusion or trapping of radicals in an unswollen, tightly coiled precipitating polymer; causes marked deviations from the usual kinetics of homogeneous radical polymerization.

trappide *Petrology.* see TRAP.

trapping *Telecommunications.* a form of radio-wave propagation in which radiated rays are greatly angled by being refracted in the lower layers of the atmosphere, thus guiding portions of the radiated energy over distances that far exceed the normal range. Also, GUIDED PROPAGATION.

trapping mode *Computer Programming.* a mode of a computer operation in which the trap or interrupt mechanism is enabled and the computer may suspend operation and respond to the trap or interrupt condition.

trash compactor *Mechanical Devices.* see COMPACTOR.

trash line *Hydrology.* an indication on a beach of the highest level of the tide, as shown by the deposit of a line of waterborne debris at this point.

trash screen *Civil Engineering.* a lattice designed to prevent the migration of debris in a stream or water course, or to keep debris from a turbine.

Traube's rule *Physical Chemistry.* the observation that the addition of a methylene substance causes the surface tension of a dilute solution to decrease threefold.

trauma [trôʹmə; trouʹmə] *Medicine.* **1.** any physical or mental shock or injury. **2.** specifically, a serious wound or injury caused by some physical action, as an automobile accident, violent assault, and so on. *Psychology.* psychological damage, or an experience that inflicts such damage. (From the Greek word for "wound.")

trauma center *Medicine.* a specialized hospital facility that deals specifically with cases of trauma; e.g., victims of automobile accidents, gunshot wounds, and so on.

traumat- or **traumato-** a combining form meaning "trauma," "wound," or "injury."

traumatic [trə matʹik] *Medicine.* **1.** relating to or caused by a trauma. **2.** used or involved in the treatment of trauma or wounds. *Psychology.* inflicting pyschological damage, especially enduring damage, such as stressful events of early childhood.

traumatic encephalopathy *Neurology.* a syndrome due to cumulative punishment absorbed in the boxing ring, characterized by the general slowing of mental functions, occasional bouts of confusion, and scattered memory loss. Also, BOXER'S ENCEPHALOPATHY.

traumatic neurosis see POST-TRAUMATIC STRESS DISORDER.

traumatic pneumonosis *Medicine.* lung disease brought on by trauma.

traumatize *Psychology.* to subject to a trauma; inflict psychological damage. *Medicine.* to injure tissue, especially by physical force.

traumatology *Surgery.* the branch of surgery dealing with wounds and injuries.

traumatopathy *Medicine.* any pathologic condition resulting from a wound or injury.

traumatotropism *Biology.* responsive growth of an organism resulting from a wound or injury.

Trauzl test *Engineering.* a test that determines the relative disruptive power of an explosive charge by measuring the change in volume of a cavity in which a set amount of charge has been placed and exploded.

trave *Architecture.* a bay formed by crossbeams.

travel *Mechanical Engineering.* the complete movement or stroke in one direction of a moving part of a machine or the distance traversed.

travel demand *Transportation Engineering.* in transport planning, the number of trips that are (or would be) made along a given route or corridor under existing (or specified) conditions.

traveling apron *Mining Engineering.* an endless belt used for conveying material of any kind.

traveling block *Mechanical Engineering.* a movable unit that is hoisted or lowered with the load in a block-and-tackle system. Also, FLOATING BLOCK, RUNNING BLOCK.

traveling charge *Ordnance.* a propelling charge that moves with the projectile through the bore as it burns. Also, LANGWEILER CHARGE.

traveling detector *Engineering.* a radio-frequency probe mounted in a slotted-line section of a waveguide; used to measure wave ratios.

traveling dune see WANDERING DUNE.

traveling gantry crane *Engineering.* a hoisting mechanism equipped with legs that run along rails at ground level and a bridgelike structure that spans the area over which it operates.

traveling grate stoker *Mechanical Engineering.* a furnace stoker with a continuously moving grate that transports coal toward the furnace.

traveling-salesman problem *Mathematics.* **1.** in a Hamiltonian graph with a length associated with each edge, the problem of finding a Hamiltonian cycle of minimum total length. **2.** any other problem that involves finding the shortest or simplest path. (From the idea that a traveling salesman would design his itinerary in this way.)

traveling-screen dryer *Chemical Engineering.* a moving belt made of screen that conveys damp substances through a heated drying zone.

traveling valve *Petroleum Engineering.* a moving discharge valve on a sucker-rod pump used in an oil well.

traveling wave *Physics.* a progressive wave in which energy is transported from one point to another, as distinguished from a standing wave in which the wave energy is stored in a confined region.

traveling-wave amplifier *Electronics.* an amplifier using a traveling-wave tube

traveling-wave antenna *Electromagnetism.* an antenna in which the current distributions are produced by waves of charges propagated only in one direction in the conductor.

traveling-wave magnetron *Electronics.* a traveling-wave tube device that operates as an oscillator in the microwave-frequency spectrum.

traveling-wave magnetron oscillations *Electronics.* oscillations in a magnetron produced by the synchronous interchange of energy back and forth between an electron stream and a traveling-wave electromagnetic field, which results in a net transfer of more energy to the field than to the stream.

traveling-wave maser *Physics.* a ruby maser that uses a slow-wave structure and yttrium iron garnet isolators; the output is in the L-band (390 to 1550 MHz).

traveling-wave parametric amplifier *Electronics.* a specialized amplifier combining the characteristics of a parametric amplifier and a traveling-wave amplifier in which the pump-, signal-, and difference-frequency waves are propagated in a traveling-wave pattern.

traveling-wave phototube *Electronics.* a sensitive phototube in which electrons emitted from the photocathode as a result of the detection of incident illumination from a laser beam become part of an electron stream that synchronously interacts with a traveling-wave electromagnetic field to produce an output signal.

traveling-wave tube *Electronics.* a specialized electron tube device that operates in the microwave frequency spectrum and whose operation depends on the synchronous interchange of energy between an electron stream and a traveling-wave electromagnetic field, resulting in a net transfer of energy from the electron stream to the field; the output is taken from a helical coil wound around the body of the tube; used chiefly as amplifiers and oscillators.

travel-time curve *Geophysics.* a plot of the various seismic waves (P-, S-, and L-wave travel times) recorded by a station used to discover the origin of a seismic disturbance.

Travers, Morris William 1872–1961, English chemist; with Sir William Ramsay, discovered krypton, neon, and xenon.

traverse *Cartography.* **1.** in surveying, the process of measuring the lengths and directions of lines that connect a series of points indicated by survey stations, in order to determine the horizontal positions of those points relative to each other. **2.** the results of such a process. **3.** the route and the series of survey stations that are to be or have been so measured. *Navigation.* the zig-zag movement of a vessel sailing into the wind by a series of tacks. Similarly, **traverse sailing.** *Ordnance.* to turn a weapon to the right or left on its mount. *Engineering.* the movement of a pivot or mounting structure. *Geology.* **1.** a line across a thin geologic section or other sample, along which mineral grains are counted or measured. **2.** a vein or fissure that cuts diagonally across a rock. *Meteorology.* a squally west wind of central France that brings snow or hail showers in winter and humid, thundery weather in summer.

traverse adjustment see BALANCING A SURVEY.

traverse hardness *Materials Science.* a profile of hardness values that provides data in hardener steels of the depth of martensite formations.

traverse level *Ordnance.* the vertical displacement above low-level air-defense systems at which aircraft can cross the area without being vulnerable to fire; it is expressed as both a height and altitude.

traverse table *Navigation.* **1.** a published table used to find the sides and course angle of a right triangle in plane or middle latitude sailings. **2.** a table created by the navigator to reduce the several courses and distances of traverse sailing to a single course and distance.

traversia *Meteorology.* a west wind that blows from the sea in western South America, especially Chile.

traversier *Meteorology.* in the Mediterranean, an often violent wind that blows directly into port.

traversing mechanism *Engineering.* the mechanism that turns a pivot or mounting structure in a horizontal plane.

travertine *Geology.* a hard, compact limestone with a white, tan, or cream color, deposited by solution from surface or ground waters. Also, CALCAREOUS SINTER, CALC-SINTER.

trawl *Engineering.* a baglike net with an open mouth that is towed beneath a ship in order to catch marine life.

trawler *Naval Architecture.* a fishing vessel that employs a trawl for catching bottom-feeding fish. Trawlers combine relatively small size with excellent seaworthiness.

tray elevator *Mechanical Engineering.* an elevator, having spaced trays attached to two vertical-mounted continuous chains, that lifts drums, barrels, or boxes.

tray tower *Chemical Engineering.* a vertical process tower containing a series of trays designed to create contact between falling liquid and rising vapor.

TRBF total renal blood flow.

TR box see TRANSMIT-RECEIVE TUBE.

tread *Civil Engineering.* the part of a step that is actually stepped on, the horizontal part of the step. *Engineering.* the part of a tire or wheel that touches the surface below, often having grooves for added traction.

treater *Chemical Engineering.* a system or vessel used for contacting a process stream with a reagent or treating chemical, as in caustic or acid treating.

treating *Chemical Engineering.* the process of contacting a fluid stream with a chemical agent to enhance the qualities of the fluid by converting, removing, or sequestering unacceptable impurities found in petroleum products, mixed gases, water, or sewage.

treble *Acoustics.* **1.** the higher portion of the audio-frequency spectrum, above about 500 hertz. **2.** notes of musical works written in the **treble clef,** which starts at middle C and progresses upward.

trebol *Materials.* the very durable, hard wood of trees of the genus *Platymiscium* of Central and South America; used for fine furniture, veneers, and musical instruments.

trebouxia *Botany.* a genus of nonmotile algae of the family Chlorococcaceae, which is a phycobiont of numerous lichens.

tree *Botany.* an imprecise term for a perennial woody plant that is larger than a bush or shrub, generally understood to describe a large growth having one main trunk with few or no branches projecting from its base, a well-developed crown of foliage, and a height at maturity of at least 12 feet. *Metallurgy.* the fulcrum or prop used as a lever in boring. *Electronics.* a network or circuit that has no closed loops. *Mathematics.* a connected graph containing no cycles; i.e., a connected forest. *Computer Programming.* a hierarchical diagram or data structure consisting of a root node and one or more levels of branch or child nodes. Each node except the root has exactly one parent node; all nodes are reachable from the root.

tree automaton *Computer Technology.* an automaton that processes input in the form of trees, usually seen in relation to context-free grammars.

tree bear see POTTO.

tree climate *Meteorology.* any climate in which many varieties of trees thrive, such as a temperate rainy climate, snow-forest climate, or tropical rainy climate; a major division in Wladimir Köppen's climatic classification, as distinguished from dry climates and polar climates.

treecreeper *Vertebrate Zoology.* see CREEPER.

tree diagram *Computer Programming.* a flow diagram of a computer program or system that has no closed paths.

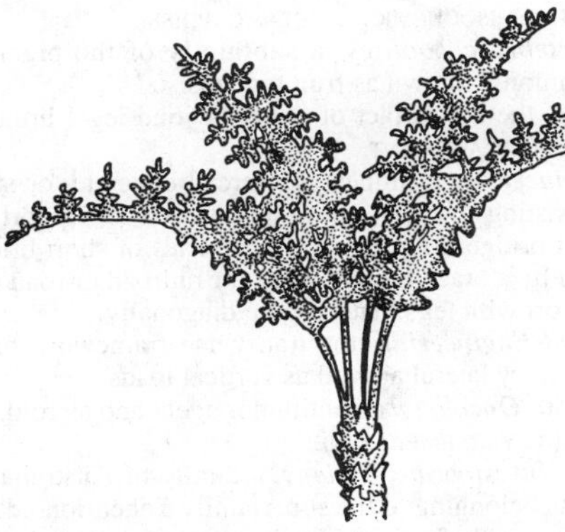

tree fern

tree fern *Botany.* the popular name for various ferns of the family Cyatheaceae that reach the size of a tree; characterized by a large woody trunk or caudex with foliage at the summit; found mostly in tropical regions.

treefish *Vertebrate Zoology.* a rockfish, *Sebastes serriceps*, marked by black bands; found in waters off southern California.

tree frog *Vertebrate Zoology.* a common name for a small, arboreal frog, especially belonging to the family Hylidae and related families, possessing expanded adhesive disks on the toes that enable it to climb trees.

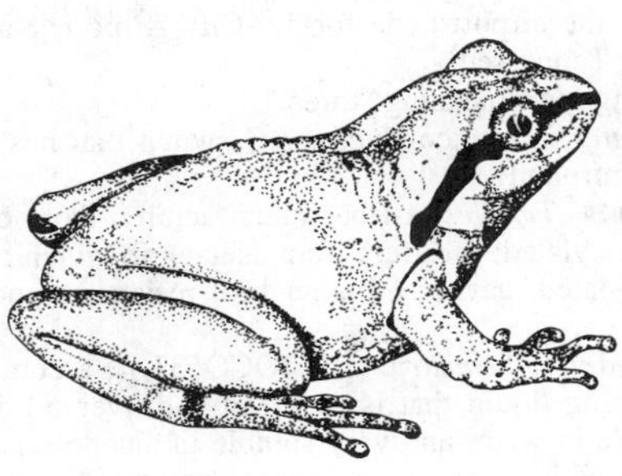

tree frog

treehopper *Invertebrate Zoology.* any of numerous hopmopterous leaping insects of the family Membracidae, having an enlarged prothorax adapted to sucking and feeding on the juices of plants.

tree line see TIMBERLINE.

treenail see TRENAIL.

tree pruning *Computer Programming.* the act of eliminating branches in a tree structure in order to reduce the number of subtrees in which to search for a record or game strategy.

tree-ring dating see DENDROCHRONOLOGY.

tree shrew *Vertebrate Zoology.* any of several insectivorous, arboreal, squirrel-like mammals of the family Tupaiidae, having a long snout; found in southeastern Asia and adjacent islands.

tree snail *Invertebrate Zoology.* any tree-living gastropod of the family Bulimulidae, characterized by a brightly colored shell with distinct whorls; found in tropical and temperate regions.

tree sort see HEAP SORT.

tree sparrow *Vertebrate Zoology.* a North American finch, *Spizella arborea*, having a reddish-brown crown and a dark spot on the breast; found commonly in the winter in the northern U.S.

tree swallow *Vertebrate Zoology.* a swallow, *Iridoprocne bicolor*, having a bluish-green back and white breast, that nests in tree cavities; found in North America.

trega- a combining form meaning 10^{12}.

trehalose *Biochemistry.* $C_{12}H_{22}O_{11}$, a disaccharide of glucose present in the hemolymph of a variety of insects and stored in certain bacteria and fungi instead of starch.

trellis(ed) drainage pattern *Hydrology.* a pattern of natural stream courses in which the main streams and elongated secondary tributaries run parallel and are intersected at more or less right angles by the primary tributary streams. Also, GRAPEVINE DRAINAGE PATTERN, ESPALIER DRAINAGE PATTERN.

Tremadocian *Geology.* a European geologic stage of the Lower Ordovician period, occurring after the Upper Cambrian period and before the Arenigian.

Trematoda *Invertebrate Zoology.* the flukes, a class of parasitic flatworms in the phylum Platyhelminthes, hermaphrodites with a spiny cuticle, suckers, and a complex life cycle involving several hosts; many infest various human organs, generally from the ingestion of undercooked fish, crustaceans, and vegetation.

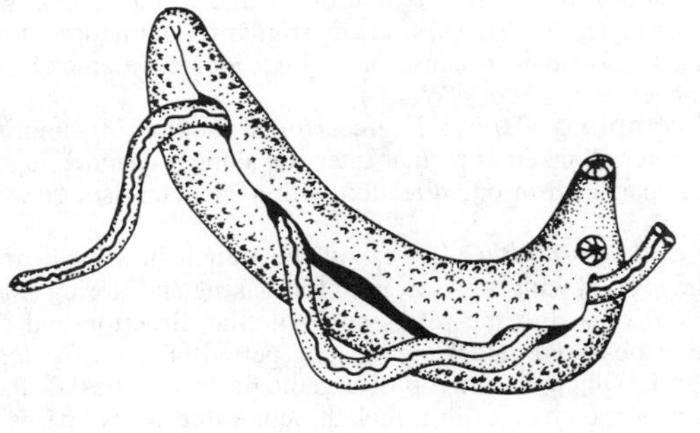

Trematoda

trematode *Invertebrate Zoology.* an organism belonging to the Trematoda; a fluke.

trematodiasis *Medicine.* an infection with trematodes.

Trematosauroidea *Paleontology.* a superfamily of labyrinthodont amphibians in the extinct order Temnospondyli; widespread in deposits of the Lower Triassic; many were almost completely marine; disappeared near the end of the Triassic.

trembles *Veterinary Medicine.* a toxic condition of cattle and sheep in the western and central United States, resulting from ingestion of white snakeroot; characterized by weakness and body tremors.

Tremellaceae *Mycology.* a family of fungi belonging to the order Auriculariaceae, which includes some edible mushrooms and occurs on decaying plant material in tropical to temperate regions.

tremellales see JELLY FUNGUS.

trémie *Engineering.* a device that is used to place concrete underwater; it consists of a hollow tube that is closed with a valve at the submerged end.

tremolite *Mineralogy.* $Ca_2(Mg,Fe^{+2})_5Si_8O_{22}(OH)_2$,$Mg/(Mg+Fe^{+2})$ $=1.0–0.9$, a gray, white, or colorless monoclinic mineral of the amphibole group occurring as bladed crystals and fibrous aggregates, having a specific gravity of 2.9 to 3.2 and a hardness of 5 to 6 on the Mohs scale; forms a series with actinolite and ferro-actinolite; found in metamorphosed dolomitic limestones and ultrabasic rocks.

tremor *Geophysics.* a minor earth movement, usually preceding or following a larger movement or a volcanic eruption. *Neurology.* a rhythmic, repetitive involuntary movement, appearing as trembling or quivering, often of constant amplitude and frequency.

Trempealeauan *Geology.* a geologic stage of the uppermost Cambrian period, occurring after the Franconian and before the Lower Ordovician.

tremulous *Neurology.* experiencing tremors; shaking, trembling, or quivering.

trenail *Mechanical Devices.* a hardwood pin used to secure a mortise and tenon. Also, TREENAIL, TRUNNEL.

trench *Geography.* a long, narrow, often U-shaped valley between two mountain ranges. *Geology.* **1.** a long narrow excavation in the earth's surface produced either artificially or naturally by erosion or tectonic movements. **2.** a steep-sided, elongated depression in the deep-sea floor, occurring between the continental margin and the abyssal hills. Also, DEEP-SEA TRENCH, OCEANIC TRENCH, SEA-FLOOR TRENCH. **3.** a narrow, steep-sided depression eroded by a stream.

trench excavator *Mechanical Engineering.* a self-propelled digging machine, usually mounted on crawler tracks, that is designed for digging trenches or ditches. Also, **trencher, trenching machine.**

trench fever *Medicine.* an infectious disease caused by the rickettsia *Rochalimaea quintana*, transmitted by body lice, and characterized in part by intermittent fever, headache, rash, and leg pains, especially severe in the shins; it was first recognized during the trench warfare of World War I and was a major problem during World War II, but is now rare. Also known by various other names, such as **five-day fever, quintan fever, Meuse fever, shinbone fever, Volhynia fever,** and **His-Werner disease.**

trench foot *Medicine.* a disease similar to frostbite, affecting the skin, blood vessels, and nerves of the feet due to prolonged exposure to cold and wet, common among soldiers serving in trenches.

trenching *Archaeology.* an excavation technique in which a site is penetrated with long, narrow trenches that reveal the vertical dimension of the area.

trench mouth *Medicine.* **1.** a contagious, painful bacterial infection of the mouth, characterized by sores and ulcers of the lining of the gums, cheeks, and tongue, often extending to the throat. Also, VINCENT'S ANGINA, VINCENT'S INFECTION. **2.** any similar inflammation of the mouth and gums. (So-called because of its high incidence among soldiers in the trenches during World War I.)

trench sampling *Mining Engineering.* a method of sampling ore in which material, taken at regular intervals along a channel, is spread out flat and channeled in one direction with a shovel. Also, CHANNEL SAMPLING.

trench silo *Agriculture.* an excavated trench in a bank or hillside, sometimes lined with concrete, used for making and storing silage.

trend *Statistics.* in time-series analysis, the direction and degree of change of observed values over a long period of time. *Geology.* **1.** the direction in which an outcrop of a geologic feature lies. **2.** the general direction of the changes in a rock structure during its phases of metamorphism. *Ordnance.* the deviation in the fall of shots, such as might be caused by incorrect speed settings on a fire support ship.

trend line *Meteorology.* a plot of monthly mean climatological values.

trennschaukel apparatus *Engineering.* an instrument that is used to determine the thermal diffusivity of a gas; it consists of a series of tubes, across which a temperature gradient is placed.

Trentepohliaceae *Botany.* a family of green algae of the order Ulotrichales, characterized by irregularly branched filaments and banded or reticulate chloroplasts that reproduce by zoospores or isogametes.

Trentepohliales *Botany.* a monofamilial order of green algae of the class Chlorophyceae, characterized by branched uniseriate filaments that form a loose tuft or a prostrate disk and by the presence of hematochrome, which colors the thallus yellow-green, yellow, orange, or red and sometimes gives off the scent of violets.

Trentonian *Geology.* a North American geologic stage of the Middle Ordovician period, occurring after the Wilderness and before the Edenian, or after the Blackriverian and before the Cincinnatian.

trepan *Mechanical Engineering.* **1.** a cutting tool, consisting of a circular tube with teeth fixed in the end, that is rotated and fed axially into a workpiece to cut a narrow grooved surface in it. Also, **trepanning tool.** **2.** to cut circular disks from stock using a trepan. *Surgery.* to operate using a trephine.

trephination *Surgery.* a surgical procedure employing a trephine. Also, **trepanation.**

trephine [trə fīn´; trə fēn´] *Surgery.* **1.** a surgical saw used to remove a disk from bone, such as the skull. **2.** an instrument for removing a circular area of cornea for transplant operations. **3.** to operate and remove a disc of bone or hard tissue by means of such a device.

trepidation *Neurology.* an involuntary trembling movement, as in a tremor.

Treponema *Bacteriology.* a family of Gram-negative, microaerophilic, motile, spiral-shaped bacteria of the family Spirochaetaceae that are found in the oral, intestinal, and genital mucosa of humans and animals; contains some pathogenic species, including *T. pallidum,* the causative agent of syphilis in humans.

Treponema pallidum hemagglutination assay *Pathology.* a test for the agent responsible for venereal, nonvenereal, and congenital syphilis in humans.

Treponemataceae *Bacteriology.* in former classifications, a family of spiral-shaped, primarily parasitic bacteria of the order Spirochaetales, which are now assigned to the family Spirochaetaceae.

treponematosis *Medicine.* any infection with the microorganism treponema, a spirochete bacteria of the family Spirochaetaceae; treponema are responsible for a wide variety of diseases including syphilis, yaws, bejel, and pinta. Also, **treponemiasis, treponemosis.**

Trepostomata *Paleontology.* a Paleozoic order of colonial bryozoans of uncertain class; characterized by long, curved, partitioned calcareous tubes; extant from the Ordovician to the Permian.

treptomorphism see ISOCHEMICAL METAMORPHISM.

Treroninae *Vertebrate Zoology.* a subfamily of the pigeon family Columbidae, commonly known as fruit pigeons.

trespass *Science.* the act or fact of going beyond legal limits. *Mining Engineering.* see ENCROACHMENT.

trestle *Civil Engineering.* a timber, reinforced concrete, or steel structure, usually consisting of many short spans, used to support a temporary or permanent bridge. *Engineering.* **1.** a series of short bridge spans that are supported by a braced tower, used for railroad or road crossings. **2.** a portable support with legs that open up diagonally.

trestle bent *Civil Engineering.* the transverse framework of a trestle structure; used to carry lateral as well as vertical loads.

trestolone acetate *Oncology.* an antitumor agent and steroid.

tretamine see TRIETHYLENEMELAMINE.

Tretothoracidae *Invertebrate Zoology.* a family of Australian beetles, coleopteran insects belonging to the superfamily Tenebrionoidea.

Treubariaceae *Botany.* a family of freshwater green algae belonging to the order Chlorococcales, having solitary, planktonic cells, a cell wall with a thin inner layer, and a multipart outer layer, often with spines or wings.

Treubiaceae *Botany.* a monogeneric family of liverworts belonging to the order Metzgeriales, characterized by large, leafy, prostrate plants without secondary pigments, having a thick mucilaginous covering on the ventral surface, and unique dorsal scales associated with succubously inserted lateral leaves.

Trevithick, Richard 1771–1833, English engineer and inventor; a pioneer in high-pressure steam engines; built the first steam railway locomotive.

TRF thyrotropin releasing factor; T-cell-replacing factor.

trf. tuned-radio frequency.

TRH thyrotropin-releasing hormone.

TRI *Aviation.* the airport code for Tri-City Airport, Kingsport-Bristol-Johnson City, Tennessee.

tri- a combining form meaning "three."

triac *Electronics.* a semiconductor AC switch that has three terminals and is gate controlled

triacetate fiber *Textiles.* a fiber manufactured from cellulose that is completely acetylated; it differs from acetate, in which cellulose is only partially acetylated, having a higher heat resistance and requiring less care.

triacetin *Organic Chemistry.* $C_3H_5(OCOCH_3)_3$, a combustible, colorless bitter-tasting liquid that is found in cod-liver oil and butter; it is slightly soluble in water and very soluble in alcohol and ether; boils at 258–260°C; used in perfumes, cosmetics, and plasticizers, and as a food additive and solvent. Also, GLYCERYL TRIACETATE.

triacylglycerol *Biochemistry.* triglyceride; a compound derived from the esterification of a glycerol molecule with three fatty acid molecules; a lipid in the blood.

triacymethane *Organic Chemistry.* a class of compounds having the formula *R*COCH(CO*R'*)CO*R* and resulting from the condensation of acyl halides with the sodium salts of 1,3-diketones.

triad a group of three items; specific uses include: *Computer Programming.* **1.** a group of three bits, bytes, or characters that form a single data item. **2.** a form of intermediate code used by compilers, consisting of an operator and two operands. Also, TRIPLE. *Electronics.* **1.** a three-dot group of phosphor particles on the inside face of a color-picture tube; the dots are arranged in a triangular pattern, and each dot emits one of the three primary colors when stimulated by an electron beam. **2.** a group of three interrelated pulses or data bits. **3.** a group of three widely separated radio direction-finder stations that are operated as a group for determining the position of aircraft or ships through direction-finding triangulation techniques. *Navigation.* a group of three transmitting stations operating as a synchronous system to provide signals that may be used by a craft to determine its position. Also, TRIPLET.

triaene *Invertebrate Zoology.* a type of sponge spicule with one long and three short rays.

triage [tré´äzh´] *Medicine.* **1.** originally, a practice used in World War I to allocate medical resources and personnel in extreme combat situations, in which wounded soldiers were classified into one of three groups: those who could be expected to live without medical care; those who would likely die even with care; those who could survive if they received care. **2.** a similar sorting out and classification of patients according to the gravity of injuries, urgency of treatment, and place for treatment, carried out as an emergency procedure during a sudden influx of large numbers of victims of war or other disaster. *Science.* any similar allocation of scarce resources based on a priority ranking.

Triakidae *Vertebrate Zoology.* the hound sharks, a family of harmless, common, cosmopolitan sharks of the order Carcharhiniformes.

trial-and-error behavior *Behavior.* a series of responses that are tested and eliminated in a problem situation until a solution is reached. Also, **trial-and-error learning.**

trial-and-error method *Crystallography.* a method that involves postulating a structure (that is, assuming locations of the atoms in the unique part of the unit cell), calculating structure factors, F_c, and comparing their magnitudes with the scaled observed values, $|F_0|$. Since the number of trial structures that must be tested increases with the number of parameters, such methods have generally been applied in solving only the simplest structures.

trial batch *Engineering.* a test sample of concrete that is used to determine the proper water-cement ratio necessary to produce the required slump and compressive strength.

trial fire *Military Science.* deliberate fire that is laid on an aiming point or target in order to adjust firing data prior to the commencement of fire for effect.

trialistic theory *Hematology.* the theory that the blood cells arise from three distinct types of primitive cells, the myeloblasts, lymphoblasts, and monocytes.

triallelic *Genetics.* of or relating to a polyploid organism that possesses three different alleles at any given locus.

trial pit *Mining Engineering.* a shallow hole with a diameter of 2–3 feet, excavated in order to test shallow minerals or determine the nature and thickness of superficial deposits and depth to bedrock.

trial shots *Engineering.* experimental shots fired into a drill site to determine the best drilling pattern.

trial structure *Crystallography.* a possible structure for a crystal, which is tested by a comparison of calculated and observed structure factors and by the results of an attempted refinement of the structure.

triamcinolone *Organic Chemistry.* $C_{21}H_{27}FO_6$, a white crystalline steroid; insoluble in water and slightly soluble in organic solvents; melts at 269–271°C; used in medicine as an anti-inflammatory agent.

triamylamine *Organic Chemistry.* $(C_5H_{11})_3N$, a combustible, colorless to yellow liquid; insoluble in water and soluble in gasoline; boils between 225 and 245°C; used in insecticides and as a corrosion inhibitor. Also, TRIPENTYLAMINE.

triamyl borate *Organic Chemistry.* $(C_5H_{11})_3BO_3$, a combustible, colorless liquid; soluble in alcohol and ether; the tri-*n* isomer of which boils at 274–276°C; used in varnishes.

triandrous *Botany.* describing a plant with three stamens.

triangle *Mathematics.* a plane polygon with three sides; equivalently, a plane polygon with exactly three (noncollinear) vertices.

triangle inequality *Mathematics.* any inequality of the form $\phi(x + y) \leq \phi(x) + \phi(y)$, where ϕ is some function or operator. ϕ is said to satisfy the triangle inequality, or to be subadditive, if the triangle inequality is true for all x and y in the domain of definition of ϕ. For example, the absolute value function satisfies the triangle inequality; i.e., $|x + y| \leq |x| + |y|$ for all real or complex numbers x and y.

triangle of forces *Mechanics.* a graphical technique used to determine the resultant of two forces; the two forces are drawn as the two sides of a triangle whose third side will represent the resultant.

triangle of velocities *Navigation.* a vector diagram used in dead-reckoning; the three vectors represent heading and true airspeed, track and ground speed, and wind speed and wind direction.

triangular *Mathematics.* having three angles; shaped like a triangle.

triangular bone see TRIQUETRUM.

triangular compass *Mechanical Devices.* a three-legged compass, having one leg attached to a double joint; used to transfer three points between drawings.

triangular face *Mathematics.* in a plane embedding of a planar graph, a face having exactly three edges on its boundary, a cut edge being counted twice. Thus, for example, if one draws a path of length 2 in the plane, and then adds a loop on the middle vertex of the path such that the loop encloses exactly one of the terminal vertices of the path, the resulting plane drawing is a triangulation of the plane.

triangular file *Mechanical Devices.* a file, triangular in cross section and tapering to a point at one end, used mostly in metal work and for sharpening saws. Also, THREE-SIDED FILE.

triangular matrix *Mathematics.* a square matrix M is said to be upper triangular if $a_{ij} = 0$ for $j < i$ and lower triangular if $a_{ij} = 0$ for $i < j$. M is said to be strictly upper triangular if $a_{ij} = 0$ for $j \leq i$ and strictly lower triangular if $a_{ij} = 0$ for $i \leq j$. The process of transforming a matrix into triangular form is called **triangularization.**

triangular method *Mining Engineering.* a method of estimating ore reserves in which it is assumed that a linear relationship holds between the grade difference and the distance between all drill holes.

triangular-notch weir *Civil Engineering.* a flow measuring device; usually a V-shaped channel.

triangular pulse *Electronics.* a pulse shape in which the voltage waveform ramps linearly to a peak value and then ramps downward at the same rate to the original level.

triangular wave *Electronics.* a waveform consisting of periodically repeating triangular pulses.

triangulation *Cartography.* a surveying method for extending horizontal control, in which the stations representing points on the ground are associated with the vertices of a network of triangles; the angles and the length of one side of the triangles are obtained instrumentally from direct field measurements, permitting the lengths of the other sides to be derived from trigonometric computation. *Navigation.* a method of determining position by obtaining the bearings of a ship or aircraft relative to two fixed radio stations separated by a known distance, thus establishing one side and three angles of a navigational triangle. *Mathematics.* **1.** the decomposition of an oriented manifold X^n into a union of adjacent n-dimensional elementary domains with orientation compatible with that of X^n. In particular, all compact manifolds can be triangulated (i.e., are *triangulable),* using only a finite number of domains. In addition, if C is the domain of integration representing the triangulation of X^n and ω is an n-form on X^n, then $\int_X n\omega = \int_C \omega$. **2.** a homeomorphism of a topological space onto a polyhedron whose points belong to simplexes of some simplicial complex. A space for which such a homeomorphism exists is said to be triangulable; the problem of determining the existence of a triangulation for a given space is called a **triangulation problem. 3.** a drawing of a planar graph in the plane is a plane triangulation if every face is a triangular face.

triangulation baseline *Cartography.* in surveying using the triangulation method, the side of the triangle selected to be measured for use in computing the lengths of the other sides; the precision and accuracy with which the triangulation baseline is measured, expressed as the probable error per fraction of the measured length of the line, determine the classification of the survey: first-order ≤ 1/1,000,000; second-order ≤ 1/500,000, third-order ≤ 1/250,000.

triangulation method see INTERSECTION METHOD.

triangulation reconnaissance *Cartography.* in surveying, a preliminary field observation of the area to be surveyed in order to determine the most likely locations to establish stations and the most productive triangulation scheme.

triangulation station *Photogrammetry.* a point on the surface of the earth whose position is determined by triangulation from other, known points, rather than by direct observation.

triangulation system *Cartography.* the main scheme or network of primary surveying stations, and any auxiliary stations, estabished to extend horizontal control, that is tied at several points into previously established triangulation stations of equal or higher order.

Triangulum *Astronomy.* the Triangle, a small constellation visible in northern autumn that lies between the constellations of Pisces and Perseus; it includes the spiral galaxy M33.

Triangulum Australe *Astronomy.* the Southern Triangle, a southern constellation lying southeast of Alpha Centauri.

Triangulum Nebula *Astronomy.* a little-used misnomer for M33, a large open-armed spiral galaxy in Triangulum that belongs to the Local Group.

Triassic *Geology.* **1.** the earliest period of the Mesozoic era, occurring after the Permian period of the Palezoic era and before the Jurassic period, from about 245 to 108 million years ago. **2.** describing the rocks formed during that time. Also, **Trias.**

triatomic *Chemistry.* **1.** consisting of three atoms. **2.** consisting of molecules having three atoms each.

Triatominae *Invertebrate Zoology.* blood-sucking true bugs, a subfamily of hemipteran insects in the family Reduviidae, including transmitters of Chagas' disease.

triaxial ellipsoid *Geodesy.* a spheroid that has three unequal axes, the shortest of which is the axis of revolution, and the other two lie in the plane of the equator.

triaxial pinch *Fluid Mechanics.* a device for heating a confined plasma, in which a discharge in the space between two concentric cylindrical conductors forms a cylindrical sheet of plasma; it is then confined and compressed by magnetic fields produced by currents flowing in the axial direction in the discharge and in the two conductors.

triaxon *Invertebrate Zoology.* a six-rayed sponge spicule.

1,2,3-triazole *Organic Chemistry.* $C_2N_3H_3$, a hygroscopic, crystalline solid; soluble in water, ether, and acetone; melts at 23°C and boils at 203°C; used in organic synthesis. Also, OSOTRIAZOLE.

1,2,4-triazole *Organic Chemistry.* $C_3N_3H_3$, a crystalline solid that is soluble in water and alcohol; melts at 120°C and boils at 260°C; important in organic synthesis; used in photocopying systems. Also, PYRRODIAZOLE.

tribal society or **tribal culture** *Anthropology.* a political system under which several bands that share cultural traits, language, and territory are formally or informally organized, often for seasonal exchanges of food and marriage partners.

tribasic *Chemistry.* **1.** describing an acid that has three hydrogen atoms which can be replaced by basic atoms or radicals. **2.** containing three atoms or groups, each having a valence of one.

tribasic calcium phosphate *Organic Chemistry.* $Ca_3(PO_4)_2$, a white amorphous powder, odorless and tasteless crystals; soluble in acids, insoluble in water and alcohol; melts at 1670°C; used in fertilizers and for various applications in the food industry. Also known by various other names, such as TERTIARY CALCIUM PHOSPHATE and TRICALCIUM PHOSPHATE.

tribasic magnesium phosphate *Inorganic Chemistry.* $Mg_3(PO_4)_2 \cdot 8H_2O$, a soft, bulky white powder; very slightly soluble in water; loses $5H_2O$ at 150°C and $8H_2O$ at 400°C; used as an antacid, adsorbent, food additive, and dietary supplement. It is also known by various other names, such as NEUTRAL MAGNESIUM PHOSPHATE and TRIMAGNESIUM PHOSPHATE.

tribasic sodium phosphate *Inorganic Chemistry.* $Na_3PO_4 \cdot 12H_2O$, toxic colorless crystals; soluble in water; decomposes at 75°C; used in water softeners, textiles, paper manufacture, and as a food additive. Also known by various other names, such as TRISODIUM PHOSPHATE, TRISODIUM ORTHOPHOSPHATE, and SODIUM ORTHOPHOSPHATE.

tribe *Anthropology.* a basic social unit consisting of a group of people descending or thought to descend from common ancestors, united under common leadership, and having a common language, territory, and cultural tradition; refers mainly to historical groups or contemporary nonliterate or nontechnological groups. *Systematics.* a taxonomic rank below a family and above a genus; in botany, such a name ends in *-eae*, in zoology, it ends in *-ini*. *Physics.* a body that rubs or causes friction with another body.

tribo- a combining form meaning "friction."

triboelectric series *Electricity.* a series of substances so arranged that any one can become positively charged when friction is applied downstream from it, or negatively charged when friction is applied upstream from it.

triboelectrification *Electricity.* the separation of electrical charges through surface friction.

tribology *Physics.* a study in mechanical phenomena concerned with frictional processes, lubrication, and wear on surfaces.

triboluminescence *Atomic Physics.* the glow or emission of light that results from friction or mechanical pressure.

tribonema *Botany.* a genus of unbranched, filamentous, freshwater, yellow-green algae of the family Tribonemataceae, characterized by asexual spores formed within cells that are freed when filaments break into H-shaped pieces.

Tribonemataceae *Botany.* a family of freshwater and marine yellow-green algae of the order Tribonematales, having unbranched, uniseriate filaments with no mucilaginous cover.

Tribonematales *Botany.* an order of freshwater and marine yellow-green algae of the class Xanthophyceae, characterized by cells with a filamentous organization.

tribromoethanol *Pharmacology.* $C_3H_3Br_3O$, a white crystalline powder used in an amylene-hydrate solution as a narcotic and general anesthetic for veterinary animals. Also, ETHOBROM.

tribromophenol see BROMOL.

tributary *Hydrology.* **1.** a stream that feeds, joins, or flows into a lake or into a larger stream at any point along its course. Also, SIDESTREAM, FEEDER. **2.** a valley that contains a tributary stream.

tributary glacier *Hydrology.* a glacier that merges with a larger glacier.

tributary station *Telecommunications.* any station other than the control station.

tributoxyethyl phosphate *Organic Chemistry.* $[CH_3(CH_2)_3O(CH_2)_2O]_3PO$, a combustible, yellow liquid; soluble in most organic liquids; boils between 215 and 228°C and freezes at –70°C; used as a plasticizer and flame retardant, and in floor waxes.

tributyl borate *Organic Chemistry.* $(C_4H_9)_3BO_3$, a combustible, water-white liquid; miscible with common organic liquids; boils at 232.4°C and freezes below –70°C; used in welding fluxes and organic synthesis, and as a textile flame retardant.

tributyl phosphate *Organic Chemistry.* $(C_4H_9O)_3PO$, a combustible, toxic, colorless liquid; soluble in water and miscible with most solvents and diluents; boils at 292°C and freezes below –80°C; used as a pigment grinding assistant, antifoam agent, heat-exchange medium, and solvent.

tributyltin acetate *Organic Chemistry.* $(C_4H_9)_3SnOOCCH_3$, a toxic, white crystalline solid that melts at 86–87°C; used as a fungicide and bactericide.

tributyltin chloride *Organic Chemistry.* $(C_4H_9)_3SnCl$, a toxic, colorless liquid; insoluble in cold water, but hydrolyzes in hot water and is soluble in alcohol; boils from 171 to 173°C (25 torr); used as a rodent poison and an intermediate.

TRIC trachoma inclusion conjunctivitis.

tricaine *Organic Chemistry.* $C_{10}H_{15}NO_5S$, a crystalline solid; soluble in water; melts at 149–150°C; used as an anesthetic for fish and aquatic amphibians.

tricalcium phosphate see TRIBASIC CALCIUM PHOSPHATE.

tricarboxylic acid cycle see KREBS CYCLE.

triceps brachii *Anatomy.* the large extensor muscle along the back of the arm that extends the forearm and adducts and extends the arm.

triceps surae *Anatomy.* the gastrocnemius and soleus muscles of the calf considered together.

Triceratops *Paleontology.* a late genus of ornithischian dinosaurs about 7 meters long and weighing 9 tons; notable for its perpetually replaced teeth, which were quickly worn down by its highly fibrous diet; like all other horned ceratops, both males and females bore horns; known only from the Upper Cretaceous of North America.

trich- or **tricho-** a combining form meaning "hair."

Trichaceae *Mycology.* a family of fungi belonging to the order Trichales that are characterized by brightly colored fruiting bodies and occur primarily on decaying trees.

trichalcite see TYROLITE.

Trichechidae *Vertebrate Zoology.* the manatees, a monogeneric family of aquatic, nocturnal mammals of the order Sirenia, found in rivers of North and South America and Africa.

Trichechiformes *Vertebrate Zoology.* a suborder of aquatic mammals of the order Sirenia, containing the manatees, sea cows, and dugongs.

trichesthesia *Physiology.* the feeling experienced with the touching of one of the hairs of the skin.

Trichiaceae *Mycology.* a family of fungal slime molds belonging to the order Trichiales.

Trichiales *Mycology.* an order of fungi belonging to the class Myxomycetes.

Trichinella spiralis *Invertebrate Zoology.* the nematode roundworm that causes trichinosis; transmitted in undercooked pork, the eggs hatch and larvae develop in the host's gut, then reproduce and encyst throughout the body, causing disease and death in humans.

trichinosis [trik´i nō´sis] *Medicine.* a serious and sometimes fatal infection, caused by the presence of the parasitic roundworm *Trichinella spiralis*, that results from ingesting the larvae in undercooked meat, especially pork, often resulting in diarrhea, fever, abdominal pain, muscle pain, swelling of the eyes, and affliction of the lungs, nervous system, and heart in more advanced cases.

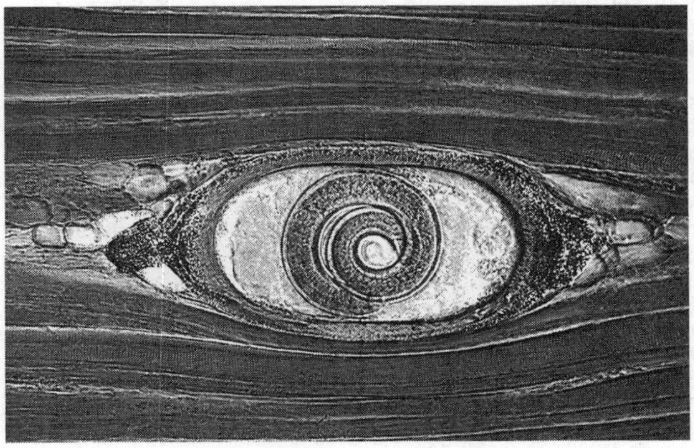

trichinosis (enlarged in muscle)

trichite *Petrology.* a thin filament or hairlike crystallite, usually black, occurring either singly or in irregular or radiating groups.

Trichiuridae *Vertebrate Zoology.* the cutlass fishes, a family of marine fishes of the teleost suborder Scombridae, characterized by an elongate body and large fanglike teeth; found in tropical and temperate seas worldwide.

trichloroacetaldehyde see CHLORAL.

trichloroacetic acid *Organic Chemistry.* Cl_3CCOOH, a nonflammable, toxic, deliquescent, colorless crystalline solid; soluble in water, alcohol, and ether; melts at 57.5°C and boils at 221°C; used in organic synthesis and in pharmaceuticals, medicine, and herbicides.

1,2,3-trichlorobenzene *Organic Chemistry.* $Cl_3C_6H_3$, combustible, toxic, white crystals; insoluble in water, slightly soluble in alcohol, and soluble in ether; melts at 53°C and boils at 221°C; used as an organic intermediate.

1,2,4-trichlorobenzene *Organic Chemistry.* $Cl_3C_6H_3$, a combustible, toxic, colorless liquid; insoluble in water and miscible with most organic solvents and oils; melts at 17°C and boils at 213°C; used as a solvent and in insecticides, dielectric fluids, lubricants, and synthetic transformer oils.

1,3,5-trichlorobenzene *Organic Chemistry.* $Cl_3C_6H_3$, a needlelike solid that is soluble in ether, benzene, and ligroin; melts at 63°C and boils at 208°C.

1,1,1-trichloroethane *Organic Chemistry.* CH_3CCl_3, a nonflammable, colorless liquid; insoluble in water and soluble in alcohol and ether; boils at 74°C and freezes at –38°C; used as a solvent. Also, METHYL CHLOROFORM.

2,2,2-trichloroethanol *Pharmacology.* $C_2H_3Cl_3O$, a hygroscopic liquid formerly used as a hypnotic and anesthetic.

trichloroethylene *Organic Chemistry.* $ClCH{=}CCl_2$, a nonflammable, toxic, colorless, photoreactive liquid; slightly soluble in water and miscible with common organic solvents; boils at 86.7°C and freezes at –85°C; used in organic synthesis, dry cleaning, and textile processing, and as a solvent and fumigant.

trichlorofluoromethane *Organic Chemistry.* CCl_3F, a noncombustible, colorless liquid; boils at 24°C and freezes at –111°C; used in organic synthesis and fire extinguishers, and as a solvent, refrigerant, and aerosol propellant. Also, FLUOROCARBON-11.

trichloroisocyanuric acid *Organic Chemistry.* $C_3Cl_3N_3O_3$, a white powder; slightly soluble in water and acetic acid; melts at 249–251°C; used as a chlorinating agent, disinfectant, and industrial deodorant. Also, SYMCLOSENE.

trichloromethane see CHLOROFORM.

trichloromethanesulfenyl chloride *Organic Chemistry.* $ClSCCl_3$, a toxic, foul-smelling, yellow oil; insoluble in water; boils with some decomposition at 148°C; used as a fumigant and dye intermediate, and in organic synthesis.

trichloromethyl chloroformate *Organic Chemistry.* $ClCOOCCl_3$, a noncombustible, toxic, colorless liquid; decomposed by hot water and soluble in alcohol and ether; boils at 128°C and freezes at –57°C; used in organic synthesis and as a military poisonous gas. Also, DIPHOSGENE.

trichloronitromethane see CHLOROPICRIN.

2,3,4-trichlorophenol *Organic Chemistry.* $Cl_3C_6H_2OH$, a needlelike solid; soluble in alcohol and ether; melts at 92–93°C; used as a fungicide and herbicide.

2,4,5-trichlorophenol *Organic Chemistry.* $Cl_3C_6H_2OH$, nonflammable gray flakes; soluble in alcohol and ether; melts at 68–70°C; used as a fungicide and bactericide.

2,4,6-trichlorophenol *Organic Chemistry.* $Cl_3C_6H_2OH$, nonflammable yellow flakes; soluble in alcohol and ether; melts at 69°C and boils at 248–249°C; used as a defoliant, herbicide, and fungicide.

2,4,5-trichlorophenoxyacetic acid *Organic Chemistry.* $C_6H_2Cl_3OCH_2CO_2H$, a light tan solid; insoluble in water and soluble in alcohol; melts at 151–153°C; used as a defoliant, herbicide, and plant hormone.

1,2,3-trichloropropane *Organic Chemistry.* $CH_2ClCHClCH_2Cl$, a toxic, combustible, colorless liquid; slightly soluble in water and soluble in alcohol and ether; boils at 156.17°C and freezes at –14°C; used as a paint and varnish remover, solvent, and degreasing agent.

1,1,2-trichloro-1,2,2-trifluoromethane *Organic Chemistry.* CCl_2FCClF_2, a colorless, volatile liquid; soluble in alcohol and ether; boils at 47°C and freezes at –35°C; used as a dry cleaning solvent and a refrigerant, and in fire extinguishers. Also, **trichlorotrifluoroethane.**

trichobothrium *Invertebrate Zoology.* an erect, bristlelike sensory hair or group of hairs on some arthropods and other invertebrates.

trichobranchiate gill *Invertebrate Zoology.* in some decapod crustaceans, such as crayfish, a gill with filamentous branches in series around an axis, growing from the skin on the legs or body. Also, **trichobranchia.**

Trichobranchidae *Invertebrate Zoology.* a family of deep-sea, tube-dwelling polychaete annelid worms belonging to the grouping Sedentaria.

trichocardia *Cardiology.* a former term for fibrinous pericarditis (an acute inflammation of the pericardium with fibrinous exudate).

Trichocephalida *Invertebrate Zoology.* a superfamily of nematode worms, disease-causing parasites of vertebrates.

trichocercous cercaria *Invertebrate Zoology.* a fluke larva with a spiny tail.

Trichocoleaceae *Botany.* a family of dioecious liverworts of the order Jungermanniales, often appearing as an intricate mass of interwoven hairs; occurs in mossy rain forests.

Trichocomaceae *Mycology.* a family of fungi belonging to the order Eurotiales and found in soil, dung, foods, or plant debris; it is composed of species that, in their asexual state, include *Aspergillus* and *Penicillum.*

trichocyst *Invertebrate Zoology.* in some protozoans, one of many tiny structures that when touched releases a thread that hardens on contact with water, probably for defense.

Trichodactylidae *Invertebrate Zoology.* a family of tropical freshwater crabs, decapod crustaceans in the section Brachyura.

Trichoderma *Mycology.* a genus of imperfect fungi belonging to the class Pyrenomycetes that occur on wood and are characterized by brightly colored ascocarps; includes both parasitic and nonparasitic types.

Trichodesmium *Bacteriology.* a genus of filamentous cyanobacteria that form red, purple, or brown colonies in marine environments and contain gas vacuoles for buoyancy.

trichoepithelioma *Medicine.* a skin tumor originating in the lanugo (fine hair) follicles.

trichogen cell *Cell Biology.* a cell from which hair or a hairlike process arises.

Trichogrammatidae *Invertebrate Zoology.* a family of tiny wasps, hymenopteran insects in the superfamily Chalcidoidea; the larvae are parasites in eggs of insects, including pests.

trichogyne *Botany.* a single or multicelled, hairlike receptor organ projecting from the female sexual apparatus of a number of simple plants, such as the red algae, lichens, and ascomycetes, that accepts the male gamete.

trichoid *Medicine.* resembling hair; hairlike.

Tricholomataceae *Mycology.* a family of fungi belonging to the order Agaricales, composed of mushrooms, some of which are edible.

trichome *Botany.* any of variously formed outgrowths from the epidermal cells of plants, such as root hairs, hooks, and spines.

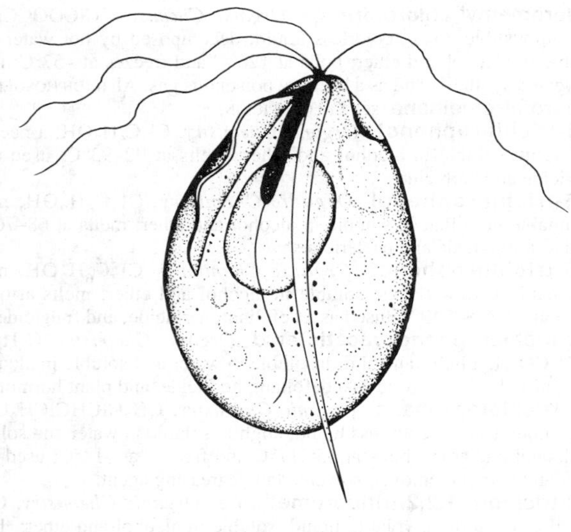

trichomonad

trichomonad *Invertebrate Zoology.* any organism that is a member of the genus *Trichomonas.*

Trichomonadida *Invertebrate Zoology.* an order of colorless flagellated protozoans in the subclass Zoomastigina.

Trichomonadidae *Invertebrate Zoology.* a family of parasitic flagellated protozoans in the order Trichomonadida.

Trichomonas *Invertebrate Zoology.* a genus of pear-shaped protozoans with five flagella and an undulating membrane; parasitic or commensal in the gut and reproductive system of many animals, causing trichomoniasis in humans.

trichomoniasis *Medicine.* an infestation with the parasitic protozoa, *Trichomonas,* especially a vaginal infection characterized by itching, burning, and yellow or green vaginal discharge, or an infection of the male genital tract which may be asymptomatic or may produce a mild urethritis; it is transmitted by sexual contact. Other forms of trichomoniasis cause serious infections in animals and birds.

Trichomycetes *Mycology.* a class of fungi belonging to the subdivision Zygomycotina, which occurs in the digestive tract of terrestrial and aquatic arthropods and insects.

trichomycin *Microbiology.* an antibiotic produced by the bacteria *Streptomyces hachijoensis* and *S. abikoensis* that is effective against yeast and fungal infections.

Trichoniscidae *Invertebrate Zoology.* a family of isopod crustaceans in the suborder Oniscoidea, found in damp places.

Trichonympha *Invertebrate Zoology.* a genus of flagellated protozoans; symbiotic in termite and roach guts, where they digest cellulose.

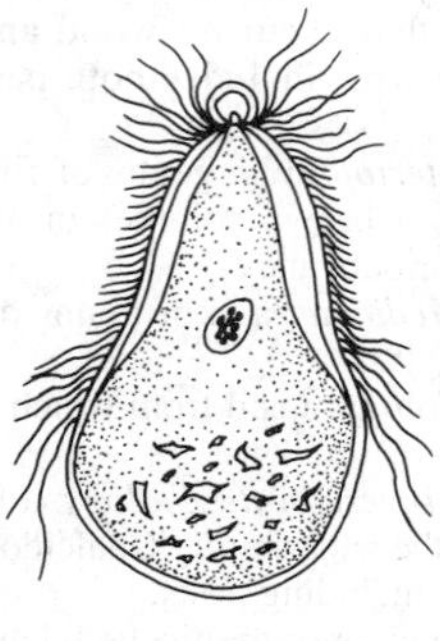

Trichonympha

trichopathophobia *Psychology.* an irrational fear of hair or concern with one's own hair.

trichopathy *Medicine.* any disease of the hair.

trichophagy *Medicine.* the habit or practice of eating hair.

Trichophilopteridae *Invertebrate Zoology.* a family of biting lice in the order Mallophaga; ectoparasites of lemurs.

trichophytin *Immunology.* a group antigen that is used in skin tests to determine the past or present occurrence of the parasitic or pathogenic fungi known as dermatophyte; occurs in the skin, hair, and nails of individuals.

Trichoptera *Invertebrate Zoology.* the caddis flies, an order of insects with two pairs of hairy, scaled wings and long antennae and legs; the wormlike aquatic larvae build cases of sand and other particles glued with secretions.

Trichopterygidae see PTILIIDAE.

trichorrhexis *Medicine.* a condition characterized by the breaking off of the hair.

Trichosida *Invertebrate Zoology.* an order of filose amoebas, common in shallow coastal and estuarine waters of most continents, that feed on algae and bacteria.

trichospore *Mycology.* in fungi belonging to the order Harpellales, a spore case that holds one spore and opens at maturity.

Trichosporon *Mycology.* a genus of fungi belonging to the class Hyphomycetes; some species are found on human skin and cause the disease white piedra.

trichosporosis see WHITE PIEDRA.

Trichostomatida *Invertebrate Zoology.* an order of ciliated protozoans in the subclass Holotrichia, having no mouth cilia.

Trichostrongylidae *Invertebrate Zoology.* a family of hookworms, nematodes in the superfamily Strongyloidea, endoparasites of domestic animals.

trichothecenes *Biochemistry.* a group of mycotoxins formed from various imperfect fungi, including those of the genera *Fusarium, Myrothecium,* and *Trichothecium*; ingestion can cause gastric hemorrhaging, and contact with skin can cause irritations.

Trichothyriaceae *Mycology.* a family of fungi belonging to the order Chaetothyriales that are parasitic to plants and other fungi, occurring in temperate to tropical environments.

trichotomous *Botany.* branching into three parts or giving off three shoots.

trichotomy property *Mathematics.* let S be a partially ordered set with order $<$. The partial ordering $<$ is said to satisfy the trichotomy property if exactly one of the following is true for every a and b in S: $a < b$, $a = b$, or $b < a$. If $<$ satisfies both the trichotomy property and the transitive property, then $<$ is a linear ordering.

trichroism *Optics.* the property of certain crystals that show a different color on each of three pairs of parallel faces when illuminated with white light.

trichromatic theory *Optics.* the theory that any color can be replicated by a suitable combination of three primary colors.

Trichuroidea *Invertebrate Zoology.* a group of slender nematode worms with a thickened posterior portion; parasites in vertebrates; includes *Trichinella spiralis,* the cause of trichinosis.

trickle charge see FLOATING CHARGE.

trickle filter *Biotechnology.* a bed of broken rock or other coarse aggregates (generally 50 to 100 mm in diameter) onto which sewage or industrial waste is sprayed intermittently; as the waste water trickles through, it is purified. Also, TRICKLING FILTER.

trickle hydrodesulfurization *Chemical Engineering.* a fixed-bed, petroleum refining method used for desulfurizing gas, oils, and middle distillates, using a catalyst of cobalt molybdenum on alumina, over which the oil trickles down against the upflowing stream of hydrogen.

trickle irrigation see DRIP IRRIGATION.

trickling *Computer Programming.* the migration of temporarily unneeded data from main memory to a secondary storage device; used when available main storage is limited.

trickling filter see TRICKLE FILTER.

triclad *Invertebrate Zoology.* any member of the order Tricladida, planarian flatworms of the class Turbellaria.

Tricladida *Invertebrate Zoology.* planarians, an order of large, free-living flatworms in the class Turbellaria, having a ventral mouth and three-branched gut; any injured part, even the head, can regenerate.

triclinic unit cell *Crystallography.* a unit cell in which there is no rotational symmetry; as a result there are no restrictions on axial ratios or interaxial angles. Also, **triclinic crystal.**

Tricomi's equation *Fluid Mechanics.* the fundamental equation of motion for transonic flow.

tricone bit *Mining Engineering.* a rock bit with three toothed, conical cutters, each mounted upon a friction-reducing bearing.

triconodont *Vertebrate Zoology.* **1.** relating to a tooth possessing three principal cusps or cones. **2.** an animal having triconodont teeth.

Triconodonta *Paleontology.* an order of primitive mammals of uncertain affinities, having a wide holarctic distribution; probably predators, they ranged from shrew-size to cat-size; extant from the Triassic to the Cretaceous.

Tricornaviridae *Virology.* a proposed family of plant viruses having isometric particles that contain tripartite ssRNA genomes.

tricosane *Organic Chemistry.* $CH_3(CH_2)_{21}CH_3$, combustible leaflets; insoluble in water and soluble in alcohol; melts at 48°C and boils at 199–200°C (3 torr); used in organic synthesis.

tricot *Textiles.* a warp-knit fabric with fine vertical wales on the face and crosswise ribs on the back; an important commercial fabric, usually made from rayon or nylon.

tricresyl phosphate *Organic Chemistry.* $(CH_3C_6H_4O)_3PO$, a combustible, toxic, colorless liquid; insoluble in water and miscible with common solvents and vegetable oils; boils at 265°C (10 torr); used as a plasticizer, gasoline additive, air filter medium, and fire retardant; commonly called **TCP.**

tricrotic *Cardiology.* relating to an arterial pulse in which there are three waves.

tricrotism *Cardiology.* the condition of having three waves representing one pulse beat.

Trictenotomidae *Invertebrate Zoology.* a family of beetles, coleopteran insects in the superfamily Tenebrionoidea.

tricuspid *Anatomy.* **1.** having three points or cusps. **2.** relating to the tricuspid valve of the heart. **3.** a tricuspid part, as a tooth.

tricuspid incompetence *Cardiology.* imperfect closure of the tricuspid valve, allowing regurgitation of blood into the right atrium during systole.

tricuspid stenosis *Cardiology.* an abnormal narrowing of the opening of the tricuspid valve.

tricuspid valve *Anatomy.* the atrioventricular valve that separates the right atrium from the right ventricle and prevents backward flow of blood.

tricycle landing gear *Aviation.* a three-wheel landing gear, usually consisting of two main wheels just aft of the center of gravity and an auxiliary wheel forward under the nose of the aircraft.

Tridacnidae *Invertebrate Zoology.* a family of marine bivalve mollusks in the subclass Eulamellibranchia; includes the giant clams.

Tridactylidae *Invertebrate Zoology.* the pygmy moles or sand crickets, a family of burrowing insects in the order Orthoptera.

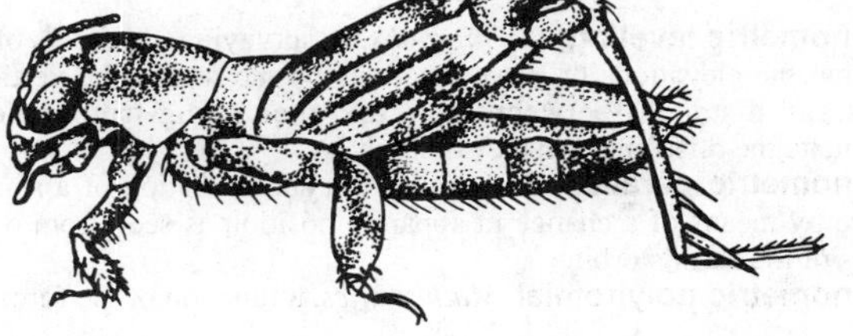

Tridactylidae

tridactylous *Zoology.* having three fingers on one hand or foot.

tridecane *Organic Chemistry.* $CH_3(CH_2)_{11}CH_3$, a combustible, colorless liquid; insoluble in water and soluble in alcohol; boils at 234°C and freezes at –5.5°C; used in organic synthesis and as a distillation chaser.

tridecyl alcohol *Organic Chemistry.* $C_{12}H_{25}CH_2OH$, a combustible white solid; soluble in alcohol and ether; melts at 32°C and boils at 155–156°C; used in perfumes, detergents, and synthetic lubricants. Also, **1-tridecanol.**

Trident *Ordnance.* a sea-based strategic weapons system consisting of the nuclear-powered Trident submarine, long-range Trident ballistic missiles, and the facilities required to support the submarine, missiles, and associated personnel. (From the *trident,* a three-pronged spear held by Neptune, or Poseidon, the god of the sea in classical mythology.)

Trident I *Ordnance.* a three-stage, solid-propellant ballistic missile that can be launched from a Trident submarine in submerged or surfaced mode; it can deploy multiple nuclear warheads to separate targets over a range of 4000 miles; officially designated **UGM-96A.** The **Trident II** is a larger, more accurate version of this missile.

tridentate ligand *Inorganic Chemistry.* a ligand that is attached to a metal ion at three different points in a coordination compound.

trident of Newton *Mathematics.* the graph in the plane of the equation $xy = ax^3 + bx^2 + cx + d$, where the coefficients a, b, c, and d are real constants and $a \neq 0$.

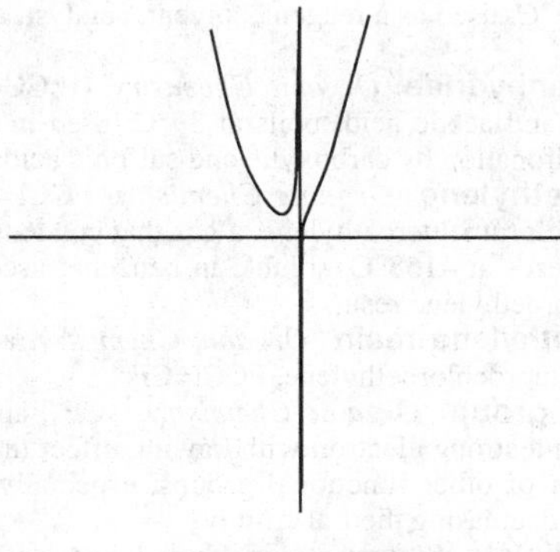

trident of Newton

tridermoma *Oncology.* a teratoma composed of fetal remains that develops during the period of primary segmentation of blastomeres in the fertilized ovum and that contains all three germ layers of the embryo.

tridiagonal matrix *Mathematics.* a square matrix in which the only nonzero entries occur on the main diagonal and the two adjacent diagonals.

triductor *Electricity.* the tripling of a power line frequency by a particular arrangement of iron-core transformers and capacitors.

tridymite *Mineralogy.* SiO_2, a colorless or white monoclinic pseudohexagonal or triclinic mineral occurring in minute, commonly twinned, tabular crystals and crystal aggregates in acidic volcanic rocks and stony meteories, having a specific gravity of 2.26 and a hardness of 7 on the Mohs scale; polymorphous with coesite, cristobalite, quartz, and stishovite.

triethanolamine *Organic Chemistry.* $(HOCH_2CH_2)_3N$, a combustible, hygroscopic, colorless liquid that is miscible with water and alcohol and slightly soluble in ether; melts at 21.2°C and boils at 335°C or 190–193°C (15 torr); used in dry-cleaning solutions, cosmetics, detergents, textile processing, wool scouring, and as a corrosion inhibitor and pharmaceutical alkalizing agent. Also, TRIS(2-HYDROXYETHYL)AMINE.

triethanolamine stearate see TRIHYDROXYETHYLAMINE STEARATE.

triethoxymethane see TRIETHYL ORTHOFORMATE.

triethylamine *Organic Chemistry.* $(C_2H_5)_3N$, a flammable, toxic, colorless liquid; soluble in water and alcohol; boils at 89.7°C and freezes at –115.3°C; used as a catalytic solvent, rubber accelerator, corrosion inhibitor, and propellant, and in the curing and hardening of polymers.

triethylborane *Organic Chemistry.* $(C_2H_5)_3B$, a toxic, flammable, colorless liquid; immiscible with water and miscible with most organic solvents; boils at 95°C and freezes at –93°C; used as a jet and rocket fuel, a fuel additive, a catalyst, and an intermediate. Also, **triethylborine.**

triethylene glycol *Organic Chemistry.* $HOCH_2CH_2OCH_2CH_2OCH_2CH_2OH$, a combustible, hygroscopic, colorless liquid; soluble in water and immiscible with gasoline; boils at 287°C or 165°C (14 torr) and freezes at –7.2°C; used as a solvent, humectant, and plasticizer, and in the dehydration of natural gas.

triethylenemelamine *Organic Chemistry.* $NC[N(CH_2)_2]NC[N(CH_2)_2]NC[N(CH_2)_2]$, a highly toxic, white crystalline powder; soluble in water and alcohol; polymerizes at 139°C; used in medicines and as an insecticide and sterilizing agent. Also, TRETAMINE.

triethylenetetramine *Organic Chemistry.* $H_2NCH_2CH_2NHCH_2CH_2NHCH_2CH_2NH_2$, a combustible, moderately viscous, yellowish liquid; soluble in water; melts at 12°C and boils at 277.5°C; used in detergents, dyes, and pharmaceuticals.

triethyl orthoformate *Organic Chemistry.* $(C_2H_5O)_3CH$, a flammable, toxic, colorless liquid; decomposes in water; soluble in alcohol and ether; boils at 145.9°C; used in organic synthesis and pharmaceuticals. Also, TRIETHOXYMETHANE.

triethyl phosphate *Organic Chemistry.* $(C_2H_5O)_3PO$, a combustible, colorless liquid; miscible in water and soluble in most organic solvents; boils at 216°C and freezes at –56.4°C; used as a solvent and plasticizer, and in the manufacture of pesticides.

trifacial neuralgia see TRIGEMINAL NEURALGIA.

trifid *Biology.* divided into three narrow parts or tubes.

Trifid Nebula *Astronomy.* a two-part nebula in Sagittarius that has a hydrogen emission region (intersected by three dark lanes of dust) and a dust reflection nebula around a hot young star; designated **M20.**

trifluoroacetic acid *Organic Chemistry.* F_3CCOOH, a nonflammable, hygroscopic, colorless liquid; very soluble in water; boils at 72.4°C and freezes at –15.25°C; used as a reagent, solvent, catalyst, and strong nonoxidizing acid.

trifluoroacetic anhydride *Organic Chemistry.* $(F_3CCO)_2O$, a liquid; soluble in ether and acetic acid; boils at 39°C; used in bringing about acylation and sulfonation by carboxylic and sulfonic acids.

trifluorochloroethylene *Organic Chemistry.* $FCCl{=}CF_2$, the legal label name for chlorotrifluoroethylene, a gas that condenses and boils at –28.4°C and freezes at –158°C; soluble in benzene; used as a base for polychlorotrifluoroethylene resin.

trifluorochloroethylene resin *Organic Chemistry.* a resin polymer derived from trifluorochloroethylene, $FCCl{=}CF_2$.

trifluoromethyl group *Organic Chemistry.* $-CF_3$, an important organic group with a strong electron withdrawing effect that markedly alters the behavior of other functional groups, especially by increasing their acidity and decreasing their alkalinity.

trifoliate or **trifoliolate** *Biology.* having three lobes, leaflets, or foils.

trifoliin A *Biochemistry.* $C_{22}H_{22}O_{11}$, a lectin present at the root surface of white clover.

trifoliosis *Veterinary Medicine.* acute photosensitization in light-skinned animals that have ingested certain leguminous plants, especially clover, resulting in swelling and inflammation of the skin and mucous membranes of the mouth.

triformol see *sym*-TRIOXANE.

trig. trigonometry; trigonometric; trigonometrical.

trigamous *Botany.* having pistillate, staminate, and perfect flowers in one head.

trigeminal *Anatomy.* of or relating to the trigeminal nerve. (From a Latin word meaning "triple.")

trigeminal nerve *Anatomy.* the fifth cranial nerve, a large mixed nerve that ramifies throughout the mouth and face.

trigeminal neuralgia *Neurology.* sudden, severe facial neuralgia affecting one or more branches of the trigeminal nerve, caused by degeneration of or pressure on the trigeminal nerve and the gasserian ganglion, and characterized by paroxysms of flashing or stabbing pain along the course of the nerve. Also, PROSOPALGIA, TRIFACIAL NEURALGIA, TIC DOULOUREUX.

trigeminy *Physiology.* any condition occurring in threes, such as three pulse beats in rapid succession.

trigger *Ordnance.* a metal projection in the firing mechanism of a firearm or similar weapon that, when pulled or squeezed by the finger, releases the firing pin and discharges the weapon. *Electronics.* **1.** the signal or stimulus that initiates a trigger circuit. **2.** to apply such a signal. *Computer Technology.* a button on a mouse or joystick used in computer games. *Computer Programming.* to transfer control to the first instruction of a program, often by intervention from the external environment. *Artificial Intelligence.* **1.** to cause a procedure to be executed. **2.** a condition, specified with a knowledge source or procedure, such that the procedure is to be executed when the condition becomes true.

trigger circuit *Electronics.* a circuit that depends on the application of an input initiating signal or stimulus to perform a specific predetermined function; examples include bistable multivibrators, monostable multivibrators, time delays, and sweep waveform generators.

trigger control *Electronics.* the on/off control of thyratrons and other gas-filled electron tubes.

trigger diode see THREE-LAYER DIODE.

triggered spark gap *Electricity.* a fixed spark gap with a trigger mechanism that receives low-power pulses from a discharge passing between two electrodes and that is activated by an auxiliary electrode.

triggerfish *Vertebrate Zoology.* any of various tropical marine fishes of the genus *Balistes,* having a compressed body and an anterior dorsal fin with three stout spines.

trigger guard *Ordnance.* a curved metal element that protects the trigger of a firearm or rocket launcher.

triggering *Electronics.* the initiation of a specific predetermined function in a circuit or device through the application of an initiating signal or stimulus.

trigger level *Electronics.* **1.** the lowest amplitude of interrogation signal that is capable of causing a reply signal from a transponder. **2.** the instantaneous amplitude of a trigger pulse that causes a circuit or device to perform a specific predetermined function.

trigger pull *Ordnance.* **1.** the pressure applied to the trigger in order to fire a weapon, especially abrupt pressure, as opposed to the gradual pressure of a trigger squeeze. **2.** the mechanical linkage of the trigger and other parts that directly fire the cartridge in a firearm.

trigger pulse *Electronics.* an electrical pulse used as a trigger.

trigger squeeze *Ordnance.* a method of firing in which a gradual pressure is placed upon the trigger by contraction of the whole hand.

trigger switch *Electricity.* a switch that is operated by pulling a trigger mounted in a pistol grip handle.

trigger tube *Electronics.* a cold-cathode gas-filled electron tube that contains one or more auxiliary electrodes to start ionization, which then continues as long as there is voltage applied to the tube.

trigistor *Electronics.* a PNPN semiconductor that has bistable characteristics and is used as a fast-acting switching device.

Triglidae *Vertebrate Zoology.* the sea robins, a family of benthic marine perciform fish characterized by an elongate body with a spiny or ridged head; found in tropical and temperate seas worldwide.

triglyceride [trī glis′ə rīd′] *Organic Chemistry.* any of a number of naturally occurring lipids formed when three fatty acids replace the three hydrogen atoms in the hydroxyl groups of glycerol, having the general formula $CH_2(OOCR_1)CH(OOCR_2)CH_2(OOCR_3)$ and representing the chief constituent of fats and oils, including the fat cells in the human body.

Trigonalidae *Invertebrate Zoology.* a family of tiny, stingless, parasitic wasps, hymenopteran insects in the superfamily Proctotrupoidea, that lay eggs on leaves, from which they are ingested by the host caterpillar.

trigonal system *Crystallography.* a crystal system in which the crystals have three-fold symmetry; the group is sometimes considered to be part of the hexagonal system

trigone *Anatomy.* **1.** the first three cusps of an upper molar tooth. **2.** a triangular area formed at the base of the urinary bladder by the orifices of the ureters and the urethra. *Botany.* a thickened angle of a cell, occurring where three or more plant cells meet.

trigonite *Mineralogy.* $Pb_3Mn^{+2}(As^{+3}O_3)_2(As^{+3}O_2OH)$, a yellow or dark-brown monoclinic mineral occurring in tabular, triangle-shaped crystals, having a specific gravity of 6.1 to 7.1 and a hardness of 2 to 3 on the Mohs scale; found in crevices in dolomite at Långban, Sweden.

trigonium urogenitale see UROGENITAL DIAPHRAGM.

trigonometric [trig′ə nə met′rik] *Mathematics.* of or relating to trigonometry.

trigonometric functions *Mathematics.* the sine and cosine functions, and any function that can be obtained from them by algebraic operations (e.g., secant, cosecant, tangent, and cotangent). Also, **trigonometric ratios.**

trigonometric leveling *Cartography.* in surveying, a process of determining the elevations of survey stations by measuring the vertical angles and distances between them and then using trigonometry to calculate the differences in elevations.

trigonometric parallax *Astronomy.* the determination of a star's distance by means of a change in apparent position as seen from opposite sides of the earth's orbit.

trigonometric polynomial *Mathematics.* a function of the form

$$p(\theta) = \sum_{k=0}^{n} (a_n \cos n\theta + b_n \sin n\theta),$$

where a_n and b_n are constants for each n. According to the Stone-Weierstrass theorem, any continuous function on the circle can be uniformly approximated by trigonometric polynomials.

trigonometric series *Mathematics.* an infinite series of the form

$$t(z) = \sum_{k=0}^{\infty} (a_n \cos nz + b_n \sin nz),$$

where a_n and b_n are constants for each n .

trigonometry [trig′ə näm′ə trē] *Mathematics.* the study of trigonometric functions, triangles, and solutions of plane or spherical triangles.

Trigonopsis *Mycology.* a genus of yeast fungi belonging to the class Hyphomycetes; the shape of its cells depends on growth conditions such as temperature and carbon and nitrogen sources; it includes only one species, *T. variabilis,* found in beer and grape must.

Trigonostylopoidea *Paleontology.* an extinct suborder of South American ungulates of uncertain order classsification; related to the astrapotheres; extant in the Paleocene and Eocene.

trigonous *Biology.* having three prominent angles, as in a plant stem or ovary.

trihedral *Mathematics.* any configuration of three noncoplanar lines intersecting in a point; e.g., the coordinate axes in Euclidean 3-space. A polyhedral angle of three faces is called a **trihedral angle.**

1,3,5-trihydroxybenzene *Organic Chemistry.* $(HO)_3C_6H_3$, a crystalline solid that is soluble in alcohol, ether, and benzene; boils at 218–221°C; an important reagent used for the estimation of furfural and of pentoses. Also, PHLOROGLUCINOL.

trihydroxyestrin see ESTRIOL.

trihydroxyethylamine stearate *Organic Chemistry.* $(HOCH_2CH_2)_3N \cdot HOOCC_{17}H_{35}$, a combustible, waxy, cream-colored solid; dispersible in hot water and soluble in methanol, mineral oil, and vegetable oil; melts at 42–44°C; used as an emulsifier in cosmetics and pharmaceuticals. Also, TRIETHANOLAMINE STEARATE.

triiodothyroacetic acid *Biochemistry.* $C_{14}H_9I_3O_4$, an agent that is unaffected by ionizing radiation; used to produce contrast in cholecystrography. Also, IOPHENOXIC ACID.

triiodothyronine *Endocrinology.* $C_{15}H_{12}I_3NO_4$, a thyroid hormone produced by the coupling of diiodotyrosine and monoiodotyrosine in thyrogloblin in the thyroid gland or by the 5'-monodeiodination of thyroxine in peripheral tissue; it is less abundant but more potent than thyroxine and acts via the same receptor as thyroxine to sustain the same physiological functions. *Pharmacology.* a preparation of this hormone used to regulate the general rate of metabolism, especially to treat hypothyroidism; similar to but much more potent than thyroxine. Also, **3,3',5-triiodothyronine.**

triisobutylene *Organic Chemistry.* $(C_4H_8)_3$, a combustible liquid that boils at 175.5–178.9°C; used in making resins, lubricating oil additives, and motor fuels.

trilateration *Cartography.* a method of land surveying that uses triangulation to measure the distance between a series of points on the earth's surface.

trilinear surveying *Cartography.* a method of determining the position of a point by making observations from that point to three points of known location and plotting the intersection of those lines of direction.

trill *Vertebrate Zoology.* **1.** a rapid succession or alternation of sounds made by a bird or insect. **2.** to make such sounds. *Crystallography.* a cyclic twin composed of three individual crystals.

trillion *Mathematics.* **1.** in the United States and France, 10^{12}. **2.** in Great Britain and Germany, 10^{18}.

Trilobita *Paleontology.* an extinct class of marine arthropods; stratigraphically very important for the Early to Middle Paleozoic era, especially the Cambrian and Ordovician; extant in the Cambrian to Permian.

Trilobita

trilobite *Paleontology.* any member of the extinct class Trilobita, the earliest known arthropods; characterized by a three-lobed body and serially repeated biramous limbs; they ranged in length from a centimeter to a few feet, but were generally small.

Trilobitoidea *Paleontology.* a term formerly used to identify as a class a small group of unusual genera found in strata from the Cambrian to the Devonian, but known mainly from the Burgess shale of the Middle Cambrian; the species are now reclassified in the group Trilobita.

trilobitomorph *Paleontology.* any three-lobed arthropod of the subphylum Trilobidomorpha.

Trilobitomorpha *Paleontology.* a subphylum of the phylum Arthropoda, containing the trilobites.

trilocular *Biology.* having three compartments or chambers, as in the capsule of a plant or the heart of a reptile.

trilogy *Medicine.* a combination of three elements, such as three concurrent symptoms.

trilogy of Fallot *Cardiology.* the threefold condition of atrial septal defect, right ventricular hypertrophy, and pulmonic stenosis.

trim *Aviation.* **1.** a condition in which an aircraft maintains a static balance in pitch and fixed attitude with regard to all wind axes. **2.** to adjust or balance an aircraft so that it is in trim. **3.** of an action or component, designed to achieve or maintain trim, such as **trim speed** or a **trim tab.** *Naval Architecture.* **1.** the way that a ship floats in the water, with emphasis on the relationship between the draft of the bow and that of the stern; a difference in draft is often a sign of leakage or improper stowage, but may be a deliberate measure or design feature. **2.** to shift ballast in order to produce the desired trim. Also, **trim ship. 3.** to adjust the sails to their optimum position in relation to the wind. Also, **trim sails**. *Graphic Arts.* to cut the folded edges from the signatures of a bound book. *Electronics.* to make a fine adjustment of resistance, inductance, or capacitance in a circuit. *Building Engineering.* any finished woodwork such as a window sash, baseboard, or decorative molding.

trim by head or **trimmed by the head** *Naval Architecture.* a condition in which a ship's bow lies abnormally low in the water. Also, DOWN BY THE HEAD.

trim by stern or **trimmed by the stern** *Naval Architecture.* a condition in which a ship's stern lies abnormally low in the water. Also, DOWN BY THE STERN.

trim die or **trimming die** *Mechanical Devices.* a die used to cut flash from a casting, forging, or stamping.

trimellitic acid *Organic Chemistry.* $C_9H_6O_6$, colorless crystals that are slightly soluble in water, partially soluble in alcohol, and insoluble in benzene; used in the organic synthesis of polymers and plasticizers. Also, 1,2,4-BENZENETRICARBOXYLIC ACID.

Trimeniaceae *Botany.* a family of dicotyledonous trees and climbing shrubs of the order Laurales, commonly accumulating aluminum and producing volatile oils; native to New Guinea, New Caledonia, Fiji, and eastern Australia.

Trimenoponidae *Invertebrate Zoology.* a family of biting lice in the order Mallophaga; ectoparasites of South American rodents.

trimer *Organic Chemistry.* a polymer consisting of three monomer molecules.

trimercuric *Chemistry.* containing three atoms of bivalent mercury.

Trimerellacea *Paleontology.* a superfamily of Paleozoic inarticulate brachiopods, characterized by lamellae and a strong calcite shell; extant in the Ordovician and Silurian.

trimeric *Organic Chemistry.* relating to or characteristic of a trimer.

Trimerophytopsida *Paleontology.* an extinct class of Devonian plants in the division Tracheophyta; descended from the Rhyniopsida, they are ancestral to several more advanced vascular plants that arose in the Devonian; e.g., Sphenopsida and Cladoxylales. Also, **Trimerophytatae.**

trimerous *Botany.* describing plant parts, usually flowers or parts of flowers, that grow in threes or multiples of three. *Entomology.* having three segments.

Trimerus *Paleontology.* a genus of Middle Paleozoic, especially Silurian, trilobites in the order Phacopida and the suborder Calymenina; characterized by small eyes and an anteriorly tapering glabella.

trimetallic plate *Graphic Arts.* a durable printing plate having an aluminum or steel base and plated with two other metals, usually copper and chromium or zinc. Also, **trimetal plate.**

trimethadione *Pharmacology.* $C_6H_9NO_3$, an anticonvulsant occurring as white crystalline granules, administered orally to control petit mal seizures.

trimethoprim *Organic Chemistry.* $C_{14}H_{18}N_4O_3$, an antibacterial, administered orally in combination with a sulfonamide in the treatment of urinary tract infections.

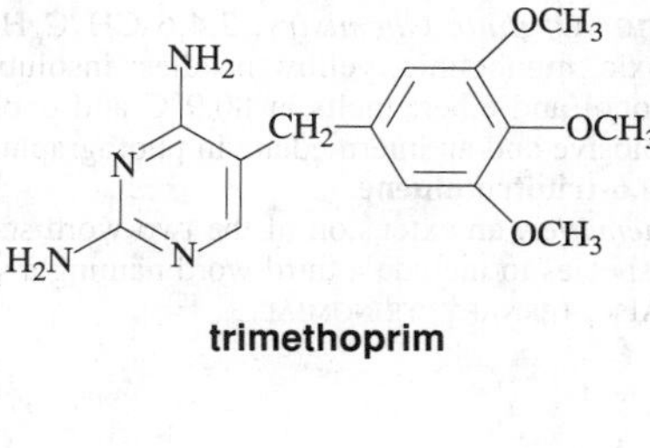

trimethoprim

trimethylamine *Organic Chemistry.* $(CH_3)_3N$, a flammable, toxic, fishy-smelling, colorless gas; soluble in water, alcohol, and ether; boils at 4°C and freezes at –117.1°C; used to odorize natural gas, and as a flotation agent and an insect attractant.

1,2,4-trimethylbenzene see PSEUDOCUMENE.

trimethyl borate *Organic Chemistry.* $(OCH_3)_3B$, a flammable, water-white liquid; miscible with ether; decomposes in presence of water; boils at 67–68°C and freezes at –29°C; used as a solvent, dehydrating agent, fungicide, and catalyst, and in neutron scintillation counters.

2,2,3-trimethylbutane see TRIPTANE.

trimethylchlorosilane *Organic Chemistry.* $(CH_3)_3SiCl$, a flammable, colorless liquid, soluble in ether and benzene, that boils at 57°C and freezes at –40°C; used as an intermediate for silicone fluids, a chain terminating agent, and water repellant. Also, CHLOROTRIMETHYLSILANE.

trimethylene see CYCLOPROPANE.

***sym*-trimethylenetrinitramine** see CYCLONITE.

trimethylol see TROMETHAMINE.

trimethylolethane *Organic Chemistry.* $CH_3C(CH_2OH)_3$, combustible, hygroscopic, colorless crystals; soluble in water and alcohol; melts at 204°C; used in the manufacture of varnishes and drying oils, and as a conditioning agent. Also, PENTAGLYCERINE.

2,2,4-trimethylpentane $(CH_3)_2CHCH_2C(CH_3)_3$, a flammable liquid; soluble in alcohol and ether; boils at 98–99°C and freezes at –109.4°C; used as 100-octane standard for motor fuel and as an azeotropic distillation entrainer. Also, ISOOCTANE.

trimethylsilyl group *Organic Chemistry.* the group $-Si(CH_3)_3$, whose resonance effects are important in many organic syntheses.

trimmer a person or thing that trims; specific uses include: *Forestry.* a machine used to trim timber or rough-finish lumber. *Building Engineering.* a wall or floor tile used to finish an angle or edge.

trimmer potentiometer see POTENTIOMETER.

trimmer resistor *Electricity.* a small rheostat, usually connected to adjust the value of a fixed resistor.

trimming *Graphic Arts.* the process of cutting the folded edges from the signatures of a bound book. *Metallurgy.* in forging or die-casting, the process of shearing to remove the parting-line flash and gates from the part.

trimorphous *Biology.* having or occurring in three different forms or castes. Also, **trimorphic.**

trim size *Graphic Arts.* the final size of a printed page or book after binding and trimming.

trimuon event *Particle Physics.* an inelastic collision of particles that produces three muons.

trindle *Mechanical Devices.* a wheelbarrow wheel. *Graphic Arts.* either of two metal bookbinding plates used to flatten a book's backbone and boards so that its front edge can be trimmed.

Trinil fauna *Paleontology.* the fauna associated with the hominid remains found at Trinil, Java; dated as Middle Pleistocene, around 1 million years Before Present; it overlies the Early Pleistocene deposits that contain the Djetis fauna and is overlain by the Late Pleistocene Ngandong strata, which contain the "Solo man" remains.

Trinitian *Geology.* a North American Gulf Coast stage of the Lower Cretaceous period, occurring after the Nuevoleonian and before the Fredericksburgian.

1,3,5-trinitrobenzene *Organic Chemistry.* $(NO_2)_3C_6H_3$, yellow crystals; insoluble in water and soluble in alcohol and ether; melts at 122°C; used as an explosive.

2,4,7-trinitrofluorenone *Organic Chemistry.* $C_{13}H_5N_3O_7$, a yellow solid that is soluble in acetone, benzene, and chloroform; melts at 176°C; used to form crystalline indole complexes for identification by mass spectroscopy.

trinitromethane *Organic Chemistry.* $CH(NO_2)_3$, white crystals; soluble in water; melts at 15°C, decomposes above 25°C, and explodes at 47°C; used in the manufacture of propellants and explosives and in mass spectroscopy.

2,4,6-trinitrophenol see PICRIC ACID.

2,4,6-trinitroresorcinol see STYPHNIC ACID.

trinitrotoluene *Organic Chemistry.* 2,4,6-$CH_3C_6H_2(NO_2)_3$, TNT; flammable, toxic, monoclinic, yellow needles; insoluble in water and soluble in alcohol and ether; melts at 80.9°C and explodes at 240°C; used as an explosive and an intermediate in photographic chemicals and dyes. Also, **2,4,6-trinitrotoluene.**

trinomen *Systematics.* an extension of the two-word scientific name or binomen of a species to include a third word naming a subspecies, variety, or form. Also, TERNARY, TRINOMIAL.

trinomial *Systematics.* **1.** see TRINOMEN. **2.** of or relating to a trinomen. *Mathematics.* a polynomial having exactly three terms.

trinomial distribution *Statistics.* a multinomial distribution with three separate outcomes that are mutually exclusive and collectively exhaustive.

trinucleid *Paleontology.* a small, blind ptychopariid trilobite of the suborder Trinucleina and family Trinucleidae; extant in the Ordovician and Silurian.

trinucleotide *Organic Chemistry.* a polymer made up of three mononucleotides.

Trinucleus *Paleontology.* a genus of blind trilobites in the family Trinucleidae; characterized by long genal spines and by a highly convex glabella with a prominent, swollen anterior lobe; about 3 centimeters long; extant in the Ordovician.

triode *Electronics.* a vacuum tube containing three elements: a cathode, a control grid, and an anode. Also, THERMIONIC TRIODE.

triode laser *Electronics.* a gas laser device containing a control grid that is used for modulating the light output of the laser.

triode transistor *Electronics.* a three-terminal transistor.

Trionychidae *Vertebrate Zoology.* the soft-shelled turtles, a family of aquatic reptiles of the order Testudines, characterized by fully webbed feet and an elongated snout with fleshy lips; found mainly in fresh waters of North America, northern Africa, and Asia.

triose *Biochemistry.* a sugar with three carbon atoms in each molecule.

***sym*-trioxane** *Organic Chemistry.* $C_3H_6O_3$, white crystals that are soluble in water, alcohol, and ether; melts at 64°C and sublimes at 115°C; used in organic synthesis, and as a disinfectant and fuel. Also, TRIFORMOL, METAFORMALDEHYDE.

trip *Transportation Engineering.* a single act of traveling between two points, from departure to arrival. *Engineering.* to release or set into motion a lever, mechanism, or circuit. *Ordnance.* in some firearms, the part of the mechanism that is released by the trigger.

tripalmitin *Organic Chemistry.* $C_3H_5(OOCC_{15}H_{31})_3$, a combustible, white crystalline powder; insoluble in water and soluble in ether and chloroform; melts at 65.5°C; used in soaps and leather dressings.

tripara *Medicine.* a female who has had three pregnancies resulting in three viable offspring.

triparanol *Pharmacology.* $C_{27}H_{32}ClNO_2$, a crystalline substance formerly used to inhibit cholesterol biosynthesis.

triparanol syndrome *Toxicology.* a group of disease conditions resulting from use of triparanol.

tripartite *Science.* having or divided into three parts.

tripentylamine see TRIAMYLAMINE.

trip hammer or **triphammer** *Mechanical Engineering.* a power hammer operated by a tripping mechanism that causes the hammer to drop.

triphenylmethane dye *Organic Chemistry.* any of a class of dyes derived from the $(C_6H_5)_3CH$ molecule, usually by substituting OH, NH_2, or HSO_3 for one of the phenyl hydrogens; includes many coal tar dyes.

triphenylmethyl radical *Organic Chemistry.* the organic group $(C_6H_5)_3C-$. Also, TRITYL RADICAL.

triphenyl phosphate *Organic Chemistry.* $(C_6H_5O)_3PO$, a combustible, toxic, colorless crystalline powder; insoluble in water and soluble in most solvents and oils; melts at 50°C and boils at 245°C (11 torr); used as a fire-retarding agent and plasticizer.

triphenylphosphorus *Organic Chemistry.* $(C_6H_5)_3P$, a combustible, white crystalline solid, insoluble in water, slightly soluble in alcohol, that melts at 79–82°C and boils above –360°C; used in organic synthesis, primarily as a polymerizing agent. Also, **triphenylphosphine.**

triphenyltetrazolium chloride *Organic Chemistry.* $C_{19}H_{15}N_4Cl$, a crystalline solid; soluble in water, alcohol, and acetone; used as a sensitive reagent for reducing sugars. Also, RED TETRAZOLIUM, TETRAZOLIUM SALT.

triphosphopyridine nucleotide *Enzymology.* the former name for nicotinamide adenine dinucleotide phosphate (NADP), a major component in enzymatic systems that participate in oxidation-reduction. Also, **triphosphopyridine dinucleotide.**

triphosphoric monoester hydrolase *Enzymology.* an enzyme that catalyzes the hydrolysis of triphosphoric acid monoesters.

triphylite *Mineralogy.* $LiFe^{+2}PO_4$, a rare, blue-gray or gray-green, orthorhombic mineral, occurring in cleavable masses and rare prismatic crystals, having a specific gravity of 3.423 to 3.565 and a hardness of 4 to 5 on the Mohs scale; forms a series with lithiophilite; found in granite pegmatites as a primary mineral.

tripinnate *Botany.* describing a pinnate leaf having three pinnate divisions; triply pinnate.

triple *Science.* having three parts, being of three types, being three times as great, and so on. *Computer Programming.* see TRIAD, def. 2.
triple bond *Physical Chemistry.* a union of two atoms in which they share three pairs of electrons between them. Also, **triple bonding, triple covalent bond.**
triple collision *Physics.* any collision involving three bodies that interact simultaneously.
triple-mode Cepheid *Astronomy.* a beat Cepheid that displays three simultaneous pulsation modes.
triple point *Physical Chemistry.* the temperature and pressure at which the gaseous, liquid, and solid phases of a given substance (e.g., water) coexist at equilibrium. *Meteorology.* a junction of three distinct tropical air masses, thought to be an ideal point of origin for a tropical cyclone.
triple precision *Computer Programming.* of or relating to processing data that has been stored into three machine words in order to increase precision. Also, **triple length.**
triple response *Physiology.* a progression of three reactions occurring at the site of trauma to the skin. The first is a dilatation of the capillaries causing a circle of redness; the second is an uneven red patch called a flare; the third is a raised area known as a weal at the site resulting from fluid oozing from the capillaries. Also, WEAL-FLARE RESPONSE.
triple screw *Mechanical Devices.* a screw having three parallel threads.
triplet *Developmental Biology.* any of three individuals produced at the same birth following coextensive gestational periods. *Genetics.* a unit of three successive bases in DNA or RNA that codes for a specific amino acid. *Engineering.* any combination of three objects or entities acting together. *Navigation.* specifically, a group of three transmitting stations operating as a synchronous system to provide signals that may be used by a craft in determining position by triangulation. Also, TRIAD. *Optics.* a lens made up of three elements.
tripletail *Vertebrate Zoology.* a large edible fish of the family Lobotidae that inhabits the tropical Atlantic and Indo-Pacific oceans; named for the appearance of the tail between back-positioned anal and dorsal fins.
triple thread *Design Engineering.* a multiple screw with three threads equally spaced around it, with the lead being three times the pitch.
triplet state *Atomic Physics.* a condition in which an atom has two unpaired electrons. *Quantum Mechanics.* a multiplet for which the spatial wavefunction is antisymmetric.
triple vaccine *Immunology.* an antigen preparation that is administered to infants to produce active immunity against the diseases diptheria, tetanus, and whooping cough; composed of diptheria toxoid, tetanus toxoid, and pertussis vaccine.
triplexer *Electronics.* in a two-receiver radar system waveguide, an automatic switching device that connects the radar transmitter to the antenna when the former is producing an output.
triplite *Mineralogy.* $(Mn^{+2},Fe^{+2},Mg,Ca)_2(PO_4)(F,OH)$, a brown to pink monoclinic mineral, massive in habit, having a specific gravity of 3.55 to 3.87 and a hardness of 5 to 5.5 on the Mohs scale; a primary mineral in granite pegmatites.
triploblastic *Developmental Biology.* of an embryo, having three germ layers or blastodermic membranes. *Zoology.* of an animal, having a three-layered body wall consisting of an ectoderm, mesoderm, and endoderm.
triploid *Genetics.* describing a cell, tissue, or organism having three haploid chromosome sets.
triploidy *Developmental Biology.* the state of having three full sets of chromosomes; often found in aborted human fetuses.
tripod *Mechanical Devices.* a stand or support for an instrument or device with three legs, often extended telescopically, and upon which is a small platform or table for holding a surveying device such as a theodolite, a camera, or other device.
tripod drill *Mechanical Engineering.* a three-legged, steam, or compressed-air driven reciprocating rock drill.
tripoli *Geology.* **1.** a light-colored, porous sedimentary rock that is friable, rough-feeling, and earthy or powdery, formed by the weathering of chert or siliceous limestone; often used for polishing metals and stones. **2.** an incompletely silicified, carbonate-leached limestone.
tripolite see DIATOMACEOUS EARTH.
tripper *Electricity.* a mechanical or electromagnetic device that is used to open a circuit breaker or starter mechanism under abnormal electrical operation conditions. *Mechanical Devices.* a device or mechanism that causes the load on a conveyor to be discharged. Also, **tripping device.**
trippkeite *Mineralogy.* $Cu^{+2}As_2^{+3}O_4$, a soft, blue-green tetragonal mineral, occurring as short prismatic crystals, having a specific gravity of 4.80 and an undetermined hardness; found in copper deposits in Chile.
triprene *Organic Chemistry.* $C_{18}H_{32}O_2S$, a yellow liquid used as a growth regular for commercial crops.
-tripsy *Surgery.* a suffix meaning "surgical crushing."
triptane *Organic Chemistry.* C_7H_{16}, a flammable, colorless liquid that is insoluble in water and soluble in alcohol; boils at 81°C and freezes at –24°C; used in organic synthesis and for aviation fuel. Also, 2,2,3-TRIMETHYLBUTANE.
trip time *Transportation Engineering.* the total duration of travel from boarding a vehicle at the origin to alighting at the destination.
tripton see ABIOSESTON.
tripuhyite *Mineralogy.* $Fe^{+2}Sb_2^{+5}O_6$, a greenish-yellow to brown, tetragonal mineral,of the ferrotapiolite group, occurring in microcrystalline aggregates, having a specific gravity of 5.6 to 5.82 and a hardness of 7 on the Mohs scale; found with other antimony minerals and in placer deposits with cinnabar. Also, FLAJOLOTITE.
Tripylina *Invertebrate Zoology.* a subdivision of radiolarians, protozoans in the order Oculosida. Also, PHAEDORINA.
triquetrum *Anatomy.* the bone in the proximal row of carpal bones lying between the lunate and pisiform bones. Also, TRIANGULAR BONE.
TRIS *Biochemistry.* $C_4H_{11}NO_2$, an amine base occurring as a white crystalline powder; used intravenously as an alkalizer for the correction of metabolic acidosis. Also, **tris(hydroxymethyl)aminomethane.**
trisaccate pollen *Botany.* a grain of pollen having three pores and a triangular appearance in cross section.
trisaccharide *Biochemistry.* a carbohydrate that results from the condensation of three monosaccharides.
trisamine or **tris amino** see TROMETHAMINE.
tris buffer or **TRIS buffer** see TROMETHAMINE.
tris[2-(2,4-dichlorophenoxyl)ethyl] phosphite *Organic Chemistry.* $C_{24}H_{21}Cl_6O_6P$, a dark liquid; boils at 200°C (0.1 torr); used as a pre-emergence herbicide for corn, peanuts, and strawberries.
trisect *Mathematics.* to divide into three equal parts.
trisecting the angle *Mathematics.* the problem of constructing, with compass and straightedge alone, an angle having a measure exactly one-third that of a given angle. It is one of a collection of impossible constructions attempted by the ancient Greeks; others include duplicating the cube and squaring the circle.
trisectrix *Mathematics.* the locus of points P such that the slope of the line from the origin to P is one-third the slope of the line from $(2a, 0)$ to P. The equation in Cartesian coordinates is $x^3 + xy^2 + ay^2 - 3ax^2 = 0$.

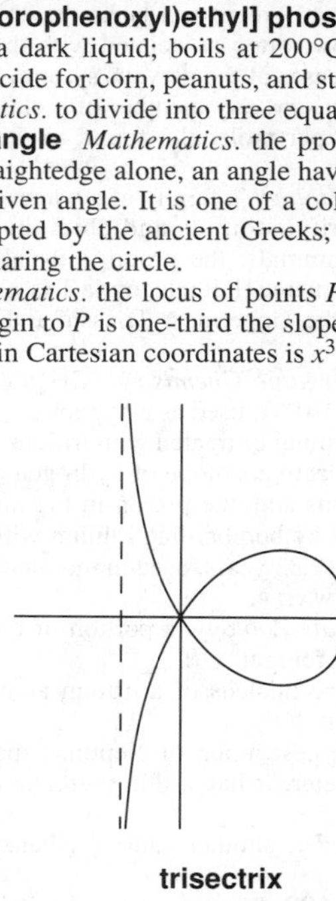

trisectrix

tris(2-hydroxyethyl)amine see TRIETHANOLAMINE.
trisistor *Electronics.* a high-speed semiconductor PNP switching device whose switching characteristics resemble those of a thyratron and are in the nanosecond range.
triskaidekaphobia *Psychology.* an irrational fear of the number 13.
triskelion *Biochemistry.* the three-appendaged, bent-propeller shape of clathrin.
trismic *Neurology.* relating to or of the nature of trismus.
trismus *Neurology.* a motor disturbance of the trigeminal nerve, particularly a tonic spasm of the masticatory muscles, causing difficulty in opening the mouth; normally an early manifestation of tetanus but also associated with dental disease and with hysteria. Also, LOCKJAW.
trisodium phosphate or **trisodium orthophosphate** see TRIBASIC SODIUM PHOSPHATE.
trisomic *Genetics.* describing a diploid cell, tissue, or organism in which one chromosome is represented three times rather than the normal two times.

trisomy *Genetics.* the state of being trisomic; a chromosomal anomaly in which a cell has an extra (third) chromosome in the normal chromosome pair, or an individual has an extra (third) chromosome in any of the chromosome pairs.

trisomy syndrome *Medicine.* any of various syndromes associated with trisomy, the state of possessing an extra chromosome.

trisomy 8 syndrome *Medicine.* a syndrome caused by a congenital trisomic condition of the eighth chromosome; characterized by mental retardation of varying severity, unusual appearance of the head and face, and chronically bent fingers. Also, **trisomy C syndrome.**

trisomy 13 syndrome *Medicine.* a syndrome caused by a congenital trisomic condition of the thirteenth chromosome; characterized by general weak health in the newborn infant, severe mental retardation, with seizures and facial deformity. Also, **trisomy D syndrome.**

trisomy 18 syndrome see EDWARDS SYNDROME. Also, **trisomy E syndrome.**

trisomy 21 syndrome see DOWN SYNDROME.

tristearin see STEARIN.

tristimulus colorimeter *Optics.* an instrument that uses tristimulus values to measure a given color stimulus.

tristimulus values *Optics.* the strength of three standardized color stimuli necessary to match a particular color of light.

trisulfide *Chemistry.* a compound containing three sulfur atoms in each molecule, such as iron trisulfide, Fe_2S_3.

trit *Mathematics.* a place value in a ternary number; i.e., one of the digits 0, 1, or 2 in the base 3 representation of a number. It is analogous to a bit in a base-2 system.

tritanopia *Medicine.* a rare form of color blindness in which the person sees only two hues, red and green, and lacks the ability to distinguish the colors blue and yellow; often associated with drug administration, retinal detachment, or diseases of the nervous system. Also, **tritanopsia.**

triterpene *Organic Chemistry.* any member of a class of terpenes containing 30 carbon atoms per molecule; found in the sap and resin of many plants.

Tritheledontidae *Paleontology.* a family of advanced carnivorous cynodonts, similar to the tritylodontids and chiniquodontids and very closely related to the mammals; the group is usually called the Ictidosauria, and the ordinal name Tritheledontia has also been proposed for the group; extant in the Uppermost Triassic and during the beginning of the Jurassic.

trithioacetaldehyde *Organic Chemistry.* $(CH_3)C_3H_3S_3$, a colorless crystalline solid; melts at 101°C; used as a hypnotic.

tritiated *Chemistry.* containing or treated with tritium.

tritium *Chemistry.* a radioactive isotope of hydrogen having the symbol T or ^3H, with two neutrons and one proton in the nucleus and thus an atomic mass of 3; formed by bombarding lithium with low-energy neutrons, it has a half-life of 12.33 years; used in mononuclear research.

tritium oxide see HEAVY WATER.

tritocerebrum *Invertebrate Zoology.* a portion of the insect brain that innervates the labrum and foregut.

triton *Nuclear Physics.* the nucleus of a tritium atom. *Organic Chemistry.* see TRINITROTOLUENE.

Triton *Astronomy.* the largest moon of Neptune, discovered in 1846; 2700 kilometers in diameter, it has a thin methane atmosphere and a methane ice polar cap.

triton B *Organic Chemistry.* another name for benzyltrimethylammonium hydroxide.

Triton X-100 or **triton X-100** *Biochemistry.* a nonionic detergent of the polyoxyethylene *p-t*-octylphenol series that is often used to make membrane proteins soluble in their naturally active state. Also, **iso-octylphenoxypolyethoxyethanol.**

tritonymph *Entomology.* the third stage in mite and tick development.

triturate *Science.* to rub to a powder. Thus, **triturated.** *Pharmacology.* a triturated substance, especially a drug rubbed up with a milk sugar.

trituration *Science.* **1.** the reduction of a solid body to a powder by continuous rubbing. **2.** the creation of a homogeneous whole by mixing. *Pharmacology.* the production of a triturate.

tritylation *Organic Chemistry.* the production of triphenylmethyl ethers by the reaction of triphenylmethyl chloride with the primary alcohol groups of sugars, in the presence of pyridine.

trityl radical see TRIPHENYLMETHYL RADICAL.

triuranium octoxide *Inorganic Chemistry.* U_3O_8, olive green to black crystals; soluble in acid; insoluble in water; decomposes at 1300°C; a radioactive poison; used in nuclear technology and preparation of uranium compounds. Also, URANOUS-URANIC ACID, URANYL URANATE.

Triuridaceae *Botany.* a family of monocotyledonous plants of the order Triuridales, characterized by scale leaves, and bisexual and unisexual flowers having no perianth and growing alone or in racemes.

Triuridales *Botany.* an order of small, mycotrophic herbs of the subclass Alismatidae, characterized by a lack of chlorophyll, endospermous seeds, distinct or nearly distinct carpels, and vascular bundles in the stem. Also, **Triuridates.**

trivacancy *Crystallography.* a point defect consisting of a cluster of four neighboring vacant lattice sites that form a triangular shape such as a tetrahedron.

trivalent *Chemistry.* having a valence of three. *Immunology.* having three binding sites. *Molecular Biology.* describing an aggregation of three or more homologous chromosomes in meiosis.

trivial *Mathematics.* describing a case or example that contributes no useful information because it is universally present or its existence is obvious. For example, a **trivial subgroup** is a subgroup consisting of the identity element or the entire group.

trivial equilibrium *Genetics.* an apparent condition of genetic equilibrium in a population due to the presence of only one type of allele at a certain locus.

trivial name *Organic Chemistry.* the popular name for an organic compound, usually based on the name of its natural source or method of procurement, and not reflecting its molecular structure.

trivium *Invertebrate Zoology.* in starfish, the three arms opposite the madreporite.

TRK *Oncology.* a rearranged proto-oncogene that is found in carcinoma of the thyroid.

tRNA or **t-RNA** transfer RNA.

tRNA isoacceptor see ISOACCEPTOR tRNA.

trocar *Surgery.* a cylindrical instrument consisting of a sharp-tipped rod within a hollow metal tube (cannula), used to pierce a body cavity or structure.

Trochacea *Paleontology.* an extinct superfamily of aspidobranch marine gastropods in the order Archaeogastropoda; includes the Paleozoic platycerates; extant from the Triassic to recent.

trochal disk *Invertebrate Zoology.* in rotifers, a flattened anterior disk with two rings of cilia, used for locomotion and food ingestion. Also, CORONAL DISK.

trochanter *Anatomy.* either of two bony processes formed at the upper end of the femur, just distal to its neck.

Trochidae *Invertebrate Zoology.* a family of snails, gastropod mollusks belonging to the order Aspidobranchia, characterized by conical top shells with a flat base.

Trochidae

Trochili *Vertebrate Zoology.* a suborder of the bird order Apodiformes that includes the hummingbirds.

Trochilidae *Vertebrate Zoology.* the hummingbirds, a large family of tiny, brightly colored, nectar-feeding birds of the order Apodiformes, found in North and South America.

Trochiliscales *Paleontology.* the oldest undoubted order of charophytic algae in which the gyrogonites are dextrally coiled; extant from the Upper Silurian to Lower Carboniferous.

trochlea *Anatomy.* **1.** a ring of fibrous material, through which the tendon of a muscle passes, enabling a muscle to pull a structure in a direction other than its own orientation. **2.** the pulley-shaped medial portion of the distal end of the humerus, for articulation with the semilunar notch of the ulna. Also, **trochlear eminence.** **3.** a smooth articular surface on a bone. (From a Latin word meaning "pulley.")

trochlear *Anatomy.* relating to a trochlea, or to any other pulley-shaped part or structure.

trochlear nerve *Anatomy.* the fourth cranial nerve, innervating the superior oblique muscle of the eye.

trochoblast *Invertebrate Zoology.* a ciliated cell of a trochophore larva.

Trochodendraceae *Botany.* a monogeneric family of dicotyledonous evergreen trees of the order Trochodendrales, characterized by simple, alternate leaves that grow in clusters at the ends of its branches, and bisexual flowers clustered in racemes.

Trochodendrales *Botany.* an order of dicotyledonous trees and shrubs of the division Hamamelidae, characterized by the vessels used and unique, elongate idioblasts found in its leaves.

trochoid *Anatomy.* resembling a pivot or a pulley. *Mathematics.* the locus of points in the plane traced by a fixed point P on the radius of a circle that rolls without slipping on a straight line. If the straight line is the x-axis, the circle has radius a, and P is at a distance b from the center of the circle, then parametric equations for the trochoid are given by $x = a\phi - b \sin \phi$ and $y = a - b \cos \phi$.

trochoidal mass analyzer *Physics.* a mass spectrometer whereby ion beams are made to follow trochoidal paths through a region that has mutually perpendicular electric and magnetic fields.

trochoidal wave *Fluid Mechanics.* a deep-water wave whose free surface and lines of equipressure are trochoids.

trochophore *Invertebrate Zoology.* in many invertebrates, a tiny free-swimming larva that is pear-shaped, translucent, and ciliated. Also, **trochosphere.**

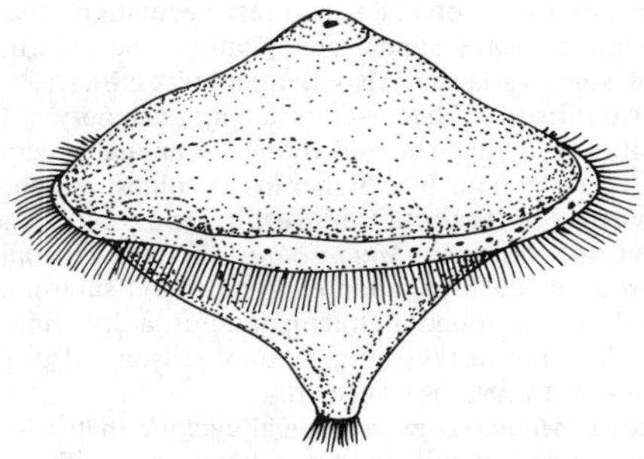

trochophore

trochus *Invertebrate Zoology.* in rotifers, the inner band of cilia on a trochal disk.

troctolite *Petrology.* a gabbro consisting primarily of calcic plagioclase and olivine, with little or no pyroxene. Also, FORELLENSTEIN.

trögerite *Mineralogy.* a rare, strongly radioactive, lemon-yellow, probably tetragonal mineral of the autunite group, with the approximate formula $(UO_2)_3(AsO_4)_2 \cdot 12H_2O$, occurring as thin tabular crystals, having a specific gravity of 3.55 and a hardness of 2 to 3 on the Mohs scale. Also, **troegerite.**

trogon *Vertebrate Zoology.* any of several brightly colored birds of the genus *Trogon* in the family Trogonidae.

Trogonidae *Vertebrate Zoology.* the trogons and quetzals, the sole family of the bird order Trogoniformes, having a long tail and short legs.

Trogoniformes *Vertebrate Zoology.* a monofamilial order of brilliantly colored birds found in tropical and subtropical regions of the Americas.

troilite *Mineralogy.* FeS, a grayish-brown, opaque, metallic, hexagonal mineral, occurring as granular masses and thin veins in iron meteorites and sheared serpentine rocks, having a specific gravity of 4.67 to 4.82 and a hardness of 3.5 to 4.5 on the Mohs scale.

Trojan horse *Computer Programming.* in computer crime, a set of instructions that will perform illegitimate functions in a system without disrupting normal operations, and thus without alerting other users. (From the Greek legend of the *Trojan horse,* a seemingly innocent large wooden horse placed outside the gates of the walled city of Troy; Greek soldiers were hidden inside the horse, and when it was brought inside the gates by the Trojans, the Greeks gained entrance to the city.)

Trojan planet *Astronomy.* **1.** an asteroid that orbits at Jupiter's distance from the sun, but 60° of orbital longitude ahead or behind it. **2.** more generally, any body that orbits in such a relationship to a larger one.

troland *Optics.* see PHOTON.

Troland and Fletcher theory *Physiology.* any of a set of theories of hearing postulating that the time nature of a sound stimulus affects the sensation of pitch.

trolley *Mechanical Engineering.* **1.** a grooved metallic wheel or pulley at the end of a pole (**trolley pole**) that picks up current from an electric wire (**trolley wire**) and carries it to a transit vehicle. **2.** any vehicle that draws power from such a mechanism. Thus, **trolley car, trolley bus, trolley locomotive, trolley vehicle. 3.** any of various small vehicles operating on a track, as in a factory, mine, or railroad yard.

Trombiculidae *Invertebrate Zoology.* the chiggers, a family of blood-sucking mites in the suborder Trombidiformes; nymphs and adults are parasites of arthropods; some larvae are disease-carrying parasites of vertebrates including humans.

Trombidiformes *Invertebrate Zoology.* a suborder of mites in the order Acarina, with a respiratory opening at the base of piercing chelicerae.

trombone *Electromagnetism.* a U-shaped coaxial line or waveguide section whose length is adjustable for impedance matching purposes; thought of as resembling the slide tube of a trombone.

trombone-action see PUMP-ACTION.

tromethamine *Organic Chemistry.* $C_4H_{11}NO_3$, a crystalline solid that is soluble in water and alcohol; it melts at 171–172°C and boils at 219–220°C (10 torr); used in pharmaceuticals, rubber accelerators, surface-active agents, and as a titrimetric standard. Also, TRIMETHYLOL, TRISAMINE.

trommel *Mining Engineering.* **1.** a revolving cylindrical screen, used in the grading of coarsely crushed ore; after material is fed into the screen at one end, the fine material drops through the holes while the coarse travels to and is delivered at the other end. Also, **trommel screen. 2.** to separate coal into various sizes by means of a trommel.

Trommsdorff-Norrish effect *Materials Science.* an increase in the molecular weight of a polymer chain, occurring in some vinyl polymerization reactions.

trona *Mineralogy.* $Na_3(CO_3)(HCO_3) \cdot 2H_2O$, a vitreous, colorless to white, gray, or yellow monoclinic mineral, occurring in fibrous or columnar masses and thick plates, having a specific gravity of 2.11 to 2.14 and a hardness of 2.5 to 3 on the Mohs scale; found in saline lake deposits and as an efflorescence; the most important of the natural sodas. Also, URAO.

troops *Military Science.* a collective term for uniformed military personnel; it is usually not applied to shipboard naval personnel.

troop safety *Military Science.* an element that defines a distance from the proposed shellburst location beyond which personnel will be safe to the degree of risk established by the commander.

trop- or **tropo-** a prefix meaning: **1.** to stimulate. **2.** to turn.

Tropaeolaceae *Botany.* a family of dicotyledonous prostrate or climbing herbaceous plants of the order Geraniales, characterized by highly irregular, bisexual, solitary flowers.

tropaeolin *Materials.* any of several azo dyes that are derived from nasturtia and used as biological stains and acid-base indicators. Also, **tropaeoline, tropeolin, tropeoline.**

Tropept *Geology.* a suborder of the soil order Inceptisol, which develops in humid tropical regions and is characterized by a moderately dark surface layer containing some organic matter overlying a brown to reddish layer.

troph- a combining form meaning "food" or "nourishment."

trophallaxis *Ecology.* the process by which the organisms of a certain species exchange food among themselves; occurring especially among social insects. Thus, **trophallactic.**

trophectoderm *Cell Biology.* the outermost layer of cells surrounding the blastocoel during the blastocyst stage of mammalian embryonic development.

trophi *Invertebrate Zoology.* mouthparts or a feeding organ, such as the set of minute jaws in rotifers, or the mouthparts of an arthropod; plural of trophus.

trophic [träf'ik; trōf'ik] *Biology.* of or having to do with nutrition or the nutritive process.

trophic cascade *Ecology.* in a food web, a trophic structure through which significant direct and indirect effects on population dynamics pass from top predators, down successively from one lower trophic level to the next.

trophic flow *Ecology.* the transfer of energy in a food chain or food pyramid.

trophic level *Ecology.* **1.** any of the series of steps on a food chain or food pyramid from producers to primary, secondary, and tertiary consumers. **2.** any of the levels of nutrients, especially the nitrate and phosphate, that exist in a body of water.

trophic structure *Ecology.* the organization of a community based on the feeding relationships between species.

trophic unit *Ecology.* a group of individuals of one or more species on the same level of a food pyramid.

tropho- a combining form meaning "food" or "nourishment."

trophobiosis *Ecology.* a relationship between two species in which both feed each other; e.g., in an ant colony, where aphids feed on plant roots inside the ants' nest and the ants then milk the aphids for their honeydew.

trophoblast *Developmental Biology.* a layer of extraembryonic ectodermal tissue on the outside of the blastocyst that attaches the embryo of mammals to the uterine wall and supplies nutrition to the embryo.

trophocyte *Entomology.* a nutrient-storing fat cell in insects. *Cell Biology.* any lower type of cell that furnishes nourishment to a higher type of cell or tissue.

trophogenic *Ecology.* **1.** relating to or resulting from food or feeding behavior. **2.** of or relating to the upper region of a lake, in which photosynthesis occurs.

trophology *Nutrition.* the study of nutritional processes in the body.

tropholytic *Ecology.* relating to the deeper region of a lake, in which the primary energy source is organic matter.

trophoneurosis *Neurology.* any trophic disorder that occurs without neurologic disease, usually attributed to nerve injury or emotional dysfunction. Thus, **trophoneurotic.**

trophonosis *Pathology.* any disease or disorder having nutritional causes.

trophopathy *Medicine.* any derangement or abnormal state of nutrition. Thus, **trophopathic.**

trophophase *Microbiology.* the early exponential stage during which a culture of microorganisms is most actively involved in growth.

trophosome *Invertebrate Zoology.* **1.** collectively, the feeding, nonsexual zooids of a hydroid colony. **2.** in certain marine worms, an organ that is colonized by bacteria serving as the host's food source.

trophotaenia *Vertebrate Zoology.* any of a number of extensions from the rectum of the embryo of certain live-bearing fish, serving to establish a functional placental relationship with the ovary of the mother.

trophozoite *Invertebrate Zoology.* the adult, feeding stage of a sporozoan protozoan.

trophus *Invertebrate Zoology.* any of the minute jaws of rotifers.

-trophy a word termination denoting a relationship to nutrition or growth.

tropia see STRABISMUS.

tropic [träp´ik] *Geography.* **1.** either of two parallels of latitude 23°27' north and south of the equator, being the Tropic of Cancer and the Tropic of Capricorn. **2. the tropics.** the region bounded by these latitudes, corresponding to the Torrid Zone. *Astronomy.* either of two circles on the celestial sphere lying on the same plane as the boundaries of the terrestrial tropics.

-tropic [träp´ik, trō´pik] a word termination meaning: **1.** stimulating. **2.** turned toward. **3.** changing.

tropical *Geography.* relating to, characteristic of, inhabiting, or occurring in the tropics. *Meteorology.* hot and humid.

tropical acne *Medicine.* severe acne occurring in hot, humid weather; characterized by large and painful cysts, nodules, and pustules.

tropical air *Meteorology.* an air mass whose characteristics are developed over low latitudes. **Maritime tropical air,** produced over tropical and subtropical seas, is very warm and humid; **continental tropical air,** produced over subtropical arid regions, is hot and very dry.

tropical climate *Meteorology.* a climate characterized by high temperatures and heavy precipitation during at least part of the year; typical of equatorial and tropical regions.

tropical cyclone *Meteorology.* a cyclone that originates over a tropical ocean, usually moving west and slightly poleward, then sometimes "recurving" back toward the east; one of the most violent storms on earth, characterized by torrential rains and winds over 65 knots (sometimes over 175 knots or 200 mph).

tropical depression *Meteorology.* a tropical cyclone having winds up to 34 knots (or up to 27 knots in some classifications).

tropical disturbance *Meteorology.* a mild tropical cyclone, with only slight surface circulation.

tropical easterlies *Meteorology.* a shallow trade wind that displays a strong vertical wind shear, occurring in the poleward margins of the tropics in summer and covering much of the tropical belt in winter. Also, SUBTROPICAL EASTERLIES.

tropical finish *Engineering.* a coating applied to electronic equipment to protect it from the high heat and humidity of the tropics.

tropical front see INTERTROPICAL FRONT.

tropical grassland *Ecology.* an area that is intermediate between a tropical rain forest and a true grassland; characterized by constant high temperature, a long dry season, and poorly fertile soil, and dominated by grasses with only occasional trees. Also, TROPICAL SAVANNA.

tropicalize *Engineering.* to apply tropical finish to a piece of electronic equipment.

tropical life zone *Ecology.* a section of the earth's surface characterized by broad-leafed evergreen forest, in which palms and mangroves are commonly found.

tropical meteorology *Meteorology.* a branch of meteorology dealing with the tropical atmosphere, with its own distinct forecasting techniques and theories.

tropical monsoon climate *Meteorology.* the type of tropical rainy climate that is warm and wet enough to produce tropical rain forest vegetation but is distinguished from the rain forest climate by a dry winter season.

tropical month *Astronomy.* the month defined by two successive passages of the Moon through the vernal equinox: 27.321582 days.

tropical rain forest *Ecology.* a region found widely in equatorial lowlands throughout the world, with heavy year-round rainfall and very little temperature variation from season to season; characterized by heavy tree growth that admits little light to the forest floor, so that there is minimal undergrowth. Tropical rain forests have a greater range of plant and animal life than any other habitat; roughly half the world's total species live in such areas.

tropical rain forest climate *Meteorology.* the type of tropical rainy climate that produces tropical rain forest vegetation; characterized by unbroken warmth, heavy annual precipitation, and high humidity, with very little seasonal variation. Also, **tropical wet climate.**

tropical rainy climate *Meteorology.* a major category of Köppen's climatic classification, characterized by heavy annual precipitation and a mean temperature of 64.4°F or higher in the coldest month.

tropical savanna or **tropical savannah** see TROPICAL GRASSLAND.

tropical savanna climate *Meteorology.* the type of tropical rainy climate that produces the vegetation of tropical and subtropical savannas; characterized by year-round high temperatures, a dry winter season, and a relatively short but heavy summer rainy season. Also, SAVANNA CLIMATE, TROPICAL WET AND DRY CLIMATE.

tropical storm *Meteorology.* a tropical cyclone that is stronger than a tropical depression but milder than a hurricane, with winds stronger than 34 knots but less than 65 (or 27–66 knots in some classifications).

tropical typhus see SCRUB TYPHUS.

tropical wet and dry climate see TROPICAL SAVANNA CLIMATE.

tropical woodland *Ecology.* an area similar to a true forest, except that it has wider spaces between trees and less undergrowth.

tropical year *Astronomy.* the year defined by two successive passages of the sun through the vernal equinox: 365.242191 days. Also, SOLAR YEAR, ASTRONOMICAL YEAR, EQUINOCTIAL YEAR.

tropical year of 1900 *Horology.* the year used as the fundamental unit of ephemeris time: 365 days, 5 hours, 48 minutes, and 46 seconds.

tropic bird or **tropicbird** *Vertebrate Zoology.* any of several tropical sea birds of the family Phaethontidae, having white plumage with black markings and two very long central tailfeathers.

tropic higher-high-water interval *Oceanography.* the lunitidal interval relating to the mean higher high waters at the time of tropic tides.

tropic high-water inequality *Oceanography.* the average vertical difference between the two high waters of a tidal day at the time of tropic tides.

tropic lower-low-water interval *Oceanography.* the lunitidal interval relating to the mean lower low waters at the time of tropic tides.

tropic low-water inequality *Oceanography.* the average difference between the two low waters of a tidal day at the time of tropic tides.

Tropic of Cancer *Astronomy.* the parallel of latitude 23°27' north of the equator on earth where the sun lies in the zenith at the summer solstice, around June 22.

Tropic of Capricorn *Astronomy.* the parallel of latitude 23°27' south of the equator on earth where the sun lies in the zenith at the winter solstice, around December 22.

tropics *Meteorology.* any region of the earth having a tropical climate, especially between the Tropic of Cancer and Tropic of Capricorn.

tropic tidal currents *Oceanography.* tidal currents associated with tropic tides, when the diurnal inequality is greatest.

tropic tide *Oceanography.* a tide that occurs twice monthly when the effect of the moon's maximum declination north or south of the equator is greatest; the diurnal range increases at these times.

tropic velocity *Oceanography.* the greater flood or ebb velocity of tropic tidal currents.
-tropin a suffix indicating an affinity for the structure or thing named in the stem, as in *gonadotropin.*
tropine *Organic Chemistry.* $C_8H_{15}NO$, a white, crystalline alkaloid that smells like tobacco; soluble in water; derivable from atropine and from various plants.
Tropiometridae *Invertebrate Zoology.* the feather stars, a family of crawling, free-living echinoderms in the class Crinoidea.
tropism [trōp´iz əm] *Biology.* the movement response of an organism to an external stimulus such as light or heat.
tropo- a combining form meaning "turn," "reaction," or "change;" also used to represent *troposphere,* as in *tropopause.*
tropocollagen *Biochemistry.* **1.** a protein that contains three polypeptide strands fashioned into a three-stranded helix; serves as the basic unit of collagen fibrils. **2.** fetal collagen.
tropomyosin *Biochemistry.* a protein in muscle fibers that controls the interaction of actin and myosin.
troponin *Biochemistry.* a complex of three polypeptide chains that mediates calcium's effect on muscle contraction.
tropoparasite *Ecology.* an organism that is parasitic for only part of its life cycle.
tropopause *Meteorology.* the upper limit of the troposphere, a layered band of atmosphere lying just below the stratosphere (about 15 to 20 kilometers high at the equator, about 10 kilometers high at the poles), marked by an abrupt decrease of lapse rate.
tropopause fold *Meteorology.* a phenomenon in which ozone-rich stratospheric air protrudes into the troposphere in a conelike fashion under conditions of low relative humidity.
tropopause inversion *Meteorology.* the characteristic decrease in lapse rate that occurs at the tropopause. Also, UPPER INVERSION.
tropopause leaf *Meteorology.* any of the discrete, overlapping layers that constitute the tropopause.
tropophilous *Botany.* of a plant, adapted to or thriving in a climate characterized by extremes of temperature and precipitation.
tropophyte *Botany.* any tropophilous plant. Also, **tropophytia.**
troposphere [träp´ə sfēr´; trōp´ə sfēr´] *Meteorology.* the lowest layer of the atmosphere, lying between the earth's surface and the tropopause, characterized by a steady drop in temperature with altitude; the part of the atmosphere where weather conditions exist and nearly all cloud formations occur.
tropospheric *Meteorology.* relating to or taking place in the troposphere.
tropospheric scatter *Telecommunications.* the propagation of radio waves by scattering as a consequence of discontinuities or irregularities in the physical characteristics of the troposphere; used for transhorizon communications at frequencies ranging from 350 to 8400 MHz. Also, **troposcatter.**
tropospheric superrefraction *Geophysics.* a property of the troposphere in which radio waves are bent back and returned to earth.
tropospheric wave *Telecommunications.* a radio wave propagated from a region of abrupt change in the dielectric constant or its gradient in the troposphere.
trough [trôf; träf] *Agriculture.* a long, narrow receptacle resembling an open crate, used for feeding farm animals. *Meteorology.* a long, narrow area of relatively low atmospheric pressure, which may contain one or more low-pressure systems; the opposite of a ridge. *Geology.* **1.** a small, linear, submerged depression oriented parallel to the shoreline, located just offshore along the bottom of a body of water and on the landward side of a longshore bar. **2.** the axis of a syncline or the line forming the lowest points of a fold. Also, **trough line. 3.** any narrow, elongated depression in the earth's surface. **4.** an elongated depression on the sea floor with gently sloping sides, broader and shallower than a trench.
trough cross-bedding *Geology.* a type of sedimentary rock layer that lies at an oblique angle to the main bedding and is characterized by lower bounding surfaces that are smoothly curved as a result of erosion. Also, CRESCENT-TYPE CROSS-BEDDING.
trough fault *Geology.* a break in a rock structure that occurs at the boundary of a structural depression or graben.
trough line *Meteorology.* the central axis of a trough.
trough plane see TROUGH SURFACE.
trough reef see REVERSE SADDLE.
trough surface *Geology.* a plane surface that joins the troughs along the series of beds in a syncline. Also, SYNCLINAL AXIS, TROUGH PLANE.
trough valley see U-SHAPED VALLEY.
troupial *Vertebrate Zoology.* **1.** a South and Central American oriole, *Icterus icterus,* of the family Icteridae, named for the sound it makes and characterized by its orange and black plumage. **2.** generally, any member of the family Icteridae, including American blackbirds, grackles, and orioles.
trout *Vertebrate Zoology.* any of several edible freshwater fish of the family Salmonidae, especially of the genera *Salmo* (the **true trout**) and *Salvelinus* (the **chars**), resembling salmon but generally much smaller. Common North American species include the **brook trout,** *Salvelinus fontinalis*; **lake trout,** *S. namaycush*; **rainbow trout,** *Salmo gairdnerii*; **cutthroat trout,** *S. clarkii*; and **brown trout**, *S. trutta.*

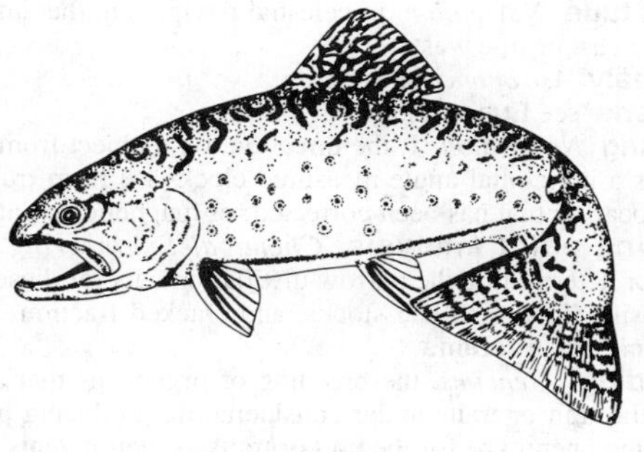

brook trout

Trouton's rule *Physical Chemistry.* the observation that the molar heat of a vapor divided by the normal boiling point of its liquid phase provides a rough average value of 21; the values for a number of common substances range from 17 to 27.
trout-perch *Vertebrate Zoology.* a North American freshwater fish, *Percopsis omiscomayaca,* belonging to the perch family (Percopsidae) but also resembling the trout.
trowel *Mechanical Devices.* any of various hand tools consisting of a broad, scooped, flat, or curved blade attached to a wooden handle; used for applying or spreading loose or plastic material or for gardening.
troy *Metrology.* **1.** of or relating to the troy system of weights. **2.** a troy weight. (From the city of *Troyes,* France, the site of a noted medieval fair that developed its own system of weights.)
troy ounce *Metrology.* a measure of weight equal to 20 pennyweights or 480 grains. Twelve troy ounces equal one troy pound.
troy pound *Metrology.* a measure of weight equal to 12 troy ounces or 240 pennyweights.
troy system *Metrology.* a system of weights used for gems and precious metals, based on a pound of 12 ounces. Thus, **troy weight**.
Trp or **trp** tryptophan.
Trp gene *Genetics.* any of a series of genes in *Escherichia coli* that participate in the synthesis of tryptophan.
Trucherognathidae *Paleontology.* a family of conodonts characterized by platelike, fibrous teeth that do not clasp the jaw ramus or the jaw tip, but simply rest on the jaw ramus; extant in the Early Paleozoic.
truck *Mechanical Engineering.* **1.** an automotive vehicle designed to tow trailers or other heavy equipment or to haul merchandise in its rear compartments. **2.** any of various wheeled frameworks used to transport heavy objects.
truck crane *Mechanical Engineering.* a crane that may be transported on a truck bed.
truck crop *Agriculture.* a vegetable crop that is raised on an acreage under intensive methods for sale at market.
truck farm *Agriculture.* a farm that produces vegetable crops that are sold at market. (From an older word *truck,* meaning "to exchange or barter.")
truck tractor *Mechanical Engineering.* a truck having a cab but no body; designed to attach to and haul a trailer.
Trudeau, Edward Livingston 1848–1915, American physician; a pioneer in tuberculosis research; founded the Trudeau Sanitarium.
true *Science.* **1.** conforming to the facts or to reality; actual or correct; not false. **2.** having characteristics that are typical of a species or group; meeting all criteria establishing its identity. *Mechanical Engineering.* precisely formed, fitted, or aligned. *Cartography.* **1.** of or relating to geographic or polar north, not magnetic north. **2.** describing a value obtained from a field survey that has been corrected for all possible errors.

true airspeed *Aviation.* the actual airspeed of an aircraft as it moves through the air mass. Because an airspeed indicator indicates true airspeed only under standard sea level conditions, true airspeed is usually calculated by adjusting indicated airspeed according to conditions (temperature, density, and pressure).

true-airspeed indicator *Aviation.* an aircraft instrument designed to calculate true airspeed. Also, **true-airspeed meter.**

true air temperature *Aviation.* ambient air temperature as measured by a thermometer aboard a flight vehicle and corrected to account for the heat of compression generated by the rapid motion of the thermometer through the air.

true altitude *Aviation.* the height of an aircraft as measured from mean sea level. *Meteorology.* see CORRECTED ALTITUDE.

true amplitude *Navigation.* in celestial navigation, the amplitude relative to true east or true west.

true anomaly *Astronomy.* see ANOMALY.

true bacteria see EUBACTERIA.

true bearing *Navigation.* **1.** the direction to an object from a point, expressed as a horizontal angle measured clockwise from true north. **2.** a compass bearing that has been corrected for magnetic deviation.

true-boiling-point analysis *Chemical Engineering.* a standard method for determining the narrow distillation cuts for diesel fuel, gasoline, kerosine, lube distillate stocks, and cracked fractions. Also, **true-boiling-point distillation.**

true breeding *Genetics.* the breeding of organisms that are homozygous for the trait or traits under consideration, producing progeny having the same phenotype for the trait or traits as their parents.

true cholinesterase see ACETYLCHOLINESTERASE.

true condensing point *Physical Chemistry.* the temperature at which a sublimed solid recondenses; a process that corresponds to liquid distillation. Also, CRITICAL CONDENSATION TEMPERATURE.

true course *Navigation.* a course relative to true north. Similarly, **true azimuth, true heading.**

true crater *Geology.* an original surface depression formed by impact or explosion, before it has been modified by or infilled with the ejected debris. Also, PRIMARY CRATER.

true dip *Geology.* see DIP, def. 1.

true direction *Navigation.* horizontal direction expressed in terms of an angular distance from true north.

true equator *Navigation.* the great circle on the earth that is equally distant from the poles, as opposed to a fictitious equator that is part of a fictitious graticule on a map or chart.

true fold see BUCKLE FOLD.

true formation resistivity *Geophysics.* the electrical resistivity of a clean porous reservoir formation containing hydrocarbons and formation water.

true freezing point *Physical Chemistry.* the temperature at which the liquid and solid phases of a substance reach equilibrium at a specific pressure, usually one standard atmosphere.

true homing *Navigation.* the process of following a course in such a manner that the true bearing relative to the target or other objective remains constant.

true horizon *Optics.* **1.** the center of perspective in a lens. **2.** the boundary of a horizontal plane as it passes through a point of vision.

true mean temperature *Meteorology.* a monthly or annual mean of the air temperatures recorded in a series of hourly weather observations at a given location.

true-motion radar *Electronics.* a radar tracking the absolute motion of all targets and the radar-carrying vehicle on the radar screen; used especially for navigation and piloting.

true north *Navigation.* **1.** the direction from an observer's position to the geographic North Pole. **2.** the north direction of any geographic meridian. *Cartography.* see ASTRONOMIC NORTH.

true place *Astronomy.* the coordinates of a star on the celestial sphere as it would be observed from the center of the sun, with respect to the true equator and equinoxes at the given time in question.

true plasma *Hematology.* blood plasma drawn directly from the blood without any change of its gas content.

true rake *Mechanical Engineering.* the angle between a plane containing the tooth face and the axial plane through the tooth point in the direction of chip flow.

true rib *Anatomy.* any of the upper seven ribs on each side, which directly join the sternum.

true seal see EARLESS SEAL.

true skin see DERMIS.

true soil see SOLUM.

true solution *Physical Chemistry.* the fully dispersed mixture of the molecules or ions of one or more substances in one or more other substances.

true strain *Mechanics.* the strain calculated by integrating the differential strains divided by the current length over a large strain; differs slightly from engineer's strain. Also, LOGARITHMIC STRAIN.

true stress *Metallurgy.* the load per unit area of a deforming specimen, continuously calculated as the area decreases.

true sun *Astronomy.* the sun as it appears in the sky as distinguished from the mean sun.

true-to-scale *Cartography.* of or relating to the condition in which the planimetric measurements on a map coincide exactly with the map's stated scale; because all two-dimensional maps involve some variation of the scale, not all measurements on a map can be true-to-scale.

true width *Mining Engineering.* the width or thickness of a vein or stratum as measured perpendicular to the dip and the strike; it is always the least width.

true wind *Meteorology.* a wind that blows relative to a fixed location on the earth's surface.

true-wind direction *Meteorology.* the direction from which the wind is blowing, as measured in relation to true north; distinguished from magnetic wind direction, which is relative to magnetic north.

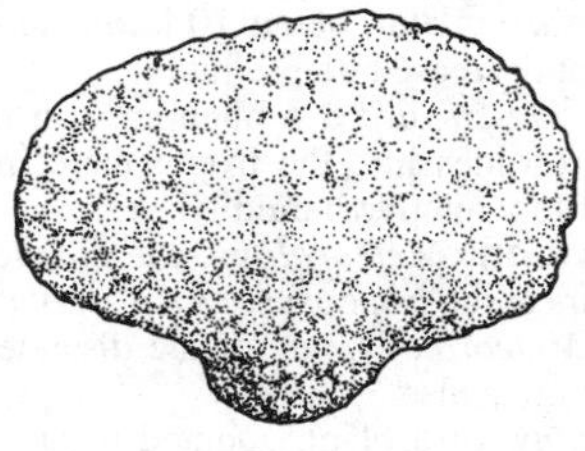

truffle

truffle *Botany.* the underground fruiting body of various members of the Tuberaceae, especially of the genus *Tuber,* grown as a food crop.

truing *Mechanical Engineering.* **1.** the process of cutting a grinding wheel so that its surface is concentric with the axis. **2.** the process of aligning a wheel so that it is concentric and in one plane.

trumpet creeper *Botany.* any of a group of woody, climbing plants of the genus *Campsis* of the bignonia family, especially *C. radicans,* having pinnate leaves and large, scarlet, trumpet-shaped flowers; found in the southern U.S. Also, **trumpet vine.**

trumpeter *Vertebrate Zoology.* **1.** a South American pheasantlike bird of the family Psophiidae, especially *Psophia crepitans,* related to the cranes and having a long neck, soft black plumage, and a loud, prolonged cry. **2.** a marine spiny fish of the perciform family Latridae that is silvery with olive stripes.

trumpeter swan *Vertebrate Zoology.* a large, pure white, wild swan, *Cygnus buccinator,* characterized by a shrill, clear cry; found in western North America; once near extinction, now recovering in numbers.

trumpeter swan

Space

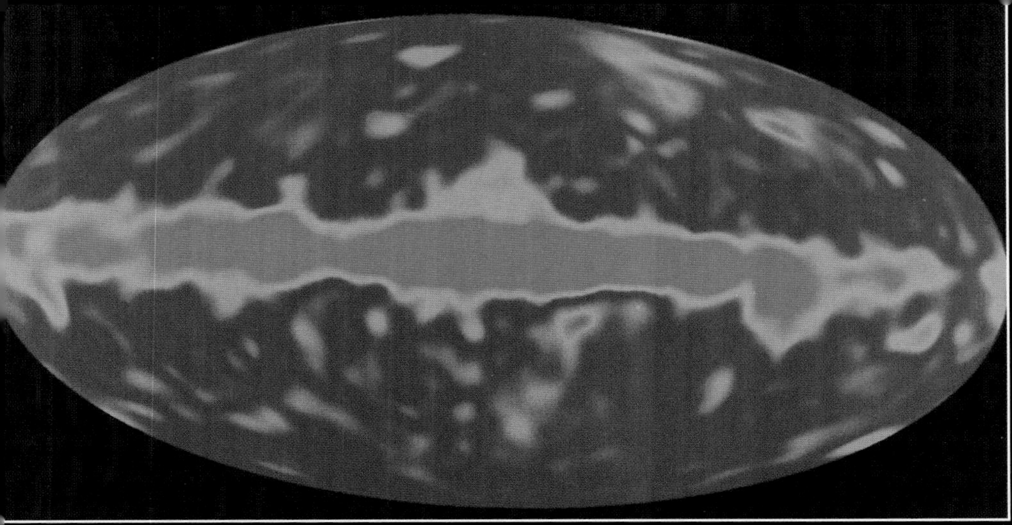

icrowave map of the sky derived from one year of data obtained by NASA's COBE satellite. Bluish regions
'e slightly cooler, and reddish ones slightly warmer, than the average background temperature of 2.736 K.
ourtesy of Lawrence Berkeley Laboratory.

Horsehead Nebula, a dark cloud or dust cloud in a hot young nebula, in Orion. Courtesy of NASA.

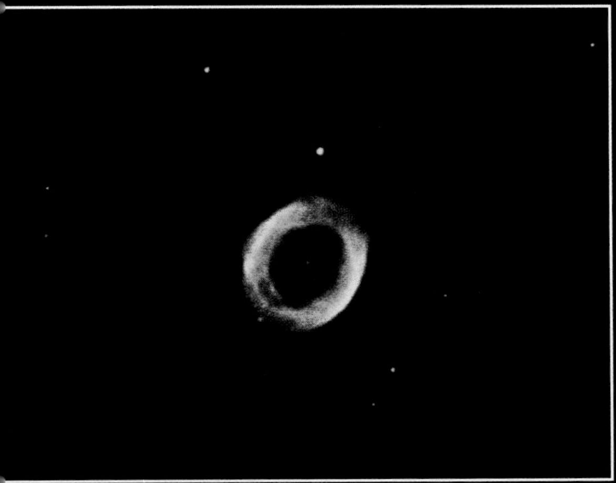

ing Nebula, an old planetary nebula, in Lyra. Courtesy of NASA.

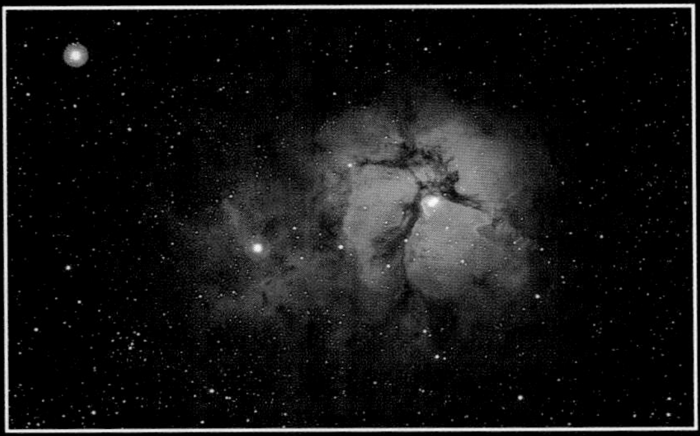

Trifid Nebula, a young nebula and massive star cluster, in Sagittarius. Courtesy of the Association of Universities for Resea... in Astronomy, Inc. The Kitt Peak National Observatory

Surface of Venus. Courtesy of NASA/JPL and The Planetarium, Armagh, N. Ireland.

Saturn as photographed by NASA's Hubble Space Telescope. Courtesy of NASA.

enter and bulge of the Milky Way Galaxy in Sagittarius with a

trumpetfish *Vertebrate Zoology.* a long, slender, scaly marine fish of the family Aulostomidae, having an elongated snout, numerous dorsal spines, and a skin pattern that acts as camouflage in reef areas; noted for the habit of orienting vertically in the water and capturing its prey from that position; found in tropical and subtropical waters of the Atlantic, Indo-West Pacific, and Pacific Oceans.

truncate *Biology.* **1.** blunt, as if cut off. **2.** having no apex. *Mathematics.* to terminate an infinite sequence or series after a finite number of terms, or to terminate a finite sequence or sum before the last term; e.g., rounding a decimal representation of a real number, forming a partial sum, or considering the first *n* terms of a sequence for some integer *n*. The computation error introduced by this process is called a **truncation error.** *Robotics.* the stopping of a robotic process before it has been completed.

truncated paraboloid *Electromagnetism.* an antenna reflector in which certain portions are removed in order to broaden the width of the major lobe.

truncus arteriosus *Developmental Biology.* the arterial trunk, leading forward from the heart. In mammals, it subdivides into the aortic and pulmonary arteries.

trunk *Anatomy.* **1.** the major part of the body, excluding the head and limbs. **2.** the undivided base of a nerve, blood vessel, or lymphatic vessel. *Botany.* the bole or main stem of a tree. *Telecommunications.* **1.** a single message circuit between two points, both of which are switching centers or individual message distribution points. **2.** in telephony, a circuit between two central offices or switching devices that is used in providing telephone connections between subscribers. Also, **trunk circuit.**

trunk exchange *Telecommunications.* a central office exchange whose principal function is to interconnect trunks. Also, **tandem switching system, toll switching system.**

trunk feeder *Electricity.* a special type of electric power transmission line used to connect generating stations, critical substations, and electrical distribution networks.

trunk stream see MAIN STREAM.

trunnion *Design Engineering.* a pin or pivot mounted on bearings so as to rotate or tilt a particular device. *Ordnance.* one of two cylindrical projections on either side of a cannon that serves to support the weapon on its carriage.

truss *Civil Engineering.* a planar structural component composed of a combination of chords and web members in a triangular arrangement that forms a rigid framework; often used to support a roof or bridge. *Medicine.* a pad or bandage supported by a belt that is used to maintain a hernia in a reduced state. *Botany.* a compact cluster or head of flowers growing upon one stalk, such as an umbel.

truss rod *Civil Engineering.* **1.** a diagonal steel reinforcement in a wooden beam. **2.** an iron or steel rod serving as a tension member.

trusting or **trust stage** *Psychology.* the earliest of Erik Erikson's eight stages of cognitive development, during the first year of life, when infants develop the sense that parents and other adults are capable of caring for them.

trust vs mistrust *Psychology.* the conflict that can arise during the trusting stage of development, if the infant cannot form the sense that parents and others in the world can be trusted to provide adequate, consistent care.

truth *Science.* the fact of being true. *Particle Physics.* see TOP.

truth function *Mathematics.* a function that assigns to each statement the truth value of that statement.

truth-maintenance system *Artificial Intelligence.* a system for keeping track of the facts, rules, and assumptions upon which a derived conclusion depends, and for withdrawing conclusions whose support has become invalid. Also, REASON MAINTENANCE SYSTEM.

truth quark see TOP QUARK.

truth serum *Organic Chemistry.* a popular name for thiopental sodium; a rapidly acting barbiturate, used as an anesthetic. (Because of the idea that a person given this drug will be uninhibited and will freely disclose repressed or secret information.)

truth table *Mathematics.* given a mathematical statement formed from constituent statements through the use of connectives such as "and," "or," conditionals, negations, and so on, a chart that displays all possible truth-value assignments for the constitutent statements and the corresponding truth values for the statement. *Computer Programming.* a set of truth values of a logical expression, one for each combination of truth values of its variables.

truth value *Mathematics.* an assignment of True or False that is made to a statement.

Tryblidium *Paleontology.* a genus of cap-shaped Paleozoic mollusks in the extinct class Monoplacophora; characterized by a shell with an outer calcitic and an inner aragonitic layer; extant in the Silurian.

Trypanochloridaceae *Botany.* a monospecific family of yellow-green algae of the order Mischococcales that occurs in the outer layers of the shells of certain terrestrial gastropods.

Trypanorhyncha *Invertebrate Zoology.* an order of tapeworms in the subclass Cestoda, endoparasites of sharks and rays.

Trypanosoma *Invertebrate Zoology.* a genus of flagellated protozoans in the family Trypanosomatidae, parasites in the blood of vertebrates, including those causing Chagas' disease and sleeping sickness, transmitted by blood-sucking arthropods.

Trypanosomatidae *Invertebrate Zoology.* a family of elongated protozoans with one flagellum and an undulating membrane, parasites with a complex life cycle in many hosts, causing serious disease in humans and animals. Also, **Trypanosomidae.**

trypanosome *Invertebrate Zoology.* any member of the genus *Trypanosoma.*

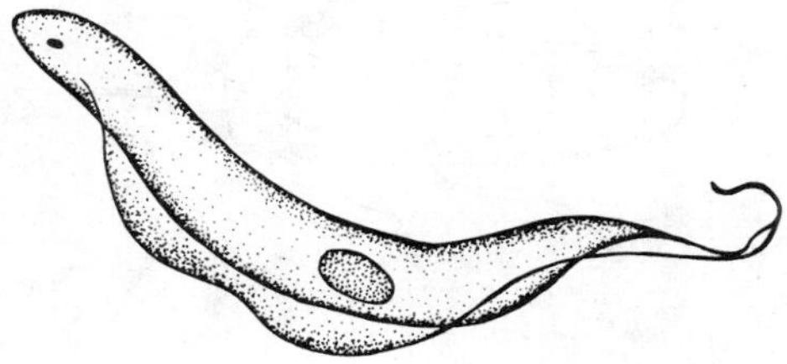

trypanosome

trypanosomiasis *Medicine.* an infestation with *Trypanosoma* parasites, which may result in various diseases, such as African sleeping sickness, Rhodesian trypanosomiasis, and Chagas' disease.

Trypetheliaceae *Mycology.* a family of fungi belonging to the order Melanommatales, composed primarily of species that live as lichens in warm temperate regions.

trypomastigote *Cell Biology.* any of the cell forms assumed during the life cycle of certain protozoa, including members of the genus *Trypanosoma,* in which the slender elongate cell has a kinetoplast and basal body located at the posterior end and a flagellum running anteriorly along an undulating membrane to become a free-flowing structure.

trypsin *Biochemistry.* an enzyme that catalyzes the hydrolysis of peptide bonds adjacent to lysine and arginine residues.

trypsinogen *Biochemistry.* the almost inactive precursor of trypsin.

tryptone *Biochemistry.* a tryptophan-rich peptone formed by the action of trypsin, as in pancreatic digestion of casein.

tryptophan *Biochemistry.* $C_{11}H_{12}N_2O_2$, one of the twenty amino acids that make up proteins; it oxidizes easily in strong acid solutions, readily absorbs ultraviolet radiation, and is essential to the mammalian diet.

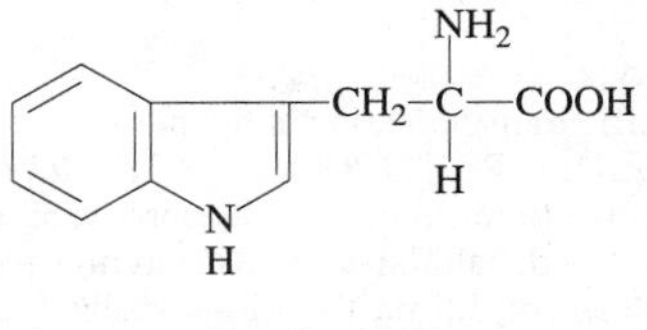

tryptophan

tryptophan hydroxylase *Enzymology.* an enzyme that catalyzes the replacement of a hydrogen atom with a hydroxyl group on the amino-acid tryptophan to form serotonin.

tryptophan synthetase *Enzymology.* an enzyme that catalyzes the formation of tryptophan by uniting serine and indole.

try square *Engineering.* a device consisting of two straightedges fastened together to form a right angle; used to draw right angles or to determine whether a work is square.

ts tensile strength.

TSA tumor-specific antigen.

Tsai-Hill criterion *Materials Science.* a criterion or description of conditions under which failure occurs in a unidirectional, fiber-reinforced composite. Also, MAXIMUM WORK CRITERION.

Tschermak, Erich 1871–1962, Austrian botanist; independently rediscovered Mendel's work and used it to confirm Darwin's theory.

tschermakite *Mineralogy.* $Ca_2(Mg,Fe^{+2})_3Al_2(Si_6,Al_2)O_{22}(OH)_2$ $Mg/(Mg+Fe^{+2})=0.5–1.0$, a dark-green to black monoclinic mineral of the amphibole group occurring as short prismatic crystals and in massive form, having a specific gravity of 3.14 to 3.35 and a hardness of 5 to 6 on the Mohs scale; found in igneous and metamorphic rocks.

T-S diagram *Thermodynamics.* a plot of states, in which temperature is plotted on the vertical and entropy is plotted on the horizontal, that serves to illustrate a thermodynamic cycle or process.

tsetse fly [tet´sē; tēt´sē] *Invertebrate Zoology.* the popular name for any fly of the genus *Glossina*, blood-sucking dipteran insects that transmit sleeping sickness.

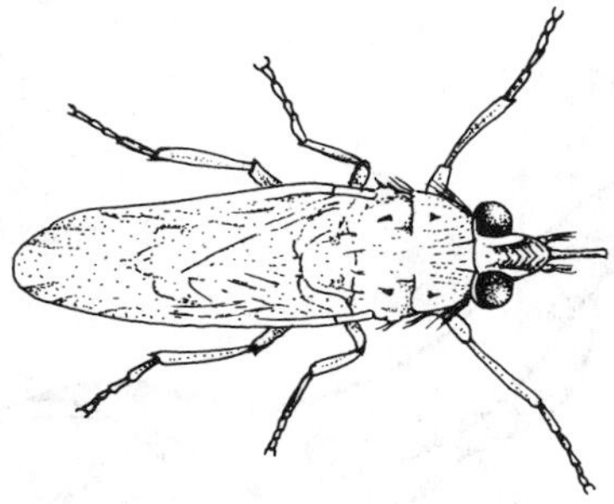

tsetse fly

TSH thyroid-stimulating hormone (thyrotropin).

TSH-RF thyroid-stimulating hormone-releasing factor.

Tsiolkovsky, Konstantin 1857–1935, Russian educator; made fundamental contributions to the theory of rocket propulsion.

T slot *Design Engineering.* in the table of a machine tool, a recessed slot in the form of an inverted T that receives the squared head of a T-bolt or slot bolt.

T-slot cutters *Design Engineering.* a type of milling cutter used to form T-slots in work tables by enlarging the bottom of the slot after it has been cut on a milling machine.

T space *Mathematics.* a topological space X that satisfies one of the five separation axioms. See SEPARATION AXIOM. Therefore: **T_0 space.** a topological space that satisfies the T_0 separation axiom. **T_1 space.** a topological space that satisfies the T_1 separation axiom. **T_2 space.** a topological space that satisfies the T_2 separation axiom. Also, HAUSDORFF SPACE. **T_3 space,** a topological space that satisfies the T_3 separation axiom. Also, REGULAR SPACE. **T_4 space,** a topological space that satisfies the T_4 separation axiom. Also, NORMAL SPACE.

TSPP tetrasodium pyrophosphate.

T-square *Mechanical Devices.* a long straightedge with a cross piece at one end that fits along the edge of a drawing board; used to draft straight lines and as a base for a triangle device or other drafting instrument.

TSS see TOXIC SHOCK SYNDROME.

TSTA tumor-specific transplantation antigen.

tsumebite *Mineralogy.* $Pb_2Cu(PO_4)(SO_4)(OH)$, a brilliant, transparent, emerald-green monoclinic mineral of the brackebuschite group occurring as small, twinned, tabular crystals, having a specific gravity of 6.133 and a hardness of 3.5 on the Mohs scale; found as a secondary mineral at Tsumeb, Namibia.

tsunami [tsü näm´ē] *Oceanography.* a long-period (usually 15–60 minutes) wave caused by a large-scale movement of the sea floor, from a volcanic eruption, submarine earthquake, or landslide; although usually barely noticeable at sea, its velocity may be as high as 400 knots, so that it travels great distances and in shoal water may reach heights of around 15 meters. (From a Japanese term for this; literally, "harbor wave.")

Tsushima Current *Oceanography.* the branch of the Kuroshio Current that flows through the Korea Strait into the Sea of Japan and then along the Japanese coast to the northern end of Honshu, where it flows eastward through the Tsugaru Strait at speeds of up to 7 knots, rejoining there the main branch of the Kuroshio.

tsutsugamushi disease *Medicine.* an acute infectious disease found in Japan, Sumatra, and Guinea, caused by the microorganism, *Rickettsia tsutsugamushi*, carried by mites and transmitted through the larva; characterized by ulceration at the bite, continual fever, colored patches on the skin, and enlargement of the lymph nodes near the bite. The fever is continuous and may last for three weeks. Also, SCRUB TYPHUS, ISLAND FEVER.

tsx protein *Biochemistry.* an outer-membrane receptor protein encoded by the tsx gene in *Escherichia coli* that may assist in the uptake of glycine, phenylalanine, serine, etc.

Tsytovich effect *Electromagnetism.* an effect in which the index of refraction of a medium is much less than unity, causing the phase velocity of electromagnetic waves in a medium to exceed the speed of light.

TTD list see THINGS-TO-DO LIST.

t test see STUDENT'S *t* TEST.

TTT diagram *Materials Science.* a graph representing the relationship between the time, temperature, and transformation of a material; used mostly for steels, showing the phases present at a given temperature after a given time and allowing predictions about heat treatment required to achieve certain structures. Also, ISOTHERMAL TRANSFORMATION DIAGRAM.

T tube *Surgery.* a drainage tube in the shape of a T, used especially in draining the common bile duct.

TTY *Computer Technology.* **1.** originally, an abbreviation for a Teletype teleprinter or for any teletypewriter. **2.** more generally, any user terminal or the user interface (keyboard, mouse, and window in which input is echoed) on a workstation.

TU transmission unit; tuberculin unit.

tuatara *Vertebrate Zoology.* a large, spiny, sluggish, lizardlike reptile, *Sphenodon punctatum*; the only living member of the family Sphenodontidae; found on the islands off New Zealand. Also, **tuatera.**

tuba *Meteorology.* the funnel-shaped cloud that descends from a cloud base in a tornado or waterspout. Also, FUNNEL CLOUD, PENDANT CLOUD, TORNADO CLOUD.

tubal bladder *Vertebrate Zoology.* a urine storage organ in fish.

tubal insufflation *Medicine.* the opening of the uterine tubes by blowing carbon dioxide gas through the passages; Rubin test.

tubal ligation *Medicine.* the surgical sterilization of a female by permanently severing and tying the fallopian tubes to prevent the egg cells from descending to the uterus.

tubal pregnancy *Medicine.* the development of an impregnated ovum in the oviduct; ectopic pregnancy.

tubatorsion *Medicine.* the twisting of the uterine tube.

tubatoxin see ROTENONE.

tube a hollow, cylindrical body or passage; specific uses include: *Biology.* a cylindrical organ or part that functions as a passage. *Botany.* the lower part of a gamopetalous corolla or a gamosepalous calyx. *Geology.* a generally cylindrical, smooth-sided passageway in a cave. *Optics.* the part of a microscope between the eyepiece and the objective. *Ordnance.* the hollow metal cylinder that contains the bore of a weapon; *tube* generally refers to artillery, while *barrel* refers to firearms. *Electronics.* **1.** an electron tube. **2.** the glass or metal envelope of an electron tube.

tube cell *Botany.* the nongenerative cell in the nucleus of pollen cells that gives rise to the pollen tube. Also, **tube nucleus.**

tube coefficient *Electronics.* any of the various attributes and specifications that define the operation of an electron tube, such as transconductance, dynamic plate resistance, or amplification factor.

tube foot *Invertebrate Zoology.* any of the small, fluid-filled tubes projecting from the body of echinoderms, controlled by pressure variance in the water-vascular system; used in feeding, locomotion, clinging, and respiration, and as a tactile sensory organ.

tubemaking *Materials Science.* the production of seamless metal tubes, a process usually involving the piercing of a solid billet followed by hot elongation and reduction in diameter. Tubes may also be produced by piercing and extrusion.

tube mill *Mechanical Engineering.* a long revolving cylindrical mill filled with a grinding medium; used for fine grinding certain ores before further treatment.

tube noise *Electronics.* electronic noise signals that are caused by the random thermal motion of charges and electron impacts in electron tubes.

tubenose fish *Vertebrate Zoology.* a marine fish belonging to the family Aulorhynchidae, characterized by a long cylindrical body with a long snout and hinged upper jaw; found along the coasts of the North Pacific. Also, **tubesnout.**

tube planting *Forestry.* the setting out for growing of young trees in narrow, open-ended cylinders, made of such materials as bamboo, veneer, and polythene.
tube punch *Mechanical Devices.* a hand punch with a hollow tube or tubes gripped like pliers; used to punch snaps or eyelids in leather.
tuber *Botany.* an enlarged, underground stem in which nutrients are stored; usually a fleshy, oblong or rounded thickening or outgrowth, as the potato, bearing minute scalelike leaves with buds or eyes in their axils. *Anatomy.* a rounded swelling or protuberance. *Neurology.* a lesion that characterizes tuberous sclerosis, appearing as a nodular glial brain lesion that generally develops in the cerebral hemispheres, cerebellum, medulla, and spinal cord.

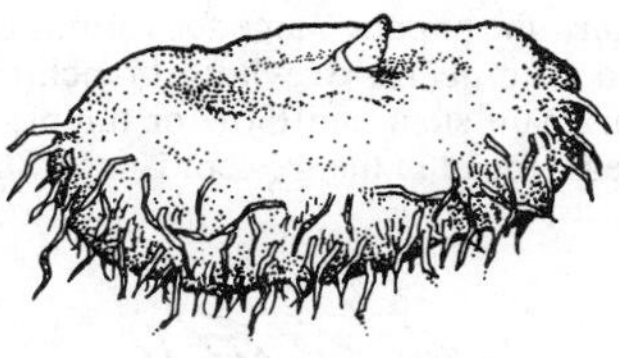

tuber

Tuberaceae *Mycology.* a family of fungi belonging to the order Tuberales that occur in forest soils, living in a symbiotic relationship with trees.
Tuberales *Mycology.* an order of fungi belonging to the class Discomycetes, including species that grow underground, such as those belonging to the genus *Tuber.*
tuber cinereum *Anatomy.* a swelling of gray matter located at the base of the hypothalamus.
tubercle *Biology.* a small, round protrusion, such as the small round projections on the roots of leguminous plants, or on the skin or bones of animals.
tubercular *Medicine.* relating to or resembling tubercules or nodules.
Tuberculariaceae *Mycology.* a family of fungi belonging to the order Tuberculariales, including species that are nonparasitic and live off dead organic matter and species that are pathogenic to plants, such as the genus *Fusarium.*
Tuberculariales *Mycology.* an order of fungi belonging to the class Hyphomycetes, characterized by a stationary sporodochium.
tuberculate *Biology.* characterized by or having tubercules.
tuberculation *Metallurgy.* the knoblike mounds of corrosion products that are found scattered over the surface of corroded materials.
tuberculin *Immunology.* a protein extract, derived from *Mycobacterium tuberculosis,* that is used in tuberculin tests to determine if an individual has a delayed hypersensitivity to tubercle baccilli.
tuberculin test *Immunology.* a test, used to determine if an individual has a delayed hypersensitivity to the tubercle bacillus, in which tuberculin is injected into or deposited in the outer layer of the skin.
tuberculosis [tə burˊkyə lōˊsis] *Medicine.* any of the infectious diseases of animals and humans caused by the microorganism *Mycobacterium tuberculosis*, characterized by the formation of tubercles and caseous necrosis, especially in the lung and bone. Also, TB.

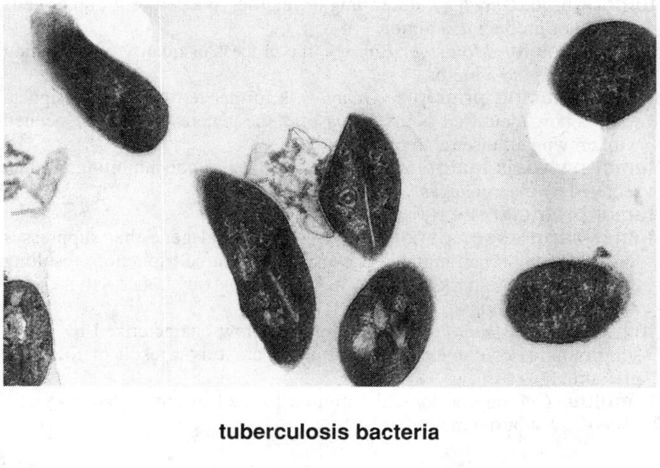

tuberculosis bacteria

tuberculous ulceration see OPEN TUBERCULOSIS.
tuberosity *Anatomy.* a rounded prominence or elevation, especially on a bone.
tuberous organ *Physiology.* an integumentary electroreceptor found in electric fish that is sensitive to high-frequency electric signals.
tuberous root *Botany.* a true root that is thickened and resembles a tuber but does not bear buds; usually one of a cluster, as in the dahlia.
tuberous sclerosis *Medicine.* a hereditary disease characterized by epilepsy and progressive mental deterioration; tumors on the surface of the lateral ventricles and hardened areas on the brain's surface are manifested, and the tumors sometimes spread to various internal organs such as the heart and kidneys, and to the eyes. Also, BOURNEVILLE'S DISEASE, EPILOIA.
tubesnout *Vertebrate Zoology.* a slender fish, *Autorhynchus flavidus*, having a long, tubelike snout; found in coastal waters from southern California to Alaska. Also, **tubenose.**
tube tester *Electronics.* test equipment used for testing electron tubes; most tube testers perform only static checks of electron tubes by indicating the amount of DC plate current on a meter.
Tubeufiaceae *Mycology.* a family of fungi belonging to the order Pleosporales that either live off decaying plant material or are parasites to scale insects or other fungi.
tube voltage drop *Electronics.* the anode voltage of an electron tube, measured with respect to the cathode, when the tube is conducting.
Tubicola *Invertebrate Zoology.* an order of tube worms, polychaete annelids that live in a secreted calcareous tube or one built of found particles; includes most of the Sedentaria.
tubicolous *Invertebrate Zoology.* living in a self-produced tube, secreted or built.
Tubifex *Invertebrate Zoology.* a common genus of small, sessile, freshwater oligochaete worms, red or brown, used to feed aquarium fish.
tubing *Engineering.* a material formed into a cylindrical, hollow body.
tuboplasty *Surgery.* the surgical repair of a tubular body structure, such as a tube of the uterus or the ear.
tubotoxin see ROTENONE.
Tubulanidae *Invertebrate Zoology.* a family of ribbonworms in the order Palaeonemertini.
tubular bowl centrifuge *Biotechnology.* a centrifuge in which a bowl is suspended upside down and hangs free; feed enters at the bottom and is jetted upward into the bowl where it forms a layer on the inside wall.
tubular capacitor *Electricity.* a paper or electrolytic capacitor in the form of a cylinder with leads or lugs projecting from one or both ends.
tubular film extrusion *Materials Science.* a process used to convert a polymer into a flattened tube of thin sheet or film. The polymer is delivered by an extruder through the annular gap in a crosshead or spider die; it emerges as a tube of molten material which undergoes a process of stretching and cooling to form the film. When cool enough to prevent mutual adhesion, the stretched tube is collapsed by guides to a layflat for later processing.
tubular gland *Anatomy.* an exocrine gland in which the secretory portion is tubular.
tubular lock *Mechanical Devices.* a projecting rim lock of a door with the tumblers set into a tube.
tubular membrane module *Biotechnology.* a membrane commonly used to filter liquids with high particulate contents; particularly useful in microfiltration and ultrafiltration processes.
tubular rivet *Mechanical Devices.* a rivet with a tube-shaped shank created by the removal of its center, leaving a thin wall; used with a punch to set its own hole in soft materials.
tubular saw *Mechanical Devices.* a crown saw with a blade longer than it is wide.
tubular scaffolding *Mechanical Devices.* a scaffold constructed of steel tubes clamped by steel-collared workpieces with screw fixings.
tubule *Anatomy.* a small canal or tube.
Tubulidentata *Vertebrate Zoology.* a monofamilial order of African mammals that includes the aardvark; the smallest order of living mammals.
tubulin *Cell Biology.* the constituent protein of microtubules; found in most eukaryotic cells.
tubulin gene *Genetics.* any of a number of genes that code for subunits of the protein tubulin, the principal component of microtubules in cells.
tubuloalveolar gland *Anatomy.* an exocrine gland that contains both tubular and saclike secretory portions.
Tucana *Astrophysics.* the Toucan, a faint southern constellation located southwest of the star Achernar; it contains the Small Magellanic Cloud.

tuckahoe *Botany.* the edible, underground, food-storage body (sclerotium) of the fungus *Poria cocos*, found on the roots of trees in the southern U.S. (From a Powhatan word meaning "it is found.") Also, INDIAN BREAD.

tuck pointing *Building Engineering.* in masonry, the finishing of joints, after cleaning and filling, with an ornamental fillet of putty, lime, or chalk which projects from the mortar joint. Also, **tuck-and-pat-pointing, tuck pointer, tuck joint pointing.**

Tudor *Architecture.* the final period of English Gothic architecture (from about 1485 to 1550), at the end of the Perpendicular phase.

Tudor arch *Architecture.* a low, wide, four-centered pointed arch common in Tudor architecture.

tufa *Geology.* a soft, porous limestone rock or deposit that is formed from solution by springs or percolations. Also known by various other names, such as CALCAREOUS TUFA, CALC-TUFA, PETRIFIED MOSS, and TOPHUS.

tufaceous *Geology.* describing material that is related to tufa or similar in appearance to tufa.

tuff *Geology.* a compacted or consolidated deposit of pyroclastic material composed of volcanic ash and dust.

tuffaceous *Geology.* describing sediments in which tuff constitutes up to half of the total composition.

tuff lava see WELDED TUFF.

tuft *Geology.* see MOUND, def. 1.

tufted titmouse *Vertebrate Zoology.* a gray titmouse, *Parus bicolor*, having bluish-gray with white plumage, rust underparts, and a crested head; found in the eastern and midwestern United States.

tugboat *Naval Architecture.* a small vessel equipped with a powerful motor that is out of proportion to its size; used to tow or push other (usually larger) vessels.

Tukey's lemma *Mathematics.* a proposition stating that each nonempty family of finite character has a maximal member, equivalent to the axiom of choice, the Hausdorff maximality principle, Zorn's lemma, the well-ordering theorem, etc.

Tukon tester *Engineering.* an instrument that uses a diamond indenter to test the microhardness of a metal.

TUL *Aviation.* the airport code for Tulsa International Airport.

tularemia [too´lə rē´mē ə] *Veterinary Medicine.* a disease of rodents, lagomorphs, certain birds, and sometimes humans, due to infection caused by the microorganism *Pasteurella tularensis* and transmitted by fleas and ticks; characterized by fever, headache, muscle pain, and nodule formations in the liver, spleen, and lymph nodes.

Tulasnellaceae *Mycology.* a family of fungi belonging to the order Metatremellales that occur on decaying organic matter in tropical to temperate regions.

Tulasnellales *Mycology.* a former term for an order of fungi belonging to the subclass Holbasidiomycetidae; they have been reclassified within the order Metatremellales.

tulip *Botany.* **1.** the common name for any of the spring-blooming members of the genus *Tulipa* of the lily family, having lance-shaped leaves and large, cup-shaped flowers in a variety of colors; propagated from bulbs and widely grown as an ornamental plant. **2.** the flower of such a plant.

tulip tree *Botany.* the common name for *Liriodendron tulipifera*, a tree of the Magnoliaceae that is native to the eastern U.S. and widely grown as an ornamental for its flowers, which resemble tulips; the state tree of Indiana, Kentucky, and Tennessee. Also, **tulip poplar.**

tulipwood *Materials.* **1.** the soft, light-colored wood of the tulip tree, used for cabinetwork. **2.** any of various striped or variegated woods of other trees.

tulle *Textiles.* a fine, sheer, net fabric with a tiny hexagonal mesh; made of silk, cotton, or synthetic fibers.

tullibee *Vertebrate Zoology.* a deep-bodied whitefish, *Coregonus artedi tullibee*, found in the Great Lakes and commercially important. (From a French-Canadian word meaning "mouth water," because of its watery flesh.)

Tulostomataceae *Mycology.* a family of fungi belonging to the order Tulostomatales that occur on sandy soil and on rotting wood in forests.

Tulostomatales *Mycology.* an order of fungi belonging to the class Gasteromycetes, composed of stalked puffballs and characterized by a roundish fruiting body supported by a well-developed stipe.

tumble *Space Technology.* of a spacecraft, probe, or other object, to rotate end-over-end about its center of mass as it continues its flight or fall.

tumble axis see TOPPLE AXIS.

tumblebug *Invertebrate Zoology.* the popular name for any of various dung beetles that roll balls of dung in which they lay their eggs and in which the young develop.

tumble home *Naval Architecture.* an inward curve of a vessel's side above the waterline, resulting in a narrower width on the upper deck than at the waterline. Rarely seen today, tumble home was a common feature of sailing warships and larger merchantmen.

tumbler *Engineering.* **1.** a locking mechanism that must be moved to the correct position before a bolt may be opened. **2.** a device in which objects are tumbled.

tumbler feeder see DRUM FEEDER.

tumbler gears *Mechanical Engineering.* a system of idler gears in a lathe gear train that is used to reverse the rotation of the lead screw or feed rod.

tumbleweed *Botany.* the popular name for various plants, such as *Amaranthus albus* and *A. graecizans*, whose branching parts, when dry, break off from the main stem and blow or tumble about in the wind, scattering their seeds; found in the western United States.

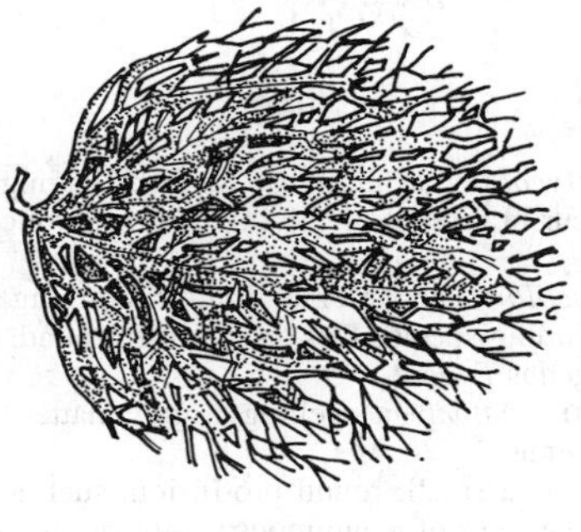

tumbleweed

tumbling *Engineering.* a process by which surface irregularities are removed by rotating an object in a tumbler filled with polishing compounds. *Computer Programming.* in computer graphics, the rotation of a display image about an axis.

tumbling barrel *Mechanical Engineering.* a revolving drum, in which small manufactured objects, especially of metal, are polished by the friction of shaking them about with abrasives. Also, **tumbling box.**

tumescence *Volcanology.* the swelling of a volcanic mountain caused by a buildup of magma in its reservoir; often followed by an eruption. *Medicine.* a swollen part or tumor.

tumid *Biology.* **1.** describing a swollen or distended organ or part. **2.** having a bulging shape.

tumor *Medicine.* **1.** an abnormal growth that arises from normal tissue, but that grows abnormally both in rate and structure and serves no physiological function. Also, NEOPLASM. **2.** a swelling or enlargement; one of the cardinal signs of inflammation.

tumoraffin *Oncology.* exhibiting an affinity for tumor cells.

tumor antigen see T ANTIGEN.

tumoricide *Oncology.* an agent or process that is destructive to cancer cells. Thus, **tumoricidal.**

tumorigenic *Oncology.* describing or relating to a cell or group of cells capable of producing a tumor.

tumorigenicity *Molecular Biology.* the process or quality of producing or giving rise to a tumor.

tumor-inducing principle *Genetics.* a former term used to designate what is now identified as the Ti plasmid, the genetic element associated with crown gall disease in plants.

tumor necrosis factor *Molecular Biology.* a tumor-inhibiting product secreted by macrophages.

tumor promoter see PHORBOL ESTER.

tumor-suppressor gene *Oncology.* a normal gene that suppresses tumor growth; when mutated it causes a loss in cell function, resulting in tumor formation. Also, GROWTH SUPPRESSOR GENE, RECESSIVE ONCOGENE, ANTI-ONCOGENE.

tumuli lava *Volcanology.* a type of lava flow characterized by small, solid mounds or domes that form on the crust as a result of localized pressure.

tumulus *Geology.* a domelike mound formed in congealed lava. *Archaeology.* a burial mound or barrow.

tuna *Vertebrate Zoology.* any of several large, pelagic, commercially important food and game species of the fish family Scombridae, closely related to the mackerel and having coarse oily flesh; found in temperate and tropical seas; common species include the **albacore**, **bluefin tuna**, and **yellowfin tuna.**

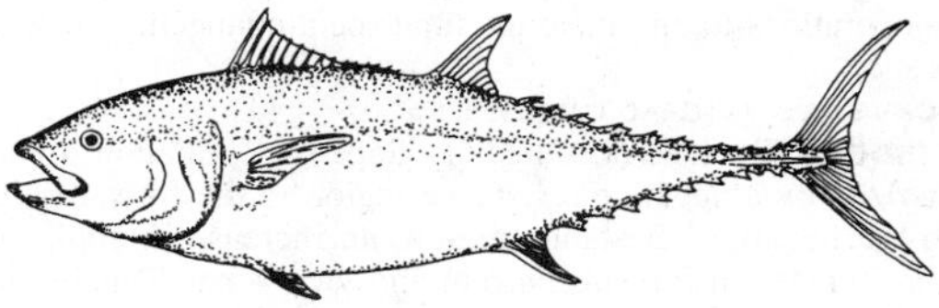

bluefin tuna

tunable filter *Electronics.* a filter device that can be tuned to the proper pass or reject frequency by adjusting one or more of the filter's components.
tunable laser *Optics.* a laser, such as a spin-flip laser, whose output frequency is adjustable across a wide range.
tunable magnetron *Electronics.* a magnetron that can be tuned to a specific frequency within a narrow range of frequencies; magnetrons are tuned either electrically or mechanically, as by adjusting the volume of a resonant cavity.
tuna clipper *Naval Architecture.* a fairly large, long-range vessel used for tuna fishing, usually equipped with brine tanks. Also, **tuna boat.**
tundish *Metallurgy.* a rectangular-shaped trough with five clay refractories, with one or more refractory nozzles at its base; used between the ladle and ingot molds in the process of teeming steel.
tundra *Ecology.* **1.** a vast, nearly level, barren, treeless region located in the Arctic; characterized by very low winter temperatures, short, cool summers, and vegetation consisting of various grasses, rushes, perennial herbs, lichens, and dwarf woody plants. **2.** see ALPINE TUNDRA. (From a Russian word for such an area; going back to the language of the Lapp people of northernmost Europe.)
tundra climate *Meteorology.* a climate that produces tundra vegetation, lacking a permanent cover of snow or ice but too cold for tree growth.
tundra swan *Vertebrate Zoology.* a swan, *Cygnus columbianus*, having a blackbill with a yellow spot at the base; found nesting in tundra regions of the New and Old Worlds.
tune *Electronics.* to adjust the resonant frequency of a circuit.
tuned amplifier *Electronics.* an amplifier, typically a radio-frequency amplifier, that can be adjusted to amplify only signals at and near a particular frequency in a given range and to attenuate signals at all other frequencies.
tuned-anode oscillator *Electronics.* an electron tube oscillator whose operating frequency is determined by the adjustment of a resonant tank circuit in the anode circuit of the tube. Also, **tuned-plate oscillator.**
tuned-base oscillator *Electronics.* a transistor oscillator whose operating frequency is determined by the adjustment of a resonant tank circuit in the base circuit of the transistor.
tuned circuit *Electronics.* a circuit containing capacitive and inductive circuit elements that operate only within the narrow band of frequencies to which the circuit is resonant; the resonant frequency is adjustable over the tuning range of the circuit.
tuned-collector oscillator *Electronics.* a transistor oscillator whose operating frequency is determined by the adjustment of a resonant tank circuit in the collector circuit of the transistor.
tuned filter *Electronics.* a filter composed of tuned circuits.
tuned-grid oscillator *Electronics.* an electron tube oscillator whose operating frequency is determined by the adjustment of a resonant tank circuit in the grid circuit of the tube.
tuned-grid tuned-anode oscillator *Electronics.* an electron tube oscillator whose operating frequency is determined by the adjustment of resonant tank circuits in both the grid and anode (plate) circuits of the tube. Also, **tuned-plate tuned-grid oscillator, tuned-grid tuned-plate oscillator.**
tuned-radio-frequency transformer *Electronics.* a transformer that operates with capacitive circuit elements to form resonant circuits in the transformer's primary and/or secondary windings; adjustment of the resonant frequency is achieved by adjusting either the capacitance or the inductance of the winding; used to provide resonant coupling between stages in radio receivers or transmitters.
tuned relay *Electricity.* a relay that has a mechanical or other resonating configuration to limit the response to currents at a designated frequency.
tuned resonating cavity *Electromagnetism.* a resonant cavity that is used in conjunction with a waveguide; the cavity length is an integral multiple of the half-wavelength of the wave in the guide.
tuned transformer *Electricity.* a transformer that is composed of an interstage coupling, in which one or both windings are tuned to resonate with the signal frequency.
tuner *Electronics.* **1.** any device or apparatus used for selecting and controlling the operating frequency of a circuit or equipment, such as the channel selector in a television receiver. **2.** a unit that selectively accepts radio or television broadcast signals, amplifying and converting them to produce intermediate-frequency or audio-frequency signals to drive other equipment, such as television monitors or audio amplifiers.
tune-up *Engineering.* a series of checks and adjustments on parts of a motor to keep it in efficient operating condition.
tungar tube *Electronics.* a gas-filled thermionic electron tube having a tungsten filament that contains a small quantity of thorium oxide and serves as a directly heated cathode and a graphite disk-shaped anode; used chiefly as a rectifier in battery charger applications.
tung nut *Botany.* the common name for the seed of the tree *Aleurites fordii,* from which tung oil is obtained, an important ingredient in certain varnishes and paints.
tung oil *Materials.* a poisonous, yellow drying oil derived from the tung nut; used in varnishes.
tungstate *Mineralogy.* any of a group of minerals that contain the radical $(WO_4)^{-2}$, such as scheelite. *Inorganic Chemistry.* a salt of tungstic acid, H_2WO_4.
tungsten *Chemistry.* a very hard, highly conductive metallic element having the symbol W, the atomic number 74, an atomic weight of 183.85, a melting point of 3410°C (the highest of all metals), and a boiling point of 5927°C; it is a white or gray, brittle metallic element that is a member of the chromium family, found combined in wolframite and scheelite; used in steel alloys, as filaments for electric light bulbs, in magnets and cemented carbides, and as heating elements in furnaces. (From a Swedish term meaning "heavy stone.") Also, WOLFRAM.
tungsten boride *Inorganic Chemistry.* WB_2, silvery octagonal crystals; insoluble in water; melts at 2900°C; used as a refractory.
tungsten carbide *Inorganic Chemistry.* WS, black hexagonal crystals; insoluble in cold water; melts at approximately 2870°C and boils at 6000°C; the strongest structural material known; used in dyes, cutting tools, and electrical resistors.
tungsten filament *Electricity.* a filament used in incandescent lamps and in thermionic vacuum tubes and other tubes requiring an incandescent cathode; smaller filaments are used in a vacuum while those used in larger lamps require an inert gas at a specific pressure.
tungsten hexachloride *Inorganic Chemistry.* WCl_6, dark blue cubic crystals; decomposes in hot water and soluble in alcohol; melts at 275°C and boils at 346.7°C; used as a coating on base metals and glass and as a catalyst.
tungstenite *Mineralogy.* WS_2, a dark lead-gray, metallic, hexagonal mineral, occurring as fine scaly or feathery aggregates, having a specific gravity of 7.74 to 7.75 and a hardness of 2.5 on the Mohs scale; found as a replacement deposit in limestone. Also, **tungstenite-2H.** *Inorganic Chemistry.* dark gray hexagonal crystals; insoluble in cold water and alcohol; decomposes at 1250°C; used as a solid lubricant. Also, **tungsten disulfide.**
tungsten lake see PHOSPHOTUNGSTIC PIGMENT.
tungsten oxychloride *Inorganic Chemistry.* $WOCl_4$, red needles; decomposed by water; melts at 211°C and boils at 227.5°C; used in incandescent lighting.
tungsten steel *Metallurgy.* a steel once widely used for cutting tools and magnets but now replaced by high-speed or hot-work steels, which contain tungsten with other alloying elements.
tungsten trioxide see TUNGSTIC OXIDE.
tungstic acid *Inorganic Chemistry.* H_2WO_4, a toxic yellow powder; insoluble in water; used in textiles, in plastics, and to make metallic tungsten products. Also, WOLFRAMIC ACID, ORTHOTUNGSTIC ACID.
tungstic anhydride see TUNGSTIC OXIDE.
tungstic oxide *Inorganic Chemistry.* WO_3, toxic, yellow rhombic crystals or yellow-orange powder, insoluble in water and acid; melts at 1473°C; used in alloys, in fireproofing, and as a pigment. Also, TUNGSTIC ANHYDRIDE, TUNGSTEN TRIOXIDE, ANHYDROUS WOLFRAMIC ACID.
tungstic trioxide see TUNGSTIC OXIDE.

tungstite *Mineralogy.* $WO_3 \cdot H_2O$, a yellow or yellowish-green, transparent, orthorhombic mineral occurring in earthy masses and microscopic platy crystals, having a specific gravity of 5.50 to 5.52 and a hardness of 1 to 2.5 on the Mohs scale; an alteration product of other tungsten minerals.

tungstophosphoric acid see PHOSPHOTUNGSTIC ACID.

tung tree *Botany.* the common name for *Aleurites fordii,* a tree of the order Euphorbiales, from whose fruit tung oil is derived. Also, **tung-oil tree.**

tunic *Anatomy.* a membranous sheath enveloping or lining an organ of the body. *Invertebrate Zoology.* the secreted cuticular covering of tunicates. *Botany.* an integument, as that covering a seed. Also, **tunica.**

tunica intima see INTIMA.

tunica media *Anatomy.* the middle coat of a blood vessel, consisting of transverse elastic and muscle fibers. Also, MEDIA.

Tunicata *Invertebrate Zoology.* the sea squirts, a subphylum of Chordata, having a nonsegmented cylindrical or globular body, a secreted covering, a pharynx perforated with gill slits, and no skeleton; sessile, floating, or free-swimming, solitary or colonial, hermaphrodites that also reproduce by asexual budding.

tunicate *Botany.* made up of a number of layers, such as the onion. *Invertebrate Zoology.* any member of the subphylum Tunicata.

tuning *Electronics.* the act of adjusting the resonant frequency of a circuit to the desired frequency or performance level. *Computer Technology.* the act of adjusting system controls to make a system perform optimally for the current workload.

tuning capacitor *Electricity.* a variable capacitor used to tune an LC circuit, series-resonant or parallel-resonant.

tuning coil *Electricity.* a fixed or variable inductor used to tune an LC circuit, series-resonant or parallel-resonant.

tuning core *Electromagnetism.* a core of powdered iron that slides or screws in and out of a coil for inductance adjustment. Also, **tuning slug.**

tuning fork *Engineering.* a two-pronged device designed to vibrate at a single frequency when struck; used to tune musical instruments and for acoustical experiments.

tuning indicator *Electronics.* any indicating device that shows whether or not a tuned circuit is resonant to the frequency of the current signal; tuning indicators usually take the form of meters or cathode-ray tube "tuning eyes."

tuning range *Electronics.* the range of frequencies over which a given tuned circuit or equipment can be adjusted.

tuning screw *Electromagnetism.* a screw that governs the penetration depth of a rod into a waveguide and is adjusted for purposes of tuning or impedance-matching.

tuning stub *Electromagnetism.* a short section of transmission line connected to a main transmission line and short-circuited at its free end; used for matching impedance.

tuning wand see NEUTRALIZING TOOL.

tunnel *Engineering.* an underground passage that is open at both ends. *Zoology.* the burrow of an animal.

immersed railway tunnel

tunnel blasting *Engineering.* a method of blasting used in rock-tunneling operations, in which a large heading is driven into the rock and filled with explosives.

tunnel borer *Mechanical Engineering.* a machine that is used to bore a tunnel, such as the massive borers used for the English Channel tunnel.

tunnel carriage *Mechanical Engineering.* a machine that combines a drill carriage with a manifold containing air and water; used in tunnel-drilling operations to eliminate the time spent connecting or waiting for drill steels.

tunnel cave see NATURAL TUNNEL.

tunnel diode *Electronics.* a heavily doped PN-junction diode that is characterized by a negative-resistance region in the forward-bias direction; in the negative-resistance region, an increase in applied voltage causes a decrease in forward current and vice versa. Tunnel diodes are used in ultrahigh-frequency and microwave oscillator and amplifier circuits. Also, ESAKI DIODE.

tunnel erosion *Hydrology.* see PIPING.

tunneling *Quantum Mechanics.* a phenomenon, impossible according to classical physics, in which a particle can penetrate and cross a small region where the potential is greater than the particle's available energy. Also, TUNNEL EFFECT, BARRIER PENETRATION, ELECTRON TUNNELING, QUANTUM BARRIER PENETRATION.

tunneling cryotron *Electronics.* a cryotronic superconducting switching device consisting of two electrodes separated by a thin film and acting similarly to a switch contact; the switch on-off state is controlled by a small current through a control line located in proximity to the main electrodes, and switching time is in the picosecond range.

tunnel kiln *Materials Science.* a type of continuous kiln used in the firing of ceramics, in which the product is moved through the length of the tunnel.

tunnel rectifier *Electronics.* a low-peak, current-rated tunnel diode used in computer memory applications.

tunnel resistor *Electronics.* a semiconductor device that combines the characteristics of a resistor and a tunnel diode; resistive material is plated across a tunneling junction, effectively connecting the resistor and the diode in parallel.

tunnel triode *Electronics.* a semiconductor device that combines a tunnel diode with a conventional transistor; the base-emitter junction is a tunneling junction, and the base-collector junction is that of a conventional triode.

tunnel vault see BARREL VAULT.

Tupaiidae *Vertebrate Zoology.* the tree shrews, a family of small, arboreal or terrestrial, insectivorous, Asian mammals characterized by a long and usually bushy tail and a long pointed snout.

tupelo *Botany.* **1.** the common name for trees of the genus *Nyssa* of the family Nyssaceae, characterized by alternate, entire leaves and purple, berrylike fruit; found in swampy regions of the eastern, southern, and midwestern U.S. **2.** the soft, light wood of these trees; used for furniture, pulpwood, crates, and boxes.

tuple *Computer Programming.* **1.** in a relational database, a single row of data elements in a relation, equivalent to a record. **2.** a set of several mathematical objects in a formal description of a system. Often preceded by a count, such as *4-tuple* or *n-tuple*.

turaco or **turakoo** see TOURACO.

turanite *Mineralogy.* a rare, radioactive, olive-green, probably orthorhombic mineral with an approximate formula of $Cu_5^{+2}(VO_4)_2(OH)_4$, having a hardness of 5 on the Mohs scale and an undetermined specific gravity; found as crusts and fibrous concretions with uranium and vanadium minerals in limestone cavities in the Turkestan.

Turbellaria *Invertebrate Zoology.* a class of free-living, ciliated flatworms in the phylum Platyhelminthes, having a ventral mouth; hermaphrodites that also reproduce by fission.

turbidimeter *Optics.* an instrument that measures the transmission and scattering of light in a fluid with suspended solids. *Biotechnology.* an instrument that estimates the growth of microorganisms by analyzing the turbidity of a sample; a light beam is passed through the prepared culture and the decrease in absorbance is measured.

turbidimetric analysis *Analytical Chemistry.* the determination of concentrations of particles in cloudy solutions by the measurement of the loss in intensity of a beam of light passing through the solution. Also, **turbidimetry.**

turbidimetric titration *Analytical Chemistry.* a titration with the end point indicated by the turbidity of the titrated solution.

turbidite *Geology.* a sediment or rock deposited from a turbidity current.

turbidity *Analytical Chemistry.* a measure of the degree of clarity of a solution. *Materials Science.* a cloudy or hazy appearance in a naturally clear liquid caused by the suspension of fine solids or colloids in the liquid. *Meteorology.* a condition in which the atmosphere becomes less transparent to radiation, especially when this reduction in transparency is caused not by cloud cover but by suspensoids such as smoke, dust, and haze. *Astronomy.* the disklike appearance of a star in a long-exposure photo caused by light scattering inside the emulsion.

turbidity coefficient *Optics.* the factor in a light-absorption calculation that indicates the quantity of light absorbed or scattered by a fluid.

turbidity current *Hydrology.* a current that transports sediments along and in contact with the bottom, as in a stream or lake. *Oceanography.* a dense current that carries large amounts of clay, silt, and sand through sea water that is less dense.

turbidity factor *Geophysics.* a measure of the rate at which the atmosphere transmits incident solar radiation.

turbidostat *Microbiology.* a device designed to maintain a bacterial culture at a constant turbidity.

turbinate *Botany.* having an inverted conical shape, attached at the point. *Anatomy.* of or relating to certain scroll-like, spongy bones of the nasal passages in humans and other vertebrates.

turbine [tur´bin; tur´bīn´] *Mechanical Engineering.* a rotary power-generating device driven by a moving fluid, such as water, steam, gas, or wind, that converts kinetic energy into mechanical energy.

turbine generator *Electricity.* an electric generator that is driven by a steam, gas, or hydraulic turbine coupled to it for electric power production.

turbine pump see REGENERATIVE PUMP.

Turbinidae *Invertebrate Zoology.* the turban snails, a family of marine gastropod mollusks that grow up to eight inches across.

turbo- *Mechanical Engineering.* a combining form indicating a machine coupled to a turbine that drives it, such as a **turbogenerator;** or a machine that is a turbine, such as a **turbomotor.**

turboalternator *Electricity.* an alternator that is driven by a steam turbine.

turboblower *Mechanical Engineering.* a centrifugal or axial-flow compressor.

turbodrill *Petroleum Engineering.* a rotary drill used in oil or gas drilling operations, driven by a turbine motor located inside the well.

turbofan *Aviation.* **1.** a ducted-fan jet engine. **2.** more narrowly, the fan used in such an engine.

turbofighter *Aviation.* a fighter aircraft propelled by a turbojet engine.

turboglyph see FLUTE CAST.

turbojet *Aviation.* **1.** a flight vehicle propelled by a turbojet engine. **2.** a turbojet engine.

turbojet engine *Aviation.* a jet engine propelled by the simplest form of gas turbine; it utilizes a compressor, combustion chamber, and turbine, with the turbine drawing just enough energy from the gas flow to drive the compressor.

turbomilling *Food Technology.* a process for the secondary milling and sorting of high-protein flour, using high-speed turbo grinders and centrifugal force to separate finer high-protein particles from the coarser starchy fraction.

turbopause *Meteorology.* the upper and lower boundaries of the turbosphere in which rapid changes in air pressure and uneven air motions occur.

turboprop *Aviation.* **1.** a turbo-propeller engine. **2.** an airplane equipped with two or more turbo-propeller engines.

turboprop

turbo-propeller engine *Aviation.* a jet engine having a turbine-driven propeller whose thrust is derived principally from the propeller, but augmented by the hot exhuast gases that issue in a jet.

turbosphere *Meteorology.* the atmospheric region characterized by turbulence.

turbostratic structures *Materials Science.* a subcrystalline microstructure characteristic of LTI, ULTI, and glassy carbons. Turbostratic structures exhibit no order between the layers, as there is in graphite, so the crystal structure of these carbons is two-dimensional.

turbosupercharger *Mechanical Engineering.* a centrifugal air compressor driven by a gas turbine, used to increase the induction pressure in an internal combustion engine.

turbot [tur´bət] *Vertebrate Zoology.* a European flatfish found especially in Atlantic coastal waters, *Scophthalmus maximus;* valued as a food fish.

turbulence a state of disorder, disarray, or agitation; specific uses include: *Meteorology.* an irregular motion of the atmosphere, as manifested by wind gusts and lulls. Also, **turbulency.** *Hydrology.* the secondary motion caused by eddies in a moving flow of water. *Chaotic Dynamics.* **1.** when describing evolution in time, another term for chaos. **2.** when describing complex evolution in space and time (sometimes called **fully developed turbulence**), it is applied where spatially localized behavior is chaotic and behavior at spatially separated points is decreasingly correlated as the separation increases. In fluids this appears as vortices and eddies of all sizes and at all length scales. Though originally used to describe fluid complexity, the term is now applied to any spatiotemporal complexity of this type. *Fluid Mechanics.* see TURBULENT FLOW.

turbulent *Science.* being in a state of turbulence.

turbulent boundary layer *Fluid Mechanics.* the fluid particles in the boundary layer that move in random paths but in which viscous stresses are dominant.

turbulent burner *Engineering.* an atomizing burner that combines fuel with air to generate turbulent flow.

turbulent flow *Fluid Mechanics.* the fluid properties at most points in the flow that exhibit random fluctuations with time. Also, TURBULENCE.

turbulent heat conduction *Oceanography.* the transfer of heat within a water mass through irregular, random eddy diffusion, laterally and vertically.

turbulent Lewis number *Physics.* a dimensionless quantity given by the eddy mass diffusivity divided by the eddy thermal diffusivity; used in studies involving turbulent mass and thermal transfer.

turbulent Prandtl number *Physics.* a dimensionless number given by the eddy viscosity divided by the eddy thermal diffusivity; used in turbulent heat transfer studies.

turbulent Schmidt number *Fluid Mechanics.* a dimensionless ratio of the eddy viscosity to the eddy mass diffusivity.

turbulent shear stress *Fluid Mechanics.* the viscous stresses that arise from the cross-products of the fluctuating velocity components in a turbulent flow. Also, **turbulent shear force.**

Turdidae *Vertebrate Zoology.* the thrushes, a diverse, cosmopolitan family of passeriform birds of the suborder Oscines.

turgid *Medicine.* swollen and distended.

turgor *Botany.* the distension of plant cells, resulting from the pressure exerted by the cell contents on the cell walls.

turgor movement *Botany.* the temporary displacement in the position of plant parts due to turgor pressure.

turgor pressure *Botany.* the stress put upon the cell wall by the fluid-saturated protoplasm of a plant cell.

Turing, Alan Mathison 1912–1954, English mathematician; a pioneer in the theory and development of computers; proposed the Turing machine.

Turing computable function *Mathematics.* a function that can be computed by a Turing machine. Every known notion of computability has been shown to be equivalent to Turing computability.

Turing machine *Computer Technology.* an early, abstract computing device consisting of a control unit, a storage tape, and a read/write head; many variants have been described.

Turing test *Artificial Intelligence.* a proposal, due to A. M. Turing, that if an interrogator, examining either a person or a program remotely via a terminal, cannot tell whether the examinee is a person or a program, then the program must be regarded as intelligent.

turion *Botany.* **1.** a small shoot or bud of a water plant that detaches and remains inert during the winter, giving rise to a new plant in the spring. **2.** any scaly shoot or sucker developed from an underground bud.

wild turkey

turkey *Vertebrate Zoology.* **1.** the popular name for both wild and domestic varieties of a large, gallinaceous bird of the family Meleagridae, especially *Meleagris gallopavo*, typically having bronze plumage, a naked head, and a fanlike tail in the male; native to North and South America and domesticated in most parts of the world. **2.** the flesh of this bird used as food.

turkey oak *Botany.* an oak, such as *Quercus laevis* and *Q. incana*, that grows on dry, sandy barrens; found in the southern U.S.

turkey vulture *Vertebrate Zoology.* a widespread New World vulture, *Cathartes aura*, belonging to the family Cathartidae, having a wingspan of about six feet and a naked, red head and black feathers. Also, **turkey buzzard.**

turmeric *Botany.* the common name for *Curcuma longa*, a plant of the Zingiberaceae, or ginger family, grown commercially for its rhizome, from which a spice and an orange-reddish dye is derived.

turmeric

turn-and-bank indicator see BANK-AND-TURN INDICATOR.

turnaround *Engineering.* a time during which a unit is shut down for repair and maintenance after a normal run, before it is returned to operation. *Transportation Engineering.* **1.** the time elapsed between arriving at a point and leaving the same point, as for aircraft. **2.** a round-trip aircraft crew duty assignment which is completed in a single day, thus not requiring a crew layover.

turnaround document *Computer Programming.* **1.** a document containing information that can be read with or without a computer, then be reread by the computer as evidence of a completed transaction. **2.** an output of one process that is input to a subsequent phase.

turnaround system *Computer Technology.* in character recognition, a system in which the input data being read has been printed by a computer from the same system or an associated system.

turnaround time *Industrial Engineering.* the time elapsed between the submission of a job to a service center or supplier for processing and the completion of the job.

turnback *Transportation Engineering.* the return of an aircraft to a departure gate after its scheduled departure from the gate.

turnbuckle *Mechanical Devices.* a fastener consisting of a metal loop screwed onto a threaded metal rod on one end and a swivel on the other, or with a long nut screwed internally with a left-hand thread at one end and a right-hand thread at the other; used to tighten stayrods and other tension connectors. Also, SCREW SHACKLE.

Turnbull's blue *Inorganic Chemistry.* a blue pigment made by the reaction of a ferrous salt with potassium ferricyanide; used in blueprints.

Turneraceae *Botany.* a family of dicotyledonous herbs, shrubs, and occasionally trees of the order Violales, characterized by perfect red or yellow flowers, and often producing alkaloids but lacking ellagic acid; native mainly to tropical and warm-temperate parts of America, Africa, and Madagascar.

Turner's syndrome *Medicine.* a rare chromosomal disorder of females marked by the absence of one X chromosome, resulting in short stature and retarded sexual characteristics, low rear hair line, deformities of the spine, chest, and extremities, and cardiac defects. (Named after Henry Hubert *Turner*, 1892–1970, American endocrinologist.) Also, GONADAL DYSGENESIS, BONNEVIE-ULLRICH SYNDROME.

Turnicidae *Vertebrate Zoology.* the button quails and hemipodes, a family of Old World, quail-like, terrestrial birds belonging to the order Gruiformes.

turning *Mechanical Engineering.* the process of producing cylindrical, flat, or tapered workpieces on a lathe.

turning basin *Civil Engineering.* an area of water in a harbor or channel in which a ship can be turned.

turning chisel *Mechanical Devices.* a chisel used to shape work on a lathe.

turning circle *Navigation.* the circular track followed by the pivot point of a vessel when circling with the rudder angled as far as possible.

turning movement *Military Science.* an offensive tactic in which the attacking force passes around or over the enemy's principal defensive positions to secure objectives deep in the enemy's rear.

turning path *Transportation Engineering.* the path followed by a given point on a vehicle structure as the vehicle makes a turn; the combined turning paths of all parts of the vehicle are a determinant of required turning clearances.

turning radius *Transportation Engineering.* the minimum radius in which an automobile, aircraft, or other vehicle can make a turn.

turnip *Botany.* the popular name for *Brassica rapa* or *B. campestri*, a plant of the family Brassiaceae, that is grown commercially for its edible leaves and root.

turnip cabbage see KOHLRABI.

turnkey *Computer Programming.* of or relating to a computer system that is delivered to the customer in a ready-to-run condition, specialized for the customer's application, with no additional equipment or modifications required. Thus, **turnkey system.**

turnkey contract *Industrial Engineering.* a contract in which the contractor agrees to complete a project for a fixed price.

turnoff mass *Astronomy.* the least mass of a star (in a group) that has had time to begin evolving away from the main sequence.

turnoff point *Astronomy.* for an H-R diagram of a star cluster, the place where stars turn off the main sequence and begin to evolve toward the giant branch.

turn-off time *Electronics.* the amount of time required for current in a switching device to fall to zero after being gated off.

turn-on time *Electronics.* the amount of time required for current in a switching device to rise to its normal value after being gated on.

turnout *Engineering.* **1.** a switching device that allows a train to change rails. **2.** a railroad siding. *Transportation Engineering.* an area along a steep grade where a slower-moving vehicle can move to the side.

turnover *Biology.* the rate at which something is replaced, specifically the atoms or molecules in a living system. *Ecology.* **1.** the ratio of the number of individuals lost from a population, as by death or emigration, to the number replaced within a given time. **2.** a process by which a species becomes extinct in an area and is replaced by another species. *Hydrology.* the circulation of water in deep temperate lakes to equalize the temperature throughout the water column. *Robotics.* a measure of how quickly a process is able to turn raw material into a finished product. *Graphic Arts.* a line of type that continues onto the next page or line. Also, RUNOVER.

turnover cartridge *Acoustical Engineering.* a phonograph cartridge that is designed to play both large-groove and fine-groove records, having a stylus for each type mounted on its opposite sides; the desired stylus is selected by pushing a lever down.

turnover number *Chemical Engineering.* the quantity of substrate or feed converted by a measured amount of catalyst. *Molecular Biology.* the number of molecules of a substrate that are transformed per minute by a single enzyme molecule when the enzyme is working at its maximum rate. Also, MOLAR ACTIVITY.

turnover rate *Chemical Engineering.* a value that corresponds to the turnover number during a specified unit of time.

turnover time *Ecology.* **1.** the time it takes for a complete cycle to occur within a system. **2.** the life span of an individual organism, from birth to death.

turnpike *Civil Engineering.* **1.** a former term for a public highway. **2.** a limited-access highway, especially one on which tolls are charged; in modern usage the term is largely limited to the eastern United States.

turnscrew *Mechanical Devices.* any device used for driving screws such as a screwdriver or wrench.

turns ratio *Electricity.* the ratio of the number of primary turns to the number of secondary turns in a transformer.

turnstile antenna *Electromagnetism.* an antenna having two dipole elements perpendicular to each other that intersect at their midpoints; used in ultrahigh-frequency transmitters to obtain an omnidirectional radiation pattern.

turnstone *Vertebrate Zoology.* a migratory shorebird of the genus *Arenaria* of the family Scolopacidae, characterized by long legs and bill; named for its practice of turning over stones in search of food.

turntable *Acoustical Engineering.* a round, motor-driven platform on top of a record player and below the stylus arm, upon which one or more records are placed for playing as the turntable rotates at a constant speed.

turntable rumble *Acoustical Engineering.* the low-frequency noise introduced into the sound picked up by a record-player cartridge, due to vibrations in the turntable.

Turonian *Geology.* a European geologic stage of the Middle or Upper Cretaceous period, occurring after the Cenomonian and before the Coniacian.

turpentine *Materials.* **1.** a sticky, viscous oleoresin obtained from coniferous trees, especially pine trees. Also, GUM TURPENTINE. **2.** a similar oleoresin obtained from the small European tree, the terebinth. Also, CHIAN TURPENTINE. **3.** a clear, volatile, essential oil distilled from these substances and used in paints, varnishes, and medicines. Also, OIL OF TURPENTINE, SPIRITS OF TURPENTINE. **4.** a similar essential oil obtained from the carbonization of pine wood. Also, WOOD TURPENTINE. **5.** to extract or apply turpentine.

turpentine camphor see TERPENE HYDROCHLORIDE.

turquoise [tur´kwōz´] *Mineralogy.* $Cu^{+2}Al_6(PO_4)_4(OH)_8 \cdot 4H_2O$, a semitranslucent, blue to green triclinic mineral of the turquoise group occurring as cryptocrystalline reniform masses and in veinlets and crusts, having a specific gravity of 2.6 to 2.8 and a hardness of 5 to 6 on the Mohs scale; formed as a secondary mineral, usually in arid regions, by the weathering of aluminous igneous rocks; used as a gemstone.

turret *Architecture.* a small tower, usually at the corner of a building and often beginning some distance above the ground. *Ordnance.* **1.** a dome-shaped or cylindrical armored structure that houses a gun or guns and usually revolves horizontally; turrets are mounted on fortifications, ships, airplanes, tanks, and other combat vehicles. Thus, **turret gun, turret mount. 2.** on a rifle scope, a part that holds the adjustable device for windage and elevation.

turret coal cutter *Mining Engineering.* a coal cutter with a horizontal jib that can be adjusted vertically to allow cutting at different levels in a seam.

turret lathe *Mechanical Engineering.* a high-production lathe in which the turret holds several tools and is mounted on a saddle that slides on the lathe bed; used for long workpieces.

turret robot *Robotics.* a tower-shaped robot for which the working envelope is a circle around the base of the robot.

turret tuner *Electronics.* a television tuner composed of a rotating cylindrical device on which resonant circuit elements are mounted; as the channel selector knob is turned, the turret tuner rotates to switch in the correct circuit element for the selected channel frequency.

Turrilites *Paleontology.* a helically coiled ammonoid in the superfamily Turrilitaceae; characterized by a gastropodlike appearance, prominent ribs, and some tubercles; extant in the Cretaceous.

turtle *Vertebrate Zoology.* **1.** any of numerous reptile species of the order Testudines, characterized by a bony or leathery shell enclosing the body and a toothless, horny beak. **2.** specifically, in popular use, an aquatic species of this type, as opposed to a terrestrial one (a tortoise). *Computer Programming.* a triangular cursor or physical plotting device used by the LOGO children's programming language, having somewhat the form of a turtle's shell. The users write programs consisting of commands to the turtle, such as "move forward" or "turn," causing it to draw pictures.

turtle

turtledove *Vertebrate Zoology.* any of several graceful, small to medium-sized doves of the genus *Streptopelia*, especially *S. turtur*, having a long, graduated tail and noted for its soft, cooing call; found in Europe.

turtlehead *Botany.* any of several perennial herbs belonging to the genus *Chelone* of the figwort family, characterized by opposite, serrated leaves and spikes of purple or white flowers; found in North America. (So-called because of the appearance of the flower.)

turtle stone see SEPTARIUM.

turunda *Surgery.* see TENT. *Pharmacology.* see SUPPOSITORY.

TUS *Aviation.* the airport code for Tucson, Arizona.

tusche *Graphic Arts.* a greasy liquid used to touch up images on lithographic press plates; this process is called **tusching.**

tusk *Vertebrate Zoology.* a long, pointed, projecting tooth, often one of a pair, as in the elephant and the walrus, but also occurring singly, as in the narwhal. *Building Engineering.* a sloping shoulder at the end of a timber for strengthening a tenon.

tussock *Ecology.* **1.** a thick, compact clump of grass or other vegetation. **2.** a mound of relatively solid ground in an otherwise swampy area, formed by plant roots that have bound together.

tutorial *Computer Programming.* a series of lessons, classes, or demonstrations that are contained within a computer system or in its documentation.

Tutte's f-factor theorem *Mathematics.* let f be an assignment of a nonnegative integer to each vertex v of an undirected graph G such that $f(v) \le \deg(v)$. For each vertex v of G, define $\hat{f}(v) = \deg(v) - f(v)$. For any subsets X and Y of $V(G)$, let $V(X, Y)$ denote the set of all edges of G having one end in X and the other end in Y. Then G has an f-factor if and only if for every two disjoint subsets X and Y of $V(G)$, the number of those components H of $G - (X \cup Y)$ for which $f(V(H)) + |V(X,Y)|$ is odd does not exceed $f(X) + \hat{f}(Y) - |V(X,Y)|$, where $f(X)$ is the sum of the values of f on the vertices in X.

Tutte's Hamiltonian theorem *Mathematics.* the theorem that every 4-connected planar graph is Hamiltonian.

Tutte's matching theorem *Mathematics.* let $c_o(H)$ be the number of odd components of the graph H, where a component is odd if it has an odd number of vertices, and let $|X|$ be the cardinality of set X. A graph with vertex set V has a perfect matching if and only if $c_o(G - S) \le |S|$ for all $S \subseteq v$.

tuya *Volcanology.* a steep-sided, flat-topped volcano.

tuyere *Metallurgy.* a refining process in which air is blown through a tube or opening in a metallurgical furnace.

TV *Telecommunications.* a shorter term for a television system or television set. See TELEVISION and related entries.

Tv virtual temperature.

TVA Tennessee Valley Authority.

TV camera scanner *Computer Technology.* an input device that uses a TV camera to optically scan characters or images, converting the image to binary, gray-scale, or color pixel arrays.

TVI television interference.

TVM *Ordnance.* track via missile, an antimissile system that tracks an enemy missile on the basis of the energy received both by the ground-based radar unit and by the antimissile missile itself.

TV set see TELEVISION RECEIVER.

TW *Aviation.* the airline code for Trans World Airlines (TWA).

TW terawatt.

TWA traveling-wave amplifier.

Twaddell scale *Engineering.* a scale used for the measure of specific gravity.

tweed *Textiles.* a rough-surfaced woolen material in a plain, twill, or herringbone twill weave. Yarn for tweed is usually dyed before weaving and then woven in two or more colors. (Originally handwoven in the area near the *Tweed* River, which separates England from Scotland.)

tweeter *Acoustical Engineering.* a small loudspeaker designed to produce high-frequency sound; often mounted in a speaker housing with an equal number of woofers.

twilight *Astronomy.* **1.** the interval between sunset or sunrise and the moment when the sun is 6° beneath the horizon (**civil twilight**), 12° (nautical twilight), or 18° (**astronomical twilight**). **2.** the soft diffuse light in the sky when the sun is below the horizon either at daybreak or at nightfall. In both the above meanings *twilight* more commonly refers to sunset than to sunrise.

twilight arch see BRIGHT SEGMENT.

twilight compass see SKY COMPASS.

twilight correction *Astronomy.* the difference in time between sunrise and when a sunshine recording device begins to record sunshine; a similar correction is applied at sunset under conditions of a clear horizon.

twilight phenomenon *Optics.* any of various phenomena that occur in the atmosphere during twilight, such as the green flash, crepuscular rays, and the anti-twilight arch.

twilight state *Neurology.* a state of disordered consciousness characterized by the performance of normal acts of which there is no subsequent recollection; this generally occurs in temporal lobe epilepsy and more rarely in delirium.

twilight zone *Astronomy.* the region of earth over which it is twilight at any given moment. *Electromagnetism.* anything resembling the twilight zone of the earth, as in a four-course radio range station in which one signal is barely heard above the monotone on-course signal.

twill *Textiles.* any fabric or garment constructed in the twill weave.

twill weave *Textiles.* one of the three basic weaves (the others being plain and satin weave), characterized by a diagonal rib, generally running upward from left to right. The rib is formed by a weaving process in which an end thread passes over two consecutive picks before weaving under, with the point of intersection moving one outward on each succeeding end.

twin *Biology.* either of two offspring produced by the same mother in the same pregnancy and developed either from one ovum (**monozygotic** or **identical twins**) or from two ova that are fertilized at the same time (**dizygotic** or **fraternal twins**). *Crystallography.* a composite crystal built from two or more crystal specimens that are grown together in a specific manner so that there is at least one plane and a direction perpendicular to it that are related in the same manner to the crystallographic axes of both parts of the twin.

twin arithmetic units *Computer Technology.* a feature whereby the arithmetic sections of a computer are virtually duplicated.

twin axial cable *Telecommunications.* a transmission line composed of two coaxial cables in a single sheath, each transmitting in the same direction.

twin axis *Crystallography.* an axis about which reflection can be observed for a pair of twin crystals having rotational symmetry between them.

twin band *Metallurgy.* the section that runs though a twin and parent crystal on a polished surface.

twin boundary *Crystallography.* the interface between the twin and the matrix.

twin-cable ropeway *Mechanical Engineering.* an aerial ropeway with parallel track cables on which carriers run in opposite directions.

twin check *Computer Technology.* a method of checking a computer operation in which hardware is duplicated and the results are automatically compared.

twin crystal *Crystallography.* a crystal composed of two or more individuals, either in contact or intergrown, in a systematic crystallographic orientation with respect to one another. Also, **twinned crystal.**

twine *Materials.* a relatively strong string or cord composed of two or more strands twisted together. *Botany.* to climb by twisting around objects for support.

twin entry *Mining Engineering.* a system of two parallel entries, one for the intake air flow and the other for the return; rooms may be worked from both entries.

twiner *Botany.* a plant that climbs by twisting its pliable stem around objects for support.

twinge *Neurology.* a brief, sharp pain.

twin-geared press *Mechanical Engineering.* a crank press with the drive gears attached to both ends of the crankshaft.

twinkling *Astronomy.* a phenomenon that is caused by turbulence in the earth's atmosphere, in which constant fluctuations in size, brightness, and color can be seen in stars.

twin law *Crystallography.* a definite pattern or arrangement in terms of the axes, angles, and planes of two or more crystals of a given crystalline material.

twinning *Crystallography.* the intergrowth of two or more crystals of a given crystalline material according to a twin law.

twinning deformation *Materials Science.* in some metals under certain conditions, a plastic deformation process in which atoms are displaced to form a mirror image of the undeformed region.

twin plane *Crystallography.* the crystallographic plane of symmetry between a twin and a matrix. Also, **twinning plane.**

twin-roller method *Materials Science.* a method for rapid quenching of alloy melts by rolling between two chilled rollers to produce a ribbon. Heat is abstracted from both sides of the splat at the same time, as the rollers constantly provide a new chill-block suface.

Twins *Astronomy.* see GEMINI.

twin-screw compounding system *Materials Science.* a system for mixing polymers with additives such as pigments, lubricants, stabilizers, cross-linking agents, and plasticizers, using twin screws to accomplish parallel extrusion and mixing often by mass transport involving volatile removal or chemical reactions. The screws may be corotating (both screws turning in the same direction) or counterrotating (two screws turning in opposite directions) intermeshed or not.

twin-screw extrusion see MULTISCREW EXTRUSION.

twin studies *Psychology.* research carried out on twins, with the general purpose of examining the issue of heredity vs. environment with respect to human behavior and personality traits.

twin-T filter see PARALLEL-T NETWORK.

twist *Design Engineering.* the number of turns per inch (or other unit length) that a strand (of yarn, rope, cable, and so on) takes about its axis. *Aviation.* an airfoil configuration having varying blade angles or angles of attack along its span or length. With *washin,* twist is positive; with *washout,* it is negative. *Robotics.* a degree of freedom based on rotational displacement around an axis. *Electromagnetism.* a section of waveguide whose cross section has a gradual rotation about its longitudinal axis.

twist boundary *Materials Science.* a small-angle grain boundary composed of an array of screw dislocations.

twist drill *Mining Engineering.* a drill having at least one, but often two, longitudinally cut spiral cutting edges capable of clearing cut material up and away from a borehole or applying coolant to such debris.

twisted pair *Electricity.* a two-wire signal cable in which the wires are twisted around each other the entire length of the cable; often used for communications.

twister *Meteorology.* an informal term for a tornado or cyclone. *Solid-State Physics.* a piezoelectric crystal that generates a voltage when subjected to twisting.

twisting number *Molecular Biology.* in relaxed closed-circular DNA, the number of base pairs in the molecule divided by the number of base pairs per revolution of the helix.

twist-lock connector *Electricity.* a safeguard designed to prevent a power plug from accidentally disengaging; the plug is twisted after insertion and locked in place.

twist of rifling *Ordnance.* the relationship of the helical grooves to the axis of the bore; it is expressed as the number of inches or calibers of length in which the rifling makes one complete spiral. Also, PITCH.

twistor *Computer Technology.* a nonvolatile computer memory element consisting of magnetic wire wound onto a short piece of nonmagnetic wire; used especially in telephone switching equipment. Also, **twistor memory.**

twitch *Neurology.* a brief contractile response of a skeletal muscle elicited by a single maximal volley of impulses in the motor neurons supplying it. Thus, **twitching.**

Twitchell reagent *Organic Chemistry.* a sulfonated petroleum or naphthalenestearosulfonic acid used as a catalyst for the acid hydrolysis of fats.

twitching motility *Microbiology.* an uncoordinated, jerky form of passive motility, exhibited by certain bacterial cells in thin films of surface liquid.

two-address code *Computer Programming.* a machine code that uses two-address instructions.

two-address instruction *Computer Programming.* an instruction that contains one operation code and two address parts, usually indicating the memory locations of two operands, either directly or indirectly.

two-beam interference *Physics.* any interference phenomena occurring between two beams or waves.

two-body force *Physics.* any force that acts between two particles or bodies and is not influenced by the presence of any other entities in the region of interest.

two-body problem *Mechanics.* a problem that deals with the motion of two mutually interacting bodies; it can be solved by reformulating the equations of motion into a single equation using the reduced mass.

two-by-four *Forestry.* an untrimmed timber measuring 2 by 4 inches in cross section. *Building Engineering.* a board trimmed from such a timber, measuring 1-5/8 by 3-5/8 inches.

two-carrier theory *Solid-State Physics.* a theory of conduction in a semiconductor that considers the simultaneous contributions of both electrons and holes to current flow.

two-color printing *Graphic Arts.* nonprocess color printing in which the second color is superimposed, usually over black.

two-component model *Fluid Mechanics.* a dynamical model of a gas flow in which the gas has two components of different temperatures.

two-component neutrino theory *Particle Physics.* a theory stating that both neutrinos and antineutrinos have zero rest mass and that a neutrino's spin is always antiparallel to its motion whereas an antineutrino's spin is always parallel to its motion.

two-cycle engine *Mechanical Engineering.* an internal-combustion engine in which only two strokes of the piston, corresponding to one revolution of the crankshaft, complete a single cycle.

two-degrees-of-freedom gyro *Mechanics.* a gyroscope whose axis of spin can freely rotate about two orthogonal axes.

two-dimensional *Science.* having the dimensions of length and width, but not that of depth. Thus, **two-dimensionality.**

two-dimensional chromatography *Analytical Chemistry.* a separation technique in which the sample is resolved by standard procedures on chromatography paper that is turned at right angles in a second solvent and resolved again.

two-dimensional electron gas *Solid-State Physics.* an array of free electrons confined to two dimensions. These occur at (a) an interface region in a heterostructure, the junction of two semiconductors of differing band gaps; (b) the oxide-semiconductor interface of a metal-oxide-semiconductor junction; or (c) the surface of liquid helium when excess electrons are deposited on the surface.

two-dimensional electrophoresis *Physical Chemistry.* the separation of molecular mixtures into fractions, due to the influence of an electrical field and with the separation conducted in a geometry that is very thin in one dimension.

two-dimensional flow *Fluid Mechanics.* the idealized flow of particles in parallel planes along identical paths in each of these planes; there is no flow and no change in fluid properties normal to these planes.

two-dimensional gel electrophoresis *Biotechnology.* a flat-bed electrophoretic technique in which the compounds are first electrophoretically separated in one direction, rotated by 90°, and then separated in a second direction.

two-dimensional storage *Computer Technology.* a direct-access storage device in which all the contents of a single file need not be stored in physically adjacent areas.

two-factor theory *Psychology.* the theory that two factors affect the degree of competence for the performance of any test or mental task, one a generalized mental capacity and the other a specific ability for the test or task.

two-finger gripper *Robotics.* a common type of end effector capable of grasping objects with two opposing digits.

two-fluid cell *Physical Chemistry.* any cell that has two different electrolytic solutions, one positive and one negative.

two-gap head *Computer Technology.* a head on a magnetic tape device with two gaps, one for reading and the other for writing. This allows reading what was just written for error checking.

two-hop transmission *Telecommunications.* a form of radio wave propagation in which waves are reflected from the ionosphere, then reflected from the earth, and then again reflected by the ionosphere before arriving at the receiver.

two-input subtractor see HALF-SUBTRACTOR.

two-layer ocean *Oceanography.* an ideal ocean model in which a surface layer of uniform density overlies a deep layer of uniformly greater density.

two-level mold *Engineering.* a mold that places one molding cavity on top of another, rather than alongside it, to reduce the clamping force required.

two-lip end mill *Mechanical Engineering.* an end milling cutter that has two cutting edges and straight or helical flutes.

two-man rule *Military Science.* a system to prohibit access by an individual to nuclear weapons or designated components by requiring the presence of two authorized persons at all times. Also, **two-man concept, two-man policy.**

two-out-of-five code *Computer Programming.* a weighted decimal code in which decimal digits are represented by five binary digits, two of which must be 1's and the other three 0's (or vice versa).

two-part code *Telecommunications.* a randomized code that consists of an encoding section in which plain text groups are arranged in a significant order coupled with their code groups in random order, and a decoding section in which code groups are arranged in significant order coupled with their meanings from the encoding section.

two-pass assembler *Computer Programming.* an assembler that passes the input program twice; the first pass resolves symbols and macros, and the second pass translates the instructions.

two-phase *Physics.* describing a system having two oscillating signals that have equal frequencies and are out of phase by 90°.

two-phase circuit *Electricity.* describing a device or circuit having two supply voltages or currents, one of which leads the other by 1/4 cycle of 90°.

two-phase five-wire system *Electricity.* a two-phase four-wire system with the fifth wire for the neutral points of each phase; the neutral is usually grounded.

two-phase flow *Fluid Mechanics.* any combination of two distinct phases under flow conditions: gas-liquid, liquid-liquid, liquid-solid particles, or gas-solid particles. *Physics.* **1.** the flow of helium II when it is assumed to be composed of two fluid components: a superfluid component (having no viscosity) and a normal fluid component. **2.** the electric current flow of electrons in a superconductor when it is assumed to be composed of two components: a superconductive component (having no electrical resistance) and an ordinary component of conduction electrons.

two-phase four-wire system *Electricity.* a two-phase system in which each phase is provided on a separate pair.

two-phase material *Materials Science.* any material composed of two phases.

two-phase three-wire circuit see THREE-WIRE SYSTEM, def. 1.

two-phase three-wire system *Electricity.* a two-phase line having a neutral which is often connected to ground.

two-piece set *Mining Engineering.* a timber set composed of a single post and a cap.

two-plus-one address instruction *Computer Programming.* an instruction that contains three address parts, the first two being for operands and the last being the address of the next instruction to be executed.

two-point cross *Genetics.* a cross involving two pairs of linked genes; it may be between a double mutant and a wild type for the genes under consideration, or between two single mutants, each carrying the wild type gene of the mutant gene carried by the other.

two-point press *Mechanical Engineering.* a mechanical press having a slide that is activated by two connections.

two-port junction *Electromagnetism.* any waveguide junction having two openings, a discontinuity and an obstacle in a waveguide.

two-port system *Control Systems.* a system that only has one input and one output.

two-pulse canceler *Electronics.* a circuit used in the moving-target indicator of a radar set; permits comparison of two successive returns from the same target to determine if the target is moving with respect to the radar set.

two-quadrant multiplier *Computer Technology.* in analog computers, a multiplier in which operation is restricted to a single sign of one input variable only.

two-range Decca *Navigation.* a Decca radio navigation system that provides circular lines of position; distinguished from the standard Decca system, which provides hyperbolic lines of position.

two-sample test *Statistics.* a test of differences between features of two populations based on a comparison of samples from each of the populations.

two's complement *Mathematics.* given a number x expressed in base 2 with n digits to the left of the radix point, the (base 2) number $2^n - x$.

two-sided ideal *Mathematics.* an ideal that is both a left ideal and a right ideal; that is, a subring I of a ring R such that both ri and ir are in I for every $r \in R$ and $i \in I$.

two-sided test see TWO-TAILED TEST.

two-source frequency keying *Telecommunications.* a frequency change signaling process in which the change from one signaling condition to another is accompanied by decay in amplitude of at least one frequency and by the heightening of amplitude of at least one other frequency. Also, FREQUENCY EXCHANGE SIGNALING.

two-stage design *Statistics.* a design in which a pilot test of small size is conducted to determine the design of a subsequent full-scale study.

two-stage digester *Biotechnology.* an anaerobic digester that separates acid-forming reactions and methanogenic reactions.

two-stage experiment *Statistics.* an experiment having two stages, where the results from the first stage determine the procedure to be carried out in the second stage.

two-stage rocket *Space Technology.* a rocket with two propulsion systems, each having its own housing and fuel source; the upper stage fires after the lower stage burns out and separates.

two-stage sampling *Statistics.* a type of sampling based on basic and secondary units. The basic units (e.g., states of the United States) are first sampled, then subsampled with respect to their secondary or subordinate units (e.g., counties).

two-state Turing machine *Computer Programming.* a variation on the Turing machine model in which the number of states has been restricted to two.

two-step grooving system *Design Engineering.* a method of winding wire or rope in which for one-half turn the rope is parallel to the drum's flanges and then it crosses over to begin the next wind. Also, counterbalance system.

two-step rocket *Space Technology.* **1.** a rocket that is boosted by a rato unit before its main propulsion unit is activated. **2.** see TWO-STAGE ROCKET.

two-step test *Medicine.* a test for coronary sufficiency in which the subject is required to take two steps up repeatedly for 90 seconds; the number of steps taken are compared against the norm for the age group, sex, and weight.

two-stroke cycle *Mechanical Engineering.* an internal-combustion engine cycle, completed in two strokes, in which the piston compresses the fuel mixture on one side and receives the thrust of the previously compressed gases on the other side during the first stroke, and draws in a fresh charge on one side while expelling burnt gasses on the other during the second stroke.

two-symbol Turing machine *Computer Programming.* a variation on the Turing machine model in which the number of symbols has been restricted to two.

two-tailed test *Statistics.* a test in which the rejection of the hypothesis may take place at either extreme of the range of values of the statistic. Also, DOUBLE-TAILED TEST, TWO-SIDED TEST.

two-tone diaphone *Acoustical Engineering.* a loudspeaker that is designed to produce only two pitches at high intensities, used for warning or distress signals.

two-tone keying *Telecommunications.* a type of keying in which the modulating wave causes the carrier to be modulated with one tone for the marking condition and with a different tone for the spacing condition.

two-tone modulation *Telecommunications.* a type of modulation in which two different carrier frequencies are utilized for the two signaling conditions; in teletypewriter operation, the transition between the two frequencies is abrupt, resulting in phase discontinuities.

two-up *Graphic Arts.* of a form, printed in duplicate on a single sheet.

two-valued logic *Mathematics.* logic in which statements can only be either true or false.

two-way shape memory *Materials Science.* a spontaneous reversible shape change that can be "trained into" certain alloys, associated with heating and cooling within the transformation temperature range. As the alloy is cycled between two temperatures, the reversible shape change might be twisting and untwisting, or bending and unbending.

two-way valve *Mechanical Engineering.* a valve that allows fluid to flow in either direction.

two-wire circuit *Electricity.* a circuit consisting of two devices interconnected by two wires. In communications, a two-wire circuit allows only one direction of transmission at a time.

two-wire electric arc metallizing *Materials Science.* a technique for producing a sprayed metallic coating for surface protection against corrosion: two wires are fed from supply spools to form the electrodes of an electric arc; the wires melt in the arc and are atomized by a blast of compressed air.

two-wire repeater *Telecommunications.* in a telephone system, a piece of equipment that allows bidirectional transmission over a two-wire line.

TWT traveling-wave tube.

TWX machine see TELETYPEWRITER.

TWX service see TELETYPEWRITER EXCHANGE SERVICE.

Twyman-Green interferometer *Optics.* a Michelson interferometer that uses a point light source, rather than an extended light source.

Twystron *Electronics.* a high-power, wide-band electron tube that incorporates elements of a traveling-wave tube and a klystron in the same envelope; the Twystron has a very high operating efficiency. (A shortened form of traveling-wave klystron tube.)

ty type.

tychite *Mineralogy.* $Na_6Mg_2(CO_3)_4(SO_4)$, a colorless to white cubic mineral occurring in small transparent octahedrons, having a specific gravity of 2.549 and a hardness of 3.5 to 4 on the Mohs scale; forms a series with ferrotychite; found in saline lake deposits.

Tycho [tī´kō] *Astronomy.* a lunar crater 85 kilometers in diameter that is the origin for a bright ray pattern conspicuous around the full moon.

Tychonic system [tī kän´ik] *Astronomy.* a noted early model for planetary motion devised by the Danish astronomer Tycho Brahe, in which the earth is stationary while the moon and the sun (around which the planets revolve) orbit the earth.

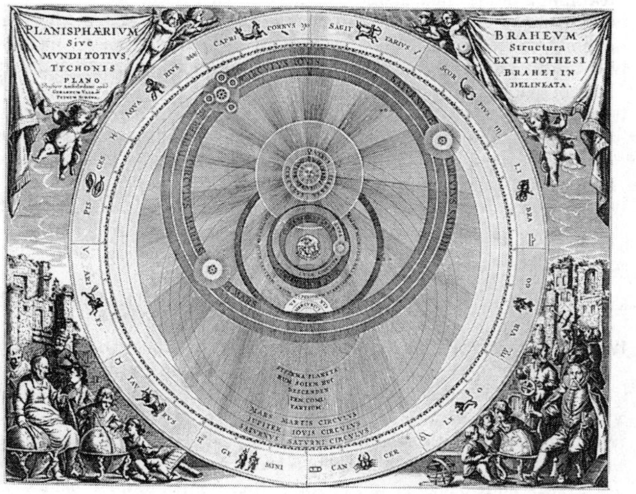

Tychonic system

Tychonoff theorem *Mathematics.* the theorem that the topological product of any family of compact spaces is compact. The theorem is true even if the family has an uncountable number of members.

Tycho's supernova *Astronomy.* a star that exploded November 11, 1572, in Cassiopeia; the remnant of the supernova is known to emit radio noise and X-rays. Also, **Tycho's star.**

tyfon see TYPHON.

tying bar see TOMBOLO.

Tylenchida *Invertebrate Zoology.* an order of nematodes in the subclass Rhabdita, having a protrusible stylet; includes free-living soil dwellers, parasites of insects, and juice-sucking parasites of plants, some of which are serious pests.

Tylenchoidea *Invertebrate Zoology.* a superfamily of nematode roundworms in the order Tylenchida.

Tylenol *Pharmacology.* the trade name for preparations of acetaminophen, a widely distributed pain reliever. See ACETAMINOPHEN.

Tyler screen *Chemical Engineering.* a standard for the number of openings in screenlike media founded on the number of meshes per linear inch.

Tyler standard screen scale *Engineering.* a scale used to classify particles by correlating the particle's size with the mesh per inch of a screen.

Tylopoda *Vertebrate Zoology.* an infraorder of mammals of the order Artiodactyla that includes the camels, llamas, and their extinct relatives.

Tylor, Edward Burnett 1832–1917, English anthropologist; often described as the founder of cultural anthropology.

tylosis *Plant Pathology.* an abnormal growth plugging the xylem vessel of a plant, and subsequently restricting the flow of water; may be caused by disease, injury, or adverse environmental conditions. Thus, **tylotic.**

tylostyle *Invertebrate Zoology.* a sponge spicule with a point at one end and a knob at the other.

tylote *Invertebrate Zoology.* a sponge spicule having a knob at each end.

Tymovirus group *Virology.* a genus of plant viruses having isometric particles containing linear ssRNA genomes; they are transmitted mechanically and by beetles, with infection causing characteristic clumping of the chloroplasts in most cell types. (From turnip yellow mosaic virus, the type member for this genus.)

tympanic cavity [tim pan´ik] *Anatomy.* a technical name for the middle ear. See MIDDLE EAR.

tympanic membrane *Anatomy.* a technical name for the eardrum. See EARDRUM.

tympanites see METEORISM.

tympanoplasty *Surgery.* any of various surgical procedures performed on the eardrum or middle ear to restore or improve hearing in patients with conductive deafness.

tympanum [tim pan´əm] *Anatomy.* **1.** see TYMPANIC MEMBRANE. **2.** see TYMPANIC CAVITY. *Architecture.* **1.** in a classical pediment, the recessed triangular surface enclosed by the cornices. **2.** the space between a doorway lintel and a surmounting arch. (Going back to a Greek word meaning "drum.")

Tyndall, John 1820–1893, British physicist and natural philosopher; discovered the Tyndall effect; examined *Penicillium.*

Tyndall cone *Optics.* the visible cone of light in a medium caused by the Tyndall effect.

Tyndall effect *Optics.* the generally bluish, visible scattering of light in a medium that contains extremely small particles.

Tyndall figure *Hydrology.* a small figure resembling a snowflake, found within an ice crystal. Also, **Tyndall flower, Tyndall star.**

Tyndallization *Engineering.* heat sterilization through the application of repeated short interval amounts of steam heat.

TYO *Aviation.* the airline code for the city of Tokyo, Japan.

type a kind or class of person, thing, condition, and so on; specific uses include: *Agriculture.* **1.** the characteristics of an animal or breed that make it desirable for a given purpose. **2.** a breed, strain, or variety of animal, or an individual belonging to such a group. *Systematics.* see NOMENCLATURAL TYPE. *Archaeology.* a classification based on the shared attributes of groups of artifacts or features, such as pottery types, projectile point types, or house types. *Hematology.* see BLOOD GROUP.

type *Graphic Arts.* **1.** a rectangular metal block engraved with a letter or other character, an impression of which is used to produce printed text. **2.** an assemblage of such blocks forming one or more lines. **3.** an impression made from such assemblages. **4.** printed text produced by a phototypesetting machine. **5.** any text material on a printed page, as opposed to illustrations or other design elements.

type A behavior *Psychology.* behavior exhibited by individuals who are intensely competitive, aggressive, and impatient, with a sense of time urgency and a strong need for recognition and advancement.

type-ahead buffer *Computer Technology.* **1.** in many computer systems, a buffer that allows the user to enter upcoming commands or other keystrokes via the keyboard prior to a prompt by the system. **2.** in a word processor, a temporary storage device that is automatically accessed when the operator types characters faster than they can be displayed on the screen.

type-alpha leader *Geophysics.* a type of stepped leader that shows little branching and has dimly lit, short steps. Also, TYPE-X LEADER.

type 1 antigen *Immunology.* a histocompatibility antigen encoded in humans by specified loci (A, B, and C). Also, CLASS I ANTIGEN.

type 2 antigen *Immunology.* a histocompatibility antigen encoded in humans by specified loci (generally DR, DP, and DQ). Also, CLASS II ANTIGEN.

type A particles see A-TYPE VIRUS PARTICLES.

type A personality *Psychology.* a personality pattern characterized by type A behavior traits, such as excessive drive and aggressiveness, competitiveness in both work and leisure activities, and a need for control; thought to correlate with a predisposition toward heart disease and other stress-related disorders.

type area or **type page** *Graphic Arts.* a rectangular area containing all the printed elements on a page, usually measured in picas. Also, TEXT AREA.

type A wave *Telecommunications.* a continuous wave.

type A1 wave *Telecommunications.* a continuous wave that is unmodulated and keyed.

type A2 wave *Telecommunications.* a continuous wave that is modulated and keyed.

type A3 wave *Telecommunications.* a continuous wave that is modulated by speech, music, or other sounds.

type A4 wave *Telecommunications.* in facsimile systems, a superaudio-frequency-modulated continuous wave.

type A5 wave *Telecommunications.* in television transmission, a superaudio-frequency-modulated continuous wave.

type A9 wave *Telecommunications.* a continuous wave in a composite transmission or situation that is different from type A1, A2, A3, A4, or A5 waves.

type B behavior *Psychology.* behavior that is characteristic of individuals who are more relaxed, patient, and self-fulfilled than those who exhibit type A behavior. Thus, **type B personality.**

type-beta leader *Geophysics.* a type of stepped leader in which the upper portion has longer and brighter steps than the lower portion.

Type B Oncovirus group *Virology.* a genus of the family Retroviridae, subfamily Oncovirinae, that includes mouse mammary tumor virus, characterized by the unique intracytoplasmic particles (Type A) that mature to Type B particles after budding.

type B wave *Telecommunications.* a damped, keyed wave.

typecase *Graphic Arts.* a compartmentalized wooden tray designed to hold metal type.

type C1 carbonaceous chondrite *Geology.* a variety of carbonaceous chondrite having a relatively low density, strong magnetic properties, and containing approximately 3.5% carbon and some sulfates.

type C2 carbonaceous chondrite *Geology.* a variety of carbonaceous chondrite having a relatively high density and low water content, and generally composed chiefly of olivine.

type C3 carbonaceous chondrite *Geology.* a variety of carbonaceous chondrite low in magnetic properties, and containing approximately 2.5% carbon and some sulfur which generally occurs as free sulfur.

type coercion *Computer Science.* the automatic conversion of data from its existing type into the type required for an operation.

Type C Oncovirus group *Virology.* a genus of the family Retroviridae, subfamily Oncovirinae, that includes avian-leukosis-sarcoma, murine leukemia and sarcoma viruses, and avian reticuloendotheliosis virus; characterized by the assemblage of virions beneath the cell plasma membrane.

type culture *Microbiology.* the culture strain that is designated to represent a newly named species of microorganism.

Type D retrovirus group *Virology.* a group of enveloped ssRNA-containing virus particles found near the plasma membrane. Also, D-TYPE VIRUS PARTICLES.

type drum *Computer Technology.* in a drum printer, a rotating steel cylinder containing 128 to 144 bands, each of which carries a complete type set.

Type I error *Statistics.* in hypothesis testing, the rejection of the null hypothesis when it is actually true. Also, FALSE REJECTION.

Type II error *Statistics.* in hypothesis testing, the acceptance of the null hypothesis when it is actually false. Also, FALSE ACCEPTANCE.

typeface *Graphic Arts.* a set of type characters having a distinct and common design, such as Times Roman or Helvetica.

type font *Graphic Arts.* see FONT.

type form *Graphic Arts.* see FORM.

type fossil *Archaeology.* a particular artifact form used to define a specific period or culture, such as an Acheulian handaxe. Also, FOSSIL DIRECTEUR.

type gauge *Graphic Arts.* a ruler used to measure lines of type, usually marked in points, picas, agates, and inches. Also, LINE GAUGE, PRINTER'S RULE.

type genus *Systematics.* a genus that is determined to be typical of a higher classification to which it belongs.

type height *Graphic Arts.* the distance from the base of a type character to its printing surface (a standard 0.918 inches in the U.S.); any object this height is said to be **type-high.**

type hierarchy *Computer Science.* **1.** in an object-oriented system, the hierarchy of data types formed by the class-superclass relationships. **2.** in general, a hierarchy of data types formed by containment by higher types; e.g., integers are a subset of reals, which are a subset of complex numbers.

type locality *Geology.* **1.** the site where a particular stratigraphic unit typically occurs and from which it derives its name. **2.** the site where a geologic feature was first identified and studied. *Systematics.* the geographic location from which a type specimen or series was collected.

type metal *Metallurgy.* any of various alloys used to make printing type, containing 2–20% tin, 2–28% antimony, and 54–95% lead.

type I reaction (immediate) *Immunology.* a hypersensitive reaction that is manifested in allergic asthma, hay fever, and excema, developing within minutes after exposure to an antigen.

type II reaction (antibody-mediated) *Immunology.* a hypersensitive reaction that is caused by an antibody reaction to cell surface antigens; occurring when red blood cells are destroyed in transfusion reactions.

type III reaction (immune complex mediated) *Immunology.* a hypersensitive reaction that is caused by the deposit of antigen-antibody complexes in tissue and blood vessels.

type IV reaction (delayed) *Immunology.* an antibody-independent hypersensitive reaction that is associated with T lymphocytes and results in the production of a lymphocyte-macrophage granuloma.

type V reaction (stimulatory) *Immunology.* a hypersensitive reaction in which autoantibodies stimulate a host tissue.

Type I rehearsal see MAINTENANCE REHEARSAL.

Type II rehearsal see ELABORATION.

type section see STRATOTYPE.

type series *Systematics.* the collection of specimens taken from the type locality and from the original description of the species. Also, **type material.**

typesetting *Graphic Arts.* the process of setting characters in type. Also, COMPOSITION.

type size *Graphic Arts.* the size of a font as measured in picas, usually from the top of the highest character to the bottom of the lowest.

type species *Systematics.* the species designated as the nomenclatural type of a genus or subgenus; i.e., the species determining the genus or subgenus name.

type-specific antigen see GROUP-SPECIFIC ANTIGEN.

type specimen *Systematics.* the single specimen or one of a series of specimens chosen from the type series to represent a newly described species or subspecies, and upon which the valid description of the species or subspecies is based. Also, PROTOTYPE.

type I superconductor *Physics.* a superconductive material that has a single critical magnetic field strength, below which there cannot exist any magnetic flux within the interior of the substance and above which the flux penetrates the material and thus reverts it to its normal state.

type II superconductor *Physics.* a superconductive material that has two critical magnetic field strengths: below the lower strength, the substance behaves as a superconductor (all magnetic flux is excluded from the interior); between the two critical field strengths, flux vortices may exist in the material; and above the higher critical state the material behaves as an ordinary substance.

type I supernova *Astronomy.* a supernova with an absolute magnitude of –19 that occurs in an elliptical or spiral galaxy; its spectrum is characterized by broad absorption lines and a gas ejection velocity of about 10,000 kilometers per second.

type II supernova *Astronomy.* a supernova with an absolute magnitude of –17 that occurs only in the disk of a spiral galaxy; its spectrum shows the same abundance of elements as the Sun and a gas ejection velocity of about 5000 kilometers per second.

type-variety analysis or **type-variety system** *Archaeology.* a method of classifying artifacts, such as pottery, according to both their type and variety. Similarly, **type-variety-mode analysis.**

typewriter *Mechanical Devices.* a machine that prints characters mechanically (i.e., by impact) on paper. *Graphic Arts.* a typeface, such as Courier, whose characters resemble those of a typewriter.

typewriter terminal *Computer Technology.* a terminal that can either be used with a keyboard and a printer for telegraphic communications and data entry or stand alone as an office typewriter.

type-X leader see TYPE-ALPHA LEADER.

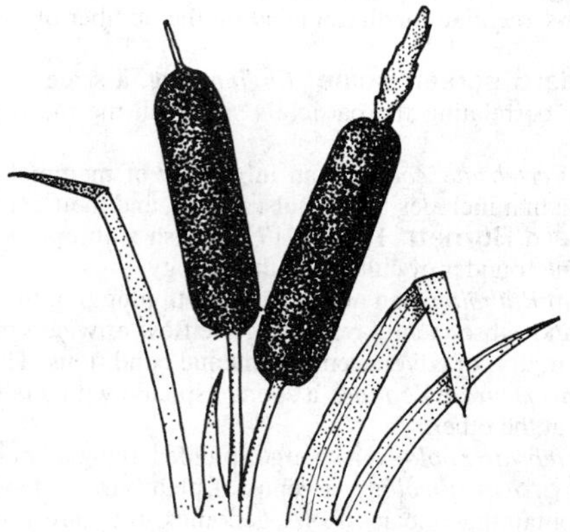

Typhaceae (cattail)

Typhaceae *Botany.* a family of monocotyledonous perennial herbs of the order Typhales, living in wet marshy places and characterized by a creeping rootstock, naked flowers in spikes, and long, linear parallel veined leaves.

Typhales *Botany.* an order of marsh or aquatic herbs of the subclass Commelinidae, characterized by rhizomes, linear leaves, and reduced, unisexual flowers found in clusters or spikes.

Typhlopidae *Vertebrate Zoology.* a family of small, blind, burrowing snakes of the suborder Serpentes; found in the tropics and subtropics worldwide.

Typhloscolecidae *Invertebrate Zoology.* a family of polychaete annelid worms in the group Errantia.

typhlosole *Invertebrate Zoology.* in various invertebrates, a longitudinal infolding of the intestinal wall that increases the digestive surface.

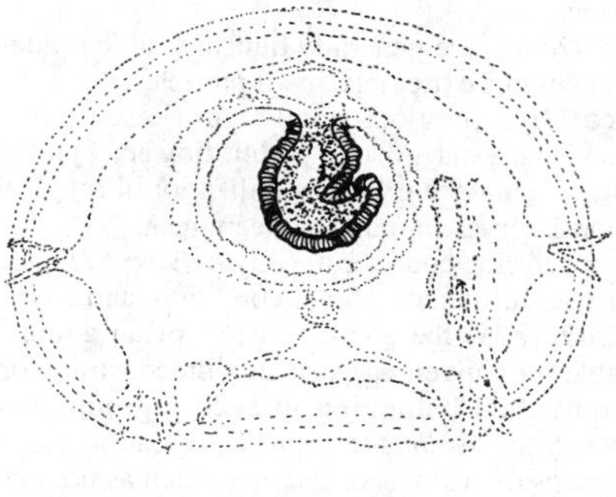

typhlosole

typhoid *Pathology.* **1.** see TYPHOID FEVER. **2.** resembling typhus.

typhoid fever *Pathology.* an acute, systemic illness caused by the bacteria *Salmonella typhi,* usually spread in contaminated food and water; characterized by inflammation and ulceration of lymph structures in the small intestine, enlargement of the spleen and certain lower abdominal lymph glands, and catarrhal inflammation of the intestinal mucous membrane; symptoms include prolonged hectic fever, skin rash, abdominal pain, delirium, and leukopenia. Also, ENTERIC FEVER.

typhoid vaccine see TAB VACCINE.

typhon *Acoustical Engineering.* a loudspeaker horn that is actuated by compressed air or steam. Also, TYFON.

Typhon *Ordnance.* a surface-to-air missile designed for shipboard deployment against high-performance aircraft and short-range tactical missiles; it carries a nuclear or nonnuclear warhead.

typhoon *Meteorology.* a severe tropical cyclone of the western Pacific Ocean and China Seas; a hurricane.

typhoon wind see HURRICANE WIND.

typhus *Pathology.* any of a group of infectious diseases caused by *Rickettsia,* transmitted by lice, fleas, mites, or ticks carried by infected rodents, and characterized by fever, chills, a dark red skin rash, severe headaches, and often delirium; various forms differ in the intensity of symptoms and severity. Also, **typhus fever.**

typhus degenerativus amstelodamensis see DE LANGE'S SYNDROME.

typical *Systematics.* relating to or serving as a nomenclatural type.

typical fiber *Mathematics.* the space to which all fibers belonging to a given bundle are homeomorphic.

typing *Systematics.* the process of determining the type category to which an entity belongs. *Microbiology.* any procedure by which two related strains of a given microorganism can be differentiated.

typogenesis *Evolution.* an explosive evolutionary period in which new forms are produced rapidly.

typographical error *Graphic Arts.* a misspelled word or similar error in printed text, made by the compositor; commonly called a **typo.**

typography *Graphic Arts.* **1.** a plan for the use of type in a printed piece. **2.** broadly, the typeface or other apparent qualities of typeset material. **3.** the art or process of using type. Thus, **typographical.**

typological method *Archaeology.* the classification of artifacts into types to compare artifacts or features across time and space, or to determine relative dates for sites.

typology *Archaeology.* **1.** a set of related types, such as a set of types of water jars. **2.** see TYPOLOGICAL METHOD.

typolysis *Evolution.* an evolutionary period preceding the extinction of a form.

typostatis *Evolution.* a comparatively static evolutionary phase in which few new forms are produced.

Typotheroidea *Paleontology.* an extinct suborder of South American ungulates in the order Notoungulata; extant from the Paleocene to Pliocene.

Tyr tyrosine.

tyramine *Pharmacology.* $C_8H_{11}NO$, a product of tyrosine decarboxylation that is structurally similar to epinephrine and neophrine, thus producing similar actions; found in decayed animal tissue, aged cheese, beer, and ergot.

Tyranni *Vertebrate Zoology.* a suborder of neotropical songbirds of the order Passeriformes, including the cotingas, plant cutters, manakins, tyrant flycatchers, and sharpbills.

Tyrannidae *Vertebrate Zoology.* the tyrant flycatchers, a diverse family of American songbirds characterized by a usually broad, flattened, hooked bill.

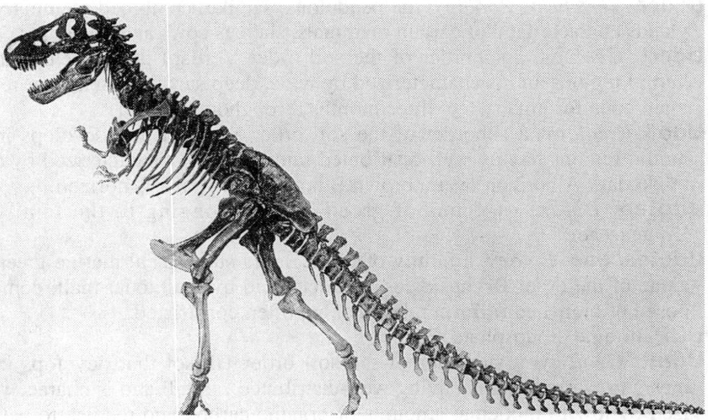

Tyrannosaurus

Tyrannosaurus [tī´ran ə sôr´əs] *Paleontology.* **1.** a genus of carnivorous saurischian dinosaurs in the family Dinodontidae; they attained a length of about 50 feet, carrying their large head about 15 feet above the ground; the most specialized of the carnivorous dinosaurs, and the largest known land predator; found in North America and Asia; extant in the Upper Cretaceous. **2.** also, **Tyrannosaur.** a member of this genus.

tyrant flycatcher *Vertebrate Zoology.* any flycatcher of the family Tyrannidae. See FLYCATCHER.

tyrocidine *Microbiology.* an antibiotic produced by the soil bacterium *Bacillus brevis*; effective against certain protozoa, bacteria, and fungi.

tyrolite *Mineralogy.* $CaCu_5^{+2}(AsO_4)_2(CO_3)(OH)_4 \cdot 6H_2O$, a green to blue, vitreous, orthorhombic mineral, occurring as fanshaped, foliated aggregates and reniform masses, having a specific gravity of 3.18 and a hardness of about 2 on the Mohs scale; found as a secondary mineral in the oxidation zone of copper deposits.

tyrosamine see TYRAMINE.

tyrosinase *Enzymology.* an enzyme that catalyzes the oxidation of monophenols and diphenols to form quinones. Also, MONOPHENOL MONOOXYGENASE.

tyrosine *Biochemistry.* $C_9H_{11}O_3N$, a crystallizable amino acid found in most proteins and synthesized metabolically from phenylalanine; a precursor of thyroid hormones, melanin, and catecholamines.

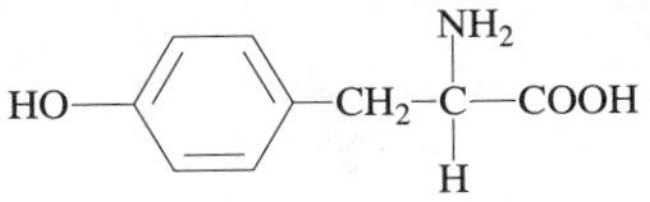

tyrosine

tyrosine aminotransferase *Enzymology.* an enzyme of the transferase class that catalyzes the breakdown of tyrosine; its defective production results in elevated blood levels of tyrosine and is associated with mental retardation and neuropsychiatric dysfunction.

tyrosine hydroxylase *Enzymology.* an enzyme that catalyzes the replacement of a hydrogen atom with a hydroxyl group on the amino acid tyrosine to form L-dopa.

tyrosine kinase *Enzymology.* an enzyme that catalyzes the transfer of a phosphoryl group from ATP to tyrosine.

tyrosinemia *Medicine.* **1.** a hereditary disorder caused by an error of metabolism of the amino acid tyrosine resulting from an enzyme deficiency; it is characterized by cirrhosis, renal rickets, renal glycosuria, and mental retardation. **2. neonatal tyrosinemia.** a transient condition of newborn infants, especially premature ones, caused by an anomaly in amino acid metabolism and characterized by an excess of tyrosine in the blood and urine; it is controlled by diet and vitamin C therapy.

tyrosinosis *Medicine.* a condition of faulty tyrosine metabolism in which an intermediate product, *para*-hydroxyphenyl pyruvic acid, appears in the urine and gives it an abnormal reducing power.

tyrosinuria *Medicine.* the presence of tyrosine in the urine.

tyrosyl *Biochemistry.* the univalent acyl radical of tyrosine.

tyrothricin *Microbiology.* an antibiotic mixture produced by the soil bacterium *Bacillus brevis;* effective against Gram-positive bacteria, and used locally for mouth and skin infections.

tyrotoxicon *Toxicology.* $C_6H_5N(:N)OH$, a poisonous crystalline compound that sometimes occurs in stale dairy products.

tyrotoxicosis *Toxicology.* a pathological condition caused by the ingestion of tyrotoxicon; symptoms may include vertigo, headache, vomiting, chills, cramps, collapse, and death.

TYS *Aviation.* the airport code for Knoxville, Tennessee.

Tyson's gland *Anatomy.* any of numerous small sebaceous glands located on the inner surface of the prepuce and around the proximal edge of the glans penis.

Tytonidae *Vertebrate Zoology.* the barn owls, a family of carnivorous nocturnal raptors of the order Strigiformes, characterized by soft plumage and a heart-shaped facial disk; found almost worldwide.

tyuyamunite *Mineralogy.* $Ca(UO_2)_2V_2O_8 \cdot 5\text{–}8H_2O$, a yellow, orthorhombic secondary mineral, having a specific gravity of 3.3 to 3.6 and a hardness of about 2 on the Mohs scale; found as crusts and coatings on limestone and sandstone in the western United States.

tyvelose *Biochemistry.* an unusual sugar isolated from the lipopolysaccharides of *Salmonella typhi.*

Tyzzeria *Invertebrate Zoology.* a genus of protozoa in the order Eucoccidiida; some species parasitize domestic ducklings and other animals.

Tyzzer's disease *Veterinary Medicine.* a disease of laboratory animals that is caused by *Bacillus piliformis* and is characterized by severe diarrhea, humped back, poor coat, and necrosis of the liver and heart; the disease has been associated with stress, unsanitary conditions, and cortisone injections.

TZP *Materials Science.* a fine-grained, fully dense, single-phase ceramic material used to make such objects as cutting tools and golf club inserts. (An abbreviation of tetragonal zirconia polycrystal.)

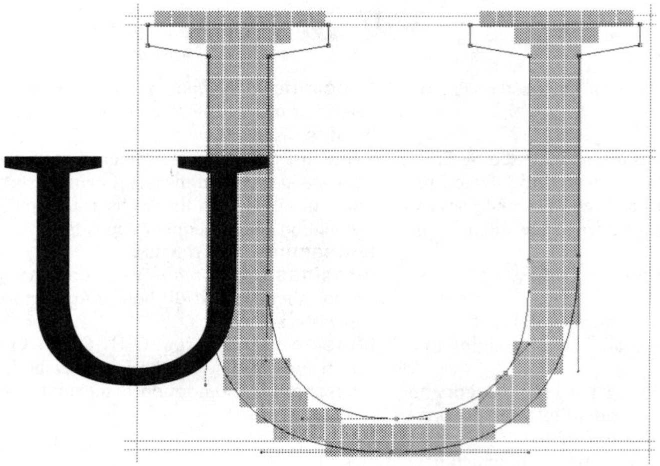

U the chemical symbol for uranium. For uranium-235, uranium-238, see URANIUM.
U uridine.
U. uniform; unit.
u uridine; uracil; an enzyme unit.
u. unit; unsymmetrical; upper.
U-2 *Aviation.* a U.S. high-altitude reconnaissance aircraft made famous by having been shot down over the Soviet Union in May, 1960.
UA *Aviation.* the airline code for United Airlines.
U/A urinalysis.
^{235}U actinouranium *Nuclear Physics.* an earlier name for the radioactive uranium isotope of mass 235 that occurs naturally, has a half-life of 7.0×10^8 years, and emits alpha particles when it decays; now called uranium-235.
UAM underwater-to-air missile.
UART universal asynchronous receiver transmitter.
UAS or **uas** upstream activation site; upstream activator sequence.
ubac *Ecology.* the shady northern side of an alpine mountain, characterized by a lower timberline and lower snow line than the sunny southern side.
u-band *Optics.* the ultraviolet absorption that occurs in a crystal from impurities in alkali halides.
U-bend die *Mechanical Engineering.* a die machined horizontally, having a rectangular cross-sectional opening that provides two edges over which metal is drawn into a channel shape.
ubiquinone *Enzymology.* another name for coenzyme Q, an enzyme closely related to vitamin K that functions as an electron carrier in the electron transport of cells.

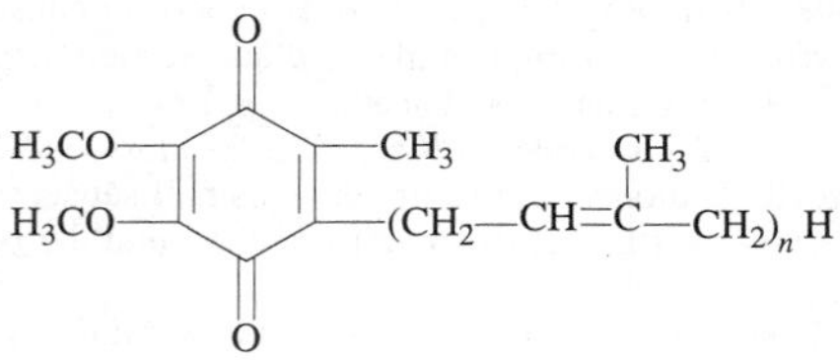

ubiquinone

ubiquitin *Biochemistry.* a polypeptide, consisting of 76 amino acids, that is present in the cells of higher eukaryotes and in *Saccharomyces cerevisiae*; it attaches to a protein prior to the selective proteolysis of that protein in the cell.
U blade *Mechanical Devices.* a curved blade of a bulldozer; used to increase motion of tractor equipment.
U-boat *Naval Architecture.* a German submarine, especially of the World War I or World War II era. (Short for a German term meaning literally "undersea boat.")
U-bolt *Mechanical Devices.* a U-shaped metal bar, threaded on both arms and used as a fastener or clip for receiving nuts.
U-bomb *Ordnance.* a uranium bomb.
ubonate *Biology.* having a conical or rounded protuberance, as found on the beak of a brachiopod shell or on the surface of many pelecypod mollusk shells.
UBV photometry *Astronomy.* a widely used method of photoelectric photometry that measures stellar brightnesses with an RCA 1P21 photomultiplier tube and a specific set of color filters.
UBV system *Astronomy.* a system of stellar magnitude determination that measures a star's apparent magnitude through three color filters: ultraviolet (U) at 3600 angstroms wavelength, blue (B) at 4200 angstroms, and visual (V) at 5400 angstroms; it is defined so that for an A0 star, B-V and U-B both equal 0.0. Also, **UBV system of stellar magnitudes.**
UC undercharge.
U center *Crystallography.* a color-center point defect in an ionic crystal lattice, due to an impurity.
U Cephei *Astronomy.* a binary star system consisting of a type B main sequence star and a type G giant, which strongly ejects gaseous matter from time to time; the gas collects in a temporary accretion ring around the B star before it falls onto the B star's surface. The system is also an eclipsing binary with 2.5-day period and brightness variation of 2.2 magnitudes.
U-clamp *Mechanical Devices.* a U-shaped metal clamp used to secure workpieces to planer beds.
U coefficient *Quantum Mechanics.* in quantum mechanics, a coefficient entering into the transformation for recoupling three or more angular momenta; it is closely related to the six-*j* symbol.
UCR unconditioned response.
UCS unconditioned stimulus.
Udalf *Geology.* a suborder of the soil order Alfisol that develops in mesic to warm regions having well-distributed rainfall and is characterized by a brown to reddish coloration.
udder *Vertebrate Zoology.* the pendulous, saclike, combined mammary glands characteristic of certain mammals, such as cows and goats.
Udert *Geology.* a suborder of the soil order Vertisol that develops in humid regions and is characterized by wide, deep surface cracks that remain open for only two to three months throughout the year.
Udoll *Geology.* a suborder of the soil order Mollisol that develops in humid regions having well-distributed rainfall and is characterized by a thick, dark A horizon over a brown B horizon and paler C horizon.
udotea *Botany.* a genus of green algae belonging to the family Udoteaceae.
Udoteaceae *Botany.* a family of tropical and subtropical marine green algae of the order Bryopsidales, characterized by multiaxial thalli composed of branched filiform tubes that are often constricted.
UDP uridine diphosphate.
Udult *Geology.* a suborder of the soil order Ultisol that develops in mesic to warm regions having well-distributed rainfall and is characterized by low to moderate amounts of organic carbon and reddish to yellowish argillic horizons.
UF urea-formaldehyde.
UFO *Astronomy.* a shorter term for an unidentified flying object. See UNIDENTIFIED FLYING OBJECT.
U format *Computer Programming.* a record format that is treated as unknown and unpredictable by the input/output control system.
UG urogenital.
U Geminorum stars *Astronomy.* a class of dwarf nova variable stars, characterized by abrupt, irregular outbursts with about four magnitude difference in brightness. Also, SS CYGNI STARS.
UGI upper gastrointestinal.

UHF *Telecommunications.* **1.** ultrahigh frequency; any radio frequency located in the band between 300 and 3000 megahertz. **2.** the over-the-air television channels 14 through 83, which carry signals with these frequencies.

Uhlbricht sphere *Optics.* a sphere with a white, diffuse, hollow interior, used in an integrating photometer.

Uhlenbeck, George 1900–1988, Indonesian-born American physicist; with S. A. Goudsmit, discovered electron spin.

uhligite *Mineralogy.* a black, cubic mineral with an approximate formula of $Ca_3(Ti,Al,Zr)_9O_{20}$, occurring as octahedral crystals, and having a specific gravity of 4.15 and a hardness of 5.5 on the Mohs scale; found in nepheline sygenite at Lake Magadi, Kenya.

UHT ultrahigh temperature.

U index *Geophysics.* the difference between successive daily mean values of the horizontal component of the geomagnetic field. Also, **U figure.**

uintahite see GILSONITE.

Uintan *Geology.* a North American geologic stage of the Middle Eocene epoch, occurring after the Bridgerian and before the Duchesnian.

Uintatheriidae *Paleontology.* a family of large herbivorous mammals belonging to the extinct order Dinocerata; extant in the Paleocene and Eocene.

Uintatherium *Paleontology.* a genus of large, primitive mammals in the extinct family Uintatheriidae, characterized by many cephalic knobs; restricted to the Middle and Upper Eocene.

Ul RNA *Genetics.* an abundant species of small nuclear RNA, having a chain composed of 165 bases; functions in excising introns and splicing exons of mRNA.

UL Underwriters' Laboratories; a label used on electrical equipment to indicate an approval rating for safety that is given by this organization.

ULA uncommitted logic array.

Ulatisian *Geology.* a North American geologic stage of the Middle Eocene epoch, occurring after the Penutian and before the Narizian.

ulcer *Medicine.* **1.** a craterlike lesion of the surface of the skin or of a mucous membrane, produced by the sloughing off of inflammatory necrotic tissue. **2.** specifically, a condition of this type occurring in the digestive system; a peptic ulcer.

ulceration *Medicine.* the fact of forming an ulcer, or a condition under which an ulcer forms.

ulcerative *Medicine.* relating to or affected by an ulcer or ulcers. Also, **ulcerous.**

ulcerative colitis *Medicine.* a chronic ulceration and inflammation of the colon, characterized by watery diarrhea, pain in the abdomen, and rectal bleeding.

ulcerative endocarditis *Cardiology.* a rapidly fatal form of acute infectious endocarditis, marked by ulcerated lesions of the endothelial lining of the heart. Also, MALIGNANT ENDOCARDITIS.

ulcerative lymphangitis *Veterinary Medicine.* a mildly contagious, chronic infection of the lymph nodes in horses and cattle that is caused mainly by organisms of the genus *Corynebacterium*, which gain access to the animal through abrasions or wounds, especially those located on the leg; ulceration appears at the sight of the wound. Also, **ulcerative cellulitis.**

ulectomy *Surgery.* **1.** the surgical excision of scar tissue. **2.** a gingivectomy.

ulegyria *Neurology.* a defect of the cerebral cortex, characterized by the weakening or sclerosis of cerebral gyri; may be congenital or may result from scarring.

ulexite *Mineralogy.* $NaCaB_5O_6(OH)_6 \cdot 5H_2O$, a colorless to white triclinic mineral occurring as rounded masses of extremely fine acicular crystals, having a specific gravity of 1.955 and a hardness of 2.5 on the Mohs scale; found in saline lake deposits. Also, BORONATROCALCITE, COTTON BALL.

uliginous *Botany.* growing in swamps or in muddy places. Also, **uliginose.**

ullage *Engineering.* the difference between the total volume of a container and the volume of the material it is presently holding.

ullage rocket *Space Technology.* a small rocket used to accelerate a liquid rocket's fuel-tank system, so that the liquid propellants flow smoothly into the pumps or combustion chamber.

ullmannite *Mineralogy.* NiSbS, a steel-gray to silver-white triclinic, pseudocubic mineral of the cobaltite group occurring as granular masses and cubic crystals with a metallic luster, having a specific gravity of 6.65 and a hardness of 5 to 5.5 on the Mohs scale; found in hydrothermal veins with other nickel minerals. Also, NICKEL-ANTIMONY GLANCE.

Ullman reaction *Organic Chemistry.* the coupling of aryl halides into diaryls with the removal of halogens, by the use of copper compounds; a modification of the Fittig synthesis in which copper powder is substituted for sodium.

Ulloa's ring see BOUGUER'S HALO.

Ullrich-Turner syndrome see TURNER'S SYNDROME.

Ulmaceae *Botany.* a family of dicotyledonous trees and shrubs of the order Urticales, characterized by simple, alternate leaves, and bisexual or unisexual flowers found in cymes.

ulmin *Geology.* a dark brown to black, amorphous, gel-like constituent of coal derived from the decay or degradation of vegetable matter. Also, HUMIN, CARBOHUMIN, HUMOGELITE, FUNDAMENTAL JELLY, GÉLOSE, VEGETABLE JELLY.

ulna *Anatomy.* the medial and larger bone of the forearm on the side opposite that of the thumb that articulates with the humerus and with the head of the radius at its proximal end, and with the radius and bones of the carpus at its distal end.

ulnoradial *Anatomy.* relating to the ulna and radius.

Ulothrix *Botany.* a genus of green algae of the family Chaetonhoraceae, consisting of simple filaments with band-shaped green chloroplasts; common in ponds.

Ulotrichaceae *Botany.* a family of green algae belonging to the order Ulotrichales, characterized by a single, parietal chloroplast, and having both attached and floating filamentous species that reproduce both sexually and asexually.

Ulotrichales *Botany.* an order of green algae of the Chlorophyta, characterized by freshwater, uninucleate members having a single, laminate chloroplast.

Ulotrichineae *Botany.* a suborder of the order Ulotrichales whose members have short cylindrical cells.

Ulsterian *Geology.* a North American provincial series of the Lower Devonian period, occurring after the Cayugan of the Silurian period and before the Erian.

ult. ultimate; ultimately; ultimo. Also, **ulto.**

ultimate analysis *Analytical Chemistry.* the determination of the percentage of elements contained in a chemical substance.

ultimate elongation *Metallurgy.* the percentage of permanent deformation remaining after a tensile rupture.

ultimate factor *Ecology.* the environmental factor that is most instrumental in causing a certain ecological event, or most essential to the survival of an organism.

ultimate lines *Astronomy.* the special spectral lines used to identify a chemical constituent in the sun or another star.

ultimate-load design *Design Engineering.* the design of a beam so it can carry the design load multiplied by the safety factor, i.e., its ultimate load capacity. Also, LIMIT-LOAD DESIGN.

ultimate recovery *Petroleum Engineering.* the estimated total volume of oil or gas that a reservoir will yield during its productive lifetime.

ultimate set *Engineering.* the ratio of the length of a specimen prior to testing to its length at the moment of fracture, expressed as a percentage.

ultimate strength *Mechanics.* the highest stress that a material can withstand before it fractures or fails.

ultimate tensile strength *Materials Science.* a result obtained from a tensile test; defined as the maximum load that the specimen withstands, expressed as $UTS = P_{max}/A_0$, where P_{max} = the maximum force and A_0 = the initial uniform cross-sectional area along the gauge length.

ultimobranchial body *Developmental Biology.* an embryonic derivative of the fifth pharyngeal pouch that migrates along with the parathyroid glands and that is incorporated in the thyroid gland. Also, POSTBRANCHIAL BODY, TELOBRANCHIAL BODY.

Ultisol *Geology.* an order of generally moist soil that develops in subtropical to tropical climates and is characterized by an argillic horizon, low in bases, overlaying a clay-rich illuvial horizon.

ultra- a combining form meaning "beyond."

ultra-audion circuit *Electronics.* a regenerative detector in which an electron tube serves as the active element, a tank circuit provides positive feedback between the control grid and the anode, and a trimmer capacitor between the anode and cathode controls the amount of regeneration.

ultra-audion oscillator *Electronics.* an electron-tube, ultra-audion circuit that resembles a Colpitts oscillator and uses interelectrode capacitances within the tube to sustain oscillations.

ultrabasic *Petrology.* of or relating to an igneous rock characterized by having a lower silica content, usually less than 45%, than a basic rock.

ultracentrifuge *Engineering.* a centrifuge with the ability to develop centrifugal fields of up to 100,000 times that of the gravitational field.

ultra-cold neutron *Physics.* a neutron with a kinetic energy of less than 10^{-7} eV.

ultradian *Biology.* of or relating to biological rhythms that recur more than once a day.

ultrafilter *Mathematics.* a filter $F(X)$ on a set X with the property that every finer filter on X is identical to $F(X)$; i.e., a filter that is not contained in any other filter.

ultrafiltration *Chemical Engineering.* the process of separating colloidal or molecular particles by filtration, using suction or pressure, by means of a colloidal filter or semipermeable membrane. *Food Technology.* a process used to separate milk proteins by passing milk at high pressure through a very fine membrane. Also, MICROFILTRATION.

ultrafiltration fermenter *Biotechnology.* a continuous fermenter that is run in conjunction with a continuous ultrafiltration system; the result is a reduction of feedback inhibition as low-molecular-weight product is continuously removed.

ultragravity waves *Fluid Mechanics.* gravity waves with periods between 0.1 and 1.0 second.

ultrahigh frequency see UHF.

ultrahigh-strength polymer *Materials.* a molecularly oriented polymer, such as gel-spun polyethylene, or a specific structure polymer, such as the copolyester of *p*-hydroxybenzoic acid, biphenol, and terephthalic acid, that is melt processable at 400°C; used in high-load applications.

ultrahigh-strength steel *Materials.* steel having a tensile strength greater than about 30 μPa.

ultrahigh vacuum *Physics.* a region in space where the pressure is on the order of 10^{-10} mm mercury or less.

ultralarge crude carrier *Naval Architecture.* an oil tanker of the largest class, having a gross register tonnage in excess of 250,000 tons. Also, SUPERTANKER.

ultralight aircraft *Aviation.* a very small, lightweight aircraft, especially one not exceeding 1000 pounds; designed for recreation or aerodynamic research.

ultralow-carbon steel *Materials.* steel containing 0.3–0.6% carbon, which has low strength but high ductility; it cannot be hardened but is machined and forged easily.

ultramafic *Petrology.* describing or relating to igneous rocks that are composed principally of mafic minerals.

ultramarine blue *Inorganic Chemistry.* a blue pigment that is made by heating a mixture of kaolin, sulfur, soda ash, and charcoal.

ultrametamorphism *Geology.* metamorphism during which the temperature of a rock exceeds its melting point.

ultramicrobalance *Engineering.* a differential weighing apparatus that is accurate to one microgram.

ultramicroscope *Optics.* a microscope that reveals extremely small particles, such as smoke, by illuminating them with an arc-lamp.

ultramicrotome *Engineering.* a microtome that uses a glass or diamond knife, enabling the dissection of cells as small as 30 nanometers in thickness.

ultraphotic rays *Electromagnetism.* electromagnetic radiation outside of the visible part of the spectrum.

ultrasonic *Acoustics.* having to do with very high frequencies of acoustic energy, above the upper range of human hearing (approximately 20,000 hertz). Also, SUPERSONIC.

ultrasonic atomizer *Mechanical Engineering.* an atomizer that forces the flow of liquid over a surface vibrating at ultrasonic frequencies.

ultrasonic camera *Electronics.* a device used in medicine to detect anomalies in living tissue; ultrasonic waves are directed through the area under study and are converted to equivalent instantaneous voltage values, which are then fed to a monitor to intensity-modulate the electron beam of a cathode-ray tube for viewing.

ultrasonic cardiography *Radiology.* a method of graphically recording the position and motion of the heart walls or the internal structures of the heart and nearby tissue, by the echo obtained from beams of ultrasonic waves directed through the chest wall. Also, ECHOCARDIOGRAPHY.

ultrasonic cell disintegration *Biotechnology.* a method used to disrupt cells by bombarding them with ultrasonic waves.

ultrasonic cleaning *Engineering.* a cleaning technique in which ultrasonic vibrations are used to remove debris from the surfaces of items immersed in a solvent.

ultrasonic coagulation *Physics.* a process in which ultrasonic waves are used to bond small particles into larger clusters.

ultrasonic communication *Telecommunications.* communication through water by keying the sound output of echo-ranging sonar.

ultrasonic delay line *Acoustical Engineering.* the passage of sound through a material such as barium titanate or mercury to induce a delay in signal processing.

ultrasonic depth finder *Engineering.* a device that measures the time delay between the emission of an ultrasonic signal and the reception of the signal's echo, and converts this delay into a measure of distance.

ultrasonic drilling *Mechanical Engineering.* a vibration-drilling technique in which the compression and extension of a core of electromagnetic material in a rapidly alternating electromagnetic field generates ultrasonic vibrations. Thus, **ultrasonic drill.**

ultrasonic echo *Acoustics.* the echo of an ultrasonic wave at a crack or flaw in a tested metal that reflects back to the initial source and can therefore be used to detect cracks, defects in casting, and other internal defects in metal objects.

ultrasonic flaw detector *Acoustical Engineering.* a device that processes time differences between transmission of ultrasonic waves and reception of their echoes to locate minute flaws in a metal surface or casting.

ultrasonic generator *Acoustical Engineering.* an electroacoustical device that generates frequencies above about 20 kilohertz.

ultrasonic grating constant *Optics.* the distance between the diffracting centers of an ultrasonic wave forming a particular light diffraction spectrum.

ultrasonic imaging see ACOUSTIC IMAGING.

ultrasonic imaging device *Acoustical Engineering.* a device used in medicine for the three-dimensional imaging of a human body or portion of the body, using transmission and reception characteristics of sound waves to form an image.

ultrasonic leak detector *Engineering.* an instrument that detects leaks by measuring the ultrasonic energy caused by gas flow through an opening.

ultrasonic light modulator *Optics.* a device that modulates light by ultrasonically affecting a transparent fluid through which the light passes.

ultrasonic machining *Mechanical Engineering.* the removal of material with an ultrasonically vibrated abrasive suspension.

ultrasonic medical tomography *Acoustics.* the use of ultrasonic sound transmissions in the human body and reception of reflected sound to form a visual image of parts of the body, using techniques such as A mode, B mode, and C mode.

ultrasonic microscope *Optics.* a microscope that uses ultrasonic radiation to magnify an object.

ultrasonic radiation *Acoustics.* high-frequency acoustic waves that can be created by disturbances of a magnetostrictive or piezoelectric transducer for a specific purpose, such as medical tomography or acoustic microscopy.

ultrasonics *Acoustics.* the study and use of sound waves having frequencies beyond the normal limit of human hearing (approximately 20,000 hertz).

ultrasonic sealing *Engineering.* the sealing of a plastic film using heat developed by vibratory pressures at ultrasonic frequencies.

ultrasonic sensor *Robotics.* a proximity sensor based on the detection of reflected sound waves.

ultrasonic testing *Engineering.* the detection of structural flaws in a material using high-frequency mechanical energy.

ultrasonic thickness gauge *Engineering.* a device used to determine the thickness of a material; the time required for an ultrasonic wave to pass through a material is measured and then converted into a distance.

ultrasonic tomography *Radiology.* the visualization of a cross section of a predetermined plane of the body by linear scanning with an ultrasonic probe across the desired site; the image is displayed on a B-scan.

ultrasonic transducer *Acoustical Engineering.* an electroacoustical device, usually piezoelectric or magnetostrictive, that converts ultrasonic waves into electromagnetic waves, or electromagnetic waves into ultrasonic waves.

ultrasonic transmitter *Acoustical Engineering.* an electroacoustic device that converts electromagnetic energy into high-intensity ultrasonic waves; typically used to track marine animals.

ultrasonic welding *Materials Science.* a deformation welding technique in which the load that forces the two surfaces together is partly introduced by a very high-frequency vibration, and bonding is achieved with very little total deformation.

ultrasonography *Radiology.* the visualization of deep body structures by recording the echoes of ultrasonic waves directed into the tissues; the image produced is an **ultrasonogram.** Also, SONOGRAPHY. *Medicine.* see ECHOGRAPHY.

ultrasonometry *Radiology.* the measurement of certain physical properties of biologic fluids by using ultrasound.

ultrasonoscope *Engineering.* an instrument that displays an echosonogram.

ultrasound *Acoustics.* generally, any acoustic wave above the normal range of human hearing, i.e., above 20,000 hertz; in practice the term usually refers to a much higher frequency used for a specific application, as about 500,000 hertz for the ultrasonic echo testing of metal objects. Ultrasound is used extensively in medical diagnosis, materials testing, industrial and medical cleaning, and for various other purposes.

ultrasound diathermy *Medicine.* the therapeutic generation of heat in body tissue through the application of ultrasound.

ultraspherical polynomials see GEGENBAUER POLYNOMIAL.

ultrastructure *Molecular Biology.* **1.** an arrangement of the smallest elements that make up a substance or organism. **2.** the structure of something that is beyond the resolution power of a light microscope.

Ultrasuede *Textiles.* a brand of synthetic, washable fabric that resembles suede.

ultraviolet *Electromagnetism.* relating to radiation having wavelengths in the range of about 4 to about 400 nanometers, beyond the violet end of the visible spectrum (and thus beyond the range of human vision).

ultraviolet absorber *Optics.* a material that absorbs ultraviolet light and reradiates it at a longer wavelength; used to reduce light sensitivity in plastics and rubbers.

ultraviolet absorption *Optics.* the absorption of light at ultraviolet wavelengths, as by a material undergoing spectroscopic testing.

ultraviolet absorption spectrophotometry *Spectroscopy.* the photometric measurement of the wavelengths of ultraviolet radiation that are absorbed by molecules and correspond to electron transitions from a bonding to an antibonding orbital.

ultraviolet astronomy *Astronomy.* the study of extraterrestrial objects at ultraviolet wavelengths.

ultraviolet-bright star *Astronomy.* a star that is brighter than a horizontal-branch star and bluer than a giant-branch star.

ultraviolet catastrophe *Physics.* the failing of the Rayleigh-Jeans law for blackbody radiation in which it is predicted that the total radiated energy would be infinite due to the short wavelength contribution.

ultraviolet degradation *Materials Science.* a phenomenon caused by sunlight and oxygen, which is manifested in deterioration of physical properties and appearance of polymers. The near ultraviolet portion of the sun's spectrum that reaches the earth includes wavelengths with sufficient energy to induce photochemical reactions in many polymers when absorbed by suitable chromophores.

ultraviolet densitometry *Spectroscopy.* an ultraviolet spectrophotometric technique used for measuring colors produced by thin-layer chromatography.

ultraviolet-erasable programmable read-only memory *Computer Technology.* a memory chip whose stored data can be erased by exposure to high-intensity shortwave ultraviolet light, and then can be redefined.

ultraviolet imagery *Electromagnetism.* the production of an image by the detection of ultraviolet radiation reflected from a target.

ultraviolet lamp *Electronics.* a lamp that is specifically designed to produce radiation in the ultraviolet range.

ultraviolet light *Electromagnetism.* electromagnetic radiation having wavelengths in the range of about 4 to about 400 nanometers; there is an overlapping of this range with the wavelengths of the long X-rays. Also, **ultraviolet radiation.**

ultraviolet microscope *Optics.* a microscope that achieves increased resolution by illuminating the subject with ultraviolet light.

ultraviolet photoemission spectroscopy *Spectroscopy.* a spectroscopic technique in which a surface is bombarded with photons in the ultraviolet energy range and the resulting electronic emission spectrum is analyzed for information about molecular orbital characteristics or the occupied and unoccupied density of states in metals and alloys. Also, **ultraviolet photoelectron spectroscopy.**

ultraviolet spectrometer *Spectroscopy.* a device calibrated to measure ultraviolet wavelengths absorbed by a sample.

ultraviolet spectrophotometry *Spectroscopy.* the photometric measurement of the wavelengths of ultraviolet radiation absorbed by specific molecules in liquids or gases.

ultraviolet spectroscopy *Spectroscopy.* the study of the ultraviolet radiation characteristically absorbed by molecular substances.

ultraviolet spectrum *Electromagnetism.* the graphical display of the intensity of ultraviolet radiation emitted or absorbed by a body.

ultraviolet stabilization *Materials Science.* the use of additives to protect polymers against ultraviolet degradation. Most plastics contain stabilizers to protect against environmental effects at the end use. Some additives, including such ultraviolet absorbers as hindered amines, nickel compounds, benzoate esters, trivalent phosphorous compounds, hindered phenols, and thiosynergists, function as chemical stabilizers; others, such as carbon black, function simply as light screens.

ultraviolet star *Astronomy.* a very hot star, often the central star in a planetary nebula, that radiates primarily in the ultraviolet range of the electromagnetic spectrum.

ultraviolet telescope *Optics.* a telescope that operates primarily in the ultraviolet region of the spectrum.

ultraviolet/visible spectroscopy *Spectroscopy.* the study of emission and absorption spectra over wavelengths in the ultraviolet and visible region of the electromagnetic spectrum.

ultraviolet visualization *Biotechnology.* a method that detects the position of nucleic acids and other compounds that absorb ultraviolet light on paper, gel, or thin-layer chromatograms.

ultravulcanian *Geology.* describing a volcanic eruption typified by violent, gaseous explosions of rock dust and angular blocks, with generally no incandescent scoria.

Ulvaceae *Botany.* a large family of green marine algae of the order Ulvales, having a two-cell thick layered thallus and reproducing both sexually and asexually.

Ulvales *Botany.* an order of green algae of the Chlorophyta, having uninucleate cells and a parenchymatous thallus that may be a sheet or a tube.

Ulvellaceae *Botany.* a family of green algae belonging to the order Ctenocladales, characterized by pseudoparenchymatous thalli with prostrate branched filaments; usually found affixed to or immersed in aquatic organisms or other solid substrates.

umangite *Mineralogy.* Cu_3Se_2, a blue-black tetragonal mineral with reddish tint that tarnishes violet; massive in habit and as small grains or fine granular aggregates, and having a specific gravity of 6.44 and a hardness of 3 on the Mohs scale; occurs with other selenides in hydrothermal veins.

umbel *Botany.* a racemose inflorescence in which the flowers are borne on undivided pedicels originating from a single node, forming a flat or rounded flower cluster.

Umbellales *Botany.* an order of dicotyledonous, herbaceous, or woody plants of the subclass Rosidae, characterized by an inferior ovary and flowers with separate petals. Also, APIALES.

Umbelliferae *Botany.* a large family of dicotyledonous shrubs and herbs of the order Umbellales, characterized by alternate, mostly compound leaves, usually bisexual flowers found in umbels, and an inferior ovary with two carpels. Also, APIACEAE.

umbilical artery *Developmental Biology.* one of two arteries carrying blood from the internal iliac to the placenta for reoxygenation.

umbilical connection *Space Technology.* see UMBILICAL CORD, defs. 1 and 3.

umbilical cord *Developmental Biology.* the flexible structure connecting the umbilicus of the embryo and fetus with the placenta, and giving passage to the umbilical arteries and vein. *Space Technology.* **1.** any of the supply lines running from an umbilical tower to a launch vehicle. **2.** also, **umbilicals.** an entire network of such lines. **3.** a line that tethers and provides life support and communications for an astronaut performing an extravehicular task in space.

umbilical duct see VITELLINE DUCT.

umbilical hernia *Medicine.* an abdominal hernia in which the bowel or omentum protrudes through the umbilical ring. Also, ANNULAR HERNIA.

umbilical tower *Space Technology.* a tower designed to service and support a rocket vehicle prior to launch, supplying it with electrical power, control signals, data links, propellant loading, high pressure gas transfer, and air conditioning. Also, **umbilical mast.**

umbilical vein *Developmental Biology.* either of the paired veins that carry blood from the chorion to the sinus venosus and heart in the early embryo, later fusing into a single vessel in the umbilical cord that carries blood from the placenta to the ductus venosus of the fetus.

Umbilicariaceae *Botany.* a family of lichens of the order Lecanorales that are attached at a central point by an umbilicus; characterized by horizontal spreading structures and a large, circular thallus.

umbilicus *Anatomy.* the round, generally depressed scar that lies in the median line of the abdomen and marks the site where, during fetal life, the umbilical vessels passed from the placenta into the fetus. Also, NAVEL.

umbo *Anatomy.* **1.** a knoblike or rounded projection on the side of a bone or a tumor. **2.** a bulge outward from the surface. *Invertebrate Zoology.* a dorsal protuberance on each valve of a bivalve mollusk, the oldest part of the shell around which the rest is built.

umbra *Optics.* a shadow that obscures all light rays from a specific source, such as the conical section in the shadow of a celestial body. *Astronomy.* the dark section of a sunspot.

umbrella antenna *Electromagnetism.* an antenna having radiating elements that are connected to a central pole or tower and are bent downward to the ground, resembling the ribs of an open umbrella.

umbrella bird *Vertebrate Zoology.* an elaborately feathered bird of the genus *Cephalopterus* in the family Cotingidae, having a heavy crest of feathers that is spread over the head during breeding in the manner of an umbrella; native to South and Central America.

umbrella bird

umbrella pine *Botany.* a tall evergreen tree of the taxodium family, characterized by an umbrellalike shape, having tufts of stiff, leaflike petioles that resemble pine needles; native to Japan, but grown worldwide. Also, STONE PINE.

Umbrept *Geology.* a suborder of the soil order Inceptisol that develops in cold to temperate regions and is characterized by high acidity, crystalline clay minerals, and a deep, dark A horizon overlying a brown B horizon and slightly paler C horizon.

Umbriel *Astronomy.* Uranus II, a moon of Uranus discovered in 1851; it has a 586-kilometer diameter and a distance from Uranus of 266,300 kilometers.

umkehr effect *Optics.* a change in the brightness of the zenith at sunset when measured at two wavelengths, one of which is affected by atmospheric ozone.

Umklapp process *Solid-State Physics.* a process in which a collision between phonons involves momentum conservation laws such that the summation of the wave vectors of the phonons before and after the collision is equal to a vector in the reciprocal lattice. Also, **U-process.**

UMP uridine monophosphate.

umwelt *Behavior.* a German term for those aspects of the environment that influence the behavior of an organism or are affected by that organism.

unaka *Geology.* **1.** a large residual structure rising above a peneplain that sometimes shows remnants of an older peneplain on its surface. **2.** a group of monadnocks spread out near the headwaters of a stream system in an area that has not yet been reduced to a peneplain.

unakite *Petrology.* an altered, granitic igneous rock consisting of epidote, pink orthoclase, and quartz, with minor opaque oxides, apatite, and zircon.

unambiguous name *Computer Programming.* a specific data item name on a computer system.

unamplified back bias *Electronics.* the negative feedback voltage developed in an amplifier's fast time constant circuits.

unarmed *Military Science.* the state of a fuse or other firing device that is not ready to function, because it has been neutralized with a safety device or because the necessary steps to place it in a functional state have not been taken.

unary *Mathematics.* relating to a function whose domain is a given set and whose range is contained in that set.

unary operation *Computer Programming.* an operation involving one operand, such as negation or complementation. Also, MONADIC OPERATION.

unary operator *Computer Science.* an operator that takes only a single argument, such as NOT or MINUS.

unattended operation *Computer Technology.* an operation, such as the transmission and reception of data, that occurs automatically without the intervention or assistance of an operator.

unattended time *Computer Programming.* a period of time in which a computer is running but has no operator present.

unavailable energy *Thermodynamics.* a portion of energy involved in an irreversible thermodynamic process that becomes nonusable for performing mechanical work.

unavoidable delay *Industrial Engineering.* a delay in a production process that is not within the worker's control and that cannot be prevented by better work technique or management; e.g., machine downtime due to a general external power failure.

unavoidable-delay allowance *Industrial Engineering.* an allowance that is built into a work schedule to allow for unavoidable delays.

unavoidable set of configurations *Mathematics.* a set of graphs having the property that at least one member of the set is a subgraph of any given planar graph; used in the Appel-Haken proof of the four-color theorem.

unbalanced line *Electricity.* a line whose current-carrying conductors are asymmetrical with respect to ground, causing earth currents.

unbalanced output *Electricity.* an output that is grounded on one side. Also, SINGLE-ENDED OUTPUT.

unbalanced translocation *Genetics.* a translocation that involves a harmful effect or is accompanied by a loss of chromosomal material.

unbalanced wire circuit *Electricity.* a wire line whose two sides are electrically dissimilar.

unbiased estimator *Statistics.* a sample statistic whose expectation is equal to the underlying population parameter to be estimated; for example, the mean of a random sample is an unbiased estimator of the population mean.

unblanking pulse *Electronics.* a pulse applied to a cathode-ray tube to activate the cathode-ray beam in order to make the trace on the screen visible during the time of the sweep.

unbonded strain gauge *Engineering.* a grid of wires strung under tension to a diaphragm; movement of the diaphragm causes a change in the resistivity of the wires that may then be used as a measure of strain.

unbounded manifold *Mathematics.* a manifold that has no boundary.

unbounded set *Mathematics.* a subset S of a normed space (with norm $\| \ \|$) such that for every real number M, there exists an element x of S such that $\|x\| > M$. For example, an unbounded set S of real numbers has the property that for every real number M, there is an element of S with the property that $|x| > M$; that is, either $x < -M$ or $x > M$. For example, all odd integers form an unbounded set.

unbounded wave *Physics.* any wave that propagates through a medium that has no boundaries and is homogeneous and nondissipative.

unbound state *Quantum Mechanics.* the state of a particle whose wavefunction remains finite at infinity in some direction and at some time.

unbreakable cipher *Telecommunications.* a cipher that cannot be solved through cryptoanalysis, regardless of computational power or time.

unbundling *Computer Technology.* the practice of marketing, pricing, and selling system hardware and software separately, rather than as a package.

uncage *Engineering.* to unlock the caging mechanism of a gyroscope.

uncatalog *Computer Technology.* the process of removing an entry from the system catalog, making the file named inaccessible.

uncertain reasoning see REASONING UNDER UNCERTAINTY.

uncertainty *Artificial Intelligence.* a measure of the confidence factor an expert system assigns to conclusions that cannot be determined definitely. *Quantum Mechanics.* the width, such as the full width at half maximum, of the probability distribution of some observable.

uncertainty principle *Quantum Mechanics.* see HEISENBERG UNCERTAINTY PRINCIPLE.
uncertainty relation *Quantum Mechanics.* the inequality expressing the fact that the product of the uncertainties of any two conjugate observables, such as position, Δx, and momentum, Δp_x, has an irreducible uncertainty, written as $\Delta x \Delta p_x \geq \{\hbar\}$.
uncharged *Electricity.* having no electrical charge.
uncharged demolition target *Ordnance.* a target that has been prepared to receive the demolition agent, and for which the necessary quantities of the agent have been calculated, packaged, and stored in a safe place with installation instructions prepared.
uncharged species *Chemistry.* an electrically neutral species.
uncinate *Neurology.* of or relating to the uncus. Also, **uncinal.** *Botany.* hooked; bent at the end like a hook.
uncinate process *Vertebrate Zoology.* a hook-shaped, bony process on certain ribs of birds, projecting backward and overlapping the succeeding rib, strengthening the thorax.
uncinate seizure *Neurology.* an attack of focal epilepsy, originating in the uncal area of the temporal lobe and characterized by dysfunction of the senses of smell and taste.
uncinate trophus *Invertebrate Zoology.* in rotifers, a trophus with a hooked or curved uncus, for shredding food.
uncinus *Invertebrate Zoology.* a hooked seta in some polychaetes or a hooked tooth on some gastropod radulas. *Meteorology.* a cloud species that is unique to the genus cirrus; characterized by a lack of gray parts and often having its parts arranged in the form of a comma with a hooked top.
uncipressure *Surgery.* the arrest of hemorrhage by applying pressure with a blunt hook.
unc mutant *Microbiology.* any bacterial cell that contains a mutation in the coupling between the proton motive force in chemiosmosis and proton ATPase activity.
uncoating *Virology.* the removal of outer layers of a virus particle, thus releasing the viral nucleic acid from the capsid or exposing it sufficiently to allow replication.
uncoiling of chromomeres *Molecular Biology.* a phenomenon in which the DNA molecule of a chromomere (chromatin granule) spins out; occurs during the creation of loops by lamp-brush chromosomes.
uncommitted logic array *Robotics.* an integrated circuit with logic elements that can be connected in a number of ways to produce different results.
uncompetitive enzyme inhibition *Enzymology.* an interaction of an inhibitor with an enzyme-substrate complex that prevents the enzymatic reaction from occurring.
unconcentrated flow see OVERLAND FLOW.
unconditional *Behavior.* see UNCONDITIONED.
unconditional convergence *Mathematics.* absolute convergence.
unconditional inequality *Mathematics.* a statement of inequality that holds for all possible values of the variables involved.
unconditional jump *Computer Programming.* an instruction that always transfers program control to another sequence of instructions. Also, **unconditional branch.**
unconditional positive regard *Psychology.* respect or affection for another person that is not dependent on the individual's behavior. Also, **unconditional love.**
unconditional probability *Statistics.* the probability that a given event will occur, regardless of whether another event occurs. Also, MARGINAL PROBABILITY.
unconditioned *Behavior.* **1.** not produced by or dependent on the conditioning of the individual. **2.** natural or innate.
unconditioned reflex *Behavior.* in classical conditioning, the behavioral response that is elicited by an unconditioned stimulus. Also, **unconditioned response.**
unconditioned reinforcer *Behavior.* a stimulus that increases the probability of a response without necessitating the learning of the value of the reinforcer. Also, PRIMARY REINFORCER.
unconditioned stimulus *Behavior.* in classical conditioning, a stimulus that elicits and may reinforce a behavioral response before conditioning occurs.
unconfined explosion *Engineering.* an explosion that occurs in the open air.
unconfined groundwater see NONARTESIAN WATER.
unconformable *Geology.* relating to rock strata that do not correspond as part of a continuous whole with the older underlying rocks in position, fit, or immediate order of age.
unconformity *Geology.* **1.** any break or gap in the geologic record in which one rock unit underlies another that is not next in stratigraphic succession. **2.** a lack of continuity in deposition between strata of different ages corresponding to or representing a period of nondeposition, weathering, or erosion that occurred before the younger strata were deposited, and usually indicated by a difference in dip or strike between the older and younger strata.

unconformity

unconformity iceberg *Oceanography.* an iceberg composed of more than one kind of ice, such as water-formed ice and the more granular firn; characterized by many silt-bands and crevasses.
unconscious *Neurology.* **1.** of or relating to a loss of awareness and the inability to perceive sensory stimuli. **2.** describing a person who is in this state of lack of awareness and sensation. *Psychology.* **1.** the part of the mind that is not readily accessible to conscious awareness, but that may manifest its existence in symptom formation, in dreams, or under the influence of certain drugs. In Freudian psychology, the term refers to the repository of not-conscious, yet mentally present and active memories, thoughts, and feelings, including certain mechanisms that influence behavior; in Jung's psychology, the term has a broadened meaning, implying the vast range of all individual and collective past, present, and potential experience of which one is not immediately aware. **2.** of or relating to this area of the mind; describing thoughts, memories, and feelings that are beyond an individual's awareness but that may affect behavior nonetheless. Thus, **unconscious motive.**
unconsciousness *Neurology.* a physical and mental state in which one is generally immobile, lacks awareness of surrounding events and circumstances, and does not respond to sensory stimuli.
unconsolidated material *Geology.* any loose aggregate of soil material or loosely arranged or unstratified sediment whose particles are not cemented together.
unconstrained optimization problem *Mathematics.* a nonlinear optimization problem having no constraint functions.
uncontrolled airport *Transportation Engineering.* an airport at which no traffic control service is available.
uncontrolled airspace *Transportation Engineering.* any airspace in which air traffic control service is not available.
uncontrolled fragments *Ordnance.* pieces of an explosive munition that are propelled outward by the force of detonation without prior design engineering to control their shape or size.
uncontrolled mosaic *Cartography.* in aerial photography, a mosaic constructed from uncorrected photographs, with overlapping details matched from print to print without reference to ground control or any other orientation.
unconventional warfare *Military Science.* a broad term for military or paramilitary operations conducted in hostile or politically sensitive territory; it includes guerilla warfare, subversion, sabotage, and other low visibility, covert, or clandestine operations that are usually carried out by predominantly indigenous personnel with external support and direction.
uncorrecting *Navigation.* a conversion of one type of directional reading into another; it includes the conversion of true direction into magnetic, compass, or gyro direction and the conversion of magnetic into compass direction.
uncorrelated random variables *Statistics.* two variables whose linear correlation (measured by the coefficient of correlation) is 0; distinguished from independence in that independent random variables are uncorrelated, but the converse does not hold.

uncountable set *Mathematics.* a set whose elements cannot be put into one-to-one correspondence with (any subset of) the natural numbers; equivalently, a set of cardinality greater than $\aleph_0$.
uncouple *Engineering.* to disengage or release.
uncoupling agent *Biotechnology.* a compound that breaks the electron transport chain during various key biological processes such as phosphorylation and photosynthesis.
uncoupling phenomena *Spectroscopy.* in diatomic molecules, deviation from predicted spectra resulting from an increase in the magnitude of the angular momentum.
unctuous *Materials.* fatty, oily, or greasy.
uncus *Neurology.* the hook-shaped end of the parahippocampal gyrus, located on the basomedial surface of the temporal lobe. *Invertebrate Zoology.* in rotifers, a hooked or toothlike part of the trophi.
undamped *Physics.* of constant or increasing amplitude.
undamped wave *Physics.* **1.** a wave that is continuously produced at some source with no change to its average amplitude. **2.** a wave that does not suffer a decrease in its amplitude as a result of dissipation.
undecanal *Organic Chemistry.* $CH_3(CH_2)_9CHO$, a colorless liquid having a sweet odor; soluble in oils and alcohol, and insoluble in water; boils 115°C (5 torr); used in flavoring and in the manufacture of perfumes. Also, *n*-UNDECYLIC ALDEHYDE, HENDECANAL.
***n*-undecane** *Organic Chemistry.* $CH_3(CH_2)_9CH_3$, a colorless liquid; freezes at –26°C and boils at 195.6°C; used in petroleum research, distillation, and organic synthesis. Also, HENDECANE.
undecanoic acid *Organic Chemistry.* $CH_3(CH_2)_9COOH$, colorless crystals; insoluble in water and soluble in alcohol and ether; melts at 28.5°C and boils at 228°C (60 torr); used in organic synthesis. Also, HENDECANOIC ACID.
2-undecanone see METHYL NONYL KETONE.
undecylenic acid *Organic Chemistry.* $CH_2{=}CH(CH_2)_8COOH$, a light-colored liquid having a fruity odor; almost insoluble in water, and soluble in alcohol and ether; boils at 137°C (2 torr) and freezes at 22–23°C; used in perfumes, flavoring, plastics, and medicinally as an antifungal.
undecylenic alcohol *Organic Chemistry.* $CH_2{=}CH(CH_2)_8CH_2OH$, a colorless liquid with a citrus odor that freezes at –3°C and boils at 133°C (15 torr); soluble in 70% alcohol; used in perfumes.
undecylenyl acetate *Organic Chemistry.* $CH_3COO(CH_2)_9CH{=}CH_2$, a colorless liquid with a floral or fruity odor; soluble in alcohol; boils at 125–127°C (7 torr); used in perfumes and flavoring.
***n*-undecylic aldehyde** see UNDECANAL.
under- a prefix used to represent *under,* meaning: **1.** located below. **2.** lesser in amount, degree, or extent. **3.** lower in quality or grade. **4.** insufficient.
underaged *Materials Science.* describing a precipitation-hardening alloy that has not been heat-treated to attain the maximum strength or hardness possible at the aging temperature.
underbead crack *Metallurgy.* a crack that occurs in a heat-affected zone but does not extend to the surface.
underbed crack *Metallurgy.* a subsurface crack in the base metal found near the weld.
underbrush *Botany.* the bushes, shrubs, and small trees that grow under groups of larger trees, as in a forest or wood.
underbunching *Electronics.* in a velocity-modulation electron tube, whose operation depends on the phenomenon of electron-bunching, the condition in which the quantity of electrons within the bunches is less than that required for best operation.
undercarriage *Mechanical Engineering.* the supporting framework of a vehicle, located beneath the chassis. *Aviation.* an aircraft landing gear. *Ordnance.* a base on which the top carriage of a weapon moves; it may be fixed or movable.
undercast *Meteorology.* a ten-tenths (1.0) cloud cover as viewed from above; used in reporting in-flight weather conditions. *Mining Engineering.* an intersection of airways such that one is deflected to pass below the other.
underchain haulage *Mining Engineering.* a mine haulage system in which chains are placed below the cars at certain intervals; suitable hooks attached to the chains thrust against the axles and propel the cars.
underclay *Geology.* a layer of fine-grained sedimentary material, usually clay, immediately underlying a coal bed or forming the floor of a coal seam. Also, COAL CLAY, ROOT CLAY, SEAT CLAY, THILL, WARRANT.
underclay limestone *Geology.* a variety of thin, dense, nodular, relatively fossil-free freshwater limestone underlying coal beds.
undercliff *Geology.* a terrace or minor cliff formed along a coast out of erosional material derived from the major cliff above.
undercoat *Vertebrate Zoology.* a growth of shorter hair or fur beneath a longer growth, as in many canine species. *Materials Science.* **1.** a coat of paint, varnish, or sealant applied beneath a finishing coat. **2.** the paint, varnish, or sealant designed for such use.
underconsolidation *Geology.* a consolidation of sedimentary material that is less than normal in relation to the existing overburden.
undercooling *Thermodynamics.* see SUPERCOOLING.
undercroft *Architecture.* a vaulted or timbered church cellar.
undercurrent *Oceanography.* a current flowing beneath a surface current, usually in the opposite direction; sometimes it flows in the same direction as the surface current but at a different speed.
undercurrent relay *Electricity.* a relay that is actuated when coil current falls below a predetermined level.
undercut *Forestry.* **1.** a notch that is cut in a tree below the level of the major cut and on the side or direction of which the tree will fall. **2.** to cut such a notch. *Mining Engineering.* **1.** to cut away a horizontal section or kerf from the bottom of a block of coal so that it may more readily fall. **2.** to cut beneath or in the lower part of a coal bed. *Engineering.* a recess that is cut into the underside of an object, leaving a lip on the top. *Metallurgy.* an unfilled groove that is melted into the base metal, adjacent to the toe of a weld. *Anatomy.* the portion of a tooth lying between the survey line and the gingivae.

undercut

undercutting *Geology.* a process of erosion in which material is removed from the base of a steep slope, cliff, or other exposed rock by the action of moving water, waves, or wind. *Chemical Engineering.* a method of removing products coming off a distillation tower at a lower temperature than the ultimate boiling point range, in order to avoid contamination of the products with the heavier compound, which would distill immediately above this ultimate boiling range. *Robotics.* the machining of a part to less than the required specification.
underdamping *Physics.* weak damping that requires many oscillations to transform the initial kinetic and potential energies of an oscillator into work done against the damping mechanism.
underdeck spray *Space Technology.* in a pad deluge, the upward spraying of water into a rocket engine.
underdeck tonnage *Naval Architecture.* the enclosed volume of a vessel below its tonnage deck, expressed in register tons.
underdevelopment *Graphic Arts.* a failure to develop a photosensitive emulsion completely, usually caused by shortened development time or a decrease in temperature of the developing solution and resulting in a loss of contrast and density in the image.
underdraft *Metallurgy.* in sheet-rolling, a downward curving of metal leaving the rolls occurring when the upper roll moves faster than the lower.
underdrive press *Mechanical Engineering.* a press in which the driving mechanism is contained within or under the bed.
underexposure *Graphic Arts.* a reduction in the exposure of a photosensitive emulsion to light, either in the camera or during enlargement. Thus, **underexpose, underexposed.**

underfit stream *Hydrology.* a stream that appears to be too small to have eroded the valley in which it is flowing.
underfloor raceway *Building Engineering.* a routing system beneath a floor, used to run electrical wires around a building for various appliances.
underflow *Computer Programming.* a condition in which a computation results in a quantity smaller than can be represented.
underflow conduit *Geology.* a deposit underlying a stream channel that is permeable enough for water to pass through.
underground *Science.* occurring or situated beneath the surface of the earth.
underground burst *Ordnance.* the underground explosion of a nuclear weapon. Similarly, **underwater burst.**
underground geology see SUBSURFACE GEOLOGY.
underground ice see GROUND ICE.
underground milling *Mining Engineering.* a mining technique using several levels of glory holes, with haulage from below; used to mine a large deposit having a strong roof. Also, **underground glory-hole method.**
underground stem *Botany.* any of various stems that grow below the surface soil, such as corms, tubers, rhizomes, or bulbs.
underground stream *Hydrology.* a body of water having a definite current and flowing beneath the ground surface within a distinct channel.
underground water see GROUNDWATER.
undergrowth see UNDERBRUSH.
underhand stoping *Mining Engineering.* any of various techniques whereby ore is mined downward or from an upper level to a lower.
underhand work *Mining Engineering.* a term for mining consisting of downward picking or drilling.
underhole *Mining Engineering.* to mine out or cut away a portion of the bottom of a coal seam so as to leave the top unsupported and ready to be worked.
underlap *Telecommunications.* **1.** in facsimile transmission, the gap between the recorded elemental area in one recording line and the neighboring elemental area in the next recording line, when these areas are below normal size. **2.** in facsimile transmission, the total number by which the effective height of the scanning spot misses the nominal width of the scanning line.
underlay *Graphic Arts.* on a printed page, an area printed in light color, over which other copy will be surprinted; similar to a tint block, but not as closely tied spatially to the surprinted copy.
underlay shaft *Mining Engineering.* a shaft that is sunk in the footwall and follows the dip of a vein, sometimes extending below the ore. Also, UNDERLIER, FOOTWALL SHAFT.
underlie *Geology.* to occur or lie under. *Engineering.* to form the foundation of; support. Thus, **underlying.**
underlier see UNDERLAY SHAFT.
underlying concept *Artificial Intelligence.* in conceptual dependency theory, an abstract concept used in the representation of a verb; e.g., *run* would have PTRANS (physical transfer) as an underlying concept.
underlying set *Mathematics.* the elements of a given algebraic object, such as a group or vector space, considered merely as members of a set without regard to algebraic or topological structure.
undermelting *Hydrology.* the melting of any floating ice from beneath.
undermine *Mining Engineering.* **1.** to mine out or cut away the earth below, especially so as to weaken the formation above and cause it to fall. **2.** to form a mine below.
undermining *Geology.* see SAPPING.
undernutrition *Nutrition.* a condition of substandard bodily health resulting from energy intake that is below the level recommended for the maintenance of good health; may be caused by inadequate food supply or by failure to ingest, assimilate, or utilize necessary food elements.
underpinning *Civil Engineering.* **1.** the process of strengthening the foundation of a structure, either to protect it from earth movement or to allow the structure to be added to or enlarged. **2.** the system of supports constructed in such a process. *Mining Engineering.* the building up of the wall of a mine shaft so that it can join the one above it.
underpitch vault *Architecture.* a vault formed by the intersection of two unequal barrel vaults. Also, WELSH VAULT.
underplanting *Forestry.* a pattern of planting in which young trees are set out or seeds are sown under an existing stand.
underplate *Mechanical Devices.* an unfinished plate fastened to an armored mortise lock casing.
underream *Engineering.* to enlarge a drill hole beneath the casing.
undersaturated *Petrology.* describing igneous rocks without free silica and containing minerals that normally do not crystallize in the presence of excess silica, such as olivine or nepheline. *Physical Chemistry.* see UNSATURATED.
undershoot *Ordnance.* to deliver a projectile short of or beneath its target. *Aviation.* to land short of a runway. *Control Systems.* the amount by which the response of a system to a change in input fails to achieve the desired result.
undershoot area *Aviation.* a strip of land that has been cleared of obstructions so as to minimize the hazards to aircraft landing short of the runway.
undershot wheel *Mechanical Engineering.* a vertical waterwheel that is driven by water pushing against the blades set into its circumference.
undershrub *Botany.* a plant having a shrublike base, such as heather.
undersize *Materials Science.* **1.** describing material that is smaller than normal. Also, **undersized. 2.** describing screened material that is small enough to pass through a certain mesh.
underspin *Mechanics.* **1.** a spinning motion imparted to a ball or other projectile, causing it to rotate backward; the opposite of topspin or overspin. **2.** the condition occurring when a projectile has an insufficient rate of spin to produce proper stabilization.
understory *Forestry.* the shrubs and plants that grow beneath the main canopy of a forest.
understressing *Metallurgy.* in fatigue testing, the process of cycling at a stress level lower than the level used at the end of the test.
underthrow distortion *Telecommunications.* a distortion that occurs in facsimile transmission when the maximum signal amplitude is insufficient.
underthrust *Geology.* a fault structure in which a lower mass of rock has been moved beneath an upper, passive mass of rock, or a large-scale, low-angle fault in which the footwall appears to have moved under the hanging wall.
undertow *Oceanography.* **1.** the seaward return of water carried up the shore by breaking waves. **2.** a supposed subsurface current near shore that pulls swimmers downward and out to sea; a mistaken description of rip currents in the surf. **3.** any current moving beneath and in a different direction from surface water.
undervoltage protection *Electricity.* the automatic disconnection of a load device from its driving source when the operating voltage falls below a predetermined level; it is sometimes automated by a Zener diode with a breakdown voltage equal to the threshold voltage in series with a disconnect relay.
undervoltage relay *Electricity.* a relay that is actuated when a voltage drop falls below a predetermined value.
underwater *Science.* situated or occurring beneath the surface of a body of water. *Engineering.* designed to be used underwater.
underwater acoustics *Acoustics.* the study of acoustic energy in a water environment, normally the ocean or a simulation of an oceanographic area.
underwater camera *Optics.* a camera enclosed in a watertight housing that permits it to be used underwater and often including a supplementary lens to correct for the difference in refractive power of water relative to that of air.
underwater demolition *Military Science.* the destruction or neutralization of underwater obstacles; it is normally accomplished by the use of explosive charges placed by underwater demolition teams.
underwater demolition team *Military Science.* a group that is specially trained and equipped to perform underwater operations including reconnaissance of approaches to prospective landing beaches, locating and improving usable channels, demolition of obstacles, and clearing mines.
underwater ice see ANCHOR ICE.
underwater mine *Ordnance.* a mine that is designed for underwater deployment; it may be actuated by contact, remote control, or various types of influence.
underwater mine fairing *Ordnance.* an aerodynamic device that may be mounted on an underwater mine in order to launch it from an aircraft.
underwater navigation *Navigation.* navigation of a submerged vessel. Also, SUBMARINE NAVIGATION.
underwater obstacle *Navigation.* a natural or artificial object that is at least partly submerged and located seaward of the high-water line and may obstruct navigation.
underwater ordnance *Ordnance.* any munition that is designed to be deployed underwater; for example, a torpedo, depth charge, or underwater mine.

underwater sound *Acoustics.* the production, transmission, and reception of sound in water, used extensively for communication, navigation, and warfare, especially in subsurface vessels employing sonar and other sound systems that use electrostrictive or magnetostrictive elements in hydrophones, projectors, and transducers.

underwater sound projector *Acoustical Engineering.* an electroacoustical element that converts electromagnetic waves into acoustic energy for underwater transmission.

underwater telephone *Telecommunications.* a type of voice communication that uses underwater sound, rather than wire signals, for transmission.

underwater television *Telecommunications.* the process of using controlled television equipment beneath the surface of the water.

underwater transducer *Acoustical Engineering.* an electroacoustical element that converts acoustic energy in water into electromagnetic energy, or converts electromagnetic energy into acoustic energy for underwater transmission.

underway *Navigation.* **1.** the state of a vessel when the anchor has been raised or the lines cast off from the dock; such a vessel may be moving or stationary in the water. **2.** the state of any moving craft, especially at the beginning of movement after a standstill.

underway sampler *Oceanography.* a device consisting of a scoop and a tube, used to collect sediment samples from the ocean floor. Also, **underway bottom sampler.**

underwinding see DNA TWISTING.

underwing *Entomology.* either of a pair of hindwings on moths and certain other insects. *Invertebrate Zoology.* the common name for forest moths of the genus *Catocala,* characterized by brightly colored hindwings.

Underwood chart *Chemical Engineering.* a graphical solution of mass balances for each equilibrium stage in computing an operation for the extraction of a solvent.

Underwood distillation method *Chemical Engineering.* a procedure for computing liquid separations from binary distillation systems conducted at partial reflux.

undetected error rate *Telecommunications.* the number of bits, blocks, characters, or other elements that are received but not detected or corrected by error-control equipment, divided by the total number of such elements that are transmitted. Also, RESIDUAL ERROR RATE.

undifferentiated cell leukemia see STEM CELL LEUKEMIA.

undifferentiation *Oncology.* a condition within tumor cells in which there is a loss of normal cell differentiation, organization, and function.

undistorted wave *Telecommunications.* a periodic wave in which both attenuation and velocity of propagation are equal for all the sinusoidal elements, and in which no sinusoidal element exists at one location that does not exist for all locations.

undisturbed *Materials Science.* describing a sample of material that has been subject to very little motion and therefore reflects the material's true strength, consolidation, and permeability.

undisturbed one output *Computer Programming.* in a digital computer magnetic core memory, the "1" output of a memory cell that has not been disturbed by the application of a partial read pulse train since it was last written into.

undisturbed zero output *Computer Programming.* in a digital computer magnetic core memory, the "0" output of a memory cell that has not been disturbed by the application of a partial write pulse train since data was last read from it.

undo *Computer Science.* to reverse the effects of an operation that turned out to be undesirable.

undulant fever see BRUCELLOSIS.

undulate *Microbiology.* of a colony, having a wavy, curved border.

undulating light *Navigation.* a nautical light whose illuminating apparatus is raised and lowered, thus creating an occulting effect for an observer on the sea or ground and allowing it to be seen by aircraft. *Aviation.* an aeronautical light having a continuous beam that is augmented by cyclic brighter flashes.

undulating membrane *Invertebrate Zoology.* in some protozoans, a thin rippling sheet of cytoplasm used in locomotion or feeding.

undulation *Physics.* **1.** a wave. **2.** the characteristic motion of waves. **3.** a wavelike motion in any medium.

undulation of the geoid see GEOID HEIGHT.

undulator *Electromagnetism.* an experimental device used to impart an undulating course to a stream of relativistic electrons in a linear accelerator by passing between pairs of magnetic poles of alternating polarity, causing them to emit microwave radiation. Also, MOTZ UNDULATOR.

undulatory extinction *Optics.* the dark areas that appear in the field of view of a microscope when the stage holding the sample is turned.

unequal crossing-over *Molecular Biology.* a recombination event that results in one of a pair of chromatids having a single copy of a segment, while the other has three copies.

unexploded explosive ordnance *Ordnance.* an explosive munition that has been prepared for actuation and deployed in such a manner that it creates a hazard to operations, personnel, installations, or material, and has not exploded, due to malfunction, design, or any other cause.

unfair *Computer Science.* of a scheduling algorithm, not guaranteed to eventually service every request.

unfavorable current *Navigation.* a current that slows the speed of a vessel. Similarly, **unfavorable winds.**

unfinished *Engineering.* of a surface or material, not treated with polish, paint, or the like. *Textiles.* **1.** of woven cloth, not sheared after looming. **2.** of worsted, having a slight nap.

unfinished bolt *Mechanical Devices.* a bolt having an unfinished hexagonal head and a finished thread.

unfired pressure vessel *Chemical Engineering.* a pressure vessel or container not directly exposed to a heating flame.

unfired tube *Electronics.* a transmit-receive tube in a radar set with no visible glow discharge to indicate that it is firing.

unformatted file *Computer Programming.* a collection of data that has not been arranged in any specific or predefined order.

unfreezing *Geology.* a process in which stones move up to the surface as a result of alternating cycles of freezing and thawing of the soil.

ung. ointment. (From *unguentum.*)

ungemachite *Mineralogy.* $K_3Na_8Fe^{+3}(SO_4)_6(NO_3)_2 \cdot 6H_2O$, a colorless to yellowish trigonal mineral, occurring as thick, tabular crystals, having a specific gravity of 2.287 and a hardness of 2.5 on the Mohs scale; found at Chuquicamata, Chile.

ungual *Vertebrate Zoology.* of or relating to a nail, claw, or hoof.

unguent [ung´gwənt] *Medicine.* an ointment, salve, or cerate.

unguentum [ung gwen´təm] *Pharmacology.* ointment (in prescriptions).

unguiculate *Zoology.* having claws. *Biology.* resembling a claw.

unguided missile *Space Technology.* a missile that is aimed at launch but not guided during flight.

unguis *Botany.* a narrow, clawlike base attaching certain petals to a receptacle. *Anatomy.* see NAIL. *Zoology.* see FANG.

ungula *Vertebrate Zoology.* the hoof of an animal.

Ungulata *Vertebrate Zoology.* a former classification including all the hoofed mammals, now separated into two orders, Perissodactyla (odd-toed animals such as the horse and the rhinoceros) and Artiodactyla (even-toed animals such as the pig and the giraffe).

ungulate [ung´gyə lāt´; ung´gyə lət] *Vertebrate Zoology.* **1.** any hoofed mammal. **2.** a member of the order Perissodactyla or Artiodactyla.

ungulicutate *Vertebrate Zoology.* having nails or claws.

unguligrade *Vertebrate Zoology.* characterized by standing or walking on hoofs.

Unh unnilhexium.

uni- a prefix meaning "one."

unialgal *Botany.* relating to a single algal cell or individual.

uniaxial crystal *Crystallography.* a crystal having one direction along which the wave-normal velocity for monochromatic light is constant, so that no double refraction results. Such a crystal has two principal refractive indices and is tetragonal, hexagonal, or rhombohedral.

uniaxial indicatrix *Crystallography.* the ellipsoid illustrating the variation of the refractive indices of a uniaxial crystal for light waves of a given wavelength in their direction of vibration. One axis represents the refractive index of the extraordinary ray, and the other two axes perpendicular to it represent the refractive indices of the ordinary ray. Also, **uniaxial indicators.**

uniaxial stress *Mechanics.* a state of stress in which only one of the principal stresses is nonzero.

unicameral *Anatomy.* having only one cavity or compartment.

Unicast process *Materials Science.* a ceramic mold precision casting process in which molten metal is poured into a ceramic mold cavity and allowed to solidify.

unicellular *Biology.* describing organisms that consist of a single cell; characteristic of protozoans and bacteria and also many algae and fungi.

uniceps *Anatomy.* of a muscle, having a single head or origin.

unicostate *Botany.* having a single large vein running down the center, as a midrib on a leaf.

unicuspid *Anatomy.* a tooth with only one cusp.

unicuspidate *Anatomy.* having only one cusp.

unidentified flying object *Astronomy.* a general term for any of various objects reportedly sighted in the sky and not positively identifiable as an aircraft, artificial satellite, or other man-made vehicle, or as a star, planet, meteor, or other celestial body or occurrence; typically described as having the appearance of a glowing disk and often popularly assumed to have come from outer space. Also, UFO.

unidirectional *Physics.* having a single direction along a straight path. *Fluid Mechanics.* specifically, of a fluid, flowing in only one direction. *Acoustical Engineering.* having high directivity in one direction, as compared to all directions. Thus, **unidirectional microphone.**

unidirectional antenna *Electromagnetism.* a highly directional antenna having a single, well-defined direction of maximum gain.

unidirectional coupler see DIRECTIONAL COUPLER.

unidirectional log-periodic antenna *Electromagnetism.* a directional antenna having a broad band and essentially constant impedance and radiation pattern for all frequencies.

unidirectional pulse-amplitude modulation *Telecommunications.* pulse-amplitude modulation in which all pulses rise in the same direction. Also, SINGLE-POLARITY PULSE-AMPLITUDE MODULATION.

unidirectional pulses *Electronics.* a set of electrical pulses that all have the same pulse polarity.

unidirectional replication *Genetics.* the movement of a single replicating fork from a replication bubble.

unidirectional solidification *Materials Science.* a process for growing single crystals; used mostly to produce superalloys for turbine airfoils, where single crystals are preferred because grain boundaries decrease the resistance to creep.

unidirectional transducer *Electronics.* a transducer that transmits signals in one direction only.

uniface *Archaeology.* a stone tool having only side or surface flaked, rather than both. Also, **uniface tool.**

unifacial *Archaeology.* of a stone tool, having flakes chipped from only one side or surface. Such a tool is **unifacially flaked.**

unification algorithm *Artificial Intelligence.* an algorithm for finding a substitution of terms for variables in a set of predicate calculus literals that will make the literals the same.

unified field theory *Physics.* **1.** any theory that combines two or more field theories; e.g., James Clerk Maxwell unified the field theories of electricity and magnetism by developing the theory of the electromagnetic field on a mathematical basis. **2.** specifically, the effort by Albert Einstein and others to unify gravitational force and electromagnetic force with a single set of laws and, more generally, to provide a geometrical interpretation for all physical interactions.

unified screw thread *Design Engineering.* a screw thread having three series of threads embedded in it: coarse, fine, and extra fine.

unifier *Artificial Intelligence.* a set of substitutions of terms for variables that, when applied to a set of predicates, will make them all identical.

unifilar suspension *Engineering.* the suspension of an object using a single thread, wire, or strip.

uniflow engine *Mechanical Engineering.* a steam-driven engine in which steam enters the cylinder through valves and exits through holes that are uncovered as the piston completes its stroke; it is designed to eliminate initial condensation.

unifocal *Optics.* relating to or derived from a single focus.

unifoliate *Botany.* having a single leaf.

unifoliolate *Botany.* having or being a leaf that is structurally compound, but that has only one leaflet, as in the orange.

uniform bound *Mathematics.* a sequence $\{f_n\}$ of functions or mappings between normed spaces X and Y is said to possess a uniform bound M (or to be **uniformly bounded**) if, for every $x \in X$ and every n,

$$\|f_n(x)\|_Y < M,$$

where $\|\ \|_Y$ is the norm in Y.

uniform boundedness theorem *Mathematics.* **1.** let f_n be a sequence of real-valued continuous functions defined on a complete metric space X. If the sequence $\{f_n(x)\}$ is bounded for each $x \in X$, then the sequence $\{f_n\}$ is uniformly bounded on some open subset of X. **2.** more generally, let $T_n: X \to Y$ be a sequence in the space $L(X,Y)$ of continuous linear mappings between Fréchet spaces X and Y. Suppose that for each $x \in X$, the sequence $\{\|T_n(x)\|_Y\}$ is uniformly bounded, where $\|\ \|_Y$ is the norm in Y. Then the sequence $\{\|T_n\|_{L(X,Y)}\}$ is also uniformly bounded, where $\|\ \|_{L(X,Y)}$ is the norm in $L(X,Y)$. Also, **uniform boundedness principle.**

uniform circular motion *Mechanics.* the motion of an object around a fixed point or axis at a constant angular speed, and at constant radius.

uniform continuity *Mathematics.* a function f is said to have the property of uniform continuity (or to be **uniformly continuous**) on a set D if for each $\varepsilon > 0$, there is a $\delta > 0$ such that $|f(x_1) - f(x_2)| < \varepsilon$ for all $x_1, x_2 \in D$ with $|x_1 - x_2| < \delta$. The choice of δ is independent of x_1 and x_2.

uniform convergence *Mathematics.* **1.** a sequence $\{a_n(x)\}$ of functions defined on an interval $[a,b]$ is said to have the property of uniform convergence (or to **converge uniformly**) to $A(x)$ in $[a,b]$ if, for each $\varepsilon > 0$, there is an integer $N > 0$ so that $|a_n(x) - A(x)| < \varepsilon$ for all $n > N$ and all x in $[a,b]$. The choice of N is independent of x. **2.** the infinite series $\sum_{n=1}^{\infty} a_n(x)$ of functions is said to converge uniformly to $S(x)$ in $[a,b]$ if the sequence of partial sums converges uniformly to $S(x)$ in $[a,b]$. **3.** the integral $\int_a^{\infty} f(x,t)\,dx$ is said to converge uniformly to $g(t)$ for $t \in [a,b]$ if, for each $\varepsilon > 0$, there is a number $M > 0$ so that

$$\left|\int_a^u f(x,t)\,dx - g(t)\right| < \varepsilon$$

for all $u > M$ and all t in $[a,b]$. The choice of M is independent of t. Uniform convergence always implies convergence.

uniform cost search *Artificial Intelligence.* a search that examines operations in order of increasing cost from the start state.

uniform distribution *Statistics.* a probability distribution having constant density; a statistical distribution in which each value has an equal probability of occurring. Also, RECTANGULAR DISTRIBUTION.

uniform field *Physics.* a field (vector or scalar) that, at a given instant in time, has the same value at all points within a specified region of interest.

uniform homeomorphism *Mathematics.* a uniformly continuous homeomorphism between metric spaces whose inverse is also uniformly continuous.

uniformitarianism *Geology.* a principle stating that ancient geological events can be explained or understood in terms of the geologic processes, phenomena, and forces that occur in the present; based on the belief that current geologic processes act in the same manner and with the same intensity as they did in the past.

uniformity *Mathematics.* given a topological space X, a family F of subsets of the Cartesian product $X \times X$ with the following properties: (a) for every point $x \in X$, (x,x) is an element of every set of F; (b) for each set V in F, $V^{-1} = \{(y,x) : (x,y) \in V\}$ is also in F; (c) for each set V in F, there is a set V^* in F such that $V\ \{(x,z) : (x,y) \in V^*$ and $(y,z) \in V^*$ for some y in $X\}$; (d) F is closed under finite intersections; and (e) any subset of $X \times X$ that contains a member of F is itself a member of F. Used to define a uniform topology on X. Also, **uniform structure on X.**

uniform line *Electricity.* a line in which the electric properties of any short element of length are identical throughout its extent.

uniform load *Mechanics.* a uniformly distributed load over part of or the whole of a length or surface.

uniform luminance *Optics.* the luminance that occurs when the brightness of an extended light source is exactly the same at all points of its surface.

uniformly diffuse reflector see PERFECTLY DIFFUSE REFLECTOR.

uniformly most powerful test *Statistics.* the test of a hypothesis that maximizes the power over all possible values of the unknown parameter in the alternative hypothesis.

uniform norm *Mathematics.* let f be an element of the vector space $C(X)$ of all bounded continuous real-valued functions on a topological space X. The uniform norm of f is

$$\|f\|_{C(X)} = \sup_{x \in X} |f(x)|;$$

that is, the supremum over X of $|f(x)|$. The space $C(X)$ is complete with the uniform norm since the uniform limit of a sequence of bounded continuous functions is itself bounded and continuous.

uniform plane wave *Electromagnetism.* a plane wave whose electric and magnetic field intensities are constant in amplitude over an entire equiphase surface; such a wave can be found only in free space and at an infinite distance from the source.

uniform property *Mathematics.* any property of a topological space that is preserved under uniform homeomorphisms, such as the properties of completeness and of Cauchy convergence.

uniform rate hypothesis *Evolution.* a theory stating that any two lineages in a phylogenetic tree diverge at a constant rate with respect to each other.

uniform space *Mathematics.* a topological space X on which a uniform topology can be defined. If X is Hausdorff and F has a countable base, then X is metrizable; conversely, every metric space is uniform.

uniform topology *Mathematics.* given a uniformity F on a topological space X, the topology whose open sets are the subsets U of X with the property that there is a set V in F such that the set $\{y : (x,y) \in V$ for some $x \in U\}$ is a subset of U. If a uniform topology exists for X, then X is said to be a *uniform space*.
unijunction transistor *Electronics.* a three-terminal transistor containing an emitter and two bases, and possessing negative-resistance characteristics; used mainly in switching circuits.
unilateral *Science.* of or relating to only one side.
unilateral anesthesia see HEMIANESTHESIA.
unilateral conductivity *Electronics.* the theoretical property of a circuit element to conduct electrical current in one direction only; in practice, even the best diode allows some reverse current to flow.
unilateral hermaphroditism *Zoology.* the presence of gonadal tissue typical of both sexes on one side, and of an ovary or testis on the other.
unilateralization *Electronics.* a technique similar to neutralization in which both the resistive and reactive components of feedback in a device are completely neutralized by external feedback to prevent undesired oscillations; the result is a device that effectively has no feedback.
unilateral surface *Mathematics.* a nonorientable two-dimensional manifold (e.g., the Möbius strip or Klein bottle) that has only one side.
unilateral tolerance method *Design Engineering.* a tolerance methodology applied to dimensioning where the tolerance is either a plus or a minus from the stated dimension; the dimension is the size or location nearest the critical condition of the material and the tolerance is either a plus or minus, whichever allows variation away from the critical point.
unilateral transducer see UNIDIRECTIONAL TRANSDUCER.
unilineal descent *Anthropology.* a practice in which one is related through a group of kin solely on the basis of one sex or the other.
unilinear cultural evolution see CULTURAL EVOLUTION.
unilugate *Botany.* of a pinnacle leaf, having only one pair of leaflets.
unimodular matrix *Mathematics.* a square matrix with determinant 1.
unimolecular *Physical Chemistry.* involving only a single molecule, rather than interaction between two separate molecules. Thus, **unimolecular reaction.**
unineme hypothesis *Molecular Biology.* a theory stating that a chromatid has only a single DNA duplex, which reaches from one end of the chromatid to the other.
uninhibited neurogenic bladder *Medicine.* a dysfunction of the bladder due to a lesion in the region of the upper motor neurons, marked by urgency and frequent involuntary voiding.
unintentional radiation exploitation *Military Science.* the operational use of non-information-bearing elements of electromagnetic energy unintentionally emanated by targets of interest.
uninterruptible power system *Electricity.* a power system that provides protection against primary power failure and variations in power-line frequency and voltage; commonly consists of a storage battery, battery charger, solid-state inverter, and solid-state switching circuit, as applied in computer systems; it may be used on-line between power line and load to provide voltage regulation and suppressed transients, or off-line and switched in only when utility power fails. Also, **uninterruptible power source** or **supply.**
union the process of uniting, or something that unites; specific uses include: *Surgery.* the process of healing; the adhesion or joining of structures or tissues, such as the edges of a wound or interfaces of a fractured bone. *Mechanical Devices.* any of a number of devices used to connect machine pipes and other parts. *Design Engineering.* a flanged or screwed pipe coupling around the outside of a joint, usually in the form of a ring fitting. *Computer Programming.* **1.** the logical OR of two propositions resulting in a True value if one of the propositions is True. **2.** the joining of two or more files or other objects. **3.** a set union formed from two sets of data. *Mathematics.* **1.** a binary set operation; denoted $\cup$ and read "union," "join," or "cup." The union $A \cup B$ of two sets A and B is the set of all objects that are elements of at least one of A and B. **2.** the union U of a collection $\{X_\alpha\}$ of sets is the set having the following properties: (a) if x is an element of X_α for some α, then x is an element of U; and (b) if x is an element of U, then x is an element of X_α for some α. Also called a *join* or *sum*; often denoted $\cup_\alpha X_\alpha$, or similar notation. **3.** the union $U = G_1 \cup G_2$ of two graphs G_1 and G_2 is a graph with $V(U) = V(G_1) \cup V(G_2)$ such that two vertices are joined by an edge in at least one G_i.
union dyeing *Textiles.* the process of dyeing fabrics made from more than one yarn type in one bath to produce either a uniform shade or a mottled effect caused by the different dyeing properties of the fibers.
union fabric *Textiles.* a fabric made from more than one fiber type, especially fabric made with a cotton warp and a wool filling.
Unionidae *Invertebrate Zoology.* a large family of freshwater mussels, bivalve mollusks in the subclass Eulamellibranchia; the outer shell surface is rough and brown or black; the inner surface is pearly and used for buttons; larvae are fish ectoparasites.
union joint *Mechanical Devices.* a threaded-pipe coupling used to connect two threaded pipes, which allows for disconnection without the need to rotate other pipe lengths.
union of languages *Computer Science.* a language whose sentences are members of any of its component languages.
unipara *Medicine.* a female who has had one pregnancy that resulted in a viable fetus, regardless of whether the infant was alive at birth, and regardless of whether it was a single or multiple birth.
uniparental inheritance *Genetics.* a form of inheritance in which the offspring possess the phenotype of only one parent.
unipolar *Electricity.* having a single pole or polarity or direction. *Anatomy.* of a nerve cell or other structure, having but a single pole or process. *Psychology.* describing a mood disorder in which only depressive episodes occur.
Unipolarina *Invertebrate Zoology.* a suborder of parasitic sporozoan protozoans in the order Myxosporida.
unipolar transistor *Electronics.* **1.** a transistor in which the main conduction path is through P-type or N-type material, but not both, and charge carriers are of only one polarity. **2.** see FIELD-EFFECT TRANSISTOR.
unipolar transmission see NEUTRAL TRANSMISSION.
unipole *Electromagnetism.* a hypothetical antenna that receives or transmits uniformly in all directions.
uniport *Biochemistry.* the transmembrane movement of one kind of solute by a carrier not linked to another solute transport across the same membrane.
unipotency *Developmental Biology.* the ability of a part to develop in one manner only. *Cell Biology.* the ability of a cell to develop into just one type of cell. Thus, **unipotent, unipotential.**
unipotential cathode see INDIRECTLY HEATED CATHODE.
unipotential electrostatic lens *Electronics.* an electrostatic lens having three apertures, with the same voltage difference between the central aperture and each of the outer apertures.
unique DNA *Molecular Biology.* any sequence of nucleotides in DNA that occurs only one time in a genome.
unique factorization domain *Mathematics.* an integral domain in which: (a) every nonzero nonunit element can be written as the product of a finite number of irreducible elements or factors; (b) any two such factorizations are composed of exactly the same number of factors; and (c) the factors in any factorization can be permuted so that corresponding factors are associates of one another.
unique factorization theorem *Mathematics.* a theorem stating that any positive integer can factored in exactly one way; follows from the fact that the integers form a unique factorization domain.
uniserial *Biology.* arranged in a single row or a series of rows along only one side of an axis.
uniserrate *Biology.* having one row of serrations on a single edge.
unisexual *Biology.* having the reproductive organs and capabilities of one sex only; male or female rather than hermaphroditic.
unit *Physics.* **1.** a quantity that is used and referred to as an accepted standard. **2.** relating to the quantity of one. **3.** any type of quantity that qualifies a physical measurement, such as a velocity measurement, which may be qualified by cm/sec, km/hr, and so on. *Engineering.* any device that can operate independently, such as a radio receiver or an oscilloscope. *Military Science.* **1.** any military organization that is structured according to competent authority, especially one that is part of a larger organization. **2.** the organizational title of a subdivision of a task force group. *Industrial Engineering.* see UNIT OF ISSUE. *Mathematics.* **1.** the multiplicative identity of a ring, field, etc. The phrases "ring with unit" or "ring with identity" are commonly used. Also, **unit element, unity. 2.** an element of a ring that has both a left and right multiplicative inverse. **3.** having length or norm equal to 1. For example, a **unit vector** is a vector of length one. A **unit binormal** (normal, tangent) is a unit vector in the same direction as a given binormal (resp. normal, tangent) vector. The **unit interval** is the interval (0, 1) or [0, 1]. **4.** centered at the origin and having radius 1. For example, the **unit circle** and **unit *n*-sphere** are the circle of radius 1 and the sphere of radius 1 centered at the origin in the plane and in n-space, respectively. The **unit ball** is the (open) interior of the unit circle or unit n-sphere; sometimes the unit ball is defined to be the (closed) union of the unit sphere and its interior.

unitarian theory see MONOPHYLETIC THEORY.

unitarity condition *Particle Physics.* a condition in which the scattering matrix for a process must be unitary because the probability for the system to end in some final state must be unity.

unitary air conditioner *Mechanical Engineering.* a small, self-contained air conditioner that is used to cool a specified space.

unitary coupling *Electromagnetism.* a perfect coupling between two coils in which all magnetic flux produced by one coil passes entirely through the other coil.

unitary decuplet *Particle Physics.* a configuration of ten hadrons arranged so that their hypercharge values and isospin values form a symmetrical pattern. Similarly, **unitary octet.**

unitary group *Mathematics.* the group of unitary transformations on a given complex finite-dimensional vector space.

unitary matrix *Mathematics.* a square invertible matrix (usually with complex entries) whose inverse equals its adjoint.

unitary module *Mathematics.* a left (resp. right) unitary module is a left (resp. right) module A over a ring R with unit 1_R such that $1_R a = a$ (resp. $a1_R = a$) for every element a in A. If, in addition, R is a division ring, then A is called a left (resp. right) vector space. The distinction between left and right is dropped in the commutative case.

unitary operator *Quantum Mechanics.* a matrix or linear operator that transforms real linear operators into themselves, so that the resulting unitary transformation leaves unchanged the normalization and orthogonality properties of a set of wavefunctions.

unitary space *Mathematics.* an inner product space.

unitary spin *Particle Physics.* a quantum number that indicates the supermultiplet (singlet, octet, and so on) to which a particle belongs, as indicated by SU_3 symmetry.

unitary structure factor *Crystallography.* the ratio of the structure amplitude, $|F(hkl)|$, to its maximum possible value, that is, the value it would have if all atoms scattered exactly in phase; denoted by U.

$$U(hkl) = |F(hkl)|/|\Sigma f(hkl)_j|,$$

where $f(hkl)_j$ contains a temperature factor expression.

unitary symmetry *Particle Physics.* one of the internal symmetry laws obeyed by the strong interactions of elementary particles predicting that as far as strong interactions are concerned, elementary particles can be grouped into multiplets containing, say, 1, 8, 10, or 27 particles and that the particles in each multiplet may be regarded as different states of the same particle. Also, COLOR SU_3, SU_3 SYMMETRY.

unitary transformation *Mathematics.* a linear transformation on a vector space V over the real or complex numbers that preserves norms and inner products; the collection of such transformations forms a unitary group. A linear transformation is unitary if and only if its matrix representation is unitary.

unit cell *Crystallography.* the basic building block of a crystal; the smallest unit of the lattice for a given crystal that displays the symmetry of the lattice. It is characterized by three vectors, *a, b,* and *c,* that form the edges of a parallelepiped and the angles between them, α, β, and γ (γ between *a* and *b*).

unit character *Genetics.* a trait or characteristic, usually controlled by a single gene or a group of inseparable genes, that is transmitted as a unit in heredity.

unit clause *Artificial Intelligence.* a logical formula consisting of only a single literal.

unit construction *Building Engineering.* the fabrication of a portion of a building, consisting of two or more walls and floor or ceiling construction, off its site location.

unit delay *Electronics.* the delay of a signal from the input to the output of a transducer or amplifier by a time interval, which is related to the operation of the circuit; for example, in a digital process controlled by a clock, a delay corresponding to one clock style.

unit die *Metallurgy.* a die block that has several cavity inserts for making different kinds of castings.

United States airways code *Meteorology.* an aviation weather-observation communication system with a unique synoptic code. Also, AIRWAYS CODE.

United States standard dry seal thread *Design Engineering.* a modified pipe thread used in pressure sealing applications in which assembly is accomplished without lubrication or sealer in refrigeration pipes, automotive and fuel line fittings, or gas and chemical shells.

unitegmic *Botany.* describing an ovule with a single coat or integument.

uniterm *Computer Programming.* a word, symbol, or number that is used as a descriptor to retrieve information from a collection; especially such a descriptor used in a coordinate indexing system.

uniterm system *Computer Technology.* a data-recording system that is based on classification of keywords in a coordinate indexing system; used by libraries.

unit heater *Mechanical Engineering.* a forced-convection heating device consisting of an encased heating surface with a fan and motor.

unitized body *Engineering.* an automotive body in which the body and the frame are contained in one unit. Thus, **unitized-body construction.**

unit length *Telecommunications.* in message transmission, a basic time element that is used to determine code speeds.

unit loading *Military Science.* the loading of troop units with their equipment and supplies into the same aircraft, ships, or vehicles.

unit magnetic pole *Electromagnetism.* a standard unit of magnetic pole strength equivalent to that exisitng when two equal magnetic poles are placed 1 centimeter apart in a vacuum and experience a repulsive force of one dyne.

unit mold *Engineering.* a plastics mold consisting of a single cavity; used for the molding containers that are difficult to blow-mold.

unit of fire *Ordnance.* a measure of ammunition supply expressed in a specific number of rounds per weapon.

unit of issue *Industrial Engineering.* the number of individual items packaged together for sale or distribution; e.g., if light bulbs are packed in pairs, the unit of issue is two bulbs.

unit operations *Chemical Engineering.* the major and distinct physical operations of chemical engineering, such as evaporation, distillation, extraction, fluid transport, heat and mass transfer, filtration, crystallization, mixing, drying, crushing and grinding, size separation, and conveying.

unit operator *Mathematics.* the identity operator on any space or algebraic object X. Denoted as 1_X or other similar notation, where $1_X(x) = x$ for all x in X.

unit power *Metallurgy.* the net amount of horsepower per cubic inch per minute required to remove a unit column of metal in unit time.

unit preference *Artificial Intelligence.* a resolution-theorem-proving strategy of preferring to resolve unit clauses, or clauses with the fewest literals.

unit process *Chemical Engineering.* a stage in a plant operation that includes a chemical conversion or change, such as ester formation in chemical manufacturing.

unit record *Computer Programming.* a record of information that is kept as a whole.

unit replacement *Ordnance.* a repair method in which a group of parts is replaced by a new group of parts.

unit state *Electrical Engineering.* a unit condition that is important to the procedures used to collect performance data.

unit strain *Mechanics.* the effect per unit length on a material that is subject to stress; e.g., the lengthening per unit area for a tensile strain.

unit strength *Military Science.* an assessment of a unit's power based on the number of personnel, the quantity of supplies, the armament equipment and vehicles, and the total logistic capabilities; it may be applied to friendly or hostile forces.

unit stress *Mechanics.* the stress on a structure at a certain place, expressed in units of force per unit of cross-sectional area.

unit string *Computer Programming.* a character string consisting of only one character.

unit substation *Electrical Engineering.* a substation consisting of transformers that are mechanically and/or electrically connected to switchgear or motor control assemblies.

unit system *Physics.* a collection of standard physical quantities that are consistent and convenient in the measurement or description of other physical quantities. *Graphic Arts.* on a rotary press, a design configuration in which each plate cylinder has its own impression cylinder.

unit test *Computer Programming.* the actual execution of an individual component of a program to check the accuracy of its performance.

unit train *Transportation Engineering.* a freight train that operates as a single unit, without uncoupling or reassembling of cars, to deliver a bulk commodity to a specific destination and return to the point of origin for another load. *Mining Engineering.* an efficient system of coal delivery in which a distinctively marked string of cars, loaded to full visible capacity, journeys without service frills and without stops.

unitunicate ascus *Mycology.* a spore sac or ascus possessing a single wall which may contain layers that remain together when the spores are ejected.

unity *Mathematics.* the number one; a quantity regarded as one.

unity gain bandwidth *Electronics.* the bandwidth of an amplifier, shown on its frequency-response curve as the range between the upper and lower frequencies at which the gain of the amplifier has a value of 1 (unity).

unity power factor *Electricity.* a power factor equal to 1.0 obtained when current and voltage are exactly in phase; a condition in which the active (or real) power is equal to the apparent power.

uniugate *Botany.* of or relating to a pinnacle leaf with only one pair of leaflets.

univ. universal; universally; university.

univalent *Chemistry.* having a valence of one. Also, MONOVALENT. *Genetics.* describing a single unpaired chromosome at meiosis, when the other chromosomes present are paired and homologous.

univalent function *Mathematics.* a one-to-one analytic function that maps the unit disk into itself. Families of univalent functions are often normalized so that $f(0) = 0$ and $f'(0) = 1$ for each f in the family. Also, SCHLICHT FUNCTION.

univalve *Invertebrate Zoology.* having a shell consisting of one piece, as a mollusk.

univariant system *Thermodynamics.* a system that, by the phase rule, is restricted to one degree of freedom; i.e., $C + 2 - P = F = 1$, where C is the number of components, P is the number of phases, and F is the number of degrees of freedom.

univariate analysis *Statistics.* the study of the distribution of a single random variable.

universal *Astronomy.* of or relating to the universe. *Science.* relating to the whole or to what is generally true; comprehensive. *Mechanical Engineering.* see UNIVERSAL JOINT.

universal algebra *Mathematics.* that area of category theory that deals with the generalization of theorems concerning one algebraic object to theorems concerning other algebraic objects.

universal antidote *Toxicology.* an activated charcoal preparation that adsorbs many chemicals.

universal asynchronous receiver transmitter *Computer Technology.* a peripheral interface device designed for converting serial data to parallel data and vice versa, thus enabling communication between a word-parallel data terminal and a bit-serial network.

universal chuck *Engineering.* a lathe chuck having concentric jaws; used in cylindrical work. *Mechanical Engineering.* see SELF-CENTERING CHUCK.

universal coupling see UNIVERSAL JOINT.

universal covering space *Mathematics.* a (topological) covering space that is simply connected.

universal cultural feature *Anthropology* a basic quality that is present in all groups of people, such as housing, childrearing practices, sex and age status, and so on.

universal developer *Graphic Arts.* any of various photographic developing solutions that can be used, with adjustments to the dilution, to process either films or papers.

universal donor *Hematology.* a person having group O blood (International Classification), who can consequently donate blood to persons of all blood types; such blood (blood cells are preferred, rather than whole blood) is sometimes used in emergency transfusion because it offers the least risk of incompatibility.

universal gas constant see GAS CONSTANT.

universal grammar *Linguistics.* the theory that all existing languages are basically the same in their underlying nature although expressed in different surface manifestations.

universal grinding machine *Mechanical Engineering.* a machine tool with a grinding head whose spindles are parallel with a reciprocating table that carries a workhead and support tailstock, allowing for a wide range of plain, tapered, and surface grinding.

universal gripper *Robotics.* a general-purpose robotic gripper that can grasp almost any kind of object.

universality *Physics.* a hypothesis claiming that for substances within a broad class having various characteristics, the critical exponents associated with the substances are the same and, furthermore, depend only on the microscopic symmetries. *Chaotic Dynamics.* behavior that seems to be generic, or characteristic of many different systems.

universality class *Physics.* a class of substances whose members have a common critical exponent relating to the hypothesis of universality.

universal joint *Mechanical Engineering.* a geared joint placed between two shafts that allows the shafts to turn or swivel at an angle. Also, UNIIVERSAL COUPLING.

universal language *Linguistics.* a language that can be understood and used by speakers of many different languages; this may be an artificial language, e.g., Esperanto, or a widely taught natural language, e.g., English. *Computer Programming.* a programming language that can express any program that could be executed on a Turing machine.

universal mill *Metallurgy.* a rolling mill that allows rolls with a vertical axis to roll the edges of the metal between some of the passes through the horizontal rolls.

universal motor *Electricity.* a series-wound motor designed to operate on both DC current and on a single-phase AC current; commonly used in small appliances and motorized tools.

universal object *Mathematics.* an object I in a category C is said to be universal if for each object C of C there exists one and only one morphism $I \to C$. Any two universal objects in a given category are equivalent. Also, REPELLING OBJECT, INITIAL OBJECT.

universal output transformer *Acoustical Engineering.* a transformer with output taps on its secondary winding so that it can be impedance-matched to nearly any speaker electrically attached to it.

universal plotting sheet *Navigation.* a plotting chart with either the latitude or the longitude lines omitted, thus allowing the navigator to draw them in as appropriate for any part of the world.

universal polar stereographic grid *Cartography.* a grid system based on the polar stereographic map projection and used by the U.S. military for its maps of polar regions north of 84° north and south of 80° south latitude.

universal potassium alum *Mineralogy.* $KAl(SO_4)_2 \cdot 12H_2O$, a colorless cubic mineral occurring as an efflorescence in clay-rich rocks, brown coals, fumaroles, and solfataras, having an astringent taste, a specific gravity of 1.76, and a hardness of 2 on the Mohs scale.

universal product code *Computer Technology.* a ten-digit code appearing as bars of varying widths, used for labeling retail products; it can be read by an optical scanner and input to a computer. Also, UPC.

universal recipient *Hematology.* an individual having type AB blood, who consequently can receive a blood transfusion from any blood type.

universal robot *Robotics.* an all-purpose manipulator that can be programmed to perform any desired task.

universal set *Mathematics.* any set U such that, in a given context, all sets under discussion are subsets of U. Also, UNIVERSE.

universal sketchmaster *Photogrammetry.* an instrument used to superimpose an image of a photograph, which may be vertical or oblique, on a map or map manuscript for cartographic purposes; the universal sketchmaster may or may not rectify the image.

universal stage *Optics.* a platform on a polarizing microscope with multiple axes that allows different crystals to be inserted into the illuminating beam of light. Also, FEDOROX STAGE, U STAGE.

Universal Time *Horology.* Greenwich Mean Time, the same time as mean time at the meridian of Greenwich, England.

Universal Time 0 *Horology.* a system of time measurement based on the diurnal transit of a star across the observer's meridian, uncorrected for polar or seasonal variations in the earth's rotation.

Universal Time 1 *Horology.* Universal Time 0 corrected for variation in the earth's polar motion.

Universal Time 2 *Horology.* Universal Time 1 corrected for seasonal variations in the earth's rotation.

Universal Time Coordinated *Horology.* Universal Time adjusted to differ by an integer number of seconds from International Atomic Time. Also, COORDINATED UNIVERSAL TIME.

universal transmission function *Geophysics.* a mathematical model describing the propagation of infrared radiation in the atmosphere.

universal transverse Mercator grid *Cartography.* a grid system based on the transverse Mercator map projection and used in the U.S. for maps of areas between the latitudes of 84° north and 80° south.

universal tube *Biotechnology.* a cylindrical shoulderless screw-cap glass vial with a volume of about 25 milliliters.

universal Turing machine *Computer Programming.* a Turing machine that can be programmed to simulate any other Turing machine.

universal veil *Mycology.* in mushroom fungi belonging to the family Agaricaceae, the tissue layer that covers the fruiting body or basidioma during its development.

universal vise *Engineering.* a vise that may be rotated in all directions and fixed in any position. Also, TOOLMAKER'S VISE.

universal wavelength function *Optics.* any of four mathematical equations that permit the refractive index of some material to be calculated when refractive indices have been determined for four standardized wavelengths.

universe *Astronomy*. the total sum of all that exists; the physical system that encompasses all known space, matter, and energy, either existing now, having existed in the past, or postulated to exist in the future. *Statistics*. the aggregated population that is the target of statistical study. Also, STATISTICAL UNIVERSE. *Mathematics*. see UNIVERSAL SET. (From a Latin term meaning literally, "all into one" or "all together as one.")

univoltine *Biology*. producing a single set of offspring in one year or breeding season.

UNIX [yoo´niks] *Computer Programming*. a general-purpose, multiprogramming, timesharing operating system developed at Bell Laboratories in the early 1970s. It features a tree-structured file system, the ability to connect separate programs via pipes, the ability to write programs at the operating system shell level, on-line documentation, and a large collection of utility programs.

unknown *Mathematics*. a symbol, usually x, y, or z, representing an unknown quantity. *Military Science*. an unidentified target.

unlimited ceiling *Meteorology*. a ceiling condition in which the total sky cover is less than 0.6, or the total transparent sky cover is at least 0.5, or surface-based obscuring phenomena obscure 0.9 or less of the sky and no layer aloft is reported as broken or overcast.

unload *Computer Programming*. **1.** to remove a large quantity of data from memory. **2.** to rewind and dismount a magnetic tape. *Ordnance*. to remove the charge from a firearm.

unloaded Q *Electronics*. the ratio of the reactance of a device or circuit to its resistance when there is no external load. Also, INTRINSIC Q.

unloader *Mechanical Engineering*. a device that removes bulk materials from freight cars or trucks, usually by electric power.

unloading *Chemical Engineering*. **1.** the emptying or depressuring of a process unit. **2.** a failure of a filter medium with release of system pressure. **3.** the release of an isolated contaminant downstream.

unloading amplifier *Electronics*. an amplifier having a sufficiently high input impedance that it appears as a negligible load to the source of the signal being amplified.

unloading circuit *Computer Technology*. a computing element or combination of elements that is capable of reproducing or amplifying voltage while drawing negligible current from the voltage source in an analog computer.

unloading device *Computer Technology*. any device, such as a floppy disk or disk pack, that holds data which has been dumped or unloaded from a computer system.

unlock *Computer Science*. to reset a lock variable, indicating that the data protected by the lock may be used by another process.

unmyelinated *Histology*. lacking a myelin sheath, such as nerve fibers.

unnilhexium *Chemistry*. a provisional name for the transuranic element having atomic number 106.

unnilpentium *Chemistry*. a provisional name for the transuranic element having atomic number 105. Also, HAHNIUM.

unnilquadium *Chemistry*. a provisional name for the transuranic element having atomic number 104. Also, RUTHERFORDIUM.

unnilseptium *Chemistry*. a provisional name for the transuranic element having atomic number 107.

unobserved fire *Military Science*. fire for which the points of impact or burst are not observed.

Unp unnilpentium.

unpack *Computer Programming*. to restore data from a packed (space-efficient) format to its original, more accessible but less space-efficient format.

unperturbed dimensions *Materials Science*. the square of the actual chain dimension in a random-coil polymer in solution, in the absence of long-range interactions.

unpolarized light *Optics*. a light whose electric vector of vibration is randomly oriented.

unpolarized particle beam *Physics*. a beam of particles whose spins are randomly oriented.

unprotect *Computer Programming*. to remove safeguarding measures designed to limit access to areas of memory, display fields, data files, programs, or networks.

unprotected reversing thermometer *Engineering*. a reversing thermometer for measuring seawater temperature that is not protected against hydrostatic pressure.

Unq unnilquadium.

unrestricted visibility *Meteorology*. a horizontal visibility of 7 miles or greater.

Uns unnilseptium.

uns- see UNSYM-.

unsatisfiable *Artificial Intelligence*. of a logical formula, false under every interpretation; inconsistent or contradictory.

unsaturated *Physical Chemistry*. of a solution, able to dissolve additional solute; not having the maximum possible amount of a substance. Also, UNDERSATURATED. *Chemistry*. of an organic compound, having double or triple bonds and thus able to add additional atoms. *Nutrition*. of or relating to unsaturated fats. *Mineralogy*. describing a mineral, such as nepheline or corundum, that cannot crystallize in the presence of free silica.

unsaturated bond *Materials Science*. the double or triple covalent bond joining two atoms in an organic molecule; when a single covalent bond replaces the unsaturated bond, polymerization can occur.

unsaturated compound *Chemistry*. a compound in which the maximum valency has not been reached and therefore may form more bonds with additional atoms.

unsaturated fat *Nutrition*. a fat that contains unsaturated fatty acid; unsaturated fats are found in large quantities in vegetable oils, such as corn oil and safflower oil; they are thought to help reduce the amount of cholesterol in the blood and are thus contrasted with saturated fats, which are thought to contribute to elevated cholesterol levels.

unsaturated fatty acid *Nutrition*. an organic compound of carbon, hydrogen, and oxygen that combines with glycerol to form fat; it contains one or more double bonds and is liquid at room temperature.

unsaturated hydrocarbon *Organic Chemistry*. one of a class of hydrocarbons that have available valence bonds, usually meaning a double or triple bond is present in the chain; this double or triple bond may or may not be in a ring structure. An unsaturated hydrocarbon is more reactive than a saturated hydrocarbon; e.g., acetylene and ethylene.

unsaturated solution *Physical Chemistry*. a solution that does not hold as much dissolved solute as would be in equilibrium with undissolved solute, and thus can still dissolve more solute.

unsaturated standard cell *Electricity*. a Weston standard cell or battery that is used for voltage calibrations that do not require accuracy greater than 0.01%.

unsaturated zone see ZONE OF AERATION.

unsaturation *Physical Chemistry*. the condition of a solution in which it has not taken in the maximum possible amount of a solute and can accept more. *Organic Chemistry*. a state in which an organic compound has valence bonds available, typically resulting in a double or triple bond; this usually applies to carbon compounds.

unscheduled DNA synthesis *Molecular Biology*. DNA synthesis that occurs in some stage other than the S phase of the cell cycle.

unscheduled maintenance *Industrial Engineering*. maintenance provided on a contingency basis (e.g., following a breakdown), rather than according to a schedule.

unscrambler *Industrial Engineering*. a device on a filling line that arranges cartons or boxes in the proper position for filling. *Telecommunications*. an electronic device that converts scrambled messages into an intelligible form by a systematic tuning of the receiver to the transmitting frequencies.

unscrolling *Computer Science*. a method of program optimization in which a loop is expanded at compile time by duplicating the contents of the loop for each value taken by the loop index and compiling the result as straight-line code. The result may take more memory but run faster.

unseen fire *Military Science*. continuous fire that is aimed by radar at the successive future positions of an aircraft; as distinguished from seen fire, which is aimed by visual observation.

unsettled *Meteorology*. of fair weather, threatening or likely to become cloudy, rainy, or stormy at any time.

unsigned *Computer Programming*. relating to a numeric field in which the most significant bit is used for the representation of the number, rather than the algebraic sign.

Unsin engine *Mechanical Engineering*. a rotary engine with two circular rotors instead of the trochoidal rotors of an eccentric rotor engine.

unsolicited message *Computer Programming*. an incoming communication message not expected by a process, program, or user.

unsprung axle *Mechanical Engineering*. a rear axle in an automobile that houses the right and left rear-axle shafts, with the wheels mounted at the outer end of each shaft.

unsprung weight *Mechanical Engineering*. the weight of the parts of an automobile that are not carried on a spring.

unstable *Physics*. a condition in which a system is subject to spontaneous changes, such as an isotope of some radioactive element. *Chemistry*. describing a compound that decomposes readily or changes into other compounds or into elements.

unstable equilibrium *Mechanics.* the condition of a system in equilibrium, such that any net force or torque that the system experiences will tend to move it farther away from equilibrium.

unstable isotope see RADIOACTIVE ISOTOPE.

unstable mutation *Genetics.* a mutation that frequently reverts to its original form.

unstable particle *Particle Physics.* a particle that is subject to spontaneous change in terms of its mass, charge, or other physical properties.

unsteady flow *Fluid Mechanics.* a flow whose properties and values are functions of time.

unsteady state *Materials Science.* a system of polyphase diffusion in which temperature or composition is changing as a function of time.

unsteady-state flow *Fluid Mechanics.* **1.** a fluid flow in which the volumetric ratios of two or more phases vary along the course of flow. **2.** a single-phase flow whose rate varies periodically or randomly with time.

unsym- *Organic Chemistry.* a combining form designating an unsymmetrical structure; in organic compounds, it denotes that the substituents are disposed unsymmetrically with respect to the carbon skeleton, or to a double or triple bond. Also, UNS-.

untranscribed spacer *Genetics.* a spacer sequence in a DNA molecule that lies between two coding genes and is not transcribed by RNA polymerase.

untuned *Electricity.* not tuned to any specified frequency.

unweathered *Geology.* see FRESH.

unwind *Mechanical Engineering.* to reverse the direction of rotation of a threaded device. *Computer Programming.* **1.** to enumerate explicitly all the instructions involved in the execution of a loop: unscroll. **2.** to remove all unnecessary or redundant operations from a computer system. **3.** to perform the actions necessary to properly exit from a series of subroutine calls, e.g., in the event of an error.

unwinding protein *Biochemistry.* a single-strand binding protein. *Molecular Biology.* any protein that has the capacity to unwind and separate the strands of the DNA double helix.

UP uroporphyrin; underproof.

up *Engineering.* of a device, functioning or performing completely.

up. upper.

upas *Botany.* **1.** a large tree, *Antiaris toxicaria*, of the mulberry family, yielding a poisonous milky sap; native to tropical Asia, Africa, and the Philippine Islands. **2.** the poisonous sap of this tree, used for arrow poison.

UPC *Computer Technology.* see UNIVERSAL PRODUCT CODE.

upcast *Mining Engineering.* **1.** the passageway through which the return air current ascends en route to its removal from the mine; the opposite of downcast or intake. **2.** an air current passing upward through a shaft or opening. **3.** material hurled upward, as by digging.

upcasting *Materials Science.* a process for continuous casting of nonferrous metals in the production of bars and rods, in which solidification occurs inside a partly submerged die that is held vertically in the molten metal.

up-converter *Electronics.* **1.** any circuit that produces an output signal at a higher frequency than the input signal. **2.** a light-emitting diode that emits visible light when stimulated by infrared radiation.

update *Computer Programming.* **1.** to modify an existing data item, record, or file. **2.** to enhance a program or software system with new capabilities or information.

update mark *Computer Programming.* a single bit designating that the data item or file in memory has been accessed or modified and needs to be updated on the storage medium.

update service *Computer Technology.* a service guaranteeing installation of program updates as they become available.

up-doppler *Acoustics.* a phenomenon in which the frequency of sound increases due to the decrease of wavelengths caused by the relative motion of the sound source, the listener, and the medium carrying the sound waves.

updraft carburetor *Mechanical Engineering.* a carburetor in which the fuel mixture is drawn upward against the force of gravity.

updraft furnace *Mechanical Engineering.* a furnace that is supplied with air from beneath the fuel bed.

updrift *Oceanography.* a shoreward movement of littoral material, opposite to its general seaward movement.

upflash *Science.* a type of lightning bolt that occurs during violent storms, flashing between a storm cloud and the air above; thought to result from the interaction of the positive charge at the top of the cloud with falling electrons from the ionosphere.

upflow *Chemistry.* solutions that enter at the bottom and leave from the top of an ion-exchange device.

upflow sludge blanket *Biotechnology.* an anaerobic digester in which the microorganisms form flocs that are kept in suspension in the lower part of the reaction tank.

upgrade *Computer Programming.* **1.** to acquire more sophisticated hardware or software than one already has. **2.** to enhance an existing product.

uphole *Mining Engineering.* **1.** a borehole that is collared in an underground working place and directed above the horizontal plane of the swivel head of the drill machine. **2.** a shothole drilled in an upward direction through rock.

uphole time *Geophysics.* the time it takes for a wave created during seismic prospecting to travel from a given depth in the shothole to the surface.

upholsterer's hammer *Mechanical Devices.* a specialty hammer with a long, narrow, rounded striking head, often magnetic, for driving nails and tacks into furniture or for use in other close spaces.

upland *Geography.* **1.** an elevated plain as distinguished from a valley plain or lowland. **2.** an area of land lying above the level where water flows or where flooding occurs. **3.** a hilly or mountainous region. Also, **uplands.**

upland cotton *Botany.* a cotton plant, *Gossypium hirsutum*, having a short staple, that is the chief commercial cotton crop in the U.S.; found in warm regions of the New World.

upland plover *Vertebrate Zoology.* a large sandpiper, *Bartramia longicauda*, that frequents upland fields and pastures; now protected and increasing in numbers; found in eastern North America. Also, **upland sandpiper.**

uplift *Geology.* a structurally elevated area within the earth's crust that was produced by deformational movements that raise or upthrust the rocks.

uplink *Telecommunications.* **1.** a communications path used to transmit signals upward from a station on the earth to a communications satellite or aircraft. **2.** to establish or utilize such a connection.

upload *Computer Programming.* the act of transferring information from a microcomputer or workstation to a remote central computer system.

uplock *Aviation.* a locking device designed to hold a retracted landing gear in place.

upmilling *Mechanical Engineering.* the process of milling an object by rotating the cutter against the feed direction.

up mutation *Genetics.* any mutation in the promotor region of a gene that stimulates the initiation of transcription.

Upper or **upper** *Geology.* **1.** describing a later or final epoch or series of the period or series named, as in Upper Cambrian. **2.** relating to rocks or strata that occur above or after earlier formations of the same subdivision of rocks.

upper air *Meteorology.* the part of the atmosphere that lies above the lower troposphere.

upper-air chart *Meteorology.* a synoptic chart of weather conditions in the upper air. Also, UPPER-LEVEL CHART.

upper-air disturbance see UPPER-LEVEL DISTURBANCE.

upper-air observation *Meteorology.* **1.** an observation of weather conditions aloft, above the range of normal weather observation. Also, **upper-air sounding. 2.** a weather observation made at a mountain station that is located at very high elevation.

upper anticyclone see UPPER-LEVEL ANTICYCLONE.

upper arm *Anatomy.* the part of the arm between the shoulder and the elbow. *Robotics.* the part of a robot's arm between the shoulder and the elbow or wrist.

upper atmosphere *Meteorology.* the part of the atmosphere that lies above the troposphere.

upper atmosphere dynamics *Meteorology.* the atmospheric phenomena that characterize motions of the atmosphere above the level of 500 km, such as internal gravity waves, sound waves, turbulence, and large-scale circulation.

upper bound *Mathematics.* an upper bound (if it exists) for a nonempty subset B of a partially ordered set A (with ordering $\leq$) is an element c of A such that $b \leq c$ for every $b \in B$. Note that upper bounds are not necessarily unique; however, if $c \leq d$ for every other upper bound d of B, then c is a **least upper bound** of B. Also, SUPREMUM.

upper branch *Geodesy.* the half of a complete celestial meridian that stretches from the north to south pole while passing through the observer's zenith.

upper bright band *Meteorology.* a level above the bright band of the melting level, at which radar echo is enhanced by the growth into snow pellets of ice crystals in a supercooled cloud. Also, **upper band.**

Upper Cambrian *Geology.* the final epoch of the Cambrian period, beginning approximately 510 million years ago.

Upper Carboniferous *Geology.* the European equivalent of the North American Pennsylvanian period.

Upper Cretaceous *Geology.* the final epoch of the Cretaceous period, beginning approximately 90 million years ago.

upper critical solution temperature *Materials Science.* the maximum temperature for phase separation in polymer binary systems such as solutions containing the two liquid phases, both of which contain amorphous polymer and one or more solvents.

upper culmination see UPPER TRANSIT.

upper cyclone see UPPER-LEVEL CYCLONE.

Upper Devonian *Geology.* the final epoch of the Devonian period, beginning approximately 365 million years ago.

upper front *Meteorology.* a front that occurs in the upper air but does not reach the earth's surface.

upper half-power frequency *Electronics.* the standard upper frequency-response limit of an amplifier; the upper frequency at which the power output of a power amplifier falls to one-half its midrange value. In the case of voltage or current amplifiers, the half-power point is the higher of the two half-power frequencies of an amplifier or filter.

upper high see UPPER-LEVEL ANTICYCLONE.

upper integral *Mathematics.* let f be a real-valued function that is bounded on an interval $[a,b]$, and let

$$a = x_0 < x_1 < x_2 < \cdots < x_{n-1} < x_n = b$$

be a partition of the interval into n parts. The **upper Riemann integral** of $f(x)$ on $[a,b]$ is

$$\inf \{ \sum_{k=1}^{n} y_k(x_k - x_{k-1}) \},$$

where y_k is the maximum value of f on the interval $[x_{k-1}, x_k]$, and where the infimum is taken over all possible (finite) partitions of $[a,b]$.

upper inversion see TROPOPAUSE INVERSION.

Upper Jurassic *Geology.* the final epoch of the Jurassic period, beginning about 155 million years ago.

upper-level anticyclone *Meteorology.* an anticylone that occurs in the upper air and is more pronounced than the anticyclonic circulation below it. Also, HIGH ALOFT, HIGH ANTICYCLONE, UPPER HIGH, UPPER-LEVEL HIGH.

upper-level chart see UPPER-AIR CHART.

upper-level cyclone *Meteorology.* cyclonic circulation that occurs in the upper atmosphere above an area having relatively little cyclonic circulation. Also, LOW ALOFT, UPPER CYCLONE, UPPER LOW, UPPER-LEVEL LOW.

upper-level disturbance *Meteorology.* a disturbance of the mean circulation pattern of the upper air, especially one that is more intensely developed aloft than near the ground. Also, UPPER-AIR DISTURBANCE.

upper-level high see UPPER-LEVEL ANTICYCLONE.

upper-level low see UPPER-LEVEL CYCLONE.

upper-level ridge *Meteorology.* a pressure ridge that occurs in the upper air, especially one that is stronger aloft than it is nearer the earth's surface. Also, RIDGE ALOFT, HIGH-LEVEL RIDGE, UPPER RIDGE.

upper-level trough *Meteorology.* a pressure trough that occurs in the upper air, especially one that is stronger aloft than it is nearer the earth's surface. Also, TROUGH ALOFT, HIGH-LEVEL TROUGH, UPPER TROUGH.

upper-level winds see WINDS ALOFT.

upper limb *Astronomy.* the upper edge of a celestial object.

upper low see UPPER-LEVEL CYCLONE.

upper mantle *Geology.* the part of the earth's mantle that extends above a depth of about 1000 km and is equivalent to the B and C layers. Also, OUTER MANTLE, PERIDOTITE SHELL.

Upper Mississippian *Geology.* the final epoch of the Mississippian period.

upper mixing layer *Meteorology.* the layer of the mesosphere that extends from about 50 to about 80 kilometers in altitude (30 to 50 miles), characterized by a rapid temperature decrease with height and considerable turbulence, with the latter probably enhanced by convection from the ozonosphere below.

Upper Ordovician *Geology.* the final epoch of the Ordovician period, beginning approximately 440 million years ago.

Upper Paleolithic *Archaeology.* the final part of the Paleolithic period, from about 35,000 years ago to about 10,000 years ago; characterized by the development of bladed stone tools, the hunting of large herd animals, and the appearance of cave paintings and other art forms.

Upper Pennsylvanian *Geology.* the final epoch of the Pennsylvanian period.

Upper Permian *Geology.* the final epoch of the Permian period, beginning about 245 million years ago.

upper punch *Metallurgy.* in powder metallurgy, the part of a die assembly that moves down into the die body to carry pressure to the powder in the die cavity.

upper ridge see UPPER-LEVEL RIDGE.

upper semicontinuous decomposition *Mathematics.* let P be a partition of a topological space X, and let P be any element of P. P is called an upper semicontinuous decomposition of X if, for every open set U containing P, there exists an open set V contained in U such that V contains P and V is the union of elements of P.

upper semicontinuous function *Mathematics.* **1.** an extended real-valued set function μ defined on a class C of sets is said to be upper semicontinuous or continuous from above at a set S in C, if for every decreasing sequence $\{S_n\}$ of sets in C for which $|\mu(S_n)| < \infty$ for at least one value of n and for which $\lim_{n\to\infty} S_n = S$, we have

$$\lim_{n\to\infty} \mu(S_n) = \mu(S).$$

It can be shown that if such a set function μ is finite, nonnegative, and additive on a ring R, and μ is continuous from above at 0 (the additive identity of R), then μ is a measure on R. **2.** a real-valued function f defined on a domain D in R^n is said to be upper semicontinuous at a point c of D if

$$f(c) \geq \lim_{x\to c} \sup f(x).$$

f is said to be upper semicontinuous on D if it is upper semicontinuous at every point of D.

upper sideband *Telecommunications.* the higher of two bands of frequencies produced by amplitude modulation.

Upper Silurian *Geology.* the final epoch of the Silurian period.

upper transit *Astronomy.* the transit of a circumpolar celestial object across the section of meridian lying above the pole. Also, SUPERIOR TRANSIT, UPPER CULMINATION.

upper-triangular matrix *Mathematics.* a square matrix whose (i,j)th entry is zero if $i > j$. If, in addition, the diagonal entries $(i = j)$ are zero, the matrix is said to be strictly upper-triangular.

Upper Triassic *Geology.* the final epoch of the Triassic period, beginning about 200 million years ago.

upper trough see UPPER-LEVEL TROUGH.

upper variation *Mathematics.* let μ be a signed measure on a measure space (X,A,μ), and let the sets A and B form the Hahn decomposition of X with respect to μ. The upper variation of μ, denoted μ^+, is the set function defined for every measurable set E by $\mu^+(E) = \mu(E \cap A)$. The **total variation** of μ is the sum of the upper and lower variations of μ.

upper winds see WINDS ALOFT.

upright silo *Agriculture.* a vertical cylindrical structure for storing silage, usually of concrete blocks or sheet metal. Also, VERTICAL SILO.

uprush *Meteorology.* the abrupt upward rush of air that occurs during a rapid development of a cumulus cloud, as before a thunderstorm. Also, VERTICAL JET. *Oceanography.* see SWASH.

UPS ultraviolet photoemission (or photoelectron) spectroscopy; uninterruptible power system (or supply); universal polar stereographic grid.

upset *Engineering.* to increase a rock drill's diameter by blunting its head. *Metallurgy.* the part of a welding cycle during which pressure is applied. *Mining Engineering.* a narrow heading driven in inclined coal to connect two levels.

upsetted moraine see PUSH MORAINE.

upset welding *Metallurgy.* in resistance welding, the process of applying pressure to the entire area of abutting surfaces or progressively along a joint before the heating is started and maintaining it throughout the heating period.

upsilon [yoop´sə län; up´sə län] the 20th letter of the Greek alphabet, written as Υ, υ.

upsilon particle *Particle Physics.* a heavy, short-lived particle consisting of a bottom quark bound to its antiquark, thus, the fifth-quark analog to the J particle; eleven additional particles with this quark structure have been identified thus far.

upslope fog *Meteorology.* ground fog that develops when air flows up a slope and is thus adiabatically cooled to or below its dewpoint.

upslope time *Metallurgy.* in resistance welding, the time associated with current increase, using slope control.

upstream *Hydrology.* toward the source of a stream. Similarly, **upriver.** *Chemical Engineering.* that section of a process stream that has yet to flow into the specific unit or system, as in proceeding upstream to a distillation column or to a refinery. *Molecular Biology.* the direction directly opposite that of the synthesis of a protein molecule or nucleic acid, e.g., from the right (or 3') end toward the left (or 5') end of messenger RNA.

upstream activation site *Genetics.* a DNA sequence that is involved in regulating transcription in a manner similar to an enhancer, but that becomes nonfuntional when located downstream of a promoter. Also, **upstream activator sequence.**

upstream face *Civil Engineering.* the face of a dam facing upstream, against which water is imponded.

uptake *Engineering.* a pipe that is used to convey exhaust gases from a boiler to a smokestack.

upthrow *Geology.* **1.** the side of a fault that has been displaced upward with respect to the other side. Also, **upthrow side, unthrown side. 2.** the measure of the upward vertical displacement on a fault.

up time or **uptime** *Industrial Engineering.* **1.** the amount of time a machine or system, such as a computer, is operating or available for operation. **2.** the time that an employee is actually working.

Upupidae *Vertebrate Zoology.* the hoopoes, a monotypic family of birds belonging to the order Coraciiformes, characterized by boldly patterned plumage and a long crest; found in Eurasia and Africa.

upward compatibility *Computer Technology.* the ability of a newer, bigger, or better computer system to run programs from a smaller or older computer system.

upward migration *Archaeology.* the movement of previously deposited objects to the soil surface as a result of plowing, bulldozing, or other such nonarchaeological processes.

upwarp *Geology.* a broad region of the crust that has been formed by differential uplift.

upwelling *Oceanography.* the movement to the surface of deep, cold, nutrient-laden water, usually caused by coastal currents that pull surface water away from the coast or by wind divergence of equatorial currents. *Volcanology.* a mild, gradual volcanic eruption.

upwind *Meteorology.* toward or against the wind or the direction from which it flows.

upwind effect *Meteorology.* the effect of a mountain in producing orographic precipitation to the windward of its base, which occurs when the airflow is forced upward before reaching the mountain slope.

UR unconditioned response.

urachus *Developmental Biology.* a fetal canal that connects the bladder with the allantois, and that persists throughout life as the median umbilical ligament.

uracil *Biochemistry.* $C_4H_4N_2O_2$, a pyrimidine component, one of the four bases that appear as nucleotide residues in RNA.

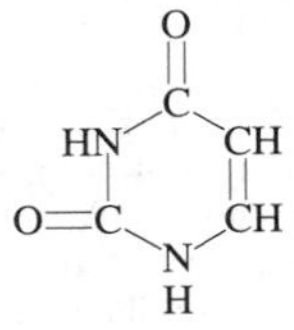

uracil

Uralian or **Uralean** *Geology.* a geologic stage of the uppermost Carboniferous period in Russia, occurring after the Gzhelian and before the Sakmarian of the Permian period.

uralite *Mineralogy.* a green, often fibrous or acicular variety of secondary amphibole, pseudomorphous after minerals of the pyroxene group.

uralitization *Geology.* a magmatic or metamorphic process in which uralite or amphibales are produced by the alteration of pyroxenes in igneous rocks.

Urals *Geography.* a mountain system in west-central Russia; regarded as the traditional boundary between Europe and Asia.

uranate *Inorganic Chemistry.* any compound made up of a metal cation with an anion of uranium and oxygen in which the uranium is in a +6 valence state.

urania or **uranic oxide** see URANIUM DIOXIDE.

uranic chloride see URANIUM TETRACHLORIDE.

Uraniidae *Invertebrate Zoology.* a family of moths, lepidopteran insects in the superfamily Geometroidea, including some brightly colored, diurnal butterflylike species.

uranine *Organic Chemistry.* $Na_2C_{20}H_{10}O_5$, an orange-red, hygroscopic, odorless powder; soluble in water and sparingly soluble in alcohol; used in dyeing silk and wool yellow, marking water for air-sea rescue and subterranean tracing, and in analytical test solutions. Also, **uranine yellow, uranin.**

uraninite *Mineralogy.* UO_2 (but usually partially oxidized), a black, brown, gray, or greenish, strongly radioactive, cubic mineral, forming a series with thorianite, often found with impurities that may include thorium, the rare earths, lead, radium, and other elements, occurring as cubic crystals and in massive and granular form, having a specific gravity of 6.5–10.0 (10.95 for pure UO_2) and a hardness of 5 to 6 on the Mohs scale; found in hydrothermal veins, sedimentary beds, and pematites; the principal ore of uranium; forms a series with thorianite.

uranite see PITCHBLENDE.

uranium *Chemistry.* a metallic element having the symbol U, the atomic number 92, an atomic weight of 238.029, a melting point of 1132°C, and a boiling point of 3818°C; a dense silvery solid with three naturally occurring radioactive isotopes: the nonfissionable **uranium-238,** composing 99% of natural uranium; **uranium-234** (0.006%), which is separated by extraction and used in nuclear research; and **uranium-235** (0.7%), which is used in nuclear fields to enrich natural uranium and was the source of energy in the original atomic bomb. (Named in 1789 after the recently discovered planet of *Uranus.*)

uranium acetate see URANYL ACETATE.

uranium age *Geology.* the age, in years, of a geological sample based on the number of ionium atoms present in the sample originally, currently, and when equilibrium with uranium is established.

uranium carbide *Inorganic Chemistry.* UC_2, radioactive and highly toxic, metallic-appearing crystals; decomposes in water and insoluble in alcohol; melts at 2350–2400°C and boils at 4370°C; used in nuclear reactor fuels. Also, **uranium dicarbide.**

uranium dating *Archaeology.* a method of dating based on measuring the rate of decay of uranium isotopes in bone and other organic remains. Also, **uranium series dating, uranium series disequilibrium dating.**

uranium dioxide *Inorganic Chemistry.* UO_2, radioactive, brown to black rhombic crystals or cubes; insoluble in water; melts at approximately 2878°C; used in nuclear fuel rods. Also, URANIA, YELLOWCAKE.

uranium enrichment *Nucleonics.* the increase in the isotopic abundance of uranium-235 relative to uranium-238 above that found in natural uranium (about 0.7% uranium-235) using methods based on isotopic mass differences (gaseous diffusion, gas centrifuge, and mass spectrograph) and isotopic spectral differences (laser excitation).

uranium hexafluoride *Inorganic Chemistry.* UF_6, colorless, monoclinic crystals; deliquescent; decomposes in cold water and alcohol; melts at approximately 64.5°C and boils at 56.2°C; used in the separation of uranium isotopes.

uranium hydride *Inorganic Chemistry.* UH_3, toxic black powder or cubes; used in separation of hydrogen isotopes and as a reducing agent.

uranium-lead dating *Geology.* the determination of the age, in years, of a geologic sample by measuring the rate of radioactive decay of uranium-238 to lead-206 and of uranium-235 to lead-207.

uranium nitrate or **uranium oxynitrate** see URANYL NITRATE.

uranium reactor *Nucleonics.* a nuclear reactor in which natural or enriched uranium is distributed in a lattice within the graphite moderator (heterogeneous reactor) or dissolved in water, the solvent being the moderator (homogenous reactor); use of enriched uranium greatly reduces the size of core required to sustain a chain reaction.

uranium series *Nuclear Physics.* a series of three naturally occurring transformation series of radioactive isotopes that form a sequence of parent-daughter relationships through alpha and beta decay, beginning with uranium-238 and ending in the stable isotope of lead-206. Also, **uranium decay series, uranium-radium series.**

uranium sulfate see TRIURANIUM OCTOXIDE.

uranium tetrachloride *Inorganic Chemistry.* UCl_4, toxic, dark green crystals, hygroscopic; soluble in water, alcohol, and acid; melts at approximately 590°C and boils at 792°C.

uranium tetrafluoride *Inorganic Chemistry.* UF_4, toxic, radioactive, green triclinic needles; very slightly soluble in water; melts at approximately 960°C; used as an intermediate in making uranium metal. Also, GREEN SALT.

uranium trioxide *Inorganic Chemistry.* UO_3, a toxic, radioactive, yellowish-red powder; insoluble in water; decomposes at its melting point; used in ceramics, pigments, and uranium refining. Also, URANYL OXIDE.

uranium yellow see SODIUM DIURANATE.

uranocircite *Mineralogy.* $Ba(UO_2)_2(PO_4)_2 \cdot 12H_2O$, a strongly radioactive, yellow-green, secondary tetragonal mineral of the autunite group, occurring in aggregates of thin tabular crystals, having a specific gravity of 3.46 to 3.53 and a hardness of 2 to 2.5 on the Mohs scale.

uranography *Astronomy.* the branch of astronomy concerned with mapping the heavens and plotting the positions of stars, star groups, star clusters, interstellar gas patches, galaxies, and the like.

uranometry *Astronomy.* the measurement of the positions of stars.

uranophane *Mineralogy.* $Ca(UO_2)_2[SiO_3(OH)]_2 \cdot 5H_2O$, a strongly radioactive, yellow or orange-yellow, monoclinic secondary mineral, occurring as crusts of acicular crystals and in microcrystalline masses, having a specific gravity of 3.83 and a hardness of about 2.5 on the Mohs scale; an alteration product of uraninite.

uranopilite *Mineralogy.* $(UO_2)_6(SO_4)(OH)_{10} \cdot 12H_2O$, a rare, strongly radioactive, soft, yellow, monoclinic secondary mineral, occurring as encrustations and aggregates of minute lathlike crystals, having a specific gravity of 3.96 and an undetermined hardness.

Uranoscopidae *Vertebrate Zoology.* the stargazers, a family of tropical and warm-temperate marine fish of the order Perciformes, characterized by a heavy broad body, eyes on top of the head, and fringed lips; found burrowing in sand with only the upper head exposed.

uranosphaerite *Mineralogy.* $Bi_2U_2O_9 \cdot 3H_2O$, a rare, orange, secondary, radioactive, monoclinic mineral, occurring in small aggregates of minute prismatic crystals, having a specific gravity of 6.36 and a hardness of 2 to 3 on the Mohs scale; found with other secondary minerals in a vein carrying uraninite, bismuth, and cobalt-nickel arsenides.

uranospinite *Mineralogy.* $Ca(UO_2)_2(AsO_4)_2 \cdot 10H_2O$, a rare, strongly radioactive, yellow to green, tetragonal secondary mineral of the autunite group, occurring in tabular crystals, having a specific gravity of 3.45 and a hardness of 2 to 3 on the Mohs scale.

uranothallite see LIEBIGITE.

uranothorite *Mineralogy.* a variety of thorite containing uranium.

uranous-uranic acid see TRIURANIUM OCTOXIDE.

uranpyrochlore *Mineralogy.* $(U,Ca,Ce)_2(Nb,Ta)_2O_6(OH,F)$, a radioactive, yellow to brown or black cubic mineral of the pyrochlore group, occurring in massive form or as octahedral crystals with a resinous luster, having a specific gravity of 4.509 to 4.90 and a hardness of 4 to 5 on the Mohs scale; found in pegmatites.

Uranus [yoo rān´əs; yůr´ə nəs] *Astronomy.* the seventh planet from the sun; it has an equatorial diameter of 51,100 kilometers, a mean solar distance of 2.88 billion kilometers, a revolution period of 84 years, and 15 known moons. (Named for *Uranus,* the god of the heavens in classical mythology; it was also for a time known as *Herschel,* for its discoverer, the English astronomer William Herschel; he himself called it "the Georgian planet," after England's King George.)

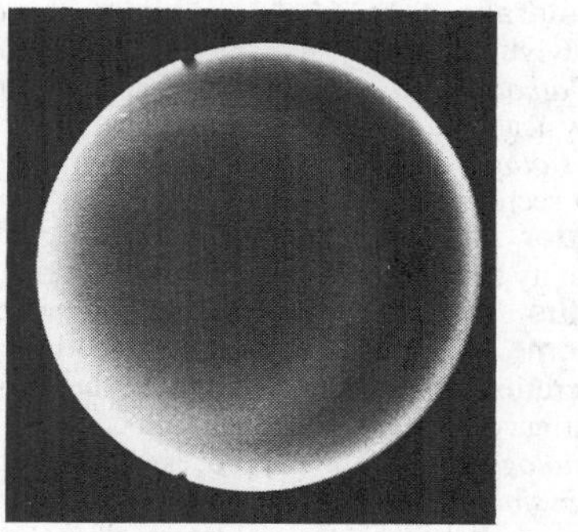

Uranus

uranyl acetate *Organic Chemistry.* $UO_2(C_2H_3O_2)_2 \cdot 2H_2O$, yellow crystals; used medicinally in the treatment of coryza.

uranyl nitrate *Inorganic Chemistry.* $UO_2(NO_3)_2 \cdot 6H_2O$, yellow rhombic crystals; deliquescent; miscible with water and very soluble in alcohol; melts at 60.2°C and boils at 118°C; used as a source of uranium dioxide and in uranium extraction. Also, URANIUM OXYNITRATE, YELLOW SALT.

uranyl oxide see URANIUM TRIOXIDE.

uranyl salt *Inorganic Chemistry.* any salt containing the UO_2^{+2} ion.

uranyl sulfate *Inorganic Chemistry.* $UO_2SO_4 \cdot 3H_2O$, toxic, radioactive, yellow-green crystals; soluble in water; decomposes at 100°C; used as an analytical reagent. Also, URANIUM SULFATE.

uranyl uranate see TRIURANIUM OCTOXIDE.

urao see TRONA.

urate calculi *Pathology.* a small, smooth, yellowish-brown concretion made up of the salt of uric acid; found for the most part in newborns or infants.

urate oxidase see URICASE.

urbacid *Organic Chemistry.* $C_7H_{15}AsN_2S_3$, a colorless, crystalline compound; insoluble in water; melts at 144°C; used to control disease in apple and coffee trees.

urban *Geography.* **1.** of or relating to a city or town. **2.** of a settlement, having a working population of which the majority is engaged in nonagricultural occupations. **3.** of a landscape, predominantly artificial.

urban anthropology *Anthropology.* the study of the cultural adjustments of those who have come to live in urban settings from other areas.

urban geography *Geography.* the geographical study of cities, especially of physical elements such as site, structure, and growth patterns.

urban heat island *Meteorology.* an urban area that exhibits temperatures higher than those in surrounding rural areas due to certain effects of development, such as the absence of vegetation and construction of reflective surfaces.

urbanization *Geography.* the process by which an area's population density grows to urban levels. *Civil Engineering.* the development of an urban-type infrastructure.

urban renewal *Civil Engineering.* the extensive demolition by public authority of decayed areas in a city to permit new construction, usually associated with economic development plans.

urban typhus *Pathology.* a mild form of murine typhus occurring mostly among indoor workers in the Mediterranean region and Malaya.

Urca process *Astrophysics.* a series of fast-working nuclear reactions that involve primarily iron-group elements and that generate neutrinos in high numbers, possibly leading to stellar collapse.

urceolate *Biology.* **1.** urn-shaped or pitcher-shaped; as applied to the calyx or corolla of a flower. **2.** having a pitcher-shaped structure, as characterized by the external tube of certain rotifers.

Urd uridine.

urea *Organic Chemistry.* $CO(NH_2)_2$, white crystals or powder, with a saline taste; occurs as a natural product of protein metabolism, and was, in 1824, the first organic compound to be synthesized; soluble in water and alcohol and almost insoluble in chloroform; melts at 132.7°C and decomposes before boiling; used in fertilizer, animal feed, plastics, flameproofing, adhesives, and cosmetics, and as a stabilizer in explosives and medicinally as a diuretic. Also, CARBAMIDE.

urea dewaxing *Chemical Engineering.* a continuous process in petroleum refining in which low-pour-point oils are manufactured; a filterable solid complex is formed by urea with the straight-chain wax paraffins in the stock.

urea-formaldehyde resin *Organic Chemistry.* a thermosetting resin derived by reacting urea with an aldehyde such as formaldehyde; such resins are commercially important because of their water-white color and translucency, and because they were the first plastics with the ability to be made into white, pastel, or delicately colored products; used in dinnerware, interior plywood, flexible foams, and insulation. Also, **urea resin.**

urea nitrate *Organic Chemistry.* $CO(NH_2)_2 \cdot HNO_3$, colorless crystals; soluble in alcohol and slightly soluble in water; decomposes at 152°C; a dangerous explosive and fire hazard; used in explosives and the manufacture of urethane.

urea peroxide *Organic Chemistry.* $CO(NH_2)_2 \cdot HOOH$, white crystals or crystalline powder; soluble in water and alcohol; decomposes at 75–78°C and is decomposed by moisture at around 40°C; a strong oxidizer and dangerous fire hazard; used as a source of water-free hydrogen peroxide, as a bleaching disinfectant, and in cosmetics, pharmaceuticals, and the modification of starches. Also, CARBAMIDE PEROXIDE.

urease *Enzymology.* an enzyme that hydrolyzes urea to carbon dioxide and ammonia; present in jack beans, and used to establish urea concentrations.

Urechinidae *Invertebrate Zoology.* a family of sea urchins, echinoid echinoderms in the order Holasteroida, characterized by an ovoid test.

urechitoxin *Toxicology.* $C_{13}H_{20}O_5$, a poisonous glycoside found in some plants of the genus *Urechites.*

Uredenales *Mycology.* an order of fungi of the class Teliomycetes, composed of rust fungi that parasitize vascular plants and crop plants and are sometimes pathogenic, causing leaf and stem infections. Also, RUST FUNGI, RUSTS.
uredinium *Mycology.* in parasitic fungi, a spore-bearing structure that grows on the stem of the host plant. Also, **uredium.**
urediospore *Mycology.* in rust fungi, a thin-walled spore formed by the genus *Uredinia.* Also, **urediniospore, uredospore.**
ureilite *Geology.* an achondritic stony meteorite composed chiefly of olivine and clinobronzite, with small amounts of nickel-iron, troilite, diamond, and graphite.
uremia *Pathology.* **1.** the presence of excessive amounts of urea, creatinine, and other nitrogenous waste products in the blood, as occurs in renal failure. **2.** the condition marked by the symptoms of chronic renal failure, including such symptoms and manifestations as nausea, vomiting, anorexia, pruritus, uremic frost on the skin, neuromuscular disorders, hypertension, edema, and electrolyte imbalances.
ureotelic *Biology.* describing animals that excrete the major part of their waste nitrogen as urea.
ureter *Anatomy.* either of the two narrow fibromuscular tubes, about 16–18 inches long, that carry urine from the kidneys to the urinary bladder; the ureter begins with the pelvis of the kidney, a funnel-like dilatation, and empties into the base of the bladder.

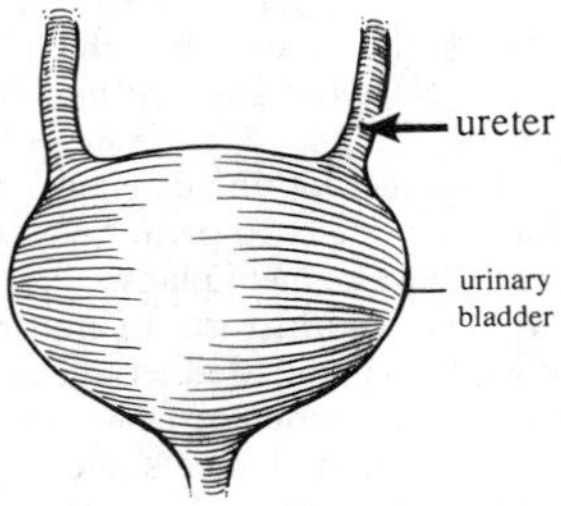

ureter

ureteral *Anatomy.* of or relating to the ureter. Also, **uretal, ureteric.**
ureteralgia *Medicine.* pain in the ureter.
ureterectomy *Surgery.* the surgical removal of all or a part of a ureter.
ureteritis *Medicine.* an inflammation of a ureter.
ureterocele *Medicine.* a hernia of the ureter into the bladder.
ureterolith *Medicine.* a stone that is lodged or has formed in the ureter.
ureterolithiasis *Medicine.* the formation of stones in a ureter.
ureteropathy *Pathology.* any disease of the ureter.
ureteroplasty *Surgery.* the surgical repair of the ureter.
ureteropyelitis *Medicine.* an inflammation of a ureter and the renal pelvis.
ureterorrhaphy *Surgery.* suture of the ureter.
ureterostenosis *Medicine.* a stricture of a ureter. Also, **ureterostegnosis.**
ureterostomy *Surgery.* the surgical formation of an opening through which a ureter may discharge its contents.
ureteroureterostomy *Surgery.* the end-to-end joining of the two portions of a severed ureter, allowing passage of fluid.
ureterovaginal *Anatomy.* relating to or communicating with a ureter and the vagina.
urethane *Organic Chemistry.* $H_2NCO_2C_2H_5$, colorless or white crystals with a saltpeterlike taste; soluble in water, alcohol, and ether, and slightly soluble in oil; melts at 49–50°C and boils at 182–184°C; used as an intermediate for pharmaceuticals, pesticides, and fungicides, and in biochemical research. Also, ETHYL CARBAMATE, ETHYL URETHANE.
urethra *Anatomy.* the membranous canal carrying urine from the bladder to the outside of the body.
urethral *Anatomy.* of or relating to the urethra.
urethralgia *Medicine.* pain in the urethra.
urethral gland *Anatomy.* the small, tubular mucus glands in the mucous membrane of the urethra.
urethrectomy *Surgery.* total or partial removal of the urethra.
urethritis *Medicine.* an inflammation of the urethra.
urethrocele *Medicine.* in women, a herniation of the urethra into the vagina.
urethrocystitis *Medicine.* an inflammation of the urethra and the bladder.
urethrography *Radiology.* X-ray visualization of the urethra after the injection of a contrast medium into the urethral orifice.
urethrorectal *Anatomy.* relating to or communicating with the urethra and the rectum.
urethrorrhagia *Medicine.* a hemorrhage from the urethra.
urethrorrhaphy *Surgery.* suture of the urethra.
urethrorrhea *Medicine.* an abnormal discharge from the urethra.
urethroscope *Medicine.* an instrument used to examine the interior of the urethra.
urethrospasm *Medicine.* a spasm of the urethral muscular tissue.
urethrotomy *Surgery.* an incision of the urethra.
urethrovaginal *Anatomy.* relating to or communicating with the urethra and the vagina.
Urey, Harold Clayton 1893–1981, American chemist; awarded the Nobel Prize for isolating deuterium; isolated isotopes of uranium, oxygen, nitrogen, carbon, hydrogen, sulfur, and boron.
ureyite see KOSMOCHLOR.
urge *Behavior.* another term for a drive. See DRIVE.
URI upper respiratory tract infection.
uric acid *Biochemistry.* $C_5H_4N_4O_3$, a compound in which amino nitrogen is excreted following amino acid metabolism; the end product of purine catabolism in primates, present in the urine.

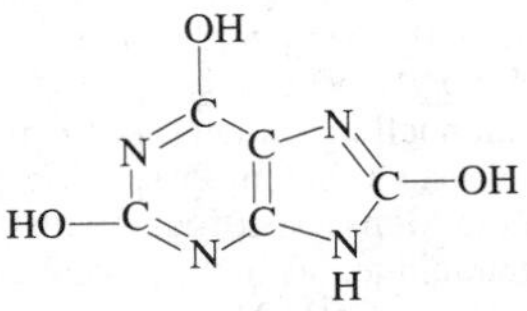

uric acid

uricase *Enzymology.* an enzyme, found in most mammals but not in primates, that converts uric acid to allantoin. Also, URATE OXIDASE.
uricosuria *Physiology.* the excretion of uric acid in the urine.
uricotelism *Physiology.* the release of uric acid as the final output of nitrogen metabolism in some animals, such as birds and reptiles.
uridine *Biochemistry.* the nucleoside, composed of uracil and D-ribose, that is a component of RNA.
uridine diphosphate *Biochemistry.* $C_9H_{14}N_2O_{12}P_2$, the transferring coenzyme in carbohydrate metabolism.
uridine diphosphoglucose *Biochemistry.* a compound that functions as a key in the transformation of glucose to other sugars.
uridylic acid *Biochemistry.* $C_9H_{13}N_2O_9P$, a compound, uridine 5'-phosphate, formed by a nuclease that hydrolyzes RNA. Also, **uridine monophosphate, uridine phosphoric acid.**
uridylylation *Biochemistry.* the transfer of a 5'-UMP group from UTP as part of the complex control of the enzyme glutamine synthetase function in *Escherichia coli.*
uridylyl transferase *Enzymology.* any enzyme that catalyzes the transfer of the uridylyl group.
urinalysis *Pathology.* an examination of a urine sample, including its chemical, physical, and microscopic aspects.
urinary *Physiology.* **1.** relating to the urine or the urinary system. **2.** containing or secreting urine.
urinary bladder *Anatomy.* a hollow muscular organ found in the lower abdominal cavity that serves as the reservoir for urine.
urinary meatus *Anatomy.* the external urethral orifice.
urinary system *Anatomy.* the organs involved in the production, storage, and excretion of urine, consisting of the two kidneys, the two ureters, the urinary bladder, and the urethra.
urinate *Physiology.* to carry on the process of urination.
urination *Physiology.* the mechanism by which urine is released from the body, accomplished by the voluntary relaxation of the sphincter muscles and the involuntary contraction of the bladder muscles.
urine *Physiology.* the liquid emitted by the kidneys, accumulated in the bladder, and passed through the urethra to the outside of the body.
uriniferous *Anatomy.* relating to structures that carry urine, such as the tubules of the kidneys.
uriniferous tubule *Anatomy.* any of the numerous collecting tubes located inside a kidney.
urkaryote *Cell Biology.* a theoretical precursor to the eukaryotes.
Urkaryote *Biology.* a proposed urkingdom containing all eukaryote organisms (i.e., plants, animals, and fungi).

urkingdom *Systematics.* a taxonomic rank in the proposed classification scheme that uses only three kingdoms, Eubacteria, Archaebacteria, and Urkaryote, rather than the five-kingdom scheme devised by Whittaker (Monera, Protista, Fungi, Plantae, and Animalia).

urn *Archaeology.* a large and decorative vase, especially one having an ornamental base. *Botany.* the theca or capsule of a moss, so named because it has a shape like that of an urn.

u1RNA an abundant form of RNA that is a component of snRNP in mammalian nuclei; used in excising introns and splicing exons of mRNA.

urobilin *Biochemistry.* an amorphous, brownish bile pigment that is an oxidized form of urobilinogen; found in the feces and sometimes in urine left standing in the air.

urobilinogen *Biochemistry.* a colorless compound formed in the intestines by the reduction of bilirubin; some is excreted in feces, some is reabsorbed and reexcreted in the bile, or at times in the urine.

urocanic acid *Biochemistry.* a compound formed from histidine by loss of ammonia.

Urochordata see TUNICATA.

urochrome *Biochemistry.* a pigment that contains a peptide and uroblin or urobilinogen, the main pigment in urine.

Urodela *Vertebrate Zoology.* an equivalent name for the tailed amphibian order Caudata, including the salamanders, newts, and efts.

urogastrone *Endocrinology.* a polypeptide that is secreted by the salivary glands and by Brunner's glands and is an inhibitor of gastric acid secretion; similar or possibly identical to epidermal growth factor.

urogenital *Anatomy.* relating to the urinary and genital apparatus.

urogenital diaphragm *Anatomy.* the tissue stretching between the ischiopubic rami, formed by the sphincter of the urethra and the deep transverse perineal muscles. Also, TRIGONIUM UROGENITALE.

urogenital system *Anatomy.* the organs concerned in the production and excretion of urine, together with the organs of reproduction.

uroglena *Botany.* a unicellular colonial genus of plantlike flagellates of the family Ochromonadaceae, having many cells united in a spherical gelatinous colony.

urogomphus *Entomology.* a projection on the terminal segment of certain insect larvae. Also, PSEUDOCERCUS.

urography *Radiology.* the radiography of the urinary tract, or a portion of the tract, after injecting a substance to render the tract opaque.

urokinase *Enzymology.* an enzyme produced in the kidney and excreted in the urine; it catalyzes the conversion of plasminogen to plasmin by preferential cleavage at an arginine-valine bond. Also, PLASMINOGEN ACTIVATOR.

urol. urology; urological.

urolithiasis *Medicine.* the formation or presence of urinary calculi.

urologic or **urological** *Anatomy.* relating to the urinary system or to urology.

urology *Medicine.* the branch of medicine dealing with the male and female urinary tracts and the male genital tracts. Thus, **urologist.**

uronema *Botany.* a green algae genus belonging to the family Chaetophoraceae, and characterized by unbranched filaments.

uronic acid *Organic Chemistry.* any of a class of compounds similar to sugars, but with the exception that the terminal carbon has been oxidized from an alcohol to a carboxyl group; examples are galacturonic acid and glucuronic acid.

uropathy *Pathology.* any disease of the urinary system.

uropepsinogen *Biochemistry.* a compound occurring in urine as the end product of the secretion of pepsinogen into the blood by gastric cells.

Urophora *Invertebrate Zoology.* a genus of gallflies of the family Trypetidae, small two-winged flies whose larvae create large round plant galls.

uropod *Invertebrate Zoology.* a flattened posterior abdominal appendage in some crustaceans; in crayfish it forms most of the tail fan, in amphipods three pairs are used for jumping, and in isopods one pair forms a tactile organ.

uroporphyrin *Biochemistry.* a porphyrin that has a carboxymethyl and a carboxymethyl substituent on each pyrrole ring.

Uropygi *Invertebrate Zoology.* the whip scorpions, an order of nocturnal arachnids having powerful anterior pincers and a whiplike tail that can eject poison at an enemy. Also, SCHIZOMIDA.

uropygial gland *Vertebrate Zoology.* a gland located at the base of the tail in most birds that secretes an oil the bird uses in preening its feathers; in waterfowl, it adds water resistance to the feathers.

urosome *Invertebrate Zoology.* the posterior portion of an arthropod.

urostyle *Vertebrate Zoology.* **1.** the solid posterior bony extension from the vertebral column in frogs and toads, representing several fused vertebrae. **2.** the hypural bone in a fish.

Urostylidae *Invertebrate Zoology.* a family of true bugs, hemipteran insects in the superfamily Pentatomoidea, having five-segmented antennae.

Ursa Major *Astronomy.* the Greater Bear, a prominent northern constellation which contains seven stars that form the Big Dipper.

Ursa Major cluster *Astronomy.* **1.** a cluster of galaxies lying at a distance of 880 million light-years in the constellation of Ursa Major. **2.** a loose star cluster to which five of the stars in the Big Dipper belong.

Ursa Minor *Astronomy.* the Lesser Bear, a small constellation in which the north celestial pole is located; its brightest stars form the Little Dipper asterism, at one end of which is Polaris.

Ursa Minor system *Astronomy.* a very faint dwarf elliptical galaxy that belongs to the Local Group and lies at a distance of 230,000 light-years.

Ursidae *Vertebrate Zoology.* the bears, a family of generally very large, terrestrial, carnivorous mammals; includes the black, brown, and sun bears, the spectacled bears, and various extinct species.

Ursidae (black bear)

Ursids *Astronomy.* a meteor shower, visible around December 22 each year, that appears to come out of the constellation of Ursa Major.

ursine *Vertebrate Zoology.* of or relating to the family Ursidae, the bears.

ursolic acid *Biochemistry.* $C_{30}H_{48}O_3$, a carboxylic acid present in a wide variety of plants in the form of a free acid or an aglycone of triterpene saponins.

urstromthal *Geology.* a wide, shallow depression formed by the erosive action of an intermittent meltwater stream flowing parallel to the advancing edge of a continental ice sheet. Also, **urstromtal, pradolina.**

Urticaceae *Botany.* a family of dicotyledonous herbs, shrubs, and trees of the order Urticales, characterized by opposite or alternate, stipulate leaves, and small flowers found in cymes or clusters.

Urticales *Botany.* an order of herbs, shrubs, and trees of the subclass Hamamelidae, having bisexual or unisexual flowers, and simple, usually stipulate leaves.

urticaria *Medicine.* a skin eruption characterized by the development of wheals of varying shapes as a result of dilatation of the capillaries, caused by a reaction to various stimuli, including allergens, stress, insect bites, food, or drugs. Also, HIVES.

urticarial *Medicine.* relating to, characterized by, or of the nature of urticaria. Also, **urticarious.**

urticate *Medicine.* **1.** marked by the presence of wheals. **2.** to produce urticaria.

urticating hairs *Entomology.* irritating hairs found on some caterpillars; hollow and with poison glands at their bases, with a barbed tip that when broken releases the poison.

Uruguay *Geography.* a river in southeastern South America, rising in southern Brazil and flowing west and south about 1000 miles into the Rio de Plata.

Urysohn metrization theorem *Mathematics.* a theorem stating that every normal space that satisfies the second axiom of countability is metrizable.

Urysohn's lemma *Mathematics.* let A and B be disjoint closed subsets of a normal space X. Then there exists a continuous real-valued function $f: X \rightarrow [0, 1]$ such that $f(x) = 0$ for $x \in A$ and $f(x) = 1$ for $x \in B$.

US *Aviation.* the airline code for USAir.

US unconditioned stimulus.

USA United States Army.

usable *Computer Programming.* a term used to describe devices, memory, programs, etc., that are easy to use by a knowledgeable user.

usable rate of fire *Ordnance.* the normal rate at which a gun is actually fired, expressed in shots per minute; it is substantially less than the maximum rate based on optimum mechanical operation.

USAF United States Air Force.

U Sagittae *Astronomy.* an eclipsing binary star of the Algol type; the period is 3.4 days and the change in brightness is 3.6 magnitudes.

U.S. airways code *Meteorology.* a code used to report synoptic observations of aviation weather conditions.

U.S.C. & G.S. United States Coast and Geodetic Survey.

USCG United States Coast Guard.

USDA or **U.S.D.A.** United States Department of Agriculture.

use *Industrial Engineering.* in work-motion studies, an elemental motion in which an object is controlled by one or both of the hands during a work cycle.

use-dilution test *Microbiology.* the process of testing a disinfectant on a surface that other cleaning did not eliminate, often using test organisms such as *Staphylococcus aureus* and *Salmonella choleraesuis.*

useful beam *Radiology.* that portion of primary radiation that is allowed to emerge from the tubehead assembly of an X-ray machine, as limited by the tubehead opening and accessory collimating devices. Also, PRIMARY BEAM.

useful service life *Electrical Engineering.* the length of time an item will operate or the time until the probability of failure exceeds the desired operating requirements.

uselife or **use life** *Archaeology.* the length of time for which a tool or other artifact is used before it is discarded.

user *Computer Programming.* **1.** a person who uses the services of a computer system. **2.** a process, device, program, or system that uses the services of another process, device, program, or computer system. *Telecommunications.* **1.** a person or group that employs the services of a telecommunications system for the purpose of transferring information. **2.** in a signaling system, a telecommunications service that utilizes a signaling network for the purpose of transferring information.

user-defined function *Computer Programming.* a subroutine defined by the user and invoked in the same manner as a built-in function of the computer language or system.

user-defined type *Computer Programming.* a data type in which the user provides the name, scope, and properties of that type.

user exit *Computer Programming.* a point in a vendor-supplied program at which the user can insert a special routine to take control of the processing.

user-friendly *Computer Programming.* describing a computer device or program that is relatively easy to operate for users with little or no previous experience with it. *Industrial Engineering.* in general, describing any device or system that is relatively easy for a novice to operate.

user group *Computer Programming.* an organization of users of a particular computer system or software product; the members of the group share information and programs; often influences vendors in product improvement.

user identification *Computer Programming.* a string of characters that uniquely identifies a user in a computer system. Also, USER ID.

user interface *Computer Technology.* any hardware, software, or both that allows a user to interact with a computer system.

user mode *Computer Programming.* a state assigned by the operating system denoting that the program currently being executed is a user program and thus can be interrupted by a supervisor program and is prohibited from executing privileged instructions.

user program *Computer Programming.* a program written by or being used by a user. Also, **user-written program, user-written code.**

user-programmable memory *Computer Programming.* the memory in a microcomputer that can be accessed by a user or user program.

user's guide *Computer Programming.* a document prepared by the developer or vendor to assist a user in learning to operate a software product or computer system.

user-to-user service *Telecommunications.* a type of switching that allows direct user-to-user connection but has no provision for message store-and-forward service.

use-wear analysis *Archaeology.* the study of the edges of stone tools to determine the type of wear they have experienced and thus the tasks for which they were used. Also, EDGE-WEAR ANALYSIS.

U-shaped abutment *Civil Engineering.* a bridge abutment, the wings of which extend back parallel to the roadway rather than projecting at an angle.

U-shaped valley

U-shaped valley *Geology.* a broad, steep-walled, relatively flat-floored valley, usually carved by glacial erosion, whose cross section resembles the letter U.

USM underwater-to-surface missile.

USMC United States Marine Corps.

USN United States Navy.

Usneaceae *Botany.* a family of lichens of the order Lecanorales, characterized by a conspicuous and varied fruticose structure, that can be ligulate, filamentous, or pendulous.

usnic acid *Biochemistry.* $C_{18}H_{16}O_7$, an antibacterial organic acid found in the lichen *Usnea barbata,* which attacks Gram-positive bacteria, such as *Mycobacterium tuberculosis,* and certain fungi.

U.S.P. or **USP** United States Pharmacopeia. Also, **U.S. Pharm.**

USPHS or **U.S.P.H.S.** United States Public Health Service.

USRDA U.S. Recommended Daily Allowances.

U.S. Recommended Daily Allowances *Nutrition.* official estimates of safe and adequate daily intake of nutrients mandatory for the maintenance of good health.

U.S. Roland *Ordnance.* an air defense artillery surface-to-air missile system under development by the U.S. Army; it is designed for short-range, low-altitude, all-weather operation and is based on the Franco-German Roland III missile system.

USR test *Medicine.* the unheated serum reagin test, a variation of the VDRL test using unheated serum, used principally for screening for syphilis.

USS or **U.S.S.** United States ship.

U stage see UNIVERSAL STAGE.

Ustalf *Geology.* a suborder of the soil order Alfisol that develops in warm, subhumid to semiarid regions, and characterized by a brown to red color.

Ustert *Geology.* a suborder of the soil order Vertisol that develops in temperate or tropical regions, and characterized by wide, deep surface cracks that may remain open for several months during the year.

ustilagic acids *Biochemistry.* any of a group of insoluble, extracellular glycolipids that are produced by plant parasites and are toxic to the infected plants; used by perfumers to create macrocyclic musks.

Ustilaginaceae *Mycology.* a family of smut fungi of the order Ustilaginales, characterized by the formation of basidiospores.

Ustilaginales *Mycology.* an order of fungi of the class Teliomycetes, most species of which are parasites of flowering plants. Also, SMUT FUNGI, SMUTS.

ustilaginism *Toxicology.* poisoning due to ingestion of corn infected with the smut *Ustilago maydis*; the symptoms may be similar to those of ergotism.

Ustilaginomycetes *Mycology.* a former class of fungi of the subdivision Basidiomycotina, now reclassified within the class Teliomycetes.

Ustilago *Mycology.* a genus of smut fungi of the family Ustilaginaceae causing diseases to grasses and cereals, such as oat and barley.

Ustoll *Geology.* a suborder of the soil order Mollisol that develops in mesic or warm regions that are generally dry or have well-distributed rainfall, and characterized by a calcic, petrocalcic, or gypsic horizon over an accumulation of lime.

Ustox *Geology.* a suborder of the soil order Oxisol that develops in warm to tropical areas of well-distributed rainfall, and characterized by low to moderate amounts of organic matter and reddish coloration.

Ustult *Geology.* a suborder of the soil order Ultisol that develops in warm to tropical regions having well-distributed rainfall with a pronounced dry season, and characterized by a low to moderate organic-carbon content and a brown to reddish coloration.

USW or **usw** ultrashort wave.

UT Universal Time.

UTC Universal Time Coordinated (coordinated universal time).

Ut dict. or **ut dict.** as directed. (From Latin *ut dictum.*)

Utend. to be used. (From Latin *utendus.*)

uterine [yoot´ə rin] *Anatomy.* of or relating to the uterus. *Medicine.* occurring in or affecting the uterus. Thus, **uterine cancer.**

uterine contraction *Medicine.* contraction of the uterus during labor.

uterine insufficiency *Medicine.* weakness of the contractile power of the uterus due to lack of adequate muscle tone.

uterine tube see FALLOPIAN TUBE.

uteritis see METRITIS.

uterometer *Medicine.* an instrument for measuring the uterus.

uterus [yoot´ə rəs] *Anatomy.* the hollow muscular organ in female mammals in which the fertilized ovum usually becomes embedded, and in which the developing embryo and fetus are nourished.

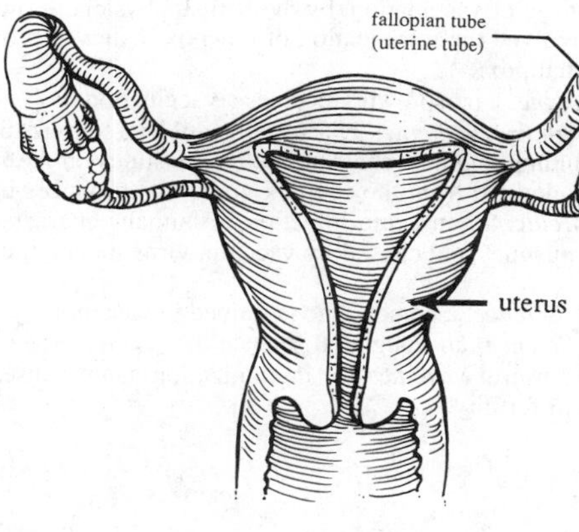

uterus

uterus bicornis *Anatomy.* a uterus that has two lateral horns or components, usually due to abnormal development.

UTI urinary tract infection.

utile *Materials.* the handsome, moderately durable wood of the African tree *Entandrophragma utile*; used for interior decoration and plywood.

utilitarian *Archaeology.* describing an artifact that has a practical as opposed to a ceremonial function, such as a cooking pot for household use.

utility *Engineering.* **1.** a municipal or regional facility such as a water works, power plant, or telephone system. **2.** a support facility for a manufacturing plant, such as steam, water cooling, and power. *Computer Programming.* see UTILITY ROUTINE. *Statistics.* in decision theory, the numerical evaluation of the consequences of alternative actions, for various values of the unknown random variables; the negative of loss.

utility branch *Robotics.* a branch in a robot's program that is triggered by an external signal.

utility knife *Mechanical Devices.* a commonly used knife consisting of a thin, sharp-edged metal blade fitted with a handle, used for cutting soft, pliable objects, especially solid foods such as meat and vegetables.

utility routine *Computer Programming.* a frequently needed standard subroutine or program that has been predefined and built into the computer language or system; examples are memory dump, sort/merge, file-handling routines, and debugging aids. Also, **utility program.**

utilizable *Nutrition.* describing the extent to which the chemical components of a given food can be used by the body for its growth, reproduction, maintenance, and repair.

utilization factor *Electricity.* in power distribution, the proportion of the maximum permitted currents to the rated capacity of the system.

utilization rate *Aviation.* a measure of average flying time for aircraft in a fleet or squadron, expressed in hours per period per aircraft. Also, FLYING HOUR RATE.

utisol *Agronomy.* one of the 10 major soil classifications; characterized by reddish-brown to reddish-yellow color and high moisture; tends to develop in warm to tropical climates.

UTM universal transverse Mercator (grid).

UTP uridine triphosphate, uridine 5'-triphosphate.

utriculus *Anatomy.* **1.** a membranous sac of the membranous labyrinth that connects to the semicircular canal of the ear and contributes to the sense of equilibrium. **2.** a structure in the male homologous to the vagina and uterus formed from the distal ends of the paramesonephric duct and located within the proslate. Also, **utricle.**

UTS ultimate tensile strength.

U-tube heat exchanger *Chemical Engineering.* a heat exchange unit that consists of a group of U-shaped tubes encased by a shell, in which one fluid flows through the tubes and the other flows through the shell.

U-tube manometer *Engineering.* a U-shaped tube filled with a liquid of known specific gravity, used to determine a pressure differential across the tube's ends. Also, LIQUID-COLUMN GAUGE.

Uukuvirus *Virology.* a genus of viruses of the family Bunyaviridae that infect a wide range of vertebrates, including birds and rodents, and that are transmitted by ticks.

UUM underwater-to-underwater missile.

UV or **U.V.** ultraviolet.

uvala *Geography.* a valleylike depression in karst country, usually formed by the linear coalescence of several sinkholes.

U valley see U-SHAPED VALLEY.

uvanite *Mineralogy.* a rare, strongly radioactive, brownish-yellow, probably orthorhombic mineral, with the approximate formula $U_2^{+6}V_6^{+5}O_{21}\cdot 15H_2O$, occurring as microcrystalline masses and coatings, having an undetermined specific gravity and hardness; found in asphaltic sandstone of the chinle formation in Utah and Arizona.

uvarovite *Mineralogy.* $Ca_3Cr_2(SiO_4)_3$, a green cubic mineral of the garnet group, occurring as dodecahedral or trapezohedral crystals and in massive form, having a specific gravity of 3.4 to 3.8 and a hardness of 6.5 to 7.5 on the Mohs scale; found in serpentine with chromite, and with chromite, in metamorphosed limestone. Also, UWAROWITE.

UVBY system *Astronomy.* a four-color photometric system that uses measurements made at ultraviolet, violet, blue, and yellow wavelengths.

UV Ceti star *Astronomy.* a faint, cool red star on the main sequence that erupts unpredictably with violent flares, occasionally brightening by two magnitudes over several seconds. Also, FLARE STARS.

uvea *Anatomy.* the vascular, pigmented middle tunic of the eye, consisting of the choroid, the ciliary body, and the iris.

uveitis *Medicine.* inflammation of all or part of the uvea, commonly involving the other tunics (the sclera and cornea, and the retina).

UV endonuclease *Enzymology.* an *E. coli* enzyme that cleaves a DNA molecule near the 5' and 3' ends of a thymine dimer; a procedure used by the cell in repairing DNA containing thymine dimers produced by ultraviolet radiation.

uviol glass *Materials Science.* a type of glass through which ultraviolet rays pass easily.

UVR ultraviolet rays.

uvr genes *Genetics.* a class of genes that are involved in the excision repair of damaged DNA.

UV stabilizer *Chemistry.* a compound that is admixed with thermoplastic resin to absorb ultraviolet rays, thereby retarding the degradation of the polymers.

uvula *Anatomy.* a small, fleshy mass of tissue hanging from the midposterior free edge of the soft palate.

uvulectomy see STAPHYLECTOMY.

uwarowite see UVAROVITE.

UXB unexploded bomb.

uxorilocal residence *Anthropology.* a residence practice in which a married couple comes to live in the wife's home.

Uzi [ooz´ē] *Ordnance.* a widely used 9-mm blowback-operated submachine gun designed in Israel; it has selective automatic/semiautomatic fire, a firing rate of 650 rounds per minute, and a muzzle velocity of 1310 feet per second. (From its designer, Uziel Gal.)

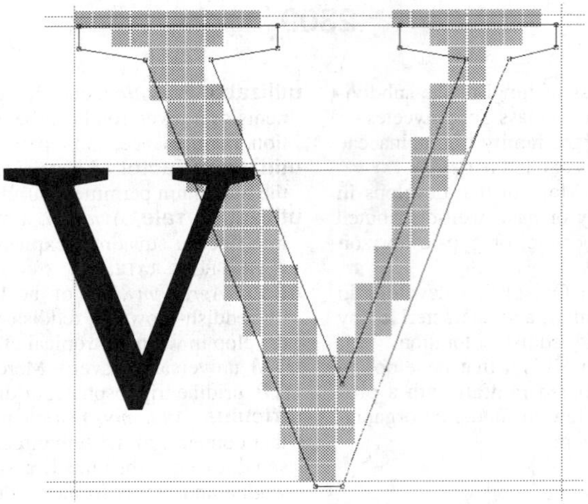

V the chemical symbol for vanadium.
V volt; valley.
V see. (From Latin *vide.*)
V. valve; vocative.
v *Computer Science.* see P AND V.
v velocity; volume.
v. vein; ventral; voltage; volume.
V-1 *Ordnance.* a robot bomb used by Germany against Britain in World War II; it was equipped with wings, stabilizers, elevators, a rudder, and a back-mounted, pulse-jet engine; it was 25.33 feet long and 2 feet 7 inches in diameter, and had a range of 150 miles. Also, BUZZ BOMB.
V-2 *Ordnance.* a rocket-powered ballistic missile used by Germany against England in World War II; it was 46 feet long and 5 inches in diameter, and had a range of 200 miles. Also, A-4.
VA vice admiral.
VA or **V.A.** Veterans Administration.
VA visual acuity; volt-ampere. Also, **va.**
VAB vehicle assembly building.
VAC *Oncology.* an acronym for a cancer chemotherapy regimen that includes the drugs vincristine, dactinomycin, and cyclophosphamide.
vac. vacuum.
vacancy *Materials Science.* a point imperfection in a crystal lattice in which an atom is missing from a site normally occupied by an atom. *Solid-State Physics.* see SCHOTTKY DEFECT.
vacancy diffusion *Materials Science.* a diffusion process in which an atom interchanges position with a vacancy.
vaccinal *Immunology.* relating to or caused by vaccine or vaccination.
vaccinate *Immunology.* to carry out a process of vaccination.
vaccination *Immunology.* the administration of treated microorganisms into humans or animals to induce an immune response, and thus provide them with resistance to disease or infection.
vaccination encephalitis *Medicine.* encephalitis that occurs following vaccination; usually following smallpox vaccination.
vaccine *Immunology.* a preparation, consisting of killed, pretreated, or living microorganisms or molecules derived from them, that is used in vaccination. *Virology.* specifically, a suspension of attenuated or killed viruses that are administered to vertebrates to produce immunity to that viral infection. (Ultimately derived from the Latin word for "cow;" the first significant use of vaccination, by the British physician Edward Jenner in 1796, involved the introduction of cowpox matter to provide immunity from smallpox.)
vaccinee *Medicine.* a person who receives a vaccination.
vaccinia *Veterinary Medicine.* **1.** a usually mild infectious disease in mammals, including humans, characterized by pustules on the skin. **2.** a virus, probably derived from the cowpox virus, that produces immunity to smallpox. *Medicine.* an acute infection in humans characterized by localized skin pustules and caused by vaccinia virus inoculation against smallpox.
vaccinologist *Science.* a person who is trained in vaccinology.
vaccinology *Science.* an integrated science that is concerned with the preparation and use of vaccines for immunization against diseases and for the control of fertility.

Vaccinology

"Vaccinology" refers to an integrated science concerning phenomena relevant to the preparation and use of "vaccines" (1) for prophylactic immunization for prevention of infectious diseases, (2) for therapeutic immunization for treatment of neoplastic and autoimmune disease, and (3) for the control of fertility. The principles of vaccinology, for the development and the use of vaccines for these purposes, derive from an understanding of the etiologic factors, pathogenic mechanisms, host immune responses, and epidemiological considerations relevant to each of the different uses to which vaccines are put.

Thus, vaccinology may be seen as a hybrid science, drawing on knowledge from several related disciplines pertaining to the knowledge required for inducing the desired effects of active immunization. For immunization requiring the presence of specific humoral antibody and/or memory cells, such effects are inducible using either noninfectious antigens, as in "killed" virus vaccines, or infectious agents, as in attenuated "live" virus vaccines. Using the former, the level of antibody response and the degree of immunological memory initially induced are dose related, the effect of which can be greatly enhanced by use of an appropriate adjuvant. These considerations also apply to the development and use of effective vaccines against cell-associated viral infections, such as HIV, EBV, and HSV, which require the induction of cellular immune mechanisms which then act by destroying or limiting the population of virus-infected cells, or of neoplastic cells or cells involved in autoimmune phenomena.

Thus, principles that apply generally, with variations that pertain to the specific agents and diseases in question, have emerged in the course of emergence of the integrated science of vaccinology.

Jonas Salk, M.D.
Founding Director and
Distinguished Professor in International Health Sciences
The Salk Institute for Biological Studies

Vacuolariaceae *Botany.* a family of flagellates belonging to the order Rahidomonadales, having chloroplasts containing chlorophylls a and c; occurring mostly in fresh water, but also in marine and brackish waters.

vacuolating virus *Virology.* a virus involving several viral families in which vacuoles are formed in infected cells.

vacuole *Cell Biology.* a large fluid-filled vesicle found in the cytoplasm of a plant cell or a protozoan. *Geology.* see VESICLE.

vacuum [vak´yoom; vak´yoo əm] *Physics.* **1.** ideally, a region of space that is completely devoid of any form of matter; a totally empty space. **2.** in practical terms, an enclosed region of space in which the pressure has been reduced (below normal atmospheric pressure) sufficiently so that processes occurring within the region are unaffected by the residual matter. *Quantum Mechanics.* according to Dirac's hole theory, the space housing an infinite sea of electrons having negative energy, which is observed normally as empty space. (From the Latin word for "empty.")

vacuum activity *Behavior.* the spontaneous occurrence of a fixed-action pattern in the apparent absence of its usual releaser.

vacuum aspiration *Medicine.* the removal of the uterine contents by applying a vacuum with the use of a hollow curet or a cannula inserted into the uterus.

vacuum atomization *Materials Science.* a process for commercial production of nickel-based superalloy powders, in which the molten metal is supersaturated under pressure with a gas, such as hydrogen, then suddenly exposed to a vacuum.

vacuum back *Photogrammetry.* see LOCATING BACK.

vacuum brake *Mechanical Engineering.* an air brake in which low pressure is maintained in the actuating cylinder, which is opened on one side to allow atmospheric pressure to apply the brake.

vacuum breaker *Engineering.* a device that decreases the vacuums formed in water supply lines, which could cause backflow. Also, BACKFLOW PREVENTER.

vacuum capacitor *Electronics.* a capacitor in which the metal plates or cylinders are enclosed in a vacuum envelope in order to increase the breakdown voltage.

vacuum casting *Metallurgy.* a process of casting metals in a vacuum. Also, SUCTION CASTING.

vacuum circuit breaker *Electricity.* a circuit breaker in which a pair of contacts is hermetically sealed in a vacuum envelope to minimize sparking.

vacuum cleaner *Mechanical Engineering.* an electrically powered mechanical device that removes and collects dry dirt and dust from a surface by means of a suction fan that creates a partial vacuum within the machine, so that the resulting rush of outside air sucks up the waste material into a bag or container.

vacuum concrete *Civil Engineering.* concrete poured into a form fitted with a vacuum mat that sucks out excess water and air, resulting in a dense, strong, fast-setting concrete.

vacuum condensing point *Chemistry.* the temperature at which a vaporized solid condenses in a vacuum during a sublimation process.

vacuum correction *Physics.* a correction imposed on the reading of a mercury barometer that is necessary because of the pressure of water vapor and air.

vacuum crystallizer *Chemical Engineering.* a crystallizer consisting of a closed vacuum vessel into which a warm, saturated solution is fed, then evaporates and cools adiabatically, resulting in crystallization.

vacuum degassing *Metallurgy.* a process for removing gases from a metal, in which two refractory tubes from a vacuum chamber dip below the surface of the molten steel, allowing vacuum jets to raise the liquid steel to a certain level in the tubes so that argon or other gases can be fed through one tube.

vacuum deposition *Metallurgy.* the condensation of thin metal coatings on a cool work surface in a vacuum.

vacuum distillation *Chemical Engineering.* a distillation performed below atmospheric pressure, but not as low as that for molecular distillation; used for distilling high-boiling-point and heat-sensitive materials such as the heavy distillates in petroleum.

vacuum drying *Engineering.* the removal of liquid from a solid material while in a vacuum system, to lower the temperature at which evaporation takes place and thus prevent heat damage to the material.

vacuum envelope *Electronics.* an airtight glass envelope used, for example, to hold the electrodes in an electron tube.

vacuum evaporation *Engineering.* **1.** the deposition of a thin film of some material on a substance by evaporation from a boiling source in a hard vacuum. **2.** in a vacuum, the loss of molecules from the surface of a solid body.

vacuum evaporator *Engineering.* a device used to coat a specimen with evaporation from metals and spectrographic carbon for use in electron spectroscopic analysis or electron microscopy.

vacuum fermentation *Biotechnology.* a fermentation system that was developed for the continuous production of ethanol and other volatile materials; reduced pressure allows the volatile material to leave the reactor continuously in the form of exhaust gas.

vacuum filter *Engineering.* a filter in which a liquid-solid slurry is applied to the high-pressure side; the liquid is pulled through the filter by the vacuum and a cake of solids is formed on the outside.

vacuum filtration *Engineering.* the introduction of a partial vacuum in order to cause an increase in fluid flow through a filter, thus separating liquid from solid.

vacuum flashing *Chemical Engineering.* the intense vaporization or flashing of a heated liquid when released to a lower pressure or vacuum.

vacuum fluorescent lamp *Electronics.* an evacuated display tube having anodes that are coated with a phosphor that glows when struck by electrons from the cathode to create a display.

vacuum forming *Materials Science.* plastic-sheet forming in which a thermoplastic sheet is clamped to a holder and then heated and drawn into the mold by a vacuum.

vacuum frame *Graphic Arts.* **1.** a mechanical device designed to hold a printing plate and a copy negative together in close, uniform contact during exposure. **2.** a similar device used to make contact photographic prints. Also, **vacuum printing frame.**

vacuum freeze dryer *Engineering.* an indirect batch dryer used to dry materials that cannot tolerate temperatures above the freezing point or that would be destroyed by the loss of volatile ingredients.

vacuum gauge *Engineering.* an instrument that indicates the absolute gas pressure in a vacuum system.

vacuum gripper *Robotics.* an end effector that uses a suction cup and a vacuum source to manipulate objects.

vacuum heating *Mechanical Engineering.* a steam-heating system in which a vacuum pump induces a pressure gradient to start a return flow of air in the return pipes.

vacuum induction melting *Materials Science.* a process in which heat is provided by an induction coil under vacuum; used in the production of nickel base and nickel-iron base superalloys, and some specialty steels.

vacuum mat *Civil Engineering.* a linen-filter screen with a vacuum pump behind it; used in a concrete form to produce vacuum concrete.

vacuum measurement *Engineering.* the measurement of a fluid pressure that is less than the atmospheric pressure.

vacuum metallurgy *Metallurgy.* the processing of metals at high temperatures by induction heating in a vacuum.

vacuum mold casting *Materials Science.* a process for casting metal in which unbonded sand is held in place in the mold by a vacuum. The pattern is covered by a heat-softened, tightly conforming thin sheet of plastic that is applied by vacuum, and a box or flask is placed over the plastic-coated pattern and filled with free-flowing sand. When the flask is evacuated, the vacuum "hardens" the sand so that the pattern can be withdrawn.

vacuum molding *Materials Science.* a method used to fabricate plastic sheet and polymer matrix composites, using a vacuum bag to exert pressure on the sheet.

vacuum-pack *Food Technology.* to pack and seal food in a container, such as a can or jar, removing as much air as possible before sealing, to preserve freshness.

vacuum pan *Biotechnology.* a sealed apparatus that controls the crystallization of solids from liquids by lowering the pressure within the sealed container.

vacuum pan salt *Chemical Engineering.* a salt produced by boiling salt brine in a triple-effect evaporator at reduced pressure.

vacuum pencil *Engineering.* a pencil-shaped piece of tubing that is attached to a vacuum pump and used to pick up semiconductor slices or chips during the manufacture of solid-state devices.

vacuum phototube *Electronics.* a phototube that has been evacuated so that its electrical characteristics are essentially unaffected by gaseous ionization.

vacuum plating *Metallurgy.* a process in which a metallic coating is spread onto the base metal by heat and under a vacuum.

vacuum polarization *Quantum Mechanics.* a process in which a vacuum-electromagnetic field produces electron-positron pairs that, in turn, exert electromagnetic fields of their own, in a manner analogous to classical dielectric polarization.

vacuum pump *Engineering.* a device with which a partial vacuum can be produced. *Robotics.* a compressor used for removing air and creating a vacuum in some component, such as a gripper.
vacuum relay *Electricity.* a relay with contacts mounted in an evacuated glass housing to minimize sparking.
vacuum relief valve *Engineering.* a device that permits liquid to enter a pressure vessel in order to prevent extreme internal pressure.
vacuum seed meter *Agriculture.* a planting mechanism that can be set to place seeds at the proper space and depth to ensure maximum growth, increasing planting speed and accuracy.

vacuum seed meter

vacuum shelf dryer *Engineering.* an indirect batch dryer consisting of a vacuum-tight chamber containing heated supporting shelves, a vacuum source, and a condenser; used to dry pharmaceuticals, temperature-sensitive materials, and easily oxidizable materials.
vacuum support *Mechanical Engineering.* the part of a rupture disk device that prevents disk deformation in case of a rapid pressure change.
vacuum switch *Electricity.* a switch in which the contacts are contained in an evacuated envelope to minimize sparking.
vacuum thermobalance *Analytical Chemistry.* a test apparatus consisting of a precision balance and a furnace; used to record changes in weight as a function of temperature.
vacuum tube *Electronics.* an electron tube that has been evacuated so that its electrical characteristics are essentially unaffected by the presence of residual gas or vapor.
vacuum-tube amplifier *Electronics.* an amplifier in which one or more vacuum tubes are used.
vacuum-tube clipping circuit *Electronics.* a clipping circuit that utilizes a vacuum tube to produce the clipping.
vacuum-tube electrometer *Electronics.* an electrometer in which a special vacuum triode, having an input resistance above 10 gigohms, amplifies the ionization current in an ionization chamber.
vacuum-tube modulator *Electronics.* a modulator utilizing a vacuum tube as the modulating element for impressing an intelligence signal on a carrier.
vacuum-tube oscillator *Electronics.* a circuit utilizing a vacuum tube to convert direct current to alternating current at the desired frequency.
vacuum-tube rectifier *Electronics.* a circuit that converts alternating current to direct current using one or more vacuum tubes.
vacuum ultraviolet radiation *Electromagnetism.* ultraviolet radiation of wavelengths less than about 200 nanometers; so-called because at shorter wavelengths the radiation is strongly absorbed by most gases.
vacuum ultraviolet spectroscopy *Spectroscopy.* a spectroscopic technique for observing ultraviolet radiation with wavelengths less than 1900 angstroms, in which evacuated equipment is required for observing spectra because of interference from absorption by oxygen.
vadose water see SOIL WATER.
vadose zone see ZONE OF AERATION.
vaesite *Mineralogy.* NiS_2, a tin-white opaque, metallic cubic mineral of the pyrite group occurring as octahedral or cubic crystals, having a specific gravity of 4.45 and a hardness of about 3.5 on the Mohs scale; forms a series with cattierite.
vagile *Ecology.* relating to or showing the ability to change location over time.
vagility *Ecology.* the ability or tendency of an organism to change its location or distribution over time.
vagina [və jī´nə] *Anatomy.* the canal in the female extending from the vulva to the cervix.
vagina fibrosa tendinis *Anatomy.* the fibrous sheath that surrounds a tendon, usually confining it to an osseous groove.
vaginal [vaj´ə nəl] *Anatomy.* of or relating to the vagina. *Medicine.* occurring in or affecting the vagina. Thus, **vaginal cancer.**
vaginal hysterectomy *Medicine.* the removal of the uterus through the vagina. *Surgery.* see COLPOHYSTERECTOMY.
vaginal touch *Medicine.* digital exploration of the vagina.
vaginate *Biology.* invested with a sheath or cover of a plant or animal body or body part.
vaginismus *Medicine.* an intense or painful spasm of the vagina caused by involuntary contraction of the vaginal and perineal muscles; the cause may be organic or psychogenic.
vaginitis [vaj´ə nī´tis] *Medicine.* an inflammation of the vagina, characterized by pain and purulent discharge.
vaginofixation *Medicine.* suture of the vagina to the abdominal wall.
vaginometer *Medicine.* an instrument for measuring the dimensions of the vagina.
vaginoplasty see COLPOPLASTY.
vagitus *Medicine.* the crying of an infant.
vagotomy *Surgery.* the cutting of the vagus nerve.
vagotonia *Neurology.* hyperexcitability of the vagus nerve, inducing vasomotor instability, muscle spasms, sweating, and constipation.
vagotonine *Biochemistry.* a pancreatic hormonal extract that heightens vagal tone, triggers bradycardia, and encourages glycogen storage in the liver.
vagrant *Ecology.* not permanently attached; moving by its own ability.
vagus *Anatomy.* either of the tenth pair of cranial nerves, consisting of motor fibers and extending from the brain to the heart, lungs, stomach, and other organs. Also, **vagus nerve.**
Vahlkampfia *Invertebrate Zoology.* a genus of naked, lobose amoebas that eat bacteria; widespread in freshwater, marine, and soil habitats.
VAL *Robotics.* a programming language developed by Unimation, Inc., and used for their robotic systems.
Val valine.
valais wind *Meteorology.* a strong wind that originates around the upper end of Lake Geneva in the Swiss canton of Valais, and blows down the Rhône Valley in south-central France.
Valangianian *Geology.* a European geologic stage of the Lower Cretaceous period, occurring after the Berrisian and before the Hauterivian.
valence *Chemistry.* a number indicating the combining power of one atom with others; that is, the number of other atoms with which it can combine. *Biochemistry.* the quantity of binding sites on a molecule that are able to react with ligands. *Immunology.* specifically, the number of antigen binding sites possessed by an antibody molecule or the number of antigenic determinants possessed by an antigen. *Behavior.* the degree of attractiveness of a stimulus. *Mathematics.* the degree of a vertex in a graph; i.e., the number of edges incident to the vertex, with loops counted twice.
valence angle see BOND ANGLE.
valence band *Materials Science.* a range of energy states filled by electrons in semiconductors and dielectrics at absolute zero; contains the electrons used in bonding. *Solid-State Physics.* the highest energy band in the spectrum of a solid crystal semiconductor or insulator that is occupied by electrons.
valence bond *Physical Chemistry.* a bond that forms between electrons from two or more different atoms.
valence-bond method *Physical Chemistry.* a method of calculating various molecular behaviors and properties according to the valence-bond theory.
valence-bond theory *Physical Chemistry.* a theory stating that the phenomenon of bond formation is the result of two adjacent molecules having one or more shared electron pairs and a suitable interaction or overlap of specific electron orbitals.
valence electron see CONDUCTION ELECTRON.
valence number see OXIDATION NUMBER.

valence transition *Physical Chemistry.* a change that takes place in the orbits of rare-earth or actinide atoms under certain conditions.

valencianite *Mineralogy.* a variety of adularia from a silver mine at Valencia, Mexico.

valency *Immunology.* the number of specific antibodies in a polyvalent antiserum or the number of distinct antigens in a polyvalent vaccine.

Valentian see LLANDOVERIAN.

valentinite *Mineralogy.* Sb_2O_3, a colorless to white, yellowish, or reddish orthorhombic mineral, dimorphous with senarmontite, occurring as tabular or prismatic crystals and in massive form, and having a specific gravity of 5.76 and a hardness of 2.5 to 3 on the Mohs scale; found as an alteration product of other antimony minerals.

***n*-valeraldehyde** *Organic Chemistry.* $CH_3(CH_2)_3CHO$, a colorless liquid; slightly soluble in water and soluble in alcohol and ether; boils at 102–103°C; used as a flavoring agent and in rubber accelerators.

valeramide *Organic Chemistry.* $CH_3(CH_2)_3CONH_2$, colorless crystals; partially soluble in water and soluble in alcohol; melts at 114–116°C. Also, PENTANAMIDE.

Valerianaceae *Botany.* a nearly cosmopolitan family of malodorous herbs and, rarely, shrubs, belonging to the order Dipsacales; characterized by ethereal oils in scattered oil cells and dicotyledonous embryos.

valerianic acid see VALERIC ACID.

valerian oil *Materials.* a combustible yellowish or brownish oily liquid with a penetrating odor; soluble in alcohol and ether; distilled from the roots of the plant species *Valeriana officinalis* and used in tobacco, perfume, and flavoring, and as an industrial odorant.

valeric acid *Organic Chemistry.* $CH_3(CH_2)_3COOH$, a colorless liquid with a penetrating odor and taste; soluble in water and alcohol; boils at 185.4°C; used as a perfume and flavor intermediate, a plasticizer, a vinyl stabilizer, and in pharmaceuticals. Also, VALERIANIC ACID, *n*-PENTANOIC ACID.

γ-valerolactone *Organic Chemistry.* $C_5H_8O_2$, a colorless liquid; soluble in water and most organic solvents; boils at 205–206.5°C; used in dye baths as a coupling agent, in cutting oils, in brake fluids, and as a solvent for adhesives, insecticides, and lacquers.

valid *Artificial Intelligence.* of a logical formula, true under every interpretation. *Systematics.* of an accepted taxonomic name, complying with the provisions of a nomenclatural code.

validation *Computer Programming.* the procedures involved in checking data or programs for correctness, compliance with standards, and conformance with the requirement specifications. Also, **validity check.**

valid name *Systematics.* the correct name for a zoological taxon.

valid program *Computer Programming.* a program in which each statement is syntactically correct and can be translated into machine code.

valine *Biochemistry.* $C_5H_{11}NO_2$, one of the twenty amino acids that make up proteins.

vallate papilla *Anatomy.* one of a group of elevated papillae at the back of the tongue.

Vallée-Poussin, Charles de la 1866–1962, Belgian mathematician; with Jacques-Salomon Hadamard, proved the prime number theorem.

valleriite *Mineralogy.* $4(Fe,Cu)S{\cdot}3(Mg,Al)(OH)_2$, a very soft, bronze-yellow, opaque, metallic hexagonal mineral occurring in massive form, having a specific gravity of 3.09 to 3.14 and an undetermined hardness; found in copper-bearing dunites, carbonatites, and serpentinized ultramafic rocks.

valley *Geology.* a generally broad, flat land area lying between hills, mountains, or other stretches of high ground and often containing a river or stream. *Oceanography.* a broad, shallow low-relief depression of the sea floor, having gently sloping sides. *Building Engineering.* the resultant depression or angle formed when two inclined sides of a roof meet. *Electrical Engineering.* the dip of a curve or waveform that occurs between two adjacent peaks.

valley attenuation *Electronics.* the attenuation at the bottom of the ripples (the higher of the two limits of attenuation) in a filter with the attenuation in the pass-band varying in a ripple fashion between an upper and a lower limit.

valley braid see ANABRANCH, def. 1.

valley breeze *Meteorology.* a gentle wind that blows up a valley or a mountain slope as a result of warming along an adjacent mountainside and valley floor during daytime solar heating in otherwise calm, fair weather conditions. Also, **valley wind.**

valley line *Hydrology.* see THALWEG, def. 1.

valley point *Electrical Engineering.* the lowest point of finite current on a current-voltage response curve.

valley-side moraine see LATERAL MORAINE.

valonia *Botany.* a marine green algae genus of the family Valoniaceae, characterized by a simple thallus that is a single oval or a long cylindrical multinucleate cell.

Valoniaceae *Botany.* a family of green algae belonging to the order Siphonocladales, distinguished by the formation of tiny lenticular segments inside the vesicle wall.

valuation *Mathematics.* a function ϕ from a field K into an ordered field such that, for all a and b in K: (a) $\phi(a) > 0$ for $a \neq 0$ and $\phi(0) = 0$; (b) $\phi(ab) = \phi(a)\,\phi(b)$; and (c) $\phi(a + b) \leq \phi(a) + \phi(b)$. Examples: (1) the "trivial" valuation $\phi(a) = 1$ for $a \neq 0$, $\phi(0) = 0$. (2) if K is an ordered field, then the absolute value function is a valuation on K. (3) if K is the field Q of rational numbers and p is a fixed prime number, then every nonzero rational number a can be written in the form $a = sp^n/t$, where s and t are integers not divisible by p. Then the function $\phi_p(a) = p^{-n}$, $\phi_p(0) = 0$ defines a valuation on Q known as the ***p*-adic valuation.**

valuation domain *Mathematics.* an integral domain with the property that given any two elements of the domain, one is a divisor of the other.

valuation ring *Mathematics.* a (discrete) valuation ring is a principal ideal domain that has exactly one nonzero prime ideal.

valval or **valvar** *Anatomy.* relating to a valve. Also, **valvular.**

valvate *Botany.* **1.** having organs that meet but do not overlap, as the parts of some buds. **2.** opening as if by valves, as do the capsules of certain dehiscent fruits and some anthers.

Valvatida *Invertebrate Zoology.* an order of starfish, echinoderms in the subclass Asteroidea.

valve *Mechanical Devices.* any of a number of devices that allow, halt, or regulate the passage of a fluid through a pipe or tube, either manually or automatically. *Anatomy.* a membranous fold or other structure that controls the flow of a fluid, such as the valve of the heart. *Botany.* the part of a fruit or seed wall that opens at dehiscence. *Invertebrate Zoology.* either half of the silicified shell of a diatom.

valve chamber *Mechanical Devices.* an assembly or casing attached to a pump or cylinder in which the valve is found and operates. Also, **valve box, valve chest.**

valve gear *Mechanical Devices.* a mechanism in an engine that controls the valve motion by actuating rods which contract slot links in forward or reverse, regulating intake and exhaust to and from the cylinders.

valve grinder *Mechanical Devices.* any of various tools used to reseat or resurface automotive valves.

valve rod *Mechanical Devices.* the metal arm in a steam engine that transmits motion received from the eccentric rod to the valve. Also, **valve spindle.**

valve seat *Mechanical Devices.* a fixed ring of heat-resistant metal upon which a plug valve or lift face rests when closing off.

valve spring *Mechanical Devices.* **1.** the spiral spring in a lift valve that enables it to close promptly, thereby efficiently regulating the flow to a gasoline engine or pump. **2.** a spiral spring of a sliding valve used to force packing rings against the valve facing.

valvula *Anatomy.* a small valve, especially a cusp of the aortic valve.

valvulitis *Medicine.* an inflammation of a valve or valvula, especially a valve of the heart.

VAMP *Oncology.* a cancer chemotherapy regimen that includes the drugs vincristine, amethopterin, 6-mercaptopurine, and prednisone.

vampire bat *Vertebrate Zoology.* an American bat of the family Phyllostomatidae that is adapted to a diet of blood, having four sharp canines it uses to slit the skin, a tubular tongue, and a clotting inhibitor in its saliva. (From the Eastern European legend of the *vampire,* an animated corpse that returns to life at night to suck the blood of the living.)

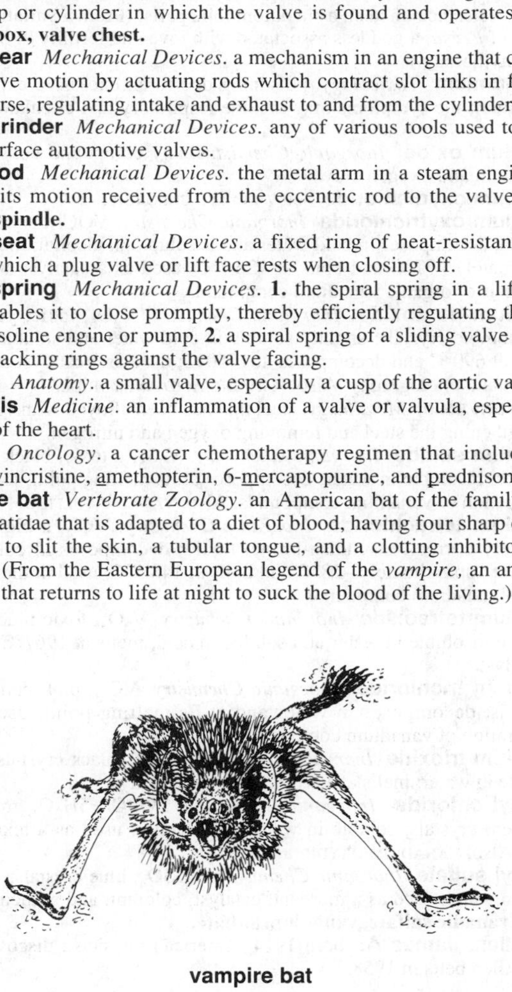

vampire bat

Vampirovibrio *Bacteriology.* a genus of Gram-negative, aerobic bacteria that occur as small, motile, vibrioid cells, each possessing a single polar flagellum; predatory on green algae of the genus *Chlorella.*

Vampyromorpha *Invertebrate Zoology.* an order of small, black, deep-sea cephalopod mollusks containing a single species, the **vampire squid,** *Vampyroteuthis infernalis.*

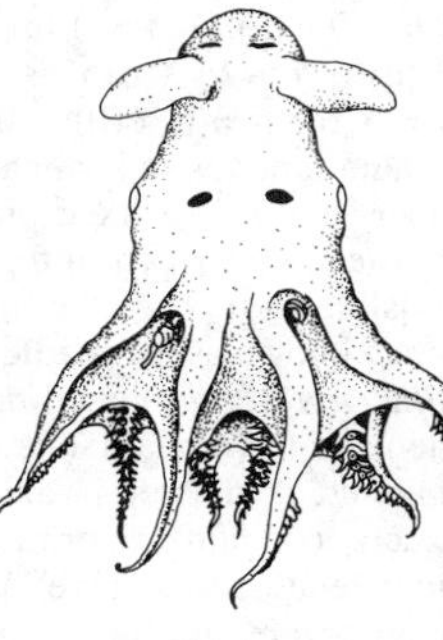

Vampyromorpha

vanadic acid anhydride see VANADIUM PENTOXIDE.

vanadic sulfide see VANADIUM SULFIDE.

vanadinite *Mineralogy.* $Pb_5(VO_4)_3Cl$, yellow, brown, or greenish crystals; an ore of lead and vanadium. Also, ENDLICHITE.

vanadium [və nā′dē əm] *Chemistry.* a rare element having the symbol V, the atomic number 23, an atomic weight of 50.942, a melting point of 1900°C, and a boiling point of about 3000°C; occurring in certain minerals and obtained as a light-gray powder with a silvery luster or as a ductile metal; used in making alloys that toughen and increase the shock resistance of steel, as in automobile parts. (From a variation of the name of *Freya,* a goddess associated with love and fertility in Scandinavian mythology.)

vanadiumism *Toxicology.* poisoning due to an excessive exposure to vanadium; the symptoms may include respiratory and circulatory irritation.

vanadium oxide *Inorganic Chemistry.* any compound consisting of vanadium and oxygen only.

vanadium oxydichloride see VANADYL CHLORIDE.

vanadium oxytrichloride *Inorganic Chemistry.* $VOCl_3$, a yellow liquid that decomposes in cold water and is soluble in alcohol; melts at approximately −77°C and boils at 126.7°C; used as a catalyst and in the synthesis of organic vanadium compounds.

vanadium pentoxide *Inorganic Chemistry.* V_2O_5, toxic, yellowish-red rhombic crystals; slightly soluble in cold water and soluble in acid; melts at 690°C and decomposes at 1750°C; used as a catalyst and in ceramics and photography. Also, VANADIC ACID ANHYDRIDE.

vanadium steel *Metallurgy.* steel that has been alloyed with vanadium, strengthening the steel and removing oxygen and nitrogen.

vanadium sulfide *Inorganic Chemistry.* V_2S_5, a toxic, black to green powder; insoluble in water; decomposes at its melting point; used in the preparation of vanadium compounds. Also, VANADIC SULFIDE.

vanadium tetrachloride *Inorganic Chemistry.* VCl_4, a highly toxic and very corrosive, reddish-brown liquid; decomposes in cold water; melts at approximately −28°C and boils at 148.5°C; used in the preparation of vanadium compounds.

vanadium tetraoxide *Inorganic Chemistry.* V_2O_4, toxic blue crystals that are insoluble in water and soluble in acid; melts at 1967°C; used as a catalyst.

vanadium trichloride *Inorganic Chemistry.* VCl_3, pink deliquescent crystals; decomposes in water and at its melting point; used in the preparation of vanadium compounds.

vanadium trioxide *Inorganic Chemistry.* V_2O_3, black crystals that are soluble in water; melts at 1970°C; used as a catalyst.

vanadyl chloride *Inorganic Chemistry.* $2VOCl_2{\cdot}5H_2O$, green deliquescent crystals; soluble in water and alcohol; used as a textile mordant. Also, VANADIUM OXYDICHLORIDE.

vanadyl sulfate *Inorganic Chemistry.* $VOSO_4$, blue crystals; very soluble in water; used as a mordant, catalyst, colorant, and reducing agent. Also, **vanadic sulfate, vanadium sulfate.**

Van Allen, James A. born 1914, American physicist; discovered the Van Allen belts in 1958.

Van Allen belt *Geophysics.* either of two regions surrounding the earth and located in the upper atmosphere, containing high concentrations of charged particles that have been captured by the earth's geomagnetic field; also known (especially formerly) as the **Van Allen radiation belt.**

van Arkel-de-Boer process *Materials Science.* a process for refining titanium and zirconium using thermal decomposition of halides.

Van Atta array *Electromagnetism.* an antenna array having pairs of reflectors equidistant from the center of the array that are connected by low-loss transmission lines, so as to reflect the signal back to the source in a narrow beam for signal enhancement without amplification.

vancomycin *Microbiology.* an antibiotic produced by the soil bacterium *Streptomyces orientalis,* used to treat severe staphylococci infections not responding to other antibiotics; inhibits the synthesis of that part of the Gram-positive bacterial cell wall called the peptidoglycan.

Van Deemter rate theory *Analytical Chemistry.* the theory that HETP (height equivalent of theoretical plate) depends on flow velocity.

Van de Graaff, Robert 1901–1967, American physicist; the inventor of the Van de Graff generator.

Van de Graaff accelerator *Electronics.* a particle accelerator that receives the required high voltage from a Van de Graaf generator.

Van de Graaff generator *Electronics.* an electrostatic generator in which high voltages are produced by a moving belt of insulating material that collects electric charges by induction and discharges them inside a large spherical electrode; used for accelerating electrons, protons, and other nuclear particles.

van den Bergh reaction *Pathology.* a method of ascertaining the presence of bilirubin by contrasting diazotized serum to a standard solution of diazotized bilirubin.

vandenbrandeite *Mineralogy.* $Cu^{+2}(UO_2)(OH)_4$, a dark-green, strongly radioactive, triclinic, secondary mineral, occurring in fibrous rosettes, lamellar masses, and as small crystals, and having a specific gravity of 5.03 to 5.08 and a hardness of 4 on the Mohs scale; formed by the oxidation of uraninite and associated copper minerals.

van der Meere, Simon born 1924, Dutch physicist; shared the Nobel Prize for the discovery of W(−), W(+), and Z subatomic particles.

Vandermonde, Alexandre 1735–1796, French mathematician; he worked on permutation groups and solvability of algebraic equations.

Vandermonde determinant *Mathematics.* the determinant of a Vandermonde matrix *A*; i.e.,

$$\det A = \prod_{i,j=1}^{n} ;\ i > j\ (x_i - x_j).$$

A Vandermonde matrix is nonsingular if and only if the x_i are all distinct.

Vandermonde matrix *Mathematics.* an $n \times n$ matrix whose (i,j)th entry is $(x_i)^{j-1}$, where the x_i are indeterminates or elements of the field of matrix entries and $x_i^0 = 1$. The Vandermonde matrix arises in the interpolation problem of finding a polynomial

$$p(x) = a_{n-1}x^{n-1} + a_{n-2}x^{n-2} + \cdots + a_1x + a_0,$$

of degree at most n–1, having coefficients a_j in a given field F, and where the x_i and y_i are given elements of F such that $p(x_i) = y_i$ for $i = 1, 2, \ldots, n$. These n interpolation conditions form a system of n equations in the n unknown coefficients a_j. The matrix associated with the system is the Vandermonde matrix, with the ith row equal to: $1, x_i, x_i^2, x_i^3, \ldots, x_i^{n-1}$. The system always has a solution if the x_i are all distinct.

Van der Pol oscillator *Electronics.* a relaxation oscillator that uses a pentode and generates a sawtooth wave. *Physics.* any vibrational system whose equation of motion is of the form

$$d^2x/dt^2 + [-dx/dt + (1/3)(dx/dt)^3] + x = 0.$$

van der Waals, Johannes 1837–1923, Dutch physicist; explained critical temperature and the van der Waals forces; awarded the Nobel Prize for developing the van der Waals equation.

van der Waals adsorption *Physical Chemistry.* adsorption that arises from van der Waals forces; it generally takes place at relatively low temperatures, involves gases that liquefy easily, and reverses itself once pressure is reduced.

van der Waals attraction see VAN DER WAALS FORCE(S).

van der Waals bond *Materials Science.* in bonding, a weak electrostatic attraction between polar molecules or noble-gas atoms where the number of electrons varies with location; the location with the deficiency attracts that with the excess of electrons.

van der Waals covolume *Physical Chemistry.* the constant in the van der Waals equation that corrects for the nonnegligible size of the molecules.

van der Waals equation *Physical Chemistry.* an equation of state that describes the behavior of real or nonideal gases by using two corrective terms that take into account the size of the molecules and the forces between them: the term representing the repulsive forces is satisfactory only in the limit of low densities and overestimates their effect at high densities; the term representing the attractive forces is well founded in theory.

van der Waals force(s) *Physical Chemistry.* a general term for those forces of attraction between atoms or molecules that are not the result of chemical bond formation or simple ionic attraction; i.e., the relatively brief and weak interactions that neutral, chemically saturated molecules experience, such as dipole-dipole forces. Also, VAN DER WAALS ATTRACTION, VAN DER WAALS-LONDON INTERACTION(S).

van der Waals-London interaction(s) see VAN DER WAALS FORCE(S).

van der Waals surface tension formula *Thermodynamics.* a relationship between the surface tension of a liquid g and the temperature T: $g = Kp_c^{2/3}T_c^{1/3}(1 - T/T_c)^n$ where K is a constant, p_c is the critical pressure, T_c is the critical temperature, and n is approximately 1.23.

Van Dorn sampler *Engineering.* a sediment sampler that has a plexiglass cylinder sealed at each end with rubber force cups.

vandyke see BROWNLINE.

vandyke brown see BLACK EARTH.

vane *Mechanical Engineering.* a flat or curved surface that moves when exposed to a flow of fluid and rechannels the flow. *Aviation.* any of a wide variety of thin, relatively flat objects designed to align with an airflow; used, for example, to direct airflow, to detect airflow and communicate with a control system, to provide stability for an aircraft, or to direct the flow in an engine, pump, or compressor. *Navigation.* **1.** a long, narrow piece of cloth that indicates the direction of the wind; on smaller vessels it is usually attached to the top of the masthead, while on larger vessels it is often mounted on the side of the ship. **2.** on an azimuth circle or pelorus, a sight that is used to observe the bearing.

Vane, John Robert born 1927, English pharmacologist; shared the Nobel Prize for work in prostaglandins.

vane anemometer *Engineering.* a portable instrument, in which a number of vanes radiate from a shared shaft, which rotates when facing the wind; used to measure low air and wind speeds in large ducts.

vane-anode magnetron *Electronics.* a magnetron containing plane-parallel walls between adjacent cavities.

vane-type instrument *Engineering.* an instrument in which the pointer is moved by either the force of repulsion between fixed and moveable magnetized vanes or the force between a coil and a pivoted vane-shaped piece of soft iron.

Vaneyellidae *Invertebrate Zoology.* a family of sea cucumbers, holothurian echinoderms in the order Dactylochirotida, with fingerlike tentacles.

Vanguard *Space Technology.* any of a series of three geodesical satellites launched by the United States in 1958 and 1959 as a part of the International Geophysical Year program.

Vanguard

van Helmont, Johann Baptista 1579–1644, Belgian alchemist; coined the term *gas*; isolated oxides from air; proposed pathogenic theory of disease.

vanilla *Botany.* a tropical orchid of the genus *Vanilla,* especially *Vanilla planifoli*, that bears a podlike fruit. *Food Technology.* **1.** the fruit of this plant. Also, **vanilla bean**. **2.** an extract made with dried, fermented vanilla beans; used in flavoring foods and in perfumes. *Computer Programming.* **1.** relating to a computer system that is very general in its application or relatively uncomplicated. **2.** a version of a computer system or program prior to enhancements or upgrades.

vanillin *Organic Chemistry.* $C_8H_8O_3$, sweet-smelling, white crystalline needles; somewhat soluble in water and alcohol; melts at 81–83°C; used in perfumes, flavorings, pharmaceuticals, as a reagent, and as a source of L-dopa. Also, **vanillic aldehyde.**

vanish *Mathematics.* a function or operator that equals zero at some point of its domain is said to vanish at that point.

vanishing line *Photogrammetry.* the unique straight line on an aerial photograph that contains all the vanishing points of all the systems of parallel lines that are parallel to a single plane.

vanishing point *Mathematics.* in a perspectivity, a point at which the lines joining corresponding points intersect. *Photogrammetry.* the point in the plane of a photograph at which a system of parallel lines in the object space converge.

vanishing tide *Oceanography.* a tide in which the water level remains at a stand for several hours instead of rising or falling continuously in the normal pattern, so that the tide seems to disappear.

vannus *Entomology.* the fanlike posterior or anal lobe of an insect's wing, differentiated from the rest of the wing by a furrow.

vanoxite *Mineralogy.* a rare, weakly radioactive, inadequately described, black mineral with an approximate formula of $V_4^{+4}V_2^{+5}O_{13} \cdot 8H_2O$, occurring as microscopic crystals and in massive form; found replacing wood and cementinc sandstone in the Colorado plateau area.

van pool *Transportation Engineering.* a work-oriented ride-sharing service, often using a company van and an employee as a regular driver.

V antenna *Electromagnetism.* a directional antenna having a pair of radiating elements forming a V and fed at the apex.

van't Hoff, Jacobus [van täf′] 1852–1911, Dutch chemist; developed the theory of stereochemistry; awarded the Nobel Prize for laws of dilute solutions and chemical thermodynamics.

van't Hoff equation *Physical Chemistry.* an equation that accounts for the effect of temperature on the equilibrium constant in a gaseous reaction when pressure remains constant. Also, **van't Hoff isochore.**

van't Hoff factor *Physics.* the proportionality constant that appears between the observed osmotic pressure and the pressure predicted by ideal conditions; given by the osmotic pressure divided by the quantity of the gas constant times the temperature divided by the volume.

van't Hoff formula *Organic Chemistry.* the formula of calculation stating that the number of stereoisomers of a sugar molecule is equal to 2^n, with n equal to the number of asymmetric carbon atoms.

van't Hoff isotherm *Physical Chemistry.* an equation that shows the change in free energy during a chemical reaction.

vanthoffite *Mineralogy.* $Na_6Mg(SO_4)_4$, a colorless monoclinic mineral occurring in massive form, having a specific gravity of 2.694 and a hardness of 3.5 on the Mohs scale; found in oceanic salt deposits.

van't Hoff's law *Physics.* a law stating that a dissolved substance exerts the same osmotic pressure as it would if it were an ideal gas that occupied the same volume as the container.

Van Vleck, John H. 1899–1980, American physicist; worked in magnetism; awarded the Nobel Prize for his research on electron correlation.

Van Vleck equation *Quantum Mechanics.* an equation giving the paramagnetism of a magnetically susceptible substance.

Van Vleck paramagnetism *Quantum Mechanics.* the paramagnetism of a collection of ions, atoms, or molecules as predicted by quantum theory.

vapor *Physical Chemistry.* a gas whose temperature is less than the critical temperature, so that it may be liquefied or solidified by compression at constant temperature; i.e., a dispersion in air of molecules of a substance that is a liquid or a solid in its normal state, such as water vapor.

vapor-compression cycle *Mechanical Engineering.* a complete cooling cycle in which the refrigerant is first made to boil, producing a cooling effect, and then recompressed into its liquid state.

vapor concentration see ABSOLUTE HUMIDITY.

vapor cycle *Thermodynamics.* a thermodynamic cycle that operates on a substance that is at all times in its vapor phase, or at least passes through the vapor phase during a portion of the cycle.

vapor degreasing *Engineering.* the cleansing of a metal surface by the condensation of solvent vapors condensed upon it.
vapor density see ABSOLUTE HUMIDITY.
vapor deposition *Materials Science.* a method by which a material is vaporized and condenses on a substrate; used to produce thin film and to dope silicon wafers in the production of integrated circuits.
vapor-dominated hydrothermal reservoir *Geology.* a geothermal system that chiefly produces dry steam.
vapor-filled thermometer *Engineering.* a thermometer consisting of a vapor-filled chamber that contracts and expands in relation to temperature and thus causes pressure changes in the device.
vapor growth *Materials Science.* a method used to fabricate a metallic or nonmetallic filamentary single crystal of ultrahigh strength; a liquid is supersaturated and crystal growth occurs by precipitation at the solid-liquid interface.
vaporimeter *Engineering.* an instrument that measures vapor pressure in a liquid, especially in order to ascertain its alcohol content.
vaporization *Engineering.* the rapid change of water into steam, especially in a boiler.
vaporization cooling *Engineering.* the cooling of hot electronic equipment by spraying a nonflammable liquid with a low boiling point and high dielectric strength onto the equipment, where it vaporizes, transporting the heat to the enclosure walls, radiators, or heat exchangers.
vaporize *Physical Chemistry.* **1.** to cause to change into a vapor. **2.** to become converted into vapor.
vaporizer *Chemical Engineering.* a vessel or still in which liquid is heated until it vaporizes; the heat may be direct, as with hot gases or submerged combustion, or indirect, as with heat-transfer fluid or steam.
vapor line *Meteorology.* on a thermodynamic diagram, an isopleth representing a moisture variable such as a saturation mixing ratio.
vapor-liquid separation *Chemical Engineering.* the process of removing liquid droplets from a flowing stream of vapor or gas by cyclonic action, impingement, and adsorption or absorption procedures.
vapor lock *Fluid Mechanics.* **1.** an interruption of a liquid flow by the formation of vapor or gas bubbles in the conduit. **2.** specifically, an obstruction of this type in the flow of gasoline to the engine in a motor vehicle.
vapor-phase axial deposition *Engineering.* a technique in which fine particles of silicon dioxide and germanium dioxide are synthesized and dehydrated to fabricate graded index optical fibers.
vapor-phase epitaxy *Solid-State Physics.* the formation of a thin solid film on a crystal surface as the result of a thermochemical vapor-solid reaction, in which the deposited film crystal orientation is determined by the substrate crystal orientation.
vapor-phase reactor *Chemical Engineering.* a heavy vessel of steel construction, used for conducting chemical reactions on an industrial level when reliable control over a vapor phase is required, as in a process of oxidation.
vapor pressure *Meteorology.* the pressure exerted by water vapor in the atmosphere, independent of any other gases or vapors; the partial pressure of water vapor. Also, VAPOR TENSION. *Thermodynamics.* the pressure associated with the vapor in equilibrium with a condensed phase.
vapor-pressure lowering *Physical Chemistry.* the effect of a reduction in vapor pressure on the properties of a system, expressed as the difference between the vapor pressure of a pure solvent and the total vapor pressure over a solution of a nonvolatile solute.
vapor-pressure osmometer *Analytical Chemistry.* an instrument that determines molecular weight by measuring the decrease in vapor pressure of a solvent when a soluble sample is introduced.
vapor rate *Chemical Engineering.* the rate of vapor flowing through a contacting device, such as upward through a distillation column.
vapor-recovery unit *Engineering.* a unit that separates, recovers, and recycles gases and vaporized gasoline from a number of other petroleum refining processes.
vapor tension see VAPOR PRESSURE.
vapor trail see CONDENSATION TRAIL.
vapor volume equivalent *Petroleum Engineering.* the volume of vapor that would be equivalent to a specific volume of liquid under designated standard conditions, from which the specific gravity of fluids recovered from gas-condensate wells may be determined.
var or **VAr** volt-ampere reactive.
var. variable; variant; variation; variety.
VAR visual-aural range.
varactor *Electronics.* a P-N semiconductor diode whose variable capacitance depends upon the applied voltage; used as a variable-reactance tuning element in oscillator and amplifier circuits, including parametric amplifiers. Also, **varactor diode.**
varactor tuning *Electronics.* the process of using an inductor and a varactor to obtain the desired frequency when tuning a circuit or adjusting the frequency of an oscillator.
Varanidae *Vertebrate Zoology.* the monitor lizards, a monogeneric family of large reptiles found in Australia, Africa, Asia, and Malaysia; most are terrestrial, but some are arboreal.
vardar *Meteorology.* a cold fall wind that blows down the Vardar Valley in Greece to the Gulf of Salonika, usually persisting for two or three days. Also, **vardarac.**
var hour *Electricity.* a unit of reactive power per unit time; its magnitude is equal to 3600 joules.
var hour meter *Engineering.* an instrument that gauges and records the reactive power in a circuit over time.
variability the fact of being variable; changeability; specific uses include: *Statistics.* the dispersion or spread of data; measured by a variety of quantities including variance, interquartile range, and mean deviation. *Military Science.* the manner in which the probability of damage to a specific target decreases with the distance from ground zero; in damage assessment, it is a mathematical factor used to average the effects of target orientation, minor shielding, and uncertainty of target response.
variable changing or likely to change; specific uses include: *Science.* in a scientific experiment, a factor that changes in order to test a hypothesis. *Mathematics.* a formal symbol that represents an unknown or arbitrary quantity in a mathematical expression or statement. *Computer Programming.* a quantity that can assume different values at different times during the execution of a program. *Artificial Intelligence.* in a formula or pattern, a symbol that can be replaced by other symbols or data structures. *Meteorology.* relating to changing conditions, such as a variable ceiling or variable cloudiness.
variable-area exhaust nozzle *Aviation.* a jet engine exhaust nozzle whose exit opening can be adjusted, thereby changing the aircraft's velocity.
variable-area meter *Engineering.* a flowmeter in which a variable restrictor in the stream is forced into a position that allows the required flow-through.
variable-area recording *Acoustical Engineering.* a photographic recording technique in which the recording density is held constant while the exposure area is varied according to sound variations.
variable-area/variable-depth gravure *Graphic Arts.* a gravure color printing process in which separated ink cells of various sizes and depths are created on a plate according to desired tonal qualities.
variable attenuator *Electronics.* an attenuator whose attenuation can be varied in steps or in a continuous manner; the term usually implies that the iterative impedance of the attenuator remains substantially constant for all attenuation values.
variable-bandwidth filter *Electronics.* a filter whose bandwidth can be changed without changing the center frequency.
variable-block *Computer Programming.* describing data in which the number of words in a block of memory is determined by the programmer.
variable capacitor *Electricity.* a capacitor whose value can be altered by varying its useful plate area, as in a rotary capacitor, or by altering the distance between the plates, as is the case in some trimmer capacitors.
variable ceiling *Meteorology.* a ceiling that swiftly increases and decreases during the course of a weather observation; under such conditions, an average of observed values may be recorded as the reported ceiling, often with the notation *V* for variable.
variable click track *Acoustical Engineering.* an audio click track in which the time intervals between the clicks may vary in order to provide a changing tempo or serve some other purpose.
variable cloudiness *Meteorology.* a cloud cover that swiftly increases and decreases during the course of a weather observation.
variable connector *Computer Programming.* **1.** a flowchart symbol representing a sequence connection that is not fixed but is varied by the flowchart procedure. **2.** an instruction in a program that can cause a transfer of control to one of several alternative paths. Also, PROGRAMMED SWITCH.
variable costs *Industrial Engineering.* production costs that vary according to the quantity of the item produced, such as labor and materials.

variable coupling *Electricity.* a coupling that is adjustable, as in an inductively coupled circuit, where the distance or angle between the coils is usually adjustable, or in a capacitively coupled circuit, where the coupling capacitance is adjustable.

variable-cycle engine *Aviation.* a jet engine characterized by its ability to alter the path of its working fluid by means of shutters and valves; its performance is similar to that of a turbojet at supersonic cruise and a turbofan for takeoff and subsonic operation; this variable flow capability results in a significant reduction in both specific fuel consumption and noise.

variable-density soundtrack *Acoustical Engineering.* a recording technique in which the recording area is held constant while the recording density is varied.

variable-depth sonar *Engineering.* sonar in which the projecting and receiving transducers can be lowered below a vessel, to a depth best suited to minimize thermal effects when detecting underwater targets.

variable diode function generator *Electronics.* a circuit that can be adjusted to produce a piecewise-linear approximation to a given nonlinear function that relates output and input levels. The circuit is made of diodes and variable resistors of potentiometers that allow independent setting of the break points and the slopes of the linear portions.

variable field *Physics.* any field (vector or scalar) that changes with time. *Computer Programming.* see VARIABLE-LENGTH FIELD.

variable flow *Fluid Mechanics.* a flow whose features, such as depth or velocity, vary both from section to section and with time.

variable-focus condenser *Optics.* a condenser that produces a wide field of illumination by varying the focal length of one of the lenses.

variable force *Mechanics.* a force that over time varies in direction or magnitude, or in both.

variable-inductance accelerometer *Engineering.* an accelerometer in which a center coil is energized by an external alternating current while two end coils joined in series generate an alternating current proportional to the mass passing through the three coils. An AC output is proportional to the mass.

variable inductor *Electromagnetism.* an inductor whose inductance value may be adjusted. Thus, **variable inductance**.

variable-interval schedule (of reinforcement) *Behavior.* in operant conditioning, an intermittent reinforcement schedule in which responses are reinforced after a variable time interval from the previous reinforcement.

variable-length field *Computer Programming.* a data field having a varying number of characters, in which the length of the field is stored within the field itself. Also, VARIABLE FIELD.

variable-length operation *Computer Programming.* an operation in which the length of the operands is allowed to vary.

variable-length record *Computer Programming.* a record whose length can take multiple values.

variable-mu tube *Electronics.* an electron tube whose amplification factor varies in a predefined manner with control-grid voltage.

variable nebula *Astronomy.* a nebula whose shape, size, or brightness changes; e.g., Hubble's Variable Nebula in the constellation Monoceros.

variable parameter *Physics.* a parameter that can be varied over a certain range but is normally fixed during some process or transformation.

variable-pitch propeller *Mechanical Engineering.* a propeller whose blade angle can be set to any angle within a given range.

variable point *Computer Programming.* a representation in which a real number is represented by a series of digits with the radix point represented by a special character at its explicit location.

variable-ratio schedule (of reinforcement) *Behavior.* in operant conditioning, an intermittent reinforcement schedule in which responses are reinforced upon completion of a variable number of responses counted from the preceding reinforcement.

variable recoil *Military Science.* a recoil system that varies the length of recoil inversely with the elevation of the gun, to prevent the rear of the weapon from hitting the ground when fired at a high elevation.

variable region *Immunology.* the N-terminal half of light chains and the N-terminal half of Fd fragments of heavy chains in immunoglobulin molecules.

variable-reluctance pickup *Acoustical Engineering.* a type of magnetic pickup cartridge in which magnetic flux is produced by movement of an iron needle in a record groove as part of a magnetic circuit.

variable-reluctance stepper motor *Electricity.* a stepper motor containing a soft iron rotor with teeth and stator coils that are independently energized in sequence. The rotor aligns itself with the stator coil that is currently energized to provide rotation.

variable-reluctance transducer *Electromagnetism.* a transducer in which a particular quantity of magnetic material is moved between two coils by the displacement being monitored; this changes the reluctance of the coil, causing the change of impedance.

variable-resistance accelerometer *Engineering.* any accelerometer operating on the principle that the electrical resistance of a conductor is a function of its dimensions.

variable-sequence robot *Robotics.* a programmable robot whose instructions can be changed by external signals.

variable speech control *Electronics.* the removal at regular intervals of small portions of speech from a tape recording, so that the remaining sounds may be stretched to fill the gaps and the playback speed may be increased without pitch changes or significant loss of intelligibility.

variable-speed drive *Mechanical Engineering.* a mechanism that transmits motion from one shaft to another, allowing the velocity ratio of the shafts to be varied continuously.

variable-speed generator *Electricity.* a generator whose speed is adjustable within certain parameters in a regulated manner so as to deliver a constant voltage, irrespective of load.

variable-speed scanning *Electronics.* a scanning method in which the deflection speed of the scanning beam in the cathode-ray tube of a television camera is determined by the optical density of the film being scanned.

variable star *Astronomy.* a star that exhibits changes in brightness, from a few thousandths of a magnitude to over 20 magnitudes, often with accompanying physical changes.

variable-transconductance circuit *Electronics.* a circuit that uses a simple differential transistor pair; one transistor controls the device's gain and the other amplifies the first's input.

variable transformer *Electricity.* an autotransformer with an adjustable output voltage from zero to a maximum value; the adjustment is generally accomplished by means of a contact arm which moves along exposed turns of a secondary winding.

variable valence cations *Materials Science.* cations of transition metals that can take on different valences due to the closeness of energy levels of adjacent orbitals.

variable visibility *Meteorology.* prevailing visibility that fluctuates markedly in the course of a weather observation; under such conditions, an average of observed values may be recorded as the reported visibility, often with the notation *V* for variable.

Variable Zone see TEMPERATE ZONE.

variance *Statistics.* a measure of dispersion around the mean, given by the arithmetic average of squared deviations; the square of the standard deviation. Also, MEAN SQUARE DEVIATION.

variance ratio test *Statistics.* a test of the hypothesis that the variance of one normal distribution is less than the variance of another; often can be put in the form of an *F*-distribution. Also, F-TEST.

variant something that changes; specific uses include: *Microbiology.* a population or strain of a given microorganism that differs from related strains in at least one way. *Telecommunications.* **1.** any of two or more code symbols that possesses the same plain text equivalent. **2.** any of two or more plain text meanings symbolized by a single code group.

variant record *Computer Programming.* a record that may contain different fields of various lengths, based on the value of a certain field within the record.

variant surface glycoprotein *Biochemistry.* a substance that produces the surface coat of electrons on a cell by periodically changing the method of synthesis present in certain species of *Trypanosoma*.

variate difference method *Statistics.* a method of time series analysis of data containing one systematic and one random factor, for the purpose of isolating the random factor and estimating its variance.

variation an instance of change, or something that changes; specific uses include: *Genetics.* **1.** an individual's deviation in characteristics from those typical of the group to which it belongs. **2.** an offspring's deviation in characteristics from the characteristics of its parents. *Astronomy.* a deviation from the normal orbit of a celestial body. *Mathematics.* let f be a real-valued function that is defined on an interval $[a, b]$, and let $a = x_0 < x_1 < x_2 < \cdots < x_{n-1} < x_n = b$ be a partition P of the interval into n parts. The variation of $f(x)$ on $[a,b]$ with respect to P is

$$\sum_{k=1}^{n} |f(x_k) - f(x_{k-1})| = V(f,P).$$

The **total variation** of f on $[a, b]$ is the supremum of $V(f,P)$ over all possible partions P of $[a,b]$; if this number is finite, then f is said to be of **bounded variation** on $[a,b]$.

variational inequality *Astronomy.* a perturbation in the rotation of the moon that arises from the tangential attractive force by the sun.
variational method *Quantum Mechanics.* an approximate method for finding the ground state energy and wavefunction of a system, by minimizing the integral of the Hamiltonian applied to a trial wavefunction with one or more variable coefficients.
variational principle *Mechanics.* any of various principles formulated to account for the observation that when a physical process takes place, certain important quantities tend to have little variation from a minimum or maximum value; e.g., Hamilton's principle. Also, **variation principle.**
variation diagram *Petrology.* a diagram showing the variations in chemical composition of the members of an igneous rock series; used to identify genetic relationships and processes that influenced the series.
variation of latitude *Geophysics.* the small, periodic changes in the astronomical latitude of points on the earth caused by polar wandering.
variation of parameters *Mathematics.* a technique for solving a nonhomogeneous differential equation in which the constants for the complementary function are considered to be variables in order to determine a particular solution.
variation per day *Geophysics.* the changes in any of the various geophysical quantities that occur over the period of one day. Similarly, **variation per hour, variation per minute.**
varicella see CHICKENPOX.
varicella zoster virus see CHICKENPOX VIRUS.
varico- or **varic-** a combining meaning "twisted and swollen" or denoting a relationship to a varix.
varicocele *Medicine.* a varicose condition of the veins of the spermatic cord, forming a baglike swelling and accompanied by a dull or dragging pain in the scrotum. Also, **varicole.**
varicose [vâr´i kōs´] *Medicine.* abnormally and permanently swollen, as in varicose veins.
varicose vein *Medicine.* an abnormally enlarged, dilated vein, usually found in the legs and feet.
varicosity *Medicine.* **1.** the state or condition of being varicose. **2.** an instance of this; a varix or varicose vein.
variegated *Biology.* **1.** describing an organism that is a mosaic of phenotypically normal and mutant portions. **2.** composed of patches of different colors. Also, **variegate.**
variegated position effect *Genetics.* a position effect that generally involves the suppression of activity of a wild-type gene when it is placed in contact with heterochromatin due to a chromosome aberration; the gene may escape suppression, but the final phenotype may be variegated with patches of normal and mutant tissue.
variegation *Plant Pathology.* a condition of plant foliage in which the leaves either lack color or are multicolored, due to the absence or localized distribution of pigments.
variety the fact of being different, or one of a range of different types; specific uses include: *Systematics.* a botanical taxonomic rank below subspecies, denoting an organism that can interbreed with the subspecies or species, but is distinguished by some minor morphological character; varieties are usually developed in cultivated or horticultural species. *Archaeology.* a group of artifacts within one type that have other more specific attributes in common; for example, pottery made over several generations by the same family. *Mathematics.* let F be an algebraically closed extension of a field K, and let S be a system of polynomial equations in n unknowns. The set of all simultaneous solutions or zeros of S is a set of n-tuples in F^n and is called the affine K-variety or algebraic set in F^n defined by S; denoted $V(S)$.
Varignon's theorem *Mechanics.* the statement that the moment of the sum of a set of concurrent forces about a point is equal to the sum of the moments of each individual force about that point.
varindor *Electromagnetism.* a saturable-core reactor, the inductance of which is varied by a direct current flowing through an auxiliary winding.
variocoupler *Electromagnetism.* a radio-frequency transformer whose mutual impedance between the primary and the secondary windings is adjustable by means of rotating one of the coils with respect to the other, while the self-impedance remains constant.
variograph *Engineering.* a device that records vector changes in the earth's magnetic field.
variola see SMALLPOX.
variola virus see SMALLPOX VIRUS.
variolate *Medicine.* **1.** relating to or resembling smallpox. **2.** to inoculate with a smallpox virus.
variole *Petrology.* a small, pea-sized spherical body in igneous rock, usually composed of fanlike sprays of plagioclase or pyroxene crystals radiating outward from a central point.
variolitic *Petrology.* of basic igneous rock, having a fine-grained texture characterized by varioles.
variolosser *Electricity.* a variable attenuator, wherein loss is controllable by voltage or current.
variometer *Electromagnetism.* an inductor possessing two or more coils that are connected in series (or one inside the other) and that are movable with respect to each other in order to achieve variable inductance; used to detect and indicate changes in one of the components of the terrestrial magnetic field vector.
varioplex *Electricity.* a time-sharing approach to transmit and receive wire telegraph signals, designed to allow the optimal use of available lines.
Variscan geosyncline *Geology.* a principal area of sediment accumulated during the Devonian period, and found in northern Africa, as well as south-central and southern Europe. Also, HERCYNIAN GEOSYNCLINE.
Variscan orogeny *Geology.* a Late Paleozoic crustal deformation that occurred in central and western Europe. Also, ARMORICAN OROGENY, HERCYNIAN OROGENY.
varistor *Electronics.* a two-electrode resistor made of semiconductor material and having voltage-dependent nonlinear resistance that drops with an increase in applied voltage. (From voltage-variable resistor.)
varix *Medicine.* a swollen and therefore overstressed vein, artery, or lymph vessel.
Varley loop test *Electricity.* a Wheatstone bridge test designed to detect and locate faults in telephone or telegraph signal lines.
var measurement *Electricity.* the measurement of reactive power in a circuit.
varmeter *Engineering.* an instrument that gauges reactive power, expressed in volt-amperes reactive (vars). Also, REACTIVE VOLT-AMPERE METER.
Varmus, Harold born 1936, American microbiologist; with J. Michael Bishop, awarded the Nobel Prize for research in oncogenes.
varney jar *Biotechnology.* a jar used for anaerobic experiments or techniques in which oxygen is removed by the combustion of phosphorus.
varnish *Materials.* a resinous solution that is spread on wood or metal surfaces to provide a hard, lustrous, generally transparent coating for protection against water and wear damage.
varnish tree *Botany.* the common name for a tree *Rhus vernicifera,* of the family Anacardiaceae, grown for its milky juice from which types of varnish and lacquer are produced.
varulite *Mineralogy.* $(Na,Ca)Mn^{+2}(Mn^{+2},Fe^{+2},Fe^{+3})_2(PO_4)_3$, a dull olive-green, monoclinic mineral of the alluaudite group, occurring in granular masses, having a specific gravity of 3.58 and a hardness of 5 on the Mohs scale; forms a series with hagendorfite; found in granite pegmatites in Sweden.
varve *Geology.* a sedimentary layer or sequence of layers, consisting of coarser and finer materials deposited within a glacial meltwater lake or other body of still water during one year's time.
varved clay *Geology.* a distinctly banded clay-rich sediment deposited in a lake or other body of still water during one year's time. Also, **varve clay.**
vas *plural,* **vasa.** *Anatomy.* any canal for carrying a fluid.
vasal *Anatomy.* relating to a vas or to a vessel.
vasa vasorum *Anatomy.* the tiny arteries and veins that supply nutrients and remove waste products from the tissues in the walls of larger blood vessels.
vascular *Anatomy.* of or relating to the blood vessels. *Hematology.* denoting a copious blood supply. *Botany.* of or relating to a vessel that carries sap or other fluid.
vascular agglutination see SLUDGED BLOOD.
vascular bundle *Botany.* a strand of conducting tissue composed of phloem and xylem.
vascular cambium *Botany.* a lateral meristem that gives rise to secondary xylem and phloem.
vascular nevus *Medicine.* a reddish patch or swollen spot on the skin caused by the hypertrophy of the skin capillaries.
vascular poison *Toxicology.* any poison that attacks the vascular system.
vascular prosthesis *Cardiology.* an artificial replacement or substitute for a diseased or nonfunctional heart valve.
vascular ray *Botany.* the ribbonlike sheets of parenchyma that extend radially through the xylem, cambium, and phloem.

vascular retinopathy *Medicine.* a disorder of the eye without swelling, resulting from changes in the blood vessels of the retinal blood vessels.

vascular-streak dieback *Plant Pathology.* a disease of the cacao plant caused by the pathogen *Oncobasidium theobromae.*

vascular system *Anatomy.* the vessels of the body, especially the blood vessels.

vascular tissue *Botany.* the specialized conducting tissue found in higher plants, and made up mostly of xylem and phloem.

vascular wilt *Plant Pathology.* any plant disease, generally fungal, in which the attacking agent is confined to the vascular plant system and causes the plant to wilt.

vasculitis *Medicine.* an inflammation of a vessel. Also, ANGIITIS.

vas deferens *Anatomy.* the excretory duct of the testis that carries sperm from the epididymis to the ejaculatory duct at the base of the penis. Also, DUCTUS DEFERENS.

vasectomized *Surgery.* having undergone the removal of the vas deferens or a portion of it by surgical means.

vasectomy *Surgery.* the surgical removal or interruption of the vas deferens, generally done for the purpose of intercepting the sperm flow, and thereby achieving a condition of male sterility.

vashegyite *Mineralogy.* $Al_{11}(PO_4)_9(OH)_6 \cdot 38H_2O$ or $Al_6(PO_4)_5(OH)_3 \cdot 23H_2O$, a white, yellow, or rust-brown orthorhombic mineral, occurring in masses and spherical fibrous aggregates, having a specific gravity of 1.934 to 2.005 and a hardness of 2 to 3 on the Mohs scale; found associated with variscite.

VASI visual approach slope indicator.

vaso- or **vas-** a combining form meaning "vessel" or "duct."

vasoactive *Physiology.* affecting the caliber of the blood vessels.

vasoactive intestinal polypeptide *Endocrinology.* a polypeptide hormone of the gut and nervous system, having a range of biological activities that includes vasodilation, relaxation of bronchial and gastrointestinal smooth muscle, and suppression of gastric acid secretion; it may also act as a neurotransmitter in the central nervous system.

vasoconstriction *Physiology.* a narrowing of the lumen or interior space of the blood vessels, causing a decrease in blood flow.

vasoconstrictor *Physiology.* an agent (a chemical substance or motor nerve) that causes narrowing of the diameter of blood vessels by stimulating a contracting response.

vasodilation *Physiology.* a widening of the lumen or interior space of the blood vessels, increasing blood flow.

vasodilator *Physiology.* an agent (a chemical substance or motor nerve) that causes a dilation of the blood vessels; e.g., nitroglycerine. Also, **vasodepressor.**

vasogenic shock *Medicine.* shock that is caused by a vasodilator, such as a toxin.

vasomotor *Physiology.* describing the nerves that affect the narrowing or widening of the lumen of the blood vessels.

vasomotor center *Physiology.* any of several regions in the medulla and hypothalamus concerned with the regulation of heart rate, blood pressure, and peripheral blood flow.

vasomotor paralysis *Neurology.* any cessation of vasomotor control.

vasoneuropathy *Neurology.* any disease or dysfunction involving both the vascular system and the nervous system.

vasoparesis *Neurology.* partial paralysis of effector organs controlled by vasomotor nerves.

vasopressin *Biochemistry.* a peptide hormone, secreted by the hypothalamus and stored in and released from the posterior pituitary, that increases blood pressure and the rate at which the kidneys absorb water; used as an antidiuretic.

vasopressin see ARGININE VASOPRESSIN.

vasorrhaphy *Surgery.* suture of the vas deferens.

vasospasm *Medicine.* a spasm of the blood vessels, causing a decrease in their diameter.

vasostomy *Surgery.* the surgical formation of an opening into the vas deferens.

vasotocin *Biochemistry.* a hormone that increases permeability in amphibian skin and bladder; found in the posterior pituitary of nonmammalian vertebrates.

vasotomy *Surgery.* an incision into or cutting of the vas deferens.

vasotribe see ANGIOTRIBE.

vasotripsy see ANGIOTRIPSY.

vasovasostomy *Surgery.* end-to-end anastamosis of the two portions of a severed vas deferens, performed in order to restore fertility in vasectomized males.

vaspar *Microbiology.* a mixture of vaseline and paraffin wax that is used to seal anaerobic culture containers.

vat *Engineering.* a tank, tub, or other large container for storing liquids; e.g., wine. *Textiles.* a preparation or a receptacle containing a vat dye.

vat dye *Textiles.* any of a number of water-insoluble dyes that can be reduced to a soluble form and used for dyeing textiles; after oxidation, the dye becomes waterfast and fade-resistant.

VATE versatile automatic test equipment.

vaterite *Mineralogy.* $CaCO_3$, a rare, relatively unstable, colorless hexagonal mineral, trimorphous with aragonite and calcite, having a specific gravity of 2.54 and a hardness of about 3 on the Mohs scale; found in hydrogel pseudomorphs after larnite.

vat printing assistant *Textiles.* in printing fabrics with vat dyes, a mixture of gums and reducing and wetting agents that carry the dye and help it penetrate the fabric.

vaucheria *Botany.* a genus of branched green algae of the family Vaucheriaceae, characterized by oogamous reproduction and commonly forming large tangled growths on mud or wet soil or growing submerged.

Vaucheriaceae *Botany.* a family of marine, freshwater, and terrestrial algae of the order Vaucheriales, with plants consisting of branched coenocytic filaments that form extensive growths on moist soil or grow submerged.

Vaucheriales *Botany.* an order of yellow-green algae of the class Xanthophyceae, distinguished by nuclear division that occurs without cell wall formation during vegetative growth.

vaudaire *Meteorology.* a foehn that blows south from Lake Geneva. Also, **vauderon.**

Vaughania *Paleontology.* a genus of tabulate corals in the superfamily Favositicae and family Vaughanidae; notable for a small, discoid corallum and the absence of tabulae; extant in the Lower Carboniferous.

vault *Architecture.* an arched masonry cover, usually forming a ceiling or roof. *Surgery.* **1.** any arched or domelike body structure. **2.** the longest palatal border obtainable through a coronal section of the maxilla. **3.** a prepared cavity in bone to receive an implant.

Vauquelin, Louis Nicolas 1763–1829, French chemist; discovered and isolated chromium, beryllium, and their compounds.

vauquelinite *Mineralogy.* $Pb_2Cu^{+2}(CrO_4)(PO_4)(OH)$, a green to brown, monoclinic mineral occurring as aggregates of small wedge-shaped crystals and in fibrous or granular form, having a specific gravity of 5.986 to 6.16 and a hardness of 2.5 to 3 on the Mohs scale.

vauxite *Mineralogy.* $Fe^{+2}Al_2(PO_4)_2(OH)_2 \cdot 6H_2O$, a blue triclinic mineral, occurring as aggregates of small tabular crystals, having a specific gravity of 2.40 and a hardness of 3.5 on the Mohs scale: found in deposits in Bolivia.

V band *Electromagnetism.* a frequency range of 46 to 56 gigahertz that corresponds to wavelengths in the frequency range 0.652 to 0.536 centimeter.

V-beam radar *Electromagnetism.* a radar system employing two fan-shaped beams that sweep about a vertical axis; one fan is vertical while the other fan is inclined to the vertical and both fans intersect at ground level; the system can determine target elevation as well as range and bearing.

V belt *Mechanical Engineering.* an endless belt having a V-shaped cross section; used with an extension pulley.

V-bend die *Mechanical Engineering.* a die that has a triangular cross-sectional opening over which bending occurs; commonly used in press-brake forming.

V block *Engineering.* a rectangular steel block in which a 90° V-shaped groove has been cut through its center. It is usually used with a vise to hold circular shaped pieces.

V-bucket carrier *Mechanical Engineering.* a conveyor with two strands of roller chain separated by V-shaped buckets to carry nonabrasive materials.

VC vital capacity.

VCG vectorcardiogram.

VCO voltage-controlled oscillator.

V coefficient *Quantum Mechanics.* a coefficient used in coupling two angular momenta in quantum mechanics; it differs from the 3-*j* number by at most a sign.

vcp vacuum condensing point.

VCR *Electronics.* videocassette recorder: a videorecorder designed for use with a standard television receiver, storing television signals on videocassettes and playing them back through the television monitor; most VCRs can also play back prerecorded tapes of motion pictures.

V cut *Mining Engineering.* a cut in which the material blasted out is roughly V-shaped; created by drilling six or eight holes, with half the holes at an acute angle to the other half.

VD or **V.D.** venereal disease; vapor density.

VDAC voltage-dependent anion channel.

VDEL venereal disease experimental laboratory.

V.D.H. valvular disease of the heart.

v-DNA see V GENE.

VDRL Venereal Disease Research Laboratory.

VDRL test *Medicine.* the Venereal Disease Research Laboratory test, a serologic flocculation test for syphilis, also positive in other treponemal diseases, such as yaws; false positive and false negative results may occur.

VDT video display terminal.

VDU visual display unit.

veal *Agriculture.* **1.** the meat of a calf, especially one that is younger than three months old and that has been fed with milk. **2.** also, **vealer.** a calf raised for its meat. (From a Latin word meaning "little calf.")

veatchite *Mineralogy.* $Sr_2B_{11}O_{16}(OH)_5 \cdot H_2O$, a colorless to white monoclinic mineral, occurring as platy to prismatic crystals and in fibrous form, having a specific gravity of 2.62 to 2.66 and a hardness of 2 on the Mohs scale; found in borate deposits.

Veblen, Oswald 1880–1960, American mathematician; made fundamental contributions to projective geometry and geometry of paths.

Vectian see APTIAN.

vectopluviometer *Engineering.* a measuring apparatus containing one or more rain gauges, used to determine the inclination and direction of falling rain.

vector *Mathematics.* **1.** a mathematical quantity that has both magnitude and direction. **2.** a one-dimensional array. *Physics.* a physical quantity that has both a magnitude and a direction and that adds like displacement; velocity, acceleration, and force are prime examples. *Computer Programming.* a data structure representing a one-dimensional array that uses an index or subscript to locate an item. *Robotics.* a set of values for the operation of a robot in which magnitude is represented by the length of a line and direction by the orientation of the line. *Aviation.* **1.** a heading issued to an aircraft to provide navigational guidance by radar. **2.** to guide a pilot, aircraft, or missile by direct communication. *Military Science.* in air intercept, close air support, and air interdiction usage, a code meaning, "Alter heading to magnetic heading indicated." *Medicine.* the carrier of an infectious agent, which acts to transfer an infection from one host to another. *Molecular Biology.* any DNA molecule that can incorporate foreign DNA and transfer it from one organism to another. Also, VEHICLE. (Going back to a Latin word meaning "to carry.")

vector analysis *Mathematics.* the study of vectors and finite-dimensional vector spaces.

vector calculus *Mathematics.* the study of vector functions between vector spaces by means of differential and integral calculus; applications include differential geometry and physics.

vectorcardiogram *Cardiology.* a graphic representation of a closed curve traced out by the resultant electrical manifestations through a complete cycle of a heartbeat.

vectorcardiography *Cardiology.* the accounting of the cardiac vectors of direction and magnitude through a complete cycle.

vector component *Mathematics.* the scalar coefficients of the basis vectors when a given vector is expressed in that basis. For example, if $\{e_i\}$ is a basis for a vector space V, and $v = \Sigma_i c^i e_i$, the scalars $\{c^i\}$ are the components of v (the superscript in c^i is not an exponent, and sometimes subscripts are used instead). If v is a vector field, then the components of v are scalar-valued functions.

vector coupling coefficient *Quantum Mechanics.* any coefficient used in transforming bases from the set of eigenfunctions of one angular momentum coupling scheme for three or more angular momenta to those of another scheme.

vector current *Particle Physics.* a current that, under a Lorentz transformation, behaves as a vector rather than as an axial vector.

vectored attack *Military Science.* an attack in which a weapon carrier not holding contact on the target is guided to the delivery point by a unit holding contact on the target.

vectored interrupt *Computer Programming.* an interrupt handling scheme in which the interrupting device number indicates the address of the appropriate service routine.

vector equation *Mathematics.* any equation involving vectors.

vector feedrate *Robotics.* a value that represents the feedrate of a tool.

vector field *Physics.* any field whose components are vectors. *Mathematics.* a function on a manifold M that assigns to each point P of M a tangent vector tangent to M at P; e.g., a function that assigns a vector to each point of n-dimensional Euclidean space. If such a vector field is differentiable, it is equivalent to a system of ordinary differential equations.

vector function *Mathematics.* a function of a scalar variable whose value at each point in its domain is a vector. Also, **vector-valued function.** *Physics.* a function of space and/or time that is a vector quantity.

vector graphics *Computer Programming.* a computer graphics technique in which the displayed images consist of lines and curves drawn point-to-point based on Cartesian coordinates.

vector gunsight *Ordnance.* a gunsight that computes the vector necessary for a bullet to strike a moving target.

vectorial group translocation *Molecular Biology.* the translocation of one or more segments across a membrane as a result of the occupancy of fixed pathways through the membrane.

vectorial structure see DIRECTIONAL STRUCTURE.

vector impedance meter *Engineering.* an instrument designed to measure both the ratio between voltage and current (magnitude of impedance) and the phase difference between the two (phase angle of impedance).

vector loop *Robotics.* a set of vectors that describe each link in a robotic arm.

vector meson *Particle Physics.* any particle of unit spin, such as the W boson, the photon, or the rho meson.

vector multiplication or **vector-product** see CROSS PRODUCT.

vector operator *Quantum Mechanics.* a mathematical operational procedure, as distinguished from a function or ordinary number, that acts to change only a vector's length, but not its direction.

vector potential *Electromagnetism.* a vector function whose curl is equal to the magnetic induction.

vector power *Electricity.* the vector representation of the power: $P_{vector} = P_{active} + jP_{reactive}$, where $j = \sqrt{-1}$. The magnitude of vector power is apparent power. The angle is the $\cos^{-1}$ (power factor). The units of measure are volt-amps.

vector-power factor *Electricity.* a proportion of the active power to the vector power; in sinusoidal quantities, it equals the power factor.

vector processor see ARRAY PROCESSOR.

vector space *Mathematics.* a module over a division ring.

vector space associated with a graph *Mathematics.* let E be the power set of the edge set of a simple graph G; i.e., the set of all subsets of the set of edges of G. If E and F are members of E, define $E + F$ to be the set of edges of G that lie in either E or F but not in both. Define scalar multiplication by elements of the field of integers modulo 2: $(E) \cdot 1 = E$ and $(E) \cdot 0 = \varnothing$, the empty set. The vector space formed in this manner is known as the vector space associated with the graph G.

vector steering *Space Technology.* a method of steering a rocket or spacecraft by changing the thrust vector (that is, the direction of the thrust force), usually by means of gimbal-mounted thrust chambers.

vector superposition map *Crystallography.* a method of analyzing the Patterson map that involves setting the origin of the Patterson map in turn on the positions of certain atoms whose positions are already known and suitably combining the superposed maps.

vector voltmeter *Engineering.* a voltmeter that employs a two-channel high-frequency sampler to measure phase in addition to voltage of two inputs of the same frequency.

Vedalia *Invertebrate Zoology.* the predatory Australian ladybird beetle, which was introduced into many countries to control the cottony cushion scale threatening citrus crops.

VEE Venezuelan equine encephalomyelitis.

vee antenna see V ANTENNA.

vee path *Engineering.* in ultrasonic testing, the path of an angle beam from an ultrasonic search unit that gives the appearance of the letter V as it bounces off the opposite surface of a test piece and back to the examination surface.

veer *Navigation.* **1.** to change direction in a clockwise sense; the opposite of *back.* **2.** to shift astern. *Naval Architecture.* to slacken or let out a line.

veering *Meteorology.* a change in wind direction, especially in a clockwise sense.

Vega *Astronomy.* Alpha (α) Lyrae, a 1st-magnitude star that lies at a distance of 26 light-years; it defines spectral type A0.

vegan *Nutrition.* a strict vegetarian who subsists on a diet of only vegetables and fruit.

Vegard's law *Metallurgy.* a law stating that the lattice parameters of substitutional solid solutions vary linearly between the values for the components and where the composition is expressed in atomic percentage.

vegetable *Agriculture.* **1.** any plant whose fruit, seeds, roots, tubers, bulbs, stems, leaves, or flower parts are used as food. **2.** the edible part of such a plant. *Botany.* of or relating to plants. *Materials Science.* describing a substance that is derived from a plant. Thus, **vegetable dye, vegetable fat, vegetable glue, vegetable wax.**

vegetable black *Materials.* any form of carbon resulting from the incomplete combustion or destructive distillation of vegetable matter.

vegetable fibers *Textiles.* any textile fiber of vegetable origin, including cotton, flax, hemp, jute, sisal, etc.

vegetable ivory *Materials.* the hard, white endosperm of the ivory nut, used in making buttons and for decorative purposes.

vegetable jelly see ULMIN.

vegetable kingdom see PLANT KINGDOM.

vegetable oil *Materials.* any oil, often edible, that is extracted from the seeds, fruit, or nuts of a plant; used in foods, as drying oils in paints, as rubber softening agents, and as pesticide carriers.

vegetable parchment *Materials.* a strong, grease-resistant and water-resistant paper made from a waterleaf base and used in packaging

vegetable tanning *Engineering.* leather tanning that utilizes plant extracts.

vegetarian *Nutrition.* an individual who subsists on a diet of vegetables and fruits but whose diet may also include milk, cheese, and eggs.

vegetarianism *Nutrition.* the practice or policy of eating a diet that consists of vegetables and fruit and that excludes meat, and often also excludes animal by-products such as cheese and eggs.

vegetate *Biology.* to grow as or like a plant.

vegetation *Ecology.* the plant life growing in a given area. *Pathology.* any plantlike fungoid growth or neoplasm.

vegetational plant geography *Ecology.* a field of study that maps vegetation regions and analyzes them in terms of environment or ecological influences.

vegetation and ecosystem mapping *Cartography.* the drawing of maps that reflect various aspects of the plant life of a given geographic area.

vegetation zone *Ecology.* **1.** a large area of the earth's surface, often spanning continents, in which vegetation with similar characteristics can be found. **2.** a plant community that forms regional patterns based on the area's geography, geology, and history.

vegetative *Biology.* **1.** of or relating to growth. **2.** of or relating to asexual reproduction. *Nutrition.* of or relating to nutrition. *Botany.* relating to or characteristic of plants. *Physiology.* **1.** functioning involuntarily or unconsciously, as in the vegetative nervous system. **2.** broadly, passive or inactive; resting. *Cell Biology.* describing the part of a cell cycle during which the cell is not involved in replication.

vegetative cell see SOMATIC CELL.

vegetative eunuch *Ecology.* a plant that completes its life cycle without producing a seed.

vegetative nervous system *Anatomy.* an older term for the autonomic nervous system.

vegetative nucleus *Invertebrate Zoology.* see MACRONUCLEUS.

vegetative propagation *Botany.* a form of asexual reproduction in which plant parts are used to give rise to an entire new plant.

vegetative state *Virology.* of or relating to the growth phase of a virus. *Physiology.* in popular use, a state or condition characterized by inactivity; torpor.

vegitofossil see TRACE FOSSIL.

vehicle [vē´ə kəl] *Mechanical Engineering.* **1.** a self-propelled machine that is designed to transport passengers or goods over land, such as an automobile, van, truck, bus, or railway car. **2.** loosely, any conveyance that transports people, such as a bicycle or sled. *Aviation.* a machine, device, or structure designed to carry passengers or goods through air or space, such as an aircraft, spacecraft, balloon, or dirigible. Also, FLIGHT VEHICLE. *Space Technology.* a rocket designed to carry a payload. Also, SPACE VEHICLE, ROCKET VEHICLE. *Graphic Arts.* in printing ink, the liquid, usually an oil or varnish, in which ink pigments are suspended. *Pharmacology.* see EXCIPIENT. *Neurology.* any medium through which an impulse travels. *Microbiology.* any means by which a microorganism, usually pathogenic, can be transmitted to a susceptible host.

vehicle clearance envelope *Transportation Engineering.* the total space needed to accomodate a moving vehicle; it equals the dynamic vehicle envelope plus a margin of safety.

vehicle control system *Aviation.* a system designed to adjust and maintain the altitude and heading of a flight vehicle, following commands received from a guidance system, by means of control surfaces or other devices; distinguished from a guidance system by its ability to change the vehicle's direction of flight. Also, FLIGHT CONTROL SYSTEM.

vehicle crush capacity *Transportation Engineering.* a vehicle's maximum passenger capacity, with all seats taken and all permissible standing room occupied.

vehicle envelope *Transportation Engineering.* the space or path occupied by a moving vehicle.

vehicle follower control *Transportation Engineering.* a type of asynchronous network control in which headways are maintained by adjusting the speed of the following vehicle.

vehicle management *Transportation Engineering.* any manual or programmed demand-responsive control of vehicle deployment.

vehicle mass ratio *Space Technology.* the ratio of a rocket vehicle's final mass to its initial mass.

vehicular telephony *Telecommunications.* in automobiles or trucks, the transmission of speech signals back and forth from radio receiver-transmitters installed in the vehicle.

veil *Biology.* a covering body part or membrane, such as the outgrowth of plant epiderm covering and protecting a sorus (as in ferns), an outgrowth hanging from the top of stipe in certain fungi, the cuplike fringe of hairs surrounding a stigma, or an insect larva case. *Meteorology.* a cloud through which objects can be seen owing to its thin aspect.

Veillonella *Bacteriology.* a genus of small, Gram-negative, anaerobic, coccoid bacteria of the family Veillonellaceae, found as nonpathogenic parasites in the mouth and alimentary tract of humans and animals.

Veillonellaceae *Bacteriology.* a family of Gram-negative, nonmotile, anaerobic bacteria found in the mouth and intestines of humans and other animals.

Veillon tube *Biotechnology.* a glass tube that is about 25 cm in length and 1 cm in diameter, and can be sealed at both ends; used for various cell cultures.

vein *Anatomy.* a vessel through which blood passes from various organs or parts back to the heart. Also, VENA. *Botany.* a branching vascular bundle found in a leaf. *Entomology.* any of the ribs stiffening and supporting the wing of an insect. *Geology.* a thin, tabular or sheetlike deposit of minerals filling a fracture or joint in a host rock.

vein

vein-banding *Plant Pathology.* a symptom of certain plant diseases, generally those caused by viruses, in which the veins in the plant leaves become prominent due to the absence of color or darkening of the tissue which borders the leaf veins.

vein-clearing *Plant Pathology.* a symptom of certain plant diseases, generally those caused by viruses, in which the veins in the plant leaves become prominent due to the translucence of the tissue which borders the leaf veins.

veined gneiss *Petrology.* a composite gneiss characterized by irregular layering.

veining *Metallurgy.* a subboundary structure that has a greater-than-average amount of precipitate of solute atoms.

veinite *Petrology.* a variety of migmatite, characterized by mobile portions formed by internal rock secretions.
vein quartz *Petrology.* a rock consisting primarily of interlocking, sutured crystals of quartz of widely variable size.
vel. velocity.
Vela *Astronomy.* the Sails, a bright constellation of the southern Milky Way; formerly considered part of the constellation of Argo Navis, Jason's Ship.
velamen *Botany.* a many-layered covering of dead cells occurring on the aerial roots of certain plants, especially the epiphytic orchid, that provides protection for the root and may aid in water retention and absorption.
Vela pulsar *Astronomy.* a pulsar with a short period (89 milliseconds) associated with the Vela supernova remnant; it lies approximately 1500 light-years distant and is one of the few pulsars detectable optically. Also, VELA X-2.
velar *Mycology.* in fungi, relating to or forming the veil-like structure called the velum.
velarium *Invertebrate Zoology.* in some medusae, a narrow peripheral membrane on the umbrella, to which tentacles attach.
velar sound *Linguistics.* a sound produced by pressing the tongue against the velum, such as the sounds made by [k, g, ng].
Vela supernova remnant *Astronomy.* the expanding cloud of gaseous debris left over after the Vela supernova exploded about 10,000 years ago. Also, GUM NEBULA.
Vela X *Astronomy.* a nonthermal radio source, part of the Vela pulsar but located approximately 0.7° away.
Vela X-1 *Astronomy.* an eclipsing X-ray binary pulsar with an orbital period of 8.96 days in the constellation Vela; it is a particularly intense emitter of hard X-rays.
Vela X-2 see VELA PULSAR.
veld or **veldt** *Ecology.* the extensive grassland region of eastern and southern Africa, characterized by a flat terrain and a mixture of trees and shrubs. (From a Dutch word meaning "field.")
veliger *Invertebrate Zoology.* in mollusks, a free-swimming larval stage developed from the trochophore.
Veliidae *Invertebrate Zoology.* the water-striders and ripple-bugs, a family of small, short-legged predatory aquatic insects in the subdivision Amphibicorisae that seem to glide on calm, freshwater surfaces, occasionally diving for prey.
Velloziaceae *Botany.* a family of monocotyledonous xerophytic shrubs of the order Liliales, characterized by persistent leaf bases at the top of the stem and adventitious roots at the lower part, leaves and flowers at the branch tips only, and resinous or gummy vessels; found in South America, Africa, Madagascar, and southern Arabia.
vellum *Materials.* **1.** a fine-grained animal skin, especially that of a lamb, kid, or calf, that has been prepared for writing. **2.** a strong, cream-colored paper that resembles natural vellum.
vellus *Anatomy.* the soft, downy hair that succeeds the lanugo and covers most of the human body.
velocimeter *Engineering.* an instrument that measures velocity, as that of sound in water or air.
Velocipedidae *Invertebrate Zoology.* a family of true bugs, hemipteran insects in the superfamily Cimicoidea.
velocity *Mechanics.* the time rate at which a body changes its position vector; velocity is a vector quantity whose magnitude is expressed in units of distance over time, such as miles per hour. (From the Latin word for "speed.")
velocity analysis *Mechanical Engineering.* a graphical technique to determine the velocities of the parts of a mechanical linkage or a mechanism.
velocity coefficient *Fluid Mechanics.* the ratio of the actual velocity of gas to the velocity calculated under ideal conditions for a compressible fluid discharged from a nozzle.
velocity constant *Control Systems.* a measure that compares the rate of change for the input command signal with the steady-state error in a control system where the two quantities are proportional.
velocity curve *Astronomy.* a plot of the changes in the line-of-sight velocity seen in the spectrum of a spectroscopic binary star.
velocity dispersion *Physics.* the separation of radiation into its components of velocity, such as in the sorting of electron velocities as electrons are passed through a magnetic field.
velocity-distance relation *Astronomy.* the relation that governs the expanding universe: the more distant a galaxy, the greater its recessional velocity.
velocity distribution function *Physics.* a function that gives the number of particles (per unit velocity range) in a system of particles having a range of velocities.
velocity filter *Electronics.* in radar practice, a storage tube device that blanks out all targets that move less than one radar cell with a specified number of antenna scans.
velocity fire *Ordnance.* fire to establish the velocity for a specific combination of weapon, propellant, projectile, and fuse.
velocity gradient *Fluid Mechanics.* the rate of change of velocity over distance normal with the course of flow.
velocity head *Fluid Mechanics.* a pressure head corresponding to the kinetic energy per unit volume of the fluid; equal to $v^2/2g$, where v is the flow velocity and g is the gravitational acceleration.
velocity-head tachometer *Engineering.* a tachometer in which the object whose speed is to be measured drives a pump or blower that in turn causes a fluid flow, which is then converted to a pressure that can be measured.
velocity hydrophone *Acoustical Engineering.* an underwater transducer that uses the instantaneous velocity of the water elements for the conversion of acoustic energy to electromagnetic energy.
velocity level *Acoustics.* the velocity of a particle expressed in decibels with respect to a reference velocity of 10^{-8} meters per second.
velocity microphone *Acoustical Engineering.* a magnetic microphone, such as a ribbon microphone, that uses a lightweight diaphragm and a moving conductor and depends on the velocity of air molecules for its operation.
velocity-modulated oscillator *Electronics.* an electron tube device that uses velocity modulation on an electron stream as it passes through a resonant cavity; the subsequent high-intensity energy is extracted from the stream as it passes through another resonant cavity, and the resulting feedback sustains oscillations.
velocity modulation *Electronics.* modulation in which the input signal of an electron tube alters the velocity of the electrons in a constant current electron beam in sympathy with the input signal.
velocity-of-light radius *Astronomy.* the distance from a spinning neutron star at which any plasma carried along by the rotation would have a velocity equal to that of light. Also, **velocity-of-light cylinder.**
velocity potential *Fluid Mechanics.* a scalar function whose gradient is the velocity vector of the fluid; the function is denoted by F, and the adopted convention is such that the value of F decreases in the direction of the velocity.
velocity profile *Fluid Mechanics.* a plot of the fluid velocity as a function of position.
velocity ratio *Mechanical Engineering.* the ratio of the input velocity of a system to its output velocity. *Oceanography.* the ratio of a tidal current's velocity at a subordinate station to the corresponding velocity at a reference station.
velocity resonance see PHASE RESONANCE.
velocity sensor *Electricity.* a device that contains two principal elements: a coil of wire and a magnetic field; it generates a voltage proportional to the velocity between the two elements. Also, **velocity pickup.**
velocity servomechanism *Control Systems.* a servomechanism that generates a signal representing the value of the velocity of the output shaft. Also, RATE SERVOMECHANISM.
velocity-shaped canceler see CASCADED FEEDBACK CANCELER.
velocity spectrograph *Physics.* a mass spectrograph that selects ions of a specified velocity and then sorts the ions by their mass as they pass through a magnetic field.
velocity-type flowmeter *Engineering.* an instrument in which fluid flow actuates the turbine-type impeller, producing a volume-time reading. Also, CURRENT METER, ROTATING METER.
velour [və loor′] *Textiles.* a soft, heavy, closely woven fabric with a velvetlike nap. (From a Latin word meaning "hairy.")
velour paper *Materials.* a paper having a velvetlike texture, produced by depositing short wool, cotton, rayon, or nylon fibers on an adhesive-coated paper.
velox [vē′läks] *Graphic Arts.* a screened photographic print that is made on a chloride printing paper. (Originally *Velox,* a trade name copyrighted by the Eastman Kodak Co.)
velum [vel′əm] *Biology.* any veil-like membrane or structure. *Invertebrate Zoology.* **1.** the ciliated swimming membrane on certain mollusk larvae. **2.** in certain jellyfish, the ringlike membrane projecting inward from the margin of bell. *Anatomy.* see SOFT PALATE. *Meteorology.* a thin cloud of great horizontal extent that is draped over or penetrated by cumuliform clouds. (A Latin word meaning "sail" or "covering.")

velvet *Textiles.* a fabric with a short, thick, soft warp pile surface and a plain, tightly woven back. The pile and backing are often made of different fibers.

velvet ants *Invertebrate Zoology.* any of several brightly colored wasps of the family Mutillidae whose females resemble ants with fine velvety hairs; they live in the nests of bumblebees, sharing their food and laying their eggs inside the larvae.

ven- a combining form meaning "vein."

vena [vē´nə] *plural,* **venae** [vē´nē]. *Anatomy.* see VEIN.

vena cava *Anatomy.* the major superior and inferior veins that carry blood from the body tissues into the right atrium of the heart.

vena contracta *Fluid Mechanics.* the section of the greatest contraction in the cross-sectional area of a jet of liquid emitting from an opening in a container.

venation *Botany.* the distribution and arrangement of the veins of a leaf. *Entomology.* the arrangement of branched and unbranched veins in an insect's wing; important in the identification of species.

vendaval *Meteorology.* a stormy southwest wind that blows from late autumn to early spring along the southern coast of Spain and in the Strait of Gibraltar, often bringing violent squalls and thunderstorms.

veneer [və nēr´] *Materials.* **1.** a thin layer of material, especially a thin sheet of expensive wood laid over a base of cheaper wood in order to improve the outward appearance of the cheaper wood. **2.** to apply such a layer or facing. **3.** any of the layers that compose a sheet of plywood. *Building Engineering.* **1.** a facing of material laid over a different material, such as a facing of stone on a wooden building. **2.** to apply such a facing.

venenosalivary see VENOMOSALIVARY.

venenosity *Toxicology.* the state or condition of being poisonous.

venenous see VENOMOUS.

Venera program *Space Technology.* a series of Soviet Venus probes initiated in 1961; produced first successful entry of Venutian atmosphere and first television pictures from the planet's surface.

venereal [və nēr´ē əl] *Physiology.* of or relating to sexual intercourse. *Pathology.* **1.** of a disease, transmitted by sexual contact. **2.** of a patient, infected with such a disease.

venereal disease *Pathology.* any disease that is transmitted through sexual contact, such as syphilis or gonorrhea.

venereal wart *Medicine.* a wart on the genitals. Also, CONDYLOMA ACUMINATUM.

Veneroida *Invertebrate Zoology.* an order of chiefly marine, actively burrowing bivalve mollusks.

Venetian red *Inorganic Chemistry.* a pure red pigment containing varying proportions of ferric oxide and calcium sulfate.

Venetz, Ignatz 1788–1859, Swiss geologist; proposed and investigated theory of former widespread Alpine glaciation.

V-engine *Mechanical Devices.* a type of automotive engine with opposite facing cylinders set at an angle that forms a V-shape in cross section, so as to promote efficiency and ease of maintenance; used to describe engines in terms of the number of cylinders: **V-6, V-8,** etc.

venipuncture *Medicine.* the surgical puncture of a vein.

venite *Geology.* a type of banded migmatite whose vein material was derived from minerals within the rock itself.

Venn diagram *Mathematics.* a pictorial representation of set union, intersection, complementation, and inclusion; used to simplify or illustrate relationships that may exist between subsets of a given universal set.

veno- a combining form meaning "vein."

venogram see PHLEBOGRAM.

venography see PHLEBOGRAPHY.

venom *Toxicology.* any of various toxins produced by certain animals, such as snakes or insects; usually injected by bite or sting.

venom hemolysis *Toxicology.* the rupture and destruction of red blood cells by a venom.

venom leukocytolysis *Toxicology.* the rupture and destruction of white blood cells by a venom.

venomosalivary *Toxicology.* of or relating to the secretion of venomous saliva. Also, VENENOSALIVARY.

venomous *Toxicology.* describing an animal that secretes venom. Also, VENENOUS.

venous [vē´nəs] *Anatomy.* of or relating to the veins.

venous blood *Hematology.* deoxygenated blood, found in the systemic veins, the right chambers of the heart, and the pulmonary arteries; it is dark red in color.

venous pressure *Physiology.* the force of the blood against the walls of the veins.

vent *Engineering.* any opening designed to allow air, water, or pressure to enter or escape from a confined space, as in a building or mechanical system. *Geology.* an opening at the surface of the earth through which volcanic material is ejected or the conduit through which such material passes. Also, VOLCANIC VENT. *Vertebrate Zoology.* any body opening or outlet, especially the anal excretory or cloacal opening in a bird, fish, or amphibian. *Medicine.* an opening from which pus is discharged. *Ordnance.* the opening in the breech of a gun that transmits fire to the explosive powder.

vent da Mut *Meteorology.* a strong wind that blows off Lake Garda in northern Italy.

vent des dames *Meteorology.* a southwest sea breeze that blows inland from the Mediterranean coast of southeastern France nearly every day during summertime.

vent du midi *Meteorology.* a warm, moist wind that blows from the south across the Massif Central of southwestern France, often followed by a southwest wind with heavy rains.

vented battery *Electricity.* a battery composed of nickel-cadmium or similar materials that is unable to recombine gases produced as a byproduct during normal operation; as such, these gases must be vented to the atmosphere to avoid rupture to its cell case.

vented extruder *Materials Science.* a type of extruder having a barrel in which there is a vent at a point along the screw; the vent can be opened or evacuated in order to extract volatiles from the polymer melt.

venter *Anatomy.* **1.** the abdominal surface of the body. **2.** the fleshy part of a muscle. *Invertebrate Zoology.* **1.** the undersurface of the abdomen of an arthropod. **2.** the outer, convex part of a gastropod or cephalopod shell. *Botany.* the thickened portion at the base of an archegonium that contains the egg.

ventifact *Geology.* any stone or pebble whose a shape or surface has been altered in any way by the abrasive action of windblown sand. Also, GLYPTOLITH, RILLSTONE, WIND-CUT STONE.

ventilation *Engineering.* the circulation and purification of air in an enclosed space. *Meteorology.* the process of exposing a weather-observation instrumentation to a flow of air.

ventilator *Mechanical Devices.* **1.** any of various devices used to produce an air flow or circulate air currents, such as a fan or blower. **2.** a device equipped with an adjustable aperture to control the flow of air.

vento di sotto *Meteorology.* a breeze that blows up Lake Garda in northern Italy. (An Italian phrase meaning "wind from below.")

vent pipe *Mechanical Devices.* a pipe used for the discharge of pressure from vessels, reactors, and procuring equipment.

ventral *Anatomy.* **1.** of or relating to the belly or venter; abdominal. **2.** situated more toward the belly surface than toward some other object of reference; anterior. *Botany.* describing the underside of a dorsiventral plant part.

ventral aorta *Vertebrate Zoology.* a major arterial route conducting blood from the heart to the first aortic arch in lower vertebrates and in the embryos of higher vertebrates.

ventral decubitus *Surgery.* the position of an individual lying on the abdomen on a horizontal surface.

ventral hernia *Medicine.* a hernia of the abdominal wall.

ventral light reflex *Invertebrate Zoology.* a method of orientation used by shrimp and other aquatic invertebrates that swim belly-up toward the light.

ventral rib *Vertebrate Zoology.* any of the ribs in fish found lying in a septum that divides the muscles of the trunk.

ventri- a combining form denoting a relationship to the belly, or to the anterior aspect of the body.

ventricle [ven´trə kəl] *Anatomy.* **1.** either of the two large muscular chambers that pump blood out of the heart. **2.** any of several cavities in the brain that contain cerebrospinal fluid.

ventricose *Biology.* swelling out in the middle or unequally, as in corolla, spores, stipe, and shells.

ventricular *Anatomy.* relating to or affecting a ventricle.

ventricular fibrillation *Cardiology.* arrhythmia in which there is contraction of ventricular muscle, caused by repetitive stimulation of myocardial fibers without a coordinating contraction of the ventricle.

ventricular flutter *Cardiology.* a condition of the heartbeat and pulse in which there are very rapid yet uniform and regular beats of 250 or more per minute.

Ventriculites *Paleontology.* a cone-shaped genus of sponges; restricted to the Cretaceous.

ventriculitis *Medicine.* an inflammation of a ventricle, especially a cerebral ventricle.

ventriculo- a combining form denoting a relationship to a ventricle.
ventriculogram *Radiology.* an X-ray image of the cerebral ventricles or of the ventricles of the heart.
ventriculography *Radiology.* the process of making a ventriculogram.
ventriculus *Zoology.* the digestive organ of insects and birds, equivalent to the mammalian stomach.
ventro- a combining form denoting a relationship to the belly, or to the anterior aspect of the body.
ventrocystorrhaphy *Surgery.* the stitching of a cyst, or of the bladder, to the abdominal wall.
ventrofixation *Medicine.* the procedure used to suspend the uterus to the wall of the abdomen.
ventromedial nucleus *Anatomy.* a group of cells in the hypothalamus of the brain.
ventromedian *Anatomy.* both ventral and median.
vent stack *Building Engineering.* the portion of a soil stack that protrudes above the highest fixture in a building.
venturi or **Venturi** *Engineering.* a specially designed constriction in a pipe that causes a pressure drop when fluid flows through it. (Named for Giovanni B. *Venturi,* 1746–1822, Italian physicist.) Also, **venturi tube.**
Venturia *Mycology.* a genus of fungi of the order Pleosporales causing plant diseases such as apple scab, pear scab, and cherry scab.
Venturiaceae *Mycology.* a family of fungi of the order Pleosporales containing species that either live off decaying organic matter or are parasites of the leaves and stems of flowering plants and conifers; some are plant pathogens.
Venturian *Geology.* a North American geologic stage of the Middle Pliocene epoch, occurring after the Repettian and before the Wheelerian.
venturi flume or **Venturi flume** *Engineering.* a flume that is constricted at one section to allow measurement of the rate of flow by comparing water levels at the constriction and at a point upstream.
venturi meter or **Venturi meter** *Engineering.* an instrument that measures rate of flow in terms of pressure drop across a venturi (or tapered throat) in a pipe.
venturi scrubber *Chemical Engineering.* a device used for gas cleaning in which liquid injected in the throat of the venturi is used to scrub mist and dust from gas passing through the apparatus.
venular *Anatomy.* relating to, composed of, or affecting venules.
venule *Anatomy.* any of the small vessels that collect blood from the capillary plexuses and join to form veins. Also, **venula.**
Venus *Astronomy.* the second planet from the sun; it has a diameter of 12,104 kilometers, a mean orbital distance of 109 million kilometers, and a retrograde rotation period of 243 days. (Named for *Venus,* the goddess of love, beauty, and vitality in Roman mythology.)

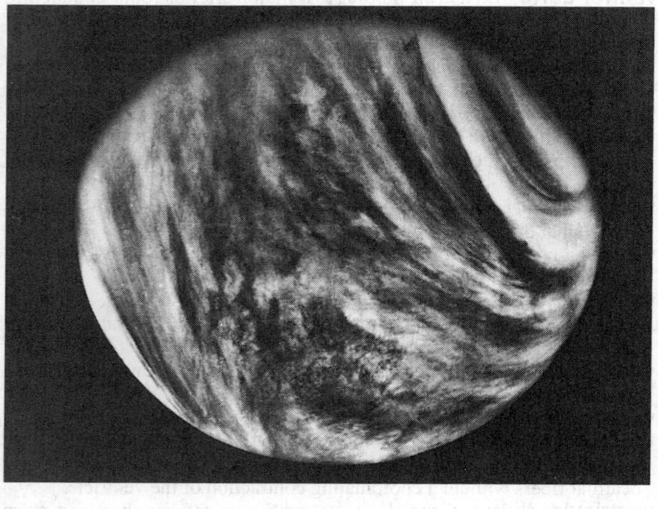

Venus

venus *Archaeology.* see VENUS FIGURINE.
Venus figurine *Archaeology.* a type of small statue characteristic of the Upper Paleolithic period, widely distributed in Europe; typically having an undefined head and exaggerated breasts and stomach.
Venus flytrap *Botany.* the common name for *Dionaea muscipula,* a plant of the order Sarraceniales, so named for its ability to trap and eat insects; native to the bogs of North and South Carolina.

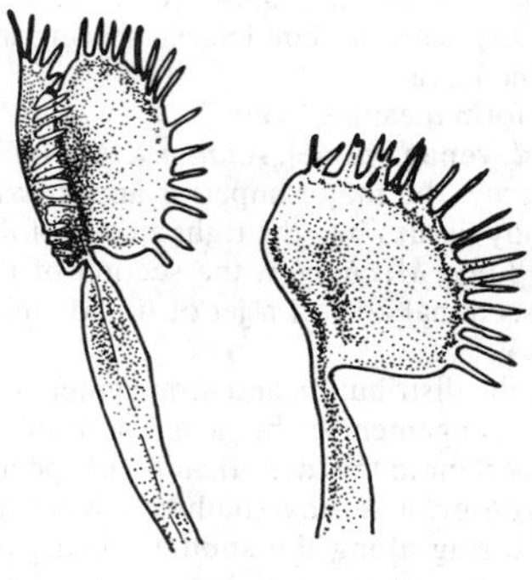

Venus flytrap

veranda or **verandah** *Architecture.* a wide, covered open gallery or porch, used for sitting.
veranillo *Meteorology.* a spell of hot, dry weather that occurs in the midst of the summer rainy season in western Mexico and Central America. (A Spanish word meaning "little summer.")
verano *Meteorology.* in Mexico and Central America, a season of relatively dry weather that occurs from November through April.
verapamil *Biochemistry.* $C_{27}H_{38}N_2O_4$, a white, crystalline, water-soluble, coronary vasodilator, occurring with veratrine in the seeds of the sabadilla; used especially in its hydrochloride form as a calcium blocker in treating angina and some arrhythmias.
veratrine *Materials.* a toxic, colorless to grayish-white alkaloid mixture that is extracted from the seed of the sabadilla plant; slightly soluble in water and soluble in alcohol and ether; melts at about 150°C; formerly used in treating rheumatism and neuralgia. Also, **veratria.**
verbal conditioning *Behavior.* in behavior therapy, the use of operant conditioning techniques in order to guide and increase the verbalized statements of patients who can learn that speaking will not result in disapproval or punishment.
verbal learning *Psychology.* the act of learning verbal material.
verbal memory *Psychology.* an individual's ability to recall something written or spoken that was learned at an earlier time, such as a poem.
Verbeekinidae *Paleontology.* a family of late, highly specialized foraminiferids belonging to the extinct suborder Fusulinina of the superfamily Fusulinacea; characterized by flat septa and related to the schwagerinids; extant in the Upper Permian.
Verbenaceae *Botany.* a family of dicotyledonous herbs, shrubs, and trees of the order Lamiales, including the beautyberry, lantana, teak, verbena, and vervain; characterized by usually opposite or whorled leaves, clusters of bisexual, often fragrant flowers found in spikes or racemes, and fleshy or dry fruit.
verdant zone see FROSTLESS ZONE.
Verde, Cape [vurd] *Geography.* a peninsula on the Atlantic coast of Senegal; the westernmost point on the African continent.
Verdet constant *Optics.* a constant in the Faraday effect that matches the angle of rotation of plane-polarized light in a magnetized material divided by the product of the length of the light path in the substance and the power of the magnetic field.
verdigris *Metallurgy.* a green or bluish coating, usually of a carbonate of copper, that forms on brass, copper, or bronze surfaces exposed to the atmosphere for long periods of time. *Chemistry.* $C_4H_6CuO_4$, a bluish-green poisonous compound used especially in the manufacture of synthetic rubber and textiles. (From an Old French word meaning literally "green from Greece.")
verdin *Vertebrate Zoology.* a small, yellow-headed titmouse, *Auriparus flaviceps,* which builds a compact, spherical nest of thorny twigs; found in the southwestern United States and Mexico.
verdohemin *Hematology.* a compound that is analogous to hemin, in which one porphyrin ring is open.
verdohemochromogen *Hematology.* a compound that is analogous to hemochromogen, in which one porphyrin ring is open.
verdohemoglobin *Hematology.* a compound that is analogous to hemoglobin, in which one porphyrin ring is open.

Verdoorniaceae *Botany.* a monotypic family of large liverworts native to New Zealand and belonging to the order Metzgeriales, characterized by a thick, fleshy thallus with polystratose wings, mucilage production, and unicellular spores.

verdoperoxidase see MYELOPEROXIDASE.

verge *Building Engineering.* the edge along a sloping roof that projects over a gable wall. *Architecture.* the shaft of a column. *Horology.* the lever in the escapement of a watch with pallets on the end that lock and release the scape wheel.

verge escapement *Horology.* an early type of escapement mechanism in which a heavy bar is mounted on a vertical spindle with a verge suspended by a torsion spring; as the bar oscillates it turns the verge, which engages a crown-shaped escapewheel, releasing it one tooth at a time, thus providing a timekeeping impulse to the works of the clock.

vergence *Geology.* the direction in which a fold is inclined or overturned. *Medicine.* the turning motion of the eyeballs toward or away from each other.

Verhulst-Pearl equation *Ecology.* a mathematical expression for a particular sigmoid curve of population growth, in which the proportional rate of increase decreases in a linear fashion as population size increases; first suggested by P. Verhulst and rediscovered by R. Pearl. Also, LOGISTIC EQUATION.

verification *Computer Programming.* **1.** the act of checking the results of an operation to see if it has been completed correctly. **2.** the process of proving mathematically that a system meets formal requirements.

verify *Computer Programming.* to determine that an operation has been performed correctly. A device that does this is a **verifier.** *Telecommunications.* to ensure that the meaning of a transmitted message is received correctly. *Ordnance.* a request from an observer or spotter of a fire-control agency to reexamine firing data and report the results.

Vermes *Invertebrate Zoology.* **1.** a former classification that included all invertebrates except the arthropods. **2.** a former term for the soft-bodied, vermiform invertebrates.

vermicule *Cell Biology.* **1.** an ookinete of *Plasmodium.* **2.** a uninucleate, motile cell of *Babesia* or *Theileria* that is formed in the gut of the tick vector.

vermiculite *Mineralogy.* any of various platy to micaceous monoclinic, phyllosilicate minerals, having the general formula $(Mg,Fe^{+2},Al)_3(Al,Si)_4O_{10}(OH)_2 \cdot 4H_2O$, and characterized by marked expansion when heated; generally formed by the weathering of micas, especially biotite; found with ultramafic igneous rocks, and used in the expanded form as insulating material and as an aggregate in concrete and plaster.

vermiform *Biology.* shaped like a worm; used to describe certain Protista and also numerous body structures, especially the appendix.

vermiform appendix *Anatomy.* the appendix, a narrow sac about three to six inches long, having no known function, that extends from the cecum of the large intestine. Also, **vermiform process.**

vermiform larvae *Entomology.* the common larvae type among the Diptera, usually with no appendages and maggotlike in appearance.

vermifuge *Pharmacology.* any agent that expels worms or intestinal animal parasites; an anthelmintic.

Vermilingua *Vertebrate Zoology.* another name for the family Myrmecophagidae, the true anteaters of South America.

vermilion see MERCURIC SULFIDE, def. 2.

vermiphobia see HELMINTHOPHOBIA.

vermis *Anatomy.* **1.** a wormlike structure. **2.** the narrow middle lobe of the cerebellum of the brain.

vernal *Geophysics.* of, relating to, or occurring in spring.

vernal equinox *Astronomy.* the moment when the sun passes northbound through the celestial equator, on or around March 21; also the place on the sky where this happens. Also, SPRING EQUINOX.

vernalization *Botany.* a method of expediting the flowering of plants by exposing seeds or seedlings to a period of cold.

vernal pool *Ecology.* a temporary pool of water formed during a spring thaw.

vernation *Botany.* the arrangement of leaves within a bud.

Verneuil flame-fusion process *Crystallography.* a method for preparing synthetic crystals by mixing compounds in a vertical, inverted blow-pipe type of furnace. Fusion at the high temperatures so generated leads to gemstones. Also, **Verneuil process.**

Vernier, Pierre [vurn yā´] 1580–1637, French mathematician and author; invented the vernier scale.

vernier [vurn´ē ər] *Engineering.* a small moveable auxiliary scale that slides in contact with the main scale to permit accurate fractional reading of the least division on the main scale. Also, **vernier scale.**

vernier capacitor *Electricity.* a low-capacitance variable capacitor connected in parallel with a higher-capacitance fixed or variable capacitor, for precision of total capacitance.

vernier compass *Cartography.* a compass on a transit that has a vernier for adjusting magnetic bearings to read as true bearings.

vernier dial *Engineering.* a fine-tuning dial whose control knob must rotate 180° to cause the slightest rotation on the main shaft.

vernier engine *Space Technology.* a small auxiliary rocket engine that is ignited or remains in operation after shutdown of a space vehicle's main propulsion, in order to refine the vehicle's speed or trajectory. Also, **vernier rocket.**

vernier sextant *Navigation.* a sextant equipped with a vernier scale to refine the angular readings of the main scale.

vernix caseosa *Developmental Biology.* an oily substance of sebum and desquamated epithelial cells that covers the skin of the fetus.

vernolate *Organic Chemistry.* $C_{10}H_{21}NOS$, an amber liquid that is used as a herbicide for sweet potatoes, peanuts, soybeans, and tobacco.

Vero cell *Virology.* an established cell line derived from the kidney of an African green monkey; used in the propagation of viruses and the study of their replication. Also, **Vero.**

Verpa *Mycology.* a genus of fungi belonging to the order Pezizales that includes edible species and grows under temperate conditions, particularly in early spring.

verrou see RIEGEL.

verruca *Medicine.* a technical term for a wart. See WART. *Biology.* **1.** a wartlike projection. **2.** a blisterlike outgrowth of body wall in some sea anemones. **3.** a protuberance tufted with bristles, as in larval insects.

verruca acuminata see VENEREAL WART.

verruca peruviana *Medicine.* the nodular eruption of Carrion's disease, appearing on the face and extremities, and evolving into bleeding tumors.

Verrucariaceae *Botany.* a family of lichens of the order Pyrenulales that grow primarily on rocks found along a coastline.

Verrucariales *Mycology.* an order of fungi belonging to the class Loculoascomycetes that are parasitic to lichens or that adapt a lichen mode of existence.

verruca vulgaris *Medicine.* the common wart.

verrucologen *Biochemistry.* a mycotoxin, generated by certain *Penicillium* species, that can cause tumors to form.

Verrucomorpha *Invertebrate Zoology.* a suborder of short-stalked barnacles, crustaceans in the order Thoracica.

verrucose *Biology.* covered with wartlike projections.

verrucous *Biology.* of or relating to a wart, or characterized by warts.

verrucous endocarditis *Medicine.* endocarditis from other than bacterial factors, in which fibrin shreds form on the ulcerated heart valves.

vers versine.

versatile *Botany.* attached at the middle and capable of swinging freely on the support, as an anther. *Invertebrate Zoology.* **1.** turning forward or backward, as the head of certain insects. **2.** freely moving up and down or from side to side, as antennae.

versine *Mathematics.* the versine of an angle θ, denoted vers θ, is the quantity $1 - \cos\theta$; i.e., the difference between the radius and the cosine of an angle constructed in a unit circle. Also, **versin, versed sine.**

version space *Artificial Intelligence.* a machine-learning technique in which a most general description and most specific description of a set of training instances are constructed until the two descriptions converge.

verso *Graphic Arts.* **1.** the left-hand, even-numbered pages in a book or other printed work; the opposite of *recto.* **2.** the back side of a printed sheet.

Verson-Wheelon process *Materials Science.* a process for rubber forming sheet metal, in which one of the dies in the conventional male and female metal die pairs is replaced by a rubber membrane; pressure is transmitted through a rubber bag under hydraulic pressure, forcing the sheet metal into the die.

vert. vertebrate; vertical.

vertebra [vurt´ə brə] *plural,* **vertebrae** [vurt´ə brē´; vurt´ə brā´]. *Anatomy.* any one of the 33 bones in the spinal column.

vertebral column see SPINE, def. 1.

vertebral foramen see SPINAL FORAMEN.

Vertebrata [vurt´ə brät´ə] *Vertebrate Zoology.* a subphylum of the phylum Chordata that includes all animals possessing a backbone, including fish, amphibians, reptiles, birds, and mammals.

vertebrate [vurt´ə brāt´; vurt´ə brət] *Vertebrate Zoology.* **1.** an organism that is a member of the subphylum Vertebrata; i.e., an animal with a backbone. **2.** of or relating to the Vertebrata.

Vertebrate Zoology

Of all the kinds of animals that exist, vertebrate animals have the most direct and intimate impact upon human life and thought. The vertebrates are those animals that have a backbone, the axial skeletal support for the body that is composed of vertebrae. This feature distinguishes them from the vast majority of animals, known as invertebrates, that have either an external skeleton or no hard skeleton at all. Fishes such as sharks, lamprey, and trout, amphibians such as frogs and salamanders, reptiles such as turtles, lizards, and snakes, birds from hawks to doves, and mammals including mice, humans, and elephants are all vertebrates.

Looking at vertebrates, we humans see our closest animal relatives and interest in them reaches back to the earliest emergence of humankind. Knowledge of the habits of vertebrates in the wild proved important for hunting and fishing. The emergence of agriculture and pastoralism in early human history led to the domestication of many animals, all of them vertebrates.

The scientific study of vertebrate animals emerged as a discipline during the early 1800s and initially flourished largely through the investigation of their comparative anatomy and embryology. In this way, vertebrates as diverse as sharks, snakes, and sloths became recognized as part of a definable group based upon common evolutionary ancestry and basic body plan.

Vertebrate zoology is traditionally divided into subdisciplines based upon the classes of animals concerned: ichthyology (fishes), herpetology (amphibians and reptiles), ornithology (birds), and mammalogy. Modern vertebrate zoology encompasses all aspects of the biology of vertebrate animals, including their behavior, natural history, evolution, ecology, anatomy, embryology, physiology, and genetics.

David M. Green
Assistant Professor and Curator of Herpetology
Redpath Museum
McGill University

vertebrate zoology *Zoology.* the branch of zoology that is concerned with the study of animals having backbones.

vertebratus *Meteorology.* a cloud variety in which the arrangement of elements resembles human vertebrae; this occurs mainly in the genus cirrus.

vertebrectomy *Surgery.* the surgical excision of a vertebra.

vertebrochondral *Anatomy.* relating to a vertebra and the costal cartilage of one of the upper three false ribs.

vertebrocostal *Anatomy.* relating to a vertebra and rib.

vertebrosternal *Anatomy.* relating to the vertebrae and the sternum.

vertex [vur´teks] *plural,* **vertices** [vurt´ə sēz´]. *Astronomy.* **1.** a point on the celestial sphere toward which a moving star group is aimed. Also, APEX. **2.** the highest elevation a celestial object attains. Also, CULMINATION. **3.** the point on any celestial great circle that lies closest to the pole. *Optics.* the point at which an axis terminates the curve in a lens. *Mathematics.* **1.** in graph theory, an element of one of two (usually finite) sets of elements that determine a graph; i.e., an element of the vertex set. The other set is called the **edge set**; each element of the edge set is determined by two elements of the vertex set. **2.** the point at which two lines or rays intersect to form an angle. **3.** a point at which two sides of a polygon meet or at which three or more faces of a polyhedron intersect. **4.** a point on a curve at which the curvature has a maximum or minimum; e.g., the vertex of a conic section. **5.** the point at which the lines generating a conic surface intersect. (From a Latin word meaning "the turning point.")

vertex cut *Mathematics.* a set S of vertices of a graph G such that $G - S$ is not connected. A vertex cut with only one element is called a **cut vertex**. The connectivity of a graph G with n vertices is the cardinal number of the smallest vertex cut of G if such a set exists; otherwise the connectivity is $n - 1$.

vertex-induced subgraph *Mathematics.* an induced subgraph; the term is used when it is necessary to distinguish the subgraph from an edge-induced subgraph.

vertex power *Optics.* the power that is the reciprocal of the back focal length in a lens.

vertical *Science.* perpendicular to the horizontal; upright. *Geodesy.* **1.** the single line that is perpendicular to the surface of the geoid at any point. **2.** the direction in which the force of gravity acts.

vertical alignment *Transportation Engineering.* the precise location of a roadway in the vertical plane, expressed in terms of feet of elevation above mean sea level.

vertical anemometer *Engineering.* an instrument designed to record the vertical component of wind speed.

vertical angles *Mathematics.* either of two pairs of congruent, nonadjacent angles formed by two intersecting lines; i.e., angles lying on opposite sides of the vertex or point of intersection.

vertical antenna *Electromagnetism.* an antenna that is mounted or supported vertically rather than horizontally, commonly consisting of a single wire or rod.

vertical axis *Aviation.* an axis passing through an aircraft from top to bottom, usually through the center of gravity. Also, NORMAL AXIS. *Naval Architecture.* the vertical line formed by a vessel's axis of yaw.

vertical axis propeller *Naval Architecture.* a propeller, having a vertical axis and spadelike blades, that is set flush in a vessel's bottom and rotates in the horizontal plane.

vertical band saw *Mechanical Engineering.* a band saw whose blade operates in the vertical plane.

vertical blanking *Electronics.* the blanking of a cathode-ray tube during the vertical retrace.

vertical boiler *Mechanical Engineering.* a fire-tube boiler that has vertical tubes placed between the top head and the tube sheet, connected to the top of an internal furnace.

vertical bomb *Ordnance.* a wingless bomb, as distinguished from a glide bomb or robot bomb; it may be equipped with control surfaces in the tail.

vertical boring machine *Mechanical Engineering.* a boring machine with a circular table that revolves about a vertical axis, so that the work-securing surface is horizontal.

vertical broaching machine *Mechanical Engineering.* a machine that shapes metal or plastic by pulling a vertically mounted broach across the surface of the object.

vertical centering control *Electronics.* a centering control that allows the vertical shifting of an image on a television receiver or cathode-ray oscilloscope.

vertical center keelson *Naval Architecture.* a fore-and-aft structural member that fits between a flat-plate keel and a rider plate.

vertical circle *Astronomy.* a great circle on the celestial sphere that passes through an observer's zenith and nadir and stands perpendicular to the horizon.

vertical compliance *Acoustical Engineering.* a measure of the ease with which the stylus can move vertically when playing a disk recording; usually measured in dynes/cm.

vertical control *Cartography.* in surveying, the determination of the elevations of survey stations with respect to a common vertical datum, usually mean sea level, providing accurate reference for the vertical positions of other features to be mapped.

vertical-control datum *Cartography.* any level horizontal surface, such as mean sea level, used as a surface of reference from which to measure elevations.

vertical coordinate *Cartography.* any value given to a point's position above or below a reference datum.

vertical-current recorder *Engineering.* an instrument that measures and records the vertical electric current in the atmosphere.

vertical curve *Civil Engineering.* a curve in the vertical plane between sections of a roadway having different gradients.

vertical definition see VERTICAL RESOLUTION.

vertical deflection oscillator *Electronics.* an oscillator that uses vertical synchronizing signals to produce a sawtooth voltage waveform, which is then amplified and fed to the vertical deflection coils on a television-receiver picture tube. Also, VERTICAL OSCILLATOR.

vertical deviation *Ordnance.* in antiaircraft artillery gunnery, the distance between the target and the point of burst, measured on a plane defined by the target's line of position and a line perpendicular to the lateral deviation.
vertical differential chart *Meteorology.* a synoptic chart, such as a thickness chart, that shows the differing manifestations of a single meteorological element at two atmospheric levels.
vertical dip slip see VERTICAL SLIP.
vertical drop *Mechanics.* the drop of an object in its trajectory, measured from its line of departure.
vertical earthrate *Navigation.* the rate at which a gyroscope must be rotated on its vertical axis to compensate for the rotation of the earth.
vertical envelopment *Military Science.* a tactical maneuver in which troops, either air-dropped or air-landed, attack the rear and flanks of a force in order to cut off or encircle it.
vertical exposure *Archaeology.* the excavating of a site to reveal its vertical extent, with relatively little breadth.
vertical field balance *Engineering.* an instrument that determines the vertical aspect of a magnetic field, by measuring the torque the field places on a horizontal permanent magnet.
vertical field-strength diagram *Electromagnetism.* a diagram indicating the intensity as a function of position of an antenna signal, measured in a vertical plane which contains the antenna.
vertical fin *Aviation.* a vertical stabilizer on an airplane or airship.
vertical firing *Mechanical Engineering.* the discharge of fuel and air perpendicular to the burner in a furnace.
vertical form index see RIPPLE INDEX.
vertical guide idlers *Mechanical Engineering.* the idler rollers placed on both sides of a belt conveyor to keep the belt in line.
vertical gyro *Aviation.* a gimbal-mounted gyro that rotates about two orthogonal axes (usually pitch and roll) and maintains its vertical spin axis by means of output signals produced by angular displacement from the orthogonal axes.
vertical heart *Cardiology.* a clockwise rotation of the heart's electrical axis, normally seen in asthenia and infancy.
vertical hold control *Electronics.* the hold control that steadies the picture on a television receiver in the vertical direction by changing the free-running period of the vertical deflection oscillator.
vertical illuminator *Optics.* a microscope that illuminates the surface of an object with a beam of light perpendicular to the object; used to view the surface of opaque materials, such as metals.
vertical-incidence transmission *Electromagnetism.* the vertical transmission of a signal from an antenna to the ionosphere whereupon the signal is reflected back to earth.
vertical instruction *Computer Programming.* a machine language instruction that performs a single operation or a time-ordered series of operations of a fixed number and type on a single set of operands.
vertical integration *Industrial Engineering.* an industrial economic structure in which the same firm conducts two or more stages in the material production chain, as when one corporation mines, refines, and shapes a metal into a final product.
vertical intensity *Geophysics.* the strength of the vertical component of the magnetic field.
vertical intensity variometer *Engineering.* a variometer in which a large magnet and very fine steel knife-edges (pivots) atop agate planes (saddles) are arranged to keep the instrument's magnetic axis horizontal. Also, Z VARIOMETER.
vertical interval *Ordnance.* in gunnery, the difference in altitude between two specified points or locations; e.g., the battery, firing ship, or observer and the target or height of burst.
vertical interval reference *Electronics.* a reference signal that is used in television transmitters and video recorders to provide a reference for maintaining color, brightness, and contrast at exact specifications; inserted into a television program signal every 1/60 seconds, in line 19 of the vertical blanking period between television frames.
vertical jet see UPRUSH.
vertical lathe *Mechanical Devices.* a machine tool with a horizontally revolving work table and work tool, which is controlled in the horizontal or vertical plane. Also, **vertical boring mill.**
vertical launch *Ordnance.* the launch of a missile or similar vehicle from a vertical position.
vertical lead *Ordnance.* the vertical angle that a gun must be aimed above a target's line of position in order for the trajectory of the projectile to pass through the target; it is the sum of the vertical deflection, the vertical pointing correction, and the superelevation.
vertical-lift bridge *Civil Engineering.* a bridge span that can be lifted vertically above the level of its approaches to allow vessels to pass underneath.
vertical linearity control *Electronics.* a control that gives linearity in the vertical direction on a television picture tube so that circular objects appear truly circular; accomplished through the narrowing or expanding of the height of the image on the upper half of the screen.
vertical magnetic intensity *Geophysics.* the strength of the vertical component of the magnetic field.
vertical metal oxide semiconductor technology *Electronics.* a semiconductor technology in which V-shaped grooves are etched into four diffused layers in silicon, and metal is then deposited in the groove to form the gate electrode.
vertical obstacle sonar *Engineering.* a type of active sonar used to determine the heights of obstacles in a submersible vehicle's path.
vertical oscillator see VERTICAL DEFLECTION OSCILLATOR.
vertical photography *Cartography.* the process of taking aerial photographs with the camera axis as close as possible to a true vertical position.
vertical polarization *Telecommunications.* the transmission of radio waves in such a way that the electric lines of force are in a vertical direction and the magnetic lines of force are in a horizontal direction.
vertical recording *Acoustical Engineering.* a recording technique used in the production of a master record by a disk phonograph recorder, in which the orientation of the cutter is perpendicular to the disk surface at all times, and vibrations are recorded by the up-and-down motion of the stylus.
vertical resolution *Electronics.* the number of horizontal lines that can be discerned in the reproduced image of a television or facsimile test pattern.
vertical retrace *Electronics.* in a cathode-ray tube, the return of the electron beam to its starting point at the top of the screen after completely transversing the screen from top to bottom.
vertical saw *Mechanical Devices.* **1.** a circular saw operating in a vertical plane. **2.** any saw with vertical frame guides.
vertical scale *Design Engineering.* the ratio of the vertical dimension of a laboratory model compared to those of its natural prototype.
vertical section *Meteorology.* any graph that shows the vertical distribution of a meteorological element or other phenomenon.
vertical seismograph *Engineering.* an instrument that measures the vertical component of an earthquake.
vertical separator *Petroleum Engineering.* a vertical cylindrical tank used to separate gas and oil.
vertical separation *Aviation.* the controlled vertical distance in altitude or flight level between two aircraft in flight. *Geology.* the vertical component of the dip slip, or the distance in a fault between two parts of a displaced marker, measured in a vertical direction.
vertical silo see UPRIGHT SILO.
vertical slip *Geology.* the vertical component of the net slip in a fault, equal to the vertical separation. Also, VERTICAL DIP SLIP.
vertical speed indicator see RATE-OF-CLIMB INDICATOR.
vertical stability *Meteorology.* see STATIC STABILITY.
vertical stabilizer *Aviation.* a fin mounted parallel to an aircraft's plane of symmetry to provide stability; the rudder, when present, is attached to this member.
vertical stretching *Meteorology.* a phenomenon in which, as altitude increases, air ascends at a faster rate and descends at a slower rate.
vertical stroke *Robotics.* the movement of a robot's arm up or down.
vertical sweep *Electronics.* **1.** in a cathode-ray tube, the sweep of the scanning beam from top to bottom of the screen. **2.** the circuit that controls such a sweep.
vertical synchronizing pulse *Electronics.* in a video signal, the pulse that synchronizes the vertical component of scanning in a television receiver with that of the transmitter and triggers vertical retrace and blanking.
vertical tab *Computer Programming.* the movement of a cursor or print head to the next line.
vertical tail *Aviation.* a component of an aircraft tail assembly, usually consisting of a moveable rudder attached to a fixed vertical stabilizer.
vertical takeoff and landing see VTOL.
vertical transmission *Genetics.* the transmission of genetic material from parent cells or organisms to their progeny, utilizing natural hereditary mechanisms such as mitosis, meiosis, and fertilization, as contrasted with induction and infection between organisms or cells of the same generation.

vertical traverse *Mechanical Engineering.* the angle through which a robot arm is free to swing in the vertical plane.

vertical turret lathe *Mechanical Devices.* a vertical lathe with a cross-rail mounted above the work table supporting a turret whose tools can be indexed and fed horizontally or vertically downward.

vertical visibility *Meteorology.* the distance an observer can see upward into the atmosphere through surface-based obscuring phenomena such as fog or rain; the visibility from the ground up to the ceiling height.

vertical vorticity *Fluid Mechanics.* the vertical component of the vorticity vector.

verticil *Biology.* a whorl or circle of body parts around a central point, such as flower petals attached around an axis.

verticillate *Botany.* relating to a plant part arranged in whorls.

Verticillium *Mycology.* a genus of fungi belonging to the class Hyphomycetes, which causes a variety of plant wilt diseases and human infections.

verticillium wilt *Plant Pathology.* a widespread and destructive disease of plants caused by fungi of the genus *Verticillium,* which causes the shriveling, drooping, and browning of plant parts.

vertiginous *Neurology.* characterized by or inducing vertigo.

vertigo [vurt´i gō] *Neurology.* **1.** an illusion or subjective sensation of movement, especially of rotation, either of one's self or of one's surroundings, resulting from diseases of the inner ear, or disturbances of the bestibular centers or pathways in the central nervous system. **2.** broadly, any form of dizziness.

Vertïsol *Geology.* one of the 10 major soil classifications that develops in regions having pronounced dry seasons; characterized by at least 30% clay and the presence of deep, wide surface cracks when dry.

vervain *Botany.* a plant of the verbena family, characterized by elongated spikes of small white, bluish, or purple flowers; found in Europe and America.

vervet *Vertebrate Zoology.* a monkey, *Cercopithecus aethiops pygerythrus,* a guenon allied to the green monkey and the grivet, but distinguished by a rusty patch at the root of the tail; found in Africa.

very close pack ice *Oceanography.* sea ice that covers almost the whole surface, with very little open water showing.

very high frequency *Telecommunications.* of a device or system, operating or able to receive signals at frequencies between 30 and 300 megahertz. Thus, **very-high-frequency oscillator, very-high-frequency radar, very-high-frequency tuner.**

very-high-frequency omnidirectional radio range *Navigation.* an air navigational radio aid that provides radial lines of position in 360°; the airborne receiving equipment determines bearings through phase comparison of a ground transmitted signal.

very large crude carrier *Naval Architecture.* an oil tanker of the class just below the largest class, with a tonnage between 100,000 and 250,000 tons; commonly classed with ultralarge crude carriers as supertankers.

very-large-scale integration *Electronics.* the fabrication of multiple complete systems, such as computers, onto a single integrated circuit chip.

very-long-baseline interferometry *Electronics.* a technique used to improve angular resolution in radio source observation.

very-long-range materiel requirements *Ordnance.* items projected as necessary for operations more than ten years in the future.

very-long-range radar *Electronics.* a radar system having a maximum line-of-sight range greater than 800 miles for a target whose area is one square meter perpendicular to the radar beam.

very low frequency *Telecommunications.* any radio frequency located on the band between 3 and 30 kilohertz; a wavelength between 10,000 and 100,000 meters.

very open pack ice *Oceanography.* floes of sea ice that cover between 10% and 40% of the surface.

very-short-range radar *Electronics.* radar having a maximum line-of-sight range less than 50 miles for a target whose area is one square meter perpendicular to the radar beam.

Vesalius, Andreas 1514–1564, Flemish anatomist, lived in Italy; called the founder of modern anatomy; wrote the first complete human anatomy; overthrew many of Galen's misconceptions.

Vesic. a blister. (From Latin *vesicula vesicatorium.*)

vesicant *Pharmacology.* **1.** producing blisters. **2.** a drug or agent that causes blistering.

vesication *Medicine.* **1.** the process of blistering. **2.** a blistered spot or surface.

vesicle [ves´i kəl] *Biology.* **1.** a small cavity or sac usually containing fluid, as a blister. **2.** small globular or bladderlike air spaces in tissues. **3.** one of three primary cavities of the brain. *Geology.* a small cavity in lava formed by the entrapment of a gas bubble during solidification of lava. Also, VACUOLE.

vesicula opththalmica see OPTIC VESICLE.

vesicular [ve sik´yə lər] *Science.* of, relating to, or resembling a vesicle or bladder.

vesicular exanthema *Veterinary Medicine.* a highly infectious viral disease of swine and sometimes of cattle, which is marked by fever and the formation of blisters on the snout, lips, tongue, and feet; clinically indistinguishable from foot-and-mouth disease in swine.

vesicular stomatitis *Veterinary Medicine.* an acute disease of cattle, swine, and numerous other species, which is caused by a vesiculovirus and characterized by vesicles on the mouth and feet; in cattle, it resembles foot-and-mouth disease.

vesicular stomatitis virus *Virology.* a virus of the *Vesiculovirus* genus of the Rhabdoviridae family that causes vesicular stomatitis in animals and can affect humans, producing mild influenzalike symptoms.

vesicular structure *Petrology.* a structure typical of igneous rocks, especially lava, characterized by an abundance of vesicles formed by expansion of gases during solidification of the magma. Thus, **vesicular lava.**

vesicular lava

Vesiculovirus *Virology.* a genus of viruses of the family Rhabdoviridae, type species *vesicular stomatitis virus,* that infects a wide range of animals.

Vespertilionidae *Vertebrate Zoology.* a large, widely distributed family of insectivorous bats of the suborder Microchiroptera, characterized by a long slender tail and well-developed ears.

vespertine *Behavior.* of or relating to an animal that is active in the evening.

Vespidae *Invertebrate Zoology.* a family of social wasps, including the hornets, yellowjackets, and paper wasps; they are hymenopteran insects belonging to the superfamily Vespoide that build papery nests; they are characterized by powerful stingers and red or yellow coloring with brown or black markings.

Vespoidea *Invertebrate Zoology.* a superfamily of wasps and hornets, hymenopteran insects in the suborder Clistogastra.

Vespucci, Amerigo [ves pyooch´ē] 1454–1512, Italian navigator; discovered the mouth of the Amazon River; described South America as a separate continent.

vessel *Naval Architecture.* any vehicle or structure designed to float on or in the water, including ships, submarines, boats, and barges. *Engineering.* a hollow container in which materials are processed or stored. *Anatomy.* a tube or duct containing or conveying blood or some other body fluid, as a vein or artery. *Botany.* any of the tubelike structures in the xylem, composed of connected cells that have lost their intervening partitions, that conduct water and mineral nutrients.

vessel member *Botany.* one of the cellular components of a vessel. Also, **vessel element, vessel segment.**

Vesta *Astronomy.* asteroid 3, belonging to type U, discovered in 1807; it has a diameter of 576 kilometers, a basaltic surface composition; the third largest and one of the four brightest asteroids. (Named for the ancient Roman goddess of the hearth.)

vestibular apparatus *Anatomy.* the sensory organs in the sacculus and utriculus of the inner ear that respond to changes in the positioning and movement of the head.

vestibular membrane see REISSNER'S MEMBRANE.

vestibular nerve *Anatomy.* the portion of the eighth cranial nerve that carries the sensation of balance and equilibrium from the inner ear to the brain.

vestibular reflexes *Physiology.* the stimulation of the receptor cells of the inner ear, resulting in communication to the muscles of the neck as to the appropriate position needed to maintain the head upright.

vestibule *Architecture.* a porch, lobby, or anteroom that leads into a larger room. *Building Engineering.* an enclosed passage, hall, or chamber located between the outer and interior doors or rooms of a building. *Anatomy.* a space or cavity at the entrance to a canal, as the vestibule of the vagina or the vestibule of the mouth.

Vestibulifera *Invertebrate Zoology.* a subclass of ciliated protozoans. Also, **Vestibuliferia.**

vestibulospinal tract *Anatomy.* a bundle of motor fibers that extends down the ventral and lateral portion of the spinal cord and controls muscle tone.

vestige [ves′tij] *Biology.* a degenerate or imperfectly developed organ or part that may have been complete and functional in some ancestor.

vestigial [ve stij′əl] *Biology.* **1.** remaining as a vestige. **2.** small and imperfectly developed.

vestigial sideband *Telecommunications.* the transmitted part of an amplitude-modulated sideband that is greatly attenuated by a transducer, while the other sideband is transmitted with minimal attenuation; used for the composite picture and synchronizing signal in television transmissions to conserve bandwidth and transmitted power. Thus, **vestigial-sideband transmission.**

vestigial-sideband filter *Electronics.* a filter inserted between a transmitter and its antenna to suppress a portion of one of the sidebands.

vesuvian see VESUVIANITE.

vesuvianite *Mineralogy.* $Ca_{10}Mg_2Al_4(SiO_4)_5(Si_2O_7)_2(OH)_4$, a white, purple, brown, yellow, or green tetragonal mineral occurring as short prismatic crystals and in massive form, having a specific gravity of 3.33 to 3.45 and a hardness of 6 to 7 on the Mohs scale; found in contact-metamorphosed limestones, nepheline syenites, and ultrabasic rocks. Also, IDOCRASE, VESUVIAN.

Vesuvian-type eruption see VULCANIAN-TYPE ERUPTION.

veszelyite *Mineralogy.* $(Cu^{+2},Zn)_3(PO_4)(OH)_3 \cdot 2H_2O$, a greenish-blue monoclinic mineral occurring as short prismatic, tabular, or equant crystals and granular aggregates, having a specific gravity of 3.4 and a hardness of 3.5 to 4 on the Mohs scale.

vet. veterinarian; veterinary.

Vetaformaceae *Botany.* a monotypic family of leafy liverworts in the order Jungermanniales that are subisophyllous, erect, pale green, and native to southern South America.

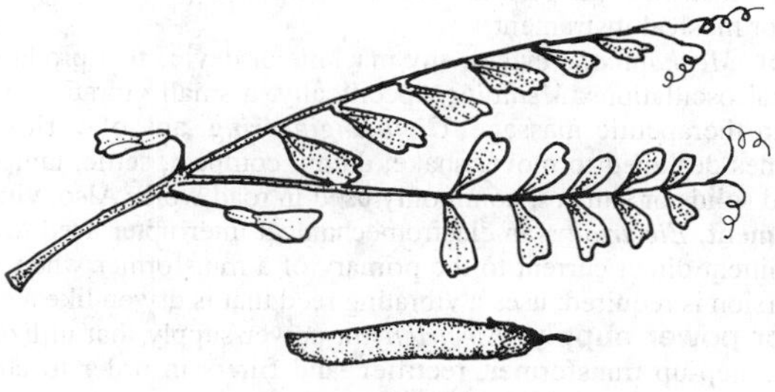

vetch

vetch *Botany.* **1.** the common name for members of the genus *Vicia* of the legume family, characterized by pinnatte leaves ending in tendrils and bearing pealike flowers, especially *V. sativa,* grown as forage and to improve soil. **2.** any of various related plants grown for their edible seeds and for forage. **3.** the beanlike seed or fruit of such plants.

veterinarian *Medicine.* a person who is professionally trained and licensed to practice veterinary medicine.

Veterinary Medicine

Veterinary medicine encompasses all aspects of biology and medicine that contribute to or deal with the nature, control, and treatment of diseases in all species of animals. With much of the world's population now suffering from chronic protein malnutrition and hunger, contributions of the veterinary profession to animal protein production is truly a humanitarian activity. Veterinary medicine as a profession existed throughout large areas of Africa and Asia as early as 2000 BC; ancient Egyptian literature includes detailed monographs on both animal and human diseases. Evidence for the parallel development of veterinary and human medicine is also found in the early writings of both Hippocrates and Aristotle. Serious outbreaks of cattle plague in Europe in the late 18th century resulted in the founding of modern schools throughout the world.

Although veterinary medicine originally developed in response to the needs of pastoral and agricultural human existence, its goals have been recently broadened to meet societal needs. Primary responsibilities now include provision for the health of all livestock; domestic poultry and wild birds; pet and companion animals; wildlife; fur-bearing animals; zoo animals; and aquatic mammals and fish. Another major responsibility is the control of over 200 diseases of animals that are transmittable to humans.

The profession also provides answers regarding the etiology and nature of animal disorders, many of which are comparable to counterpart diseases in humans. Such comparative medical research on animals has provided many clues and breakthroughs important to understanding and ultimately preventing many human diseases. Veterinary medical subdisciplines are now common and parallel the organ system specialties in human medicine; in addition, there is specialization in the husbandry and diseases of specific animal species. The role of animals in providing companionship to the lonely and aged is now recognized as another important contribution to human mental and physical well-being.

Charles E. Cornelius
Emeritus Professor of Veterinary Medicine
University of California, Davis

veterinary medicine *Medicine.* the field of medicine involving the treatment or prevention of disease or injury in animals.

vetiver oil *Materials.* an essential oil with a violetlike odor that is derived from the East Indian grass *Vetiveria zizanioides*; used in perfumery.

VF video frequency; voice frequency.

V.F. vocal fremitus.

v.f. field of vision.

V factor *Microbiology.* a growth factor needed by certain parasitic bacteria of the genus *Haemophilus* and found in human red blood cells.

V flume *Mining Engineering.* a flume in the shape of a V, supported by trestlework and sometimes miles in length; used to bring timber and wood down from high mountains; the water in the flume is also used for mining purposes.

V format *Computer Programming.* a data set format in which logical records are of varying length and each physical record begins with a length indicator.

VFR Visual Flight Rules.

VFR between layers *Aviation.* a flight condition in which a cloud ceiling exists, but an aircraft may operate between two layers of clouds according to modified Visual Flight Rules, as visibility permits.

VFR flight see VISUAL FLIGHT.
VFR on top *Aviation.* a flight condition in which a cloud ceiling exists, but modified Visual Flight Rules are in effect if the aircraft travels above the cloud layer.
VFR terminal minimums *Aviation.* the minimum conditions of ceiling and visibility under which Visual Flight Rules may be used at an airport.
VFR weather *Aviation.* a route or terminal weather condition of aircraft operation under Visual Flight Rules.
v.g. for example. (From Latin *verbi gratia.*)
VGA *Computer Technology.* a standard video display device for IBM PCs, having graphics resolution up to 1024 × 760 pixels in 256 colors. (Short for video graphics array.)
v gene *Genetics.* the gene that codes for the variable region of an immunoglobulin chain. Also, V-DNA, V STRAND.
V guide *Mechanical Engineering.* a V-shaped groove that guides a wedge-shaped element, such as a belt or chain.
VHF or **V.H.F.** very high frequency. Also, **vhf, v.h.f.**
VHN Vickers hardness number.
VHS *Telecommunications.* the most common format for videocassette recording and playback, using cassettes of half-inch videotape. (Short for video home system.)
VI volume indicator; viscosity index.
v.i. see below. (From Latin *vide infra.*)
viability *Developmental Biology.* **1.** the ability to live after birth. **2.** specifically, the ability of a fetus to live outside the uterus under normal conditions.
viability count *Cell Biology.* the number of cells in an in vitro cell population that are capable of growing and proliferating; usually determined by the number of visible colonies that develop on a plated culture inoculated from a highly diluted suspension. Also, **viable count.**
viable *Developmental Biology.* **1.** able to live, especially after birth. **2.** specifically, describing a fetus that has reached such a state of development that it can live outside the uterus. *Botany.* describing a plant that is able to live and grow under normal conditions.
viaduct *Civil Engineering.* **1.** a bridge, usually long and multispan, that carries a road or railroad over a river, a valley, or an urbanized area. **2.** any bridge, as over a road, channel, or the like.
vial *Science.* a small tubular glass bottle, commonly used for blood samples or for the growth of cultures of microorganisms.
Vianaidae *Invertebrate Zoology.* a family of true bugs, hemipteran insects in the superfamily Tingoidea.
Vi antigen *Immunology.* a polysaccharide microcapsular antigen that occurs on certain bacteria, such as *Salmonella typhi,* and *Salmonella paratyphi.*
via point *Robotics.* a point halfway between the starting position and stopping position of a robot's tool tip, through which it passes without stopping. Also, WAY POINT.
vibraculum *Invertebrate Zoology.* a modified bryozoan zooid in the form of a whiplike filament, used to sweep debris and parasites from the surface of the colony.
vibrate *Mechanics.* to undergo a process of vibration.
vibrating *Mechanics.* oscillating about a position of equilibrium; undergoing or characterized by vibration.
vibrating capacitor *Electricity.* a capacitor in which the potential on the electrode of the capacitor is varied by mechanical oscillation in order to convert a steady applied potential to an alternating potential that is more easily amplified.
vibrating needle *Engineering.* a needle on a compass that determines the relative intensity of the horizontal aspects of both the earth's magnetic field and the magnetic field surrounding the compass.
vibrating pebble mill *Mechanical Engineering.* a size-reducing mill in which the action of vibrating pebbles grinds the feed.
vibrating reed *Mechanical Engineering.* a mechanism that is used in various measuring instruments, consisting of a set of steel reeds that are energized by an electromagnet transporting alternating current. Thus, **vibrating-reed electometer, vibrating-reed frequency meter, vibrating-reed magnetometer, vibrating-reed tachometer.**
vibrating screen *Mechanical Engineering.* an oscillating screen powered by a solenoid, by magnetostriction, or mechanically by eccentrics or unbalanced spinning weights; used to separate fine and coarse material. Thus, **vibrating-screen classifier.**
vibrating-wire transducer *Engineering.* an instrument that measures ocean depth by noting the frequency at which a tungsten wire stretched in a magnetic field vibrates.
vibration *Mechanics.* a process of oscillation (variation from one limit to another) about a position of equilibrium; an alternating or reciprocating motion.
vibrational circular dichroism *Spectroscopy.* the differential absorption of left- and right-circularly polarized light for optically active samples, as measured by observation of different vibrational modes.
vibrational energy *Physical Chemistry.* the difference between energy generated by a theoretical diatomic molecule whose rotational energy is set at zero, and one whose rotational energy is generated by stopping the nuclei from vibrating without slowing the motion of its electrons.
vibrational level *Physical Chemistry.* the level of energy in a molecule with more than one nucleus (diatomic or polyatomic) that is distinguished by a particular vibrational energy.
vibrational quantum number *Physical Chemistry.* a quantity that represents the vibrational energy describing the motion of nuclei in a molecule.
vibrational spectroscopy *Spectroscopy.* a classification of spectroscopic techniques including infrared and Raman spectroscopy whereby molecular vibrations are detected by observing changes in the electric dipole moments of the molecules; such techniques require that the specimen not be heated so that molecular rotations can clearly be distinguished from vibrational modes.
vibrational spectrum *Spectroscopy.* a molecular spectrum produced by energy transitions between vibrational levels.
vibrational sum rule *Spectroscopy.* a rule stating that the sum of the band strengths of all emission bands having the same upper state or all absorption bands having the same lower state is proportional to the number of molecules in the upper state or lower state, respectively.
vibration damping *Mechanical Engineering.* the process of reducing the amount of vibration in a mechanical system by increasing the amount of damping present in the system. Also, **vibration suppression.**
vibration drilling *Mining Engineering.* a drilling technique in which a vibrating drill head is used to fracture rock.
vibration galvanometer *Engineering.* an alternating-current galvanometer in which the current being measured is equal to the natural oscillation frequency of the moving elements.
vibration isolation *Mechanical Engineering.* in structures and vehicles, the isolation of those motions that are classified as mechanical vibration.
vibration machine *Mechanical Engineering.* a mechanism that can create a controlled, reproducible mechanical vibration at a desired amplitude and frequency.
vibration magnetometer *Engineering.* an instrument that determines the horizontal magnetic field strength of a needle by measuring how long it vibrates.
vibration meter see VIBROMETER.
vibration parameters *Crystallography.* see DISPLACEMENT PARAMETERS.
vibration pickup *Electricity.* an electromechanical transducer, such as a microphone or crystal, that is used to transform the oscillatory motion of a surface, such as mechanical vibrations, into an electrical voltage or current.
vibration separation *Mechanical Engineering.* the separation of material using a vibrating screen.
vibrato *Acoustics.* the periodic variations in pitch produced by a human voice or musical instrument.
vibrator *Mechanical Devices.* any machine or device that produces mechanical oscillations. *Medicine.* specifically, a small vibrating machine used in therapeutic massage. *Civil Engineering.* any of various large machines designed to move, shake, dump, compact, settle, tamp, pack, or feed solids or slurries; commonly used in road work. Also, **vibratory equipment.** *Electricity.* an electromechanical interrupter used to supply intermittent direct current to the primary of a transformer when voltage conversion is required; uses a vibrating reed that is driven like a buzzer.
vibrator power supply *Electricity.* a power supply that utilizes a vibrator, step-up transformer, rectifier, and filters in order to change a low-DC voltage into a high-DC voltage.
vibrator-type invertor *Electricity.* a device that utilizes a vibrator in an associated transformer or other inductive type of device to change DC input power into AC output power.
vibrator-type rectifier *Electricity.* a vibrator that connects one terminal of a secondary winding of a transformer to an output terminal each time that terminal is positive; when the negative half-cycle appears at the transformer terminal, the vibrator is open such that the AC output of the transformer is rectified into DC.

vibratory [vī′brə tôr′ē] *Mechanics.* **1.** experiencing vibrations; vibrating. **2.** causing or capable of causing vibrations.

vibratory centrifuge *Mechanical Engineering.* a high-speed centrifugal device used to remove moisture from pulverized solids.

vibratory hammer *Mechanical Engineering.* a pile hammer that uses eccentric cams to vibrate a pile into position.

vibratory milling *Materials Science.* a fast and efficient method of raw material processing that reduces particles to a smaller size distribution than is usually obtained by other milling processes.

vibratory saw *Surgery.* a hand-held surgical saw designed to cut through hard materials such as bone or a cast, leaving soft tissue such as skin or muscle untouched.

vibratron *Electronics.* a triode having an anode that can be moved or vibrated by an external force, thus varying the anode current in proportion to the amplitude and frequency of the applied force.

Vibrio *Bacteriology.* a genus of Gram-negative bacteria of the family Vibrionaceae that occur as motile, rod-shaped cells with one or more polar flagella; the species include the causative agents of human cholera and many other animal diseases.

vibrio *Bacteriology.* an organism of the genus *Vibrio.*

vibriocins *Bacteriology.* any bacteriocins, produced by species of the bacterial genus *Vibrio,* that are effective against certain enterobacteria as well as certain mammalian cells.

Vibrionaceae *Bacteriology.* a family of Gram-negative, rod-shaped, motile, anaerobic bacteria; usually found in saltwater or freshwater environments, and sometimes pathogenic in humans and other animals.

vibrissa *plural,* **vibrissae.** *Vertebrate Zoology.* **1.** any of the erectile, tactile hairs found on the face of most mammals except humans. Also, WHISKER, SINUS HAIR. **2.** any of the feathers near the mouth of many birds that may help in keeping insects caught as food from escaping.

vibro- a combining form meaning "vibration."

vibroenergy separator *Mechanical Engineering.* a vibratory screen that classifies or separates grains of solids by a gyratory motion and a vibration caused by balls bouncing against the underside of the screen.

vibrograph *Engineering.* an instrument that records mechanical vibration.

vibrometer *Engineering.* an instrument designed to measure the displacement, velocity, or acceleration of a vibrating solid.

vibrothermography *Materials Science.* a nondestructive evaluation technique used to locate and identify regions of discontinuities in a variety of materials; it is especially useful for detecting delaminations in advanced composite materials.

vic. vicinity.

vic- *Organic Chemistry.* a prefix meaning "vicinal" (i.e., adjoining or neighboring), used to designate a substituent position on a carbon ring or chain.

Vicalloy *Materials.* the trade name for an alloy of iron, cobalt, and vanadium that is a permanent magnet and can be drawn into thin tape or wire.

vicariance *Ecology.* a condition in which closely related species originally forming a common ancestral population are separated into different geographical regions by a natural barrier, such as a mountain range or an ocean. Also, **vicarism.**

vicariants *Ecology.* closely related animal species that were isolated from each other following the development of a natural barrier, such as the European reindeer and the North American caribou. Also, **vicariads.**

vicarious *Ecology.* relating to or characterized by vicariance. Thus, **vicarious species.**

vicarious learning *Behavior.* the acquisition of knowledge or ability through indirect experience and observation, rather than direct experience or practice.

vicarious reinforcement *Behavior.* reinforcement that results from an individual's observation of another being rewarded or punished for a certain behavior.

Vicat temperature *Materials Science.* the temperature at which a 1-millimeter penetration of the specimen occurs in a Vicat test.

Vicat test *Materials Science.* a hardness test for thermoplastics that have no definite melting point; it involves the penetration of a plastic specimen with an indentor of 1 square millimeter cross section (a **Vicat needle**) at a specified heating rate.

vicinal *Organic Chemistry.* denoting adjoining or neighboring positions on a carbon ring or chain.

vicinal faces *Crystallography.* abnormal faces that modify normal crystal faces, and whose indices cannot be expressed in small whole numbers.

Vickers *Ordnance.* **1.** a main battle tank developed by the British company, Vickers, Ltd., in 1958–1963; it is equipped with a 105-mm gun, as well as .50-caliber and .30-caliber machine guns. **2.** any of various machine guns built by Vickers. *Aviation.* any of various aircraft built by Vickers and used during World War I.

Vickers hardness test *Materials Science.* a test that measures the hardness of a material as a function of its resistance to penetration. A diamond pyramid is pressed into the sample, and the diagonals of the square impression are measured; the **Vickers hardness number** is then read from a chart for a given diagonal length and applied load. Also, DIAMOND-PYRAMID HARDNESS TEST.

Vicksburgian *Geology.* a North American Gulf Coast stage of the Oligocene epoch, occurring after the Jacksonian and before the Chickasawhay.

Victaulic coupling *Mechanical Devices.* a coupling in which a groove is cut around each end of pipe aligned with a fitted, rubber ring around the pipe joint; the joint is drawn together with two bolts, which provides for a watertight seal.

Victoria, Lake or **Victoria Nyanza** *Geography.* a large lake (area: 26,828 sq. mi.) between Kenya, Uganda, and Tanzania.

Victoria blue *Organic Chemistry.* $C_{33}H_{31}N_3 \cdot HCl$, a bronze crystalline powder; soluble in hot water, alcohol, and ether; used to dye silk, wool, and cotton, and as a biological stain.

Victoria Falls *Geography.* a waterfall (height: 355 ft) on the Zambesi River between Zambia and Zimbabwe.

Victorian *Architecture.* denoting British and American architecture during the reign of Queen Victoria (1837–1901).

vicuna [vī koon′ə] *Vertebrate Zoology.* the popular name for *Lama vicugna,* a camel-like ruminant native to the Andes mountains that is prized for its soft yet strong and resilient wool; recently classified as a rare species and now protected. *Textiles.* a fabric or garment made of this wool. Also, **vicuña** [vī koon′yə].

vicuna

vidarabine *Pharmacology.* $C_{10}H_{13}N_5O_4$, an antiviral agent produced by the bacterium *Streptomyces antibioticus* and used in immunosuppressed patients to treat serious herpesvirus infections. Also, ADENOSINE ARABINOFURANOSIDE, VIRA-A.

videcon *Robotics.* the electron tube in a television camera that converts reflected light into electronic signals.

video *Electronics.* **1.** the component of a television signal that contains the picture information. **2.** of or relating to the bandwidth and spectrum position of a radar or television scanning signal. **3.** a videotape recording, especially one that has been commercially produced. (From a Latin word meaning "to see.")

video amplifier *Electronics.* an amplifier providing wide-band operation in the frequency range of approximately 15 cycles to 5 megacycles per second.

video camera *Electronics.* a camera designed to record moving pictures directly onto videotape, which can then be played back, usually without an intervening developing process, on a VCR or similar device.

videocassette *Electronics.* a cassette holding magnetic videotape, designed for use with a VCR.

videocassette recorder see VCR.

video correlator *Electronics.* a radar circuit that aids in locating and identifying a target with high precision.
video data digital processing *Telecommunications.* in television, a type of digital processing in which a computer compares each scanned line with neighboring lines and removes the aberrations caused by electromagnetic interference.
video discrimination *Electronics.* the use of a radar circuit or device to reduce the frequency band of the video-amplifier stage in which it is used.
videodisk *Electronics.* a magnetic or optical disk that is used to record images. Also, **videodisc.**
videodisk player *Electronics.* a device used to play back videodisks through a television monitor.
videodisk recorder *Electronics.* a videorecorder that records television signals on a magnetic or optical disk.
video frequency *Telecommunications.* a frequency in the bandwith required for a video signal, ranging from approximately 0 to 5 megahertz.
videognosis *Medicine.* medical diagnosis based on the interpretation of televised images.
video home system see VHS.
video integration *Electronics.* the use of the redundancy of repetitive signals to sum successive video signals to improve the output signal-to-noise ratio.
video masking *Electronics.* a form of radar signal processing that makes the desired targets more readily visible by removing ground clutter and other unwanted echoes.
videomicroscopy *Optics.* a technique in which television cameras are used to illuminate magnified images too dim for the human eye.
video monitor *Computer Technology.* a video display device similar to a television screen that displays an image from a video signal generated by an input/output controller of the computer.
videophone *Telecommunications.* a telephone that enables a caller to see and hear the subscriber at the other end of the circuit by means of a small television camera and screen. Also, VIDEO TELEPHONE, VIEWPHONE.
video player *Electronics.* a device that converts a recorded television program, such as a videodisk or videotape, into signals that are capable of driving a home television receiver.
video recorder *Electronics.* a device that records television signals and stores them on a videotape or videodisk.
video replay *Electronics.* **1.** a procedure in which a television program is recorded and rebroadcast at a later time. Also, **videotape replay. 2.** see INSTANT REPLAY.
video sensing *Computer Technology.* a technique in optical character recognition in which a document image is reflected by light onto a video camera.
video signal *Telecommunications.* the part of a television signal that contains all of the information required to establish a visual television image, including intensity, color, synchronization, and blanking.
videotape *Electronics.* **1.** a magnetic tape designed to record and play back a composite black-and-white or color television signal. **2.** the record of a television broadcast made on such magnetic tape.
videotape recorder *Electronics.* a VCR (videocassette recorder) or other device that records material on videotape.
videotape recording *Electronics.* **1.** a method of recording television picture and sound signals on tape. **2.** a videotape containing a record of a television broadcast.
video telephone see VIDEOPHONE.
videotex *Telecommunications.* a European information service that uses a modified domestic television receiver coupled with a regular public telephone line to provide an interactive data retrieval service for public or private usage. Also, **videotext, viewdata.**
video transformer *Electronics.* a transformer whose bandwidth is adequate for the transfer of picture information signals from one circuit to another.
vidicon *Electronics.* a television camera tube in which an electron beam scans a charged-density pattern that has been formed and stored on the surface of a photoconductor.
VIE *Aviation.* the airport code for Vienna, Austria (Schwechat).
Vienna definition language *Computer Programming.* a metalanguage used to define the syntax and semantics of a programming language operationally in terms of the computations to be performed during execution.
Viete, Francois (Franciscus Vieta) 1540–1603, French mathematician; called the father of modern algebra; established the use of letters for unknowns; improved methods of solving degree equations.
view camera *Optics.* a camera that can be focused in any direction so that a photographer can control the shape of the subject in the image; a ground glass at the back of the camera also permits the photographer to view the image before recording it.
viewfinder *Optics.* a device that displays an optically or electronically formed image corresponding to that seen by the associated lens.
viewing time *Electronics.* the period of time during which an image remains on the viewing screen of a storage tube.
viewport *Computer Technology.* **1.** an area on a display screen through which an operator can view all or a portion of a graphical or textual image. **2.** a process by which a user can display a specific picture at a selected location on the screen.
VIG vaccinia immune globulin.
vigia *Navigation.* a possible obstruction or navigational danger, such as a rock, shoal, or island, that has been reported but has not been identified by survey; it is placed on the charts with a note (vigia), indicating its lack of positive identification.
vigilance *Behavior.* the state of alertness of an animal to changes in the environment, especially to the possible presence of a predator. Also, **vigilant behavior.**
vignafuran *Biochemistry.* a substance produced by the cow pea (genus *Vigna*) and used to counteract disease.
vignette [vin yet′] *Graphic Arts.* a halftone in which the background gradually fades out toward the edges so that the lightest tones are about the same as the tone of the paper on which it is printed.
vignetted *Graphic Arts.* of an image, having edges that fade out gradually.
vignetting *Optics.* a decrease in brightness on the fringes of an optical devices's field of view that arises when the sides of the aperture block out light rays. *Graphic Arts.* the use or technique of making a vignette.
vigor *Biology.* a combination of attributes of a living organism, usually expressed by rapid growth, high fertility and fecundity, large size, and long life.
Vigreaux column *Analytical Chemistry.* an apparatus formerly used in fractional distillation, consisting of a long glass tube with a thermometer at the top and a side arm attached to a condenser.
Viking *Space Technology.* a series of two U.S. probes that in 1976 orbited and landed on Mars, conducting search-for-life experiments and extensively photographing the planet both from orbit and from its surface.

Viking II on Mars

villa *Architecture.* **1.** a Roman or Renaissance Italian country estate. **2.** a modern suburban or rural house with extensive grounds that is considered to be pretentious.
Villafranchian *Geology.* **1.** a European stage of the Upper Pliocene and Lower Pleistocene epoch, roughly equal to the Astian of the Pliocene epoch. **2.** in France and Italy, the terrestrial equivalent of the marine Calabrian.

Villari effect *Physics.* an observed change in the magnetic induction of a ferromagnetic material resulting from mechanical stresses applied to the material.
Villari reversal *Physics.* the changing over of expansion to contraction in some ferromagnetic materials that are subjected to an increasing magnetic field.
villiaumite *Mineralogy.* NaF, a red or yellow, cubic mineral, occurring in granular masses and as minute crystals, having a specific gravity of 2.79 and a hardness of 2 to 2.5 on the Mohs scale; found in alkalic igneous rocks.
villin *Biochemistry.* a protein in intestinal microvilli that can bind to and cleave actin filaments.
villoma see PAPILLOMA.
villous *Biology.* **1.** covered with soft hairs; shaggy. **2.** covered with villi. Also, **villose.**
villous adenoma *Medicine.* a slow-growing but potentially malignant polyplike tumor of the epithelium occurring on the mucous membrane of the large intestine; characterized by bleeding and large amounts of mucoid diarrhea in the stool.
villous tenosynovitis *Medicine.* a tuberculous infection of the tendon sheaths and bursa in which the sheaths are shaggy, with a hairlike surface.
villus *plural,* **villi.** *Anatomy.* a small process or protrusion, especially such a protrusion from the free surface of a mucous membrane. *Botany.* any of the soft, long hairs covering some fruits, flowers, and other plant parts. (A Latin word meaning "tuft of hair.")
vimen *Botany.* a long, flexible plant shoot; a twig.
vimentin *Cell Biology.* a protein that is found in certain intermediate filaments; common in many eukaryotic cells including muscle cells.
vimineous *Botany.* producing long shoots or twigs.
Vim-Silverman needle *Surgery.* a needle used in needle biopsy. (Named for Irving *Silverman,* born 1904, American surgeon.)
vinasse *Materials Science.* the liquid residue resulting from the distillation and fermentation of alcohols and liquors.
vinblastine *Pharmacology.* $C_{46}H_{58}N_4O_9$, an antineoplastic alkaloid extracted from the woody herb *Vinca rosea;* used in the treatment of Hodgkin's disease and other lymphomas.
Vincent's infection *Medicine.* a bacterial infection in which several agents may act in symbiosis, ulcerating and giving rise to gangrene in the tonsils, and possibly also affecting the mouth and gums and causing trench mouth. Also, **Vincent's angina, Vincent's gingivitis.** (Named for Henri *Vincent,* 1862–1950, French physician.)
vincristine *Pharmacology.* $C_{46}H_{56}N_4O_{10}$, an antineoplastic alkaloid extracted from the herb *Vinca rosea;* used in the treatment of acute lymphocytic leukemia.
vinculin *Biochemistry.* a fibrous protein thought to help anchor actin microfilaments to other cell components.
vindesine *Oncology.* $C_{43}H_{55}N_5O_7$, an antineoplastic agent.
Vindobonian *Geology.* a European geologic stage of the Lower Miocene epoch.
vine *Botany.* a pliable plant stem that is unable to support itself upright, twining around other objects for support.
vinegar *Materials.* a sour liquid containing acetic acid, derived from the fermentation of diluted alcoholic beverages such as cider and wine.
vinegar eel *Invertebrate Zoology.* a minute nematode worm, *Anguillula aceti,* found in vinegar, fermenting paste, and the like. Also, **vinegar worm.**
vine snake *Vertebrate Zoology.* a common name for any of several very slender species of venomous, rear-fanged members of the genus *Oxybelis* of the family Colurbridae; inhabits the tropics worldwide and is easily camouflaged as a vine or branch.

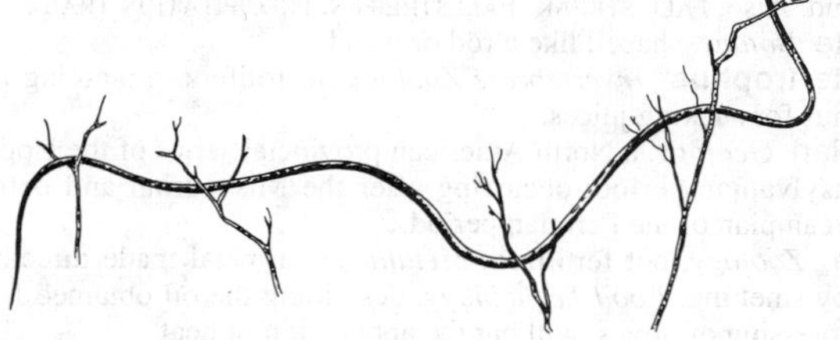

vine snake

viniculture *Agriculture.* the growing of grapes, especially for winemaking.
viniferins *Biochemistry.* stilbene oligomer disease-fighting agents produced by *Vitis vinifera* (the grapevine).
vintage *Food Technology.* **1.** the product of grapes or wine from a particular vineyard, winery, or region during a single harvest. **2.** the calendar year in which a particular wine is produced.
vintage wine *Food Technology.* an often superior wine of a particular type, year, and origin.
vinyl *Materials.* **1.** any resin that is formed by the polymerization of a compound that contains the vinyl group. **2.** also, **vinyl plastic.** a plastic that is made from a vinyl resin. *Textiles.* a synthetic fiber that is composed of at least 85% vinyl alcohol and acetyl units, with the vinyl alcohol units constituting at least 50% of the total weight.
vinyl acetal resin *Organic Chemistry.* $[CH_2CH(OC_2H_5)]_x$, a colorless, odorless thermoplastic that is light-stable, soluble in alcohol, and insoluble in water; used in lacquers, coatings, and molded objects.
vinyl acetate *Organic Chemistry.* $CH_3COOCH{=}CH_2$, a colorless liquid; soluble in most organic solvents and sparingly soluble in water; boils at 73°C; used in safety glass interlayers and the production of polymers and copolymers such as polyvinyl resins.
vinyl acetate resin *Organic Chemistry.* $[-CH_2-CHO-COCH_3-]$, an odorless thermoplastic derived from polymerization of vinyl acetate; insoluble in water and oils and soluble in alcohol and chlorinated hydrocarbons; used in coatings, lacquers, and molded products.
vinylacetylene *Organic Chemistry.* $H_2C{=}CHC{\equiv}CH$, a colorless gas or liquid that boils at 5°C; a combustible dimer of acetylene used in the manufacture of neoprene, as an intermediate, and in organic synthesis.
vinyl alcohol *Organic Chemistry.* $CH_2{=}CHOH$, an unstable taritomer of acetaldehyde that is found only in ester form or as the polymer, polyvinyl alcohol. Also, ETHENOL.
vinylation *Chemistry.* the process of forming vinyl derivatives by adding acetylene to hydroxyl, amino, amido, carboxylic, or mercapto groups.
vinylbenzene *Organic Chemistry.* an alternative name for styrene, $C_6H_5CH{=}CH_2$, an oily, aromatic liquid used in making polystyrene plastics and rubbers.

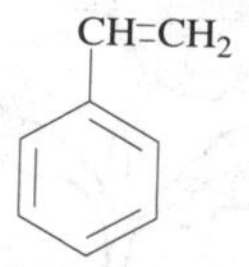

vinylbenzene

vinyl chloride *Organic Chemistry.* $CH_2{=}CHCl$, a gas with an ethereal odor; slightly soluble in water and soluble in alcohol and ether; boils at −13.9°C; a very important vinyl monomer that is extremely toxic and a carcinogen; used in polyvinyl chloride and copolymers, organic synthesis, and plastic adhesives. Also, CHLOROETHENE, CHLOROETHYLENE.
vinyl chloride resin *Organic Chemistry.* $(CH_2CHCl)_x$, a white powder derived from polymers of vinyl chloride; used to make chemical-resistant pipes when unplasticized, and bottles when plasticized.
vinyl ether *Organic Chemistry.* $CH_2{=}CHOCH{=}CH_2$, a colorless, light-sensitive liquid; slightly soluble in water and soluble in alcohol and ether; boils at 28°C; used as a copolymer for plastic products and medicinally as an anesthetic.
vinyl ether resin *Organic Chemistry.* any of a variety of vinyl ether polymers, such as polyvinyl ethyl ether or polyvinyl butyl ether.
vinylethylene see 1,3-BUTADIENE.
vinyl group *Organic Chemistry.* $CH_2{=}CH-$, a radical of ethylene; highly reactive, polymerizes easily; used as the base of many important plastics.
vinylidene chloride *Organic Chemistry.* $CH_2{=}CCl_2$, a colorless liquid; insoluble in water; readily polymerizes; boils at 32°C; used in copolymers with vinyl chloride or acrylonitrile to form various kinds of saran.
vinylidene resin *Organic Chemistry.* a polymer with a monomer structure of $(-H_2CCX_2-)$, with X being a chlorine, fluorine, or cyanide radical. Also, POLYVINYLIDENE RESIN.
vinylog *Organic Chemistry.* a compound that is related to another by addition or removal of a vinylene linkage, $(-CH{=}CH-)$; e.g., ethyl crotonate is a vinylog of both ethyl acetate and the next highest vinylog, ethyl sorbate.

vinyl polymerization *Organic Chemistry.* addition polymerization with the unsaturated monomer containing a CH_2=CH– group.

vinylpyridine *Organic Chemistry.* $C_5H_4NCH=CH_2$, a colorless liquid; soluble in water, dilute acids, alcohol, ketones, esters, and hydrocarbons; boils at 159°C; used in elastomer production and pharmaceuticals.

2-vinylpyridine *Organic Chemistry.* $C_5H_4NCH=CH_2$, a colorless liquid; soluble in water, dilute acids, and alcohol; an irritant to skin and eyes; boils at 79–82°C (29 torr); used in elastomer production and in pharmaceuticals.

1-vinyl-2-pyrrolidone *Organic Chemistry.* C_6H_9OH, a colorless liquid; boils at 148°C (100 torr) and melts at 13.5°C; an irritant and narcotic; used in polyvinylpyrrolidone and in organic synthesis.

***N*-vinyl-2-pyrrolidone** *Organic Chemistry.* C_6H_9OH, a colorless liquid that boils at 148°C (100 mm Hg); used in polyvinylpyrrolidone and in organic synthesis.

vinyl resin *Materials.* any of a group of thermoplastic, synthetic resins produced by polymerizing a compound that contains the vinyl radical, CH_2=CH–; used in adhesives, floor coverings, toys, and records.

vinylstyrene see DIVINYLBENZENE.

vinyltoluene see 3-METHYLSTYRENE.

vinyltrichlorosilane *Organic Chemistry.* $CH_2=CHSiCl_3$, a colorless-to-pale yellow liquid that is soluble in most organic solvents and that polymerizes easily; boils at 90.6°C; used in silicone and adhesives. Also, **trichlorovinylane.**

vinyon *Textiles.* an artificial fiber composed of a long-chain synthetic polymer that is at least 65% by weight of vinyl chloride units.

Violaceae *Botany.* a family of dicotyledonous herbs and shrubs of the order Violales, including the pansy and violet, characterized by alternate, simple or compound leaves, bisexual regular or zygomorphic flowers and seeds containing endosperm.

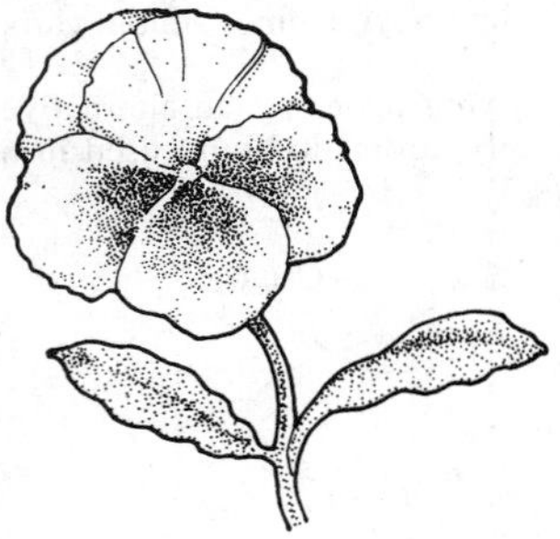

Violaceae (pansy)

Violales *Botany.* an order of dicotyledonous plants of the subclass Dilleniidae, characterized by a compound ovary and mostly parietal placentation.

violarite *Mineralogy.* $Fe^{+2}Ni_2^{+3}S_4$, a violet-gray cubic mineral of the linnaeite group, massive in habit and occurring as small nodules, having a specific gravity of 4.5 to 4.8 and a hardness of 4.5 to 5.5 on the Mohs scale; found with other sulfides of hydrothermal origin.

violet *Botany.* **1.** any low, stemless or leafy-stemmed plant of the genus *Viola*, characterized by purple, blue, or white flowers. **2.** the flower of any native, wild species of violet. *Optics.* one of the distinctive colors of this flower, reddish-blue; the color sensation that corresponds to a wavelength in the range of 390 to 455 nanometers.

violet layer *Astronomy.* a haze layer in the upper Martian atmosphere that renders the atmosphere relatively opaque at blue and violet wavelengths; it is subject to occasional "blue clearings" during which the effect lessens.

violle *Optics.* a unit of brightness that equals 20.17 candelas.

viologen display *Electronics.* an electrochromic display consisting of an electrolyte made up of an aqueous solution of dipositively charged organic salt, containing a colorless cation undergoing a one-electron reduction process producing a purple radical cation when negative potential is applied to the electron.

viomycin *Microbiology.* an antibiotic produced by the bacterium *Streptomyces puniceus*; due to its toxicity, it is used to treat tuberculosis only when other antibiotics are not effective; may cause neurological and kidney damage.

VIP vasoactive intestinal polypeptide.

viper *Vertebrate Zoology.* **1.** generally, any of several poisonous snakes that belong to the family Viperidae, characterized by long fangs that fold against the roof of the mouth when the jaws close. **2.** specifically, *Vipera berus*, a common European venomous snake with dark markings.

Viperidae *Vertebrate Zoology.* a family of fanged venomous snakes of the suborder Serpentes that are generally heavy-bodied with large heads, nocturnal, and terrestrial; found in Eurasia and Africa.

vira-A see VIDARABINE.

viral *Virology.* of, relating to, caused by, or characteristic of viruses or of a specific virus.

viral Eastern equine encephalitis see EASTERN EQUINE ENCEPHALOMYELITIS.

viral encephalomyelitides *Medicine.* a viral disease common to horses and contagious to humans, causing inflammation of the brain and nervous system, and resulting in convulsions, fever, and eventually coma.

viral gastroenteritis *Medicine.* an infectious viral disease of the stomach and intestines, with inflammation of the gastrointestinal mucous membranes. Also, **viral enteritis.**

viral hepatitis *Pathology.* a chronic virus of global proportions, predominantly affecting children and young adults; primarily transmitted through ingestion of contaminated food, but sometimes through blood transfusions; the disease is known to be passed, in the early contagious stages, through the stool of infected patients. Symptoms include fever, general malaise, nausea, abdominal and gastrointestinal discomfort, vomiting, and anorexia. Also, INFECTIOUS HEPATITIS, HEPATITIS A.

viral pneumonia *Medicine.* pneumonia in which the infecting agent is a virus.

viral shedding *Virology.* the release of an infectious virus from cells into media or from an organism into the environment, irrespective of cytopathology or signs and symptoms of disease.

viral-specific enzyme *Enzymology.* an enzyme produced in a host cell coded for by viral genes that have infected the host genome.

viral Western equine encephalitis see WESTERN EQUINE ENCEPHALITIS.

virazon *plural,* **virazones.** *Meteorology.* a strong westerly or southwesterly sea breeze of South America or the Iberian peninsula.

Virchow, Rudolf 1821–1902, German pathologist and anthropologist; the founder of cellular pathology.

viremia *Medicine.* the presence of a virus or viruses in the blood.

vireo *Vertebrate Zoology.* any of numerous New World birds of the family Vireonidae, characterized by a plump body with usually green plumage and a stout, hook-tipped bill with a bristled base.

vireo

Vireonidae *Vertebrate Zoology.* the vireos, a family of small, New World songbirds of the passerine suborder Oscines, characterized by green plumage and a short hooked bill.

virga *plural,* **virga.** *Meteorology.* a curved streak of water droplets or ice particles that falls from a cloud but evaporates before reaching the ground. Also, FALL STREAK, FALLSTREIFEN, PRECIPITATION TRAIL.

virgate *Botany.* shaped like a rod or wand.

virgate trophus *Invertebrate Zoology.* in rotifers, a piercing toothed trophus for sucking juices.

Virgilian *Geology.* a North American provincial series of the Uppermost Pennsylvanian period, occurring after the Missourian and before the Wolfcampian of the Permian period.

virgin *Zoology.* not fertilized. *Metallurgy.* a metal made directly from ore by smelting. *Food Technology.* describing the oil obtained from the first pressing of olives, without the application of heat.

Virgin *Astronomy.* the constellation Virgo.

virgin flow *Hydrology.* a stream flow that is not affected by any artificial influences, such as dams, reservoirs, or other constructions.
virgin medium *Computer Technology.* a storage medium on which data has never been written.
virgin neutron *Nucleonics.* a neutron, from any source, that has not yet undergone a collision and therefore retains its initial energy.
virgin tape see RAW TAPE.
virgin wool *Textiles.* new wool directly clipped from a sheep, or wool that has never been used in any kind of fabric manufacturing process.
Virgo *Astronomy.* the Maiden, a zodiacal constellation visible during northern spring between Leo and Libra; it contains the white 1st-magnitude star Spica.
Virgo A *Astronomy.* a radio source associated with the peculiar galaxy M 87, from which an optical jet, which is a strong source of synchrotron emission, extends.
Virgo cluster *Astronomy.* a cluster of galaxies in Virgo and Coma Berenices approximately 50 million light-years away that has about 2500 members.
Virgo X-1 *Astronomy.* an X-ray source associated with the same radio galaxy (M 87) as Virgo A.
virial coefficients *Thermodynamics.* the temperature-dependent coefficients $a, b, c, \ldots$, appearing in the most general form of the equation of state

$$pV = RT(1 + a/V + b/V^2 + c/V^3 + \cdots +),$$

where p is the pressure, V is the volume of one mole of the gas, R is the gas constant, and T is the absolute temperature.
virial of a system *Physics.* a quantity derived for a closed system of particles given by the expression $-(1/2)\ \Sigma\ (r_iF_i)$, where r_i is a vector from the origin to the ith particle and F_i is the force acting on the ith particle; the summation is averaged over a long time interval and, by the virial theorem, is the average kinetic energy of the system.
virial theorem *Quantum Mechanics.* a theorem stating that the average kinetic and potential energies of a system are related to the system's total energy and internuclear separation; the relation between the binding energy and kinetic energy of a system in equilibrium. *Physics.* see CLAUSIUS VIRIAL THEOREM.
virial-theorem mass *Astronomy.* the mass of a star cluster or cluster of galaxies that is derived from the mean motions.
viridans streptococci *Microbiology.* a group of pathogenic and saprophytic streptococci bacteria that do not have well-defined group antigens. Also, α-HEMOLYTIC STREPTOCOCCI.
Viridivelleraceae *Botany.* a monotypic family composed of a small moss of the order Dicranales, known only in Queensland, Australia, that lacks vegetative leaves, accomplishing photosynthesis by means of a persistent, filamentous protonema.
viridobufagin *Toxicology.* $C_{23}H_{34}O_5$, a toxic substance derived from the skin glands of the toad, *Bufo viridis,* that has a digitalislike effect on the heart.
virilism *Medicine.* **1.** the physical or behavioral characteristics associated with the male; masculinity. **2.** the development of secondary male sexual characterstics in a female. **3.** the premature development of masculine characteristics in a male. *Behavior.* the development in a female of behavioral patterns associated with males.
virilocal residence *Anthropology.* a residence practice in which a married couple comes to live in the husband's home.
virino *Microbiology.* a theoretical infectious entity composed of protein and a short nucleic acid sequence that does not encode any proteins.
virion *Virology.* a complete viral particle, composed of the nucleoid and the capsid.
Virmel engine *Mechanical Engineering.* an engine controlled by a gear and crank system in which the pistons stop and restart when a chamber reaches the spark plug.
virogene *Genetics.* a viral gene that is integrated in a chromosome of a eukaryotic host cell.
virogenetic *Virology.* of viral origin; caused by a virus.
virogenic *Virology.* virus-producing.
virogenic stroma *Virology.* in insect virus infections, a complex of viroplasm and chromatinlike material that forms the nuclear site of viral DNA synthesis.
viroid *Virology.* any of a class of infectious agents consisting of a small strand of RNA that is not associated with a protein, does not code for proteins, is not translated, and is apparently replicated by host cell enzymes; known to cause various plant diseases.

Virology

Viruses were discovered a century ago as infectious agents that were small enough to pass through filters that retained all bacteria. Virology is the study of how viruses replicate and cause diseases in their animal, plant, and bacterial hosts. Among the numerous diseases of humans caused or initiated by viruses are measles, mumps, chicken pox, the common cold, influenza, infectious mononucleosis, shingles, rabies, poliomyelitis, smallpox, hepatitis, cold sores, genital herpes, AIDS, and some cancers and immunological disorders. Viruses also cause a variety of diseases in plants and animals, including ones of economic importance to humans.

Since the 1950s, great advances have been made in understanding how viruses work and in controlling diseases caused by viruses. Nevertheless, much remains to be learned if we are to win the war against viruses such as influenza, herpes, leukemia, and immunodeficiency viruses. The development of antiviral agents for treating diseases such as AIDS, herpes sores, and the common cold is a fairly new, rapidly growing field. The present AIDS epidemic and our continuing discovery of new viruses shows pointedly how tentative our control over viruses really is.

Virology has and continues to contribute greatly to other fields such as the molecular biology of gene expression. Viruses and viral-based vectors are used extensively by the burgeoning biotechnology and pharmacological industries for the isolation, characterization, and manufacture of numerous economically important genes and their products. Insect viruses can serve as species-specific insecticides for the treatment of some crop diseases. Lastly, viral vectors are beginning to be used to genetically engineer plants and animals; in the near future, they may enable us to cure genetic diseases in people as well.

Janet E. Mertz
Professor of Oncology
University of Wisconsin Medical School

virology *Microbiology.* the branch of microbiology that deals with viruses and viral infections. Thus, **virologist.**
virolysis *Virology.* holes or lysis in the membrane of enveloped viruses, caused by interaction with antibodies.
viropexis *Virology.* the surrounding of virus particles by the cell plasma membrane or by particles bound to cell-surface receptors by a nonspecific phagocytic process.
viroplasm *Virology.* a type of inclusion body in which virus replication or assembly occurs. Also, FACTORY AREA.
virosis *Virology.* the condition resulting from a virus infection.
virosome *Virology.* a liposome that incorporates viral proteins at the surface.
virostatic *Virology.* of or relating to the ability to minimize or prevent viral replication.
Virtanen, Artturi 1895–1973, Finnish biochemist; studied agricultural chemistry; awarded the Nobel Prize for developing a method of fodder conservation.
virtual *Science.* being of a certain nature in effect or implications, though not literally of this nature.
virtual address *Computer Programming.* an address to a location in virtual storage, not a physical location; must be converted to a real address prior to access.
virtual cathode *Electronics.* between the electrodes of an electron tube, any area that contains an accumulation of electrons.
virtual center see INSTANTANEOUS CENTER.

virtual direct-access storage *Computer Technology.* a storage system in which data are retrieved from mass storage (via batch processing) and transferred onto a conventional disk drive prior to use, then automatically re-stored at completion.
virtual displacement *Mechanics.* an infinitesimal displacement of a system, actual or theoretical, that is consistent with all of the system's constraints.
virtual entropy *Thermodynamics.* the entropy of a system neglecting the contribution to the entropy by nuclear spins; this contribution remains constant in almost all chemical processes.
virtual gravity *Meteorology.* the force of gravity on a parcel of air.
virtual height *Geophysics.* the height of a layer of the atmosphere, measured by the time it takes for a pulse traveling at the speed of light to reach the layer and return to earth.
virtual image *Optics.* an image that forms at a point from which divergent light rays appear to emanate, but from which, in fact, they do not emanate.
virtual laboratory *Computer Science.* an arrangement in which laboratory instruments are connected to computers that are connected to networks; requests for experiments and experimental results can be communicated via the networks, allowing geographically separated teams of scientists to collaborate on a single problem. Also, COLLABORATORY.
virtual leak *Engineering.* the portion of a leak in a vacuum system that is caused by the slow desorptive release of gas at an unpredictable rate.
virtual level *Nuclear Physics.* the state of energy produced in a compound nucleus in a virtual state, characterized by having a lifetime that greatly exceeds the time it takes a nucleon with the same level of energy to penetrate it.
virtual machine *Computer Technology.* **1.** a computer system in which hardware and software are designed to make use of virtual memory techniques. **2.** the combined system resources in a multiprogramming environment that are dedicated to a single program. **3.** a machine that is simulated by hardware and software.
virtual memory *Computer Programming.* the memory in a computer system that includes both internal memory and external storage, yet can be treated by the programmer as a single memory, because the computer translates the virtual address to a physical location and handles movement of data between external and internal memory. Also, VIRTUAL STORAGE.
virtual meridian *Navigation.* in a gyrocompass affected by a speed-course-latitude error, the meridian in which the spin axis settles.
virtual orbital *Physical Chemistry.* a region around the nucleus in which no electrons are found when the energy level of an atom is at ground state.
virtual pressure *Meteorology.* the atmospheric pressure of a parcel of moist air that has the same density as a parcel of dry air at the same temperature.
virtual process *Quantum Mechanics.* an intermediate process involving the emission and reabsorption of a virtual quantum, during which energy conservation is violated provided that the energy imbalance, dE, and its duration, dt, satisfy the Heisenberg uncertainty principle, and the product of dE and dt is less than Planck's constant.
virtual quantum *Quantum Mechanics.* a photon or other particle that exists only temporarily (for a time $\{\hbar\}/mc^2$) while being exchanged between other particles as they are emitted and absorbed as part of a virtual process, and that cannot exist independently of the process. Also, **virtual particle.**
virtual reality *Artificial Intelligence.* a computer simulation of a system, either real or metaphorical, that allows a user to perform operations on the simulated system and shows the effects in real time; e.g., a system for architects might allow the user to "walk" through a proposed building design, displaying how the building would look to someone actually inside it. Also, ARTIFICIAL REALITY. *Computer Technology.* any computerized simulation that is regarded as very faithful to reality, as in a video game, interactive learning program, and so on.
virtual source *Acoustics.* **1.** an imaginary image positioned where a sound source could be if there were no reflective or absorptive barrier. **2.** an imaginary source that creates ambiguity in determining the real source.
virtual storage see VIRTUAL MEMORY.
virtual temperature *Meteorology.* the temperature of a parcel of dry air that has the same density and pressure as the surrounding air of a moist-air system; calculated from actual temperature by the formula $Tv = (1 + 0.61q)T$, where T is actual temperature and q is relative humidity.
virtual unemployment *Robotics.* the jobs in an industry that have been eliminated through the introduction of robots and other automated equipment.
virtual work *Mechanics.* work done by a force through a virtual displacement.
virucidal *Virology.* able to cause the inactivation of a virus.
virulence *Virology.* the disease-producing ability of a microorganism. *Microbiology.* the degree of this ability, generally indicated by the severity of the infection in the host and the ability of the virus to invade the host's tissues.
virulent *Microbiology.* **1.** of, relating to, or characterized by virulence. **2.** exceedingly pathogenic, noxious, or deleterious.
virulent phage *Virology.* a phage that causes lysis of its host cell and cannot induce a lysogenic state.
virus *Virology.* a noncellular microbial entity that consists of a core of RNA or DNA enclosed in an outer coat of protein and, in some forms, a protective outer membrane, and that can live and reproduce only in susceptible host cells. Viruses infect bacteria, plants, and animals, and more than 200 types have been identified as capable of causing diseases in humans, such as influenza, the common cold, measles, smallpox, and herpes. *Computer Programming.* an illicit program that instructs a computer to change or destroy information, to display a satirical message, or to otherwise disrupt processing, including making copies of itself; often hidden in apparently normal software and becoming active at a given time or after a specific sequence of events; so-called because of its ability to "infect" other programs. (From the Latin word for "poison.")
virus assembly *Virology.* the merging of constituent parts of a virus particle and the resulting formation of that particle; morphogenesis.
viruslike particle *Virology.* a structure that resembles a virus particle, but has not been demonstrated to be pathogenic.
virusoid *Genetics.* a small species of ssRNA found within the virions of certain plant viruses; composed of approximately 300–400 nucleotides.
virus particle *Virology.* the morphological structure of a virus, usually consisting of a genome engulfed by a protein capsid and sometimes a lipid envelope.
virus receptor *Virology.* the configuration of or structure on a host cell membrane that acts as an attachment site for a surface protein of the virus.
virus replication *Virology.* the formation of progeny virus from input virus induced by the replication of the viral genomic nucleic acid and merging of progeny virus particles.
visa point *Robotics.* a point, midway along a line connecting the starting and stopping positions of a robot's tool tip, through which the tip passes without stopping.
vis breaking see VISCOSITY BREAKING.
Viscaceae *Botany.* a cosmopolitan family of dicotyledonous brittle evergreen shrublets belonging to the order Santalales that are hemiparasitic on tree branches and sometimes stimulate the host to produce witches' brooms; includes the traditional Christmas mistletoe.
viscacha *Vertebrate Zoology.* any of several burrowing South American rodents of the genera *Lagostomus* and *Lagidium* in the family Chinchillidae.
viscera [vis´ə rə] *Anatomy.* the soft internal organs of the body, including the lungs, heart, digestive, excretory, reproductive, and circulatory organs; the plural of *viscus.*
visceral [vis´ə rəl] *Anatomy.* relating to the viscera, especially those in the abdomen. *Behavior.* relating to instinctive or appetitive drives.
visceral arch see BRANCHIAL ARCH.
visceralgia *Medicine.* pain in the viscera or in any bodily organ.
visceral leishmaniasis *Medicine.* a serious chronic disease that is fatal if untreated, caused by infection with the protozoan parasite *Leishmania donovani*, transmitted by the bite of a sand fly, and characterized by fever, vomiting, anemia, inflammation of the liver and spleen, dropsy, and wasting away; it is prevalent in the Mediterranean, parts of Asia, Africa, South and Central America, and Brazil. Also, KALA-AZAR.
visceral mass *Invertebrate Zoology.* **1.** in mollusks, a dorsal enlargement of the body containing the visceral organs, leaving the ventral part for use in locomotion. Also, **visceral hump. 2.** in some parasitic crustaceans, a mass of tissue containing the degenerate rudiments of internal organs.
visceral pericardium see ENDOCARDIUM.
visceral peritoneum *Anatomy.* the portion of the thin lining of the abdominal cavity that covers the abdominal organs.
visceral pleura *Anatomy.* the thin layer of pleura that surrounds and coats the lungs.

visceratonia *Psychology.* a personality type associated with the endomorphic body build, characterized by love of food, enjoyment of physical relaxation and comfort, sociability, and affection.

visceroptosis *Medicine.* a prolapse of the viscera, the major internal organs, especially of the abdomen.

viscid [vis´id] *Botany.* of a plant surface, covered with a sticky or glutinous substance.

viscoelastic *Mechanics.* exhibiting or characterized by viscoelasticity.

viscoelasticity *Mechanics.* a condition of a liquid or solid that exhibits viscosity but also memory of past deformation, with the ability to store energy elastically and to dissipate energy due to viscosity of the medium; exhibiting features of hysteresis, relaxation, and creep. *Materials Science.* the typical manifestation of this quality in a polymer; the shear viscosity is expressed by the relaxation time, which is the ratio of the viscous to the elastic characteristics in the determinant property.

viscometer *Engineering.* an instrument that measures the viscosity of a fluid. Also, **viscosimeter.**

viscometric analysis *Fluid Mechanics.* the determination of flow properties by viscometry.

viscometry *Engineering.* **1.** the study of the behavior of viscous fluids. **2.** the technology of measuring viscosities of fluids. Also, **viscosimetry.**

viscose [vis´kōs] *Fluid Mechanics.* see VISCOUS. *Chemical Engineering.* a form of cellulose acetate used in the production of regenerated cellulose fibers *Textiles.* see VISCOSE RAYON.

viscose process *Chemical Engineering.* a process for making regenerated rayon by converting cellulose to the soluble xanthate, which can then be spun into fibers and reconverted to cellulose by an acid treatment.

viscose rayon *Textiles.* the most common of three types of rayon, made by chemically treating cellulose sheets, shredding and aging the sheets, forming them into a viscose solution, and extruding the solution through a spinneret to form fibers.

viscosity [vis käs´i tē] *Fluid Mechanics.* the internal friction of a fluid; the resistance to flow exhibited by a liquid or gas subjected to deformation; viscosity is expressed by a coefficient, in units of poise (1 poise = 0.1 Pa-sec).

viscosity breaking *Petroleum Engineering.* a method in petroleum refining for lowering or breaking the viscosity of high-viscosity residuum through thermal cracking of molecules at rather low temperatures. Also, VIS BREAKING.

viscosity coefficient *Fluid Mechanics.* the coefficient m in the equation relating shear stress t to velocity gradient du/dy: $t = m(du/dy)$; qualitatively, it is the force per unit area required to maintain unit velocity gradient.

viscosity conversion table *Chemical Engineering.* a chart or table that gives the conversion, at the same temperature, of kinematic viscosity in centistokes to Saybolt viscosity in seconds.

viscosity curve *Fluid Mechanics.* a plot of viscosity against temperature.

viscosity-gravity constant *Petroleum Engineering.* an index of the chemical components of crude oils, given by an empirical relationship between Saybolt Universal viscosity and specific gravity; the constant is high for naphthenic crude oils and low for paraffinic crude oils.

viscosity index *Petroleum Engineering.* an arbitrary scale using an empirical formula to indicate the effect of temperature changes on the viscosity changes in lubricating oils; shown by comparing Pennsylvania oil that measures 100 on the scale to an asphaltic oil that measures zero.

viscosity-temperature chart *Petroleum Engineering.* a chart used to determine the kinematic or Saybolt viscosity of a petroleum oil at any temperature, within a limited scope, if viscosities at two temperatures are known.

viscous [vis´kəs] *Fluid Mechanics.* characterized by a high degree of friction between component molecules as they slide by each other.

viscous damping *Mechanical Engineering.* the dissipation of energy that occurs when a particle in a vibrating system is resisted by a force whose magnitude is a constant, independent of displacement and velocity, and whose direction is opposite to the direction of the velocity of the particle.

viscous dissipation function *Fluid Mechanics.* a quadratic function of spatial derivatives of velocity components that gives the rate at which mechanical energy is converted to irrecoverable thermal energy through the action of viscous forces. Also, DISSIPATION FUNCTION.

viscous drag *Fluid Mechanics.* the rearward force exerted on an object due to fluid molecules adhering to the surface when the object passes through a viscous fluid.

viscous filler *Mechanical Engineering.* a packaging machine that fills cartons with a viscous liquid.

viscous flow *Fluid Mechanics.* **1.** a flow in which adjacent parallel fluid layers slide against each other due to viscosity. **2.** a duct flow in which the mean free path of the fluid particles is small compared to the smallest transverse section of the duct.

viscous flow *Materials Science.* the inelastic rate of deformation of liquids, e.g., glassy material generally at elevated temperatures.

viscous fluid *Fluid Mechanics.* a fluid whose viscosity is large enough that the viscous forces are not negligible in the total force field.

viscous force *Fluid Mechanics.* the tangential frictional forces or internal shearing forces of a viscous fluid.

viscous impingement filter *Engineering.* a filter consisting of a loosely arranged medium coated with an adhesive that causes the airstream to change directions frequently as it passes through the filter.

viscous lubrication see COMPLETE LUBRICATION.

viscous remanent magnetization *Geophysics.* the remanent magnetization manifested by certain rock particles that, due to either size or composition, were unable to retain the remanent magnetization acquired at the time of formation and took on the new orientation of the earth's magnetic field after a reversal of that field.

viscus *plural,* **viscera.** *Anatomy.* any large interior organ situated in any of the three great cavities of the body, especially in the abdomen.

vise *Mechanical Devices.* any of various instruments used to hold and grip work, consisting of two serrated jaws, one or both of which can be moved by a screw, lever, or cam, the entire mechanism capable of being clamped or bolted onto a work table.

Viseán *Geology.* a European geologic stage of the Lower Carboniferous period, occurring after the Tournaisian and before the Namurian. Also, **Visian.**

vise clamps *Mechanical Devices.* angle strips of brass, lead, or copper that are used as false jaws in a vise over the existing jaws to protect a secured workpiece. Also, **vise claws.**

vise grips *Mechanical Devices.* a plierslike tool operated by hand with serrated jaws and a lock mechanism to firmly hold a part in its jaws; the lock is released by pulling one jaw away from the other.

visibility *Optics.* the clarity with which an object can be seen. *Meteorology.* as a measure of weather conditions, the greatest distance in a given direction at which a standard object can be seen and identified with the unaided eye.

visibility factor *Electronics.* the ratio of the minimum input-signal power detectable by ideal instruments connected to the output of a receiver to that which is detectable by a human operator through a display connected to the same receiver. Also, DISPLAY LOSS.

visibility function *Optics.* see LUMINOSITY FUNCTION.

visibility meter *Optics.* a photometer that artificially diminishes the visibility of an object to the point where it can barely be seen and then gauges the reduction on a standard scale. *Engineering.* any instrument, such as a transmissometer, used to measure visual range in the atmosphere or the physical characteristics of the atmosphere that affect visual range.

visible *Optics.* capable of being seen; perceptible to the sight.

visible absorption spectrophotometry *Spectroscopy.* the photometric measurement of the wavelengths of visible radiation absorbed by a sample and which correspond to electron transitions from the ground state to an excited state.

visible bearing *Navigation.* a bearing determined by visual observation with an azimuth circle or pelorus.

visible horizon *Astronomy.* a projection of the local horizon outward to intersect with the celestial sphere. Also, APPARENT HORIZON.

visible radiation *Optics.* see LIGHT.

visible spectrophotometry *Spectroscopy.* the photometric measurement of the wavelengths of visible radiation absorbed by a sample that has been irradiated by a tungsten lamp having an emission spectrum of 380–780 nanometers.

visible spectrum *Spectroscopy.* a spectrum having wavelengths in the visible region of the electromagnetic spectrum.

vision *Physiology.* **1.** the process by which objects outside the body are perceived; the act of seeing. **2.** see VISUAL ACUITY. **3.** a subjective sensation of vision that is not elicited by actual visual stimuli; an apparition. *Robotics.* the ability of a robot to detect an object and react to it because of feedback from a video camera or other vision sensor.

vision optical system *Robotics.* a system of visual feedback based on various devices, such as video cameras, photo cells, or other apparatus, allowing a robot to recognize objects or measure their characteristics.

vision sensor *Robotics.* a device, such as a camera, that uses reflected light to find and identify an object and sends the information, in the form of electronic signals, to a controller.
vision slit *Ordnance.* a narrow aperture in a tank or armored vehicle to allow the crew to see outside the vehicle.
visna virus *Virology.* the prototype of the subfamily Lentivirinae, which causes chronic pnuemonitis and encephalitis in sheep.
visor tin *Mineralogy.* a variety of cassiterite occurring in twinned crystals with a characteristic notch, resembling the visor of an iron helmet.
visual *Physiology.* relating to vision or sight.
visual achromatism *Optics.* a system that gauges the removal of chromatic aberrations or chromatic differences of magnification that might be detected by the human eye. Also, OPTICAL ACHROMATISM.
visual acuity *Physiology.* the crispness or sharpness of a visual image.
visual aid to navigation *Navigation.* an object on which bearings may be taken visually, either with the naked eye or with optical instruments; its position is indicated on government-approved navigational charts.
visual angle *Optics.* the angle that subtends the nodal point of the viewer's eye and the object.
visual anthropology *Anthropology.* a branch of anthropology that studies the visual elements of a culture in the form of artwork and utititarian objects; involving issues such as why a culture chooses one form over another or one material type over another.
visual apraxia see OPTIC APRAXIA.
visual-aural range *Navigation.* an earlier VHF radio range system that displayed reciprocal lines of position as aural code signals; it has been replaced by VOR.
visual binary *Astronomy.* a binary star system that is resolvable into its two components by telescopic observation. Also, **visual double.**
visual bombing *Military Science.* bombing in which a visible aiming point is used for sighting.Thus, **visual bombsight.**
visual colorimetry *Analytical Chemistry.* the comparison of the color of an unknown solution to color-tinted disks for the purpose of identification.
visual comparator see OPTICAL COMPARATOR.
visual fix *Navigation.* a fix based on visual observation of fixed objects. Similarly, **visual line of position.**
visual flight *Aviation.* flight in which the pilot uses direct vision, rather than instruments, to determine attitude and position. Instruments are used in visual flight to determine altitude, airspeed, compass direction, and so on. Also, VFR FLIGHT.
Visual Flight Rules or **visual flight rules** *Aviation.* a set of regulations established by the U.S. Civil Aeronautics Board to govern visual flight, including minimum visibility standards (relating to cloud separation, haze, precipitation, and so on), that are more stringent than those for instrument flight.
visual Herschel effect see HERSCHEL EFFECT.
visualization *Psychology.* a behavioral technique for improving performance, in which the individual is encouraged to create a mental picture of the successful execution of the task. *Medicine.* the process of viewing, or of achieving a complete visual representation of, an object, as in roentgenography. *Computer Programming.* any presentation of data using visual images.
visually coupled display see HELMET-MOUNTED DISPLAY.
visual magnitude *Astronomy.* the magnitude of a celestial body in a wavelength range approximating the human eye.
visual meteor *Astronomy.* a meteor bright enough to be visible to the naked eye.
visual mine firing indicator *Ordnance.* a device used with exercise mines to indicate that the mine would have fired had it been poised.
visual photometer *Optics.* a photometer that allows an observer to compare the luminance of two surfaces.
visual photometry *Optics.* **1.** the use of the human eye to measure intensity of light, as distinguished from photoelectric photometry. **2.** the use of a visual photometer to compare the luminance of two surfaces.
visual pigment *Biochemistry.* any of the photosensitive pigments of vertebrate or invertebrate photoreceptors that contain an opsin and a vitamin A aldehyde.
visual projection area *Physiology.* the region of the cortex of the occipital lobe that is functionally involved in vision.
visual purple see RHODOPSIN.
visual radio range *Navigation.* any radio range designed to provide course lines that may be flown by visual reference to a course-deviation indicator.
visual range see DAYTIME VISUAL RANGE.
visual righting reflex *Physiology.* a repositioning of the neck and limbs in order to realign the head in response to a visual image of the surroundings.
visual scanner *Electronics.* a device that scans written or printed data and produces the corresponding signals.
visual seizure *Medicine.* an epileptic seizure, occurring alone or as an aura at the beginning of a seizure, in which the subject experiences flashes of light, or red or other colors, or other visual hallucinations due to neural discharges from the visual projection or visuopsychic areas of the nervous system.
visual servoing *Robotics.* the use of feedback from a solid-state camera on the end of a robot's end effector.
visual telephony *Telecommunications.* television transmission over telephone lines.
visual transmitter *Electronics.* the radio equipment used expressly for the transmission of picture signals.
visual violet see IODOPSIN.
visual yellow *Biochemistry.* a substance formed in the retina when rhodopsin is exposed to light.
visuognosis *Physiology.* the recognition and interpretation of visual impressions.
visuomotor (behavioral) rehearsal *Psychology.* see VISUALIZATION.
visuopsychic area *Anatomy.* the area of the cerebral cortex concerned with visuognosis.
visuosensory *Physiology.* relating to the perception of stimuli giving rise to visual impressions.
Vitaceae *Botany.* a family of dicotyledonous woody vines of the order Rhamnales, characterized by climbing tendrils, small bisexual or unisexual flowers, and a fruit that is a berry.

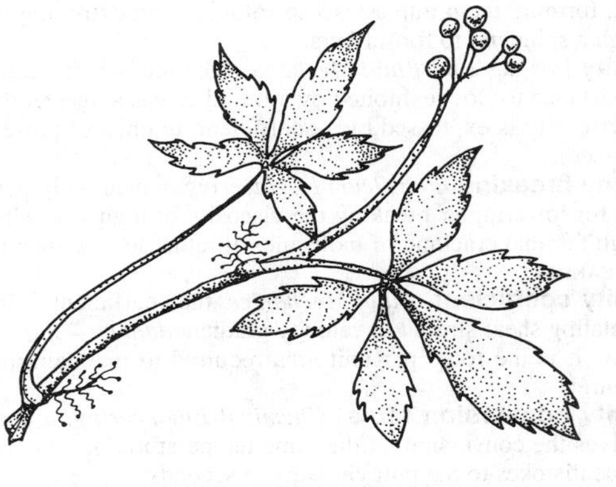

Vitaceae (Virginia creeper)

vital *Biology.* necessary to life or relating to life.
vital area *Military Science.* a designated area or installation to be defended by air defense units.
vital capacity *Physiology.* the amount of air that can be forced out of the lungs after a maximum inhalation.
Vitallium *Materials.* the trade name for a ductile metal and cobalt alloy, usually containing chromium, molybdenum, manganese, and iron; used in prostheses.
vital point *Neurology.* the area of the medullary respiratory center that, if damaged, results in respiratory arrest and death.
vital staining *Microbiology.* a method of coloring viable cells with certain biological dyes, such as methylene blue.
vital statistics *Statistics.* statistics regarding the principal events in the lives of individuals, such as birth, death, and marriage.
vitamin *Biochemistry.* any of various organic substances that occur in many foods in small amounts and are also produced synthetically, and that are necessary in trace amounts for the normal physiologic and metabolic functioning of the body. (From the Latin word for "life" plus *amine*; they were originally believed to be amino acids.)
vitamin A *Biochemistry.* $C_{20}H_{30}O$, a fat-soluble, water-insoluble compound, structurally related to carotenes and essential for normal growth and metabolism, especially vision; found in liver oils and carotenoids.
vitamin A aldehyde see RETINAL.
vitamin B_1 see THIAMINE.
vitamin B_2 see RIBOFLAVIN.
vitamin B_3 see PANTOTHENIC ACID.

vitamin B_6 *Biochemistry.* pyridoxine and related water-soluble substances, including pyridoxal and pyridoxamine; found in most foods, especially meats, liver, vegetables, whole-grain cereals, and egg yolk; function as coenzymes in the metabolism of amino acids.

vitamin B_6 hydrochloride see PYRIDOXINE HYDROCHLORIDE.

vitamin B_{12} see COENZYME B_{12}.

vitamin B complex *Biochemistry.* the broad group of water-soluble vitamins, including thiamine, riboflavin, niacin, pyridoxine, pantothenic acid, inositol, biotin, folic acid, and vitamin B_{12}, that occur in foods such as cereal germ, liver, and yeast.

vitamin C *Biochemistry.* $C_6H_8O_6$, a water-soluble vitamin that is found in numerous frutis and vegetables and is necessary for human growth; its absence leads to scurvy and poor wound repair. Also, ASCORBIC ACID.

vitamin C

vitamin D *Biochemistry.* steroids with cholecalciferol biological activity; either of two fat-soluble sterol-like compounds with antirachitic properties; found in egg yolk, fatty fish, and enriched milk; not required in the diet if sunlight exposure is sufficient.

vitamin D_2 see ERGOCALCIFEROL.

vitamin D_3 see CHOLECALCIFEROL.

vitamin D milk see ENRICHED MILK.

vitamin D-resistant rickets *Pathology.* a hereditary human disease having symptoms resembling those of rickets, but actually involving defective transport of phosphorus by the kidneys rather than deficiency of vitamin D; thus it does not respond to the prescribed treatment for rickets, i.e., megadoses of vitamin D, sunlight, and a healthy diet. It is characterized by two different anomalies: in **hypophosphatemic vitamin D-resistant rickets,** hypophosphatamia is the main element, while in **hypocalcemic vitamin D-resistant rickets,** phosphate levels in the blood are not of concern, but the low levels of calcium concentration are. Also, PSEUDODEFICIENCY RICKETS.

vitamin E *Biochemistry.* any of eight related fat-soluble compounds, found in vegetable oils, that act as antioxidants; necessary for normal reproduction, muscle development, and resistance to red blood cell hemolysis.

vitamin G see RIBOFLAVIN.

vitamin H see BIOTIN.

vitaminization *Nutrition.* the feeding of vitamins to make an individual vigorous or to purportedly prevent or cure disease.

vitamin K *Biochemistry.* any of a group of three fat-soluble compounds required for the synthesis of blood-clotting factors, such as prothrombin.

vitamin K_1 see PHYTONADIONE.

vitamin P see BIOFLAVONOID.

vitamin requirements *Nutrition.* organic substances needed in very small quantities by the body for normal growth and maintenance of life, and which play a catalytic and regulatory role in the body's metabolism.

vitellarium *Invertebrate Zoology.* in flukes and tapeworms, an accessory genial gland that secretes the yolk and shell for the fertilized egg. Also, **vitelline gland.**

vitellin *Biochemistry.* a phosphoprotein found in the yolk of eggs.

vitelline *Developmental Biology.* relating to or resembling the yolk of an egg or ovum. Also, OMPHALOMESENTERIC.

vitelline artery *Developmental Biology.* an artery in vertebrate embryos that passes from the yolk sac to the primitive aorta. Also, **omphalomesenteric artery.**

vitelline duct *Developmental Biology.* the narrow tube connecting the yolk sac with the midgut of the mammalian embryo. Also, **umbilical duct, omphalomesenteric duct, yolk stalk.**

vitelline membrane *Developmental Biology.* a protective coat, usually composed of protein fibers, surrounding the egg of many invertebrates. Also, **vitelline layer.**

vitelline vein *Developmental Biology.* any of the veins that return blood from the yolk sac to the primitive heart of the early embryo. Also, **omphalomesenteric vein.**

vitellogenesis *Developmental Biology.* the production and development of an egg yolk, especially a hen's egg yolk.

vitellogenin *Biochemistry.* a protein from birds and amphibia that is the precursor of the proteins phosvitin and lipovitellin when stimulated by female sex hormones.

vitiligo *Medicine.* a benign, usually progressive, chronic skin disorder of unknown origin, characterized by irregular patches of skin without pigment that may be surrounded by a hyperpigmented border.

Viton *Materials Science.* an elastomer that is a copolymer of vinylidene fluoride and hexafluoropropylene; used in high-temperature gaskets, the material is resistant to oil, steam, solvents, and water.

vitrain *Geology.* a coal lithotype characterized by a brilliant, glassy appearance, jet-black color, cubic cleavage, and conchoidal fracture; a component of bonded coal. Also, PURE COAL.

Vitreoscilla *Bacteriology.* a genus of gliding bacteria belonging to the family Beggiatoaceae of the order Cytophagales, consisting of colorless filaments that metabolize organic material.

Vitreoscillaceae *Bacteriology.* formerly classified as a family of gliding bacteria of the order Beggiatoales, now assigned to the families Beggiatoaceae and Cytophagaceae.

vitreous *Materials Science.* relating to, characteristic of, or resembling glass. *Anatomy.* glasslike or hyaline; often used alone to denote the vitreous body.

vitreous body *Anatomy.* the transparent gel that occupies the inner portion of the eyeball between the lens and the retina. Also, **vitreous humor, vitreum.**

vitreous chamber *Anatomy.* the large space between the lens of the eye and the retina, which is filled with a clear, gelatinous material.

vitreous enamel *Materials Science.* a glass coating that is applied to metal by spreading a glass powder over the metal surface and then fusing it to the metal at temperatures ranging from 600 to 1000°C; used for cloisonné jewelry and art objects; sinks, bathtubs, and home appliances; and industrial storage tanks, silos, reaction vessels, and smokestacks. Also, PORCELAIN ENAMEL.

vitreous luster *Optics.* a luster that resembles glass.

vitreous silica *Materials.* a transparent glass consisting mainly of silica that is chemically stable and heat-resistant.

vitreous state *Solid-State Physics.* an amorphous state of a solid in which there is no regular crystalline structure, but the crystalline state may eventually set in after a long period of time.

vitric *Geology.* describing any glassy, pyroclastic material containing at least 75% glass.

vitric tuff *Geology.* a tuff consisting primarily of fragments of volcanic glass.

vitrification *Geology.* a process in which a glassy or noncrystalline material is formed by fusion under conditions of intense heat. *Materials Science.* the progressive fusion of a material during the firing process; as it proceeds, glassy bonding increases and the porosity of the fired product decreases.

vitrified wheel *Mechanical Devices.* a grinding wheel whose abrasive surface is bound by glass or porcelain.

vitrinite *Geology.* an oxygen-rich maceral group in coal derived from decayed wood or bark, and associated with peat formation.

vitrinoid *Geology.* a type of vitrinite found in bituminous caking coal having a reflectance of 0.5–2%.

vitrophyre *Petrology.* any porphyritic igneous rock having a pronounced glassy groundmass. Also, GLASS PORPHYRY.

vittate *Botany.* having lengthwise ridges or stripes.

viuga *Meteorology.* a cold storm that blows across the Russian steppes from the north or northeast.

vivax malaria *Medicine.* a form of malaria caused by the parasite *Plasmodium vivax* and characterized by febrile paroxysms varying from daily to three days apart, but typically occurring every other day. Although rarely fatal, it is the most common form of malaria, the most difficult to cure, and the one in which relapses are most likely to occur. Also, BENIGN TERTIAN MALARIA.

Viverridae *Vertebrate Zoology.* the mongooses, civets, and genets, a family of carnivorous mammals of the suborder Feliforma, characterized by a long slender body with short legs, a long furry tail, and well-developed perianal scent glands; found in Asia and Australia.

vivi- a combining form meaning "life" or "alive."

vivianite *Mineralogy.* $Fe_3^{+2}(PO_4)_2 \cdot 8H_2O$, a blue monoclinic mineral of the vivianite group, occurring in clusters of prismatic to tabular crystals, as fibrous masses, and in earthy crusts, having a specific gravity of 2.68 and a hardness of 1.5 to 2 on the Mohs scale; found as a secondary mineral in some metallic veins and pegmatites, and as concretions in clays. Also, BLUE IRON EARTH, BLUE OCHER.

Viviparidae *Invertebrate Zoology.* a family of freshwater snails, gill-breathing gastropod mollusks in the suborder Pectinibranchia.

viviparous *Physiology.* bearing live young; having offspring that develop within the body and are born alive, rather than producing an egg that develops outside the body; e.g., humans and most other mammals.

vixen file *Mechanical Devices.* a flat file having curved cutting edges, used to file soft metals.

Vizing's theorem *Mathematics.* a theorem stating that if G is a simple graph, then the edge-chromatic number of G is one of Δ or $\Delta + 1$, where Δ is the maximum degree in G.

V jewels *Mechanical Devices.* a set of jewel bearings used in sensitive electronic measuring equipment, located at the apex of a conical recess of a conical pivot.

V-joint *Mechanical Devices.* an angular, hollow, mortar joint used in masonry work.

Vlasov equation *Physics.* a differential equation that applies to a hot plasma where collisions are insignificant and the primary source of particle interaction is their collective electromagnetic interaction.

VLDL very-low-density lipoprotein.

VLF or **vlf** very low frequency.

VMC visual meteorological conditions.

V.M.D. Doctor of Veterinary Medicine. (From Latin *Veterinariae Medicinae Doctor.*)

V-notch weir see TRIANGULAR-NOTCH WEIR.

v-number see ABBÉ NUMBER.

vocal cords *Anatomy.* either of two pairs of folds of mucous membrane in the larynx that are involved in speech production. The **true vocal cord,** or **vocal fold,** covers the vocal muscle and forms the inferior boundary of the ventricle. The **false vocal cord** separates the ventricle from the vestibule. The stretch and shape of the vocal cords, controlled by the vocal muscle, determines the pitch of the voice.

vocal mimicry see MOCKING.

vocal muscle *Anatomy.* the muscle in the larynx that shortens and stretches the vocal cords.

vocal sac *Vertebrate Zoology.* either of a pair of distensible pockets of skin in the throat region of many male frogs and toads that resonates the voice.

Vochysiaceae *Botany.* a family of dicotyledonous mostly woody plants of the order Polygalales, characterized by simple or opposite whorled leaves, a single fertile stamen, and bisexual flowers.

vocoder see VOICE CODER.

VODAS *Electronics.* a system that prevents the overall voice-frequency singing of a two-way telephone circuit by blocking one direction of transmission at all times. (An acronym for voice-operated device anti-sing.)

VODER *Electronics.* in telephony, an electronic device that artificially produces voice sounds, using electron tubes connected to electrical filters controlled through a keyboard. (An acronym for voice-operated demonstrator.)

VOGAD *Electronics.* in telephony, a device used to give a virtually constant volume output for a wide range of inputs. (An acronym for voice-operated gain-adjusting device.)

Vogel, Hermann Carl 1841–1907, German astronomer; found Doppler shifts on the star Algol; modified Secchi's spectral classification.

Vogel-Fulchen relation *Materials Science.* an equation used in thermal analytical investigation of polymeric materials; it allows for the kinetic treatment of differential scanning calorimetry data.

vogesite *Petrology.* a syenite lamprophyre consisting of hornblende phenocrysts embedded in a groundmass of orthoclase and hornblende.

Voges-Proskauer test *Microbiology.* any of four dye tests using indole methyl red Voges-Proskauer citrate (IMVC) to distinguish between the colon group of bacteria and the aerogenes group of bacteria; the identifying characteristic is the bacteria's ability to produce the chemicals acetoin and diacetyl.

voglite *Mineralogy.* a very rare, strongly radioactive, green, monoclinic mineral with the approximate formula $Ca_2Cu^{+2}(UO_2)(CO_3)_4 \cdot 6H_2O$, occurring in secondary scaly coatings on sandstone and as an alteration product of uraninite, having a specific gravity of 2.8 to 3.06 and an undetermined hardness.

Vogt-Russell theorem *Astrophysics.* a theorem stating that if the values of temperature, density, and chemical composition govern the pressure, opacity, and rate of energy generation in a star, then mass and chemical composition are sufficient to describe the structure of the star.

voice *Physiology.* the sounds produced by a human or by an animal, especially the type of sounds characteristic of a given person.

voice-activated *Electronics.* of a device, activated or operating in response to a voice or other sound signals. Also, VOICE-OPERATED.

voice call sign *Telecommunications.* a call sign used almost exclusively for voice communications.

voice channel *Telecommunications.* a communication channel with sufficient bandwidth to carry understandable voice frequencies.

voice coder *Electronics.* a device that converts speech input into a digital form that can be enciphered for secure transmission. At the receiver, the transmitted digital signals are deciphered and converted back to speech. Also, VOCODER.

voice coil *Acoustical Engineering.* a speaker coil, attached to the diaphragm of a loudspeaker, that moves freely in the air gap of speaker pole pieces.

voice/data system *Telecommunications.* a communications system that transmits both voice and digital data information.

voice digitization *Electronics.* the conversion of analog voice signals into digital signals.

voice frequency *Telecommunications.* in telephony, an audio frequency in the band that allows transmission of clearly intelligible speech, ranging from 200 to 3400 hertz. Also, COMMERCIAL SPEECH BAND, SPEECH FREQUENCY.

voice-frequency dialing *Telecommunications.* a telephone dialing system in which DC pulses are converted into voice-frequency AC pulses.

voice-frequency telegraph system *Telecommunications.* a telegraph system that allows the use of a number of telegraph channels on one voice grade channel or circuit for further processing through a metallic or radio network. Also, **voice-frequency carrier telegraphy.**

voice-grade channel *Telecommunications.* a channel that has a bandwidth large enough to handle the transmission of voice-frequency signals.

voice mail *Telecommunications.* a form of electronic mail in which spoken messages are stored electronically and later replayed upon command of the recipient.

voice-operated see VOICE-ACTIVATED.

voiceover *Acoustical Engineering.* a voice that is heard on a film or videotape without the speaker's being seen on camera, as that of the narrator of a documentary film or the announcer for a television program or commercial.

voiceprint or **voice print** *Acoustical Engineering.* a sound spectrogram of a particular voice, having specific characteristics that can be used to identify the person from whose voice the spectrogram is created.

voice-recognition unit *Telecommunications.* a computer device that recognizes a limited number of words and then transposes them into equivalent digital signals that may be used as computer input.

voice response *Computer Technology.* computer output in the form of a humanlike voice, either recorded and replayed or artificially generated.

voice synthesizer *Electronics.* any device that simulates speech by assembling a language's elements or phonemes under digital control.

voicing sounds *Linguistics.* the formation of some sounds using the vocal cords and a certain amount of air; by opening the vocal cords and letting air pass through unobstructed, which is called **voiceless sound** (as in [p]), or by constricting air and vibrating the vocal cord, which is called **voiced sound** (as in [b]).

void *Astronomy.* a large region of the universe containing few or no galaxies. *Physiology.* to discharge a bodily waste. *Materials Science.* an empty space that can vary in size from that of one atom to a microscopically visible area within a group of atoms that can be filled with different atoms; voids are important in ferrimagnetic ceramics because magnetic strength is directly related to the way in which atoms are placed in these sites. Also, INTERSTITIAL SITES. *Computer Programming.* **1.** the inadvertent omission of ink in a character image. **2.** a null or invalid pointer or address. **3.** to nullify a previous operation or set of data.

void channels *Materials Science.* the open channels of a porous or packed medium that allow the flow of liquid or gas.
void coefficient *Nucleonics.* a rate of change in reactivity, expressed as the partial derivative of reactivity with respect to a void formed by the removal of material (such as the formation of steam bubbles in a water reactor system) at a specified location within a reactor.
void ratio *Geology.* the ratio of the amount of empty space to the amount of solid material in substances such as soil samples, sediments, or rocks that contain both void space and solid material.
void set *Mathematics.* the empty set; denoted $\varnothing$.
Voigt, Woldemar [fōkt] 1850–1919, German mathematical physicist; developed the equations later known as the Lorentz transformations.
Voigt body or **Voigt-Kelvin solid** see KELVIN BODY.
Voigt effect *Optics.* the birefringence that occurs when light travels through an isotropic gas under the influence of a magnetic field perpendicular to the direction of light propagation. Also, MAGNETIC DOUBLE REFRACTION.
Voigt model *Materials Science.* the representation of the reaction of polymer chains to stress by a parallel combination of elasticity represented by a spring and viscosity represented by a dashpot; the oldest two-dimensional model for viscoelastic behavior of a polymer.
Voigt notation *Mechanics.* in elasticity theory, a notation by which the numbers 1, 2, 3, 4, 5, 6 are used to label elastic constants and elastic moduli instead of using the pairs of letters xx, xy, xz, yy, yz, zz.
voile *Textiles.* a sheer, crisp fabric made from hard-twist cotton, silk, worsted, rayon, or acetate yarns in a plain weave.
vol. volume; volcano.
Volans *Astronomy.* the Flying Fish, a small dim constellation of the Southern Hemisphere lying between Carina and the pole. Also, PISCIS VOLANS.
volar *Anatomy.* relating to the palm of the hand or sole of the foot.
volatile *Chemistry.* of a substance, readily vaporizable at a relatively low temperature. Thus, **volatile fluid.**
volatile component *Geology.* a constituent of a magma having a sufficiently high vapor pressure that it comes out of solution in the gaseous phase. Also, **volatile, volatile flux.**
volatile file *Computer Programming.* a data file that is frequently updated.
volatile-oil reservoir *Petroleum Engineering.* a bubble-point oil reservoir with conditions leading to volatized oil (high temperature and low liquid density) and thus lower recovery.
volatile storage *Computer Technology.* a storage device that loses its contents when the power source is removed. Also, **volatile memory.**
volatile transfer *Volcanology.* a process in which a gaseous phase (or volatile component) separates from a magma, moves along with it, and releases dissolved material back into the magma, usually in an area of reduced air pressure. Also, GASEOUS TRANSFER.
volatility *Thermodynamics.* the property of a liquid having a low boiling point and a high vapor pressure at ordinary pressures and temperatures.
volatility product *Chemistry.* the product of the equilibrium pressures of gases produced by the vaporization of a liquid or solid.
volatilization *Thermodynamics.* the vaporization of a substance by a heat transfer or the lowering of the pressure.
volborthite *Mineralogy.* $Cu_3^{+2}V_2^{+5}O_7(OH)_2 \cdot 2H_2O$, a dark-green, yellow, or brown monoclinic mineral occurring as masses, scaly crusts, and honeycombed to fibrous aggregates, having a specific gravity of 3.42 and a hardness of 3.5 on the Mohs scale; found as a secondary mineral with carnotite in sandstone; a principal ore of vanadium.
volcanello see SPATTER CONE.
volcanic *Volcanology.* relating to or produced by a volcano.
volcanic ash *Volcanology.* see ASH.
volcanic bomb *Volcanology.* see BOMB.
volcanic breccia *Petrology.* a pyroclastic rock containing angular volcanic fragments larger than 64 millimeters in diameter, in the presence or absence of a matrix.
volcanic chain *Volcanology.* a line of volcanic mountains, often along a fault.
volcanic cone *Volcanology.* a cone-shaped hill formed by volcanic discharges around a volcanic vent, composed of lava or pyroclastic materials.
volcanic dome *Volcanology.* a steep-sided, rounded mass of viscous lava forming a dome-shaped or bulbous mass over a volcanic vent.
volcanic dust *Volcanology.* see ASH.
volcanic eruptive see EXTRUSIVE.
volcanic gases *Geology.* the volatile materials formerly dissolved in a magma that are released during a volcanic eruption.
volcanic glass *Geology.* a natural glass produced when molten lava suddenly cools and solidifies before crystallization can take place.
volcaniclastic rock *Petrology.* a clastic rock containing volcanic material.
volcanic mud *Geology.* a mixture of water and volcanic ash formed either on land or at the time of eruption. Thus, **volcanic mudflow.**

volcanic neck

volcanic neck *Volcanology.* a mass of hardened volcanic magma that has been thrust upward into a volcanic cone by subterranean pressure, remaining upright after the surrounding material has been stripped away by erosion. *Geology.* see PIPE.
volcanic rift zone *Geology.* in Hawaii, a belt of features associated with volcanic activity that overlies dike assemblages.
volcanic rock *Geology.* a finely crystalline or glassy igneous rock formed as a result of volcanic activity at or near the earth's surface.
volcanics *Petrology.* igneous rocks that reached or approached the earth's surface prior to solidification.
volcanic sink *Volcanology.* see SINK.
volcanic theory *Astronomy.* a now-superseded theory that most lunar formations such as craters are volcanic in origin.
volcanic vent *Geology.* see VENT.
volcanic water *Volcanology.* magmatic water that is on or just under the earth's surface.
volcanism or **vulcanism** *Geology.* any of the processes in which magma and its associated gases rise up from the earth's interior and are discharged onto the surface and into the atmosphere. Also, **volcanicity.**
volcano *Geology.* **1.** a vent or fissure in the earth's surface through which magma and its associated materials are expelled. **2.** the generally conical structure formed by the expelled material.

volcano

Volcanology

Volcanology may be defined as that branch of science that deals with the eruption of magma (molten material plus its gaseous content) upon the surface of the earth or its rise into levels near the surface. Although volcanology is closely related to geology and may be considered a phase of geology, it is also intimately related to seismology, geophysics, and geochemistry.

The history of volcanology is interwoven with superstition and mythology. In Roman mythology, Vulcan was the god of fire, the blacksmith of the gods. Poets identified Vulcan's workshop with various active volcanoes in the belief that the "smoking" mountain was the chimney of Vulcan's forge. In ancient mythology Vulcan's forge was located on the island of Vulcano, one of the Lipari islands off the coast of Sicily. From this association the name "volcano" was applied to all mountains which give off "smoke and fire."

Perhaps the first objective account of a volcanic eruption was the description by Pliny the Younger of the AD 79 eruption of Vesuvius in which his uncle, Pliny the Elder, lost his life. The account of the eruption was contained in two letters, written by Pliny the Younger to the Roman historian Tacitus, giving the details of his uncle's death. These letters may well be the earliest contribution to the science of volcanology. With the acceptance of volcanoes as a natural phenomenon, volcanology began to be organized as a body of knowledge. However, no real progress was made until the development of geology. In fact the development of geology was, to a large extent, fostered by the controversies that surrounded volcanoes.

The eruption of Mt. Pelee in 1902 and the resulting death of nearly 30,000 inhabitants of St. Pierre focused attention on volcanology. The tragedy drew the sympathy and attention of the entire world, and scientists from many countries came to study the eruption. All agreed that information must be made available to warn inhabitants vulnerable to possible eruptions, and studies in an effort to predict volcanic eruptions are of top priority today. Another event that had a profound influence on volcanology was the U.S.A. program to land a man on the moon. It had long been known that there were craters on the moon, but whether they were the results of volcanic activity or due to the impact of meteorites was the subject of a lively controversy. Certainly if we were going to land a man on the moon, it was imperative that we know the character of the lunar surface. The intensive investigation of the earth's volcanoes in an effort to determine the character of the lunar surface dramatically advanced the knowledge of volcanology.

Fred M. Bullard
Emeritus Professor of Geological Sciences
The University of Texas at Austin

volcanology or **vulcanology** *Geology.* the study of the causes and phenomena associated with volcanism.

vole *Vertebrate Zoology.* a small mouselike rodent of the genus *Microtus* and related genera of the family Cricetidae, characterized by short legs, small ears, and a blunt nose; found chiefly in Palearctic regions.

vole

Volga *Geography.* a river in western Russia, rising in the Valdai Hills near Moscow and flowing about 2300 miles east and south to the Caspian Sea.

Volhynia fever see TRENCH FEVER.

volley *Military Science.* **1.** the simultaneous discharge of a number of weapons; in modern usage, it is usually applied to a salute fired by a group of riflemen. Similarly, **volley bombing**. **2.** the bullets, projectiles, or missiles discharged in such a manner. **3.** in artillery gunfire support, a method in which the individual pieces fire a given number of rounds as rapidly and accurately as possible, without synchronizing fire with the other pieces. Also, **volley fire.**

volt *Electricity.* the SI unit of potential difference or electromotive force; equivalent to the potential difference between two points requiring one joule of work to move one coulomb of electricity from the point of lower potential to the point of higher potential; alternately, one volt is equivalent to the electromotive force that will send a current of one ampere through a resistance of one ohm. (Named for Alessandro *Volta.*)

Volta, Count Alessandro 1745–1827, Italian physicist; invented the voltaic pile (voltaic cell) and the capacitor; improved the electroscope.

Volta *Geography.* a river system in western Africa; the **Black Volta** and **White Volta** both rise in Upper Volta and flow south to form the Volta, which flows through the artificial **Lake Volta** into the Gulf of Guinea.

voltage *Electricity.* the potential difference or electromotive force as measured in volts.

voltage amplification *Electronics.* the amplification of an input-signal voltage to produce a higher output-signal voltage.

voltage amplifier *Electronics.* any amplifier designed to increase a signal's voltage.

voltage clamp technique *Biotechnology.* a technique for studying the effects of membrane potentials; microelectrodes are placed in a cell and adjusted so that the membrane potential is held at a specific voltage.

voltage coefficient *Electricity.* the ratio of the change in a value to the change in voltage, for components whose values vary with voltage; it indicates the extent to which a value changes under the influence of voltage and is generally expressed in percent per volt or in parts per million per volt (ppm/V).

voltage contrast *Materials Science.* an image produced by a scanning electron microscope in which one area of a specimen appears brighter than another; it occurs when there are voltage differences between adjacent sections of a specimen; used to find shorts or open circuits in microchips.

voltage-controlled oscillator *Electronics.* an oscillator having an output frequency that varies with the applied direct current control voltage.

voltage corrector see VOLTAGE REGULATOR.

voltage-current dual *Electricity.* the replacement of elements of one circuit in a circuit pair by their dual elements in another circuit, according to the duality principle; for example, currents replaced by voltages, or capacitances by resistances.

voltage-dependent anion channel *Biochemistry.* a porin channel found in the outer membrane of the mitochondria.

voltage divider *Electricity.* a tapped resistor, potentiometer, or two fixed resistors in series, connected across a voltage source; the output voltage (from the center terminal to one side of the source) is less than the source voltage. Also, POTENTIAL DIVIDER.

voltage doubler *Electronics.* a rectifier circuit that rectifies each half cycle of an alternating voltage separately and adds the two rectified voltages to produce a direct voltage having approximately twice the peak amplitude of the original applied alternating voltage.

voltage drop *Electricity.* **1.** the voltage difference between two points in a circuit. **2.** the voltage difference across an element in a circuit. Also, POTENTIAL DROP.

voltage feed *Electromagnetism.* the excitation point on a transmitting antenna at which the voltage is applied at an antinode or voltage loop.

voltage gain *Electronics.* the ratio of an output-signal voltage level to the input-signal voltage level, usually expressed in decibels.

voltage gating *Biotechnology.* the regulating of permeability of a membrane by opening a channel in response to a change in membrane potential.

voltage generator *Electronics.* a two-terminal circuit element whose terminal voltage is independent of its current.

voltage gradient *Electricity.* the voltage difference per unit length in a specified path, such as a resistor or other conductive path, or in free space. Also, POTENTIAL GRADIENT.

voltage multiplier *Electricity.* a specialized rectifier circuit that delivers a DC output voltage as a multiple of the peak value of the AC input voltage, thus allowing a voltage step-up without a transformer.

voltage-multiplier circuit *Electricity.* a circuit utilized to obtain a high DC potential of two or more times the peak value of the AC input supply; it is effective only when the load current is negligible, as in an anode supply to a CRT; most useful for high-voltage, low-current supplies.

voltage node *Electromagnetism.* in a system that supports a standing wave, a point at which the voltage is zero, as at the midpoint of a half-wave antenna.

voltage phasor *Electricity.* the length of a line that represents the magnitude of a sinusoidally varying voltage, and whose phase is represented by an angle with the positive x-axis.

voltage quadrupler *Electronics.* a rectifier circuit that uses four diodes to produce a DC voltage that is four times the peak of the AC input voltage.

voltage rating *Electricity.* **1.** the maximum voltage that an electronic device can carry or withstand without failing. Also, WORKING VOLTAGE. **2.** the specified output voltage of a generator.

voltage reflection coefficient *Electromagnetism.* in a system in which currents or fields may be reflected, the ratio of the complex electric field strength or voltage of the reflected signal to that of the incident signal.

voltage-regulating transformer *Electromagnetism.* a transformer that maintains a constant AC output by means of a resonant circuit; used in conjunction with a saturable core.

voltage regulation *Electricity.* the percentage variation of an output voltage of a power supply for either a supply voltage (line) variation or an output current (load) variation.

voltage regulator *Electronics.* a circuit or device that produces a nearly constant voltage output, even though the voltage input (line) and current output (load) vary widely. Also, VOLTAGE CORRECTOR.

voltage-regulator diode *Electronics.* a diode in which the voltage is independent of the current for a considerable range of current values, such as a Zener diode. The diode may be used as the main component of a simple voltage regulator, or as a constant voltage reference for a more elaborate voltage regulator circuit.

voltage-regulator tube *Electronics.* a glow-discharge tube that maintains an essentially constant tube voltage drop over the operating range of current.

voltage saturation see ANODE SATURATION.

voltage transformer *Electricity.* **1.** a transformer utilized to transform voltage with little or no current. **2.** a small step-up transformer used for increasing the sensitivity of an AC voltmeter. Also, POTENTIAL TRANSFORMER.

voltaic cell *Electricity.* a primary cell that consists of two electrodes made of differing metals immersed in an electrolyte; used to generate a DC voltage. Also, GALVANIC CELL.

voltaic pile *Electricity.* a rudimentary primary battery that consists of a series of disks made of two different metals, such as zinc and copper, stacked alternately with electrolytes, soaked cloth, or paper disks.

voltaic vertigo *Neurology.* the inclination of the head toward the stimulated side upon application of an electric current to the vestibular fibers of the eighth nerve. Also, GALVANIC VERTIGO.

voltaite *Mineralogy.* $K_2Fe_5^{+2}Fe_4^{+3}(SO_4)_{12}\cdot 18H_2O$, a green to black cubic mineral occurring as octahedral and dodecahedral crystals and in massive form, having a specific gravity of 2.7 and a hardness of 3 on the Mohs scale; found in mines with other secondary sulfates.

voltametry *Physical Chemistry.* a technique for measuring current passing through an electrolytic solution while a series of voltages is applied.

voltammeter *Electricity.* an instrument that can measure current and voltage.

volt-ampere *Electricity.* the SI unit of apparent power, equal to the product of 1 volt and 1 milliampere.

volt-ampere hour *Electricity.* the SI unit for the integral of apparent power over time, equal to the product of 1 volt-ampere and 1 hour.

volt-ampere reactive *Electricity.* the SI unit of reactive power. Also, REACTIVE VOLT-AMPERE.

Volterra, Vito 1860–1940, Italian mathematician; founder of functional analysis.

Volterra dislocation *Solid-State Physics.* a type of dislocation in which a ring of solid material is cut, the cut surfaces are displaced slightly, and the ends are then rejoined.

Volterra integral equations *Mathematics.* **1.** the Volterra integral equation of the first kind is $f(x) = \int_a^x K(x,y)\phi(y)dy$, where the functions $K(x,y)$ (the kernel) and $f(x)$ (the perturbation function) are continuous in x and y on some interval $[a,b]$, and the function $\phi(x)$ is to be determined. Also, **Volterra integral equation of type 1. 2.** the Volterra integral equation of the second kind has the unknown function $\phi(x)$ outside the integral; it is $f(x) + \lambda\int_a^x K(x,y)\phi(y)dy = \phi(x)$. Also, **Volterra integral equation of type 2.**

voltmeter *Engineering.* an instrument that measures the potential difference between two points.

voltmeter sensitivity *Electricity.* the ratio of a voltmeter's total resistance to its full-scale reading in volts; expressed in ohms per volt.

volt-ohm-milliammeter *Engineering.* an instrument that measures direct current for voltage, resistance, and current. Also, CIRCUIT ANALYZER, MULTIMETER.

voltzite *Mineralogy.* a reddish or yellowish mixture of the mineral wurtzite and an organometallic zinc compound.

volume *Mathematics.* **1.** a measure of the amount of space that is enclosed by a closed surface in Euclidean 3-space. For example, the volume of a cube having edges of length e is e^3. **2.** the volume of any bounded set S in Euclidean n-space is determined as follows: The volume of an n-dimensional cube having edges of length e is e^n. Let S_a be the total volume of a given collection of nonoverlapping n-cubes that completely cover S, and let α denote the infimum of all such volumes S_a. Let S_b be the total volume of a given collection of nonoverlapping n-cubes that are completely contained in S, and let β denote the supremum of all such volumes S_b. If $\alpha = \beta$, then the set S is said to have volume α(or β); if $\alpha = \beta = 0$, then S has volume zero; and if $\alpha \neq \beta$, then the set S does not have volume. **3.** the volume of an unbounded set S is determined as follows: For a given n-cube C, let S_C denote the volume of $S \cap C$, computed in the sense of definition 2. Then the volume γ of S is the supremum, if it exists, of all such volumes S_C . If this supremum does not exist, then S is said to have infinite volume (if the supremum tends to infinity); or S does not have volume. *Acoustics.* a measure of the loudness or intensity of a sound or signal. *Acoustical Engineering.* this measure as indicated in volume units (VU) on an appropriate meter. *Transportation Engineering.* the total flow of traffic past a given point or along a stretch of road, normally expressed in vehicles per hour or per day. *Computer Technology.* a physical unit of storage medium that can be read or written with a single access mechanism, such as a tape reel or a floppy disk.

volume compressor *Acoustical Engineering.* an electroacoustic circuit that automatically limits the volume range of a sound transmitter to 30 to 40 decibels, thus allowing more modulation and a higher signal-to-noise ratio.

volume control *Acoustical Engineering.* a control used to adjust the loudness of a receiver or amplifier, normally by adjusting a variable resistance.

volume control system *Acoustical Engineering.* a device that keeps the volume of input from an amplifier, receiver, or any other electroacoustical device within a specific decibel range.

volume diffusion *Materials Science.* the diffusion of atoms through the interiors of grains.

volume dose *Radiology.* the total amount of energy absorbed by a human or other living object exposed to radiation, symbolized in gram-rads (100 ergs). Also, INTEGRAL DOSE. *Nucleonics.* see INTEGRAL DOSE.
volume expander *Acoustical Engineering.* an electroacoustical circuit that increases the volume range of a receiver by providing automatic gain control to amplify stronger signals and attenuate weak signals; one use for such a circuit is to counter the action of a volume compressor used at a broadcasting or recording station.
volume flow rate *Fluid Mechanics.* the volume of fluid passing through a unit area per unit time.
volume form *Mathematics.* a nondegenerate differential n-form on an n-manifold.
volume indicator *Acoustical Engineering.* a device on a sound system that indicates the amplitude, in decibels, of the electromagnetic wave at one point in the sound system. Also, VOLUME UNIT METER.
volume integral *Mathematics.* the volume of a domain D in space, computed as the triple integral (if it exists) of the function $f(x,y,z) = 1$ over D, with respect to the measure $dx\, dy\, dz$.
volume lifetime *Solid-State Physics.* the average amount of time that a minority carrier exists in a bulk semiconductor, from the instant of its generation to the instant of recombination.
volume-limiting amplifier *Electronics.* an amplifier equipped with an automatic device that functions when the input volume exceeds a specified level, then reduces the gain in such a way that the output volume remains essentially constant despite further input volume increases.
volume range *Acoustical Engineering.* the dynamic range of volume, expressed in decibels, for a sound system or an electroacoustic signal.
volume resistivity *Electricity.* **1.** the electrical resistance of a specific volume of material, based on the resistance between opposite faces of a one centimeter cube of the given material; expressed in ohm-centimeters. Also, SPECIFIC INSULATION RESISTANCE. **2.** the ratio of the potential gradient parallel to the current in a material to the current density.
volume shift see FIELD SHIFT.
volume-speed-density relationship *Transportation Engineering.* the relationship of total traffic volume to speed and density. Volume is proportional to density times average speed; at high densities, however, further increases of density produce lower speeds rather than greater volume.
volume strain *Mechanics.* a strain that causes a change in the volume of a substance, as opposed to one causing a change only in its shape.
volume stress *Mechanics.* a measure of the external force causing a change in the volume of a substance, per unit area of the substance.
volume target *Electromagnetism.* a radar target composed of a large number of objects too close together to be resolved.
volumeter *Engineering.* an instrument that measures the flow of a gas, liquid, or solid, either directly or indirectly.
volume transport *Oceanography.* the amount of water that a current transports between two points of reference, expressed in cubic meters per second; determined by multiplying the current speed times the cross-sectional areal limits of the current.
volumetric analysis *Analytical Chemistry.* the quantitative measure of a solution of known volume but unknown concentration, determined by adding reagents of known concentration until the reaction is complete.
volumetric efficiency *Thermodynamics.* the ratio of the actual mass of gas compressed to the mass of gas occupying piston displacement at the inlet pressure and temperature; used as a measure of engine efficiency.
volumetric flask *Analytical Chemistry.* a flask with only one calibration, used to measure and prepare definite fixed volumes of liquids.
volumetric loading rate *Biotechnology.* the rate of addition of raw materials to a fermenter or to an aerobic digester, expressed in terms of weight of material per unit volume per unit time.
volumetric pipet *Analytical Chemistry.* an open-ended graduated glass tube used to measure different quantities of liquids.
volumetric radar *Engineering.* radar that displays three-dimensional data on a variety of targets.
volumetric strain see VOLUME STRAIN.
volumetric stress see VOLUME STRESS.
volume unit *Acoustical Engineering.* an amplitude unit on a sound level meter, used to monitor an electroacoustic signal; it is based on the measurement by a volume unit meter calibrated for a sensitivity of 0 VU at 4 decibels (1.228 volts in a 600-ohm circuit) for U.S. standards.
volume unit meter see VOLUME INDICATOR.
volume velocity *Acoustics.* the rate of flow of a fluid medium that is driven by a sound wave, across a specified boundary.
voluntary association *Anthropology.* a group that is open to members of different age and kin groups and in which membership is by choice.
voluntary muscle *Physiology.* a muscle controlled by conscious will.
volunteer plant *Botany.* a term for a plant that grows without direct human supervison, control, or cultivation, especially one growing spontaneously from seeds lost from a previous crop.
volute *Architecture.* a spiral or scroll-shaped ornament, as on an Ionic or Corinthian capital. *Mechanical Devices.* any machine part having a spiral form or rotary action.
volute pump *Mechanical Engineering.* a centrifugal pump housed in a spiral casing.
volute spring *Mechanical Engineering.* a conical-shaped coil spring that extends in the direction of the axis of the coil.
Volutidae *Invertebrate Zoology.* a family of marine snails, gill-breathing gastropod mollusks in the order Neogastropoda.
volutin *Biochemistry.* a basophilic substance found as granules in the cytoplasm of yeast, cyanobacteria, and other microorganisms.
volvent nematocyst *Invertebrate Zoology.* an unarmed, coiled-tube nematocyst that is closed at the end.
Volvocaceae *Botany.* a biflagellate family of the order Volvocales, characterized by cells embedded in a gelatinous matrix, coenobia that are curved disks or are spherical to ellipsoidal, and always an even number of cells, from 2 to 50,000; it contains the well-known genus *Volvox,* and is abundant in enriched bodies of fresh water.
Volvocales *Botany.* an order of green algae of the Chlorophyta, characterized by motile members that are unicellular or colonial.
Volvocida *Invertebrate Zoology.* an order of green flagellated protozoans in the class Phytomastigophorea, sometimes classified as algae. Also, **Volvocina.**
Volvox *Botany.* the type genus of aquatic green flagellates of the family Volvocaceae, found in rotating spherical colonies propelled by flagella.

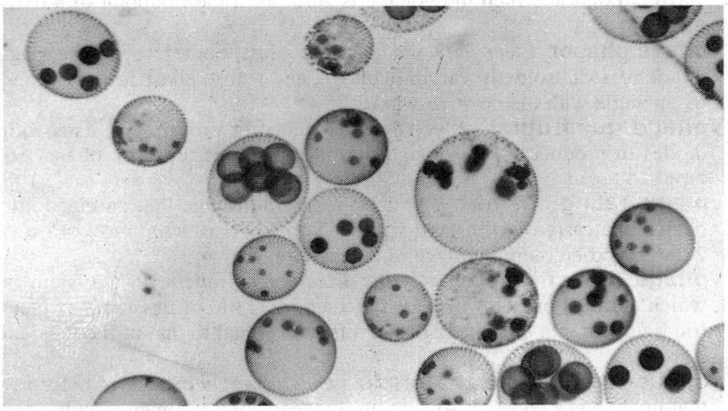

Volvox

volvulosis see ONCHOCERCIASIS.
volvulus *Medicine.* a knotting or twisting of the intestine, causing an obstruction.
VOM volt ohm meter.
Vombatidae *Vertebrate Zoology.* the Australian wombats, a family of marsupial mammals of the order Diprotodontia that are usually nocturnal, solitary, and semifossorial.
vomer *Anatomy.* the narrow bone forming the lower and posterior half of the nasal septum.
vomeronasal cartilage *Anatomy.* the two narrow pieces of cartilage that project anteriorly from the bony segments of the lower portion of the nasal septum.
vomica *Pathology.* **1.** a pus-filled cavity, usually in the lungs. **2.** the pus content of such a cavity.
vomit *Medicine.* to forcefully eject the contents of the stomach through the mouth; regurgitate.
vomitus *Medicine.* **1.** the act of vomiting. **2.** vomited matter.
von Békésy, Georg 1889–1972, Hungarian-born American physicist; awarded the Nobel Prize for the physiology of stimuli within the cochlea.
von Braun, Wernher [vän brän´] 1912–1977, German and American rocket engineer; played a major role in developing the German V-2 and the American Explorer I and Saturn rockets.

von Economo's encephalitis *Medicine.* an epidemic encephalitis characterized by increasing languor, apathy, and drowsiness, passing into lethargy; originally described by von Economo and observed in various parts of the world between 1915 and 1926.

von Gierke's disease *Medicine.* a glycogen storage disease resulting from a congenital error of metabolism, in which an excessive amount of glycogen is deposited in the kidneys and liver, characterized by hypoglycemia, ketoacidosis, and hyperlipemia. Also, GLYCOGEN STORAGE DISEASE, GLYCOGENOSIS.

von Hippel-Lindau disease *Pathology.* one of the phacamatosis group of congenital diseases, often characterized by multiple hemangiomas of the retina and hemangioblastomas of the cerebellum with attendant lesions and cysts on the spinal cord, pancreas, kidneys, and other viscera. Also, RETINOCEREBRAL ANGIOMATOSIS.

von Kármán, Theodore 1881–1963, Hungarian-born American physicist; worked in aerodynamic theory; analyzed Kármán's vortex street.

von Mohl, Hugo 1805–1872, German botanist; studied and named protoplasm; proposed cell division; explained osmosis.

von Neumann, John 1903–1957, American mathematician; worked in game theory, quantum mechanics, and high-speed computers.

von Neumann bottleneck *Computer Programming.* an inherent limitation in any von Neumann machine, arising because most computer time is spent moving data between storage and the central processing unit.

von Neumann machine *Computer Programming.* a sequential computer in which instructions and data are stored in the same storage medium.

von Recklinghausen's disease see NEUROFIBROMATOSIS.

von Willebrand's disease *Hematology.* a congenital hemorrhagic condition, characterized by prolonged bleeding time and deficiency of coagulation, associated with increased bleeding after trauma or surgery, menorrhagia, and postpartum bleeding. Also, PSEUDOHEMOPHILIA.

von Willebrand's factor *Hematology.* the attribute necessary for the adhesion of platelets to vascular elements; deficiency results in prolonged bleeding time as seen in von Willebrand's disease.

VOR *Navigation.* an air-navigational radio aid that provides radial lines of position in 360°; the airborne receiving equipment determines bearing through phase comparison of a ground transmitted signal. (An acronym for very-high-frequency omnidirectional radio range.)

Vorce diaphragm cell *Chemical Engineering.* a cylindrical, electrolytic cell with an asbestos-covered cathode and graphite anodes; used for producing chlorine.

vorobievite *Mineralogy.* a pink, reddish, or purplish-red variety of beryl containing cesium; used as a gemstone. Also, MORGANITE.

VORTAC *Navigation.* an air navigational ground station that combines VOR and tacan systems, providing VOR bearing with tacan bearing and distance.

vortex *Fluid Mechanics.* **1.** a flow field in which fluid particles move in concentric paths. **2.** see VORTEX TUBE.

vortex amplifier *Engineering.* a device in which the supply flow is introduced at the circumference of a hollow cylinder and the vortex developed can either reduce or increase flow.

vortex burner *Engineering.* a burner that uses a vortex to mix the combustion air and the fuel.

vortex distribution method *Fluid Mechanics.* an analytic method used in the study of aerodynamics that neglects the thickness of the profile (usually an airfoil) being studied, such as the theory of thin airfoils.

vortex filament *Fluid Mechanics.* a straight line connecting all point vortices that theoretically have infinite vorticity. Also, VORTEX LINE.

vortex line *Fluid Mechanics.* **1.** a line showing the instantaneous direction of rotation of the flow. **2.** see VORTEX FILAMENT.

vortex precession flowmeter *Engineering.* an instrument that measures gas flow by noting the rate of precession of vortices generated by radial vanes obstructing the flow. Also, SWIRL FLOWMETER.

vortex ring *Fluid Mechanics.* a line vortex in which the vortex filament forms a closed curve.

vortex shedding *Fluid Mechanics.* the periodic creation and leaving behind of vortices downstream from objects that are in the fluid path.

vortex-shedding meter *Engineering.* a flowmeter that determines velocity by noting the frequency at which vortices are generated by flow obstructions.

vortex sheet *Fluid Mechanics.* a surface composed of an infinite number of vortex filaments, each of infinitesimal strength and each extending to infinity or to a flow boundary.

vortex street *Fluid Mechanics.* the series of vortices shed systematically downstream from a body in a rapid fluid flow, Also **vortex trail, vortex train.**

vortex thermometer *Engineering.* a thermometer that adjusts for adiabatic and frictional temperature increases by generating a vortex in the air as it passes the sensing element; used in aircraft.

vortex tube *Fluid Mechanics.* a tubular surface formed by vortex lines passing through a closed curve.

Vorticella *Invertebrate Zoology.* a genus of one-celled animals, characterized by a ciliate, bell-shaped body on a slender retractile stalk; often found attached to a plant or other object underwater.

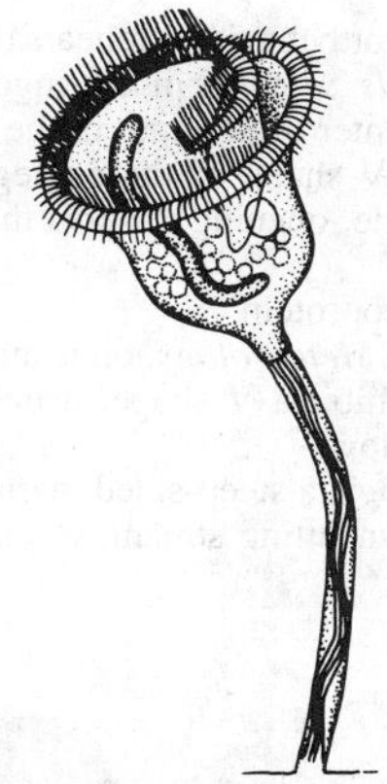

Vorticella

vorticity *Fluid Mechanics.* the curl of the velocity vector; a measure of the circulation of a fluid.

vorticity equation *Fluid Mechanics.* an equation describing the horizontal circulation of fluid particles about a vertical axis: $dw/dt = -(w \times \mathrm{div}_h q)$, where w is the absolute vorticity and $\mathrm{div}_h q$ is the horizontal divergence of the velocity vector q.

vorticity-transport hypothesis *Fluid Mechanics.* a theory stating that fluid vorticity, instead of momentum, is conserved in turbulent eddy flows.

VOS vertical obstacle sonar.

v.o.s. dissolved in yolk of egg. (From Latin *vitello ovi solutus.*)

Voskhod *Space Technology.* a series of Soviet manned space flights launched in 1964–65; included the first extravehicular activity in space.

Vostok *Space Technology.* a series of Soviet manned artificial satellites launched in 1961–63; included the first manned space flight, by Yuri Gagarin in *Vostok 1* (April 12, 1961).

vougesite see VOGESITE.

voussoir *Architecture.* one of the wedge-shaped stones or bricks used to form an arch or vault.

VOX voice-activated; voice-activated transmission.

Voyager *Space Technology.* a series of U.S. space probes that flew by Jupiter, Saturn, Uranus, and Neptune, en route to deep space, taking pictures and sending back data to scientists on earth.

voyeur [voi yür′] *Psychology.* a person who engages in voyeurism; one who derives pleasure from observing the sexual activities of others.

voyeurism [voi′yür iz əm] *Psychology.* the fact of achieving sexual pleasure from observing the sexual activity of other people, or observing people in the nude or in the act of undressing. Also, SCOPOPHILIA.

VP variable pitch; vent pipe.

VP-16 see ETOPOSIDE.

V particle *Particle Physics.* an unstable particle that decays so as to produce the characteristic V-shaped track in a cloud chamber.

VPg *Virology.* a virus-coded protein connected by an amino acid to the terminal end of the virion nucleic acid of picorna viruses. An abbreviation for a poliovirus virion protein, genome-linked.

VPI vapor phase inhibitor.

VPP virus pneumonia of pigs; variable-pitch propeller.

VPRC volume of packed red cells.

VPR chart *Navigation.* a chart used with the virtual plan position indicator reflectoscope.

VR virtual reality.

V.R. vocal resonance.

vrbaite *Mineralogy*. $Tl_4Hg_3Sb_2As_8S_{20}$, a gray-black, opaque, orthorhombic mineral occurring as groups of small, thick tabular or pyramidal crystals, having a specific gravity of 5.3 and a hardness of 3.5 on the Mohs scale; found at Allchar, Yugoslavia.

V region see VARIABLE REGION.

vriajem see FRIAGEM.

VS or **V.S.** veterinary surgeon; volumetric solution.

V's *Mechanical Devices*. the V-shaped longitudinal guides or ways of a planer, lathe bed, or the like, as a ridge or recess along the surface for moving a carriage or table.

v.s. see above. (From Latin *vide supra*.)

v.s. vibration seconds.

VSB vestigial sideband.

VSBR or **vsbr** visuomotor behavioral rehearsal.

V-segregates *Materials Science*. in the ingot-casting of steel, segregates that form in the center of an ingot in the final stages of solidification; the characteristic V shape of the segregates occurs as equiaxed dendrites slump and settle, opening fissures that become filled with enriched liquid.

VSG variant surface glycoprotein.

V-shaped depression *Meteorology*. on a surface weather chart, a low about which isobars exhibit a V shape, usually extending toward the equator from the parent low.

V-shaped valley *Geology*. a steep-sided, narrow-bottomed valley, usually excavated by a downcutting stream, whose cross profile resembles the letter V.

V-shaped valley

VSI vertical speed indicator.

V/STOL vertical short takeoff and landing.

v strand see V GENE.

VT vacuum tube; variable time; voice tube.

VT fuse *Ordnance*. a variable time fuse, a type of proximity fuse designed to detonate a charge in the vicinity of a target when activated by an external influence other than contact.

V-thread *Mechanical Devices*. a screw thread with a thread angle of 60° in the form of a cut around a cylinder surface.

VTOL *Aviation*. a nonrotary airplane capable of taking off and landing vertically, especially one able to clear a 50-foot obstacle within 50 feet of takeoff. Also used to form compounds such as **VTOL aircraft.** (An abbreviation of vertical takeoff and landing.)

VTR videotape recorder; videotape recording.

VU volume unit.

vug or **vugh** *Petrology*. a small cavity in a vein or rock, usually lined with crystals of different mineralogical composition from the enclosing rock.

Vulcan *Astronomy*. a planet once hypothesized to be orbiting the sun inside the orbit of Mercury to account for the advance of Mercury's perihelion, which has since been explained by relativity. *Ordnance*. an Army air defense artillery gun designed for low-altitude air defense and direct fire against surface targets; it has six 20-mm, rotary-fired barrels. (From *Vulcan*, the Roman god of fire.)

Vulcanian eruption *Geology*. a volcanic eruption that is characterized by violent periodic explosions of viscous lava, volcanic bombs, and ash clouds.

vulcanization *Materials Science*. **1.** generally, a chemical reaction that causes cross-linking of rubber with sulfur. **2.** a term for other cross-linking reactions of polymers, as those that occur in some silicone rubbers.

vulcanized fiber *Materials Science*. a fibrous material that is obtained by treating paper pulp with zinc chloride; used for electrical insulation.

vulnerant *Surgery*. **1.** causing a wound or inflicting injury. **2.** an agent that so acts.

vulnerary *Surgery*. **1.** of or relating to wounds or their healing. **2.** an agent that stimulates the healing of wounds.

Vulpecula *Astronomy*. the Fox, a small constellation of summer and autumn lying in the Milky Way between Cygnus and Aguila.

vulsella *Surgery*. forceps with sharp, clawlike hooks at the ends, used to grasp tissue.

vulture *Vertebrate Zoology*. the popular name for several species of large carrion-feeding or predatory birds in the family Cathartidae, usually having a large body, a naked head, and somewhat weak claws; includes some of the largest flying birds.

turkey vulture

vulva *Anatomy*. a general term for the external female genitalia, including the labia majora, labia minora, clitoris, and vestibule of the vagina.

vulval or **vulvar** *Anatomy*. relating to the vulva.

vulval gland *Anatomy*. a mucus-secreting gland that empties into the vulval area and provides lubrication during sexual intercourse.

vulvectomy *Medicine*. surgical removal of the vulva.

vulvitis *Medicine*. inflammation of the vulva.

vulvovaginitis *Medicine*. inflammation of the vulva and vagina or of the associated mucous glands.

vuthan *Meteorology*. an intense storm of southeastern South America.

vv. veins. (From Latin *venae*.)

v/v volumn (solute) per volume (of solvent).

v-value see ABBÉ NUMBER.

VV Cephei stars *Astronomy*. a class of long-period, eclipsing binary stars with M supergiant primaries and blue (usually type B) supergiant or giant secondaries.

V.W. vessel wall.

V-W transition *Microbiology*. a form of reversible phase variation found in certain bacterial strains of *Citrobacter*, in which the V form synthesizes the VI antigen, and the W form does not.

Vycor *Materials Science*. the trade name for a type of high-silica glass made by chemically leaching all the ingredients except silica from alkali-borosilicate glass, and then consolidating the resultant porous glass; used for laboratory and industrial glassware because of its durability and resistance to heat.

VZIG varicella zoster immune globulin.

VZV varicella zoster virus.

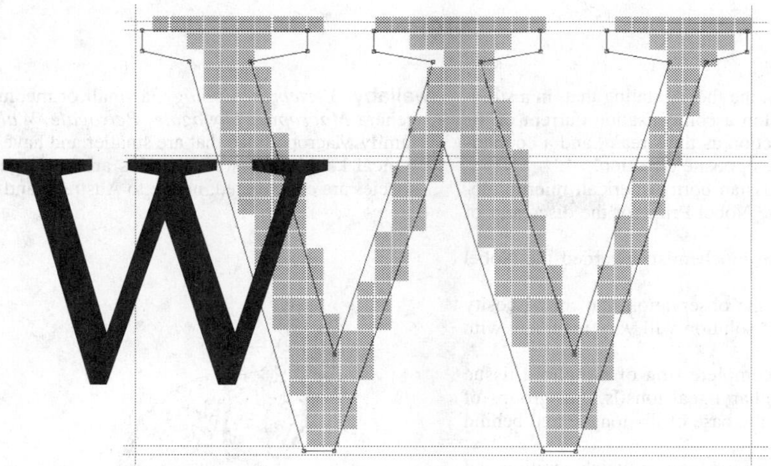

W the chemical symbol for tungsten. (From German *Wolfram.*)

W or **W.** watt; watts.

w or **w.** watt; watts.

w. weight; wide; width.

w/ with.

wacke [wak´ə] *Petrology.* sandstone consisting of a mixture of angular, poorly sorted or unsorted mineral and rock fragments in an abundant groundmass of clay and fine silt.

Wacker process *Chemical Engineering.* a procedure used to oxidize ethylene to acetaldehyde, by contacting with oxygen in the presence of cupric chloride and palladium chloride.

wad *Ordnance.* **1.** a cloth or cardboard plug that was originally used in muzzle-loading weapons to hold the shot or projectile against the powder; in some modern cartridges it is used to hold the propellant in place. Also, **wadding. 2.** any of the cardboard, fiber, or plastic devices used in shotgun shells to contain propellant gases and protect the shot charge, or to act as filler with smaller charges; the combination of such devices in a particular shell is called the **wads, wadding,** or **wad column.** *Mineralogy.* a general term for soft, dark-brown to black massive manganese oxides of low specific gravity that are not specifically identified, usually also containing other metallic oxides; often found in marshy areas as a result of the decomposition of manganese minerals.

wadi [wäd´ē] *Geology.* **1.** a channel or steep-sided valley or ravine in certain desert regions that carries water only during the rainy season. Also, **wady, waddy. 2.** the intermittent stream that flows through such a channel. (From the Arabic word for such a feature.)

wading bird *Vertebrate Zoology.* a general term for any of the usually long-legged birds of the order Ciconiiformes that wade in water in search of food, including various shorebirds and inland water birds such as herons, egrets, storks, and ibises. Also, **wader.**

Wadsorth mounting *Optics.* a system used to reduce astigmatism and spherical aberration in diffraction gratings; consists of a concave mirror and a plate holder mounted normal to the grating.

wafer *Engineering.* a small, flat disk, such as a washer or piece of insulation. *Electronics.* a flat semiconductor slice on which microcircuits or individual microcircuit devices can be fabricated. *Medicine.* a thin sheet of dry paste used to enclose a powder to be swallowed.

wafering *Materials Science.* the machining of crystalline materials such as silicon into thin slices for use in electronic devices.

wafer lever switch *Electronics.* a lever switch that engages one or more contacts on a movable wafer segment; actuated by the lever with contacts on one or both sides of the wafer.

wafer socket *Electronics.* an electron tube or transistor socket consisting of one or two plastic or ceramic wafers with spring-metal clips that grip the terminal pins of the tube or transistor.

waffle weave *Textiles.* a fabric with a textured surface, resembling a honeycomb, that is produced in the weaving process.

Wagner ground *Electricity.* a pair of impedances with a common point grounded, used in an AC bridge network to neutralize the effect of stray capacitance errors when high impedances are measured. Also, **Wagner earth connection.**

wagnerite *Mineralogy.* $(Mg,Fe^{+2})_2(PO_4)F$, a translucent to opaque, yellow, red, or greenish monoclinic mineral occurring as prismatic crystals and in massive form, having a specific gravity of 3.15 and a hardness of 5 to 5.5 on the Mohs scale; found in a variety of environments.

Wagner's reagent *Analytical Chemistry.* an aqueous solution of potassium iodide and iodine; used in the microscopic analysis of alkaloids. Also, **Wagner's solution.**

Wagner von Jauregg, Julius 1857–1940, Austrian neurologist; awarded the Nobel Prize for developing malarial inoculation for dementia paralytica.

wagon drill *Mechanical Devices.* a pneumatic, percussive-type rock drill mounted vertically on a three-wheeled or four-wheeled wagon with steel or rubber tires.

wagon vault see BARREL VAULT.

wagtail *Vertebrate Zoology.* a chiefly Old World perching bird of the family Motacillidae, characterized by a trim, slender body and a long tail that it habitually wags up and down.

Wahl correlation *Petroleum Engineering.* a correlation based on assumed data for pressure, volume, and temperature; used to estimate total recovery from a solution-gas-drive oil reservoir.

Waidner-Burgess standard *Optics.* a unit of luminous intensity equal to the luminous intensity of 1 square centimeter of a blackbody at the melting point of platinum; equivalent to 60 candelas.

wainscot *Architecture.* a decorative and protective paneling on the lower portion of an interior wall.

wairakite *Mineralogy.* $CaAl_2Si_4O_{12}\cdot 2H_2O$, a colorless to white, brittle monoclinic mineral of the zeolite group occurring as subhedral crystals, having a specific gravity of 2.26 and a hardness of 5.5 to 6 on the Mohs scale; found in tuffs and tuffaceous rocks from numerous volcanic regions.

WAIS *Psychology.* Wechsler Adult Intelligence Scale, a widely used intelligence test for adults that measures both verbal and performance abilities and yields an IQ score.

waist *Engineering.* the center portion of a vessel, whose diameter is smaller than adjacent areas.

wait *Computer Science.* **1.** to suspend progress of a program, e.g., by busy waiting, suspending a process pending an interrupt, or use of a semaphore, until some event external to the program occurs. **2.** an operation on a semaphore, used, for example, to wait until a resource is available or until some data is available to be processed.

waiting line *Industrial Engineering.* see QUEUE.

waiting time *Industrial Engineering.* **1.** the time during which orders are awaiting transportation, a part of manufacturing lead time. **2.** see STANDBY TIME. *Transportation Engineering.* the delay time due to waiting in a queue; the time actually spent in the queue minus the time it would take to pass through the queue zone if traffic were moving freely. *Computer Science.* time spent waiting for some external event; e.g., waiting for a disk access to be completed. Also, **wait time.**

wait state *Computer Programming.* a state in which a computer program is unable to use the central processing unit due to an unfinished I/O request or interleaving of other programs.

wake *Navigation.* the visible trail of waves left behind by a vessel as it moves through the water. *Fluid Mechanics.* the region downstream from the separation point where the adverse pressure gradient causes fluid backflow near the boundary.

wake flow *Fluid Mechanics.* a turbulent eddy flow in the wake region.

wake gain *Naval Architecture.* the increase in thrust of a propeller due to the forward motion of the water around the propeller that is caused by the vessel's forward movement.

wake stream theory *Oceanography.* the theory stating that, in a stratified ocean, a wake stream will develop a compensation current on its right side that flows in the same direction as the stream, and a countercurrent on its left side that flows in the opposite direction.

Waksman, Selman 1888–1973, Russian-born American microbiologist; studied soil bacteria; awarded the Nobel Prize for the discovery of streptomycin.

Wald, George born 1906, American biochemist; awarded the Nobel Prize for the discovery of retinene.

Walden's rule *Physical Chemistry.* the observation that the viscosity and the conductance in an electrolytic solution will vary inversely with each other.

Waldeyer's Ring *Anatomy.* the incomplete ring of lymphoid tissue formed by the faucial, lingual, and pharyngeal tonsils, and groups of smaller lymphatic follicles located at the base of the tongue and behind the posterior pillars of the soft palate.

Waldhof fermenter *Biotechnology.* a fermenter in which aeration and agitation are provided by a hollow, bladed, wheel-type impeller.

Waldseemüller, Martin 1470?–1521?, German cartographer; the first to show the New World as being distinct from Asia; proposed the name "America" in honor of Amerigo Vespucci.

wale *Textiles.* **1.** in knit fabric, the lengthwise ribs produced by a series of loops formed by the action of one needle. **2.** in woven fabric, any one rib, cord, or raised portion running lengthwise with the warp.

walk *Mathematics.* a finite, nonempty, alternating sequence of vertices and edges $v_0, e_1, v_1, e_2, \ldots, v_{k-1}, e_k, v_k$ in a graph G such that v_{i-1} and v_i are the end vertices of the edge e_i, $1 \leq i \leq k$. v_0 and v_k are the end or terminal vertices of the walk; all other vertices are internal vertices. Edges and vertices may appear more than once. A walk is open if $v_0 \neq v_k$; otherwise it is closed. Also, PATH.

walkdown *Electronics.* a failure in magnetic core storage, in which successive drive pulses cause charges in magnetic flux that remain after the magnetic field is removed.

walker *Medicine.* a device that aids in walking, especially a lightweight four-legged metal frame, about waist-high, used for support in walking by the infirm or disabled.

walkie-talkie *Telecommunications.* a radio-transmitter/receiver that is small enough to be carried in one hand or on the back. Also, PACK UNIT.

walking beam *Mechanical Engineering.* a rigid member whose ends rest on supports that move up and down independently, and whose center is hinged to the load it carries.

walking catfish *Vertebrate Zoology.* a catfish, *Clarias batrachus,* of the family Clariidae, distinguished by an auxillary breathing apparatus that allows it to survive on land for extended periods of time and a tail and spiny fins with which it can crawl over ground; native to Asia and introduced to Florida.

walking dragline *Mechanical Engineering.* a large-capacity shovel mounted on moving feet that are mechanically worked to maneuver as required.

walking fern *Botany.* a fern, *Camptosorus rhizophyllus,* characterized by triangular fronds that taper into a slender prolongation that frequently roots at the tip. Also, **walking leaf.**

walking machine *Mechanical Engineering.* a transportation device that uses four operator-controlled mechanical legs as a means of locomotion.

walking plow *Agriculture.* the simplest form of plow, pulled by a draft animal with the plowman walking behind.

walkingstick or **walking stick** *Invertebrate Zoology.* the popular name for any of various wingless insects of the family Phasmidae, noted for having a long, slender body with the appearance of a stick. The **Northern walkingstick,** *Diapheromera femorata,* is widely found in the eastern half of the United States.

walk-off *Graphic Arts.* the failure of an image to stick to a lithographic plate, such that part of the image disappears during a press run.

walkthrough *Computer Programming.* a step-by-step review during the design phase of a computer program or system, conducted by a group of peer programmers and analysts to check for errors. Also, **walkthrough method.**

walk-up *Architecture.* an apartment building with no elevator.

wall *Engineering.* a vertical structure that forms an enclosure. *Geology.* **1.** the side of a passage in a cave. **2.** the surface of country rock that laterally bounds a mineral or ore vein. **3.** the mass of rock forming a particular side of a fault. *Mining Engineering.* **1.** the side of a drift, level, or entry of a mine. **2.** the face of a longwall stall or working, often termed the **coal wall. 3.** a rib of solid coal that stands between two rooms.

wallaby *Vertebrate Zoology.* a small or medium-sized kangaroo of the genera *Macropus, Thylogale, Petrogale, Wallubus,* and others in the family Macropodidae that are smaller and have a different dentition than typical kangaroos; some species are no larger than rabbits and some species are endangered; native to Australia and Tasmania.

wallaby

Wallace, Alfred Russel 1823–1913, English naturalist; independently formulated the theory of natural selection; the founder of biogeography.

Wallace effect *Evolution.* the process of developing and reinforcing reproductive isolation between populations of elementary biological species by selection against hybrids with relatively lower fitness. Also, **Wallace hypothesis.**

Wallace's line *Paleontology.* an imaginary line proposed by Alfred Russel Wallace, between the different faunal assemblages characteristic of Southeast Asia and of Australia during the Cenozoic; the line runs just southeast of Java, Borneo, and the Philippines; there has been relatively little faunal mixing across the line, although in periods of great ice formation, as from 30,000 to 10,000 years ago, the straits through which the line runs have been very narrow, so that increased faunal migration was possible.

Wallach, Otto 1847–1931, German chemist; awarded the Nobel Prize for his work in alicyclics and terpenes.

Wallach transformation *Organic Chemistry.* an acid-catalyzed conversion of azoxybenzene into a *para*-hydroxyazobenzene, accomplished with concentrated sulfuric acid.

wall anchor *Building Engineering.* a steel or metal anchor 1.5 inches by 0.5 inch by 2 feet long, with bent-up ends fastened to every second or third common joist about 4 feet apart vertically; built into brickwork of a wall to provide lateral support. Also, JOIST ANCHOR.

wall-attachment amplifier *Fluid Mechanics.* a bistable fluidic device using two walls set back from the supply jet port, control ports, and channels to define two downstream outputs.

wallboard *Materials.* fibrous material that is produced in sheets for use on walls and ceilings.

wall cake *Petroleum Engineering.* the deposit of solids along a borehole due to the filtration of drilling-mud liquids into the formation; can lead to serious drilling problems. Also, FILTER CAKE.

wall crane *Mechanical Engineering.* a light crane that is pivoted on bearings and affixed to the wall of a building.

wall drill *Mechanical Devices.* a fixed or radial drilling machine bolted to the wall of a workshop.

wall effect *Electronics.* an increase in ionization caused by the liberation of electrons from the walls of an ionization chamber during the ionization process.

wall energy *Solid-State Physics.* the energy per unit area existing between two ferromagnetic domains that have different orientations.

Waller's law *Neurology.* a law stating that if sensory fibers are separated near the ganglion of the spinal cord, the fibers between the lesion and the spinal cord degenerate while those between the lesion and the ganglion remain intact.

walleye *Medicine.* the popular name for exotropia, a condition in which one eye deviates outward relative to the other. *Vertebrate Zoology.* **1.** a large game fish, *Stizostedion vitreum,* having large, staring eyes. Also, **walleyed pike. 2.** any of various other fishes with large, staring eyes.

Walleye *Ordnance*. a guided air-to-surface glide bomb designed for the stand-off destruction of large, semihard targets; it uses a contrast-tracking television guidance system.

Walley engine *Mechanical Engineering*. an engine with four elliptical rotors that turn clockwise and lead to excessively high rubbing velocities.

wall-forming body *Cell Biology*. a granular structure found in certain protozoans that forms a part of the oocyst wall.

wall friction *Fluid Mechanics*. the drag on a flow due to viscous adherence of the fluid molecules to the wall of the conduit.

Wallis, John 1616–1703, English mathematician; introduced the ∞ symbol, the concept of limit, negative and fractional exponents, and the graphic representation of complex numbers.

Wallis formulas *Mathematics*. formulas giving the values of the definite integrals from 0 to $\pi/2$ of the functions $\sin^n(x)$, $\cos^n(x)$, and $\cos^m(x)\sin^n(x)$ for any positive integers m and n. Also, **Wallis' theorem.**

Wallis product *Mathematics*. the infinite product

$$\pi/2 = (2/1)(2/3)(4/3)(4/5)(6/5)(6/7)\cdots = \prod_{k=1}^{\infty} (2k)^2/(2k-1)(2k+1).$$

wall niche see MEANDER NICHE.

wall off *Engineering*. to seal cracks in a borehole using cement, mud cake, or the like.

wall outlet *Electricity*. an electric outlet mounted in a wall, usually in a protective box with a cover that allows a plug to be easily inserted.

wall plate *Building Engineering*. a horizontal member which is built into or laid along the top of a wall so as to distribute the load-bearing pressure from joists or rafters.

wall ratio *Design Engineering*. **1.** the ratio of the outside radius of a tube or other cylindrical device as it relates to its inside radius. **2.** the ratio of the corresponding diameters.

wall reef *Geology*. a long, narrow, steep-sided coral reef built up on a reef wall.

wall rock *Geology*. the rock enclosing or adjoining a fault, mineral vein, or igneous intrusion.

wall-rock alteration *Geology*. the physical or chemical changes in the wall rock adjoining a hydrothermal ore deposit that result from the action of the fluids that formed the deposit itself.

wall-sided glacier *Hydrology*. a glacier on a steep slope, usually unconfined laterally by valley walls.

wall tie *Building Engineering*. in a cavity wall, a sturdy corrosion-resistant metal or steel tie which is fitted into bed joints; they inhibit moisture penetration.

walnut *Botany*. **1.** any of a number of related trees of the genus *Juglans*, of the North Temperate Zone, that bear a roundish, edible nut with a two-lobed seed. **2.** the nut of such a tree. *Materials*. the dark brown to black wood of trees of the genus *Juglans*, known for its strength, elasticity, and durability; it is used for cabinetry and furniture.

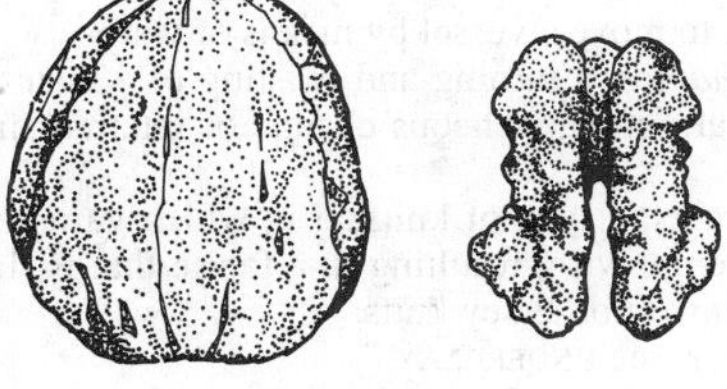

walnut

walpurgite *Mineralogy*. $Bi_4(UO_2)(AsO_4)_2O_4 \cdot 2H_2O$, a colorless to yellow, transparent to transluscent, rare, radioactive, triclinic mineral, occurring in aggregates of lathlike crystals, often twinned and appearing like gypsum, and having a specific gravity of 5.95 and a hardness of 3.5 on the Mohs scale; found with other secondary uranium minerals.

walrus *Vertebrate Zoology*. a large carnivorous marine mammal, *Odobenus nosmarus* of the family Otariidae, possessing canine teeth that are enlarged into large tusks, flippers, and a tough, wrinkled skin; found mainly in the shallow waters of the arctic regions of the Atlantic and Pacific Oceans.

Walther *Ordnance*. one of several automatic pistols produced by the Walther company beginning in 1929; the most famous is a 9-mm, 8-shot pistol used by the German military during WWII; officially designated **Walther 9-mm automatic pistol** or **P-38.**

Walton, Ernest Thomas Sinton born 1903, Irish physicist; with John D. Cockcroft, shared the Nobel Prize for discovering transmutations of atomic nuclei.

Waltz filtering *Artificial Intelligence*. an algorithm for finding consistent label sets for a labeled graph structure; e.g., interpretations of a line drawing of a polyhedron. Given a graph structure in which each node has a set of possible labels and nodes are connected by edges that represent constraints, the label sets are reduced by requiring arc consistency and by constraint propagation.

wand *Computer Technology*. a long, slender device connected to a terminal used in merchandising systems to read bar codes or other magnetic information from merchandise tickets, credit cards, and other documents. Also, **wand reader.**

wandering dune *Geology*. a sand dune that is slowly moved as a whole in the leeward direction of the prevailing winds, due to insufficient anchoring vegetation. Also, MIGRATORY DUNE, TRAVELING DUNE.

wandering of the poles *Astronomy*. the small variation of earth's rotation axis relative to the surface, caused by the varying gravitational perturbations from the moon and sun.

wandering star *Astronomy*. a pretelescopic name for a planet, because of its starlike appearance, but movement against a background of known, fixed stars.

waning moon *Astronomy*. lunar phases that occur after the full moon but before the new moon; so-called because its illuminating area is decreasing.

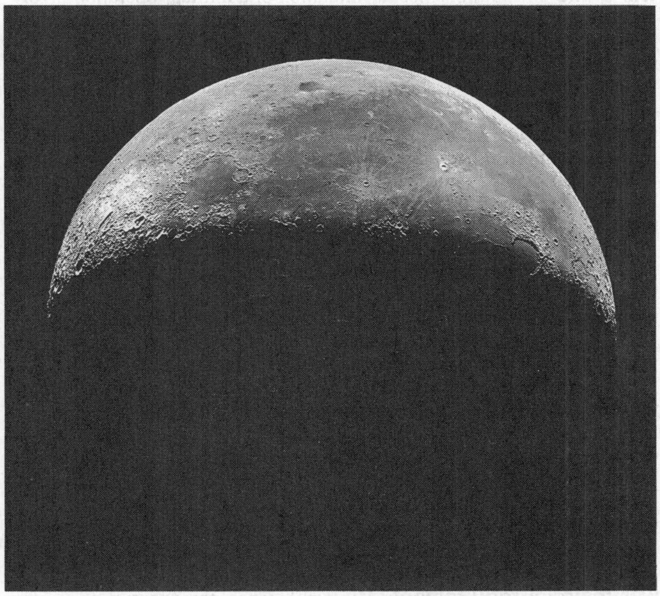

waning moon

Wankel engine [wäng´kəl] *Mechanical Engineering*. a rotary internal-combustion engine developed by the German engineer Felix Wankel, having a three-lobe rotor that turns in a close-fitting chamber, so that the power stroke is applied to each of the three faces of the rotor in turn as they pass a single spark plug.

Wannier function *Solid-State Physics*. a Fourier transform of the momentum eigenfunction (Bloch function) for a crystal.

Wannier-Mott exitons *Materials Science*. in a crystal insulator or semiconductor, an exiton (electron-hole pair) in which the average distance between electron and corresponding electron hole is larger than the lattice constant.

wapiti [wäp´i tē] *Vertebrate Zoology*. another name for the North American elk, *Cervus elaphus*; taken from a Native American name for the animal; used to distinguish this animal from the European elk. See ELK, def. 2.

warbler *Vertebrate Zoology*. any of numerous small, active, insectivorous songbirds of the diverse family Sylviidae, characterized by drab coloring but a beautiful song; most found in the Old World, with a few in America; often includes other birds resembling the Sylvidae, such as the Parulidae.

Warburg, Otto 1883–1970, German physiologist; studied the metabolism of cancer cells; awarded the Nobel Prize for the discovery of the yellow enzyme.

Warburg apparatus *Biotechnology.* a manometer that is used for studying the cellular respiration of tissues or cells by measuring oxygen uptake and carbon-dioxide evolution.

Warburg effect *Biochemistry.* **1.** the inhibition of photosynthesis due to high levels of oxygen. **2.** the high cellular levels of lactic acid in tumors.

Warburg impedance *Physical Chemistry.* the element in the equivalent circuit on an electrochemical cell that represents diffusional processes.

Warburg's old yellow enzyme see NADPH HYDROGENASE.

Wardiaceae *Botany.* a monotypic South African family of robust dark mosses of the order Isobryales that live attached to rocks in running streams; characterized by rigid stems with irregular distal branches and terminal sporophytes.

wardite *Mineralogy.* $NaAl_3(PO_4)_2(OH)_4 \cdot 2H_2O$, a colorless, white, or light blue-green, vitreous, transparent to translucent tetragonal mineral occurring as symmetrically developed pyramidal crystals and as granular to fibrous aggregates, having a specific gravity of 2.76 to 2.81 and a hardness of 5 on the Mohs scale; found in pegmatites and in variscite deposits.

Ward-Leonard speed-control system *Control Systems.* a system that uses the armature voltage of a separately driven direct-current motor, controlled by a motor-generator set, to control the speed of a direct-current motor.

war game *Military Science.* a simulation of a military operation involving two or more opposing forces and using rules, data, and procedures designed to depict an actual or assumed real life situation.

war gas *Ordnance.* see GAS.

Warhawk *Aviation.* a popular name for the P-40 pursuit (fighter) aircraft.

warhead *Ordnance.* the part of a missile, projectile, torpedo, rocket, or other munition that contains the system or material intended to inflict damage; it may be a nuclear or thermonuclear system, a high explosive system, chemical or biological agents, or inert material.

warhead delivery *Microbiology.* a process by which a substance that does not normally gain entry into a cell, such as an antimicrobial agent, is transported into the cell attached to a carrier that is normally endocytosed, such as a peptide.

warhead installation *Ordnance.* a warhead with the additional items necessary for mating it to the carrier.

warhead mating *Ordnance.* the act of attaching the warhead to a rocket or missile body, torpedo, airframe, motor, or guidance section.

warhead section *Ordnance.* a completely assembled warhead including skin sections and related components.

warm-air heating *Mechanical Engineering.* the process of heating by circulating warm air by means of a direct-fired furnace surrounded by a bonnet through which the air circulates.

warm air mass *Meteorology.* an air mass whose temperature is warmer than that of the surrounding air or warmer than the surface over which it is moving.

war materiel requirement *Military Science.* the quantity of an item required to equip and support approved forces during the period prescribed for war materiel planning purposes; the particular forces depend upon current Secretary of Defense guidance.

warm-blooded *Biology.* describing animals, such as mammals and birds, whose blood ranges from about 98° to 112°F, remaining constant in spite of fluctuations in the surrounding air or water temperature.

warm boot *Computer Programming.* **1.** to restart a computer system without having to reinitialize all of the data. **2.** to reinitialize a computer system without turning off the power. Also, **warm start.**

warm braw *Meteorology.* a warm, dry foehn that blows for 4–8 days during the east monsoon in the islands north of New Guinea.

warm-core ring *Oceanography.* a ring formed on the north side of the Gulf Stream that consists of a core of warm water surrounded by colder continental slope water.

warm front *Meteorology.* any nonoccluded front (or part of a front) that advances in such a manner that colder air is replaced by warmer air.

warm high *Meteorology.* any high that is warmer at its center than at its outer edges. Also, **warm anticyclone, warm-core high, warm-core anticyclone.**

warm low *Meteorology.* any low that is warmer at its center than at its outer edges. Also, **warm cyclone, warm-core low, warm-core cyclone.**

warm occlusion *Meteorology.* any occluded front (or part of a front) that advances in such a manner that colder air is replaced by warmer air. Also, **warm occluded front, warm-type occlusion.**

warm pool *Meteorology.* a limited area of relatively warm air that is surrounded by colder air, especially in high latitudes when a cutoff high is formed; identified on a thickness chart as a thickness maximum. Also, **warm drop, warm-air drop.**

warm sector *Meteorology.* the area where the warm air of a wave cyclone is found, between the cold front and the warm front of the storm; it steadily diminishes as a result of occlusion.

warm spring *Hydrology.* any spring whose water temperature is significantly above the environmental temperature of its locale, but below human body temperature.

warmth *Acoustical Engineering.* a qualitative characteristic of room acoustics opposite to that of brilliance; a warm room is an enclosure whose reverberation time for low frequencies is greater than the reverberation time for high frequencies.

warm tongue *Meteorology.* a mass of warm air that extends poleward from a tropical cyclone.

warm-tongue steering *Meteorology.* the steering effect that an upper-level warm tongue exerts on a tropical cyclone.

warm-water sphere *Oceanography.* a general term for the warm (generally over 8°C), upper layers of the ocean in middle and low latitudes; also called, by analogy with the atmosphere, the **ocean troposphere,** whereas the cold-water sphere is also called the **ocean stratosphere.**

warning behavior *Behavior.* behavior performed in response to the appearance of a predator that serves to alert others to the presence of the danger. Thus, **warning call, warning signal, warning coloration.**

warning device *Computer Technology.* a visible or audible alarm that informs the operator of a malfunction or error condition.

warning net *Telecommunications.* a communication system that distributes warning information of enemy movements.

warning-receiver system *Electronics.* an electronic countermeasure in which a pilot is notified when his aircraft is being illuminated by radar signals above certain power thresholds.

warning stage *Hydrology.* the elevation of a river at which it is necessary to issue flood warnings if adequate precautions are to be taken.

Warnowiaceae *Botany.* a family of marine dinoflagellates of the order Gymnodiniales, which are planktonic, nonthecate, nonphotosynthetic, and phagotrophic; distinguished by the presence of a complex light receptor composed of crystalline elements.

warp *Textiles.* **1.** the yarns running lengthwise in a woven fabric and interwoven with the weft; warp yarns are stretched in parallel lines on a loom to form a foundation, and the weft or filling threads are woven through them by means of a shuttle. **2.** the yarns running lengthwise in a warp-knit fabric. **3.** of or relating to the selvage way of a fabric. *Geology.* **1.** a slight, generally broad or regional upward or downward bend in the earth's crust. **2.** a sedimentary layer that was deposited by water. *Navigation.* **1.** a hawser that is used in moving a ship in harbor; it is attached at one end to a buoy, anchor, or fixed point on shore and hauled in at the ship end. **2.** a rope or wire by which a trawl is set and recovered by a trawler. **3.** to move a vessel by means of warps.

warpage *Mechanics.* a turning and twisting of a material out of shape, due to a gradual, inhomogeneous change of internal stress or mechanical properties.

warp knit *Textiles.* a type of knitting in which yarns interlock to form rows running lengthwise, resulting in a fabric that is flatter, closer, and less elastic than weft or jersey knits.

warrant *Geology.* see UNDERCLAY.

warrant officer *Military Science.* an officer ranking above noncommissioned officers and below commissioned officers; there are four grades.

Warren truss *Civil Engineering.* a truss having diagonal bracing only, without vertical members.

war reserve *Ordnance.* nuclear weapons stockpiled by the Department of Energy or transferred to the Department of Defense, and intended for employment in the event of war.

war reserve materiel requirement *Military Science.* the portion of the war materiel requirement required to be on hand on D-day; it is equal to the war materiel requirement minus the sum of the peacetime assets assumed to be available on D-day and the war materiel procurement capability.

war reserves *Ordnance.* stocks of materiel amassed in peacetime to meet the increase in military requirements upon an outbreak of war; such reserves are intended to provide the interim support necessary to sustain operations until resupply can be effected.

war surplus *Ordnance.* an item of military materiel that is obsolete, unserviceable, or unnecessary to meet current and reserve requirements.

wart *Medicine.* **1.** a benign epidermal lesion with a horny surface, caused by a human papillomavirus, transmitted by contact or by autoinoculation, and usually occurring on the fingers and hands. **2.** any of various nonviral, epidermal proliferations with similar characteristics. Also, VERRUCA.

Wartenberg's symptom *Neurology.* **1.** itching of the nostrils and tip of the nose in cases of cerebral tumor. **2.** flexion of the thumb when attempting to flex the four fingers against resistance, exhibited as a sign of corticospinal tract lesion.

Warthin's tumor see ADENOLYMPHOMA.

warthog or **wart hog** *Vertebrate Zoology.* a large wild pig, *Phacochoerus aethiopicus*, of the family Suidae; characterized by backcurved, heavy tusks protruding from a thick snout, sparse hair, and a shaggy mane; found on the savannas of Africa. (So named because of the large, wartlike growths on its snout.)

warthog

warwickite *Mineralogy.* $(Mg,Ti,Fe^{+3},Al)_2(BO_3)O$, a dark brown to dull-black, subtranslucent, orthorhombic mineral occurring as slender prismatic crystals with rounded ends, having a specific gravity of 3.35 and a hardness of 3.5 to 4 on the Mohs scale; found in crystalline limestone with magnetite and ilmenite near Warwick, New York.

WAS *Aviation.* the airline code for the city of Washington, D.C.; principal airports in the Washington Metropolitan Area are Dulles International (IAD), National (DCA), and Baltimore/Washington International (BWI).

Wasatchian *Geology.* a North American geologic stage of the Eocene epoch, occurring after the Clarkforkian of the Upper Paleocene epoch and before the Bridgerian.

Wasatch winds *Meteorology.* strong, easterly winds that blow out of the canyons of the Wasatch Mountains onto the plains of Utah.

wash to apply water or another liquid, as in cleaning; specific uses include: *Geology.* **1.** any loose, unconsolidated or eroded surface material carried and deposited by running water. **2.** an area of land that is intermittently washed over by a sea or river. **3.** the dry bed of an intermittent stream, usually at the bottom of a canyon. Also, DRY WASH, WASHOUT. **4.** see ALLUVIAL CONE. *Engineering.* **1.** to clean material out of a borehole by the jetting and buoyant action of a flow of water or mud-laden liquid. **2.** the erosion of core or drill string equipment by rapidly flowing water or mud-laden liquid. *Aviation.* the wake of turbulent air behind an aircraft in flight. *Metallurgy.* a coating that is applied to the face of a mold before casting.

washability *Mining Engineering.* the amenability of a coal to quality improvement by means of cleaning.

wash-and-strain ice foot *Oceanography.* an ice foot formed from ice casts and slush, having a rough and irregular surface; although it may become as solid and fast as other types of ice foot, travel on it is more difficult than on smoother types.

wash cone see OUTWASH CONE.

washed clot *Hematology.* a blood clot that is composed of fibrin and platelets. Also, WHITE CLOT.

washer *Mechanical Devices.* a ring, usually of metal but also of rubber and other materials, placed between moving parts or on fasteners to fill space, reduce friction, and secure joints.

washer cutter *Mechanical Devices.* a fixed center device with one or two adjustable cutting points for cutting washers from soft materials, such as leather.

washer thermister *Electronics.* a washer-shaped thermister having a center hole for a mounting screw or a stacking rod.

washin or **wash-in** *Aviation.* **1.** a method of increasing lift by increasing the angle of incidence on the outer part of an airplane wing; used to counteract the effects of torque; the opposite of *washout.* **2.** a twist built into a wing to accomplish this effect.

washing a process of cleaning by means of the flow of a liquid; specific uses include: *Analytical Chemistry.* the removal of impurities from precipitates by adding solvents to the precipitate, decanting, and then repeating the process until the desired purity is reached. *Chemical Engineering.* the process of cleaning a solids bed or cake with a liquid in which the solid is not soluble.

washing plant *Mining Engineering.* a plant using washing, tumbling, or scrubbing processes to remove slimes from coarse ore or to disintegrate diamondiferous gravel to produce sized screen fractions.

washing soda *Inorganic Chemistry.* a popular name for sodium carbonate decahydrate, $Na_2CO_3 \cdot 10H_2O$, a substance that is often used in cleansers and for washing and bleaching. See SODIUM CARBONATE DECAHYDRATE.

Washitan *Geology.* a North American Gulf Coast stage of the Upper Cretaceous period, occurring after the Fredericksburgian and before the Woodbinian.

wash load *Geology.* **1.** the part of a total stream load that is composed of fine particles derived from bank erosion or some source upstream. **2.** a suspended load.

wash metal *Metallurgy.* molten metal that is used to wash out a furnace, ladle, or other containers.

Washoe zephyr *Meteorology.* a chinook that blows along the eastern slopes of the Sierra Nevada in northern California.

washout *Aviation.* **1.** a method of decreasing lift of an airplane wing by decreasing the angle of attack toward the tip; the opposite of *washin.* **2.** a warp built into a wing to accomplish this effect. *Engineering.* **1.** an enlarged well bore caused by the solvent and erosional action of the drilling fluid. **2.** a fluid-cut opening caused by leaking fluid. *Mining Engineering.* **1.** sandstone or shale found within a channel in a coal seam cut by flowing water. Also, CUTOUT, HORSE, HORSEBACK, ROLL, SWELL, SYMON FAULT. **2.** a barren, thin, or jumbled portion of a coal seam without actual disruption or vertical displacement of the coal and strata.

washover *Geology.* **1.** the sedimentary material deposited by wave overwash, especially a small delta produced by storm waves on the landward side of a barrier. Also, STORM DELTA, WAVE DELTA. **2.** the process by which a washover is formed.

wash plain **1.** see ALLUVIAL PLAIN. **2.** see OUTWASH PLAIN.

wash water *Chemical Engineering.* water that has contacted gas or liquid process streams, filter cakes, or packed beds to dissolve or flush away unclean substances and impurities.

wasp *Invertebrate Zoology.* any of numerous social or solitary hymenopterous insects of the Vespidae and Sphecidae families, or other allied families, having a slender body, a narrow waist, and a powerful sting; they usually feed on other insects, spiders, and decaying animal matter.

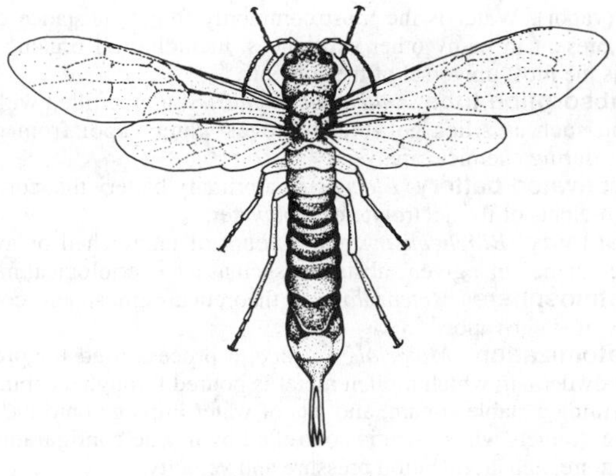

horntail wasp

Wassermann, August von [wäs´ər mən; väs´ər mən] 1866–1925, German physician and bacteriologist; he devised the **Wassermann test,** a blood test formerly in wide use for diagnosing syphilis.

Wassermann reagin see REAGIN, def. 2.

waste *Engineering.* **1.** any building or residential trash. **2.** dirty water resulting from mining use, industrial use, or other industrial wastes. **3.** the quantity of excavated material that exceeds fill. *Mining Engineering.* **1.** in a mine, the barren rock or a portion of an ore deposit that is not currently economically viable, but that could be profitably treated at a later time. **2.** the unpacked, unsupported region behind a working face. **3.** the refuse resulting from the dressing and smelting of ore, or the fine coal resulting from the mining and preparation of coal.
waste bank *Mining Engineering.* a bank of earth made parallel to the ditch from which the earth was dug.
waste filling *Mining Engineering.* waste material, such as rock, mill tailings, sand and gravel, and smelter slag, used as support in heavy ground and in large stopes to prevent the collapse of rock walls and to control subsidence, thus allowing the extraction of pillars left behind during earlier phases of mining.
waste flake *Archaeology.* a piece of stone occurring as a waste product during the manufacture of a stone tool; either a larger piece flaked off from the original stone (**primary waste flake**) or a smaller piece removed during finishing (**secondary waste flake**).
waste heat *Engineering.* detectable, noncombustible heat in gases that will be used in processes downstream in the system.
waste-heat boiler *Chemical Engineering.* a unit for recovering heat that uses hot oil or gas by-products from chemical processes; used for producing steam in a boiler-type system. Also, GAS-TUBE BOILER.
waste lubrication *Engineering.* a method for delivering lubrication to a bearing surface through the wicking movement of cloth waste.
waste plain see ALLUVIAL PLAIN.
waste raise *Mining Engineering.* an excavation in which barren rock and other economically valueless material is broken up for use in filling a stope.
waste rock *Mining Engineering.* rock that cannot be economically treated but must be fractured and removed in order to access and excavate ore.
waste water see RETURN FLOW.
wasting disease *Pathology.* any disease marked by progressive weakness and emaciation.
WAT word association test.
watch *Horology.* a portable timepiece small enough to be carried in the pocket or strapped to the wrist. *Navigation.* a work shift during which part of a ship's crew is on duty, usually alternating every four hours.
watch crystal *Horology.* see CRYSTAL.
watch error *Horology.* the discrepancy in the time indicated by a watch from correct time as indicated by a timekeeping standard; usually denoted by seconds fast or slow per day.
watching mine *Ordnance.* a naval mine that is secured to its mooring but observable on the surface, possibly only under certain tidal conditions.
watch rate *Horology.* a measurement of the accuracy of a watch based on the amount of time that it regularly gains or loses each day.
water *Chemistry.* H_2O, a colorless, odorless, tasteless liquid having a melting point of 0°C and a boiling point of 100°C at standard atmospheric pressure, and having the allotropic forms of ice (solid) and steam (vapor). Water is the most commonly found substance on earth and is present in many other substances, including all organic tissues, and it is the most universal of the solvents.
water-absorption tube *Analytical Chemistry.* a tube filled with an absorbent, such as silica gel, that removes water vapor from gaseous streams during chemical analysis.
water-activated battery *Electricity.* a primary battery that contains all the ingredients of its electrolyte except water.
water activity *Biochemistry.* the amount of unattached or available water contained in a given substance, such as a bacteriological medium.
water atmosphere *Meteorology.* a theoretical atmosphere composed entirely of water vapor.
water atomization *Materials Science.* a process used for producing metal powders, in which molten metal is poured through a ceramic nozzle, forming a stable stream, and jets of water impinge onto the stream, forming droplets whose size is controlled by nozzle configuration, pour temperature, and atomization pressure and velocity.
water ballast *Naval Architecture.* water that is taken aboard a vessel deliberately as ballast. Thus, **water-ballast tank.**
water-base mud *Petroleum Engineering.* drilling mud having water as the liquid base, used in oil-well drilling operations.
water-base paint *Materials.* any paint, such as latex paint, in which the vehicle or binder is dissolved in water. Also, WATER PAINT.
water bath *Microbiology.* a container in which water is maintained at a constant temperature for the incubation of microbial culture vessels.
water bear see TARDIGRADA.
Water Bearer see AQUARIUS.
water-bearing strata *Geology.* a sedimentary formation lying directly below the standing water level, so that it contains or yields water.
water biscuit see ALGAL BISCUIT.
water block *Petroleum Engineering.* an accumulation of water-oil emulsion near the lower end of an oil-well borehole, retarding fluid movement through the formation and toward the borehole.
water boatman *Invertebrate Zoology.* any of various aquatic insects of the family Corixidae, characterized by long, oarlike hind legs by which they propel themselves through the water.
water-boiler reactor *Nucleonics.* a reactor employing a modification of the pressurized-water concept to produce steam directly, by using fission heat to boil water within a reactor core rather than in an external heat exchanger.
waterborne *Science.* conveyed by water.
waterborne disease *Pathology.* any disease carried by water and transmitted to the host by water ingestion, such as typhoid fever.
water brake *Mechanical Engineering.* an absorption dynamometer that measures the power output of an engine shaft in which the mechanical energy is converted to heat in a centrifugal pump, with a free casing where turning moment is measured.
water break *Metallurgy.* a break in the water film found on a metal when it is removed from a bath.
water budget or **water balance** see HYDROLOGIC BUDGET.
water buffalo *Vertebrate Zoology.* a massive ox, *Bubalis bubalis,* of the family Bovidae, standing about 6 feet tall and having large splayed hooves and huge, ridged and upcurving horns that spread up to 7 feet across; domesticated in India and Southeast Asia as a beast of burden since about 3000 BC, and gradually introduced worldwide.

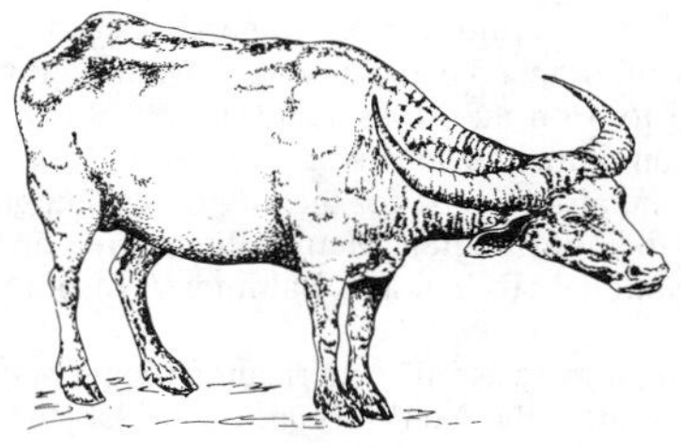

water buffalo

waterbug *Invertebrate Zoology.* a popular term for any of various aquatic insects, such as the water boatmen, the back swimmers, or the giant waterbugs of the family Belostomatidae.
water calorimeter *Engineering.* an instrument that measures radio-frequency power by noting the rise in temperature of water absorbing the radio-frequency energy.
water clock *Horology.* an early timekeeping device based on the uniform flow of water out of or into a graduated vessel. Also, CLEPSYDRA.
water cloud *Meteorology.* a cloud composed entirely of liquid water drops; a cloud of above-freezing temperatures.
watercolor *Materials.* any pigment for which water, rather than oil, is used as the vehicle.
water conservation *Ecology.* the process or methods of protecting water resources through efficient development and management policies.
water content *Hydrology.* **1.** the quantity of water present within a porous sediment or sedimentary rock, expressed as the ratio of the weight of water in the sediment to the weight of the sediment when dried. **2.** see FREE-WATER CONTENT.
water-cooled *Mechanical Engineering.* describing an engine, machine gun, or other device that is cooled by water circulating in pipes or a water jacket. Thus, **water-cooled engine, water-cooled reactor,** and so on.
water-cooled condenser *Mechanical Engineering.* a condenser in which cooling water is used to remove the latent or sensible heat of hot fluids; the heat is transferred from the condenser tube to water circulating in passageways around it, while the water is air-cooled separately in a radiator or cooling tower before recirculation.

water-cooled furnace *Mechanical Engineering.* a fuel-fired furnace with tubes through which water circulates to provide insulation, generate steam, and limit heat loss.

watercourse *Hydrology.* **1.** a natural or artificial channel through which water flows or may flow intermittently or continuously. **2.** any stream or current of water.

watercress *Botany.* a small, perennial creeping plant, *Nasturtium officinale*, growing in watery places; its leaves have a pungent flavor and are eaten as a salad and used in flavoring and garnishing.

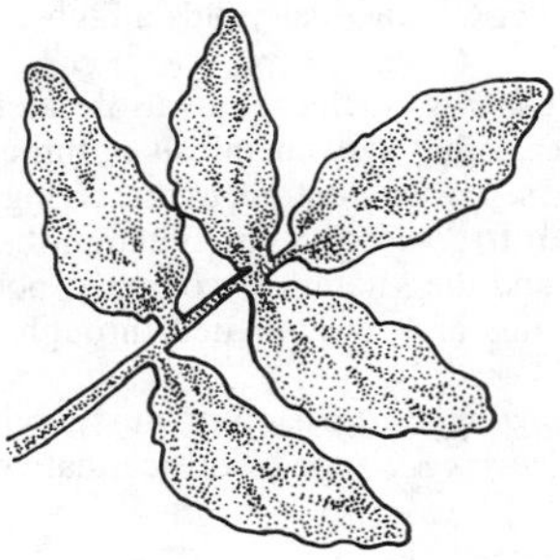

watercress

water crop see WATER YIELD.

water cupola *Volcanology.* a bulge in the surface of the ocean directly above an underwater volcanic explosion.

water curb see GARLAND, def. 1.

water cycle see HYDROLOGIC CYCLE.

water demineralizing *Chemical Engineering.* the process of removing minerals, such as calcium, sodium, and magnesium, from water by using distillation, ion-exchange, or chemical procedures.

water-drive reservoir *Petroleum Engineering.* a reservoir into which water is injected to control the pressure as oil or gas is extracted.

water equivalent *Meteorology.* a depth assigned to water from the equivalent of melting snowpack or icepack in which the water equivalent of a new snowfall equals the amount of precipitation for that same snowstorm.

water exchange *Oceanography.* the volume and rate of water exchange between bodies of water in a given location, controlled by tides, winds, river discharge, and currents; also, the volume and rate of water exchange between a body of water and the atmosphere.

waterfall *Hydrology.* a place in a stream where the course is interrupted suddenly, causing the water to descend more or less vertically without support for a significant distance.

waterfall lake see PLUNGE POOL, def. 2.

waterfall method *Materials Science.* in commercial ceramic tile production, a process for applying a glaze coating in which the glaze is run freely over the face of the tile.

water-flow pyrheliometer *Engineering.* an absolute pyrheliometer in which the radiation-sensing element is a blackened cylinder enclosed in a chamber through which water flows at a constant rate; the water temperature is monitored by thermometers as it enters and exits, to determine radiation intensity.

water fountain see FOUNTAIN.

waterfowl *Vertebrate Zoology.* any aquatic bird of the order Anseriformes, especially ducks, geese, and swans.

water front *Geography.* the land margin of a body of water, especially a harbor.

water gap *Geology.* a deep, narrow pass across a ridge, mountain, or other area of high elevation through which a river or stream flows.

water gas *Materials Science.* a flammable, explosive, toxic mixture of gases consisting of carbon monoxide, hydrogen, carbon dioxide, and nitrogen; it is prepared from coke, air, and steam and is used for fuel, illumination, and organic synthesis. Also, BLUE GAS.

water gas reaction *Chemical Engineering.* a procedure for producing carbon monoxide by passing steam above hot coal or coke at temperatures of 600–1000°C.

water gauge *Engineering.* a gauge that displays the water level in a boiler by means of a vertical glass tube connected to the water and steam spaces in the boiler.

water glass see SODIUM SILICATE.

water hammer *Fluid Mechanics.* a pressure increase in a pipeline due to a sudden retardation or acceleration of the flow.

water-hammer pulse see CORRIGAN'S PULSE.

water-harden *Metallurgy.* to cool steel rapidly by quenching, thus forming a shock-resistant core and a hard exterior that can be ground off. Thus, **water-hardening, water-hardened (tool) steel.**

water heater *Mechanical Engineering.* **1.** a perforated nozzle that heats water within a vessel by blowing steam into it. **2.** any system, industrial or domestic, used to heat water.

waterhole *Hydrology.* a small natural depression that contains water, especially in an arid or semiarid region.

water horizon see AQUIFER.

Waterhouse-Frederickson syndrome *Medicine.* the malignant or fulminating form of epidemic cerebrospinal meningitis, characterized by fever, coma, hemorrhages of the skin and mucous membranes, adrenal hemorrhage, and cyanosis.

water hyacinth *Botany.* a floating aquatic plant, *Eichhornia crassipes,* of the family Pontederiaceae, that grows very rapidly on the surface of tropical and subtropical rivers, lakes, and swamps.

water influx *Petroleum Engineering.* **1.** the natural flow or injection of water into petroleum reservoirs. **2.** an oil-recovery method using water to displace the reservoir fluids and move them toward the borehole.

water jacket *Mechanical Engineering.* a casing filled with circulating water, used in water-cooling an engine, machine gun, or other device.

waterjet *Naval Architecture.* a propulsion system consisting of a pump-driven jet of water directed from the rear of the vessel.

water-knockout drum *Petroleum Engineering.* an apparatus used to remove water from the hydrocarbon fluids in an oil well. Also, **water-knockout trap, water-knockout vessel.**

waterleaf *Materials Science.* paper that has not been sized with rosin prior to tub sizing.

water level *Hydrology.* see WATER TABLE.

water lily *Botany.* any of several aquatic plants in the genera *Nymphaea* and *Nuphar* of the family Nymphaecea, having large floating leaves and showy flowers on long stalks.

waterline or **water line** *Geology.* **1.** the temporal line of contact, at any given time, between the land and any body of standing water. **2.** see SHORELINE, def. 1. *Hydrology.* see WATER TABLE. *Naval Architecture.* **1.** the vertical level at which a vessel floats. **2.** a line or lines on the hull indicating the level and trim at which a vessel should float under various loading conditions.

water load *Electromagnetism.* a waveguide termination whereby electromagnetic energy is dissipated into a specified amount of water; the initial and final temperatures of the load of water can aid in determining the amount of microwave power.

water main *Civil Engineering.* a public water-supply line, such as that under a street, feeding into individual users' supply lines.

watermark *Graphic Arts.* a design, often the name or logo of the manufacturer, that is pressed into wet paper during manufacturing.

water mass *Oceanography.* a body of water that can be identified by a particular set of physical and chemical characteristics, often consisting of two or more water types.

watermelon *Botany.* a large, round or oblong fruit, *Citrullus lanatus*, of the family Cucurbitaceae, having a hard, green rind and juicy, pink or red pulp containing many seeds.

water meter *Engineering.* an instrument that measures the quantity of water flowing past a given point in a piping system.

water microbiology *Microbiology.* the branch of microbiology that concerns microorganisms existing in water environments.

water moccasin *Vertebrate Zoology.* a venomous, semiaquatic pit viper, *Agkistrodon piscivorous*, that inhabits swamps in the southern and southeastern United States. Also, COTTONMOUTH.

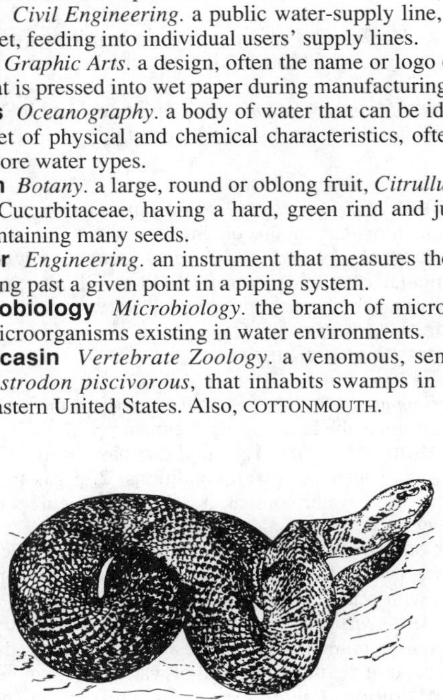

water moccasin

water-moderated reactor *Nucleonics.* a reactor in which either ordinary (light) water or heavy water (deuterium oxide) is used as a moderator; in some reactors, water is used both as coolant and as moderator.

water noise *Acoustics.* hydrodynamic noise; that is, noise resulting from the flow of water, as in a river or stream, or especially through a restricted volume such as a resonant pipe, producing knocking sounds and other turbulent flow noises.

water of constitution *Chemistry.* water that is chemically combined in a compound.

water of crystallization *Chemistry.* water that is a physical constituent of certain crystals or hydrated salts; can be removed by heating to leave an anhydrous salt. Also, **water of hydration.**

water opal see HYALITE.

water opening *Oceanography.* see OPENING.

water paint *Materials.* **1.** see WATER-BASE PAINT. **2.** any paint that can be diluted with water.

water path *Engineering.* the distance between an ultrasonic search unit and the test piece in an immersion-type ultrasonic examination.

water pig see CAPYBARA.

water plane *Naval Architecture.* the plane through a vessel's hull in which its waterline lies.

water plane area *Naval Architecture.* the area in vertical-projected cross section of a vessel's water plane.

water plane coefficient *Naval Architecture.* the ratio of a vessel's water plane area to that of the rectangle formed by its waterline length and beam.

water pollution *Ecology.* a general term for the presence in water of substances, usually chemicals or waste matter introduced by humans, that are harmful to organisms living in the water or to those that drink from it or are otherwise exposed to it.

water power *Engineering.* any power that is generated by the force of moving water; for example, in contemporary use, electric power generated through hydraulic turbines; historically, mechanical power from a waterwheel.

waterproof *Engineering.* **1.** impervious to water. **2.** to make impervious to water by treating with a waterproofing agent.

waterproofing *Materials.* a waterproofing agent. *Engineering.* the process of making something waterproof.

waterproofing agent *Materials.* **1.** a substance, such as paint or wax, used to form a water-repellent film on various kinds of surfaces such as cement, metal, or textiles. **2.** a metal salt or other chemical that penetrates textile fibers, making them water-repellent but allowing the passage of air through the fabric.

water purification *Civil Engineering.* the process of purifying drinking water supplies to ensure safety before the water enters the distribution system.

water putty *Mechanical Devices.* a woodwork powder that, when mixed with water, is used as a filler for cracks, nail holes, or other crevices.

water reducer *Materials Science.* an admixture agent in concrete, such as lignosulfonate, that provides good workability at lower water-to-cement ratios.

water regimen *Hydrology.* see REGIMEN.

water-repellent *Engineering.* resistant but not impervious to water. *Materials.* a material having this quality.

water requirement *Agriculture.* the total amount of water needed to grow a particular crop under normal field conditions, including both available environmental water and irrigated water.

water-resistant see WATER-REPELLENT.

water rheostat see LIQUID RHEOSTAT.

water ring see GARLAND.

waters *Developmental Biology.* a popular name for the amniotic fluid, a liquid that surrounds the fetus during pregnancy.

water saturation *Chemistry.* **1.** a solid that has absorbed its maximum amount of water under the given conditions. **2.** a gas that reaches its dew point due to its water content. **3.** a liquid that enters into a second liquid phase after the introduction of more water.

water scrubber *Chemical Engineering.* a system or device used to wash away traces of water-soluble components of a gas stream by contacting gases with water through bubbling or spraying.

water seal *Materials.* any waterproof sealant. *Engineering.* the seal formed in a trap, a plumbing fixture designed to prevent the flow of air and gases; a bend in the pipe stays full of water, creating the seal.

water-sealed holder *Engineering.* a low-pressure gas holder having cylindrical sections that telescope into a tank of water.

watershed *Hydrology.* **1.** the point of high ground dividing two different drainage systems. **2.** the region drained by a specific water system. *Forestry.* a forest that serves as a source of water for rivers and streams by soaking up rain and melted snow in its soil.

water sky *Meteorology.* a term used primarily in the polar regions to denote the dark appearance of the underside of a cloud layer occurring over open water; it is darker than land sky.

water smoke see STEAM FOG.

water snake *Vertebrate Zoology.* any of various snakes living in or around water, particularly any of several harmless snakes of the genus *Natrix.*

Water Snake see HYDRUS.

water snow *Hydrology.* snow that yields a high content of water when melted.

water softening *Chemistry.* the removal of calcium and magnesium ions from hard water, either with chemicals or through ion exchange.

water-soluble *Chemistry.* susceptible to dissolving in water.

water-soluble vitamin *Nutrition.* any vitamin that dissolves in water, such as vitamin C and the vitamin B complex; not stored in the body, water-soluble vitamins are disseminated through blood and body tissues.

waterspout *Meteorology.* a tornado or dust devil that occurs over a water surface. *Engineering.* a pipe or orifice that conveys or discharges water.

waterspout

water swivel *Mechanical Devices.* a device that connects a water hose to a drill rod, allowing the drill to operate while shavings from the borehole are washed out and the drill bit is cooled. Also, SWIVEL NECK.

water system see RIVER SYSTEM.

water table *Hydrology.* the upper boundary of the zone of saturation, being the surface of a body of unconfined groundwater at a pressure equal to the pressure of the atmosphere. Also, GROUNDWATER TABLE, PHREATIC SURFACE. *Building Engineering.* a projecting string course that is used to divert rainwater from a structure or building.

water-table cement or **water-table mound** see GROUNDWATER CEMENT.

watertight *Engineering.* designed, fitted, or secured so as to be impervious to water. Thus, **watertight compartment.**

watertight subdivision *Naval Architecture.* the subdivision of a vessel into a number of separate watertight compartments, allowing leakage or flooding to be limited to one portion of the vessel.

water tower *Civil Engineering.* an elevated water tank that supplies a water distribution system under a gravity pressure head.

water-transport number see TRANSFERENCE NUMBER.

water trap *Geology.* a section of a cave in which the roof drops below the water level, thus filling the area with water. *Civil Engineering.* see TRAP.

water treatment *Biotechnology.* the process of treating sewage or other effluent water sufficiently to permit it to be released into the environment by use of filtration, aeration, bacterial treatment, and so forth.

water-tube boiler *Mechanical Engineering.* **1.** a boiler in which water is contained in a series or sections of tubes. **2.** a steam boiler in which a number of closely spaced water-cooling tubes are connected to drums that act as water pockets and steam separators, an arrangement yielding rapid water circulation and quick steaming. Also, INTEGRAL-FURNACE BOILER.

water tunnel *Aviation.* an open-throated or closed-throated chamber or apparatus designed for aerodynamic testing, similar to a wind tunnel but using water instead of air as its working fluid.

water type *Oceanography.* seawater of a particular temperature and salinity.

water vapor *Physical Chemistry.* water that is in its gaseous state, especially when it is in this gaseous state below the boiling point of water. *Meteorology.* atmospheric water in vapor form; one of the most important of all components of the atmosphere, serving as the raw material for clouds and rain, as a vehicle for energy transport, and as a regulator of planetary temperatures via the greenhouse effect. Also, AQUEOUS VAPOR.

water-vapor laser *Optics.* a laser in which a water-vapor substance produces a beam of infrared radiation at specific wavelengths.

waterwall *Mechanical Engineering.* a water-cooling system used in boilers, in which a network of tubes carrying circulating water forms or lines one wall of the boiler.

waterway *Navigation.* **1.** a navigable channel, especially a dredged channel. **2.** a channel formed along the outboard edge of a weather deck which leads water to the scuppers.

water well *Civil Engineering.* any well from which groundwater is drawn. Depending on the depth of the water table, such wells may be dug, driven by means of a pointed pipe, bored by means of an auger, or drilled by means of a water-well drill.

water-well drill *Civil Engineering.* a specialized power-drilling apparatus used to tap very deep groundwater reservoirs.

water-well drill

water-wettable *Chemistry.* of a substance, able to absorb or adsorb water.

waterwheel *Mechanical Engineering.* a wheel arranged with floats or buckets so that it can be turned by flowing water; used as a source of energy to drive machinery or raise water.

water white *Chemistry.* a standard of color for liquids with the appearance of clear water.

waterworks *Civil Engineering.* **1.** a municipal or regional water distribution plant, typically consisting of a pumping station and purification facilities. **2.** an entire public water supply system, including reservoirs, pipelines, and other facilities.

water year *Hydrology.* any twelve-month period used as a basis for hydrologic data; in the United States, the period beginning October 1 and ending September 30.

water yield *Hydrology.* the amount of runoff from a drainage basin, equal to the amount of precipitation minus the amount lost by evapotranspiration. Also, RUNOUT.

WATS [wäts] *Telecommunications.* a telephone service that allows a subscriber to direct-dial an unlimited number of calls to one or more of six regions in the continental United States for a flat monthly charge. (An acronym for wide-area telephone service.)

Watson, James Dewey born 1928, American biochemist; with Crick, discovered the double helix and constructed the Watson-Crick model.

Watson, John Broadus 1878–1958, American psychologist; founder of the behaviorist school of psychology.

Watson-Crick model *Molecular Biology.* the double-helical molecular structure of DNA propounded by James D. Watson and Francis Crick.

Watson equation *Physical Chemistry.* an empirical equation that is used to calculate heat of vaporization data for organic compounds to within 10–15°C of their critical temperature.

Watson-Sommerfeld transformation *Mathematics.* a device for converting a series whose kth term is the product of the kth Legendre polynomial and a coefficient a_k into the sum of a contour integral of $a(k)$ and terms involving the poles of $a(k)$, where $a(k)$ is a meromorphic function such that $a(k) = a_k$ at integer values of k.

Watson-Watt, Sir Robert 1892–1973, Scottish electrical engineer; developed radar.

Watt, James 1736–1819, Scottish engineer and inventor; invented the condenser and the first practical steam engine.

watt *Physics.* a unit of power in the SI or MKS system of units, equivalent to one joule per second; in electrical power it is equal to the current in amperes multiplied by the electrical potential in volts. (Named for James *Watt.*)

Watt's law *Thermodynamics.* a law stating that the sum of the latent heat of vaporization of a sample of water at any temperature T and the heat transfer required to raise the same amount of water from 0°C to T is constant.

wattage *Electricity.* the power output or consumption of an electrical device, expressed in watts.

wattage rating *Electricity.* **1.** the recommended output power of an electronic device. **2.** the recommended power dissipation of an electronic device.

watt current see ACTIVE CURRENT.

wattevillite *Mineralogy.* a white, silky, orthorhombic or monoclinic mineral, with the approximate formula $Na_2Ca(SO_4)_2 \cdot 4H_2O$, occurring as aggregates of minute, hairlike or acicular, water-soluble crystals with a sweetish to astringent taste, having a specific gravity of 1.81 and an undetermined hardness; found in pyrite-rich lignite near Bischofsheim, Germany.

watt-hour *Electricity.* the practical unit of energy, equal to the power of 1 watt absorbed continuously for 1 hour.

watt-hour capacity *Electricity.* the number of watt-hours a storage battery can deliver safely under specific operating conditions including temperature, pressure, and other parameters.

watt-hour meter *Electrical Engineering.* an integrating meter that measures the total electric energy consumed in a circuit over time.

wattle *Building Engineering.* an interlacing of twigs or tree branches used for fence, wall, roof, or framing construction. *Vertebrate Zoology.* a flap or fold of flesh that hangs down from the throat of certain animals, such as the turkey.

wattle-and-daub *Archaeology.* denoting a type of construction in which a framework of flexible wooden sticks (wattle) is coated with mud (daub).

wattless current see REACTIVE CURRENT.

wattless power see REACTIVE POWER.

wattmeter *Engineering.* an instrument that measures electric power in terms of watts.

watt-second *Physics.* a unit of energy equivalent to one joule.

wave *Physics.* a uniformly advancing disturbance in a medium, in which the moved parts undergo a double oscillation; a collective disturbance that propagates at a definite speed. Also, WAVE MOTION. *Fluid Mechanics.* any disturbance or sudden change in conditions that moves through a flow. *Oceanography.* such a disturbance occurring on the surface of the sea or other body of water, taking the form of a moving swell or ridge of water. *Military Science.* a formation of forces, landing ships, craft, amphibious vehicles, or aircraft that is required to beach or land at approximately the same time; it may be classified according to type, function, or landing order as follows: assault wave, boat wave, helicopter wave, numbered wave, on-call wave, or scheduled wave.

wave acoustics *Acoustics.* the study of acoustic energy behavior, based on the premise that sound travels in waves, a theory supported by transmission phenomena such as refraction, reflection, and diffraction as well as phenomena such as the Doppler effect, which can be related to wavelength and frequency.

wave amplitude *Physics.* the maximum value of a waveform.

wave angle *Electromagnetism.* the angle of transmission or reception of an antenna with respect to some reference direction.

wave antenna *Electromagnetism.* an antenna consisting of several horizontal, parallel radiating elements of lengths 1/2 to several wavelengths long, each of which is terminated to ground at its far end through its characteristic impedance.

wave band *Telecommunications.* see BAND.

wave base *Hydrology.* the depth of water beyond which suspended sediments are not agitated by wave action.

wave-built terrace *Geology.* a gently sloping surface built up at the seaward margin of a wave-cut platform from sediments that drifted along the shore or across the platform and were deposited in the deeper water beyond.

wave celerity see PHASE VELOCITY.

wave converter *Electromagnetism.* any of various waveguide devices that change a wave from one type to another; e.g., baffle plates, gratings, and sheath-reshaping converters.

wave crest *Physics.* the position at which a waveform has a maximum; crests propagate with the wave through the medium.

wave-cut bench *Geology.* a level or slightly inclined erosional surface or platform that is carved by the action of waves and projects outward from the base of a sea cliff. Also, BEACH PLATFORM, HIGH-WATER PLATFORM.

wave-cut cliff *Geology.* a cliff formed along a coast by the action of waves cutting horizontally and landward into rock.

wave-cut notch *Geology.* a small, V-shaped depression cut out at the base of a sea cliff by the erosive action of waves.

wave-cut platform *Geology.* a gently sloping erosional surface or ledge carved by wave erosion that extends from the base of a wave-cut cliff into the sea for some distance. Also, **wave-cut terrace, wave-cut plain.**

wave cyclone *Meteorology.* a cyclone that forms and moves along a front, producing by its circulation a wavelike deformation of the front. Also, **wave depression.**

wave delta see WASHOVER.

wave depth see WAVE BASE.

wave disturbance *Meteorology.* a partially or poorly developed wave cyclone.

wave duct *Electromagnetism.* **1.** a channel set up in the atmosphere, through which electromagnetic waves of certain frequencies may pass easily. **2.** a waveguide having tubular boundaries capable of sustaining and concentrating electromagnetic wave propagation.

wave equation *Physics.* a partial differential equation of space and time given by

$$\nabla^2\psi - (1/c^2)\partial^2\psi/\partial t^2 = 0,$$

where ψ is the disturbance that is propagating, c is the propagation velocity, ∇^2 is the Laplacian operator, and t is the time coordinate.

wave filter *Electricity.* a transducer or other circuit generating device that allows different levels of attenuation for signals of different frequencies.

wave forecasting *Oceanography.* the prediction of future wave conditions from observed or calculated ocean and meteorological data.

waveform *Physics.* a graphical representation of the shape of a wave for a given instant in time over a specified region in space.

waveform analysis *Physics.* the resolution of a complex waveform into a sum of simply periodic waveforms, usually by mathematical or electronic means.

waveform analyzer *Acoustical Engineering.* an electroacoustical device that measures various characteristics of complex electroacoustic waves, using a graphic display.

wavefront *Physics.* a surface of constant phase in a wave's motion.

wavefront reversal see OPTICAL PHASE CONJUGATION.

wavefront splitting *Optics.* any technique that produces two beams of light from a single source and then recombines them in order to produce interference.

wave function see SCHRÖDINGER WAVE FUNCTION.

wave gait *Robotics.* the oscillating movement of a multilegged robot.

wave gauge *Engineering.* an instrument that measures the height of electromagnetic or radio waves and the time it takes for them to complete one cycle of oscillation.

wave group *Physics.* several waves propagating together but having slightly different wavelengths.

waveguide *Electromagnetism.* a hollow metal pipe, usually of rectangular cross section, that is designed to transmit microwave energy along its longitudinal axis.

waveguide assembly *Electromagnetism.* a complete system of waveguide devices and components arranged so as to perform a specific function with microwave energy.

waveguide attenuation *Electromagnetism.* the decrease from one point of a waveguide to another in the power carried by an electromagnetic wave in a waveguide.

waveguide bend *Electromagnetism.* a waveguide section having a smooth and continuous change in the direction of its longitudinal axis. Also, **waveguide elbow.**

waveguide cavity *Electromagnetism.* a section of waveguide that is enclosed by a pair of irises, which form shunt susceptances.

waveguide critical dimension *Electromagnetism.* a cross-sectional dimension of a waveguide that determines the cutoff frequency.

waveguide cutoff frequency *Electromagnetism.* a limiting frequency that bounds a range of frequencies that can be propagated through a waveguide.

waveguide filter *Electromagnetism.* a device composed of waveguide components that is used to alter the amplitude-frequency response characteristics of a waveguide system.

waveguide hybrid *Electromagnetism.* a waveguide junction of four arms in which microwave energy entering through one arm will be divided two ways and exit the junction through the adjacent arms, but not through the opposite arm.

waveguide propagation *Telecommunications.* the process and techniques of long-range communications in the 10–35 kHz frequency range using the waveguide characteristics of the atmospheric duct created by the ionospheric D-layer and the surface of the earth; also used in microwave optical communications systems.

waveguide shim *Electromagnetism.* a thin metallic sheet that serves to provide electrical continuity between waveguide components. Also, **waveguide gasket.**

waveguide slot *Electromagnetism.* a narrow slot cut into the wall of a waveguide, used for coupling or the insertion of probes or other measuring devices into the guide.

waveguide switch *Electromagnetism.* a movable section of waveguide that is designed to mate with one of several other waveguide components.

wave-height correction *Navigation.* a correction applied to the altitude of a celestial sight in order to account for the elevation of the visible horizon by ocean waves.

wave impedance *Electromagnetism.* in a waveguide, the ratio of the transverse component of the electric field to the transverse component of the magnetic field.

wave intensity *Physics.* a measure of the rate of wave energy flow; the wave energy per unit area per unit time.

wavelength *Physics.* the spatial distance between adjacent points of equal phase.

wavelength constant see PHASE CONSTANT.

wavelength shifter *Electronics.* **1.** a frequency shifter whose performance is indicated in units of wavelength. **2.** a photofluorescent compound that is used with a scintillator material to emit photons of a longer wavelength than those absorbed; used to increase the efficiency of a photocell or phototube.

wavelength standards *Spectroscopy.* known wavelengths emitted by specific light sources that are used for comparison purposes to interpolate unknown wavelengths in spectra.

wave line *Oceanography.* a line or ridge indicating the farthest advance of waves as they break on the shore, formed by the accumulation of sand, bits of seaweed, debris, and the like at this point.

wavellite *Mineralogy.* $Al_3(PO_4)_2(OH,F)_3 \cdot 5H_2O$, a colorless, white, green, yellow, or brown, transparent to translucent, orthorhombic mineral occurring as crusts and spherical aggregates of radiating acicular crystals, having a specific gravity of 2.36 and a hardness of 3.5 to 4 on the Mohs scale; found in aluminous, low-grade metamorphic rocks and in limonite and phosphate-rock deposits.

wavemark *Oceanography.* see WAVE LINE.

wave mechanics see SCHRÖDINGER'S WAVE MECHANICS.

wavemeter *Engineering.* an instrument that measures the geometrical spacing in an electromagnetic wave between a series of surfaces with the same phase.

wave microphone *Acoustical Engineering.* a microphone having a directivity pattern that depends on the phenomenon of wave interference, such as a line microphone or a reflector microphone.

wave motion *Physics.* see WAVE.

wave motor *Mechanical Engineering.* a motor whose energy is derived from the undulating motion of sea waves.

wave normal *Physics.* a unit vector that is perpendicular to a wave front.

wave number *Physics.* **1.** the reciprocal of the wavelength of a wave. **2.** sometimes, 2π times the reciprocal of the wavelength.

wave optics *Optics.* the branch of optics that analyzes radiant energy in terms of its wave characteristics.

wave packet *Physics.* the superposition (addition) of several waves of different wavelengths and phases to produce a pulselike or localized traveling waveform.

wave-particle duality *Quantum Mechanics.* the principle that photons and other particles behave like a wave motion when being propagated, and behave like particles when interacting with matter; thus descriptions of their behavior solely as waves or solely as particles are inadequate. Also, DUALITY PRINCIPLE.

wave period *Physics.* the time that elapses between the passage of adjacent points of equal phase through a fixed point in space; given by the reciprocal of the frequency for periodic waves.

wave plate see RETARDATION PLATE.

wave refraction *Physics.* the change in direction of wave propagation when a wave passes from one medium to another.

wave setdown *Oceanography.* a decrease in the mean water level in the zone of wave break, caused by a pressure field.

wave setup *Oceanography.* an increase in mean water level between the zone of wave break and shore, caused by the momentum of the waves.

wave shaper *Engineering.* in explosives, an insert containing either inert material or explosives with varying detonation rates, used to alter the shape of the detonation wave.

wave-shaping circuit *Electronics.* a circuit producing an output waveshape that differs from the input waveshape; for example, converting a square wave into a ramp.

wave speed see PHASE VELOCITY.

wave system *Oceanography.* a group of waves of similar height, direction, and length; generally, ocean waves consist of several superimposed wave systems.

wave tail *Electronics.* the part of a wave that begins at the initial decay and ends where the wave amplitude is zero.

wave theory of cyclones *Meteorology.* a theory of cyclogenesis that is founded upon the concept of wave formation at the interface between two fluids, as along a weather front.

wave theory of light *Optics.* the theory that light travels as a wave, originally stated by Huygens (1690) and supported by the experiments of Young (1801) and Fresnel (1814–1815), as opposed to the conflicting view that it is a stream of tiny particles, proposed by Newton (1704); modern quantum theory describes light as having both wave and particle characteristics.

wave tilt *Electromagnetism.* a slight tilt forward in the propagation of the electric field vector of a radio wave that is radiated by a vertical antenna at the surface of the earth, due to its proximity to the ground.

wave train *Physics.* a continuous series of waves.

wave trap *Electronics.* a resonant circuit that prevents a wave of particular frequency from entering a receiver. *Civil Engineering.* a series of small breakwaters designed to break the force of ocean waves entering a harbor.

wave trough *Physics.* the position at which a waveform has a minimum; troughs propagate with the wave and are found between adjacent wave crests.

wave vector *Physics.* a vector whose direction is the direction of propagation and whose magnitude is given by the wave number, $2\pi/\lambda$, where λ is the wavelength.

wave-vector space *Solid-State Physics.* the space of wave vectors of the state functions in a system; used in calculations of energies of electrons, X-ray diffraction, and thermal vibrations in a crystal lattice. Also, K-SPACE, RECIPROCAL SPACE.

wave velocity see PHASE VELOCITY.

WAW *Aviation.* the airport code for Warsaw, Poland.

wax *Materials.* **1.** a solid, yellowish substance secreted by bees for use in building the honeycomb; it is composed of esters, cerotic acid, and hydrocarbon and is easily molded when warm. Also, BEESWAX. **2.** any of a wide variety of similar substances, including paraffin, cerumen, spermaceti, and vegetable wax; used for coatings, candles, and polishes. *Engineering.* to apply or use such materials.

waxbill *Vertebrate Zoology.* any of numerous seed-eating tropical Old World songbirds of the family Estrildidae, characterized by brilliantly colored plumage with striking markings and a waxy white to reddish bill.

wax D *Biochemistry.* a peptidoglycolipid in the cell wall of a species of *Mycobacterium.*

wax fractionation *Chemical Engineering.* a lube-oil refining process that involves continuous solvent recovery and crystallization; used for the manufacture of waxes with low oil content from wax concentrates.

wax gland see CERUMINOUS GLAND.

waxing *Engineering.* the process of applying a wax to any surface or material.

waxing *Astronomy.* of the moon, increasing the extent of its illumination.

waxing moon *Astronomy.* the lunar phases from the new moon through the full moon.

wax manufacturing *Chemical Engineering.* a method of petroleum refining that is much like wax fractionation, used to produce oil-free waxes by chilling and crystallization from a solvent.

wax paper *Materials.* a type of paper that is coated with wax to make it liquid repellent; used as a sanitary wrapping for food and medical supplies. Also, **waxed paper.**

wax tailings *Materials Science.* a brown, sticky, semiasphalt substance that is obtained just before the formation of coke from the destructive distillation of petroleum; used as a preservative for wood and in roofing paper.

waxwing *Vertebrate Zoology.* any of three bird species of the genus *Bombycilla* in the family Bombycillidae, found in the Northern Hemisphere; characterized by a plump body, a short bill, a tail and wings having waxy red or yellow tips, and a showy crest.

waxwing

waxy *Materials.* **1.** covered with or made of wax. **2.** resembling wax. *Mineralogy.* describing a type of mineral luster that resembles the luster of wax.

way point *Navigation.* a specific point on the course between departure and destination that a craft can use to check its progress. *Robotics.* see VIA POINT.

ways *Mechanical Engineering.* **1.** the smooth longitudinal guides on the top surface of a lathe bed on which the carriage and tailstock slide. **2.** the timbers from which a ship slides when being launched.

wayside *Transportation Engineering.* located along a transport path or guideway.

wayside control *Transportation Engineering.* a command and control system operating from a guideway.

WB or **W.B.** water ballast; weather bureau.

Wb weber.

WBC or **W.B.C.** white blood cell; white blood count.

W boson *Particle Physics.* a particle that, along with the Z boson (or Z particle), carries the weak force in particle interactions. Also, **W particle.**

W chromosome *Genetics.* the female-determining sex chromosome in certain organisms, such as chickens and moths, in which the female is the heterogametic sex; the males are ZZ and the females are WZ.

weak acid *Chemistry.* an acid in an aqueous solution that does not ionize to an appreciable extent and yields few hydrogen ions.

weak convergence *Mathematics.* the sequence $\{x_n\}$ of elements of a topological vector space X converges weakly to x (and x is the **weak limit** of x_n) if $\lim_{n\to\infty}(x_n) = f(x)$ for every continuous linear functional f that is defined on X.

weak coupling *Particle Physics.* an interaction of four fermion fields resulting in an interaction that is many orders of magnitude less than that of electromagnetic or strong interaction. *Electricity.* see LOOSE COUPLING.

weak direct product *Mathematics.* **1.** the subgroup (subring, and so on) of a direct product $\Pi_\lambda S_\lambda$ consisting of those elements having all but a finite number of components equal to the additive identity is called the **external weak direct product.** If the index set is finite, then the weak direct product is the same as the direct product. If all the S_λ of an external weak direct product are additive Abelian groups (rings, and so on), then the weak direct product is usually called the **(external) direct sum** and denoted $\Sigma_\lambda S_\lambda$, and the S_λ are called **summands. 2.** a group G that is the union of a family of its normal subgroups $\{N_\lambda\}$ indexed by some set Λ, such that each N_λ intersects the union of the others only at the identity element, is said to be the **internal weak direct product** of the family $\{N_\lambda\}$; this is also called the **internal direct sum** if G is additive Abelian. Since G is isomorphic to the external weak direct product of the family $\{N_\lambda\}$, the words *external* and *internal* are often omitted when the context is clear.

weak electrolyte *Physical Chemistry.* a compound that experiences only a small degree of dissociation in an aqueous solution; e.g., acetic acid or ammonium hydroxide.

weaker topology *Mathematics.* a topology T is weaker than a topology S if every open set in T is also open in S.

weakfish *Vertebrate Zoology.* a spiny-finned saltwater food fish of the genus *Cynoscion,* such as *C. regalis,* having a tender mouth and found along the Atlantic coast of the United States.

weak flour *Food Technology.* any wheat flour having a low gluten content and loose chemical structure, such as cake flour; generally grown under more humid conditions than strong flours.

weak forces see WEAK INTERACTIONS.

weak ground *Mining Engineering.* underground roofs and walls in danger of collapse without suitable support.

weak interactions *Particle Physics.* a class of forces between particles that are approximately 10^{14} times weaker than the strong forces that hold the nucleus together; e.g., the neutrino, a weakly interacting particle, will travel through approximately 10^{16} cm of iron before scattering, whereas the neutron, a strongly interacting particle, will travel through only about 10 cm before scattering. Also, WEAK FORCES.

weakly bonding orbital *Physical Chemistry.* a relationship between an electron and a nucleus in which the electron's energy decreases slightly as it draws closer to the nucleus, causing them to have a weak interaction. Also, **weakly bonding molecular orbital.**

weak method *Artificial Intelligence.* a problem-solving technique, such as search, that may apply to a variety of problems but is not guaranteed to find a solution or to be efficient.

weak topology *Mathematics.* the weakest topology on a vector space that makes linear functionals continuous.

weak* topology *Mathematics.* the weakest topology on the dual space X^* of a topological vector space X in which the linear functionals on X^* (induced by evaluation of elements of X^* at points of X) are continuous.

weal-flare response see TRIPLE RESPONSE.

wean *Biology.* to accustom a child or young animal to food other than the mother's milk.

weaning *Biology.* the process in which an animal's dependence on its mother for food and protection comes or is brought to an end.

weanling *Biology.* a child or animal that has only recently been weaned from the mother's milk.

weapon *Ordnance.* any instrument, device, or system designed for use against an enemy in combat; it may range from a knife or rifle to a bomb or nuclear warhead.

Weapon Alpha *Ordnance.* a U.S. Navy rocket-propelled depth charge with a range of 1000 yards; officially designated **RUR-4.**

weapon and payload identification *Military Science.* **1.** the determination of the type of weapon being used in an attack. **2.** the discrimination of a reentry vehicle from penetration aids being used with the vehicle.

weapon debris *Ordnance.* the residue of a nuclear weapon after its explosion; it includes fission products, unexpended plutonium or uranium, casing materials, and other components of the weapon.

weapon delivery *Ordnance.* the process of acquiring a target, releasing the weapon in the proper direction, and guiding it toward the target if necessary.

weapon engagement zone *Ordnance.* in air defense, a defined airspace within which a particular weapon system is responsible for engagement.

weapons-grade uranium *Nucleonics.* uranium enriched to at least 50% uranium-235.

weapons of mass destruction *Ordnance.* an arms-control term for weapons that are capable of a high order of destruction or of killing large numbers of people; they may be nuclear, chemical, biological, or radiological; the term does not include the means of transporting or propelling the weapon if such means is separable or divisible from the weapon.

weapons readiness state *Military Science.* the degree of readiness of air defense weapons that can become airborne or be launched to carry out an assigned task; it may be expressed in numbers of weapons and numbers of minutes.

weapons recommendation sheet *Military Science.* a sheet or chart that defines the intention of an attack and recommends the nature of weapons and resulting damage expected, as well as the tonnage, fusing, spacing, desired mean points of impact, and intervals of reattack.

weapons system *Ordnance.* a combination of one or more weapons with all related equipment, materials, services, personnel, and means of delivery and deployment required for self-sufficiency. Also, **weapon system.**

weapon-target line *Ordnance.* an imaginary straight line from a weapon to a target.

wear *Engineering.* the deterioration of a surface caused by repeated contact with another part.

wearability *Textiles.* the durability of a fabric under normal use.

wearing course *Civil Engineering.* the exposed, uppermost layer of a road surface.

wear tables *Ordnance.* tables providing estimated data on the decrease in muzzle velocity due to firing a given number of rounds; such data are used to correct calibration data at various points in a weapon's use.

weasel *Vertebrate Zoology.* a small, nocturnal, carnivorous mammal of the genus *Mustela* of the family Mustelidae, characterized by a long body, short legs, a reddish coat, and white underside; found in terrestrial, arboreal, and aquatic habitats in all parts of the world except Australia; the **long-tailed weasel,** *M. frenata,* is found in most of North America.

weasel

weather *Meteorology.* **1.** the short-term state of the atmosphere, as distinguished from the long-term conditions of climate; this includes temperature, humidity, precipitation, wind, visibility, and other factors, chiefly considered in terms of their effects on life and human activity. **2.** in a surface weather observation, a category of atmospheric phenomena that are used to identify a localized atmospheric state at the time the observation is taken; this includes conditions such as rain, sleet, snow, tornado, and thunderstorm.

weatherboard *Naval Architecture.* the side of a vessel toward the wind.

weather central *Meteorology.* an organization that acts as a principal source in processing meteorological data information for a given area, including collecting, collating, evaluating, and disseminating weather information.

weathercock *Meteorology.* a weather vane with the figure or shape of a rooster (cock) on it. *Aviation.* of a flight vehicle, to head into the wind; to align, or attempt to align, its longitudinal axis with the direction of the wind. This process is commonly called **weathercocking** or **weathervaning.**

weathercock stability see DIRECTIONAL STABILITY.

weather deck *Naval Architecture.* a partly or wholly uncovered deck, open to the elements; often the upper deck or main deck of a vessel.

weathered iceberg *Oceanography.* an iceberg of irregular shape, resulting from extensive ablation; some specific types are arch, drydock, pinnacled, pyramidal, and overturned icebergs.

weathered layer *Geophysics.* an earth layer characterized by low seismic-wave velocities and found just under the surface of the earth.

weather flight *Aviation.* a flight whose primary mission is the collection of meteorological data.

weather forecast *Meteorology.* a forecast highlighting changes in meteorological elements, thus characterizing the state of the atmosphere at a given time, such as 24 hours from the current time.

weathering *Geology.* the natural processes by which the actions of atmospheric and other environmental agents, such as wind, rain, and temperture changes, result in the physical disintegration and chemical decomposition of rocks and earth materials in place, with little or no transport of the loosened or altered material. Also, DEMORPHISM, CLASTATION. *Materials Science.* the exposure of a material to environmental conditions; e.g., solar or ultraviolet light, temperature extremes, oxygen, humidity, or airborne biological and chemical agents.

weathering correction *Geophysics.* a correction applied to travel times of reflected or refracted seismic waves to account for the effect of the weathered layer.

weathering-potential index *Geology.* a measurement of a given rock or mineral's susceptibility to weathering, calculated by chemical analysis.

weathering steel *Materials.* a type of steel with a relatively low alloy content, which forms a corrosion-resistant coating upon atmospheric exposure; used mainly in the building of bridges.

weathering test *Petroleum Engineering.* a standard test used to isolate the heavy components of a liquid-petroleum gas sample by evaporating off the volatile components. Also, BOILAWAY TEST.

weathering velocity *Geophysics.* the speed at which a P wave travels through the weathered layer of the earth's crust.

weather map *Meteorology.* a graphical representation on a synoptic chart in which the state of the atmosphere and general circulation pattern, with associated weather elements, are indicated after a thorough study of weather observations over a network of reporting stations.

weather minimum *Meteorology.* an aviation weather operation term used to indicate the limit of inclement weather conditions during which aviation operations may be active, under either visual or instrument flight rules, with respect to minimum ceiling, visibility, or specific flight hazards.

weather modification *Meteorology.* an effort to alter the natural weather phenomena of the atmosphere by artificial means, such as cloud seeding used to enhance precipitation during a storm event.

weather observation *Meteorology.* an evaluation that is used to describe the atmospheric state at a prescribed altitude or at the earth's surface, highlighting one or more meteorological elements.

weatherometer *Engineering.* an apparatus used to accelerate the weathering effects on an object, especially by exposing it to ultraviolet rays or brine.

weather pit *Geology.* a small, shallow indentation on the generally flat summit of exposed granite or granitic rocks formed by the intense, local solvent action of impounded water.

weatherproof *Materials.* describing a material that can withstand the effects of all kinds of weather.

weather radar *Engineering.* a type of radar designed to detect and differentiate precipitation, ranging from light showers to heavy storms.

weather resistance *Engineering.* the ability of a material to maintain its integrity and appearance against exposure to severe weather conditions, such as rain or cold.

weather satellite *Meteorology.* an artificial satellite designed to collect data used in weather forecasting. Also, METEOROLOGICAL SATELLITE.

weather shore *Navigation.* describing the direction of the shore relative to the observation made from a vessel at sea, when considering the prevailing wind direction shoreward.

weather side *Navigation.* the side of a maritime vessel that is exposed to the prevailing weather characteristics over a water surface.

weather signal *Meteorology.* the indication of a weather forecast via a visual signal.

weather station *Meteorology.* a place or facility, such as the U.S. Weather Bureau, at which operations are performed for the observation, recording, measurement, and transmission of meteorologic data to the public or other business, government, and industry interests.

weather strip or **weatherstrip** *Building Engineering.* a weatherproofing material made out of a narrow wood, metal, or rubber strip that is fastened to the joints of a window or door sash to minimize the effect of external weather conditions, such as wind, rain, or dust. Also, **weather stripping** or **weatherstripping.**

weather type *Meteorology.* a graphical representation on a series of general synoptic weather conditions, including typical atmospheric pressure patterns, utilized to extend a forecast time range. Also, **weather-map type.**

weather vane or **weathervane** *Meteorology.* a wind vane consisting of a rodlike device to which a freely rotating pointer is attached to indicate the wind direction. Also, WIND VANE, VANE.

weather window *Engineering.* an opportune time interval during which weather conditions are favorable for some procedure that cannot otherwise be undertaken.

weave *Textiles.* to interlace threads, yarns, or strips to form a fabric. *Materials Science.* as applied to composites, any of a variety of one-, two-, or three-dimensional patterns in which carbon fibers reinforced in a carbon matrix are woven around each other; in a **plain weave,** one warp yarn runs over and then under one fill yarn.

weaverbird *Vertebrate Zoology.* any of numerous small, seed-eating birds of the family Ploceidae, characterized by a short tail and conical bill and named for their extraordinary large and complex nests, many of which are hanging flasks or tunnels; found in Africa and Asia. Also, **weaver finch.**

weaving *Textiles.* the process of creating fabric by interlacing lengthwise warp threads and crosswise filling threads. The three fundamental weaves are plain, satin, and twill. *Transportation Engineering.* the crossing of highway traffic streams by the merging and diverging of individual vehicles.

web *Invertebrate Zoology.* the thin, delicate, silken material that is spun by spiders and some larvae; a cobweb. *Vertebrate Zoology.* **1.** the integumentary membrane stretching between the fingers or toes in various animals. **2.** the series of barbs on each side of the shaft of a feather. *Textiles.* a woven fabric. *Graphic Arts.* a continuous roll of paper. *Civil Engineering.* the vertical plate between the flanges of a girder. *Mechanical Engineering.* **1.** the plated or central portion of a structure. **2.** the plate in a beam or girder that connects the upper and lower flat plates. *Metallurgy.* the thin section of metal found at the bottom of a cavity or depression. *Optics.* see WIRE.

Weber, Wilhelm Eduard [vā´bər] 1804–1891, German physicist; developed the electrodynamometer and the centimeter-gram-second system of electromagnetic measurement.

weber [vā´bər] *Electromagnetism.* the meter-kilogram-second unit of magnetic flux, equivalent to the flux (in a single-turn coil) that produces an electromotive force of 1 volt as it is reduced uniformly to zero over 1 second. (Named for Wilhelm Eduard *Weber.*)

Weber-Christian disease *Medicine.* an inflammatory reaction of subcutaneous fat, characterized by recurring fever and the formation of painful nodules on the lower extremities or the trunk. (Named for Friedrich Eugen *Weber,* 1832–1891, German otologist, and Henry Asbury *Christian,* 1876–1951, American physician.) Also, NODULAR NONSUPPURATIVE PANNICULITIS.

Weber differential equation *Mathematics.* the differential equation

$$u'' + (n + 1/2 - z^2/4)u = 0;$$

this is equivalent to a particular case of the confluent hypergeometric equation. The solutions are parabolic cylinder functions.

Weber-Fechner law see FECHNER'S LAW.

Weberian apparatus *Vertebrate Zoology.* a chain of four (sometimes three) small bones derived from the anterior vertebrae of certain fishes that act as a sense organ by transmitting water-borne vibrations from the swim bladder to the inner ear. Also, **Weberian ossicle.** (Named for Ernst Heinrich *Weber,* 1795–1878, German anatomist and physiologist.)

weberite *Mineralogy.* Na_2MgAlF_7, a pale gray, vitreous, translucent orthorhombic mineral occurring as irregular grains and fine-grained masses, having a specific gravity of 2.96 and a hardness of 3.5 on the Mohs scale; found associated with cryolite.

Weber number 1 *Fluid Mechanics.* a dimensionless ratio of inertial forces to surface tension; We = v^2rL/s, where v is the flow velocity, r the fluid density, L a characteristic length, and s the surface tension.
Weber number 2 *Fluid Mechanics.* the square root of Weber number 1.
Weber number 3 *Chemical Engineering.* a dimensionless number equal to the surface tension times the acceleration of gravity divided by the product of the liquid density, the depth of liquid on the tray being studied, and the square of the fluid velocity; used in calculating the interfacial area of liquid droplets in distillation equipment.
Weber's law *Physiology.* a constant escalation of a stimulus in which the least amount of increase in the stimulus that can be perceived remains constant, if the proportion of the escalation of the stimulus to the whole stimulus stays unchanged.
web-fed press *Graphic Arts.* a high-speed, rotary printing press fed by a continuous roll of paper. Also, WEB PRESS.
web frame *Naval Architecture.* a metal frame with a flanged or otherwise stiffened edge, which provides additional reinforcement every sixth frame.
Webley *Ordnance.* any of several pistols that were manufactured by Webley & Scott, Ltd. and used by the British military during World Wars I and II; the Webley .455 six-chambered revolver was adopted in 1887 and used through six models until 1947.
web press see WEB-FED PRESS.
Wechsler Adult Intelligence Scale see WAIS.
Wechsler Intelligence Scale for Children see WISC.
Weddell Current [wed′əl; wə del′] *Oceanography.* an ocean current flowing eastward at around 50°S in the South Atlantic, north of the Antarctic Circumpolar Current. (Named for James *Weddell,* 1787–1834, a British navigator who explored this area.)
weddellite *Mineralogy.* $Ca(C_2O_4)\cdot 2H_2O$, a colorless, white, or yellowish-brown tetragonal mineral occurring as minute, isolated pyramidal crystals, having a specific gravity of 1.94 to 2.02 and a hardness of about 4 on the Mohs scale; found in mud at the bottom of the Weddell Sea, Antarctica.
Wedderburn theorems *Mathematics.* **1.** the theorem that the following conditions on a left Artinian ring R are equivalent: (a) R is simple; (b) R is primitive; (c) R is isomorphic to the indomorphism ring of nonzero finite dimensional vector space V over a division ring D; and (d) for some positive integer n, R is isomorphic to the ring of all $n \times n$ matrices over a division ring. **2.** the theorem that the following conditions on a ring R are equivalent: (a) R is a nonzero semisimple left Artinian ring; (b) R is a direct product of a finite number of simple ideals, each of which is isomorphic to the endomorphism ring of a finite dimensional vector space over a division ring; and (c) there exist division rings $D_1, \ldots, D_t$ and positive integers $n_1, \ldots, n_t$ such that R is isomorphic to the ring $M_1 \times \cdots \times M_t$, where M_k is ring of the $n_k \times n_k$ matrices over D_k for $k = 1, 2, \ldots, t$. Each theorem is also known as the **Wedderburn-Artin theorem** or the **Wedderburn structure theorem.**
wedeloside *Biochemistry.* an analog occurring in the Australian weed, *Wedelia asperrima,* and capable of inhibiting the ADP/ATP carrier system in mitochondria.
wedge *Mechanical Devices.* **1.** the acute angular shape of most cutting instruments, consisting of two inclined planes that merge in the front to form an edge. **2.** an instrument having this shape, usually made of metal or wood, used to exert pressure for cutting, rending, propping up, or spreading apart. *Design Engineering.* a piece of resistant material whose two major surfaces form an acute angle. *Electromagnetism.* a sample of dissipative material (such as carbon) with a tapered length; used as a waveguide termination. *Engineering.* a device that channels waves of ultrasonic energy into a test object at an angle; used in ultrasonic testing. *Optics.* a prism having plane surfaces inclined toward one another at a small angle and tending to divert light toward their thicker sections. *Telecommunications.* in a test pattern, a convergent pattern of equally spaced black and white lines used to indicate resolution. *Meteorology.* an elongated area of high atmospheric pressure.
wedge-base lamp *Electricity.* a small indicator lamp with wire leads folded over opposite sides of a flat glass base, so that it can be inserted into a socket with a wedge-shaped spring contact.
wedge bit *Mechanical Devices.* a noncoring drill bit, V-shaped in cross section, used to ream out a bore hole in conjunction with a steel deflecting wedge. Also, BULL-NOSED BIT.
wedge bonding *Engineering.* a thermocompression bonding technique in which a wedge-shaped tool presses a small section of lead wire onto the bonding pad of an integrated circuit.
wedge core lifter *Mechanical Engineering.* a core-gripping device consisting of a series of serrated-face, tapered wedges in recesses cut into the inner surface of a lifter case that is threaded to the inner tube of a core barrel; the barrel is raised to pull the wedges tight and grip the core.
wedge cut see DOUBLE-WEDGE CUT.
wedge filter *Nucleonics.* a radiation filter used to increase the uniformity of radiation dosage in certain types of treatment by continuous or stepwise variations in thickness or transmission characteristics from one edge to the other. *Optics.* a filter that gradually attenuates a light beam as it moves through progressively denser sections of the filter.
wedge ice see FOLIATED ICE.
wedge-out see THIN-OUT.
wedge photometer *Engineering.* a photometer in which a wedge of absorbing material is pushed into the brighter of two beams until the luminous flux densities of the two light sources is equal.
wedge product see EXTERIOR PRODUCT.
wedge spectrograph *Spectroscopy.* a spectrograph in which an optical wedge is used as an attenuator to vary the intensity of incident radiation.
wedging *Engineering.* **1.** the removal of a block of stone by driving a wedge into a longitudinal crack created by a row of drill holes. **2.** the use of a deflecting wedge near the bottom of a deep diamond-bore hole to either restore direction or obtain samples. **3.** the logging of multiple wedge-shaped pieces of core inside a core barrel. **4.** the material used to tighten a shaft lining.
WEE western equine encephalitis.
weed *Botany.* a subjective term for any plant not valued for use or beauty, growing wild, and regarded as hindering the growth of useful vegetation or otherwise negatively affecting the landscape; a plant regarded as a weed in one context may not be so regarded in another.
week *Astronomy.* a period of seven successive days, generally considered to begin with Sunday and end with Saturday.
Weeksiaceae *Botany.* a family of red algae in the order Cryptonemiales, having been recently separated from the family Dumontiaceae; characterized by thalli of various morphologies and spermatangia occurring singly.
weep hole *Civil Engineering.* a pipe through a retaining wall, through which ground water may flow to reduce the pressure behind the wall.
weeping core *Petroleum Engineering.* a core cut that has been brought to the surface with tearlike drops of fluid, indicating that the deposit will likely yield little oil.
weeping spring see SEEPAGE SPRING.
weeping willow *Botany.* a willow tree, *Salix babylonica,* characterized by long, slender, drooping branches; native to Asia and cultivated in Europe and America.
weevil *Invertebrate Zoology.* the popular name for any of various insects of the coleopteran family Curculionidae; also called **snout beetles**; they are noted as highly destructive crop pests; e.g., the boll weevil.
weft *Textiles.* the yarns that run crosswise on a fabric, or from selvage to selvage at right angles to the warp; it is woven through the warp threads by means of a shuttle. Also, FILLING, WOOF, PICK.
weft knit *Textiles.* a type of knitting in which yarns interlock to form rows running crosswise. Weft knits can be produced on both flat and circular knitting machines and are the most common type of knitting.
Wegener, Alfred Lothar 1880–1930, German geologist; formulated the theory of continental drift; with Wladimir Peter Köppen, wrote a noted work on climatology.
Wegener's granulomatosis *Medicine.* a rare multisystem disease of unknown cause, characterized by the formation of nodules or tumorlike masses in the upper and lower respiratory tracts, necrotizing vasculitis, and glomerulonephritis; symptoms include sinus pain, bloody or purulent nasal discharge, cough, chest pain, anorexia, and skin lesions.
wehrlite *Petrology.* a variety of peridotite that consists primarily of olivine and clinopyroxene, with traces of opaque oxides.
Weibull distribution *Statistics.* a probability distribution that resembles the gamma and exponential distributions except that it has three parameters, which allows more flexibility of shape for curve fitting.
weibullite *Mineralogy.* $Pb_6Bi_8(S,Se)_{18}$, a steel gray, opaque, metallic, very brittle, orthorhombic mineral with dark gray streaks, occurring as foliated or fibrous masses, and having a specific gravity of 6.97 to 7.145 and a hardness of 2 to 3 on the Mohs scale.
Weibull modulus *Materials Science.* the flaw size distribution or data scatter of a given volume of ceramic under a uniform stress; a factor used in analytical design approaches to ceramics.

Weibull statistics *Materials Science.* a series of probabilistic relationships often used in designing ceramic components for applications in which high stresses or complex stress distributions are present and trial-and-error approaches are inadequate.

Weierstrass, Karl [vī´ər shträs´] 1815–1897, German mathematician; developed the modern theory of analytical functions.

Weierstrass approximation theorem *Mathematics.* the theorem that for every real-valued function f that is continuous on $[a,b]$ and any $\varepsilon > 0$, there exists a polynomial P such that $|f(x) - P(x)| < \varepsilon$ for every x in $[a,b]$. That is, f may be uniformly approximated by polynomials.

Weierstrass elliptic function *Mathematics.* a doubly periodic meromorphic function of order two with its two poles coalesced into a single double pole; it is uniquely defined by this requirement up to an affine change of variables in its domain and range. It satisfies the complex ordinary differential equation

$$[p'(z)]^2 = 4p^3(z) - g_2p(z) - g_3,$$

where g_2 and g_3 are constants depending on the choice of periods. Also, **Weierstrass *p* function, Weierstrass *pe* function.**

Weierstrass factorization theorem see FACTORIZATION THEOREM.

Weierstrass functions *Mathematics.* functions that determine solutions to the Euler-Lagrange equation and Jacobi's condition while maximizing a given integral; used in the calculus of variations.

Weierstrass M test *Mathematics.* let $\{f_n\}$ be a sequence of functions such that, for each n, $|f_n(x)| \le M_n$ for all x on the interval (a,b). If $\sum_{n=1}^{\infty} M_n$ converges, then $\sum f_n$ converges uniformly on (a,b).

Weierstrass transform *Mathematics.* the function given by

$$F(t) = \int_{-\infty}^{\infty} (4\pi t)^{-1/2} \exp[-(x-y)^2/4] f(y)dy,$$

where $f(y)$ is a given real function; used in studying the heat equation.

Weierstrass zeta function see ZETA FUNCTION.

weighing rain gauge *Engineering.* a recording rain gauge consisting of a funnel that directs the rain into a bucket on a scale; the rainwater weight is converted into inches of precipitation.

weight *Mechanics.* **1.** the gravitational force experienced by a body on the earth's surface or in some other gravitational field. *Weight* and *mass* are not synonyms; *weight* is relative to force and thus can change according to position. For example, an object in outer space will have less weight than it does on earth, because of its distance from the earth's gravitational field, but its mass will not be affected. **2.** see APPARENT WEIGHT. *Metrology.* a system for determining the mass of objects; e.g., avoirdupois weight. *Engineering.* **1.** a body of known mass used on a scale as a comparison in weighing objects. **2.** an object used for its heaviness, as to hold something in opposition or to maintain a condition of balance. *Statistics.* **1.** a measure of the relative importance of an item in a population. **2.** to assign a constant to a single item in a population as a measure of its importance.

weight and balance *Aviation.* in the loading of an aircraft, the monitoring and control of weight distribution in order to ensure that the aircraft will maintain its essential flight characteristics.

weight-and-balance sheet *Aviation.* a log in which the weight and location of items composing an aircraft load are recorded, in order to maintain proper weight and balance.

weight barometer *Engineering.* a barometer that measures atmospheric pressure in terms of the weight of the mercury in a column.

weight density *Physics.* the weight of an object divided by the volume that the object occupies.

weighted *Statistics.* adjusted to some representative value.

weighted area masks *Computer Technology.* a set of characters that theoretically renders all input specimens unique; each character consists of weighted points in the reader.

weighted average *Statistics.* an average of values of a random variable that have been weighted, or adjusted in value, because of their frequency of occurrence or perceived importance.

weighted character *Systematics.* in phenetics and numerical taxonomy, a character that is given a higher value or more importance than others in establishing phyletic relationships.

weighted code *Computer Programming.* a decimal code in which each bit position has a different value or weight, so that the value of the digit is the sum of each bit multiplied by its weight.

weight function *Mathematics.* a nonnegative function used to define an inner product on a space of functions via an integral over the common domain of the functions.

weighting *Engineering.* the correction of measurements to account for unusual factors. *Textiles.* a finishing process in which a material, especially silk, is chemically or mechanically treated to give it more weight or body.

weighting network *Acoustical Engineering.* in an electronic device such as a sound system, a processing network that assigns relative importance to various signals used in processing.

weightless *Mechanics.* experiencing a condition of weightlessness, as in the interior of a spacecraft.

weightlessness *Mechanics.* a condition in which there is no detectable acceleration due to gravity or any other force. Also, ZERO GRAVITY.

weightlessness switch see ZERO-GRAVITY SWITCH.

weight-loaded regulator *Engineering.* a pressure regulator valve that is preloaded by counterbalancing weights to either open or close at predetermined limits of a pressure range.

weight of measurement *Crystallography.* a number assigned to express the relative precision of each measurement; in least-squares refinement, the weight should be proportional to the reciprocal of the square of the estimated standard deviation of the measurement. Other weighting schemes are frequently used.

weight zone *Ordnance.* any of a number of defined weight ranges that are sometimes used with projectiles of 75 mm and larger, as an aid in choosing the optimum projectile for a particular situation; the weight zone is marked on the projectile.

Weigle reactivation *Microbiology.* a phenomenon in which certain ultraviolet-irradiated bacteriophages containing damaged DNA have a greater survival rate in host cells that have also been irradiated prior to bacteriophage infection.

Weil-Felix reaction [vīl] *Immunology.* the production of nonspecific agglutinins in the sera of individuals having typhus or other rickettsial infections.

Weil-Felix test *Immunology.* a test used to determine the presence of typhus or other rickettsial infections in an individual, in which proteus bacteria serve as an antigen and the production of agglutinins is observed to determine whether or not an infection exists.

Weil's disease *Medicine.* the most severe form of leptospirosis, usually caused by infection with *Leptospira interrogans*; characterized by jaundice, nephritis, fever, enlargement of the liver and spleen, and muscular pains. Also, LEPTOSPIRAL JAUNDICE.

Weinberg, Steven [wīn´burg] born 1933, American physicist; awarded the Nobel Prize for independently establishing the unity of electromagnetism and the weak force.

Weinberg-Salam theory *Particle Physics.* a unified theory of weak and electromagnetic interactions between particles that predicts the behavior of the W and Z particles as the agents for the weak interaction. Also, ELECTROWEAK THEORY.

Weingarten formulas *Mathematics.* given a submanifold L of a manifold M, the Weingarten formulas express the covariant derivative of a vector field normal to L with respect to a vector tangent to L, as a sum of vectors tangent and normal to L in M.

weinschenkite *Mineralogy.* $YPO_4 \cdot 2H_2O$, a colorless, white, or gray, transparent, vitreous monoclinic mineral occurring as rosettes, crusts, and spherulites of minute, lathlike elongated crystals, having a specific gravity of 3.265 and a hardness of 3 on the Mohs scale; found in limonite deposits with other phosphates; now known as **churchite-(Y).**

weir [wēr´] *Civil Engineering.* a low dam designed to permit water to overflow across its entire length. A **measuring weir** is a spillwaylike device used to measure flow through a water channel.

weir tank *Petroleum Engineering.* an oil-field storage tank using weir boxes (low- and high-level) and controls to meter the liquid content.

Weismann, August [vīs´män´] 1834–1914, German biologist; formulated the germ-plasm theory of heredity; described meiosis.

Weismannism *Genetics.* the generally accepted concept that phenotypic characters acquired in one generation are not inherited, and that only changes in the germ cells are inherited by offspring. Also, **Weismann's germ plasm theory.**

Weissenberg effect [vīs´ən burg] *Materials Science.* the manifestation of elasticity in a viscoelastic material exemplified by the tendency of a polymic liquid solution to climb the walls of the cup or cylinder in which it is rotating, or on a shaft rotating in a cup; used in rotational viscometric studies to measure normal stresses.

Weissenberg method *Solid-State Physics.* a method of mounting a crystal in preparation for diffraction analysis, in which the incident beam strikes the crystal perpendicularly to one of its real axes rather than one of its reciprocal lattice axes.

Weissenberg photograph *Crystallography.* a distorted image of a given layer of the reciprocal lattice. It is an oscillation photograph in which the camera moves in a particular manner as the crystal rotates; a narrow band of diffracted radiation normal to the rotation axis (a layer line) is selected. The position of a spot on the film is related to the angle of rotation of the crystal at the instant of diffraction for that particular spot.

weissite *Mineralogy.* Cu_5Te_3, a bluish-black, opaque, metallic, shiny hexagonal, pseudocubic mineral, massive in habit, and having a specific gravity of about 6 and a hardness of 3 on the Mohs scale; found in hydrothermal veins with other tellurium minerals.

Weiss magneton *Atomic Physics.* a unit of magnetic moment mathematically represented by 1.853×10^{-21} erg/oersted, approximately 1/5 of the Bohr magneton as experimentally derived from magnetic moments of certain molecules approximating integral multiples of this property.

Weiss molecular field *Solid-State Physics.* a magnetic field postulated by the Weiss theory of ferromagnetism, in which atomic magnetic moments within a domain tend to align; the alignment of these atomic moments subsequently generates this magnetic field.

Weiss theory *Solid-State Physics.* a theory of ferromagnetism claiming that below the Curie temperature, a ferromagnetic material is believed to consist of molecular domains that are subject to spontaneous magnetization in the presence of an external magnetic field.

Weisz ring oven *Analytical Chemistry.* an apparatus used in microanalytic techniques to vaporize solvent from filter paper, leaving the solute behind in a circular ring.

Weizmann, Chaim [vīts´män´; wīts´mən] 1874–1952, Russian-born Israeli political leader and chemist; improved methods of synthesizing acetone and butyl alcohol.

Weizsäcker's theory *Astronomy.* a theory for the origin of the solar system proposing that the planets formed from turbulent eddies which became permanent and self-gravitating.

Weizsäcker-Williams method *Quantum Mechanics.* an approximate method for computing bremsstrahlung in a collision between two particles having kinetic energies that are large compared with their rest energies; the problem is considered in the rest frame of one particle, in which virtual photons representing the Coulomb field of the other are Compton scattered.

weld *Metallurgy.* **1.** to carry on a process of welding. **2.** a union made by welding.

weldability *Metallurgy.* the capacity of a metal to be welded under specified conditions and to perform well in the intended service.

weld bead *Metallurgy.* a deposit of filler metal from a single welding pass.

weld decay *Materials Science.* a corrosion attack at or adjacent to a weld as the result of galvanic action resulting from structural differences in the weld or between it and the adjacent metal.

weld delay time *Metallurgy.* the time that current is delayed when starting a forge delay timer for synchronizing the forging pressure and the welding heat.

welded tuff *Petrology.* a banded or streaky pyroclastic rock, indurated by the combined action of the heat retained by the particles, the weight of overlying material, and hot gases. Also, TUFF LAVA.

welder *Metallurgy.* **1.** also, **welding machine.** a machine used in welding. **2.** a person qualified for and capable of performing a welding operation. *Robotics.* an end-of-arm tooling used for welding.

welder's hammer *Mechanical Devices.* a device used for joining pieces of suitable metals by striking an electric arc between an electrode and a metal rod.

weld gauge *Engineering.* a device used to monitor the size and shape of welds.

welding *Metallurgy.* the joining of two metal surfaces that have been heated, melted, and fused together. *Geology.* a process in which sediments are consolidated under pressure exerted either by the weight of superincumbent material or by earth movements.

welding current *Electricity.* a strong electric current that is passed through a workpiece to produce the needed heat for welding.

welding cycle *Metallurgy.* the steps it takes to make a resistance weld.

welding electrode *Metallurgy.* filler metal that is in the form of a wire or rod, through which current is conducted between the electrode holder and the arc.

welding generator *Electricity.* a generator that supplies the heavy electric current used to weld a workpiece.

welding rod *Metallurgy.* a rod or wire filler metal that is used in welding.

welding schedule *Metallurgy.* a record of welding-machine settings and a list of the machine's capabilities for a given material, size, and finish.

welding sequence *Metallurgy.* the order of welding in which various parts of a structure or weldment are made.

welding stress *Metallurgy.* stress that is caused by localized heating and cooling during welding.

welding tip *Engineering.* the replaceable nozzle on a gas torch.

welding torch *Engineering.* a gas mixing and burning tool that welds metal.

welding transformer *Electricity.* a special ultrahigh-current stepdown transformer used in electric welding applications.

weld interval *Metallurgy.* the total heat and cool times in making one multiple-impulse weld.

weld-interval timer *Engineering.* a device that regulates weld intervals.

weld line *Metallurgy.* a line at which the weld metal and the base metal meet, or at which the base-metal parts meet when filler metal is not used.

weldment *Engineering.* an assembly whose parts are joined by welding.

weld metal *Metallurgy.* the part of a weld that is melted during the welding process.

weld time *Metallurgy.* the time that the welding current is applied to the work.

Welge method *Petroleum Engineering.* a method used to estimate anticipated production from a gas-cap-drive oil reservoir.

well *Engineering.* a hole that is dug in the earth to gain access to water, oil, brine, gas, or the like. *Building Engineering.* a vertical open area or shaft in a building that extends vertically from the floor to the roof to accommodate stairs or an elevator.

wellbore see BOREHOLE.

wellbore hydraulics *Petroleum Engineering.* the study of the behavior of fluids (water, gas, and oil) in the tubing, casing, or annulus between tubing and casing in a wellbore.

well completion *Petroleum Engineering.* the operation of sealing off a depleted well.

well conditioning *Petroleum Engineering.* **1.** the operation of controlling production rate and associated pressure drawdown in preparation for well sampling. **2.** the removal of solid deposits from the sides of a wellbore or the breakage of a water block to increase recovery rate.

well core *Engineering.* a cylindrical sample of rock or soil taken from a borehole.

well-deck vessel *Naval Architecture.* a former merchant vessel with two sunken weather decks, the first between the forecastle and a full-width bridgehouse amidships, and the second between the bridgehouse and the poop deck.

well-defined *Mathematics.* of a function, having a value that is independent of any choices permitted by its definition.

Weller, Thomas Huckle born 1915, American virologist; shared the Nobel Prize with John F. Enders and Frederick C. Robbins for the isolation and cultivation of poliomyelitis; he also isolated and cultivated chicken pox, mumps, and herpes zoster.

well-formed formula *Mathematics.* a finite sequence of symbols that is syntactically correct for a given set of syntactic rules. *Artificial Intelligence.* a formula composed of symbols that are arranged in a way that is allowed under the rules of the language.

wellhead *Civil Engineering.* the top level of a well, from which water pumped from the well is allowed to flow freely. *Hydrology.* the source of a stream or the place where a spring emerges from the ground.

wellhole *Mining Engineering.* **1.** a large-diameter, vertical hole in quarries and opencast pits, used to accept heavy charges during blasting. **2.** the sump in which water collects, located in a shaft below the point at which skips are caged at the bottom.

well injectivity *Petroleum Engineering.* an injection-well characteristic that defines its ability to receive injected water or gas; can be influenced negatively by a retarding factor such as a water block.

well logging *Engineering.* the recording and analysis of characteristics relating to formations penetrated by a drill hole.

well-ordering *Mathematics.* an ordering of a set S is a well-ordering if every subset of S has a minimal element under the given ordering. In that case, S is said to be a **well-ordered set.**

well-ordering principle *Mathematics.* the assumption that every set can be given an order so that it becomes a well-ordered set; logically equivalent to the axiom of choice, Tukey's lemma, Zorn's lemma, the Hausdorff maximality principle, and so on.

well performance *Petroleum Engineering.* actual well production as compared to anticipated production capacity, pressure drop, or flow rate.

wellpoint *Mechanical Devices.* a long, hollow rod or pipe, consisting of a ball valve, screen, and jet tip as part of a wellpoint system, which is inserted into the ground to pump out ground water, thereby lowering the water table and preventing flooding during construction.

wellpoint system *Civil Engineering.* a row or rows of wellpoints used to protect a construction excavation.

well-posed problem *Mathematics.* a problem with a uniquely determined solution that depends continuously on the initial conditions.

well-regulated system *Control Systems.* a system in which a regulator works in conjunction with the physical environment to produce a stable system.

well shooting *Engineering.* the firing of an explosive charge in the bottom of a well to increase the flow of water, oil, or gas.

well-sorted *Geology.* describing a sorted sediment whose constituent particles are nearly uniform in size and have a sorting coefficient of less than 2.5.

well spacing *Petroleum Engineering.* the geographic pattern of producing oil or gas wells in an oil field, regulated by government authority and established in consideration of maximum ultimate production per reservoir. Also, SPACING PATTERN.

well-type manometer *Engineering.* a double-leg, glass-tube manometer in which the diameter of one leg is very small while the other is a reservoir for the liquid.

Welsbach, Baron Carl von 1858–1929, German chemist; invented Welsbach mantle; isolated neodymium and praseodymium.

Welsh vault see UNDERPITCH VAULT.

Welwitschiales *Botany.* a rare plant located in the southwestern coast of Africa; the most bizarre plant of all the gymnosperms, it is protected by law. (Named for the Austrian botanist Friedrich *Welwitsch* in 1860.)

Welwitschiidae *Botany.* one of three subclasses of the gymnosperm subdivision Gneticae in some classification systems, composed of the single monotypic family Welwitschiaceae and native to arid desert regions of Angola and southwest Africa; characterized by short evergreen dioecious shrubs, a woody inverted conelike stem, and only two leaves that grow throughout the life of the plant, giving the shrub the appearance of a large wooden carrot; some plants are estimated to be between 1500 and 2000 years old.

wen *Medicine.* a benign encysted tumor of the skin, especially of the scalp that is formed when fatty matter secreted by the sebaceous gland collects inside the gland.

wenge *Materials.* the hard, coarse-textured wood of the African tree, *Millettia laurentii*; used for cabinets, flooring, and decorative pieces. Also, PANGA PANGA.

Wenlockian *Geology.* a European geologic stage of the Middle Silurian period, occurring after the Taranian and before the Ludlovian.

Wentworth classification *Geology.* a method of classifying sedimentary particles according to size by providing diameter limits for the size classes.

Wentworth scale *Geology.* a grade scale for defining the range of grain-size clastic sedimentary rocks from particles to boulders, using a constant ratio of 1:2 to relate the size classes.

Wentzel-Kramers-Brillouin approximation *Quantum Mechanics.* an approximate method of solving Schrödinger's equation for slowly varying potentials, in which the logarithm of the wave function is expanded in powers of Planck's constant. Also, WKB METHOD.

Wenzel, Karl Friedrich 1740–1793, German chemist; determined many rates of chemical reactions and weights of compounds.

Werdnig-Hoffman disease *Medicine.* an inherited, usually fatal disorder characterized by progressive muscular atrophy, beginning in infancy or early childhood.

Werfenian stage see SCYTHIAN STAGE.

Werner, Abraham Gottlob 1749–1817, German geologist; devised the first systematic mineral classification.

Werner, Alfred 1866–1919, French-born Swiss chemist; awarded the Nobel Prize for formulating a comprehensive coordination theory of valences.

Werner band *Spectroscopy.* a spectral band in the ultraviolet region of the hydrogen spectrum, extending from 116 to 125 nanometers.

Werner's syndrome *Medicine.* a hereditary syndrome in which there is premature aging in the adult, characterized by sclerodermal skin changes, cataracts, subcutaneous calcification, muscular atrophy, a tendency to diabetes mellitus, baldness, and a high incidence of neoplasm.

Wernicke's encephalopathy or **syndrome** [vur′ni kēz′] *Medicine.* a neurological disorder caused by thiamine deficiency and most often associated with alcohol abuse, characterized by confusion, apathy, paralysis of the eyes or double vision, and decreased mental function as a result of hemorrhagic lesions of the gray matter of the brain. (Named after Karl *Wernicke,* 1848–1905, German neurologist.) Also, POLIOENCEPHALITIS HAEMORRHAGICA.

Wessel, Caspar 1745–1818, Norwegian mathematician; introduced the geometric representation of complex numbers; devised the Argand diagram.

West African subregion *Ecology.* a distinct zoogeographical region that includes the area of western Africa south of the Sahara.

West Australia Current *Oceanography.* the eastern segment of the South Pacific gyre, flowing northward off the western coast of Australia from around 45° to 15°S; it is a strong current, but its speed and depth vary considerably with the seasons. Also, **West Australian Current.**

westerlies *Meteorology.* the dominant west-to-east motion of the atmosphere resident over the middle latitudes of both hemispheres, principally extending between 35° and 55° latitude. Also known by various other names, such as **circumpolar westerlies, mid-latitude westerlies, middle-latitude westerlies, polar westerlies, subpolar westerlies, subtropical westerlies, temperate westerlies, zonal westerlies,** and **zonal winds.**

westerly wave *Meteorology.* an atmospheric wave disturbance that is entrained in the mid-latitude westerly circulation.

western blotting see PROTEIN BLOTTING.

western boundary currents *Oceanography.* the strong, permanent, poleward-moving currents found in all five major oceanic gyres: the Brazil, East Australian, Kuroshio, and Agulhas currents, and the Gulf Stream; the East Africa Coast Current is a western boundary current in the small gyre of the North Indian Ocean during the northern hemisphere summer, but it virtually disappears in winter.

Western diamondback *Vertebrate Zoology.* a large, extremely venomous rattlesnake, *Crotalus atrox*, of the southwestern United States and Mexico.

Western diamondback

Western Equatorial Countercurrent *Oceanography.* a weak system of drift currents sometimes found flowing eastward in winter in the Western Atlantic near the equator.

western equine encephalitis *Medicine.* a viral disease common to horses and contagious to humans, causing inflammation of the brain and nervous system and resulting in convulsions, fever, and eventually coma. It is the less severe of the two forms of viral encephalomyelitis; the other is eastern equine encephalitis. Also, VIRAL WESTERN EQUINE ENCEPHALITIS.

Western Hemisphere *Geography.* the half of the earth that lies west of the Atlantic Ocean, including North and South America; usually charted westward from 20°W to 160°E.

West Greenland Current *Oceanography.* the current that flows northward along the western coast of Greenland into Baffin Bay; it consists of water from the East Greenland Current, joined by some warmer Atlantic water, the latter keeping the coast relatively ice-free.

West Indian satinwood *Materials.* the high-density, nondurable, fine-textured wood of the Central and South American tree, *Zanthoxylum flavum*; used for inlay work, furniture, and cabinets.
westing *Navigation.* the distance a craft makes good to the west.
West Nile fever *Medicine.* an acute illness caused by a mosquito-borne virus, manifesting in headaches, lymph disorder and swelling of the lymph nodes, myalgia, and rash.
Weston standard cell see STANDARD CELL.
Westphal balance *Engineering.* a balance in which the relative density is measured by suspending a solid from the balance beam into a liquid of known density.
Westphalian *Geology.* a European geologic state of the Upper Carboniferous period, occurring after the Namurian and before the Stephanian.
west point *Astronomy.* the point at which the celestial equator meets the horizon. *Cartography.* the point on the horizon, to the left of an observer facing north, intersected by the prime vertical plane.
westward intensification *Oceanography.* a hypothetical explanation of the intensification of currents on the western sides of the oceans, derived from a mathematical model of the effects of internal friction and zonal wind stress at the surface.
west-wind surface drift *Oceanography.* the surface layer of the Antarctic Circumpolar Current. Also, **West Wind Drift.**
wet *Chemistry.* describing a substance that has accepted water or another liquid on its surface or within it. *Physics.* a condition in which the contact angle between a liquid and a solid is between 0° and 90°; in such a case the solid is said to be **wetted** by the liquid. *Meteorology.* **1.** characterized by heavy or frequent showers, mist, and so on. **2.** damp or rainy weather. *Aviation.* of air or other fluid, to come in contact with and flow across an airfoil or any other body, surface, or area.
wet ashing *Organic Chemistry.* the decomposition and conversion of an organic compound into ash by exposure to nitric or sulfuric acid.
wet assay *Mining Engineering.* the use of liquid processes, such as solution or flotation, to determine the amount of a desired constituent in an ore, metallurgical residue, or alloy.
wet avalanche see WET-SNOW AVALANCHE.
wet-bag isostatic pressing *Materials Science.* in the forming of advanced ceramics, a method of compacting powdered materials into a solid homogeneous mass before machining or sintering; used in the production of high-voltage insulators, rocket nose cones, and tubes.
wet battery *Electricity.* a battery containing a number of wet cells.
wet blasting *Mining Engineering.* the process or technique of detonating a shot in a wet hole.
wet-bulb depression *Meteorology.* the differential in degrees between the wet- and dry-bulb air temperatures.
wet-bulb temperature *Meteorology.* **1.** the lowest temperature possible on a wet-bulb thermometer in a given air parcel, if such a parcel is cooled adiabatically to saturation at constant pressure by evaporation of water into it with all latent heat supplied by the parcel. **2.** a temperature as read from a wet-bulb thermometer and identified with isobaric wet-bulb temperature.
wet-bulb thermometer *Meteorology.* a thermometer whose bulb is encased by a piece of muslin saturated with water; used to measure the evaporation of water into the air.
wet cell *Electricity.* an electric cell containing a liquid electrolyte.
wet-cell caplight *Mining Engineering.* a headlamp powered by rechargeable batteries, which are usually worn on the belt.
wet classifier *Engineering.* a device that separates solid particles from a liquid into coarse and fine or heavy and light fractions without employing a screen.
wet climate see TROPICAL RAIN FOREST CLIMATE.
wet contact *Electricity.* a term for any contact through which direct current flows.
wet cooling tower *Mechanical Engineering.* a tower that cools water by atomization into a circulating air stream so that heat is lost through evaporation.
wet criticality *Nucleonics.* the criticality achieved in a reactor with the coolant present.
wet drill *Mechanical Engineering.* a percussive drill with a water feed through the machine or by means of a water swivel that suppresses the dust produced by drilling.
wet electrolytic capacitor *Electricity.* an electrolytic capacitor that utilizes a liquid electrolyte; its leakage current is higher than that of the dry electrolytic type, but it is self-healing during momentary voltage breakdowns.
wet emplacement *Space Technology.* a rocket launch pad designed to provide a pad deluge as a cooling procedure during launch.
wet engine *Mechanical Engineering.* a term for an engine that contains oil, coolant, and fuel.
wet flashover voltage *Electronics.* the voltage necessary to break down the air surrounding a clean, wet insulator between electrodes.
wet grinding *Mechanical Engineering.* **1.** the milling of materials in water or other liquid. **2.** the process of applying a coolant to the work and wheel to facilitate the grinding process.
wether *Agriculture.* a castrated male sheep.
wet hole *Mining Engineering.* a borehole that crosses a water-bearing formation whose flow is strong enough to fill the hole.
wetland or **wetlands** *Ecology.* any section of low-lying land that is periodically submerged or whose soil contains a great deal of moisture.
wet mill *Mechanical Engineering.* **1.** a mill in which the material to be ground is mixed with liquid. **2.** a mill powered by a liquid stream.
wet pressing see PRESSURE FILTRATION.
wetproof see WATERPROOF.
wet-reed relay *Electricity.* a reed-type relay having mercury at its contacts to minimize arcing.
wet scrubber *Engineering.* a device in which a gas stream is drawn through a liquid to remove pollutants from the gas.
wet season see RAINY SEASON.
wet sleeve *Mechanical Engineering.* a cylinder lining that has more than 70% of its surface exposed to coolant.
wet snow *Meteorology.* snow that has fallen to the ground and is largely composed of liquid water.
wet-snow avalanche *Hydrology.* one of the basic types of snow avalanche, caused by a sudden thaw that releases a large mass of wet, heavy snow; it is the slowest moving type of avalanche. Also, SLUSH AVALANCHE.
wet spinning *Materials Science.* the process of spinning fibers from dissolved polymers and then extruding to remove the solvent while the spinneret is immersed in a liquid.
wet storage *Ordnance.* a storage method in which large-caliber ammunition is placed in racks surrounded by nonflammable liquid; it is a safety precaution used in transporting such ammunition in combat vehicles.
wet strength *Materials Science.* the ability of paper to resist tearing when it is wet, due to the addition of resins to the stock during manufacturing. Thus, **wet-strength paper, wet-strong paper.**
wettability *Chemistry.* the ability of a solid to interact with or accept a liquid that is spread over it.
wet tabling *Mining Engineering.* a method of washing a pulp consisting of two or more minerals by flowing it across an inclined, riffled table, shaking it longwise, and water-washing it crosswise.
wetted *Chemistry.* of a substance, having accepted liquid spread on its surface.
wetted perimeter *Geology.* the length of wetted contact between a stream and its channel, measured along the cross section perpendicular to the direction of flow.
wetted surface *Naval Architecture.* the surface area of a vessel's underwater body to the deep load line.
wetted-wall column *Chemical Engineering.* a vertical column, such as a spinning-band column, that functions with inner walls wetted from the processed liquid; used in theoretical studies of analytical distillations and for determining mass transfer rates.
wet-test meter *Engineering.* an instrument that counts the revolutions of a shaft mounted with water-sealed gas-carrying cups of specific size, in order to determine gas flow.
wetting *Electronics.* a term for the process of coating a contact surface with an adherent film of mercury.
wetting agent *Chemical Engineering.* a compound added to a liquid in small quantities in order to enhance the spread of the liquid on a surface or the penetration of the liquid into a material. *Graphic Arts.* see DAMPENING SOLUTION.
wetting angle *Fluid Mechanics.* the acute contact angle between a solid and a liquid surface such as the angle between the water surface in a capillary tube and the tube wall.
wetting phase *Petroleum Engineering.* in a formation where the reservoir fluid is two-phase (oil and water), the phase (water) that wets the pore surfaces of the formation.
wetware *Artificial Intelligence.* a description of the brain, viewed as a kind of computer hardware that is "wet," as opposed to an electronic computer, which is dry.

wet well *Mechanical Engineering.* a chamber used to store liquids. *Nucleonics.* in a boiling-water reactor, an annular chamber that retains water between it and the primary containment structure's steel cylinder, thus providing a means for pressure suppression in a loss-of-coolant accident.

Weyl, Hermann [vīl] 1885–1955, German mathematician; made major contributions in differential geometry and group theory.

Weyl group *Mathematics* given a root system Φ embedded in a Euclidean space E, the Weyl group of Φ is the group of automorphisms of E generated by reflections through hyperplanes perpendicular to the roots in Φ. The resulting group is a subgroup of the permutation group on the roots of Φ. The automorphism that takes each root to its negative is not an element of the Weyl group.

Weyl tensor *Mathematics.* the Weyl tensor C is a four-argument tensor field on a manifold M which vanishes identically if M is conformally flat. On a (pseudo-)Riemannian manifold, C is invariant under conformal mappings of M into itself, coincides with the Riemann curvature tensor in Ricci-flat ($R_{ij} = 0$) spaces, and has trace zero ($C^{is}_{ks} = 0$). In component form,

$$C^{ij}_{km} = R^{ij}_{km} - (R^i_k \delta^j_m - R^j_k \delta^i_m + R^j_m \delta^i_k - R^i_m \delta^j_k)/(n-2) + R\,\varepsilon^{ij}_{km}/(n-1)(n-2),$$

where n is the dimension of M. Also, **conformal curvature tensor.**

wff [wůf] *Artificial Intelligence.* well-formed formula.

W.G. or **w.g.** water gauge; wire gauge.

WH or **wh** watt-hour.

whale *Vertebrate Zoology.* any large, air-breathing marine mammal of the orders Mysticeta and Odotoceta (or, in some classifications, of the order Cetacea), characterized by a thick torpedo-shaped body, anterior paddle-shaped limbs, a flat horizontal tail for propulsion, and a blowhole on the highest part of the head for respiration; included is the largest animal ever to have lived on earth, the **blue whale.**

blue whale

Whale *Astronomy.* another name for the constellation Cetus.

whaleback dune *Geology.* a smooth, elongated mound or ridge of desert sand with a rounded crest, having a form similar to the back of a whale. Also, SAND LEVEE.

whaleboat *Naval Architecture.* **1.** a light, fast rowing boat carried by a whaling ship, and used in the actual pursuit of whales. **2.** a double ended, high-ended ship's boat of highly seaworthy design.

whalebone see BALEEN.

whalebone whale see BALEEN WHALE.

whale oil *Materials.* an oil that is derived from whale blubber; used formerly in the manufacture of soap and candles, and as fuel for lamps.

whale shark *Vertebrate Zoology.* a harmless shark, *Rhinocodon typu,* of the family Rhiniodontidae; the largest living fish, reaching up to 45 feet long; it inhabits all tropical seas and is characterized by tiny teeth, fine gill rakers, and a white-speckled brown back with light stripes.

wharf *Civil Engineering.* a ship's berth lying parallel to the waterfront.

wheal *Medicine.* a smooth, slightly elevated area of edema on the body surface, usually but not always accompanied by severe itching, that tends to change its size or shape or disappear within a few hours. Also, **hive, welt.**

wheal and flare *Immunology.* a skin reaction, associated with immediate hypersensitivity, that is characterized by a sudden elevation of the skin surface (wheal) encircled by a red inflamed patch (flare).

wheat *Botany.* any of a number of related cereal grasses of the genus *Triticum,* having spikes filled with seeds; the spikes in some species have awns and in others are bare. *Food Technology.* the kernels of such a plant ground or otherwise processed for bread, pasta, breakfast cereal, and other food products.

wheat germ *Botany.* the fat-rich embryo of the wheat kernel, milled out as an oily flake and used to enrich breads and cereals.

wheat germ agglutinin *Cell Biology.* a lectin derived from wheat that is of biological use due to its ability to bind oligosaccharides containing *N*-acetylglucosamine.

wheat germ oil *Materials.* a pale-yellow oil that is soluble in fats and is derived from wheat germ; used as a dietary supplement.

Wheatstone, Charles 1802–1875, English physicist; invented the "magic lyre," Wheatstone's bridge, and (with William Cooke) the electric telegraph.

Wheatstone bridge or **Wheatstone network** *Electricity.* a four-arm balancing circuit device that measures an unknown resistance in terms of a standard resistance, using a null indicator consisting of two parallel resistance branches, with each branch containing two resistances in series. Also, RESISTANCE BRIDGE.

wheel *Mechanical Engineering.* **1.** any of a wide variety of circular frames or disks designed to revolve around a central axis. **2.** any of various machines, devices, or the like characterized by a revolving circular frame or disk. **3.** see STEERING WHEEL. *Mathematics.* a graph formed from a cycle by adding one new vertex and an edge from the new vertex to each vertex of the cycle. If the cycle has n vertices, the resulting graph is called an n-wheel and is denoted W_n.

wheel-and-belt method *Materials Science.* a process for the continuous casting of nonferrous metal rods or bars, using a casting wheel; solidification occurs along the periphery of the wheel mold as it rotates.

wheelbarrow *Engineering.* a small, hand-powered vehicle consisting of two long handles supporting a boxlike body on a single wheel; used for transporting various materials. Also, BARROW.

wheelbase or **wheel base** *Design Engineering.* the distance in the direction of travel from the front to the rear wheels of a vehicle; it is measured between the ground contact centers of each wheel.

wheel conveyor *Mechanical Engineering.* a conveyor trough containing wheel units that are set at a downward slope so that objects are conveyed by gravitational force. Also, GRAVITY WHEEL CONVEYOR.

wheel dresser *Engineering.* a tool designed to clean, resharpen, and restore the cutting faces of grinding wheels.

wheeled crane *Mechanical Engineering.* a self-propelled crane, supported by large rubber tires, that uses one engine for both hoisting and propulsion.

Wheeler, Sir Mortimer 1890–1976, British archaeologist; improved excavation and recording methods.

Wheeler-Feynman theory *Physics.* a relativistic theory that accounts for radiation damping by positing the existence of enough bodies in the universe to absorb all effects radiated by any charged particle.

Wheelerian *Geology.* a North American geologic stage of the Upper Pliocene epoch, occurring after the Venturian and before the Hallian.

wheelhouse see PILOT HOUSE.

wheel printer see DAISY-WHEEL PRINTER.

wheel sleeve *Mechanical Devices.* an adapter for a precision grinding machine that allows the arbor of the grinder to fit the hub of the grinding wheel.

wheel spats see SPAT.

wheel sprinkler *Mechanical Devices.* an apparatus used as part of a moveable irrigation system, typically consisting of water pipes cut with small holes through which water is pumped.

wheel sprinkler

wheel static *Electronics.* interference in an automobile radio caused by a buildup of static charge on the wheel cylinders developed by friction between the tires and the road.

wheeze *Medicine.* a whistling sound made during respiration.

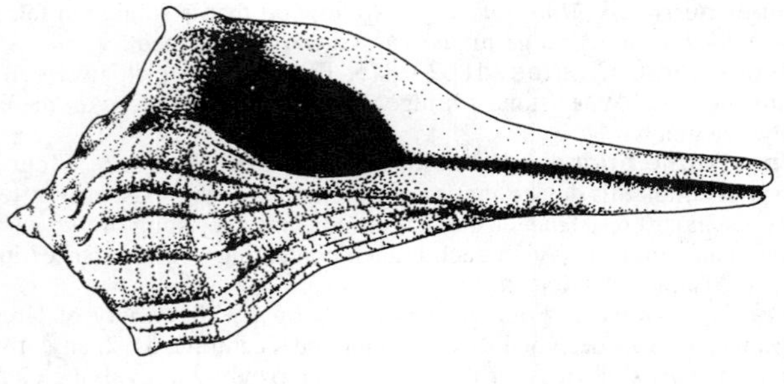

whelk

whelk *Invertebrate Zoology.* any of several marine gastropods of the family Buccinidae, characterized by a spiral shell and used for food in some cultures.

whelp *Vertebrate Zoology.* **1.** to give birth to; said especially of a female dog. **2.** an unweaned mammal, especially a puppy or other canine or large cat. *Design Engineering.* a tooth on a sprocket wheel.

wherryite *Mineralogy.* $Pb_4Cu(CO_3)(SO_4)_2(Cl,OH)_2O$, a light green to pale yellow monoclinic mineral occurring as fine-grained to fibrous masses and aggregates of equidimensional crystals, having a specific gravity of 6.45 and an undetermined hardness.

whet *Mechanical Engineering.* to sharpen a knife or other such edged tool.

whetstone *Mechanical Devices.* a hard, fined-grained, abrasive stone, usually of a siliceous slate, used for sharpening knives and other edge tools. *Computer Programming.* a unit of measure of a computer's throughput; used to compare competitive systems on scientific problems.

Whewell equation *Mathematics.* an equation relating the length of a plane curve to the angle of inclination of the tangent to the curve.

whewellite *Mineralogy.* $CaC_2O_4 \cdot H_2O$, a colorless, white, brownish, or yellowish, transparent to translucent, monoclinic mineral occurring as masses of short prismatic crystals, having a specific gravity of 2.21 to 2.23 and a hardness of 2.5 to 3 on the Mohs scale; found with lignite or coal and in septarian nodules.

whey *Food Technology.* the residue from milk after removal of the casein and most of the fat.

whey cheese *Food Technology.* any cheese made from whey, such as ricotta or Mysost.

whiffletree *Mechanical Devices.* a pivoted crossbar to which the traces of a harness are attached for pulling a plow, cart, and so on.

whiffletree switch *Electronics.* a multiposition electronic switch composed of gates and flip-flops; used in computers. (The term is derived from the shape of the switch's circuit diagram, which resembles a whiffletree.)

WHILE statement *Computer Programming.* a statement that is found in various programming languages, in which a statement or a block of statements is repeatedly executed, as long as a specified condition is continually met.

whip antenna *Electromagnetism.* a vertical telescoping antenna commonly used in mobile or portable communication systems.

whipcord *Textiles.* a durable twill-weave fabric or knitting of hard-twisted worsted, cotton, or synthetic fibers, having a sharply defined warp twill.

whipgraft *Botany.* a graft in which the scion and stock, both cut on a long slant with a slit in each cut surface, are fitted together by inserting the tongue of one into the slot of the other. Thus, **whipgrafting.**

whiplash *Medicine.* a popular term for an acute cervical sprain, especially one sustained in a rear-end auto collision. Also, **whiplash injury.**

Whipple, George Hoyt 1878–1976, American pathologist; with Minot and W. Murphy, awarded the Nobel Prize for the use of liver to treat pernicious anemia.

Whipple's disease *Pathology.* a rare intestinal lypodystrophy marked by diarrhea, steatorrhea, lymphadenopathy, fever, and arthritis with attendant lesions attached to the central nervous system, and in which the lamina propria is invaded by PAS-positive bacillary microorganisms. (Named for George Hoyt *Whipple.*)

Whipple's triad *Oncology.* the three essential clinical features of insulin-producing tumors: spontaneous hypoglycemia, central nervous or vasomotor system symptoms, and relief of symptoms by the administration of glucose. (Named for George Hoyt *Whipple.*)

whippoorwill [wip′ər wil′] *Vertebrate Zoology.* a nocturnal goatsucker bird found in North America, *Caprimulgus vociferus* of the family Caprimulgidae, known for its vigorous call heard at nightfall or just before dawn and characterized by a round body, a wide mouth, speckled brown plumage, and a light crescent mark on the throat. Also, **whip-poor-will.** (Its name is an imitation of the sound of its call.)

whippoorwill

whippoorwill storm see FROG STORM.

whipsnake *Vertebrate Zoology.* any of several thin, long, arboreal snakes, including the tropical green vine snake, the coachwhip of the Americas, and the copperhead of Australia.

whipstock *Petroleum Engineering.* a wedge introduced into an oil-well borehole during drilling procedures to crowd the drill to the side of the hole and deflect it away from some obstruction.

whipworm disease *Medicine.* infection with the parasitic intestinal whipworm, *Trichuris trichiura.*

whirlpool *Hydrology.* an eddy or vortex of water moving rapidly in a circular path and producing a depression in the center, into which floating objects may be drawn.

Whirlpool Galaxy *Astronomy.* a large spiral galaxy in Ursa Major, lying at a distance of about 20 million light-years; designated **M 51.**

whirlpool separator *Biotechnology.* a hydroclone that is used for the separation and recycling of cells from the effluent of fermentation.

whirly *Meteorology.* a localized violent storm with a diameter of 3–100 yards, frequently occurring over Antarctica at the time of the equinoxes.

whisker *Materials Science.* a metallic or nonmetallic, single-crystal fiber with ultrahigh strength and a micrometer-size width; created when crystals are grown under conditions in which the growth is rapid in one direction, and often used in aerospace applications. *Vertebrate Zoology.* see VIBRISSA. *Robotics.* a type of contract sensor composed of thin wires protruding from an end effector.

whiskey *Food Technology.* a grain spirit distilled from a mash of barley, rye, or other cereal grain that is malted and fermented. (From Gaelic *uisgebeatha,* "the water of life.")

whistler *Geophysics.* a radio signal generated by some lightning discharges capable of traveling great distances through the earth's magnetic field; so-called because of the characteristic whistlelike sound that is picked up by radio receivers. *Vertebrate Zoology.* see THICKHEAD.

whistling meteor *Electromagnetism.* a radio meteor when a spherical system for detection is used in which the meteor presence causes a rapid change in audio-frequency radar signal.

white *Optics.* **1.** a color having no hue, at the end of the scale of grays opposite to black. **2.** having this color, reflecting all or nearly all the rays of the spectrum.

White, Charles 1728–1813, English physician; promoted hospital hygiene; described gangrene; wrote on comparative anatomy.

White, Leslie 1900–1974, American anthropologist; revived the idea of cultural evolution.

White, Paul Dudley 1886–1973, American physician; a leading authority on heart disease.

White, Stanford 1853–1906, American architect; designed Washington Arch (New York City) and Newport (Rhode Island) Casino.

white arsenic see ARSENIC TRIOXIDE.

white blood cell see LEUKOCYTE.

whitecap *Oceanography.* the froth of bubbles in the crest of a breaking wave.

white cast iron *Materials.* iron that is formed when most of the carbon in a molten cast iron forms iron carbide instead of graphite upon solidification. (So called because it fractures to produce a "white" or bright crystalline fractured surface.)

white caustic see CAUSTIC SODA.

white cell see LEUKOCYTE.

white cement *Materials.* a cement made from pure calcite limestone or limestone containing iron treated with fluorspar (natural calcium fluoride); ground finer than ordinary cement to give superior working qualities, but having physical properties similar to those of ordinary cement.
white clay see KAOLIN.
white clot see WASHED CLOT.
white coal see TASMANITE.
white coat *Building Engineering.* the final coating applied to plaster.
white compression *Telecommunications.* in television or facsimile transmission, the reduction of picture signal gain at levels equal to light areas, compared to the gain at the level for midrange light values; used to lessen the contrast in picture highlights.
white corpuscle see LEUKOCYTE.
white damp *Mining Engineering.* carbon monoxide, a poisonous gas that may be found in the afterdamp of a gas or coal-dust explosion or among the gases given off by a mine fire.
white dwarf star *Astronomy.* a small, hot, but intrinsically faint star, essentially the leftover and exposed core of a red giant star which has puffed off its outer layers to form a planetary nebula.
white-eye *Vertebrate Zoology.* any of numerous tropical bird species of the family Zosteropidae, known primarily for the white rings around their eyes; characterized by a short tail, short wings, and a pointed bill; found in Africa, Asia, New Guinea, Australia, and New Zealand.
whiteface or **white face** *Agriculture.* a popular name for Hereford cattle, whose head and facial markings are uniformly white.
whitefish *Vertebrate Zoology.* **1.** any of several freshwater fishes of the family Salmonidae, especially of the genus *Coregonus*, that are prized as food fish. **2.** any of various fishes of other families that resemble the true whitefish, including some species in the families Coregonidae, Lactariidae, and Sparidae.
white fox see ARCTIC FOX.
white garnet see LEUCITE.
whitehead *Medicine.* a whitish accumulation under the skin due to the secretion of an oil gland duct. Also, MILIUM.
Whitehead, Alfred North 1861–1947, English mathematician and philosopher; the cofounder (with Bertrand Russell) of mathematical logic.
white ice *Hydrology.* ice that has a whitish color with no discoloration, as when it forms from snow falling on lake or glacier ice.
white infarct *Medicine.* necrosis of an area of tissue due to a sudden failure of circulation in a blood vessel. Also, ANEMIC INFARCT, PALE INFARCT.
white iron *Metallurgy.* a hard cast iron that is chilled in a metallic mold.
white lead *Materials.* any basic lead carbonate of variable composition that appears as a white, amorphous powder; used as a pigment in lead paint.
white level *Telecommunications.* in television or facsimile transmission, the carrier signal that corresponds to maximum picture brightness.
white light *Optics.* light containing any mixture of monochromatic radiation in such proportions that it is perceived as colorless.
white light hologram *Optics.* a hologram that is illuminated with ordinary light.
white-light interference *Optics.* interference between beams of light having a broad spectral width, such as the light emitted by a tungsten bulb.
white mahogany see PRIMAVERA.
white metal *Metallurgy.* **1.** any of a group of white-colored metals having low melting points such as lead, antimony, tin, or zinc. **2.** any of the usually tin-based alloys made from these metals.
white mica see MUSCOVITE.
white mineral oil see MINERAL OIL.
white nickel see RAMMELSBERGITE.
White Nile *Geography.* a tributary of the Nile, flowing about 2300 miles north from Lake Victoria and joining the White Nile at Khartoum to form the Nile's trunk stream.
whitening filter *Electronics.* a filter used with a noise source, with a frequency response that makes the output noise approximate white noise. Also, PREWHITENING FILTER.
white noise *Acoustics.* random or thermal noise that typically has a wide frequency range and also constant energy per unit bandwidth; sound that is analogous to white light in that it is uniformly distributed throughout the complete audible sound spectrum. *Acoustical Engineering.* inoffensive background noise, such as the recorded sound of falling rain or of ocean waves, that is intended to mask the sound of more disturbing environmental noises. Also, **white sound.**
white oak *Botany.* any of several oak trees of the genus *Quercus* having white to light-gray bark, especially *Q. alba* of eastern North America and *Q. garryana* and *Q. lobata* of western North America.
white object *Optics.* an object that reflects all wavelengths of light nonselectively, at a high level, and with a high degree of diffusion.
white oil *Materials.* any of a number of combustible oils that are derived from paraffinic hydrocarbons; used as a textile finisher and lubricant and as a plastics modifier.
whiteout *Meteorology.* an atmospheric optical phenomenon resident in the polar regions in which a uniform white glow envelops an observer such that sense of depth, orientation, and clouds are lost. Also, MILKY WEATHER. *Graphic Arts.* **1.** a white fluid used in correcting printed copy and in touching up photographic negatives or other illustrations. **2.** to delete copy or other material by means of this fluid.
white paraffin oil see MINERAL OIL.
white phosphorus *Chemistry.* one of the main forms of the element phosphorus. See PHOSPHORUS.
white phosphorus grenade *Ordnance.* a grenade containing a main charge of white phosphorus that, when scattered by a smaller explosive charge, creates a smoke screen with an incendiary effect.
white piedra *Medicine.* piedra caused by *Trichosporon cutaneumis.* It is manifested as white to light brown nodules that are softer than those seen in black piedra, and usually affects the hair of the beard, axilla, and groin. It is most common in tropical regions. Also, TRICHOSPOROSIS.

white pine forest

white pine *Botany.* a large pine tree, *Pinus strobus,* of eastern North America, having irregular branches and light-gray bark. *Forestry.* the soft, light wood of this tree, widely used in various building and manufacturing applications.
white potato see POTATO.
whiteprint see DIAZO PAPER.
white radiation *Crystallography.* any radiation, such as X rays or sunlight, with a continuum of wavelengths.
white rainbow see FOGBOW.
white scours see CALF SCOURS.
white shark *Vertebrate Zoology.* any of various light-colored predatory sharks of tropical and warm waters, such as the great white shark.

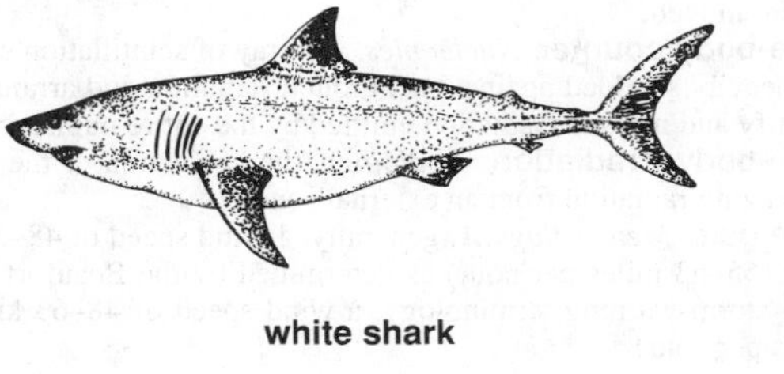

white shark

white signal *Telecommunications.* in FAX transmission, the signal that is produced by the scanning of a minimum density area of the subject copy.

white squall *Meteorology.* a sudden squall condition that develops over tropical or subtropical waters; its arrival is accompanied not by the usual squall cloud but by a whitish line of broken water or whitecaps over the sea surface.

white stochastic process *Mathematics.* a stochastic process with no correlation (including autocorrelation) between any of its (vector) components at different times; i.e., a stochastic process with a power spectrum uniformly distributed over all frequencies.

white substance of Schwann *Neurology.* see MYELIN.

white tellurium see KRENNERITE.

white-to-black amplitude range *Telecommunications.* **1.** in a facsimile system that uses positive-amplitude modulation, the ratio of the signal voltage or current for picture white to the signal voltage or current for picture black at any point in the system. **2.** in a facsimile system that uses negative-amplitude modulation, the ratio of the signal voltage or current for picture black to the signal voltage or current for picture white at any point in the system.

white-to-black frequency swing *Telecommunications.* in a frequency-modulated facsimile system, the numerical difference between the signal frequencies corresponding to picture white and picture black at any point in the system.

white-to-black transmission *Telecommunications.* **1.** in an amplitude-modulated facsimile system, a form of transmission in which the maximum transmitted power corresponds to the minimum density of the subject copy. **2.** in a frequency-modulated facsimile system, a form of transmission in which the lowest transmitted frequency corresponds to the minimum density of the subject copy.

whitewash *Materials.* a mixture of lime and water or of whiting, size, and water that is used to paint or whiten such surfaces as walls, fences, and woodwork; not usually water resistant.

white water *Hydrology.* a popular term for the rough water in ocean breakers or river rapids.

white whale *Vertebrate Zoology.* a large Arctic whale of the family Monodontidae, characterized by a rounded body, a slight beak, and brown skin at birth that lightens to white. Also, BELUGA.

whitleyite *Geology.* an achondritic stony meteorite that contains black chondrite fragments.

whitlockite *Mineralogy.* $Ca_9(Mg,Fe^{+2})H(PO_4)_7$, a colorless, white, gray or pinkish, transparent to translucent, trigonal mineral, having a specific gravity of 3.13 and a hardness of 5 on the Mohs scale; occurs as a secondary mineral in granite pegmatites and in phosphate rock deposits, and has been identified in lunar rocks.

whitlow *Medicine.* an inflammation affecting the finger or toe, especially toward the end or around the nail. Also, **felon.**

Whitney, Eli 1765–1825, American inventor; noted for the development of the cotton gin and of the first mechanisms with standardized interchangeable parts.

Whittaker, Edmund Taylor 1873–1956, English mathematician and physicist; made major contributions in analytical dynamics, functions of complex variables, and special functions.

Whittaker differential equation *Mathematics.* the differential equation $u'' + (-1/4 + k/z + (1/4 - m^2)/z^2)u = 0$; equivalent to the confluent hypergeometric equation. The solutions are called **Whittaker functions.**

Whitworth, Sir Joseph 1803–1887, English mechanical engineer; devised a standardized system of gauges and of screw threads (**Whitworth screw thread**); improved method of milling and testing plane surfaces.

WHO or **W.H.O.** World Health Organization.

whole-arm fusion *Molecular Biology.* a break in the very short arms of two acrocentric chromosomes, succeeded by the fusion of the long parts into one chromosome. Also, CENTRIC FUSION.

whole blood *Hematology.* blood from which none of the elements have been removed.

whole-body counter *Nucleonics.* an array of scintillation counters that are heavily shielded against background radiation and arranged so as to identify and measure radiation emitted by the entire human body.

whole-body irradiation *Radiology.* the subjection of the entire body to ionizing radiation from an external source.

whole gale *Meteorology.* **1.** generally, a wind speed of 48–55 knots per hour (55–63 miles per hour) as determined by the Beaufort wind scale. **2.** in storm-warning terminology, a wind speed of 48–63 knots (55–72 miles per hour).

whole human blood *Hematology.* blood that has been drawn from a selected donor under strict aseptic conditions, containing citrate ion or heparin as an anticoagulant, and used as a blood replenisher.

whole milk *Food Technology.* milk that contains all of its butterfat, usually 4 percent by volume.

whole mount *Microbiology.* an entire specimen placed on a microscope slide for examination.

whole range point *Aviation.* the point on the surface of the earth directly below an aircraft at the moment of impact of a bomb that was released by the aircraft; the surface distance the aircraft travels between the moment of release and the moment of impact.

whole step *Acoustics.* an interval in musical pitch that has a frequency ratio of 9:8 in the tempered scale. Also, **whole tone.**

wholistic see HOLISTIC.

whooping cough *Medicine.* the popular name for pertussis, a highly contagious respiratory disease. See PERTUSSIS.

whooping crane

whooping crane *Vertebrate Zoology. Grus americana*, a large, rare North American wading bird of the family Gruidae, characterized by a trumpetlike call, long legs and neck, and plumage that is all white except for black markings on the face and bright red skin on the head.

WHO plate *Immunology.* an instrument used in testing small volumes of serum in serial dilutions, composed of a rectangular block of plastic having small indentations.

Whorf or **Whorfian hypothesis** *Anthropology.* the concept advanced by the American linguist Benjamin Lee Whorf (1897–1941), that a culture's native language exerts an influence over its members' perceptions and thought processes, which leads to variations in behavior from culture to culture. Also, SAPIR-WHORF HYPOTHESIS.

whorl *Botany.* an arrangement of several identical parts in a circle around a single point on a plant, such as the calyx, the corolla, the stamens, and the pistils. *Anatomy.* **1.** a turn in the spiral cochlea of the ear. **2.** a spiral twist of the muscle fibers at the apex of the heart.

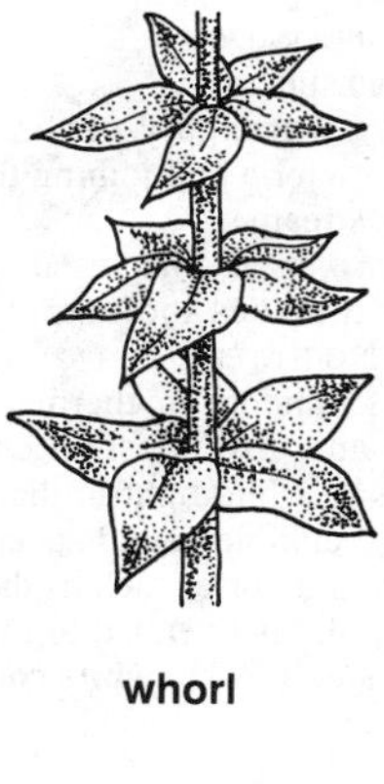

whorl

why explanation *Artificial Intelligence.* a facility provided by an expert system that can explain why it requires certain information from the user.

WI wrought iron.

WIA wounded in action.

wiborgite see RAPAKIVI.

wichtisite see TACHYLITE.

wicking *Engineering.* solder flow under the insulation of covered wire.

Widal reaction *Immunology.* an agglutination reaction to the Widal test indicating the presence of enteric infections by *Salmonella.*

Widal test *Immunology.* a test, used to detect the presence of enteric infections, in which serum antibodies are put in contact with species of the bacteria *Salmonella* and the subsequent agglutination is observed.

wide-angle lens *Optics.* a lens having a focal length substantially shorter than the diagonal dimension of the image it forms.

wide-area data service *Telecommunications.* an automatic teletypewriter data exchange service that uses leased commercial telephone lines to transmit data.

wide area telephone service see WATS.

wide band *Acoustical Engineering.* a large range of frequencies, typically greater than 1/3 of an octave, used for electroacoustical processing. *Electronics.* **1.** capable of transmitting a broad range of signal frequencies intelligibly. **2.** having a bandwidth greater than the voice range (6 kHz). *Telecommunications.* a bandwidth characterized by a data transmission speed of 10,000 to 500,000 bits per second; such speed requires more than one voice-frequency band for operation.

wide-band amplifier *Electronics.* an amplifier that passes a wide band of frequencies without attenuation.

wide-band communications *Telecommunications.* **1.** a communications system that provides a number of channels that are virtually invulnerable to interruption by countermeasures or natural phenomena, including multichannel telephone cable, tropospheric scatter, and a multichannel line-of-sight radio system such as microwave. **2.** a communications systems for signals substantially greater than voice band, typically referring to 1.54 or 2.048 Mbps PCM carrier systems.

wide-band ratio *Telecommunications.* in a particular communication system, the ratio of the occupied frequency bandwidth to the intelligence bandwidth.

wide-band repeater *Electronics.* a repeater that operates over a large band of input and output frequencies; commonly used in satellite communications and low-altitude aircraft, both of which require data transmission beyond line of sight.

wide-band switching *Electronics.* switching circuitry that is capable of operating over a wide range of frequencies.

wide-band transformer *Electricity.* a transformer that operates over a wide range of frequencies.

wide berth *Navigation.* ample clearance; **to give a wide berth to** means "to keep well away from."

widebody *Aviation.* a jet airliner whose fuselage is wide enough to accommodate several hundred passenger seats, which are usually divided by two aisles running the length of the airplane.

widebody jet (Boeing 747)

wide-open *Electronics.* having the ability to respond to a wide range of frequencies and the inability to discriminate against any of them.

widgeon *Vertebrate Zoology.* any of several freshwater ducks of the genus *Anas* of Eurasia and the Americas, having metallic-green wing feathers with a white patch, and a white or buff-colored forehead.

widget *Computer Technology.* an informal term for any small interactive displayed graphical object, such as a menu or a simulated push-button, along with related software, that can be used as a component in constructing a graphical user interface.

Widmanstatten pattern *Geology.* a triangular pattern of intersecting bands on the cut, polished, and acid-etched surface of an iron meteorite. Also, **Widmanstatten figure.**

Widmanstatten structure *Materials Science.* the precipitation of a second phase from the matrix when there is a fixed crystallographic relationship between the precipitate and matrix crystal structures, often involving the formation of needlelike or platelike structures.

widow *Graphic Arts.* a break in the continuity of typeset text resulting in a page that begins with the last word or last few words of a paragraph; considered poor typography. (From the idea that the isolated word or words are left alone like a woman who is a widow.)

width *Science.* the horizontal measurement of something; the extent from side to side; breadth. *Telecommunications.* the elapsed time of a pulse.

width coding *Telecommunications.* the modification of the pulses emitted from a transponder for recognition in the display, using a preset code.

width control *Electronics.* the ability to adjust the amplitude of the horizontal sweep and consequently the width of an image, as in a television receiver or oscilloscope.

width of sheaf *Ordnance.* in artillery support, the lateral interval between center of flank bursts or impacts.

Wiedemann, Gustave Heinrich 1826–1899, German physical chemist; formulated the Wiedemann-Franz law (with Rudolf Franz) and Wiedemann's rule.

Wiedemann effect *Electromagnetism.* a twist incurred in a current-carrying conductor when subjected to a magnetic field parallel to the current.

Wiedemann-Franz law *Solid-State Physics.* a law stating that for all metals the ratio of the thermal conductivity to the electrical conductivity is proportional to the absolute temperature. Thus, **Wiedemann-Franz ratio.**

Wiedemann's rule *Physical Chemistry.* the rule that a mixture's magnetic susceptibility in mass is the sum of the susceptibilities in weight of each of its components. Also, **Wiedemann's additivity law.**

Wiegand effect *Electricity.* the ability of a stressed ferromagnetic wire to recognize the rapid switching of magnetization when subject to a direct-current magnetic field.

Wiegand module *Electricity.* the apparatus consisting of a Wiegand wire, two small ferromagnets, and a pickup coil; used to generate an electrical pulse following the Wiegand effect.

Wiegand wire *Electricity.* a mechanically stressed wire whose magnetic permeability is much greater at the surface than at the center.

Wieland, Heinrich Otto 1877–1957, German chemist; awarded Nobel Prize for research on structure of bile acids and poisons.

Wien, Wilhelm [vēn] 1864–1928, German physicist; awarded Nobel Prize for formulating the Wien laws of blackbody energy.

Wien bridge *Electronics.* a frequency-sensitive four-arm bridge in which two adjacent arms are resistances and the other two adjacent arms are RC combinations.

Wien bridge oscillator *Electronics.* an oscillator that employs a Wien bridge to provide phase shift and frequency determination via feedback.

Wien capacitance bridge *Electricity.* a four-arm AC frequency-sensitive bridge, in which two adjacent arms are resistances and the other two adjacent arms are RC combinations; useful as a simple audio-frequency meter since it can be balanced at only one frequency at a time; also used for capacitance and resistance measurements.

Wien constant *Physics.* the constant in Wien's displacement law; it is approximately equal to 0.2897 centimeter-kelvin.

Wien displacement law *Physics.* a law stating that the product of the absolute temperature T and the wavelength of maximum intensity λ_m from a blackbody is equal to a constant: $\lambda_m T = s$.

Wien distribution law *Physics.* an empirical law stating that the blackbody energy distribution of radiation is related to the absolute temperature T and wavelength λ by the relation $dE_\lambda = A\lambda^{-5}e^{-B/\lambda T}d\lambda$, where A and B are experimentally determined constants.

Wien effect *Physical Chemistry.* the property that causes an electrolyte to increase its conductance when a high voltage is applied.

Wiener, Norbert [wēn′ər] 1894–1964, American mathematician; the founder of cybernetics; helped develop high-speed computers; worked in probability theory and Fourier analysis.

Wiener experiment *Optics.* an experiment in which a mirror covered with a photographic emulsion was exposed to a light beam perpendicular to its surface; following development, stationary waves were found in the emulsion whose nodes synchronized with the nodes of the electric vector rather than those of the magnetic vector.

Wiener-Hopf method *Mathematics.* a technique for solving certain integral and differential equations with boundary conditions; based on the observation that a complex function, analytic in the strip $\alpha < \mathrm{Im}(z) < \beta$ and decaying to zero at infinity, can be represented uniquely as the sum of two functions, analytic and decaying to zero on the half-planes $\mathrm{Im}(z) > \alpha$ and $\mathrm{Im}(z) < \beta$, respectively.

Wiener-Khintchine theorem *Mathematics.* the theorem that the (auto)variance function of a given stochastic process and the power spectral density of the process are Fourier cosine transforms of each other.

Wiener process *Mathematics.* a stochastic process $x(t)$ such that: (a) $x(t)$ has stationary independent increments; (b) every increment is normally distributed; (c) the mean value function $E[x(t)]$ is zero for all values of t; and (d) $x(0) = 0$. The conditional probability distribution of the Wiener process, given by $f\{x(t) \mid x(t_0) = x_0\}$ for $t < t_0$, satisfies the diffusion equation. It is used in the study of Brownian motion, quantum mechanics, and thermal noise in electric circuits; it is also called the **Wiener-Lévy process.**

Wien frequency bridge *Electricity.* a Wien capacitance bridge that is modified to measure varying frequencies.

Wien inductance bridge *Electricity.* a modified Wien capacitance bridge in which inductors are substituted for capacitors; used to measure unknown inductances.

Wien radiation law *Physics.* a relationship between the differential emissive power of a blackbody and the differential wavelength interval evaluated at the wavelength of maximum intensity: $dE_m = CT^5 d\lambda$, where C is a constant.

Wierl equation *Electronics.* a formula for the intensity of an electron beam being scattered through a predetermined angle by diffraction from gas molecules.

Wiese formula *Engineering.* an observable relationship between motor fuel antiknock values over 100 with respect to performance numbers.

Wiesel, Torsten born 1924, Swedish-born American physiologist; with Hubel, awarded the Nobel Prize for work on processing of visual information.

Wiffley table see SHAKING TABLE.

Wigner, Eugene Paul born 1902, Hungarian-born American physicist; awarded the Nobel Prize for improving and extending methods of quantum mechanics; formulated symmetry principles.

Wigner coefficients see CLEBSCH-GORDAN SERIES.

Wigner-Eckart theorem *Quantum Mechanics.* a theorem, from the quantum theory of angular momentum, stating that each matrix element of an operator may be factored into a vector coupling coefficient and another quantity that is independent of the magnetic quantum numbers.

Wigner effect *Nucleonics.* the changes in physical and chemical properties of graphite resulting from the displacement of lattice atoms under irradiation by fast (high-energy) neutrons and other energetic particles in a reactor; the resulting buildup of stored energy leads to changes in crystal lattice dimensions and overall bulk size.

Wigner energy *Nucleonics.* the excess potential energy stored within the crystal lattice of graphite as a result of the Wigner effect.

Wigner force *Nucleonics.* a nonelectromagnetic (nonexchange) force acting over very short distances between nucleons in an atom.

Wigner 6-j symbol see SIX-J SYMBOL.

Wigner nuclides *Nuclear Physics.* a class of nuclides whose respective nucleons mirror each other and consist of pairs of isobars in which the atomic number and the neutron number differ by one.

Wigner release *Nucleonics.* in a graphite-moderated reactor, a controlled annealing procedure for relieving thermal-induced and radiation-induced stresses (**stored Wigner energy**) by increasing the reactor temperature above its normal working level for a very short time.

Wigner's theorem *Quantum Mechanics.* **1.** a theorem stating that in a collision of the second kind, total spin angular momentum will be conserved. **2.** a theorem stating that an eigenfunction of the Hamiltonian may be multiplied by a symmetry element of the Hamiltonian to yield another eigenfunction having the same eigenvalue.

Wijs' iodine monochloride solution *Analytical Chemistry.* a reagent that is used to determine iodine values of fats and oils; composed of a mixture of iodine trichloride and iodine in glacial acetic acid. Also, **Wijs' special solution.**

Wilcoxian *Geology.* a North American geologic stage equivalent to the Sabinian.

Wilcoxon's rank-sum test *Statistics.* a nonparametric test for significant differences between paired observations in which the numerical value of the differences between the paired characteristics is determined and these differences are ranked by their absolute values; similar to the Mann-Whitney test.

Wilcoxon's signed rank test *Statistics.* a nonparametric test used to determine the location of a symmetric population.

wild boar *Vertebrate Zoology.* an Old World wild pig, *Sus scrofa*, of the family Suidae, characterized by enlarged canine teeth on both jaws and a heavy body covered with stiff bristles.

wild card *Computer Programming.* a character that can be used to represent any single character or set of characters in a search argument or file specification.

wildcat *Vertebrate Zoology.* a popular term for any of several small or medium-sized cats, including the jungle cat, lynx, ocelot, European wildcat, and kaffir cat. *Naval Architecture.* the recessed drum of an anchor windlass. *Mining Engineering.* **1.** a prospect well for oil or natural gas drilled in an unproved area. **2.** to drill one or more such boreholes.

Wildcat *Aviation.* a popular name for the F-4F fighter.

Wildcat (F-4F)

wildcat drilling or **wildcatting** *Mining Engineering.* the speculative drilling of boreholes in unproved areas.

wild cinnamon see BAYBERRY.

wildebeest see GNU.

wilderness *Ecology.* any uncultivated region that is not inhabited by humans.

Wilderness *Geology.* a North American geologic stage of the Middle Ordovician period, occurring after the Porterfield and before the Trentonian.

Wild fence *Engineering.* an enclosure, typically 16 feet square and 8 feet high, built around a precipitation gauge in order to minimize eddies around the gauge and ensure accuracy.

wildflower *Botany.* any flowering plant that normally grows without cultivation, in a meadow, field, forest, and so on.

wildflysch *Geology.* a flysch facies characterized by irregular twisted and contorted beds and irregularly sorted blocks and boulders.

wildlife *Zoology.* **1.** a general term for all undomesticated animals that live free in nature. **2.** relating to, abounding in, or preserved for such animals. Thus, **wildlife refuge, wildlife reserve.**

wildness *Metallurgy.* a term for a condition occurring during cooling, when molten metal evolves gas and becomes violently agitated, causing metal to eject from the mold.

wild shot *Ordnance.* **1.** any shot that is not within the normal dispersion pattern. **2.** specifically, in artillery support, a shot whose impact is more than four probable errors and six firing table probable errors from the center of impact. **3.** any shot that is fired without aiming.

wild snow *Meteorology.* snow that has been deposited under calm conditions and low air temperature, having a fluffy nature that renders it rather unstable.

wild strain *Virology.* a virus isolate formed by a natural infection, as opposed to a strain created under laboratory conditions.

wild type *Genetics.* **1.** the standard gene, or the standard genotype or phenotype of an individual, that is found in a wild or natural population. **2.** a form that is arbitrarily designated as such. *Virology.* the original parental strain of a virus.

wilkeite *Mineralogy.* **1.** a rose-red or yellow variety of apatite in which the phosphate is partly replaced by silicate and sulfate. **2.** a phosphatian variety of fluorellestadite.

Wilkins, Maurice born 1916, British biophysicist; for X-ray diffraction studies of DNA, shared the Nobel Prize with Watson and Crick.

Wilkinson, Geoffrey born 1921, British chemist; with E. O. Fischer, awarded Nobel Prize for molecular studies relating to pollution control.

Willans line *Mechanical Engineering.* the line on a graph that indicates the steam consumption versus the power output for a steam engine or turbine, and that can be extended to show fuel consumption for an entire power plant.

willemite *Mineralogy.* Zn_2SiO_4, a colorless to white, green, yellow, or reddish-brown trigonal mineral, occurring as prismatic crystals and in massive form, having a specific gravity of 3.89 to 4.19 and a hardness of 5.5 on the Mohs scale; found in crystalline limestones and as a secondary mineral in the oxidized zone of zinc deposits; a minor ore of zinc.

Williams, Daniel Hale 1856–1931, American surgeon; a pioneer in open-heart surgery.

Williams-Hazen formula see HAZEN-WILLIAMS FORMULA.

Williams-Landel-Ferry relation *Materials Science.* the temperature dependency of melt viscosity expressed in terms of glass transition temperatures and universal constants; a major result of the theory of free volume.

Williamson, William Crawford 1816–1895, English naturalist; wrote on structure and distribution of fossil plants.

Williamsoniaceae *Paleontology.* a family of plants in the order Cycadeoidales; they grew to heights of 20–35 feet, with a crown of fronds and a slender trunk; related to the Bennettitaceae; extant in the Upper Triassic and Jurassic.

Williamson synthesis *Organic Chemistry.* a method of preparing ethers, utilizing the interaction of an alkyl halide and a sodium alcoholate or phenolate.

Williams refractometer *Optics.* a refractometer that splits a beam of light from a single slit into two beams by passing it through a pentagonal prism.

Williams tube *Electronics.* a cathode-ray tube that stores information as a pattern of the charges controlled by the tube's electron beam.

Willis, Thomas 1621–1675, English physician; wrote descriptions of typhoid fever, diabetes, and the human nervous system.

williwaw *Meteorology.* a violent squall that occurs in near-polar areas such as Alaska and the Strait of Magellan, especially during the winter season.

willow *Botany.* any tree or shrub of the genus *Salix* in the family Salicaceae, having narrow leaves, tassel-like spikes of flowers, and usually flexible twigs. *Materials.* the brownish-yellow wood of such trees, known for its durability and nonshrinkability and used in furniture making.

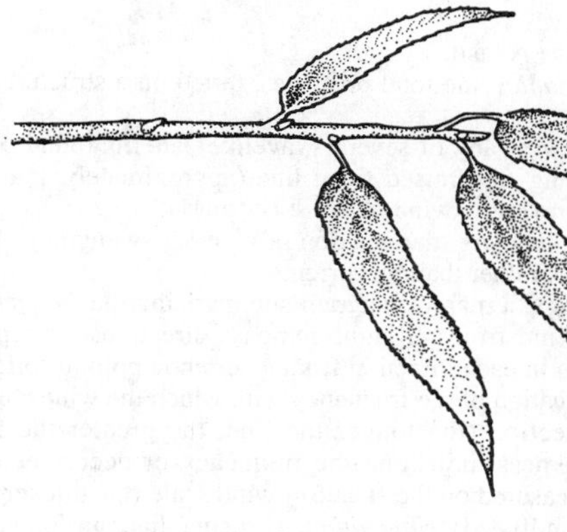

willow

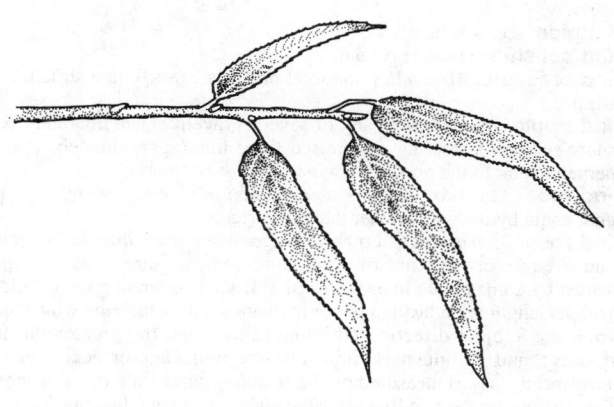

Willstätter, Richard 1872–1942, German chemist; awarded the Nobel Prize for research on chlorophyll and other plant pigments.

willy-willy *Meteorology.* a severe tropical cyclone occurring over Australia.

Wilms' tumor *Oncology.* a rapidly developing malignant tumor of the kidneys, composed of embryonal elements; usually affects children before the fifth year. (Named for Max *Wilms,* 1867–1918, German surgeon.)

Wilson, Charles Thomson Rees 1869–1959, Scottish physicist; awarded the Nobel Prize for invention of the Wilson cloud chamber.

Wilson, Kenneth G. born 1936, American physicist; awarded the Nobel Prize for his method of analyzing changes in matter at phase transitions.

Wilson, Robert Woodrow born 1936, American astrophysicist; shared the Nobel Prize (with Arno Penzias and Pyotr Kapitsa) for the study of cosmic microwave radiation.

Wilson-Bappu effect *Astronomy.* a relationship between the absolute magnitudes of late-type stars and the width of the K2 emission core in the resonance line of ionized calcium.

Wilson cloud chamber *Nucleonics.* an apparatus for delineating the tracks of high-speed charged particles as they pass through a chamber filled with a saturated vapor that, when supercooled by adiabatic expansion, condenses to form droplets of liquid around ions left in the particles' path.

Wilson effect *Astronomy.* the apparent displacement of the penumbra in a sunspot as the spot rotates toward the sun's limb.

Wilson electroscope *Electricity.* a leaf-type electroscope that uses a single leaf hanging vertically, thus maximizing its sensitivity.

Wilson experiment *Electromagnetism.* an experiment in which a hollow dielectric cylinder coated on its inner and outer surfaces with a conductive material is rotated about its longitudinal axis parallel to a magnetic field; a voltage is detected across the inner and outer surfaces, thus verifying that which is predicted by electromagnetic theory.

Wilson plot *Crystallography.* in crystal structure determination by X-ray diffraction, a plot of averages of observed intensities in ranges of $\sin^2\theta/\lambda^2$ with the theoretical values of a unit cell composed of randomly arranged atoms. From this plot, a scale factor and an average temperature factor may be derived.

Wilson's disease *Genetics.* a human disease that is inherited as an autosomal recessive allele; homozygous individuals are unable to process and excrete copper adequately, and damage to the brain and liver results from its accumulation in the tissues. Also, HEPATOLENTICULAR DEGENERATION.

Wilson's theorem *Mathematics.* the theorem that $(n - 1)! + 1$ is divisible by n if n is prime. The converse is also true: $(n - 1)! + 1$ is divisible by n if and only if n is prime.

wilt *Botany.* to become limp; droop. *Plant Pathology.* any of several diseases of plants (such as potatoes, apples, and bananas) characterized by plugging of the conductive tissue by a fungus, resulting in wilting.

wilting capacity *Ecology.* the amount of water remaining in the soil when a plant is permanently wilted.

Wimshurst machine *Electricity.* a machine that produces high-voltage static electricity; used, for example, to charge Leyden jars for storage.

winch *Mechanical Devices.* **1.** any of various devices in which the rotation of a drum winds or unwinds a rope, chain, or cable to raise or lower a load. **2.** to hoist by means of such a device.

Winchester *Ordnance.* any of the firearms produced by the Winchester Repeating Arms Company founded by Oliver F. Winchester; the most famous are the Winchester lever-action repeating rifles, such as the Winchester '73 and the Winchester '94.

Winchester disk *Computer Technology.* a nonremovable, moving-head disk unit using Winchester technology.

Winchester technology *Computer Technology.* innovations in disk drive technology that use nonremovable, sealed disk units, a very light read/write head, a lubricated disk surface, and magnetic orientation of particles on the disk surface to produce a high-density disk.

wind *Meteorology.* air that is in motion, especially horizontally, in relation to the earth's surface.

wind *Electronics.* the manner in which magnetic tape is wound onto a reel. In **A wind,** the coated surface faces the hub; in **B wind,** the coated surface faces away from the hub. Any wind that has a flat radial side free of protrusions is called a **uniform wind.**

windage *Mechanics.* **1.** the amount of deflection of a projectile due to wind. **2.** a correction in aiming or sighting made to compensate for such deflection.

windage scale *Ordnance.* a table or scale for adjusting the sight to allow for the effect of wind on a bullet or other projectile. Also, WIND GAUGE.

windage yaw *Ordnance.* in aerial gunnery, the deviation caused by the interaction of the gun and the air.

Windaus, Adolf 1876–1959, German chemist; awarded the Nobel Prize for his work on sterols; discovered and synthesized vitamin D_3.

windbreak *Agronomy.* an area of trees or bushes serving to shield crops or livestock from the effects of wind. *Engineering.* any device designed to interfere with wind flow in order to provide protection from its effects.

wind burn *Medicine.* skin injury caused by excessive exposure to wind; manifestations include dryness, redness, soreness, and itching. *Botany.* injury to plant foliage from hot, dry winds.

wind charger *Electricity.* a wind-powered machine, with blades often resembling airplane propellers, that converts wind power to rotary shaft power and finally into electrical power by means of a generator.

wind chill *Meteorology.* **1.** the sensation of the combined effects of cold air and wind on exposed surfaces of the body. **2.** see WIND-CHILL FACTOR. *Medicine.* loss of body heat caused by exposure to wind.

wind-chill factor *Meteorology.* an expression of the combined effects of cold air and wind, derived by subtracting a factor of the wind speed from the air temperature. For example, if the temperature is 10°F and the wind speed is 20 miles per hour, the wind-chill factor is –9, which is regarded as equivalent to a temperature of –9°F. Also, WIND CHILL, CHILL-WIND FACTOR.

wind-chill index *Medicine.* the net cooling effect expressed by a value of kilograms per hour per square meter of heat loss by a body through the effects of air temperature and wind speed, with variations per body size, composition, and metabolic rate.

wind cone *Engineering.* a cone-shaped piece of fabric pivoted at its opened larger end and used to indicate wind direction. Also, WIND SOCK, WIND SLEEVE.

wind correction *Engineering.* any adjustment that is made to compensate for the effects of wind, as on projectile flight, sound reception, and the like.

wind crust *Hydrology.* a snow crust formed by the packing of previously deposited snow into a hard layer as a result of wind action.

wind current *Meteorology.* a quasipermanent, large-scale wind system such as the westerlies, the trade winds, or the polar easterlies.

wind-cut stone see VENTIFACT.

wind deflection *Mechanics.* deflection of the trajectory of a projectile in flight due to wind.

wind direction *Meteorology.* the prevailing direction from which the wind blows at a given moment in time.

wind-direction shaft *Meteorology.* a graphical representation on a synoptic weather chart by a straight line drawn upward from a weather station circle; wind speed barbs are attached as perpendicular lines to the end of the shaft.

wind divide *Meteorology.* a semipermanent feature of the atmospheric circulation on opposite sides of which the prevailing wind directions differ greatly; usually a high-pressure ridge.

wind drift *Acoustics.* the apparent shift in the position of a sound source or target due to the motion of the medium carrying the sound, especially the motion of air that is caused by wind.

wind-driven current see DRIFT CURRENT.

winder *Building Engineering.* in a staircase, the wedge-shaped step whose tread is wider at one end than at the other, and which changes direction for stairs laid above it.

wind erosion *Geology.* an erosion process by which topsoil is detached, carried away, and deposited by the action of wind.

wind-fire angle *Ordnance.* the clockwise horizontal angle between the plane of fire and the direction in which the ballistic wind is blowing.

wind gap *Geology.* a shallow notch or pass in the upper section of a mountain ridge, normally marking the site of a former water gap. Also, AIR GAP, WIND VALLEY.

wind gauge *Engineering.* see ANEMOMETER. *Ordnance.* see WINDAGE SCALE.

winding *Electricity.* **1.** a conductive path consisting of one or more turns of wire forming a continuous coil; used in transformers, relays, and other electric devices. **2.** the manner in which wire is wound in a given device.

winding number *Mathematics.* the number of times that a closed curve in the plane passes around a designated point of the plane in a positive (counterclockwise) direction.

windkill *Ecology.* the loss of plant life from strong winds.

windlass *Mechanical Engineering.* a simple hauling or lifting device, composed of a horizontal cylindrical component upon which is wound a rope or cable that is attached, directly or indirectly, to the object to be moved; the cylinder may be turned by a hand crank or a motor.

wind measurement *Meteorology.* the determination of an air parcel's size, speed, and direction of motion.

windmill *Mechanical Engineering.* a device that converts wind energy to mechanical energy by means of wind-propelled vanes or sails radiating about a horizontal shaft.

windmill anemometer *Engineering.* a rotation anemometer in which the axis of rotation is horizontal and the relation between wind speed and angular rotation is nearly linear.

wind noise *Acoustics.* a significant source of noise in certain environments or at certain times due to the rapid movement of air around or through objects.

window *Architecture.* **1.** an opening to admit light and usually air into a building, consisting of a framework or sash and a sheet of glass called a windowpane. **2.** see WINDOWPANE. *Computer Programming.* a division of a display screen in which a set of information is displayed. *Aviation.* the area of skin of an aircraft used for optics or infrared instrumentation. *Ordnance.* a confusion reflector consisting of chaff, wire, or bars cut to resonate at enemy radar frequencies; it is dropped from aircraft or delivered by shells or rockets. Thus, **window rocket.** *Space Technology.* a brief period of time when a spacecraft can be launched from a particular site into a selected orbit. *Electronics.* the time period during which a digital circuit is open to signal sampling. *Geophysics.* **1.** a wavelength interval in the electromagnetic spectrum that is capable of penetrating the earth's atmosphere. **2.** any of various devices placed in the upper atmosphere in order to create a radar echo; used in tracking airborne devices or the wind. *Hydrology.* a part of a river that remains unfrozen during all or part of the winter although it is completely surrounded by river ice. *Geology.* an area in which erosion has effected a thrust or recumbent fold, exposing the underlying rock. *Nucleonics.* **1.** a thin sheet (often of mica) covering the end of a Geiger counter through which low-energy radiation is received and measured. **2.** an aperture or thin portion of wall in a nuclear reactor for passage of particles or radiation. **3.** an arbitrarily narrow energy range of relatively high transparency in the total neutron cross section of a material; such windows, of importance in neutron shielding, arise from interference between potential and resonance scattering in elements of intermediate atomic weight. **4.** a material, generally opaque to the passage of radiation, that acts as a window when it selectively transmits a specific small range of radiation.

window editor *Computer Programming.* an editor program that incorporates the use of windows that allow the user to view or act upon the data being edited.

window ice *Hydrology.* a thin deposit of ice crystals that forms when water condenses on the interior of a cold window surface and then freezes. Similarly, **window frost.**

windowing *Computer Programming.* **1.** the act of dividing a display screen into separate areas in which different information can be viewed simultaneously. **2.** the act of selecting a portion of a large drawing to be displayed in detail on the screen.

windowpane *Architecture.* the sheet of glass that admits light through a window.

windpipe see TRACHEA.

wind polish see DESERT POLISH.

wind pressure *Mechanics.* the total pressure exerted on a structure by wind.

wind ripple *Meteorology.* any of several wavelike patterns found on a snow surface, appearing as a raised ridge line (approximately 1 inch) perpendicular to the prevailing wind. Also, SNOW RIPPLE.

wind rode *Naval Architecture.* the position of a vessel swung at its anchor chain by the wind, rather than by currents.

wind rose *Meteorology.* a map or diagram summarizing the frequency and strength of the wind from all eight compass directions, as represented by a line drawn in each direction from a common point of origin, with its length an indication of the frequency with which the winds blow from the subject direction (the longer the line, the greater the frequency), and its thickness indicating the frequency of occurrence of windspeed class as measured on the Beaufort wind scale (the thicker the line, the higher the windspeed). *Navigation.* **1.** such a diagram found on certain charts, such as pilot charts, showing the average winds for a certain location, consisting of a circle with a number of lines or arrows radiating from it. The arrows fly with the wind. Their length indicates the percentage of the time a wind blows from that direction. On American charts the number of feathers indicates the average speed of the wind in Beaufort numbers. The center may include a number that represents the percentage of calm. **2.** on ancient Mediterranean charts, from the time before the invention of the magnetic compass, the wind rose showed the eight named winds that mariners would presumably be able to identify and use as guides to steering. Also, **wind star.**

windrow *Agriculture.* a row of hay or straw that has been cut and laid out for drying and collection. *Geology.* a low accumulation of material formed naturally by air or water currents, or artificially on or near a construction site.
winds aloft *Meteorology.* the windspeed and direction aloft at varying atmospheric levels as determined by any winds-aloft observation. Also, UPPER-LEVEL WINDS, UPPER WINDS.
winds-aloft observation *Meteorology.* the analysis of wind speed and direction at various levels in the atmosphere.
wind scoop *Meteorology.* the resulting depression in a snow surface when the wind is deflected by obstructions in a local area such as trees, houses, or buildings, thus creating an eddy effect that changes the snow surface ahead of the obstruction.
wind-scoured basin *Geology.* see DEFLATION BASIN.
wind shear *Meteorology.* the rate at which the windspeed varies at different points in its general direction of motion; it is particularly hazardous to aircraft operations, especially near the ground.
windshield *Engineering.* a glass screen that insulates the interior of a vehicle from the elements.
windshield wiper *Mechanical Devices.* a device used to remove moisture from a vehicle's windshield, consisting of a rubber blade attached to a pivoting mechanical arm.
wind-shift line *Meteorology.* a narrow zone of abrupt wind direction change.
wind ship [wind] *Naval Architecture.* a ship equipped with sails to augment its engine.
wind ship [wīnd] *Navigation.* to swing a ship by hauling in and paying out anchor or mooring lines.
wind slab or **windslab** *Hydrology.* a layer of snow that is compacted tightly by the wind as the snow is falling.
wind-slab avalanche *Hydrology.* a type of avalanche initiated by the breaking off of a large wind slab from the underlying snow.
wind sock or **wind sleeve** see WIND CONE.
wind speed or **windspeed** *Meteorology.* the rate at which air moves.
windstorm *Meteorology.* a low-pressure condition in which the strength of the winds is the predominant characteristic.
wind stress *Meteorology.* a drag or tangential force imposed on the earth's surface by the motion of an adjacent body of air.
wind tee *Engineering.* a T-shaped device designed to indicate the direction of surface wind at an airfield. Also, LANDING TEE.
wind tide *Oceanography.* the rise in the still-water level on the leeward side of a large body of water, caused by wind stress on the surface; expressed as the difference between the still-water level on the leeward side and that on the windward side. Also, **wind setup.**
wind triangle see TRIANGLE OF VELOCITIES.
wind tunnel *Engineering.* a chamber, duct, or apparatus in which a high-speed movement of air or other gas is generated, often by a fan, and within which objects (or models of objects) such as engines, aircraft, airfoils, and rockets are placed in order to study the airflow about them and the aerodynamic forces acting upon them.
wind-tunnel balance *Fluid Mechanics.* see AERODYNAMIC BALANCE.
wind-tunnel instrumentation *Engineering.* any mechanical, electrical, or optical measuring devices used in connection with wind-tunnel tests.
wind turbine *Electricity.* a wind-driven machine containing curved vanes or blades inside a wheel set vertically on a revolving shaft; wind or air pressure against the vanes turns the wheel, and the rotating shaft may then drive a dynamo to produce electrical power.
windup *Mechanical Engineering.* the slight twisting of a shaft under a torsional load.
wind valley see WIND GAP.
wind vane *Engineering.* an asymmetrical object that is balanced on a rotating axis and indicates the direction the wind is blowing.
wind velocity *Meteorology.* the speed and direction in which the wind blows.
windward *Meteorology.* toward the wind, or toward the direction from which it blows.
wind wave *Oceanography.* a wave caused by wind.
wine *Food Technology.* an alcoholic beverage (usually 12–14% alcohol by volume) that is made by fermenting the juice of grapes; the main types of wine are **sparkling wines** (e.g., champagne); **red table wines** (e.g., Bordeaux, Burgundy, Chianti); **white table wines** (e.g., Chablis); **rosé wines**; **appetizer wines** (e.g., sherry); and **dessert wines** (e.g., port).
wine gallon *Metrology.* see GALLON, def. (a).
wing *Zoology.* an organ of flight, such as either of two anterior appendages of a bird or any of the paired lateral extensions of the body wall of an insect. *Aviation.* an airfoil structure that is similar in form to the wing of a bird, designed to develop lift for an aircraft, either extending on one side of the aircraft and separated from its mate by the fuselage (but sometimes considered as a unit, as in the calculation of wing area) or extending uninterrupted on both sides of the aircraft. Used to form numerous compounds such as **wing chord, wing flap,** and **wing lift.** *Architecture.* a large but subordinate part of a building that projects from the building's main or central part. *Geology.* **1.** the forward, projecting section of a dune. Also, HORN. **2.** see VESICLE. *Military Science.* **1.** a military unit composed of the necessary organization to perform a specific mission. **2.** the part of a military unit on either side of its main body.
wing area *Aviation.* the surface area of an aircraft's wing, measured on the plane of its chords from the projection of the plan-view outline. For purposes of this calculation, the wing is considered to include the ailerons and to extend without interruption through the fuselage and nacelles.
wing axis *Aviation.* the axis of a wing about which all its aerodynamic forces are centered.
wing dam *Civil Engineering.* see GROIN.
wing divider *Mechanical Devices.* a divider or compass with a curved piece or wing projecting from one leg through the slot of the other and a setscrew to fix its position. Also, **winged dividers.**
winged *Zoology.* describing an organism that has wings. *Design Engineering.* having wings or winglike parts.
winged headland *Geography.* a promontory with lower projections of land on either side.
Winged Horse see PEGASUS.
winged missile *Ordnance.* a missile having wings, as distinguished from wingless missiles such as bullets and projectiles.
winged rocket *Space Technology.* a rocket vehicle with aerodynamic surfaces.
wing fence *Aviation.* a barrier or other surface on top of an aircraft wing, designed to retard spanwise airflow.
wing gun *Ordnance.* a fixed gun that is mounted on the wing of an airplane.
wingless abutment *Civil Engineering.* a bridge abutment with no wing walls extending back from it.
wing load *Aviation.* the load imposed on a wing by the weight of an aircraft or its contents.
wing loading *Aviation.* the ratio of the total weight of an airplane to its wing area.
wing nut or **wingnut** *Mechanical Devices.* a nut having projecting metal flanges used for hand tightening or loosening. Also, BUTTERFLY NUT, HAND NUT.
wing panel *Aviation.* **1.** a part of a wing that can be removed or constructed independently. **2.** any readily identifiable part of a wing. **3.** the two mated wings on either side of an airplane, when considered as a unit as in the calculation of wing area.
wing photograph *Photogrammetry.* any photograph taken through the side or wing lenses of a multilens camera, usually at an oblique angle.
wing pod *Aviation.* a streamlined nacelle carried beneath an airplane wing, usually housing a jet engine.
wing point *Cartography.* in aerial photography, three points or notches produced by the camera along each edge of the photograph, used to extend radial control in making controlled mosaics.
wing profile *Aviation.* the outline of a wing cross section.
wing rib *Aviation.* the primary structure of a wing that forms its airfoil shape and joins the leading edge to the trailing edge.
wing root *Aviation.* the base of an airplane wing, where it joins the fuselage.
wing section *Aviation.* on a wing or other airfoil, any cross section that is parallel to the plane of symmetry or to some other specified plane.
wingspan or **wing span** *Aviation.* on a fixed-wing aircraft, the span or straight-line distance between one wingtip and the other, including any projecting ailerons. (The term *wingspan* always refers to the tip-to-tip span of an entire wing, never to a wing half on one side of a fuselage.) Also, **wingspread.**
wing spot generator *Electronics.* electronic circuitry that causes wings to appear on a radar G display video target. The wing size is inversely proportional to the target range.
wing tank *Aviation.* **1.** a primary fuel tank built into the wing of an aircraft. **2.** see WINGTIP TANK.

wingtip or **wing tip** *Aviation.* **1.** the extreme outer edge of a wing. **2.** relating to or situated along this edge; used to form compounds such as **wingtip float** and **wingtip vortex.**

wingtip rake *Aviation.* the slant of a wingtip as seen in plan view; considered positive when the trailing edge of the wing is longer than the leading edge.

wingtip tank *Aviation.* an auxiliary fuel tank mounted on an airplane wingtip. Also, WING TANK.

wink *Industrial Engineering.* a basic division on a microchronometer, equivalent to 1/2000 (0.0005) minute.

wink counter see MICROCHRONOMETER.

Winkler titration *Analytical Chemistry.* a method for estimating the dissolved oxygen in seawater; involves the addition of manganous hydroxide to the sample with the subsequent liberation of iodine, which is then titrated with a sodium thiosulfate standard solution.

winning *Mining Engineering.* **1.** the process of excavating, loading, and removing valuable material from the ground, subsequent to development; mining. **2.** a new mine opening. **3.** the part of a coal field that is laid out and ready for working.

winnow *Agriculture.* **1.** to separate grain from its husks by means of air or wind currents. **2.** a device used in this process.

winnowing gold *Mining Engineering.* a method of extracting gold from dry auriferous powder by tossing the material in the air and recovering the heavier particles not blown away.

winter *Astronomy.* the season beginning at the winter solstice, approximately December 22, and lasting until the vernal equinox, about March 21. *Agriculture.* of grain or other crops, growing in winter; planted in autumn and harvested in late spring or early summer. Thus, **winter wheat, winter rye,** and so on.

Winteraceae *Botany.* a family of dicotyledonous plants in the order Magnoliales, characterized by hypogynous flowers, exstipulate leaves, and usually laminar stamens; occurring in Madagascar, Australia, New Guinea, the southwestern Pacific, and Latin America.

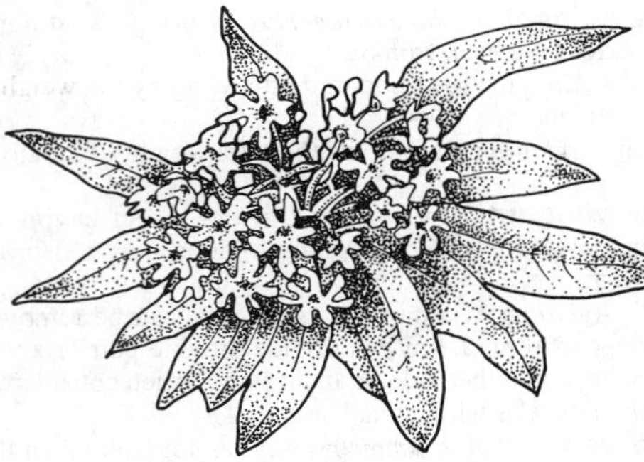

Winteraceae (drimys)

winter balance *Hydrology.* the change in mass of a glacier from its minimum value of a given year to the maximum value of that year.

winter buoy *Navigation.* a small, usually lighted buoy that replaces another buoy during the winter, in areas where the regular buoy is liable to ice damage.

wintergreen *Botany.* a small, creeping evergreen shrub, *Gaulteria procumbens*, of the family Ericaceae, of North America, whose leaves afford a fragrant volatile oil consisting of methyl salicylate.

wintergreen oil see METHYL SALICYLATE.

winter ice *Oceanography.* sea ice in the stage following young ice; i.e., it is less than a year old but is unbroken and solid, between 15 centimeters and 2 meters thick.

winterize *Engineering.* to prepare (equipment, a building, pipes, and so on) for winter weather conditions, as by adding antifreeze to a car's cooling system. Thus, **winterization.**

winter load line *Naval Architecture.* the deepest safe loading line for winter sea conditions; one of several standard load lines marked on merchant vessels.

winter solstice *Astronomy.* the moment at which the greatest southern declination of the sun occurs.

winter-talus ridge *Geology.* a wall-like curved ridge of boulders and coarse debris on the flow of a cirque, formed by frost action that resulted in the breaking apart of cliffs above a snowbank-occupied cirque.

winze *Mining Engineering.* **1.** a vertical or inclined opening or excavation in a mine to connect two levels; it is distinguished from a raise in that it is sunk underhand while a raise is put up overhand. **2.** a subsidiary shaft, sunk in an ore body, that begins and ends underground and generally serves to connect two levels.

wipe *Metallurgy.* to apply filler metal (such as solder) in a semifluid state and distribute mechanically. Thus, **wiped joint.**

wipe-on plate *Graphic Arts.* a lithographic plate designed to be treated before exposure by having photosensitive emulsion applied to its surface.

wiper *Electricity.* a mechanical arm having one end mounted on the rotating shaft of a switch, stepping relay, or variable resistor and the other end sliding or moving over a conducting member of a contact. *Mechanical Devices.* see WINDSHIELD WIPER.

wiping *Metallurgy.* the process of activating a metal surface by mechanical rubbing or wiping in order to bring out enhancements such as a phosphate coating. Thus, **wiping effect.** *Navigation.* the process of reducing the permanent magnetism in a steel vessel by surrounding it with a coil and moving the coil up and down while it is energized.

wiping contact *Electricity.* a contact that slides to make or break a connection.

wire *Electrical Engineering.* **1.** a long, slender strand of drawn conductive metal, either bare or covered with insulation. **2.** a group of such strands, clustered together and either left bare or covered with a common insulator. **3.** to connect electrical circuits using a system of such strands. *Metallurgy.* any continuous length of metal drawn from a rod. *Telecommunications.* **1.** to send by telegraph. **2.** to connect to a television cable system. *Optics.* an element, generally made of a thin metal wire, attached to the field of view on a telescope's eyepiece as a reference. Also, WEB. *Surgery.* **1.** a slender, flexible structure of metal, with applications in surgery and dentistry. **2.** to insert such material into a body structure to promote healing, for example, to immobilize a fracture or promote clot formation in an aneurysm.

wire board *Computer Technology.* see BOARD.

wire bonding *Electricity.* the process of providing a low-resistance path when joining two wires, either by direct contact or with an interconnecting tie. Permanent wire bonding for electrical continuity is obtained by fusing the junction, usually by soldering.

wire cloth *Materials.* a fabric made from woven wire.

wired-program *Computer Technology.* **1.** describing a device whose control program is hard-wired. **2.** an obsolete technology in which the placement of wires on a control panel determined the sequence of instructions that the computer executed.

wired-program computer see FIXED-PROGRAM COMPUTER.

wire drag *Oceanography.* a buoyed wire that is towed at a desired depth under the ocean surface in order to detect obstructions in areas where normal sounding methods are insufficient.

wiredrawing *Metallurgy.* the process of reducing a cross section by pulling a wire through a die.

wire facsimile system *Telecommunications.* a facsimile (FAX) system in which messages are transmitted over wires rather than by radio.

wire flame spray gun *Engineering.* a device that uses heat from a gas flame and a wire or rod to perform a flame-spraying operation.

wire frame *Robotics.* a mesh or grid used in a finite analysis program.

wire fusing current *Electricity.* a current that is sufficiently high to heat a wire to its melting point, causing it to fuse.

wire gauge *Engineering.* a gauge used to measure the diameter of wire or the thickness of sheet metal. *Design Engineering.* a standard series of sizes, arbitrarily indicated by numbers, to which the diameter of wire or the thickness of sheet metal is usually made.

wire grating *Electromagnetism.* a waveguide grating composed of wires arranged in such a way so as to filter certain modes of propagation.

wireless *Science.* having no wires. *Electronics.* a former term for radio.

wire line *Mechanical Devices.* a rope or cable composed of strands of metal or metallic alloy.

wire-line coring *Petroleum Engineering.* the use of a wire line to secure rock samples from a reservoir formation during oil-well drilling operations.

wire-link telemetry *Telecommunications.* telemetry in which signals are transported over transmission lines, rather than being sent over radio. Also, HARD-WIRE TELEMETRY.

wire mile *Electricity*. a unit of measurement used to express the length of a two-conductor wire required between two selected points. The total number of wire miles for any given system is obtained by multiplying the basic unit distance by the number of circuits to be wired.

wire nail see FINISHING NAIL.

wirephoto *Telecommunications*. a photograph sent over telegraph or telephone lines.

wire recorder *Acoustical Engineering*. an earlier recording device, an electromagnetic recorder using a stainless steel recording medium rather than magnetic tape.

wire recording *Acoustical Engineering*. an early technique in magnetic recording, such as recording on stainless steel.

wire rod *Metallurgy*. coiled metal that has been hot-rolled and then cold-drawn into wire.

wire rope *Engineering*. steel rope consisting of strands of twisted steel wires wound about a central core.

wire saw *Mechanical Engineering*. a saw consisting of thin wire cables running as a belt over pulleys, creating an abrasive action to cut out quarry stone.

wiresonde *Meteorology*. an atmospheric sounding instrument that is suspended by a balloon and tied to the ground by a wire cable; used to obtain temperature and humidity data from the ground up to a height of a few kilometers.

wire stripper *Mechanical Devices*. a hand tool that is designed to remove the insulation around a wire without causing damage to the wire inside.

wiretap *Telecommunications*. **1.** a secretly established, concealed connection to a telephone circuit or other wiring system, used to monitor conversations. **2.** to make such a connection.

wire telegraphy *Telecommunications*. telegraphy in which communications are transmitted over wires rather than by radio.

wire train *Mechanical Engineering*. an assembly used to coat wire with resin, consisting of an extruder, a crosshead and die, cooling means, and wire feed and take-up spools.

wireway *Mechanical Engineering*. a hinged, covered trough lined with sheet metal, used to house electrical conductors or cables.

wire-weight gauge *Engineering*. a device used to measure a river's height by lowering a weight suspended on a wire from a point of known elevation to the river's surface.

wire-wound cryotron *Physics*. a device (operated at cryogenic temperatures) consisting of a superconductive insulated wire wound by a control coil; when current is passed through the coil, the magnetic field that is produced destroys the superconductivity.

wire-wound potentiometer *Electricity*. a wire-wound, variable resistor with three terminals: the two ends of the winding, which provides the resistance; and a movable contact (tap), which varies the resistance by tapping off a selected segment of the winding.

wire-wound resistor *Electricity*. a resistor made of wire that is wound on an insulating form.

wire-wound rheostat *Electricity*. a wire-wound, variable resistor with two terminals: one end of the winding, which provides the resistance (the other end of the winding is not connected); and a movable contact (tap), which varies the resistance by tapping off a selected segment of the winding.

wire-wrap connection *Electricity*. an electrical connection in which a wire is wrapped several turns around a square terminal, usually by a motorized tool, with such precision and tightness that no other means (such as soldering) is required to maintain the connection.

wiring *Electricity*. **1.** all the wires that connect electrical circuits together into an operating system. **2.** the process of connecting electrical circuits together with wire. *Surgery*. the process of fixing into position (e.g., segments of fractured bone) by means of wire.

wiring harness *Electricity*. an assembly of wire and cables configured so that it can be inserted, connected, or removed as a unit.

Wirsung's duct *Anatomy*. the excretory duct of the pancreas that extends through the gland from the tail to the head where it empties into the duodenum at the duodenal papilla. (Named for Johann Georg *Wirsung*, 1600–1643, German physician.) Also, PANCREATIC DUCT.

WISC *Psychology*. a widely used intelligence test for children ages 6 to 16, adapted from the WAIS test, measuring both verbal and performance abilities, and yielding an IQ score. (An acronym for Wechsler Intelligence Scale for Children.)

Wisconsian *Geology*. in North America, the fourth and final stage of the Pleistocene epoch, occurring after the Sangamon interglacial stage. Also, **Wisconsin.**

wisent *Vertebrate Zoology*. a large European bison, *Bison bonasus*, of the family Bovidae, characterized by long horns and brown hair with a shaggy mane and face.

wisent

wish fulfillment *Psychology*. **1.** in psychoanalytic theory, the gratification of id impulses which, because of interference from the superego, often have to be redirected through such means as dreams, fantasy, or sublimation. **2.** any behavior that serves to gratify an individual's desires, especially when not explicitly directed toward those desires.

Wiskott-Aldrich syndrome see ALDRICH SYNDROME.

wisper wind *Meteorology*. a cold wind that blows overnight in the Wisper Valley in Germany during otherwise clear weather conditions.

witch *Anthropology*. a person, usually a woman, who practices witchcraft; a man who does this is usually called a **wizard** or **warlock.**

witchcraft *Anthropology*. the use of sorcery or magic in an effort to harm or affect another person. Also, **witchery.**

witch doctor *Anthropology*. a popular term for a shaman.

witch hazel *Botany*. a shrub, *Hamamelis virginiana*, of the family Hamamelidaceae of eastern North America, having jointed, twisting branches; toothed, ovoid leaves; and small, yellow flowers. *Pharmacology*. a liquid extracted from the bark and leaves of this plant, used as an astringent and a soothing liniment.

witch of Agnesi *Mathematics*. a name for the curve, symmetric about the y-axis and asymptotic to the x-axis in both directions, given by the equation $x^2y = 4a^2(2a - y)$. (Named for the Italian mathematician Donna Maria Gaetana *Agnesi*, 1718–1799, who discussed it in an early calculus textbook; *witch* is a pun derived from an Italian name for the curve.)

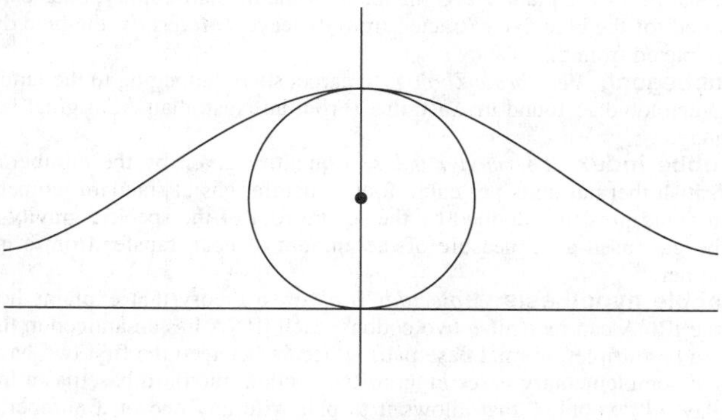

witch of Agnesi

withdrawal operation *Military Science*. a planned operation involving the disengagement of a force in contact with the enemy.

withdrawal symptoms or **withdrawal syndrome** *Medicine*. the delirium that occurs when a patient addicted to drugs or alcohol withdraws from the substance. Also, ABSTINENCE SYMPTOMS.

Withering, William 1741–1799, English physician and botanist; introduced digitalis treatment for dropsy; classified vegetables.

witherite *Mineralogy*. $BaCO_3$, a colorless, white, or gray, transparent to translucent, orthorhombic mineral of the aragonite group occurring as twinned dipyramidal crystals and in massive form, having a specific gravity of 4.291 and a hardness of 3 to 3.5 on the Mohs scale; fluoresces and phosphoresces blue-white under shortwave ultraviolet light; a low-temperature mineral in hydrothermal veins.

withers *Vertebrate Zoology*. the top of the shoulders of a horse.

witness mark *Cartography.* in surveying, a marker placed in a known and easily accessible location to aid in recovering a property corner or survey station. Also, **witness point.** *Materials Science.* a localized mark on a material, such as a scuff mark on a ceramic or a smear on the impacting material, indicating that a slight impact occurred.

Witte-Margules equation *Oceanography.* an expression of the slope of the boundary layer between two water masses of differing density and velocity; the equation takes into account the effects of the earth's rotation.

Wittgenstein, Ludwig [vit´gən shtīn´] 1889–1951, Austrian philosopher; independently introduced mathematical truth tables for truth-value analysis.

wittichenite *Mineralogy.* Cu_3BiS_3, a steel-gray to tin-white, metallic, opaque orthorhombic mineral occurring as prismatic or thick tabular crystals and in massive form, having a specific gravity of 6.01 to 6.19 and a hardness of 2 to 3 on the Mohs scale; found in silver ores and veins with copper-iron sulfides and other bismuth minerals.

Wittig, Georg 1897–1987, German chemist; awarded the Nobel Prize for discovering the Wittig rearrangement.

Wittig rearrangement *Organic Chemistry.* the rearrangement of benzyl and alkyl ethers in the presence of a methylating agent, giving secondary and tertiary alcohols via a 1,2 shift. Also, **Wittig ether rearrangement.**

wittite *Mineralogy.* $Pb_9Bi_{12}(S,Se)_{27}$, a lead-gray, opaque, metallic monoclinic mineral, occurring in massive form, having a specific gravity of 7.12 and a hardness of 2 to 2.5 on the Mohs scale; found in amphibolite and garnet tactite rocks.

Wittrockiellaceae *Botany.* a monospecific family of green algae of the order Incertae Sedis, noted for forming a thin brown stratum on mud in brackish waters amid various blue-green algae; characterized by short, erect, sparsely branched filaments that are attached by rhizoids and covered by tough mucilage.

wizzy-wig see WYSIWYG.

WKB method see WENTZEL-KRAMERS BRILLOUIN APPROXIMATION.

W.L. or **w.l.** wavelength; waterline.

WLF equation *Materials Science.* the temperature dependency of melt viscosity expressed in terms of glass transition temperatures and universal constants; a major result of the theory of free volume.

wm wattmeter.

WMA or **W.M.A.** World Medical Association.

WMO or **W.M.O.** World Meteorological Organization.

WN *Aviation.* the airline code for Southwest Airlines.

W/O water-in-oil.

woad *Botany.* a plant, *Isatis tinctoria,* of the mustard family, once cultivated for the blue dye extracted from its leaves. *Materials.* the blue dye extracted from this plant.

wobbegong *Vertebrate Zoology.* a carpet shark belonging to the family Orectolobidae; found in Australia. (From an Australian Aboriginal language.)

Wobbe index *Thermodynamics.* a quantity given by the number of British thermal units per cubic foot of natural gas at standard temperature and pressure, divided by the square root of the specific gravity of the gas; used as a measure of the amount of heat transfer from a gas burner.

wobble hypothesis *Molecular Biology.* a theory that explains how one tRNA can recognize two codons: each tRNA has an anticodon that is a base triplet; normal base-pairing occurs between the first two bases and complementary bases in an mRNA codon; the third base has a free play, or "wobble," that allows it to pair with any one of a number of bases that are in the third position of different codons.

wobble-wheel roller *Mechanical Engineering.* a skip body mounted on oscillating, smooth rubber tires; used for compaction and fine rolling of soil.

wobbulator *Electronics.* a signal generator whose frequency is automatically swept periodically over a defined range; often used with an oscilloscope in determining a circuit's frequency response.

Wöhler, Friedrich 1800–1882, German chemist; isolated aluminum and beryllium; first to synthesize an organic compound (urea); with von Liebig, classified organic compounds.

Wohlwill process *Materials Science.* an electrolytic process for refining gold, using anodes of crude gold, cathodes of pure gold, and an electrolyte solution of gold chloride-hydrochloric acid at 70°C; the resulting cathode deposit is about 99.5% gold. Silver is recovered from the slimes as silver chloride; and copper, platinum, and palladium accumulate in the electrolyte.

Wolbachia *Bacteriology.* a genus of Gram-negative bacteria belonging to the tribe Wolbachieae of the family Rickettsiaceae, which occur as coccoid or rod-shaped cells growing intracellularly or extracellularly in arthropod hosts.

Wolbachieae *Bacteriology.* a tribe of bacteria of the family Rickettsiaceae that occur either parasitically or symbiotically within the cells of arthropods.

wolf *Vertebrate Zoology.* a predatory, carnivorous mammal belonging to the genus *Canis* of the dog family Canidae, resembling and closely related to the domestic dog (of a breed such as the German shepherd), and having erect ears, a long muzzle, a long, bushy tail, and nonretractile claws; includes the **gray** or **timber wolf,** *Canis lupus,* and the **red wolf,** *C. rufus* (and also the coyote, *C. latrans*). Wolves were formerly common throughout the Northern Hemisphere, but now are chiefly restricted to the more remote and unpopulated parts of their range.

gray wolf

Wolf, Maximilian 1863–1932, German astronomer; used a moving camera and long exposure to discover the asteroid Brucia.

Wolf 359 *Astronomy.* a star of absolute magnitude 16.6 that lies 7.7 light-years from the sun; it flares up in brightness and emits radio noise.

wolfberry *Botany.* a shrub, *Symphoricarpos occidentalis,* of the honeysuckle family, characterized by gray, hairy, egg-shaped leaves, pink bell-shaped flowers, and white berries, for which it is sometimes grown; found in North America.

Wolfcampian *Geology.* the North American provincial stage of the Lowermost Permian period, occurring after the Virgilian of the Pennsylvanian period and before the Leonardian.

wolf eel *Vertebrate Zoology.* a long, slender marine fish of the family Anarhichatidae; distinguished by its ferocity and strong teeth; found along the western North American coast and the North Pacific and Atlantic oceans. Also, **wolffish.**

Wolfe graft or **Krause-Wolfe graft** *Surgery.* a skin graft using the full thickness of the skin. (Named for Fedor *Krause,* 1857–1937, German surgeon, and John *Wolfe,* 1824–1904, Scottish ophthalmologist.)

wolfeite *Mineralogy.* $(Fe^{+2},Mn^{+2})_2(PO_4)(OH)$, a reddish-brown to dark-brown, transparent to translucent monoclinic mineral occurring in fibrous aggregates and as prismatic crystals, having a specific gravity of 3.79 and a hardness of 4.5 to 5 on the Mohs scale; dimorphous with satterlyite; found in granite degmatites.

Wolff, Caspar Friedrich 1734–1794, German anatomist and physiologist; formulated a theory of epigenesis; anticipated the concept of embryonic layers.

Wolffian body see MESONEPHROS.

Wolffian duct see MESONEPHRIC DUCT.

Wolff-Kishner reduction *Organic Chemistry.* the reduction reaction in which a ketone or aldehyde is converted into the hydrazone; this derivative is then heated in a sealed tube with sodium ethoxide in absolute ethanol to give nitrogen and the corresponding hydrocarbon.

wolf herring *Vertebrate Zoology.* a carnivorous fish, *Chirocentrus dorab,* of the family Chirocentridae, having a voracious appetite and fanglike teeth in the jaws; found in the Indian and Pacific Oceans.

Wolf-Lundmark-Melotte galaxy *Astronomy.* a dwarf, irregular galaxy (like the Magellanic Clouds) that is part of the Local Group and lies 5 million light-years from earth.

Wolf number *Astronomy.* a number indicating the degree of sunspot activity on the sun. Also, RELATIVE SUNSPOT NUMBER, ZURICH NUMBER.

wolf plant *Agronomy.* a term for an apparently palatable plant that grazing animals will leave largely uneaten.

wolfram *Chemistry.* an alternate name for tungsten, used especially outside the United States. See TUNGSTEN.

wolframic acid see TUNGSTIC ACID.

wolframite *Mineralogy.* $(Fe^{+2},Mn^{+2})WO_4$, an intermediate member of the series hübnerite-ferberite.

Wolf-Rayet star *Astronomy.* a young, short-lived class of stars with extremely hot surface temperatures (100,000 to 350,000 kelvin) and broad emission lines in the spectra.

wolfsbane *Botany.* any of several species of aconite genus *Aconitum,* including *A. lycoctonum,* which bears stalks of hood-shaped, purplish-blue flowers; *A. napellus,* also called **monkshood,** which yields a poisonous alkaloid used medicinally; and other various-colored plants.

wolf spider *Invertebrate Zoology.* any of various ground spiders of the family Lycosidae, including the tarantula, *Lycosa taretula,* having keen eyesight and strong legs and jaws; noted for stalking prey rather than using a web.

Wolinella *Bacteriology.* a genus of Gram-negative, anaerobic bacteria of the family Baccteroidaceae that occur as helical, curved, or straight rod-shaped cells, motile with a polar flagellum; found in cattle rumen and the human oral cavity.

Wollaston, William Hyde 1766–1828, English scientist; isolated rhodium and palladium; invented the camera lucida, the reflecting goniometer, and the **Wollaston lens**; discovered spectral lines.

wollastonite *Mineralogy.* $CaSiO_3$, a colorless white, grayish, or greenish, transparent to translucent, triclinic mineral occurring in fibrous to granular masses and as tabular twinned crystals, having a specific gravity of 2.87 to 3.09 and a hardness of 4.5 to 5 on the Mohs scale; found in contact-metamorphosed crystalline limestones.

Wollaston polarizing prism *Optics.* a device composed of two wedge-shaped prisms whose optic axes are perpendicular to one another; incident light emerges from the prism as two colored, oppositely polarized beams.

Wollaston wire *Engineering.* an extremely fine wire, usually platinum, formed by encasing one metal in another, drawing the two metals together, and then dissolving the outer metal.

Wolman's disease *Genetics.* a human disease that is inherited as a recessive allele on chromosome 4; homozygous individuals are deficient in lipase production.

wolverine *Vertebrate Zoology.* the largest weasel, *Gulo gulo,* of the carnivorous mammalian family Mustelidae, having a black body with light stripes on the sides and a light forehead; noted for its strength and voraciousness and now found mostly in northern forests.

womb *Anatomy.* another term for the uterus, especially in older use. See UTERUS.

wombat *Vertebrate Zoology.* a semifossorial marsupial of the family Vombatidae, characterized by a sturdy, stocky body with short legs; found in Australia.

womera *Archaeology.* a spear-throwing device used among the Australian aborigines.

wood *Botany.* the hard lignous substance, composed primarily of xylem, that is found under the bark of the trunks and branches of trees and shrubs. *Materials.* **1.** such a substance that has been cut and prepared for use in building, carpentry, fire-making, etc. **2.** relating to, composed of, or derived from wood. Thus, **wood preservative, wood block, wood filler, wood turpentine,** and so on. *Ecology.* a thick growth of trees that is larger than a grove and smaller than a forest.

Wood, Robert Williams 1868–1955, American physicist; worked on the spectrum analysis of metallic vapor; a pioneer in color photography.

wood alcohol see METHYL ALCOHOL.

woodbine *Botany.* any of several climbing honeysuckles, such as *Lonicera periclynum* of Europe, having fragrant, pale-yellow flowers, or the Virginia creeper of North America.

Woodbinian *Geology.* the North American Gulf Coast stage of the Upper Cretaceous period, occurring after the Washitan and before the Eaglefordian.

woodchuck *Vertebrate Zoology.* a thick-set burrowing rodent, *Marmota monax,* having short legs and a bushy tail, that hibernates in the winter. Also, GROUNDHOG.

woodcreeper *Vertebrate Zoology.* any of various tropical songbirds of the family Dendrocolaptidae, characterized by stiffened tail feathers.

woodcut *Graphic Arts.* **1.** an engraved block of wood from which prints are made. **2.** a print or impression made from such a block.

Wood effect *Optics.* the transparency exhibited by alkali metals when they are exposed to ultraviolet light.

wood ether see DIMETHYL ETHER.

wood flour *Metallurgy.* a pulverized wood product used in a foundry to overcome sand expansion in the mold, increase flowability, and improve the casting finish.

wood frog *Vertebrate Zoology.* a common brown frog, *Rana sylvatica,* characterized by a dark, masklike marking on the head; found in damp woodlands of northern and eastern North America.

wood frog

woodhoopoe *Vertebrate Zoology.* any of several tropical birds of the family Phoeniculidae, characterized by metallic, blackish plumage and a slender, curved bill; found in Africa.

woodhouseite *Mineralogy.* $CaAl_3(PO_4)(SO_4)(OH)_6$, a colorless, white, or flesh-pink, vitreous, transparent to translucent trigonal mineral of the beudantite group occurring as small pseudocubic and tabular crystals, having a specific gravity of 3.012 and a hardness of 4.5 on the Mohs scale; found in veins with quartz, lazulite, and topaz cutting an andalusite deposit in the White Mountains of California.

wood hyacinth see BLUEBELL.

wood lily *Botany.* a lily, *Lilium philadelphicum,* having orange-red flowers; found in eastern North America.

woodman's axe *Mechanical Devices.* an axe used in the felling of trees, having a double-bitted head on a long handle.

woodpecker *Vertebrate Zoology.* any one of over 200 different species of the family Picidae, small to medium-sized birds of stocky build, having straight, hard-pointed bills adapted to hacking and chiseling the wood of tree trunks and long, protruding tongues for extracting insects from holes; found mainly in wooded regions and savannas of North, Central, and South America, Africa, and Eurasia.

woodpecker

wood pulp *Materials.* wood reduced to the form of pulp by means of various chemical or mechanical processes; used in the manufacture of paper.
wood rat see PACK RAT.
wood ray see XYLEM RAY.
Woodruff key *Mechanical Devices.* a semicircular machine key used with shafts no more than 2-1/2 inches in diameter.
wood screw *Mechanical Devices.* a metal fastener used for wood, usually having a flat, slotted head, a pointed shank, and a coarse thread.
Wood's glass *Materials.* a glass having high ultraviolet but low visible light transmission.
Wood's lamp *Microbiology.* an electric lamp equipped with a filter so that it emits light of only 365-nm wavelength; used in the diagnosis of certain diseases, such as ringworm. Also, **Wood's light.**
Wood's metal *Metallurgy.* one of a group of alloys that have melting points below that of tin (449°F) and are named after their inventors; Wood's metal is 50% bismuth, 25% lead, 12.5% tin, and 12.5% cadmium and has a melting point of 154°F.
woodstave pipe *Mechanical Devices.* a pipe composed of thin, narrow wood strips that are placed side by side and banded with a metal wire and collared joints; used in agricultural or mining irrigation operations.
woodstone see PETRIFIED WOOD.
wood tin *Mineralogy.* a brownish, reniform or nodular variety of cassiterite with a concentric structure of radiating fibers, which resembles dry wood in appearance.
wood turpentine see TURPENTINE, def. 4.
wood vinegar see PYROLIGNEOUS ACID.
wood warbler *Vertebrate Zoology.* any of numerous small birds of the family Parulidae, characterized by variable plumage; the northern species are highly migratory; found in the New World.
Woodward, John 1665–1728, English geologist and naturalist; pioneer in stratigraphy; cataloged the fossils of England.
Woodward, Robert Burns 1917–1979, American chemist; awarded the Nobel Prize for synthesizing chlorophyll and other organic compounds.

Robert Burns Woodward

Woodward-Hoffmann rule *Organic Chemistry.* a rule that predicts the stereochemical course of certain one-step organic reactions, by analyzing the symmetry of the interacting molecular orbitals.
woodwardite *Mineralogy.* an inadequately described, greenish-blue, translucent mineral, having an approximate formula of $Cu_4^{+2}Al_2(SO_4)(OH)_{12}\cdot 2\text{–}4H_2O$, occurring as tiny fibrous concretions, and having a specific gravity of 2.38 and an undetermined hardness; first found at Cornwall, England.
Woodward's Reagent K see *N*-ETHYL-5-PHENYLISOXAZOLIUM 3'-SULFONATE.
Wood-Werkman reaction *Biochemistry.* the carboxylation of pyruvate or any similar carboxylation that also produces oxaloacetate.
woodworker *Mechanical Engineering.* a carpenter, joiner, cabinetmaker, or other worker in wood.
woodworker's vise *Mechanical Devices.* a vise used by carpenters, designed to grip the work but not damage it.
woody structure *Metallurgy.* a macrostructure that shows elongated surfaces of separation when it fractures; usually found in wrought iron and extruded rods of aluminum alloys.
woof *Textiles.* **1.** the threads or yarn running from side to side on a loom. **2.** fabric; cloth.
woofer *Acoustical Engineering.* a large loudspeaker that is designed to produce low-frequency sound; often part of a speaker housing with an equal number of tweeters.
woof yarns *Textiles.* a term used mostly in Great Britain for the crosswise yarns on a fabric; usually called the filling or weft yarns in the United States.
wool *Vertebrate Zoology.* the soft, often thick and curly undercoat of fur in various mammals, especially the fur of the domestic sheep, but also that of goats, camels, and other related mammals. *Textiles.* the fleece of sheep, characterized by its insulating value and moisture absorbency; widely used in clothing and carpeting.
woolen *Textiles.* **1.** made or consisting of wool. **2.** cloth of carded wool yarn whose fibers vary in length. **3. woolens.** wool clothing or cloth. Also, **woollen.**
woolen yarn *Textiles.* a yarn that is at least 95% wool and is made by carding but not combing fibers, then spinning them into yarn; generally, the staples used are shorter than those used in worsteds.
wool fat or **wool wax** see LANOLIN.
Woolley, Sir Leonard 1880–1960, English archaeologist; excavated Ur in Sumeria (Iraq).
woolpack see BALLSTONE.
wool-sorters' disease see PULMONARY ANTHRAX.
wool wax see LANOLIN.
word *Linguistics.* a distinguishable unit of speech or writing that has a recognized meaning and that is generally separated from other units of this type, in speech by a pause before and after and in writing by a space before and after; one of the items making up the vocabulary of a language. *Computer Programming.* a set of bits, bytes, or characters that the processing unit treats as a fundamental unit of storage. Also, MACHINE WORD. *Mathematics.* let $\{G_i\}$ be a family of groups, where i is in some index set I. It may be assumed, by relabeling as necessary, that the G_i are mutually disjoint. Let $X = \cup_{i \in I} G_i$, and let $\{1\}$ be a one-element set disjoint from X. A **word on X** is any sequence $(a_1, a_2, \ldots, a_i)$ such that $a_i \in X \cup \{1\}$ and for some integer $n \geq 1$, $a_i = 1$ for all $i \geq n$.
word blindness see ALEXIA.
word boundary *Computer Programming.* a byte address that is also the address of a full word.
word concatenation *Acoustical Engineering.* the audible construction of recognizable words from syllables produced by a computer that is engaged in voice response.
word deafness see KOPHEMIA.
word format *Computer Programming.* an arrangement determining what data is contained in each position or group of positions within a word.
word length *Computer Programming.* the number of bits contained in a machine word.
word mark *Computer Programming.* a marker bit in a variable-word-length computer used to signal the beginning or ending of a word.
word-oriented *Computer Technology.* an early-style computer in which memory is addressable only by words, and bits within words can be addressed only by manipulating the data, using additional instructions.
word processing *Computer Programming.* **1.** a general term for programs that facilitate the creation, modification, formatting, display, and printing of text files. **2.** the use of such a program.
word processor *Computer Programming.* a hardware or program package that is specifically designed to process text.
word rate *Computer Technology.* the amount of time elapsed from the beginning of the transmission of one word to the beginning of the transmission of the next word.
words per minute *Telecommunications.* in telegraphy, the measure of the speed with which information can be transmitted.
word vision *Neurology.* the ability to perceive printed or written words.
word wrap *Computer Programming.* the ability of a word processing system to automatically move a word that extends into the right margin to the beginning of the next line. Also, WRAP MODE.

work *Mechanics.* the transfer of energy to a body by the application of a force, so as to cause the body to move a certain measurable distance in the same direction as the applied force. The work done by a constant force is equal to the product of the component of force in the direction of the displacement, times the magnitude of this displacement. Work is not accomplished unless there is a measurable displacement; the application of force alone does not constitute work. *Industrial Engineering.* **1.** any activity, physical, mental, or a combination of both, that has some practical value; especially, an activity that produces revenue or accomplishes some desired goal. **2.** material in any stage of manufacture or production, as a machine tool. **3.** to operate for productive purposes, as a mine or a farm. *Mechanical Engineering.* to move irregularly so as to become out of gear.

workable *Agronomy.* of soil, able to be broken up easily for planting.

work-and-back see SHEETWISE.

work-and-turn *Graphic Arts.* a common press imposition in which the sheets are printed on one side through half the press run, then turned so that they can be printed on the reverse side; the two sides can thus be printed in one impression.

work angle *Metallurgy.* in arc welding, the angle between the electrode and one of the joints.

work assembly *Computer Programming.* the activities related to organizing collections of data records and programs.

work coordinates *Robotics.* a coordinate system used in machining operations.

work cycle *Industrial Engineering.* a single complete sequence of a repetitive work process.

worked-out *Mining Engineering.* describing a coal seam or ore deposit that is depleted or exhausted.

worked penetration *Engineering.* the penetration of a lubricating grease sample immediately after it has been raised to a specified temperature and subjected to a specified number of strokes by a worker.

work-energy theorem see WORK-KINETIC ENERGY THEOREM.

worker *Industrial Engineering.* any person who carries on work, especially someone who does physical work. *Invertebrate Zoology.* **1.** a sexually underdeveloped, nonreproductive bee that works for the community in collecting food and maintaining the hive. **2.** a similar class of ants, wasps, or termites.

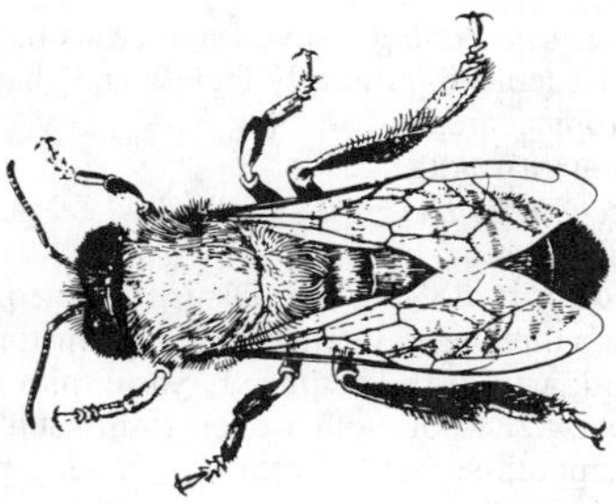

worker bees

work factor *Industrial Engineering.* in time-motion studies, an index of the time required in addition to basic time, to allow for the variables of manual control and weight.

work file see TRANSACTION DATA.

work flow layout *Industrial Engineering.* see LAYOUT.

work function see HELMHOLTZ FREE ENERGY.

workhardening or **work hardening** *Mechanics.* the increase in rigidity of a material, such as a metal, as a result of previous strain it has experienced.

working *Mining Engineering.* **1.** the operation of mining or exploiting. **2.** a shaft, quarry, stope, level, or other mining excavation; often used in the plural. **3.** describing ground, rock, or a coal seam under great pressure and emitting creaking noises. **4.** the entire strata excavated in the mining of a seam. *Navigation.* the progress of a vessel, particularly a sailing vessel, in the desired direction of travel, especially under unfavorable conditions; e.g., working to windward in a gale. *Telecommunications.* the carrying on of radio communications by means of telephony, telegraphy, or facsimile for a purpose other than calling.

working aperture *Optics.* the maximum aperture at which a lens will provide a sharp image.

working cell *Robotics.* a work station equipped with a microcomputer or minicomputer that is specialized for graphics, computer-aided design and engineering, or scientific applications. Also, **work cell.**

working cell

working character type *Psychology.* a character type that exhibits objective reasoning, unselfish love, and productive work habits.

working electrode *Physical Chemistry.* an electrode that is used to test for corrosion in an electrochemical cell.

working envelope *Robotics.* a rectangular, spherical, cylindrical, or irregular set of points that represent the maximum reach of a robot's wrist in all directions, excluding the tool tip. Also, **working profile.**

working face *Mining Engineering.* see FACE, def. 2.

working load *Engineering.* the heaviest load that a given structural member can support.

working memory *Psychology.* the immediate contents of consciousness, in which certain items from long-term memory may be held temporarily and readily brought back into awareness. *Artificial Intelligence.* in a production system, the memory that represents the data about the current problem; analogous to human short-term memory.

working pressure *Engineering.* the maximum amount of pressure under which a pressurized vessel or conduit can operate.

working program *Computer Programming.* a program that can be translated into machine language and is thought to produce correct results when executed.

working Q see LOADED Q.

working range *Robotics.* the range of variables that describe every point within a robot's working envelope.

working set *Computer Programming.* the smallest collection of data and instruction pages that must be loaded into memory for efficient processing without the excessive overhead associated with paging.

working-set window *Computer Programming.* the time interval in which a program's working set is referenced.

working space volume *Mechanical Engineering.* the volume enclosed by the working envelope.

working storage *Computer Technology.* **1.** the area in the COBOL programming language used to define and initialize data fields. **2.** see TEMPORARY STORAGE.

working substance *Mechanical Engineering.* **1.** any substance that undergoes a physical change, as in temperature or pressure, as part of a process of work. **2.** specifically, a gas or liquid that is used to exchange heat and perform work in a heat engine; e.g., water is the working substance in a steam turbine. Also, **working fluid.**

working voltage see VOLTAGE RATING, def. 1.

work-kinetic energy theorem *Mechanics.* a theorem stating that the change in kinetic energy of a particle in motion is equal to the work done by all the forces acting on the particle. Also, WORK-ENERGY THEOREM.

work lead *Metallurgy.* the electrical arc that connects the source of the welding current to the work.

work life *Chemical Engineering.* the period of time that an adhesive or liquid resin stays usable after it is mixed with a catalyst, solvent, or other compounding element.

Workman-Reynolds effect *Geophysics.* the process of electric-charge separation in slightly impure, freezing water.

work measurement *Design Engineering.* an analytical technique used independently or in conjunction with cost accounting so that data on actual hours and production work units can be collected and the relationship between the labor hours and work performed can be calculated.

work module *Industrial Engineering.* a set of interrelated tasks that are assigned to one worker.

work of adhesion see ADHESIONAL WORK.

work of rupture see FRACTURE SURFACE ENERGY.

workpiece *Industrial Engineering.* the output product of a given work task. *Robotics.* the item being manipulated or machined by a robotic system.

workpiece program *Control Systems.* a program that carries out the machining of a component under computer control.

workpiece removal rate *Design Engineering.* the number of workpieces removed in a machining operation over a set period of time.

work sampling *Industrial Engineering.* the analysis of a job cycle by intermittent observation of the work process, in order to determine the proportion of total time devoted to each of the activities that constitute the job.

worksoftening *Mechanics.* the decrease in rigidity of a material, such as vinyl, as a result of previous strain.

workspace *Computer Programming.* **1.** the amount of memory needed by a computer program to execute over and above the amount required to store the program and data. **2.** the memory containing the data currently being processed. *Robotics.* the envelope of space in which a robot does its work.

work specialization *Industrial Engineering.* the subdivision of an organizational task into individual jobs. Also, DIVISION OF LABOR.

work standardization *Industrial Engineering.* the division of a task into standardized elements or subtasks to allow greater prediction of productivity, easier training of workers, and the like.

workstation or **work station** *Industrial Engineering.* the space in which a worker carries out a specific operation. *Computer Technology.* a single-user computer system with inherent processing capability; may also be a terminal to a central computer. The term usually refers to a computer that is more powerful than a personal computer.

workstation independence *Robotics.* a robotic or numerical control program that is not dependent upon the workstation that is using it.

work tape see SCRATCH TAPE.

work task *Industrial Engineering.* the work function performed by an individual worker, machine, or group.

work volume see WORKING ENVELOPE.

world *Science.* the universe or a complex whole conceived as resembling the universe. *Physics.* a term describing space-time in four dimensions as distinguished from three-dimensional space.

world calendar *Astronomy.* a proposed calendar that divides the year into four equal quarters, with January, April, July, and October beginning on a Sunday and having 31 days each, while the other months have 30 days; the one day left at the end of December belongs to no month, and leap-year days follow June 30th.

world coordinates *Robotics.* a robotic coordinate system that uses the earth as its reference point.

World Geographic Reference System see GEOREF GRID.

world line *Physics.* a line visualized in four-dimensional space-time.

world modeling *Robotics.* a method of robotic programming that allows the system to carry out complex tasks based on data stored in its memory.

world rift system *Geology.* a system of mid-oceanic ridges and their associated valleys, along which tensional splitting and the upwelling of magma resulting in seafloor spreading are believed to occur.

world view *Anthropology.* the manner in which a society views reality and the workings of the world; often unexpressed but implicit in the modal personality and approach to life of the culture.

worm *Invertebrate Zoology.* **1.** any of numerous small, slender, soft-bodied invertebrates with no backbone, including the flatworms, roundworms, acanthocephalans, nemerteans, gordiaceans, and annelids. **2.** any of numerous small creeping or crawling animals with slender, elongated bodies that resemble the true worm, including earthworms, tapeworms, insect larvae, and adult forms of some insects. *Mechanical Devices.* **1.** any mechanical device that is spiral or vermiculate in form. **2.** a revolving screw whose threads mesh with the teeth of a rack or worm wheel. **3.** a condensing tube used in distilling. **4.** a tool, in the shape of a double corkscrew, used to extract the shot and powder from a muzzle loading gun. *Metallurgy.* an irregularity caused when gas evolution forces the sweat from molten metal through the top crust of solidifying metal. *Computer Programming.* **1.** a copy-protection algorithm applied to proprietary data or software that destroys data whenever an illegal copy is detected. **2.** an unauthorized program that runs within a computer system, may cause damage, and may send copies of itself to other computers over networks.

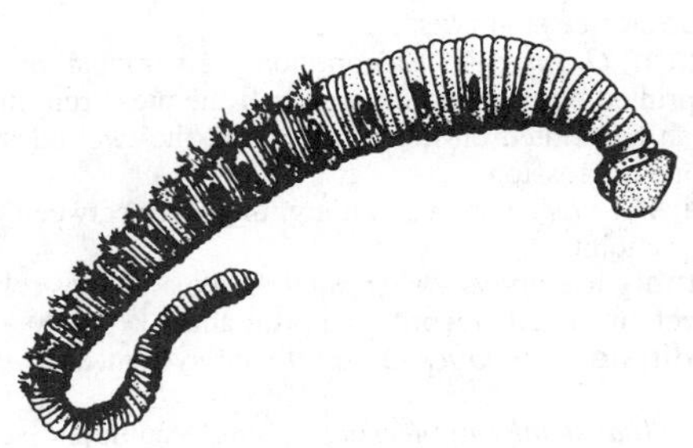

worm

WORM *Computer Technology.* a nonreusable, high-capacity, optical storage device used for archival storage. (An acronym for write once read many times.) Also, **WORM storage.**

worm drive *Mechanical Devices.* a propulsion or drive assembly made up of a worm in conjunction with an operative driver gear.

worm eel *Vertebrate Zoology.* any of the small, burrowing, tropical eels of the family Ophichtidae of the Indo-Pacific and Atlantic Oceans, resembling an elongated worm.

worm gear *Mechanical Devices.* a mechanism consisting of a worm in operation with a worm wheel. *Robotics.* a gear, used to connect nonparallel, nonintersecting shafts, that has teeth cut on an angle and is driven by a worm.

worm lizard *Vertebrate Zoology.* any of numerous burrowing lizards of the family Amphisbaenidae, primarily legless and shaped like an earthworm; found in tropical areas.

worm snake see BLINDSNAKE.

worm wheel *Mechanical Devices.* a toothed wheel gearing that has the threads of a worm.

Worsaae, Jens Jacob 1821–1885, Danish archaeologist; further developed Thomsen's three-age system; wrote a prehistory of Denmark.

worsted [wûr′stid; wŭs′tid] *Textiles.* **1.** yarn spun from long-staple combed wool fibers. **2.** wool cloth woven from such yarns, having no nap and a hard, smooth surface. (From *Worstead,* a parish in England that was a noted early site for the manufacture of this cloth.)

wort *Food Technology.* the rich, liquid fraction that is extracted from malted barley and other cereals through the process of mashing and that, after fermentation, becomes beer or ale.

wound *Surgery.* an injury induced by physical means, with disruption of normal anatomical continuity. *Botany.* an injury to plant tissue due to an external agency other than disease.

woundwort *Botany.* any of several plants of the genus *Stachys* of the mint family, characterized by hairy stems and leaves and clusters of small, red flowers; formerly used for treating wounds.

woven fabric *Textiles.* any fabric composed of warp and filling threads interlaced with each other at right angles.

wow *Acoustical Engineering.* a low-frequency sound produced during the playback of a record, caused either by off-center placement of the hole in the record or by variations in the speed of the turntable.

WP wettable powder; white phosphorus; word processing; word processor.

WPA or **W.P.A.** Western Pine Association.

WPC watts per candle.

wpm or **WPM** words per minute.

wpn weapon.

W.R. Wassermann reaction.

wrap *Graphic Arts.* to run type characters around an illustration.

wrap-around hanger *Petroleum Engineering*. in an oil well, a tubing hanger with two hinged halves, featuring a resilient sealing element situated between two steel mandrels.

wraparound plate *Graphic Arts*. a metal printing plate that is flexible enough to be used on a cylinder for rotary or offset printing.

wrap forming see STRETCH FORMING.

wrap mode see WORD WRAP.

wrapper sheet *Mechanical Engineering*. **1.** the outer casing of a firebox in a fire-tube boiler. **2.** a plate that forms the sides and top of a locomotive boiler.

wrap test *Computer Programming*. a diagnostic test in which a data pattern is transmitted, echoed back, and compared to the original.

wrasse *Vertebrate Zoology*. any of various spiny-finned marine fishes of the family Labridae, especially of the genus *Labrus*, characterized by thick, fleshy lips, powerful teeth, and usually a brilliant coloration; certain species are valued as food fish; found in warm seas.

wrecking ball *Engineering*. a large, heavy iron or steel ball that is swung or dropped by a derrick to demolish old buildings, compact waste stone, or break up hard substances.

wrecking bar *Mechanical Devices*. a metal bar, one to three feet long, having a chisel edge on one end and a claw on the other; used to pull or pry nails.

wren *Vertebrate Zoology*. any one of the approximately 60 species of the family Troglodítae of small, extremely quick, active birds, characterized by brown or gray plumage that is striped, streaked, or spotted; short, rounded wings; and sharp, pointed, slightly downcurved bills; found in North, Central, and South America, Eurasia, and the extremes of North Africa.

wren

Wren, Christopher 1632–1723, English architect, astronomer, and mathematician; designed St. Paul's Cathedral, London.

wrench *Engineering*. a lever with adjustable openings, used to turn or hold a bolt or pipe. *Mechanics*. the combination of a couple and a force parallel to the torque exerted by the couple. Any force-couple system in which the couple is not perpendicular to the force vector can be reduced to a wrench.

wrench fault *Geology*. a fault along which the strike separation is nearly vertical. Also, BASCULATING FAULT, TORSION FAULT.

wrench-head bolt *Mechanical Devices*. a square- or hex-shaped bolt head designed to be used with a wrench.

wrentit *Vertebrate Zoology*. the only New World babbler, *Chamaea fasciata*, of the family Timaliidae, or in some schemes, composing the family Chamaeidae; characterized by a gray-brown back and white underbelly, a long tail, and short round wings; found west of the Rockies.

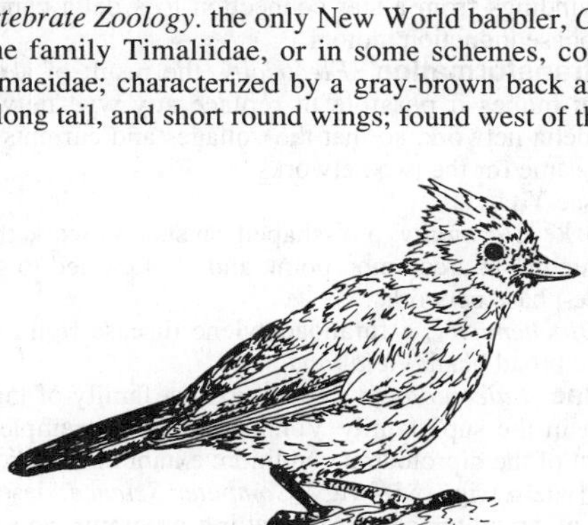

wrentit

Frank Lloyd Wright's Johnson Wax Building

Wright, Frank Lloyd 1869–1959, American architect; developed the prairie style and open planning; designed the Guggenheim Museum.

Wright, Orville 1871–1948 and **Wilbur** 1867–1912, American inventors; pioneers in airplane design and construction.

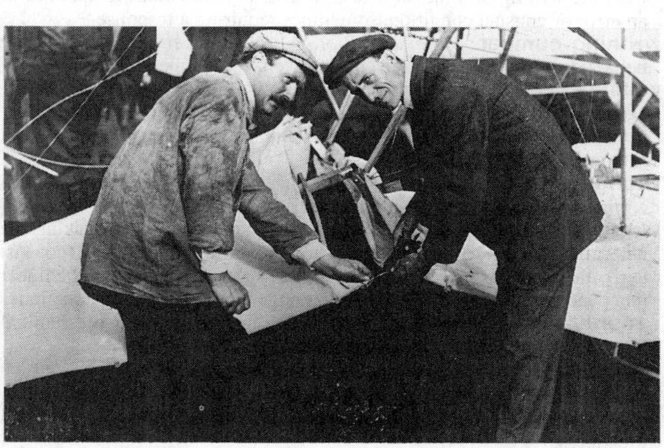

Wright brothers

Wright, Sewall 1889–1988, American geneticist; made contributions to evolutionary theory.

Wright's inbreeding coefficient *Genetics*. **1.** the percentage of loci at which an individual is homozygous. **2.** the probability that any two allelic genes in an individual have the same ancestral origin common to both its parents.

Wright's phenomenon *Astronomy*. the apparent larger diameter of Mars as viewed at ultraviolet wavelengths than when viewed at infrared wavelengths.

Wright telescope *Optics*. a variation of the Schmidt telescope in which an ellipsoidal mirror replaces the spherical primary mirror.

wringing fit *Design Engineering*. a fit of machine parts in which there is zero-to-negative allowance or tolerance.

wrinkle ridge *Astronomy*. a long, low, prominent ridge, perhaps offset by faults, which is seen on all the lunar maria.

wrinkling *Metallurgy*. a wavy conditon in metal that is caused by drawing, when an area of the metal passes over the draw radius; may also be caused by unbalanced compressive forces.

wrist *Anatomy*. the part of the upper limb between the forearm and hand, in the region of the wrist joint. Also, CARPUS. *Robotics*. the set of joints to which the end effector or impulse connection of a robot is attached. Also, **wrist socket.**

wristdrop *Medicine.* a condition resulting from paralysis of the extensor muscles of the hand and fingers. Also, CARPOPTOSIS.

wrist force sensor *Robotics.* a contact sensor used in Cartesian-coordinate robotic systems that measures the forces affecting a wrist joint.

wrist pin see GUDGEON PIN.

wrist turnover *Robotics.* one axis of movement for a pick-and-place robot.

writable control store *Computer Technology.* a portion of control store of a central processing unit into which microprograms can be loaded or dynamically altered.

write *Computer Programming.* to record information onto a storage medium; to copy information into a memory location.

write error *Computer Programming.* a condition in which data is incorrectly transferred to a storage medium, usually due to dust, damage, or malfunction.

write head *Electronics.* a device that records individual bits of a digital signal onto a magnetic tape, drum, or disk.

write protection *Computer Programming.* a procedure taken to prevent overwriting of an area of memory or storage medium by unauthorized users or programs, or by mistake; examples include a disk write-protect notch and a tape write-protect ring.

write-protect ring *Computer Technology.* a removable plastic ring that is attached to the hub of a reel of magnetic tape and allows the tape drive to write on the tape; if the ring is removed, the tape cannot be written on. Also, **write-enable ring**, **write-inhibit ring, write ring.**

writer *Computer Programming.* a program that writes data from an output queue to the output device.

write time *Computer Technology.* the amount of time required to transfer data onto a storage medium.

write-to operator *Computer Programming.* a service routine that allows the writing of a message to inform the system console operator of an error or unusual condition, sometimes requiring a response.

writhing number *Molecular Biology.* the number of times that the helix axis of a DNA molecule crosses itself while supercoiling.

writing speed *Electronics.* **1.** the rate at which successive charge-storage elements can be written to inside a charge-storage tube. **2.** the maximum speed that an electron beam can travel and still produce a satisfactory trace.

wrong-reading *Graphic Arts.* of a print or film image, reading as if in a mirror, with characters running from right to left.

Wronskian *Mathematics.* a differential invariant formed by taking the determinant of the $n \times n$ matrix whose ith row is the list of $(i-1)$st derivatives of a set of functions $f_1, f_2, \ldots, f_n$; used to establish the linear independence of solutions of linear homogeneous differential equations.

wrought alloys *Metallurgy.* alloys that are suitable for mechanical forming at below melting-point temperatures.

wrought carbon steel see PLAIN-CARBON STEEL.

wrought iron *Metallurgy.* a commercial iron that contains less than 0.3% carbon and containing 1.0 or 2.0% slag, giving it ductility and toughness.

wrought-iron piling *Materials Science.* a wrought-iron production method used during the Roman Empire period for weapons and high-quality tools.

WT watertight; wireless telegraphy.

wt. weight.

WT1 *Oncology.* a tumor-suppressor gene found to be altered in Wilms' tumor.

Wu, Chien-Shiung born 1912, American physicist, born in China; confirmed Lee and Yang's work disproving the law of conservation of parity.

Wuchereria *Medicine.* a white threadlike worm found in warm, humid regions of the world; some strains are carried by mosquitoes and cause disease in humans.

Wuchereria bancrofti *Medicine.* the form of *Wuchereria* roundworm carried by the *Culex* mosquito, which passes into the blood in immature form and later affects the lymph, interfering with its circulation and resulting in lymphangitis, chyluria, and elephantiasis.

wuchereriasis *Medicine.* an infection with the filarial nematode, *Wuchereria.*

wulfenite *Mineralogy.* $PbMoO_4$, an ore of molybdenum, a yellowish-orange, brown, pink, or blue tetragonal mineral with a resinous to adamantine luster, occurring as square tabular crystals and in massive form, and having a specific gravity of 6.5 to 7 and a hardness of 2.75 to 3 on the Mohs scale; found as a secondary mineral in the oxidized zone of ore deposits.

Wulff process *Chemical Engineering.* the production of acetylene by treating a hydocarbon gas with superheated steam in a regenerative refractory furnace at 1148–1371°C.

Wullenweber antenna *Electromagnetism.* an electronically steerable antenna used in ground-to-air communication.

Wundt, Wilhelm 1832–1920, German psychologist and physiologist; founded the first laboratory for experimental psychology.

Wurdemanniaceae *Botany.* a monospecific family of red algae of the order Nemaliales, recently separated from the family Gelidiaceae, about which little is known.

Würm *Geology.* **1.** the European geologic stage of the uppermost Pleistocene epoch, occurring after the Riss and before the Holocene epoch. **2.** the fourth and final glaciation stage of the Pleistocene epoch in the Alps, occurring after the Riss-Würm interglacial stage.

W Ursae Majoris stars *Astronomy.* a contact binary system whose two stars orbit in close proximity, sharing a common gaseous envelope and eclipsing each other's light, so that the system's brightness varies almost continuously.

Wurtz, Charles Adolphe 1817–1884, French chemist; discovered the amines of ammonia; devised the Wurtz reaction.

Wurtz-Fittig reaction *Organic Chemistry.* a modified Wurtz reaction in which the alkyl halide reacts with an aromatic halide in the presence of metallic sodium and an anhydrous solvent; the end product is an alkylated aromatic hydrocarbon.

wurtzilite *Geology.* a black, asphaltic pyrobitumen derived from the metamorphosis of petroleum.

wurtzite *Mineralogy.* (Zn,Fe)S, a brownish-black to orange-brown, translucent hexagonal or trigonal mineral occurring as hemimorphic pyramidal or short prismatic to tabular crystals and as fibrous crusts, having a specific gravity of 3.98 to 4.08 and a hardness of 3.5 to 4 on the Mohs scale; found chiefly in hydrothermal vein deposits; trimorphous with matraite and sphalerite.

Wurtz reaction *Organic Chemistry.* a method of synthesizing hydrocarbons by treating alkyl halides with metallic sodium in ethereal solution; the reaction usually involves alkyl iodide; for example, $2CH_3I + 2Na \rightarrow CH_3CH_3 + 2NaI$.

wüstite *Mineralogy.* FeO, a cubic mineral of the periclase group, metallic gray in reflected light, having a specific gravity of 5.74 to 5.88 and a hardness of 5 on the Mohs scale; a component of fusion crusts on meteorites, and also found in mid-ocean ridge basalt with native iron.

w/v weight per volume.

W Virginis stars *Astronomy.* a type of Cepheid variable star with a characteristic period 10 to 30 days long and a brightness "bump" on the declining light curve. Also, TYPE II CEPHEIDS.

WWV *Telecommunications.* the call letters of a radio station under the supervision of the National Bureau of Standards that provides technical services, such as radio propagation disturbance warnings.

WWVH *Telecommunications.* the call letters of the National Bureau of Standards radio station at Maui, Hawaii.

Wyandotte *Agriculture.* a breed of chicken that lays brown eggs and also produces meat; a number of different varieties have been developed in the U.S. from several other breeds.

wye *Electricity.* a three-phase star connection. *Engineering.* a pipe branching into two directions from the main run at 45° angles. Also, Y.

wye-delta switching starter *Electricity.* a motor starter that switches the stator windings from a star connection to a delta connection; used with three-phase induction motors.

wye-delta transformation *Electricity.* the result of a mathematical process that makes it possible to replace any wye network with an equivalent delta network, so that the voltages and currents at the terminals are the same for the two networks.

wye-level see Y LEVEL.

wye network *Electricity.* a Y-shaped or star network that has three branches meeting at a common point, and is often tied to ground; often used in three-phase networks.

wyerone *Biochemistry.* a furanoacetylene disease-fighting agent produced by the broad bean, *Vicia faba.*

Wynyardiidae *Paleontology.* a small, extinct family of marsupials that may belong in the superfamily Vombatoidea; an example of the early development of the diprotodont condition; extant in the Miocene.

WYSIWYG [wiz′ə wig′; wiz′wig′] *Computer Science.* describing a program, such as an editor or text-formatting program, whose screen display is a direct reflection of the state of the underlying document or data. (An acronym for the phrase "what you see is what you get.") Also, WIZZY-WIG.

X

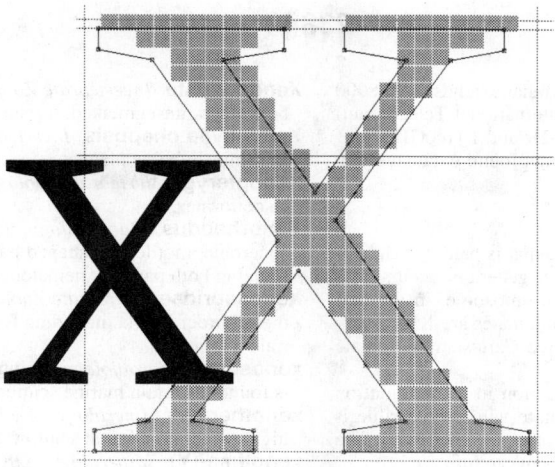

X a chemical symbol for an electronegative atom.
x a chemical symbol for mole fraction.
x or **X** *Mathematics*. **1.** an unknown quantity or a variable. **2.** the *x*-axis.
xanth- or **xantho-** a combining form meaning "yellow."
xanthan gum *Organic Chemistry*. a natural heteropolysaccharide consisting of D-glucose, D-mannose, and D-glucuronic acid residues; a water-soluble gum with a molecular weight of between five and ten million; produced by the pure culture fermentation of glucose with *Xanthomonas campestris*.
xanthelasma *Medicine*. a soft yellow plaque or spot usually occurring in groups on the eyelids.
xanthene *Organic Chemistry*. $CH_2(C_6H_4)_2O$, yellowish, crystalline leaflets; very slightly soluble in water, slightly soluble in alcohol, and soluble in ether; melts at 100.5°C and boils at 310–312°C; the central structure of fluoroscein, eosin, and rhodamine dyes, and used in organic synthesis and as a fungicide.
xanthene dye *Organic Chemistry*. any of a group of dyes related to xanthene, with the chromophore consisting of (C_6H_4) groups.
Xanthidae *Invertebrate Zoology*. the largest crab family, small, nonswimming, shallow water decapod crustaceans in the section Brachyura, with a strong carapace and claws; includes edible stone and coral crabs.
Xanthidium *Botany*. a genus of spiny desmids of the family Desmidiaceae, including some common plankton forms.
xanthin *Botany*. the yellow coloring matter of flowers.
xanthine *Organic Chemistry*. $C_5H_4N_4O_2$, a purine base that occurs in the blood and urine and in some plants; a yellowish-white powder; insoluble in water and soluble in mineral acids; sublimes with partial decomposition; toxic by ingestion; used in medicine and organic synthesis. Also, DIOXOPURINE.
xanthine oxidase *Enzymology*. an enzyme that catalyzes the oxidation of certain purines to urate. Also, SCHARDINGER ENZYME.
xanthinuria *Pathology*. **1.** the production of abnormally large amounts of xanthine in the urine. **2.** a rare inherent malady caused by an insufficiency of the enzyme xanthine oxidase, resulting in the overproduction of xanthine in place of uric acid in the urine, and in some cases generating xanthine calculi.
xanthism *Biology*. a condition in which the coloring of an animal, especially its skin, is marked by an abnormal amount of yellow pigment. Also, **xanthochroism.**
Xanthobacter *Bacteriology*. a genus of Gram-negative, aerobic, coccoid or rod-shaped bacteria of uncertain affiliation that occur as nitrogen-fixing organisms in soil and aquatic habitats.
xanthochromia *Medicine*. any yellow discoloration, as of the skin.
xanthochromic *Neurology*. yellow-colored; applied specifically to the cerebrospinal fluid. Also, **xanchromatic, xanthochromatic.**
xanthoconite *Mineralogy*. Ag_3AsS_3, a dark-red, dull-orange, or brown, monoclinic mineral occurring as tabular crystals and reniform masses, having a specific gravity of 5.54 and a hardness of 2 to 3 on the Mohs scale; found in low-temperature hydrothermal veins with ruby silvers and other silver minerals; dimorphous with proustite.
xanthoderma *Medicine*. a yellowish discoloration of the skin.
xanthoma *Medicine*. a benign condition caused by the deposit of lipids on the skin and characterized by small yellow lesions in the subcutaneous layer of skin, often around tendons.
xanthoma *Oncology*. a tumor composed of lipid-laden foam cells.
xanthomatosis *Medicine*. an imbalance in lipid metabolism causing the accumulation in the body of excess lipids and characterized by the formation of lipoid tumors.
xanthomegnin *Biochemistry*. a fungal liver toxin produced by species of *Aspergillus, Penicillium,* and *Trichophyton.*
Xanthomonas *Bacteriology*. a genus of Gram-negative, aerobic, rod-shaped bacteria of the family Pseudomonadaceae that produce a yellow pigment and are generally pathogenic for plants.
xanthomycin *Microbiology*. an antibiotic produced by certain *Streptomyces* bacteria, active against a variety of Gram-positive organisms.
xanthone *Organic Chemistry*. $CO(C_6H_4)_2O$, a tricyclic compound that occurs in some plant pigments; white needles or crystalline powder; insoluble in water and soluble in alcohol; melts at 173°C and boils at 350°C; used as a larvacide, and as an intermediate for dyes, perfumes, and pharmaceuticals.
xanthophore *Cell Biology*. a yellow chromatophore or pigment cell.
xanthophose *Neurology*. the subjective perception of the color yellow.
Xanthophyceae *Botany*. one of the three major classes within the Chrysophyta, including common filamentous alga; found in the phytoplankton of fresh waters and the oceans.
xanthophyll *Biochemistry*. $C_{40}H_{56}O_2$, a group of yellow pigments that are oxygen derivatives of the carotenes; found in green leaves, avian egg yolk, and human plasma.
Xanthophyllaceae *Botany*. a monogeneric family of small, dicotyledonous trees of the order Polygalales, often accumulating aluminum and native to the Indo-Malaysian region.
xanthopsis *Oncology*. a yellow pigmentation found in cancers.
Xanthoria *Botany*. a genus of arborescent, foliaceous, yellow or orange lichens.
Xanthorrhoeaceae *Botany*. a family of stout shrubs and herbs of the order Liliales, characterized by secondary stem growth of the monocotyledonous type, perennial pungent leaves, and small trimerous flowers; native to Australia, Tasmania, New Guinea, and New Caledonia.
xanthosarcoma *Oncology*. a giant cell sarcoma of the tendon sheath.
xanthosis *Medicine*. a yellowish discoloration, especially of the skin.
xanthosomes *Cell Biology*. particulate bodies found in certain foraminiferan cells and thought to be waste products.
xanthous *Medicine*. yellow or yellowish.
xanthoxenite *Mineralogy*. $Ca_4Fe_2^{+3}(PO_4)_4(OH)_2 \cdot 3H_2O$, a yellow triclinic mineral occurring as platy masses and fibrous aggregates, having a specific gravity of 2.8 to 2.97 and a hardness of about 2.5 on the Mohs scale; found as an alteration mineral in granite pegmatites.
xanthurenic acid *Biochemistry*. $C_{10}H_7NO_4$, a metabolite of L-tryptophan present in the urine.
x-axis *Mathematics*. in R^3(3-dimensional Cartesian space), the line that is the intersection of the planes given by $y = 0$ and $z = 0$, usually represented by a line from the upper right through the origin toward the lower left to indicate that it comes directly out of the paper toward the viewer. It is positive to the lower left and negative to the upper right in a right-handed coordinate system. In R^2, it is generally represented as a horizontal line, with the right half positive and the left half negative. *Cartography*. **1.** the line on a map or chart on which distances to the east or west of the prime reference line (meridian) are determined. **2.** in aerial photography, a line joining opposite fiducial marks in the direction roughly parallel to the line of flight.

X band *Telecommunications.* a microwave band that extends from 8000 to 12,000 megahertz (8 to 12 gigahertz); the International Telecommunications Union specifies a range between 8.5 GHz and 10.68 GHz.

X bodies *Virology.* inclusion bodies or structures found exclusively in tobacco-mosaic-virus infected cells.

X boson see SUPERHEAVY BOSON.

X_c *Electricity.* the symbol for capacitive reactance.

X chromosome *Genetics.* the sex chromosome that is paired in the homogametic sex of most animal species, and that generally occurs with another sex chromosome, designated the Y chromosome, in the heterogametic sex; thus female mammals are XX and males are XY.

x coordinate *Mathematics.* the first of the three Cartesian coordinate functions in R^3.

x-direction *Cartography.* in surveying, a direction in a triangulation figure that is observed and given an approximate value; this value is then used during the mathematical process of correcting the figure. The corrected value for an *x*-direction is called the **x-correction.**

Xe the chemical symbol for xenon.

xen- or **xeno-** a combining form meaning "strange" or "foreign."

Xenarthra *Vertebrate Zoology.* a suborder of the Edentates (order Edentata), including the infraorders Pilosa (sloths), Vermilingua (anteaters), and Cingulata (armadillos).

xenia *Botany.* the action of pollen on a seed that, in the same generation, shows characteristics which result from pollination.

xenoantigen *Immunology.* an antigen found in organisms of more than one species.

xenoblast *Mineralogy.* a mineral that has grown during metamorphism without developing its characteristic crystal faces.

Xenococcus *Bacteriology.* a genus of Cyanobacteria, the blue-green algae, occurring as single, rounded, vegetative cells of variable size.

xenocryst *Crystallography.* a crystal foreign to the rock in which it is found.

xenodiagnosis *Medicine.* a technique of animal inoculation using laboratory-bred insects and animals to diagnose certain parasitic infections that cannot be demonstrated in blood films; originally used to diagnose *Trypanosoma cruzi* (Chagas' disease), and also used with *Trichinella spiralis* infections.

xenogamy *Botany.* a cross-pollination of the stigma of a flower with a flower on another plant.

xenogeneic *Immunology.* relating to substances that originate from different species.

xenograft see HETEROPLASTIC GRAFT.

xenol see PHENYLPHENOL.

xenolith *Petrology.* an inclusion of foreign material in a rock.

xenomorphic *Geology.* relating to the texture of an igneous rock in which the crystals are not bounded by their own crystal faces and whose form is impressed upon them by adjacent preexisting crystals. Also, ALLOTRIOMORPHIC, ANIDIOMORPHIC, LEPTOMORPHIC.

xenon [zē´nän] *Chemistry.* an element of the noble gas group of the periodic table, having the symbol Xe, the atomic number 54, an atomic weight of 131.3, a melting point (liquid) of –112°C, and a boiling point (liquid) of –107°C; a colorless, odorless gas (or liquid) that is chemically unreactive but not inert, reacting only under highly oxidizing conditions; used in lamps, luminescent tubes, lasers, and bubble chambers. (From a Greek word meaning "strange.")

xenon-135 *Nuclear Physics.* a radioactive isotope of xenon with a half-life of 9.1 hours, produced in a nuclear reactor.

xenon-absorption effect *Nucleonics.* in a nuclear reactor, a buildup of xenon-135 from radioactive (beta) decay of iodine-135, which may delay the restarting of a reactor after shutdown.

xenon arc lamp *Electricity.* a xenon-gas-filled tube in which a brilliant artificial light is produced by a prolonged electric arc passing between two electrodes and through the xenon gas; often used in photography. Also, **xenon flash lamp.**

xenon poisoning *Nucleonics.* in a nuclear reactor, the buildup of xenon-135 from radioactive (beta) decay of iodine-135, a fission product; until the accumulation of xenon-135, which has the highest known cross section for thermal neutron capture, has largely decayed away, it may delay by a day or more the restarting of a reactor after shutdown.

Xenopeltidae *Vertebrate Zoology.* the sunbeam snakes, a monospecific family of harmless, burrowing iridescent snakes; native to southeastern Asia.

xenophobia [zen´ə fō´bē ə; zēn´ə fō´bē ə] *Psychology.* **1.** an irrational fear of strangers or strange places. **2.** an irrational fear or hatred of nations, cultures, or societies other than one's own. Thus, **xenophobic.**

Xenopneusta *Invertebrate Zoology.* an order in the phylum Echiurida, burrowing, unsegmented, marine worms with a proboscis and no brain.

Xenopsylla cheopsis *Invertebrate Zoology.* the common rat flea that transmits bubonic plague.

Xenopterygii *Vertebrate Zoology.* another name for the fish order Gobiesociformes.

Xenorhabdus *Bacteriology.* a genus of Gram-negative, facultatively anaerobic, motile, rod-shaped bacteria of the family Enterobacteriaceae, found in both parasitic nematodes and the nematode hosts, insect larvae.

Xenosauridae *Vertebrate Zoology.* a family of small, stocky lizards of the suborder Sauria, including four rare species found in China and Central America.

xenosome *Microbiology.* an intracellular symbiotic microorganism that is found in certain marine scuticociliates.

xenothermal *Mineralogy.* of a hydrothermal mineral deposit formed at high temperature but at shallow depth; also, of that environment.

xenotime-(Y) *Mineralogy.* YPO_4, a moderately radioactive tetragonal mineral occurring in variously colored aggregates of prismatic or pyramidal crystals, having a specific gravity of 4.4 to 5.1 and a hardness of 4 to 5 on the Mohs scale; forms a series with chernovite-(Y); found in acidic igneous rocks, in pegmatites, and as a detrital mineral.

xenotropic *Virology.* describing a genetically transmitted retrovirus that cannot replicate in its host but can infect and replicate in hosts of another species.

Xenungulata *Paleontology.* a monogeneric order of large South American ungulates containing a single genus, *Carodnia*, that is similar to the embrithopods, uintatheres, and pantodonts of the other continents; extant in the Paleocene.

xenyl *Organic Chemistry.* the functional biphenyl group $C_6H_5C_6H_4^-$.

xer- or **xero-** a combining form meaning "dry."

Xeralf *Geology.* a suborder of the soil order Alfisol that develops in mediterranean climates; characterized by a reddish or brownish color and good drainage.

xerarch succession *Ecology.* an ecological sequence that begins in a dry area.

Xerert *Geology.* a suborder of the soil order Vertisol that develops in Mediterranean climates; characterized by wide, deep surface cracks that open and close annually.

xeric *Ecology.* **1.** describing a location or habitat with very little moisture. **2.** describing an organism that lives in such an environment.

xericulture *Agriculture.* **1.** the growing of plants with a minimal use of water. **2.** gardens, landscapes, or the like that require little water.

xeriscaping *Agriculture.* landscaping using minimal amounts of water, as by growing drought-tolerant plants, using drip irrigation, and so on.

xeroderma *Medicine.* a disease characterized by rough, dry skin.

xeroderma pigmentosum *Medicine.* a rare inherited skin disease, inherited as autosomal recessive alleles, in which there is extreme photosensitivity to ultraviolet light, characterized by early development of senile changes in the skin, keratoses, papillomas, malignancies, and severe ophthalmologic abnormalities.

xerogel *Chemistry.* a gel from which all volatile liquid has been removed.

Xerography *Graphic Arts.* a trade name for electrostatic printing, copyright by the Xerox Corporation, which developed this technique.

Xeroll *Geology.* a suborder of the soil order Mollisol that develops in Mediterranean climates; characterized by a calcic, petrocalcic, or gypsic horizon, or a duripan.

xeromorphic *Botany.* describing plant forms adapted to an arid environment.

xerophilous *Ecology.* living or thriving in dry habitats; drought-tolerant. Thus, **xerophile, xerophily.**

xerophthalmia *Medicine.* conjunctivitis caused by vitamin A deficiency in which the eye is dry, without discharge, and atrophies. Also **xerosis conjunctivae.**

xerophyte *Ecology.* a plant that lives in an area with little moisture. Thus, **xerophytic.**

xerosere *Ecology.* a sere that occurs on dry soil.

xerosis *Medicine.* any abnormal dryness, as of the eye, skin, or mouth.

xerostomia *Pathology.* a dryness of the mouth due to diminished functioning of the salivary glands caused by aging, disease, drug reaction, and so on.

Xerothermic *Geology.* a postglacial interval characterized by a relatively warmer, drier climate.

xerothermic *Meteorology.* describing or relating to a climate characterized by dry and hot weather conditions. Also, **xerothermal.**

Xerox *Engineering.* the brand name for a copying machine that reproduces printed, written, or graphic material by xerography. (Taken from a Greek phrase meaning literally "dry copy.")

original Xerox machine

Xerult *Geology.* a suborder of the soil order Ultisol that develops in Mediterranean climates and is characterized by a brown to reddish color and a low to moderate organic carbon content.

X factor *Microbiology.* hemin, a growth factor required by certain species of *Haemophilus* undergoing aerobic growth.

x-height *Graphic Arts.* in a type font, the height of lowercase letters without ascenders or descenders, such as the x.

xi [zī] the fourteenth letter of the Greek alphabet, written Ξ, ξ.

xid mouse *Immunology.* a mouse bearing a Y-linked immune deficiency gene.

xi hyperon *Particle Physics.* an isotopic spin multiplet of quasi-stable baryons with positive parity, spin 1/2, isotopic spin 1/2, hypercharge –1, and approximate mass 1318 MeV. Also, **xi particle.**

xi-minus particle *Particle Physics.* a xi hyperon that carries a unit negative charge.

X inactivation *Genetics.* the random condensation and inactivation of one of the two X chromosomes during the early embryonic development of a female mammal, thereby compensating for the double dose of genes carried on these chromosomes; the condensed chromosome is often visible as a Barr body.

xiphi- or **xipho-** a combining form meaning "sword."

Xiphiidae *Vertebrate Zoology.* the swordfishes, a monotypic family of elongated fishes having a long flattened snout, a high dorsal fin, a blue-black back, gray-blue sides, and a white belly; found in tropical and temperate parts of all oceans, and the Mediterranean and Black Seas.

xiphisternum *Anatomy.* the xiphoid process of the sternum.

Xiphodontidae *Paleontology.* an extinct family of mammals in the suborder Tylopoda; the xiphodonts resembled camels in some ways, but exact affinities have not been established; restricted to Europe in the Late Eocene and Early Oligocene.

xiphoid *Anatomy.* shaped like a sword.

xiphophyllous *Botany.* bearing sword-shaped leaves.

Xiphydriidae *Invertebrate Zoology.* a family of wood wasps or horntails, a small family of hymenopteran insects in the suborder Chalastogastra; the larvae bore into deciduous trees.

xi-zero particle *Particle Physics.* a xi hyperon that carries no charge.

X_L *Electricity.* the symbol for inductive reactance.

x-line *Graphic Arts.* the upper limit of x-height.

X linkage *Genetics.* the most common form of sex-linkage, in which the loci of certain genes are found only on the X chromosome and not on the Y chromosome.

XMP xanthosine 5'-monophosphate.

XMP pathway *Biochemistry.* a cyclic metabolic channel for yeasts growing on methanol to assimilate formaldehyde.

XO *Genetics.* an abnormal chromosome type in a heterogametic organism characterized by the presence of a single X chromosome without a second X or a Y chromosome; in humans this results in the sexual phenotype known as Turner's syndrome.

xonotlite *Mineralogy.* $Ca_6Si_6O_{17}(OH)_2$, a colorless to white, pink, or gray monoclinic and triclinic mineral occurring as compact to fibrous masses, having a specific gravity of 2.71 and a hardness of 6 to 6.5 on the Mohs scale; found as veining in serpentine rocks or contact zones.

X organ *Invertebrate Zoology.* in some crustaceans, a small ductless gland in the eyestalk, producing a secretion that affects molt.

x-press *Biotechnology.* a device used to disrupt microbial cells by freezing them and then pressing a frozen paste of the cells through a small hole.

x-radiation see X-RAY.

X-ray or **X ray** *Radiology.* **1. X-rays** or **X rays**. energy beams of very short wavelengths (0.1 to 1000 angstroms) produced by high-velocity electrons that encroach on various materials, particularly heavy metals. They are commonly created by passing a current of high voltage through a Coolidge tube, they can penetrate most substances, and they are often used for inner-body examination and treatment. Also, ROENTGEN RAYS, X-RADIATION. **2.** a device that uses X-rays to produce inner-body images. **3.** an image created by this device. Also, **x-ray** or **x ray**. (From *X* in the sense of "unknown," so called by their discoverer, Wilhelm Roentgen, because of their seemingly mysterious nature.)

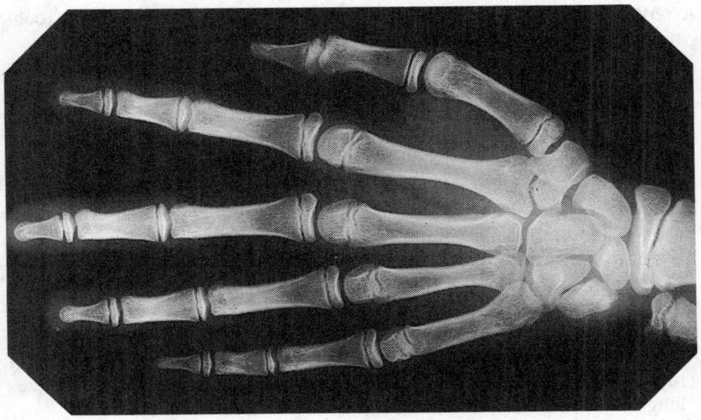

X-ray view of the human hand

X-ray absorption *Electromagnetism.* the absorption of X-ray frequencies in a medium through which X-ray radiation is passing.

X-ray analysis *Physics.* the application of X-rays to study crystal structure, atomic composition, and heavy elements in the presence of lighter elements.

X-ray astronomy *Astronomy.* the branch of astronomy that studies the X-ray emission from celestial objects.

X-ray background *Astronomy.* diffuse X-ray emission from many unresolved sources beyond the immediate solar system.

X-ray binary *Astronomy.* an X-ray source that is observed to be a member of binary systems; e.g., Cygnus X-1, Centaurus X-3, and X Persei.

X-ray burster *Astronomy.* a celestial source that emits X-rays in sharp, short bursts.

X-ray camera *Crystallography.* a device for holding film in an appropriate manner to intercept an X-ray diffraction pattern.

X-ray cluster *Astronomy.* a cluster of galaxies with detectable X-ray emission from a hot diffuse medium pervading the cluster.

X-ray crystallography *Crystallography.* the analysis of the arrangement of atoms in a crystal by an analysis of the diffraction pattern of that crystal.

X-ray crystal spectrometer *Spectroscopy.* a device that is capable of producing an X-ray spectrum of radiation and measuring the wavelengths of the component frequencies by means of X-ray diffraction analysis, using a crystal of well-known lattice spacing.

X-ray dermatitis *Radiology.* an inflammatory reaction of skin resulting from biologically effectual radiation exposure. Also, RADIODERMATITIS.

X-ray diffraction *Physics.* the scattering of X-rays from a crystal, producing a characteristic interference pattern whose information is useful in determining the crystal structure. Also, **X-ray microdiffraction.**

X-ray diffraction analysis *Crystallography.* the analysis of diffraction patterns produced when X-rays impinge on a crystalline material, leading to a determination of the three-dimensional arrangement of atoms in the crystal.

X-ray film *Radiology.* a thin, flexible, transparent sheet of plastic that is coated with a radiant-energy sensitive emulsion; used in roentgenography. *Graphic Arts.* photographic film that is coated with a photosensitive emulsion that is sensitive to X-rays.

X-ray fluorescence *Atomic Physics.* the emission of the characteristic X-ray line spectrum of a substance on exposure to X-rays. Also, **X-ray emission.**

X-ray fluorescence analysis *Analytical Chemistry.* a nondestructive method of chemical analysis that uses an intense X-ray beam to irradiate a sample. The subsequent fluorescence is used to determine the elements in the sample from the wavelengths of the spectral lines; concentration is based on intensity of these lines. Also, **X-ray fluorometry.**

X-ray fluorescent emission spectrometer *Spectroscopy.* a crystal spectrometer device that is used to observe and measure fluorescent X-rays in which bent transmitting or reflecting crystals are used to concentrate low-intensity X-rays into a beam.

X-ray generator *Electronics.* a metal that emits large amounts of X-rays when bombarded with high-velocity electrons.

X-ray goniometer *Engineering.* a scale that determines the angle between incident and refracted beams; used in X-ray diffraction analysis.

X-ray hardness *Electromagnetism.* a measure of the ability of X-rays to penetrate a substance.

X-ray holography *Electromagnetism.* the use of holographic techniques to image objects at X-ray wavelengths.

X-ray image spectrography *Spectroscopy.* a technique that is used in X-ray fluorescent emission analysis in which X-rays from the source undergo Bragg scattering from a cylindrically bent crystal, which produces a slightly enlarged image on a photographic film.

X-ray laser *Electromagnetism.* a laser that is capable of producing X-ray radiation in a collimated beam.

X-ray machine *Radiology.* the X-ray tube and associated equipment used to generate X-ray photographs.

X-ray microscope *Engineering.* any instrument that utilizes X-radiation for magnification of 100–1000 diameters or for chemical analysis; including contact and projection microradiography, reflection X-ray microscopy, or X-ray image spectrography.

X-ray monochromator *Engineering.* an apparatus that diffracts X-rays from a crystal to produce a beam with a narrow range of wavelengths.

X-ray nebulae *Astronomy.* a cloud of gas hot enough to emit X-rays; often remnants of supernovae.

X-ray nova *Astronomy.* a short-lived X-ray source that appears suddenly in the sky and dramatically increases in strength over a period of a few days and then decreases, with an overall lifetime of a few months. Also, TRANSIENT X-RAY SOURCE.

X-ray optics *Electromagnetism.* the aspects of X-ray physics that are similar to properties exhibited by visible light.

X-ray photoelectron spectroscopy *Spectroscopy.* a technique in which a specimen is irradiated with X-rays and the energies of the photoelectrons emitted from the specimen's surface are determined.

X-ray photogrammetry *Photogrammetry.* the theories and techniques of obtaining reliable measurements by means of radiographs produced from X-rays, rather than from photographs.

X-rays see X-RAY.

X-ray sickness see RADIATION SICKNESS.

X-ray spectrograph *Spectroscopy.* a device that is equipped to achieve X-ray diffraction from a crystal specimen and record the diffraction pattern on a photographic plate or film.

X-ray spectrometry *Spectroscopy.* the measurement and analysis of X-ray diffraction produced by crystals. Thus, **X-ray spectrometer.**

X-ray spectrum *Spectroscopy.* a graphic display of the X-ray intensities as a function of wavelength produced by any process in which X-rays are emitted, such as in bremsstrahlung, in which sharp lines are superimposed on a continuous distribution.

X-ray target *Electronics.* in an electron tube, the metal body with which high-velocity electrons collide to produce X-rays.

X-ray telescope *Engineering.* a device that can detect X-rays emanating from space and resolve the X-rays into an image.

X-ray therapy *Medicine.* radiation therapy by the application of X-rays, often of higher energy than diagnostic X-rays; used especially in the treatment of cancer.

X-ray tube *Electronics.* a specialized, high-voltage diode in which X-rays are produced through the acceleration of electrons to a high velocity by an electrostatic field and their sudden stop in a collision with a target.

X-ray unit *Radiology.* **1.** a measure of X-ray exposure equaling the quantity that causes redness of the skin. Also, KIENBOCK UNIT. **2.** a machine that produces X-rays for diagnostic or therapeutic purposes. *Spectroscopy.* see SIEGBAHN.

X server *Computer Science.* software that accepts commands in the form specified by the X protocol and produces corresponding displays on a graphical display device.

x unit *Physics.* a unit formerly used to express the wavelength of X-rays and gamma rays, equal to approximately 10^{-11} cm.

XXY trisomy *Genetics.* see KLINEFELTER SYNDROME.

Xyelidae *Invertebrate Zoology.* a small family of primitive sawflies, hymenopteran insects in the suborder Chalastogastra; the larvae eat the leaves of trees.

xyl- or **xylo-** a combining form meaning "wood."

xylan *Biochemistry.* any of several polysaccharides consisting primarily or completely of xylose residues; found in woody tissues, straw, and corncobs.

xylanases *Enzymology.* any enzyme that catalyzes the breakdown of xylans, polysaccharide components of wood, cereals, or bran.

xylem [zī´ləm] *Botany.* a woody tissue in vascular plants, consisting of traceal tissue and parenchyma, that helps to provide support and conducts water and nutrients upward from the roots.

xylem ray *Botany.* a vascular ray located in the xylem. Also, WOOD RAY.

xylene *Organic Chemistry.* $C_6H_4(CH_3)_2$, an isomeric mixture of *o*-, *m*-, and *p*-xylene; a clear liquid, with various grades having different boiling points; insoluble in water and soluble in alcohol and ether; used in aviation gasoline, coatings, lacquers, rubber cements, organic synthesis, and polyester resin manufacture. Also, DIMETHYLBENZENE.

***m*-xylene** *Organic Chemistry.* 1,3-$C_6H_4(CH_3)_2$, a clear, colorless liquid; soluble in alcohol and insoluble in water; boils at 139.3°C; used as a solvent, as an intermediate for dyes and organic synthesis, and in insecticides and aviation fuel.

***o*-xylene** *Organic Chemistry.* 1,2-$C_6H_4(CH_3)_2$, a clear, colorless liquid; soluble in alcohol and insoluble in water; boils at 144°C; used in phthalic anhydride manufacturing, in vitamin synthesis, and in dyes, insecticides, and motor fuels.

***p*-xylene** *Organic Chemistry.* 1,4-$C_6H_4(CH_3)_2$, a colorless liquid; soluble in alcohol and insoluble in water; boils at 138.5°C and freezes at 13–14°C; used in terephthalic acid synthesis for polyester resins and fibers such as Dacron and Mylar, in vitamin and pharmaceutical syntheses, and in insecticides.

xylenol *Organic Chemistry.* $(CH_3)_2C_6H_3OH$, a white, crystalline solid; soluble in most organic solvents; slightly soluble in water; melts between 20° and 76°C due to differences among the various (six) isomers; used in disinfectants, solvents, pharmaceuticals, insecticides and fungicides, wetting agents, and dyestuffs. Also, DIMETHYLPHENOL.

xylidine *Organic Chemistry.* $(CH_3)_2C_6H_3NH_2$, a liquid; soluble in alcohol and ether and slightly soluble in water; boils at 213–226°C, varying among the various (six) isomers; used in organic synthesis, in pharmaceuticals, and as a dye intermediate.

xylinite *Geology.* a variety of provitrinite composed of woody tissue.

xylitol *Organic Chemistry.* $CHOH(CHOH)_3CH_2OH$, a pentahydric alcohol, melting at 95–97°C, that is derived from xylose. Also, **xylite.**

Xylocaine *Pharmacology.* a trademark name for lidocaine preparations. See LIDOCAINE.

xylocarp *Botany.* a hard, woody fruit.

Xylocopidae *Invertebrate Zoology.* the carpenter bees, a family of hymenopteran insects belonging to the suborder Clistogastra; stout, hairy bees that burrow into dry wood timbers.

xylogen *Botany.* wood or xylem in an early or formative stage.

xyloid *Botany.* of or relating to wood or like wood.

xylophagous *Science.* feeding on wood, as certain larvae and insects. **2.** destroying timber, as certain mollusks and fungi.

xylophilous *Science.* growing or living on wood.

xylose *Biochemistry.* $C_5H_{10}O_5$, a five-carbon, pentose sugar found in woody materials that is used as a nonnutritive sweetener and in dyeing and tanning.

xylulose *Biochemistry.* $C_5H_{10}O_5$, a five-carbon ketose sugar formed from xylose that plays a role in carbohydrate metabolism, particularly in the hexose monophosphate cycle.

X-Y servo *Robotics.* a peripheral enhancement with the ability to place a load in any position that can be reached by a robot.

xyster see RASPATORY.

xystus or **xyst** *Architecture.* **1.** in classical architecture, a long, open portico used for exercise in bad weather. **2.** a tree-lined walk.

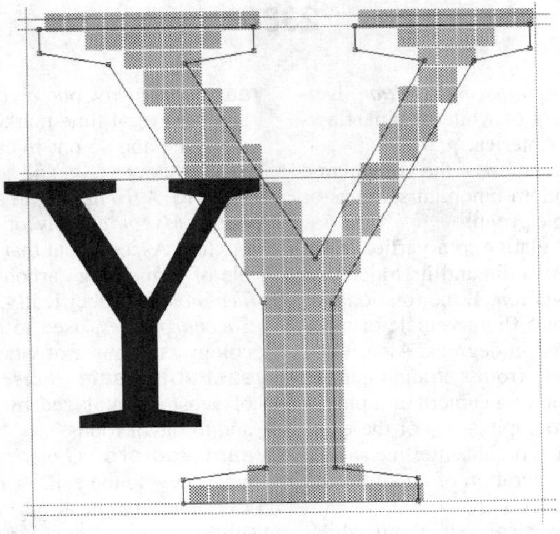

Y *Engineering*. see WYE.

Y the chemical symbol for yttrium.

y or **Y** *Mathematics*. **1.** an unknown quantity or a variable, especially a second one in a statement where *x* represents the first one. **2.** the *y*-axis.

y. or **y** yard(s).

YACC [yak] *Computer Science*. a widely used context-free, parser-generator program for producing syntax-directed translators such as compilers. (An acronym for yet another compiler-compiler.)

yacht *Naval Architecture*. **1.** a small, fast sailing ship, historically of Dutch origin, adopted for use as a small naval vessel and (especially) as a pleasure craft. **2.** a relatively large, seagoing racing or pleasure craft. In British usage, it usually implies a sailing vessel, while in American usage it implies a luxurious motor-powered vessel.

YA-7F *Aviation*. a single-seat U.S. Air Force aircraft developed beyond the experimental stage, but produced in limited numbers for service testing.

YAG *Materials*. a synthetic garnet of yttrium and aluminum oxide; used to generate laser beams and as a gemstone.

Yagi-Uda antenna *Electromagnetism*. an end-fire antenna array constructed of several unconnected, equally spaced dipole elements lying in a common horizontal plane, with one element fed by a transmission line. (Named for Hidetsugu *Yagi*, 1886–1976, the Japanese electrical engineer who invented it.) Also, **Yagi antenna.**

YAG laser see YTTRIUM-ALUMINUM-GARNET LASER.

yak *Vertebrate Zoology*. the large, stocky, domesticated ox, *Bos grunniens* of the family Bovidae, having short, smooth hair on the back and long hair falling in fringes around the underside and ankles; raised for its milk, meat, and hair and used as a beast of burden; found in the high plateaus and mountains of Tibet and Central Asia.

yak

yalca *Meteorology*. a local term for a violent snowstorm with squall-like winds blowing over the Andes Mountain passes in Peru.

Yale Cross-Cultural Survey *Anthropology*. a former name for the Human Relations Area File.

Yalow, Rosalyn born 1921, American biophysicist; awarded the Nobel Prize for developing a method of measuring hormone levels in the body.

yam *Botany*. a large, edible, starchy, tuberous root of any of various plants of the genus *Dioscorea*, growing in tropical climates and cultivated in warm regions.

yamase *Meteorology*. a cool, onshore easterly wind that blows shoreward toward the Senriku district of Japan in the summer season.

yampee *Botany*. a vine, *Dioscorea trifida*, having large leaves and edible tubers; found in South America. Also, CUSH-CUSH.

Yang, Chen Ning born 1922, Chinese-born American physicist; with Tsung Dao Lee, shared the Nobel Prize for disproving the law of conservation of parity.

Yang-Mills theory *Particle Physics*. a field theory originally based on the hypothesis that nuclear forces between nucleons can be derived by imposing local isospin invariance, implying that the interaction must occur through the exchange of three massless vector bosons.

Yangtze [yang´sē] *Geography*. the longest river in Asia, flowing about 3600 miles from the Tibetan plateau across central China to the East China Sea.

Y antenna see DELTA MATCHED ANTENNA.

yard *Architecture*. **1.** an open area, often planted, adjacent to a building. **2.** the grounds of a public building such as an inn or a college. *Metrology*. a traditional unit of length equal to 36 inches or 3 feet; equivalent to 0.9144 meter. *Naval Architecture*. **1.** a horizontal spar set across a mast, and used to support a square sail. Nonsailing vessels may be fitted with yards to support signal halyards or electronic equipment. **2.** a shipbuilding facility; shipyard.

yardage *Metrology*. an amount expressed in yards. *Mining Engineering*. **1.** a price paid per yard, rather than per ton, for mining or cutting coal. **2.** extra compensation paid to a miner working in deficient coal or in a narrow place, based on the number of yards advanced.

yardarm *Naval Architecture*. **1.** one of the outer ends of a yard; it is a common attaching point for halyards and other hoists. **2.** a short yard high on a modern ship's foremast; used for signal hoists.

yard maintenance *Engineering*. the level of maintenance that includes the complete reconstruction of parts, subassemblies, and components.

yardstick *Mechanical Devices*. a stick one yard long, commonly marked with subdivisions; used for measuring.

yare *Navigation*. describing a ship that is easily steered or maneuvered.

Yarkovsky effect *Astronomy*. the anisotropic emission of radiation from a rotating object orbiting the sun, due to uneven heating.

Yarmouthian interglacial *Geology*. the second interglacial stage of the Pleistocene epoch in North America, occurring after the Kansan and before the Illinoian glacial stages. Also, **Yarmouth.**

yarn *Textiles*. a continuous strand of two or more fibers twisted loosely together, or a single filament of a natural or synthetic fiber; used for weaving or knitting. Yarns are categorized as either spun yarn or continuous filament yarn.

yarn count *Textiles*. see COUNT.

yarn dyeing *Textiles*. the process of dyeing yarn before weaving or knitting.

yarrow *Botany.* a common composite plant, *Achillea millefolium*, having finely divided leaves and close, flat clusters of white or pink flowers; native to Europe and naturalized in North America.

Yates's chi-squared test *Statistics.* a modification of the chi-squared test for significance, used to compare two random binomial samples or matched observations of a single random sample group.

yaw *Mechanics.* **1.** the angular displacement, relative to a vertical axis, between the axis of symmetry of a slender projectile and its trajectory. **2.** to rotate or oscillate about a vertical axis. *Aviation.* **1.** the rotational or oscillatory movement about the vertical axis of a flight vehicle or other body. **2.** the extent of such movement measured in degrees. Also, **yawing.** *Navigation.* of a ship, to deviate temporarily from a straight course. *Cartography.* in aerial photography, the rotation of a camera or a photograph coordinate system about either the photograph z-axis or the exterior z-axis. *Robotics.* wrist movement around a vertical centerline.

yaw acceleration *Mechanics.* the angular acceleration of an aircraft, a rocket, a ship, or a similar body about a vertical axis.

yaw axis or **yawing axis** *Mechanics.* the vertical axis about which yawing occurs. *Aviation.* specifically, the vertical axis through a flight vehicle or other body, about which the body yaws.

yaw damper *Aviation.* any device or control system designed to reduce yaw in a flight vehicle.

yawhead *Aviation.* on a flight vehicle, a device projecting into the field of flow and designed to detect any change in yaw.

yaw in bore *Ordnance.* the maximum angle between the longitudinal axis of a projectile and the axis of the bore through which it is fired.

yawl *Naval Architecture.* **1.** a two-masted rig, having a small mizzenmast stepped abaft of the rudderpost; as distinguished from a ketch in which the mizzenmast is forward of the rudderpost. **2.** a vessel equipped with such a rig.

yawmeter *Aviation.* a yawhead or other device designed to indicate the degree of yaw of a vehicle in flight. Also, **yaw indicator.**

yaws *Medicine.* an endemic, infectious tropical disease caused by *Treponema pertenue*, spread by direct contact, and characterized by chronic, ulcerating sores anywhere on the body with eventual tissue and bone destruction; it may be effectively treated with penicillin B. Also, THYMIOSIS.

yaw simulator *Control Systems.* a test instrument that is used to study probable aerodynamic behavior in a controlled environment under specific conditions.

y-axis *Mathematics.* in R^3 (3-dimensional Cartesian space), the line that is the intersection of the planes given by $x = 0$ and $z = 0$; usually chosen to be represented by a horizontal line through the origin, positive toward the right and negative to the left. In R^2, it usually is represented as a vertical line, with the upper half positive and the lower half negative. *Cartography.* the line on a map or chart on which distances to the north or south of a reference line (such as the equator) are determined.

yazoo or **yazoo stream** *Hydrology.* a tributary flowing along the base of a natural levee formed by the main stream before joining it at a considerable distance downstream. (From a noted example of this type of tributary, the *Yazoo* River, which joins the Mississippi River at Vicksburg.)

Yb the chemical symbol for ytterbium.

y base *Biochemistry.* a greatly modified guanine residue in tRNA that possesses a distinct fluorescence.

Y-block *Metallurgy.* a test casting in the shape of the letter Y that is used to appraise cast iron and other low-shrinkage alloys.

Y-branch *Mechanical Devices.* a pipe fitting with three openings, two in a line and one at an angle, used to unite three pipes coming from different directions.

Y chromosome *Genetics.* the sex chromosome that is found only in the heterogametic sex of most animal species; in mammals, the Y chromosome is male-determining and carries few structural genes.

Y circulator *Electromagnetism.* a waveguide junction that has three arms, symmetrically arranged, and a ferrite post or wedge located at the center; energy entering through one arm will be transmitted to only one of the other two.

Y connection see Y NETWORK.

y coordinate *Mathematics.* the second of the three Cartesian coordinate functions in R^3.

yd yard; yards.

Y-delta transformation *Electricity.* the result of a mathematical process that makes it possible to replace any Y network with an equivalent delta network. The transformation requires that an equal resistance between equivalent terminals of the Y and delta networks be maintained.

year *Astronomy.* one revolution of the earth around the sun reckoned by any of several time markers, such as vernal equinox passage, relative to the stars, and so on; in common usage, 365.25 days. See also TROPICAL YEAR, LUNAR YEAR, and so on.

yearling *Agriculture.* an animal that is between one and two years old.

yeast *Mycology.* any of various very small, single-celled fungi of the phylum Ascomycota that reproduce by fission or budding and are capable of fermenting carbohydrates into alcohol and carbon dioxide. *Food Technology.* specifically, any of several such organisms of the genus *Saccharomyces*, used to ferment beer, to leaven bread, and in pharmacology as a source of vitamins and proteins.

yeast autolysate *Biotechnology.* the end product of the decomposition of yeast as catalyzed by intracellular enzymes; used in culture media and to flavor foods.

yeast vectors *Genetics.* yeast cells that contain plasmids carrying genes new to the cell.

YEG *Aviation.* the airport code for Edmonton (Alberta) International.

yellow *Optics.* the color sensation that corresponds to a wavelength in the range of 577 to 597 nanometers, lying between red and green in the color spectrum.

yellow-bellied sapsucker *Vertebrate Zoology.* a type of woodpecker, *Sphyrapicus varius*, that feeds on the sap from trees, having black and white plumage with a yellowish belly and a red throat and forehead; found in eastern North America.

yellow-billed cuckoo *Vertebrate Zoology.* a cuckoo, *Coccyzus americanus*, noted for constructing its own nest and rearing its own young, unlike most cuckoos; characterized by a yellow bill; found in North America.

yellow birch *Botany.* a birch, *Betula allehaniensis*, having yellowish or silvery gray bark and yielding a strong, light-brown wood used in construction and furniture; found in North America.

yellow-breasted chat *Vertebrate Zoology.* a large warbler, *Icteria virens*, having an olive-green back and yellow breast; noted for imitating the songs of other species; found in North America.

yellow cake or **yellowcake** *Mining Engineering.* the final precipitate produced during the milling of uranium ores. *Inorganic Chemistry.* a popular name for uranium dioxide, UO_2. See URANIUM DIOXIDE.

yellow coal see TASMANITE.

yellow cross *Toxicology.* see MUSTARD GAS.

yellow fat cell *Histology.* an adipocyte containing a single large, yellowish fat droplet and little cytoplasm.

yellow fever *Medicine.* an acute arborviral infection transmitted by certain mosquitoes; historically a major epidemic threat in the Caribbean and still found in West Africa and tropical America; characterized by headache, fever, jaundice, vomiting, and bleeding.

yellow fever virus *Virology.* a virus of the genus *Flavivirus*, family Flaviviridae, endemic to Africa and South America; transmitted primarily by certain mosquitoes and causing fatal hepatitis in humans and other primates.

yellowfin tuna *Vertebrate Zoology.* an important food fish, *Thunnus albacares*, often growing to 6 feet long and weighing about 300 pounds; found in the Atlantic and Pacific Oceans. Also, **yellowfin.**

yellow-green alga *Botany.* a freshwater and terrestrial alga, including unicellular, colonial, filamentous, and siphonaceous forms; characterized by disklike yellow-green plastids, and known for accumulating oil and leucosin rather than starch.

yellowjacket *Invertebrate Zoology.* any of several small, black paper wasps of the family Vespidae, having bright yellow markings.

yellow lead oxide see LITHARGE.

yellow mercurous iodide see MERCUROUS IODIDE.

yellow mud *Geology.* a type of mud characterized by a yellow color resulting from the presence of iron compounds.

yellow precipitate see MERCURIC OXIDE, def. 2.

yellow print *Graphic Arts.* a positive print made from a blue color-separation negative.

yellow prussiate of soda see SODIUM FERROCYANIDE.

yellow pyoktanin see AURAMINE.

Yellow River *Geography.* a river that flows about 2900 miles eastward across northern China to the Yellow Sea. Also, HUANG HE, HWANG HO.

yellow salt *Inorganic Chemistry.* a popular name for uranyl nitrate, $UO_2(NO_3)_2 \cdot 6H_2O$. See URANIUM TRIOXIDE.

Yellow Sea *Geography.* an inlet of the East China Sea between Korea and China.

yellow snow *Hydrology.* snow that has a yellowish tint, as from the presence of certain yellow-green algae.

yellowtail *Vertebrate Zoology.* any one of various fishes with yellow tails, especially *Seriola lalandei*, a California game fish, or *Ocyurus chrysurus*, a small West Indian snapper.

Yenisei *Geography.* a river that flows about 2500 miles north through Siberia to the Arctic Ocean.

Yerkes, Robert Mearns 1876–1956, American psychologist and psychobiologist; a pioneer in the study of the great apes.

Yerkes-Dodson law or **principle** *Psychology.* the principle stating that performance of a task is enhanced by moderate levels of arousal and reduced by either very high or very low levels, depicted as an inverted U-shaped function.

Yersin, Alexandre 1863–1943, Swiss bacteriologist; discovered (independently) the agent of bubonic plague and also developed a serum against it.

Yersinia *Bacteriology.* a genus of Gram-negative, facultatively anaerobic, rod-shaped to ovoid bacteria of the family Enterobacteriaceae; species include *Y. pestis,* the causative agent of bubonic plague.

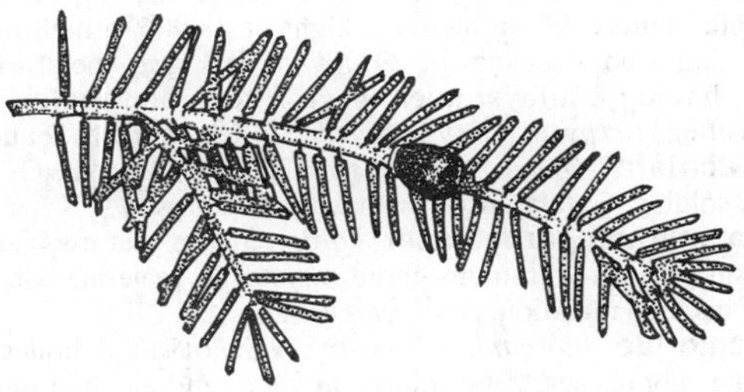

yew

yew *Botany.* **1.** a cone-bearing evergreen tree of the genus *Taxus,* related to the pines, having dark green leaves and red berries. **2.** the tough, elastic wood of these trees.

YF-23 *Aviation.* a U.S. Air Force delta-winged, single-seat, twin-jet tactical fighter, adaptable for use as a long-range interceptor or strategic reconnaisssance plane.

YF-23 tactical fighter

Y factor *Petroleum Engineering.* an empirical relationship of oil-reservoir bubble-point pressure and formation volume, used to smooth out the curve representing solution-gas/oil-ratio data.

Y gun *Ordnance.* a Y-shaped antisubmarine weapon with two barrels, designed to throw depth charges on either side of a surface vessel.

YHZ *Aviation.* the airport code for Halifax (Nova Scotia) International.

yield *Agronomy.* the quantity of products raised or grown at a given time or in a given area, such as the amount of grain harvested per acre. *Ordnance.* see NUCLEAR YIELD. *Engineering.* the product of a reaction or process. *Mechanics.* the plastic deformation that occurs when a solid is strained beyond its elastic limit.

yielding *Mechanics.* beginning to deform plastically.

yielding arches *Mining Engineering.* steel arches placed in underground openings to support any loads that may result from ground movement or faulted or fractured rocks during ground removal.

yielding floor *Mining Engineering.* a soft floor that, when subjected to heavy pressures due to packs and pillars, heaves and flows into open spaces.

yielding prop *Mining Engineering.* a steel support of adjustable length that features a sliding or flexible joint; used when the roof pressure exceeds a certain value.

yield-pillar system *Mining Engineering.* a means of maintaining natural roof strength by relieving pressure in working areas and transferring load to abutments in areas that are clear of workings and roadways.

yield point *Mechanics.* **1.** the stress point at which strain continues to increase without any further application of stress. **2.** see YIELD STRESS. *Engineering.* the load at which stress is no longer proportional to strain; i.e., the stress-strain "curve" ceases to be a straight line.

yield strength *Mechanics.* the stress level at which a material attains a specified strain, for example, a 0.2% offset.

yield stress *Mechanics.* **1.** the lowest stress that will lead to yield. Also, PROOF STRESS; YIELD POINT. **2.** see FLOW STRESS.

yield surface *Mechanics.* a surface in a yielding solid on which the yield is localized; certain solids will yield as a series of slipping slabs rather than with uniform strain.

yield temperature *Engineering.* the temperature at which a fusible material in a plug device melts and dislodges the plug from its hold.

YIG or **yig** *Electronics.* a synthetic yttrium-iron garnet, used in filters and amplifiers.

YIG device *Electronics.* any device that uses an yttrium-iron garnet crystal along with a variable magnetic field to attain wideband tuning in microwave circuits.

YIG filter *Electronics.* an electromagnetically tuned filter consisting of a YIG crystal set in a magnetic field produced by a permanent magnet and a solenoid. Variable current through the solenoid is used to tune the filter; the magnet minimizes the need for current by tuning the filter to the center of the band.

YIG-tuned oscillator *Electronics.* a microwave oscillator that achieves precisely controlled wideband tuning by varying the current through a tuning solenoid acting on a YIG filter in the tunnel-diode oscillator circuit. Also, **YIG-tuned tunnel-diode oscillator.**

YIG-tuned parametric amplifier *Electronics.* a parametric amplifier that uses a continuously varying direct current through the solenoid of a YIG filter for tuning.

Y junction *Electromagnetism.* a three-arm waveguide junction whose longitudinal axes all lie in a common plane and form a Y.

ylang-ylang oil see ILANG-ILANG OIL.

ylem *Astrophysics.* the initial substance (primordial matter) of the universe from which all subsequent matter (formation of chemical elements) is said to be derived.

Y level *Engineering.* a level with Y-shaped rests, used to support a surveyor's telescope. Also, WYE-LEVEL.

ylide *Organic Chemistry.* a compound containing two adjacent atoms bearing formal positive and negative charges: a carbanion and a heteroatom with the positive charge; both atoms have an octet of electrons.

Y linkage *Genetics.* a form of sex linkage in which the loci of certain genes are found only on the Y chromosome and not on the X chromosome; such genes are therefore expressed only in the heterogametic sex of the species.

ylium ion see ENIUM ION.

Y-5 molars *Anthropology.* a cusp pattern of five cusps separated by grooves that seem to form the letter Y; present on the molars of dryopithecines and recognized as a pattern of all hominoids, including humans.

YMQ *Aviation.* the airport code for Montreal, Quebec.

Y network *Electricity.* a Y-shaped or star network that has three branches meeting at a common point. The configurations of the Y-connection network in power (transformer) circuits and the T-connection network in filter circuits are physically the same but schematically different. Also, Y CONNECTION.

Ynezian *Geology.* a North American geologic stage of the lower Paleocene epoch, occurring after the Upper Cretaceous period and before the Bulitian.

YOB year of birth.

yogurt or **yoghurt** *Food Technology.* a fermented milk product made by inoculating whole, low-fat, or nonfat milk with a lactic acid culture containing *Lactobacillus bulgaricus* and *Streptococcus thermophilus.*

yoke *Agriculture.* originally, a frame for harnessing a pair of oxen or other draft animals, consisting of a crosspiece linking two bow-shaped headpieces. *Mechanical Devices.* **1.** a similar frame or clamp that holds machine parts or devices in place. **2.** a two-branched water pipe that joins two separate water streams, such as hot and cold. *Mechanical Engineering.* a slotted crosshead used in place of a connecting rod in some steam engines. *Naval Architecture.* a crosspiece on a boat rudder. *Electromagnetism.* a piece of ferromagnetic material that connects two or more magnetic cores. *Electronics.* **1.** the system of coils used for magnetic deflection of the electron beam in a cathode-ray tube. **2.** the ferromagnetic ring or cylinder that holds the pole pieces of a dynamo-type generator, acting as part of the magnetic circuit. *Computer Technology.* a group of rigidly fastened heads that move together to read or write two or more tracks on a storage medium.

yoked basin see ZEUGOGEOSYNCLINE.

yolk *Biochemistry.* the yellow spherical mass of nutrient material stored in the ovum. (From the Old English word for "yellow.")

yolk glands see VITELLARIUM.

yolk sac *Developmental Biology.* the extraembryonic membrane that connects with the midgut; in humans, expanding at the end of the fourth week of development into the umbilical vesicle connected to the body of the embryo by the yolk stalk.

yolk stalk see VITELLINE DUCT.

Y organ *Invertebrate Zoology.* in crustaceans, one of a pair of anterior glands that regulate molting.

Yorkian *Geology.* a European geologic stage of the Lower Upper Carboniferous period, after the Lamarkian and before the Staffordian.

Yorkshire [yôrk´shər; yôrk´shēr´] *Agriculture.* a breed of hog that has white hair, a long narrow body, and erect ears.

young *Geology.* **1.** describing a geologic or topographic structure in the first stage of the cycle of erosion. Also, **youthful.** **2.** to face toward or in the same direction as a younger surface or formation. Also, FACE.

Young, John Wesley 1879–1932, American mathematician; the co-founder (with Veblen) of projective geometry.

Young, Thomas 1773–1829, English physicist and physician; revived the wave theory of light; formulated the principle of interference and the Young-Helmholtz theorem; described astigmatism.

Young construction *Optics.* a graphical technique that traces a ray of light passing through a spherical boundary between two media with different refractive indices.

Young diagram *Mathematics.* given a partition of a positive integer n with the summands n_i listed in decreasing order, a Young diagram for the partition consists of n boxes arranged in rows such that the ith row has n_i boxes in it. Also, **Young shape, Young frame.** If the boxes in the diagram are filled in with quantities such as the integers from 1 to n, then the configuration is called a **Young tableau.**

Young-Helmholtz theory *Physiology.* the supposition that the phenomenon of color vision relies on three different groups of retinal fibers, responding to red, green, and violet.

young ice *Hydrology.* newly formed, floating, flat sea ice in the stage following nilas. Also, FRESH ICE.

Younginiformes *Paleontology.* a suborder of small, primitive diapsid reptiles in the extinct order Eosuchia, known mainly from **Youngina,** which lived in southern Africa in the Late Permian and is a possible ancestor of all lizards and rhynchocephalians; Late Permian and Triassic.

Young's modulus *Mechanics.* the ratio between tensile or compressive stress and elongation of a solid stressed in one direction.

Young's two-slit interference *Optics.* a method used by Thomas Young to disprove Newton's corpuscular theory of light; a beam of light from a single source was split into two beams by closely spaced parallel slits and the resulting interference was observed on a screen.

youth *Geology.* **1.** the first stage in the cycle of erosion in the topographic development of a landscape or area; characterized by the dominance of the original surface or structural features. Also, TOPOGRAPHIC YOUTH. **2.** the stage characterized by an ungraded profile of equilibrium in the development of a shore or coast. *Hydrology.* the stage at which the erosion of the stream channel has just begun in the development of a stream; characterized by increasing vigor and by the ability to carry a greater load than the stream is already carrying.

YOW *Aviation.* the airport code for Ottawa, Ontario.

y-parallax *Cartography.* in aerial photography, the difference between the perpendicular distances of the two images of a point, on a pair of photographs, and the vertical plane containing the air base.

y parameter *Electronics.* a device or network parameter expressed as an admittance.

Yponomeutidae *Invertebrate Zoology.* a family of small moths, lepidopteran insects in the superfamily Tineoidea, including ermine moths; some larvae are fruit pests.

Ypresian *Geology.* a European geologic stage of the Lowermost Eocene epoch, occurring after the Thanetian of the Paleocene epoch and before the Cuisian. Also, LONDONIAN.

yrast state *Nuclear Physics.* a state in which the energy level of a nucleus is less than the other energy levels in the same spin.

Y-suppressed lethal gene *Genetics.* a recessive lethal gene identified in *Drosophila* that causes death in XO individuals but not in normal XY males.

ytterbium [i tur´bē əm] *Chemistry.* a rare-earth element having the symbol Yb, the atomic number 70, an atomic weight of 173.04, a melting point of 824°C, and a boiling point of 1427°C; a lustrous, malleable metal used in lasers and as a dopant for garnet gemstones. (Named after *Ytterby,* a quarry in Sweden in which it was found.)

ytterbium oxide *Inorganic Chemistry.* Yb_2O_3, colorless crystals; insoluble in water; used in alloys, ceramics, lighting, glasses, and as a catalyst. Also, **ytterbia.**

yttrium [it´trē əm] *Chemistry.* a metallic element having the symbol Y, the atomic number 39, an atomic weight of 88.906, a melting point of 1500°C, and a boiling point of 2927°C; a dark-gray metal used in nuclear technology, alloys, microwave filters, and semiconductors. (Named after *Ytterby,* a quarry in Sweden in which it was found.)

yttrium acetate *Organic Chemistry.* $Y(C_2H_3O_2)_3 \cdot 9H_2O$, colorless crystals; soluble in water; used in analytical chemistry.

yttrium-aluminum-garnet laser *Optics.* a laser that uses neodymium ions in an yttrium-aluminum-garnet crystal to generate a continuous beam of infrared radiation. Also, YAG LASER.

yttrium chloride *Inorganic Chemistry.* $YCl_3 \cdot 6H_2O$, deliquescent, reddish-white rhombic crystals; soluble in water and alcohol; decomposes at 100°C; used in analytical chemistry.

yttrium oxide *Inorganic Chemistry.* Y_2O_3, colorless to yellowish cubes or powder; insoluble in water and soluble in alcohol; melts at 2410°C; used in TV phosphors, microwave filters, and as a stabilizer for refractories. Also, **yttria.**

yttrium sulfate *Inorganic Chemistry.* $Y_2(SO_4)_3$, white powder; soluble in water; decomposes at 1000°C; used as a reagent.

yttrocrasite-(Y) *Mineralogy.* $(Y,Th,Ca,U)(Ti,Fe^{+3})_2(O,OH)_6$, a very rare, radioactive, black orthorhombic mineral occurring as prismatic or thin tabular crystals, having a specific gravity of 4.80 and a hardness of 5.5 to 6 on the Mohs scale; found in granite pegmatites.

yttrotantalite-(Y) *Mineralogy.* $(Y,U,Fe^{+2})(Ta,Nb)O_4$, a moderately radioactive, black or brown orthorhombic mineral occurring as prismatic and tabular crystals and in massive form, having a specific gravity of 5.7 and a hardness of 5 to 5.5 on the Mohs scale; dimorphous with formanite-(Y); found in granite pegmatites.

Yucatan Current *Oceanography.* a rapid current that flows northward from Honduras through the Yucatan Channel between Mexico and Cuba, along the eastern edge of the Yucatan Shelf, and into the Gulf of Mexico; often considered the starting point of the Gulf Stream.

yucca *Botany.* any plant of the genus *Yucca* of the agave family, characterized by narrow, stiff, sword-shaped leaves and an upright cluster of waxy, bell-shaped white flowers; found in warmer regions of America.

yugawaralite *Mineralogy.* $CaAl_2Si_6O_{16} \cdot 4H_2O$, a colorless or white monoclinic mineral of the zeolite group occurring as flat crystals, having a specific gravity of 2.20 to 2.23 and a hardness of 4.5 on the Mohs scale; found as veins in andesitic volcanics altered by hot spring waters.

Yukawa, Hideki 1907–1981, Japanese physicist; awarded the Nobel Prize for the mathematical prediction of the meson.

Yukawa force *Nuclear Physics.* the short-range force between nucleons that holds them together and is postulated to arise from the exchange between a proton and a neutron of a finite mass particle.

Yukawa meson *Particle Physics.* a meson particle with finite rest mass that was postulated to account for the strong forces between nucleons.

Yukawa potential *Nuclear Physics.* a formula that measures the strength and range of the strong force that is believed to arise when a finite mass particle is exchanged.

Yukon *Geography.* a river formed by the confluence of the Lewes and Pelly rivers in northwestern Canada, flowing 1979 miles across Alaska to the Bering Sea.

YVR *Aviation.* the airport code for Vancouver International.

YWG *Aviation.* the airport code for Winnepeg, Manitoba.

YYC *Aviation.* the airport code for Calgary, Alberta.

YYZ *Aviation.* the airport code for Pearson International, Toronto.

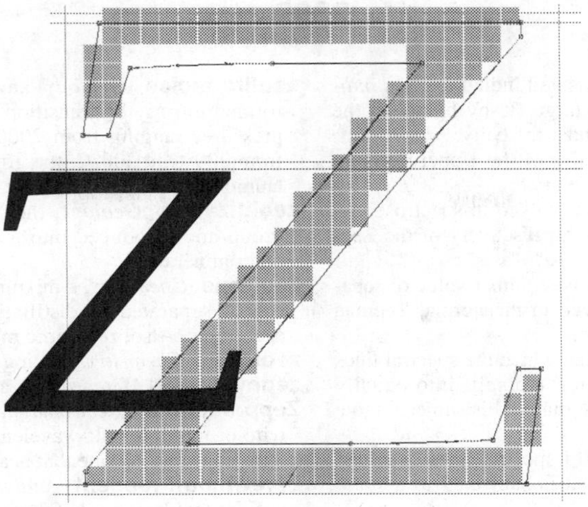

z *Mathematics.* **1.** an unknown quantity or variable, especially the third one in a statement in which x and y are the first two. **2.** the z-axis.

z. zero.

z or **Z** zone.

Z zenith distance.

Z *Chemistry.* the symbol for atomic number. *Electricity.* the symbol for impedance. *Mineralogy.* the symbol for the number of formula units per unit cell of a mineral.

Zaire River see CONGO.

Zalambdalestidae *Paleontology.* a family of small, primitive mammals of Mongolia; extant in the Cretaceous.

Zambesi or **Zambezi** *Geography.* a river flowing about 1700 miles south and east from northwest Zambia into the Mozambique Channel.

Zanclidae *Vertebrate Zoology.* the moorish idols, a family of perciform fishes of the Indo-Pacific region; in some classification systems, this group is ranked as a subfamily of Acanthuridae and called **Zanclinae.**

Zannichelliaceae *Botany.* a cosmopolitan family of monocotyledonous, submerged, rhizomatous herbs of the order Najadales, characterized by a single axial vascular bundle with vessel-less xylem surrounded by phloem; found in fresh, alkaline, or brackish water.

Zanzibar *Geography.* an island off the eastern coast of Africa (area: 640 square miles), belonging to Tanzania.

zap *Computer Programming.* **1.** a command in some database management and spreadsheet packages that erases all the data in a file or a spreadsheet at once. **2.** to accidentally delete a file, clear a display, and so on. **3.** to quickly correct a problem.

Zapodidae *Vertebrate Zoology.* a family of rodents in the suborder Myomorpha, characterized by very long hindlegs and tail; found in the Northern Hemisphere.

zaratite *Mineralogy.* $Ni_3(CO_3)(OH)_4 \cdot 4H_2O$, an emerald-green, cubic mineral with vitreous luster, occurring as mamillary incrustations or as compact masses, having a specific gravity of 2.57 to 2.69 and a hardness of 3.5 on the Mohs scale; found as a secondary mineral in serpentine and basic igneous rocks.

Zaunkonig torpedo *Ordnance.* a German acoustic torpedo used as an antishipping weapon during World War II; officially designated **T-5.** Also, **GNAT** (German Naval Acoustic Torpedo).

z-axis *Mathematics.* in R^3 (three-dimensional Cartesian space), the line that is the intersection of the planes given by $x = 0$ and $y = 0$; usually chosen to form a right-handed coordinate system in the order x, y, z. It is generally a vertical line, positive upward, and negative downward.

Z-bar *Building Engineering.* a Z-shaped steel bar used in construction.

Z^0 boson *Particle Physics.* a particle that, along with the W boson, mediates the neutral-current weak force in particle interactions. Also, Z PARTICLE.

Z Camelopardalis star *Astronomy.* a type of dwarf nova variable star that erupts occasionally and whose decline in brightness may contain protracted and unpredictable pauses at a constant brightness.

Z chromosome *Genetics.* a sex chromosome in certain organisms, such as chickens and moths, in which the female is the heterogametic sex; males are ZZ and females are WZ.

z coordinate *Mathematics.* the third of the three Cartesian coordinate functions in R^3.

ZD zero defects; zenith distance.

Z DNA or **zDNA** see Z FORM DNA.

Zea *Botany.* a genus of grasses of the family Gramineae, including all wild and crop species of corn.

Zealand *Geography.* the largest island of Denmark (area: 2709 square miles), on which Copenhagen is located.

zearalenone *Biochemistry.* a fungal toxin that contaminates damp grain-based foodstuffs, causing disease in sows, poultry, and cattle.

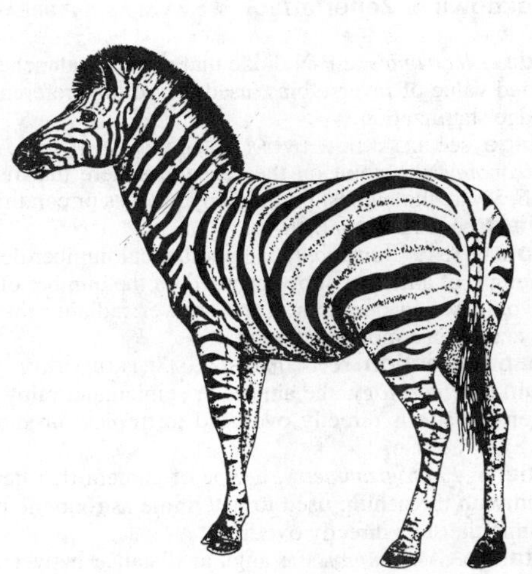

zebra

zebra *Vertebrate Zoology.* any of three species of black-and-white striped, fleet-footed equine African mammals of the genus *Equus* in the family Equidae.

zebra plant *Botany.* any of various plants, such as those of the families Acanthaceae and Marantaceae, having foliage characterized by contrasting stripes.

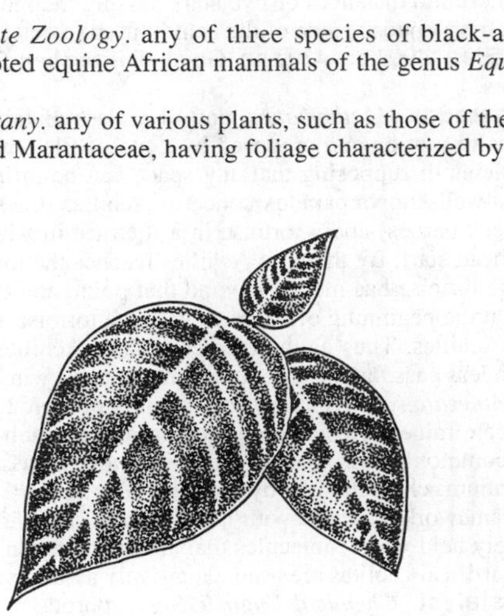

zebra plant (Acanthaceae)

zebu *Vertebrate Zoology.* a variety of domesticated Indian ox, *Bos indicus*, characterized by long drooping ears, a large fleshy hump on the shoulders, and loose skin forming a dewlap under the belly.

Zechstein *Geology.* a European geologic series of the Upper Permian period, occurring after the Rotliegende.

Zeeman, Pieter 1865–1943, Dutch physicist; worked in spectroscopy; shared the Nobel Prize with Hendrik Lorentz for discovery of the Zeeman effect.

Zeeman displacement *Spectroscopy.* the wave number value of separation between adjacent spectral lines observed in the normal Zeeman effect when there is a unit magnetic field present.

Zeeman effect *Spectroscopy.* a phenomenon in which the spectral lines associated with atomic or molecular radiation are split into equally spaced levels (with small separation) when the material is subjected to a static magnetic field.

Zeeman split *Spectroscopy.* a fine splitting of spectral lines observed in the emission spectra of a substance when it is subjected to a static magnetic field.

Zeidae *Vertebrate Zoology.* the dories, a family of marine fish of the order Zeiformes, characterized by a compressed body and a deep mouth; found in temperate zones of the Atlantic and Pacific Oceans.

Zeiformes *Vertebrate Zoology.* an order of bony fish characterized by highly compressed deep bodies and an extremely protractile mouth; includes the dories, oreos, boarfish, and other families; usually found in small schools in tropical and temperate seas. Also, **Zeoidea, Zeomorphi.**

Zemorrian *Geology.* a North American geologic stage of the Oligocene and Miocene epochs, after the Refugian and before the Saucesian.

Zener breakdown or **Zener effect** see AVALANCHE BREAKDOWN, def. 2.

Zener diode *Electronics.* a P-N diode that shows avalanche breakdown at a specified value of reverse bias; used for voltage reference purposes or for voltage stabilization.

Zener voltage see BREAKDOWN VOLTAGE, def. 1.

zenith *Astronomy.* the point on the celestial sphere that lies vertically overhead. *Science.* the highest point in any process or condition.

zenithal chart see AZIMUTHAL CHART.

zenithal hourly rate *Astronomy.* a hypothetical number describing the relative strength of a meteor shower, based on the number of meteors an observer would see in one hour if the shower radiant remained in the observer's zenith on a moonless night.

zenithal map projection see AZIMUTHAL MAP PROJECTION.

zenithal rain *Meteorology.* the annual or semiannual rainy season that occurs when the sun is directly overhead in tropical or subtropical regions.

zenith camera *Photogrammetry.* a type of camera that has its optical axis aligned with the zenith; used to determine astronomic positions by photographing the stars directly overhead.

zenith distance *Astronomy.* the angular distance between the zenith and a celestial object. Also, **zenith angle.**

zenith telescope *Optics.* a telescope that measures the difference between the zenith distances of two stars passing the meridian at the same time but from opposite side of the zenith; used to determine latitude.

Zeno of Elea 490? BC–430? BC, Greek philosopher; formulated Zeno's paradoxes.

Zeno's paradox *Mathematics.* any of a class of fallacious paradoxes dealing with motion and proposed by Zeno of Elea, which demonstrate the problems in supposing that any space can be infinitely subdivided. The most well-known paradox concerns Achilles (the swiftest of the ancient Greek heroes) and a tortoise in a footrace in which the tortoise is given a head start. By the time Achilles reaches the tortoise's initial position, the tortoise has moved beyond that point, and the situation is the same as at the beginning of the race, with the tortoise still having a head start on Achilles. Thus as the race continues, Achilles can never catch up, much less pass the tortoise, and so the tortoise wins the race.

zeolite *Mineralogy.* a group, or any member of a group, of hydrous aluminosilicate minerals containing sodium and calcium as major cations or, less commonly, barium, beryllium, lithium, potassium, magnesium, and strontium; characterized by the ratio (Al+Si):O=1:2, an open tetrahedral framework structure with large cations capable of ion exchange, and loosely held water molecules that allow reversible dehydration; natural and artifical zeolites are used extensively as water softeners.

zeolite catalyst *Chemical Engineering.* a porous material, natural or synthetic, consisting of hydrated silicates of aluminum and sodium or calcium; used as a catalyst in petroleum cracking.

zeolite facies *Geology.* any of a set of metamorphic facies that are formed during the transition from diagenesis to metamorphism under pressures ranging from 2000 to 3000 bars and temperatures ranging from 200°C to 300°C; the group includes analcime, heulandite, stilbite, laumontite, and wairakite.

zeolitization *Geology.* the process by which minerals of the zeolite group are introduced into a sediment or rock, or alter or replace the other minerals.

zeotrope *Chemistry.* a mixture of liquids whose individual components can be separated by distillation even though the mixture itself is not azeotropic. Also, **zeotropic mixture.**

zeotropic *Chemistry.* relating to or being a zeotrope.

zephyr [zef´ər] *Meteorology.* a soft, gentle breeze.

Zepp antenna *Electromagnetism.* a horizontal antenna consisting of a wire of multiple half-wavelengths fed at one end by one wire of a two-wire transmission line that is a multiple of quarter-wavelengths long.

Zerewitinoff reagent *Analytical Chemistry.* a solution of methylmagnesium iodide and *n*-butyl ether that reacts with water and oxygen; used to determine the presence of water, alcohols, and amines in inert solvents.

Zermelo, Ernst 1871–1953, German mathematician; formulated basic axioms of set theory.

Zernike, Frits 1888–1966, Dutch physicist; awarded the Nobel Prize for development of the phase-contrast microscope.

zero *Mathematics.* **1.** the additive identity of the real or complex numbers; denoted 0. **2.** more generally, the additive identity of a ring. Also, **zero element. 3.** let f be a function defined in a neighborhood of a point x_0 and suppose that

$$\lim_{x \to x_0} |f(x)/(x - x_0)s| = c,$$

where s and c are nonzero real numbers. If $s > 0$, then x_0 is said to be a **zero of order** ***s***. (If $s < 0$, x_0 is a pole of order s.) Also, **zero of the function** ***f***. **4.** another term for a root of a function. *Engineering.* the point at which a graduation begins, as on a thermometer scale.

Zero *Aviation.* a single-engine Japanese fighter plane of World War II.

zero-access storage *Computer Technology.* a storage device in which the latency is negligible.

zero-address instruction see ADDRESSLESS INSTRUCTION (FORMAT).

zero adjuster *Engineering.* a device used to calibrate an instrument to record zero when the measured quantity is zero.

zero-age main sequence *Astronomy.* a line on the Hertzsprung-Russell diagram showing the location of main sequence stars that have not yet begun to evolve; often abbreviated **ZAMS.**

zero-base *Industrial Engineering.* describing a calculation or forecast that is made solely on the basis of presently existing conditions, without reference to earlier conditions or practices. Thus, **zero-base budgeting, zero-base planning,** and so on.

zero beat *Electricity.* describing the result from combining two signals having exactly the same frequency.

zero-beat reception see HOMODYNE RECEPTION.

zero-bevel gear *Mechanical Devices.* a bevel gear whose curved teeth have a zero-degree spiral angle.

zero bias *Electronics.* a state in which the control grid and cathode of an electron tube are at the same DC voltage. Thus, **zero-bias tube.**

zero branch *Spectroscopy.* a spectral band corresponding to a Fortrat parabola whose vertex lies very close to the wave number axis.

zero charge potential *Physical Chemistry.* the potential of an electrode where there is no excess electrical charge on the electrode surface. Also, POTENTIAL OR POINT OF ZERO CHARGE.

zero compression *Computer Programming.* any of a number of techniques used to avoid storing insignificant leading zeros during data processing.

zero condition *Computer Programming.* the state representing binary 0 in a storage bit. Also, ZERO STATE.

zero curtain *Hydrology.* the layer of ground above the permafrost, characterized by a nearly constant temperature of 0°C for a considerable time during the freezing and thawing of the overlying active layer.

zero defects *Industrial Engineering.* a quality-control philosophy based on the ideal goal of entirely defect-free output.

zero deflection *Ordnance.* the adjustment of a gunsight so that it is parallel to the axis of the gun bore.

zero divisor *Mathematics.* a nonzero element a of a ring R for which there exists a nonzero element b of R such that $ab = 0$.

zero error *Electronics.* **1.** a lack of error, or an inconsequentially low error, in instruments or measurements. **2.** the inherent delay in the transmitter and receiver circuits in a radar system.

zero fill *Computer Programming.* the process of filling unused storage locations with the character representation of zero or with binary 0 bits.

zero flag *Computer Programming.* an indicator that is set to binary 1 if the register being referenced has all bits set to zero or an operation results in a zero value.

zerogel *Chemistry.* a gel that has dried to form a solid.

zero geodesic see NULL GEODESIC.

zero gravity see WEIGHTLESSNESS.

zero-gravity switch *Electricity.* a switch that operates when apparent weightlessness or zero gravity is reached.

zero height of burst *Ordnance.* a term applied to a group of shots fired under the same conditions in which there are an equal number of air bursts and ground bursts.

zero in *Ordnance.* **1.** to adjust the sights of a weapon using calibrated firing results. **2.** to synchronize one device to another.

zero layer *Oceanography.* a reference level in the ocean at which there is practically no horizontal motion.

zero-length *Space Technology.* designed to launch a rocket or other vehicle directly into the air, without providing guidance.

zero-length launcher *Ordnance.* a short, mobile launcher that holds a weapon in position for launching but does not provide guidance; it is normally used with a rocket-powered weapon with sufficient thrust to launch it directly into the air. Thus, **zero-length rocket.**

zero-length launching *Ordnance.* a technique in which the first movement of a missile or aircraft removes it from the launcher.

zero level *Acoustical Engineering.* a reference level against which sound intensity measurements are compared; for the decibel scale, an audio reference level of 0.006 watt is equivalent to zero decibels.

zero-level address see IMMEDIATE ADDRESS.

zero-level addressing see FIRST-LEVEL ADDRESSING.

zero-lift angle *Aviation.* the angle of attack of a wing that produces a zero lift.

zero-lift chord *Aviation.* a reference line between the trailing edge and the relative wind of an airfoil at zero-lift angle.

zero offset *Robotics.* the elimination of less-than-significant digits on either side of a decimal point.

zero-order hold *Control Systems.* a device that takes sampled output and holds it at that value until the next output sample is taken.

zero-order reaction *Physical Chemistry.* a chemical reaction in which the rate of reaction does not depend on the concentration of the individual reactants, e.g., a photochemical reaction in which light determines the rate of reaction.

zero output *Electronics.* **1.** the absence of output signal or power, sometimes disregarding noise output. **2.** the normal condition of no signal pulse at a particular output terminal in a flip-flop. **3.** the output obtained when a zeroed magnetic cell is read.

zero phase *Electricity.* in an AC circuit, a condition in which there is no phase difference between the current and voltage, or in which a similar relationship exists between two other quantities.

zero-phase sequence relay *Electricity.* a relay that is actuated during zero phase.

zero point *Ordnance.* the location of the center of burst of a nuclear weapon at the instant of detonation; it may be above, on, or below the surface of land or water.

zero-point energy *Physics.* the nonthermal energy of a system, most evident as temperature approaches absolute zero.

zero-point entropy *Physics.* the limiting value of the entropy of a system as its temperature approaches absolute zero.

zero-point pressure *Physics.* the limiting pressure of a system as its temperature approaches absolute zero.

zero-point vibration *Physics.* the nonthermal vibrational motion in a crystal lattice, due to quantum effects.

zero potential *Electricity.* a neutral charge without potential energy. The ground or earth, which is used as a reference, is considered to have a zero potential; regardless of the number of electrons added to or removed from the ground, the charge is negligible and the potential remains zero.

zero-power reactor *Nucleonics.* an experimental nuclear reactor having very low neutron flux density and power level, thus requiring no forced cooling; the insignificant buildup of fission products allows the fuel elements to be handled after use for reactor physics studies.

zero state see ZERO CONDITION.

zero subcarrier chromaticity *Telecommunications.* in color television transmission, the chromaticity that is intended to be shown when the subcarrier amplitude is 0.

zero-sum game *Mathematics.* a finite game in which the sum of the payoffs of the players is zero or constant in every situation. If the game has n players and the sum of the payoffs of the players is a constant c in every situation, then the game can be considered as a game of $n + 1$ players in which the payoff sum is always zero.

zero suppression *Computer Programming.* the act of replacing leading zeros with blanks during a printing operation.

zeroth law of thermodynamics *Thermodynamics.* a law stating that when two bodies are both equal in temperature to a third body, they in turn are equal in temperature to each other; this law provides the necessary basis for temperature measurement.

zero time reference *Electronics.* the time reference of the schedule of events during one cycle of radar operation.

zero twist *Ordnance.* rifling with no twist; it is sometimes used at the origin of rifling.

zero vector *Mathematics.* **1.** the element of a vector space having all components equal to the zero element of the scalar field. **2.** any element of a vector field having length or magnitude zero.

zeta the sixth letter of the Greek alphabet, written Z, ζ.

Zeta Aurigae star *Astronomy.* a variable star consisting of a binary star in which a supergiant primary of spectral class K and a main-sequence secondary eclipse each other.

zeta function *Mathematics.* **1.** see RIEMANN ZETA FUNCTION. **2.** an odd, doubly quasi-periodic, meromorphic function $\zeta(z)$ that satisfies the ordinary differential equation $d\zeta/dz = -p(z)$, where $p(z)$ is the Weierstrassian elliptic function. Also, WEIERSTRASS ZETA FUNCTION.

Zeta Geminorum stars *Astronomy.* a specialized group of Cepheid variable stars with a quasi-bell-shaped light curve.

zeta potential *Physical Chemistry.* a potential gradient that arises across the interface between the boundary liquid layer in contact with a solid and the movable diffuse layer in the body of the liquid. Also, ELECTROKINETIC POTENTIAL.

zeugite *Cell Biology.* a cellular structure within which the dikaryotic life-cycle stage of certain fungi is terminated by the fusion of the nuclei.

zeugmatography *Medicine.* a technique of imaging that combines nuclear magnetic spectroscopy with methods of focusing and scanning radio waves to produce cross-sectional images of the human body which help distinguish tumors from normal tissue.

zeugogeosyncline *Geology.* a parageosyncline with an adjacent uplifted area that receives clastic sediments. Also, YOKED BASIN.

zeunerite *Mineralogy.* $Cu^{+2}(UO_2)_2(AsO_4)_2 \cdot 10\text{–}16H_2O$, a green, radioactive, tetragonal secondary mineral of the autunite group, occurring as tabular crystals, having a specific gravity of 3.39 and a hardness of 2.5 on the Mohs scale; found in the oxidized zone of deposits containing uranium and arsenic-bearing minerals.

z.f. zero frequency.

Z form DNA *Genetics.* the form of DNA in which the helix is pitched in a left-handed rather than a right-handed direction (as it is with the A and B forms). Also, Z DNA, zDNA.

ZI zone of interior.

Ziegler, Karl 1898–1963, German chemist; with Giulio Natta, received the Nobel Prize for improvements in catalysis of polymerization.

Ziegler process *Chemical Engineering.* a catalytic process for the linear polymerization of ethylene and the stereospecific polymerization of propylene that uses low pressure to create synthetic rubber and high-density polymers.

Ziehl-Neelsen stain *Microbiology.* a method of staining bacteria that are resistant to other staining methods, such as those of the genus *Mycobacterium*; the primary chemical ingredient of the stain is carbolfuchsin. Also, **Ziehl-Neelsen carbolfuchsin.**

Ziemann's dots *Microbiology.* tiny dots that are sometimes found in malaria-infected erythrocytes upon treatment with Romanowsky stains.

Zieve syndrome *Toxicology.* poisoning due to excessive ingestion of ethyl alcohol by a person having preexisting liver disease; symptoms may include hemolytic anemia and hyperlipidemia.

ziggurat *Architecture.* an ancient Mesopotamian pyramidal temple tower rising in diminishing stages linked by ramps.

zigzag lightning *Geophysics.* a cloud-to-ground discharge that appears to have a single, irregular channel.

zigzagplasty *Surgery.* a surgical procedure by which the visual effect of a linear scar is minimized by dividing it into irregular segments situated at right or acute angles to one another.

zigzag reflections *Electromagnetism.* multiple reflections of a radio signal occurring within the ionosphere.

zigzag rule *Mechanical Devices.* a sectioned measuring rule that can be folded up for easy carrying.

Zimm plot *Analytical Chemistry.* a graph of the root-mean-square end-to-end distances of coiled polymer molecules as determined through scattered-light photometry.

zinc *Chemistry.* a metallic element having the symbol Zn, the atomic number 30, an atomic of weight of 65.38, a melting point of 419°C, and a boiling point of 907°C; a lustrous, bluish-white transition metal found in ores and used in alloys (such as bronze and brass), galvanized metals, electroplating, dry-cell batteries, and nutritional supplements. (Probably derived from a Greek term meaning "point" or "prong;" from the shape assumed by the crystals in smelting.)

zinc-65 *Nuclear Physics.* a radioactive isotope of zinc with a half-life of 244 days, used in alloy-wear tracer studies and in metabolism investigations.

zinc acetate *Organic Chemistry.* $Zn(C_2H_3O_2)_2 \cdot 2H_2O$, white, pearly, monoclinic crystalline plates with an astringent taste; soluble in water and alcohol; dehydrates at 100°C; melts with decomposition at 200°C; used as an astringent, wood preserver, reagent, polymer cross-linking agent, dietary supplement, and feed additive.

zincalism *Toxicology.* zinc poisoning; the symptoms include fever, chills, vomiting, and headache.

zincaluminite *Mineralogy.* $Zn_6Al_6(SO_4)_2(OH)_{26} \cdot 5H_2O$, a white to light-blue, probably hexagonal mineral, occurring as tufts and crusts of minute platy crystals, having a specific gravity of 2.26 and a hardness of 2.5 to 3 on the Mohs scale; found in zinc mines at Laurium, Greece.

zinc arsenite *Inorganic Chemistry.* $Zn(AsO_2)_2$, a toxic white powder; insoluble in water; used as a timber preservative and insecticide.

zincate *Inorganic Chemistry.* any compound containing the ZnO_2^{2-} ion, derived from the reaction of zinc hydroxide and a strong base.

zinc blende see SPHALERITE.

zinc bloom see HYDROZINCITE.

zinc borate *Inorganic Chemistry.* $3ZnO \cdot B_2O_3$, white, triclinic crystals or amorphous powder; soluble in cold water; melts at 980°C; used in medicine, fireproofing, mildew inhibitors, and ceramic flux.

zinc bromide *Inorganic Chemistry.* $ZnBr_2$, colorless, rhombic crystals, hygroscopic; soluble in water and very soluble in alcohol; melts at 394°C and boils at 650°C; used in photography, rayon manufacture, and as a radiation shield.

zinc carbonate *Inorganic Chemistry.* $ZnCO_3$, colorless, trigonal crystals; soluble in acid; melts at 300°C; used in ceramics, fireproofing, cosmetics, pharmaceuticals, and medicine.

zinc chill see SPELTER'S FEVER.

zinc chloride *Inorganic Chemistry.* $ZnCl_2$, white, hexagonal deliquescent crystals that are soluble in water, alcohol, and ether; melts at 283°C and boils at 732°C; used as a catalyst and antiseptic, and in fireproofing, textiles, and medicine.

zinc chromate *Inorganic Chemistry.* $ZnCrO_4$, lemon-yellow prisms; insoluble in cold water and soluble in acid; used as a pigment.

zinc cyanide *Inorganic Chemistry.* $Zn(CN)_2$, toxic, colorless, rhombic crystals; insoluble in water and alcohol; decomposes at 800°C; used in metal plating and as a chemical reagent and insecticide.

zinc fluoride *Inorganic Chemistry.* ZnF_2, colorless, monoclinic or triclinic crystals; soluble in water; melts at 872°C and boils at approximately 1500°C; used in phosphors, ceramics, wood preserving, electroplating, and fluorination.

zinc formate *Organic Chemistry.* $Zn(CH_2O)_2 \cdot 2H_2O$, white crystals that dehydrate at 140°C; soluble in water and insoluble in alcohol; toxic by ingestion; used as a catalyst for methanol production, in waterproofing, in textiles, and as an antiseptic.

zinc fume fever see SPELTER'S FEVER.

zinc halide *Inorganic Chemistry.* any compound composed of zinc and an element from the halogen group, such as fluorine or chlorine.

zinc hydroxide *Inorganic Chemistry.* $Zn(OH)_2$, colorless rhombic crystals that are very slightly soluble in cold water and soluble in acid; decomposes at 125°C; used as an intermediate, in medicine, and in rubber manufacture.

zincite *Mineralogy.* $(Zn,Mn^{+2})O$, a yellow to orange or red hexagonal mineral occurring as hemimorphic pyramidal crystals and in massive form, having a specific gravity of 5.67 to 5.684 and a hardness of 4 on the Mohs scale. Also, RED ZINC ORED, RUBY ZINC.

zinckenite see ZINKENITE.

zinc metaarsenite see ZINC ARSENITE.

zinc naphthenate *Organic Chemistry.* $Zn(CH_3C_5H_5COO)_2$, an amber, viscous liquid or solid; the liquid contains 8–10% zinc, while the solid is 16% zinc; very soluble in acetone; used as a drier and wetting agent in paints, varnishes, and resins, and as an insecticide, fungicide, wood preservative, and textile waterproofer.

zinc oxide *Inorganic Chemistry.* ZnO, white, hexagonal crystals, soluble in acid and insoluble in water and alcohol; melts at 1975°C; widely used in skin ointments and protectants, plastics, ceramics, cosmetics, photography, and pigments. Also, ZINC WHITE, CHINESE WHITE.

zinc phosphate *Inorganic Chemistry.* $Zn_3(PO_4)_2$, a white powder; insoluble in water; melts at 900°C; used in dentistry, phosphors, and metal coating; also occurs in several hydrated forms, such as $4H_2O$ and $8H_2O$. Also, **zinc orthophosphate.**

zinc phosphide *Inorganic Chemistry.* Zn_3P_2, a gritty dark-gray powder; melts above 420°C and boils at 1100°C; decomposes in water; reacts with acids to give the toxic and flammable phosphine gas, and reacts violently with oxidants; extremely toxic; used as a rodenticide.

zinc point *Thermodynamics.* the normal melting point of zinc at standard atmospheric pressure: 419.58°C; a fixed-point temperature of the International Practical Temperature Scale of 1968.

zinc selenide *Inorganic Chemistry.* ZnSe, yellowish to red crystals; insoluble in water; melts above 1100°C; a fire hazard in contact with water or acids; used in phosphors and infrared optics.

zinc sulfate *Inorganic Chemistry.* $ZnSO_4 \cdot 7H_2O$, colorless crystals; loses water to the air or on heating to 280°C; soluble in water; melts at 100°C; used in animal feeds and as a mordant and reagent. Also, **zinc vitriol.**

zinc sulfide *Inorganic Chemistry.* ZnS, a yellowish-white powder; insoluble in water; exists in two forms and transforms from the beta to the alpha form at 1020°C; used in glass, rubber, and plastic manufacture, in pigments and dyes, and for various other purposes.

zinc telluride *Inorganic Chemistry.* ZnTe, reddish crystals; insoluble in water; melts at about 1240°C; used in semiconductor research and as a photoconductor.

zinc white *Inorganic Chemistry.* a popular name for zinc oxide, ZnO.

Zinfandel or **zinfandel** *Botany.* a dark purple wine grape developed and grown in California. *Food Technology.* a spicy, fruity wine made mostly or entirely from this grape.

Zingiberaceae *Botany.* a family of tropical monocotyledonous plants of the family Zingiberales, characterized by highly aromatic rootstocks; includes ginger and turmeric.

Zingiberales *Botany.* an order of monocotyledonous herbs or shrubs of the subclass Commelinidae, characterized by pinnately veined leaves and irregular flowers with well-differentiated sepals and petals, an inferior ovary, and one to five functional stamens; found primarily in tropical or subtropical climates.

Zingiberidae *Botany.* a subclass of terrestrial or ephiphytic herbs or small trees belonging to the class Liliopsida, characterized by a simple, unbranched trunk.

Zinj *Anthropology.* short for *Zinjanthropus boisei.*

Zinjanthropus boisei *Anthropology.* an earlier name for *Australopithecus boisei.*

zinkenite *Mineralogy.* $Pb_9Sb_{22}S_{42}$, a steel-gray, opaque hexagonal mineral with metallic luster, massive in habit, as fibrous aggregates, or as slender prismatic crystals, having a specific gravity of 5.36 and a hardness of 3 to 3.5 on the Mohs scale; found in hydrothermal veins with tin and base metal sulfides and sulfosalts. Also, ZINCKENITE.

zinnwaldite *Mineralogy.* $KLiFe^{+2}Al(AlSi_3)O_{10}(F,OH)_2$, a pale-violet, dark-gray, yellowish-brown, or green monoclinic mineral of the mica group, occurring as short prismatic or tabular crystals and in scales or scaly aggregates, having a specific gravity of 2.9 to 3.3 and a hardness of 2.5 to 4 on the Mohs scale; found in granite pegmatites, greisens, and high-temperature quartz veins.

Zinsser, Hans 1878–1940, American bacteriologist; developed immunization against typhus.

Ziphiidae *Vertebrate Zoology.* the beaked whales, a family of the order Odontoceta, characterized by a lack of teeth in the upper jaw and a beaklike projection on the front of the head.

zip mode *Computer Technology.* a mode of plotter operation in which the input commands can either increase or decrease the speed relative to one or both axes.

zipp *Veterinary Medicine.* a paste that is applied to wounds in veterinary surgery; made by grinding together one part zinc oxide, two parts iodoform, and two to three parts liquid paraffin. (An acronym for zinc-iodoform-paraffin-paraffin.)

zippeite *Mineralogy.* $K_4(UO_2)_6(SO_4)_3(OH)_{10}\cdot 4H_2O$, a very rare, strongly radioactive, yellow to reddish-orange, orthorhombic mineral occurring as coatings and microcrystalline aggregates, having a specific gravity of 3.66 and a hardness of 2 on the Mohs scale; found in pitchblende ores as an efflorescence.

zipper *Engineering.* a slide fastener that meshes interlocking teeth to provide a closure for adjacent pieces of a material.

zipper conveyor *Mechanical Engineering.* a conveyor belt whose zipperlike teeth mesh to form a closed tube; used to transport fragile or easily degradable items. Also, CLOSED-BELT CONVEYOR.

zircon *Mineralogy.* $ZrSiO_4$, a tetragonal mineral that is the chief ore of zirconium, ranging from colorless to yellow, green, brown, and red, occurring as prismatic or dipyramidal crystals or irregular grains, having a specific gravity of 4.6 to 4.7 and a hardness of 7.5 on the Mohs scale; found in igneous rocks, as a detrital mineral, and in meteorites and lunar rocks; colorless varieties are used as gemstones. Also, HYACINTH, ZIRCONITE.

zircon

zirconia see ZIRCONIUM OXYCHLORIDE.

zirconia anhydride see ZIRCONIUM OXIDE.

zirconite see ZIRCON.

zirconium *Chemistry.* a metallic element having the symbol Zr, the atomic number 40, an atomic weight of 91.22, a melting point of 1850°C, and a boiling point of 4377°C; used in coatings for nuclear fuel rods, corrosion-resistance alloys, and laboratory ware. (Probably from an Arabic word describing its color.)

zirconium-95 *Nuclear Physics.* a radioactive isotope of zirconium with a half-life of 64.0 days; decays by beta emission and is used to trace petroleum pipeline flows.

zirconium anhydride see ZIRCONIUM OXIDE.

zirconium boride *Inorganic Chemistry.* ZrB_2, toxic, gray metallic crystals or powder; melts at about 3200°C; used in cutting tools and in various high-temperature electrical applications. Also, **zirconium diboride.**

zirconium carbide *Inorganic Chemistry.* ZrC, a very hard, gray crystalline solid; insoluble in water; melts at 3540°C and boils at 5100°C; ignites spontaneously in air; used as an abrasive, in cutting tools, and in various high-temperature electrical applications.

zirconium chloride **1.** see ZIRCONIUM OXYCHLORIDE. **2.** see ZIRCONIUM TETRACHLORIDE.

zirconium dioxide see ZIRCONIUM OXIDE.

zirconium halide *Inorganic Chemistry.* any compound made up of zirconium and an element from the halogen group (periodic table Group VIIa), such as fluorine or chlorine.

zirconium hydride *Inorganic Chemistry.* ZrH_2, a gray-black metallic powder that is a dangerous fire hazard, especially in contact with oxidants; used in metallurgy, as a source of hydrogen, and as a catalyst and reducing agent.

zirconium hydroxide *Inorganic Chemistry.* $Zr(OH)_4$, a white amorphous powder; insoluble in water; loses $2H_2O$ at 550°C; used to make zirconium compounds and glass colorants.

zirconium lamp *Electronics.* a high-intensity point-source lamp that utilizes a zirconium-oxide cathode in an argon-filled bulb.

zirconium nitride *Inorganic Chemistry.* ZrN, a very hard, brass-colored powder; melts at about 2980°C; used to make crucibles, cermets, and refractories.

zirconium oxide *Inorganic Chemistry.* ZrO_2, a heavy white amorphous powder; insoluble in water; melts at 2700°C and boils at 5000°C. The fused form is reported to be 11.0 on the Mohs scale and thus harder than diamond. It is used in heat-resistant fibers, special glasses, high-frequency induction coils, and for a variety of other industrial purposes. Also known by various other names, such as ZIRCONIA ANHYDRIDE and ZIRCONIUM DIOXIDE.

zirconium oxychloride *Inorganic Chemistry.* $ZrOCl_2\cdot 8H_2O$, white crystals; loses $6H_2O$ on heating to 150°C and $8H_2O$ at 210°C; soluble in water; used in textiles, cosmetics, antiperspirants, and for various other purposes. Also, ZIRCONIA.

zirconium phosphate *Inorganic Chemistry.* $ZrO(H_2PO_4)_2\cdot 3H_2O$, a white amorphous powder that is soluble in acids and insoluble in water; decomposes on heating; used as a chemical reagent. Also, **zirconium orthophosphate.**

zirconium tetrachloride *Inorganic Chemistry.* $ZrCl_4$, irritant, white, lustrous crystals; decomposes in water and sublimes on heating to 331°C; used in analytical chemistry and for various industrial purposes. Also, ZIRCONIUM CHLORIDE.

zirconyl chloride see ZIRCONIUM OXYCHLORIDE.

zirkelite *Mineralogy.* $(Ca,Th,Ce)Zr(Ti,Nb)_2O_7$, a radioactive, black, monoclinic, pseudocubic mineral, occurring as commonly twinned pseudo-octahedral crystals, having a specific gravity of 4.706 to 4.741 and a hardness of 5.5 on the Mohs scale; dimorphous with calciobetafite.

zI test *Statistics.* a significance test adapted from the zM test, applying to instances rather than measurements. (From z test for Instances.)

zitterbewegung *Quantum Mechanics.* the intrinsic, extremely high-frequency oscillatory motion of electrons predicted by the Dirac theory in the presence of a potential barrier.

Z line *Histology.* the line formed by the attachment of actin filaments between two sarcomeres.

zM test *Statistics.* a significance test that compares a random sample of 1 or more measurements to a large population whose mean and standard deviation are known. (From z test for Measurements.)

Zn the chemical symbol for zinc.

zo- a combining form meaning "animal."

Zoantharia *Invertebrate Zoology.* one of two subclasses of anthozoan coelenterates, including sea anemones and most corals.

Zoanthidea *Invertebrate Zoology.* an order of anthozoans, coelenterates in the subclass Zoantharia, with no skeleton or basal disk; most are colonial, with polyps joined at the base resembling a sea anemone, attached to sponges, rocks, and the shells of hermit crabs.

Zoanthinaria *Invertebrate Zoology.* an order of anthozoan coelenterates resembling sea anemones, but typically colonial or social, mostly found in warm shallow seas.

Zoarcidae *Vertebrate Zoology.* the eelpouts, a family of shallow water marine fish composing the suborder Zoarcoidei of the order Gadiformes, characterized by spineless fins and wide gill openings; found in cold portions of the Atlantic and Pacific Oceans and in the Arctic and Antarctic.

zod. zodiac.

zodiac *Astronomy.* **1.** the band of sky with twelve constellations that extends 8° to either side of the ecliptic and through which the sun, moon, and planets apparently pass during the year; it is divided into twelve equal parts, each represented by one of the twelve constellations (**signs of the zodiac**). **2.** a chart or diagram in which these twelve constellations are represented by drawings or figures traditionally associated with each constellation.

zodiacal constellation *Astronomy.* any of the twelve ancient figures along the zodiac: Aries (Ram), Taurus (Bull), Gemini (Twins), Cancer (Crab), Leo (Lion), Virgo (Virgin), Libra (Balance, or Scales), Scorpio (Scorpion), Sagittarius (Archer), Capricorn (Goat), Aquarius (Water Bearer), and Pisces (Fish); widely believed in ancient times (and still regarded by some in modern times) as significantly influencing a person's character and personality and the course of human affairs.

zodiacal light *Geophysics.* a phenomenon, much like airglow, in which a cone of light appears to spread upward from the horizon to the ecliptic; seen in the west just after sunset or in the east just before sunrise. (So called because it is seen against the zodiacal constellations.)

zodiacal pyramid *Geophysics.* the luminous, pyramid-shaped cone formed by zodiacal light. Also, **zodiacal cone**

zoea *Invertebrate Zoology.* in decapod crustaceans, especially marine crabs, an early larval stage with long spines on the cephalothorax. Also, **zoaea.**

zoetic *Biology.* of or relating to life.

zoisite *Mineralogy.* $Ca_2Al_3(SiO_4)_3(OH)$, a vitreous or pearly orthorhombic mineral of the epidote group, occurring in a range of colors, massive in habit or as prismatic crystals, having a specific gravity of 3.3 to 3.55 and a hardness of 6 to 7 on the Mohs scale; dimorphous with clinozoisite; found in metamorphic and basic igneous rocks, altered granitic rocks, and pegmatites.

zoite *Cell Biology.* any of a number of life-cycle stages in certain protozoans.

Zollinger-Ellison syndrome *Oncology.* an uncommon tumor of the pancreas, characterized by severe peptic ulceration, gastric hypersecretion, elevated serum gastrin, and gastrinoma of the pancreas or the duodenum, seen most often in people between twenty and fifty years of age.

zonal *Science.* relating to or characteristic of a zone or zones. *Meteorology.* denoting a latitudinal aspect (from east to west) as distinguished from a longitudinal aspect (from north to south).

zonal flow *Meteorology.* the east to west component of wind flow along a global latitude. Also, **zonal circulation.**

zonal harmonics *Mathematics.* axially symmetric spherical harmonics; i.e., spherical harmonics that do not depend on azimuth.

zonal index *Meteorology.* the determination of the strength of the mid-latitude westerlies by computing the horizontal pressure differential (or, the geostrophic wind) between 35 and 55 degrees latitude.

zonal kinetic energy *Meteorology.* the kinetic energy in the mean zonal wind flow, determined by averaging the zonal wind component about a fixed latitude circle.

zonal soil *Geology.* any soil having well-developed characteristics that reflect the influence of soil-forming agents. Also, MATURE SOIL.

zonal theory *Mineralogy.* a theory of mineral-deposit formation and mineralogical sequence patterns based on the changes undergone by a mineral-bearing fluid as it moves upward from its magmatic source to the surface.

zonal wind *Meteorology.* **1.** the wind component referred to along a local parallel of latitude. **2. zonal winds.** see WESTERLIES.

zonal wind-speed profile *Meteorology.* a graphical representation in which the speed of local flow is one coordinate and the latitude is the other.

zona pellucida *Histology.* the thick, solid elastic envelope of the ovum. (From a Latin term meaning literally "transparent zone.")

zonation *Geology.* a condition in which geological materials are arranged or formed in zones.

Zond *Space Technology.* a series of Soviet space probes launched between 1964 and 1970; **Zond 5** was the first probe to orbit the moon and return to earth.

zonda *Meteorology.* a hot wind of Argentina.

zone a continuous, relatively distinct area having certain identifiable limits or boundaries; specific uses include: *Geography.* a latitudinal band around the earth sharing general characteristics, such as the north frigid zone or the temperate zones. *Geology.* **1.** any horizontal, vertical, concentric, or otherwise oriented belt, layer, band, or strip of earth materials characterized by some particular property or content. **2.** a stratigraphic unit of any kind that can be distinguished from its surroundings by certain general characteristics. *Paleontology.* a biostratigraphic area or belt of strata of any thickness or lithology that is characterized by fossils of two or more species with different but restricted vertical ranges, and by the absence of species characteristic of the zones on either side. *Military Science.* any important tactical area, particularly near the front; it is usually used with a term describing the specific area; e.g., a combat zone. *Ordnance.* in howitzer and mortar fire, an area in which projectiles will fall as the weapon's elevation is varied between minimum and maximum, using a given propelling charge. *Analytical Chemistry.* a defined area such as the band through which a solute spreads within a series of tubes in a liquid-liquid extraction procedure. *Mathematics.* the portion of a sphere lying between two parallel planes that intersect the sphere. If one of the planes is tangent to the sphere, the zone is said to have one base; if neither plane is tangent to the sphere, the zone is said to have two bases. *Crystallography.* a group of crystal faces whose intersections are parallel lines.

zone axis *Crystallography.* an axis that runs through the center of a crystal and is parallel to the faces of a crystal zone.

zone charge *Ordnance.* with semifixed ammunition, the amount of propellant necessary to reach a particular zone of fire; e.g., zone charge three consists of three units of propellant.

zone control *Engineering.* the division of a process or building into zones to allow independent temperature regulation in each zone.

zone description *Navigation.* a number from 1 to 12 that must be applied to the zone time to obtain Greenwich Meridian time. The number is positive in west longitude and negative in east longitude.

zone electrophoresis *Physical Chemistry.* a form of electrophoresis in which there is a separation of the components of a molecular mixture into distinct zones, as when the solution moves through a porous or semisolid medium, such as filter paper.

zone fire *Military Science.* artillery or mortar fires that are delivered in a constant direction at several quadrant elevations.

zone indices *Crystallography.* the three numbers U,V,W that define a crystal zone. If the indices of two faces defining the zone axis are $h_1k_1l_1$ and $h_2k_2l_2$, then $W = k_1l_2 - l_1k_2$, and so on.

zone law *Crystallography.* a law stating that $hu + lv + kw = 0$, where h, k, and l are the Miller indices of a plane lying in a crystal zone and u, v, and w are the zone indices.

zone melting crystallization *Chemical Engineering.* a process for purifying crystalline solids in which a sample is placed in a narrow column and heated so that a molten zone passes through and entrains impure substances.

zone meridian *Astronomy.* the meridian of longitude (always a multiple of 15°) used to define the mean time for a time zone.

zone noon *Astronomy.* mean solar noon as seen from the meridian of a given time zone.

zone I (nuclear) *Military Science.* a circular area with ground zero as the center and minimum safe distance I as the radius, from which all ground forces are to be evacuated; two surrounding circular rings at greater distances from ground zero require maximum and minimum personnel protection, respectively. Thus, **zone II (nuclear), zone III (nuclear).**

zone of accumulation or **zone of illuviation** see B HORIZON.

zone of aeration *Hydrology.* the area immediately below the ground surface in which the interstices are filled with water at a pressure less than the pressure of the atmosphere, and air at atmospheric pressure. Also, UNSATURATED ZONE, VADOSE ZONE, ZONE OF SUSPENDED WATER.

zone of avoidance *Astronomy.* the region along the Milky Way (and a short distance away from it) in which reduced numbers of galaxies are observed because of obscuration from interstellar dust and gas.

zone of cementation *Geology.* the area of the earth's crust underlying the zone of weathering, in which loose sediments are cemented by dissolved minerals leached down from the overlying layers by percolating waters. Also, BELT OF CEMENTATION.

zone of fire *Military Science.* an area within which a designated ground unit or fire support ship delivers or is prepared to deliver fire support.

zone of maximum precipitation *Meteorology.* the point of greatest annual precipitation identified by a particular mountain elevation region or belt.

zone of optimal proportion *Immunology.* a zone, occurring when an antibody and antigen are mixed, in which both the antigen and the antibody are totally combined.

zone of saturation *Hydrology.* the subsurface zone in which all interstices are filled with water at a pressure greater than the pressure of the atmosphere. Also, SATURATED ZONE, PHREATIC ZONE.

zone of silence see ANACOUSTIC ZONE.

zone of subduction see SUBDUCTION ZONE.

zone of suspended water see ZONE OF AERATION.

zone plan *Transportation Engineering.* a graphic representation of the fare zones in a transit system.

zone plate *Optics.* a plate in which alternate transparent and opaque rings obstruct every other Fresnel half-period zone, so that light traveling through the plate forms an intense point image similar to one formed by a lens.

zone-position indicator *Engineering.* a radar set used to display the general position of an object on another radar set with a narrower field.

zone refining *Metallurgy.* the process of refining germanium and silicon to produce the elements used in making transistors. Also, **zone purification.**

zonesthesia *Neurology.* a subjective sensation that a portion of the body, particularly the trunk, is being constricted, as by a girdle.

zone time *Astronomy.* the mean time observed throughout a designated time zone, regardless of what the actual local solar time is.

Zoology

Zoology is the study of animals. It is a subdivision of biology, which is the study of all living things. Zoology encompasses all aspects, from the molecular level, to the organ, to that of the whole animal, and from unicellular organisms to whales. There are many subdisciplines of zoology (e.g., genetics, biochemistry, parasitology, physiology, ethology, ecology, systematics, ichthyology, ornithology, mammology, and so on.), with further subdivisions (e.g., physiological ecology, neurophysiology, respiratory physiology, and so on).

The study of zoology has always been part of the life of humans because of the intimate relationship between us and other animals for food, assistance, and companionship. This was true when we were hunters, then farmers, and is just as true in the information age. Now that we have the capacity to radically change the biosphere through our own actions, either we can do it blindly or we can try to manage the change, understanding the effect it has on the animals with whom we share the planet. This requires knowledge of animals, and this knowledge constitutes zoology.

Initial studies of zoology concentrated on describing the anatomy and categorizing animals (systematics), and this goes on today. Not all animals are named and described (especially insects), and there is a continuing argument as to the most appropriate grouping of animals (classification). More recently, interest has developed into the molecular aspects of zoology, made possible by techniques developed in chemistry and physics but applied to biological systems. There are many similarities at the molecular level between animals and plants, and consequently the subject is usually referred to as Molecular Biology rather than Molecular Zoology or Botany.

David Randall
Professor of Zoology
University of British Columbia

zoning the fact of being in zones or having a relationship to zones; specific uses include: *Building Engineering.* a set of local regulations limiting the type of buildings, building densities, and so on, that can occur in a specified area. *Geology.* the relationship among areas characterized by a particular mineral or suite of minerals, thus reflecting conditions of formation and composition. *Crystallography.* an imperfect reaction in crystal growth between the solid and liquid phases, resulting in variations in the chemical composition within the crystal.

zonite *Invertebrate Zoology.* a millipede body segment. *Geology.* see RANGE ZONE.

zonula adherens *Cell Biology.* a specialized cellular adhering junction into which are inserted microfilaments that apparently interact with other microfilaments to bind adjacent cells, giving strength to a tissue.

zoo- a combining form meaning "animal."

zooagglutinin *Toxicology.* a substance found in animal venoms that causes agglutination of red blood cells.

zooarchaeology *Archaeology.* the study of animal remains from archaeological sites.

zoochory *Botany.* the dispersal of fruit or seeds by animals, especially birds.

zooecium *Invertebrate Zoology.* in colonial bryozoans, the body wall, a generally secreted gelatinous, chitinoid, or calcareous covering of individual zooids.

zooecology *Ecology.* the study of the relationship between animals and their environment. Thus, **zooecological.**

zooflagellate *Invertebrate Zoology.* any of various flagellated protozoans of the subclass Zoomastigina, lacking plantlike characteristics.

zoogenesis *Ecology.* **1.** the origin of animal life on earth. **2.** the origin and development of a particular animal species.

zoogenic *Ecology.* relating to or produced by the activities of animals.

zoogeographic *Ecology.* describing the relationship between geography and animal life, especially the effect of geographical barriers such as deserts, mountain ranges, or oceans on the type of animal life found in various areas. Also, **zoogeographical.**

zoogeographic(al) region *Ecology.* any of the major geographical areas into which the earth is divided on the basis of distinct forms of animal life; e.g., the Australian region includes marsupial and monotreme forms not found elsewhere. Also, **zoogeographic(al) realm.**

zoogeography *Geography.* a branch of geography dealing with the relationships between geography and animal populations, especially with the effect of geography on species distribution.

zoogloea *Microbiology.* a gelatinous mass of bacterial cells produced by certain genuses such as *Zoogloea ramigera*; used to treat sewage.

zooid *Invertebrate Zoology.* an individual forming part of a colony; a polyp.

zool. zoology; zoological.

zoological [zō′ə läj′ə kəl] *Zoology.* of or relating to zoology.

zoologist [zō äl′ə jist] *Zoology.* a person who studies or is an expert in zoology.

zoology [zō äl′ə jē] *Biology.* **1.** the branch of biology dealing with animals. **2.** the sum of collective knowledge that is known about animals. **3.** the animal life of a given area.

zoom *Optics.* **1.** to alter the size of an image by bringing it closer, without causing it to become out of focus. **2.** any instrument or device used in zooming. *Computer Programming.* to uniformly increase or decrease the size and scope of the displayed image in computer graphics.

Zoomastigina *Invertebrate Zoology.* a subclass of flagellated protozoans in the class Mastigophora, without plantlike characteristics; usually symbiotic or parasitic with asexual reproduction.

Zoomastigophorea *Invertebrate Zoology.* a protozoan class of the subphylum Mastigophora, including choanoflagellates, retortamonads, trichomonads, oxymonads, diplomonads, and hypermastigids.

zoom lens *Optics.* an optical system having a variable focal length and fixed focal plane, thus allowing variable magnification of the image.

zoomorph *Archaeology.* a design element that depicts an animal, such as those found in cave paintings.

zoomorphic *Behavior.* relating to or characterized by zoomorphism; regarding human behavior as equivalent to animal behavior.

zoomorphism *Behavior.* the viewing of human behavior in terms of the behavior of animals, especially the principle that human actions are entirely the result of biological and instinctual drives rather than reason or emotion. *Anthropology.* the fact of conceiving or representing a deity as having animal form.

zoonomy *Zoology.* the physiology of animals.

zoonosis *Pathology.* any disease of animals that can be transmitted to humans under natural conditions, such as rabies or brucellosis.

zoophilic *Microbiology.* describing a pathogenic organism that prefers animal hosts to human ones.

zoophilous *Botany.* describing a plant that is pollinated by animals other than insects.

zoophobia *Psychology.* an irrational fear of animals.

zoosemiotics *Behavior.* the study of the methods by which animals use signaling as a form of communication.

zoosporangium *Botany.* a spore case in which zoospores are produced.

zootechnics *Zoology.* the science of animal husbandry.

zootoxin *Toxicology.* any toxin produced by an animal.

zootrophotoxism *Toxicology.* poisoning due to ingestion of contaminated food of animal origin.

Zooxanthellaceae *Botany.* a family of marine flagellates constituting the order Zooxanthellales.

zooxanthellae *Invertebrate Zoology.* plantlike flagellate protozoans in the order Dinoflagellida, photosynthesizing symbiotes inside the cells of sponges, corals, and others.

Zooxanthellales *Botany.* a monofamilial order of mostly intrafamilial symbiont flagellates of the class Dinophyceae, noted for excreting large amounts of photosynthate as glycerol and aiding in the calcification of corals.

Zorn's lemma *Mathematics.* let X be a nonempty partially ordered set with the property that every totally ordered subset of X has an upper bound in X. Then X contains a maximal element; i.e., there is an element x of X such that if $x \leq a$ for some element a of X, then $a = x$.

Zoroasteridae *Invertebrate Zoology.* a family of deepwater starfish, echinoderms in the order Forcipulatida, with a domed disk, long arms, inconspicuous arm plates, and highly specialized pedicellariae.

Zosterapidae *Vertebrate Zoology.* the white-eyes, a family of small green birds of the passerine suborder Oscines, characterized by a conspicuous white eye ring; found in Africa, Asia, New Guinea, Australia, and New Zealand.

Zosterophyllopsida *Paleontology.* a class of early terrestrial vascular plants in the division Propteridophyta; leafless and rootless; lived from the end of the Silurian to the Late Devonian; probably ancestral to the lycopods, which first appeared in the Early Devonian.

Zosterophyllum *Paleontology.* a genus of primitive vascular plants in the extinct class Zosterophyllopsida, characterized by lateral spiked sporangia; extant in the Devonian.

zoysia *Botany.* any of several low-growing grasses of the genus *Zoisia,* especially *Z. matrella*; native to tropical Asia and widely used in the United States as a lawn grass.

Z parameter *Electronics.* a device or network parameter expressed as an impedance.

Z particle see Z^0 BOSON.

Zr the chemical symbol for zirconium.

ZRH *Aviation.* the airport code for Zurich, Switzerland.

Zsigmondy, Richard 1865–1929, German chemist; awarded the Nobel Prize for research on the nature of colloidal solutions.

Zsigmondy filter *Chemistry.* any of a series of graded semipermeable membranes that are used to separate colloidal particles by size from a colloidal suspension.

Zsigmondy gold number *Chemistry.* the number of milligrams of a colloidal suspension that are needed to prevent 10 ml of gold sol from coagulating when 0.5 ml of 10% NaCl solution is added to the solution.

z-transform *Mathematics.* given a sequence $\{a_n\}$, the series $\sum_{n=1}^{\infty} a_n z^{-n}$ (**one-sided z-transform**) or $\sum_{n=-\infty}^{\infty} a_n z^{-n}$ **(two-sided z-transform),** where z is complex. The z-transform can be viewed as the Fourier or Laplace transform of a function f whose values are known (sampled) only at the integers; i.e., $f(n) = a_n$ for each integer n.

Z-twist *Textiles.* a right-hand yarn twist in which the spiral formed by twisting slants in the same direction as the center diagonal of the letter Z, as distinguished from an S-twist. Also, RIGHT TWIST, REGULAR TWIST.

Zuloagan *Geology.* a North American Gulf Coast stage of the Upper Jurassic period, occurring before the LaCasitan, and equivalent to the Oxfordian stage.

Zulu time *Horology.* another term for Greenwich mean time.

Z variometer see VERTICAL INTENSITY VARIOMETER.

zwischenbirge see MEDIAN MASS.

Zworykin, Vladimir 1889–1982, Russian-born American physicist; developed the iconoscope, kinescope, and electron microscope.

Zygaenoidea *Invertebrate Zoology.* a superfamily of moths, lepidopteran insects in the suborder Heteroneura, with a rudimentary proboscis.

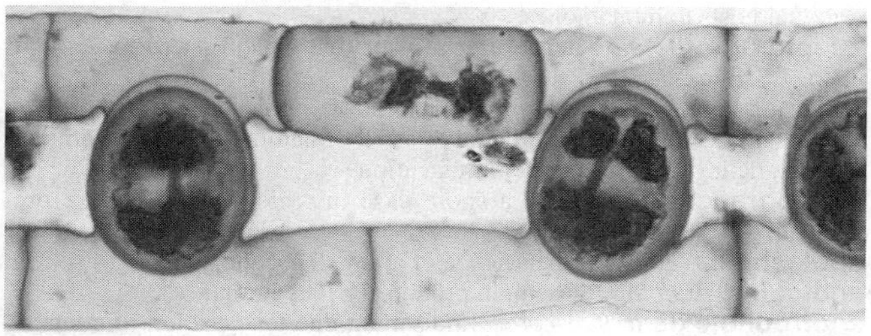

Zygnema

Zygnema *Botany.* a common filamentous green algae genus of the family Zygnemataceae, characterized by a pair of stellate chloroplasts and isogamous or anisogamous conjugation.

Zygnemataceae *Botany.* a family of filamentous plants belonging to the order Conjugales, differentiated by spiral, bandlike, or cushionlike chloroplast.

Zygnematales *Botany.* an order of freshwater and subaerial green algae of the class Chlorophyceae, distinguished by sexual reproduction by conjugation of amoeboid or passive gametes, the lack of flagella at any stage, and occurrence as unicells or as uniseriate filaments.

zygo- or **zyg-** a combining form meaning "yoke," "paired," or "joined."

zygodactyl *Vertebrate Zoology.* having a toe arrangement of two in front and two behind; used to describe the feet of certain birds.

Zygodiscaceae *Botany.* a monospecific family of flagellate algae of the order Coccosphaerales, for which coccospheres are unknown but with one known genus having a series of parallel crossbars on a long, narrow coccolith.

zygogenesis *Developmental Biology.* **1.** the production of a zygote. **2.** gametic reproduction.

zygoma *Anatomy.* a long, slender, strong process arising from the lower portion of the squamous part of the temporal bone, passing forward to join the zygomatic bone, and thus forming the zygomatic arch.

zygomatic arch *Anatomy.* the arch formed by the articulation of the broad temporal processs of the zygomatic bone and the slender zygomatic process of the temporal bone.

zygomatic bone *Anatomy.* the bone that forms the prominence of the cheek.

zygomorphic *Botany.* having or characterized by bilateral symmetry.

zygophore *Botany.* any of several specialized branches of hyphae bearing isogametes that unite to produce a zygospore.

zygophyte *Botany.* a plant that reproduces by the fusion of two similar gametes of zygospores.

Zygoptera *Invertebrate Zoology.* the damselflies, a suborder of dragonflies in the order Odonata, with slender, often orange or bright blue bodies and long, narrow, transparent wings held upright at rest.

zygosity *Developmental Biology.* **1.** the condition of a cell or individual with regard to the alleles determining a specific character, whether identical (homozygosity) or different (heterozygosity). **2.** in twinning, the fact of developing from one zygote (monozygosity) or two (dizygosity).

zygospore *Botany.* a spore formed by conjugation of two similar gametes.

zygote *Developmental Biology.* the cell resulting from the union of a male and female gamete, until it divides; the fertilized ovum.

zygotene *Developmental Biology.* the second of five sequential stages of meiotic prophase, during which the synapsis of homologous chromosomes occurs. Also, **zygonema.**

zygotic *Developmental Biology.* **1.** of or relating to a zygote. **2.** of or relating to zygosity.

zygotic induction *Virology.* a lysogenic process in which a prophage is transmitted to a recipient cell not having a phage repressor protein.

zygotic lethal gene *Genetics.* a gene that allows the normal functioning of gametes, but has lethal effects during embryonic or larval development.

zygotic meiosis *Developmental Biology.* a meiotic process that occurs after formation of the zygote in certain protozoa and fungi.

zymase *Enzymology.* a complex of enzymes, found in yeast extract, that catalyze glycosis and other reactions of alcoholic fermentation.

zymo- or **zym-** a combining form meaning "enzyme" or "fermentation."

zymogen *Enzymology.* the inactive precursor of an enzyme that becomes active by proteolysis.

zymogen granules *Enzymology.* proenzyme granules, found in gland cells, that store and secrete the zymogens synthesized by the ribosomes in endoplasmic reticulum.

zymogenic *Microbiology.* **1.** causing fermentation. Also, **zymogenous. 2.** of or relating to fermentation.

zymogram *Microbiology.* a representation of enzymatically active parts that are separated by electrophoretic means. *Physical Chemistry.* a recording of the extent of fermentation.

Zymomonas *Bacteriology.* a genus of Gram-negative, facultatively anaerobic bacteria that occur as motile, rod-shaped cells with polar flagella in fermenting beverages and plants.

zymophore *Enzymology.* the active part of an enzyme.

zymosan *Biochemistry.* a mixture of polysaccharides, proteins, and ash, derived from the cell walls or the entire cell of yeast; used in assaying properdin.

zymosis *Microbiology.* another term for fermentation. *Pathology.* any infectious or contagious disease.

zymosterol *Biochemistry.* an unsaturated yeast sterol, occurring in fungi and molds, that yields cholesterol on hydrogenation.

ZZ Ceti star *Astronomy.* any of a class of variable white dwarf stars whose brightness variations average 0.3 magnitude.

Symbols and Units

Greek Alphabet

Α	α	alpha	Ι	ι	iota	Ρ	ρ	rho
Β	β	beta	Κ	κ	kappa	Σ	σ ς	sigma
Γ	γ	gamma	Λ	λ	lambda	Τ	τ	tau
Δ	δ ∂	delta	Μ	μ	mu	Υ	υ	upsilon
Ε	ϵ	epsilon	Ν	ν	nu	Φ	ϕ φ	phi
Ζ	ζ	zeta	Ξ	ξ	xi	Χ	χ	chi
Η	η	eta	Ο	ο	omicron	Ψ	ψ	psi
Θ	θ ϑ	theta	Π	π	pi	Ω	ω	omega

International System of Unit Prefixes

Prefix	Symbol	Multiple	
exa	(E)	10^{18}	(quintillions)
peta	(P)	10^{15}	(quadrillions)
tera	(T)	10^{12}	(trillions)
giga	(G)	10^{9}	(billions)
mega	(M)	10^{6}	(millions)
kilo	(k)	10^{3}	(thousands)
hecto	(h)	10^{2}	(hundreds)
deka	(da)	10^{1}	(tens)
deci	(d)	10^{-1}	(tenths)
centi	(c)	10^{-2}	(hundredths)
milli	(m)	10^{-3}	(thousandths)
micro	(μ)	10^{-6}	(millionths)
nano	(n)	10^{-9}	(billionths)
pico	(p)	10^{-12}	(trillionths)
femto	(f)	10^{-15}	(quadrillionths)
atto	(a)	10^{-18}	(quintillionths)

International System Derived Units

Quantity	Name	Symbol	Dimension
Capacitance	farad	F	$A^2\ s^4\ kg^{-1}\ m^{-2}$
Charge	coulomb	C	A s
Conductance	siemens	S	$kg^{-1}\ m^{-2}\ s^3\ A^2$ or Ω^{-1}
Energy	joule	J	$kg\ m^2\ s^{-2}$
Force	newton	N	$kg\ m\ s^{-2}$
Magnetic flux	tesla	T	$kg\ s^{-2}\ A^{-1}$
Polarizability		α	$A^2\ s^4\ kg^{-1}$
Potential	volt	V	$kg\ m^2\ s^{-3}\ A^{-1}$
Power	watt	W	$kg\ m^2\ s^{-3}$
Pressure	pascal	Pa	$kg\ m^{-1}\ s^{-2}$
Resistance	ohm	Ω	$kg\ m^2\ s^{-3}\ A^{-2}$

Fundamental Physical Constants

Recommended values for fundamental physical constants; adapted from E. R. Cohen and B. N. Taylor from *CODATA Bulletin*, number 63, November 1986, and *Journal of Research of the National Bureau of Standards*, volume 92, number 2, pages 85–95, 1987.
Values in bold are exact.
In parentheses is the one-standard-deviation uncertainty of the final digit of a given value.

General Constants

Symbol	Quantity	Value	Units	Relative uncertainty (ppm)
	Universal Constants			
G	Newtonian constant of gravitation	6.672 59 (85)	10^{-11} m^3 kg^{-1} s^{-2}	128
μ_0	permeability of vacuum	$\mathbf{4\pi \times 10^{-7}}$	N A^{-2}	
		≈12.566 370 614	10^{-7} N A^{-2}	
ϵ_0	permittivity of vacuum	$\mathbf{1/\mu_0 c^2}$		
		≈8.854 187 817	10^{-12} F m^{-1}	
h	Planck constant	6.626 0755 (40)	10^{-34} J s	0.60
	in electron volts, $h/\{e\}$	4.135 6692 (12)	10^{-15} eV s	0.30
$\hbar$	$h/2\pi$	1.054 572 66 (63)	10^{-34} J s	0.60
	in electron volts, $\hbar/\{e\}$	6.582 1220 (20)	10^{-16} eV s	0.30
l_P	Planck length, $\hbar/m_P c = (\hbar G/c^3)^{1/2}$	1.616 05 (10)	10^{-35} m	64
m_P	Planck mass, $(\hbar c/G)^{1/2}$	2.176 71 (14)	10^{-8} kg	64
t_P	Planck time, $l_P/c = (\hbar G/c^5)^{1/2}$	5.390 56 (34)	10^{-44} s	64
c	speed of light in vacuum	**2.997 924 58**	10^8 ms^{-1}	
	Electromagnetic Constants			
μ_B	Bohr magneton, $e\hbar/2m_e$	9.274 0154 (31)	10^{-24} J T^{-1}	0.34
	in electron volts, $\mu_B/\{e\}$	5.788 382 63 (52)	10^{-5} eV T^{-1}	0.089
	in hertz, μ_B/h	1.399 624 18 (42)	10^{10} Hz T^{-1}	0.30
	in kelvins, μ_B/k	0.671 7099 (57)	K T^{-1}	8.5
	in wavenumbers, μ_B/hc	46.686 437 (14)	m^{-1} T^{-1}	0.30
e	elementary charge	1.602 177 33 (49)	10^{-19} C	0.30
$2e/h$	Josephson frequency–voltage ratio	4.835 9767 (14)	10^{14} Hz V^{-1}	0.30
Φ_0	magnetic flux quantum, $h/2e$	2.067 834 61 (61)	10^{-15} Wb	0.30
μ_N	nuclear magneton, $e\hbar/2m_P$	5.050 7866 (17)	10^{-27} J T^{-1}	0.34
	in electron volts, $\mu_N/\{e\}$	3.152 451 66 (28)	10^{-8} eV T^{-1}	0.089
	in hertz, μ_N/h	7.622 5914 (23)	MHz T^{-1}	0.30
	in kelvins, μ_N/k	3.658 246 (31)	10^{-4} K T^{-1}	8.5
	in wavenumbers, μ_N/hc	2.542 622 81 (77)	10^{-2} m^{-1} T^{-1}	0.30
e^2/h	quantized Hall conductance	3.874 046 14 (17)	10^{-5} S	0.045
R_H	quantized Hall resistance, $h/e^2 = \mu_0 c/2\alpha$	25,812.8056 (12)	Ω	0.045

Atomic Constants

Symbol	Quantity	Value	Units	Relative uncertainty (ppm)
a_0	Bohr radius, $\alpha/4\pi R_\infty$	0.529 177 249 (24)	10^{-10} m	0.045
α	fine-structure constant, $\mu_0 ce^2/2h$	7.297 353 08 (33)	10^{-3}	0.045
α^{-1}	inverse fine-structure constant	137.035 9895 (61)		0.045
E_h	Hartree energy, $e^2/4\pi\epsilon_0 a_0 = 2R_\infty hc$	4.359 7482 (26)	10^{-18} J	0.60
	in eV, $E_h/\{e\}$	27.211 3961 (81)	eV	0.30
R_∞	Rydberg constant, $m_e c\alpha^2/2h$	10,973,731.534 (13)	m^{-1}	0.0012
	in eV, $R_\infty hc/\{e\}$	13.605 6981 (40)	eV	0.30
	in hertz, $R_\infty c$	3.289 841 9499 (39)	10^{15} Hz	0.0012
	in joules, $R_\infty hc$	2.179 8741 (13)	10^{-18} J	0.60
$h/2m_e$	quantum of circulation	3.636 948 07 (33)	10^{-4} m^2 s^{-1}	0.089
h/m_e		7.273 896 14 (65)	10^{-4} m^2 s^{-1}	0.089

Fundamental Physical Constants

Atomic Constants (*continued*)

Symbol	Quantity	Value	Units	Relative uncertainty (ppm)
	Electron			
r_e	classical electron radius, $\alpha^2 a_0$	2.817 940 92 (38)	10^{-15} m	0.13
λ_C	Compton wavelength, $h/m_e c$	2.426 310 58 (22)	10^{-12} m	0.089
$\lambda\!\!\!^{-}{}_C$	$\lambda_C/2\pi = \alpha a_0 = \alpha^2/4\pi R_\infty$	3.861 593 23 (35)	10^{-13} m	0.089
m_e/m_α	electron–α-particle mass ratio	1.370 933 54 (3)	10^{-4}	0.021
m_e/m_d	electron–deuteron mass ratio	2.724 437 07 (6)	10^{-4}	0.020
g_e	electron g-factor, $2(1 + a_e)$	2.002 319 304 386 (20)		1×10^{-5}
μ_e	electron magnetic moment	928.477 01 (31)	10^{-26} J T^{-1}	0.34
μ_e/μ_B	in Bohr magnetons	1.001 159 652 193 (10)		1×10^{-5}
μ_e/μ_N	in nuclear magnetons	1838.282 000 (37)		0.020
a_e	electron magnetic moment anomaly, $\mu_e/\mu_B - 1$	1.159 652 193 (10)	10^{-3}	0.0086
m_e	electron mass	9.109 3897 (54)	10^{-31} kg	0.59
	in atomic mass units	5.485 799 03 (13)	10^{-4} u	0.023
	in electron volts, $m_e c^2/\{e\}$	0.510 999 06 (15)	MeV	0.30
$M(e)$, M_e	electron molar mass	5.485 799 03 (13)	10^{-7} kg/mol	0.023
μ_e/μ_μ	electron–muon magnetic moment ratio	206.766 967 (30)		0.15
m_e/m_μ	electron–muon mass ratio	4.836 332 18 (71)	10^{-3}	0.15
μ_e/μ_p	electron–proton magnetic moment ratio	658.210 6881 (66)		0.010
m_e/m_p	electron–proton mass ratio	5.446 170 13 (11)	10^{-4}	0.020
$-e/m_e$	electron specific charge	−1.758 819 62 (53)	10^{11} C kg^{-1}	0.30
σ_e	Thomson cross section, $(8\pi/3)r_e^2$	0.665 246 16 (18)	10^{-28} m^2	0.27
	Muon			
m_μ/m_e	muon–electron mass ratio	206.768 262 (30)		0.15
g_μ	muon g-factor, $2(1 + a_\mu)$	2.002 331 846 (17)		0.0084
μ_μ	muon magnetic moment	4.490 4514 (15)	10^{-26} J T^{-1}	0.33
μ_μ/μ_B	in Bohr magnetons	4.841 970 97 (71)	10^{-3}	0.15
μ_μ/μ_N	in nuclear magnetons	8.890 5981 (13)		0.15
a_μ	muon magnetic moment anomaly, $[\mu_\mu/(e\hbar/2m_\mu)] - 1$	1.165 9230 (84)	10^{-3}	7.2
m_μ	muon mass	1.883 5327 (11)	10^{-28} kg	0.61
	in atomic mass units	0.113 428 913 (17)	u	0.15
	in electron volts, $m_\mu c^2/\{e\}$	105.658 389 (34)	MeV	0.32
$M(\mu)$, M_μ	muon molar mass	1.134 289 13 (17)	10^{-4} kg/mol	0.15
μ_μ/μ_p	muon–proton magnetic moment ratio	3.183 345 47 (47)		0.15
	Proton			
σ_{H_2O}	diamagnetic shielding correction for protons in pure water, spherical sample, 25°C, $1 - \mu_p'/\mu_p$	25.689 (15)	10^{-6}	
$\lambda_{C,p}$	proton Compton wavelength, $h/m_p c$	1.321 410 02 (12)	10^{-15} m	0.089
$\lambda\!\!\!^{-}{}_{C,p}$	$\lambda_{C,p}/2\pi$	2.103 089 37 (19)	10^{-16} m	0.089
m_p/m_e	proton–electron mass ratio	1836.152 701 (37)		0.020
γ_p	proton gyromagnetic ratio	26,752.2128 (81)	10^4 s^{-1} T^{-1}	0.30
$\gamma_p/2\pi$		42.577 469 (13)	MHz T^{-1}	0.30
γ_p'	uncorrected (H_2O, sph., 25°C)	26,751.5255 (81)	10^4 s^{-1} T^{-1}	0.30
$\gamma_p'/2\pi$		42.576 375 (13)	MHz T^{-1}	0.30
μ_p	proton magnetic moment	1.410 607 61 (47)	10^{-26} J T^{-1}	0.34
μ_p/μ_B	in Bohr magnetons	1.521 032 202 (15)	10^{-3}	0.010
μ_p/μ_N	in nuclear magnetons	2.792 847 386 (63)		0.023
m_p	proton mass	1.672 6231 (10)	10^{-27} kg	0.59
	in atomic mass units	1.007 276 470 (12)	u	0.012
	in electon volts, $m_p c^2/\{e\}$	938.272 31 (28)	MeV	0.30
$M(p)$, M_p	proton molar mass	1.007 276 470 (12)	10^{-3} kg/mol	0.012
m_p/m_μ	proton–muon mass ratio	8.880 2444 (13)		0.15
e/m_p	proton specific charge	9.578 8309 (29)	10^7 C kg^{-1}	0.30
μ_p'	shielded proton moment (H_2O, sph., 25°C)	1.410 571 38 (47)	10^{-26} J T^{-1}	0.34
μ_p'/μ_B	in Bohr magnetons	1.520 993 129 (17)	10^{-3}	0.011
μ_p'/μ_N	in nuclear magnetons	2.792 775 642 (64)		0.02

Fundamental Physical Constants

Atomic Constants (*continued*)

Symbol	Quantity	Value	Units	Relative uncertainty (ppm)
		Neutron		
$\lambda_{C,n}$	neutron Compton wavelength, h/m_nc	1.319 591 10 (12)	10^{-15} m	0.089
$\bar{\lambda}_{C,n}$	$\lambda_{C,n}/2\pi$	2.100 194 45 (19)	10^{-16} m	0.089
μ_n/μ_e	neutron–electron magnetic moment ratio	1.040 668 82 (25)	10^{-3}	0.24
m_n/m_e	neutron–electron mass ratio	1838.683 662 (40)		0.022
μ_n	neutron magnetic moment	0.966 237 07 (40)	10^{-26} J T^{-1}	0.41
μ_n/μ_B	in Bohr magnetons	1.041 875 63 (25)	10^{-3}	0.24
μ_n/μ_N	in nuclear magnetons	1.913 042 75 (45)		0.24
m_n	neutron mass	1.674 9286 (10)	10^{-27} kg	0.59
	in atomic mass units	1.008 664 904 (14)	u	0.014
	in electron volts, $m_nc^2/\{e\}$	939.565 63 (28)	MeV	0.30
$M(n)$, M_n	neutron molar mass	1.008 664 904 (14)	10^{-3} kg/mol	0.014
μ_n/μ_p	neutron–proton magnetic moment ratio	0.684 979 34 (16)		0.24
m_n/m_p	neutron–proton mass ratio	1.001 378 404 (9)		0.009
		Deuteron		
μ_d/μ_e	deuteron–electron magnetic moment ratio	0.466 434 5460 (91)	10^{-3}	0.019
m_d/m_e	deuteron–electron mass ratio	3670.483 014 (75)		0.020
μ_d	deuteron magnetic moment	0.433 073 75 (15)	10^{-26} J T^{-1}	0.34
μ_d/μ_B	in Bohr magnetons	0.466 975 4479 (91)	10^{-3}	0.019
μ_d/μ_N	in nuclear magnetons	0.857 438 230 (24)		0.028
m_d	deuteron mass	3.343 5860 (20)	10^{-27} kg	0.59
	in atomic mass units	2.013 553 214 (24)	u	0.012
	in electron volts, $m_dc^2/\{e\}$	1875.613 39 (57)	MeV	0.30
$M(d)$, M_d	deuteron molar mass	2.013 553 214 (24)	10^{-3} kg/mol	0.012
μ_d/μ_p	deuteron–proton magnetic moment ratio	0.307 012 2035 (51)		0.017
m_d/m_p	deuteron–proton mass ratio	1.999 007 496 (6)		0.003

Physico-Chemical Constants

Symbol	Quantity	Value	Units	Relative uncertainty (ppm)
m_u	atomic mass constant, $\frac{1}{12}m(^{12}C)$	1.660 5402 (10)	10^{-27} kg	0.59
	in electron volts, $m_uc^2/\{e\}$	931.494 32 (28)	MeV	0.30
N_A, L	Avogadro constant	6.022 1367 (36)	10^{23} mol^{-1}	0.59
k	Boltzmann constant, R/N_A	1.380 658 (12)	10^{-23} J K^{-1}	8.5
	in electron volts, $k/\{e\}$	8.617 385 (73)	10^{-5} eV K^{-1}	8.4
	in hertz, k/h	2.083 674 (18)	10^{10} Hz K^{-1}	8.4
	in wavenumbers, k/hc	69.503 87 (59)	m^{-1} K^{-1}	8.4
F	Faraday constant	96,485.309 (29)	C mol^{-1}	0.30
R	molar gas constant	8.314 510 (70)	J mol^{-1} K^{-1}	8.4
N_Ah	molar Planck constant	3.990 313 23 (36)	10^{-10} J s mol^{-1}	0.089
N_Ahc		0.119 626 58 (11)	J m mol^{-1}	0.089
V_m	molar volume (ideal gas), RT/p			
	$T = 273.15$ K, $p = 101{,}325$ Pa	22.414 10 (19)	L/mol	8.4
n_0	Loschmidt constant, N_A/V_m	2.686 763 (23)	10^{25} m^{-3}	8.5
V_m	$T = 273.15$ K, $p = 100$ kPa	22.711 08 (19)	L/mol	8.4
c_1	radiation constant, first, $2\pi hc^2$	3.741 7749 (22)	10^{-16} W m^2	0.60
c_2	radiation constant, second, hc/k	0.014 387 69 (12)	m K	8.4
	Sackur–Tetrode constant (absolute entropy constant), $\frac{5}{2} + \ln\{(2\pi m_ukT_1/h^2)^{3/2}kT_1/p_0\}$			
S_0/R	$T_1 = 1$ K, $p_0 = 100$ kPa	−1.151 693 (21)		18
	$p_0 = 101{,}325$ Pa	−1.164 856 (21)		18
	Stefan–Boltzmann constant,			
σ	$(\pi^2/60)k^4/\hbar^3c^2$	5.670 51 (19)	10^{-8} W m^{-2} K^{-4}	34
b	Wien displacement law constant, $b = \lambda_{max}T \approx c_2/4.965\ 114\ 23$	2.897 756 (24)	10^{-3} m K	8.4

Fundamental Physical Constants

Standard Values/Maintained Units

Symbol	Quantity	Value	Units	Relative uncertainty (ppm)
u	atomic mass unit (unified), $1\ u = m_u = \frac{1}{12}m(^{12}C)$	1.660 5402 (10)	10^{-27} kg	0.59
eV	electron volt, $(e/C)J = \{e\}$ J	1.602 177 33 (49)	10^{-19} J	0.30
g_n	standard acceleration of gravity	**9.806 65**	ms^{-2}	
atm	standard atmospherre	**1.013 25**	10^5 Pa	
Maintained Electrical Units				
A_{BI85}	BIPM maintained ampere,	$1 - 6.03\ (30) \times 10^{-6}$	A	
	$A_{BIPM} = V_{76-BI}/\Omega_{69-BI}$	= 0.999 993 97 (30)	A	0.30
Ω_{BI85}	BIPM maintained ohm, Ω_{69-BI}	$1 - 1.563(50) \times 10^{-6}$	Ω	
	$\Omega_{BI85} \equiv \Omega_{69-BI}$(1 Jan. 1985)	= 0.999 998 437 (50)	Ω	0.050
$\frac{d\Omega_{69-BI}}{dt}$	Drift rate of Ω_{69-BI}	−0.0566 (15)	μΩ/a	
V_{76-BI}	BIPM maintained volt,	$1 - 7.59(30) \times 10^{-6}$	V	
	$V_{76-BI} \equiv 483{,}594$ GHz $(h/2e)$	= 0.999 992 41 (30)	V	0.30
X-Ray Standards				
Å*	Å*: $\lambda(WK\alpha_1) \equiv 0.209\ 100$ Å*	1.000 014 81 (92)	10^{-10} m	0.92
$xu(CuK\alpha_1)$	Cu x-unit: $\lambda(CuK\alpha_1) \equiv 1537.400$ xu	1.002077 89 (70)	10^{-13}m	0.70
a	lattice spacing of Si (in vacuum, 22.5°C),	0.543 101 96 (11)	nm	0.21
d_{220}	$d_{220} = a/\sqrt{8}$	0.192 015 540 (40)	nm	0.21
$V_m(Si)$	molar volume of Si, $M(Si)/\rho(Si) = N_A a^3/8$	12.058 8179 (89)	cm^3/mol	0.74
$xu(MoK\alpha_1)$	Mo x-unit : $\lambda(MoK\alpha_1) \equiv 707.831$ xu	1.002 099 38 (45)	10^{-13} m	0.45

Measurement Conversion

Values in bold are exact.

Length

Units	in.	ft	yd	mi	cm	m
1 inch	**1**	0.08333333	0.02777778	0.00001578283	**2.54**	**0.0254**
1 foot	**12**	**1**	0.3333333	0.0001893939	**30.48**	**0.3048**
1 yard	**36**	**3**	**1**	0.0005681818	**91.44**	**0.9144**
1 mile	**63,360**	**5280**	**1760**	**1**	**160,934.4**	**1609.344**
1 centimeter	0.3937008	0.03280840	0.01093613	0.000006213712	**1**	**0.01**
1 meter	39.37008	3.280840	1.093613	0.0006213712	**100**	**1**

Area

Units	$in.^2$	ft^2	yd^2	mi^2	cm^2	m^2
1 square inch	**1**	0.006944444	0.0007716049	0.0000000002490977	**6.4516**	**0.00064516**
1 square foot	**144**	**1**	0.1111111	0.00000003587006	**929.0304**	**0.09290304**
1 square yard	**1296**	**9**	**1**	0.0000003228306	**8361.2736**	**0.83612736**
1 square mile	**4,014,489,600**	**27,878,400**	**3,097,600**	**1**	**25,899,881,103.36**	**2,589,988.110336**
1 square centimeter	0.1550003	0.001076391	0.0001195990	0.00000000003861022	**1**	**0.0001**
1 square meter	1550.003	10.76391	1.195990	0.0000003861022	**10,000**	**1**

Volume

Units	$in.^3$	ft^3	yd^3	cm^3	dm^3	m^3
1 cubic inch	**1**	0.0005787037	0.00002143347	**16.387064**	**0.016387064**	**0.000016387064**
1 cubic foot	**1728**	**1**	0.03703704	**28,316.846592**	**28.316846592**	**0.028316846592**
1 cubic yard	**46,656**	**27**	**1**	**764,554.857984**	**764.554857984**	**0.764554857984**
1 cubic centimeter	0.06102374	0.00003531467	0.000001307951	**1**	**0.001**	**0.000001**
1 cubic decimeter	61.02374	0.03531467	0.001307951	**1000**	**1**	**0.001**
1 cubic meter	61,023.74	35.31467	1.307951	**1,000,000**	**1000**	**1**

Mass

Units	avdp oz	avdp lb	short ton	long ton	kg	metric ton
1 avdp ounce	**1**	**0.0625**	**0.00003125**	0.00002790179	**0.028349523125**	**0.000028349523125**
1 avdp pound	**16**	**1**	**0.0005**	0.0004464286	**0.45359237**	**0.00045359237**
1 short ton	**32,000**	**2000**	**1**	0.8928571	**907.18474**	**0.90718474**
1 long ton	**35,840**	**2240**	**1.12**	**1**	**1016.0469088**	**1.0160469088**
1 kilogram	35.27396	2.204623	0.001102311	0.0009842065	**1**	**0.001**
1 metric ton	35,273.96	2204.623	1.102311	0.9842065	**1000**	**1**

Energy

	J	kg	m^{-1}	Hz	K	eV	u	hartree
1 joule	**1**	$1.11265006 \times 10^{-17}$	5.0341125×10^{24}	$1.50918897 \times 10^{33}$	7.242924×10^{22}	6.2415064×10^{18}	6.7005308×10^{9}	2.2937104×10^{17}
1 kilogram	$8.987551787 \times 10^{16}$	**1**	4.5244347×10^{41}	$1.35639140 \times 10^{50}$	6.509616×10^{39}	5.6095862×10^{35}	6.0221367×10^{26}	2.0614841×10^{34}
1 $meter^{-1}$	$1.9864475 \times 10^{-25}$	$2.2102209 \times 10^{-42}$	**1**	299,792,458	0.01438769	$1.23984244 \times 10^{-6}$	$1.33102522 \times 10^{-15}$	$4.5563352672 \times 10^{-8}$
1 hertz	$6.6260755 \times 10^{-34}$	$7.3725032 \times 10^{-51}$	$3.335640952 \times 10^{-9}$	**1**	4.799216×10^{-11}	$4.1356692 \times 10^{-15}$	$4.43982224 \times 10^{-24}$	$1.5198298508 \times 10^{-16}$
1 kelvin	1.380658×10^{-23}	1.536189×10^{-40}	69.50387	2.083674×10^{10}	**1**	8.617385×10^{-5}	9.251140×10^{-14}	3.166829×10^{-6}
1 electron volt	$1.60217733 \times 10^{-19}$	$1.78266270 \times 10^{-36}$	806,554.10	$2.41798836 \times 10^{14}$	11,604.45	**1**	$1.07354385 \times 10^{-9}$	0.036749309
1 unified atomic mass unit	$1.49241909 \times 10^{-10}$	$1.6605402 \times 10^{-27}$	$7.51300563 \times 10^{14}$	$2.25234242 \times 10^{23}$	1.0809478×10^{13}	931.49432×10^{6}	**1**	3.42317725×10^{7}
1 hartree	$4.3597482 \times 10^{-18}$	$4.8508741 \times 10^{-35}$	21,947,463.067	$6.5796838999 \times 10^{15}$	3.157733×10^{5}	27.2113961	$2.92126269 \times 10^{-8}$	**1**

Chemistry

Periodic Table of the Elements[a]

New notation → Previous IUPAC form → CAS version

1 Group IA	2 IIA	3 IIIA IIIB	4 IVA IVB	5 VA VB	6 VIA VIB	7 VIIA VIIB	8	9 VIIIA VIII	10	11 IB	12 IIB	13 IIIB IIIA	14 IVB IVA	15 VB VA	16 VIB VIA	17 VIIB VIIA	18 VIIIA
1 H 1.00794																	2 He 4.00260
3 Li 6.941	4 Be 9.01218											5 B 10.811	6 C 12.011	7 N 14.0067	8 O 15.9994	9 F 18.9984	10 Ne 20.1797
11 Na 22.9898	12 Mg 24.3050											13 Al 26.9815	14 Si 28.0855	15 P 30.9738	16 S 32.066	17 Cl 35.4527	18 Ar 39.948
19 K 39.0983	20 Ca 40.078	21 Sc 44.9559	22 Ti 47.88	23 V 50.9415	24 Cr 51.9961	25 Mn 54.9380	26 Fe 55.847	27 Co 58.9332	28 Ni 58.69	29 Cu 63.546	30 Zn 65.39	31 Ga 69.723	23 Ge 72.61	33 As 74.9216	34 Se 78.96	35 Br 79.904	36 Kr 83.80
37 Rb 85.4678	38 Sr 87.62	39 Y 88.9059	40 Zr 91.224	41 Nb 92.9064	42 Mo 95.94	43 Tc (98)	44 Ru 101.07	45 Rh 102.906	46 Pd 106.42	47 Ag 107.868	48 Cd 112.411	49 In 114.82	50 Sn 118.710	51 Sb 121.75	52 Te 127.60	53 I 126.905	54 Xe 131.29
55 Cs 132.905	56 Ba 137.327	57 La ★ 138.906	72 Hf 178.49	73 Ta 180.948	74 W 183.85	75 Re 186.207	76 Os 190.2	77 Ir 192.22	78 Pt 195.08	79 Au 196.967	80 Hg 200.59	81 Tl 204.383	82 Pb 207.2	83 Bi 208.980	84 Po (209)	85 At (210)	86 Rn (222)
87 Fr (223)	88 Ra 226.025	89 Ac ▲ 227.028	104 (261)	105 (262)	106 (263)	107 (262)	108 (265)	109 (267)									

★ Lanthanide series	58 Ce 140.115	59 Pr 140.908	60 Nd 144.24	61 Pm (145)	62 Sm 150.36	63 Eu 151.965	64 Gd 157.25	65 Tb 158.925	66 Dy 162.50	67 Ho 164.930	68 Er 167.26	69 Tm 168.934	70 Yb 173.04	71 Lu 174.967
▲ Actinide series	90 Th 232.038	91 Pa 231.036	92 U 238.029	93 Np 237.048	94 Pu (244)	95 Am (243)	96 Cm (247)	97 Bk (247)	98 Cf (251)	99 Es (252)	100 Fm (257)	101 Md (258)	102 No (259)	103 Lr (260)

Source: Adapted from *Chemical Engineering News*, page 27, February 4, 1985.

[a] Atomic masses shown here are adapted from the 1985 IUPAC values (maximum of six significant figures).

Atomic Weights[a]

Element	Symbol	Atomic number	Atomic weight	Element	Symbol	Atomic number	Atomic weight	Element	Symbol	Atomic number	Atomic weight
Actinium	Ac	89	227.0278	Gadolinium	Gd	64	157.25	Potassium	K	19	39.0983
Aluminum	Al	13	26.981539	Gallium	Ga	31	69.723	Praseodymium	Pr	59	140.90765
Americium	Am	95	(243)	Germanium	Ge	32	72.61	Promethium	Pm	61	(145)
Antimony	Sb	51	121.75	Gold	Au	79	196.96654	Protactinium	Pa	91	231.03588
Argon	Ar	18	39.948	Hafnium	Hf	72	178.49	Radium	Ra	88	226.0254
Arsenic	As	33	74.92159	Helium	He	2	4.002602	Radon	Rn	86	(222)
Astatine	At	85	(210)	Holmium	Ho	67	164.93032	Rhenium	Re	75	186.207
Barium	Ba	56	137.327	Hydrogen	H	1	1.00794	Rhodium	Rh	45	102.90550
Berkelium	Bk	97	(247)	Indium	In	49	114.82	Rubidium	Rb	37	85.4678
Beryllium	Be	4	9.012182	Iodine	I	53	126.90447	Ruthenium	Ru	44	101.07
Bismuth	Bi	83	208.98037	Iridium	Ir	77	192.22	Samarium	Sm	62	150.36
Boron	B	5	10.811	Iron	Fe	26	55.847	Scandium	Sc	21	44.955910
Bromine	Br	35	79.904	Krypton	Kr	36	83.80	Selenium	Se	34	78.96
Cadmium	Cd	48	112.411	Lanthanum	La	57	138.9055	Silicon	Si	14	28.0855
Calcium	Ca	20	40.078	Lawrencium	Lr	103	(262)	Silver	Ag	47	107.8682
Californium	Cf	98	(251)	Lead	Pb	82	207.2	Sodium	Na	11	22.989768
Carbon	C	6	12.011	Lithium	Li	3	6.941	Strontium	Sr	38	87.62
Cerium	Ce	58	140.115	Lutetium	Lu	71	174.967	Sulfur	S	16	32.066
Cesium	Cs	55	132.90543	Magnesium	Mg	12	24.3050	Tantalum	Ta	73	180.9479
Chlorine	Cl	17	35.4527	Manganese	Mn	25	54.93805	Technetium	Tc	43	(98)
Chromium	Cr	24	51.9961	Mendelevium	Md	101	(258)	Tellurium	Te	52	127.60
Cobalt	Co	27	58.93320	Mercury	Hg	80	200.59	Terbium	Tb	65	158.92534
Copper	Cu	29	63.546	Molybdenum	Mo	42	95.94	Thallium	Tl	81	204.3833
Curium	Cm	96	(247)	Neodymium	Nd	60	144.24	Thorium	Th	90	232.0381
Dysprosium	Dy	66	162.50	Neon	Ne	10	20.1797	Thulium	Tm	69	168.93421
Einsteinium	Es	99	(252)	Neptunium	Np	93	237.0482	Tin	Sn	50	118.710
Element 104 (Unnilquadium)		104	(261)	Nickel	Ni	28	58.69	Titanium	Ti	22	47.88
Element 105 (Unnilpentium)		105	(262)	Niobium	Nb	41	92.90638	Tungsten	W	74	183.85
Element 106 (Unnilhexium)		106	(263)	Nitrogen	N	7	14.00674	Uranium	U	22	238.0289
Element 107 (Unnilseptium)		107	(262)	Nobelium	No	102	(259)	Vanadium	V	23	50.9415
Element 108		108	(265)	Osmium	Os	76	190.2	Xenon	Xe	54	131.29
Element 109 (Unnilennium)		109	(267)	Oxygen	O	8	15.9994	Ytterbium	Yb	70	173.04
Erbium	Er	68	167.26	Palladium	Pd	46	106.42	Yttrium	Y	39	88.90585
Europium	Eu	63	151.965	Phosphorus	P	15	30.973762	Zinc	Zn	30	65.39
Fermium	Fm	100	(257)	Platinum	Pt	78	195.08	Zirconium	Zr	40	91.224
Fluorine	F	9	18.9984032	Plutonium	Pu	94	(244)				
Francium	Fr	87	(223)	Polonium	Po	84	(209)				

[a] The atomic weights given are adapted from the 1985 IUPAC Atomic Weight table. Values in parentheses represent the mass number of the isotope of longest known half-life.

"Stable" Particles

Class	Symbol	Spin, parity	*I*-spin T	T_3	Strangeness	Mass (MeV/c²)	Mean life (sec)	Common decay modes	(percent)
Photon	γ	1				0			
Lepton	ν_e	$\frac{1}{2}$				<0.00006			
	ν_μ	$\frac{1}{2}$				<0.57			
	$e^\pm$	$\frac{1}{2}$				0.5110034 ± 0.0000014	$>10^{29}$		
	$\mu^\pm$					105.65946 ± 0.00024	$(2.19712 \pm .000077) \times 10^{6}$	$e\nu\nu$	(100)
	τ^-	$\frac{1}{2}$				1782 ± 4	$(4.6 \pm 1.9) \times 10^{-13}$	$\mu\nu\nu$	(18)
								$e\nu\nu$	(17)
								hadrons + neutrals	(65)
Meson	$\pi^\pm$	0^-	1	±1	0	139.5669 ± 0.0012	$(2.6030 \pm .0023) \times 10^{-8}$	$\mu\nu$	(100)
	π^0	0^-	1	0	0	134.9626 ± 0.0039	$(0.828 \pm .057) \times 10^{-16}$	$\gamma\gamma$	(98.8)
								γe^+e^-	(1.2)
	$K^\pm$	0^-	$\frac{1}{2}$	$\pm\frac{1}{2}$	±1	493.669 ± 0.015	$(1.2371 \pm .0026) \times 10^{-8}$	$\mu\nu$	(63.5)
								$\pi\pi^0$	(21.2)
								$\pi\pi\pi$	(5.6)
								$\pi\pi^0\pi^0$	(1.7)
								$\mu\pi^0\nu$	(3.2)
								$e\pi^0\nu$	(4.8)
	K^0	0^-	$\frac{1}{2}$	$-\frac{1}{2}$	+1	497.67 ± 0.13	50% K_S, 50% K_L		
	$K_S(K_1)$						$(0.8923 \pm .0022) \times 10^{-10}$	$\pi^+\pi^-$	(68.6)
								$\pi^0\pi^0$	(31.4)
	$K_L(K_2)$						$(5.183 \pm .040) \times 10^{-8}$	$\pi^0\pi^0\pi^0$	(21.5)
								$\pi^+\pi^-\pi^0$	(12.4)
								$\pi\mu\nu$	(27.0)
								$\pi e\nu$	(38.8)
								$\pi^+\pi^-$	(0.20)
	$D^\pm$	0^-	$\frac{1}{2}$	$\pm\frac{1}{2}$	0 (Charm = ±1)	1868.3 ± 0.9	$(2.5 {+3.5 \atop -1.5}) \times 10^{-13}$		
	D^0	0^-	$\frac{1}{2}$	$-\frac{1}{2}$	0 (Charm = +1)	1863.1 ± 0.9	$(2.3 {+0.8 \atop -0.5}) \times 10^{-13}$		
Baryon	p	$\frac{1}{2}$	$\frac{1}{2}$	$+\frac{1}{2}$	0	938.2796 ± .0027	$>10^{37}$		
	n	$\frac{1}{2}$	$\frac{1}{2}$	$-\frac{1}{2}$	0	939.5731 ± .0027	917 ± 14	$pe^-\nu$	(100)
	Λ	$\frac{1}{2}$	0	0	−1	1115.60 ± 0.05	$(2.632 \pm .020) \times 10^{-10}$	$p\pi^-$	(64.2)
								$n\pi^0$	(35.8)
	Σ^+	$\frac{1}{2}$	1	+1	−1	1189.36 ± 0.06	$(0.800 \pm .004) \times 10^{-10}$	$p\pi^0$	(51.6)
								$n\pi^+$	(48.4)
	Σ^0	$\frac{1}{2}$	1	0	−1	1192.46 ± 0.08	$(5.8 \pm 1.3) \times 10^{-20}$	$\Lambda\gamma$	(100)
	Σ^-	$\frac{1}{2}$	1	−1	−1	1197.34 ± 0.05	$(1.482 \pm .011) \times 10^{-10}$	$n\pi^-$	(100)
	Ξ^0	$\frac{1}{2}$	$\frac{1}{2}$	$+\frac{1}{2}$	−2	1314.9 ± 0.6	$(2.90 \pm 0.1) \times 10^{-10}$	$\Lambda\pi^0$	(100)
	Ξ^-	$\frac{1}{2}$	$\frac{1}{2}$	$-\frac{1}{2}$	−2	1321.32 ± 0.13	$(1.641 \pm .016) \times 10^{-10}$	$\Lambda\pi^-$	(100)
	Ω^-	$\frac{3}{2}$	0	0	−3	1672.22 ± 0.31	$(0.82 \pm 0.03) \times 10^{-10}$	ΛK^-	(68.6)
								$\Xi^0\pi^-$	(23.4)
								$\Xi^-\pi^0$	(8.0)
	Λ_c^+	$\frac{1}{2}$	0	0	0 (Charm = +1)	2273 ± 6	$\sim 7 \times 10^{-13}$		

The Solar System

The Planets[a]

Planet	Mass (10^{24} kg)	Equatorial radius (km)	Density (10^3 kg m^{-3})	Escape velocity (km sec^{-1})
Mercury	0.3303	2439	5.43	4.25
Venus	4.870	6051	5.25	10.4
Earth	5.976	6378	5.518	11.2
Mars	0.6421	3393	3.95	5.02
Jupiter	1900	71,398	1.332	59.6
Saturn	568.8	60,330	0.689	35.5
Uranus	86.87	26,200	1.18	21.3
Neptune	102.0	25,225	(1.54)	23.3
Pluto	(0.013)	1145	1.84	1.3

[a] Values in parentheses are uncertain.

Planetary Orbits

Planet	Semimajor axis (10^6 km)	Orbital period (yr)	Eccentricity	Inclination[a] (deg)
Mercury	57.9	0.24085	0.206	7.003
Venus	108.2	0.61521	0.0068	3.394
Earth	149.6	1.00004	0.0167	0.000
Mars	227.9	1.88089	0.0933	1.850
Jupiter	778.3	11.86223	0.0483	1.309
Saturn	1427.0	29.45774	0.0559	2.493
Uranus	2869.6	84.018	0.047	0.772
Neptune	4496.6	164.78	0.0087	1.779
Pluto	5900.1	248.4	0.247	17.146

[a] Inclination is measured with respect to the orbital plane of the earth.

Planetary Spins and Magnetic Fields

Planet	Rotation period	Obliquity of spin axis[a] (deg)	Magnetic moment (G cm^3)	Surface field[b] (G)
Mercury	58.65 days	2 ± 3	2.4×10^{22}	0.002
Venus	243.01 days	177.3	$<4 \times 10^{21}$	<0.00002
Earth	23.9345 hr	23.45	7.98×10^{25}	0.3
Mars	24.6299 hr	23.98	2.5×10^{22}	0.0006
Jupiter	9.841 hr (equator) 9.925 hr (interior)	3.12	1.5×10^{30}	4
Saturn	10.233 hr (equator) 10.675 hr (interior)	26.73	4.6×10^{28}	0.2
Uranus	17.24 hr (interior)	97.86	4.1×10^{27}	0.2
Neptune	18.2 ± 0.4 hr	(29.56)	—	—
Pluto	6.387 days	(118.5)	—	—

[a] Values in parentheses are uncertain.
[b] Measured at the magnetic equator.

Geological Timetable

Era	Period	Epoch	Millions of years ago	Important geological and biological events	Temporal distribution of life forms
Cenozoic	Quaternary	Recent		Abnormally high provincialism and few epicontinental seas	
			0.011		
		Pleistocene		Great Ice Age (glacial and interglacial periods)	
			2.5		
	Tertiary	Pliocene		Renewed uplift of Rocky Mountains; Grand Canyon, Cascades begin to form	Modern humans
			7.0		
		Miocene			
			26	Antarctica freezes	
		Oligocene			
			38	Major uplift of Himalayan Mountains as India collides with Asia	
		Eocene			
			54		
		Paleocene			
			65	Marine and terrestrial mass extinction	
Mesozoic	Cretaceous				
			136	Major uplift of Rocky Mountains	Angiosperms
	Jurassic				
			190		
	Triassic			Break-up of Pangea begins	Birds, mammals
			225	Marine mass extinction	

Paleozoic	Permian	280	Completion of uplift of Appalachian Mountains: supercontinent Pangea assembled; glaciation	Reptiles
	Carboniferous	345	Extensive coal swamps	Amphibians Terrestrial invertebrates Gymnosperms
	Devonian	395		? Mosses, liverworts (probably older than fossil record shows) Ferns and fern-allies
	Silurian	435	Glaciation	
	Ordovician	500	Second marine diversification	Fishes (including ostracoderms)
	Cambrian	570	Major marine diversification	Marine invertebrates
Proterozoic (Precambrian)			Glaciation; widespread continental rifting	
		1000		Simpler eukaryotic organisms; protozoa, algae, fungi
		2000	Free oxygen appears in atmosphere	
		3500	Oldest known fossils	Bacteria and blue-green algae
		3700	Oldest known terrestrial rocks	
		4600	Oldest known lunar rocks (? age of solar system)	

Source: Adapted from Eugene N. Kozloff (1990). "Invertebrates." Saunders College Publishing, Philadelphia.

Five-Kingdom Classification of Organisms

Kingdom Monera

Prokaryotes (cells lack a true nucleus and membrane-bound organelles); mostly unicellular; some occur in filaments or clusters.

Phylum Schizophyta: Bacteria; about 2,500 species including eubacteria (true bacteria), rickettsias, mycoplasmas, and spirochetes; mostly heterotrophs; some photosynthetic and chemosynthetic autotrophs; reproduction usually asexual by binary fission.

Phylum Cyanophyta: Blue-green algae or cyanobacteria; about 200 species of photosynthetic autotrophs with chlorophyll *a* and accessory pigments; no chloroplasts; mostly filamentous; some unicellular; reproduction asexual by binary fission or fragmentation

Phylum Prochlorophyta: Protosynthetic autotrophs; contain chlorophyll *a* and *b*, xanthophylls, and carotenes.

Kingdom Protista

Diverse group of unicellular and simple multicellular eukaryotes (cells having a true nucleus and membrane-bound organelles).

Phylum Euglenophyta: Euglenoids; about 800 species of unicellular, photosynthetic/heterotrophic organisms with chlorophyll *a* and *b*; usually having a single flagellum; reproduction asexual.

Phylum Mastigophora: Flagellates; about 2,500 species; mostly parasitic; includes *Trypanosoma* and *Trichonympha*.

Phylum Sarcodina: Sarcodines; about 11,500 species that move by means of pseudopodia; includes amoebas.

Phylum Ciliophora: Ciliates; about 7,200 species; locomotion by cilia, or sessile; includes paramecia and stentors.

Phylum Sporozoa: Sporozoans; about 6,000 species of nonmotile parasites; includes *Plasmodia,* the cause of malaria.

Phylum Chrysophyta: Golden algae; about 12,000 species of photosynthetic autotrophs with chlorophylls *a* and *c* and carotenes, xanthophylls, and fucoxanthins; most are unicellular and aquatic; includes diatoms.

Phylum Pyrrophyta: Fire algae; about 1,100 photosynthetic species with chlorophylls *a* and *c* and xanthophyll; unicellular; major component of marine phytoplankton; includes dinoflagellates.

Phylum Chlorophyta: Green algae; about 7,000 photosynthetic species with chlorophylls *a* and *b* and carotenoids; includes unicellular, colonial, and multicellular species; probable ancestor of modern land plants.

Phylum Phaeophyta: Brown algae; about 1,500 photosynthetic species with chlorophylls *a* and c and fucoxanthin; includes kelps.

Phylum Rhodophyta: Red algae; about 4,000 photosynthetic species with chlorophylls *a* and *d*, carotenes, and phycobilins; includes filamentous, multicellular seaweeds.

Kingdom Fungi

Eukaryotic heterotrophs that obtain food by absorption; includes saprophytes and parasites; most are multicellular, composed of intertwined filaments (hyphae); cell wall of most species contains chitin.

Phylum Myxomycophyta: Slime molds; about 600 species; body consists of multinucleate plasmodium that creeps by amoeboid movement; plasmodium separates into funguslike sporangia during reproduction; forms spores.

Phylum Eumycophyta: True fungi; about 81,500 species; mostly filamentous with chitinous cell walls; reproduction may include sexual and asexual stages.

Class Oomycetes: Aquatic fungi; some motile cells in certain stages of life cycle; approximately 200 species.

Class Zygomycetes: Terrestrial molds; about 600 species; reproduce asexually by spores and sexually by the fusion of nuclei from the tips of different mating strains.

Class Basidiomycetes: Club fungi; about 25,000 species of terrestrial fungi that produce spores on basidia; includes mushrooms, rusts, and smuts.

Class Ascomycetes: Sac fungi; about 30,000 terrestrial and aquatic species; spores form in an ascus (little sac) which results from the sexual combination of two gametes or strains; includes yeasts that reproduce asexually by budding, powdery mildews, morels, and truffles.

Class Deuteromycetes: Imperfect fungi; about 25,000 species that either do not reproduce sexually or have sexual life cycles that are not fully understood; includes *Penicillium.*

Kingdom Plantae

Multicellular, eukaryotic autotrophs that carry out photosynthesis in chloroplasts; chlorophyll *a* is the photoreactive pigment; chlorophyll *b* and various carotenoids serve as accessory pigments; primarily terrestrial; cell walls contain cellulose; body has distinct tissues; life cycle of alternating sporophyte and gametophyte generations.

Phylum Bryophyta: Bryophytes; about 15,600 species that lack vascular tissues and true roots, stems, and leaves; obtain nutrients by osmosis and diffusion; found chiefly in moist habitats; sperm must swim to eggs; gametophyte is dominant form in life cycle.

Class Muscopsida: Mosses; about 9,500 species with small, leafy gametophytes; sporophyte nonphotosynthetic, attached to and dependent upon gametophyte.

Class Hepaticopsida: Liverworts; about 6,000 extremely tiny species; gametophytes generally leafy, with scaly upper surface; sporophyte nonphotosynthetic, attached to and dependent upon gametophyte.

Class Antherocerotopsida: Hornworts; about 100 species; gametophyte not differentiated into roots, stems, or leaves; stomata on sporophyte; sporophyte grows from basal meristem.

All of the following plant phyla are tracheophytes with true roots, stems, and leaves in the dominant sporophyte; gametophyte is greatly reduced.

Phylum Psilophyta: Whisk ferns; a few species of seedless plants lacking roots and leaves; sperm must swim to eggs.

Phylum Sphenophyta: Horsetails; about 15 species of seedless plants with hollow, siliceous stems; sperm must swim to eggs.

Phylum Lycophyta: Club mosses; about 1,000 diverse species of seedless plants with leafy sporophytes; some have only one type of spore; others produce microspores and megaspores; sperm must swim to eggs.

Phylum Pterophyta: Ferns; about 12,000 diverse species of seedless plants; sporophyte bears haploid spores on underside of fronds; spores germinate into free-living gametophytes; sperm must swim to eggs.

Phylum Cycadophyta: Cycads; about 100 species of palmlike plants with slow cambial growth; gymnosperms.

Source: Adapted from Harvey D. Goodman *et al.* (1986). "Biology." Harcourt Brace Jovanovich, Orlando.

Five-Kingdom Classification of Organisms

Phylum Ginkgophyta: Ginkgo; one species only; fan-shaped leaves; gymnosperms; separate sexes.

Phylum Gnetophyta: Seed plants similar to angiosperms; about 70 species; motile sperm; xylem with vessels.

Phylum Coniferophyta: Conifers; about 550 species of gymnosperms; gametophyte much reduced; male pollen is dispersed by wind; female gametophytes remain on seed cone, where fertilization and seed development take place; leaves needlelike or scalelike; most species are evergreens.

Phylum Anthophyta: Angiosperms; about 235,000 species of plants that produce enclosed seeds; reproductive structures are flowers; mature seeds are enclosed in fruits; gametophyte is much reduced in size; double fertilization involves two sperm nuclei.

Class Monocotyledones: Monocots; angiosperms that produce seeds with one cotyledon (seed leaf); flower parts usually in threes; leaves with parallel veins; almost all are herbaceous.

Class Dicotyledones: Dicots; angiosperms that produce seeds with two cotyledons; flower parts usually in fours or fives; leaves usually with nonparallel veins; herbaceous and woody plants.

Kingdom Animalia

Multicellular, eukaryotic heterotrophs that obtain food by ingestion; most are motile; reproduction is predominantly sexual.

Phylum Porifera: Sponges; about 5,000 aquatic, mostly marine, species of asymmetrical animals that lack distinct tissues and organs; body consists of two layers supported by a stiff skeleton; sponges are aquatic, mostly marine, and sessile; reproduction sexual or asexual.

Class Calcarea: Simple shallow-water sponges; calcium carbonate spicules; includes *Grantia*.

Class Hexactinella; Deep water sponges; silica spicules; includes Venus's flower basket.

Class Demospongiae: Large sponge; spicules of spongin and silica; freshwater and marine species; includes bath sponges and finger sponges.

Class Sclerospongiae: Spicules of calcium carbonate, silica, and spongin.

Phylum Coelenterata (Cnidaria): Coelenterates; about 9,000 aquatic species; radially symmetrical; digestive cavity with only one opening; two body layers separated by jellylike mesoglea; tentacles armed with stinging cells; two body forms; vase-shaped polyp and bell-shaped medusa; live singly or in colonies often made up of specialized individuals; reproduction sexual or asexual.

Class Hydrozoa: Hydras and related animals; freshwater hydras occur singly with only polyp form; other hydrozoans are colonial and have a life cycle with both polyps and medusae; reproduction sexual or asexual.

Class Scyphozoa: Jellyfish; marine coelenterates; dominant medusa form reproduces sexually; free-swimming larvae become sessile polyps that form new medusae asexually.

Class Anthozoa: Sea anemones, corals, and related ''flower animals;'' marine coelenterates with no medusa stage; sea anemones are solitary; corals are colonial organisms with either an internal or external skeleton.

Phylum Ctenophora: Sea walnut and comb jellies; about 90 species; gelatinous marine animals with eight bands of cilia; often bioluminescent.

Phylum Platyhelminthes: Flatworms; about 13,000 species; bilaterally symmetrical with three germ layers; digestive cavity has only one opening; no circulatory or respiratory systems; no coelom or pseudocoelom.

Class Turbellaria: Planarians and related free-living, carnivorous flatworms; simple nervous system with sense receptors; sexual reproduction by mutual fertilization; asexual reproduction by fission.

Class Trematoda: Flukes; parasites covered by protective cuticle; most are endoparasites; may have a complex life cycle involving more than one host.

Class Cestoda: Tapeworms; parasitic species that live as adults in the intestines of vertebrates; no digestive system; food absorbed through body surfaces.

Phylum Nematoda: Roundworms; about 12,000 mostly parasitic species; tubular and bilaterally symmetrical body form; digestive tract has two openings; pseudocoelom; reproduction is sexual.

Phylum Nematomorpha: Horsehair worms; bodies very slender, up to one meter (3.3 ft) long; adults are free-living; larva parasitic.

Phylum Acanthocephala: Spiny-headed worms; about 500 species; parasitic with no digestive tract; head with recurved spines for attachment to host.

Phylum Rotifera: Rotifers or ''wheel'' animals; wormlike or spherical; complete digestive tract; crown of cilia on anterior end resembling a wheel.

Phylum Bryozoa: ''Moss'' animals; microscopic, aquatic organisms that form branching colonies; U-shaped row of ciliated tentacles for feeding (lophophore).

Phylum Brachiopoda: Lamp shells; about 250 species, 30,000 extinct; two shells, one dorsal and one ventral; adults sessile; feed by lophophore.

Phylum Mollusca: Mollusks; about 47,000 species of soft-bodied animals with a true coelom, a head-foot, a visceral mass, and a mantle; most are aquatic; many have one or more shells.

Class Pelecypoda: Bivalves; mollusks with two shells and a hatchet-shaped foot; no head; open circulatory system; many species sessile; includes clams, scallops, and oysters.

Class Gastropoda: Snails and slugs; aquatic and terrestrial mollusks with a locomotor belly-foot; open circulatory system; adults have an asymmetrical body twisted by torsion; most have a coiled shell.

Class Cephalopoda: Octopuses, squids, and nautiluses; marine mollusks with a large head and a foot divided into tentacles; closed circulatory system; highly developed nervous system; shell may be internal, external, or absent.

Class Scaphopoda: Tooth shells; about 350 species; marine mollusks with tubular shells.

Class Polyplacophora (Amphineura): Chitons; closely resemble ancestral mollusk; eight dorsal shell plates with reduced head and elongated body.

Phylum Annelida: Segmented worms; about 9,000 species; body has a true coelom; longitudinal and circular muscles; fairly complex circulatory, respiratory, and nervous systems; elimination by nephridia.

Class Oligochaeta: Earthworms and related species; earthworms have paired setae, a closed circulatory system, and a cerebral ganglion; hermaphroditic; sexual reproduction by mutual fertilization.

Class Hirudinea: Leeches; mainly freshwater annelids with posterior and anterior suckers; either free-living or parasitic; reproduction sexual, usually by mutual fertilization of hermaphrodites.

Class Polychaeta: Marine worms; body with many bristles; about 6,000 species; includes sandworms.

Phylum Arthropoda: Arthropods; at least 1 million species; segmented body; paired, jointed appendages; exoskeleton; open circulatory system; complex nervous system with two ventral nerve cords and a brain.

Class Merostomata: Horseshoe crabs; four species; aquatic with fangs and book gills.

Class Arachnida: Arachnids; about 57,000 species; body has two divisions, eight legs, and paired chelicerae and pedipalps; respiration by tracheae, book lungs, or both; includes spiders, ticks, and mites.

Class Crustacea: Crustaceans; about 25,000 mostly aquatic species; paired mandibles and thoracic appendages; usually two pairs of maxillae; respiration by gills; includes lobsters and barnacles.

Class Chilopoda: Centipedes; about 3,000 terrestrial species; mandibles; many body segments, each with one pair of legs.

Class Diplopoda: Millipedes; about 7,500 terrestrial species; mandibles; many body segments, each with two pairs of legs.

Class Insecta: Insects; over 750,000 terrestrial and freshwater species; three body divisions; three pairs of legs; head has antennae, compound eyes, and mouthparts; thorax has legs and, in many species, wings; most species undergo incomplete or complete metamorphosis; respiration by tracheae.

Order Protura: Proturans; piercing-sucking or chewing mouthparts; lacking wings, antennae, and eyes; front legs adapted as organs of touch; no metamorphosis.

Order Thysanura: Bristletails and silverfish; chewing mouthparts; two or three tails; wingless; no metamorphosis.

Order Collembola: Springtails; chewing mouthparts; wingless; compound eyes reduced or absent; no metamorphosis.

Order Ephemeroptera: Mayflies; chewing mouthparts that are reduced; membranous wings (usually two pairs); nymphs are aquatic; adults do not feed at all during their brief existence; incomplete metamorphosis.

Order Ondonata: Dragonflies and damsel flies; two pairs of transparent wings; chewing mouthparts; not able to walk; incomplete metamorphosis.

Order Plecoptera: Stone flies; chewing mouthparts which are reduced in many species; two pairs of membranous wings; nymphs are aquatic; incomplete metamorphosis.

Order Orthoptera: Crickets and grasshoppers; two pairs of wings or wingless; chewing mouthparts; incomplete metamorphosis.

Order Dermaptera: Earwigs; chewing mouthparts; two pairs of wings or wingless; incomplete metamorphosis.

Order Embioptera: Web spinners; chewing mouthparts; wingless with the exception of some males; spin silk with special organs on forelegs; incomplete metamorphosis.

Order Isoptera: Termites; social insects that have chewing mouthparts; only the reproductive males and females have wings; incomplete metamorphosis.

Order Mallophaga: Chewing lice; wingless; chewing mouthparts; parasites of mammals and birds; incomplete metamorphosis.

Order Anoplura: Sucking lice; wingless; piercing-sucking mouthparts; parasites of mammals; incomplete metamorphosis.

Order Corrodentia: Bark lice, book lice; chewing mouthparts; wingless or two pairs of membranous wings; incomplete metamorphosis.

Order Hemiptera: Bugs; piercing-sucking mouthparts; two pairs of wings or wingless; incomplete metamorphosis.

Order Homoptera: Aphids, scale insects, and cicadas; wingless or winged (one or two pairs); piercing-sucking mouthparts; incomplete or complete metamorphosis.

Order Thysanoptera: Thrips; piercing-sucking mouthparts; wingless or two pairs of wings with long hairs attached to them; incomplete or complete metamorphosis.

Order Mecoptera: Scorpion flies; chewing mouthparts; wingless or two pairs of membranous wings; in some males, tip of abdomen is curved, resembling the tail of a scorpion; complete metamorphosis.

Order Neuroptera: Helgramites (dobson fly); Larva aquatic with large mandibles.

Order Trichoptera: Caddis flies; chewing mouthparts which are reduced; two pairs of wings; larvae are aquatic; adults do not feed extensively; complete metamorphosis.

Order Lepidoptera: Moths and butterflies; two pairs of broad, scaly wings; tubelike, sucking mouthparts; complete metamorphosis.

Order Diptera: Flies and mosquitoes; transparent front wings; hind wings reduced to knobby balancing organs; lapping and piercing mouthparts; complete metamorphosis.

Order Siphonaptera: Fleas; wingless; flattened body; piercing-sucking mouthparts; parasites of birds and mammals; complete metamorphosis.

Order Coleoptera: Beetles; winged (two pairs of wings, front pair serves as horny cover for membranous hind pair) or wingless; chewing mouthparts; complete metamorphosis.

Order Strepsiptera: Strepsipterans; chewing mouthparts which are reduced; females are wingless; males have two pairs of wings, with the front pair greatly reduced; complete metamorphosis.

Order Hymenoptera: Ants, bees, wasps; winged (two pairs that interlock in flight) or wingless; chewing or lapping mouthparts; some social species; complete metamorphosis.

Phylum Echinodermata: Echinoderms; abut 6,000 marine species; calcium endoskeleton, tube feet, and a water-vascular system; adults radially symmetrical; includes starfish, sand dollars, and sea urchins.

Class Crinoidea: Sea lily; five rays; tube feet without suckers; most have stalklike body.

Class Asteroidea: Starfish; five rays with two rows of tube feet in each ray; eyespots.

Class Ophiuroidea: Brittle stars; five slender, delicate rays.

Class Echinoidea: Sea urchin, sand dollar; rays lacking; body spherical or oval.

Class Holothuroidea: Sea cucumber; long, thick body with tentacles; no rays.

Phylum Hemichordata: Acorn worms; about 80 species; hydrostatic skeleton similar to starfish's water vascular system; dorsal and ventral nerve cords, dorsal cord hollow in some species.

Phylum Chordata: Chordates; about 43,000 species that at some stage have a notochord, dorsal nerve cord, pharyngeal gill slits, and tail.

Subphylum Urochordata: Tunicates; about 1,300 marine species; adults sacklike; usually sessile; larvae free-swimming.

Subphylum Cephalochordata: Lancelets; about 30 species of thin, fishlike marine animals; basic chordate features throughout life.

Subphylum Vertebrata: Vertebrates; about 41,700 species; long or cartilaginous endoskeleton includes vertebrae; characterized by distinct cephalization and a closed circulatory system; excretion by kidneys; most have two sets of paired appendages.

Class Agnatha: Jawless fishes; about 60 species of lampreys and hagfishes; have suckerlike mouths; scaleless, cylindrical bodies without appendages; skeleton of cartilage; two-chambered heart.

Class Chondrichthyes: Sharks, skates and rays; about 625 mostly marine species of jawed fishes; cartilaginous skeleton; teeth replaced continuously; skin covered with tooth-derived scales; eggs fertilized internally; two-chambered heart.

Class Osteichthyes: Bony fishes; about 20,000 species of jawed fishes; bony skeletons; found in marine and freshwater habitats; body has bone-derived scales, an air-filled swim bladder, and a two-chambered heart; fertilization external in most species.

Class Amphibia: Amphibians; about 2,500 species characterized by a gill-breathing larval stage and a lung-breathing adult stage; adults have moist skin and three-chambered heart with double circulatory system; eggs usually laid in water.

Order Anura: Frogs and toads; adults are tailless with rear legs adapted to jumping; fertilization external, in water.

Order Urodela: Salamanders; adults have tails and four legs of about equal size; fertilization internal.

Order Apoda: Caecilians; legless, usually blind amphibians that live underground.

Class Reptilia: Reptiles; about 6,000 terrestrial species; fertilization occurs internally; embryo protected by a shelled, amniotic egg; skin is waterproof, covered with horny plates or overlapping scales; double circulatory system.

Order Rhyncocephalia: Tuatara; primitive four-legged species; skin covered by plates; three-chambered hearts.

Order Chelonia: Turtles and tortoises; shelled, toothless species with four limbs and a three-chambered heart.

Order Squamata: Lizards and snakes; reptiles with overlapping scales and a three-chambered heart; most lizards have four legs; snakes are legless.

Order Crocodilia: Crocodiles, alligators, and related species; broad-bodied, water-dwelling carnivores with powerful tails and jaws; skin covered by bony plates; four-chambered heart.

Class Aves: Birds; about 8,600 species; winged vertebrates with feathers; endothermic (warm-blooded); double circulation and a four-chambered heart; fertilization internal; amniotic eggs usually incubated in nest; respiratory system with lungs and air sacs.

Order Rheiformes: Rheas; large, flightless South American birds.

Order Struthioniformes: Ostriches; large, flightless African birds.

Order Procellariiformes: Albatrosses, shearwaters, petrels, and other species of related seabirds with tubelike nostrils.

Order Sphenisciformes: Penguins; flightless swimming birds with wings modified as paddles.

Order Gaviiformes: Loons; diving birds with three webbed toes.

Order Pelecaniformes: Pelicans, boobies, and related water birds with four-toed, webbed feet.

Order Ciconiiformes: Storks, herons, and related long-legged wading birds.

Order Gruiformes: Cranes, coots, limpkins, rails, gallinules,and related species.

Order Charodriiformes: Auks, gulls, sandpipers, and related shore and water birds.

Order Anseriformes: Ducks, geese, swans, screamers; short-legged water birds.

Order Falconiformes: Falcons, eagles, hawks, condors, and related diurnal birds of prey.

Order Galliformes: Chickens, turkeys, pheasants, grouse, and related fowl-like birds.

Order Columbiformes: Pigeons, doves, sandgrouse, and related species.

Order Psittaciformes: Parrots; hook-billed birds that eat seeds, fruit, or nectar.

Order Cuculiformes: Cuckoos, roadrunners, and related species.

Order Strigiformes: Owls; nocturnal birds of prey.

Order Caprimulgiformes: Goat-suckers, whippoorwill, night hawk, and related species.

Order Apodiformes: Swifts and hummingbirds; weak-legged birds with strong wings.

Order Coraciiformes: Kingfisher; fishing birds.

Order Piciformes: Woodpeckers, toucans, barbets, and related birds.

Order Passeriformes: Perching birds, including all songbirds.

Five-Kingdom Classification of Organisms

Class Mammalia: Mammals; about 4,500 species; young are nourished by mother's milk and receive extensive parental care; endothermic and covered by hair; double circulation; four-chambered heart; respiration by lungs; highly developed, complex brain; fertilization internal.

Order Monotremata: Monotremes, spiny anteaters and duck-billed platypus; the only mammals that lay eggs; incomplete control of body temperature; milk secreted from abdominal sweat glands.

Order Marsupialia: Marsupials, including kangaroos, wombats, and opossums; young are born in an immature state and develop further attached to a nipple in the mother's marsupium (pouch).

The following orders are placental mammals in which young undergo substantial prenatal development within the uterus. During gestation, exchange of nutrients, gases, and wastes occurs through the placenta.

Order Insectivora: Insectivores; small, chiefly nocturnal mammals that feed largely on insects.

Order Chiroptera: Bats; the only flying mammals.

Order Artiodactyla: Hoofed mammals with two or four toes; large herbivores including deer, pigs, cattle, sheep, goats, camels, antelopes, giraffes, and hippopotamuses; most are ruminants.

Order Perissodactyla: Hoofed mammals with one or three toes; large herbivores including horses, tapirs, and rhinoceroses.

Order Proboscidea: Elephants; enormous herbivores with trunk.

Order Carnivora: Carnivores; bears, cats, dogs, seals, and related predators.

Order Cetacea: Whales, porpoises, and dolphins; marine mammals with streamlined body; forelimbs adapted as flippers; no hindlimbs; nostril (blow-hole) at top of head.

Order Sirenia: Sea cows; aquatic herbivores with short, paddlelike forelimbs and no hindlimbs.

Order Rodentia: Rodents; small, plant-eating mammals; one upper and one lower curving, chisel-like incisors that grow continuously.

Order Lagomorpha: Rabbits, hares, and pikas; rodentlike mammals with two pairs of upper and one pair of lower incisors.

Order Edentata: Armadillos, anteaters, and tree sloths; insect eaters with few or no teeth.

Order Primates: Primates, including lemurs, monkeys, apes, and humans; mammals with features adapted to arboreal life.

Chronology of Modern Science

1403 The first quarantine is imposed in Venice, Italy, in an attempt to stem the tide of the Black Death.

1408 The windmill is first used in Holland to carry inland waters out to coastal areas.

1436 Italian artist Leon Battista Alberti publishes a book on perspective, a forerunner of projective geometry, which is developed 400 years later.

c. 1450 The harquebus, the first gun that is small enough to be carried and fired by one person, is invented in Spain, marking the beginning of "small arms" (although it is actually so heavy that it needs supports when fired).

1451 The German scholar Nicholas of Cusa constructs spectacles with concave lenses to aid the nearsighted; prior to this only convex lenses for farsightedness had been used.

1454 After more than 20 years of experimentation, the German inventor Johannes Gutenberg prints 300 copies of a Latin Bible having 1282 pages; it is known today as the Gutenberg Bible, and it inaugurates the era of movable type. In the next 50 years, approximately 10 million books are printed using this system.

1464 The German astronomer Johann Miller, better known as Regiomontanus, writes *De triangulis omnimodis (On Triangles of All Types)*, summarizing the law of sines and other aspects of geometry. However, the book is not published until 1533.

1472 Regiomontanus documents a comet's path for the first time, overcoming superstitions and irrational terror previously connected with comets. Centuries later, this comet comes to be known as Halley's Comet.

1474 William Caxton prints the first book in the English language, a history of ancient Troy.

1480 Twenty-eight-year-old Leonardo da Vinci describes a workable parachute.

1484 *Triparty en la sciences des nombres*, by Nicolas Chuquet, is published. He uses negative numbers as exponents, although not as solutions to equations.

1490 Leonardo da Vinci describes capillary action, noting that liquid in small-diameter tubes crawls upward.

1492 German geographer Martin Behaim constructs the first terrestrial globe, omitting the Americas and the Pacific Ocean.

1492 On his first voyage to the Americas, Italian navigator Christopher Columbus notes that the magnetic compass directional arrow varies somewhat as longitude changes, a principle later known as magnetic declination.

1492 Leonardo da Vinci diagrams a flying machine.

1494 Italian mathematician Luca di Pacioli publishes his influential algebra book *Summa de arithmetica, geometria, proportioni et proportionalita*, introducing double-entry bookkeeping and including a study of cubic equations.

1497 Nicolaus Copernicus records the occultation (eclipse) of a star by the moon.

1500 Jakob Nufer performs the first recorded Caesarean operation on a living woman in Switzerland.

1502 Peter Heinlein of Nuremberg builds the first pocket watch, which has a spring-driven mechanism and is known as the "Nuremberg Egg."

1502 Amerigo Vespucci publishes an assertion that the land discovered by Columbus is a "new world," completely separate from Asia. He further claims that Asia lies farther west, beyond a second ocean.

1507 German mapmaker Martin Waldseemüller publishes 1000 copies of a map of the new continent discovered by Columbus and calls it "America" after Amerigo Vespucci.

1512 Copernicus states that the earth and the other planets revolve around the sun, rather than the earth being at the center of the universe, as was commonly believed at the time.

1513 Vasco Nuñez de Balboa crosses the Isthmus of Panama and is the first European to view the Pacific Ocean, although he probably did not realize it was the second great ocean about which Amerigo Vespucci had speculated eleven years earlier.

1514 The plus (+) and minus (–) signs are used in algebraic expression by the Dutch mathematician Vander Hoecke.

1518 The first smallpox epidemic attacks the Americas, concentrated among the Indians of Hispaniola. Two years later, an epidemic devastates the Aztecs, aiding in Cortés' conquest.

1519 Ferdinand Magellan of Portugal sets sail from Spain on the first voyage to circumnavigate the world. Only one of five ships makes it back to Spain almost exactly three years later, and Magellan himself is killed in the Philippine Islands nineteen months into the journey. Nevertheless, the journey provides information on the size of the earth and the independent structure of the continents.

1520 Scipione del Ferro partially solves cubic equations (involving x to the third power), but tells only his pupil Antonio Maria Fiore, just before his death in 1526.

1528 Swiss physician Philippus Paracelsus publishes the first surgery manual, *Die kleine Chirurgia.*

1535 Niccolo Fontana (better known as "Tartaglia") reveals to another Italian mathematician, Geronimo Cardano, a general method for solving certain types of cubic equations, which were not known to del Ferro or to Fiore.

1537 Tartaglia's *Della nova scientai (Of the New Science)* introduces the science of ballistics.

1538 Girolamo Fracastoro is the first European to observe that a comet's tail always points away from the sun, a fact also known in China since at least 635 AD. Two years later, Peter Bennewitz (also known as Peter Apian) independently arrives at the same conclusion and publishes the first scientific drawing of a comet.

1540 Ether is first produced by combining sulfuric acid and alcohol.

1540 Michael Servetus discovers pulmonary circulation.

1541 Niccolo Tartaglia and Antonin Maria Fiore engage in a mathematical "duel" of cubic equations, which Tartaglia wins.

1542 French doctor Jean François Fernel publishes a description of appendicitis and of the transporting contractions of the digestive system known as peristalsis.

1543 Andreas Vesalius publishes *De fabrica corporis humani (Concerning the Structure of the Human Body)*, a book that corrects over 200 errors of the ancient Greeks, revolutionizes medicine, and founds modern anatomy.

1543 A few months before his death, Copernicus publishes *De revolutionibus orbium coelestium (On the Revolution of Celestial Bodies)*, although the manuscript had been circulating among scientists for years. In the book, Copernicus argues that, contrary to the geocentric system developed by Hipparchus and Ptolemy, the earth and other planets revolve around the sun in a circular path. The book, which was banned by the Catholic Church for almost 300 years, causes a scientific revolution.

Events of 1544-1596

1544 Sebastian Münster publishes the first compendium of world geography, *Cosmographia generalis*.

1544 Michael Stifel publishes *Arithmetica integra*, summarizing all known arithmetic and algebra.

1545 Geronimo Cadano publishes *Ars magna*, which contains Tartaglia's solution to cubic equations (published against his wishes), the solution for quartic equations involving an unknown raised to the fourth power developed by Cardano's pupil Luigi Ferrari, practical application of negative numbers, and references to complex numbers.

1545 Ambroise Paré, surgeon to Henry II of France, publishes a book on surgery, reporting the success of soothing ointments rather than burning oils on wounds and of tying off arteries rather than cauterizing them.

1546 In *De contagione*, Gerolamo Fracastoro advances the idea that infections and epidemic diseases are caused by transferable seedlike entities.

1546 The Flemish cartographer Gerardus Mercator discovers the earth's magnetic pole.

1551 Georg Joachim Rheticus publishes trigonometric tables, prepared in connection with the research of his teacher, Copernicus, into planetary motion; Rheticus is the first to relate ratios to angle size rather than to circle arcs.

1551 Erasmus Reinhold prepares a set of planetary tables that refine Copernicus' determinations of planetary motions.

1552 In *Tabulae anatomicae*, Bartolommeo Eustachio describes the tube connecting the ear and throat (now known as the eustachian tube), the adrenal glands, and the detailed structure of the teeth. However, the book is not published until 1714.

1553 Michael Servetus is burned at the stake for heresy found in his *Christianismi restitutio*, which includes his theory that blood circulates from the heart to the lungs and back.

1555 In *L'historie de la nature des oyseaux (History of the Nature of Birds)*, Pierre Belon classifies 200 species and describes homologies in the skeletons of all vertebrates.

1556 *De re metallica (Concerning Metallic Things)* by Georgius Agricola is published posthumously; it summarizes a wide range of mineralogy.

1559 Realdo Colombo publishes *De re anatomica (Of Anatomical Matters)*, in which he postulates that blood circulates from the right chamber of the heart, through the lungs, and to the left chamber of the heart, from which it then is pumped to the rest of the body.

1560 The first established scientific society is founded in Naples by Giambattista della Porta. Although it is later suppressed by the Inquisition, it leads to other, more persistent scientific societies.

1561 Gabriel Fallopius publishes *Observationes anatomicae*, including descriptions of the inner ear and the female reproductive system.

1568 Gerardus Mercator introduces cylindrical projection in maps, now known as mercator projection. The cover of his book of maps shows the Greek Titan Atlas balancing the world on his shoulders, a picture that gives the name *atlas* to books of maps.

1572 Tycho Brahe discovers what he calls a nova or new star (which is now established as a supernova) in the constellation Cassiopeia, and watches it for about 15 months. The star is also documented by Chinese astronomers.

1572 Rafael Bombelli publishes *Algebra*, applying complex numbers to solve equations and continuing fractions to approximate roots.

1577 From his observatory built two years before in Denmark, Tycho Brahe observes the great comet of 1577 and tries to determine its distance using parallax; he shows that the comet is at least four times as far away as the moon, disproving the Greek belief that comets are atmospheric phenomena.

1579 Franciscus Vieta publishes *Canon mathematicus*, extending Rheticus' trigonometric tables to each second of a degree and demonstrating the decimal representation of numbers.

1581 While sitting in church, seventeen-year-old Galileo compares the swaying of a chandelier to his pulse and discovers that the time taken for a complete swing is independent of the width of the swing, but depends solely on the length of the pendulum itself; he further verifies this conclusion in home experiments. His discoveries provide the groundwork for the use of pendulums in measuring time, an idea developed 70 years later.

1582 Catholic Europe replaces the Julian calendar with the Gregorian calendar, named after the current pope, Gregory XIII, and developed by Christoph Clavius. In the reformation, the 10 days between October 4 and 15 are dropped, making 1582 the shortest year on record with only 354 days. Under the new system, the year 1600 would be a leap year, but 1700, 1800, and 1900 would not be and 2000 would, making a total of 97 leap years every 400 years. The Gregorian calendar was not adopted by Great Britain for 200 years, nor by Russia for 350 years.

1583 Mathematician Simon Stevin demonstrates the laws of hydrostatics, showing that the surface pressure of a liquid is related to the height of the liquid above the surface and the area of the surface, but not to the shape of the vessel holding the liquid.

1586 Stevin demonstrates the use of decimal fractions to simplify computations involving fractions. He may also have performed an important gravitational experiment, by noting that two different weights dropped at the same time will hit the ground at the same time. (It is said that this same experiment was also performed by Galileo from the top of the Leaning Tower of Pisa.)

1589 Galileo performs a series of experiments with balls of various weights on an inclined plane, determining that weight does not affect the pull of gravity, and that under the constant pull of gravity an object will experience constant acceleration. The latter discovery showed that without friction, an object will remain in constant motion. His discoveries are published two years later in *De motu (On Motion)*.

1589 The Reverend William Lee of England invents the first knitting machine that can knit faster than human hands, using a stocking frame.

1590 Spectacle-maker Zacharias Janssen constructs the first microscope using two convex lenses and a tube.

1591 In *Isagoge in artem analyticam*, Franciscus Vieta is the first person to use letters of the alphabet to symbolize constants and unknowns in mathematical equations.

1592 Galileo develops the first primitive thermometer, known as a thermoscope, using air in a tube instead of liquid.

1592 An Italian engineer, Domenico Fontana, uncovers the ruins of the Roman city Pompeii, buried in lava in 79 AD. This incident is often viewed as the beginning of modern archaeology.

1596 Ludolf van Ceulen extends the value of π from the three decimal places established by Archimedes to twenty places, and several years later to thirty-five places.

1596 *Opus palatinum de triangulis* by Rheticus is published posthumously; it contains trigonometric tables for the six standard trigonometric functions.

Events of 1597–1645

1597 Galileo states in a letter to Johannes Kepler that he believes the Copernican theory of the solar system to be true.

1597 Andreas Libavius publishes *Alchemia*, the first true chemical textbook, in which he describes the preparation of hydrochloric acid, sulfuric acid, and tin tetrachloride.

1600 William Gilbert publishes *De magnete (Concerning Magnets)*, presenting a model of the solar system based on magnetism, explaining gravity in electromagnetic terms, and noting magnetic dip in a compass. Though agreeing with Copernicus that the stars are stationary and the earth rotates on its axis, Gilbert retains the belief that the sun revolves around the earth.

1600 In Rome, the Italian metaphysical scholar Giordano Bruno is burned at the stake as a heretic for believing that the earth revolves around the sun.

1602 Tycho Brahe's *Astronomiae instauratae progymnasmata (Introduction to the New Astronomy)* is published posthumously by Johannes Kepler. It gives positions for 777 stars and describes the supernova of 1572.

1603 In his *Uranometria*, Johann Bayer develops the modern system of star nomenclature based on the Greek and Latin alphabets; the system is still in use today.

1603 Hieronymus Fabricius publishes a pioneering anatomical study of valves and veins later developed by his student William Harvey.

1604 Fabricius publishes an important study of embryology, *De formata foetu (On the Formation of the Fetus)*, containing a description of blood circulation in the umbilical cord.

1604 Kepler describes how the eye focuses light and calculates light intensity as the square of the distance from the source, in his book *Ad vitellionem paralipomena quibus astronomiae pars optica traditor (The Optical Part of Astronomy)*.

1608 Hans Lippershey of the Netherlands builds the first practical telescope.

1609 In *Astronomia nova,* Kepler formulates his first two laws of planetary motion—the elliptical orbits of the planets and the alteration of planetary speed according to distance from the sun.

1609–1611 Galileo improves the telescope and pioneers its use for astronomical observation, discovering mountains on the moon, many new stars, four satellites of Jupiter, the phases of Venus, and the composition of the Milky Way. His findings, which provide overwhelming evidence supporting Copernican theory, are published in *Sidereus nuncius (Starry Messenger)*, and make him famous all over Europe.

1610 Lippershey and Zacharias Jansen invent the compound microscope, which Galileo then uses to study insect anatomy.

c. 1611 Thomas Harriot, Fabricius, Father Scheiner, and Galileo discover sunspots at about the same time. By observing the changing position of sunspots, Galileo demonstrates the rotation of the sun in 1613.

1614 John Napier introduces logarithms and publishes tables and rules for their use.

1614 Santorio Santorio conducts experiments involving his own body weight, wastes, perspiration, pulse, and temperature, introducing the study of metabolism.

1616 The Catholic Church bans Copernican theory and warns Galileo not to support it.

1619 Kepler formulates his third law of planetary motion, involving a calculation to determine the distance of a planet from the sun, and his theories of harmony in nature. He also publishes a defense of the Copernican system, and it is immediately banned by the Catholic Church.

1620 In his influential *Novum organum scientiarum*, Francis Bacon champions scientific observation and the inductive method over postulation and *a priori* reasoning, describing what is known today as the scientific method.

1621 Willebrod Snell discovers that in the refraction of a ray of light there is a constant relationship between the sines of the angles rather than between the angles themselves, as Ptolemy had postulated. This finding comes to be known as Snell's law.

c. 1622 William Oughtred invents the slide rule.

1623 In his book *Pinax theatri botanici*, Gaspard Bauhin introduces binomial names for genera and species in botanical nomenclature.

1624 Jan Baptista van Helmont coins the word "gas," from the Flemish word for "chaos," to describe compressible fluid.

1626 Santorio measures human temperature using a thermometer for the first time.

1627 Kepler publishes his *Rudolphine Tables*, which is a listing of planetary motions based on elliptical orbits and Napier's logarithms. It also includes a star map based on Tycho's work, but expanded to include over 1000 stars.

1628 William Harvey publishes his important research into the circulation of blood in *Exercitatio anatomica de motu cordis et sanguinis (Anatomical Treatise on the Movement of the Heart and Blood)*. He compares the heart to a pump and claims that blood leaves the right ventricle for the lungs and returns to the left ventricle, from which it goes to the body generally, then returns to the right ventricle. Though not received well at first, his theories form the foundation of modern physiology.

1632 Galileo risks publishing *Dialogo sopra i due massimi sistemi del mundo (Dialogue on the Two Chief World Systems)*, containing a Ptolemy-supporter (a thinly veiled description of the Pope), a Copernicus-supporter, and a truth-seeker (Galileo himself).

1633 The Catholic Inquisition arrests Galileo and forces him to renounce Copernican theory. It is said that at the end of his recantation, he muttered, *"E pur se muove,"* or "Nevertheless, it (the earth) moves."

1637 In *Discours de la méthode*, René Descartes champions deductive reasoning from mathematical laws; famous appendices to this immensely influential work contain important advances in physics, optics, meteorology, and mathematics (including analytic geometry, exponential notation, and the theory of equations).

1638 Galileo publishes his major work on mechanics, including descriptions of experiments on acceleration, friction, inertia, and falling bodies that correct many of Aristotle's errors.

1640 The French mathematician Blaise Pascal publishes his theory of conics.

1640 Pierre de Fermat discovers Fermat numbers and formulates new theorems in number theory.

1640 In *De motu gravium*, Evangelista Torricelli applies Galileo's laws of motion of fluids, establishing the science of hydrodynamics.

1643 Torricelli invents the mercury barometer and, in the process, makes the first artificially created vacuum known to science.

1644 In *Principia philosophia*, Descartes outlines his vortex theory of the origin and state of the solar system, advances Galileo's theory of inertia, and further develops the scientific method.

1645 Otto von Guericke builds the first vacuum pump and takes the first measurement of the density of air.

Events of 1648-1686

1648 To demonstrate that air pressure changes with altitude, Pascal sends his brother-in-law up a mountain carrying some barometers. The barometric mercury columns drop from 30 to 27 inches after he has climbed one mile up the mountain, leading Pascal to conclude that the atmosphere can have only finite height.

1650 Giovanni Battista Riccioli is the first to detect a double star as he views the middle star of the Big Dipper's handle.

1651 In his map of the moon, Riccioli introduces many names of lunar features still in use today, including the names of his heroes—Tycho, Ptolemaeus, Copernicus, and Kepler.

1653 Another set of bodily vessels, in addition to the veins and arteries, is discovered by Olof Rudbeck, which he calls lymphatics. He demonstrates their existence to Queen Christiana of Sweden, using a dog.

1654 In response to questions from a French gambler, Pascal and Fermat together work out mathematical techniques to determine probability.

1654 Guericke begins public demonstrations of the power of air pressure in a vacuum, including experiments in which teams of horses are unable to separate a sphere that is held together by a vacuum.

1656 Using improved telescope lenses created by Benedict Spinoza, Christiaan Huygens observes the rings on Saturn, seen but not understood by Galileo in 1612, as well as Saturn's moon Titan and the Orion Nebula.

1656 Huygens builds the first pendulum clock based on the cycloid, much like a modern grandfather clock. It is the first clock to keep highly accurate time.

1658 Jan Swammerdam first observes red blood cells under a microscope.

1659 Huygens is the first to observe surface features on Mars, a dark, triangular marking he names Syrtis Major (a Latin term meaning "great bog").

1660 Marcello Malpighi extends Harvey's research into circulation and discovers the capillaries that provide the passage between arteries and veins.

1660 Guericke demonstrates the principle of static electricity, using a globe of sulfur.

1661 Robert Boyle publishes *The Skeptical Chymist*, in which he defines the properties of an element, discusses alkalis and acids, and refutes Aristotle's and Paracelsus' ideas regarding the chemical composition of matter.

1662 Boyle discovers that air can be compressed and formulates Boyle's law, which states that in an ideal gas of constant temperature, pressure and volume vary inversely.

1662 King Charles II of England charters the Royal Society, one of the outstanding groups of scientists of all time.

1662 John Graunt, a draper by trade, publishes the first book of statistics, in which he outlines birth rates and life expectancy in London.

1664 Giovanni Alfonso Borelli calculates that the orbit of a comet is a parabola rather than a circle, ellipse, or line.

1665 Giovanni Cassini determines the rotational speeds of Jupiter, Mars, and Venus.

1665 While examining some cork under a microscope, Robert Hooke is the first to identify cells. He publishes his discovery in *Micrographia*, in which he also compares light to waves of water.

1665 *Physicomathesis de lumine* by Francis Grimaldi, published posthumously, explains the diffraction of light using his theory that light is a wave phenomenon.

1665–1666 Isaac Newton conducts a series of experiments, discovering that white light is composed of different colors that can be separated by a prism, formulating his first law of universal gravitation, and inventing the first form of calculus.

1668 John Wallis suggests the law of conservation of momentum, according to which the total momentum of a closed system does not change.

1668 Theories of spontaneous generation of living organisms are disproved by Francesco Redi in a controlled experiment involving meat and maggots.

1668 Newton invents the reflecting telescope; all previously developed telescopes had been refracting telescopes.

1669 Nicolaus Steno correctly describes the process by which living things become fossils.

1669 In his quest to create gold from another material, Hennig Brand discovers a new, faintly glowing element he names "phosphorus." This is the first discovery of an element that can be attributed to a specific person and date.

1669 Erasmus Bartholine discovers, but does not fully understand, the principle of double refraction as he examines a piece of transparent crystal known as Iceland spar.

1669 In his famous study of silkworms, Marcello Malpighi gives the first detailed description of an invertebrate's anatomy.

1670 Thomas Willis discovers the relationship of sugar in the urine to diabetes.

1672 Giovanni Cassini calculates the distance from the earth to the sun to be 87,000,000 miles, a figure accurate to within 6,000,000 miles of modern estimates. Scientists begin to realize just how large the known solar system must be.

1675 In connection with his study of Jupiter's moons, Olaus Roemer estimates the speed of light to be about 141,000 miles per second, about 75% of what we now believe to be its actual speed. Before Roemer, many scientists believed light had infinite speed.

1675 Gottfried Leibniz develops differential and integral calculus.

1676 or **1677** Looking at pond water through a microscope of his own construction, Anton van Leeuwenhoek is the first person to view what he called "animalcules," known today as protozoa or microorganisms.

1678 Edmond Halley publishes the findings of two years of study conducted in the south Atlantic Ocean, cataloguing 341 southern stars previously unknown to astronomers.

1680 The flightless dodo bird of the Raphidae family becomes extinct due to slaughter by settlers on its native island of Mauritius near Madagascar; this is the first noted extinction of a species in historic times.

1682 Halley first views the comet later named for him, but which he calls "the great comet." In 1705 he correctly predicts that it will reappear in 1758.

1682 Sexual reproduction in plants is described by Nehemiah Grew in *The Anatomy of Plants*, in which he also describes differing tissues in a plant's roots and stems.

1683 Leeuwenhoek views what are now known to be bacteria through finely ground lenses of his own making. No one else would be able to see bacteria for another century.

1683 Count Enrenfried von Tschirnhous develops important transformations to simplify polynomial equations by using new methods to solve equations of the third and fourth degrees.

1686 John Ray publishes the first of three volumes describing and classifying 18,600 plant species, laying important groundwork for the later efforts of Linnaeus.

Events of 1687-1740

1687 Isaac Newton publishes what is often described as the most important single book of science ever written, *Philosophiae naturalis principia mathematica (The Mathematical Principles of Philosophy)*, better known as *Principia Mathematica* or simply *Principia*. In it Newton establishes his three laws of motion and the law of universal gravitation and theorizes on the oblate spheroid shape of the earth.

1693 John Ray follows up his classification of plants with a book logically classifying animals according to their hooves, toes, and teeth. He correctly identifies whales as mammals.

1698 The first practical steam-powered machine is built by Thomas Savery to pump water from coal mines. He names it "The Miner's Friend."

1701 Giacomo Pylarini gives the first smallpox inoculations to three children in Constantinople.

1704 Newton publishes his important work on the nature of light, entitled *Optics*. He proposes that light is particulate in nature and closes with a series of what he considers to be very important unanswered questions.

1706 Francis Hauksbee builds a static-electricity generator using a glass sphere turned by a crank. Interest in experimentation with electricity greatly increases.

1707 English physician John Floyer develops a "pulse watch," which runs for exactly one minute per winding. It is used to count the number of heartbeats in a minute and is a precursor to watches with second hands.

1709 British iron maker Abraham Darby develops the use of coke in iron smelting, enabling factories to produce iron at a faster rate and preparing Great Britain for the Industrial Revolution.

1709 Irish philosopher George Berkeley publishes *New Theory of Vision*, a significant contribution to psychological thought.

1710 Jacob Christoph Le Blon develops three-color printing.

c. 1710 The Pennsylvania Dutch invent the first modern rifle, later known as the Pennsylvania (Kentucky) long rifle, with two or three times the range of a musket and much greater accuracy.

1712 Thomas Newcomen devises a new kind of steam engine with a piston-and-cylinder construction that utilizes ordinary low-pressure steam to pump water out of coal mines.

1714 Gabriel Fahrenheit develops a mercury thermometer and the Fahrenheit scale of measurement, based on three fixed temperatures: 0° for the freezing point of ice, water, and ammonium chloride; 32° for the freezing point of water, and 96° for normal human body temperature (which is shown later to actually be 98.6°).

1716 Edmond Halley invents and demonstrates a diving bell with an air-refreshment system.

1717 Lady Mary Wortley Montagu brings the practice of inoculation against smallpox to England from Turkey and has her own two children vaccinated.

1718 By comparing star measurements recorded by early Greek astronomers with current data, Halley concludes that the stars are not fixed, but move at varying speeds and in varying directions.

1723 Jacob Leupold publishes his nine-volume *Theatrum machinarum generale (General Theory of Machines)*, which is the first detailed and systematic exposition of mechanical engineering; it includes a design for a noncondensing, high-pressure steam engine similar to the ones that would be built 200 years later.

1726 Stephen Hales is the first to measure blood pressure.

1728 James Bradley notes the periodic shift of the stars' positions during a year and explains it by the aberration of light caused by the earth's orbit. He uses his findings to recalculate the speed of light, which he places within 5% of what we now believe it to be.

1728 Pierre Fauchard publishes the first book of dental science, explaining how to fill decayed areas in the teeth with tin, lead, or gold, as well as the use of artificial dentures and crowns.

1728 English instrument-maker John Harrison invents a highly accurate clock that can be used at sea, called a chronometer, which makes the precise measurement of longitude possible.

1728 Dutchman Vitus Bering is commissioned by Peter I of Russia to explore the area between Siberia and North America to determine if there is a land connection. Two hundred thirty-six years after Columbus first discovered the New World, it is finally proven that North America stands completely apart from Asia.

1729 Through a series of experiments, Stephen Gray discovers that some materials conduct electricity better than others and that static electricity travels on the surface of objects rather than through the interior.

1731 John Hadley and Thomas Godfrey independently invent the reflecting quadrant, a precursor to the modern sextant, for use in measuring distances at sea.

1733 Chester Moor Hall invents the achromatic telescope using two types of lenses to filter out distorting colors, thus making it possible to create a shorter, more convenient telescope.

1733 Charles-Francois de Disternay du Fay divides electrical charges into two types; he states that materials such as glass are charged with *vitreous* electricity and are attractive to materials such as amber, which are charged with *resinous* electricity.

1733 Englishman John Kay patents the flying shuttle loom, enabling one person to do much more work with less effort.

1735 Carolus Linnaeus publishes *Systema naturae (Systems of Nature)*, introducing a system of plant classification still in use today, which he extends in later volumes to also include animals.

1735 Charles-Marie de La Condamine measures the equator from Peru while Pierre-Louis Moreau de Maupertuis measures degrees of latitude from Lapland. Their results verify Newton's theory that the earth is an oblate spheroid with an equatorial bulge. On the way back to France, La Condamine explores the Amazon River and brings the first rubber and curare back to Europe.

1735 Georg Brandt discovers that a blue ore similar to copper ore and known to miners is actually a new element. He names it "cobalt," after *kobolds*, or earth spirits. This is the first new metal to be discovered since the ancients and the first new element discovered since phosphorus in 1669.

1736 Claudius Aymand performs the first successful appendectomy.

1736 Leonhard Euler writes the first book on mechanics based on differential equations, *Mechanica sive motus scientia analytice exposita*.

1738 Daniel Bernoulli develops his kinetic theory of gases, according to which pressure within a closed container of gas increases as temperature rises and decreases as temperature goes down.

1740 Swiss naturalist Abraham Trembley discovers the freshwater hydra and shows it to be a very primitive animal that possesses some properties of plants, such as the ability to grow into two organisms when cut in two or to merge two organisms into one.

Events of 1742-1771

1742 Anders Celsius suggests replacing the Fahrenheit scale with his Centigrade scale, which places the freezing point of water at 0 degrees and the boiling point at 100 degrees. Later the name of this scale is changed to *Celsius scale* by international agreement.

1742 Christian Goldbach proposes that every even number greater than 2 can be expressed as the sum of two prime numbers. This theory, known as Goldbach's conjecture, has never been proved or disproved by any mathematician.

1743 Jean le Rond d'Alembert expands on Newton's laws of motion in *Traité de dynamique*, proposing that actions and reactions in a closed system of moving bodies are balanced. He applies this principle to several mechanical problems.

1745 Pieter van Musschenbroek and Ewald Georg von Kleist independently invent the capacitor, an electricity-storing device that came to be called the Leyden jar after Musschenbroek's employer, the University of Leyden.

c. 1746 Euler determines the mathematics of light refraction, based on Huygens' theory of 1690 that light is composed of different colors of varying wavelengths.

1747 German chemist Andreas Marggraf detects that there is sugar in beets, a discovery that lays the foundation for the European sugar industry.

1748 Euler publishes *Introductio in analysin infinitorum (Introduction to Infinitesimal Analysis)*, outlining pure analytical mathematics.

1748 Jean Nollet findss that a pig's bladder acts as a semipermeable membrane, leading him to discover osmotic pressure.

1748 The Spanish scientist Antonio de Ulloa reports his discovery of platinum while on a trip to South America. He also notes that it is denser, higher-melting, and less reactive than gold.

1749 George-Louis Leclerc de Buffon advances a new theory of evolution in his first of 44 volumes of *Natural History*. He also suggests that the earth was formed by a collision of a massive comet with the sun some 75,000 years ago.

1750 Nicolas de Lacaille travels to the Cape of Good Hope, where he determines solar and lunar parallaxes and collects information for an extensive catalog of 2000 southern stars.

1751 In his *Philosophia botanica*, Linnaeus rejects any ideas of evolution.

1751 Axel Cronstedt isolates nickel and discovers that, like iron, it is attracted by a magnet.

1752 Benjamin Franklin conducts his famous experiment with a kite in a lightning storm, showing that lightning is a form of electricity. Based on his findings, he invents the lightning rod, which becomes popular almost overnight in America and Europe.

1752 Following experimentation with the digestive processes of a hawk, René-Antoine Ferchault de Réaumur discovers the role of gastric juices, proving that digestion is a chemical as well as a mechanical process.

1754 John Dolland invents a dual-image telescope called a heliometer, which can be used to determine angular distances, find planetary diameters, and determine distances between two heavenly objects.

1754 Joseph Black discovers "fixed air," or carbon dioxide, when he heats limestone. He determines that carbonates are compounds of a gas and a base.

1755 Immanuel Kant submits that the sun, earth, and Milky Way are part of a lens-shaped galaxy and that some nebulas, such as one in Andromeda, comprise other galaxies.

1758 Marggraf develops the flame test in chemistry for distinguishing one substance from another according to the color of flame it produces when burned.

1758 Cronstedt publishes *Essay on the New Mineralogy*, dividing minerals into four classes: earths, bitumens, salts, and metals. This work establishes the classification of minerals by chemical structure.

1759 In a pioneering work in embryology, Kaspar Wolff shows that rather than being contained in miniature in an egg or sperm, specialized organs develop out of unspecialized tissue.

1759 In his *Tentamen theorieae electricitatis et magnetismi (Examination of a Theory of Electricity and Magnetism)*, Franz Aepinus rejects mechanical theories of electricity and suggests that ordinary matter repels itself in the absence of electricity, a theory of electrostatics similar to Newton's law of gravity and supporting Benjamin Franklin's one-fluid theory of electricity.

1760 John Mitchell proposes that earthquakes are caused by the buildup of pressure within the earth and produce waves that travel at measurable speeds. He further submits that the center of these waves can be determined and that this center is the point of origin of the earthquake.

1761 Joseph Black discovers that ice absorbs heat without changing temperature when melting. He calls this property "latent heat," a quality later important in the improvement of the steam engine.

1761 Giovanni Batista Morgagni founds the science of pathological anatomy with his book, *De sedibus et causis morborum per anatomen indagatis (On the Causes of Diseases)*, a study of 640 postmortems he had conducted during his career.

1761 Mikhail Vasilevich discovers the atmosphere of Venus during an unsuccessful attempt to determine the parallax of the planet during its transit across the sun.

1763 Joseph Kohlreuter carries out landmark experimentation into the pollination of plants by animal pollen-carriers.

1764 While repairing a Newcomen steam engine, James Watt invents the condenser to improve the efficiency of the machine, using his friend Joseph Black's theories of latent heat.

1764 James Hargreaves first introduces his new invention, called a "spinning jenny," which can spin eight threads at once.

1766 Henry Cavendish discovers and outlines the properties of a highly flammable substance he calls "fire air," now known as hydrogen.

1766 Albrecht von Haller shows that nervous stimulation controls muscular movement and that all nerves lead to the spinal cord or brain.

1767 In *The History and Present State of Electricity*, Joseph Priestley offers an explanation for rings formed by an electrical charge on metal (now known as Priestley's rings) and proposes that electrical forces follow the same inverse square law that gravity does.

1767 Although he is now completely blind, Euler dictates *Vollstandige Anleitung zur Algebra (Complete Instruction in Algebra)*, a book that outlines the basic principles of algebra that are still in use today. The following year he publishes the first of three volumes of *Institutiones calculi integralis*, which establishes the basic theories of integral calculus and also contains Euler's own differential equations.

1769 In one of the key developments leading to Britain's industrial revolution, Richard Arkwright invents a spinning frame that produces cotton threads for textile manufacture.

1771 In experiments with plants in carbon dioxide-filled chambers, Priestley discovers that plants somehow convert the gas breathed out by animals and created by fire back into oxygen.

Events of 1772–1798

1772 Daniel Rutherford, Joseph Priestley, Carl Scheele, and Henry Cavendish independently discover nitrogen, although Rutherford is usually given the credit as being the first.

1774 Priestley publishes his discovery of oxygen, although Scheele had previously discovered it in 1772 but did not publish his findings until 1777, and Lavoisier actually named the element.

1774 Carl Scheele discovers manganese, chlorine, and barium, although he is not given credit for discovering any of them. However, he does not recognize chlorine to be an element, thinking it to be a combination of oxygen and some other substance. It is not rediscovered for about thirty years.

1775 James Watt patents a new steam engine he had developed in 1765, one that is six times as efficient as the old Newcomen engine. Its condenser is separated from the cylinder, allowing the steam to act directly on the piston.

1776 In the first orderly classification of human beings, Johann Friedrich Blumenbach divides humanity into five races based on superficial characteristics—Caucasians, Mongolians, American Indians, Malayans, and Ethiopians.

1776 The first blast furnace is created by John Wilkinson, using a steam engine.

1777 Charles-Augustin de Coulomb demonstrates how his torsion balance can be used to measure tiny weights with more accuracy than a balance scale.

1778 Scheele and Lavoisier independently conclude that air is composed mainly of oxygen and nitrogen.

1779 Jan Ingenhousz discovers the two distinct plant cycles that comprise photosynthesis: one cycle in the daytime during which plants consume carbon dioxide and produce oxygen, and one cycle at night during which oxygen is consumed and carbon dioxide is produced.

1780 During experiments with oxidation of organic gases, Johann Wolfgang discovers that platinum is a catalyst and that there are certain similarities of reaction among elements. The latter discovery forms the basis of the periodic table.

1781 After viewing what he believes to be a comet, William Herschel realizes it is actually a new planet, later named Uranus.

1781 Priestley "creates" water by igniting hydrogen in oxygen.

1781 Watt expands on his steam engine invention to create a mechanical attachment with a rotary movement that enables the engine to power a wider variety of machines.

1781 Arkwright builds the first modern factory to house his water frame for spinning.

1782 John Goodricke, a 17-year-old deaf-mute, suggests that a variable star actually has a dim companion that orbits around it and periodically eclipses it.

1783 Herschel explains the apparent movement of stars together and apart from each other by arguing that our sun itself is moving toward some stars and away from others, changing their relative positions.

1783 From his experiments with the amount of heat and carbon dioxide produced by a guinea pig, Lavoisier concludes that respiration is a form of combustion.

1783 Joseph and Étienne Montgolfier launch the first hot-air balloon, the earliest successful air flight in history. The second flight carries a cargo of animals, and the third flight carries two men. The same year, Jacques Charles creates and flies in the first hydrogen balloon.

1784 George Atwood develops a formula for determining the acceleration of a free-falling body.

1784 Benjamin Franklin invents bifocal lenses to solve his own vision problems.

1784 Ironmaster and inventor Henry Cort invents both the puddling method of manufacturing wrought iron from coke-smelted iron and the rolling mill with grooved rollers.

1785 In *On the Construction of the Heavens*, Herschel gives a reasonable description of the lenslike shape of the Milky Way, although he underestimates its size and places the sun in the center instead of near the edge. He also determines that some of what were previously thought to be nebulas are actually "globular clusters" of thousands of stars, and he is the first to detect what are now known as black holes.

1785 The theory of uniformitarianism, or the very gradual and regular evolution of the earth's structure, is introduced in *Theory of the Earth*, by James Hutton. According to his ideas, the earth is so old that it is almost impossible to determine when it began.

1785 Adrien Legendre publishes his first statement of the law of quadratic reciprocity and begins his important work on elliptic integrals and functions.

1787 Jacques Charles demonstrates that different gases expand at the same rate for a given temperature. This principle, later known as Charles's law, had actually been discovered by Guillaume Amontons as early as 1699.

1787 John Fitch builds and launches the first steamboat on the Delaware River. Although a technological success, it is a commercial failure.

1789 Herschel builds a technologically advanced telescope with a 49-inch lens and a 40-foot focal length. He discovers two satellites of Uranus and two more of Saturn.

1789 In the important chemistry book *Traite élémentaire de chimie (Elementary Treatise on Chemistry)*, Lavoisier introduces the law of conservation of mass that he had developed years before, and establishes a new system of chemical nomenclature.

1790 Samuel Slater builds the first American factory, constructed entirely from his memory of blueprints he had seen earlier in England, and ushers in the Industrial Revolution in the United States.

1793 Eli Whitney invents the cotton gin to pull the cotton threads off the seeds in cotton bolls. The machine is so successful that it transforms the economy of the southern U.S., where cotton becomes the main crop, and encourages the growth of slavery to supply the labor needed to keep up with the new machine.

1794 Legendre publishes *Elements de géométrie*, a simplification of Euclid's work that is the standard geometry textbook for the next 100 years.

1796 Using fluid from a cowpox sore on a milkmaid's hand, Edward Jenner inoculates an eight-year-old boy. Two months later he inoculates the boy with smallpox, but the boy does not get the disease. The process, which is called vaccination after the vaccinia virus that causes cowpox, soon is in use all over Europe.

1796 Pierre-Simon Laplace advances his "nebular hypothesis" of the origin of the world, which states that our solar system was formed by the condensation of a nebula of dust and gas, a theory still well established today.

1798 Henry Cavendish accurately determines the mass of the earth by determining the gravitational constant, the unknown G in Newton's equations. Cavendish calculates the mass to be about 6,600,000,000,000,000,000,000 tons, or 5.5 times the density of water.

Events of 1798-1821

1798 Eli Whitney is awarded a government contract for 10,000 muskets. In manufacturing these weapons, he develops methods for producing standardized interchangeable parts, a procedure important to the Industrial Revolution.

1799 Joseph-Louis Proust demonstrates that elements in a compound always combine in definite proportions, thereby differing from mixtures. This is sometimes known as Proust's law.

1799 Engineers of Napoleon's army excavate a large black stone near Rosetta, Egypt, on which the same message is engraved in three different languages—Greek and two ancient Egyptian languages. The Rosetta Stone, as it comes to be called, makes the deciphering of Egyptian hieroglyphics possible.

1799 In his doctoral dissertation, Karl Friedrich Gauss proves that every algebraic polynomial equation has a solution, the fundamental theorem of algebra.

1800 After experimenting with bowls of salt solution connected by metal arcs, Alessandro Volta constructs the first electric battery using wire and round plates of copper, zinc, and felt moistened with salt water.

1800 Less than two months after Volta's battery is publicized, William Nicholson and Anthony Carlisle build a battery and use it to perform electrolysis with water, separating it for the first time into hydrogen and oxygen.

1800 While investigating the heat of different colors in a spectrum, Herschel discovers that an invisible light beyond the red end of the spectrum that produces the most heat. He publishes his findings on infrared radiation in *An Investigation of the Powers of Prismatic Colours to Heat and Illuminate Objects.*

1800 Humphry Davy discovers nitrous oxide and experiences firsthand its power to influence the emotions and sensations (thus its popular name, "laughing gas"). He suggests its use as an anesthetic during surgery.

1801 Astronomer Giuseppe Piazzi accidentally discovers the first known asteroid, Ceres, between Mars and Jupiter. Its orbit is calculated by Karl Gauss, using a law developed by Johann Bode in 1772 when he predicted that a planet might exist where Ceres was found. In the next few years, many more asteroids are discovered along a belt between Mars and Jupiter.

1801 Lamarck becomes the first to classify invertebrates, in *Systeme des animaux sans vertebres (System for Animals without Vertebras)*, in which he also includes the beginnings of his theory of evolution.

1801 Building on Herschel's discovery of radiation beyond the red end of a spectrum, Johann Ritter experiments with silver chloride and detects radiation beyond the violet end as well, now known as ultraviolet radiation.

1801 Thomas Young discovers that light passing through two narrow slits produces an interference pattern, further support for the wave theory of light.

1803 The English chemist John Dalton becomes the first to propose a formal theory for the atomic composition of matter, claiming that atoms (a word he took from Democritus's writings of 440 BC) explain why chemicals combine only in specific proportions. Dalton also pioneers the concept of atomic weight, although many of his original values are later found to be far from the actual weight.

1803 William Henry proposes the law named after him, which states that the mass of a gas dissolved in a liquid is proportional to the pressure.

1803 In *Essai de statique chimique (Essay on Static Chemistry)*, Claude Berthollet applies physics to chemistry to show that reaction rates depend on the amounts of reacting substances as well as on affinities.

1807 Using the process of electrolysis, Humphry Davy discovers two new elements in one week in October—potassium and sodium. A year later, using the same process, he discovers barium, strontium, and calcium, and infers the presence of magnesium (isolated in 1829).

1807 The *Clermont*, a 133-foot-long steam-powered paddleboat built by Robert Fulton, makes its maiden voyage up the Hudson River. It proves to be a huge commercial success.

1808 Joseph Louis Gay-Lussac publishes his discovery that gases combine in definite proportions, just as elements do, when forming compounds.

1808 Étienne Malus discovers the polarization of light by reflection.

1809 In *Philosophie zoologique*, Lamarck states his beliefs that animals evolved from simpler forms, that over time unused body parts will degenerate or new body parts develop as needed for survival, and that such acquired characteristics can be inherited.

1811 Amadeo Avogadro suggests that at a given temperature and pressure, equal volumes of any gas will always contain equal numbers of molecules. This eventually comes to be known as Avogadro's hypothesis but is not widely recognized for another 50 years.

1811 While manufacturing potassium nitrate for gunpowder, Bernard Courtois accidentally produces dark crystals he has never seen before. He passes the crystals on to another chemist, Humphry Davy, who determines the crystals are composed of a new element, which he names iodine.

1814 Joseph von Fraunhofer manufactures the best lens and prism to date and uses it to view solar spectra. He notes hundreds of dark lines in the spectrum where wavelengths are missing. After measuring and mapping the most prominent lines, he determines that they always fall in the same place.

1814 George Stephenson builds the first practical steam locomotive in England, capable of hauling heavy loads much faster than is possible by horse-drawn wagons.

c. 1816 René Laënnec invents the stethoscope.

1818 Augustin-Jean Fresnel proposes that a transverse motion of light waves would explain polarized light and could still account for reflection, refraction, and diffraction.

1818 Johann Franz Encke works out the orbit of a comet discovered the previous year by Jean-Louis Pons. With an orbit of only three and a third years, Encke's Comet is the first short-period comet discovered and remains to this day the comet with the smallest known orbit.

1819 Pierre-Louis Dulong and Alexis-Thérèse Petit demonstrate that the specific heat of an element is inversely proportional to its atomic weight. The Dulong-Petit law makes it much easier to compute atomic masses.

1820 Hans Christian Oersted publishes his discovery of electromagnetism, based on experiments with a compass and an electric current. André Marie Ampère then expands on these experiments and finds that wires carrying current in the same direction attract each other and wires carrying current in opposite directions repel. Also expanding on Oersted, Johann Schweigger invents the galvanometer to measure the strength of a current.

1821 Thomas Seebeck discovers that heating certain metals to a given point produces electricity, a phenomenon now known as thermoelectricity.

1821 Michael Faraday publishes the results of his research into electromagnetic rotation and uses his findings to create an electricity-powered motor.

Events of 1822-1848

1822 In *Grundriss der Mineralogie (Outline of Mineralogy)*, German mineralogist Friedrich Mohs sets up a scale of mineral hardness, now known as the Mohs scale, and presents a mineral classification system.

1822 Jean Baptiste Fourier proposes that any continuous function is the sum of sine and cosine curves, that units on two sides of an equation have to balance just as numbers do, and that the basic set of units should be mass, length, and time.

1822 With the publication of *Traité des propriétés projectives des figures (Treatise on the Properties of Figures)*, Jean-Victor Poncelet inaugurates projective geometry.

1823 Gay-Lussac notes that two compounds containing silver have the same molecules in the same proportions, but differ widely in their properties. He and his colleague Jons Jakob Berzelius determine that the arrangement of the molecules must have some bearing. Berzelius calls these compounds with the same formula but different properties *isomers*.

1823 Michael Faraday is able to liquefy chlorine by a systematic use of cold and pressure. Later he uses the same technique to liquefy other gases.

1823 William Sturgeon invents the electromagnet, which is improved by Joseph Henry eight years later to the point where it can hold a ton of iron suspended.

1824 Niels Abel proves that the general quintic equation cannot be solved by algebraic means.

1825 The first successful steam locomotive to carry both passengers and freight, invented by George Stephenson, makes its first run.

1825 Oersted isolates aluminum, using a procedure so difficult that aluminum remains a precious metal for about sixty years until a simpler method is invented.

1826 Nikolai Lobachevsky develops a system of non-Euclidean geometry that he presents in a paper to the Kazan University in Russia. Others, including Janos Bolyai and Karl Friedrich Gauss, had worked out the same ideas earlier, but had not made them public.

1827 Gauss introduces differential geometry in *Untersuchungen zur Differentialgeometrie*.

1827 Robert Brown discovers the constant motion of microscopic particles that comes to be known as Brownian motion. Eighty years later this motion forms the basis for proof of the existence of atoms.

1827 In *Die galvanische kette, mathematisch bearbeitet*, Georg Simon Ohm states that an electrical current is equal to the ratio of the voltage to the resistance (Ohm's law).

1830 Charles Lyell publishes the first volume of his landmark *Principles of Geology*, advancing the uniformitarian theory that is still generally accepted today.

1831 Electromagnetic induction, a principle involved in electric generators, is discovered independently by Michael Faraday and Joseph Henry. Henry goes on to build the first practical electric motor.

1831 Charles Sauria develops the first practical friction match, using phosphorus on the tip.

1831 Thomas Graham announces that the rate of gas diffusion is inversely proportional to the square root of its molecular weight, a formulation now known as Graham's law.

1832 Faraday introduces the laws of electrolysis, the study of passing electric current through a liquid or a solution.

1833 Sugar manufacturer Anselme Payen isolates and names the first enzyme, diastase, in an effort to speed the conversion of starch to glucose.

1834 Benoit-Pierre Clapeyron presents a formulation of the second law of thermodynamics, with the statement that entropy tends to increase in a closed system; this actually predates the formal statement of the first law of thermodynamics.

1834 Cyrus McCormick patents a mechanical reaping machine that is one of a series of inventions revolutionizing farm work and food production.

1836 Theodor Schwann isolates the first animal enzyme, pepsin, which acts to break down food in the stomach.

1837 Louis Agassiz advances the idea of an Ice Age, a period during which huge glaciers covered most of the earth's surface.

1838 Friedrich Bessel determines the distance to a star other than the sun by using the parallax of 61 Cygni.

1838 Matthias Jakob Schleiden discovers that all living plant tissue is made up of cells. The following year, Theodor Schwann announces that the same is true for animal tissue. Their discoveries are known as the Schleiden-Schwann cell theory.

1839 Photography is born with the invention by Louis Jacques Daguerre of a process for making a silver image on a copper plate, known as a daguerrotype.

1839 In a laboratory accident, Charles Goodyear invents vulcanized rubber, a more stable and much more useful substance than plain rubber.

1842 Ether is used as an anesthetic in surgery for the first time by Dr. Crawford Williamson Long, although credit for its discovery is often given to William Morton, who publishes his findings about its use in 1846, three years before Long.

1842 Christian Doppler discovers that the wave frequency of a moving sound changes as the source of the sound moves toward or away from the observer. This becomes known as the Doppler effect.

1843 James Prescott Joule determines the mechanical equivalent of heat, or the quantity of work required to produce a unit of heat.

1843 William Rowan Hamilton devises a noncommutative mathematical system of hypercomplex numbers.

1844 Samuel F. B. Morse demonstrates his federally funded telegraph by sending the message, "What hath God wrought?" from Baltimore to Washington, D.C.

1845 With a huge telescope of his own making, William Parsons discovers a nebula having a spiral shape. Later he finds fourteen more spiral nebulas and discovers the first crab nebula.

1846 Based on calculations made independently by John Couch Adams and Urbain Leverrier, Johann Galle discovers the planet Neptune. Later that year, William Lassell discovers Neptune's very large satellite, which he names Triton.

1846 Elias Howe patents a sewing machine small enough to be used in a home. Five years later I. M. Singer improves it significantly.

1847 Hermann von Helmholtz publishes *Uber die Erhaltung der Kraft (On the Conservation of Energy)*, establishing the law stating that the amount of energy in the universe remains constant, that energy can be neither created nor destroyed, that energy is also constant within any given closed system, and that energy can be converted from one form to another. This becomes known as the first law of thermodynamics.

1848 William Thomson (Baron Kelvin) proposes a temperature of –273°C as the "absolute zero" point, or least possible temperature, since at that point the energy of molecules in a gas would reach zero. He proposes a new temperature scale (now known as the Kelvin scale) based on absolute zero, eliminating the need for negative numbers to express low temperatures.

Events of 1849-1879

1849 The speed of light is accurately calculated by Armand Fizeau using a turning disk and a mirror placed five miles away. His assistant, Jean Bernard Léon Foucault, improves the technique in 1850 and determines the speed of light to be 185,000 miles per second, which is only 0.7% off the figure accepted today.

1850 Rudolf Clausius establishes the second law of thermodynamics, which states that energy converted into heat cannot be reconverted into any other kind of energy, and that the amount of this unavailable or "useless" kind of energy in the universe is continuously increasing.

1851 Jean Foucault suspends a huge pendulum from a church ceiling to demonstrate the rotation of the earth.

1852 James Prescott Joule and Baron Kelvin demonstrate that an expanding gas with no outside energy source will become cooler; this is now known as the Joule-Thomson effect.

1854 John Snow definitely establishes the long-suspected link between poor hygiene and disease during the cholera epidemic in London. When he bars access to a water pump drawing water from a well a few feet from a sewer pipe, disease in the area decreases dramatically.

1854 Bernhard Riemann proposes a second type of non-Euclidean abstract geometry that a century later would be considered to present a truer picture of the universe than Euclidean geometry.

1855 Heinrich Geissler designs a mercury-powered air pump and uses it to develop the most effective vacuum tubes to date, called Geissler tubes, which eventually lead to the discovery of electrons.

1856 In *Recherches sur la putrefaction (Researches on Fermentation)*, Louis Pasteur advances the idea that microorganisms, not chemical action, produce fermentation. He further notes that gentle heating will kill yeast and prevent diseases in liquids, a process now called pasteurization.

1856 Henry Bessemer develops a new way to remove carbon from cast iron in steelmaking by sending a blast of air through molten iron in special furnaces. This production method is now known as the Bessemer process.

1856 Workers near the Neander River in Germany discover the first known skeleton of a Neanderthal man in a limestone cave, which becomes an important first step in studying the evolution of the human species.

1858 The first telegraph cable to connect the United States and Europe is laid across the Atlantic Ocean.

1859 Charles Darwin publishes *On the Origin of Species,* a book outlining his theory of evolution by natural selection. It causes great controversy and becomes the foundation of a completely new biology.

1859 The first oil well, funded by Edwin Laurentine Drake, is drilled in Titusville, Pennsylvania.

1859 Gustav Kirchhoff and Robert Bunsen experiment with spectral analysis and find that each element gives off a distinct frequency of spectral lines. In the next few years, this leads to the discovery of several new elements and to the realization that sodium, hydrogen, and several other elements exist in the sun and its atmosphere.

1859 Jean Lenoir builds the first working internal-combustion engine, fueled by coal gas. The following year he incorporates it in a "horseless carriage."

1862 Louis Pasteur's experiments with bacteria conclusively disprove the theory of spontaneous generation and advance the germ theory of disease, revolutionizing medical research and treatment.

1862 Julius von Sachs discovers bodies that are later identified as chloroplasts in plant cells and determines that starch is produced by photosynthesis.

1864 James Clerk Maxwell publishes *A Dynamical Theory of the Electromagnetic Field*, outlining his famous equations that describe the unified phenomena of electricity, magnetism, and light.

1865 After years of research with pea plants, Gregor Mendel publishes his theories of genetics in an obscure publication. His ideas are generally unknown for over thirty years.

1865 Friedrich August Kekulé realizes that benzene is made up of carbon and hydrogen atoms in a stable hexagonal ring, not a chain. This changes the way chemists view molecular structure.

1866 Alfred Nobel tames the volatility of nitroglycerin by enclosing it in kieselguhr, a packing material, making a much safer explosive he terms *dynamite*.

1869 Dmitri Ivanovich Mendeleev is the first of several scientists to publish a periodic table of the elements, with elements arranged in order of increasing atomic weight. In his revised edition two years later, he leaves three empty spaces where he believes newly discovered elements will be placed in the future. The spaces are filled in by 1885.

1869 Thomas Andrews suggests that there is a "critical temperature" for every gas, at which it can no longer be liquefied by pressure alone.

1871 Darwin publishes *The Descent of Man,* in which he applies his principles of evolution to the human race.

1873 Following the description given in Homer's *Iliad*, Heinrich Schliemann locates the site of the ancient city of Troy in Turkey. The discovery stimulates great interest in the science of archaeology.

1873 Johannes van der Waals refines Boyle's and Charles's laws of gases, introducing modified equations that allow for a wider range of molecular behavior. Thirty-seven years later he receives the Nobel Prize in physics for this work.

1873 Charles Hermite identifies *e* as the first transcendental number, a number that cannot solve any algebraic equation with rational coefficients.

1876 Alexander Graham Bell transforms the world of communication when he patents the first telephone. It is improved almost immediately by Edison, who adds carbon powder to the mouthpiece, allowing it to carry more current.

1876 In *The Geographical Distribution of Animals,* English naturalist Alfred Wallace establishes biogeographical support for the evolutionary theory.

1876 Robert Koch isolates and cultivates the bacterium that causes anthrax in cattle, the first step toward finding a cure for bacterial diseases.

1877 Louis Paul Cailletet and Raoul-Pierre Pictet independently liquefy oxygen.

1877 Edison invents the phonograph, his favorite invention, just one year after opening his research laboratory at Menlo Park, New Jersey.

1879 Limitations on human activity caused by the darkness of night are conquered as Edison in America and Joseph Swan in England independently produce carbon-thread electric lamps. Edison's first bulb burns for forty continuous hours, and only a year later he opens the first electric generation station in London, used primarily for lighting.

1879 While synthesizing organic compounds in the laboratory, Ira Remsen and Constantine Fahlberg accidentally discover saccharin when the latter puts his fingers to his lips.

Events of 1879-1904

1879 While working with bacteria, Pasteur accidentally discovers that they can be weakened, and that in this weakened state they will impart immunity without provoking the disease. This is the foundation of immunizations. Pasteur then uses this information to develop the first artificially produced vaccine, against anthrax, in 1881.

1880 John Milne invents a seismograph for measuring waves caused by earthquakes, founding the science of seismology.

1880 Charles Louis Alphonse Laveran discovers that a protist, not a bacterium, causes malaria. This is the first time an organism other than a bacterium is identified as causing disease.

1881 Albert Abraham Michelson invents the interferometer, which measures distance by observing interference patterns of a split light beam.

1882 Walther Flemming publishes his discovery of chromosomes and mitosis in *Cell Substance, Nucleus, and Cell Division.* This discovery fits in with Mendel's research into genetics, which is still not being recognized at this time.

1882 Robert Koch isolates the bacterium that causes tuberculosis, but is unable to find a cure for the disease. The following year he discovers the bacterium causing cholera.

1882 Ferdinand Lindemann proves that π is a transcendental number and that the centuries-old problem of squaring a circle with a straightedge and compass cannot be done.

1884 Svante Arrhenius is given only a mediocre grade for his doctoral dissertation outlining his theories of ionic dissociation, in which ions detach into positively and negatively charged ions when in a solution. Less than twenty years later, he receives the Nobel Prize for the same paper.

1884 Ottmar Mergenthaler patents the Linotype, a machine that can set a whole line of type at one time, greatly speeding up the production of printed matter, which is especially useful to the newspaper industry.

1885 Louis Pasteur develops a rabies vaccine and tests it on young Joseph Meister, who was bitten by a rabid dog. The vaccine saves the boy's life.

1885 A Westinghouse Company employee, William Stanley, invents the transformer for shifting the voltage and amperage of alternating current, enabling electricity to be transported over long distances with relatively little loss.

1885 Carl Friedrich Benz designs the first working automobile, which has an internal-combustion, gasoline-burning engine but only three wheels.

1887 In his studies of air and sound, Ernst Mach notes that air molecules behave irregularly when pushed faster than the speed of sound.

1887 An experiment by Albert Michelson and Edward Morley seems to indicate that the earth has no detectable motion in relation to the ether, a hypothetical medium whose existence was then thought to be necessary for the motion of light waves. This finding, called the *null result,* anticipates Einstein's later description of light, which does not require the existence of the ether.

1887 Heinrich Hertz observes that light can affect electrical phenomena, the so-called photoelectric effect. This leads him to discover the existence of radio waves, originally called Hertzian waves, the following year.

1888 George Eastman designs the first camera for nonprofessional use, which he arbitrarily names *Kodak.*

1890 Emil von Behring discovers antitoxins and uses them to develop tetanus and diphtheria vaccines. For this, he receives the first Nobel Prize in physiology and medicine in 1901.

1890 William Stewart Halsted introduces the practice of wearing rubber gloves during surgery and also performs the first radical mastectomy to treat breast cancer.

1893 Karl Benz builds the first four-wheeled car with a combustion engine, and Henry Ford builds his first automobile.

1893 Wilhelm Wien discovers the inverse proportion of wavelength and absolute temperature. Wien's law is used to establish star temperatures and indirectly leads to the quantum theory.

1895 While investigating fluorescence in cathode rays, Wilhelm Roentgen discovers X-rays and begins a period of revolutionary discoveries. The first diagnostic X-ray is taken less than a year later; he is awarded the first Nobel Prize in physics in 1901.

1896 Basing his research on Roentgen's work, Henri Becquerel discovers radioactivity emitted by a uranium compound.

1896 Working under the tutelage of Hendrik Antoon Lorenz, Pieter Zeeman finds that a magnet can change the spectral lines of a light source, a discovery later used to determine fine details of atomic structure and the composition of stars.

1897 Joseph John Thomson discovers the electron, the first known subatomic particle, and determines its mass.

1898 Marie and Pierre Curie isolate and analyze radium and polonium and their radioactive properties. For their work on radioactivity, they share a Nobel Prize in physics in 1903, and Marie wins a second award in chemistry in 1911, after Pierre's death.

1899 Ernest Rutherford notes that radiation given off by uranium takes two forms; he names the positively charged radiation alpha rays and the negatively charged radiation beta rays.

1900 Gregor Mendel's work with genetics is rediscovered by three biologists who are working independently in separate countries. They finally give publicity and credence to Mendel's theories.

1900 Sigmund Freud publishes *Die Traumdeutung (The Interpretation of Dreams)*, which explores the role of the subconscious and unconscious mind in human behavior.

1900 Paul Ulrich Villard observes gamma rays, proving them to be a form of radiation more powerful than X-rays.

1900 The quantum theory is formulated by Max Planck, showing that energy is given off in pieces that are sized in inverse proportion to the wavelength. His work marks the beginning of modern physics and earns him a Nobel Prize in 1918.

1900 The first dirigible to make a successful flight is built by Count Ferdinand von Zeppelin.

1900 Walter Reed shows that yellow fever is not caused by a bacterium or by human contact, but by mosquitoes carrying what he later shows to be a virus. His work reduces the yellow fever death rate so rapidly that the last epidemic in the U.S. is in 1905.

1901 The first radio waves are broadcast across the Atlantic from England to Newfoundland by Guglielmo Marconi, using an antenna he invented in 1895.

1903 The first heavier-than-air machine is flown for 59 seconds over 850 feet by Wilbur and Orville Wright at Kitty Hawk, North Carolina.

1904 Beltram Borden Boltwood proves that uranium changes into radium through the process of radioactive decay. He later expands his theory, showing that radium decays into lead and that, in general, radioactive substances eventually decay into nonradioactive substances. This theory is then used in radioactive dating of inorganic substances.

Events of 1905-1927

1905 Albert Einstein formulates his light-quantum theory, which finally explains the photoelectric effect (for which he is awarded the 1921 Nobel Prize in physics); he calculates how molecular movement causes Brownian motion and publishes his Special Theory of Relativity. This contains the revolutionary equation $E=mc^2$ that establishes the law of mass-energy equivalence.

1905 Alfred Binet devises the first test to determine the "intelligence quotient" of children.

1906 Frederick Hopkins suggests the presence of trace substances in foods that are not carbohydrates or proteins but that are essential to life. These substances come to be known as vitamins.

1906 The first broadcast of words and music by means of amplitude modulation (AM) radio waves is sent by Reginald Fessenden.

1906 Charles Barkla discovers that each element has a characteristic X-ray, and that the penetration degree of these X-rays is related to the atomic weight of the element. This discovery proves to be the key to atomic number.

1906 In his third law of thermodynamics, Walther Nernst proposes that absolute zero is a point that can be closely approached, but never actually reached.

1907 Paul Ehrlich develops the earliest form of chemotherapy, using a specially formulated dye to kill protozoa in cells that cause sleeping sickness, but leaving the rest of the cell intact.

1907 Thomas Hunt Morgan begins genetic experiments with chromosome-simple fruit flies that prove the function of chromosomes, demonstrate the role of mutation in heredity, and illustrate sex-linked characteristics.

1907 Ivan Petrovich Pavlov conducts his famous experiments with a dog that salivates when it hears the ringing of a bell that it associates with food, demonstrating the law of reinforcement or conditioned responses and greatly influencing the development of behavioral psychology.

1908 Ernest Rutherford and his assistant Hans Geiger develop the Geiger counter to detect radiation levels.

1908 Henry Ford revolutionizes the manufacturing process when he sets up the first assembly line in his Model-T auto factory.

1908 Hermann Minkowski publishes *Space and Time,* proposing time as a fourth dimension for study of the universe.

1909 Robert Peary and Matthew Hensen become the first men to reach the North Pole.

1911 Rutherford announces his theory of a positively charged atomic nucleus surrounded by negative electrons, and it is almost immediately accepted.

1911 Roald Amundsen becomes the first man to reach the South Pole.

1911 Heike Kamerlingh Onnes discovers superconductivity—the loss of electrical resistance in mercury near absolute zero temperatures. It is soon discovered that other metals and alloys become superconductors at very low temperatures.

1912 Using electroscopes on balloon flights, Victor Hess discovers that the ionization of air increases with altitude, indicating the presence of what he names cosmic radiation.

1912 Studies of Cepheid variables in Magellanic Clouds by Henrietta Swan Leavitt lead to the period-luminosity law, a key discovery that replaces the method of parallax in determining huge distances outside the Milky Way.

1913 Frederick Soddy announces a theory of isotopes based on the radioactive displacement law.

1913 Niels Bohr applies the quantum theory to the atom and decides that electrons have fixed orbits around the nucleus and that they emit or absorb specific amounts of energy, or quanta, when they jump into another orbit.

1913 In the famous "oil drop experiment," Robert Andrews Millikan measures the charge of a single electron to be sixteen-quintillionths of a coulomb.

1914 Rutherford suggests that the positive nucleus of atoms contains protons that balance the charge of electrons but have a much greater mass.

1914 Through studies of earthquake waves, Beno Gutenberg determines that the earth has at its center an inner core and an outer core, which together measure about 4200 miles in diameter, with a sharp dividing line between the mantle and the outer core that is now known as the Gutenberg discontinuity.

1916 Einstein publishes his General Theory of Relativity, or Theory of Gravitation, equating gravitational and inertial mass and showing space to be curved.

1916 Using Einstein's theory of relativity, Karl Schwarzschild postulates on the existence in space of phenomena that are later named black holes.

1916 Gilbert Newton Lewis and Irving Langmuir independently formulate a system of chemical bonding based on the number of electrons in each shell of an atom.

1917 Carl Gustav Jung publishes *Psychology of the Unconscious,* which postulates that the unconscious mind consists of both a personal and a collective unconscious.

1918 Using a new 100-inch telescope at Mt. Wilson Observatory near Pasadena, California, Harlow Shapley determines that rather than being at the center of our galaxy, our solar system is about 20,000 light years from one end of the galaxy and 80,000 light years from the other.

1919 Observations of a total eclipse of the sun show a gravitational deflection of light that supports Einstein's General Theory of Relativity.

1921 Frederick Grant Banting isolates insulin from degenerating dog pancreases and begins experiments in treating diabetes, first administering it to a human patient the following year.

1923 Arthur Holly Compton illustrates the dual wave and particle nature of X-rays and gamma rays by showing how their wavelengths increase when they collide with electrons.

1925 Wolfgang Pauli proposes a fourth quantum number for energy levels of atomic electrons and theorizes that no two electrons with the same quantum number can possibly occupy the same atom. This theory accounts for the chemical properties of elements.

1925 In the "Scopes Monkey Trial" in Dayton, Tennessee, John T. Scopes is prosecuted for teaching evolution to his high school science class.

1925 Scottish inventor John L. Baird transmits the first television signal—a human face.

1926 The first talking motion picture, *The Jazz Singer*, begins a new era in popular entertainment.

1926 Robert Goddard launches the first rocket propelled by liquid fuel.

1926 Erwin Schrödinger introduces the theory of wave mechanics, viewing the electron as a wave rather than a particle. Max Born applies probabilistic interpretations to the theory, giving rise to quantum mechanics.

1927 George Lemaître proposes the earliest version of what is now known (and generally accepted) as the "Big Bang" theory of the origin of the universe.

Events of 1927-1953

1927 Werner Heisenberg presents the "uncertainty principle," which states that the position and the momentum of a subatomic particle cannot both be determined simultaneously.

1927 In *Coming of Age in Samoa*, Margaret Mead describes life stages in a primitive society.

1928 Station WGY of Schenectady, New York, schedules the first television broadcast.

1928 Alexander Fleming accidentally discovers a mold that destroys bacteria, which he isolates and names penicillin. Its use in clinical medicine does not begin for over a decade.

1929 Edwin Hubble announces Hubble's law, stating that other galaxies are receding from ours at a rate proportional to their distance from us. This law gives support to the expanding universe theory.

1929 The first particle accelerator, originally known as an "atom-smasher," is devised by John Douglas Cockcroft and Thomas Sinton Walton. The following year Ernest O. Lawrence expands on their idea to create the cyclotron.

1929 Deoxyribonucleic acid, better known as DNA, is discovered by Phoebus Levene.

1930 The planet Pluto is discovered by Clyde William Tombaugh.

1930 The first analog computer is invented by Vannevar Bush to solve differential equations.

1932 After years of speculation, the existence of the neutron in an atom is demonstrated by James Chadwick, for which he is awarded the Nobel Prize three years later. Based on this discovery, Werner Heisenberg proposes that an atom nucleus is composed of protons and neutrons, explaining the existence of isotopes.

1932 The positron, a positively charged electron and the first form of antimatter to be discovered, is detected in cosmic-ray tracks by Carl David Anderson.

1932 Ernst August Ruska builds the first primitive electron microscope capable of a 400× magnification. By the following year he has already improved it to 1200×. By 1937 it can magnify 7000×.

1933 Philo T. Farnsworth, often called the father of modern television, develops the dissector tube, making electronic television possible.

1934 The husband-and-wife team of Frédéric and Irène Joliot-Curie develop the first artificially radioactive element, an isotope of phosphorus; Irene is the daughter of Pierre and Marie Curie.

1935 Charles Richter invents a scale to measure the strength of earthquakes.

1935 The first isotope of uranium, U-235, is discovered by Arthur Jeffrey Dempster.

1935 Robert Alexander Watson-Watt uses microwaves to devise a radio detection and ranging system, better known as radar.

1937 Carl David Anderson discovers the muon, a subatomic particle larger than an electron but smaller than a proton.

1937 Theodor Dobzhansky publishes *Genetics and the Origin of the Species*, thus finally linking Darwin's evolutionism and Mendel's theory of mutation in genetics.

1938 Hans Bethe and Carl von Weizsacker independently determine that the nuclear fusion of hydrogen into helium produces the energy of the sun and other stars.

1938 Otto Hahn and Lise Meitner of Germany are the first to split a uranium atom, although they hesitate to call their success nuclear fission.

1939 Paul Hermann Müller develops the organic insecticide DDT, which proves especially effective in fighting typhus, but is later found to have numerous dangerous side effects.

1939 The first workable single-rotor helicopter is produced in the United States by Igor Sikorsky.

1939 Edwin Armstrong invents frequency modulation (FM) for the almost static-free transmission of radio waves.

1941 The first plane powered by a jet engine, developed in 1930 by Frank Whittle, is successfully flown.

1942 Enrico Fermi and a team of researchers at the University of Chicago create the first controlled nuclear chain reaction in a pile of uranium and graphite.

1944 Based on experiments with pneumococci, Oswald T. Avery determines that DNA is material carrying genetic properties in nearly all living organisms.

1945 The first two nuclear fission bombs are created by a team of scientists in a vast effort code-named "The Manhattan Project" in Los Alamos, New Mexico, headed by J. Robert Oppenheimer. One bomb is uranium-235-based and the other is plutonium-based.

1946 John W. Mauchly and John P. Eckert construct ENIAC (electronic numerical integrator and computer), the first entirely electronic computer. It weighs 30 tons and takes up 1500 square feet of space.

1947 Willard Frank Libby introduces a method of dating archaeological objects by determining the concentration of radioactive carbon-14 in the item.

1947 Air Force test pilot Chuck Yeager flies an X-1 rocket plane faster than the speed of sound, breaking the sound barrier for the first time.

1947 Cecil Frank Powell discovers the pion, a particle with mass between that of electrons and protons that interacts with neutrons and protons in the atomic nucleus. It is believed to be the meson-particle predicted by Hideki Yukawa twelve years earlier.

1948 William Bradford Shockley, Walther Houser Brattain, and John Bardeen invent the transistor, using small, solid semiconductors instead of delicate vacuum tubes to transmit current. By 1952, transistors are being used in radios and hearing aids.

1948 Norbert Weiner publishes *Cybernetics*, a detailed account of the mathematics of communication and a forerunner of computer science.

1948 George Gamow, Ralph Alpher, and Robert Herman formulate the Big Bang theory, explaining the origin of the universe in terms of a huge explosion producing a burst of energy.

1951 Mauchly and Eckert create UNIVAC I, the first electronic computer to store data on magnetic tape and to be mass-produced.

1952 Jonas Salk develops a polio vaccine using dead viruses. Inoculations begin in 1954; the Salk method is replaced in 1957 by a live-virus vaccine developed by Albert Sabin.

1952 A group led by Edward Teller develops and explodes the hydrogen bomb, the first thermonuclear device working by nuclear fusion. It has 500 times the power of the atomic bomb dropped on Hiroshima in 1945.

1953 Francis H. Crick and James Dewey Watson suggest the double helix model for DNA, giving a logical explanation for how genetic traits are transferred.

1953 The discovery of the Great Global Rift in the Atlantic Ocean by Maurice Ewing and Bruce Charles Heezen establishes the study of plate tectonics and revolutionizes geology.

Events of 1954–Present

1954 The first successful kidney transplant is performed between identical twins in Boston, Massachusetts.

1954 The first nuclear reactor built for the peacetime production of power begins operation in the Soviet Union, and the first nuclear-powered submarine, the *Nautilus*, is launched in the United States.

1956 George Emil Palade determines the protein-manufacturing function of ribosomes in the cell and Jacques-Lucien Monod and Francois Jacob determine the roles of messenger-RNA and transfer-RNA.

1957 The Space Age begins with the launching of the first artificial satellite, Sputnik I, by the Soviet Union. It takes the United States another year to make a successful satellite launch.

1959 The first man-made object lands on the moon, the Soviet Lunik II rocket.

1960 Theodore Harold Maiman designs the first laser (light amplification by stimulated emission of radiation). Within two years, the laser is being used in eye surgery.

1960 The first semiconductors are etched with transistor circuits, making integrated circuits that eventually lead to smaller, cheaper, and more powerful modern computers.

1961 The Soviets send the first human being into space, Yuri Gagarin, who orbits the earth in a 1.8-hour mission aboard the Vostok I.

1961 Murray Gell-Mann and others propose a system for classifying subatomic particles which Gell-Mann describes as "the eightfold way." This system is the foundation of the quark theory.

1963 Maarten Schmidt detects a huge redshift of a so-called "radio star," which turns out to be a quasar, an extremely distant galaxy with a very active center that enables it to be seen.

1965 Space technology advances rapidly, with the Soviets making the first space walk in March, the United States launching the first communications satellite in April, the United States completing the first space rendezvous of two satellites in December, and the Soviets sending out a space probe that lands on Venus.

1965 Using a computer, Hugh Williams, R. A. German, and C. R. Zanke are able to answer the "cattle of the sun" problem that was first posed by Archimedes 2000 years earlier; their solution is a number with 206,545 digits.

1967 Dr. Christiaan Barnard performs the first human heart transplant in a hospital in Cape Town, South Africa. The recipient lives for eighteen days.

1967 Jocelyn Bell discovers the first pulsar, a fast-moving neutron star that emits regular bursts of microwaves.

1969 The first person to walk on the moon is Neil A. Armstrong of the United States, followed by fellow crew member Edwin "Buzz" Aldrin, with the third crew member Michael Collins orbiting the moon in the Apollo 11 spacecraft.

1970 The first widebody jet airliner, the Boeing 747, begins making regular flights across the Atlantic.

1970 Floppy disks for computer data storage are introduced.

1970 Hamilton O. Smith and Daniel Nathans discover an enzyme that can cut up a DNA strand, which can then be recombined to form new genes. Recombinant DNA techniques later become important in genetic engineering.

1970 Scientists at the University of Wisconsin, led by Har Gobind Korana, succeed in synthesizing a gene without using any existing genes as a template, as was the practice in prior syntheses.

1971 Texas Instruments introduces the first pocket calculator, weighing about 2.5 pounds and costing about $150.

1971 The United States launches Pioneer 10, a space probe that will travel beyond Mars. It passes through the magnetic field of Jupiter in 1973. In 1983, it passes beyond Neptune.

1974 Howard Georgi and Sheldon Glashow develop the first Grand Unified Theory (GUT) to unite strong, weak, and electromagnetic forces.

1975 The development of microchips makes possible the first personal computer, the Altair 8800, with 256 bytes of memory.

1976 The Four-Color Theorem, a centuries-old conjecture that a map in a plane could be colored with four colors so that no two countries sharing a border are the same color, is solved by a computer programmed by K. Appel and W. Haken.

1977 Donald Johanson discovers "Lucy," a four-million-year-old female hominid fossil.

1977 The last recorded case of smallpox occurs in Somalia, and the first known case of Acquired Immune Deficiency Syndrome (AIDS) occurs in New York, although it is not officially recognized as a disease until 1981.

1978 In Great Britain, the first baby is born as a result of fertilizing an egg in a laboratory test tube, rather than in the mother's body.

1981 The *Columbia*, a manned, reusable space shuttle is successfully launched into space and then returned to earth.

1981 A team at the University of Wisconsin, led by Joseph Cassinelli, discovers R136a, a star 100 times brighter than the sun and with a mass 2500 times as large; it is the most massive star known to date.

1981 Chinese scientists successfully clone a golden carp.

1983 Kay Davies and Robert Williamson pinpoint the DNA marker for Duchenne muscular dystrophy—the first marker for a genetic disease that is indisputably identified.

1984 Alec Jeffreys devises a genetic fingerprinting method based on DNA sequences that are unique to each individual.

1984 Louis deBranges proves the Bieberbach Conjecture, a very important, ages-old theorem in complex analysis.

1986 In its flight past Uranus, Voyager 2 detects ten more satellites of the planet in addition to its already-known five, and sends back additional information about the planet itself.

1987 The superconductivity of ceramic materials at warmer temperatures near 30 K is discovered by Alex Muller and Georg Bednorz.

1987 The first supernova since 1604 in or near our own galaxy is observed by Ian Shelton in the Large Magellanic Cloud in the galaxy next to ours.

1992 A team led by astrophysicist George Smoot of the University of California, Berkeley reports that NASA's COBE satellite (Cosmic Background Explorer) has detected slight temperature variations in the cosmic microwave background; this is described as relic radiation from the Big Bang and is regarded as providing the strongest support to date for the Big Bang theory.

Photo Credits

abacus: HBJ Photo; **aborigine:** Courtesy Dept. of Library Services, American Museum of Natural History; **accelerator:** Fermi National Accelerator Laboratory; **acetobacter:** HBJ Photo; **Acropolis:** Scala/Art Resource; **Adélie penguin:** Dietkich/U.S. Navy; **adjustable wrench:** Courtesy of J.H. Williams Industrial Products, Inc.; **agitator:** Courtesy of Greerco Corp.; **Airacomet:** Bell Aircraft Corporation; **aircraft carrier:** U.S. Navy; **air pollution:** AERO Service, Litton Industries; **algae:** Electro-Nucleonics Laboratories; **Allen wrench:** Courtesy of Holo-Krome Co.; **altocumulus:** News Bureau, General Electric Company; **ameba:** General Biological Supply House, Chicago; **amethyst:** Courtesy Dept. of Library Services, American Museum of Natural History; **amphibian:** W.P. Taylor/U.S. Fish and Wildlife Service; **amphibious assault ship:** U.S. Navy; **amphitheater:** B.W. Muir/U.S. Forest Service; **anaphase:** Carolina Biological Supply Company; **aneroid barometer:** U.S. Weather Bureau; **antenna:** NASA; **anticlinal fold:** U.S. Geological Survey; **anti-missile missile:** Courtesy of Raytheon Co; **antiproton:** Brookhaven National Laboratory; **Apollo 11:** NASA; **arch bridge:** Tennessee Valley Authority; **Archimedes:** HBJ Photo; **arete:** U.S. Geological Survey; **argillite:** Courtesy Dept. of Library Services, American Museum of Natural History; **arthritis:** Arthritis Foundation; **asphalt paver:** Courtesy of Ingersoll-Rand Co.; **astronomical camera:** NASA; **atomic bomb:** Atomic Energy Commission; **atomic clock:** National Bureau of Standards; **attapulgite:** Mineral Constitution Laboratories, Penn State U.; **aurora borealis:** Courtesy Dept. of Library Services, American Museum of Natural History; **Australopithecus boisei:** Smithsonian Institution; **automated board assembly:** Courtesy of WT Automation, Inc.; **automatic focusing system:** Courtesy of Polaroid Corporate Archives; **awl:** Courtesy of CooperTools; **axes:** Courtesy of CooperTools; **Babbage:** The Mansell Collection, London; **backfire:** U.S. Forest Service; **bacteriophage:** G. Peuso/Institute of Sanitation, Rome; **Badlands:** Publicity Division, South Dakota Dept. of Highways; **baler:** Courtesy of John Deere & Co.; **ball-peen hammer:** Courtesy of CooperTools; **battleship:** U.S. Navy; **beam network:** Courtesy of Weyerhaeuser Co.; **Becquerel:** Courtesy Dept. of Library Services, American Museum of Natural History; **bedding:** U.S. Geological Survey; **Bell:** Courtesy of AT&T; **Bernard:** National Library of Medicine, Washington, D.C.; **beryl:** HBJ Photo; **Bevatron:** Lawrence Radiation Laboratory, Berkeley, California; **biomass:** Courtesy of Georgia-Pacific Corp.; **biplane:** HBJ Photo; **black-capped chickadee:** Courtesy Dept. of Library Services, American Museum of Natural History; **blades:** Courtesy of Peabody Museum, Harvard University; **blasthole drill:** Courtesy of Ingersoll-Rand Co.; **block cutting:** U.S. Forest Service; **blood cells:** General Biological Supply House, Chicago; **blossom:** A.W. Schoof/HBJ Photo; **blue crab:** General Biological Supply House, Chicago; **bluffs:** U.S. Geological Survey; **Bohr:** Black Star; **Bothriolepsis:** Field Museum of Natural History, Chicago; **bougainvillea:** Smithsonian Institution; **box wrench:** Courtesy of J.H. Williams Industrial Products, Inc.; **Brahe:** HBJ Photo; **bridge:** U.S. Geological Survey; **bronze ax:** Courtesy Dept. of Library Services, American Museum of Natural History; **brown bat:** George A. Smith; **bubble memory:** HBJ Photo; **Burgess Shale:** Smithsonian Institution; **butte:** U.S. Geological Survey; **C-17:** Courtesy of LTV Aircraft Products Group; **cable skidder:** Courtesy of John Deere & Co.; **calcite:** Mario Fanten, 1973/Photo Researchers; **calculator:** Courtesy of IBM; **Callisto:** NASA; **canal:** Bureau of Reclamation, U.S. Department of the Interior; **cancer cell:** Wide World Photos; **Canis Major/Canis Minor:** Courtesy Dept. of Library Services, American Museum of Natural History; **Cannon:** Harvard College Observatory; **canyon:** Union Pacific Railroad; **Carson:** New Bedford Standard-Times; **cartilage:** General Biological Supply House, Chicago; **Carver:** The Bettmann Archive; **caterpillar:** George A. Smith; **C clamp:** Courtesy of J.H. Williams Industrial Products, Inc.; **chambered nautilus:** HBJ Photo; **cheek cells:** Photo Researchers; **chevron fold:** U.S. Geological Survey; **chiseling tool:** Courtesy Dept. of Library Services, American Museum of Natural History; **cholla:** U.S. Forest Service; **ciliated protozoa:** HBJ Photo; **cirque:** U.S. Geological Survey; **cirrocumulus:** News Bureau, General Electric Company; **cirrus:** Helen Faye/HBJ Photo; **citrine:** Courtesy Dept. of Library Services, American Museum of Natural History; **Closterium:** Courtesy Dept. of Library Services, American Museum of Natural History; **cloud chamber:** University of California, Berkeley; **cloud formation:** HBJ Photo; **cluster:** Lick Observatory, University of California, Santa Cruz; **cocoon:** Courtesy Dept. of Library Services, American Museum of Natural History; **coelenterate:** Carolina Biological Supply Company; **cold-rolled steel:** HBJ Photo; **colloid mill:** Courtesy of Greerco Corp.; **Columbia:** NASA; **Columbus:** Brown Brothers; **columnar joint:** U.S. Geological Survey; **combination wrenches:** Courtesy of J.H. Williams Industrial Products, Inc.; **combine:** Courtesy of John Deere & Co.; **communications satellite:** Courtesy of Bell Telephone Laboratories; **Comstar:** Courtesy of AT&T; **cone crater:** HBJ Photo; **conglomerate:** Smithsonian Institution; **coniferous forest:** Standard Oil Co.; **connector blocks:** Courtesy of TW Comcorp.; **constant-velocity star:** Mount Wilson and Palomar Observatories; **container ship:** Courtesy of Matson Navigation Company; **contour plowing:** U.S. Soil Conservation Service; **control panel:** Courtesy of Piper Aircraft Corporation; **Copernican system:** Courtesy Dept. of Library Services, American Museum of Natural History; **coral:** Courtesy Dept. of Library Services, American Museum of Natural History; **coronagraph:** High Altitude Observatory, Climax, Colorado; **Cousteau:** Paris Match/Archive Photos; **crab:** Douglas P. Wilson/Marine Biological Laboratory; **crater:** U.S. Geological Survey; **crawler drill:** Courtesy of Ingersoll-Rand Co.; **crevasses:** U.S. Geological Survey; **Cro-Magnon:** Courtesy Dept. of Library Services, American Museum of Natural History; **cross-bedding:** U.S. Geological Survey; **crown:** L.C. Goldman/Bureau of Sport Fisheries & Wildlife; **crustacean:** Courtesy Dept. of Library Services, American Museum of Natural History; **crystallization:** Hoffman-La Roche & Co; **cultivator:** Courtesy of John Deere & Co.; **Curies:** The Bettmann Archive; **Cygnus:** Courtesy Dept. of Library Services, American Museum of Natural History; **d'Alembert:** HBJ Photo; **dam:** Las Vegas News Bureau; **datolite:** Courtesy Dept. of Library Services, American Museum of Natural History; **Davenport electric motor:** Smithsonian Institution; **deep-submergence rescue vehicle:** U.S. Navy; **defense-suppression missile:** Courtesy of Northrop Corp.; **deforestation:** Soil Conservation Service; **Delphinus:** Courtesy Dept. of Library Services, American Museum of Natural History; **delta:** NASA; **dendritic drainage:** U.S. Geological Survey; **deposition:** U.S. Geological Survey; **Descartes:** Smith Collection, Columbia University Libraries; **desert soil:** Arabian American Oil Co.; **De Vries:** Science Service; **diamond:** Indiana Wire Die Company; **difference engine:** Courtesy of IBM; **dig:** Field Museum Of Natural History, Chicago; **Dione:** NASA; **diphenylamine:** Hoffman-La Roche & Co.; **disk harrow:** Courtesy of John Deere & Co.; **dive bomber:** Courtesy of McDonnell Douglas; **docking station:** Courtesy of WT Automation, Inc.; **dome:** U.S. Geological Survey; **downy woodpecker:** National Audubon Society/Photo Researchers; **drone:** HBJ Photo; **drought:** Bureau of Reclamation/U.S. Department of the Interior; **drumlin:** U.S. Geological Survey; **Dubos:** Simpson Kalisher; **duckbill pliers:** Courtesy of J.H. Williams Industrial Products, Inc.; **durum wheat:** Publicity Division, South Dakota Dept. of Highways, Pierre; **Earth:** NASA; **earthfill dam:** Corps of Engineers; **Eastern mixed forest:** U.S. Forest Service; **echinoderm:** Carolina Biological Supply Company; **eclipse:** NASA; **Eddington:** HBJ Photo; **Edison:** Edison National Historic Site, U.S. Department of the Interior; **Einstein:** Wide World Photos; **embryo:** Martin M. Rotker/Taurus Photos; **emerald:** Courtesy Dept. of Library Services, American Museum of Natural History; **emphysema:** National Tuberculosis & Respiratory Disease Association; **engraving:** Art Resource; **ENIAC:** Univ. of Pennsylvania; **epithelium:** Ward's Natural Science Establishment, Inc., Rochester, N.Y.; **Epstein-Barr virus:** University of Bristol, Dept. of Pathology; **eruption:** Pan American World Airways; **etching:** HBJ Photo; **Europa:** NASA; **excavation:** HBJ Photo; **exfoliation:** U.S. Geological Survey; **extravehicular activity:** NASA; **Faraday:** Harper's Weekly; **fault:** U.S. Geological Survey; **FAX:** Courtesy of Ricoh Corporation; **feller-buncher:**

Photo Credits

Courtesy of John Deere & Co.; **Fermi:** HBJ Photo; **ferromanganese:** Bureau of Mines; **fetus:** HBJ Photo; **fibrin:** Keith R. Porter/Rockefeller Institute for Medical Research; **files:** Courtesy of CooperTools; **fjord:** U.S. Geological Survey; **fledgling:** Florida State News Bureau, Tallahassee; **Fleming:** British Information Services; **flint:** Courtesy Dept. of Library Services, American Museum of Natural History; **flycatcher:** U.S. Forest Service; **folded rock:** Courtesy Dept. of Library Services, American Museum of Natural History; **forward-looking infrared system:** Courtesy of Northrop Corp; **fossil:** U.S. Geological Survey; **fracture:** Courtesy Dept. of Library Services, American Museum of Natural History; **framing:** Courtesy of Weyerhaeuser Co.; **front suspension:** Courtesy of Ford Motor Co.; **fuel-injection engine:** Courtesy of Ford Motor Co.; **fungus:** U.S. Department of Agriculture; **galaxy:** Hale Observatories; **Galilean telescope:** Courtesy Dept. of Library Services, American Museum of Natural History; **gantry robot:** Courtesy of CIMCORP, Inc.; **Ganymede:** NASA; **garden snail:** Hugh Spencer/Photo Reseachers; **Gatling gun:** National Archives; **Gemini VII:** NASA; **Giant's Footprint:** NASA; **glaciation:** Oregon State Highway Commission; **glacier:** U.S. Geological Survey; **glasswing butterfly:** Courtesy Dept. of Library Services, American Museum of Natural History; **gneiss:** U.S. Geological Survey; **Goddard:** Esther C. Goddard Photo; **gold:** Bureau of Mines; **gonorrhea virus:** HBJ Photo; **Grand Canyon:** Union Pacific Railroad Photo; **granitic rock:** Field Museum of Natural History, Chicago; **grapple skidder:** Courtesy of John Deere & Co.; **Great Red Spot:** NASA; **gripper:** Courtesy of Advanced Manufacturing Systems, Inc.; **ground beetle:** Roche; **guided-missile ship:** U.S. Navy; **gully erosion:** U.S. Geological Survey; **guppy:** Courtesy Dept. of Library Services, American Museum of Natural History; **gypsum:** Courtesy Dept. of Library Services, American Museum of Natural History; **hair follicles:** Hoppock Associates; **Hale telescope:** Mount Wilson and Palomar Observatories; **halite:** Bureau of Mines; **Halley's Comet:** Hale Observatories; **hanging valley:** U.S. Geological Survey; **Harpoon missile:** U.S. Navy; **harvester:** Courtesy of Georgia-Pacific Corp.; **hatching stage:** HBJ Photo; **headframe:** Jim Van Gundy Photography/Courtesy of Hecla Mining Co.; **heavier-than-air craft:** Smithsonian Institution; **hermit crab:** General Biological Supply House, Chicago; **herons:** Florida State News Bureau; **Herschel:** HBJ Photo; **hibiscus:** Smithsonian Institution; **hieroglyphics:** HBJ Photo; **hillock:** U.S. Geological Survey; **hogback:** U.S. Geological Survey; **homogenizer:** Courtesy of Greerco Corp.; **Hopper:** HBJ Photo; **horned owl:** U.S. Forest Service; **hornfels:** Courtesy Dept. of Library Services, American Museum of Natural History; **horst:** U.S. Geological Survey; **hummingbird:** HBJ Photo; **hurricane:** Providence Journal; **Huxley:** The National Portrait Gallery, London; **hydrofoil:** U.S. Navy; **iceberg:** U.S. Navy; **ichthyosaur:** Courtesy Dept. of Library Services, American Museum of Natural History; **impact crater:** NASA; **incubator:** Courtesy of Lyon Electric Co.; **influenza:** WHO Photo; **interceptor:** U.S. Navy; **internal-combustion engine:** Courtesy of Ford Motor Co.; **intestinal cells:** University of Iowa; **Intruder:** U.S. Navy; **Io:** NASA; **Irish moss:** Hugh Spencer/Photo Reseachers; **irrigation:** Standard Oil Co.; **jackhammer:** Courtesy of Ingersoll-Rand Co.; **jellyfish:** Robert C. Hermes/Photo Reseachers; **joint:** U.S. Geological Survey; **Jovian moons:** Jet Propulsion Laboratory/California Institute of Technology; **Jupiter:** Jet Propulsion Laboratory/California Institute of Technology; **Jurassic fossils:** Courtesy Dept. of Library Services, American Museum of Natural History; **kame:** Wisconsin Conservation Department; **karst plain:** U.S. Geological Survey; **Koch:** ACME Newspictures, Inc.; **Kohlberg:** Newsweek; **kunzite:** Courtesy Dept. of Library Services, American Museum of Natural History; **Lagoon Nebula:** Hale Observatories; **LaPlace:** HBJ Photo; **lava:** U.S. Geological Survey; **lava fountain:** U.S. Geological Survey; **Lavoisier:** Library of Congress; **Lawrence:** University of California, Berkeley; **leaf and dart:** HBJ Photo; **Leakeys:** National Geographic Society; **Lee and Yang:** The Bettmann Archive; **leech:** Carolina Biological Supply Company; **Leeuwenhoek:** New York Academy of Medicine; **lichens:** U.S. Forest Service; **lightning:** Herald Tribune Photo; **limestone:** Bureau of Mines; **Linotype:** Farm Security Administration; **Lister:** Science Service; **lithophysae:** Courtesy Dept. of Library Services, American Museum of Natural History; **live oak:** Standard Oil Co.; **locks:** U.S. Army Photograph; **locking-jaw pliers:** Courtesy of CooperTools; **locomotive:** Association of American Railroads; **loess:** U.S. Geological Survey; **logic board:** Courtesy of Apple Computer, Inc.; **log loader:** Courtesy of John Deere & Co.; **longleaf pine:** U.S. Department of Agriculture; **lunar rover:** NASA; **Lyell:** Courtesy Dept. of Library Services, American Museum of Natural History; **lymphocyte:** HBJ Photo; **Magdalenian artifact:** HBJ Photo; **Magellan:** Courtesy Dept. of Library Services, American Museum of Natural History; **mammary gland:** HBJ Photo; **marble:** Bureau of Mines; **Marconi:** HBJ Photo; **Mars:** NASA; **mast-mounted sight:** Courtesy of Northrop Corp.; **Maxwell:** The National Portrait Gallery, London; **McClintock:** Nik Kleinberg/Picture Group; **McCormick:** International Harvester/Photo News; **meander:** U.S. Geological Survey; **measles:** Courtesy of Chas. Pfizer & Co., Inc.; **Mendel:** Courtesy Dept. of Library Services, American Museum of Natural History; **merino:** Wool Bureau, Inc.; **mesophyll:** L.K. Shumway, College of Eastern Utah; **metaphase:** Carolina Biological Supply Company; **meteor crater:** Courtesy Dept. of Library Services, American Museum of Natural History; **meteorite:** HBJ Photo; **Michelson:** Yerkes Observatory; **microchip shield:** Courtesy of General Electric Co.; **micrograph:** General Biological Supply House, Chicago; **milling machine:** Courtesy of Ingersoll-Rand; **Minoan artifacts:** The Metropolitan Museum of Art, Fletcher Fund; **Miocene fossils:** National Science Foundation; **missile launcher:** Courtesy of LTV Missiles and Electronics Group; **mitosis:** HBJ Photo; **moldboard plow:** Courtesy of John Deere & Co.; **monadnock:** U.S. Geological Survey; **monarch butterfly:** Courtesy Dept. of Library Services, American Museum of Natural History; **monazite:** Bureau of Mines; **monoplane:** Courtesy of Piper Aircraft Corp.; **moon:** NASA; **moored mine:** U.S. Navy; **moraine:** U.S. Geological Survey; **morel:** HBJ Photo; **Morgan:** Science Service; **moropus skeleton:** Courtesy Dept. of Library Services, American Museum of Natural History; **mower-conditioner:** Courtesy of John Deere & Co.; **mud crack:** U.S. Geological Survey; **multiple birth:** The Bettman Archive; **multirole aerial target:** Courtesy of Northrop Corp.; **mushroom:** HBJ Photo; **mutation:** Brookhaven National Laboratory; **Mycobacterium:** Dept. of Health, Education and Welfare; **natural arch:** U.S. Geological Survey; **Neanderthal skull:** Smithsonian Institution; **nebula:** Harvard College Observatory; **needlenose pliers:** Courtesy of J.H. Williams Industrial Products, Inc.; **neurons:** Manfred Kage, Peter Arnold, Inc.; **Newton:** HBJ Photo; **nippers:** Courtesy of J.H. Williams Industrial Products, Inc.; **normal erosion:** U.S. Geological Survey; **nuclear submarine:** U.S. Navy; **obsidian:** Smithsonian Institution; **oceanographic submersible:** U.S. Navy; **octopus:** Courtesy Dept. of Library Services, American Museum of Natural History; **offshore drilling:** Texas Gas Transmission Corp.; **oil well:** Texas Mid Continental Oil & Gas Assoc./American Petroleum Institute; **opal:** Courtesy Dept. of Library Services, American Museum of Natural History; **open-hearth process:** U.S. Steel; **open-pit mine:** Courtesy of Hecla Mining Co.; **ophicalcite:** Courtesy Dept. of Library Services, American Museum of Natural History; **Opuntia cactus:** Courtesy Dept. of Library Services, American Museum of Natural History; **Orion:** Mount Wilson and Palomar Observatories; **outwash plain:** U.S. Geological Survey; **overdrive:** Courtesy of Ford Motor Co.; **overstory:** U.S. Forest Service; **overthrust:** Standard Oil Co.; **pahoehoe lava:** U.S. Geological Survey; **Panther helicopter:** Courtesy of LTV Aircraft Products Group; **Pascal:** Art Resource; **passive sensors:** Courtesy of Northrop Corp.; **Pasteur:** Brown Brothers; **Patriot missile:** Courtesy of Raytheon Co.; **Pauli:** American Institute of Physics; **pectolite:** Courtesy Dept. of Library Services, American Museum of Natural History; **Pei:** Evelyn Hofer/HBJ Photo; **Peking man:** HBJ Photo; **periglacial area:** U.S. Geological Survey; **periscope:** Courtesy of Optical Technology Devices, Inc.; **Perseus:** National Optical Astronomy Observatories; **personal computer:** HBJ Photo; **pewee:** U.S. Department of Agriculture; **phases of the moon:** Courtesy Dept. of Library Services, American Museum of Natural History; **Phillips screwdriver:** Courtesy of CooperTools; **photocopier:** Courtesy of Ricoh Corporation; **phyllite:** Courtesy Dept. of Library Services, American Museum of Natural History; **Piaget:** Wide World Photos; **pickerel frog:** Leonard Lee Rue III; **piedmont:** U.S. Geological Survey; **pine:** U.S. Forest Service; **pipeline mixer:** Courtesy of Greerco Corp.; **Piper Cub:** Courtesy of Piper Aircraft Corp.; **Pithecanthropus:** Courtesy Dept. of Library

Photo Credits

Services, American Museum of Natural History; **plate tectonics:** U.S. Geological Survey; **playa:** U.S. Geological Survey; **Pliohippus:** Courtesy Dept. of Library Services, American Museum of Natural History; **plumb bobs:** Courtesy of CooperTools; **Polaris:** U.S. Navy; **Polaroid:** Courtesy of Polaroid Corporate Archives; **polecat:** Bureau of Sport Fisheries & Wildlife; **polio:** R.C. Williams/National Foundation; **pollination:** Treat Davidson/Photo Researchers; **polyp:** Courtesy Dept. of Library Services, American Museum of Natural History; **porphyry:** Smithsonian Institution; **Portuguese man-of-war:** Douglas P. Wilson/Marine Biological Laboratory; **pothole:** U.S. Geological Survey; **power saw:** Courtesy of Dremel; **pre-Columbian sculpture:** HBJ Photo; **prehistoric arrow points:** Courtesy of Museum of the American Indian, Heye Foundation; **press:** Courtesy of Magnat Machinery, Inc.; **printer:** Courtesy of Ricoh Corporation; **proceratops:** Courtesy Dept. of Library Services, American Museum of Natural History; **promontory:** Courtesy Dept. of Library Services, American Museum of Natural History; **prophase:** Carolina Biological Supply House; **Protozoa:** General Biological Supply House, Chicago; **Pseudomonas fluorescens:** HBJ Photo; **Ptolemaic system:** Courtesy Dept. of Library Services, American Museum of Natural History; **pueblo:** Arizona State Museum; **punch card:** HBJ Photo; **pyrite:** Courtesy Dept. of Library Services, American Museum of Natural History; **quartz:** Courtesy Dept. of Library Services, American Museum of Natural History; **radar antenna:** Courtesy of Vantage Associates; **radial tire:** Courtesy of Goodyear Tire & Rubber Co.; **radio telescope:** NASA; **ragweed:** Smithsonian Institution; **rain erosion:** U.S. Geological Survey; **Ramapithecus:** Courtesy Dept. of Library Services, American Museum of Natural History; **ramjet:** NASA; **rangefinder:** Courtesy of Azimuth Corporation; **ratcheting box wrench:** Courtesy of J.H. Williams Industrial Products, Inc.; **rat flea:** U.S. Department of Agriculture; **reaper:** International Harvester Company/Photo News; **rear suspension:** Courtesy of Ford Motor Co.; **recording thermometer:** Courtesy of Wahl Instruments, Inc.; **red blood cells:** Richard Baker/USC Dept. of Microbiology; **Redstone:** NASA; **redwoods:** Redwood Empire Association Photo; **reentry vehicle:** NASA; **relief sculpture:** Art Resource; **rescue/salvage ship:** U.S. Navy; **rhizopods:** Courtesy of General Biological Supply House, Chicago; **rills:** U.S. Geological Survey; **ring laser gyro:** Courtesy of Northrop Corp.; **rings of Saturn:** Mount Wilson and Palomar Observatories; **ripper:** Courtesy of John Deere & Co.; **rippled sandstone:** Ward's Natural Science Establishment, Inc., Rochester, N.Y.; **ripple mark:** U.S. Geological Survey; **rockfall:** U.S. Geological Survey; **roller:** Courtesy of Ingersoll-Rand Co.; **Rorschach:** HBJ Photo; **roseate spoonbill:** National Audubon Society/Photo Researchers; **rotifers:** Courtesy Dept. of Library Services, American Museum of Natural History; **Russell:** The Bettmann Archive; **Rutherford:** Cavendish Laboratory, University of Cambridge; **Saarinen:** HBJ Photo; **Sabin:** National Foundation; **Sagittarius:** Mount Wilson and Palomar Observatories; **saguaro cactus:** U.S. Department of Agriculture; **sal ammoniac:** Courtesy Dept. of Library Services, American Museum of Natural History; **Salk:** National Foundation for Infantile Paralysis; **salt crystals:** Morton Salt Company; **sand dunes:** Florida State News Bureau; **sandstone:** U.S. Geological Survey; **Sargassum:** General Biological Supply House, Chicago; **satellite:** NASA; **Saturn:** NASA; **sawmill:** Courtesy of Weyerhaeuser Co.; **scaling hammer:** Courtesy of CooperTools; **scarab:** Maslowski & Goodpaster/Photo Reseachers; **scarp:** U.S. Geological Survey; **sea arch:** U.S. Geological Survey; **sea horse:** Douglas P. Wilson/Marine Biological Laboratory; **sea ice:** Photo by L. Shapiro; **seaplane:** Free Lance Photographers Guild; **Sea Sparrow:** U.S. Navy; **sedimentary rock:** Field Museum of Natural History, Chicago; **servo-controlled robot:** Courtesy of GMFanuc Robotics; **set screw wrench:** Courtesy of J.H. Williams Industrial Products, Inc.; **Shapley:** HBJ Photo; **shatter cone:** Photo by George Williams; **ship-based missile:** Courtesy of LTV Missiles and Electronics Group; **Siamese twins:** HBJ Photo; **sickled red blood cell:** Richard Baker/USC Medical School; **silver:** Courtesy Dept. of Library Services, American Museum of Natural History; **sinkhole:** U.S. Geological Survey; **skin:** General Biological Supply House, Chicago; **Skinner:** HBJ Photo; **skull:** Percy Brooks/New York Hospital; **Skylab:** NASA; **sledgehammer:** Courtesy of CooperTools; **slip-joint pliers:** Courtesy of J.H. Williams Industrial Products, Inc.; **slump:** U.S. Geological Survey; **Smilodon:** Courtesy Dept. of Library Services, American Museum of Natural History; **soil compactor:** Courtesy of Ingersoll-Rand Co.; **solar energy unit:** Courtesy of RCA; **somatic mutation:** Brookhaven National Laboratory; **space shuttle:** NASA; **sperm:** HBJ Photo; **spherulites:** Courtesy Dept. of Library Services, American Museum of Natural History; **spinning machine:** Science Museum; **spiracles:** Courtesy Dept. of Library Services, American Museum of Natural History; **spiral galaxy:** Hale Observatories; **sporophytes:** Hugh Spencer/Photo Reseachers; **sprouting:** Carolina Biological Supply Company; **stalagmites:** HBJ Photo; **Staphyloccocus aureus:** Eli Lilly and Company; **starch:** HBJ Photo; **Stealth bomber:** Courtesy of LTV Aircraft Products Group; **steamboat:** The Bettmann Archive; **step trench:** Arkansas Archeological Survey; **stibnite:** Courtesy Dept. of Library Services, American Museum of Natural History; **Stonehenge:** HBJ Photo; **stony meteorite:** Courtesy Dept. of Library Services, American Museum of Natural History; **stratified rock:** Courtesy Dept. of Library Services, American Museum of Natural History; **striae:** U.S. Geological Survey; **strip cropping:** U.S. Department of Agriculture; **submarine rescue ship:** U.S. Navy; **suction hose:** Courtesy of Northrop Corp.; **sugar maple:** General Biological Supply House, Chicago; **Sullivan (Carson Pirie Scott Building):** Chicago Architectural Photographing Co.; **sundial:** HBJ Photo; **Superfortress:** Wide World Photos; **surface mount assembly:** Courtesy of Adept Technology, Inc.; **surface-to-surface missile:** Courtesy of LTV Missiles & Electronics Group; **Surveyor III:** NASA; **sweat gland:** Medichrome Division, Clay-Adams Co.; **syncline:** U.S. Geological Survey; **synoptic terrain photography:** NASA; **syphilis:** HBJ Photo; **tadpole:** HBJ Photo; **talcose slate:** Courtesy Dept. of Library Services, American Museum of Natural History; **target docking vehicle:** NASA; **Taurus:** Courtesy Dept. of Library Services, American Museum of Natural History; **teeming steel:** Ewing Galloway; **telophase:** Carolina Biological Supply Company; **Telstar:** Courtesy of Bell Telephone; **tepee:** HBJ Photo; **terminal:** Courtesy of Digital Equipment Corporation; **thrust fault:** U.S. Geological Survey; **thunderhead clouds:** National Oceanic and Atmospheric Administration; **timber harvesting:** Courtesy of Georgia-Pacific Corp.; **tongue-and-groove pliers:** Courtesy of CooperTools; **tornado:** U.S. Navy; **torpedo:** U.S. Navy; **tourmaline:** Courtesy Dept. of Library Services, American Museum of Natural History; **tracking satellite:** NASA; **tractor:** Courtesy of John Deere & Co.; **transverse dunes:** U.S. Geological Survey; **trichinosis:** HBJ Photo; **trilobite:** Courtesy Dept. of Library Services, American Museum of Natural History; **trumpeter swan:** U.S. Fish and Wildlife Service; **tuberculosis bacteria:** HBJ Photo; **tunnel:** Standard Oil Co.; **turboprop:** Courtesy of Piper Aircraft Corp.; **turtle:** HBJ Photo; **Tychonic system:** Courtesy Dept. of Library Services, American Museum of Natural History; **Tyrannosaurus:** Courtesy Dept. of Library Services, American Museum of Natural History; **unconformity:** U.S. Geological Survey; **undercut:** Washington State Historical Society, Tacoma; **Uranus:** NASA; **U-shaped valley:** U.S. Geological Survey; **vacuum seed meter:** Courtesy of John Deere & Co.; **Vanguard:** U.S. Navy; **vein:** U.S. Geological Survey; **Venus:** NASA; **vesicular lava:** Courtesy Dept. of Library Services, American Museum of Natural History; **Viking II:** NASA; **vitamin C:** Mortimer Abramowitz; **volcanic neck:** U.S. Geological Survey; **volcano:** U.S. Geological Survey; **volvox:** W.J. Garnett/HBJ Photo; **V-shaped valley:** U.S. Geological Survey; **waning moon:** Lick Observatory, University of California, Santa Cruz; **waterspout:** Science Service; **waterwell drill:** Courtesy of Ingersoll-Rand Co.; **Western diamondback:** Courtesy Dept. of Library Services, American Museum of Natural History; **wheel sprinkler:** U.S. Department of Agriculture; **white pine forest:** U.S. Forest Service; **whooping crane:** Luther C. Goldman/Bureau of Sport Fisheries & Wildlife; **widebody jet:** Courtesy of LTV Aircraft Products Group; **Wildcat:** U.S. Navy; **wood frog:** Harold V. Green/HBJ Photo; **Woodward:** Harvard University; **worker bee:** HBJ Photo; **working cell:** Courtesy of WT Automation, Inc.; **Wright Brothers:** The Bettmann Archive; **Wright, Frank Lloyd (Johnson Wax Building):** S.C. Johnson & Son, Inc.; **X-ray:** HBJ Photo; **Xerox:** HBJ Photo; **YF-23:** Courtesy of Northrop Corp.; **zircon:** Courtesy Dept. of Library Services, American Museum of Natural History; **Zygnema:** General Biological Supply House, Chicago.